33rd European Photovoltaic Solar Energy Conference and Exhibition (EU PVSEC 2017)

Amsterdam, The Netherlands
25-29 September 2017

Volume 1 of 4

Editors:

A. Smets
P. Helm
N. Taylor

ISBN: 978-1-5108-5397-3

Printed from e-media with permission by:

Curran Associates, Inc.
57 Morehouse Lane
Red Hook, NY 12571

CURRAN ASSOCIATES INC. proceedings.com

Some format issues inherent in the e-media version may also appear in this print version.

Printed by Curran Associates, Inc. (2018)

WIP – Renewable Energies
Sylvensteinstr. 2
81369 Munchen
Germany

Phone: +49 89 72012735
Fax: +49 89 72012791

wip@wip-munich.de

Additional copies of this publication are available from:

Curran Associates, Inc.
57 Morehouse Lane
Red Hook, NY 12571 USA
Phone: 845-758-0400
Fax: 845-758-2633
Email: curran@proceedings.com
Web: www.proceedings.com

33rd European Photovoltaic Solar Energy Conference and Exhibition

Proceedings of the International Conference held in Amsterdam, The Netherlands

25 September – 29 September 2017

Edited by:

A. SMETS
Delft University of Technology
Professor Solar Energy
Delft, The Netherlands

N. TAYLOR
European Commission Joint Research Centre
Ispra, Italy

P. HELM
WIP Renewable Energies
Munich, Germany

Coordination of the Technical Programme:
European Commission Joint Research Centre
Via E. Fermi 1
21020 Ispra (VA)
Italy

Institutional Support:
European Commission
UNESCO – United Nations Educational, Scientific and Cultural Organization, Natural Sciences Sector
WCRE – World Council for Renewable Energy

Supporting Organisations:
ARE – Alliance for Rural Electrification
EREF – European Renewable Energies Federation
ESA – European Space Agency
EUFORES –The European Forum for Renewable Energy Sources
EUREC – The Association of European Renewable Energy Research Centres
European Technology & Innovation Platform PV
IEA PVPS – International Energy Agency, Photovoltaic Power Systems Programme
REN21 – Renewable Energy Policy Network for the 21st Century
TKI Urban Energy - Topsector Energie

Institutional PV Industry Cooperation:
SolarPower Europe
SOLARUNITED

Supporting Associations:
AIE – European Association of Electrical Contractors
AREA – African Renewable Energy Alliance
CABA – Continental Automated Buildings Association
EPIC – European Photonics Industry Consortium
GENSED - Turkish Solar Energy Industry Association
MPIA - Malaysian Photovoltaic Industry Association
SASIA - Saudi Arabia Solar Industry Association

EU PVSEC realised by:
WIP
Sylvensteinstr. 2, 81369 München, Germany
Tel: +49 89 720 12 735, Fax: +49 89 720 12 791
Email: pv.conference@wip-munich.de
www.photovoltaic-conference.com
www.wip-munich.de

Proceedings produced and published by:
WIP
Sylvensteinstr. 2, 81369 München, Germany
Tel: +49 89 720 12 735, Fax: +49 89 720 12 791
Email: pv.conference@wip-munich.de
www.photovoltaic-conference.com
www.wip-munich.de

ISSN 2196-0992

33rd EUROPEAN PHOTOVOLTAIC SOLAR ENERGY CONFERENCE AND EXHIBITION
25 SEPTEMBER – 29 SEPTEMBER 2017
AMSTERDAM, THE NETHERLANDS

EU PVSEC COMMITTEES

INTERNATIONAL SCIENTIFIC ADVISORY COMMITTEE (ISAC)

Chair
P. Szymanski, European Commission, JRC, Director of Energy, Transport and Climate, Petten, The Netherlands

Secretary
N. Taylor, European Commission, JRC, Institute for Energy and Transport, Ispra, Italy

Committee Members
P. Frankl, Head of the Renewable Energy Division, International Energy Agency, France
M. Getsiou, European Commission, DG RTD, Brussels, Belgium
S.W. Glunz, Head of Division "Solar Cells – Development and Characterization", Fraunhofer ISE, Freiburg, Germany
D. Habermann, Chief Innovation Officer, Meyer Burger Technology, Gwatt (Thun), Switzerland
P. Malbranche, General Director, CEA INES, France
P. Menna, European Commission, DG ENER, Brussels, Belgium
S. Nowak, Managing Director of NET Nowak Energy & Technology, St. Ursen, Switzerland
W.C. Sinke, Program Development Manager, ECN Solar Energy, The Netherlands
A.H.M. Smets, Professor Solar Energy at Delft University of Technology, The Netherlands
M. Topič, Chairman ETIP PV, Head of Laboratory of Photovoltaics and Optoelectronics of the University of Ljubljana, Slovenia
P. Verlinden, Chief Scientific and Vice Chair, State Key Lab. TRINA Solar, China
A. Grassi, Managing Director, ETA-Florence Renewable Energies, Florence, Italy
P. Helm, Scientific & Managing Director, WIP - Renewable Energies, Munich, Germany

CONFERENCE EXECUTIVE COMMITTEE

Conference General Chairman
A.H.M. Smets, Professor Solar Energy at Delft University of Technology, The Netherlands

Technical Programme Chairman
N. Taylor, European Commission, JRC, Institute for Energy and Transport, Ispra, Italy

Committee Members
H. Ossenbrink, Former European Commission, Italy
P. Helm, Scientific & Managing Director, WIP - Renewable Energies, Munich, Germany
W.C. Sinke, Program Development Manager, ECN Solar Energy, The Netherlands
S. Nowak, Managing Director of NET Nowak Energy & Technology, St. Ursen, Switzerland
A. Grassi, Managing Director, ETA-Florence Renewable Energies, Florence, Italy

2017 SCIENTIFIC COMMITTEE
Topic Organisers and Paper Review Experts

Topic 1 – New Materials and Concepts for Photovoltaic Devices

N.J. Ekins-Daukes, Imperial College London, United Kingdom
A. Martí Vega, UPM, Madrid, Spain
J. Poortmans, imec, Leuven, Belgium
J.A.M. van Roosmalen, ECN, Petten, The Netherlands

V. Bukauskas, Center for Physical Sciences and Technology, Vilnius, Lithuania
J. Govaerts, imec, Leuven, Belgium
J.F. Guillemoles, CNRS, Chatou Cedex, France
M.A. Hernandez-Fenollosa, UPV, Valencia, Spain
J.C. Hummelen, University of Groningen, The Netherlands
D. König, UNSW Australia, Sydney, Australia
I. Konovalov, University of Applied Sciences Jena, Germany
C. Lévy-Clément, CNRS, Thiais, France
K.R. McIntosh, PV Lighthouse, Coledale, Australia
Y. Okada, University of Tokyo, Japan
B. Rech, HZB, Berlin, Germany
F. Roca, ENEA, Portici, Italy
M. Rusu, HZB, Berlin, Germany
V. Saly, Slovak University of Technology, Bratislava, Slovakia
R.E.I. Schropp, Utrecht University, Driebergen, The Netherlands
H. Shirai, Saitama University, Saitama-City, Japan
W. Warta, Fraunhofer ISE, Freiburg, Germany

Topic 2 – Silicon Photovoltaics

O. Anspach, PV Crystalox Solar, Erfurt, Germany
D.L. Bätzner, Meyer Burger Research, Hauterive, Switzerland
R. Brendel, ISFH, Emmerthal, Germany
J. Cárabe, CIEMAT, Madrid, Spain
F. Ferrazza, eni spa, San Donato Milanese, Italy
S.W. Glunz, Fraunhofer ISE, Freiburg, Germany
G. Hahn, University of Konstanz, Germany
D. Muñoz, CEA, Le Bourget du Lac, France
S. Riepe, Fraunhofer ISE, Freiburg, Germany
M. Topic, University of Ljubljana, Slovenia
A.W. Weeber, ECN, Petten, The Netherlands
P. Wohlfart, Singulus Technologies, Kahl am Main, Germany

A.G. Aberle, SERIS, Singapore
N. Azkona, UPV/EHU, Zamudio, Spain
J. Bagdahn, Anhalt University of Applied Sciences, Köthen, Germany
R. Barinka, Fill Factory, Rožnov pod Radhoštm, Czech Republic

P.A. Basore, PV Specialist Services, Solana Beach, USA
S. Bowden, Arizona State University, Tempe, USA
T.M. Bruton, Progress in Photovoltaics, Woking, United Kingdom
A. Danel, CEA, Le Bourget du Lac, France
B. Dauksher, Arizona State University, Tempe, USA
M. Debucquoy, imec, Leuven, Belgium
P. Delli Veneri, ENEA, Portici, Italy
T. Desrues, CEA, Le Bourget du Lac, France
D. Dimova-Malinovska, Bulgarian Academy of Sciences, Sofia, Bulgaria
S. Dubois, CEA, Le Bourget du Lac, France
T. Dullweber, ISFH, Emmerthal, Germany
P. Fath, RCT-Solutions, Konstanz, Germany
S. Gall, HZB, Berlin, Germany
I. Gordon, imec, Leuven, Belgium
J. Haschke, EPFL, Neuchâtel, Switzerland
K. Hesse, Wacker Chemie, Burghausen, Germany
J. Horzel, CSEM, Neuchâtel, Switzerland
J.-R. Huang, Motech Industries, Taoyuan, Taiwan
J. John, imec, Leuven, Belgium
D.-W. Kang, Cheongju University, Republic of Korea
S.D. Kim, San Jose, USA
A. Meeder, Geo-En Energy Technologies, Berlin, Germany
J. Meier, Meier Technologies, Berlingen, Switzerland
A. Metz, Alzenau, Germany
C. Molpeceres, UPM, Madrid, Spain
J.W. Müller, Hanwha Q CELLS, Bitterfeld-Wolfen, Germany
Y. Ohshita, Toyota Technological Institute, Nagoya, Japan
S. Peters, Hanwha Q CELLS, Bitterfeld-Wolfen, Germany
K. Petter, Hanwha Q CELLS, Bitterfeld-Wolfen, Germany
R. Preu, Fraunhofer ISE, Freiburg, Germany
J. Rentsch, Fraunhofer ISE, Freiburg, Germany
E. Sauar, Brighterlite, Oslo, Norway
H. Schlemm, Meyer Burger, Hohenstein-Ernstthal, Germany
J. Schmidt, ISFH, Emmerthal, Germany
M.C. Schubert, Fraunhofer ISE, Freiburg, Germany
B. Terheiden, University of Konstanz, Germany
M. Vetter, IPHT, Jena, Germany
G.P. Willeke, Fraunhofer ISE, Freiburg, Germany
D. Wood, Hemlock Semiconductor, Hemlock, USA
A. Yamada, Tokyo Institute of Technology, Japan
D.L. Young, NREL, Golden, USA
M. Zeman, Delft University of Technology, The Netherlands
J. Zhao, Meyer Burger, Hohenstein-Ernstthal, Germany

Topic 3 – Thin Film Photovoltaics

C.J. Brabec, University of Erlangen-Nuremberg, Germany
A.N. Tiwari, EMPA, Dübendorf, Switzerland
S.C. Veenstra, ECN, Eindhoven, The Netherlands
W. Wischmann, ZSW, Stuttgart, Germany

G. Bardizza, European Commission JRC, Ispra, Italy
R. Birkmire, University of Delaware, Newark, USA
D. Cheyns, imec, Leuven, Belgium
S. Cros, CEA, Le Bourget du Lac, France
T. Dalibor, AVANCIS, Torgau, Germany
B. Dimmler, Manz CIGS Technology, Schwäbisch Hall, Germany
K. Durose, Stephenson Institute for Renewable Energy, Liverpool, United Kingdom
S. Huang, UNSW Australia, Sydney, Australia
S.J.C. Irvine, Swansea University, St. Asaph, United Kingdom
C.-F. Lin, NTU, Taipei, Taiwan
M. Lira-Cantú, CIN2, Barcelona, Spain
M.Ch. Lux-Steiner, HZB, Berlin, Germany
M. Meuris, imec, Leuven, Belgium
J. Parisi, University of Oldenburg, Germany
R. Po, eni spa, Novara, Italy
M. Powalla, ZSW, Stuttgart, Germany
A. Redinger, University of Luxembourg, Luxembourg
A. Romeo, University of Verona, Italy
V. Sittinger, Fraunhofer IST, Braunschweig, Germany
A. Virtuani, EPFL, Neuchâtel, Switzerland

Topic 4 – Concentrator and Space Photovoltaics

G. Flamand, imec, Leuven, Belgium
C. Signorini, European Space Agency, Noordwijk, The Netherlands

I. Antón, UPM, Madrid, Spain
A.W. Bett, Fraunhofer ISE, Freiburg, Germany
R. Campesato, CESI, Milan, Italy
V. Khorenko, Azur Space, Heilbronn, Germany
R. King, Arizona State University, Tempe, USA
F. Rubio Berenguel, Soitec Solar, Freiburg, Germany
G. Siefer, Fraunhofer ISE, Freiburg, Germany
G. Timò, RSE, Piacenza, Italy

Topic 5 – Performance, Reliability and Sustainability of Photovoltaic Modules and Balance of System Components

M. de Wild-Scholten, SmartGreenScans, Groet, The Netherlands
R. Gottschalg, Loughborough University, United Kingdom
N.M. Pearsall, Northumbria University, Newcastle upon Tyne, United Kingdom
M. Perrin, CEA, Le Bourget du Lac, France
T. Sample, European Commission JRC, Ispra, Italy
K. Wambach, Wambach-Consulting, Aindling, Germany

K. Baert, KU Leuven, Heverlee, Belgium
G. Beaucarne, Dow Corning, Seneffe, Belgium
R. Einhaus, Apollon Solar, Lyon, France
G. Friesen, SUPSI, Canobbio, Switzerland
V. Fthenakis, Columbia University, New York, USA
W.J. Gambogi, DuPont, Wilmington, USA
L. Gomes de Freitas, Federal University of Uberlândia, Brazil
G. Graditi, ENEA, Portici, Italy
Y. Hishikawa, AIST, Tsukuba, Japan
U. Jahn, TÜV Rheinland Energy, Cologne, Germany
S. Krauter, University of Paderborn, Germany
A.R. Lagunas, CENER, Sarriguren-Navarra, Spain
C. Monokroussos, TÜV Rheinland, Shanghai, China
M.A. Muñoz-García, UPM, Madrid, Spain
H. Nagel, Fraunhofer ISE, Freiburg, Germany
H. Nussbaumer, ZHAW, Winterthur, Switzerland
S. Oktik, Sisecam, Istanbul, Turkey
G. Oreski, PCCL, Leoben, Austria
W. Palitzsch, Loser Chemie, Zwickau, Germany
J.N. Roy, IIT Kharagpur, India
R. Sinton, Sinton Instruments, Boulder, USA
M. Van Iseghem, EDF R&D, Moret-sur-Loing, France
E. Voroshazi, imec, Leuven, Belgium
A. Wade, First Solar, Mainz, Germany

Topic 6 – PV System Performance and Integration

F.P. Baumgartner, ZHAW, Winterthur, Switzerland
F. Bonemazzi, ENEL, Rome, Italy
F. Frontini, SUPSI, Canobbio, Switzerland

P. Lechner, ZSW, Stuttgart, Germany
O. Mayer, GE Global Research, Garching, Germany
F. Montanari, ENEL Green Power, Rome, Italy
C. Protogeropoulos, EEPS, Athens, Greece
A. Scognamiglio, ENEA, Portici, Italy
W.G.J.H.M. van Sark, Utrecht University, The Netherlands
I. Weiss, WIP - Renewable Energies, Munich, Germany

C. Alonso-Tristán, UBU, Burgos, Spain
J. Binder, ZSW, Stuttgart, Germany
P. Blanc, MINES ParisTech, Paris, France
P. Bonomo, SUPSI, Canobbio, Switzerland
F. Bourry, CEA, Le Bourget du Lac, France
I. Da Silva, Strathmore University, Nairobi, Kenya
B. Gaiddon, Hespul, Lyon, France
M. Grottke, WIP - Renewable Energies, Munich, Germany
I.B. Hagemann, Architekturbüro Hagemann, Aachen, Germany
C. Helmke, Clake Energy, Erfurt, Germany
T. Huld, European Commission JRC, Ispra, Italy
E. Kaijuka-Okwenje, Rural Electrification Agency, Kampala, Uganda
S. Koopman, CSIR, Pretoria, South Africa
P.A. Leicester, Loughborough University, United Kingdom
P. Malbranche, CEA, Le Bourget du Lac, France
G. Mütter, Alternative Energy Solutions, Vienna, Austria
C. Nyman, Soleco, Porvoo, Finland
K. Peter, ISC Konstanz, Germany
T. Reijenga, BEAR-iD, Gouda, The Netherlands
J. Remund, Meteotest, Bern, Switzerland
K.T. Roro, CSIR, Pretoria, South Africa
M. Schroedter-Homscheidt, German Aerospace Center, Wessling, Germany
M. Sengupta, NREL, Golden, USA
D. Stellbogen, ZSW, Stuttgart, Germany
J. Sutterlueti, Gantner Instruments, Schruns, Austria
S. Tselepis, Pikermi, Athens, Greece
X. Vallvé, Trama TecnoAmbiental, Barcelona, Spain
M. Zehner, Rosenheim University of Applied Sciences, Germany

Topic 7 – PV Economics, Markets and Policies

C. Breyer, Lappeenranta University of Technology, Finland
M. Getsiou, European Commission DG RTD, Brussels, Belgium
G. Masson, Becquerel Institute, Brussels, Belgium
T. Nordmann, TNC Consulting, Feldmeilen, Switzerland
S. Nowak, NET Nowak Energy & Technology, St. Ursen, Switzerland
E. Perezagua, Consultores de Energía Fotovoltaica, Madrid, Spain

S. Caneva, WIP - Renewable Energies, Munich, Germany
I. Kaizuka, RTS, Tokyo, Japan
D. Mayer, MINES ParisTech, Paris, France
P. Mazzucchelli, EUREC, Brussels, Belgium
S. Szabó, European Commission JRC, Ispra, Italy

SUBJECT INDEX

Topic Code

1 New Materials and Concepts for Photovoltaic Devices
2 Silicon Photovoltaics
3 Thin Film Photovoltaics
4 Concentrator and Space Photovoltaics
5 Performance, Reliability and Sustainability of Photovoltaic Modules and Balance of System Components
6 PV System Performance and Integration
7 PV Economics, Markets and Policies

Session Type

P = Plenary Session
O = Oral Session
V = Visual Session

Day Codes

A = Monday, 25 September 2017
B = Tuesday, 26 September 2017
C = Wednesday, 27 September 2017
D = Thursday, 28 September 2017
E = Friday, 29 September 2017

e.g. 3CP.1 ⇒ 3=Thin Film Photovoltaics, C=Wednesday, P=Plenary Session, 1=Session 1

TABLE OF CONTENT
of EU PVSEC 2017 Proceedings Papers

Silicon Photovoltaics

Oral PRESENTATIONS 2CO.9 Production Technologies for Silicon Solar Cells

Oral PRESENTATIONS 2CO.10 c-Si Solar Cell Process Technology

Oral PRESENTATIONS 2CO.11 c-Si Homojunction Cells

Visual PRESENTATIONS 2AV.3 Heterojunction Solar Cells

Thin Film Photovoltaics

Visual PRESENTATIONS 3DV.2 CI(G)S, CdTe and Related Thin Film Solar Cells and Modules (II) / Perovskite, Organic and Dye-Sensitised Devices

Concentrator and Space Photovoltaics

Oral PRESENTATIONS 4DO.4 III-V-Based Devices for Terrestrial and Space Applications

Visual PRESENTATIONS 4CV.4 III-V-Based Devices for Terrestrial and Space Applications

Performance, Reliability and Sustainability of Photovoltaic Modules and Balance of System Components

Plenary SESSION 5DP.1 Performance, Reliability and Sustainability of Photovoltaic Modules and Balance of System Components

Oral PRESENTATIONS 5BO.5 Backsheet and Encapsulant Materials

Oral PRESENTATIONS 5BO.6 Electrical Characterisation of PV Devices

Oral PRESENTATIONS 5CO.5 Performance Enhancing Coatings and Outdoor Performance

Oral PRESENTATIONS 5CO.6 Potential Induced Degradation, Light & Elevated Temperature Induced Degradation and Partial Shading Effects on PV Modules

Oral PRESENTATIONS 5EO.1 Sustainability and Recycling

Visual PRESENTATIONS 5BV.4 PV Module Performance and Reliability (I)

PV System Performance and Integration

Plenary SESSION 6DP.2 PV System Performance and Integration

Oral PRESENTATIONS 6BO.7 Advanced Field Performance Estimation

Oral PRESENTATIONS 6BO.8 Failure Modes and Degradation

Visual PRESENTATIONS 6BV.3 Solar Resource and Forecasting / Building, Infrastructure and Landscape Applications / Grid and Energy System Integration

PV Economics, Markets and Policies

Oral PRESENTATIONS 7DO.8 Global PV Economics and Market Trends

Oral PRESENTATIONS 7DO.9 Innovative National PV Market Economics Business Cases

Oral PRESENTATIONS 7EO.3 PV-Related Policies, Strategies and Societal Issues

Visual PRESENTATIONS 7DV.1 PV Economics and Markets / PV-Related Policies, Strategies and Societal Issues

FOREWORD

The EU PVSEC 2017 in Amsterdam concluded with the strong message that the photovoltaic revolution is just getting started. Research and innovation across the sector, together with fast deployment and numerous application options, are paving the way to putting PV at the centre-stage of sustainable energy systems. This year 2,516 participants from 89 countries took part in this most inspiring platform. It provided a vital forum for information exchange in the field of PV solar energy research, demonstration and applications. In the exhibition 64 companies from all parts of the world welcomed visitors and presented their products and services.

The EU PVSEC 2017 Proceedings give a comprehensive overview of the photovoltaic solar sector, its current status and future prospects in science, research, innovation, development and deployment on more than 4,500 pages. Selection for inclusion in the conference was made by paper review experts and topic organisers (see the listing on pages III-V), to whom we express our sincere gratitude for their comprehensive review work and overall contribution to the success of the conference. In addition to the 619 submitted papers, the proceedings include 70 presentations (slides) shown during the plenary and oral presentations as well as 232 poster files of the visual presentations. In total this amounts to 921 publications.

The 2017 EU PVSEC Proceedings maintain our commitment to providing quick and open access to high quality scientific results. They constitute a powerful tool for targeted and quick information search and retrieval, enabling you to search by keyword, paper title and authors. We are sure that these features will help to simplify the use and exploit the full potential of this extensive source of information.

The Conference Proceedings are published as downloadable DVD file and are also fully accessible online. A DOI code (Digital Object Identifier) has been assigned to each paper. This ensures unequivocal and permanent identification and full citability. The papers can be viewed and downloaded in a full free open access from the EU PVSEC Proceedings website www.eupvsec-proceedings.com.

We are confident that these Proceedings will play an important role in providing a comprehensive overview of the current actors and activities in the global PV sector and that they will disseminate information on the state-of-the-art of technologies and applications. This can generate further research, add momentum to innovation and promote interest in PV worldwide.

We would like to cordially thank all authors and participants of the EU PVSEC 2017 for their contributions and look forward to welcoming you in 2018 at the 35th EU PVSEC 2018 in Brussels, Belgium.

The Editors

EU PVSEC 2017 25 - 29 September 2017

33rd European PV Solar Energy Conference and Exhibition

RAI Convention & Exhibition Centre • Amsterdam • The Netherlands

Selected Highlights

Nigel Taylor, Technical Programme Chair

.... with help from the session chairs and others

European Commission

Silicon

European Commission

2nd Fabrication Run of n-Type HP mc Solar Cells
IV Results

Fraunhofer ISE

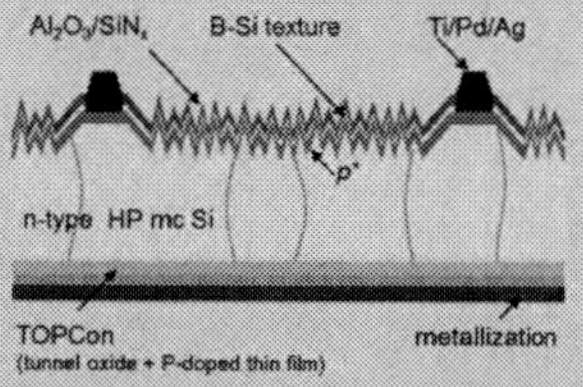

V_{oc} [mV]	J_{sc} [mA/cm²]	*FF* [%]	*pFF* [%]	*η* [%]
674.2	41.1	80.5	81.6	22.3*

Cell area 2×2 cm², aperture area (*) certified by Fraunhofer ISE CalLab

- 2nd cell batch
- Applying of „lessons learned" in cell fabrication
- Same material as previous solar cells

→ New record efficiency of 22.3% for a mc silicon solar cell

2DO.3.1 Approaching 22% Efficiency with Multicrystalline n-Type Silicon Solar Cells
J. Benick et al, Fraunhofer ISE, Freiburg, Germany

European Commission

Q.ANTUM technology on p-Cz substrates in mass production

Hanwha Q CELLS

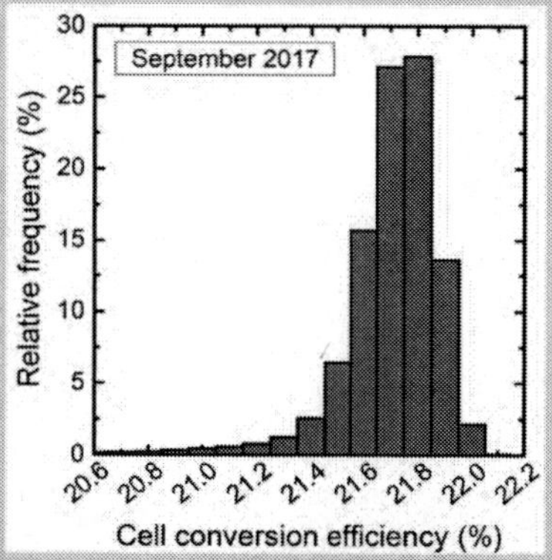

Platform change to Q.ANTUM[1] secures:

- Efficiency increase of +0.6%abs. per year
- Evolutionary development process[2] and upgrade of existing tool set
- Silver consumption: -50%
- Throughput increase
- **Avg. cell efficiency of 21.7% in production (max. >22%)**

2CO.9.6 Mass Production of Q.ANTUM Solar Cells and Modules on p-Type Cz Silicon Substrates
F. Fertig et al, Hanwha Q CELLS, Bitterfeld-Wolfen, Germany

European Commission

24.13% efficiency IBC cell

Trinasolar

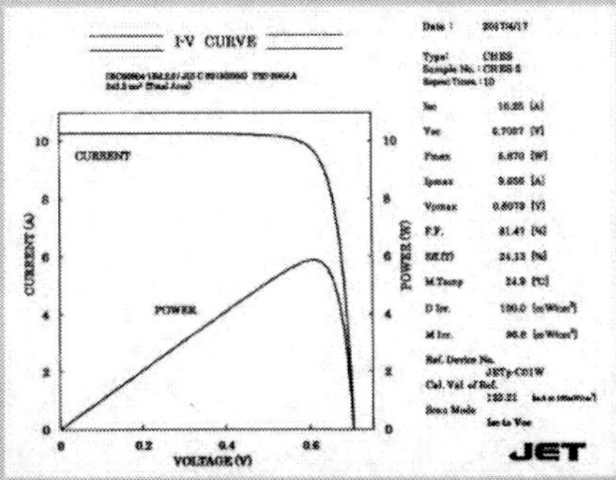

Compared to 23.5% cell (2016):

- 10 mV gain in V_{oc} by using selective emitter structure;
- J_{sc} is also improved by using less absorption in SiN_x;
- Higher FF is due to the reduced R_s, mainly from lower ρ_c on emitter.

Area	V_{oc}	I_{sc}	V_{mpp}	I_{mpp}	FF	Eff
[cm²]	[mV]	[A]	[mV]	[A]	[%]	[%]
244.9	702.7	10.25	607.9	9.66	81.47	24.13

* Independently measured at Japan Electrical Safety & Environment Technology Laboratories (JET)

2CO.11.5 Large-Area (6 Inch) Screen-Printed IBC Solar Cells with Efficiency Approaching 24% without Passivated Contacts

G. Xu et al, Trina Solar Energy, Changzhou, China

European Commission

Advantages of a-Si/c-Si heterojunction (HJ) back contact (BC) Si solar cell

a-Si/c-Si HJ : High V_{oc} and FF by excellent a-Si/c-Si interface

Back contact : High J_{sc} and FF by small optical and resistive loss

Si wafer-based: Reasonable production cost

KANEKA

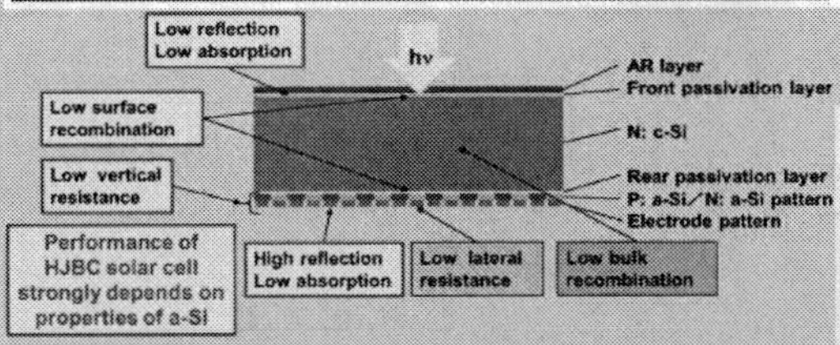

High efficiency c-Si solar cell; PV parameters

*da: designated area *ta: total area *ap: aperture area

	V_{OC} (V)	J_{SC} (mA/cm²)	FF (%)	Eff (%)	Area* (cm²)	Cell type
Panasonic	0.740	41.8	82.7	25.6	143.7 (da)	HJBC [1]
SunPower	0.737	41.3	82.7	25.2	153.5 (ta)	IBC [2]
Fraunhofer ISE	0.725	42.5	83.3	25.7	4.017(da)	TOPCon [3]
Kaneka	0.738	40.8	83.5	25.1	151.9 (ap)	HJ [4]
Kaneka	0.744	42.3	83.8	26.3	180.4 (da)	HJBC [5]
Kaneka	0.740	42.5	84.7	26.6	179.7 (da)	HJBC [6]
Kaneka	0.738	42.7	84.9	26.7	79.0 (da)	HJBC
Theoretical limit	0.753	43.6	89.2	29.3		150μm-thick wafer [7]
Ratio (26.7%/limit)	0.980	0.979	0.952	0.911		

- High V_{oc}, J_{sc} and FF obtained by excellent PECVD passivation and structure design
- Challenge for improvement of FF is a key to achieve >27% efficiency.

[1] K. Masuko et al., IEEE Journal of Photovoltaics, 4, 1433 (2014). [2] "SunPower 2015 Analyst Day" (2015). [3] A. Richter et al., Silicon PV proceedings (2017). [4] D. Adachi et al., Applied Physics Letters, 107, 233506 (2015). [5] K. Yoshikawa et al., Nat. Energy 2, 17032 (2017) [6] K. Yoshikawa et al., Silicon PV proceedings (2017) [7] A. Richter et al., IEEE J. Photovolt. 3, 1184 (2013).

2BP1.1 Record-Breaking Efficiency Back-Contact Heterojunction Crystalline Si Solar Cell and Module

K. Yamamoto et al, Kaneka, Osaka, Japan
J.L. Hernández, Kaneka, Westerlo-Oevel, Belgium

European Commission

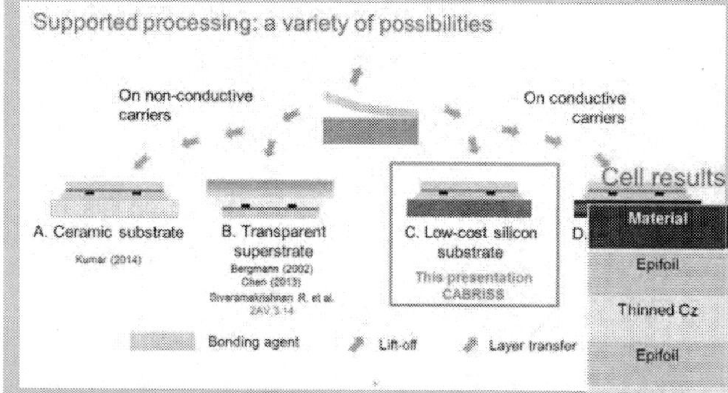

imec

TU WIEN · SINTEF · PyroGenesis · CABRISS · INKRON

Cell results

Material	Configuration		J_{sc} [mA/cm²]	V_{oc} [mV]	FF [%]	Eff. [%]	
Epifoil	Bonded to p⁺ Cz substrate	Best Avg.	33.2 32.8	681 676	69.2 69.1	15.7 15.3	*This work*
Thinned Cz	Bonded to p⁺ Cz substrate	Best Avg.	33.2 32.7	738 739	73.0 72.1	17.9 17.3	*This work*
Epifoil	Freestanding	Best Avg.	33.5 32.9	693 656	73.2 72.1	17.0 15.6	*Bearda et al. [2AV.3.26]*
Thinned Cz	Freestanding	Best	34.0 33.9	744 723	74.4 71.6	18.8 17.6	*Bearda et al. [2AV.3.26]*
Comparison with literature (cells bonded to conductive substrates)							
Epifoil (18 µm)	Bonded to steel	Best	34.5	632	77.2	16.8	*Amberwave/UNSW*
Slim-cut (25 µm)	Bonded to steel	Best	33.6	580	76.7	14.9	*Astrowatt*

2CO.12.2 Solar Cells on < 50µm Thick Epitaxial Foils Conductively Bonded to Low-Cost Si Carrier

H. Sivaramakrishnan Radhakrishnan et al, imec, Leuven, Belgium
N. Bednar & N. Adamovic, Vienna University of Technology, Austria
R. Roozeman & J. Heikkinen, INKRON, Espoo, Finland
A. Ulyashin & M. Syvertsen, SINTEF, Oslo, Norway

European Commission

Thin Film

European Commission

Key techniques implemented into the record sub(mini)modules

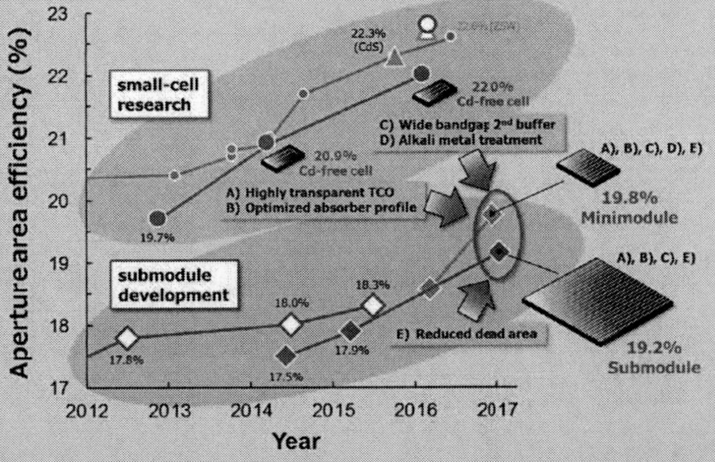

3CP.1.1 Present Status of Solar Frontier Cu(In,Ga)(Se,S)2 Record Efficiencies and Overall Progress

V. Bermudez et al, Solar Frontier, Atsugi, Japan

European Commission

HIGH EFFICIENCY MODULES

- Current in-house 30×30cm² champion module with 18.2% efficiency
- high Ga/high S absorber, wide bandgap buffer - highly Na doped In_xS_y:Na/i-$Zn_{0.85}Mg_{0.15}O$

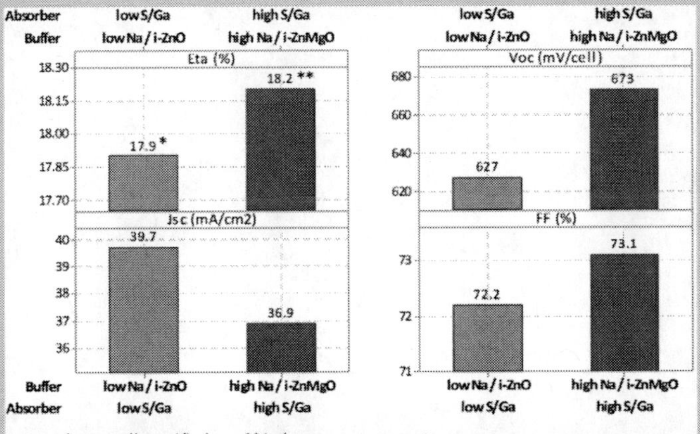

* externally certified ** in-house measurement

3BO.9.1 Wide Bandgap Sequential Absorber with Tunable Buffer Bandgap for CIGSSe Solar Modules at 18% Efficiency

R. Lechner et al, Avancis, Munich, Germany

European Commission

SOLIBRO Hi-Tech
A Hanergy Company

FULL SIZE: CHAMPION MODULE GEN 3

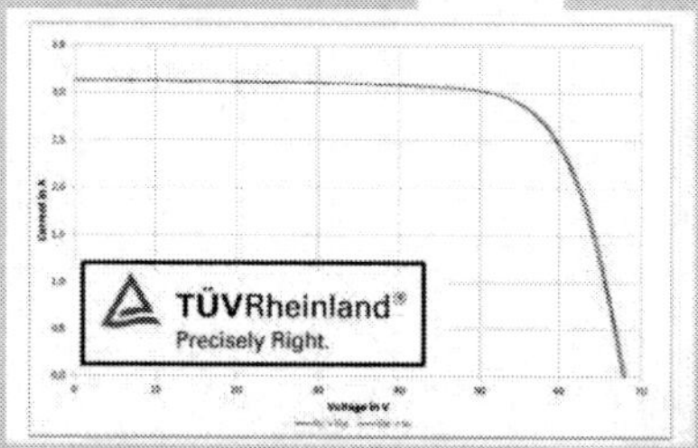

P_{mpp} [W]	159.4
V_{oc} [V]	67.9
I_{sc} [A]	3.12
FF [%]	75.2
η	17.0 %

- 17.9% Aperture Area Efficiency
- Externally verified
- Processed in SL2 production
- Module with future target process GEN3

17.0% @ 0.94m²
totale area efficiency
New world record

CELL AND MODULE DESIGN

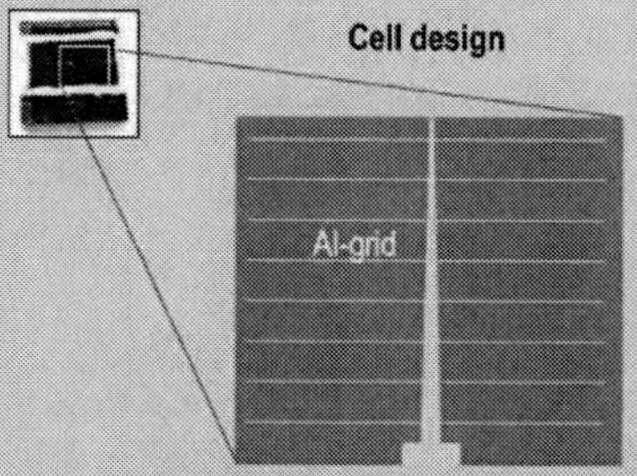

- Sheet resistance: 70 Ohms/sq
- Cell area: 10x10 mm²
- Dead area loss: ~ 3%
- Encapsulation: MgF_2

3CP.1.2 **17% Total Area Efficiency at Commercial Size CIGS Module**
P. Kratzert et al, Solibro Research, Uppsala, Sweden

Double Vision

Light Trapping
IV Results of III-V//Si Solar Cells

- Increase in J_{sc}
  - 11.6 → 12.7 mA/cm²
- Efficiency: 33.3%
- Record efficiency for 2T multi-junction solar cell

	V_{oc} [V]	J_{sc} [mA/cm²]	*FF* [%]	η [%]
Gen 1				
	3.046	11.7	87.5	31.3
Gen 2				
	3.125	11.6	86.5	31.4
Gen 3				
	3.127	**12.7**	83.8	**33.3**

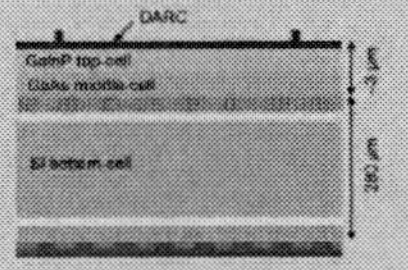

1AP.1.3 Monolithic III-V//Si Multi-Junction Solar Cell Exceeding an Efficiency of 31%
Fraunhofer ISE, Freiburg, Germany

Surplus collection of carriers at 3rd terminal injected from top cell

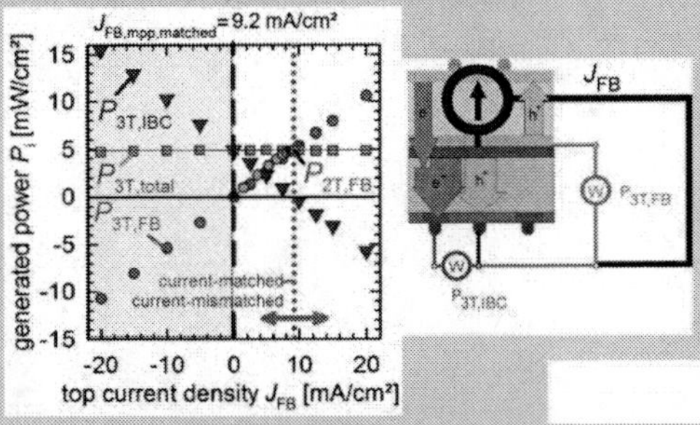

1AP.1.2 : Maximum Power Extraction Enabled by Monolithic Tandems Using Interdigitated Back Contact Bottom Cells with Three Terminals
ISFH and NREL

European Commission

22.7% Perovskite-CIGS thin film tandem solar cells

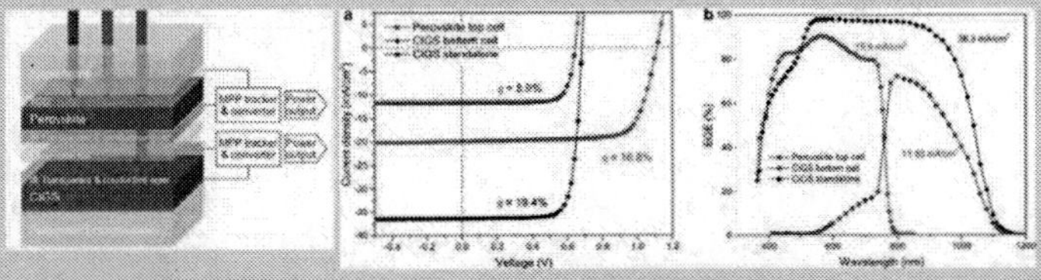

Solar Cells	V_{oc} (V)	J_{sc} (mA/cm²)	FF (%)	η (%)	MPP (%)	Cell area (cm²)	Efficiency gain
CIGS (stand alone)	0.684	36.4	78.2	19.4	19.4	0.213	
CIGS bottom cell	0.645	11.6	78.1	5.9	5.9	0.213	
Perovskite top cell	1.116	19.9	75.7	16.8	16.8	0.273	
Perovskite-CIGS 4-terminal tandem					22.7	0.213	3.3

Further improvement in Perovskite-CIGS thin film tandem solar cells via bandgap engineering, optical coupling etc.

Laboratory for Thin Films and Photovoltaics 13

3DO.7.4 Efficient and Stable NIR-Transparent Perovskite Solar Cells Prepared by Partial Ion Ex-Change Method for All-Thin-Film Tandem Applications
F. Fu, S. Pisoni, T. Feurer, A. Wäckerlin, S. Nishiwaki, A.N. Tiwari & S. Buecheler
EMPA, Dübendorf, Switzerland

European Commission

Performance

European Commission

~~MWp~~ MWh/MWp

Energy rating uncertainty

Total-order sensitivity analysis

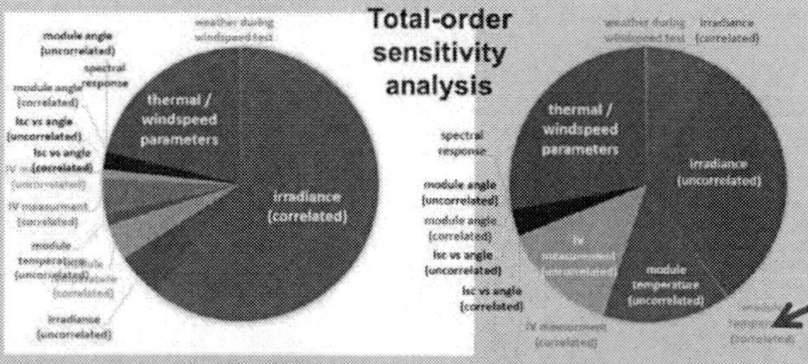

Uncertainty of $E_{mod,year}$ 2.16 % (95% conf.)

Uncertainty of *CSER* 1.84 % (95% conf.)

Simulation of 1000 modules, mixed types.
Uncertainty model based on survey of accredited test labs.

IEC 61853 Energy rating standard

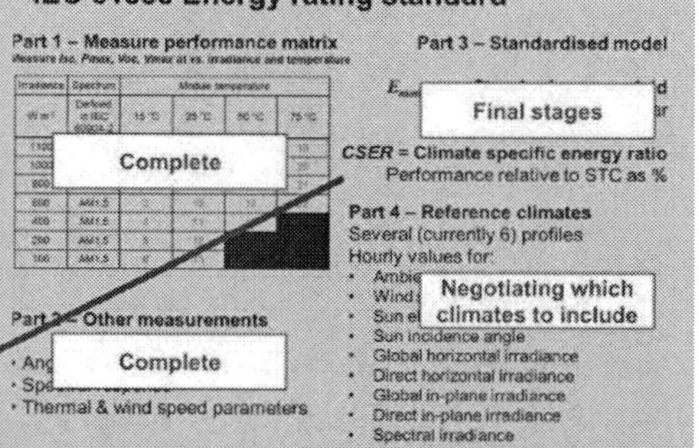

PhotoClass PV-Enerate

5CO.7.4 Progress in Energy Rating Standards: Accuracy and Optimisation

J.C. Blakesley, NPL, United Kingdom

T. Huld & H. Müllejans, European Commission JRC, Ispra, Italy

European Commission

Bifacial impact in the field

Exceptional service in the national interest — Sandia National Laboratories

Bifacial Single Axis Tracker (NM)

- Daily Potential Bifacial Energy Gains were estimated from front and back irradiance data using reference cells.
- Potential gains increase when tracker is off-track, but yield decreases

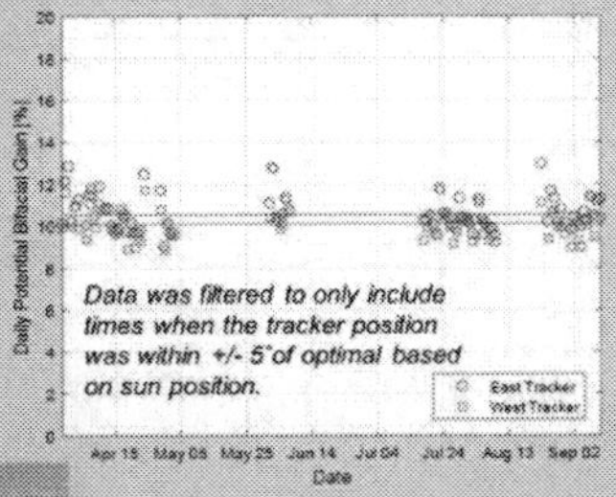

6CO.13.3 Outdoor Field Performance from Bifacial Photovoltaic Modules and Systems

J.S. Stein et al, Sandia National Laboratories, Albuquerque, United States
C. Deline, NREL, Golden, United States
F. Toor, University of Iowa, United States

European Commission

Light and elevated Temperature Induced Degradation

Piercing the Puzzle Together

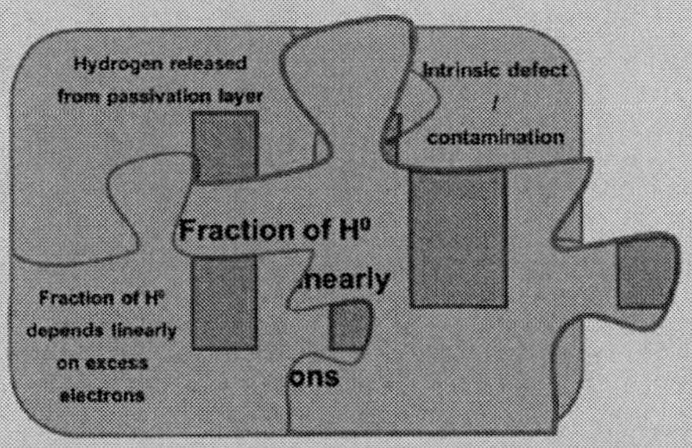

Model
Degradation

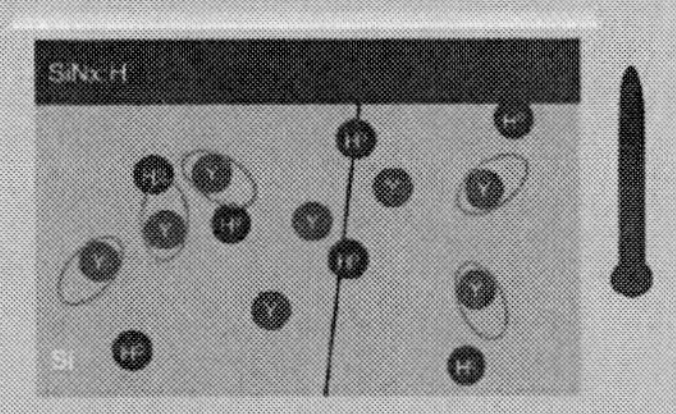

- LeTID observed in all p-type Si materials
- Defect activation only possible after fast firing of SiN_x:H-layer
- LeTID defect involves one fast-diffusing species
- Defect activated by inducing a change in charge state

→ Key to controlling LeTID: Manipulation of hydrogen

- Promising route for investigations: Float-zone silicon

2BP.1.4 Understanding Light-Induced Degradation in Multicrystalline Silicon: Possible Complex Formation Mechanisms

F. Schindler et al, Fraunhofer ISE, Freiburg, Germany

European Commission

Big data can help (1)

More than one million PV systems < 30 kWp have been installed in Germany.

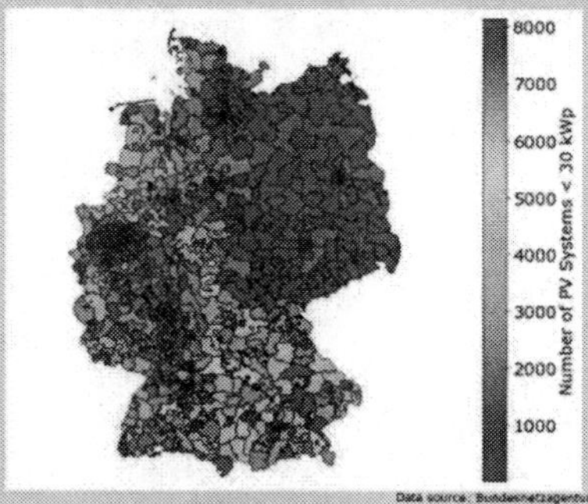

There is an annual loss of €450m to €500m by residential roof-top PV systems in Germany.

Year	Annual electric energy production in TWh	Annual yield loss in TWh	Monetary loss in million EUR
2014	36.1	1.46	€460m
2015	38.7	1.59	€489m
2016	37.6	1.58	€460m

Source for mean Feed-In Tariff: BMWI, EEG in Zahlen. Vergütungen, Differenzkosten und EEG-Umlage 2000 bis 2017.

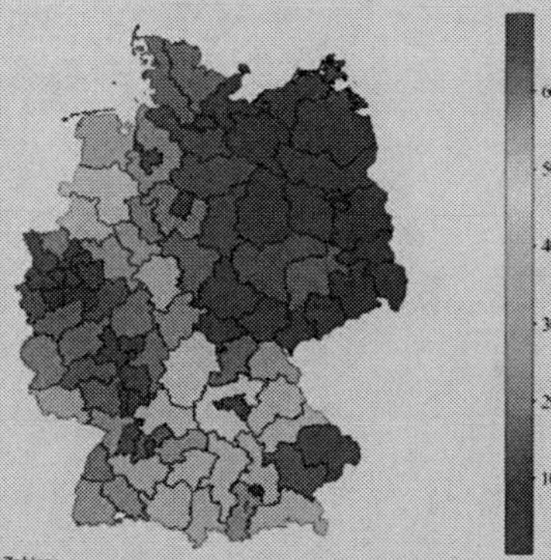

6BO.7.1 Survey on Yield of PV Systems in Germany 2014 to 2016

H. te Heesen & M. Rumpler
Trier University of Applied Science, Neubrücke, Germany
V. Herbort Ulm University of Applied Sciences, Germany

European Commission

Big data can help (2)

Model development: Model Architecture

Two parallel approaches

- **Generic Fault Prediction (Supervision-Diagnosis Model):** *predict inverter deviations from nominal behavior*
- **Specific Fault Prediction (Short-Term Fault Prediction Model):** *predict a specific fault class*

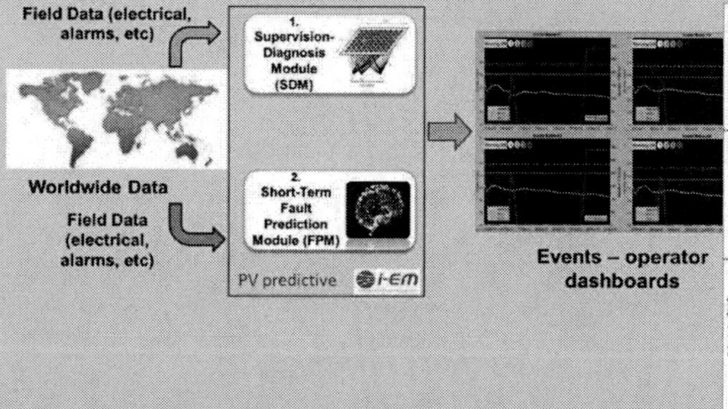

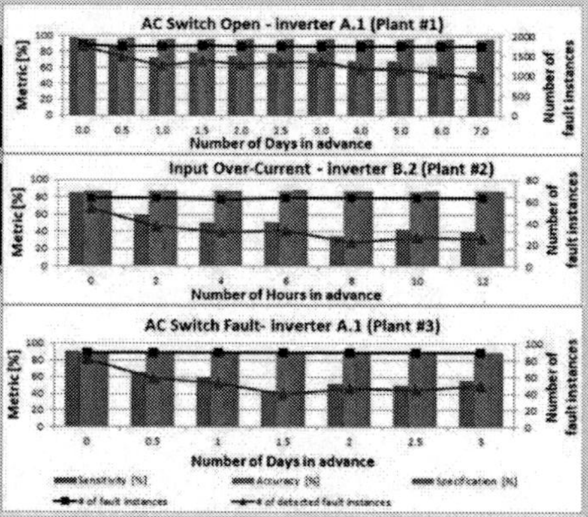

6DP.2.4 Predictive Maintenance in Photovoltaic Plants with a Big Data Approach

Alessandro Betti et al, i-EM (Livorno, ITALY), Enel Green Power (Rome, Italy)

European Commission

Non-destructive testing increasingly quantitative

Indication for PID

PID
- Homogeneous T increase
- Located at one string end
- Trend of PID-ratio increase towards the end
- < 50% of modules affected
- Incubation time
- PID-cells close the frame

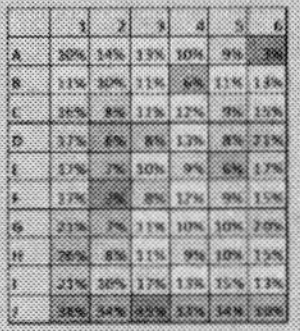

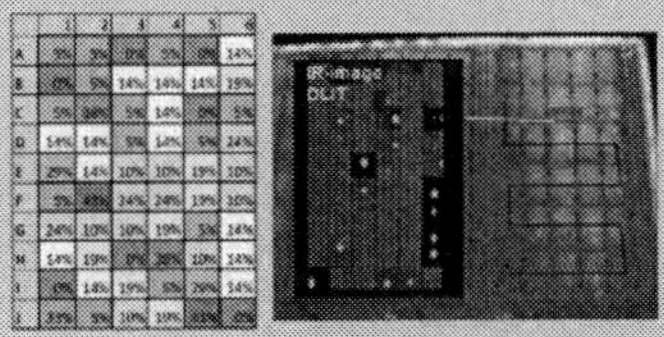

Pseudo-PID
- Homogeneous T increase
- Arbitrarily distributed
- many modules may be affected
- Close to commissioning
- Affected cells everywhere

6BO.8.2 Quantitative Study of Potential Induced Degradation of a Roof-Top PV-Installation with IR-Imaging

C. Buerhop-Lutz et al, ZAE Bayern, Erlangen, Germany
C. Zetzmann Rauschert, Pressig, Germany
C.J. Brabec University of Erlangen-Nuremberg, Germany

European Commission

Which module is an insurance case?

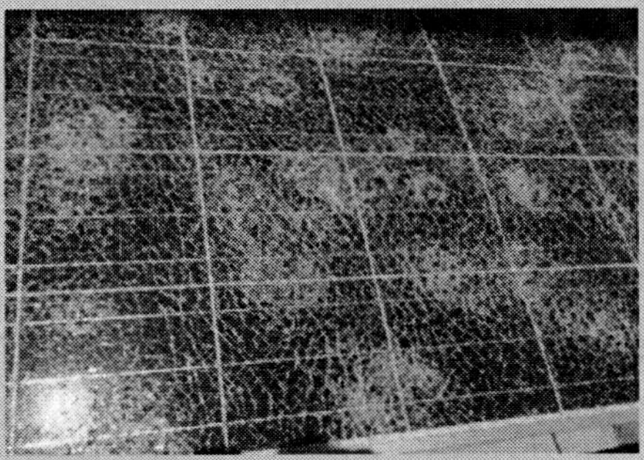

6BO.8.3 Scientific Investigation of a PV Generator After Hail

W. Mühleisen et al, CTR, Villach, Austria
M. Spielberger PVSV, Guttaring, Austria
H. Sonnleitner ENcome, Klagenfurt, Austria
Y. Voronko OFI, Vienna, Austria

European Commission

Product Environmental Footprint Category Rules (PEFCR) for PV

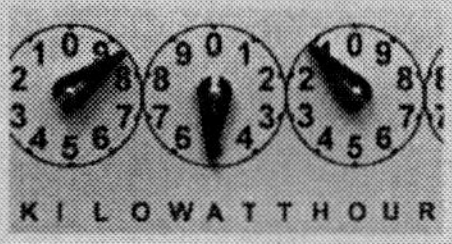

The PEFCR provides a harmonized set of rules to evaluate the environmental impacts of electricity produced from photovoltaic modules installed in the region of the EU28 and EFTA countries.

Circular Economy

Is it feasible to aggregate the complex results of a Product Environmental Footprint evaluation into a single score?

Communication Vehicles (CVs)

Performance label	Loyalty schemes	Campaigns targeting user behaviour
Performance label + QR code	Leaflets, catalogues, etc.	Marketing campaigns/ advertising
Performance improvement label	Instruction manuals	Declaration/ product passport
Barcode	Websites (producer, vendor)	On invoices
Pictogram, POS product advertisement	Websites (3rd party)	PEF external communication report
Consumer receipt – basket comparison	App based on performance	Performance tracking report.

Communication vehicles tested during the PEF Pilot Phase for the product category of PV electricity production

5DP.1.4 The Product Environmental Footprint (PEF) of Photovoltaic Modules – Lessons Learned from the Environmental Footprint Pilot Phase on the Way to a Single Market for Green Products in the European Union

A. Wade, First Solar, Mainz, Germany
P. Stolz & R. Frischknecht Treeze, Uster, Switzerland
G. Heath, NREL, Golden, United States
P. Sinha, First Solar, Tempe, United States

European Commission

Product Innovation

European Commission

BIPV / VIBV / IIPV

Solar module for automotive applications

TOYOTA

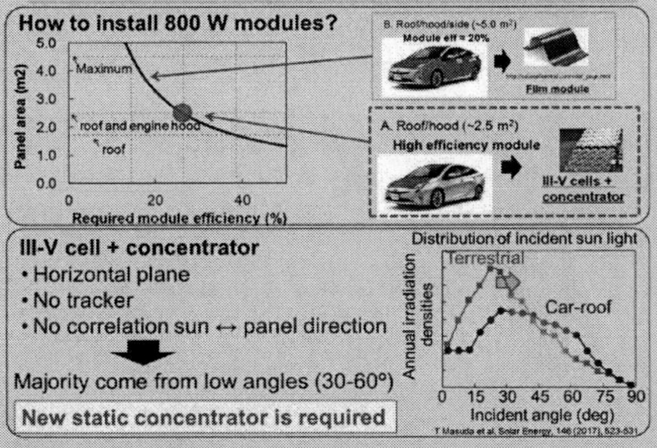

- New static low concentrator was proposed for the automotive application.

- Prototype module was fabricated by using cutting tool for lens processing.

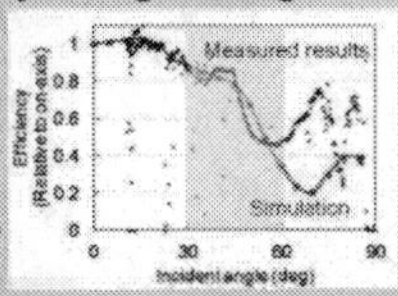

- Characteristics of lens is in good agreement with calculated results.

4DO.4.6 Solar Powered Vehicles with Static Concentrator Photovoltaics

T. Masuda et al, Toyota, Susono, Japan
K. Araki & M. Yamaguchi, TTI, Nagoya, Japan

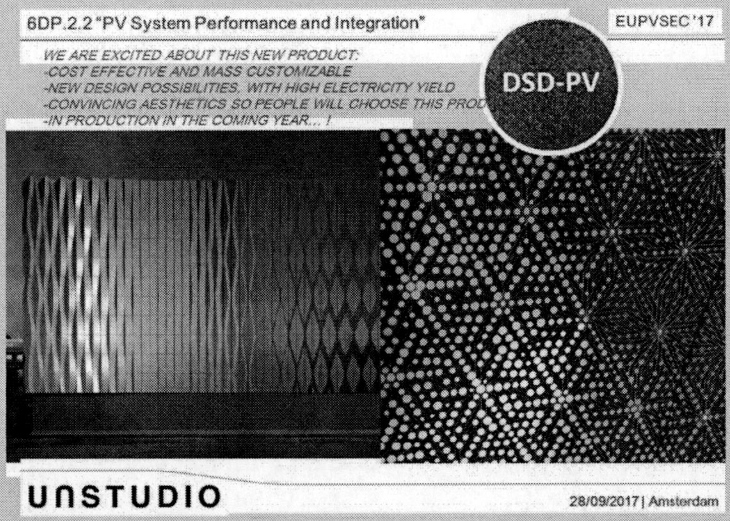

6DP.2.2 Aesthetics and Performance of PV

T. Minderhoud, UNStudio, Amsterdam, Netherlands

Terrawatt

European Commission

Different Phases of VRE Integration – Towards Sector Coupling

Phase	Description	Country Examples
1	VRE capacity is not relevant at the all-system level	Most countries, incl. Mexico, Indonesia, South Africa
2	VRE capacity becomes noticeable to the system operator	Brazil, China, India, Sweden, Texas
3	Flexibility becomes relevant with greater swings in the supply/demand balance	Italy, Germany, Portugal, Spain, UK
4	Stability becomes relevant. VRE capacity covers nearly 100% of demand at certain times	Ireland, South Australia, Denmark
5	Structural surpluses emerge; electrification of other sectors becomes relevant	
6	Bridging seasonal deficit periods and supplying non-electricity applications; seasonal storage & synthetic fuels	

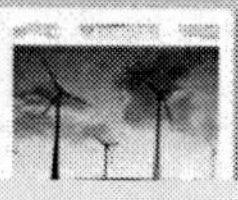

Electricity only accounts for around one fifth of total final energy demar
renewables will require multiplying their uses in buildings, indus

Acceptance
Landscapes
Clean Energy for All

EUPVSEC 2017, Opening, Paolo Frankl, IEA

European Commission

Open access proceedings

Thanks!

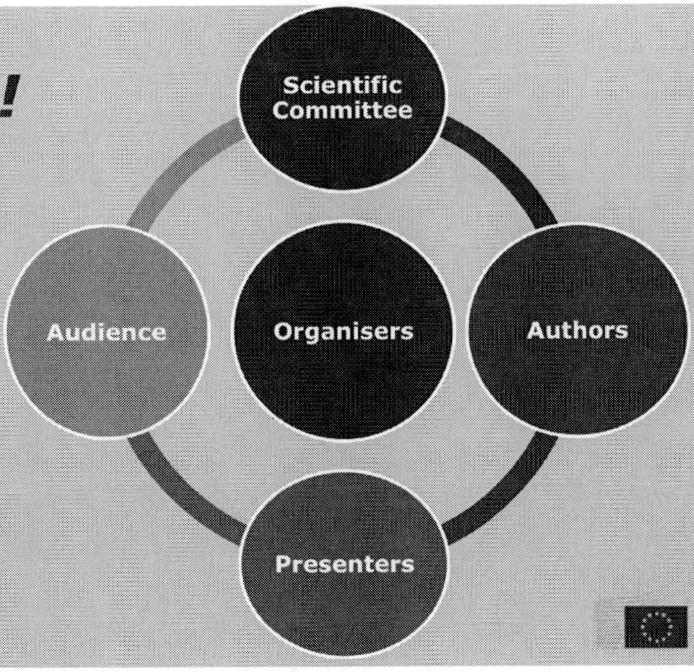

LIST OF EXHIBITORS
(in alphabetical order)

3D-Micromac	DE
Abet Technologies	US
Advanced Silicon Group	US
ASYS	DE
Bentham Instruments	UK
BERGER Lichttechnik	DE
Berlin-Brandenburg Energy Network	DE
CSEM SA	CH
Delft University of Technology	NL
Dutch Energetics Management	NL
ECN	NL
Engineered Materials Systems	US
ETA Florence Renewable Energies	IT
EUREC	BE
European Commission, DG JRC	BE
FME	NL
greateyes	DE
h.a.l.m. elektronik	DE
Heraeus Photovoltaics	DE
Holland Solar	NL
HydroPV Technologies	NL
IEA PVPS	CH
InnoLas Solutions	DE
Ionx Cleaning Facilities	NL
ISC Konstanz	DE
IZOVAC Technologies	BY
Jonas & Redmann	DE
JRT Metallization Lines	DE
Kipp & Zonen	NL
KOPEL / KYOSHIN ELECTRIC	JP
Luvata Pori Oy	FI
Newport Spectra-Physics	DE
pv magazine group	DE
pv-tools	DE
Quantum Zurich	CH
RENA Technologies	DE
Semilab	HU
SINGULUS TECHNOLOGIES	DE
Sinton Instruments	US
Sisecam Flat Glass	TR
Solar Swiss Connect	CH
SOLARC Innovative Solarprodukte	DE
SOLARUNITED	DE
Solaxess SA	CH
Solibro	DE
Solliance	NL
Stäubli Electrical Connectors	CH
Sunprojects	NL
SUPSI ISAAC	CH
SVCS	CZ
Tempress Systems	NL
TFSC-Instrument	FR
TKI Urban Energy	NL
University of Ljubljana, LPVO	SI
Valentin Software	DE
Van der Valk Solar Systems	NL
Vela Solaris	CH
VITRONIC	DE
VON ARDENNE	DE
Wiley	US
WIP Renewable Energies	DE
ZSW Baden-Württemberg	DE

We thank the EU PVSEC 2017 Sponsors

BERGER Lichttechnik

LONGi Solar

RENA

SINGULUS

MECHANICALLY STACKED DUAL-JUNCTION AND TRIPLE-JUNCTION III-V/SI-IBC CELLS WITH EFFICIENCIES OF 31.5 % AND 35.4 %

M. Rienäcker[1], M.Schnabel[2], E. Warren[2], A. Merkle[1], H. Schulte-Huxel[2], T.R. Klein[2], M.F.A.M. van Hest[2], M.A. Steiner[2], J. Geisz[2], S. Kajari-Schröder[1], R. Niepelt[1], J. Schmidt[1,3], R. Brendel[1,3] , P. Stradins[2], A. Tamboli[2] and R. Peibst[1,4]

[1] Institute for Solar Energy Research Hamelin (ISFH), Am Ohrberg 1, D-31860 Emmerthal, Germany
[2] National Renewable Energy Laboratory, 15013 Denver West Parkway, Golden CO 80401, USA
[3] Institute of Solid State Physics, Leibniz Universität Hannover, 30167 Hannover, Germany
[4] Institute of Electronic Materials and Devices (MBE), Leibniz Universität Hannover, 30167 Hannover, Germany

ABSTRACT: The theoretical efficiency limit of 29.4 % for single-junction crystalline Silicon (c-Si) solar cells is an insurmountable barrier that is being steadily approached within the last decades. Combining the Si cell with a second absorber material on top in a dual junction tandem or triple junction solar cell is an attractive option to surpass this limit significantly. We demonstrate a mechanically stacked GaInP//Si dual-junction cell with an in-house measured efficiency of 31.5 % and a GaInP/GaAs//Si triple-junction cell with a certified efficiency of 35.4±0.5 %

Keywords:

1 Introduction

Steady improvements over decades have led to exceptional success of market-dominating silicon-based photovoltaics, which recently demonstrated an outstanding record efficiency of 26.7 % for an interdigitated back contact solar cell [12].Stacking of multiple junctions has proven to be a viable route to increase the cell efficiency further beyond the theoretical limit of 29.4% for silicon single-junction solar cells[6, 2]. Increasing research interest is currently dedicated to tandem concepts based on Si bottom cells for terrestrial, non-concentrated applications [2 and Ref. therein, 5, 10 and Ref. therein]. Here we prepared a mechanically stacked GaInP//Si dual-junction cell and a GaInP/GaAs//Si triple-junction cell, achieving energy conversion efficiencies of 31.5 % and 35.4 %, respectively. With this result, we demonstrate the potential of silicon based multi-junction solar cells for high efficiencies.

2 Experimental

2.1 Fabrication of POLO-IBC bottom cell

We prepare ion-implanted and inkjet-patterned interdigitated back contacted solar cells with **POL**y-Si on **O**xide (POLO) junctions for both polarities (POLO-IBC cells) [7, 8] with an active area of 7.6 mm × 15 mm (to match the top cell area) on saw-damage etched 156 mm × 156 mm *n*-type Czochralski silicon wafers with a base resistivity of 4 Ωcm and a final thickness of 155 µm.
After growing a thin thermal silicon dioxide layer in a tube furnace, undoped amorphous silicon (*a*-Si) is deposited on both sides by using LPCVD (E2000 from Centrotherm). Hereafter, the front side of the wafer receives a blanket phosphorus implantation, followed by masked phosphorus and blanket boron implantation on the rear side. For the latter, the phosphorus locally overcompensates the boron, and an interdigitated pattern with a pitch of 952 µm is formed. For the masking of the implantations, we use a sputtered dielectric layer, which is patterned by inkjet-printed hotmelt wax (Pixdro LP50 from Meyer Burger) and subsequent wet-chemical etching.

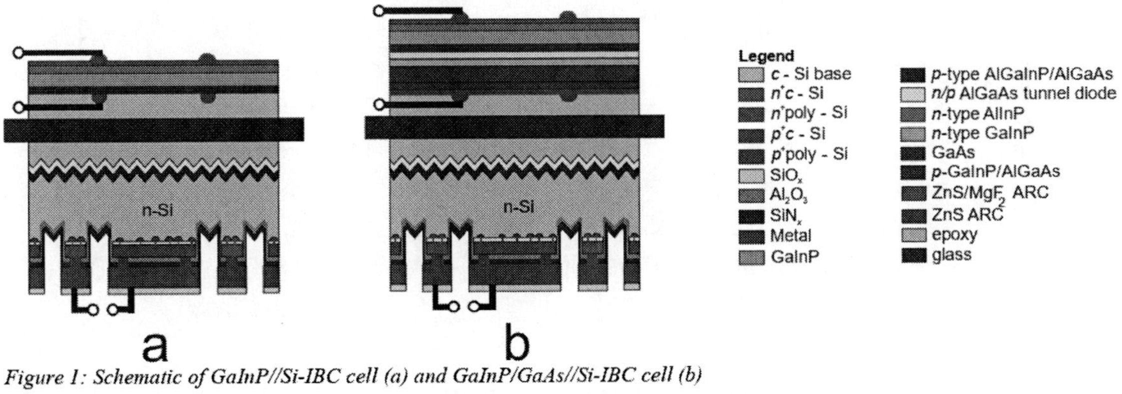

Figure 1: Schematic of GaInP//Si-IBC cell (a) and GaInP/GaAs//Si-IBC cell (b)

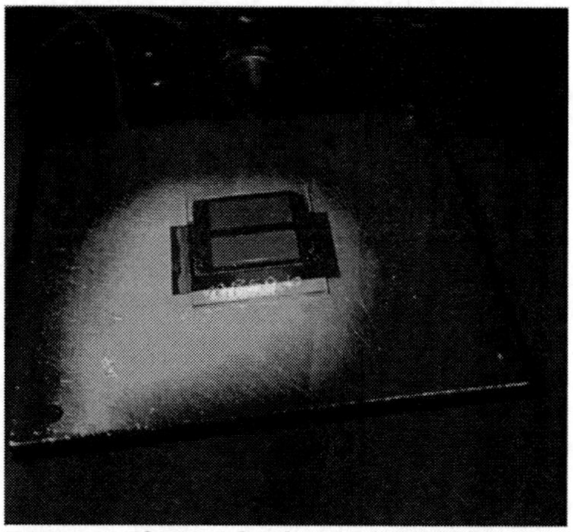

Figure 2: Photograph of GaInP/GaAs//Si-IBC cell with an efficiency of 35.4% illuminated with blue LED light

After the removal of dielectric implant masks, a high temperature treatment for the formation of the POLO junctions with an annealing duration of 60 minutes at 1050°C is performed. The junction formation annealing leads to recrystallization of the a-Si, redistribution and activation of dopants and a pinhole creation within the interfacial SiO_2 layer [11] for efficient carrier transport between c-Si and poly-Si [7]. Furthermore gettering of the bulk material with n^+ POLO is achieved [4].

During the junction formation step, a thick silicon dioxide layer is grown on top of the poly-Si by wet thermal oxidation.

After the oxidation and junction formation process, the grown silicon dioxide layer is again patterned via inkjet printing and wet-chemical etching on the rear, and at the same time removed on full area from the front side of the wafer. The remaining SiO_2 on the rear acts as an etching barrier for a subsequent texturization process, which yields a textured front side and a separation of n^+ POLO BSF and p^+ POLO emitter regions by a textured trench of a width of 100 µm. The trench isolation significantly improves the overall recombination behaviour of the cells [8] and furthermore facilitates contact separation via the RISE process [1]. After removing the SiO_2 by HF etching from the rear side, the cell precursors are passivated with an aluminum oxide/silicon nitride stack on the rear side and a silicon nitride on the front side.

An additional SiO_2 layer is deposited by PECVD on the front side in order to improve the optics for one sun application rather than tandem cells. This leads to non-ideal optical coupling between silicon and III-V cells and represents an area for future improvements.

We use a picosecond UV laser to locally ablate the dielectric layers and realize via contact openings with an opening fraction of about 4.5% with respect to the total cell area.

The solar cell precursors are metalized by vacuum evaporation of a 10 µm thick aluminum layer on the rear side in a high-throughput tool from Applied Materials and the contact separation is performed via the RISE process [1]. Finally, the POLO-IBC bottom cells, as shown in Fig. 1, are laser-scribed and cleaved from the wafer.

2.2 Fabrication of GaInP single junction cell

The GaInP cell was grown inverted, with a rear heterojunction, on a GaAs substrate with a 2° miscut towards <111> by metalorganic chemical vapor deposition (MOCVD), using the reactor, precursors, and process window described in Ref.[3]. Cell processing of the MOCVD-grown stack begins with a rear-side Au grid prepared by photolithographically patterned Au electroplating, followed by evaporation of ZnS as a rear-side antireflection coating. The cell is then bonded to glass using epoxy, and the GaAs substrate removed by etching. Finally, the front side is processed: a front grid is prepared by photolithographically-patterned Ni/Au electroplating, a photolithographically-patterned mesa etch is performed to define the cell area, and a MgF_2 / ZnS double-layer antireflection coating is evaporated.

2.3 Fabrication of GaInP/GaAs dual-junction cell

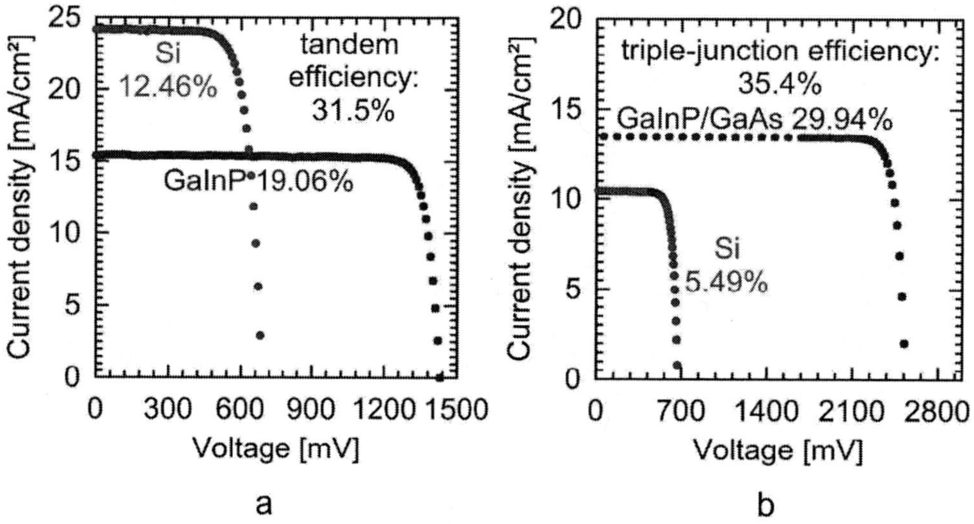

Figure 3: IV data of GaInP/Si-IBC cell (a) and GaInP/GaAs/Si-IBC cell

The process sequence for the GaInP/GaAs dual-junction cell is the same as for the single junction GaInP cell, except that both cells are grown in sequence during the MOCVD step. The GaInP cell is the same, and is grown first, followed by a $Al_{0.3}Ga_{0.7}As$/GaAs tunnel junction and then the GaAs cell, as described in ref. [3].

2.4 Stacking of the III-V cell on the Si cell

The III-V cell on glass is bonded to the front side of the Si cell using epoxy. Stacking and bonding is done under an infrared camera in order to align the top cell area with the IBC grid fingers of the bottom cell. The final four-terminal triple-junction cell is shown in Fig. 2.

2.5 *IV* measurement

For the GaInP/Si IBC cell, the quantum efficiency and specular reflectance of both cells were measured on a custom-built instrument. The *IV* measurement of the GaInP/Si IBC cell was taken on a class A adjustable solar simulator with a primary calibration reference cell used to set the intensity. The two cells were contacted separately and measured sequentially. For each cell, the the illumination was set such that each cell was at an illumination level, which is the equivalent of AM1.5G and the filtered equivalent for the bottom cell. A shadow mask was used to ensure no photocarrier generation outside the 1 cm^2 active cell area. The GaInP cell was measured first, and during measurement of the Si cell, a resistor was used to hold the GaInP at its maximum power point (mpp). This is necessary because under open circuit conditions, GaInP luminescence coupling to the Si cell is higher than at mpp, artificially inflating its performance [2]. Luminescent coupling from Si to GaInP is negligible because Si luminescence is weak and not absorbed by the GaInP cell.
The GaInP/GaAs//Si-IBC cell was measured using the a calibrated measurement with same procedure at NREL's certified device performance laboratory.

3 Results and discussion

3.1 GaInP//Si dual-junction solar cell

The resulting in-house measured *IV* data is show in Fig. 3 a. After stacking, the GaInP top solar cell has a single junction efficiency of 19.1 % (V_{OC}=1430 mV, J_{SC}= 15.4 mA/cm², *FF*=86.6 %) and the Si bottom cell yields an efficiency of 12.5 % (V_{OC}=687 mV, J_{SC}= 24.2 mA/cm², *FF*=75.1 %, GaInP cell at mpp). A cumulative efficiency of 31.5 % is obtained, which exhibits, compared to the recently published III-V/Si record efficiency of 32.5 % [2], an 1 $\%_{abs}$ lower efficiency. This is attributed to the about 1 $\%_{abs}$ lower efficiency of the GaInP top cell. The Si bottom cell has nearly the same performance, but the V_{OC} of our POLO-IBC cell is 7 mV lower, which can be mainly explained by degradation during the stacking procedure as well as lower passivation quality of the SiN_x passivation at the front side and degraded passivation quality of the trench passivation at the rear side compared to excellent passivation by amorphous Si used in Ref. [2]. Further optimization of the back-end processing and minimizing perimeter recombination will lead to an increase in V_{OC} above 700 mV.
The low *FF* of the silicon cell is tentatively attributed to non-ideal heat-sinking during the measurement.
The absence of a metal grid, the high transparency of the front side and decent reflectivity of the rear side of our POLO-IBC bottom cell is responsible for the high short circuit current of 24.2 mA/cm² (1.1 mA/cm² higher than for the Si bottom cell in Ref [2]).Since the antireflection coating applied is optimized for the non-filtered solar spectrum, further improvements are anticipated by tuning the antireflection coating on the Si cell to the relevant infrared wavelength range.
Taking possible improvements into account and using a 20 % efficient GaInP cell and a filtered efficiency for our POLO-IBC cell above 13 %, a GaInP/Si tandem efficiency above 33 % is within reach.

3.2 GaInP/GaAs//Si triple-junction solar cell

For the GaInP/GaAs//Si triple-junction cell a calibrated *IV* measurement was performed at NREL's certified device performance laboratory.
The current-matched GaInP/GaAs dual-junction achieved an efficiency of 29.94±0.42 % (V_{OC} = 2535 mV, J_{SC} = 13.43 mA/cm², *FF* = 87.9 %) and the POLO-IBC bottom cell added another 5.49±0.08 % (V_{OC} = 670 mV, J_{SC}= 10.4 mA/cm², *FF*=78.8 %, GaInP/GaAs cell at mpp) filtered efficiency to achieve an overall four-terminal, triple-junction efficiency of 35.43±0.5 %. As with the GaInP/Si cell, improvements in Si J_{SC} are anticipated with optimized antireflection coatings.
This excellent efficiency is the second-highest after and close to the current world record of 35.9 % for III-V/Si triple-junction cells [2].
On its own, the efficiency of this type of Si wafer cell with POLO junctions reaches 25 %. Thus, the addition of the GaInP/GaAs top cell yields a total efficiency improvement of about 10 $\%_{abs}$. The efficiency of the four-terminal GaInP/GaAs//Si cell is even close to the current world record of a series-connected 37.9 % for a non-Si triple-junction cell [9].

4. Conclusion

We have demonstrated lab scale dual-junction and triple-junction four-terminal III-V//Si solar cells using POLO junctions on the Si IBC cells for both polarities with efficiencies of 31.5 % and 35.4 %, respectively. The latter result is very close to the world record for III-V//Si and all-III-V triple-junction solar cells and therefore shows that silicon wafer cells, which provide the mature and cheap basis for over 90 % of today's photovoltaic (PV) devices, are well suited for tandem applications.

Acknowledgements

Funding at ISFH was provided by the German State of Lower Saxony, the German Federal Ministry for Economics and Energy (BMWi) within the research project "EASi" (FKZ0324040) and EU's FP7 within the research project "HERCULES" (608498). Funding for this work at NREL was provided by DOE through EERE contract SETP DE-EE00030299 and under Contract No. DE- AC36-08GO28308.

We are grateful to M. Young, W. Olavarria and J. Carapella for the III-V growth and processing at NREL, S. Essig for developing the stacking procedure, and to H. Kohlenberg, T. Friedrich, M. Tatarzyn, S. Schmidt, A.

Raugewitz, R. Winter, N. Wehmeier, F. Heinemeyer, J. Hensen, D. Sylla, T. Neubert for sample processing at ISFH.
We appreciate G. Glowatzki, B. Koch and J. Krügener form Institute of Electronic Materials and Devices (MBE) at Leibniz Universität Hannover for helping with LPCVD deposition and ion implantation.
The authors would like to thank M. Pickrell from Sun Chemical Ltd. for providing the hotmelt wax.

References

[1] P Engelhart, NP Harder, T Neubert, H Plagwitz, B Fischer, R Meyer, and R Brendel. Laser processing of 22% efficient back-contacted silicon solar cells. In *21st European Photovoltaic Solar Energy Conference*, pages 773–776, 2006.
[2] Stephanie Essig, Christophe Allebé, Timothy Remo, John F Geisz, Myles A Steiner, Kelsey Horowitz, Loris Barraud, J Scott Ward, Manuel Schnabel, Antoine Descoeudres, David L. Young, Michael Woodhouse, Matthieu Despeisse, Christophe Ballif, and Adele Tamboli. Raising the one-sun conversion efficiency of III–V/Si solar cells to 32.8% for two junctions and 35.9% for three junctions. *Nature Energy*, 6:17144, August 2017.
[3] John F Geisz, Myles A Steiner, I Garcia, Sarah R Kurtz, and Daniel J Friedman. Enhanced external radiative efficiency for 20.8% efficient single-junction gainp solar cells. *Applied Physics Letters*, 103(4):041118, 2013.
[4] Jan Krügener, Felix Haase, Michael Rienäcker, Rolf Brendel, H Jörg Osten, and Robby Peibst. Improvement of the srh bulk lifetime upon formation of n-tpe polo junctions for 25% efficient si solar cells. *Solar Energy Materials and Solar Cells*, 2017.
[5] Agnes Merkle Talysa R. Klein Nikhil Jain Stephanie Essig Henning Schulte-Huxel Emily Warren Maikel F.A.M. van Hest John Geisz Jan Schmidt Rolf Brendel Robby Peibst Paul Stradins Adele Tamboli Manuel Schnabel, Michael Rienäcker. Iii-v/si tandem cells utilizing interdigitated back contact si cells and varying terminal configurations. In *Proceedings of 43rd IEEE PVSC*, 2017.
[6] Armin Richter, Martin Hermle, and Stefan W. Glunz. Reassessment of the Limiting Efficiency for Crystalline Silicon Solar Cells. *IEEE Journal of Photovoltaics*, 3(4):1184–1191, oct 2013.
[7] Michael Rienäcker, Marcel Bossmeyer, Agnes Merkle, Udo Römer, Felix Haase, Jan Krügener, Rolf Brendel, and Robby Peibst. Junction resistivity of carrier-selective polysilicon on oxide junctions and its impact on solar cell performance. *IEEE Journal of Photovoltaics*, 7(1):11–18, 2017.
[8] Michael Rienäcker, Agnes Merkle, Udo Römer, Heike Kohlenberg, Jan Krügener, Rolf Brendel, and Robby Peibst. Recombination behavior of photolithography-free back junction back contact solar cells with carrier-selective polysilicon on oxide junctions for both polarities. *Energy Procedia*, 92:412–418, 2016.
[9] T. Takamoto, H. Washio, and H. Juso. Application of ingap/gaas/ingaas triple junction solar cells to space use and concentrator photovoltaic. In *2014 IEEE 40th Photovoltaic Specialist Conference (PVSC)*, pages 0001–0005, June 2014.
[10] Jérémie Werner, Bjoern Niesen, and Christophe Ballif. Perovskite/Silicon Tandem Solar Cells: Marriage of Convenience or True Love Story? - An Overview. *Advanced Materials Interfaces*, page 1700731, September 2017.
[11] T Wietler, D Tetzlaff, J Krügener, M Rienäcker, F Haase, Y Larionova, R Brendel, and R Peibst. Pinhole density and contact resistivity of carrier selective junctions with polycrystalline silicon on oxide. *Applied Physics Letters*, 110(25):253902, 2017.
[12] Kunta Yoshikawa, Hayato Kawasaki, Wataru Yoshida, Toru Irie, Katsunori Konishi, Kunihiro Nakano, Toshihiko Uto, Daisuke Adachi, Masanori Kanematsu, Hisashi Uzu, and Kenji Yamamoto. Silicon heterojunction solar cell with interdigitated back contacts for a photoconversion efficiency over 26%. *Nature Energy*, 2:17032, mar 2017.

MULTI-SEGMENT PHOTOVOLTAIC LASER POWER CONVERTERS AND THEIR ELECTRICAL LOSSES

Rok Kimovec[1], Henning Helmers[2], Andreas W. Bett[2], Marko Topič[1]
[1]University of Ljubljana, Faculty of Electrical Engineering, Tržaška cesta 25, 1000 Ljubljana, Slovenia
Tel: +386 1 4768 276, Fax: +386 1 4768 276, E-mail: rok.kimovec@fe.uni-lj.si
[2] Fraunhofer Institute for Solar Energy Systems ISE, Heidenhofstraße 2, 79110 Freiburg, Germany

ABSTRACT: The electrical power losses of small area six-segment monolithic interconnected GaAs laser power converter are analyzed based on a variety of measurements and an experimentally validated numerical 2.5-dimensional distributed electrical model. Input parameters for the developed model were extracted from a wide-range of independent measurements performed on the manufactured devices. The complex geometry of the device is implemented in the model with digital images of photolithographic masks. Validity of the built model is confirmed by a good agreement between measured and simulated spatial electroluminescence response and a good agreement between measured and simulated dark and illuminated *J-V* curves is observed. It is shown that distributed ohmic losses limit the efficiency of the studied device above the optimal ($G_{optimal}$=13.2 W/cm^2) monochromatic irradiance (λ_0=809 nm). At high irradiance (G_{high}=83.1 W/cm^2), the photovoltaic conversion efficiency is decreased by 17%$_{abs.}$, due to combined ohmic losses, where more than 88%$_{rel.}$ of the losses are caused by Joule heating in the lateral conduction layer (which is specific to this kind of multi-segment device). Below the optimal irradiance, minority carrier recombination at the perimeter of the *pn*-junction is identified as the limiting factor. At G_{low}=1.8 W/cm^2 perimeter recombination causes 5.5%$_{abs.}$ decrease of the conversion efficiency. Additionally, photo-induced conductivity in the semi-insulating GaAs substrate leads to a leakage current between adjacent segments which reduces the efficiency for all studied irradiances equally by 1.2%$_{abs.}$

Keywords: photovoltaics, monolithic interconnection, gallium arsenide based cells, III-V semiconductors, simulation, distributed model, electroluminescence, electrical losses, power-over-fiber, power-by-light

1 INTRODUCTION

Laser Power Converters (LPC) [1–3] are photovoltaic (PV) devices converting monochromatic light into electrical energy typically used in a "power-over-fiber" (alternatively "power-by-light") electrical power supplies [4–6]. Utilizing a nonconductive energy transmission medium, (usually a multi-mode optical fiber) such power supply systems are inherently immune to electromagnetic radiation and provide a galvanic isolation between the power source and powered electronic circuits or devices.

The output voltage of a single-junction single-segment LPC made of GaAs in the maximum power point is around 1 V [7]. Since this voltage is often too low to power electronic devices directly, multi-segment LPCs are used to increase the output voltage of the PV device (also known as monolithic interconnected modules or MIMs) [2, 8]. In a multi-segment LPC, one vertical *pn*-junction is grown on a semi-insulating substrate and a wet chemical treatment is then applied to electrically separate individual segments directly on the chip. Thus, a galvanic and spatial separation of the *pn*-junctions is achieved. With the means of series interconnections of adjacent segments, a mini-module with an output voltage around N_{seg}*1V is realized, where N_{seg} is the number of segments [9].

In this work we present a detailed electrical power loss analysis of a six-segment monolithically interconnected GaAs laser power converter based on an experimentally validated electrical distributed model. The majority of the input parameters for the model were extracted from independent measurements performed on the manufactured device. Furthermore, digital images of photolithographic masks were used to maintain exact geometrical features of the multi-segment LPC. Measured and simulated illuminated and dark *J-V* curves are in good agreement, showing us that developed model is able to reproduce the measured data. To ensure the validity of the model, we used completely independent characteristic of the device presented by measured and simulated electroluminescence images. With the validated model we investigate and quantify the contribution of various electrical loss mechanisms (i.e. Joule heating, recombination and shunt losses) to overall electrical power losses in the manufactured device.

2 SPECIMEN DESCRIPTION

In this work we study and present electrical loss mechanisms in a small area radial pie-shaped six-segment monolithic interconnected GaAs laser power converter manufactured at Fraunhofer ISE. A picture of the device is shown in Figure 1. The diameter of the circular area measured between the busbars is 2.08 mm, resulting in an inner device area of 3.39 mm^2. In the studied device, 11.1% of the area of an individual segment (0.63 mm^2) is covered with a cathode busbar and additionally 7.2% is covered by metal front grid. This metal coverage results in a segment active area of approximately 0.52 mm^2 exposed to the impinging light.

FIGURE 1: A macro photograph of the six-segment GaAs MIM LPC investigated in this work. Radius of inner circular area between the busbars is R=1.04 mm.

The epitaxial structure, shown in Figure 2, was grown by metalorganic vapor phase epitaxy (MOVPE) on a 4" semi-insulating (S.I.) GaAs substrate and is based on a GaAs *np*-junction (3650 nm thick absorber). On top of the *np*-junction is a 400 nm thick *n*-type GaInP window layer (FSF) [10] which is transparent for the monochromatic light and reinforces lateral current flow between grid fingers. Below the *np*-junction is a *p*-type GaInP back surface field (BSF). A highly doped 5 μm thick *n*-type GaAs lateral conduction layer (LCL) is used for lateral interconnection of the individual PV cell segments. A tunnel diode provides ohmic connection between the *p*-type BSF and the *n*-type LCL. A highly doped *n*-type GaAs cap layer above the window provides a low-ohmic front side metal-semiconductor contact. After the MOVPE growth of the structure devices on the wafer are fabricated with standard micro-structuring photolithographic processes, wet etching, polyimide insulation, metal evaporation and a two layer anti-reflection (AR) coating [11]. Further details on the device structure and processing can be found elsewhere [2, 12].

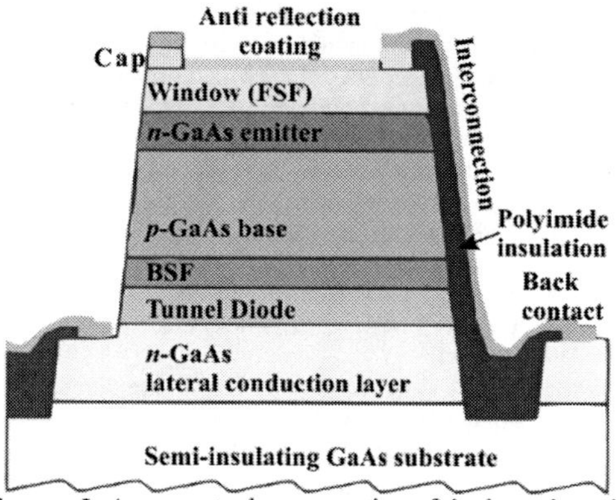

FIGURE 2: A conceptual cross-section of the investigated device. (not to scale)

3 NUMERICAL DISTRIBUTED MODEL OF SIX-SEGMENT GaAs MIM LPC

A complex geometry and epitaxial structure result in a complex current flow in the investigated MIM device. Therefore, a distributed numerical model based on a real device structure is required to accurately simulate electrical losses. In this section, we provide basic description of the investigated six-segment GaAs MIM LPC model implemented in two and a half dimensional (2.5D) PV simulation tool PVMOS [13]. It should be pointed out that most of the electrical input parameters for the model were extracted from a variety of measurements performed on the manufactured device, as described in the next section, and remaining few were obtained from literature [12]. Such modeling approach enables us to implement a stack of two dimensional simulation electrodes, where each electrode represents a specific device layer, described by realistic electrical parameters. Planar electrodes are vertically interconnected with each other with so called elementary units, to form a geometrically accurate three dimensional electrical representation of the device.

The used modeling approach mimics the manufacturing procedures, i.e. the MOVPE growth, micro-structuring, polyimide insulation, and metal evaporation as described above. First we defined a stack of four resistive simulation electrodes, where each simulation electrode presents the sheet resistances of the modeled layer (i.e. S.I.-substrate, LCL + base, emitter + window and metallization). Between the nodes of the electrodes, vertical connections were established with elementary units. The role of the elementary units depends on its position in the device; standard two diode electrical PV model describe the bulk photoactive region, whereas simple resistive elementary units describe the rest of the device. The resistance of each resistive elementary unit is set to the appropriate value, based on its position in the device.

Afterwards the definition of geometrical features of the device is set. In the areas, defined by the digital images of photolithographic masks, the properties of simulation electrodes and elementary units were changed to represent the role of the explicit area in the device. For example, mesa etch sheet resistances in mesa etch regions were set to high values and all elementary units in these regions were set to high resistances to suppress the simulated current flow in those regions. Further, to model the photo-induced conductivity in the S.I.-substrate [12, 14], its sheet resistance was lowered at the isolation trenches where the substrate is exposed to incident light. Increased minority carriers recombination were specifically implemented at the perimeter of the *pn*-junctions by increasing the dark saturation current of the second diode J_{02} in the two diode PV model of the photo active area [12].

In similar manner, all specific device features were taken into account and implemented accurately. Further details and explanations can be found in Ref. [12].

4 PARAMETRIZATION OF THE SPECIMEN

A variety of measurements was performed on the investigated specimen to evaluate the device performance and extract electrical parameters of individual device features.

First a dark current-voltage (*I-V*) curve and a series of illuminated *I-V* curves were measured at constant temperature T=25 °C. The dark *I-V* curve, measured with a precision source-measurement unit, was divided by segment area to obtain the dark current density-voltage (*J-V*) curve. From this curve, in reverse polarity bias, we extracted the area-weighted shunt resistance. This value presents a *pn*-junction shunting and was accordingly modeled as a parallel conductance in two-diode electrical PV model of the active area. Additionally, the combined (i.e. depletion region and perimeter recombination) dark saturation current density for the second diode in two diode PV model (J_{02}) was extracted from the dark *J-V* curve. Bulk component J_{02B} (depletion region) of the combined J_{02} was obtained from literature [15] and the perimeter J_{02P} component was obtained from following equation:

$$J_{02} = J_{02b} + \frac{P}{A} \cdot J_{02p},$$

where perimeter to area ratio (P/A) of the investigated device is 57 cm^{-1}.

Illuminated *I-V* curves were measured under transient monochromatic (λ_0=809 nm) illumination at various irradiances G, utilizing a four probe measurement technique [16]. Again, measured *I-V* curves were divided by segment area to obtain *J-V* curves, from which current density-open circuit voltage (J_{SC}-V_{OC}) pairs were extracted. These pairs were further used to obtain saturation current density for the first diode (J_{01}) in two diode PV model (neutral region recombinations). Ideality factors (n), for the two diode electrical PV model, were preset to 1 and 2 for J_{01} and J_{02B}, J_{02P} respectively.

Furthermore, sheet resistances of individual epitaxial layers were measured with a transmission line method (TLM) [17] on the same wafer as the manufactured device. The same method was used to measure the specific metal-semiconductor contact resistance at the cap layer (front contact) as well as at the lateral conduction layer (back contact). Also the sheet resistance of metallization was measured on the test structure, manufactured on the same wafer. The area weighted resistance for vertical current flow through the structure was calculated from nominal doping levels and layers thicknesses and literature values for mobility of majority carriers [18].

Electrical parameters of the semi-insulating substrate were obtained independently. The sheet resistance of the substrate in the dark was calculated from nominal thickness and literature resistivity [19]. Additionally, illuminated sheet resistance of the substrate was obtained iteratively, for each irradiance separately, by fitting the modeled slope of the *J-V* curve at J_{SC} to the measured slope of the *J-V* curve again at J_{SC}.

With described procedures a full set of electrical parameters, used as input parameters for the model, was obtained. For quantitative values and details see [12].

4 VALIDITY OF THE MODEL

The complexity of the device results in a complex numerical model. Comparison of the measured and simulated dark and illuminated *J-V* curves as seen in Figure 3 were used to confirm that the established model is able to reproduce the device behavior. An illuminated *J-V* curve presented in this plot was obtained under monochromatic (λ_0=809 nm) irradiance G=4.4 W/cm^2. At this irradiance very good agreement between simulation and measurement is observed. For higher irradiance slight underestimation of ohmic losses in our model is presented [12], which is also observed in the dark *J–V* curve at high voltages (>7 V); voltage region where the device is limited by series resistance losses.

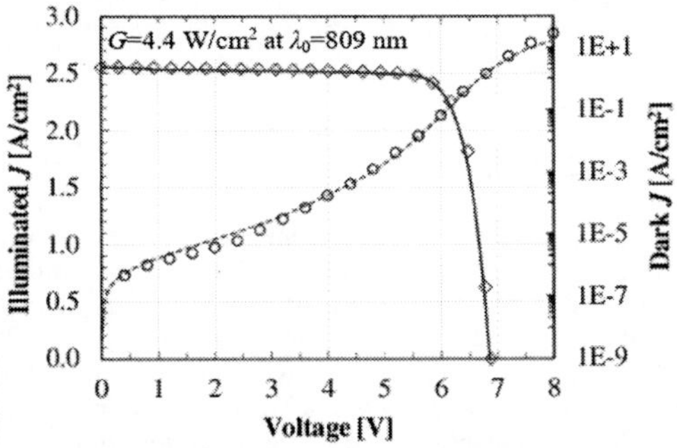

FIGURE 3: Simulated (symbols) and measured (lines) illuminated and dark *J-V* curves of the six-segment MIM GaAs LPC.

The good agreement between measured and simulated *J-V* curves is not sufficient to confirm the validity of the model, since various sets of input parameters of such complex model, could result in similar simulated *J-V* characteristics. Also, some of the input parameters were extracted from the same measured *J-V* curves, and therefore good agreement between the measurements and simulations is anticipated. To really validate the model, good agreement between simulation and measurement for an independent characteristic of the device, which has not been used for the parametrization, must be established. For that purpose, we used measured and simulated electroluminescence (EL) images. Very good agreement between the measured and simulated normalized EL, as plotted in Figure 4, ensure us that our model well represents the device in detail, is able to predict its behavior, and consequently can be used to study different loss mechanisms in the specimen.

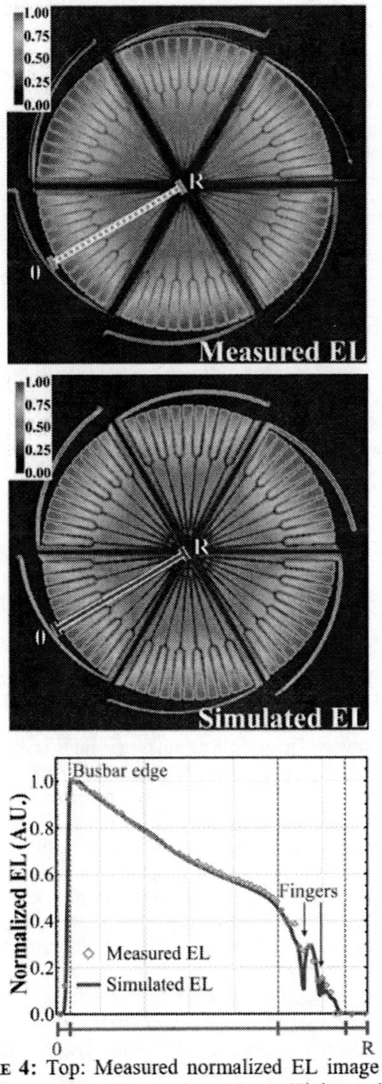

FIGURE 4: Top: Measured normalized EL image of the six-segment MIM LPC at I_{bias}=10mA. Higher values on the color scale present higher EL emission. Middle: Simulated normalized EL image of the six-segment MIM LPC at I_{bias}=10mA. Bottom: Line scans from the edge towards center as marked with the lines in the EL images. Further explanations are given in the text.

The measured EL image was obtained with a scientific CCD camera, fitted with a microscope objective, at forward current bias I_{bias}=10 mA (V_{bias}≈6.8 V) and at controlled temperature T=25 °C [20]. Dark background was subtracted from the raw EL image and then the image was normalized to the maximum value. The simulated EL image was obtained from the simulated spatial map of the junction voltage V_j (at I_{bias}=10 mA, V_{bias}≈6.8 V, T=25 °C) as [21]:

$$EL = const. \cdot \exp\left(\frac{q \cdot V_j}{k \cdot n_{EL} \cdot T} - 1\right),$$

where kT/q is the thermal voltage, and n_{EL}=2 was obtained from the measured $EL(V_{bias})$ slope.

Investigation of the spatial EL data and a representative line scan (bottom of Figure 4) from the edge (0) towards the center (R) reveal operating characteristics of the studied device. At the origin of the line scan, EL emission from the device is completely blocked by the busbar metallization. Further towards the center of the specimen, the negative slope of the EL signal originates from the voltage drop related to the lateral flow of the injected current through semiconductor layers. Primarily, the EL emission drop is caused by the current flow through LCL; thus, the drop increases with increasing distance from the rear contact. In this region, two slopes of the EL signal drop can be observed. The transition of the slopes occurs where two grid fingers merge into one; indicating that the drop of the junction voltage is not caused only by the LCL, but also by the lateral current flow in the finger grid and window layer. At the edges of the *pn*-junctions the EL emission severely drops because of the increased recombination at the perimeter. In the line scan this is observed as the pronounced drop of the EL signal at the center, while distinct dips in the EL emission relate to the metal fingers.

5 RESULTS AND DISCUSION

The major electrical losses in MIM LPC are caused by three distinct loss mechanisms.

- Inherently long distances for the lateral current flow through the structure with a finite conductance results in a severe ohmic losses related Joule heating. This effect is severely pronounced in the lateral conduction layer, where no metallization is present to provide very low ohmic lateral current path and all photo generated current must flow through this semiconductor layer.
- Small area and segmentations of the specimen results in a large perimeter to area ratio. Due to imperfections in the crystal structure at the perimeter, probability for minority carrier recombination is elevated in this region. Therefore, perimeter recombination is the major recombination loss mechanism in small area MIMs.
- Unique shunting mechanism in studied MIMs is caused by the leakage current flow through semi-insulating substrate between adjacent segments. This effect is especially influential for the device performance at very low irradiance [22] or when the substrate is illuminated by the impinged light [14]; as is the case for the studied device.

The influence of a distinct loss mechanism on the conversion efficiency of the studied device, illuminated with homogeneous monochromatic (λ_0=809 nm) light, is shown in Figure 5. The individual efficiency curves were obtained by individually excluding the distinct loss mechanisms from the model; e.g. for the curve with excluded Joule heating, all resistive components were set to zero. In the following paragraphs, each loss mechanism and its effect on device performance is described and quantified. For more detailed explanations of the presented effects refer to [12].

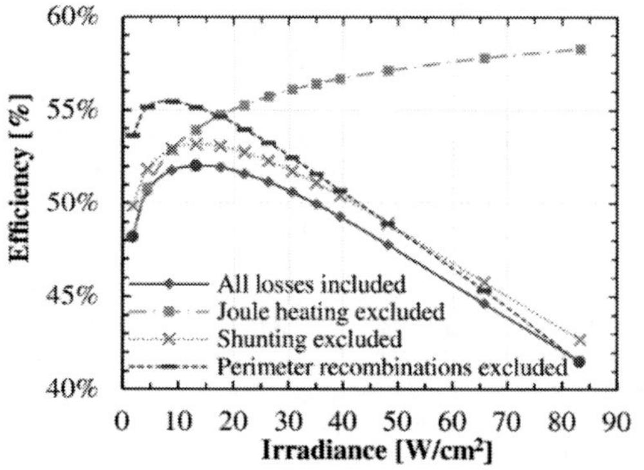

Figure 5: The influence of distinct electrical loss mechanism on the monochromatic conversion efficiency (λ_0=809 nm) versus irradiance for the studied six-segment MIM LPC. Red circles on the curve with all losses included mark irradiances for further analysis. Namely: G_{low}=1.8 W/cm^2, $G_{optimal}$=13.2 W/cm^2, G_{high}=83.1 W/cm^2. Modified from [12].

5.1 Joule heating – distributed ohmic losses

Joule heating losses in the studied device increase significantly with increasing irradiance. They are the major cause for impaired device performance at high irradiance. The conversion efficiency is deteriorated by Joule heating by 2%$_{abs.}$ at $G_{optimal}$, but this drop elevates to almost 17%$_{abs.}$ at G_{high}.

The most of the Joule heating is produced by current flow in the lateral conduction layer below the PV segments. This effect accounts for 83%$_{rel.}$ of the overall Joule heating at G_{low} and is further elevated to 88%$_{rel.}$ at G_{high}. Reduction of the LCL sheet resistance to 50% of the current value, would increase conversion efficiency at G_{high} by 6.5%$_{abs.}$

5.2 Recombination losses

Minority carrier recombination losses in the studied device are most severe at low irradiance, as seen in Figure 5. The multi-segment design and small area of the studied device result in a high perimeter to area ratio (P/A=57 cm^{-1}). In combination with GaAs [23, 24] this results in a domination of the perimeter recombination to overall recombination losses for up to very high irradiances. The reason for that is an increased probability for the non-radiative minority carrier recombination at the perimeter, due to either the adsorption of impurities at the perimeter or dangling bonds at the edge of the crystal lattice [24, 25].

To study the effect of perimeter recombination on the efficiency of the device we excluded it from the model by setting J_{02P} to the value of J_{02B}. Conversion efficiency is reduced by 5.5%$_{abs.}$ at G_{low} and by 3.5%$_{abs.}$ at $G_{optimal}$ due to increased recombination at perimeter. Additionally, the peak efficiency of the device without increased perimeter recombination shifts to a lower irradiance (8.8 W/cm^2). At G_{high} perimeter recombinations become saturated and recombinations in the neutral region determines the recombination losses [12].

Noteworthy, the effect of perimeter recombination can also be clearly observed in the EL images as seen in Figure 4. As discussed above, the pronounced drop in the EL signal is observed all around the perimeter of the *pn*-junctions. Such behavior is expected in small area GaAs PV cells, due to increased non-radiative recombinations in these regions. With proper passivation of the edges of the *pn*-junctions, this loss effect can be mitigated [26]. Therefore, analysis of the spatial EL images can be used to quickly evaluate the performance of the passivation of the GaAs small area cell's edges.

5.3 Shunting losses

The effect of the junction shunting (as present in conventional PV cells) in the studied MIM LPC, is overshadowed by the shunting between adjacent LPC PV segments due to a leakage current through the substrate. We have shown recently, that light impinging on the semi-insulating GaAs substrate at the isolation trenches, causes a significant reduction of the substrate resistivity [14]. This phenomenon leads to a significant leakage current between adjacent segments in the studied device. Due to a reciprocal relationship between the irradiance and the substrate resistivity the conversion efficiency due to this unique phenomenon is impaired by $1.2\%_{abs.}$ independently of irradiance.

6 SUMMARY AND CONCLUSION

In this work we studied the influence of various electrical loss mechanisms (e.g. Joule heating, perimeter minority carrier recombination and shunting through the semi insulating-substrate) on the performance of the small area six-segment monolithically interconnected (MIM) GaAs laser power converter (LPC) based on a distributed 2.5 dimensional model. We paid special attention to the parametrization of the manufactured device, from which we obtained a full set of realistic electrical parameters for the model. Additionally, the complex geometry of the device was preserved in the built model by digital images of the photolithographic masks used in the manufacturing of the device.

Using the model good agreement between measured and simulated illuminated and dark *J-V* curves was achieved. Moreover, the model was independently verified with spatial maps of measured and simulated electroluminescence emission which also showed very good agreement. Thus, the model accurately represents the manufactured device and can be used for analysis of individual electrical loss mechanisms.

Based on the validated model, we studied the influence of the individual electrical loss mechanism on the conversion efficiency of the device, illuminated with various monochromatic (λ_0=809 nm) irradiances. At low irradiance (G_{low}=1.8 W/cm^2), below the peak conversion efficiency, the performance of the device is limited by increased minority carrier recombination at the edges of the *pn*-junction. Here at G_{low}, perimeter recombination accounts for $5.5\%_{abs.}$ decrease in the conversion efficiency. For irradiances above the peak conversion efficiency, Joule heating dominates the losses. At G_{high}=83.1 W/cm^2, conversion efficiency is decreased by $16.8\%_{abs.}$ due to combined ohmic losses. The dominant contribution to Joule heating losses stems from the lateral conduction layer, which accounts for $88\%_{rel.}$ of the overall ohmic losses at G_{high}. Additional $1.2\%_{abs.}$ decrease of conversion efficiency was observed for all irradiances, due to photo-induced leakage current through the semi-insulating GaAs substrate. This effect was identified as a major shunting mechanism in the studied device and completely diminishes the effect of conventional junction shunting.

7 ACKNOWLEDGEMENTS

The authors thank Kasimir Reichmuth for fruitful discussions. R.K. thanks Bart Pieters, author of the PVMOS tool, for the support regarding the simulation tool used in this work. Kasimir Reichmuth and Alexander Wekkeli provided esthetically pleasing photograph of the specimen. The financial support from the Slovenian Research Agency is gratefully acknowledged (program P2-0197 and PhD funding for R.K.).

8 REFERENCES

[1] V. Andreev, V. Khvostikov, V. Kalinovsky, V. Lantratov, V. Grilikhes, V. Rumyantsev, M. Shvarts, V. Fokanov, A. Pavlov, Proceedings of 3rd World Conference on Photovoltaic Energy Conversion (2003) 761–764.

[2] J. Schubert, E. Oliva, F. Dimroth, W. Guter, R. Loeckenhoff, A. W. Bett, IEEE Transactions on Electron Devices 56 (2009) 170–175.

[3] M. York, S. Fafard, Journal of Physics D: Applied Physics 50 (2017) 173003.

[4] R. Pena, C. Algora, I. R. Matías, M. López-Amo, Applied optics 38 (1999) 2463–2466.

[5] G. Bottger, M. Dreschmann, C. Klamouris, M. Hubner, M. Roger, A. W. Bett, T. Kueng, J. Becker, W. Freude, J. Leuthold, IEEE Photonics Technology Letters 20 (2008) 39–41.

[6] K. Worms, C. Klamouris, F. Wegh, L. Meder, D. Volkmer, S. P. Philipps, S. K. Reichmuth, H. Helmers, A. Kunadt, J. Vourvoulakis, A. W. Bett, C. Koos, W. Freude, J. Leuthold, W. Stork, Wind Energy 20 (2017) 345–360.

[7] T. Shan, X. Qi, Infrared Physics & Technology 71 (2015) 144–150.

[8] R. Pena, C. Algora, I. Anton, Proceedings of 3rd IEEE World Conference On Photovoltaic Energy Conversion, Osaka, Japan, Vol.1 (2003) 228–231.

[9] H. Helmers, L. Wagner, C. E. Garza, K. S. Reichmuth, E. Oliva, S. P. Philipps, D. Lackner, A. W. Bett, Proceedings SENSOR 2015, Nürnberg, Germany (2015) 519–524.

[10] E. Oliva, F. Dimroth, A. W. Bett, Prog. Photovolt: Res. Appl. 16 (2008) 289–295.

[11] R. Kimovec, M. Topic, Facta universitatis – series: Electronics and Energetics 30 (2017) 93–106.

[12] R. Kimovec, H. Helmers, A. W. Bett, M. Topič, Submitted for publication (2017).

[13] B. E. Pieters, Proceedings of IEEE 40th Photovoltaic Specialist Conference (PVSC) Denver, CO, USA (2014) 1370–1375.

[14] R. Kimovec, H. Helmers, A. W. Bett, M. Topič, Submitted for publication (2017).

[15] M. Steiner, S. P. Philipps, M. Hermle, A. W. Bett, F. Dimroth, Prog. Photovolt: Res. Appl. 19 (2011) 73–83.

[16] S. K. Reichmuth, H. Helmers, C. E. Garza, D. Vahle, M. de Boer, L. Stevens, M. Mundus, A. W. Bett, G. Siefer, Proceedings of 32nd European Photovoltaic Solar Energy Conference (EU PVSEC), Munich, Germany (2016) 5–10.

[17] D. K. Schroder, Semiconductor Material and Device Characterization, Wiley-IEEE Press, Piscataway, NJ (2006) 1–59.

[18] M. Sotoodeh, A. H. Khalid, A. A. Rezazadeh, Journal of Applied Physics 87 (2000) 2890–2900.
[19] D. C. Look, in Semiconductors and Semimetals, Elsevier (1983) 75–170.
[20] R. Kimovec, H. Helmers, A. W. Bett, M. Topič, Informacije MIDEM – Journal of Microelectronics, Electronic Components and Materials 46 (2016) 142–148.
[21] U. Rau, Phys. Rev. B 76 (2007) 085303.
[22] Y. Wang, N. Chen, X. Zhang, Y. Bai, Y. Wang, T. Huang, H. Zhang, H. Shi, Science China Technological Sciences 53 (2010) 1240.
[23] T. B. Stellwag, P. E. Dodd, M. S. Carpenter, M. S. Lundstrom, R. F. Pierret, M. R. Melloch, E. Yablonovitch, T. J. Gmitter, Proceedings of the 21st IEEE Photovoltaic Specialists Conference, Kissimmee, FL, USA, Vol 1. (1990) 442–447.
[24] P. Espinet-González, I. Rey-Stolle, M. Ochoa, C. Algora, I. García, E. Barrigón, P rog. Photovolt: Res. Appl. 23 (2015) 874–882.
[25] A. Belghachi, S. Khelifi, Solar Energy Materials and Solar Cells 90 (2006) 1–14.
[26] M. T. Sheldon, C. N. Eisler, H. A. Atwater, Advanced Energy Materials 2 (2012) 339–344.

UNCERTAINTY PROPAGATION OF SPECTRAL MATCHING RATIOS MEASURED USING A CALIBRATED SPECTRORADIOMETER

Diego Pavanello[1]*, Roberto Galleano[1], Robert Kenny[1] and Harald Müllejans[1]

[1]European Commission, DG Joint Research Centre, Directorate C – Energy, Transport and Climate
Via E. Fermi 2749, 21027 Ispra (VA), Italy
*phone: +39 0332 786449; e-mail: diego.pavanello@ec.europa.eu

ABSTRACT: The international standard IEC62670-3 "Photovoltaic Concentrators (CPV) Performance Testing – Part 3 - Performance Measurements and Power Rating" sets the guidelines for power measurement of a CPV device, both under indoor and outdoor conditions. When measuring under outdoor conditions, the acquired data have to be filtered "a posteriori", in order to select only those points measured with ambient conditions close to the Concentrator Standard Operating Conditions (CSOC).

The most stringent requirement to be met is related to the three Spectral Matching Ratios (SMR), which have all to be within the limit of 1.00 ± 0.03. SMR are usually determined by the ratio of the currents of component reference cells set to monitor the outdoor spectral ratio conditions during the CPV device power measurements. Alternatively the SMR can be calculated from measured spectral irradiance if the spectral responsivities of the junctions are known. Experience demonstrates that obtaining real world data meeting these strict conditions is very difficult in practise. However, increasing the acceptable range would make the entire filtering process less appropriate from a physical point of view.

Given the importance of correctly measuring the SMR, an estimate of their associated measurement uncertainties is needed to allow a proper assessment of the validity of the 3% limit. In this study a Monte Carlo simulation has been used, to allow a preliminary estimation of the propagation of uncertainties for SMR measured with a calibrated spectroradiometer.

The method consists of applying both random and wavelength correlated errors to the measured spectra and to the spectral responsivities of the three CPV cell junctions, accordingly to the measurement uncertainties of the European Solar Test Installation (ESTI). The experimental data used in this study have been acquired during clear sky conditions in May 2016, at ESTI's facilities in Ispra, northern Italy (45°49'N 8°37'E). Several considerations are then inferred in the case of SMR measured with component reference cells.

KEYWORDS: concentrator photovoltaics, measurement uncertainty, standardization, Monte Carlo simulation, component reference cells

1. Introduction

Concentrator photovoltaics (CPV) occupies a small niche of the global photovoltaic (PV) market. However, several power plants with capacity of more than 30 MWp have been realized, mainly in China and in the US, with a total installed capacity of about 360 MWp already grid-connected [1]. Even though manufacturing capacity has recently decreased worldwide, CPV technology remains very promising: the highest efficiency values for concentrator multi-junction cells reached 46.0 ± 2.2% at 508 suns [2] and the scientific outlook is robust. However, in order to be competitive with c-Si technology CPV has to increase system efficiency to at least 40%, and concentration ratios up to 1000X [3]. Given that the price depends on the rated power, as is the case for flat-plate modules, reliable measurements of maximum power (and consequently of the efficiency) are crucial. In 2017 the international standard IEC62670-3 "Photovoltaic concentrators (CPV) – Performance testing – Part3: Performance measurements and power rating" [4] was published; it specifies different methods to measure the electrical characteristics of CPV devices both indoor and outdoor. In order to explore the standard's applicability, several measurements on a triple-junction CPV device have been performed at the European Solar Test Installation (ESTI) of the European Commission's Joint Research Centre in Ispra (45°49'N 8°37'E). The approach for outdoor measurements is to acquire performance data of the CPV device under clear sky conditions, together with the relevant meteorological quantities at ground level and the direct normal spectral irradiance with simultaneous acquisitions throughout the day. Subsequently a set of filters is applied to the acquired data in order to reduce the data set only to the points near CSOC (Concentrator Standard Operating Conditions) as defined in [5]. The most stringent requirement is related to the spectral match between the reference spectrum AM1.5D and the measured spectrum, which is determined by calculating the three (in the case of a triple junction device) spectral matching ratios (SMR_{ik}, $i,k = 1\ldots3$, $i \neq k$), which must lie in the interval 1.00 ± 0.03. This stringent acceptance limit makes sense only if the measurement uncertainty of each SMR_{ik} is much less than the limit itself. In the present study the uncertainty sources of the SMR_{ik} quantities are considered, using real data from the ESTI calibration chains and data acquired in-the-field, in order to estimate the uncertainties of each SMR_{ik} by a Monte Carlo simulation. Each run of the algorithm consisted in 100000 iterations, in order to generate a sufficient number of random errors which are added to the measured functions. The resulting values are then compared to the acceptance threshold prescribed in the standard. Several considerations about the uncertainties of the three SMRs measured making use of component reference cells are presented at a preliminary stage, analyzing outdoor data taken from ESTI measurements on a triple junction component cell.

2. Measurement method

The spectral matching ratios SMR_{ik} as function of time t are defined as the ratio of short circuit current J between the two junctions i,k normalized to the short circuit currents produced under the reference condition AM1.5D:

$$SMR_{ik}(t) = \frac{J_i(t)}{J_k(t)} \frac{\hat{J}_k}{\hat{J}_i} \quad (1.)$$

where $J_{i,k}(t)$ are the measured short circuit currents of i-th and k-th junctions, and $\hat{J}_{i,k}$ are the short circuit currents of the same junctions under reference spectral conditions.

The calculated $SMR_{ik}(t)$ (SMR for simplicity) are functions of time because the solar spectrum changes during the day, affecting the first term, while the second term is a function only of the spectral responsivities of the two junctions and of the reference spectrum, so it is constant. As a ratio of four currents, the SMR is a dimensionless quantity.

There are two methods to measure the SMR: by making use of a calibrated spectroradiometer or by using a calibrated component reference cell set, composed of three single-junction cells, each having spectral responsivities representative of the respective junction in multi-junction devices.

The latter approach is simpler in terms of operation, because it does not need a complex calibrated (and hence expensive) spectroradiometer system to continuously measure the spectral irradiance. Moreover the three spectral responsivities of the CPV device under test do not have to be measured. However, a good result is guaranteed only if the CPV device under test and the component cells have sufficiently similar spectral responsivities, and if the cells are properly calibrated, otherwise this method only furnishes an approximate measurement [1].

Using the spectroradiometer method considered in the present study the short circuit currents are calculated by

$$J_i(t) = \int \Phi_i(\lambda) G(t,\lambda) d\lambda \quad (2.)$$

$$\hat{J}_i = \int \Phi_i(\lambda) \hat{G}(\lambda) d\lambda \quad (3.)$$

where Φ is the spectral responsivity, G is the spectral irradiance, $\hat{G}$ the reference spectrum and λ the wavelength. Substituting Eq. (2)-(3) into Eq. (1) we obtain

$$SMR_{ik}(t) = \frac{\int \Phi_i(\lambda) G(\lambda,t) d\lambda}{\int \Phi_k(\lambda) G(\lambda,t) d\lambda} \frac{\int \Phi_k(\lambda) \hat{G}(\lambda) d\lambda}{\int \Phi_i(\lambda) \hat{G}(\lambda) d\lambda} \quad (4.)$$

Eq. (4) highlights the remarkable symmetry property of the SMR with respect to all the four functions $\Phi_i(\lambda)$, $\Phi_k(\lambda)$, $G(\lambda,t)$ and $\hat{G}(\lambda)$: a linear transformation of one or more of the four functions (name generically $f(\lambda)$) of the type

$$f(\lambda) \Rightarrow kf(\lambda) \qquad k \in \mathbb{R}_0 \quad (5.)$$

has no impact on the SMR (the SMR does not depend on the absolute values of the measured quantities).

3. Sources of Uncertainty

The uncertainties associated with the three SMR depend on the uncertainties of the quantities in the four integrals in Eq. (4). All of them are experimentally measured quantities except for the reference spectrum Ĝ, which is tabulated in [2] with no associated uncertainty. Given that the quantities are measured at different wavelengths, depending on the instrument and on the method, a suitable interpolant (usually a polynomial) is used to define all of them at the same wavelengths. In the present study a piecewise cubic Hermite polynomial has been used as interpolant, because it offers good control of the first derivative of the function. In general, the interpolation error is limited by [3]:

$$\|f(x) - \mathbb{H}[f(x)]\|_\infty \le C\delta x^{n+1} \|f^{n+1}(x)\|_\infty \quad (6.)$$

where the left term is the norm of the error between a generic measured data set and the corresponding interpolant (in this specific case a Hermite piecewise cubic polynomial), which is proportional and limited by some constant C to the *(n+1)*-th order derivative and the distances between adjacent points (in this case the measured wavelengths).

In this study the interpolation error has been evaluated negligible, and therefore not used.

In the implementation of the Monte Carlo method, it is essential to first associate the random error to all the measured points individually, and only afterwards perform the interpolation; associating the errors to the interpolated points would lead to incorrect results.

3.1 Uncertainty budget associated with the spectral responsivity

The spectral responsivity of multi-junction devices is performed at ESTI with suitable bias light to saturate the junction(s) not under measurement and narrow band light at wavelengths where the junction under measurement responds. The narrow band light is obtained from a broadband Xenon light source filtered by interferential bandpass filters with bandwidth from 10-20 nm full-width-half-maximum (FWHM), covering the wavelength range from 300 to 1900 nm. The quasi-monochromatic light is then chopped at a fixed frequency, and the short-circuit current signal is measured by lock-in amplifiers, which enable extraction of the signal due to the quasi-monochromatic light from the static component due to the bias lights [4]. The narrow band light simultaneously illuminates the device under test and the reference device; the spectral responsivity of the device under test is calculated from the ratio of the two device currents.

The following components mainly contribute to the uncertainty of the spectral responsivity:

- Electrical uncertainty (lock-in amplifiers, current-voltage converters, experimental noise)
- Temperature uncertainty (measurement conditions at 25.0 ± 1.0 °C and thermometers calibrations)
- Optical uncertainty (the spatial non-uniformity of quasi-monochromatic light on the measurement plane, alignment of device under test and reference device, area of device under test and reference)
- Reference cell uncertainty (calibration and drift)

The electrical standard uncertainty due only to the lock-in amplifiers and the current-voltage converters is ±0.56%, excluding the electrical noise contribution, which is measured during each acquisition by the standard deviation of the signal ratios as each wavelength point is measured a number of times. This value varies with wavelength and device.

The temperature of the device and of the reference cell is controlled during the measurement by two separate base plates, controlled in temperature by a water circuit and by a Peltier element respectively. The temperature standard uncertainty contribution, considering both items, amounts to ± 0.06% and is the smallest of the four uncertainty components. The two devices are kept physically as close as possible, but non-uniformity of the incident quasi-monochromatic light is unavoidable, due to the non-uniformity of the Xenon light source itself and of the filters; each filter generates a different uniformity pattern on the test area. Considering the worst case, the standard non-uniformity uncertainty is ± 1.16% for small area reference cells (typically 400 mm^2). Given that the two devices are placed on two different plates, a tilt alignment error within 2° is assumed, corresponding to a standard uncertainty of ±0.03%. The last component of the optical uncertainty budget is due to the area of the devices, which enters in the calculation of the ratio of the short-circuit current ratio, and which has to be normalized to the unit area. The area is measured by a calibrated electronic scale ruler equipped with a magnification camera, traceable to SI Units with a standard uncertainty of ±0.07%. The area contribution is excluded in the present exercise because it does not influence the SMR (see Eq. (5)).

The measured spectral responsivity curves of the device used in this study with associated standard uncertainties are shown in Figure 1.

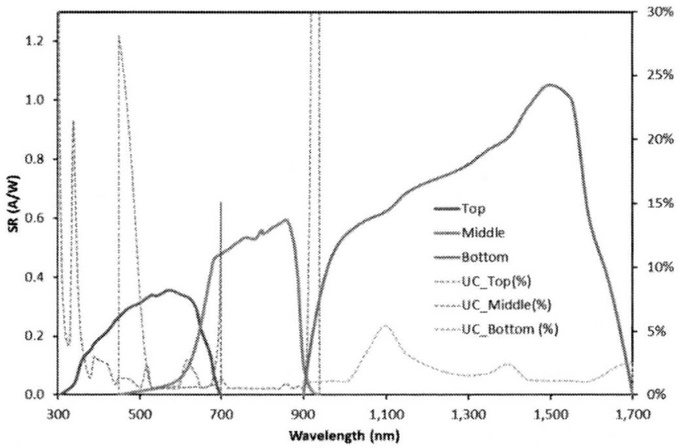

Figure 1: Spectral responsivity curves of the three junctions (continuous lines, left axis) with their wavelength-dependent associated standard uncertainties (dotted lines, right axis); the peaks are due to the fact that high random relative errors occur at very low currents and due to poor signal/noise ratio. The shown curves are those interpolated with a 1 nm step.

3.2 Uncertainty budget associated with the spectrum measurement

The uncertainty budget of the spectrum measurement is composed of the following items:

- Standard lamp calibration uncertainty and drift
- Internal calibration transfer uncertainty
- Temperature control of the detectors under outdoor conditions

As already explained, the SMR depends only on the shape of the measured curves, so systematic effects affecting uniformly all the wavelengths do not impact the SMR uncertainty; for example, since all the devices are placed on the same solar tracker, possible tracking errors have not to be taken into account in the uncertainty budget.

The spectroradiometer used in the present study has been internally calibrated against a standard lamp, calibrated at NPL. Uncertainty arises mainly from calibration, including uncertainty of calibration laboratory and expected lamp calibration drift. Average NPL calibrated standard lamp uncertainty is ±0.70%, plus a calibration transfer uncertainty to spectroradiometer accounting for another ±0.55%. During 50 hours of operation, the lamp drift is estimated to be ±1.00 %; if this is considered the limit of a rectangular distribution, the associated standard uncertainty is ±0.59%. The uncertainty of the standard lamp is then transferred to the spectroradiometer via periodic internal recalibrations. During the process the main uncertainty components are stray light, lamp to spectroradiometer entrance slit distance, lamp current and map orientation. Stray light contribution is estimated to be ±0.29%, even with blackened walls and surfaces, and no object in the room being at a distance less than 1.5 meters from the optical table. The distance lamp-spectroradiometer is set by a 500 mm calibrated gauge block; a residual uncertainty in the lamp distance is estimated to be within ±0.5 mm, leading to a standard uncertainty in the irradiance spectral intensity of ±0.10%. The spectral distribution of the lamp is affected by the current injected into the lamp during calibration of the spectroradiometer. Uncertainty of the power supply is due to short and long-term drift; however these two combined terms account for a standard uncertainty contribution of only ±0.03%.

During outdoor operations the two detectors (c-Si for the range 300-1150 nm and InGaAs until 1700 nm) are temperature controlled by distinct Peltier elements inside the instrument case. However, a temperature gradient between the temperature read by the feedback sensor and the true junction temperature might be possible, depending upon the position of the temperature sensors inside the instruments. If the real temperature differs from the temperature at which the instrument was calibrated, the current produced by the detectors changes, due to the change of spectral responsivity of crystalline silicon and InGaAs materials with temperature. This contribution is modelled in the Monte Carlo method, by multiplying the portion of measured spectral irradiance using the following average values [5]:

$$\begin{cases} 0.20\ \%/_{°C} & 850 \leq \lambda \leq 1150\,nm \\ -0.24\ \%/_{°C} & 1500 \leq \lambda \leq 1700\,nm \end{cases} \quad (7.)$$

Finally, as an empirical check of the input uncertainties estimation, the expanded uncertainty (k=2) of the spectral irradiance measurement have been compared with the deviations with respect to the average spectrum of 14 consecutive acquisitions of a steady-state solar simulator Xenon lamp. Approximately 95.45% of the points were within the uncertainty limits as expected providing a verification of the uncertainties.

4. Results

The standard uncertainties of the three SMRs have been calculated as the standard deviations of the associated probability density functions. The sensitivity analysis of the various sources of error was performed running the Monte Carlo simulation on four different spectra measured during the 6th May 2016 at Ispra, ranging from the early morning to the early afternoon. Four spectra have been chosen for the analysis, and they are plotted in figure 2 together with the AM1.5D reference spectrum. Three of the spectra result in SMR values within the 3% acceptance limit, while the first acquired spectrum of the day would clearly give results outside the limit, because of its low UV and visible content (figure 3). It was included in this analysis to check that the results are applicable also to spectra acquired further from AM1.5 conditions. The uncertainty wavelength-dependent function is also shown.

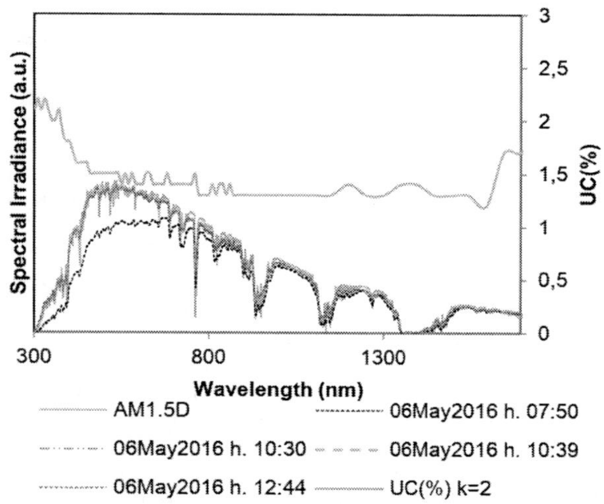

Figure 2: Measured spectra and reference spectrum AM1.5D (IEC 60904-3 ed.3) with the expanded uncertainty function. The spectrum acquired at 07:50 would lead to SMR values outside the 3% limit, due to the low content in the UV and the visible. However, it has been included in the analysis to check that the results are applicable also to data far from AM1.5D conditions

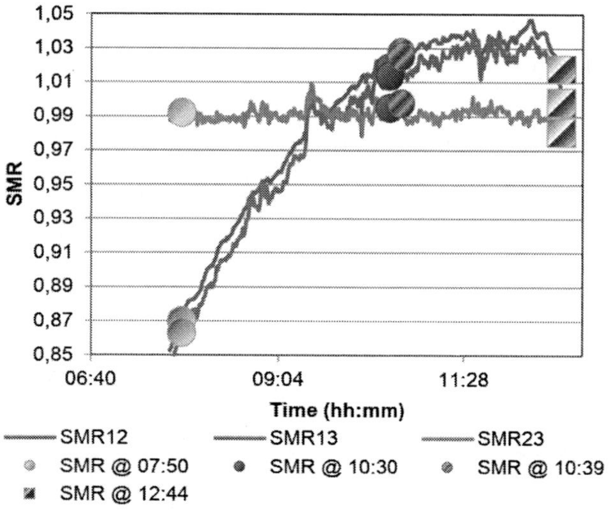

Figure 3: Spectral matching ratios during the day, calculated using the measured spectra and spectral responsivities without any added error. The four triplets of spectral matching ratios given by the spectra used in this study are shown

4.1 Contribution of spectral responsivity errors

The errors on the spectral responsivity of a junction affects only the SMR in which the junction is involved in; as an example, errors on the top junction do not affect SMR_{23}. Moreover, the relative impact on the two SMR is exactly the same. For each graph shown in Figure 4, only one spectral responsivity is considered with errors, while the other two have been used without uncertainty, in order to isolate all the single contributions to the final budget. The simulation has run for all the four spectra, and the average value of the errors obtained with the four spectra is calculated (Table 1).

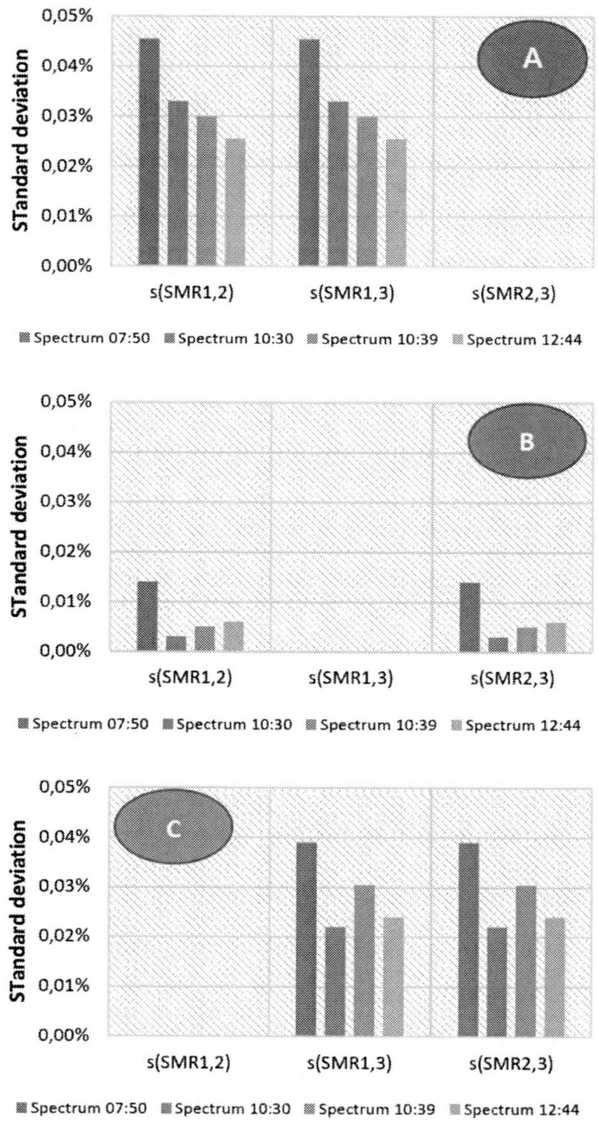

Figure 4: Errors on the three SMR generated by the input errors on the spectral responsivities of the top cell (A), middle cell (B) and bottom cell (C) considered separately and calculated for all the four spectra.

Table 1: Average SMRs variations due to spectral responsivity measurement uncertainties

Junction	$\sigma(SMR_{12})$	$\sigma(SMR_{13})$	$\sigma(SMR_{23})$
Top	±0.033%	±0.033%	0
Middle	±0.008%	0	±0.008%
Bottom	0	±0.028%	±0.028%

4.2 Contribution of measured spectra random errors

In the previous section the contributions of Top, Middle and Bottom junctions spectral responsivities have been analysed. The other source of uncertainty comes from the measured spectrum $G(t, \lambda^{(G)})$, which, in the case of this study, is measured by a grating spectroradiometer with two detectors having different spectral ranges (crystalline silicon and InGaAs). The uncertainties listed in Section 4.2 act as random variables, for which there is no correlation between their values at adjacent wavelengths. The effect of the integral operator is then to smooth the variability of the output integral quantity. Moreover, the SMR function is invariant in respect with multiplication, so absolute errors acting on all the wavelengths by the same amount do not affect the final value.

As can be seen in Figure 5 the impact of random errors on spectrum is evidently smoothed by the integral operator: all the values show a very low variability, and are comprised between 0.05% and 0.06% approximately. As a conservative approach, the value of 0.06% can be considered for the random spectral errors contribution to the overall final budget.

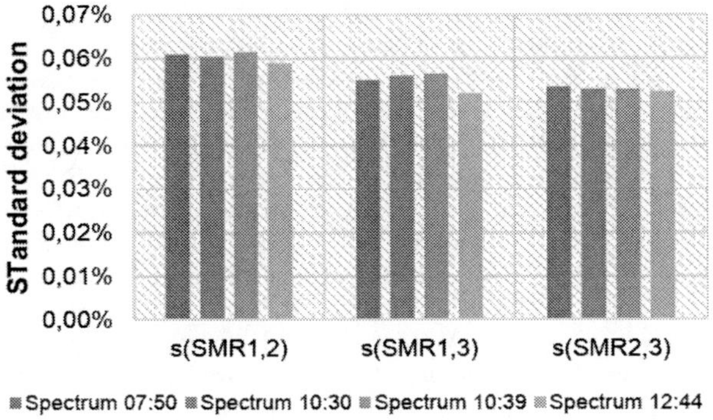

Figure 5: Impact of random errors of measured spectrum on the three SMR.

4.3 Contribution of temperature errors on measured spectrum

As already described, the direct normal spectral irradiance is measured using a calibrated grating spectroradiometer with two detectors. The measurement has to be performed maintaining both detectors at the same temperature at which the calibration of the spectroradiometer was performed. The sensors have their temperature controlled by two separate Peltier elements, but the measured temperature differs from the effective junction temperature, due to the position of the temperature sensor with respect to the detector, to the irradiance variability during the measurement, and to the exposition of the spectroradiometer body to the outdoor conditions. The effect of this discrepancy has to be estimated; in the present study it is modelled with a Gaussian distribution with $2\sigma = \pm 1.5$ °C. The uncertainty associated with the temperature of the two sensors does not affect all the wavelengths, but only the intervals specified in Eq. (7) [5]. The algorithm to adjust the measured spectra is

$$G\left(t,\lambda;T_{c-Si}^{(i)},T_{InGaAs}^{(i)}\right) = p(\lambda;T_{c-Si}^{(i)},T_{InGaAs}^{(i)})G(t,\lambda) \quad (8.)$$

where the function p varies only in the regions $850 \leq \lambda \leq 1150$ nm and $1500 \leq \lambda \leq 1700$ nm, otherwise it is unity. The index *(i)* in Eq. (8) refers to the trial number, because the function changes at each iteration following the Gaussian of all the possible values of temperature errors. The two Gaussian curves modelling crystalline silicon and InGaAs temperature errors are independently generated. An example is shown in Figure 6.

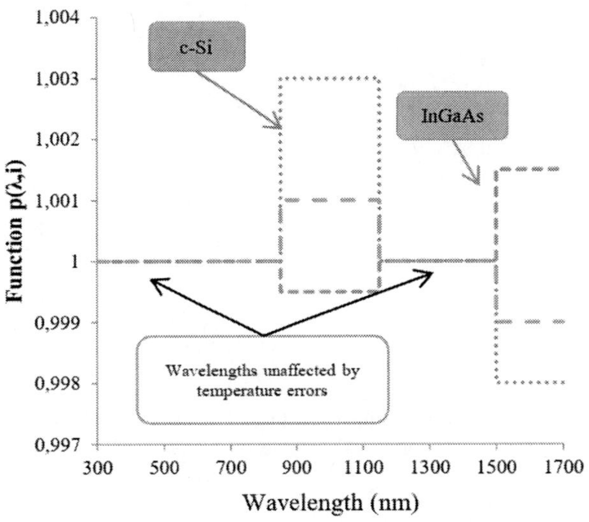

Figure 6: Weight function generated in three different Monte Carlo runs to adjust the spectrum in the InGaAs and c-Si regions independently. The weight function is always equal to one except in the interested wavelengths, where the spectrum is scaled up or down depending on the random number sampled from the Gaussian distribution

In order to study the uncertainty component associated to each of the two temperature errors separately, the weights of c-Si and InGaAs have been used alternately, varying the amplitude of the Gaussian probability and assuming the two contributions are statistically independent:

c-Si temperature: $p(\lambda;T_{c-Si}^{(i)},T_{InGaAs}^{(i)}) \rightarrow p(\lambda;T_{c-Si}^{(i)})$

InGaAs temperature: $p(\lambda;T_{c-Si}^{(i)},T_{InGaAs}^{(i)}) \rightarrow p(\lambda;T_{InGaAs}^{(i)})$

The results are shown in Figure 7 for c-Si and InGaAs separately, considering the different spectra.

The c-Si temperature error affects all the three SMR, because it is present in both the Top and Middle junctions; the entity of the variation of the three SMR depends on the values of spectral irradiance and spectral responsivities in the interval. [850;1150] nm. Regarding the variation associated with error in InGaAs, only the bottom junction responds in the wavelength range from 1500 to 1700 nm and therefore, only SMR_{13} and SMR_{23} are affected by the same amount.

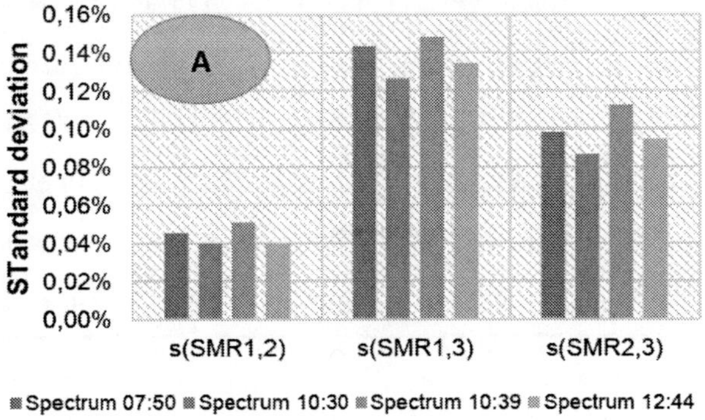

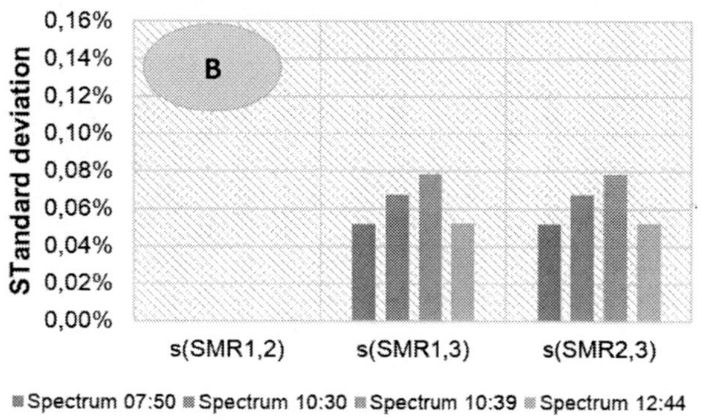

Figure 7: Uncertainties associated with temperature errors of spectrum measurement, hypothesizing a 1.5 °C error in the temperature controller: (A) only the c-Si detector and (B) only InGaAs detector.

Considering the estimated $\sigma = \pm 1.5$ °C, the corresponding output standard deviations are summarised in Table 2.

Table 2: Variations associated with spectroradiometer temperature errors for the case input σ=±1.5 °C.

($\sigma =$	σ(SMR$_{12}$)	σ(SMR$_{13}$)	σ(SMR$_{23}$)
c-Si	0.04%	0.14%	0.10%
InGaAs	0	0.06%	0.06%

4.4 Overall Uncertainties with all contributions

The overall uncertainty is due to the superposition of all the contributions separately analysed in the previous sections, considered all together in a single Monte Carlo run. The simulation is then performed for all the different measured spectra, to assess the stability of the result upon different irradiance levels and spectral distributions. The results of the Monte Carlo method with all the different degrees of freedom are summarized in Table 3.

Table 3: overall uncertainties obtained with all the input components considered simultaneously. The sign "±" is omitted for clarity.

(k=1)	σ(SMR$_{12}$)	σ(SMR$_{13}$)	σ(SMR$_{23}$)
Spectrum 07:50	0.10%	0.15%	0.21%
Spectrum 10:30	0.09%	0.17%	0.23%
Spectrum 10:39	0.08%	0.14%	0.17%
Spectrum 12:44	0.09%	0.15%	0.21%
Average	*0.09%*	*0.15%*	*0.21%*
Expanded Uc (k=2)	**0.18%**	**0.30%**	**0.42%**

5. Consideration about SMR measured with component cells

In the previous sections the SMR were derived by calculation from the spectral irradiance measured with a spectroradiometer and the three spectral responsivities. The second method described in the IEC 62670-3 standard foresees the use of component cells, which are three single junction cells made by the same technologies as of the junctions composing a triple junction device. There are two possibilities prescribed in this case: 1) to use cells matched with the device under test or 2) to use standard component cells, which are representatives of the most common triple junction devices, but not necessarily of the specific device under test. In the former case, the technology of the device under test must be known, or the three spectral responsivities have to be measured by the laboratory, but the final spectral match is going to be much more reliable than in the latter case. The use of standard component cell gives the coarsest estimation of the spectral match, and the error on the measurement is likely to be very large due to the different spectral responsivity between the cells and the device under test. The use of matched component cells partially solves this issue, and the error on the SMR measurement is mainly due to the calibration uncertainty of the component cells. At ESTI, primary and secondary reference cells calibration can be performed either indoor on various solar simulators or outdoor using the Global Sunlight Method (DSM), the Direct Sunlight Method (DSM) or the Simplified Global Sunlight Method (SGSM). Measuring indoor the bottom cell very large mismatch corrections are necessary to correct to STC, in the order of 50% even using a dual source continuous solar simulator. For this reason indoor calibration is not considered in this study. Here we consider only the outdoor DSM method, for which the expanded uncertainty at ESTI is ±0.48% (k=2). In this case the mismatch correction is about 3-4% even for the bottom cell against a c-Si reference cell (secondary calibration) or a cavity (primary calibration). The uncertainty of the three SMR is mainly due to the uncertainty of the three ratios of the calibration values, which is ± 0.68% (k=2).

6. Conclusions

Accordingly to the International Standard IEC 62670-3 the data of power performance of CPV devices measured outdoor have to be filtered in order to restrict the set only to those points closer to the Concentrator Standard Operating Conditions. The strictest requirement is that all the three Spectral Matching Ratios (SMR) must be in the range 1.00 ± 0.03, to eliminate points acquired with spectral conditions too far from the reference spectrum. The present study calculated the uncertainties associated with the three SMR measured using a calibrated spectroradiometer and the spectral responsivities of the triple-junction CPV device, using a Monte Carlo approach. The values of the expanded (k=2) uncertainties are one order of magnitude less than the limit, with the highest value of ±0.42% for SMR$_{23}$, and hence the 3% value is meaningful for this measurement method. However, the uncertainties of spectrum and spectral responsivity measurements at ESTI are representative of only a few top-class laboratories in the world, and hence typical uncertainties on SMR are expected to be generally higher. It is advisable for each laboratory dealing with outdoor power measurements of CPV devices to declare also the specific uncertainties of the SMR used to filter the acquired data. The uncertainties of SMR calculated using component cells has been considered only partially in this study, but the fact that the component cell device sub-junctions are generally different than the particular device under test leads to an approximate method, with higher uncertainty even with matched component cells and very precise calibration.

References

[1] S. Philipps, K. Horowitz, S. Kurtz, Current technology status of concentrator photovoltaic (CPV), Freiburg, 2016. https://www.ise.fraunhofer.de/content/dam/ise/de/documents

/publications/studies/2016_02_09_CPV_Report_ISE_NREL _Version_1_2.pdf (accessed March 6, 2017).
[2] M.A. Green, K. Emery, Y. Hishikawa, W. Warta, E.D. Dunlop, Ho-Baillie, D.H. Lev, A.W. Y., Solar cell efficiency tables (version 49), Prog. Photovolt Res. Appl. 25 (2017) 3–13. doi:10.1002/pip.
[3] N.J. Ekins-daukes, P. Sandwell, J. Nelson, A.D. Johnson, G. Duggan, N.J. Ekins-daukes, P. Sandwell, J. Nelson, D. Andrew, E. Herniak, What does CPV need to achieve in order to succeed?, AIP Publ. Artic. 20004 (2016). doi:10.1063/1.4962072.
[4] IEC, International Standard IEC62670-3 Edition1.0 2017-02, IEC. (2017).
[5] IEC, International Standard IEC 62670-1 Edition1.0 2013-09, IEC. (2013). doi:10.1109/IEEESTD.2007.4288250.
[6] M. Norton, R. Galleano, G. Siefer, M. Steiner, R. Kenny, W. Zaaiman, Analysis of the output of two isotype cell sets compared to a precision spectroradiometer, AIP Conf. Proc. 1679 (2015). doi:10.1063/1.4931529.
[7] J. Bureau International des Poids et Measures (BIPM), Evaluation of measurement data - Guide to the expression of Uncertainty in Measurement, 2008.
[8] E.R. Woolliams, Determining the uncertainty associated with integrals of spectral quantities, 2013. www.m4ssl.npl.co.uk.
[9] Y. Ma, Y. Zhang, Y. Gu, X. Chen, S. Xi, B. Du, H. Li, Tailoring the performances of low operating voltage InAlAs/InGaAs avalanche photodetectors, Opt. Express. 23 (2015) 19278. doi:10.1364/OE.23.019278.
[10] IEC, International Standard IEC60904-3 Edition 3.0 2016-04, 2016.
[11] Q. Alfio, S. Riccardo, S. Fausto, Matematica Numerica, Springer, 2000.
[12] IEC, International Standard IEC60904-8 Ed.3 2014-05, 2014.
[13] IEC, International Standard ISO/IEC 17043 Edition 1.0, 2010.

ANALYSIS OF BACKSHEET AND REAR COVER REFLECTION GAINS FOR BIFACIAL SOLAR CELLS

Max Mittag, Alex Grünzweig, Martin Wiese, Nabil Mahmoud, Alexandra Schmid, Martin Heinrich
Fraunhofer Institute for Solar Energy Systems ISE
Heidenhofstr. 2, 79110 Freiburg
max.mittag@ise.fraunhofer.de

ABSTRACT: Bifacial solar cells are known to increase module power and performance. Due to their active rear side additional gains are possible from internal reflections. Existing models to analyze cell-to-module (CTM) gains need to be extended. We analyze reflection within modules with bifacial cells and establish a system and a nomenclature for gains resulting from internal reflection. Transmission through the cell, subsequent reflection on module cover layers and a second absorption of light in the solar cell leads to additional gains in the range of 0.5 – 0.8% for bifacial cells in modules with reflecting rear cover materials (i.e. white backsheets) under STC conditions. We present models to calculate gains and perform measurements on bifacial modules with different backsheets and covers. Cover coupling gains for bifacial cells are higher compared to monofacial cells (20%$_{rel}$ in modules with white backsheets).

Keywords: Photovoltaic Module, Optical Gains, Simulation, CTM, Cell-to-Module, Bifacial, Backsheet Coupling

1 INTRODUCTION

Today, photovoltaic modules mainly use monofacial solar cells [1] that are only capable of converting irradiance from the front side into electrical power. Bifacial solar cells are a promising technology, which allows additional gains from rear irradiance [2]-[6]. By introducing bifacial cells into photovoltaic modules, existing models for cell-to-module (CTM) efficiency analysis or yield prediction [7] are no longer sufficient due to additional optical effects within the photovoltaic module such as additional relevant internal reflections [8].

Conventional solar modules with monofacial cells are known to profit from reflections from the modules rear cover ("backsheet gain") [9]. Previous research shows that these power gains are in the range of 1 to 3% [7][10] for common module setups and components. Since monofacial cells are only capable of using light irradiant from the front side, only a fracture of possible gains from internal reflection of light can be realized. Bifacial cells have additional electrically active surfaces on the cell rear side and therefore higher reflective gains are possible [11]. Results regarding the magnitude of reflection gains have been presented [9][12].

We conduct an analysis of factors that allow reflection gains from the module covers for bifacial cells and present a nomenclature as well as models for calculation. We manufacture several modules of different setups and measure the effects of front side irradiance on modules with bifacial solar cell.

2 GAINS BY INTERNAL REFLECTION

Monofacial solar cells in modules with a reflecting (opaque) rear cover profit from reflection of light within the module. Rays incident on the opaque backsheet may be reflected onto the front glass and afterwards may again be reflected on active solar cell area (Figure 1). We will use the description "k_{11}" for this effect in accordance to the methodology of Hädrich et al. [7] to analyze the cell-to-module (CTM) ratio and influencing effects.

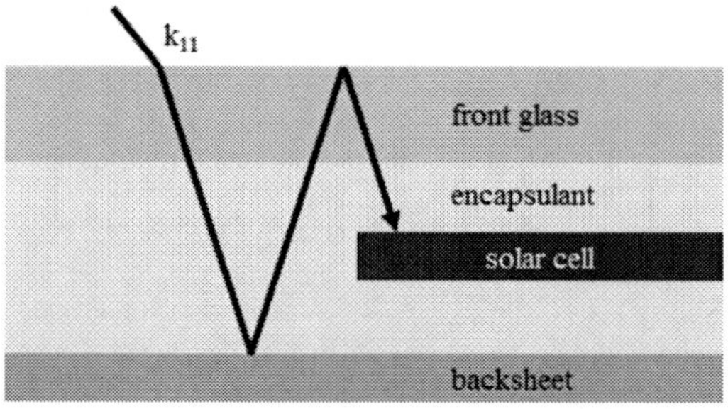

Figure 1: schematic drawing of backsheet reflection gains in modules with monofacial cells and opaque rear cover [13]

By introducing transparent backsheets and double-glass-modules an extension of the nomenclature is necessary. We therefore rename the k_{11} gain factor to "cover coupling" and extend it by using additional indices to allow a further distinction between different separate contributors. The first index letter describes the origin of the incident light (module front "f" or module rear "r") and the second letter names the cell side receiving the ray. Figure 2 shows the extended model for transparent rear cover materials (monofacial cell).

The introduction of transparent module rear covers not only affects the internal gains and losses but also module power characterization. Reflections from measurement chucks or surrounding equipment have to be considered for cell and module measurements [14][15]. In this work we do not consider irradiance from the module rear side as models for front side irradiance may be applied accordingly.

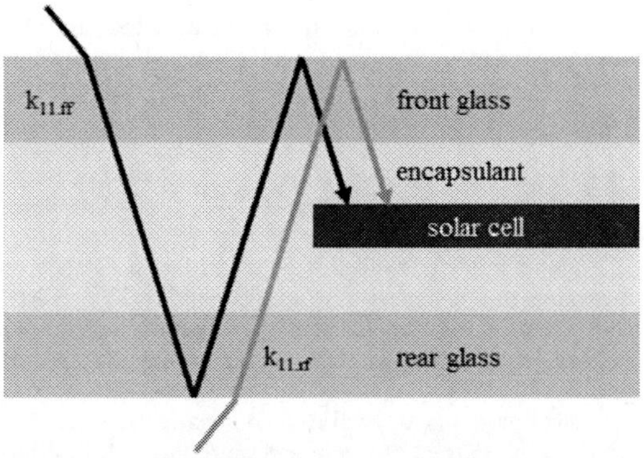

Figure 2: Schematic drawing of cover reflection gains in modules with monofacial cells and transparent rear cover

An additional extension is necessary if bifacial cells are introduced into modules since a second active cell surface has to be considered. Figure 3 shows the additional gain mechanisms $k_{11.fr}$ and $k_{11.rr}$.

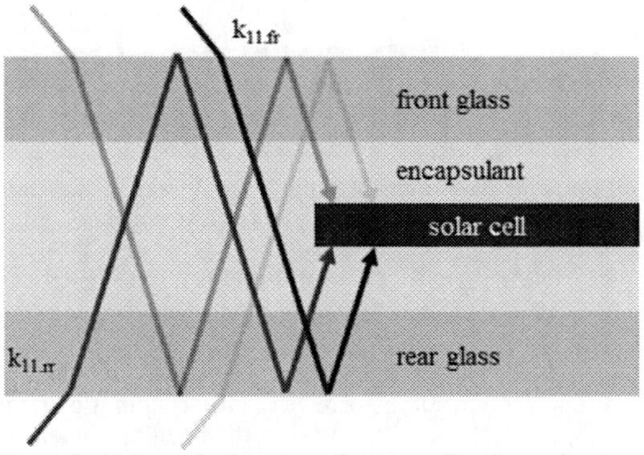

Figure 3: Schematic drawing of cover reflection gains in modules with bifacial cells and transparent rear cover

Additional reflectors in cell and string spacing areas have been presented to increase the module power in bifacial modules [8][16]. Our system and nomenclature can also be applied to categorize these gains (Figure 4).

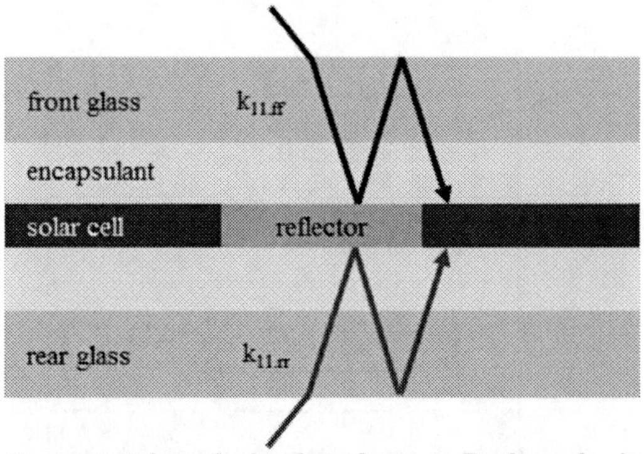

Figure 4: Schematic drawing of cover reflection gains in modules with bifacial cells, transparent rear cover and additional reflectors in the cell/string spacing area

3 GAINS BY TRANSMISSION THROUGH SOLAR CELLS AND SUBSEQUENT REFLECTION

Another possible gain mechanism is described by Singh et al. [10]. Bifacial cells are translucent in wavelengths > 1000 nm. Light transmitted through the solar cell may be afterwards reflected at a module cover and subsequently be absorbed by the solar cell (Figure 5).

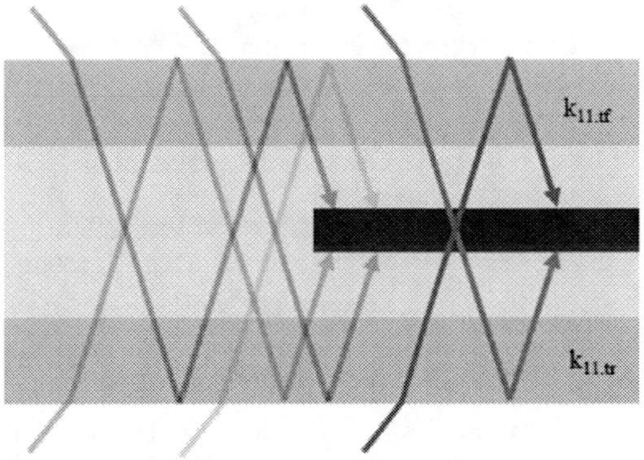

Figure 5: Schematic drawing of cover reflection gains resulting from transmission of light through the solar cell and subsequent reflection

We create a calculation model and perform optical measurements (Figure 6, Figure 7) to quantify transmission gain effects ($k_{11.tx}$).

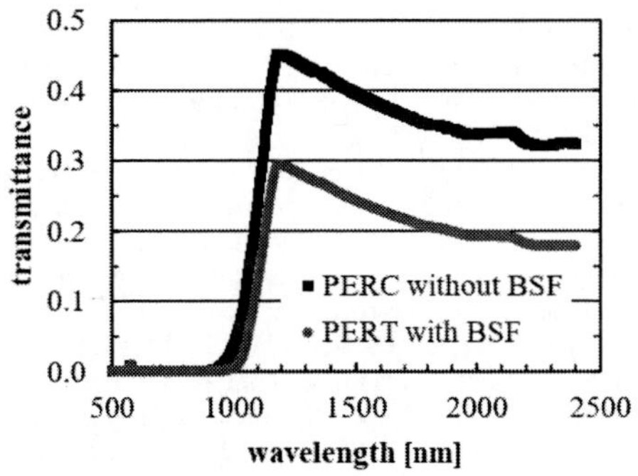

Figure 6: Measured transmittance of different bifacial solar cells

Neglecting multiple reflections or total reflection, we assume light irradiant from the module front has an AM1.5g spectrum and is partially reflected at the front glass. It passes through EVA three times in total and is being transmitted through the cell. Reflection on a rear cover material (such as a backsheet or a glass) is considered. If the rear cover is transparent (i.e. a glass) we include another pass through the transparent cover and a reflection on the outer module interface (i.e. $k_{11,tr}$). The additional irradiation on the cell rear side can be estimated using the following equation:

$$E_{cellrear} = E_{AM1.5} \cdot (1 - r_{glassfront}) \cdot t_{glassfront} \cdot (3 \cdot t_{EVA}) \cdot t_{solarcell} \cdot r_{rearcover}$$

The cell features a spectral response (SR) which has to be considered in the calculation of electrical gains.

We perform a measurement of all parameters. Selected results are displayed in Figure 7. Transmission gains are caused by light with a wavelength between approx. 1000 and 1200 nm (below blue and red curve).

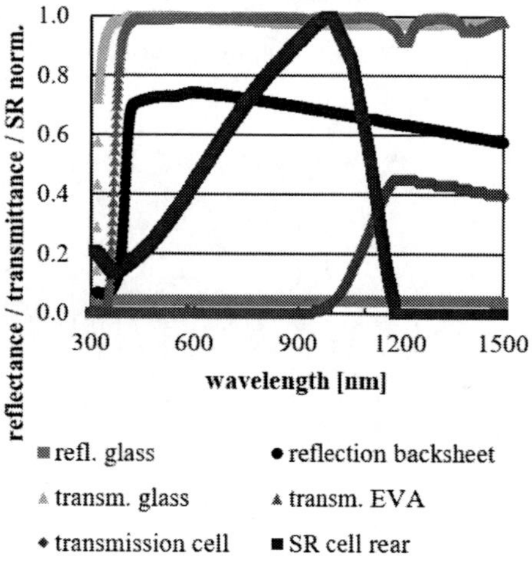

Figure 7: Measured optical properties of module materials and a bifacial solar cell

We calculate the transmission gain to be 0.4% for an exemplary monofacial module setup using bifacial solar cells, low-iron glass without anti-reflective coating and a white TPT-backsheet. We calculate the transmission gains of a double-glass module as well as a module with black backsheet and find them to be neglectable (0.03%). Multiple reflections, total reflection or additional effects have not been considered.

To estimate the influence of the transmission gains, power measurements on samples using commercial bifacial cells and white as well as highly absorbing backsheets are measured. The initial cell power is measured and module results are corrected according to differences in initial cell power. We find the difference in I_{SC} measurements between modules with black and white backsheets to be 0.5 to 0.8% which can be associated to transmission gains. We therefore conclude that transmission gains are a relevant factor for monofacial modules with bifacial solar cells.

Considering these results we further extend the CTM model [17][18] and nomenclature for gains resulting from transmission of light through the solar cell, reflection on module cover interfaces and consequent absorption. Gains after transmission are marked as $k_{11.tx}$ with x describing the direction of incidence on the solar cell after internal reflections (Figure 5).

4 MODELLING OF COVER COUPLING GAINS

Cover coupling gains depend on:

- Cell and string spacing
- Optical properties of module materials (i.e. reflection, absorption)
- Module setup (i.e. layer thickness)
- Cell edge length and cell size

An empirical model to describe backsheet gains has been presented by Haedrich et al. [7] which is based on measurements and an exponential fit to describe the coupling gain as a function of the cell distance *d*:

$$I_{SC\,gain} = a \cdot \exp\left(\tfrac{d}{b}\right) + c = \frac{I_{SC}(0mm) - I_{SC}(d)}{I_{SC}(0mm)}$$

Gains modeled with this equation that are based on measurements using shading masks (Figure 8) or LBIC (Laser Beam Induced Current) [9] already include the optical properties of module materials and the module setup.

This model is also valid for bifacial solar cells. The short circuit current at 0 mm and any other cell distance is higher for bifacial cells due to transmission gains and therefore $I_{SC\,gain}$ is changed for bifacial cells. If an estimation of the transmission gains is performed a correction of I_{SC}(0mm) may be performed.

Effects of varying cell edge lengths (i.e. for halved cells or shingles) can be considered based on measurements on full-size cells and a linear correction [19].

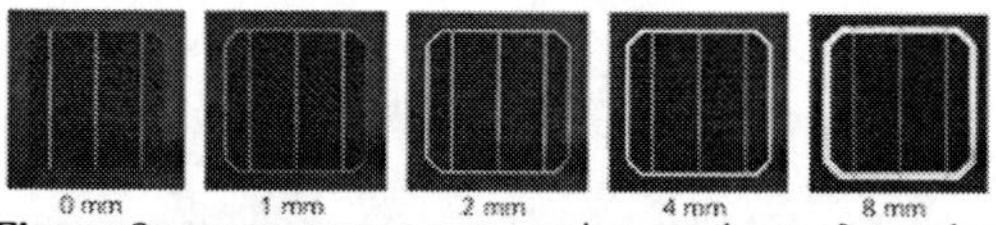

Figure 8: measurement setup using masks to determine cover coupling gains for different cell spacings

5 INFLUENCE OF DIFFERENT COVERS ON MODULE POWER

We build and measure single-cell modules of different setup to compare the cover coupling gains for different rear cover materials. All modules contain the same commercial EVA, glass and interconnector ribbons. Commercially available monofacial as well as bifacial cells are used. Modules are measured at Fraunhofer ISE Module-TEC and coupling gains for different cell/string spacing are determined. The modules contain different rear cover materials. We use a flash sun simulator with IV-curve measurement system and shading masks to vary cell and string spacing (Figure 8).

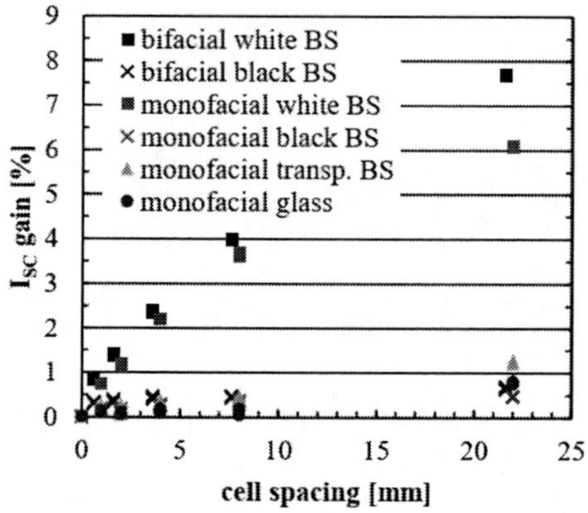

Figure 9: cover coupling gains of modules with different rear cover materials at different cell/ string distances, mono- and bifacial cells, mask measurement

Coupling gains of black or transparent rear cover materials are small compared to reflective covers such as white backsheets. Figure 9 shows increased coupling gains for reflective rear covers both for monofacial and bifacial cells. Measurements are performed with front side irradiance only. Bifacial cell cover gains are higher compared to monofacial cells. We find them to be approx. 20% higher compared to monofacial cells in modules using white backsheets.

It can be concluded that the power of modules with bifacial cells can be increased with reflecting rear covers if only front side irradiation is considered. Gains from additional irradiation from the module rear side have not been considered within this study.

To confirm the increased cover coupling gains for bifacial cells in monofacial module setups with reflecting rear covers (Figure 9), we build additional 4-cell modules (2 mm cell spacing, 600x400 mm², one sample per setup) with black, white and transparent backsheets as well as with a rear glass cover and measure the short circuit current I_{SC} at Fraunhofer ISE CalLab PV Modules. Results prove the increased optical gains for bifacial solar cells (Figure 10).

The strong influence of the rear cover on module power proves the necessity to consider the background in the measurement of bifacial modules (if a single-side-irradiation measurement setup is used).

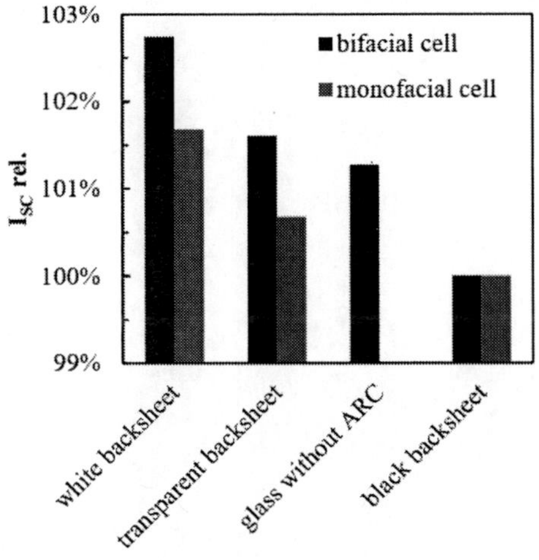

Figure 10: Short circuit current gain of monofacial and bifacial solar cells (different manufacturer) in 4-cell-modules with different rear cover materials; 2 mm cell spacing; front side irradiation only, mask used, normalized to I_{SC} measured with black blacksheet

Due to the lost bifaciality of the module itself, monofacial module setups with opaque rear covers might not be preferred. Applying reflective elements in the cell spacing area but not behind the cell provides a solution [9]. Bifaciality is preserved and cover coupling gains can be observed.

We build additional 4-cell modules with reflectors in the cell spacing area using a white polymer coating or a white tape on the inside of the rear glass (4 mm cell spacing) and measure the change in short circuit current compared to an uncoated glass.

The I_{SC} of a bifacial module increases by 1.5% compared to an uncoated glass without reflective elements (Figure 11, front irradiation only). The white tape that is usually used to cover string interconnector ribbons raises the I_{SC} by 1.1%.

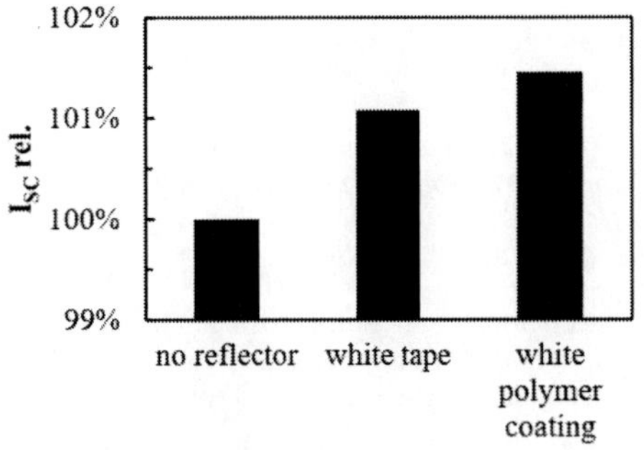

Figure 11: Short circuit current of bifacial solar cells in modules with different reflectors compared to an untreated rear glass (no reflector), 4 mm cell spacing, front irradiance only, mask used

6 SUMMARY

Module efficiency analysis and cell-to-module calculations require new models for bifacial cells due to additional optical gain mechanisms related to the second active cell side. We analyze possible gain mechanisms and identify three additional light paths compared to monofacial cells.

Bifacial cells are partially transparent in wavelengths between 1000 and 1200 nm. Transmission, subsequent reflection on module cover layers and a second absorption of light in the solar cell leads to additional gains in the range of 0.5 – 0.8% for bifacial cells in modules with reflecting rear cover materials (i.e. white backsheets). Transmission gains in modules with absorbing or transparent rear covers can be neglected.

We model cover coupling gains and perform measurements on modules of different setup and confirm the predicted additional gains for bifacial cells.

7 ACKNOWLEDGEMENTS

We would like to thank the German Ministry of Economic Affairs and Energy ("Kokobi" FKZ 0325875) for their funding.

We would like to thank Matthieu Ebert for his valuable contributions to this work and the photovoltaics module department of Fraunhofer ISE.

8 REFERENCES

[1] *"International Technology Roadmap for Photovoltaic (ITRPV)"*, 8th edition, 2017

[2] Woehrle, N., et al, *"Solar cell demand for bifacial and singulated-cell module architectures"*, Photovoltaics International, vol. 36, 2017

[3] Kraus, K. et al. 2016, *"biPERC silicon solar cells enabling bifacial applications for industrial solar cells with passivated rear sides"*, physica status solidi (a), Vol. 213, No. 1, pp. 68–71.

[4] Woehrle, N., et al, "*Understanding the rear-side layout of p-doped bifacial PERC solar cells with simulation driven experiments*", Energy Procedia, 2017, [forthcoming]

[5] Fellmeth, T., et al, "*Co-diffused bi-facial PERT solar cells*", Energy Procedia, 2017, [forthcoming]

[6] Lohmueller, E., et al, "*Bifacial p-type PERL solar cells with screen-printed pure Ag metallization and 89% bifaciality*", 33rd European PV Solar Energy Conference and Exhibition, 2017

[7] Haedrich, I., et al, "*Unified methodology for determining CTM ratios: Systematic prediction of module power*", Solar energy materials and solar cells 131 (2014), pp.14-23, ISSN: 0927-0248, SiliconPV, 2014

[8] Guerrero-Lemus, R., et al, "*Bifacial solar photovoltaics – A technology review*", Renewable and Sustainable Energy Reviews, Vol. 60, 2016

[9] Van Aken, B., et al, "*White bifacial modules – improved STC performance combined with bifacial energy yield*", 31st European PV Solar Energy Conference and Exhibition, 2016

[10] Singh, J., et al, "*Comparison of Glass/Glass and Glass/Backsheet PV Modules Using Bifacial Silicon Solar Cells*", IEEE Journal of Photovoltaics, Vol. 5, 2015

[11] Koentopp, M., et al, "*Optimized Module Design: A Study of Encapsulation Losses and the Influence of Design Parameters on Module Performance*", IEEE Journal of Photovoltaics, Vol. 3, 2013

[12] Ponce-Alcántara, S., et al, "*The importance of optical characterization of PV backsheets in improving solar module power*", Photovoltaics International, Vol. 26, 2014

[13] Sánchez-Illescas, P.J., et al, "*Performance of Photovoltaic Modules with White Reflective Back Sheets*", 23rd European PV Solar Energy Conference and Exhibition 2008

[14] Hohl-Ebinger, J., et al, "*Bifacial Solar Cells in STC Measurement*", 25th European PV Solar Energy Conference and Exhibition, 2010

[15] Rauer, M., et al, "*Monofacial IV Measurement of bifacial silicon solar cells in an inter-laboratory comparison*", 32nd European PV Solar Energy Conference and Exhibition, 2016

[16] Witteck, R., et al, "*Optimized Interconnection of Passivated Emitter and Rear Cells by Experimentally Verified Modeling*", IEEE Journal of Photovoltaics, Vol. 6, 2016

[17] Mittag, M., et al, "*Systematic PV module optimization with the cell-to-module (CTM) analysis software*", Photovoltaics International, vol. 36, 2017

[18] Mittag, M., et al, "*Electrical and Thermal Modeling of Junction Boxes*", 33rd European PV Solar Energy Conference and Exhibition, 2017

[19] Mittag, M., et al, "*Cell-to-Module (CTM) Analysis for Photovoltaic Modules with Shingled Solar Cells*", 44th IEEE Photovoltaic Specialists Conference, 2017

[20] Holst, H., et al, "*Application of a new ray tracing framework to the analysis of extended regions in Si solar cell modules*", Energy Procedia 38, SiliconPV, 2013

BENEFIT OF TEXTURED CIGS CELLS FOR LOW REFLECTING NANOGRID APPLICATION

Joop van Deelen[1], Marco Barink[2]
[1]Solliance/TNO, High Tech Campus 21, 5656 AE, Eindhoven, The Netherlands
[2]Holst Centre/TNO, High Tech Campus 31, 5656 AE, Eindhoven, The Netherlands

ABSTRACT: Metallization of the front contact can increase the conductivity. We show the optical benefit by combining nanowire with a texture of the thin film stack. The parallel nanowires are placed in crevices of a texture in the optical model. Texturization reduces the reflection of both the stack and the nanowire. For a period of 500 nm, the reflection is virtually independent of the nanowire width for a texture height/period ratio of only 1. For a texture height of 1000 nm, the reflection is minimized with maximum of 3% (between 600 nm and 750 nm wavelength) and this would mean that the surface would appear as black, even with a surface metal coverage of 30%. Moreover, outside this wavelength range, the reflection is below 1% and marks a significant increase in current density up to 38 mA/cm^2 compared to the flat layer stack (27 mA/cm^2).
Keywords: lightmanagement, metallization, grid

1 INTRODUCTION

The front contact of thin film solar cells need to be transparent and conducting. A special class of materials has been developed dedicated to these two combined functionalities: the transparent conductive oxides, such as SnO_2:F and ZnO:Al. However, transmittance and conductivity are mutually contradicting material qualities and result in a limited conductivity with a certain transparency [1]. Addition of metallic grid has been shown to give a boost in conductivity with a relatively small drop in transparency [2,3]. We have previously shown results on the benefit for solar cells by enhancing the front contact with a metallic micro grid [4-7]. However, increased conductivity of the front contact comes with an optical penalty as the metallic grid reflects the light proportionally to the surface coverage. Nanowires could be an option to further improve the transparency as function of the conductivity [8]. Unfortunately, a nanowire grid does not conduct as well as a micro grid for a similar surface coverage, but has the potential for a much lower reflectivity. For this reason, a new design is proposed by introducing a nanogrid-texture combination with strongly reduced reflection, which opens a new array of possibilities [9,10].

We show results from optical modeling of a metallic grid and texture with a spacing of 500 nm. Both the texture height and the nanowire width was changed systematically. The optical model gives absorption and reflection spectra and from these maximum current densities were derived.

2 METHODS

The optical behaviour of nanogrid on top of Cu(In,Ga)Se_2 (CIGS) cells for both flat and for textured layers was modeled with FEM using COMSOL solver software. The basic texture and grid design is shown in fig. 1. Note that the top medium here is glass as to mimic the actual panel situation, in contrast to most experimental work on thin film panels that have a front contact/air interface. The Ag nanowire width is denoted as LF and the nanowire height was set to a fixed value of 50 nm. Translation to current density was done by combining calculated absorption spectra of CIGS and the current density spectrum (fig. 1d). This way, the data shows directly the potential impact on PV performance. Although the spectra do not include the glass/air interface, for calculation of the current density, a 4% front interface reflection and no absorption was included. The IQE was set at 100%.

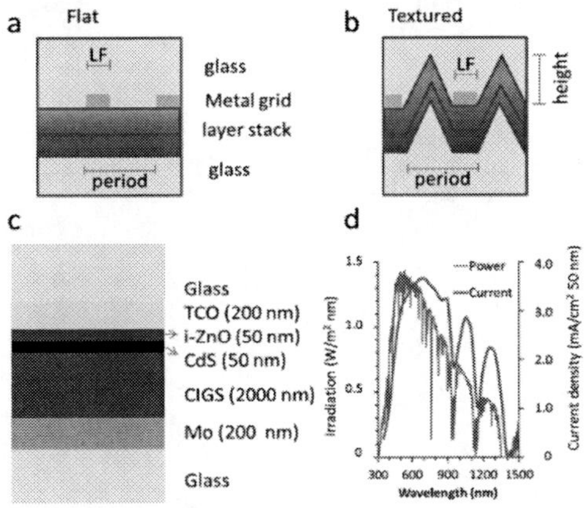

Figure 1: Design of nanogrid on top of flat (a) and textured (b) cells tacks, detailed cell stack (c) and power and current density spectra (d) [9].

3 RESULTS

3.1 Overall spectra

Fig. 2a shows that for a flat layer stack, the nanowire of 25 nm wide (with pitch of 500 nm) results in significant reflection. For textured layer stacks, the reflection is greatly reduced (see figs, 2b) and almost no reflection is visible for a texture with height of 500 nm an height/period aspect ratio of 1 (figure 2c). It is seen that the spectral distribution of the reflection and the absorption changes strongly with texture height (H), so instead of functioning only as an 'anti-reflection' texture, there are many optical phenomena occurring, which give surprising outcomes that can be used to tune spectral distribution of light in optoelectronic devices. Widening the wires width to 150 nm (30% surface coverage) increases the reflection and also shows enlarged changes in spectral distribution as shown in figures 2d and 2e. For such a high surface coverage, there is still reflection present at a high texture of 500 nm, although it is much less than for the flat layer stack.

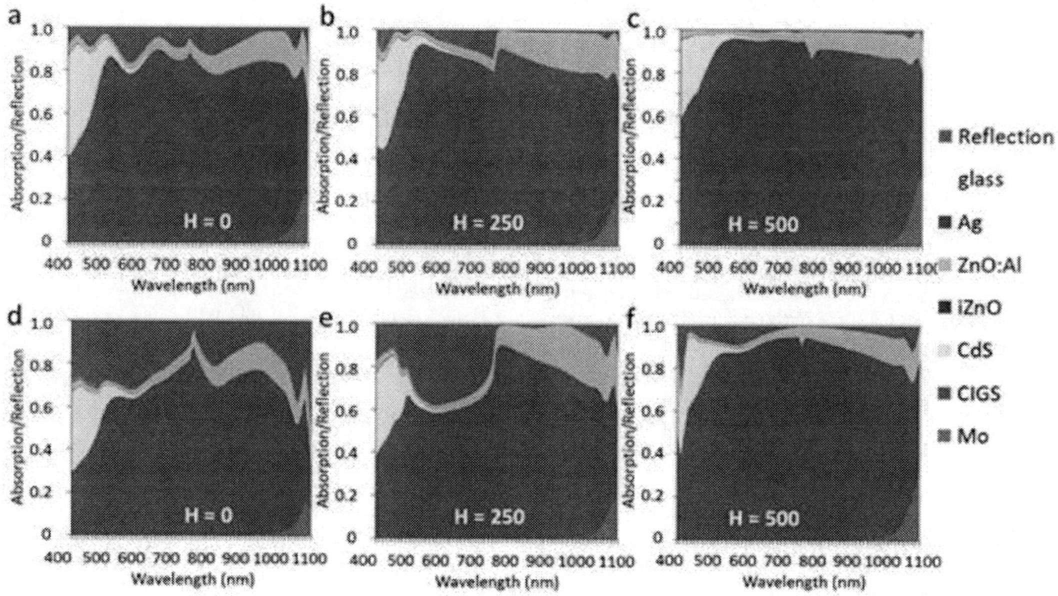

Figure 2: Absorption and reflection spectra for various texture heights (H, in nm). the legend shows the meaning of the colors. The wires width of figures a,b,c is 25 nm and the wire width of figures d,e,f is 150 nm.

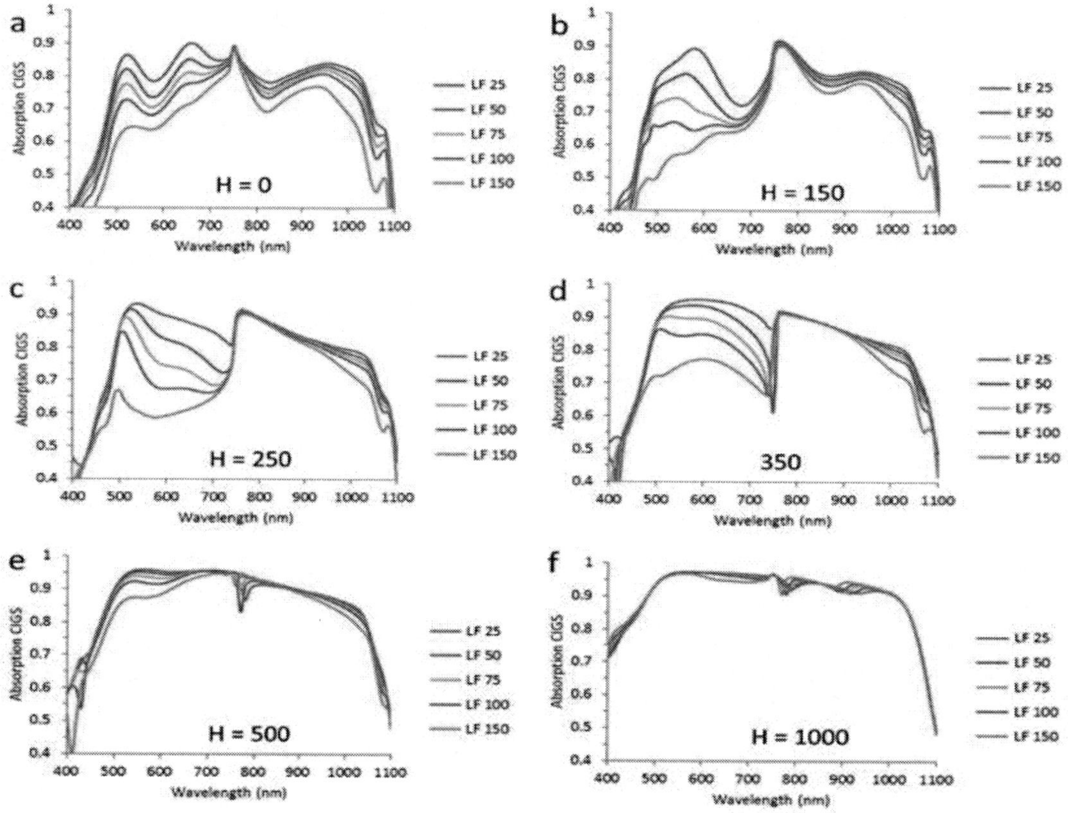

Figure 3: CIGS absorption spectra for various texture heights (H, in nm)) and wires widths (LF, in nm).

3.2 CIGS absorption and total reflection spectra.

The spectra in figure 2a-f show that the texture induced increase of CIGS absorption is mainly related to the reduced reflection. To increase the insight in the precise relations between spectral distribution of the CIGS absorption, the nanowire width (LF) and the texture height (H), these were plotted as shown in figures 3a-f.

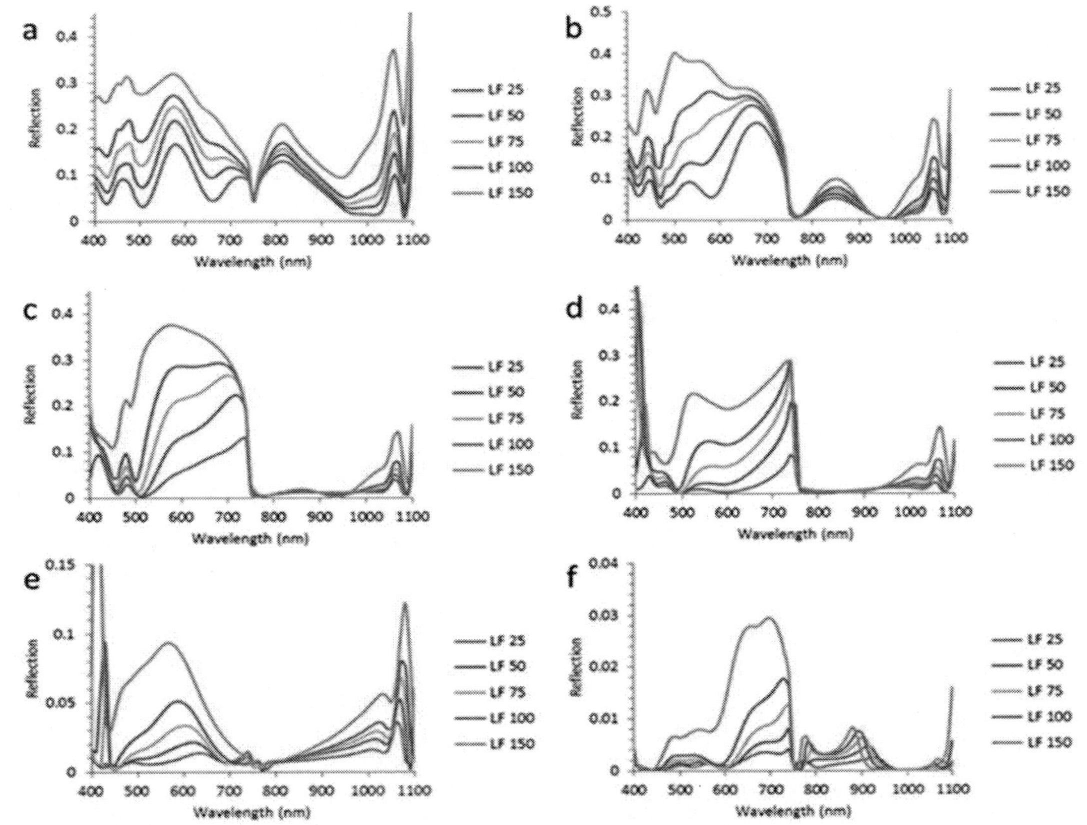

Figure 4: The total reflection in the CIGS device stack as function of the wavelength for a period of 500 nm and a texture height of (a) 0 nm, (b) 150 nm (c) 250 nm, (d) 350 nm, (e) 500 nm, (f) 1000 nm. The width of the nanowire (LF) varies between 25 nm and 150 nm, as indicated in the legend.

The spectral distribution of the CIGS absorption flat layer stacks show a pattern (see figure 3a) and as the wire width increases, the pattern changes to some extent. However, the absorption maximum of 0.9 at 740 nm, does not change. In contrast, texture heights of 150 nm, 250 nm and 350 nm) have a dramatic change in spectral distribution as the wire-width is increased from 25 nm to 150 nm. For very high textures of 500 nm and 1000 nm, the CIGS absorption is high over the whole spectrum and increase of wire-width has only small (at 500 nm texture height) or virtually no (at 1000 nm texture height) impact. This suggests that the optical losses are minimal for these textures, regardless of the wire width.

Reflection data as shown in figure 4 confirms the impression from the total spectra in figure 2; the main loss is reflection and the trends seen in the CIGS absorption are also observed in the reflection spectra. Note that in this case, the y-axis is different for the different figures and give more detailed insight in the spectral distribution of the reflection.

For all texture heights, the main pattern of the spectral distribution of the reflection is the same. There is significant reflection up to a wavelength of approximately 750 nm and very low reflection between 750 nm and 1050 nm. However, notice that the maximum reflection depends on the texture height and varies from 3% for the 1000 nm high texture (fig. 4f) to 40% for the lower texture heights of 150 nm and 250 nm.

3.3 Current density

The more general trends were made visible by showing the CIGS absorption in terms of current density (CIGS current). Figure 5 shows the CIGS absorption, the reflection and the silver absorption in current densities for all the investigated texture heights and nanowire widths for a period of 500 nm and a nanowire height of 50 nm. For this period, the CIGS absorption and the reflection are not depending on the nanowire width for texture heights above 600 nm. Moreover, the overlap is stronger and the height at which this occurs is smaller than observed for 1000 nm for equivalent surface coverages (see ref [10] for data on 1000 nm period). Note that also the total current loss associated with the nanowires is smaller. This is partially attributed to the fact that the nanowires are smaller and thereby less prone to give reflection, as can already be seen spectra in figure 4. According to the optical model, the current without nanowires or texture would be 33.85 mA/cm^2.

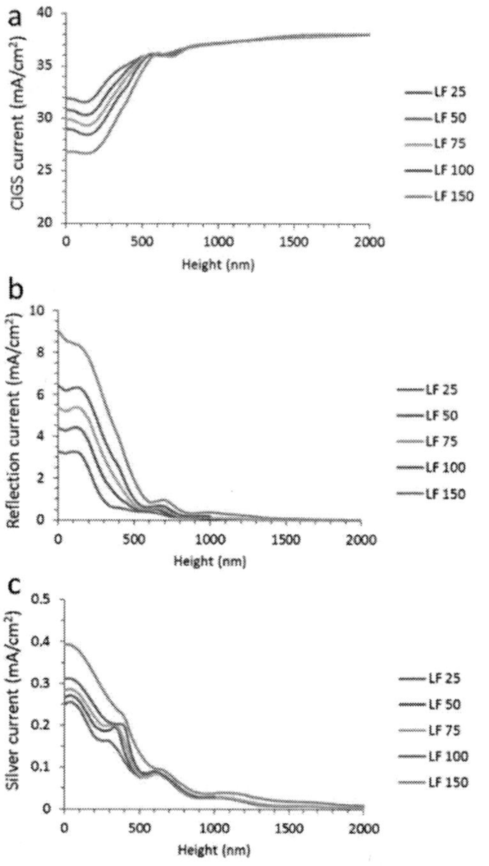

Figure 5: (a) The CIGS absorption of a textured device with a period of 500 nm expressed in current density with metal nanowires as function of the texture height for the different nanowire widths (LF) between 25 nm and 150 nm as indicated in the legend. (b) The total reflection expressed in current density as derived from the spectra as described in the text as function of the texture height for the different nanowire widths (LF) between 25 nm and 150 nm as indicated in the legend. (c) The absorption of the silver nanowires expressed in current density as derived from the spectra as described in the text as function of the texture height for the different nanowire widths (LF) between 25 nm and 150 nm as indicated in the legend.

The absorption in the CIGS expressed in current density increases after some threshold in the texture height and as the reflection reaches near-zero values (see figure 5b), the CIGS reaches a maximum current density above 38 mA/cm^2 as derived from the spectra, i.e. without taking into account only the optical characteristics. The spectra themselves show how the evolution of these overall trends with texture height are wavelength specific. It is foreseen that in addition to the trends discussed in this work, variation in shape of both the texture and the nanowire will induce different effects and future work could be performed in this direction to find a balance between optimum optical design and manufacturability.

4 CONCLUSIONS

The concept of parallel nanowires to improve the conductivity of the front contact is combined with a device texture as to reduce optical losses. Optical modeling was performed for such design as to investigate the critical parameters with respect to optical effects and find texture dimensions that minimize reflection losses. It was found that texture increases the CIGS absorption, mainly due to reduction of reflection losses. For a period of 500 nm, the reflection is virtually independent of the nanowire width for a height/period ratio of only 1. For a texture height of 1000 nm, the surface would appear as black with 3% reflection in the visible, even with a surface metal coverage of 30%. Therefore, a significant increase in current density up to 38 mA/cm^2, which represents an increase of 11mA/cm^2 compared to the flat layer stack.

5 REFERENCES

[1] A. Illiberi, B. Kniknie, J. van Deelen, H.L.A.H. Steijvers, D. Habets, P.J.P.M. Simons, A.C. Janssen, E.H.A. Beckers, Solar Energy Mater. and Solar Cells 95 (2011) 1955.

[2] J. van Deelen, L.A. Klerk, M. Barink, H. Rendering, P. Voorthuizen, A. Hovestad, Thin Solid Films 555 (2014) 159.

[3] J. van Deelen, H. Rendering, A. Hovestad, Proceedings Mater. Res. Soc. Symp 1323 (2011) 17

[4] J. van Deelen, M. Barink, L. Klerk, W.P. Voorthuijzen, A. Hovestad, Prog. Photovolt. Res. Appl. 23 (2015) 498.

[5] J. van Deelen C. Frijters, M. Suijlen, M. Barink, Thin Solid Films 594 (2015) 225.

[6] J van Deelen, Y. Tezsevin, M. Barink, Materials 9 (2016) 96.

[7]. J. van Deelen, C. Frijters, Solar Energy 143 (2017) 93.

[8] J van de Groep, D. Gupta, M.A. Verschuuren, M.M. Wienk, R.A.J. Janssen, A. Polman Scientific Reports 5 (2015) 11414.

[9] V.K. Narasimhan, T.M. Hymel, R.A. Lai, Y. Cui, ACS Nano 9 (2015) 10590.

[10] J. van Deelen, A. Omar, M. Barink, Materials 10 (2017) 392.

HIGHLY RELIABLE LOW CONCENTRATION INGAP/GAAS/SI 3-JUNCTION SOLAR CELLS WITH SMART STACK TECHNOLOGY

Kikuo Makita[1], Hidenori Mizuno[1], Ryuji Oshima[1], Takeshi Tayagaki[1], Masaaki Baba[2], Noboru Yamada[2], Hidetaka Takato[1], and Takeyoshi Sugaya[1]
[1]National Institute of Advanced Industrial Science and Technology (AIST) , Japan
[2]Nagaoka University of Technology, Japan
AIST Tsukuba, Central 2, Umezono 1-1-1, Tsukuba, Ibaraki, 305-8568, JAPAN
Phone: +81-29-861-3263, Fax: +81-29-861-5615, E-mail: kikuo-makita@aist.go.jp

ABSTRACT: Low concentration multi-junction (MJ) solar cells have a practical solution to consist with high efficiency and low cost. We have demonstrated a low concentration InGaP/GaAs/Si 3-junction solar cell with mechanical stacking method. In this development, our key technology is the direct bonding using conductive nanoparticle alignment, which is named "Smart Stack" technology. Using this technology, we fabricated an InGaP/GaAs/Si 3-junction solar cell and observed the maximum efficiency of 23.74% at 8.0 suns (AM1.5g). In addition, we discuss the reliability of InGaP/GaAs/Si 3-junction solar cells showing the results of accelerated aging test and thermal cycling test. It was confirmed that our solar cells have high long-term reliability under sever conditions. Appling general lifetime estimation method with Arrhenius plot, the life time under 10 suns was estimated to be more than 1.5×10^5 hours at 60℃. These results strongly suggested the high potential of our Smart Stack technology to achieve low cost and highly reliable mechanically stacked MJ solar cells for practical use.
Keywords: Silicon, III-V Semiconductors, Nanoparticles, Multijunction Solar Cell

1 INTRODUCTION

Multi-junction (MJ) solar cells have enabled very high efficiencies over 30% due to the effective utilization of the sun's energy. The MJ solar cells using mechanical stacking technology have been particularly interested as next generation solar cell [1, 2]. The mechanical stacking enables high efficiency and low-cost because of flexible and most appropriate combination of different cells. Among them, GaAs/Si MJ solar cell is a useful combination. GaAs-based cell essentially has high efficiency. Si cell has high sensitivity for long wavelength region and holds the promise of low cost bottom cell. According to our theoretical prediction, the current-matched InGaP/GaAs/Si 3-junction solar cells have a potential to reach more than 30% efficiency. In addition, low concentration technique is a practical solution to obtain cost effective system. Low concentration system does not require a tracking system. According to our cost model, low concentration systems below 10 suns realize a competitive cost level.

Against this background, we have developed a unique mechanical staking method using a Pd nanoparticle alignment, which is named "Smart Stack" technology [3-6]. We successfully reported an InGaP/GaAs/InGaAsP/InGaAs 4-junction solar cell with 31.6% efficiency [7] and an InGaP/GaAs/CuInGaSe 3-junction solar cell with 24.2% [5]. Recently, we obtained an InGaP/GaAs/Si 3-junction solar cell with 25.1% [8]. In this paper, we examined the low concentration InGaP/GaAs/Si 3-junction solar cells with Smart Stack technology and addressed the long-term reliability under sever conditions as an initial inspection.

2 DEVICE APPROACH

Figure 1 shows a schematic diagram of the fabricated low concentration InGaP/GaAs/Si 3-junction solar cell. As the top cell, two-junction cell consisting of p-InGaP (Eg-1.89 eV) and p-GaAs (Eg-1.42 eV) absorption layers connected with a tunneling layer were used. As the bottom cell, the simplest and widely-used Si cell with Al-back surface field-type was used. Si (100) wafer was a single-side mirror polished with the thickness of 625 μm. In here, our key technology is the semiconductor bonding with Smart Stack technology, which is the integration of conductive nanoparticle (Pd) alignments at the bonding interface. Pd nanoparticles were aligned on a bottom Si cell through the use of self-assembled block copolymer templates [3,6]. A top GaAs-based cell was separated from the growth substrate by Epitaxial Lift-off (ELO) technique. Finally, two cells were stacked under the light load (5 N/cm^2). Using this technology, low bonding resistance (< 1 Ωcm^2) as well as low interfacial optical loss (<2 %) was reported.

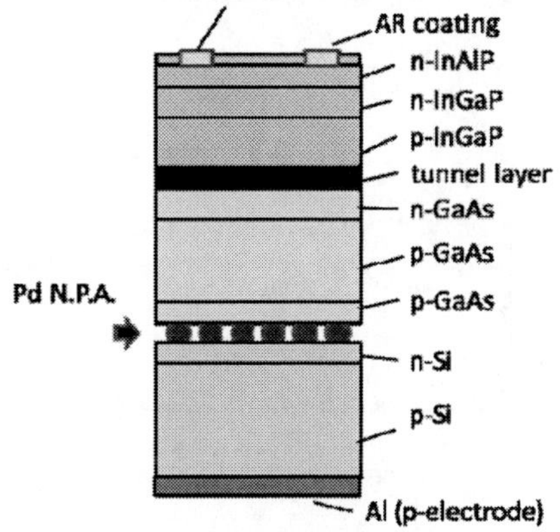

Figure 1: Schematic diagram of the fabricated InGaP/GaAs/Si 3-junction solar cells with Smart Stack technology.

3 EXPERIMENT AND DISCUSSION

3.1 Device Experiment

The current-voltage characteristics were measured under the concentrated AM1.5g spectrum. Figure 2 shows the efficiency in dependence of the concentration factor. Under 1sun, the total efficiency (η) was 19.26% with open-circuit voltage (Voc), short-circuit current density (Jsc) and fill factor (FF) of 2.59V, 11.59mA/cm^2

and 0.64, respectively. The efficiency increases for increasing concentration factor. Maximum efficiency was 23.74% at 8.0 suns. Figure 3 shows the open circuit voltage and the fill factor for concentration factor. The improvement of efficiency is concerned in proper increase of open circuit voltage. The efficiency decreases along with the deterioration of fill factor in concentration factor 10 or more. This degradation is due to the occurrence of negative resistance, which is originated in the tunneling layer of top cell. In the same experiment for InGaP/GaAs 2-junction solar cells, similar tendency was also observed for high current density of over 100mA/cm^2. Therefore, the main cause depends on the insufficient characteristic of top GaAs cell. However, as a result, this performance is practical level for low concentration system below 10 suns. This result is the useful demonstration as low concentration examination for GaAs/Si MJ solar cell with Smart Stack technology.

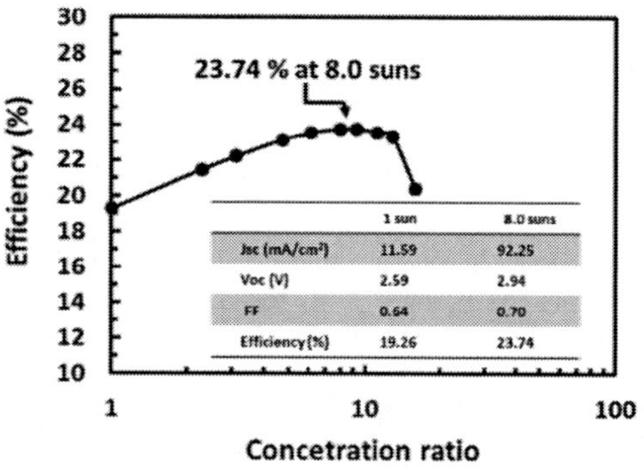

Figure 2: Efficiency of the InGaP/GaAs/Si 3-junction solar cells against the concentration factor.

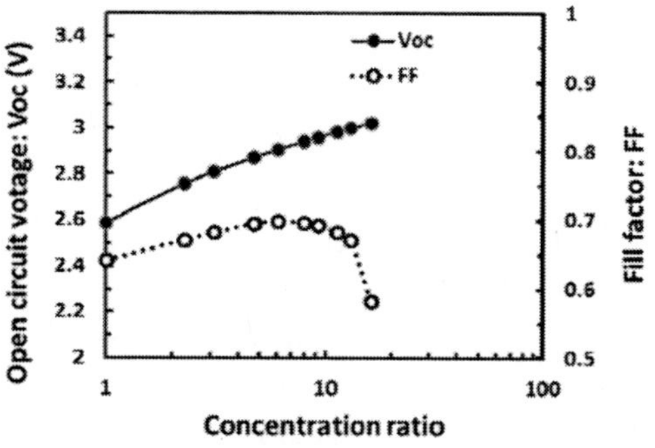

Figure 3: Open circuit voltage and fill factor of the InGaP/GaAs/Si 3-junction solar cells against the concentration factor.

3.2 Reliability Experiment

For practical use, it is essential to achieve solar cells with high reliability. From this point of view, we performed accelerated aging test and thermal cycling test. The solar cells used here were InGaP/GaAs/Si 3-junction solar cells bonded by the Pd nanoparticles alignment (with the AR coating). The accelerated aging test was conducted at 150 ℃ under the forward current for 100 hours. The forward current was set to be 100mA/cm^2, which was an equivalent current to use for the low concentrated sunlight (10 suns). After the designated time, dark J-V measurements of the cells were carried out at room temperature. In Fig. 4, the diode resistivity (at 100 mA/cm^2), which was determined by numerically calculating dJ/dV, are shown as a parameter for aging time. The diode resistivity includes the cell resistance and the bonding resistance. According to our simulation, if it is deterioration less than 2 Ωcm^2, the relative efficiency deterioration at 10 suns is held down to less than 10%. In this study, the deterioration standard was set to be 2 Ωcm^2. In Fig.4, the 100-hour data was judged with lower than the deterioration standard experimentally. In addition, Figure 5 shows the result of light J-V measurements before and after the aging test at 1 sun. It was confirmed that the relative degradation rate of the conversion efficiencies was less than 2%. The thermal cycling test was also carried out under the temperature of -40 and +85℃. Figure 6 is the result of light J-V measurements before and after the 200 cycles test. Obvious resistivity degradations was not observed after 200 times, and the conversion efficiencies were thus maintained almost the same.

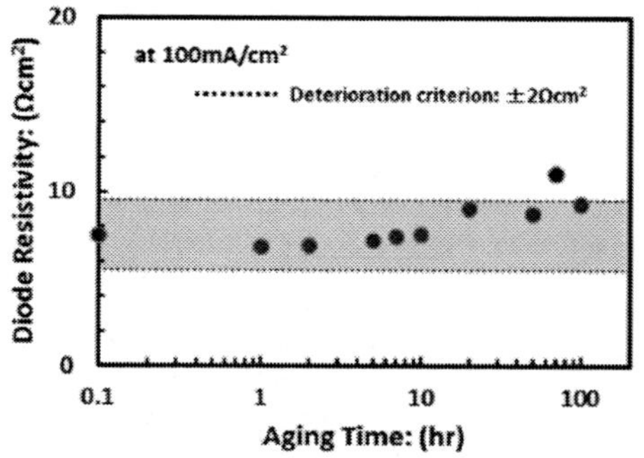

Figure 4: Change in the diode resistivity (at 100 mA/cm^2) with aging times obtained by the accelerated aging test.

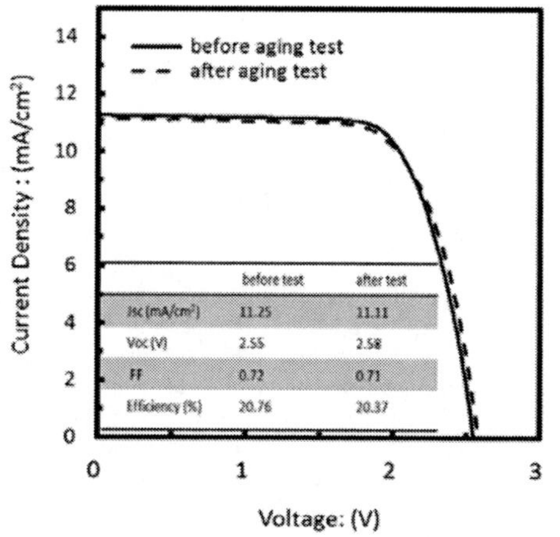

Figure 5: Light J–V characteristics of the InGaP/GaAs/Si 3-junction solar cell before and after the aging test.

The reliability tests for Smart Stack GaAs/Si multi-junction cells were performed as an initial inspection. The obtained results suggest that the bonding interface with Pd nanoparticle is highly stable for rugged environments. Appling the result of accelerated aging test to Arrhenius plot method, the life time under 10 suns was estimated to be more than 1.5×10^5 hours at 60℃ by using activation energy of 1 eV [7]. This activation energy was preliminary value estimated with GaAs/InP MJ solar cells. The stability is in a sufficient level for actual solar cell applications.

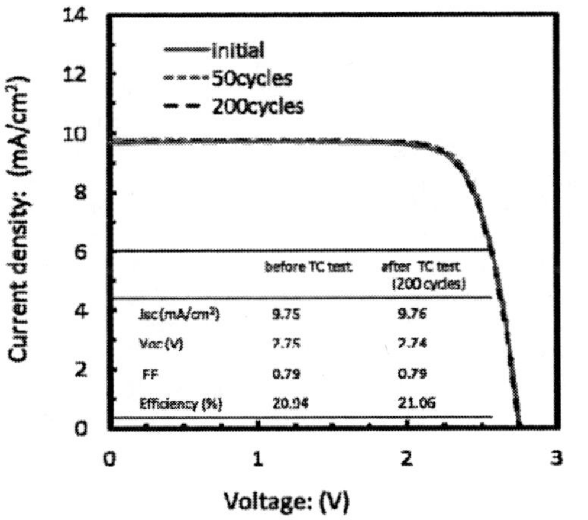

Figure 6: Light J–V characteristics of the InGaP/GaAs/Si 3-junction solar cell before and after the 200 cycles test.

4 CONCLUSIONS

We fabricated the low concentration InGaP/GaAs/Si 3-junction solar cell with Smart Stack technology, and observed the maximum efficiency of 23.74% at 8.0 suns. According to our theoretical prediction, the efficiency can be improved over 30%. In addition, we demonstrated the bonding interface is stable for a long term through the reliability tests of temperature acceleration and temperature cycling. These results suggest that Smart Stack technology is promising to obtain low cost and highly reliable low concentration InGaP/GaAs/Si MJ solar cells, as next generation solar cell. The improvement for device performance is currently ongoing in our group.

This work was supported by the New Energy and Industrial Technology Development Organization.

References

[1] P. T. Chiu, D. C. Law, R. L. Woo, S. B. Stinger, D. Bhusari, W. D. Hong, A. Zakaria, J. Boisvert, S. Mesropian, R. R. King, and N. H. Karam, "Direct semiconductor bonded 5J cell for space and terrestrial applications", IEEE Journal of Photovoltaics, Vol. 4 (2014) 493.

[2] A. W. Bett, S. P. Philipps, S. Essig, S. Heckelmann, R. Kellenbenz, V. Klinger, M. Niemeyer, D. Lackner, and F. Dimroth, "Overview about technology perspectives for high efficiency solar cells for space and terrestrial applications", 28th European Photovoltaic Solar Energy Conference, (2013) 1AP.1.1.

[3] H. Mizuno, K. Makita, and K. Matsubara, "Electrical and optical interconnection for mechanically stacked multi-junction solar cells mediated by metal nanoparticle arrays", Appl. Phys. Lett., Vol. 101 (2012) 191111.

[4] K. Makita, H. Mizuno, H. Komaki, T. Sugaya, R. Oshima, H. Shibata, K. Matsubara, and S. Niki, "Over 20% efficiency mechanically stacked multi-junction solar cells fabricated by advanced bonding using conductive nanoparticle alignments", 2013 MRS Spring Meeting, (2013) FF2.08.

[5] K. Makita, H. Komaki, H. Mizuno, H. Sai, T. Sugaya, R. Oshima, H. Shibata, K. Matsubara, and S. Niki, " A 24.2%-efficiency Ⅲ - V /CuInGaSe mechanical stacking multi-junction solar cells using semiconductor bonding method" , 29th European Photovoltaic Solar Energy Conference, , (2014) 3O.4.1.

[6] H. Mizuno, K. Makita, T. Sugaya, R. Oshima, Y. Hozumi, H. Takato, and K. Matsubara, "Palladium nanoparticle array-mediated semiconductor bonding that enables high-efficiency multi-junction solar cells", Jap. J. Appl. Phys., Vol. 55 (2016) 025001

[7] K. Makita, H. Mizuno, R. Oshima, S. Nonaka, A. Furukawa, H. Takatao, and T. Sugaya, "Development of smart stacking technology for multi-junction solar cells", 21st International Photovoltaic Science and Engineering Conference, (2015) GAS. O. 13.

[8] H. Mizuno, K. Makita, T. Tayagaki, T. Mochizuki, T. Sugaya, and H. Takato, "High-efficiency Ⅲ-V//Si tandem solar cells enabled by the Pd nanoparticle array-mediated "smart stack" approach", Appl. Phys. Express, Vol.10 (2017) 072301.

ACHROMATIC LENS CASTING NEARLY UNIFORM IRRADIANCE OVER MJ SOLAR CELLS

Marta Victoria*, Guido Vallerotto, Stephen Askins, Ignacio Antón, Gabriel Sala
Instituto de Energía Solar, Universidad Politécnica de Madrid
*marta.victoria@ies.upm.es

ABSTRACT: The Achromatic Doublet on Glass (ADG) Fresnel lens is a novel concentrator that combines a low-dispersion elastomer and a high-dispersion plastic to obtain a Fresnel lens with reduced chromatic aberration, and consequently, increased concentration capability. This article summarizes the outcomes of the ray-tracing simulations as well as the experimental results obtained within the past two years. In addition, the uniformity of the irradiance profile cast by the lens onto Multi-Junction solar cells is analyzed and compared to that of a classic Silicone on Glass (SoG) Fresnel lens used as a benchmark. The more uniform profile cast by the achromatic lens requires lower horizontal current transportation across the MJ solar cell since the photocurrent distributions for the top and middle subcells are similar. It has been experimentally shown that, although the Fill Factor (FF) of the solar cell illuminated by the SoG Fresnel lens strongly depends on the lens-to-cell distance (showing a characteristic W-shaped curve), for the ADG Fresnel lens, the FF remains high at every position. This implies a higher tolerance not only to errors in the module assembly but also to variations in the temperature of the primary lens.
Keywords: Concentrators, Characterization, Multijunction Solar Cell

1 INTRODUCTION

CPVMatch is a Horizon 2020-funded project that aims to attain 40% efficient concentrating photovoltaic (CPV) modules using monolithic Multi-Junction (MJ) solar cells and improved concentrating optics [1]. Within this framework, the IES-UPM is developing a novel achromatic lens design and manufacturing method. The Achromatic Doublet on Glass (ADG) Fresnel lens benefits from the usual advantages of refractive concentrators (high efficiency, ease of parquet assembly and low cost) while simultaneously reducing their main drawbacks (limited attainable concentration due to chromatic aberration and high thermal sensitivity).

The ADG Fresnel lens is composed of glass, plastic and elastomeric materials. The lens design takes advantage of the different dispersion of the materials to reduce chromatic aberration at the wavelengths range converted by the MJ solar cells. First, a 'bi-Fresnel' lens is obtained by plastic injection, compression molding or hot-embossing. Then, it is laminated together with the elastomer over a glass substrate (Fig. 1). The manufacturing method, completely different to previous approaches [2], is based on standard industrial processes and it is one of the key innovations of the project. In summary, the proposed process enables, for the first time, an achromatic lens affordable for the CPV market.

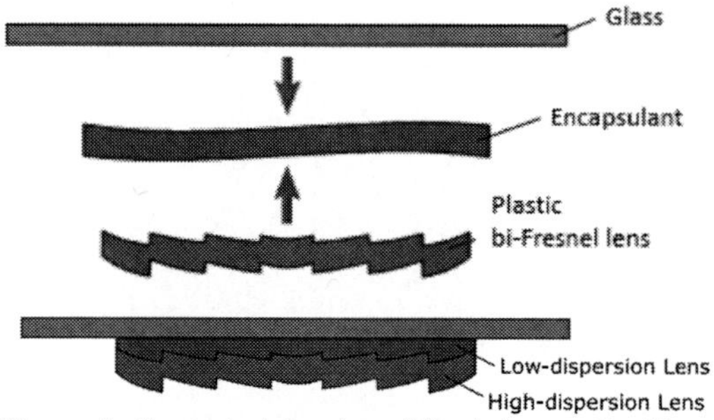

Figure 1: Conceptual drawing of the Achromat Doublet on Glass (ADG) Fresnel lens: (top) the components prior to the lamination process, (bottom) the laminated doublet.

The paper is organized as follows. The main simulation and experimental results obtained within the past two years are summarized in section 2. Then, section 3 focuses on the analysis of the irradiance profiles cast by the ADG and SoG Fresnel lenses and their impact on the performance of a MJ solar cell. Finally, the main conclusions are summarized in section 4.

2 SIMULATED AND EXPERIMENTAL RESULTS

The first step in the development of this novel concentrator comprised an extensive set of ray-tracing simulations to assess the performance of the ADG Fresnel lens that we have designed. The attainable concentration, acceptance angle, efficiency, irradiance profile cast by the lens, as well as the impact of real-world constraints (draft angles, tip rounding) and manufacturing error have been reported elsewhere [3]. In both cases, for the ray-tracing and for the experimental tests, a classic Silicone on Glass (SoG) Fresnel lens [4] with the same optical aperture (40 x 40 mm) and focal distance (approximately 75 mm) is used as a reference against which the results are evaluated.

The ray-tracing simulations showed that the reduced chromatic aberration increases the concentration-acceptance product (CAP) of the lens. Hence, either a higher concentration or a wider acceptance angle can be achieved. For example, operating at the same optical efficiently, the concentration attained by the ADG Fresnel lens is 722X while that achieved by the SoG is 267X [3].

The ray-tracing also showed that the best pair of materials was polycarbonate (PC) and ethylene-vinyl acetate (EVA) so they were chosen for the first prototypes. The rigid bifacial Fresnel lens was manufactured out of PC by Fresnel Optics GmbH according to our design. We used a custom laminator to encapsulate these lenses to small pieces of low-iron glass. After several weeks of fine-tuning the process, we were able of producing the elementary unit prototype shown in Fig. 1 which has very good transparency and it is air bubble free. Furthermore, we had also been able of obtaining arrays of lenses by one-single lamination (Fig. 2). We believe the ease with which we could proceed with lamination bodes well for the manufacturability prospects of this lens.

Figure 2: A PC-EVA ADG Fresnel lens prototype laminated at the IES-UPM.

Figure 3: Photograph of the parquet composed of ADG Fresnel lenses. The PC bi-Fresnel lenses were first manufactured by plastic injection, and then, encapsulated to EVA and a glass substrate in a single lamination at the IES-UPM.

The achromatic behavior of the ADG Fresnel lens was experimentally proven by measuring the size of the spot cast by the lens using a charge-coupled device (CCD) camera. This is one of the techniques usually employed at IES-UPM to characterize CPV primary optics and a detailed description can be found in [5]. In summary, the lens is illuminated with a flash solar simulator for CPV modules [6] and high-pass or low-pass filters are added to the CCD camera. This allows to discriminate between the spots of light that can be converted into electricity by the different subcells within a MJ solar cell. By measuring the evolution of the 'blue' and 'red' spots as the lens-to-cell distance is varied, the achromatic behavior of the lens was proven [7]. The measurements were repeated at different lens temperatures proving that the thermal sensitivity of the ADG Fresnel lens is lower than that of the reference SoG Fresnel lens. We believe that two effects are the main responsible for the reduced thermal sensitivity. First, refractive index variation as a function of temperature is lower for the plastic and elastomeric materials comprising the ADG lens than for the silicone. Second, the difference in the coefficient of thermal expansion (CTE) of the materials comprising the lens is particularly adverse in the glass-silicone pair, causing the deformation of the silicone and the change in the optics profile [8]. Conversely, the materials comprising the ADG show a more similar CTE which implies a reduced deformation of the optically active surfaces. This is a significant result since SoG Fresnel lens are known to be highly impacted by changes in the temperature of the lens [8], [9] which can reduce the energy harvested by the CPV system throughout a year [10].

The tolerance of the ADG Fresnel lens to variation in the lens-to-cell distance has also been experimentally characterized using solar cells of different sizes [11]. Additionally, the spectrally-resolved optical efficiency of the ADG Fresnel lens has been measured by using a recently introduced experimental technique described in [12]. In this experiment, the efficiency at different narrow spectral bands is measured using a MJ solar cell as light sensor and a series of band-pass filters. Additional bias light is provided to saturate the subcells whose spectral response is out of the transmittance of every band-pass filter.

3 IRRADIANCE CAST ON MULTI-JUNTION SOLAR CELLS

The chromatic aberration suffered by a SoG Fresnel lens causes an irradiance profile over the MJ solar cell that is neither spatially nor spectrally uniform. This may worsen the performance of the solar cell due to an increased effective series resistance. If the distributions of photogenerated currents are very different for the top and middle subcells, the current needs to flow horizontally across the device which increases the effective series resistance and decrease the solar cell fill factor (FF) [13], [14]. For SoG Fresnel lenses, due to the high sensitivity of silicone refractive index with temperature, the irradiance profile significantly varies during the lens outdoor operation (see Fig. 4). On the top of that, the variation in the spectral content of the incident light as a function of the time of the year does not counterbalance this effect but exacerbates it. Thus, in winter, as the temperature decreases, the silicone refractive index increases, the lens becomes more converging and 'red' light (converted by the middle subcell) is focused on the MJ solar cell entrance. At the same time, since air mass (AM) values are higher during summer, the 'blue' light is absorbed and a red-shifted spectral distribution is found at the entrance of the optical system. The opposite situation is found in summer. High temperatures reduce silicone refractive index focusing 'blue' light on the cell. Simultaneously, lower AM values found in summer also increase the 'blue' content of the input spectrum [15].

One quick approach to analyze this effect consists in recording the IV curve of a MJ solar cell illuminated by a primary lens as the lens-to-cell distance is varied. For SoG Fresnel lenses, this measurement has been previously reported by several authors [16]–[18]. When plotted against the lens-to-cell distance, the FF of the solar cell, shows a characteristic W-shaped signature. This is known to be caused by the evolution of two combined effects: the spatial concentration and the mismatch among the subcell within the MJ device [18].

As a general rule, the extent to which non-uniformity reduces the FF of a solar cell depends on both, the characteristics of the non-uniform irradiance distribution produced by a certain optics and the ability of the solar cell to smooth the uneven profile. This capability depends on the cell size and its technological characteristics (which determines the effective series resistance for the horizontal current transportation).

In our case, the outdoor measurements of the IV curve of a lattice-matched Ge-based MJ solar cell illuminated by the reference SoG Fresnel lens unveil the typical W-shaped evolution of the FF (see Fig. 6). However, when the cell is illuminated by the ADG Fresnel lens and the lens-to-cell distance is modified, the FF remains almost constant for every position. This results can be predicted by observing the evolution of the photocurrent distributions for the top and middle subcells of a Lattice Matched Germanium-based (LM-Ge) obtained from ray-tracing (Fig. 5). Details for the ray-tracing simulations are included in [3]. In the first place, the photocurrent distributions for top and middle subcells are more similar, reducing the horizontal current flow. In the second place, the ADG Fresnel lens is less sensitive to temperature changes (which are to some extent, equivalent to variations in the lens-to-cell distance).

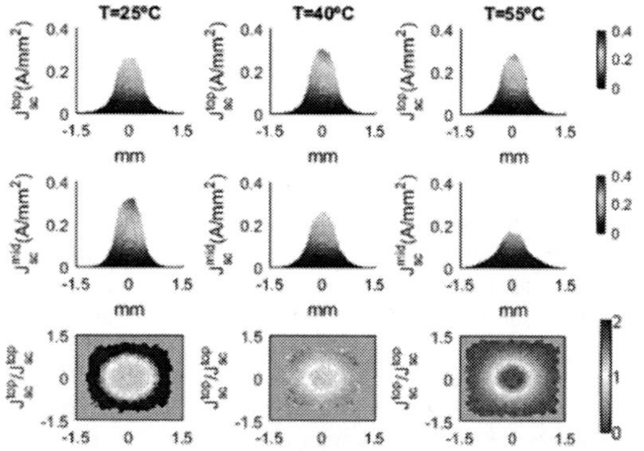

Figure 4: Results for the SoG Fresnel lens used as a benchmark. Spatial distribution of the photogenerated current for the top (first row) and middle (second row) subcells of a LM-Ge MJ solar cell at different temperatures predicted by ray-trace simulations. Contour plots (third row) show the ratio between top and middle photocurrents. Blue areas represent cell regions where there is an excess of top photogenerated current (middle limited) while red areas represent an excess of middle photogenerated current (top limited). Regions where both subcell photocurrents are matched are shown in white. Gray areas represent regions where both photocurrent values are below 0.1% of the maximum, that is, dark areas.

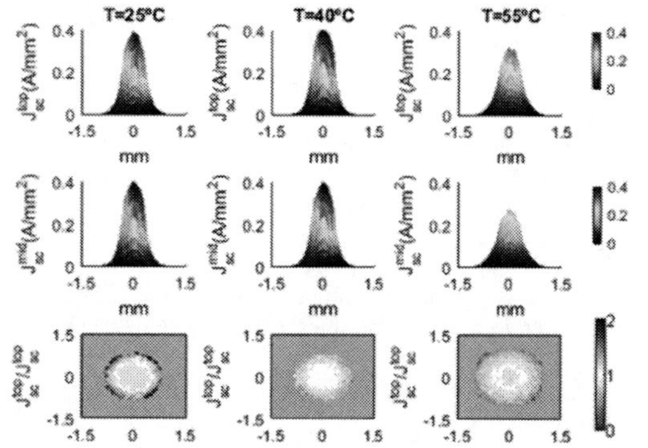

Figure 5: Results for the ADG Fresnel lens. Spatial distribution of the photogenerated current for the top (first row) and middle (second row) subcells of a LM-Ge MJ solar cell at different temperatures predicted by ray-trace simulations. Contour plots (third row) show the ratio between top and middle photocurrents. Blue areas represent cell regions where there is an excess of top photogenerated current (middle limited) while red areas represent an excess of middle photogenerated current (top limited). Regions where both subcell photocurrents are matched are shown in white. Gray areas represent regions where both photocurrent values are below 0.1% of the maximum, that is, dark areas.

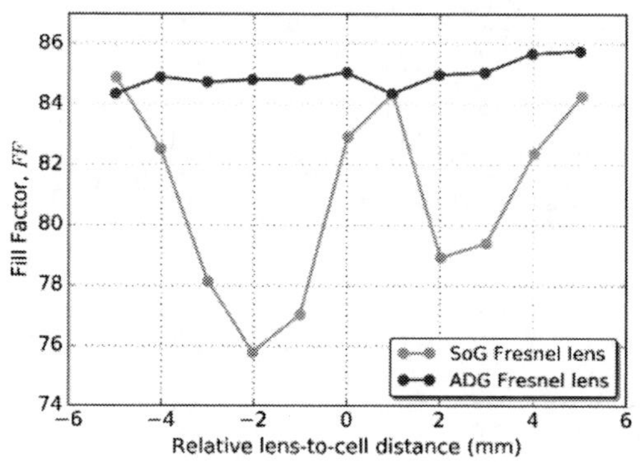

Figure 6: Fill factor *FF*, of the solar cell illuminated by the ADG Fresnel lens as a function of the relative lens-to-cell distance. The measurement was performed outdoors at Fraunhofer-ISE. For comparison, a SoG Fresnel lens with the same optical aperture and focal distance was also characterized.

4 CONCLUSIONS

The Achromatic Doublet on Glass (ADG) Fresnel lens is the first achromatic lens affordable for Concentrating Photovoltaics (CPV) thanks to the proposed manufacturing method in which a low-dispersion elastomer and a high-dispersion plastic are combined in a low-cost lamination process. Reducing the chromatic aberration allows increasing the concentration-acceptance angle product (CAP) of the ADG Fresnel lens when compared to the values attained by a classic Silicone on Glass (SoG) Fresnel lens. Elementary unit and parquet of lenses prototypes have been manufactured showing good transparency and no bubbles. The achromatic behavior of the ADG has been experimentally demonstrated in previous works.

This paper focuses on the analysis of the irradiance distribution cast by the ADG Fresnel lens onto a MJ solar cell and its effect on the cell performance. The ADG Fresnel lens produces similar photocurrent distributions for the top and middle subcells, and consequently, reduces the effective series resistance, and increases the Fill Factor (FF) of the MJ solar cell. The FF has been experimentally measured as a function of the lens-to-cell distance. While the FF corresponding to the SoG Fresnel lens used a benchmark shows a characteristic W shape, the FF for the ADG Fresnel remains constantly high for different positions proving the advantages of a more uniform illumination. Moreover, the sensitivity to changes in temperatures is lower for the ADG Fresnel lens.

5 ACKNOWLEDGEMENTS

This work has been partially supported by the Spanish

Ministry of Economy and Competitiveness under the Acromalens project (ENE2013-45229-P) and it has received funding from the European Union's Horizon 2020 research and innovation program within the project CPVMatch under grant agreement No 640873. The authors thank Maike Weisenfarth at Fraunhofer-IES for the outdoor measurements of the ADG Fresnel lens.

6 REFERENCES

[1] "CPVMatch - Concentrating photovoltaic modules using advanced technologies and cells for highest efficiencies," *AIP Conf. Proc.*, vol. 1766, no. 1, p. 060002, 2016.

[2] F. Languy *et al.*, "Flat Fresnel doublets made of PMMA and PC: combining low cost production and very high concentration ratio for CPV," *Opt. Express*, vol. 19 Suppl 3, pp. A280-294, 2011.

[3] G. Vallerotto *et al.*, "Design and modeling of a cost-effective achromatic Fresnel lens for concentrating photovoltaics," *Opt. Express*, vol. 24, no. 18, pp. A1245–A1256, 2016.

[4] E. Lorenzo and G. Sala, "Hybrid silicone-glass Fresnel lens as concentrator for photovoltaic applications," in *Sun II*, 1979, vol. 1, pp. 536–539.

[5] M. Victoria, S. Askins, R. Herrero, I. Antón, and G. Sala, "Assessment of the optical efficiency of a primary lens to be used in a CPV system," *Sol. Energy*, vol. 134, pp. 406–415, 2016.

[6] C. Domínguez, I. Antón, and G. Sala, "Solar simulator for concentrator photovoltaic systems," *Opt Express*, vol. 16, no. 19, pp. 14894–14901, 2008.

[7] G. Vallerotto, S. Askins, M. Victoria, I. Antón, and G. Sala, "A novel achromatic Fresnel lens for high concentrating photovoltaic systems," in *AIP Conference Proceedings*, 2016, vol. 1766, p. 050007.

[8] S. Askins, M. Victoria, R. Herrero, C. Domínguez, I. Antón, and G. Sala, "Effects of Temperature on Hybrid Lens Performance," *AIP Conf. Proc.*, vol. 1407, pp. 57–60, 2011.

[9] V. D. Rumyantsev, N. Y. Davidyuk, E. A. Ionova, P. V. Pokrovskiy, N. A. Sadchikov, and V. M. Andreev, "Thermal Regimes of Fresnel Lenses and Cells in 'All-Glass' HCPV Modules," *AIP Conf. Proc.*, vol. 1277, pp. 89–92, 2010.

[10] T. Hornung, M. Steiner, and P. Nitz, "Estimation of the influence of Fresnel lens temperature on energy generation of a concentrator photovoltaic system," *Sol. Energy Mater. Sol. Cells*, vol. 99, pp. 333–338, 2012.

[11] G. Vallerotto, M. Victoria, S. Askins, I. Antón, and G. Sala, "Experimental characterization of achromatic doublet on glass (ADG) Fresnel lenses," *AIP Conf. Proc.*, vol. 1881, no. 1, p. 030010, 2017.

[12] M. Victoria, C. Domínguez, N. Jost, G. Vallerotto, I. Antón, and G. Sala, "Spectrally-resolved optical efficiency using a multi-junction cell as light sensor: Application cases," *AIP Conf. Proc.*, vol. 1881, no. 1, p. 030011, 2017.

[13] S. R. Kurtz and M. J. O'Neill, "Estimating and controlling chromatic aberration losses for two-junction, two-terminal devices in refractive concentrator systems," in *Proceeding 25th IEEE Photovoltaic Specialists Conference*, 1996, pp. 361–364.

[14] P. Espinet-González, I. Rey-Stolle, C. Algora, and I. García, "Analysis of the behavior of multijunction solar cells under high irradiance Gaussian light profiles showing chromatic aberration with emphasis on tunnel junction performance," *Prog. Photovolt. Res. Appl.*, vol. 23, no. 6, pp. 743–753, 2015.

[15] M. Victoria *et al.*, "Tuning the current ratio of a CPV System to maximize the energy harvesting in a particular location," in *CPV-9*, Miyazaki, Japan, 2013, vol. 1556.

[16] H. Cotal and R. Sherif, "The effects of chromatic aberration on the performance of GaInP/GaAs/Ge concentrator solar cells from Fresnel optics," in *Photovoltaic Specialists Conference, 2005. Conference Record of the Thirty-first IEEE*, pp. 747–750.

[17] V. M. Andreev, V. A. Grilikhes, A. A. Soluyanov, N. H. Timoshina, E. V. Vlasova, and M. Z. Shvarts, "Weakening of the chromatic aberration negative effect on the performance of concentrator multi-junction solar cells," in *22nd EPVSEC*, Milan, 2008, pp. 126–131.

[18] M. Victoria, R. Herrero, C. Domínguez, I. Antón, S. Askins, and G. Sala, "Characterization of the spatial distribution of irradiance and spectrum in concentrating photovoltaic systems and their effect on multi-junction solar cells," *Prog. Photovolt. Res. Appl.*, vol. 21, no. 3, pp. 308–318, 2013.

COLORED SOLAR MODULES USING INTEGRATED PIXELATED RESONANT DIELECTRIC NANOSCATTERER ARRAYS

Verena Neder,[1,2] Stefan L. Luxembourg,[3] and Albert Polman[2]
[1]Institute of Physics, University of Amsterdam
Science Park 904, 1098 XH Amsterdam, the Netherlands
[2]Center for Nanophotonics, AMOLF,
Science Park 104, 1098 XJ, Amsterdam, the Netherlands
[3]ECN Solar Energy, PO Box 1, 1755 ZG Petten, the Netherlands

ABSTRACT: Green colored photovoltaic modules based on silicon heterojunction solar cells integrated with periodic arrays of light scattering Si nanoscatterers were made. Here, we expand this concept to multiple colors and color combinations. We show that random arrays of silicon nanoarrays can be combined to pixels to create the desired color. By adjusting the coverage of pixels composed of different nanoparticle arrays different colors can be designed and the calculated chromaticity values can be obtained from the fabricated sample. The pixelated nanoscattering concept was used to realize a solar mini-module with a scattering spectrum close to white. The calculated relative loss in short-circuit current is only 19%. The nanoparticle pixel arrays can be made using soft imprint lithography opening up an alternative route for designing low loss colored PV modules.
Keywords: nanoparticles, silicon solar cell, building integration

1 INTRODUCTION

For photovoltaics (PV) to be applied on a very large scale it is essential that PV panels are well integrated into our built environment and landscape. Typically, PV panels have a black or dark blue appearance that results from the textured surface of the solar cells in combination with a dielectric antireflection (AR) coating, designed to optimize the absorption of sunlight in the solar cell. However, for building- and landscape-integrated applications it would be beneficial to have PV panels in a variety of well-defined colors, while maintaining a high PV conversion efficiency.

Recently, we have reported a new method to create a green colored solar module using resonant light scattering from silicon nanoparticles with optical Mie resonances [1]. We exploited the use of dielectric Mie scatterers, integrated with the top cover layer of a solar panel, to selectively scatter a narrow band of the solar spectrum resulting in a bright green color. In the present paper, this concept is expanded in order to achieve a broad range of colors; and to create white solar mini-modules. The reduction in short circuit current densities is only 11% for the green colored mini-modules and 19% for the white colored mini-module.

2 DIELECTRIC SCATTERERS FOR COLORATION

Dielectric nanostructures are known to have bright scattering colors due to the excitation of Mie resonances. The resonant spectrum can be tailored by particle shape and refractive index [2]. If made from low-loss dielectrics the Mie resonance can have a relatively narrow linewidth, resulting in a bright scattering spectrum with -depending on the used dielectric- negligible absorption losses. Several authors have demonstrated colorful examples of light scattering from arrays of Mie scatterers made on planar substrates [3]-[7].

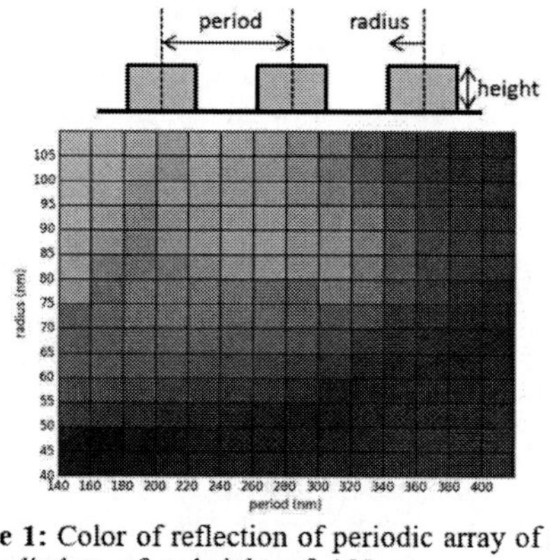

Figure 1: Color of reflection of periodic array of silicon Mie cylinders of a height of 100 nm on a sapphire substrate for different radius and period. Numerical simulations of reflection spectra were done using Lumerical FDTD and are transformed to visible color. A schematic of the geometry of the structure is shown above.

Figure 1 shows the result of numerical finite-difference-time-domain (FDTD) simulations (Lumerical) of the reflected color of square arrays of silicon nanocylinders (height 100 nm) on a semi-infinite sapphire substrate. The visual colors were obtained using the concept described in [8], using the average midday light spectrum in Western Europe (D65) as illumination source. By varying the radius and period of the Si cylinders, the Mie resonance spectrum is controlled and the color can be tuned over the entire color spectrum visible by the human eye. In Fig. 2 a photograph of a sample with six fields of silicon nanocylinder arrays of different radius and period (height ~105-110 nm) on top of sapphire is shown. The fabrication was done by electron beam lithography and reactive ion etching of a planar silicon-on-sapphire substrate. The measured scattering spectra of the fields in Fig. 2a are plotted in Fig. 2b and show that with increasing radius and period the Mie resonances are shifting to larger wavelengths. The simulated reflectance spectra (height 105 nm) plotted in the same figure match well with the experiment. The scattering bandwidth is determined by the scattering rate and absorption losses in the nanoparticles and varies over the shown spectral range.

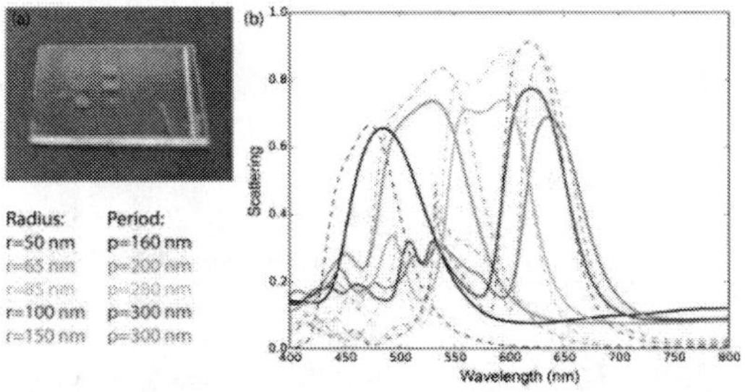

Figure 2: **(a)** Photograph of 6 fields of periodic arrays of silicon cylinders on top of a sapphire substrate with different radius and period (height 105-110 nm). Radius and periods of the simulated arrays are indicated. They match with ±10 nm the fabricated dimensions. **(b)** Measured (solid lines) and simulated (dashed lines) scattering spectra of 5 colored fields.

3 PRINTED SINGLE COLORED SOLAR CELL

To integrate the Mie scattering effect into a solar module, the nanopatterned sapphire cover slides with periodic arrays of Si cylinders (diameter 100-120 nm; height 240 nm) on the bottom, were placed on a front- and back-contacted Si heterojunction solar cell using index matching fluid. The integrated design is shown in Fig. 3. The nanoscatterer array was fabricated by substrate conformal imprint lithography (SCIL).

Figure 3: Schematic of silicon nanoscatterer arrays on bottom of a sapphire cover slide, integrated into a silicon heterojunction solar mini-module using immersion oil. The sapphire slide and the immersion oil mimic commonly used module glass and EVA.

Figure 4 (taken from [1]) shows the measured scattering spectrum of the integrated Mie scattering design. A clear resonance (peak amplitude 35-40%), due to the magnetic (550 nm) and electric (528 nm) dipole resonances is observed [9,10]. Reference measurements for a bare Si heterojunction solar cell and the unpatterned module are also shown. The bare cell shows low reflection due to the combination of texture and optimized AR coating (interference minimum at ~630 nm). The unpatterned module has a reflectivity of 7-12 % as expected given the refractive index for sapphire. The inset in Fig. 4 shows a photograph of the 12x12 mm^2 sapphire slide with the green reflecting Mie scattering design on top of the HIT solar cell. The green scattering intensity is quite insensitive to variation in incoming angle. For more details, see [1].

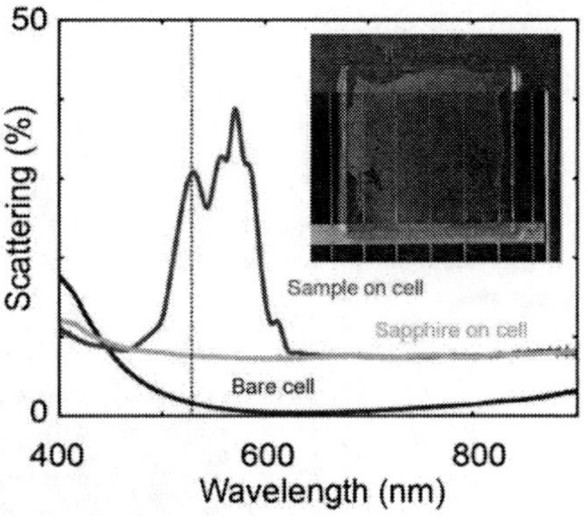

Figure 4: Measured scattering efficiency (green) of integrated design. Data for an unpatterned mini-module (light blue) and the bare Si heterojunction solar cell (dark blue) are also shown. Vertical black dashed line at λ=528 nm marks the electrical dipole resonance. Inset: Photograph of nanopatterned module shows bright green reflectance. Sapphire sample is 12x12 mm. Image taken from [1].

4 MIXED COLOR PIXELS

As described in Section 2, by variation of the size and density of the nanocylinders, the reflected color can be tuned to the desired spectral color range. To achieve mixed colors or white, we combine pixels of different colors, with each pixel composed of 50x50 μm^2 nanocylinder arrays. Figure 5 shows SEM images of the blue (radius 50 nm) and red (radius 90 nm) scattering random nanocylinder arrays. A random geometry was implemented to avoid diffraction effects for large scattering angles.

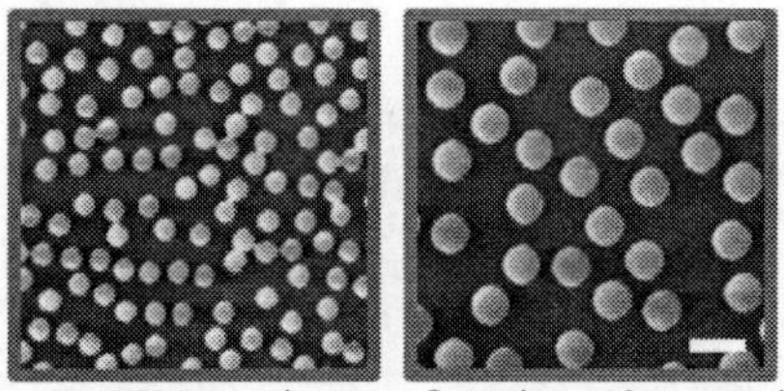

Figure 5: SEM top image of random Si nanocylinder arrays on sapphire, scattering in blue (left) and red (right). Scale bar is 500 nm.

In Figure 6, the measured scattering spectra of the blue and the red fields are shown (blue and red solid lines). The blue spectrum shows a weak resonance around 500 nm, the red spectrum shows strong first order resonances (electric and magnetic) around 800 nm and higher order resonances at 480 and 580 nm. The chromaticity values in the CIE 1931 xyY color space of the blue and red fields were determined by the concept described in Section 2 (see caption of Fig. 6).

In order to obtain a chromaticity close to white by combining red (r) and blue (b) pixels, the ratio of the red and blue pixel coverage r:b was chosen so that the xy values match the white point of the CIE 1931 color space of D65 (x_{White}=0.3127, y_{White}=0.3290). The chromaticity values x_n and y_n of the combined color can be determined by:

$$(x_n, y_n) = (x_b, y_b)\frac{bB_b}{B_n} + (x_r, y_r)\frac{rB_r}{B_n}$$

with the brightness of the blue array $B_b = X_b + Y_b + Z_b$, the red array $B_r = X_r + Y_r + Z_r$ and the combined brightness of the mixed fields $B_n = B_b + B_r$ [11]. Using the xyY values determined for the reflection spectra, we find that a pixel coverage ratio of r:b=1:3 leads to chromaticity values of x_n=0.3181 and y_n=0.3283, close to the white point.

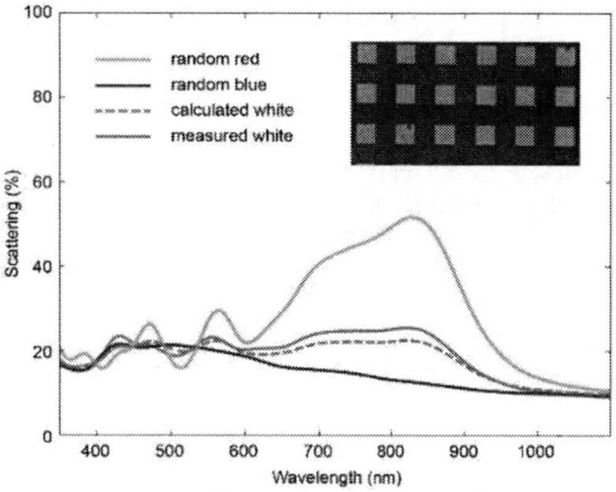

Figure 6: Measured scattering efficiency of blue and red reflecting random nanoarrays (blue and red solid lines). Spectra for calculated combined field (dashed grey line) and measured combined field (solid grey line) with ratio r:b=1:3. The chromaticity values of the plotted scattering curves are (x_r=0.3868, y_r=0.3573, Y_r=0.3160) for the red spectrum, (x_b=0.2813, y_b=0.3128, Y_b=0.1718) for the blue spectrum, (x_{cw}=0.3181, y_{cw}=0.3283, Y_{cw}=0.2078) for the calculated combined spectrum and (x_{mw}=0.3281, y_{mw}=0.3323, Y_{mw}=0.2221) for the combined measured spectrum. Inset: Optical microscope image of combined field. One red pixel and three blue pixels build a unit cell. Pixel is 50x50 µm^2.

Combining the two reflectance spectra in the ratio 1:3, and applying the method to determine the CIE color space values gives the dashed calculated line in Fig. 6. Next, the fields are combined experimentally in pixels of 50x50 µm^2 (see microscope image in inset of Fig. 6). The measured scattering spectrum (solid grey line in Fig. 6) of this combined field matches well the calculated scattering curve; resonance peaks are matching well over the entire spectral range. The reflected scattering spectrum of the mixed pattern has an average intensity of 22 % over the visible spectral range (450-800 nm)

5 TOWARDS A WHITE SOLAR CELL

Using the same approach as described in Section 4, a white colored solar mini-module was created. In these experiments the height of the nanocylinders was slightly smaller. Red to blue pixels had the optimized ratio r:b=1:9, and are shown in the microscope image in the inset of Fig. 7. The scattering spectra of the white integrated design, shown in Fig. 7, shows an average intensity of 20 % over the visible spectral range. The chromaticity values obtained from the scattering measurement of the nanopatterned module with combined red and blue fields are (x=0.3320, y=0.3491, Y=0.2152) and are close to the white point.

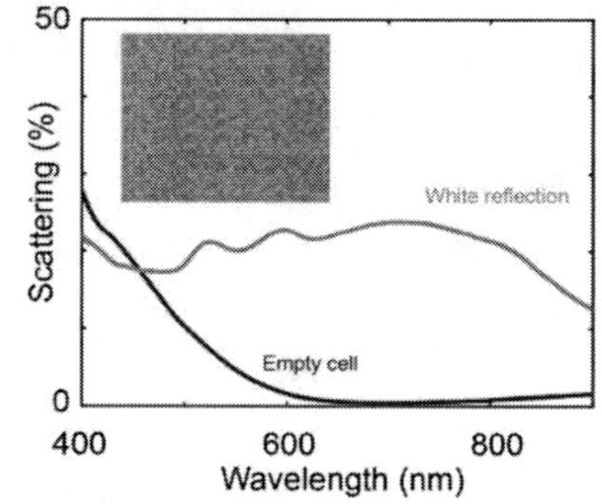

Figure 7: Measured scattering efficiency (grey) of integrated design of white field composed red and blue pixels. A spectrum for the bare Si heterojunction solar cell (dark blue) is also shown. Inset: Optical microscope image of combined field. One red pixel and 9 blue pixels are building a unit cell. Pixel is 50x50 µm^2.

Figure 8 shows the measured external quantum efficiency (EQE) of the nanopatterned mini-module and the bare Si solar cell. The EQE spectrum for the integrated design shows a reduction (15%-40%) over the entire visible spectrum up to 800 nm, resulting from light scattering and absorption by the Si nanoscatterers. The reduction observed above 800 nm is due to the reflection off the sapphire substrate. Correspondingly, the calculated J_{sc} is reduced by 25% (relative) from 40.8 to 30.7 mA/cm^2 compared to the bare Si heterojunction cell. Compared to the unpatterned module this would mean a relative reduction by 19%, taking into account 7% reflection from the sapphire slide.

One way to decrease the losses in this configuration is to tune the resonance of the red field, in order to have less reflection above 700 nm, as any reflection in that spectral range cannot contribute to the color perception, but reduces the short circuit current. A photograph of the white sample is depicted in the inset of Fig. 8. The color appearance is grey, which matches the measured spectra, as grey has the same chromaticity as white but is less luminescent. The brightness of the structure could be increased by increasing the scattering intensity of the pixels respectively.

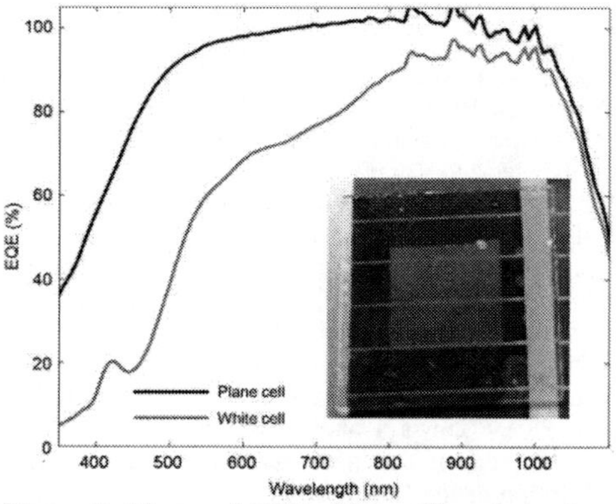

Figure 8: Measured EQE of bare silicon heterojunction solar cell (dark blue) and nanopatterned module with combined red and blue fields (grey). Inset: Photograph of nanopatterned module shows grey reflection. Sapphire sample is 12x12 mm.

6 CONCLUSION

In this paper, we expanded the technology to create efficient bright green colored solar mini-modules to additional colors and white. We demonstrated the colored fabricated solar mini-module with a white scattering surface using pixels of integrated random nanoarrays. The concept can be widened to any desired color. The colored solar modules presented here can find applications in building- and landscape-integrated photovoltaics of many different kinds.

7 METHODS

7.1 Fabrication - Electron beam lithography

Silicon-on-sapphire substrates (500 nm Si on 0.46 mm sapphire, MTI Corporation) were etched down to a thickness of 200 nm using reactive ion etching (RIE) in a HBr_2 (25 sccm) and O_2 (2 sccm) plasma. For this, a 50 nm thick layer of HSQ was spincoated on top of a silicon-on-sapphire wafer and patterned with a dose of 3400 $\mu C/cm^2$ in a Raith Voyager e-beam system. The samples were developed in TMAH (25%) for 60 seconds. Two subsequent RIE etch steps were performed, first to break through the sol-gel pattern using CHF_3 (25 sccm) and Ar (25 sccm), and second, to transfer the pattern into the silicon layer using HBr_2/O_2.

7.2 Fabrication – SCIL

The Silicon-on-sapphire substrates were etched down in the way described in section 7.1. Next, we used substrate conformal imprint lithography (SCIL) to fabricate arrays of silicon cylinders. A 45-nm-thick silica sol-gel film was spin-coated and patterned using a SCIL stamp that was made from a pre-patterned silicon master, fabricated by electron-beam lithography. To transfer the pattern in the Si, the same etching steps as described in section 7.1 were performed

7.3 Reflactance measurements

Measurements were performed with an integrating sphere in reflection configuration. A collimated beam from a supercontinuum light source was coupled in at an incidence angle of 8°. In this setup, all light that is scattered/reflected from the sample in the upper halfsphere is collected and sent to a spectrometer.

7.3 EQE measurements

The active-area EQE spectrum was measured using slightly focused light (2 mm diameter) under normal incidence from a Xenon light source sent through a monochromator. The EQE of the bare Si heterojunction cell approaches 100% in the 600-900 nm spectral range.

8 ACKNOWLEDGMENTS

We gratefully acknowledge Paula Bronsveld and her team for fabricating the Si heterojunction solar cells, Marc Verschuuren for help with substrate conformal imprint lithography (SCIL), the assistance of Dimitry Lamers and Bob Drent for cleanroom support and Mark Knight for useful discussions. This work is part of the research program of the Netherlands Foundation for Scientific Research (NWO); it is also funded by the European Research Council.

9 REFERENCES

[1] V. Neder, S. L. Luxembourg, and A. Polman, Applied Physics Letters, 111, 073902 (2017).
[2] J. Van De Groep and A. Polman, Opt. Express 21,26285 (2013).
[3] E.-H. Cho, H.-S. Kim, B.-H. Cheong, P. Oleg, W. Xianyua, J.-S. Sohn, D.-J. Ma, H.-Y. Choi, N.-C. Park, and Y.-P. Park, Opt. Express 17, 8621 (2009).
[4] L. Cao, P. Fan, E.S. Barnard, A.M. Brown, and M.L. Brongersma, Nano Lett. 10, 2649 (2010).
[5] S. Sun, Z. Zhou, C. Zhang, Y. Gao, Z. Duan, S. Xiao, and Q. Song, ACS Nano 11, 4445 (2017).
[6] J. Proust, F. Bedu, B. Gallas, I. Ozerov, and N. Bonod, ACS Nano 10, 7761 (2016).
[7] Y. Kanamori, T. Ozaki, and K. Hane, Opt. Express 22, 25663 (2014).
[8] J. Henrie, S. Kellis, S.M. Schultz, and A. Hawkins, Opt. Express 12, 1464 (2004).
[9] A.I. Kuznetsov, A.E. Miroshnichenko, Y.H. Fu, J. Zhang, and B. Luk'yanchuk, Sci. Rep. 2, 1 (2012).
[10] A.I. Kuznetsov, A.E. Miroshnichenko, M.L. Brongersma, Y.S. Kivshar, and B. Luk'yanchuk, Science (80). 354, (2016).
[11] SemanticScholar, "Perceiving Color 5.1 the Color Stimuli" (2008).

SHINGLING TECHNOLOGY FOR CELL INTERCONNECTION: TECHNOLOGICAL ASPECTS AND PROCESS INTEGRATION

D. Tonini, G. Cellere, M. Bertazzo, A. Fecchio, L. Cerasti, and M. Galiazzo[1]
Applied Materials Italia Srl,
Via Postumia Ovest 244, Olmi di San Biagio di Callalta (TV) Italy

ABSTRACT:
Shingling technology is a promising approach to improve photovoltaic module efficiency and optimize its energy yield, because it allows an increase of module output power without introducing substantial modification in the cell fabrication process and in the module bill of materials (BOM).
In a shingled module, c-Si PV cells are cut in smaller stripes (usually 3 to 6 stripe per cell, according to the layout) called "shingles" and assembled in strings by connecting the front of the leading edge of each shingle to the back of the tail edge of next one, in a similar way to tiles on a rooftop. Electrically conductive adhesives (ECAs) are typically used to make the shingles interconnection while the bussing ribbons can be attached using ECA as well as traditional soldering.
In this work, we analyze the technological aspects of shingling module production from cell to assembled string, describe the most relevant process steps and provide proof that shingled modules deliver power advantages compared to standard cell-based modules
Keywords: C-Si, Silicon solar cell, laser processing, manufacturing and processing, PV module.

1 PURPOSE AND APPROACH OF THE WORK

Shingling technology for cell interconnection in a module is not new in PV: in fact, it was one of the first method to create the series between the strings, for example it had been adopted in early space applications [1]. In recent years, there was a revamping of this technology due to increased need for differentiation among module manufacturers, specifically aiming at higher power densities [2], and availability of suitable Electrically Conductive Adhesives (ECAs) [3] and equipment. The shingling approach provides several advantages compared to standard modules:

- Lower ohmic losses: overall string current is lower than traditional ribbon-connected strings because of the smaller area of the shingles connected in series. This greatly reduces the ohmic losses.
- Better area utilization: since there is no need for cell-to-cell gape, denser cell packing can be obtained in the same area.
- Lower T processing: ECAs curing temperature is lower than ribbon soldering reducing residual cell stress and bowing.
- The lower current results in lower operating temperature – hence increase energy generation in real world operations.
- The absence of visible busbars (BB) and ribbons results in improved all-black aesthetics.

In this work we demonstrate the performance gain obtained with shingling interconnection technology in terms of module output power and efficiency. Moreover, we describe the technological challenges for each step in the shingling assembly process flow.

2 TECHNOLOGICAL ASPECTS

In the shingled module scheme, a solar cell is cut in 3 to 6 stripes (the shingles) that are subsequently assembled in strings by connecting the front of each shingle to the back of the next one (Figure 1). Shingled strings are made of several cells and can be up to 2 meters long according to the module design; strings are then connected by bussing ribbons in series or parallel arrangement and processed using the traditional module making procedure.

Due to the small area of the shingles, the overall current flowing in a string when in operation is significantly lower than a standard one made of full cells. This dramatically reduces the ohmic losses, providing significant advantages in terms of power output of the module and lowering the operating temperature. Moreover, the dense packing of the shingles and the absence of visible BBs leads to better area utilization, increasing the efficiency of the module compared to a standard one of the same area.

Shingles are attached one to the other using ECAs, which cure at low temperature and are flexible enough to accommodate the thermal stresses during processing and thus contributes to improving the reliability of the device.

Since the shingling approach impacts only the way that cells are interconnected to form strings, it can be applied to any cell architecture as long as contacts are on the opposite sides of the cell. It can be applied with almost no process modifications to BSF, PERC, bifacial-PERC, n-PERT/PERL, HJ, among others.

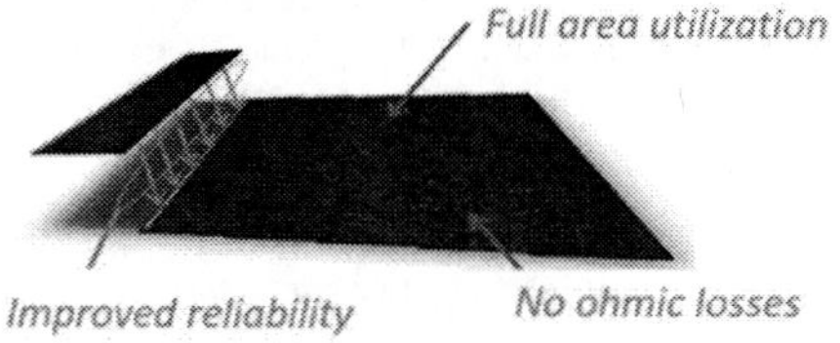

Figure 1: shingling connection scheme

3 PROCESS INTEGRATION

The production of shingled modules requires the substitution of traditional stringing/tabbing with four additional process steps and the modification of the cell grid layout in order to allow for the subsequent separation of the stripes. The module BOM and lamination process are not affected and do not require changing.

In this section, we analyze the major process steps changes, highlighting the effects on device performances and yield.

3.1 Cell layout

Moving from a standard 3-4 BBs layout to a shingling one requires some grid modifications. The fingers must be interrupted to guarantee a good separation of the cell into shingles, while the front and back pseudo-BBs layout must be changed in accordance with the shingles attachment scheme (each shingle has a pseudo-BB on the front at one end and one in the back at the other end - as in Figure 1).

In the shingled cell layout, fingers are attached to pseudo-BBs only on one end, so the path current has to flow through is longer compared to a standard 4 BB layout. Hence the number of fingers and their aspect ratio require some optimization to keep satisfactory FF at cell and shingle level; moreover, having interruption-free fingers becomes more important to avoid resistance losses.

3.2 Singulation

Separation of the cell into shingles is done using a laser scribing and break approach. To do this, a pulsed infrared laser is used to induce a crack at the separation position and a subsequent mechanical action completes the singulation process. Laser power, repetition rate and scanning speed have been carefully selected to minimize the amount of damage to the cell. Scribing can be performed from the back of the front side of the cell. In order to study the possible impact of the side the laser is fired to, we ran a comparison test based on half cut cells. We measured the saturation current density or J0 of half-cut cells before laser scribing, after scribing and after the complete separation to understand how the reverse behavior is affected by laser processing.

As shown in Figure 2, laser scribing does not create relevant bulk defects (J01 is not changing), but affects the junction quality when scribing on the frontside (J02 increases). This justifies the adoption of backside scribing throughout the remaining of the work.

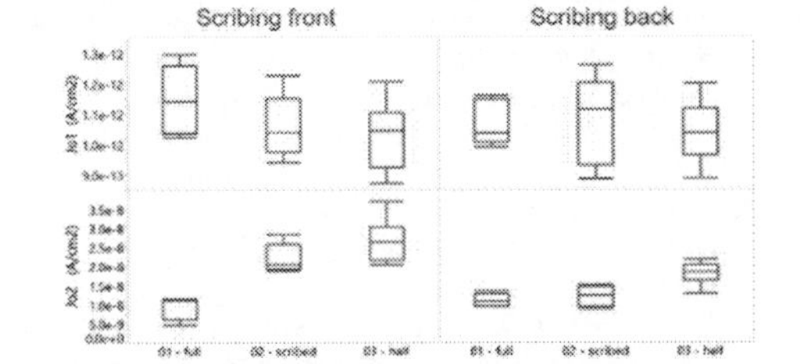

Figure 2: J0 measurement on half-cut cells: scribing on the front side (left) and on the backside (right). Backside scribing does not damage the junction, J02 left unvaried

Laser accuracy is fundamental to obtain uniform shingles, so alignment precision, consistency of pitch between shingles, and scribing depth repeatability were carefully tested. Our experiments show that alignment precision <±50μm, pitch repeatability better than ±10μm, and scribing depth variation range of ±10μm, are attainable.

3.3 ECA printing

The most important technological task in the shingling approach is the connection between the shingles, which must provide electrical continuity along the strings and have enough mechanical strength to sustain the stress a module goes through both during testing and operating life. ECAs offer a viable solution based on proven reliability in other fields; however, they have not been widely used in photovoltaic technologies.

There are two competing ways to deposit ECA on a shingle with the required accuracy, precision, and weight control— dispensing and screen printing. Between them, we believe screen printing is the most suitable solution because it is a proven technology with a decades-long history of production in c-Si photovoltaic. In fact, it is a simple, robust and fast process, capable of delivering with high throughput fine control of the ECA laydown with no issues like nozzle clogging and cumbersome cleaning.

Tests done using specifically designed screens allowed depositing less than 10mg/cell of ECA without compromising the electrical performances and thus greatly contributing to reduce the total cost of ownership of shingling process and, ultimately, the cost per Watt of the final module.

3.4 String assembly

Individual shingles must be arranged together in a string. Typical length of a string must match that of a module characteristic dimension, so strings up to 2 meters long (the long side of a 72-cell module) are possible. On the other side, accuracy in shingle-to-shingle alignment and in the total string length are important in order to achieve uniform power output, reliability and aesthetic consistency among all the strings. High alignment accuracy is needed to avoid contacting issues like those in the bottom EL image of Figure 3, where alignment mismatch causes an electrical interruption in some shingles (black areas)

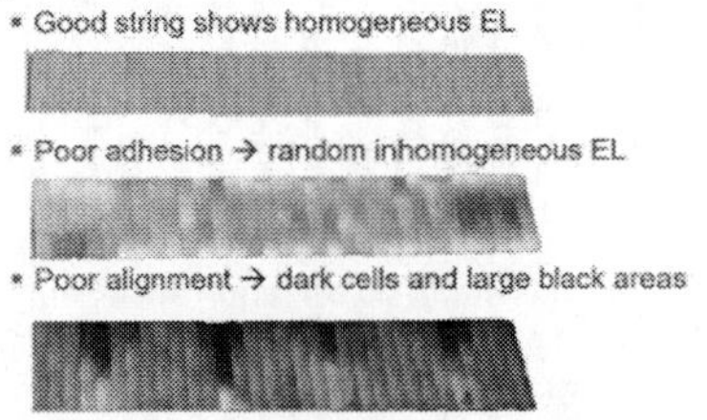

Figure 3: Electroluminescence imaging of a good string (top), a string with ECA adhesion issue (middle), and a string with alignment mismatch with non-electrically connected areas (bottom)

3.5 ECA curing

ECAs tested so far are thermally cured and require process temperature lower than 200°C. ECAs differ from traditional silver pastes for the front side cell grid because they do not have (or have very limited) solvents that evaporate. Instead they contain an organic matrix which reticulates upon the application of heat providing mechanical strength while forcing the silver particles to be in contact with one another to create a percolation path for current flow. The curing reaction is exothermic and can be easily studied by Differential Scanning Calorimetry (DSC). DSC provides a way to determine if the reaction went to completion: we analyzed the fresh ECAs and compared them to thermally treated samples, as shown in Figure 4. The exothermic peak completely disappear in the cured samples, meaning the curing reaction is completed in the oven and no residual uncured ECA is left, avoiding any out-gassing effect during lamination or module service Non complete curing during thermal processing can lead to adhesion issues in shingling interconnection as in Figure 3 (middle).

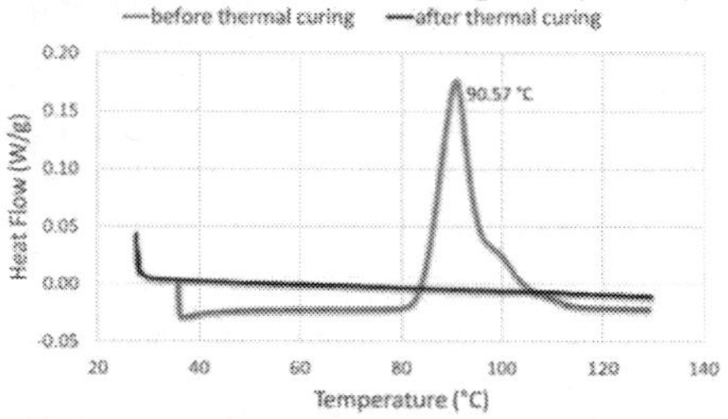

Figure 4: DSC analysis of an ECA before and after the thermal curing. As shown the exothermic peak of curing reaction disappears after the thermal treatment, meaning the reaction was already completed in the oven.

4 DEVICE PERFORMANCES

4.1 Module power

The first assessment of module power gain was conducted on 60 cell equivalent modules. Eighteen modules were assembled at a customer site and then tested. Baseline was a conventional commercial module made with the same starting material (mono PERC cell); all the shingled modules outdid the baseline power value, giving an average gain of 9.6%, with four modules exceeding 10% of power gain (Figure 5).

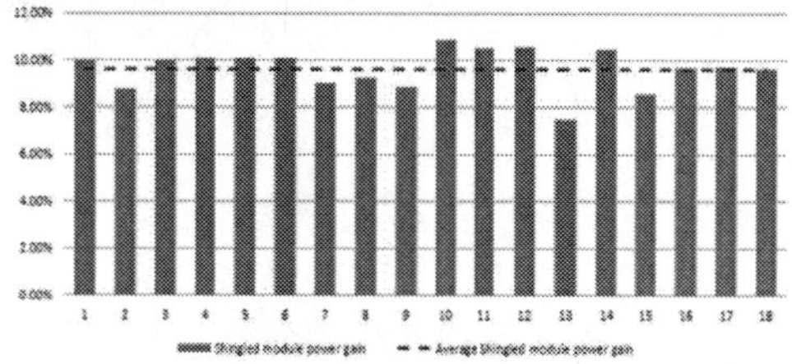

Figure 5: Test results on 18 60 cell equivalent shingled modules: all the samples surpassed the baseline, giving an average power gain of 9.6%

A second test was performed on 68 shingled modules.

Since starting cells were pseudo-square, two kind of modules were assembled: 1) those made by the central shingles of the cell, thus being rectangular in shape (called full square in Fig.6), and 2) those made by the first and last shingle of each cell, thus having rounded corners (called pseudo square in Fig.6). As expected, modules made by pseudo square shingles present a smaller power gain compared to full rectangular ones, due to the smaller active area. Figure 6 shows the power gain for the two kind of modules, in the case of full area shingles a gain up to 8% can be achieved.

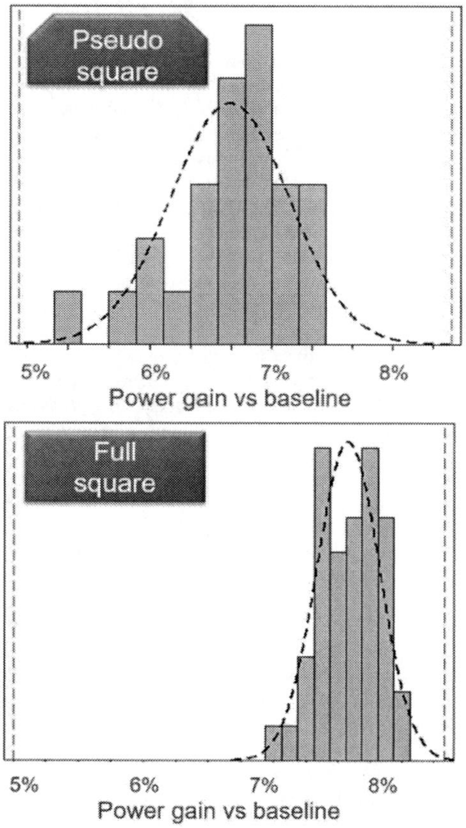

Figure 6: power gain for 72 cells equivalent shingled modules. Rounded (top) and full rectangular (bottom) shingles. As expected the power gain is lower for modules made by rounded shingles, because of the reduced active area.

4.2 Effect of ECA deposit on module power

In a shingled module, the amount of ECA contributes significantly to the overall CoO, so we tested the effect of the amount of ECA on the overall performances, adjusting the print layout to minimize the deposit.

Tests on small sized modules (1-cell equivalent) showed that a dramatic reduction of the ECA laydown is possible without affecting the electrical performances. The graphs in Figure 7 show the power and FF reduction for two different ECAs.We obtained a power loss lower than 2% reducing the wet deposit from 25 to 4 mg/cell. In particular, the second ECA (lower two graphs) allows getting to extremely low deposit without any significant performance degradation.

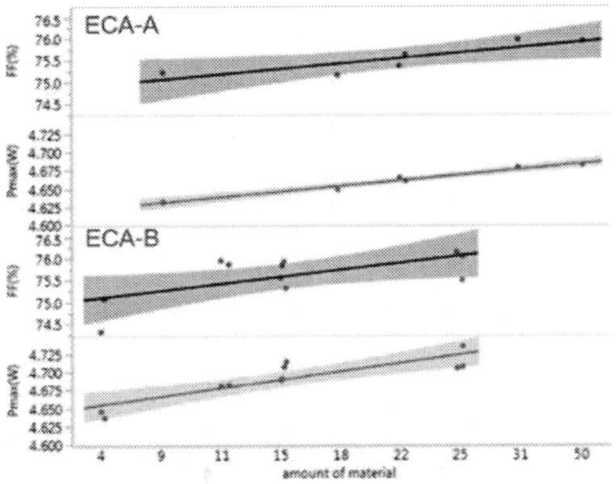

Figure 7: power losses Vs ECA laydown on minimodule testing for two different ECAs.

4.3 Reliability test

Shingling modules do not have an extensive history of reliability data, so a deep analysis on this aspect is needed. First results of thermal cycling are promising: three modules have been tested up to TC800 showing power losses lower than 3% (Figure 8) with no significant damage revealed by electroluminescence imaging. Test is still ongoing aiming to a complete characterization up to 1000 thermal cycles.

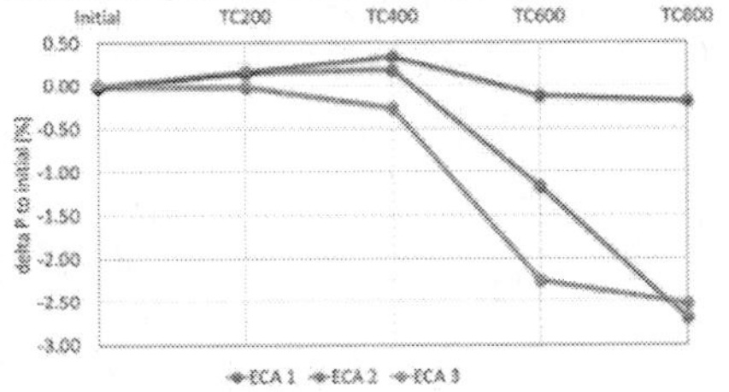

Figure 8: thermal cycling test for three modules showing power losses <3% after TC800.

5 CONCLUSIONS

Shingled modules can be realized with almost all the existing cell technologies (BSF, PERC, bifacial-PERC, n-PERT/PERL, HJ) without major process modifications, allowing higher power density better energy yield vs. standard modules and with high reliability and.

In this work we analyzed the main technological changes and the required process integration steps needed to mass produce shingled modules. Sample modules in two separate tests showed an increase of power per module up to 10%. Keeping cost per Watt at the same level or even lower than conventional modules looks possible by better material utilization and ECA laydown reduction. While ECA adoption is definitely not widespread in solar industry, thermal cycling tests showed limited power losses (<3%) after TC800, assessing the reliability of this kind of approach.

Overall, this work demonstrates the potential of the technology from a cost, performance, and reliability standpoint. Of course, introducing the technology in HVM manufacturing will require a number of further assessments and developments.

6 REFERENCES

[1] Chapin, D.M., Fuller, C.S., Pearson, G.L . 1954, "A new silicon p-n junction photocell for converting solar radiation into electrical power", J. Appl. Phys., Vol. 25, No. 5, p. 676

[2] SunPower Corporation 2017, "SunPower introduces new solar panel: The performance series" [https://us.sunpower.com/blog/2015/11/12/sunpoweri ntroduces-performance-seriessolar-panel/]

[3] Beaucarne, G. 2016, Materials Challenge for shingled cells interconnection, Energy Procedia, Vol. 98, pp. 115–124.

FEM-BASED DEVELOPMENT OF NOVEL BACK-CONTACT PV MODULES WITH ULTRA-THIN SOLAR CELLS

Andreas J. Beinert[1, 2], Roman Leidl[3], Paul Sommeling[4], Ulrich Eitner[1] and Jarir Aktaa[2]
[1]Fraunhofer Institute for Solar Energy Systems ISE, Heidenhofstraße 2, 79110 Freiburg, Germany
[2]Karlsruhe Institute of Technology (KIT), Institute for Applied Materials, Hermann-von-Helmholtz-Platz 1, 76344 Eggenstein-Leopoldshafen, Germany
[3]AIT Austrian Institute of Technology GmbH, Giefinggasse 2, 1210 Vienna, Austria
[4]Energy Research Centre of the Netherlands (ECN), Westerduinweg 3, 1755 LE Petten, The Netherlands
Corresponding author: Andreas J. Beinert | Phone: +49 (0)761 4588 5630 | E-mail: Andreas.Beinert@ise.fraunhofer.de

ABSTRACT: With the availability of ultra-thin back contact solar cells, the question arises if they can be integrated into PV modules. Particularly the single-side metallization and joint architecture of back contact solar cells may cause critical stress. We develop a three-dimensional finite element model of a frameless 60-cell module with an electrically conductive backsheet to simulate the cell stress in terms of mechanical push load. The FEM model is validated by mechanical load tests of frameless modules. With the validated model we perform a variation of the cell and encapsulant thickness from 80 µm to 180 µm and 200 µm to 460 µm, respectively. Moreover we compare three different encapsulant materials, one Ethylene-vinyl acetate copolymer (EVA), one thermoset Polyolefin elastomer (POE-TS), one thermoplastic POE (POE-TP) and two different electrically conductive adhesives. The combination of 460 µm thick EVA and 180 µm cells shows the lowest first principal stress σ_I which, however, is far above the measured critical fracture stress of 134 MPa for the MWT cells used. We then modify the FEM model to simulate a framed glass-foil as well as a framed and frameless glass-glass module with the same material combination, 80 µm cells and 200 µm encapsulant. The framed glass-foil module shows a by 54% reduced deflection and a stress of 114 MPa, which is much smaller than without a frame (296 MPa). The lowest value is achieved for the framed glass-glass module with 66 MPa. An interesting finding from the thickness variation is, that for cells below about 110 µm thickness the stress increases with increasing encapsulant thickness, while for thicker cells the stress decreases.
Keywords: thin solar cells, thermomechanical stress, FEM modeling, material characterization, PV module design

1 INTRODUCTION

Since the development of novel module designs is an expensive, time-consuming process, virtual prototyping by the finite element method (FEM) is a convenient way for identifying robust PV module designs. [1] From a mechanical point of view, the mechanical load is a crucial test in the IEC61215 [2] sequence for novel module designs. In this work we investigate the potential of reducing the encapsulant and solar cell thickness in PV modules with back contact solar cells connected by electrically conductive adhesives (ECA) and an electrically conductive backsheet. This concept has been investigated before with a focus on the thermomechanical challenges due to temperature cycling e.g. [3–5]. In this work, we develop a 3-dimensional FEM-model which is capable of resolving the stress which arises from lamination and mechanical load in great detail. The FEM-model is validated by a mechanical load test. With the validated FEM-model we compare six different material combinations as well as different layer thicknesses of the back contact solar cells and encapsulant.

2 METHOD

The work consists of three interacting parts (figure 1). The main part is the FEM-modelling, which gets input from the material characterization. The results of the FEM-simulation are compared with experimental results from 4-line cell bending tests in terms of cell fracture probability and validated by mechanical load tests on full-size modules.

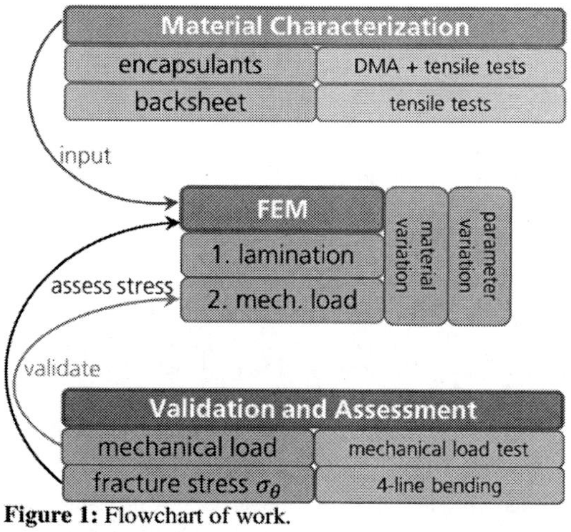

Figure 1: Flowchart of work.

2.1 Material characterization

Since the material parameters of the encapsulants are crucial for the thermomechanical behavior of the laminate, we measure the Young's modulus and Poisson's ratio. We use a dynamic mechanical analyzer (DMA) to measure the Young's modulus for the temperature range of -100 °C...+160 °C. The Poisson's ratio is measured by a tensile tester using an optical extensometer at ambient temperature. Additionally we measure the backsheets Young's modulus at ambient temperature by a tensile tester. We investigate one Ethylene-vinyl acetate copolymer (EVA), one thermoset Polyolefin elastomer (POE-TS), one thermoplastic POE (POE-TP) and one electrically conductive backsheet. The corresponding results are shown in table 1 and figure 2.

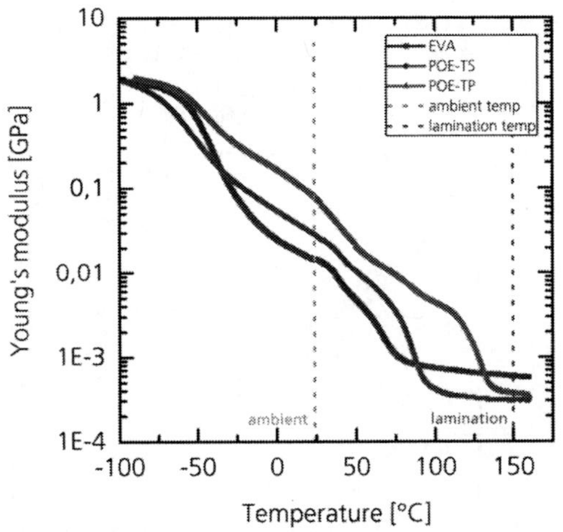

Figure 2: Measured temperature dependent Young's modulus of the three different encapsulants.

2.2 Finite Element model

Because of the large aspect ratio of the solar cell edge length to its thickness, mechanical modelling of PV modules is a challenging task. In order to minimize the computational effort, we exploit the symmetry of the PV laminate by modelling a quarter laminate. Additionally we apply the sub-modelling method, which allows separating the model into two problems: the computation of the displacement in the global-model and the computation of the stress in the sub-model with a minimized geometry and finer mesh, more details in [6]. The global-model consists of a 60-cell unframed PV laminate with monocrystalline full-square 156x156 mm² back contact solar cells. The overall dimension of the PV laminate is 1.61x0.98 m². The target thickness of each layer and the materials used are shown in table 2.

Table 1: Material properties used in the FEM-model. *: provided by manufacturer; $^+$: measured.

Material	Density	Young's modulus	Poisson's ratio	CTE
	ρ [kg/m³]	E [GPa]	ν [-]	α [10^{-6} K^{-1}]
Solar glass	2530*	74*	0.24*	9.5*
EVA	820$^+$	*T*-dep.$^+$	0.22$^+$	270 [7]
POE-TS	941$^+$	*T*-dep.$^+$	0.31$^+$	270 [7]
POE-TP	865$^+$	*T*-dep.$^+$	0.26$^+$	270 [7]
Cz-silicon	2329 [7]	Elasticity Matrix [7]		*T*-dep. [8, 9]
ECAI	3950*	*T*-dep.$^+$	0.3*	46*
ECAII	3600*	*T*-dep.$^+$	0.3*	*T*-dep.*
Back sheet	1315*	6.36$^+$	0.29 [7]	50.4 [7]
Al	2700 [10]	70 [10]	0.33 [10]	23 [10]
Rubber inlay	67*	0.0074*	0.3*	769*

The sub-model consists of the cell matrix embedded in the encapsulants and 31 ECA contact joints per cell. The result of the global model is transferred as boundary and initial conditions to the sub-model. A rectangular mapped mesh is used. In the sub-model the number of 180 elements per cell edge is increased by 10%. Also the amount of elements per layer height is increased for each layer individually.

The model is computed in two successive steps. First the lamination process by cooling down from 150 °C to 23.8 °C. Secondly the mechanical push load of 2400 Pa is applied onto the thermally stressed module at 23.8 °C. In the first step the laminate is not mounted, while in the second step six laminate clamps are used. For this purpose, the clamps are implemented by an additional sub-model to the global-model.

From a FEM pre-study we find that the load application pattern has a significant impact on the module deflection. Therefore we implement the used vacuum cups of the test stand by a two-dimensional spatially resolved Gaussian distribution, shown in figure 3. We assume the load magnitude to be equal for each vacuum cup. We validate the deflection by mechanical load tests on full-size modules. With the validated model we compare six different material combinations (three encapsulants and two ECA's). Moreover we conduct a parameter variation study of the thickness of the solar cell and the encapsulant for all combinations. For the material combination with the least stress we modify the FEM model to simulate a framed module as well as a glass-glass setup with and without frame.

Table 2: PV module materials used in the FEM model with corresponding thickness (target thickness in bold).

Layer	Material	Thickness [µm]
Front glass	solar glass	3200
Encapsulant	EVA, POE-TS, POE-TP	100, **200**, 300, 460
Solar cells	monocrystalline silicon	**80**, 120, 180
ECA	ECAI, ECAII	200
Backsheet	Cu-PA-PET-PVF	380
Frame	Aluminum with rubber inlay	1500

2.3 Validation

For the validation, we measure the deflection of three frameless PV-modules with 180 µm thick MWT-cells during several mechanical load cycles. In pre-tests we find that for higher loads the laminate slips out of the laminate clamps. Therefore the test load is limited to 1000 Pa. The deflection is measured by four laser sensors with a measurement resolution of 1 mm. Figure 3 shows the positions of the laser sensors.

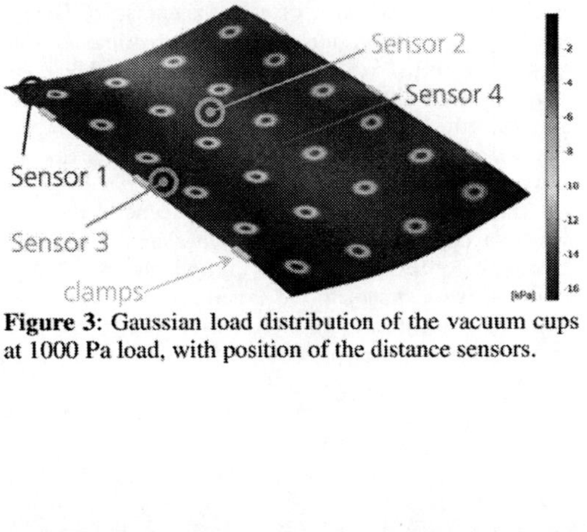

Figure 3: Gaussian load distribution of the vacuum cups at 1000 Pa load, with position of the distance sensors.

3 RESULTS

3.1 Fracture stress by 4-line-bending

In order to assess the criticality of the results from the FEM-model, we determined the characteristic fracture stress on the rear side of ten industrial MWT-cells, used in the validation experiment by a four-line-bending test. The results are presented in figure 4 as a Weibull plot [11, 12]. We obtain a characteristic fracture stress of 134 MPa. Since the fracture stress of silicon is a tensile stress, we evaluate in the FEM study the first principal stress σ_I, which is the tensile stress equivalent. More precisely, the highest stress values arise at the contact joint of the contact pad and the solar cell. Thus we evaluate the mean first principal stress $\bar{\sigma}_I$ over this contact area at the contact joint with maximum first principal stress σ_I.

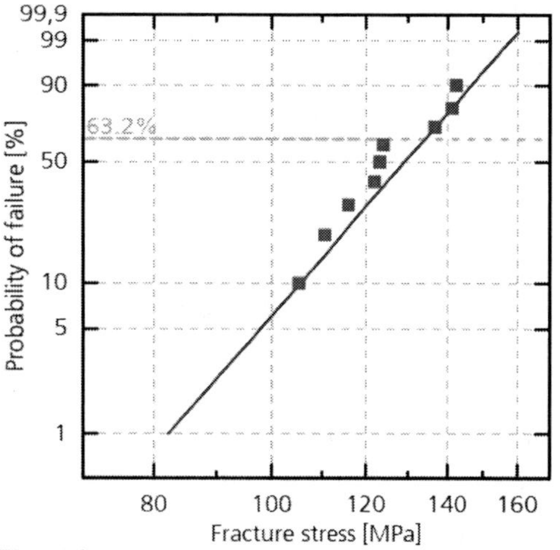

Figure 4: Weibull-plot for 4-line bending fracture tests of ten MWT cells. The dashed yellow line is the value at which the characteristic fracture stress is determined.

3.2 FEM-model validation

Due to their position close to the clamps, sensor 1 and 3 show minor deflection values, consequently the uncertainty of the laser sensor measurment is fairly large compared to the deflection. Therefore we evaluate only sensor 2 and 4 for the validation. The deflection measured by sensor 4 in dependence on the applied load is shown in figure 5 together with the result of the FEM-model. A similar graph is obtained for sensor 2 (not shown). The FEM results lie within the upper third of the experimental data for both sensors. This slight underestimation of the deflection might be due to the assumption of equal load of each vacuum cup. Overall speaking, the FEM results match the experimental data reasonably well. Therefore an equal load of each vacuum cup is considered to be a valid assumption for this work. The deflection values at 0 Pa load represent the deflection due to gravity. According to the FEM results the gravitational deflection is about 3 mm.

The experimental values of deflection are only available up to 1000 Pa, but the mechanical load test according to IEC61215 [2] is carried out at 2400 Pa, therefore, we extrapolate the experimental data using a linear fit of the data of all modules. This is a reasonable approximation since previous results show a linear behavior of the deflection in this range [13]. At 2400 Pa the FEM results deviate by about 10 mm from the extrapolated data, which corresponds to about 10 % and is equal to the scattering of the experimental data.

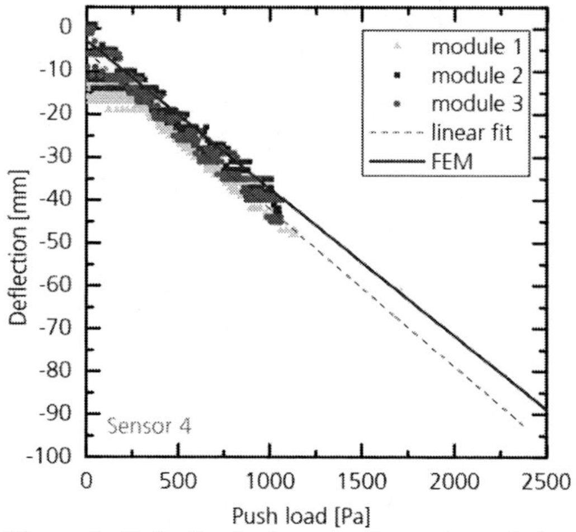

Figure 5: Deflection-load curve of experimental data (symbols) with extrapolated linear fit (dashed line) and FEM results (solid line) for the distance sensor 4.

We investigate the tested modules after different load levels for new cracks by electroluminescence (EL) imaging. The obtained crack pattern coincides very well with the stress pattern obtained from the FEM simulation. Figure 6 shows exemplarily the EL-image of validation module 3 after 1000 Pa push load overlaid with the corresponding FEM result. We therefore consider the FEM model to be sufficiently valid.

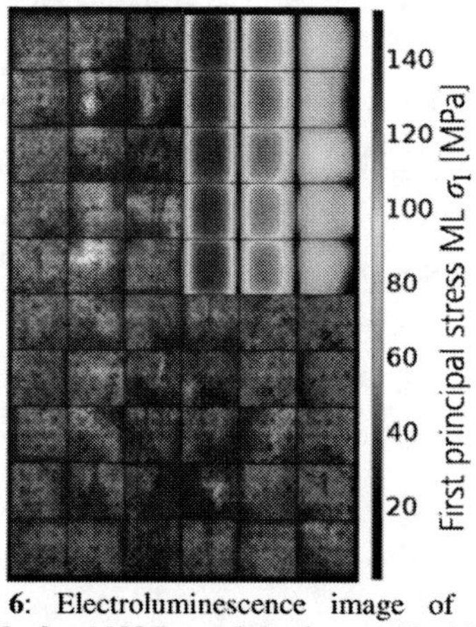

Figure 6: Electroluminescence image of validation module 3 after 1000 Pa push load overlaid with the FEM result at the same load.

3.3 Lamination

Due to the mismatch of the coefficients of thermal expansion of the different layers within the PV laminate (see table 1), the cooling down after lamination leads to thermomechanical stress. We set the lamination temperature of 150 °C as the stress-free temperature. Therefore the stress increases with decreasing temperature. With the backsheet contracting stronger than the front glass, the module slightly bends, showing a convex shape from the sunny side. Accordingly, the solar cell deformation is also convex and the front side is in tensile mode. Thus, the mean first principal stress $\bar{\sigma}_I$ is evaluated on the front side by taking the mean stress value over the contact pad area at its position on the front side. Figure 7 shows the maximum mean first principal stress $\bar{\sigma}_I$ for all six material combinations with the target material layer thicknesses given in table 2. All variations are far below the characteristic fracture stress σ_θ, hence we do not expect any cell cracking after lamination.

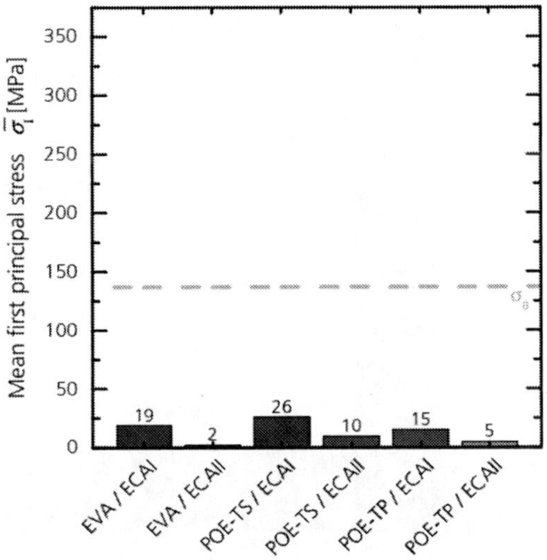

Figure 7: Mean first principal stress $\bar{\sigma}_I$ on the front side of the solar cells after lamination. The dashed yellow line indicates the characteristic fracture stress σ_θ.

3.4 Mechanical load

The bending of the PV laminate due to 2400 Pa mechanical pressure load leads to tensile stress on the rear side of the solar cells. Exemplarily, figure 8 shows the first principal stress σ_I on the rear side of the solar cells at 2400 Pa pressure load for the combination of EVA with ECAII with the target thickness. The pattern of the contact joints is clearly visible due to the higher stress values, because the contact pads apply an additional strain to the solar cell.

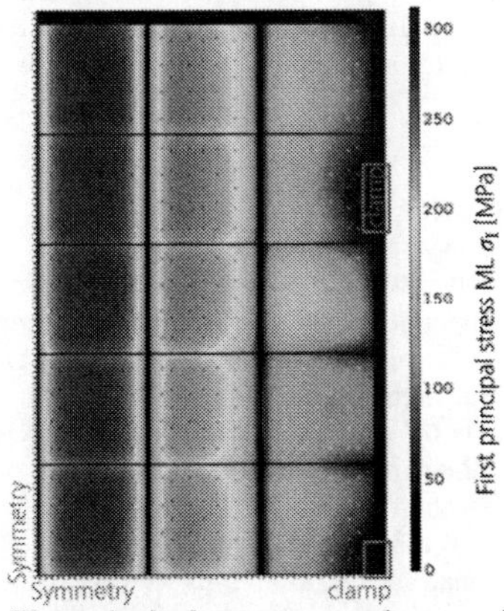

Figure 8: First principal stress σ_I on the rear side of the solar cells at 2400 Pa mechanical push load for the combination of EVA with ECAII with target thickness.

The dependence of the mean first principal stress $\bar{\sigma}_I$ on the cell and encapsulant thickness is depicted in figure 9 and figure 10, respectively. For the cell thickness variation we use a EVA thickness of 200 µm. For all material combinations the deflection decreases between 8 - 10%, depending on the combination, with increasing cell thickness (not shown). The same behavior is obtaind for the stress decreases. The lowest stress is achieved by the combination of EVA (200 µm) with ECAI and 180 µm cell thickness. However with 260 MPa, the value is about two times above the characteristic fracture stress σ_θ of 134 MPa. Comparing this to standard solar cells, which have a higher characteristic fracture stress σ_θ of about 200 MPa [14], the vealue is still in a critical range. This corresponds well with previous findings for a clamped frameless common PV module with 460 µm EVA [6]. Decreasing the cell thickness to the target thickness of 80 µm increases the stress between 4% and 12%, depending on the material combination. EVA with ECAI shows the largest increase.

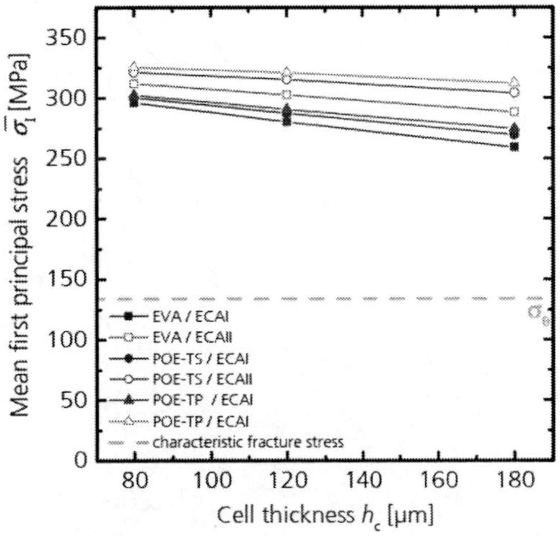

Figure 9: Mean first principal stress $\bar{\sigma}_I$ on rear side of the solar cells at 2400 Pa pressure load for different solar cells thickness with an encapsulant thickness of 200 µm.

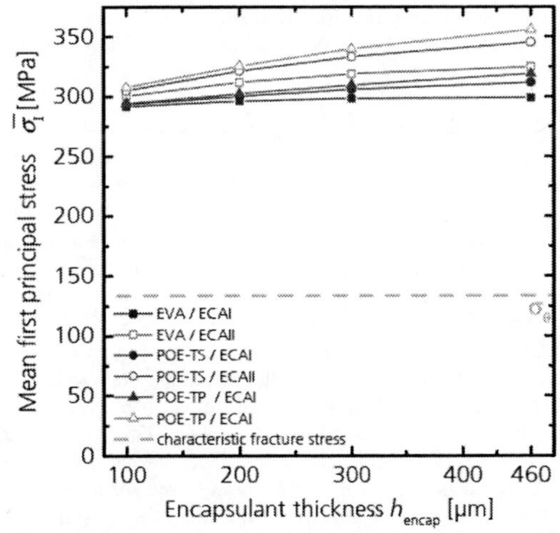

Figure 10: Mean first principal stress $\bar{\sigma}_I$ on the rear side of the solar cells at 2400 Pa pressure load for different encapsulant thickness with 80 µm thin solar cells.

Surprisingly, the mean first principal stress $\bar{\sigma}_I$ increases with increasing encapsulant thickness for 80 µm thick cells. The reason for this behavior lies in the coupling of the cells to the backsheet through the contact pads. So the strain in the solar cells is directly coupled to the strain in the backsheet. Moreover, the strain in the backsheet due to push load increases with increasing module thickness, assuming a constant bending radius on the glass surface, i.e. the bending radius of the backsheet increases for an increasing module thickness. Now, with increasing encapsulant thickness, the module thickness increases and so does the strain in the backsheet. Simultaneously the difference between the strain of the solar cells and the strain of the backsheet increases. On the other hand, the module bend decreases by 4 - 6% for an increasing encapsulant thickness due to higher strain in the encapsulant and a higher laminate bending stiffness, which causes a decreasing stress. According to the FEM results, the cell stress is mainly influenced by the strain in the backsheet for thin cells. For thicker solar cells (>110 µm), the influence of the backsheet vanishes, because the stiffness of the solar cell increases with its thickness. The turning point is visible by the nearly constant stress level between 100 µm and 110 µm cell thickness in figure 11.

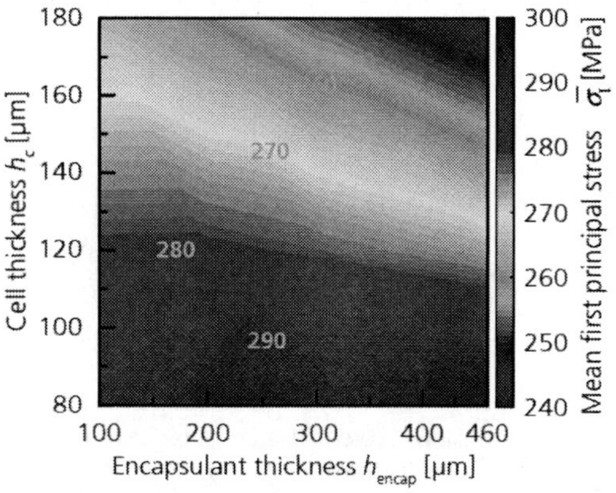

Figure 11: Mean first principal stress $\bar{\sigma}_I$ on rear side of the solar cell of the model with EVA with ECAI at 2400 Pa pressure load for different encapsulant and solar cells thickness. All values are above the characteristic fracture stress σ_θ.

However, all obtained stress values are far above the characteristic fracture stress σ_θ. From prior studies it is known that a frame reduces the stress in the solar cells significantly [6]. Therefore we replace the clamps in the FEM model by a frame fixed at four positions, according to [6]. With this modified model we simulate for 200 µm encapsulant and 80 µm cell thicknesses the material combination with the lowest stress, namely EVA with ECAI. Additionally we investigate the potential of a glass-glass setup to reduce the stress by replacing the backsheet with a glass. For this setup the front glass has a thickness of 2 mm and the rear glass 2.035 mm, taking the electrically conductive copper layer into account. We simulate this setup with and without frame for EVA with ECAI. Due to the alteration of the mounting system, the FEM model of framed modules is not validated. However, since the mounting system is implemented in a sub-model of the laminate, the laminate itself, particularly the mesh, is identical for both FEM models. Therefore, the modified FEM model can be considered as reasonable and the results are presented in figure 12.

As expected, the frame reduces the deflection of the module significantly by 54%. Also the mean first principal stress $\bar{\sigma}_I$ in the solar cells is significantly reduced to 114 MPa, which is below the characteristic fracture stress. A further decrease is achieved by the glass-glass setup. Here the influence of the frame is not significant anymore with stress values of 67 MPa and 66 MPa for frameless and framed modules respectively. So the main impact is a reduction of the strain in the rear layer of the module.

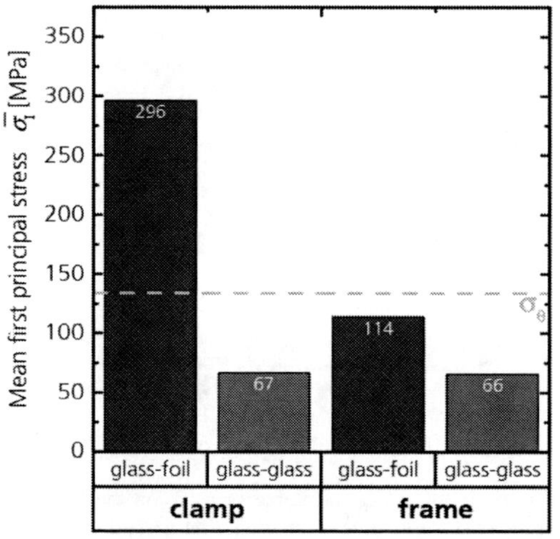

Figure 12: Mean first principal stress $\bar{\sigma}_I$ on rear side of the solar cell (80 µm) of the model with 200 µm EVA with ECAI at 2400 Pa pressure load for different module designs. The dashed yellow line indicates the characteristic fracture stress σ_θ.

In conclusion ultra-thin back contact solar cells can be integrated in modules using thin encapsulants if either a glass-glass setup is chosen or a frame is used. In this study the characteristic fracture stress σ_θ of the MWT-cells is rather low, comparing to standard solar cells with about 200 MPa [14]. So keeping in mind that the fracture stress might be higher for other back contact solar cell concepts, like IBC, ultra-thin solar cells can also be integrated into a framed glass-foil module using thin encapsulants.

4 CONCLUSION

This paper presents a validated FEM model of a frameless 60-cell back contact module. We compare six different material combinations according to their impact on the tensile stress in the solar cells. We measure the temperature dependent Young's modulus of three encapsulants by a dynamic mechanical analyzer in order to capture their material bevahiour properly. Additionally we measure their Poisson's ratio as well as the Young's modulus of the backsheet with a tensile testing machine. For all six material combinations we vary the thickness of the solar cells and the encapsulant in the FEM simulations. We obtain the lowest stress value (260 MPa) for 200 µm thick EVA with ECAI and 180 µm cells. Depending on the material combination, the stress increases between 4% and 12% if the cell thickness is reduced to 80 µm. Due to the electrical connection concept used, the stress dependency on the encapsulant thickness is different for cells below or above about 110 µm thickness. For thin cells (< 110 µm), the stress increases with increasing encapsulant thickness, while for thick cells the stress decreases.

For the combination of 200 µm EVA, ECAI and 80 µm cells, the stress is significantly reduced (296 MPa to 114 MPa) by using a frame instead of laminate clamps. A glass-glass setup reduces the stress even further to 67 MPa for clamped and 66 MPa for framed modules. With this the FEM simulations we show that it is possible to integrate ultra-thin back contact solar cells in a PV module.

5 ACKNOWLEDGEMENT

This project has received funding from the European Union's Seventh Programme for research, technological development and demonstration under grant agreement No 609788, acronym CHEETAH and from the Cusanuswerk by a PhD-scholarship. The authors like to thank Li Carlos Rendler and Pascal Romer (both Fraunhofer ISE) for the fruitful discussions.

6 REFERENCES

[1] B. Klein, *FEM - Grundlagen und Anwendungen der Finite-Element-Methode im Maschinen- und Fahrzeugbau,* 9th ed. Wiesbaden, Germany: Vieweg+Teubner Verlag, 2012.

[2] *IEC 61215: 2005-04. Crystalline silicon terrestrial photovoltaic (PV) modules - Design qualification and type approval*, 2005.

[3] M. Meuwissen *et al,* "Simulation assisted design of a PV module incorporating electrically conductive adhesive interconnects," in *Proc. of the 21st European Photovoltaic Solar Energy Conference and Exhibition*, 2006, pp. 2485–2490.

[4] F. Kraemer, S. Wiese, E. Peter, and J. Seib, "Mechanical problems of novel back contact solar modules," *Microelectronics Reliability*, vol. 53, no. 8, pp. 1095–1100, 2013.

[5] I. Bennett and N. Loiseaux, "Measurement of Strains in MWT Modules During Manufacture," *Proceedings of the 2nd International Conference on Crystalline Silicon Photovoltaics SiliconPV 2012*, vol. 27, no. 0, pp. 697–702, 2012.

[6] A. J. Beinert, M. Ebert, U. Eitner, and J. Aktaa, "Influence of Photovoltaic Module Mounting Systems on the Thermo-Mechanical Stresses in Solar Cells by FEM Modelling," in *Proc. of the 32nd European Photovoltaic Solar Energy Conference and Exhibition*, München: WIP, 2016, pp. 1833–1836.

[7] U. Eitner, S. Kajari-Schroeder, M. Koentges, and H. Altenbach, "Thermal stress and strain of solar cells in photovoltaic modules," in *Advanced Structured Materials*, vol. 15, *Shell-like Structures: Non-classical Theories and Applications*, H. Altenbach and V. A. Eremeyev, Eds, Berlin/Heidelberg: Springer, 2011.

[8] R. B. Roberts, "Thermal expansion reference data: silicon 80-280K," *Journal of Physics D: Applied Physics*, vol. 15, no. 9, pp. L119, 1982.

[9] R. B. Roberts, "Thermal expansion reference data: silicon 300-850 K," *Journal of Physics D: Applied Physics*, vol. 14, no. 10, pp. L163, 1981.

[10] W.M. Haynes, Ed, *CRC handbook of chemistry and physics*: CRC Press, 2014.

[11] W. Weibull, "A statistical distribution function of wide applicability," *Journal of Applied Mechanics*, vol. 18, pp. 293–305, 1951.

[12] W. Weibull, *A Statistical Theory of the Strength of Materials*. Stockholm: Generalstabens Litografiska Anstalts Förlag, 1939.

[13] M. Sander, A. Dietrich, M. Pander, M. Ebert, and J. Bagdahn, "Systematic investigation of cracks in encapsulated solar cells after mechanical loading," *Solar Energy Materials and Solar Cells*, vol. 111, pp. 82–89, 2013.

[14] F. Kaule, W. Wang, and S. Schoenfelder, "Modeling and testing the mechanical strength of solar cells," *Solar Energy Materials and Solar Cells*, vol. 120, Part A, pp. 441–447, 2014.

TPEDGE: PROGRESS ON COST-EFFICIENT AND DURABLE EDGE-SEALED PV MODULES

Max Mittag[1], Ulrich Eitner[1], Tobias Neff[2]
[1] Fraunhofer Institute for Solar Energy Systems ISE
Heidenhofstr. 2, 79110 Freiburg, Germany
[2] Bystronic glass, c/o Bystronic Lenhardt GmbH,
Karl-Lenhardt-Straße 1-9, 75242 Neuhausen-Hamberg, Germany
max.mittag@ise.fraunhofer.de

ABSTRACT: With TPedge we present an advanced frameless, polymer free encapsulation concept for silicon solar cells which addresses disadvantages and cost factors related to conventional solar modules. TPedge is a gas-filled, edge sealed, glass-glass module without polymeric encapsulation foils. The cost calculation indicates 5.7% lower costs of ownership for TPedge compared to the conventional module production due to savings for encapsulation foils and frame. Results from successful and extended module testing (i.e. 4000 hours damp-heat, ΔP_{MPP} = -1.3%) and 4 years of outdoor exposure prove good reliability of the concept. We perform a cell-to-module (CTM) analysis of TPedge and other concepts and calculate gain and loss factors. We find TPedge to have increased reflection losses compared to conventional modules and a lower CTM_{power} ratio (0.906). We combine CTM-analysis and cost calculation and find the specific costs (€/Wp) for TPedge to be 1.2% lower compared to the conventional module concept. Weight analysis shows a lower weight of TPedge modules compared to conventional modules and glass-glass laminates (-0.4 kg). A Carbon footprint analysis performed for different module concepts and important module materials shows a lower CO_2-footprint of the materials used in TPedge modules compared to conventional modules.

Keywords: CTM, Cell-to-Module, Module Manufacturing, Durability, Reliability, Cost reduction, Weight Analysis

1. INTRODUCTION

Conventional photovoltaic modules use polymeric foils like ethylene-vinyl acetate (EVA) as solar cell encapsulation [1]. Several effects are known that cause failure or power loss of the solar module and are directly related to the encapsulation material or incomplete protection of the solar cells from environmental influences [2]-[6]. A lamination process is needed for module production to encapsulate the cells in polymer foils. This process takes 8-15 minutes [7] and is known to represent a bottle neck in solar module production. The costs of foils and aluminum frames are a significant cost factor to PV module production [8] for conventional modules. Various innovations have been proposed such as the introduction of ultra-fast cure material, non-curing thermoplastics or other material groups [9] as well as multi-stage laminators for faster processing times or glass-glass-laminates for improved aging stability. None of these measures were able to completely eliminate the intrinsic disadvantages of the encapsulation foils at competitive cost levels.

In previous work we presented the "TPedge" module concept; a gas-filled glass-glass module with an edge sealing [10]-[13] that targets the disadvantages of conventional laminates. This work reveals the progress in the development of this module concept and its maturity by presenting additional results from accelerated aging (4000 hours damp-heat) and outdoor exposure. We compare TPedge with other module concepts such as conventional glass-foil-laminates as well as with glass-glass-laminates.

We perform a weight analysis of module materials and evaluate TPedge modules using 2 mm thin glass on reliability and weight reduction.

We perform a detailed analysis of the cell-to-module (CTM) power ratio of the TPedge concept in comparison to other photovoltaic modules. A cost analysis for the TPedge module concept is performed. We combine results of the CTM-analysis with the cost calculation and calculate specific costs (€/Wp).

2. THE TPEDGE MODULE CONCEPT

The "TPedge"-module concept applies an edge sealing process, well known from the manufacturing of double glazing insulation windows. The edge sealing consists of a thermoplastic spacer (TPS) filled with drying silicates and a silicone which renders the mechanical stability of the module. The glass spacing is filled with air. A double side front glass with anti-reflective coating as used in solar thermal collectors is used to minimize reflection losses.

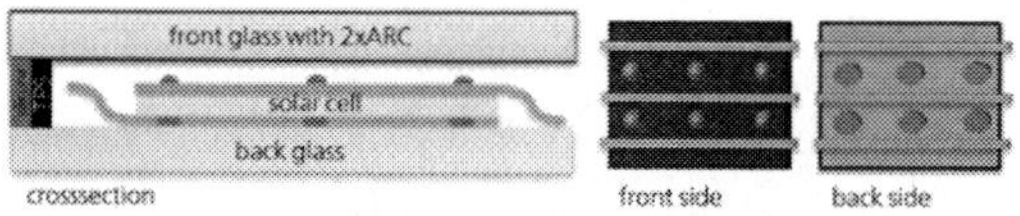

Figure 1: TPedge-module sketch with position of adhesive pins and double layer edge sealing [11]

Small pins, consisting of an UV-curing adhesive, glue the solar cells to the rear side glass pane. Glass spacing is provided by a set of transparent distance pins on the front side of the solar cells that cover approximately 0.02% of the cell area and provide additional mechanical stability.

Metal frames or similar additional supporting constructions are not necessary.

Figure 1 shows a schematic drawing of the cross section of a TPedge module with the positions of the adhesive pins on the front and backside of the solar cell. As a comparison the standard module architecture is shown in Figure 2.

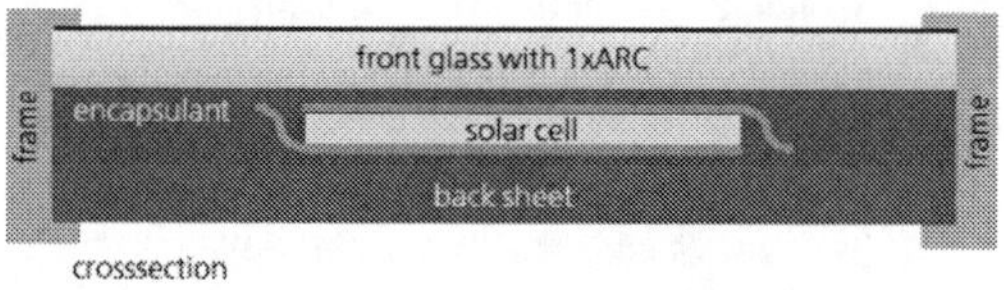

Figure 2: Conventional solar module schematics

3. RELIABILITY TESTING

a. Damp-Heat-Testing

Damp-heat tests according to IEC 61215 and IEC 61730 are performed on seven modules at Fraunhofer ISE TestLab PV Modules, see Table I. Three TPedge modules, three conventional glass-foil-modules and one glass-glass-laminate are tested. One TPedge and one conventional module (B) include a different set of cells and 20 cells per module only. Commercially available module materials are used and modules are manufactured to achieve comparability (same materials, same production parameters etc.). After every 1000 hours power measurements and EL-inspections are performed.

Table I: Power measurement results before and after damp-heat tests (85 °C, 85% r.h.)

	Initial Power [Wp]	After 4000 h damp-heat [Wp]	number of cells
TPedge A1	262.97	259.55	60
TPedge A2	262.62	260.67	60
TPedge B2	81.81	81.54	20
Conventional A1	274.87	217.59	60
Conventional A2	274.49	159.94	60
Conventional B2	86.06	84.06	20
Glass-Glass A1	272.34	263.59	60

Test results show only minor damp-heat-related effects on TPedge modules (ΔP_{MPP} < 1.3%). The conventional glass-backsheet modules using cells of manufacturer A fail after 4000 hours with a power loss of 27% and 41%. The smaller module using cells of manufacturer B is measured with a power loss below the 5% fail criteria (ΔP_{MPP} = -2.3%).

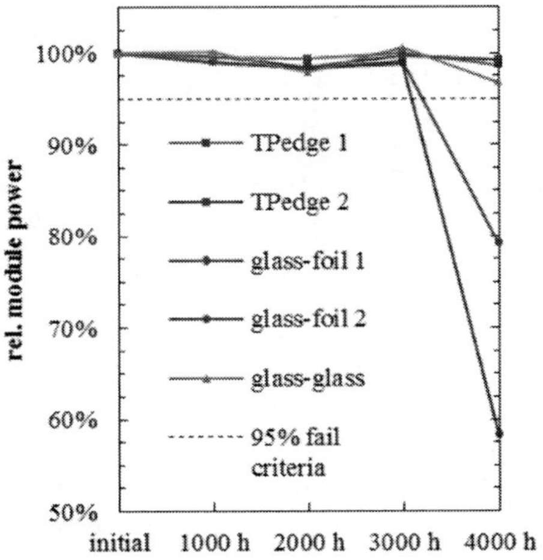

Figure 3: Power measurement results of full sized modules of different setups after several damp-heat-measurement cycles

Aging of the full size modules is clearly visible in the electroluminescence-images (Figure 4). The glass-glass-laminate withstands the extended test (ΔP_{MPP} = -3.2%) but EL shows damp-heat induced degradation in cells close to the module edge (Figure 4).

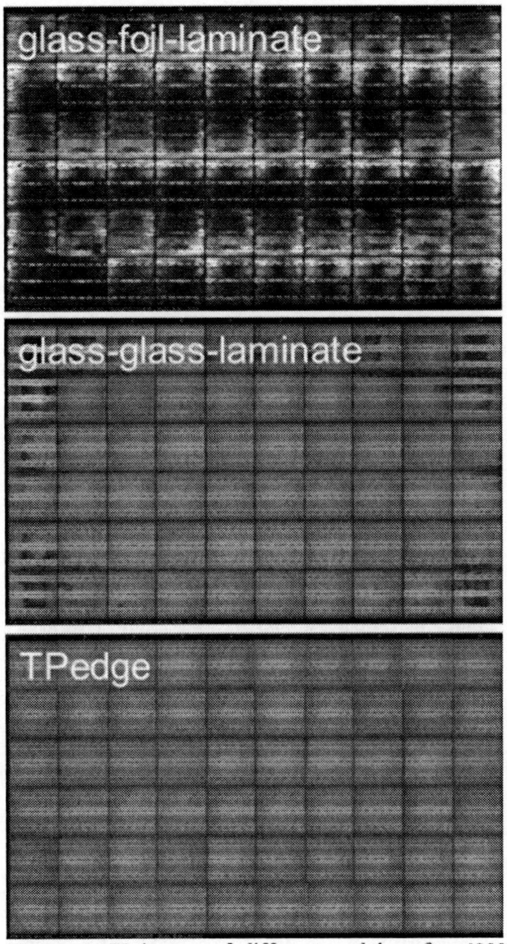

Figure 4: EL-images of different modules after 4000 hours damp-heat-testing

b. Outdoor Testing

Since August 2013 ten TPedge BiPV modules are operating in a vertical façade at Fraunhofer ISE, Freiburg (Figure 5) [14]. The modules contain MWT back-contact cells processed at Fraunhofer ISE PV-TEC [14] that are soldered to strings at Fraunhofer ISE ModuleTEC using the technology of shaped interconnectors [15].

Figure 5: Façade application of 70 TPedge BiPV modules at Fraunhofer ISE, Freiburg

In 2015 60 additional modules of two different sizes (42 and 14 cells) have been added to the façade. During installation of the additional modules an inspection of a previously installed module has been performed. No damage has been discovered using electroluminescence (Figure 6). A power measurement shows a ΔP_{MPP} of -1.9% of the inspected module (measurement uncertainty of initial and final measurement is 1.6%).

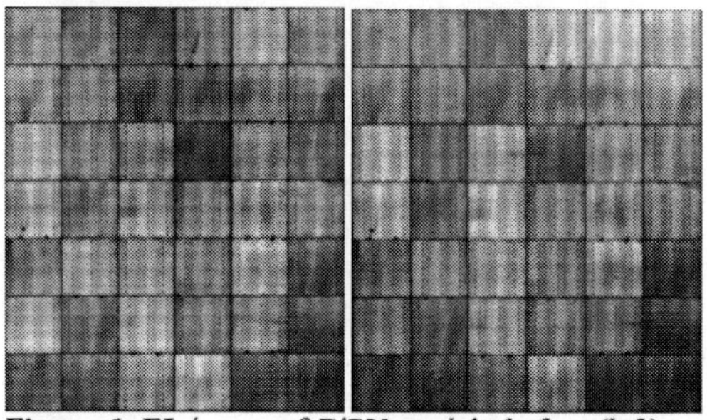

Figure 6: EL-image of BiPV module before (left) and after (right) two years of outdoor exposure. prototype MWT back-contact solar cells

Monitoring results show no decrease in the performance ratio of the installed modules (Figure 7).

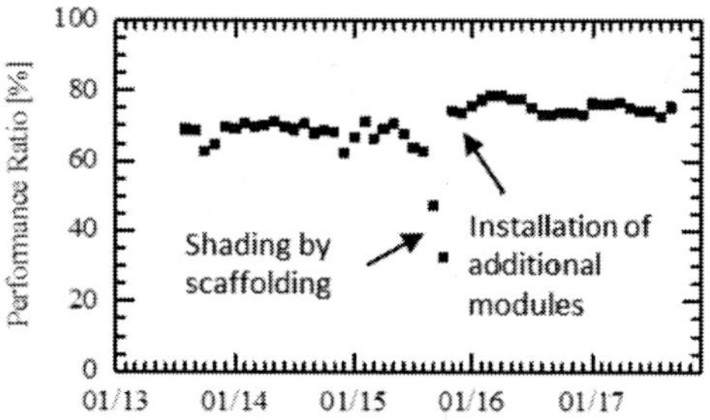

Figure 7: Performance ratio of TPedge BiPV application (modules facing 30° south, vertically installed)

Additional outdoor testing has been performed on nine full sizes modules of different setups. Four TPedge modules, three conventional modules and two glass-glass modules are tested for 14 months in Freiburg. Modules are similar to the modules tested for damp-heat-resistance. Cells of two different manufacturers and a TPedge module using 2 mm thin glass are tested.

The modules show no signs of degradation or power loss ($\Delta P_{MPP} < 0.9\%$) after 14 months of outdoor exposure. Electroluminescence shows no change between initial images and images after testing.

Additional test results regarding TPedge can be found in previous publications. Results of extended thermal cycling, mechanical load testing, PID-stability, hail stability have been published earlier [10][11][13].

4. WEIGHT REDUCTION

TPedge is a double glass module so that a higher weight is expected for a TPedge setup using two 3 mm glass panes compared to a conventional module. We measure the weight of module components and analyze weight reduction potentials.

Table II: Specific weight of module components

Component	Weight	
Glass (3 mm)	7.5	kg/m²
Glass (2 mm)	5.0	kg/m²
Encapsulant	0.4	kg/m²
Backsheet	0.45	kg/m²
Solar cells	11.0	g/pcs
String ribbon	17.0	g/m
Cell ribbon	2.6	g/m
Edge sealing	0.1	kg/module
Aluminum frame	2.5	kg/module
Junction box + cables	1.9	kg/pcs

Analysis shows that weight reduction of 30% is possible by using 2 mm thin glass for TPedge instead of 3 mm glass. Also TPedge using thin glass features the lowest weight compared to a conventional module (3 mm glass) and a glass-glass-laminate (2x 2 mm glass) (Figure 8).

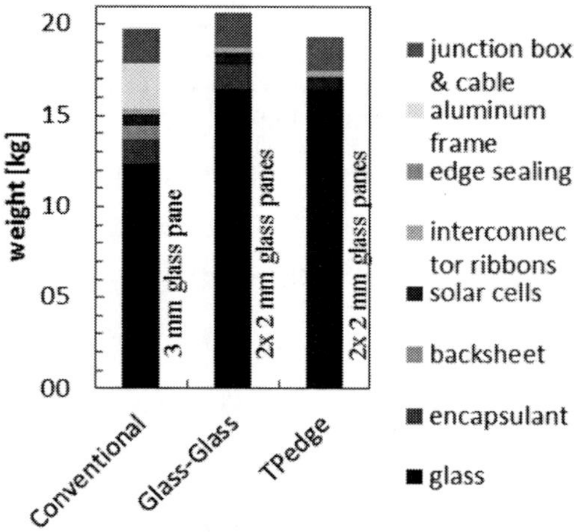

Figure 8: Weight analysis of different module concepts; edge sealing not visible due to small contribution to module weight

We have reported in previous publications [11][13] that the reduction of weight for the TPedge concept is possible by using 2 mm thin glass. Previously reported laboratory reliability tests cover mechanical load, damp-heat as well as hail tests. Outdoor test results with thin glass TPedge modules have also been presented. We demonstrated that the weight of TPedge modules can be successfully reduced without compromising on reliability.

5. CTM-ANALYSIS

We perform a cell-to-module analysis of different module concepts to analyze estimated losses from the additional air-glass interface in TPedge modules. Analysis is performed using Fraunhofer ISE SmartCalc.CTM [17]-[20]. The bill of materials of the analyzed setups is shown in Table III.

Table III: Bill of materials of analyzed module concepts

Component	Specification
Front glass	2.0 mm for glass-glass-laminate (1x ARC) and TPedge (2x ARC); 3.0 mm for conventional modules (1x ARC)
Rear glass	2.0 mm for TPedge and glass-glass laminates (no ARC)
Backsheet	White Tedlar-PET backsheet, 0.35 mm for conventional module
Encapsulant	EVA, 460 µm for laminates; air for TPedge
Solar cells	4.77 Wp, mono-crystalline, pseudo-square, 3 busbars (tapered)
String ribbon	6.0x0.3 mm copper core with SnPbAg coating
Cell ribbon	1.5x0.2 mm copper core with SnPbAg coating

Analysis confirms the additional reflection losses on position 2 (inner interface) of the TPedge front glass (Figure 9). Also coupling gains from module encapsulation reduce the module power of TPedge modules compared to conventional laminates. Reduction of interconnector losses is caused by lower currents of TPedge modules due to decreased irradiance on cell level. While TPedge features only small coupling gains, losses from encapsulant absorption become negligible due to the thin (< 1 mm) air filling of the glass spacing. The CTM_{power}-ratio of the analyzed conventional module is calculated to be 0.954. The CTM_{power} for TPedge is 0.906. Another analysis is performed for a glass-glass-laminate. CTM_{power} of 0.947 for the analyzed glass-glass setup is lower compared to the conventional module due to reduced cover coupling gains of the transparent rear glass [21].

Simulation results are compared with power measurement. Differences between measurement and simulation are below 0.3% for all three concepts.

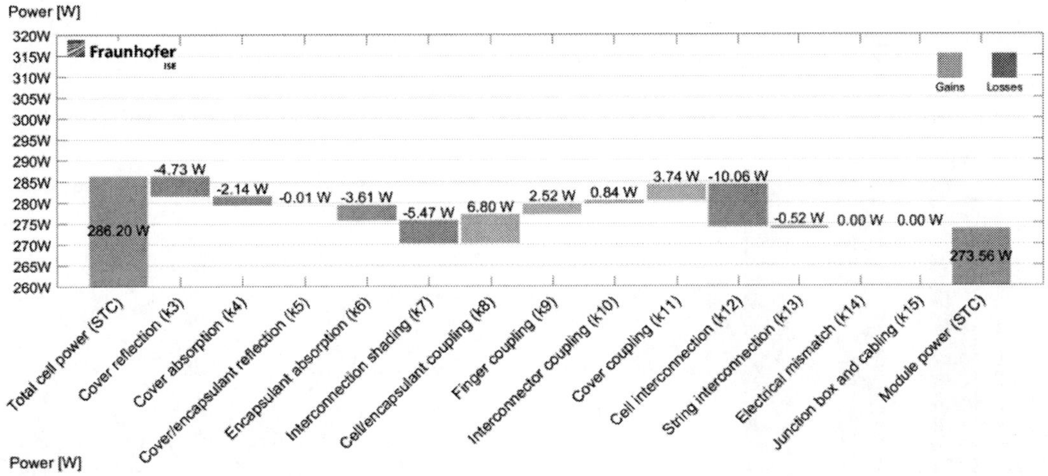

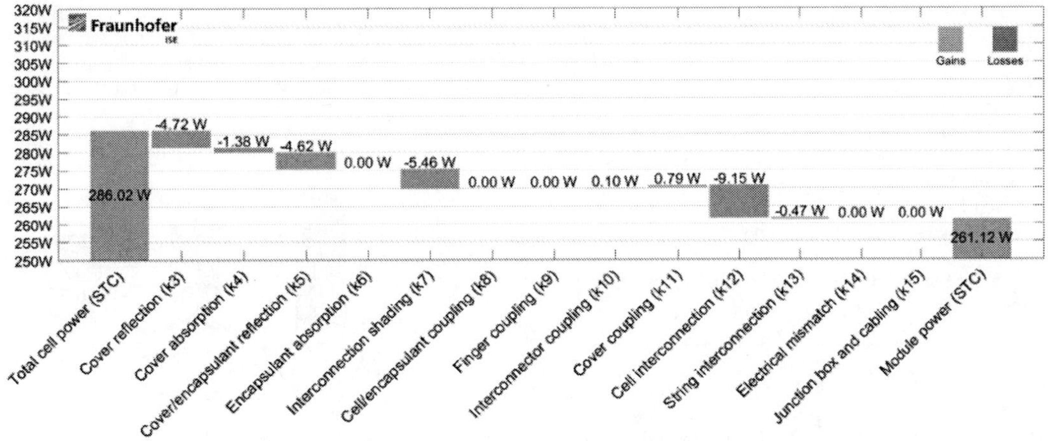

Figure 9: Cell-to-module analysis of a conventional module (top) and a TPedge solar module (bottom)

6. COST ANALYSIS

We perform a cost of ownership calculation for three module concepts using material and process costs. Calculation is performed with a cost analysis tool developed by Fraunhofer ISE considering material and process costs [16]. Important material costs are shown in Table IV.

Table IV: material prices as input for cost analysis

Component	Price	
Safety glass, 0x ARC	4.50	€/m²
Safety glass, 1x ARC	5.40	€/m²
Safety glass, 2x ARC	5.90	€/m²
Encapsulant	1.30	€/m²
Backsheet	2.00	€/pcs
Solar cells	0.225	€/Wp
String ribbon	0.04	€/m
Cell ribbon	0.24	€/m
Pin adhesive	200	€/l
Aluminum frame	6.00	€/pcs
Junction box + cable	4.00	€/pcs

We find TPedge to have lower Cost of Ownership than the other module concepts (Figure 10). While costs for solar cells, junction boxes, labeling etc. are the same for all concepts costs for glass, encapsulants, edge sealing and the aluminum frame are different. Saving the frame is the main factor for the glass-glass-modules price advantage compared to the conventional module. TPedge is cheaper due to the missing polymer foils.

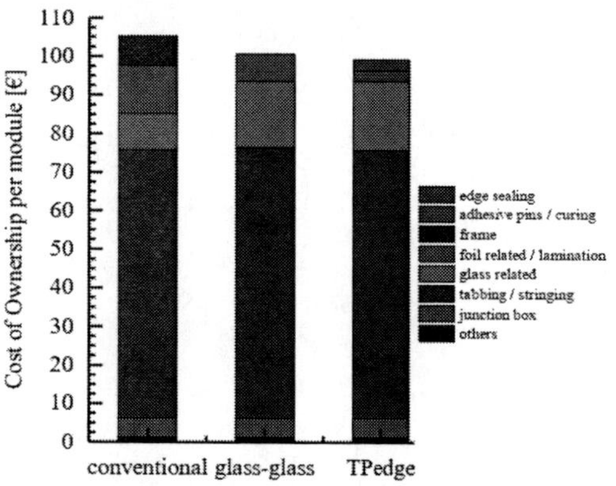

Figure 10: Cost of Ownership of different modules concepts. The cell costs are included in the tabbing/stringing part.

Costs of the analyzed setups are shown in Table V. We find the TPedge module to save 5.7% compared to the conventional module setup and to have the lowest costs of all compared module concepts.

Table V: Cost of Ownership of different module concepts

Module	Price	
Conventional module	105.32 €	100%
Glass-glass laminate	100.69 €	95.6%
TPedge	99.28 €	94.3%

Combining the results of CTM-analysis with the cost analysis we are able to calculate the specific costs (€/Wp). Modules in the cost calculation use 4.77 Wp solar cells as those were used for prototyping and CTM-analysis. Results are displayed in Table VI.

Table VI: Specific costs of different module concepts

	Conventional Module	Glass-glass Laminate	TPedge
COO	105.32 €	100.69 €	99.28 €
CTM	0.954	0.947	0.906
Specific Costs	0.385 €/Wp	0.382 €/Wp	0.390 €/Wp
	100%	96.8%	98.7%

We find the conventional module setup to have the highest specific costs and TPedge to save 1.4%. In the analyzed setups the frameless glass-glass laminate has the lowest specific costs (-3.5% compared to the conventional setup).

Additional costs resulting from logistics (e.g. packaging) or system integration (e.g. mounting) have not been considered.

7. CARBON FOOTPRINT

We perform an analysis of the carbon footprint of the materials that are used for module production. We consider important materials that are characteristic for the module concepts. Materials that are the same in all analyzed concepts (solar cells, junction boxes etc.) are ignored (ceteris paribus). Data is obtained from literature [22]-[33].

Table VII: CO_2 footprint of important module materials

Component	CO_2-footprint	
Glass (3 mm)	14.0	kg $_{CO2}$/m²
Glass (2 mm)	9.3	kg $_{CO2}$/m²
Encapsulant	0.8	kg $_{CO2}$/m²
Backsheet	0.7	kg $_{CO2}$/m²
Aluminum frame	19.4	kg $_{CO2}$/module

The glass-glass-laminate and TPedge use two 2 mm thin glasses. Modules have a size of 1.65 x 1.00 m² and – if applicable – use two layers of encapsulant and one backsheet. Analysis shows that materials used in TPedge feature a lower CO_2-footprint compared the other concepts (Figure 11). Additional effects from energy consumption during production (e.g. lamination) have not been considered.

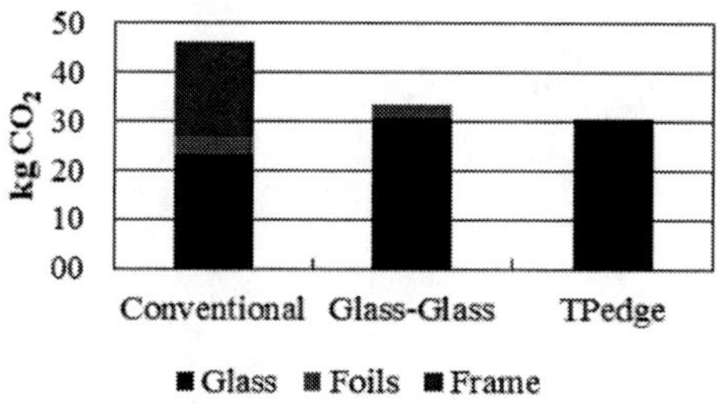

Figure 11: Combined CO_2-footprint of materials used for module production

8. SUMMARY

TPedge is successfully tested in an extended damp-heat test (4000 hours, 85 °C, 85%r.h.) with a change in P_{MPP} < 1.3%. Electroluminescence imaging reveals no defects in TPedge modules due to testing. We find damp-heat induced aging effects in the manufactured reference modules (conventional glass-backsheet and glass-glass configuration).

Cell-to-module analysis of TPedge, glass-backsheet and glass-glass modules shows increased reflection losses of TPedge. CTM_{power} ratio for TPedge (0.906) is lower than for conventional (0.954) and glass-glass-laminates (0.957).

Cost calculation shows lower cost of ownership (COO) of the TPedge module compared to the conventional module (94.3%) and the glass-glass-laminate (95.6%).

We calculate the specific costs (€/Wp) using results of COO- and CTM-analysis and find TPedge to have lower specific costs than the conventional module (98.8%).

We measure the weight of important module components and analyze weight reduction potentials for TPedge. We find a weight reduction of 30% possible by switching from 3 mm to 2 mm glass panes. Frameless thin-glass TPedge modules are lighter that framed glass-backsheet modules. Modules using 2 mm glass panes successfully perform in laboratory and outdoor testing so that weight reduction does not compromise the excellent reliability properties of TPedge concept.

We calculate the carbon footprint of module materials for different module concepts and find TPedge to have a 34% lower CO_2-footprint considering important module materials.

9. ACKNOWLEDGEMENTS

We would like to thank the German Ministry of Economic Affairs and Energy (FKZ 0325591 A) and the German Ministry for the Environment, Nature Conservation, Building and Nuclear Safety (FKZ 0325081) for their funding.

Also we would like to thank Bystronic Lenhardt GmbH for their partnership during the development of the TPedge module concept.

We would like to thank Matthieu Ebert for his valuable contributions to SmartCalc.CTM and the photovoltaics module department of Fraunhofer ISE.

10. REFERENCES

[1] SEMI Group, "*8th International Technology Roadmap for Photovoltaic*", 2017

[2] Peike, C., et al, "*Research on decrease of cell to module loss for crystalline silicon photovoltaic module*", Proceedings of the 3rd International Conference on Crystalline Silicon Photovoltaics (SiliconPV), 2013

[3] Ferrara, C., et al, "*Why do PV modules fail?*", International Conference on Materials for Advanced Technologies, 2011

[4] Oreski, G., et al, "*Damp heat induced physical aging of PV encapsulation materials*", 12th IEEE Intersociety Conference on Thermal and Thermomechanical Phenomena in Electronic Systems (ITherm), 2010

[5] Peike, C., et al, "*Towards the origin of photochemical EVA discoloration*", 39th IEEE Photovoltaic Specialists Conference, 2013

[6] Kraft, A., et al, "*Investigation of acetic acid corrosion impact on printed solar cell contacts*", IEEE Journal of Photovoltaics Vol.5 No. 3, 2015

[7] Wirth, H., et al, "*PV module technology and reliability - status and perspectives*", Green, Vol. 2, No. 4, pp.159-169, 2012

[8] Powell, D.M., et al, "*Crystalline silicon photovoltaics: a cost analysis framework for determining technology pathways to reach baseload electricity costs*", Energy & Environmental Science, 2012

[9] Poliskie, M., et al, "*Solar Module Packaging – Polymeric Requirements and Selection*", Energy & CRC Press, 2011

[10] Haedrich, I., et al, "*Encapsulant-free, edge-sealed module process technology*", PHOTON's 8th PV Production Equipment Conference, Shanghai (China), 2012

[11] Mittag, M., et al, "*TPedge": Qualification of a gas-filled, encapsulation free glass-glass photovoltaic module concept*", 29th European PV Solar Energy Conference and Exhibition, 2014

[12] Wirth, H. et al, "*Photovoltaic Modules: Technology and Reliability*", ISBN 978-3-11-034827-9, 2016

[13] Mittag, M., et al, "*TPedge": Glass-glass Photovoltaic Module for BiPV-Applications*", Engineered Transparancy / glasstec, 2014

[14] Drews, A., et al, "*HIP-MWT solar cells–pilot-line cell processing and module integration*", 27th European PV Solar Energy Conference and Exhibition, 2012

[15] Eitner, U., et al, "*Interconnector-based module technology for thin MWT-cells*", 27th European PV Solar Energy Conference and Exhibition, 2012

[16] Nold, S., et al, "*Cost modeling of silicon solar cell production innovation along the PV value chain*", 27th European PV Solar Energy Conference and Exhibition, 2012

[17] Haedrich, I., et al, "*Unified methodology for determining CTM ratios: Systematic prediction of module power*", Solar energy materials and solar cells 131 (2014), pp.14-23, ISSN: 0927-0248, SiliconPV, 2014

[18] www.cell-to-module.com

[19] Mittag, M., et al, "*Cell-to-Module (CTM) Analysis for Photovoltaic Modules with Shingled Solar Cells*", 44th IEEE Photovoltaic Specialists Conference, 2017

[20] Mittag, M., et al, "*Systematic PV module optimization*
with the cell-to-module (CTM) analysis software", Photovoltaics International, vol. 36, 2017

[21] Mittag, M., et al, "*Analysis of Backsheet and Rear Cover Reflection Gains for Bifacial Solar Cells*", 33rd European PV Solar Energy Conference and Exhibition, 2017

[22] Teenou, R.Y., "*Energy and CO2 emissions associated with the production of Multi- glazed windows lls*", Master Thesis, University of Östersund, Sweden, 2012

[23] www.aluwatch.org

[24] www.carbonfootprint.at

[25] https://ecrm.marketgate.com/help/519/842.htm

[26]
http://www.nsg.com/en/sustainability/glassandclimatechange/embodiedc02infloatglass

[27] https://iea-etsap.org/E-TechDS/PDF/I10_AlProduction_ER_March2012_Final%20GSOK.pdf

[28]
https://ec.europa.eu/clima/sites/clima/files/ets/allowances/docs/bm_study-glass_en.pdf

[29] http://de.saint-gobain-glass.com/de/umwelt/energy-co2.htm

[30] British Glass, "*Glass Recycling - Life Cycle Carbon Dioxide Emissions A Life Cycle Analysis Report*", 2003

[31] Bundesverband Glasindustrie e.V., "*Stellungnahme Staatliche Beihilfe SA.42393 (2016/C) (ex 2015/N) – Deutschland Förderung der Kraft-Wärme-Kopplung in Deutschland*", 2016

[32] de Wild-Scheuten, M.J., et al, "*Environmental Life Cycle Analysis Of Dye Sensitized Solar Devices; Status And Outlook*", 22nd European Photovoltaic Solar Energy Conference, 2007

[33] van Sark, W.G.J.H.M.., et al, "*CO2 Emissions of PV in the Perspective of a Renewable Energy Economy*", 22nd European Photovoltaic Solar Energy Conference, 2007

HOW CELL TEXTURING IMPACTS ANNUAL YIELD OF SOLAR MODULES AND THE ROLE OF MODULE EMBEDDING

Ingrid Haedrich[1], Marco Ernst[1], Andrew Thomson[1], Peiting Zheng[2], Xinyu Zhang[2], Hao Jin[2], Daniel McDonald[1]
[1] Australian National University ANU, Canberra 2601 ACT, Australia
[2] R&D Center Jinko Solar, Haining, Zhejiang, China

ABSTRACT: Two emerging trends in multicrystalline silicon cell texturing are plasma texturing and metal catalyzed chemical etching. Both process roughen silicon surfaces increasing light absorption. These processes are attractive as they are applicable to diamond-wire sawn wafers. In this work, the optical properties of these surfaces, and other conventionally textured surfaces, are investigated for cells in air and after encapsulation, with direct irradiance over a large range of angles of incidence. We find that the angular optical performance in air varies strongly with cell texture, and that these variations are mitigated significantly once embedded. This is especially notable for plasma etched, black silicon cell structures, which suffer comparatively from poorer index matching and light recycling. This behavior causes low optical cell to module ratios of 96.5% at normal and 87.5%d at 70° angle of incidence, compared to cell to module ratios of 97.5-99.5% at normal and 90%-97.5% at 70° angle of incidence for other chemically etched cells. The impact of angular optical performance on module yield for different cell surfaces was computed. Annual yield losses attributed to the module embedding were 2% to 11%; the highest losses are caused by embedding the black silicon cell structures.

Keywords: black silicon, cell to module ratio, solar module, embedding, cell texture, solar module yield

1 MOTIVATION

A key paradigm of PV cell module development is higher efficiency and power, measured under STC with normal irradiance. However, fielded performance strongly impacts the true cost of PV electricity, where most radiation is typically non-normal. Clearly different front cell textures impact angular light absorption [1-10]. However, it is the angular behavior of a cell embedded in a module sandwich, and how the module embedding impacts the cell angular behavior, that impacts field performance. Hence to truly reduce costs by manipulation and design of the cell front surface, the performance after embedding the cell into a module should be considered. In this work we investigate cell types with various front textures including chemical and plasma etched front structures and compare them to common acidic/ISO- and random textured wafers. We measure their angular light capturing performance in air and within a module configuration. Based on these measurements we calculate their impact on the module annual yield and before on the cell to module ratio. We focus our work on recently developed black silicon cell structures; particularly chemically etched textures applied on wire sawed wafer material and plasma etching on previously ISO textured wafers. We quantify the cell to module losses for different angles of incidence and we define a cell to module yield ratio, which is a parameter indicating losses explicitly caused by module embedding. We furthermore quantify the impact of anti-reflective glass coatings on the optical losses in comparison to non-coated glass for all investigated cell types.

2 APPROACH

We investigated seven solar cells from five different manufacturers (see Table 1). We laser-cut the solar cells into samples in size 20 x 17.5 mm^2. After measuring the samples against air, we embedded them into an industrial common module setup using a UV through EVA and a 3.2 mm thick low iron glass without anti-reflective coating (see Figure 2). To consider multiple light bouncing effects within the module, e.g. from reflected light at the finger surface, it was necessary apply mirrors on the edge of the small module samples. The specular mirrors redirect light that would otherwise be lost back in to the sample. The global hemispherical reflectance of cells in air and inside a module are measured inside an integrating sphere using a center mount sample holder. The light spot has a size of around 10 x 2 mm2 at 8° and of about 10 x 10 mm2 at 80° AoI. The reflectance of all samples is measured at an angle of incidence of 8°, 20°, 40°, 60°, 70°,75° and 80° at a wavelength range from 250 nm to 1500 nm in 10 nm steps.

Table 1: Overview of the material and design characteristics of the investigated solar cell structures including scanning electron microscope images (SEM). The given reflectance is measured with a spectrophotometer using an integrating sphere at an angle of incidence of 8°.

name	*photo*	*wafer*	*front surface texture*	*R at 8° in air (%)*	*SEM images*
A.1		mc-Si	*metal catalyzed* chemical etching (MCCE) directly applied on diamond sawn wafer	8.7	
A.2		mc-Si	same as A 1 *without front metallization*	6.6	
B		mc-Si	plasma texture (RIE) on acidic textured diamond sawn wafer	5.4	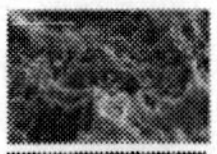
C.1		mc-Si	acidic texture (isotexture)	9.2	
C.2		mc-Si	same as C.1 *without front metallization*	6.8	
D		mc-Si	random pyramids (RP)	7.2	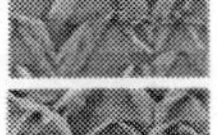
E		Cz-Si	random pyramids (RP)	7.4	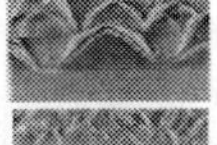
F		Cz-Si	random pyramids (RP)	5.7	
G		Cz-Si	random pyramids (RP)	6.9	

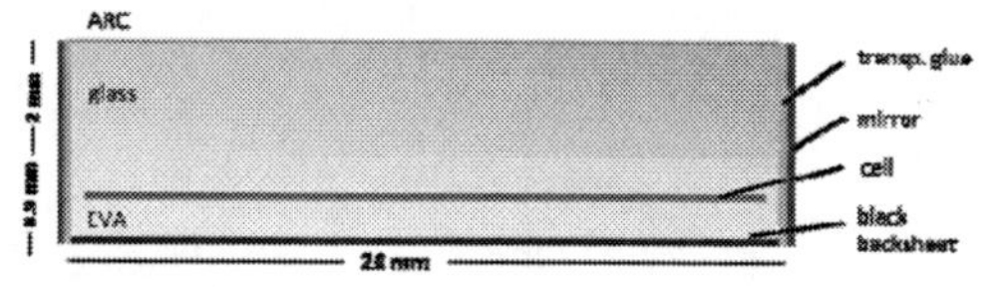

Figure 1: Illustration of the module setup including dimensions

3 RESULTS

Angular spectral absorbance

By weighting the measured absorption curves with the AM1.5 spectrum (in Wm-2nm-1) and the cells internal spectral response (in A/W) we determine the effective angular absorption (see **Error! Reference source not found.**). The internal spectral response is taken from the internal quantum efficiency (IQE) of a high efficient IBC solar cell. For the samples measured in air the black silicon sample shows the highest absorption over all AoIs, starting from 94.6% at 8° down to 89.3% at 70°. Isotextured and the MCCE on diamond cut wafers showed the lowest absorption reaching from 90.8% / 91.3% to 84.8% / 78.91% (A.1/ C.1) (see graph (a)). By looking to the normalized absorption curves in air (see **Error! Reference source not found.**d) the black silicon cells (B) and the isotextured sample (C.2) show the least decrease in absorption by retaining 94.5% (for B) of their initial performance at 8° AoI. The worst angular performance is found for the two MCCE textured diamond cut wafers (A.1 and A.2). Their normalized absorption decreases down to 86.4% and 87.7% respectively at 70°.

By putting the cell samples into a regular module setup, the spreading of the angular absorption between the cell types is significantly reduced for all angles. For example, whereas the cell samples vary in their absorbance in air at 70° by an absolute 10.4% (A.1 vs. B), this reduces to 2.2 points for module samples without AR coating and 2.5 percentage points for modules with AR coating (see **Error! Reference source not found.**b). This behavior can be explained by the photon recycling effects during embedding which help cells with an initial high reflectance loss to couple relatively more light into the cell after embedding whereas cells with an initially low reflectance loss in air do not experience such a large coupling gain. This effect is further described in [10].

The ability to absorb angular light inside a module configuration is mainly driven by the reflectance of the glass/air interface. In a module setup without AR coating the cells absorb 74% to 76% absolute at 70° (see **Error! Reference source not found.**b) or, in other words, they can hold 83% to 84% of their performance shown at 8° AoI. Given that at an AoI of 70° the air/ glass reflectance losses for uncoated glass are known to reduce the absorbance by about 18%, the other 6% to 8% are caused by reflection and absorption losses within the module structure. Further it needs to be considered that the angle the light hits the module surface at is reduced when the light enters the embedding materials, due to their higher refractive index (n_{glass}= 1.52 and n_{eva}= 1.47) (Fresnel law [11]). For example, an AoI of 70° will lead to an incoming angle at the cell surface of only 38°. At this new effective AoI the cells still absorb more than 97% relative to perpendicular irradiance. Note that applying an anti-reflective coating on the glass surface changes the amount of light entering a module, but not the angle at which the light hits the cell surface, since this is determined by the refractive index of the material which directly surrounds the cell.

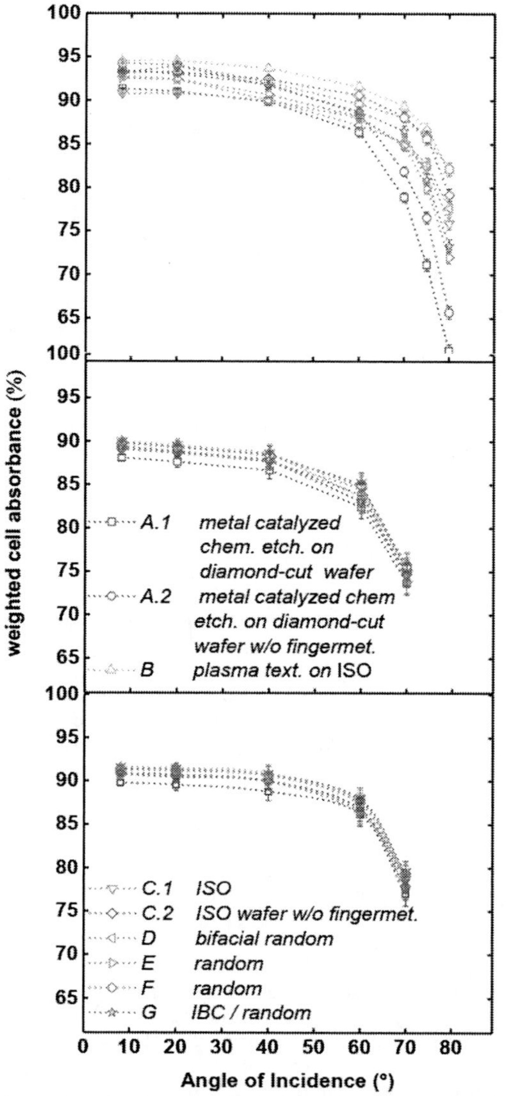

Figure 2: Absolute weighted absorption curves for all angles and all investigated samples in air (*top*) and embedded in the module configurations with (*centre*) and without antireflective coating (ARC) (*bottom*).

Angular cell to module (CTM) ratio

Focusing more closely on the optical losses caused by module embedding, we use the CTM ratio, which relates the cell absorption within a module to the absorption measured in air at the same AoI (see **Error! Reference source not found.**).

For modules without an antireflective coating we determine CTM ratios between 0.95 for B to 0.98 for C.1 at an AOI of 8°. In other words, the reflection losses at the air/glass surface and the absorption in glass and EVA cause 5% reduction in power in the case of black silicon solar cells, but only 2% for the isotextured solar cells having the same front metallization. This reflects the fact that the index matching gain impacts the isotextured cells more than the plasma etched surfaces.

Following the CTM ratios over increasing AoIs it is obvious that the black silicon cells show the lowest CTM ratios over all angles. At 70° AoI the CTM ratio for this cell type B is 0.85 whereas for the plasma texturing on acidic textures A.1 it is still 0.93 (0.91 for A.2). The latter can be explained by their initial high absorption losses at high AoIs in air which do not decrease much further by embedding them.

Applying an AR coating on top of the glass increases the CTM ratios on average by 0.02 (for 8°, 20° and 40°) and by 0.04 at 70° AoI. This enables CTM ratios above 1 to be achieved, meaning that the module embedding causes a gain compared to the samples measured in air. This is particularly the case for the isotextured samples C.1 and the bifacial samples D.

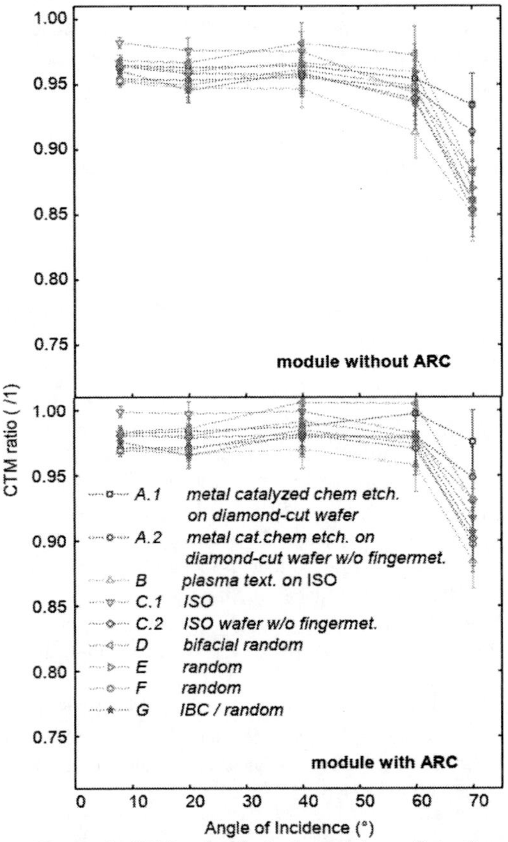

Figure 3: CTM ratios for both module configurations for all investigated samples. The CTMs tend to decrease for higher angles, with some exceptions. The CTM ratios for the samples embedded with AR coated glass are on average 2% higher than the same samples embedded without AR coating.

Annual optical yield (AOY)

We calculate the angular resolved irradiance received on a tilted module plane using the software SUNCALCULOR as a function of the module tilt. Based on annual irradiances we calculate the theoretical annual yield of a cell exposed directly in air and then inside a module. We chose two Australian locations, one with a high diffuse to global irradiance ratio (Melbourne) and one with a high direct to global irradiance ratio (Alice Springs). We use solar irradiance data provided by the Australian Bureau of Meteorology (BOM). We calculate the annual optical yield of a cell in air / module by summing the angular resolved spectral irradiance multiplied with the absorption spectra over the AoI [6].

For the absorption in air we measured up to an AOI of 80° and then used a linear interpolation fit. The absorption at an AoI of 90° was set to 0. For the absorption in a module we fitted the measured curves using the Martin Ruiz module [12].

As a reference, we further calculated the annual optical yield for an ideal cell with an assumed spectral absorption of 1, representing a cell with no optical or electrical losses. Figure 4 normalizes the annual yield of all cell and module configurations to this reference.

Further we calculated the theoretical annual yield for all configurations assuming an absorption equal to 8° for all angles of incidence. With this, yield calculations based on standard testing conditions, so without considering any angular absorption behavior, are represented. The difference between the STC and realistic conditions are an indicator for annual losses caused by specifically angular effects.

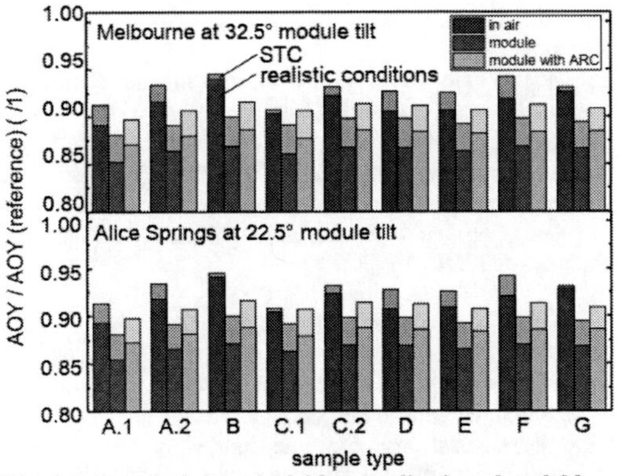

Figure 4: optical annual yield normalized to the yield an ideal solar cell (absoption =100% for all angles of incidence) would achieve (dark colours). Further the annual yield assuming a constant angular absorption (8°) is displayed (bright colors) for modules installed in Alice springs and Melbourne in optimum tilt angle.

We find that the highest annual yield will be achieved by the plasma textured solar cells no matter if calculated for air or being embedded in a module. Even though this cell type shows the highest angular embedding losses (7%) the total annual yield still accedes the yield of all other investigated cell textures. The lowest annual yield we predict for cell type A.1, so for the MCCE textured samples. For this cell type the losses attributed to module embedding are only 4%.

Further we find that an anti-reflective coating would increase the annual yield by in average over all samples by a maximum of 2%.

Comparing the STC based calculations to the realistic conditions (see Figure 5) we find that by not considering any angular effects we would overestimate the annual yield by 3% for a module being installed in its optimum tilt and up to 9% if installed at a 90° tilt angle. In other words, angular effects cause up to 9% of the annual yield losses.

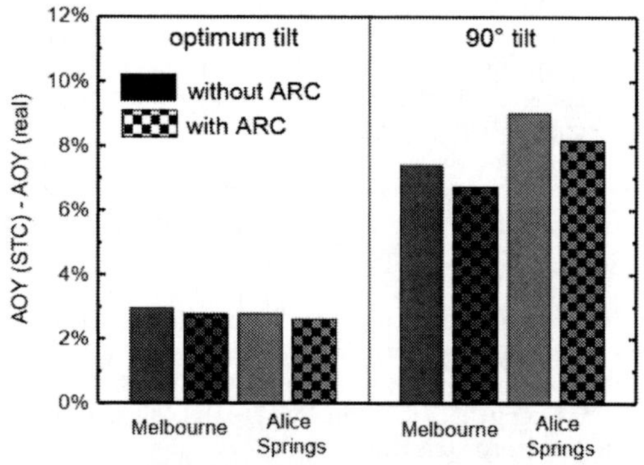

Figure 5: annual yield losses attributed to angular effects shown for two locations and for two different tilt angles.

4 SUMMARY

We investigated 9 different cell samples and determined their angular absorption in air and embedded in module. We find the highest absorption for the plasma etched cell structure (B) and the lowest for cell type A.1, MCCE structure.

In general, the wide spread in the angular absorption measured in air (Δ=3.8% at 8°, Δ=10% at 70°) is significantly reduced when the cells are embedded in a module structure (Δmax=2.3% at 60°). Therefore, the optical advantages which are seen for example for cell type B (plasma on acidic texture) are significantly mitigated at the module level.

Focusing particularly on losses caused by module embedding, described with help of the CTM ratio, we find values between 0.95 for the plasma etched cells and 0.98 for isotextured cells at perpendicular irradiance. The CTM ratio decreases slowly before reaching an AoI of 60°, after this it decreases more drastically down to 0.85 (B) or 0.93 at best (cell type A.1).

We find that the highest annual yield will be achieved by the plasma textured solar cells no matter if calculated for air or being embedded in a module. Even though this cell type shows the highest angular embedding losses (7%) the total annual yield still accedes the yield of all other investigated cell textures.

Calculating the annual yield by not considering any angular losses (STC) would overestimate the annual yield by 3% for a module being installed in its optimum tilt and up to 9% if installed at a 90° tilt angle.

Acknowledgements

The authors acknowledge the Australian Renewable Energy Agency funding of this work through Grant 2014/RND008

References

1. Wirth, H., K.-A. Weiß, and C. Wiesmeier, Photovoltaic Modules - Technology and Reliability. 2016, De Gruyter

2. Andrew Thomson, M.E., Ingrid Haedrich, Jiadong Qian, Impact of PV module configuration on energy yield under realistic conditions. Optical and Quantum Electronics, 2017. 49(2): p. 82.

3. Parretta, A., et al., Angle-dependent reflectance measurements on photovoltaic materials and solar cells. Optics Communications, 1999. 172(1): p. 139-151.

4. Xiaogang Liu ab, P.R.C.c., Marius Peters b, Bram Hoex b, Jacqueline M. Cole *ad and Derek J. Fray c, Black silicon: fabrication methods, properties and solar energy applications. 2014.

5. Khoo, Y.S., et al., Method for quantifying optical parasitic absorptance loss of glass and encapsulant materials of silicon wafer based photovoltaic modules. Solar Energy Materials and Solar Cells, 2012. 102(0): p. 153-158.

6. Haedrich, I., et al., Unified methodology for determining CTM ratios: Systematic prediction of module power. Solar Energy Materials and Solar Cells, 2014. 131: p. 14-23.

7. Peters, I.M., K. Yong Sheng, and T.M. Walsh, Detailed Current Loss Analysis for a PV Module Made With Textured Multicrystalline Silicon Wafer Solar Cells. Photovoltaics, IEEE Journal of, 2014. 4(2): p. 585-593.

8. Gjessing, J. and E.S. Marstein, Optical performance of solar modules. Energy Procedia, 2013. 38: p. 348-354.

9. Schneider, A., R. Harney, and S. Koch, Impact of Cell Texturing Quality on Cell to Module Losses. Energy Procedia, 2015. 77: p. 414-419.

10. Geisemeyer, I., et al., Angle Dependence of Solar Cells and Modules: The Role of Cell Texturization. IEEE Journal of Photovoltaics, 2017. 7(1): p. 19-24.

11. Krauter, S. and P. Grunow. Optical Modelling and Simulation of PV Module Encapsulation to Improve Structure and Material Properties for Maximum Energy Yield. in Photovoltaic Energy Conversion, Conference Record of the 2006 IEEE 4th World Conference on. 2006.

12. Martin, N. and J.M. Ruiz, Calculation of the PV modules angular losses under field conditions by means of an analytical model. Solar Energy Materials and Solar Cells, 2001. 70(1): p. 25-38.

NEW CHEMICAL FUNCTIONALIZATION CONCEPT FOR ANTI-REFLECTIVE AND ANTI-SOILING FRONT GLASS OF PV MODULES BASED ON SURFACE STRUCTURING AND MODIFICATION

Charlotte Pfau[1,2*], Klemens Ilse[1,3], Jens Schneider[1], Marko Turek[1], Paul-T. Miclea[1,2], Christian Hagendorf[1],
Przemysław Ząbek[4], Wiesław Doros[4]
[1]Fraunhofer-Center for Silicon-Photovoltaics CSP, Otto-Eißfeldt-Str. 12, 06120 Halle (Saale), Germany
[2]Centre for Innovation Competence SiLi-nano, Martin Luther University of Halle-Wittenberg
Karl-Freiherr-von-Fritsch-Str. 3, 06120 Halle (Saale), Germany
[3]Anhalt University of Applied Sciences
Faculty EMW, Bernburger Str. 55, D-06366, Koethen, Germany
[4]DAGlass, ul. Warszawska 55, 35-205 Rzeszów, Poland
*Corresponding author: +49 345 5589 5411, charlotte.pfau@csp.fraunhofer.de

ABSTRACT: This works presents a new functionalization concept for front glasses of PV modules and an analysis of their anti-reflective and anti-soiling properties. The described functionalization technology is based on surface structuring and modification obtained by a low cost chemical treatment. The functionalized, nanostructured surface is part of the original glass and no additional coating is required, which makes it more reliable. Nanostructured glasses are compared to conventional anti-reflective coated and untreated reference glasses. The optical and self-cleaning properties of the glasses are investigated by spectrally resolved optical analysis and by a newly developed anti-soiling test method, respectively. Additionally, the performance of 1-cell modules equipped with these glasses is simulated on the basis of optical data and measured electrically in solar simulators. 60-cell modules with different glasses are compared for their electrical performance. It is shown that the anti-reflective behavior of a glass with nanostructured surface without any additional anti-reflective coating is as good as conventional solar glass with standard anti-reflective coating. Furthermore, it is shown that loss in integral transmission after soiling process is significantly reduced (25 % relative) by a proper wet chemical surface treatment. This reduces solar module operation costs in arid regions.
Keywords: PV Module, Module Technology, Solar glass, anti-reflective coating, anti-soiling coating

1 INTRODUCTION

DAGlass' low cost, chemical surface modification technology is a new concept of anti-reflective (AR) as well as anti-soiling (AS) glass treatment. The functionalized surface is part of the original glass and no additional coating is required which makes it much more reliable. The oldest glass installations (by greenhouses) with AR treatment of D.A. Glass are about 4-6 years old. The function of the treatment remains on the constant level without any significant decrease. Beside classical coating technologies, *DAGlass* develops various own technologies for structuring the glass surface on different length scales reaching from mm down to about 100 nm. The combination of anti-reflective and anti-soiling effects, i.e. by superimposing different surface structures without deposition of additional coatings, on PV module front glass has a high potential for PV electricity generation cost reduction. Especially the cost reduction potential of anti-soiling effects is not yet being fully utilized. Dust deposition on PV modules can cause severe problems, especially in desert applications. Cumulative energy losses can reach more than 1 % per day [1], also causing tremendous financial impact on operating and maintenance costs. Thus, it is of high interest to scientifically assess this technology and to explore its opportunities for the PV market.

Various concepts of glass surface functionalization are reported in literature **[2]**. To improve anti-reflective properties, usually a single anti-reflective layer is introduced by magnetron sputtering or more commonly by the Sol-gel technology. A popular AR layer for glasses is nanoporous SiO_2. The optimum (effective) refractive index and optimum thickness of the coating reduces the reflection at the desired wavelength by destructive interference. Instead of deposition of a nanoporous layer with effective optical properties DA glass selectively etches the glass surface and generates a nanoporous structure in the glass itself. Therefore no additional interface is present between the glass and quasi-AR layer. Such a structure might be less costly and more reliable which is subject to further testing.

Considering anti-soiling functionality, often self-cleaning concepts with (super-) hydrophobic or (super-) hydrophilic surface wettability are targeted [3]. Especially in desert areas, the self-cleaning ability could not always be provided since most of these concepts require water, which typically is rare and expensive in arid and semi-arid regions. Furthermore, outdoor exposure tests in desert environments show that hydrophobic/hydrophilic coating functionalization gets often lost within a short period of time due to coating degradation and severe contamination. For example, cementation processes due to typical dew events during night and morning hours can dramatically increase surface adhesion [4]. A promising anti soiling strategy for desert locations is to increase natural self-cleaning by wind. This could be reached by introducing surface roughness and surface chemistry to lower Van der Waals and/or capillary forces between glass and particle surfaces, also minimizing initial dust adhesion.

The main objective is to develop functionalized glasses for PV applications with the targets of increasing solar module power and long term reliability leading to a maximal energy yield. This objective is pursued by improving the glasses in an iterative process based on process modifications and subsequent PV applicability evaluation of differently functionalized anti-reflective and anti-soiling surfaces.

In this work, the assessment of the first batch of glasses, processed after a first optimization step, is presented.

For this purpose, spectrally resolved measurements and optical simulations as well as indoor anti-soiling tests are performed on differently treated glasses. Additionally, 1-cell modules built with *DAGlass'* glasses are simulated and measured electrically under solar simulators.

2 EXPERIMENTAL DETAILS

2.1 Glasses and their functionalization

Table 1 gives an overview on the analyzed samples and their basic parameters. *DAGlass* has used low iron solar float *Pilkington optiwhite S* in 2.05 mm thickness as a base glass (Ref). *DAGlass* has prepared 4 differently functionalized glasses (AR1-AR4); they all have random craters with structural size of about 150 nm on the surface (see AFM image in **Figure 2**).

The crater structure was introduced by chemical etching in liquid bath in which the cations are etched from the glass surface based on DA Glass technology. Subsequently, for the samples AR2 and AR4, an intrinsic SiO_2 coating was introduced by a chemical after-treatment process in a chemical liquid bath. The achieved thickness of the SiO_2 layer can be controlled by the treatment time and is about 10-15 nm. In the last step, a TiO_2 layer with a thickness of about 20 nm was deposited on samples AR3 and AR4 by magnetron sputtering.

Table 1: Overview on samples from *DAGlass* (Ref., AR1-4). In all cases the surface structure is composed of random craters with structural size of about 150 nm. All glass used is low iron solar float base glass Ref and S1 for functionalization, respectively. The double side treated glasses (AR1-AR4) are treated similarly on both sides.

Sample	Surface structure	Surface coating	Sample thickness in mm	Single or double side treatment
Ref	No	No	2.05	No
AR1	Yes	No	2.05	Double
AR2	Yes	SiO_2	2.05	Double
AR3	Yes	TiO_2	2.05	Double
AR4	Yes	SiO_2+TiO_2	2.05	Double

For technological reasons the glasses AR1-AR4 are treated similarly on both sides, although a single side surface treatment is more meaningful for PV applications. In the next optimization step a single side treated glass will be prepared.

Figure 2: AFM image of a nanostructured surface of glass sample AR1 obtained by *DAGlass'* functionalization technology.

2.2 Analysis methods

Spectroscopy

In a first step, global hemispheric transmittance and reflectance spectra of all glasses and used module materials are measured using a *Perkin Elmer Lambda 1050* UV/VIS/NIR spectrophotometer equipped with a Spectralon integrating sphere. The Fresnel reflection coefficients at the air-glass interface and further reflection and absorption losses in the module stack are calculated from this spectrally resolved data. A spectral weighting with solar spectrum and spectral response of the cells is carried out; it allows for the calculation of reflection loss factors enabling an evaluation of the PV applicability and a direct comparison of the anti-reflective properties of the different glasses. Additionally, similar to our approach described in Ref. **[5]** effects of the glass treatment on solar modules short circuit currents densities are calculated from optical simulations

Optical characterization was also used to determine soiling losses by performing measurements of total hemispheric transmittance before and after dust deposition. The transmission loss was calculated as

$$T_{Loss} = \frac{\int_{400\,nm}^{1200\,nm} T_{initial}(\lambda) - T_{dust}(\lambda) d\lambda}{\int_{400\,nm}^{1200\,nm} 1\, d\lambda}$$

Laboratory soiling tests

The anti-soiling behavior of the different glass samples/coatings (see Table1) was evaluated by using a specially designed soiling test setup which is capable of controlling sample temperature, relative humidity as well as the airborne dust concentration. The test chamber was built to simulate outdoor conditions [6] will be described in detail in a forthcoming publication. Two different test procedures were applied to the samples:

Test 1 (dry): Samples were cooled down to 15 °C at 60 % rH allowing them to stabilize for 10 min. Arizona A2 fine test dust suspended in dry air was blown into the chamber for 1 min. The resulting airborne dust concentration was about 1-2 g/m³ at 50 ± 5% rH. The dust was allowed to settle on the sample surfaces for 30 min without any air disturbance (gravitational settling). Afterwards, the samples exposed to wind blowing (wind speed ≤ 5 m/s).

Test 2 (wet/dew): the test parameters were analogous to the ones of test 1 except for the sample temperature of 5 °C at 60 % rH, resulting in dew forming on the glass surfaces.

To evaluate the surface properties of the different samples, water contact angle measurements were performed using the measurement tool *Dataphysics OCA35* applying the sessile drop method.

Light microscopy

In addition to the determination of the transmission loss, light microscopy (dark field imaging) was used to analyze surface coverage C with a Zeiss Axio Scope.A1 Vario (20x objective). C is defined as quotient of area covered by particles divided by total area of the acquired microscope image. For this, 10x10 pictures were collected and stitched for 8 different positions per sample. Image analysis was performed with ImageJ software using auto threshold function.

Mini modules and electrical characterization

In PV modules, there is a spectral interaction of the optical properties of all components such as the glass, the encapsulant, the solar cell, and the backsheet [7]. Hence, a systematic assessment of the impact of the glass on the module performance requires the setup of fully functional modules. Regarding the optical properties, one cell mini-modules are sufficient to give a quantitative indication of the glass properties on the module performance. Therefore, each glass type has been tested by preparing 3 single cell mini-modules made of identical materials except the front glass. All cells have been subject to an electrical I-V-characterization before the lamination to ensure a homogeneous distribution of cell parameters for the entire set of mini-modules. Additionally, the spectral properties have been determined by measuring the external and internal quantum efficiency of the solar cell. Subsequent to the preparation of the mini-modules, the I-V-curve of each module has been measured. All measurements have been performed using the LOANA measurement system by pv tools. Special focus has been set on the comparison of the short circuit current changes of the modules compared to the solar cell values.

3 RESULTS AND DISCUSSION

3.1 Anti-reflective behavior

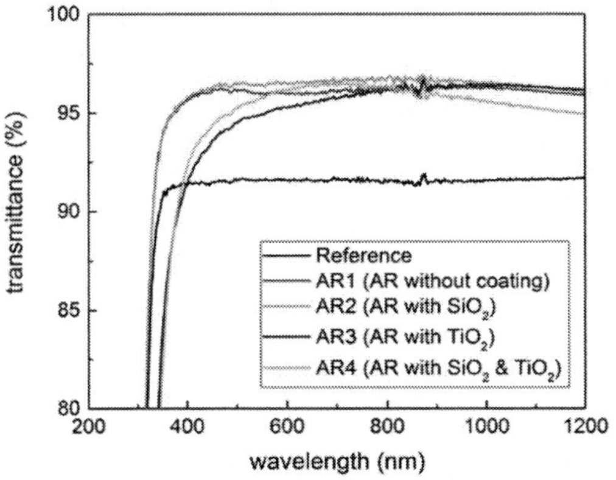

Figure 3: Hemispheric transmittance spectra of different anti-reflective glasses from *DAGlass* (AR1 has a purely nano-structured surface, whereas AR2-AR4 additionally have SiO_2 and/or TiO_2 coating), corresponding untreated references (ref). Note that transmittance measurements are shown for free standing glasses with double side surface treatment .

The measured hemispheric transmittance spectra, which are plotted in **Figure 3**, show strong effects of the different glass surface treatments summarized in **Table 1**. Note that the glasses are structured on both sides leading reducing the reflection on the two surfaces and therefore doubling the effect. With optical data from all components in the modules and the quantum efficiency data from the solar cells, the expected module currents and optical loss mechanisms can be calculated (see Fig.4) [5-8]. A short circuit current for the reference glass of $J_{SC,ref}$=38mA/cm² and for the AR1 glass of $J_{SC,AR1}$=39mA/cm² is expected. This corresponds to an increase of about +1mA/cm² or +2.6%$_{rel}$

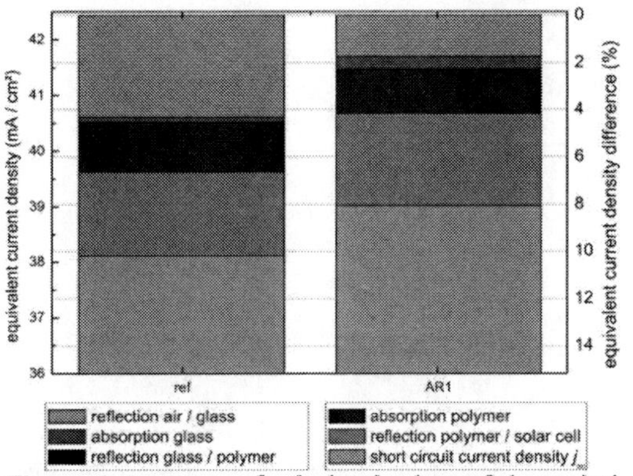

Figure 4: Summary of calculated values of the optical losses and estimated current density for 1-cell-modules with front glass sample ref and AR1 at AM1.5g sun radiation.

The calculated enhancement is in good agreement with the j_{sc} enhancement measured electrically with the LOANA solar simulator for the 1-cell modules. This confirms the expected positive anti-reflective effect of the glass treatment for module applications (see Fig.5).

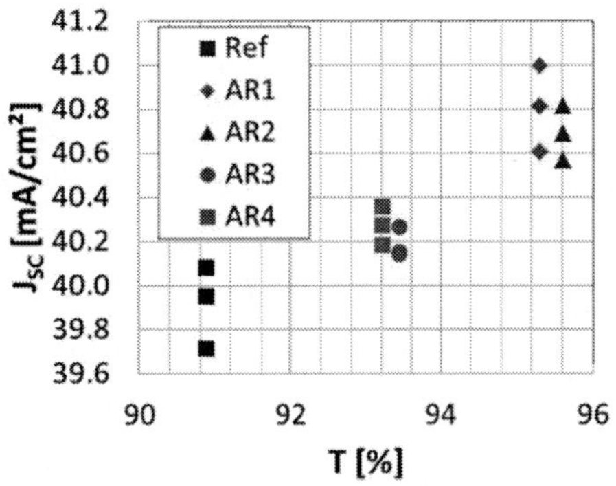

Figure 5: Short circuit current versus transmittance for the difference glasses. An increase of close to 1mA/cm² is obtained for the best glass. Please note that large perimeters in mini modules produce higher short circuit current densities than modelled or expected for large modules.

The results are also confirmed on large scale modules. Performance results show an increase of short circuit current density J_{SC} by +0,93mA/cm² or +2,5%$_{rel}$. The efficiency is increased by +0.35%$_{abs}$ or +2.1%$_{rel}$. This is comparable to other ARC data published earlier [7].

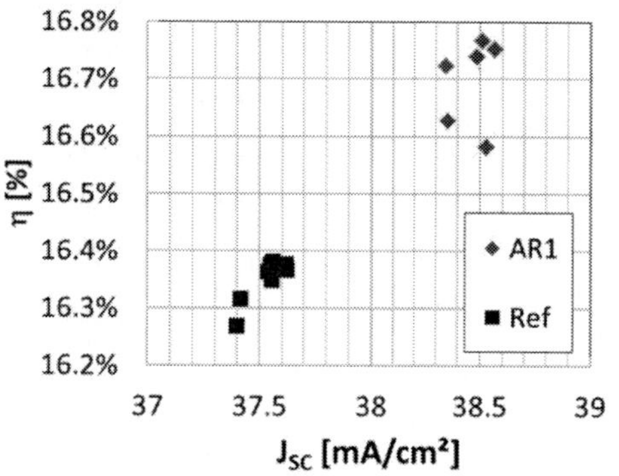

Figure 6: Module full area efficiency versus short circuit current density for 60-cell solar modules with AR1 and Reference glass..

3.2 Anti-soiling behaviour

The results of water contact angle measurements are shown in Table 2, indicating a super-hydrophilic behavior for the structured samples with and without SiO_2 coating and more hydrophobic behavior for the samples with TiO_2.

Table 2: Overview on water contact angles of samples from *DAGlass* which were used for soiling tests.

Sample	Ref	AR1	AR2	AR3	AR4
Water Contact Angle [°]	48 ± 9	< 10	< 10	86 ± 9	81 ± 3

Fig. 7 shows results of surface coverage and transmission loss for test 1 (15 °C cooling temperature, no dew). Both characterization methods show the same trend for the different samples types, with coating AR2 and AR4 outperforming the other samples: the comparison of the loss in integral transmission after soiling process of untreated reference glass and surface treated glasses shows that a reduction of transmission loss from 3.2 % to 2.4 % (0.8 % absolute) is obtained by proper surface treatment (see Fig. 7). This corresponds to a relative improvement of about 25 %.

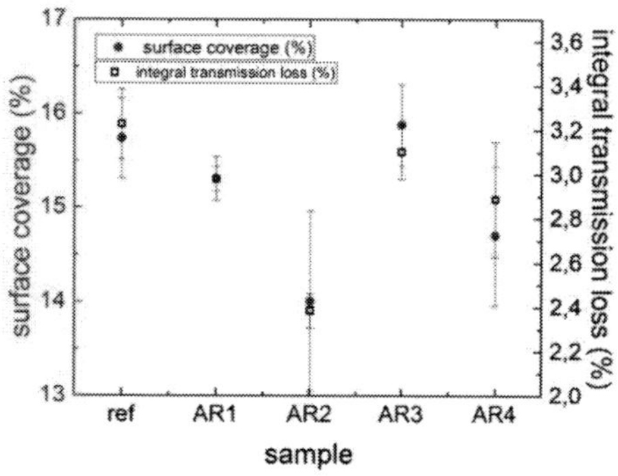

Figure 7: Surface coverage and loss in transmission after soiling process for differently treated anti-reflective (AR) from DAGlass in comparison to the untreated reference glass (top).

Figure 8 shows results of soiling tests for test 2 (5 °C cooling temperature, dew formation). In this case, only 4 samples were tested. Again, surface coverage and transmission losses show similar ranking for different sample types, with AR2 and AR4 outperforming the Ref and AR3. In comparison with Test 1, the results of Test 2 show higher levels of surface coverage and transmission loss. This can be attributed to wet surface due to dew which leads to higher capillary forces between particles and surfaces. Furthermore, particle caking effects due to rearrangement and compaction during dew events also increase of particle adhesion [6].

The dew also leads to higher standard deviation since homogeneity of deposited dust is lower according to different dew patterns (see Fig. 9). The more hydrophobic coatings AR3 and AR4 (TiO_2 coating) tend to concentrate particles in droplet shape structures, whereas the hydrophilic Ref and AR2 show homogenous distribution of dust particles.

The comparison of the anti-soiling behavior of the different coatings shows that the anti-soiling functionality cannot directly be attributed to the water contact angle, since the best working coatings AR2 and AR4 are hydrophobic and hydrophilic, i.e. the surface chemistry is not the decisive factor in these experiments (compare results of AR3 and AR4). On the other hand, the SiO_2 deposition seems to promote anti soiling behavior. We assume that this can be attributed to structural effects resulting from the deposition method, but will be investigated in detail in forthcoming studies.

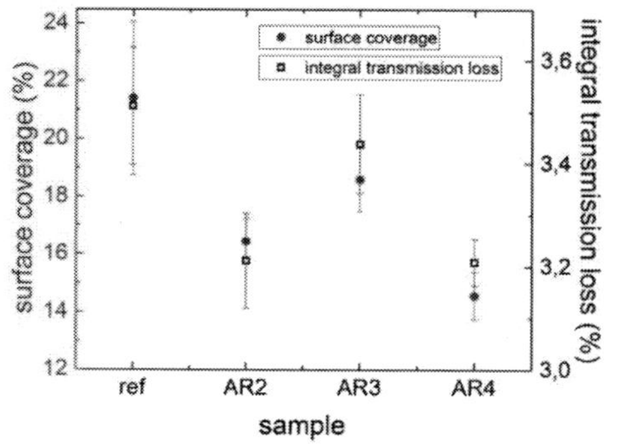

Figure 8: Surface coverage and loss in transmission after soiling process for differently treated anti-reflective (AR) from DAGlass in comparison to the untreated reference glass. The test parameters used are the same as in Figure 7 except from the sample temperature which is set to 5 +/-1 °C in order to increase dew formation.

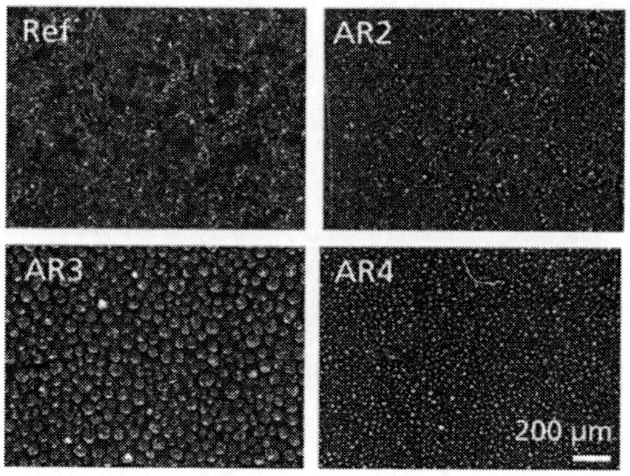

Figure 9: light microscopy images (dark field operation mode) for differently treated glasses after laboratory soiling test 2

Improved anti-soiling behavior of coating AR2 and AR4 may also result in a significant reduction of solar module operation costs. Lowering soiling rate by 25% would also reduce operating costs for cleaning by a similar factor, since cleaning frequency can be reduced. But: indoor soiling test provide only an estimation which coating may perform better under specific conditions. Comparability to outdoor exposure tests was shown on a qualitative level [6], but quantitative behavior has still to be evaluated by real outdoor experiments. Nevertheless, the laboratory soiling tests can reduce R&D costs for coating optimization since preselection of potential coatings reduces efforts in outdoor testing.

4 CONCLUSION AND OUTLOOK

We found that the anti-reflective behavior of a glass with nanostructured surface is comparable to conventional solar glass with standard anti-reflective coating. Concerning, the self-cleaning properties, it is shown that *DAGlass* has various possibilities to change the surface properties being visible in significant changes of water contact angles. The comparison of the loss in integral transmission after soiling process of untreated reference glass and surface treated glasses shows that a reduction of transmission loss from 3.2 % to 2.4 % (0.8 % absolute) for proper surface treatment with SiO_2. This corresponds to a relative improvement of about 25 %, which indicates a possible reduction for solar module operation costs in dusty environments.

Further testing is planned to improve the microstructural understanding of the anti-soiling behavior. Furthermore outdoor testing qualifying module performance under real conditions is planned.

5 ACKNOWLEDGEMENTS

We thank the Federal Ministry of Economic Affairs and Energy (BMWi) of the German Government for the financial support of this work within the "FunGlass" project (0324021A).

Charlotte Pfau wants to thank the German Ministry for Education and Research ('Bundesministerium für Bildung und Forschung') for their financial support within the Centre for Innovation Competence SiLi-nano (Project Number 03Z2HN11).

6 REFERENCES

[1] A. Sayyah, M. N. Horenstein, and M. K. Mazumder, Sol. Energy 107, 576–604 (2014), DOI: 10.1016/j.solener.2014.05.030

[2] G. Seifert. et al. 2015, "Light management in solar modules",in Wehrspohn, R.B. et al. (Eds), Photon Management in Solar Cells. Weinheim, Germany: Wiley-VCH.7.

[3] Illya Nayshevsky, QianFeng Xu, Gil Barahman, Alan M. Lyons "Anti-reflective and anti-soiling properties of KleanBoost(TM) a superhydrophobic nano-textured coating for solar glass", Washington D.C., 2017

[4] K. Ilse et al., Microstructural analysis of the cementation process during soiling on glass surfaces in arid and semi-arid climates, physica status solidi (RRL) 10 (7), 509–572 (2016).

[5] Krauter, S. & Grunow, P.,"Optical modelling and simulation of PV module encapsulation to improve structure and material properties for maximum energy yield", *Proc. IEEE 4th WCPEC*, Waikoloa, Hawaii, USA. 2006

[6] Haedrich, I. et al. 2014, "Unified methodology for determining CTM ratios: Systematic prediction ofmodule power", *Sol. Energy Mater. Sol. Cells*, pp. 14–23

[7] J. Schneider, M. Turek, M. Dyrba, I. Baumann, B. Koll, and T. Booz. „Combined effect of light harvesting strings, anti-reflective coating, thin glass, and high ultraviolet transmission encapsulant to reduce optical losses in solar modules" *Prog. Photovolt: Res. Appl.* 2014; 22:830-837.

[8] Hanifi, H. et. al. 2016, "Investigation of cell-to-module (CTM) ratios of PV modules by analysis of loss and gain mechanisms", *PV International 32*, p. 89

[9] K. Ilse, J. Rabanal, L. Schönleber, M. Khan, V. Naumann, C. Hagendorf, J. Bagdahn "Comparing indoor and outdoor soiling experiments for different glass coatings and microstructural analysis of particle caking processes" IEEE Photovoltaic Specialists Conference - IEEE PVSC 44 Washington D.C. 2016

COMBINED INTERCONNECTION AND LAMINATION OF BIFACIAL BUSBARLESS CELLS THROUGH WOVEN WIRING

Tom Borgers[1] Tom.Borgers@imec.be, Jonathan Govaerts[1] Jonathan.Govaerts@imec.be, Jan D'Haen[2] jan.dhaen@uhasselt.be, Philippe Nivelle[2,3] Philippe.Nivelle@imec.be, Eszter Voroshazi[1] Eszter.Voroshazi@imec.be, Jozef Szlufcik[1] Jozef.Szlufcik@imec.be, Jef Poortmans[1,3,4] Jef.Poortmans@imec.be, Michaël Daenen[2,3] Michael.Daenen@UHasselt.be
[1]imec vzw, Kapeldreef 75, 3000 Leuven, Belgium [2]imomec [3]UHasselt [4]KULeuven

ABSTRACT: In this work we present a new technology for simultaneously interconnecting and encapsulating bifacial busbarless solar cells, using a multi-wire approach combined with weaving technology. A background discussion on existing multi-wire interconnection technologies allows to compare this new approach, which potentially improves performance and reliability and reduces process complexity and costs. After a more in-depth review of the weaving aspects, an extensive experimental campaign has been performed to optimize the combined interconnection/lamination process and obtain homogeneous solder joints. Preliminary thermal cycling tests have been executed, showing the stability and reliability of the solder joint contacts on 1-cell glass-glass laminates. Finally, the realization of a 2x2-cell-module proof-of-concept demonstrates the technical viability and performance of the technology.
Keywords: Interconnection, bifacial, multiwire

1 INTRODUCTION

Nowadays, the most common PV module fabrication technology involves stringing of 2-side contacted photovoltaic cells. The generated electrical current is collected through distributed metal fingers across the cell into typically two or more busbars. By soldering tinned copper ribbons to these busbars, cells are electrically connected in series to form cell strings. The size of these ribbons is a compromise between shadowing on the illuminated surface of the cells and resistive losses. The individual cell strings are connected with ribbons and laminated into a module.
For both improving electrical performance and reducing optical losses, a trend toward an increasing amount of (more narrow) busbars is becoming a reality [1]. This trend evolves ultimately to multi-wire interconnection technologies, with the additional advantage that busbars are no longer needed, reducing the cost of the silver metallization on cell level. Two of these multi-wire interconnection technologies are in a very advanced stage of development. The first approach effectively mimics the standard technology by soldering wires on finger solder pads, replacing the busbar. As in standard production, this step is then followed by a separate encapsulation process [2]. Such an approach requires controlling wire expansion during the soldering process and accurate alignment of the wires to the dedicated finger solder pads. A second approach applies a contact foil directly onto the metallized cell followed by a lamination process [3]. The contact foil integrates low-temperature-solder-coated copper wires on an optically transparent supporting film with an adhesive layer. During lamination the wires of the contact foil are soldered directly to the metal fingers of the cell. The use of low-temperature solder reduces stress between the wire-to-finger contact points on the cell. Silver metallization on cell level is further reduced as no dedicated finger solder pads are needed. The contact foil has to be engineered to allow the wires to contact neighboring bifacial cells on opposite sides of the supporting film. Similarly as for the first approach, thermomechanical stress considerations may require some attention for compensating differences in thermal expansion between the cells and the wires. Moreover, a layer of encapsulant material is still needed for the subsequent module lamination step.
Building further on these evolutions, the replacement of the contact foil with a woven interconnection sheet would allow for a combination of both interconnection and encapsulation processes into a single lamination step, without introducing additional materials (see Table I). Such a woven sheet can be made by interweaving encapsulant ribbons and metal wires and simultaneously integrating thermal stress relief features [4].

Table I: *Comparison of Multi-Wire interconnection technologies*

Approach	Thin cells	Stress relief	No added materials	Assembly
Wire stringing [2]	✗	✗	✗	
Contact foil stringing [3]	✓	✗	✓	
Woven inter-connect [4]	✓	✓	✗	

In this work we present our progress on this technology development towards a simplified and reliable way for contacting bifacial solar cells.

2 NEW INTERCONNECTION AND LAMINATION PROCESS

2.1 Woven interconnection and encapsulation weave

A weave consists of yarns, wires or ribbons that are joined together at right angles to each other, to form a two-dimensional structure. In this case, metal wires are interwoven perpendicularly with encapsulant ribbons. Figure 1 demonstrates such a hybrid weave in which the metal wires extend on both sides of the weave and thus can be also contacted electrically.

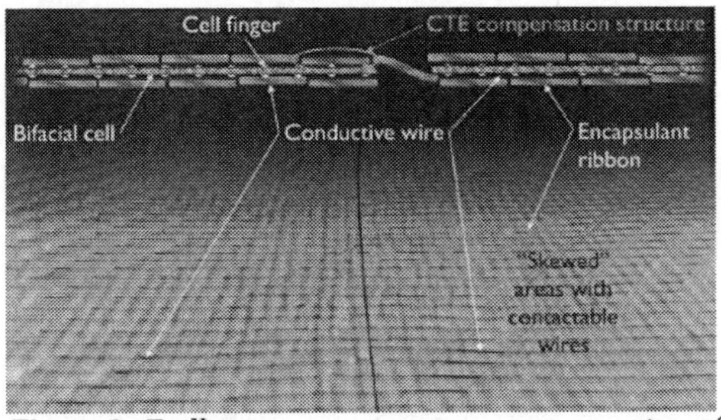

Figure 1: *Twill interconnection weave, cross-section of cell lay-up*

This intertwining of wires and encapsulant provides additional interesting opportunities:

- the undulating shape of the woven wires (alternatingly contacting the cell metallization and floating above the cell) reduces thermomechanical stress generated after soldering or during thermal cycling
- the thickness and structural and dimensional stability of the weave can be tuned, while keeping the freedom in wire diameter, pitch and contacting ratio, i.e. how much the wires protrude on the cell side vs. the glass side.

For our application we choose a twill weave for a number of reasons. Most importantly, this weave pattern is simple, low-cost and at the same time allows for

- sufficient amount of encapsulant material, dimensional stability, thickness and covering due to fewer intertwining
- a good and reliable electrical connection of the wires to the cell fingers due to the exposed extended wire sections between two consecutive intersections (also creating two different fabric sides)
- finger contact enhancement due to a diagonal progression of intertwining of the wires along the weave (twill lines)

2.2 Simplified module lay-up

During the module lay-up phase, the weave is brought into contact with the metal finger contacts on the cell surface. Orientation of the weave is such that the conductive wires in the weave are at an angle with those fingers, allowing a soldered interconnection at the intersections during lamination. Since any orientation of the weave relative to the fingers of the cell can be used, a diagonal finger orientation (i.e. an angle of 45°) on (at least one side of) the cell would avoid the use of end-bussing bars between two cell strings. This would considerably reduce the amount of Cu consumption in module manufacturing and avoid time-consuming and production-yield restricting process steps.

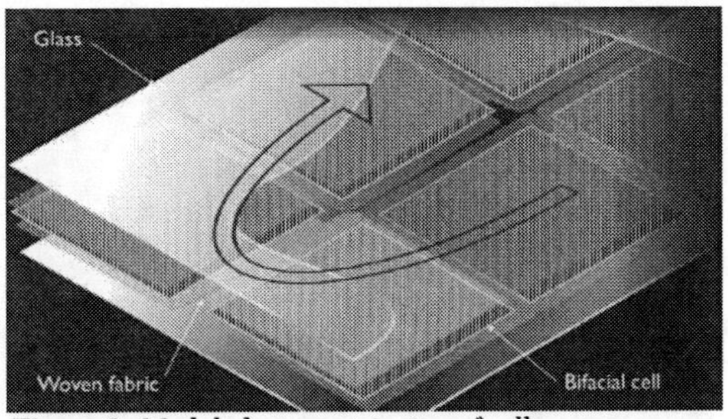

Figure 2: *Module lay-up: stringing of cells*

Lay-up of the solar module could start immediately on a module glass plate by placing a woven contact sheet on the glass, with the contact side facing up (i.e. the side where wires mostly protrude from the weave). The first (bifacial) cell is laid on this side. A second sheet is laid on the cell, with the first contact-side half facing down. The next cell is laid on the second half of this contact sheet (i.e. the half where the contact side is facing up). In the same way, the first cell chain is built up. By placing the last contact sheet of the cell string perpendicularly, a connection to the next cell string is created; the electrical connection of the foil with the cell is not affected due to the adapted co-design of the cell (having a diagonal finger pattern). After placing the first cell of the next cell string, the entire module may be assembled with cells and sheets in the same way, without the need for end-bussing of the cell strings. As a final step a sealing glass plate (or backsheet) is positioned. No additional encapsulation material is needed. This layup is schematically illustrated in Figure 2.

2.3 Combined cell interconnection and lamination

After layup, the module stack is laminated, wherein the conductive wires of the weave are soldered to the metal fingers of the cells. Initially the polymer material is not yet fully liquefied to avoid penetration of encapsulant material between the cell finger and the wires, which might lead to bad contacts. Further increase in temperature causes melting of the thermoplastic encapsulant ribbons, transforming them into a uniform encapsulation layer.

3 EXPERIMENTAL SETUP

3.1 Process optimization

Two solar grade polyolefin types were used for the encapsulant ribbons: Type A, and Borealis Quentys^{TM} BPO882F (further indicated as Type B), both transmitting wavelengths between 240 nm and 1200 nm. The electrically conductive wires with a copper diameter of 200 µm are coated with a SnBiAg-based solder alloy with a low-temperature melting point of 139 °C. The woven contact sheet size was adapted to the glass size of the laminate. A total of 20 wires were evenly integrated in the weave to create a uniform distributed contact area in accordance to the contact surface of the solar cell.

For initial process optimization, single-cell laminates were made using glass on the front- and backside. Because of the wide availability, all used cells were full-sized (156 x 156) mm^2 semi-square monocrystalline Si-cells with Al-BSF and Ag-printed front grid with two busbars (Table II). Preliminary tests showed improved solder joint formation when solder flux was applied on the wires before lamination. A very thin PTFE release film was added between the woven contact sheet and the front side glass plate, and the cell backside and backside glass plate. This lay-up facilitates the removal of the glass plates after lamination for cross-section sample preparation and SEM-EDX solder joint analysis.

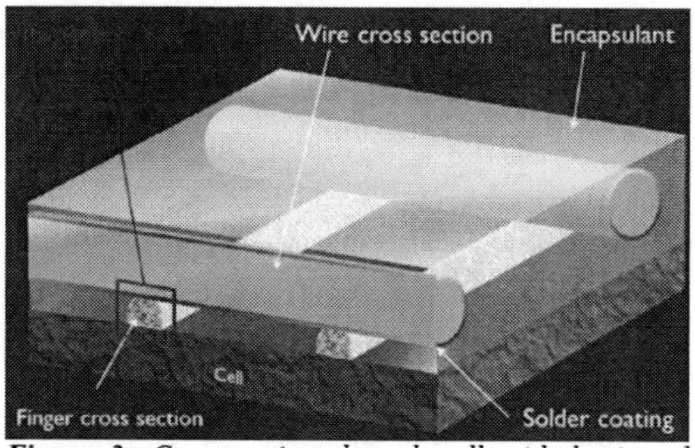

Figure 3: *Cross-section through cell with laminated contact sheet*

Initial lamination recipes were provided by the encapsulant manufacturers. A baseline lamination and interconnection recipe was derived from a test matrix with varying temperatures and laminator membrane pressure. Interconnection quality was evaluated with electroluminescence (EL) imaging and IV-measurements.

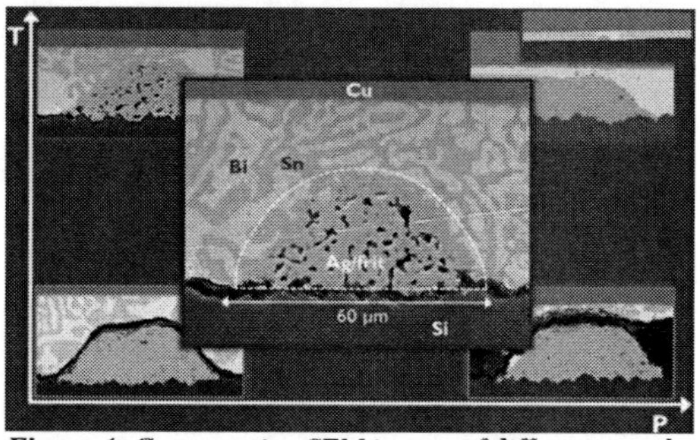

Figure 4: *Cross-section SEM images of different samples in function to the lamination pressure and temperature*

Solder joint quality was evaluated with cross-sections at the wire-finger interface, as illustrated in Figure 3, focusing on intermetallic formation. Based on this quality assessment, optimized lamination conditions were extrapolated from the test matrix. As can be seen in the cross-section SEM images in Figure 4, intermetallic joints were formed at light pressure and higher temperatures. Segregation of Bi and Sn in the proximity of the cell fingers was seen at high pressure and high temperature conditions. The final conditions were selected from a second iteration with a refined test matrix: at intermediate temperatures and adapted pressure, very smooth and homogeneous solder joints were formed without encapsulant flowing in between the interface of the cell fingers and the wire (except on places where the wire was separated from the cell surface by an encapsulant ribbon).

3.2 Initial solder contact reliability tests

Preliminary thermal cycling tests were initiated on glass-glass single-cell laminates to evaluate the stability and reliability of the solder joint contacts. Backside interconnection of the cells was performed by soldering SnPbAg-coated copper ribbons perpendicular to the orientation of the backside busbars to avoid residual stress and bowing of the cells after soldering. The soldered cells were laminated with a wired contact sheet at the front side and an encapsulant sheet at the back side, as illustrated in Figure 5. Thick bussing ribbons were placed in the laminate to contact the wires during the lamination process.

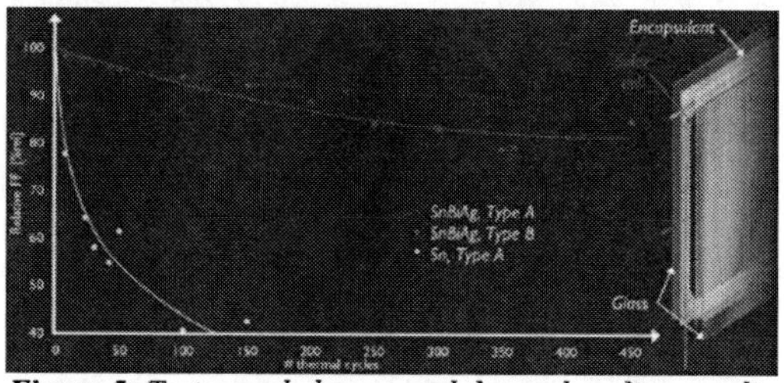

Figure 5: *Test sample lay-up and thermal cycling results*

One reference sample was prepared without solder contacts by using high-melting point Sn solder-coated wires in the woven contact sheet, to test to what extent this sample would be outperformed by the samples with soldered contacts. The samples were measured after lamination; an average fill factor (FF) of 77.1 % was achieved, compared to 67.1 % for the sample without solder contacts. The samples were exposed to up to 450 thermal cycles between -40 °C and +85 °C. An evolution of the FF is plotted in Figure 5. Compared to the steep degradation of the reference sample (reaching a FF of 39.3% after 150 cycles), only a slight continuous degradation is observed. Potentially part of this (initial) degradation is caused by the backside ribbon pressure contact with the overall Al backside of the cell, which is not a soldered contact and therefore it is gradually degenerating during thermal cycling.

However, to remove any doubt about this potential degradation mechanism, the experiment will be repeated with cells having appropriate backside contacting possibilities.

3.3 Proof of concept 4-cell module

To further prove the interconnection principle, a 2x2-cell glass-glass module was fabricated with 4 pieces of a weave as demonstrated in Figure 6.

Figure 6: *4-cell proof of concept, lay-up scheme and EL-image*

Wires were interrupted between 2 cells in one weave to create a series interconnection of the cells. No additional encapsulant layers were added. At the time of this experiment, no bifacial cells were available; Therefore monofacial cells were used. Contact sheets were placed perpendicular to the backside busbars, to enable – limited – backside solder joint formation.

Table II: *4-point IV-measurements on 4-cell module*

Cell	V_{oc} [mV]	I_{sc} [A]	P [mW]	FF (%)
1	635,6	8,794	4294	76,8
2	634,8	8,788	4444	79,7
3	631,9	8,733	4398	79,7
4	629,7	8,718	4139	75,4
12*	1261,5	8,875	8565	76,5
23*	1258,8	8,742	8641	78,5
34*	1253,2	8,794	8367	75,9
4-cell	2486,0	8,570	16627	78,0

The laminator recipe for this 4-cell laminate is the one that was optimized from the test matrix. During the lamination process, an electrical soldered connection is formed between the wires and the different cells; as such, an electrical series connection is established and a uniform encapsulated module was formed (Figure 6). Due to the specific configuration of the side collection busbars, all individual cells and cell strings can be assessed with 4-point IV-measurements. Table II provides an overview of the cell string performance in the module. Note that values related to 2- and 4-cell strings are indicative and recalculated due to a limited exposure area of the solar cell simulator we used for these measurements. Compared to the FF of the bare cell, the measured values are very high, indicating the potential of the technology. This is partly due to the 4-point measurement layout (for cells 2 and 3), and partly due to the fact that the cells' front side finger (and busbar) metallization is oversized for this technology. The fingers are dimensioned to carry the current from all over the cell to the 2 busbars, whereas this technology limits the current path through the fingers due to the closely distributed wires. Through cell and interconnect co-design the cell metallization should be reduced with minimal performance losses hence leading to considerable gains in metallization cost.

4 CONCLUSIONS

A flexible and low mechanical stress method is presented for electrically contacting and interconnecting busbar-free bifacial solar cells, simultaneously providing encapsulation of the cells. The method was supported by experiments, showing a hybrid woven weave that combines conductive wires and encapsulant ribbons. The combined soldering and lamination process has been optimized to establish reproducible and reliable metallurgical solder connections between solar cells and the wires of the woven sheet, with supporting evidence by SEM imaging. First thermal cycling tests confirm the reliable solder joint (further experiments are in progress). Finally an efficient 4-cell laminate demonstrates the potential technological advantages and performance.

5 ACKNOWLEDGEMENTS

This project receives the support of the European Union, the European Regional Development Fund ERDF, Flanders Innovation & Entrepreneurship and Province of Limburg. The authors like to acknowledge the funding agencies and the industrial partners of the imec Industrial Affiliation Program in SiPV, and the EFRO-SALK SolSThore project, for their support and for providing the materials used in this work.

6 REFERENCES

[1] International Technology Roadmap for Photovoltaic (ITRPV) 2015. Giorgio Cellere, et al. Seventh Edition, October 2016. pp. 14-15, 27-28.

[2] S. Braun, G. Hahn, et al., Multi-Busbar Solar Cells and Modules: High Efficiencies and Low Silver Consumption, s.l. : Energy Procedia, vol: 38, pp. 334–339, 2013, pp. vol: 38, pp. 334–339.

[3] A. Faes, M. Despeisse, et al., SmartWire Solar Cell Interconnection Technology, Amsterdam : s.n., 2014. 29th EU-PVSEC.

[4] T. Borgers, J. Szlufcik, Electrically contacting and interconnecting photovoltaic cells. WO2016156276 (A1), 2016.

CO-EXTRUSION OF A NOVEL MULTILAYER PHOTOVOLTAIC BACKSHEET BASED ON POLYAMIDE-IONOMER ALLOY SKIN LAYERS

Christopher Thellen, Andreas Rothacker, David Santoleri
Tomark-Worthen LLC
11 McComb Avenue Glen Mills, PA 19342 USA

ABSTRACT: Through the use of co-extrusion technology, Tomark-Worthen has produced a durable 5-layer backsheet material based on a proprietary polyamide alloy with no interlayer adhesives and no PET or fluoropolymers. The use of co-extrusion processes has allowed for the production of this material in a single step, eliminating multiple steps common in typical laminated backsheet manufacturing processes. The polyamide alloy material is being evaluated as a replacement for fluorine-based polymers as the skin-layers of the backsheet and is highly stable and resistant to UV light and moisture while being cost-effective in the PV market. This backsheet has excellent barrier to oxygen and moisture barrier which was achieved through the use of low permeability polymers in the core layers of the 5-layer construction. Accelerated UV light testing of the alloy skin layers has shown that after 4000 hours of testing, the alloy material exhibited a LAB Δe value of only 0.64, indicating no noticeable difference in color. Long term damp heat testing indicates no hydrolysis of the material, which is common in PET backsheet and is a common mode of failure in those materials. Reflectivity testing of this alloy resulted in a reflectivity value of 91.2%. Relative thermal index (RTI) of this material has been evaluated and achieved a value of 125°C. Environmentally speaking, this material is made without volatile organic compounds (VOC) and trim can be recycled back into the production process given that all polymers in the construction are thermoplastic in nature.

Keywords: PV Materials, Backsheet, Stability, Environmental Effect, Characterization

1 INTRODUCTION

Photovoltaic (PV) backsheet is a critical component of a solar module that provides insulation and environmental protection to the system. In order for solar modules to perform for 25+ years in various environments, the backsheet must provide excellent weatherability, electrical insulation, mechanical strength, and adhesion to the PV encapsulant. To this day, lamination has been the major process used to construct PV backsheet consisting of several different materials. Recently, manufacturers have looked toward producing multi-layer PV backsheet through a one-step co-extrusion of thermoplastic materials as an alternative manufacturing process to lamination. Co-extrusion is the simultaneous extrusion of two or more polymers through a single die where the polymers are joined together such that they form distinct, well-bonded layers forming a single extrusion product [1]. Co-extrusion is an industrial process, which is widely used to form multi-layered sheets or films that are suitable for various products ranging from food packaging materials to reflective polarizers [2]. A depiction of the co-extrusion process is presented in Figure 1 [3].

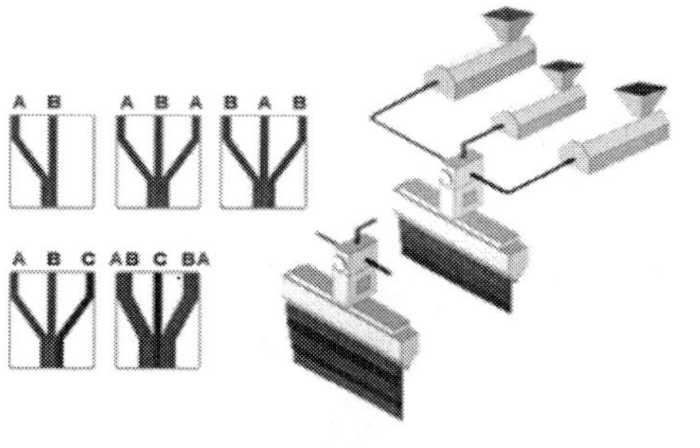

Figure 1: Schematic of the sheet co-extrusion process

Continuous co-extrusion of multilayer PV backsheet has both advantages and disadvantages to conventional lamination techniques. Some advantages of the co-extrusion method include efficient manufacturing, elimination of interlayer adhesives, absence of hazardous emissions or volatile organic compounds, and removal of the oven drying step in lamination processes. As with any process, there are challenges which must carefully be weighed against the potential advantages of the method. These challenges include rheological limitations of material combinations, properly designed extrusion equipment for materials used, material compatibility and layer to layer uniformity.

In general, the most common backsheet failures observed on PV panels are cracking due to UV-light exposure and tension forces as well as pitting/porosity in the backsheet due to degradation. Example images of these failures are shown in Figure 2 below [4].

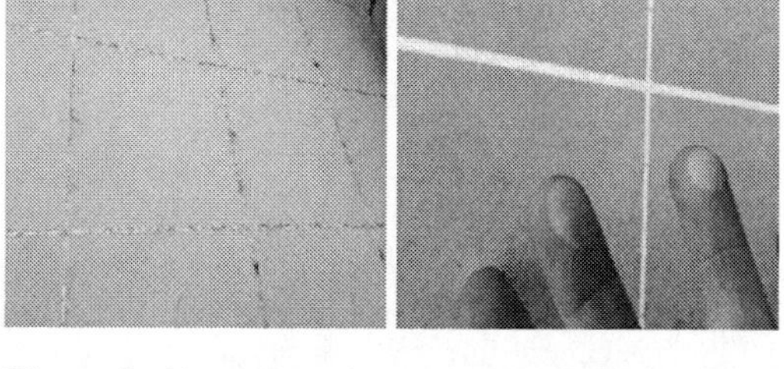

Figure 2: Examples of common backsheet failures including cracking (left) and porosity (right).

According to data provided by the Rinovasol Group, a group of companies specializing in refurbishing, repairing and recycling solar modules, approximately 65% of failed modules that they receive have failed backsheet. Of these failures, 75% were due to cracking from UV-light exposure. Cracks formed from tension/stresses in the backsheet accounted for 10% of failures and porosity in the backsheet was observed in 15% of failed backsheet. Rinovasol receives approximately 200,000 panels per month globally with various defects. Along with their partner PV Cycle, they refurbish failed PV panels due to hot spots, delamination and broken backsheet.

This paper will address the technological advantages and challenges associated with producing PV backsheet through the co-extrusion method and will highlight a novel multi-layer backsheet construction that utilizes a proprietary polyamide-ionomer alloy as the environmentally-resistant skin-layers of the backsheet.

2 CO-EXTRUSION VS. LAMINATION

2.1 Advantages of Co-Extrusion over Lamination

2.1.1 Elimination of Interlayer Adhesives

Typical lamination processes to produce multilayer backsheet use interlayer adhesives to bond the multiple layers together that make up the solar backsheet. These adhesives act as the sole mechanism for preventing the backsheet components to separate during the life of the product. Adhesive failure is a common mode of failure in which delamination of the layers can lead to reduced optical transparency, heat dissipation, and protection from outside elements [5]. In co-extruded structures, materials that are compatible with each other are extruded together in the melt form and interlayer adhesion is dependent on a combination of the chemical reaction between polymers in adjacent layers and inter-diffusion of polymer chains across the interfaces. Given that these bonds are not solely dependent on an adhesive to maintain integrity, co-extruded structures will have longer-lasting bonds than laminated structures.

2.1.2 Ability to Recycle

Unlike laminated backsheet, co-extruded backsheet is usually made with thermoplastic materials, meaning that they can be collected and re-processed back into the manufacturing stream. Current technology does not allow for the separation of backsheet from existing solar modules, but trim and scrap materials from the backsheet manufacturing step can be collected and re-processed back into one or several of the backsheet layers given that the collected trim is compatible with the individual layer in the backsheet into which it is being recycled.

2.1.3 No Hazardous Materials

Adhesives used in lamination of solar backsheet layers are often solvent-based and/or toxic in nature. These typically require the use of proper engineering controls, personal protective equipment and hazardous wastes that are expensive to dispose of. Co-extrusion processes do not require solvent-based adhesives and therefore eliminate the need for many of these safety and environmental controls.

2.2 Challenges Associated with Co-Extrusion

2.2.1 Material Rheological Differences

Analyzing the flow of coextruded polymer melts through dies can be difficult due to the complex three-dimensional flow patterns that exist. Rheological effects such as viscous encapsulation or elastic rearrangement cause a flow perpendicular to the main flow direction that can distort the coextruded structure [6]. These distortions are also often visible as visual defects known as interfacial instability. When co-extruding different polymer materials into a multilayer structure, it is critical to maintain even flow rates and stresses exiting the extrusion die to avoid these defects. Proper structure design, material selection and process engineering will help to negate defects caused by this instability.

2.2.2 Thermal Limitations

Although co-extruded structures are produced using separate extruders for each material in the multilayer structure, they must still combine in a feed-block or multi-manifold die before exiting as a single structure. Therefore, polymers that are highly sensitive to temperature or materials with vastly different melt flow properties at a given temperature will be more difficult to co-extrude. Laminations are typically done at temperatures lower than the melting point of the polymers and therefore are not subject to this limitation.

3 CO-EXTRUDED MULTILAYER BACKSHEET

3.1 Materials and Structure

A high-performance co-extruded (HP-CoEx) backsheet was developed by Tomark-Worthen LLC with a trade name of PhotoMark® Reflections™ as a fluoropolymer and polyethylene terephthalate (PET) free backsheet. PET-based backsheet is known to suffer from hydrolysis-based degradation which leads to yellowing and severe losses in mechanical strength and properties. Reliability testing of PV components has shown that PET degradation is present in most PET-based backsheet, regardless of the protective layers [7]. The HP-CoEx backsheet is also produced without the use of fluoropolymer materials that are common in laminated backsheet. The absence of fluoropolymers in this structure reduces the raw material cost and increases the recyclability of this material in the manufacturing process. This novel co-extruded backsheet is comprised of outer skin layers made from a proprietary polyamide-ionomer alloy, a functional polyolefin core layer and polyamide-based tie layers as illustrated in Figure 3.

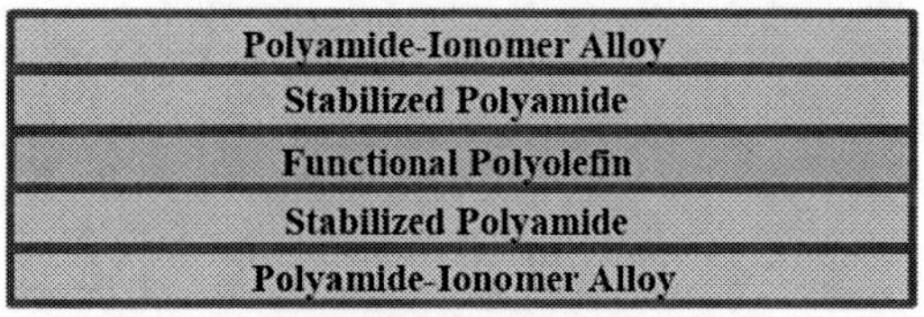

Figure 3: Cross-sectional schematic of co-extruded backsheet structure

3.2 Physical Properties

3.2.1 Interlayer Adhesion Strength

Interlayer adhesion strength of the co-extruded backsheet was measured using a 90-degree T-Peel test using a Comten tensile testing machine. Test samples were created by first laminating the backsheet on either side of a 15580P ethylene vinyl acetate (EVA) encapsulant sheet at a temperature of 148°C. Samples were conditioned for 24 hours at 50% RH prior to conducting the T-Peel test. Results of the PhotoMark® Reflections ™ co-extruded backsheet, along with a TPT-type (Polyvinyl

Fluoride/Polyester/Polyvinyl Fluoride) control laminated backsheet, is presented in Table 1.

Table 1: Adhesion Strength

Backsheet Type	Interlayer Adhesion Strength (N/cm)	Adhesion Strength to EVA (N/cm)
PhotoMark® Reflections™	No Peel	72.4
TPT Control	87.6	87.6

The data presented in Table 1 indicates that the adhesion strength of the backsheet to the EVA encapsulant was similar in both samples, with the control having a 17% higher adhesion. Important to note is that the internal layers of the PhotoMark® Reflections™ backsheet did not delaminate at all during the test indicating a strong inter-layer bond, which is expected with a co-extruded product vs. a laminated product.

3.2.2 Resistance to UV Exposure

The majority of photovoltaic module manufacturer's offer a 25-year warranty for their products and some of them are even in the stage of extending their warranty for longer terms. The reliability and durability of the solar modules are highly dependent on the UV stability of the encapsulation system and the protecting backsheet system [8]. In order to evaluate the resistance of the backsheet to UV degradation, 25mm x 50mm samples were subjected to UV exposure using an Atlas Suntest CPS+ xenon test instrument at conditions of 500W/50°C. A BYK Gardner color spectrophotometer was used to measure the color difference (Δe) in the backsheet samples at exposure intervals of 500 hours from 0 to 4000 hours. The results of this testing are presented in Figure 4 below. In this figure Tedlar-Polyester-Polyolefin (TPE) backsheet and Kynar-Polyester-Polyolefin (KPE) backsheets are used as controls. These controls are high-performance, fluoropolymer-based backsheet used in the industry today. Both the fluoropolymer side (Face) and the polyolefin side (Back) were tested. The PhotoMark® Reflections™ backsheet is designated as HP-CoEx. Only one curve is shown for this backsheet as both the face and back are identical materials. It is commonly recognized that Δe values greater than 1 indicate a color difference that is recognizable by the human eye.

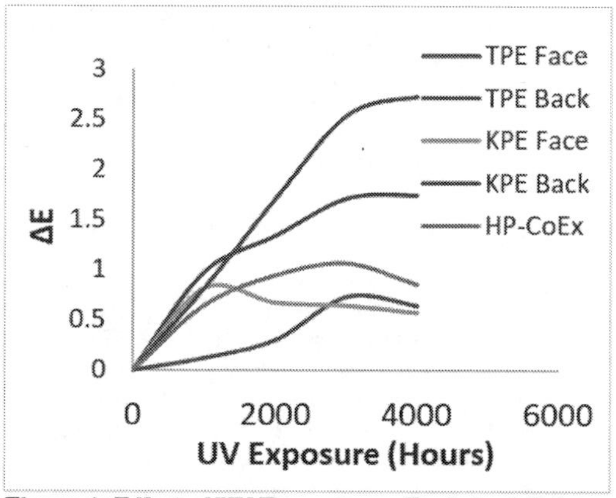

Figure 4: Effect of UV Exposure on Backsheet Color

The data from Figure 4 indicates that, as expected, the polyolefin side of the backsheet shows greater yellowing and Δe values exceeding 1.0 in less than 2000 hours of UV exposure. The fluoropolymer side of these samples show much greater UV resistance with only minor color changes after 4000 hours of exposure. The non-fluoropolymer containing HP-CoEx sample demonstrated similar performance to the Tedlar and Kynar samples with Δe values less than 1.0 after 4000 hours of exposure.

3.2.3 Pressure Cooker Test (PCT)

A pressure cooker test is commonly used within the photovoltaic community to measure hydrolytic stability of backsheet materials with time as a variable. A PCT was carried out on the PhotoMark® Reflections™ backsheet (HP-CoEx) along with stabilized photovoltaic grade PET as controls. Testing was conducted for a total of 135 hours with samples taken and tested at intervals of 28, 56, 83, 110, and 135 hours. Tensile testing was carried out post-PCT on the samples according to ASTM D882, "Standard Test Method for Tensile Properties of Thin Plastic Sheeting". Percent elongation results with respect to time in the pressure cooker test are presented in Table 2 below for the PhotoMark® Reflections™ backsheet along with two stabilized PET controls from Polyplex. Polyplex results are taken from the Polyplex technical brochure using test results from test conditions of 121°C, 100%RH, 2.1bar.

Table 2: PCT Results (% Elongation)

Hours	HP-CoEx	PET HZ100	PET HZ600
0	235	130	155
28	86	120	150
56	52	30	105
83	43	No test	20
110	39	No test	No Test
135	22	No test	No Test

The data from Table 2 indicates that the % elongation decreases as a function of time in all tested samples. PET data was not available for PET samples after approximately 83 hours of testing. Additionally, mechanical testing was conducted on the PhotoMark® Reflections ™ samples after PCT testing to evaluate the effect of PCT testing on tensile strength and tensile modulus. These results are presented below in Figure 5.

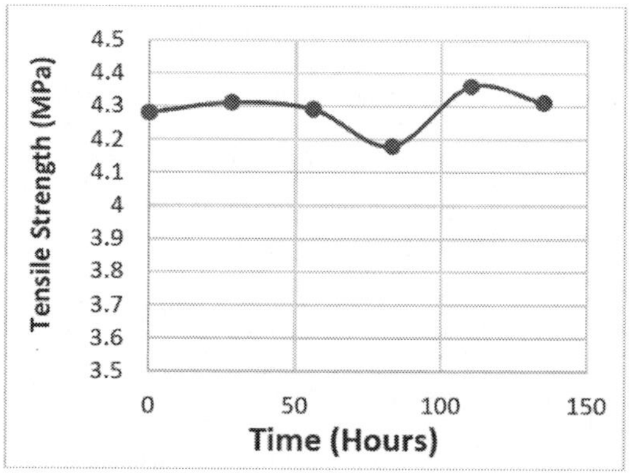

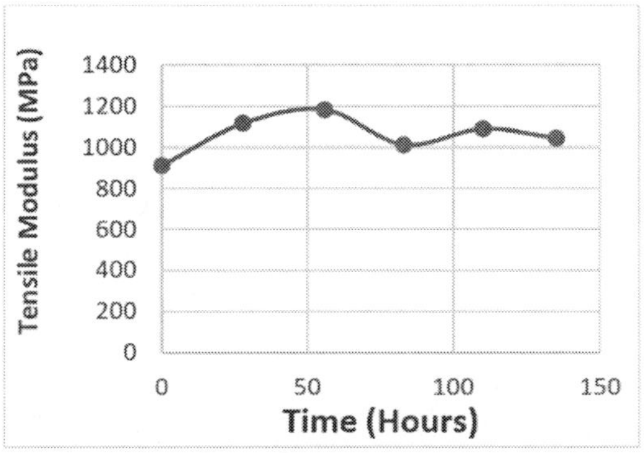

Figure 5: Tensile strength (top) and tensile modulus (bottom) of PhotoMark® Reflections™ backsheet after PCT

The data presented in Figure 5 indicates that the tensile strength and the tensile modulus of the PhotoMark® Reflections™ backsheet stays consistent throughout the PCT and no major loss or gain of strength or stiffness is observed through 135 hours of testing.

3.2.4 UV/Mechanical Properties and Stability

The determination of the degradation of mechanical properties as a result of UV exposure is critical in predicting durability of backsheet after many years of environmental exposure. Accelerated UV-aging testing was accomplished using an Atlas Ci35 Xenon-Arc Weatherometer. Test samples were cut into dog-bone shape according to ASTM D882 and exposed to an irradiance of 340nm at 0.55W/m²/nm at chamber conditions of 70°C and 50% relative humidity. Tensile properties of the test samples were determined using a Comten tensile testing machine according to ASTM D882. Tensile tests were carried out on samples prior to testing and at intervals of 1000 and 2000 hours of UV exposure. Test results for the PhotoMark® Reflections™ backsheet along with a TPT-type backsheet control are presented in Tables 3 and 4.

Table 3: Tensile Strain (%) After UV Exposure

Sample	0 Hours UV	1000 Hours UV	2000 Hours UV
TPT Control	58	63	71
PhotoMark® Reflections	308	351	345

Table 4: Tensile Strength (MPa) After UV Exposure

Sample	0 Hours UV	1000 Hours UV	2000 Hours UV
TPT Control	131	121	120
PhotoMark® Reflections	51	44	42

The mechanical properties presented in Tables 3 and 4 indicate that both backsheets demonstrated increases in tensile strain as a result of the test. Given that the test is conducted at a relative humidity of 50%, it is believed that the hydroscopic components of the backsheet absorbed moisture which caused a plasticization of the material. The control and the PhotoMark® Reflections™ backsheet demonstrated only small losses in tensile strength as a result of the UV exposure.

4. SUMMARY

Pressure on the PV industry to manufacture solar modules with 25+ years of efficient service life at reduced costs has driven research and development of new PV materials and production methods. Co-Extrusion of PV backsheet, as an alternative to lamination processes, has achieved recognition as an efficient method for producing such backsheet. Advantages of co-extrusion manufacturing over lamination include the ability to recycle, efficient manufacturing and the elimination of interlayer adhesives and toxic solvent-based processes. Challenges with co-extrusion are mainly associated with thermal limitations and rheological properties of polymeric components, which limit material selection.

Tomark-Worthen LLC has developed a high-performance co-extruded backsheet with the trade name of PhotoMark® Reflections™ that contains no PET or fluoropolymers. This material has demonstrated excellent interlayer adhesion strength, as well as adhesion to EVA encapsulant material. Data suggests that color shifts of this new backsheet as a direct results of UV exposure is comparable to fluoropolymer-type backsheet (TPE, KPE). UV exposure testing (50% RH) has shown that PhotoMark® Reflections™ material maintains ductility after 2000 hours of exposure, while losing 18% of its tensile strength in the same time period. Further evaluations of this backsheet are currently underway which include sequential environmental testing, backsheet stress testing and module evaluations.

5. ACKNOWLEDGEMENTS

The authors would like to thank Rinovasol GmbH and PV Cycle for the data contributions to this paper. As a leader in the renovation of defective panels, RINOVASOL's worldwide reach is fast expanding with additional operations set up recently in Australia, United States, Canada, and India. Rinovasol is producing new PV panels under the name Suntronic and is evaluating Tomark-Worthen's advanced backsheet for their modules. Additionally, the authors would like to thank Specialized Technology Resources (STR) for testing and evaluating the backsheet samples through UV exposure and peel strength testing. STR's expertise in the area of encapsulant technology has proven invaluable to the success of this work.

4.3 References

[1] Rauwendaal, C. Polymer Extrusion, Hanser Publishers, New York, 1994.

[2] Co-Extrusion of Multilayer Structures: Interfacial Phenomenon. Encyclopedia of Polymer Science and Technology, John Wiley and Sons Publishers, 2013.

[3] Won Jee Plastics Machinery Website http://www.ueng-jyi.com.tw/en/pl-2.html Accessed 11-29-2016.

[4] Personal communication with Rinovasol on September 17, 2017.

[5] Julien, S.; Miller, D.; Gu, X.; Wan, K. "Characterization of the Adhesion Integrity in Photovoltaic Panels using Various Thermomechanical Blister Tests" www.NIST.gov; Accessed 08/21/2017.

[6] Kim, H.; Koppi, K.; Wrisley, R. "The Effect of Viscous Encapsulation on Layer Uniformity and Rheology in Multilayer Co-Extrusion", SPE ANTEC 2016, 734-740.

[7] Knausz, M.; Oreski, G.; Eder, G.; Voronko, Y.; Duscher, B., Koch, T.; Pinter, G.; Berger, K. "Degradation of Photovoltaic Backsheets: Comparison of the Aging Induced Changes on Module and Component Level", Journal of Applied Polymer Science, 132, 24, 2015.

[8] Yang, S.; Whitfield, K.; Rhee, E. "UV Degradation Study of Encapsulate Backsheet Systems for PV Modules Using Metal Halide Lamp", NREL PVMRW, Feb 25-26, 2014.

3D CYLINDRICAL APPROACH TO DETERMINE THE EXCESS MINORITY CARRIERS' DENSITY OF AN N^+/P SOLAR CELL UNDER CONSTANT MONOCHROMATIC ILLUMINATION

Assane Diouf[1, 2], Serigne Ndiangue Leye[1, 2], Senghane Mbodji[1, 2], Amadou Diao[2] and Gregoire Sissoko[2]
[1]Research team in Renewable Energies, Materials and Laser of Department of Physics, Alioune DIOP University of Bambey, Bambey, Senegal;
[2]Laboratory of Semiconductors and Solar Energy, Department of Physics, Faculty of Science and Technology, Cheikh Anta Diop University, Dakar, Senegal
E-mail:* *senghane.mbodji@uadb.edu.sn; zoura42@gmail.com
Phone:00221775220101

ABSTRACT: The columnar cylindrical model is used for the characterization of an N^+/P polycrystalline silicon solar cell in steady state under monochromatic illumination. From this model and using both the Bessel functions and the dynamic junction recombination velocity, the continuity equation has been solved. A new expression of the excess minority carriers' density in the base depending to the radius (r) and the depth (z) at a position of the grain is obtained. This mathematical modeling allowed us to study the effects of grain size (R) and grain boundaries recombination velocity (Sgb) on the behavior of excess minority carriers' density in the base of the solar cell.
Keywords: Columnar cylindrical model, grain boundaries recombination velocity, grain radius (r), grain size (R).

1 INTRODUCTION

Many works were done to improve the solar cells' efficiency. Thus, some methods concerning the manufacturing technologies [1] and operating conditions that help to minimize effects limiting the performance of solar cells [2-3] were developed. To do so, simulation works based on the resolution of the continuity equation for excess minority carrier's density can be adopted.
These approaches are done in steady state [4] or in dynamics ones [5] and, usually, have considered the 1D orientation [6] or the 3D columnar cubic representation [3-7] which permits to take into account the bulk recombination noted at the grain boundary of the polycrystalline solar cell.

These two orientations have permitted to determine the lifetime, the effective back side surface recombination velocity and diffusion length, intrinsic junction recombination velocity, the short-circuit and open circuit operating conditions, the photocurrent, the phototention, the power, the series and shunt resistances, efficiencies of the solar cell .
For a good consideration of the film thickness, the intragrain material quality, and the grain boundary properties of the polycrystalline silicon solar cell, works done in columnar cubic orientation must be extended to three-dimensional cylindrical model as tried by S. Elnahwy [8] and N. Adeeb and A. Trabelsi et al.[9].

S. Elnahwy and N. Adeeb used three-dimensional cylindrical model for a polycrystalline solar cell to study the effects of the grain size and the grain boundary recombination velocity on the spectral response and the short-circuit current of N/P junction silicon solar cells with different values for the cell thickness. A. Trabelsi et al. showed that three types of recombination such as the bulk recombination, the surface recombination and the recombination at the grain boundaries have an important effect on the cell parameters. They also did the comparison between an elementary cylindrical grain solar cell and an elementary cubic grain one.

What would be missing for the improvement of these previous work [8-9] is the use of the dynamic junction recombination velocity (Sf). Hence, our study is based to Sf and proposed an analytical expression of the excess minority carriers' density in the base of an N^+/P ploycristalline solar cell in steady state and under monochromatic illumination. We studied the effects of the dynamic junction recombination velocity (Sf), the grain size (R), the grain boundaries recombination velocity (Sgb), the backside surface recombination (Sb) on the behavior of excess minority carriers in the base of the photovoltaic cell.

2 THEORY

Here, we use a three-dimensional cylindrical (3Dcyl) model for an N^+/P polycrystalline solar cell as shown in fig.1.

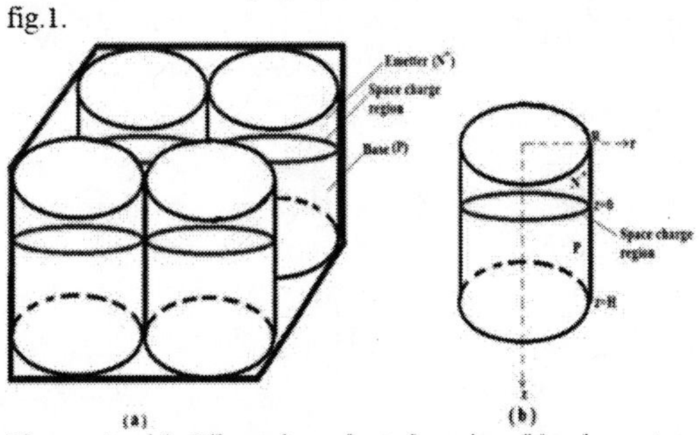

Figure 1: (a) Fibrously oriented grain, (b) elementary cylindrical grain solar cell.

2.1 Assuption

This model is treated considering the following assumptions [8-9-10]:

- for reasons of symmetry, the angle θ is not taken into account;
- the solar cell is illuminated by its front face;
- the thickness of the emitter is neglected; we will then have a generation rate dependent only on the base depth (z).

2.2 Continuity equation

The continuity equation in the base of the solar cell is presented as [8-9-10]:

$$\frac{\partial^2\delta(\mathrm{r},\mathrm{z})}{\partial r^2}+\frac{\partial^2\delta(\mathrm{r},\mathrm{z})}{\partial z^2}+\frac{1}{r}\frac{\partial\delta(\mathrm{r},\mathrm{z})}{\partial r}-\frac{1}{L^2}\delta(\mathrm{r},\mathrm{z})=-\frac{G(z)}{D} \quad (1)$$

δ(r, z) is the excess minority carriers density in base of the solar cell; D and L represent the coefficient and the diffusion length, respectively; G (z) is the electron-hole pairs generation rate [10-11]:

$$G(z)=\alpha \mathrm{I}_0(1-R')\exp(-\alpha z) \quad (2)$$

The excess minority carriers' density is given by [6-7-8]:

$$\delta(\mathrm{r},\mathrm{z})=\sum_{k\geq 1} f_k(\mathrm{r})\sin(\mathrm{C}_k\mathrm{z})+\mathrm{K}_k \quad (3)$$

Inserting equation (3) into (1), after derivation and arrangement we obtain:

$$\sum_{k\geq 1} f_k''(r)\sin(C_k z)+\frac{1}{r}\sum_{k\geq 1} f_k'(r)\sin(\mathrm{C}_k\mathrm{z})-\sum_{k\geq 1}\frac{1}{L_k^2}.f_k(r)\sin(C_k z)=-\frac{G(z)}{D}+\frac{K_k}{L^2} \quad (4)$$

With:

$$L_k=\left(\frac{1}{L^2}+C_k^2\right)^{-\frac{1}{2}} \quad (5)$$

Using the orthogonally condition of the sinus function on the length of the base [8-9-10], we get a differential equation in f_k (r):

$$f_k''(r)+\frac{1}{r}f_k'(r)-\frac{1}{L_k^2}f_k(r)=\frac{2G_0}{HD}\frac{C_k}{C_k^2+\alpha^2}(1+(-1)^{k+1}\exp^{-\alpha H})+\frac{2K}{L^2H}\frac{1-\cos(C_k\mathrm{H})}{C_k} \quad (6)$$

The resolution of the differential equation without a second member gives:

$$f_k(\mathrm{r})=\mathrm{A}_k.r \quad (7)$$

Hence, the general solution of the radial function is:

$$f_k(r)=A_k r+\frac{2G_0}{HD}\frac{L_k^2C_k}{C_k^2+\alpha^2}(1+(-1)^{k+1}\exp^{-\alpha H})-\frac{2K_k}{L^2H}\frac{1-\cos(C_k\mathrm{H})}{C_k} \quad (8)$$

The excess minority carriers' density in the base is then expressed as:

$$\delta(r,\mathrm{z})=\sum_{k\geq 1}\left[\begin{array}{l}A_k.r+\frac{2G_0}{HD}\frac{L_k^2C_k}{C_k^2+\alpha^2}(1+(-1)^{k+1}\exp^{-\alpha H})\\-\frac{2L_k^2K_k}{L^2H}\frac{1-\cos(C_k\mathrm{H})}{C_k}\end{array}\right]\sin(C_k.z)+\mathrm{K}_k \quad (9)$$

2.3 Boundary conditions

For the determination of the coefficients A_k, K_k and C_k we admit the following boundary conditions [8,9, 12,13]:

2.3.1 At the junction (z=0) [12,13]:

$$\left.\frac{\partial\delta(r,z)}{\partial z}\right|_{z=0}=\frac{S_f}{D}\delta(\mathrm{r},0) \quad (10)$$

Sf represents the dynamic junction recombination velocity [12, 14] and is imposed by the external load and losses at the junction.

Sf is the sum of two terms: $Sf=Sf_0+Sf_j$. Sf_0 corresponds to leakage current at the junction and Sj is due of electric current flow through the junction.

2.3.2 At grain boundary (r=R) [8,9]:

$$\left.\frac{\partial\delta(r,z)}{\partial r}\right|_{r=R}=-\frac{S_{gb}}{D}\delta(R,\mathrm{z}) \quad (11)$$

Sgb is the grain boundary recombination velocity [13].

2.3.3 At back side surface of the solar cell (z=H)[8,9,12,13]:

$$\left.\frac{\partial\delta(r,z)}{\partial z}\right|_{z=H}=-\frac{S_b}{D}\delta(r,H) \quad (12)$$

Sb is the recombination velocity which evaluates the rate of losses of minority carriers at the back side surface of the solar cell [9, 10]. After combination and arrangement of equations (10), (11), (12) and (9) we obtain:

$$\tan(C_kH)=-\frac{D}{S_b}C_k(1-\frac{S_b}{S_f\cos(C_kH)}) \quad (13)$$

$$\delta(\mathrm{r},\mathrm{z})=\sum_{k\geq 1}\frac{2G_0L^2C_k}{D^3(C_k^2+\alpha^2)}\{1+(-1)^{k+1}\exp^{-\alpha H}\}\left[\begin{array}{l}\frac{1}{2\frac{D}{S_{gb}}L_k^2(1-\cos(C_k\mathrm{H}))+L^2HS_f(\frac{1}{S_{gb}}-\frac{R}{D})}\\\left\{\begin{array}{l}(S_fL_k^2r+\frac{2\frac{D}{S_{gb}}L_k^2(1-\cos(C_k\mathrm{H}))}{L^2H})\sin(C_kz)\\+2L_k^2C_k(\frac{D}{S_{gb}})\end{array}\right\}\\-\frac{D}{H}\sin(C_k\mathrm{H})\end{array}\right] \quad (14)$$

Equation (13) is a transcendental equation. We use the numerical method "Newton-Raphson" [14] to obtain the solutions C_k of equation (13).

3 SIMULATION RESULTS AND DISCUSSIONS

The distribution of excess minority carriers' density in the base of an N^+/P polycrystalline silicon junction solar cell as a function of the base depth (z), the dynamic junction recombination velocity (Sf), the grain size (R), the grain boundaries recombination velocity (Sgb) and the back side surface recombination velocity (Sb) is studied by Matlab software.

Figures 2, 3 and 4 show the effects of the grain size (R), the dynamic junction recombination velocity (Sf) and the grain boundary recombination velocity (Sgb) on the distribution of the density of the carriers, respectively.

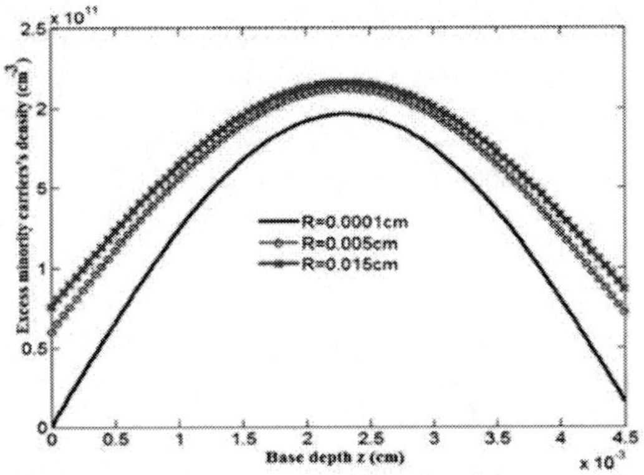

Figure 2: Excess minority carriers' density versus the base depth (z); $D=26cm^2.s^{-1}$, $Sf= 4.10^4 cm.s^{-1}$, $Sb=10^4$ $cm.s^{-1}$, $Sgb=5000$ $cm.s^{-1}$, $\lambda=1000$ nm, L = 0.01cm, H=130 μm.

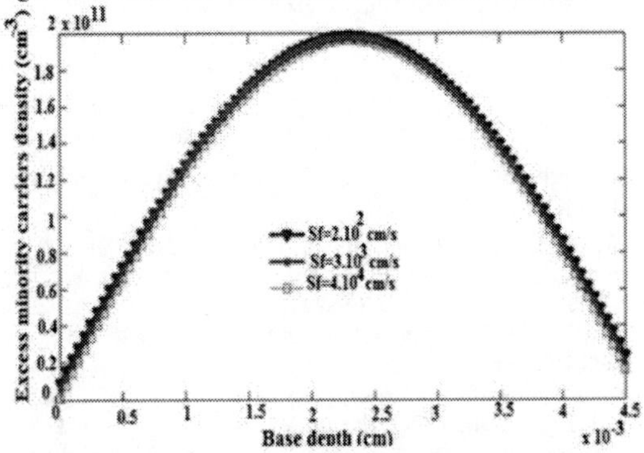

Figure 3: Excess minority carriers' density versus the base depth (z); $D=26cm^2.s^{-1}$, $Sb=10^4$ $cm.s^{-1}$, $Sgb=5000$ $cm.s^{-1}$, $\lambda=1000$ nm, L = 0.01cm, H=130 μm.

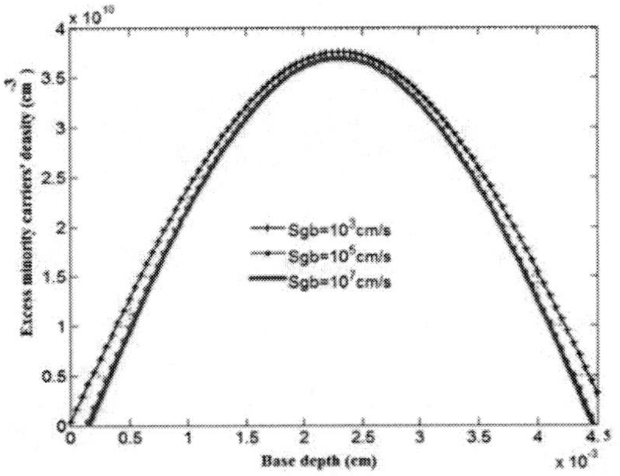

Figure 4: Excess minority carriers' density versus the base depth (z); $D=26cm^2.s^{-1}$, $Sf= 4.10^4 cm.s^{-1}$, $Sb=10^4$ $cm.s^{-1}$, $\lambda=1000$ nm, L = 0.01cm, H=130 μm.

Analysis of these figures showed that the excess minority carriers' density in the base has a maximum along the thickness of the base. As shown by *Mbodji et al.* [15] region with a positive slope corresponds to minority carriers' collection area. Lost minority carriers in the bulk are in region with negative slope [15]. As the grain size (R) increases, we note an increase of the excess minority carriers density. This is plausible because as shown by Mbodji et al., when the grain size increases the recombination at grain boundary decreases corresponding to an increase of photocurrent and phototension.

When recombination at grain boundaries increases, we have a decrease of the excess carrier's minority density. The grain boundaries are considered to be passing areas of impurity atoms which reduces their rate in the crystallite as well and these atoms are distributed disorderly because of the large number of defects due to incomplete atomic bonds [16].

When the dynamic junction recombination velocity increases (Sf), the carrier density decreases. (Sf_0), the recombination velocity of the intrinsic junction, related to the losses that occur through the junction, is not taken into account.

In Figure 5, we plotted the excess minority carriers density versus both the base depth (z) and the grain size (R).

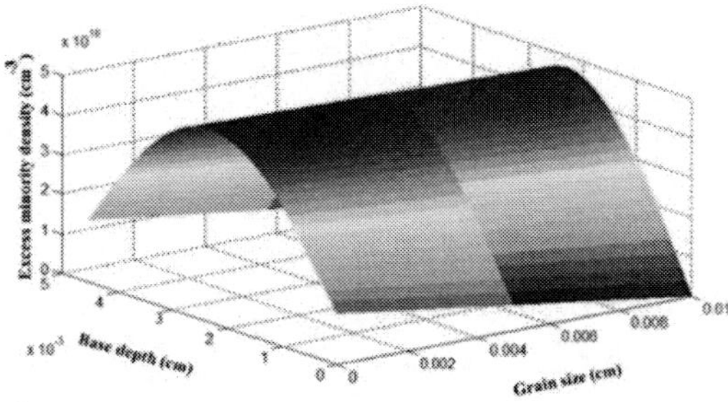

Figure 5: Excess minority carriers' density versus both the base depth and grain size: $D=26cm^2.s^{-1}$, $Sf= 4.10^4 cm.s^{-1}$, $Sb=10^4 cm.s^{-1}$, $Sgb=5000\ cm.s^{-1}$, $\lambda=1000$ nm, L = 0.01cm, H=130 µm.

Figure 5 confirms the analyzes of the preceding results. We note that recombination at the grain interfaces have no effect on the central region of the grain as shown by Mbodji et al. and [15].

In figure 6 we showed the evolution of the excess minority carriers density versus the dynamic junction recombination velocity (Sf) and the back side surface recombination velocity (Sb), respectively, for different grains boundary recombination velocity (Sgb).

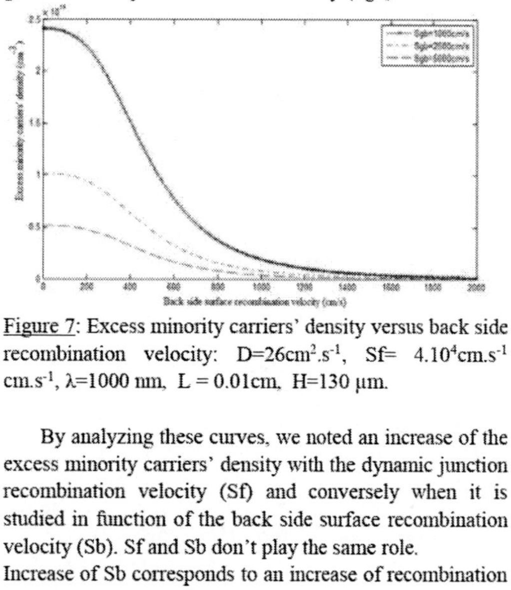

Figure 7: Excess minority carriers' density versus back side recombination velocity: $D=26cm^2.s^{-1}$, $Sf= 4.10^4 cm.s^{-1}$ $cm.s^{-1}$, $\lambda=1000$ nm, L = 0.01cm, H=130 µm.

By analyzing these curves, we noted an increase of the excess minority carriers' density with the dynamic junction recombination velocity (Sf) and conversely when it is studied in function of the back side surface recombination velocity (Sb). Sf and Sb don't play the same role.

Increase of Sb corresponds to an increase of recombination at the back side surface of the solar. But increase of Sf means that the external load is important because the space charge is enlarged leading to an increase of electrons' flow crossing the junction as shown by Mbodji et al [15]

4 CONCLUSION

A 3D columnar cylindrical approach for the resolution of the continuity equation of an N^+/P polycrystalline silicon solar in steady state and under monochromatic illumination conditions is presented. It allowed us to establish a new expression of the excess minority carriers' density in the base of the solar cell. This expression is based on the dynamic junction recombination velocity parameter. The effects of the grain size (R), the grain boundaries recombination velocity (Sgb), the back side surface recombination velocity (Sb) and the dynamic junction recombination velocity (Sf) were studied. We concluded that the excess minority carriers density increases with Sf, and R but decreases with Sb and Sgb as shown by the previous works.

ACKNOWLEDGMENT

The authors would like to thank Alioune DIOP University of Bambey and the "Centre d'Excellence en Mathématiques, Informatique et TIC (CEA-MITIC)" in Gaston Berger University, Senegal. They both sponsored the participation of the authors in EUPVSEC 2017.

5 REFERENCES

[1] D. A Neamen, 2003 Semiconductor Physics and Devices Basic Principle, New York.

[2] Sinton, R.A. and P.M. Swanson, 1987 An Optimization Study of Si Point Contact Concentration 15th I.E.E.E. Photon. Spect. Conf. USA, pp: 1207-1208.

[3] W. Shockley Electrons and Holes in Semiconductors Van Nostrand, Princeton, NJ, 1959, pp. 318-325.

[4] S. Mbodji, I. Ly, H. L. Diallo, M.M. Dione, O. Diasse and G. Sissoko Modeling Study of N+/P Solar Cell Resistances from Single I-V Characteristic Curve Considering the Junction Recombination Velocity (Sf). Research Journal of Applied Sciences, Engineering and Technology.

[5] A. G. Camara, N.Thiam, I. Gaye, S.Gokan and G. Sissoko. 3D Modeling of a Vertical Junction Polychristalline Silicon Solar Cell Under monochromatic Illumination in Frequency Modulation. Current Trends in Technology and Science

[6] H. Ly Diallo, B. Dieng, I. Ly, M.M. Dione, M. Ndiaye, O.H. Lemrabott, Z.N. Bako, A. Wereme and G. Sissoko. Determination of the Recombination and Electrical Parameters of a Vertical Multijunction Silicon Solar Cell. Research Journal of Applied Sciences, Engineering and Technology ISSN: Maxwell Scientific Organization, 2012

[7] S. Mbodji, M. Dieng, B. Mbow, F.I. Barro, G. Sissoko Influence of Grain Size and Grain Boundary Recombination Velocity on the Diffusion Capacitance of a Polycrystalline Bifacial Silicon Solar Cell. Journal des Sciences pour l'Ingénieur N°11 (2009) 64 - 69

[8] S. Elnahwy and N. Adeeb Exact Analysis of a Three-dimensional Cylindrical Model for a Polycriystalline Solar Cell. Journal of Applied Physics 64, 5214 (1988)

[9] A. Trabelsi, A. Zouari and A. B. Arab Modeling of Polycrystalline N^+/P Junction Solar Cell with Columnar Cylindrical Grain. Revue des Energies Renouvelables Vol. 12 N°2 (2009) 279 – 297.
[10] A. Trabelsi, M. Krichen, A. Zouari and A. B. Arab Effect of Quasi-monocrystalline Porous Silicon at the Backside on the Photovoltaic Parameters of a Polycrystalline Silicon Solar Cell. Revue des Energies Renouvelables Vol. 11 N°3 (2008) 395 – 405.
[11] Sze, S.M. and K.N. Kwok. Semiconductor Device Physics and Design. Springer Physics of Semiconductors Devices 2008.
[12] H. L. Diallo, A. S. Maiga, A. Wereme and G. Sissoko, J. Appl. Phys., Vol. 42, (2008), pp. 203- 211.
[13] G.Sissoko, C. Museruka, A. Correa, I. Gaye and A. L. Ndiaye, Proc. of World Renewable Energy Congress, pp. 1487 – 1490, 1996, Denver, USA.
[14] M. Goyal, Computer Based Numerical & Statistical Techniques, Infinity Science Press, LLC, 2007.
[15] S. Mbodji, B. Mbow, F. I. Barro and G. Sissoko. A 3D Model for Thickness and Diffusion Capacitance of Emitter-base Junction Determination in a bifacial Polycrystalline Solar Cell Under Real Operating Condition. Turk J Phys 35 (2011), 281 – 291.
[16] J. Y. Seto, J. Appl. Phys. Vol. 46. No2, 5247 - 52.54, (1975).

TEMPERATURE EFFECT ON THE CAPACITANCE OF THE BIFACIAL SOLAR CELL CONSIDERING THE COLUMNAR CYLINDRICAL ORIENTATION.

Serigne Ndiangue LEYE[1], Assane DIOUF[1], Amadou DIAO[1], Senghane MBODJI[1,2] and Gregoire SISSOKO[1]
[1]Laboratory of Semiconductors and Solar Energy, Department of Physics, Faculty of Science and Technology, Cheikh Anta Diop University, Dakar, SENEGAL;
[2]Research team in renewable energies, materials and laser of Department of Physics, Alioune DIOP University of Bambey, Bambey, Senegal.

E-mail: senghane.mbodji@uadb.edu.sn; ndiangueleye@hotmail.com
Adresse: PO. Box 30, Bambey

ABSTRACT: In this paper, temperature's effect on the transient capacitance's new expression of the bifacial solar cell, considering the cylindrical model of the grains tis proposed. The temperature effects are verified and it is noted that the transient capacitance decreases with temperature.
Keywords: Polycrystalline, Solar Cell, Junction Recombination Velocity, Capacitance, Temperature

INTRODUCTION

In dynamic state, the P-N junction is the seat of the transient capacitance or the associated junction and the diffusion capacitance due to ionized dopants and to the the charges injected in the neutral regions, respectively. The transient capacitance is simply associated with the charge Q contained in the junction and the diffusion capacitance translates the delay between voltage and current [1].

In steady state, the transient capacitance is predominant and the solar cell space charge region can be assimilated to a plane capacitor with two identical plane electrodes separated by a thickness [2,3,4] which depends to the Sf that appears to be very important in determining the extension region width [2,3,4], the back side surface field [2,3,4], the grains size [2], the grain boundary recombination velocity [2], the magnetic [3] field, the wavelength [2], the irradiation level [5] and the temperature[6] [2].

These results are, only, got at one (1D) dimension [2,3,4] and in three (3D) dimensions [2] studies. All 3D modeling studies are based on columnar approach using an isolated grain of the solar cell. .

The polycrystalline solar cell is made of many small individual grains with random orientation which can be modeled by columnar [2], cylindrical [7, 8] and spherical orientations. For cylindrical studies [7, 8] we noted that works are limited to determine the absolute spectral response, short-circuit current density, the effects of the emitter-base reverse saturation current density, the cell thickness and the grain radius on the photovoltaic parameters. These studies don't address the capacitances and temperature of the solar cells and are done using two real operating condition: the open circuit and short-circuit.

That is why, taking into account the concept of dynamic junction velocity (Sf) [9] and using an cylindrical orientation, this paper aims to study the temperature effect on the transient diffusion capacitance of an N^+P-P^+ bifacial solar cell.

First, we present the theorical backgrounds which help us to resolve the continuity equation. Then the work is extended to determine the solar cell's excess minority carriers density and transient capacitance. Finaly, the different mechanics driving the temperature effects on the transient capacitance.

1 Theoretical Background.

We present in figure 1 an isolated cylindrical grain of the bifacial silicon solar cell under monochromatic illumination.

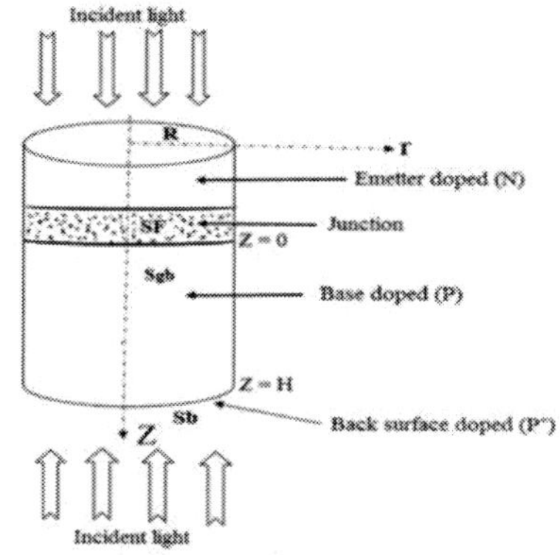

Figure 1: an isolated cylindrical grain.

The bifacial solar cell is assumed to have four zones: the emitter doped N^+ the base doped P, the junction N^+-P which is setted between the emitter and the base, and the back contact where another junction PP^+ is created by a thin film strongly doped P+. This second junction is the seat of the very intense electric field which returns the minority carriers towards the base [10-12]. This electrical field limits the recombination at the back side of this type of solar cell which are more efficient than the conventional solar cell [10-12].

According to [7], the transfer phenomena in the base of the solar cell are modeled by the continuity equation in cylindrical coordinates:

$$\frac{\partial^2\delta(r,\theta,z)}{\partial r^2}+\frac{\partial^2\delta(r,\theta,z)}{\partial z^2}+\frac{1}{r^2}\cdot\frac{\partial^2\delta(r,\theta,z)}{\partial\theta^2}+\frac{1}{r}\cdot\frac{\partial\delta(r,\theta,z)}{\partial r}-\frac{\delta(r,\theta,z)}{L^2(T)}=-\frac{G(z)}{D(T)} \quad (1)$$

As, we have an azimuthal symmetry, the angle θ is not processed. Therefore the continuity equation becomes:

$$\frac{\partial^2\delta(r,z)}{\partial r^2}+\frac{\partial^2\delta(r,z)}{\partial z^2}+\frac{1}{r}\cdot\frac{\partial\delta(r,z)}{\partial r}-\frac{\delta(r,z)}{L^2(T)}=-\frac{G(z)}{D(T)} \quad (2)$$

$\delta(r,z)$: minority carrier's density;
L(T) is the electron diffusion length in the base and τ the lifetime;
D(T) is the electron diffusion coefficient in the base given by the well-known Einstein relation temperature dependant and expressed as:

$$L^2(T)=\tau\cdot D(T) \quad (3)$$

And

$$D(T)=\mu(T)\cdot\frac{k_b}{q}\cdot T \quad (4)$$

μ (T) is the coefficient mobility [13] for electrons, given as:

$$\mu(T)=1.43\cdot10^9\cdot T^{-2.42}cm^2\cdot V^{-1}\cdot S^{-1} \quad (5)$$

k_b is the Boltzmann constant ;
q is the elementary charge of an electron ;
G(z) is the electron-hole pairs generation expressed as [14]:

$$G(z)=\alpha\cdot I_0\cdot(1-R)\cdot\exp(-\alpha\cdot z) \quad (6)$$

The coe□cient α denotes the absorption of light for wavelength λ; and I_0 is the incident photon flux [15].
While proceeding by the separation method of the variables used by [16], we put:

$$\delta(r,z)=\sum_{k\geq1}^{\infty}f_k(r)\cdot\sin(c_k\cdot z)+K_k \quad (7)$$

The solution of the equation is given by:

$$\delta(r,z)=\sum_{k\geq1}^{\infty}\left[\begin{array}{l}A_k\cdot r+\frac{2\cdot\alpha\cdot(1-R)}{H\cdot D(T)}\cdot\frac{L_k^2\cdot c_k}{c_k^2+\alpha^2}\cdot G(H)-\\ \frac{2K_k}{L^2(T)\cdot H}\cdot\frac{1-\cos(c_k\cdot H)}{c_k}\end{array}\right]\cdot\sin(c_k\cdot z)+K_k \quad (8)$$

Coefficients A_k and K_k are obtained from the boundaries conditions of the solar cell:

- At the junction (z=0) [2]

$$\left.\frac{\partial\delta(r,z,T)}{\partial z}\right|_{z=0}=\frac{Sf}{D(T)}\cdot\delta(r,T,z=0) \quad (9)$$

- At the back side of the solar cell [2]:

$$\left.\frac{\partial\delta(r,z,T)}{\partial z}\right|_{z=H}=-\frac{Sb}{D(T)}\cdot\delta(r,T,z=H) \quad (10)$$

- At the grain boundary (r=R) [7,8]:

$$\left.\frac{\partial f(r,T)}{\partial r}\right|_{z=R}=-\frac{Sgb}{D(T)}\cdot f(R,T) \quad (11)$$

As shown by [2,17,18], Sf is dynamic junction velocity and both traduces losses occurring across at the junction and current flow to the external load.
Sb is the back side surface recombination velocity. It quantifies the rate at which excess minority carriers are lost at the back side surface of the cell [2,19]

Sgb is the grain boundary recombination velocity on the back of excess charge minority carriers in the base [2, 17,20].

2 CAPACITANCE

The bifacial solar cell's transient diffusion is given by [14]:

$$C(Sf,Sb,Sgb,\lambda,T)=\frac{q}{V_T}\cdot\left[\begin{array}{l}\delta(z=0,Sf,Sgb,Sb,\lambda,T)\\ +\frac{n_i^2(T)}{N_b}\end{array}\right] \quad (12)$$

N_b is the doping rate and $n_i(T)$is the intrinsic minority charge carriers density [21] expressed as :

$$n_i(T)=A\cdot T^{3/2}\cdot\exp(-\frac{E_g}{2\cdot k_b\cdot T}) \quad (13)$$

E_g is the energy gap, it corresponds to the difference between the energies of the conduction band (E_c) and the valence band (E_v).
In our case, $E_g=1.12\cdot1.6\cdot10^{-19}J$ and A is a constant value: $A=3.87\cdot10^{16}cm^{-3}.K^{-3/2}$

3 RESULTS AND DISCUSSIONS

In figures 2, we plotted the minority charge carriers density versus the depth in the base for different temperature values. Figure 2 shows an increase of the excess minority carriers density with the temperature increase as shown by [6].

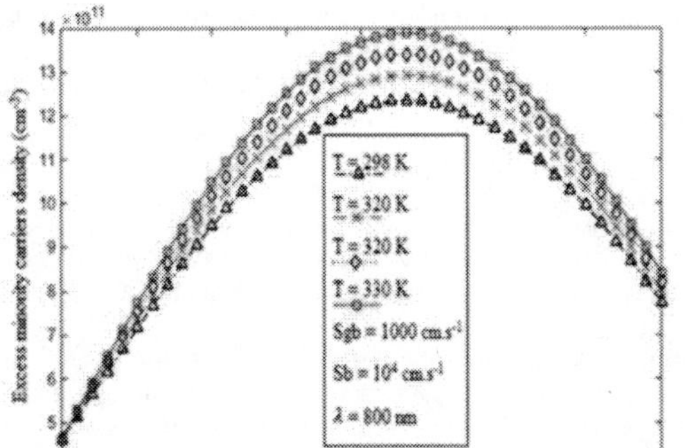

Figure 2: Excess minority carriers density versus base depth for different temperature values and when the solar is illuminated by its front side: $Sf=4.10^4$ cm.s^{-1}

We plotted in figure 3 the transient diffusion capacitance versus the junction recombination velocity (SF) and in figure 4 the logarithm of the transient capacitance of the solar cell versus the logarithm of the temperature (T).

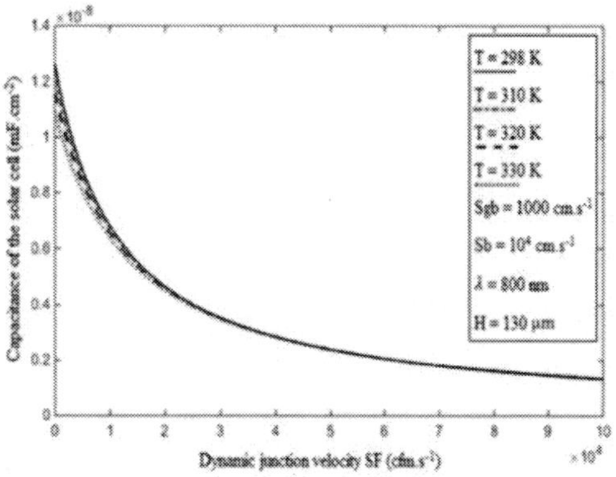

Figure 3: Transient capacitance versus SF for various Temperature (T) when the solar cell is illuminated by its front side:R = 0.01 cm

The figure 3 shows a decrease of the capacitance of the solar cell with SF and when the temperature increases.

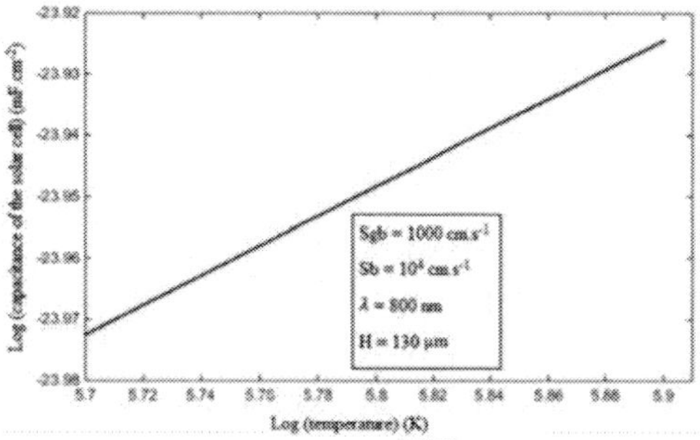

Figure 4: Logarithm of the transient capacitance versus logarithm of the temperature (T) when the solar cell is illuminated by its front side: SF = 4.10^4 $cm.s^{-1}$, R= 0.01 cm.

CONCLUSION

In this paper, we have determined the transient capacitance using the cylindrical orientation of solar cell's grains.

We have shown that, as for the 1D study and the columnar isolated grain model, the transient capacitance decreases when the temperature of the solar cell increases.

ACKNOWLEDGMENT

The authors would like to thank Alioune DIOP University of Bambey and the "Centre d'Excellence en Mathématiques, Informatique et TIC (CEA-MITIC)" in Gaston Berger University, Senegal. They both sponsored the participation of the authors in EUPVSEC 2017.

References

[1] S. Valkov, Electronique analogique, Cours avec problèmes résolus, Educalivre, (1995)

[2] Mbodji, S., Mbow, B., Barro, F. I. and Sissoko, G, Turkish Journal of Physics, Vol. 15, Issue 3, 281 (2011)

[3] S. Madougou, F. Made, M. S. Boukary and G. Sissoko, Advanced Materials Research, Vol. 18, Issue 19, 303 (2007)

[4] G. Sissoko, B. Dieng, A. Correa, M. Adj, D. Azilino, Proceedings of the 5 th Word Renewable Energy Congress., Vol. 3, 1852 (1998), Florence, Italy

[5] M. S. Ndoye, B. Seibou, I. Ly, M. S. Diouf, M. Wade, S. Mbodji and G. Sissoko, International Journal of Innovative Technology and Exploring Engineering (IJITEE), Vol. 6, Issue 3, (2016)

[6] I. Diatta, I. Ly, M. Wade, M. S. Diouf, S. Mbodji and G. Sissoko, World Journal of Condensed Matter Physics, Vol. 6, 261 (2016)

[7] A. Trabelsi, A. Zouari and A. Ben Arab, Revue des Energies Renouvelables, Vol. 12, N° 2, 279 (2009)

[8] S. Elnahwy and N. Adeeb, Journal of Applied Physics Vol. 64, 5214 (1988); doi: 10.1063/1.342435.

[9] S. Mbodji, M. Zoungrana, I. Zerbo, B. Dieng and G. Sissoko, World Journal of Condensed Matter Physics, Vol. 5, 284 (2015)

[10] G. Sissoko, A. Correa, E. Nanema, M. N. Diarra, A. L. Ndiaye and M. Adj, Proc. World Renewables Energies Congress, Vol. 3, 1856 (1998)

[11] S. Mbodji, B. Dieng, M. B. Mbow, F. I. Barro and G. Sissoko, Journal of Applied Science and Technology (JAST), Vol. 15, N° 1 & 2, 109 (2010)

[12] S. Mbodji and G. Sissoko, Int. J. Pure Appl. Sci. Technol., Vol. 6, N° 2, 103 (2011)

[13] M. Kunst and A. Sanders, Semiconductor Science and Technology, Vol 7, 51 (1992)

[14] S. Mbodji, M. Dieng, B. Mbow, F. I. Barro and G. Sissoko, Journal of Applied Science and Technology (JAST), Vol. 15, 109 (2010)

[15] M. A. Green, and M. Keevers, Progress in Photovoltaics., Vol. 3, 189 (1995)

[16] J. Dugas, Solar Energy Materials and Solar Cells., Vol. 32, 71 (1994)

[17] S. Mbodji, M. Zoungrana, I. Zerbo, B. Dieng and G. Sissoko, World Journal of Condensed Matter Physics, Vol. 5, 284 (2015)

[18] Grégoire Sissoko and Senghane Mbodji, Int. J. Pure Appl. Sci. Technol., Vol. 6, Issue 2, 103 (2011)

[19] S. Mbodji, A. S. Maiga, M. dieng, A. Wereme and G. Sissoko, Global Journal of Pure and Applied Sciences., Vol. 15, 125 (2009)

[20] S. Elnahwy and N. Adeeb, Journal of Applied Physics, Vol. 64, N⁰. 10, 5214 (1988)

[21] C. D. Thurmond, Journal of the Electrochemical Society, Vol. 122, 133 (1975).

PARAMETER EXTRACTION OF OXIDIZED Ni/Au AND Ni-ONLY TRANSPARENT CONDUCTING OXIDES (TCOs) ON N-TYPE GaN SCHOTTKY BARRIER DIODE WITH BIAS DEPENDENCE BARRIER HEIGHT AND IDEALITY FACTOR AT DIFFERENT TEMPERATURES

Ali.T. Hajjiah[1]*, Asmaa.A. Alkhabbaz[1] , Noah.P.Allen[2], Louis.J.Guido[2]
[1] Kuwait University, College of Engineering and Petroleum, Electrical Engineering Department, P.O. Box 5969, Safat 13060-Kuwait
[2] Virginia Polytechnic Institute and State University, Bradley Department of Electrical and Computer Engineering, Blacksburg, VA, 24060
*Corresponding author: phone: +965 24985829, fax: +965 24817451, e-mail: ali.hajjiah@ku.edu.kw

ABSTRACT: The forward current-voltage (I-V) characteristics of oxidized Ni-Au and Ni-only Schottky contacts on n-type GaN were measured in the temperature range 160 K-400 K. The Schottky barrier diode whose barrier height (Φ_b) and ideality factor (β) are both bias dependent has been studied as a function of temperature using a technique which makes use of voltage-dependent differential slope curve $\alpha(V) = [d\ln(I)/d\ln(V)]$. The extracted parameters include the saturation current (zero bias barrier height), the voltage dependence of the barrier height, the voltage dependence of the ideality factor as well as the series resistance. The deviation of the characteristics from the ideal thermionic behavior are more pronounced with both temperature and voltage. The results obtained indicate the presence of other current transport mechanisms as temperature and voltage are changed. Moreover, the increase in barrier height with increasing temperature and voltage has been explained as an effect of barrier inhomogeneity. Furthermore, the series resistance (Rs) values estimated from three different methods (Cheung's, Direct, and $[dV/d\ln(I)]$) vary from a value of 58.54Ω at 160K and 31.25Ω at 400K for Ni-only and a value of 735.8Ω at 160K to 139.94Ω at 400K for oxidized Ni-Au and are seen to be strongly temperature dependent.
Keywords: TCO Transparent Conducting Oxides, III-V Semiconductors, n-type, Schottky diode.

1 INTRODUCTION

Metal-semiconductor (MS) structures are important research tools in the characterization of new semiconductor materials, and at the same time, the fabrication of these structures plays a vital role in constructing some useful devices in technology [1]. There is a current interest in developing ohmic and Schottky contacts on gallium nitride (GaN) for applications in optoelectronic devices and high power/high frequency devices such as light emitting diodes (LEDs) [2], laser diodes (LDs), photo detectors [3], hetero-junction field effect transistors (HFETs) [4], metal oxide field effect transistors (MOSFETs) [5], high electron mobility transistors (HEMTs) [6] and Schottky rectifiers [7]. However, to take further advantage of thermal stability and chemical inertness of GaN, it is necessary to understand the behavior of different metal contacts to GaN at different temperatures. Electrical characteristics of metal contacts to n-GaN under different temperature conditions have been studied by many research groups [11,12,13]. Published reports indicate a barrier height of between 0.53-1.03 eV which is dependent on the types of metals used. Guo et al [12] reported a room temperature barrier height of 0.66 eV on Ni Schottky contacts using the IV method, while a barrier height of 0.84 eV was also measured by Hacke et al [13] on Au Schottky contacts by using the IV technique. Dogan et al [11] performed measurements on Au/Ni/n-GaN Schottky contacts in the 40-320 K temperature range. Their results indicate the variation in the IV characteristics in three different regions, with thermionic emission being the dominant transport mechanism. From their report, they revealed the variation of the Richardson constant within these regions. Donoval et al [14] also studied the IV characteristics of GaN Schottky contacts in the 125-300 K temperature range while Schmidtz et al [15] studied the IV characteristics of the Pd/GaN Schottky contacts up to 500 °C. The effort has been made by several groups to form Schottky contacts on n-type GaN using a number of metals e.g., Au [8], Pt [8], Pd [8], Ni [8], Ti [8], and Cr [9]. From the studies on the n-GaN Schottky contacts, it has been found that the Schottky barrier height (Φ_b) depends on the metal work function due to the ionic nature of GaN [10]. However, there are still wide variations in reported values of Schottky barrier height (Φ_b), even when popular metals such as Pt, Ni, and Au and Pt are used. In all their reports, they have reported the dependence of the barrier height and ideality factor on temperature. However, reports on the variation of the observed leakage currents with temperature and the voltage dependence of the ideality factor with temperature for oxidized Ni-Au TCO has been very little (if any). This paper is going to study the I-V-T characteristics of both Ni and oxidized Ni-Au TCOs on n-type GaN and determine the voltage dependence of the ideality factor (β) for both. This is important because it helps to identify the exact current flow mechanisms and to extract accurate parameters of Schottky diodes such as the saturation current (I_S) which gives the zero-bias barrier height (Φ_{bo}), and the voltage dependence of Φ_b in the presence of series resistance (R_S).

2 EXPERIMENTAL PROCEDURES

Schottky diode test structures were made using n-type GaN material grown by metalorganic chemical vapor deposition (MOCVD) at Virginia Tech Micron lab. The test structure starts with a 4um thick unintentionally doped GaN nucleation layer grown on a 2 inch diameter C-plane sapphire (Al_2O_3) substrate at low temperature. To achieve good structural quality, a 1 µm thick undoped GaN buffer layer was then grown at high temperature on top of the GaN nucleation layer. Finally, a 1 um thick silicon doped GaN test layer which serves as the device platform in contact with both the Schottky and ohmic electrodes was grown at high temperature with $(N_D - N_A) = 1 \times 10^{17}$ cm^{-3} as established by capacitance-voltage measurements at room temperature. Next, to form the ohmic contacts, a thick 30 nm Ti / 400 nm Al metallic bilayer was deposited by electron-beam evaporation and annealed at 600 °C in a hydrogen (5%) / argon (95%) gas mixture. Finally, the Schottky electrode material was then deposited by electron-beam evaporation using Kurt Lesker PVD under high vacuum (2.5×10^{-6} Torr) on top of the GaN layer. The deposited Schottky material consists of 5 nm of Ni followed by 5 nm of Au. A standard photoresist lift-off process was used to pattern the Schottky electrodes. Test structures with Ni/Au bilayers were annealed at 500 °C for 10 minutes in air to "oxidize" the Schottky electrodes. Finally, a thick 50 nm Ni / 400 nm Ag pads were deposited on the Schottky electrodes for probe testing

3 RESULTS AND DISCUSSION

3.1 Current-Voltage Characteristic Curves

The I-V relation for a metal/semiconductor diode, based on the thermionic emission (TE) theory and including the bias dependency of the ideality factor can be expressed as

$$I = I_s \exp\left(\frac{qV}{\beta(V)kT}\right)\left[1-\exp\left(\frac{-qV}{kT}\right)\right] \quad (1\text{-a})$$

and when considering $V > 3kT/q$

$$I = I_s \exp\left(\frac{qV}{\beta(V)kT}\right) \quad (1\text{-b})$$

Here I is the measured current, V is the applied voltage, q is the electron charge, $\beta(V)$ is the ideality factor as a function of applied voltage which describes the deviation from the ideal diode equation, k is Boltzmann's constant, T is the absolute temperature in Kelvin, I_s is the saturation current derived from the straight line intercept of the *ln I-V* plot at zero-bias and is given by

$$I_s = AA^*T^2 \exp\left(-\frac{q\phi_{bo}}{kT}\right) \quad (2)$$

Where A is the Schottky contact area, A^* is the effective Richardson constant and Φ_{bo} is the apparent barrier height at zero-bias.

Fig.1 shows the I-V characteristics of (a) Ni and (b) oxidized Ni-Au Schottky contact on n-type GaN in the temperature range of (160 - 400 K) with a step of 10 K. The figure shows that there is an increase in the saturation current as the temperature is increased. For Ni-only the saturation current is found to vary from 2.00E-21 A at 160 K to 3.18E-08 A at 400 K. On the other hand, for oxidized Ni-Au Schottky contact, the saturation current values vary from 4.00E-19 A at 160 K to 2.00E-12 A at 400K. This shows that the Ni material is more temperature sensitive than the oxidized Ni-Au material.

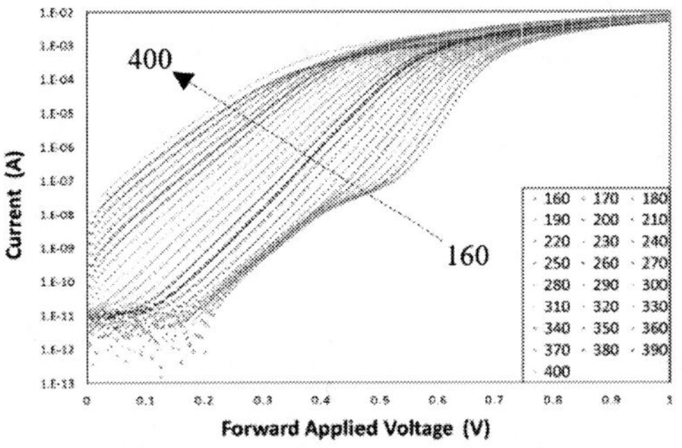

(a)

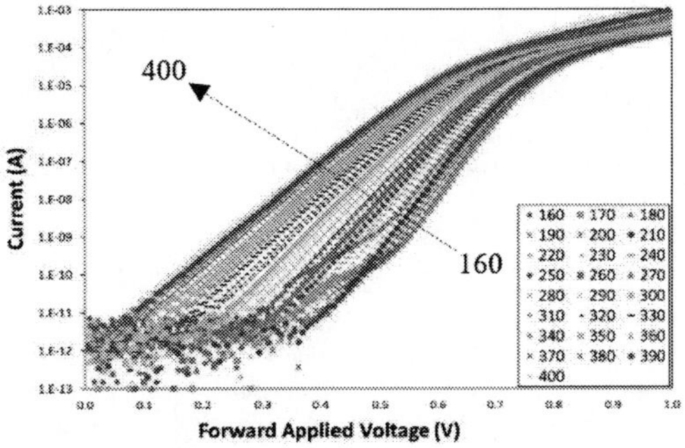

(b)

Fig.1: Semi-logarithmic forward bias current-voltage characteristics of **(a)** Ni-only and **(b)** oxidized Ni-Au on n-type GaN Schottky barrier diode in the temperature range of 160K-400K.

3.2 Temperature and Bias Dependence Schottky Diode Parameters

The semi-logarithmic forward bias current-voltage characteristics (left) and power exponent-voltage (right) of both Ni only and oxidized Ni-Au on n-type GaN is presented in **Fig.2.** The power exponent α is the slope of the logarithmic values of the I-V characteristic and is presented as follow [16,17]

$$\alpha = \frac{d(\ln I)}{d(\ln V)} \quad (3)$$

The α parameter is calculated from **Eq. (3)** and plotted in **Fig. 2**. It can be seen from **Fig.2** that the α-V curve(s) increase linearly reaching a maximum α_m before decreasing and stabilizing at α=1. The influence of the series resistance is insignificant at low bias levels, and therefore, the curve is a measure of the bias independence of β and Φ_b at this level. The peak of the α-V curve marks the point which the series resistance (R_s) has a large effect. Furthermore, the α-V curve is ohmic in the regime where α=1 since the voltage drop across R_s dominates. The usefulness of transforming the experimental I-V characteristics in accordance with **Eq.(3)** was demonstrated in Refs. [21-22]. This formalism enables to trace changes in the current flow mechanisms and to extract device parameters of a single and double injection system as well as metal-semiconductor and P-N diodes whose ideality factor is voltage dependent.

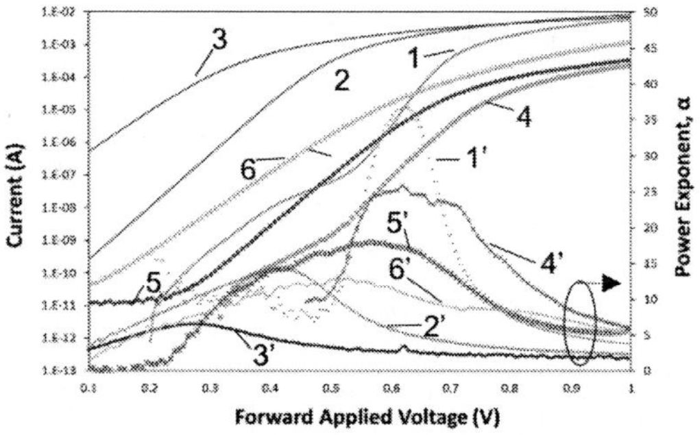

Fig.2: Semi-logarithmic measured I-V characteristics (left) and α-V (right) of both Ni-only [brown (1,1'): 160K, green (2,2'): 300K, dark blue (3,3'): 400K] and oxidized Ni-Au [blue (4,4'): 160K, red (5,5'): 300K, yellow (6,6'): 400K] on n-type GaN Schottky barrier device.

The dependence of γ on voltage was extracted from α-V curves by **Eq.(4)** to construct **Fig.3** for both Ni-only and oxidized Ni-Au TCOs on n-type GaN sample. It can be seen that at relatively large voltages, there is a super-linear increase in the γ-V curve(s) while in the case of a pure thermionic emission theory it equals 1. In the case of including the effect of Rs, it can be seen that the γ-V curve decreases with increasing Rs. The characteristic behavior of γ can be used to distinguish between the different current mechanisms as well.

$$\gamma = \frac{d(\ln \alpha)}{d(\ln V)} \tag{4}$$

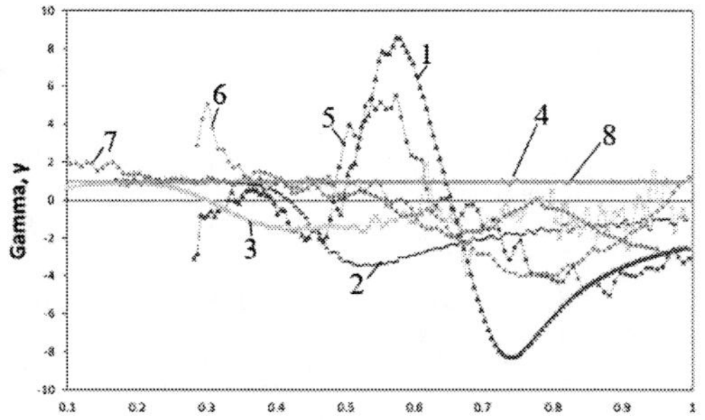

Fig.3: γ-V plot for both Ni-only [dark blue (1): 160K, red (2): 300K, yellow (3): 400K] and oxidized Ni-Au (purple (5): 160K, green (6): 300K, light blue (7): 400K). The constant lines (grey 4, pink 8) represent (Rs=0).

Fig.4 shows a plot of ideality factor with respect to the bias voltage at different temperatures. The voltage dependence ideality factor can be calculated using one of the following equations **Eq.(5a)** or **Eq.(5b)** [16,17]

$$\beta = \frac{1}{1 - 3\alpha \dfrac{kT(1 - \dfrac{qV}{\alpha kT})^2}{qV(\gamma - 1)}} \tag{5-a}$$

Or,

$$\beta = \frac{qV}{kT}(1 - \frac{IR_s}{V}) \frac{1}{\int_{V_o = 3(kT/q)}^{V} \frac{\alpha}{V} dV} \tag{5-b}$$

It is clearly shown from the figure that the ideality factor is changing as a function of voltage at different temperatures and it is not equal to 1. One explanation for this deviation of β from unity is due to image forces acting on the barrier height Φ_b and interface states or trap localized in the unavoidable native oxide covering the semiconductor which are populated or released with applied voltage, thereby, modifying the effective barrier height.

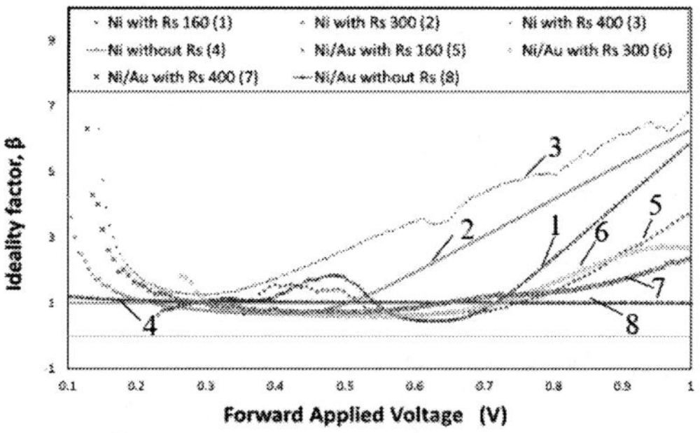

Fig.4: Plot of the ideality factor for both Ni-only [(1): 160K, (2): 300K, (3): 400K, (4): Rs=0] and oxidized Ni-Au [(5): 160K, (6): 300K, (7): 400K, (8): Rs=0] Schottky contact on n-type GaN at differnet temperatures as a function of voltage.

Next, the dependence of saturation current on both bias voltage and temperature can also be seen in **Fig.5**. The figure was constructed using **Eq.(6)**. From **Fig.5**, the saturation current is almost constant for the low-moderate voltage ranges while it starts to decrease as the bias is increased. This may be due to a decrease in the recombination process due to the fact that the series resistance is dominating [17].

$$I_s = \frac{I}{\exp \dfrac{qV}{kT} \left[1 - 3\alpha \dfrac{kT(1 - \dfrac{qV}{\alpha kT})^2}{qV(\gamma - 1)} \right]} \tag{6}$$

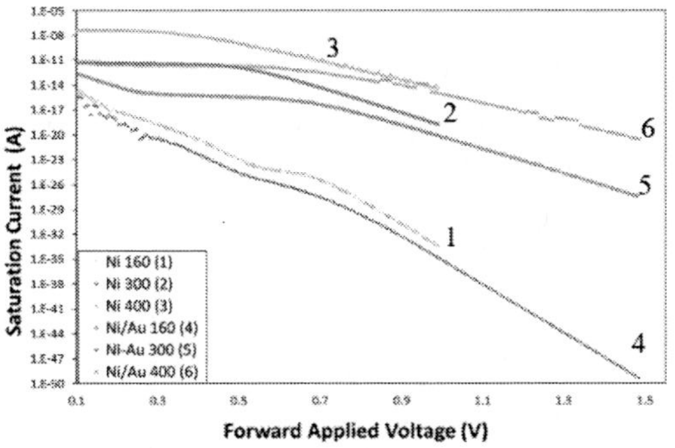

Fig.5: Bias dependence of Saturation current for Ni-only [grey (1): 160K, blue(2): 300K, yellow(3): 400K] and oxidized Ni-Au [red(4): 160K, brown(5): 300K, green(6): 400K] on GaN Schottky diode.

Furthermore, the zero-bias barrier height can be calculated using the following equation [17].

$$\phi_{bo} = \text{kT} \ln\left(\frac{\text{I}_s}{\text{A}^* A T^2}\right) \qquad (7)$$

The barrier dependence on both bias voltage and temperature is shown in **Fig.6**. From the figure, it is clear that the zero-bias barrier height is increasing with the voltage. This may be due to the inhomogeneity of the barrier height.

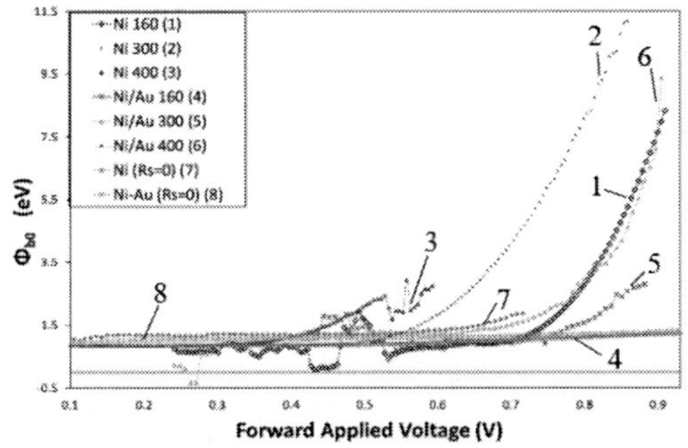

Fig.6: Plot of zero-bias barrier height for both Ni-only [(1): 160K, (2): 300K, (3): 400K, (4): Rs=0] and oxidized Ni-Au [(5): 160K, (6): 300K, (7): 400K, (8): Rs=0] Schottky contacts on n-type GaN at differnt temperatures as a function of voltage.

3.3 The Effect of Temperature on the Series Resistance

Three series resistance extraction methods were used to investigate the effect of temperature on the series resistance of both Ni-only and oxidized Ni-Au Schottky contacts on n-type GaN diode. The first method is called the Direct method where the series resistance was extracted from the measured semi-log forward bias I-V assuming thermionic emission. The I-V relationship of the Schottky diode with series resistance can be expressed as [18]:

$$I = I_s \exp\left[\frac{q(V - IR_{s,direct})}{\eta K T}\right] \qquad (8)$$

Where the $IR_{s,direct}$ term is the voltage drop across series resistance of the device. **Eq.(9)** was used as the second method for extracting the series resistance from the plot of dV-dLn(I). The equation can be expressed as follows [18].

$$\frac{dV}{dLn(I)} = IR_{s-derivative} + \eta\left(\frac{KT}{q}\right) \qquad (9)$$

To further verify the previous two methods, Cheung's method was used as a third method for extracting our series resistance and investigating the effect of temperature on the series resistance, see **Eq.(10)** and **Eq.(11)**

$$H(I) = V - \eta\left(\frac{KT}{q}\right) Ln\left(\frac{I}{AA^*T^2}\right) \qquad (10)$$

$$H(I) = IR_{s-cheung} + \eta\phi_{BO} \qquad (11)$$

It is clearly shown in **Fig.7**. that the value of series resistance is decreasing with the increase of temperature for both Ni-only and oxidized Ni-Au on n-type GaN and a good match between the three extraction methods can clearly be seen in the figure. For Ni-only on top of n-type GaN, a value of 58.54Ω at 160K and 31.25Ω at 400K was obtained. However, for oxidized Ni-Au on top of n-type GaN, a value of 735.8Ω at 160K and 139.94Ω at 400K was obtained. These values were obtained from the Direct method.

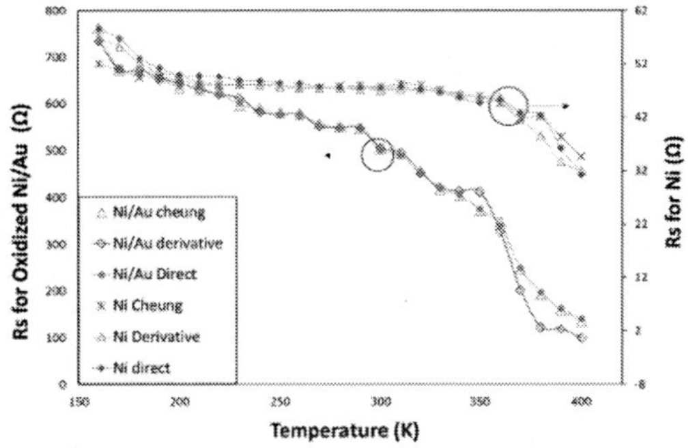

Fig.7: Temperature dependence of the extracted series resistance for both Ni-only and oxidized Ni-Au on n-type GaN Schottky diode in the temperature range of 160-400 K using three different extraction methods.

Table 1: Series resistance values extracted using direct, H(I), and dV/dln(I) methods for values of 160 K and 400 K.

Material \ Method		Direct (Ω)	H(I) (Ω)	dV/dln(I) (Ω)
Ni	160 K	58.54	52.04	56.951
	400 K	31.255	34.63	31.98
Ni/Au	160 K	735.8	763.11	734.98
	400 K	139.94	134.57	99.08

Table.1 shows the extracted values of the series resistance at the temperature boundaries used in this article (160K and 400K). Values are almost matched for the three methods used in the extraction of the resistance R_s. As temperature increases, it is obvious that the series resistance for the oxidized Ni-Au Schottky contact decreases more rapidly than the series resistance of Ni-only.

4 CONCLUSIONS:

The forward current-voltage (I-V) characteristics of both materials oxidized Ni-Au and Ni-only Schottky contacts on n-type GaN were investigated in the temperature range 160 K-400 K with a step of 10 K. A technique that makes use of the voltage-dependent differential slope curve $\alpha(V) = [d\ln(I)/d\ln(V)]$ has been applied to the sample to check the dependency of the diode parameters on the bias voltage. Through this method, the diode parameters would be extracted more properly than other methods. The extracted parameters include the saturation current (zero bias barrier height), the voltage dependence of the barrier height and of the ideality factor as well as series resistance. The ideality factor (β) changed with respect to the voltage and it was not equal to 1. low-moderate voltage values led to a constant saturation current that started to deviate as the series resistance began to dominate. The inhomogeneity of the barrier height resulted in an increase in the barrier with respect to the bias voltage. Furthermore, the series resistance (R_s) values estimated from three different methods (Cheung's method, Direct method, and $[dV/d\ln(I)]$ method are seen to be strongly temperature dependent.. As temperature increases, the series resistance for the oxidized Ni-Au Schottky contact decreases more rapidly than the series resistance of Ni-only.

5 REFERENCES

[1] E. H. Rhoderick and R. H. Williams, "Metal-Semiconductor Contacts," 2nd Edition, Clarendon Press, Oxford, 1988.

[2] L.C. Chen, C. Y. Hsu, W. H. Lan, S. Y. Teng-Solid State Electron. 47, 1843 (2003).

[3] L. B. Flannery, I. Harrison, D. E. Lacklison, R. I.Dykeman, T. S. Cheng, C. T. Foxon-Mater. Sci. Eng. B 50, 307 (1997).

[4] J. Moon, M, Micovic, A. Kudoghlian, P. Janke, P.Hashimoto, W. Wong, L. McCray, C. Nguyen-IEEE Electron. Devices Lett. 23, 637 (2002).

[5] D. Mistele- Mater. Sci. Eng. B 93, 107 (2002).

[6] H. Amano, M. Kito, X. Hiramatsu, I. Akasaki - Jpn. J.Appl. Phys. 28, L2112 (1998).

[7] R. Khanna, S. J. Pearton, F. Ren, I. Kravchenko, C. J. Kao, G. C. Chin- Appl. Phys. Lett. 87, 052110 (2005).

[8] A. C. Schmitz, A. T. Ping, M. Asif Khan, Q. Chen, J.W. Yang, I. Adesida- Semicond. Sci. Technol. 11, 1464 (1996).

[9] E. V. Kalinina, N. J. Kurnestov, V. A. Dmitreiev, K.G. Irvine, C. H.Carter-Jr., J. Electron. Mater. 25, 831 (1996).

[10] J. S. Foresi, T. D. Moustakas-Appl. Phys. Lett. 62, 2859 (1993).

[11] S. Dogan, S. Duman, B. Gurbulak, S. Tuezemen, H. Morkoc, Phys. E. 41, 646-651 2009.

[12] J. D. Guo, F. M. Pan, M. S. Feng, R. J. Guo, P. F. Chou, C. Y. Chang, J. Appl. Phys. 80(3) 1996.

[13] P. Hacke, T. Detchprohm, K. Hiramatsu, and N. Sawaki, Appl. Phys. Lett. 63 (2676)1993.

[14] D. Donoval, V. Kulikov, P. Beno, and J. Racko, ASDAM 2002 Smolenice Castle, Slovakia, October 2002.

[15] A. C. Schmitz, A. T. Ping, M. Asif. Khan, Q. Chen, J. W. Yang, and I. Adesida, Electron. Lett. 32 (19) 1996.

[16] V. Mikhelashvili & G. Eisenstein-J. Appl. Phys. 86, 6965 (1999).

[17] V. Mikhelashvili, G. Eisenstein & R. Uzdin, Solid-State Electronics, 45(2001), 143-148.

[18] S.K. Cheung and N.W. Cheung-Appl. Phys. Lett. **49**, 85 (1993).

[19] M. K. Hudait, P. Venkateswarlu & S. B. Krupanidhi Solid-State Electronics, 45(2001), 133-141.

[20] S. Chand, J. Kumar- J. Appl. Phys. 80, 288 (1996)

[21] Zyuganov AN, Svechnikov SV. Injection contact phenomena in semiconductors. Kiev: Naukova Dumka; 1981 [in Russian].

[22] Mikhelashvili V, Eisenstein G, Garber V, Fainleib C, Bahir G, Ritter D, Orenstein M, Peer A. J Appl Phys 1999; 85:6873±83.

SELF-CONSISTENT EVALUATION OF OPTICAL PATH LENGTH FACTOR, Z, IN SI SOLAR CELLS

Luigi Abenante and Massimo Izzi
ENEA, Italian National Agency for New Technologies, Energy, and Sustainable Economic Development
Via Anguillarese 301, 00123, Roma, Italy

ABSTRACT: An expression for cell collection efficiency, η_c, as a function of rear-surface internal reflectance, R_B, is presented, which derives from the Basore expression for internal quantum efficiency, *IQE*, in Si solar cells. Using the expression for η_c together with the Rand and Basore expression for effective optical path-length factor, *Z*, provides a calculation approach for a self-consistent evaluation of R_B, η_c, and *Z* in textured cells, where η_c and *Z* are determined by R_B. The calculation approach is applied in two reported Si solar cells and checked by using the numerical simulation program PC1D.
Keywords: Light Trapping, Simulation, Solar Cell.

1 INTRODUCTION

Light trapping in solar cells extends the optical path length of weakly absorbed light beyond the physical thickness, *w*, of the device. In [1], a wavelength-dependent figure of merit for light trapping has been presented, which is measured in multiples of *w*. This figure of merit is called effective optical path-length factor, *Z* [1].

In [1], a general analytical expression for this quantity is presented, which includes the ratio IQE/η_c. *Z* therefore depends on external reflectance and internal optical parameters. Internal optical parameters are: propagation angle of internal light with respect to the device normal, θ, rear-surface internal reflectance, R_B, and front-surface internal reflectance, R_{FI}.

In [2], a sensitivity analysis is performed on a complicate expression for *IQE*, which includes multiple values for internal optical parameters, which reveals that only θ_n and $R_{B,n}$, which are θ and R_B at the third and successive passes of light across a device, are important to determine *IQE*. Based on this result, we have set the aforementioned expression for *IQE* at constant internal optical parameters and obtained an expression for η_c as a function of *IQE*, θ, and R_B to be substituted in the aforementioned general expression for *Z* [1]. Since $\theta=\theta_n=60°$ [2], we find that *Z* can be calculated in textured cells at all wavelengths by only evaluating R_B, provided that curves of external reflectance and *IQE* are available. Arbitrary assignments to R_B are avoided because R_B is determined such that, at weak absorption, η_c is independent of wavelength, as expected physically [1-2].

We have evaluated analytically effective R_B, η_c, and *Z* in two textured Si solar cells reported in the literature [3-4] by using digitized data of *IQE* and external reflectance. The effective R_B-values and η_c-curves obtained analytically are subsequently confirmed numerically with the simulation program PC1D [3].

In conclusion, in this work, a calculation approach for a reliable self-consistent and physically consistent simultaneous evaluation of R_B, η_c, and *Z* in textured Si solar cells is presented, where the evaluation of R_B is crucial.

2 THEORY

After [1], at each wavelength, λ, *IQE* can be expressed as

$$IQE = \eta_c[1 - \exp(-\alpha Zw)] \times \frac{1 - R_f}{1 - R} \tag{1}$$

where α is the band-to-band absorption coefficient, *R* is the total external reflectance, and R_f is the extrapolated external front-surface reflectance [1].

From (1) one obtains the general expression for *Z* [1]

$$Z = -\frac{1}{\alpha w}\ln\left[1 - \frac{1 - R}{1 - R_f} \times \frac{IQE}{\eta_c}\right] \tag{2}$$

In [2], an analysis of the sensitivity of *IQE* to parameters is performed exploiting a complicate expression for *IQE*, which accommodates multiple values for internal optical parameters [2]. The conclusion of the analysis is that *IQE* mostly depends on constant values of R_B and θ. These values are those assigned at the third and successive optical passes across the device [2]. This implies that $\theta=60°$.

At constant values for internal optical parameters the complicate expression for *IQE* of [2] reduces to

$$IQE = \eta_c \frac{(1 - T)(1 + R_B T)}{1 - R_B T^2} \tag{3}$$

where $T=\exp(-\alpha w/\cos\theta)$ is the internal optical transmittance. From (3) we obtain

$$\eta_c = IQE \times \frac{1 - R_B T^2}{(1 - T)(1 + R_B T)} \tag{4}$$

Substituting (4) into (2) shows that *Z* only depends on θ and R_B. As a result, using (2) together with (4) allows calculating *Z* at all λ-values in a textured cell, whose curves for *IQE*, *R*, and R_f are available, by only assigning values to R_B.

The right member of (4) is the product of two functions with opposite monotony. R_B can be consequently determined such that (4) gives η_c~constant when αw approaches zero, as required in [1-2].

3 APPLICATION

3.1 Experimental

Effective R_B, η_c, and *Z* were evaluated by applying the approach outlined in the previous Section in two reported Si solar cells: a 170-µm $p^+/n/n^+$ PERT cell A [3] and a 180-µm $n^+/p/p^+$ PERC cell B [4]. Cell A has CZ-Si bulk resistivity, ρ_B, equal to 3 Ωcm and emitter sheet resistance, ρ_{SHEET}, equal to 55 Ω/sq [3]. Cell B has mc-Si ρ_B between 1 and 3 Ωcm and ρ_{SHEET}=50 Ω/sq [4].

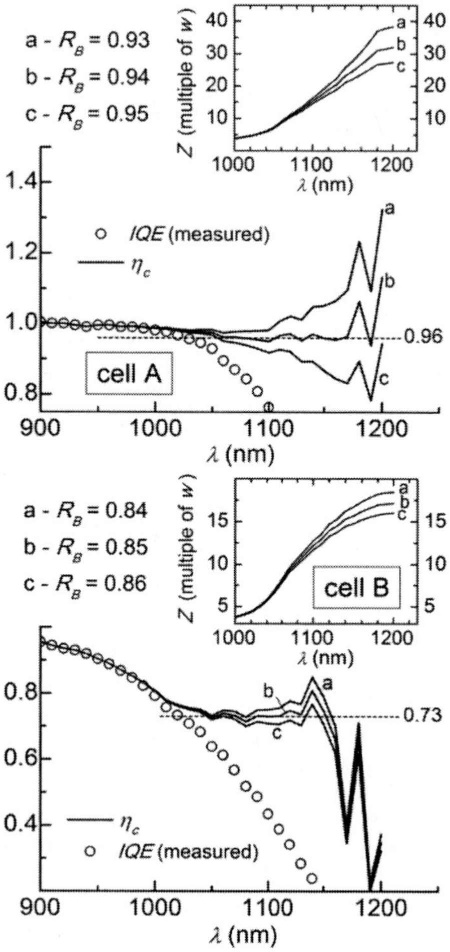

Figure 1: Examples of physically consistent evaluation of effective R_B, η_c, and Z in actual textured Si solar cells A and B. As can be seen, the near-bandgap monotony of the η_c-curves varies with R_B. In each cell, an R_B-value can be found, for which the η_c-curve can be fitted by the shown near-bandgap η_c-value. η_c-curves are calculated with (4) and Z-curves with (2) using the reflectance curves in Fig. 2 and the relevant measured IQE-curve shown in both this figure and Fig. 2.

Consistent with the previous Section, calculations were performed at θ=60°. Results are shown in Fig. 1 together with portions of measured IQE-curves. The whole measured curves of IQE of the two cells are reported in Fig. 2 together with the measured R-curves and relevant extrapolated R_f-curves.

In Fig. 1, it is shown that, notwithstanding the perturbations introduced by the measured IQE-curves, for each cell, an effective R_B-value can be determined (R_B-value b), for which (4) gives, as expected, η_c-constant at most near-bandgap wavelengths (η_c-curves b). At these wavelengths, in fact, as reported in Fig. 1, η_c-curves b can be fitted by constant η_c-values. Consequently, according to the criterion proposed in the previous Section, in each cell, light trapping is evaluated non-arbitrarily by the Z-curve b alone.

It is worth remarking that Z-curves in Fig. 2 are not perturbed because, after substituting (4), (2) is only a function of θ and effective R_B.

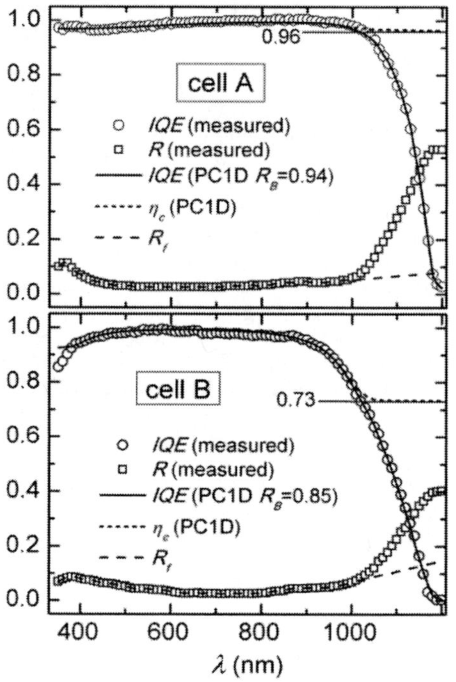

Figure 2: Measured data of IQE and R for cells A and B as digitized from [3] and [4], respectively, and relevant extrapolated R_f-curves are shown along with fit-curves to IQE and relevant η_c-curves calculated with PC1D by assigning the reported R_B-values.

3.2 Theoretical

Eq. (3) was obtained with the IQE-formula in [2], which is relevant to T_{inc}=1–R_f and R_{FI}≤1. It can be verified that Eq. (3) can be obtained as well with

$$IQE = \eta_c \frac{A}{1-R} \qquad (5)$$

by using the Basore formula for absorptance A [2] at T_{inc}=1–R,

θ and R_B constant, and R_{FI}=1 at all optical passes. This allows fitting quantum efficiency without adjusting R_{FI}.

4 CHECK

The application in subsection 3.2 was implemented in PC1D v. 5.9 [5] by assigning R_{FI}=1 at all optical passes and T_{inc}=1–R. In addition, the surface-texture facet-angle was adjusted so as to impose θ=60° at all passes of trapped light. Under these conditions, the b-values of R_B reported in Fig. 1 were assigned and IQE-data were fitted by adjusting transport parameters and emitter doping profiles. While, for cell A, a Gaussian emitter doping profile could be assigned having peak doping of 2×10^{19} cm^{-3}, junction depth of 2.72 μm, and ρ_{SHEET}=55 Ω/sq; for cell B, it was necessary assigning a 0.3 μm-deep uniform doping profile of 5.162×10^{19} cm^{-3} having ρ_{SHEET}=50 Ω/sq. The resulting PC1D fit-curves and IQE-data are shown with full lines in Fig. 2. In the same Fig. 2, the corresponding η_c-curves calculated with PC1D are shown with short-dashed lines. As can be seen, η_c-curves calculated with PC1D and η_c-curves calculated with (4) agree.

5 CONCLUSIONS

Modeling light trapping in a textured solar cell at internal optical parameters that do not vary with optical passes across the device is consistent with the sensitivity analysis performed in [2]. This has allowed expressing η_c and Z as functions of constant R_B.

A self-sufficient and self-consistent calculation approach has been consequently set up that can be reliably used to calculate Z, and hence, evaluate light trapping in a given textured Si solar cell by assigning non-arbitrary values to R_B provided that data of external reflectance and *IQE* are available.

References

[1] J. A. Rand, P. A. Basore, Conference Record of the Twenty Second IEEE Photovoltaic Specialists Conference, vol. I (1991) 192.

[2] P. A. Basore, Conference Record of the Twenty Third IEEE Photovoltaic Specialists Conference (1993) 147;
M. A. Green, Silicon Solar Cells: Advanced Principles & Practice, University of New South Wales, (1995) Appendix G.

[3] Jian Sheng, Wei Wang, Shengzhao Yuan, Wenhao Cai, Yun Sheng, Yifeng Chen, Jiangning Ding, Ningyi Yuan, Zhiqiang Feng, Pierre J. Verlinden, Sol. Energy Mater. Sol. Cells, vol. 152 (2016) 59.

[4] W. Deng, D. Chen, Z. Xiong, P. J. Verlinden, S. Member, J. Dong, F. Ye, H. Li, H. Zhu, M. Zhong, Y. Yang, Y. Chen, Z. Feng, and P. Altermatt, IEEE Journal of Photovoltaics, vol. 6, no. 1 (2016) 3.

[5] P. A. Basore, D. A. Clugston, PC1D version 5.9, University of New South Wales (2003).

ON THE EFFECT OF In, P SURFACTANTS ON THE GaAs PV CELL FORMATION

A.S. Vlasov[1*], L.B. Karlina[1], B.Y. Ber[1], F.E. Komissarenko[2], D.Yu. Kazantsev[1], N.Kh. Timoshina[1], M.M. Kulagina[1], A.B. Smirnov[1]
[1] Ioffe Institute, 26 Polytechnicheskaya, St.Petersburg, Russia
[2] ITMO University, 49 Kronverkskiy, St.Petersburg, Russia
*e-mail: vlasov@scell.ioffe.ru

ABSTRACT: Influence of Indium and Phosphorous on the GaAs photovoltaic cells has been studied. In, P diffusion was made from the gas phase in hydrogen flow atmosphere. The samples were characterized by secondary ion mass spectroscopy (SIMS), Raman spectroscopy, scanning electron microscopy (SEM) and atomic force microscopy (AFM). The p-n junction was formed by Zn diffusion from the gas phase. SERS experiment with Au nanoparticles revealed a strong reduction of As cluster concentration in subsurface region after annealing in In, P vapours. At the same time In-rich clusters are found in the annealed samples. A comparative study of PV cells prepared with and without In, P treatment was performed. It was found, that despite the higher Zn concentration in the emitter, surfactant-induced modification of the surface results in higher external quantum efficiency at short wavelength region. Also it is found, that p-n junction formed in the presence of surfactants reveals higher open circuit voltage and fill factor values of the I-V curve. As a result, 15-20% relative PV cell efficiency increase was observed with the In, P diffused samples.

Keywords: Diffusion, gallium arsenide based cells, characterisation

1 INTRODUCTION

Zn diffusion in III-V materials is still used for solar and laser PV converters [1-3]. Usually optimization of the diffusion process is focused on the source selection (e.g. use of Ga+Zn source [4]), impact of oxide layers on the diffusion [5], surface roughness [6] etc.

Diffusion method suggested by our group earlier [7] simplifies the Zn diffusion process: it allows one-stage processing, reduces nonradiative recombination in the bulk of the p-emitter and on the surface even at high doping levels. No "dead" layer etching is necessary. Emitters in PV structures should meet several conditions. First, the p-n junction depth must provide necessary lateral conductance without degradation of the spectral photoresponse function. Recombination decrease at high doping level allows to obtain Ohmic contacts directly on the emitter surface. In this report we provide a comparative study of the PV structures characterization and describe the possible origin for the observed modifications, caused by annealing in In, P vapours.

2 EXPERIMENTAL

Figure 1 presents SIMS profiles of three elements: In, P and Zn. The samples are: epi-ready GaAs substrate (fig. 1, a) and the same substrate with preliminary In, P treatment (fig. 1, b). Zn-Sn and In-P-Sn melts were used sources of Zn and In, P vapours. Diffusion was done for 1 hour at 670 ^{0}C. Zn diffusion was done simultaneously. Epi-ready Si-doped (n=2x10^{16}-3x10^{18} cm^{-3}) 6^0 inclined GaAs substrates were used.

It can be seen from figure 1, that the maximum Zn concentration and diffusion rate differ inspite of the fact that both samples were treated at the same conditions. Previously we have shown, that In controlls the Zn diffusion rate, while P is responsible for the effective surface passivation [7]. It should be noted, that In is observed only in the subsurface region, i.e. it is responsible for the diffusion mechanism, but there's no direct impact of neither In nor P on the p-n junction properties. It was found that Zn diffusion rate depends also on the substrate doping level (fig. 2), however the influence of In, P treatment appears to be similar: the diffusion rate in In, P doped samples was approx. 2-3 times higher (diffusion conditions for the samples in figure 2 are the same as those, shown in figure 1).

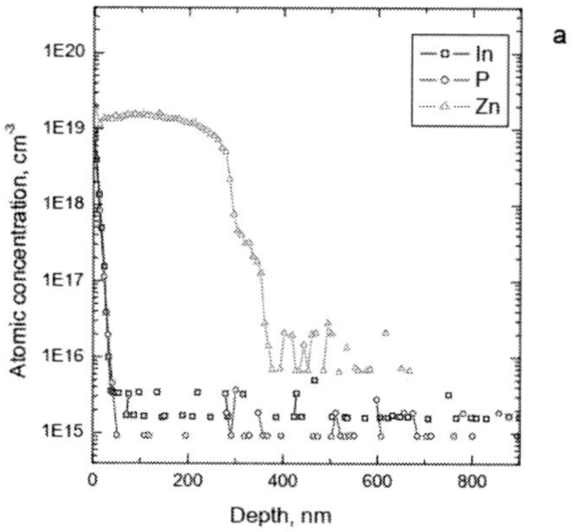

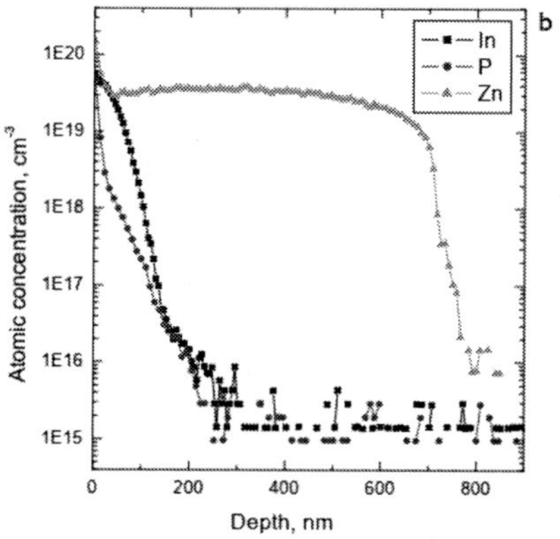

Fig.1 SIMS profiles of Zn (triangles), In (squares) and P (circles) after Zn diffusion in n-GaAs (n=3x10^{16} cm^{-3}).

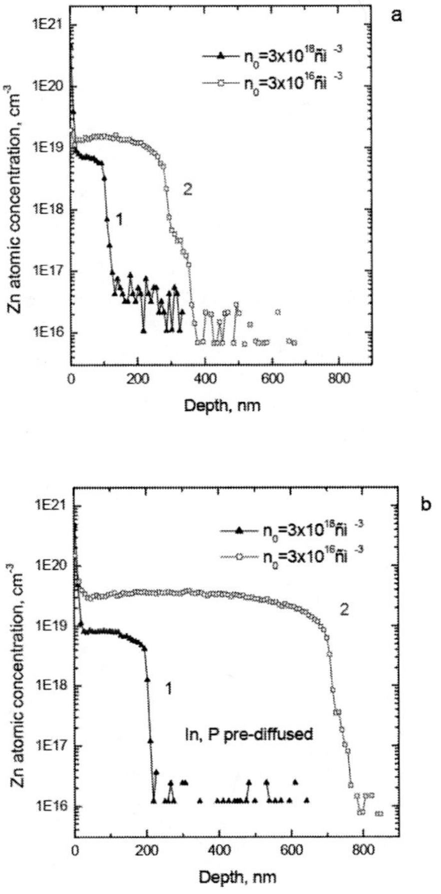

Fig. 2. SIMS profiles of Zn in GaAs with different doping level: triangles - n=3x10^{18} cm^{-1}, squares – n=3x10^{16} cm^{-3}. a – epi-ready substrates, b – after annealing in In, P vapours.

GaAs surface before and after annealing was examined by a SERS experiment. The samples were covered by Au nanoparticles and then the Raman spectra were recorded. Nanoparticles were obtained by e-beam evaporation of gold at low sweep rate (which means high local temperature under the beam). Au particles of few nanometers in diameter interact with incoming laser beam and produce secondary electro-magnetic waves, which intensity appears to be extremely strong in the vicinity of such a small particle. Thus, the intensity of Raman signal of vibrations of atoms situated close to Au particles might be enhanced. The RS spectra (shown in fig. 3) were measured in backscattering geometry in two configurations: z(xy)z (black) and z(xx)z (red).

In figure 3, a one can see the spectra of a reference (untreated) GaAs sample covered with Au particles. Dotted line presents the Raman spectrum of the same sample without Au. Initially the Raman spectrum represents only longitudinal mode at 268 (phonon-plasmon mode) and 292 cm^{-1} (depletion layer mode). Au particles activate As-As bond type vibration originating from As clusters. These arsenic clusters have trigonal R-3m structure. The modes at 200 and 257 cm^{-1} correspond to E_g type (active in (xy) configuration) and $A1_g$ type (active in (xx) configuration) vibrations. After annealing in In, P vapors (the sample is annealed without Au particles and Au particles are inserted simultaneously in the same process as the reference sample) the intensity of As-As bond type vibration is strongly decreased (fig 3, b) evidencing that the amount of As clusters is reduced during the annealing process. At some annealing time-temperature conditions this mode can totally disappear.

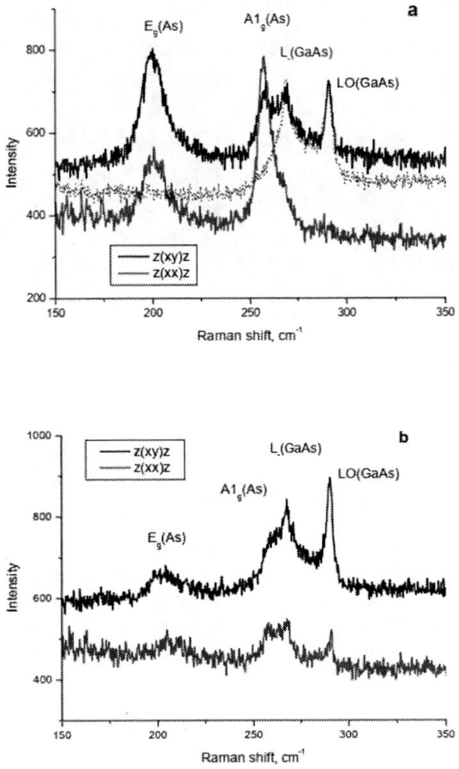

Fig.3. Raman spectra of Au nanoparticles coated GaAs. (a) – reference substrate, (b) – after In, P diffusion.

Figure 4 presents an SEM image of the annealed sample obtained by the backscattered electron (in lens) detector. SEM image was recorded on the Zeiss Neon 40 EsB setup working with accelerating voltage of 5kV. The penetration depth of electrons at such a voltage is ~15 nm. The image in backscattered electrons is sensitive to chemical composition: heavier atoms reflect the incident electrons more efficient. It can be seen, that there are bright spots of about 10 nm in diameter distributed randomly along the surface. According to the experiment conditions, bright spots can be associated with In or Sn aggregation. SIMS measurements show that In concentration at the surface is at least an order of magnitude higher than Sn one. Thus, we can conclude that bright spots in the SEM image can be identified as In-rich clusters. Since in the Raman scattering experiment we observe the reduction of As cluster concentration, we conclude that during the annealing process these clusters might be the centers for In incorporation. Initially As clusters cannot be seen in SEM since the atomic numbers of As and Ga are close. We can also suppose that these clusters and their content are responsible for the observed differences in Zn diffusion rates. Analysis of the Raman spectra of the samples after annealing reveal considerable shifts of the depletion layer mode, which might indicate on high In concentration (up to several percent) in the subsurface layer.

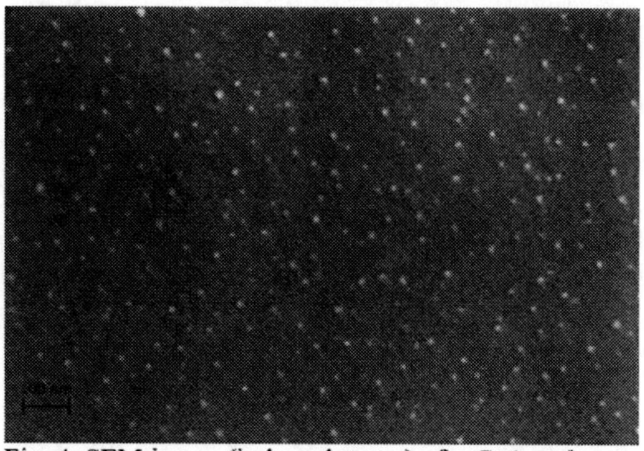

Fig. 4. SEM image (in lens detector) of a GaAs substrate after annealing.

Images of the same surface obtained by the Everhart-Thornley detector do not exhibit any change in the morphology. This means that these adatoms are incorporated in the lattice. To check this additionally, we have examined the surface by an atomic force microscopy (AFM). In all cases, the annealing leaded to a slight flattening of the surface: the RMS (root mean squared) roughness of all the samples was registered to be 0.2-0.3 nm. Thus AFM measurements confirmed, that GaAs surface do not degrade after annealing, no new structures are formed, and Indium atoms are incorporated in the crystal lattice.

PV cells

It is well known, that p-n junction depth is responsible for the spectral sensitivity, lateral conductance and to some extent I-V curve shape. One always has to find a compromise between these parameters when configuring high power (concentrator or laser power converter) devices. Figure 5 presents spectral characteristics of two devices obtained in the same conditions of Zn diffusion with and without In, P treatment. Inspite of the fact, that presence of Indium on the GaAs surface increases the diffusion length (i.e. the p-n junction is deeper), it is seen, that short wavelength side of the photoresponce function is higher for the In, P treated sample. This indicates on higher mobility and good surface passivation in this case. Surface passivation was also registered in photoluminescence spectra of the In, P treated samples.

I-V curves of the devices with In, P diffusion also reveal better performance in all studied current density range (see figure 6). Open circle voltage and Fill factor are higher with rhese devices. It should be noted that current research was made without any additional treatment (e.g. surface passivation, antireflection catings) to exclude any uncertainty effect, so that only influence of In, P treatment on the Zn diffusion and resulting device properties is considered .

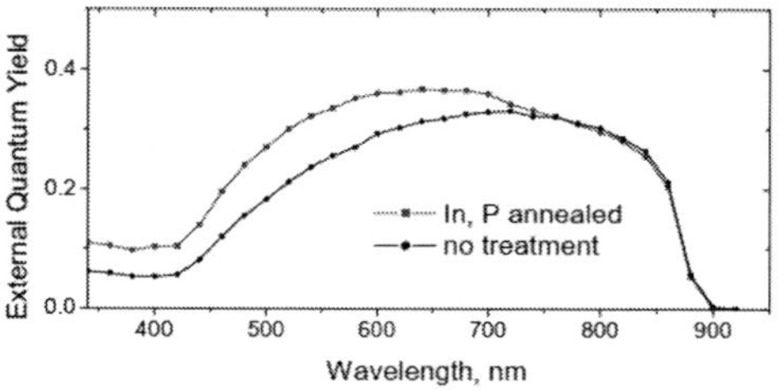

Fig 5. External quantum yield spectral dependence og GaAs PV cells. Squares – reference sample d=150 nm, I_{sc}=8,17 mA/cm^2 (AM1.5D 320-940 nm), p=1x10^{19} cм$^{-3}$. Circles – In, P diffused sample d=240 nm, I_{sc}=9.45 mA/cm^2 p=5x10^{19} cм$^{-3}$.

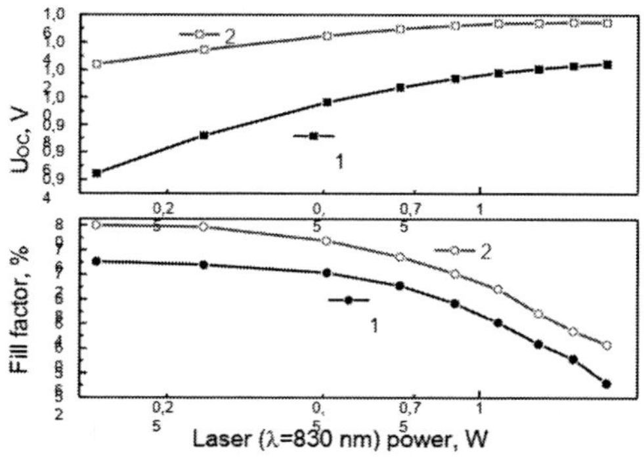

Fig. 6. I-V curve parameters of the cells without (line 1) and with (line 2) surfactant treatment.

7 SUMMARY

A comparative study of the impact of In and P surfactants on the Zn diffusion in GaAs has been performed. Strong dependence of the Zn diffusion rate on both surfactant treatment and substrate doping level was found via SIMS measurements. Raman scattering revealed decrease of the number of As clusters followed by appearance of In-rich clusters on the In, P treated surface. Sufficient increase of the PV characteristics of the In, P treated samples was found.

REFERENCES

[1] [1.] S. P.Chang, Advanced Materials Research, **684**, 312, (2013).

[2] N.L. Dmitruk et al., Solar Energy Materials & Solar Cells , **137**,124 (2015)

[3] L. Tang et al., Solar Energy Materials & Solar Cells, **122,** 94 (2014).

[4] H. Ye and Q. Ni , Journal of Materials Science **51**, 15, (2016).

[5] T. Schlegl et al., AIP Conf. Proc. 738, 396 (2004);

[6] A.A. Akopyan et al., Semiconductors , **43** (3) 368 (2009).

[7] L.B. Karlina et al. Journal of Crystal Growth **432** 1338 (2015)

Effect of Cu deficiency on the optical properties of dual ion beam sputtered CZTSSe thin film

Brajendra S. Sengar [1], Vivek Garg [1], Vishnu Awasthi [1], Shailendra Kumar [2] and Shaibal Mukherjee [1,*]
[1]Hybrid Nanodevice Research Group (HNRG), Electrical Engineering, Indian Institute of Technology (IIT) Indore, Indore-453552, India
[2]Raja Ramanna Center for Advanced Technology, Indore-452013, India.
* E-mail: shaibal@iiti.ac.in

ABSTRACT: The near-surface region of thin-film polycrystalline CZTSSe is considered influential because it is the region where the electrical junction forms in a CZTSSe photovoltaic device. Spectroscopic Ellipsometry measurements of polycrystalline CZTSSe films reveal that there is a thin layer at the surface which has different optical and electronic properties from those of the bulk film. This surface layer of thin-film CZTSSe has a larger bandgap. These properties indicate that the surface layer is more Cu deficient than the bulk in the nearly stoichiometric thin-film CZTSSe used in photovoltaic devices. Cu deficiency at the surface is also confirmed by using x-ray photoelectron spectroscopy (XPS). This work provides an insight into the importance of surface layer engineering for photovoltaic device design.
Keywords: Thin Film Solar cell, Kasterite, Sputtering, CZTSSe

1 INTRODUCTION

Thin-film Kasterite polycrystalline $Cu_2ZnSn(S,Se)_4$ (CZTSSe) is used as the absorber layer for high efficiency photovoltaic solar cells. The efficiency of laboratory thin-film CZTSSe solar cells has surpassed 13%. High efficiency solar cells always use slightly Cu-poor. The surface layer of CZTSSe is considered important because it is the region where the electrical junction forms in photovoltaic devices and it plays a crucial role in the device performance. There have been studies on the properties of the surface of CIS/CIGS/CZTSSe materials [1], [2]. It has been found that the surface region and the bulk of CIGS are structurally similar, with no ordered defect chalcopyrite structure observed, but the chemical composition of the surface is slightly different than the composition of the bulk. It has also been reported that the surface of thin CIGS films has wider bandgap than the bulk [3]. An angle resolved x-ray photoelectron spectroscopy study on epitaxial thin CIS films showed that a severe Cu depletion exists on the top of one to two layers [4]. Energy dispersive x-ray spectroscopy and x-ray photoelectron spectroscopy studies on thin-film CIS showed that CIS layers have a Cu-deficient surface composition. Grazing incidence x-ray diffraction analysis showed that there exists a Cu-depleted surface layer in thin CIS films, which is expected to be due to the reconstruction of the surface of slightly Cu-deficient CIS film[5]. However surface analysis of CZTSSe thin films are very rare. In the present work, we have analyzed the surface of dual ion beam sputtered CZTSSe thin films.

2 EXPERIMENTAL TECHNIQUE

CZTSSe thin films with thicknesses of 400 nm are grown on Mo coated Soda Lime Glass substrate at 300°C, by Elettrorava dual ion beam sputtering (DIBS) system. The background pressure in the deposition chamber is $\sim 1\times10^{-8}$ mbar. The deposition of thin films are carried out in Ar atmosphere with a chamber working pressure of 2.4×10^{-4} mbar and with RF power of 68 W. The CZTSSe films growth is carried out using a 4N (99.99%) pure 4-in-diameter stoichiometric single CZTSSe target. More details on the fabrication process can be found elsewhere [6]. XPS measurements are performed by utilizing the beam line of INDUS-1 synchrotron source.

Spectroscopic Ellipsometry (SE) measurement is carried out using J.A. Woolam M-2000D Ellipsometry. Further, data analysis is performed using a three-layer optical model consisting of a Mo as a substrate, CZTSSe layer, and a top surface layer.

3 RESULTS AND DISCUSSIONS

3.1 Composition analysis

Energy dispersive X-ray spectroscopy (EDX) measurement was carried out for the detailed investigation of the elemental ratios of CZTSSe films .The measurement data thus obtained from EDX analysis, as in Table 1. EDX results presented in the table are average values of the EDX spectra measured at different points on the sample. These points were taken over a large area (200 μm × 200 μm), since it was essential to have the data at various points in order to achieve an accurate chemical composition of the samples.

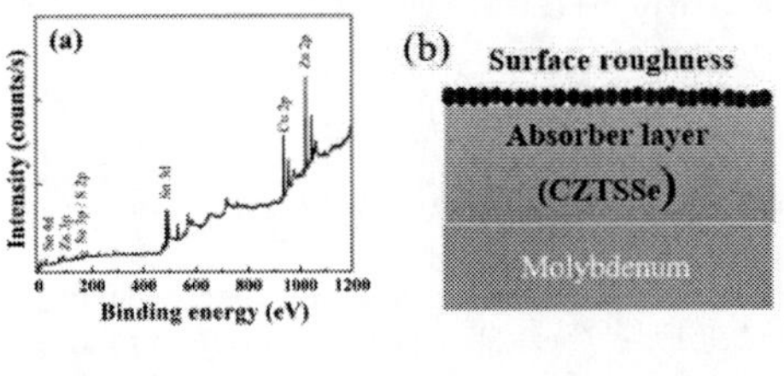

Fig. 1: (a) XPS survey spectra of the CZTSSe Thin film, (b) Multilayer model with a Bruggeman effective medium approximation layer for SE analysis.

We now focus on a comparison of bulk and surface composition of the CZTSSe film. The relative bulk composition as determined by EDX of the original film immediately after growth is 24.15% Cu, 15.30% Zn, 13 .11% Sn, and 49.24% S+Se, and the corresponding Cu/(Zn + Sn) and Zn/Sn ratios are (0.85) and (1.16), respectively. The investigated CZTSSe bulk is thus slightly Cu-poor and Zn-rich in comparison to stoichiometric CZTSSe (Cu:Zn:Sn:S+Se = 2:1:1:4).The surface composition was determined by quantification of the XPS data(as shown in Fig. 1(a)). The XPS derived relative cation compositions are compared to the bulk sensitive EDX data in Table 1. Comparing the XPS with the EDX data, a significant deviation in the relative cation composition can be identified. While the bulk composition derived by EDX is similar to the expected nominal Cu:Zn:Sn:S+Se = 2:1:1:4 stoichiometry of CZTSSe , the XPS data indicates a considerably Cu-poor and (Zn + Sn)-rich chemical surface structure. With respect to the bulk sensitive EDX data and the nominal 2:1:1:4 composition, XPS shows the CZTS surface to contain more Zn and Sn and less Cu.

Table I: Chemical composition, as measured by EDX and XPS of CZTSSe thin film

	Cu	Zn	Sn	S + Se
EDX Atomic Composition (%)	24.15	13.50	13.11	49.24
XPS Atomic Composition (%)	19.30	14.77	16.75	49.18

3.2 Spectroscopic Ellipsometry (SE) Measurement

In order to determine accurate optical functions of CZTSSe thin film, we adopt an effective medium approximation layer to account for the imperfect interface between the sample and the ambient. The optical properties of this surface roughness layer are determined by mixing the optical constants of CZTSSe layer and void [7], [8]. Fig. 1(b) depicts the mathematical schematic model used in the ellipsometric analysis in this study. The model used to calculate ψ and Δ consists of three layers. The first layer is the molybdenum layer because the light source cannot penetrate to the soda-lime glass. The second layer is the CZTSSe, whose model dielectric function is represented by a sum of Tauc Lorentz oscillators. Finally, there is a layer of surface roughness, which is represented using a Bruggeman effective medium approximation (EMA) composed of a mixture of the underlying CZTSSe and void [9], [10].

Fig. 2(a) shows the fit results of ψ and Δ at the incidence angles of 65°, 70° and 75° in the spectral region of 1.2 – 6.4 eV for CZTSSe thin film. We find that the value of mean squared error in the spectral range of 1.2–6.4 eV is improved from 12.12 to 2.09 by replacing the standard surface layer depending on a underlying bulk layer optical constant with a surface layer independent of the bulk optical constants. Fig. 2(b) shows the difference between the calculated fit and the experimental data in the spectrum at angle of 70°. The fit is significantly improved, as shown in Fig. 2(b), when an independent surface layer is used in the surface layer compared, when we used effective medium approximation with the underlying CZTSSe layer with void. Standard deviation is also decreased. Without an independent surface layer, the fit was satisfactory; however, with that, the difference between calculated fits and experimental data is reduced to even within a resolution in ψ (< .1°) in a fundamental bandgap region. Light starts to be absorbed from the bandgap and due to transparency caused by low absorption below the bandgap, Ellipsometry parameters ψ and Δ stay oscillatory with decreasing amplitude until there is no more back reflection in the sample.

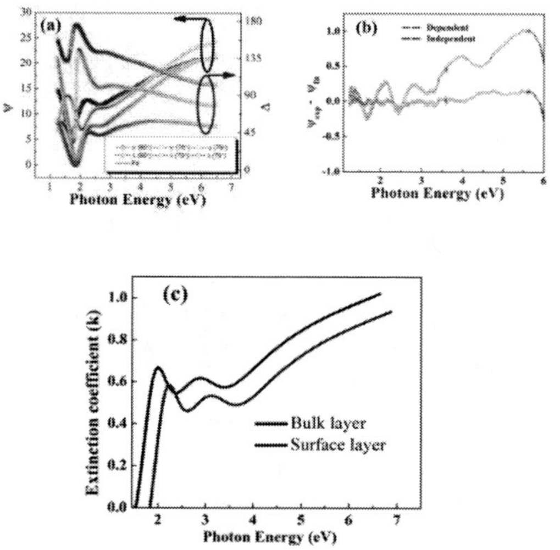

Fig. 2: (a) The fitted experimental ψ and Δ spectra of CZTSSe thin films for different angle of incidence (b) Difference in ψ between the data and the fit with and without an independent surface layer dielectric, (c) Optical constants k of bulk layer and surface layer.

Fig. 2(c) compares two optical functions. The black and blue solid lines represent the optical functions of the bulk layer and surface layers in the CZTSSe thin-film, respectively. In particular, surface layer exhibit a reduction in the extinction coefficient k in 1–2.8 eV relative to that for bulk CZTSSe. This has been shown to be a clear indicator of Cu deficiency. Depression of the extinction coefficient k of Cu-poor CZTSSe is observed in the 1–3 eV spectral range may be due to a reduction in repulsive interaction between Cu $3d$ and (S,Se) $4p$ states caused by Cu deficiency at the valence band maximum similar to CIGS. This implies that Cu is more deficient at surface region than in the bulk region of thin CZTSSe films, which has already been verified by EDX and XPS.

The optical properties of the surface layer for the CZTSSe thin-film in Fig. 2(c) suggest that the reduction in the extinction coefficient is greater in the 1.12–1.8 eV range compared with the full range of 1- 2.8 eV CZTSSe bulk layer. This is also confirmed by bandgap analysis (Surface layer Bandgap = 1.79 eV,Bulk layer Bandgap = 1.51 eV).The increase of the bandgap for Cu-poor CZTSSe Surface layer can be explained. It has been found that change in bandgap is due to p-d hybridization between Cu d-levels and Se p-levels on the basis of band structural calculations in the case of ternary chalcopyrite compounds [11]. In a similar manner, the decrease in the bandgap with increase in Cu content in case of CZTSSe films may be attributed to changes in the extent of p-d hybridization between Cu d-levels and (S, Se) p-levels.

4 CONCLUSIONS

In summary, we have investigated the chemical structure of CZTSSe thin films by XPS and EDX .We provide spectroscopic evidence for Cu deficiency in the surface layer of the nearly stoichiometric thin-film CZTSSe. The reduction in absorption in the spectral region of 1–2.8 eV and the increase in the bandgap at the surface layer are found for CZTSSe thin-film. We find that the surface layer is more Cu deficient than a bulk layer in the CZTSSe thin-film. This work presents the possibility of probing the optical and electronic properties of a surface layer in the thin-film polycrystalline semiconductors by adopting a material layer with voids using effective medium. This work is also of significance in engineering of the solid-state electronic and optoelectronic device design near the junction interface.

ACKNOWLEDGMENT

This work is partially supported by Board of Research in Nuclear Sciences (BRNS), Department of Atomic Energy (DAE), Government of India and Clean Energy Research Initiative (CERI), Department of Science and Technology (DST), Government of India. We are thankful to DIBS and EDX facility equipped at Sophisticated Instrument Centre (SIC) at IIT Indore. The authors (Brajendra S. Sengar, Vivek Garg) acknowledge CSIR and UGC, India for their fellowships. Prof. Shaibal Mukherjee is thankful to Deity YFRF, Government of India award. We are thankful to Dr. D. M. Phase, Dr. R. J. Chaudhary and Mr. A. Wadikar for using AIPES beam line Indus-1 facility at RRCAT.

REFERENCES

[1] M. Bär, B. a. Schubert, B. Marsen, S. Krause, S. Pookpanratana, T. Unold, L. Weinhardt, C. Heske, and H. W. Schock, "Impact of KCN etching on the chemical and electronic surface structure of Cu2ZnSnS4 thin-film solar cell absorbers," *Applied Physics Letters*, vol. 99, no. 15, pp. 26–29, 2011.
[2] M. Bär, B. A. Schubert, B. Marsen, S. Krause, S. Pookpanratana, T. Unold, L. Weinhardt, C. Heske, and H. W. Schock, "Native oxidation and Cu-poor surface structure of thin film Cu 2ZnSnS4 solar cell absorbers," *Applied Physics Letters*, vol. 99, no. 11, pp. 54–57, 2011.
[3] M. J. Romero, K. M. Jones, J. AbuShama, Y. Yan, M. M. Al-Jassim, and R. Noufi, "Surface-layer band gap widening in Cu(In,Ga)Se 2 thin films," *Applied Physics Letters*, vol. 83, no. 23, pp. 4731–4733, 2003.
[4] D. Liao and A. Rockett, "Cu depletion at the CuInSe[sub 2] surface," *Applied Physics Letters*, vol. 82, no. 17, p. 2829, 2003.
[5] I. M. Kötschau and H. W. Schock, "Depth profile of the lattice constant of the Cu-poor surface layer in (Cu2Se)1−x(In2Se3)x evidenced by grazing incidence X-ray diffraction," *Journal of Physics and Chemistry of Solids*, vol. 64, no. September 2003, pp. 1559–1563, 2003.
[6] B. S. Sengar, V. Garg, V. Awasthi, S. Kumar, C. Mukherjee, M. Gupta, and S. Mukherjee, "Growth and characterization of dual ion beam sputtered Cu 2 ZnSn (S , Se) 4 thin films for cost-effective photovoltaic application," *Solar Energy*, vol. 139, pp. 1–12, 2016.
[7] V. Awasthi, V. Garg, B. S. Sengar, S. K. Pandey, Aaryashree, S. Kumar, C. Mukherjee, and S. Mukherjee, "Impact of sputter-instigated plasmonic features in TCO films: For ultrathin photovoltaic applications," *Applied Physics Letters*, vol. 110, no. 10, pp. 1–6, 2017.
[8] V. Garg, B. S. Sengar, V. Awasthi, and P. Sharma, "RSC Advances Localized surface plasmon resonance on Au nanoparticles: tuning and exploitation for performance enhancement in ultrathin photovoltaics," *RSC Advances*, vol. 6, pp. 26216–26226, 2016.
[9] S. K. Pandey, V. Awasthi, B. S. Sengar, V. Garg, P. Sharma, and S. Kumar, "Band alignment and photon extraction studies of Na-doped MgZnO / Ga-doped ZnO heterojunction for light-emitter applications Band alignment and photon extraction studies of Na-doped MgZnO / Ga-doped ZnO heterojunction for light-emitter applications," vol. 165301, 2015.
[10]V. Awasthi, S. K. Pandey, V. Garg, B. S. Sengar, P. Sharma, S. Kumar, C. Mukherjee, and S. Mukherjee, "Plasmon generation in sputtered Ga-doped MgZnO thin films for solar cell applications," *Journal of Applied Physics*, vol. 119, no. 23, p. 233101, 2016.
[11]J. E. Jaffe and A. Zunger, "Theory of the band-gap anomaly in ABC2 chalcopyrite semiconductors," *Physical Review B*, vol. 29, no. 4, pp. 1882–1906, 1984.

AN ALTERNATIVE METHODOLOGY TO INVESTIGATE PROPERTIES OF MINORITY CARRIERS: EFFECTS OF N-DOPANT SPECIES

Hassanet Sodabanlu,[1,*] Amaury Delamarre,[1] Kentaroh Watanabe,[1] Masakazu Sugiyama,[1,2] Yoshiaki Nakano[2]
[1] Research Center for Advanced Science and Technology, the University of Tokyo
[2] Department of Electrical Engineering and Information systems, School of Engineering, the University of Tokyo
*Corresponding author: sodabanlu@hotaka.t.u-tokyo.ac.jp

ABSTRACT: We propose in this communication a simplified and straightforward method for determining the lifetime of minority carrier in the materials for solar cells. This technique consists in placing the layer to be investigated between a p-n junction in the light-incident side, and measuring the extracted photocurrent for various illumination wavelength or layer thickness. Our strategy is close to extracting the diffusion length by fitting the EQE of a solar cell [1], but is more precise since the wavelength range carrying the relevant information is larger. As a support for this proposal, we investigated the effects of n-dopant species, Te and S, on the properties of GaAs solar cells. The preliminary study showed that the carrier extraction was degraded with increasing layer thickness, and this effect was more pronounced in S-doped solar cells. The increasing doping concentration also deteriorated photo-current, and, in this case, influence of Te doping was more than that of S. In summary, this proposed methodology has been proved useful to investigate the properties of minority carriers in semiconductors.
Keywords: III-V Semiconductors, Doping, Gallium Arsenide Based cells, Lifetime, Photovoltaic

1 INTRODUCTION

A solar cell principally operates with the drift currents of minority carriers, both electron and hole, generated by sunlight. Hence, the properties of minority carriers including diffusion length, velocity and lifetime are important factors toward the development and improvement of solar cells. Although the transport properties of majority carriers are rather straightforward to obtain for example by Hall effect measurement, the properties of minority carriers require more sophisticated techniques including time-resolve photoluminescence [2-3], electron beam induced current [4] and moving grating techniques [5]. However, these techniques do not fully consider the minority carriers in solar cells under practical operation. The photo-current measurement [6-7], investigating the steady-state current generated by a known carrier generation rate, is one of available methods to characterize the minority carrier lifetime under a practical condition. Adopted from this measurement technique, we proposed an alternative, simplified, and straightforward method to characterize minority carriers based on p-i-n/n-i-p solar cell structures. The layer to be investigated is insert in the junction of a p-on-n or n-on-p solar cell corresponding to the type of minority carriers (electron or hole). For example, in case of hole minority carrier, the investigated n-doped layer is inserted in the junction of n-on-p solar cell, and vice versa. This situation is similar to a n-i-p solar cell of which the absorber layer (i-layer) is intentionally doped in order to generate a flat band diagram, so that the light generated hole minority carriers can be confined in this layer. By fitting the collectable photocurrent with the device simulation, the minority carrier lifetime can be theoretically extracted.

Among the donor impurities for GaAs and III-V semiconductors, Si, S and Te are the most widely used dopants. Each dopant has its own unique properties, for example, incorporation coefficient, dopant energy level, and atomic size. These properties can affect the characteristics of hole minority carrier. Hence, in this paper, we provided the prove of concept of our proposed methodology by investigating effects of n-dopant species on minority holes by n-n-p solar cells. The thickness and carrier concentration (doping level) of n-GaAs absorber layer will be varied and the performance of GaAs solar cells will be examined and compared.

2 EXPERIMENTAIL DETAILS

The crystal growth in this work was carried out in a planetary MOVPE reactor using standard precursors with H_2 carrier gas including trimethylgallium (TMGa), trimethylaluminium (TMAl), trimethylindium (TMIn), tertiarybutylarsine (TBAs), and tertiarybutylphosphine (TBP). While dimethylzinc (DMZn) was only one p-type dopant, there were two n-type dopants, diethyltelluride (DETe) and hydrogen sulfide (H_2S), utilized for investigating the influence of dopant species on the properties of minority carrier lifetime and solar cell performance. All epitaxial growths were done under the reactor temperature and pressure of 610 °C and 100 mbar, respectively. The experiment was started by studying the incorporation efficiency into GaAs of both Te and S dopants by varying the ratio between partial pressures of dopant gas and TBAs. The impurity concentrations were evaluated by secondary ion mass spectrometry (SIMS). The electrochemical capacitance-voltage (ECV) profiling technique was employed to measure the activated free carrier in doped GaAs layers. After that n-n-p GaAs solar cells schematically depicted in Fig. 1(a) were fabricated. The test solar cell structure consists of n-GaAs absorber in which hole minority carriers can be systematically analyzed. There are two series of solar cell samples fabricated in this research. The first series consists of n-n-p cells with 3-μm thick n-GaAs absorbers of which free carrier concentrations are varied, $5x10^{16}$, $1x10^{17}$, and $1x10^{18}$ cm^{-3}. The second one is varied the n-GaAs absorber thickness, 1, 3, and 5 μm with a carrier concentration of $1x10^{17}cm^{-3}$. The effects of thickness and doping concentration of absorber layer on behaviors and properties of holes can be observed by I-V under AM1.5G illumination and EQE characteristics of solar cells. Because of a limitation in our MOVPE reactor, the gas dopants, DETe and H_2S, cannot be switched during the growth. Hence, the dopant in n-InGaP window and n-GaAs emitter is the same as that in n-GaAs absorber. The band diagram of n-n-p structure in Fig. 1(b) was simulated assuming the thickness and carrier density of 1 μm and $1x10^{17}$ cm^{-3}, respectively. The experiment data will later be compared and fitted with the simulation results.

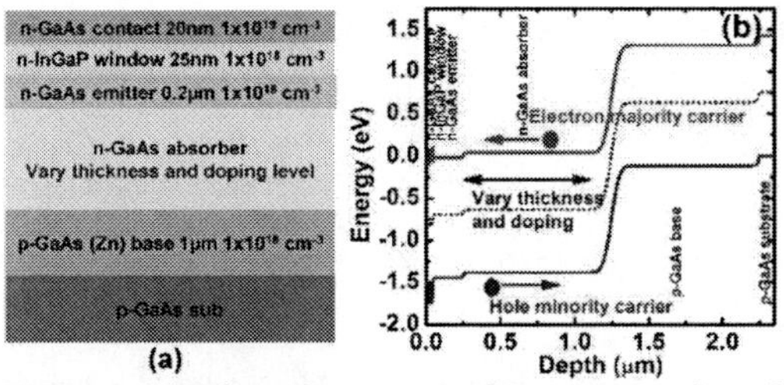

Figure 1: (a) Schematic structure of GaAs n-n-p solar cell investigated in this work. (b) Energy band diagram of the GaAs n-n-p solar cell.

3 SIMULATION MODEL

Figure 2(a) shows a simple n-n-p-p structure which the dependences of short circuit current density (J_{sc}) under AM1.5G (100 mW/cm^2) illumination on the hole lifetime were simulated using PC1D program [8]. The thickness of 1x10^{17} cm^{-3} n-doped GaAs absorber was varied from 1 to 5 μm. Each curve was normalized with its saturate J_{sc} with 10-μs hole lifetime. Obviously, the carrier density decreased with shorter hole lifetime because of recombination loss. This behavior was more pronounced with thicker absorber layer. By fitting experimental data with these tendencies, characteristics of minority carriers including lifetime, diffusion length and mobility could be potentially extracted. Note that, the electron lifetime of 0.25 μs, the Shockley-Read-Hall recombination rate at each surface of 1x10^6 cm/s, the contact resistance of 0.015 Ω, the shunt resistance of 1x10^4 Ω, and the surface reflectance of 30% were assumed for the simulations in this work.

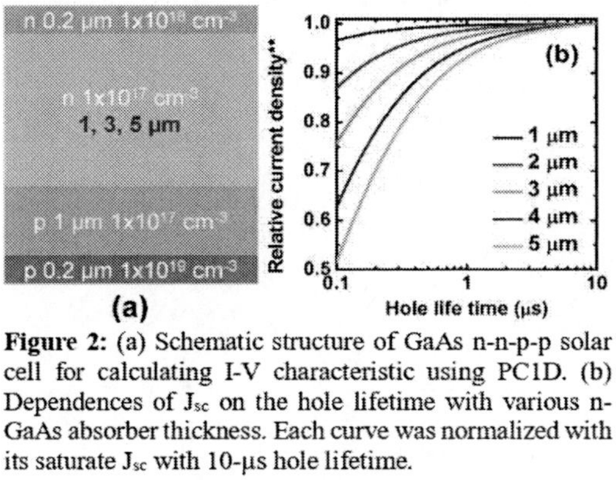

Figure 2: (a) Schematic structure of GaAs n-n-p-p solar cell for calculating I-V characteristic using PC1D. (b) Dependences of J_{sc} on the hole lifetime with various n-GaAs absorber thickness. Each curve was normalized with its saturate J_{sc} with 10-μs hole lifetime.

4 RESULTS AND DISCUSSIONS

SIMS and ECV results in Figs. 3 respectively show the dependences of impurity and free electron concentrations in GaAs layers on partial pressures of dopant gas with respect to that of TBAs. Obviously, Te atoms could incorporate into GaAs approximately two orders better than S atoms. The Te and S concentrations could be expressed in terms of the ratio of dopant and TBAs partial pressures as given by Eqs. (1) and (2), respectively.

$$[Te] = 9.89 \times 10^{22} \left(\frac{[DETe]}{[TBAs]}\right)^{1.0078} \quad (1)$$

$$[S] = 2.58 \times 10^{20} \left(\frac{[H_2S]}{[TBAs]}\right)^{0.9366} \quad (2)$$

Without any post-annealing process, about 30-40% of Te atoms could be electrically activated with a saturated free electron concentration of approximately 10^{19} cm-3. On the other hand, almost all S atoms could be activated, and a maximum carrier concentration of 4x10^{18} cm^{-3} was realized as indicated in Fig. 3. For a low S doping range, the difference between the free electron and S concentrations was most likely due to carrier compensation by background p-type doping including carbon. The activation of S impurities in this work is in good agreement with the data reported in Ref. 9. The carrier concentrations in n-n-p GaAs solar cells in this work were thoroughly adjusted regarding these doping profiles.

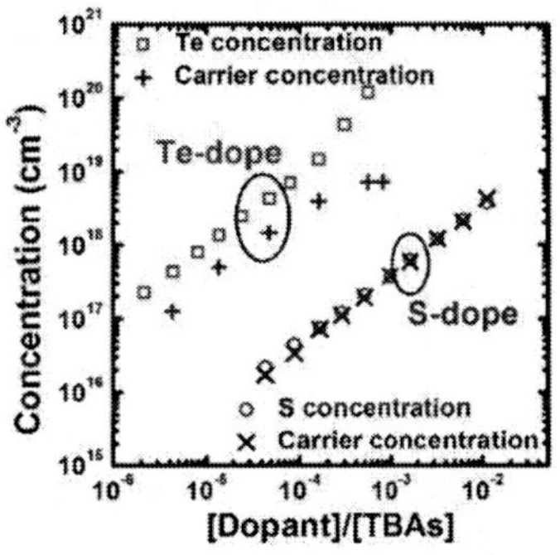

Figure 3: Dependences of impurity and free carrier concentrations of Te- and S-doped GaAs on dopant partial pressures.

Figure 4 shows dependences of J_{sc} of n-n-p GaAs solar cells without anti-reflection coating under AM1.5G (100 mW/cm^2) illumination with (a) various n-GaAs absorber thicknesses using the carrier density of 1x10^{17} cm^{-3} and (b) different carrier concentrations with the absorber thickness of 3 μm. In Fig. 4(a), a thick absorber layer caused a difficulty in carrier extraction, especially holes, and resulted in degradation of current density. S-doped samples were influenced by this effect more than Te-doped cells. The reason might be related to thermal inter-diffusion of S and Zn atoms near n-p junction. A thicker absorber layer necessitated a longer deposition time, resulting in deeper diffusion and worse solar cell efficiency. The Te-doped samples did not exhibit this trouble possibly because Te atoms have a smaller diffusion coefficient than S atoms. Dot-lines in Fig. 4(a) represented simulation results of J_{sc} as functions of absorber thickness as described in Fig. 2. The hole lifetime was varied from 0.01 to 0.12 μs. The simulation lines are not well fitted with the experimental data. The influences from InGaP window layer including light absorption and reflection as well as the SRH recombination at the rear interface especially in the samples with a thin absorber might cause this inconsistency between the measurement and the simulation data. Hence, the further improvement of simulation model is necessitated in order to correctly extract parameters of minority carrier. The solar cell structure should also be optimized to minimize undesirable loss mechanisms, for example insert a high bandgap back surface filed layer. In case of increasing the free carrier density of n-GaAs absorber in Fig. 4(b), the recombination loss via defects and dopant impurities became a dominant mechanism and diminished photocurrent. In contrast with the thickness dependency, this effect was manifest in Te-doped GaAs. The explanation is most likely related to the fact that only 30% of Te could be electrically active. Therefore, in order to obtain a similar free electron concentration, Te concentration would be much larger than S concentration. Such a highly inactive Te atoms could act as carrier traps and degrade solar cell performance.

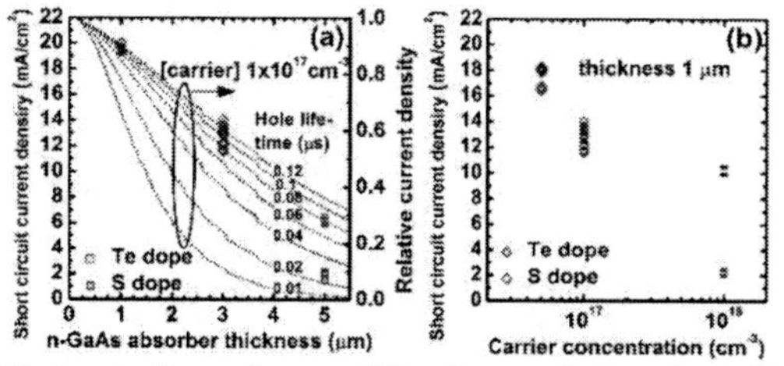

Figure 4: Dependences of J_{sc} of n-n-p GaAs solar cells without anti-reflection coating under AM1.5G (100 mW/cm^2) illumination with (a) different n-GaAs absorber thickness using carrier density of $1x10^{17}$ cm^{-3} and (b) different carrier concentration with the absorber thickness of 3 μm.

Figure 5 shows EQE spectra of the n-n-p GaAs solar cells with different thickness and carrier concentration in n-GaAs absorber corresponding to the J_{sc} in Fig. 4. The degradation of current density is most likely ascribed to recombination loss of minority hole in n-GaAs absorber as indicated by decreasing of EQE. However, we could not clearly observe EQE degradation in short wavelength range corresponding to recombination of hole generated near surface vicinity. An explanation should be ascribed to the influences of InGaP window and the reflectivity of incident light. With increasing either thickness or carrier concentration of the absorber, the degradation of hole extraction was clearly observed. For the first case, this effect was more noticeable in S-doped samples, while Te-doped cells exhibit more pronounced tendency in the latter case.

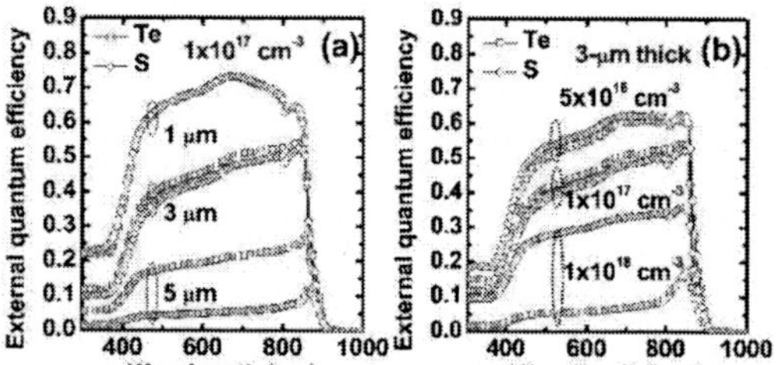

Figure 5: EQE spectra of n-n-p GaAs solar cells with different thickness and carrier concentration in n-GaAs absorber corresponding to the J_{sc} in Fig. 4.

5 CONCLUSIONS

A simplified and straightforward method for determining the lifetime of minority carrier in the materials for solar cells was proposed in this study. With this technique, the layer to be investigated is placed between a p-n or n-p junction, and the collectable photocurrent for various illumination wavelength or layer thickness is measured. As a support for this proposal, we investigated the effects of n-dopant species, Te and S, on the properties of GaAs solar cells. The results showed that, with increasing either thickness or carrier concentration of the absorber, the degradation of hole extraction was clearly observed. Further improvement of both device structure and simulation model are necessitated. In addition, the wavelength dependency of photocurrent should be also carried out. Nevertheless, this methodology is evidently promising for investigating the quality of semiconductor via its minority carrier properties.

6 ACKNOWLEDGEMENT

Part of this study was supported by the NEDO project: the research and development of ultrahigh-efficiency and low-cost III-V compound semiconductor solar cell modules.

7 REFERENCES

[1] P. A. Basore, Proceedings 23rd IEEE Photovoltaic Specialists Conference (1993) 147.
[2] R. K. Ahrenkiel, Solid-State Electronics 35 (1992) 239.
[3] D. Kiliani, G. Micard, B. Steuer, B. Raabe, A. Herguth, and G. Hahn, J. Appl. Phys. 110 (2011) 054508.
[4] L. Chernyak, A. Osinsky, H. Temkin-H, J. W. Yang, Q. Chen, and M. A. Khan, Appl. Phys. Lett. 69 (1996) 2531.
[5] U. Haken, M. Hundhausen, and L. Ley, Appl. Phys. Lett. 63 (1993) 3066
[6] J. Müller, H. Bernt, and H. Reichl, Sol-State Elec. 21 (1978) 999.
[7] D. J. Roulston, N. D. Arora, and S. G. Chamberlain, IEEE Trans. Elec. Devices ED-29 (1982) 284.
[8] D. A. Clugston and P. A. Basore, Proceedings 26th IEEE Photovoltaic Specialists Conference (1997) 207.
[9] Ph. Bove, J. Ch. Garcia, Ph. Maurel, and J. P. Hirtz, Appl. Phys. Lett. 58 (1991) 1973.

LIGHT TRAPPING SIMULATED BY FAST AND ACCURATE TRANSFER MATRIX APPROACH

J. Holovsky[1,2,*], R. Nevyhosteny[1]

[1] CTU Faculty of Electrical Engineering Technická 2, 166 27 Prague, Czech Republic

[2] Institute of Physics ASCR, v. v. i., Cukrovarnická 10, 162 00 Prague, Czech Republic

*Phone: +420 224 352 167; E-mail: jakub.holovsky@fel.cvut.cz

ABSTRACT: Today's photovoltaic technologies are optimized to conditions that hardly happen at the same time: 1000W/m^2 normal irradiance and 25°C cell temperature. Partially, this makes sense for a concept of solar parks generating and selling electricity at a fixed price. This is however an outdated concept when the generation in the middle of a sunny day had a priority. The future concepts will on the other hand take into account also the electricity demand, thus making the optimization process much more complicated. Mathematical tools that can quickly calculate energy generation at a point in time with given irradiance, angle and temperature will be highly desirable in device optimization. Obviously, technology of crystalline silicon will stay involved. There, we cannot forget the chase for higher efficiency which pushes down the crystalline silicon wafer thicknesses while maintaining the optical absorption. Concept of light scattering on a nano-rough surfaces might then be applied together with usual geometrical approaches. Another popular concept will be a tandem of crystalline silicon and hybrid perovskite which is usually a layer with notable light scattering at nano-porous TiO layer. In all these cases the proper light management will be a crucial optimization field. It will inevitably combine geometrical ray-tracing in thick wafer together with thin-film Fresnel optics in contacts or top thin-film sub-cell, but also light scattering. The concept of light scattering and light trapping used to be inevitable for maintaining the technology of amorphous silicon profitable for long time. This strong light trapping concept will again rise in importance for current technologies, but fast methods of simulations will have priority in more comprehensive optimization over the time-consuming methods such as finite element or FDTD method. In this paper we present a novel approach of optical simulation of multilayer with accounting surface scattering that is based on *complex* transfer matrix concept, similarly as the non-scattered Fresnel component. This allows treatment of light trapping and effects of evanescent waves also for the scattered light. The only condition is to include phase randomization present in the case of light scattering.

Keywords: optical simulations, transfer matrix, scattering, light trapping, solar cells

1 STANDARD TRANFER MATRIX APPROACH

Elegant matrix approach exists for coherent treatment of electric fields propagating in positive and negative direction E^+,E^- inside a smooth multilayer structure. This is solved by a product of L matrices 2×2 describing interfaces, labelled I_{2x2} together with L–1 matrices 2×2 describing propagation, labelled P_{2x2}, where L is a number of interfaces [1, 2].

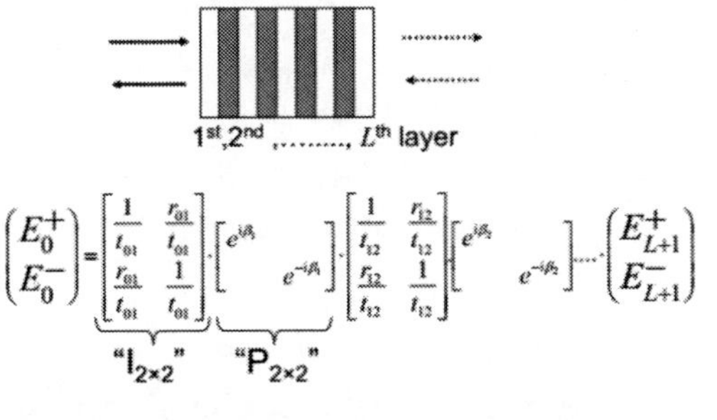

Figure 1: Transfer matrices for calculating light propagation perpendicularly at smooth interfaces in the coherent approach.

Matrices I_{2x2} are composed of standard coherent Fresnel coefficients r_{ij} and t_{ij}, that are complex, calculated from complex refractive index $N=n+ik$, where $k=\alpha\lambda_0/4\pi$ is extinction coefficient, λ_0 is vacuum wavelength and α is absorption coefficient. Under oblique incidence, Fresnel coefficients are different for **s** and **p** polarization and take into account cosine factor $\cos\varphi$ that can also be complex where φ is the angle of propagation that can also be complex. Matrices P_{2x2} are composed of the exponential term, where $\beta_i = 2\pi d_i N_i \cos\varphi_i / \lambda_0$, see Fig. 1. For propagation under oblique incidence the 2×2 matrices are duplicated and distinguished as I^S_{2x2} for **s** polarized and I^P_{2x2} for **p** polarized light, constituting larger matrix I_{4x4} for interface P_{4x4} for propagation, see Fig. 2.

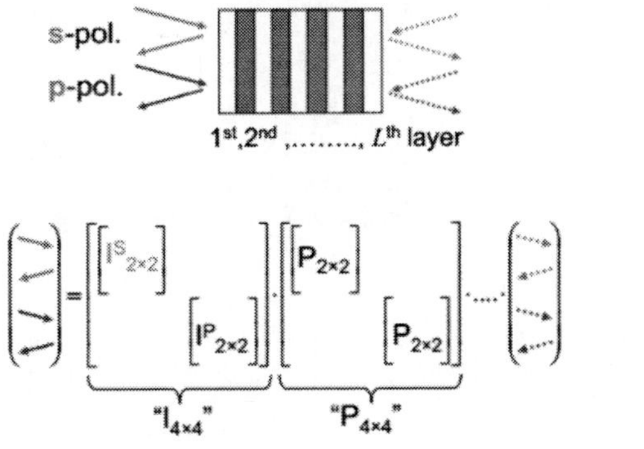

Figure 2: Transfer matrices for calculating light propagation at an oblique angle to the smooth interfaces in the coherent approach.

2 TREATMENT OF LIGHT TRAPPING

Similar matrix approach can be done also for incoherent (real) intensities that do not contain information about a phase. Therefore, one of the appealing way of accounting scattering could represent light by a vector of intensities propagating in discrete angles (matrices $M \times M$, where M is the number of discrete angles). This concept has been used in literature [3,4] already. Matrices contain a separate intensity coefficients for each discrete angle. Unfortunately, light trapping is in principle non-linear effect because in the condition of total reflection the intensity of light inside the layer has no linearly linked component outside the structure (in air). Therefore it is not possible to describe the light trapping in the framework of such series of $2L$-1 incoherent matrices $M \times M$. One of the solution [3] is to build one large matrix with dimension $(LM)^2$, where each angle for each layer is represented by one line/column, and use Gauss elimination or make an inversion matrix [4]. Complexity of calculations increase quadratically with number of layers L and is therefore more time-consuming than the approach of sequencing linear matrices. Still, this approach is unable to account for evanescent waves that, in a structure containing ultrathin layers, should not be neglected. We therefore stick to complex formalism where we represent such (evanescent) waves by a complex angle φ according to the equation (1). We allow product $n\sin\varphi$ to be in a range 0 – 5 and discretize it in arbitrary M steps.

$$\cos\varphi_i = \sqrt{1-\sin^2\varphi_i} \in \mathrm{C} \quad , \quad n \cdot \sin\varphi \in (0\,,\,5) \qquad (1)$$

Extending the matrices I_{4x4} and P_{4x4} for M angles results in matrices I_{4Mx4M} for interfaces and P_{4Mx4M} for propagation, see Fig 3. These matrices are composed all only from 2×2 matrices at diagonal, so that up to this point the vectors elements representing different angles and polarizations are calculated independently and could, in principle, be calculated separately.

3 TREATMENT OF LIGHT SCATTERING

We include the scattering by inserting scattering interfaces into a medium. Bulk scattering is represented by such interface placed in middle of the layer, surface scattering is represented by placing two such scattering interfaces on both sides of the surface. Scattering is the effect connecting together different angles and different polarizations. The matrices I_{4Mx4M} and P_{4Mx4M} are duplicated reserving one part for specular and one part for scattered component, see Fig. 4.

Intensity lost due to scattering from specular (non-scattered) part "leaks" into scattered part that is represented by the same formalism. The reduction of intensity and the "leakage" is represented by scattering matrices S_{4Mx4M} that are defined element-wise as follows:

$$\mathrm{S}_{ij} = \underbrace{s(\lambda n \rho)\delta_{ij}\mathbf{U}_{4x4}}_{\mathrm{S}'_{ij}} + \underbrace{ADF(\lambda n \sigma \varphi_i)[1-s(\lambda n \sigma)]\,\mathbf{W}_{4x4}}_{\mathrm{S}''_{ij}} \qquad (2)$$

$$\sum_{j=1}^{M} ADF(n\lambda\sigma\varphi_{ij}) = 1 \qquad (3)$$

$$\mathbf{U}_{4x4} = \begin{bmatrix} 1 & & & \\ & 1 & & \\ & & 1 & \\ & & & 1 \end{bmatrix} \qquad \mathbf{W}_{4x4} = \frac{1}{2}\begin{bmatrix} 1 & & 1 & \\ & 1 & & 1 \\ 1 & & 1 & \\ & 1 & & 1 \end{bmatrix} \qquad (4)$$

Loss of intensity in specular part is expressed by the first term in (2) that affects only diagonal elements and contains

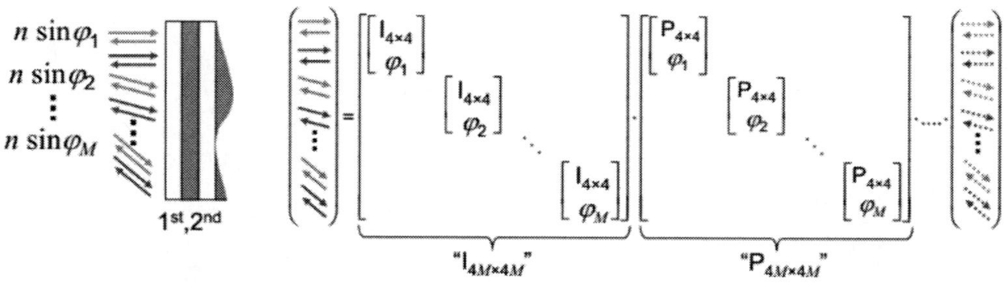

Figure 3: Transfer matrices for calculating multiple angles at interfaces in the complex coherent approach. Green colour represent **s** polarization and blue colour represent **p** polarization.

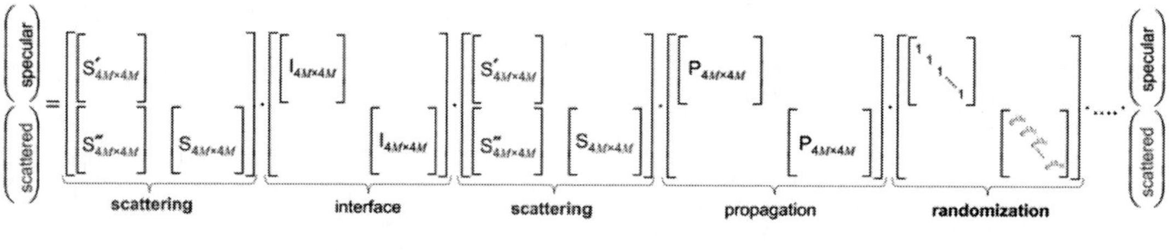

Figure 4: Complete representation of matrix formalism containing specular part in top line and scattered part in bottom line. Non-diagonal matrices are responsible for transfer from specular to scattering part.

scattering factor $0 < s(n\lambda\sigma) < 1$ defined by scalar scattering theory [5]. *ADF* denotes angular distribution function of scattered light, that has to fullfil normalization (3). Because scattering also mixes polarization of scattered part, this effect is represented using matrix $\mathbf{W}_{4\times4}$. in the second part of the equation (2). This second part affects also non-diagonal elements.

4 TREATMENT OF RANDOMIZATION

Finally, some strategy of randomization of a loss of coherence must be used. In nature, the coherence loss is caused by summation of many random contributions. This is a cumbersome mathematic solution that however is also being used sometimes. This work employs for the purpose of de-coherence a simple and fast approach published recently [6]. As all the matrices I_{4Mx4M} and P_{4Mx4M} are composed of 2×2 matrices, it means that in the calculation, there are allways superposed only 2 waves aeia $ae^{i\alpha}$ and $be^{i\beta}$. Basically, we want to get the same result as with summing intensities – the standard approach for incoherent calculations. See equations We do it by adding a "randomization" matrix containing phase manipulation term $be^{i\Delta}$.

$$ae^{i\alpha},\ be^{i\beta}e^{i\Delta} \qquad (5)$$

$$\left|ae^{i\alpha} + be^{i(\beta+\Delta)}\right|^2 \qquad (6)$$

$$a\cos\alpha + b\cos(\beta+\Delta) + i\left(a\sin\alpha + b\sin(\beta+\Delta)\right) \qquad (7)$$

$$a^2 + b^2 + 2ab\cos(\alpha - \beta - \Delta) \qquad (8)$$

If the phase factor d is carefully selected in one of the summand, so that $\Delta \to \beta - \alpha + \pi/2$ then the second term in equation (8) vanishes.

5 SUMMARY, OUTLOOK

We have presented a preliminary work on our novel optical model. We presented strategies to treat light trapping, light scattering and randomization fully within the complex formalism of the transfer matrix approach. This unique approach allows to correctly treat the evanescent waves that some other approaches such as ray-tracing [7] do not allow and that are necessary in the presence of ultrathin layers such as tunnelling oxide. Moreover this approach is promising to be very fast compared to other approaches due to the fact that the complexity of the calculation increases linearly with number of layers and linearly with required accuracy. This is very important for including this model into more complex calculation or optimization algoritms.

5.1 Acknowledgment

This work was supported by Czech Science Foundation grant 16-10429J, Czech Ministry of Education, Youth and Sports project CZ.02.1.01/0.0/0.0/15_003/0000464 "Centre of Advanced Photovoltaics".

5.2 References

[1] B. Harbecke, Appl. Phys. B 39 (1986) 165
[2] R. YEH, P., 1988. Optical Waves in Layered Media. New York: Wiley
[3] R. Santbergen, R.J.Ch. van Zolingen, Energy Conversion and Management 47 (2006) 3572
[4] F. Leblanc, J. Perrin, J. Appl. Phys. 75 (2), 15 (1994) p. 1074
[5] Beckmann, P., Spizzichino, A., 1963. Scattering of Electromagnetic Waves from Rough Surfaces. Pergamon, London.
[6] A. Campa, J. Krc, M. Topic, Progress In Electromagnetics Research, 137 (2013) p. 187
[7] J. Springer, A. Poruba, M. Vanecek, Journal of Applied Physics, 96 (2004) 5329

Study of the SPR tunability of Graphene Coated Metal Nano-Spheres utilizing Graphene-Perovskite interaction: Application in Photovoltaic

Shivani Bhardwaj* and R. P. Sharma
Centre for Energy Studies, Indian Institute of Technology, Delhi-110016, India
*Email: sshivanni@gmail.com

ABSTRACT: Graphene and Perovskite (methyl ammonia lead halide: $CH_3NH_3PbI_3$) interaction plays a vital role to couple the electromagnetic (EM) field to sub wavelength structures and this field further merged with thin films for the Photovoltaic. Perovskite illustrates a tunable behaviour like organic and inorganic materials while dielectric function of graphene can be tuned by varying its thickness. The study of such coupled system is useful to understand the physics of plasmonic behaviour of metal-dielectric interface. In present work, we have studied the coupling of EM field with noble Metal Nano-particles (MNPs), that supports localized surface plasmon resonance (LSPR). This interaction controls the resonance band width and full width half maxima (FWHM) of the resonance band under the quasi-static approximation for the core-shell spherical nano-geometry. In this geometry, core is metallic (Cu) and shell is graphene mono-layer (GML) embedded in two different media of air and perovskite matrix. The optical parameters such as extension efficiency and SPR tunability were studied from ultra-violet to infrared domain (310nm -1400 nm), which steer a wide range of AM1.5 spectrum. There is a blue shift in SPR peak as we increase the core radii from 11 nm to 19 nm. The tunability of SPR peaks of Cu-GML nano-geometry is strongly dependent of the thickness of GML (varies from 0.1 nm to 0.6 nm). As the size of NPs or core-shell increases, there is a red shift in the resonance peak. We have also analysed that extinction efficiency for Cu-GML is higher in perovskite medium than air. Without graphene, LSPRs of NPs do not show appreciable shift in resonance wavelengths, which demonstrates the role of graphene as an electromagnetic field coupling agent.

Keywords: Core/shell geometry, Graphene shell, Perovskite, Noble metal nanoparticle, Surface plasmon resonance and extinction efficiency

1 INTRODUCTION

The optical properties of the metal nano-particles (MNPs) exhibit interesting fundamental significance in nano-optics[1]. MNPs support localized surface plasmon resonances (LSPRs) [1-3]. LSPRs play a vital role in the research and industrial applications to enhance efficiency of the optoelectronic as well as the photonic devices such as photovoltaic cells[4], enhanced Raman spectroscopy [5],light-emitting diodes[6], optical sensing [7], ultrafast optical switching[8], optical filters[9], biological imaging and medical applications [10]. In the recent years,there are the various type of nano-geometries such as coated MNPs are used to enhance and tune the optical responses[10-14].The tunability of plasmonic response basically depends on size of core, shell thickness and physical environment of host medium in which MNPs are embedded. Core/shell materials like metal/semiconductor metal/metal and semiconductor/metal allow the wide range of SPR tunability in UV-visible-IR region[10-12]. Recently, many group of researchers used Graphene as a coating material that having high potential to boost and tune the plasmon modes due to its unique optical properties[12].Graphene's optical properties depend on its conductivity ,transparency and its strength. Graphene itself is a plasmonic material having high conductivity with charge-carrier mobility $>10^5 cm^2 V^{-1} s^{-1}$ at room temperature. Graphene plasmonic behavior can be tuned by doping as well as the gating method [15-17].
Methylammonium lead iodide ($CH_3NH_3PbI_3$) also known as perovskite having a great potential in photovoltaics with direct optical band gap (Eg ~ 1.55 eV), high absorption coefficient and large charge carrier diffusion. Perovskite is a very low cost material, also posseses organic–inorganic dual features. Due to these features, it is the highly efficient to convert the sun light into the electricity[18-20]. The interaction of incident light with perovskite provides a broadband absorption in perovskite materials without increasing the device complexity [19, 20].

In this paper we have studied the tunability of LSPR and full width at half maxima (FWHM) of coated nano sphere. These optical features are studied from UV to IR region for graphene (as shell) coated Cu (as Core) NPs embedded in air and Perovskite matrix. The extinction efficiency and FWHM for this core/shell nano-geometry has been calculated for different core radius and different graphene shell thicknesses. This study has been performed under electro-static approximation in which the particle size is very small in comparison to the wavelength of incident radiation. The coated nano geometry plays the most important role to tune the optical modes in comparition to the non-coated NPs. All the known parameters has been taken from the literature[2, 21, 22]. The optical parameters of perovskite material has been taken from the literature [19, 20]

2 THEORETICAL MODELING

2.1 Optical-properties of metal nano-spheres

The electric field inside and outside of the spherical core-shell nano particles can be calculated by solving the Laplace equation or the potentials by applying the boundary conditions[23]. In ideal dipole approximation, the analytical expression for the absorption and scattering cross-sections can be written as [3, 23]

$$C_{absa} = k\,\mathrm{Im}(\alpha) = \frac{8\pi^2 \varepsilon_3^{1/2} R_2^{\ 3}}{\lambda}\,\mathrm{Im}\left|\frac{\varepsilon_2\varepsilon_{R_1} - \varepsilon_3\varepsilon_{R_2}}{\varepsilon_2\varepsilon_{R_1} + 2\varepsilon_3\varepsilon_{R_2}}\right| \quad (1)$$

$$C_{scat} = \frac{k^4}{6\pi}|\alpha|^2 = \frac{128\pi^5}{3\lambda^4}\varepsilon_3^{\ 2} R_2^{\ 6}\left|\frac{\varepsilon_2\varepsilon_{R_1} - \varepsilon_3\varepsilon_{R_2}}{\varepsilon_2\varepsilon_{R_1} + 2\varepsilon_3\varepsilon_{R_2}}\right|^2 \quad (2)$$

Where $k=(2\pi/\lambda)n$ and α is the polarizability, ε_1, ε_2 are dielectric function of core and shell, $\varepsilon_{R1}$and ε_{R2} are the effective dielectric functions, R_1 and R_2 are the radii of core and core+shell respectively [24]. Further for a single NP the value of extinction efficiency Q_{exn} in term of C_{scat} and C_{abs} is expressed as[2, 3, 23]

$$Q_{extn} = (C_{abs} + C_{scat})/\pi R_2^{\ 2} \quad (3)$$

The size dependent dielectric function $\varepsilon(\omega)$ of MNPs are obtained by using Drude and Lorentz model [2, 3, 24]

$$\varepsilon(\omega) = \varepsilon_{bulk}(\omega) + \frac{\omega_p^2}{\omega^2 + j\omega\gamma_{bulk}} - \frac{\omega_p^2}{\omega^2 - j\omega\gamma} \quad (4)$$

Where ω is frequencyof incident light,bulk plasmon frequency is ω_p, γ is the relaxation frequency of the effective MNPs and the relaxation frequency for bulk material is γ_{bulk}.

2.2 Optical properties of graphene

The graphene is used as coating material in core@shell nano-geometry, where the thickness of shell is very crucial parameter to tune the SPR wavelength. The Kubo formula is used to understand the thickness dependent optical properties of graphene in terms of conductivity [25-27]. The conductivity of thin graphene sheet depends on the frequency of incident light. This frequency-dependent surface conductivity can be quantitatively described as the summation of intra-conductivity (σ_{intra}) and inter-conductivity (σ_{inter}) and expressed as [12, 16, 27, 28]

$$\sigma_{2D}(\omega) = \sigma_{intra}(\omega) + \sigma_{inter}(\omega) \quad (5)$$

where

$$\sigma_{intra}(\omega) = -i\frac{e^2 k_B T}{\pi\hbar^2(\omega - i2\Gamma)} \times \left[\frac{\mu_c}{k_B T} + 2\ln\left(\exp\left(-\frac{\mu_c}{k_B T}\right) + 1\right)\right] \quad (6)$$

$$\sigma_{inter}(\omega) = -i\frac{e^2}{4\pi\hbar}\ln\left[\frac{2|\mu_c| - \hbar(\omega - i2\Gamma)}{2|\mu_c| + \hbar(\omega - i2\Gamma)}\right] \quad (7)$$

Where ω is the incident light frequency, Γ is relaxation process rate having a value $\Gamma \leq 1\,meV$,T is temperature, μ_c is the chemical potential of graphene lies between 50 to 1000 meV, and ħ is reduced Planck's constant. Further the relative permittivity (ε_g) can be found in terms of complex-valued surface conductivity by the coupling of ohm's law and Maxwell's equation as[25, 29]

$$\varepsilon_g(\omega) = \varepsilon_0\left(1 + i\frac{\sigma_{2D}(\omega)}{\varepsilon_0\,\omega\,d}\right) \quad (8)$$

Where d is the thickness of graphene and ε_0 is dielectric constant of vacuum and the Graphene shell thickness is d $= R_1 - R_2$, as shown in fig.1.

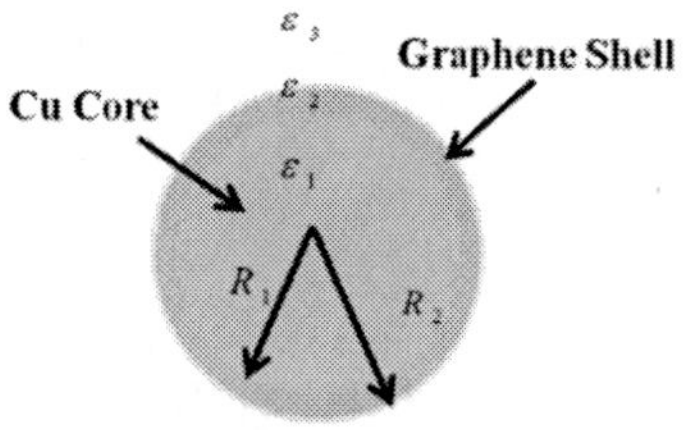

Figure 1: Schematic diagram of Core/shell nanosphere.

3 RESULTS AND DISCUSSION

We have studied the tunability of the SPR wavelengths for Graphene coated spherical NPs in term of the Extinction efficiency (Qext). The extinction efficiency is the function of wavelength of the incident light. The core radius variation responses for the Graphene coated Copper (Cu) core, core/shell NP embedded in the host media of the air (N=1.05) and perovskite matrix (N is wavelength dependent) as shown in fig. 2 and fig.3, respectivily. It is evident that, there is only one plasmonic peak lies from visible to IR domain (from 521nm-1334nm) with high tunability of SPR approximately 813 nm for Cu/Graphene(Core/shell) spherical NP as shown in fig. 2.

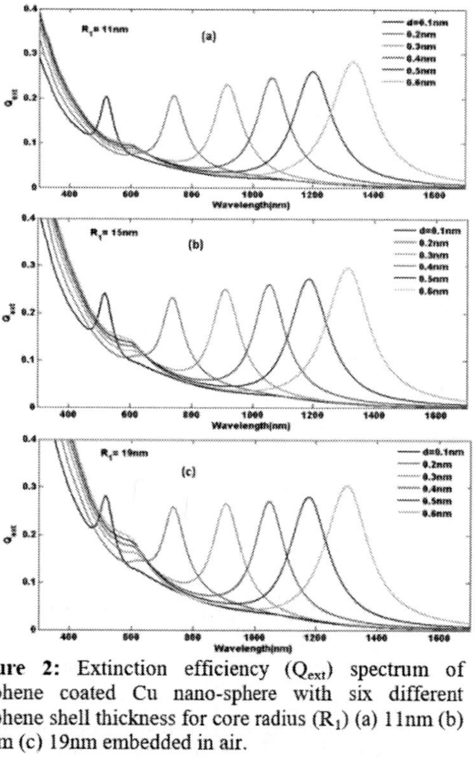

Figure 2: Extinction efficiency (Q_{ext}) spectrum of graphene coated Cu nano-sphere with six different graphene shell thickness for core radius (R_1) (a) 11nm (b) 15nm (c) 19nm embedded in air.

It was found that all plasmonic peaks follow the same trend as graphene shell thickness increases from 0.1nm to 0.6nm. SPR peaks are blue shifted as core radii (R_1) is increased from 11 nm to 19 nm for a fixed Graphene shell thickness.

The advantage of introducing perovskite can be observed in fig 3 as Cu/Graphene NP exhibits dual plasmonic peaks, embedded in the perovskite matrix. From this figure, we found that the SPR tunability for first plasmonic peak for Cu/Graphene spherical NP is very low around 7nm that lies in UV domain (385 nm to 392nm). The second plasmonic peak lies from visible to IR region (from 522nm to1420 nm), is having wide range of SPR tunability approximatly 883 nm with the increasing amplitude of the extinction efficiency as the graphen shell thickness increase from 0.1nm to 0.6nm at fixed core radius (R_1). Both plasmonic peaks are red shifted with increase in the graphene shell thickness up to 0.6nm.The magnitude of the extinction efficiency (Qext) is higher for the Cu/Graphene NP embedded in the Perovskite medium than Air. Extinction efficiency spectrum can be calculated by using "(3)".

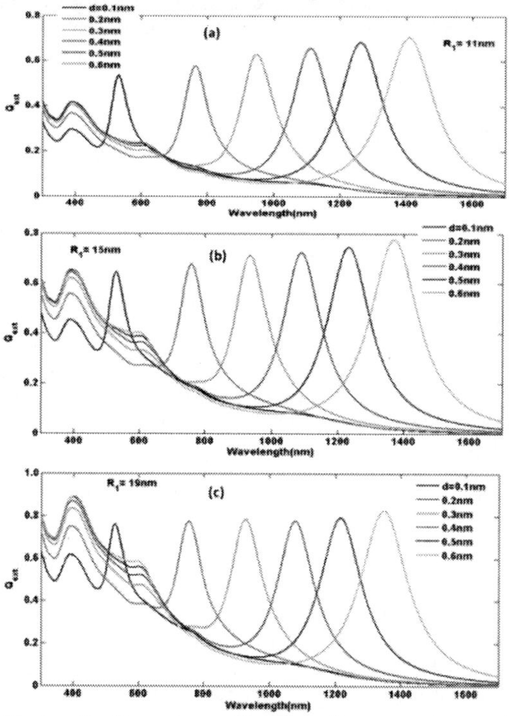

Figure 3: Extinction efficiency (Q_{ext}) spectrum of graphene coated Cu nano-sphere with six different graphene shell thickness for core radius (R_1) (a) 11nm (b) 15nm (c) 19nm, embedded in Perovskite matrix.

Fig.4 shows the variation in FWHM as the function of graphene shell thickness. It is clear that the enhancement in FWHM follws identically linear pattern as the shell thickness increses from 0.1nm to 0.6nm for both media (Air and Perovskite). As the shell thickness increases from its minimum to maximum value, FWHM also increase for a fixed core radius (R_1). As the core radii increses from 11nm to 19 nm there is a blue shift observed in the FWHM.

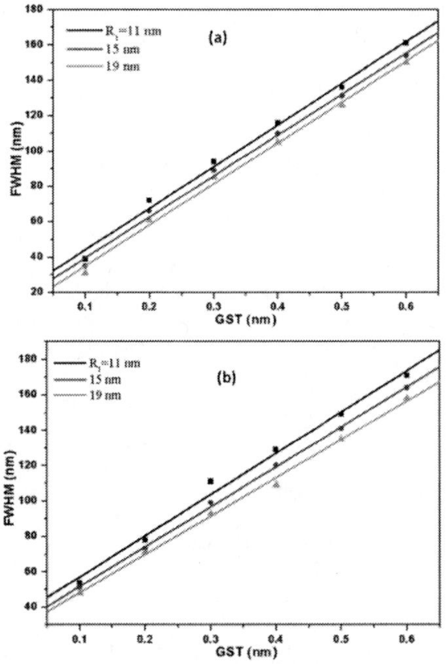

Figure 4: Variation of FWHM of graphene coated Cu NPs with Graphene shell thickness (GST) for three different core radius (R_1) embedded in (a) air (b) Perovskite matrix.

Due to the same trand of FWHM increment for all core radius and different host media we have chosen 15nm core radius at 0.4nm graphene shell thickness for discussion. The FWHM is found to be 110nm and 120 nm for air and Perovskite respectively. Cu/Graphene composition is more suitable in Perovskite medium to enhance the the FWHM in the comparison to the air which can be observed from fig. 4(a) and fig. 4(b).
The non coated Cu nano spheres does not show the appropriate SPR tunablility in the host media of Perovskite as shown in fig.5.

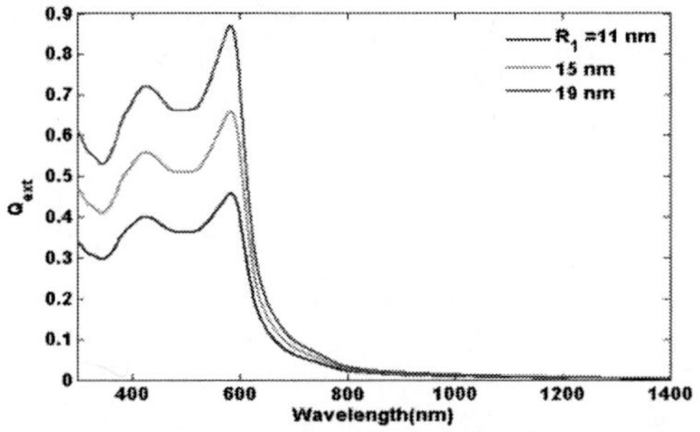

Figure5 Wavelength-dependent extinction efficiency spectra of non-coated spherical Cu nano-particles for three different core radius (R_1)= 11nm, 15nm, and 19nm embedded in Perovskite medium.

4 CONCLUSION

In this work, we have studied the graphene coated Cu NS (Nano sphere), having a wide range of the SPR tunability (in the comparison to the non-coated Cu NP) from Visible to IR (517nm-1329nm) and UV to IR (380nm -1420 nm) wavelength domain embedded in air and perovskite matrix, respectively. Enhancement of the extinction efficiency and broadening of FWHM depends on the host media (Air and Perovskite). It can be seen that Cu/Graphene NP shows high extinction efficiency and high FWHM for Perovskite than air. The SPR tunability is very sensitive parameter that depends on the size of core radii and thickness of the Graphene shell. This geometry has high potential to couple of incident light into optical modes of other materials and applicable in Plasmonic and Photovoltaics.

References

[1] Noguez C 2007 Surface Plasmons on Metal Nanoparticles: The Influence of Shape and Physical Environment *The Journal of Physical Chemistry C* **111** 3806-19
[2] Kreibig U and Vollmer M 2013 *Optical properties of metal clusters* vol 25: Springer Science & Business Media)
[3] Maier S A 2007 *Plasmonics: fundamentals and applications*: Springer Science & Business Media)
[4] Atwater H A and Polman A 2010 Plasmonics for improved photovoltaic devices *Nat Mater* **9** 205-13
[5] Moskovits M 2005 Surface-enhanced Raman spectroscopy: a brief retrospective *Journal of Raman Spectroscopy* **36** 485-96
[6] Liang Z, Sun J, Jiang Y, Jiang L and Chen X 2014 Plasmonic Enhanced Optoelectronic Devices *Plasmonics* **9** 859-66
[7] McFarland A D and Van Duyne R P 2003 Single silver nanoparticles as real-time optical sensors with zeptomole sensitivity *Nano Letters* **3** 1057-62
[8] Inouye H, Tanaka K, Tanahashi I, Hattori T and Nakatsuka H 2000 Ultrafast optical switching in a silver nanoparticle system *Japanese Journal of Applied Physics* **39** 5132
[9] Dirix Y, Bastiaansen C, Caseri W and Smith P 1999 Oriented pearl-necklace arrays of metallic nanoparticles in polymers: A new route toward polarization-dependent color filters *Advanced Materials* **11** 223-7
[10] Jain P K, Lee K S, El-Sayed I H and El-Sayed M A 2006 Calculated absorption and scattering properties of gold nanoparticles of different size, shape, and composition: applications in biological imaging and biomedicine *J. Phys. Chem. B* **110** 7238-48
[11] Ma Y, Li W, Cho E C, Li Z, Yu T, Zeng J, Xie Z and Xia Y 2010 Au@ Ag core− shell nanocubes with finely tuned and well-controlled sizes, shell thicknesses, and optical properties *ACS Nano* **4** 6725-34
[12] Bhardwaj S, Uma R and Sharma R 2016 A Study of Metal@ Graphene Core–Shell Spherical Nano-Geometry to Enhance the SPR Tunability: Influence of Graphene Monolayer Shell Thickness *Plasmonics* 1-9
[13] Rai A, Chaudhary M, Ahmad A, Bhargava S and Sastry M 2007 Synthesis of triangular Au core–Ag shell nanoparticles *Materials Research Bulletin* **42** 1212-20
[14] Liu M and Guyot-Sionnest P 2004 Synthesis and optical characterization of Au/Ag

core/shell nanorods *The Journal of Physical Chemistry B* **108** 5882-8

[15] Grigorenko A N, Polini M and Novoselov K S 2012 Graphene plasmonics *Nat Photon* **6** 749-58

[16] Lu H, Cumming B P and Gu M 2015 Highly efficient plasmonic enhancement of graphene absorption at telecommunication wavelengths *Opt. Lett.* **40** 3647-50

[17] Koppens F H L, Chang D E and García de Abajo F J 2011 Graphene Plasmonics: A Platform for Strong Light–Matter Interactions *Nano Letters* **11** 3370-7

[18] Lin Q, Armin A, Nagiri R C R, Burn P L and Meredith P 2015 Electro-optics of perovskite solar cells *Nature photonics* **9** 106-12

[19] Lu Z, Pan X, Ma Y, Li Y, Zheng L, Zhang D, Xu Q, Chen Z, Wang S and Qu B 2015 Plasmonic-enhanced perovskite solar cells using alloy popcorn nanoparticles *RSC Advances* **5** 11175-9

[20] Löper P, Stuckelberger M, Niesen B, Werner J, Filipic M, Moon S-J, Yum J-H, Topič M, De Wolf S and Ballif C 2014 Complex refractive index spectra of $CH_3NH_3PbI_3$ perovskite thin films determined by spectroscopic ellipsometry and spectrophotometry *The journal of physical chemistry letters* **6** 66-71

[21] Palik E D 1998 *Handbook of optical constants of solids* vol 3: Academic press)

[22] Johnson P B and Christy R W 1972 Optical Constants of the Noble Metals *Physical Review B* **6** 4370-9

[23] Bohren C F and Huffman D R 2008 *Absorption and scattering of light by small particles*: John Wiley & Sons)

[24] Lv W, Phelan P E, Swaminathan R, Otanicar T P and Taylor R A 2013 Multifunctional core-shell nanoparticle suspensions for efficient absorption *Journal of Solar Energy Engineering* **135** 021004

[25] Wang B, Zhang X, Yuan X and Teng J 2012 Optical coupling of surface plasmons between graphene sheets *Applied Physics Letters* **100** 131111

[26] Bao Q, Zhang H, Wang B, Ni Z, Lim C H Y X, Wang Y, Tang D Y and Loh K P 2011 Broadband graphene polarizer *Nat Photon* **5** 411-5

[27] Zhao B, Zhao J M and Zhang Z M 2014 Enhancement of near-infrared absorption in graphene with metal gratings *Applied Physics Letters* **105** 031905

[28] Bhardwaj S, Pathak N K, Ji A, Uma R and Sharma R P 2017 Tunable Properties of Surface Plasmon Resonance of Metal Nanospheroid: Graphene Plasmon Interaction *Plasmonics* **12** 193-201

[29] Chew W C 1995 *Waves and fields in inhomogeneous media* vol 522: IEEE press New York)

SPUTTER-INSTIGATED PLASMONIC FEATURES IN TCO FOR ULTRATHIN PHOTOVOLTAICS: A CASE STUDY OF GA-DOPED ZNO

Vivek Garg[1], Brajendra S. Sengar[1], Vishnu Awasthi[1], Shailendra Kumar[2], and Shaibal Mukherjee[1, *]
[1]Hybrid Nanodevice Research Group (HNRG), Electrical Engineering, Indian Institute of Technology (IIT) Indore, Indore-453552, India
[2]Raja Ramanna Center for Advanced Technology, Indore-452013, India
*shaibal@iiti.ac.in

ABSTRACT: Electrical and optical properties of dual ion-beam sputtered (DIBS) Ga-doped ZnO (GZO) film is analyzed. Usage of assist ion source in DIBS precisely controls excitation of plasmonic features in GZO film. The plasmonic features observed in GZO are due to the formation of metal and metal-oxide nanoclusters. Such plasmon generation is verified by electron energy loss spectra as obtained by ultraviolet photoelectron spectroscopy, spectroscopic ellipsometry, and field emission scanning-electron microscopy measurements. This is promising in terms of increasing the efficiency of ultrathin solar cells by increasing optical path length in the absorbing layer. Keywords: DIBS, UPS, plasmons, GZO

1 INTRODUCTION

Cost-effective and efficient thin film solar cells (SCs) fabrication is a challenge for contemporary researchers. The leading approaches to overcome these challenges are: (1) to use cost-effective and more efficient thin film materials, and (2) by using cost-effective nanoclusters in thin films to compensate for the reduction in physical layer thickness by enhancing optical layer thickness. In recent years, focused research on ZnO films has been gaining momentum and is now becoming the most established alternate as a transparent conducting oxide (TCO) material due to its higher transparency in visible region, high band gap tunability, high electro-chemical stability, and large exciton binding energy (60 meV). A number of ZnO based TCO materials are investigated such as MgZnO, GaZnO, AlZnO and i-ZnO etc. The incorporation of Mg in ZnO helps in the band gap tunability. Moreover, incorporation of both Ga and Al in ZnO matrix adds a flexibility to enhance the carrier concentration.

In order to improve the performance of thin film SCs and reduce their manufacturing cost simultaneously, the concept of plasmonic is introduced. Enhanced absorption coefficient by the excitation of plasmons greatly influences the performance of SCs with reduced physical layer. The usage of nanoclusters of noble metal such as Au, Pt and Ag is quite common as plasmonic layer [1-3]. Even though these nanoclusters offer higher carrier concentrations and limited chemical interactions, the realization of such noble metal nanoclusters is costly due to material cost and their incorporation in thin films SCs is a complicated procedure. Therefore, there is a requirement of cost-effective alternatives for Plasmon excitation mechanism in ultrathin SCs. Here, we also report a straightforward and single-step procedure for formation of Ga, Zn, Mg and their oxide nanoclusters in GZO film, which act as TCO and buffer layer in SCs. It has been reported earlier that these metallic or metal-oxide nanoclusters in semiconducting films are created due to difference in the sputtering rates of different constituent elements of semiconductors.

2 EXPERIMENTAL TECHNIQUE

GZO thin film with thicknesses of 200 nm is grown separately on Si substrate by Elettrorava DIBS system. The background pressure in the deposition chamber is $\sim 1\times10^{-8}$ mbar. The deposition of individual thin films is carried out in Ar atmosphere with a chamber working pressure of 2.4×10^{-4} mbar and with RF power of 68 W. GZO film growth is carried out using a 4N (99.99%) pure

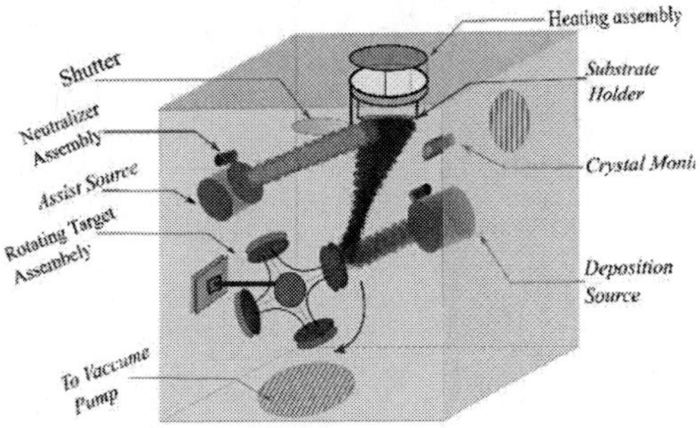

Fig.1: Schematic illustration of DIBS deposition system.

4-in-diameter 1 at. %, Ga-doped-ZnO (99.99%) target respectively. Further, details of the deposition system are reported elsewhere[4, 5].UPS measurements are performed by utilizing the beam line of INDUS-1 synchrotron source.

SE measurement is carried out using J.A. Woolam M-2000D Ellipsometry. Further, data analysis is performed using a three-layer optical model consisting of a Si substrate, GZO layer, and a top roughness layer. Theoretical spectra by using fitting parameters of the General Oscillator (Gen-Osc) model is in good agreement with experimental data with low value of mean square error (MSE) of 9.64 confirms the suitability and reliability of the model. Tauc-Lorentz (T-L) oscillators in GenOsc model is utilized to model the complex dielectric function, which is occurring due to excitonic transitions.[6, 7]

3 RESULTS AND DISCUSSION

3.1 Electrical and Optical Properties

GZO thin films have been studied for their electrical properties. For this, a four-probe Hall-effect measurement along with 2612A Keithley sourcemeter is used. Hall measurement has revealed n-type conductivity in GZO film with carrier density of 1.06×10^{21} cm^{-3}, resistivity of 4.11×10^{-3} Ω-cm, and mobility of 14.83 $cm^2V^{-1}s^{-1}$ at room temperature.

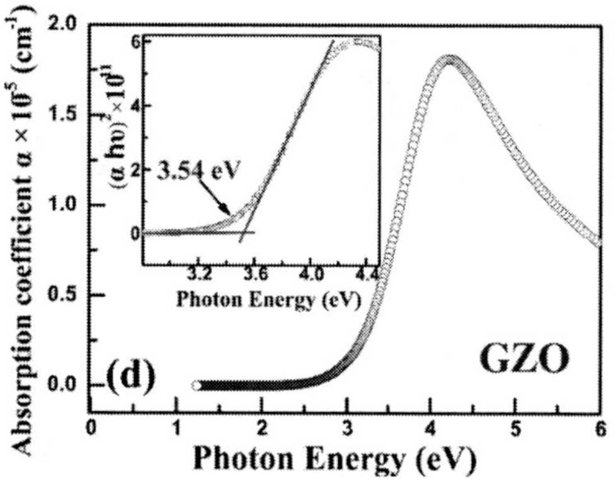

Fig.2: Transmission spectra of GZO. Inset shows respective Taucs plot.

The transmission spectra of GZO thin film grown on sapphire substrate are shown in Fig. 2. GZO film is highly transparent. The transmittance of GZO film is better than 96% in 400-800 nm. Band gap value for GZO, as evaluated from Taucs plot obtained from transmission spectra, is 3.55 eV.

3.2 Spectroscopic Ellipsometry (SE) Measurements

Fig. 3(a) depicts the spectral variation of the real part (ε_1) and the imaginary part (ε_2) of the complex dielectric function, $\varepsilon(E)=\varepsilon_1(E)+i\varepsilon_2(E)$, in the photon energy range of 1.5-6 eV. The values of ε_1 and ε_2 have significant role in designing optical and optoelectronic devices, since they are closely associated to the electronic polarizability of ions and the localized fields inside materials [8-10]. The origin of ε_1 peaks in GZO is excitonic and mainly due to exciton-phonon complex transitions, as reported elsewhere [11, 12]. Fig. 3(a) shows the maximum value of ε_1 is 4.79 (at ~3.54 eV. In general, the broadening parameter of ε_1 peak is inversely proportional to corresponding excitonic lifetime. Therefore, the results indicate good excitonic stability in GZO.

Fig. 3(a) depicts that onset of ε2, which corresponds to the absorption at the fundamental band gap (E_g), observed at 3.53 eV. There is an unusual peak in ε2 spectra of GZO at photon energy larger than Eg (>4 eV). This peak is due to additional absorption in films, which might be caused by particle plasmons produced by metal or metal oxide nanoclusters in GZO thin films. Additionally, spectral variation in refractive index (*n*) and extinction coefficient (*k*) is plotted in Fig. 3(b). Maximum value of n is 2.2 and the value of k is nearly zero below 2.75 eV, which indicates that these films are completely transparent below 2.75 eV. The onset in absorption spectra above Eg ~3.5 eV, is observed in the film, which is caused by the interband transition. The particle plasmon features can be observed in the spectra of GZO in the form of a peak at 4.22 eV, which might correspond to plasmon resonance peak of nanoclusters formed in corresponding thin film.

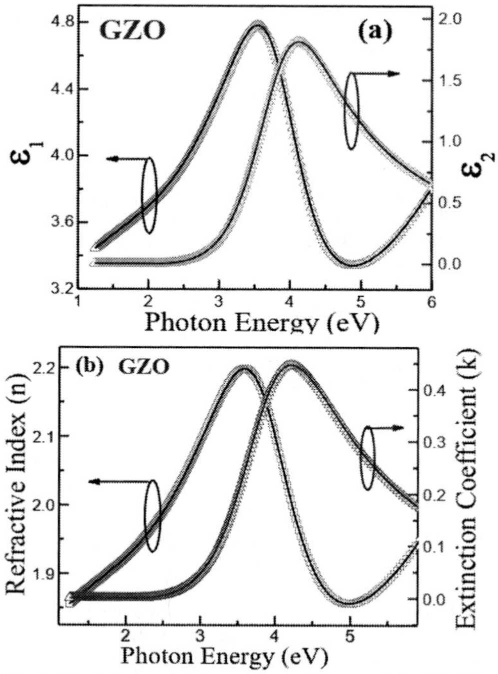

Fig.3: Variations of (a) $\varepsilon_1$and ε_2, (b) *n* and *k* of GZO thin film with incident photon energy.

3.3 Electron Energy Loss Spectra

Some broad plasmon peaks are detected in the UPS spectra of GZO thin film. Fig. 4(a-c) shows the UPS spectra of GZO film before and after 5 and (5+5) minute

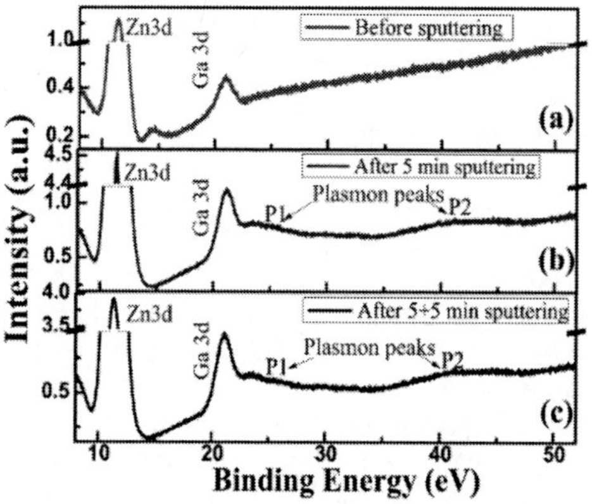

Fig.4: UPS spectra of GZO (a)before, (b)after 5 minute, and(c) after (5+5) minute.

of Ar+ sputtering. Broad plasmon peaks (P1 and P2) are observed at 23.5 and 41 eV in the UPS spectra of both the 5 and (5+5) minute sputtered films. However, the unsputtered

GZO film does not exhibit existence of any such broad plasmon peaks. Plasmon peaks in sputtered film are generally formed due to kinetic energy loss of Zn-3d core photoelectrons after scattering with valence plasmons as a result of inelastic scattering mechanism[13, 14]. These broad plasmon peaks can be the valence bulk plasmon (VBP) and valence surface plasmons (VSP) of Zn, ZnO, Ga, GaO nanoclusters embedded within GZO matrix [13]. Calculated energy values of VBPs, VSPs and particle plasmon resonance in ZnO and air medium are tabulated in Table 1. Additionally, theoretical calculation of the VBP, VSP and particle plasmon resonance energy (PPRE) of different nanoclusters in ZnO, and in air medium are performed by using following equations, and tabulated in Table 1.

Theoretically the bulk plasmon energy, $\hbar\omega_b$, of particle material in the free electron approximation is evaluated by equation (1) [7, 15, 16].

$$(\hbar\omega_b)^2 = (4\pi\hbar^2 e^2 N_v/m^*) \quad (1)$$

Where, m^* is the effective mass of electrons, and N_v is the density of valence band electrons, which can be calculated by

$$N_v = N_a * \rho * \frac{\Sigma x_i n_i}{\Sigma x_i A_i} \quad (2)$$

Where, ρ is density of material, N_a is the Avogadro number, x_i, n_i, and A_i, represents number of atom present in molecule, the number of valance electron, and atomic weight of ith element, respectively. In order to determine the energy of VSPs (VSPE), equation (3) is employed [17, 18]:

$$VSPE = VBPE/\sqrt{2} \quad (3)$$

Where, VBPE is the energy of VBP. In addition, the particle plasmon resonance energy ($\hbar\omega_{pp}$) for different nanoclusters can be calculated by using the following equation [17, 18].

$$\omega_{pp}^2 = \frac{\omega_b^2}{\varepsilon_\infty + 2\varepsilon_a} \quad (4)$$

Where ε_∞ is the high-frequency dielectric constant, ε_a (for ZnO ~8.59 and air ~1), is the dielectric constant of surrounding medium of a nanoclusters. The values of different constants required for the calculation of equations (1-4) are given in Table I. Due to the lack of information available in the literature, the value of ε_∞ is considered to be ~3 for Zn and Ga, nanoclusters. Here, VBP, VSP and PPRE of nanoclusters embedded in GZO, are calculated and the data is populated in Table II. Plasmonic peak positions, the distance of those plasmonic peaks from the Ga 3d core-level peak position, and their corresponding energy values for the VBPs and VSPs of Ga, Zn, GaO, and ZnO nanoclusters in sputtered GZO thin films are given in Table III.

Table I: Parameters for plasmon energy calculation.

Clusters	Density (gm cm-3)	No. of valence electrons	Atomic weight	High frequency Dielectric Constant
Ga	5.91	3	69.72	3
Zn	7.14	2	65.38	3
GaO	6.44			3.57
ZnO	5.61			3.76

Table II: Calculated energy values of VBPs, VSPs, and particle plasmon resonance in ZnO and in air medium.

Clusters of elements and compounds	Energy of VBP (eV)	Energy of VSP (eV)	Particle plasmon resonance energy (eV) (in ZnO medium)	Particle plasmon resonance energy (eV) (in air medium)
Ga	14.53	10.27	3.82	6.49
Zn	13.46	9.52	3.54	6.02
GaO	23.69	16.75	6.10	10.04
ZnO	7.31	5.17	1.87	3.04

These results are in good agreement with FESEM micrograph of GZO is shown in Fig. 5(a), and it confirms the formation of nanoclusters in GZO thin film. Clusters present in the GZO thin films are analyzed by further

Table III: Calculation of plasmon energies of different peaks and their indexing with different plasmonic nanoclusters.

Peak	Distance of plasmon peak from Ga 3d peak position (eV) 20.06	Corresp onding energy for VSP of nanoclus ters (eV)	Corresp onding PPR Energy (eV) (in ZnO medium)	Corresp onding PPR Energy (eV) (in air medium)
P_{a1}	3.14	----	Ga, Zn	ZnO
P_{a2}	20.94	(Ga+Zn), (GaO+ ZnO)	----	(GaO+ ZnO+Zn), (ZnO + Ga+ GaO)

processing of FESEM images using imageJ software. From image processing, calculated average size of different nanoclusters present in GZO thin film is in the range of 20-35 nm.

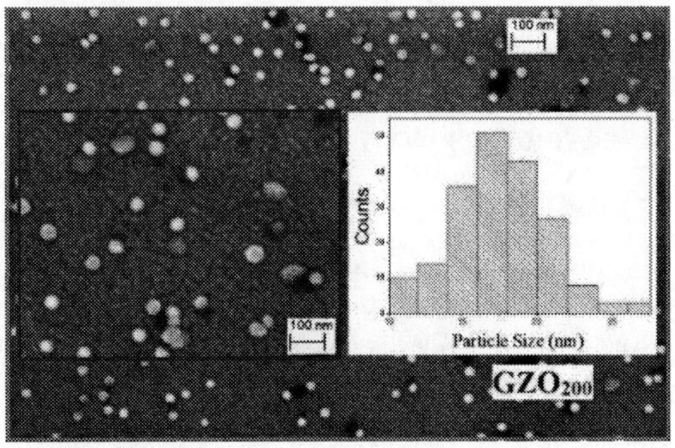

Fig.5: FESEM micrograph of GZO thin film.

The particle plasmon resonance energy is in the range of ~1.87-10.04 eV for different nanoclusters in both ZnO and air medium. The energy range of nanoclusters falls in the ultraviolet-visible-infrared range of solar spectrum and can certainly improve the absorption cross-section by scattering mechanism, where the effective optical path in the photoactive medium increases and hence ultimately enhance the performance of the solar cell.[19-25] Currently, high-efficiency and ultra-thin solar cells use precious metals like Au and Ag with particle plasmon energy of 2.8 and 4.5 eV,[26-30] respectively which adds to the cost of device fabrication. Nevertheless, these precious and noble metals can be replaced by the nanoclusters generated by a short-duration Ar+ sputtering of DIBS deposited films for realizing economical and high-performance photovoltaics.

4 CONCLUSIONS

The electrical and optical properties are studied for the thin film of GZO. The spectroscopic Ellipsometry reveals the favorable optical constants of both films for solar cell application and confirms plasmon generation. Moreover, the broad plasmon peaks observed in UPS spectra due to metal and metal oxide nanoclusters. The particle plasmon resonance energy in GZO are in the range of ~1.87-10.04 eV for different nanoclusters are in both ZnO and air medium. Moreover, FESEM confirms generation of nanoclusters in the individual thin films. From this, we can conclude that, the DIBS system can be used for one-step generation of metal and metal oxide nanoclusters in TCO thin films. These nanoclusters are extremely promising to enhance the optical scattering and trapping of the incident light, which increases the optical path length in the absorber layer for achieving all sputtered cost-effective ultra-thin SCs and eventually increases its efficiency.

5 ACKNOWLWDGEMENTS

This work is partially supported by Board of Research in Nuclear Sciences (BRNS), Department of Atomic Energy (DAE), Government of India, Clean Energy Research Initiative (CERI), Department of Science and Technology (DST), Government of India, and UGC-DAE CSR. Brajendra S. Sengar and Vivek Garg are also thankful to CSIR and UGC, respectively for the award of fellowship. Dr. Shaibal Mukherjee is thankful to DeitY YFRF, Government of India award. We are also thankful to DIBS facility equipped at Sophisticated Instrument Centre (SIC) at IIT Indore.

5 REFERENCES

1. S Pillai, M Green. Solar Energy Materials and Solar Cells. 2010;94(9):1481-6.
2. HA Atwater, A Polman. Nature materials. 2010;9(3):205-13.
3. V Garg, BS Sengar, V Awasthi, A Shree, P Sharma, C Mukherjee, et al. RSC Advances. 2016.
4. A Kumar, M Das, V Garg, BS Sengar, MT Htay, S Kumar, et al. Applied Physics Letters. 2017;110(25):253509.
5. SK Pandey, V Awasthi, BS Sengar, V Garg, P Sharma, S Kumar, et al. Journal of Applied Physics. 2015;118(16):165301.
6. V Awasthi, V Garg, BS Sengar, SK Pandey, Aaryashree, S Kumar, et al. Applied Physics Letters. 2017;110(10):103903.
7. V Awasthi, SK Pandey, V Garg, BS Sengar, P Sharma, S Kumar, et al. Journal of Applied Physics. 2016;119(23):233101.
8. BS Sengar, V Garg, V Awasthi, Aaryashree, S Kumar, C Mukherjee, et al. Solar Energy. 2016;139:1-12.
9. Y Zhang, C Yam, GC Schatz. J Phys Chem Lett. 2016;7(10):1852-8.
10. S Law, C Roberts, T Kilpatrick, L Yu, T Ribaudo, EA Shaner, et al. Phys Rev Lett. 2014;112(1):017401.
11. J Kim, GV Naik, AV Gavrilenko, K Dondapati, VI Gavrilenko, SM Prokes, et al. Physical Review X. 2013;3(4).
12. V N'Tsame Guilengui, L Cerutti, JB Rodriguez, E Tournié, T Taliercio. Applied Physics Letters. 2012;101(16):161113.
13. L Kövér, G Moretti, Z Kovács, R Sanjinés, I Cserny, G Margaritondo, et al. Journal of Vacuum Science & Technology A: Vacuum, Surfaces, and Films. 1995;13(3):1382-8.
14. S Kumar, C Mukherjee, VK Dixit, SD Singh, SN Jha. Applied Surface Science. 2011;258(1):143-6.
15. P Yadav, R Gupta, M Modi, S Kumar. Solid State Communications. 2013;156:12-5.
16. RF Egerton. Reports on Progress in Physics. 2009;72(1):016502.
17. T Oates, H Wormeester, H Arwin. Progress in Surface Science. 2011;86(11):328-76.
18. A Vishnu, P Sushil Kumar, K Shailendra, C Mukherjee, G Mukul, M Shaibal. Journal of Physics D: Applied Physics. 2015;48(48):485305.
19. A Novitsky, A Uskov, C Gritti, I Protsenko, B Kardynał, AV Lavrinenko. Progress in Photovoltaics: Research and Applications. 2014;22(4):422-6.
20. M Schmid, J Klaer, R Klenk, M Topič, J Krč. Thin Solid Films. 2013;527:308-13.
21. VE Ferry, JN Munday, HA Atwater. Advanced materials. 2010;22(43):4794-808.
22. S Fahr, C Rockstuhl, F Lederer. Photonics and Nanostructures-Fundamentals and Applications. 2010;8(4):291-6.
23. F Beck, A Polman, K Catchpole. Journal of Applied Physics. 2009;105(11):114310.
24. K Catchpole, A Polman. Applied Physics Letters. 2008;93(19):191113.
25. K Catchpole, A Polman. Optics express. 2008;16(26):21793-800.
26. X Li, T Chen, Y Liu, K Leong. Optics express. 2014;22(5):5124-32.
27. S Peters, S Peredkov, M Neeb, W Eberhardt, M Al-Hada. Surface Science. 2013;608:129-34.
28. D Gaspar, A Pimentel, T Mateus, J Leitao, J Soares, B Falcao, et al. Scientific reports. 2013;3.
29. S Hayashi, T Okamoto. Journal of Physics D: Applied Physics. 2012;45(43):433001.
30. S Zhu, T Chen, Z Cen, E Goh, S Yu, Y Liu, et al. Optics express. 2010;18(21):21926-31.

EFFECT OF SPUTTERING AND ANNEALING PARAMETERS ON PROPERTIES OF SILICON QUANTUM DOTS

Mihir Kumar Sahoo[1,a], J.P. Kar[2,b], P.G. Kale[1,c]
1 Department of Electrical Engineering, NIT Rourkela, 769 008, Odisha, India
2 Department of Physics, NIT Rourkela, 769 008, Odisha, India
[a]kumarmihir1992@gmail.com, [b]karjp@nitrkl.ac.in, [c]pareshkale@nitrkl.ac.in

ABSTRACT: Fabrication of the third generation photovoltaic (PV) solar cell uses thin film deposition technique to deposit dielectrics such as SiO_2, Si_3N_4, SiC, and SiO_x. The layer works as a barrier layer, capping layer, tunneling medium, and an intermediate i- layer in the p-i-n structure solar cell. Control of physical parameters and morphology of thin film layer is essential for the synthesis of the quantum dots PV solar cell. The presented work use sputtering as deposition technique to fabricate SRO/SiO_2 multilayers on controlling the deposition parameters such as RF sputtering power, the proportion of sputtering gas flow rate, and sputtering deposition time. The fabricated device is examined for physical and optical properties using spectroscopic techniques such as a surface profiler, EDS-SEM, XRD, FTIR, UV-Vis, and PL.

Keywords: RF Sputtering; Annealing; Si-QDs; i-layer; SRO layer; Deposition rate; O_2 gas flow; Ar gas flow

1 INTRODUCTION

Si-QDs tune the effective bandgap (E_g) of PV solar cell to absorb a broad range of solar spectrum [1, 2]. Si-QDs embedded with SiO_2 dielectric matrix gives quantum confinement effect, which exhibits higher E_g as compared to bulk Si for the same solar spectrum [3]. When the QDs are close to each other, carriers can tunnel between them to form quantum dots superlattice [4]. The choice of deposition method and the annealing environment decide the properties of Si QDs matrix. Among many other techniques e.g. CVD, evaporation, sputtering is also used for fabricating the third-generation solar cell structures. In general, sputtering offers relatively high deposition rate of 4.4 nm/min to 10 nm/min at low operating temperatures between 27 ^{0}C to 300 ^{0}C. Various types of sputtering techniques reported in the literature are RF, Magnetic, and DC.

RF sputtering offers the advantage of using both conducting and non-conducting target to deposit the dielectric matrix with high deposition rate and at low temperature [5, 6]. The paper reports use of RF sputtering for the fabrication of p-i-n structured QDs solar cell device. RF sputtering method is used to deposit multiple alternate layers of SiO_2-SRO functioning as intermediate layer [7]. Upon annealing, Si QDs are precipitated in SRO layer ($2SiO_x \rightarrow (2-x)\, SiO_2 + x\, Si$) [8]. The Si-QDs size and density vary with the variation of annealing temperature of the P-i-N structure solar cell [8].

2 EXPERIMENTATION

2.1 Fabrication of Si-QDs

N-type phosphorus-doped Si wafer (1-10 Ω-cm resistivity, single side polished, thickness 275 ± 25 μm, <100> orientation, 2-inch diameter) is used as the substrate. Si wafers are cut into smaller pieces according to the requirement. All the Si wafers are rinsed with DI H_2O (5 min) followed by sonication in acetone (5 min) and repeat the same once again followed by sonication in IPA (10 min). Sonicated samples are again sonicated with DI H_2O (5 min). The wafers are then put in H_2SO_4 (98.4%) (3 min) and the same proportion of H_2O_2 (1:1) for some time to nullify the reaction. Then rinsed with DI H_2O (15 min) followed by (HF) (5%) (5 min) followed by DI H_2O (10 min). Spin coating machine dries the wet wafers.

Keep the dry wafers in the substrate holder. The substrate holder rotates at a speed of 10 rpm for uniform thin film deposition. A Si target (99.9% purity, 3-inch diameter and 3 mm in thickness with the copper backing plate) is placed in the target holder of the sputtering system. The sputtering system is an assembled one with a 10 cm of separation between target and substrate. Two types of pumps are used to create a vacuum in the sputtering chamber. The rotary pump creates vacuum up to $4*10^{-3}$ mbar pressure, and diffusion pump creates a vacuum of $9.7*10^{-7}$ mbar pressure. For the deposition of thin films, the base pressure is $9.7*10^{-7}$ mbar. The Ar and O_2 gas flow are controlled by a rotameter with proper proportion to deposit SRO layer. The working pressure is kept constant at $4*10^{-3}$ mbar. Pre-sputtering is done by supplying 10 Watt of RF power (5 min) followed by 20 Watt (5 min) and similarly reach up to 50 Watt (15 min). Plasma appears inside the chamber during the pre-sputtering period.

The p-i-n structured solar cell consists of i-layer. The i-layer consists of SiO_2 – SRO – SiO_2 matrix structure. The SRO layer will produce Si QDs due to phase separation in an annealing process. The size of Si-QDs depends on the thickness of the SRO layer. The thickness of SRO layer depends on the deposition rate of the sputtering deposition technique at the specified deposition parameters. The value of (x) in SiO_x should be to the requirement of the density of Si QDs.

2.1.1 Variation in sputtering deposition time

Sputtering power supply was kept constant at 50 Watt. Both the target and substrate shutters are opened to deposit SRO film. Six samples of are fabricated for different deposition time. The samples are named as 5_min, 10_min, 15_min, 30_min, 45_min, 60_min to distinguish the samples according to their deposition time as 5, 10, 15, 30, 45, 60 min respectively.

Stylus surface profiler (Veeco Dektak 150) on the optically flat Si substrates measures the thickness of the deposited layer. The thickness of sample 5_min and 10_min sample could not be determined as the minimum thickness measured by it is 10 nm. The deposition rate (nm/min), increases with the increase in deposition time in a non-linear manner for a constant gas flow rate, as in Fig.1. The deposition rate is considered to be around 0.6 nm/min for a lower thickness of the deposited thin film.

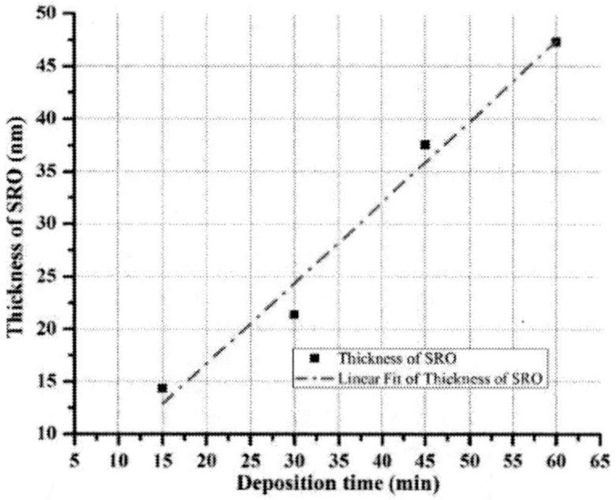

Figure 1: SRO thickness at different deposition time for a constant Ar: O_2 proportion

2.1.2 Variation in Proportion of Gas flow rate
The ratio of Ar and O_2 is varied to get the proper value of (x) of SiO_x (i.e., SRO) layer. The experiment is carried out with a reduced O_2 percentage for a constant deposition time of 30 min. The stylus surface profiler measures the thickness of the deposited sample. Stylus surface profiler (Veeco Dektak 150) on the optically flat Si substrates measures the thickness of the deposited layer. **Error! Reference source not found.** shows the changes in the thickness of the deposited sample with the change in the proportion of sputtering gasses. As the O_2 percentage increases from 7.5 sccm to 13.5 sccm, SRO thickness decreases with deposition rate (slope) of nearly 0.9 nm/min, as in Fig. 2.

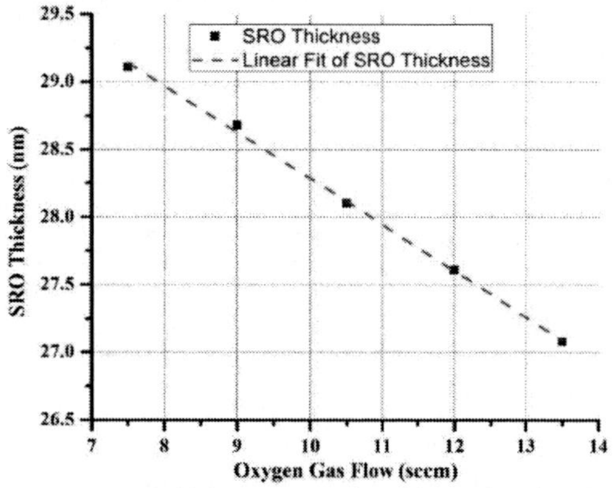

Figure 2: SRO thickness at 30 min of deposition time for variable proportion of Ar: O_2 ratio

2.1.3 Variation in sputtering power
The increase in the sputtering power increases the thickness of the SRO sample. And the increase in the proportion of the gas flow rate gives a higher value of (x) of the SRO layer. The sputtering power is kept at 75 watts and the gas flow rate of Ar: O_2 is maintained at 1:1. For a deposition time of 60 min, the stylus surface profiler measures the thickness of the deposited sample to be 70.4 nm. So, the deposition rate is 1.1 nm per minute.

2.2 Characterization of Si-QDs
The SEM-EDS spectroscopy determines the value of (x) in SRO layer. The value of (x) is nearly 0.01 for the sample deposited at a sputtering power of 50 watts, and at a lower value of oxygen gas. The value of (x) increases with the increase in the sputtering power of 75 watts and increase the oxygen gas flow rate of (Ar: O_2:: 1:1). The SEM characterization of the sample is carried out at 150X magnification and for that of EDS is an area of the sample. The value of (x) is 0.73 for the experiment carried out at 75 watts.

2.2.1 XRD Analysis
The XRD spectroscopy gives the peaks at various values of 2θ. The peaks centered around 56.28 (i.e., 2θ) attributable to Si (311) Bragg peaks. This diffraction peak indicated the crystalline precipitates and formed after annealing; the Si QDs size is 2.82 nm. The experiment is carried out with scan range (2θ) from 10^0 to 80^0, scan width of 0.02 nm, and at a scan rate of 20^0 per minute. There is a systematic nature in the peaks of the graph of the as-deposited samples and the annealed samples, as in Fig. 3. There are systematic peaks at some specified point, both for the as-deposited sample and the annealed sample, when the samples with different sputtering power, as in Fig. 4 , and a different proportion of the sputtering gas flow rate are plotted, as in Fig. 5.

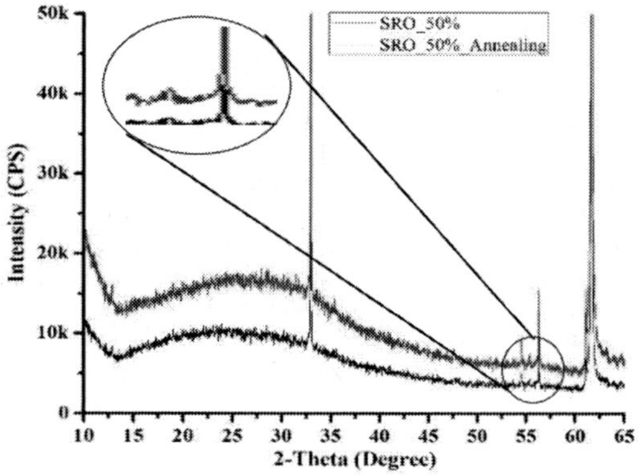

Figure 3: XRD peaks of the sample SRO_50% and SRO_50%_Annealing

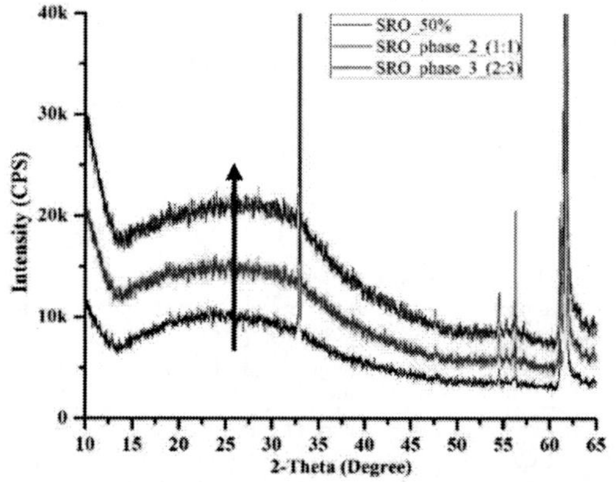

Figure 4: XRD peaks for different sputtering power and gas flow rate of the as-deposited sample

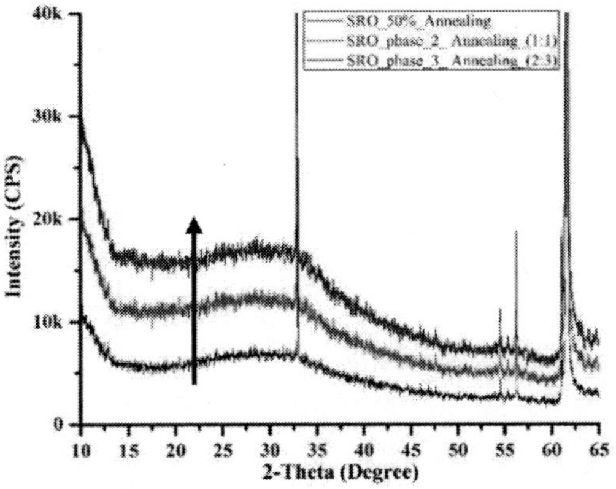

Figure 5: XRD peaks for different sputtering power and gas flow rate of the annealed sample

2.2.2 FTIR Analysis

The FTIR spectrophotometer (IRAffinity-1S, Shimadzu) do the characterization of all the samples. The transmittance of infrared (IR) spectra changes randomly for both the samples deposited at a different time of deposition, as in Fig. 6, and the samples deposited for the various proportion of gas flow rate, as in Fig. 7, respectively, for 50 watts of sputtering power. FTIR spectroscopy determines the composition, bonding configuration, and passivation state of as-deposited and annealed sample. The peak appears at 1100 cm^{-1} confirms the Si-O vibrational bond present in both the as-deposited sample and the annealed sample, as in Fig. 6 and Fig. 7. Fig. 8 exhibits the FTIR spectra of the as-deposited and annealed sample for 75 watts of sputtering power. The peaks appear at 1200 cm^{-1} shows the bridging vibrational Si-O bond, as in Fig. 8. The FTIR transmittance peak shift of the as-deposited and annealed sample suggests the Si atoms are bonded to more oxygen atom. It indicates that a near stoichiometric SiO_2 matrix is formed gradually and the phase separation of Si and SiO_2 occurs.

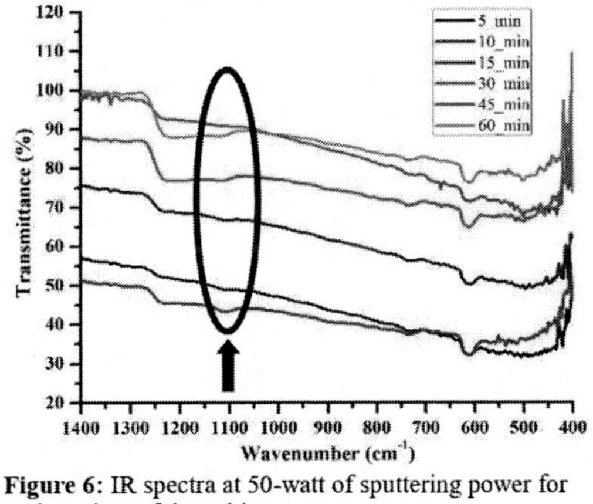

Figure 6: IR spectra at 50-watt of sputtering power for various time of deposition

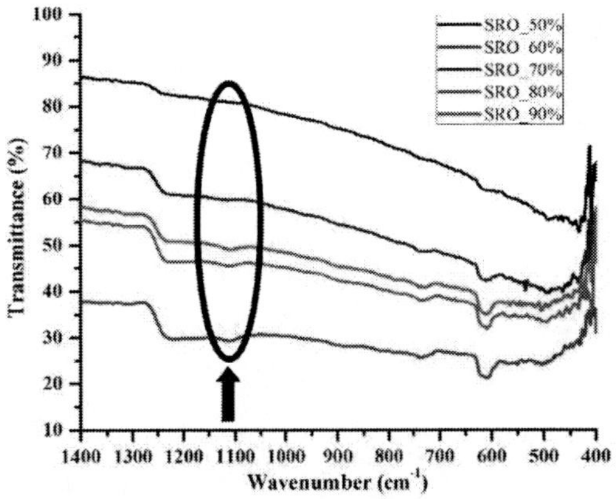

Figure 7: IR spectra at 50-watt of sputtering power for various proportion of gas flow rate (Ar: O_2)

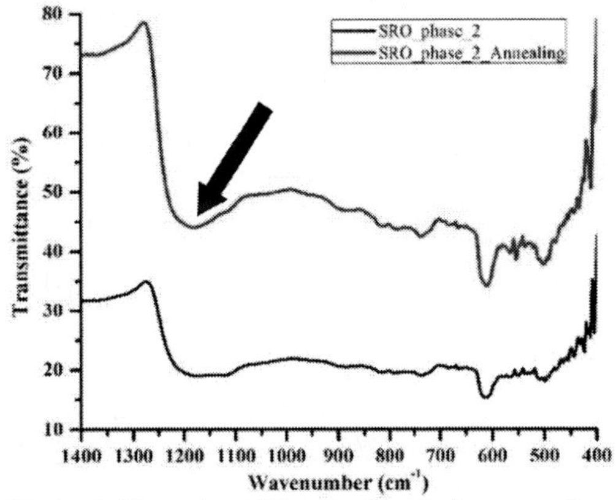

Figure 8: IR spectra at 75 watts of sputtering power for the 1:1 (Ar: O_2) proportion of gas flow

2.2.3 UV-Vis Analysis

The UV-Vis absorption spectra, as in Fig. 9, shows that the Si QDs in SRO layer absorb between 250 to 550 nm corresponding to a bandgap of 2.25-4.96 eV. The peak edge of the Si QDs at around 360 nm is 3.4 eV. The UV-Vis spectra indicate a blue shift from the as-deposited sample to the annealed sample, as in Fig. 9. Blue shift indicates a change in both the distance between the Si QDs and the refractive index of surroundings i.e. SRO layer. The increase in absorption for the deposited thin film layer, observed at short wavelength, indicates the absorption from the encapsulated Si QDs. The absorption increases with increase in the proportion of gas flow rate for both the as-deposited sample, as in Fig. 10, and the annealed sample, as in Fig. 11.

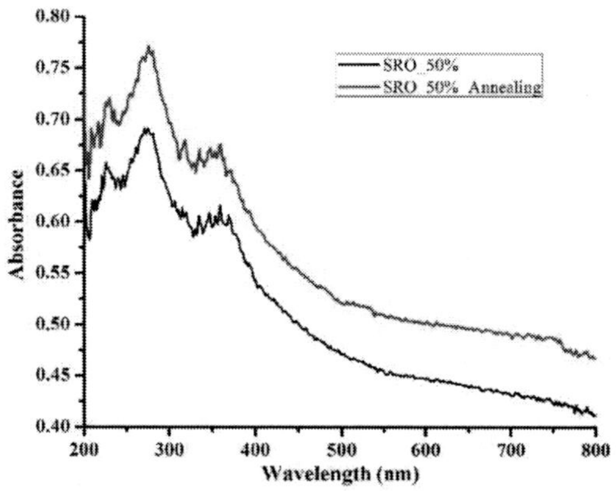

Figure 9: UV-Vis peaks of the sample SRO_50% and SRO_50%_Annealing

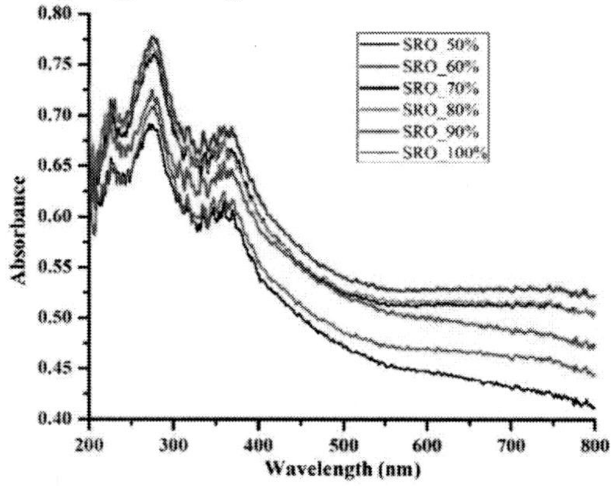

Figure 10: UV-Vis peaks at 50-watt sputtering power for as-deposited samples

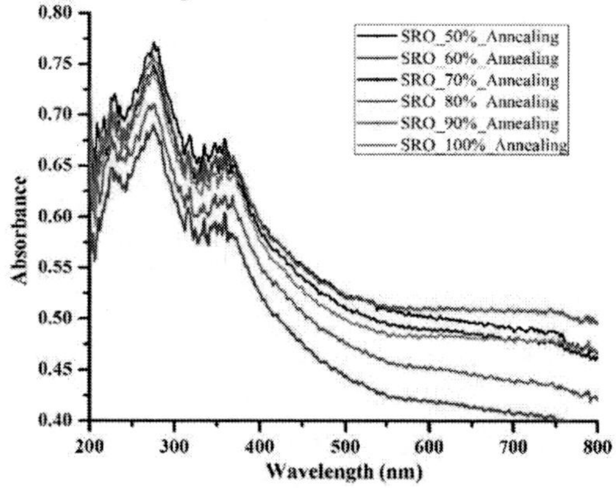

Figure 11: UV-Vis peaks at 50-watt sputtering power for annealed samples

2.2.4 Photoluminescence Analysis

There is a significant increase in PL intensity as the QD size decreases. The blue shift of emission band occurs for 0.01 of x-value of layer SiO_x, as in Fig. 12. PL upon excitation at a wavelength of 210 nm showed emission as a broad peak 550-700 nm with a maximum at 584 for the SRO_50%_Annealing sample. PL response is limited in blue region independent of the size distribution of Si QDs. PL properties depend on the quantum confinement and surface structure [9]. The surface passivation dangling bonds of Si QDs, with hydrogen, affect the PL properties [9]. The blueshift of PL and rapid change of intensity level after annealing process confirms the quantum confinement effect [10]. Due to enhanced confinement, the red shift of PL spectra occurs in the post deposition annealing. The Si nanocrystal size and the phase separation of Si/SiO_2 affect the PL spectra.

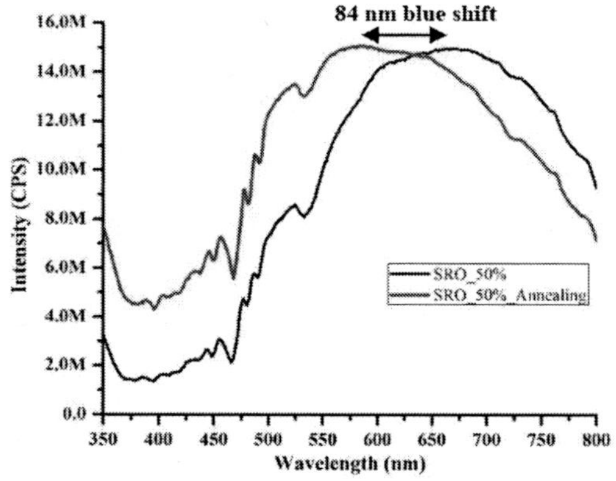

Figure 12: PL spectra of SRO layer (annealed at 950 ^{0}C, 60 min) deposited onto Si substrate

3 CONCLUSION

Si QDs are fabricated successfully by using the sputtering deposition method. The size of the Si QDs varies from 3 nm to 5 nm for various sputtering parameters. This Si QDs can be employed for the third-generation PV solar cell in which, the p-i-n structured PV solar cell has more efficiency than that of the previous generations' solar cell due to the wide range of solar spectrum absorption.

4. References

[1] E. Cho, S. Park, X. Hao, D. Song, G. Conibeer and M. A. Green, "Silicon quantum dot/crystalline silicon solar cell," *Nanotechnology*, vol. 19, no. 24, p. 245201, 2008.

[2] I. Perez-Wurfl, X. Hao, A. Gentle, D. Kim, G. Conibeer and M. A. Green, "Si nanocrystal p-i-n diodes fabricated on quartz substrates for third generation solar cell applications," *Applied Physics Letters*, vol. 95, no. 15, p. 153506, 2009.

[3] P. Kale and C. Solanki, "Silicon quantum dot solar cell using top-down approach," *International Nano Letters*, vol. 5, pp. 61-65, 2015.

[4] M. Green, G. Conibeer, D. Konig, E. Cho, D. Song and e. al, "Progress with all-silicon tandem cells based on silicon quantum dots in a dielectric matrix," *Proceedings 21st EU PVSEC, Dresden*, 2006.

[5] F. Rebib, E. Tomasella, L. Thomas, J. Cellier, T. Sauvage and M. Jacquet, "Study of Reactive RF Sputtering Process of Silicon Target Under Ar-O2-N2 Atmospheres," *Plasma Processes and Polymers*, vol. 4, no. S1, pp. S986-S990, 2007.

[6] M. Vila, C. Prieto, J. Garcia-Lopez and M. Respaldiza, "Influence of the target and working gas on the composition of silicon nitride thin films prepared by reactive RF-sputtering," *Nuclear Instruments and Methods in Physics Research*, vol. 211, no. 2, pp. 199-205, 2003.

[7] H. Seifarth, R. Grotzschel, A. Markwitz, W. Matz, P. Nitzsche and L. Rebohle, "Preparation of SiO2 films with embedded Si nanocrystals by reactive r.f. magnetron

sputtering," *Thin Solid Films,* vol. 330, no. 2, pp. 202-205, 1998.

[8] X. Hao, A. Podhorodecki, Y. Shen, G. Zatryb, J. Misiewicz and M. Green, "Effects of Si-rich oxide layer stoichiometry on the structural and optical properties of Si QD/SiO2 multilayer films," *Nanotechnology,* vol. 20, no. 48, p. 485703, 2009.

[9] M. H. Balci, J. Maria, F. Vullum-Bruer and e. ai, "Synthesis of Monodisperse Silicon Quantum Dots Through a K-Naphthalide Reduction Route," *Journal of Cluster Science,* vol. 23, no. 2, pp. 421-435, 2012.

[10] D. Lockwood, Z. Lu and J. Baribeau, "Quantum Confined Luminescence in Si/SiO2 Superlattices," *Physical Review Letters,* vol. 76, no. 3, pp. 539-541, 1996.

BOTH SURFACE TEXTURED GLASS: A NEW, INNOVATIVE AND EFFECTIVE APPROACH TO IMPROVE THE PERFORMANCES OF SUPERSTRATE TYPE THIN FILM SOLAR CELLS

Gourab Das*[a], Jayasree R Sharma [a], Sukanta Bose [a], Sumita Mukhopadhyay [a], Chandan Banerjee [b], Asok K Barua [a]
[a]Centre of Excellence for Green Energy and Sensor Systems, Indian Institute of Engineering Science & Technology, Shibpur, Howrah 711103, India
[b]National Institute of Solar Energy, Gwal Pahari, Gurgaon –122003, India

ABSTRACT: Light trapping is one of the major criteria to improve the solar cell performances. In the case of silicon based thin film solar cell texturization of front transparent conducting oxides (TCOs) is one of the practical ways for better light trapping by enhancing optical diffused transmittance and minimizing the reflection losses. A new method is adopted here; glass surface(s) is/are texturized instead of texturing the front TCO and is used in the fabrication of silicon based thin film solar cell. We used two different textured patterns for two different sides of the glass. This textured glass is found to be highly effective in terms of their optical properties. Over 12% and 13% improvement of short circuit current density is noticed in the cell fabricated single junction microcrystalline and amorphous silicon solar cell respectively on both surface textured glass as compared to non-textured glass. Single junction microcrystalline silicon solar cell (with simple p-i-n structure) and amorphous silicon solar cell with initial efficiency of 8.24% and 9.24% respectively are fabricated on this type of glass substrate without using back reflector.
Keywords: a-Si:H; μc-Si:H; Texturisation; TCO Transparent Conducting Oxides; Light Trapping; Thin Film Solar Cell

1 INTRODUCTION

Light trapping is one of the primary concerns among the researchesr in recent years to improve the solar cell performance. As in the superstrate type silicon based thin film solar cell, light enters the cell through the glass, texturization of front transparent conducting oxides (TCOs) coating is one of the regular ways for better light management by enhancing optical diffused transmittance and minimizing the reflection losses [1]. Light scattering in thin-film solar cells is traditionally achieved by using a textured transparent conductive oxide (TCO) layer, such as commercially available SnO_2:F with a randomly textured surface or textured ZnO:Al deposited either by sputtering followed by a post-deposition hydrochloric acid etching or by low pressure chemical vapor deposition [2-6] But due to the unoccasional spikes in textured front TCO, thickness of the window layer should be thick enough to avoid shorting of the cell through puncturing of the p-layer. To eliminate this problem we first textured the surface of the glass onto which TCO is deposited and avoid texturization of the front TCO. This will enable us to reduce the thickness of the window layer, without any shorting of the cell which in turns allows more light to the active part of the cell.

It was thought that sharp edges and occasional spikes on the textured TCO surface are responsible for cell shunting [7]. We texture the optically bottom side of the glass (used for superstrate type solar cell) on which TCO is coated in lieu of texturing the TCO to avoid shunting problem of the cell without compromising light trapping phenomena. In doing so, it was noticed that there is subsequent amount of reflection from the top surface of the glass. For this reason we also textured the top surface of the glass to cut down the reflected light and enhance the diffused part of the sunlight to the active part of the cell.

We have succeeded to maximize the total transmittance by using double side textured glass and optimize after using many different textured patterns. It also helps to lower the window layer thickness of the cell without shorting of the cell by puncturing through of the spike of TCO.

2 EXPERIMENTAL DETAILS

Texturization of the glass is carried out in reactive ion etching (RIE) unit using gas mixture of SF_6 and O_2 under various process conditions and optimized for solar cell application. 2 mm Schott glass are used here. Various process parameters for optimization of texturization process are shown in Table 1. Structural properties were investigated by Field Emissive Scanning Electron Microscopy (FESEM) supplied by Carl Zeiss and Atomic Force Microscopy (AFM, NT-MDT). Optical studies i.e. total transmission; diffuse transmittance and optical haze were performed by UV-VIS-NIR spectrophotometer (Shimadzu-UV1700).

Textured glass surface was covered by optimized 800 nm of ZnO:Al by DC sputtering (front TCO). Single junction μc-Si and a-Si solar cell have been fabricated onto these ZnO:Al coated textured glass substrate. We have also fabricated same μc-Si:H solar cells on the ZnO:Al coated non-textured glass and chemically etched ZnO:Al on non-textured (flat glass) and a-Si solar cell on standard Ashai-U SnO2:F and ZnO:Al coated non-textured glass for reference. Schematic diagram of the fabricated solar cells are shown in Fig. 1.

Table 1: Optimized parameters of the RIE process

Textured glass side	Gas ratio (SF_6:O_2)	Power density (mW/cm^2)	Pressure (Torr)	Time (Min)
Top side	5:1	200	0.15	15
Bottom side				40

The best texturization occurs at the chamber pressure of 150 mTorr, power density of 200 mW/cm^2. We optimized the texture profile for the both surface for obtaining highest total transmission. Top surface was processed for 15 min whereas bottom surface was textured for 40 minutes to obtained desired results.

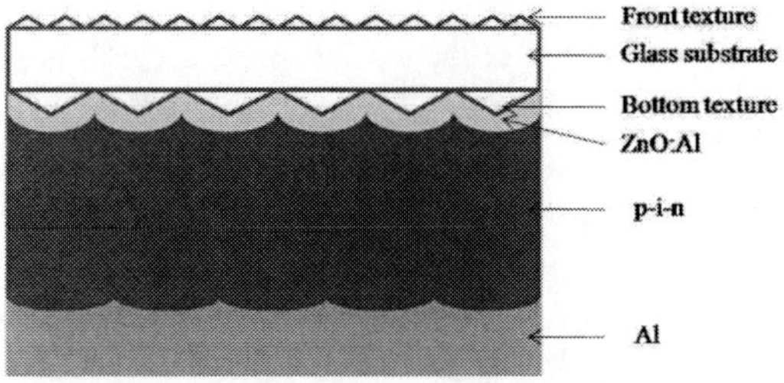

Fig. 1: Schematic diagram of thin film solar cell on both side textured glass substrate.

3 RESULTS AND DISCUSSIONS

Root mean square (R.M.S) roughness and average roughness of the bare Schott glass is 6.271 nm and 2.217 nm respectively. We textured two sides of glass surface differently. R.M.S roughness and average roughness of the textured glass of the top surface is 29.610 nm and 23.439 nm. Textured pattern of larger feature size is used for bottom surface. RMS roughness and average roughness of bottom surface texturization is 82.119 nm and 70.890 nm respectively.

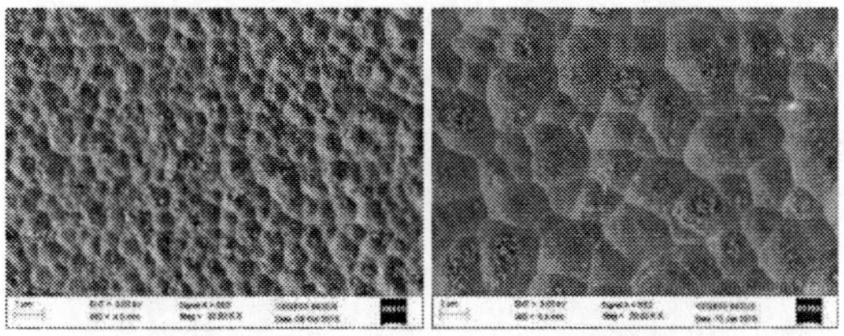

Fig. 2: FESEM images of textured of top surface of the glass (left) textured of bottom surface of the glass (right).

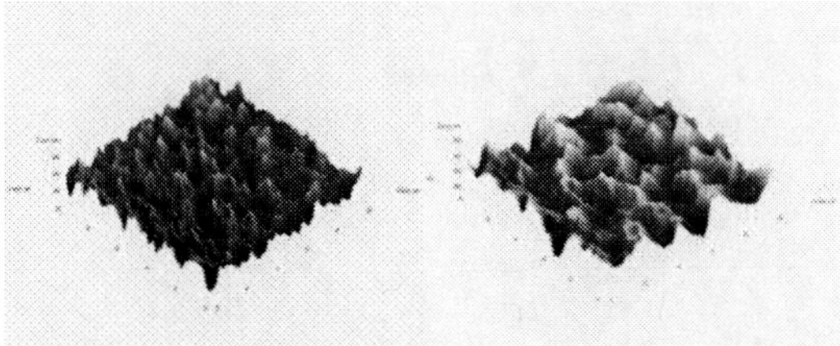

Fig. 3: AFM images of textured of top surface of the glass (left) textured of bottom surface of the glass (right).

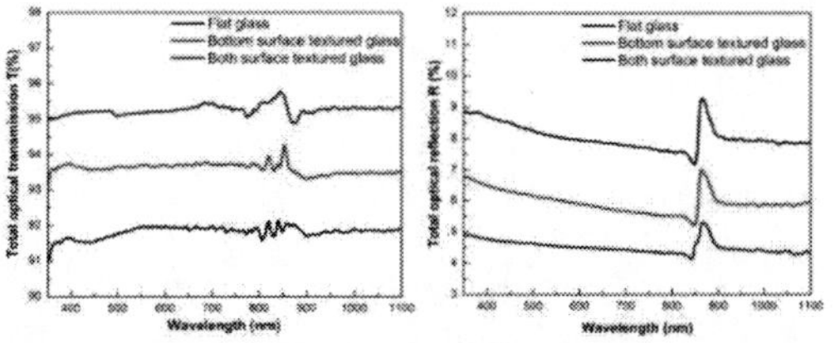

Fig. 4: Total optical transmittance (left) and total optical reflection (right) flat glass, bottom and both surface textured glass

We have measured the total optical transmittance and reflectance of the non-textured, bottom side textured and both side textured glasses. It is evident from Fig. 4 that texturing the both surface of the glass enhance the total light transmittance as well as the diffused transmittance. Total transmittance of the both side textured glass is enhanced by ~ 4% (absolute) compared to non-textured flat glass. It is also observed that reflection from the both side textured glass is found minimum with ~ 4.5% where as it is 8.05% and 5.95% in case of non-textured and only bottom side textured glass respectively. It is noticed that optical haze at 550 nm is increased to 51% for both side textured glass from 33% in only bottom surface texture and total transmittance is found to be 95.26% in average where it was 93.62% in case of only bottom surface textured glass.

Single junction μc-Si solar cell with initial efficiency (η) = 7.41% was obtained with J_{sc} = 23.79 mA/cm^2, V_{oc} = 0.48 V and F.F. = 64.9% in the flat substrate. I-V characteristics of the fabricated solar cells are depicted in Fig. 5. In the chemically textured ZnO:Al substrate, η = 7.75% was observed with J_{sc} = 25.00 mA/cm^2. In case of textured glass substrate we have found highest initial efficiency of 8.01% which is 8.1% higher than the solar cell on flat substrate and 3.4% higher than the chemically textured ZnO:Al coated substrate. It is noticed that J_{sc} was mainly enhanced in the textured glass substrate due to its high diffused and total transmittance. All the PV parameters of the μc-Si soalr cells are shown in Table 2.

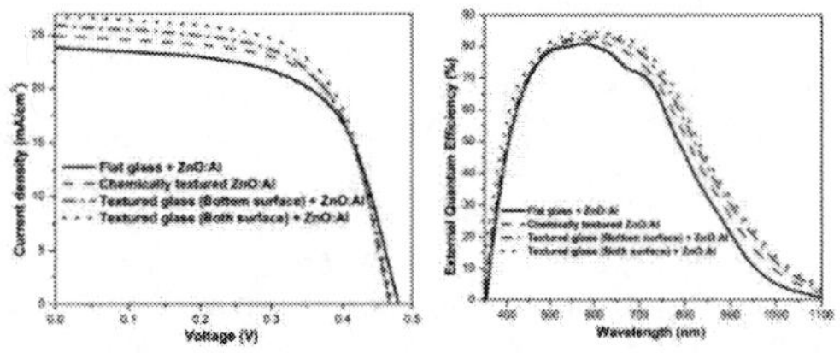

Fig. 5: I-V characteristics (left) EQE curve (right) of the μc-Si soalr cell on flat glass, bottom side, both side textured and conventional glass substrate.

Table 2: PV parameters of μc-Si solar cells fabricated on different substrates.

Type of substrate	J_{sc} (mA/cm^2)	V_{oc} (V)	FF	η (%)
Flat glass	23.79	0.48	0.649	7.41
Chemically textured ZnO:Al	25.00	0.47	0.660	7.75
Textured glass (Bottom surface)	25.90	0.47	0.658	8.01
Textured glass (Both surface)	26.73	0.47	0.653	8.24

From the EQE spectra (Fig. 5) it is observed that response increases after 500 nm of the solar spectrum in the solar cell on textured glass substrate. It can be noticed from the I-V curve (Fig. 5) and Table 2 that short circuit current density (J_{sc}) is enhanced by 3.2% in the solar cells fabricated on both side textured glass substrate than that of only bottom surface textured glass substrate without affecting the all other PV parameters resulting an improvement in initial efficiency (η) by 2.9%. From the EQE spectra it is observed that there is enhancement in the spectral response mainly in lower wavelength region. This is may be due to less reflection from the top glass surface of the textured glass.

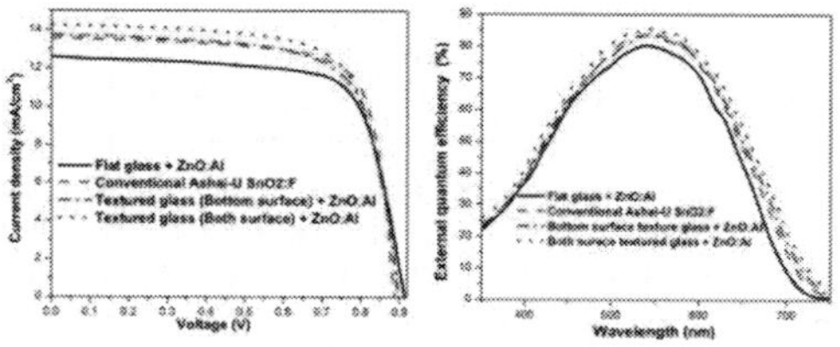

Fig. 6: I-V characteristics (left) EQE curve of the μc-Si solar cell on flat glass, bottom side, both side textured and conventional glass substrate.

Table 3: PV parameters of a-Si solar cells fabricated on different substrates.

Type of substrate	J_{sc} (mA/cm^2)	V_{oc} (V)	FF	η (%)
Flat glass	12.60	0.91	0.74	8.48
Conventional Ashai-U	13.57	0.89	0.749	9.05
Textured glass (Bottom surface)	13.74	0.90	0.724	8.95
Textured glass (Both surface)	14.26	0.90	0.72	9.24

Single junction a-Si solar cell with initial efficiency (η) = 7.41% was obtained with J_{sc} = 12.60 mA/cm^2, V_{oc} = 0.91 V and F.F. = 74.0% in the flat substrate with ZnO:Al as TCO. I-V characteristics of the fabricated solar cells are depicted in Fig. 6. In the conventional Ashai-U SnO_2:F substrate initial efficiency, η = 9.05% was observed with J_{sc} = 13.57 mA/cm^2. In case of only bottom side textured glass substrate we have found the initial efficiency of 8.95% which is 5.5% higher than the solar cell on flat substrate but 1.1% lower than the conventional Ashai-U FTO coated substrate as the fill factor was dropped by 3.3% due to lower sheet resistance of the in house ZnO:Al compared to Ashai-U FTO. All the PV parameters of the a-Si soalr cells are shown in Table 3

From the EQE spectra (Fig. 6) it is observed that response increases after 460 nm of the solar spectrum in the solar cell on textured glass substrate compared to the solar cell on flat glass substrate. It can be also noticed that the response in the higher wavelength region of the solar cells on only bottom side textured glass substrate and on Ashai-U glass substrate is more or less same.

It is noticed from the Table 3 that short circuit current density (J_{sc}) is enhanced by 3.8% in the solar cells fabricated on both side textured surface than that of only bottom surface textured glass substrate without affecting the other PV parameters (only a very little reduction in fill factor is noticed) by resulting an improvement in initial efficiency (η) by 3.24%. From the EQE spectra it is observed that there is a slight enhancement in the spectral response in whole of the wavelength region.

CONCLUSIONS

Texturization of both surfaces of the glass has a superior effect in terms of light management in silicon based thin film solar cell. Size, shape and distribution of the craters are well controlled by RIE process. Texturization with large feature size in the bottom surface helps to promote the total transmittance enhancing the diffused part and small feature size in the top surface of the glass helps additionally to cut down the reflection. So texturing the both surface of the glass is a simple, innovative and very effective method to improve the performance of superstrate type silicon thin film solar cell.

REFERENCE(S)

1. M. Mizuhashi, Y. Gotoh and K. Adachi, Japanese Journal of Applied Physics, 27 (1988) 2053.
2. S. Fernández, O. de Abril, F.B. Naranjo, J.J. Gandía, Solar Energy Materials and Solar Cells 95 (2011) 2281–2286.
3. Y-H. Hu, Y-C. Chen, H-J. Xu, H. Gao, W-H. Jiang, F. Hu, Y-X. Wang, Texture ZnO Engineering 2 (2010) 973-978.
4. X. Yan, S. Venkataraj, A. G. Aberle, International Journal of Photoenergy, Volume 2015 (2015), Article ID 548984.
5. W. W. Wenas, A. Yamada, M. Konagai and K. Takahashi, Japanese Journal of Applied Physics, Volume 30 (1991) L441.
6. S. Nicolay, M. Despeisse, F-J. Haug, C. Ballif, 95 (2011) 1031–1034.
7. A. Chowdhury, D-W. Kang, M. Isshiki , T. Oyama, H. Odaka, P. Sichanugrist, M. Konagai, Solar Energy Materials & Solar Cells 140 (2015) 126–133.

EFFICIENT LIGHT HARVESTING IN SURFACE BARRIER SOLAR CELLS WITH QUASIPERIODICAL MICRORELIEF AND METAL NANOWIRES

N.L.Dmitruk, A.V.Korovin, O.Yu.Borkovskaya, I.B.Mamontova, S.V.Mamykin, N.V.Kotova, V.R.Romanyuk
Institute for Physics of Semiconductors, National Academy of Sciences of Ukraine
45 Nauky Prospect, Kyiv 03028, Ukraine

ABSTRACT: Effect of increasing of Au(ITO)/GaAs surface barrier solar cells efficiency due to improving of electrical characteristics and light harvesting by quasiperiodical microrelief (quasigratings) formation and metal nanowires deposition has been studied. The technology of quasigrating preparation on GaAs substrate and deposition of Au nanowires has been elaborated. Theoretical modelling and experimental investigation of optical, photoelectric and electric characteristics of these structures shows that besides of regular light trapping the physical mechanisms of the photocurrent enhancement effect are: 1) excitation of surface plasmons/plasmon polaritons in metal nanowires; 2) improving of electrical characteristics due to decrease of the interface oxide layer thickness; 3) increase of photocarriers collection area due to dense net of Au nanowires on the surface of GaAs substrate.
Keywords: nanowires, surface plasmon (polariton), interfaces

1 INTRODUCTION

Surface texturing is known as an effective method to enhance solar cell (SC) efficiency via reduction of optical losses caused by the light reflection and absorption in metal contact electrodes. For polar semiconductors III-V (GaAs, InP) the technologies of various type surface microrelief (quasigrating, dendritic, pyramidal) preparation by wet chemical anisotropic etching in corresponding solutions have been elaborated [1]. The elaborated technology [2] allowed us to prepare quasiperiodical microrelief (quasigrating) with intended parameters variation at (100) surface of n-GaAs. An alternative method of the optical losses reduction is to use metal nanoparticles or periodical array of nanowires (NW) for excitation of surface plasmons (SP) or evanescent waves-surface plasmon polaritons (SPP). To investigate the possibilities of the solar cell photocurrent (and efficiency) enhancement by joint effect of these methods we use the formerly determined method of forming plasmonic metal (Au) nanowires on the ridges of GaAs surface microrelief during the deposition of Au from corresponding salt solution [3,4]. Here the surface microrelief of quasigrating-type has been manufactured on *n*-GaAs (100) with 10^{17} cm^{-3} doping impurities concentration by chemical anisotropic etching. The surface plasmons/plasmon polaritons (SP/SPP) excitation in periodic 1D array of metal nanowires can lead to essential increase of the light transmittance into semiconductor and corresponding photocurrent enhancement in a certain spectral range [5]. The investigated surface barrier solar cells with accordingly prepared GaAs surface were fabricated by evaporation deposition of Au in the form of circular diode or contact grid for Au/GaAs structures or by cathode evaporation of ITO layer forming ITO/GaAs structures. Microrelief morphology was investigated by AFM technique.

2 EXPERIMENTAL DETAILS

The quasiperiodical microrelief (quasigrating) was obtained on (100) surface of n-GaAs wafer with 10^{17} cm^{-3} concentration of doping impurities by chemical anisotropic etching in $1H_2SO_4{:}2HF{:}1.5H_2O_2$ mixture. Parameters of microrelief (period, the depth of grooves) were controlled by etchant temperature (22-24°C) and etching duration (40-130 s) (Fig.1).

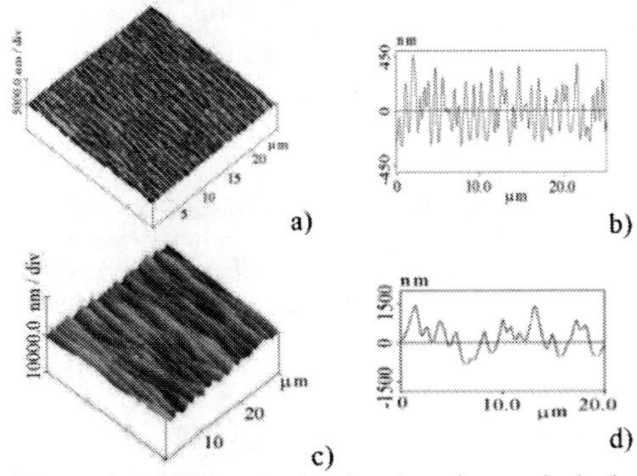

Figure 1: AFM images (a, c) and section analysis (b, d) of quasigrating microrelief surfaces with $N_d \cong 10^{17}$ cm^{-3} of n-GaAs having various predominating periods of gratings: $L_1 \cong 0.76$ μm (a, d), $L_2 \cong 1.9$ μm (c, d).

The increase of the mean period and depth of grooves was obtained by increase of the etchant temperature and process duration. Moreover, the microrelief parameters were also dependent on the semiconductor doping [6].

Au nanoparticles or nanowires were grown (Fig.2) on the flat and microrelief surfaces by photoinduced chemical deposition from aqueous $AuCl_3$ salt solution [7]. Au nanoparticles deposited on a flat surface by this method do not display preferred directions or specific arrangements and usually are randomly distributed.

In the case of the non-flat surface, gold nanoparticles are localized on ledges of the microrelief (Fig.2b). Isolated aggregates of particles thus placed on a substrate in such a manner as peaks of surface microrelief do. In the case of microrelief of the quasi-grating type, metal nanoparticles are localized at the ledges of the relief; thus 1D percolation of metal nanoparticles takes place along the edges. This means that the gold coating is a system of approximately parallel nanowires.

Investigated diode structures Au/GaAs and ITO/GaAs were fabricated in two ways: 1) by vacuum evaporation of semitransparent Au layer (~30 nm thickness) through the mask in the form of circular diodes with different surface area (1.33 and 0.5 mm^2); 2) by deposition of 100 nm ITO film (cathode sputtering) with following deposition of Au contact grid of SC type.

Surface topology was characterized with help of AURA100 SEM (SERON technology) and by AFM measurements which were carried out in the tapping mode using a Nanoscope IIIA (Digital Instruments, USA) with a silicon nitride tip.

Spectral dependencies of the photocurrent were measured in short circuit regime under choppered monochromatic polarized light illumination.

a)

b)

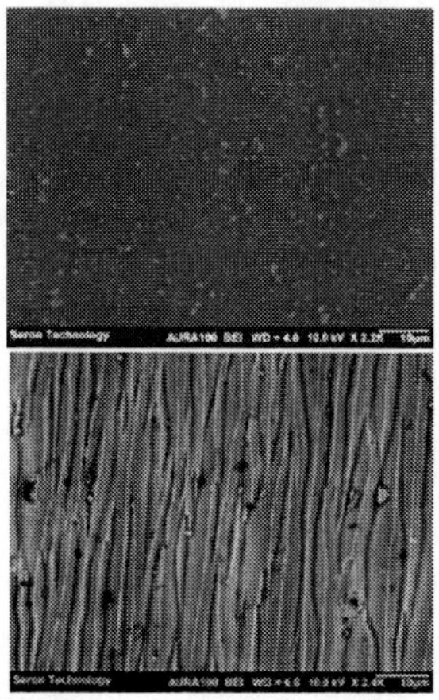

Figure 2: SEM images of flat (a) and quasigrating type (b) GaAs substrate after Au nanoparticles (a) and Au nanowires (b) deposition by photoinduced chemical growth method from aqueous $AuCl_3$ salt solution.

3 EXPERIMENT AND DISCUSSION

Measurement of the SC short-circuit photocurrent spectra (Fig.3) and the light current-voltage characteristics under AM0 simulated illumination showed that photocurrent enhancement caused by microrelief with predetermined parameters can reach from 2 to 10 (Fig.4) in different spectral range and it can be increased in addition by a factor of 2 (or ≥10 in the ITO/GaAs case Fig.5) due to Au nanowires deposition.

3.1 Photoelectric characterization of investigated structures

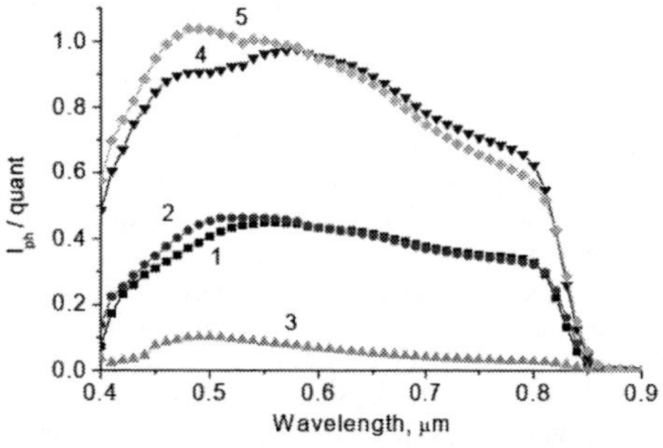

Figure 3: Spectra of short-circuit photocurrent, normalized to the diode surface area, for Au/GaAs structures with flat surface (3) and with quasigrating-type microrelief (1,2,4,5) without (1,2) and with Au nanowires (4,5) under *p*- (1, 4) and *s*- (2, 5) polarized light.

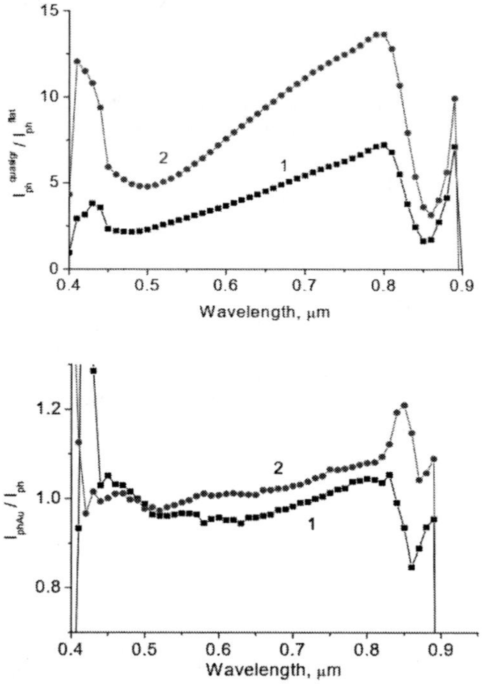

Figure: 4. a) Spectra of the photocurrent enhancement due to GaAs surface texturing (curve 1) and joint effect of Au nanowires and texturing (curve 2) for Au/GaAs diode structure.

b) Spectra of the ratio of photocurrent enhancement caused by Au nanowires, under *p*-polarization of light to that at the *s*-polarization for Au/GaAs structures with the mean period of quasigrating of 1.9 μm (curve 1) and 0.76 μm (curve 2).

3.2 Photovoltaic characteristics

The light current-voltage characteristics of the investigated Au(ITO)/GaAs structures with self-organized microtexturing of quasigrating type and Au NW on the ridges of relief, measured under simulated AM0 illumination, are shown in Fig. 6 and Fig. 7.

The forward dark current-voltage dependencies for Au/GaAs diode structures with flat and microtextured interfaces without Au and with Au nanoparticles or nanowires on the textured interface, are shown in Fig. 8.

Analysis of the obtained characteristics allowed us to determine the main mechanisms of joint effect of SC photocurrent enhancement caused by elaborated technology of Au nanowires formation on quasiperiodical microrelief of GaAs interface. It was shown that besides the optical effects a significant contribution into the enhancement effect gives the change of SC electrical characteristics, specifically the increase of the collection area for photocarriers, generated outside the barrier contact and decrease of the oxide layer thickness at the surface barrier interface.

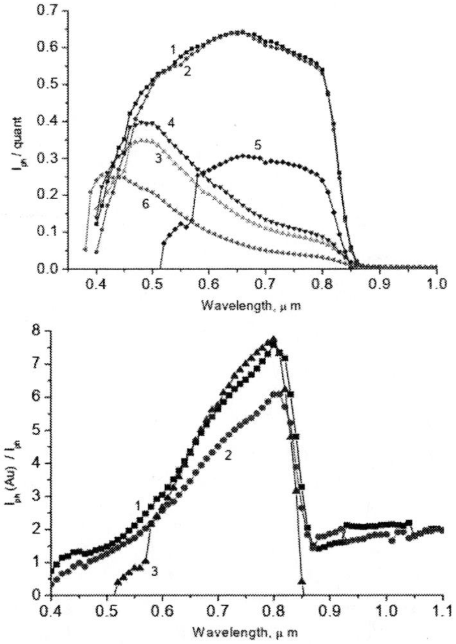

a)

b)

Figure: 5. a) Spectra of the short-circuit photocurrent for ITO/GaAs structures with flat surface of GaAs (curves 6, 5) and with interface microrelief of quasigrating-type (1-4) without (6, 3, 4) and with Au nanowires (1, 2), with Au nanoparticles (5) at *p*- (3, 1) and *s*- (4, 2) polarization of light.

b) Spectra of the photocurrent enhancement for ITO/n-GaAs structures due to Au nanoparticles on the flat interface (3) and Au nanowires on the microtextured one (1, 2) under *p*- (1) and *s*- (2) polarization of incident light.

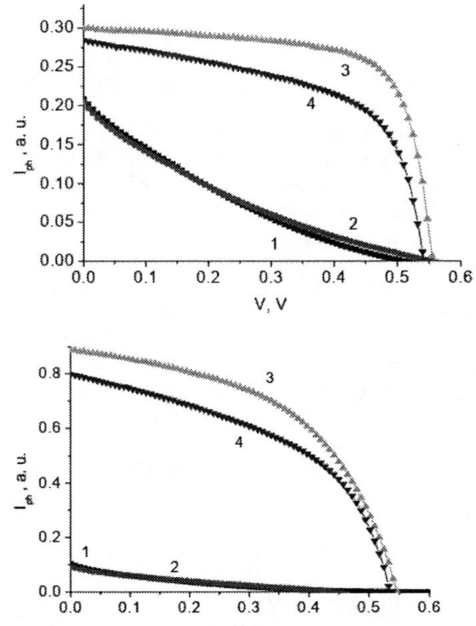

a)

b)

Figure 6: The light current-voltage characteristics under simulated AM0 illumination for Au/GaAs structures with circular diode (with diameter of ~0.8 mm) (a) and contact grid form of Au contact (b) and interface microrelief of quasigrating type without (1, 2) and with Au nanowires (3, 4). The mean period of quasigrating is equal to 1.9 μm (1,3) and 0.76 μm (2,4).

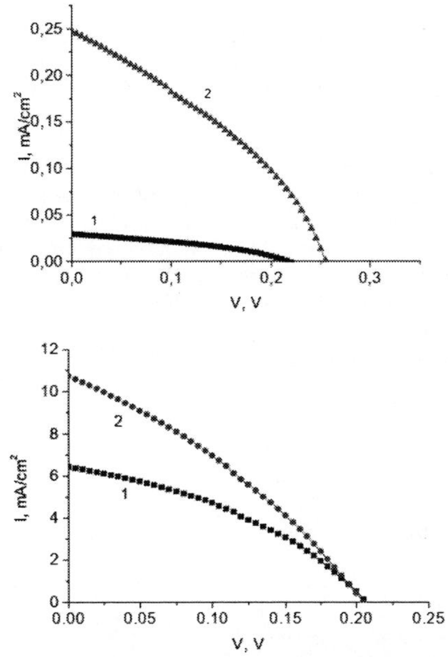

a)

b)

Figure 7: The light current-voltage characteristics of ITO/n-GaAs structures with flat (1) and microtextured (2) interfaces without Au deposition – (a), and with Au nanoparticles (b) on the flat interface (1) and Au nanowires on the textured one (2) – (b).

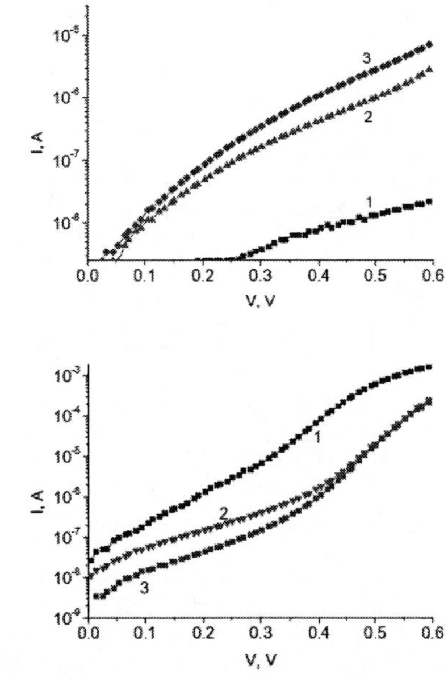

a)

b)

Figure 8. The forward dark current-voltage dependencies for Au/GaAs diode structures with flat (1) and microtextured (2, 3) interfaces without Au – (a), and with Au nanoparticles (1) or nanowires (2, 3) on the textured interface (2, 3) with mean period of quasigrating of 1.9 μm (2) and 0.76 μm (3) – (b).

4 THEORETICAL MODELING

Parameters of microrelief (mean period and depth of grooves, Fig.9) were varied by a change of the technological regime and were determined by AFM technique. These parameters were used in the theoretical modeling [4] and calculation of optical (transmittance, reflectance and absorption) spectra for *p*- and *s*-polarized light (Fig.10).

Theoretical modeling of optical properties for quasigrating GaAs surface with Au nanowires, located on relief edges, were performed using modified differential formalism method for solution of Maxwell's equations [8]. Using curvilinear coordinate transformation at the boundaries of the periodic array of one-dimensional (1D) nanowires deposited on a semiconductor substrate (GaAs, Si), the spatial distribution of electric and magnetic fields of s- and p-polarized light has been calculated.

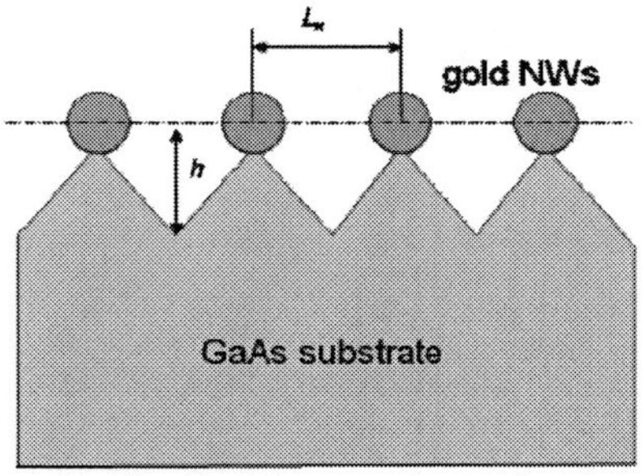

Figure 9: Schematic diagram of simplified model used in calculations on optical response of nanowire ensemble deposited on semiconductor substrate with QG microrelief. Here *h* is the depth substrate texturization and L_x is the nanowires period.

Based on electric and magnetic fields distribution the transmittance was calculated (Fig.10) at the position of the metal/semiconductor (substrate) interface with averaging over the plane of interface. Calculation were made for geometrical parameters shown on Fig.10 which is roughly correspond to our experimenatl samples. Quasigrating period were 0.76 μm (Fig.10a) and 1.92 μm (Fig.10b) with varied modulation depth. The optical parameters of a bulk semiconductor and gold were interpolated using the tabular data given from [9]. For practical usage of periodical ensembles of plasmon-carrying wires in SCs, the natural light is utilized. Therefore, the light transmittance was averaged over both polarization states $(T_p+T_s)/2$. Thus, in Fig. 10, we observe both increase and decrease of light transmittance in various spectrum ranges. And these peculiarities are especially strongly manifested for p-polarized light in contradiction to s-polarization, and there are clearly observed peculiarities for p-polarized light (arround 760 nm, Fig.10a) that correspond to the diffraction orders according to the momentum conservation law for SPP's excitation. Also, there are some dispersionless peculiarities (mainly at 600 and 800 nm, Fig.10a) in the transmittance of p-polarized light that could be explained in terms of local surface plasmon excitations across the single NW.

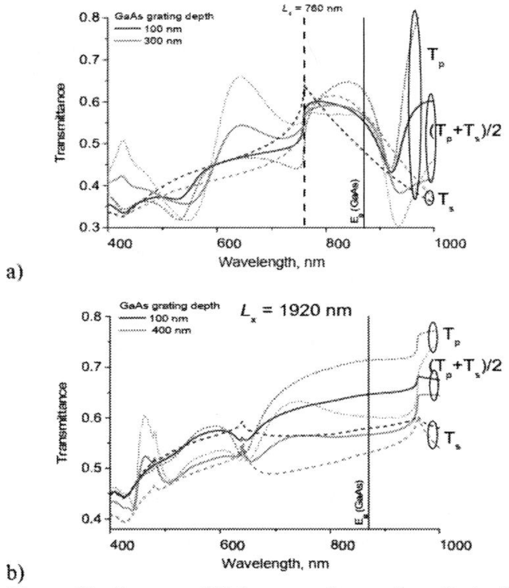

Figure 10: Spectra of light transmittance into GaAs for structures with the grating period of 760 nm (a) and 1920 nm (b) under illumination of *p*- and *s*-polarized and nonpolarized $(T_p+T_s)/2$ light.

CONCLUSIONS

It is found the possibility to obtain joint effect of the Au(ITO)/GaAs SC photocurrent (and efficiency) enhancement using two methods of optical losses decrease – the quasiperiodical microtexturing of GaAs surface and deposition of Au nanowires on the microrelief ridges causing excitation of surface plasmons/plasmon polaritons (SP/SPP) in them. The elaborated self-organized technology allowed us to realize such modification of interface in surface barrier structures.

Theoretical modelling and experimental investigation of optical, photoelectric and electrical characteristics of these structures allowed us to determine physical mechanisms of the photocurrent enhancement effect and its dependence on geometrical parameters of microrelief and nanowires controlled with technological regime.

It was determined that introduction of plasmonic metal (Au) nanowires in Au(ITO)/GaAs SC structures by means of elaborated technology increases the photocurrent enhancement effect caused by microtexturing by a factor of ~2 (or ~10 in ITO/GaAs). Besides the optical causes the mechanism of enhancement effect includes changes of electrical characteristics due to increase of photocarrier collection area with Au nanowires and decrease of the interface oxide layer thickness.

ACKNOWLEDGEMENTS

This work was supported by the NATO SPS grant NUKR.SFPP984617 "Nanostructured metal-semiconductor thin films for efficient solar harvesting".

6 REFERENCES

[1]. N.L. Dmitruk, O.Yu. Borkovskaya, I.B. Mamontova, S.V. Mamykin, Solar Energy Materials and Solar Cells 60 (2000) 379.

[2]. N.L. Dmitruk, O.Yu. Borkovskaya, I.N. Dmitruk, I.B. Mamontova, Solar Energy Materials and Solar Cells 76 (2003) 625.

[3]. N. Dmitruk, T. Barlas, A. Dmytruk, A. Korovin, V. Romanyuk, J. Nanoscience and Nanotechnology 8 (2008) 564.

[4]. N. L. Dmitruk, O. Y. Borkovskaya, I. B. Mamontova, S. V. Mamykin, S. Z. Malynych, and V. R. Romanyuk, Nanoscale research letters, 10 (2015).

[5]. N.L. Dmitruk, A.V. Korovin, Semiconductor Science and Technology, 28 (5) (2013) 055013.

[6]. N.L. Dmitruk, A.V. Korovin, O.Yu. Borkovskaya, A.M. Dmytruk, I.B. Mamontova, S.V. Mamykin. Proceedings 27th European Photovoltaic Solar Energy Conference (2012) 408.

[7]. Barlas, T. R., Dmitruk, N. L., Kotova, N. V., Mayeva, O. I., & Romanyuk, V. R. (2005). Superlattices and Microstructures, 38(2), 130-141.

[8]. A.V. Korovin, JOSA A 25 (2008) 394.

[9]. Palik E 1985 Handbook of Optical Constants of Solids (Orlando, FL: Academic)

FEATURES OF SI^+ IMPLANTED N-GASB (100) PHOTOSENSITIVE STRUCTURE

R.V.Ghita[1], D.Pantelica[2], C.Logofatu[1], C.C.Negrila[1], P.Cristea[3], L.Fara[4]

[1]National Institute of Materials Physics, P.O.Box MG-7, Magurele, Bucharest, ROMANIA
[2]National Institute for Physics and Nuclear Engineering-"Horia Hulubei", P.O.Box MG-6, Magurele, Bucharest, ROMANIA
[3]University of Bucharest, Faculty of Physics, P.O.Box MG-11, Magurele, Bucharest, ROMANIA
[4]University "Politehnica" Bucharest, Faculty of Applied Sciences, Splaiul Independentei, No.313, Bucharest, ROMANIA
E-mail:ghitar@infim.ro

ABSTRACT

Gallium antimonide (GaSb) is an III-V semiconducting compound with a distinct lattice and band gap energy structure, enabling innovative application in optoelectronic devices. As substrate material in device technology, it allows operation on a wide spectral range i.e. 0.8÷ 4.3μm. Consequently, GaSb is an important competitor for solar energy conversion e.g. high efficiency thermo-photovoltaic (TPV) cells. Nowadays, GaSb technology is devoted to the development of high quality single crystals, efficient surface passivation techniques or competitive design of active *p-n* junction. The present work is dedicated to the study of Si^+ ion implantation in *n-GaSb*, with the aim of defining an active *p-n* junction at a precise level of local doping. The *n-GaSb* substrates were chemically prepared by cleaning and selective etching, and afterwards were exposed to Si^+ at E=1 MeV. The technological procedure included a recovery treatment and for defining the device structure, the metals were vacuum deposited in order to realize ohmic contacts i.e. PdGeAu/*n-GaSb* and Ag/*p-GaSb*. The characteristics of the sample in different technological stages were investigated by AFM analysis and selective XPS analysis together with electrical and optical measurements. The result of this experiment was the obtaining of photosensitive structure with a maximum photo-current response at 1.5μm.

Keywords: GaSb, Doping, Photoelectric Properties

1. INTRODUCTION

In this period of modern industry development that requires different energy sources, a solution in good viability with environmental protection is the use of photovoltaic cells for direct conversion of Sun energy. The Gallium Antimonide (GaSb) is the basis of the most PV cells and modern TPV (thermophotovoltaic) systems. In the development of PV and TPV applications there are three vectors involved, namely: efficiency, temperature resistance and cost. Regarding devices perspectives, GaSb has shown applications in laser diodes with low threshold voltage [1], photodetectors with high quantum efficiency [2], highly frequency devices [3] or in high efficiency thermophotovoltaic (TPV) cells [4]. GaSb is an III-V semiconductor compound with zinc blende crystal structure and direct band gap energy of 0.726 eV and is worth to mention that the structure GaAs/GaSb has set a record for solar cell efficiency of 35% opening a new era for photovoltaics applications [5]. In this view, we can say that the GaSb photosensitive structures offers the possibility of an almost total conversion of sun energy from visible to IR spectrum range where heat is transformed in electricity by TPV effect. We remark that the absorption wavelength of different industrial toxic gases are in the IR range where GaSb devices are recommended due to band gap structure and also, the near IR regime is suitable for biological and medical applications.

It is remarked [6] that GaSb is one of optimum substrate for some long-wavelength lasers and photodetectors and in this perspective, ion implantation is an attractive method to achieve a local layers doping in these devices. Ion implantation at 30-200 keV energy range is a well-established process in semiconductor industry [7]. The optimization of ion implantation in MeV regime requires, as in the case of low-energy process, a multiple characterization of implanted samples by different techniques as: ion channelling, TEM, electronic transport measurements, optical spectroscopy [8]. When the energy of implants is in the MeV range, implantation effects can extend several micro-meters deep into the sample. Generally speaking, when the defects are introduced in the lattice by implantation, the translation order is disrupted and for further purposes, as defining a device structure, the implanted sample is subjected to a recovery thermal treatment. The present work is dedicated to the presentation of some characteristics of Si^+ implanted GaSb substrate and to the related technology that finally conduced to an IR photosensitive structure.

2. EXPERIMENTAL

In present work, the *n*-GaSb (100) substrate surfaces were chemically prepared by cleaning in organic solvents and selective etching in HF:H_2O (DIW) (1:1) (t=10 sec) and HCl:H_2O (DIW) (1:1) (t=10 sec) at room temperature, in order to remove the native oxides. The as prepared samples were exposed to Si^+ using the 3 MeV ion source of the Tandetron accelerator at IFIN-HH (Magurele-Romania). The main ion beam was operated at E=1MeV, an implantation fluence $\Phi= (10^{13}\div10^{14})$ cm^{-2} with a beam current in the range I= (10÷15) nA. For the implanted *p-n* structure was calculated the main characteristics (as penetration depth; 0.9μm) by the use of SRIM software [9]. The samples were subjected to a recovery treatment at T= 600°C, t= 30 sec, in N_2 vapor flux (an RTA procedure). The sample was investigated regarding surface morphology before and after RTA, using an Atomic Force Microscopy facilities offered by a MFP 3D SA, Asylum Research microscope operated in the intermittent contact mode under ambient atmosphere. The tip used had a radius of curvature of 10 nm, spring constant k = 3.4 N/m and resonance frequency about 107 kHz.

Contact processing technology consists in vacuum deposition of ohmic contacts, namely PdGeAu/*n-GaSb* and Ag/*p-GaSb*. PdGeAu contact was obtained by successive deposition of Pd ~ 30 nm), Ge (~60nm), Au (~ 40 nm) from a W boat in HOCHVAKUUM Dresden installation at a medium pressure p~ $5\cdot10^{-5}$ torr. After deposition, the back side ohmic contact was subjected to an annealing process in low vacuum. Ag contact (~ 200 nm) was obtained by evaporation from a Mo boat in HOCHVAKUUM installation in medium vacuum condition. The front side ohmic contact was annealed in low vacuum T~450^{0}C. The characteristics of PdGe base contact, namely the depth profile were obtained by a controlled Ar^+ ion etching in an XPS system, followed by a XPS analysis. In this analysis was implied XPS a SPECS Spectrometer based on a Phoibos analyser with monochromatic X-rays emitted by an anti-cathode of Al (1486.7 eV). The hemispherical analyzer operated in the constant energy mode with a pass energy of 5 eV, giving an energy resolution of 0.4eV, which was established as FWHM of the Ag 3d 5/2 peak. The analysis chamber was maintained as specified at ultrahigh vacuum conditions (~10^{-9} torr). The recorded spectra were processed using Spectral Data Processor v 2.3 (SDP) software.

The electrical evaluation of ohmic contacts was established by the use of a Keithley 6517 a electrometer and the spectral response of the photoactive device was recorded in an optical dedicated set-up.

3. RESULTS AND DISCUSSION

Some data regarding the Si^+ implantation process are extracted from the SRIM simulation, as they are presented. In Figure 1(a,b,c,d). As can be observed, the penetration depth is in the range of 0.9µm, the collision events have an increasing trend to the 1µm limit (b) and there are extended damage defects by vacancies, replacement collisions or displacements as a function of Si^+ ion energy loss in the GaSb lattice (c,d)

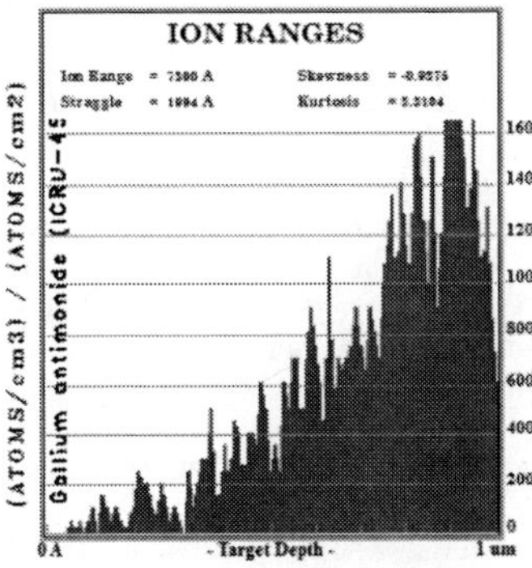

(a)

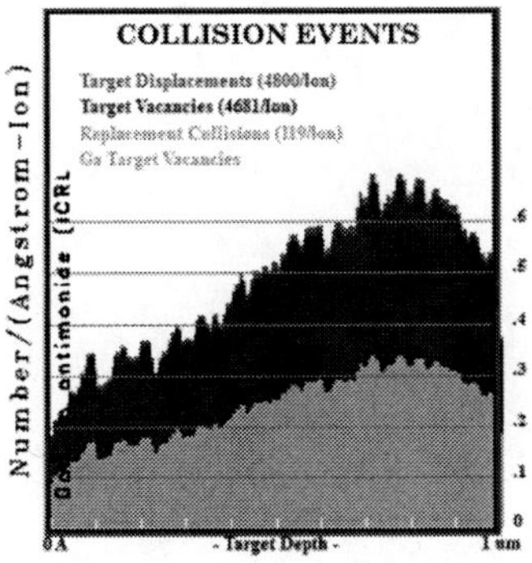

(b)

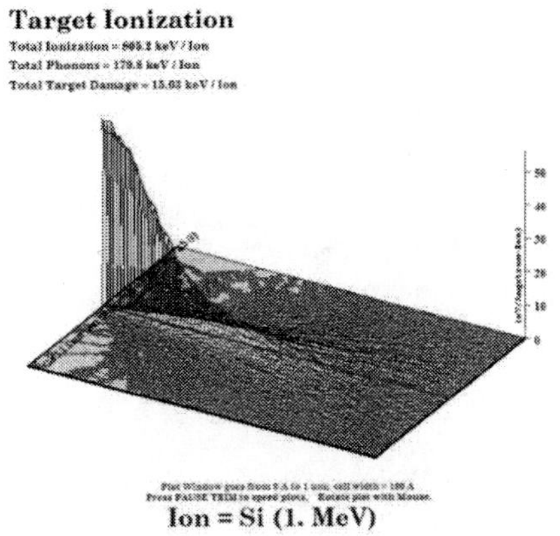

(c)

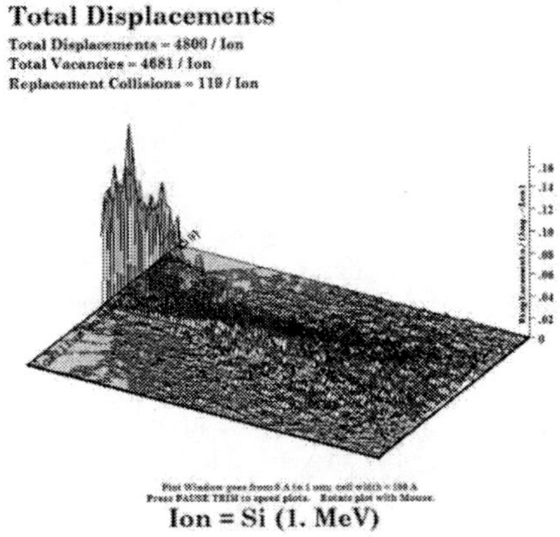

(d)

Figure 1 (a,b,c,d) Implantation effects in *n-GaSb*

General aspect of the implanted surface before and after recovery treatment is presented in Figure 2 (a,b), as AFM images. The surface details are in the same observations characteristics as presented in literature [9], the effect of thermal treatment being visual observed. We notice that there exist a structural modifications after ion implantation due to radiation damage, some of them being lattice defects. The effect of recovery treatment was also putted into evidence by the modification in photo-current in the sense of its increase, from I= 2µA, to I= 3 mA

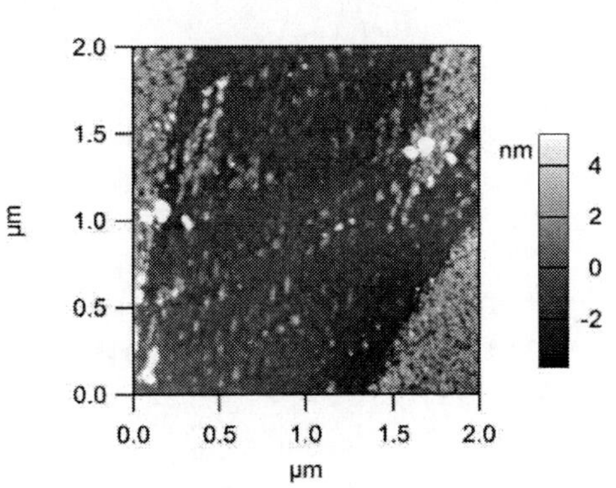

(a)

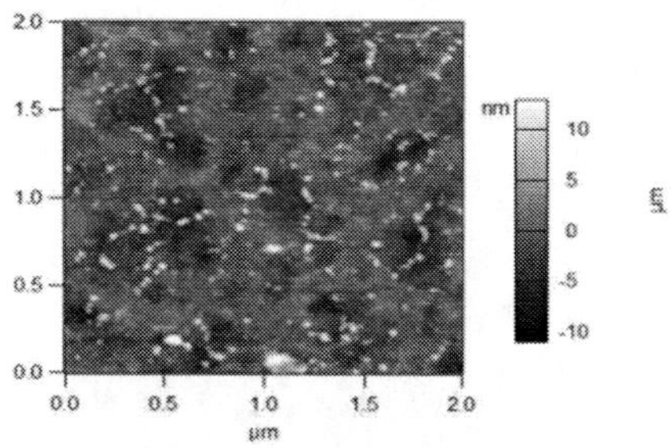

(b)

Figure 2 (a,b) AFM image of $Si^{+}/n\text{-}GaSb$ for as implanted sample (a) and after recovery treatment (b)

The active structure is subjected to a vacuum deposition procedure in order to obtain ohmic contacts on p-n junction, as PdGeAu/n-GaSb and Ag/p-GaSb. From XPS spectra, we notice in Figure 3 (a) that A-A doublet is the signal Sb 3d related to metallic Sb, in alloy or in a compound GaSb. The B-B doublet is the signal Sb 3d related to Sb oxide in the form Sb2O3 (ratio Sb: Sb2O3→43%:57%). In Figure 3(b), we notice that A-peak (29.61 eV) represents Ge in its metallic form, probably in alloy with the other metallic elements. The B-peak (32.28 eV) is related to oxidized Ge in the form Ge2O3. The doublet C-F (31.83eV; 33.16 eV) represents the line Sb 4d related to metallic Sb. The doublet D-E (34.32 eV; 35.74 eV) represents Sb oxide.

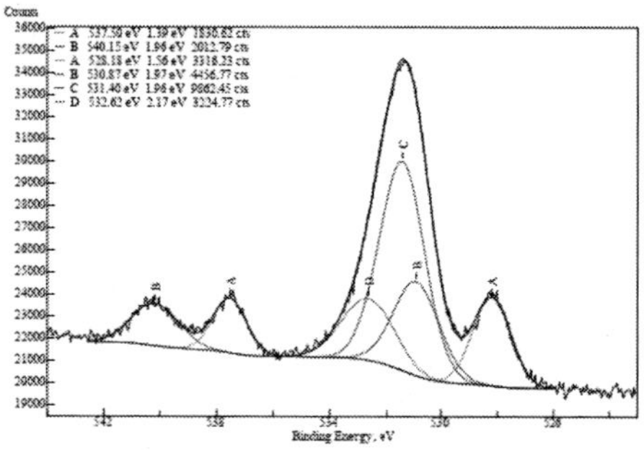

(a)

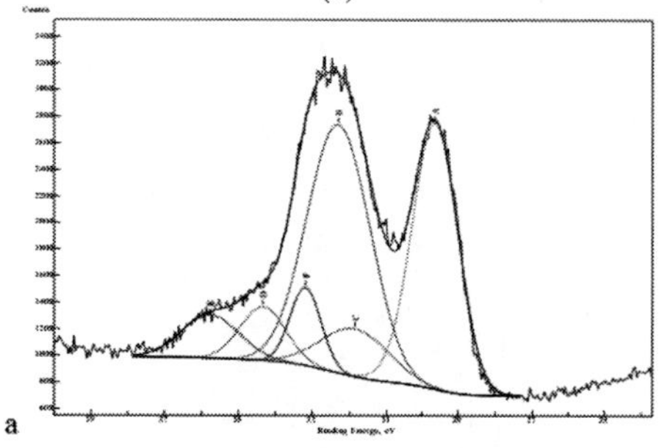

(b)

Figure 3 (a,b) XPS spectrum for Sb 3d on as deposited layer (a) and (b) XPS spectrum of Ge3d-Sb4d on as deposited sample

In Figure 4 (a) we present the Au 4f spectrum that indicates a single form of Au where the Biding Energy (BE) is higher in respect to pure Au and is characteristic to Au alloy, and the lines are characterizen an uniform surface. In Figure 4(b) the Ga 3d line (A-peak) is related to metallic Ga in alloy as in GaSb form; the B-peak is related to Ga oxide in the form Ga2O3 (Ga: Ga2O3→ 39%:61%). The C and D peaks represents the components of O 2s detected in O spectrum (metallic oxide-C peak and contamination oxides- peak D).

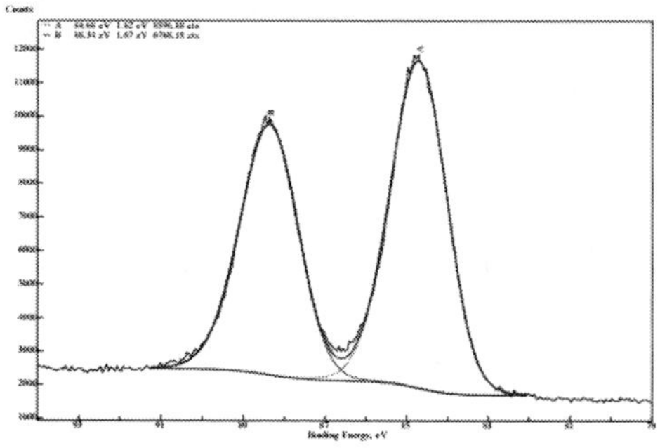

(a)

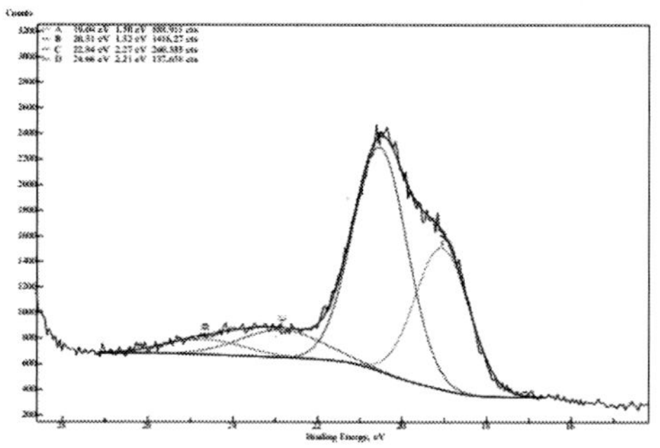

(b)

Figure 4 (a,b) XPS spectrum of Au 4f (a) and Ga 3d (b) lines on as deposited sample

The effect of controlled depth profiling can be observed in Figure 5 in the form of evolution of atomic ratios from the surface to the inner of semiconductor. The surface atomic composition indicates some benefic characteristics for ohmic contacts, namely: 1)- Pd and Ge are forming stable alloys and they are present at the interface and 2)- Ga is extracted from GaSb and generates Ga vacancies that can be replaced by Ge and in this way are leading to the increase of carriers number (n+doping of GaSb).

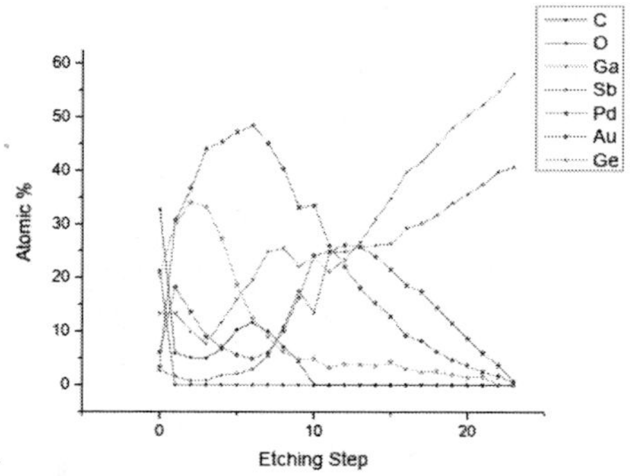

Figure 5 Evolution of atomic ratio of elements during controlled etching

The PdGeAu ohmic contact characteristic is presented in Figure 6 (a), where I –V curve slope is RS= 14.4 Ω. The same quality of Ag/p-GaSb ohmic contact extracted from I-V, indicates a curve slope RS = 10 Ω as is seen in Figure 6 (b).

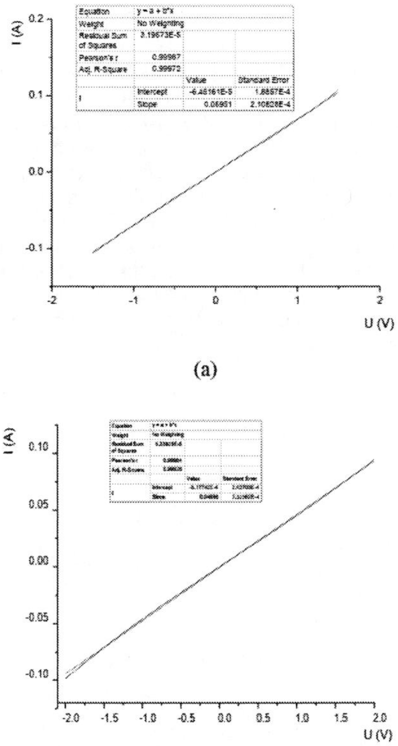

(a)

(b)

Figure 6 (a,b) I-V curves for PdGeAu/*n-GaSb* (a) and Ag/*p-GaSb* (b)

The spectral response of the photo-current I (λ) is presented for the active device in Figure 7. As can be observed the photo-current has a maximum at 1.5μm, and its responsivity as an IR sensor is similar to different devices on the market.

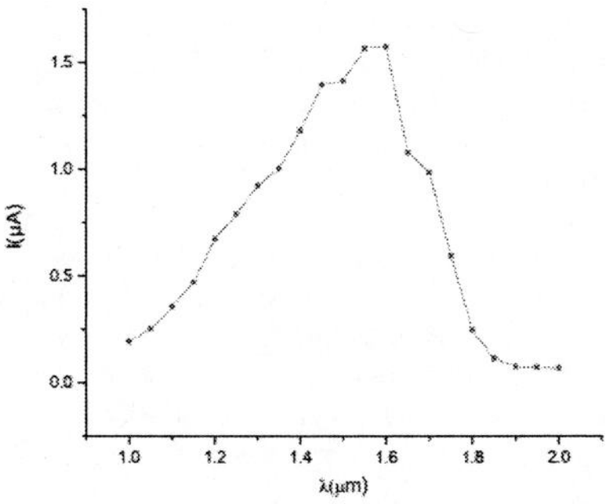

Figure 7 Spectral response of photo-current for *p-GaSb/n-GaSb* Si^+ implanted structure

4. CONCLUSIONS

Controlled doping of GaSb by Si+ ion implantation, conduced to the possibility of defining a new structure for a photosensitive device in near IR region. The technological steps in p-GaSb/n-GaSb device design implies: recovery treatment; selective chemical etching; vacuum deposition of PdGe base metals and Ag and their low vacuum annealing. The way to an active IR device implied also different characterization techniques, namely: AFM, XPS, electrical and optical measurements in dedicated set-ups. The photo-sensitive structure response in IR region indicates as regarding responsivity characteristics the presence of a competitive device on the market.

ACKNOWLEDGEMENTS

The authors thank for the financial support provided by the Ministry of Education- Romania-UEFISCDI Project No.68/2014-Partneship.

REFERENCES

[1] G.Motosugi and T.Kagawa, *Jpn.J.Appl.Phys,* 19, (1980), 2303.

[2] O.Hildebrand, W.Kuebant and M.H.Pilkuhn, *Appl.Phys.Let t* 37, (1980), 80.

[3] C.Hilsum and H.D.Rees, *Electron. Lett,* 6, (1970), 277.

[4] L.M.Fraas, G.R.Girard, J.E.Avery, B.A.Arau, V.S.Sundaram, A.G.Thomson and J.M.Gee, *J.Appl.Phys,* 66, (1989), 3866.

[5] L.M.Fraas, J.E.Avery, V.S. Sundaram, V.T Dinh, T.M. Davenport, J.M.Yerkes, „Photovoltaic Specialists Conference 1990", Conference Record of the Twenty First IEEE, Vol.1, 21-25 May, Kissimmee, Fl. USA, (1990), 190.

[6] R.Callec, P.N.Favennec, M.Salvi, H.L'Haridon and M.Gauneau, *Appl.Phys.Lett.* 59, (1991), 1872.

[7] M.D.Giles in *VLSI Technology*, 2nd ed., Edited by S.M.Sze, McGraw-Hill, New York, (1988), 327.

[8] G.Braunstein, D.Tuschei, S.Chen and S.T.Lee, *J.Appl.Phys.*66, (1989), 3515.

[9] J.F.Ziegler, M.D.Ziegler and J.P.Biersack, *Nucl.Instrum.Methods B*, 268, (2010), 1818.

[10] P.Kluth, J.Sullivan, W.Li, R.Weed,C.S. Schnohr,R.Giulian, L.L.Araujo, W.Lei, M.D.Rodriguez, B.Afra, T.Bierschenk,R.C.Ewing and M.C.Ridgway, *Appl.Phys. Lett.*, 104, (2014), 023105.

QUANTUM DOT LUMINESCENT SOLAR CONCENTRATOR: OPTIMIZATION OF CONCENTRATION AND THICKNESS

M. Rafiee*, S. Chandra, H. Ahmed and S. J. McCormack
Dept. of Civil, Structural and Environmental Engineering, Trinity College Dublin, Dublin 2, Ireland
* Corresponding author. Email: rafieem@tcd.ie Tel: +353 1 896 2671

ABSTRACT: The absorption coefficient of a Luminescent Solar Concentrator (LSC) can non-linearly affect the total optical efficiency (η_{opt}) and solar concentration ratio (C_P) of the device. Absorption is determined by two critical design parameters; the concentration of the luminescent material and the thickness of LSC plate. This paper presents a theoretical approach using a mathematical model based on a Ray Tracing algorithm to optimize these parameters to obtain the best LSC configuration design. A 60 × 60 × 3 mm LSC of CdSe/ZnS quantum dots (QDs) doped in epoxy resin was modelled and the model results were validated by comparing them to experimental results. To achieve the optimum thickness and luminescent material concentration of the LSC, a sensitivity analysis was carried out to predict the behavior of the LSC under different solar radiation. The LSC optical efficiency under standard AM1.5 global solar radiation was found to have reached a maximum (3.6%) at QD concentration of 0.55 wt%. Above this concentration, the optical efficiency was found to decrease due to re-absorption losses. Then, the thickness of the LSC was varied from 0.5 to 30 mm while the QD concentration was kept constant at the optimum (i.e. 0.55 wt%). Under these circumstances, the total optical efficiency has been observed to increase from 1.2% to 13.2% while the C_P was significantly decreased from 1.19 to 0.22 due to a decrease in the geometric gain.

Keywords: Luminescent Solar Concentrator, Ray Tracing, Modeling, Concentration, thickness, Quantum Dot, Emission, Absorption, Sensitivity Analysis

1 INTRODUCTION

Reducing spectral losses and enhancing efficiencies in Photovoltaic (PV) cells can be implemented by non-imaging optics based solar concentrating technology, such as LSC (Chatten et al., 2011, Chandra et al., 2012, S. Chandra, 2017). LSCs, illustrated in Fig. 1, consist of luminescent materials doped in a host material where the PV cell is attached to the edge of the device. As is seen, a portion of the solar radiation at the top surface is absorbed by the luminescent species and emitted at a single wavelength, where PV cells have higher efficiency. Then, through Total Internal Reflection (TIR), the emitted light is wave-guided to one of the LSC edges where the PV cell is located (Weber and Lambe, 1976, Goetzberger and Greube, 1977, Goetzberger, 1978). This improves the solar panel efficiency, as well as reducing the required PV area; thus, decreasing the final cost of the generated energy. LSC devices are able to convert and concentrate both diffuse and direct light; therefore, it is an adequate technology where diffuse solar radiation is dominant, such as in northern European countries where over 50% of light is diffuse (van Sark et al., 2008). They are low-cost and can be fabricated at small and large scales as transparent clear or coloured windows (Aste et al., 2015, Earp et al., 2004a). These features make them a preferred choice for use in Building Integration Photovoltaic (BIPV) systems and façades of buildings which brings us closer to the goal of constructing buildings with zero carbon energy consumption and buildings which are able to cover their required energy by renewable energies (Aste et al., 2011, Pagliaro et al., 2010, Debije and Verbunt, 2012). The absorption of LSC can non-linearly affect the performance and the total optical efficiency of the LSC device. From Beer Lambert law (Klampaftis et al., 2009, Şahin and Ilan, 2013, Abderrezek et al., 2013) absorbance is given as follows:

$$A = log_{10} T \tag{1}$$

Where T is the value of transmittance of LSCs. Absorbance is determined by two important design parameters; the concentrations of the luminescent material and the thickness of LSC plate.

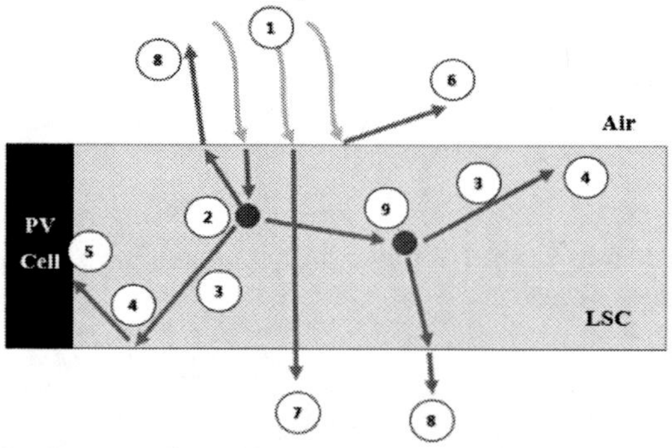

Figure 1: Configuration of LSC which shows: 1- solar ray is irradiated to the LSC, 2- absorbed by the luminescent material, then 3- re-emitted at longer wavelength and 4- wave-guided by Total Internal Reflection (TIR) and 5- reaches the PV cell. Losses include: 6- front surface reflection 7- the fraction of light which is lost through the bottom and 8- other surfaces (escape cone loss) and 9- the emitted light which is reabsorbed by another luminescent molecules and its energy is decreased

Modeling methods are required in order to design and optimize the LSC configuration. LSC mathematical models are implemented by either Thermodynamic or Ray Tracing methods (Batchelder et al., 1979, Carrascosa et al., 1983, Glassner, 1989, Barnham et al., 2000, Chatten et al., 2004, Earp et al., 2004b, Burgers et al., 2005, Kennedy et al., 2007, Kerrouche et al., 2014, Parel et al., 2015, Hughes et al., 2015). In a ray tracing model, each ray of light with specific wavelength and direction is traced until it leaves the system. The accuracy of both models were reported to be similar (Farrell et al., 2005) ; though, the ray tracing algorithm is more complex and time consuming in comparison with thermodynamic model (Glassner, 1989). The advantage of ray tracing is that it is more flexible and can simulate different geometries and shapes of LSC with multiple luminescent species under diffuse and direct light (van Sark et al., 2008).

In this paper, LSC doped with CdSe/ZnS QD is modelled using a ray tracing algorithm. Modeling results were compared to experimental results to validate the ray tracing algorithm. Afterwards, a sensitivity analysis was undertaken to study and estimate the behavior of the LSC while changing the thickness of LSC and the QD concentrations under different solar radiation intensities. The modeling results were used to achieve the best and optimized configuration for the LSC.

2 RAY TRACING IMPLEMENTATION AND VALIDATION

The LSC ray tracing algorithm is based on a set of ray intersection calculation processes in an iterative loop due to the phenomena such as reflection, refraction, absorbing, scattering, attenuating and TIR events inside the LSC device. The process is continued till the fate of the ray is detected and then the same algorithm is executed for the other incident rays. The occurrence probability of each event at each stage of the ray tracing algorithm is determined according to parameters such as wavelength and angle of the incident ray, characteristics of the host and luminescent material, dimensions and configuration of the LSC. After implementing the iterative loop for all input rays, the optical efficiency of the LSC device is calculated by:

$$\eta_{opt} = \frac{\int_{\lambda_{min}}^{\lambda_{max}} E_{P_{OUT}}(\lambda).d(\lambda)}{\int_{\lambda_{min}}^{\lambda_{max}} E_{P_{IN}}(\lambda).d(\lambda)} \qquad (2)$$

Where $E_{P_{OUT}}(\lambda)$ is the output energy spectrum detected at the edge of the LSC and $E_{P_{IN}}(\lambda)$ is the energy spectrum for the input solar radiation. Having known the optical efficiency of the LSC, the solar concentration ratio, C_P can be obtained by:

$$C_P = \Gamma \times \eta_{opt} \qquad (3)$$

Where Γ is the geometric gain of the LSC and it is known as the aperture surface ratio (A_{APR}) to the total PV cell area (A_{PV}) which can be calculated as follows:

$$\Gamma = \frac{A_{APR}}{A_{PV}} \qquad (4)$$

The ray tracing algorithm was used to model a 60 × 60 × 3 mm CdSe/ZnS QD LSC plate (Kennedy, 2010, Rowan, 2007, Kennedy et al., 2008, Kennedy et al., 2009) in which the PV solar cell was attached to one of its edges while the other edges have mirrors with reflectivity of ~1. The emission and absorption spectra of the LSC plate are shown in Fig.2.

The model was run for 100,000 input rays and the final results can be seen in Table I. Around 3.9% of the irradiated rays were reflected back from the top surface of the LSC device. The rest of the rays (96.1%) were refracted inside the LSC plate. Around 5.2% of rays were lost as heat inside the LSC which were either attenuated by the host material or absorbed by the QD and lost due to the non-unity quantum yield of the QD. Around 89% of the rays exited the LSC plate and were lost due the escape cone. Only a small portion of rays (~ 1%) were wave-guided to the edge of the LSC and detected by PV solar cell. The optical efficiency was calculated and found to be 1.6% which was in close agreement with the reference optical efficiency (1.63%).

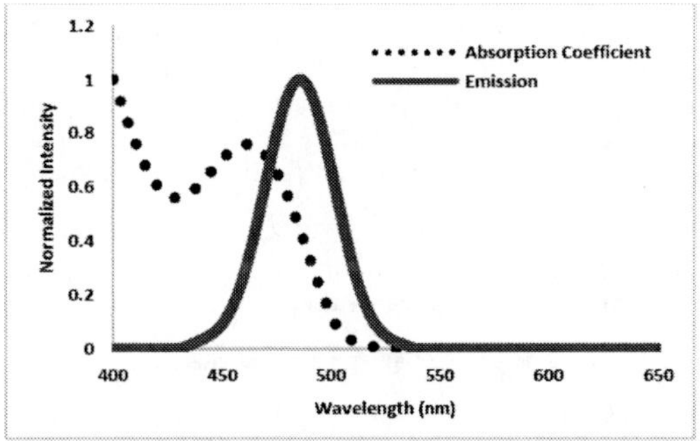

Figure 2: Normalized absorption coefficient and emission spectra for CdSe/ZnS QD LSC plate of 60×60×3 mm

Table I: Comparison and validation of the results obtained by the ray tracing model

Fate of the Rays	Value (%)
Reflected Light	3.90
Refracted Light	96.10
Lost Light	95.08
Light Reached PV	1.02
η_{opt} (REF=1.63% (Kennedy, 2010, Rowan, 2007, Kennedy et al., 2008, Kennedy et al., 2009))	1.6
C_P	32

3 SENSITIVE ANALYSIS AND DISCUSSION

In the second stage of the study, a sensitive analysis was undertaken to optimise the thickness of the QD LSC plate and the concentration of the QD under the standard AM1.5 solar radiation (Global, Direct and Diffuse). The QD concentration was varied from 0.1 to 1.2 wt% while the thickness of the LSC plate was 3 mm (Γ = 20). The modelling results are shown in Fig 3. Under global solar radiation, it has been observed that, the optical efficiency is linearly increasing as the QD concentration increased and reached its maximum (0.036) at concentrations between 0.55 - 0.6 wt%. Higher concentrations than 0.6 wt%, the optical efficiency is dropped and a saturation point is reached. The same trend was observed for the C_P as presented in Fig.3.b where it reached a maximum value of 0.7 and then started to decrease. The main reason for such behaviour is that by increasing the QD concentration, the re-absorption loss probability is increased. Moreover, by increasing the QD concentration, the emission peak was observed to be red-shifted. This is believed due to the fact that the photon loses some part of its energy in each re-absorption cycle. Fig 3.c shows the wavelength of the emission peak resulting from LSCs with different QD concentration values. As can be seen, for QD concentration in the ranges of 0.1 to 0.3 wt%, the emission peak is red-shifted sharply from around 495 nm to 504 nm. It remains constant up to QD concertation of 0.5 wt% and then it is slightly increased again.

Under direct and diffuse solar radiation, the behaviour of the LSC has a similar trend to LSC under global radiation. The maximum η_{opt} and C_P for the LSC under diffuse solar radiation were respectively 0.043 and 0.86, at a QD concentration of 0.5 wt%. Under direct solar radiation, the η_{opt} and C_P were found to be 0.016 and 0.32, respectively at a QD concentration of 0.55 wt%.

The optimum QD concentration is therefore suggested to be between 0.5 - 0.6 wt% where the highest optical efficiency can be achieved under different solar radiation intensities.

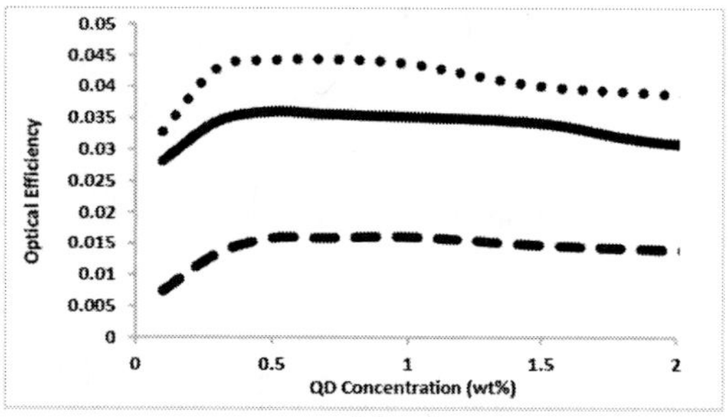

(a)

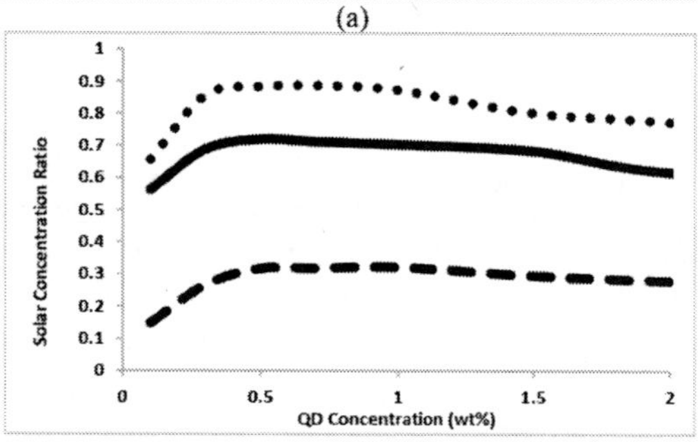

(b)

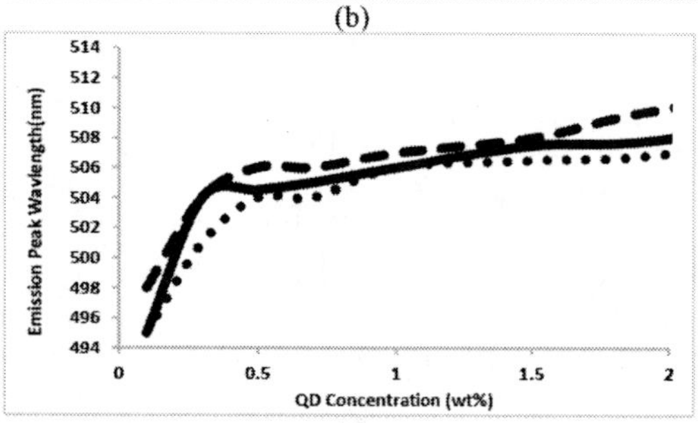

(c)

Figure 3: Modelling results under different solar radiations (Line: AM1.5 Global, Dash: AM1.5 Direct, Dots: AM1.5 Diffuse): (a) optical efficiency, (b) solar concentration ratio and (c) emission peak wavelength vs QD LSC concentrations

In the next step, the optimum concertation of the QD (0.55 wt%) was used to optimise the thickness, *d*, of the LSC plate in which *d* was varied from 0.5 to 30 mm while the value of QD concentration was kept constant. Fig.4 shows the optical efficiency for different LSC thicknesses.

It has been found that, the η_{opt} under global and diffuse solar radiation are in close agreement. The total optical efficiency increased non-linearly from around 0.01 to 0.13 for thicknesses between 0.5 to 30 mm. However, C_P is significantly decreased from around 1.7 to 0.5 for thicknesses between 0.5 to 5 mm and for thickness above 6 mm, C_P is slowly decreased to 0.22 (in thickness of 30 mm).

Under direct solar radiation, η_{opt} increased from 0.009 to 0.037 while the C_P decreased from 0.73 to 0.074.

This analysis shows that increasing the LSC thickness results in more light being detected at the edge of the LSC and increasing the optical efficiency; though, it reduces the geometric gain of the LSC which decreases the final solar concentration ratio, C_P.

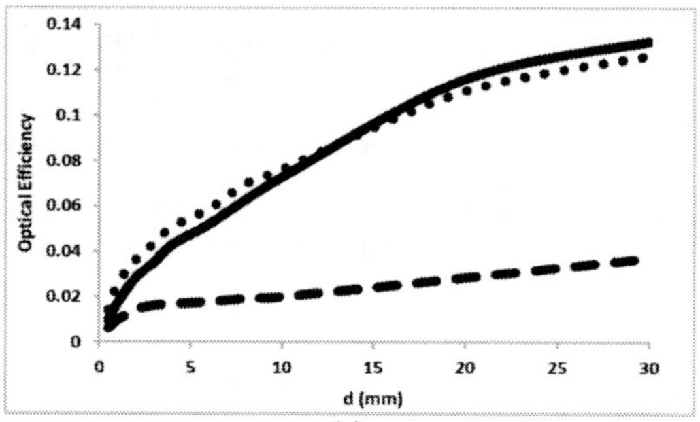

(a)

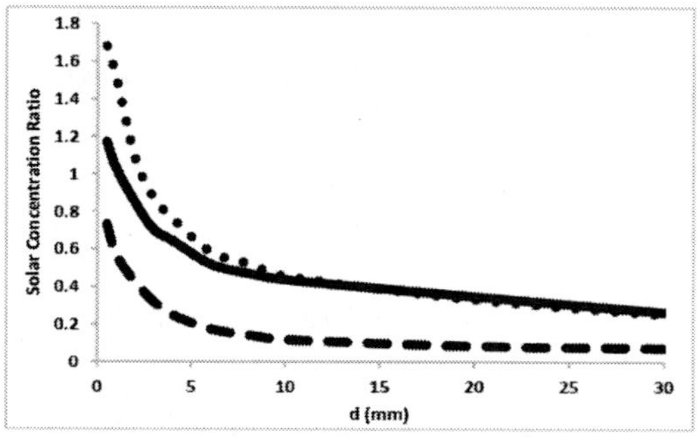

(b)

Figure 4: Modelling results under different solar radiations (Line: AM1.5 Global, Dash: AM1.5 Direct, Dots: AM1.5 Diffuse): (a) optical efficiency and (b) solar concentration ratio vs LSC thicknesses

4 CONCLUSION

In this paper, a ray tracing algorithm was used in order to model CdSe/ZnS QD LSC doped in epoxy resin polymer. The model was validated by comparing the simulation and measured results which were found to be in excellent agreement. Then, the validated model was used to study and predict the behavior of the LSC under different solar radiation intensities, LSC thicknesses and QD concentrations. The results showed that with QD concentration in the range of 0.5 - 0.6 wt%, the best η_{opt} and C_P can be obtained for the LSC plate under different solar radiation intensities investigated. It was found that, below a thickness of 6 mm the C_P was significantly increased due to the increase in geometric gain.

5 ACKNOWLEDGEMENT

The authors would like to acknowledge the funding from the European Research Council grant entitled PEDAL: Plasmonic enhancement of advanced luminescent solar devices (13379: 203889) and funding from Science Foundation Ireland (SFI). In addition, we also thank Dr. Manus Kennedy and Dr. Brenda Rowan whose thesis and research were used for validation of the model in this paper.

6 References

ABDERREZEK, M., FATHI, M., DJAHLI, F. & AYAD, M. 2013. Numerical Simulation of Luminescent Downshifting in Top Cell of Monolithic Tandem Solar Cells. *International Journal of Photoenergy,* 2013.

ASTE, N., ADHIKARI, R. & DEL PERO, C. Photovoltaic technology for renewable electricity production: Towards net zero energy buildings. Clean Electrical Power (ICCEP), 2011 International Conference on, 2011. IEEE, 446-450.

ASTE, N., TAGLIABUE, L. C., PALLADINO, P. & TESTA, D. 2015. Integration of a luminescent solar concentrator: Effects on daylight, correlated color temperature, illuminance level and color rendering index. *Solar Energy,* 114, 174-182.

BARNHAM, K., MARQUES, J. L., HASSARD, J. & O'BRIEN, P. 2000. Quantum-dot concentrator and thermodynamic model for the global redshift. *Applied Physics Letters,* 76, 1197-1199.

BATCHELDER, J., ZEWAI, A. & COLE, T. 1979. Luminescent solar concentrators. 1: Theory of operation and techniques for performance evaluation. *Applied Optics,* 18, 3090-3110.

BURGERS, A., SLOOFF, L., KINDERMAN, R. & VAN ROOSMALEN, J. Modelling of luminescent concentrators by ray-tracing. Presented at the 20th European Photovoltaic Solar Energy Conference and Exhibition, 2005. 10.

CARRASCOSA, M., UNAMUNO, S. & AGULLO-LOPEZ, F. 1983. Monte Carlo simulation of the performance of PMMA luminescent solar collectors. *Applied Optics,* 22, 3236-3241.

CHANDRA, S., DORAN, J., MCCORMACK, S., KENNEDY, M. & CHATTEN, A. 2012. Enhanced quantum dot emission for luminescent solar concentrators using plasmonic interaction. *Solar Energy Materials and Solar Cells,* 98, 385-390.

CHATTEN, A., BARNHAM, K., BUXTON, B., EKINS-DAUKES, N. & MALIK, M. 2004. Quantum dot solar concentrators. *Semiconductors,* 38, 909-917.

CHATTEN, A. J., FARRELL, D. J., BOSE, R., DIXON, A., POELKING, C., GÖDEL, K. C., MAZZER, M. & BARNHAM, K. W. Luminescent and geometric concentrators for building integrated photovoltaics. Photovoltaic Specialists Conference (PVSC), 2011 37th IEEE, 2011. IEEE, 000852-000857.

DEBIJE, M. G. & VERBUNT, P. P. 2012. Thirty years of luminescent solar concentrator research: solar energy for the built environment. *Advanced Energy Materials,* 2, 12-35.

EARP, A. A., SMITH, G. B., FRANKLIN, J. & SWIFT, P. 2004a. Optimisation of a three-colour luminescent solar concentrator daylighting system. *Solar Energy Materials and Solar Cells,* 84, 411-426.

EARP, A. A., SMITH, G. B., SWIFT, P. D. & FRANKLIN, J. 2004b. Maximising the light output of a Luminescent Solar Concentrator. *Solar Energy,* 76, 655-667.

FARRELL, D., CHATTEN, A., JERMYN, C., THOMAS, P. & BARNHAM, K. Thermodynamic Modelling of the Luminescent Solar Concentrator. INTERNATIONAL SOLAR ENERGY SOCIETY UK SECTION-CONFERENCE-C, 2005. 179.

GLASSNER, A. S. 1989. *An introduction to ray tracing,* Elsevier.

GOETZBERGER, A. 1978. Fluorescent solar energy collectors: operating conditions with diffuse light. *Applied physics,* 16, 399-404.

GOETZBERGER, A. & GREUBE, W. 1977. Solar energy conversion with fluorescent collectors. *Applied Physics,* 14, 123-139.

HUGHES, M. D., WANG, S.-Y., BORCA-TASCIUC, D.-A. & KAMINSKI, D. A. 2015. Analysis of ultra-thin crystalline silicon solar cells coupled to a luminescent solar concentrator. *Solar Energy,* 122, 667-677.

KENNEDY, M. 2010. *Monte-Carlo Ray-Trace Modelling of Quantum Dot Solar Concentrators.* PhD, Dublin Institute of Technology.

KENNEDY, M., MCCORMACK, S., DORAN, J. & NORTON, B. 2008. Ray-trace Modelling of Quantum Dot Solar Concentrators.

KENNEDY, M., MCCORMACK, S., DORAN, J. & NORTON, B. 2009. Improving the optical efficiency and concentration of a single-plate quantum dot solar concentrator using near infra-red emitting quantum dots. *Solar Energy,* 83, 978-981.

KENNEDY, M., ROWAN, B., MCCORMACK, S., DORAN, J. & NORTON, B. 2007. Ray-trace modelling of Quantum Dot Solar Concentrators and comparison with fabricated devices.

KERROUCHE, A., HARDY, D. A., ROSS, D. & RICHARDS, B. S. 2014. Luminescent solar concentrators: From experimental validation of 3D ray-tracing simulations to coloured stained-glass windows for BIPV. *Solar Energy Materials and Solar Cells,* 122, 99-106.

KLAMPAFTIS, E., ROSS, D., MCINTOSH, K. R. & RICHARDS, B. S. 2009. Enhancing the performance of solar cells via luminescent down-shifting of the incident spectrum: A review. *Solar Energy Materials and Solar Cells,* 93, 1182-1194.

PAGLIARO, M., CIRIMINNA, R. & PALMISANO, G. 2010. BIPV: merging the photovoltaic with the construction industry. *Progress in Photovoltaics: Research and Applications,* 18, 61-72.

PAREL, T. S., PISTOLAS, C., DANOS, L. & MARKVART, T. 2015. Modelling and experimental analysis of the angular distribution of the emitted light from the edge of luminescent

solar concentrators. *Optical Materials,* 42, 532-537.
ROWAN, B. 2007. The development of a quantum dot solar concentrator. *Doctoral*, 12.
S. CHANDRA, H. A., M. RAFIEE AND S. J. MCCORMACK 2017. Plasmonic Quantum Dot Solar Concentrator. *SPIE Photonics West 2017.*
ŞAHIN, D. & ILAN, B. 2013. Radiative transport theory for light propagation in luminescent media. *JOSA A,* 30, 813-820.
VAN SARK, W. G., BARNHAM, K. W., SLOOFF, L. H., CHATTEN, A. J., BÜCHTEMANN, A., MEYER, A., MCCORMACK, S. J., KOOLE, R., FARRELL, D. J. & BOSE, R. 2008. Luminescent Solar Concentrators-A review of recent results. *Optics Express,* 16, 21773-21792.
WEBER, W. & LAMBE, J. 1976. Luminescent greenhouse collector for solar radiation. *Applied Optics,* 15, 2299.

Performance Characterization of Crystalline Silicon Solar Cells Based on Combination of Plasmonics Silver Nanoparticles and Luminescent Downshifting Eu-Doped Phosphor-Particles

Bang-Jin You, Wen-Jeng Ho*, Sheng-Kai Feng, Zong-Xian Lin, and Jheng-Jie Liu
Department of Electro-Optical Engineering, National Taipei University of Technology
No. 1, Sec. 3, Zhongxiao E. Rd., Taipei 10608, Taiwan, R.O.C.
*e-mail: wjho@ntut.edu.tw

ABSTRACT:The photovoltaic-performance characterizations of crystalline silicon (C-Si) solar cells based on the combination of plasmonics scattering of silver nanoparticles (Ag NPs) and luminescent downshifting (LDS) of Eu-doped phosphor-particles are experimentally demonstrated in this work. The surface morphologies and plasmonics scattering of Ag NPs of various dimensions are characterized by the measurements of scanning electron microscopy (SEM), optical reflectance and absorbance, and external quantum efficiency (EQE). Furthermore, the LDS characteristics of Eu-doped phosphor-particles silicate layer coated by spin-on film technique over the surface of Ag NPs are revealed on the spectrum of optical reflectance and EQE. The photovoltaic-performance improvements of the C-Si solar cells with the combination of plasmonics scattering and LDS are confirmed by photovoltaic current-voltage (I-V) measurement under one sun AM 1.5G illumination. In this study, the efficiency enhancement of 21.95% for the cell with a antireflective coating (ARC) of SiO_2, 23.62% for the cell with Ag NPs embedded in a pure SiO_2 ARC, and 26.49% for the cell with Ag NPs embedded in a Eu-doped phosphor-SiO_2 ARC were obtained, respectively, compared to the reference bare solar cells.
Keywords: Antireflective coating (ARC), crystalline silicon solar cells, Eu-doped phosphor-particles, luminescent downshifting (LDS), plasmonics scattering

1.Introduction

Photovoltaic energy is a clean form of energy and it promises to be a major component of energy technology in the future. How to enhance the conversion efficiency and reduce the manufacturing cost of the solar cells are very important issues for silicon-based solar cells. The pyramidal textured and antireflection (AR) coating are mostly used technologies for improving the conversion efficiency of wafer-based single-crystal Si solar cells [1]. However, the new device structure [2] and novel concepts still need to research and develop to reach more higher efficiency and much lower device cost. Recently, the study shows that the nano-scale metallic particles have unique optical properties, such as the appearance of surface plasmons (SPs), and the metallic nanoparticles (NPs) have been applied in photovoltaic devices to enhance the conversion efficiency [3-5]. On the other hand, the application of a luminescent downshifting (LDS) layer on photovoltaic devices has been proposed as a method for improving the poor spectral response of solar cells in short-wavelength lights [6-12]. However, scant studies have reported that the combination of plasmonics scattering and spectral conversion to enhance the conversion efficiency of photovoltaic devices.

In this work, the combination of plasmonics scattering of silver NPs (Ag NPs) and LDS of Eu-doped phosphor-particles to enhance photovoltaic performance is the first demonstrated. The surface morphologies and plasmonics scattering of Ag NPs of various dimensions are characterized by the measurements of scanning electron microscopy (SEM), optical reflectance and absorbance, and external quantum efficiency (EQE). The LDS characteristics of Eu-doped phosphor-particles silicate layer coated by spin-on film technique over the surface of Ag NPs are revealed on the spectrum of optical reflectance and EQE. Furthermore, the photovoltaic-performance improvements of the crystalline silicon (C-Si) solar cells with the combination of plasmonics scattering and LDS are also confirmed by photovoltaic current-voltage (I-V) measurement under one sun AM 1.5G illumination. The effectiveness of the combination of plasmonics scattering and spectral conversion on C-Si solar cells is demonstrated for the cells having much high efficiency.

2. Experiments

2.1 Plasmonics scattering of Ag NPs of various dimensions

A 20 nm thick SiO_2 space layer was deposited on the bare silicon substrates (glass substrates) using e-beam evaporation. Next, 3, 5, and 7 nm thick silver films were deposited on the substrates with a 20 nm SiO_2 space layer by e-beam evaporation and subsequently annealed at 200 °C for 30 mins under ambient H_2 to form Ag NPs of various dimensions. Then, a 75 nm thick SiO_2 film was deposited on the surface of substrates with Ag-NPs layer using spin-on film process, which the Ag-NPs was embedded in SiO_2 layer. A silicon substrate coated with a 95 thick SiO_2 layer (a SiO_2 antireflective coating; a SiO_2 ARC) was also prepared for comparing. The surface morphologies and plasmonics scattering of Ag NPs of various dimensions are characterized by the measurements of scanning electron microscopy (SEM), optical reflectance, and absorbance.

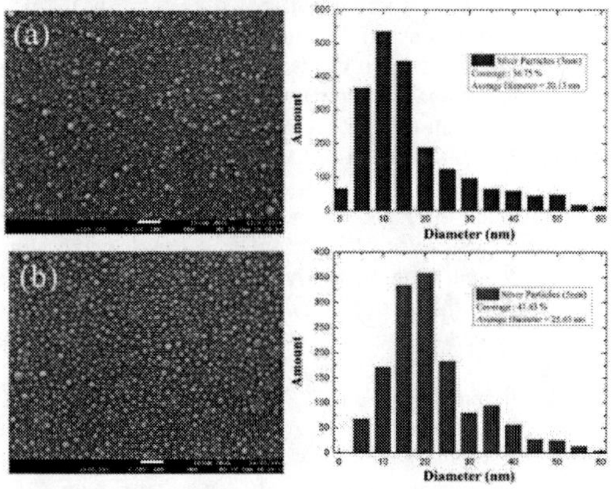

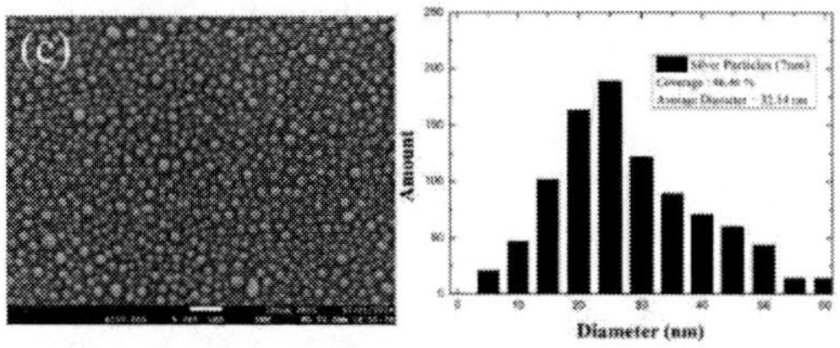

Fig.1:Top-view SEM image and particles distribution profile of Ag NPs through e-beam deposition of (a) 3, (b) 5, and (c) 7 nm thick.

2.2 Luminescent downshifting of Eu-doped phosphor-particles

In order to understand the optical property of Eu-doped phosphor, the Silicafilm solution (Emulsitone Company, USA) mixed with 3 wt% Eu-doped phosphor powder (InteMatix Company, USA; O6040™) was coated firstly on the silicon substrates using spin-on technique at 3000 rpm for 40 s and then baked at 150℃ for 1 min under an air atmosphere. The photoluminescence (PL) emission of the sample with a Eu-doped phosphor silica film was characterized at 610 nm wavelength. Then, the samples with Ag NPs of various dimensions were deposited a Eu-doped phosphor silica film and characterized by optical reflectance measurement. Both abovementioned characterizations are used to examine the LDS effects of Eu-doped phosphor-particles.

2.3 Combination of the luminescent downshifting and plasmonics scattering

To characterize the LDS and plasmonics-scattering effects on the C-Si solar cells, the bare C-Si solar cells were fabricated firstly. The p-type (~5 Ω·cm), CZ, (100), 525 μm-thick C-Si substrates were firstly cleaned by a standard RCA-clean process. Then, the n^+-emitter layer was formed by a spin-on film process using the Phosphorosilicafilm (Emulsitone Company, USA) source and rapid thermal annealing (RTA) at 800 °C for 3 min in an ambient N_2. After n^+-Si emitter layer diffusion, the remaining phosphorous silicate glass on the front-side surface was removed by a buffered oxide etchant (BOE). Next, 250 nm-thick Al film and Ti(20-nm)/Al(250-nm) films were deposited on the rear-side electrode and front-side finger-electrode, respectively, of all samples using a e-beam evaporator and RTA at 450 °C for 30 min in an ambient N_2, which obtaining the bare-type C-Si solar cells as reference.

Then, a 20 nm thick SiO_2 space layer was deposited on the bare-type C-Si solar cells using an e-beam evaporator. Next, 3, 5, and 7 nm thick silver films were deposited on the surface of the bare solar cells with a 20 nm SiO_2 space layer by e-beam evaporation and subsequently annealed at 200 °C for 30 min under ambient H_2 to form Ag-NPs of various dimensions. Then, a 75 nm thick SiO_2 film was deposited on the surface of Ag-NPs layer using spin-on film process, which Ag-NPs was embedded in SiO_2 layer. Finally, we prepared silicate solution mixing with $(Sr_{1-x}Ba_x)_2SiO_4:Eu^{2+}F$ phosphors-powder of 3 wt.% instead of pure SiO_2 solution and spun on the cell with Ag-NPs. For comparison, we also produced a C-Si solar cell coated with the same Silicafilm layer (SiO_2) of 95 nm thick but without the phosphor particles. The optical reflective spectra and external quantum efficiency (EQE) response of these cells are also measured and compared to examine LDS and plasmonics scattering effects in a wavelength range of 300 to 1000 nm. Furthermore, the photovoltaic current-voltage (J–V) characteristics of these cells were measured and compared to confirm the contribution of LDS and plasmonics scattering.

3. Results and Discussion

Figure 1 shows the top-view SEM image and particles distribution profile of Ag NPs through e-beam deposition of (a) 3, (b) 5, and (c) 7 nm thick. The averaged dimension and coverage of Ag NPs were larger as the Ag film deposited thickness is higher. Fig. 2 shows the optical reflectance of the Si sample coated with a SiO_2 ARC and the Si sample with Ag-NPs of various dimensions embedded in SiO_2 ARC. The reflectance of the Si sample with Ag-NPs embedded in SiO_2 ARC was lower than the cell with an only SiO_2 ARC one due to plasmonics scattering. In addition, the plasmonics scattering of the large dimension particles was higher than that of the smaller dimension particles because of exhibited a lower reflectance. Moreover, the optical absorbance of the samples with Ag-NPs of various dimensions embedded and non-embedded in SiO_2 ARC is shown in Fig. 3. The peaks of absorbance signals indicated the surface plasmon resonance (SPR) of Ag-NPs, which shows that SPR was depended on the dimension and environment dielectric of Ag NPs. Fig. 4 shows the optical reflectance of the samples with Ag-NPs of various dimensions embedded in a Eu-doped phosphor silica ARC. A lower reflectance at the wavelengths below 450 nm and a higher reflectance at the wavelength of about 610 nm are attributed to the LDS of Eu-doped phosphor particles. Fig. 5 shows the EQE response of the Si solar cell coated with a SiO_2 ARC and the Si solar cells with Ag-NPs of various dimensions embedded in SiO_2 ARC. The EQE values of the Si sample with Ag-NPs embedded in SiO_2 ARC was lower than the cell with an only SiO_2 ARC one due to plasmonics Ag NPs self-absorption at the wavelengths below 550 nm. However, the EQE values were enhanced in the wavelength beyond 550 nm because of the plasmonics scattering of Ag NPs.

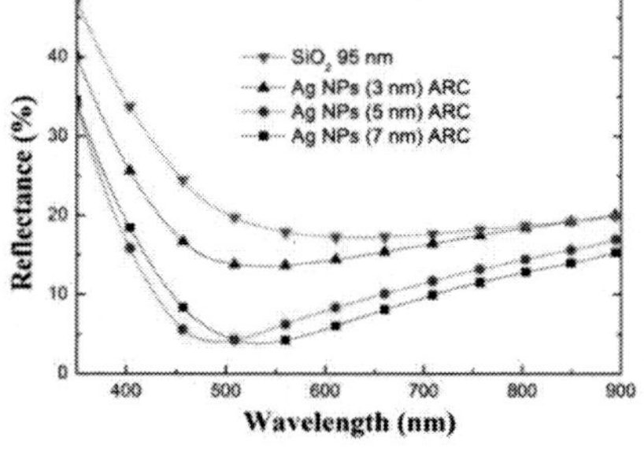

Fig.2:Optical reflectance of the Si sample coated with a SiO_2 ARC and the Si sample with Ag-NPs of various dimensions embedded in SiO_2 ARC.

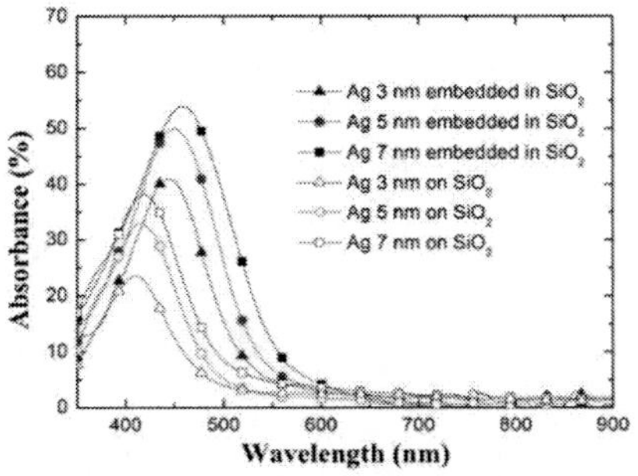

Fig.3:Optical absorbance of the samples with Ag-NPs of various dimensions embedded and non-embedded in SiO_2 ARC.

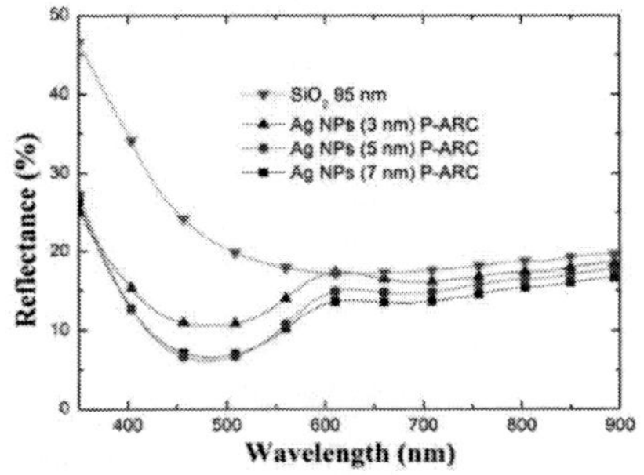

Fig.4:Optical reflectance of the samples with Ag-NPs of various dimensions embedded in a Eu-doped phosphor silica ARC.

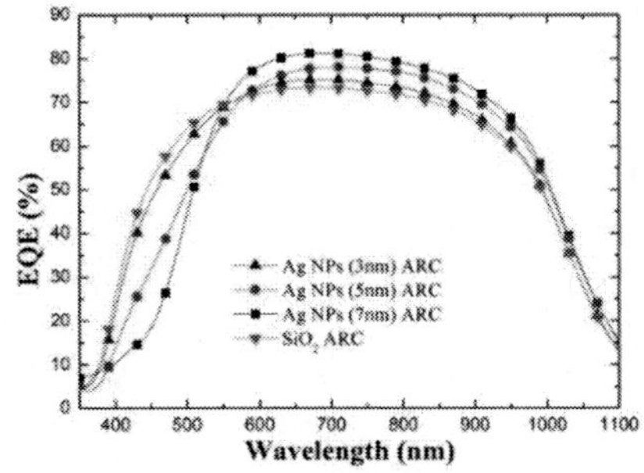

Fig. 5:EQE response of the Si solar cell coated with a SiO_2 ARC and the Si solar cells with Ag-NPs of various dimensions embedded in SiO_2 ARC.

Figures 6 and 7 show the EQE response and photovoltaic current-density versus voltage (J-V) characteristics of the Si solar cell coated with a SiO_2 ARC and the Si solar cells with Ag-NPs (the deposited thickness of 3 nm) embedded in a Eu-doped phosphor silica ARC (P-ARC), respectively. The photovoltaic performances were summarized in Table I. The EQE values were enhanced in the wavelength beyond 600 nm due to the LDS of Eu-doped phosphor and plasmonics scattering of Ag NPs. Especially, the EQE enhancement for the cell with Ag-NPs embedded in a P-ARC was impressive improved at short wavelength range (350-550 nm) due to LDS effect. In general, the short-circuit current density (J_{SC}) of the cell is proportional to the EQE response. The J_{SC} of 34.07 mA/cm^2, open-circuit voltage (V_{OC}) of 527.27 mV, fill factor (FF) of 71.10%, and conversion efficiency (η) of 12.77% were obtained when the cell with Ag-NPs embedded in SiO_2 ARC, compared to J_{SC} = 33.40 mA/cm^2, V_{OC} = 516.90 mV, FF = 73.43%, and η = 12.50% from the cell coated with a SiO_2 ARC without Ag NPs. These results show that the increase in J_{SC} of 0.67 mA/cm^2 and in η of 0.27% are attributed to the plasmonic scattering of Ag NPs embedded in SiO_2 ARC and it is well agreed to the enhancement of EQE. Moreover, the additional increase in J_{SC} to 34.36 mA/cm^2 and η to 13.56% for the Si solar cell with Ag-NPs embedded in a Eu-doped phosphor silica ARC (P-ARC) are attributed to the contribution of the combination of LDS of Eu-doped phosphor particles and plasmonics scattering of Ag NPs.

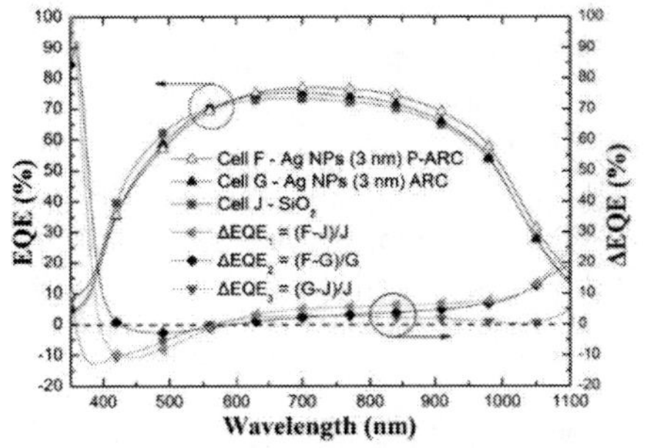

Fig. 6:EQE response and EQE enhancement of the Si solar cell coated with a SiO_2 ARC and the Si solar cells with Ag-NPs (the deposited thickness of 3 nm) embedded in a Eu-doped phosphor silica ARC (P-ARC).

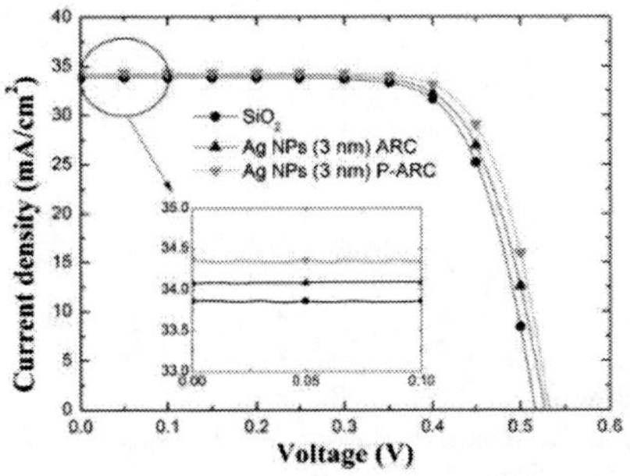

Fig.7:Photovoltaic current density-voltage (J-V) characteristics of the Si solar cell coated with a SiO_2 ARC and the Si solar cells with Ag-NPs (the deposited thickness of 3 nm) embedded in a Eu-doped phosphor silica ARC (P-ARC).

Solar Cell	J_{sc} (mA/cm^2)	V_{oc} (mV)	FF (%)	η (%)	ΔJ_{sc} (%)	Δη (%)
Bare SC	26.40	518.29	74.67	10.22		
Cell with SiO_2 ARC	33.40	516.90	72.43	12.50	26.51	21.95
Bare SC	26.49	528.97	73.71	10.33		
Cell with Ag NPs ARC	34.07	527.27	71.10	12.77	28.61	23.62
Bare SC	26.44	532.73	75.95	10.69		
Cell with Ag NPs P-ARC	34.36	532.79	74.09	13.56	29.95	26.49

Table 1 :Photovoltaic performances of the bare C-Si solar cell, cell with Ag NPs embedded in SiO_2 ARC, and cell with Ag-NPs embedded in Eu-doped phosphor silica ARC (PARC).

4.Conclusion

The photovoltaic performances of C-Si solar cells depending on the plasmonic scattering of Ag NPs of various dimensions embedded in SiO_2 ARC and the combination of LDS of Eu-doped phosphor particles and plasmonics scattering of Ag NPs were experimentally demonstrated. It was found that larger Ag-NPs exhibited a better plasmonics scattering beyond the wavelength of 550 nm in this study. Besides, the combination of LDS of Eu-doped phosphor particles and plasmonics scattering of Ag NPs applied on the C-Si solar cells to enhance photovoltaic performances are also confirmed by EQE response and photovoltaic performances. An additional increasing in J_{SC} from 33.40 to 34.36 mA/cm^2 and η from 12.50% to 13.56% for the Si solar cell with Ag-NPs embedded in a Eu-doped phosphor silica ARC are obtained, compared to the cell with a SiO_2 ARC.

Acknowledgement

The authors would like to thank the Ministry of Science and Technology of the Republic of China for financial support under Grant MOST 103-2221-E-027-049-MY3.

References

[1]. Y. Matsumoto, et al., ***38th IEEE Photovoltaic Specialists Conference***, 258 (2012).
[2]. J. Caiab and L. Qi, ***Mater. Horiz. 2***, 37 (2015).
[3]. M. A. Garcia, ***J. Phys. D: Appl. Phys. 44***, 283001 (2011).
[4]. H. A. Atwater and A. Polman, ***Nature Mater. 9***, 205 (2010).
[5]. W. J. Ho, Y. Y. Lee, and S. Y. Su, ***Nanoscale Res. Lett.* 9: 483**, (2014).
[6]. E. Klampaftis, D Ross, K. R. McIntosh, and B. S. Richards, ***Sol. Energy Matter. Sol. Cells 93***, 1182 (2009).
[7]. T. Trupke, A. Shalav, B.S. Richards, P. Wurfel, M. A. Green, ***Sol. Energy Matter. Sol. Cells 90***, 3327 006).
[8]. C. K. Huang, Y. C. Chen, W. B. Hung, T. M. Chem, K. W. Sun, and W. –L. Chang, ***Prog. Photovoltaics 21***, 1507 (2013).
[9] J. Liu, K. Wang, W. Zheng, W. Huang, C. –H. Li, and X. –Z You, ***Prog. Photovoltaics 21***, 668 (2013).
[10] J. –Y Chen, C. K. Huang, W. B. Huang, K. W. Sun, and T. M. Chen, ***Sol. Energy Matter. Sol. Cells 120***, 168 (2014).
[11] G. Shao, C. Lou, and D. Xiao, ***J. Luminescence 157***, 344 (2015).
[12] B. M. van der Ende, L. Aarts, A. Meijerink, ***Phys. Chem. Chem. Phys. 11***, 11081 (2009).

EFFECTS OF SULFURIZATION TIME ON MOS_2 ABSORBER LAYER FOR THIN FILMS SOLAR CELLS APPLICATIONS

H. Rashid[1], K.S. Rahman[1], M.I. Hossain[3], F.H. Alharbi[3,4], N. Tabet[3,4], M.A. Islam[5] and N. Amin[1,2]*
[1]Department of Electrical, Electronic and Systems Engineering, Faculty of Engineering & Built Environment, The National University of Malaysia, 43600, Bangi, Selangor, Malaysia
[2]Solar Energy Research Institute (SERI), The National University of Malaysia, 43600, Bangi, Selangor, Malaysia
[3]Qatar Environment and Energy Research Institute, Hamad Bin Khalifa University, Doha, Qatar
[4]College of Science and Engineering, Hamad Bin Khalifa University, Doha, Qatar
[5]Graduate School of Material Science, Nara Institute of Science and Technology, Ikoma, Nara 630-0101, Japan
*Corresponding author: nowshad@ukm.edu.my

ABSTRACT: Thin films of MoS_2 were prepared on top of soda lime glass by using RF magnetron sputtering technique and further sulfurized under different timing profiles with static gas pressure of Nitrogen in growth chamber set at 370 torr at room temperature (RT). Four different time intervals i.e. 60 min, 90 min, 120 min and 150 min were considered at 597 °C. The effects of this parameter change were studied on absorber layer properties for photovoltaic solar cell applications. The structural, topographical and morphological properties as well as cross sectional thickness of MoS_2 layers were analysed. MoS_2 films were grown at RT at 100W RF power by sputtering. As-deposited films exhibited two XRD peaks corresponding to $(100)_{HEX}$ and $(110)_{HEX}$ planes at $2\theta = 33.4°$ and 59.5° respectively, whereas post-deposition treatment (sulfurization) revealed four diffraction peaks corresponding to $(002)_{HEX}$, $(100)_{HEX}$, $(103)_{HEX}$ and $(110)_{HEX}$ at 14.3°, 33.7°, 39.9° and 59.4° respectively. Thickness of sulfurized films was measured within the range of 446.6 nm to 595.5 nm during FESEM while as-deposited film was measured 419.5 nm and surface was observed to be rough for films sulfurized for 60 min and 90 min. EDX results showed S/Mo ratio as 3.01 for films sulfurized for 120 minutes whereas as-deposited film was with 2.44 ratio. Film morphology was found to be very smooth with a low average surface roughness value of 0.98 nm for films sulfurized for 150 min. Enriched ratio of S/Mo, XRD peaks validation and surface morphological results encouraged its potential suitability for the use of MoS_2 as absorber layer material for solar photovoltaic applications.
Keywords: MoS_2 thin films, RF Sputtering, Absorber Layer, FESEM, XRD, AFM, Sulfurization.

1 INTRODUCTION

The prosperous segregation of graphene from graphite has directed to its widespread study in physics, materials and nano engineering due to its extraordinary electrical and mechanical properties. Yet, a lack of a band gap limits its potential for electronic device applications and has encouraged the exploration of other two dimensional (2D) layered materials. Among them, transition metal dichalcogenides (TMDCs) such as MoS_2, $MoSe_2$, WS_2 and WSe_2 are the utmost studied materials. TMDC materials possess an X-M-X (X are chalcogen atoms and M is transition metal atoms) crystal structure. In which, sandwich layers are coupled by weak van der Waals (vdW) forces while the intra-layer bonding between M and X atoms have strong covalent bonding.

The optoelectronic study of exfoliated flakes has formed the basis to understand their unique optical and electrical properties [1-4]. First principle density functional theory (DFT) calculations accord with these studies and show that the band structure of TMDC change dramatically with material exfoliation owning to strong quantum confinement effects [5]. Specifically, the bulk TMDC convert from an indirect bandgap semiconductor to direct bandgap semiconductor when thinned to a single X-M-X layer, which results in a drastic enhancement of enhanced photoluminescence [5-8]. It has also been shown that single or few layer TMDC exhibit superior light harvesting characteristics compared with their bulk counterparts, due to strong light matter interactions. For example, a single $MoSe_2$ monolayer, 0.4 nm thick was estimated to generate a maximum 4.6 $mA.cm^{-2}$ photocurrent under standard illumination conditions, which is equivalent to the photocurrent produced by 50 nm thick Si or 15 nm GaAs [9]. In addition, an ultrathin 2D TMDC layer does not suffer from the disadvantageous anisotropic electronic properties intrinsic to their bulk counterparts (i.e. charge carrier transport across the layers is three orders of magnitude slower than the transport along the layer plane). Finally, given the inertness of the chalcogenide surface, atomic layer heterojunctions (or vdW heterojunctions) can be constructed without lattice strain that can lead to defects and recombination [10].

Sulfurization is a post depostion process of thermal heating utilized for re-crystallization of deposited films to improve the films quality as well as solar cell performance parameters [11]. In this study, thin films of MoS_2 were deposited on top of soda lime glass (SLG) by RF magnetron sputtering technique and further sulfurized under different timing profiles with static gas pressure of Nitrogen in growth chamber. The structural, topographical and morphological properties as well as cross sectional thickness of MoS_2 layers were analysed.

2 EXPERIMENTAL SETUP

Thin films of MoS_2 were deposited by using RF magnetron sputtering technique on top of SLG substrates by using 99.99% pure MoS_2 target, purchased from Applied Science Corporation, Republic of South Korea. The samples were cleaned in an ultrasonic bath within the fume hood for 10 minutes in methanol, acetone, then again in methanol and finally in deionized (DI) water for 20 minutes. Dry nitrogen (N_2) gas was further used to ensure no marks of any liquid remain on the surface of cleaned substrate.

During sputtering, the Argon (Ar) flow rate was maintained at 4.0 SCCM. The working pressure was observed 3.20 × 10^{-3} kPa throughout the deposition. In order to achieve substantial growth of MoS_2 thin films, the deposition was carried out for 60 min. The sputtering conditions used are itemized in Table I.

Table I: Summary of parameters of MoS_2 thin films

Parameters	Conditions
Base Pressure (Torr)	4.2E-6
Operating Pressure (Torr)	2.4E-2
Deposition Temperature (°C)	RT
Deposition Time (min)	60
Deposition Power: RF (W)	100
Argon Gas Flow Rate (SCCM)	3.5

Subsequently, the sulfurization processes were performed in the presence of pure N_2 gas (99.99%) in the chamber with working pressure set at 370 torr. Summary of sulfurization process parameters is shown in Table II.

Table II: Summary of sulfurization parameters

Parameters	Conditions
Base Pressure (Torr)	0.11
Operating Pressure (Torr)	370
Temperature (°C)	580
Ramp Rate (°C/min)	15
Background Gas	N2
Sulfur (mg)	250
Sulfurization Time (min)	60, 90, 120, 150

3 RESULTS & DISCUSSION

In this section, the effects of sulfurization process or deposition time has been discussed to realize the structural, morphological and compositional properties of MoS_2.

3.1 Structural Analysis

The structural analysis was performed by using XRD method and the patterns were obtained in the range of 10° to 80° of 2θ diffraction angle. Fig. 1 illustrates XRD patterns for MoS_2 thin films sulfurized for various timing profiles i.e. 60 min, 90 min, 120 min and 150 min. XRD patterns revealed that the first peak of (002) orientation around 14° is becoming sharper with the increase of sulfurization process time. Two new peaks (103) and (110) were introduced and become vigilant for the samples sulfurized for 150 min. As well as, the peak of (100) around 33° was turning more intense with the increase of sulfurization processing time. It is well-known that heating plays a vital role in the growth of phases from amorphous to polycrystalline, which has been observed in this study during the transition from as-sputtered films to the increase in hold time. Films sulfurized for 150 min demonstrated high crystallinity as compare with as-sputtered films. There were total four peaks observed such as (002), (100), (103) and (110), maintained hexagonal structure of crystals of MoS_2 with confirmation of JCPDS card no. 37-1492.

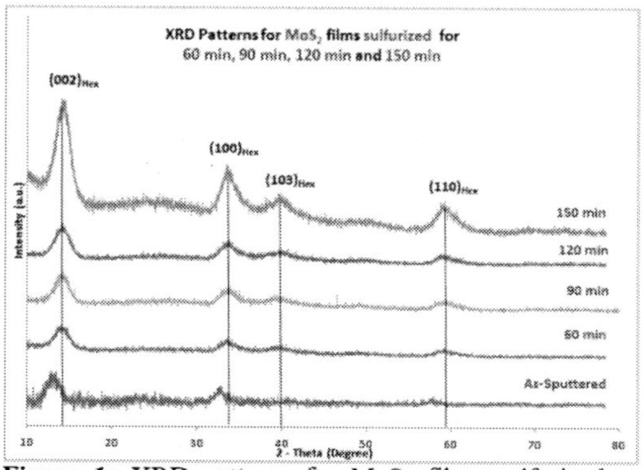

Figure 1: XRD patterns for MoS_2 films sulfurized at various processing times

3.2 Surface Topology Analysis

The AFM analysis was performed to calculate average surface roughness and RMS roughness data.

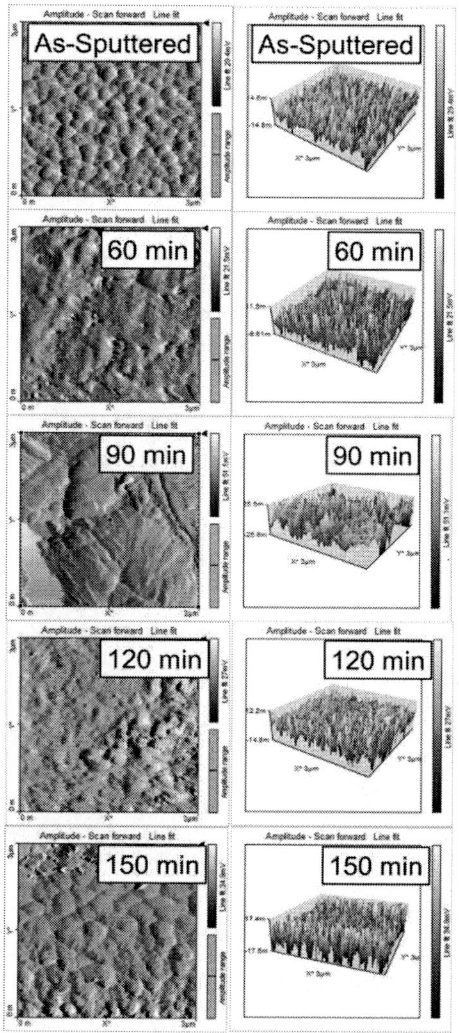

Figure 2: AFM results obtained from sulfurization of MoS_2 films under different hold time and as-sputtered

The average surface roughness was observed to be in the range of 6 nm to 9 nm with an increasing trend as a function of sulfurization time. As sputtered films have RMS roughness value of 8.02 nm whereas the highest value was observed for films sulfurized for 150 min. The obtained topological images are shown in Fig. 2. The summary of measured values of roughness is presented in Table III.

Table III: Summary of parameters of MoS_2 thin films

Sulfurization Process Time	Ava. Roughness (Sa) – nm	RMS Roughness (Sq) - nm
As-Sputtered	6.4443	8.0289
60 min	6.7637	8.5360
90 min	6.8632	9.1434
120 min	7.7396	10.7500
150 min	9.0012	12.7370

3.3 Morphological Analysis

Fig. 3 is signifying FESEM images (surface and cross-section) of as-sputtered and sulfurized films of MoS_2 with different process times.

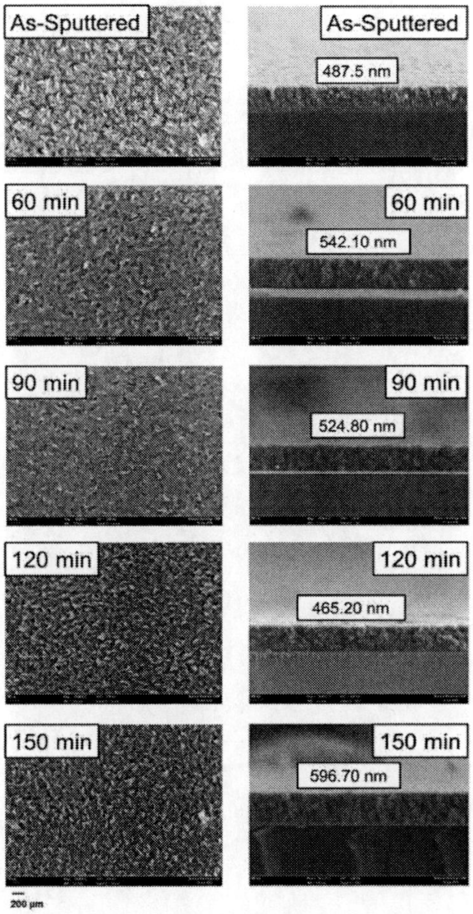

Figure 3: FESEM images (surface and cross section) of MoS_2 films, as-sputtered, sulfurized for 60, 90, 120 and 150 min

The surface of films was observed to be rough and 'nano-flower' like structure was noticed. The thickness was observed to be 487.5 nm for as-sputtered film and 542.1 nm, 524.8 nm, 465.20 nm and 596.70 nm for sulfurization time of 60, 90, 120 and 150 min respectively. No particular trend was observed in MoS_2 films evolution with respect to processing time of sulfurization.

3.4 Compositional Analysis

The compositional ratio of Mo and S was calculated by EDX for sulfurized and as-sputtered thin films of MoS_2, the normalized values are summerized in Table IV.

It can be seen that S/Mo ratio has an increasing trend with sulfurization processing time. However, the ratio was found to be increasing only till 120 min of sulfurization, afterwards its decreasing drastically. This could be due to more sulfurization time, causing S elimination. In general, all the films found non-stoichiometric with S/Mo ratio of x: 1, where x is more than 2 in number.

Sulfurization	Normalized		Ratio
Process Time	Mo	S	S/Mo
As-Sputtered	29.11	70.89	2.44
60 min	28.56	71.44	2.50
90 min	26.62	73.38	2.76
120 min	24.95	75.05	3.01
150 min	29.20	70.80	2.43

4 CONCLUSION

Thin films of MoS_2 were deposited on top of SLG by means of RF magnetron sputtering and subsequently sulfurized under different time profiles i.e. 60, 90, 120 and 150 min in the presence of static gas pressure of N_2 in the growth chamber for re-crystallization. XRD results revealed four diffraction peaks corresponding to (002), (100), (103) and (110) planes with hexagonal structure, proving MoS_2 presence. The peaks were found to be intense as the sulfurization time increased; this is obvious in satisfaction of re-crystallization. The films sulfurized for 150 min have an intense and dominant peak corresponding to (002) plane.

The average surface roughness was found to have an increasing trend during AFM analysis and the values observed in the range of 6 nm to 9 nm. The thickness of films was witnessed 487.5 nm for as-sputtered film and 542.1 nm, 524.8 nm, 465.20 nm and 596.70 nm for sulfurized films of 60, 90, 120 and 150 min respectively. EDX results indicated non-stoichiometric films with normalized S/Mo ratio more than 2. Films sulfurized at 150 min hold time are found to have intense and dominant peaks, better re-crystallinity, bigger grains, 'nano-flower' like structure and S/Mo ratio more closer to MoS_2 compositional structure, which is evident for its suitable use as absorber layer for thin film photovoltaic applications.

5 ACKNOWLEDGEMENT

This work was supported thankfully by Qatar National Research Fund (QNRF) by the grant code of NPRP 6-931-2-382.

6 REFERENCES

[1] Q. H. Wang, K. Kalantar-Zadeh, A. Kis, J. N. Coleman, and M. S. Strano, "Electronics and optoelectronics of two-dimensional transition metal dichalcogenides," *Nature nanotechnology,* vol. 7, no. 11, pp. 699-712, 2012.
[2] R. Ganatra, and Q. Zhang, "Few-layer MoS2: a promising layered semiconductor," *ACS nano,* vol. 8, no. 5, pp. 4074-4099, 2014.
[3] X. Huang, Z. Zeng, and H. Zhang, "Metal dichalcogenide nanosheets: preparation, properties and applications," *Chemical Society Reviews,* vol. 42, no. 5, pp. 1934-1946, 2013.
[4] B. Peng, P. K. Ang, and K. P. Loh, "Two-dimensional dichalcogenides for light-harvesting applications," *Nano Today,* vol. 10, no. 2, pp. 128-137, 2015.
[5] A. Splendiani, L. Sun, Y. Zhang, T. Li, J. Kim, C.-Y. Chim, G. Galli, and F. Wang, "Emerging photoluminescence in monolayer MoS2," *Nano letters,* vol. 10, no. 4, pp. 1271-1275, 2010.
[6] M. Fontana, T. Deppe, A. K. Boyd, M. Rinzan, A. Y. Liu, M. Paranjape, and P. Barbara, "Electron-hole transport and photovoltaic effect in gated MoS2 Schottky junctions," *Scientific reports,* vol. 3, 2013.
[7] H. Rashid, K. Rahman, M. Hossain, N. Tabet, F. Alharbi, and N. Amin, "Prospects of molybdenum disulfide (MoS2) as an alternative absorber layer material in thin film solar cells from numerical modeling," *Chalcogenide Letters,* vol. 11, no. 8, pp. 397-403, 2014.
[8] H. Rashid, K. Rahman, M. Hossain, N. Tabet, F. Alharbi, and N. Amin, "Deposition and Characterization of MoS2 as an Alternative Absorber Layer for High Efficiency Solar Cells by PVD Technique."
[9] M. Bernardi, M. Palummo, and J. C. Grossman, "Extraordinary sunlight absorption and one nanometer thick photovoltaics using two-dimensional monolayer materials," *Nano letters,* vol. 13, no. 8, pp. 3664-3670, 2013.
[10] E. Khestanova, F. Guinea, L. Fumagalli, A. Geim, and I. Grigorieva, "Universal shape and pressure inside bubbles appearing in van der Waals heterostructures," *Nature communications,* vol. 7, 2016.
[11] M. Islam, K. Rahman, F. Haque, M. Akhtaruzzaman, M. Alam, Z. Alothman, K. Sopian, and N. Amin, "Properties of low temperature vacuum annealed CZTS thin films deposited on polymer substrate," *Chalcogenide Letters,* vol. 11, no. 5, pp. 233-239, 2014.

Novel $Zn_xSn_{1-x}Se$ Absorber For Use in Thin-Film Solar Cells

T.M.Razykov[1], E. Artegiani[2], G.S. Boltaev[3], A.Bosio[4], B.Ergashev[1], K.M.Kouchkarov[1], N.K. Mamarasulov[1]
A.A. Mavlonov[1], A.Romeo[2], N.Romeo[4], R.Yuldoshov[1]

[1]Physical-Technical Institute, Uzbekistan Academy of Science,
Bodomzor Yoli 2B, Tashkent 100084, Uzbekistan
Phone: +998-71-235-4103, Fax: +998-71-235-4291, E-mail: razykov@uzsci.net
[2]Universita' di Verona, Ca' Vignal 2- Strada Le Grazie 15, 37134 Verona, Italy
[3]Ion Plasma and Laser Technologies, Uzbek Academy of Sciences, Tashkent 100125, Uzbekistan
[4]University of Parma, G.P Usberti 7/A - 43124 Parma, Italy

ABSTRACT
$Zn_xSn_{1-x}Se$ (x=0) thin films were fabricated by chemical vapor deposition (CVD) using polycrystalline tin selenide (SnSe) precursors. The morphology, structure, optical and electrical properties of films were studied as a function of the composition of precursors and the substrate temperature. Results obtained have shown that Se rich films were fabricated at low substrate temperatures, despite the different compositions of the SnSe precursor during the synthesis. In this case, the grain sizes of the films vary in the range of 8-20 μm, depending on the substrate temperature. X-ray diffraction (XRD) analysis of the samples showed that the films have a orthorhombic crystalline structure. The optical measurements have revealed that the samples have optical bandgap of 1.21 eV and the absorption coefficient of ~10^5 cm^{-1}. The samples performed p-type conductivity being in the range of 10-10^2 $(Ohm\times cm)^{-1}$ depending on the deposition conditions.

Keywords: $Zn_xSn_{1-x}Se$, CVD, thin film, bandgap.

1. INTRODUCTION

Today the leading materials in the world photovoltaic market are Si wafers, $Cu(In,Ga)Se_2$ (CIGS) and CdTe thin films. Currently, remarkable efficiencies were achieved on these cells, i.e. Si (26.0 %), CIGS (22.6 %) and CdTe (22.1 %). However, these materials have serious disadvantages: Si has low absorption coefficient and wafer form, CIGS contains high cost materials (In and Ga) and CdTe has a toxicity issue of Cd and limited resources of Te. While novel, very promising $Zn_xSn_{1-x}Se$ thin films have ideal properties for fabrication of low cost and high efficiency thin film solar cells for large scale terrestrial application: 1) high absorption coefficient of ~10^5 cm^{-1}, 2) band gap of 0.9-2.7 eV which allows fabrication of single (1.46 eV) and double junction (1.0 eV and 1.7 eV) cells, and 3) the conductivity can be varied in very wide range of 10-10^2 $(Ohm\times cm)^{-1}$ which means that doping to reduce the series resistance is not necessary, since the conductivity can be varied by the controlling of the intrinsic point defects. Moreover, it is formed by abundant and environmental-friendly elements. Variety of techniques have been used for the deposition of SnSe films, such as vacuum evaporation [1], chemical bath deposition [2], electro-deposition [3], solid state reactions [4], solvothermal [5], molecular beam epitaxy [6], pulsed laser deposition [7] and chemical vapor deposition [8].

Moreover, this material is promising for applications in many fields, such as solar cells [9–11], thin film light emitting diodes (LED) [11], low threshold lasers, optical amplifiers for telecommunication networks [12], guided wave devices, chemical and biosensors [13].

In this work, effect of the stoichiometric, Sn rich and Se rich precursors on structural, morphological, optical and electrical properties of SnSe films is discussed for the first time.

2. EXPERIMENTAL

The stoichiometric, Sn rich and Se rich granules with 99.999 % purity were used as precursors. The temperature of the precursors was varied in the range of 850℃-950℃. The temperature of the substrate was varied between 450℃ - 550℃. The flow rate of the carrier gas was ~ 20 cm^3/min. The duration of the deposition process depends on the required film thickness and ranges between 30-60 min. All SnSe thin films have been grown on borosilicate glasses. The crystal structure and the phase of the material compositions were studied by XRD using a "Panalytical Empyrean" diffractometer (Cu Kα radiation, λ = 1.5418 Å) with a wide-angle measurements of 2θ in the range of 20°-80° and a step of 0.010. The experimental data were analyzed using the Joint Committee on Powder Diffraction Standard (JCPDS). Morphological studies were carried out using a scanning electron microscope (SEM-EVO MA 10), and film compositions were determined using an energy-dispersive X-ray spectroscopy (EDX, Oxford Instrument - Aztec Energy Advanced X-act SDD). Optical parameters of selected SnSe thin films were determined from the transmission spectra obtained in a wide spectral range between 400-2500 nm using the HR 4000 Ocean Optic spectrometer with a resolution of 2 nm. Silver was used as ohmic contact for the p-type films. The type of conductivity of the samples was determined by thermoelectric effect. The thickness of the films was determined by using microinterferometer MII-4, and the method of precision micro-weighting (on FA 120 4C scales with an accuracy of 0.1 mg).

3. RESULTS AND DISCUSSIONS

According to EDX data, the obtained films have no any impurity elements within the sensitivity of the method (Fig. 1). The deposited films from the Sn rich SnSe precursor have

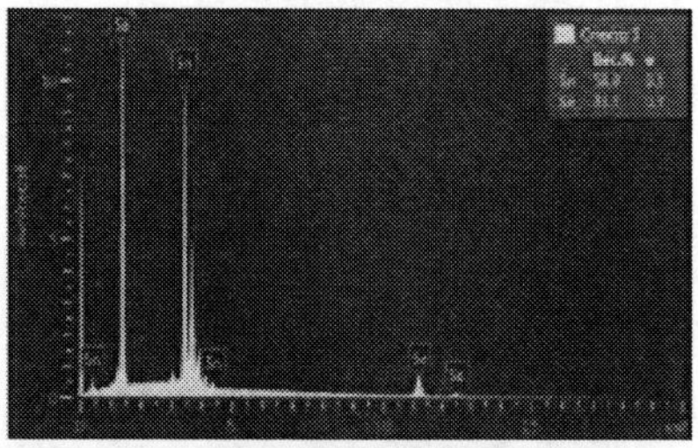

$Sn_{0.98}Se$ T=450°C

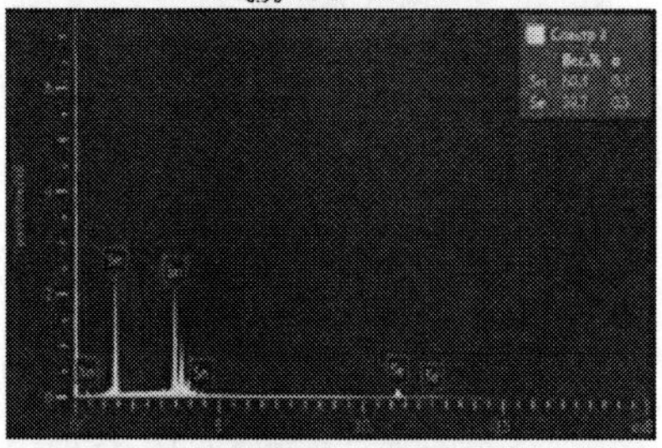

$Sn_{1.01}Se$ T=500°C

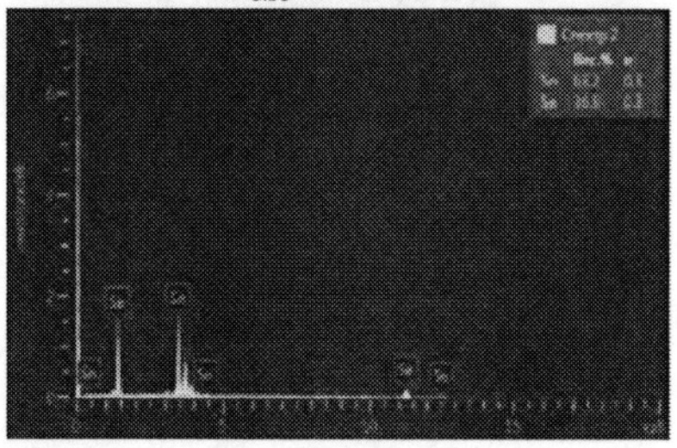

$Sn_{1.14}Se$ T=550°C

Figure 1 Results of energy dispersive X-ray spectroscopy analysis for SnSe thin films deposited from Sn enriched precursor at indicated substrate temperatures.

almost similar as stoichiometric precursor at substrate temperatures of 450℃ and 500℃. On the other hand, the film grown at 550℃ has Sn rich composition. This can be explained by the fact that when the substrate temperature increases, Se re-evaporates from the substrate, which results in enrichment of the Sn in the films. All films deposited from the SnSe precursors, i.e. Se enriched and with stoichiometric composition, have Se enriched composition at different temperatures (Table 1). All SnSe films studied in this work have a smooth surface without cracks and pores on the surface of the substrate. Figure 2 shows the XRD patterns of SnSe films deposited from three different SnSe precursors, i.e. Sn enriched, Se enriched and with stoichiometric composition, for three different substrate temperatures. According to XRD analysis, all samples have single-phase, the orthorhombic structure of SnSe and there are no other crystalline phases of Sn, Se, Sn_2Se_3, $SnSe_2$, SnO_2, etc. The main peaks are corresponding to the (400) and (111) preferential orientation of SnSe (Fig. 2). A similar results were also noted in several papers devoted to SnSe films obtained by thermal evaporation and chemical vapor deposition [8].

The authors conducted this, presence of both (400) and (111) preferential orientations to the growth conditions, e.g. substrate temperature, the distance between the substrate and the target. In this work, (111) plane was observed only for SnSe thin films deposited from Sn enriched SnSe precursor at 550℃ (Fig. 2a). Analyzing the XRD data of all samples with a database (JCPDS: 01-089-0233) showed that all films have an orthorhombic structure.

The lattice parameters of the crystal for all samples were calculated using the following formula: $1 / d^2 = h^2 / a^2 + k^2 / b^2 + l^2 / c^2$, where d is the distance between the planes, h, k, l are Miller indices and a, b, c are lattice constants. The lattice constants for thin films deposited at different substrate temperatures from Sn and Se enriched and stoichiometric SnSe precursors have the following values: a = 11.52 Å, b = 4.16 Å, c = 4.43 Å and a = 11.48 Å, b = 4.17 Å, c = 4.39 Å, respectively. These values are in good agreement with the JCPDS database, as well as the data presented in the literature for SnSe thin films grown by various methods [14, 15].

Fig. 3 illustrates SEM images for all samples grown at different substrate temperatures from the Sn and Se enriched as well as stoichiometric SnSe precursors. Although the microcrystals for all SnSe films are uniformly distributed over the film surface, the microstructure (shape and grain size) of the samples depends on the substrate temperature and the precursor composition. With increasing the substrate temperature, the shape of the grains changes, whereas the grain sizes do not change significantly for the films deposited from Sn enriched SnSe precursor (Fig. 3a), and vice versa for the samples obtained from Se enriched and stoichiometric SnSe precursors. Furthermore, the grain shapes of all the samples have a flattened appearance for all substrate temperatures, except for films obtained from Sn enriched in SnSe compound at a substrate temperature of 550℃, which has the form of a parallelepiped.

Table 1 Results of energy dispersive X-ray spectroscopy analysis for SnSe thin films deposited from Se enriched and stoichiometric SnSe precursors

$T_{п}$(°C)	SnSe thin films deposited from stoichiometric SnSe precursor			SnSe thin films deposited from Se enriched SnSe precursor		
	Composition of the film	Sn%	Se%	Composition of the film	Sn%	Se%
450	$Sn_{0.91}Se$	57.8	42.2	$Sn_{0.89}Se$	57.3	42.7
500	$Sn_{0.91}Se$	57.8	42.2	$Sn_{0.96}Se$	59.2	40.8

550	$Sn_{0.92}Se$	58.2	41.8	$Sn_{0.95}Se$	58.7	41.3

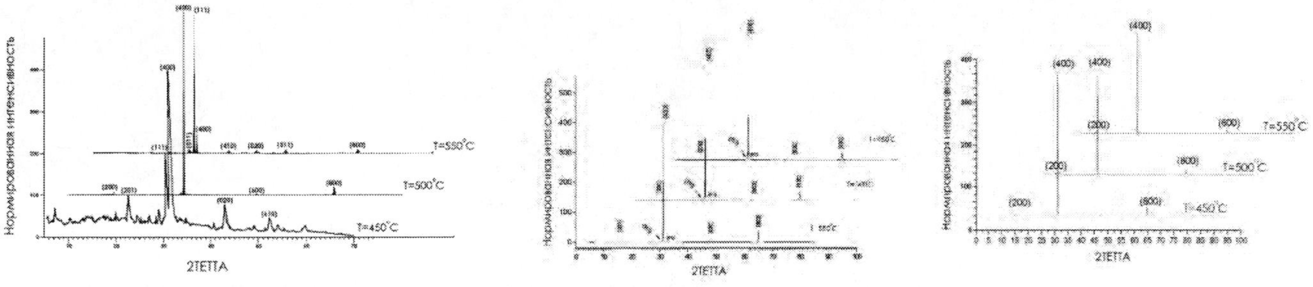

a) b) c)

Figure 2. Wide-angle X-ray data of SnSe films obtained from a) Sn enriched, b) stoichiometric and c) Se enriched SnSe precursors at three different substrate temperatures.

Table 2. Electrical parameters of SnSe films. T_s-substrate temperature, σ-electrical conductivity, E_{ac}-activation energy.

T_s (°C)	SnSe thin films deposited from Sn enriched SnSe precursor				SnSe thin films deposited from stoichiometric SnSe precursor				SnSe thin films deposited from Se enriched SnSe precursor			
	Composition of the film	σ (Ohm·cm)$^{-1}$	E_{ac} (eV)	Type of cunductivity	Composition of the film	σ (Ohm·cm)$^{-1}$	E_{ac} (eV)	Type of cunductivity	Composition of the film	σ (Ohm·cm)$^{-1}$	E_{ac} (eV)	Type of cunductivity
450	$Sn_{0.98}Se$	90	0.0023	*p*	$Sn_{0.91}Se$	6,5	0.0023	*p*	$Sn_{0.89}Se$	5,5		*p*
500	$Sn_{1.01}Se$	70	0.0024	*p*	$Sn_{0.91}Se$	24	0.0025	*p*	$Sn_{0.96}Se$	20		*p*
550	$Sn_{1.14}Se$	15	0.0406	*p*	$Sn_{0.92}Se$	27	0.0026	*p*	$Sn_{0.95}Se$	24		*p*

The samples obtained from Sn enriched precursor have similar grain sizes of 8-10 µm and have a polycrystalline structure for all substrate temperatures. However, the grain sizes of the films deposited from Se enriched and stoichiometric SnSe precursors increased, i.e. 8-20 µm, at the substrate temperature of 550℃ and the structure become more densely packed. Moreover, disappearance of vertically deposited grains on the surface of all SnSe films was observed with increasing the substrate temperature.

The optical properties of selected SnSe film, i.e. deposited from stoichiometric SnSe precursor at 450℃, were studied to determine the absorption coefficient (α) and the optical bandgap (Fig. 4 and 5). As shown, SnSe thin film has the absorption coefficient of ~10^5 cm^{-1} and the optical bandgap of 1.21 eV. These values are in good agreement with [15].

The electrical parameters of the SnSe films, e.g. conductivity σ, activation energy E_{ac} and the type of conductivity, are given in Table 2. As shown, all SnSe thin films illustrate *p*-type conductivity. The electrical conductivity of SnSe films deposited from Sn enriched SnSe precursor decreases with increasing the substrate temperature. This is explained by increasing the Sn content in the film deposited at high substrate temperature. Unlike, the electrical conductivity of those samples obtained from Se enriched and stoichiometric SnSe precursors increases with increasing the substrate temperature. This improvement is associated with an increase in the grain size and a decrease in the grain boundary density of the films grown at higher substrate temperatures.

4. SUMMARY

We have studied the morphology, structure, optical and electrical properties of SnSe films fabricated at different substrate temperatures and compositions of precursors. It has been found that: 1) at low substrate temperatures, SnSe thin films are Se enriched, i.e. independent of the initial precursor composition; 2) with increasing the substrate temperature, foreign particles disappear from the film surfaces and the films are more closely packed polycrystalline structure; 3) the films have an orthorhombic structure and are characterized by preferential (400) and (111) plane orientations; 4) the electrical conductivity of the SnSe films deposited from Sn enriched precursor decreases with increasing the substrate temperature, whereas increases in SnSe films obtained from the Se enriched and stoichiometric SnSe precursors. Optical measurements showed that the SnSe thin films have a direct band gap of 1.21 eV and a high absorption coefficient of ~ 10^5 cm^{-1}.The samples performed *p*-type high conductivity of

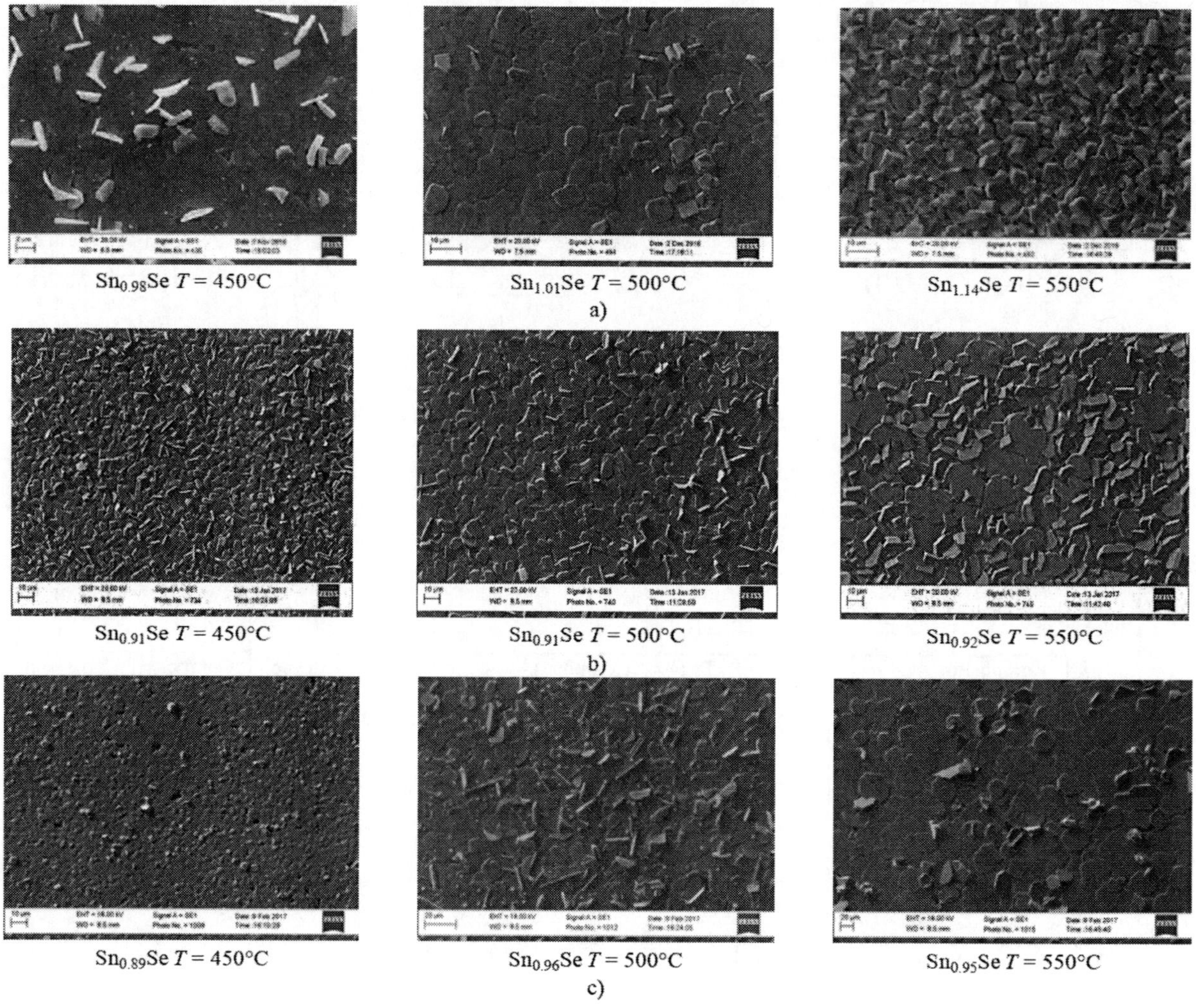

Figure 3. SEM images of SnSe thin films deposited from a) Sn enriched, b) stoichiometric and c) Se enriched SnSe precursors at indicated substrate temperatures.

10-10^2 (Ohm×cm)$^{-1}$ depending on the deposition conditions. Our next step will be fabrication of $Zn_xSn_{1-x}Se$ films in the whole range of compositions of $0 \leq x \leq 1$.

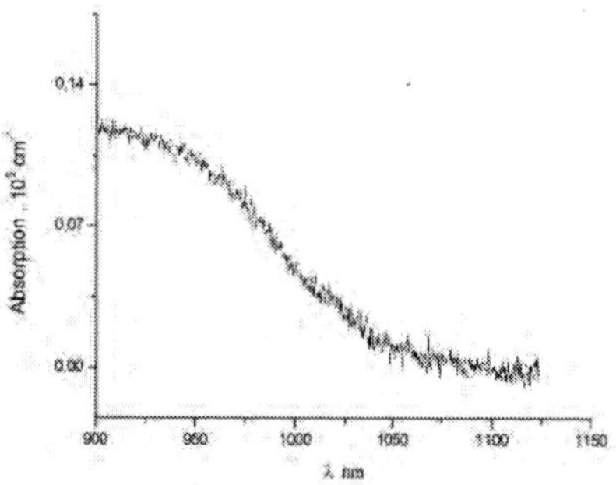

Figure 4. Absorption coefficient as a function of wavelength for SnSe thin film deposited from stoichiometric SnSe precursor at 450 °C.

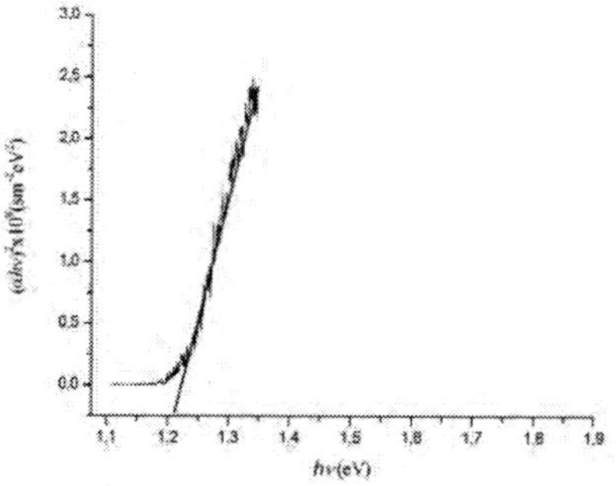

Figure 5. Absorption coefficient versus photon energy for SnSe thin film deposited from stoichiometric SnSe precursor at 450 °C.

5 REFERENCES

[1] D. Pathinettam Padiyan, A. Marikani, K.R. Murali, Electrical and Photoelectrical Properties of Vacuum Deposited SnSe Thin Films, Cryst. Res. Technol. 35 (2000) 949-957.
[2] Z. Zainal, N. Saravanan, K. Anuar, M.Z. Hussein, W.M.M. Yunus, Chemical bath deposition of tin selenide thin films, Mater. Sci. Eng., B, Solid-State Mater. Adv. Technol. 107 (2004) 181-185.
[3] B. Subramanian, T. Mahalingam, C. Sanjeeviraja, M. Jayachandran, M.J. Chockalingam, Electrodeposition of Sn, Se, SnSe and the material properties of SnSe films, Thin Solid Films 357 (1999) 119-124.
[4] D. Tran Quan, SnSe Thin Films Synthesized by Solid State Reactions, Thin Solid Films 149 (1987) 197-203.
[5] Y. Xie, H. Su, B. Li, Y. Qian, A direct solvothermal route to nanocrystalline selenides at low temperature, Mater. Res. Bull. 35 (2000) 459-464.
[6] I.R. Nuriev, E.Y. Salaev, A.K. Sharifova, Ser. Fiz.-Tekh. Mat. Nauk 3 (1982) 91.
[7] R. Teghil, A. Santagata, V. Marotta, S. Orlando, G. Pizzella, A. Giardini-Guidoni, A. Mele, Characterization of the plasma plume and of thin film epitaxially produced during laser ablation of SnSe, Appl. Surf. Sci. 90 (1995) 505-514.
[8] N.D. Boscher, C.J. Carmalt, R.G. Palgrave, I.P. Parkin, Atmospheric pressure chemical vapour deposition of SnSe and SnSe 2 thin films on glass, Thin Solid Films 516 (2008) 4750–4757.
[9] S. Sagadevan, Semiconductor Nanomaterials, Methods and Applications: A Review, Nanosci. Nanotechnol. 3(3), (2013) 62-74.
[10] S.M. Ali, J. Muhammad, S.T. Hussain, S.A. Bakar, M. Ashraf, Study of microstructural, optical and electrical properties of Mg dopped SnO thin films, Mater. Sci. Mater. Electron. 24(7), (2013) 2432-2437.
[11] Y. Moon, S. Kim, J. Park, The influence of substrate temperature on the properties of aluminum-doped zinc oxide thin films deposited by DC
magnetron sputtering, J. Mater. Sci.: Mater. Electron. 17, (2006) 973-977.
[12] NanoScience and Technology: Nanostructured Materials and Their Applications, Ed. S. Logothetidis, (Springer, Berlin, 2012).
[13] Z.L. Wang, W. Wu, Nanotechnology-enabled energy harvesting for self-powered micro-/nanosystems, Angew. Chem. Int. Ed. 51 (2012) 11700-11721.
[14] N.R. Mathews, Electrodeposited tin selenide thin films for photovoltaic applications, Solar Energy 86 (2012) 1010–1016.

[15] F. K. Butt, C. Cao, W. S. Khan, Z. Ali , R. Ahmed, F. Idrees, I. Aslam, M. Tanveer, J. Li, S. Zaman, T. Mahmood, Synthesis of highly pure single crystalline SnSe nanostructures by thermal evaporation and condensation route, Materials Chemistry and Physics 137 (2012) 565-570.

RECORD-LIGHT WEIGHT C-SI MODULES BASED ON THE SMALL UNIT COMPOUND APPROACH – MECHANICAL LOAD TESTS AND GENERAL RESULTS

Hartmut Nussbaumer*, Markus Klenk, Nico Keller, Peter Ammann, Jan Thurnheer
ZHAW, Zurich University of Applied Sciences, SoE, Institute of Energy Systems and Fluid Engineering
Technikumstrasse 9, CH-8401 Winterthur, Switzerland
*phone: +41 (0) 58 934 4799; e-mail: hartmut.nussbaumer@zhaw.ch

ABSTRACT: The "small unit compound" (SUC) concept [1] allows the preparation of very light weight c-Si solar laminates utilizing thin glass or plastic materials. A main difference to the common module type is the replacement of the frame at the laminates fringe by a lattice-like structure at the rear or at the rear and front side. Due to the small distances between the mechanical supporting elements which prevent the dishing of the laminate, the stiffness of the laminate itself can be reduced to a minimum. This enables the use of thin glass or alternative materials such as polymer foils. In this paper we present results of mechanical tests with such modules and discuss general results concerning the module lay-out. The paper has a focus on glass/glass and glass/backsheet laminates with glass thicknesses of 0.8, 1.1, 1.6 and 2 mm. A 60 cell glass/backsheet module 0.8 mm front glass with a weight of as low as 6.3 kg (without junction box and cables) has been fabricated. According to our knowledge this is the lowest weight for 60-cell c-Si modules with front glass ever reported. First hail resistance tests show that these modules may surpass the IEC 61215 norm. Only slightly lower weights are possible if alternative materials are used instead of glass for the front side cover.

Keywords: PV module, new concept, compound, new material, construction

1 INTRODUCTION

For some PV applications the weight of c-Si PV modules is an obstacle. Particularly roofs of commercial and industrial buildings are sometimes not suited for conventional rooftop solar installation without costly structural improvement. In addition some special PV systems, e.g. PV elements for foldable parking roofs or greenhouses need lightweight modules to enable cost effective mounting structures [2].

In general, a lower module weight may also result in more lightweight and thus cheaper mounting solutions. This fact is increasingly relevant as the share of cost for mounting the PV modules has increased significantly in the last years due to the falling prices of PV modules. Thin light weight modules without frame may also drastically enhance the packing density and therefore have a great potential to lower the transport cost.

The segment of light weight modules can be subdivided in thin film and crystalline silicon modules. Light weight thin film modules are mainly flexible structures intended for a flat installation on roof tops. However, a horizontal positioning on the ground is unfavorable with regard to energy harvest, soiling and residual water, which can lead to faster degradation of the modules. Most light weight concepts for c-Si however use alternative materials to glass, e.g. ETFE as transparent medium at the front side, supported by a rigid material such as glass fibre reinforced plastic at the laminates rear [6]. However, there are also approaches for light weight c-Si PV modules on the basis of glass/backsheet and glass/glass laminates.

Standard, 60 cell, crystalline silicon glass/backsheet modules typically have a weight in the range of 18- 22 kg, depending on the thickness of the glass and the frame. This results in a specific weight of about 12.5 kg/m^2. The mechanical rigidness is due to the glass and a circumferential aluminum frame.

Glass/glass modules are gaining market share, due to a presumed longer lifetime and other benefits such as the potential bifaciality. Glass/glass modules may be mounted without an additional frame depending on the thicknesses of the glasses and the sub construction used. Light weight glass/glass modules have been reported from Fujipream achieving a weight of 8.2 kg using glass thickness as low as 1.1 mm and an ionomer– based encapsulant [3, 4]. The size of the 215 Wp module is not given in the references, therefore a specific weight cannot be calculated. Currently heat strengthened solar glass is commercially available with a thickness in the range of 2 mm. We have included thermal strengthened glass with a thickness of 1.6 mm. Thinner glasses are, to our knowledge, chemically strengthened and by far more expensive. However there are ongoing research activities, which aim at thermally strengthened glass with 1 mm thickness [5].

Modules using alternative materials to glass are commercially available and achieve specific weights of as low as 2.74 kg/m^2 [6]. These semi-flexible modules use a polymeric material as front sheet and glass-fibre reinforced plastics to obtain sufficient mechanical stability.

The basic concept of standard c-Si PV modules has not changed significantly in the last couple of years and has proven its long-term reliability. However, the concept also has some inherent structural drawbacks. Due to the laminates dimension and weight there are considerable forces which have to be considered in the module design, particularly because of the large open laminate area [7].

The stiffness of a standard c-Si module is basically depending on the frame and the front glass; both components are also the heaviest parts of the module structure. Reducing the dimensions and weight of either frame or glass directly results in a lower mechanical stability. Standard glass/backsheet modules with circumferential frame also have a large unsupported central laminate area, for glass/glass modules there are similar issues leading to a considerable dishing of the central laminate area if a mechanical load is applied, which may result in a damage of the module. Using significantly thinner glass as a replacement in the conventional structure is therefore not a suitable means to obtain reliable modules with reduced weight. Even though inexpensive, thermally hardened 2 mm solar glass is readily available it cannot replace the typical 3.2 mm glass in a standard glass/backsheet layout. Obviously, the use of glass with thickness below 2 mm or alternative materials such as polymers or GRP is even more problematic.

One option to achieve sufficient mechanical bending strength is to use a rigid supporting structure on the back of the module. A different approach to prevent dishing for light weight c-Si laminates uses a lattice instead of a backing plate [1]. Compared to a standard module the laminate area is now subdivided into smaller units. This results in reduced lever arms between supporting points and suppressed dishing. Due to the subdivision of the modules in smaller mechanical segments the modules are named Small Unit Compounds (SUC) modules. The beams can be positioned on the modules rear side or on the rear and the front side. The latter lay-out may be of particular interest for glass/glass modules or other concepts with symmetrical lay-out, because the cell matrix within the laminate is then along the neutral fibre.

Figure 1: SUC modules with conventional glass/backsheet laminate and lattice structure at the rear.

In this approach, drawing a dividing line between module laminate and sub-construction cannot be done so easy any more. The function of the sub construction is normally to fix the rigid module onto the building or mounting structure. In the new approach, the sub-construction has two functions: enhance the mechanical rigidness of the module and fixing the module to building or mounting structure.

This aspect has to be considered when weights are compared. In this respect it makes more sense to compare weights of complete installed systems (module together with sub-construction) rather to compare the modules / laminates itself. However, when comparing complete systems, one has to consider the different prerequisites. PV modules installed on a flat roof with a tilted angle for example, need additional mechanical elements comparing to a modules installed flat on a tilted roof.
Also for measuring the mechanical stability it makes sense to measure the complete system consisting of the relevant part of the sub-construction together with the module laminate.

Furthermore the approach of assigning functionality of the mechanical rigid module to the sub-construction opens the door to transport extreme lightweight PV laminates and increase the packing density considerably by reducing the module height due to the missing frame. Increasing the packing density will decrease the transportation costs of PV modules.

2 MECHANICAL LOAD TESTS

Different SUC designs with lattices and beams (mainly aluminium, but also GRP) were tested, mainly in form of rear side support structures with several laminate types. For each laminate/lattice structure combination the bow was determined as a function of load. Loads up to 285 kg per m^2 (480 kg in total) were applied, which is above the 240 kg/m^2 as required by the IEC 61215. In order to enable a comparison with the state-of-the-art a standard glass/backsheet module (3.2 mm glass/frame) was tested in the same manner; also the weights of the respective structures are compared.

The experiments with different set-ups focused on two groups of designs. The first group of trials was carried out with a comparatively conventional 60 cell glass/backsheet laminate with 2 mm glass thickness. Beams with different arrangements, diameters, weight and material were applied to this laminate type in order to reveal general trends and to measure the resulting bow when exposed to mechanical load.

In the second group of trials different laminate structures with thinner glass and alternative material were respectively tested with two sets of rear side beams (same dimensions but with two different arrangements). Glass/glass and glass/backsheet laminates (60 cells) were prepared with 0.8, 1.1, 1.6 and 2 mm glass thickness; also a glass/backsheet laminate with 3.2 mm glass and a flexible laminate based on GRP were included in the experiments. Again the respective bow was determined for different loads up to 2854 Pa. The 0.8 mm and 1.1 mm glass is a non-standard, chemically hardened solar glass since heat strengthening is not possible for this low solar glass thickness. Latest tests have been performed on glass/backsheet laminates using a 1.6 mm thermally strengthened glass.

The mechanical tests have been performed by using a test rig on which the tested module is loaded with sand sacks with a weight of 12 kg each. The sand sacks were positioned in a defined sequence in order to achieve a homogeneous load.

Different module lay-outs were tested and already presented earlier [1]. It was found that a real lattice structure is not superior to simpler lay-outs with supporting beams parallel or perpendicular to the length axis of the laminate, see figure 2. When installed, the respective ends of the small beams at the rear side are connected to two perpendicular oriented main beams of the mounting. For the lay-out with beams parallel to the length axis (figure 32) an additional central support is beneficial to obtain the best results with regard to weight and mechanical stiffness. The lowest weight of the total system was obtained with the shorter beams as depicted in figure 3a. In order to carry out mechanical load tests the mounting system is replaced by the test rig.

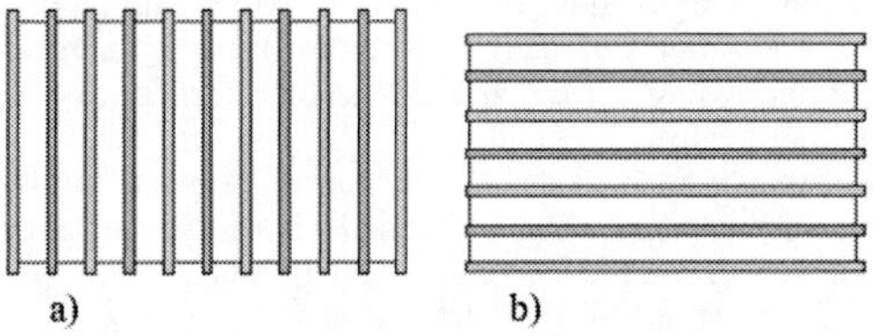

a) b)

Figure 2: Positioning of the supporting beams; either perpendicular (a) or parallel (b) to the length axis of the laminate.

Even though some experiments were carried out with beams made of glass fibre reinforced plastics (GRP) aluminium beams were used as standard. Using GRP instead of aluminium may be beneficial in several regard, particularly if a real lattice structure has to be applied, as intersections could be easier realised than with aluminium beams. Using GRP leads however also to additional issues, such as the connection to the laminate, UV stability and cost. Since very low weights in combination with sufficient mechanical rigidness's were obtained with layouts as shown in figure 3a and 3b using aluminium beams, we focused on these approaches for the moment.

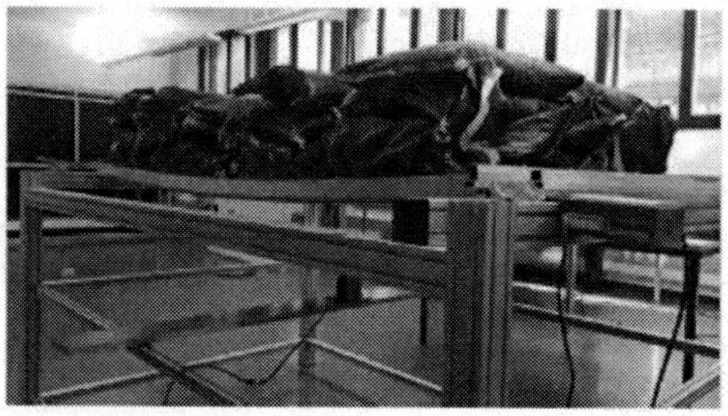

Figure 3: Laminate with applied mechanical load in form of sandbags on the back side of the laminate.

3 EXPERIMENTAL RESULTS

When using very thin glass, hail testing is an obvious issue. First hail tests according to the IEC 61215 norm had been performed. Deviating from the norm, the velocity of the ice bullet was not measured in situ. The setting was adjusted in order to ensure a velocity of the ice bullets with a diameter of 25mm to be higher than 25 m/s. Up to now we found no indication (IV-measurement, EL) for a significant damage even for laminates with 0.8 mm glass (glass/backsheet laminates with 0.8 mm glass still have to be tested).

By comparing the dishing of laminates with different glass thicknesses it was found that the influence of the glass thickness on the mechanical rigidness is negligible for glass thicknesses of 2 mm and below [1]. The mechanical stiffness is only determined by the amount, the cross section and the orientation of the supporting beams.

Figure 5 shows the mechanical stiffness of designs with differing supporting beam lay-outs (amount of beams, cross-section, and material) obtained by mechanical load tests. The general lay-out, which was used in this particular series, corresponds to figure 3a. The stiffness is reflected by the bow of the central laminate area (dishing), which is caused by the applied load. It is not surprising that the stiffness is increased with increasing amount and increasing thickness of walls of the beams. From figure 5 it can be seen, that configurations using 6 or 11 supporting beams result in a less pronounced dishing compared to the reference module (3.2mm glass/backsheet, Al-frame).

It is assumed that the measured dishing is an appropriate means to appraise the potential damage to the cell matrix within the laminate. While the assumption may not be perfectly valid on all regards, e.g. due to local bending effects, it probably gives a very good indication if the IEC 61215 can be passed.

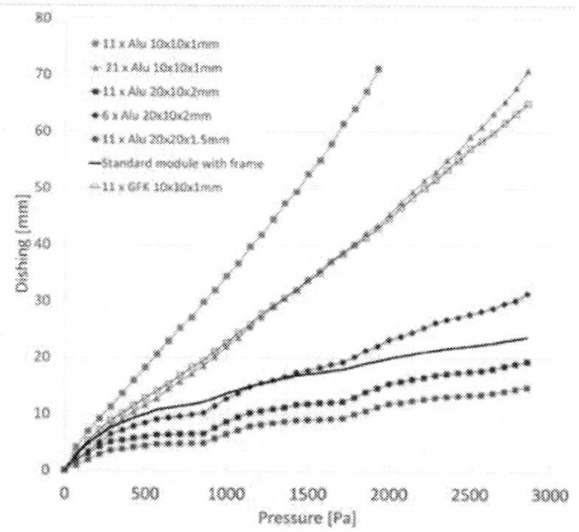

Figure 4: Maximum dishing as a function of the mechanical load for different types of supporting beams. The supporting beams are orientated parallel to the short edge of the laminate (see figure 3a). The supporting beams exhibit different cross sections and wall thicknesses as indicated. All beams are made out of aluminum except one built out of glass fiber reinforced plastic (GRP). The measurements were done on glass/backsheet laminates using a 2 mm thick front glass. A standard glass/backsheet module (framed, 3.2 mm glass) is shown as a reference.

A final approval of the system can only be achieved, if the whole mounting system has been designed and tested. This aspect makes the tests more complicated. Corresponding tests were also carried out; an example is shown in figure 5.

Figure 5: PV laminate and sub-construction under load demonstrating the mechanical stability of the construction.

Testing of the specific example shown in figure 5 showed that the total system, including the sub-construction, is mechanically stable Dishing could be limited to values below the ones observed for standard modules with frame.

4 DISCUSSION, CONCLUSION AND OUTLOOK

The replacement of the module frame by a lattice like structure opens the path towards the use of glass with thickness equal or below 2 mm or the use of very thin and flexible non-glass laminates. It could be shown, that with this approach, the dishing of modules under mechanical load can be limited to similar values of standard modules at considerably lower weight. Also first hail tests indicate that such light weight modules with thin glass could be an alternative to standard modules.

Figure 6 shows a system using a Tritec sub-construction which was adapted by us to mount SUC modules as presented in this paper. The Tritec sub-construction is designed for the installation of framed standard modules on lightweight corrugated metal roofs, e.g. on industrial buildings [8]. In this design the modules are in landscape position, as mainly used for flat roofs and east/west orientation. The tilt angle is adjustable in the low angle range around 15°. The basic layout of the beams corresponds to the one shown in figure 3 b. For this layout it was found to be beneficial to insert an additional central mounting beam, parallel to the two mounting beams at the short sides. The module layout is shown in figure 7 in more detail.

Figure 6: Modified sub-construction based on a Tritec mounting system for corrugated metal roofs. The sub-construction is adapted to the mounting of a SUC-module with rear side aluminum beams. In this example a frameless glass/backsheet with 0.8 mm glass thickness has been applied.

Figure 7: Assembly drawing of the laminate with seven aluminum profiles parallel to the long side of the module and three crosswise oriented bars.

Even though the shown example was not optimized with regard to the system weight a considerable weight reduction to the standard could be obtained. (see table 1) The new approach results in a weight reduction of 5.1 kg (module and sub-construction) using standard aluminum profiles available on the market. According to our calculations a further weight reduction of 3.4 kg would be possible by increasing the aspect ratio of the aluminum bars.

Table 1: The total weight of the designed system of figure 6 compared to a standard system.

System	Module weight + sub con. [kg]	Gross weight [kg]	Gross weight per area [kg/m^2]
Construction see fig. 6	6.3 + 11.1	17.4	10.5
Standard	19.7 + 2.8	22.5	14.1

This would, by keeping the mechanical stiffness, lead to a gross weight of only 14 kg and a weight per area of only 8.8 kg/m^2 for the whole system.

From BenQ a light weight glass/backsheet module is commercially available. The module has a weight of 10.5 kg [9]. Our module laminate itself has a weight of only 6.3 kg (3.9 kg/m^2). To our knowledge, this is the lowest ever reported value for a glass/backsheet 60 cell module. Including the cross bars the weight increased to about 9.5kg (5.9 kg/m^2), depending on the orientation and cross section of the bars. It has to be mentioned however, that the module of BenQ has already a mechanical supporting structure resulting in a mechanical stiffness in all directions without additional sub-construction. In our approach the perpendicular cross bars of the mounting need to be added in order to reach similar mechanical stiffness summing up to similar weights.

Using a supporting beam structure in combination with an ultra-light weight PV module can be an approach for serving a niche market. Even though the extremely thin, chemically hardened solar glasses with 0.8 mm and 1.1 mm did pass the mechanical load and hail tests in our experiments, their price is an obstacle. Thermally strengthened glasses are commercially available (Lisec), making a cost effective light weight glass/backsheet module more likely to be transferred to mass production. The approach would also be suited for ultra-light weight modules using GRP [6].

The SUC laminate design could allow a high packaging density during transportation. In addition the new approach also opens the path for using larger modules, as the mechanical stiffness is no longer depending on the dimension of the frame. Whether the use of large light weight modules could lead to a cost reduction of PV systems by reducing the installation cost needs to be proven. Using large laminates could also be of interest for special applications such as carports.

5 ACKNOWLEDGEMENT

The work is done in the European project U-Light: Ultra lightweight PV modules, in the solar-era.net program. The development is financially supported by the Swiss Federal Office of Energy under the contract SI/501280-01. The project is carried out with the partners ISC Konstanz, Polymer Competence Center Leoben PCCL and Isovoltaic.

Tritec and Schweizer Metallbau are acknowledged for their free of charge contribution of mounting parts for flat roof installations. Meyer Burger AG is acknowledged for their free contribution of solar cell strings. Lisec Austria GmbH is acknowledged for donating thermally strengthened glass plates with a thickness of 1.6 mm.

6 REFERENCES

[1] H. Nussbaumer, M. Klenk, N. Keller, SMALL UNIT

COMPOUND MODULES: A NEW APPROACH FOR LIGHT WEIGHT PV MODULES, Proceedings of the 32nd European Photovoltaic Solar Energy Conference and Exhibition, Munich, 2016

[2] F. Baumgartner, A. Büchel, F. Carigiet, T. Baumann, R. Epp, A. Wirtz, A. Huegeli, U. Graf F. "URBAN PLANT" light-weight Solar System FOR PARKING AND Other URBAN DOUBLE USE APPLICATIONS, Proceedings of the 28th European Photovoltaic Solar Energy Conference and Exhibition, Paris, 2013

[3] pv magazine, News 1st of March 2012, "PV Expo Japan: solutions for a unique market"

[4] PhotonicsOnline, News October 13th, 2010, New DuPont Encapsulant and NSG Thin Glass Enable Fujipream To Build World's Thinnest Crystalline Silicon Solar Panel

[5] LIMES, Light Innovative Materials for Enhanced Solar Efficiency, http://www.limes.nu/

[6] DAS Energy GmbH, Data sheet: Module DAS-Serie 240 – 250P

[7] A. J. Beinert et al. "Influence of photovolataic module mounting systems on the thermo-mechanical stresses in solar cells by FEM Modeling", EUPVSEC, Munich, 5BV.1.14, 2016

[8] Tritec subconstruction system "TriStand" http://www.tritec-energy.com/de/montagesysteme-komponenten/tri-stand-montagezubehoer-c-142/

[9] BenQ Solar claims title for world's lightest crystalline silicon PV module
by Nitin Nampalli on March 13, 2014 in Solar Choice News, http://www.solarchoice.net.au/blog/news/benq-solar-claims-title-worlds-lightest-crystalline-silicon-pv-modules-130314g

OUTDOOR AND INDOOR TESTING OF TRANSPARENT ANTISOILING COATING BASED FLUORINE-DOPED TIN OXIDE FOR SOLAR ENERGY APPLICATIONS

Kawtar Belrhiti Alaoui*[1, 2], Ahmed Alami Merrouni[1], Badr Ikken[1], Abdelkader Outzourhit[2]

[1]Research Institute for Solar Energy and New Energies (IRESEN), 16, street Amir Sidi Mohamed souissi - Rabat, Morocco

Tel: +212 (0) 537 682 236 | Fax : +212 (0) 537 688 852

[2]Nanomaterials Laboratory for Energy and Environnement (NLEE), Faculty of Sciences Semlalia, Cadi Ayyad University, PO Box: 2390, Marrakech 40000, Morocco

**belrhiti@iresen.org, alami@iresen.org, ikken@iresen.org, , aoutzour@uca.ma*

ABSTRACT: The installation of photovoltaic modules in the MENA region is very beneficial if we consider the high amount of solar irradiation received by this area. Nevertheless, the presence of soiling sources mainly from dust and sand storm reduces, significantly, the performances of this technology and increases the Operation and Maintenance (O&M) costs. For this reason, it's crucial to find solution for dust removal, to enhance the modules optical efficiency, thus the electricity production. Super-hydrophobic coatings can be a good solution for this issue. In this paper, we evaluated the soiling effect on three different PV glazing samples under the semi-arid climate of Benguerir in Morocco. Furthermore, the coating performances were evaluated by comparing the transmittance drop of two glass samples with two different coating layers based on SnO_2 and SnO_2F.

The results show that soiling is an immense problem in Morocco where the transmittance drop can reach 38% after 13 weeks of exposition. Furthermore, adequate coatings on glass can be a good anti-soiling solution. It is found that the SnO_2:F coating on glass enhances the transmittance by 3.22% in comparison to 2.99% for SnO_2 on glass.

Keywords: coatings, soiling, outdoor testing

1 INTRODUCTION

Species in nature have developed great functions and capabilities, and pushed scientists to mimic biological systems. Several successful materials have already been described in scientific literature such dye-sensitized solar cell that mimics the photosynthetic process of plants, photonic crystal mimics the wings of a butterfly, and super-hydrophobic and self-cleaning surfaces mimics the surface of a lotus plant [1]

A comparative study of photovoltaic uncoated and coated modules with self-cleaning properties in arid climate was elaborated in Spain. The self-cleaning coating enhanced the performances of the modules [2]. The dirt losses have a negative influence on the performance ratio of the photovoltaic systems. These are mainly due to dust particles in the module surface that reduce the performance of the photovoltaic systems in dry climate to 15% [3]. Consequently, the modules should be cleaned very often, which may be very expensive especially in large PV fields. Therefore, the key to saving water and energy is to install photovoltaic modules coated with anti-soiling films.

Several surface treatments are currently being tested or produced in the Green Energy Park (GEP) [4] in order to develop a technology that is adapted to arid and semi-arid climates, and to reduce the influence of the deposition of dirt on the surfaces of glass, and thus the cost of cleaning,. Tests are performed outdoor in the semi-arid region of Benguerir (Morocco).

In this work the influence of fluorine-doped tin oxide the (SnO_2:F) coatings on the antisoiling properties of the glass surfaces under Moroccan climate was investigated.

2 OBJECTIVE

The decrease of solar cells power caused by the influence of soiling on the glass surface has attracted great intention at the research scale, especially in arid climate. In this work, we will investigate the influence of the SnO_2:F coating on the quantity of dirt resulting from the sand storms that characterize the region during the summer. The study will aid in deciding to include this coatings as an additional treatment for future research on surface treatments adapted to Morocco and MENA region.

This work is an experimental study, which focuses on a comparative study of tree types of glasses, one with SnO_2:F coating, one with SnO_2 coating, both deposed by CVD technique, and a simple bare soda lime glass. The objective is to evaluate the effect of the soiling on these glasses.

3 MATERIALS AND METHOD

As mentioned above, this study has two main objectives: (i) understanding the soiling effect on PV glass samples during the dry period of the year, (ii) evaluating the performance of two anti-soiling coatings. For this purpose, three sets of glass samples with a dimension of 10x10cm were exposed at Green Energy Park research facility (32°13'18.2"N 7°55'44.0"W) from May the 5th until August the 17th 2017. South facing exposure rack (Fig.1) is inclined by 30° to fit the latitude of the study area.

Figure 1: The exposition rack installed at GEP.

The exposed glass samples were subdivided into three groups: B, C_1 and C_2. B refers to the bare glass samples. These samples are regular glass with a maximum transmittance of 91.12% and they are used as reference in this study to evaluate the coatings performances. The C_1 group contains the samples coated with an SnO_2 layer. Their maximum transmittance value is 84.27%. The samples of the 3rd group are coated by an SnO_2:F layer and they will be referred to as C_2. The maximum transmittance of the C_2 samples is 87.73%.

In this work, transmittance is the key element to evaluate the effect of soiling and the coating performance. To measure the transmittance, we used Jenway 6850UV/Vis Spectrophotometer (figure 2). This device measures the transmittance from 190 to 1100 nm, with a variable spectral bandwidth, ranging from 5 nm down to 0.5nm, using tungsten halogen and deuterium light sources.

Figure 2: Jenway spectrophotometer.

Another important parameter for the glass surface characterization is the wettability. For PV glazing, the surface wettability can describe the degree of soiling removal. Indeed, by measuring the contact angle, it may be possible to determine the surface tension of the material and its attraction force with respect to soil as well [5, 6, and 7]. In this study, the Drop Shape Analyzer – DSA100 (Figure 3) was used to measure the contact angle of our glass samples.

Figure 3: Kruss, drop Shape Analyzer DSA 100.

To evaluate the effect of soiling on the glass transmittance, and to analyze the efficiency of the two coatings, five sets of glass were exposed for 15 weeks from May the 5th to August the 17th. Each set contains 2 samples from each group (B, C_1 and C_2). Before the samples exposition, the transmittance of the clean state was measured and the average transmittance curve for each glass type is presented in Figure 4.

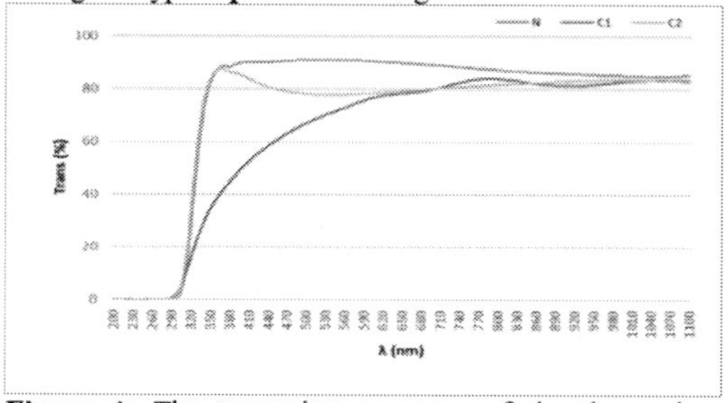

Figure 4: The transmittance curve of the three glass samples.

During the exposure period, the five sets were left to get naturally soiled. After each period, one set of samples was removed for measurement. Table I, represents the measurements time as well as the average transmittance values recorded.

Table I: Measurements period and transmittance values

	5-May	*19-May*	*2-Jun*	*16-Jun*	*1-Aug*	*17-Aug*
N samples	75.62	66.48	56.91	53.49	48.66	57.98
C_1samples	63.48	61.35	46.97	46.06	45.32	47.81
C_2samples	70.36	66.54	55.64	50.06	43.45	59.43

3 RESULTS AND DISCUSSION

The effect of dust on the transmittance drop is shown in Fig.5. As it can be seen, the transmittance drop of the regular glass samples (B) is the highest especially for the first week of exposure where the transmittance drop reached around 10%. This is not the case for the coated systems, where the transmittance drop was around 2% and 3.5% for C_1 and C_2 respectively. After the first month of exposure, the transmittance drop became steadier, especially for the B and C_1 samples and it was respectively around 3% and 1%.

To better evaluate the transmittance drop -or in other words the soiling effect- and to allow a clear comparison between the three glass samples, we determined the Soiling Rate (SR). This parameter is calculated using the following equation:

$$SR = \left(1 - \left(\frac{\int_{\lambda=200}^{\lambda=1100} \tau}{\int_{\lambda=200}^{\lambda=1100} \tau C}\right)\right) x\ 100$$

where τ is the transmittance at the soiled state and τ_c at the clean state.

(a)

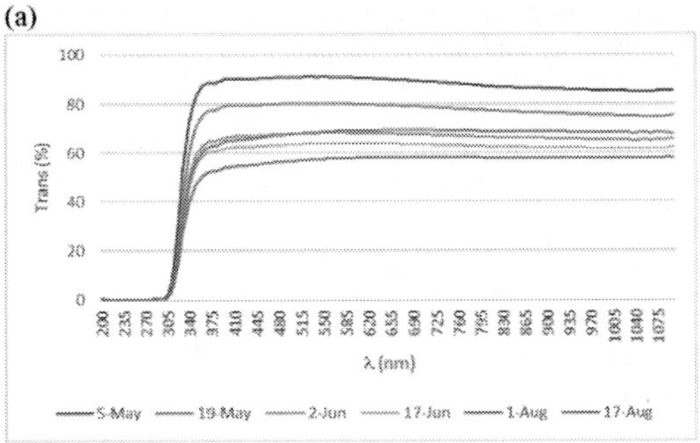

(b)

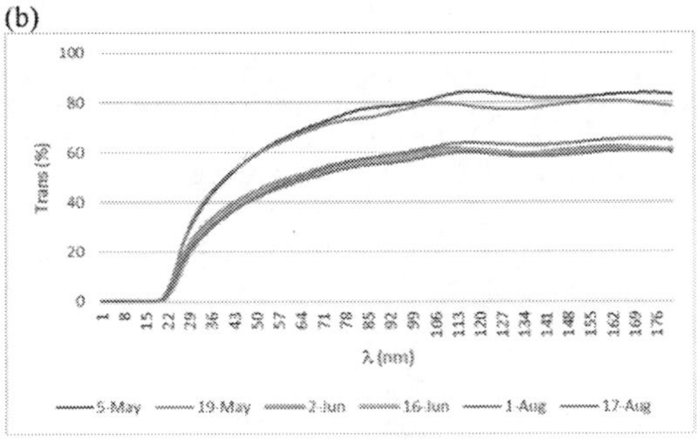

(c)

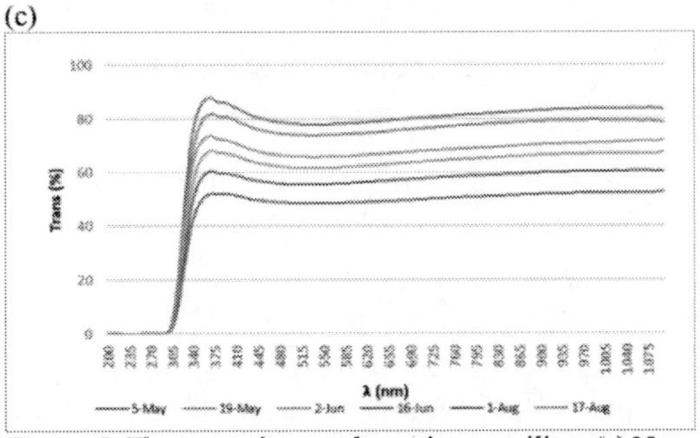

Figure 5: The transmittance drops due to soiling. (a) Non-coated samples, (b) C_1 coated samples, (c) C_2 coated samples.

The higher the SR is, the higher is the soiling effect on the transmittance drop and vise-versa. The soiling rate values of the three glass samples are plotted in Fig.6. As described above, the SR of the B samples is the highest after the first week of exposure. This parameter keeps increasing to reach about 36% after 13 weeks of exposure (1st of August). The Soiling rate of the C_1 samples was the lowest during the first week (around 3%). This parameter increased to 13% after two weeks, but, it stabilized after 13 weeks at about 30%.

Regarding the C_2 samples, the SR after the first week of exposure was lower than the one of the uncoated samples; nevertheless, it was lower than that of the C_1 samples. The SR kept increasing linearly to reach 38% after 13 weeks of exposure.

This situation changed completely after the 1st of August where a drop on the SR values was observed, i.e. an enhancement on the transmittance. Indeed, due to a rain event on August the 6th (Fig.8) the SR decreased to 18%, 4% and 12% for the C_2, C_1 and the bare samples respectively.

This leads us to conclude that during the dry period (from May the 5th to August the 1st) the C_1coating is more performant than the C_2. However, with the presence of natural cleaning agents, like rain, humidity and dew, the C_1 coating shows a high performance in comparison to C_2. This is due to the high hydrophobicity of C_2 samples in comparison to the other samples.

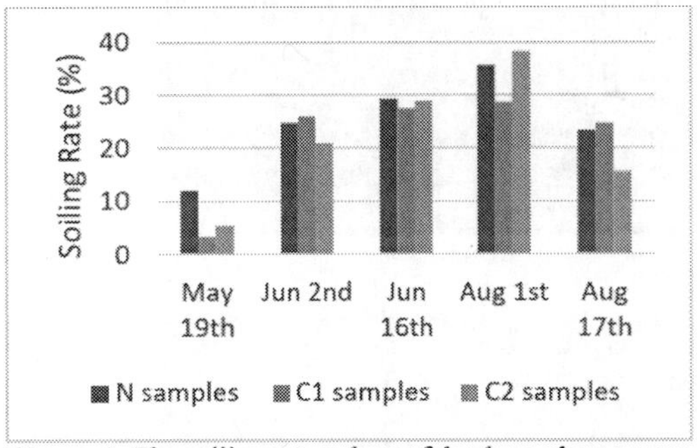

Figure 6: The soiling rate values of the three glass groups

Indeed, as illustrated in figure 7, the average contact angle of the C_2 glass samples is 67.1° compared to 52.6 and 32. 6° for the C_1 and the uncoated samples respectively.

(a)

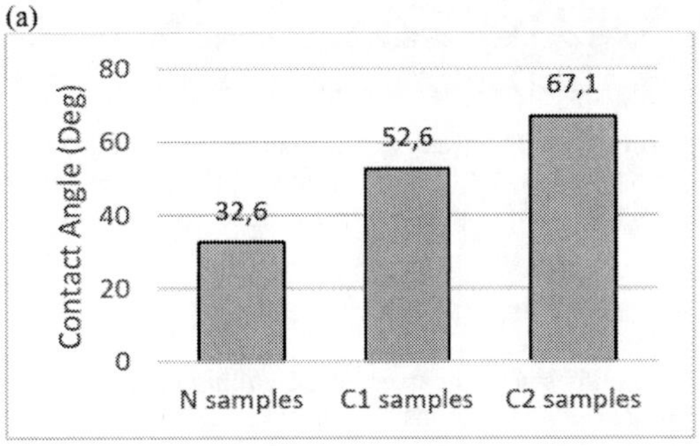

(b)

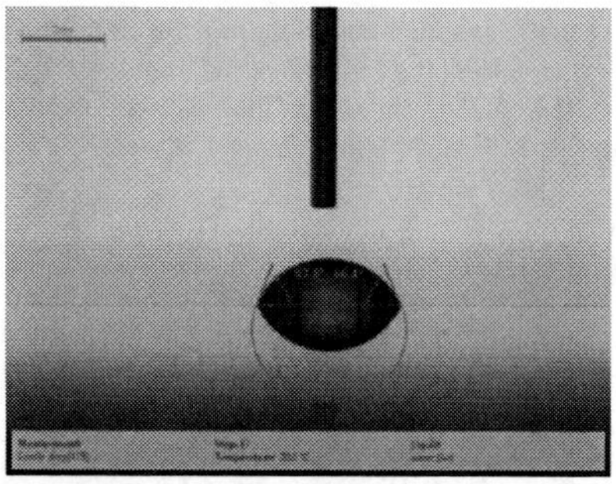

(c)

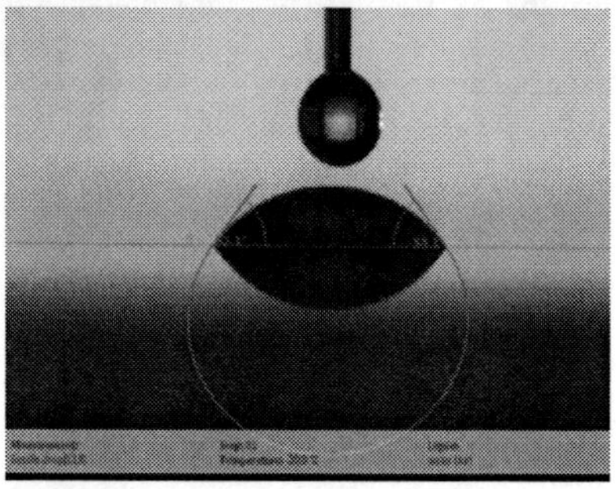

(d)

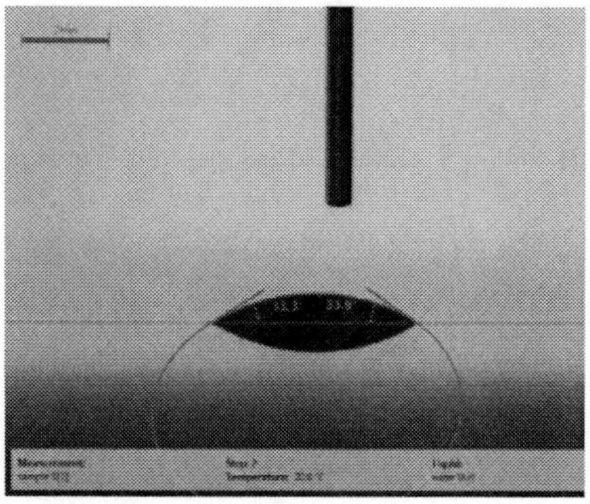

Figure 7: (a) The average contact angle values of the three glass samples. (b,c & d) screenshot from the DSA 100 for the C_2, C_1 and N samples respectively.

4 CONCLUSION

In this study, we assessed the soiling effect on three different PV glazing samples: bare (B) glass, and samples with two different coatings C_1 and C_2, during the dry period of the year at Green Energy Park research facility in Morocco. Results show that soiling is an efficiency limiting factor for PV operation. The soiling rate due to the dust accumulation on the glass samples can reach 38% after 13 weeks of exposure. Coatings based on SnO_2 and SnO_2F can be a solution to enhance the optical properties of the PV modules especially in the first week of exposition.

In order to compare the two coatings used in this study we can divide the exposure period on two parts:

- From May the 5th to August the 1st: this period was completely dry, with high temperature, low humidity values and no rainfall.
- From August the 1st to August the 17th: This period was still hot and dry, but we experienced a rainfall event on the 6th (Figure 8).

During the first period, the C_1 coatings were more efficient on preventing the dust deposition and enhancing the optical efficiency as the SR was around 28% at the end of this period in comparison to 36% and 38% for the B and the C_2 samples.

Due to some rainfall recorded on the 6th of August, the SR values of the C_2 decreases significantly in comparison to the samples coated with C_1 (18% and 4% respectively). This suggests that C2 coatings are more efficient in the presence of natural cleaning agents and they will require less cleaning efforts than samples coated with C1. This is due to their higher wettability/contact angle.

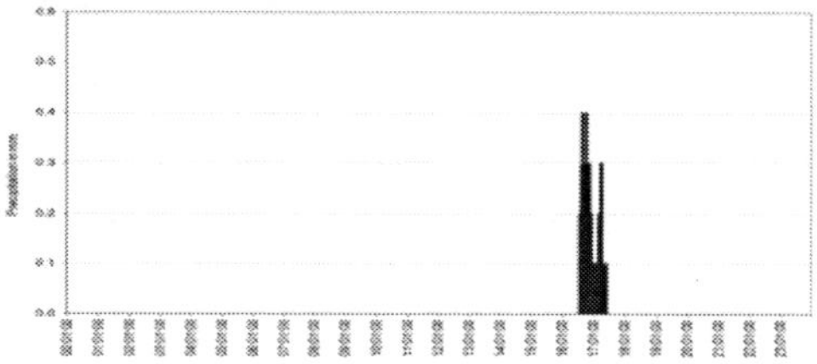

Figure 8: Rainfall records on August the 6th.

Generally, if we average the soiling rate over all the experiment period we can say that the coating C_2 is more efficient than C_1 and it can enhance the transmittance by 3.22% in comparison to 2.99% by C_1, see figure 9.

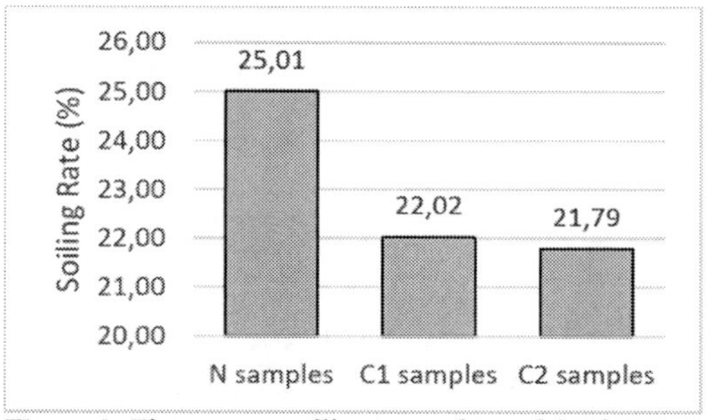

Figure 9: The average soiling rate values of the three glass groups

5 ACKNOWLEDGEMENT

Authors would like to warmly thank Mr. Abdelatif Bouaichi and the Faculty of Sciences and Technologies Errachidia for conducting the transmittance measurements in there laboratories.

6 REFERENCES

[1] Xiong, J., Das, S. N., Shin, B., Kar, J. P., Choi, J. H., & Myoung, J. M. (2010). Biomimetic hierarchical ZnO structure with superhydrophobic and antireflective properties. Journal of colloid and interface science, 350(1), 344-347.

[2] Piliougine, M., Cañete, C., Moreno, R., Carretero, J., Hirose, J., Ogawa, S., & Sidrach-de-Cardona, M. (2013).

Comparative analysis of energy produced by photovoltaic modules with anti-soiling coated surface in arid climates. Applied energy, 112, 626-634.
[3] Piliougine, M., Carretero, J., Sidrach-de-Cardona, M., Montiel, D., & Sánchez-Friera, P. (2008, September). Comparative analysis of the dust losses in photovoltaic modules with different cover glasses. In Proceedings of 23rd European Solar Energy Conference (Vol. 2008, p. 269).
[4] GREEN ENERGY PARK Platform. URL: http://www.greenenergypark.ma/
[5] Nakajima, A., Fujishima, A., Hashimoto, K., & Watanabe, T. (1999). Preparation of transparent superhydrophobic boehmite and silica films by sublimation of aluminum acetylacetonate. *Advanced Materials*, *11*(16), 1365-1368.
[6] Faustini, M., Nicole, L., Boissiere, C., Innocenzi, P., Sanchez, C., & Grosso, D. (2010). Hydrophobic, antireflective, self-cleaning, and antifogging sol– gel [s: an example of multifunctional nanostructured materials for photovoltaic cells. *Chemistry of Materials*, *22*(15), 4406-4413.
[7] Ganesh, V. A., Raut, H. K., Nair, A. S., & Ramakrishna, S. (2011). A review on self-cleaning coatings. *Journal of Materials Chemistry*, *21*(41), 16304-16322.

EFFECTS OF MULTIPLE ORGANIC DYES FOR LUMINESCENT DOWN-SHIFTING LAYERS APPLICATION

S. Gilligan*, H. Ahmed, S. Chandra and S.J. McCormack
Dept. of Civil, Structural, & Environmental Engineering, Trinity College Dublin, Dublin 2, Ireland
*Corresponding Author. E-mail Address: gilligs3@tcd.ie Tel: +353 1 896 2671

ABSTRACT: High energy photons in the UV region are not efficiently used in solar cells and limit the overall efficiency of the cell. Luminescent Down-Shifting (LDS) layers can be used to convert the high energy photons to a lower energy before it reaches the cell, which is more efficiently absorbed. In this study, an investigation of the optics of polymer thin-films LDS layer when comprising of mixed organic dyes, against single dye thin-films was investigated. The effect of creating a layer comprising of an equal ratio of BASF Lumogen Yellow 083, Lumogen Orange 240, and Lumogen Violet 570 in a polymethyl methacrylate (PMMA) polymer was investigated and compared to the experimental results from each single dye. This aided in determining whether using multiple organic dyes in a single film was beneficial for use as an LDS material. The interaction of the dyes with each other was investigated. It was determined that the combination of dyes created a larger absorbance range following the trend of each of the dyes it contained. The resulting emission of the mixed layer has an emission peak at the wavelength matching the longest emission peak wavelength of the comprising dyes. Further tests containing only two of the Lumogen dyes agreed with this finding.

Keywords: Luminescent Down-Shifting, Emission, Absorbance, Thickness, Organic Dyes, Lumogen

1. INTRODUCTION

With world population reaching 7 billion in 2011 and continuing to grow, there is a large strain on traditional fuel sources such as coal, oil and natural gas. These fuel sources also are placing the world's ecosystem in danger due to pollution. To combat this, the development of an energy source which is both sustainable and renewable is vital. Solar cells are photovoltaic (PV) devices which harness solar energy and convert it into electrical energy via the photovoltaic effect. The technology in PV devices has rapidly developed since the 1950s due to the strain a growing population has placed on the world. In Ireland in 1990, only 1.8% of the gross amount of energy used was renewable, which has increased to 8.3% in 2015 [1].

While PV cells are a renewable energy source, solar cells suffer large energy losses due to various factors. A portion of losses can be attributed to spectral losses in the conversion process, in which high energy photons are not efficiently absorbed by the solar cell [2,3]. This prevents the maximum efficiency from being reached. A process known as Luminescent-Down Shifting (LDS) is a method in which high energy photons are converted to a lower energy using a luminescent planar sheet on top of the existing solar cell [4,5]. It was first proposed in the 1970s by Hovel et al. as a method to increase the number of photons being absorbed. It is a desirable method as the planar sheets can be fitted to existing solar cells with ease and is an inexpensive process [6]. As the photons are shifted, they are absorbed in the photoconductive layer, creating more electron-hole pairs and increasing the external quantum efficiency. The incident photons, which have been converted, are more efficiently absorbed as they are at a lower energy, thus increasing the efficiency of the solar cell [7,8].

Different luminescent materials used have varying absorbance and emission peaks so they can be chosen more specifically depending on the material of the solar cell being used. For a LDS material to be beneficial or not the majority of photons down-shifted must not be reabsorbed by the LDS layer [9]. So far, the question of whether having multiple organic dyes in a single layer as a possible LDS material has not been fully addressed. This paper aimed to fill this gap in knowledge. In this paper, various organic dyes from Lumogen (BASF) will be used to create thin-films both separately and combined to see what effect multiple dyes combined will have on each other.

2. METHOD

A host material consisting of PMMA and toluene in a ratio of 7:3 was magnetically stirred for 90 minutes at a low speed to prevent bubbles. This ratio was chosen as it would consistently produce thin-film layers of 50 μm thickness on a substrate when fabricated using a spin coater(Specialty Coating Systems –G3P-12), as it has a thick viscosity without solidifying too quickly on contact with the air. The layers were coated at 1000 RPM for 240 seconds. Using the host material, 4 different solutions of 0.1 wt% concentration were mixed on the magnetic stirrer for 60 minutes. The 4 different sample solutions were Lumogen Yellow 083, Lumogen Orange 240, Lumogen Violet 570, and an equal amount of each dye in one solution, known as the mixed layer. Thin-films were then created using the four types solutions. The films were dried after spin coating, however were not yet easily removed from the substrate in a single film. To easily remove the thin-films, the samples were left in an oven at 40°C for 5 days, and in room temperature for a day further to allow for easy removal from the substrate. The resultant thin-films can be seen in figure 1 below.

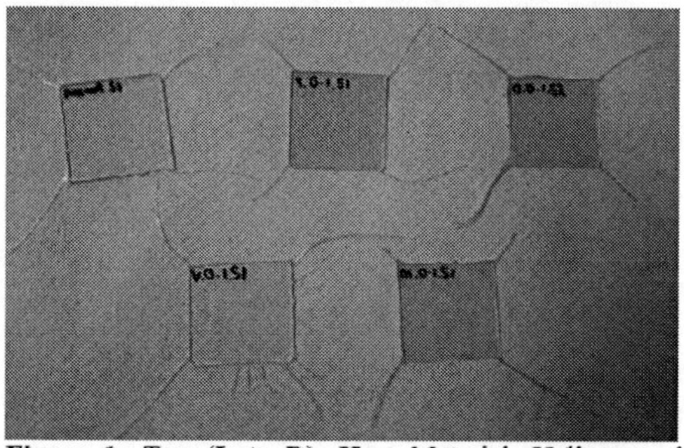

Figure 1: Top (L to R): Host Material, Yellow, and Orange. Bottom (L to R): Violet and Mixed thin-films

The absorbance and emission measurements of the thin-films were taken using a LabSphere Integrated Sphere and the SpectraWiz software. For absorbance measurements, the setup consists of deuterium and halogen light sources to simulate the solar spectrum.

The emission measurements used the Follow Spot 1200 as a solar light source which was focused at different filters depending on the absorbance range of each Lumogen being tested. This was then collimated into an optic fibre which connected to an integrating sphere. Inside the integrating sphere is a holder which the thin-film is clamped to. The integrating sphere is then connected to a detector, and data accessed using SpectraWiz software. From these results, the absorbance and emission for each sample thin-film could be analysed.

3. RESULTS

Uniform LDS layers were successfully fabriated at an average thickness of 50 μm. The layers are flexible and easily removed from the substrate. As shown in figure 2 below, each Lumogen dye absorbs over a different region. As a result of mixing the dyes together in an equal ratio, the absorbance region of the mixed solution is over a much larger range. Yellow has a peak absorbance at 475 nm, Orange peaks at 526 nm, and Violet peaks at 381 nm, leaving the Mixed layer to peak at 480 nm.

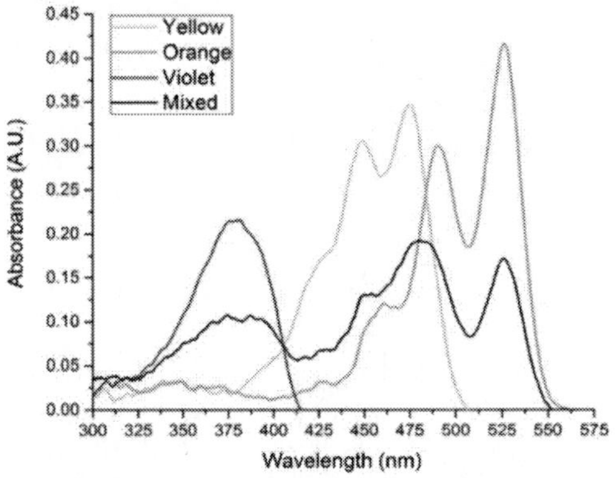

Figure 2: Absorbance measurements of sample thin-films containing organic dyes

The emission measurements show the shift in peak wavelength, which can be seen in figure 3 below. Yellow emits at a peak of 552 nm (a shift of 77 nm), Orange emits at 582 nm (shift of 56 nm), Violet emits at 434nm (shift of 54 nm), and finally Mixed emits at 581 nm (the largest shift of 101.5 nm).

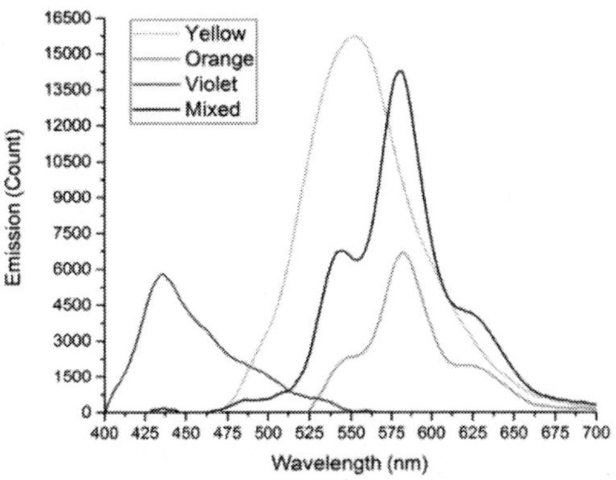

Figure 3: Emission measurements of sample thin-films containing Lumogen Dyes

4. DISCUSSION

From the absorbance measurements of the samples, it is clear that the Mixed sample has a larger range, but as a consequence the height of the absorbance peaks are not as large as the organic dyes when not combined with each other.

Table 1: Table containing the absorbance and emission peak and range of wavelengths

	Absorbance		Emission	
	Peak (nm)	Range (nm)	Peak (nm)	Range (nm)
Yellow	475	300 - 506	552	470 - 700
Orange	526	400 - 560	582	525 - 700
Violet	378	315 - 416	434	400 - 600
Mixed	480	300 - 550	581	468 - 700

To determine if the absorbance curve attained for the Mixed layer is as expected, a curve was calculated using the results from the layers individually. To determine this curve, the absorbance results from each layer were added together and divided by 3 as they were combined in equal quantities. This produced the curve as shown below.

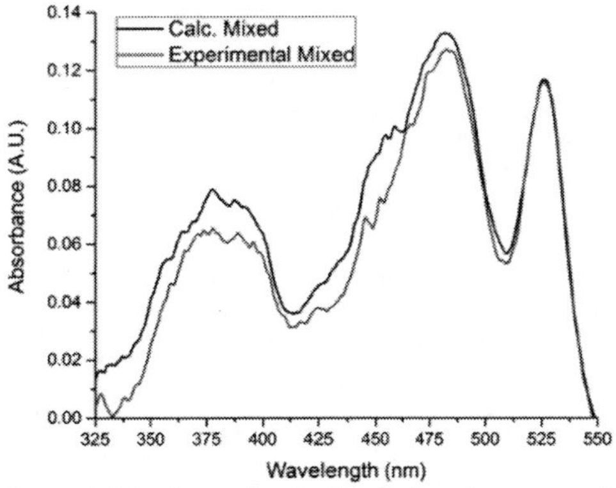

Figure 4: Mixed sample experimental results compared against calculated.

The emission peak of the Mixed layer is blue shifted 1 nm than the emission peak ofLumogen Orange. From this it can be gathered that when a sample consists of materials which have various emission peaks, the resultant peak will be formed at the wavelength which is longest. Further tests performed included fabricating thin-films consisting of only two dyes – a solution of Violet and Orange, and a solution of Violet and Yellow. When the characteristics were measured, they too had emission peaks matching the dye which has the longest wavelength.

4. CONCLUSION

LDS polymer thin-films were fabricated with various organic Lumogen Dyes and investigated. The layers were uniform at a thickness of 50 μm due to optimising the ratio of PMMA to toluene in the host material. This was futher optimised by testing different recipies on the Spin Coater to consistently create this thickness. Through this investigation, it was determined that when different dyes, Lumogen Yellow 083, Lumogen Orange 240, and Lumogen Violet 570 in this case, with different absorbance and emission ranges or peaks are combined, the resulting emission of the layer will take on the peak of the dye which emits at the longest wavelength. The absorbance of the resulting thin-film is also over a much larger range, encompassing each of the dyes which are present in the thin-film.

5. ACKNOWLEDGEMENTS

The authors would like to acknowledge the funding from the European Research Council (Starter Grant PEDAL Project Number 639760).

6. REFERENCES

1. Holland, M., and Howley, M. (2016). Renewable Electricity in Ireland 2015. 32. Available at: http://www.seai.ie/Publications/Statistics_Publications/Renewable_Energy_in_Ireland/Renewable-Electricity-in-Ireland-2015.pdf.
2. Klampaftis, E., Ross, D., McIntosh, K.R., and Richards, B.S. (2009). Enhancing the performance of solar cells via luminescent down-shifting of the incident spectrum: A review. Sol. Energy Mater. Sol. Cells *93*, 1182–1194. Available at: http://www.sciencedirect.com/science/article/pii/S0927024809000762.
3. Rothemund, R., Kreuzer, S., Umundum, T., Meinhardt, G., Fromherz, T., and Jantsch, W. (2011). External quantum efficiency analysis of Si solar cells with II-VI nanocrystal luminescent down-shifting layers. Energy Procedia *10*, 83–87. Available at: http://dx.doi.org/10.1016/j.egypro.2011.10.157.
4. Hovel, H.J., Hodgson, R.T., and Woodall, J.M. (1979). The effect of fluorescent wavelength shifting on solar cell spectral response. Sol. Energy Mater. *2*, 19–29.
5. Van Sark, W.G.J.H.M., Meijerink, a., Schropp, R.E.I., Van Roosmalen, J. a M., and Lysen, E.H. (2005). Enhancing solar cell efficiency by using spectral converters. Sol. Energy Mater. Sol. Cells *87*, 395–409.
6. Danos, L., Parel, T., Markvart, T., Barrioz, V., Brooks, W.S.M., and Irvine, S.J.C. (2012). Increased efficiencies on CdTe solar cells via luminescence down-shifting with excitation energy transfer between dyes. Sol. Energy Mater. Sol. Cells *98*, 486–490. Available at: http://dx.doi.org/10.1016/j.solmat.2011.11.009.
7. Ahmed, H., Sc, B., and Sc, M. (2014). Materials Characterization and Plasmonic Interaction in Enhanced Luminescent Down- Shifting Layers for Photovoltaic Devices.
8. De JONG, M. (2015). Deposition of luminescent thin films for solar energy applications .
9. Ross, D., Klampaftis, E., Fritsche, J., Bauer, M., and Richards, B.S. (2012). Increased short-circuit current density of production line CdTe mini-module through luminescent down-shifting. Sol. Energy Mater. Sol. Cells *103*, 11–16. Available at: http://dx.doi.org/10.1016/j.solmat.2012.04.009.

IMPROVING THE MATHEMATICAL MODEL FOR LUMINESCENT DOWN-SHIFTING LAYERS BY INVESTIGATING THEIR LOSS MECHANISMS

M. Rafiee*, H. Ahmed, S. Chandra and S. J. McCormack
Dept. of Civil, Structural and Environmental Engineering, Trinity College Dublin, Dublin 2, Ireland
* Corresponding author. Email: rafieem@tcd.ie Tel: +353 1 896 2671

ABSTRACT: A Luminescent Down-Shifting (LDS) layer is an optical approach to improve the solar cell's optical response by using luminescent species doped in a polymer matrix material to form a thin layer to be deposited on top of the PV solar cell. In this article, the performance of different polymer matrix materials with various loss mechanisms (attenuation and scattering losses) are mathematically modeled based on a ray tracing algorithm. The simulation is carried out with and without counting the loss mechanisms in the mathematical model. First, the model studied the performance of LDS layer (12.49 × 12.49 × 0.015 cm) of Europium complex (Eu) doped in a Polyvinyl Acetate (PVA) film and deposited on top of monocrystalline Silicon (c-Si) under standard AM1.5 global solar radiation. The comparison of the modelling and experimental results proved that the model's discrepancy was improved from 27% to 2% when the loss mechanisms of the matrix materials are counted in the model. Optical efficiency of 80% was obtained for the LDS layer and an excellent agreement has been achieved between the experimental and model results. Subsequently, the model was used to study the performance of other matrix materials such as glass, epoxy resin polymer, and Poly Methyl Methacrylate (PMMA) polymer for their loss mechanisms. The results have shown that although these polymer are transparent above 300 nm, their optical response significantly changes in the region between 300 to 400 nm hence the optical efficiency and response of the LDS layer.

Keywords: Luminescent Down Shifting, Ray Tracing, Matrix Material, Loss Mechanisms, Optical Efficiency, Emission, Absorption

1 INTRODUCTION

In all Photovoltaic (PV) technologies developed to date, the poor optical response for short wavelength light represents a fundamental limit to the maximum efficiency achieved by the solar cell. In this region (300-500 nm), the high energy photons are not used efficiently and their energy is lost through thermalisation (Rothemund, 2014). The potential exists to increase solar cell efficiency by making better use of short wavelength solar radiation. As is seen in Fig. 1, LDS layer is an optical approach to improve the solar cell's optical response by using luminescent species doped in a polymer matrix material to form a thin layer to be deposited on top of the PV solar cell (Ahmed, 2014, Hovel et al., 1979).

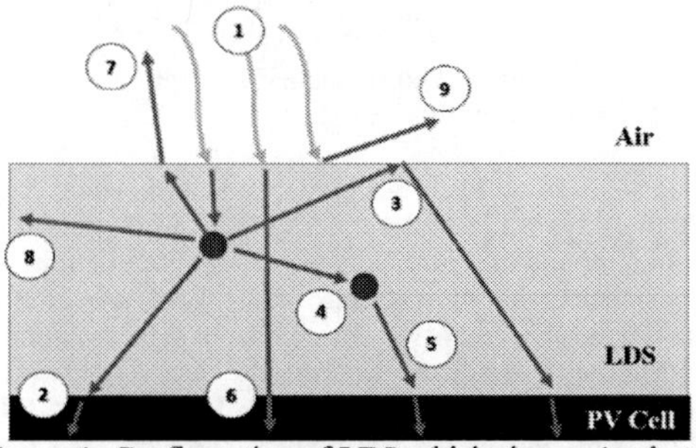

Figure 1: Configuration of LDS which shows: 1- photon enters the LDS absorbed by the luminescent material. Then, 2- re-emitted at longer wavelength and reaches the PV cell 3- or it is wave-guided to the PV cell by Total Internal Reflection (TIR) 4- or re-absorbed by another luminescent molecule and 5- re-emitted with less energy and then reaches the PV cell. 6- Some photons directly reach the PV cell without red-shifting. The other losses include: 7- escape cone loss 8- edge losses and 9- front surface reflection. Note that, the light may also be scattered or attenuated by the matrix material which is not shown here

LDS layers can convert both diffuse and direct solar radiation; therefore, it is an adequate technology where diffuse solar radiation is dominant, such as in northern European countries where over 50% of light is diffuse (van Sark et al., 2008). These features make them a preferred choice for use in Building Integration Photovoltaic (BIPV) systems and façades of buildings which brings us closer to the goal of constructing buildings with zero carbon energy consumption and buildings which are able to cover their required energy by renewable energies (Aste et al., 2011, Pagliaro et al., 2010, Debije and Verbunt, 2012).

In LDS layers, the luminescent species absorbs the short wavelength solar radiation and emits at longer wavelengths where the external quantum efficiency (EQE) of the solar cell is higher. EQE is defined as the ratio of the number of collected electrons to the number of incident photons (Wenham, 2012, Solanki, 2015, Radziemska, 2006, Wen et al., 2012, Abderrezek et al., 2013, Şahin and Ilan, 2013):

$$EQE(\lambda) = \frac{me(\lambda)}{mp(\lambda)} = IQE(\lambda) \times (1 - R_{PV}) \quad (1)$$

Where $me(\lambda)$ is the number of electrons generated by the PV, $mp(\lambda)$ is the number of photons striking the PV at each wavelength, IQE is the internal quantum efficiency which is the ratio of the collected electrons to the absorbed photons by the PV and R_{PV} is the probability of the reflectance by the PV (Yang et al., 2008). When solar radiation is incident on the PV solar cell, the absorption probability is rapidly reduced due to two main loss sources: optical and recombination losses (Rothemund, 2014). This significantly decreases the External Quantum Efficiency (EQE) of the solar cell (Wilson and Richards, 2009).

It is cost effective to use modeling methods to design and optimize the LDS configuration before fabricating actual LDS/PV. Thermodynamic or Ray Tracing methods (Kerrouche et al., 2014, Batchelder et al., 1979, Burgers et al., 2005, Parel et al., 2015, Barnham et al., 2000, Hughes et al., 2015, Glassner, 1989, Kennedy et al., 2007, Chatten et al., 2004, Earp et al., 2004, Carrascosa et al., 1983) can be used to model LDS while the accuracy of both models were similar (Farrell et al., 2005). Ray tracing algorithms are more time consuming in comparison with thermodynamic models (Glassner, 1989). However, the ray trace model is more flexible and can simulate different geometries and shapes with multiple luminescent species under diffuse and direct light (van Sark et al., 2008).

In this paper, LDS layer with EU^{3+} complex doped in PVA matrix was modelled using a ray tracing algorithm with and without counting the PVA matrix loss mechanisms (attenuation and scattering losses). The modeling and experimental results (Liu et al., 2013) were compared to validate the ray tracing algorithm. Subsequently, the model was run for LDS layers with different matrix materials such as glass, epoxy resin, and PMMA of the same size and configuration. The results were compared to investigate the loss mechanisms effect on the LDS layer efficiency and optical response.

2 MATHEMATICAL LDS MODELLING USING A RAY TRACING ALGORITHM

The LDS ray tracing algorithm is based on a set of ray intersection routines due to phenomena such as reflection, refraction, absorbing, scattering, attenuating and TIR events occurs inside the device. Each ray of light with specific wavelength and direction is traced until it leaves the system (Glassner, 1989). The routines are applied in an iterative loop for each ray irradiated to the device based on the exposed light spectrum. When a ray strikes the top surface of the LDS layer, it is either reflected or refracted according to the light incident angle as Snell's Law (Glassner, 1989). After refraction, the ray may be scattered or/and attenuated by the matrix material, lost as escape cone loss, absorbed and then emitted by the luminescent materials or wave-guided by TIR. The routine is continued until the fate of the ray is determined and then the same process is executed for the other rays. The occurrence probability of each event at each stage of the ray tracing algorithm is determined according to parameters such as wavelength, angle of the illuminated ray, characteristics of the matrix/ luminescent material, dimensions and configuration of the LDS layer. After implementing the iterative loop for all input rays, the optical efficiency of the device is calculated by:

$$\eta_{opt} = \frac{\int_{\lambda_{min}}^{\lambda_{max}} E_{P_{OUT}}(\lambda).d(\lambda)}{\int_{\lambda_{min}}^{\lambda_{max}} E_{P_{IN}}(\lambda).d(\lambda)} \quad (2)$$

Where $E_{P_{OUT}}(\lambda)$ is the energy spectrum of the output photons detected by the PV and $E_{P_{IN}}(\lambda)$ is energy spectrum of the input solar radiation.

To evaluate the performance of the LDS/PV device, the EQE of the PV with and without the LDS layer is modelled by (Rothemund, 2014):

$$EQE_{LDS}(\lambda) = (\eta_{opt}.ESM.f_{abs}(\lambda)) + (EQE_{PV}(\lambda).f_{trans}(\lambda)) \quad (3)$$

Where EQE_{PV} is the EQE of the solar cell without the LDS. $f_{abs} = 1 - e^{-\alpha d}$ is the absorption function of the luminescent material which gives the number of photons absorbed (α is the absorption coefficient in 1/mm and d is thickness of the LDS in mm). $f_{trans} = 1 - f_{abs}$ is the transmission function of the luminescent material which gives the number of photons transmitted through the layer. *ESM* (Emission Spectral Matching) is a factor that is used to normalize emission spectra and also match it to the EQE_{PV}:

$$ESM = \frac{\int_{\lambda_{min}}^{\lambda_{max}} PL(\lambda).EQE_{PV}(\lambda).d(\lambda)}{\int_{\lambda_{min}}^{\lambda_{max}} PL(\lambda).d(\lambda)} \quad (4)$$

Where *PL* is the photoluminescence (or emission spectrum) of the luminescent material.

3 RESULTS, VALIDATION AND DISCUSSION

In this section, the experimental results of LDS doped with EU^{3+} complex (Liu et al., 2013) are used to validate the mathematical ray tracing modelling. The absorption and emission spectra of the luminescent material is shown in Fig. 2. The size of the LDS layer is 156 cm^2 with 0.15 mm thickness, deposited on top of c-Si PV. The radiation used in the experiment and modelling was AM 1.5 global solar radiation whose information obtained from ASTM G173-03 Reference Spectra. The concentration of the EU^{3+}complex was 1%.

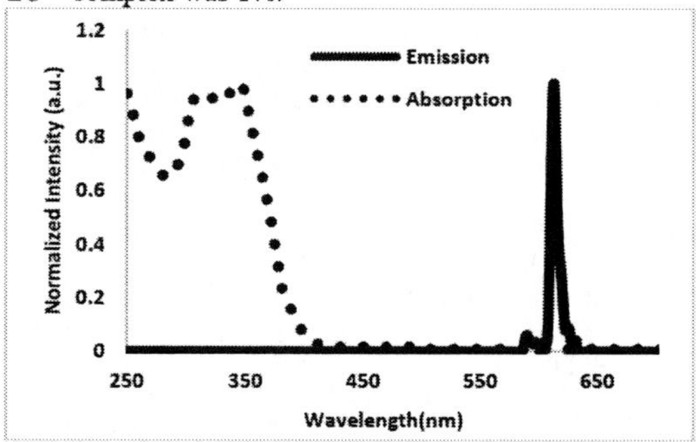

Figure 2: Normalized emission and absorption spectra for EU^{3+} complex (Liu et al., 2013)

The LDS based PVA layer was simulated under 250,000 rays while all loss mechanisms were considered in the modelling. Fig. 3.a shows the simulation and experimental output spectrum of the LDS (in arbitrary units) which are compared with the input solar radiation of the LDS layer. As it can be seen, the output spectrum of the LDS layer has a peak at around 610 nm which is due to the emission peak of the EU^{3+} complex. The input irradiation is red-shifted 60 nm (from 300 nm to 360nm) by the LDS layer. Fig.3.b shows the simulation and the experimental EQE_{LDS} curves in comparison with the EQE of uncoated PV. As can be seen, the EQE of the PV is significantly improved by around 46% in the wavelength range of 300-365nm due to the presence of the LDS layer. The achieved outputs of the model can be seen in Table I. In the PVA modelling, around 17% of rays is reflected after striking the top surface of the LDS layer due to the mismatching in refraction index of air (1) and the PVA film (1.49). It has been found that, 2.7% of rays were lost as heat which is due to either attenuation/scattering loss of the PVA matrix material or non-unity quantum yield of the luminescent material. Majority of rays (~80%) reached the PV cell at the bottom layer which resulted in approximately 79% optical efficiency and a 31.39 $mA.cm^{-2}$ current density. It is observed that by counting the loss mechanisms, the model is valid and the simulation results are in an excellent agreement with experiments (maximum 2.2% discrepancy error in peak).

In the next step, the LDS based PVA layer was simulated (250,000 rays) but without counting the loss mechanisms of the PVA matrix material. As it can be seen in Fig.4, the modelling and experimental spectra are not well-matched in this case. In addition, as is seen in Table. 1, in the case where the loss mechanisms were not counted, the number of rays lost as heat decreased from 2.69% (in PVA modelling) to 1.01% (Without loss modelling) and that is due to the non-unity quantum yield of the luminescent material. This results in the final optical efficiency of 81% and gives a current density of 31.41 $mA.cm^{-2}$ without considering the losses in the model. As is seen, by considering loss mechanisms, the discrepancy error in the peak is enhanced from around 27% (Without loss modelling) to 2% (PVA modelling). This clearly shows the effect of counting material losses in the LDS modeling.

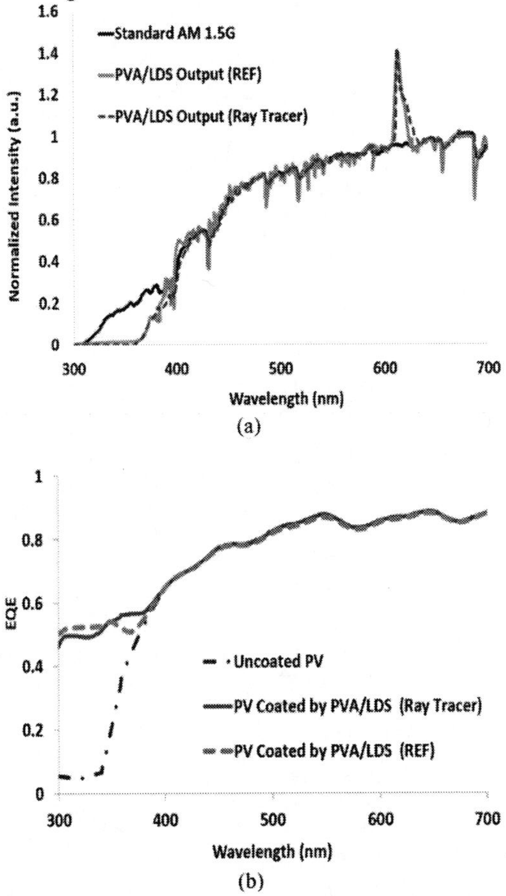

(a)

(b)

Figure 3: (a) Output spectrum of the LDS based PVA layer obtained from modeling (red) and experimental (blue) in comparison with the input solar radiation spectrum (black) and (b) Modeling (red) and experimental (blue) EQE_{LDS} curves in comparison with the EQE of uncoated c-Si PV cell (black).

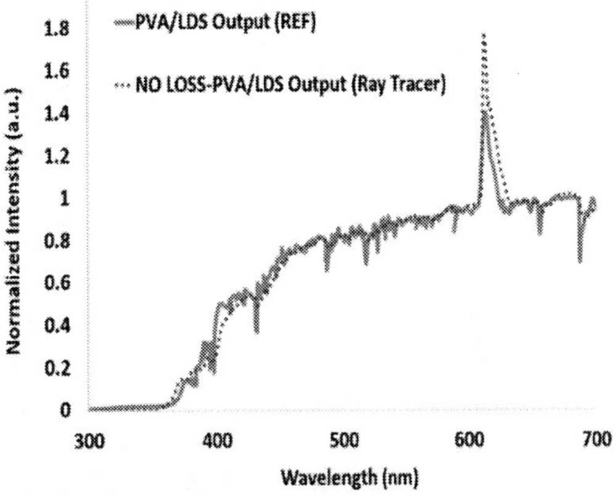

Figure 4: Output spectrum of the LDS based PVA layer obtained from modeling (red and no loss mechanisms is counted) and experimental (blue)

Table I: Statistic results obtained by the ray tracing model for LDS devices modelled with different matrix materials

Modelled Matrix Material	PVA	Without Loss	Glass	Epoxy Resin	PMMA
Number of Input Rays	250,000				
Reflected (%)	17.03	16.86	17.05	16.99	16.92
Refracted (%)	82.97	83.14	82.95	83.01	83.08
Lost as Heat (%)	2.69	1.01	2.44	0.87	1.02
Escape Cone Loss (%)	0.01	0.01	0.003	0.01	0.01
Reached to PV (%)	80.27	82.12	80.50	82.13	82.05
Optical Efficiency (%)	79.45	81.31	79.85	81.52	81.43
I_{sc} ($mA.cm^{-2}$)	For Uncoated PV : 31.19 $mA.cm^{-2}$				
	31.39	31.41	31.39	31.41	31.40
Red-Shifting (nm)	In REF: 63nm				
	60	53	48	15	18
Discrepancy Error in Peak Point in comparison with the REF PVA Layer(%)	2.2	27.34	13.77	19.57	18.84

In the next step, LDS based Epoxy, Glass and PMMA were simulated under same conditions for the LDS layer (250,000 rays). As can be seen in Table 1, by using resin epoxy as the matrix material of the LDS, the least rate of lost as heat (0.87%) was achieved in comparison with other host materials. This resulted in the highest amount of optical efficiency (81.52%) and current density (31.41 $mA.cm^{-2}$). The resulted output spectra of the modelled devices are shown in Fig.5. As can be seen, the trend and the position of the peaks in all layers are similar; however, the red-shifting and the discrepancy error in the peak are different. This is due to the different absorption spectra of the matrix materials resulting in different loss mechanisms over wavelength which can significantly change the optical response of the LDS layer.

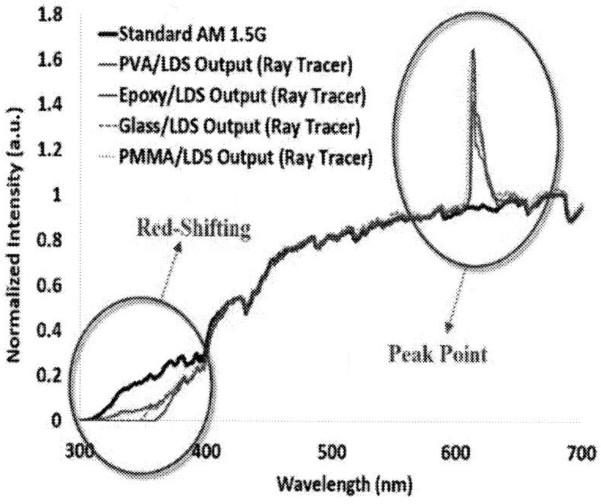

Figure 5: Output spectrums of LDS layers based, epoxy, glass and PMMA compared to the solar radiation spectrum. The LDS PVA layer emission is also shown.

4 CONCLUSION

In this paper, different matrix polymer materials with various loss mechanisms were modeled in LDS layer encapsulation by using a ray tracing algorithm. First, the developed model simulated the performance of an LDS layer of EU^{3+} doped in PVA film on top of c-Si PV cell. It was found that, when loss mechanisms (attenuation and scattering) are counted in the model algorithm, the model's discrepancy in comparison with the experimental results, is significantly improved from around 27% to 2%. By considering the losses, an excellent agreement has been achieved between the experimental and the model results. By using the developed model, the LDS layer achieved 79% optical efficiency and 31.39 $mA.cm^{-2}$ current density. In addition, the EQE of the PV is significantly improved by around 46% in the wavelength range of 300-365nm.

Second, the model was used to simulate the performances of LDS based glass, epoxy and PMMA layers to see the effect of the loss mechanisms for these matrix materials. The best results were achieved by LDS based epoxy layer with around 82% optical efficiency and 31.41 $mA.cm^{-2}$ current density. Different amounts of red-shift was observed in the modelled LDS layers when compared to the solar radiation spectrum. Therefore, the selection of the matrix material is very important when designing the LDS layer. The matrix material must be a suitable environment for the luminescent species and should exhibit a high transmittance and low scattering over the LDS operating wavelength, 300-500 nm. Also, the matrix materials should be highly transparent in the region where the PV cell is efficient so that it does not attenuate the incident rays in this region. It has been observed that, all used matrix materials are transparent after ~ 300nm; however, their optical response in the region between 300-400nm can significantly change the response and optical properties of LDS layer.

5 ACKNOWLEDGEMENT

The authors would like to acknowledge the funding from the European Research Council grant entitled PEDAL: Plasmonic enhancement of advanced luminescent solar devices (13379: 203889) and funding from Science Foundation Ireland (SFI).

6 References

ABDERREZEK, M., FATHI, M., DJAHLI, F. & AYAD, M. 2013. Numerical Simulation of Luminescent Downshifting in Top Cell of Monolithic Tandem Solar Cells. *International Journal of Photoenergy*, 2013.

AHMED, H. 2014. *Materials Characterization And Plasmonic Interaction In Enhanced Luminescent Down-Shifting Layers For Photovoltaic Devices*. PhD, Dublin Institute of Technology.

ASTE, N., ADHIKARI, R. & DEL PERO, C. Photovoltaic technology for renewable electricity production: Towards net zero energy buildings. Clean Electrical Power (ICCEP), 2011 International Conference on, 2011. IEEE, 446-450.

BARNHAM, K., MARQUES, J. L., HASSARD, J. & O'BRIEN, P. 2000. Quantum-dot concentrator and thermodynamic model for the global redshift. *Applied Physics Letters*, 76, 1197-1199.

BATCHELDER, J., ZEWAI, A. & COLE, T. 1979. Luminescent solar concentrators. 1: Theory of operation and techniques for performance evaluation. *Applied Optics*, 18, 3090-3110.

BURGERS, A., SLOOFF, L., KINDERMAN, R. & VAN ROOSMALEN, J. Modelling of luminescent concentrators by ray-tracing. Presented at the 20th European Photovoltaic Solar Energy Conference and Exhibition, 2005. 10.

CARRASCOSA, M., UNAMUNO, S. & AGULLO-LOPEZ, F. 1983. Monte Carlo simulation of the performance of PMMA luminescent solar collectors. *Applied Optics*, 22, 3236-3241.

CHATTEN, A., BARNHAM, K., BUXTON, B., EKINS-DAUKES, N. & MALIK, M. 2004. Quantum dot solar concentrators. *Semiconductors*, 38, 909-917.

DEBIJE, M. G. & VERBUNT, P. P. 2012. Thirty years of luminescent solar concentrator research: solar energy for the built environment. *Advanced Energy Materials*, 2, 12-35.

EARP, A. A., SMITH, G. B., SWIFT, P. D. & FRANKLIN, J. 2004. Maximising the light output of a Luminescent Solar Concentrator. *Solar Energy*, 76, 655-667.

FARRELL, D., CHATTEN, A., JERMYN, C., THOMAS, P. & BARNHAM, K. Thermodynamic Modelling of the Luminescent Solar Concentrator. INTERNATIONAL SOLAR ENERGY SOCIETY UK SECTION-CONFERENCE-C, 2005. 179.

GLASSNER, A. S. 1989. *An introduction to ray tracing*, Elsevier.

HOVEL, H., HODGSON, R. & WOODALL, J. 1979. The effect of fluorescent wavelength shifting on solar cell spectral response. *Solar Energy Materials*, 2, 19-29.

HUGHES, M. D., WANG, S.-Y., BORCA-TASCIUC, D.-A. & KAMINSKI, D. A. 2015. Analysis of ultra-thin crystalline silicon solar cells coupled to a

luminescent solar concentrator. *Solar Energy,* 122, 667-677.
KENNEDY, M., ROWAN, B., MCCORMACK, S., DORAN, J. & NORTON, B. 2007. Ray-trace modelling of Quantum Dot Solar Concentrators and comparison with fabricated devices.
KERROUCHE, A., HARDY, D. A., ROSS, D. & RICHARDS, B. S. 2014. Luminescent solar concentrators: From experimental validation of 3D ray-tracing simulations to coloured stained-glass windows for BIPV. *Solar Energy Materials and Solar Cells,* 122, 99-106.
LIU, J., WANG, K., ZHENG, W., HUANG, W., LI, C. H. & YOU, X. Z. 2013. Improving spectral response of monocrystalline silicon photovoltaic modules using high efficient luminescent down-shifting Eu3+ complexes. *Progress in Photovoltaics: Research and Applications,* 21, 668-675.
PAGLIARO, M., CIRIMINNA, R. & PALMISANO, G. 2010. BIPV: merging the photovoltaic with the construction industry. *Progress in Photovoltaics: Research and Applications,* 18, 61-72.
PAREL, T. S., PISTOLAS, C., DANOS, L. & MARKVART, T. 2015. Modelling and experimental analysis of the angular distribution of the emitted light from the edge of luminescent solar concentrators. *Optical Materials,* 42, 532-537.
RADZIEMSKA, E. 2006. Effect of temperature on dark current characteristics of silicon solar cells and diodes. *International Journal of Energy Research,* 30, 127-134.
ROTHEMUND, R. 2014. Optical modelling of the external quantum efficiency of solar cells with luminescent down-shifting layers. *Solar Energy Materials and Solar Cells,* 120, 616-621.
ŞAHIN, D. & ILAN, B. 2013. Radiative transport theory for light propagation in luminescent media. *JOSA A,* 30, 813-820.
SOLANKI, C. S. 2015. *Solar photovoltaics: fundamentals, technologies and applications*, PHI Learning Pvt. Ltd.
VAN SARK, W. G., BARNHAM, K. W., SLOOFF, L. H., CHATTEN, A. J., BÜCHTEMANN, A., MEYER, A., MCCORMACK, S. J., KOOLE, R., FARRELL, D. J. & BOSE, R. 2008. Luminescent Solar Concentrators-A review of recent results. *Optics Express,* 16, 21773-21792.
WEN, C., FU, C., TANG, J., LIU, D., HU, S. & XING, Z. 2012. The influence of environment temperatures on single crystalline and polycrystalline silicon solar cell performance. *Science China Physics, Mechanics and Astronomy,* 55, 235-241.
WENHAM, S. R. 2012. *Applied photovoltaics*, Routledge.
WILSON, L. & RICHARDS, B. 2009. Measurement method for photoluminescent quantum yields of fluorescent organic dyes in polymethyl methacrylate for luminescent solar concentrators. *Applied optics,* 48, 212-220.
YANG, W., MA, Z., TANG, X., FENG, C., ZHAO, W. & SHI, P. 2008. Internal quantum efficiency for solar cells. *Solar Energy,* 82, 106-110.

LOW TEMPERATURE DEPOSITION OF CONDUCTING INDIUM OXIDE FILMS FOR SOLAR CELL APPLICATIONS

Y. Vygranenko[1*], M. Fernandes[1,2], M. Vieira[1,2], G. Lavareda[1,3], C. Nunes de Carvalho[3,4], P. Brogueira[4,5], A. Amaral[4,5]
[1]CTS-UNINOVA, Campus da Caparica, 2829-516 Caparica, Portugal
[2]Departamento de Engenharia Electrónica e Telecomunicações e de Computadores, Instituto Politécnico de Lisboa, 1959-007 Lisboa, Portugal
[3]Departamento de Ciência dos Materiais, Faculdade de Ciências e Tecnologia, Universidade Nova de Lisboa, 2829-516 Caparica, Portugal
[4]CeFEMA and [5]Departamento de Física, Instituto Superior Técnico, Universidade de Lisboa, Av. Rovisco Pais 1, 1049-001 Lisboa, Portugal
*Corresponding author (Tel.: +351 21 831 7289, e-mail: yvygranenko@deetc.isel.ipl.pt)

ABSTRACT: In this contribution, we report on low temperature deposition of conducting indium oxide (In_2O_3) films by a radio-frequency plasma enhanced reactive thermal evaporation technique. The films were deposited on glass and flexible polyethylene terephthalate substrates without intentional heating of the substrate and at temperatures up to 150 °C. The material stoichiometry was accurately controlled by adjusting the deposition conditions including oxygen flow, process pressure, pumping speed, and rf-power. The Hall effect measurements were performed on the fabricated samples to determine the relation between the deposition conditions and the electrical properties of the films. The resistivity of 4×10^{-4} Ω-cm was reached under optimized deposition conditions. Coatings on flexible plastic substrates with 16 Ω/sq sheet resistance show an 82% peak value of transmittance in the visible spectral range. The film morphology is also analyzed by scanning electron microscopy (SEM).
Keywords: Antireflection Coating, ITO, Flexible Substrate, Electrical Properties, Optical Properties.

1 INTRODUCTION

Commercial polymer films with indium-tin oxide (ITO) coatings are considered to be the most promising substrates for flexible thin-film solar cells [1]. Substrate materials such as polyethylene terephthalate (PET) or polyethylene naphthalate (PEN) are most attractive due to their high transparency and low cost [2]. The technological challenge is that ITO coating should be deposited at low temperatures, desirably not exceeding the glass transition point for these plastics. Several vacuum techniques such as dc and rf sputtering, ion beam-assisted evaporation, and arc-discharge ion plating have been used for deposition of ITO on polymeric substrates [3-6]. However, the properties of low temperature ITO on these plastics are substantially inferior in comparison to ITO grown on glass substrates at high temperatures. To overcome the limitation on the deposition of ITO films with satisfactory electrical and optical properties on plastic substrates, we have developed a radio-frequency plasma-enhanced reactive thermal evaporation (rf-PERTE) technique [7]. This work reports on electrical, optical, and structral properties of indium oxide films deposited on glass and PET substrates by the rf-PERTE technique. The deposition conditions have been optimized to reach the best performance characteristics of In_2O_3 coatings on flexible PET substrates.

2 EXPERIMENTAL DETAILS

2.1 Deposition technique

The rf-PERTE system is based on a bell jar type vacuum chamber with a pumping group including diffusion and mechanical pumps. The typical configuration for thermal evaporation is used, with a distance between the tungsten boat and the sample holder of 32 cm. For plasma assisting, an rf-electrode in the form of a copper ring is placed in the half-way between them. An electrically-driven shutter is placed about 15 mm below the sample holder to shield the plastic substrate from oxygen plasma before the deposition starts. The oxygen injection into chamber is controlled by a SmartTrak 100 Series mass flow controller. A Genesys™ series programmable regulated power supply and a Cesar™ Generator, Model 136, are used as dc and rf power sources, respectively. All electronic units and electrical parts are linked to a microprocessor based control unit either through analog, digital or communication (RS-232) ports. The developed software enables the full control over critical deposition conditions such as mass flow of oxygen, process pressure, current flowing through crucible, rf-power, and substrate temperature. The details on the deposition system have been reported elsewhere [8].

2.2 Measurement techniques

A Dektak 3 profilometer was used to measure the film thickness. The sheet resistance of films on glass substrates was measured using a Veeco FPP-5000 four point probe. For coatings on rigid and flexible substrates, the Hall effect and resistivity measurements were performed using 4-terminal van der Pauw geometry. The magnetic field strength was ± 1 T. The transmittance of films in the UV/visible region was measured using a Shimadzu UV-2501PC spectrophotometer using the integrating sphere configuration.

3 RESULTS AND DISCUSSION

3.1 Optimization of deposition conditions

The metallic atoms evaporated from the heated crucible may undergo collisions in the oxygen ambient during their transit to the substrate, leading to the formation of metal oxide molecules in the gas phase. The surface oxidation reaction on the growing layer also takes place due to intense oxygen ion bombardment. The deviation from stoichiometry δ in $In_2O_{3-\delta}$ can be varied in wide range by adjusting the deposition conditions such as pressure, gas flow rate, rf-power, and evaporation rate. The conducting and semiconducting films have been produced under various deposition conditions [9, 10].

To produce transparent and highly-conductive $In_2O_{3-\delta}$ films, an optimal balance between the evaporated metal mass and absorbed volume of oxygen should be reached. This balance can be evaluated by measuring the pressure at steady oxygen flow, pumping speed, and rf-power. Fig. 1 shows that the pressure in the chamber decreases, when the evaporation of indium starts, and then stabilizes when the evaporation rate becomes constant. The deviation from stoichiometry in $In_2O_{3-\delta}$ is related to the difference between the initial pressure, p_i, and deposition pressure, p_d, that is determined by the evaporation rate. The evaporation rate depends on multiple factors such as the dc power applied to the crucible, process pressure, amount of metal still in the crucible, etc. To stabilize the evaporation rate, the dc current through the crucible is automatically adjusted to keep the deposition pressure constant. This approach enables the reproducible deposition of metal oxide films with required stoichiometry simply by setting the value of the differential pressure $\Delta p = p_i - p_d$.

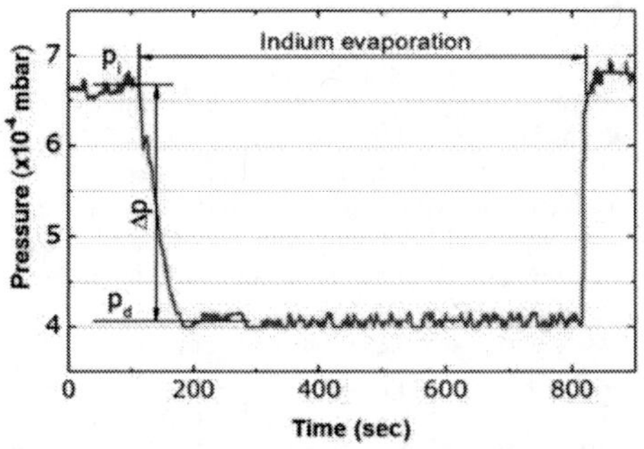

Figure 1: Pressure in the vacuum chamber during the deposition process.

To optimize the deposition conditions, two series of $In_2O_{3-\delta}$ films were deposited on unheated glass substrates by varying the differential process pressure or rf-power. For the first series, Δp was varied in the range of $(1.6 - 2.7)\cdot10^{-4}$ mbar at an oxygen flow rate of 15 sccm and an rf-power of 70W. The second series was obtained at 15 sccm flow rate and $\Delta p = 2.5\cdot10^{-4}$ mbar, varying the rf-power in the range of 30 – 150W. The thickness of $In_2O_{3-\delta}$ layers was ranging from 98 to 117 nm at 10 min deposition time. Fig. 2 shows the resistivity variation with Δp and rf-power. The lowest resistivity value of $9\cdot10^{-4}$ Ω·cm was achieved at $\Delta p = 2.5\cdot10^{-4}$ mbar and 70 W rf-power. For this sample, the Hall mobility and electron density are 26 $cm^2/V\cdot s$ and $2.7\cdot10^{20}$ cm^{-3}, respectively. At larger Δp values, the Hall mobility drops down to 8.6 $cm^2/V\cdot s$ at $\Delta p = 2.7\cdot10^{-4}$ mbar indicating a growing number of atomic-scale defects as a result of oxygen deficiency. The same occurs at low rf-powers.

To further optimize the deposition conditions, a series of $In_2O_{3-\delta}$ films was deposited at various substrate temperatures up to 190 °C. Fig. 3 shows variation of Hall mobility and electron density with substrate temperature. The Hall mobility monotonically increases with substrate temperature reaching a value of 56 $cm^2/V\cdot s$ at 190 °C.

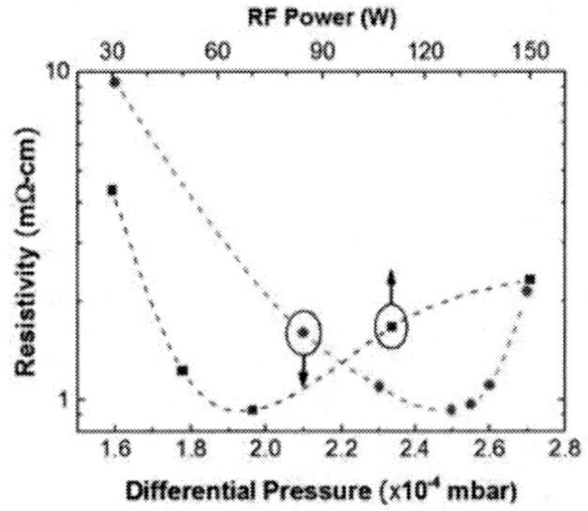

Figure 2: Variation of resistivity with differential process pressure and rf-power for $In_2O_{3-\delta}$ films.

The highest charge carrier density of $4.6\cdot10^{20}$ cm^{-3} combined with the lowest resistivity of $3.9\cdot10^{-4}$ Ω·cm are observed in the film deposited at 150 °C. At higher temperatures, the oxygen vacancy generation is limited by the increasing metal oxidation rate.

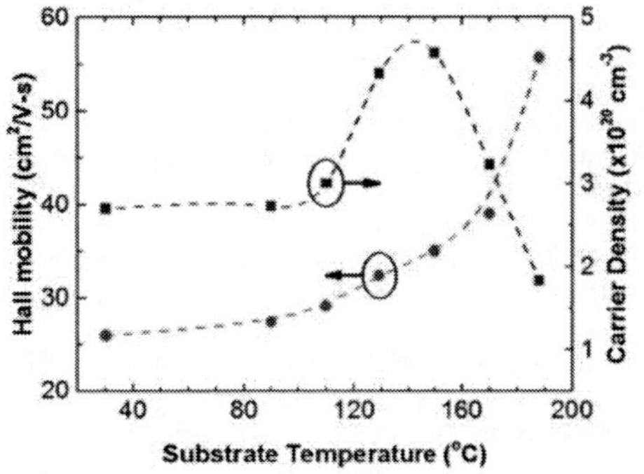

Figure 3: Variation of Hall mobility and charge carrier density with substrate temperature.

3.2 In_2O_3 films on plastic substrates

To demonstrate the large area deposition capability, In_2O_3 films were deposited on 100-µm-thick 10 × 10 cm^2 PET substrates. The depositions were performed without intentional substrate heating (sample 1) and at 150 °C substrate temperature (samples 2 and 3). The thickness of coatings was adjusted targetting the sheet resistance values of 60, 25, and 15 Ω/sq for samples 1, 2, and 3, respectively. The fine-tuning of the deposition parameters was performed using the experimental data discussed above. The measured performance parameters of the In_2O_3 films are summarized in Table I. The Hall mobility and resistivity values of films on plastic correspond to that for films on glass substrates. The achieved electrical properties in terms of sheet resistance and conductivity

Table I: Electrical properties of In_2O_3 films on PET.

	#1	#2	#3
Substrate temperature, °C	30	150	150
Thickness, nm	175	154	250
Sheet resistance, Ω/sq.	61	23.4	16.4
Hall mobility, $cm^2/V\cdot s$	26	34	35
Carrier density, $\times10^{20}$ cm^{-3}	2.3	4.9	4.5
Resistivity, $\times10^{-4}$ Ω·cm	10.7	3.6	4.1

are better than that for ITO coated PET substrates by sputtering reported elsewhere [11].

Fig. 4 shows the transmittance spectra of the In_2O_3 coated PET substrates. The spectrum of PET film is also shown as a reference. For samples 1 and 2, the first order interference maxima are 86.8% at 706 nm and 86.9% at 662 nm, respectively. For sample 3 the second order interference maximum is 82% at 498 nm. The transmittance peak positions are in good agreement with thickness values reported in Table 1. In the UV region the transmittance is limited by fundamental absorption in In_2O_3.

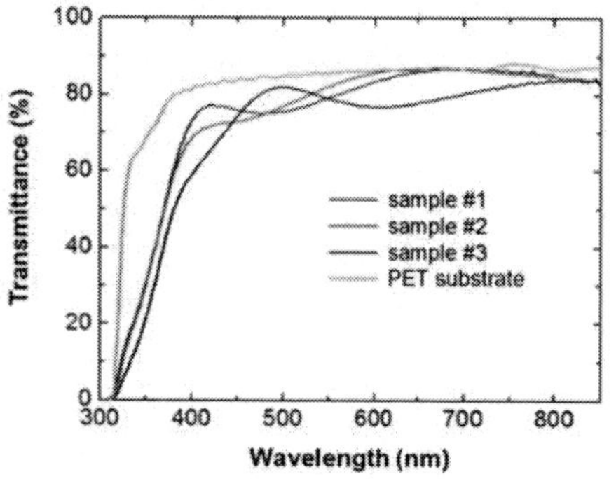

Figure 4: Transmittance spectra of In_2O_3 films on PEN substrates. The sample numbering corresponds to that in Table I.

Fig. 5 shows the SEM images of films deposited on the unheated substrate and at 150 °C. The fine grained structure is observed in the film deposited on the unheated substrate, while the larger (~20 nm) grain size is observed in the 150 °C sample. The increase in material crystallinity leads to an improvement in the carrier mobility (see Table I).

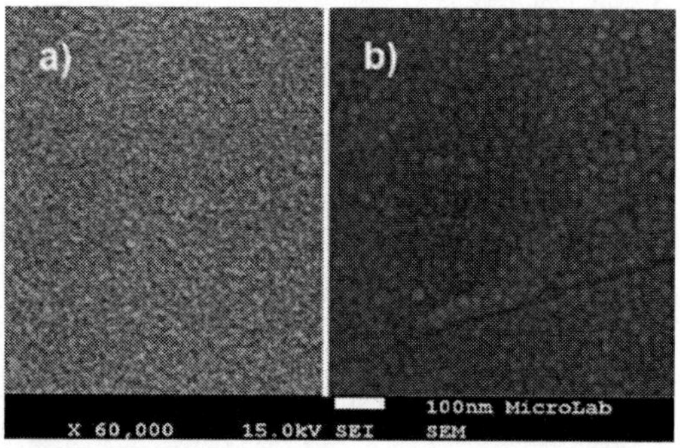

Figure 5: SEM images of In_2O_3 films deposited on plastic at (a) room temperature, and (b) 150 °C substrate temperature.

4 CONCLUSIONS

Transparent conducting indium oxide coatings have been deposited on glass and PET substrates at low (≤150 °C) temperatures using the rf-PERTE method. It is demonstrated that the electrical properties of $In_2O_{3-\delta}$ can be controlled by adjusting the differential process pressure Δp that is a function of the metal evaporation rate at a constant oxygen flow rate. The Hall mobility increases with substrate temperature due to the improving film crystallinity. The achieved resistivity values are $9 \cdot 10^{-4}$ and $3.6 \cdot 10^{-4}$ Ω·cm for films deposited without intentional heating of the substrate and at 150 °C temperature, respectively. $In_2O_{3-\delta}$ films on PET substrates with a sheet resistance of 16 Ω/sq show 82% peak value of transmittance in the visible spectral range. The electrical and optical performance characteristics of developed In_2O_3 coated substrates meet the requirements for flexible organic and a-Si:H solar cells.

ACKNOWLEDGEMENTS

The authors are grateful to the Portuguese Foundation of Science and Technology through fellowship SFRH/BPD/102217/2014 and to Instituto Politécnico de Lisboa (ID&CA project-SolWin) for financial support of this research.

REFERENCES

[1] Cheng C, Wagner S., Overview of flexible electronics technology. In: Wong,WS, Salleo A. editors. Flexible Electronics: Materials and Applications, New York: Springer; 2009. p. 1–28.
[2] MacDonald MK, Looney et al. Latest advances in substrates for flexible electronics. Journal of the SID 15 (2007)1075.
[3] Tseng K-S, Lo Y-L. Appl. Surf. Sci. 285 (2013) 157.
[4] Kim D-H et al. Appl. Surf. Sci. 253 (2006) 409.
[5] Meng Y, et al. J. Vac. Sci. Technol. A 20 (2002) 288.
[6] Niino F, Hirasawa H, Kondo K. Thin Solid Films 411 (2002) 28.
[7] C. Nunes de Carvalho, G. Lavareda, E. Fortunato, A. Amaral. Thin Solid Films 427 (2003) 215.
[8] M. Fernandes et al. Energy Procedia 102 (2016) 96.
[9] G. Lavareda, C. Nunes de Carvalho, E. Fortunato, A. R. Ramos, E. Alves, O. Conde, A. Amaral. J Non-Cryst Solids, 352 (2006) 2311.
[10] C. Nunes de Carvalho, G. Lavareda, A. Amaral, Conde O, A.R. Ramos. J Non-Cryst Solids, 352 (2006) 2315.
[11] C. Guillén, J. Herrero, Proceedings 24th European Photovoltaic Solar Energy Conference, Vol. I (2009) 579.

PHASE TRANSFER AND POLYMER COMPATIBILITY OF GOLD AND GOLD CORE SILVER SHELL NANORODS FOR LUMINESCENT SOLAR DEVICES

A.Sethi[a,]*, S. Chandra[a], H. Ahmed[a], S. Gilligan[a] and S. J. McCormack[a]
[a]Dept. of Civil, Structural and Environmental Engineering. Trinity College Dublin, Dublin 2, Ireland.
*Corresponding author Email: sethia@tcd.ie Tel: + 353 1 896 2671

ABSTRACT: Luminescent solar concentrator (LSC) device is based on energy down shifting phenomenon with a primary goal of trapping and guiding light via total internal reflection. The trapped light propagates and is concentrated at the edge where photovoltaic cells absorb and convert it to electricity. Luminescent Down shifting (LDS) devices are a special case of LSC where geometric gain is unity with its main function to improve the spectral response of the photovoltaic cells in short wavelength solar radiation. Low quantum efficiencies of luminescent species have limited the optical efficiency of the aforementioned devices. Plasmonic properties of metal nanoparticles (MNPs) such as gold (Au) and silver (Ag) nanorods can be used for the enhancement of fluorescence of quantum dots, rare earth complexes and organic dyes in LSC and LDS devices. Au Nanorods (NRs) were synthesized using the seed mediated wet chemical process, with CTAB (hexadecyltrimethylammonium bromide) as the capping agent. To make use of localized plasmonic resonances due to silver, (Au) core (Ag) shell NRs also labelled as Au_Ag NRs were fabricated using the synthesized Au nanorods as seeds and Silver Nitrate ($AgNO_3$) as the precursor. The nanorods obtained have a high shape yield. Purification and separation of NRs from the by-products was performed by multiple centrifugation steps. Functionalization of the nanorods was done at room temperature using thiolated PEG molecule. Thiol group has a stronger affinity for Au/Ag surface than the surfactant and binds to the ends of the NRs. Phase transfer to chloroform was successfully carried out for Au and Au_Ag NRs in which methanol acts as a phase transfer agent. Nanorods in organic solvent were stable for more than a month. Compatibility of Au NRs with PMMA and silicone polymer encapsulate was investigated.

Keywords: Core shell nanorods, Gold Nanorods, Phase Transfer, Polymer film, luminescent solar concentrator

1 INTRODUCTION

Luminescent solar concentrators can be used to concentrate both direct and diffuse sunlight without the need of expensive solar tracking systems [1]. Luminescent down shifting indirectly increases the efficiency of solar cells by conversion of one photon of high energy light, to one photon with an energy that can be efficiently absorbed by the photovoltaic material [2]. Re-absorption of the photons due to an overlap of the absorption and emission spectra of fluorophores increases the escape cone and non-unity quantum yield losses for the LSC and LDS [3].

Plasmonic interaction between metal nanoparticles (MNPs) and the fluorophores can help reduce these losses thus enhancing optical efficiency of LSC/LDS. Surface plasmon resonance (SPR) frequency is primarily the function of the size and shape of the MNPs. The longitudinal surface plasmon resonance (LSPR) results in a strong electric field enhancement that decays exponentially from the surface of the nanorods [4]. When the fluorophores are placed in the range of enhanced field, the plasmonic interaction occurs and consequently there is an enhancement in fluorescence.

Au and Au_Ag nanorods have been synthesized using a previously reported research [5]. Silver has a higher interband transition energy than gold and has localized plasmon resonances in the near ultraviolet spectral range. Moreover, the electric field enhancements are larger for silver than gold. Unfortunately, anisotropic Ag nanorods are not as good as Au nanorods, in terms of the shape and size uniformity [6]. Synthesis of Ag nanorods often lead to a high production of triangles as by- product thus making it difficult to separate the nanorods from the triangles due to similar sized nanoparticles. Silver nanorod synthesis required a very precise control of temperature and even a slight change in temperature changed the aspect ratio of the nanorods. Therefore, control of the aspect ratio of Ag nanorods and thus their longitudinal plasmon wavelengths has remained difficult. Therefore, to utilize the plasmonic properties of silver, Ag shell was grown on Au nanorods.

The limitations of the host material in LSC/LDS and the environment require the nanorods to be dispersed in a solvent that is compatible with the host material and does not change the shape, size of the nanorods while retaining their optical properties. This work focuses on the phase transfer of the nanorods to an organic solvent. It is important that the phase transfer should not lead to agglomeration of the nanorods. These changes can affect the optical properties of the NRs. Moreover, the organic solvent should be compatible with the host material (PMMA, epoxy, silicone encapsulate) without changing the optical properties of the nanorods. To ensure a successful phase transfer the nanorods need to be capped by thiols or amines [7] to improve the solubility of the nanorods in organic media. CTAB forms a strong bilayer on the nanorods [8] to remove the excess CTAB/CTAC it is required to perform multiple centrifugation steps soon after the synthesis as the excess can cause the nanorods to change their shape.

2 MATERIALS AND METHODS

2.1 Materials

Hydrogen tetrachloroaurate trihydrate ($HAuCl_4.3H_2O$; 99%), sodium borohydride ($NaBH_4$; 99%), Cetyltrimethylammoniumbromide (CTAB), L-ascorbic acid (AA), silver nitrate ($AgNO_3$; 99%), Cetyltrimethylammoniumchloride (CTAC), O-[2-(3-Mercaptopropionylamino)ethyl]-O′-methylpolyethylene glycol (MW: 5,000) were obtained from Sigma Aldrich. E1184 optical encapsulate from Dow Corning. Plexit 55 Rohn Acryl Resin from Carl Roth Germany. Epoxy. HPLC grade Methanol (CH_3OH), Chloroform ($CHCl_3$)

and Toluene were purchased from the Sigma Aldrich Deionized water was used for all synthesis.

2.2 Synthesis of Au and Au_Ag Nanorods

Gold nanorods were chemically synthesized using a seed mediated growth method previously described in [4]. A seed solution of gold nanospheres was synthesised by adding freshly prepared ice cold 0.01 M $NaBH_4$ solution to an aqueous solution of 0.05 M $HAuCl_4$ and 0.1 M CTAB under vigorous stirring at 1300rpm. Growth solution of NRs was prepared by mixing the gold precursor with CTAB at 27-30°C to form a complex between the Au salt and CTAB. After the addition of mild reducing agent 0.1 M AA, the growth solution changed from dark yellow to colourless due to the reduction of Au(III) to Au(I). 40 µl and 100 µl of 0.005 M $AgNO_3$ solution was added to this growth solution to help in the formation of single crystalline rods, with the latter used as seeds for the synthesis of Au_Ag NRs. 120µl of seed solution was added to this growth media to form Au NRs. The solution was kept at 27°C in a water bath for half an hour to complete the nanorod growth.

Silver overgrowth was performed similar to the work published in [9]. The Au NRs with 100 µl of 0.005 M $AgNO_3$ solution were used as the seeds for the synthesis of the core shell nanorods. As-prepared Au NRs were centrifuged to remove the excess CTAB and additives from the solution. NRs were then transferred into a 0.08 M CTAC solution and washed to ensure a full surfactant exchange. Five aliquots were prepared containing 2ml of Au NRs as seeds, 0.04, 0.12, 0.24, 0.4, 0.56 ml of 0.01 M $AgNO_3$ was subsequently added to the aliquots followed by the addition of 0.1 M AA. They have been named as samples 1-5 for further reference in this paper. The volume of AA solution is half the volume of the $AgNO_3$ in the aliquots. The resultant solution is kept in an oven at 65°C for 4.5hrs.

2.3 Characterisation of Au, Au_Ag Nanorods and Host Material layers

UV/Vis/NIR spectra were recorded using a Steller Net spectrometer for the optical characterisation of the nanoparticles at room temperature. PMMA cuvettes with a path length of 1 cm were used for characterising all the samples. SEM images were obtained using Carl Zeiss Ultra, a low concentration of the NRs was used for imaging the nanoparticles. The nanorod solution was drop cast on the surface of a silicon wafer that had a high p doping to avoid charge deposition. InLens detector was used with a low accelerating voltage between 3-5kV keeping in mind that a higher accelerating voltage causes charge deposition and makes it difficult to image the nanoparticles. ImageJ software was used to do the size evaluation. An integrating sphere was used to measure the absorbance/transmission spectra of the Au NRs embedded in host material layers.

2.4 Purification of Au and Au_Ag Nanorods

Shape yield obtained after the seeded growth mechanism is usually very low to be sufficient for the targeted application. Spheres, cubes and triangles are the inevitable by-products in the wet chemical synthesis of Au and Au_Ag NRs. The as-grown Au NRs and Au_Ag NRs were first centrifuged at high centrifugal force (12,000g) for 30 minutes to remove excess CTAB/CTAC and additives [10]. The pellet of Au NRs was re-dispersed in water and centrifuged at a lower force (3,500g) for 15 minutes. The resulting supernatant containing the higher aspect ratio NRs as evident by the UV/Vis spectrum in Figure 1 (a) was used for phase transfer process. For Au_Ag NRs, samples 1 and 2 were centrifuged at 4,000g for 15 minutes whereas samples 3, 4 and 5 were centrifuged at 5,000g for 15 minutes. The Au_Ag NRs in the pellet were re-dispersed in deionized water and used for the phase transfer.

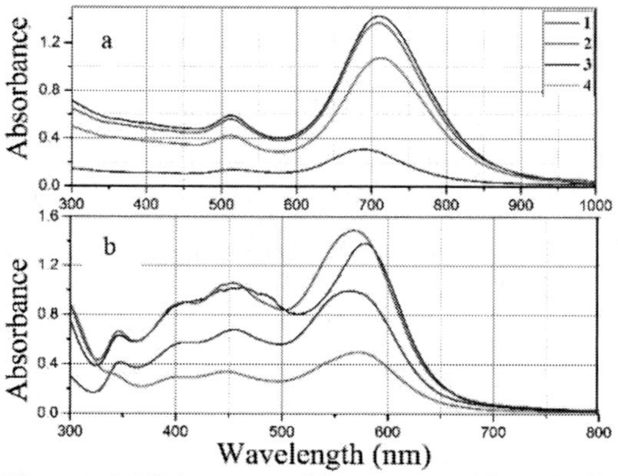

Figure 1: UV/Vis spectra of (a) Au NRs with 40µl of 5 mM $AgNO_3$ in the growth solution. No.4 spectrum is the supernatant after second centrifugation cycle with LSPR peak at 711 nm, whereas the No.3 spectrum is of the pellets with the LSPR peak at 689 nm. (b) Au_Ag NRs with 0.4 ml of 10 mM $AgNO_3$ in the growth solution. No.4 spectrum denotes the supernatant with an absence of a prominent peak 4 at 346 nm, this peak is clearly visible in the pellet solution denoted by No.3. There is also a blue shift as expected for peak 1 in the pellet solution. The LSPR peak is at 565 nm for the pellet solution and at 573 nm for supernatant solution.

2.5 Functionalization of Au and Au Ag Nanorods

Wet chemical synthesis results in the formation of a strongly bound bilayer of CTAB on the Au NR. It is imperative for the nanoparticles to be soluble and stable in an organic solvent with no change in the shape or size of the particles. Phase transfer of the Au and Au Ag NRs from aqueous media to organic solvents was carried out using thiolated poly ethyl glycol as a phase transfer agent [11]. 0.5 mM PEG-SH solution was added to 4 ml of 104 ppm nanorod solution and stirred at room temperature for 18 hours. The PEGylated NRs were then added to chloroform forming two immiscible layers, with the NRs in the top layer. Appropriate amount of methanol was added to the solution and the resultant solution was swirled to initiate the phase transfer. The NRs completely transferred to chloroform after 15 minutes.

2.6 Dispersion of Au Nanorods in host material layers

The organic solvent should be compatible with the host materials to avoid any change in the NRs after dispersing them into the polymer. Two host materials PMMA and silicone encapsulate were chosen. PMMA and toluene were mixed in 75:25 wt% and the resultant mixture was stirred for 90 minutes. The silicone encapsulate was mixed in 1:1 ratio. Appropriate amount of Au NRs dispersed in chloroform were added to the host materials to form layers of 3mm thickness.

3 RESULTS AND DISCUSSION

An optical spectrum can give a wealth of information about the quality of the material. The position of the plasmon band can give us an estimate of the aspect ratio of the nanorods, whereas the full width half maximum and the shape of the longitudinal LSPR band can tell us about the NR size dispersion. The high ratio between the maximum of the longitudinal and the transverse peak and the absence of a shoulder on the transverse band is an indication of the absence of by-products in the Au NR solution. Despite this, purification of the parent solution is important to ensure the removal of extra surfactant which hinders further functionalization of the NRs. The UV/Vis spectra shown in Figure 1 (a) gives a comparison of the as-grown Au NRs and Au NRs after each centrifugation step with a blue shift in the LSPR peak for the pellet, thus, indicating smaller aspect ratio rods in the pellet.

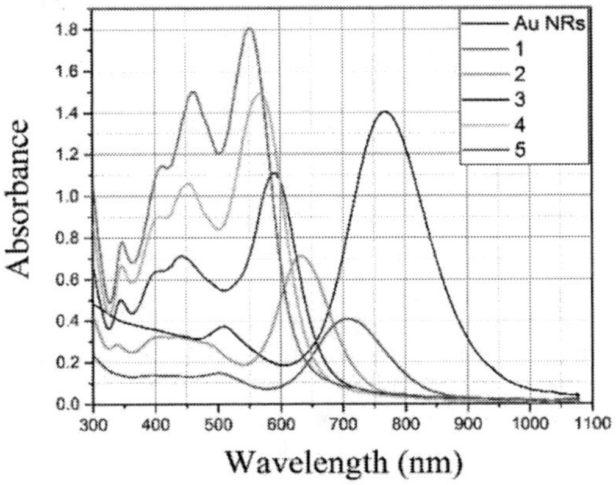

Figure 2: Optical spectra of Au NR and as obtained core (Au) shell (Ag) NRs. Samples 1-5 were obtained by increasing the amount of Ag precursor in the solution, hence, there is an increase in the intensity of the plasmon bands.

The extinction spectra of Au_Ag NRs have four plasmon bands, the peaks can be labelled as 1-4 from low to high energies respectively. These plasmon bands become stronger with an increase in the $AgNO_3$ precursor as can be seen in Figure 2. The dilution caused by the AA and AgNO3 solution has been corrected for the optical spectra of the core shell NRs, therefore, the particle concentration is the same for all the samples.

Purification is an important step for the core shell nanorods. The removal of additional CTAC is crucial for the functionalization of the Au_Ag NRs. Figure 1 (b) depicts the UV-Vis spectra of the as-grown sample 4 Au_Ag NRs and the supernatant containing the Au_Ag NRs after second centrifugation indicating an absence of peak 4 due to the lack of a prominent silver shell on the Au_Ag NRs.

Peak 1 which is the LSPR peak in the core shell nanorods blue shifts from 707 nm for sample 1 to 552 nm for sample 5. This blue shift is not only due to the decrease in the aspect ratio of the nanorods but also because of the increasing effect of Ag in the optical properties. The plasmon resonance wavelength due to Ag is shorter than Au. Peak 2 at 503 nm initially blue shifts then red shifts to 461 nm when the $AgNO_3$ volume is 0.24 ml, further increase in the $AgNO_3$ volume increases the intensity of the transverse peak such that the longitudinal and transverse are of comparable intensities. Peak 3 and 4 become prominent when $AgNO_3$ volume reaches 0.24 ml in the solution. The appearance of these peaks indicates that the plasmonic properties of the silver shell screen the effect of gold on the plasmon bands.

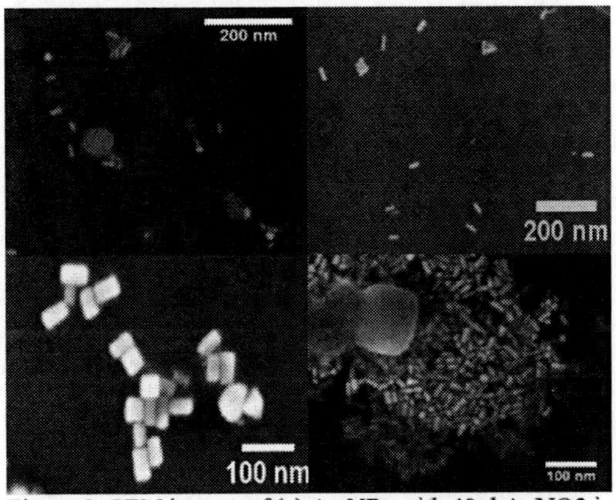

Figure 3: SEM images of (a) Au NRs with 40ul AgNO3 in solution, AR ~ 2.9±0.4 (b) Au NRs with 100ul of AgNO3 in solution, AR ~4±0.5 (c) Au_Ag core shell NRs with 0.56 ml $AgNO_3$ solution (d) Au NRs coagulated on the Si wafer.

SEM images (Figure 3) of the samples were taken after the purification of the parent solution. Note that the nanorods have a curved surface at the ends. The average aspect ratio of the Au NR samples with 40 and 100 µl of 5 mM $AgNO_3$ solution was ~2.9±0.4 and ~4±0.5 respectively. Sample 5 of the core shell nanorods was chosen for SEM to illustrate the difference in the shape and size compared to the Au NRs. The decrease in the aspect ratio of the core shell nanorods is mainly due to the non-uniform thickness of the Ag shell on the sides and ends of the nanorods. Therefore, they look more like cuboids than nanorods. The average length, diameter and aspect ratio of the core shell nanorods for sample 5 was 45±3 nm, 25±3 nm, 1.8±0.3 respectively.

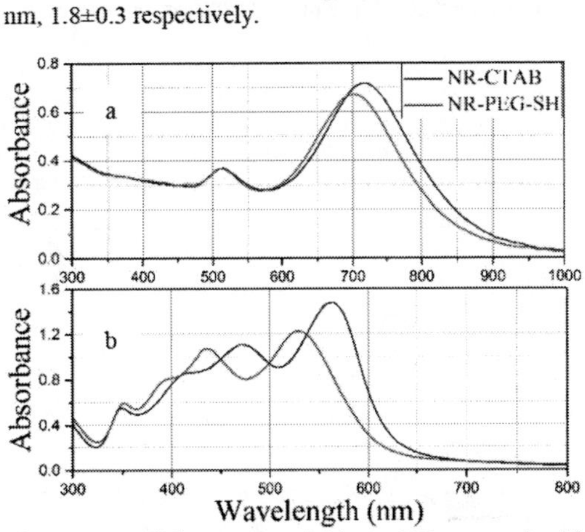

Figure 4: UV/Vis spectra of (a) Au NRs capped with CTAB (black) and functionalized with PEG-SH (red), a blue shift of LSPR peak is observed for PEG-SH Au NRs. (b) Au_Ag NRs in sample 5 capped with CTAB (black) and PEG-SH (red), a strong blue shift is observed for the longitudinal plasmon peak.

CTAB/CTAC are very important as capping agents to prevent nanoparticle aggregation, CTAB also helps in the anisotropic growth of the NRs above a critical micelle concentration by forming a bilayer that preferentially binds to the longitudinal surface of the Au NRs [12]. After washing the nanorods twice it was relatively easy to load PEG-SH molecule onto the nanorods. Figure 4 shows the optical spectrum taken before and after PEGylation of Au and Au_Ag nanorods. A blue shift of 5±3 nm is observed in the LSPR peak of the Au NRs. The blue shift is more pronounced 10±3 nm for the core shell nanorods. This might be due to excess CTAC in the solution. The shift in the LSPR peak was found to be independent of the final concentration of the PEG-SH in the NRs solution.

Phase transfer into chloroform was performed efficiently with the help of methanol. There is no shift in the LSPR wavelength of the Au nanorods as can be seen in Figure 5. For the Au_Ag nanorods there is a blue shift in chloroform but it is consistent with the blue shift observed after the PEGylation of the Au_Ag NRs. Nanorods in chloroform are stable for months at room temperature.

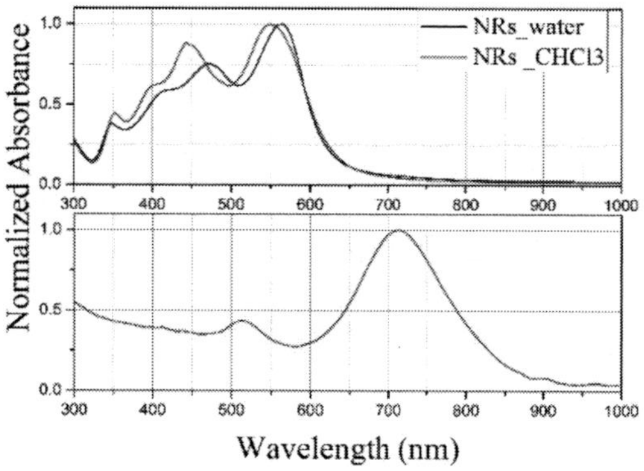

Figure 5: UV/Vis spectra of (a) Au_Ag NRs in water (black) and after phase transfer in chloroform (red). (b) Au NRs in water and chloroform spectra overlap completely without any shift in the wavelength.

1 ml of Au NRs in chloroform were used to form host material layers. It was easier to form a homogeneous solution for Silicone polymer as compared to PMMA. The nanorods retain their optical properties in the host material.

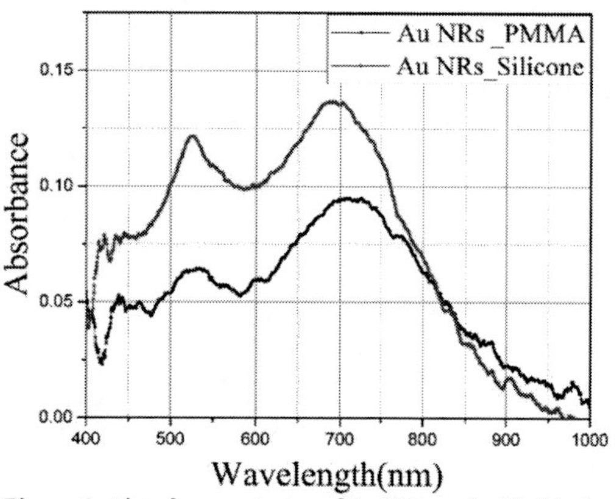

Figure 6: Absorbance spectra of Au NRs embedded in the host material Silicone encapsulate (red) and PMMA (black).

4 CONCLUSIONS

Au nanorods and Au_Ag core shell nanorods were repeatedly and efficiently synthesized giving a high shape yield. Purification and shape separation of the nanorods was carried out with the help of centrifugation. The functionalization of the nanorods was carried out at room temperature using thiolated PEG molecule. A blueshift was observed for the LSPR wavelength which was more pronounced for the Au_Ag nanorods this might be due to the extra CTAC present in the solution. Phase transfer to chloroform was efficient and the nanorods were stable for months. In this work, we also checked the compatibility of the Au nanorods in chloroform with the host material PMMA and silicone encapsulate. There was no observable shift in the plasmon peaks of the nanorods. For the same thickness of the layers a very well-defined transverse and longitudinal peak was observed for the silicone encapsulate as compared to PMMA. It is easier to handle silicone and it is a non-toxic material.

5 ACKNOWLEDGEMENTS

The authors would like to acknowledge the funding from the European Research Council grant entitled PEDAL: Plasmonic enhancement of advanced luminescent solar devices (13379: 203889) and support funding from Science Foundation Ireland (SFI).

5 REFERENCES

[1] Weber, W. H., and John Lambe. "Luminescent greenhouse collector for solar radiation." Applied Optics 15.10 (1976): 2299-2300.

[2] Klampaftis, Efthymios, et al. "Enhancing the performance of solar cells via luminescent down-shifting of the incident spectrum: A review." Solar Energy Materials and Solar Cells 93.8 (2009): 1182-1194.

[3] Van Sark, Wilfried GJHM, et al. "Luminescent Solar Concentrators-A review of recent results." Optics Express 16.26 (2008): 21773-21792.

[4] Thomas, Reshmi, et al. "Optical effects near metal nanostructures: towards surface-enhanced spectroscopy." Current Science (2012): 85-96.

[5] Nikoobakht, Babak, and Mostafa A. El-Sayed. "Preparation and growth mechanism of gold nanorods (NRs) using seed-mediated growth method." Chem. Mater 15.10 (2003): 1957-1962.

[6] Pietrobon, Brendan, Matthew McEachran, and Vladimir Kitaev. "Synthesis of size-controlled faceted pentagonal silver nanorods with tunable plasmonic properties and self-assembly of these nanorods." ACS nano 3.1 (2008): 21-26.

[7] Wang, Xinnan, et al. "A rapid phase transfer method for nanoparticles using alkylamine stabilizers." Journal of colloid and interface science 348.1 (2010): 24-28.

[8] Murphy, Catherine J., et al. "Anisotropic metal nanoparticles: synthesis, assembly, and optical applications." (2005): 13857-13870.

[9] Jiang, Ruibin, et al. "Unraveling the evolution and nature of the plasmons in (Au core)–(Ag shell) nanorods."

Advanced Materials 24.35 (2012).
[10] Scaletti, Federica, et al. "Rapid purification of gold nanorods for biomedical applications." MethodsX 1 (2014): 118-123.
[11] BaniáYaseen, A. I. "Facile phase transfer of gold nanoparticles from aqueous solution to organic solvents with thiolated poly (ethylene glycol)." RSC Advances 4.95 (2014): 52676-52679.
[12] Casas, Justin, et al. "Replacement of cetyltrimethylammoniumbromide bilayer on gold nanorod by alkanethiol crosslinker for enhanced plasmon resonance sensitivity Biosensors and Bioelectronics 49 (2013): 525-530.

ELECTRO-OPTICAL MODELING OF A ZNO/CU$_2$O SUBCELL IN A SILICON-BASED TANDEM HETEROJUNCTION SOLAR CELL

Ørnulf Nordseth[1], Laurentiu Fara[2,3*], Raj Kumar[4], Sean Erik Foss[1], Constantin Dumitru[2], Vlad-Florin Muscurel[2], Florin Drăgan[2], Dan Crăciunescu[2], Kristin Bergum[4], Halvard Haug[1], Paul Sterian[2,3], Irinela Chilibon[5], Cristina Vasiliu[5], Laurentiu Baschir[5], Dan Savastru[5], Edouard Monakhov[4], Bengt Gunnar Svensson[4]
[1]Institute for Energy Technology, Instituttveien 18, 2007 Kjeller, Norway
[2]University Politehnica of Bucharest, Spl. Independentei 313, RO-060042, Bucharest, Romania
[3]Academy of Romanian Scientists, Spl. Independentei 54, RO-030167, Bucharest, Romania
[4]Department of Physics/Center for Materials Science and Nanotechnology (SMN), University of Oslo, P.O. Box 1048, Blindern, NO-0316 Oslo, Norway
[5]National Institute of Research and Development for Optoelectronics (INOE-2000), Bucharest-Magurele, Str. Atomiştilor 409, Măgurele RO-077125, Romania.
*Corresponding author: lfara@renerg.pub.ro

ABSTRACT: In this work, the optical and electrical characteristics of a ZnO/Cu$_2$O subcell in a Si-based tandem heterojunction solar cell configuration were investigated. Highly Al-doped ZnO and undoped Cu$_2$O thin films were prepared on quartz substrates by magnetron sputter deposition. Thermal annealing of the Cu$_2$O layer at 900°C enhances the electrical properties and reduces optical absorption, presumably as a result of increased grain size. Hall effect measurements show that the majority carrier (hole) mobility increases from 10 to 50 cm^2/V·s and the resistivity decreases from 560 to 200 Ω·cm after annealing. Numerical modeling was carried out to investigate the effect of incorporating a buffer layer between the ZnO emitter layer and the Cu$_2$O absorber layer on the optical and electrical characteristics of the ZnO/Cu$_2$O subcell. The analysis suggests that the performance of the ZnO/Cu$_2$O subcell can be improved by the incorporation of a ZnS, TiO$_2$, or Ga$_2$O$_3$ buffer layer as a result of reduced conduction band offset.

Keywords: Metal oxides, tandem, silicon heterojunction, electro-optical modeling

1 INTRODUCTION

The semiconducting metal oxides ZnO and Cu$_2$O are promising candidate materials for implementation in photovoltaic applications, since they are earth-abundant, non-toxic and have the potential for low production cost [1]. The theoretical maximum efficiency for a ZnO/Cu$_2$O heterojunction solar cell is 19-20% [2]. However, the highest conversion efficiency currently achieved is about 8% [3], with the potential to reach 14% [4]. For such heterojunction solar cells, the properties of the ZnO/Cu$_2$O interface are highly important in order to obtain a high power conversion efficiency. Reducing the defect density at the heterojunction interface to a minimum is critical in order to avoid recombination losses. In addition, ZnO features a conduction band misalignment of approximately 0.9 eV with Cu$_2$O, which results in a reduced open circuit voltage. Further, a few nm thick intermediate layer of CuO has been reported to form at the ZnO/Cu$_2$O interface which also reduces the open circuit voltage [5]. For these reasons, different buffer layer materials, which are inserted between the ZnO and Cu$_2$O layers, have been investigated with the aim of improving the energy band alignment. Among the materials that have been proposed are Zn$_{1-x}$Ge$_x$O, Ga$_2$O$_3$, TiO$_2$, ZnS, and Zn(O,S) [3, 6, 7].

The objective of this work is to evaluate the performance of a ZnO/Cu$_2$O subcell in a Si-based tandem heterojunction solar cell configuration. In general, the direct bandgap of approximately 2.1 eV makes Cu$_2$O attractive as a top cell absorber layer for tandem solar cells. Figure 1 shows a schematic device design of a four-terminal tandem solar cell, combining a conventional crystalline silicon subcell with a ZnO/Cu$_2$O subcell in a mechanical stack of independently connected cells, i.e. a four-terminal configuration. The ZnO/Cu$_2$O subcell is realized by reactive magnetron sputter deposition of metal oxides on a transparent glass substrate (quartz), enabling low-energy photons to be transmitted through the ZnO/Cu$_2$O top subcell for subsequent absorption in the c-Si bottom subcell. The ZnO/Cu$_2$O subcell, could in this architecture serve as the module glass encapsulating the c-Si subcell, requiring insignificant changes to the design of the bottom cell. In this paper, electro-optical modeling was carried out in order to investigate the performance of the ZnO/Cu$_2$O subcell for different buffer layer materials inserted between the ZnO emitter and Cu$_2$O absorber layers.

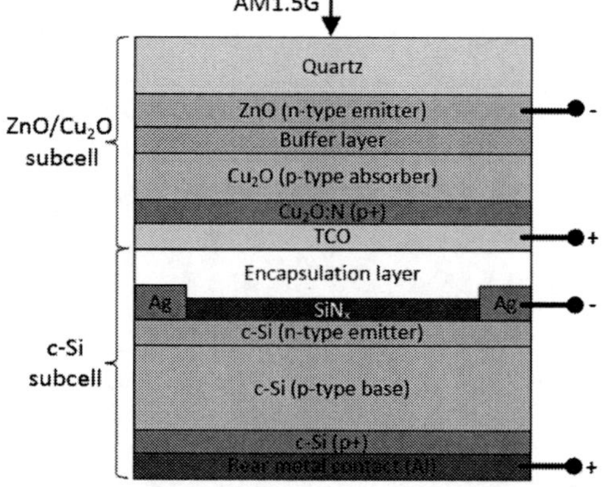

Figure 1: Schematic layout of a four-terminal tandem solar cell consisting of a ZnO/Cu$_2$O heterojunction subcell on top of a conventional wafer based crystalline Si subcell.

2 EXPERIMENTAL DETAILS

Cu$_2$O and Al-doped ZnO (AZO) thin films were deposited on 10 x 10 x 0.5 mm^3 quartz substrates using a direct current/radio frequency (DC/RF) magnetron sputtering system (Semicore Triaxis). Cu$_2$O was deposited by reactive sputtering of a 99.999% Cu target in O$_2$/Ar (6/49 sccm) with a substrate temperature of 400 °C. The power density was fixed at 2.2 W/cm^2. As-grown

Cu_2O films were annealed at 900 °C for 3 minutes in vacuum (oxygen partial pressure ~10^{-1} Torr). AZO was deposited by co-sputtering of a 99.99% pure ZnO ceramic target at 50 W and a 99.999% Al target at 3 W in Ar with a substrate temperature of 400 °C, resulting in an aluminum content of approximately 4 wt% in the deposited thin films. During the magnetron sputtering deposition, the base pressure was below 3.0 x 10^{-7} Torr. The optical properties and surface morphology of the Cu_2O and AZO thin films were analyzed using a Horiba Jobin Yvon UVISEL spectroscopic ellipsometer and a QUANTA INSPECT F 50 scanning electron microscope (SEM), respectively. The optical transmittance spectrum in the wavelength range from 300 nm to 1000 nm was measured using a setup with spectrophotometers, a deuterium-halogen light source, and an integrating sphere. Room temperature Hall effect measurements (LakeShore 7604) using the van-der Pauw configuration were carried out to determine the carrier mobility and concentration, and resistivity of the AZO and Cu_2O thin films.

3 RESULTS

3.1 Optical analysis

Figure 2a shows the optical transmittance spectra for a 500 nm thick as-grown and annealed Cu_2O thin film, as well as a 200 nm thick AZO thin film, on quartz. Based on the measured transmittance spectra, a Tauc plot analysis was carried out to determine the optical band gap of the AZO and Cu_2O (as-grown and annealed) thin films [8]. The resulting Tauc plots are shown in Figure 2b and suggest that the optical band gap is increased from E_g = 2.06 eV for the as-grown Cu_2O film to E_g = 2.19 eV after annealing. The widening of the optical band gap after annealing at 900°C might be due to partial elimination of defects states [9] and more phase-pure Cu_2O films with negligible contribution of the CuO phase. For the AZO thin film the optical band gap was estimated to E_g = 3.61 eV, evidencing the so-called Burstein-Moss effect.

Figure 3a shows the calculated percentage reflectance, absorptance, and transmittance as a function of wavelength for the ZnO/Cu_2O subcell depicted in Figure 1. In this simulation, an 80 nm thick front AZO layer, a 2 μm thick Cu_2O layer, and a 75 nm thick bottom AZO layer (transparent conductive oxide) were implemented in the model. The complex refractive indices for the AZO and Cu_2O layers were obtained from spectrometric ellipsometry measurements and implemented in the optical analysis [10]. The absorber layer thickness of 2 μm was chosen mainly due to experimental constraints, i.e. the experimentally feasible thickness range is typically a few micrometers as a result of the relatively low deposition rate for the sputtered

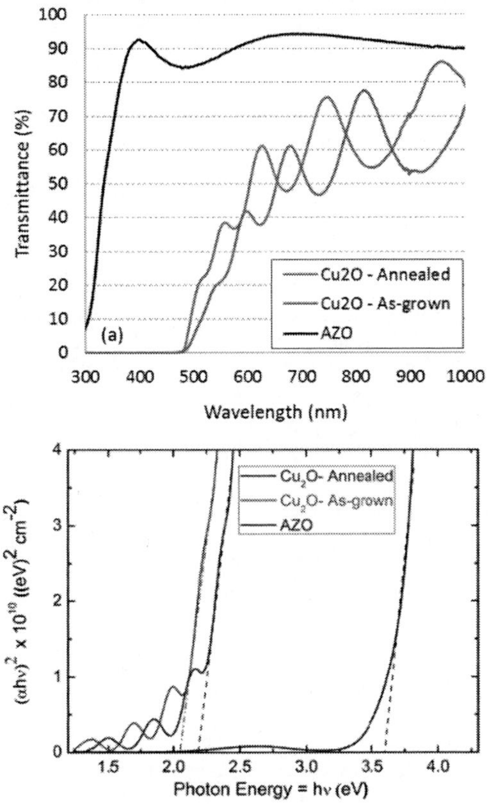

Figure 2: (a) Optical transmittance spectra and (b) Tauc plot for the as-grown and annealed Cu_2O and AZO thin films deposited on quartz. The optical band gap values are estimated from extrapolation to the abscissa (dashed lines).

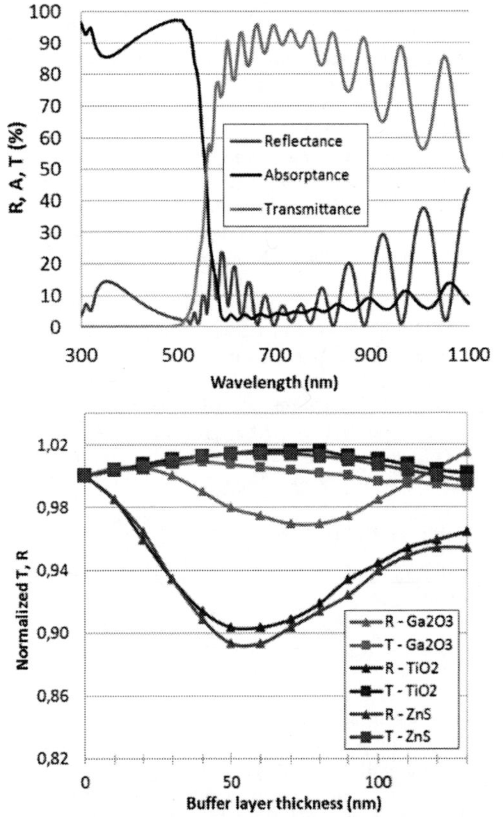

Figure 3: (a) Percentage reflectance, absorptance and transmittance as a function of wavelength for the ZnO/Cu_2O subcell without a buffer layer. (b) Normalized average reflectance and transmission for the ZnO/Cu_2O subcell as a function of buffer layer thickness for various materials.

Cu_2O thin film. Furthermore, the AZO layer thickness of ~80 nm was chosen based on optimization of the optical performance [10], i.e., 80 nm thickness provides a minimum front reflectance and maximum long-wavelength transmittance for the ZnO/Cu_2O subcell. Planar surfaces were adopted in the model. The calculated curves shown in Figure 3a suggest that photons are absorbed in the top ZnO/Cu_2O subcell in the wavelength range below ~600 nm and transmitted in the wavelength range above ~600 nm. It can also be seen that in the long-wavelength range, the AZO layers contribute to parasitic absorption, caused by free carrier absorption. The average optical reflection for the ZnO/Cu_2O subcell without a buffer layer is about 10 % in the wavelength range from 300 – 1100 nm, which suggests that there is room for improving the optical characteristics for example by introducing a more advanced anti-reflection coating in combination with texturing of the front surface.

The optical characteristics for the ZnO/Cu_2O subcell were also analyzed for different buffer layers inserted between the AZO emitter layer and the Cu_2O absorber layer. Figure 3b shows the normalized average optical reflectance and transmission as a function of Ga_2O_3, TiO_2, and ZnS buffer layer thickness. Tabulated values for the refractive index of the materials were used in the calculations [11, 12, 13]. The data shows that the optical performance of the ZnO/Cu_2O subcell is affected by the incorporation of a buffer layer and that minimum reflectance and maximum transmission is achieved for a buffer layer thickness around 50 – 70 nm.

3.2 Electrical analysis

The majority carrier mobility, film resistivity and majority carrier concentration for Cu_2O (as-grown and annealed) and AZO thin films deposited on quartz, derived from room temperature Hall effect measurements, are presented in Table I. The data in the table suggests that the electrical properties for the Cu_2O thin film are improved after annealing, i.e., the majority carrier (hole) mobility increases from 10 to 50 $cm^2/V{\cdot}s$ and the resistivity decreases from 560 to 200 $\Omega{\cdot}cm$ after annealing. These values are comparable to those reported previously for sputter-deposited polycrystalline Cu_2O thin films on quartz [14, 15], suggesting that the annealed Cu_2O thin films are well suited for photovoltaic applications. The increase in carrier mobility after annealing can, at least partly, be attributed to the increase in grain size and reduced grain-boundary scattering. The average grain size for a 500 nm thick Cu_2O thin film on quartz increases from about 70 nm for the as-grown film to about 600 nm for the annealed film, as revealed from SEM images of the thin film surface [10].

Table I: Majority carrier mobility, film resistivity, and majority carrier concentration for AZO and Cu_2O thin films on quartz, derived from room temperature Hall effect measurements. (*) denotes the Cu_2O thin film after annealing at 900 °C.

Parameter	AZO	Cu_2O	Cu_2O*
Resistivity (Ohm·cm)	5e-4	560	200
Mobility ($cm^2/V{\cdot}s$)	20	10	50
Carrier concentration (cm^{-3})	3e20	3e15	1e15

The electrical performance of the ZnO/Cu_2O subcell with incorporation of different buffer layers was evaluated based on device modeling in Silvaco Atlas. The model assumes the heterojunction as several layers for which the physical properties were based on experimental and theoretical results previously reported in the literature [4]. Table II summarizes the parameters for each layer implemented in the model. The thickness of the AZO transparent electrode was set to 250 nm in order to provide sufficient lateral transport of charge carriers which is critical for the electrical performance of a solar cell. Typically, a thickness of about 200 nm for the AZO layer has been reported for Cu_2O-based heterojunction solar cell devices [3]. An interface defect layer (IDL) and a defective Cu_2O (d-Cu_2O) layer was included between the AZO and Cu_2O layers in order to represent interface defect states, which significantly influences the electrical characteristics of the ZnO/Cu_2O heterojunction. A Gaussian defect distribution with a midgap level and a width of 0.1 eV was assumed for all layers [4].

Figure 4 shows the calculated current-voltage (I-V) characteristics for the ZnO/Cu_2O subcell. The I-V curve for the ZnO/Cu_2O subcell without a buffer layer is shown along with the I-V curve for the ZnO/Cu_2O subcell with a 50 nm thick idealized ZnO buffer layer inserted between the AZO emitter layer and the 2 μm thick Cu_2O absorber layer, i.e., the value for the electron affinity was chosen to optimize the electrical performance of the ZnO/Cu_2O subcell under AM1.5G illumination.

Table II: Material parameters and their values implemented in the Silvaco model for the ZnO/Cu_2O subcell.

Parameter	Al:ZnO	Buffer layer	IDL	d-Cu_2O	Cu_2O	Cu_2O:N
Thickness (nm)	250	50	1	5	2000	50
Electron affinity (eV)	4.4	Varied	3.2	3.2	3.2	3.2
Band gap (eV)	3.35	Varied	2.1	2.1	2.1	2.1
Relative permittivity	9	Varied	7.6	7.6	7.6	7.6
Acceptor concentration (cm^{-3})	0	0	1e15	1e15	1e15	1e21
Donor concentration (cm^{-3})	3e20	1e19	0	0	0	0
Hole mobility ($cm^2/V{\cdot}s$)	5	5	50	50	50	50
Electron mobility ($cm^2/V{\cdot}s$)	10	10	100	100	100	100
Effective density of states in conduction band (cm^{-3})	2.2e18	2.2e18	2.4e19	2.4e19	2.4e19	2.4e19
Effective density of states in valence band (cm^{-3})	1.8e19	1.8e19	1.3e19	1.3e19	1.3e19	1.3e19
Capture cross section of holes (cm^2)	1e-15	1e-15	1e-13	1e-15	1e-15	1e-15
Capture cross section of electrons (cm^2)	1e-12	1e-12	1e-13	5e-13	5e-13	5e-13
Defect density (cm^{-3})	1e18	5e17	1e20	3.5e19	1e13	1e13

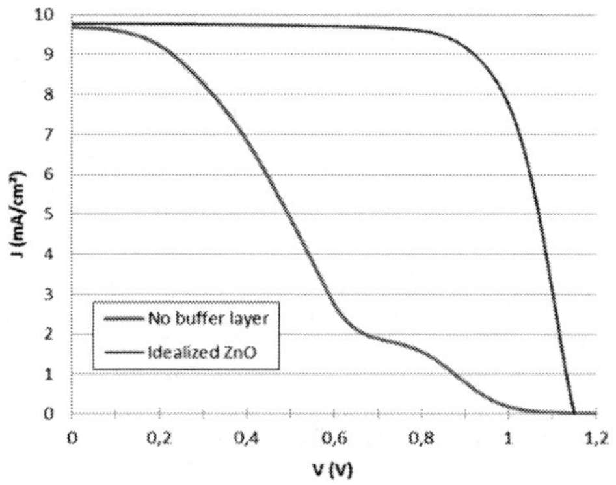

Figure 4: I-V curve for the ZnO/Cu_2O subcell without a buffer layer and with a 10 nm thick idealized ZnO buffer layer.

For the ZnO/Cu_2O subcell the highest performance was achieved for an electron affinity of 3.65 eV for the idealized ZnO buffer layer, yielding a fill factor (FF) of 74 % and a power conversion efficiency of 8.3 %, assuming an electron affinity of 3.2 eV for Cu_2O. However, the ZnO/Cu_2O subcell without a buffer layer displays S-shaped I–V characteristics, limiting the FF to around 28 %. Such distortion of the I-V curve has been previously reported for Cu_2O heterojunction solar cells [4]. The origin of the S-shaped deformation might be space charge accumulation caused by restricted charge transport over the ZnO/Cu_2O interface, e.g., the conduction band offset can result in an energy barrier that limits the electron current flowing across the interface. Previously reported simulation results for heterojunction solar cells have suggested that such distortion of the I-V curve may appear for conduction band offsets of more than 0.4 - 0.5 eV [16].

Table III summarizes the I-V parameters for the ZnO/Cu_2O subcell for five different buffer layer materials. The data shows that the conversion efficiency can be increased by several percent absolute by the incorporation of a buffer layer. Based on these simulation results, ZnS, TiO_2, and Ga_2O_3 seem to be suitable candidate materials for enhancing the electrical performance of the ZnO/Cu_2O subcell, as compared to a cell with no buffer layer incorporated.

Table III: I-V parameters for the ZnO/Cu_2O subcell for selected buffer layer materials. J_{sc} and V_{oc} denote the short circuit current density and open circuit voltage, respectively.

Buffer layer material	J_{sc} mA/cm^2	V_{oc} V	FF %	Eff. %
No buffer layer	9.7	1.00	28.3	2.7
ZnO (idealized)	9.8	1.16	74.0	8.3
ZnS	9.8	1.15	58.4	6.6
TiO_2	9.8	1.14	52.2	5.8
Ga_2O_3	9.7	1.14	51.7	5.8
Zn(O,S)	9.7	1.14	47.0	5.2

Figure 5 shows the corresponding external quantum efficiency (EQE) spectrum for the ZnO/Cu_2O subcell for different buffer layers. The EQE curves for TiO_2 and Ga_2O_3 are almost identical and have been represented by the same curve in the plot. For the ZnO/Cu_2O subcell with an idealized ZnO buffer layer, the EQE is approximately 90 % in the wavelength range from 400 to 500 nm, whereas for the ZnO/Cu_2O subcell without a buffer layer the EQE is below 20 % in the same wavelength range.

Figure 6 shows a plot of the percentage conversion efficiency of the ZnO/Cu_2O subcell versus electron affinity for the different buffer layers investigated in this work. The electron affinity adopted for Cu_2O, ZnS, TiO_2, Ga_2O_3, and Zn(O,S) was 3.2, 3.9, 4.0, 4.0 and 4.05 eV, respectively, assuming a composition of $ZnO_{0.42}S_{0.73}$ for Zn(O,S) [7]. The data shown in Table III suggests that the performance of the ZnO/Cu_2O subcell decreases with increasing difference in electron affinity between the Cu_2O absorber and the buffer layer, i.e., for a conduction band offset of more than ~0.5 eV the FF is significantly reduced due to distortion of the I-V curve.

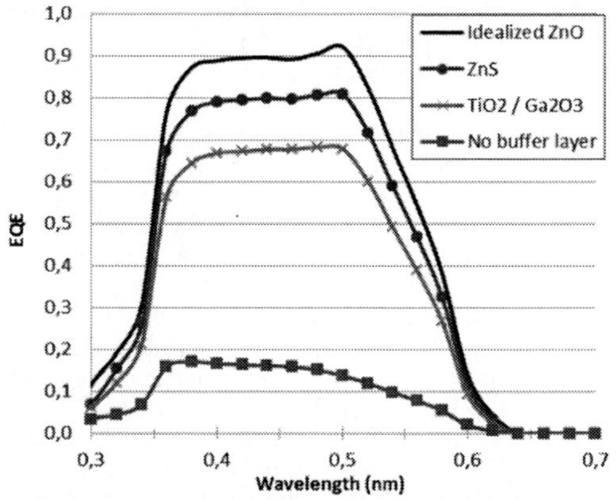

Figure 5: External quantum efficiency spectrum for the ZnO/Cu_2O subcell for different buffer layer materials.

The relatively low electron affinity of Cu_2O ($\chi = 3.2$ eV) makes it challenging to find material combinations with suitable energy band alignments. In order to develop a high performance ZnO/Cu_2O heterojunction solar cell with a sufficiently low conduction band offset, manipulation of the energy band alignment at the heterojunction interface will most likely be required, for instance by adjusting the deposition conditions, in addition to the incorporation of a buffer layer.

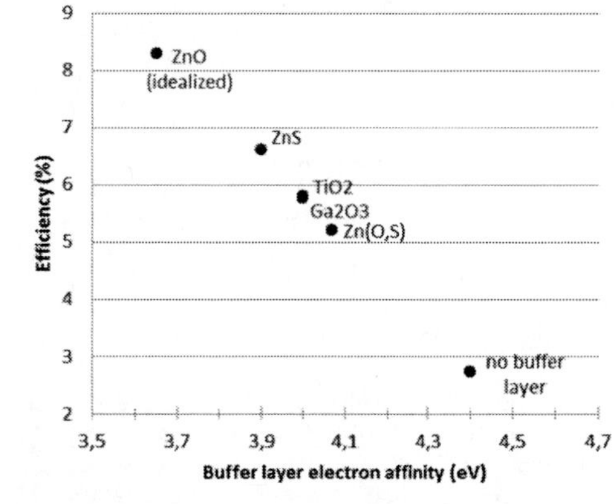

Figure 6: Percentage conversion efficiency for the ZnO/Cu_2O subcell vs. electron affinity for different buffer layer materials.

The concentration and charge transition energies of bulk defects add restrictions to the position of the Fermi energy level in the materials, which affect the energy band alignment at the heterojunction interface [17]. Thus controlling the bulk defects during the thin film growth could be essential for the performance of the heterojunction solar cell. The influence of the defects on the electrical characteristics of the ZnO/Cu_2O subcell was not analyzed in this work, but will be the subject of further investigation.

In general, the thickness of the absorber layer will affect the photocurrent of a solar cell. For a monolithic (two-terminal) tandem solar cell, current matching of the subcells is necessary, which implies that the thickness of the absorber layer becomes an important design parameter. Figure 7 shows the calculated conversion efficiency and short circuit current density for the ZnO/Cu_2O subcell for different thickness of the Cu_2O absorber layer. In this case, an idealized ZnO buffer layer was used. The maximum performance of the device is obtained for an absorber layer thickness of approximately 5 µm. By increasing the thickness from 1 µm to 5 µm the efficiency is increased from 7.3 % to 9.0 %, with J_{sc} saturating at approximately 11.1 mA/cm^2. In comparison, the current density for a conventional c-Si solar cell is approximately 38 mA/cm^2 under 1 sun illumination. Placing a ZnO/Cu_2O subcell on top of a c-Si solar cell in a tandem configuration (cf. Figure 1) will result in a current density of about 22 mA/cm^2 for the c-Si bottom subcell, as a result of the short-wavelength photons being absorbed in the top subcell. The rather large current mismatch between the two subcells in such tandem configuration favors the arrangement of a mechanical stack of independently connected cells, i.e. a four-terminal configuration.

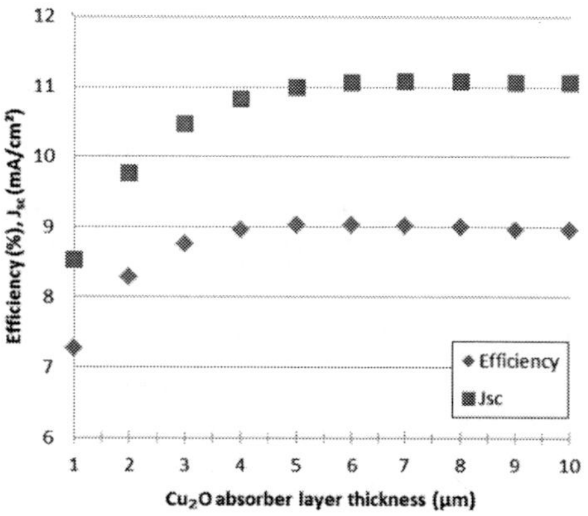

Figure 7: Percentage conversion efficiency and J_{sc} for the ZnO/Cu_2O subcell with an idealized ZnO buffer layer vs. Cu_2O absorber layer thickness.

4 CONCLUSION

The sputter-deposited Cu_2O and Al-doped ZnO thin films presented in this work show good potential for application in a ZnO/Cu_2O subcell in a c-Si based tandem solar cell configuration. After annealing at 900°C, the Cu_2O thin films display high carrier mobility (50 cm^2/V·s) and low resistivity (200 Ohm·cm). Device modeling showed that the I-V curve for the ZnO/Cu_2O subcell is distorted due to the conduction band offset which limits the electron current flowing across the heterojunction interface. However, the conduction band alignment can be improved by incorporation of a ZnS, TiO_2, or Ga_2O_3 buffer layer, enhancing the electrical performance of the ZnO/Cu_2O subcell. The maximum performance of the ZnO/Cu_2O subcell with an idealized ZnO buffer layer is obtained for an absorber layer thickness of approximately 5 µm, yielding a conversion efficiency around 9 % and a short circuit current density of about 11 mA/cm^2. The relatively large current mismatch between the ZnO/Cu_2O and c-Si subcells favors a four-terminal tandem solar cell configuration.

ACKNOWLEDGEMENTS

This work was conducted under the research project "High-performance tandem heterojunction solar cells for specific applications (SOLHET)", project 251789, financially supported by the Research Council of Norway (RCN) and the Romanian Executive Agency for Higher Education, Research, Development and Innovation Funding (UEFISCDI) through the M-Era.net program. RCN is also acknowledged for the support to the Norwegian Micro- and Nano-Fabrication Facility, NorFab, project number 245963/F50.

REFERENCES

[1]. L. Fara and M. Yamaguchi (editors), "Advanced Solar Cell Materials, Technology, Modeling, and Simulation", IGI Global Publishing House (2013).

[2]. W. Shockley and H.J. Queisser, "Detailed Balance Limit of Efficiency of PN Junction Solar Cells". J. Appl. Phys. 32 (1961) 510-519.

[3]. T. Minami, Y. Nishi and T. Miyata, "Efficiency enhancement using a $Zn_{1-x}Ge_x$-O thin film as an n-type window layer in Cu_2O-based heterojunction solar cells", Appl. Phys. Express 9 (2016) 052301.

[4]. Y. Takiguchi and S. Miyajima, "Device simulation of cuprous oxide heterojunction solar cells", Jpn. J. Appl. Phys. 54 (2015) 112303.

[5] J. Gan, S. Gorantla, H. N. Riise, Ø. S. Fjellvåg, S. Diplas, O. M. Løvvik, B. G. Svensson, E. V. Monakhov, and A. E. Gunnæs. "Structural properties of Cu_2O epitaxial films grown on c-axis single crystal ZnO by magnetron sputtering", Appl. Phys. Lett. 108 (2016) 152110.

[6]. T. Minami, T. Miyata and Y. Nishi, "Cu_2O-based heterojunction solar cells with an Al-doped ZnO/oxide semiconductor/thermally oxidized Cu_2O sheet structure", Solar Energy 105 (2014) 206-217.

[7]. R.E. Brandt, M. Young, H.H. Park, A. Dameron, D. Chua, Y.S. Lee, G. Teeter, R.G. Gordon and T. Buonassisi, "Band Offsets of N-Type Electron-Selective Contacts on Cuprous Oxide (Cu_2O) for Photovoltaics", Applied Physics Letters 105 (2014) 363901.

[8] J. Gan, V. Venkatachalapathy, B. G. Svensson and E. V. Monakhov, "Influence of Target Power on Properties of Cu_xO Thin Films Prepared by Reactive Radio Frequency Magnetron Sputtering". Thin Solid Films 594 (2015) 250-255.

[9]. Y. Wang, P. Miska, D. Pilloud, D. Horwat, F. Mücklich and J. F. Pierson, "Transmittance enhancement

and optical band gap widening of Cu_2O thin films after air annealing". J. Appl. Phys. 115 (2014) 073505.

[10] Ø. Nordseth, R. Kumar, K. Bergum, L. Fara, S. E. Foss, H. Haug, F. Drăgan, D. Crăciunescu, P. Sterian, I. Chilibon, C. Vasiliu, L. Baschir, D. Savastru, E. Monakhov and B. G. Svensson, "Optical Analysis of a ZnO/Cu_2O Subcell in a Silicon-Based Tandem Heterojunction Solar Cell", Green and Sustainable Chemistry 7 (2017) 57-69.

[11] M. Rebien, W. Henrion, M. Hong, J. P. Mannerts, M. Fleischer, "Optical properties of gallium oxide thin films", Appl. Phys. Lett. 81 (2002) 250.

[12] J. Cui, T. Allen, Y. Wan, J. McKeon, C. Samundsett, D. Yan, X. Zhang, Y. Cui, Y. Chen, P. Verlinden and A. Cuevas, "Titanium oxide: A re-emerging optical and passivating material for silicon solar cells", Solar Energy Materials & Solar Cells 158 (2016) 115–121.

[13] J. Siquerios, R. Machorro and L. Regalado, "Determination of the optical constants of MgF_2 and ZnS from spectrophotometric measurements and the classical oscillator method", Applied Optics 27 (1988) 2549–2553.

[14]. S. Ishizuka, T. Maruyama and K. Akimoto, "Thin-film deposition of Cu_2O by reactive radio-frequency magnetron sputtering", Jpn. J. Appl. Phys. 39 (2000) L786-L788.

[15]. Y. S. Lee, M. T. Winkler, S. C. Siah, R. Brandt and T. Buonassisi, "Hall mobility of cuprous oxide thin films deposited by reactive direct-current magnetron sputtering", Appl. Phys. Lett. 98 (2011) 192115.

[16] A. O. Pudov, A. Kanevce, H. A. Al-Thani, J. R. Sites, and F. S. Hasoon, "Secondary barriers in CdS-$CuIn_{1-x}Ga_xSe_2$ solar cells", J. Appl. Phys. 97 (2005) 064901.

[17] S. Siol, J. C. Hellmann, S. D. Tilley, M. Graetzel, J. Morasch, J. Deuermeier, W. Jaegermann and A. Klein, "Band alignment engineering at Cu_2O/ZnO heterointerfaces", ASC Appl. Mater. Interfaces 8 (2016) 21824.

CELL-TO-MODULE CONVERSION LOSS SIMULATION FOR SHINGLED-CELL CONCEPT

Jorge Rabanal-Arabach[1*], Dominik Rudolph[1], Ingo Ullmann[1], Andreas Halm[1], Andreas Schneider[1], Thomas Fischer[2]
[1] International Solar Energy Research Center (ISC), Rudolf-Diesel-Str. 15, 78467 Konstanz, Germany
[2] Teamtechnik Maschinen und Anlagen, Planckstraße 40, 71691 Freiberg, Germany
[*] Corresponding author: Phone: +49 7531 36183-341; Fax: +49 7531 36183-11
Email: jorge.rabanal@isc-konstanz.de

ABSTRACT: The "shingled solar cell interconnection concept" for c-Si solar cells has been investigated since more than 26 years [1]. Since 2013, several articles about high voltage module applications have been published with main focus on accelerated aging tests, showing promising results [2-4]. The present work focuses on a theoretical study carried out to carefully determine the cell-to-module losses based on experimentally embedded stripes of commercial c-Si solar cells (originally 6-inch, full square Al-BSF) into a module interconnected by the shingled-cell concept. Results show an acceptable cell-to-module calculation accuracy. The simulated active area efficiency is up to 3%rel higher than for a module made out of traditional front-to-back cell-to-cell interconnection. By splitting a cell in five stripes (S 1/5 concept), the FF improves in average by 2%abs mainly due to lower resistance loss hence lower electrical power loss compared to a module made of 5 busbar solar cells connected by the traditional front-to-back interconnection concept (S 1/5 REF in the article).
Keywords: Design, PV modules, polycrystalline, shingled-cell.

1 INTRODUCTION.

The "shingle" solar cell interconnection technique cannot be called a novel concept or technological approach since it was first published already in 1990 [1]. At that time, the authors demonstrated a cell to module conversion (C2M) loss of 0.5% in terms of efficiency, achieving a module efficiency of 17.3% with 0.96 packing density. Standard industrial solar modules based on soldering that time had a packing density factor of 0.89 for mc-Si solar cells, and nowadays of 0.93 and higher. A similar concept was employed in the fabrication of a high voltage module by J. Summhammer et al. [2-4] in the years of 2013 to 2016, with positive results achieved from accelerated aging tests. The authors estimated a saving on cell's metallization of 70% to 90% if compared to a typical industrial module made out of 60 c-Si cells (6-inch full square) comprising soldering for cell interconnection [3].

The shingled-cell concept is of great interest at industrial level not only from a technical point of view but also for an aesthetical one. The former refers to the attempt to diminish the C2M electrical (ohmic) power losses mainly by applying a lower current per string and at the same time also allowing for increasing the packing factor –resulting in higher module efficiency. It is also important to point out, that the possibility to obtain high module voltage without compromising module efficiency due to the length of the string exists. The improvement in aesthetics is linked to the absence of ribbons on top of each cell and the space between one cell and the immediately consecutive in series connected one. This all results in a very homogeneous optical appearance and also enhances the packing density and thus the module efficiency. Currently, only few proto-type PV modules out of this type exist in the market and scientific articles related to this topic are difficult to find. The aim of this work is to contribute with in depth studies on defining, calculating and evaluating C2M losses for manufacturing of photovoltaic (PV) modules made out of mc-Si solar cells by means of the "shingle concept". In the present work, cell-to-stripe-to-module losses are calculated and compared to experimental results for 3, 4 and 5 cell stripes obtained per wafer. PV modules consisting out of 1-cell and equivalent power are manufactured to experimentally determine the C2M losses. To complement the study, the experimental results are extrapolated to industrial size PV modules by using two different scenarios of module optimizations: scenario 1) focus on module $P_{MPP,}$ and scenario 2) focus on module area.

2 DESIGN OF EXPERIMENT

2.1 The concept

The shingled-cell concept corresponds to a full size c-Si solar cell which is reduced (cut) in area by laser cutting parallel to the busbar (on the busbar or close to it), resulting in cell stripes (*pcs*) with electrical interconnections at the long top and bottom edges only. By overlapping the back of one cell to the front side connection of the next one the stripes are later directly interconnected in series, forming this way a shingled string. Figure 1 a) shows a schematic lateral view of traditional front-to-back cell-to-cell interconnection and in Figure 1 b) a shingle concept made out of 4 cell stripes.

For module manufacturing, conductive glue is used to establish a low resistive and stable electrical contact and providing a sufficient mechanical support level. The concept is assumed to be implemented in a fully automated module line. Therefore, a feasible laser cutting process and a least overlapping cell area are employed.

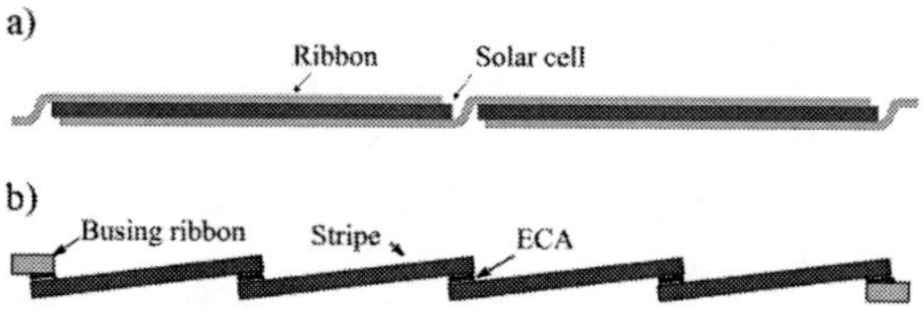

Figure 1: Schematic lateral view of: a) traditional front-to-back cell-to-cell interconnection, and b) shingle concept made of 4 cell stripes (S 1/4).

2.2 Cell-to-module conversion calculations

To calculate the C2M gain/loss, first the shingled cell power is simulated. The previous is done by calculating the power loss contribution by taking the electrical resistance of the emitter resistance, the electrical contact of the silver to the emitter and the line resistance of fingers and busbar of each *pcs*. This is made by adapting the formulas explained in section 1.2.2 of A. Mette PhD thesis [5] to the geometry of the shingled-type stripe with busbar printed on both sides of a full-square Al-BSF solar cell. The electrical power loss ($P_{loss,\ pcs}$) is assumed to affect in first order the V_{MPP} rather than the I_{MPP} of the stripe, and it is calculated as following:

$$P_{loss,pcs} = R_{eq}\left[\frac{I_{MPP\,pcs}}{l_{pcs}\,w_{pcs}}\left(l_f + \frac{o_{pcs}}{2}\right)\frac{s_f}{2}\right]^2 n_f + R_{bb}\, I^2_{MPP\,pcs} \quad (1)$$

$$R_{eq} = \frac{(R_e + R_{e,c} + R_f)}{2} \quad (2)$$

Where R_e, $R_{e,c}$, R_f and R_{bb} are the electrical resistance of the emitter, the emitter contact, the finger and the busbar, l_{pcs} and w_{pcs} correspond to the length and width of the solar cell stripe, o_{pcs} is the overlapping between cells (or busbar width at the edge), and n_f , l_f and s_f are the number of fingers, finger length and spacing between fingers, respectively.

$$R_e = \frac{1}{6} r_{sheet}\left(\frac{s_f - w_f}{l_f}\right) \quad (3)$$

$$R_{e,c} = \frac{\sqrt{r_{sheet}\rho_c}}{l_f} coth\left(w_f\sqrt{\frac{r_{sheet}}{\rho_c}}\right) \quad (4)$$

$$R_f = \frac{2}{3}\rho_f \frac{l_f}{A_f} \quad (5)$$

To calculate the optical power loss (caused by reflection and absorption of glass, cell anti-reflection coating, and encapsulant) and the electrical (ohmic) power loss (caused by cell-to-cell connectors and busing ribbons) due to the module assembly, the same procedure and equations are used as proposed in [6] and [7]. Changes in J_{SC} during the manufacturing process are considered as fit parameters for the optical power loss. It's important to clarify that the C2M power loss due to cell mismatching is not included in the simulation.

2.3 Experimental setup

Several mc-Si solar cells (without metallization) were printed with 4 different metal grid layouts. Each layout is optimized for the next individual string interconnection concepts: i) asymmetrical 4BB H-pattern, front-to-back cell-to-cell interconnection, ii) shingle stripes made of 1/4 of a solar cell (S 1/4), iii) asymmetrical 5BB H-pattern, front-to-back cell-to-cell interconnection, and iv) stripes made of 1/5 of a solar cell (S 1/5). The optimization corresponds to optimum number of fingers due to cell size, its shading factor and the cost of the silver paste to print the respective amount of fingers. The design of the metallization grid is discussed in [8]. Examples of module samples for groups S 1/4 REF, S 1/4 and S 1/5 are presented in Figure 2.

In addition, 3 stripes are obtained from each cell of a batch of commercial mc-Si cells (6-inch, full square Al-BSF, 3 busbar, 92 fingers). The average measured maximum power point (P_{MPP}) of the batch is 4.13 ± 0.01 W at standard test conditions (STC: AM 1.5G, 1000 W/m² and 25°C). The measured P_{MPP} for those stripes is 1.20 ± 0.01 W at STC, meaning a power drop of 12.8%$_{rel}$ in average occurs. In the case of the cells with optimized metallization grid, the measured power drop is approximately 6.1%$_{rel}$ and 4.5%$_{rel}$ for S 1/4 and S 1/5, respectively. Due to the lowest power drop from cell-to-stripe achieved with the optimized metallization grid, several modules with 1-cell equivalent power are fabricated based on stripes from the 4 groups previously presented (see Table I). All modules are manufactured by using 3.2 mm textured glass, EVA and transparent backsheet. ECA is employed to ensure electrical connection and mechanical stability [9]. The manufactured modules are characterized by EL and I-V measurements. Modules without cracks and less manufacture failure were selected for data analysis.

Table I. Overview on the group samples.

Group	Description	Overlap (mm)	Sample quantities
S 1/4 REF	4 busbars front-to-back	-	3
S 1/4	4 *pcs* per wafer	1.5	4
		2.0	4
S 1/5 REF	5 busbars front-to-back	-	3
S 1/5	5 *pcs* per wafer	1.5	4
		2.0	4

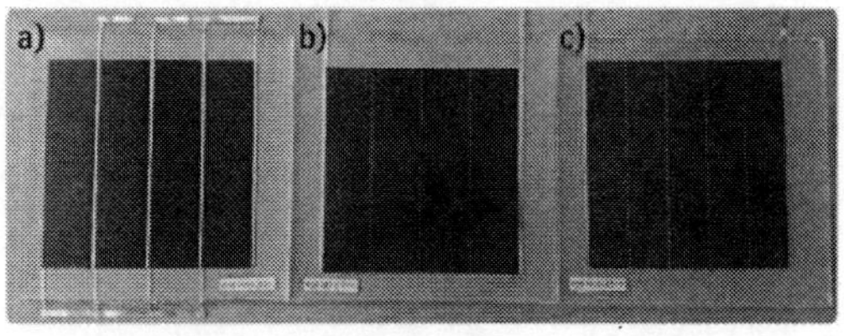

Figure 2: Example of manufactured modules for groups: a) S 1/4 REF, b) S 1/4, and c) S 1/5.

3 RESULTS AND DISCUSSIONS

Figure 3 shows the relative change in electrical parameters for cell samples embedded into a 1-cell equivalent module. The relative drop in power is not improved by the application of the shingled-cell concept but fill factor losses are strongly reduced. The power loss is mainly driven by the decrease in current density, which itself increases with the overlap distance. The improvement in *FF* can be attributed to the increase in V_{MPP}:V_{OC} ratio (see Figure 4). Two reasons could drive this improvement: i) low voltage drop in the metallization grid (ohmic loss) due to short finger lengths, low I_{MPP} or the combination of both, and ii) small drop in V_{OC} (less than 1%$_{rel}$) due to edge recombination caused by caused by the larger edge length to cell area ratio if compared to a larger cell. The J_{MPP} to J_{SC} ratio seen in Figure 4 reveals an enhancement for S 1/5 only, reason why the *FF* shows higher improvement for S 1/5 than for S 1/4. Nevertheless, no differences have been noticed in J_{MPP} to J_{SC} ratio for the different overlap employed in the manufacture of S 1/4 or S 1/5.

With the experimental results on hand, parameters to calculate the optical and electrical losses could be fitted.

Figure 5 shows a) the simulated P_{MPP} of each stripe (*pcs*) for a certain type of stripe obtained from 1 wafer (pcs_per_wafer), and b) the electrical parameters of the in-house manufactured stripes with optimized grid. The experimental measurements of stripes show better V_{MPP} than the theoretical calculated.

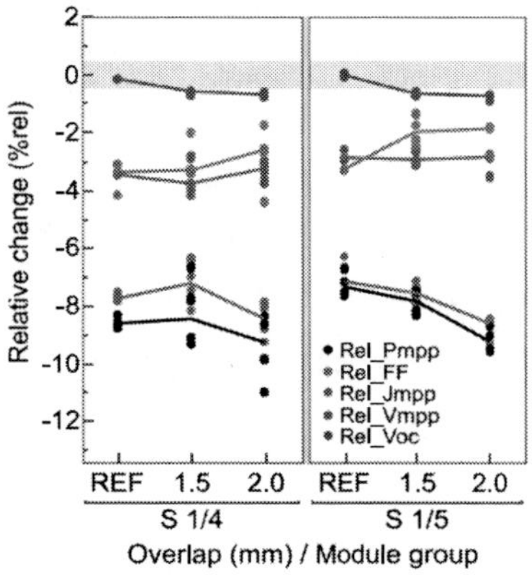

Figure 3: Relative change of C2M electrical parameters. Calculations based on experimental results. The relative change of J_{SC} has a similar trend and proportion as J_{MPP}, and it's therefore not plotted. The grey band denotes to measurement and calculation accuracy.

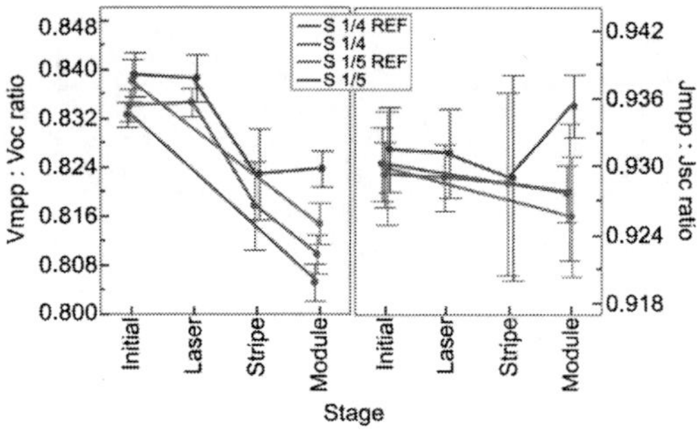

Figure 4: V_{MPP} to V_{OC} ratio and J_{MPP} to J_{SC} ratio for various manufacturing stages.

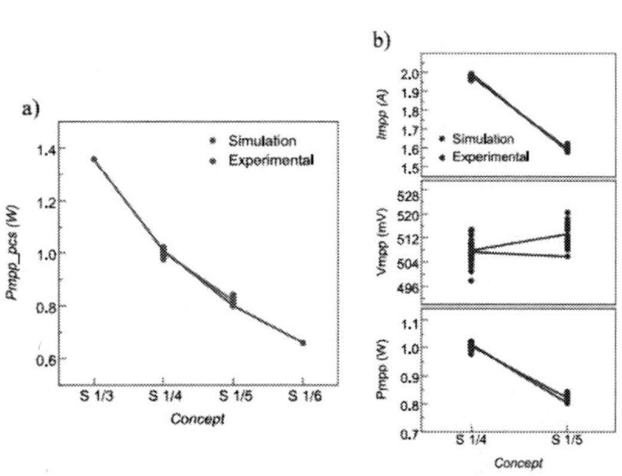

Figure 5: a) Max power of shingle stripes with different amount of fingers per wafer, b) Electrical parameters at MPP for shingle stripes with metallization grid optimized to each shingle concept.

In Figure 6 the power obtained from the C2M calculation is compared with experimental data. 1-cell equivalent power is considered for the calculations. An acceptable agreement between simulation and measurements can be stated. The low power experimentally obtained for group S 1/4 with overlap of 2.0 mm can be attributed to technical difficulties in the production line, such as misleading cell stripe alignment during shingle-string assembly, causing it an overlap higher than the targeted and/or provoking the ECA to be glued on non desirables areas. Other explanation for this deviation is linked to simulation assumptions and simplifications, meaning that more experiments need to be addresses before state a strong conclusion about functionality of the code and equations employed to run the simulation.

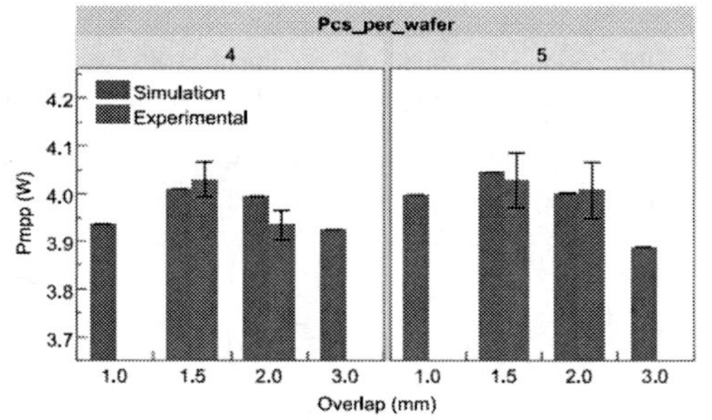

Figure 6: P_{MPP} of 1-cell equivalent shingled strings plotted vs. the overlap for the C2M simulation and experimental data.

4 FULL SIZE AREA MODULE EXTRAPOLATION

To complement the study, two different module optimizations are carried out: focus on module P_{MPP} (scenario 1) and module area (scenario 2). For both scenarios, the inner area is considered as the sum of the active area (area covered by solar cells) and the area of the space between cell to cell and string to string.

4.1 Module power (scenario 1)

In this scenario a module power of 300 W is targeted. The amount of cells per string (connected in series) and strings in parallel are calculated considering the module values of V_{MPP} and I_{MPP} at STC to be in the range of commercial PV modules with similar characteristics, meaning 42 V and 8 A. The amount of strings in parallel shall be proportional to the amount of shingle stripes obtained from the full size solar cell, i.e. if the solar cell is divided in 4 shingle stripes, 4 strings will be connected in parallel in order to obtain an output current equivalent to a traditional module interconnection. Both, electrical losses by cell interconnection and optical losses by shading due to overlapping, encapsulating material and glass, are considered.

In Figure 7 the results for the power optimization scenario are summarized, i.e., the module power achieved by using 4 or 5 *pcs* per wafer for different overlap. REF stands for a module made out of traditional front-to-back cell-to-cell interconnection. The module efficiency of shingled-cell modules is mainly driven by the overlap. The module power is always higher than 300 W due to the fact that the results for the amount of *pcs* per string and strings in parallel are rounded up to the next integer, thereby avoiding a portion of a stripe or string as a result.

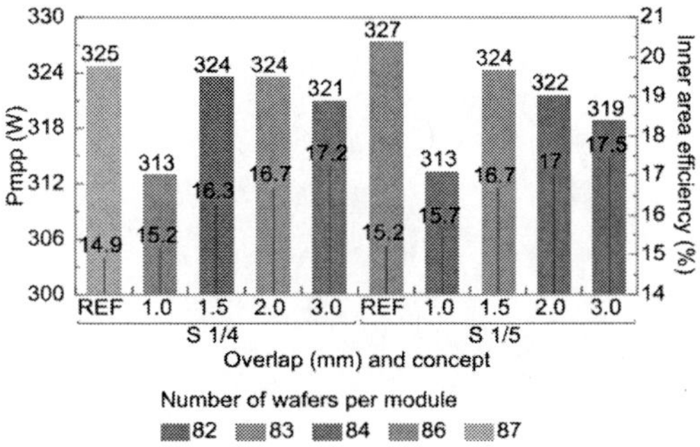

Figure 7: Module power (P_{MPP}) and inner area efficiency for power optimization plotted vs. overlap distance and the shingled concept. The color of the bars relates for each calculation to the amount of wafer needed per module to achieve this power.

4.2 Module area (scenario 2)

For module area optimization an equivalent 60-cell module is used as reference (REF). This module is considered to be made out of full size poly c-Si solar cells (as previously mentioned), and assembled employing the traditional cell interconnection (H pattern and front-to-back interconnection by soldered ribbons). Its measured P_{MPP} is 235 W at STC. The inner and outer module areas are 1.46 m² and 1.50 m² respectively. Therefore, its inner and total area efficiencies are 15.9% and 15.7%, respectively. Its packing density is 0.97.

As in the power optimization scenario, both, electrical losses by cell interconnection and optical losses by overlapping, encapsulating material and glass, are taken into account.

The results for the equivalent area optimization are shown in Figure 8. The inner area efficiency is always higher than for the reference module due to the higher packing density of the shingle concept. It's clear that a module power higher than the reference module (REF) is not possible to reach on 1.50 m² of module area but the efficiency can be up to 15%$_{rel}$ higher than for the REF by using 1/5 or 1/6 with 1 mm overlap (see Figure 9).

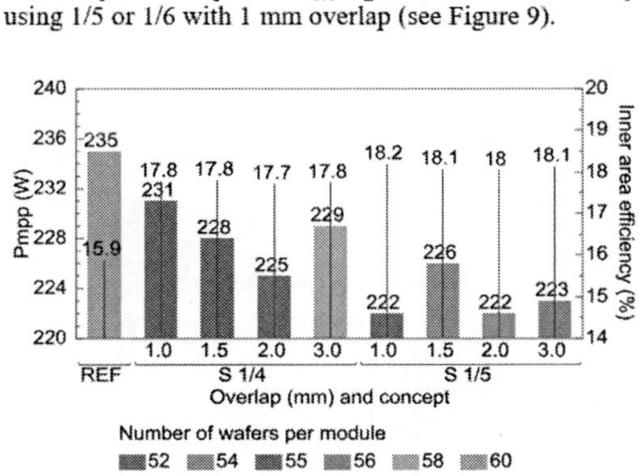

Figure 8: Module P_{MPP} for the different interconnection concepts. REF stands for the 60-cells module as previously described.

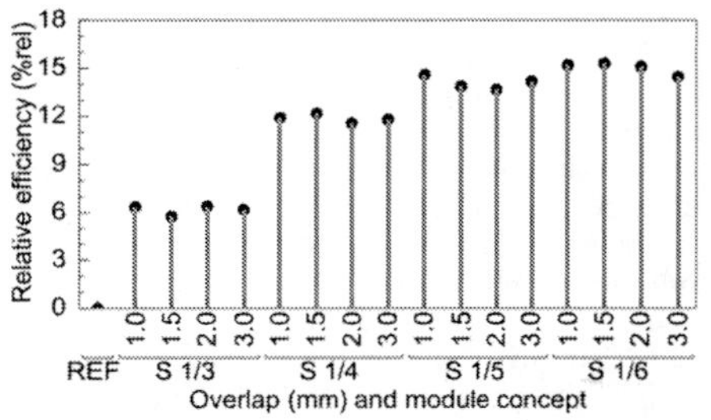

Figure 9: Relative difference in inner area efficiency for the different interconnection concepts regarding to the 60-cell module used as reference (REF).

5 CONCLUSIONS

In the present article, we have shown a comparison of the cell-to-stripe-to-module loss and gain between simulation and experimental results for shingled solar cells. The later shows that if a commercial mc-Si solar cell (6", full-square, Al-BSF, 3 busbar) is converted to a shingled-cell stripe, it exhibits cell-to-stripe loss up to approx. 13%$_{rel}$ in power. If the metallization grid is optimized for the shingle concept, this difference in cell-to-stripe power loss decreases to 6%$_{rel}$ and 4%$_{rel}$ for S 1/4 and S 1/5, respectively.

Results for C2M calculations show in the power optimization scenario, that a module P_{MPP} of 320 W is reachable using 83 shingled cells per string and 4 strings in parallel, thus 82 wafers. The inner area efficiency is lower than initially expected. This can be attributed to the fact of a lack in module architecture optimization. For the module area scenario, the higher efficiency regarding to REF is attributed to the compact area of the shingled-cell modules. To achieve higher power, solar cells with V_{MPP} higher than the reference of this work needs to be employed. Nevertheless, the simulation and experiment show the possibility to manufacture modules with improved efficiency as long as module materials of the right fitting are available in the market.

Although experimental and simulated power losses are very well in agreement, more experiments needs to be carried out to complement the current findings for other shingle concepts as 1/3 and 1/6, for several overlap distances.

6 ACKNOWLEDGEMENTS

This work was supported by the German Federal Ministry of Economic Affairs and Energy (BMWi) under contract number FKZ 03241098 (Schindel - Neuartige Verbindungskonzepte für Solarmodule)

7 REFERENCES

[1] W. Schmidt and K. D. Rasch, "New Interconnection Technology for Enhanced Module Efficiency", IEEE Transactions on Electron Devices 37, no. 2, 1990, pp. 355-357.

[2] J. Summhammer and Z. Halavani, "High-voltage PV-modules with crystalline silicone solar cells", 28th EU-PVSEC (2013), pp. 3119-3122.

[3] Z. Halavani, J. L. Lang and J. Summhammer, "Results of pressure-only cell interconnections in high voltage PV-Modules", 29th EU-PVSEC (2014), pp. 64-68.
[4] J. Summhammer and Z. Halavani, "Cell interconnection without gluing or soldering for crystalline Si photovoltaic modules", *EPJ Photovoltaics*, vol. 7, 2016.
[5] Ansgar Mette, "New concepts for front side metallization of industrial silicon solar cells", PhD thesis, Albert-Ludwigs-Universität, Freiburg in Breisgau, 2007.
[6] I. Haedrich, U. Eitner, M. Wiese, and H. Wirth, "Unified methodology for determining CTM ratios: Systematic prediction of module power", *Sol. Energy Mater. Sol. Cells*, vol. 131, no. 7, pp. 14–23, 2014.
[7] S. Guo, J. P. Singh, M. Peters, A. G. Aberle, and J. Wong, "Two-dimensional current flow in stringed PV cells and its influence on the cell-to-module resistive losses", *Sol. Energy*, vol. 130, pp. 224–231, 2016.
[8] D. Rudolph, et al. "Cell design optimization for shingled modules", 33rd EU-PVSEC (2017).
[9] Guy Beaucarne, "Materials challenge for shingled cells interconnection", *Energy Procedia*, vol. 98 pp. 115–124, 2016.

ANALYSIS OF PEEL AND SHEAR FORCES AFTER TEMPERATURE CYCLE TESTS FOR ELECTRICAL CONDUCTIVE ADHESIVES

Stephan Hoffmann, Torsten Geipel, Michel Meinert, Achim Kraft
Fraunhofer Institute for Solar Energy Systems ISE
Heidenhofstraße 2, 79110 Freiburg, Germany

ABSTRACT: Soldering of copper ribbons on busbars is the standard process for the series interconnection of solar cells in industrial PV modules. As an alternative interconnection approach electrical conductive adhesives (ECAs) can be used for cell interconnection. ECAs are lead free and can be processed at temperatures below 200°C. However, the higher costs and the lower adhesive forces between interconnector and solar cell compared to established solder-based interconnection prevented a larger market share until now. We present results of 90° peel-tests according to DIN EN 61189-2:2006 and shear tests according to DIN EN 1465 for seven commercially available ECAs and a comparison to the electrical performance of corresponding 3-cell modules in the temperature cycle test according to IEC 61215. The results show that all samples pass the temperature cycle test with a loss of power below 5%. The measured peel forces decrease after TC600 whereas the shear forces remain at their initial level.
Keywords: PV Module, Interconnection Technologies, Electrical Conductive Adhesives, Reliability

1 INTRODUCTION

In the PV-industry, the mechanical quality of the interconnection between a solar cell and a ribbon is usually determined by a peel test. The recommended peel force is 1 N per mm ribbon width [3]. As an alternative to soldering, electrically conductive adhesives (ECAs) can be used for interconnecting the solar cells. The main advantages are that ECAs are lead free and the curing temperature is typically in a range between 140°C and 180°C. This makes ECAs a natural choice for temperature- or stress-sensitive cell structures (e.g. HJT). The main disadvantages are the costs, because usually ECAs are filled with silver, and the low adhesive forces achieved in the peel test [1, 2]. In this work we compare seven different ECAs in terms of the achieved adhesive forces in case of a 90° peel test. In addition, we introduce a shear test based on testing standard DIN EN 1465 [4]. The motivation of the shear test is that – once laminated - no loads perpendicular to the interconnection joint appear in a PV module. The results of the peel and shear tests are compared with the electrical performance of three cell modules which were subjected to the temperature cycle test [5]. The temperature cycle test is chosen due to thermomechanical stress that results from different CTEs of the materials (silicon, copper) and which directly affect the interconnection joint. For our analysis, samples for peel and shear test are also subjected to the temperature cycle test and characterized.

2 EXPERIMENTAL

2.1 Sample preparation

For all samples, commercially available multicrystalline 6" full square 3BB solar cells with an Al-BSF are used. The ribbon is a silver covered copper ribbon (1.5 mm x 0.2 mm). In case of the soldered reference for the thermal cycle test, the copper ribbon is covered with Sn62Pb36Ag2 solder and Kester 952s is used as flux. All ribbons are stretched by 0.5% prior to gluing. The curing conditions of the seven different ECAs are listed in Table I. The laminates which were assigned to the temperature cycle test are build up with a glass and a layer of EVA at the front, followed by the cell matrix and a layer of EVA and a back sheet at the rear side. In case of the peel and shear test samples an additional layer of non-adhesive foil is added in order to remove them from the laminate after TC600.

Each ECA is cured according to the data sheet. The curing time and temperature is given in Table I.

Table I: Curing conditions of the ECAs used

ECA	temperature	curing time
A	160°C	3 min
B	180°C	0,5 min
C	180°C	10 min
D	180°C	5 min
E	180°C	10 min
F	180°C	10 min
G	180°C	10 min

The different sample designs are chosen depending on the characterization that is done and are shown in Figure 1. In case of the peel test samples, the ribbon is placed over the complete busbar. For each ECA, one cell is prepared. For the shear test, an ECA area with a length of about 10 mm and a width according to the busbar (1.5 mm) is made. For each ECA, six interconnection joints are made. The modules for IV measurement and electroluminescence characterization are made of a string with three interconnected cells. For each ECA, two electrically independent strings are made.

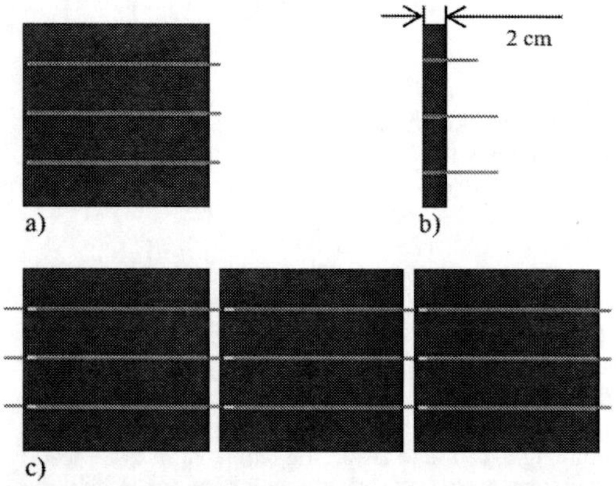

Figure 1: Schematic drawing of the samples. a) sample for peel tests, b) sample for shear test, c) sample for temperature cycle test

The curing of the ECAs and the soldering is done under reproducible conditions on a semi-automatic contact soldering station. The samples are subjected to the temperature cycle test following IEC 61215. The samples for the peel and shear test are used to analyze the forces for the initial state and after 600 temperature cycles.

2.2 Peel and shear test

In the PV industry the 90° peel test is an established method to characterize the mechanical properties of an interconnection joint. It is recommended to achieve a peel force of 1 N/mm joint width. A schematic drawing is shown in Figure 2. The sample is fixed on a substrate and the ribbon is pulled in a 90° angle.

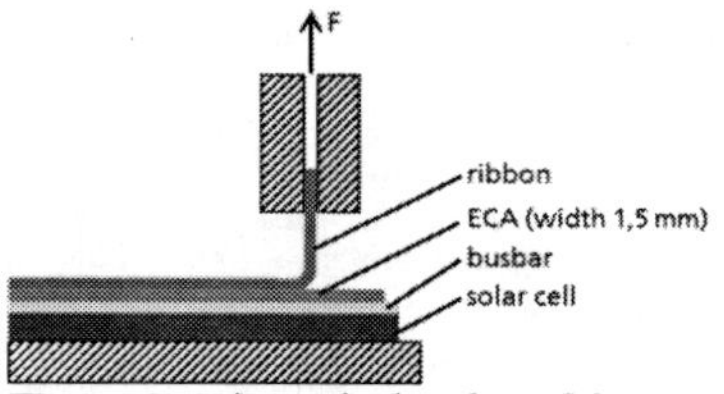

Figure 2: Schematic drawing of the 90° peel test

As the mechanical loading of cell and ribbon in laminated modules is expected to be dominated by shearing modes we introduce a 0° lap-shear test for the mechanical properties of the interconnection joint. Figure 3 shows a schematic drawing. The sample is fixed on a substrate and the ribbon is pulled in a 0° angle.

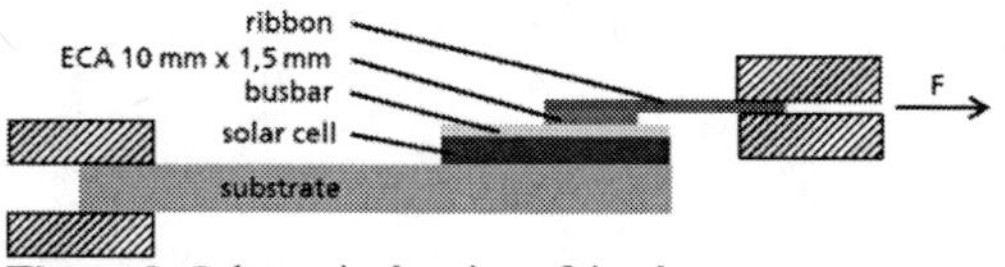

Figure 3: Schematic drawing of the shear test

3 RESULTS

3.1 Peel forces

Figure 4 shows the results for the peel test at initial state and after TC600. Only ECA D fulfills the criterion of 1 N/mm in the 90° peel-test at initial state.

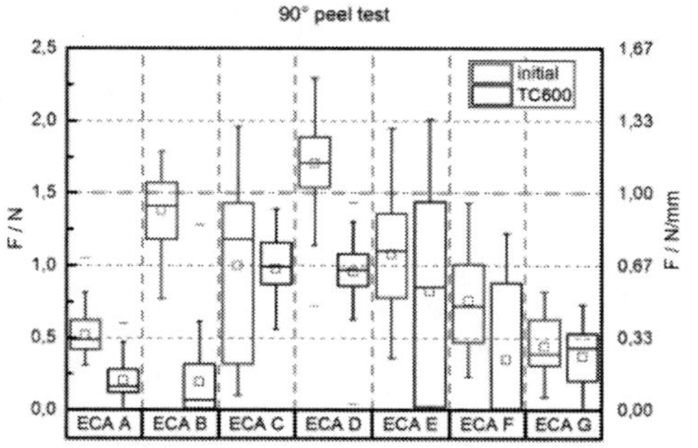

Figure 4: Results of the 90° peel forces in initial state and after TC600.

After the temperature cycle test, the measured peel forces drop for every ECA tested. In case of ECA E and ECA F the values show a wider spread. This is due to low peel forces at the edges of the interconnection joint and values in the middle that are at a comparable level of the initial state.

3.2 Shear forces

The result of the 0° shear test is shown in Figure 5. Besides ECA A, the results show a similar trend to the 90° peel test which is not necessarily expected because the different applied loading mode.

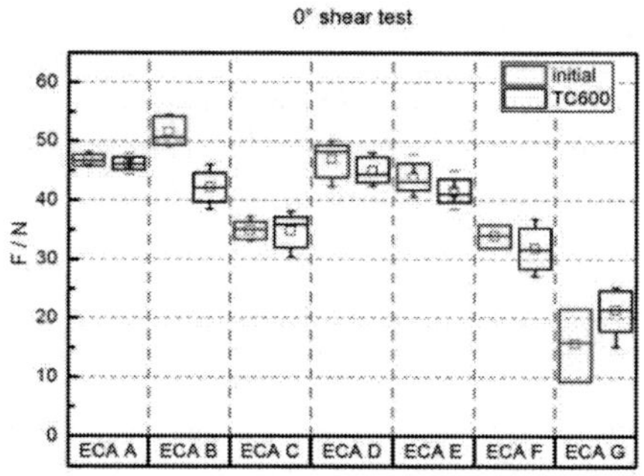

Figure 5: Results of the 0° shear forces in initial state and after TC600.

Besides ECA B, the measured shear forces after the temperature cycle test remain at their initial level.

A comparison of the measured adhesive force in case of the peel test and the shear test shows three adhesives which will be investigated in more detail: ECA A, which shows poor adhesive forces in terms of the peel test but performs well in terms of the shear test. ECA D, which performs well in both tests and ECA G, which shows the worst results of the ECAs tested.

3.3 IV measurement

The temperature cycle test according to IEC61215 is considered a highly relevant test for the reliability of interconnection joints as it addresses the thermomechanics of joints and module. Figure 6 shows the normalized results of the module efficiency. For each ECA, two independent cell strings are measured. All the samples pass 600 temperature cycles of the ongoing temperature cycle test. In this state, the measured adhesive forces in terms of the peel and shear test do not show an influence on the performance of the ECAs in a PV-module.

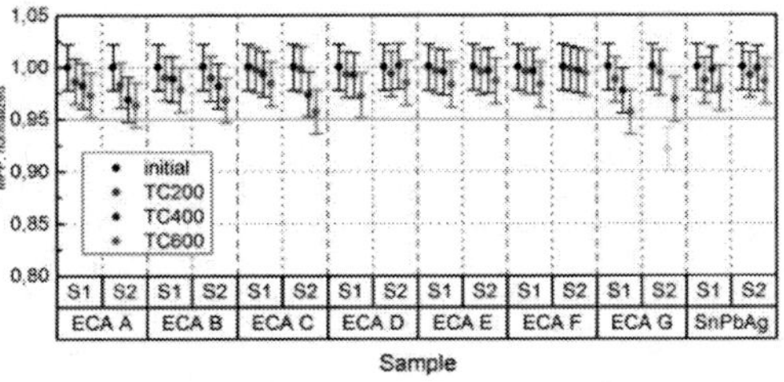

Figure 6: Normalized efficiency for each ECA and string. After one round of tempereture cycle test according to IEC61215, all samples have passed the test with less than 2% loss in efficiency. For each ECA and the soldered reference, two strings are tested (S1 and S2)

The IV-data show that the losses in efficiency result from FF-losses and are thus directly linked to the interconnection quality (**Fehler! Verweisquelle konnte nicht gefunden werden.**). The values for short circuit current and open circuit voltage remain stable.

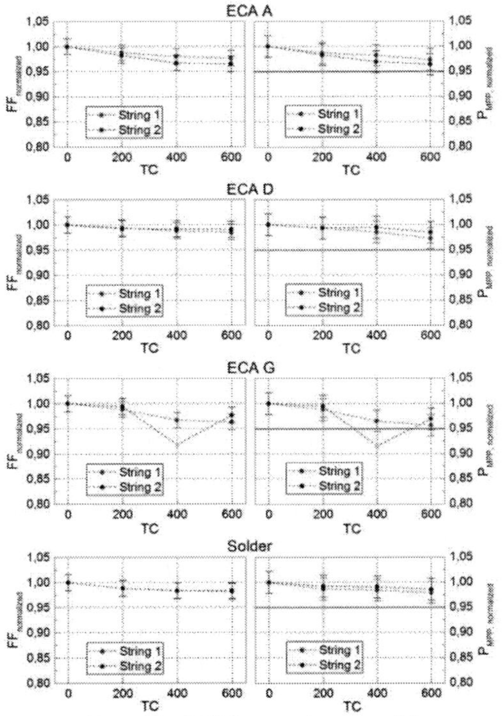

Figure 7: Normalized fill factor and power of three ECAs and the soldered reference.

3.4 Electroluminescence imaging

In addition to the IV-measurement, electroluminescence imaging is performed to get a better understanding of the occurred failure mode. Images are taken with 4 A forward bias current. After 600 temperature cycles only the module with ECA A shows suspicious areas.

Table II: Comparison of electroluminescence images of ECA A, D and G, as well as the soldered reference S.

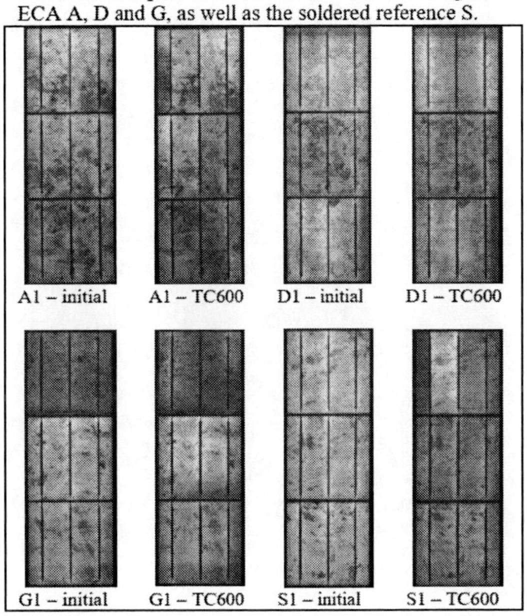

4 SUMMARY

We present results of a comparison between a 90° peel test and a 0° shear tests of solar cells interconnected with seven different commercial available ECAs. Six of the seven ECAs in the test show a correlation between the measured initial peel and shear forces. The comparison of the electrical performance of 3-cell modules after the temperature cycle test according to IEC 61215 shows that all modules pass 200 temperature cycles with a power loss < 2% and 600 cycles (3x standard) with a power loss < 5%, although in the initial state only one ECA passes the 1 N/mm criterion in terms of the 90° peel test. The results clearly prove that adhesive forces below 1 N/mm for solar cells interconnected by conductive adhesives do not necessarily result in a failure in the temperature cycle test. In fact, we show that the glued joints withstand extended reliability testing.

5 ACKNOWLEDGEMENTS

This project was funded by the German Ministry for Economic Affairs and Energy within the research project "KleVer" under contract number 0325833B.

6 REFERENCES

[1] T. Geipel et al., Reliable interconnection of the front side grid fingers using silver-reduced conductive adhesives", Energy Procedia, vol. 55, pp. 336-341, 2014

[2] T. Geipel and U. Eitner, Electrically conductive adhesives: An emerging interconnection technology for high-efficiency solar modules, Photovoltaics International, vol. 11, no. 21, pp. 27-33, 2013

[3] Solar cells - Datasheet information and product data for crystalline silicon solar cells, EN 50461, Beuth Verlag GmbH, 2007

[4] DIN EN 1465:2007-07

[5] IEC 61215-2 terrestrial photovoltaic (PV) modules – Design qualification and type approval – Part 2: test procedures

POLYMERIC MICROLENSES FOR PHOTOVOLTAIC MICROCONCENTRATOR APPLICATIONS: PROTOTYPE CHARACTERIZATION AND SIMULATION

Fausta Loffredo*, Fulvia Villani, Giuseppe Nenna, Riccardo Miscioscia, Carla Minarini, Franco Roca
ENEA Portici Research Center,
P.le Enrico Fermi 1, 80055 Portici (NA), Italy

ABSTRACT: Poly(methylmethacrylate)-based lenses with an area around $4*10^{-6}$ m^2 were manufactured on glass (covered with a hydrophobic layer) by single-drop casting process. The morphology of the polymeric optical structures were studied by interferometric and non-contact profilometric analyses and the effect of a double-step annealing treatment on the lens shape was investigated. The results confirm that the used method could represent a new route for a low-cost lenses production. Moreover, the simulation analysis demonstrated that these structures can be employed with common commercial trackers utilized with traditional modules. This study opens new perspectives in the field of solar micro concentration.
Keywords: lens, PMMA, microconcentrator, casting, polymer

1 INTRODUCTION

In order to improve the performances of photovoltaic devices the use of microconcentrator system could be a further advantage to enhance the efficiency thanks to the miniaturization of the active area and the consequent reduction of defects percentage in the fabrication of solar cells. In this study we investigated the possibility to manufacture by solution process lenses (millimeter size) useful for microconcentrator system (similar to that schematized in **Figure 1**) consisting of a lens coupled with a microcell (dimension of tens of micrometers). Concerning the concentrating optics component, in literature, several fabrication techniques have been proposed for the realization of differently shaped lenses characterized by low aberration and good uniformity of the focal length, like photolithography, reactive ion etching, laser assisted techniques, e-beam lithography, photopolymerization, printing techniques [2-6].

In this wok we developed a process based on a single-step casting to fabricate a lens prototype with diameter up to 2 mm. These optical structures of this dimension could be useful to produce an optical microconcentrator system including micro-solar cells having linear dimensions range from 50 to 200 μm. Indeed, in a solar micro-concentrator system the lens base area (A_{lens}) has to be in the right ratio with respect to the solar cell area ($A_{solar\ cell}$). In particular, in order to reach an effect comparable to that of the traditional concentrator systems A_{lens} / $A_{solar\ cell}$ has to be about 500 [7].

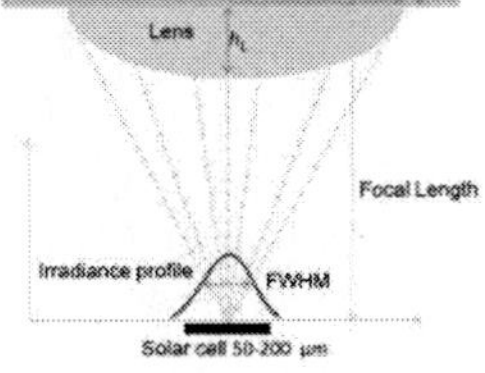

Figure 1. The schematized system of the micro-concentrator system investigated in these study.

Among the most commonly used polymers for the fabrication of lightweight optical components, PMMA was selected as low cost material, with high optical grade, good mechanical and thermal properties [5]. The polymer based solution was deposited onto glass substrates covered by high hydrophobic tetraethyl orthosilicate 1H,1H,2H,2H- perfluorodecyl-triethoxysilane (TEOS/PFTEOS) buffer layer [6]. After deposition, annealing processes were employed to optimize the lenses, both in terms of shape and optical quality. A complete optical and geometrical characterization of the manufactured lenses was carried out by interferometric and non-contact profilometric analyses to investigate the optical properties. Finally, in order to study the possibility to utilize the as-fabricated optical components in microconcentrator systems raytracing simulations of the manufactured lenses were performed through a dedicated software downstream of the interferometric analysis.

2 EXPERIMENTAL PART

2.1 Materials and method to fabricate the PMMA lenses

Commercial glasses (from Schott), employed as substrates, were cleaned by sonication through sequential steps in deionized water, acetone and isopropanol, followed by drying with nitrogen flow and finally in oven at 130 °C for 1 h.

The substrates were coated by a TEOS/PFTEOS film to give highly hydrophobic character to the glass surface where the optical microstructures have been successively fabricated [6].

A polymeric solution was prepared by dissolving PMMA (Mw 120000, Aldrich) in NMP (Aldrich) at 80 mg/mL concentration. Hemispherical lenses were manufactured by casting deposition of PMMA solution on the hydrophobic surface of the TEOS/PFTEOS film.

The droplet depositions (1.1 μL of single drop volume) were performed by Gilson pipette (2μL of total volume) in different points of the substrate in order to obtain an array of lenses (**Figure 2**). During the casting process and for the successive 6h, the substrate temperature was kept at 20 °C, then a double-step thermal treatment was carried out in air on hot-plate. In detail the PMMA structures were heated at 90 °C for 1h and subsequently at 150°C for 7 min.

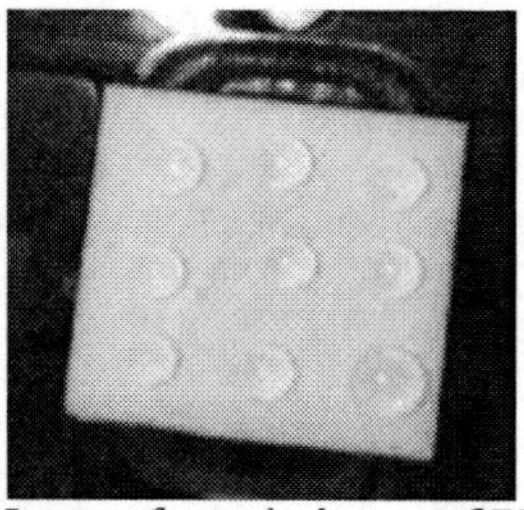

Figure 2. Image of a typical array of PMMA optical structures deposited at 17°C on top of a hydrophobic glass.

2.2 Characterization of the optical microstructures

The morphology and the profiler of the PMMA-microstructures were investigated after each thermal treatment by surface optical profiler (Taylor Hobson CCI4K) while the optical properties of the lenses were studied by a Mach-Zehnder interferometer in confocal configuration [5,6,8,9]. The interferometric images were recorded by means of a EOS Canon camera, removing the owner optical objective, and using directly the CCD camera array to capture the images.

3 RESULTS

3.1 Morphological characterization

Arrays of PMMA lens were deposited on glass (**Figure 2**) by casting (1.1µL per drop) of the polymer solution performed by a chemical pipette. In order to reduce as much as possible the splashing of the liquid during the deposition, before casting, the glass substrate was covered with a hydrophobic layer and the casting process was carried out at 17°C. Moreover, as already mentioned, in order to dry the drop and to improve the roughness and the spherical shape of obtained polymeric structures, the lenses were heated first at T= 90° C for 1h and then at 150° C for 7min.

The effect of these thermal treatments on the geometrical and optical properties of the PMMA-microstructures were investigated by an optical profilometer and a Mach-Zehnder interferometer, respectively. In particular, in Figures 3 we report the optical profiles of a typical PMMA lens obtained after thermal treatments at 90°C (top) and 150°C (bottom).

From **Figures 3** it is possible to see the shape of the realized lens and its structural imperfections are mainly due to not perfect centering of the maximum. This problem resulted even more critical when we used volume of solution bigger than 1.1µL. This suggests that we are in the limit condition of work. For the amount of solution deposited the effect of gravity on the structure is much more dominant and appears to be in strong competition with the forces due to the surface energy of the deposited drop.

The comparison between the profiles reported in **Figures 3**, measured after the first and the second thermal treatment, suggests that there is not significant widening by increasing the temperature up to 150°C. Indeed, the base diameter of the lens is about constant (about 2.3 mm). On the other hand, it is possible to observe an improvement of the lens shape due to re-distribution of PMMA as a consequence of heating at T=150°C (see **Figure 4** where the extrapolated error profile, calculated by subtracting the perfect spherical shape, is reported).

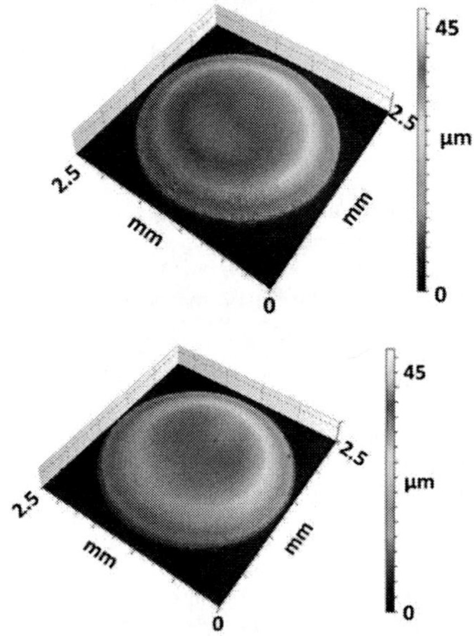

Figures 3: we report the optical profiles of a typical PMMA lens obtained after thermal treatments at 90°C (top) and 150°C (bottom).

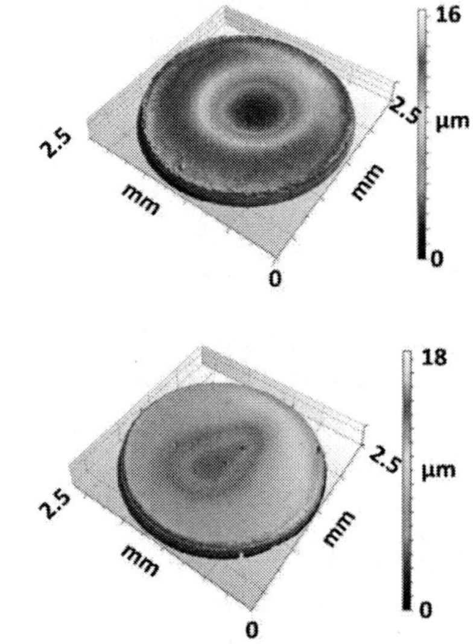

Figure 4: The error profiles of the fabricated optical structure related to the lens obtained after the first (top, T=90°C) and second (bottom, T=150°C) annealing.

The base diameter of the fabricated lens, measured by profiler analysis, was around 2.32 mm (A_{lens} = 4.22*10^{-6} m^2) useful to be coupled with a circular solar cell of ~ 100 µm of diameter ($A_{solar\ cell}$ = 7.85*10^{-9} m^2).

Considering the geometrical parameters of the double-step annealed lens (the best optical structure) we can estimate the focal length *f* of about 45 mm through formula 1 and 2 [10]:

$$f = \frac{ROC}{n-1} \text{ (1), with } \quad ROC = \frac{h_L}{2} + \frac{r^2}{2h_L} \text{ (2),}$$

where ROC is the radius of curvature, n is the refractive index of PMMA (approximately 1.49 at 543 nm), h_L and r are the height and the base radius of the microlens (in this case $h_L \approx 30$ µm and $r \approx 1150$ µm).

3.2 Optical characterization

The optical structures were analyzed by means of a Mach-Zehnder system [5,6,8,9].

In **Figure 5** are reported the optical micrographs in confocal position of two different PMMA structures annealed at 90°C (a,c) and at 150°C (b,d).

After the 1st thermal annealing (T=90°C)	After the 2nd thermal annealing (T=150°C)
a) 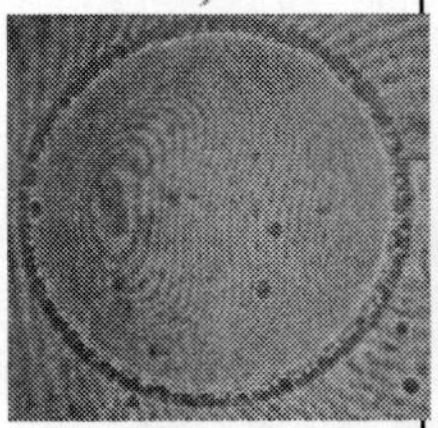	b) 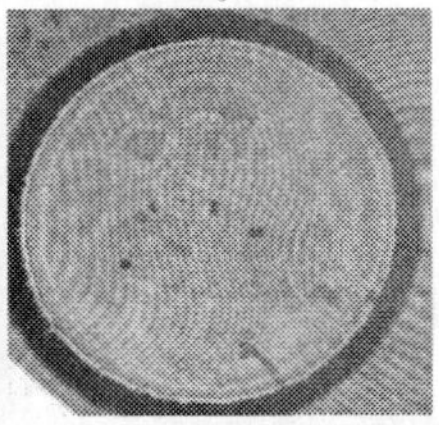
c) 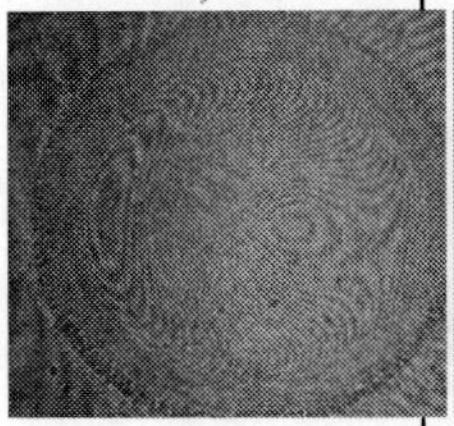	d) 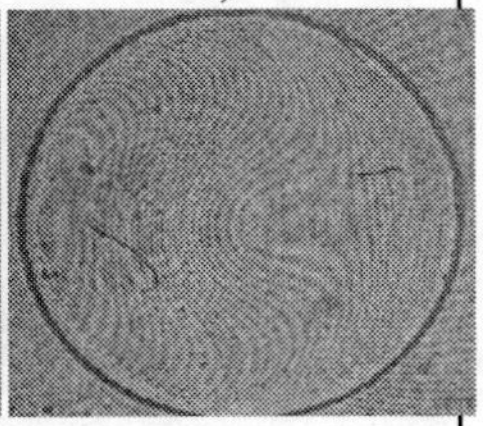

Figure 5: Interferometric fringes of two manufactured optical PMMA structures. For each structure the image was detected after heat treatment at 90°C (a,c) and 150°C (b,d).

In accordance with profilometric results, for both PMMA structures the final shape improved significantly by increasing the temperature up to 150°C (**Figure 5**). Indeed the lenses are better centered and the interference fringes are much straighter.

However, the realization of optimized lenses (**Figure 5a,5b**) is not always reproducible and many optical structures having appreciable misalignment of the maximum and an irregular distribution of the polymer are produced (**Figure 5c**). For these samples the higher temperature treatment only induces a toroidal distribution of the polymer at the edges of the structure (**Figure 5d**). Probably, the process may become more easily replicable and the toroidal distribution can be avoided by making confinement forms by means of photolithography process on a compatible polymeric film previously deposited on the substrate or through more sophisticated treatments [11].

3.2 Simulation results

The experimental results demonstrated that the employed casting-based process is well suited to manufacture lenses sized for solar micro-cells having linear dimensions range from 50 to 200 µm. Taking into account these geometrical parameters, we performed optical simulations of the system (**Figure 6**) by considering a target (micro-cell) with radius of 100 µm, fixed at the distance of 45 mm from the lens, and where the incident angles of light beam (with respect to the optical axis of the lens) are 0° and 0.1°. In **Figure 7** the images of the focused light spot when the source is perfectly aligned with the lens (top) or tilted with an angle of 0.1° (bottom) are reported. In the same figure the intensity profiles along orthogonal axes (blue line for x, green line for y) are shown, too.

The intensity profiles clearly highlight the peak shift in the two examined cases. In particular, if the source is tilted more than 0.1° the focused light spot moves out of the active area of the considered micro-cell (green circle). In this case, however, we are perfectly aligned with the possibilities which have commercial trackers [12].

The simulation results, obtained by varying the h_L of lenses from 38 µm to 24 µm, indicate that the focal lengths change from 30 mm up to 46 mm. These values agree with the focal length f of the manufactured lenses.

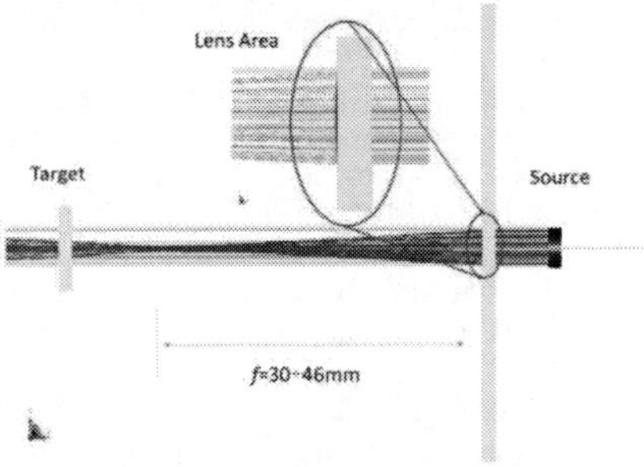

Figure 6: Ray-tracing simulation related to the realized optical structures.

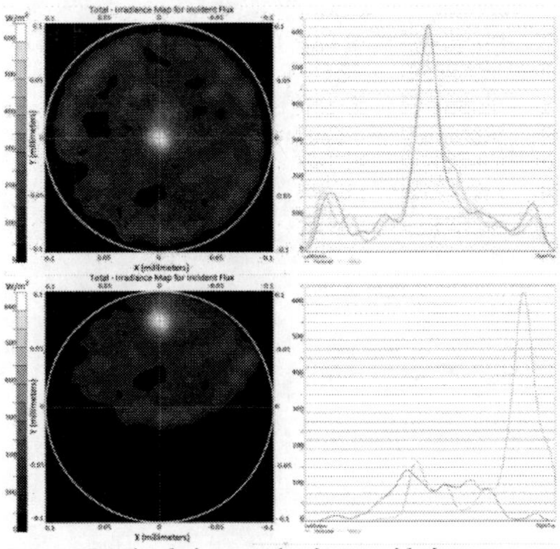

Figure 7: Simulation results by considering a target (micro-cell) where the incident angles of light beam are

0° (top) and 0.1° (bottom).

4 CONCLUSIONS

Large polymeric lenses were manufactured by a single-step solution-processable deposition method. In detail, a casting-based process for PMMA optical structures with a base area around $4x10^{-6}$ m^2 was developed. The lenses are demonstrated to have a spherical shape choosing the right deposition process parameters and post-processing annealing temperature (T=150°C). In particular, the temperature treatment improved the lenses morphology in terms of roughness and optical quality. The results confirm that the described method could represent a new route for low cost lenses production, opening new perspectives in the field of solar microconcentration.

5 ACKNOWLEDGMENT

This research is part of the EU CHEETAH project and has received funding from the European Union's Seventh Programme for research, technological development and demonstration under grant agreement No. 609788.

6 REFERENCES

[1] O. Fidaner, F.A. Suarez, M. Wiemer, V. A. Sabnis, T. Asano, Appl. Phys. Lett. 104, (2014) 103902.

[2] M. He, X.- C. Yuan, N. Q. Ngo, J. Bu, S. H. Tao, J. Opt. A: Pure Appl. Opt., 6 (2004) 94.

[3] J.- R. Ho, T.- K. Shih, J.- W. J. Cheng, C.-K. Sung, C.- F. Chen, Sens. and Act. A, 135 (2007) 465.

[4] L. Jacot-Descombes, V. J. Cadarso, A. Schleunitz, S. Grützner, J. J. Klein, J. Brugger, H. Schift, G. Grützner, Optics Express, 25 (2015) 25365.

[5] I. A. Grimaldi, S. Coppola, F. Loffredo, F. Villani, C. Minarini, V. Vespini, L. Miccio, S. Grilli, P. Ferraro, Optics Letters, 37 (2012) 2460.

[6] I. A. Grimaldi, S. Coppola, F. Loffredo, F. Villani, G. Nenna, C. Minarini, V. Vespini, L. Miccio, S. Grilli, P. Ferraro, Appl. Opt., 52 (2013) 7699.

[7] A. A. M. Sayigh, Energy and the Environment: Into the 1990s, proceedings of the 1st world renewable energy congress, UK, 23-28 Sept 1990.

[8] T. Miyashita, Jap. J. of Appl. Phys., 46 (2007) 5391.

[9] T. Mancebo, S. Bara, J. of Modern Opt., 45 (1998) 1029.

[10] Ph. Nussbaum, R. V.¨Olkel, H. P. Herzig, M. Eisner, S. Haselbeck, Pure Appl. Opt., 6 (1997) 617.

[11] Wei-Chih Chen, Tsung-Ju Wu, Wen-Jong Wu and Guo-Dung J Su, J. Micromech. Microeng. 23 (2013) 065008.

[12] F. Pascarella, G. Leanza, A. Citarella, "Dispositivo per la Misura dell'Angolo d'Incidenza dei Raggi Solari su una Superficie Piana Mobile, in Particolare per Eliostati", RM2003A000561, 1343703 (2003)

ANALYSIS FOR DIFFERENT MATERIALS USED AS UP CONVERTERS WHEN INCORPORATED IN BIFACIAL SILICON SOLAR CELLS USING THE PROGRAM PC1-D

Aline Cristiane Pan, Leandro Santos Grassi Cardoso, Fernando Soares dos Reis
Solar Energy Technology Nucleus (NT-Solar), Pontifical Catholic University of Rio Grande do Sul
Av. Ipiranga, 6681, Porto Alegre, 90619-900, RS, Brazil, e-mail: aline.pan@pucrs.br

ABSTRACT: A complete and detailed description of the physical processes occurring in the up conversion phenomena is very difficult to accomplish with complete accuracy. On the one hand, it would be necessary to know how it is distributed luminescent centers in different energy sublevels; and secondly, how does the propagation of light in the converter, and how to deal with this phenomenon. Considering these factors, the objective of this work is to obtain comparative results (I x V) for different materials used as up converters (UC) when incorporated in bifacial silicon solar cells using the unidimensional program PC1-D, in order to get the best candidates for use as UC. Up conversion phenomena consists generally of a non-absorbent matrix containing luminescent centers at a given concentration. The absorption and emission resulting from the luminescence were simplified considering the luminescent center as a series of three energy levels. With the mathematical approaches developed, was obtained tables with photon density per unit for all ranges of wavelengths, between 300 and 2100 nm. The simulation was performed assuming a bifacial silicon solar cell already experimentally characterized and contrasted with theoretical data. Forms of illuminations have been modified for this solar cell considering the information of UC, where changed the intensities constants (transmitted sunlight) and conversion efficiency. The simulations of I x V curves demonstrated that UC which absorb at wavelengths between 1800 and 2000 nm have a better potential in relation to others, the efficiencies found (~20 %) for solar cells with UCs implemented are shown below the limit values calculated (~36 %). However, approaches the values found in practice (~16%) and modeled.
Keywords: Bifacial, silicon solar cell, simulation, modelling.

1 INTRODUCTION

Actually meeting the ever increasing demand for energy has been one of the greatest challenges of modern times, estimate that in the last 10,000 years mankind has gone from producing 100 W of energy supplied from human labor to 1.5 GW produced by a steam turbine , an increase of approximately 15 million times, of this growth estimated that about 99% occurred in the 20th century. Only in the last 50 years oil consumption is equivalent to about 80% of all oil already consumed by humanity [1]. At the same time, contrasting with this reality, the main sources of energy used have been quite disputed. Both because of environmental issues and because some of these energy sources such as oil, natural gas and coal are finite resources.

In 2010, the International Energy Agency announced that if consumption were to remain the same as in 2006, coal would reach its peak of production by 2070, natural gas will peak around 2030 and fossil fuels by 2028. For nuclear fission, estimated that uranium will peak in the period between 2015 and 2035. It is due to this future shortage of non-renewable energy sources, and the growing concern about the preservation of the environment, without crippling the productivity of society that has been encouraging the search for new sources of renewable energy that do not cause degradation to the environment and are both efficient and cheap. In this context, solar photovoltaic energy presents itself as a good alternative because it is an inexhaustible source of energy, and because it does not degrade the environment. The investment in this technology itself attests to this fact. From 2010 to 2014 the solar capacity in the world went from 39 GW to 180 GW [2]. According to data from the European Photovoltaic Industry Association (EPIA) since 1998 average annual growth has remained above 35%.

Many of the recent efforts devoted to improve the efficiency of the solar cells are based on the study of the so-called "third generation" concepts [3], which aims at a better use of the solar spectrum to surpass the theoretical limits calculated by Shockley and Queisser [4], in the range of 30% efficiency for a conventional solar cell. One of these third-generation concepts is photon conversion. The photon converters try to take advantage of the photons with less energy than those of the semiconductor bandgap ($E < Eg$), or of those of greater energy ($E > Eg$), and are referred to as "up-converters" or "down-converters", respectively [5]. The up-converters can take advantage of the transmitted energy, while the down-converter reduces the energy lost through thermalization.

A complete and detailed description of the physical processes occurring in the up conversion phenomena is very difficult to accomplish with complete accuracy. On the one hand, it would be necessary to know how it is distributed luminescent centers in different energy sublevels; and secondly, how does the propagation of light in the converter, and how to deal with this phenomenon. Considering these factors, the objective of this work is to obtain comparative results (I x V) for different materials used as up converters (UC) when incorporated in bifacial silicon solar cells using the unidimensional program PC1-D, in order to get the best candidates for use as UC.

2 CHARACTERISTICS OF UP CONVERTER

In this work commercial UC and quantum dots (QD) by companies Phosphors Tecnology, Evident Tecnoloy and Maxmax were used. All analyzed UC were distributed as a powder and the PQ in liquid form. The method used to introduce them into silicon bifacial silicon solar cell was during the encapsulation to form the photovoltaic module, by means of a silicone gel [6].

Twenty-four samples were prepared with different concentrations of UC and QD on a silicone base for characterization optics. The two parts of the elastomer (3 ml of base + 0.3 ml) of curing agent were mixed according to the orientation different masses of UC and different volumes of PQ. A more detailed description of the preparation of samples of the PV module assembly can be found in Pan *et al.* [6]. In Tab. 1 we can see the chemical composition for each sample.

Table I: Chemical composition and measure of reflectance and transmittance at the wavelength of 1500 nm,

Sample	Concentracion (UC and QD)	Reflectance (%)	Transmitance (%)
1	3,3 mL silicone	5,6 ± 4,1	87,4 ± 1,9
2	UC 545/F 0,5g	63,6 ± 1,0	11,1 ± 4,2
3	UC 545/F 1,0g	70,5 ± 0,8	11,0 ± 1,3
4	UC 545/UF 0,5g	70,8 ± 1,0	4,7 ± 6,9
5	UC 545/UF 1,0g	83,0 ± 0,8	2,0 ± 3,7
6	UC 545/F 0,5g + 0,2mL QD	75,8 ± 0,9	7,8 ± 7,3
7	UC 545/F 1,0g + 0,2mL QD	67,6 ± 0,3	6,5 ± 6,7
8	UC 545/UF 0,5g + 0,1mL QD	60,2 ± 2,8	2,8 ± 16,6
9	UC 545/UF 1,0g + 0,1mL QD	8,6 ± 2,1	1,1 ± 1,9
14	UC 545/F 0,5g + 0,1mL QD	60,9 ± 3,5	14,6 ± 7,4
15	UC 545/F 1,0g + 0,1mL QD	68,3 ± 2,2	8,7 ± 3,2
16	UC 545/UF 0,5g + 0,2mL QD	73,2 ± 2,1	8,4 ± 2,5
17	UC 545/UF 1,0g + 0,2mL QD	80,9 ± 1,0	2,7 ± 6,2
18	UC 545/UF 1,0g + 0,3mL QD	81,8 ± 1,0	2,2 ± 4,1
19	UC 545/UF 1,5g + 0,2mL QD	83,58 ± 1,23	0,97 ± 2,60
20	UC 545/UF 1,5g + 0,3mL QD	83,1 ± 1,7	1,3 ± 6,1
21	UC 545/UF 0,5g + UC 545/F 0,5g + 0,1mL QD	77,0 ± 3,4	3,5 ± 0,1
22	UC 545/F 1,0g + 0,2mL QD	69,4 ± 0,7	7,3 ± 7,8
23	UC 545/F 1,0g +0,3mL QD	69,7 ± 0,7	5,4 ± 5,0
24	UC 545/F 1,5g + 0,2mL QD	73,9 ± 0,9	3,0 ± 5,9
25	UC 545/F 1,5g + 0,3mL QD	71,3 ± 0,9	2,4 ± 3,5
26	UC R 0,5g	60,7 ± 0,8	20,6 ± 5,2
27	UC G 0,5g	74,4 ± 0,8	8,4 ± 4,5
28	UC SPG 0,5g	43,9 ± 1,7	32,5 ± 3,4

The measurements of reflectance and transmittance were made using the LAMBDA 950 UV / Vis / NIR spectrometer, Perkin Elmer, for optical analysis of samples. The range from 300 nm to 2100 nm was used for all analyzes.

3 MATHEMATICAL MODEL

A complete and detailed description of the physical processes occurring in the converter is very difficult to accomplish with complete accuracy. On the one hand, it would be necessary to know how it is distributed luminescent centers in different energy sublevels; and secondly, how does the propagation of light in the converter, and how to deal with this phenomenon [7]. Considering these factors, this study sought to find greed in the density of photons of a solar cell when incorporated into UC.

To obtain the power density for each wavelength interval is calculated first the energy in this range by Eq. (1):

$$E = h\nu = \frac{hc}{\lambda} \qquad (1)$$

where h is Planck's constant, v is the frequency, c is the speed of light and wavelength λ.

For the calculation of the photon density (Φ) on the average wavelength interval (Δλ), Eq. (2), it was using the energy value already calculated and irradiance values (F) for the solar spectrum tabulated AM1 .5G [10]. With the unit adjustments result was obtained at n/cm²s:

$$\Phi = (\frac{F}{E}) * \left(\frac{\Delta\lambda}{2}\right) \qquad (2)$$

The next step was to calculate the power density in the same wavelength range, according to Eq. (3) W/cm²:

$$H = \Phi * E \qquad (3)$$

With Eq. (4) there was the power density constant for all wavelength intervals by the sum of all irradiances of photons arranged:

$$C_H = \sum H \qquad (4)$$

Considering the insertion of UC, was calculated again for the density of photons of wavelength ranges, but considering the radiation that is transmitted. That is, the optical transmittance was measured (T) for a bifacial silicon solar cell, and added to these values for Eq. (2), making it in Eq. (3):

$$\Phi = (\frac{F}{E}) * \left(\frac{\Delta\lambda}{2}\right) * T \qquad (5)$$

There was also again the calculation of power density for each wavelength range, Eq. (3), considering the UC. According to Eq. (6):

$$H_{UC} = \Phi_{UC} * E \qquad (6)$$

And finally, the constant of power density was obtained by the addition of UC according to Eq. (7):

$$C_{UC} = \sum H_{UC} \quad (7)$$

In the calculations can vary the conversion efficiency using a conversion factor of Eq. (7) and thus vary UC efficiencies. For all it considered that the emission of UC would be exactly half of its absorption. Because theoretically to occur UC phenomenon, it is necessary that two or more lower energy photons to recombine in a higher energy. So we used this relationship for this model, and it was considered UC of 1000 nm to 2200 nm. As, for example, for a UC who absorbs two photons of 1000 nm that will emit a photon of 500 nm to the sum of all densities from 1000 nm. This is just an initial simplification, it is known that the reality of each iones trivalent rare earth ions (TR^{3+}) and your host material will be another.

4 RESULTS AND DISCUSSION

With the mathematical approaches developed, was obtained tables with photon density per unit (W/cm^2) for all ranges of wavelengths, between 300 and 2100 nm, where those data were entered into the software which simulates dimensional solar cells [8]. The simulation was performed assuming a bifacial silicon solar cell already experimentally characterized and contrasted with theoretical data [9]. Forms of illuminations have been modified for this solar cell considering the information of UC, where changed the intensities constants (transmitted sunlight) and conversion efficiency. With these simulations were achieved important results, such as the current curves versus potential difference (I x V) for these cells when incorporated into the UC. For all the results the UC of 2000 nm, it is, that emits two photons in 1000 nm is what gives better responses when incorporated in the solar cell.

Fig. 1 shows a comparison of the curves I x V of UC of 2000 nm for different conversion efficiencies and for two sample with different concentration. Where maintaining the fixed power density of photons in 0.00208 W/cm^2, it is, considering that all solar radiation transmitted (non-absorbed) by the solar cell is absorbed and re-emitted by the UC again to the solar cell. This value was used considering the thermodynamic limits of a silicon solar cell [4], where 20.8 % of the solar radiation spectrum AM1,5G is transmitted. There is a variation of approximately 40 % on electrical current compared to concentrations of 25 and 100 %, and this response does not follow linearity. When the variation in the concentration of 25 % compared to 75 % is observed, senses a variation in current of approximately 32 %, and when compared to 50 % the difference is 22 %. To a 75 % concentration factor compared with 100 % and 50 % are obtained by this variation between 6 % and 10 %, respectively. Finally, comparing the concentrations of 100 %, and 50 % reaches 16 % of variation in electric current. In addition, it can be seen that the results obtained with the samples (*A4* and *A5*) produced are less than a 25% of the conversion.

In Fig. 2 used the irradiance constant calculated for each converter and different UC with the same concentracion (*A2, A4, A26, A27* and *A28*) for 2000 nm. Five different UCs were analyzed with the same concentration. And it is possible to note that only sample *A28* has a higher current. However, this increase is too small to be observed.

In order to analyze the results of the emitted radiation found for UC and that calculated for AM 1.5G [4] obtained the comparative graph which is shown in Fig. 1 (c).

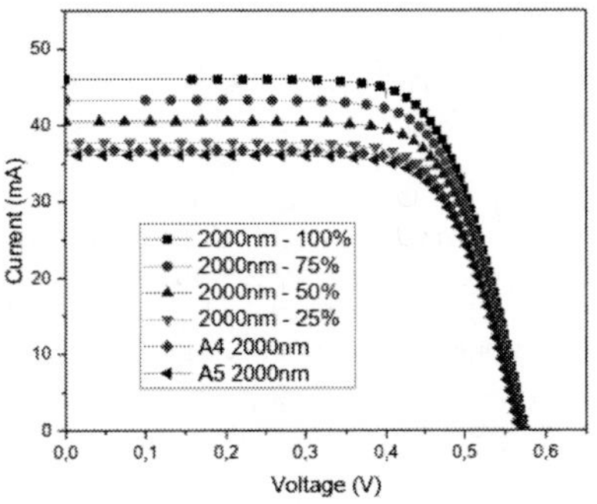

Figure 1: I x V curves obtained by PC1D using variations of concentrations of 2000 nm UC (theorical) and for two samples used (A4 and A5).

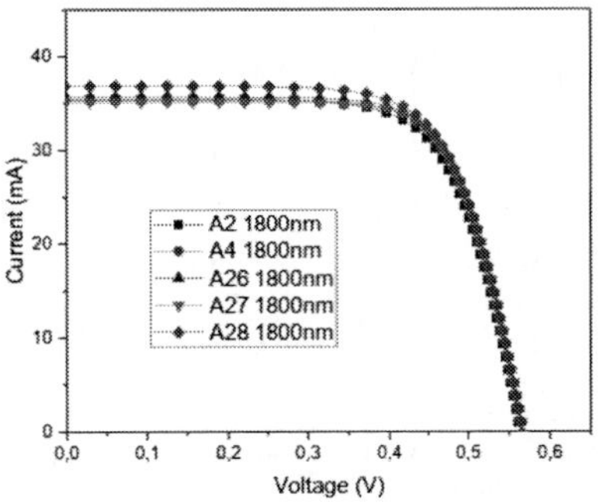

Figure 2: I x V curves obtained by PC1D using variations of constants of irradiation calculated for different UC with the same concentracion for 1800 nm.

In order to analyze the results of the emitted radiation found for UC and that calculated for AM 1.5G [4], considering 20.8% of the transmission radiation, obtained the comparative graph which is shown in Fig. 3.

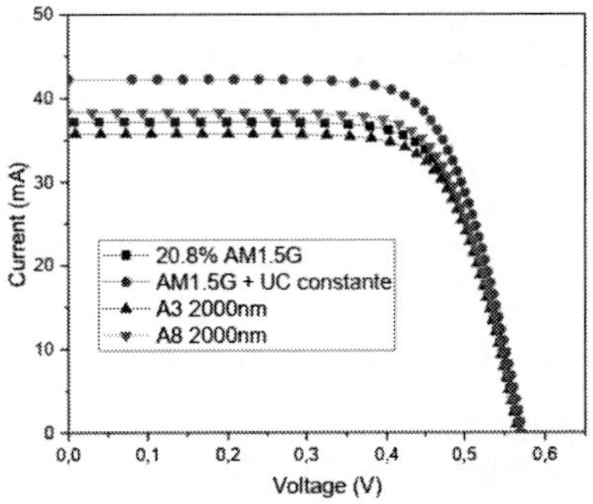

Figure 3: I x V curves obtained by PC1D using variations of irradiance for the 2000 nm UC-theorical, two samples and AM1.5G for 2000 nm.

It used to the UC for 1800 and 2000 nm for this analysis, as this always has the best efficiencies. It is evident that when it is assumed that all the solar radiation which was transmitted by the solar cell is absorbed by the UC (20.8 %) the electric current increases excessively, more than 50 % compared with when it is considered AM1.5G and the constant 0.0107 W/cm^2 concerning of own UC of 2000 nm, which would be a more realistic comparison. Considering the configuration in which it used to constant calculated for the converter and the AM 1.5G with constant of converter note to an electric current difference of 14 %.

The simulations of I x V curves demonstrated that UCs which absorb at wavelengths between 1800 and 2000 nm have a better potential in relation to others, the efficiencies found (~20 %) for solar cells with UCs implemented are shown below the limit values calculated [11] (~36 %). However, approaches the values found in practice (~16 %) [12] and modeled [13].

5 CONCLUSIONS

The absorption and emission resulting from the luminescence were simplified considering the luminescent center as a series of three energy levels. With the mathematical approaches developed, was obtained tables with photon density per unit for all ranges of wavelengths, between 300 and 2100 nm.

The simulation with PC1D was performed assuming a bifacial silicon solar cell already experimentally characterized and contrasted with theoretical data. Forms of illuminations have been modified for this solar cell considering the information of UC, where changed the intensities constants (transmitted sunlight) and conversion efficiency.

The simulations of I x V curves demonstrated that UC which absorb at wavelengths between 1800 and 2000 nm have a better potential in relation to others, the efficiencies found (~20%) for solar cells with UCs implemented are shown below the limit values calculated (~36 %). However, approaches the values found in practice (~16%) and modeled.

ACKNOWLEDGEMENTS

This work was performed as part of the Universal Project 2014, n° 458145/2014-9, of the National Council for Scientific and Technological Development (CNPq). In same way, by Research Program for Students of PUCRS (BPA/2017).

REFERENCES

[1] A. García-Olivares, J. Ballabrera-Poy. Tecnological Forecasting & Social Change, Vol. 90, (2015) 587-98.
[2] B. Cointe. Energy Research & Social Science, Vol. 8 (2015) 151-161.
[3] A. Martí, A. Luque, Next Generation Photovoltaics, Institute of Physics Publishing Bristol and Philadelphia, (2004).
[4] W. Shockley, H. Queisser, Journal of Applied Physics, Vol. 32-3 (1961) 510.
[5] B. Richards, A. Shalav. IEEE Transactions on Electron Devices. Vol 54-10 (2007) 2679.
[6] Pan, A. C. *et al.* Análise Óptica da Implementação de Conversores Fotônicos e Pontos Quânticos em Células Solares Bifaciais de Silício, V Congresso Brasileiro de Energia Solar, Recife (2014).
[7] Pan, A. C. Convertidores Fotónicos para Células Solares Bifaciales de Silicio, Tese de Doutorado, IES, UPM, Madri (2009).
[8] PC1D. Software for modelling a solar cell, Download in: http://www.engineering.unsw.edu.au/energy-engineering/pc1d-software-for-modelling-a-solar-cell (2008).
[9] A. C. Pan, C. Canizo, A. Luque. 22nd European Photovoltaic Solar Energy Conference and Exhibition, Milão (2007) 1438-1442.
[10] L. M. Fraas, L. D. Partain. Solar Cells and Their Applications, Wiley, 2ª edition (2010).
[11] T. Trupke, M. Green, P. Würfel. Journal Applied Physics, Vol. 92, n. 7 (2002) 4117-4122.
[12] S. Fisher *et al.* Solar Energy Materials and Solar Cells, Vol. 140 (2015) 217–223.
[13] J. Gordon *et al.*, 2005. Journal of the Optical Society of America B (2005) 1654–1659.

HARD AND TRANSPARENT DLC COATING AS A PROTECTIVE LAYER FOR SOLAR CELLS

Azzeddine DEHBI-ALAOUI*, Brahim SEQUAQ, Amine OBAIDI, Mustapha Ait IBBA
Laboratoire des Couches Minces et Traitement de Surface par Plasma.
Ecole Normale Supérieure, Université Sidi Mohamed Ben Abdellah,
FES, 30050, Morocco
(*) adehbi_alaoui@yahoo.fr, Tel 212662011434

ABSTRACT: Solar cells are subject to sand wind, they need a hard and transparent layer on the front surface. Diamond-like carbon films (DLC) coatings are known for their high hardness but they are not completely transparent in the visible. In this work we will show a method, using a Fast Atom Beam (FAB) source, in a special configuration that allowed us to deposit hard and transparent (clear) 100nm DLC films on glass substrates at room temperature and at pressure around 10^{-2} Pa. The coating is deposited from the reflected FAB beam from an Aluminum target which was not sputtered. The energy of the reflected FAB beam particles has been reduced due to collisions with the reflected particles from the substrates. Infrared spectroscopy showed that these films are not hydrogenated this is logical because only Argon was used to sputter carbon from the internal FAB source graphite walls and were accelerated out of the FAB source. Physical properties have been measured using different characterization techniques. The results showed that this coating is a tetrahedral (ta-C) type of DLC film and can be used as a protective layer for photovoltaique solar cells.

Keywords: Diamond Like carbon films (DLC), transparent and hard coating, solar cells protection, optical properties.

1 INTRODUCTION

Hard and transparent coatings are needed as protective films against harsh environment for solar cells. In general diamond like carbon (DLC) films are very hard , transparent in the infrared but in the visible they are not transparent ,clear like glass and their reflectance can be more than 20% even if the film is very thin (100nm). If the carbon source is a hydrocarbon gas it is hydrogenated and called amorphous hydrogenated carbon films (a-C:H) and if it produced from graphite without adding hydrogen to the process it is not hydrogenated and called amorphous carbon films (a-C) or tetrahedrique carbon films (ta-C). This late type has around 90% Sp^3 C-C bonding and is the closest type of DLC to Diamond. Many plasma based techniques have been used [1,2,3] to deposit DLC. In this work we are interested in FAB source [4] because it is able to produce DLC films at room temperature, however the transparency of DLC produced by FAB source in the visible is moderate (around 60% for very thin films and only 30% for thick films (more than 2 microns). In an earlier work [5] we have been able to improve FAB source transmittance by adding hydrogen to the hydrocarbon gas used (1% butane and 99% H_2) ; Transmittance increased to more than 70% but this was not high enough. It is known [6] that in order to increase transmittance and optical band gap of DLC films, we have to follow three rules: 1) increase the plasma degree of ionization, 2) reduce the carbon content in the plasma relatively to the gas used (H2, HnCm, or Argon) , 3) reduce the impact energy of particles reaching the samples. In this work we have used a new configuration in the FAB source method in order respect these three rules see Fig.1.

2 EXPERIMENTAL DETAILS

2.1 Approach used

The deposition technique used is FAB 110 source [7]. The extraction voltage can be chosen between 800V and 2000V. The maxim discharge current is 50mA. The vacuum chamber pressure is 10^{-4}Pa.

In our approach we did not use a hydrocarbon gas, instead, only Argon was used. When FAB source is in operation the internal walls which are made of graphite are sputtered and the beam that emerges from the source contain only argon and a small proportion of carbon particles. One way of reducing the beam particles energy is by reducing the beam extracting voltage. However below 800V the beam cannot be extracted from the source and the glow discharge inside the source stops.

In order to reduce the particles energy we have set an Aluminum Target at an angle of 45° in front of the beam. This target was not used to be sputtered but was put to receive the first beam high impact energy and reflect the particles to the samples holder with a lower impact. We have verified that no aluminum was sputtered from the target. This is due to the low voltage (800V) used for the FAB anode. The samples were set in front of the reflected beam from the target (Fig 1).

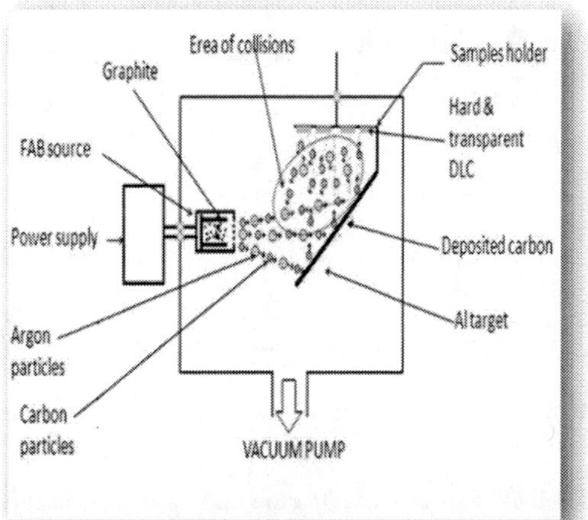

Figure 1: Configuration of the technique used to deposit hard and transparent DLC film

This configuration has created an area situated between the target and the samples holder, in which collisions occurs between the reflected particles from the target, the

FAB beam particles and the particles reflected from the samples which were situated at a distance of 5cm from the target. These collisions contributed to a decrease in impact energy of particles reaching the samples. Therefore we have created a situation where carbon content in the beam is low, degree of ionization is high inside the FAB source and the impact energy is low.

2.2 Measurements

The thickness was measured using the AFM technique. Single-beam digital Fourier transmission spectroscopy and was used for the optical evaluations. The optical band gap was assessed from visible wavelength absorption spectra, using the Tauc [8] relationship, and the refractive index from transmission and reflection spectra using PUMA software [9] using Swanepoel method [10]. Raman Spectroscopy was used to identify the type of DLC obtained. Vickers micro hardness was used to see the effect of the DLC obtained on the total hardness of the coated glass substrate.

3 RESULTS AND DISCUSSION

100 nm thick DLC films were deposited in one hour. This new configuration has allowed us to deposit for the first time a completely transparent FAB source film. We tried to increase this thickness by repeated coatings but, for reasons which are not known to us, further growth did not occur.

3.1. Optical evaluations

3.1.1. Visible regime

Measurement of transmittance and reflectance spectra in the visible regime from the obtained carbon films on glass are shown in Figure 2. The average reflectance is around 7% and the transmittance is more than 90% from 600nm to 900nm.

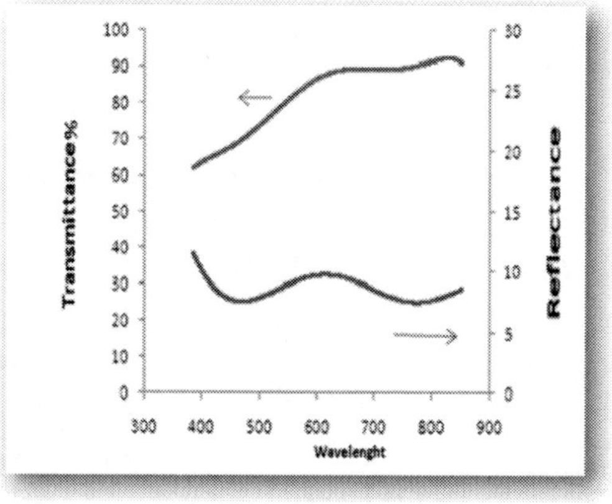

Figure 2: Transmittance and reflectance of the obtained DLC films

From these spectra the refractive index and the extinction coefficient are calculated using PUMA software [9] and are shown in Figure 3. These values are typical of reported DLC films optical constants [11]; n and k decreases with wave length from 2.2 to 1.6 and from 0 to 0.1, respectively.

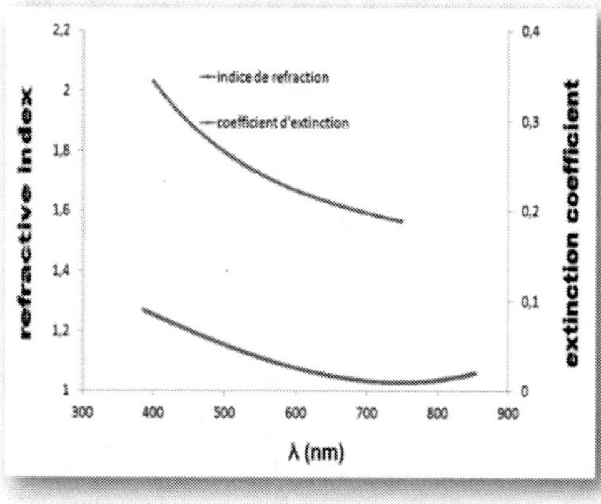

Figure 3: Refractive index and extinction coefficient vs wave length calculated using PUMA software .

The band gap of the prepared films was estimated from the plot of the squared $(\alpha h\nu)^2$ versus photon energy ($h\nu$). (Fig4) where α is the absorption coefficient, by extrapolating the straight line of the plot to intercept the photon energy axis at $h\nu = 0$ eV. Eg is found equal to 2.4 eV for these transparent films. This value is higher than the Eg obtained in normal direct FAB deposition using hydrocarbon gases where the optical band gap is close to 1eV [12].

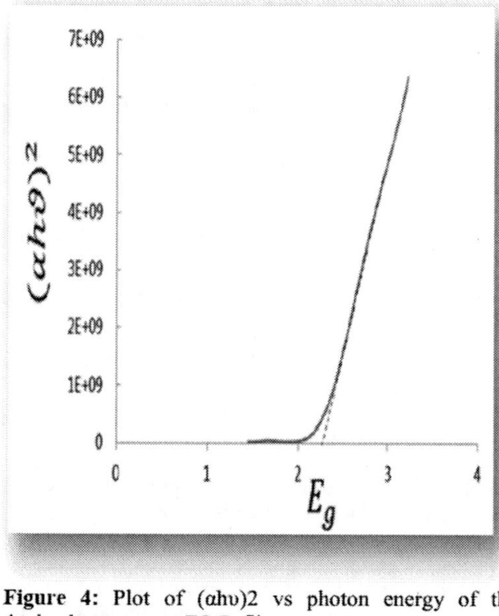

Figure 4: Plot of (αhυ)2 vs photon energy of the obtained transparent DLC film

3.1.2 Infrared Regime

The spectrum in Infrared regime is shown in Figure 5. No C-H bands are present around 3000cm-1 which confirms that the film is not hydrogenated.

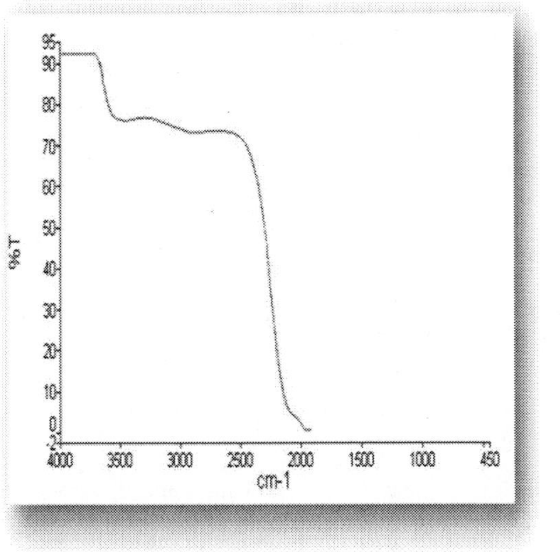

Figure 5: Infrared transmittance of transparent DLC film deposited on glass

3.1.3 Raman spectroscopy

Bonding state of carbon in DLC coating was investigated by Raman spectroscopy. Raman spectrum in Fig.6 shows that the D band at 1282 cm-1 is small relatively to the G band at 1541. This is a typical spectrum of ta-C films [13]

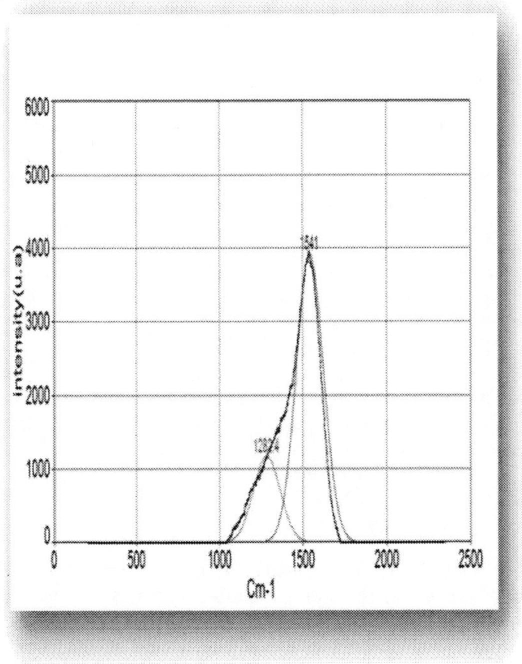

Figure 6: Raman spectra of the obtained transparent DLC film

4. Hardness:

Vickers micro hardness at 25g load of non coated and coated glass substrate with the obtained DLC is found equal respectively to 887Hk And 1017Hk.

4 CONCLUSION

For the first time hard and transparent carbon film has been deposited at room temperature using FAB source. The results showed that reducing the impact energy and the FAB beam carbon content contributed to a deposition of a hard and transparent DLC on glass substrates. Infrared and Raman spectra indicates that is a a non hydrogenated ta-C type of DLC. We think that it can be used as a protective layer for PV and CPS cells and for many other applications whenever hardness and transparence are needed. The low energy used will not harm semi-conductor cells and can be deposited on soft polymeric cells. In the next phase this coating will be deposited on PV cells and its performance will be studied.

4.1 References

[1] A.Dehbi-Alaoui, Diamond and Related Materials, (1992) (445-449)
[2] J. C. Angus, P. Koidi and S. Domitz, in J. Mort and F. Jansen (eds.), *Plasma Deposited Thin Films,* CRC Press, Boca Raton, FL, 1986, p. 89.
[3] Grill, A. Thin Solid Films 355–356, 189–193 (1999)
[4] Franks, J., *Vacuum, 34 (1984) 259-61.*
[5] A. Dehbi Alaoui, B. Ollivier and A. Matthews, Le Vide les Couches Minces. Supplement N°261, 221 (March/April 1992
[6] A. Dehbi-Alaoui, PhD Theses, Hull University, 1994.UK.
[7] A. A. Voevodina, J. M. Schneider, C. Caperaa, StevensonC & A. Matthewsc *Surface and Coatings Technology,* 47(1991) 327—335.
[8] J. Tauc, Optical Properties of Non-crystalline Solids, in F. Abels (ed), Optical Properties of Solids,North-Holland, Amsterdam, 1972, pp. 277—313.
[9] Ernesto G. Birgin, et al, Applied Numerical Mathematics 47 (2003) 109–119.
[10] R Swanepoel J.Phys.E : Sc.Instrum., Vol. 17, 1984, (896-903);
[11] Y. Bounouh and M. L. Theye (Paris) A. Dehbi-Alaoui and A. Matthews(Hull) , J. P. Stoquert (Strasbourg Physical Review B.Avril 1995, Volume 51, Number15/9597(9).
[12] A. Dehbi-Alaoui, and A. Matthews, Surface Coating & Technology, 47 (1991) 722 -729.
[13] J.Robertson, Materials Science and Engineering R 37 2002 (129-281).

STRESS-FREE FABRICATION OF PHOTOVOLTAIC MODULES USING ROOM TEMPERATURE INTERCONNECTION

Hae Wook Chung[a],Hyung-Jun Song[b], Jeong In Lee[b], Eun-Hye Park[a], Dong-Youn Shin[a,*]

[a)] Department of Graphic Arts Engineering, Pukyong National University,
365, Sinseon-ro, Nam-gu, Busan, 48547, Republic of Korea
[b)] Photovoltaic Laboratory, Korea Institute of Energy Research,
Daejeon, 34101, Republic of Korea

[*] Corresponding Author: Dong-Youn Shin
Phone: +82 (0)51 629 6384, E-mail: dongyoun.shin@gmail.com

ABSTRACT: This study demonstrates a novel route to fabricate photovoltaic modules using a highly conductive liquid metal. To validate the reliability of the proposed interconnection method using a liquid metal, we compares the efficiencies of the unit modules, which were fabricated using the conventional soldering and the proposed interconnection methods, after 500 hours of thermal cycling. It is found that there is no significant difference in efficiency, 17.69% for the best unit module interconnected using the conventional soldering method and 17.66% for the other best unit module interconnected using a liquid metal. Since the proposed interconnection method is free of thermomechanical stress, it is able to cope with thin solar cells, which otherwise are apt to be broken in the interconnection process of photovoltaic modules.
Keywords: Silicon Solar Cell, Ribbons, Module Manufacturing, Thermomechanical Stress

1 INTRODUCTION

In order to further dwindle the materials cost of crystalline silicon solar cells, the as-cut thickness of a solar cell wafer is expected to be reduced from 180 μm as of now to 140 μm by the year of 2021, which results in the more production of solar cell wafers from a silicon ingot [1].

However, as the thickness of a solar cell wafer reduces, it becomes far more vulnerable to mechanical and thermal stresses. When the crystalline silicon solar cells are interconnected with the metal ribbons using a conventional soldering method at a temperature of 200°C and above, the mechanical stress induced by the thermal mismatch between the solar cell wafer and the metal ribbon is strong enough to bend a thin crystalline silicon solar cell [2]. The thermal shock in the mid of soldering could also lead to the breakage of a thin crystalline silicon solar cell by developing micro-cracks inside a solar cell wafer. Therefore, an alternative to the conventional soldering method has been sought for [3].

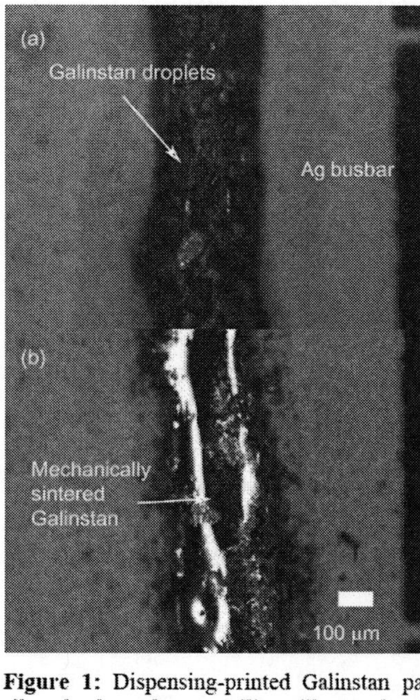

Figure 1: Dispensing-printed Galinstan paste on the silver busbar of a crystalline silicon solar cell, (a) as-printed and (b) mechanically sintered.

2 APPROACH

Monocrystalline silicon solar cells with three front busbars were used for experiments. As a conductive interconnection material, a liquid metal alloy, *i.e.*, Galinstan was chosen. It was mixed with the in-house developed carrier vehicle and then loaded in a syringe. After manually syringe-dispensing Galinstan paste on the busbars of a solar cell, it was dried on a hot plate for a while to remove the solvent ingredients of Galinstan paste. Then, metal ribbons were placed on them and temporarily bonded using a UV curable adhesive. This temporally bonded solar cell was placed between two EVA sheets, on the top and the bottom of which glass and white back sheet were subsequently placed. The fabrication of this sandwiched photovoltaic module was completed in a vacuum-laminator.

3 RESULTS AND DISCUSSIONS

Galinstan is a non-toxic alloy of Ga (68.5 wt%), In (21.5 wt%) and Sn (10%), which takes a liquid form at room temperature [4]. Albeit its electrical conductivity is as high as 3.46×10^6 S/m [5], its high surface tension of 534.6 mN/m renders Galinstan difficult to be harnessed for practical uses [4]. We have succeeded in harnessing Galinstan by embedding it in a liquid carrier vehicle with a mild yield stress. The characteristic Bingham pseudoplastic behaviour of the carrier vehicle rendered Galinstan droplets stably dispersed and printable using a syringe dispenser, as shown in Figure 1(a).

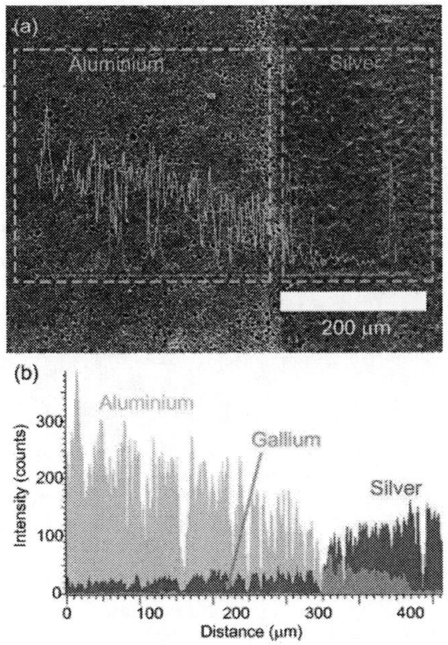

Figure 2: (a) SEM image and (b) EDS line-scan profiles of Al, Ag and Ga on the aluminium and silver electrodes.

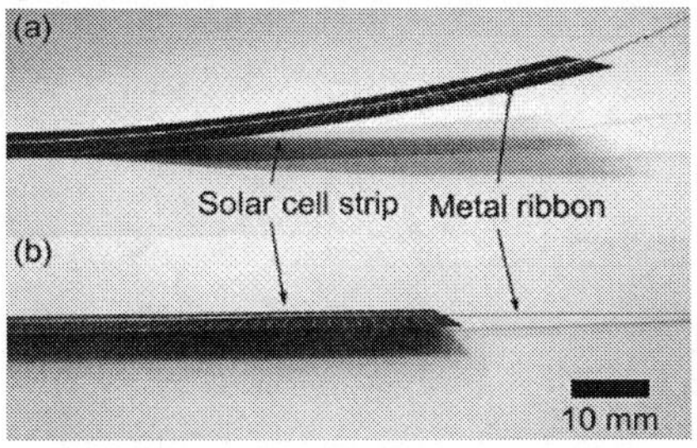

Figure 3: Comparison of bending, (a) conventional soldering method and (b) the proposed interconnection method using a liquid metal.

Although the Galinstan paste is non-conductive at this as-printed state, the application of a mild mechanical pressure turns it into a highly conductive state by squashing and merging Galinstan droplets, as shown in Figure 1(b) [6]. This unique trait of Galinstan paste even enables the interconnection of crystalline silicon solar cells with the metal ribbons at room temperature, which eventually liberates photovoltaic modules from the thermomechanical stress in the fabrication process.

In order to check the adverse influence of Galinstan on the aluminium and silver electrodes of a crystalline silicon solar cell, Galinstan paste was applied on the centre of the silver busbar and then placed on a hot plate at 100°C for 200 hours. As can be seen in Figure 2, Gallium was hardly detected on the aluminium electrode, proving that Galinstan does not diffuse across the silver busbar and corrode the aluminium electrode.

Solar cell strips were prepared using the conventional soldering and the proposed interconnection methods for the comparison purpose. Figure 3 clearly shows that the proposed interconnection method using a liquid metal does not exhibit bending at all since it is completely free of thermomechanical stress.

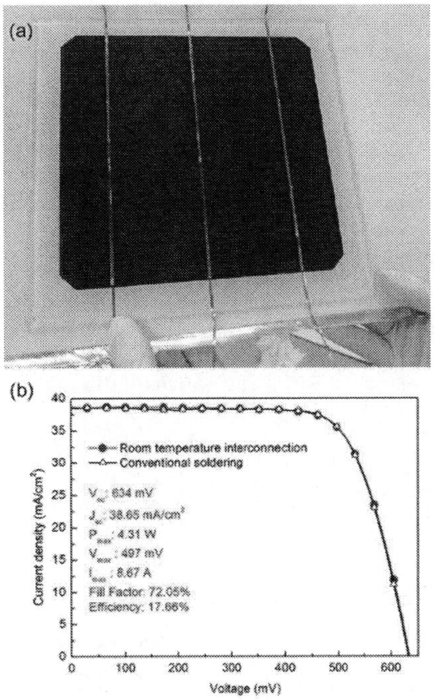

Figure 4: (a) Unit module interconnected using a liquid metal and (b) comparison of module efficiencies after thermal cycling tests for 500 hours.

Unit modules were also fabricated using the conventional soldering and the proposed interconnection methods, as shown in Figure 4(a), and their module efficiencies were measured after 500 hours of thermal cycling. As shown in Figure 4(b), the best unit modules interconnected using each method did not show any noticeable difference in efficiency, 17.66% for the best unit module interconnected using a liquid module and 17.69% for the best unit module interconnected using the conventional soldering method. However, it was also found that the liquid state reaction of Galinstan and the silver busbar increased series resistance of the unit module. Preventing the formation of a resistive intermetallic compound is left as future works.

4 CONCLUSIONS

In summary, we have introduced a new interconnection method that can fabricate silicon photovoltaic modules at room temperature using a liquid metal. The difficulty of a liquid metal in printing on a substrate was successfully resolved by incorporating it in a carrier vehicle with a mild yield stress. The best unit module fabricated using the developed Galinstan paste showed no noticeable difference in conversion efficiency, compared to the other best one fabricated using the conventional soldering method after 500 hours of thermal cycling. Although proper measures to prevent the formation of a resistive intermetallic compound through the liquid state reaction between the silver busbar and Galinstan have to be taken, the intermediate experimental results presented in this study suggest this novel interconnection method could be an alternative to conventional interconnection methods for the fabrication of photovoltaic modules with thin silicon wafers.

5 ACKNOWLEDGEMENTS

This work was supported by the Basic Science Research Program through the National Research Foundation of Korea (NRF), which is funded by the Ministry of Education, Science and Technology (Grant No. 2015R1D1A3A01016204), and the New & Renewable Energy Technology Development Program of the Korea Institute of Energy Technology Evaluation and Planning (KETEP) funded by the Korean Ministry of Knowledge Economy (Grant No. 20163020010890).

6 REFERENCES

[1] SEMI PV Group, International Technology Roadmap for Photovoltaic (ITRPV.net), Results 2016, 2017. <http://www.itrpv.net>

[2] L.C. Rendler, J. Walter, T. Geipel, M. Volk, C. Ebert, U. Eitner, Modelling and Verification of Mechanical Stress Induced By Soldering of Wires for Multi Busbar Interconnection, EU PVSEC, Hamburg, Germany, 2015.

[3] M.T. Zarmai, N.N. Ekere, C.F. Oduoza, E.H. Amalu, A review of interconnection technologies for improved crystalline silicon solar cell photovoltaic module assembly, Appl. Energy 154 (2015) 173-182.

[4] T. Liu, P. Sen, C.J. Kim, Characterization of Nontoxic Liquid-Metal Alloy Galinstan for Applications in Microdevices, J. Microelectromech. Syst. 21 (2) (2012) 443-450.

[5] A. Fassler, C. Majidi, Liquid-Phase Metal Inclusions for a Conductive Polymer Composite, Adv. Mater. 27 (2015) 1928-1932.

[6] J.W. Boley, E.L. White, R.K. Kramer, Mechanically Sintered Gallium-Indium Nanoparticles, Adv. Mater. 27 (2015) 2355-2360.

RECORD-BREAKING EFFICIENCY BACK-CONTACT HETEROJUNCTION CRYSTALLINE SI SOLAR CELL AND MODULE

Kenji Yamamoto, Daisuke Adachi, Kunta Yoshikawa, Wataru Yoshida, Toru Irie, Katsunori Konishi, Takahisa Fujimoto, Hayato Kawasaki, Masanori Kanematsu, Hirotaka Ishibashi, Toshihiko Uto, Yuji Takahashi, Toru Terashita, Gensuke Koizumi, Naoaki Nakanishi, Masashi Yoshimi
KANEKA Corporation, Settsu, Osaka 566-0072, Japan

ABSTRACT: Record-breaking efficiencies of crystalline Si (c-Si) solar cell and module using a back-contact heterojunction (BCHJ) technology are reported. Importance of a surface passivation technology for c-Si wafer by amorphous Si layer on improvements of conversion efficiency is demonstrated by both-side-contacted HJ c-Si solar cells. Utilizing our surface passivation technology of c-Si by amorphous Si and metallization technology, we have so far achieved a conversion efficiency of 26.7% in BCHJ c-Si solar cell and 24.4% in module. Energy loss analysis revealed a 27%-efficiency is reachable using our technologies. These results show the advantage of the BCHJ technology and provide a path to access new high-efficiency photovoltaic market for the c-Si solar cell.
Keywords: Heterojunction, Silicon Solar Cell, High-Efficiency, Reliability

1 INTRODUCTION

In the last years, several high conversion efficiency c-Si solar cells were reported. There are mainly two technologies for these progresses; passivated contact technology including heterojunction (HJ) technology and interdigitated back contact (BC) technology [1]. Especially, the HJ c-Si solar cell with amorphous (a-Si) passivation layers on c-Si wafers have been recognized as one of the most important technologies to improve the power generation per unit area of c-Si solar cell, due to their high conversion efficiency and small power temperature coefficient [2]. The BC structure is known as a most ideal structure for high efficiency solar cell by means of reduced optical loss due to the structure with electrode formed only on rear side [3]. A combination of the HJ technology and the BC technology is promising as it was reported in 2014 that a record-breaking conversion efficiency of 25.6% was achieved by a back contact HJ (BCHJ) c-Si solar cell [4].

We have so far focused our attentions to the HJ technology [5]–[10], and we reported a high efficiency both-side-contacted HJ c-Si solar cell with a record cell conversion efficiency of 25.1% in 2015 [9, 10]. For the back-contact technology, we reported a 23%-efficiency c-Si solar cell using doped a-Si thin film as a precursor for the doping of Si wafer in 2011 [11] and a ~25%-efficiency 6-inch BCHJ c-Si solar cell in 2016 [12].

In this paper, we report our record-breaking efficiencies of c-Si solar cell (26.7%) and module (24.4%). Also, we report a result of energy loss analysis of the high efficiency BCHJ c-Si solar cell and briefly discuss the path for a cell conversion efficiency of 27%.

2 BASE TECHNOLOGIES OF BCHJ C-SI SOLAR CELL

2.1 Advantage of thin-film Si solar cell technology

As the surface of the c-Si wafer of the BCHJ c-Si solar cell is covered by a-Si layers [4], the performance of the BCHJ c-Si solar cell should be changed by the properties of the a-Si layers, such as optical property (Isc), electrical resistivity (FF, series resistance), and passivation property (Voc and FF). Therefore, in the development of the high efficiency BCHJ c-Si solar cell, one of the biggest challenges is a good balancing of such properties of the a-Si layers. To overcome the difficulty, application of the thin-film Si technology to the development of the BCHJ c-Si solar cell was indispensable, which was same as the development of the both-side-contacted HJ c-Si solar cell as we reported in the past EUPVSEC [9]. For example, very common technologies in thin-film Si solar cell, such as gap-controlling of a-Si layer, buffer layer between i-layer and doped layer, and contacting layer to TCO layer, could be applicable to the c-Si solar cell with the HJ. We have developed optimal a-Si layers for the HJ c-Si solar cell using the thin-film Si technology.

2.2 Impact of surface passivation on cell performance

We have developed the optimal a-Si layers by investigating the impact of the passivation properties to the cell performance using both-side-contacted HJ c-Si solar cells. The both-side-contacted HJ c-Si solar cells with different preparation conditions of a-Si layers were fabricated using 6-inch semi-square n-type Czochralski Si wafers as described elsewhere [7]. Illuminated current–voltage (I–V) characteristics were determined by a conventional I–V tester at 25 °C under illumination conditions of AM1.5 and 100 mW/cm^2. All I–V measurements are performed at an independent laboratory (Fraunhofer ISE CALLAB). Effective minority carrier lifetime (τ_e) measurements with different injection level for the finished cells were carried out using a Sinton lifetime tester (WCT-120) with the 'Suns-Voc' mode.

To realize high conversion efficiency in c-Si solar cell, surface passivation of c-Si wafer surface is an indispensable technology. We had so far developed the surface passivation technology of c-Si by amorphous Si thin film using a both-side-contacted HJ c-Si solar cell. To evaluated surface passivation of c-Si wafer, we have analyzed τ_e of our 23.8%[Voc: 739 mV, Isc: 39.9 mA/cm^2, FF: 80.8%]-, 24.5%[Voc: 741 mV, Isc: 40.1 mA/cm^2, FF: 82.6%]-, and 25.1%[Voc: 738 mV, Isc: 40.8 mA/cm^2, FF: 83.5%]-HJ c-Si solar cells which were reported in 2014~2015[8, 9]. The τ_e was analyzed using a model reported by Olibet *et al.* [13]. By this analysis, the carrier recombination center density at a-Si/c-Si interface (Ns) are also estimated by a curve-fitting of the $\tau_e(\Delta n)$ curves as shown in Fig. 1.

Figure 2 shows the relative-Voc and -FF of the HJ c-Si solar cell as a function of Ns. A clear dependence of Ns on FF is observed, whereas no significant change in Voc is found in Fig. 2. This result is thought to be caused by the fact that the τ_e at Voc point (excess carrier density

(Δn) of ~2×10^{16} cm^3) is dominantly determined by intrinsic recombination of Si wafer rather than the carrier recombination at the a-Si/c-Si interface. On the other hand, at Pmax point, a τ_e of the carrier recombination at the a-Si/c-Si interface ($\tau_{surface}$) (2 ~4.4 ms) is smaller than a τ_e of the bulk recombination in c-Si wafer (τ_{bulk}) (9~10 ms) and τ_e is dominated by $\tau_{surface}$, which indicates that the τ_e at Pmax point is mainly attributed to the carrier recombination at the a-Si/c-Si interface, and an increase in $\tau_{esurface}$ directly reflects the increase in τ_e, resulting a monotonical increase in FF with Ns. These results indicate that the reduction of carrier recombination centers at the a-Si/c-Si interface should be one of the most crucial issues for the further improvement of FF even in the HJ c-Si solar cells with efficiency over 25%.

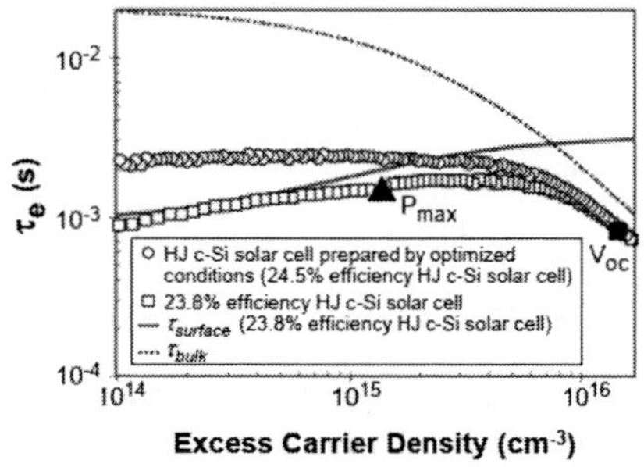

Figure 1 Measured $\tau_e(\Delta n)$ curves of 23.8% efficiency HJ c-Si solar cell and HJ c-Si solar cell prepared by optimized process conditions (24.5% efficiency HJ c-Si solar cell) [10].

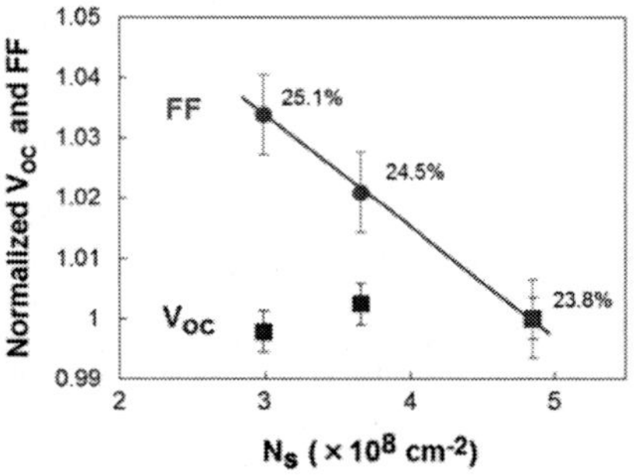

Figure 2: Normalized-Voc and -FF as function of Ns. Voc and FF of 23.8% efficiency HJ c-Si solar cell (Ns of 4.9×10^8 cm^{-2}) are set to unity, respectively. Error bars indicate measurement error [10].

3 BACK-CONTACT HETERO-JUNCTION C-SI SOLAR CELL

3.1 Experimental

BCHJ c-Si solar cells are fabricated using n-type Czochralski crystalline 6 inch Si wafer. Figure 5 shows a typical schematic image of cross section of the BCHJ c-Si solar cell [15]. Rear side passivation is done by P^+ HJ layer stack (i:a-Si layer deposition followed by p:a-Si layer deposition) and N^+ HJ layer stack (i:a-Si layer deposition followed by n-type thinfilm Si layer deposition) by a PECVD method. The front-side (light receiving side) passivation is done by a-Si layer followed by anti-reflection (AR) layer deposition. The P^+ and N^+ HJ layer stacks are patterned into interdigitated design on the rear surface of the Si wafer. The electrode structure is formed on both p:a-Si and n:a-Si layer.

I–V measurement of the HJBC solar cell reported in this paper was performed at an independent laboratory. To estimate electrical losses in the BCHJ c-Si solar cell, we have analyzed the I–V characteristics and properties of τ_e on different Δn which is carried out by a Sinton lifetime tester with the 'Suns-VOC' mode.

A module with industrially-feasible structure using BCHJ c-Si solar cells is prepared. I–V measurement of the best performing module reported in this paper was also performed at an independent laboratory.

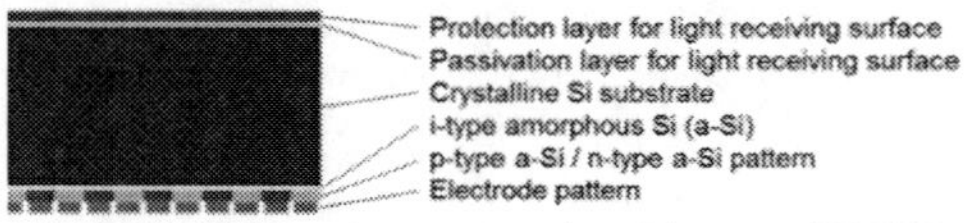

Figure 5 Schematic cross-sectional image of BCHJ c-Si solar cell.

3.2 Results and discussions

To improve the conversion efficiency in BCHJ c-Si solar cell, we have improved our HJ technology with high-quality a-Si and a low resistance electrode technology for a back-contact structure [12]. By a combination and an optimization of the technologies, we have developed a BCHJ solar cell with conversion efficiency of 26.3% [Voc: 744 mV, Isc: 42.3 mA/cm^2, FF: 83.8%] in 2016, which is a record-breaking efficiency for Si solar cell [16].

Energy loss factors in the 26.3% BCHJ c-Si solar cell are investigated as follows. To estimate the series resistance loss, the I–V curve and 'SunsV_{OC} curve' which is generated from the V_{OC} measurements at different incident light intensities are compared, since the 'SunsVOC curve' is expected to be unaffected by the voltage drop due to series resistance [17]. As a result, the series resistance loss is estimated to be absolute efficiency of ~0.5% by a comparison of the value of 'psedo-FF (pFF)' (85.6%) obtained by the 'SunsVOC curve' and the value of FF (83.8%).

Other dominant energy loss factor is an extrinsic recombination loss which includes surface recombination loss at the HJ interface and Shockley-Read-Hall recombination in the c-Si wafer. By an analysis of the dependence of effective minority carrier lifetime on different excess carrier density, the extrinsic recombination loss is assigned to be absolute efficiency of ~1.2%. Optical loss is calculated to be absolute efficiency of ~1.0% from the EQE measurement. The result of the energy loss analysis of the 26.3%-efficiency BCHJ solar cell is summarized in Fig. 7 [15].

In our preliminary experiments of the development of passivation technology, we have confirmed that a further improvement of effective minority carrier lifetime is possible, which reduces the extrinsic recombination loss to be absolute 0.7% from 1.2%. Small reductions of series resistance loss and optical loss are also possible by a tiny optimization. Therefore, it is reasonable to expect that conversion efficiency of 27% is achievable by using our state-of-art technology [15].

Targeting the conversion efficiency of >27%, we have improved our HJBC technologies. Now, we reached a conversion efficiency of 26.7% [1]. Figure 6 shows I– and Power (P)–voltage curves of the 26.7%-efficiency BCHJ c-Si solar cell under standard test condition of

25°C and an illumination of AM1.5 spectrum and 100 mW/cm^2 in designated area measurement.

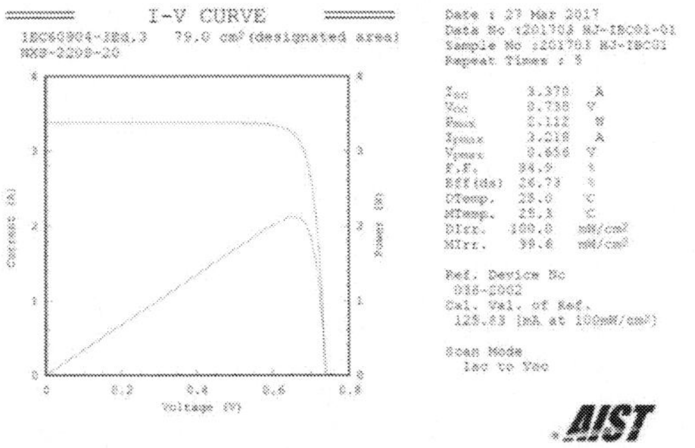

Figure 6: Current and power–voltage curves of 26.7%-efficiency BCHJ c-Si solar cell under standard test condition in designated area measurement, measured by the National Institute of Advanced Industrial Science and Technology (AIST).

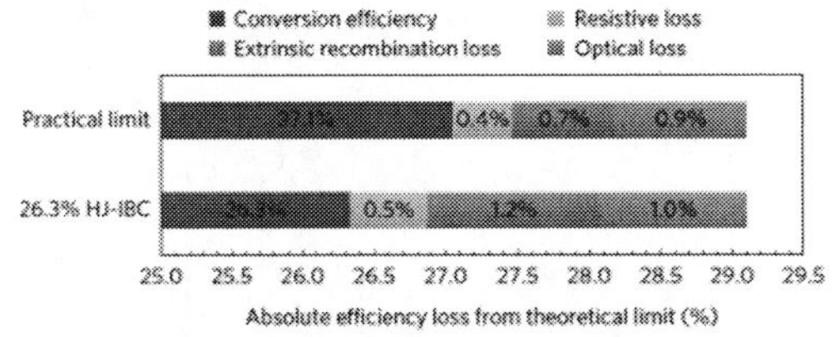

Figure 7: Energy loss factors of 26.3%-efficiency BCHJ c-Si solar cell [15].

Besides the cell, using high efficiency BCHJ c-Si solar cells and newly developed technologies to minimize the resistive loss of inter-cell connection in the module and to raise the collection efficiency of the light radiated into the module, we achieved a record breaking conversion efficiency in c-Si solar cell module of 24.4% in 2016 [Voc: 79.5 V, Isc: 5.04 A, FF: 80.1%, Area: 13,177 cm^2] (Fig. 8) [18]. Further improvement of module conversion efficiency can be expected by the optimization and the development of the technologies for the cell and module described above. These results show advantages of the back-contact heterojunction technology, and will provide bases to c-Si solar cell to access high-efficiency PV market, besides the reduction of the cost of PV electricity by the improvement of conversion efficiency.

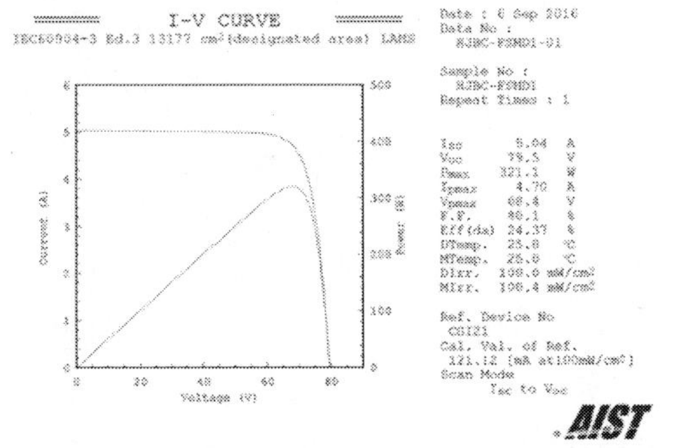

Figure 8: Current and power–voltage curves of 24.4%-efficiency module using BCHJ c-Si solar cells under standard test condition in designated area measurement, confirmed by AIST.

4 CONCLUSIONS

KANEKA's record-breaking efficiencies of crystalline Si (c-Si) solar cell and module using a back-contact heterojunction technology are reported. These results was achieved an application of both-side-contacted HJ c-Si solar cell technology such as passivation technology by a-Si layers.

For the BCHJ c-Si solar cell, we have achieved a record-breaking conversion efficiency of 26.7% in a c-Si solar cell, by means of a combination of our surface passivation technology, low resistance electrode technology and back-contact structure. Energy loss factors in the 26.3%-efficiency BCHJ c-Si solar cell revealed that 27%-efficiency is achievable using our technology. In addition, we have also achieved another record-breaking efficiency of 24.4% in c-Si solar cell module. These results show advantages of the back-contact heterojunction technology and provide bases to c-Si solar cell to access high-efficiency solar cell market.

ACKNOWLEDGEMENTS:

This work was supported in part by the New Energy and Industrial Technology Development Organization (NEDO) under the Ministry of Economy, Trade and Industry of Japan.

REFERENCES:

[1] For example, M.A. Green, K. Emery, Y. Hishikawa, W. Warta, E.D. Dunlop, D.H. Levi, and A.W.Y. Ho-Baillie "Solar cell efficiency tables (version 50)", *Progress in Photovoltaics*, vol. 25, pp.668-676, 2017.

[2] C. Ballif, S. De Wolf, A. Descoeudres, and Z.C. Holman, "Amorphous silicon/crystalline silicon heterojunction solar cells", in: G.P. Willeke, E.R. Webe (Eds.), Advances in photovoltaics: Part 3, Academic Press, California, pp.73-120, 2014.

[3] R.M. Swanson, S.K. Beckwith, R.A. Crane, W.D. Eades, Young Hoon Kwark, R.A. Sinton, and S.E. Swirhun. "Point-contact silicon solar cells". IEEE Trans. Electron Devices, vol.31, pp. 661-664, 1984.

[4] K. Masuko, M. Shigematsu, T. Hashiguchi, D. Fujishima, M. Kai, N. Yoshimura, T. Yamaguchi, Y. Ichihashi, T. Yamanishi, T. Takahama, M. Taguchi, E. Maruyama, and S. Okamoto, "Achievement of more than 25% conversion efficiency with crystalline silicon heterojunction solar cell", IEEE J. Photovoltaics, vol.4, pp. 1433-1435, 2014.

[5] J.L. Hernández, K. Yoshikawa, A. Feltrin, N. Menou, N. Valckx, E. Van Assche, J. Poortmans, D. Adachi, M. Yoshimi, T. Uto, H. Uzu, T. Kuchiyama, C. Allebé, N. Nakanishi, T. Terashita, T. Fujimoto, G. Koizumi, and K. Yamamoto, "High efficiency silver-free heterojunction silicon solar cell", Jpn J. Appl. Phys. vol. 51, pp. 10NB04-1-10NB04-3, 2012.

[6] J.L. Hernández, D. Adachi, K. Yoshikawa, D. Schroos, E. Van Assche, A. Feltrin, N. Valckx, N. Menou, K. Vandersmissen, H. Philipsen, J. Poortmans, M. Yoshimi, T. Uto, H. Uzu, M. Hino, K. Nakano, R. Mishima, T. Kuchiyama, G. Koizumi, C. Allebé, T. Terashita, M. Hiraishi, N. Nakanishi, and K. Yamamoto, "High efficiency copper electroplated heterojunction solar cells", in 27th European Photovoltaic Solar Energy Conference and Exhibition, pp. 655-656, 2012.

[7] J.L. Hernández, D. Adachi, D. Schroos, N. Valckx, N. Menou, T. Uto, M. Hino, M. Kanematsu, H. Kawasaki, R. Mishima, K. Nakano, H. Uzu, T. Terashita, K.

Yoshikawa, T. Kuchiyama, M. Hiraishi, N. Nakanishi, M. Yoshimi, and K. Yamamoto, "High efficiency copper electroplated heterojunction solar cells and modules – the path towards 25% cell efficiency", in 28th European Photovoltaic Solar Energy Conference and Exhibition. pp. 741-743, 2013.
[8] K. Yamamoto, D. Adachi, H. Uzu, M. Ichikawa, T. Terashita, T. Meguro, N. Nakanishi, M. Yoshimi, and J.L. Hernández, "High-efficiency heterojunction crystalline Si solar cell and optical splitting structure fabricated by applying thin-film Si technology", Jpn. J. Appl. Phys. vol. 54, pp. 08KD15-1-08KD15-5, 2015.
[9] K. Yamamoto, D. Adachi, H. Uzu, T. Uto, T. Irie, M. Hino, M. Kanematsu, H. Kawasaki, K. Konishi, R. Mishima, K. Nakano, T. Terashita, K. Yoshikawa, M. Ichikawa, T. Kuchiyama, T. Suezaki, T. Meguro, N. Nakanishi, M. Yoshimi, D. Schroos, N. Valckx, N. Menou, and J.L. Hernández, "Progress & challenges in thin-film silicon photovoltaics: heterojunctions & Multijunctions", in 31st European Photovoltaic Solar Energy Conference and Exhibition, pp. 1003-1005, 2015.
[10] D. Adachi, J.L. Hernández, and K. Yamamoto, "Impact of carrier recombination on fill factor for large area heterojunction crystalline silicon solar cell with 25.1% efficiency", Appl. Phys. Lett., vol. 107, pp. 233506-1-233506-3, 2015.
[11] K. Yoshikawa, A. Feltrin, J.L. Hernández, N. Menou, N. Valckx, E. Van Assche, C. Allebé, K. Yamamoto, B. Pawlak, J. Robbelein, M. Debucquoy, M. Aleman, S. Singh, T. Janssens, P. Verlinden, N. Posthuma, I. Gordon, K. Baert, and J. Poortmans, "Over 22.5% efficiency interdigitated back contact Cell using PECVD thin film layer as doping precursors" in 21st international photovoltaic science and technology conference (PVSEC-21), 5A-1O-07, 2011.
[12] K. Yoshikawa, H. Kawasaki, W. Yoshida, K. Konishi, K. Nakano, T. Uto, D. Adachi, T. Irie, M. Kanematsu, H. Uzu, T. Terashita, T. Meguro, M. Yoshimi, and K. Yamamoto, "6 inch High Efficiency Back Contact Crystalline Si Solar Cell Applying Heterojunction and Thinfilm Technology" in 43th IEEE Photovoltaic Specialist Conference, 2016.
[13] S. Olibet, E. Vallat-Sauvain, and C. Ballif, "Model for a-Si:H/c-Si interface recombination based on the amphoteric nature of silicon dangling bonds", Phys. Rev. B vol. 76, 035326, 2007.
[14] D. Adachi, T. Terashita, T. Uto, J.L. Hernández, and K. Yamamoto, "Effects of SiO_x barrier layer prepared by plasma-enhanced chemical vapor deposition on improvement of long-term reliability and production cost for Cu-plated amorphous Si/crystalline Si heterojunction solar cells" Sol. Energy. Mater. Sol. Cell., vol. 163, pp.204-209, 2017.
[15] K. Yoshikawa, H. Kawasaki, W. Yoshida, T. Irie, K. Konishi, K. Nakano, T. Uto, D. Adachi, M. Kanematsu, H. Uzu and K. Yamamoto, "Silicon heterojunction solar cell with interdigitated back contacts for a photoconversion effciency over 26%" Nat. Energy, vol. 2, 17032, 2017.
[16] M.A. Green, K. Emery, Y. Hishikawa, W. Warta, E.D. Dunlop, D.H. Levi, and A.W.Y. Ho-Baillie "Solar cell efficiency tables (version 49)", *Progress* in Photovoltaics, vol. 25, pp.3-13, 2017.
[17] R. A. Sinton and A. Cuevas. "A quasi-steady-state open-circuit voltage method for solar cell characterization." in: 16th European Photovoltaic Solar Energy Conference. pp. 1152–1155, 2000.
[18] News release of NEDO and Kaneka (October 27, 2016), "World's highest conversion efficiency of 24.37% achieved in a crystalline silicon solar cell module – Represents significant progress toward power generation cost targets"

PILOT LINE RESULTS OF N-TYPE IBC CELL PROCESS IN MASS PRODUCTION ENVIRONMENT

N. Guillevin[1], I. Cesar[1], A.R. Burgers[1], P. Venema[2], Z. Wang[3], J.Y. Zhai[3], D. Liu[3]

[1] *Energy research Centre of the Netherlands, P.O.Box 1, 1755 ZG Petten, The Netherlands,*
[2] *Tempress Systems B.V., Radeweg 31, 8171 MD Vaassen, NL*
[3] *Yingli Energy (China) Co., Ltd. 3399 North Chaoyang Avenue, Baoding, China, 071051*

Together with Yingli Solar and Tempress Systems, ECN accelerates the development of the Mercury IBC technology by implementing its low-cost cell process in a mass production environment. The Mercury solar cell concept based on a bifacial IBC cell design with a front floating emitter (FFE) can now be fully processed on Yingli's pilot line. Pilot processing in an industrial environment offers a great opportunity for the Mercury technology to gain in maturity through an effective bottom-up approach. With a cell process using similar specifications as the commercial n-PERT process, the implementation of the Mercury IBC technology went fast reaching cell efficiencies close to 21% in less than a year. The FFE which renders the cell resilient to electrical shading and reduces the demands on feature size, allowing a high tolerance cell processing. This key feature of the Mercury cell concept plays a major role when it comes to high throughput industrial manufacturing up to module level. Before the end of the year, a set of optimisations will be applied to the rear-side diffusion pattern and to the metallisation with the objective the reach cell efficiencies of 22%. Manufacturability aspects, lab performance and the results of piloting the IBC Mercury technology are described in this paper. Further efficiency gain towards 24% based on the use of polysilicon is also discussed in this paper.

Keywords: interdigitated back-contact; n-type silicon; pilot production

1 INTRODUCTION

The recent international technology roadmap for photovoltaics predicts an increasing market share of back-contact module technology and n-type material for the coming decade [1]. This forecast is supported by market trends towards higher performance, lower costs/kWh, and by the interest in improved aesthetics of PV systems. Connected to this forecast, a relatively new trend is the appearance of bifacial modules in the market. With the incentive of developing state-of-the-art photovoltaic technologies for the industry, ECN has designed a low-cost and industrial bifacial n-type Interdigitated Back Contact (IBC) homo-junction solar cell known as the Mercury IBC technology currently implemented in mass production environment.

In contrast with the well-known IBC cells with front surface field (FSF), the ECN's IBC cell technology employs a conductive p+-doped front floating emitter (FFE). A schematic cross-section of the Mercury cell is shown in Figure 1. The FFE, key feature of ECN's Mercury IBC cells, enhances lateral transport properties for minority carriers (holes) at the front. In addition, it induces a "pumping effect" which transports holes from regions above the BSF to the rear emitter. The additional collection pathway limits the effect of electrical shading of the BSF areas at the rear that conventional FSF IBC cells suffer from.

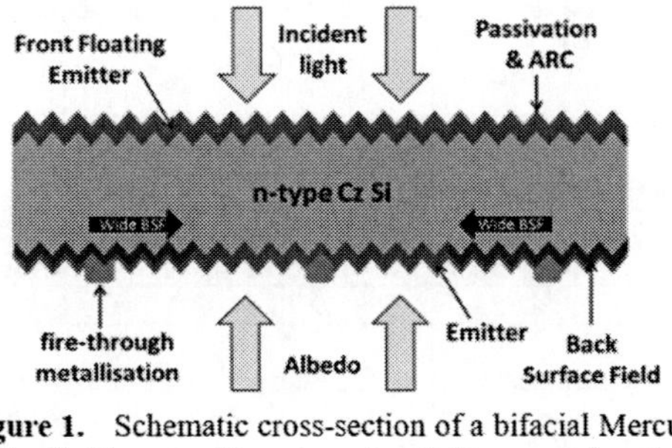

Figure 1. Schematic cross-section of a bifacial Mercury IBC cell. The rear side can receive light, except for where the metallization is present.

With proper tuning of conductance and J_0, the FFE can be applied as an effective means to increase the BSF width with marginal loss in cell performance while assuring process simplification and cost reduction [2].

The integral IBC Mercury solar cell and module concept is currently being piloted in mass production environment at Yingli Solar. This step towards industrialisation is the perfect opportunity for a fast development of the Mercury IBC concept and a precious gain of knowledge on manufacturability of such back-contact solar cell and module technology.

After a review of the manufacturability aspects and lab performance, the pilot line results of the IBC Mercury cell process are discussed. Cell performance progression and pathway to reach the cell efficiency target of 22% at Yingli before the end of 2017 are shown.

Finally, ambitions to raise efficiency of the IBC Mercury cell technology towards 24% keeping an industrial and cost effective process will be discussed.

2 MANUFACTURABILITY AND PERFORMANCE

a. A cost effective industrial process

Despite a different cell architecture, ECN's Mercury IBC process technology remains close to our industrial n-PERT technology in production at Yingli. Similar process equipment as well as process parameters are used without increasing the number of major manufacturing steps making the Mercury process compatible with an industrial scale production and throughput. Cells are processed on commercially available 6 inch n-Cz wafers. The cell structure comprises an interdigitated boron-doped emitter and a phosphorous-doped Back Surface Field (BSF) on the rear-side and a boron-doped FFE on the front-side. Doped regions are created by means of tube diffusion processes and are designed to be suitable for industrial throughput, i.e. lower cycle time and high load density. The FFE and the rear emitter are formed in a single Boron diffusion step. Structuring of the rear-side diffusions is based on conventional screen printing process and does not involve the creation of a gap between the emitter and the BSF. This patterning approach greatly simplifies processing of the device and reduces manufacturing costs compared to complex and costly high resolution patterning techniques such as lithography or LASER ablation processes. In addition, this approach offers a great flexibility in implementing different diffusion pattern designs and matching metallisation grid designs. Front-side and rear-side surface passivation and anti-reflecting coatings are applied by industrial ALD and PECVD equipment respectively. The metallisation consists of a firing-through Ag paste deposited in a single step, for both emitter and BSF, by screen-printing and features an open grid design suitable for thin wafers and bifacial applications. The currently used metallisation grid design at ECN includes busbars and 62 interconnections pads as shown in figure 2.

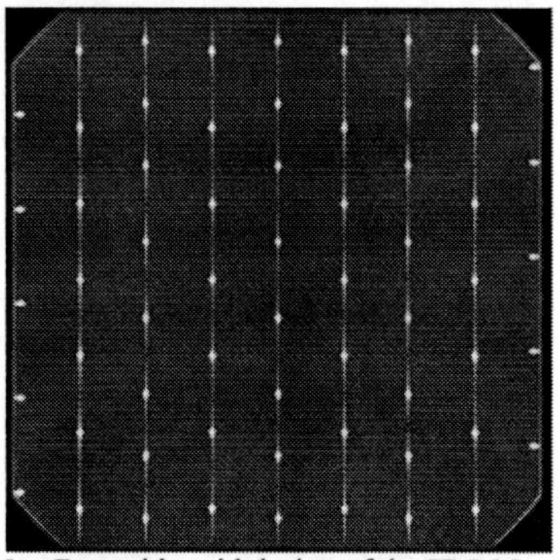

Figure 2. Rear-side grid design of the IBC Mercury cell as processed at ECN.

A major objective in the development of the Mercury IBC technology was to change the market perception that an IBC process is laborious and expensive. Great efforts dedicated to simplify the manufacturing process resulted in placing the process complexity of Mercury cells next to that of p-PERC cells. The process cost, excluding the cost for wafer and silver paste, reached a cost level comparable to that of p-PERC [2]. With an attractive module and system cost, the Mercury IBC technology is ready to enter the market of low-cost PV technology.

Flexibility of the diffusion pattern and metallisation grid designs offers freedom when it comes to the choice of module interconnection technology. Both diffusion and metallisation patterns can be easily modified to be made compatible with many interconnection technologies. Based on the current metallisation grid design which includes interconnection pads, the cells can be readily processed into modules using ECN's foil-based interconnection technology [4-5]. ECN's module manufacturing technology is based on an interconnection foil with integrated conductor layer (e.g. copper or aluminium), on which the cells are electrically contacted using a conductive adhesive. Compared to a tabbed interconnection technology, the interconnection foil allows reduction of the module series resistance by using more interconnect metal (more cross-sectional area) and thereby reduces the cell to module FF loss. Also, the module manufacturing based on integrated back-foil can be done with higher yield and reduced interconnection-process-related stress, allowing use of (much) thinner cells and therefore offering additional cost reduction possibilities. The interconnection technology by itself has also potential for improvement and further cost reduction, as reported in [6]. So far, 2x2 cells IBC Mercury laminates successfully passed damp heat and thermal cycle tests as described by the IEC 61215 standard. The IBC cells that are free of front metallisation give the module an aesthetical appearance as illustrated in figure 3. Based on the same interconnection technology, a first 60-cell module is in preparation to monitor cell-to-module losses.

Figure 3. Photographic image of an all-black 2x2 Mercury IBC module

The metallisation grid design can be easily modified to be made compatible with for instance the multi-wire interconnection technology [7]. This alternative type of interconnection will allow the technology to benefit from the full bifaciality potential of the Mercury IBC cells for increased annual energy yield when installed in the field.

b. Cell performance processed on ECN's pilot line

We have been able to steadily increase the efficiency of our cells on ECN's pilot line, and reported our best efficiency to date of 21.1% [2]. The I/V parameters of this cell are listed in Table 1. The I/V parameters were obtained with an in-house measurement using a class AAA solar simulator. The measurement chuck was especially designed for our Mercury cells, with current and voltage probes only contacting the module interconnection points, and a reflective but non-electrically-conductive chuck surface, representative for the situation in a module. The J_{sc} was corrected for spectral mismatch. The measurement was calibrated with a Fraunhofer-ISE-calibrated front-and-rear contact cell measured with a different chuck. Both for calibration and for IBC cell measurements, the chuck surface outside of the cell was masked with black tape, to avoid calibration errors due to variation in the chuck area and chuck reflectance.

Table 1. I-V parameters of the best Mercury IBC cell measured at ECN. Short circuit current is corrected for spectrial mismatch

Area	J_{sc} (mA/cm^2)	V_{oc} (mV)	FF (%)	Eta (%)	Bifaciality factor
239	41.2	653	78.4	21.1	83%

Performance of our Mercury cells can be further improved, in particular in V_{oc} and FF. By improving the rear-side diffusion pattern and metal contact properties that enable smaller contact areas and less recombination, efficiency improvement beyond 22% is within reach.

Work in this direction is being carried out and will be discussed in the next sections.

The bifaciality factor of our Mercury IBC cells was also measured on ECN's I/V tester. To allow a proper rear-side I/V measurement, ribbons were connected to the contact pads of the cell metallisation grid by soldering. Front and rear-side I/V measurements were carried out by contacting the cell to the soldered ribbons. A high bifaciality factor of 83% was measured illustrating the great potential for higher annual energy yield when the cells will be interconnected in a bifacial module configuration.

3 MERCURY IBC OPTIMISED IN MASS PRODUCTION ENVIRONMENT

a. An effective bottom-up technology development

Only a few 6 inch wafer based IBC pilot lines exist in the world and information on the manufacturability of this cell type is scarce because of that. From a long-lasting cooperation with the Chinese PV manufacturer Yingli Solar and with the support of Dutch equipment supplier Tempress, ECN has the chance to implement the cost effective Mercury IBC process in a mass production environment to accelerate development of the technology. This collaboration towards high performance industrial cell and module technology is the perfect opportunity to ease development aiming at reaching a technology readiness level suitable for a successful introduction to the market.

Rapidly after setup of the Mercury IBC pilot line at Yingli, implementation of the IBC cell process started in the third quarter of 2016 as shown in figure 4. In only a quarter of a year, Yingli had processed their first batch of Mercury IBC cells with a top efficiency of 19.3% illustrating the compatibility of ECN's IBC Mercury technology with mass production equipment. After couple of month of process stabilisation, a cell efficiency of 20.5% was reached just before the end of 2016. With this important milestone completed, the consortium could focus on improving further the cell efficiency through a step by step optimisation of the different key design and process aspects.

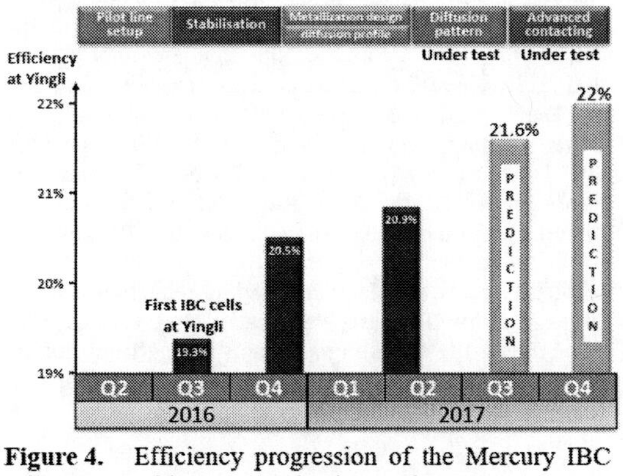

Figure 4. Efficiency progression of the Mercury IBC cells processed in mass production environment

b. Technology optimization – achieved results

All processing is performed on Yingli's PANDA production line offering high throughput pilot process for a fast turnover of experiments. As shown in figure 4, the first quarter of 2017 was dedicated to the metallisation grid design improvement. To evaluate performance of the cells, design of the metallisation grid has to be compatible with the contacting chuck used to measure I/V parameters. In order to process the first measurable batch of IBC cells as early as possible, the metallisation grid was designed to match Yingli's measurement chuck operational at this time which included 31 contact points. This suboptimal design leads to significant fill factor (FF) losses estimated at 1.1% absolute. A new chuck design with 81 contact points became available at Yingli at the beginning of 2017 offering the opportunity to redesign the metallisation grid to reduce resistive losses. The earlier metallisation grid design (from 2016) and current metallisation grid design are shown in figure 5. From modelling, the FF would increase by around 0.7% absolute by increase the number of contact points from 31 to 81, without increasing of the metal paste fraction. The modelled fill factor loss breakdown is shown in table 2.

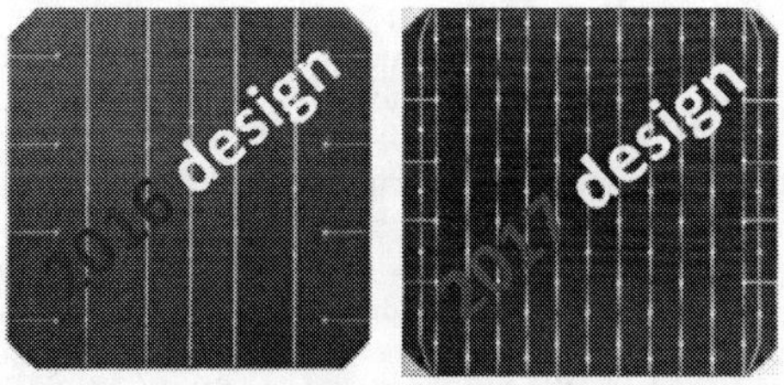

Figure 5. Mercury IBC former metallisation grid design with 31 contact pads (2016 design, left picture) and current metallisation grid design with 81 contact pads (2017 design, right picture) as processed at Yingli

Table 2. Modelled FF loss breakdown for the 2 different metallisation grid designs featuring 31 and 81 contact pads

Absolute FF loss	31 contact pads	81 contact pads
In metal fingers	0.60%	0.20%
In metal Busbars	0.50%	0.15%
Total FF loss	**1.10%**	**0.35%**

Surface passivation of IBC cells is of high importance, firstly for high J_{sc} due to the long path lengths that minority carriers need to travel before being collected, and secondly to build up a substantial carrier density in order to achieve high V_{oc}. Therefore, in parallel to the metallisation grid design optimisation, emitter, FFE and BSF diffusion profiles were improved for better surface passivation in order to further increase the V_{oc} and J_{sc} of our IBC cells. Improvement of the surface passivation was achieved by reducing dopant concentration at the surface of the three doped layers (FFE, emitter and BSF). In order to keep the overall cell process industrial and cost-effective, the adjustment of the three diffusion profiles is integrated within an already existing process step. Therefore the number of major manufacturing process steps remains the same. Also, the improvement of the dopant profiles consists of a uniform reduction of dopant concentration over the wafer surface. No additional patterning step is involved to create selectivity in the diffusion profiles commonly used to maintain high dopant concentration under the metal contacts to prevent FF loss. In consequence, this optimisation step is a trade-off between improving surface passivation and maintaining good contact resistance to prevent FF loss.

Thanks to the high throughput of Yingli's pilot line, fast processing of batches of hundreds of cells can be processed improving results statistics. Also, test structures were included in each process run to easily measure valuable parameters such as J_0 of the surface. Three different optimisation levels were tested. For each optimisation level, J_0 surface of the emitter, of the FFE and of the BSF was measured. The J_0 surfaces of the original and most performant doping profiles are reported in table 3.

Table 3. Surface dark saturation current densities (J_0) of the original [D1] and most performant [D3] emitter, FFE and BSF profiles, measured on test structures.

	[D1]	[D3]
Emitter & FFE J_0 surface	65 fA/cm^2	50 fA/cm^2
BSF J_0 surface	290 fA/cm^2	200 fA/cm^2

Means and confidence intervals of the open circuit voltage (V_{oc}) and fill factor (FF) of the IBC cells processed with the three different optimisation levels are shown in figure 6. The experiment is based on batches of more than hundred cells. For each optimisation level of the doping profiles, a firing optimisation was performed to adjust the fire-through contact formation to the new dopant surface concentration. A significant increase of around 6 mV in V_{oc} is reached with the third optimisation level [D3] without FF loss nor J_{sc} loss. This V_{oc} gain is consistent with the J_0 surface reduction of the different diffusions reported in table 3.

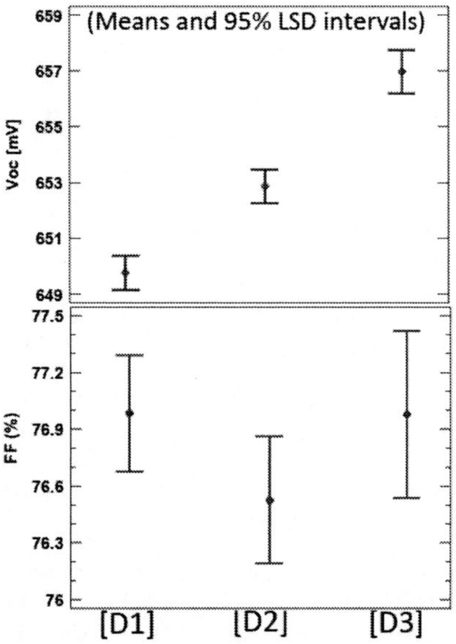

Figure 6. Open circuit voltage (V_{oc}) and fill factor (FF) (means and confidence intervals LSD: Least Significant Difference $p<0.05$) of the IBC cells processed with the three optimisation levels of the doping profiles [D1], [D2] and [D3].

The optimised diffusion profiles were tested together with the new metallisation grid design in a run carried out at Yingli at the beginning of the second quarter of 2017. From this new run, a significant efficiency increase of 0.4% absolute was obtained on the 2016 champion cells with the V_{oc} and FF gain expected from the metallisation design and diffusion process improvements. To confirm I/V measurement accuracy, these champion cells were also measured at ECN. The I/V data comparing the champion cell of 2016 and the champion cell of 2017 are shown in table 4.

Table 4. I/V characteristics of Yingli's champion IBC cells (2016 and 2017) measured at ECN. J_{sc} is corrected for spectral mismatch, cell area=242.8cm2

Date	Cell features	J_{sc} [mA/cm^2]	V_{oc} [mV]	FF [%]	Eta [%]
December 2016	31 contacts [D1] profile	40.4	657	77.4	20.5
April 2017	81 contacts [D3] profile	40.5	663	78.0	20.9

In less than a year, efficiency of 20.9% were reached on 6 inch n-type Cz wafers fully processed at Yingli which demonstrate an effective bottom-up efficiency development in a mass manufacturing environment that is hard to achieve at an R&D facility.

c. Towards 22% efficiency

The current performance of the Mercury IBC cells is currently limited by the open circuit voltage, due to high recombination dominated by the emitter contacts and the heavily doped BSF area. Therefore, reducing both the BSF area and the emitter contact fraction is a route to decrease the recombination in the cell and therefore enhance the cell performance.

Depending on the contact width and the screen printing tolerances, a minimum width of the passivated BSF area is required, which is typically more than 300 µm. In a one-dimensional interdigitated finger design, the only option to reduce the BSF area fraction further is then to increase the emitter width, but this induces large transport losses. Therefore, we reduced the BSF length within the unit cell, and in this way we created "islands" of BSF surrounded by the rear side emitter. Unit cell schematic representations of the standard interdigitated and "BSF islands" diffusion patterns are shown in figure 7. The performance of the "BSF islands" diffusion pattern together with improvement of the FFE passivation has been already reported by ECN on a previous publication [8]. Similar point-contact structures for the diffused areas have been studied before for IBC cells [9]-[10], but in these cases, the BSF islands were mainly created to study the electrical shading reduction benefits. In the Mercury IBC case, electrical shading is not a major issue due to the collecting and transporting front floating emitter. Therefore, the BSF area reduction will mainly improve the passivation of the cell.

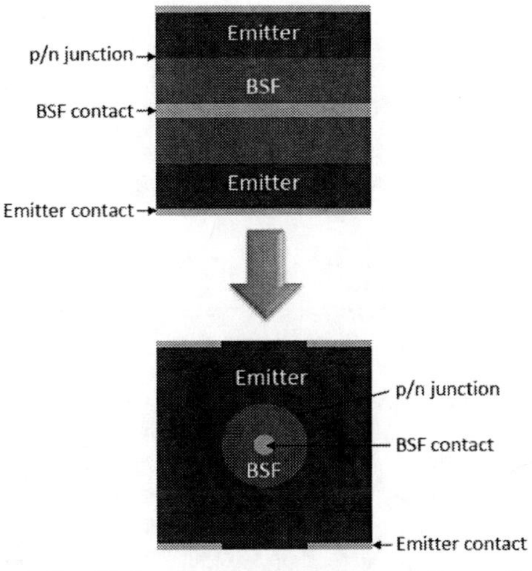

Figure 7. Unit cell schematic representations of the standard interdigitated (top) and "BSF islands" (bottom) diffusion patterns

The "BSF islands" design was tested in the latest ECN's IBC baseline run together with improvement of the FFE passivation. The ECN's IBC baseline run is meant to regularly control ECN's equipment and process stability. The baseline process does not include the latest improvements required to reach efficiencies in the range of the champion cells. Therefore the baseline group reaches average efficiencies around 20.5%. In this test, the baseline reference group is process with an interdigitated diffusion pattern with 37% BSF fraction and 6% emitter metal contact fraction. The "BSF islands" diffusion pattern has 20% BSF fraction and 4% emitter metal contact fraction. As shown in figure 8, the group processed with the "BSF islands" diffusion pattern outperforms the baseline group processed with the standard interdigitated diffusion pattern. The 0.5% absolute efficiency gain obtained thanks to the "BSF islands" diffusion design combined with improved front side passivation is in fair agreement with the simulation results based on Quokka previously reported in [8]. The "BSF islands" diffusion pattern is currently being tested at Yingli.

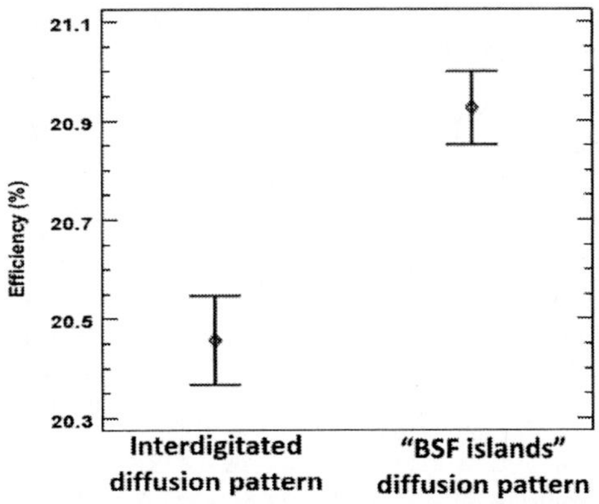

Figure 8. Mercury IBC cell efficiencies results (means and confidence intervals) from the latest ECN's baseline run. The reference group is processed with the interdigitated emitter and BSF diffusion pattern. The test group is processed with the "BSF islands" diffusion pattern.

Short circuit current of Yingli's IBC cells can yet be further improved by simple tuning of the SiNx refractive index to match performance of ECN's champion IBC cells. In combination with the 0.5% absolute efficiency gain from "BSF islands" diffusion pattern, Mercury IBC cells are predicted to reach efficiencies of 21.6% before the end of the third quarter of 2017 as marked in figure 4.

Additional gains in open circuit voltage and fill factor are required to reach the 22% efficiency target before the end of the year. As mentioned earlier, the current Mercury cell process involves screen-printed firing through (FT) metallization. The contacting mechanism of FT metallization is a trade-off between the depth to which the paste has to etch into the doped silicon to obtain contact resistance as low as possible and the recombination resulting from the etching into the diffusion. Current FT pastes tested on our Mercury IBC cells do not allow contact resistances below 5 mOhm.cm^2. A commercial screen printable non-fire through (NFT) metal paste was recently tested at ECN as alternative to conventional FT pastes. Contact resistance as low as 0.5 mOhm.cm^2 was measured on test structures processed with the NFT paste without increasing recombination at the contacts compared to conventional FT paste. Based on these results, the NFT paste will offer the opportunity to further reduce contact area for additional V_{oc} increase while improving the cell fill factor. Efficiency gain potential was simulated with the 2D modelling program Quokka based on Yingli efficiencies. Simulation inputs are shown in table 5 and simulation results are summarized in the table 6.

As predicted by the simulation run 3, thanks to its excellent contacting abilities, the NFT paste allows a V_{oc} increase of 4mV and a 1% absolute FF gain while reducing further the contact area by 1%. Combined with the "BSF islands" design, the used of the screen printable NFT paste will allow efficiency of the Mercury cells to reach 22%.

Table 5. Inputs parameters used in Quokka 2D modelling program

Simulation runs	Diffusion pattern & BSF fraction	Emitter contact area	Contact resistance [mΩ.cm^2]
1	Interdigitated 37% BSF	6%	5
2	BSF islands 20% BSF	4%	5
3	BSF islands 20% BSF	3%	0.5

Table 6. Output I/V parameters obtained from Quokka 2D modelling program based on inputs listed in table 6. J_{sc} increase of run 2 and 3 are related to the front SiNx optical properties optimisation.

Simulation runs	J_{sc} [mA/cm^2]	V_{oc} [mV]	FF [%]	Eta [%]
1	40.5	663	78	20.9
2	41.2	671	78	21.6
3	41.2	675	79	22.0

Implementation of the NFT metallisation process will shortly begin at Yingli with the objective to reach 22% before the end of 2017.

4 BEYOND 22% EFFICIENT MERCRY IBC CELLS

As discussed previously, the performance of our Mercury IBC cells is currently limited by recombination at doped surfaces and at the metal contacts. Carrier-selective contacts based on the combination of a tunnel oxide and doped polysilicon (poly-Si) layer are therefore of a great interest to further reduce contact recombination and increase cell performance. Passivating contacts based on polysilicon are currently studied at ECN for the so-called PERPoly (Passivated Emitter Rear PolySilicon) solar cell concept [11]. Excellent surface and contact passivation quality of p and n doped poly-Si was recently obtained at ECN. To keep the cell process industrial and cost-effective, ECN focuses on screen printed firing through metallisation process for which contact properties to the poly-Si doped layers was also tested. Table 7 summarizes the dark saturation current density J_0 measured on test structures of contacted (J_0,c) and not contacted (J_0,nc) poly-Si layers for both n- and p-type diffusions with a sheet resistance of respectively 60 Ohm/sq and 200 Ohm/sq. Firing-through metallisation contact resistance (R_{co}) measured on test structures are also reported in table 7.

Table 7. Passivation and contact properties of n- and p-doped poly-Si layers processed at ECN as measured on test structures.

parameters	Units	n doped poly-Si	n doped poly-Si
R_{sheet}	[Ω/sq]	60	200
J_0,c	[fA/cm^2]	200	500
J_0,nc	[fA/cm^2]	5	15
R_{co}	[mΩcm^2]	2	2

Based on these results, several IBC cell architectures were modelled using the Quokka 2D simulation program. Schematic cross sections of the cell architectures simulated are summarized in figure 9. The first study case involves an hybrid poly-Si IBC cell consisting of a conventional Mercury IBC cell with n-doped poly-Si as BSF. The second IBC cell architecture modelled features a full poly-Si rear-side with interdigitated n- and p- doped poly-Si emitter and BSF regions. In the two first study cases, the firing-through metal fingers are 80 microns wide. Finally the third study case is similar to the second study case with interdigitated n- and p- doped poly-Si at the rear-side but the firing-through metal fingers are set to a width of 40 microns. In all study cases, the front side remains the standard p-doped FFE homojunction specific to the Mercury IBC cells.

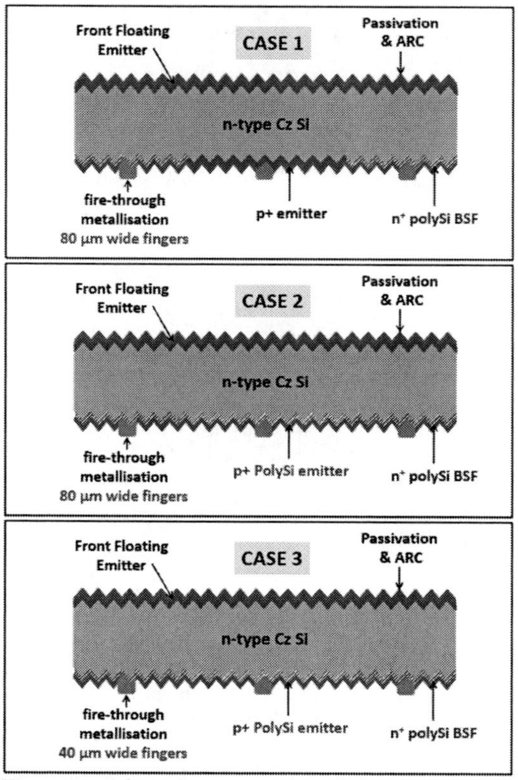

Figure 9. Schematic cross-section of the Mercury IBC cells modelled in Quokka.

Results of the modelling are presented in figure 10 under the form of efficiency contour plots. Efficiency responses are drawn as a function of emitter and BSF half widths. For each contour plot, a red dot indicates the maximum efficiency point with the associated emitter and BSF width. For each study case, a wafer resistivity of 9 Ohm.cm and a bulk lifetime of 2 milliseconds are set as input parameters.

The efficiency contour plot resulting from the simulation of the hybrid poly-Si IBC cell (case 1) shows a maximum efficiency at 22.7% with a wide BSF of 700 microns and a wide emitter of 620 microns as optimum. From the efficiency contour plot of the second case, an almost 1% absolute increase on the maximum efficiency is predicted by the modelling. With a BSF of 560 microns width and an emitter of 600 microns width, a peak efficiency of 23.5% is within reach according to the model predictions. Finally, by reducing metal finger width by half and choosing for narrower BSF and emitter (340 microns and 460 microns respectively) a maximum efficiency of 23.7% is predicted by the modelling. With a 2-3% lower metal fraction compared to the cell structures studied in the two first cases, the cell structure studied in the third case will benefit from an increase bifaciality factor in addition to higher front-side efficiency potential.

The cell design specifications such as emitter and BSF width predicted by the modelling to reach the highest efficiencies remain compatible with high throughput industrial processes such as screen printing.

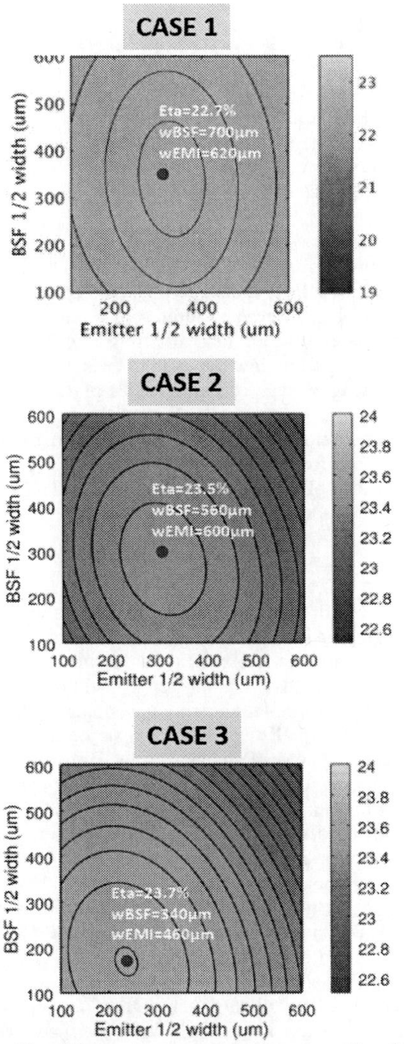

Figure 10. Efficicency contour plots as a function of BSF and emitter half width . These 3 contour plots are the resutls of Quokka 2D simulation for the 3 different IBC architectures showed in figure 9.

5 CONCLUSION

ECN recently teamed up with industry to accelerate the development of its integral bifacial solar cell and module concept based on an interdigitated back-contact cell structure with a front floating emitter: the Mercury IBC technology. Engineered based on a highly tolerant cell process at a cost level of a p-PERC cell, the Mercury IBC technology is currently being piloted on Yingli's production line which offers an ideal environment to bring the concept to high technology readiness level, usually hard to achieve at an R&D facility. The implementation of the process in mass production environment took only a quarter of a year and efficiencies of 20.9% were reached in less than a year. This fast progress illustrated the compatibility of the ECN's IBC Mercury process flow with mass production and shows the high potential for rapid improvements. Based on optimisation of the diffusion pattern and improved metallisation contact scheme, Mercury cells with efficiency of 22% are expected to be manufactured at Yingli by the end of the year. Further efficiency improvement towards 24% become possible when involving passivated contacts as already developed by ECN for the PerPoly cell concept.

With such efficiencies, a high bifaciality factor of 83% and a flexible choice of module interconnection methods, the ECN's IBC Mercury technology clearly becomes a remarkable competitor on the photovoltaic market.

6 ACKNOWLEDGEMENTS

ECN acknowledges the cooperation with Tempress Systems and Yingli solar.

7 REFERENCES

[1] http://www.itrpv.net/Reports/Downloads/ - ITRPV Eighth Edition 2017- Maturity Report 20170906.pdf – page 36

[2] Ilkay Cesar et al., "Enablers for IBC: integral cell and module development and implementation in PV industry", *7th International Conference on Silicon Photovoltaics, SiliconPV 2017.*

[4] Bennett IJ, Eerenstein W, Rosca V. Reducing the cost of back-contact module technology. *Energy Procedia 2013; 38:329-33.*

[5] Bende EE, Van Aken BB. The effect of reduced silver paste consumption on the cost per wp for tab-based modules and conductive-foil based modules. *Energy Procedia 2015;67:163-74.*

[6] Goris MJAA, Kikkert BWJ, Kroon JM, Rozema K, Bennett IJ, Verlaak J. Production of low cost back contact based PV modules, *32nd European Photovoltaic Solar Energy Conference, Munich, Germany, p. 99-104, 2016.*

[7] Johann Walter et al, Multi-wire Interconnection of Busbar-free Solar Cells, *Energy Procedia Volume 55, 2014, Pages 380-388*

[8] A. Mewe, BSF islands for reduced recombination in IBC cells, *IEEE PVSEC-44, Washington DC, June 25-30 2017*

[9] C. Reichel, F. Granek, M. Hermle and S. W. Glunz, "Back-contacted silicon solar cells with insulating thin films", *NPV workshop, 2011*

[10] J. Haschke, N. Mingirulli, and B. Rech, "Progress in point contacted rear silicon heterojunction solar cells", *Energy Procedia 27, 2012, p. 116*

[11] M. Stodolny, Material properties of LPCVD Processed n-type Polysilicon Passivating Contacts and its Application in PERPoly Industrial Bifacial Solar Cells, *7th International Conference on Silicon Photovoltaics, SiliconPV 2017*

Simultaneous Fabrication of n & p Contacts for Bifacial Cells by a Novel Co-plating Process

R. Russell[1,*], L. Tous[1], E. Cornagliotti[1,2], D. Hendrickx[1], F. Duerinckx[1],J. Szlufcik[1],
[1] Imec, Kapeldreef 75, 3001 Leuven, Belgium
[2]Now at Kaneka Belgium, Nijverheidsstraat 16, 2260 Westerlo-Oevel, Belgium
*corresponding author: Richard.Russell@imec.be, Tel: +32-16-28 7668,

ABSTRACT: This paper describes a n & p surface Ni/Ag co-plating process for the metallization of bifacial cells suitable for multi-wire module integration. The process, based on electroless Ni and immersion Ag plating, mitigates typical problems of achieving simultaneous electroless Ni plating onto n & p surfaces by the use of a novel silicon selective activation step. Key attractive features of this metallization process include low cost from low Ag usage (typically < 8 mg/cell) and low stress batch processing with no cell contacting using low Capex high throughput tools. The plating process is applied to metallize bifacial nPERT+ cells achieving average cell efficiencies of 22.4% with typical cell bifacialities above 95%. Limited 1 cell laminate reliability data is given with IEC61215 thermal cycling and damp heat test criteria being met. Finally we present cost-of-ownership estimates at the cell level ($/cell), module level ($/Wp) and system level LCOE ($/kWh) for multi-wire modules made from nPERT+ cells using the co-plating process and compare the results to conventional modules with screen printed pPERC+ cells. Reductions in the LCOE of ≥ 5% are estimated for co-plated nPERT+ cell multi-wire modules for albedo values ≥10%.
Keywords: Metallization, Contact, Bifacial

1 INTRODUCTION

Interest in bifacial solar cells is steadily growing in the PV industry and this is reflected in the 2017 ITRPV [1] which predicts a 30% market share by 2027. Large installations do already exist such as Yingli's 50MWp system in China. EDF has an even larger 90MWp project on-going in Mexico. Current total capacity for bifacial solar cell production is estimated to be circa 2GWp. [2] Growth in bifacial cell/module production is due to the fact this technology can offer an effective way of increasing module power without significantly increasing costs thus lowering the levelized-cost-of-electricity (LCOE), a main market driver.

Currently, industrial bifacial solar cells are either typically based on a passivated emitter and rear structure (PERC) using p-type wafers or passivated emitter and rear totally diffused (PERT) or silicon heterojunction (SHJ) structures on n-type wafers [3]. The metallization of such cells is technologically more challenging than mono-facial cells as shading losses are required to be low on both sides and many techniques applicable to one side are not directly applicable to the other.

Metallization by screen-printing is widely used but has certain disadvantages depending on the cell structure. For pPERC+ cells (+ sign indicating bifaciality) an Al paste is used for contact fingers on the p-side whilst an Ag paste is used for the n-side. This makes optimal co-firing for both Ag and Al pastes difficult. Furthermore, fine-line printing (with sufficient conductivity) of Al pastes is generally not possible, resulting in increased p-side shading limiting cell bifaciality.

For nPERT+ cells an AgAl paste is normally used for the p-side (emitter) allowing finer lines to be printed. However, now both sides use Ag containing pastes, increasing costs and the AgAl pastes must not spike through the emitter which constrains the emitter doping profile to non-optimal conditions.

Silicon heterojunction bifacial cells cannot withstand the firing temperatures of conventional pastes. Therefore they require low temperature Ag pastes which are more expensive and also have the additional disadvantage of having lower conductivity.

Metallization of bifacial cells by plating is also challenging. Although it is possible to simultaneously plate Ni from electroless nickel solutions onto n and p silicon, without an activation step it is not possible using stable long lasting solutions and/or at similar rates on both surfaces [4]. A palladium activation step can be used but is relatively expensive and tends to activate areas as well where plating is not desirable (e.g. on silicon nitride) creating unwanted ghost plating.[5]

Light induced plating (LIP), another common technique used in mono-facial cells, can only be used to plate onto the n-side. One option would be to plate separately the n & p sides one after the other using two different techniques. The n-side being plated by LIP and the p-side by field-induced plating (FIP) [6]. During FIP the n-side is electrically contacted but must not be in contact with the plating solution. This approach is sequential, hence slow, high in Capex and requires relatively complex cell contacting and manipulation.

Another approach used mainly with high efficiency SHJ cells is the blanket sputtered seed layer + mask + electro-plating approach as described in [7]. Due to masking, very narrow and high aspect ratio fingers are achievable but this process introduces significant additional material (PVD seed layer, mask deposition/removal) and Capex costs (3 tools) and does not therefore provide a lower cost alternative to screen-printing.

2 IMEC'S BIFACIAL PLATING PROCESS

At imec we have tried to address some of the shortcomings outlined above in current screen-printing or plating techniques used to metallize bifacial cells. Our metallization approach has three main objectives i) low Ag usage to reduce material costs ii) the tool set must be simple and capable of high throughput to reduce Capex costs and iii) it must enable stable high cell/module efficiencies.

Our metallization approach is based on electroless and immersion plating to enable simultaneous, contact free, co-plating of n & p contacts. A multi-wire interconnection approach relaxes the finger line conductivity to achieve good fill factors so that sufficient line conductivity can be achieved by nickel plating capped with a thin immersion Ag layer. The metallization sequence being developed is shown in table I. As only two thin metal layers are required the processing time is relatively short and being a batch plating process throughput can be high and easily scaled-up.

Table I: The co-plating sequence developed at imec

Process Step (n & p surfaces)	Chemistry	Time	Function
Surface activation	imec iNi+	2min	Silicon selective surface activation
Electroless Ni	MacDermid	~15min	Ni plating on n & p surfaces
Immersion Ag	MacDermid	2-6min	For line conductivity and capping
Sinter 350°C	Low [O_2] atmosphere	~4 min	Reduces line & contact resistance

The key attractive features of this metallization route are:

- Self-aligned plating (no masking) with line widths < 20µm achievable.
- Batch process with no contacting to cells, low stress.
- N & p surfaces are co-plated simultaneously.
- Low capex equipment with tank size determining throughput.
- Low consumable costs, enabled by stable plating baths and < 8 mg Ag needed per cell.
- Excellent contacting properties of Ni enabling n & p diffusions with high implied Voc.

Succesful bifacial plating by this approach is primarily due to the surface activation step. The surface activation step deposits a thin layer of Ni by a modified immersion plating process. This process is silicon selective depositing Ni only on exposed silicon (thus avoiding ghost plating problems often seen with paladium activation). As a thin Ni layer is deposited during activation the subsequent electroless Ni solution deposits Ni onto an existing Ni layer with similar rates on both sides, which is much less challenging than simultaneously plating onto n and p silicon. Figure 1 shows scanning electron microscope (SEM) pictures of n and p fingers before and after surface activation clearly showing a conductive (Ni) film being deposited.

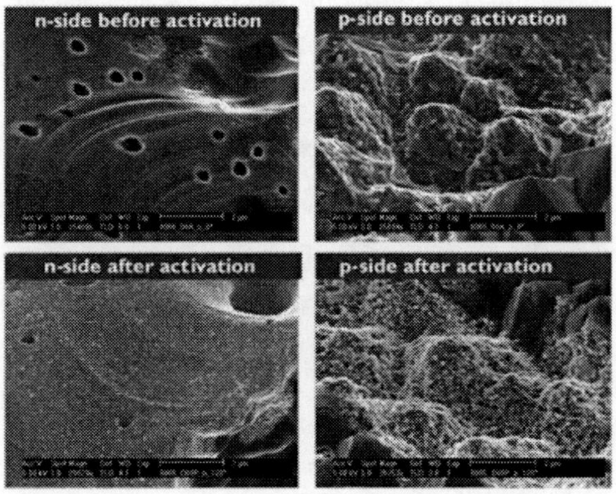

Figure 1. Laser doped (locally flattened) n-surface and laser ablated p-surface before and after the activation step prior to electroless Ni plating. Note that the pits in the laser doped n-surface are an artefact of the doping process.

A SEM picture of a plated finger after electroless Ni plating and immersion Ag shows the high selectivity of the activation and plating processes with no ghost plating occurring (Figure 2).

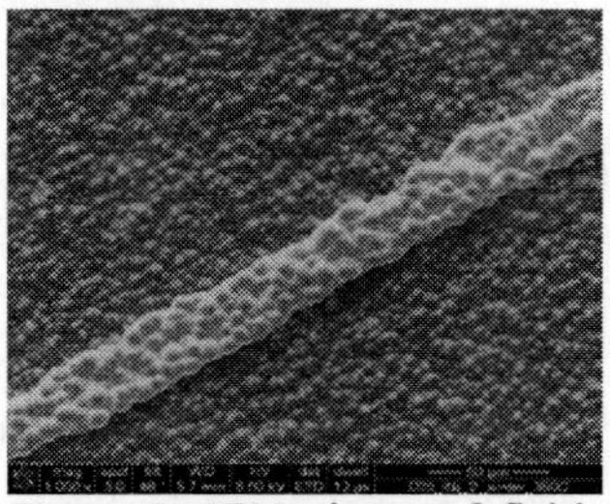

Figure 2. SEM picture of finished fully plated finger showing no spurious ghost plating.

3 CELL RESULTS

The co-plating process outlined in table I was used to metallize bifacial Passivated Emitter Rear Totally diffused cells on n-type wafers (nPERT+). The cell structure is schematically shown in figure 3 and the processing sequence is outlined in figure 4.

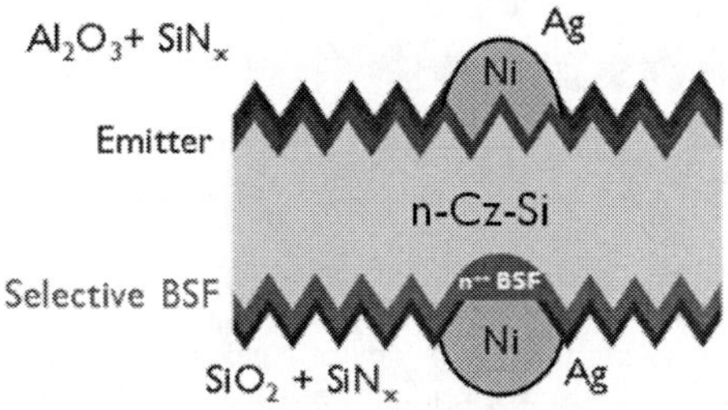

Figure 3. Schematic of nPERT+ cell structure used

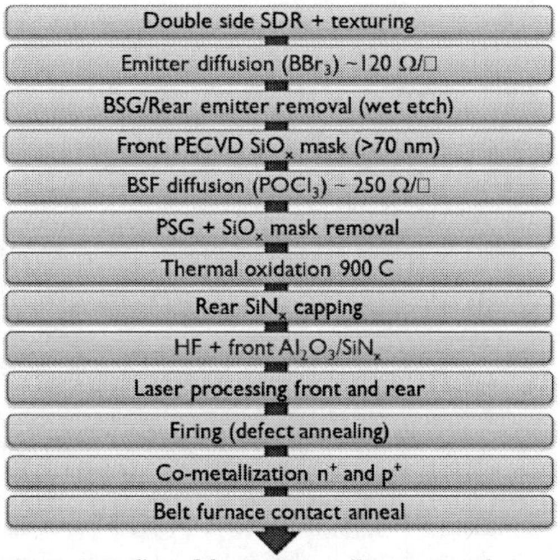

Figure 4. Outline of the nPERT+ cell process used

To demonstrate the feasibility of high throughput cassette plating the wafers were co-plated using 25 wafers/cassette. The cell I-V data of all 44 finished cells is given in table II. The I-V data was obtained with a Grid^{TOUCH} measurement system from Pasan (with 30 wires for current extraction plus 5 wires for voltage measurement on the front and rear) on a low reflection black chuck [8]. J_{sc} is corrected to remove wire shading from the I-V measurement. As finger line resistance is relatively high in these cells there is a small but significant voltage offset between the voltage and current wires in our Grid^{TOUCH} system which causes cell fill factors to be overestimated. The data in table II shows the downwards corrected fill factor values from re-plotting the IV curve taking this effect into account knowing the finger line resistances (n-side 21.4Ω/cm, p-side 20.0 Ω/cm in this case). Typical bifaciality values as measured for a representative nPERT+ cell are also shown in Table II.

Table II. IV data on 44 cells measured using a Pasan Grid^{TOUCH} contact system on a WACOM I-V solar simulator (Cell area 239cm^2 (M0-sized wafers) , wafer resistivity: 5Ωcm, wafer thickness: 180μm, I-V measurement based on calibrated reference cell from Fhg-ISE Callab).

44 cell Result	Jsc [mA/cm^2]	Voc [mV]	FF [%]	η [%]
Average	40.4	691.2	80.3	22.4
Std. deviation	0.1	1.6	0.6	0.2
Best cell	40.5	694.2	81.1	22.8
Bifaciality	97.4%	99.9%	99.9%	97.2%

Effective co-plating is demonstrated by the high average FF and Voc values providing evidence for good plating uniformity, low contact and line resistances suitable for multi-wire cell interconnection whilst maintaining low junction and contact recombination currents.

4 ADHESION AND RELIABILITY DATA

4.1 Adhesion

As Ni/Ag nPERT+ cells have no busbars, adhesion measurements can only be made on individual metallized fingers. We have used the "stylus drag" method to do this where the lateral force applied by a stylus to a finger is measured until failure and the maximum value recorded. [9] There are typically two failure mechanisms: i) where a finger is dislodged before breaking (typical for Ni/Cu/Ag contacts) and ii) where the stylus cuts through the finger removing a section of finger (typical for screen printed contacts). Type ii) failures were observed with Ni/Ag contacts.

Adhesion measurements normalized to finger contact width for cells with different finger line resistances (achieved by depositing differing amounts of Ag) are shown in Figure 5. Process A and process B refer to modifications in the plating process. Figure 5 shows that finger adhesion is fairly independent of line conductivity and can be influenced by plating process conditions. Although these values are promising, field and accelerated module testing are required to judge if co-plating nPERT+ cells can be integrated with SWCT to produce reliable modules.

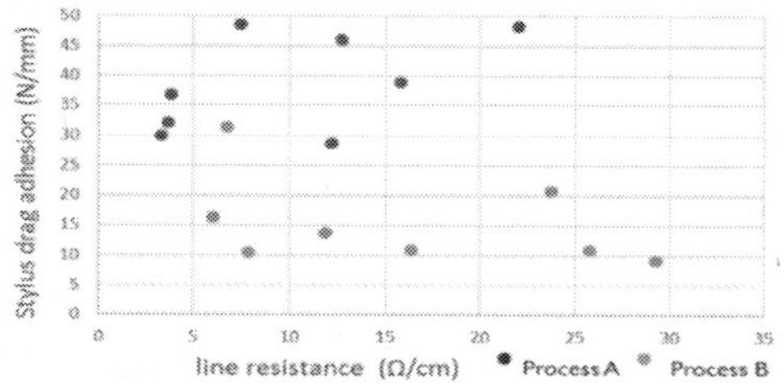

Figure 5. Normalized finger adhesion data versus line resistance for two plating conditions

4.2 Reliability

The proposed plating sequence providing relatively low finger line conductivity is designed for multi-wire interconnection. It is essential therefore that a low and stable contact resistance between the wires and plated fingers can be achieved.

Evidence that uniform low Rs contacts can be made between Ni/Ag fingers and smart wires (18 wires of 300μm diameter) is given in Figure 6.

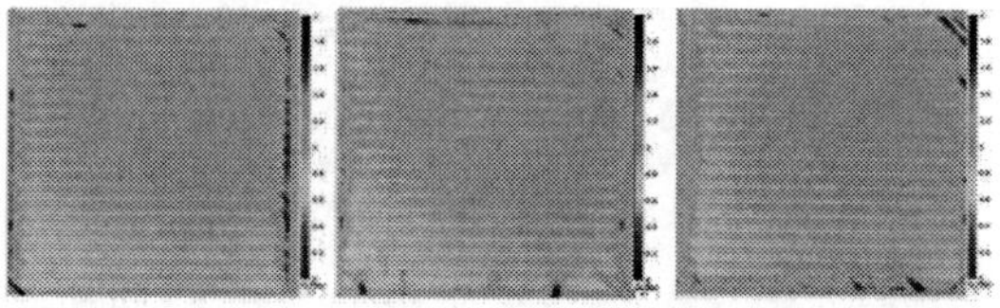

Figure 6. PL-Rs maps of 3 single cell Ni/Ag nPERT+ cell SWCT laminates. Full scale red-black is 0-2 Ωcm^2

Damp heat and thermal cycling testing according to IEC61215 has been performed on single cell Ni/Ag nPERT+ SWCT glass/glass laminates. The results are presented in Figures 7 and 8. Although this initial reliability data is limited to single cell laminates it is encouraging with IEC61215 damp heat and thermal cycling test criteria being met. Further module optimization is being made and more extensive testing planned.

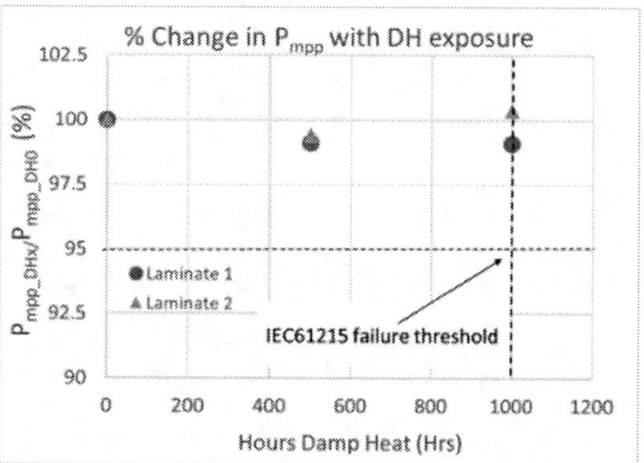

Figure 7. Damp heat results for two single cell laminates

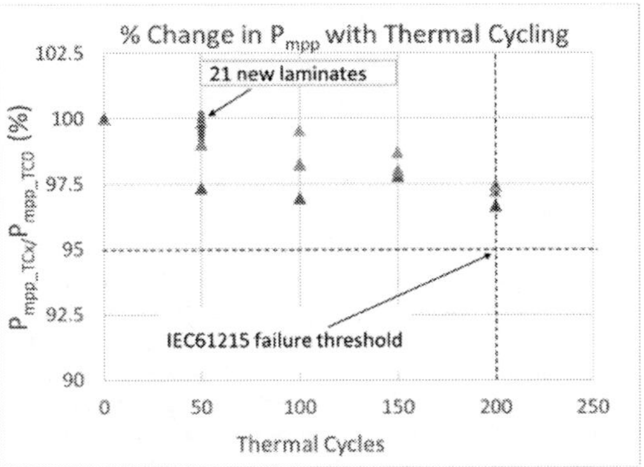

Figure 8. Thermal cycling results on single cell laminates

5 CO-PLATING COST ESTIMATES

One key objective of imec's co-plating process is that it should be cheaper than alternative bifacial metallization routes. To judge this, cost estimates have been made at three levels: at the cell level (cost of metallization/cell in \$/cell), at the module level (module in \$/Wpe) and at the system level (levelized-cost-of electricity in \$/kWh) and compared to two alternative cell structures.

The two alternative screen print based bifacial cell structures selected for comparison are (i) pPERC+ using five busbars and conventional module technology and (ii) nPERT+ cells with minimal Ag usage enabled by Smart Wire Connection Technology (SWCT) as proposed by Meyer Burger. [10]. The cell structures compared are depicted in Figure 9.

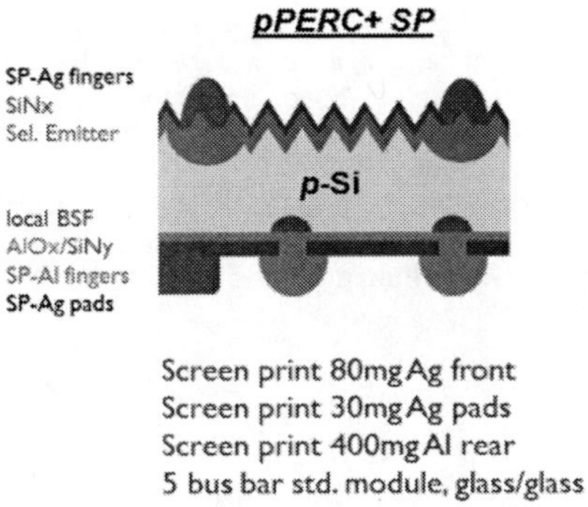

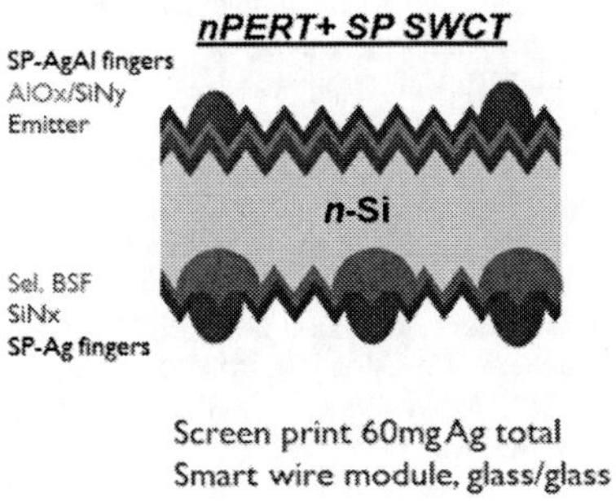

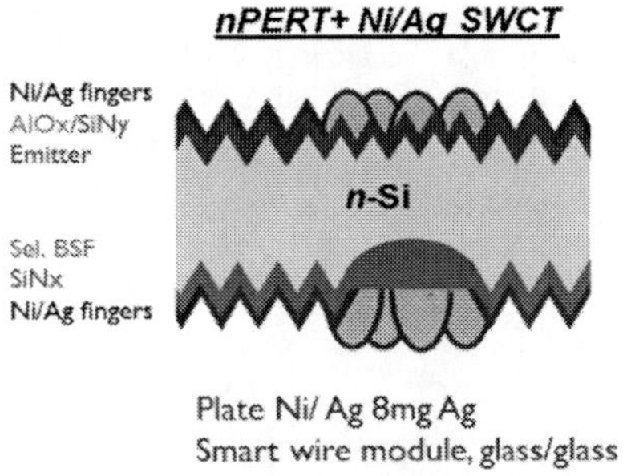

Figure 9. Cell structures and module type used in cost comparisons

5.1 Cost of metallization (\$/cell)

The process steps included in estimating the cost of metallization per cell for the three cell scenarios depicted in Figure 9 are shown in Table III. The costs estimates not only include material costs but also costs related to depreciation, labour, waste, utilities, floor space and yield losses are depicted in Figure 10. Some key assumptions in the cost estimates are: equipment depreciation over 7years, labour costs typical for Asia, Ag paste 800\$/Kg, Al paste 45\$/Kg, Ag pad paste 600\$/Kg, waste disposal costs for contaminated rinse water and plating chemicals included for the plated contacts.

Table III. The process steps included in the cost of metallization for screen printing and Ni/Ag co-plating.

Screen-print (SP)	Ni/Ag
4500 wfr/hr	6000 wfr/hr
• Screen print front • Dry • Screen print rear • Dry • Fast-firing • Test/sort	• Laser ablate front • Laser dope rear • Fast-firing anneal • Ni/Ag plating • N_2 anneal • Test/sort

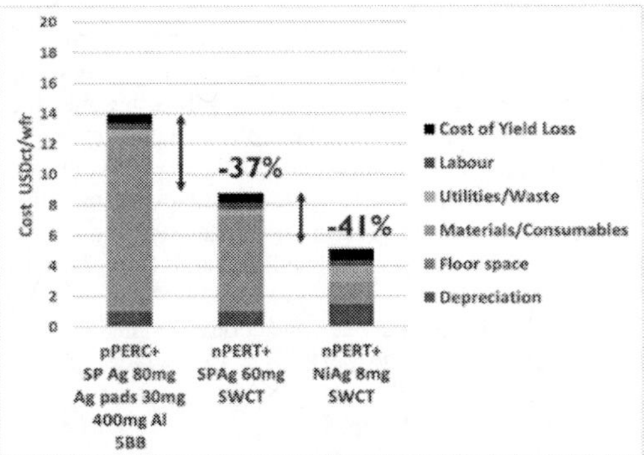

Figure 10. Cost breakdown for the metallization of the cell structures outlined in Figure 9.

The estimated cost per wafer for metallization using Ni/Ag is ~40% cheaper than even the best screen print scenario using a total of only 60mg Ag/cell (suitable for SWCT). This cost reduction is driven solely by a reduction in consumable costs.

5.2 Module level costs (\$/Wpe)

Estimated module level costs are given in Figure 11 for three cell/module technology scenarios being considered. It is assumed that all modules are of a glass/glass construction and the STC front side 72 cell module power used in the estimate is as shown in Figure 11. Note that 375Wp for the Ni/Ag nPERT+ case is equivalent to a bare cell efficiency of 22%.

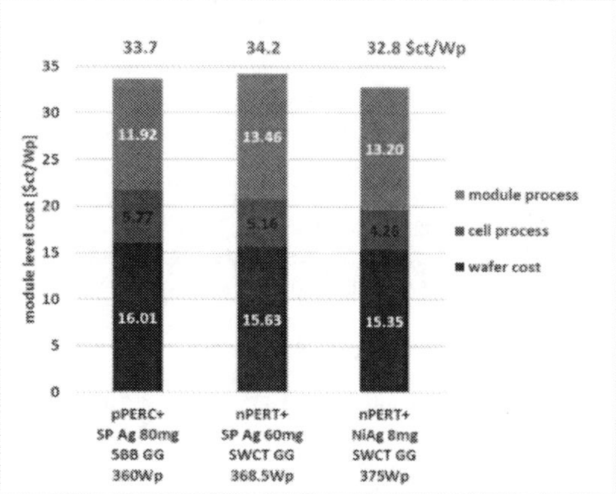

Figure 11. Module level costs ($ct/Wp) for the cell structures and module technologies outlined in Figure 9.

Figure 11 shows that although SWCT module costs are slightly higher, Ni/Ag nPERT+ is still competitive with conventional module pPERC+ due to lower metallization costs and higher power.

5.3 System level costs ($/kWh)

In estimating the system level levelized-cost-of-electricity (LCOE) the main calculation assumptions were: i) a 30 year system lifetime, ii) a 2% degradation in year 1 and a 0.7%/year degradation for all subsequent years and iii) all system installation and running costs (except module costs) have the same $/m^2 cost (independent of module technology). The system installation and running cost $/m^2 value used was calculated from ITRPV 2017 published data. [1].

LCOE values for bifacial cells depend on insolation levels, module bifaciality and albedo (module rear side light intensity as a percentage of module front side light intensity). Table IV shows the module bifacialities used in estimating LCOE for two levels of insolation/albedo. Note that as the insolation level increases absolute values for LCOE decrease but the *percentage* difference between the three module scenarios remains the same (not shown in table IV). Conversely, as albedo increases the *percentage* difference in LCOE between the reference pPERC+ and the nPERT+ modules increases due to the latter's higher bifaciality. Taking pPERC+ SP module as a reference reductions of $\geq$ 5% in LCOE appear feasible with nPERT+ Ni/Ag SWCT modules.

Table IV. Estimated LCOE for the cell structures and module technologies outlined in figure 9 for two values of albedo and insolation.

Cell type	Bifaciality %	Albedo %	Insolation kWh/m^2/yr	LCOE $ct/kWh	LCOE %Δ
pPERC+ SP 5BB	75	10	2000	2.97	Ref
nPERT+ SP SWCT	95	10	2000	2.88	-3.3%
nPERT+ Ni/Ag SWCT	95	10	2000	2.81	-5.4%
pPERC+ SP 5BB	75	30	1000	5.22	Ref
nPERT+ SP SWCT	95	30	1000	4.90	-6.1%
nPERT+ Ni/Ag SWCT	95	30	1000	4.79	-8.2%

6 SUMMARY

Screen-printing or conventional plating techniques have certain disadvantages for the metallization of bifacial cells. They are either relatively expensive, complex or compromise performance. We introduce a relatively simple co-plating approach to try to mitigate these issues which is designed for multi-wire module integration.

A fine line (<20μm) Ni/Ag batch plating process is described with very low Ag usage enabled by a novel silicon selective activation step prior to Ni plating.

Efficiencies up to 22.8% are reported on nPERT+ bifacial cells using this metallization process with an average of 22.4% over 44 cells with high typical cell bifaciality (>95%).

Limited initial 1 cell laminate reliability data is so far encouraging with IEC61215 thermal cycling and damp heat test criteria being met.

Driven by very low Ag usage and low Capex/high throughput tools, costs for this metallization process are low. Cost-of-ownership estimates at the cell level ($/cell), module level ($/Wp) and system level LCOE ($/kWh) for multi-wire modules made from co-plated nPERT+ cells are reported. The LCOE is estimated to be $\geq$ 5% lower for co-plated nPERT+ cell multi-wire modules compared to 5 busbar pPERC+ modules for albedo values $\geq$10%.

7 ACKNOWLEDGEMENTS

The authors gratefully acknowledge the financial support of the partners of Imec's Industrial Affiliation Program for Photovoltaics (IIAP-PV) and MacDermid inc for the adhesion measurements.

This work receives the support of the European Union, the European Regional Development Fund ERDF, Flanders Innovation & Entrepreneurship and the Province of Limburg.

8 REFERENCES

[1] www.itrpv.net
[2] Kopecek R. 2017, Quo vadis bifacial PV? Photovoltaics inyernational , 35th edn, p89-96
[3] Dullweber, T. 2016, PERC+: industrial PERC solar cells with rear Al grid enabling bifaciality and reduced Al paste consumption, PIP, Vol 24, pp 1487-1498
[4] Guo. J.H. 2005, Metallization improvement on frabrication of interdigitated backside and double sided buried contact solar cells, SOLMAT, vol 86, pp 485-498
[5] Karmalkar. S. 2004, Effects of nickel and palladium activations on the adhesion and IV characteristics of as plated electroless nickel deposits on polished silicon wafers, J. Electrochem . Soc.Vol 151, pp C554-C558
[6] Wang. X. 2014, Laser-doped metal plated bifacila silicon solar cells, SOLMAT, vol 131, pp 37-45
[7] Geissbuhler. J. 2014, Silicon heterojunction solar cells with copper-plated grid electrodes: status and comparison with silver thick-film techniques. IEEE J. Photovolt, vol 4 No.4, pp1055-1062

[8] Fakhfouri. V. 2014, Measurement accuracy and precision:the Grid^{TOUCH} contacting system, Meyer Burger-white paper, 2014
[9] Chen. R. 2015, Failure modes identified during adhesion testing of metal fingers on silicon solar cells. Energy Procedia 67 194-201
[10] Soderstrom T. 2013, Smart wire connection technology, 28th EUPVSEC conference pp495-499

This is a pre-peer review version of the paper submitted to the journal Progress in Photovoltaics.
It has been selected by the Executive Committee of the EU PVSEC 2017 for submission to Progress in Photovoltaics.

UNDERSTANDING THE LIGHT-INDUCED DEGRADATION AT ELEVATED TEMPERATURES: SIMILARITIES BETWEEN MULTICRYSTALLINE AND FLOATZONE P-TYPE SILICON

T. Niewelt[1,2], F. Schindler[1,2], W. Kwapil[1,2], R. Eberle[1], J. Schön[1,2], M.C. Schubert[1]
[1]: Fraunhofer Institute for Solar Energy Systems ISE, [2]: Freiburg Materials Research Centre FMF
[1]: Heidenhofstraße 2, 79110 Freiburg, Germany, [2]: Stefan-Meier-Straße 21, 79104 Freiburg, Germany

ABSTRACT: This paper discusses degradation phenomena in crystalline silicon. We present new investigations of the light- and elevated temperature induced degradation (LeTID) of multicrystalline silicon. The investigations provide insights into the defect parameters as well as the diffusivity and solubility of impurity species contributing to the defect. We discuss possible defect precursor species and can rule out several metallic impurities. We find that an involvement of hydrogen in the defect could explain the characteristic observations for LeTID. Furthermore, we demonstrate analogies to the light-induced degradation mechanisms at elevated temperatures observed in floatzone silicon, where several experimental results also indicate an involvement of hydrogen in the defect. Based on the similarities between multicrystalline and floatzone silicon we suggest that both degradation phenomena might be caused by the same or similar defects. As we do not expect large concentrations of cobalt in floatzone silicon, we suggest that complexes of intrinsic lattice defects and hydrogen might cause both degradation phenomena.
Keywords: Silicon; Degradation; Defects

1 INTRODUCTION

Light-induced degradation (LID) in silicon material has been subject to extensive research efforts for several decades. The most prominent example of LID is the degradation due to the activation of boron-oxygen (BO) defects in Czochralski-grown (Cz) silicon addressed to as BO-degradation or BO-LID [1, 2] (and references therein). More recently, silicon with copper contamination was also shown to cause a significant lifetime decrease under illumination referred to as Cu-LID [3] (and references therein). Furthermore, the light-induced dissociation of iron-boron (or other Fe-Acceptor) pairs in contaminated material (e.g. [4, 5]) is often mentioned in the context of LID. These phenomena have a common behaviour in that they can be observed upon illumination at room temperature and that the crucial components of the defects have been identified.

In 2012, *Ramspeck* et al. extended the investigation of multicrystalline (mc) silicon PERC solar cells to conditions relevant for photovoltaic module operation in the field, i.e. to illumination at elevated temperatures above ~60°C for long timescales [6]. As a result, they observed a severe drop in the cell efficiency that exceeded effects that could reasonably be explained with iron contamination or BO defects. The effect was termed Light- and elevated Temperature-Induced Degradation (LeTID) later on [7]. Several distinct differences such as the degradation kinetics, its injection dependence and the absence of direct correlation to the concentrations of boron or interstitial oxygen indicate that LeTID is not related to BO-degradation, FeB dissociation or Cu-LID, e.g. [6, 8, 9].

In the recent months and years, many different experimental observations concerning LeTID by several groups have been published which – in conjunction – yield a set of clues which eventually may help to identify the responsible "LeTID defect". This contribution merges results from literature and findings of the authors and discusses the possible conclusions which can be drawn. Section 2 focuses on the multicrystalline materials usually associated with LeTID, while section 3 discusses implications of similar observations on monocrystalline silicon. Finally, we present a hypothesis for defect components and complex formation mechanisms underlying LeTID, which explains the degradation mechanisms in multi- and monocrystalline silicon.

2 MULTICRYSTALLINE SILICON

2.1 Experimental Insights

The LeTID effect has been investigated by several groups on various different sample types (including different solar cell concepts) and under different conditions. After initial suggestions that the degradation might arise from the degradation of dielectric surface passivation layers [8], many results indicate that the degradation arises from a recombination-active bulk defect [7, 10, 11]. The studies suggest a great variety of features of the responsible defect and factors influencing the degradation. Some of the findings recur in different studies and from different aspects, while others arise from specific experimental details. For the sake of clarity, we structure the discussion of the factors influencing the LeTID effect and features of the underlying defects into three parts. The first part introduces general features of the overall degradation effect and how they are affected by sample processing conditions. The observed transitions between the metastable defect states are discussed in the second part, and the third part discusses characteristic features observed in the degraded state.

2.1.1 General Features and Influence of Processing Conditions

There are indications that different silicon materials (e.g. wafers from different mc silicon ingots or different positions in the ingot) show a different degradation behaviour, e.g. [10, 12]. Notably, it was perceived early on that both B- and Ga-doped material are susceptible to LeTID [6], although there are indications that the degradation is slowed in Ga-doped material [13]. A detailed analysis of the underlying differences of the materials could provide interesting insights into the responsible defects. However, comparability of experiments on different materials from multiple studies is hindered by the observation that even the same starting material already features very different LeTID effects for slight variations in sample processing schemes, e.g. [14].

In particular, the severity of LeTID seems to be determined by high temperature applications such as the fast firing step. As *Bredemeier* et al. [15], *Nakayashiki* et al. [10], and *Chan* et al. [16] showed, the magnitude of LeTID scales with the peak firing temperature: stronger degradation is observed at higher peak temperature. Below a threshold sample temperature of around ~700 °C, no significant LeTID is observed. Besides the influence of peak temperature an influence of the temperature ramps is observed. We demonstrated in Ref. [14] that LeTID can be effectively suppressed by applying slower temperature ramps. It is still unclear whether the slower heating or cooling or the longer time above a critical temperature cause the suppression. Alternatively, the degradation behaviour, i.e. the magnitude and the kinetics, can be influenced by adding a subsequent temperature treatment after the firing [16, 17], which may even lead to a suppression of the degradation effect.

In addition to the firing process, the surface passivation layer seems to have a crucial impact on LeTID: if no passivating layer is present during the firing step, no LeTID is observed [18]. The presence of dielectric passivation layers (such as silicon nitride SiN_x aluminium oxide AlO_x [18] or silicon oxide SiO_x [19]) during firing gives rise to LeTID. However, the passivation scheme affects LeTID: while SiN_x and AlO_x/SiN_x-stacks processed under identical conditions lead to a very similar behaviour, samples featuring only AlO_x passivation layers show a slower and less pronounced degradation [18].

2.1.2 Defect State Transitions

The detrimental aspect of LeTID from a solar cell perspective is the **degradation** of charge carrier lifetime during illumination at elevated temperature, i.e. during the typical working conditions of photovoltaic modules. The experiments of *Ramspeck* et al. were performed at about 75 °C [6]. At such temperatures, the degradation is observed on timescales of several hundred hours of illumination, i.e. it proceeds rather slowly in comparison to other LID effects. Not surprisingly, the degradation can be significantly accelerated by using higher temperatures, as shown e.g. by *Bredemeier* et al. who determined the activation energy of the degradation process to be ~0.9 eV [20]. Below a threshold of around 50 °C hardly any degradation is observed [13].
The degradation is caused by illumination but also occurs under current injection, e.g. [7]. Thus, it is likely caused rather by excess charge carrier injection than light itself. In a recent publication we have used this feature to create defined injection conditions in solar cells susceptible to LeTID and study the defect activation kinetics [21]. We found a linear dependence of the degradation rate constant to the injected charge carrier density Δn at least in the regime of 10^{13} - 10^{15} charge carriers per cm^3. This finding is in accordance with the observation of several groups that increased illumination intensity gives rise to a faster degradation, e.g. [22]. So far, no saturation of the degradation rate for stronger illumination was reported and in the study underlying Ref. [21] we observed no saturation for strong current injection. It should be noted, that the investigation of the defect state transitions is hindered by their simultaneous occurrence, e.g. the likely superposition of degradation and recovery during illumination at elevated temperature.

A characteristic feature of LeTID is that the highly recombination-active degraded state of the underlying defect is metastable under the conditions that cause degradation. If the degradation conditions (i.e. charge carrier injection *and* elevated temperature) are maintained for a longer time, lifetime **recovery** (or regeneration) sets in. The recovery usually saturates at approximately the initial lifetime level [23]. Like the degradation, the recovery kinetics are strongly influenced by the injection conditions and temperature [22]. So far, experimental studies indicate that the resulting state is then stable to further illumination at elevated temperature, e.g. [7].
This defect state is not the only state of the defect that can be reached from the degraded state. If the sample is subjected to elevated temperature without carrier injection, a lifetime recovery can be observed as well, but the resulting state is prone to re-degradation upon subsequent degradation conditions. The process is usually addressed to as the **annealing** transition. We have observed this annealing transition at a temperature of 200 °C. To our knowledge no detailed study on its temperature dependency has been published yet. Alternatively, a similar transition to an unstable recovered state can be achieved via illumination at low temperatures (i.e. below the apparent degradation threshold temperature of 50°C), e.g. [18]. This transition is carrier-driven: by applying a constant forward bias to solar cells in the dark at a temperature of ~25°C, we observed a recovery to roughly the initial solar cell parameter level (not shown here), which could be re-degraded. At present, it is unclear whether the defect configurations are the same after dark anneal and carrier injection at room temperature.
There are indications that the second degradation after a defect anneal proceeds faster than the first [24]. This finding could be related to results published by *Chan* et al.: they showed that a dark anneal of several minutes to hours can also affect the defect, if it is performed *prior* to degradation [25]. They demonstrated that moderate temperature annealing significantly changes the subsequent degradation: annealing between ~150 °C and ~200 °C accelerates the degradation, whereas annealing performed between ~225 °C and ~275 °C causes a slower degradation. In a later work, Chen et al. [26] proved that even annealing in the dark at 175 °C induces degradation followed by regeneration, which they attributed to the thermally generated minority carriers. Interestingly, as mentioned above a (short) anneal at temperature exceeding 275 °C prior to degradation can suppress the degradation [16].

2.1.1 Degraded State

In the degraded state the defects responsible for LeTID introduce a strong recombination activity. It is a reasonable assumption that the recombination causing the lifetime degradation is a measure for the concentration of defects in this state. This allows investigating the spatial distribution of the LeTID defect via spatially resolved measurement techniques such as electroluminescence (EL) or photoluminescence (PL) imaging or laser beam induced current (LBIC). Such investigations have indicated that the LeTID defect is distributed in a mostly homogeneous pattern across mc silicon samples and solar cells.

Denuded Zones

In the state of minimum lifetime (i.e. maximum concentration of active defects) thin regions of decreased degradation around grain boundaries have been observed, e.g. [24].
These regions are addressed to as **denuded zones.** One possible origin of these denuded zones is a kinetic effect

due to the lower injection level close to grain boundaries. It could cause a local retardation of the degradation and thereby an *apparently lower* concentration at the moment of the overall degradation maximum. This source of doubt can be discarded from the investigations performed on solar cells. There, the recombination activity of grain boundaries is counteracted by compensating currents through the emitter and metal grid. Thus, the experiments we performed to study the degradation kinetics (see in Ref. [21]) were performed with laterally almost homogeneous injection density. We still observed typical denuded zone patterns around grain boundaries in the EL monitoring (see an example in Fig.1) and did not observe the grain boundaries to exhibit a retarded degradation. It should be noted that the same argumentation is true for LBIC studies of degraded solar cells (e.g. [24]), which are however, prone to effects arising from the different injection dependence in initial and degraded state.

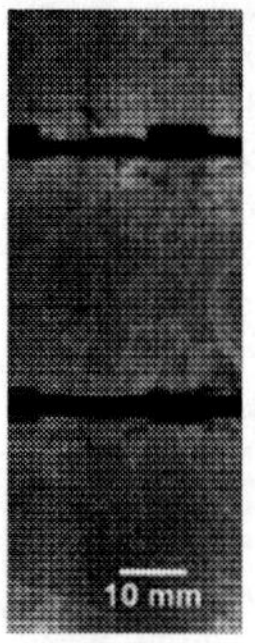

Figure 1: Fraction of an EL image of a PERC solar cell (p-type 1 Ωcm base material, processed for maximum LeTID, see Ref. [21]) at the state of maximum degradation showing denuded zones around grain boundaries. Please note that the horizontal lines of higher EL intensity arise from the local rear contacts.

A second explanation for the observation of denuded zones arises from artefacts due to the strong injection-dependence of charge carrier lifetime in multicrystalline silicon: For instance, a determination of the effective defect concentration $N_t^* = 1/\tau_{deg} - 1/\tau_{initial}$ from lifetime τ calibrated images featuring different excess charge carrier concentrations (with respect to time or spatial position) would implicate large errors on the derived effective defect concentration including lateral variations which might appear as denuded zones. To exclude such artefacts, we performed further studies that prove the existence of actual denuded zones of lower LeTID defect concentration: In a detailed investigation, we ensured that the initial and degraded state are evaluated at the same injection level Δn (here: $1\cdot10^{13}$ cm^{-3}) for every pixel. This was achieved via interpolation of injection dependent lifetime-calibrated PL image stacks [27] to create images at fixed injection (as discussed e.g. in Ref. [28]). Beforehand, the PL lifetime images were corrected for lateral charge carrier diffusion by applying a de-smearing algorithm, as suggested by *Phang* et al. [29], in order to obtain images of the real local charge carrier lifetimes. This was found to be especially important in the initial state, where high lifetimes in the grains and low lifetimes at the grain boundaries give rise to lateral current flows concealing the actual recombination mechanisms. We expect a slight remaining overestimation of N_t^* in our evaluation in regions of strong recombination (e.g. dislocation clusters) due to effects of optical blurring that affects the initial state stronger than the degraded state. The investigations confirmed that thin regions around the grain boundaries in the investigated lifetime samples feature a smaller degradation (corresponding to smaller effective defect concentrations N_t^*) than the grains themselves, which exhibit homogeneous degradation. Fig.2 exemplarily shows three images of a commercial 1 Ωcm p-type high performance mc-Si wafer from these investigations: two de-smeared bulk lifetime images in the initial (a) and the degraded state (b) (taken at an illumination intensity corresponding to ~ 1 sun) as well as an image of the effective defect concentration N_t^* (c). The denuded zones are visible both in the lifetime image in the degraded state (b) as brighter regions of increased lifetime and in the N_t^* image (c) as dark regions of reduced defect concentration. In order to determine the width of the denuded zones we performed high resolution microscopic PL investigations using a scanning confocal microscope, as discussed in [30]. The total width (i.e. in both directions from the grain boundary) of the denuded zones was found to vary in a range of 200-400µm. From this information we can draw conclusions regarding the defect species responsible for LeTID, which will be discussed in section 2.2.

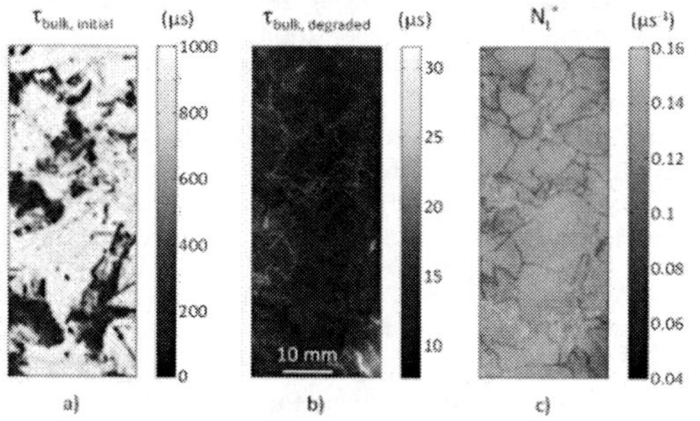

Figure 2: Degradation of a commercial 1 Ωcm p-type (boron doping) high performance mc-Si wafer: a) Fraction of a lifetime-calibrated PL image taken at 1 sun equivalent in the initial state b) Lifetime-calibrated PL image taken at 1 sun equivalent at maximum degradation after 105 hours (same sample fraction) c) Effective defect concentration N_t^* illustrating the additional recombination caused by LeTID defects evaluated at fixed injection level of $\Delta n=1\cdot10^{13}$ cm^{-3} (same sample fraction).

Defect Parameters

As discussed e.g. in Ref. [31] the Shockley-Read-Hall parameters describing the recombination activity of a defect can serve as a fingerprint for its identification. Several groups have performed analyses of injection dependent lifetime measurements to isolate the dominant defect in LeTID affected samples and study its recombination activity [10, 32-34]. As discussed in detail in Ref. [35] such investigations can determine possible pairs of the capture constant ratio k and the energy level in the band gap E_t but need further input to determine the right pair. This information can usually be derived from temperature dependent measurements of the recombination activity of the defect. However, such investigations require the defect to remain unchanged during the measurements, which is complicated for the LeTID defect due to its strong sensitivity to temperature treatments. Therefore, most studies use the assumption that E_t is at or around the middle of the band gap, where the $k(E_t)$ function usually features a plateau. Thus, *Nakayashiki* et al. [10] reported a value for k = 28.5, *Bredemeier* et al. [32] determined a value for k = 20±7, and *Morishige* et al. [33] reported on values in the range of $26 < k < 36$.

More recently *Vargas* et al. performed temperature dependent analyses and reported ranges of k and E_t [34]. In order to avoid changes of the defect during the measurements, they restricted the investigated temperature range to between -25 °C and 75 °C. They report on two possible defect energy levels of E_t - E_i = -(0.32±0.05) eV or $E_t - E_i$ = (0.21±0.05) eV in the lower and upper halves of the bandgap with corresponding values for the capture cross section ratios of k = 56±23 or k = 49±21, respectively.

a)

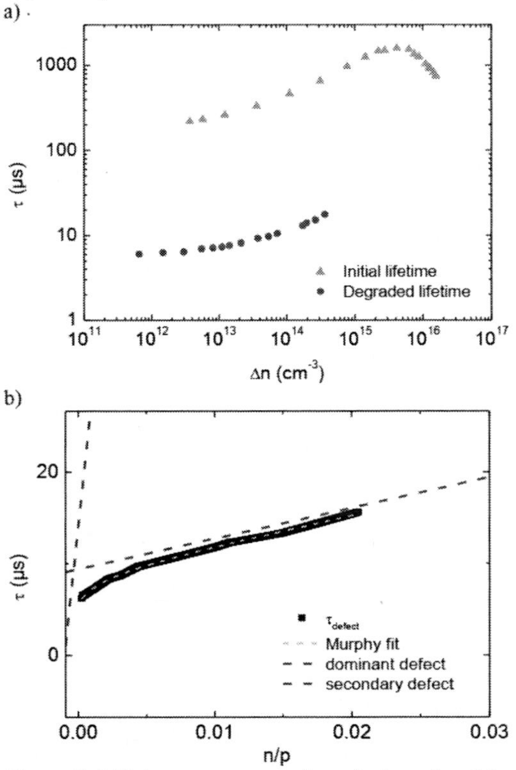

Figure 3: Lifetime spectroscopy in a single grain of the sample shown in Fig.2 (commercial 1 Ωcm p-type high performance mc-Si wafer) a) Lifetime as a function of excess carrier density before (orange triangles) and after (cyan circles) activation of LeTID defects (~1 sun equivalent, ~75 °C). b) The defect-related lifetime limitation τ_{defect} as a function of the ratio of minority and majority carrier density $X = n/p$. The dashed lines indicate that τ_{defect} can be described by superposition of two defects. The dominant defect features a value of $k \approx 35$ assuming a mid-band energy level.

All literature results reported above were obtained by making use of global measurement techniques which measure average lifetimes across several grains of the mc silicon. These measurements are thus affected by influences of lateral sample inhomogeneity (e.g. grain boundaries and dislocation clusters) that can affect the measured injection dependence. We performed an experiment to ensure that this experimental limitation does not cause structural misinterpretations. Based on our spatially resolved measurements we were able to avoid the averaging over structural defects. We thereby performed an injection-dependent lifetime spectroscopy within a single grain on a wafer. The PL image stacks were corrected for lateral carrier diffusion as indicated above and the lifetime was averaged over a homogeneous grain to reduce signal noise. Fig.3a) shows the diffusion-corrected lifetime measured and averaged in a single grain of a commercial 1 Ωcm p-type high performance mc silicon wafer (same sample as shown in Fig.2) before and after degradation (~1 sun equivalent halogen lamp illumination at ~75 °C sample temperature). The recombination introduced by LeTID defects was isolated via comparison of degraded and annealed state. The quantity $\tau_{defect} = (1/\tau_{degraded} - 1/\tau_{initial})^{-1}$ was investigated as a function of the ratio of minority and majority carrier density n/p (according to the linearized SRH approach suggested by *Murphy* et al. [36]). Thereby we identified a dominant defect which features a ratio of capture cross sections of $k \approx 35$ (*cf.* Fig.3b)) under the assumption of a deep defect level. We thus observe a comparable injection level dependence as that reported in literature. Therefore we can exclude severe influences of averaging artefacts.

2.2 Implications for the Responsible Defect

The various experimental findings allow narrowing down potential components of the defect complex responsible for LeTID. In the initial state of samples after typical solar cell (or lifetime sample) processing steps, little recombination activity due to LeTID defects is observed. Thus, the defect is not (or only weakly) recombination-active, indicating that the majority of defects are either in an inactive state or not yet formed. However, as discussed above, details of the sample process greatly influence the amount and the kinetic behaviour of LeTID. This indicates that they affect the concentration of formed defects or the concentration of available precursor species necessary to form the defect during a degradation treatment.

We have demonstrated in the previous section that spatially resolved investigations show denuded zones around grain boundaries. This indicates that the defect or an important precursor species is internally gettered in the grain boundaries. This internal gettering can occur either during crystallization, phosphorus diffusion, or during the cooling after the final firing step. The binding energies between crystal defects and impurities are too small to cause segregation at high temperatures [37]. Thus, the responsible gettering mechanism is most likely precipitation, which would allow gettering only at temperatures with a supersaturated impurity. Some of the impurity species observed in mc silicon can be ruled out from causing the denuded zones in LeTID as they have too low diffusivities to form denuded zones of >100 µm width. These are elements such as B, Ga, P, W, Ti, V, O_i, Zn, and Na, as they cannot diffuse far enough during typical crystal growth or any high temperature step during the solar cell processing. Thus, the observation of denuded zones shows that they cannot be the LeTID defect by themselves but would need to form a complex with a more mobile species. This conclusion contradicts the suggestions of *Nakayashiki* et al. and *Morishige* et al. based on their recombination activity analyses, which suggest that the LeTID defect might be interstitial titanium (Ti_i) or substitutional tungsten (W_s) [10, 33].

The assumption that the denuded zones are created during crystal growth would rule out further species, as well. Impurities that are known to diffuse fast at moderate temperatures would be redistributed homogeneously during solar cell processing (e.g. during phosphorus diffusion steps) due to a solubility below typical impurity concentrations at process temperatures. Denuded zones in solar cells and fully processed lifetime samples could therefore not have been formed by Cr, Mn, Mo, Fe, Cu, Ni

This is a pre-peer review version of the paper submitted to the journal Progress in Photovoltaics.
It has been selected by the Executive Committee of the EU PVSEC 2017 for submission to Progress in Photovoltaics.

and Co during the crystallization step.

Thus, the creation of the denuded zones related to LeTID is likely to occur during the firing steps, which is in good agreement with the observation of the strong influence of the firing step process details on the LeTID defect or its precursors. If we assume that the denuded zones are formed during the cool down in the firing process, the fast-diffusing species ruled out above become the likely candidates for defect precursors. However, for the typical times and temperatures of firing processes, even the diffusivities of Cr, Mn, Mo and Fe are too small to explain the width of the denuded zones. This leaves Cu, Ni and Co as possible candidates featuring diffusivities in a suitable range (e.g. [38]) to form denuded zones of >100 µm width during cooling down. One of these elements, Cu, has already been shown to cause light-induced degradation.However, several features of Cu-LID differ significantly from LeTID, e.g.: (i) Cu-LID occurs also at room temperature [3] as opposed to an apparent onset of LeTID at elevated temperature, (ii) Cu-LID proceeds significantly faster than LeTID [9], (iii) the capture cross section ratio k of the dominant defect in the degraded state has been determined to 1.7 - 2.6 [39] compared to k~30 for LeTID. In addition, no observation of a regeneration comparable to the complete LeTID cycle has been reported so far. Therefore, we exclude the possibility that Cu-LID and LeTID are identical phenomena, although a participation of Cu in LeTID cannot be dismissed in this line of argument. Yet, both copper and nickel are unlikely candidates for a different reason: The solubility of Cu is, similar to Ni, quite high and all Cu complexes should dissolve already at temperatures well below typical T_{peak} of the firing process. In that case the strong dependency of the LeTID on T_{peak} cannot be explained.

In contrast, in the interesting peak temperatures range the solubility of Co rises from $S_{Co} \approx 2 \times 10^{11}$ cm^{-3} at 700 °C to $S_{Co} \approx 1 \times 10^{12}$ cm^{-3} at 750 °C, which is in the range of typical Co concentrations in as-grown mc silicon originating from crucible coatings [40, 41]. For lower temperatures, the solubility of Co strongly declines (e.g. $S_{Co} \approx 5 \times 10^{9}$ cm^{-3} at 600 °C), which could explain the sensitivity of LeTID to the firing step peak temperature. During typical phosphorus diffusion gettering processes the total Co concentration decreases by around two orders of magnitude [41], which is in agreement with the observation of reduced LeTID in gettered samples [12]. Together with its suitable diffusivity, this renders cobalt the most interesting candidate from the typical metal impurities for being involved in LeTID.

Additionally, a variety of different Co defects are known [38, 42]. As mentioned above, the extent of LeTID degradation clearly correlates with the temperature ramps of the firing process [14]. In this context, the reported sensitivity of Co to cooling rates after high temperature steps is remarkable: while slow cooling processes lead to a diffusion of Co to the wafer surface, fast cooling ramps can facilitate its agglomeration in precipitates in the volume [38]. It should be noted that these forms of Co were observed in intentionally contaminated monocrystalline silicon and that further species might exist in mc silicon due to the presence of additional nucleation sites. Interesting candidates for the LeTID defect in the variety of known Co-related defects are the complexes of Co and hydrogen, which are believed to be recombination-active [38].

Hydrogen is also a likely candidate for a precursor species by itself. While not a common species in freshly crystallized material, it is well known to be introduced to silicon from the wafer surfaces during firing steps, especially in the presence of hydrogen-rich dielectric passivation layers. The diffusivity of hydrogen depends very much on the specific sample and process conditions (e.g. [43]) but is high enough to penetrate to the wafer bulk. The large density of defects (e.g. dangling bonds) at grain boundaries makes them effective sinks for H, potentially creating hydrogen-denuded zones.

3 MONOCRYSTALLINE SILICON

It has been demonstrated recently that LeTID also occurs on monocrystalline p-type Cz silicon [26, 44]. The effect is observed in addition to the conventional BO degradation or after specific process steps to mitigate BO defect activation. This finding supports the idea that LID effects observed in floatzone (FZ) silicon (e.g. by *Sperber* et al. [45, 46] and the authors [47]) might arise from a similar or the same defect. We had suggested this possibility in Ref. [47] but now regard it as even more likely in the light of the reports on LeTID in Cz silicon. In the first part of this section we therefore report the experimental findings on FZ-LID that bear similarity to LeTID. For the sake of brevity our own experimental results on FZ-LID are only mentioned in summarized form in the same structure as used in section 2.1. For further details please refer to Ref. [47]. The second part discusses the observed effects and potential implications for LeTID.

3.1 Similarities of FZ-LID and LETID

3.1.1 General Features and Influences of Processing Conditions

An important similarity of LeTID and FZ-LID is that significant degradation is only observed on p-type samples after passivation with dielectric layers and subjection to a short high temperature step ("firing"). We did not observe FZ-LID on n-type samples and are not aware of reported LeTID on n-type material. The peak temperature of the firing step T_{peak} was observed to have a strong impact on the magnitude of the FZ-LID effect. The effect ranged from slight lifetime reduction for moderate T_{peak} around 700 °C to a severe τ_{eff} decay for T_{peak} of 900 °C, measured on equally processed samples. No clear degradation was observed for sister samples fired at only 650 °C or subjected to slow-rate annealing steps at 425 – 450 °C. It should be noted that we observed FZ-LID although our samples were fired with rather slow cooling ramps (in fact the exact same firing processes that caused only weak LeTID in our study presented in Ref. [14]). Besides the dependence on the firing step we observed a crucial influence of SiN_x capping layers: significant degradation was only observed when a SiN_x layer was present, either as a single film (see *Sperber* et al. [45]) or on top of the AlO_x layers that we used to passivate our samples. By contrast, at the same T_{peak} of 800 °C, the omission of SiN_x capping layers drastically reduced the degradation effect.

3.1.2 Defect State Transitions

Besides these similar preconditions of the sample processing there are parallels concerning the observed defect state transitions. We observed the FZ-LID under the same conditions as applied in many studies of LeTID, i.e. ~1 sun equivalent halogen lamp illumination at ~75 °C sample temperature. *Sperber* et al. also demonstrated the

effect to occur at higher temperatures [46]. By dark annealing FZ samples, in degraded state can be reverted to a high lifetime state that is prone to repeated degradation [45], resembling the effect of annealing of LeTID defects. We also observed a lifetime recovery of degraded FZ samples upon illumination at room temperature with a laser, resembling the light-induced annealing of LeTID defects.
When the conditions that cause degradation (i.e. illumination at elevated temperature) were maintained after complete degradation, samples affected by FZ-LID exhibited a complete lifetime recovery. The resulting state was demonstrated to be stable to further exposition to degradation conditions, as demonstrated in Ref. [48]. In summary, our experiments demonstrate that the same transitions observed in samples affected by LeTID occur in samples prone to FZ-LID. However, the processes occurred on different – i.e. significantly shorter – time scales in the FZ samples. Under the tested degradation conditions the whole cycle of degradation and complete recovery was completed in less than 40 hours compared to typically several hundred hours for LeTID.

3.1.3 Degraded State

In our study of FZ-LID we applied PL imaging to investigate the lateral distribution of the effect. In the FZ-LID degraded state the wafers featured a characteristic ring pattern. We observed fine striations in the measured PL intensity in a range of 5%. An example is given in Fig.4 for a FZ sample featuring in degraded state. These local variations are likely to be associated with a lateral variation of the concentration of the defect causing FZ-LID. Similar patterns have been shown for the variation of the doping concentration of FZ wafers, e.g. [49]. Thus, it appears that the defect responsible for FZ-LID includes either dopant atoms or a yet unknown species which is distributed alike during the FZ crystallization process. In mc silicon the strong lateral inhomogeneities caused by structural features and the superposition with other recombination channels make it difficult to observe or exclude such features. However, we observed a similar pattern in the lateral distribution of the degradation time constants [28] that might be related to lateral doping variations in mc ingots.

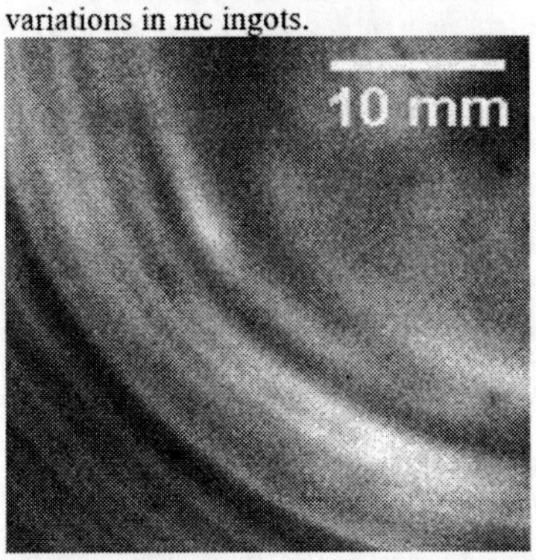

Figure 4: Fraction of a PL image of a p-type 1 Ωcm FZ sample (processed for strong FZ-LID, as discussed in Ref. [47]) in FZ-LID degradation maximum.

The additional recombination introduced by the bulk defects responsible for FZ-LID exhibits a strong injection dependence. We performed an analysis of the isolated defect-related lifetime limitation τ_{defect} similar to the one presented in section 2.1.3. Good agreement with the measurements was achieved under the assumption of a deep defect level with a capture cross section ratio $k \approx 30$. *Sperber* et al. found a similar value for k of 20±2 [45]. In our experiments samples prone to FZ-LID featured indications of this injection dependent recombination already in initial state. This supports the assumption that a fraction of the defects causing FZ-LID might be present/active directly after sample processing. If this is the case, the defects observed by *Grant* et al. in Ref. [50] might be the same, which is supported by the similar value of $k \approx 26$ they found to fit their data. Note that these values reported for FZ silicon thus show good agreement with the ranges reported for LeTID, as discussed in section 2.1.3.

3.2 Discussion and Implications

This set of similarities is no proof that the underlying defects are actually the same. However, we consider it reasonable to discuss the potential implications concerning LeTID.

The observation of shorter process durations in the FZ samples agrees well with the concept that electrons (or more general excess charge carriers) play a role in the defect state transitions. Besides the observed influence of illumination intensity on the transitions of LeTID affected samples, this concept is supported by our study presented in Ref. [21]. There, we demonstrated a direct correlation of the degradation rate constant of mc silicon based solar cells and the injected charge carrier concentration. The FZ samples we investigated featured an excellent surface passivation and low background recombination caused by contaminations or structural defects. Therefore, the same illumination conditions gave rise to a higher excess charge carrier concentration Δn than in typical mc silicon samples, which provides a good explanation for the faster degradation and recovery. An estimation of the injection level in the FZ and mc samples during illumination supports that the degradation-recovery cycles are quite comparable, if the influence of Δn is taken into account.

Thus, the kinetics of LeTID are sensitive to lifetime limitations introduced by other defects in the material or influences of surface passivation, metal contacts, etc. For mc silicon this implies that higher material purity and crystallographic quality affect the observed LeTID, potentially explaining several observations in literature, e.g. [7, 10, 12]. For Cz grown silicon a strong influence of lifetime limitations induced by BO defects can be expected. Besides arising apparent influences of the concentrations of boron and interstitial oxygen it must be kept in mind, that the transitions between different BO defect states are also known to occur at the conditions used for investigation of LeTID. Similar considerations should be true for other metastable defects such as FeB and CrB pairs or Cu-LID.

The central implication of LeTID in Cz and FZ silicon concerns the question which precursor species appear to be reasonable. In section 2.2 we have discussed several potential precursor species based on the common contaminants in mc silicon and identified the most likely ones. Given that Cz and FZ silicon are grown under different conditions concerning cleanliness and growth conditions than mc silicon, different crystal defect species are expected. The incorporation of metal contaminations (including Cobalt) during crystal growth can be expected to be orders of magnitude smaller than for typical mc silicon. Thus, potential defect precursors are most likely to be either intrinsic lattice defects such as interstitial silicon atoms Si_i or lattice vacancies V_{Si} (e.g. [51]), the dopant species boron (or Ga), nitrogen introduced during crystal growth (e.g. [50]) or species introduced from the wafer surfaces during processing.

As mentioned in section 2.2, the dopant species should not show denuded zones and can therefore not be the concentration-limiting precursor species.

A species which is known to be introduced to silicon wafers during the process sequences giving rise to LeTID and FZ-LID is hydrogen. An involvement of H in the defect formation could explain the influences of surface passivation layers and the firing conditions. Release of H from dielectric passivation layers during thermal treatments is expected and peak temperatures above 600 °C would allow its diffusion throughout the wafer thickness (e.g. [52-55] and references therein). H atoms themselves are not likely to be recombination-active defects. Also, many crystal defects can be effectively passivated upon binding with hydrogen (e.g. grain boundaries: [56, 57]). However, formation and subsequent dissolution or evolution of a recombination-active H complex would be a valid explanation for observations related to LeTID and FZ-LID. The diffusivity of hydrogen varies by orders of magnitude depending on its charge state, which can be influenced by illumination, e.g. [43, 58]. The conditions applied for degradation increase the concentration of neutral hydrogen H^0, which is very mobile and can therefore diffuse to other defect precursors. The necessity to promote recharging of hydrogen might also be the underlying mechanism causing the dependence of the degradation rate on Δn discussed in Ref. [21].

The lateral distribution of FZ-LID we have observed exhibited a characteristic striation pattern. This implies that a grown-in species in the wafers contributes to the defect causing FZ-LID. This structure could arise from slight instabilities of the FZ growth process that cause a variation of the intrinsic defect concentration. It is a reasonable assumption, that the FZ and Cz samples we and the other groups have investigated were cut from crystals grown rather vacancy-rich due to throughput considerations. However, this is speculative and the inclusion of intrinsic lattice defects in mc silicon depends on the specific growth conditions. We can therefore neither confirm nor exclude the involvement of Si_i or V_{Si} but consider them interesting candidates for future investigations.

4 CONCLUSION

In this work, we have discussed light-induced degradation effects observed in multicrystalline and FZ silicon. In the first part we excluded many typical defects from playing an important role in LeTID in mc silicon based on the occurrence of denuded zones. Species that are known to diffuse slowly are unlikely to show depletion on the length scales observed for the widths of the denuded zones. They can therefore only be precursors of the LeTID defect if they form a complex with a fast-diffusing species. Cobalt and hydrogen were identified as species that could create recombination-active defects and feature suitable diffusivities to form the observed denuded zones. Also, both species are known to be affected by fast high temperature steps. The involvement of hydrogen would further explain the influences of dielectric passivation layers.

The second part of this paper compares features of a light-induced degradation effect observed in FZ silicon samples to those of LeTID in mc silicon. We demonstrate that both effects are very similar and thus might arise from similar or even the same defect reactions. If both degradation effects are caused by the same defect, the similar extent of degradation in FZ silicon renders the involvement of metal contaminations unlikely. In this line of reasoning, LeTID is suggested to be caused by defect complexes of mobile hydrogen and an intrinsic crystal defect.

ACKNOWLEDGEMENT

This work was supported by the German Federal Ministry for Economic Affairs and Energy BMWi and by the industry partners within the research cluster SolarLIFE under contract No. 0325763A. The authors are responsible for the content.

REFERENCES

[1] K. Bothe and J. Schmidt, J Appl Phys **99**, 13701 (2006).
[2] T. Niewelt, *et al.*, IEEE J. Photovolt. **7**, 383 - 398 (2016).
[3] J. Lindroos and H. Savin, Sol Energ Mat Sol C **147**, 115-126 (2016).
[4] G. Zoth and W. Bergholz, J Appl Phys **67**, 6764-71 (1990).
[5] L.J. Geerligs and D. Macdonald, Appl Phys Lett **85**, 5227-9 (2004).
[6] K. Ramspeck, *et al.*, 861 - 865 (2012).
[7] F. Kersten, *et al.*, Sol Energ Mat Sol C **142**, 83-86 (2015).
[8] F. Fertig, *et al.*, physica status solidi (RRL) – Rapid Research Letters **9**, 41-46 (2015).
[9] H. Vahlman, *et al.*, physica status solidi (a) **214**, 1700321-n/a (2017).
[10] K. Nakayashiki, *et al.*, IEEE J. Photovolt. **6**, 860-868 (2016).
[11] T. Luka, *et al.*, 826 - 828 (2015).
[12] A. Zuschlag, *et al.*, Progress in Photovoltaics: Research and Applications **25**, 545-552 (2017).
[13] J. Fritz, *et al.*, **[submitted]**, (2017).
[14] R. Eberle, *et al.*, physica status solidi (RRL) – Rapid Research Letters **10**, 861-865 (2016).
[15] D. Bredemeier, *et al.*, AIP Adv. **6**, 035119 (2016).
[16] C.E. Chan, *et al.*, IEEE J. Photovolt. **6**, 1473-1479 (2016).
[17] P. Engelhart and F. Kersten, *Solarzellenherstellungsverfahren und Solarzellenbehandlungsverfahren Solar cells and solar cell manufacturing processes treatment method*. 2015, Google Patents: Germany.
[18] F. Kersten, *et al.*, Energy Procedia **92**, 828-832 (2016).
[19] D. Skorka, *et al.*, 643 - 646 (2016).
[20] D. Bredemeier, *et al.*, Solar Energy Materials & Solar Cells **(submitted)**, (2017).
[21] W. Kwapil, *et al.*, Solar Energy Materials & Solar Cells **(available online)**, (2017).
[22] D.N.R. Payne, *et al.*, Sol Energ Mat Sol C **158**, 102-106 (2016).
[23] M.A. Jensen, *et al.*, IEEE J. Photovolt. **7**, 980-987 (2017).
[24] T. Luka, *et al.*, Sol Energ Mat Sol C **158**, 43-49 (2016).

[25] C. Chan, *et al.*, Solar RRL **1**, 1600028-n/a (2017).
[26] D. Chen, *et al.*, Solar Energy Materials & Solar Cells **172**, 293-300 (2017).
[27] J.A. Giesecke, *et al.*, Solar Energy Materials & Solar Cells **95**, 1011-8 (2011).
[28] M. Selinger, *et al.*, Energy Procedia **92**, 867-872 (2016).
[29] S.P. Phang, *et al.*, Appl Phys Lett **103**, 192112 (2013).
[30] F.D. Heinz, *et al.*, physica status solidi (RRL) – Rapid Research Letters **9**, 697-700 (2015).
[31] L.E. Mundt, *et al.*, IEEE J. Photovolt. **5**, 1503-1509 (2015).
[32] D. Bredemeier, *et al.*, Energy Procedia **92**, 773-778 (2016).
[33] A.E. Morishige, *et al.*, IEEE J. Photovolt. **6**, 1466-1472 (2016).
[34] C. Vargas, *et al.*, Appl Phys Lett **110**, 092106 (2017).
[35] S. Rein, Lifetime spectroscopy: a method of defect characterization in silicon for photovoltaic applications, (2005, Berlin Heidelberg, Springer).
[36] J.D. Murphy, *et al.*, J Appl Phys **111**, 113709 (2012).
[37] M. Seibt, *et al.*, Materials Science and Engineering: B **159**, 264-268 (2009).
[38] K. Graff, Metal Impurities in Silicon-Device Fabrication, (2nd ed. 2000, Berlin, Springer).
[39] A. Inglese, *et al.*, J Appl Phys **120**, 125703 (2016).
[40] M.C. Schubert, *et al.*, IEEE J. Photovolt. **3**, 1250-8 (2013).
[41] J. Schön, *et al.*, Sol Energ Mat Sol C **142**, 107-115 (2015).
[42] T.M. Gibbons, *et al.*, J Appl Phys **121**, 045704 (2017).
[43] C. Sun, *et al.*, J Appl Phys **117**, 045702 (2015).
[44] F. Fertig, *et al.*, Energy Procedia **[submitted]**, (2017).
[45] D. Sperber, *et al.*, physica status solidi (RRL) – Rapid Research Letters **11**, 1600408 (2017).
[46] D. Sperber, *et al.*, IEEE J. Photovolt. **7**, 463-470 (2017).
[47] T. Niewelt, *et al.*, J Appl Phys **121**, 185702 (2017).
[48] T. Niewelt, *et al.*, IEEE J. Photovolt. **7**, 1197-1202 (2017).
[49] S.Y. Lim, *et al.*, IEEE J. Photovolt. **3**, 649-655 (2013).
[50] N.E. Grant, *et al.*, J Appl Phys **117**, 055711 (2015).
[51] R. Falster and V.V. Voronkov, Materials Science and Engineering: B **73**, 87-94 (2000).
[52] S. Wilking, *et al.*, J Appl Phys **113**, 194503 (2013).
[53] H.F.W. Dekkers, *et al.*, Appl Phys Lett **89**, 013508 (2006).
[54] B.J. Hallam, *et al.*, IEEE J. Photovolt. **4**, 88-95 (2014).
[55] D. Mathiot, Physical Review B **40**, 5867-5870 (1989).
[56] B.J. Hallam, *et al.*, Energy Procedia **77**, 799-809 (2015).
[57] P. Karzel, *et al.*, J Appl Phys **114**, 244902 (2013).
[58] P. Hamer, *et al.*, physica status solidi (RRL) – Rapid Research Letters **9**, 111-114 (2015).

This is a pre-peer review version of the paper submitted to the journal Progress in Photovoltaics.
It has been selected by the Executive Committee of the EU PVSEC 2017 for submission to Progress in Photovoltaics.

RECONSTRUCTING PHOTOLUMINESCENCE SPECTRA AT LIQUID NITROGEN TEMPERATURE FROM HEAVILY BORON DOPED REGIONS OF CRYSTALLINE SILICON SOLAR CELLS

HUITING WU[1], HIEU T. NGUYEN[1], ANYAO LIU[1], DANIEL MACDONALD[1]
[1]Research School of Engineering, The Australian National University, Canberra ACT 2601

ABSTRACT: When measured at low temperature (79K), the photoluminescence (PL) spectra from silicon wafers containing a diffused heavily-doped layer exhibit a second peak due to band gap narrowing in the diffused region. This work aims to decompose this peak into components arising from the various doping concentrations within the diffused layer. While the peak position of silicon band-to-band PL spectra changes significantly with the doping concentration in silicon, the shape of the spectra also varies strongly with doping concentrations due to the broadening effects of band-filling and band-tail states. By measuring PL spectra on a range of uniformly heavily doped wafers, we show that these changes in spectral position and shape can be accurately modelled for doping concentrations above 1×10^{15} cm^{-3} using simple parameterisations, with minimal impact of variations in excitation intensity or injection level. This allows the PL spectra for a range of arbitrary doping concentrations to be reconstructed. We then show that the PL spectra from a thermally boron-diffused wafer, in which the boron concentration changes with depth, can be reconstructed based on a superposition of PL spectra arising from the layers of different doping beneath the surface. Furthermore, the scaling factor for each layer can be accurately estimated based on the doping profile and the fraction of incident light absorbed in the layer.

Keywords: Characterization, Doping, Photoluminescence, c-Si, Boron, Diffusion.

1 Introduction

Heavily-doped regions serve important functions in silicon solar cells, including forming the p-n junction, lowering the contact resistance for metallisation, and shielding minority carriers from the surface. The most conventional method of achieving high concentrations of dopants near the wafer surface is through thermal diffusion, which results in a non-uniform doping profile. There are two standard methods used to determined doping profiles of heavily-doped layers: Secondary-Ion Mass-Spectrometry (SIMS) and Electro-Chemical Capacitance-Voltage (ECV). However, both of these methods are destructive, and are generally only suitable for relatively large areas (around 1mm^2 or larger), meaning they cannot easily be applied to micron-scale diffused regions typical in many high efficiency devices.

Various techniques for measuring dopant concentrations in silicon based on spectrally-resolved or intensity variations in PhotoLuminescence (PL) have been developed [1-7]. Such techniques are contactless, and can in principle be applied at the micron-scale via the use of confocal optics. Methods based on spectrally-resolved PL, or PL Spectroscopy (PLS), utilise the fact that the heavily-doped regions are subject to band-gap narrowing effects, reducing the energy of the PL emission from these regions. When measured at room temperature, the entire PL spectrum is thermally broadened, and the component emitted from the heavily-doped region appears as a shoulder on the PL peak from the underlying substrate, and is therefore not easily resolved. Nevertheless, room temperature PLS methods have been developed which can reveal the sheet resistance as well as doping concentration for uniformly doped or diffused regions[1, 2, 4].

In this work, we attempt to use liquid nitrogen temperature PL spectra, for which the heavily boron-doped region results in a distinct and characteristic peak[3], to probe the boron dopant profile. The approach is based on the use of PL spectra measured on a set of uniformly heavily boron-doped samples, to reconstruct the PL spectra measured on diffused samples, which contain a range of dopant concentrations as a function of depth. This differs from another recent approach based on a comparison with spectra measured at a bath temperature of 79K on reference diffused samples [6], in which the PL peak position was used to estimate the surface dopant concentration, and the peak intensity used to estimate the depth of the diffused layer, assuming a Gaussian shape for the dopant profile.

Here we aim to show that the shape of the PL peak arising from the heavily doped layer can be accurately reconstructed based on modelled PL spectra arising from the layers within the diffused region, and that the shape of this peak contains some extractable information about the dopant profile below the surface. The ultimate aim of this work is to enable dopant profiles to be estimated based on modelling of the PL spectral shape, without assuming a Gaussian dopant distribution below the surface, and without relying on the intensity of the PL peak, which is prone to artefacts [6]. This would then enable boron dopant profiles to be approximated directly from the PL spectra with micron-scale spatial resolution, allowing accurate characterisation of localised boron diffused and laser doped regions. The work presented here is a first step towards this objective.

2 Experimental Methods

PL spectra were measured on wafers with a range of uniform grown-in boron concentrations, with $N_A < 1\times10^{15}$ cm^{-3} and from 1×10^{17} to 1.1×10^{20} cm^{-3} and on boron-diffused samples with a range of boron profiles. The samples were float zone mono-crystalline silicon wafers, and were chemically polished, with no surface passivation films present. The PL spectra were measured on samples placed within a liquid nitrogen cryostat at a bath temperature of 79K. A

This is a pre-peer review version of the paper submitted to the journal Progress in Photovoltaics.
It has been selected by the Executive Committee of the EU PVSEC 2017 for submission to Progress in Photovoltaics.

Horiba LabRAM tool equipped with a confocal microscope, with excitation via a 532 nm continuous wave laser focused into a 1μm diameter spot on the samples with a 50x objective lens (numerical aperture of 0.55). The laser power incident on the sample surface was measured to be 23.5mW. This power was reduced in some cases by inserting neutral density filters. Note the sample surface temperature, when subject to the incident laser beam, might be slightly higher than the bath temperature. The optical path of the luminescence coincides with that of the excitation beam before being reflected onto a monochromator (of 150 grooves per mm, and spectral resolution of 0.25nm). The PL beam was then measured by a liquid nitrogen cooled InGaAs array detector with a photon wavelength detection range of 900 nm to 1600 nm.

3 Results and Discussion

3.1 PL Spectra for Thermally Diffused Sample

The most significant feature when comparing the PL spectra from a thermally boron-diffused sample to an undiffused sample (with $N_A < 1 \times 10^{15}$ cm^{-3}), is an additional broad peak at lower energy than the main band-to-band peak from the silicon substrate, as shown in Figure 1. Note that the smaller peak at lower energy for the undiffused sample is a phonon replica of the main peak, and is not related to the doping. The separate peak from the diffused layer occurs due to band-gap narrowing effects in the heavily doped region, and its shape and position changes depending on the doping concentrations present[3]. In order to understand the composition of the peak arising from the boron-diffused layer, PL spectra generated from samples with different levels of uniform boron doping were studied, analysed and parameterised in this work.

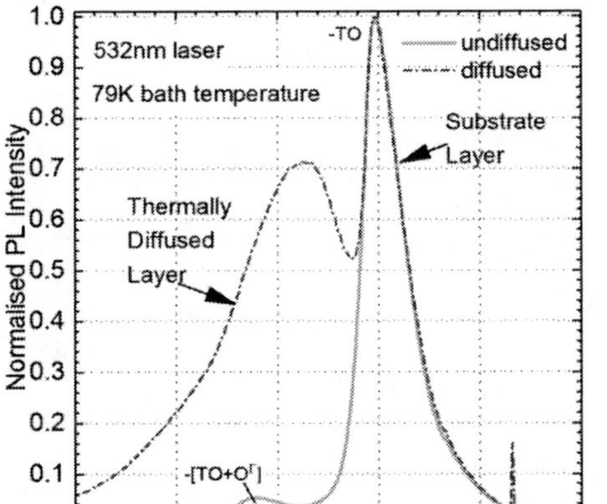

Figure 1. Comparison between the normalised PL spectra measured at 79K from a thermally boron-diffused sample ($N_{surf} = 2.4 \times 10^{20}$cm^{-3}) and an undiffused sample ($N_A < 1 \times 10^{15}$ cm^{-3}). The peak labelled $-\mathrm{TO}$ represents the transverse optical phonon emission and $-[\mathrm{TO} + \mathrm{O}^{\Gamma}]$ is the combination of transverse optical phonon and the optical zone-centre phonon [8]. Also note the spike at the position of around 1.166eV is the second harmonic generation of the laser beam and unrelated to the PL.

3.2 PL Spectra for Uniformly Doped Samples

Samples with high, uniform boron concentrations between 1×10^{17} cm^{-3} to 1.1×10^{20} cm^{-3} were collected and their PL spectra measured at 79K bath temperature, as shown in Figure 2. The central peak position shifts towards lower photon energy and the peak broadens as the doping level increases. In addition to the central transverse optical phonon emission peak ($-\mathrm{TO}$) and the emission peak of transverse optical and optical zone-centre phonons ($-[\mathrm{TO} + \mathrm{O}^{\Gamma}]$)[8], there are other subtle features evident compared to the spectrum from light doping. These will be discussed further below.

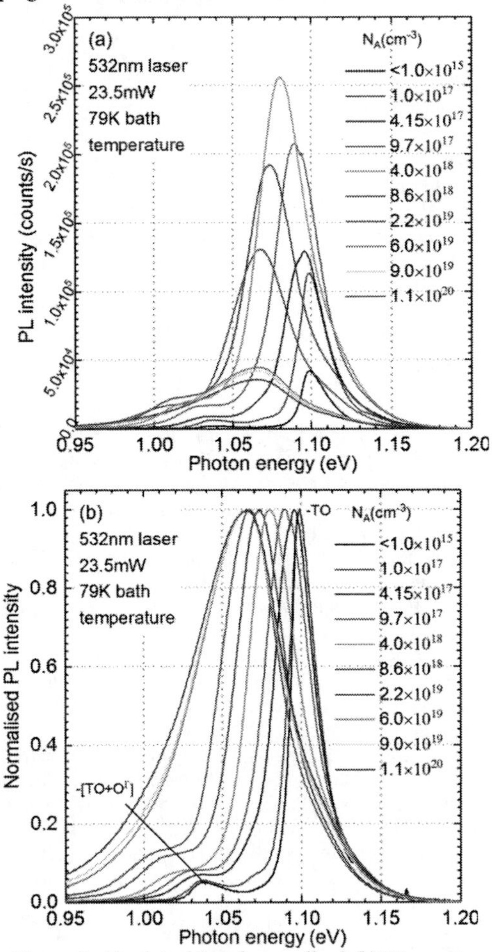

Figure 2. Absolute (a) and normalised (b) PL spectra measured at 79K on wafers with a range of uniform boron dopant concentrations N_A (cm^{-3}).

In this work the spectra measured on samples with different doping concentrations were modelled based on a PL curve that was measured on a lightly doped silicon wafer ($N_A < 1 \times 10^{15}$ cm^{-3}) which has minimum influence from the dopants at liquid nitrogen temperature [9]. Note that no model was applied to the mentioned PL spectra although in principle it can be modelled with the Maxwell-Boltzmann distribution[10]. The normalised PL spectrum for this lightly doped sample is provided in a supplementary file.

This method of modifying a measured curve was compared to that of a Gaussian-curve fitting method. It was found that given the same number of parameters in determining the central position, height and widths of the

This is a pre-peer review version of the paper submitted to the journal Progress in Photovoltaics.
It has been selected by the Executive Committee of the EU PVSEC 2017 for submission to Progress in Photovoltaics.

curves, the measured curve achieved a more accurate fit potentially due to the imbedded replicas in the measured signal. Hence a measured baseline curve is used throughout this work. However, future work will again focus on Gaussian-curve fitting method for practicality.

A shift parameter, E_{shift} (eV), accounting for the bandgap narrowing effect, was used to shift the original curve to the required energy position for each doping concentration. A broadening factor E_S (see Equation (1), where E is the photon energy, and I the PL intensity) was then used to broaden the curve to match the measured one.

$$I_{broadening}(E) \propto \int_0^{\infty} I(E')exp(-\frac{(E'-E)^2}{E_S^2})d(E') \quad (1)$$

We note that this broadening factor is different to that used by Dumke and Wagner[11-13]. Those authors separated their broadening factor E_S (attributed to incomplete ionization and band tails) from the impacts of the Fermi energy level E_F in their formulas, whereas we include all these factors in our single broadening factor. In this study, this factor is entirely determined from fitting our experimental spectra, thus allowing us to reconstruct the PL spectrum for a specific doping level using only a standard PL spectrum of lightly-doped silicon, without knowledge of Fermi levels. In the process of matching the adjusted curves with measurements, however, it was found that a single curve was insufficient to capture the shape of the various PL curves of different doping concentration with accuracy. Instead, a spectrum made up of a superposition of three such curves (referred to as the central, left and right curves) was found to provide optimal results. The model consisted initially of 8 parameters controlling independently the shift, broadening and the scaling factor of the three curves. Note that the PL spectra are normalised, thus the scaling factor for the central peak is set at a fixed value of 1. Previous authors have used combinations of Gaussian curves to fit similar PL spectra, although as a function of temperature, rather than doping[14]. In either case, the number of fit parameters required is not significantly changed.

Figure 3 illustrates how the three-curve model is required to obtain a good fit to the PL spectra for three representative boron doping concentrations. A single central curve alone is not sufficient, even when shifted and broadened. An additional curve on each side of the peak is required to accurately reproduce the measured spectra.

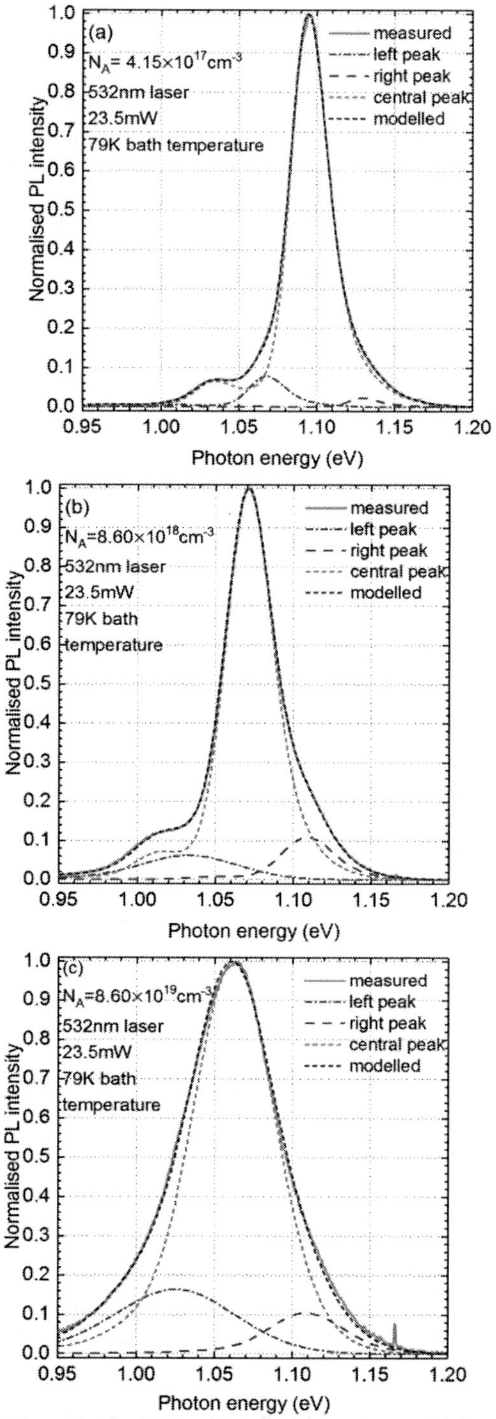

Figure 3. The fitting using the three-curve model for PL spectra for three representative boron doping concentrations. Graph (a) for $N_A = 4.15\times10^{17}$ cm^{-3}, graph (b) for $N_A = 8.60\times10^{18}$cm^{-3}, and graph (c) for $N_A = 8.60\times10^{19}$ cm^{-3}.

To further justify the use of three curves, an error analysis was performed on how the fit quality changes with the absence of either of the side peaks. The error factor was determined as the sum of the squares of the difference between the model and measurement (both normalised), multiplied by the normalised intensity of the measurement at that energy. The aim of including the second weighting factor was to give relatively higher importance to the higher intensity part of the spectrum, in order to preserve a good match for the central peak positions as well as their shapes, as these are the most important for the ultimate purpose of fitting the spectra measured on diffused samples.

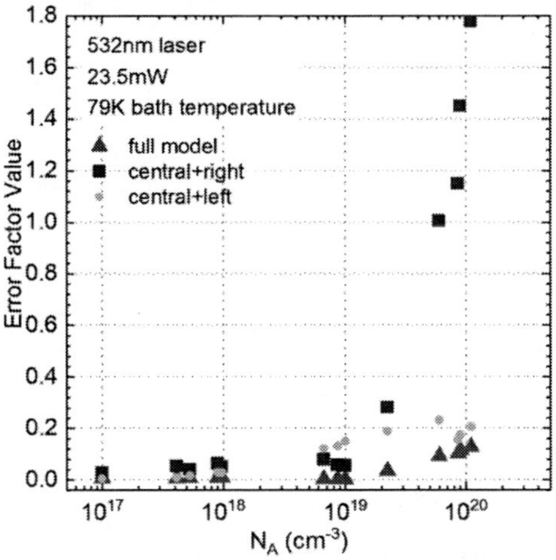

Figure 4. The error factor values across the range of doping concentration for the full model with three curves compared to fits with only two curves (central + right and central + left).

Figure 4 reveals how this error factor changes with doping concentration for the full model (3 curves) or the 2-curve models. It shows that, overall, the three-curve model provides a consistently higher accuracy in the match over the range of doping concentrations measured, and that the relevance of the sides peaks varies with the doping. The left peak, for example, is more important for doping concentrations over 1×10^{15} cm^{-3}.

Once the 3-curve model was established, the 8 fit parameters were individually optimised to model the shapes of PL spectra measured across the set of samples with different doping concentrations. Establishing the connection between the fit parameters and the doping concentration in principle allows a generalised description of PL curves for any specified boron doping within the range measured.

3.3 Parameterization of PL Spectra on uniformly-doped samples

3.3.1 Parameters

The fit parameters were obtained from the three-curve modelling for the PL spectra of the samples with uniform boron doping. The model was optimised to achieve the best fit (a majority of the samples matched with the model within 0.01 error factor whilst doping over 1×10^{15} cm^{-3} achieved an error factor in the range of 0.01~0.11).

The 8 parameters, namely E_S, E_{shift} (for broadening and horizontal shift of the central curve), left/right shift (for adjusting the position of the two side curves), left/right scaling factor (vertical scaling of the side curves), and the left/right broadening factor (for broadening the minor curves with the E_S formula) are plotted against the doping concentration, and the trends are fitted with various polynomial functions, as shown in Figure 5. The polynomial functions are listed in the Appendix. A laser power of 23.5 mW was used in all cases. Reducing the laser power to 1.2mW was found to have significant impact on the fit parameters as shown in figure 7(b). For one it affected the broadness of the main peak as well as the magnitude of the left peak, as reflected in the changes of E_S, shown in Figure 5(a), and the left peak scaling factors in Figure 5(e). For another, it reveals addition features (peaks at around 1.07eV and 1.09eV) for low injection levels PL signals that the current model does not capture. More details are provided in Section 3.4.

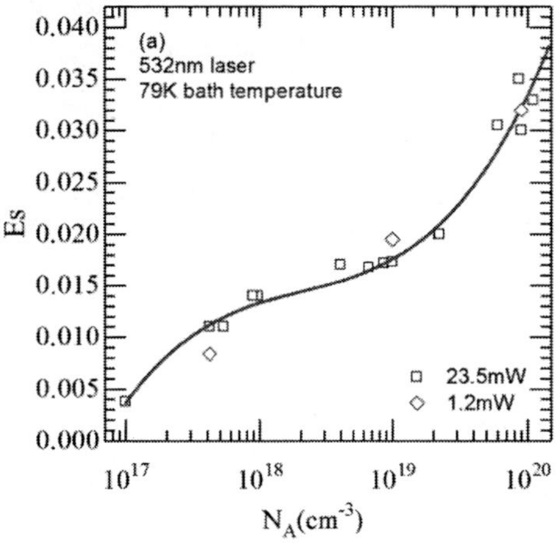

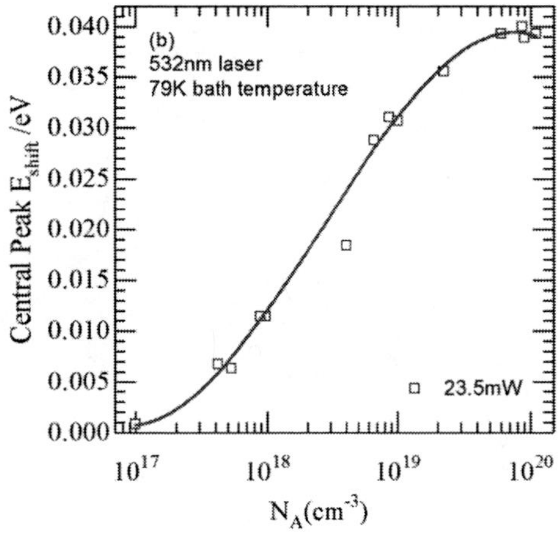

This is a pre-peer review version of the paper submitted to the journal Progress in Photovoltaics.
It has been selected by the Executive Committee of the EU PVSEC 2017 for submission to Progress in Photovoltaics.

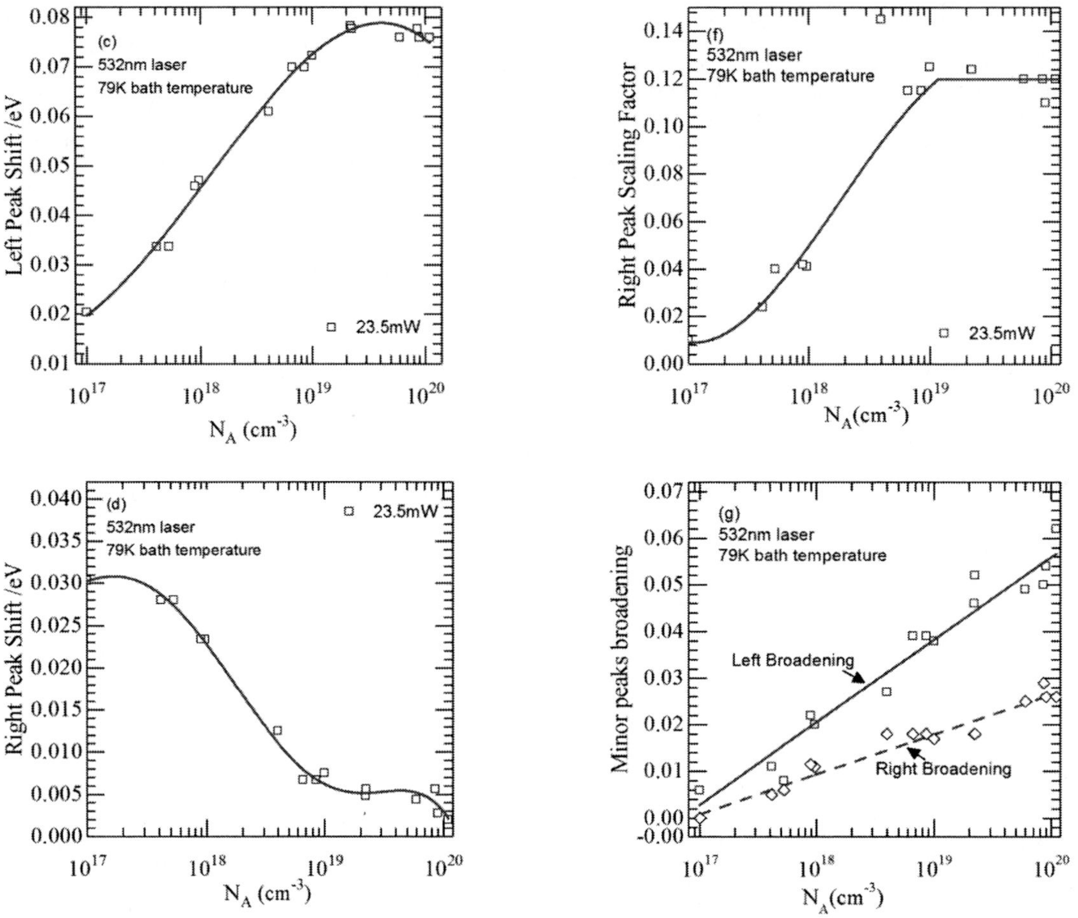

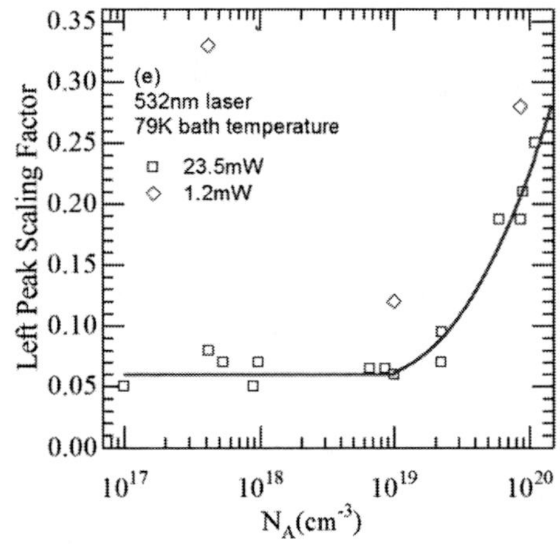

Figure 5. The fit parameter values plotted against the uniform boron doping concentration. Both the E_{Shift} and left peak shift value are defined as the relative distance to the left of the lightly-doped silicon curve. The right peak shift is defined as the distance to the right of the lightly-doped silicon curve. Simple empirical functions are used to fit the data, as shown in each plot (with fit formulae provided in the Appendix). The plots for E_S and the left peak scaling factor show the changes in the relevant parameter when the laser power was reduced to 1.2mW. For all other fit parameters, reducing the laser power had no observable impact.

The empirical fits to the parameters above provide the fundamental basis for later constructing the PL spectra of a thermally diffused wafer. The left and right broadening factors share similarities in behaviour in that the left broadening values are very close to double that of the right.

When the empirical functions for the fit parameters were used in reverse to predict the shape of the PL spectra for a representative set of the uniformly-doped samples, the error factors were found to increase roughly by a factor of 10 compared to an unconstrained fit. Figure 6 demonstrates the results, and indicates that the predicted PL spectra are still of good quality.

This is a pre-peer review version of the paper submitted to the journal Progress in Photovoltaics.
It has been selected by the Executive Committee of the EU PVSEC 2017 for submission to Progress in Photovoltaics.

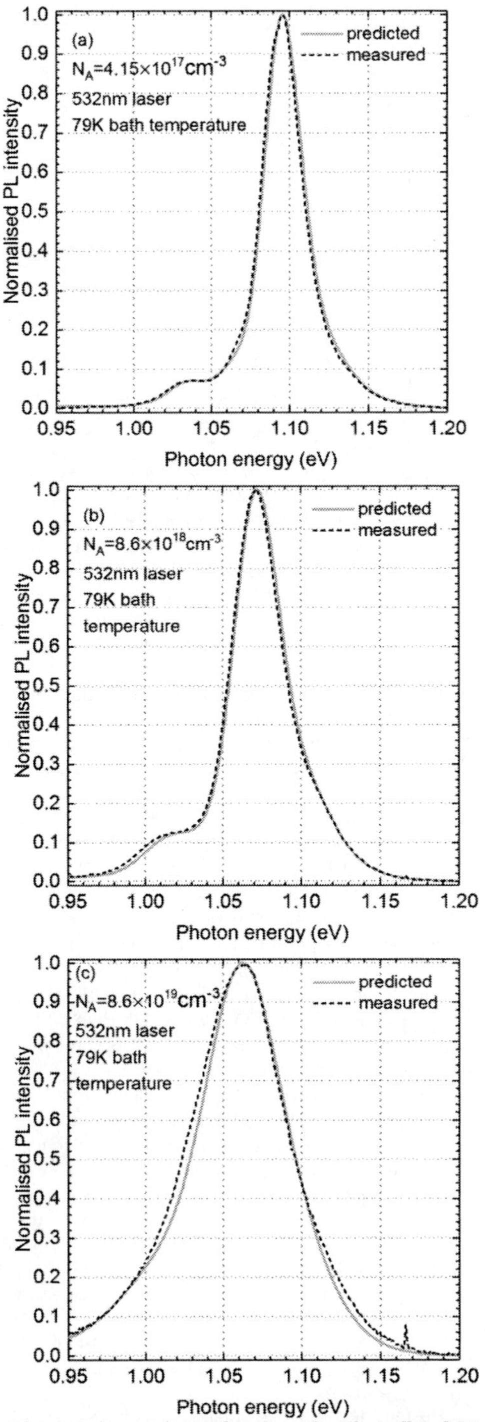

Figure 6 Comparison of measured and predicted PL spectra for three representative uniform boron dopant concentrations. The predicted spectra were constructed using the empirical functions for the fit parameters, from Figure 5.

3.3.2 Physical mechanisms underlying changes in the PL spectra

There are several established physical mechanisms that occur in doped crystalline silicon that affect the PL spectral shape. Firstly, as the doping concentration increases, more energy states are introduced into the forbidden band gap caused by the screening cloud exchange among the free carriers and the dopant atoms. This causes a rigid shift of the conduction and valence bands towards one another and reduces the band gap [15-18]. Secondly, at high dopant concentrations, band tails form and deepen due to the random distribution of the impurities in the host lattice resulting in an emission mechanism with a distribution of energies below the band gap [15, 19-23]. Thirdly, the aggregated energy states of the dopant atoms form a dopant band, which for p-type silicon wafers is located slightly above that of the valence band. The dopant band widens as a result of increased dopant interactions as their concentration increases. Fourthly, as the doping approaches the Mott transition, the dopant band approaches the band tail, before both the band tails and the dopant band eventually merge[22, 24]. And finally, the increased doping concentration results in a shift in the Fermi level towards the valence band (for p-type silicon) and ultimately merging into the valence band at degenerate doping concentrations[25].

Referring to the PL spectra plotted against photon energy levels in this work, the first mechanism shifts to lower energy all components of the standard PL spectrum captured from lightly-doped Si, with minimal effect on the shape of each individual emission component. The band tails will broaden the lower energy side of the PL spectra. The dopant band and its merging with the band tails will add to the weighting and the broadening for the low energy side of the spectra. And the shift in the Fermi level will broaden all spectral components and shift them toward the high energy side.

The three-curve model is an attempt to capture the changes in these different phenomena using an efficient number of fit parameters. Whilst all of the phenomena described may affect any of the three curves with varying degrees, each curve has a larger dependence on a specific phenomenon, as described below.

The central curve reflects the rigid shift of both conduction and valence bands, and the effects of the Fermi level. The rigid shift moves the peak position of the standard PL spectrum captured from lightly-doped Si, while the changes in the energy states distribution broadens the spectra.

The left curve is more related to the presence of the band tails, the dopant band, and the eventual merging between them. This is reflected by the drastically increasing importance of the left curve for high doping densities in Figure 4, and the escalating scaling factor for the left curve in Figure 5(e). The saturation of the left curve shift value at higher doping concentration in Figure 5(c) also relates to the description of band merging. The prominence of the left curve is also observed to be highly dependent on the injection level. Section 3.4 provides further details.

The right curve largely reflects the shifting of the Fermi level. This is indicated by the higher significance of the right curve at doping densities above 1×10^{19} cm^{-3} in Figure 4. Once the material becomes degenerate, however, further changes in the doping concentration produce less impact on the weighting of the right curve. This echoes the flat line on the right peak scaling factor in Figure 5 for doping concentrations above 1×10^{19}cm^{-3}.

3.4 Impact of injection level

The above three-curve model and parametrization is

based on measurements with the same laser power at the sample surface of 23.5mW. This means that the injection level for each PL spectrum may vary as the boron concentration changes. Therefore, it is important to investigate the possible effects of changing injection level (or laser power) on the shape of the PL spectra. For this purpose, samples with three representative doping concentrations were measured under different laser powers, and their spectra plotted in Figure 7.

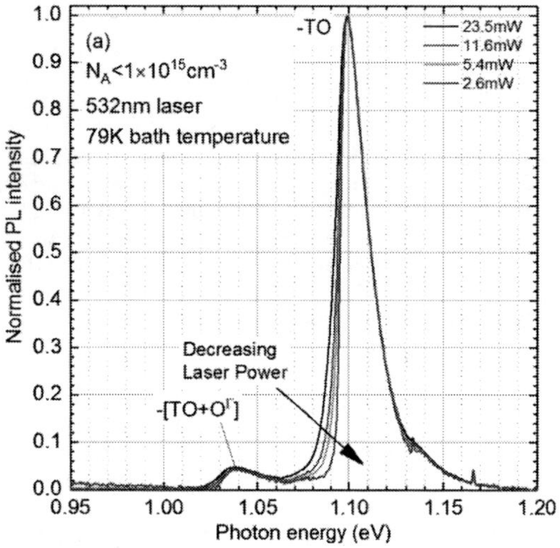

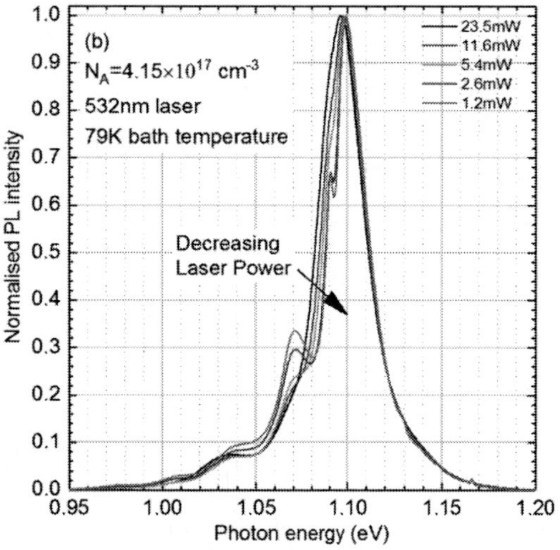

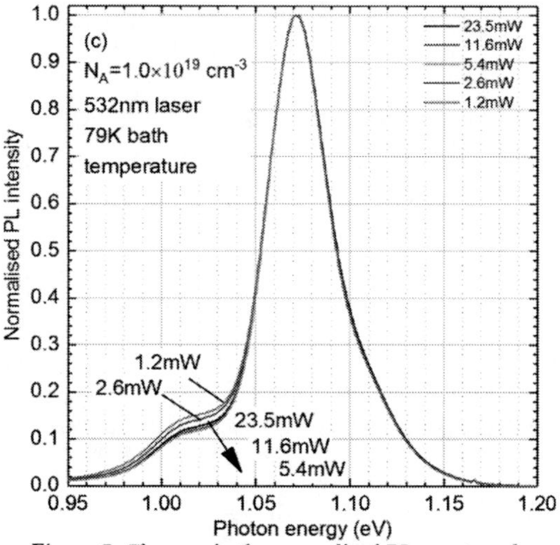

Figure 7. Changes in the normalised PL spectra when the laser power (and injection level) is reduced, for three different dopant concentrations.

As shown in Figure 7 (a), for a lightly doped sample, there is an obvious narrowing on the lower energy side of the central peak as the laser power and injection level is reduced. This is due to the receding of band tails when the carrier density is reduced. This phenomenon has been observed previously [15, 26].

For a moderately doped sample, as shown in Figure 7 (b), some other features appear at lower laser power. In addition to the general narrowing of the central peak as the laser intensity is reduced, as was observed for the lightly doped sample, two new sharp peaks appear on the low energy side of the central peak. They were first observed by Dean *et al.* [8] on moderately and heavily boron doped silicon, denoted by Dean as B_2^{TO} (around 1.07eV) and B^{TO} (around 1.09eV). The B_2^{TO}peak is due to the TO phonon side band of free-to-bound recombination involving Boron acceptors [8] whilst the B^{TO} peak is due to TO phonon sideband of ICBE (impurity-cluster-bound exciton) which occurs below the Mott transition[27].

Once the doping concentration is increased to 1×10^{19} cm^{-3} as in Figure 7(c), both the central curve narrowing effect and the distinct dopant-related peaks become much less pronounced. That is because at these higher doping levels the samples are under low injection and any changes in the minority carrier density do not significantly impact the shape of the PL spectra. A minor exception is the small residual variation in the height of the left peak on the low energy side. However, in practice, this part of the spectrum has a negligible impact on the PL spectral shape emitted from diffused regions, and we therefore do not attempt to account for it in the following sections.

Besides the minimal changes in the shapes of the normalised PL spectra in Figure 7 (c) for the heavily doped sample, a change in the PL photon flux as a function of incident laser power was also observed, with the results plotted in Figure 8 for a sample with doping of 1×10^{19} cm^{-3}. The figure suggests a linear relationship between the laser power and the PL intensity.

This is a pre-peer review version of the paper submitted to the journal Progress in Photovoltaics.
It has been selected by the Executive Committee of the EU PVSEC 2017 for submission to Progress in Photovoltaics.

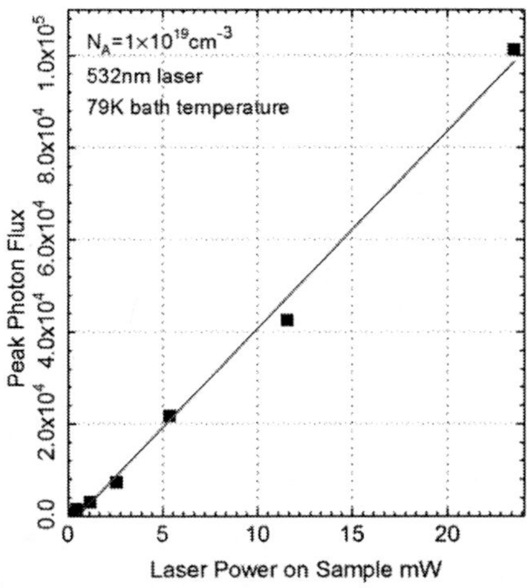

Figure 8 Relationship between the incident laser power and peak PL intensity for a sample with uniform boron doping of 1×10^{19} cm^{-3}

Summarizing these results, for higher doping concentrations above 1×10^{19} cm^{-3}, the PL spectral shape is largely insensitive to the injection level, and the PL intensity is linearly related to the incident photon flux, under the excitation conditions used in this work. Since, in practice, most boron diffusions are largely composed of boron concentrations above this value, we therefore proceed to model the PL spectra from a boron diffused sample by simply using linear combinations of spectra from the constituent dopant concentrations, without needing to take account of the effects of injection level.

Limiting the fitting to the higher doping concentrations brings another potential simplification to the subsequent attempt to model the PL spectra from thermally diffused samples. Since increased dopant concentrations reduce the carrier motilities, the effect of depth-wise carrier diffusion is less significant. This potentially means the carrier concentration in a thermally diffused sample will be more directly affected by the extent of optical absorption in each layer of the profile, and less so by the transport of carriers between layers, as discussed in more detail below.

3.5 Re-construction of PL spectra measured on diffused samples

3.5.1 Reconstruction

To enable simplified modelling of the PL spectra measured on thermally boron-diffused samples using the parameters obtained from various uniformly doped samples, the following simplifying assumptions are made. Firstly, the effect of photon reabsorption of the PL signal is ignored at bath temperature of 79K due to the low absorption coefficient [28], the short penetration depth of the incident laser light, as well as a steep attenuation of the carrier density as a function of depth due to Auger recombination [7]. Secondly, for sufficiently heavily doped regions of a thermally diffused sample, we assume a linear dependency of the PL signal intensity on the incident excitation power, as was confirmed in section 3.4. Finally, within the heavily doped region, carrier diffusion is assumed to have a relatively small impact on the local excess carrier density, which is therefore largely determined by the local generation rate.

Based on these assumptions, the PL spectra measured on a thermally boron-diffused sample can be reconstructed using modelled PL curves for a set of doping concentrations extracted from an ECV profile, as shown in Figures 9 and 10. In this case we have divided the diffused region into five equally thick layers with decreasing boron concentrations. The thickness of the layers is chosen such that the doping levels range from 2.4×10^{20} cm^{-3} down to near 1×10^{19} cm^{-3}. The PL spectra measured from a lightly-doped substrate with $N_A = 1 \times 10^{15}$ cm^{-3} is also used to construct the PL component from the substrate. By altering the scaling factor for each component curve (see Figure 12 and section 3.5.2) we can achieve a good match to the measured spectra, as shown in Figure 9(b). Note that only layers with high doping concentrations are modelled, to avoid the impact of varying injection levels on the spectral shape for the lower doping concentrations.

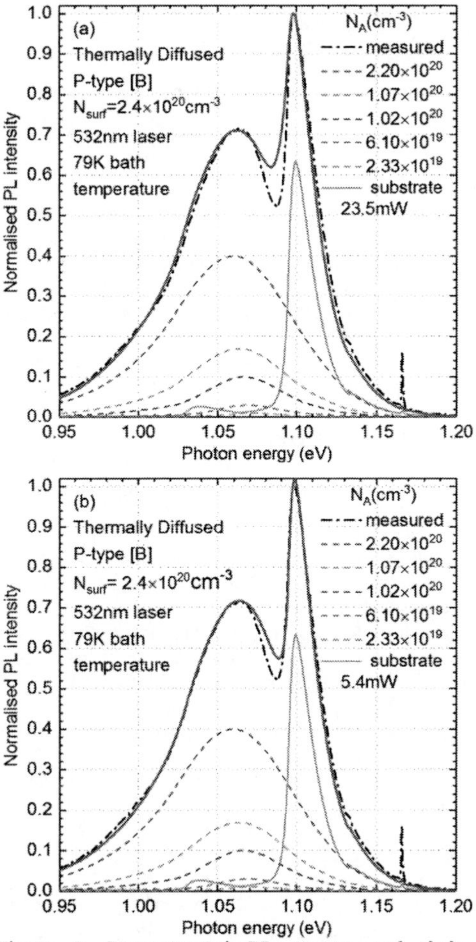

Figure 9. Reconstructed PL spectra and their components for a thermally boron-diffused sample. The left plot (a) shows a fit with a substrate spectrum based on a laser power of 23.5 mW, and on the right (b) with a reduced laser power of 5.4mW for the substrate spectrum, due to absorption of laser light in the overlying heavily doped region.

This is a pre-peer review version of the paper submitted to the journal Progress in Photovoltaics.
It has been selected by the Executive Committee of the EU PVSEC 2017 for submission to Progress in Photovoltaics.

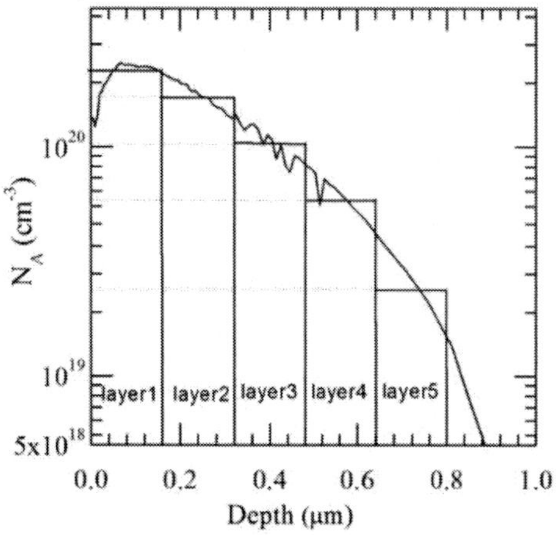

Figure 10 The doping profile measured from ECV showing the change in doping concentration up to 1μm below the surface of the thermally diffused sample. The heavily diffused region is uniformly divided to into five layers where each layer is represented by one average doping value.

Figure 9 (b) demonstrates the effect of using a PL spectrum for the substrate measured with a lower laser power (or injection level). It indicates that the measured PL signal from a lower injection level gives an overall better match to the measured spectra. This reflects the fact that the injection level plays an important role in the spectral shape of the lightly doped substrate, but not the heavily doped regions. The better fit for the lower injection spectra for the substrate is since the overlying heavily-doped region absorbs some of the laser light before it reaches the substrate, effectively reducing the generation rate and injection level there.

Nevertheless, there is still some deviation between the model in Figure 9 (b) and the measured spectra on the diffused sample, in particular in the valley between the two peaks. Using even lower injection level peaks for the substrate resulted in a poorer fit on the low energy side of the substrate peak, meaning that other effects may also affect the fitting in this part of the spectrum. This small discrepancy will be the subject of future work.

Overall, the result shows that the diffusion peak can be decomposed into PL spectra from the different doping levels existing in the diffused layer. The injection effect is significant in the substrate layer, and can be accounted for by using a substrate spectrum with reduced injection/laser power. However, injection effects can be quite safely neglected in the heavily diffused region where the doping is over 1×10^{19} cm^{-3}.

Figure 9 shows that the layer with the highest doping contributes the most to the overall PL spectrum from the diffused layer. This raises the question as to whether the deeper layers are necessary to achieve a good overall fit. Figure 11 shows an attempt to fit the spectra using only the most heavily doped layer. This results in a poor fit, in which the heavily-doped peak is too broad to match the measured spectrum. This indicates that the contributions from the deeper layers with lower doping are required to describe the measured spectrum accurately.

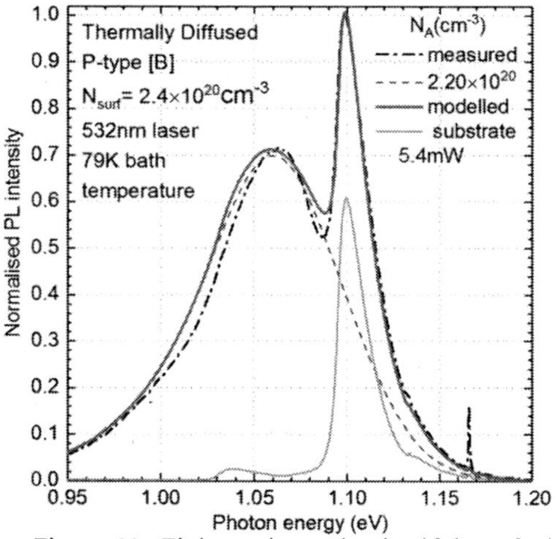

Figure 11. Fitting using only the highest doping concentration spectrum and the substrate spectrum. The PL curve from the diffused sample exhibits a narrower diffusion peak compared to that of the highest doping concentration.

3.5.2 Estimating the scaling factors of the component layers

Decomposing the diffused PL spectrum into individual PL spectra for layers of different doping suggests a simplified model for estimating the relative scaling factors of each component. We may assume that each layer of the diffused region can be represented by a region of uniformly-doped Si with the doping concentration equal to that of the central doping value, as indicated in Figure 10.

The PL intensity emitted from each layer will be determined by the local injection level. In general, this is difficult to estimate, as carriers are not only generated in a region, but can also diffuse from other regions, altering the carrier density there. However, as a first order simplification, we may assume that such diffusion currents are small, and that the injection level in a given layer is determined largely by the amount of light absorption there, as was discussed in section 3.4.

Estimating the absorption of laser light in each layer can be achieved by assuming the standard exponential decay based on an assumed absorption coefficient, combined with the thickness of each layer, which in this case is fixed. The absorption coefficient was considered constant across the diffused layers, and its magnitude was adjusted to provide a best fit to the scaling factors. A constant was used to linearly scale these values, and the result is shown in Figure 12. The result shows the amount of laser light absorbed in each layer follows the same trend as the scaling factors achieved by forcing the best fit to the overall spectrum in Figure 9. This close relationship could provide a basis for a method to estimate the dopant profile based on the measured PL spectra, as will be explored in future work.

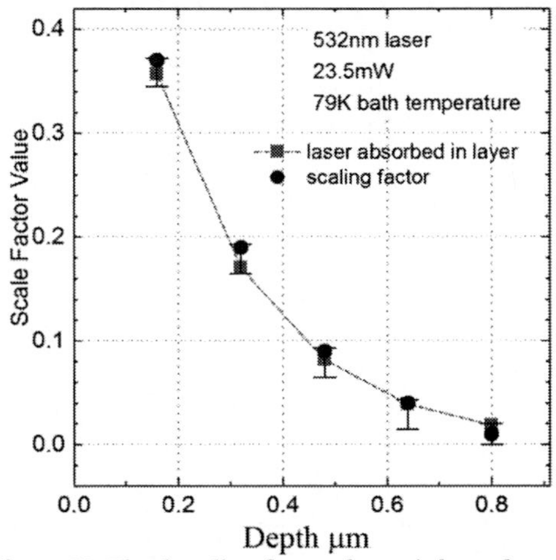

Figure 12. Fitted scaling factors for each layer from Figure 9 (circles) and estimated laser light absorption in each layer (squares).

In this case the optimum absorption coefficient was found to be 4.8 μm^{-1} which means that the laser light is attenuated to 10% of the initial value at a depth of 0.5 μm. We note that there is a discrepancy between these values and the ~25% incident laser power required for the PL spectrum from the substrate in Figure 9, which lies 1 μm below the surface. This difference may be caused by carrier diffusion effects which likely play a larger role in the more lightly doped regions. The result is that carriers migrate from the diffused region into the substrate, resulting in an increased injection level in the substrate.

In summary, carrier diffusion plays a part in determining the injection level of the substrate, but is less significant for modelling the heavily diffused region within a short distance below the surface.

4 Conclusion

At liquid nitrogen temperature, the PL spectra from different uniformly boron doped crystalline silicon samples have been measured and parameterized using 8 fit parameters with a 3-curve model. It was found that PL spectra for higher doping concentrations (above 1×10^{19} cm^{-3}) are much less affected by changes in the excitation power and injection level than regions of lighter doping. The fit parameters were used to model PL spectra from component layers within a thermally boron-diffused sample, which are linearly combined to accurately reconstruct the measured PL spectrum. Moreover, the scaling factor for each of the component layers is found to be closely related to the fraction of incident light absorbed in each layer. It was also found that the PL spectra from only the substrate and most heavily-doped surface layer is insufficient in establishing a satisfactory model for the measured PL from a thermally boron-diffused sample. This suggests the PL spectra do contain useful information of the doping profile, and that the photoluminescence-based technique can potentially be used for approximating the doping profile of thermally diffused boron doped silicon wafers, and applied with micron-scale spatial resolution.

5 References

1. P. Gundel, M.C. Schubert, and W. Warta, *Comprehensive microscopic analysis of laser-induced high doping regions in silicon.* IEEE Transactions on Electron Devices, 2011. **58**(9): p. 2874-2877.

2. H.T. Nguyen, et al, *Quantification of sheet resistance in boron-diffused silicon using micro-photoluminescence spectroscopy at room temperature.* Submitted to Solar RRL, 2017.

3. H.T. Nguyen, et al, *Micro -photoluminescence spectroscopy on heavily -doped layers of silicon solar cells.* physica status solidi (RRL)-Rapid Research Letters, 2015. **9**(4): p. 230-235.

4. A. Roigé, et al., *Microscale spatially resolved characterization of highly doped regions in laser-fired contacts for high-efficiency crystalline Si solar cells.* IEEE Journal of Photovoltaics, 2015. **5**(2): p. 545-551.

5. M. Tajima, *Determination of boron and phosphorus concentration in silicon by photoluminescence analysis.* Applied Physics Letters, 1978. **32**(11): p. 719-721.

6. Y. Han et al., *Determination of dopant density profile of p-type doped silicon from low temperature micro-photoluminescence spectroscopy.* Sumitted to IEEE Journal of Photovoltaics, 2017.

7. F.D. Heinz, et al, *Doping density in silicon and solar cells analyzed with micrometer resolution.* IEEE Journal of Photovoltaics, 2013. **3**(1): p. 341-347.

8. P.J. Dean, J.R. Haynes, and W.F. Flood, *New Radiative Recombination Processes involving Neutral Donors and Acceptors in Silicon and Germanium.* Physical review, 1967. **161**(3): p. 711-&.

9. A. Liu, H.T. Nguyen, and D. Macdonald, *Quantifying boron and phosphorous dopant concentrations in silicon from photoluminescence spectroscopy at 79 K.* physica status solidi (a), 2016. **213**(11): p. 3029-3032.

10. I. Pelant et al., *Photoluminescence assessment of B, P, and Al in Si wafers: The problem of sample heating by a laser beam.* Journal of applied physics, 1993. **73**(7): p. 3477-3481.

11. W. Dumke, *Comparison of band - gap shrinkage observed in luminescence from n+ -Si with that from transport and optical absorption measurements.* Applied Physics Letters, 1983. **42**(2): p. 196-198.

12. J. Wagner, *Photoluminescence and excitation spectroscopy in heavily doped n-and p-type silicon.* Physical Review B, 1984. **29**(4): p. 2002.

13. J. Wagner, W. Appel, and M. Warth, *Photoluminescence from heavily doped Si layers grown by liquid-phase epitaxy.* Journal of Applied Physics, 1986. **59**: p. 1305-1308.

14. W.S. Yoo, K.Kang, G.Murai, and M.Yoshimoto, *Temperature Dependence of Photoluminescence Spectra from Crystalline Silicon.* ECS Journal of Solid State Science and Technology, 2015. **4**(12): p. P456-P461.

15. H. T. Nguyen, D.Macdonald, *On the composition of luminescence spectra from heavily doped p-type silicon under low and high excitation.* Journal of Luminescence, 2017. **181**: p. pp. 223–229.

16. J.C. Inkson, *Screened exchange in semiconductor.* 1973. **Solid state physics** (6): p. L183-L185.

17. G.D. Mahan, *Energy gap in Si and Ge: Impurity dependance.* Journal of Applied Physics, 1979. **45**(2634).

This is a pre-peer review version of the paper submitted to the journal Progress in Photovoltaics.
It has been selected by the Executive Committee of the EU PVSEC 2017 for submission to Progress in Photovoltaics.

18. K.F. Berggren and B.E. Sernelius, *Band-gap narrowing in heavily doped namy-valley semiconductors.* Physical Review, 1981. **24**(4).

19. R.A. Logan, and A.G. Chynoweth, *Effect of Degenerate Semiconductor Band Structure on Current-Voltage Characteristics of Silicon Tunnel Diodes.* Physical Review, 1963. **131**(1): p. 89-95.

20. I. Kudman and T. Seidel, *Absorption Edge in Degenerate p - Type GaAs.* Journal of Applied Physics, 1962. **33**(3): p. 771-773.

21. P.A. Wolff, *Theory of the Band Structure of Very Degenerate Semiconductors.* Physical Review, 1962. **126**(2): p. 405-412.

22. J.R. Lowney, *Impurity bands and band tailing in moderately doped silicon.* Journal of Applied Physics, 1986. **59**(6): p. 2048-2053.

23. P. Van Mieghem, *Theory of band tails in heavily doped semiconductors.* Reviews of modern physics, 1992. **64**(3): p. 755.

24. P.P. Altermatt, A. Schenk, and G. Heiser, *A simulation model for the density of states and for incomplete ionization in crystalline silicon. I. Establishing the model in Si: P.* Journal of Applied Physics, 2006. **100**(11): p. 113715.

25. K. Morigaki and F. Yonezawa, *Metal-Nonmetal Transition in Doped Semiconductors.* Progress of Theoretical Physics Supplement, 1975. **57**: p. 146-155.

26. A. Liu, H.T. Nguyen, and D. Macdonald, *Photoluminescence Spectra of Moderately Doped, Compensated Silicon Si:P,B at 79-300 K.* IEEE Journal of Photovoltaics, 2017. **7**(2): p. 581-589.

27. M. Tajima et al., *Photoluminescence due to impurity-cluster-bound exciton in highly doped and highly compensated Si.* Japanese Journal of Applied Physics, 2015. **54**(11): p. 111304.

28. E. Daub and P. Würfel, *Ultralow values of the absorption coefficient of Si obtained from luminescence.* Physical review letters, 1995. **74**(6): p. 1020.

This is a pre-peer review version of the paper submitted to the journal Progress in Photovoltaics.
It has been selected by the Executive Committee of the EU PVSEC 2017 for submission to Progress in Photovoltaics.

6 Appendix

Empirical polynomial equations for the spectra fit parameters

$$x = ln(N_A)$$

	x^5	x^4	x^3	x^2	x^1	x^0
E_g			0.00023249	−0.02941740	1.24216387	−17.4891021
E_{shift}			−0.00025243	0.03210401	−1.35235539	18.88651490
left peak shift			−0.00034141	0.04256123	−1.75533273	23.9931829
right peak shift	−0.00002485	0.00520653	−0.43566432	18.19578648	−379.318347	3157.539971
left peak scaling	(for $N_A < 1 \times 10^{19} cm^{-3}$)					0.06
	(for $x \geq 1 \times 10^{19} cm^{-3}$)		0.00253212	−0.31648337	13.17413396	−182.5780022
right peak scaling	(for $N_A < 1 \times 10^{19} cm^{-3}$)		−0.00132567	0.16734028	−7.00917301	97.48026528
	(for $x \geq 1 \times 10^{19} cm^{-3}$)					0.12
left peak broadening	2 × right peak broadening					
right peak broadening					0.00365	−0.14192

EFFICIENT CARRIER INJECTION FROM AMORPHOUS SILICON INTO CRYSTALLINE SILICON DETERMINED FROM PHOTOLUMINESCENCE

A. Paduthol[1]*, M.K. Juhl[1], Z. Hameiri[1], G. Nogay[2], P. Löper[2], T. Trupke[1]
[1] The University of New South Wales, 2052, Sydney, Australia
[2] École Polytechnique Fédérale de Lausanne Lausanne (EPFL), Institute of Microengineering (IMT), Photovoltaics and Thin Film Electronic Laboratory (PV-Lab), Rue de la Maladière 71b, 2002 Neuchâtel, Switzerland

ABSTRACT: For devices with intrinsic amorphous silicon layer on a crystalline silicon substrate, the light absorbed in the amorphous silicon layer can be weakly electronically coupled into the crystalline silicon base. Such carrier injection has previously been reported using measurements on finished devices containing numerous layers. Here we use the spectral response of photoluminescence, a contactless approach that has been previously shown to provide a relative measurement of a device's internal quantum efficiency, to investigate carrier injection on simpler structures. In solar cells with intrinsic amorphous silicon front layers and a crystalline silicon base, such as common heterojunction solar cells, the effect of the absorption in the front layer should be reflected in the internal quantum efficiency. A highly absorbing front layer is expected to cause a drop in the internal quantum efficiency at short wavelengths. However, if electron hole pairs that are generated in the front layer are subsequently injected into the base, the optical losses will be reduced, resulting in a partial recovery of the internal quantum efficiency at short wavelengths. Here we quantify the efficiency of carrier injection from the intrinsic amorphous silicon front layer to the base, by measuring the spectral response of photoluminescence on heterojunction test structures and comparing it with the internal quantum efficiency expected from the absorption losses in amorphous silicon.

Keywords: Amorphous Silicon, Characterisation, Heterojunction, Photoluminescence, Spectral Response

1 INTRODUCTION

Intrinsic amorphous (aSi(i)) silicon is known to provide excellent surface passivation [1]–[4] on crystalline silicon (cSi) and is used as the passivation layer for high efficiency silicon heterojunction (SHJ) solar cells [5]. However, these layers have a strong optical absorption and can thus limit the performance of SHJ solar cells. The light absorbed in an aSi(i) layer has been previously demonstrated to be weakly electronically coupled with the cSi base, adding to the cell's photocurrent [6]. These investigations used internal quantum efficiency (IQE) and ellipsometry measurements to quantify such carrier injection. But, a complete device structure with good carrier collection probabilities and full optical modelling is required for these measurements, which inherently prevents the systematic investigation of individual layers in the absence of highly doped layers or transparent conducting oxides. Here, we first present a contactless approach that avoids the requirement of full device fabrication and full device modelling; we then report on significantly higher carrier injection from aSi(i) into cSi compared to previous reports.

The spectral response of photoluminescence (PL) provides a relative measure of the internal quantum efficiency (IQE) for a device [7], [8]. As PL is a contactless measurement technique, this method can be used on partially processed devices. For device structures with high effective lifetime, such as those studied here, the spectral response of PL (EQE_{PL}) provides a relative measure of external quantum efficiency (EQE) which, when corrected for reflection, provides IQE_{PL}.

$$IQE_{SC}(\lambda) \propto IQE_{PL}(\lambda) = \frac{I_{PL}(\lambda)}{N(\lambda)(1-R(\lambda))}$$

where IQE_{SC} is the conventional IQE, based on the short circuit current, of the device, IQE_{PL} is the relative IQE measured from PL, I_{PL} is the PL intensity per unit area, N is the photon flux the of incident light, R is the reflection, and λ is the wavelength of incident light.

For an n-type cSi device with aSi(i) passivation, it has been previously shown that the carriers generated in the aSi(i) can be partially injected into the cSi [6]. The current techniques to quantify such a carrier injection is by performing electrical IQE measurements and thus requires electrical contacts on a device. To quantify the carrier injection efficiency the following two quantities need to be determined.

1. the optical generation rate in the aSi(i)
2. the total recombination rate (representing the total carrier density) within the crystalline silicon base.

Here we determine the recombination rate in the crystalline silicon base using the spectral response of PL. The optical generation rate in the amorphous top layer is determined using a combination of spectral ellipsometry measurements and optical modelling. Comparing these two values provides the carrier injection efficiency (β) from aSi(i) to cSi, as demonstrated below.

First, we studied double-side polished cSi samples with a thermally grown silicon oxide (SiO_x) layer in between the aSi(i) layer and the cSi which prevents carrier flow from aSi(i) to cSi (β=0). For these devices we show that, as expected, the IQE_{PL} matches the optical generation in the bulk silicon. Following this, devices with just an aSi(i) layer on cSi were studied. A very efficient carrier injection from aSi(i) to cSi, with β~1.00, was observed in such devices.

2 EXPERIMENTAL SETUP

Measurement of EQE_{PL} was performed with an experimental setup that was built in-house and is described in detail elsewhere [10]. The tool has an illumination source with multiple LED based excitation wavelengths (FutureLED GmbH). In this study, seven different excitation wavelengths were used to measure PL (395 nm, 420 nm, 435 nm, 455 nm, 505 nm, 590 nm and

660 nm). The spectral width of the LEDs was confirmed to have a negligible effect on our measurements [11]. The photon flux for each wavelength is from 7×10^{15} to 10^{16} photons/cm^2/s. An InGaAs photodiode was used as the PL detector. Appropriate filtering was used to ensure that the detector was insensitive to the LED emission for all illumination wavelengths. EQE_{PL} was measured using a constant signal method wherein the incident photon flux is adjusted such that the PL signal is the same for all wavelengths and the EQE is determined from the inverse generation rate. This ensures that the device remains under the same injection level at each wavelength. To ensure that only the cSi emission was measured, the PL spectra at different excitation wavelengths were measured with a spectrophotometer (BwTek Sol 1.7). The IQE_{PL} was then determined through combination of EQE_{PL} and reflectance measurements, taken with a Universal Reflectance Accessory spectrophotometer (PerkinElmer Lambda 1020).

The absorption rate within the amorphous silicon layer was determined using a combination of optical measurements and modelling. The refractive index and thicknesses of the films were determined by spectroscopic ellipsometry measurements (VASE Ellipsometer, J.A.Woollam). The optical properties for the cSi substrate and thermally grown SiO_x were determined from Herzinger *et. al* [12], while the tauc-Lorentz oscillator model [13] was used to fit the optical model for the a-Si layers. The measured specular reflection was used to verify the obtained optical model. Excellent agreement was found between measured and simulated reflection data. The optical modelling was based on the transfer matrix method [14].

The surface recombination velocity losses for the device structures measured in this paper are assumed to be negligible. The internal quantum efficiency in the absence of parasitic absorption is thus close to unity. For a device structure with n thin film layers on the top surface, and in the absence of carrier injection from those layers into the crystalline silicon substrate the maximum achievable IQE (IQE_{Max}) is defined as:

$$IQE_{Max} \propto IQE_{PL}(\lambda)$$

With no carrier flow between the layers the measured spectral variation of IQE_{PL} is thus expected to reflect the absorption losses in the surface layers.

3 RESULTS

3.1 Carrier injection in aSi(i)-SiO_x-cSi structure

To validate the PL based approach for quantum efficiency measurements on heterojunction structures, devices with no carrier injection from the aSi(i) to the cSi bulk (β=0) were investigated. To prevent the carrier injection, a SiO_x layer (~5 nm) was thermally grown between the aSi(i) and cSi. Three thickness of aSi(i) were investigated, 8 nm, 11 nm and 17 nm. The IQE_{PL} of the devices are shown in Fig 1(a). The PL spectra from these three devices showed only the cSi band-band luminescence peak at all illumination wavelengths. As mentioned earlier, the thicknesses and the refractive index of the a-Si was determined from a fit to spectroscopic ellipsometry data with a single Tauc-Lorentz oscillator [13]. IQE_{Max} was calculated for the three aSi(i) layer thicknesses using the absorptance from the optical modelling, results are shown in Fig 1(a) as dashed lines. The IQE_{PL} data for the same samples was scaled relatively to match IQE_{Max} at the longest wavelength (660 nm). Excellent spectral agreement can be seen between IQE_{PL} and IQE_{Max} for these device structures, demonstrating that the carrier density in the base is reduced in proportion to the absorption losses occurring in the top layer, indicating negligible carrier flow between the layers.

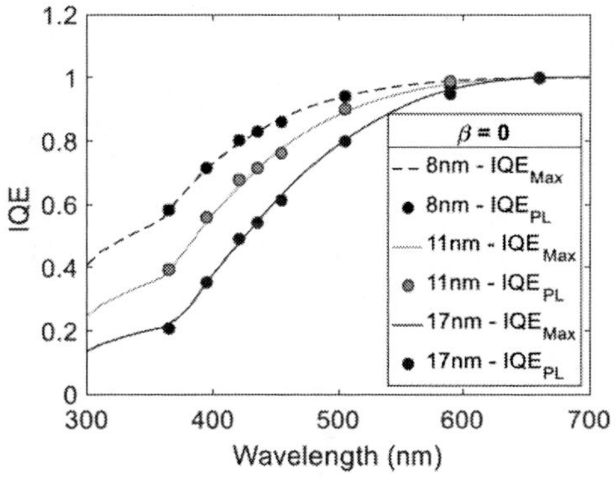

Figure 1: (a) IQE_{Max} and IQE_{PL} data for three n-type cSi samples with aSi(i) top layers with different thicknesses. All samples contained a thermally grown SiO_x blocking layer between the cSi base and the aSi(i) top layer. The IQE_{PL} matches well with the IQE_{Max} values.

3.2 Carrier injection in aSi(i)-cSi structure

Next, the carrier injection efficiency of aSi(i) passivated cSi devices is investigated. The device structures consisted of double side polished n-type Si with aSi(i) layers (8 nm and 14 nm) on both sides. These samples are largely identical to the samples investigated in the previous section, except for the fact that the thermally grown SiO_x blocking layer between the cSi base and the aSi(i) top layer, was omitted.

The PL emission from only the cSi is measured for all illumination wavelengths. The spectra for these two samples, shown in Fig 2(b), show the typical cSi emission characteristics, with no sign of any spurious luminescence affecting the data. IQE_{PL} and IQE_{Max} were determined as described above and the results are shown in Fig 2(a). The reduction in IQE_{Max} towards shorter wavelengths represents the significant absorption losses in the aSi(i) layer. While these losses were reflected in IQE_{PL} in the samples containing the oxide blocking layer (see Fig.1), the IQE_{PL} on these sample data do not reflect this trend at all. On the contrary, the results show an almost constant PL response across all wavelengths. This implies that even at short wavelengths all non-reflected incident photons ultimately contribute to photo-generation in the cSi base, as if the absorbing aSi(i) layer were not present. The explanation for this observation is that almost all the carriers generated by photon absorption in the aSi(i) layer are efficiently injected into the cSi, making the aSi(i) layer effectively transparent. This effect might explain the lower than expected current losses that are experimentally observed in interdigitated back contact (IBC) cells with a-Si front side passivation layers, in comparison to the losses expected from optical simulations [5], [15]. In these studies, on IBC cells, the a-

Si layer is covered by a silicon nitride layer which might prevent the highly efficient carrier injection as measured here.

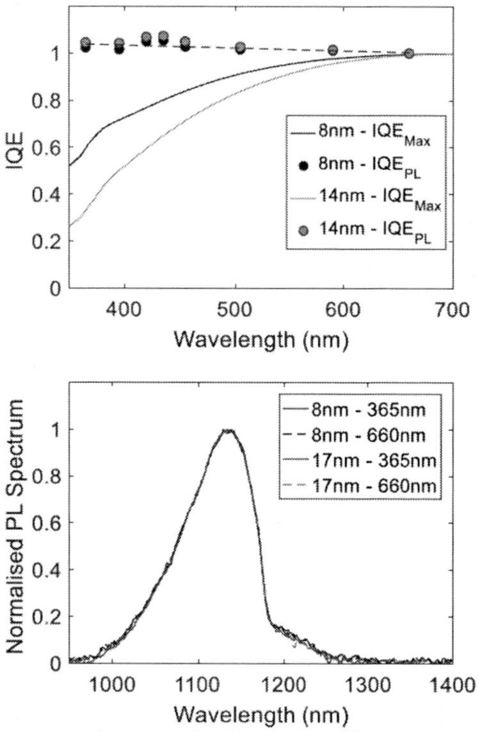

Figure 2: (a) IQE_{PL} for the sample aSi(i) passivated on cSi. (b) The PL spectrum from the 8nm and 14nm samples for the shortest and longest illumination wavelength used. The data shows only emission from the cSi is observed.

As mentioned earlier, the previous methods to measure such carrier injection require forming electrical contacts on the device. The IQE_{PL} based technique is a unique way to measure the carrier injection on devices during fabrication, just after the passivation step, with no need for additional contacting layers. This enables a much more detailed analysis of a much wider range of samples and of how processing conditions and the presence of additional layers, such as TCOs or highly doped amorphous layers, may impact the carrier injection probability from the intrinsic layer into the base. Such investigations are currently under way but beyond the scope of this study. The possibility of a close to unity injection efficiency provides an opportunity for an improved overall collection efficiency for aSi(i) passivated solar cells.

4 CONCLUSIONS

We presented a contactless spectral response of PL based technique to measure carrier injection between aSi(i) and cSi in partially processed heterojunction structures. We compared the measured IQE_{PL} with the maximum IQE predicated by optical modelling. The accuracy of the method was demonstrated on samples containing an oxide barrier layer between the amorphous silicon and the base, which therefore have no carrier injection between the layers. Next, the carrier injection of a-Si on cSi was investigated, and it was found that the carrier injection is almost 100% for two different thicknesses of the aSi(i) layer. This is much higher than previously reported carrier collection efficiency values. The PL based technique demonstrated here provides a unique way to measure such a carrier injection between different layer during fabrication and thus can be used to for further optimisation of heterojunction photovoltaic cells.

5. ACKOWNLEDGEMENTS

This program has been supported by the Australian Government through the Australian Renewable Energy Agency (ARENA) grant RND009. Responsibility for the views, information or advice expressed herein is not accepted by the Australian Government. The authors gratefully acknowledge support by the Swiss National Science Foundation (SNF) under grant No. 200021_14588/1 and No. IZLIZ2_156641 and by the Swiss Federal Office for Energy (SFOE) under grant No. SI/501253-01. The authors thanks Dr. Angus Gentle for performing the spectroscopic ellipsometry measurements.

6. REFERENCES

[1] H. Keppner, R. Flückiger, A. Shah, C. Fortmann, G. Willeke, K. Happle, and H. Kiess, "Passivation properties of amorphous and microcrystalline silicon layers deposited by VHF-GD for crystalline silicon solar cells," *Sol. Energy Mater. Sol. Cells*, vol. 34, pp. 201–209, 1994.

[2] S. Dauwe, J. Schmidt, and R. Hezel, "Very low surface recombination velocities on p- and n-type silicon wafers passivated with hydrogenated amorphous silicon films," in *Conference Record of the Twenty-Ninth IEEE Photovoltaic Specialists Conference, 2002.*, pp. 1246–1249.

[3] S. Dauwe, L. Mittelstadt, A. Metz, J. Schmidt, and R. Hezel, "Low-temperature rear surface passivation schemes for >20% efficient silicon solar cells," in *3rd World Conference on Photovoltaic Energy Conversion*, 2003.

[4] J.-W. A. Schüttauf, K. H. M. van der Werf, I. M. Kielen, W. G. J. H. M. van Sark, J. K. Rath, and R. E. I. Schropp, "Excellent crystalline silicon surface passivation by amorphous silicon irrespective of the technique used for chemical vapor deposition," *Appl. Phys. Lett.*, vol. 98, no. 15, p. 153514, Apr. 2011.

[5] K. Yoshikawa, W. Yoshida, T. Irie, H. Kawasaki, K. Konishi, H. Ishibashi, T. Asatani, D. Adachi, M. Kanematsu, H. Uzu, and K. Yamamoto, "Exceeding conversion efficiency of 26% by heterojunction interdigitated back contact solar cell with thin film Si technology," *Sol. Energy Mater. Sol. Cells*, Jun. 2017.

[6] Z. C. Holman, A. Descoeudres, L. Barraud, F. Z. Fernandez, J. P. Seif, S. De Wolf, and C. Ballif, "Current Losses at the Front of Silicon Heterojunction Solar Cells," *IEEE J. Photovoltaics*, vol. 2, no. 1, pp. 7–15, Jan. 2012.

[7] G. D. Pettit, J. M. Woodall, and H. J. Hovel, "Photoluminescent characterization of GaAs solar cells," *Appl. Phys. Lett.*, vol. 35, no. 4, pp. 335–337, Aug. 1979.

[8] M. K. Juhl, M. D. Abbott, and T. Trupke, "Relative

External Quantum Efficiency of Crystalline Silicon Wafers From Photoluminescence," *IEEE J. Photovoltaics*, pp. 1–7, 2017.
[9] M. K. Juhl and T. Trupke, "The impact of voltage independent carriers on implied voltage measurements on silicon devices," *J. Appl. Phys.*, vol. 120, no. 16, p. 165702, Oct. 2016.
[10]A. Paduthol, M. K. Juhl, and T. Trupke, "Addressing limitations of photoluminescence based external quantum efficiency measurements," *J. Appl. Phys.*
[11]A. R. Paduthol, M. K. Juhl, and T. Trupke, "Estimating the effect of LED spectra on EQE measurement in solar cells," in *2016 IEEE 43rd Photovoltaic Specialists Conference (PVSC)*, 2016, pp. 1580–1584.
[12]C. M. Herzinger, B. Johs, W. A. Mcgahan, J. A. Woollam, and W. Paulson, "Ellipsometric determination of optical constants for silicon and thermally grown silicon dioxide via a multi-sample, multi-wavelength, multi-angle investigation," *J. Appl. Phys.*, vol. 831, no. 10, pp. 1272–561, 1998.
[13]G. E. Jellison Jr. and F. A. Modine, "Parameterization of the optical functions of amorphous materials in the interband region," *Appl. Phys. Lett.*, vol. 69, no. 3, pp. 371–373, Jun. 1998.
[14]C. C. Katsidis and D. I. Siapkas, "General transfer-matrix method for optical multilayer systems with coherent, partially coherent, and incoherent interference," *Appl. Opt.*, vol. 41, no. 19, p. 3978, Jul. 2002.
[15]B. Paviet-Salomon, A. Tomasi, A. Descoeudres, L. Barraud, S. Nicolay, M. Despeisse, S. De Wolf, and C. Ballif, "Back-Contacted Silicon Heterojunction Solar Cells: Optical-Loss Analysis and Mitigation," *IEEE J. Photovoltaics*, vol. 5, no. 5, pp. 1293–1303, Sep. 2015.

ON THE DETERMINATION OF THE CONTACT RESISTIVITY FOR PASSIVATING CONTACTS USING 3D SIMULATIONS

Gamze Kökbudak[1,2], Ralph Müller[1,3], Frank Feldmann[1,3], Andreas Fell[1], Raşit Turan[2], Stefan W. Glunz[1,3]
[1]Fraunhofer Institute for Solar Energy Systems, Heidenhofstr. 2, 79110 Freiburg, Germany
[2] The Center for Solar Energy Research and Applications (GÜNAM), Middle East Technical University, Dumlupınar Bulvarı 1, 06800 , Ankara, Turkey
[3] Laboratory for Photovoltaic Energy Conversion, University Freiburg, Germany

ABSTRACT: Besides surface passivation, a low contact resistivity is one of the most important requirements of passivating contacts in order to achieve good carrier selectivity. In this work, different methods to determine the contact resistivity were performed and compared using a passivating contact structure consisting of poly silicon on a thin silicon oxide. As the traditional 1D Transmission Line Model (TLM) includes many assumptions that are not valid for complicated structures and even the application of the analytical 2D TLM is limited, especially for very low contact resistivity values, 3D numerical simulations with Quokka3 were carried out to accurately model passivating contact structures and determine the contact resistivity by comparison to electrical measurements. The investigated sample showed both, good passivation quality with implied open-circuit voltage iV_{oc} of 716 mV together with a low specific contact resistivity of 0.21 mΩ·cm^2 as extracted from the numerical simulations. Using our method, it is possible to distinguish the sources of the contact resistivity. A fraction of 0.11 mΩ·cm^2 could be attributed to the interface between metal & poly-Si layer and 0.10 mΩ·cm^2 results from the interface between poly-Si layer & bulk.
Keywords: Passivating contact, specific contact resistivity, transfer length method, numerical simulations, Quokka

1 INTRODUCTION

Passivating contacts are a key ingredient to further push the efficiency of silicon solar cells. In the past few years, there has been a growing interest in passivating contact structures which feature an ultra-thin silicon oxide layer and a heavily doped silicon layer (TOPCon [1], poly-Si [2] or POLO [3]). Recently, solar cells have achieved efficiencies greater 25% using a passivating rear contact [4]. Moreover, SunPower is using passivating contacts for their highly efficient interdigitated back contact solar cells [5].

To obtain highly efficient passivating contacts, the recombination at the contacts needs to be minimized in order to get a high internal voltage (iV_{oc}). Additionally, the requirement of carrier selectivity can be fulfilled by allowing one type of charge carrier while blocking the other type which is necessary to obtain a high external open circuit voltage (V_{oc}). To fulfill these requirements, the structure is generally created by depositing a thin interface oxide layer and a conductive thin film on the silicon wafer. So this structure has not only several layers but also several interfaces differently from traditional solar cell structures featuring only diffused regions. In order to obtain a high fill factor (FF) from such a cell, having low contact resistivity is a necessity.

The *contact resistivity* ρ_c (Ω·cm^2), sometimes called *specific contact resistance* or *specific contact resistivity* characterizes the contact independent of the contact area [6]. For traditional silicon solar cells like diffused emitter Al-BSF cells, a low contact resistivity can be usually achieved simply by a stronger doping. However, when it comes to passivating contact cell structure which has many different layers and interfaces, the contact resistivity depends on many parameters. For solar cell designs employing localized contacts, very low contact resistivity values down to 1 mΩ·cm^2 may be needed. Hence it is very important to determine the contact resistivity of passivating contact systems accurately in a wide range.

Fig.1a shows a simple diffused emitter structure with conventional "Transfer Length Method" metal pads, which can be evaluated with the simple one-dimensional (1D) Transmission Line Model (TLM) to obtain the contact resistivity values [7-8]. In contrast, Fig.1b shows a passivating contact with the metal pattern on top of it. As the passivating contact structure includes several conductive layers and a thick bulk, there are many current paths which make the analysis of the contact resistivity more complicated. There is no existing simple evaluation method for such a case. Vertical measurement from front through the bulk to the back side works by measuring the sum of the all the ρ_c values even without current crowding effect near the metal edges as can be seen in Fig 1c. However, this method is only useful for a limited ρ_c range as the bulk resistance dominates the overall resistance when it comes to very low contact resistivity values.

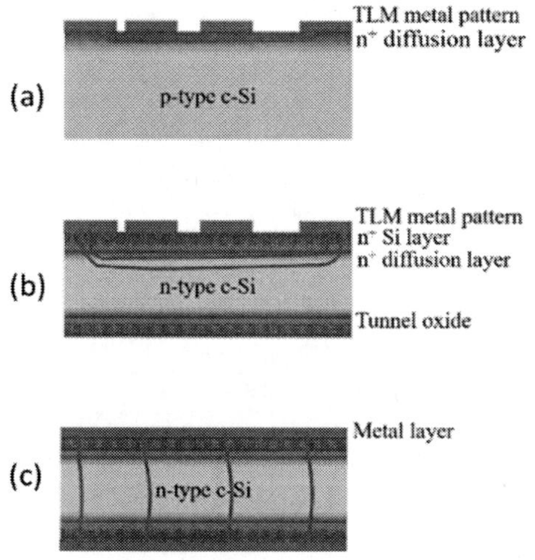

Figure 1: Schematic of the Transfer Length Method (TLM) measurement samples. Diffused emitter structure with a 1D current path (a), Passivating contact structure with many interfaces and complicated current paths (b), Passivating contact structure with vertical current flow through silicon bulk (c)

This paper focuses on the comparison of different analytical methods and uses a rigorous numerical simulation approach to accurately determine the contact resistivity accurately for complicated structures such as passivating contacts.

2 EXPERIMENTAL

n-type floating zone silicon wafers with 1 Ω·cm base resistivity and 200 µm thickness were used. After RCA cleaning, an ultra-thin silicon oxide layer was formed using the nitric acid oxidation of Si (NAOS) method on both sides. Subsequently, about/approx. 40 nm thick amorphous silicon layers were deposited by the Low Pressure Chemical Vapor Deposition (LPCVD) technique symmetrically on both sides. Subsequently, Phosphorus ion implantation was carried out with a dose of 1.5×10^{15} cm^{-2}. Then, the samples were annealed in a tube furnace at 850 °C to activate dopants and crystallize the Si layer simultaneously. Thereafter, a remote plasma hydrogen passivation (RPHP) was applied at 400°C for 30 min to further reduce interface recombination. The samples were prepared symmetrically to obtain Quasi Steady State PhotoConductance (QSSPC) and TLM measurements at exactly the same location on the same sample.

The passivation quality was characterized using the QSSPC technique. Next, this sample was prepared for contact resistivity measurements by defining a TLM pattern with evaporated Ti/Pd/Al stack via a photolithographical lift-off process. Finally, the resistance R as a function of the pad spacing d was determined via four probe measurements to carry out further evaluations of the contact resistivity.

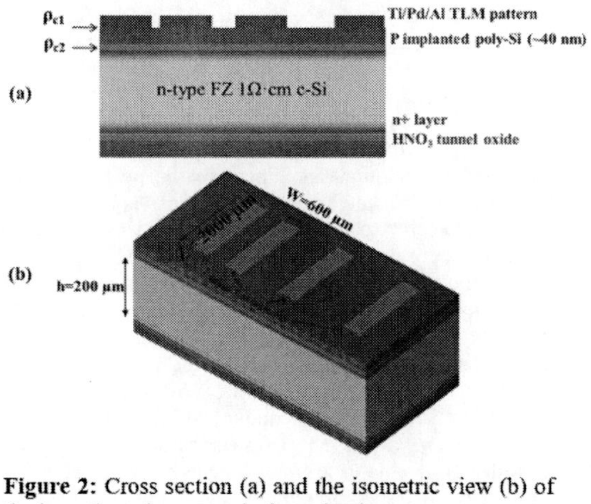

Figure 2: Cross section (a) and the isometric view (b) of the investigated symmetrical passivating contact structure

For the TLM structure, metal pad lengths of 2000 µm and widths of 600 µm were used. The spacing between metal pads increases from 20 µm to 240 µm.

For the numerical method that was proposed in this work, additional resistance obtainment via four probe measurements was carried out after implementing plasma etching process using SF_6 as a reactant gas, to remove the conductive layers in-between the metal pads on the front side of the sample (see Fig. 3).

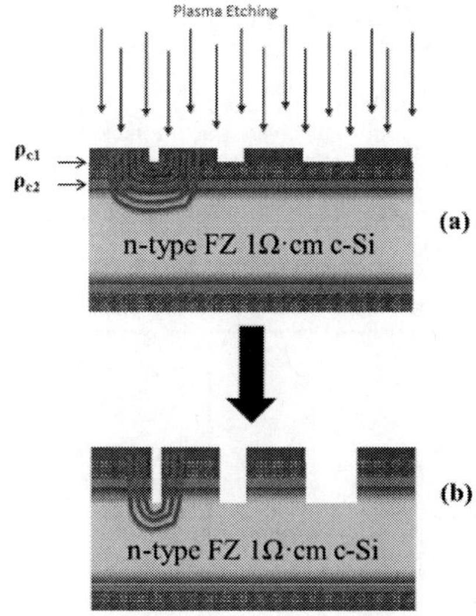

Figure 3: Passivating contact experimental structure and schematic current flow before plasma etch (a), after plasma etch (b)

3 RESULTS AND DISCUSSION

3.1 Analytical 1D-Transmission Line Model (TLM)

The contact resistivity can be obtained using the simple one-dimensional (1D) TLM, for the samples which have only one conductive layer (of very small thickness) and one interface. For this model, the total resistance is measured for various metal spacing and created as an *R(d)* plot as can be seen in Figure 5. Non-linearity occurs for very small metal pad distances due to 2D current flow in the bulk [9]. By fitting in the linear region, some parameters can be extracted such as contact resistance R_c, sheet resistance R_{sheet} and transfer length L_T using a simple analytical theory of TLM with many assumptions that are not compatible with the complicated structures such as passivating contacts. Then, the contact resistivity value can be calculated using the extracted parameters from this plot.

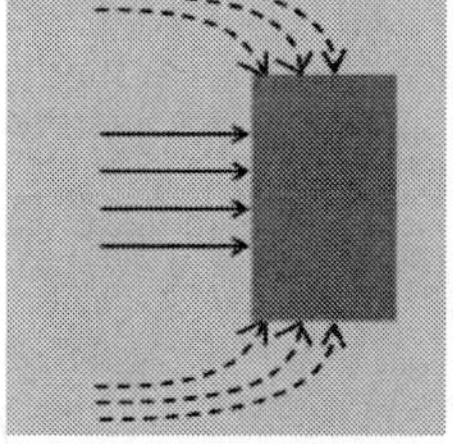

Figure 4: Current trails from the edges of the metal

The assumptions include counting only the 1D current flow by disregarding the current flowing through bulk as this model was developed only for very thin conductive layers. Current trails from the edges of the metal are also ignored (see Fig 4.) which could be avoided by cutting the sample at the edges of the metal pads. However, this may cause other problems since this technique can introduce shunts at the cut planes and needs some additional technological effort. In the 1D TLM, all conductive layers such as doped poly-Si, the n^+ diffusion from the poly-Si into the bulk, and the bulk are all regarded as one effective layer. Another assumption is coming from the influence of the interfaces since this evaluation method includes only one "effective" interface. Furthermore, the metal layer resistance is disregarded.

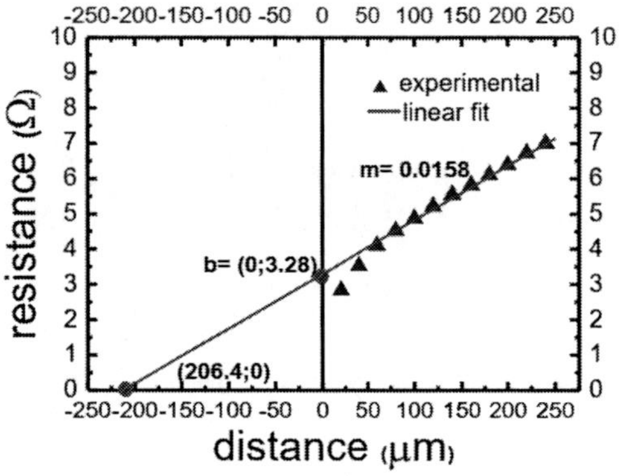

Figure 5: Plot of the measured total resistance as a function of distance between metal pads

Figure 5 shows the plot of the measured total resistance as a function of metal gap distance together with a linear fit for pad distances > 100 μm with interception points on x & y axes as well as the slope. The 1D TLM evaluation was carried out according to Equation (1) to (3). This gives the sheet resistance while transfer length L_t could be calculated from the solution of Equation (2). Here, the transfer length refers to the distance which most of the current transfers from semiconductor into metal and vice versa [10]. Finally, ρ_c could be determined as **3.45 mΩ·cm²** for this specific sample using Equation (3).

$$R_{sheet} = m \times Z = 31.6\ \Omega/\square \qquad (1)$$

$$R_C = \frac{R_{sheet}\, L_t}{Z} \coth\left(\frac{W}{L_t}\right) \qquad (2)$$

Solve for L_t:

$$L_t = 105.2\ \mu m$$

$$\rho_c = R_{sheet} \times L_t^{\,2} \qquad (3)$$

Solve for ρ_c:

$$\rho_c \approx 3.45\ m\Omega\cdot cm^2$$

3.2 2D Analytical TLM evaluation

As the 2D current transport through the bulk of the silicon is neglected in standard 1D TLM evaluation, Eidelloth and Brendel proposed an analytical expression for the extraction of the contact resistivity from TLM measurements considering the 2D current flowing through the bulk [11]. Although this method stands for correcting the 2D bulk effect, it does not include a lateral edge effect correction (see Fig. 4). Furthermore, this method is not suitable for the structures having more than one conductive layer and several interfaces. So the highly doped poly-Si and the n^+ layer below the thin oxide are not taken into account and the contact resistivity of all interfaces is lumped together. The metal layer is also assumed to be infinitively conductive as in 1D TLM. Moreover, it was already stated that this 2D model is not applicable for small contact resistivity values such as $\rho_c \ll 1$ mΩ·cm² [11].

From the linear fit to the measured data shown in Fig.5, the ratio of intersection with the y-axis and slope (b/m) was extracted. Using Equations (4), (5), (6) and (7), equation (8) was solved to extract ρ_c according to Ref [9]. This 2D TLM evaluation yields a contact resistivity value of **1.39 mΩ·cm²** which is much lower than the value 1D TLM gave.

$$G_{1D\_TLM} = \sqrt{\gamma}\coth(\sqrt{\gamma}) \qquad (4)$$

$$G_{CM} = 1 + \gamma + \gamma\frac{\delta}{\pi}\left(\ln 4 - \ln\left(e^{\frac{\delta}{\pi}} - 1\right)\right) \qquad (5)$$

$$\gamma = W^2\rho_b/(\rho_c h) \qquad (6)$$

$$\delta = h/W \qquad (7)$$

$$\rho_c \approx 1.39\ m\Omega\cdot cm^2$$

3.3 3D numerical simulations

As all the analytical models include many assumptions that are not fulfilled by investigations of complex structures, Quokka3 simulations [12] were performed to model the passivating contacts accurately. Quokka3 features a fast solver for ohmic-only carrier transport and it can account for up to two conductive layers on top of the bulk including their respective interface- (i.e. contact-) resistivities. With this, Quokka3 is well suited to simulate TLM structure on the passivating contact system. Furthermore, to account for the lateral edge effect (see Fig 4), a 3D model was used. With the 3D multi-layer model it is possible to separate $\rho_{c,1}$ and $\rho_{c,2}$ where the former stands for the contact resistivity between metal and the poly-Si layer and the latter refers to the contact resistivity between poly-Si and the n^+ layer (across the thin oxide).

The fixed simulation input parameters are bulk resistivity & thickness, the symmetrical layer structure and the metal pad geometry. The variable input parameters are the sheet resistance of the two conductive layers (poly-Si and n^+ layer) as well as the contact resistivity at each interface (metal to poly-Si and poly-Si to n^+ layer across the thin oxide). These four variable parameters were varied to find the best match between simulation and experimental data. With Quokka3 being currently restricted to two conductive layers per side, we chose to disregard the sheet resistance of the metal layer in order to be able to account for the n^+ layer below the thin oxide. This assumption is reasonable as the metal layer conductivity is very high and its effect was found to be very weak in respective simulations.

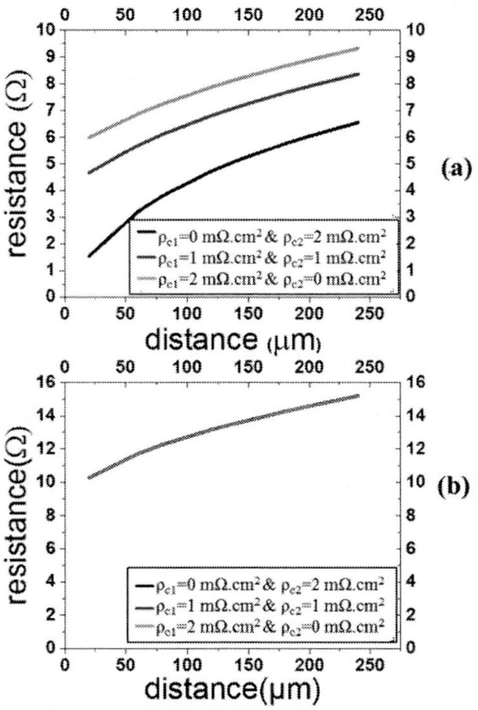

Figure 6: R(d) plots of different simulation sets created by using the fixed parameters of $R_{sheet,\ poly\text{-}Si}=10^2$ Ω/□ & $R_{sheet,\ n^+\ layer}=10^2$ Ω/□ (a), $R_{sheet,\ poly\text{-}Si}=10^{12}$ Ω/□ & $R_{sheet,\ n^+\ layer}=10^{12}$ Ω/□ (b)

Since there are four variable parameters, there may be more than one parameter set matching the experimental data. As an example, Figure 6a shows three simulations where the effective contact resistivity (sum of ρ_{c1} and ρ_{c1}) was kept constant but the contribution of ρ_{c1} and ρ_{c2} to the effective contact resistivity was varied. If the sheet resistance of the poly-Si and n^+ layer is low, the individual contributions of ρ_{c1} and ρ_{c2} have a strong influence on the total resistance (see Fig. 6a). In contrast, one cannot distinguish anymore between the contribution of ρ_{c1} and ρ_{c2} if the sheet resistances are very high (see Fig. 6b). In order to evaluate accurate values for both ρ_{c1} and ρ_{c2}, additional experimental data is needed. Therefore, plasma etching process was applied using SF_6 as a reactant gas, to remove the conductive layers in-between the metal pads on the front side (see Fig. 3) and the resistance of each structure was measured again.

Figure 3a shows the cross section of passivating contact structure and possible current paths before the plasma etching step. 4 probe resistance measurements were carried out while current is flowing from one metal pad to the second one through each of the conductive layers and through the bulk. After etching of the conductive layers between the metal pads, the current is constrained to flow from one metal contact, through all of the interfaces into semiconductor bulk and up into the second metal contact pad as shown in Figure 3b. After performing another 4 probe resistance measurement, the simulations were performed to find a unique simulation parameter set that matches both experimental data sets (before and after etch) simultaneously.

Figure 7 shows the experimental data together with the best-matching simulation parameter set.

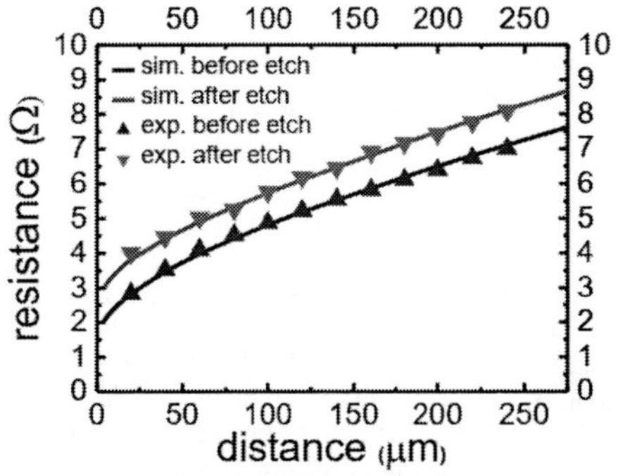

Figure 7: R(d) plot showing the best match between simulation and experimental data (ρ_{c1}=0.11 mΩ.cm², ρ_{c2}=0.10 mΩ.cm², $R_{sheet,\ poly\text{-}Si}$=15000 Ω/□, $R_{sheet,\ n^+\ layer}$=600 Ω/□)

From the simulations it was found that the investigated sample shows a contact resistivity value of **0.11 mΩ.cm²** and **0.10 mΩ.cm²** for the interfaces between metal & poly-Si (ρ_{c1}) and poly-Si & bulk (ρ_{c2}), respectively. In addition to the contact resistivity values for each interface, sheet resistances for conductive layers were found to be 15000 Ω/□ for the poly-Si layer and 600 Ω/□ for the n^+ diffused layer. Calculating the overall sheet resistance of the whole sample (parallel connection of poly-Si & n+ layer at the front and rear together with the bulk) yields 42.6 Ω/□. This value is very close to the overall sheet resistance extracted from QSSPC measurements (43 Ω/□).

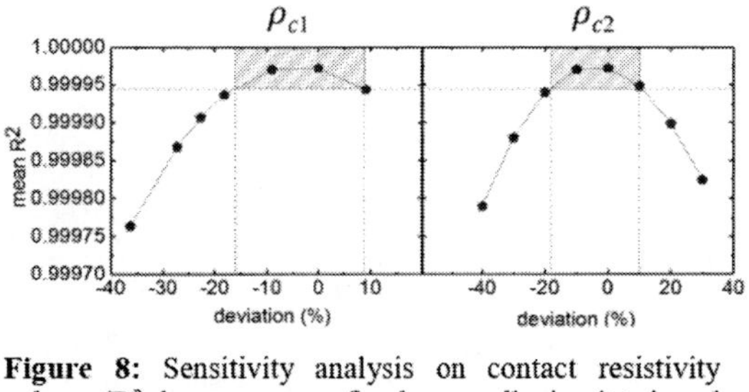

Figure 8: Sensitivity analysis on contact resistivity values (R^2 is a measure for how well simulated and experimental data match, as shown in Figure 7)

The uncertainty of the results was examined by a sensitivity analysis. A high sensitivity to ρ_c was found for both interfaces as shown in Figure 8. The shaded regions indicate the parameter range where the simulated and experimental data show a good match. For ρ_{c1}, the best match was found with 0.11 mΩ.cm² and the accuracy range is between 0.09 and 0.12 mΩ.cm². For ρ_{c2}, the best match is at 0.10 mΩ.cm² and the acceptable range was found to be between 0.085 to 0.11 mΩ.cm². Therefore, there is an uncertainty of about 10% for each extracted contact resistivity value.

For the sheet resistance of the n^+ layer, the uncertainty is less than 4% (600±20 Ω/□). Only the sheet resistance of the poly-Si layer cannot be determined accurately as its conductivity is very low. A good match was obtained in the range from 10000 to 30000 Ohm/sq.

4 SUMMARY

In order to improve the cell performance, the contact resistivity has to be low and it is required to be accurately determined especially when the cell design's contact fraction is small.

Since analytical TLM evaluations (1D & 2D) are not suitable for more than one layer and interface, a numerical simulation model using Quokka3 for exact 3D modelling of the multilayer structures was employed. The 3D model was applied to a structure consisting of several layers, i.e. passivating contact structure with LPCVD deposited and ex-situ phosphorus implanted poly-Si layer on a thin silicon oxide. Table 1 shows the comparison of different methods in terms of limitations and applicability of the methods coupled with the evaluation results for the specific sample investigated in this work. It can be seen that for the structure investigated in this work the analytical models show a large error, highlighting the usefulness and necessity of the numerical simulation approach. Notably, the speed and simplicity of the Quokka3 simulations did not impose any practical limitation on its extensive application.

Table 1: Comparison of different methods to determine the contact resistivity of passivating contact structures

Evaluation Technique	1D analytical TLM	2D analytical TLM	3D numerical simulation
Limitations for passivating contacts	• Only 1D current flow • No current through the bulk • No current flow at the edges • Only one conductive layer • Only one interface • No sheet resistance of metal layer	• No current flow at the edges • Only one conductive layer • Only one interface • No sheet resistance of metal layer	• No sheet resistance of metal layer
Applicability	• 1D pad geometry • 1 conductive layer (1D) • 1 interface	• 1D pad geometry (only with large pad distance) • 1 conductive layer (bulk 2D) • 1 interface	• 2D pad geometry • 2 thin conductive layers + 3D bulk • 3 interfaces • Back side included
Respective result: Contact resistivity [mΩ.cm²]	**3.45**	**1.39**	**0.11+0.10**

5 ACKNOWLEGMENTS

This work has partially funded by the German Federal Ministry for Economic Affairs and Energy under grant No. 0325827B "26+" and 03225877D "PEPPER". We also thank our colleagues at Fraunhofer ISE, especially A. Leimenstoll, F. Schätzle, A. Lösel, R. van der Vossen, A. Seiler, S. Seitz for their contribution to this work.

G. Kökbudak further thanks CHEETAH project under grant No. 609788. and The Scientific And Technological Research Council Of Turkey under grant No. 114F306.

6 REFERENCES

[1] F. Feldmann, M. Bivour, C. Reichel, M. Hermle, and S. W. Glunz, "Passivated rear contacts for high-efficiency n-type Si solar cells providing high interface passivation quality and excellent transport characteristics," Sol. Energy Mater. Sol. Cells, vol. 120, pp. 270–274.

[2] Yan, D., Cuevas, A., Bullock, J., Wan, Y., & Samundsett, C. (2015). Phosphorus-diffused polysilicon contacts for solar cells. Solar Energy Materials and Solar Cells, 142, 75-82. doi:10.1016/j.solmat.2015.06.001

[3] R. Peibst et al., "A simple model describing the symmetric I–V characteristics of p polycrystalline Si/n monocrystalline Si, and n polycrystalline Si/p monocrystalline Si junctions." IEEE J. Photovoltaics, vol. 4, no. 3, pp. 841–850, 2014.

[4] Richter, A. et al. (2017). Silicon Solar Cells with Passivated Rear Contacts: Influence of Wafer Resistivity and Thickness. Freiburg, Germany: SiliconPV 2017.

[5] D. D. Smith et al., "Toward the practical limits of silicon solar cells" IEEE Journal of Photovoltaics, vol. 4, no. 6, pp. 1465–1469, 2014.

[6] D. K. Schroder & D. L. Meier (1984). "Solar cell contact resistance – A review". IEEE Transactions on Electron Devices, 31(5), 637-647.

[7] W. Shockley, A. Goetzberger, and R. M. Scarlett, "Theory and experiments on current transfer from alloyed contact to diffused layer."Appendix B of "Research and investigation of inverse epitaxial UHF power transistors," Shockley Res. Lab., Palo Alto, CA USA, Tech. Rep.AFALL TDR 64-207, 1964.

[8] H. Berger, "Contact resistance on diffused resistors," in IEEE Int. Solid-State Circuits Conf., Dig. Tech. Papers, vol. 12. Feb. 1969, pp. 160–161.

[9] R. Brendel, & S. Eidelloth (2014). "Analytical Theory for Extracting Specific Contact Resistances of Thick Samples From the Transmission Line Method.", IEEE ELECTRON DEVICE LETTERS.

[10] D. K. Schroder (2006). "Contact Resistance and Schottky Barriers. In Semiconductor Material and Device Characterization" , (pp. 127-168).

[11] M. Rienäcker, R. Peibst, & R. Brendel (2016). "Junction Resistivity of Carrier-Selective Polysilicon on Oxide Junctions and Its Impact on Solar Cell Performance", IEEE JOURNAL OF PHOTOVOLTAICS.

[12] A. Fell, J. Schön, M. C.Schubert, S. W. Glunz (2017). "The concept of skins for silicon solar cell modeling" SOLMAT.

FRONT SIDE METALLIZATION OF P AND N-TYPE SI SOLAR CELLS: A PERCOLATION MODEL FOR EXPLAINING THE CURRENT PATH

Michael Pfeffer, Praveen Kumar, Marius Zehender, Benjamin Willsch, Oliver Eibl
Institute of Applied Physics
Eberhard Karls University Tuebingen
Auf der Morgenstelle 10
D-72079 Tuebingen Germany
Phone: +49 7071 2976015
Email: michael.pfeffer@uni-tuebingen.de

Screen printed mono-crystalline Si solar cells yield efficiencies that are limited by electrical losses due to the front side metallization process. Front side metallization of n-type and p-type Si solar cells with maximum efficiencies, p-type 18% and n-type 20 %, were analyzed with respect to their electrical properties and microstructure. In this work, the electrical characterization together with a percolation model is presented, the microstructural results are presented in an accompanying paper. To explain the current path, we introduced a percolation model, in which the Ag colloids constitute current filaments in the glass layer with reduced resistance. The introduced percolation model is new and different to the classical percolation model since it directly relies on the macroscopic resistance of such filaments. The chemical analysis of the glass layer shows that it is semiconducting rather than insulating. Since the contact resistance across the metallization is an important part of the series resistance of the solar cell, engineering of the glass layer becomes essential for high-efficiency Si solar cells.
Keywords: Silicon Solar Cell, Solar Cell Efficiencies, Electrical Properties, Modelling

1 INTRODUCTION

Mono-crystalline, high-efficiency Si solar cell play a key role in the future as a part of renewable, eco-friendly energy sources due to their simple and cheap production as compared to other photovoltaic devices. The most relevant technology to contact such cells is screen printing, due to its simplicity and high throughput. The electrical losses across the so formed metallization interface limit the efficiency of the cell and are summarized as the contact resistance of a cell. In high-efficiency cells, this contact resistance is in the range of few $m\Omega.cm^2$ [1-5] and is a part of the series resistance of the cell.
The formation of this contact region, with respect to its electrical and microstructural properties was review in [6]. Despite the high importance of the contact interface to the efficiency, the current transport from the bulk Si to the Ag finger is still under debate. Different hypotheses exist in the literature and were reviewed in [7]. The two most commonly discussed models are

- Nonuniform conduction in which the current flows across multiple Ag nanocrystals that are possibly in direct contact between the Si emitter and the Ag bulk [2,8]
- Current flow through the thin interfacial glass layer richly decorated with Ag nanocolloids, assisted by multistep tunneling [9,10]

In contrast to these models, we proposed a new percolation model [11]. The main assumption for this model is to treat the glass layer as a dirty semiconductor instead of an insulator. The chemical analysis shows that the glass layer is not pure SiO_2, but enriched with Pb and traces of dopants [12].
The current can then flow across the high resistive glass layer, but can take low resistance "short-cuts" across the Ag nano-colloids, thus leading to a percolative path of low resistance across the metallization interface.

2 EXPERIMENTAL

In this paper, high-efficiency, mono-crystalline Si solar cells with n- and p-type emitter were under investigation. Each type of solar cell was produced with the same production parameters and only differ in the screen printing process. More detailed parameters of these cells are given in refs. [4, 5]. As a result, the differences in efficiency and the electrical losses due to the series and contact resistance can only be attributed to the differences in the metallization interface. This enables us to perform a structure-property correlation of the interfacial layer, by linking the electrical and microstructural results on these cells.
The p- and n-type cells under investigation are labeled according to their efficiency, starting from p-T1 (n-T1) for highest to p-T3 (n-T3) for lowest efficiency. The big wafer data is provided in Table I.

Table I: Full wafer data of p- and n-type cells. The cells are labeled according to their efficiencies, starting from T1 (highest efficiency) to T3 (lowest efficiency) [13].

Solar Cell	η[%]	$R_s[\Omega\ cm^2]$	$\rho_c[m\Omega\ cm^2]$
p-T1	18.0	0.61	5
p-T2	16.9	0.87	21
p-T3	13.6	3.91	131
n-T1	19.96	0.53	4
n-T2	19.86	0.57	4
n-T3	16.09	2.79	>100

The electrical measurements were performed on small cut pieces of the solar cells ($1\times1.5cm^2$) in an optical cryostat. Starting from 80K to room temperature (RT), I-V curves of the cells were recorded and evaluated with respect to the series resistance.
Microstructural analysis was performed by SEM and TEM, both equipped with EDX to probe the chemical composition of the interfacial layer.

3 RESULTS AND DISCUSSION

3.1 Temperature dependent series resistance

The big wafer data are given in Table I and show the importance of the series and contact resistance to the efficiency of the cells. There is a clear correlation of low resistances (both series- and contact resistance) with highest efficiencies. To get a better understanding of this correlation, I-V curves of all cells were recorded at different temperatures and evaluated with respect to their series resistance [13]. The results are shown in Fig. 1.

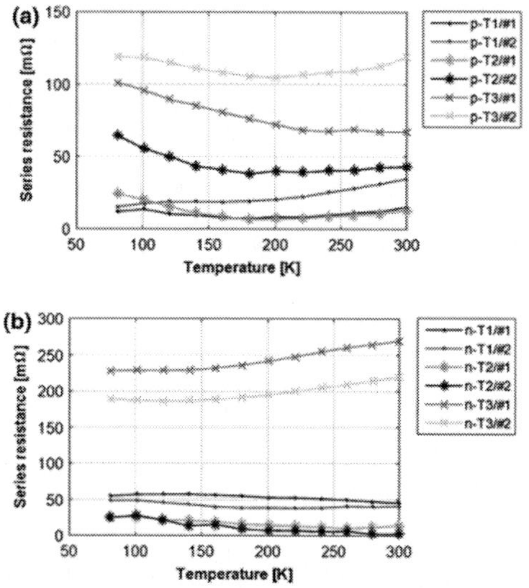

Figure 1: Temperature dependent series resistance measurements of a) p-type and b) n- type cells.

For both n- and p-type cells, the high-efficiency cells show the lowest series resistance over the whole temperature range, while the low resistance cells show a higher series resistance. A direct connection of Ag crystallites to the Ag bulk of the finger, as proposed by [2], should show metallic behavior. Both high-efficiency cells of p- and n-type show only a small temperature dependence over the whole temperature range.

3.2 Microstructural and chemical analysis

SEM investigation of more than 10 cells has been performed. Fig. 2a-b show the color coded plan-view BSE images of high-efficiency p-T1 and medium efficiency p-T2 acquired at the Si/metallization interface. The different colors could be assigned to the different phases by SEM-EDX microanalysis, i.e. Si pyramids (yellow), Ag (blue), glass phase (pink), carbon (green) and Zn-rich oxide phases (sky blue). For high-efficiency p-T1 cell we observed a homogenous wetting of the Si emitter by the glass phase, whereas a discontinuous wetting of the Si emitter was found for p-T2.

Fig. 3 shows SEM results of high and low-efficiency n- and p-type cells. The microstructure of the metallization for high-efficiency Si solar cells is similar for n-type and p-type cells: (i) pronounced wetting of the Si emitter by a glass layer, (ii) high density of Ag colloids in glass layer and (iii) thickness of glass layer < 1µm [4,5,11]. In contrast, the low-efficiency cells show much smaller densities of Ag colloids in the glass layer. There is a clear correlation visible between the density of Ag colloids with the contact resistance of the cells and by this the efficiency of the investigated cells. This shows again the importance of the Ag colloids for low contact resistances and high-efficiency cells.

Fig. 4 shows a bright-field image of the glass layer of high-efficiency cell p-T1. The glass layer contained a high density of Ag colloids. In a companion paper [12], a large number of bright-field, dark-field and energy filtered TEM (EF-) images were acquired of the glass layer of different cells. They prove the presence of metallic precipitates and Ag colloids in the glass layer and suggest that the glass layer of such cells is a dirty semiconductor rather than an insulator. Such images are important for explaining the current path through the glass layer, e.g. due to percolation.

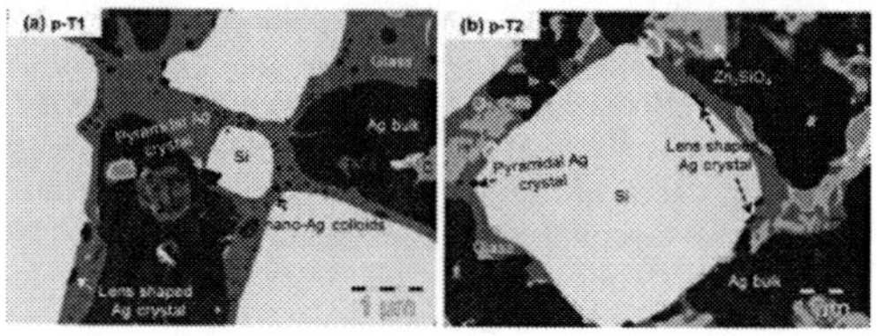

Figure 2: Plan-view false color BSE images obtained at the Si/Ag interface. (a) high-efficeincy p-T1, a continuous wetting of the glass phase and a high density of Ag nanocolloids was observed. (b) p-T2, showed a discontinuous wetting of the Si emitter by the glass phase.

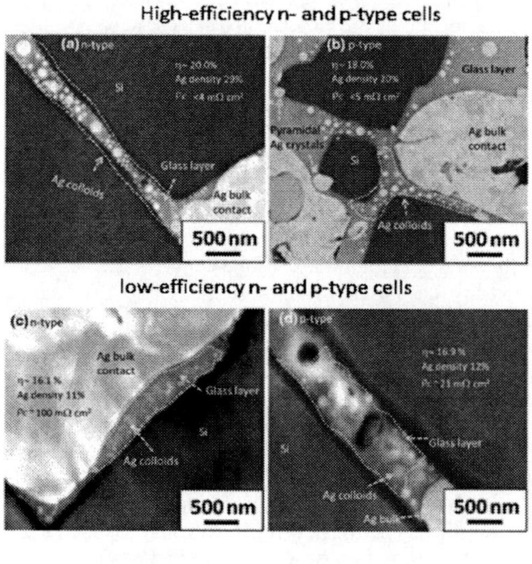

Figure 3: SEM micrographs of n- and p-type high and low-efficiency cells [11].

The chemical analysis both by TEM-EDX and SEM-EDX spectroscopy show that the glass layer of all cells does not consist of pure SiO_2, but it is rather made up of a (SiOx)Pb compound. In addition, Zn, Ti, Al, Ag P and B as minor constituents with mole fractions above the detection limit were found [12]. This shows that the glass layer is not an insulator as expected for a pure SiO_2 matrix, but is rather a dirty semiconductor with (at least) poor conductivity.

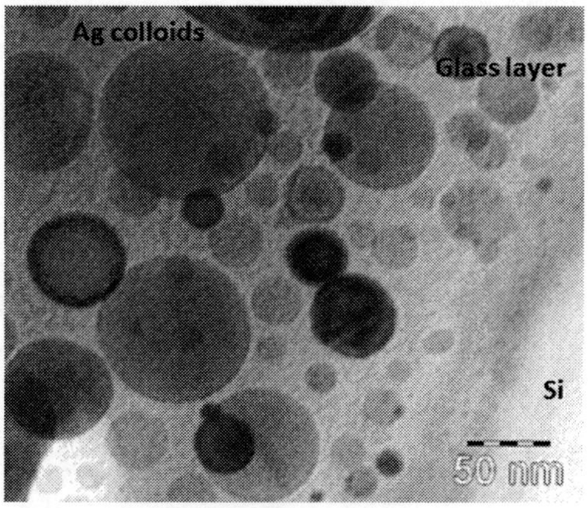

Figure 4: Bright field image of the glass layer of high-efficiency p-type cell.

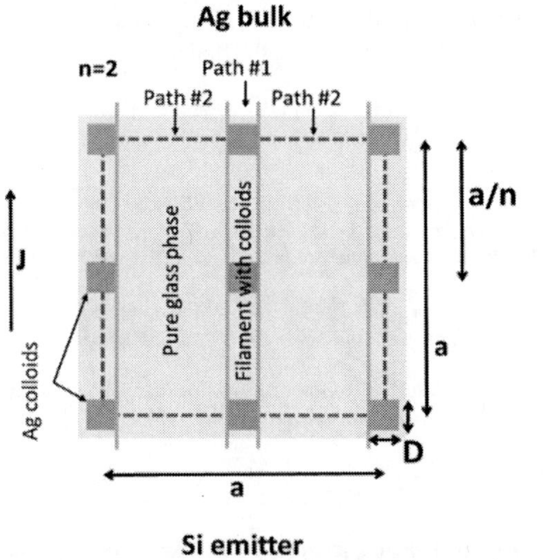

Figure 5: Ag colloid arrangement and current paths for square-shaped 2D. The colloids are distributed periodically across the glass matrix. Current flow is only allowed from bottom to top [11].

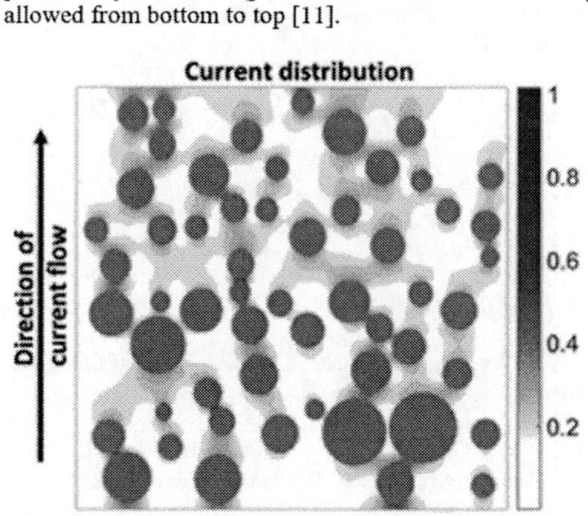

Figure 6: Current distribution across the glass layer; the maximum value is set to one. Due to the colloids, there are bands of high current across the structure, guided by the Ag colloids.

The resistance of a 2D structure of Ag colloids in a glass matrix is shown in Fig. 5. With increasing volume fraction of Ag colloids, the resistance is decreasing. Fig. 6 shows one particular simulation of a random arrangement of Ag colloids, bands of high currents emerge that are guided by the Ag colloids through the poorly conductive glass layer. The introduced percolation model explains the correlation between high Ag colloid density and high cell efficiencies and emphasizes the importance of the glass layer for the current transport [11].

3.3 Structure-property correlation: percolative current transport

To get a complete picture of the current transport across the metallization interface, the structure-property correlation has to be performed to link the electrical and microstructural results. From this, a percolation model was proposed to explain the current transport across the glass layer [11]. The main assumptions and their link to the individual measurements are:

- The chemical analysis shows that the glass matrix is not a pure SiO_2, but rather (SiOx)Pb which is considered a dirty semiconductor with at least poor conductivity [12].
- Microstructural analysis shows a clear correlation between high densities of Ag-colloids with low contact resistance. This shows, that the current path involves the highly conductive Ag colloids [4,5,11].

A simple depiction of our model is shown in Fig. 5.

To keep the starting point of the model simple, a 2D cut through the glass layer is considered (Si-emitter on bottom, Ag bulk on top), where current flow is only allowed in one dimension from the bottom to the top. In addition, only regularly arranged, square-shaped colloids with vanishing resistivity are considered. These assumptions allow the analytical calculation of the resistance of the glass layer. To make the calculation independent of actual values for the resistivity of the glass matrix, the results are divided by the resistance of the pure, colloid free glass layer, yielding the reduction of the resistance by the colloids:

$$fractional\ resistance = \left(1 + \frac{\left(\frac{nD}{a}\right)^2}{1 - \frac{nD}{a}}\right)^{-1}$$

$$= \frac{1 - \sqrt{VF}}{1 + VF - \sqrt{VF}}$$

$$with\ VF = \frac{n^2 D^2}{a^2}$$

In this formula, plotted in Fig. 7, n is the number of colloids in each dimension of the regular grid, D their size and a the thickness of the glass layer. In Fig. 7, the fractional resistance is not only shown for the presented 2D model with square-shaped colloids, but also for spherically shaped colloids in 3D. The calculation of the 3D model follows the same steps and is given in more details in [11]. The determined volume fraction of the high- and low-efficiency cells is also displayed as yellow lines. The result in Fig. 7 shows that the reduction of the resistance does only depend on the amount (i.e. the volume fraction) of the Ag colloids, but not on how they are distributed across the glass layer.

To get more generalized results, the calculation has also been performed for circularly shaped colloids in 2D, as well as cubically and spherically shaped colloids in 3D. From Fig. 7 it can be seen, how this model explains nicely the reduced contact resistance (and thus the higher efficiency) of cells with a larger density of Ag colloids. It is noteworthy, that within our model the fractional resistance depends only on the volume fraction of Ag in the glass layer.

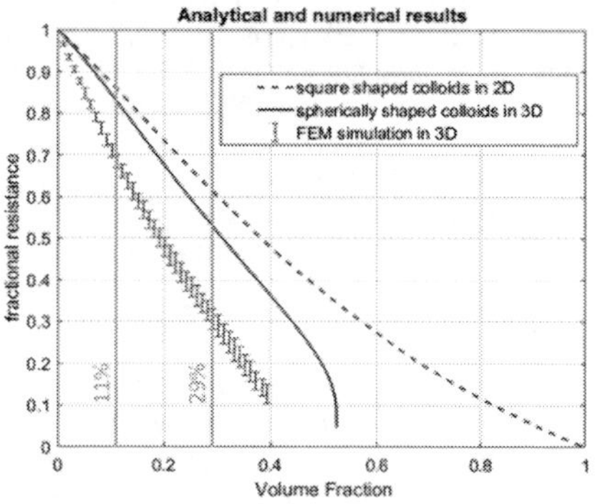

Figure 7: Fractional resistance of the glass layer in 2D (analytically) and 3D (analytically and numerically).

The strict restrictions on one-dimensional current flow and regular arrangement of same-sized colloids can be avoided by performing numerical simulations. To do so, size distributions of the Ag colloids were determined from the 2D SEM micrographs and transformed to 3D distributions according to [14]. Then, random arrangements of random-sized colloids were generated and their electrical resistance was determined using the finite-element method. The results are shown in Fig. 7.

It can be seen, that the (more demanding) simulation yields similar results as the analytical approach. For the determined VF of high- and low- efficiency cells, the simulation predicts an even stronger difference in the contact resistance as the VF is increasing. But only the simple analytical approach is able to give some insight about the exact dependencies on the model parameters as the VF.

4 CONCLUSION

Electrical and microstructural characterization of high and low efficiency, n- and p-type single crystalline Si solar cells were performed. The electrical data clearly showed the correlation of high cell efficiencies and low series- and contact resistances. In addition, the temperature dependent measurements of the series resistance showed a temperature independent behavior of the high-efficiency cells. From this it was concluded, that the current path across the metallization interface is made up of different components with different temperature-dependent behavior.
The microstructural results showed, that high-efficiency cells always contain a high density of Ag colloids in the glass layer, in contrast to low-efficiency cells with low densities of Ag colloids. In addition, chemical analysis showed that the glass layer itself seems to be a dirty semiconductor, rather than an insulator.
From the structure-property correlation of these results, a percolation model was deduced. It successfully explains why low contact resistances are found in high-efficiency solar cells with high colloid density. With this model, the glass layer becomes more important: the chemical composition of the glass layer as well as the colloid volume fraction are characteristic properties that should be measured.

5 REFERENCES

[1] R. Hoenig, D. Voessing, F. Clement, D. Biro, R. Preu, J. Wilde, *Energy Procedia 38* (2013) 737

[2] C. Ballif, D.M. Huljic, G. Willeke, A. Hessler-Wyser, *Applied Physics Letters* 82(2003)1878.

[3] S. Kontermann, G. Willeke, J. Bauer, *Applied Physics Letters* 97(2012) 191910.

[4] P. Kumar, M. Pfeffer, B. Willsch, O. Eibl, L. Koduvelikulathu, V. Mihailetchi, R. Kopecek, *Solar Energy Materials and Solar Cells* 157 (2016) 200.

[5] P. Kumar, M. Pfeffer, B. Willsch, O. Eibl, *Solar Energy Materials and Solar Cells 145* (2016) 358.

[6] M. Hörteis, T. Gutberlet, A. Reller, S.W. Glunz, *Advanced Functional Materials* 20(2010) 476.

[7] G. Schubert, F. Huster, P. Fath, *Solar Energy Materials & Solar Cells* 90 (2006) 3399.

[8] S. Kontermann, R. Preu, G. Willeke, *Applied Physics Letters* 99 (2011) 111905..

[9] M.M. Hilal, S. Sridharan, C. Khadilkar, A- Shaikh, A. Rohatgi, S. Kim, *Journal of Electronic Materials* 35 (2006) 2041.

[10] Z.G. Li, L. Liang, L.K. Cheng, *Journal of Applied Physics 105* (2009) 066102.

[11] M. Pfeffer, P. Kumar, O. Eibl, *Journal of Electronic Materials 45* (2016) 5764.

[12] P. Kumar, M. Pfeffer, O. Eibl, *Abstract submitted to EU PVSEC 2017*

[13] B. Willsch, P. Kumar, O. Eibl, *Journal of Electronic Materials 45* (2016) 2656.

[14] J. Gegner, C. Henninger, A. Öchsner, *Materialwissenschaft und Werkstofftechnik* 35 (2004) 36.

INVESTIGATION ON THE AG-AL METAL SPIKING INTO BORON-DIFFUSED P^+ LAYERS OF INDUSTRIAL BIFACIAL *N*-TYPE SILICON WAFER SOLAR CELLS BY NUMERICAL SIMULATION

Mengjie LI[1, 2, a], Fa-Jun MA[3], Rolf STANGL[1], Armin G. ABERLE[1, 2], Ganesh S. SAMUDRA[2], Bram HOEX[3]

[1] Solar Energy Research Institute of Singapore (SERIS), National University of Singapore (NUS), Singapore 117574
[2] Department of Electrical & Computer Engineering, National University of Singapore (NUS), Singapore 117576
[3] School of Photovoltaic and Renewable Energy Engineering (SPREE), University of New South Wales (UNSW), Sydney NSW 2052, Australia
[a] Corresponding author: Mengjie LI, *E-mail address*: mengjie.li@u.nus.edu

ABSTRACT: A small amount of Al is usually added to the Ag paste to facilitate a sufficiently low specific contact resistance to contact boron diffused p^+ silicon layers. However, Al can result in spiking from the metal contacts into the diffused layer leading to a pronounced recombination at the metal-silicon interface. This significantly limits the open-circuit voltages of screen printed *n*-type silicon solar cells. The aim of this work is to investigate the sensitivity of industrially feasible boron diffusion doping profiles to various spike densities/shapes using Sentaurus TCAD device simulation. Four different experimental BBr_3 diffused doping profiles were analyzed. Bifacial *n*-type front and back contacted (*n*FAB) silicon solar cells were simulated. It was found that the cell efficiency was only significantly affected when the localized spikes penetrate as deep as that they contact to a doping density below ~ 1×10^{18} cm^{-3}. Beyond that point, the defect-related recombination at the metal-Si interface starts to increase by orders of magnitude. The cell efficiency could degrade by relatively ~ 2 - 3 % to up to ~ 10 % when the spike fraction increases from 5 - 10 % to 100 %.
Keywords: Contact, Metallization, Modelling, *n*-type, Recombination, Simulation

1 INTRODUCTION

Screen printing and firing-through technology is widely applied for the contact formation of industrial high-efficiency silicon solar cells. In *n*-type silicon solar cells, the boron doped p^+ layers are usually contacted using Al containing Ag fire-through paste. The Al provides a lower contact resistance [1, 2]. However, it catalyses spikes and consequently can cause severe recombination losses [3-10]. Recently, several research groups have studied the microscopic structure at the metal-Si interface [8-12]. Local spikes after screen printing and firing are observed with SEM imaging. Various simulation models have been proposed to get a better understanding of the influence of the Ag/Al spikes on boron doped surface in silicon solar cells [6, 8, 9, 13]. In 2014, Edler *et al.* [6] first presented a two-dimensional (2D) model using Silvaco ATLAS [14] simulation to simulate the resulting reduction in open circuit voltage (V_{oc}) of their *n*-type cell with Ag/Al contacted boron doped p^+ layers. At that time, a fair agreement to experimental observations was achieved with a full area metal penetration to a depth of 200 nm above the *p-n* junction. In 2015, Koduvelikulathu *et al.* [13] introduced a more detailed 2D metal spiking model to explain the metallization induced recombination losses in Ag contacted phosphorus doped layers. They showed that the superficially grown Ag crystallites only contribute a negligible degradation in V_{oc}. In 2016, a locally presented metal spiking model was presented by Wohrle *et al.*[9] and Kiefer *et al.*[8]. Wohrle *et al.* [9] demonstrated that the dark saturation current density at the metal-Si interface (J_{0e_met}) was heavily influenced only when the spikes protrude closely towards the *p-n* junction. Additionally, the major recombination effect comes from the spike tip, while the spike flanks only contributing a minor part. A spike density of 10% had to be assumed in their simulations in order to reproduce the measured J_{0e_met}. Unlike Wohrle *et al.* [9], Kiefer *et al.* [8] presented a locally planar-shaped spike model. To effectively minimise the recombination induced by metal spiking, a deep junction doping profile is suggested in many publications [6, 8, 9, 13]. However, it is still of great interest to see how sensitive the doping profiles are in responding to various spike densities/shapes. This work builds on previous research results [6, 8, 9, 13] and provides a more comprehensive model to investigate the influence of Ag/Al spiking into boron diffused p^+ layers of industrial bifacial n-type front and back contacted (*n*FAB) silicon solar cells. We perform a 2D unit-cell simulation with localized "planar-shaped" spikes with TCAD Sentaurus Device [15] simulation. Four differently processed industrial feasible boron profiles were imported into the device simulator to calculate the cell characteristics with a variation on spike depth (d_{spike}) and spike coverage (f_{spike}). With the help of simulation, we discuss the tolerance of Ag/Al spikes of industrial boron profiles, the benefits that can be achieved with deeper profiles and suggestions for further development for high-efficiency *n*-type silicon solar cells.

2 SIMULATION APPROACH

In this work, we performed a two-dimensional (2D) unit cell simulation with localized "planar-shaped" spikes (as sketched in Figure 1 (b-2) and Figure 1 (b-3)) with TCAD Sentaurus Device [15] simulation, as an expansion of results from other groups [6, 8, 9, 13]. Instead of focusing on the prediction of the dark saturation current densities at metal-Si interface (J_{0e_met}), we focused on the solar cell device performance as a function of spike depth d_{spike} and spike coverage ratio f_{spike}. Figure 1 (b-1) depicts the simulated *n*-type bifacial front and back contacted (*n*FAB) cell structure. Figure 1 (b-2) depicts the "planar shaped" metal spike structure, describing the microscopic details at the metal-Si interface. The three input parameters to describe the metal spikes are: the spike depth d_{spike} and spike coverage ratio f_{spike} and the surface recombination velocity at the metal-Si interface S_{met}. However, difficulty was faced when simulating the structure depicted in Figure

1 (b-2) due to the reason that a boundary condition has to be defined where the metal spike flank has to contact to the p^+ layer with a concentration gradient. Additionally, as proved by Wohrle [9], the major recombination effect comes from the spike tip, while the spike flanks only contributing a minor part. Thus, we further simplified the simulated structure as shown in Figure 1 (b-3). Figure 1 (b-3) shows the equivalent simplification of Figure 1 (b-2) as used in the simulation. The red dotted rectangular box in Figure 1 (b-3) indicates the 2-D symmetry element simulated in Sentaurus TCAD device simulation.The diffusion profile under the non-spiking region was assumed to be identical to the measured electrochemical capacitance voltage (ECV) profile. Under the metal spiking region, the first few hundred nanometres of the profile was removed, which corresponds to the spike depth, d_{spike}. The generation profile was calculated using the open-source program Module ray tracer [16] with calibration of the rear internal reflectance towards the measured reflectance between the metal grid fingers of processed solar cells. It was found that the experimental reflectance curve measured on gold-plated copper chuck can be reproduced with a rear internal reflectance of ~0.85. The surface recombination parameters were derived by reproducing measured effective minority carrier lifetimes in a similar way as done in Ref. [17]. In the unit cell simulation, Q_{total_front} was set as -4×10^{12} cm^{-2}, Q_{total_rear} as $+2\times10^{12}$ cm^{-2}, and S_{pass} as 1×10^4 cm/s. The surface recombination velocity S_{met} at the metal-Si interface was set at 1×10^7 cm/s. The spiking depth is varied in steps of 100 nm up to 50 nm of the p-n junction. The metal grid had a finger width of 50 µm and a pitch of 1.8 mm.

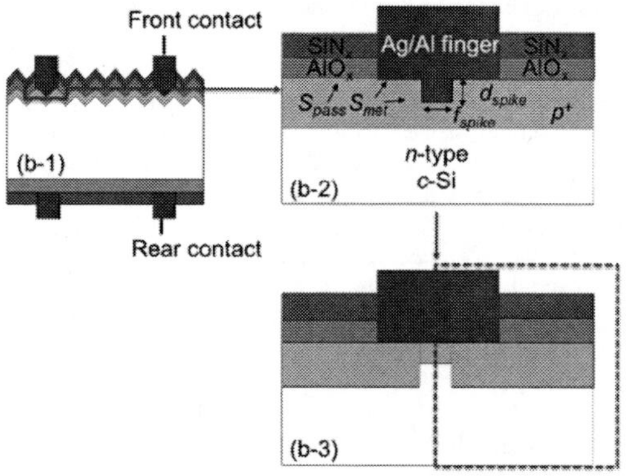

Figure 1: (b-1) Schematic illustration of the cross-sectional nFAB cell structure. (b-2) Enlargement of the "planar shaped" metal spike structure, indicated with a red box in (b-1). (b-3) The equivalent simplification of (b-2) as used in the simulation. S_{pass} is the surface recombination velocity of electron and hole at the AlO_x-Si interface, S_{met} is the surface recombination velocity at the metal-Si interface, d_{spike} is the spike depth, f_{spike} is the percentage of spike coverage over the entire metal finger. The red dotted rectangular box in (b-3) indicates the 2-D symmetry element simulated in Sentaurus TCAD device simulation.

3 EXPERIMENTS

Four differently processed boron doping profiles fabricated with BBr_3 tube diffusion (Tempress, TS81004) were investigated in this study. The doping profiles were measured with electrochemical capacitance voltage profiling (ECV, WEP-CVP21). After boron diffusion and passivation with plasma-enhanced chemical vapour deposition (PECVD, Meyer Burger, MAiA) of AlO_x/SiN_x stacks, a specially designed test structure was printed on only one side of the sample with screen printing by an industrial screen printer (Dek, PVD1200IR) and followed by firing through in a fast firing furnace (BTU International SinTerr). The details of the sample fabrication are described in Ref. [18]. The metallized test samples were used in this study to check the microscopic structure of the metal-Si interface. A scanning electron microscope (Zeiss Auriga) is used to view the cross-sectional surface of the metal-Si interface. The results are presented in the next section.

4 RESULTS

4.1 P^+ layers

Figure 2 shows the boron profiles as measured by ECV. The profile named 35 features a sheet resistance (R_{sheet}) of 35 ohms/sq. The same rule applies to the other three profiles. The 35 ohms/sq. and the 60 ohms/sq. profiles featured a higher peak concentration compared to the 40 ohms/sq. and 100 ohms/sq. profiles. The 40 ohms/sq. profile features the deepest junction across the profiles investigated in this study with a junction depth reach almost 1 µm.

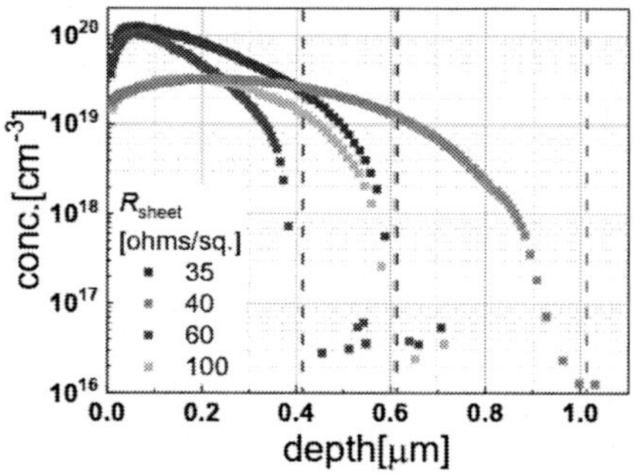

Figure 2: Experimental boron profiles as measured by ECV. The vertical dashed lines depict the position of the p-n junction of each of the doping profiles.

4.2 Metal-Si interface

Figure 3 depicts two examples of the metal spikes observed at the cross-sectional surface of the metal-Si interface. The samples were fired at optimized firing conditions. Metal spikes along the interface was observed and marked with red circles. Red arrows mark the voids formed due to Al alloying Si. The spikes were found to be distributed inhomogeneously at the interface mainly in the Al-rich regions.

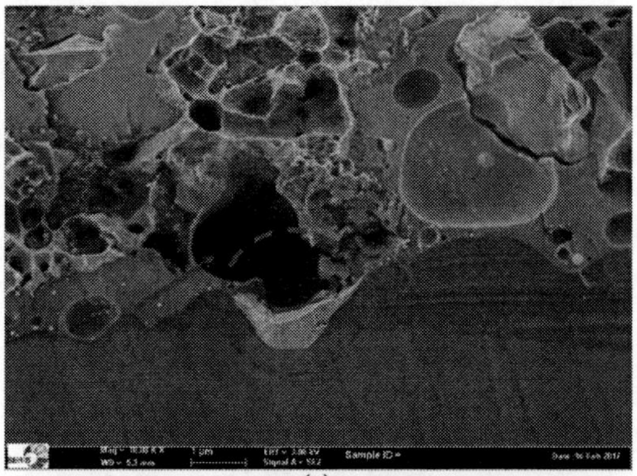

(a)

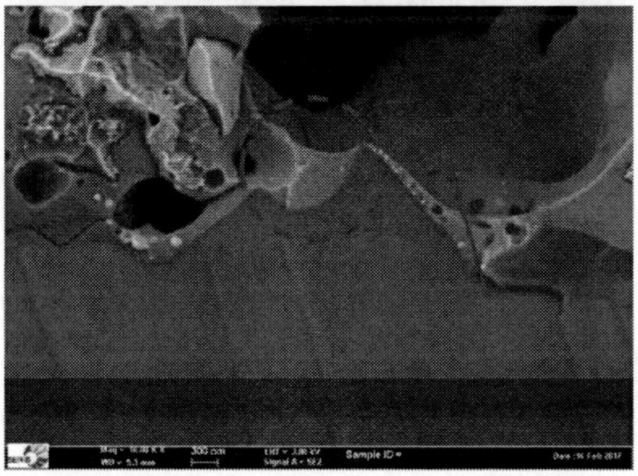

(b)

Figure 3: (a, b) Two SEM images of the cross section of the interface between silicon and the Ag/Al paste. Red circles mark the typical spikes grown into the silicon. The voids formed due to Al alloying Si are marked by red arrows.

4.3 Variation of spike depth and surface coverage

Figure 4 shows simulated interface recombination taking place at the spike tip at the metal-Si interface. The recombination rate increases by orders of magnitude when the spike is reaching the *p-n* junction. The doping profile with a R_{sheet} of 100 ohms/sq. is investigated. The vertical dashed line in the plot indicate the position of the *p-n* junction of the doping profile. The recombination mechanism at the metal-Si interface of a highly doped region was initially explained by del Alamo *et al.* [19] and further interpreted by Wohrle *et al.* [9]. When the surface recombination is pinned to the maximum, which is the case at the metal-Si interface, the recombination rate is inversely related to the doping density, as a higher doping density results in a lowered minority carrier density. Hence, the recombination activity significantly increases when the spike depth increases due the increased minority carrier concentrations at these depths.

Figure 5 shows the simulated V_{oc} as a function of the spike penetration depth d_{spike} and the spike coverage ratio f_{spike}. The V_{oc} decreases with increased d_{spike} and f_{spike}. A higher f_{spike} leads to a faster and more severe degradation of V_{oc}. For a f_{spike} of 1 %, we only see a small V_{oc} decrease of 5 mV. There are two things to be keep in mind. First, our model simulates a planar surface, whereas a real surface of a solar cell is textured. Second, the real distribution of the metal spikes is far more complicated than assumed in our simulations. The simulation was conducted with an evenly distributed single spike. Even with the simplifications in the simulation, consistent results with what has been reported in the literature [6, 8, 9, 20, 21] was obtained. This shows that simulation with localised planar-shaped spikes is a fair simplification of the reality.

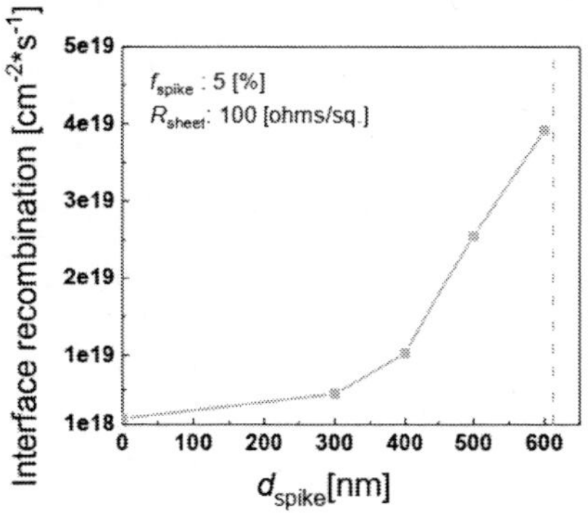

Figure 4: Simulated interface recombination rate at the spike tip at metal-Si interface as a function of spike depth d_{spike}. The doping profile with a R_{sheet} of 100 ohms/sq. is investigated. The vertical dashed line indicate the position of the *p-n* junction of the doping profile.

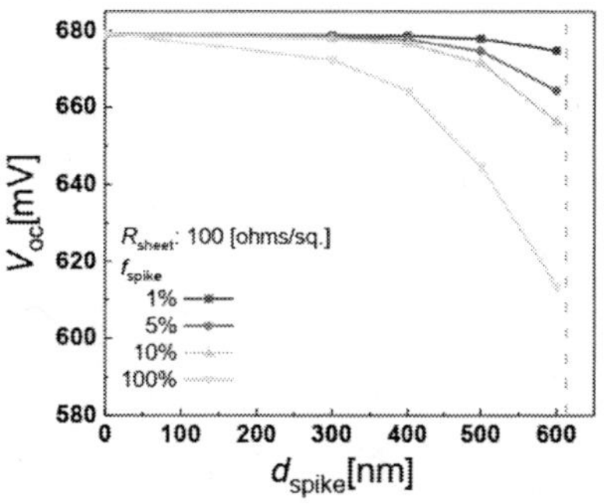

Figure 5: Simulated V_{oc} as a function of f_{spike} and d_{spike}. The 100 ohms/sq. diffusion profile in Figure 2 was used in these simulations. The vertical dashed line indicate the position of the *p-n* junction of the doping profile.

4.4 Sensitivity of the different doping profiles

Figure 6 (a) shows simulated open-circuit voltages of the *n*FAB cells with different front side boron diffused p^+ layers. The f_{spike} was assumed to be 5 %. The V_{oc} was severely degraded with the presence of spikes, especially with deep spikes reaching the *p-n* junction. Without spiking (d_{spike}=0), the V_{oc} is highly related to the surface doping density. For example, the 100 ohms/sq. and the 35 ohms/sq. profiles have the same junction depth but different surface doping concentration. The V_{oc} of the 100 ohms/sq. profile is much higher. In the presence of deep spikes, there are two crucial parameters: (1) the distance between the junction depth and the spike interface, (2) the doping density at the spike-Si interface. Significant degradation in V_{oc} was observed, when the doping density at the spike-Si interface dropped below ~ 1×10^{18} cm^{-3}. The trend on cell efficiency basically agrees with that in V_{oc}, as shown in Figure 6 (b). However, the 40 ohms/sq. profiles shows higher cell efficiency than the 100 ohms/sq. profile especially when the spike size are relatively small (not reaching the *p-n* junction). This is due to the reason that the cell with the 40 ohms/sq. front p^+ layer has a higher *FF* (not shown). Figure 7 shows the simulated relative solar cell efficiency degradation as a function of f_{spike} of the four different boron profiles investigated. For the 35 ohms/sq. and the 100 ohms/sq. profiles, a 600 nm deep spike already contacts to a doping density below 10^{18} cm^{-3}, the degradations in η increase from ~ 2 - 3 % to ~ 10% with increasing f_{spike} from 5 - 10 % to 100 %. For the 40 ohms/sq. profile, the 600 nm spike still contacts to a doping density ~ 10^{19} cm^{-3}, therefore the cell efficiency degrades by only ~ 2 %. Thus, the optimization on diffused p^+ layers towards a higher efficiency should not only focus on building deeper profiles. A focus probably also be put on developing profiles with a prolonged plateau of boron density above 1×10^{18} cm^{-3}. However, the validity of the statement requires further experimental verification.

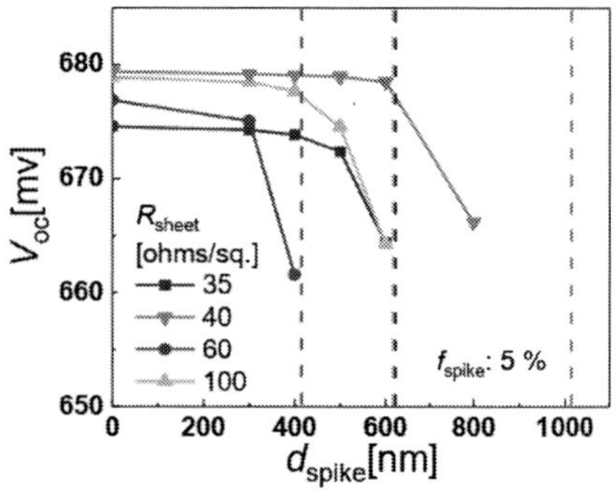

(a)

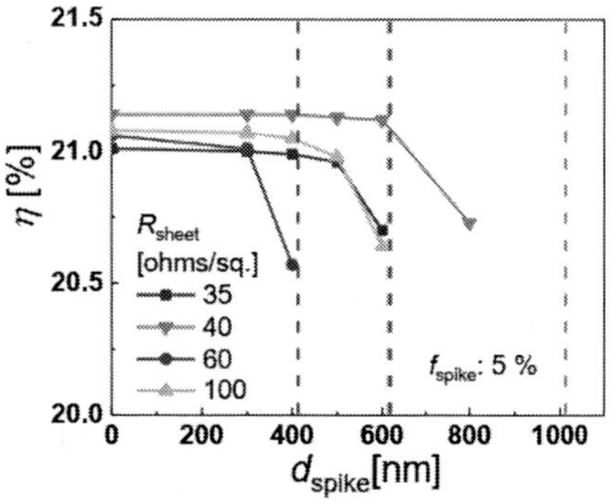

(b)

Figure 6: (a) Simulated V_{oc} as a function of d_{spike} of *n*FAB cells with the four different boron diffused p^+ layers. (b) Simulated solar cell efficiency η of four different boron doping profiles of *n*FAB solar cells. The f_{spike} was assumed to be 5 %. The vertical dashed lines depict the position of the *p-n* junction of each of the doping profiles.

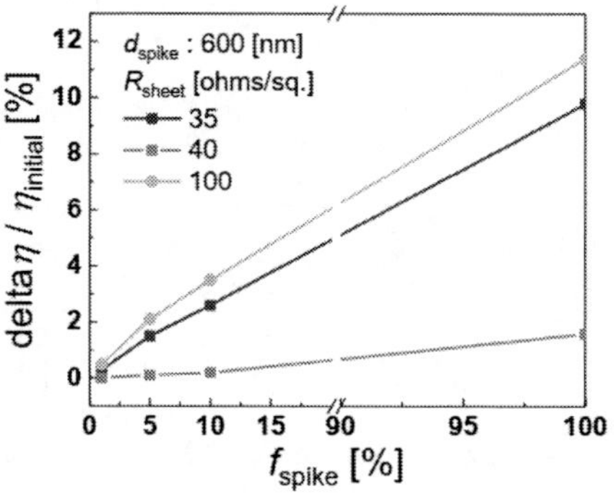

Figure 7: Simulated relative η degradation as a function of f_{spike} of different boron profiles.

5 CONCLUSION

In this work, we show that the solar cell efficiency will only be significantly affected when the localized spikes penetrate as deep that they contact to a doping density below ~1×10^{18} cm^{-3}. Beyond that point, the defect-related recombination at the metal-Si interface increases by orders of magnitude. The cell efficiency could degrade by ~ 2 - 3 % up to ~ 10 % relative when f_{spike} increases from 5-10 % to 100 %. The optimization on diffused p^+ layers towards a higher efficiency in the presence of deep spikes should not only focus on building deeper profiles. A new focus may also be put on developing profiles with a prolonged plateau of boron density above 1×10^{18} cm^{-3}. In the case of shallow spiking, the cell efficiency is highly related to the surface doping density. For cells diffused with front side p^+ layer with similar junction depth but different surface concentration, higher efficiency was observed in the ones with a relatively lower surface concentration. Thus in the case of shallow spiking, the priority should be put on developing profiles with a contactable low surface concentration.

6 ACKNOWLEDGEMENTS

The authors would like to thank their colleagues from the Silicon Materials and Cells Cluster of the Solar Energy Research Institute of Singapore (SERIS) for their assistance in sample processing and measurements. SERIS is sponsored by the National University of Singapore and Singapore's National Research Foundation through the Singapore Economic Development Board. This research is supported by the National Research Foundation, Prime Minister's Office, Singapore under its Energy Innovation Programme Office. The authors would like to thank Dr. Cangming KE from SERIS for the assistance in scanning electron microscope imaging.

7 REFERENCE

[1] S. Riegel, F. Mutter, G. Hahn, and B. Terheiden, "Influence of the dopant on the contact formation to p$^+$-type silicon," *Energy Procedia,* vol. 8, pp. 533-539, 2011/01/01 2011.

[2] R. Lago *et al.*, "Screen printing metallization of boron emitters," *Progress in Photovoltaics: Research and Applications,* vol. 18, no. 1, pp. 20-27, 2010.

[3] T. Fellmeth, A. Born, A. Kimmerle, F. Clement, D. Biro, and R. Preu, "Recombination at Metal-Emitter Interfaces of Front Contact Technologies for Highly Efficient Silicon Solar Cells," *Energy Procedia,* vol. 8, pp. 115-121, 2011.

[4] R. Hoenig, M. Glatthaar, F. Clement, J. Greulich, J. Wilde, and D. Biro, "New measurement method for the investigation of space charge region recombination losses induced by the metallization of silicon solar cells," *Energy Procedia,* vol. 8, pp. 694-699, 2011.

[5] A. E. Lejo J. Koduvelikulathu, Valentin D. Mihailetchi. Corrado Comparotto, Andreas Halm, Radovan Kopecek, Kristian Peter, "Metallization and firing process impact on V_{oc}--a simulation study " *Proceedings of 27th European Photovoltaic Solar Energy Conference and Exhibition,* 2012.

[6] A. Edler, V. D. Mihailetchi, L. J. Koduvelikulathu, C. Comparotto, R. Kopecek, and R. Harney, "Metallization–induced recombination losses of bifacial silicon solar cells," *Progress in Photovoltaics: Research and Applications,* vol. 23, no. 5, pp. 620-627, 2015.

[7] F. Kiefer, R. Peibst, T. Ohrdes, J. Krügener, H. J. Osten, and R. Brendel, "Emitter recombination current densities of boron emitters with silver/aluminum pastes," in *2014 IEEE 40th Photovoltaic Specialist Conference (PVSC)*, 2014, pp. 2808-2812.

[8] F. Kiefer, J. Krügener, F. Heinemeyer, H. J. Osten, R. Brendel, and R. Peibst, "Structural Investigation of Printed Ag/Al Contacts on Silicon and Numerical Modeling of Their Contact Recombination," *IEEE Journal of Photovoltaics,* vol. 6, no. 5, pp. 1175-1182, 2016.

[9] N. Wöhrle, E. Lohmüller, J. Greulich, S. Werner, and S. Mack, "Towards understanding the characteristics of Ag–Al spiking on boron-doped silicon for solar cells," *Solar Energy Materials and Solar Cells,* vol. 146, pp. 72-79, 3. 2016.

[10] F. D. Heinz *et al.*, "Microscopic origin of the aluminium assisted spiking effects in n-type silicon solar cells," *Solar Energy Materials and Solar Cells,* vol. 131, pp. 105-109, 12. 2014.

[11] S. Fritz, M. König, S. Riegel, A. Herguth, M. Hörteis, and G. Hahn, "Formation of Ag/Al Screen-Printing Contacts on B Emitters," *IEEE Journal of Photovoltaics,* vol. 5, no. 1, pp. 145-151, 2015.

[12] E. Lohmüller, S. Werner, R. Hoenig, J. Greulich, and F. Clement, "Impact of boron doping profiles on the specific contact resistance of screen printed Ag–Al contacts on silicon," *Solar Energy Materials and Solar Cells,* vol. 142, pp. 2-11, 11. 2015.

[13] L. J. Koduvelikulathu, V. D. Mihailetchi, S. Olibet, D. Rudolph, E. Cabrera, and R. Kopecek, "Two-Dimensional Modeling of the Metallization-Induced Recombination Losses of Screen-Printed Solar Cells," *Photovoltaics, IEEE Journal of,* vol. 5, no. 1, pp. 159-165, 2015.

[14] Silvaco, "Atlas-Device Simulation Framwork," *Available at:* http://www.silvaco.com/products/tcad/device_simulation/atlas/atlas.html

[15] Synopsys, Synopsys TCAD, Version L-2016.03, Available at: http://www.synopsys.com

[16] "Module ray tracer," *Available at: https://www.pvlighthouse.com.au/module-ray-tracer*

[17] F.-J. Ma, S. Duttagupta, M. Peters, G. S. Samudra, A. G. Aberle, and B. Hoex, "Numerical Modelling of Silicon p^+ Emitters Passivated by a PECVD AlOx/SiNx Stack," *Energy Procedia,* vol. 33, pp. 104-109, 2013.

[18] Mengjie Li, Johnson Wong, Ning Chen, Armin G. Aberle, Rolf Stangl, "Griddler AI assisted p^+ layer optimization towards lower screen printing induced recombination losses for industrially relevant n-type bifacial Si wafer solar cells," *under preparation, to be submitted to IEEE Journal of Photovoltaics,* 2017.

[19] J. A. d. Alamo and R. M. Swanson, "The physics and modeling of heavily doped emitters," *IEEE Transactions on Electron Devices,* vol. 31, no. 12, pp. 1878-1888, 1984.

[20] F. Kiefer *et al.*, "Bifacial, fully screen-printed n-PERT solar cells with BF2 and B implanted emitters," *Solar Energy Materials and Solar Cells,* vol. 157, pp. 326-330, 12. 2016.

[21] P. Kumar *et al.*, "N-type single-crystalline Si solar cells: Front side metallization for solar cells reaching 20% efficiency," *Solar Energy Materials and Solar Cells,* vol. 157, pp. 200-208, 12. 2016.

THE ROLE OF THE OXIDE IN THE CARRIER SELECTIVITY OF METAL/POLY-SI/OXIDE CONTACTS TO SILICON WAFERS

G.J.M. Janssen* (janssen@ecn.nl), M.K. Stodolny, I.G. Romijn, L.J. Geerligs
ECN Solar Energy, P.O. Box 1, NL-1755 ZG Petten, The Netherlands
*Phone: +31 88 515 4803

ABSTRACT: Simulation results are presented that elucidate the role of the interfacial oxide layer in the passivation and carrier selectivity exhibited by doped poly-Si contact layers. An approach is introduced by which the oxide is represented as a layer with limited charge carrier mobility. This approach is easy to implement in numerical simulations and directly affects the conductivity of minority as well as of majority charge carriers, which determine the carrier selectivity. Our results show that for optimal passivation and carrier selectivity the oxide should not be a perfect tunnel oxide but should show higher effective charge carrier mobility. This means that poly-Si/oxide layers with substantial in-diffusion tails can be used, which has benefits when fire-through metallization is applied.
Keywords: contact, passivation, carrier selectivity, simulation, silicon solar cell.

1 INTRODUCTION

Effective carrier selective contact layers can be made by deposition of highly-doped poly-Si on an interfacial oxide grown on the c-Si wafer [1-3]. Carrier selectivity is provided by an asymmetrical conductivity, which enhances the transport of one carrier to the contact and suppresses the other [4]. The high dopant concentrations in the poly-Si are the basis for the carrier selectivity of the poly-Si/oxide contact layers. The interfacial oxide is a diffusion barrier for dopants like phosphorous and boron, which means that in the poly-Si layer uniform dopant concentrations close to the solubility level can be obtained [3,5,6]. But, as experimental and theoretical studies have indicated, the presence of the thin interfacial oxide layer also has a function in the reduction of the contact recombination parameter J_{0c} [5-8].

It has been assumed that the oxide layer is a barrier for charge carrier transport to the metal. An effective suppression of the minority carrier transport will reduce the J_{0c}; at the same time this barrier should not significantly impede majority carrier transport, i.e. lead not to a high contact resistance ρ_c. It has been recognized that the key for a good carrier selective contact is a good balance between the recombination properties and contact resistance [4,9].

Another contribution to the contact recombination will come from the recombination at the surface of the wafer. It is well known that interfacial oxides grown either thermally or in a wet-chemical process contribute to improved surface passivation by reduction of the density of interface states (D_{it}). Hydrogenation further reduces the D_{it}. Most passivation schemes for Si-surfaces also include a field or band-bending effect, i.e. by applying fixed charges at the surface the concentration of minority carriers is reduced. In the poly-Si/oxide contacts, the highly-doped poly-Si layer presents the field effect.

In this paper we will first consider first the wafer passivation and its relation with the dopant concentration in the poly-Si, the thickness of interfacial oxide and the D_{it}, or alternatively the surface recombination parameter S_{int}, to which it is proportional.

Next, we consider in more detail the importance of the charge carrier transport by the oxide. Experimental data suggest that properties like thickness, homogeneity, impurity level of the oxide depend very much on the processing conditions [5-7,10]. Different assumptions on the nature of the oxide have already led to two very different physical models for the oxide, i.e. as a quantum-mechanical tunnel barrier [11], or as layer of pin-holes causing current constriction [6,12].

Here we will present numerical simulations in which we represent the oxide either as a quantum-mechanical tunnel or by a thin silicon layer with limited charge carrier mobility. It will be shown that for very low effective charge carrier mobility these models give similar results, but that the oxide with the optimal (J_{0c}, ρ_c) has an effective mobility higher than that of a perfect quantum-mechanical tunnel oxide. Moreover, experimental data suggest such a higher effective mobility is more representative for the carrier transport through the oxide.

2 MODELLING AND SIMULATION DETAILS

2.1 The Girisch model

The Girisch model [13] was used to estimate the band-bending induced in the poly-Si on one side of an insulating oxide and in the c-Si on the other side. The assumptions used by the Girisch model are that the effective charge induced by band-bending is the same in both semiconductors (charge neutrality), that the electrostatic potential is continuous and that the quasi - Fermi levels are constant throughout the system. This latter assumption is only valid if the recombination is not too high. Moreover, band-gap narrowing effects are not included. The oxide is represented by a change in electrostatic potential Ψ:

$$\Delta\Psi = \frac{Qqd}{\epsilon_{SiOx}\epsilon_0} \quad (1)$$

Here Q is the charge density associated with the bend-bending, d is the thickness of the oxide, ϵ_0 the vacuum permittivity, ϵ_{SiOx} the relative dielectric constant of the oxide (3.9 in this work) and q the elementary charge. Following the approach of ref. [13] the minority carrier density n_s at the oxide/c-Si interface can be calculated for a given splitting of the quasi-Fermi-levels ΔE_F; the recombination parameter J_{0p} characterizing the recombination at this interface is then obtained from:

$$J_{0p} = \frac{qn_sS_{int}}{\exp(q\Delta E_F/kT)} \quad (2)$$

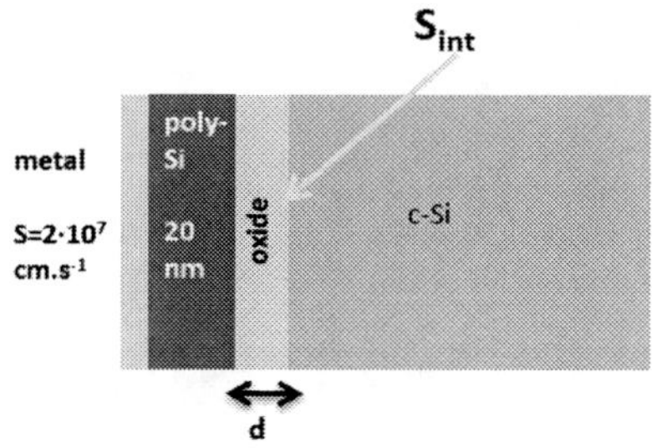

a)

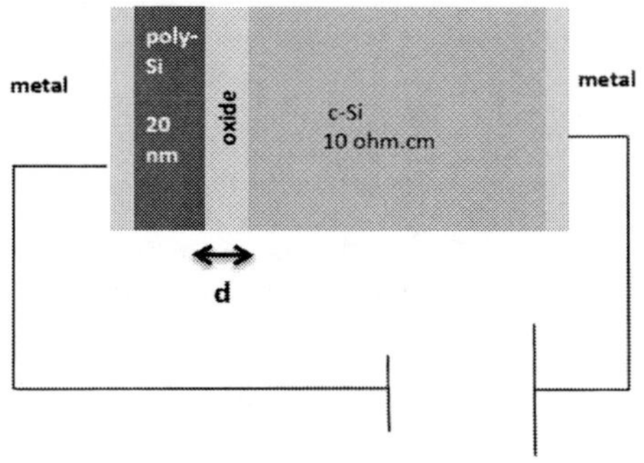

b)

Figure 1: Schematics of the structures used in Atlas simulation to calculate J_{0c} (a) and ρ_c (b).

Here S_{int} is the recombination velocity of the minority carrier at the interface, and k and T are the Boltzmann constant and temperature respectively. At strong enough band-bending (high enough Q) J_{0p} values are independent of the chosen Fermi-level splitting [14].

2.2 Atlas simulations

The schematic structures of Fig. 1 were used to calculate the J_{0c} and ρ_c, with the simulation package Atlas of Silvaco [15]. In this work the polarity of the c-Si and poly-Si was taken the same. For the calculation of J_{0c} in all cases a base dopant concentration of $5 \cdot 10^{14}$ cm^{-3} was assumed, in the calculation of the ρ_c the base resistivity was 10 ohmcm. For the calculation of J_{0c} the metal is represented by a surface with theoretical maximum recombination velocity of $2 \cdot 10^7$ cms^{-1}. The poly-Si thickness was 20 nm, i.e. sufficient to include the band-bending zone. The recombination velocity at the metal surface is so high that the actual lifetime in the poly-Si is not critical. In this work the Shockley-Read-Hall lifetime of the poly-Si was set at a rather high value of 1 μs; in reality it can be as low as 1 ps. The high recombination value assumed at the surface is in fact equivalent to recombination velocity of a 20 nm thick poly-Si layer with a lifetime of 0.1 ps. In the simulation the poly-Si lifetime will only affect the distribution of the recombination between the poly-Si volume and the surface. The charge carrier mobility in poly-Si was set to 14 cm^{2}V^{-1}s^{-1} in agreement with experimental data [3]. Also this value is not critical for the J_{0c}.

At the interface of the oxide and c-Si a minority recombination rate S_{int} was assumed; in this way J_{0c} includes J_{0p}. In some cases an in-diffusion tail in the c-Si was modelled by a Gaussian profile with a maximum concentration N_s at the oxide/c-Si interface and depth parameter a. The Auger recombination modelled according to ref. [16] and Schenk's band-gap narrowing model [17] were used for both c-Si and poly-Si.

R_{comb} is calculated by summing all recombination at the poly-Si/metal interface, in the poly-Si and at the oxide/wafer interface (R_{comb}) and any tail region present. J_{0c} is then extracted from:

$$qR_{comb} = J_{0c}\frac{\Delta n(N_D+\Delta n)}{n_i^2} \qquad (3)$$

Δn and N_D are the excess and dopant concentration in the base, respectively, and n_i is the intrinsic charge carrier concentration assumed to be $8.6 \cdot 10^9$ cm^{-3}, i.e. the value commonly used when extracting J_0 values from lifetime measurements with the WTC120 Lifetime Tester [18]. At high injection levels J_{0c} becomes independent of Δn, here the value at $\Delta n = 10 \cdot N_D$ will be reported.

The area specific junction resistance ρ_c is calculated from an IV curve of the structure in Fig.1b in the range 0-60 mV. Contact resistances between metal and poly-Si and between metal and wafer were set to zero, in practice values < 10 mΩ.cm^2 are expected for this contribution to the total ρ_c. The junction resistance ρ_c is calculated by subtracting the base resistance (0.177 Ω.cm^2 in this work) from the total resistance.

Two approaches were used to represent the oxide:

a) As a SiO_2 tunnel oxide. The effective masses for holes and electrons were set at 0.4 m_0 (m_0 is the electron rest mass). With a SiO_2 bandgap of 8.9 eV the tunnel barrier is about 3.4 eV for electrons and 4.4 eV for holes.

b) As an intrinsic Si layer with thickness 1.5 nm and an effective charge carrier mobility μ that represents the permeability for charge carriers in the oxide. In the present simulation the mobility of holes and electrons was given the same value. Usually electrons are more mobile than holes, but their mobilities are expected to be in the same order of magnitude [9].

2.3 Quokka simulations

Quokka [19] was used to calculate the maximum efficiency possible for a given combination of J_{0c} and ρ_c. To exclude all other factors affecting the efficiency, the base resistivity was set to 10^6 Ωcm and an infinite base lifetime as assumed. All other resistances and recombination parameters were set zero; the generated current density was taken as at 43.3 mAcm^{-2} [20].

3 RESULTS

3.1 Passivation effect

Calculations using the Girisch model as outlined in 2.1 showed that the field of the highly doped polysilicon is very effective in the reduction of the surface recombination of the wafer (Fig. 2). Layers of n+ or p+ poly-Si with dopant concentration > 10^{19} cm^{-3} provide fields that are equivalent to a charge density > 10^{12} cm^2, i.e. of similar magnitude as field provided by the fixed charges of AlO_x or SiN_x. At these high dopant concentrations the results are similar for $p+$ poly Si layers (not shown). The charges slightly decrease with oxide thickness d, but the effect is not very critical in the range 1-2 nm (Fig. 2b). The thickness of the polysilicon

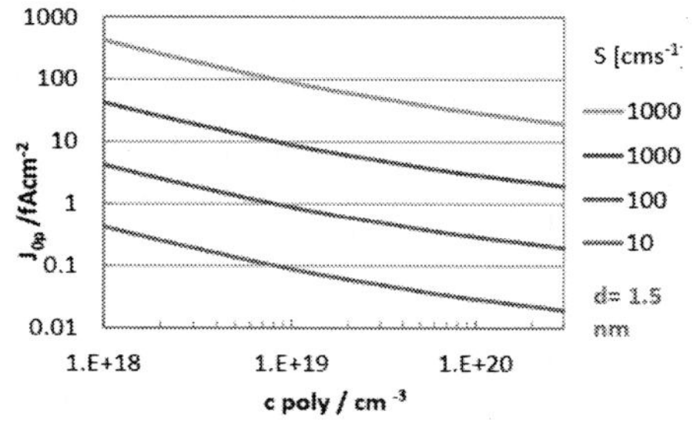

a)

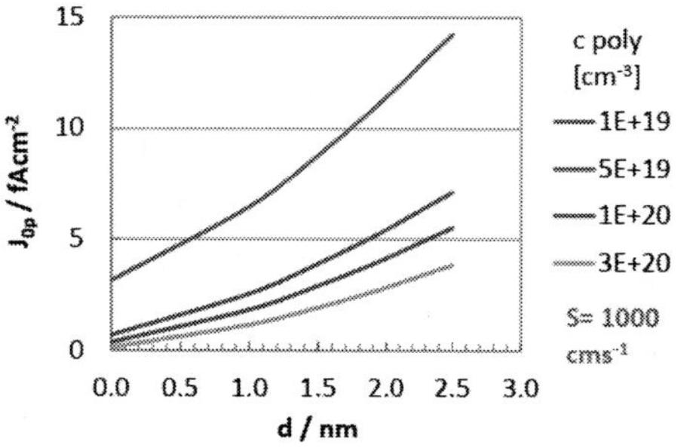

b)

Figure 2: The recombination parameter J_{0p} of the oxide/c-Si interface calculated according to Girisch model a) with a fixed oxide thickness of 1.5 nm as function of the n+-poly dopant concentration, and b) as function of the oxide thickness d.

is even less critical, but it must be larger than the width of the region where band bending occurs, i.e. a few times the Debye length λ. Using $\lambda = \sqrt{\epsilon_{Si}\epsilon_0 kT/(q^2 N_D)}$ we find that λ is $\approx$ 0.5 nm for poly-Si with doping levels of about N_D=10^{20} cm^{-3} and relative dielectric constant for Si ϵ_{Si}=11.7, which means that poly-Si layers as thin as 5 nm are sufficient to provide the field effect.

A very critical parameter is the S_{int}, which reflects the number of defects at the oxide/c-Si interface. Fig. 2a shows that, within the approximations of the Girisch model, the defect density must be such that S_{int} < 1000 cms^{-1} in order to achieve J_{0p} < 10 fAcm^{-2}.

3.2 Carrier selectivity

In agreement with the results of Steinkemper *et al.* [21], we found that a tunnel oxide is very efficient to suppress the J_{0c} of poly-Si layers (Fig. 3). In all calculations with oxides thickness between 1 and 2 nm we found that J_{0c} was < 0.5 fAcm^{-2} when we assume S_{int}= 0 cms^{-1}. Despite the higher tunnel barrier for holes than for electrons, the J_{0c} values for n-type and p-type are the same. Only at d < 1 nm we observed a significant increase in J_{0c} values for p+-poly up to 3 fAcm^{-2}, reflecting that the oxide barrier becomes more permeable for electrons.

The data in Fig. 3 also include values obtained with a more realistic S_{int} = 1000 cms^{-1}; the J_{0c} then essentially represents the J_{0p} already discussed above using the Girisch model. The increase in J_{0c} with oxide thickness d reflects the reduction of the field effect. This also determines the main difference between values for the p-type and n-type poly.

The thickness of the tunnel oxide is much more critical for the calculated values of ρ_c. Over the range 1-2

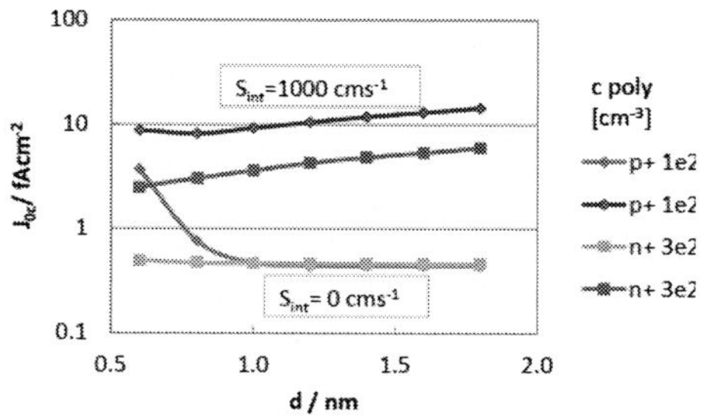

Figure 3: The recombination parameter J_{0c} of the metal/poly-Si/oxide/c-Si from Atlas simulations using the quantum-mechanical tunnel model as function of the oxide thickness d. The lines represent different polarity and values used for S_{int}.

nm these values change by 3-4 orders of magnitude (Fig. 4). Moreover, the different barrier for electron and holes now leads to very significant difference in ρ_c for p-type and n-type contacts. As pointed out by Peibst *et al.*, these features are at conflict with reported data for doped poly-Si/oxide layers [6,9,12] No strong correlation of ρ_c and oxide thickness d in the range 1-2 nm has been observed, nor a systematically larger resistance ρ_c for p+ poly-Si compared to n+ poly-Si [5-7,12]. Moreover, even for n+ poly-Si layers reported values of ρ_c are in the order 5-20 mΩcm^2 rather than the 0.1-1 Ωcm^2 consistent with a tunnel oxide [1,6,7,22]. It has also been reported that the presence of an in-diffusion tail reduces the ρ_c [11]; this is corroborated by the results in Fig. 4. But notice that these tails do not change the dependency on d, nor do they remove the very different tunnel permeability for holes and electrons.

According to recent analyses, in high efficiency cells ρ_c will be < 1 Ωcm^2 [9]. However, it has been recognized before that the oxide probably does not always behave like a perfect quantum-mechanical tunnel barrier. One cause can be that here are pin-holes in the oxide [6,12]; another reason can be that the tunnelling is assisted by the presence of traps in the oxide. Rather than to modify the parameters of the tunnel model or to extend the model, we now represent the oxide barrier by a thin (1.5 nm) layer of intrinsic silicon with an effective charge carrier mobility μ. Because the layer is so thin, the carrier density at either side of the layer determines the carrier density in the layer. By using the mobility of the charge

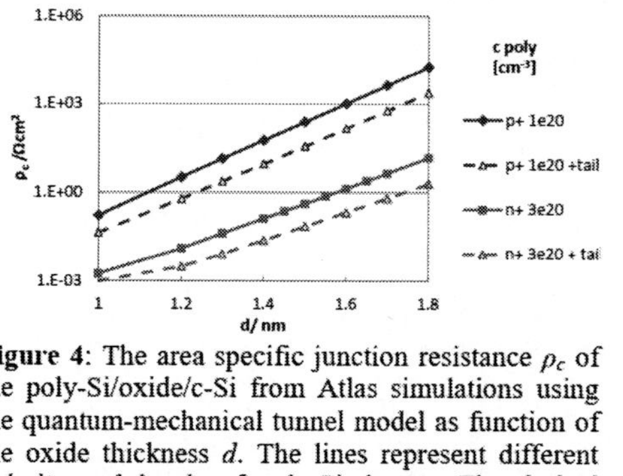

Figure 4: The area specific junction resistance ρ_c of the poly-Si/oxide/c-Si from Atlas simulations using the quantum-mechanical tunnel model as function of the oxide thickness d. The lines represent different polarity and levels of poly-Si dopant. The dashed lines were calculated using a gaussian in-diffusion tail with depth factor a=0.5 µm and N_s=1·10^{19} cm^{-3} for p+ and N_s=5·10^{19} cm^{-3} for n+, respectively.

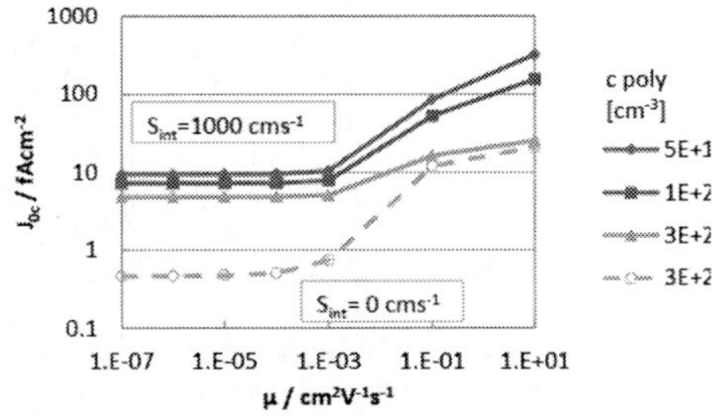

Figure 5: The recombination parameter J_{0c} of the metal/poly-Si/oxide/c-Si from Atlas simulations as function of the effective charge carrier mobility in the i-Si layer representing the oxide. The lines represent different levels of n+ poly-Si dopant.

carriers as a parameter, the conductivity of holes and electrons in the barrier layer can be further tuned, as is also the basis of a similar model proposed by Peibst *et al.* [6,9,12].

Fig.5 shows that using an effective charge carrier mobility < 10^{-3} Vcm^{-2}s^{-1} is as effective as a tunnel barrier in reducing J_{0c}. Note, that such mobility values are much smaller than in the poly-Si, or in doped crystalline silicon. The asymmetric conductivity of holes and electrons determines the selectivity, but due to the large driving force for minorities to the metal, where holes and electrons are at equilibrium, a low conductivity for minorities is needed to sufficiently suppress the minority transport to the contact [4].

At the high end of mobility values the J_{0c} is strongly dependent on the poly-dopant level as the minority concentration at the contact is now dominated by the dopant level and not by the barrier. Here, for clarity we only present results for the n+-poly layers. The results for p-type layers will only be different because of the different doping levels, and small differences in the band-bending between n+/n and p+/n systems resulting in different contributions of J_{0p}. Note, that because of the expected low lifetime in poly-Si, the reduced charge carrier mobility is also instrumental in minimizing the recombination at the non-contacted area of poly-Si layers.

As expected the ρ_c as function of charge carrier mobility has the opposite trend, i.e. it strongly decreases with the mobility (Fig.6). By comparing data of Fig. 4 at 1.5 nm to Fig. 6 we can conclude that the quantum-

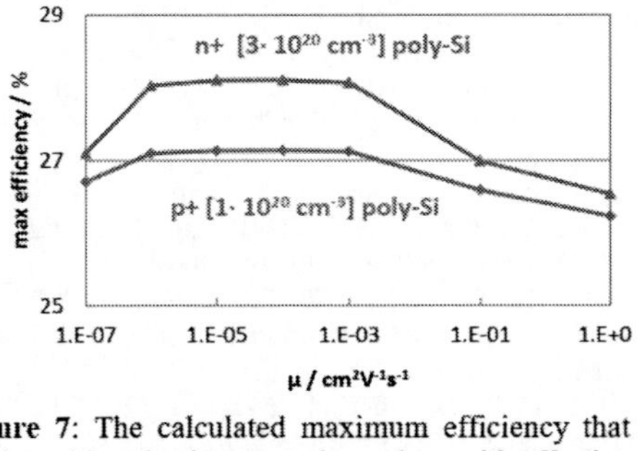

Figure 7: The calculated maximum efficiency that can be achieved using J_{0c} and ρ_c values with effective mobility charge carrier mobility μ in the i-Si layer representing the oxide.

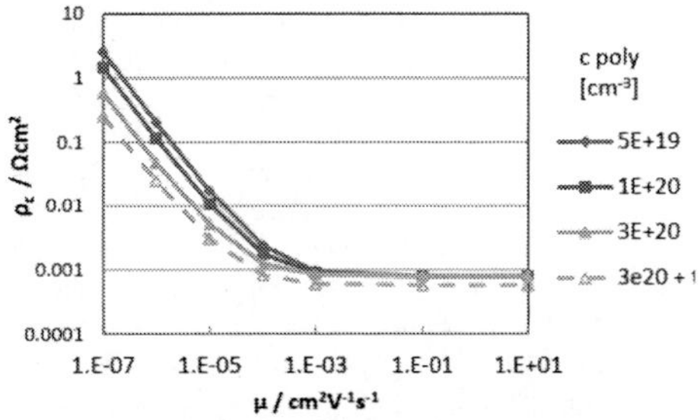

Figure 6: The area specific junction resistance ρ_c of the poly-Si/oxide/c-Si from Atlas simulations as function of the effective charge carrier mobility in the i-Si layer representing the oxide. The lines represent different levels of poly-Si dopant. The dashed line was calculated using a gaussian in-diffusion tail with depth factor 0.5 µm and N_s=5·10^{19} cm^{-3}.

mechanical tunnel mobility for electrons is in the order of < 10^{-6} cm^2 V^{-1}s^{-1} for electrons and even less for holes. But as Fig. 5 shows, such low values are not required for a low J_{0c}. The calculation of the maximum efficiency as function of the charge carrier mobility confirms this (Fig. 7). The range of mobility values that give the highest efficiency comprises a few orders of magnitude. Especially for p-type the calculated maximum efficiency can be much higher than predicted on the basis of the tunnel model.

The reduction of ρ_c due to the in-diffusion tail is also described using the model of effective charge carrier mobility. However, as Fig. 6 also shows, this effect is in absolute values not significant if the effective mobility is above the limit of the quantum-mechanical tunnel model.

3.3 Fire-through metallization

The bifacial n-type PERPoly (Passivated Emitter Rear PolySilicon) solar cells of ECN feature an n+-poly-Si layer at the rear. This n+-poly-Si layer is contacted by an industrial, screen-printed metallization pattern [23]. This configuration differs in two ways from the one depicted in Fig.1 and discussed in section 3.2 above. First, a large part of the poly-Si area is not contacted but covered by a SiN_x layer. However, due to the low lifetime in poly-Si the J_0 of this area will be similar to the J_{0c} of a metal covered area. Secondly, some of the metal will spike through the oxide, which would lead to J_0 values locally much higher than the J_{0c} calculated in section 3.2. The degree of spiking can to some extent be controlled by the poly-Si layer thickness, but the presence of in-diffusion tails can also effectively reduce the J_0 [24].

4 DISCUSSION

With the field provided by poly-layers with dopant concentrations ≥ 10^{20} cm^{-3}, a recombination velocity S_{int} = 1000 cms^{-1} at the oxide/c-Si interface is adequate to obtain J_{0c} < 10 fAcm^{-2} values. Such low values of S_{int} can be achieved by hydrogenation of the oxide/wafer interface, e.g. by deposition of SiN_x:H on the poly-Si and optional subsequent firing [3]. Since the oxide acts as a diffusion barrier, the dopant concentration in the poly can reach a uniform value close to the solubility limit over the whole poly thickness. As this limit is higher for phosphorous than boron, the field effect of n-poly can be stronger. The less good passivation often observed for p-poly is at least in part due to a weaker field although also the oxide/c-Si interface may contain more defects because of the presence of boron ions.

The simulation results presented this paper show that the oxide has a significant role in providing good carrier selectivity for the contacts. Good carrier selectivity means that the transport of the majority carrier to the contact is unimpeded but the transport of minority carriers is efficiently suppressed. As discussed in 3.2 we found that treating the oxide as a perfect quantum mechanical tunnel barrier gives low J_{0c} values, but also large ρ_c values. Moreover, experimental data also suggest that the oxide is often very different from the ideal tunnel barrier because of e.g. dopants and pinholes etc. [6,10,12].

With the effective charge carrier mobility model more justice can be done to these differences. This model is easy to use with existing simulation software, and it retains the important function of the oxide in mechanism of the carrier selectivity, i.e. the reduction of the conductivity for carrier transport. By setting the effective mobility at values < 10^{-6} $cm^2V^{-1}s^{-1}$ the tunnel behaviour can be reproduced, including the effect of in-diffusion tails. But by using larger mobility values different behaviour of charge transport can be described, such as through pinholes or trap-assisted tunnelling. Using a Si layer with an effective charge carrier mobility is therefore an easy to implement and flexible way to represent the oxide layer in numerical simulations. The charge carrier mobility values can be derived from experimental values of the junction resistance ρ_c.

The simulation result that the maximum efficiency is not obtained for a perfect tunnel oxide but rather for an oxide more permeable for charge carriers seems favourable for the development of this type of contact layers. First of all it shows that p-contacts with low J_{0c} do not necessarily have a prohibitively large ρ_c. Secondly, oxides that are damaged by high temperature treatment, or contain large concentrations of dopant can still be good enough in the suppression of minority transport.

This observation, that the interfacial oxide does not necessarily have to be a perfect tunnel oxide, is also consistent with excellent passivation results obtained at ECN with "leaky" oxides [10], which have significant in-diffusion tails. Rather than destroying the selectivity, such tails can be beneficial in providing additional lateral conductivity in the case of non-full area contacts. This enables the use of thinner poly layers, and hence less optical losses. Moreover, in the case of fire-through contacts, when metal spikes through the interfacial oxide, such tails can provide extra shielding of the contact. However, in all cases the benefits of the tail have to outweigh the additional Auger recombination.

5 CONCLUSIONS

The interfacial oxide plays an essential role both for the passivating properties and for the carrier selectiveness of poly-Si/oxide contact layers. As during processing the oxide is a diffusion barrier, high dopant levels close the solubility level are achieved, which provide a large field effect in the wafer. Hydrogenation is still required but with known oxides the interface properties provide excellent wafer passivation.

We proposed an intrinsic thin (1.5 nm) silicon layer with a low effective charge carrier mobility as a good representation of the oxide as barrier for charge carriers. The simulations showed that this barrier function is essential for the carrier selectivity, i.e. a charge carrier mobility much smaller than found in doped c–Si is consistent with low J_{0c} values. Too high values obviously impede the majority transport too much; the optimum value for the effective mobility through the barrier is not as low as would correspond to tunnel model.

The proposed method to represent the oxide in simulations is not only convenient and easy to implement, it also seems better suited to represent the oxide in simulations especially in the case of p+-type layers and for more "leaky oxides" with large in-diffusion tails. This enables a more realistic study of the benefits of these in-diffusion tails in reducing the junction resistance, providing lateral conductivity or contact shielding in the case of fire-through metallization.

REFERENCES

[1] F. Feldmann, M. Bivour, C. Reichel, M. Hermle, and S. W. Glunz, Solar Energy Materials and Solar Cells 120, Part A (2014) 270-274.

[2] U. Romer, R. Peibst, T. Ohrdes, B. Lim, J. Krugener, T. Wietler, and R. Brendel, IEEE Journal of Photovoltaics 5 (2015) 507.

[3] M. K. Stodolny, M. Lenes, Y. Wu, G. J. M. Janssen, I. G. Romijn, J. R. M. Luchies, and L. J. Geerligs, Solar Energy Materials and Solar Cells 158, Part 1 (2016) 24-28.

[4] U. Würfel, A. Cuevas, and P. Würfel, IEEE Journal of Photovoltaics 5 (2015) 461-469.

[5] D. Yan, A. Cuevas, Y. Wan, and J. Bullock, Solar Energy Materials and Solar Cells 152 (2016) 73-79.

[6] R. Peibst, U. Römer, Y. Larionova, M. Rienäcker, A. Merkle, N. Folchert, S. Reiter, M. Turcu, B. Min, J. Krügener, D. Tetzlaff, E. Bugiel, T. Wietler, and R. Brendel, Solar Energy Materials and Solar Cells 158 (2016) 60-67.

[7] D. Yan, A. Cuevas, J. Bullock, Y. Wan, and C. Samundsett, Solar Energy Materials and Solar Cells (2015)

[8] H. Steinkemper, B. Michl, M. Hermle, and S. W. Glunz, Energy Procedia 92 (2016) 225-231.

[9] R. Brendel and R. Peibst, IEEE Journal of Photovoltaics 6 (2016) 1413-1420.

[10] M. K. Stodolny, J. Anker, L. J. Geerligs, G. J. M. Janssen, B. W. H. van de Loo, J. Melskens, O. Isabella, J. Schmitz, M. Lenes, J. M. Luchies, W. M. M. Kessels, and I. G. Romijn, Energy Proc. (2017) in press.

[11] H. Steinkemper, F. Feldmann, M. Bivour, and M. Hermle, Energy Procedia 77 (2015) 195.

[12] R. Peibst, U. Römer, K. R. Hofmann, B. Lim, T. F. Wietler, J. Krügener, N. P. Harder, and R. Brendel, IEEE Journal of Photovoltaics 4 (2014) 841-850.

[13] R. B. M. Girisch, R. P. Mertens, and R. F. De Keersmaecker, IEEE Transactions on Electron Devices 35 (1988) 203.

[14] L. E. Black and K. R. McIntosh, IEEE Journal of Photovoltaics 3 (2013) 936-943.

[15] Atlas, www.silvaco.com, (2013).

[16] P. Altermatt, Journal of Computational Electronics 10 (2011) 314-330.

[17] A. Schenk, J.Appl.Phys. 84 (1998) 3684-3695.
[18] R. Sinton and D. Macdonald, *WCT-120 Photoconductance Lifetime Tester and optional Suns-VOC Stage User Manual,* Sinton Consulting, Inc, Boulder, Colorado, 2006.
[19] A. Fell, IEEE Transactions on Electric Devices 60 (2013) 733-738.
[20] A. Richter, J. Benick, F. Feldmann, A. Fell, M. Hermle, and S. W. Glunz, Solar Energy Materials and Solar Cells (2017) in press.
[21] H. Steinkemper, F. Feldmann, M. Bivour, and M. Hermle, IEEE Journal of Photovoltaics 5 (2015) 1348-1356.
[22] M. Rienäcker, M. Bossmeyer, A. Merkle, U. Römer, F. Haase, J. Krügener, R. Brendel, and R. Peibst, IEEE Journal of Photovoltaics 7 (2017) 11-18.
[23] M. K. Stodolny, M. Lenes, Y. Wu, G. J. M. Janssen, I. G. Romijn, J. R. M. Luchies, and L. J. Geerligs, Solar Energy Materials and Solar Cells 158, Part 1 (2016) 24-28.
[24] M. K. Stodolny, L. J. Geerligs, G. J. M. Janssen, R. van de Sanden, J. Melskens, R. Santbergen, O. Isabella, M. Lenes, J. M. Luchies, W. M. M. Kessels, and I. G. Romijn, 33rd EUPVSEC, Amsterdam (2017).

MULTI-WIRE INTERCONNECTION FOR MULTI-BUSBAR INTERDIGITATED BACK-CONTACT CELLS: OPPORTUNITIES AND PITFALLS IN CELL-MODULE CO-DESIGN

Jonathan Govaerts[1], Tom Borgers[1], Patrizio Manganiello[1], Maarten Debucquoy[1], Arvid van der Heide[1], Hans Goverde[1], Eszter Voroshazi[1], Jozef Szlufcik[1] and Jef Poortmans[1,2]
[1]Imec, Kapeldreef 75, 3001 Leuven, Belgium
[2]Also part-time professor at KULeuven and UHasselt

ABSTRACT: In a previous contribution, we presented a multi-wire approach for interconnecting back-contact cells. While a proof-of-concept was clearly demonstrated, some remaining fill factor (FF) losses were not fully explained and initiated the need for a more in-depth analysis. The current abstract therefore reports on this analysis.
As a first step, we explain the reasoning towards an optimized theoretical cell-module co-design for multi-busbar interdigitated back-contact cells. Indicating the opportunities, we also present the measured results of the fabricated demo modules. Then, looking closely at the remaining FF losses, we determine a hypothesis for this unforeseen pitfall and confirm it through basic circuit simulations. Relying on this hypothesis, the solution is very straightforward and could be implemented at either cell- or interconnect-level. After also experimentally verifying this phenomenon and its solution, we finally present 1-cell laminates to indicate the potential of the technology in terms of Cell-to-Module-performance and compare with previous modeling results on expected FF loss.
Keywords: back-contact module, interconnection

1 BACKGROUND

In the field of c-Si PV interconnection, a growing interest towards multi-wire approaches for 2-side contacted (increasingly bifacial and busbarless) cells is materializing [1,2,3,4]. This trend is in line with the predictions put forward in the ITRPV roadmap [5], and is attracting attention due to the potential it holds for a significantly improved trade-off between optical (finger shading) and electrical (resistive transport) losses. This benefit holds up on the cell-level and especially on the module-level performance.

Also the multi-wire approach can be very interesting for back-contact cells, as we pointed out previously [6]. In this case the technology allows an improved and interesting trade-off between electrical and financial implications, independent of optical considerations [7].

In this paper we elaborate the previously shown results on interconnecting multi-busbar IBC cells with a multi-wire woven fabric approach. Next to the potential of an advanced (tapered) interconnect design, we share our learnings on the attention points that need to be taken into account on either cell- or module-level.

2 WOVEN WIRING

In our multi-wire interconnection concept, the metal wires are woven into a fabric with glass fibre that is used for insulating both polarities of the cell, as indicated in Figure 1. While explained thoroughly in [7], it is important to note that this interconnection concept can be applied to all types of back-contact (BC) cells, and has so far been demonstrated on MWT [8] and IBC cells [6] as proof-of-concept (with very different contact pad layouts). Here we show that a thorough cell-module co-design is required to reach the full potential of the technology.

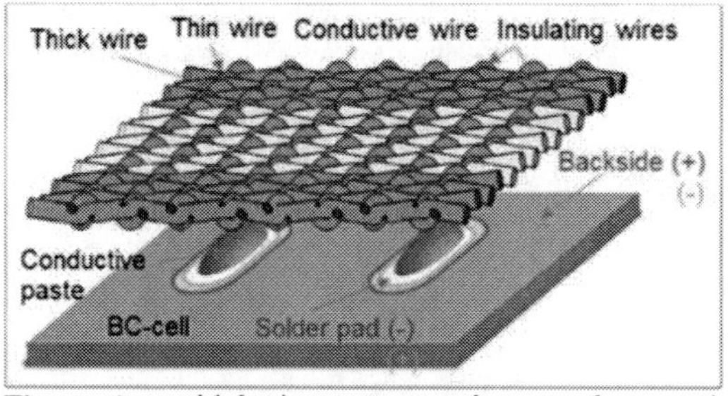

Figure 1: multiple interconnect wires are integrated inside the woven fabric in this approach

3 CELL-INTERCONNECT DESIGN

To benefit most of the multi-wire approach, the interconnection is best done perpendicular to the busbar direction to allow a wide distribution of the wires and to be independent of the amount of busbars. Furthermore, the IBC cell should be designed with asymmetric busbars (opposing polarities at opposing cell edges) to minimize resistive losses in the interconnect [7]. This leads to a layout and buildup as illustrated in Figure 2.

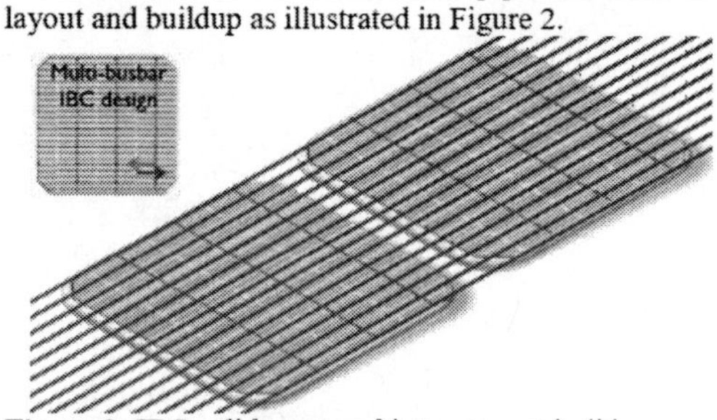

Figure 2: IBC cell layout and interconnect buildup

4 OPPORTUNITIES

Based on such a perpendicular busbar-wire buildup, the most straightforward way of interconnection would be to make an "interdigitated" wire pattern, where alternatingly wires are contacting the positive and negative busbars on the same cell (and the opposite polarities on the neighbouring interconnected cell obviously). However, significant gains can be obtained, both in electrical performance as in cost considerations, by implementing a tapered design instead:

- Resistive interconnection loss can be reduced through the tapering principle (increasing the interconnect cross-section for increasing electrical current)
- Solder paste consumption can be significantly reduced
- Cell stress is reduced as each wire is connected to that same cell at only 1 location
- The fabric has a more uniform stiffness since the wire interruptions are more distributed across the whole area

Both approaches are illustrated next to each other in Figure 3 and have been elaborated in [7].

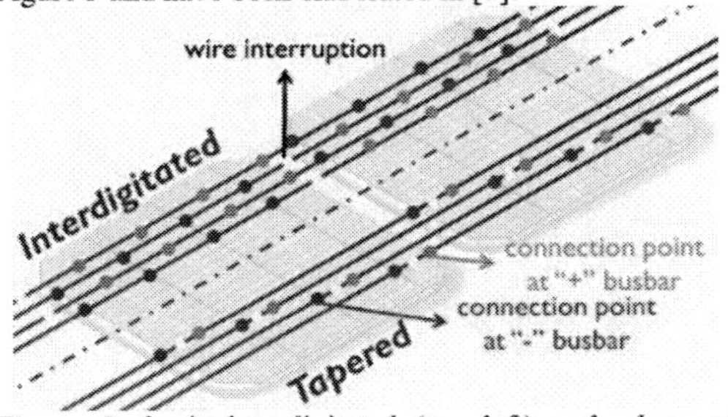

Figure 3: basic interdigitated (top left) and advanced tapered (bottom right) interconnect schemes

5 EXPERIMENTAL PROOF-OF-CONCEPT

In the experimental phase we implemented the above cell and tapered interconnect layout, with fabrication details reported in [7]. This resulted in 2 2x2-cell modules that were subsequently evaluated with IV- and EL-measurements. Figure 4 indicates how the first module exhibits some significant shunting, with a pFF drop of ~6%. This pFF drop is resolved in the second module through improved inspection and rework using IR- and EL-imaging. However, the overall FF is still low at ~71%.

	P_{mpp} [W]	I_{sc} [A]	V_{oc} [V]	FF [%]	pFF [%]	R_{sh} [Ω]	R_s [mΩ]
cell	5.231	9.657	0.683	79.3	81.9	10	2
ideal module	20.923	9.657	2.733	79.3	81.9	39	7
Module 1	17.435	9.054	2.713	71.0	76.1	2	31
Module 2	18.482	9.503	2.738	71.1	80.4	14	29

Figure 4: IV-performance of 4-cell laminates

To trace back the cause of this low FF, some clue can be found in the IV-curves. In Figure 5, the graph shows a clear kink around the maximum powerpoint, deviating from the ideal curve. A similar type of deviations has been reported earlier, e.g. [9, 10], and seems to originate from putting serially interconnected cells with non-identical performance in parallel. As can be seen in Figure 5, also the EL-picture is highly non-uniform within the cells.

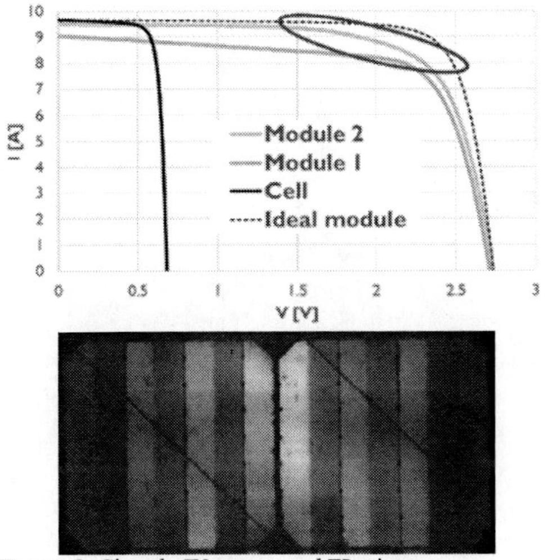

Figure 5: Clues in IV-curves and EL-picture

6 PITFALLS, LEARNINGS AND SOLUTIONS

To better explain this behaviour, Figure 6 shows how the used IBC cells can be electrically divided into several "subcells" in parallel, alternatingly sharing busbars of the same polarity.

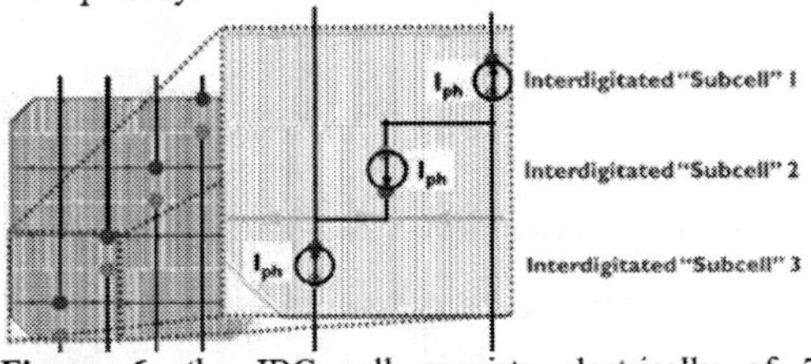

Figure 6: the IBC cell consists electrically of 7 "subcells": the magnified area shows how 3 such neighbouring subcells alternatingly share busbars of the same polarity

As schematically illustrated in Figure 7, the serial (tapered) interconnection of 2 such cells, consisting of several subcells, creates significant imbalances in current at the outer subcells: the bottom subcell of the top cell has to absorb the current of 2 subcells of the bottom cell (oversupply), while the top subcell of the bottom cell has to supply current for the 2 top subcells of the top cell. Qualitatively, this behavior is demonstrated in the EL image (Figure 5), where the subcells with an oversupply of current appear more bright. It is also indicated in Figure 7 (dashed black arrow) how this imbalance can be resolved by providing an electrical connection between the busbars with the same polarity of the outer subcells.

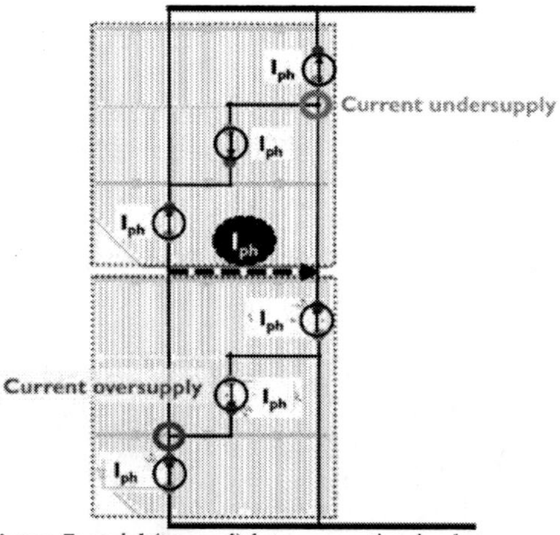

Figure 7: serial (tapered) interconnection leads to current imbalances in the outer subcells

To test this hypothesis quantitatively, circuit simulations were carried out using 7 subcells with identical 1-diode behaviour. This simplified layout and electrical schematic is shown in Figure 8. To closely mimic the actual cell design, we also implement a (resistive) path between one polarity of the outer subcells, and vary the value of its resistance R.

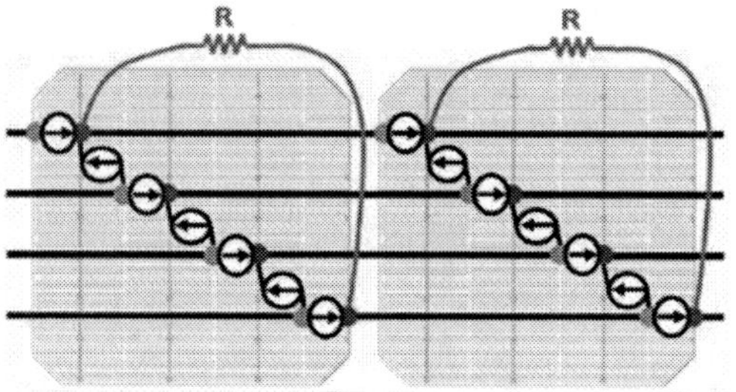

Figure 8: simplified layout and schematic for the electrical circuit simulations

Figure 9 shows the outcome of the simulations:

- If there is no redistribution path (R=∞), a current loss, equal to the current of one subcell is obtained
- If there is a lossless redistribution path (R=0), the kink is not observed (as expected)
- If there is a resistive redistribution path the kink is clearly present and can be matched to the measurement

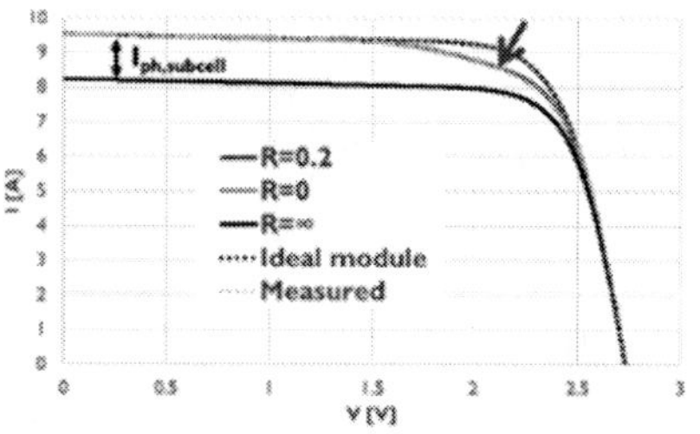

Figure 9: electrical circuit simulation results indicate the impact of a redistribution path with varying resistance

As discussed above, and confirmed in the simulations, this mechanism can be avoided by implementing a redistribution path between the outer subcells. This can be done on different levels, and is illustrated schematically in Figure 10:

(a) The cell metallization provides a common polarity on the cell by electrically connecting the busbars of one polarity

(b) An additional solder paste dot connects the same polarity of the outer subcells

(c) The woven interconnect fabric integrates a perpendicular interconnect wire to link the (outer) subcells

(d) An adjusted design immediately links the subcells with identical behaviour and avoids the underlying current imbalance

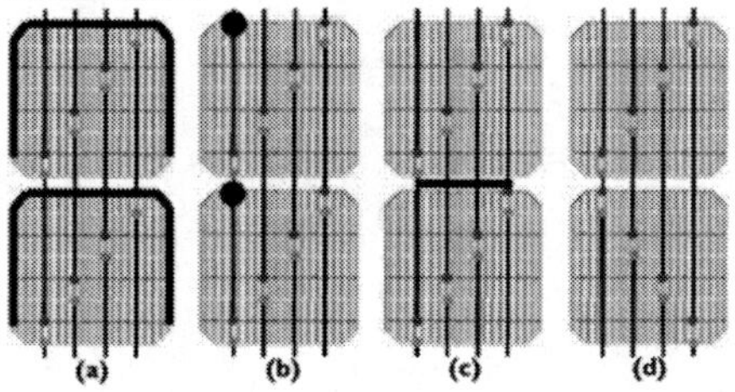

Figure 10: implementation of potential solutions (in black) to integrate a redistribution path or avoid the need for its presence

7 EXPERIMENTAL VERIFICATION

In addition, to also experimentally verify this hypothesis, we fabricated samples with a simplified structure: narrow "stripes" of IBC cells, ~16 mm wide and each similarly to the full cell containing 7 subcells, were interconnected with the tapered pattern and measured.

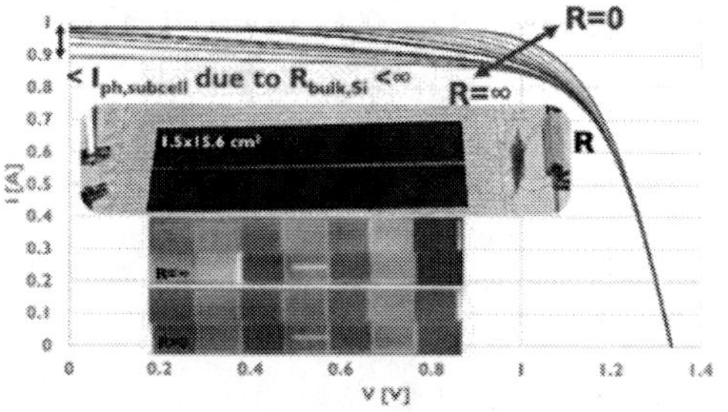

Figure 11: measured IV-curves of interconnected subcells with a redistribution path with varying resistance R

The hypothesis was then checked by implementing a variable resistance on the same string in between the 2 cells. Figure 11 indicates the resulting measurements. The trend indeed confirms the simulations: with increasing resistance between the outer subcells, first the kink in the IV-curve appears, and then further deteriorates yielding an increasing drop in current. This drop in current is not fully equal to the current of 1 subcell, due to the fact that there is a redistribution path, albeit highly resistive, through the bulk of the wafer. The EL pitures finally confirm also the IV measurements, indicating high non-uniformity if there is no low-resistive redistribution path.

Finally, to indicate the cell-to-module (CtM) performance of the technology and validate the model for determining resistive losses [7], we also fabricated 2 1-

cell laminates. The resulting samples and measurements are shown in Figure 12. The loss in FF of ~3.2% due to interconnection matches relatively well with the calculated value of 2.9%.

	P_{mpp} [W]	I_{sc} [A]	V_{oc} [V]	FF [%]
Standalone	5.358	9.851	0.684	79.5
Soldered	5.103	9.976	0.670	76.3
Laminated without ARC	5.025	9.644	0.681	76.5
CtM	93.8	97.9	99.6	96.2
Laminated with ARC	5.188	9.875	0.682	77.0
CtM	96.8	100.2	99.7	96.9

Figure 12: CtM-performance of 1-cell laminates

5 CONCLUSION AND OUTLOOK

In this paper, we report on our progress on our multi-wire interconnection approach for back-contact cells. In particular, we elaborate on our understanding of the opportunities and pitfalls, and how these can be avoided, of the tapered multi-wire interconnection technology on multi-busbar interdigitated back-contact cells. To support this understanding, we present both simulations and experimental verification. The optimized cell-module co-design is implemented in 1-cell laminates as final proof-of-concept.

6 ACKNOWLEDGMENT

The authors gratefully acknowledge imec's SiPV industrial affiliation programme and its partners, VdSWeaving for discussions and manufacturing of woven interconnect material, ECN for module measurements and DSM for supplying AR-coated glass. This project receives the support of the European Union, the European Regional Development Fund ERDF, Flanders Innovation & Entrepreneurship and the Province of Limburg. Module materials are developed within the LENO PV project funded by Flanders Innovation & Entrepreneurship under grant No155025. This project has received funding from the European Union's Horizon 2020 research and innovation programme under grant agreement No 72752.

7 REFERENCES

[1]] T. Söderström et al., EUPVSEC, 2013
[2] S. Braun et al., SiliconPV, 2013
[3] T. Borgers et al., IEEE PVSC, 2016
[4] T. Borgers et al., EUPVSEC, 2017
[5] ITRPV roadmap, 2016
[6] J. Govaerts et al., EUPVSEC, 2016
[7] T. Borgers et al., PiP, 2016
[8] T. Borgers et al., BCWorkshop, 2015
[9] R.M. Sullivan, NASA report, 1965
[10] A. van der Heide et al., PiP, 2005

CONTACTLESS DETERMINATION OF DIELECTRIC ABSORPTION FROM THE SPECTRAL RESPONSE OF PHOTOLUMINESCENCE

Mattias Klaus Juhl[1], Michael Pollard[1], Appu Rshikesan Paduthol[1], Angus Gentle[2], Thorsten Trupke[1,3], Ziv Hameiri[1]
1. The University of New South Wales, Sydney, Australia
2. University of Technology Sydney, Australia
3. BT Imaging Pty Ltd, Sydney, Australia

ABSTRACT: Light absorption within a solar cell's dielectric coatings represents a loss mechanism that proves challenging to measure, particularly at short wavelengths at which the substrate is strongly absorbing. This paper presents a method based on measurements of the spectral response of photoluminescence that determines a relative measure of the absorption in a dielectric coating. The use of the spectral response of photoluminescence is suggested as it is a contactless and non-destructive technique that can be applied to a solar cell at any stage of production. The absorptance determined from this method is compared to ellipsometry measurements on polished samples, and is then used to determine the absorptance of a silicon nitride layer on the textured surface of a screen printed solar cell.
Keywords: Spectral Response, Absorption, Characterisation, Photoluminescence

I. INTRODUCTION

To increase solar cells performance, it is helpful to determine both how and where the power is lost, such that this loss can be reduced. One loss mechanism that is difficult to determine on a finished solar cell, is the fraction of light that is absorbed within the front dielectric layer/s. This paper works towards this goal by investigating if the spectral response of photoluminescence (SR_{pl}) can be used to determine the absorption in SiN_x at short illumination wavelengths (<600 nm).

Currently there are two approaches that are used to determine the absorption in a dielectric layer in photovoltaics. One method is based on spectral ellipsometry data, while the other is based on external quantum efficiency (EQE) and reflection data.

Spectral ellipsometry provides a means of determining the complex refractive index and thickness of dielectrics on silicon samples. When this information is combined with optical modelling of the surface texture of a solar cell, the absorption in dielectrics can be determined. Unfortunately, this approach can often be ineffective for silicon solar cells as spectral ellipsometry is usually only performed on polished wafers. The most common method is thus to deposited the dielectric being investigated onto a polished sample, and ensure it experiences the same thermal processing as a finished cell. Spectral ellipsometry is then used to determine the complex refractive index. This process assumes that the film properties are identical to that on the finished cell, irrespective of it being deposited on a polished or textured wafer. This assumption is not always true, especially when the deposition is done using a dynamic (in-line) systems where the thermal mass of the wafer impacts the deposition temperature. Several methods also exist to determine the complex refractive index of dielectrics that are deposited on random pyramid textured monocrystalline [1], [2] and acid etched samples [2], [3]. While these techniques enable measurements to be performed on these structures the techniques do have limitations. These include a reduction in the information that can be obtained from the ellipsometry measurement, and requiring additional hardware such as tilting stages. Furthermore, no method currently exists that can determine the complex refractive index on a surface with a black silicon texture. When the complex refractive index of a film has been obtained from the spectral ellipsometry data, detailed optical modelling is then required to determine the absorption that occurs in the film, adding another source of potential errors.

The second approach uses a combination of EQE and reflection measurements [4]. This approach assumes that the decrease in EQE at short illumination wavelengths is a combination of (a) front surface reflection, (b) absorption in the front dielectrics, (c) surface recombination, and (d) losses in the front diffusion. By creating a device with very low losses in the front diffusion (e.g. a well passivated sample with a low resistivity diffusion) and measuring the EQE and spectral reflection, an upper limit for the absorption loss in the front dielectric can be determined. This method has been claimed to be the more accurate method for determining the absorption within front dielectrics [4], as it directly measures the losses rather than requiring modelling. One issue with this technique is that it requires specially processed solar cells.

This paper introduces a third approach to measure the absorption in the front dielectric. The method is based on SR_{pl} measurements, which are contactless, non-destructive, and can be performed on samples at any stage of production. Note that in this proposed method we monitor the photoluminescence emission of the dielectric and not of the bulk silicon. Measurements can be performed under the same conditions as EQE measurements, i.e. with the incident illumination perpendicular to the sample. It thus allows a more direct measurement of the absorption in the film. This paper presents a comparison between the proposed PL based technique with the two established approaches for determining absorption in dielectric films, that were discussed above. This paper presents a proof of concept for this new method, focusing on confirmation that SR_{PL} in fact provides a relative measure of the absorption in SiN_x. Finally, we propose a path to determine the absolute absorptance in dielectric coatings.

II. BACKGROUND

The SR_{pl} monitors the photoluminescence (PL) emission (I_{pl}) of a sample under monochromatic illumination with known photon flux (N_γ):

$$SR_{\rm pl}(\lambda) = \frac{I_{\rm pl}(\lambda)}{N_\gamma(\lambda)} \tag{1}$$

The monochromatic illumination wavelength (λ) is scanned to cover the wavelength range of interest. Note that in the context of the spectral response of PL the term $I_{Pl}(\lambda)$ refers to the spectrally integrated PL signal as a function of illumination wavelength.

The SR_{pl} has previously been used to determine different sample properties, including the optical band gap [5], [6], surface recombination [7], [8], and diffusion length [9]. The SR_{pl} has also been used to determine a relative measure of a sample's absorptance [10]–[12] and the external quantum efficiency of wafers and cells [13], [14].

Here we apply SR_{pl} to determine a relative measure of the absorptance of the dielectric layer on the front surface. The emitted PL intensity is related to the depth averaged excess carrier density in a thin film by [15]:

$$I_{\mathrm{pl}} \propto \int BB\alpha_{\mathrm{bb}}\, \overline{n}\, \overline{p}\, d\, \lambda_{\mathrm{pl}} \quad (2)$$

where BB is black body radiation, $\alpha_{\mathrm{bb}}$is the band to band absorption coefficient of the dielectric film, $\overline{n}$ and $\overline{p}$ are the average electron and hole densities in the dielectric film, and λ_{pl} is the PL emission wavelength. The excess carrier density is related to the absorbed photon flux that results in charge carrier generation ($N_{\gamma-abb}$):

$$\overline{n} = \tau_{\mathrm{eff}} N_{\gamma-\mathrm{abb}} W \quad (3)$$

where τ_{eff} is the effective lifetime, and W is the thickness of the dielectric film. Inserting (3) into (2), under the assumption that the sample is in low injection conditions results in:

$$I_{\mathrm{pl}} \propto \tau_{\mathrm{eff}} N_{\gamma-\mathrm{abb}} \int BB\alpha_{\mathrm{bb}}\, d\, \lambda_{\mathrm{pl}} \quad (4)$$

Here, the majority carrier concentration has been lumped into the proportionality sign. Noting that the integral of the PL spectrum is a function of material parameters, the integral can also be absorbed into the proportionality symbol. Similarly, τ_{eff} can be absorbed into the proportionality symbol for a sample with constant τ_{eff}, resulting in

$$I_{\mathrm{pl}} \propto N_{\gamma-\mathrm{abb}} \quad (5)$$

Substituting (5) into (1) provides

$$I_{\mathrm{pl}} \propto \frac{N_{\gamma-\mathrm{abb}}}{N_{\gamma}} \quad (6)$$

This is very close to the definition of the absorptance (A):

$$A = \frac{N_{\gamma-a}}{N_{\gamma}} \quad (7)$$

where, $N_{\gamma-a}$ is the total absorbed photon flux, which includes both transitions that lead to the generation of electron-hole pairs and parasitic absorption such as free carrier absorption. Thus, when the number of photons that result in excess carrier generation ($N_{\gamma-abb}$) is proportional to number of photons absorbed ($N_{\gamma-a}$) in the silicon nitride, an assumption that can generally be assumed to be valid at low illumination intensities, the SR_{pl} provides a relative measure of the absorptance:

$$A \propto \frac{I_{PL}}{N_{\gamma}} \quad (8)$$

This paper investigates several samples to determine whether this relationship holds true.

III. EXPERIMENTAL DEMONSTRATION

To validate the proposed method, the absorptance of several silicon nitride layers was determined and compared to SR_{pl} measurements. The two methods, discussed earlier, were used to determine the absorptance of the SiN_x being: spectral ellipsometry and a combination of EQE and reflection data on specially made cells.

Single side polished samples were used for spectral ellipsometry measurements, as both SR_{pl} and spectral ellipsometry measurements can be performed on this type of sample. The samples were p-type wafers with bulk doping of ~5×10^{15} cm^{-3}. The wafers were RCA cleaned, dipped in hydrofluoric (HF) solution, before two different SiN_x deposition using an industrial dynamic plasma enhanced chemical vapor deposition system (MAiA, Meyer Burger).

In the first experiment, spectral ellipsometry measurements were performed using an EC-400 from J.A Woollam. The measurements were performed at three different angles for the wavelength range of 300 to 1000 nm. The measured ellipsometry data was fitted using the software package WVASE [16] using a Tauc-Lorentz oscillator [17] and a graded layer. As the sample's had a polished surface the absorptance in the SiNx was calculated using the transfer matrix method.

The SR_{pl} was measured using a custom set-up, schematically shown in Figure 1. The monochromatic illumination was provided by a tunable laser. The spectral width of the excitation light was approximately 4 nm. The collimated light was passed through several short pass filters to further reject any light in the range of the expected PL signal. The light was then reflected off a beam splitter, and focused onto the sample. For each sample, the focus was adjusted to maximise the PL signal. The PL emission from the dielectric layer was collected, long pass filtered, and coupled into a CCD based spectrometer (Princeton Instruments IsoPlane 160) with a back-thinned silicon Electron Multiplying Charge Coupled Device (Princeton Instruments ProEM 1600). The intensity of the laser was monitored during the measurement by sending a fraction of the laser's intensity, reflected off the beam splitter, to a calibrated silicon photodiode. While spectral detection is not strictly necessary for this method, it was performed here to ensure that only luminescence from the dielectric was being detected.

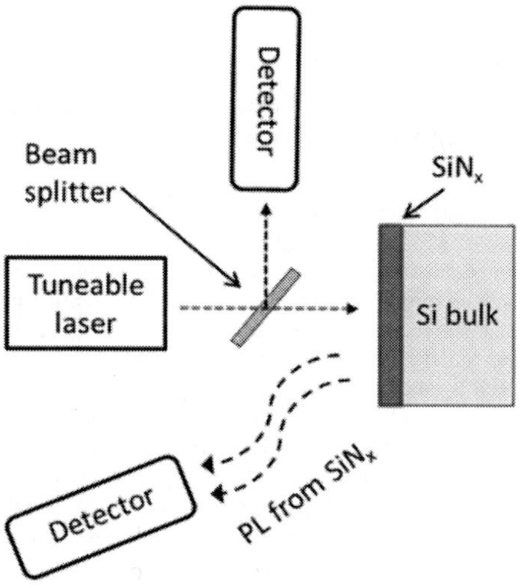

Figure 1: Schematic of the experimental set-up used to measure the spectral response of photoluminescence.

SR_{pl} was measured for excitation wavelength in the range 450 nm to 550 nm. The absorptance was also calculated from the optical parameters in the same range. The correlation between the SR_{pl} measurement and the absorptance in the SiN_x is shown in figure 2(a) for two samples. Linear fits to the data are shown as dashed lines. The fit shows good correlation between the absorptance measured from ellipsometry and SR_{pl} measurement for both SiN_x films investigated. Unfortunately, the line of best fit does not pass through the origin for either sample, with the highly absorbing SiN_x having a positive intercept and the less absorbing SiN_x having a negative intercept. The cause of this error is currently unknown, but it is believed to associated with an error in the measurement of the background value used in the photoluminescence measurements. Using the slope and intercept determined from Figure 2 (a), the SR_{pl} data is converted into an absolute absorption and shown in the more typical form, as a function of wavelength, in Figure 2 (b). This procedure represents a scaling of the relative absorption data obtained from PL to the absolute absorption data obtained from the optical modeling.

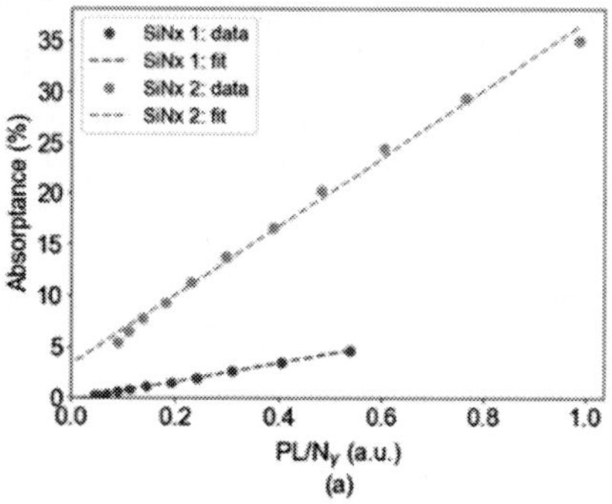

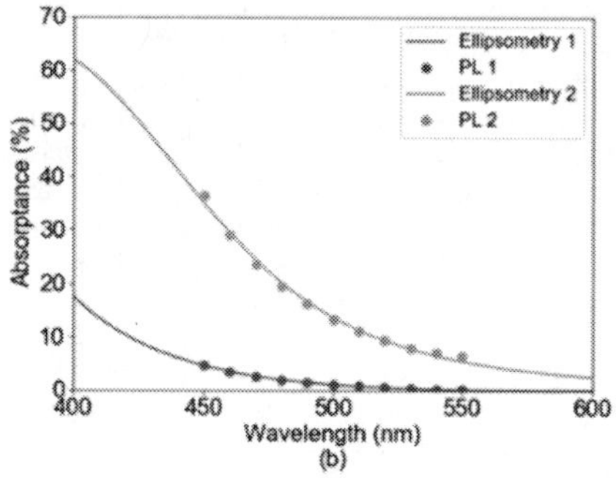

Figure 2: Compassion of the absorptance of two different SiN_x films, determined from spectral ellipsometry data and the spectral response of photoluminescence. (a) Absolute absorptance obtained from optical modelling as a function of the raw SR_{pl} and (b) spectral absorptance after scaling the PL data (see text for details).

In a second experiment the relative SR_{pl} measurement is compared to the absorptance of films determined from conventional EQE (measured using an EQ8 from PV measurements) and reflection data (measured on a Cary 500 spectrophotometer) on two screen printed solar cells. The two samples were deposited with the same SiN_x but had different diffusions. The absorption of the films was determined from the sample with the light diffusion (> 120 Ω/sq), from a combination of EQE and reflection data using

$$A_{SiN} = 1 - EQE(\lambda) - R(\lambda) \qquad (9)$$

where A_{SiN} is the absorptance of the SiN_x film, $R(\lambda)$ is the spectral reflectance of the front surface and $EQE(\lambda)$ is the external quantum efficiency. An implicated assumption in Equation 9 is that the only current loss mechanisms are the reflection and absorption in the dielectric layers. The absorptance determined from Equation 9 for the lightly doped sample is shown Figure 3(a). A linear correlation is observed, as expected.

The heavily diffused sample is not shown in 3(a) as it has additional losses in the diffused region and thus the assumptions behind Equation 9 do not hold. The SR_{pl} measurements were then converted into absolute absorption using the slope and intercept determined from the linear fit to the data shown in Figure 3(a). Since both the lightly diffused and the heavily diffused samples have the same nitride film, the same calibration constants were applied to both SR_{pl} data sets. The absolute absorption in the film is shown in Figure 3(b) for both samples. Good agreement for the absorption obtained from PL between the light and heavy diffusion is observed, demonstrating that the photoluminescence emission of the SiN_x was not affected by the diffusion.

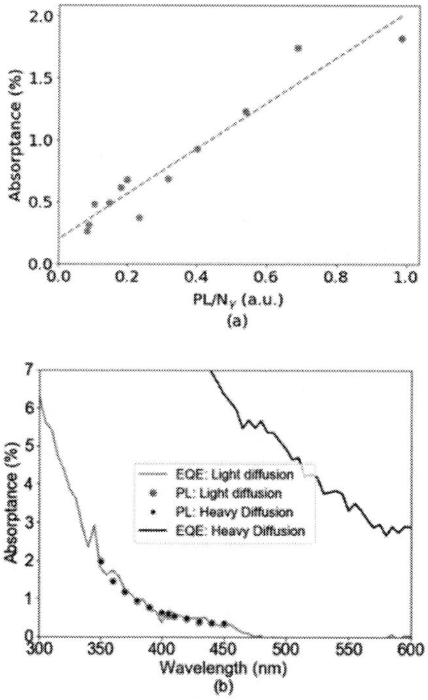

Figure 3: Comparison of the SiN_x absorptance on fully processed cells, determined from conventional EQE and reflection data and from the spectral response of photoluminescence, respectively. (a) Absolute absorptance as a function of relative SR_{pl} and (b) spectral absorptance after scaling the PL data (symbols) for the two samples using the same scaling factor (see text for details). Solid lines represent 1-$EQE(\lambda)$-$R(\lambda)$ as per Eq.9.

Figure 3(b) shows that on both samples the absorption by the nitrite causes similar losses. On the sample with light diffusion, the absorption data from PL are spectrally well aligned with values for 1-$EQE(\lambda)$-$R(\lambda)$. In contrast, the deviation between the absorption data from PL from 1-$EQE(\lambda)$-$R(\lambda)$ for the other cell [black line in Figure 3(b)] are a result of additional losses in the heavily diffused emitter. The experimental approach demonstrated here thus enables quantitative a distinction of different loss mechanisms.

IV. CALIBRATION

To be a useful technique for the characterization of arbitrary samples with unknown optical and electronic parameters, the relative absorptance of a film measured by SR_{pl} needs to be converted into an absolute absorptance, ideally without the need for calibration against other experimental data. If this is possible then the absorptance of the dielectric top layers can be determined without the need for either specially prepared samples, optical modeling of the texturing, or determination of the complex refractive index of the dielectric. One suggested method is through comparison of SR_{pl} to EQE measurements at wavelengths below 360 nm. At wavelengths below 360 nm, the absorption depth is so short (< 10 nm) that photons can be considered to be absorbed at the surface [4]. The collection efficiency is then independent of wavelength. Variations in the internal quantum efficiency or in 1-$EQE(\lambda)$-$R(\lambda)$ below 360 nm are thus associated with an absolute increase in parasitic absorption. If SR_{pl} data can be acquired within this wavelength range it may provide a means of converting the relative absorptance into an absolute absorptance. Unfortunately, our current hardware is limited to 350 nm excitation, but this approach will be evaluated in future work.

V. CONCLUSION

We have presented a method, based on the spectral response of PL, to provide a measure of the absorption of a film on a finished cell without the need for numerical simulation or preparation of special samples. The SR_{pl} was compared to the absorptance determined from ellipsometry and a combination of EQE and reflection data. Good correlation was found between the proposed method and the other techniques. In its current implementation, the SR_{pl} can only determine a relative measure of the absorption in dielectric layers. Further work will focus on validating a proposed method by which the SR_{pl} can be converted into absolute absorptance in a self-contained fashion.

VI. ACKNOWLEDGEMENTS

This program has been supported by the Australian Government through the Australian Renewable Energy Agency (ARENA) grant RND009. Responsibility for the views, information or advice expressed herein is not accepted by the Australian Government.

VII. REFERENCES

[1] J. Sun, M. F. Saenger, M. Schubert, J. N. Hilfiker, R. Synowicki, C. M. Herzinger, and J. a Woollam, "Characterizing antireflection coatings on textured mono-crystalline silicon with spectroscopic ellipsometry," in the *34th IEEE Photovoltaic Specialists Conference* , 2009.

[2] M. Saenger, J. Sun, M. Schädel, J. Hilfiker, M. Schubert, and J. Woollam, "Spectroscopic ellipsometry characterization of SiNx antireflection films on textured multicrystalline and monocrystalline silicon solar cells," *Thin Solid Films*, vol. 518, no. 7, pp. 1830–1834, Jan. 2010.

[3] S. C. Siah, B. Hoex, and A. G. Aberle, "Accurate characterization of thin films on rough surfaces by spectroscopic ellipsometry," *Thin Solid Films*, vol. 545, pp. 451–457, 2013.

[4] B. Fischer and P. P. Altermatt, "A simple emitter model for quantum efficiency curves and extracing the emitter saturation current," in the *28th European Photovoltaic Solar Energy Conference*, 2013.

[5] A. Cornet, J. R. Morante, and M. Lo, "Determination of the direct band-gap energy of InAlAs matched to InP by photoluminescence excitation spectroscopy," *Journal of Applied Physics*, vol. 81, no. 10, pp. 10–14, 1997.

[6] D. A. Evans, A. G. McGlynn, B. M. Towlson, M. Gunn, D. Jones, T. E. Jenkins, R. Winter, and N. R. J. Poolton, "Determination of the optical band-gap energy of cubic and hexagonal boron nitride using luminescence excitation spectroscopy," *Journal of Physics: Condensed Matter*, vol. 20, no. 7, p. 075233, Feb. 2008.

[7] J. M. Dishman, "Characterization of Dominant Recombination Centers in Semiconductors from the Temperature Dependence of Luminescence Excitation

Spectra: GaP(Zn,O)," *Physical Review B*, vol. 5, no. 6, pp. 2258–2267, 1972.
[8] D. Baek, J. Lee, and B. Choi, "Diffusion length and resistivity distribution characteristics of silicon wafer by photoluminescence," *Materials Research Bulletin*, vol. 58, pp. 157–163, Oct. 2014.
[9] D. Baek, J. Lee, and B. Choi, "Surface recombination evaluation of a silicon wafer by using the photoluminescence intensity ratio method for solar cell applications," *Journal of the Korean Physical Society*, vol. 65, no. 3, pp. 336–341, Aug. 2014.
[10] C. Itoh, K. Tanimura, N. Itoh, and M. Itoh, "Threshold energy for photogeneration of self-trapped excitons in SiO2," *Physical Review B*, vol. 39, no. 15, pp. 11183–11186, 1989.
[11] M. K. Juhl, T. Trupke, M. D. Abbott, B. Mitchell, and B. J. Hallam, "Spatially Resolved Light Trapping of Silicon Wafers from Photoluminescence Imaging Using Long Wavelength Excitation," in the *31st European Photovoltaic Solar Energy Conference*, 2015
[12] M. K. Juhl, T. Trupke, M. Abbott, and B. Mitchell, "Spatially Resolved Absorptance of Silicon Wafers From Photoluminescence Imaging," *IEEE Journal of Photovoltaics*, vol. 5, no. 6, pp. 1840–1843, Nov. 2015.
[13] G. D. Pettit, J. M. Woodall, and H. J. Hovel, "Photoluminescent characterization of GaAs solar cells," *Applied Physics Letters*, vol. 35, no. 4, pp. 335–337, Aug. 1979.
[14] M. K. Juhl, M. D. Abbott, and T. Trupke, "Relative External Quantum Efficiency of Crystalline Silicon Wafers From Photoluminescence," *IEEE Journal of Photovoltaics*, vol. 7, no. 4, pp. 1074–1080, 2017.
[15] M. A. Green, "Analytical expressions for spectral composition of band photoluminescence from silicon wafers and bricks," *Applied Physics Letters*, vol. 99, no. 13, p. 131112, Sep. 2011.
[16] J. A. Woollam, B. D. Johs, C. M. Herzinger, J. N. Hilfiker, R. A. Synowicki, and C. L. Bungay, "Overview of variable-angle spectroscopic ellipsometry (VASE): I. Basic theory and typical applications." *Optical Metrology*, vol. 1, pp. 3–28, 1999.
[17] G. E. Jellison and F. A. Modine, "Parameterization of the optical functions of amorphous materials in the interband region," *Applied Physics Letters*, vol. 69, no. 3, pp. 371–373, Jul. 1996.

ANGLE-DEPENDENT REFLECTANCE OF ISOTEXTURED SILICON

Aina Alapont Sabater[1], Johannes M. Greulich[1], Nico Tucher[1], Benedikt Bläsi[1], Stefan W. Glunz[1,2]
[1]Fraunhofer Institute for Solar Energy Systems ISE, Heidenhofstraße 2, 79110 Freiburg, Germany
[2] Albert-Ludwigs-University Freiburg, Laboratory for Photovoltaic Energy Conversion, Georges-Köhler-Allee 105, 79110 Freiburg, Germany
Telephone: +49 761 4588 5075, e-mail: aina.alapont.sabater@ise.fraunhofer.de

ABSTRACT: Multicrystalline silicon (mc-Si) is the workhorse of photovoltaics with a market share above 60%, being acidic textured (isotextured) silicon the standard. For optical simulations, the spherical caps model is widely applied, but it was reported that is does not correctly describe the angular dependence of reflection. In this work, our first objective is to verify this. Secondly, we want to obtain an improved model for optics of isotextured mc-Si wafers and solar cells able to describe the light trapping properties of cells and modules correctly. We investigate the reflectance of isotextured silicon by varying the angle of incidence of the light between 8° and 80°. We perform measurements of nine isotextured samples, obtaining reflectance curves with a similar shape, but they differ by up to $60\%_{abs}$ from previously published data. The first simulations are performed using the spherical caps model, which is shown to be inaccurate to describe the angle-dependent reflectance of isotextured silicon. In the second set of simulations we vary the texture morphology, reducing the deviations of the simulation from the measurement from 8% to 1.2% for incident angles smaller than 60°. We conclude that alternative morphologies are worth doing further research.
Keywords: Silicon Solar Cell, Multicrystalline Silicon, Optical Properties, Simulation, Light trapping

1 INTRODUCTION

Multicrystalline silicon (mc-Si) with an isotextured front surface, i.e. isotropically etched silicon by hydrofluoric acid (HF) and nitric acid (HNO_3) [1], has a market share in the photovoltaic industry of more than 60% and is expected to keep a share well above 40% for the next ten years [2]. Simulation is the basis for a thorough analysis of the performance of the isotexture including light trapping in the cell and especially in the module. For optical simulations of isotextured silicon solar cells, the spherical caps model was proposed [3] and is widely used. By adjusting the opening angle as characteristic parameter this model is capable of describing the hemispherical reflectance [4] and hemispherical transmittance [5] for normally incident light. However, it lacks a simple and accurate description of its angular dependence [3, 6], which is of utmost importance for the light trapping within the solar cell and within the module.

In this paper we present angle-dependent reflectance measurement results of isotextured silicon which show a different behaviour than previously published data [3, 5, 6]. We also evaluate different isotexture morphologies to look in further detail at the reflectance for different angles of incidence of the light. We present the corresponding simulations applying alternative morphologies and compare our results with the simulations using the spherical caps approach.

This work presents a new model for the optical simulation of isotextured silicon, worthy of further investigation, since the accurate description of angle-dependence would be a step forward in the solar cell simulation.

2 EXPERIMENTAL

We work with nine different mc-Si wafers, which are wet-chemically textured. Four of them were processed at Fraunhofer ISE at two different process speeds and two different acid concentrations, resulting in the formation of four different textures: "strong" (see Figure 1a), "medium" (see Figure 1b) and two "weak". Additionally, four wafers were kindly provided by the Australian National University (ANU), also with different textures, and the SCHMID Group sent us a textured wafer with "DW PreTex" treatment [7].

As representative group, we present in this paper the results of three samples with different textures: "strong" from ISE (see Figure 1a), "medium" from ISE (see Figure 1b) and "weak" from the ANU (see Figure 1c). The planar wafer, used as reference, is a polished float zone (FZ). From each wafer we cut two samples in the dimensions of 2x3.3 cm² and 3.3x4.5 cm².

A spectrophotometer with an Ulbricht sphere

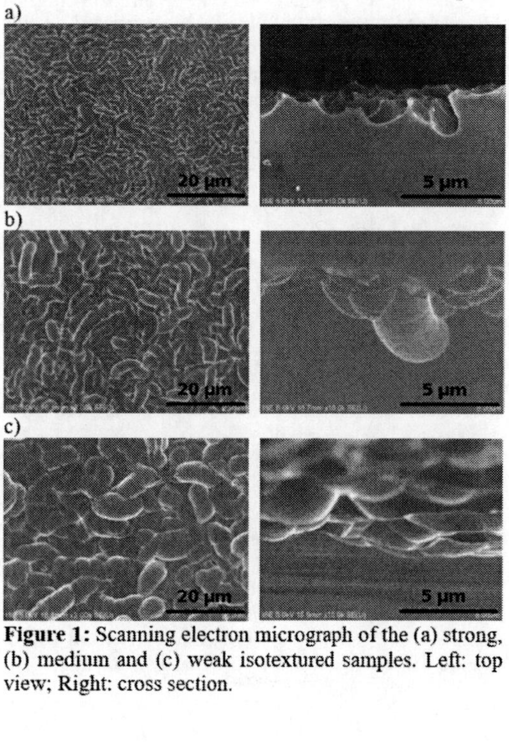

Figure 1: Scanning electron micrograph of the (a) strong, (b) medium and (c) weak isotextured samples. Left: top view; Right: cross section.

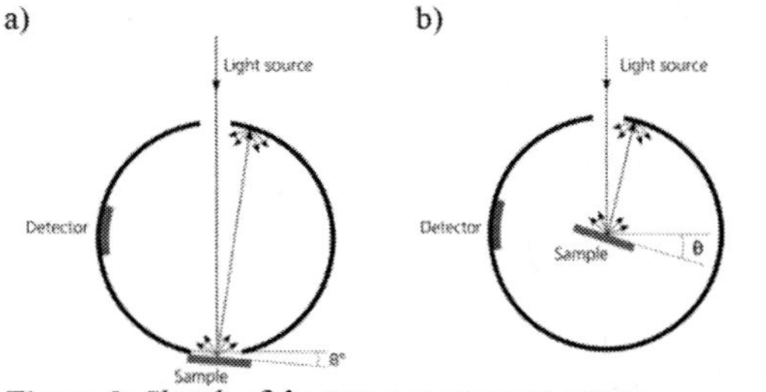

Figure 2: Sketch of the two measurement setups.

(Fourier-Spectrometer), with diameter of 20 cm is used for the measurements. The reference measurement is performed as shown in Figure 2a: sample mounted externally at the sphere at an angle of the incident light of 8°. The other measurements are performed with the sample center mounted within the Ulbricht sphere (see Figure 2b), which is rotated, thereby increasing the angle of incidence of the light θ.

In order to get the most accurate measurements, i.e. to avoid that the reflected light on the sphere strikes again the sample, the sample should be as small as possible. Meanwhile, the sample has to be large enough so that the whole light beam strikes it. Considering these two requirements, we use the small samples for incident angles $\theta \leq 60°$ and the large samples for $\theta \geq 60°$.

3 SIMULATION

We perform three different sets of optical simulations, which are described below. All of them apply 3D geometric optics (ray tracing) and the transfer matrix method. The incident angle θ is varied between 0 and 88°.

3.1 Spherical caps

The first set of simulations are performed using the Module Ray Tracer (MRT) by PV Lighthouse [8], using refractive index data for silicon from Ref. [9]. The simulations of the hemispherical reflectance are performed with 10^6 rays and wavelengths between 300 and 1000 nm in steps of 10 nm. The MRT uses the spherical caps approach. As shown in Figure 3, the morphology is described as an inverted spherical cap by the characteristic angle ω_c.

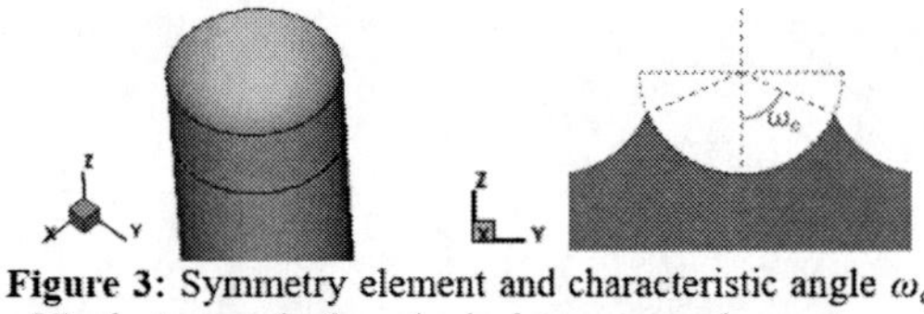

Figure 3: Symmetry element and characteristic angle ω_c of the isotexture in the spherical cap approach.

3.2 Quadratic grid

The second and third sets of numerical simulations are performed using Sentaurus Device [10], since this software allows us more flexibility, e.g. by varying the texture morphology and the ray randomization algorithms. We use in this case a spherical caps morphology, out of which a pillar with quadratic cross section is cut, resulting in the symmetry element shown in Figure 4, also described by the characteristic angle ω_c.

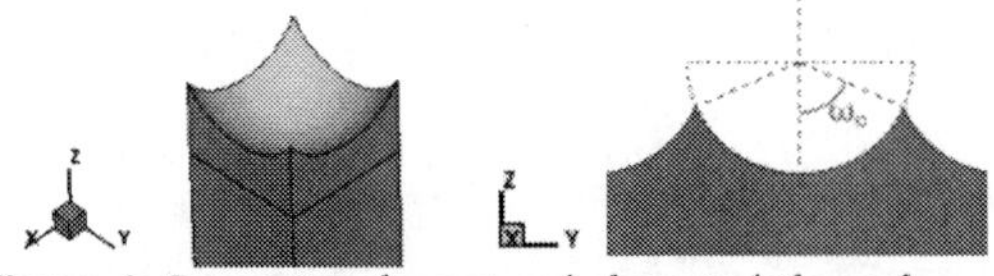

Figure 4: Symmetry element and characteristic angle ω_c of the isotexture in the quadratic grid approach. The cross section corresponds to a cut through the centre of the symmetry element.

3.3 Tub form

As depicted in scanning electron microscopy (SEM) images in Figure 1, the morphology of isotextured silicon appears to resemble a tub form. We implemented this as optical model by defining two characteristic parameters: the characteristic angle ω_c and the characteristic length L_c (see Figure 5). Furthermore, the stronger the texture is, the shorter is L_c, the shape being more rounded and the geometry more similar to a spherical cap.

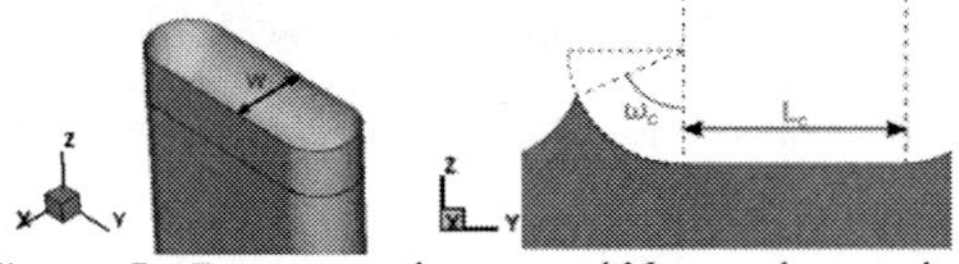

Figure 5: Symmetry element, width w, characteristic angle ω_c and characteristic length L_c of the isotexture in the tub form approach.

4 RESULTS

First, we compare our measurement results of the three samples with data from Ref. [3]. As shown in Figure 6, our measurements show a much less pronounced dependence on θ than the measurements from Ref. [3]. The same holds not only for the samples fabricated at Fraunhofer ISE, but also for our measurements at samples from the original batch of Ref. [3] that were kindly provided by the ANU. The differences in R are above $27\%_{abs}$ for angles greater than 75°. After discussing this discrepancy with the authors of that study, we conclude that an error occurred in part of the data from Ref. [3], besides the error when determining the characteristic angle ω_c of the texture that

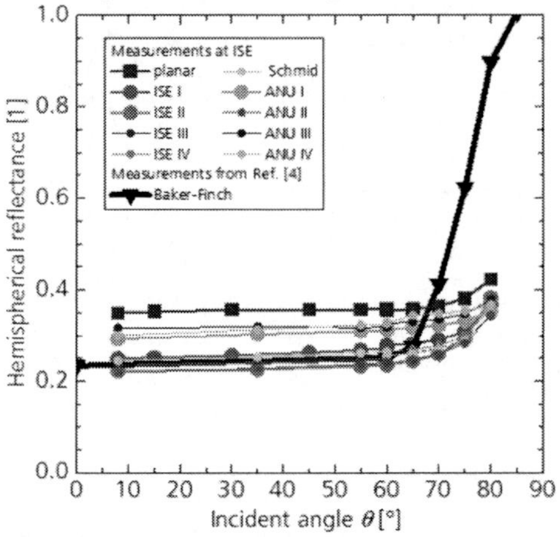

Figure 6: Reflectance R of isotextured silicon for $\lambda = 600$ nm for two cases: 1) Measurements of this work performed at ISE; 2) Measurements from Baker-Finch et al. [3].

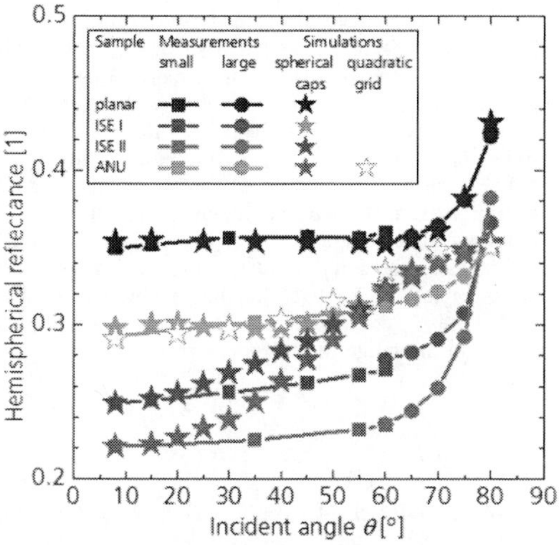

Figure 7: Simulated (stars) and measured (lines) reflectance R for $\lambda = 600$ nm for one planar and three isotextured silicon samples.

the authors recently reported [5].

Second, we focus on the spherical caps model, using the MRT by PV Lighthouse, and compare it to our measurements. The simulated and measured reflectance of the planar sample, used as reference, are in close agreement with deviations smaller than $0.9\%_{abs}$ (see Figure 7), which indicates that both the experimental setup as well as the simulation scheme work properly. We adjust ω_c in the MRT, so that the simulated hemispherical reflectance R_{sim} matches the measured reflectance R_m at $\theta = 8°$, obtaining $\omega_c^{ISE\ I} = 60°$, $\omega_c^{ISE\ II} = 66°$, $\omega_c^{ANU} = 52°$. The next simulations are carried out using this characteristic value, and varying the angle of incidence θ between 8 and 80°. The results shown in Figure 7 reveal a high discrepancy; differences in R are above $5\%_{abs}$ for $\theta = 60°$ for the sample ISE I and above $8\%_{abs}$ for $\theta = 60°$ for the sample ISE II. We also simulate the sample from the ANU with the quadratic grid approach (see Figure 7), obtaining very similar results to those using the spherical caps approach.

Third, we test the tub form model in order to analyse and understand better this new approach. To that end, we analyse the influence of the characteristic parameters on the reflectance. As shown in Figure 8, the greater the ratio L_c/w and the smaller the ω_c, the higher the reflectance R. For small ω_c ($60° < \omega_c < 70°$) the reflectance consists of two straight lines with a sharp change in the slope at $\theta = 85°$. When ω_c increases up to 88°, R is a curved line.

Furthermore, we perform another variation, this time of the azimuth φ (described in Figure 9) and the characteristic length L_c, for a fixed $\omega_c = 88°$.

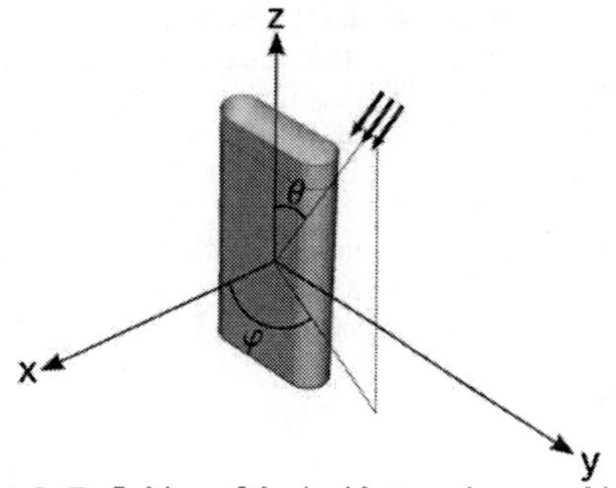

Figure 9: Definition of the incident angle or zenith θ and azimuth φ.

Like in the previous case, the reflectance is higher when the value of L_c/w increases (see Figure 10). For small φ, the reflectance only rises with the incident angle θ. However, for larger φ the reflectance rises and falls with the variation of θ.

At this point, we perform the simulations using the tub form approach and compare them to the measurements. As for the previous simulations, we adjust the characteristic parameters, obtaining the values shown in Table 1. Other combinations of L_c/w and ω_c might be possible to reach the same reflectance at 600nm for $\theta = 8$. Because isotextured silicon is irregular (see Figure 1) and every tub has a different orientation, we simulate the

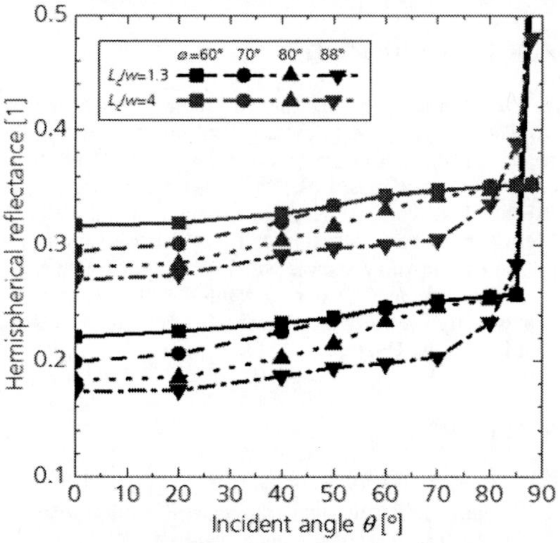

Figure 8: Simulated reflectance R for $\lambda = 600$ nm of the tub geometry including a variation of the characteristic parameters ω_c and L_c/w for fixed $\varphi = 0°$.

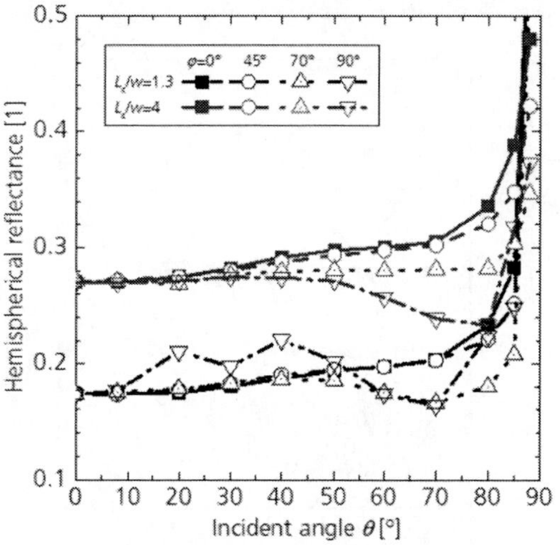

Figure 10: Simulated reflectance R for $\lambda = 600$ nm of the tub geometry including a variation of azimuth φ and characteristic length L_c for fixed characteristic angle $\omega_c = 88°$.

whole range of φ (from 0 to 90°) and calculate an averaged reflectance.

Table 1: Hemispherical reflectance and fitting ratio L_c/w and characteristic angle ω_c for the three isotextured samples for the tub geometry.

Sample	R(600nm, θ = 8°) [%]	L_c/w [-]	ω_c [°]
ISE I	24.9	3.2	88
ISE II	22.1	2.2	88
ANU	29.2	4.7	88

The results depicted in Figure 11 show a good agreement for the planar sample for all angles, as well as for the isotextured samples for incident angles of up to 60°: in this region the difference between simulated and measured results is always below 1.2%$_{abs}$. From this point the behavior of the simulated reflectance R_{sim} differs gradually from the measured reflectance R_m. To see more clearly the difference between the two approaches (spherical caps and tub form) we plot in Figure 11 again the reflectance obtained with the spherical caps model for the sample ISE II.

5 DISCUSSION

The angle-dependence is not well described by the spherical caps model, resulting in an overestimation of the reflectance (see Figure 7) of up to 8%$_{abs}$ for incident angles around 60°. For isotextured samples with poor optical properties, which exhibit a reflectance of almost 30% for an incident angle of 8°, the discrepancies between measurements and simulations with the spherical caps model are smaller (below 2%$_{abs}$). Nevertheless, it is in industry's interest to produce solar cells with good optical properties, as the samples ISE I and ISE II with only 25% and 22% reflectance for θ = 8° respectively. The same argument applies to the quadratic grid approach. This presents similar results as the spherical caps approach and consequently is also inaccurate.

In the range $0 < \theta < 60°$, the tub form approach describes quite accurately the angle-dependent

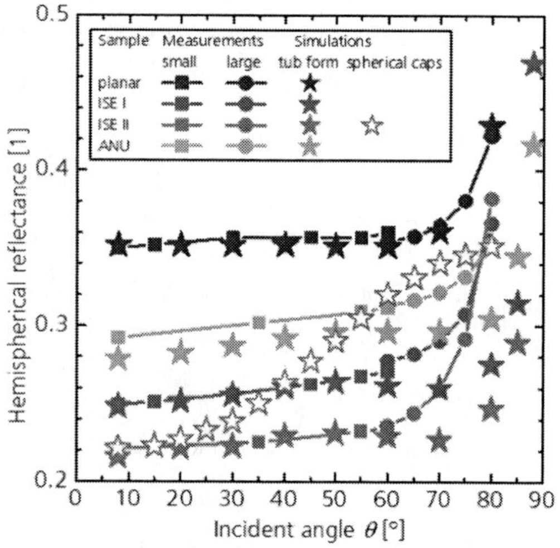

Figure 11: Simulated (stars) and measured (lines) reflectance R for $\lambda = 600$ nm for one planar and three isotextured silicon samples.

reflectance of isotextured silicon (see Figure 11). The results show discrepancies much smaller than by using the spherical caps approach. Furthermore, solar cells are not installed alone, but they are encapsulated with ethylene vinyl acetate (EVA) and glass. Due to the refractive index of these two layers, the incident angle is reduced; reaching rarely values over 50° on the solar cell. For this important range, the tub approach represents an improvement compared to the spherical caps approach. Nevertheless, the complete range of incident angles up to 90° is important for light trapping within the module. Further research is required to improve the optical models in the range of incident angles between 60° and 90°.

6 SUMMARY AND CONCLUSION

We have presented an optical analysis of isotextured silicon, studying the angle-dependent reflectance. Our work has been developed in three parts. First, we have compared our measurements of nine different isotextured samples with published data [3]. We revealed that the published data are partly inaccurate and we laid the foundations for a thorough analysis of isotextured silicon by generating reliable reflectance data.

Second, we have performed simulations by using the Module Ray Tracer by PV Lighthouse, where the spherical caps model was implemented. Comparing the simulated results with the measurements, we have demonstrated that this model is not able to describe the measured reflectance when the incident light is not perpendicular to the cell, reaching differences between simulated reflectance R_{sim} and measured reflectance R_{meas} above 8%$_{abs}$ for well-textured samples.

Third, we have analysed two more morphologies of the simulation model. We have shown a superior match between measurements and simulations using the tub morphology for incident angles from 0 to 60°. The difference R_{sim}-R_{meas} is reduced to 1.2%$_{abs}$ for incident angles θ from 0 to 60°. We conclude that the new tub form approach is promising and that further research is required for further improvements.

ACKNOWLEDGEMENTS

This work was partly funded by the German Federal Ministry for Economic Affairs and Energy within the research project "BiZePS" (contract number 0325909).

The authors would like to thank their colleagues at ISE for their work, Simeon C. Baker-Finch and Keith R. McIntosh for the discussion and the SCHMID group for providing a textured sample with DW PreTex treatment.

Aina Alapont Sabater gratefully acknowledges the support by scholarship funds from the Deutsche Bundesstiftung Umwelt.

REFERENCES

[1] Sarti, D., Le, Q. N., Bastide, S., Goaer, G., & Ferry, D., "Thin industrial multicrystalline solar cells and improved optical absorption," *Proceedings of the 13th European PV Solar Energy Conference*, pp. 25–28, 1995.

[2] "International Technology Roadmap for Photovoltaic (ITRPV): 2015 Results including maturity reports," Seventh Edition, Oct. 2016.

[3] S. C. Baker-Finch, K. R. McIntosh, and M. L. Terry, "Isotextured Silicon Solar Cell Analysis and Modeling 1: Optics," *IEEE J. Photovoltaics*, vol. 2, no. 4, pp. 457–464, 2012.

[4] J. Greulich *et al.*, "Optical Simulation and Analysis of Iso-textured Silicon Solar Cells and Modules Including Light Trapping," *Energy Procedia*, vol. 77, pp. 69–74, 2015.

[5] K. R. McIntosh, T. G. Allen, S. C. Baker-Finch, and M. D. Abbott, "Light Trapping in Isotextured Silicon Wafers," *IEEE J. Photovoltaics*, vol. 7, no. 1, pp. 110–117, 2017.

[6] K. R. McIntosh, M. D. Abbott, and B. A. Sudbury, "Ray Tracing Isotextured Solar Cells," *Energy Procedia*, vol. 92, pp. 122–129, 2016.

[7] SCHMID Group | Gebr. SCHMID GmbH, *SCHMID develops new process for the texturing of diamond-wire-cut multi-wafers: Press release, Feb 2, 2017.*

[8] *PV Lighthouse: Module Ray Tracer.* [Online] Available: https://www.pvlighthouse.com.au. Accessed on: Jun. 07 2017.

[9] M. A. Green, "Self-consistent optical parameters of intrinsic silicon at 300K including temperature coefficients," *Solar Energy Materials and Solar Cells*, vol. 92, no. 11, pp. 1305–1310, 2008.

[10] Synopsys, "Sentaurus™ Device User Guide: vI-2013.12,"

BENCHMARKING MECHANICAL STRENGTH DATA FOR NEW SOLAR CELL CONCEPTS

Felix Kaule[1,*], Sylke Meyer[1] and Stephan Schoenfelder[1,2]
[1]Fraunhofer Center for Silicon Photovoltaics CSP, Otto-Eissfeldt-Strasse 12, 06120 Halle (Saale)
[2]Leipzig University of Applied Science, Karl-Liebknecht-Str. 134, 04277 Leipzig
*corresponding author: Felix.Kaule@csp.fraunhofer.de

ABSTRACT: Unlike the efficiency of the solar cell, the mechanical strength is not a key parameter for solar cell and module manufacture, yet. Though, it is known that the metallization reduces the mechanical strength after texturization. But no systematic comparison between different cell types was performed so far in literature. In this work, a mechanical benchmark test for 10 different solar cells from 6 different manufacturers with 1525 cells is presented. Significant differences in the probability of failure were found depending on the individual cell process. The presented test procedures are powerful tools to identify the most damaging process steps and to increase the yield during cell and module manufacturing.

Keywords: solar cell, mechanical strength, 4-point bending test, characterization, metallization

1 INTRODUCTION

The mechanical strength of a material is defined as its ability to withstand an applied load without failure. For silicon as brittle material, a single defect is responsible for the breakage of the solar cell. Hence, breakage occurs if the load is too high or a defect is too large. During the manufacturing of solar cells, the strength varies from the as-cut wafer to the final solar cell [1]. The strength of the as-cut wafer depends on the sawing type and sawing parameter and is especially low for fixed diamond wire sawn wafer. After texturization the strength increases because the upper surface is removed and so the strength limiting cracks are altered or removed. After the texturization step the strength decreases due to the metallization [2].

Although there are some investigations about the influence of metallization for Al-BSF solar cells [3] [4], there is still no comparison between different cell types. Furthermore, it is not possible to compare results from different publications, because there is not standard which defines the determination of fracture strength. Especially, for solar cells with metallization it is important to use the correct silicon thickness and consider nonlinearities in the bending test [5].

This work determines the influence of the metallization of different solar cell types. Therefore, a benchmark test is presented using fracture tests with a 4-point bending setup in all possible loading configurations. The benchmark test displays significant differences in the mechanical strength for different solar cell batches and may be used in future to identify critical defects.

2 MATERIAL AND METHODS

2.1 Material

For this work 10 different solar cell types from 6 different manufacturers with two different cell types (Al-BSF and PERC) where used (see Table 1). There were four batches with multi-crystalline Al-BSF silicon cells and one batch with mono-crystalline Al-BSF silicon cells. For the PERC type, there was one batch with multi-crystalline silicon cells and four batches with mono-crystalline silicon cells, while one of this four were bifacial cells. Overall, 1525 solar cells were broken in the 4-point bending setup described below.

Table 1: overview of all used solar cells types

Batch No.	Cell Type	Si Type	Number of Cells
1	Al-BSF	multi	120
2	Al-BSF	multi	113
3	Al-BSF	multi	140
4	Al-BSF	multi	200
5	Al-BSF	mono	120
6	PERC	multi	200
7	PERC	mono	200
8	PERC	mono	180
9	PERC	mono	120
10	PERC (bifacial)	mono	152

2.2 Methods

For fracture tests the 4-point bending setup is used [6], because it loads a large area homogenously by uniaxial bending moments including the surface and edge defects of the solar cell. In this test, the load occurs as homogeneous tensile stress on the bottom surface of the solar cell. The 4-point bending configurations had an outer span (outer rollers) of 80 mm and inner load span (inner rollers) of 40 mm. The rollers are made of steel with a diameter of 10 mm. For better contact behavior and reduced friction PTFE foils were put between the rollers and the cells. The cells are loaded in four different configurations with variation of sunny and back side as well as a variation of the busbars cross and parallel to the rollers of the 4-point bending (see Figure 1).

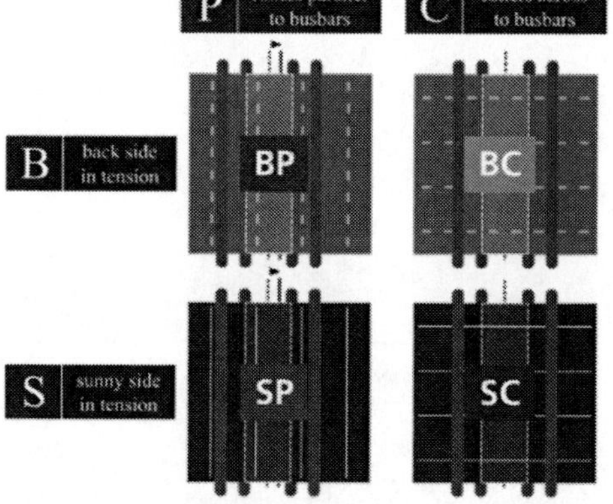

Figure 1: Overview of all four loading configurations for each solar cell type in the 4-point bending setup.

All tests were performed on a universal testing machine ZWICK 005, using a load cell of 500 N. The deflection of the sample was measured by the machine position. The load speed was defined to be 0.2 mm/s. During testing the force and deflection was recorded for each sample, beginning after reaching an initial force of 0.5 N and ending at the fracture of the wafer.

A Finite Element model was used to calculate the fracture stresses in the 4-point bending experiment for consideration of nonlinearities during the bending test [7]. Furthermore, the cell models are reduced to the silicon layer because the metallization does not have a major influence on the stiffness or bending stresses in the 4-point bending for silicon [5]. Though, the metallization governs the defect structure within the silicon layer. Hence, the facture stress in the silicon correlates to defects due to cell metallization. The fracture stresses σ_i were statistically evaluated using a two parameter Weibull distribution [8] [9]

$$P = 1 - \left[\exp\left(-\frac{\sigma_i}{\sigma_\theta}\right)^m\right] \qquad \text{(Eq. 1)}$$

where σ_Θ is the characteristic fracture stress at which 63.2 % of all samples fails and m is the Weibull modulus which represents the scattering of the fracture stresses. For each test, a certain effective area A_{eff} is calculated from the Finite Element model. A high Weibull modulus represents a small variation of fracture stresses and a low Weibull modulus represents large variation in fracture stress values. The Weibull parameters were calculated using the Maximum-Likelihood (MLE) estimation according to [10] or [11]. From the effective area A_{eff}, which is unique for each test setup, the size independent scale parameter σ_0 can be determined from the equation:

$$\sigma_0 = \sigma_\theta A_{eff}^{1/m} \qquad \text{(Eq. 2)}$$

This equation can also be used to calculate a certain characteristic fracture stress for a given effective area. For evaluation of all cells the effective area was set to A_{eff} = 9116mm², which was also used in previous publications for better comparison [2]. For a prediction of breakage rate a random load σ can be applied simultaneously on all configurations (BP, BC, SP, SC). Using the principle of independent action the probability of failure P

$$P = 1 - Exp\left[-A_{eff}\sum_{i=1}^{4}\left(\frac{\sigma}{\sigma_{0,i}}\right)^{m_i}\right] \qquad \text{(Eq. 3)}$$

with

i = BP, BC, SP, SC

can be calculated for this random load σ and effective area A_{eff}.

3 RESULTS

The Weibull parameter for all cells are shown in Figure 2 and listed in Table 2 for A_{eff}= 9116 mm². The fracture stresses for the Al-BSF and PERC solar cells are shown in Figure 3 and Figure 4. The characteristic fracture stress σ_θ is marked with a diamond symbol and for each solar cell types all broken cells for the four different loading configurations are plotted. Additionally, the predicted breakage rate is shown in grey bars with the numbers on the bottom side of the plot.

Table 2: Weibull parameter for all cells in the four configurations for A_{eff} = 9116 mm²

Batch No.	Characteristic Fracture Stress [MPa]				Weibull Modulus [-]			
	BP	BC	SP	SC	BP	BC	SP	SC
1	162	166	215	195	17.4	20.3	9.7	6.9
2	139	159	119	123	7.9	9.1	5.7	5.4
3	158	156	176	179	19.0	18.1	10.5	11.9
4	145	149	142	126	8.6	12.9	6.2	4.7
5	187	138	198	200	22.5	38.6	10.2	14.8
6	163	164	183	185	6.0	9.3	6.6	10.2
7	157	137	223	205	9.7	12.1	11.5	10.1
8	169	158	180	181	7.8	17.9	14.3	11.7
9	178	170	158	155	5.4	4.0	5.1	4.3
10	145	158	183	166	12.9	5.3	8.5	7.2

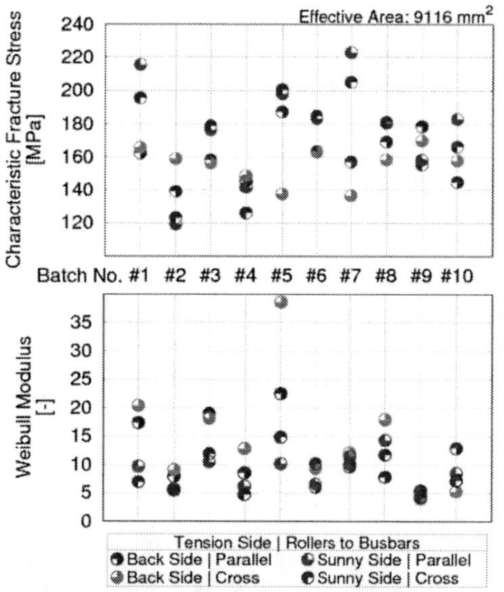

Figure 2: Weibull parameter for all cells in the four configurations for A_{eff}=9116 mm². The symbol color was chosen according to the scheme given in Figure 1.

For the Al-BSF cell batches (#1 - #5) a very differing strength behavior was observed. The back side displayed a lower scattering than the sunny side, represented by higher Weibull modulus in Figure 2. Though, no general behavior could be observed. For the batches #1, #3 and #5 the back side is weaker than the sunny side and for #2 and #4 the sunny side is weaker. Despite some exceptions the cross and parallel loading for each side (sunny/back) are in the same range, so the defects seem to be mostly isotropically distributed for each side. The predicted breakage rate for the random load of 110 MPa varies strongly. There are very high breakage rates for cells #2 and #4 and very low for the cells #1, #3 and 5. It should be noticed, that the batches with the highest scattering have the highest breakage rates.

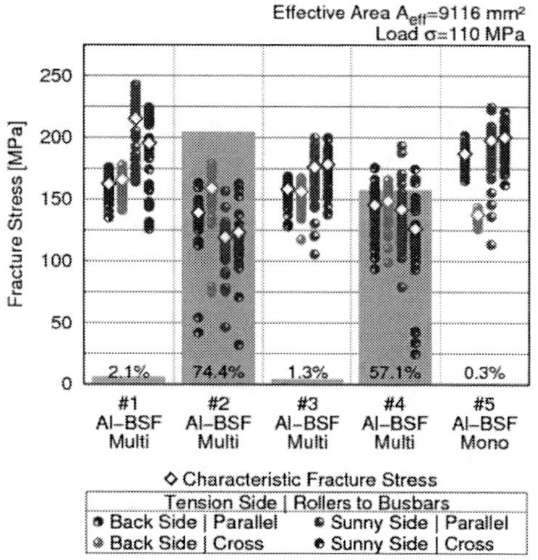

Figure 3: Fracture stresses for all Al-BSF cells in all four configurations with probability of failure for simultaneous load of σ = 110 MPa on all configurations. Circles represent single data points and the characteristic fracture stress σ_θ is marked with a diamond symbol.

For PERC cells (batches #6 - #10) also a very differing strength behavior was observed. For the batches #6, #7, #8 and #10 the back side has a lower characteristic fracture stress compared to the sunny side. For batch #9 the characteristic fracture stress is higher. Though, for the back side the scattering is also higher caused by many cells with lower fracture stresses on the back side compared to the sunny side. Like for the Al-BSF cells, with some exceptions the cross and parallel loading for each side (sunny/back) were in the same range. Thus, the defects seem to be mostly isotropically distributed for each side.

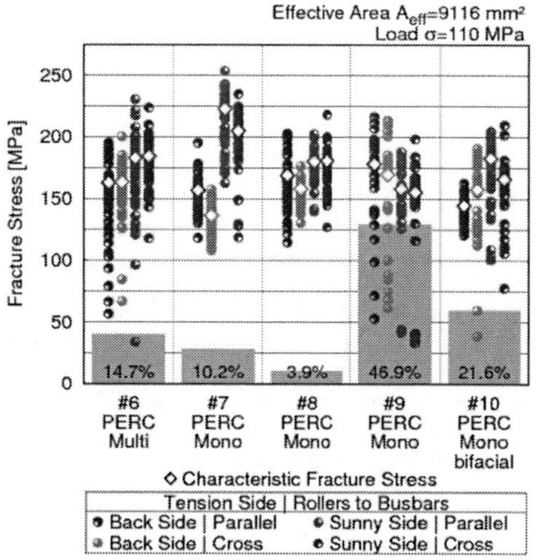

Figure 4: Fracture stresses for all PERC cells in all four configurations with probability of failure for simultaneous load of σ=110 MPa on all configurations. Symbols meaning as in Figure 3.

No major difference between the two cell architectures Al-BSF (#1 - #5) and PERC (#6 - #10) could be observed. There were only trends with minor exceptions. For example, Al-BSF cells showed mostly lower scattering on the back sides compared to the PERC cells. The predicted breakage rate of Al-BSF cells strongly varies between very high or very low values while the predicted breakage rates for PERC cells are more similar.

4 DISCUSSION

The cell type (Al-BSF/PERC) itself does not show a characteristic influence on the mechanical strength. Though, it is known that the metallization governs the strength of solar cells [5]. Hence, the composition of the metallization (pastes) and process parameter like firing temperatures seems to strongly influence the strength because of the very different results here. However, a clear correlation of metallization and strength is difficult in this work because there was no detailed process information available.

It was observed that the sunny side has higher scattering of fracture stresses compared to the back side especially for Al-BSF cells. This could be caused by different defect distribution, (i) sawing defects remained after texturing, (ii) local defects in the busbar regions. Though, also handling problems may cause single larger defects, leading to higher scattering.

The predicted breakage rate with the loading on all configurations presented here is artificially imposed and represents a theoretical worst-case scenario because all defects are stressed simultaneously. Nevertheless, it is a good way to predict the breakage rate because it is taken the hole strength behavior of the cell type into account.

The bow of the solar cells due to the CTE mismatch of the metallization and the silicon is not considered. Thus, the intrinsic stress due to the bow of the cell is the starting situation of the 4-point bending test. This means for the back side in tension all stress applied until the cell is flat should be subtracted from the final fracture stress. For the sunny side, this stress needs to be added, because the cell is already bent. This means, the strength on the back side could be overestimated and vice versa, the strength on the sunny side would be underestimated. Though, it is assumed here that the intrinsic stress can be neglected for strength analysis.

In order to evaluate the mechanical strength for manufacturers of cells and modules there is still no standard loading which represents real situation in the manufacturing. If this would be available one could define certain failure criteria and limits for fracture stresses of solar cells, i.e., breakage rates can be predicted and used as parameter for process optimization.

5 CONCLUSION

A mechanical benchmark test for different types and batches of solar cells was presented. In this work, 10 different batches from 6 different manufacturers were tested in terms of mechanical strength. There were no characteristic differences observed between Al-BSF and PERC cell types. Though, there were some trends detectable regarding fracture stresses and scattering of sunny sides and back sides between Al-BSF and PERC. However, strong differences in the predicted breakage rates were found depending on the individual cell process. This means that for each solar cell process the benchmark test must be performed because the influence is not only limited to the cell type but very likely more to process parameter and material composition of the metallization.

6 REFERENCES

[1] R. Koepge, F. Weget, S. Thormann and S. Schoenfelder, "Characterization of damage and mechanical strength of wafers and cells during the cell manufacturing process," Photovoltaics International , no. Volume 22, pp. 55-62, 2013.

[2] F. Kaule, M. Oswald, R. Koepge, C. Klute and S. Schoenfelder, "Comprehensive analysis of strength and reliability of silicon wafers and solar cells regarding their manufacturing processes," Photovoltaics International (29th), pp. 22-29, 2015.

[3] C. Kohn, R. Kuebler, M. Krappitz, G. Kleer, I. Reis, M. Retzlaff, D. Erath and D. Biro, "Influence of the metallization process on the strength of silicon solar cells," in 24th European Photovoltaic Solar Energy Conference, 2009.

[4] V. Popovich, M. Janssen and I. Bennett, "Breakage issues in silicon solar wafers and cells,," Photovoltaics International, 2011.

[5] F. Kaule, W. Wang and S. Schoenfelder, "Modeling and testing the mechanical strength of solar cells," Solar Energy Materials and Solar Cells, 2014.

[6] DIN SPEC 91351: Strength testing for photovoltaic wafers, 2017-04.

[7] S. Schoenfelder, J. Bagdahn, M. Petzold and J. Burghartz, "Mechanical Characterisation and Modelling of Thin Chips," in Ultra-thin Chip Technology and Applications, New York, Springer , 2011, pp. 195-218.

[8] W. Weibull, A Statistical Theory of the Strength of Materials, Stockholm: Generalstabens Litografiska Anstalts Färlag, 1939.

[9] W. Weibull, "A statistical distribution function of wide applicability," Journal of Applied Mechanics, pp. 293-297, 1951.

[10] DIN EN 843-5:2007-03: Hochleistungskeramik - Mechanische Eigenschaften monolithischer Keramik bei Raumtemperatur - Teil 5: Statistische Auswertung, 2007.

[11] ASTM C 1239-07: Reporting Uniaxial Strength Data and Estimating Weibull Distribution Parameters for Advanced Ceramics, 2007.

ACCURACY AND SIGNIFICANCE OF THE PROJECTIONS IN THE INTERNATIONAL TECHNOLOGY ROADMAP FOR PHOTOVOLTAIC (*ITRPV*)

P. Baliozian[1], J. Trube[2], M. Fischer[3], S. Al-Hajjawi[1], R. Yadav[4], S. Nold[1] and R. Preu[1]
[1]Fraunhofer Institute for Solar Energy Systems (ISE), Heidenhofstr. 2, 79110 Freiburg, Germany
[2]VDMA Photovoltaic Equipment, Lyoner Str. 18, 60528 Frankfurt, Germany
[3]Hanwha Q Cells GmbH, Sonnenallee 17-21, 06766 Bitterfeld-Wolfen OT Thalheim, Germany
[4]Albert-Ludwigs-Universität Freiburg, Zentrum für Erneuerbare Energien (ZEE), Tennenbacher Str. 4, 79106 Freiburg, Germany
Contact email: puzant.baliozian@ise.fraunhofer.de

ABSTRACT: With the advancement of photovoltaics (*PV*) in the energy markets and the rapid technological developments, the existence of planned PV technology roadmaps becomes beneficial. The International Technology Roadmap for Photovoltaic (*ITRPV*) has proven itself to be a leading PV technology roadmap in the crystalline silicon (*c-Si*) field. The yearly reports and the widening scope of its structured projections cover essential factors in the value chain of PV production. The main aim of having a summative accuracy study of the *ITRPV* is to assist research and development scientists, engineers and investors in taking attentive decisions. The importance of the *ITRPV* increases with the reliability and accuracy of its projections. In this work, the accuracy analysis of the *ITRPV* is performed by choosing topics and gathering the data from the yearly *ITRPV* reports. As projection accuracy measures, projection absolute deviations (*PADs*) and projection absolute percentage deviations (*PAPDs*) are considered paving the way for mean projection absolute deviation (*MPAD*) and mean projection absolute percentage deviation (*MPAPD*) deductions. In our case the *8th edition* report projecting values for six topics are taken as reference values while 2016 projected values from *2nd* till *8th editions* are taken for the *PAD* study. Using our approach, different graphical representations show the projection absolute deviations and projection absolute percentage deviations from the chosen reference values. For instance, considered multicrystalline silicon wafer thickness projections have shown an *MPAD* of 26,7 µm and an *MPAPD* of 14,8%. Considering the six projection topics, *MPAPD* as a function of projection time interval is deduced. A time dependent inaccuracy is then used to plot expected ranges for future projection values of each studied topic. From a cost perspective, for the silver amount per cell, an *MPAD* of 13 mg/cell could correspond to an inaccuracy of around 0,078 €/cell for a given silver price. *ITRPV* projection accuracy leads to enhancing cost-effective decision making.
Keywords: Photovoltaic, projection accuracy, roadmap, *ITRPV*.

1 INTRODUCTION

Roadmapping is a constructive innovation management tool supporting the development of the field it serves. The objective of a photovoltaic (PV) technology roadmap is to support in projecting the future milestones of the technology's development. The International Technology Roadmap for Photovoltaic (*ITRPV*) provides the photovoltaics community with yearly reports projecting the development of numerous crucial value chain items (materials, process parameters, and products). *ITRPV* was first introduced in 2010 by *SEMI PV group* and has been evolving ever since [1]. The latest version of the *ITRPV 8th edition, 2017)* includes data from 40 different companies and research institutes, and is processed anonymously by *VDMA (German Mechanical Engineering Industry Association)* [2]. Median values of the collected data from experts are plotted. Each contributing company or institution vote counts as one no matter how many contributors from the company are participating in the report. The *ITRPV* is by far the main PV roadmap among existing ones to focus almost entirely on technology topic projections for crystalline silicon (*c-Si*) PV [3].

According to Garcia *et al.* [4], a technology roadmapping process includes three essential phases, briefly stated as:

I. Preliminary phase of defining the leadership, scope, and boundaries of the roadmap,
II. Creating the technology roadmap report,
III. Follow-up activity including the critique and validation of the roadmap.

The *ITRPV* team has succeeded to initiate, create and review the reports yearly. The approach in this paper is to further investigate the accuracy of the reports, by using statistical methods common in forecasting/projection accuracy studies. This quantitative approach reinforces the third essential phase of roadmapping, aiming to enhance the predictions and further criticize and validate specific projections stated in the different reports.

The following section (section 2) of the paper describes the method used to obtain a quantitative measure, mainly mean projection absolute deviation and mean projection absolute percentage deviation. Section 3 shows the results obtained by adopting the method. The method was performed on six topics included in the *ITRPV* projection reports from the *2nd* till the *8th editions*. Studied topics are: multicrystalline (mc-Si) wafer thickness, silver amount per cell, emitter sheet resistance, mc-Si ingot mass, bulk recombination current density and finger width. Finally, section 4 concludes the paper and gives insight to the possibility of future studies joining cost models to the accuracy of *ITRPV* projection topics.

2 METHOD OF PROJECTION DATA ANALYSIS

The main aim of our data analysis is to create a validity (accuracy) measure for *ITRPV* by using known metrics. It is important to differentiate between the terms accuracy and reliability for purposes of clarity. Accuracy is defined as the closeness of a measurement to the 'true' value (reference value), while reliability means consistency of the results [5]. Studying the accuracy of projections, scale dependent and independent metrics could be used. To start with, the projection deviation (E) is the difference between the projected year value (P) and the reference value (Y) shown in equation 1.

$$E = P - Y \qquad (1)$$

The projection absolute deviation (*PAD*) is of interest for our purpose, which is merely the absolute value of E. Figure 1 shows an example of the difference in projections in silver amount per cell between the 5th and 7th *editions*, from projected year 2015 to 2021 (values are adapted from the two reports).

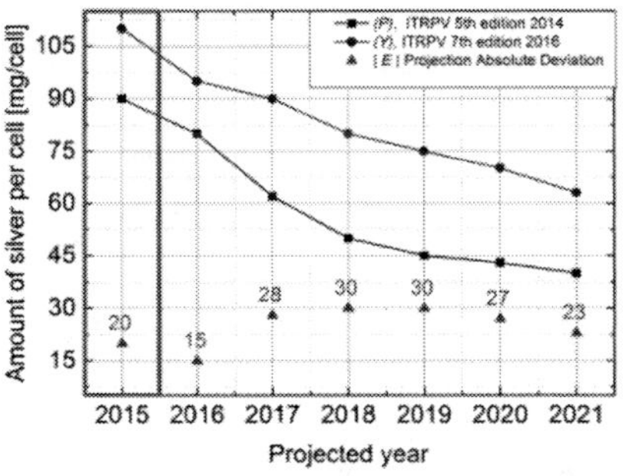

Figure 1: Amount (mass) of silver per cell from 2015 until 2021 projected in *ITRPV* 5th *edition* and 7th *edition* (point connecting line acts as a guide-to-the-eye). Some projection data points in the graph are deduced by means of linear interpolation of two report data points. (Projected values taken from Ref. [2])

To calculate the scale dependent accuracy, the mean projection absolute deviation (*MPAD*) is used as a metric (shown in equation 2) [6-8], *PAD* being the projection absolute deviation and N being the number of projections that are studied.

$$MPAD = (\sum PAD)/N \qquad (2)$$

The scale dependent accuracy metrics are understood by PV *R&D* groups, however to make it reader-friendly for other entities such as investors the scale independent metrics (unitless) are adopted. Another reason for adopting a percentage based measure is the ability to directly compare topics having different units. For instance wafer thickness in µm and silver amount in mg/cell projection deviations could be compared when a scale independent measure is chosen. For that purpose, widely known projection accuracy metric, the projection absolute percentage deviation (*PAPD*) is used (shown in equation 3). The percentage deviation reference value is taken to be the 2016 projection value of the latest released *ITRPV* 8th *edition* 2017 report.

$$PAPD = \frac{PAD}{Y} * 100 \qquad (3)$$

The mean absolute percentage deviation (*MPAPD*) is the summation of *PAPDs* divided by the number of projections *(N)* taken into account (shown in equation 4).

$$MPAPD = (\sum PAPD)/N \qquad (4)$$

In this method, data is extracted from the reports themselves for the purpose of approaching the data from a reader's point of view. An intuitive data acquisition method for each topic is created where a raster includes all the projections from the past projecting the future values. $P_{tr\text{-}tp}$ is taken to be a projected value, where *tr* indicates the report edition and *tp* indicates the value's projected year. The raster itself is named by the topic. For instance, wafer thickness $P_{3rd\text{-}2017}$ is the projection of *ITRPV* 3rd *edition* to the wafer thickness in 2017. To put this in an example, the reference value can be wafer thickness $P_{8th\text{-}2016}$ and the projected value wafer thickness $P_{5th\text{-}2016}$. It is important to note that the reference value and the projected value should be for the same specific topic and the same projected year. Consequently, $E_{5th\text{-}8th\text{-}2016}$ would be the projection deviation between the 5th *edition* report and the 8th *edition* for the year of projection 2016. $P_{8th\text{-}2016}$ value is taken as the reference for the different chosen topics. The ideal reference value would be the actual market value of the topics taken from all PV manufacturers (difficulty of attaining data from all market stakeholders). Figure 2 shows the general steps taken in this accuracy study approach.

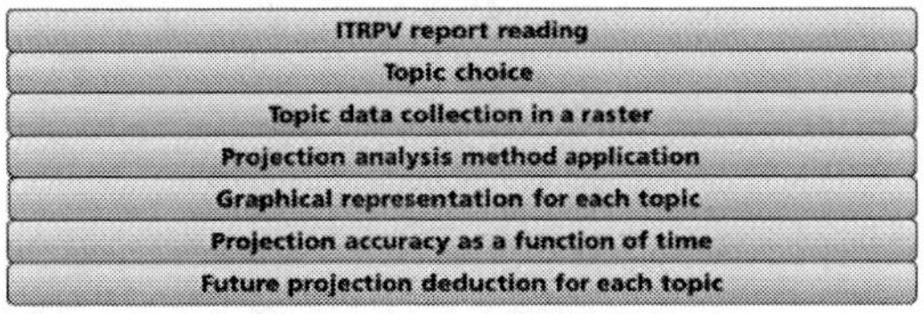

Figure 2: General flowchart of the *ITRPV* accuracy study method.

3 RESULTS

The explained method was adopted for six previously mentioned *ITRPV* projection topics. The main reason of topic choice is availability of projections in the *ITRPV* reports from the 2nd *edition* until the 8th *edition*. The *ITRPV* dynamically includes new topics in every published edition and removes/edits others. Figure 3 shows the development of the *ITRPV* in terms of number of contributing companies and plotted technology roadmap results.

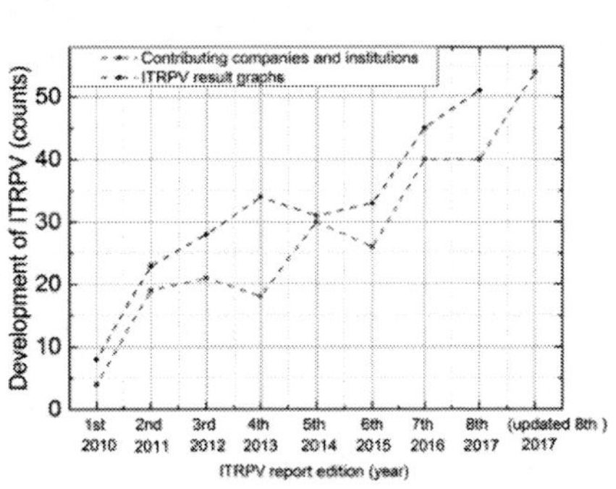

Figure 3: Development of *ITRPV* number of contributing companies and institutions, as well as roadmap technology topic result graphs.

With the advancement of editions, the pool of chosen topics is expected to be larger. The inclusion of new technologies among the already existing topics increases the complexity of the choice for the accuracy study.

However, the widening of the topic scope within the *ITRPV* and the addition of new technologies within the topics is inevitable and of vital importance for bringing up new notions. Yearly more accurate estimations require the update of the reference value (*Y*) with the advancement of the publication years. The projection accuracy study results of the studied topics are as follows:

3.1 mc-Si wafer thickness

The projection absolute deviations *PADs* are plotted by taking $P_{8th\text{-}2016}$ = 180 µm as the reference value. The mean projection absolute deviation (*MPAD*) is calculated to be 26,7 µm. The *PAD* decreases the closer it gets to the projected year. *PAPD* is plotted for the purpose of having a scale independent result (in percentage). *MPAPD* is calculated to be 14,8%. *PAD* and *PAPD* plots for mc-Si wafer thickness are plotted in Figure 4. Such an approach is taken for the rest of the chosen topics.

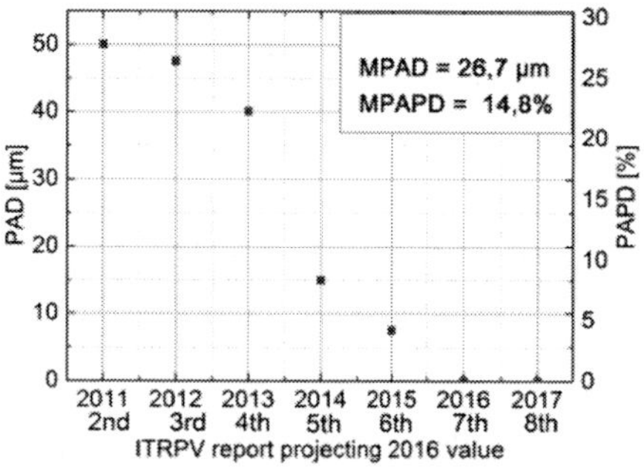

Figure 4: The mc-Si wafer thickness value projection absolute deviations with data from *ITRPV* (*2nd to 8th editions*) reports. The reference value is taken to be $P_{8th\text{-}2016}$ = 180 µm.

3.2 Silver amount per cell

For the topic of silver, the same approach has shown an *MPAD* of 13 mg/cell and an *MPADP* of 13,3%. (see Figure 5). Projection inaccuracy of silver amount can be directly linked to an inaccuracy in the cost per cell. Taking a given silver price of 600 €/kg, the *MPAD* could correspond to an inaccuracy of 0,078 €/cell.

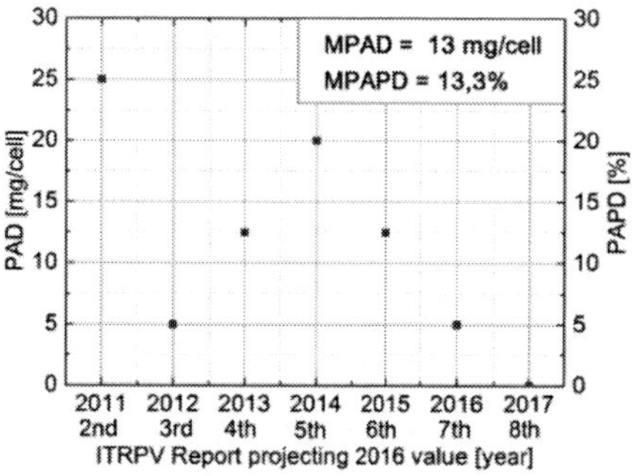

Figure 5: The silver amount per cell projection absolute deviations and projection absolute percentage deviations with data from *ITRPV* (*2nd to 8th editions*) report. The reference value is taken to be $P_{8th\text{-}2016}$ = 10 mg/cell.

3.3 Emitter sheet resistance

The reference value taken for the projection accuracy study of the emitter sheet resistance is $P_{8th\text{-}2016}$ = 96 Ohm/sq. The calculated *MPAD* is 13,3 Ohm/sq whereas the *MAPD* is 13,8% (see Figure 6).

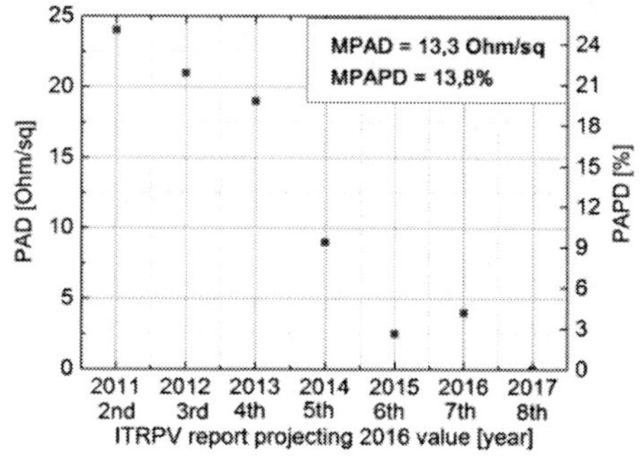

Figure 6: The emitter sheet resistance *PADs* and *PAPDs* with data from *ITRPV* (*2nd to 8th editions*) reports. The reference value is taken to be $P_{8th\text{-}2016}$ = 96 Ohm/sq.

3.4 mc-Si ingot mass

The method application on mc-Si ingot mass projections has shown an *MPAD* of 94 kg and an *MPAPD* of 13% (see Figure 7).

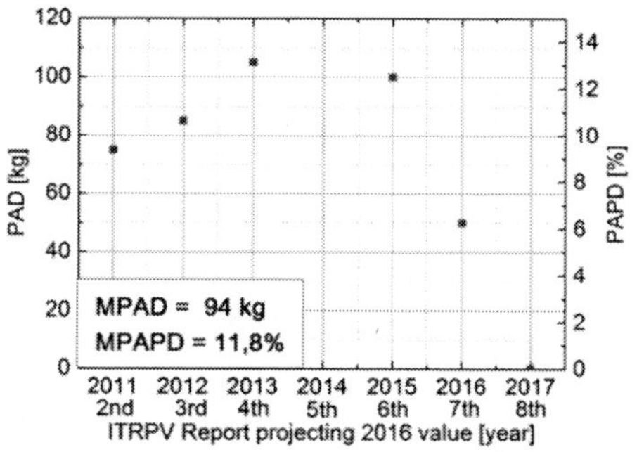

Figure 7: The mc-Si ingot mass *PADs* and *PAPDs* with data from *ITRPV* (*2nd to 8th editions*) report. The reference value is taken to be $P_{8th\text{-}2016}$ = 800 kg.

3.5 Bulk recombination current

The method was adopted for p-type mc-Si bulk recombination current density projections. The projections were taken from the *2nd to 8th ITRPV editions*. *MPAD* is shown to be 33,3 fA/cm² while *MPAPD* is 13,8% (see Figure 8).

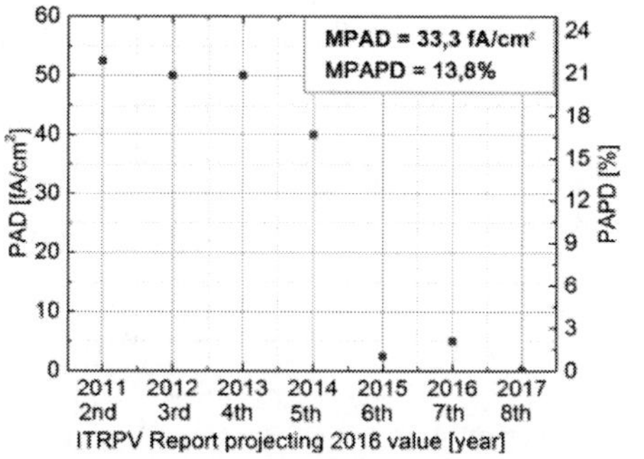

Figure 8: The bulk recombination current density for p-type mc-Si *PADs* and *PAPDs* with data from *ITRPV* (2nd to 8th *editions*) report. The reference value is taken to be $P_{8th\text{-}2016}$.= 240 fA/cm².

3.6 Finger width

The method adoption for the finger width projections showed an *MPAD* of 1,75 µm and *MPAPD* of 3,6% (see Figures 9).

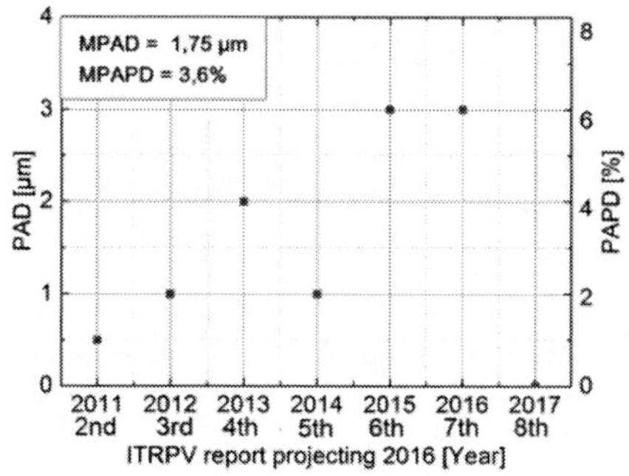

Figure 9: The finger width *PADs* and *PAPDs* with data from *ITRPV* (2nd to 8th *editions*) report. The reference is taken to be $P_{8th\text{-}2016}$ = 48 µm.

3.7 Summary of studied topics

MPAD and *MPAPD* of the studied topics are listed in Table I. Figure 10 shows a graphical summary of the projection absolute percentage deviations deduced for the six studied topics. The graph demonstrates that there is no specific trend for the deviations of each report from the reference value. For example, mc-Si wafer thickness and emitter sheet resistance topics have shown a decrease in *PAPD* the closer the projection gets to the reference year (8th *edition*). These results are different from the *PAPDs* of the topics finger width and silver amount per cell, where silver amount prediction of the 3rd *edition* has shown a relatively better accuracy than the 4th and 5th *edition* projections. Such variations in projection values might be due to the change of contributors in the different *ITRPV* reports. The dependence of projection topics on ongoing changes in the developments of the technologies and the dynamic market might be possible explanations.

Table I: *MPAD* and the *MPAPD* of the chosen topics taking *ITRPV* 8th *edition* projection of 2016 as a reference value.

Chosen topics from 2nd to 8th *editions*	MPAD	MPAPD (%)
mc-Si wafer thickness	26,7 µm	14,8
Finger width	1,75 µm	3,7
Emitter sheet resistance	13,3 Ohm/sq	13,8
Silver amount per cell	13 mg/cell	13,3
mc-Si ingot mass	94 kg	13
Bulk recombination current density	33,3 fA/cm²	13,8

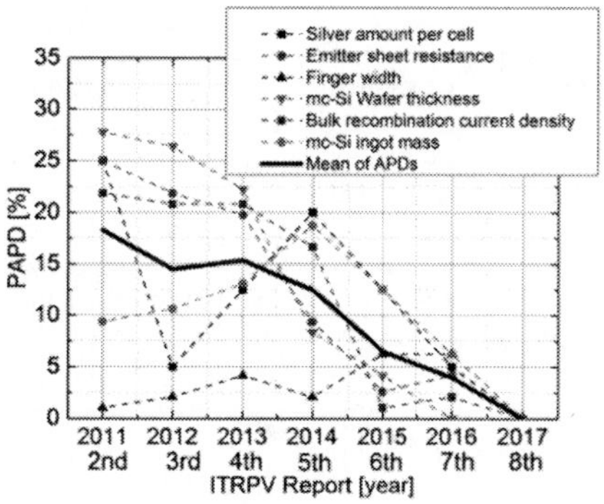

Figure 10: Summary of the projection absolute percentage deviation of projections for the six chosen topics, taking the *ITRPV* 8th *edition* projection of 2016 values as references ($P_{8th\text{-}2016}$).

3.8 Time dependent projection accuracy

From the results of *PAPDs* of the six topics, the yearly mean projection absolute percentage deviation depending on projection time interval is plotted (see Figure 11). The 4-year projection time interval *PAPD* value is higher than the 5-year time interval due to the previously mentioned silver amount projection of the 3rd edition. A linear fit is done to obtain the general deduction for the time interval dependency of the projection accuracy. The result of the fit is then used to define future projection accuracy limits; see Figure 11.

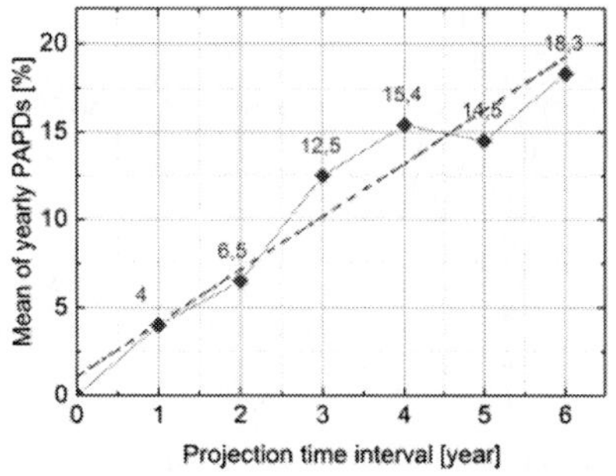

Figure 11: Mean projection absolute percentage deviation of projections of the six chosen topics depending on projection time interval. $P_{8th\text{-}2016}$ is taken as a reference for the *PAPD* calculations. As a result of the fit the slope (a) = 3,03 and the y-intercept (b) = 1,08.

3.9 Future projections within accuracy range

Figures 12-17 show the *8th edition* projections of the studied six topics. An upper and lower accuracy limit is defined by using the time dependent projection accuracy shown in section 3.8. In other words, the *PAPD* of the time interval is taken and multiplied by the projected value, to get the deviation that is then added and subtracted to the projected value, thus creating the upper and lower accuracy limits respectively.

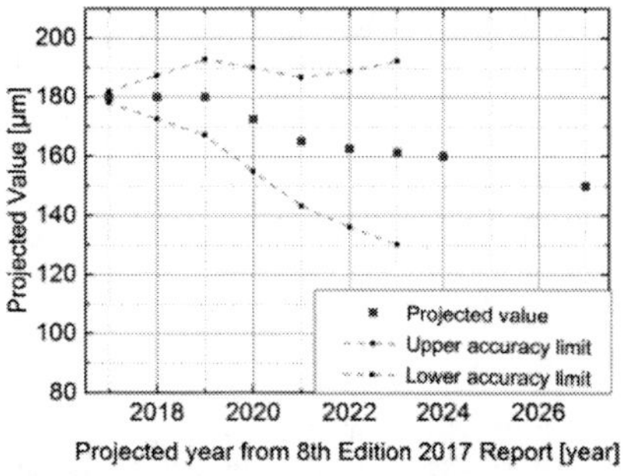

Figure 12: The future projections of mc-Si wafer thickness from the *8th edition ITRPV* within the defined upper and lower accuracy limits. (Projected values taken from Ref. [2]).

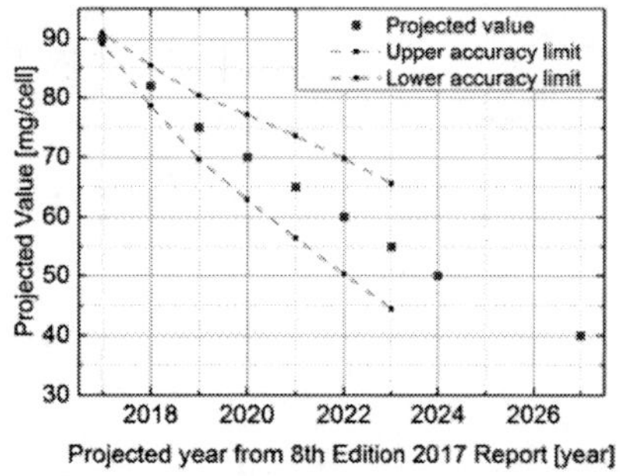

Figure 13: The future projections of silver amount per cell from the *8th edition ITRPV* within the defined upper and lower accuracy limits. (Projected values taken from Ref. [2]).

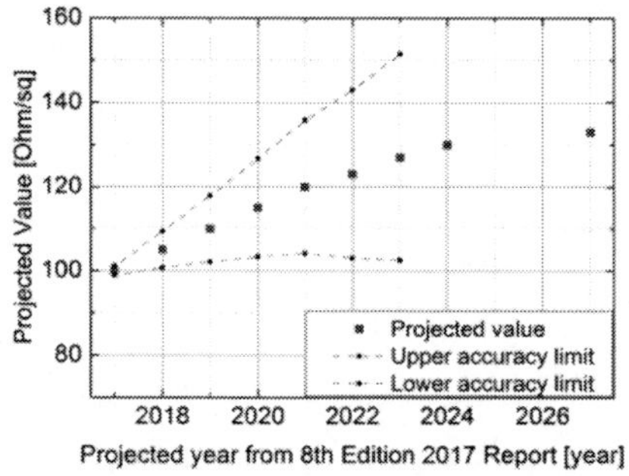

Figure 14: The future projections of emitter sheet resistance from the *8th edition ITRPV* within the defined upper and lower accuracy limits. (Projected values taken from Ref. [2]).

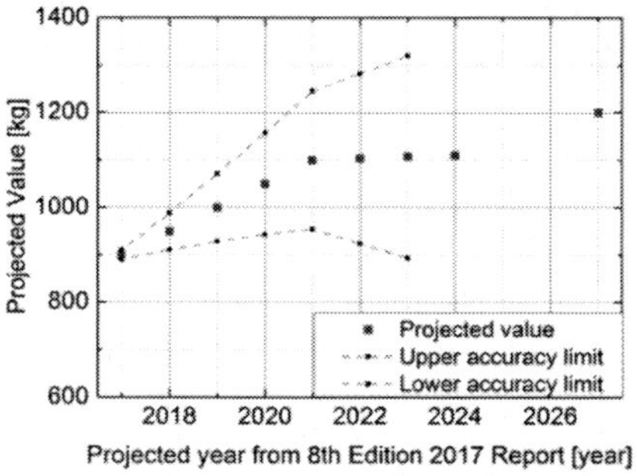

Figure 15: The future projections of mc-Si ingot mass from the *8th edition ITRPV* within the defined upper and lower accuracy limits. (Projected values taken from Ref. [2]).

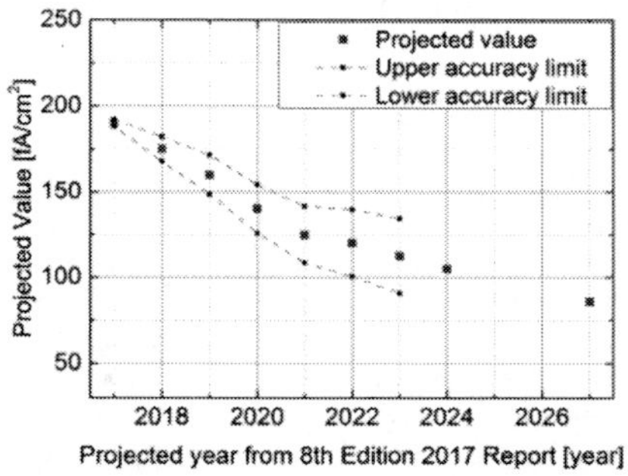

Figure 16: The future projections of the bulk recombination current density of p-type mc-Si from the *8th edition ITRPV* within the defined upper and lower accuracy limits. (Projected values taken from Ref. [2]).

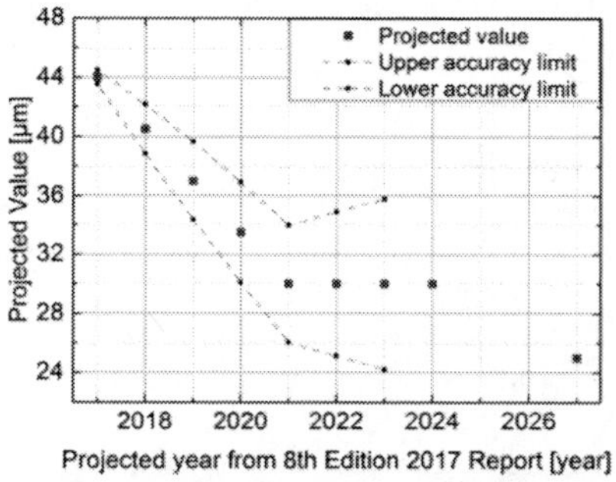

Figure 17: The future projections of the finger width from the *8th edition ITRPV* within the defined upper and lower accuracy limits. (Projected values taken from Ref. [2]).

The results are merely statistical and don't include any physical limits. Other methods can be developed to show different scenarios of future projection accuracy limits. Nevertheless, these plots show the expected range of the future projections lying within the upper and lower accuracy limits based on the deduced time-dependency of the accuracy.

4 CONCLUSION AND OUTLOOK

Knowing that *ITRPV* is the leading technology roadmap in crystalline silicon photovoltaics, and considering the importance of its continuous development, an *ITRPV* accuracy study allows the better understanding of the topics showing different projection accuracies. This quantitative analysis would assist first of all collaborating institutes in the *ITRPV* to pin-point the topics that are over/underestimated in specific time gaps, in an attempt for future projection enhancements. Moreover, differentiating accurate from less accurate projections would guide *R&D* groups to better understand the deviation from initial goals set in previous reports.

In this paper, six topics were taken into consideration. Projection absolute deviations and projection absolute percentage deviations were calculated for each chosen topic. Based on the results, future projection accuracy limits are defined and plotted for each discussed topic. Such an approach could be applied on a broader pool of topics that will be present in the future reports with the advancement of the *ITRPV* publications. The advancement of this study with the development of *ITRPV* enhances the follow-up phase of roadmapping.

Having a sensitivity analysis of the value chain topics in parallel to the advancement of the *ITRPV* can improve the understanding of cost dependence on inaccuracy. In other words, the financial risk can be defined based the projection accuracy done in the *ITRPV* or any other report. No projection can be verified before it becomes in the past, however learning from the past allows more accurate projections in the future. It is explicit that the increase in contribution of more PV *R&D* groups would allow projections closer to the market value.

ACKNOWLEDGEMENTS

The authors acknowledge the work of contributing experts from different companies and institution in the yearly published *ITRPV* reports. The authors would also like to thank all co-workers at Fraunhofer ISE.

5 REFERENCES

[1] Fischer, M. *et al.* (2012) SEMI International technology roadmap for photovoltaics – Challenges in c-Si Technology for suppliers and manufacturer, 27th European Photovoltaic Solar Energy Conference and Exhibition (EUPVSEC), Frankfurt.

[2] VDMA (2017). International Technology Roadmap for Photovoltaics - www.*ITRPV*.net (all *ITRPV* reports).

[3] Baliozian, P. *et al.* (2016), Photovoltaic Development Standardizing Based on Roadmaps and Technology Readiness Levels (*TRLs*). 32nd European Photovoltaic Solar Energy Conference and Exhibition

[4] Garcia, M.L. and Bray, O.H. (1997). Fundamentals of Technology Roadmapping. Strategic Business Development Department Sandia National Laboratories.

[5] ISO 5725-1(1994). Accuracy, trueness and precision of measurement methods and results - Part 1: General principles and definitions. p.1

[6] Ezeliora Chukwuemeka, D. *et al.*(2015) Measuring the Forecasting Accuracy for Masters Energy Oil and Gas Products, Journal of Scientific and Engineering Research, 2015, 2(2):52-70

[7] Hyndman, R. and Koehler, A. (2005). Another look at measures of Forecast Accuracy. Department of Econometrics and Business statistics Monash University and Department of Decision Sciences and Management Information Systems Miami University.

[8] Hoover, J. (2009) How to Track Forecast Accuracy to Guide Process Improvement, Foresight: The International Journal of Applied Forecasting.

TOWARD 21.4 % EFFICIENCY BY IMPLEMENTING INDUSTRIALLY FEASIBLE TECHNOLGIES IN PRINTED-ALO_X PERC TECHNOLOGY

Yung-Sheng Lin[1*], Jui-Yi Hung[2], Chen-Hao Ku[1], Jung-Ching Wang[3], Tsung-Cheng Chen[1], Cheng-Shun Hu[1], and Ching-Chang Wen[1]
1. E-ton Solar Tech Co., LTD., No.498, Sec. 2, Bentian Rd., Annan Dist., Tainan City 709, Taiwan, R.O.C.
2. New E Materials Co., LTD., 2F, No. 3, Luke 1st Rd., Luzhu Dist., Kaohsiung City 82151, Taiwan, R.O.C.
3. Eternal Chemical Co., LTD., No. 22, Changhsing Road, Luchu District, Kaohsiung 821, Taiwan, R.O.C.

ABSTRACT: In our previous work, we presented cell efficiency beyond 21 % achieved by our unique printed-AlO_x passivated emitter and rear cells (PERC) technology. In this paper, we demonstrate that the efficiency of this printed-AlO_x PERC cells can be further improved by synergizing both improvements in front passivation and rear passivation. These passivation-improvement technologies are selective emitter (SE), SiO_x/SiN_x dual layer capping (DL CAP) and improved anti-reflection coating (Imp. ARC), respectively. In DL CAP technology we developed a SiO_x paste for screen-printing deposition of SiO_x layer. Compared to our standard PERC cells, a significant average efficiency gain of 0.45 % was achieved, in which SE, DL CAP and Imp. ARC contributed to the efficiency gain of 0.16 %, 0.13 % and 0.16 %, respectively. We also demonstrate that an average efficiency of 21.44 % with the best efficiency of 21.52 % was achieved with these technologies by carefully choosing the in-coming wafer quality and well controlling each process step. Moreover, it is worthy to note that our developed DL CAP technology, i.e. implementing the printed-SiO_x layer in conventional SiN_x capping layer, can be easily integrated to existing production line with minimum additional capital cost for equipment and can be applied to other vapor-phase-deposited-AlO_x PERC cells as well.
Keywords: PERC, AlO_x, SiO_x, paste

1. INTRODUCTION

As of now, passivated emitter and rear cells (PERC) have gained a significant market share in the market of high-efficiency silicon-based solar cell. However, the efficiency of PERC cells needs to be continuously improved to sustain its competitiveness for the demand of ever-higher module output power. Several researchers have showed that emitter loss (included front passivation loss) is the top one recombination loss of today's state-of-the-art PERC cells or of maximal simulated efficiency PERC cells [1–4], implying there is still plenty of room for improvement in the front side. Selective emitter (SE) technology, i.e. heavy doping underneath fingers reduces contact recombination; meanwhile light doping in-between fingers improves blue response, is the most realistic approach to reduce the emitter loss. As for front passivation quality, improving the thermal stability of front anti-reflection coating (ARC) against high-temperature co-firing at metallization step is crucial for further reducing the front passivation loss.

In this study, we further implemented SE technology in our unique printed-AlO_x PERC cells [5] and modified the ARC process to improve the thermal stability of front passivation (ARC) layer. Moreover, we found that applying an extra thin printed-SiO_x layer on ARC layer can significantly improve the passivation quality of front side [6]. Based on the same concept, we also applied this printed-SiO_x layer into the rear-side passivation stack for improving the passivation quality of rear side. It is worthy to note that applying this printed-SiO_x into rear-side passivation stack provides an opportunity for fine tuning the internal back reflectance of rear-side dielectric stack as well [7]. In the following content, the effects of each abovementioned technology on the performance of our printed-AlO_x PERC cells will be demonstrated.

2. EXPERIMENTAL

The PERC cells shown in this study were produced in our ExcelTon™-Cell-III mass production line and the standard process flow was illustrated in our previous work [5]. In this study, three technologies were further implemented in the standard ExcelTon™-Cell-III PERC process as follows, and the standard PERC cells were used as the reference cells for comparisons.

(1) SE: Heavily doped regions (regions underneath the fingers) were formed by using laser doping technology. In this doping technology, the phosphosilicate glass (PSG) formed after conventional furnace diffusion process was used as doping source.
(2) DL CAP: An extra screen-printing SiO_x deposition, which is similar with the AlO_x deposition [5], was carried out before the SiN_x capping layer deposition, i.e. the rear-AlO_x passivation layer was capped by the SiO_x/SiN_x DL CAP layer to form an AlO_x/SiO_x/SiN_x stack layer. To obtain different thickness of SiO_x layer, different wet deposit of SiO_x paste was carried out (0.18 g/pcs – 0.36 g/pcs).
(3) Imp. ARC: A modification in ARC process was also carried out to improve the thermal stability of ARC layer.

μ-PCD lifetime mapping (WT-2000, Semilab) was used as the performance index to compare the passivation quality of PERC cells with different technologies. Spectrophotometer (U-4100, Hitachi) was used to evaluate the reflectance of PERC cells with different capping layers. Please note that the lifetime mapping measurement and the reflectance measurement were carried out after co-firing step but without opening and metallization steps. To investigate the impact of different technologies on spectral response, the quantum efficiency (QE) of cells was also determined by using QEX7 (PV Measurements, Inc) QE measurement system.

3. RESULTS AND DISCUSSION

3.1 The thickness of DL CAP layer

In Fig. 1(a), from the side view of SEM analysis on the finished cell, the thickness of SiO_x layer was about 70 nm as paste usage of 0.18 g/pcs was used. In Fig. 1(b) and Fig. 1(c), the thickness of SiO_x layer increased to 120 nm and to 340 nm by using the paste usage of 0.22 g/pcs and 0.36 g/pcs, respectively. Some voids were observed in thick SiO_x layer as shown in Fig. 1(c). The location of void was randomly distributed in the layer and the form of void was also irregular. In contrast, there were no voids observed in the case of thin SiO_x layer (Fig.1(a) and Fig.1(b)). Interestingly, if a single SiO_x layer was used as capping layer, there were no voids observed even when the thickness of SiO_x layer was as thick as 340 nm (see Fig. 1(d)). These observations indicate that not only the out gassing of the SiO_x layer itself but also the interaction between the SiO_x layer and the SiN_x layer plays a crucial role in the mechanism of void formation during the high-temperature co-firing step.

Our developed printed-AlO_x/SiO_x dielectric stack provides a low-cost and non-vapor-phase-deposited approach to accomplish the rear passivation layer of PERC cells. However, this printed-SiOx capping layer did not feature the same thermal and chemical stability as the SiN_x capping layer or as the SiO_x/SiN_x DL CAP layer did during high-temperature co-firing by using current commercial aluminum paste (not show here). We are now working on this application aiming to improve the thermal and chemical stability of this printed-SiO_x by adjusting the composition of SiO_x paste or the composition of aluminum paste or both. Taking the stability of capping layer into consideration, SiN_x was used as the outermost layer for all cells in this study.

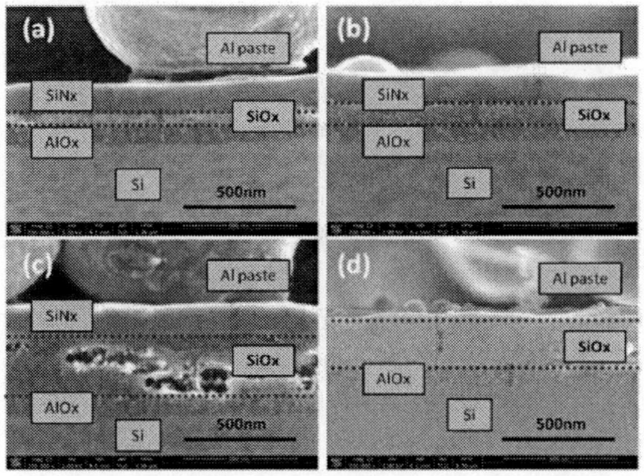

Figure 1: The cross-sectional SEM images of PERC cells with different capping layers. DL CAP layers with SiO_x paste usage of (a) 0.18 g/pcs, (b) 0.22 g/pcs and (c) 0.36 g/pcs, and (d) SiO_x capping layer with paste usage of 0.36 g/pcs.

3.2 The reflectance of DL CAP layer

Fig. 2 shows the reflectance comparison of the PERC cells with different capping layers. The benefit from the low-refractive-index printed-SiO_x layer to internal back reflectance was not observed in this study. In contrast, the reflectance at 1200 nm decreased as the SiO_x paste usage (the thickness of the SiO_x layer in DL CAP layer) increased. Compared to the reflectance at 1200 nm of the reference cell, a reflectance drop of about 4 % was observed when the-thickest-SiO_x DL CAP layer was used. This abnormal reflectance drop can be explained by more voids are formed in the SiO_x layer as higher usage of SiO_x paste are used. And these voids scatter light as a result of this reduction in the reflection at long wavelength. Taking this negative effect of voids on internal back reflectance into consideration, the usage of SiO_x paste was chosen at 0.18 g/pcs in the following DL CAP implementation.

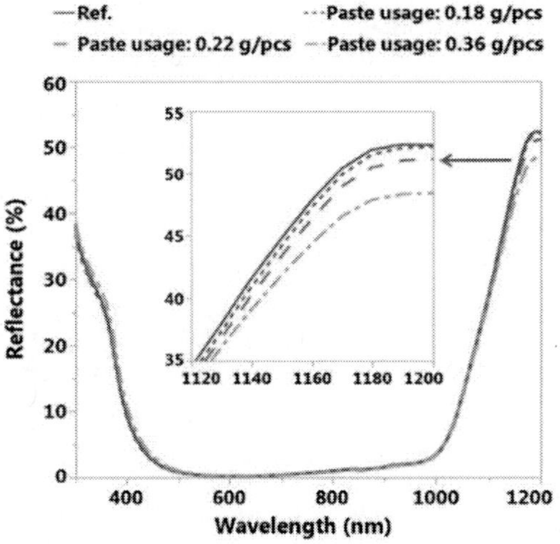

Figure 2: The reflectance comparison of PERC cells with different capping layers (DL CAP layers with different SiO_x paste usage and the reference SiN_x capping layer, respectively).

3.3 The evolution of lifetime improvement by each technology

The evolution of lifetime improvement by each technology is shown in Fig 3. As expected, the lifetime incrementally improved by implementing each technology, from the reference cell with an average lifetime of 54 µs increased to 98 µs by implementing all three technologies. A significant lifetime gain of 26 µs by implementing SE is consistent with the fact that the top one recombination loss of today's state-of-the-art PERC cells is emitter loss [1–4]. DL CAP and Imp. ARC further contributed to smaller lifetime gains of 8 µs and 10 µs, respectively, indicating rear passivation loss and front passivation loss were further improved. This result suggests that the printed-SiO_x layer in DL CAP layer plays a role similar to what it did in the application to improve the front passivation [6], i.e. this printed-SiO_x layer is a hydrogen species provider or a hydrogen species diffusion barrier or both which can help the hydrogenation passivation at the Si/AlO_x interface. Moreover, the result also confirms that our Imp. ARC process indeed improves the ARC thermal stability against high-temperature co-firing at metallization step.

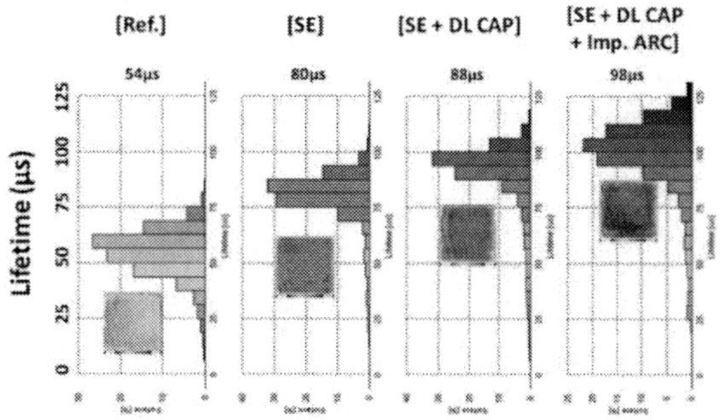

Figure 3: The influence of different technologies on lifetime after high-temperature co-firing step: the images of μ-PCD lifetime mapping and their corresponding lifetime distributions of PERC cells by implementing different technologies.

3.4 Cell performance by implementing each technology

To diminish lot-to-lot and cell-to-cell variance, we randomly picked one lot in our ExcelTon^{TM}-Cell-III mass production line and well shuffled these cells into four groups in the first test run; one reference group performed the standard PERC process and the other three experimental groups implemented different technologies, respectively. The efficiency box plots of each group (Fig. 4) and their corresponding average electrical performance (Table I) demonstrate that the efficiency was able to incrementally improve by implementing each technology. First, compared to the reference cells, an efficiency gain of 0.16 % was achieved by implementing SE. This efficiency gain can be mainly ascribed to the improvement in blue response by the light-doping emitter of SE as confirmed by QE analysis (Fig. 5). Next, an efficiency gain of 0.13 % was achieved by implementing DL CAP. Also, QE analysis, as shown in Fig. 5, confirms that this efficiency gain mainly contributes to the improvement in red response resulted from the improvement in rear passivation. Finally, an efficiency gain of 0.16 % was achieved by further implementing Imp. ARC. This synergistic effect of both improvements in front passivation and rear passivation resulted in a significant efficiency gain of 0.45 % compared to that of reference cells. Moreover, by carefully choosing the in-coming wafer quality and well controlling each process step, the efficiency can be improved further by implementing these three technologies. This result is also shown in Fig. 4 and Table I, denoted as 2nd run. An average efficiency of 21.44 % with the best efficiency of 21.52 % was demonstrated in this preliminary optimization.

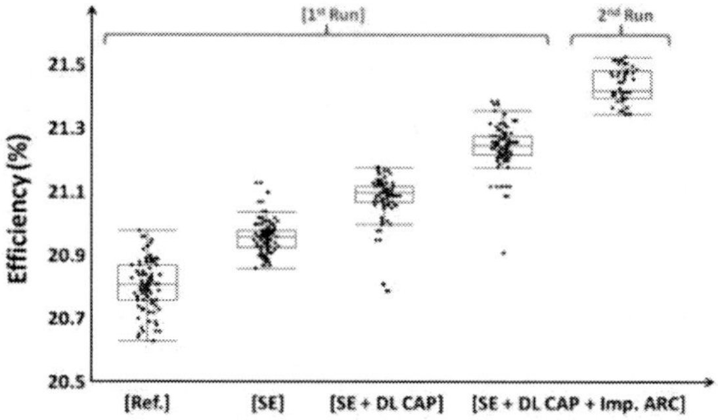

Figure 4: Incremental efficiency improvement by implementing different technologies.

Table I: The average electrical performance of PERC cells with different technologies.

Run	Technology	V_{oc}(V)	I_{sc}(A)	FF(%)	Efficiency(%)
1st	Standard(Ref.)	0.655	9.75	79.54	20.80
	SE	0.660	9.77	79.43	20.96
	SE + DL CAP	0.662	9.79	79.51	21.09
	SE + DL CAP+ Imp. ARC	0.663	9.83	79.61	21.25
2nd	SE + DL CAP+ Imp. ARC	0.667	9.87	79.51	21.44
	SE + DL CAP+ Imp. ARC (The Best Cell)	0.668	9.89	79.56	21.52

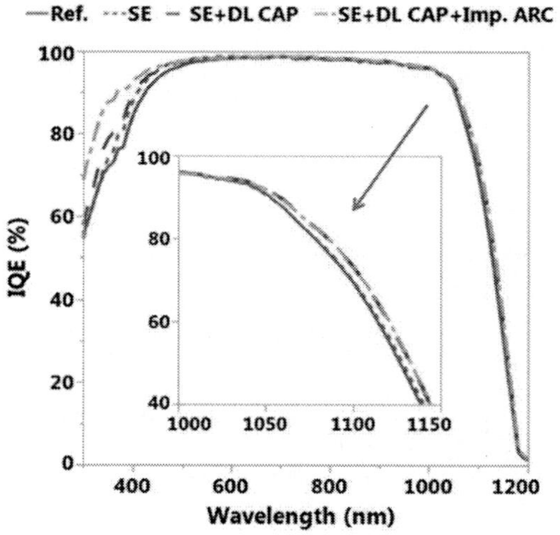

Figure 5: The IQE comparison of PERC cells with different technologies.

3.5 Cell performance in our ExcelTonTM-Cell-III mass production line

As above mentioned, the capability of DL CAP for increasing efficiency has been demonstrated. However, we think there is still plenty of room for improvement in printed-SiOx, such as void-free DL CAP or as capping layer without SiNx. Therefore, only SE and Imp. ARC were implemented into our PERC production line first. Fig. 6 illustrates the efficiency distribution of our PERC product in August, the corresponding average and champion electrical parameters are shown in Table II, respectively. The average efficiency was 21.35 % with an average Voc of 664 mV. As for the champion cell, the efficiency reached 21.70 % with an additional Voc gain of 4 mV, indicating great potential of higher average efficiency can be achieved by using ExcelTonTM-Cell-III technology. And we aim at average efficiency of 21.8 % by the end of 2018 by further implementing the printed-SiOx layer or other industrially feasible technologies.

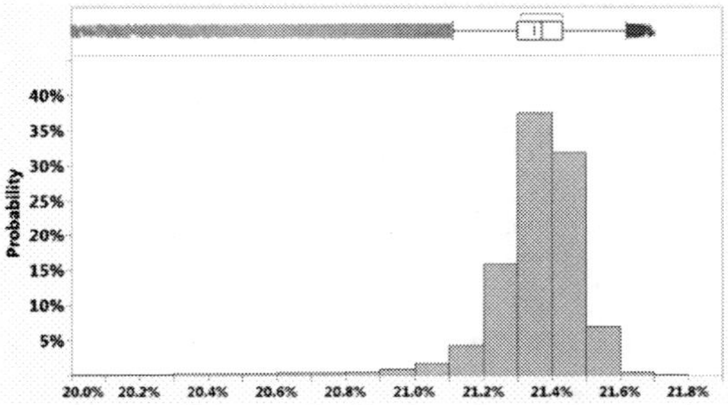

Figure 6: The efficiency distribution of our PERC product by implementing SE and Imp. ARC.

Table Ⅱ: The average electrical parameters of PERC cells in our mass production line.

ExcelTon^{TM}-Cell-III	Eta [%]	V_{oc} [V]	I_{sc} [A]	FF [%]	PCS
Average	21.35	0.664	9.87	79.61	263,025
Champion cell	21.70	0.668	9.90	80.10	1

4. CONCLUSION

An efficiency gain beyond 0.4 % was demonstrated by synergizing both improvements in front passivation and rear passivation resulted from simultaneously implementing SE, DL CAP and Imp. ARC technologies in standard PERC process. The average efficiency beyond 21.4 % was also demonstrated by a preliminary optimization of these three technologies. Among these three technologies, DL CAP still has plenty of room for further improvement. In this study, we only took the benefit of passivation improvement by using thin printed-SiO_x layer since thicker printed-SiO_x layer in DL CAP featured more voids which are harmful to internal back reflectance. The efficiency will be further improved if the advantage of the increase in internal back reflectance by lower-refractive-index SiO_x can be taken.

5. REFERENCE

[1] C. Kranz, J. H. Petermann, T. Dullweber, and R. Brendel, "Simulation-Based Efficiency Gain Analysis of 21.2 %-Efficient Screen-Printed PERC Solar Cells," *Energy Procedia*, vol. 92, pp. 109-115, 2016.

[2] D. C. Walter, B. Lim, and J. Schmidt, "Realistic Efficiency Potential of Next-Generation Industrial Czochralski-Grown Silicon Solar Cells after Deactivation of the Boron–Oxygen-Related Defect Center," *Progress in Photovoltaics: Research and Applications*, vol. 24, pp. 920-928, 2016.

[3] B. Min, H. Wagner, M. Muller, H. Neuhaus, R. Brendel, and P. P. Altermatt, "Incremental Efficiency Improvements of Mass-Produced PERC Cells UP to 24 %, Predicted Solely with Continuous Development of Existing Technologies and Wafer Materials," in *31st European Photovoltaic Solar Energy Conference and Exhibition*, 2015, pp. 473-476.

[4] D. Chen et al., "21.40 % Efficient Large Area Screen Printed Industrial PERC Solar Cell," in *31st European Photovoltaic Solar Energy Conference and Exhibition*, 2015, pp. 334-340.

[5] T.-C. Chen et al., "E-Ton's Printed-AlOx PERC Cells: Efficiencies beyond 21 % with a Next-Generation AlO_x Paste," in *32nd European Photovoltaic Solar Energy Conference and Exhibition*, 2016, pp. 680-682.

[6] Y.-S. Lin et al., "Improving the Passivation and the Reflectance Uniformity of SiN_x Anti-Reflection Coating by Using the SiO_x Passivation Paste," in *43rd IEEE Photovoltaic Specialist Conference*, 2016, pp. 2943-2945.

[7] K. O. Davis et al., "Investigation of the Internal Back Reflectance of Rear-Side Dielectric Stacks for C-Si Solar Cells," *IEEE Journal of Photovoltaics*, vol. 3, no. 2, pp. 641-648, 2013.

Industrially feasible PERC cells on diamond wire sawing multi-crystalline silicon wafers textured by RIE towards 20.78% efficiency

Wei Wang, Jianwen Dong, Quanhua Ye, Yadi Yang,Wenhao Cai, Jian Sheng, Jian Yang, Chun Zhang, Xiaoxue Zhou, Shuguang Dong

R&D Center, Solar BU, GCL System Integration Technology Co., Ltd., 288#, ChenFeng Road, YangShe Town ZhangJiaGang E&T Deve. Zone, SuZhou, JiangSu, China, 215600. wangwei6@gclsi.com

ABSTRACT: Improved cell efficiency and cost reduction are two keys to PV production processes. Passivated emitter and rear cell (PERC) due to its high efficiency and compatibility with existing p-type materials and production lines are expected to be the mainstream cell technology with the largest gain in market share over the next few years. Meanwhile, diamond wire sawing (DWS) can significantly reduce the wafer processing cost and is going to gain more market shares in multi-crystalline silicon (*mc*-Si) with advantages of texturing. Reaction ion etch (RIE) technology has been successfully applied to realize the diamond wire cutting *mc*-Si wafer surface texturing regardless of grain orientations of *mc*-Si wafers, and is helpful for efficiency improvement. Combining the high efficiency structure with low cost material, the final cell efficiency reached 20.43% in mass production and 20.78% for the best cell.

Keywords: DWS, RIE, PERC, multi-crystalline

1 INTRODUCTION

Several solar cell manufactures have introduced PERC or RIE technology into *mc*-Si solar cells mass production in recent years, but few companies combined these two technologies together. PERC technique is to improve the rear side of cell by replacing the conventional Al-BSF structure with a dielectrically passivated rear surface [1]. RIE, solves the texturing problem of diamond-wire sawing multi-crystalline silicon (DWS *mc*-Si), furthermore, it can achieve an efficiency increase of 0.4~0.6% based on p-type DWS *mc*-Si compared to the conventional Al-back surface field (Al-BSF) solar cells [2][3]. The structure of front RIE surface PERC solar cells can achieve an efficiency increase of >1.2%abs, meanwhile, the advantage of low cost diamond-wire sawing multi-crystalline silicon (DWS *mc*-Si) wafer can also be obtained. This paper presents the development of this kind of RIE-PERC *mc*-Si cells and power loss analysis is carried out to analyze the potential for higher efficiency on mass production.

2 EXPERIMENTAL DETAILS

All experiments were fabricated on 156.75×156.75 mm^2 p-type *mc*-Si wafers from GCL company, with a thickness of about 180~200μm and a bulk resistivity of about 1.6 Ω·cm. The DWS *mc*-Si wafers were first damage-etched in an HNO_3/HF mixture solution, and then the nano-texture were fabricated by RIE. The subsequent damage etch was performed in a chemical solution and the reflectivity was optimized at the same time. After a phosphorus diffusion, an emitter with an R_{sheet} of 100~110Ω/□ was formed. The Al_xO_3 and SiN_x layers were deposited by the PECVD method using MAiA equipment on the rear side. On the front, the deposition of silicon nitride (SiN_x) antireflection coating was optimized to match the nano-texture surface. The laser ablation process was applied in the rear side to open the Al_xO_3 / SiN_x layers. Finally all of the cells were performed by conventional screen printing and optimized firing.

A QSSPC lifetime tester, Sinton Instruments WCT-120, was used to monitor the passivation properties of the different FSFs and the recombination parameter J_0 of the rear side. Contact resistivity of double side metallization was evaluated using the Transfer Length Method (TLM) [4]. A quantum efficiency measurement system was used to characterize the reflectance of different samples between the wavelength of 300nm and 1180nm.

Quokka, a free and fast multi-dimensional simulator developed by Andreas Fell [5], was used to simulate the series resistance and the saturation current density. The solar cell performance was measured on a production flash tester, using reference cells calibrated by Fraunhofer ISE CalLab.

3 RESULTS AND DISSCUSION

3.1 RIE process

RIE is a promising solution for DWS *mc*-Si wafer texturing. As shown in Fig. 1 (b), 300 nm to 500 nm nanostructure were successfully fabricated, which is smaller than that of the wire slurry sawing (WSS) *mc*-Si wafer (3-4 μ m), as shown in Fig.1 (a). The nanostructure surface shows lower reflectance even post ARC deposition, as shown in Fig. 2.

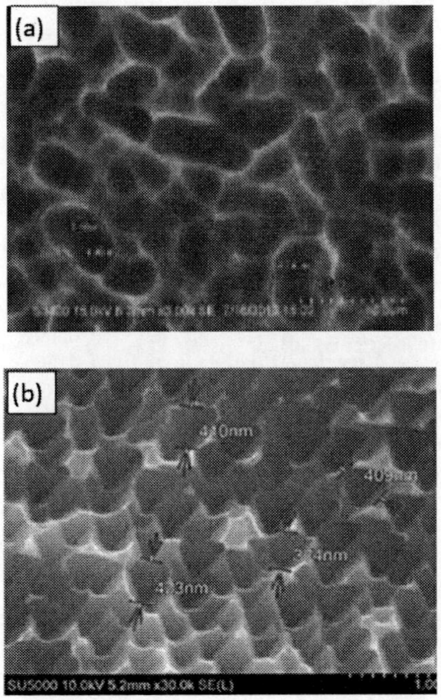

Figure 1: SEM images of (a) Regular acid texturing on WSS *mc*-Si wafer, (b) RIE on DWS *mc*-Si wafer

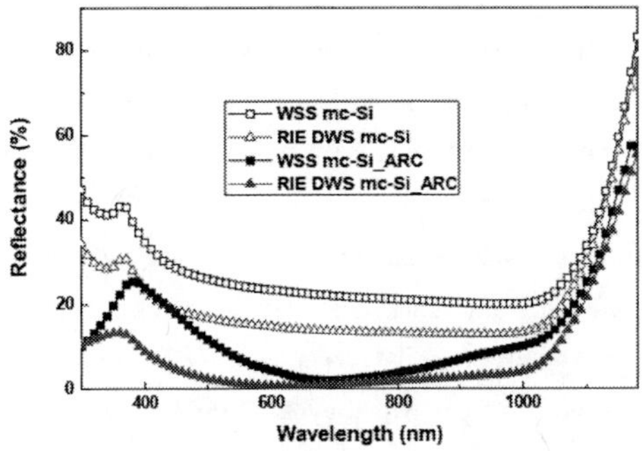

Figure 2: With and without ARC reflectance curves of regular acid texturing WSS and RIE texturing DWS *mc*-Si wafers

The front nanostructure surface passivation is a challenge for industrial cell fabrication. In this paper, around 82 fA/cm^2 surface recombination current densities J_{0e} was got through special technology on front passivation area, meanwhile the contact resistivity was acceptable. Fig. 3 showed the J_{0e} change when using different diffusion process and passivation technology. Group A is the baseline of J_{0e}, the diffusion and passivation process is normal, Group B has special passivation layer with normal diffusion, Group C is based on the Group B passivation technology, which has optimized diffusion process.

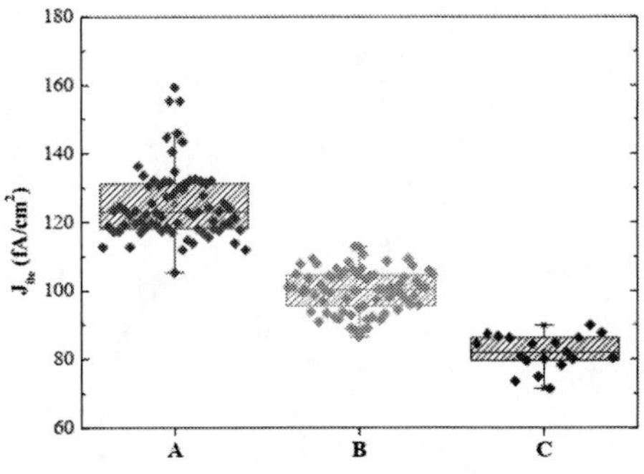

Figure 3: J_{0e} of different process: Group A is baseline; Group B with special passivation; Group C with special passivation and diffusion

3.2 Rear passivation

PERC cell aims to improve the rear side by replacing the conventional Al-BSF with a dielectrically passivated rear surface. As shown in Fig.4. The surface of raw WSS *mc*-Si wafer is much more defect rich, mounts of "worm" structure was formed after acid texturing of etching back process. For DWS *mc*-Si wafer, the front surface was single side RIE textured, the rear side was kept smooth, when saw damage removing or etching back by industrial normal process were applied. The flat rear surface of DWS *mc*-Si wafer is easy to be passivated by Al_xO_3/SiN_x layer. The rear surface recombination current densities J_{0rear} can reach to ~18*f*A/cm^2.

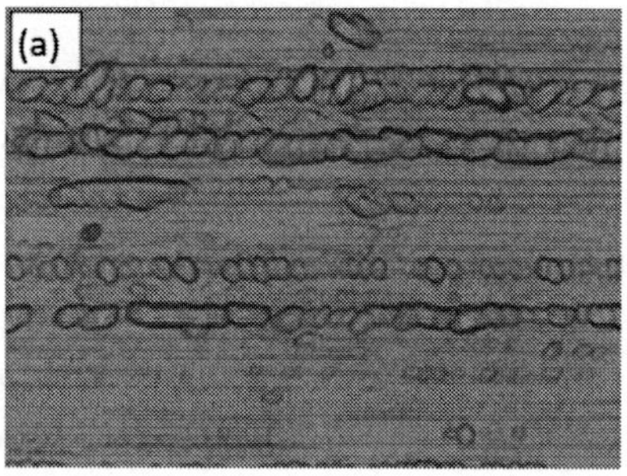

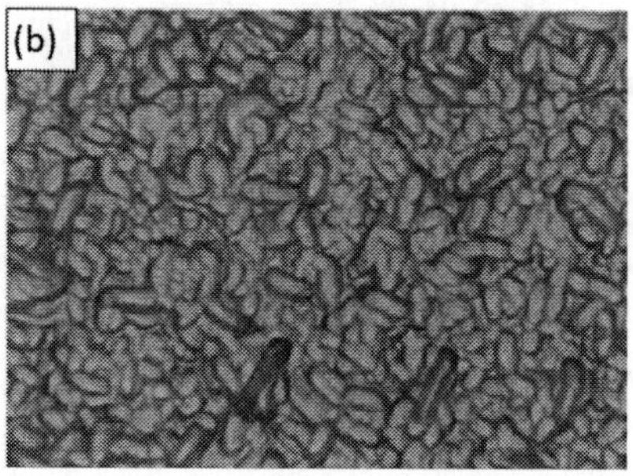

Figure 4: Surface morphology of (a) DWS and (b) WSS *mc*-Si wafers etched back by industrial normal process

3.3 Cell efficiency

Using RIE and PERC technology, the low cost material DWS *mc*-Si wafers had been successfully transferred to mass production, and the final cell efficiency reached to 20.43% in average and 20.78% for the best cell. The IV performances are shown in Table 1, the group of RIE+PERC DWS *mc*-Si cells almost 1.73% absolute higher than wire slurry sawing (WSS) *mc*-Si group.

Table1: I-V performance of solar cells with different texture

Group	Voc (mV)	Jsc (mA/cm^2)	FF (%)	Eff (%)
WSS *mc*-Si	633.5	36.81	80.18	18.70
RIE+PERC DWS *mc*-Si	656.5	39.10	79.57	20.43
The best of RIE+PERC DWS *mc*-Si	661.4	39.12	80.31	20.78

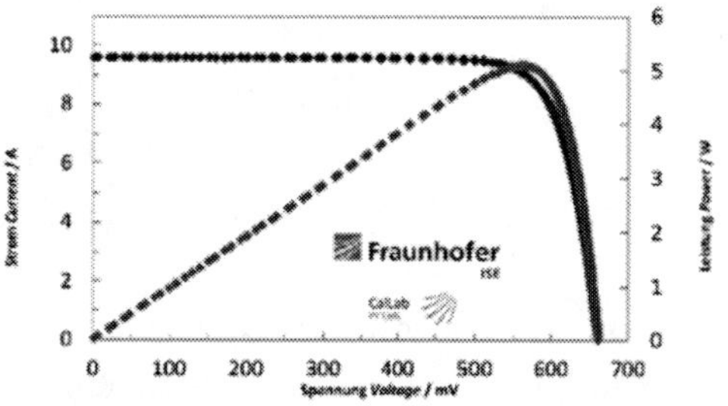

Figure 5: I-V curve of the champion 20.78% efficiency cell, which was independently confirmed at Fraunhofer ISE CalLab.

3.4 Module performance

On mass production, when the average efficiency of RIE+PERC DWS *mc*-Si solar cells is 20.2%, the module performance is showed in Fig. 6, the main power grade is 290W, now the average efficiency is up to 20.43%, the main power grade may move up to 295W.

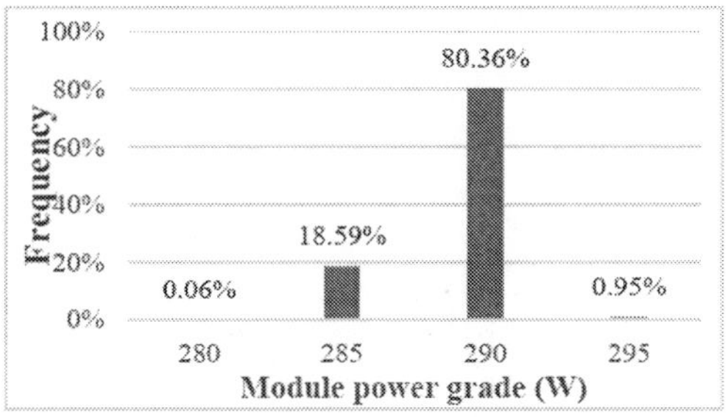

Figure 6: Module power distribution of 20.2% average efficiency RIE+PERC-DWS *mc*-Si solar cells.

4 ANALYSIS

4.1 Recombination loss

The total saturation current density, J_{0tot} was calculated from the one-diode equation below [6, 7, 8]:

$$J_{0tot} = \frac{J_{sc}}{e^{(\frac{V_{oc}q}{nKT})}} \qquad (1)$$

J_{0tot} can also be expressed as

$$J_{0tot}=J_{0e}*(1-f)+J_{0e,metal}*f+J_{0,nbulk}+J_{0rear}*(1-F_{metal})+J_{0rear,metal}*F_{metal} \qquad (2)$$

Components of J_{0tot} equation above are shown in Figure 7 below.

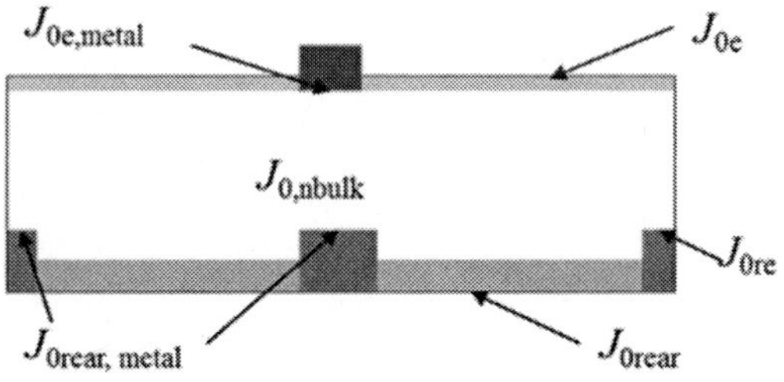

Figure 7: Components of the total recombination parameter J_{0tot}

The average J_{sc} and V_{oc} value come from the electrical performance presented in Table 1, where k, T, and q are the Boltzmann constant, temperature, and the electron charge, respectively, and *n* is assumed to be 1. The J_0 for each surface was tested by QSSPC measurements. The distribution of recombination loss is shown in Figure 8. It can be seen that the bulk and emitter, play a significant role in the J_0 make-up and limit the V_{oc} improvement. Therefore, the emitter design should be optimized, selective emitter may be used in future. The base improvement of multi-crystalline wafer is always important for *mc*-Si cells, recently, hydrogen passivation and high quality cast technology are expected.

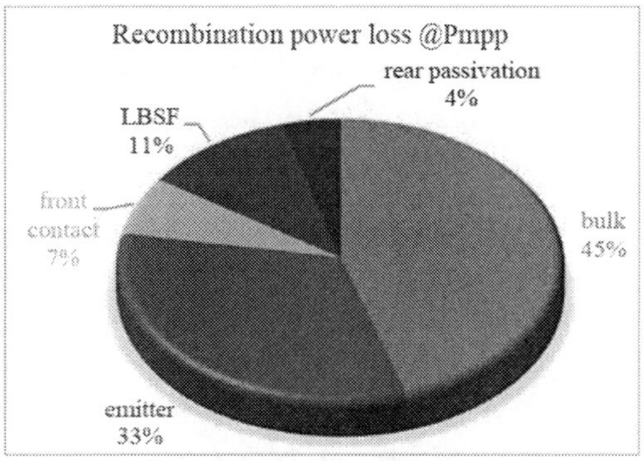

Figure 8: Recombination current density distribution.

4.2 Series resistance loss

The distribution of the series resistance losses is shown in Figure 9. The bulk and front sheet resistance (61% of the total series resistance) are the key limiting factors for the total series resistance. An increase of the front busbars would contribute to a reduction in both bulk and front sheet resistance, also would decrease the front metal resistance, but the V_{oc} would decrease as well, offsetting the benefit of the reduction in the series resistance. There is a trade-off between V_{oc} and series resistance to reach an optimum efficiency.

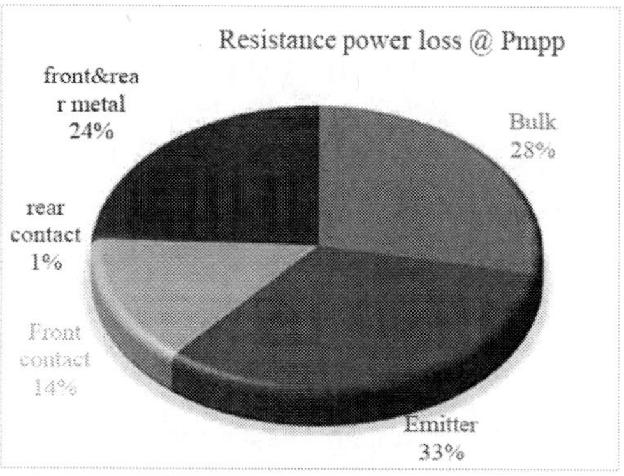

Figure 9: Components of the series resistance loss.

4 SUMMARY

RIE has been successfully applied to realize the diamond wire cutting *mc*-Si wafer surface texturing regardless of grain orientations of *mc*-Si wafers, and is helpful for efficiency improvement. Combining the high efficiency structure of PERC and RIE with low cost DWS *mc*-Si wafers, the cell efficiency has reached 20.4% in mass production and 20.78% for the best cell. In the future, the numbers of busbars, the selective emitter, and the base optimization are important direction for *mc*-Si cells further improvement.

5 ACKNOWLEDGEMENTS

The authors would like to thank Shengzhao Yuan, Yanfeng Cui, Huiming Zhang, Zhendong Shen, Xianzhao Chen, Guidong Chen, Wenchao Gu, Jiaxian Lv, Xiaoli Qian, Wenwen Wei, Yibo Zhang, Ji Li.

6 REFERENCES

[1] A.W. Blakers, A.Wang, A. M. Milne, J. Zhao, and M. A. Green, "22.8% efficient solar cell," Appl. Phys. Lett., vol. 55, pp. 1363–1365, 1989.

[2] J.Yoo, G.Yu, J.Yi, Large-area multi crystalline silicon solar cell fabrication using reactive ion etching (RIE), Sol. Energy Mater. Sol.Cells 95(2011)2–6

[3] Y.D. Yang, W. Wang, Q.H. Ye, J. W. Dong, et al. Fabricating nano-texture on diamond-wire saw multi-crystalline silicon using RIE-etch, 32th European Photovoltaic Solar Energy Conference and Exhibition. Munich, Germany, 2016, pp.667-670

[4] D. L. Meier, D. K. Schroder, "Contact resistance: its measurement and relative importance to power loss in a solar cell," IEEE Trans. Electr. Dev. 31, 647(1984).

[5] A. Fell, "A free and fast three-dimensional/two-dimensional solar cell simulator featuring conductive boundary and quasi-neutrality approximations," IEEE Trans. Electr. Dev. 60, 733(2013)

[6] Y. W. Ok, A. D. Upadhyaya, Y. G. Tao, F. Zimbardi, K. Ryu, M. H. Kang, A. Rohatgi, "Ion-implanted and screen-printed large area 20% efficient N-type front junction Si solar cells," Solar Energy Materials & Solar Cells. 123, 92(2014).

[7] J. Benick, B. Hoex, G. Dingemanns, A. Richter, M. Hermle, S. W. Glunz, "High-efficiency N-type silicon solar cells with front side boron emitter," 24th EUPVSEC, 2009, pp. 863-870.

[8] D. Suwito, U. Jäger, J. Benick, S. Janz, M. Hermle, S. W. Glunz, "Industrially feasible rear passivation and contacting scheme for high-efficiency N-type solar cells yielding a V_{oc} of 700mV" IEEE Trans. Electr. Dev. 57, 2032(2010)

ULTRASONICALLY TINNED PVD AL REAR CONTACTS ON HIGH-EFFICIENCY CRYSTALLINE SILICON SOLAR CELLS FOR MODULE INTEGRATION

H. Nagel[1,*], D. Eberlein[1], S. Hoffmann[1], M. Graf[1], B. Steinhauser[1], F. Feldmann[1], A. Kraft[1], U. Eitner[1], M. Glatthaar[1], M. Hermle[1], S. W. Glunz[1], H. Haverkamp[2], T. Fischer[3], A. Hain[4], P. Wohlfart[4], V. Mertens, J. M. Müller[5] and T. Buck[6]

[1]Fraunhofer Institute for Solar Energy Systems (ISE), Heidenhofstraße 2, 79110 Freiburg, Germany
[2]Gebr. SCHMID GmbH, Robert-Bosch-Str. 32-36, 72250 Freudenstadt, Germany
[3]teamtechnik GmbH, Planckstraße 40, 71691 Freiberg, Germany
[4]SINGULUS TECHNOLOGIES AG, Hanauer Landstrasse 103, 63796 Kahl am Main, Germany
[5]Hanwha Q Cells GmbH, Sonnenallee 17-21, 06766 Bitterfeld-Wolfen, Germany
[6]International Solar Energy Research Center Konstanz e.V., Rudolf-Diesel-Straße 15, 78467 Konstanz, Germany
*Author for correspondence: Henning Nagel, Phone: +49 (0)761-4588-5493, e-mail: henning.nagel@ise.fraunhofer.de

ABSTRACT: We investigated ultrasonic tinning of physical vapor deposited (PVD) Al rear contacts on *p*-type passivated emitter and rear cells and *n*-type back junction cells, respectively. By means of electroluminescence (EL) imaging after ultrasonic tinning we identified small local damage of the cells' rear side passivation only at the positions where the soldering irons were initially placed on the Al. Though measured interconnector peel force was < 1 N/mm and hence lower then required by the standard DIN EN 50461, we obtained power degradation < 5 % after 400 thermal cycles, 2000 h in damp heat test and 20 cycles in the humidity freeze test, respectively, which is double the load demanded by the IEC 61215 standard. EL images revealed no local degradation induced by the climate tests. The microstructure of the tinned PVD Al is investigated by secondary electron microscopy and energy dispersive X-ray spectroscopy.
Keywords: metallization, back contact, module integration

1 INTRODUCTION

At present, most industrial-type crystalline silicon solar cells feature screen-printed rear contacts consisting of large-area Al and local Ag solder pads. However, the contacts suffer from shortcomings like possible reduction of the surface passivation of the cells due to the applied high sintering temperature, increased effective surface recombination velocity in comparison with boron back surface field (BSF) and damage of the *n*-doped BSF of *n*-type cells. Hence, high-efficiency c-Si solar cells often rely on physical vapor deposited (evaporated or sputtered, PVD) Al. It makes good ohmic contact to *p*- and *n*-type Si, provides excellent back reflection of infrared light passing through the cells and features very high electrical conductivity. However, for a broad industrial utilization of PVD Al, a reliable inter-connection technology is needed. In this work, we studied local ultrasonic tinning of PVD Al, a technology that was originally developed for screen-printed Al [1, 2]. To the best of our knowledge, the application to PVD Al was not comprehensively reported before.

2 EXPERIMENTAL

Two types of 6 inch CZ Si solar cells were investigated: about 21 % efficient *n*-type back junction cells prepared by Hanwha Q Cells [3] and about 20 % efficient *p*-type front junction passivated emitter and rear cells (PERC) featuring locally laser-fired PVD Al rear contacts on top of an electrically passivating Al_2O_3/SiN stack. The Al on the rear of the PERC cells was deposited in a high-rate inline thermal evaporation vacuum system to a thickness of about 2.7 µm. The front sides of both cell types were contacted by screen-printed and fired Ag. In order to make the rear Al solderable, 3 tin busbars were simultaneously applied within 6 s per cell by ultrasonic soldering irons in a setup equipped with automated cell transport, see Figure 1. 99.9 wt% pure Sn was used without flux. During tinning, the cells were positioned on a temperature controlled chuck at 270 °C. Sn wires were fed to the tips of the irons which were heated to 380 °C. In an automated tabber-stringer 1.5 mm x 0.2 mm Cu ribbons with standard SnPb coating were soldered onto the cells by means of controlled infrared lamp heating. Because Cu interconnectors and Al were already tinned, we obtained very good solder wetting. Subsequently, the cells were interconnected to 3-cell-strings. Two strings, one of every cell type, were laminated in each of the 3 manufactured photovoltaic modules. The cross-sectional construction of the modules was soda-lime glass / ethylene vinyl acetate (EVA) / solar cells / EVA / Tedlar-PET-Tedlar backsheet. Illuminated *IV* curves of the individual strings in the modules were measured at standard test conditions (25 °C, 1000 W/m², AM1.5G). Electroluminescence (EL) images were taken under forward bias. The *IV* and EL measurements were performed before and after light-induced degradation (3 h under 1 sun halogen-lamp illumination) and after subsequent damp heat, thermal cycling and humidity

Figure 1: Photo of the ultrasonic tinning process applied to the Al rear contact on a CZ Si solar cell. Departing from this photo, the cells in the study featured 3 busbars instead of 5.

freeze tests according to the standard IEC 61215. On individual strings, which were not laminated into modules, the 90° peel force of interconnectors soldered onto the back side was measured. In order to investigate the microstructure and composition of tinned PVD Al on the rear of PERC cells, cross sections were prepared by ion milling and studied by secondary electron microscopy (SEM) as well as energy dispersive X-ray spectroscopy (EDX).

For the purpose of demonstrating the efficiency potential, a 'TopCon' solar cell [4] featuring a full-area carrier-selective contact on the back side and PVD Al as a metal electrode on the rear *n*-doped Si layer was prepared. The cell was characterized by illuminated *IV* measurement.

3 RESULTS AND DISCUSSION

3.1 Microstructure of the tinned PVD Al rear contact

We investigated the microstructure of the tinned PVD Al in order to shed light on potential failure mechanisms in case of mechanical stress. Our analysis was confined to *p*-type PERC cells. The SEM images depicted in Figure 2 show cross sections of the ultrasonically tinned PVD Al on the alkaline textured rear of the cells. After tinning, the cells were annealed with the same temperature profile as used for the soldering of interconnectors. It can be seen from the upper picture, that Al and Sn resemble the textured Si surface well without large gaps. However, a conspicuous result is that there are small voids in the Sn bulk featuring diameters of about 100 nm. We explain their formation by the well known cavitation effect: When liquids – in this case molten Sn – are subjected to ultrasound the sonic waves locally create low pressure that ruptures the liquid. During cool down the Sn solidifies fast and the formed cavities are frozen into it. The voids are also present at the interface between Sn and Al as can be seen in the middle picture. Because their number per unit area is small at the interface, we expect that the voids don't reduce the adherence of Sn on the PVD Al significantly.

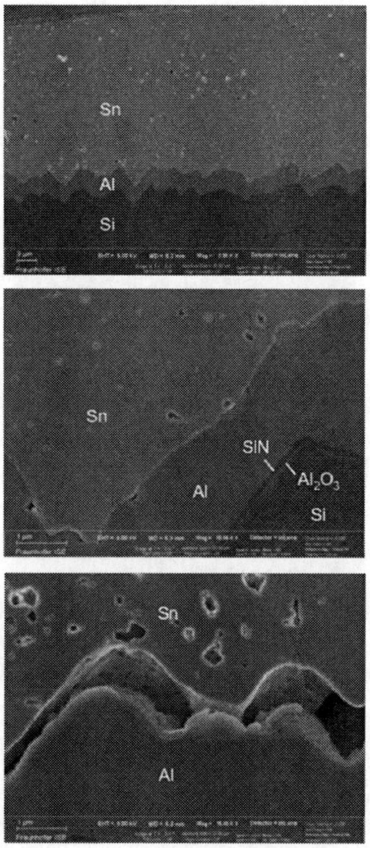

Figure 2: SEM pictures showing three cross sections of ultrasonically tinned PVD Al on the alkaline textured rear of *p*-type PERC cells.

On the microscale the Sn most often is also in intimate contact to the PVD Al, see SEM picture in the middle. However, in some areas very narrow gaps are located between Sn and Al. Obviously Sn didn't wet Al well there during ultrasonic soldering. Since the area of these spots is notable, it can be concluded that non-wetting plays a role when it comes to the mechanical adherence of the Sn layers to the Al.

After applying mechanical force, the Sn layer was detached from the Al as can be seen in the lower picture. Because we didn't observe fracture in the Sn, Al or Si bulk at any studied position, the Sn-Al interface appears to be the weakest site. This is confirmed by the interconnector pull-off tests described below.

EDX mappings of tinned PVD Al cross sections are shown in Figure 3. It can be seen from the upper picture that no extended intermetallic phases at the interface between Sn and Al are formed. This is known from the alloy phase diagram [5]. In general, it can be expected that this limits the adherence of Sn on Al. The lower picture in Figure 3 reveals a disrupted Al_2O_3 layer at the same interface which is the intended consequence of

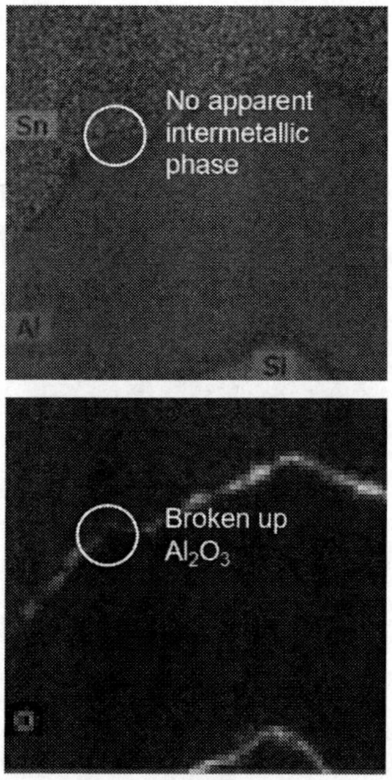

Figure 3: EDX cross-sectional view of an ultrasonically tinned PVD Al rear contact on a *p*-type PERC cell. Image width: 5.8 μm.

applying ultrasound during tinning. In some areas the Al_2O_3 seems to be still intact, but no gap between Sn and Al is visible at this position.

Note that at the bottom edge in the lower picture of Figure 3 an about 10 nm thick Al_2O_3 surface passivation layer can be seen.

3.2 Interconnector peel force

The normalised 90° interconnector peel force measured on *p*-type PERC cells after stringing is shown in Figure 4. It is 0.5 N/mm on an average. This value does not comply with the standard DIN EN 50461 (> 1 N/mm). The failing interface was between the Sn deposited by ultrasonic soldering and the PVD Al in accordance with the findings described in paragraph 3.1. We expect that the adherence can be increased e. g. by increasing the ultrasonic power during Al tinning.

Apparently, the adherence of PVD Al on the dielectric passivation layer stack is higher. It must be mentioned that it is a strong function of the top dielectric layer material. Recently, in a work on 'PassDop' solar cells we found that the adherence of thermally evaporated Al is very high on SiO_2 and very low on MgF_2 [6]. In [7] we reported high adherence of thermally evaporated Al on SiN and very high adherence on crystalline Si. Furthermore, the adherence of PVD Al on dielectric layers depends on the thermal treatment before Al deposition [8] and the substrate temperature during Al deposition [9].

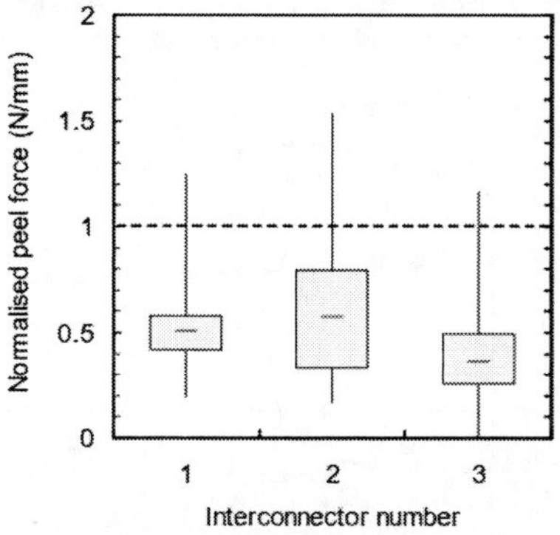

Figure 4: Measured normalised 90° peel force of interconnectors soldered on ultrasonically tinned PVD Al rear contacts of *p*-type PERC cells.

3.3 Module climate tests

Figure 5 shows the electrical power at maximum power point (P_{mpp}), the open-circuit voltage (V_{oc}), the short-circuit current density (J_{sc}) and the fill factor (*FF*) of the prepared 3-cell-strings in photovoltaic modules measured under standard test conditions. The values were normalised to the ones before light-induced degradation (LID). The strings contain the *p*-type front-junction laser-fired contact PERC cells (upper graph) and *n*-type rear-junction cells (lower graph), respectively. The measurement sequence depicted from left to right in each graph is a) illuminated *IV* measurement before LID, b) after LID, c) after 200 thermal cycles from -40 °C to 80 °C and d) after 400 thermal cycles. Power degradation is below 5 % (dotted horizontal line) and hence meets the standard IEC 61215. Note that we performed twice the number of cycles required. It is also worth to mention that in Figure 5 V_{oc} of both cell types increased after 200 and 400 cycles compared with the initial state. Obviously, defects in the bulk and/or at the surfaces were annealed during the elevated module test temperature.

Figure 6 depicts EL images taken from the *p*-type PERC strings in the thermal cycling test. The pictures reveal that the upper two cells in the string suffer from LID. The reason is the formation of BO-complexes in the CZ Si bulk under illumination. Subsequently, during

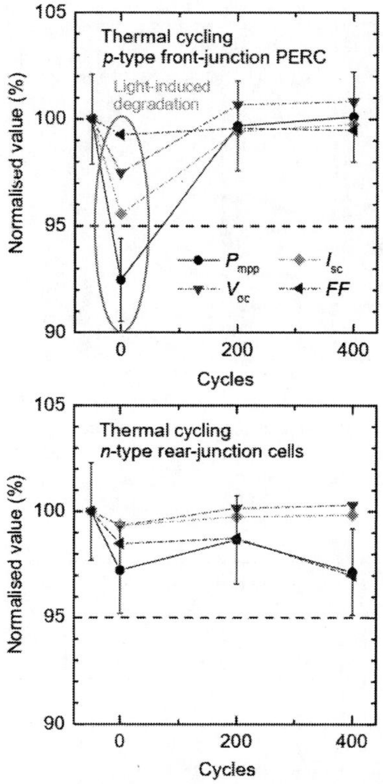

Figure 5: Electrical performance parameters of 3-cell-strings laminated in photovoltaic modules measured before and after LID and after subsequent thermal cycles between -40 °C and 85 °C.

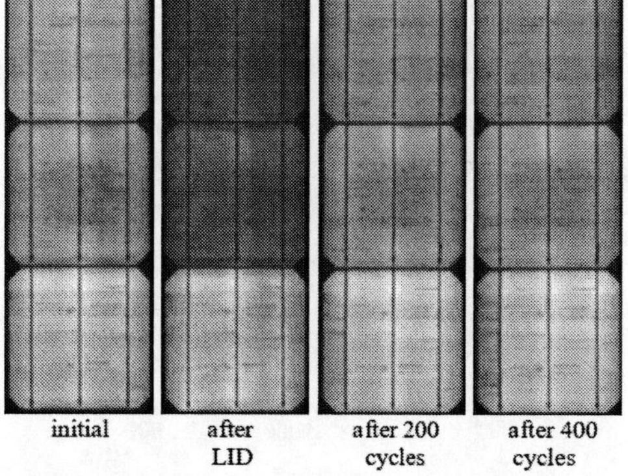

Figure 6: EL images of the 3-cell-string investigated in the upper graph in Figure 5. The pictures were taken under forward bias after consecutive LID and thermal cycling stress.

elevated temperatures in the thermal cycling test BO-complexes dissociate and both cells recover.

The EL images after 200 and 400 cycles virtually are the same as before LID. Hence, excellent stability against the applied stress is proven on the whole cell area.

Figure 7 shows the corresponding EL images of the *n*-type rear-junction cells. Again, there is no visible local degradation after thermal cycling. Additionally, the *n*-type Si cells don't suffer from LID as expected.

A close look at the electroluminescence images in

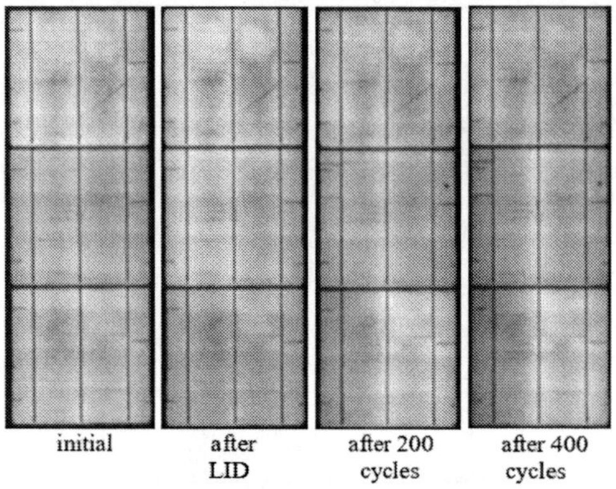

Figure 7: Forward bias EL images of the 3-cell-string investigated in the lower graph in Figure 5. The pictures were taken after consecutive LID and thermal cycling stress.

Figures 6 and 7 reveal that the cells' rear side electrical passivation is not damaged by ultrasonic tinning of the capping PVD Al layer except where the soldering iron first came in touch with the Al. Optimization of the tinning process e. g. by lowering the ultrasound power at the beginning of the tin busbars can probably eliminate the damage there.

Figures 8 and 9 show the normalised electrical performance parameters of 3-cell-modules measured before and after LID and after subsequent damp heat test (85 °C and 85 % relative humidity) and humidity freeze test (-40 °C to 85 °C at 85 % relative humidity). Again, highest power degradation was 4 % after double IEC 61215 testing. The EL images of the strings (not shown here) revealed no visible local degradation due to the climate tests.

3.4 High-efficiency TopCon cell

Table 1 shows the measured electrical performance parameters of a 10 x 10 cm² TopCon cell featuring a 1 µm thick PVD Al rear contact. An efficiency of 23.4 % was obtained. From the high open-circuit voltage of 713 mV and electroluminescence imaging (not shown here) we deduce that the PVD Al did not spike through the *n*-doped Si layer on the rear of the cell. However, the cell's series resistance is above what was expected due to a high contact resistance of about 150 mΩ cm² between Al and *n*-doped Si as measured on extra test samples. Hence, further process optimization is needed.

Note that the thermally evaporated PVD Al adheres well on the *n*-doped Si. Hence, it was possible to

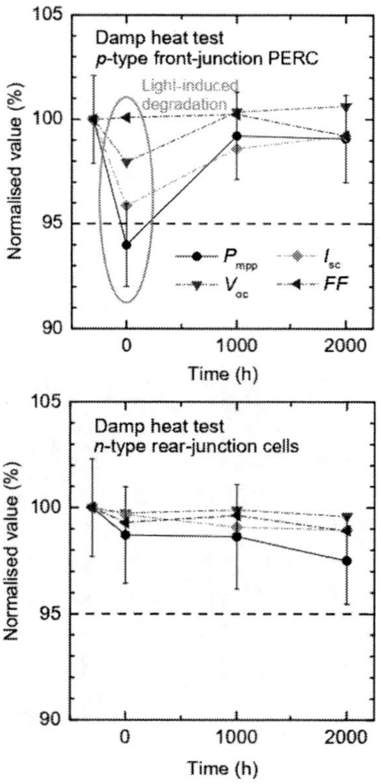

Figure 8: Electrical performance parameters of 3-cell-strings laminated in photovoltaic modules measured before and after LID and after subsequent damp heat test.

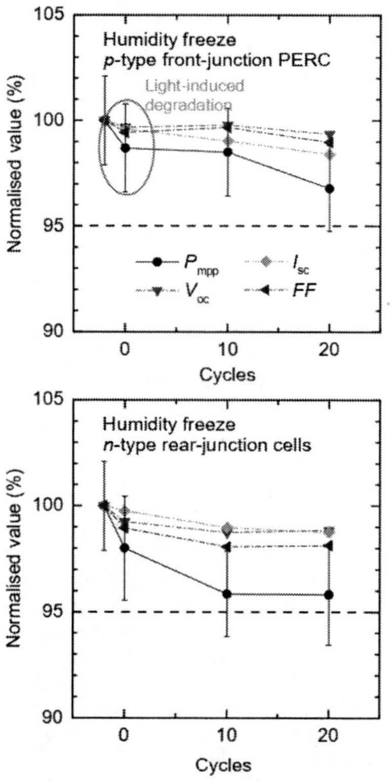

Figure 9: Measured electrical performance parameters of 3-cell-strings in modules before and after LID and after humidity-freeze cycles. Upper picture: *p*-type front junction PERC cells. Lower picture: *n*-type rear-junction cells.

ultrasonically tin the Al as described before for the *p*-type PERC and the *n*-type rear-junction cells.

Table 1: Measured illuminated *IV* parameters of a 100 cm² TopCon solar cell featuring PVD Al rear metal contact.

V_{oc} (mV)	J_{sc} (mA/cm²)	*FF* (%)	Eta (%)	R_{ser} (mΩ cm²)
713	40.3	81.5	23.4	0.59

4 SUMMARY AND CONCLUSION

Ultrasonic tinning of PVD Al was investigated for making the rear contact of high-efficiency Si solar cells solderable. Electroluminescence imaging revealed that the cells' rear electrical passivation was not damaged by ultrasonic tinning of the capping PVD Al except at very small positions where the soldering iron first came in touch with the Al. The measured average normalised 90° interconnector peel force after stringing was 0.5 N/mm on *p*-type PERC cells which does not yet comply with the standard DIN EN 50461 (> 1 N/mm). The failing interface was between ultrasonically deposited Sn and PVD Al.

A high stability of the prepared modules in the performed climate tests was found. After double thermal cycling, damp-heat and humidity freeze tests the electrical power degradation was below 5 %. No local degradation became obvious from EL images.

In order to demonstrate the performance of PVD Al as a rear contact, a 100 cm² TopCon cell was prepared reaching an efficiency of 23.4 %. The contact resistivity of the PVD Al on the *n*-doped Si layer limited the fill factor.

In conclusion, ultrasonically tinned PVD Al rear contacts are a reliable solution for manufacturing solderable high-efficiency Si solar cells. Ultrasonic tinning can easily be implemented at the end of the cell processing sequence and is characterized by a high technology readiness level. In order to increase the interconnector peel force, further optimization e. g. of the ultrasonic power is necessary.

ACKNOWLEDGEMENT

The authors thank the German Federal Ministry for Economic Affairs and Energy for funding within the project 'IdeAl' under contract number 0325889B.

5 REFERENCES

[1] H. v. Campe, S. Huber, S. Meyer, S. Reiff and J. Vietor, Proceedings of the 27th European Photovoltaic Solar Energy Conference, p. 1150, 2012.
[2] P. Schmitt, D. Eberlein, C. Ebert, M. Tranitz, U. Eitner and Harry Wirth, Energy Procedia 38 (2013) 380.
[3] V. Mertens, T. Ballmann, J. Cieslak, M. Kauert, A. Mohr, F. Stenzel, M. Junghänel, K. Suva, Ch. Klenke, G. Zimmermann and J. W. Müller, Proceedings of the 28th European Photovoltaic Solar Energy Conference, p. 714, 2013.
[4] F. Feldmann, B. Steinhauser, S. Kluska, M. Hermle and S. W. Glunz, this conference.
[5] A. J. McAlister and D. J. Kahan, Bulletin of Alloy Phase Diagrams 4 (1983) 410.
[6] B. Steinhauser, A. Büchler, H. Nagel, S. Kluska, J. Benick, P. Saint-Cast, S. Gutscher, J. Bartsch and Martin Hermle, Energy Procedia 124 (2017) 313.
[7] H. Nagel, M. Kamp, D. Eberlein, A. Kraft, J. Bartsch, M. Glatthaar and S. W. Glunz, Proceedings of the 32nd European Photovoltaic Solar Energy Conference, p. 48, 2016.
[8] J. Kumm, P. Hartmann, D. Eberlein and A. Wolf, Thin Solid Films 612 (2016) 393.
[9] J. R. Black, Proceedings of the 15th Int. Reliab. Phys. Symp. Las Vegas, p. 257, 1977.

EFFECTIVENESS OF THE IEC 60904-9 SPECTRAL MATCH CLASSIFICATION FOR INDUSTRIALLY-RELEVANT SILICON SOLAR CELLS

Harrison Wilterdink[1], Ron Sinton[1], Adrienne Blum[1], Eric Schneller[2], Kristopher O. Davis[2], Halden Field[3]
Phone: +1 (303)-945-2113, Fax: +1 (303)-945-2199, E-mail: harrison@sintoninstruments.com
[1]Sinton Instruments: 4720 Walnut St., Suite 102, Boulder, CO 80301
[2]Dept. of Materials Science and Engineering, Univ. of Central Florida: 4000 Central Florida Blvd, Orlando, FL 32816
[3]PV Measurements, Inc: Point Roberts, WA 98281

ABSTRACT: Solar simulator spectral class (determined by IEC 60904-9) does not correlate strongly to measurement uncertainty when measuring current-voltage (*I-V*) curves or rating the power of photovoltaic (PV) devices. An assessment of measurement uncertainty for these applications requires consideration of other factors besides the simulator spectrum. In this paper, we conduct a relatively simple uncertainty assessment for errors related to spectral mismatch, using methods based on IEC 60904-7 and statistical sampling theory. We demonstrate how such an assessment could be applied to a production line environment for silicon solar cells or modules.

Keywords: silicon, silicon solar cell, spectral response

1 INTRODUCTION

IEC 60904-9 classifies solar simulators on three metrics, one of which is their similarity to the integrated irradiance of the AM1.5G spectrum in six wavelength "bins." The classification is relatively simple, but is known to be ineffective in predicting measurement accuracy of simulators used for measuring current-voltage (*I-V*) curves and power rating of photovoltaic (PV) devices [1 – 4]. This ineffectiveness is at least partially due to a historical idiosyncrasy—parts of the spectral classification were adopted from ASTM E491 [5], a standard used for thermal balancing in space applications. PV absorbers "see" *photons*, whereas thermal absorbers "see" *energy* (irradiance); use of irradiance-based metrics for PV applications can lead to ambiguities in total photon flux, and thus, predicted photocurrent [6].

In comparison, IEC 60904-7 specifies a method to quantify errors due to spectral mismatch in *I-V* and power rating measurements. The method is more complex, requiring knowledge of the simulator spectrum, as well as spectral response of both the device under test and the reference device used to set the simulator's irradiance during measurement. However, the end result of the spectral mismatch calculation is a quantitative assessment of spectral mismatch error and a "spectral mismatch factor" (MMF) that can be used to correct measured data. For example, MMF = 1.01 indicates the solar cell's short-circuit current (I_{sc}) measured by the simulator is overestimated by 1 %, relative to the I_{sc} that would be measured under AM1.5G.

This paper demonstrates a straightforward application of IEC 60904-7 that can be used to estimate spectral mismatch errors that occur in production lines for wafer-based silicon PV technologies.

2 METHODS

2.1 EQE measurement of bare cells

We acquired one batch each of monocrystalline silicon passivated emitter and rear (mono-Si PERC) cells and multicrystalline silicon aluminum back surface field (multi-Si Al-BSF) cells. Each batch contained approximately 25 cells, and was manufactured using industrial processing techniques applicable to high-volume manufacturing. We measured the external quantum efficiency (EQE) of each cell at the University of Central Florida using a Flash QE system made by Tau Science [7]. For each cell, we measured EQE at 25 different points in a 5 x 5 grid across the cell, and then averaged the resulting curves. In principle, the averaging process makes the reported EQE less sensitive to the effects of individual regions of the cell with high or low optical performance, therefore allowing for more accurate prediction of photocurrent from the whole cell. Variability of the averaged curves across all cells in each sample set is shown in Figure 1.

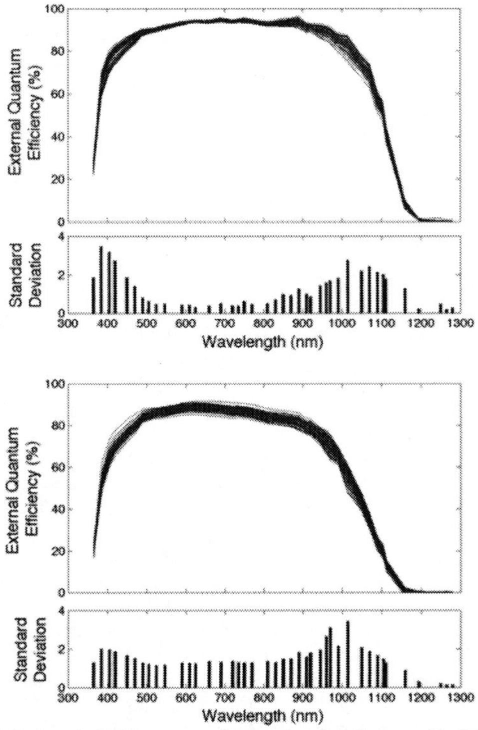

Figure 1: EQE curves from a batch of mono-Si PERC cells (top) and multi-Si Al-BSF cells (bottom) measured by the Tau Science Flash QE system [7]. The units of standard deviation are absolute percentage points.

2.2 Estimating EQE after encapsulation

The encapsulation process can decrease a device's EQE at some wavelengths (for example, by absorbing UV light) while increasing EQE at other wavelengths (for example, providing an optical interface that increases total internal reflection). The internal reflection gains can be significant, especially for acid-textured multi-Si samples.

We were unable to encapsulate the cells and measure the effect directly, so instead, we approximated the internal reflection phenomenon by measuring the EQE of a single cell while coated in ethanol. The ethanol does not absorb UV in the same way as some encapsulants, but it does have a roughly similar index of refraction ($n \approx 1.3 - 1.4$), which adequately simulates the internal reflection phenomenon.

For this experiment, EQE was measured on a bare cells and ethanol-coated cells using a QEX10 system at PV Measurements, Inc. Compared to the Flash QE system, the QEX10 measures over a wider range of wavelengths (especially in the UV) with finer wavelength resolution. For this experiment, EQE was only measured at a single point on each cell.

2.3 General MMF analysis

We calculated MMF for each cell using the measured spectral irradiance of several solar simulators that span different classes of spectral match. For each combination of simulator and cell batch, this resulted in a distribution of MMF values. For a given cell batch, the mean and standard deviation of the MMF distribution quantify two different components of the spectral mismatch error:

- Mean indicates systematic errors in measured cell current for the whole sample set.
- Standard deviation indicates random errors that arise from the interaction of each cell's EQE with the simulator spectrum. Note that EQE measurement reproducibility also factors in to the standard deviation, although we did not attempt to isolate the reproducibility effect in this work.

The methodology used here is a direct adaptation of work by Herrmann [7], who also calculated distributions of MMF values, but instead used a range statistic (the difference of maximum and minimum) to compare simulators. In addition, we compare MMF distributions when using a spectrally *matched* vs. *mismatched reference device*:

- *Matched*: the reference device is arbitrarily chosen as the first cell in the same sample set (*i.e.*, multi-Si reference for multi-Si devices).
- *Mismatched*: The reference device is arbitrarily chosen as the first cell in the opposite sample set (*i.e.*, mono-Si reference for multi-Si devices).

The theory underpinning IEC 60904-7 indicates that if the reference device is perfectly matched to the test device(s), all spectral mismatch errors should be eliminated. In reality, all devices within a given batch show some variation, so the reference device will never be a perfect match to all devices in the sample set. The matched reference device *mostly* eliminates systematic effects, but leaves random effects unchanged. For a given simulator and cell batch, comparing the two cases of matched and mismatched reference device shows the potential magnitude of systematic MMF error that might be encountered in practice, depending on the simulator itself and calibration procedures employed in its use.

2.4 Extending the MMF analysis to production lines

The calculations described in Section 2.3 can be applied to production line environments by framing the problem in the following manner: Assume that the batch of cells has been randomly sampled from a solar cell production line in which a simulator performs *I-V* measurements on every cell. Since no simulator can perfectly replicate AM1.5G, each cell on the production line will be measured with some non-zero error due to spectral mismatch (quantified by MMF).

The MMF distribution of the sample set can be used to infer details of the MMF distribution for the total population of production line cells. The appropriate analysis is readily found in applied statistics texts [9], and permits the construction of a confidence interval (CI) for both the mean (μ) and standard deviation (σ) of the production line MMF distribution. Throughout this paper, we report results using the common practice of a 95 % CI (for a more detailed description of the CI calculation, see Section 5).

3 RESULTS AND DISCUSSION

For brevity, analysis and results presented here are for the multi-Si Al-BSF sample set (results are qualitatively very similar for the mono-Si PERC sample set). Note that all values in Table I and Table II are written with precision out to the fourth decimal place *for convenience only*, in order to show differing numerical values for each simulator—this is not an indication of precision for the calculated values (for more information on measurement uncertainty, see Section 3.2)

3.1 Multi-Si Al-BSF cells

Table I and Figure 2 summarize the calculated MMF values with CIs on μ and σ for a theoretical production line. A few observations are worth noting from Table I:

First, regardless of how the simulator is calibrated (spectrally *matched* or *mismatched reference device*), the spread in MMF values, quantified by σ, is relatively constant. This is because the spread is primarily caused factors unrelated to the reference device, as discussed in Section 2.3. For all simulators shown here, the estimated σ is relatively small: on the order of 0.1 % – 0.3 % or less.

Second, the systematic measurement error, quantified by μ, strongly depends on how the simulator is calibrated. When using a spectrally *matched reference device*, all simulators have a systematic error ≤ 0.1 %. However, when using a spectrally *mismatched reference device*, there are drastic differences in performance between simulators; more specifically:

1) The continuous-illumination Xe + halogen system performs best overall. This is not surprising, considering that this simulator is used at a European test lab where high-accuracy is expected for many kinds of PV devices (not just silicon). Ideally for this application, the simulator must make accurate measurements regardless of differences in EQE between reference device and test device.
2) The pulsed simulator using five monochromatic LEDs performs the worst overall. As the choice of five LEDs perhaps indicates, it is likely this simulator was designed to have excellent adherence to the IEC 60904-9 standard for spectral match, which does not actually require that it

Table I: Summary of MMF distribution parameters calculated from multi-Si samples. Simulator spectral class is listed according to IEC 60904-9:2007, including the unofficial "A+" designation introduced by TUV Rheinland.

Simulator details			MMF-related uncertainty estimate (95 % CI)			
			Matched reference device		***Mismatched reference device***	
Operation	**Light source**	**Class**	μ	σ	μ	σ
Pulsed	Xe, unfiltered	C	1.0003 ± 0.0010	[0.0018, 0.0032]	0.9888 ± 0.0010	[0.0018, 0.0032]
Pulsed	Xe, optimized filtering for low Si MMF	C	1.0000 ± 0.0003	[0.0007, 0.0010]	0.9948 ± 0.0003	[0.0007, 0.0010]
Pulsed	Xe, optimized filtering for high class	"A+"	1.0003 ± 0.0001	[0.0002, 0.0004]	1.0103 ± 0.0001	[0.0002, 0.0004]
Pulsed	Five monochromatic LEDs	"A+"	1.0010 ± 0.0004	[0.0007, 0.0012]	1.0360 ± 0.0004	[0.0007, 0.0013]
Continuous	Xe + halogen, filtered	"A+"	1.0000 ± 0.0001	[0.0002, 0.0003]	1.0000 ± 0.0001	[0.0002, 0.0003]

produce light across all wavelengths where silicon is spectrally-active. Examination of the raw spectral irradiance of this simulator (not shown) indicates that no light is produced by the LEDs for λ < 400 nm and λ > 1000 nm. This simulator represents the clearest case where high class (according to IEC 60904-9) does not guarantee a low measurement uncertainty: for the mismatched reference case, the systematic error is 3.6 %!

3) The three pulsed Xe simulators perform similarly, despite the fact that they have very different spectral classes.
   a) Examination of the raw spectral irradiance data (not shown) indicates that the majority of spread in MMF from the unfiltered Xe simulator is due to significant light produced by atomic Xe emission peaks in the range 800 nm – 1000 nm, which is also a region of high EQE variability within the sample set.
   b) The Class "A+" Xe simulator uses shortpass filtering (cutoff ≈ 800 nm) to correct this problem. An unfortunate consequence of this approach is that the filtered light is very weak in the extended infrared (λ > 1000 nm), which is also a region of significant EQE variability within the sample set. Based on published spectral irradiance data from various simulators [2], we believe this kind of compromise (unintentional or otherwise) between high spectral class and accuracy in the infrared is quite common among pulsed Xe simulators.
   c) A slightly more sophisticated approach is to use notch filtering in order to decrease light intensity only in the 800 nm – 1000 nm range, leaving λ > 1000 nm untouched. This is the strategy employed by the Class C filtered Xe simulator (second row in Table I and Table II). The systematic error for this simulator is < 1 %, which is slightly better than the "A+" filtered Xe simulator. This again illustrates the shortcomings of the IEC 60904-9 letter grade concept.

To summarize, the MMF analysis allows us to compare systematic and random MMF errors for different simulators when measuring the same sample set. In all cases considered here, it is not possible to reliably estimate measurement uncertainty using *only* the simulator's spectral class from IEC 60904-9. This result has been well-known within the literature for some time [1 – 4, 6]. Total measurement uncertainty due to spectral mismatch effects depends not only on the simulator, but also on the spread in EQE for the devices being measured, as well as the reference device used to calibrate the simulator. One advantage of the present analysis is that it separates the errors into systematic and random effects by framing the discussion in terms a MMF distribution and its statistical parameters (μ, σ).

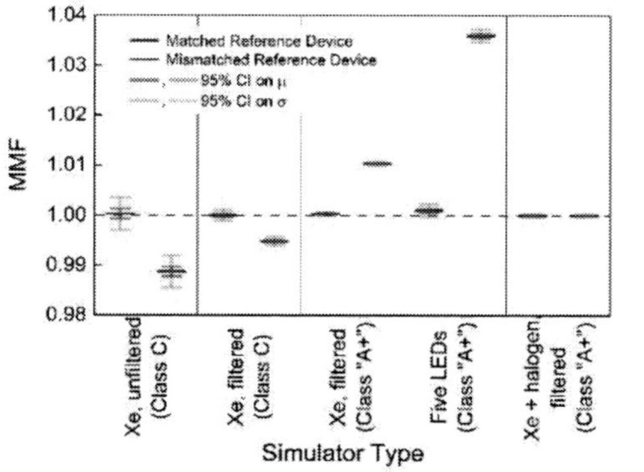

Figure 2: Graphical representation of the CI data given in **Table I**.

3.2 Estimation of MMF after encapsulation

Figure 3 shows the relative difference in EQE for one multi-Si Al-BSF cell with and without ethanol on the surface. Using the relative EQE gain observed in this cell, we estimated a "post-encapsulation EQE curve" for each cell and re-ran the MMF analysis of the previous section. Table II and Figure 4 summarize the calculated MMF values with CIs on μ and σ for a theoretical production line.

Overall, the MMF results are qualitatively very similar to that of the previous section. However, it is worth noting that magnitude of systematic error, quantified by μ, decreases slightly for the mismatched reference device

Table II: Summary of MMF distribution parameters calculated from multi-Si samples coated in ethanol. Simulator spectral class is listed according to IEC 60904-9:2007, including the unofficial "A+" designation introduced by TUV Rheinland.

Simulator details			MMF-related uncertainty estimate (95 % CI)			
			Matched reference device		*Mismatched reference device*	
Operation	Light source	Class	μ	σ	μ	σ
Pulsed	Xe, unfiltered	C	0.9982 ± 0.0014	[0.0026, 0.0045]	0.9928 ± 0.0014	[0.0026, 0.0045]
Pulsed	Xe, optimized filtering for low Si MMF	C	0.9996 ± 0.0004	[0.0008, 0.0013]	0.9946 ± 0.0004	[0.0008, 0.0013]
Pulsed	Xe, optimized filtering for high class	"A+"	1.0005 ± 0.0002	[0.0004, 0.0007]	1.0075 ± 0.0002	[0.0004, 0.0007]
Pulsed	Five monochromatic LEDs	"A+"	1.0016 ± 0.0007	[0.0012, 0.0021]	1.0302 ± 0.0007	[0.0013, 0.0022]
Continuous	Xe + halogen, filtered	"A+"	1.0001 ± 0.0001	[0.0002, 0.0003]	0.9992 ± 0.0001	[0.0002, 0.0003]

case. This is most likely due to the ethanol's effect in the "shoulder regions" of the EQE curve (approximately 400 nm – 500 nm and 900 nm – 1000 nm). For the bare cells, the shoulder regions have some of the greatest EQE variation (Figure 1), but the internal reflection phenomenon simulated by the ethanol tends to "flatten out" the shoulder regions and increase the EQE values closer to 100 %. The net effect is that every EQE curve becomes slightly more similar to the mono-Si reference, resulting in slightly less systematic MMF error.

It is also worth noting that spread of errors, quantified by σ, is slightly larger for the cells coated in ethanol. This is probably not related to the ethanol itself, but rather the difference in data collection methods used for this experiment (see Sections 2.1 and 2.2). For the ethanol dataset, EQE was measured at only one location on each cell (no spatial averaging across the total cell area). The resulting MMF values are likely affected by a few outlier measurements. Note that this is only a possible explanation, and we did not attempt to verify this quantitatively using the spatially-mapped EQE datasets.

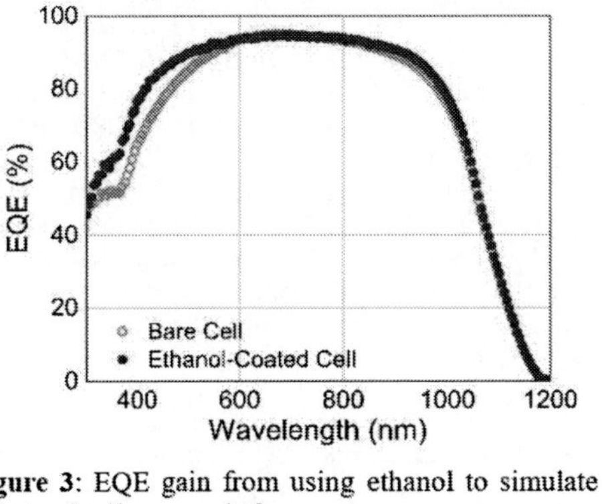

Figure 3: EQE gain from using ethanol to simulate the effects of cell encapsulation.

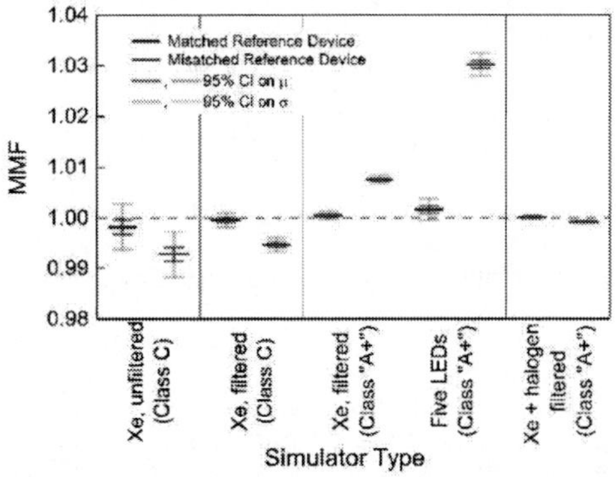

Figure 4: Graphical representation of the CI data given in **Table II**.

3.3 Uncertainty in MMF

One complication with the MMF-based approach presented here is propagation of uncertainty from measured quantities into the final result. The MMF computation of IEC 60904-7 requires three measured quantities: solar simulator spectral irradiance, spectral response of the device under test, and spectral response of the reference device. Uncertainties in all these measured quantities will affect the MMF calculation. Theoretical calculation of this uncertainty propagation from first principles is rather difficult, as the uncertainties can be wavelength-dependent. Furthermore, these values are numerically integrated in order to produce the MMF result, and the process of mathematical integration does not easily lend itself to direct calculation of propagated uncertainty. For additional information, we refer the reader to more in-depth work on the subject [10 – 12].

We can make a rough estimate of the uncertainty by comparing our theoretically computed J_{sc} values directly to measured values for the Class "A+" filtered Xe simulator (see Figure 5). The same samples and reference cell are used to obtain both the theoretical and measured J_{sc} values, the only difference is that the theoretical values are computed using measured EQE and spectral irradiance values with some unknown uncertainty. On average, the approximate difference between theoretical and measured J_{sc} is about 0.1 mA cm^{-2}, or close to 0.3 %. We take this as a rough estimate for the uncertainty in MMF, which is within the same order-of-magnitude as Monte-Carlo-based simulation results [3].

The MMF differences stated between Xe simulators in Table I and Table II should be interpreted with this 0.3 % uncertainty estimate in mind. In those tables, we recorded the MMF values with precision to the fourth decimal place, which is obviously not justified in terms of actual measurement precision. From practical point of view, an apparently large difference in simulator spectral class (C vs. "A+") does not always translate to a significant and measurable difference in simulator performance with respect to spectral mismatch errors.

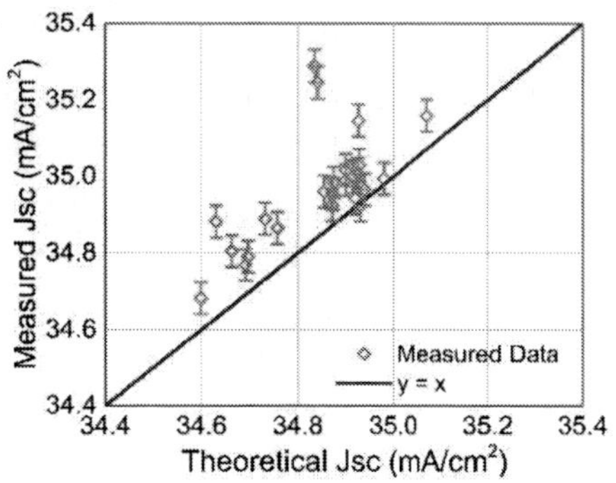

Figure 5: Measured solar cell J_{sc} vs. theoretical J_{sc}. Error bars are ± 0.12 % ($k = 2$), estimated based on measurement reproducibility for two cells in the sample set.

4 SUMMARY

When measuring *I-V* curves or power rating PV devices, the simulator spectral class per IEC 60904-9 does not generally correlate with measurement uncertainty. A MMF-based approach shows that measurement uncertainties depend not only on the simulator, but also on the variability of the samples being measured (spread of sample EQE curves) and how the simulator is calibrated (EQE of the reference device).

In the best case, when a simulator is calibrated using a spectrally *matched reference device*, the uncertainties among all simulators are quite comparable. In the worst case, when using a spectrally *mismatched reference device*, we even showed one simulator would have a systematic error of 3.6 %. The primary problem in that case was lack of full-spectrum illumination in the regions where silicon is spectrally-active, specifically a lack of light in the UV and IR. This is all despite the apparent "A+" rating of the simulator in question.

For the Xe simulators considered here when measuring silicon cells, there is no significant difference in measurement uncertainty between a simulator with Class C vs. "A+" spectral match. In the worst case, it is actually quite possible for the spectral class ranking to be inversely related to the measurement uncertainty (Class "A+" having a greater uncertainty than lower-class systems [3]).

When employing a spectrally matched reference device, spectral mismatch errors are likely to be quite small. Many other measurement uncertainties are likely to dominate the overall measurement uncertainty; for example, capacitance effects in high-efficiency silicon devices [13] and angle-of-incidence effects in simulators with a significant amount of diffuse illumination [4].

5 ADDITIONAL INFORMATION

For a sample set with mean $\bar{x}$ and standard deviation s taken from a normal distribution with unknown μ and σ, the (100 – α) % CI for μ is given by:

$$\bar{x} - \frac{s}{\sqrt{n}} t_{\alpha/2,n-1} \leq \mu \leq \bar{x} + \frac{s}{\sqrt{n}} t_{\alpha/2,n-1}$$

where n is the number of samples, and $t_{\alpha/2,n-1}$ denotes the percentage point of the t-distribution with $n - 1$ degrees of freedom, such that Prob{ $t_{n-1} \geq t_{\alpha/2,n-1}$ } = $\alpha/2$. The (100 – α) % CI for σ is given by:

$$s^2 \frac{n-1}{\chi^2_{\alpha/2,n-1}} \leq \sigma^2 \leq s^2 \frac{n-1}{\chi^2_{1-\alpha/2,n-1}}$$

where $\chi^2_{\alpha/2,n-1}$ denotes the percentage point of the chi-square distribution with $n - 1$ degrees of freedom, such that Prob{ $\chi^2_{n-1} \geq \chi^2_{\alpha/2,n-1}$ } = $\alpha/2$. Note that the CI for μ is relatively insensitive to the normality assumption, whereas the CI for σ is not [9].

6 REFERENCES

[1] W. Herrmann et al., "Advanced Intercomparison Testing of Testing of PV Modules in European Test Laboratories," in *Proc. 22nd EU PVSEC* (Milan, 2007).

[2] W. Herrmann and L. Rimmelspacher, "Uncertainty of Solar Simulator Spectral Irradiance Data and Problems with Spectral Match Classification," in *Proc. 27th EU PVSEC* (Frankfurt, 2012).

[3] C. Monokroussos *et al.*, "IEC 60904-9 Spectral Classification and Impact on Industrial Rating of c-Si Devices," in *Proc. WCPEC-6* (Kyoto, 2014).

[4] C. Monokroussos *et al.*, "Effect of Glass/EVA/Backsheet Encapsulation Package on the Industrial Rating of c-Si PV Modules," in *Proc. 31st EU PVSEC* (Hamburg, 2015).

[5] C. Osterwald, personal communication (July 3, 2017).

[6] R. Sinton *et al.*, "Critical Evaluation of the Foundations of Solar Simulator Standards," in *Proc. 44th IEEE PVSC* (Washington D.C., 2017).

[7] E. Schneller *et al.*, "Crystalline Silicon Device Loss Analysis through Spatially Resolved Quantum Efficiency Measurements," in *IEEE Journal of Photovoltaics* 2017; 7: 957 – 965. DOI: 10.1109/JPHOTOV.2017.2689160

[8] W. Herrmann, "Advances in Spectral Irradiance Analysis of Solar Simulators," in *Proc. 26th PVSEC* (Singapore, 2016).

[9] D.C. Montgomery, Introduction to Statistical Quality Control (7th Edition): John Wiley & Sons Inc., 2013.

[10] J. Hohl-Ebinger and W. Warta, "Uncertainty of the Spectral Mismatch Correction Factor in STC Measurements on Photovoltaic Devices," in *Prog. Photovolt: Res. Appl.* 2011: **19**:573 – 579. DOI: 10.1002/pip.1059

[11] H. Field and K. Emery, "An Uncertainty Analysis of the Spectral Correction Factor using Wavelength-Dependent Error Sources," in *Proc. 23rd IEEE PVSC* (Kentucky, 1993).

[12] F. Plag *et al.*, "Comprehensive Analysis of a Pulsed Solar Simulator to Determine Measurement Uncertainty Components," in *Proc. 29th EU PVSEC* (Amsterdam, 2014).

[13] R. Sinton, H. Wilterdink, and A. Blum, "Assessing Transient Measurement Errors for High-Efficiency Silicon Solar Cells and Modules," to be published in *IEEE Journal of Photovoltaics*.

Properties of multi-crystalline silicon ingot grown by self-nucleating crucible.

J. Laurent[1,a], G. Rancoule[a], E.Drode[a], C. Reimann[b,c], M. Trempa[b], C. Kranert[c], J.Friedrich[b,c], L.Teale[d], R. Dyer[d], I.Dorrity[d]

[1] *Corresponding author*
[a]*Vesuvius France, 68 rue Paul Deudon, 59750 Feignies, France*
e-mail / phone number = julien.laurent@vesuvius.com / (033) 3 27 69 71 31

[b]*Fraunhofer IISB, Schottkystr. 10, 91058 Erlangen, Germany*

[c]*Fraunhofer THM, Am St.-Niclas-Schacht 13, 09599 Freiberg, Germany*
e-mail / phone number = Christian.Reimann@iisb.fraunhofer.de / +49-(0)9131/761-272

[d]*PV Crystalox Solar, Brook House, Abingdon, Oxfordshire, UK, OX14 4SE*
e-mail / phone number = leslie.teale@pvcrystalox.com / +44 (0)1235 437175

ABSTRACT: High performance multi-crystalline silicon (HP-mc Si) wafers are usually grown by directional solidification of molten silicon from an unmelted layer consisting of silicon grains. The HP-mc Si ingots allow producing solar cells with high efficiency by decreasing the structural defects in silicon, especially dislocation clusters. Nevertheless, it has the disadvantage of losing important usable ingot length because of a remaining unmelted silicon layer and a larger bottom red zone than for standard multi-crystalline silicon ingots. Further, longer process times are needed for the careful melting process. A self-nucleating crucible where nucleation is directly governed by the crucible bottom without any silicon grain layer has been developed to overcome these issues.
The new developed Vesuvius HP-mc crucible was tested in laboratory G1 as well as in industrial G4 scale and was compared to standard crucibles. Within the study grain structure properties as well as recombination active defect content and cell efficiencies on ingots and wafers obtained by the two crucible types were explored.
The results show clearly that ingots originating from the self-nucleating crucible can reach a similar quality than HP-mc ingots grown by a silicon seeding process. It will be shown that using a self-nucleating crucible can lead to a significant lower dislocation content and higher cell efficiencies and then subsequently can contribute to a significant wafer cost reduction.

Keywords: crystallization, high performance mc-Si, crucible, cell efficiency.

1 INTRODUCTION

High performance multi-crystalline material (HP mc-Si) is actually supplanting conventional multi-crystalline silicon (mc-Si) wafers and its market share will be increasing over the next years up to 50% [1]. By this new casting technology, where small silicon seeds were used to obtain a small grained microstructure, high efficiency silicon solar cells can be generated [2-6]. But this process induces larger bottom red zone and lost area due to remaining unmelted seed. Further, it takes more time to carefully melt the silicon feedstock to remain the silicon seed layer. To reduce the cost of this process, a self-nucleating crucible has been developed which allows generating a HP-mc grain structure including its profitable properties by complete melting of the silicon feedstock. In consequence the drawbacks of higher yield loss in the bottom region and the higher time consume are eliminated.

2 EXPERIMENT

2.1 Crystallization

In this study, three different kinds of ingots were crystallized. 1) "Standard mc-Si" ingots using a RTU Vesuvius solar crucible, 2) "Reference HP-mc Si" ingots with silicon seeds put on the bottom of the RTU Vesuvius crucible, 3) "Vesuvius HP-mc Si" ingots using the Vesuvius HP-mc self-nucleating crucible. Laboratory scale (G0: 1,1 kg and G1: 15 kg) and industrial scale (G4: 280 kg) ingots are used in this study. For all experiments Wacker polysilicon feedstock doped with boron for 2 Ohm cm resistivity were used.

The reference HP-mc Si technology consists on feeding the crucible with silicon seeds on the bottom covered by silicon feedstock. Then the melting phase is crucial because the seeds must be partially melted to keep a seeding layer. By lower the temperature of the bottom crucible, the ingot growth is initiated until obtaining the so-called HP-mc Si ingot.

The Vesuvius HP-mc crucible consists on a nucleating layer on the bottom crucible which allows the ingot manufacturer to resign on any silicon grain layer and obtain an ingot with comparable quality as a standard HP-mc Si ingot but with a smaller bottom red zone.

2.2 Samples preparation and characterization

For characterization of the grain structure properties, the G0 ingots (Ø100mm, 60mm height) were vertically halved followed by horizontal cutting at 30 mm ingot height. The G1 ingots (220x220x130 mm^3) were vertically cut in the center followed by horizontal cutting of 156x156mm^2 central slices at 20 mm and 80 mm ingot height. The G4 ingots were cut into 16 bricks. The central brick was cut into wafers over ingot height.

The grain structure properties (grain size, grain orientation distribution and grain boundary type distribution) were measured with a grain detector tool in combination with the so called "Laue Scanner" [8]. The photoluminescence imaging measurements were done with an OPTECTION imaging tool.

2.3 Solar cells

Solar cells were made from single wafers over brick height from the G4 ingots by using a mc-PERC process consisting of phosphorous diffusion, rear side SiON/SiN passivation and metallization.

3. RESULTS

3.1 G0 and G1 experiments (laboratory scale)

3.1.1 Grain structure

The grain structure of grown for G0 and G1 ingots are shown in figure 1 and figure 2. In figure 2c, the silicon seeds are shown at the bottom of the ingot which leads to an increase of the red zone size of several millimeters compared to the standard mc-Si ingot. It can clearly be seen from Fig. 1b/2b that the Vesuvius HP-mc crucible leads to finer grain structure than in the standard mc ingot (1a/2a) and almost as fine as the reference H-mc ingots (1c/2c) for both type of ingot for both G0 and G1 scale.

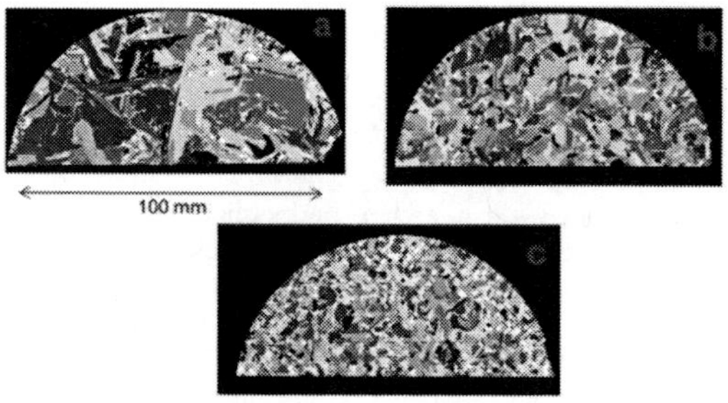

Figure 1: Grain structure images from horizontal cuts at ~30 mm ingot height (G0 ingot) for a) Standard mc-Si, b) Vesuvius HP-mc and c) Reference HP-mc Si ingots.

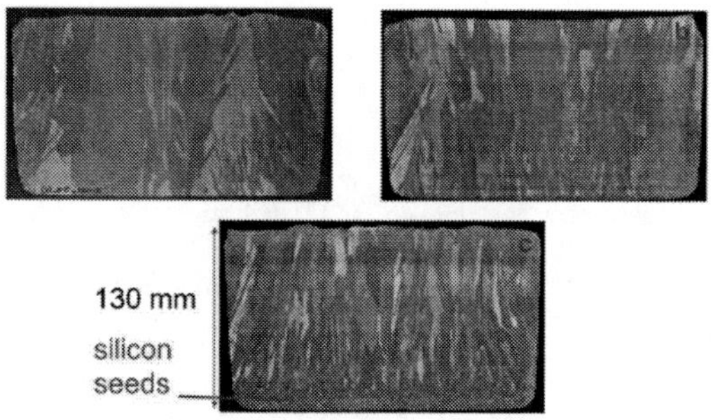

Figure 2: Grain structure images from vertical cuts (G1 ingot) for a) Standard mc-Si, b) Vesuvius HP-mc Si and c) Reference HP-mc Si ingots.

Figure 3 and figure 4 show the orientation distribution on horizontal cuts for the G0 ingots and on wafers (156×156 mm²) for the G1 ingots. We can clearly see that the Vesuvius HP-mc crucible (3b/4b) leads to a homogeneous orientation distribution similar to the reference HP-mc Si ingot (3c/4c).

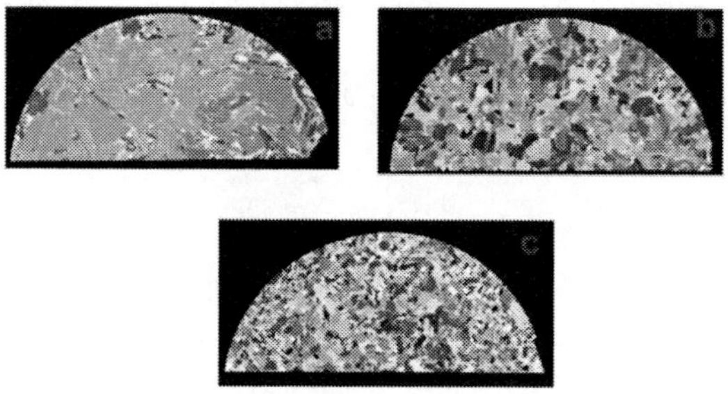

Figure 3: Orientation distribution at ~30 mm ingot height (G0 ingot) for a) Standard mc-Si, b) Vesuvius HP-mc Si and c) Reference HP-mc Si ingots.

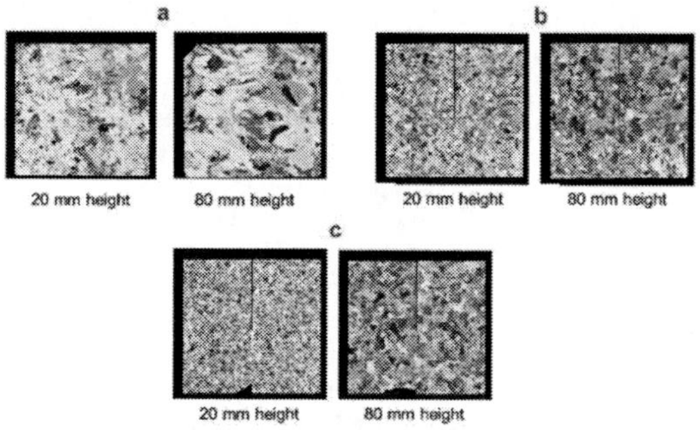

Figure 4: Orientation distribution at ~20 mm and 80 mm ingot height (G1 ingot) for a) Standard mc-Si, b) Vesuvius HP-mc Si and c) Reference HP-mc Si ingots.

The grain boundaries are then studied on the wafers from the previous G1 ingots. The results are displayed in figure 5. Vesuvius HP-mc crucible increases the so-called random grain boundaries (R) from 15/22% up to 35/55%, which is helpful to restrain the movement of dislocations through the ingot. However, the random fraction increases with crystal height which is not a typical trend for HP-mc silicon seeded on silicon [7]. Indeed, in the figure 5, the reference HP-mc ingot shows a decreasing trend for the random fraction.

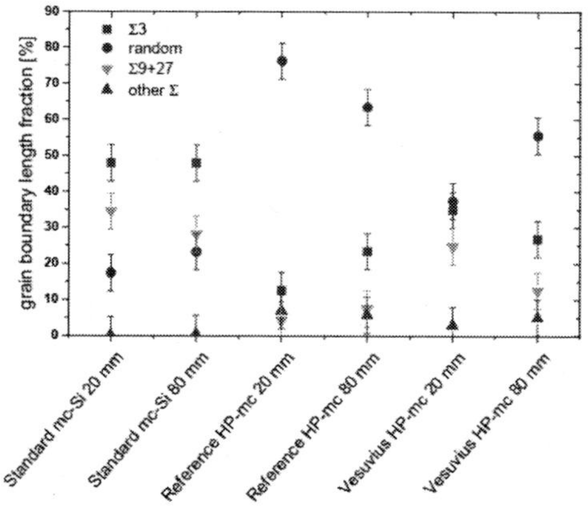

Figure 5: Grain boundary type distribution at ~20 mm and 80 mm ingot height (G1 ingot) for Standard mc-Si (left), Reference HP-mc Si (middle) and Vesuvius HP-mc Si (right) ingots.

Concerning Σ_3 grain boundaries, the values are much higher for the standard mc-Si ingot (>50%) than for the two others (between 10% and 30%). It clearly shows that the HP-mc technology exhibits lower twin content. As the twins allow dislocations to cross over them resulting in many clusters over the ingot height, it is a very positive result to lower the amount of twins by the HP-mc technology.

3.1.2 Photoluminescence measurements

Photoluminescence measurements (figure 6) are made to investigate the recombination active defects. Recombination active dislocation area seems to be indirect proportional to the length fraction of random grain boundaries. However, the recombination active area for the ingot with R=55% is still high, which pronounces that the single R-fraction value on these ingot height is not the key factor.

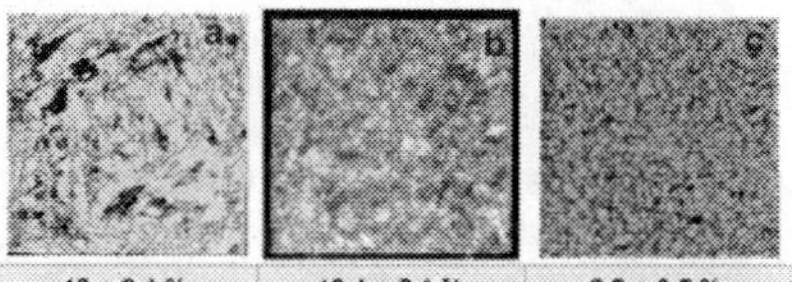

Recombination active area	12 ± 2.4 %	10.4 ± 2.1 %	2.3 ± 0.5 %
Random length fraction	23%	55%	63%

Figure 6: PL measurement and recombination active dislocation area values compared to random length fraction for a) Standard mc-Si, b) Vesuvius HP-mc Si and c) Reference HP-mc Si ingots.

In conclusion from the laboratory scale ingots, the Vesuvius HP-mc crucible results in comparable values in term of grain structure (grain size, orientation) as the Reference HP-mc ingot except the fact that the initial fraction of random grain boundaries is not as and therefore the R-fraction trend over the ingot height is increasing which is rather typical for a Standard mc ingot than for a Reference HP-mc ingot.

3.2 G4 experiments (industrial scale)

3.2.1 Grain structure

Figure 7 shows wafers from the central brick of G4 ingots in the bottom region (20 mm height) by using the Vesuvius HP-mc crucible and a standard crucible. The Vesuvius HP-mc crucible obviously exhibits smaller silicon grain size in comparison with the standard mc-Si crucible. This is also shown in figure 8 where the mean grain size of Vesuvius HP-mc Si ingot is between standard mc-Si and the reference HP-mc Si.

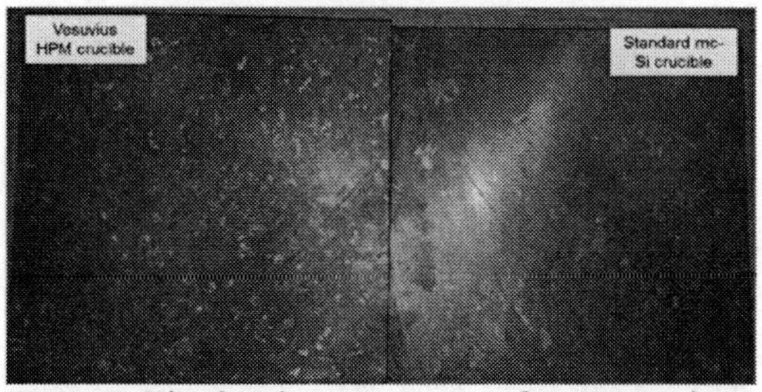

Figure 7: Visual grain structure on wafers (20 mm ingot height) for Vesuvius HP-mc crucible (left) and Standard mc-Si crucible (right).

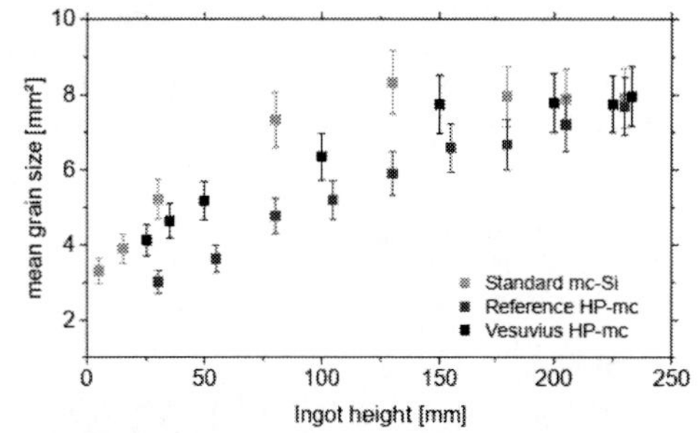

Figure 8: Mean grain size vs. ingot height for the standard mc-Si, the Vesuvius HP-mc Si and reference HP-mc Si ingots.

The grain boundary fraction is exposed in figure 9 for the three studied ingots for different ingot heights. Vesuvius HP-mc crucible leads to similar random fraction as the reference HP-mc and here a decreasing random fraction trend over ingot height analogue to the reference HP-mc ingot could be obtained.

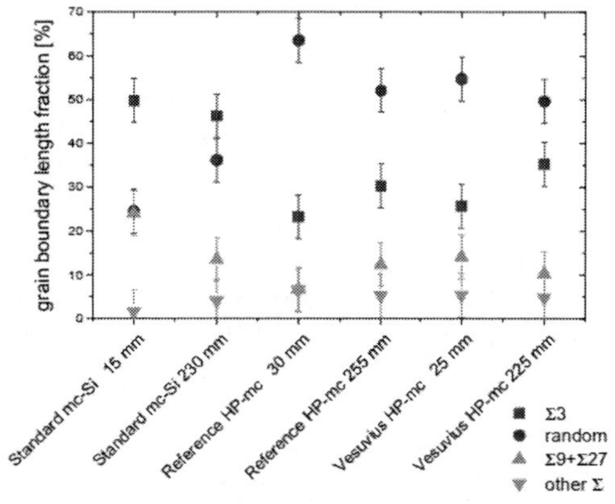

Figure 9: Grain boundary type distribution on industrial scale (G4) for Standard mc-Si (left), Reference HP-mc Si (middle) and Vesuvius HP-mc Si (right) ingots.

3.2.2 Photoluminescence measurements

Figure 10 shows the recombination active area fraction for the three G4 ingots.

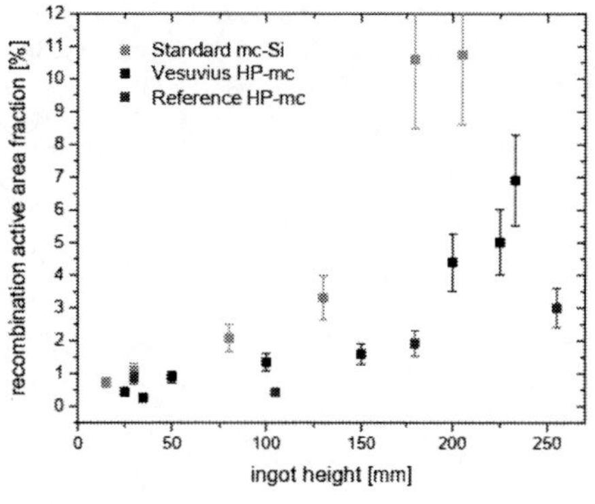

Figure 10: Recombination active area fraction vs ingot height for the standard mc-Si, the Vesuvius HP-mc and

reference HP-mc Si ingots.

The increase of recombination area over ingot height is much smoother in the case of Vesuvius HP-mc crucible compared to the Standard mc-Si ingot. And the recombination area is somewhere between Standard mc-Si and Reference HP-mc Si.

From these results, it can be stated that a sufficiently high random fraction at the ingot bottom – leading to a decreasing trend over ingot height - results in small recombination active area fractions (= low dislocation content) and therefore to high quality HP-mc Si material.

3.2.3 Cell efficiency

Figure 11 shows clearly that the Vesuvius HP-mc crucible leads to higher efficiencies (~0.1 to 0.5% abs.), especially in the upper 2/3 of the ingots. The highest random fraction and lowest recombination active area fractions obtained with the Vesuvius HP-mc crucible shows the highest efficiencies in comparison with a standard mc-Si. Unfortunately, the data for the reference HP-mc are missing but will be available later.

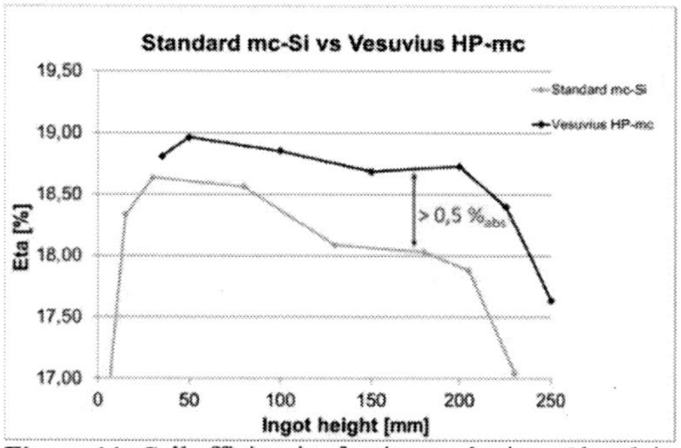

Figure 11: Cell efficiencies for the standard mc-Si and the Vesuvius HP-mc Si ingots.

4. CONCLUSION

Vesuvius developed a HP-mc crucible which permits to generate a "close-to-HPM" grain structure by applying a standard multi process including a non-time consuming complete melting process. Ingots grown with this crucible exhibit good grain structure properties and therefore shows excellent material quality comparable to HP-mc ingots seeded on silicon feedstock. The drawbacks of classical HP-mc method (larger bottom red-zone, longer process times) are solved by this technology.

Finally, by using this Vesuvius HP-mc crucible, the efficiency is increase up to 0.5%abs especially in the upper 2/3 of the ingots. Nevertheless, this crucible must be customized with every customer to fit the best with their furnace condition.

ACKNOWLEDGEMENT

The authors would like to thanks the International Solar Energy Research Center Konstanz e.V. for the solar cell processing and their characterization.

REFERENCES

[1] International Technology Roadmap for Photovoltaic, www.itrpv.net, (2017).

[2] Y.M. Yang, A. Yu, B. Hsu, W.C. Hsu, A. Yang, C.W. Lan, Development of high performance multicrystalline silicon for photovoltaic industry, Prog. Photovolt.: Res. Appl. 23 (2015) 340–351.

[3] C.W. Lan, Grain control in directional solidification of photovoltaic silicon, presented in part at the 5th International Workshop on Crystal Growth Technology, Berlin, Germany, June, 2011.

[4] C.W. Lan, W.C. Lan, T.F. Li, A. Yu, Y.M. Yang, C. Hsu, B. Hsu, A. Yang, Grain control in directional solidification of photovoltaic silicon, Journal of Crystal Growth 360 (2013) 68–75.

[5] C. W. Lan, Grain control in directional solidification of photovoltaic silicon, Presented in Part at the 5th International Workshop on Crystalline Silicon Solar Cells, Boston, USA, Nov., 2011.

[6] C.W. Lan, The emergence of high-performance multi-crystalline silicon in photovoltaics, Journal of Crystal Growth 468 (2017) 17-23.

[7] M. Trempa, I. Kupka, C. Kranert, T. Lehmann, C. Reimann, J. Friedrich, Evolution of grain structure and recombination active dislocations in extraordinary tall conventional and high performance multi-crystalline silicon ingots, Journal of Crystal Growth 459 (2017), 67-75.

[8] T. Lehmann, M. Trempa, E. Meissner, M. Zschorsch, C. Reimann, J. Friedrich, Laue scanner: A new method for determination of grain orientations and grain boundary types of multicrystalline silicon on a full wafer scale, Acta Materialia 69 (2014) 1–8

ECO-SOLAR FACTORY: MULTICRYSTALLINE SILICON INGOT CRYSTALLISATION FROM REUSABLE SILICON NITRIDE CRUCIBLES

M.P. Bellmann[1], G. Stokkan[1], A. Ciftja[2], R. Roligheten[2], J. Denafas[3], F. Buchholz[4], T. Kaden[5]

[1]SINTEF Materials and Chemistry, Trondheim, Norway;
e-mail: Martin.Bellmann@sintef.no
[2]Steuler Solar Technology AS, Porsgrunn, Norway
[3]UAB Soli Tek R&D, Vilnius, Lithuania
[4]International Solar Energy Research Center - ISC Konstanz, Konstanz, Germany
[5]Fraunhofer THM, Freiberg, Germany

ABSTRACT: Preliminary results on silicon ingot crystallisation from reusable silicon nitride crucibles are reported. High Performance multicrystalline silicon ingots (HPmc-Si) using granular polysilicon for seeding were grown. The obtained material quality was investigated at ingot, wafer and cell level. Finally, PV modules were manufactured to demonstrate the potential of such material at commercial scale. So far, the silicon nitride crucible was used in five subsequent runs. Material performance and cell efficiencies are comparable to reference material grown from high purity electronic grade quartz crucibles. Two 60 cell modules were assembled and the module power output is comparable to commercial modules of the same category.

Keywords: silicon nitride crucibles, crystallisation, HPmc-Si, crystalline silicon solar cells, reuse

1 INTRODUCTION

Silicon at high temperature reacts easily with almost everything, and due to purity and mechanical requirements, manufactures are restricted to very few crucible materials. Fused silica and fused quartz glass remains the material of choice for crucible and mould applications because it is readily available in high purity form and presents little reactive to liquid silicon. The four largest drawbacks of these materials are:

- being a thermal insulator
- oxygen contamination
- the need to employ a silicon nitride powder coating in order to prevent sticking of the crucible to the ingot
- cost related due to single use.

Fig. 1: Cracked silica (l.) and intact silicon nitride (r.) crucible after silicon ingot crystallisation.

Silica-based crucibles contribute with up to 30% to the conversion cost from silicon-feedstock to the as-grown ingot. Due to a phase transformation at high tempereratures the crucible cracks during the cooling process. After usage, the crucible is discarded and landfilled. Very high purity quartz is used for crucibles used for CZ monocrystalline silicon, but this is not economical for the crucibles used to produce HPmc-Si, and lower quality grade sand is used instead.

Crucibles based on silicon nitride are potential successors to the more traditional silica-based crucibles for growth of silicon ingots. The immediate advantages of using silicon nitride crucibles in multicrystaline silicon poduction are:

- their increased resilience, which enables them to be reused multiple times,
- very high purity (comparable to the silica crucibles for CZ silicon)
- the lower oxygen content and
- decreased thermal resistance.

Steuler Solar Technology has developed a concept for reusable crucibles based on advanced silicon nitride ceramics. Silicon nitride crucibles are produced by a novel state-of-the-art slip casting technique and subsequent nitridation process (patented) which include the following steps: pure silicon powder, water and other chemicals are mixed together to make a slurry, which then is poured into a plaster mould for slip. After water is absorbed by the plaster, the "greenware" crucible is produced and then removed from the plaster. Nitridation of the "greenware" crucible occurs at high temperatures (above 1000 °C) and produces the final silicon nitride crucible.

The technical potential of these crucibles for crystallisation of multicrystalline crystalline silicon growth has been investigated at SINTEF.

2 EXPERIMENTAL DETAILS

Crystallisation experiments have been conducted in an induction furnace (Crystalox DS250) for directional solidification of cylindrical ingots. Crucibles have been charged with 15 kg of silicon feedstock. To obtain HPmc-Si, FBR-granules were placed in the bottom of the crucible, covered by a layer of poly-Si chunks. The material has been doped with boron to achieve p-type conductivity. For the reference material fused quartz crucibles (Electronic Grade used for CZ silicon growth) were used. Both crucible types, fused quartz and silicon nitride were coated with α-Si_3N_4 and burned to prevent sticking.

The material was characterized by a range of methods including Fourier Transform Infra-Red (FTIR) to determine segregation profiles of substitutional carbon C_s, interstitial oxygen O_i and nitrogen complexes NN and NNO, Neutron Activation Analysis (NA) to determine content of metallic impurities, lifetime analysis including Microwave Photoconductance Decay (μPCD) and Quasi-Steady-State Photoconductance (QSSPC).

Conventional solar cells with full-area aluminium back surface field (BSF) were manufactured by SoliTek and assembled into monofacial glass/glass modules.

3 RESULTS

3.1 HPmc-Si ingot growth

Five ingots were grown from the same silicon nitride crucible, resulting in four times reuse. With the help of a vacuum pad the ingot was released from the crucible. The crucible remained intact. The coating was scratched during ingot removal and the crucible was cleaned and re-coated before each run.

Fig. 2: Five mc-Si ingots grown from the same silicon nitride crucible.

3.2 Impurities

Oi-concentrations (shown in Fig. 3) are relatively stable for ingots grown in fused quartz crucibles. For the silicon nitride crucibles, there is a clear reduction towards the top. At 8 cm distance from the bottom the shift suggests that silicon nitride crucibles can outperform quartz crucibles, especially for taller ingots. An explanation for the initial high oxygen might be relegated to the oxidation of the pre-coated silicon nitride crucible which is expected during burning, with formation of silicon oxynitride.

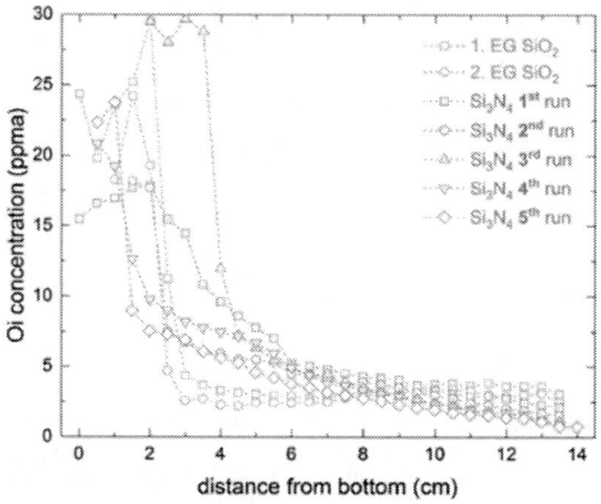

Fig. 3: Longitudinal Oi-segregation profiles for ingots grown in quartz (orange) and silicon nitride (blue) crucibles.

The material grown in a silicon nitride crucible shows higher amount of dissolved carbon (Fig. 4) than that grown in the quartz crucibles. A potential source for this is possibly the crucible itself, which contains small amounts of silicon carbide. Another possible explanation would be contamination from the CO gas produced during smelting/solidification of silicon from the reaction of SiO gas (emitted from the silicon melt and the crucible) and the carbon from the lining refractories. However, we believe the former route is unlikely to be the source of carbon, as in a previous work carried out at SINTEF it was found that when silicon nitride crucibles are employed for CZ silicon growth, it results with very low SiO evaporation (hence, very low oxygen content in the silicon crystal) but with higher carbon contamination.

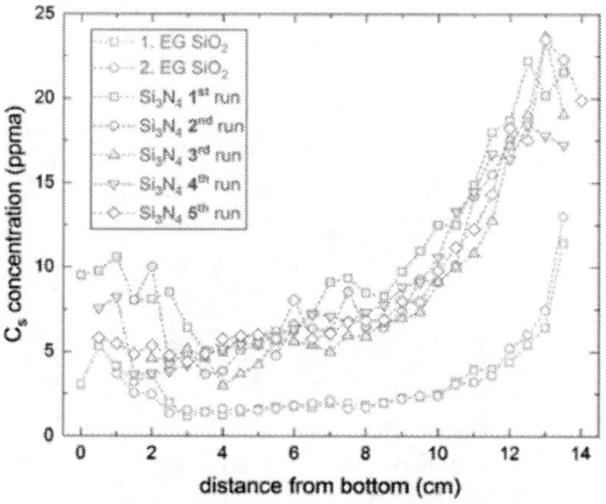

Fig. 4: Longitudinal Cs-segregation profiles for ingots grown in quartz (orange) and silicon nitride (blue) crucibles.

Nitrogen concentration levels (Fig. 5) is comparable with reference material grown from fused quartz crucibles. We would like to point out that the nitrogen concentration levels are primarily above the solid solubility limit of nitrogen (0.09 ppma). A possible explanation could be the increased nitrogen solubility due to the presence of boron, carbon and oxygen (multicomponent system).

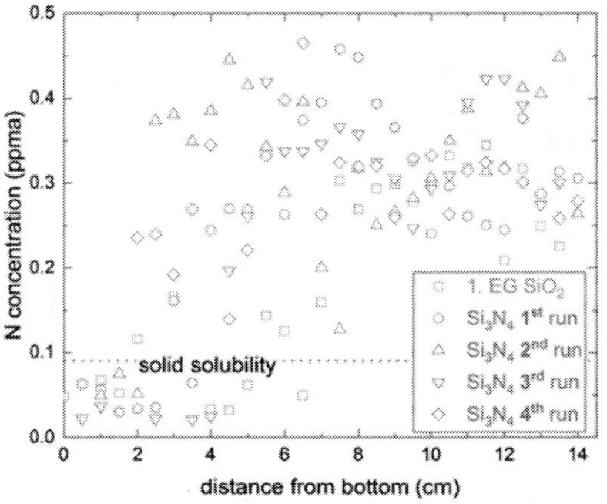

Fig. 5: Longitudinal Cs-segregation profiles for ingots grown in quartz (orange) and silicon nitride (blue) crucibles.

Comparable metallic impurity concentration levels were revealed in material grown from high purity quartz and silicon nitride crucibles. For material grown from silicon nitride crucibles, iron was even below the detection limit.

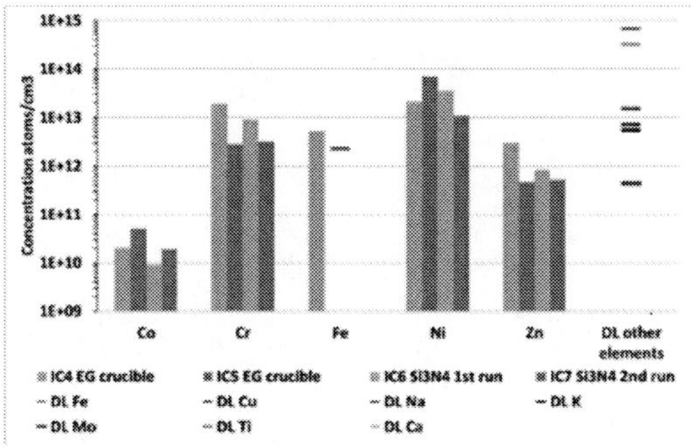

Fig. 6: Metallic impurity levels measured in material crystallised in quartz (orange) and silicon nitride (blue) crucibles.

3.3 Charge carrier lifetime

Fig. 7 shows results from QSSPC measurements. Except for the ingot which was obtained after first time of reusing the silicon nitride crucible, bulk carrier lifetimes are in the same range.

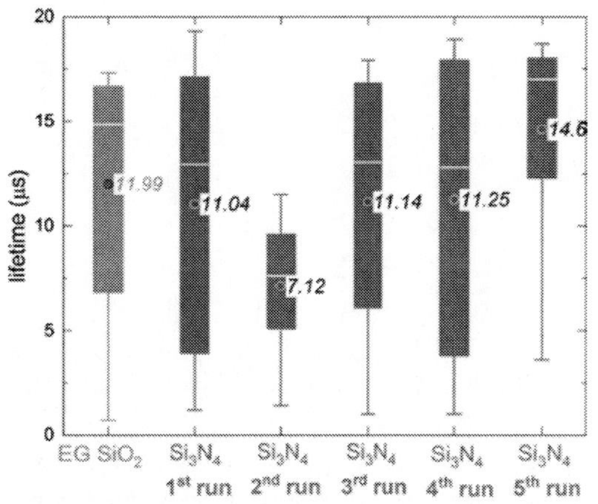

Fig. 7: Bulk lifetime for ingots grown in quartz (orange) and silicon nitride crucibles (blue). The injection levels are in the range of 1-5x10^15 cm^-3.

These results are confirmed by the lifetime maps obtained from µPCD measurements shown in Fig. 8. For the crystallisation series from silicon nitride crucible the lifetime is worst in the second run. Interestingly lifetime improves in the third run. This clearly indicates that after a "bad run", material of superior quality can be obtained in the following run.

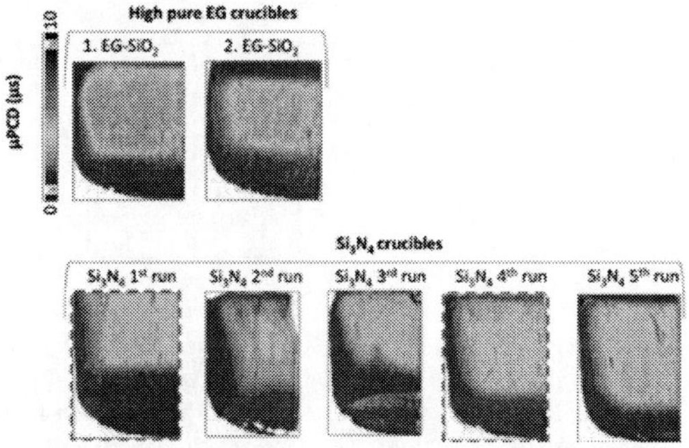

Fig. 8: Lifetime maps on cross sections for ingots grown from fused quartz crucibles (top) and silicon nitride crucibles (bottom)

3.4 Cell efficiencies

Because of relative high carrier lifetime, ingots from the first and fourth run have been chosen to saw wafers and make solar cells. Standard Al-BSF 3BB technology has been applied for cell production with very little process optimization. For the reference material both available ingots have been selected. Measured cell efficiencies are compared in Fig. 9. Cell efficiencies for both material groups are comparable and in average 17.2 to 17.6%. Cells made from bottom material (first run) show lower performance, in average 16.3%. The reason for this relies in the extended "red zone" as seen from Fig. 8 (first run) compared to the fourth run with much more reduced "red zone" area in the bottom.

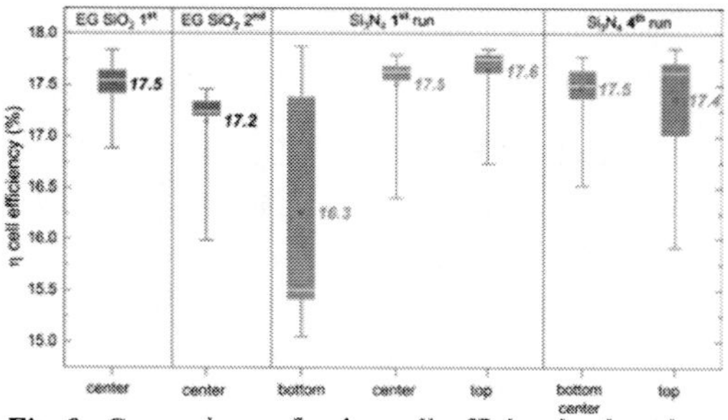

Fig 9: Comparison of solar cell efficiencies based on material grown from fused quartz and silicon nitride crucibles.

Such solar cell efficiency trend (bottom - low, center & top – high efficiency) is very similar to industrial type mc-Si wafer material used by Solitek where the efficiency difference between bottom and top ingot positions is up to 0.4 - 0.5 % abs.

3.5 Module performance

The best solar cells (17.7% – 17.9%) were sorted and carefully selected for a glass/glass module production giving a module output comparable to the industrial type solar modules.

Module	Isc (A)	Imp (A)	Voc (V)	Vmp (V)	Pmp (W)
A: Si3N4	8.95	8.45	38.53	30.90	260.93
B: commercial (17.6%-17.8%)	9.14	8.54	38.21	30.16	257.63

***Tab.: 1:** Comparison of average short circuit current (Isc), maximum power current (Imp), open circuit voltage (Voc), maximum power voltage and maximum power (Pmp) values for modules based on material grown from silicon nitride crucibles (A) compared to commercial available modules (B) from a slightly lower class (17.6-17.8% cell efficiency).*

Si_3N_4 modules have been installed in a small outdoor PV plant and real-life performance data collection is performed.

5 CONCLUSION

This study suggests that silicon nitride crucibles can outperform traditional silica crucibles in terms of metallic impurities and oxygen especially for tall ingots. At the moment of writing this paper the crucible has been used in five subsequent runs. Material performance is comparable to material grown from electronic grade fused quartz crucibles. This was also confirmed at cell and module level. Cell efficiencies are in the order of 17.5 % and module power of ~ 260 W.

7 ACKNOWLEDGEMENT

This project has received funding from the European Union's Horizon 2020 research and innovation programme under grant agreement No. 679692.

IDENTIFICATION OF DEFECT-SUPPRESSING GRAIN BOUNDARIES IN MULTICRYSTALLINE SILICON BASED ON MEASUREMENTS OF AS-CUT WAFERS USING ADVANCED IMAGE PROCESSING

T. Strauch, M. Demant, P. Krenckel, S. Riepe, S. Rein
Fraunhofer Institute for Solar Energy Systems ISE
Heidenhofstraße 2, 79110 Freiburg, Germany

ABSTRACT: The mobility and multiplication of dislocation clusters in cast Silicon depend on many factors, especially grain competition and grain boundary types. The great success of High-Performance (HP) mc-Si material in the last years is mainly based on a high occurrence of stress-absorbing grain boundaries. Where the growth of a dislocation cluster is confined by a grain boundary, the cluster displays distinct margins. Therefore cluster and grain show a mutual boundary and a corresponding vertical development. We present an image processing tool that extracts those regions with corresponding and possibly correlated development of dislocation clusters and grain structure in multicrystalline silicon (mc-Si), based on photoluminescence (PL) and optical measurements on wafers with a vertical distance in the brick of 4-8 mm. The vertical development of grain structure and PL signal is reconstructed layer by layer through the complete brick height and thus allows an insight into the whole crystal development of the whole brick and a spatially resolved macroscopic 3D-analysis on industrial scale. Grain orientations and grain boundary types were investigated within regions with corresponding development in and around dislocation clusters. Exemplary results quantitatively support the general assumption that among the dislocation-suppressing grain boundaries, random grain boundaries have the largest share with 42%.
Keywords: dislocation clusters, grain boundaries, deformation field, stress absorption

1 INTRODUCTION

Silicon wafer material quality is crucial for increasing solar cell efficiency and reducing costs per watt peak and defect development has a large impact on bulk lifetime. In previous research, theories and models have been developed with regard to the interaction between grain and defect development [3–5]. Yet, up to now, the connection between grain and defect development has not been investigated quantitatively on a statistically relevant basis due to the lack of suitable methods. Within the frame of this work, we have developed a method to identify regions automatically with specific defects (such as dislocation clusters) within the analysed wafers and to quantify the effect of grain boundaries on dislocation development as well as their development. With these means, theories and models based on few empirical data can be investigated on a statistical basis and thus a deeper insight into crystal defect development can be gained. Moreover, our method allows a complete reconstruction of the macroscopic development process of grain structure and defect development and will provide valuable feedback for crystal growth.

To investigate and quantify the effect of grain boundaries in detail, the grain boundary types and grain orientations serve as important input parameters for our evaluation. They can be determined with Laue measurements [1]. Although this method is non-destructive and faster than electron back scattering diffraction (EBSD), it is still a detailed single-wafer measurement and takes time according to the number of grains to be measured. The identification of the critical regions in the wafer, which is part of our method, makes it possible to reduce the necessary measuring coordinates considerably, thereby ensuring more relevant data in a shorter time. Moreover, our method tracks the development of dislocation clusters through brick height by fast measurements – Photoluminescence (PL) and optical measurements – on as-cut wafers and thus extends the impact of the detailed information vertically through brick height, for as far as the dislocation cluster extends and the grain relation remains the same.

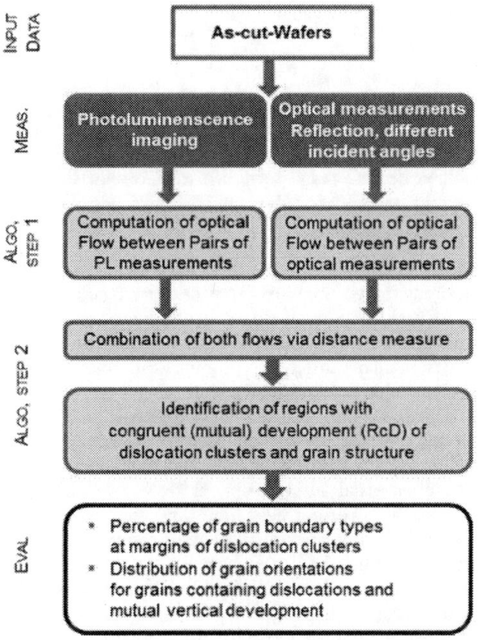

Figure 1: Outline of the method and included algorithm and evaluation steps.

2 MEASUREMENTS AND ANALYSIS

2.1 Algorithm for image registration

The outline of our method and algorithm is shown in Figure 1. Aligned PL images and optical image stacks (different illumination angles) serve as input for the so-called image registration. We track the vertical development of defect objects via vector fields that characterize the development of grain structure or PL signal from wafer to wafer (vertical distance between 4 and 8 mm). In particular, the optical flow is computed following the method of Bruhn et al. and Liu et al. [2,3]. This type of image registration methods can be based on

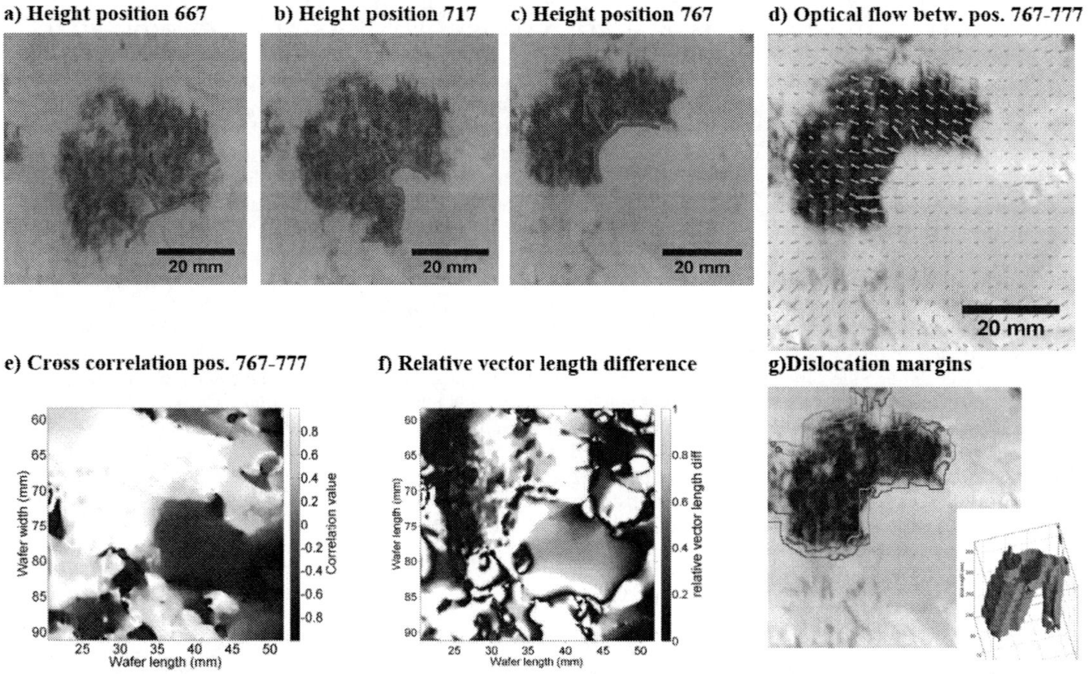

Figure 2: a)-c) Sequence of PL image sections showing the development of the same dislocation cluster over a vertical distance of 19 mm between subsequent wafers. The boundary to the overgrowing grain and its movement are marked magenta. **d)** Flow field corresponding to PL-image in c), plotted as cyan overlay arrows. **e)** Angular and **f)** length distance as intermediate results within the algorithm for flow comparison between PL and optical image. **g)** Regions at cluster margins, marked cyan for mutual flow fields and magenta otherwise. In the corner, part of the corresponding reconstructed 3D object is shown.

different image properties and criteria. In our case, the vector field is computed so that the image intensity and gradient, i.e., PL intensity or the vector of reflection values under different illumination angles remain constant along the flow. To ensure that the problem is well-defined and reflects the physical reality, other conditions (e.g. a certain flow homogeneity, called regularity) must be included in the function [4,5]. With the solver used for the optical flow calculation, it is possible to reconstruct the flow in regions with homogeneous intensity (like high-quality parts in PL images), using the parts surrounding this region that contain more information. Especially if several of the surrounding grains show a similar development, this information completion is reliable.

2.2 Evaluation and possible error sources

The vector fields resulting from the respective optical flow computations based on PL and optical measurements provide locally resolved information about grain competition and defect development in vertical direction, see for example Figure 3 a) and b).

By combining the two flow fields for every wafer pair, we identify regions where the respective vertical development of grain structure and dislocation clusters coincide and thus indicate an interaction or correlation between crystal structure and defect development. We will call these regions *regions with congruent development* (RcD). These correlations between the development of grain structure and defect clusters are investigated by computing a similarity measure between the vector fields characterizing the development of the grain structure and the defect clusters, respectively. The normalized cross correlation value, i.e., the cosine of the angle between the flow vectors, serves as measure for the angular distance. Cross correlation values near 1 signify similar flow directions and thus a corresponding vertical development of grain boundaries and dislocation clusters, see Figure 2 e). Relative differences between the flow vector lengths, visualized for the same wafer region in Figure 2 f), serve as additional distance measure. Combining those measures at the grain boundary locations gives us the RcD and thus the means to quantify the impact of the grain structure on the dislocation development. In Figure 2 g), the result can be observed as a cyan marking around dislocation clusters for the RcD, whereas the regions with magenta contours do not belong to the RcD.

The RcD are marked and grain boundaries that confine dislocation clusters are extracted. For this purpose, dislocation lines were first extracted from PL images [6] and then isolated from recombination-active grain boundaries by combining the PL feature image with the grain boundary structure extracted from the segmented optical images [7].

Laue measurements are applied on chosen samples and yield information about grain boundary types in the regions of interest and grain orientations within the dislocation cluster regions. Along with the optical flow fields and the identified regions of interest, all information is available for a quantitative evaluation of the impact of different grain boundary types on dislocation development. Thus, among the grain boundaries that confine dislocation clusters, the percentages of the respective grain boundary types are evaluated. Furthermore, the orientation of grains containing dense dislocation clusters is separated in those within the RcD and those outside. It must be remarked again that in the inner part of the grains, the development flow is completed from the surrounding regions.

As in-situ-measurements are not applicable for measuring grain growth and dislocation cluster development of large-scale bricks during growth, no reference method exists, and, to our best knowledge, no other methods with the same purpose have been published. A useful and necessary criterion to test the correctness of the vector field between a pair of wafers is that, if used as a mapping function, one of the images can be computed by warping the other image with the flow field. One of the main challenges for the computation of the vector fields is the relative lack of information within many image regions. Thus the solution is not unique, i.e., even if the deformation field reproduces the next image from the previous one perfectly, this does not mean that this is indeed the only accurate way how the object can have developed. Via further constraints in the underlying equations, the algorithm can complete missing information, yet it is not possible to prove the correctness of the identified solution unambiguously.

2.3 Material set

A large set of bricks (more than 1000 wafers from 11 bricks of different materials and manufacturers) is investigated in terms of optical flow computation. Fast 2D wafer measurements, namely PL and optical (reflection) measurements, are applied on every tenth or 20th wafer throughout the bricks. PL imaging serves to extract the recombination-active defect structures [6], the optical measurements serve to extract the grain structure [7], different illumination directions highlighting different contrasts between grains. The RcD are marked for all wafers. For three bricks (conventional mc-Si), grain orientations and grain boundary types determined via Laue measurements serve for a detailed analysis.

3 RESULTS AND DISCUSSION

3.1 General results and suitability of the analysis

Optical flow fields give direct feedback concerning relevant crystal growth properties to crystal growers. By proceeding from wafer pair to wafer pair, the whole vertical development is characterized. In the following we name the most important properties that can be recognized directly from the visualized flow vector fields as they can be seen in Figure 3 a) and b):

- Lateral growth direction (via vector angle)
- Relation between flow directions in neighbor grains or near grain boundaries
- Overgrowth angles at grain boundaries (via vector length and knowledge of wafer distance)

The color code for the optical flow fields simplifies the understanding of the growth information because it shows the development in high resolution and makes it possible to compare flows quickly with the naked eye. It can be considered as a 2D mapping which leads to a 3D understanding of the brick if observed for consecutive wafer pairs. Thus, while not directly showing or analyzing 3D objects, we get an impression of every whole brick we analyze.

Our method is suitable to give an insight into crystal growth. It fulfills the necessary criterion that the flow field, used as a mapping function can reproduce the next wafer from the previous except up to a reasonable error threshold. The flow field may be inaccurate especially at grain corners, i.e. sharp direction changes of the grain boundaries because the differential equations are based on differentiable functions.

In comparison with the expectations via naked eye, the flow fields mainly show good results, see for example Figure 2 a)-d). This concerns the developments for PL and optical measurements separately but also, in particular, for the comparison of the developments in and at dense dislocation clusters. However, we also observe inconsistencies in the flow from wafer pair to wafer pair. That means, although the flow vectors in one grain do – normally – not vary much over short vertical distances, general flow directions in a region might differ compared to the next flow. A possible explanation may be that the constraints make the algorithm react more sensitively to small changes like differences in the measurements, noise or saw marks. Yet, this does not yet explain why, in certain regions, the flow directions alternate such that for every second pair they are similar. This requires more thorough investigations.

Thus, the method allows the localization of beneficial grain boundaries. Moreover, this is possible without considering dislocation clusters as objects with sharp contours. Scientifically, this is of advantage because dislocation clusters are no compact 3D objects but consist of arranged defect lines.

3.2 Detailed evaluation of flow field correspondence

It can be observed that at the margins of dislocation clusters, the percentage of $\sum 3$ and random grain boundaries is very high, see Figure 3 e) as an example.

For random grain boundaries, the high occurrence probability at dislocation margins and in the RcD corresponds to what is expected because they absorb stress. The partly even higher $\sum 3$-percentages probably have other reasons because with the naked eye we do not observe it at sharp dislocation margins and it is known that this grain boundary type does not absorb stress. Usually, for the multiplication and mobility of dislocations, those boundaries do not constitute an impediment. As mentioned before, dislocation clusters are no solid objects. Thus, the term of dislocation margins can be ambiguous. Furthermore, twin grains, which are confined by $\sum 3$ boundaries, often develop mutually with their neighbors and are close to each other. Thus, if by chance they are situated within a RcD, their high spatial density leads to a high percentage.

For the results of the grain orientations, we observe that grains with dislocation clusters within RcF are mostly concentrated to few accumulations in the right lower corner of the inverse polefigure, i.e., near to <101>, or central in the figure towards <112> (see for example Figure 3 f), while within the non-congruent flow regions the orientations are rather scattered over the upper regions of the inverse polefigure up to <001> and <111>. As dislocation clusters are most mobile within the gliding planes, some orientations are more often decorated with dislocations. However, our data set is varied but not representative and we will investigate this on a larger data set in the future.

a) Flow field PL

b) Flow field grain structure

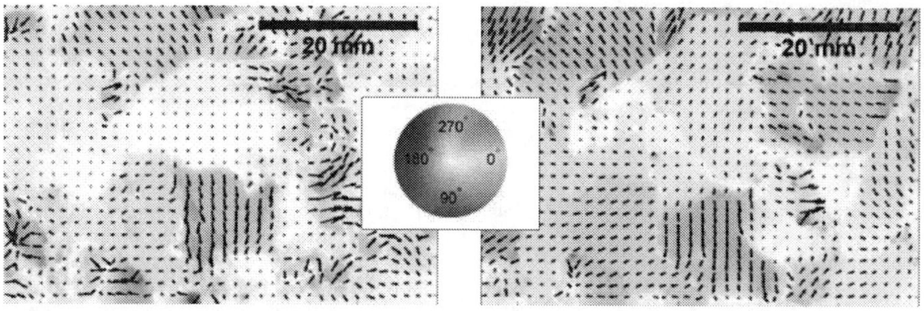

c) GB type

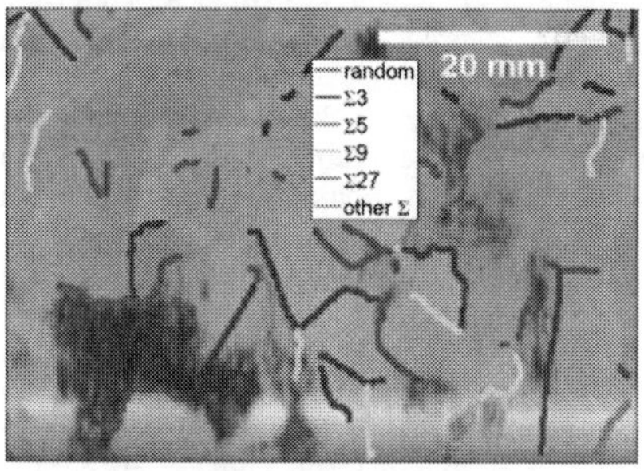

d) Grain orientation

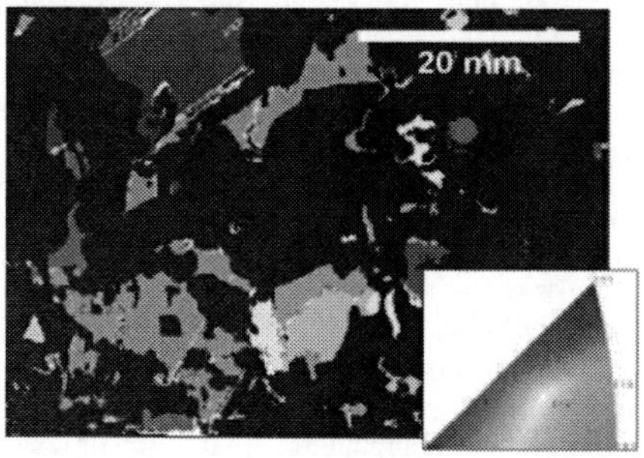

e) Percentage of grain boundary types in RcD near dislocation clusters

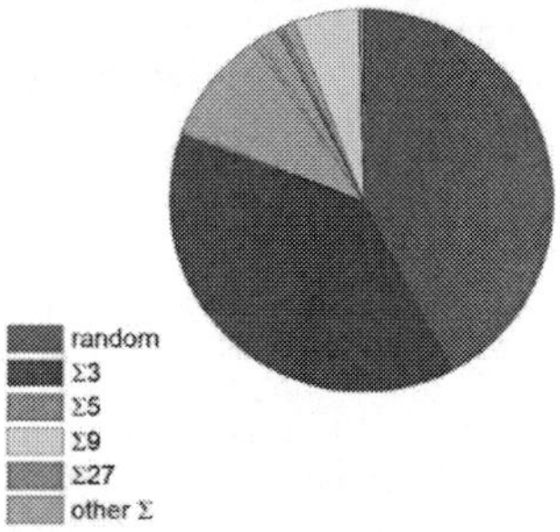

f) Distribution of grain orientations at dislocation clusters within and outside the RcD

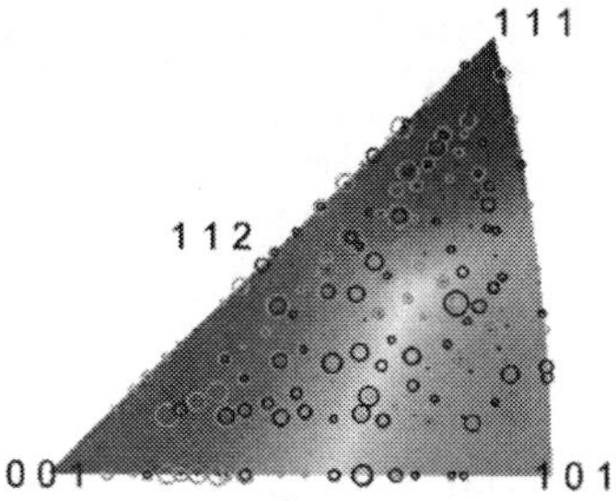

Figure 3: a) and b) Image section of the respective visualized flow fields for a) PL intensity and b) grain structure of an mc-Si wafer. The color in the regions and the direction of the black arrows encode the angle of the development in lateral direction, the saturation (on a relative scale) and length of the black arrows show the length of the vectors, which is related to the vertical overgrowth angle. **c)** Same image section for the PL image with overlayed information on the determined grain boundary types. **d)** Grain orientations in the regions with congruent development. **e)** Percentages of the respective grain boundary types (in terms of line length) in regions with congruent development at the margins of dislocation clusters. **f)** Distribution of grain orientations for those grains within (marked with black circles) and without (marked with grey circles) the RcD containing dislocations.

3.3 Outlook and possible optimization

For the optical flow computation, it has to be kept in mind that the underlying equations for the optical flow contain no physical model for grain growth, which is hardly possible. Still, certain physical properties of the growth make it necessary to find an optimal choice of the underlying model, the image properties used, the solvers, parameters and mathematical constraints. Among these varied degrees of freedom, it will take more time and data evaluation to get closer to the optimum. Especially, achieving consistency between consecutive vector fields will serve to reconstruct the whole grain growth and competition.

Furthermore, decorated dislocation lines evolve relatively fast, so that it is almost impossible to track the exact growth, and thus the crystallographic gliding planes, over the vertical distance of >4 mm. By computing the flow between smaller vertical distances, not only dislocation developments could be tracked with higher accuracy, but also the glide planes could be detected without applying an extra Laue measurement.

The choice of the distance measure between the flow

fields will be further optimized since it has a large impact on the results. Moreover, we will discern between sharp (distinct) and ambiguous (indistinct) grain boundaries, which will improve the distinction between the effects of the respective grain boundary types.

4 CONCLUSIONS

We developed a method to analyse locally resolved information about the development of grain structure and larger recombination-active defects like dislocation clusters. This was attained via optical flow computation between wafer pairs. The results consist in vector fields which designate the direction and overgrowth angle within the crystal. By computing the flow for all consecutive wafer pairs through the brick, the whole growth was reconstructed. By comparing and combining the vector fields of grain structure and defects, we gained information where they might be correlated. Those regions with mutual development were extracted and further restricted to the margins of dislocation clusters where grain boundaries may have a beneficial effect. In this way, measurements for the determination of local crystal orientation and grain boundary types can be restricted to fewer data points.

The vector fields displaying the development were in good agreement with the naked eye. They further showed low errors for the comparison between original image and the image calculated using the flow as mapping function. They contain valuable feedback for crystal growers concerning defect growth and grain structure development. With a further optimized algorithm, the flow can be assembled to a complete growth reconstruction.

Within the regions of flow congruency, we investigated the impact of the respective grain boundary types for chosen samples. Our quantitative results support the general assumption that dislocations often occur in certain grain orientations and that random grain boundaries can often suppress dislocation clusters. For a consolidation of those assumptions and a deeper understanding how to suppress defects in the most efficient way, more evaluations and larger data sets with grain orientations and grain boundary types will be carried out.

5 ACKNOWLEDGMENTS

This work was financially supported by the German Federal Ministry for the Economy and Energy within the project Q-Crystal (contract number 0324103A).

6 REFERENCES

[1] Lehmann T, Trempa M, Meissner E, Zschorsch M, Reimann C, Friedrich J. Laue scanner: A new method for determination of grain orientations and grain boundary types of multicrystalline silicon on a full wafer scale. *Acta Materialia* 2014;69:1–8.

[2] Bruhn A, Weickert J, Schnörr C. Lucas/Kanade Meets Horn/Schunck: Combining Local and Global Optic Flow Methods. *International Journal of Computer Vision* 2005;61:211–31, doi:10.1023/B:VISI.0000045324.43199.43.

[3] Liu C. *Beyond Pixels: Exploring New Representations and Applications for Motion Analysis*. Cambridge, MA, USA; 2009.

[4] Horn BK, Schunck BG. Determining optical flow. *Artificial Intelligence* 1981;17:185–203, doi:10.1016/0004-3702(81)90024-2.

[5] Lucas BD, Kanade T. An iterative image registration technique with an application to stereo vision. *IJCAI* 1981.

[6] Demant M, Rein S, Haunschild J, Strauch T, Höffler H, Broisch J, Wasmer S, Sunder K, Anspach O, Brox T. Inline quality rating of multi-crystalline wafers based on photoluminescence images. *Progress in Photovoltaics: Research and Applications* 2015, doi:10.1002/pip.2706.

[7] Strauch T, Demant M, Krenckel P, Riepe S, Rein S. Analysis of grain structure evolution based on optical measurements of mc Si wafers. *Journal of Crystal Growth* 2016;454:147–55, doi:10.1016/j.jcrysgro.2016.09.009.

GALIUM DOPED MONOCRYSTALLINE SILICON BY CONTINUOUS CZOCHRALSKI PROCESS FOR MAKING LIGHT INDUCED DEGRADATION FREE P-TYPE PERC CELLS

Yonghui Zhang, Qiao Li and Rui Zhou
Longi Green Energy Technology Corp, 401 East Chang'an Street, Xi'an, China
Han Xu*, Scott Turchetti and Sean Keohane
GT Advanced Technologies, 243 Daniel Webster Highway, Merrimack, NH 03054, USA
*Phone: +1 603 204 2838; e-mail: han.xu@gtat.com

ABSTRACT: Light induced degradation (LID) due to boron-oxygen (B-O) complexes can cause significant power loss from high performance p-type mono-crystalline PERC cells. The paper shows that replacing B with Ga as the dopant in the mono-crystalline silicon can eliminate this degradation. The paper also shows that the continuous Czochralski (CCz) process enables the production of Ga doped mono-crystalline ingot of very uniform resistivity distribution, thus, increasing the material yield and lowering the production cost.

Keywords: Gallium Doping, Monocrystalline Silicon, Light Induced Degradation, Czochralski, LID, PERC, Cz, CCz

1 INTRODUCTION

Light induced degradation (LID) due to boron-oxygen (B-O) complexes in boron (B) doped, p-type monocrystalline silicon solar cells can result in significant loss in power output [1]. The issue is more severe for the high performance cells, such as passivated emitter rear contact (PERC) cells [2]. While the relative power loss for the aluminum back surface field (Al-BSF) cell is typically 2-3%, the loss for the PERC cell is as high as 3-6%. Because the degradation is caused by the B-O complexes, the severity of the loss is related to the boron concentration in the base matereial. The loss is the highest for PERC cell of lowest base resistivity (highest boron concentration) which is preferred due to PERC cell's reduced back contact area [3,4]. Hydrogen regeneration can recover the lost power but the recovery is not 100% and the amount of recovery also depends on boron concentration [5]. Minimizing oxygen in the silicon crystal can mimize the formation of B-O complexes, thus, reducing LID, but the reduction of oxygen is limited for the traditional Czochralski (Cz) process unless much more expensive magnetic Czochralski (MCz) or float zone (FZ) technique is used.

Another approach to prevent the formation of B-O complexes is to replace the boron with the gallium (Ga) as the dopant in the monocrystalline crystals. However, because of the very low segregation coefficient of Ga in silicon, traditional Cz crystal pulling process will result in >10 times resistivity difference along the crystal length limiting the yield of usable crystal for making the solar cells. Low yield increases the wafer production cost. Continuous Czochralski (CCz) process, on the other hand, can pull monocrystalline crystals with extremely uniform resistivity distribution regardless of segregation coefficient. This is because the dopant concentration in the silicon melt can be managed by continuous feeding of silicon and dopant [6]. Furthermore, the target resistivity can be tuned easily to meet the PERC cell requirement for the best base resisvitity.

2 EXPERIMENTAL

2.1 Ingot growth and characterization

CCz process was used to pull a Ga doped mono-crystalline ingot (2m length, ϕ215mm) targeting resistivity of 0.8 Ω-cm. As-grown ingot was cut into three 650mm long ingot sections (top, middle and bottom) and 2mm thick sample slugs were cut before and after each ingot section. The ingot sections were characterized for minority carrier lifetime (MCLT), resistivity (R_s), interstitial oxygen (O_i) and substitutional carbon (C_s). MCLT was measured by the quasi-steady state photoconductance (QSSPC) method using Sinton BCT400 at 1x10^15 minority carrier density. The resistivity was measured on the sample slugs both before and after thermal donor annihilation with a 4-point probe. The interstitial oxygen and substitutional carbon concentration were measured on the slugs using FT-IR spectrometer per ASTM F121-83 standard.

2.2 PERC cell fabrication

The ingot sections were squared and sliced into 156mm pseudo square wafers (M2 size) by diamond wire saw. Wafers were separated into three groups based on which ingot section (top, middle and bottom) they came from and were processed into PERC cells on an industrial production line. Cells were tested for conversion efficiency before and after light induced degradation (LID). The LID is done by light exposure (1 Sun, 65°C) for 5 hours.

3 RESULTS AND DISCUSSIONS

3.1 Ingot characterization

Figure 1 shows resistivity measured from sample slugs at the center and edge locations before thermal donor annihilation. Figure 2 shows resistivity measured from the slugs after thermal donor annihilation. The axial resistivity distributions are very uniform, varying from 0.74 to 0.94 Ω-cm at the center and 0.85 to 1.06 Ω-cm at the edge, respectively. The radial resistivity distributions vary from 12 to 16% post thermal donor annihilation.

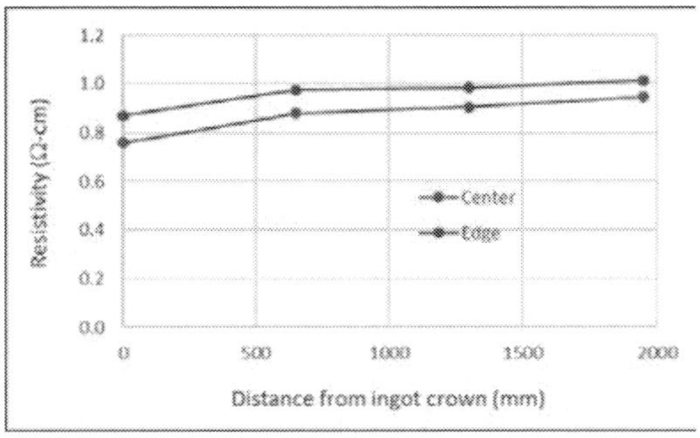

Figure 1: Resistivity measured on sample slugs at the center and edge locations before thermal donor annihilation

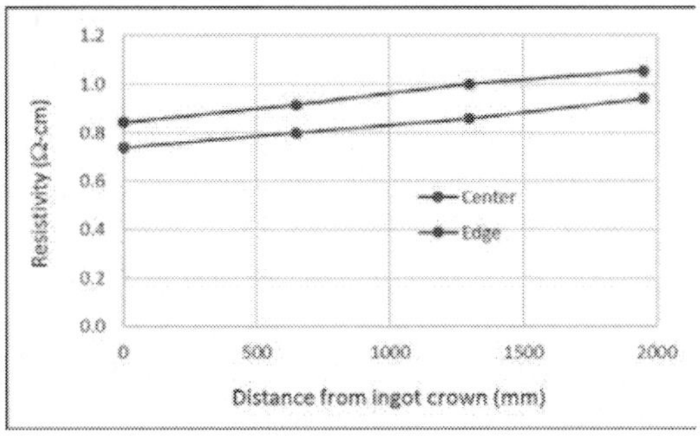

Figure 2: Resistivity measured on sample slugs at the center and edge locations after thermal donor annihilation

Figure 3 shows the minority carrier lifetime (MCLT) measured on the cross-section surfaces of each ingot section by QSSPC method. MCLT varies from 241 to 299 across the ingot length. Notice the MCLT stays quite constant throughout the entire ingot that is different from that of Cz ingot which decreases over the ingot length due to the decrease in ingot resistivity.

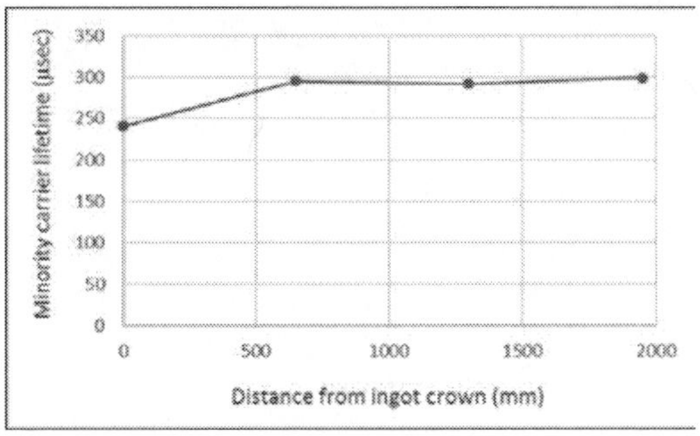

Figure 3: Minority carrier lifetime measured at the center of the cross section of ingot

Figure 4 shows the interstitial oxygen measured from the sample slugs. The interstitial oxygen varies from 13.8 to 14.8ppma at the center of slug, 11.8 to 13.4ppma at the edge of slug. The relatively low oxygen is due to the small contact area between the silicon melt and quartz crucible surface which is unique to the CCz process. Substitutional carbon concentrations are below the detection limit of 1ppma. Low interstitial oxygen is consistent with the very small observed change in resistivity before and after thermal donor annihilation.

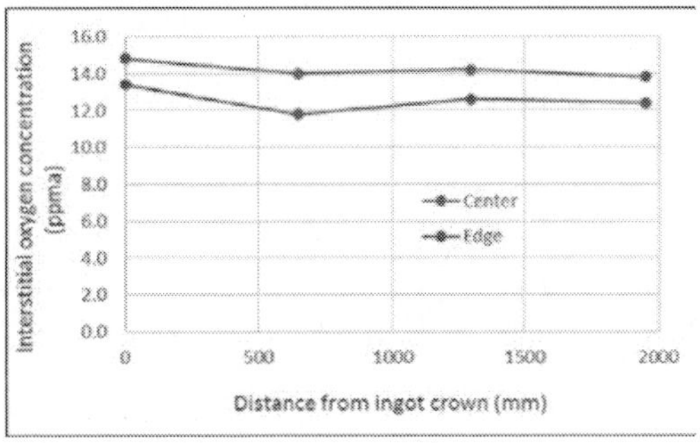

Figure 4: Interstitial oxygen measured at the center and edge of the sample slugs

3.2 PERC cell efficiency and LID

Figure 5 compares the efficiency distribution of the CCz wafers with the production baseline data collected from the same cell line around the same time when the CCz wafers were being processed. The wafers used in production were produced by conventional Cz with recharge (RCz) process. The data for wafers from the top, middle and bottom ingot sections are plotted as one group. The average efficiency of CCz wafers is nearly identical to that from the production baseline but the distribution is much tighter. The much tighter distribution is likely the result of much tighter resistivity distribution of the CCz wafers. Future work is planned to test CCz wafers of even lower resistivity to optimize the cell efficiency. Table I shows the initial cell efficiency difference from the baseline average as well as the relative cell efficiency degradation after 5 hours of light exposure (1 Sun, 65°C). The cells from the top, middle and bottom sections of the ingot are tested separately. The average efficiencies from the top and the middle sections are slightly higher than the baseline average while the average efficiency from the bottom section is slightly lower than the baseline average. The LID can be controlled to <1% relative with Ga doping. The remaining loss is suspected to be caused by a loss mechanism different from B-O complexes, for example, the copper-oxygen pairs [7]. Another thing to note is that <1% LID is achieved for cells with low base resistivity (<1 Ω-cm) which would have quite high LID if the base material is doped with B.

Table I: Average initial cell efficiency difference from production baseline and average LID

Ingot Section	Δ Eta	LID
Top	+0.07%, *abs*	0.94%, *rel*
Middle	+0.07%, *abs*	0.99%, *rel*
Bottom	-0.04%, *abs*	0.99%, *rel*

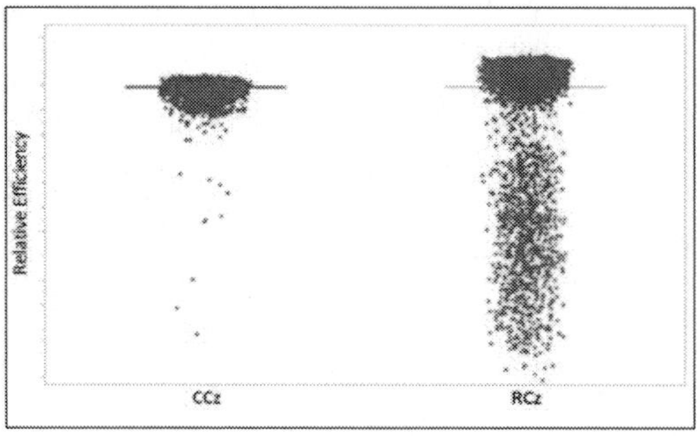

Figure 5: Relative cell efficiency distributions of CCz wafers and production baseline (RCz). The horizontal lines represent the distribution averages.

4 CONCLUSIONS

Replacing B with Ga as the dopant for monocrystalline silicon is an approach at the material level to eliminate LID due to B-O complexes from the high performance monocrystalline PERC cells. Ga doping allows high dopant concentration to be used to optimize PERC cell performance without increased LID. The CCz process enables Ga doped crystal to be produced with very uniform resistivity distribution across the entire ingot, thus, increasing the usable material, reducing the production cost and producing cells with tighter efficiency distribution.

5 REFERENCES

[1]. J. Schmidt, AG Aberle and R. Hezel, Investigation of carrier lifetime instabilities in Cz-grown silicon, Proc. 26th IEEE PVSC, 1997, p13
[2]. A. Das and A. Rohatgi, The impact of cell design on light induced degradation in p type silicon solar cells, Prof. 37th IEEE PVSC, 2011, p158
[3]. F. Wolny, T. Weber, M. Muller and G. Fisher, Light induced degradation and regeneration of high efficiency Cz PERC cells with varying base resistivity, Energy Procedia 38 (2013), p523
[4]. C. Arndt, D. Stichtenoth, N. Muller, et al, Manufacturing of 300Wp modules by industrialization of PERC solar cell technology includeing crystal, cell and module development, Proc. 29th EU PVSEC, 2014, p376
[5]. F. Wolny, T. Weber, G. Fischer, A. Herguth and S. Wilking, Enhanced stable regeneration of high efficiency Cz PERC cells, Energy Procedia, 77 (2015) p546
[6]. H. Xu, Characterization of n-type mono-crystalline silicon ingots produced by continuous Czochralski (Cz) technology, Energy Procedia, 77 (2015) p658
[7]. J. Lindroos, M. Yli-Koski, A. Haarahiltunen, M. Schubert and H. Savin, Light-induced degradation in copper-contaminated gallium-doped silicon, Phys. Status Solidi RRL, 7 (2013), p262

INFLUENCE OF DIFFERENT TRANSITION METAL CONTAMINATIONS ON DEGRADATION AND REGENERATION IN MC SI

Andreas Schmid, Annika Zuschlag, Daniel Skorka, Jakob Fritz, Clemens Winter, Giso Hahn
University of Konstanz, Department of Physics, 78457 Konstanz, Germany

ABSTRACT: Light and elevated temperature induced degradation (LeTID) affects significantly the performance of multicrystalline Silicon (mc-Si) solar cells. The underlying mechanisms of LeTID and following regeneration are still unknown and might depend on contaminations. To investigate the influence of different transition metals on the degradation and regeneration behavior, mc-Si ingots were intentionally contaminated in the Si melt with either 20 ppma Fe, Cu, Cr or a combination of Fe and Cu. Minority charge carrier lifetime (τ_{eff}) samples were processed and the degradation and regeneration behavior under illumination (0.9 ± 0.1 sun) and elevated temperature (75°C) is measured repetitively by time-resolved photoluminescence imaging (TR-PLI) at room temperature. The resulting minority charge carrier lifetime maps are analyzed and the effective defect concentration within one sample and also for differently contaminated samples were calculated. This approach leads to different defect generation rates for the differently contaminated samples, which might hint to an influence of different transition metals on the LeTID causing defect.
Keywords: Degradation, Multicrystalline Silicon, Impurities, Lifetime

1 INTRODUCTION

Multicrystalline (mc) Si PERC (passivated emitter and rear cell) solar cells, but also lifetime samples, show a strong degradation of minority charge carrier lifetimes under illumination at elevated temperatures (*e.g.* [1-4]), so-called light and elevated temperature induced degradation (LeTID). The observed degradation kinetics cannot be explained by the already known degradation mechanisms like BO-correlated defects or the dissociation of FeB pairs [1-3]. Also a regeneration process after degradation has been observed for solar cells [3] and lifetime samples [5]. The kinetics of degradation and regeneration is influenced by measurement conditions, like illumination and temperature (*e.g.* [3, 6]), while the underlying mechanisms causing the degradation and regeneration of mc-Si are still unknown. In literature, several authors speculate on possible degradation and regeneration mechanisms [4, 7-9]. For all of them, defects correlated with getterable impurities play a crucial role. While in [7] a movement of impurities towards the wafer surface is assumed as the mechanism of regeneration, in [4] it is pointed out that the degradation strength can be influenced by preceding effective gettering processes. Lifetime spectroscopy data also gave first hints on possible candidates responsible for the degradation due to the measured ratio of capture cross sections of the underlying lifetime limiting defects [8, 9]. In this study we investigate the influence of several transition metal impurities common in mc-Si materials by using mc-Si materials intentionally contaminated with different elements in the melt before crystallization.

2 EXPERIMENTAL

To investigate the influence of different transition metals on the degradation and regeneration process, four intentionally contaminated mc-Si ingots are compared. For the contaminated mc-Si materials, 20 ppma of Fe, Cu, Cr or a combination of Fe and Cu were intentionally introduced into the Si melt. The detrimental effect of transition metals on the minority charge carrier lifetime in Si materials is known and, *e.g.*, described in [10]. NAA (neutron activation analysis) results regarding the transition metal concentration of the here investigated materials are shown in [11]. In the middle of the ingots the concentration of the intentionally implemented impurities is in the range of $7x10^{14}$ cm^{-3} for the Cu content of the Cu contaminated ingot, $5x10^{13}$ cm^{-3} for the Fe concentration of the Fe contaminated ingot and $9x10^{12}$ cm^{-3} for the Cr concentration of the Cr contaminated ingot. Regarding the ingot intentionally contaminated with a combination of Fe and Cu, the resulting impurity contents are in the same range as for the material contaminated with only one of these species. Additionally, standard mc-Si material is processed and characterized for reference. To investigate samples with different contamination levels of the same transition metal impurity, mc-Si wafers from three different ingot heights are available, as impurities are unevenly distributed over the ingot height due to segregation.

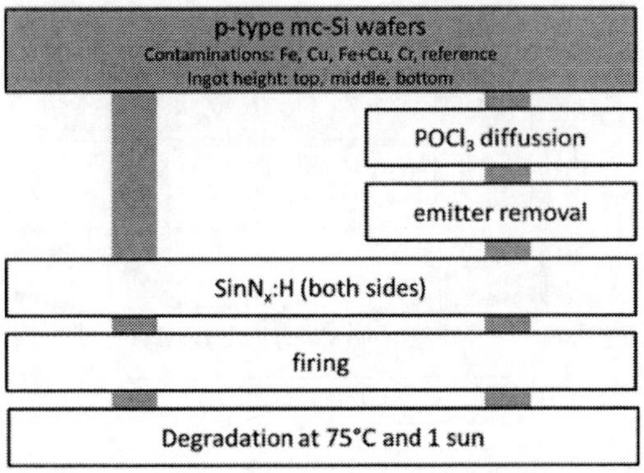

Figure 1: Process sequence of the investigated lifetime samples.

Sister wafers (only mid ingot height) of each contamination type with comparable grain and defect structure were used to compare the influence of the gettering step on the degradation and regeneration behavior of the lifetime sample. A scheme of the applied process sequences is given in Fig. 1. All samples were etched to remove the saw damage. On half of the wafers only a surface passivation was realized by firing of a PECVD (plasma-enhanced chemical vapor deposition) SiN_x:H layer. The sister wafers of these samples were gettered by a $POCl_3$ diffusion (55 Ω/□). Afterwards, the emitter was removed and the samples were also surface passivated by firing of a PECVD SiN_x:H layer.

For degradation, the samples are held at a temperature of approx. 75°C under illumination with halogen lamps (0.9 ± 0.1 sun). Minority charge carrier lifetime τ_{eff} is measured repetitively by the fast and self-calibrated TR-PLI method [12] at room temperature, resulting in a series of spatially resolved lifetime maps for each sample over degradation time. For a statistically relevant analysis of areas differing in material quality, an array of 2,500 areas (each 150x150 µm^2 in size) is distributed over the TR-PLI lifetime map of the 5x5 cm^2 mc-Si samples. Average τ_{eff} values within these areas are extracted over degradation time and further analyzed. This method allows tracking of changes in τ_{eff} of different sample areas under degradation conditions. The advantage of this approach is that the complete information of the spatially resolved τ_{eff} maps over degradation time can be easily displayed. The τ_{eff} data for each 150x150 µm^2 area on the samples is plotted over time and coded by a rainbow color bar based on the areas' lifetime values at the beginning of the degradation experiment (*cf.* [4]).

3 RESULTS

The harmonic average τ_{eff} for four different contaminated lifetime samples from middle ingot regions are shown in Fig. 2. All shown samples have been P-gettered. The ungettered samples showed initial lifetimes between 1 and 10 µs only and nearly no LeTID effect can be observed on such low lifetime levels. The P-gettering process reduces the contamination level, so that the overall minority charge carrier lifetime is significantly higher and LeTID as well as regeneration processes are easier to investigate.

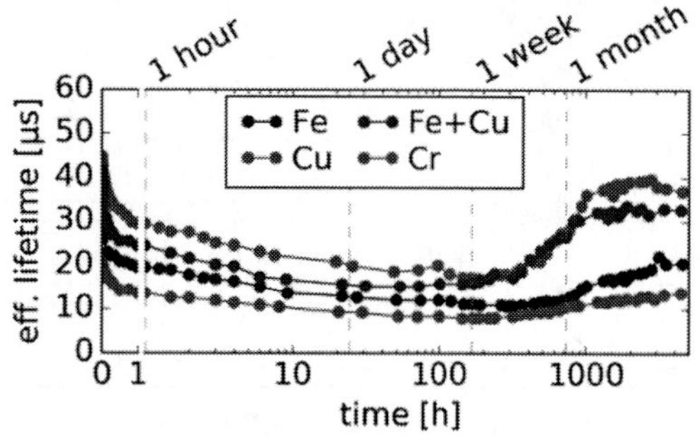

Figure 2: Harmonic average τ_{eff} data for differently contaminated mc-Si wafers. Degradation and regeneration behavior of lifetime samples can be strongly influenced by contamination type.

As expected, initial harmonic average τ_{eff} is lowest for Cr, slightly higher for Fe (both around 25 µs), and highest for Cu contaminated samples. As Cu is forming co-precipitates with Fe, the initial τ_{eff} of the Fe+Cu sample is closer to the only Cu contaminated sample (both around 45 µs) than to the only Fe contaminated one. These differences in the initial harmonic average τ_{eff} can be explained by different effects depending on the contamination type and the resulting gettering efficacy. The transition metals segregate differently during the cooling down of the Si in the ingot (internal gettering and formation of precipitates according to temperature dependent diffusivity and solubility). Furthermore, the higher the diffusion coefficient of the transition metal, the better the intentional contamination is reduced during the applied P-gettering step. Here the local grain structure of the sample becomes important for the distribution of contaminations (internal gettering).

Fig. 2 shows clearly that LeTID and regeneration is observed in all samples. Comparing the degradation behavior, the slight differences between the materials contaminated with Cu and a combination of Fe and Cu hint already on different degradation kinetics in dependence of the impurity species. Beside the degradation kinetics, also the strength of degradation is influenced by the different transition metal contents. For the regeneration effect different onsets and regeneration strengths were observed and seem to depend on contamination, too.

The disadvantage of average τ_{eff} analyses is that information on local τ_{eff} distribution measured by spatially resolved TR-PLI gets lost. Therefore, further analyses of the degradation and regeneration behavior will be exemplarily discussed in more detail based on the Fe+Cu contaminated sample shown in Fig. 3. The following description of the observed degradation and regeneration behavior holds true also for samples originating from mid ingot heights of the other intentionally contaminated samples.

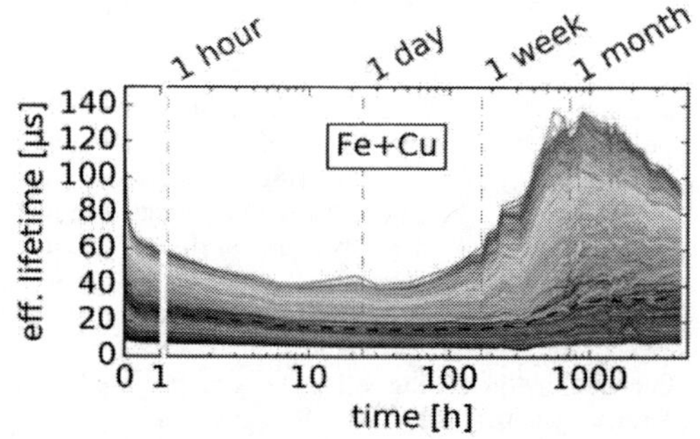

Figure 3: τ_{eff} for the Fe+Cu contaminated mc-Si wafer (x-axis: 0-1 h linear scale, 1-5,000 h log scale). Each line in the graph represents an area of 150x150 µm^2 in the spatially resolved TR-PLI lifetime measurements and is color-coded to its value at the beginning of degradation.

Each single line in Fig. 3 represents one of the 2,500 small areas on the lifetime samples as described above. The applied color code, based on the initial τ_{eff} of each sample at the beginning of the experiment, shows that the relative τ_{eff} distribution is almost kept constant within each sample during the experiment, which means that sample areas with the highest initial τ_{eff} show also the highest τ_{eff} at the maximum degradation level as well as during the regeneration process. This holds true even if the range of lifetime is narrowed down during degradation or is spread again during regeneration as already described by [4] for standard industrial mc-Si.

After approximately 50 h a regeneration process sets in, with regeneration in areas with higher initial τ_{eff} (red lines) starting significantly earlier than in areas of lower initial τ_{eff} at maximum degradation level, again in good agreement with industrial mc-Si as described in [4]. In contrast to [4], τ_{eff} of some areas, mainly areas with high initial τ_{eff}, exceeds the initial τ_{eff}. It has to be mentioned that the sample shown in Fig. 2 has a lower τ_{eff} level due to the remaining impurity content compared to samples shown in [4]. After approximately 1,000 h a second decrease in τ_{eff} sets in. As stated in [4, 13] this behavior can be attributed to a decreasing surface passivation quality which limits the τ_{eff} values.

To investigate the influence of the local defect structure on the LeTID and regeneration behavior, all sample areas with 50 and 100 µs initial τ_{eff} are extracted from the rainbow plot shown in Fig. 3. The chosen sample areas (more than 200 pixels in the TR-PLI map for each of the chosen initial τ_{eff} values) are distributed all over the lifetime sample representing areas with same initial τ_{eff} but different defect structure. As the degradation behavior of areas with the same initial τ_{eff} is very similar, an influence of the local defect structure can be excluded. Additionally, the comparable behavior of different sample areas with the same initial τ_{eff} hints on the same root-cause of degradation even if the local defect structure is different.

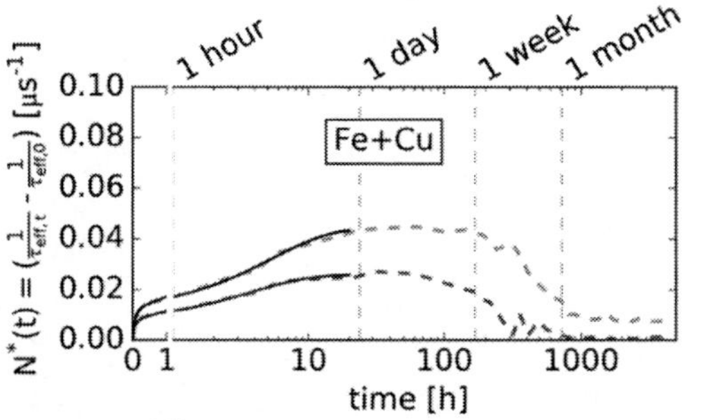

Figure 4: Time-dependent effective defect concentration of the Fe+Cu contaminated sample. The dotted red line represents all areas with 100 µs initial τ_{eff}, the dotted blue line all areas with 50 µs initial τ_{eff}. The solid lines show the exponential fit curves of the degradation.

To further quantify these observations, the harmonic average τ_{eff} of the extracted curves with initial τ_{eff} of 50 and 100 µs were calculated. Out of these data, the time-dependent effective defect concentration $N^*(t)$ is determined by

$$N^*(t) = \left(\frac{1}{\tau_{eff,t}} - \frac{1}{\tau_{eff,0}} \right)$$

and is shown in Fig 4. For degradation, it is possible to fit $N^*(t)$ within one sample with two saturating exponential functions with the same parameters. This is exemplarily shown at the Fe+Cu contaminated sample in Fig. 4, where the solid lines represent the fit-curves of the different initial τ_{eff}. Despite probably slightly different injection levels for areas with different initial τ_{eff}, this leads to comparable defect generation rates and the assumption of the same root-cause for LeTID in areas of different initial τ_{eff} within the Fe+Cu sample.

Considering $N^*(t)$ of the differently contaminated samples, it is possible to fit the degradation data with the same parameters within one sample as described above for the Fe+Cu sample. A comparison of $N^*(t)$ for areas with the same initial τ_{eff} of 50 µs of the differently contaminated samples is shown in Fig. 5. The corresponding fit parameters differ for samples with different impurity species resulting in different defect generation rates. Due to the same initial τ_{eff} in Fig. 5 and the exclusion of a dominating influence of the local defect structure, as discussed above, the calculated differences of the defect generation rates seem to be due to the varying impurities in the investigated materials.

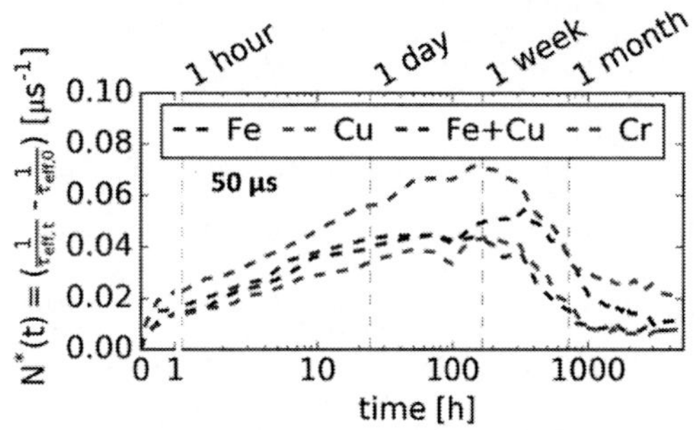

Figure 5: Time-dependent effective defect concentration of the differently contaminated samples with the initial τ_{eff} of 50 µs.

4 SUMMARY AND DISCUSSION

The influence of different transition metal contaminations on the strength and kinetics of LeTID and regeneration could be shown using data from TR-PLI. Spatially resolved τ_{eff} and harmonic average τ_{eff} have been measured and allow for a separation of degradation and regeneration kinetics in areas of varying material quality. While the lifetime of ungettered samples is so low that LeTID is difficult to observe, P-gettered samples show the known LeTID and regeneration behavior. Areas with the same initial τ_{eff} but different local defect structure show no obvious dependence of LeTID on structural defects.

A comparison of differently contaminated samples shows different LeTID and regeneration kinetics, resulting in different defect generation rates in dependence of the impurity species. The investigated impurities have been excluded as root-cause for LeTID *e.g.* by [12-14], based on their material properties in crystalline Si. It has to be mentioned that material properties like diffusion coefficient are usually investigated and determined in defect-lean mono-crystalline Si at high temperatures. Therefore, it is reasonable, that those material properties might be different in mc-Si due to its higher (structural) defect density. In defect-rich mc-Si, the diffusion of impurities is influenced by trapping. This might lead to diffusion coefficients varying by several orders of magnitude as it is, *e.g.*, the case for the diffusion of H [14]. Therefore, based on the results discussed above it is still possible that the investigated impurities might be involved in the root-cause of LeTID or at least influence the defect causing LeTID, *e.g.* by trapping H, forming complexes, or changing/influencing recombination activity or charge state.

8 REFERENCES

[1] K. Ramspeck, S. Zimmermann, H. Nagel, A. Metz, Y. Gassenbauer, B. Birkmann, A. Seidl, Proc. 27th EUPVSEC, Frankfurt/Main, Germany, 2012, 861.
[2] F. Fertig, K. Krauss, S. Rein, Physica Status Solidi RRL 9(1) (2014) 41.
[3] F. Kersten, P. Engelhart, H.-C. Ploigt, A. Stekolnikov, T. Lindner, F. Stenzel, M. Bartzsch, A. Szpeth, K. Petter, J. Heitmann, J. Müller, Solar Energy Materials and Solar Cells 142 (2015) 83.
[4] A. Zuschlag, D. Skorka, G. Hahn, Progress in Photovoltaics: Research and Applications 25(7) (2017) 545.
[5] A. Zuschlag, D. Skorka, G. Hahn, Proc. 43rd IEEE PVSC, Portland 2016, 1051.
[6] J. Fritz, A. Zuschlag, D. Skorka, G. Hahn, Energy Procedia 124 (2017) 718.
[7] D. Bredemeier, D. Walter, S. Herlufsen, J. Schmidt, AIP Advances 6 (2016) 035119-1.
[8] A.E. Morishige, M.A. Jensen, D.B. Needleman, K. Nakayashiki, J. Hofstetter, T.-T.A. Li, T. Buonassisi, IEEE Journal of Photovoltaics 6(6) (2016) 1466.
[9] K. Nakayashiki, J. Hofstetter, A. Morishige, T. Li, D. Needleman, M. Jensen, T. Buonassisi, IEEE Journal of Photovoltaics 6(4) (2016) 860.
[10] G. Coletti, P.C.P. Bronsveld, G. Hahn, W. Warta, D. Macdonald, B. Ceccaroli, K. Wambach, N. Le Quang, J.M. Fernandez, Advanced Functional Materials 21 (2011) 879.
[11] S. Riepe, I. Reis, W. Kwapil, M. Falkenberg, J. Schön, H. Behnken, J. Bauer, D. Kreßner-Kiel, W. Seifert, W. Koch, Physica Status Solidi C 8(3) (2011) 733.
[12] D. Kiliani, G. Micard, B. Steuer, B. Raabe, A. Herguth, G. Hahn, Journal of Applied Physics 110 (2011) 054508.
[13] D. Sperber, A. Graf, D. Skorka, A. Herguth, G. Hahn, IEEE Journal of Photovoltaics (2017), accepted, doi: 10.1109/JPHOTOV.2017.2755072.
[14] S. Kleekajai, F. Jiang, M. Stavola, V. Yelundur, K. Nakayashiki, A. Rohatgi, G. Hahn, S. Seren, K. Kalejs, Journal of Applied Physics 100(9) (2006) 093517.

GROWTH OF OXYGEN PRECIPITATES AND DISLOCATIONS IN CZ SILICON: IMPACT ON MINORITY CARRIER LIFETIME

F. E. Rougieux[1], H. T. Nguyen[1], D. Macdonald[1], B. Mitchell[2] and R. Falster[3]
[1]Research School of Engineering, College of Engineering and Computer Science, The Australian National University, Canberra ACT 0200, Australia
[2]School of Photovoltaic and Renewable Energy Engineering, University of New South Wales, Sydney NSW 2052 Australia
[3]SunEdison Semiconductor, via Nazionale 59, 39012 Merano, Italy

ABSTRACT: The impact of point defect created during ingot growth and Tabula Rasa on the growth of oxygen precipitates is investigated. We show that both point defect from ingot growth and point defect from Tabula Rasa affect the growth of oxygen precipitates during solar cell processing. We show that the vacancy concentration during precipitate growth affects the recombination activity of oxygen precipitates. Finally, we demonstrate the impact of non-equilibrium point defect concentrations on dislocation growth.
Keywords: Silicon, defects, precipitates, oxygen, dislocations, FTIR, photoluminescence, vacancy, interstitials.

1 INTRODUCTION

This work aims at understanding the impact of intrinsic point defects on oxygen precipitation and recombination in monocrystalline silicon. Oxygen precipitation leads to a significant volume expansion with an associated strain. This strain is released via the absorption of silicon vacancies or emission of silicon self-interstitials. Evidently vacancy-rich silicon will tend to promote oxygen precipitation and growth while interstitial-rich silicon will tend to slow oxygen precipitation and growth but promote dislocation growth. Upon heating to 1100°C vacancies and interstitials are generated in equal concentrations through Frenkel-pair generation. A nitrogen ambient also leads to vacancy surface generation, while an oxygen ambient leads to interstitial surface generation. Through controlled annealing one can thus explore the creation of extended defects in vacancy-rich and interstitial-rich environments, and their relative impact on the minority carrier lifetime.

In this paper, we perform tabula rasa treatments followed by two-step anneal treatments in order to examine the impact of dislocations and oxygen precipitates on the minority carrier lifetime. We investigate the impact of dislocations and precipitates on the minority carrier lifetime. We show that at similar precipitated concentration, the recombination activity of precipitates is significantly increased in vacancy-rich silicon. Finally, our results indicate that the cooldown rate after annealing may lead to out-of-equilibrium intrinsic point defect concentrations that contribute to the growth of oxygen precipitates or the formation of dislocations during subsequent processing

2 EXPERIMENTAL METHOD

The precipitation anneal consists of an anneal at 800°C in nitrogen for 4 hours followed by an anneal at 1000°C in oxygen for 8 hours. The 800°C step simulates the thermal impact of a diffusion. Falster *et al.* have shown that this anneal does not lead to significant nucleation of oxygen precipitates, in fact some precipitates below a certain size dissolve [4]. However the precipitates that survive grow in size (the distribution of precipitates size "drifts" to higher sizes) allowing them to survive the subsequent anneal at 1000°C [4].

The subsequent anneal at 1000°C leads to the growth of oxygen precipitates such that they become observable not only by lifetime spectroscopy [5], but also Fourier Transform Infrared Spectroscopy (FTIR), micro photoluminescence (μ-PL) and defect etch. This is a standard process for oxygen precipitate growth [6].

Micro photoluminescence spectroscopy (μ-PL) measurements were conducted using a Horiba LabRam tool equipped with a confocal microscope and a liquid nitrogen cooled InGaAs array detector. An excitation wavelength of 810 nm was chosen in order to probe deeper than the immediate surface and the measurements were performed at 80 K. The excitation absorption depth at 80 K is ~80 μm. For all measurements, a 50× objective lens was used to focus the laser beam into a spot of 1.8 μm in diameter on the slabs. FTIR measurements were conducted using a Bruker Vertex 80 system with a resolution of 4 cm^{-1}.

Photoluminescence (PL) images were captured using a BT imaging tool. Minority carrier lifetime measurements were performed using the QSSPC and transient photoconductance methods with a Sinton WCT120 system. We used the thermal oxide grown during the two-step anneal as well as PECVD silicon nitride for surface passivation (with $J_{0\text{-}SiN}$ = 3.3 $fA.cm^{-3}$). The measured lifetime was found to be the same with both films, but because the silicon nitride passivation featured surface scratches, we present only the thermally-grown oxide passivated results for all photoluminescence images and spectroscopy measurements.

3 RESULTS

3.1 Oxygen reduction

Fig. 1. shows the FTIR spectra before and after the tabula rasa and two-step anneals. In the as-grown wafer (Fig. 1a.), there is little difference between the oxygen reduction in the vacancy-rich or the interstitial-rich regions after two-step annealing. This is in agreement with the minority carrier lifetime measurements showing little difference between the interstitial-rich region and vacancy-rich region.

In the oxygen pre-annealed slab (Fig. 1b.), the interstitial oxygen concentration stays almost unchanged after the two-step annealing. This demonstrates that oxygen pre-annealing is effective at preventing oxygen precipitation.

Both the as-grown slab and the nitrogen pre-annealed slab show a significant and similar reduction of the interstitial oxygen concentration after two-step anneal. However, for the nitrogen annealed wafers, a peak appears at 1230 cm^{-1} corresponding to oxygen precipitates, in particular disc-shape oxygen precipitates [9]. It appears that the same amount of oxygen is precipitated in both the as-grown sample and the nitrogen pre-annealed sample, but only the nitrogen sample shows this oxygen precipitate peak. Possible reasons for this will be discussed below.

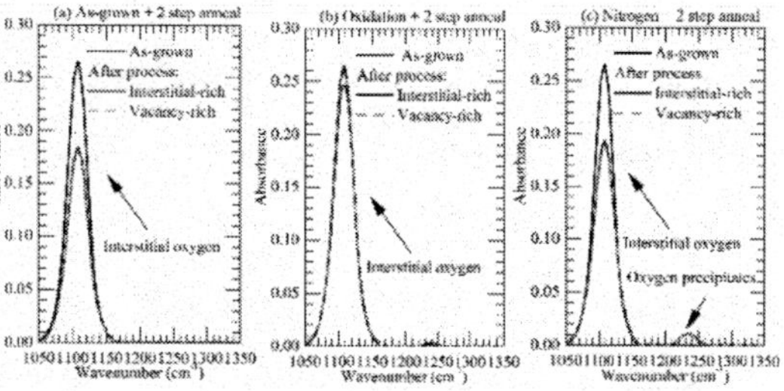

Fig. 1. FTIR spectra in the (a) as-grown slab, (b) oxygen pre-annealed slab and (c) nitrogen pre-annealed slab; after a two-step anneal. The peak at 1107 cm^{-3} is the absorption peak for interstitial oxygen, the peak at 1230 cm^{-3} is the absorption peak for oxygen precipitates.

2.2 Oxygen precipitates and dislocations

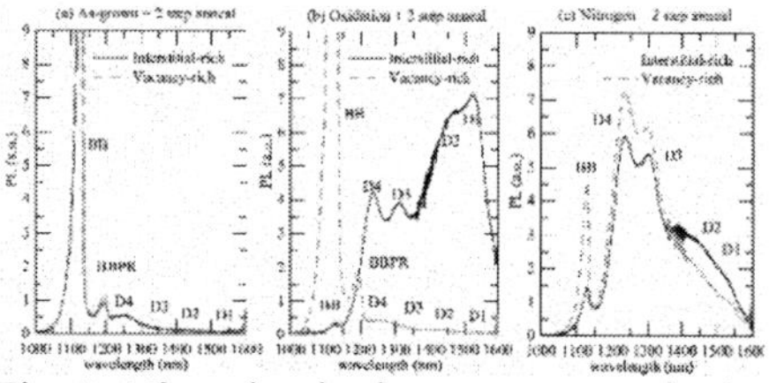

Fig. 2. Micro-photoluminescence spectra in the (a) as-grown slab, (b) oxygen pre-annealed slab and (c) nitrogen pre-annealed slab; after a two-step anneal. BB is the main band-to-band emission assisted by a transverse-optical phonon. BBPR is a phonon replica of the main band-to-band peak (band-to-band emission assisted by a transverse-optical phonon plus an optical zone centre phonon). The D lines (D1-D4) are sub-band gap luminescence from dislocations and/or oxygen precipitates.

Fig. 2. shows the micro-PL spectra in the interstitial-rich and vacancy-rich zone of all slabs. Note the sub-band-gap luminescence peaks labelled D3/D4 reflect the intrinsic properties of dislocations [10]. On the other hand, the D1/D2 peaks are usually associated with secondary defects or decoration around oxygen precipitates and dislocations in silicon, which are generally highly recombination active [10]. In the as-grown slab, the band-band luminescence intensity is similar after the two-step anneal, which is consistent with the lifetime measurements. The D-lines have low intensity in this sample.

In the oxygen pre-annealed slab, the micro-PL spectra shows a much lower band-band peak in the interstitial-rich region and a higher band-band peak in the vacancy-rich region, again in agreement with the PL image. The lower BB peak is correlated with high D-lines emissions in the interstitial-rich region, especially D1/D2, which are expected to be recombination active. This suggest that more precipitates and/or dislocations form in the interstitial rich-region of the oxygen annealed slab.

In the nitrogen pre-annealed slab, the micro-PL spectra show a very low band-band peak both in both the interstitial-rich and vacancy-rich regions. This is also correlated with high D-line peaks in both regions, although D3/D4 are dominant over D1/D2. It seems that a large density of active oxygen precipitates and/or dislocations form in both regions of the nitrogen annealed slab. A more detailed discussion can be found in [11].

2.3 Lifetime

Uncalibrated PL images after the two-step anneals are shown in Fig. 2. In the as-grown slab, the lifetime is homogeneously low in the slab after the two-step anneal, except for sections of the P-band, which are even lower. In the vacancy-rich region, void nucleation competes with oxygen precipitation. This means a lower density of oxygen precipitates form than in the P-band. The P-band has the lowest lifetime post the two-step anneal. This region is where the vacancy concentration is high but just below the concentration required for void nucleation [8]. This means during ingot cooling, vacancies are consumed by oxygen precipitates in this region [8]. These precipitates subsequently grow during the two-step anneal.

In the pre-oxidized slab, the lifetime is lower in the interstitial-rich region after the two-step anneal. From the injection dependence of the lifetime it appears that the same defect may be limiting the interstitial-rich region and the vacancy-rich region, with greater concentration in the interstitial-rich region.

In the nitrogen annealed slab, the lifetime is homogeneously low and the injection dependence of the lifetime is not measurable. It appears that high concentrations of defects and/or defects with large capture cross-sections are present throughout the samples and reduce the minority-carrier lifetime to values well below 1μs. More details can be found in [11].

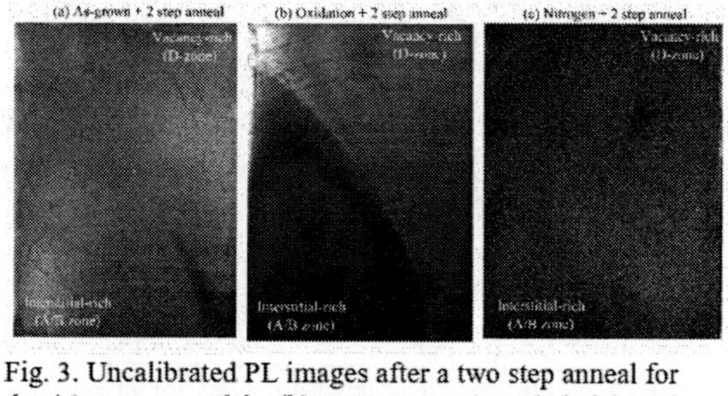

Fig. 3. Uncalibrated PL images after a two step anneal for the (a) as-grown slab, (b) oxygen pre-annealed slab and (c) nitrogen pre-annealed slab, after a two-step anneal. The contrast was adjusted allowing to see the features

4 CONCLUSIONS

In conclusion, we show both intrinsic point defects

from ingot-growth and Tabula rasa have an impact on the growth of oxygen precipitates during cell processing. We also how that precipitates grown in a vacancy-rich environment are much more recombination active or numerous. Finally, Tabula Rasa needs not only to dissolve oxygen precipitates but also to reduce the intrinsic point defect concentration to a minimum to avoid the formation of precipitates and dislocations during cell processing.

Acknowledgments

This work has been supported by the Australian Research Council (ARC) through a Discovery Early Career Researcher Award, and the Australian Renewable Energy Agency (ARENA) through project RND009.

References

[1] L. Mule'Stagno, "A Technique for Delineating Defects in Silicon," in *Gettering anf Defect Engineering in Semiconductor Technology IX*, 2001, pp. 753-758.

[2] K. H. Yang, "A preferential Etch for silicon crystals," *Semiconductor Processing, ASTM STP 850, D. C. Gupta, Ed., American Society for Testing and Materials, Philadelphia, Pa, USA,* 1984.

[3] K. H. Yang, "An Etch for Delineation of Defects in Silicon," *Journal of The Electrochemical Society,* vol. 131, pp. 1140-1145, 1984.

[4] R. J. Falster, M. Cornara, D. Gambaro, M. Olmo, and M. Pagani, "Effect of High Temperature Pre-Anneal on Oxygen Precipitates Nucleation Kinetics in Si " *Solid State Phenomena,* vol. 57-58, pp. 123-128, 1997.

[5] J. D. Murphy, R. E. McGuire, K. Bothe, V. V. Voronkov, and R. J. Falster, "Minority carrier lifetime in silicon photovoltaics: The effect of oxygen precipitation," *Solar Energy Materials and Solar Cells,* vol. 120, Part A, pp. 402-411, 2014.

[6] K. F. Kelton, R. Falster, D. Gambaro, M. Olmo, M. Cornara, and P. F. Wei, "Oxygen precipitation in silicon: Experimental studies and theoretical investigations within the classical theory of nucleation," *Journal of Applied Physics,* vol. 85, pp. 8097-8111, 1999.

[7] V. V. Voronkov and R. Falster, "Vacancy-type microdefect formation in Czochralski silicon," *Journal of Crystal Growth,* vol. 194, pp. 76-88, 1998.

[8] V. V. Voronkov and R. Falster, "Effect of vacancies on nucleation of oxide precipitates in silicon," *Materials Science in Semiconductor Processing,* vol. 5, pp. 387-390, 2002.

[9] B. Pivac, A. Borghesi, M. Geddo, A. Sassella, and A. Stella, "Stoichiometry of oxygen precipitates in silicon," *Applied Surface Science,* vol. 63, pp. 245-248, 1993.

[10] M. Tajima, Y. Iwata, F. Okayama, H. Toyota, H. Onodera, and T. Sekiguchi, "Deep-level photoluminescence due to dislocations and oxygen precipitates in multicrystalline Si," *Journal of Applied Physics,* vol. 111, p. 113523, 2012.

[11] F. E. Rougieux, H. T. Nguyen, D. Macdonald, B. Mitchell, and R. Falster, "Growth of oxygen precipitates and dislocations in Cz silicon: Impact on lifetime," *IEEE Journal of Photovoltaics,* vol. (accepted), 2017.

INVESTIGATING THE INFLUENCE OF INTERSTITIAL IRON ON THE STUDY OF BORON-OXYGEN DEFECTS

Moonyong Kim, Daniel Chen, Malcolm Abbott, Stuart Wenham, and Brett Hallam
School of Photovoltaic and Renewable Energy Engineering, University of New South Wales, Sydney, NSW 2052, AUSTRALIA

ABSTRACT: In this paper, we explore the influence of interstitial iron on studying boron-oxygen related degradation in p-type Czochralski silicon. We show that the presence of interstitial iron can potentially explain several key experimental results used to derive theories on the fundamental properties of the boron-oxygen defect. A theoretical analysis of the influence of interstitial iron and the boron-oxygen defect concentrations on the apparent capture cross section ratio indicates that at a concentration of 4×10^{11} /cm^3, interstitial iron can result in an apparent capture cross section ratio of >50 for a typical normalised boron-oxygen defect density after fast degradation. With the same interstitial iron concentration, after full degradation, the apparent capture cross section converged to the lifetime ratio with only boron-oxygen defects. We then compare the literature data of first annihilation rate of B-O defects after rapid degradation, to the kinetics of iron-boron pair association. The similar reaction rates suggest that with no iron, after either fast- or slow- degradation, only a single stage of recovery occurs. Another key potential influence of iron occurs when modulating the applied voltage used to induce B-O defect formation. Reducing the applied voltage can result in the association of Fe-B pairs and a subsequent recovery of V_{OC}. We conclude that careful attention must be paid to interstitial iron when studying boron-oxygen defects and that the current models related to boron-oxygen defects may need to be revised.
Keywords: Carrier-induced degradation, interstitial iron, boron-oxygen

1 INTRODUCTION

Carrier-induced boron-oxygen (B-O) related degradation in Czochralski silicon wafers has been widely studied over decades [1], [2]. However there is still significant debate about the structure and properties of the B-O defect [3]–[6]. One particular discrepancy in the literature is related to the capture cross-section ratio of the B-O defect. In early studies, the fast-forming recombination center (FRC) and slow-forming recombination center (SRC) were proposed to be two distinctively different defects of B-O due to the different time-constants of lifetime degradation and different capture cross section ratios ($k_{FRC,a}$ and k_{SRC}) of 100 and 10 respectively [7], [8]. The k of FRC was later revised to be 65 ($k_{FRC,b}$) [9].

It was also shown that an applied voltage of 0.4 V could suppress the fast degradation while still enabling the slow degradation. Furthermore, when the applied voltage is shifted from 0.8 V to 0.4 V, partial recovery can be observed, which then continues to degrade by B-O formation with the same degradation rate as the one that had applied voltage of 0.4 V only. Based on the observation, it was suggested that at higher applied voltage of 0.8 V, both fast and slow degradations occur while lower applied voltage of 0.4 V can only introduce the slow degradation. However, the observation is not sufficient to conclude the existence of two independent defects. The two-defect theory cannot certainly explain how improvement of V_{OC} can be observed when applied voltage is changed from 0.8 V to 0.4 V. If the defects were independently formed, the V_{OC} should have continued to decrease regardless of whether the fast degradation, as suggested in the paper, was suppressed with lower V_{OC} instead of converging to the other condition with constant applied voltage.

However, recently it was suggested that a single defect could be responsible for B-O degradation due to the measurement of identical recombination properties for fast and slow degradation with capture cross section ratio of close to 19 [6]. Two other key experimental results could also suggest the involvement of separate defects. One of the key results was a two-stage recovery of lifetime after rapid degradation [7]. This was compared to a single-stage recover of lifetime after full. The paper showed that the first stage of lifetime recovery could partially occur at close to room temperature.

In another study by Niewelt *et al.*, a revised single-defect parameterisation was presented with a k of 18 ($k_{B\text{-}O,b}$) for the donor level that was also able to describe both fast- and slow- degradation on multiple samples [10]. It should be noted that the difference in the donor-level k from earlier work ($k_{FRC,a}$) was due to the fact that Bothe *et al.* assumed a single-level defect, while in the more recent studies by Hallam *et al.* and Niewelt *et al.*, the two-level properties of B-O defects were included [6], [11]–[13], as well as the different assumptions used in modelling [14].

Furthermore, we have demonstrated the method to modulate the fraction of fast and slow degradation [15]. This observation strongly suggested that the defect that there are two recombination inactive precursors, A_1 and A_2 where A_1 is the precursor of A_2. The degradation occurs when A_2 transits to B under an illumination. In our previous work where we determined identical recombination properties for rapid and slow B-O related degradation, we eluded to the possibility of an influence of iron [16]. Iron is a common impurity in p-type Cz silicon that is influenced by carrier-injection. In the interstitial form, iron can associate with substitutional boron to form Fe-B pairs [17]. However, these pairs can be readily dissociated with carrier injection. Under illumination, the dissociation of Fe-B pairs usually occurs within 10 seconds, in contrast to fast and slow B-O related degradation, which can take up to 300 s and 48 hours, respectively [7], [16].

A peculiar behaviour of interstitial iron is when compared to the resultant lifetime in the Fe-B state, the lifetime above the cross-over point increases, and lifetime below the cross-over point decreases where the cross-over point depends on the background doping density [18]. In contrast, as B-O defects form, the lifetime in all injection levels decrease, although the effect is more pronounced at low-injection levels due to the increasing dominance of dark-saturation current density related recombination with increasing carrier concentration. Therefore, during light-soaking/dark annealing, in samples containing iron, the association/dissociation of Fe-B pairs could potentially lead to a misinterpretation of the B-O related properties.

This paper will provide a theoretical evidence, which demonstrates how the behaviour of interstitial iron can adequately provide an alternative explanation to explain the previous difference between the fast and slow degradation. The results will be compared to the key results published in the literature that has been used for the development of theories for the fundamental properties of the boron-oxygen defect, in the context of interstitial iron.

2 RESULTS AND DISCUSSION

2.1 Influence of interstitial iron and B-O defects in mono-crystalline silicon

To investigate the influence of interstitial iron in B-O defect studies, we performed a theoretical simulation to determine an apparent capture cross-section ratio (k_{app}) of Shockley-Read-Hall (SRH) recombination in the presence of both defects. We used the known recombination properties of the interstitial iron, $k_{Fei} \approx 286$ [16], and B-O defects, $k_{B\text{-}O,c} \approx 9.5$ [5], [19], where a parameterisation method [20] was used for SRH lifetime approximation with normalised defect density of B-O defects, $NDD_{B\text{-}O}$ in Figure 1.

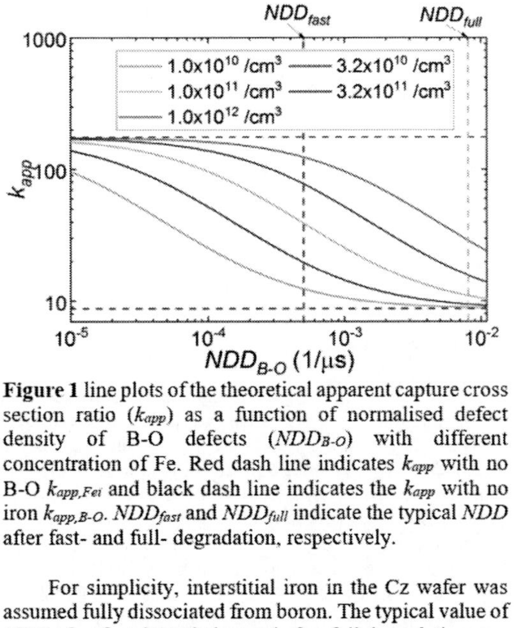

Figure 1 line plots of the theoretical apparent capture cross section ratio (k_{app}) as a function of normalised defect density of B-O defects ($NDD_{B\text{-}O}$) with different concentration of Fe. Red dash line indicates k_{app} with no B-O $k_{app,Fei}$ and black dash line indicates the k_{app} with no iron $k_{app,B\text{-}O}$. NDD_{fast} and NDD_{full} indicate the typical NDD after fast- and full- degradation, respectively.

For simplicity, interstitial iron in the Cz wafer was assumed fully dissociated from boron. The typical value of NDD after fast degradation and after full degradation was statistically estimated using the multiple measurements of identical structures with lifetime range of 60 to 90 μs. For the B-O defect, the relationship between the normalised defect density (NDD) and B-O defect concentration was estimated using Murphy's method [21]. The k_{app} was estimated by taking the ratio of the SRH lifetime at two injection levels, 10^{16} /cm³ and 10^{12} /cm³[16]. The simulated impact of interstitial iron concentration and normalised B-O defect density on the k_{app} is shown in **Fehler! Verweisquelle konnte nicht gefunden werden.**. Increasing the [Fe$_i$] increases k_{app} for a given NDD, whereas increasing the NDD for a given [Fe$_i$] decreases k_{app}. This conceptual simulation significantly highlights the possible explanation of the difference in k with different NDD with a fixed amount of interstitial iron, where it would be expected to have a higher initial k_{app} throughout the early stages of B-O degradation. The theoretical result offers a possible explanation for the higher capture cross-section ratio attributed to the 'FRC' in early studies and is in agreement with suggestions of a single defect causing both rapid and slow B-O related degradation [6], [15].

2.2 Influence of iron to the annihilation of B-O defects

The association kinetics of Fe-B pairs can have a significant impact on B-O studies during dark annealing. During dark annealing, the iron association can occur, which depends on the temperature and N_A. Therefore, if B-O defects and iron are both present in p-type silicon, you can expect to have the simultaneous reaction of Fe-B association and B-O annihilation. The Fe-B association rate [22], $\kappa_{Fe,a}$ with $N_A = 1.5 \times 10^{16}$ /cm³ was compared to the known FRC annihilation rate [7], $\kappa_{FRC,a1}$ in Figure 2. In the temperature range of 295 K and 330 K, $\kappa_{Fe,a}$ is within the errors of the reported fast recovery rate, $\kappa_{FRC,a1}$. The measurement was taken with the intensity of 10 W/cm² or 0.1 Sun, which can expect to be the injection point below the cross-over point. Below the cross-over point, Fe-B association can appear to increase the V$_{OC}$ measured. It should be noted that there was a discrepancy between experimental data from Bothe *et al.* at temperatures above 330 K and the Fe-B association rate [7]. However, there were only limited data points were used between the time frame of 1 to 100 seconds to calculate the rate in the paper[7] that it is unclear whether the number of measurements is sufficient to fit the rate of the reaction. Furthermore, the difference in kinetics of the second stage of fast defect annihilation (1.36 eV) and the annihilation rate for slow-degradation (1.32 eV) by Bothe *et al.* was very similar. Lim *et al.* also suggested the two reactions are due to same defect annihilation [23]. Therefore, the previous observation of first stage recovery of fast may have been due to the combination of Fe-B association and a single rate for B-O defect annihilation.

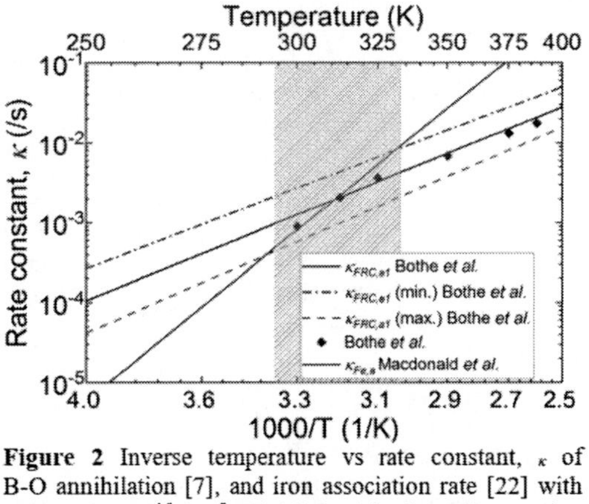

Figure 2 Inverse temperature vs rate constant, κ of B-O annihilation [7], and iron association rate [22] with $N_A = 1.5 \times 10^{16}$ /cm³. Highlighted in the yellow region represents the temperature range where iron association rate is within the range of B-O defect annihilation rate ($\kappa_{Fe,a} = \kappa_{FRC,a1}$).

2.3 Influence of iron on the formation of B-O defects

The influence of iron dissociation to the B-O defect formation study was theoretically investigated. It was known that an applied voltage can form B-O defects as well as dissociate Fe-B pairs [24], [25]. The Fe-B dissociation rate is proportional to the light intensity [26], which strongly indicates that the dissociation rate is excess carrier concentration dependent. The V_{OC} under 0.1 Sun with a different fraction of interstitial iron to the total iron concentration of 3×10^{11} /cm^3 is shown in Figure 3. At this iron concentration, this trend was opposite to that previously obtained by Schmidt *et al.* [27], where for a total iron concentration below 1×10^{12} /cm^3 an increase in the V_{OC} with Fe-B pair dissociation would decrease in the V_{OC}. However, as the V_{OC} was estimated at 0.1 Sun instead of at 1 Sun, it is expected to operate below the cross-over point [16], where dissociation of Fe-B pair would lower the effective lifetime [28], hence lower V_{OC}. The figure shows that dissociation, which can occur with the application of a bias voltage, can reduce the V_{OC}. Furthermore, if Fe-B association occurs, for example, with slower dissociation rate, V_{OC} can be increased.

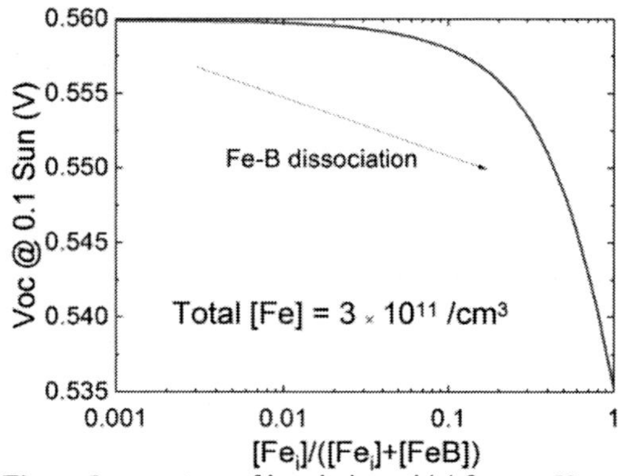

Figure 3 percentage of iron in interstitial form vs V_{OC} at 0.1 Sun with total [Fe] = 3×10^{11} /cm^3.

When the transition rate of A_2 to B is significantly reduced, typically with lower excess or hole concentrations, it can appear as a single exponential degradation with two subsequent precursor stages, A_1 and A_2. In order to demonstrate the apparent single exponential degradation, the two degradation conditions were used, where Condition 2 had a lower transition rate of A_2 to B than that of Condition 1. The normalised concentrations in each of the states $N_i(t)$, where i = A_1, A_2, B, are governed by a system of linear differential equations given by Equations 1-3, and t is the light soaking time. Here, κ_{ij} is the rate constant from state i to state j.

$$\frac{dN_{A1}}{dt} = -\kappa_{A1A2}N_{A1} \quad (1)$$

$$\frac{dN_{A2}}{dt} = \kappa_{A1A2}N_{A1} - \kappa_{A2B}N_{A2} \quad (2)$$

$$\frac{dN_B}{dt} = \kappa_{A2B}N_{A2} \quad (3)$$

Using the experimental results of fractional defect density over light soaking time with pre-dark annealing condition of 200 °C at 10 mins from Kim *et al.* [15], the formation rate of fast and slow B-O related degradation at room temperature is simulated in Figure 4. Condition 1 curve is plotted based on the fitted time constant of the results. κ_{A2B} of the Condition 2 is 20 times lower than κ_{A2B} of Condition 1. Condition 2 effectively shows a single exponential formation rate, despite that there are two reactions are occurring. This can explain the previous observation of a single formation rate, which was interpreted as the suppression of fast degradation [7]. Furthermore, the recovery of V_{OC}, when the applied voltage is changed from 0.8 V to 0.4 V is likely that the dissociation rate of Fe-B had reduced due to the lower applied voltage, which can effectively improve the V_{OC}. The applied voltage of 0.4 V may have been insufficient to dissociate Fe-B pairs such that no reduction was observed in the early stage of degradation by the constant applied voltage. The converging curves in the study by Bothe *et al.* [7] would, therefore, be adequately explained by the condition when the Fe-B dissociation rate is reduced while B-O defects continue to form at the same rate with the applied voltage.

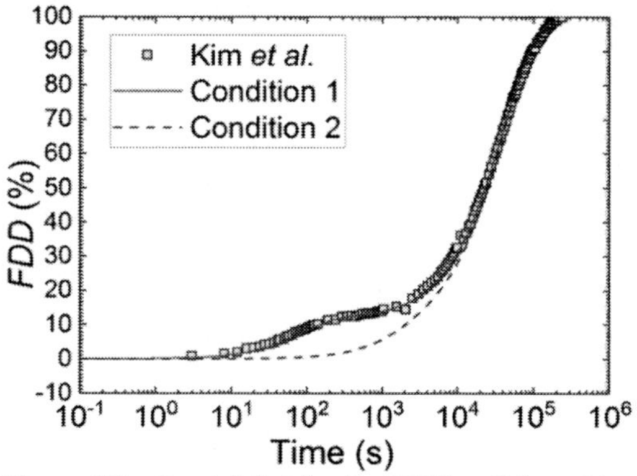

Figure 4 Fractional defect density (FDD) vs light soaking time. Condition 1 is the fitted curve. Condition 2 is the modified condition of where κ_{A2B} of the Condition 2 is 20 times lower than κ_{A2B} of Condition 1

CONCLUSION

This paper highlights the significant influence that interstitial iron can have when studying the B-O defect system. Theoretical simulation results in this paper indicate that the presence of iron can greatly influence the apparent behaviour and property of B-O defects., which could offer an alternative explanation of previous observations. The higher k of the B-O defects that is formed by the fast degradation, or namely FRC may be the combination of Fei and B-O defect. The known behaviour of iron and boron association can adequately explain a two-stage recovery of carrier lifetime. We theoretically demonstrate that the dissociation of Fe-B decreases V_{OC} at 0.1 Sun, which can adequately explain previous observation where the recovery of V_{OC} is observed when the applied voltage is reduced. Reducing the applied V_{OC} can reduce the dissociated stage of iron. The previous observation of the V_{OC} improvement with the reduced applied voltage before the degradation is observed can be explained by the new equilibrium of Fe_i with different reaction rate. The apparent single exponential degradation, which was previously interpreted as a suppression of fast degradation can be adequately explained by the slower transition rate of A_2 and B, or κ_{A2B}.

The theoretical results that are demonstrated in this study suggest that current defect models should be revised and the potential influence of iron should be considered as even in very small concentrations, can have significant impacts on studies of the B-O defect. The two independent defects that previous studies have suggested would be due to the pre-assumptions of the existence of two defects and likely the misperception of the influence of interstitial iron. Therefore, we conclude that the B-O defect is a single defect.

ACKNOWLEDGEMENTS

The authors would like to acknowledge Kyung Kim, Ly Mai, Nino Borojevic, Hongzhao Li, and the MAiA processing team who assisted with wafer processing. This Program has been supported by the Australian Government through the Australian Renewable Energy Agency (ARENA 1-A060) and the Australian Center for Advanced Photovoltaics (ARENA 1-SRI001). Brett Hallam would like to thank the Australian Research Council (ARC) for financial support through a Discovery Early Career Researcher Award (DE170100620) and the Australian Academy of Science through the J.G. Russell Prize. The views expressed herein are not necessarily the views of the Australian Government, and the Australian Government does not accept responsibility for any information or advice contained herein. The authors would like to thank the commercial partners of the ARENA 1-A060 project, and the UK Institution of Engineering and Technology (IET) for their funding support for this work through the A.F. Harvey Engineering Prize.

4.2 References

[1] J. Schmidt and K. Bothe, "Structure and transformation of the metastable boron- and oxygen-related defect center in crystalline silicon," *Phys. Rev. B*, vol. 69, no. 2, p. 24107, 2004.

[2] D. W. Palmer, K. Bothe, and J. Schmidt, "Kinetics of the electronically stimulated formation of a boron-oxygen complex in crystalline silicon," *Phys. Rev. B*, vol. 76, no. 3, p. 35210, 2007.

[3] V. V Voronkov, R. Falster, A. V Batunina, D. Macdonald, K. Bothe, and J. Schmidt, "Lifetime degradation mechanism in boron-doped Czochralski silicon," *Energy Procedia*, vol. 3, pp. 46–50, 2011.

[4] V. V Voronkov, R. Falster, K. Bothe, and B. Lim, "Light-induced lifetime degradation in boron-doped Czochralski silicon: are oxygen dimers involved?," *Energy Procedia*, vol. 38, pp. 636–641, 2013.

[5] V. Voronkov and R. Falster, "The nature of boron- oxygen lifetime- degrading centres in silicon," *Phys. Status Solidi*, vol. 13, no. 10–12, pp. 712–717, 2016.

[6] B. Hallam, M. Abbott, T. Nærland, and S. Wenham, "Fast and slow lifetime degradation in boron-doped Czochralski silicon described by a single defect," *Phys. Status Solidi - Rapid Res. Lett.*, vol. 10, no. 7, pp. 520–524, 2016.

[7] K. Bothe and J. Schmidt, "Electronically activated boron-oxygen-related recombination centers in crystalline silicon," *J. Appl. Phys.*, vol. 99, no. 1, p. 13701, 2006.

[8] V. V. Voronkov and R. Falster, "Light-induced boron-oxygen recombination centres in silicon: Understanding their formation and elimination," *Solid State Phenom.*, vol. 205, pp. 3–14, 2014.

[9] V. V. Voronkov, R. Falster, K. Bothe, B. Lim, and J. Schmidt, "Lifetime-degrading boron-oxygen centres in p-type and n-type compensated silicon," *J. Appl. Phys.*, vol. 110, no. 6, p. 63515, 2011.

[10] T. Niewelt, S. Mägdefessel, and M. C. Schubert, "Fast in-situ photoluminescence analysis for a recombination parameterization of the fast BO defect component in silicon," *J. Appl. Phys.*, vol. 120, no. 8, p. 85705, 2016.

[11] V. Voronkov and R. Falster, "The nature of boron-oxygen lifetime-degrading centres in silicon," *Phys. status solidi*, vol. 13, no. 10–12, pp. 712–717, Dec. 2016.

[12] B. Hallam *et al.*, "Recent insights into boron-oxygen related degradation: Evidence of a single defect," *Sol. Energy Mater. Sol. Cells*, 2017.

[13] T. Niewelt, J. Schön, J. Broisch, W. Warta, and M. Schubert, "Electrical characterization of the slow boron oxygen defect component in Czochralski silicon," *Phys. Status Solidi - Rapid Res. Lett.*, vol. 9, no. 12, pp. 692–696, 2015.

[14] N. Nampalli, T. H. Fung, S. Wenham, B. Hallam, and M. Abbott, "Statistical analysis of recombination properties of the boron-oxygen defect in p-type Czochralski silicon," *Front. Energy*, vol. 11, no. 1, pp. 4–22, 2017.

[15] M. Kim, M. Abbott, N. Nampalli, S. Wenham, B. Stefani, and B. Hallam, "Modulating the extent of fast and slow boron-oxygen related degradation in Czochralski silicon by thermal annealing: Evidence of a single defect," *J. Appl. Phys.*, vol. 121, p. 53106, 2017.

[16] D. Macdonald, A. CuevasJennifer Wong-Leung, A. Cuevas, and J. Wong-Leung, "Capture cross sections of the acceptor level of iron–boron pairs in p-type silicon by injection-level dependent lifetime measurements," *J. Appl. Phys.*, vol. 89, no. 12, pp. 7932–7939, 2001.

[17] D. Macdonald, "Recombination activity of interstitial iron and other transition metal point defects in p- and n-type crystalline silicon," *Appl. Phys. Lett.*, vol. 85, no. 18, pp. 4061–4063, 2004.

[18] D. MacDonald, T. Roth, P. N. K. Deenapanray, T. Trupke, and R. A. Bardos, "Doping dependence of the carrier lifetime crossover point upon dissociation of iron-boron pairs in crystalline silicon," *Appl. Phys. Lett.*, vol. 89, no. 14, p. 142107, 2006.

[19] T. Niewelt, J. Schön, J. Broisch, S. Mägdefessel, W. Warta, and M. C. Schubert, "A unified parameterization for the formation of boron oxygen defects and their electrical activity," *Energy Procedia*, vol. 92, pp. 170–179, 2016.

[20] W. Shockley, W, Read, "Statistics of the Recombinations of Holes and Electrons," *Phys. Rev.*, vol. 87, no. 5, pp. 835–842, 1952.

[21] J. D. Murphy, K. Bothe, R. Krain, V. V. Voronkov, and R. J. Falster, "The impact of oxide precipitates on minority carrier lifetime in Czochralski silicon," *ECS Trans.*, vol. 50, no. 5, pp. 137–144, 2012.

[22] D. Macdonald, T. Rothprakash, P. N. Deenapanraykarsten, K. Bothe, P. Pohl, and J. Schmidt, "Formation rates of iron-acceptor pairs in crystalline silicon," *J. Appl. Phys*, vol. 98, p. 83509, 2005.

[23] B. Lim *et al.*, "Lifetime recovery in p-type Czochralski silicon due to the reconfiguration of boron – oxygen complexes via a hole-emitting process Lifetime recovery in p -type Czochralski silicon due to the reconfiguration of boron – oxygen complexes via a hole-emitting," *Appl. Phys. Lett.*, vol. 98, p. 162104, 2011.

[24] J. Knobloch, S. W. Glunz, D. Biro, W. Warta, E. Schaffer, and W. Wettling, "Solar cells with efficiencies above 21% proccessed from Czochroalski grown silicon," in *Proceedings of the 25th IEEE Photovoltaic Specialists Conference, Washington, DC IEEE*, 1996, pp. 405–408.

[25] A. A. Istratov, H. Hieslmair, and E. R. Weber, "Iron and its complexes in silicon," *Appl. Phys. A Mater. Sci. Process.*, vol. 69, no. 1, pp. 13–44, 1999.

[26] L. J. Geerligs and D. Macdonald, "Dynamics of light-induced FeB pair dissociation in crystalline silicon," *Appl. Phys. Lett.*, vol. 85, no. 22, pp. 5227–5229, 2004.

[27] J. Schmidt, "Effect of dissociation of iron-boron Pairs in crystalline silicon on solar cell properties," *Prog. Photovoltaics Res. Appl.*, vol. 13, no. 4, pp. 325–331, 2005.

[28] D. H. Macdonald, L. J. Geerligs, and A. Azzizi, "Iron detection in crystalline silicon by carrier lifetime measurements for arbitrary injection and doping," *J. Appl. Phys.*, vol. 95, no. 3, pp. 1021–1028, 2004.

MACHINING BEHAVIOUR OF SILICON IN WIRE EDM FOR PV APPLICATIONS

Makarand M Kane* Aditya Jadhav* Mastik Kumar* S V Kulkarni S S Joshi
Indian Institute of Technology, Bombay
Mumbai, India 400 076

ABSTRACT: Silicon is the dominant material used to make photovoltaic (PV) cells today and silicon wafers are sliced from ingots by abrasive cutting processes. In this work, Wire Electric Discharge Machining (WEDM) is investigated as a promising alternative for the conventional wafer-slicing methods. Through an experimental approach, the behaviour of WEDM is characterised for silicon and compared with that for mild-steel. Parameteric analysis is done with repsect to the available voltage and current levels and switching time intervals. In particular, the voltage-current (V-I) characteristics, power consumption and material removal rate (MRR) for the two materials are compared. Significant differences are observed in the I-V characteristics, which are discussed in this paper. The power consumed in the case of silicon is observed to be 43% more than that for steel, for a certain set of parameters. To confirm the suitablity of the method for PV application, the cut surfaces of silicon are subjected to Scanning Electron Microscopy (SEM) and Inductively Coupled Atomic Emission Spectroscopy (ICPAES) for testing of surface roughness and contamination.

Keywords: Silicon Solar Cell, Manufacturing and Processing

1 INTRODUCTION

Extracting wafers from raw silicon, i.e., from feedstock to ingot and from ingot to wafers, is the most energy intensive part of the solar module manufacturing supply chain and accounts for about 33% of the module's production cost [1]. Currently, silicon wafers are sliced from ingots by two prominent abrasive cutting processes: i) Wire loose slurry method and ii) Diamond saw cutting method. Both these methods can cause microfractures and cracks as deep as 20 μm into the surface of the wafers they produce. This leads to fragile wafers with typical thickness of 180 μm [2]. Also, as much as 50% of the feedstock ingot material is lost as kerf losses [3]. Apart from that, these methods have other drawbacks like contamination of the wafers due to the slurry [4]. Hence, it is critical to look for alternative techniques for slicing wafers from silicon ingots.

Efforts to reduce material loss by using thinner wires have been made. However, the wire thickness itself is limited due to high tension required for precise cutting of the ingots and due to wear of the wire caused by friction [5]. There have been attempts to obtain *kerfless* wafers [6], [7]. However, it is likely that ingot slicing will continue to be the popular form of wafering due to its widespread acceptance and inertia in the industry.

Recently, there have been attempts to employ Wire Electric Discharge Machining (WEDM) for semiconductors in general and for slicing of wafers from ingots in particular [3], [8], [9], [10]. However, most of the literature focuses on the use of *existing* WEDM technology for slicing of silicon ingots. However, WEDM is a spark-erosion-based machining process and hence, is well suited for metals. It is not *designed* for semiconductors like silicon. Its behaviour with silicon should be quite distinct from that for metals. Very few authors have addressed this issue [11]. However, there is no literature published providing direct comparison of behaviour of WEDM while cutting silicon vis-à-vis metals. If this distinction is clearly understood, it might help to customise WEDM for silicon wafering. For the comparison, one of the most widely used metals in the industry i.e. mild steel (carbon-steel) is chosen and the behaviour of WEDM while cutting silicon is compared with that while cutting the steel.

The work reported in the paper is classified in two parts. The first part focuses on distinction between electrical characterization of the WEDM-spark for silicon and steel. This includes results obtained by recording voltage and current of the WEDM-spark while cutting of silicon ingots. The second part contains results of surface topography and contamination analysis with SEM and ICPAES. This part is included to investigate and establish the effectiveness of WEDM for silicon cutting from the perspective of contaminations and surface quality. Although similar results regarding surface topography are found in literature [11], [12], analysis of surface topography is presented here to confirm that the WEDM is indeed a useful process for ingot-cutting, particularly for PV applications.

The paper is organized in six sections. As the work focuses on current and voltage charactertistics of the WEDM-sparks, the next section is devoted to explain the typical voltage and current waveforms of WEDM-sparks. Third section describes the experimental setup and the approach used for the experimentation. Section 4 highlights major findings of the work with comments on I-V characteristics and power consumption. The fifth section contains discussion of SEM and ICPAES results. The last section presents conclusion.

2 TYPICAL WEDM CHARACTERISTICS

WEDM is a non-contact spark erosion process for cutting of electrically conductive workpieces. The sparks are generated by applying pulsed voltage between the tool, which is a wire, and the workpiece. Typical discharge current and voltage waveforms are shown in Figure 1.

There are 5 stages ('a' through 'e') in the discharge process as follows [13]:

a) **Pre-breakdown**: Voltage is applied between the work-piece and the wire but there is no breakdown across the dielectric.
b) **Breakdown**: When the two electrodes are close enough i.e. at a distance of the order of 10-100 μm,

*equal contribution

then breakdown occurs. The voltage starts decreasing and the current starts increasing via formation of plasma channel.

c) **Discharge** or **Sparking phase**: The actual cutting takes place by erosion. During this phase the plasma channel produces a lot of heat. The temperature is of the order of 8100 K.

d) **End of discharge**: During this phase, electrons and ions *implode* and hence the current and voltage drops to zero.

e) **Post-spark**: The voltage across the electrodes is deliberately made zero. This is also called as *dead time*. The flushing dielectric fluid removes debris and frozen metal particles during this time.

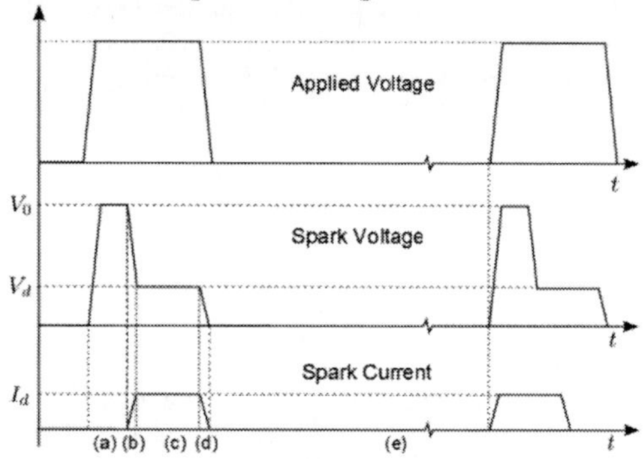

Figure 1 Ideal voltage and current waveforms of WEDM spark [13]

3 EXPERIMENTAL SETUP

The schematic of experimental setup is as shown in Figure 2. The terminal voltage and current (provided by pulsed power supply) signals are recorded with the help of a Digital Storage Oscilloscope (DSO) – Tektronix TDS 2002B with recording frequency of 60 MHz. Blocks of p-doped polycrystalline silicon and mild steel are chosen as workpieces. The apparatus contains:

- **WEDM**: Servo Control High Quality Wire Cut EDM BM400C-C available at Machine Tools Laboratory, Department of Mechanical Engineering, IIT Bombay. This will be referred to as BM4 here onwards.
- **Wire**: 180 µm Molybdenum wire
- **Dielectric Fluid**:
  - Emulsion-based dielectric fluid (EML) This is provided by the manufacturer.
  - Deionsed water (DIW)
- **Type of Silicon**: 'p'-type
- **Type of steel**: mild (carbon) steel
- **Workpiece dimensions**: 74 × 42 × 22 mm (For both, silicon and steel)

Table I Variable parameters of BM4 WEDM. Electrical characterization of spark is done by varying one parameter at a time. Around 100 sets of experiments are completed.

Sr No.	Parameter	Range
1	Pulse ON Time (t_{ON})	10-30 µs
2	Machining Period (t_T)	170-900 µs
3	Voltage level (V)	75 V and 110 V
4	Peak Current* (I_{peak})	1 to 6
5	Wire Speed	5 – 11 m/min

* Current is varied in indices 1 to 6. It is found that they correspond to a range of 10 A to 50 A.

Five distinct parameter settings in BM4 are as shown in Table I. There are more than 5 discrete steps possible for each parameter, except for voltage. Hence, total number of combinations turn out to be more than 1000. However, only one parameter is varied at a time and others are kept constant. Apart from variation of workpiece material, the dielectric used for cutting is also varied – two dielectrics are used – EML and DIW. Such variation is done because EML might introduce some impurities in the cut-wafers as it contains some emulsion.

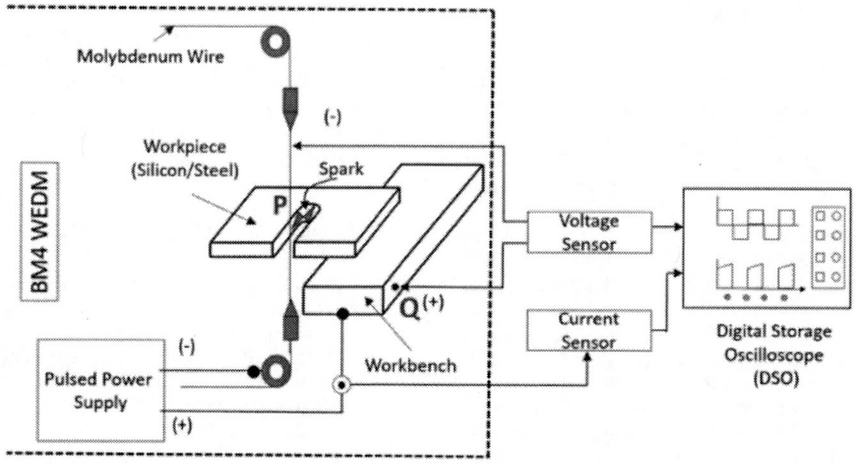

Figure 2 Schematic of setup for recording voltage and current waveform of EDM spark

Hence, DIW is also used as another dielectric to ensure lesser impurities in the cut wafers.

One such experimental *run* includes recording voltage and current of the spark for several machining cycles. The recording is done for two workpieces – for silicon as well as steel. Hence, one *set* of readings contains 4 data vectors – 2 for each workpiece material; 47 of such *sets* are completed.

4 EXPERIMENTAL RESULTS

Voltage and current waveforms recorded in the 47 *sets* are plotted using MATLAB and MsExcel. The results presented in next subsections are *representative* of observations common across all the sets.

4.1 Voltage and Current Profiles

Figure 3 and Figure 4 show the spark voltage and current profiles for steel and silicon. Figure 5 shows how this difference gets translated into I-V characteristics of the spark. The I-V characteristics in case of steel matches well with the I-V characteristics of EDM spark reported by Li et al. [14]. The experimental setup used by them is same as that used in this work. In particular, they have also used Molybdenum wire to cut carbon steel. The I-V characteristics for silicon are clearly different from steel.

4.2 Power Consumption

The power consumed for the cutting process during one machining cycle is calculated from the recorded voltage and current data. To eliminate variation of power consumption over different cutting cycles, several cycles are recorded (see Figure 6) and averaged. Power difference for silicon and steel is as shown in Figure 8. These are representative readings where, time period (t_T) and current index is varied. The power consumed by silicon is greater than steel for all the readings. In some cases, it is 43% more than that for steel. However, the powers consumed in both the cases is comparable for lower values of open circuit voltages.

4.3 Material Removal Rate (MRR)

MRR is calculated by measuring the time taken for a 10 mm straight cut, while operating with the same machine parameters for both workpieces. The MRR was found to be 20.63 mm²/min and 20.95 mm²/min for silicon and steel respectively. Thus, these preliminary investigations have shown that MRRs for steel and silicon are almost equal.

4.4 Observation and comments on results

From these observations following comments can be made:

1. The I-V characteristics of silicon and steel are clearly different. From Figure 3 through Figure 5, it can be said that four distinct stages in the spark discharge (as indicated in the ideal waveforms in Figure 1) are clearly visible in case of steel. However, in case of silicon, *breakdown* and *discharge* stages are *combined* into one single stage. Table II presents this fact clearly.

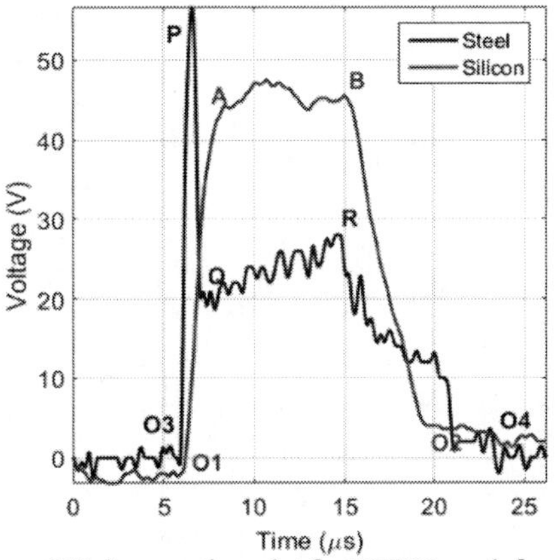

Figure 3 Voltage vs time plot for WEDM spark for silicon and steel

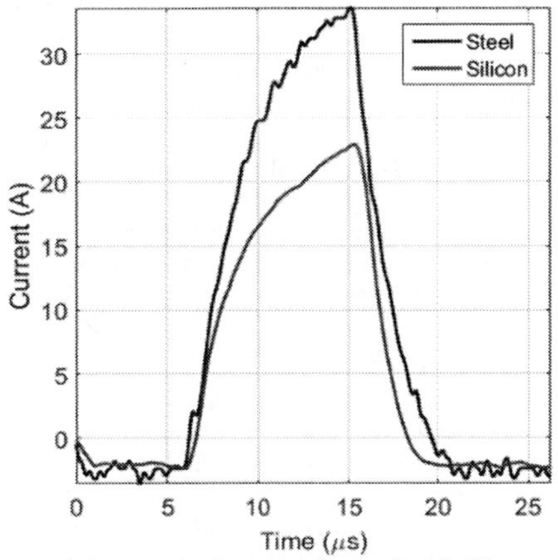

Figure 4 Current vs time plot for steel and silicon

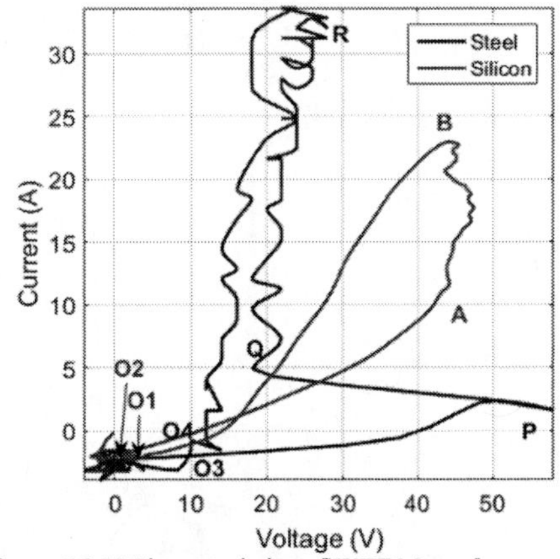

Figure 5 I-V Characteristics of WEDM spark: comparison of silicon and steel

Table II Phases of WEDM as observed from spark voltage and current for cutting of steel and silicon with WEDM (see Figure 5)

Phase of WEDM	Steel	Silicon
Pre-breakdown	O3-P	O1-A
Breakdown	P-Q	A-B
Discharge	Q-R	
End of discharge	R-O4	B-O2

2. Following relationships are found from Figure 3 through Figure 6.

$$I_P^{(Si)} < I_P^{(St)}$$
$$V_P^{(Si)} > V_P^{(St)}$$
$$P^{(Si)} < P^{(St)}$$

where, I_P is the peak current, V_P is the peak voltage and P is the power consumed.

3. One of the reasons for the two observations above is, the presence of bulk and contact resistances in the case of silicon. Resistivity of silicon is around 10-22 Ωcm whereas that for the steel is $1.45 \times 10^{-5}\Omega$cm. There is a difference of 5 orders of magnitude in the resistivities of silicon and steel. Therefore, for a silicon workpiece, the point of sparking P and the point of application of pulsed voltage to the metallic workbench, i.e., point Q, (see Figure 2) are at different potentials. A high voltage drop (*for the same current value*) across the workpiece might be responsible for the difference in the characteristics seen above.
4. Another reason could be as follows: the workpiece is kept on a metallic bench for cutting in the WEDM. The complete set of material-boundaries involved in the problem can be shown schematically as in Figure 7. The junction formed between contact of the metal and the semiconductor may be affecting the observed characteristics [12]. In short, there can be two prime reasons for the observed characteristics:
   a. Junction formed between metallic workbench and semiconductor workpiece
   b. Fundametal difference in the spark generated between metal and semiconductor

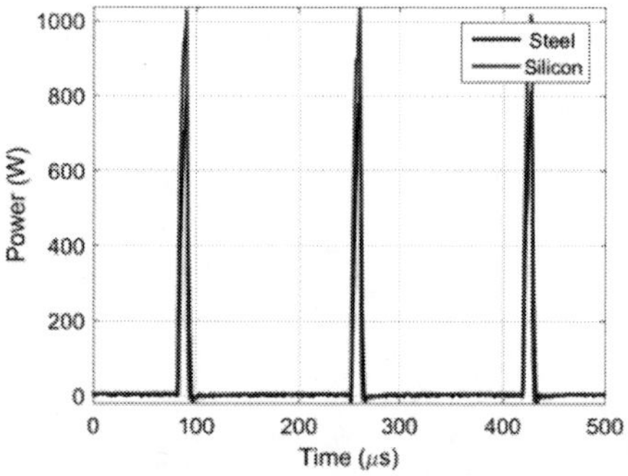

Figure 6 Instantaneous power $[p(t) = v(t) \times i(t)]$ vs time plot for several cycles. Power is calculated by averaging power for three cycles.

5. Huijun et al. [11] have investigated the distinction between the plasma formed during silicon cutting vs metal cutting. However, this explanation is primarily based only on the negative temparature coefficient of resistance and it is not adequate. More investigation needs to be done to understand the characteristics fully.
6. Effect of dielectric: The distinction between waveforms is similar irrespetive of the dielectric used. In fact the power consumed is more in case of EML than that with DIW.

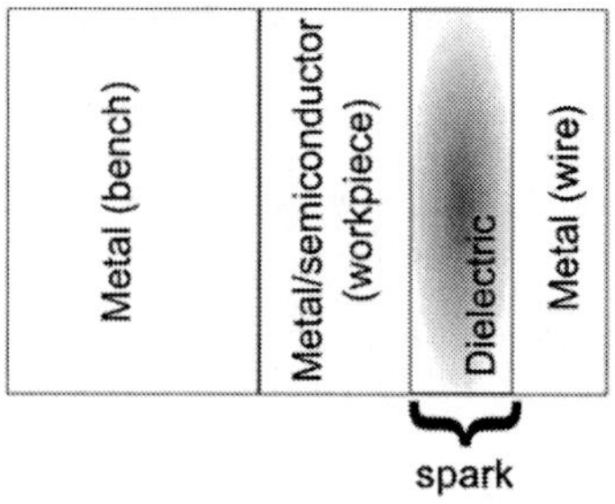

Figure 7 Schematic description of materials and boundaries involved while WEDM cutting. It is clear that, for cutting of silicon, the discharge properties will be different than those for metal cutting.

5 SURFACE TOPOGRAPHY AND CONTAMINATION ANALYSIS

SEM is widely used in literature to study the surface morphology of crystalline silicon solar cells [15]. Energy Dispersive X-ray spectroscopy (EDX) is then used as a tool to detect elemental contaminations that remain deposited on the cut surface. Whereas, ICPAES is used to quantify the percentage of contaminations in the wafers. Thus, SEM denotes the contaminations due to cutting process which are easily removable, and ICPAES denotes the contaminations which may have penetrated the workpiece (silicon) due to cutting process.

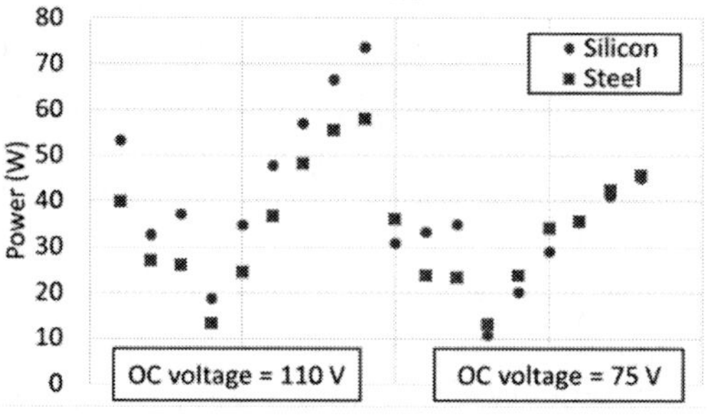

Figure 8 Difference in power consumption during EDM cutting of silicon and steel. It is seen that silicon consumes more power for OC (Open Circuit) voltage of 110 V. However, this behaviour is not seen at lower OC voltage of 75 V

The SEM image of a silicon piece in *as-cut condition* is shown in Figure 9. When it is compared with the image of a wafer cut by diamond-wire saw in literature ([16] – Figure 2), it can be observed that, the chipping and sawing marks are seen on the wafers cut by wire-saw however, they are absent in the wafers cut by WEDM. Instead, that wafer shows some resolidified layer of material after implosion of WEDM plasma.

Surface of the cut samples (10 nos) is scanned at several places using EDX to see the contaminations. The results are shown in Figure 10. It is seen that the significant contaminations are manganese, sodium and aluminium. These contaminations on surface can be removed easily by usual etching process used in wafer manufacturing.

ICPAES results of cut samples (in *as cut* conditions) are shown in Table III. It is clear that, the contaminations introduced in the samples are very less in percentage. Some of the contaminations are due to EML used as dielectric. If a WEDM is developed customized for silicon, then the contaminations like copper can be eliminated completely.

Figure 9 WEDM-cut-wafers: does not show any sawing mark magnification 1500x

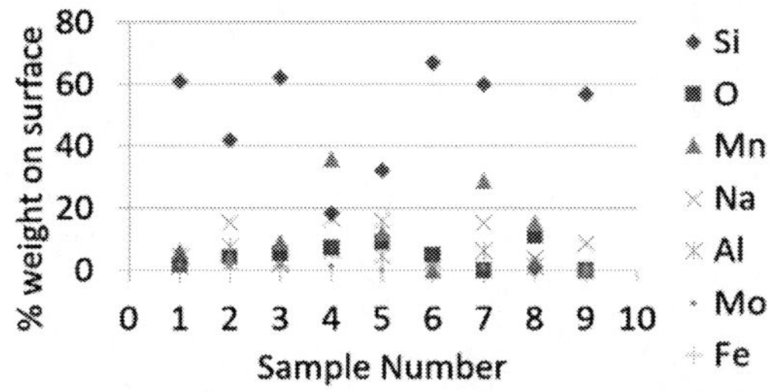

Figure 10 SEM-EDX results of surface of silicon cut by wire EDM. It is seen that, there are several metals on the surface.

6 CONCLUSION

The differences in machining of silicon and steel by WEDM are documented in this paper. Unlike steel, there is a significant voltage drop across the bulk and contact resistances in the case of silicon, which leads to differences in the voltage and current waveforms. The I-V characteristics of the spark while cutting silicon are significantly different from the ideal I-V characteristics of WEDM spark reported earlier. Also, the power consumed in the case of silicon is more than that for steel, while the MRR is nearly equal. It is now clear that, to address these differences, a new form of WEDM specifically designed for silicon wafering is required.

Table III Full elemental scan of Silicon sample machined with Molybdenum wire based WEDM

Elements	Sample (% weight)
Boron (B)	0.0041
Barium (Ba)	0.0015
Calcium (Ca)	0.3111
Chromium (Cr)	0.0036
Copper (Cu)	0.0053
Iron (Fe)	0.1489
Potassium (K)	0.0154
Magnesium (Mg)	0.025
Manganese (Mn)	0.0018
Sodium (Na)	0.0278
Strontium (Sr)	0.0004
Zinc (Zn)	0.0027
Molybdenum (Mo)	0.008
Chlorine (Cl)	1.0195
Silicon (Si)	98.4248

Thus, the impact of this work in terms of *scientific innovation and relevance* is threefold: a) It provides better understanding of the spark-erosion process in WEDM for the case of silicon. b) It will provide a blueprint for the development of WEDM specialised for silicon wafer slicing c) If the custom WEDM thus designed is proven to be faster and more efficient than the existing techniques, it will help reduce the cost of wafer production and consequently module manufacturing.

ACKNOWLEDGEMENTS

Authors thank National Centre for Photovoltaic Research and Education (NCPRE), IIT Bombay and Machine Tools Laboratory, IIT Bombay for their support in terms of equipment and manpower for this project.

7 REFERENCES

[1] B. P. Rand, J. Genoe, P. Heremans, and J. Poortmans, "Crystalline Silicon Solar Module Technology: Towards the 1€ Per Watt-Peak Goal," *Prog. Photovolt Res. Appl.*, vol. 15, no. February 2013, pp. 659–676, 2007.

[2] B. Sopori, S. Devayajanam, and P. Basnyat, "Surface characteristics and damage distributions of diamond wire sawn wafers for silicon solar cells," *Materials (Basel).*, vol. 3, no. June, pp. 669–685, 2016.

[3] G. G. Dongre, C. Vesvikar, R. K. Singh, and S. S. Joshi, "Efficient dicing of silicon ingots for photovoltaic applications," in *Conference Record of the IEEE Photovoltaic Specialists Conference*, 2010.

[4] H. J. Möller, "Wafering of Silicon," *Semicond. Semimetals*, vol. 92, no. 4, pp. 63–109, 2015.

[5] A. Bidiville, K. Wasmer, M. Van Der Meer, and C. Ballif, "Wire-sawing processes: Parametrical study

and modeling," *Sol. Energy Mater. Sol. Cells*, vol. 132, pp. 392–402, 2015.

[6] K. Nakayashiki *et al.*, "Fabrication and analysis of high-efficiency String Ribbon Si solar cells," *Solid. State. Electron.*, vol. 50, no. 7–8, pp. 1406–1412, 2006.

[7] L. Lichtensteiger, "Kerf-free wafer slicing and thinning for semiconductor applications," *2013 10th China Int. Forum Solid State Light. China SSL 2013*, pp. 85–88, 2015.

[8] H. V. B. Dominik Reynaerts, Paul Henries Heeren, "Micro-Structuring of Silicon by Electro Discharge Machining," *Sensors Actuators A Phys. J.*, vol. 60, no. 1, pp. 212–218, 1997.

[9] Y. Uno, A. Okada, Y. Okamoto, and T. Hirano, "Wire EDM slicing of monocrystalline silicon ingot," *Proc. 2000 ASPE Annu. Meet.*, vol. 22, pp. 172–175, 2000.

[10] H. Takino, T. Ichinohe, K. Tanimoto, S. Yamaguchi, K. Nomura, and M. Kunieda, "Cutting of polished single-crystal silicon by wire electrical discharge machining," *Precis. Eng.*, vol. 28, no. 3, pp. 314–319, 2004.

[11] P. Huijun, L. Zhidong, G. Lian, Q. Mingbo, and T. Zongjun, "Study of small holes on monocrystalline silicon cut by WEDM," *Mater. Sci. Semicond. Process.*, vol. 16, no. 2, pp. 385–389, 2013.

[12] Y. F. Luo, C. G. Chen, and Z. F. Tong, "Investigation of silicon wafering by wire EDM," *J. Mater. Sci.*, vol. 27, no. 21, pp. 5805–5810, 1992.

[13] A. Descoeudres, "Characterization of Electrical Discharge Machining Plasmas," vol. 3542, p. 137, 2006.

[14] C. Li, J. Bai, J. Ding, and Y. Fan, "Gap current voltage characteristics of energy-saving pulse power generator for wire EDM," *Int. J. Adv. Manuf. Technol.*, 2015.

[15] L. C. T. Markvart, Augustin McEvoy, *Solar Cells, Materials, Manufacture and Operation*. Elsevier Science, 2012.

[16] A. Kumar, S. Kaminski, S. N. Melkote, and C. Arcona, "Effect of wear of diamond wire on surface morphology, roughness and subsurface damage of silicon wafers," *Wear*, vol. 364–365, pp. 163–168, 2016.

TOWARDS MULTI MILLISECOND SPATIALLY HOMOGENEOUS CARRIER LIFETIMES FROM EPITAXIAL SILICON WAFERS GROWN ON POROUS SILICON

Sarah Kajari-Schröder[1,*], Catherin Gemmel[1], Jan Hensen[1] and Rolf Brendel[1,2]

[1] Institute for Solar Energy Research Hamelin (ISFH), Am Ohrberg 1, 31860 Emmerthal, Germany
[2] Insitute for Solid State Physics, Leibniz University Hanover, Appelstr. 2, 30167 Hanover, Germany
[*]Corresponding author, Kajari-schroeder@isfh.de

ABSTRACT: Kerfless wafering is receiving renewed attention internationally due to a) the need of PV industry to realize cost savings and to enhance resource efficiency along the value chain and b) due to the recent step by step demonstrations of key components for the implementation of kerfless wafers in mainstream photovoltaics: high electronic quality of the kerfless wafers, suitability for existing machinery (drop-in replacement), and high solar cell efficiencies. We use the porous silicon (PSI) process for the fabrication of kerfless monocrystalline silicon wafers. Here, we analyze the PSI wafer quality by analyzing the injection-dependent carrier lifetimes of the samples with infrared lifetime mappings and photo conductance decay measurements. We evaluate the reorganization of the PSI by scanning electron microscopy (SEM). We successfully lift-off 7 out of 7 epitaxial PSI wafers on an area of 10 x 10 cm^2 pseudosquare. The effective lifetimes of these layers range from 836 to 1022 µs at an injection level of n=10^{15} cm^{-3} and the lifetime is spatially homogeneous on the wafer area.

Keywords: Kerfless silicon, porous silicon, epitaxial wafer

1 INTRODUCTION

Kerfless wafering, and in particular the porous silicon (PSI) process [1,2], is receiving renewed attention internationally due to a) the need of PV industry to realize cost saving and resource efficiency along the full value chain [3] and b) due to an increasing number of reports on the demonstration of high electronic wafer quality [4–6] and high solar cell efficiencies [7–10].
The PSI wafer quality depends on the porosification, the reorganization and the actual epitaxy processes. The porosification has to ensure a low porosity surface for a reliable closing of the surface during the reorganization process prior to epitaxy, which is necessary for a high-quality homoepitaxy [2,11,12]. The epitaxy itself has to be of a high quality and should preferably exhibit a high growth rate to be suitable for industrialization. From the substrate preparation to the epitaxy process no significant contamination must be introduced into the PSI wafer, since this would lead to a lower lifetime [4,12,13].
Additionally, less obvious process parameters also influence the PSI wafer quality, such as the application of an HF-dip prior to epitaxy [14] or stresses in the reorganized silicon layer [15].
Here, we report on our progress towards fully detachable high electronic quality of the PSI wafers.

2 EXPERIMENTAL PROCEDURE

We start with round 6" p$^+$-type polished CZ silicon wafers with a resisitivity of 16.4 mΩcm as a substrate. We use an HF-based electrolyte with HF:H_2O:Ethanol=1:1:1 to anodically etch a porous double layer: a low-porosity starting layer of ~1.3 µm thickness and a high-porosity separation layer of ~0.3 µm thickness. In a first batch A, we vary the separation layer etching current densities from 100 to 150 mA /cm^2 in order to determine a process window for fault-free lift-off, which is then used for processing an optimized second batch B.

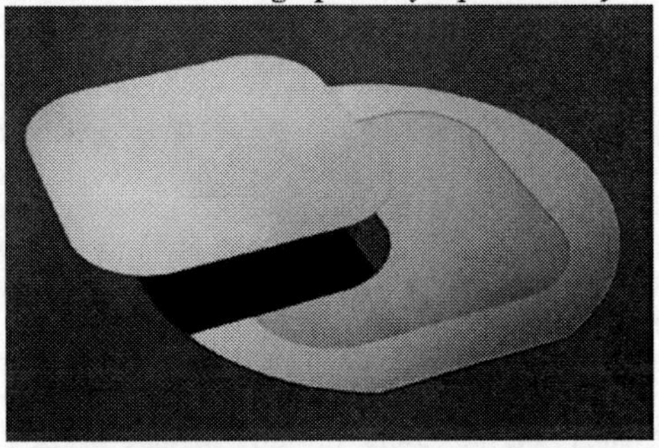

Figure 1: Photograph of a detached PSI wafer and the corresponding substrate.

We perform an HF-dip, immediately followed by reorganization for 30 min at 1100 °C in hydrogen atmosphere and n-type silicon epitaxy with a thickness of 120 µm.
We define an area of 10 x 10 cm^2 pseudosquare by laser scribing the epitaxial layer while still being attached to the substrate and subsequently perform the lift-off of the layer. Figure 1 shows a photograph of the lifted-off epitaxial wafer and the substrate. Fractions of the epitaxial layer are prepared for SEM imaging. From the lifted-off epitaxial wafers we remove the residual porous silicon in KOH and clean the samples with standard RCA. We passivate both sides with 10 nm Al_2O_3 using plasma assisted atomic layer deposition (ALD) and measure the injection

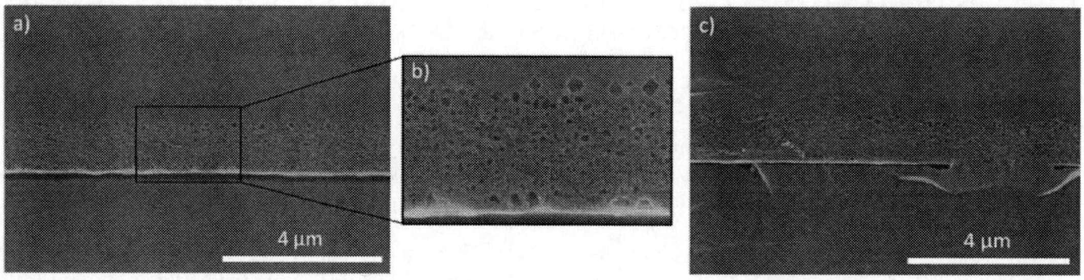

Figure 2: Cross-sectional SEM images of the reorganized porous layers from Batch A after epitaxy. Image a) corresponds to a non-detachable layer etched with 130 mA/cm^2 separation layer etching current density and shows a wide region of a completely reorganized separation layer. Image b), a zoom from image a), shows faceted pores, indicating a strong reorganization. Images c) shows another separation layer of a non-detachable layer etched with the same current density. Here, we see that the separation layer is interrupted by regions of a collapsed and resealed separation layer.

dependent lifetime by the photoconductance decay (PCD) technique and by dynamic infrared lifetime mapping (dynILM).

3 REORGANIZATION AND LIFT-OFF

The samples from batch A are either completely non-detachable, or only partially detachable. The separation layer etching current densities leading to at least large-area lift-off lie in the range of 100 to 110 mA/cm^2. Figure 2 shows cross sectional scanning electron micrographs of the PSI wafers from this batch. Figure 2 a) corresponds to a non-detachable layer etched with 130 mA/cm^2 separation layer etching current density and shows a wide region of a completely reorganized separation layer. The starting layer is also strongly reorganized, as can be seen in the zoom of Figure 2 b) by the closed and faceted pores. Figure 2 c) shows another separation layer of a non-detachable layer etched with the same current density. The large void of the separation layer is interrupted only by rare regions where the separation layer apparently collapsed and resealed, which then causes a non-detachable layer.
This issue was remedied in the follow-up batch B by adapting the etching current to values between 100 – 110 mA/cm^2. Then, we are able to perform fault-free lift-off of all samples on an area of 10 x 10 cm^2 pseudosquare in a batch of seven wafers.

4 LIFETIME ANALYSIS

Our recently published result of 4.5 ms lifetime on a small area of a 5 Ohmcm PSI wafer [4] was achieved by a sample from batch A. The strongly reorganized separation layer hindered the lift-off of most of the layers in that batch. We therefore processed a new batch B of the substrates with suitable etching parameters for fault-free lift-off of all seven samples in that batch. These layers have a resistivity of 1.7 Ωcm.
At an injection level of $n = 10^{15}$ cm^{-3} the samples show lifetimes between 836 µs (Nr 02) and 1022 µs (Nr 01). This is prior to any post-treatment such as gettering which is known to improve the lifetime of PSI wafers significantly [4,5]. Two of those wafers, Nr 01 and Nr 04, receive an additional phosphorous gettering step that increases the measured lifetime to 1195 and 1171 µs, respectively (not shown in Figure 3).
Figure 4. Spatially resolved lifetime of the seven 10 x 10 cm^2 pseudosquare PSI wafers measured with dynILM. Two of those wafers were treated with a phosphorous gettering step, which further increases and homogenizes the lifetime. shows the lifetime mappings of the lifetime samples at an illumination of 0.14 suns as measured by dynILM at a temperature of 70 °C. The lifetime is on a high level on most of the wafer area, and increases in homogeneity after gettering.

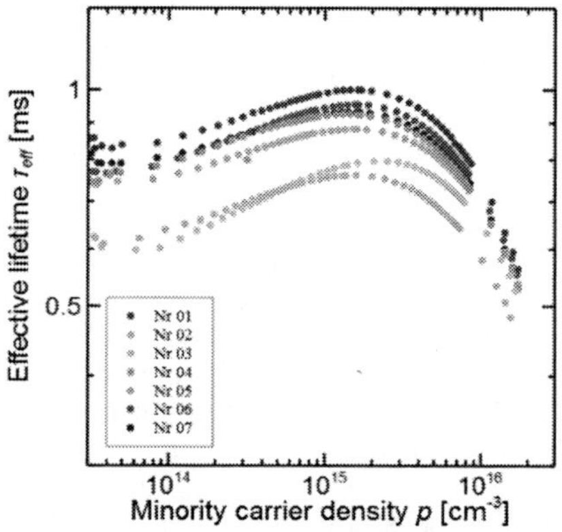

Figure 3: Effective lifetime in dependence of the injection level of all seven kerfless epitaxial wafers from one batch.

5 SUMMARY AND CONCLUSION

We have demonstrated high carrier lifetimes of 836 to 1022 µs before gettering and up to 1195 µs after gettering on the full 10 cm x 10 cm pseudosqare PSI wafers. This was achieved for all of the seven wafers of one batch, of which also all seven epitaxial PSI wafers were detachable without fault.
The key for this improvement was the optimization of the separation layer etching current density that depends on a combination of parameters such as wafer doping, electrolyte chemistry, storage time between porosification and epitaxy, pre-epi

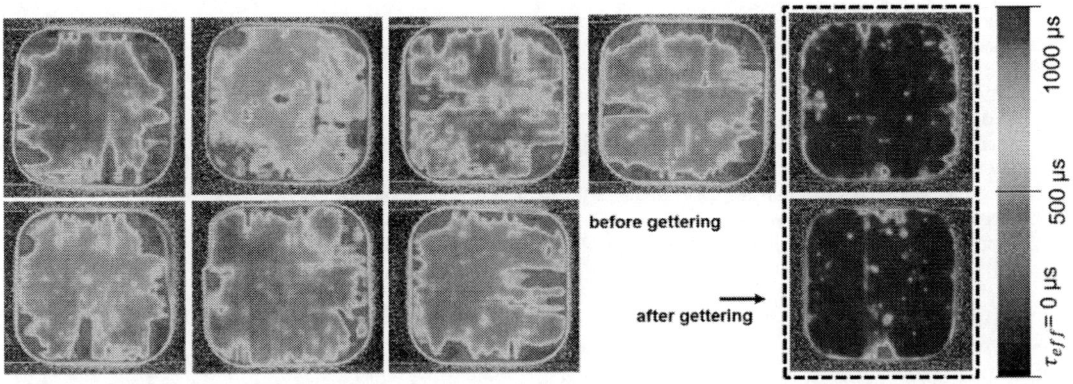

Figure 4. Spatially resolved lifetime of the seven 10 x 10 cm^2 pseudosquare PSI wafers measured with dynILM. Two of those wafers were treated with a phosphorous gettering step, which further increases and homogenizes the lifetime.

treatment and the quality of the epitaxy itself. We found that if the etching current density for the separation layer is too high, it can collapse and reseal, thus hindering a successful lift-off of the PSI wafer.

The success rate for lift-off and carrier lifetime after the optimization is promising and demonstrates that reproducible high quality PSI wafers can be fabricated. However, the narrow etching current density regime identified here requires adjustment of the etching parameters for e.g. substrate wafers with a different doping concentration. Tracking the doping concentration of substrate wafers and adjusting the corresponding etching parameters in porosification is possible and in fact in line with efforts to implement "smart factories". However, it is certain that the necessity for the implementation of fine-tuning parameters to individual substrates is disadvantageous and would better be avoided.

In the meantime, we demonstrated that a wide process window is also feasible by further optimizing the process sequence [16]. In recent experiments we achieved both, a fault-free lift-off and lifetimes before gettering of 4 ms and after gettering of up to 7 ms in combination with a wide process window for the etching current density of the separation layer [16].

6 ACKNOWLEDGEMENTS

The authors thank Jessica Strey for sample processing. This work was funded by the German Federal Ministry for Economic Affairs and Energy under Grant 0324107 (NEPSIC) and the state of Lower Saxony.

7 REFERENCES

[1] H. Tayanaka, T. Matsushita. Separation of thin epitaxial Si film on porous Si for solar cells. In: *Proc. 6th Sony Res. Forum*; 1996, p. 556.

[2] R. Brendel. A novel process for ultrathin monocrystalline silicon solar cells on glass. In: *14th European Photovoltaic Solar Energy Conference*; 1997.

[3] ITRPV. *International Technology Roadmap for PHotovoltaic (ITRPV): 2015 Results including maturity reports*; 2016. Available at: http://www.itrpv.net/Reports/Downloads/ [accessed 05.02.2017].

[4] C. Gemmel, J. Hensen, S. Kajari-Schröder, R. Brendel. *IEEE J. Photovoltaics* 2017; **7(2)**: 430–6, DOI: 10.1109/JPHOTOV.2016.2642640.

[5] Powell, D. M., Markevich, V. P., J. Hofstetter, Jensen, M. A., Morishige, A. E., S. Castellanos, B. Lai, Peaker, A. R., T. Buonassisi. *J. Appl. Phys.* 2016; **119(6)**: 65101, DOI: 10.1063/1.4940947.

[6] H. Sivaramakrishnan Radhakrishnan, R. Martini, V. Depauw, K. van Nieuwenhuysen, M. Debucquoy, J. Govaerts, I. Gordon, R. Mertens, J. Poortmans. *IEEE J. Photovoltaics* 2014; **4(1)**: 70–7, DOI: 10.1109/JPHOTOV.2013.2282740.

[7] R. Hao, T. S. Ravi, V. Siva, J. Vatus, I. Kuzma-Filipek, F. Duerinckx, M. Recaman-Payo, M. Aleman, E. Cornagliotti, P. Choulat, R. Russell, A. Sharma, L. Tous, A. Uruena, J. Szlufcik, J. Poortmans. *IEEE J. Photovoltaics* 2016; **6(6)**: 1451–5, DOI: 10.1109/JPHOTOV.2016.2601945.

[8] I. Kuzma-Filipek, M. Recaman, F. Duerinckx, R. Hao, T. S. Ravi, E. Cornagliotti, P. Choulat, A. Sharma, J. John, M. Aleman, L. Tous, R. Russel, A. Uruena, J. Szlufeik, J. Poortmans. 22.5 % n-PERT solar cells on epitaxially grown silicon wafers. In: *32nd European Photovoltaic Solar Energy Conference*; 2016, 2DO.16.1.

[9] E. Kobayashi, Y. Watabe, R. Hao, T. S. Ravi. *Prog. Photovolt: Res. Appl.* 2016; **24(10)**: 1295–303, DOI: 10.1002/pip.2813.

[10] V. Mertens, S. Bordihn, A. Mohr, K. Petter, J. W. Müller, Jeong, Daniel J. W., R. Hao, T. S. Ravi. 21.4 % Efficient Fully Screen Printed n-type Solar Cell on Epitaxially Grown Silicon Wafers with Built-In Boron Rear Side Emitter. In: *31st European Photovoltaic Solar Energy*; 2015, pp. 1000–1002.

[11] B. Terheiden. Homoepitaxy on Porous Silicon. In: Canham L, (ed). *Handbook of Porous Silicon*; 2014, pp. 567–579.
[12] Hariharsudan Sivaramakrishnan Radhakrishnan. *Silicon epitaxial layers grown on buried porous silicon templates: detailed electrical and chemical understanding*. Dissertation. Leuven, Belgien; 2014.
[13] D. M. Powell, J. Hofstetter, D. P. Fenning, R. Hao, T. S. Ravi, T. Buonassisi. *Appl. Phys. Lett.* 2013; **103(26)**: 263902, DOI: 10.1063/1.4844915.
[14] N. Milenkovic, M. Drießen, E. Gust, S. Janz, S. Reber. *Energy Procedia* 2014; **55**: 552–8, DOI: 10.1016/j.egypro.2014.08.023.
[15] M. Karim, R. Martini, H. Sivaramakrishnan Radhakrishnan, K. van Nieuwenhuysen, V. Depauw, W. Ramadan, I. Gordon, J. Poortmans. *Nanoscale Res Lett* 2014; **9(1)**: 348, DOI: 10.1186/1556-276X-9-348.
[16] C. Gemmel, J. Hensen, L. David, S. Kajari-Schröder, R. Brendel. *JJAP*: (submitted).

ORIGIN AND IMPACT OF CRYSTALLOGRAPHIC DEFECTS IN EPITAXIALLY GROWN SI WAFERS

S. Janz, D. Amiri, E Gust, S. Kühnhold-Pospischil, S. Riepe, F. Heinz, M. Drießen
Fraunhofer Institute for Solar Energy Systems (ISE)
Heidenhofstrasse 2, D-79110 Freiburg, Germany

ABSTRACT: In recent years epitaxially grown wafers (EpiWafers) made their way from small laboratory to production scale fabrication. A lot of effort has been put into upscaling of tools, improvement of material quality and in the fabrication of highly efficient solar cells. In this publication we investigate typical crystal defects appearing in EpiWafer material and correlate them with process or reactor specific features like porous templates, thick epitaxial layer growth or inline processing. We find two categories of stacking faults (SFs) which are either decorated with inclusions or not. Besides particles on the surface of the growth template contaminations with oxygen are found to be reasons for such polycrystalline inclusions. Stress leads to either delamination of the template or local lattice strain both reasons for defect growth. Parallel faults which propagate along slipping lines in the material are interacting with SFs although the correlation is not fully understood yet. Finally the recombination activity in SFs is found to be especially high where the {111} planes meet or when strong decorating occurs. This work is a first step to systematically investigate the crystallographic nature of EpiWafer material and will help to further improve processes and equipment.
Keywords: Silicon, EpiWafer, defects, kerf-less wafering, stacking faults

1 INTRODUCTION

Efficiency improvement and reduced consumption of resources are main drivers for today's PV developments. Epitaxially grown silicon wafers "EpiWafers", prepared by detachment of the silicon layer grown epitaxially on a reorganized porous silicon template, offer the possibility to reduce silicon and energy consumption and with this production costs while maintaining high efficiency potential. In recent years a lot of effort has been put in the development of equipment capable of high-throughput and low cost production in order to enable a technology transfer from semiconductor to photovoltaic industry. Today the EpiWafer is on its way to become a cost-efficient alternative for standard mono-crystalline wafers in photovoltaic applications[1]. Besides the significant progress in industrialisation of the EpiWafer technology several solar cell concepts have been successfully applied to this new material [1], [2]. However, although some research groups report on the necessity of post-treatments like gettering to exploit the full electrical potential of the epitaxially grown wafers [3], [4] systematic studies about specific defects and their origin are completely missing so far. In this publication we try to close this gap with the analysis of defects which we find in EpiWafer material and compare them to known defects in epitaxially grown Si from literature. Almost all available publications on related subjects so far focus on epitaxy on standard wafers and epitaxial layer thicknesses of below 100 µm. Only recently the interest in Si epitaxy of layers beyond 100 µm thickness used for power devices has been raised again [5]. However, so far we could not find any literature about epitaxially grown Si layers with thicknesses exceeding 100 µm grown on porous Si templates. Besides that many publications do not clearly indicate how the EpiWafers were fabricated (reactor and process details) and crystallographic information like stacking fault and etch pit density are simply missing. In this work we explain in detail how the EpiWafers have been processed, identify different classes of defects, investigate their origin in the course of the fabrication process and characterize their influence on the electrical quality of the material.

2 EXPERIMENTAL

The porosified 6-inch Cz samples for all experiments have been prepared by the Institute for Microelectronics in Stuttgart (IMS) featuring a low porosity layer on top of two high porosity layers. All wafers were cut to 10 cm edge length on one side and were cleaned in aqueous hydrofluoric acid (HF) with a concentration of 1 % before mounting in the quasi-inline RTCVD 160 furnace [6]. The reorganization processes took place at a peak temperature of 1150 °C under 100 % hydrogen atmosphere for up to 30 min. Subsequently the quasi-inline Si epitaxy process was done at the same temperature in trichlorosilane (TCS) / H_2 atmosphere with a Cl / H gas flow ratio of 12 % and process times of 80 min to achieve layer thicknesses of 150 to 200 µm. For the n-type samples the PH_3 flow was set to achieve a constant layer doping of approximately 3E15 cm^{-3}, which is a quite common value for highly efficient wafer solar cells. Some samples were just reorganized and measured afterwards with Secondary Ion Mass Spectroscopy (SIMS) at RTG Mikroanalyse GmbH to determine oxygen concentrations throughout the porous silicon layer stack. Therefore the reorganized samples were broken just before loading into the SIMS chamber in order to avoid any native oxide growth inside the porous Si structure which could distort the measurement.

Cross sections were prepared with simple sample cleaving and investigated with Scanning Electron Microscopy (SEM) and an in-situ Energy Dispersive X-ray spectrometry (EDX) setup. Before scanning the surface with an optical microscope the samples were etched in SECCO[2] solution to visualize defects. From these images stacking fault distributions were derived.

Prior to detachment 5x5 cm² areas were defined on the wafers using a dicing saw from DISCO. We detached the EpiWafers in an in-house developed lifting tool

[1] www.nexwafe.com

[2] SECCO-etch: HF + $K_2Cr_2O_7$ + H_2O in a ratio of HF : H_2O = 2 : 1 with 44 g $K_2Cr_2O_7$ dissolved in 1 l of H_2O.

which basically provides homogeneous soaking through special membrane materials over defined sample areas. The residual porous and highly doped Si layer was etched in a CP71 wet chemical bath with an etching rate of around 1.4 µm/min for around 3 min. Both sides of all samples for electrical characterization were passivated in an Atomic Layer Deposition (ALD) tool with 20 nm of Al_2O_3. The spatially resolved measurement of minority carrier lifetimes was done using micro-photoluminescence (µPL) [7].

3 RESULTS AND DISCUSSION

The first inspection of the surface with microscopy shows squared shaped stacking faults with shiny and dark surfaces. This difference in reflection originates from the surface structure which is (i) completely flat and in the same plain as the surrounding epitaxial layer for the shiny regions and (ii) coming out of the plain and structured for the dark ones. We call the shiny ones non-decorated and dark ones decorated SFs. It is known from literature that in general SFs arise from atomic defects. They originate from small mismatch between first growth area with respect to the substrate, are therefore initiated at the substrate / layer interface and grow along inclined {111} planes [8]. On a (100) oriented substrates there are four such planes which are tilted by 54.7° (in relation to the surface) leading to a squared shape structure at the intersect with the surface. If just one of such a pyramidal structure is occurring and no additional defects arise in the proximity of the SF we call it a non-decorated SF. These kinds of SFs are readily found in epitaxially grown Si layers on Si substrates. In the following we focus on decorated stacking faults and try to give some explanations for their origins especially linked to the reorganized porous layer stack which is used for EpiWafers as growth template.

3.1 Particles and oxygen background

Some of the SFs show poly crystalline inclusions of which some have their origin in particles located at the substrate / layer interface. In Fig.1 one such particle can be found in the SEM micrograph of a respective cross section. Several twin grain boundaries, intercepting crystal planes and the non-oriented structures coming out of the surface plain show the high degree of decoration. EDX measurements presented here (Fig.1, right) identify quartz particles which can be explained by carrier abrasion during movement of the samples. Besides movement related to abrasion at reactor components which are made of e.g. quartz or coated graphites, particles can also arise from gas phase nucleation or parasitic depositions on the reactor walls and then fall on the growth template. Moreover the handling of porosified wafers can be another source of particles or scratches which can influence not just the epitaxial growth but the reorganization process itself.

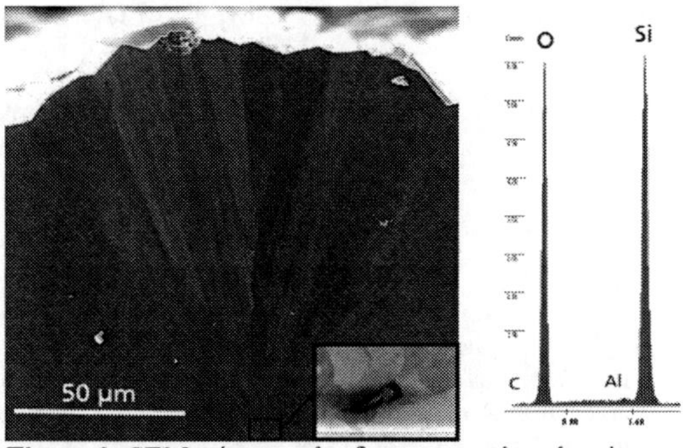

Figure 1: SEM micrograph of a cross section showing a decorated stacking fault (left) with an included particle (inset). EDX data measured at particle position (right) showing high oxygen concentrations which suggest a quartz particle.

Contaminations at the growth template can be another reason for the decoration of SFs. Especially metals and oxygen are known to cause such defects in epitaxially grown Si layers on even crystallographically perfect Si substrates [9]. Ravi *et al.* [10] report that C concentrations of E19 at./cm^3 already lead to SF growth. On the reorganized porous layers the gettering effect caused by the large internal Si surface [11] has to be considered. All impurities which are already present in the parent substrate or in the reactor ambient during reorganization are gettered and can lead to locally very high e.g. metal concentrations. Pomerantz *et al.* [12] report that high contamination levels of metals such as Cu, Fe, Ni and others lead to precipitates which are nucleation sites for SF growth. Although no such precipitates could be found we still consider metal impurities as one important origin of crystal defects. For one there are many sources of metal contaminants inside the reactor and gettering of EpiWafers shows significant improvement in minority carrier lifetime [3].

In order to investigate the oxygen concentration in the reorganized porous layer and on its surface we measured such structures with SIMS after hydrogen processing (reorganization). In Fig.2 the SIMS profile from seed wafer to surface (right to left) shows oxygen concentration of up to E21 atoms cm^{-3} in the porous layer. An unwanted oxidation of samples during preparation can be excluded as all samples were broken directly before transferring them to the SIMS chamber. The concentration at the surface is lower than in the porous bulk (E20 atoms cm^{-3}) and shows a strong dip close to the transition zone between these two. Besides the native oxide growth during shipment, which could not be prevented, the signal accuracy in the first nanometers of the profile is questionable. We therefore cannot finally conclude on the real oxygen concentration on the template. However, as the concentration inside the porous structure is very high oxygen is supposed to play a role in the first layers of Si growth. Finch *et al.* [13] report that oxide can produce steps at the surface forcing the silicon crystal to grow around such an oxide inclusion leading to the formation of stacking faults.

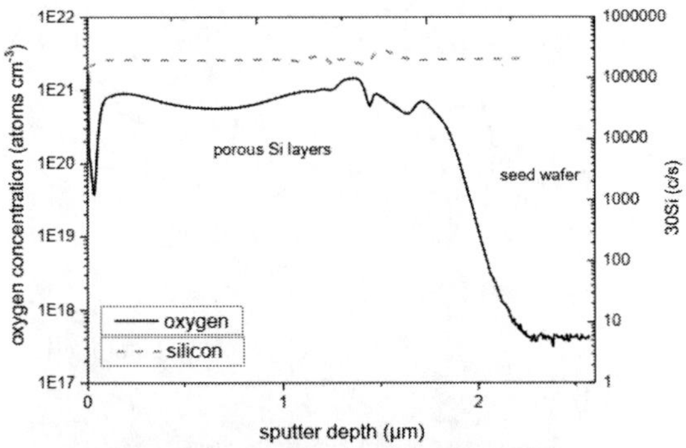

Figure 2: SIMS measurement through porous Si layers into seed wafer after reorganization showing oxygen (solid line) and silicon (dashed line) content.

3.2 Stress and template

Other reasons for defects in epitaxially grown layers can arise from structural or crystal imperfections of the template. Such imperfections can be e.g. not fully reorganized templates containing holes or uneven surfaces [14]. However, even a smooth template is strained due to stress between the seed wafer and the porous layer. In certain cases the strain can be such high that a delamination occurs. In Fig.3 the cross section of an epitaxial layer on a partly delaminated porous layer stack is shown. Originating from the delaminated part of the template a defect rich crystal growth is visible. Surprisingly the left part of the SF seems to be mainly unharmed which leads to a non-symmetric shape of the SF. We call this a partly decorated SF.

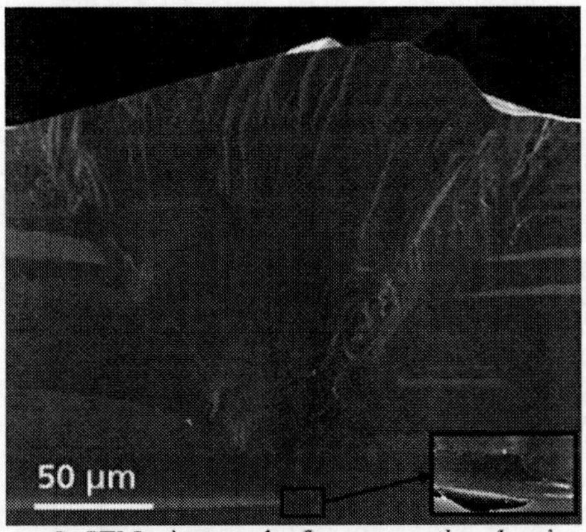

Figure 3: SEM micrograph of a cross section showing a partly decorated stacking fault with a locally detached porous Si layer stack (inset) being the origin of the polycrystalline growth.

Besides this macroscopic effect of strain on the crystal growth a more microscopic effect has to be considered. In an energetically optimized state all pores should be either spherically shaped or oriented along crystallographically preferred planes. Due to limited energy / time during reorganization and / or residual oxygen on the pore's inner surfaces preventing defect diffusion not a single such perfect pore can be found. However, the pores in proximity to the surface show additional deformations (see Fig.4, left) which look like a flattening. During reorganization the Si atoms are looking for energetically preferential sites which leads to dissolving of small pores or the growth of the larger ones [15]. The templates surface is a pore with infinite size which means that Si atoms want to diffuse there. On the other hand pores in proximity of it tend to close their surface as it is energetically preferential for them. These competition leads to the flattening of the pores which is an energetic compromise leading to microscopic lattice strains in such areas. After epitaxy this strained areas can be identified as the origin for several close parallel faults which grow in the direction on the {111} plane (Fig.4, right).

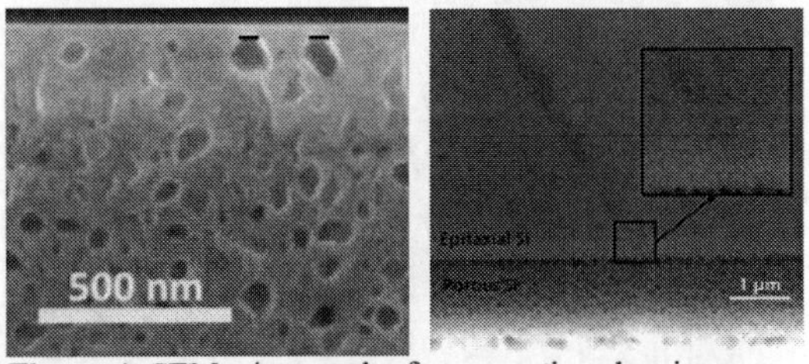

Figure 4: SEM micrograph of cross section showing a reorganized porous layer with flattened pores in close proximity to the growth template (left) and parallel faults originating from one of such pores (right).

During epitaxy the generated mismatch is compensated through these close parallel faults. In Fig.5 (left) one example for such an accumulation of defects on the fringes of a SF is shown in cross section view. Several variations with an increasing number of close parallel faults can be found in the EpiWafer material. The number of additional growth planes could be explained by different strain levels leading to increased mismatch between crystal structure of template and growing layer. In Fig.5 (right) top views of intersects of the {111} planes with the surface are shown. The increasing number of parallel faults (counterclockwise) caused by increased mismatch / strain leads to pyramidal structures with increasing height.

Figure 5: SEM micrograph of defect etched cross section showing close parallel faults along the {111} planes (left). Top view (microscope) of epitaxially grown layers with increasing number of parallel faults (right, counterclockwise).

3.3 Effect of stacking faults on electrical quality of the EpiWafer

In a first step we identified SFs without and with polycrystalline inclusions (decorated SF) in an optical microscope. On the same spot non-calibrated µPL measurements were performed. In Fig.6 the microscope pictures (top) show a non-decorated SF (left), a SF with some parallel faults (middle) and one filled with polycrystalline inclusions (right). The corresponding lifetime mappings for the first two show cross like features identifying highly recombination active regions. These cross like shapes overlap with the intersects of the {111} planes which are partial dislocations extending from the nucleation to the surface. It is not surprising though that these regions show enhanced recombination activity. The fully decorated SF seems to be highly recombination active throughout its whole volume. It is not clear so far for all these defects if the decoration with contaminants such as metals or dangling bonds / vacancies is the reason for the high recombination activity.

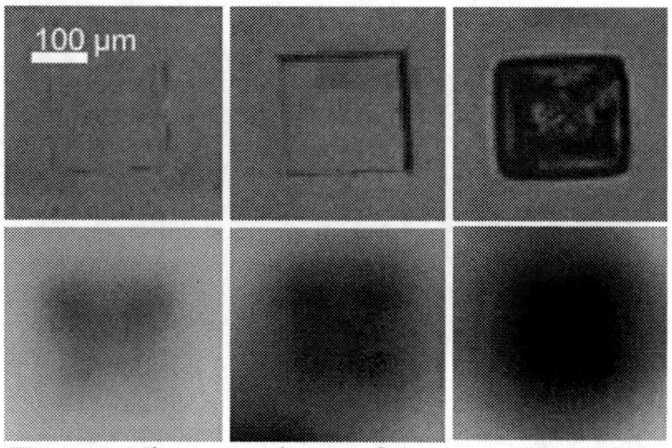

Figure 6: Microscope pictures of top view on epitaxial Si layer (top) and µPL mapping of corresponding regions (bottom).

When looking at µPL mappings of larger areas of the EpiWafer we find regions with enhanced recombination (black) following exactly the family of directions <110> (see Fig.7). Obviously these lines correspond to another crystal defect. Considering their orientation we conclude that these are line defects which glide along slip planes of the crystal. It is known that such defects occur when the temperature gradient along the wafer is in the range of 5 – 10 K cm^{-1} [16] at growth temperature. In most cases they start at wafer edges initiated by little damages. In our case we find that most line defects intersect with SFs. Especially at the crossing points of such slip lines the highly recombination active (decorated) SFs can be found. We came up with two explanations for these effects: (i) After growing an epitaxial layer of 150 µm and more the EpiWafer is exposed to a temperature gradient exceeding 5 – 10 K/cm which initiates parallel faults at existing SFs which then glide to the fringes of the wafer. (ii) During reorganization parallel faults are generated again due to thermal process inhomogeneities at micro stress regions in the porous layer itself and glide then along slip lines through the whole template. At the crossings of such line defects step like surfaces can lead to SF growth due to lattice mismatch.

Ravi *et al.* [10] find this effect on (111) wafers although without any porous layer involvement. Arguments that several crossings can be found where there is no SF visible support theory (i). Unfortunately there are not many µPL mappings available showing such features. Furthermore these mappings are limited in terms of spatial resolution which makes it difficult to verify such arguments.

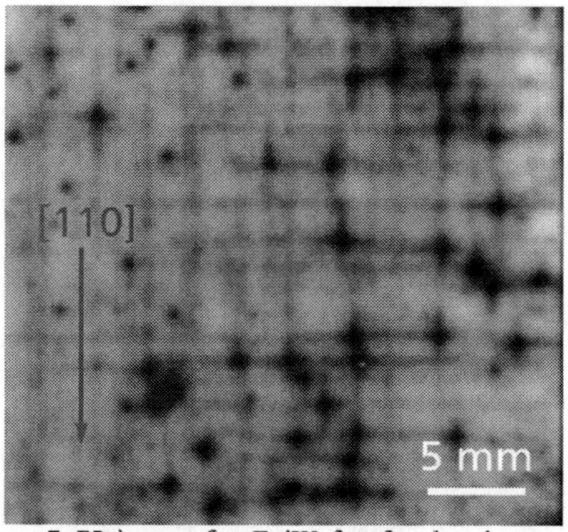

Figure 7: PL image of an EpiWafer after detachment showing SF and line defects as highly recombination active areas in black.

At the moment we can conclude that there are good arguments for both hypotheses and that further investigations using defect etching on just reorganized samples looking for initial parallel faults and µPL images with higher resolution will be necessary to finally understand the relation between parallel faults and SFs.

4 CONCLUSIONS

We identify stacking faults (SFs) with and without poly crystalline inclusions as the most detrimental crystallographic defects in EpiWafers. Decorations within such SFs seem to be initiated by either inclusions like quartz particles from the sample holder or by delamination of the porous Si layers. SIMS measurements reveal oxygen (≈1E21 at/cm^3) on the reorganized Si template as another reason for the appearance of SFs. Macroscopic and microscopic stress influences the template and can lead to delamination, a strained lattice or parallel faults throughout the whole wafer. µPL reveals that especially decorated SFs are highly recombination active which can be explained by the high concentration of defects and grain boundaries.

Although several reasons for crystal defects like metal contamination or high dopant concentration on the growth template could not be investigated and the role and origin of stress is not fully understood yet our investigations will already help to further improve the material quality and to understand related effects in the solar cells' performance like e.g. decreased fill factors and others.

5 ACKNOWLEDGEMENTS

The authors would like to express their gratitude to Philipp Barth, Astrid Seiler, Mira Kwiatkowska, Michaela Winterhalder, Karin Zimmermann, Antonio Leimenstoll and Felix Schätzle at ISE for their support and input in many valuable discussions. This work would not have been possible without the close cooperation with the NexWafe GmbH in particular with Nena Milenkovic, Kai Schillinger and Stefan Reber. Stefan Janz wants to especially thank Kris Van Nieuwenhuysen from IMEC for many valuable discussions about epitaxial growth on porous templates and for her tireless promotion of kerf-less wafering.

This work was funded by the European Commission (EC GA No. 609788) and the German Federal Ministry for Economic Affairs and Energy (FKZ 020E-41V7452).

6 REFERENCES

[1] E. Kobayashi, Y. Watabe, R. Hao, and T. S. Ravi, "High efficiency heterojunction solar cells on n-type kerfless mono crystalline silicon wafers by epitaxial growth," Appl. Phys. Lett., vol. 106, no. 22, p. 223504, 2015.

[2] R. Hao *et al.,* "Kerfless epitaxial mono crystalline Si wafers with built-in junction and from reused substrates for high-efficiency PERx cells," *IEEE J.* Photovoltaics, vol. 6, no. 6, pp. 1451–1455, 2016.

[3] D. M. Powell et al., "Exceptional gettering response of epitaxially grown kerfless silicon," J. Appl. Phys., vol. 119, no. 6, p. 65101, 2016.

[4] C. Gemmel, J. Hensen, S. Kajari-Schröder, and R. Brendel, "4.5 ms effective carrier lifetime in kerfless epitaxial wafers from the porous silicon process," IEEE J. Photovoltaics, vol. 7, no. 2, pp. 430–436, 2017.

[5] M. Künle, J. Baumgartl, K. Koren, and O. Fiedler, "Multiple epitaxial Si film deposition by APCVD for power semiconductors," in 2015 26th Annual SEMI Advanced Semiconductor Manufacturing Conference: Proceedings, Saratoga Springs, NY, 2015, pp. 455–459.

[6] S. Reber, C. Haase, N. Schillinger, S. Bau, and A. Hurrle, "The RTCVD160 -- a new lab-type silicon CVD processor for silicon deposition on large area substrates," in 3rd World Conference on Photovoltaic Energy Conversion: Proceedings, Osaka, 2003, pp. 1368–1371.

[7] F. D. Heinz, P. Gundel, C. Schubert, and W. Warta, "Mapping the quantitative carrier lifetime and doping concentration with micro resolution," in 35th IEEE Photovoltaic Specialists Conference: Proceedings of, Honolulu, HI, 2010, pp. 498–503.

[8] G. R. Booker, "Crystallographic imperfections in silicon," Discuss. Faraday Soc., vol. 38, pp. 298–304, 1964.

[9] W. C. O'Mara, "Oxygen, Carbon and Nitrogen in Silicon," in Handbook of semiconductor silicon technology, W. C. O'Mara, R. B. Herring, and L. P. Hunt, Eds., Park Ridge: Noyes, 1990, pp. 451–549.

[10] K. V. Ravi, "Crystallographic defects in epitaxial silicon films," Thin Solid Films, vol. 31, no. 1-2, pp. 171–183, 1976.

[11] R. Bilyalov et al., "Crystalline silicon thin films with porous Si backside reflector," Thin Solid Films, vol. 403-4, pp. 170–174, 2002.

[12] D. Pomerantz, "A cause and cure of stacking faults in silicon epitaxial layers," Journal of Applied Physics, vol. 38, no. 13, pp. 5020–5026, 1967.

[13] R. H. Finch, H. J. Queisser, G. Thomas, and J. Washburn, "Structure and origin of stacking faults in epitaxial silicon," Journal of Applied Physics, vol. 34, no. 2, pp. 406–415, 1963.

[14] N. Milenkovic, M. Driessen, C. Weiss, and S. Janz, "Porous silicon reorganization: Influence on the structure, surface roughness and strain," Journal of Crystal Growth, vol. 432, pp. 139–145, 2015.

[15] H. S. Radhakrishnan et al., "Kerfless layer-transfer of thin epitaxial silicon foils using novel multiple layer porous silicon stacks with near 100% detachment yield and large minority carrier diffusion lengths," Sol. Energ. Mat. Sol. Cells, vol. 135, pp. 113–123, 2015.

[16] W. C. O'Mara, R. B. Herring, and L. P. Hunt, Eds., Handbook of semiconductor silicon technology. Park Ridge: Noyes, 1990.

PRINCIPLES OF CARRIER-SELECTIVE CONTACTS BASED ON INDUCED JUNCTIONS

Martin Bivour[1], Christoph Messmer[1], Lisa Neusel[1], Florian Zähringer[1], Jonas Schön[1],
Stefan W. Glunz[1,2], Martin Hermle[1],
[1]Fraunhofer Institute for Solar Energy Systems (ISE), Heidenhofstrasse 2, 79110 Freiburg, Germany
[2]Department of Sustainable Systems Engineering (INATECH), Albert-Ludwigs University Freiburg,
Georges-Köhler-Allee 103, 79110 Freiburg, Germany

ABSTRACT: The applicability of different high (low) work function thin films for the formation of alternative hole (electron) selective contacts is currently (re-) explored for silicon solar cells. To provide some insight into contact schemes based on induced junctions their operation principles, important design parameters and losses are reviewed experimentally and with the help of numerical device simulations. Simulations with Sentaurus TCAD are used to address the importance of the work function and an efficient tunneling transport. It is highlighted that "non-selective" contacts will not obey the standard diode theory. This calls for an adapted loss analysis and one approach for this is presented for different hole contacts prepared by evaporation, sputtering, atomic layer deposition and PECVD. The results show that the "classical" electrical losses of selective contacts, like recombination and ohmic transport which are well quantified by J_0 and the ohmic contact resistance, are not sufficient for the evaluation of "non-selective" contacts.
Keywords: Heterojunction, Inversion-layer, a-Si, Simulation, Fundamentals

1 INTRODUCTION

Passivating and carrier selective contacts [1] have the potential to overcome some of the intrinsic efficiency limitations of the established homojunction contact schemes. For the latter, doping of the silicon absorber surface is applied to obtain carrier selective device regions below the unpassivated metal electrodes. The high recombination in this regions results in high J_0 and therefore moderate voltage at MPP and open-circuit conditions. Both, passivating and non-passivation carrier selective contacts typically obey the diode model described by the Shockley equations

$$J = J_{gen} - J_0\{\exp[q\,V/(n\,k_BT)] - 1\} \qquad (1)$$

One important assumption for this is that the voltage (V) corresponds to the external voltage ($V_{external}$) and that this external voltage matches the quasi-Fermi level splitting within the absorber at the edge of the contact (depletion) region. The latter voltage can be expressed by the implied voltage ($V_{implied}$). This assumption ($V_{external} = V_{implied}$) is typically fulfilled when the gradient in the majority carrier quasi-Fermi level in the contact region is negligible. This means that low-injection conditions in the contact region during operation must be maintained [2], i.e. that the majority carrier density remains well above the minority carrier density ($n_{maj} \gg n_{min}$). Or, in a more general expression, that the conductivity of majority carriers remains well above those of the minority carriers ($\sigma_{maj} = q\,\mu_{maj}\,n_{maj} \gg \sigma_{min} = q\,\mu_{min}\,n_{min}$) [3]. Basically, we define contacts to be "selective" for which $V_{external} = V_{implied}$ and hence the validity of the diode equation is fulfilled. Accordingly, we define contacts to be "non-selective" when having a significant voltage loss in the contact region during operation leading to $V_{external} < V_{implied}$.

Similar to classical homojunctions, for TOPCon [4] and poly-Si [5] related contacts obtaining carrier selectivity is typically not the main design constraint. This is explained by the high doping efficiency of the partly crystalline silicon thin films and the c-Si wafer surface. It ensures that the majority carrier density in the contact region during equilibrium ($n_{maj,0}$) is already much higher than the excess carrier density that will be added during operation ($\Delta n = \Delta p$). Accordingly,

$$n_{maj} = n_{maj,0} + \Delta p \gg n_{min} = n_{min,0} + \Delta n$$

with $\Delta n \gg n_{min,0}$, is typically fulfilled. For TOPCon and poly-Si related contacts, an intermediate passivating dielectric buffer (SiO_x) ensures that those contacts are not only selective but also passivating. Accordingly, J_0 is low and the device characteristic and the limiting losses are well described by the diode equation.

The application of different high (low) work function thin films for the formation of passivating and hole (electron) selective contacts is another currently explored approach for silicon solar cells. So far, most promising candidates are transition metal oxide based heterojunctions leading to efficiencies of about 22 % [6], [7]. However, other materials are also under investigation [8]. Work in this field can be understood as a renaissance of the induced silicon junctions from the 1970's [9] (e.g. metal-insulator-semiconductor) by taking advantage of (i) novel or better studied high (low) work function contact materials mainly adapted from organic electronics [10] and (ii) improvements of the passivating intermediate buffer layers (a-Si:H, ultrathin dielectrics). However, knowledge regarding relevant junction and material properties and their engineering, stability issues and the integration of those high (low) work function contact materials in the c-Si solar cell device architectures is still in its infancy. Basically, engineering towards sufficient selectivity seems to be of more importance compared to devices featuring n- and p-type doped carrier selective silicon regions.

Within this work the operation principles, important design parameters and losses of contact schemes based on induced junctions are reviewed experimentally and with the help of numerical device simulations. Simulations with Sentaurus TCAD are used to address the importance of the induced c-Si dark band bending, the voltage loss in the contact region, the band line-up and an efficient tunneling transport. Hole contacts prepared by evaporation, sputtering, atomic layer deposition and PECVD are screened and results for simple solar cells with MoO_x, V_2O_x and a-Si:H(p) hole contacts are presented.

2 SIMULATION RESULTS

2.1 Work function and showcase for selective and non-selective contacts.

The importance of a high work function (WF) for a

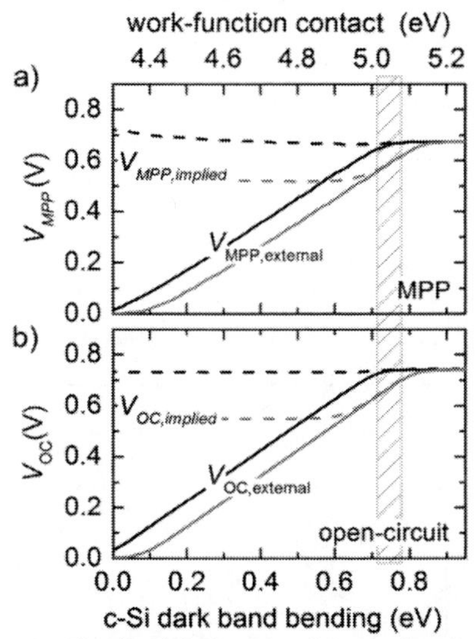

Figure 1: Simulated influence of work function and chemical c-Si passivation on the external and implied voltage for a) MPP and b) open-circuit conditions. For the grey lines poor chemical passivation is assumed (S_0= 1000 cm/s), for black lines S_0= 10 cm/s. The shaded region shows the transition between selective and non-selective contacts.

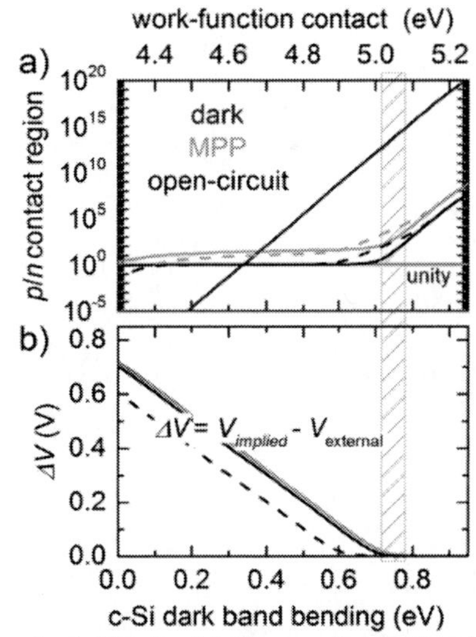

Figure 2: a) Simulated influence of work function on the ratio of hole and electron density at the metal / c-Si interface for equilibrium, MPP and open-circuit conditions. b) Corresponding difference between external and implied voltage. Solid lines show results for excellent and dashed lines for poor chemical passivation.

hole selective contact is motivated by simulating a n-type absorber with ideal electron contact and a metal / c-Si hole contact. Variables are the work function of a metal which is changed from extremely low values close to the c-Si work function (4.3 eV) to high values close above the c-Si valence band energy (5.17 eV). At the interface either excellent or poor chemical surface passivation is assumed.

The dashed lines in Figure 1 show that the implied voltage (V_{implied}) is higher for the better chemical passivation (black line) for both MPP and open-circuit condition (except for very high WF). This reflects the higher density of excess holes and electrons in the absorber. The V_{implied} provides the upper limit of the extractable voltage. The external voltage approaches this value for a high WF and induced c-Si band bending (V_{bb}) of above 5.0 eV and 0.7 eV, respectively. For lower values, power losses occur since the hole quasi-Fermi level in the hole contact region cannot be maintained. The corresponding gradient in the hole quasi-Fermi level represents the difference between the external voltage and implied voltage ($\Delta V = V_{\text{implied}} - V_{\text{external}}$). Accordingly, when analyzing only the external voltage the upper limit of the extractable voltage provided by the passivation of the contact is underestimated by its insufficient selectivity. Therefore, a sufficient high equilibrium band bending and inversion of the c-Si surface is needed to build a hole selective contact ($V_{\text{external}} \approx V_{\text{implied}}$). Good chemical passivation ensures that the voltage is on a high level.

Figure 2 explains why a work function above ~5.0 eV is needed for a hole selective contact and negligible ΔV. The blue line in Figure 2a shows how the equilibrium p/n ratio in the contact region is increased with increasing work function and induced c-Si band bending. During operation (MPP [orange], open-circuit [black]) this asymmetry is reduced by the increasing density of excess electrons and when p/n approaches unity (grey line) ΔV becomes significant since $n_{maj} \gg n_{min}$ and $\sigma_{maj} \gg \sigma_{min}$ are not fulfilled anymore.

2.2 Transition metal oxide based silicon heterojunctions. Beyond the importance of the work function.

Figure 3 shows a possible band diagram featuring an inverted c-Si surface, an a-Si:H(i) passivation layer, a high band gap and high work function transition metal oxide (TMO) and a TCO electrode. While the importance of a high hole density for the c-Si part of the junction has been motivated before, the sketched band alignment suggests that for an efficient hole extraction from the c-Si towards the TCO an efficient tunneling transport should be involved as well. To derive some basic design rules

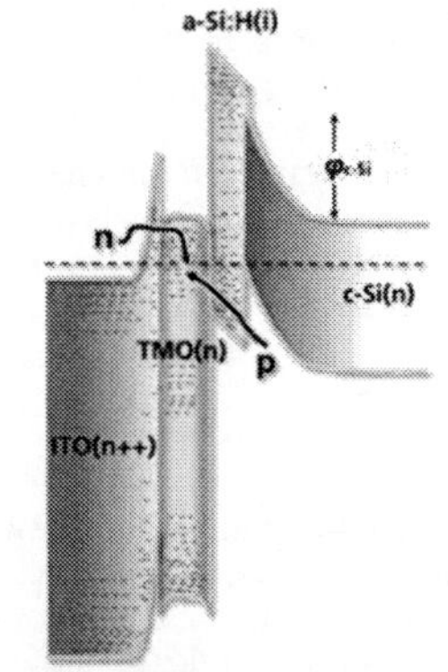

Figure 3: Sketched band diagram of a transition metal oxide based silicon heterojunction. An efficient hole transport via band-to-band tunneling is not possible since the conduction band energy of the TMO is below the valence band energy of the a-Si:H. Hence, the arrows indicate the need for trap assisted tunneling via the TMO.

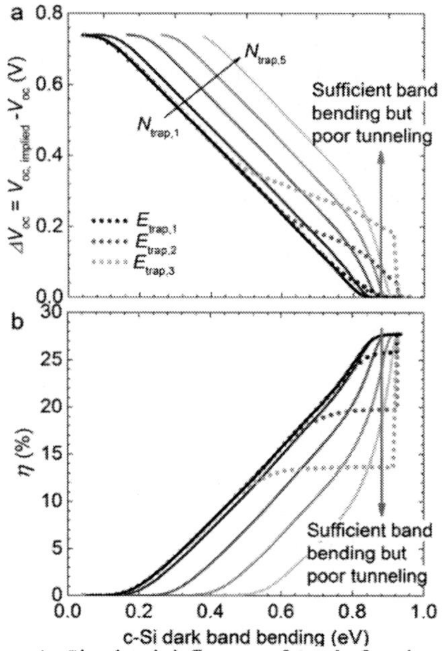

Figure 4: Simulated influence of work function and traps in the TMO on a) selectivity and b) efficiency of a contact similar to Figure 3.

and for better insight we performed more advanced numerical device simulations [11]. Figure 4 reviews some findings on the importance of an efficient tunneling via traps in the TMO. The latter is assumed to be essential for the case that the actual band alignment does not allow for band-to-band tunneling from the valance band of the a-Si:H(i) into the conduction band of the TMO. The black line in Figure 4 shows the case where trap assisted tunneling (TAT) is efficient. ΔV and the efficiency are mainly limited by the induced c-Si band bending. Assuming a sufficient high band bending (> 0.8 eV, grey arrows) it can be seen that for a reduction of the trap density (N_{trap1} [black] → $N_{trap,5}$ [orange]) the efficiency decreases due to an increased ΔV. Another important parameter is the energetic position of the traps in the TMO. If the traps within the TMO are not energetically located close to the valence band of the buffer (E_{trap1} [blue] → $E_{trap,3}$ [cyan]) the efficiency is also reduced since ΔV is increased.

Such design rules gained from simulations are very useful for the engineering of "novel" contacts. It helps to evaluate the actually limiting losses and finally the heterojunction parameters that need to be improved experimentally.

3 EXPERIMENTAL RESULTS

The limited validity and applicability of the diode equation for the loss analysis of contacts limited by an insufficient selectivity calls for a useful figure of merit for engineering such contacts towards sufficient selectivity. This is in contrast to selective contacts for which the "classical" electrical losses like recombination and ohmic transport are well quantified by J_0 and the contact resistance. In this sense, we can determine ΔV which is experimentally accessible without much effort for open-circuit conditions [12]. For further insight the

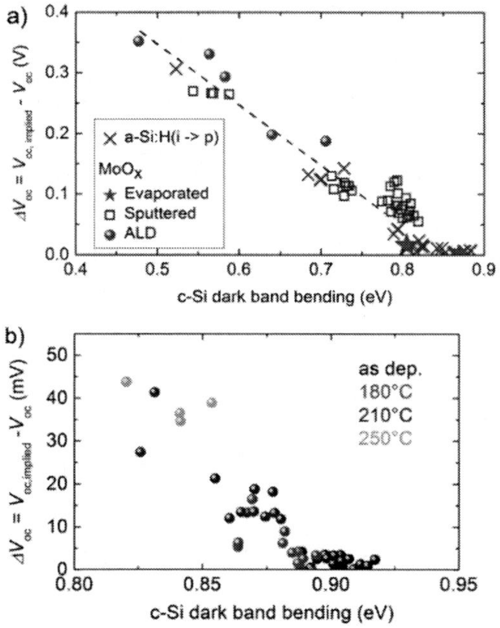

Figure 5: ΔV_{oc} for different hole contacts as a function of the induced c-Si band bending. The latter is determined by surface photo-voltage (SPV) measurements. a) Influence of fabrication conditions for a-Si:H(p) and MoO_x. b) Influence of post deposition annealing for thermally evaporated MoO_x.

induced band bending can be measured by surface photo-voltage measurements or C-V.

For the sake of completeness, it shall be pointed out that a similar way to probe the selectivity is to take advantage of the fact that the selectivity is (further) decreased by increasing the illumination intensity above 1 sun. On the one hand, this increases the sensitivity for assessing a small ΔV for 1 sun conditions which might be limited by the measurement accuracy. On the other hand the increase of ΔV with illumination causes a characteristic slope of the Suns-V_{OC} characteristic reflecting a very low or negative ideality factor of a corresponding diode equation [12], [13]. Such ideality factors lack physical meaning, substantiating the limited validity of the diode equation for non-selective contacts.

3.1 Probing the selectivity

The importance of the induced c-Si junction for a low ΔV_{oc} is shown in Figure 5a for different hole contacts: A doping variation for p-type a-Si:H [12] and MoO_x which

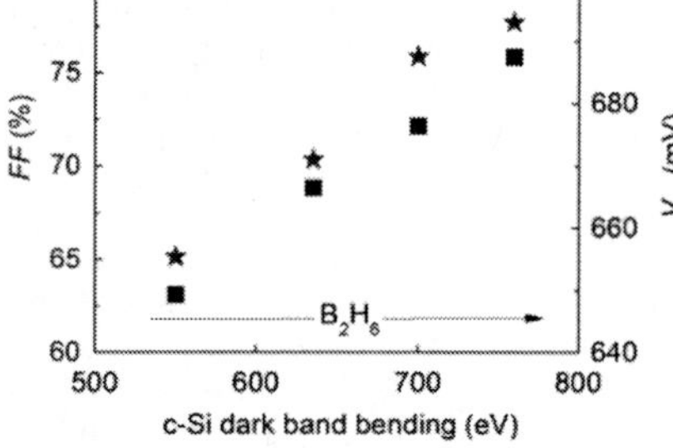

Figure 6: Influence of induced c-Si band bending for a-Si:H(p) emitter doping of silicon heterojunction cells.

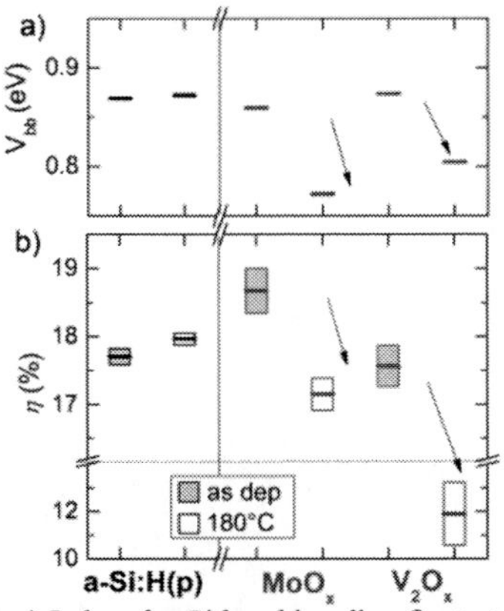

Figure 7: a) Induced c-Si band bending for test structures and b) efficiency of simple silicon heterojunction cells featuring different hole contacts before and after annealing.

is deposited by thermal evaporation [14], ALD [15] or sputtering [16]. Good qualitative agreement with the simulation results is obtained. Most important, a linear dependence of ΔV_{oc} on the induced c-Si band bending is obtained. A higher band bending and lower ΔV_{oc} results from higher doping of the p-type a-Si:H(p). For MoO_x only thermal evaporation enables a high band bending and good selectivity. However, Figure 5b shows that a post deposition annealing has a negative influence on the band bending and ΔV_{oc} for MoO_x [17]. The loss of band bending suggests that a reduction of the effective work function of the MoO_x is involved in the reduction of selectivity. This degradation of the selectivity highlights the main challenge for the application of MoO_x, but also other currently investigated materials: The successful integration of such materials in the c-Si solar cell device architectures and the compatibility with the corresponding fabrication steps.

3.2 Proof of concept solar cells

The importance of the induced c-Si band bending for the silicon heterojunction featuring an a-Si:H(p) hole contact is shown in Figure 6. With increasing gas phase doping the band bending is increased resulting in improved *FF* and V_{oc}. That these improvements are caused by an improved selectivity is reflected in the disappearance of an s-shaped *J-V* characteristic, the temperature dependence of the V_{oc} and *FF*, the ideality factor obtained from Suns-V_{OC} measurements and by the increased fulfilment of the superposition principle (overlap of dark and light *J-V* curve shifted by J_{sc}) as shown in Ref. [18]. This is pointing towards the fact that "classical" losses and hence the validity of the diode equation becomes more dominate with increasing a-Si:H(p) doping.

The basic applicability of the two high work function transition metal oxides MoO_x and V_2O_x for c-Si solar cells is shown in Figure 7b [19] and was explored by others [6], [20], [21]. Compared to the control device featuring an a-Si:H(p) hole contact encouraging efficiencies are obtained before annealing. However, annealing which is an integral part of the back-end cell processing results in a significant reduction of efficiency for MoO_x and V_2O_x which is caused by the reduction of the selectivity as shown in Ref. [19]. It can be seen in Figure 7a that this efficiency reduction comes along with a decrease of the induced c-Si band bending (V_{bb}).

4 CONCLUSIONS

It was stressed that, in the un-optimized case, novel passivating and carrier selective contacts might not obey the diode theory which means that the "classical" electrical losses like recombination and ohmic transport which are well quantified by J_0 and the contact resistance are not sufficient for contact evaluation. In such a case a useful figure of merit is needed for engineering the contact towards a sufficient selectivity and hence validity of the diode theory. In this sense, probing the voltage loss is suggested which occurs for non-selective contacts in the contact region during operation since low-injection conditions and an asymmetric majority / minority carrier conductivity are not maintained. A relatively simple experimental approach is to check if the external voltage matches the implied voltage during open-circuit conditions.

The need for a sufficient high induced c-Si band bending to obtain a selective contact was highlighted for a-Si:H(p) and two metal oxide based hole contacts.

5 ACKNOWLEDGEMENTS

K. Zimmerman and colleagues at ISE are gratefully acknowledged for experimental / technical assistance and fruitful discussions. This work was funded by the European project DISC under the European Union's Horizon 2020 research and innovation program under grant agreement N°727529.

REFERENCES

[1] R. M. Swanson, "Approaching the 29% limit efficiency of silicon solar cells," in *Proceedings of the 31st IEEE Photovoltaic Specialists Conference*, Orlando, USA, 2005, pp. 889-94.

[2] S. M. Sze, *Physics of semiconductor devices*, 2nd ed. New York: John Wiley & Sons, Inc., 1981.

[3] P. Würfel, *Physics of Solar Cells - From principles to new concepts*. Weinheim: Wiley-Vch Verlag GmbH & Co KgaA, 2005.

[4] F. Feldmann, *et al.*, "Passivated rear contacts for high-efficiency n-type Si solar cells providing high interface passivation quality and excellent transport characteristics," *Solar Energy Materials and Solar Cells*, vol. 120, pp. 270-274, 2014/01/01/ 2014.

[5] R. Peibst, *et al.*, "Building blocks for back-junction back-contacted cells and modules with ion-implanted poly-Si junctions," in *2014 IEEE 40th Photovoltaic Specialist Conference (PVSC)*, 2014, pp. 0852-0856.

[6] J. Geissbühler, *et al.*, "22.5% efficient silicon heterojunction solar cell with molybdenum oxide hole collector," *Applied Physics Letters*, vol. 107, p. 081601, 2015.

[7] X. Yang, *et al.*, "High-Performance TiO2-Based Electron-Selective Contacts for Crystalline Silicon Solar Cells," *Advanced Materials*, pp. n/a-n/a, 2016.

[8] J. Bullock, *et al.*, "Survey of dopant-free carrier-selective contacts for silicon solar cells," in *2016 IEEE 43rd Photovoltaic Specialists Conference (PVSC)*, 2016, pp. 0210-0214.

[9] R. Singh, *et al.*, "Review of conductor-insulator-semiconductor (CIS) solar cells," *Solar Cells,* vol. 3, pp. 95-148, 1981.
[10] L.-M. Chen, *et al.*, "Interface investigation and engineering - achieving high performance polymer photovoltaic devices," *Journal of Materials Chemistry,* vol. 20, pp. 2575-2598, 2010.
[11] C. Messmer, *et al.*, "Numerical Simulation of Silicon Heterojunctions Featuring Metal Oxides," *Submitted to IEEE-JPV* 2017.
[12] M. Bivour, *et al.*, "Doped Layer Optimization for Silicon Heterojunctions by Injection-Level Dependent Open-Circuit Voltage Measurements," *IEEE Journal of Photovoltaics,* pp. 566-574, 2014.
[13] S. W. Glunz, *et al.*, "Analyzing back contacts of silicon solar cells by suns-voc-measurements at high illumination densities," in *Proceedings of the 22nd European Photovoltaic Solar Energy Conference* Milan, Italy, 2007, pp. 849-53.
[14] M. Bivour, *et al.*, "Molybdenum and Tungsten Oxide: High Work Function Wide Band Gap Contact Materials for Hole Selective Contacts of Silicon Solar Cells," *Solar Energy Materials and Solar Cells,* pp. 34–41, 2015.
[15] M. Bivour, *et al.*, "Atomic Layer Deposited Molybdenum Oxide for the Hole-selective Contact of Silicon Solar Cells," *Energy Procedia,* vol. 92, pp. 443-449, 2016/08/01/ 2016.
[16] M. Bivour, *et al.*, "Sputter-deposited WO_x and MoO_x for hole selective contacts," *Energy Procedia,* 2017.
[17] L. Neusel, *et al.*, "Selectivity issues of MoOx based hole contacts," *Energy Procedia,* 2017.
[18] M. Bivour, "Silicon Heterojunction Solar Cells: Analysis and Basic Unterstanding " PhD, Faculty of Engineering University of Freiburg, 2015.
[19] M. Bivour, *et al.*, "High work function metal oxides for the hole contact of silicon solar Cells," in *2016 IEEE 43rd Photovoltaic Specialists Conference (PVSC)*, 2016, pp. 0215-0220.
[20] L. G. Gerling, *et al.*, "Transition metal oxides as hole-selective contacts in silicon heterojunctions solar cells," *Solar Energy Materials and Solar Cells,* vol. 145, pp. 109-115, 2016.
[21] M. Mews, *et al.*, "Oxygen vacancies in tungsten oxide and their influence on tungsten oxide/silicon heterojunction solar cells," *Solar Energy Materials and Solar Cells,* vol. 158, Part 1, pp. 77-83, 2016.

THERMAL STABILITY OF NOVEL HOLE-SELECTIVE CONTACTS FOR SILICON WAFER SOLAR CELLS

Chang-Yeh Lee[1], Tian Zhang[1], Kean Khoo[1], Amir Abdallah[2], Sergey Rashkeev[2], Nouar Tabet[2], and Bram Hoex[1]
[1]School of Photovoltaic and Renewable Energy Engineering, UNSW Sydney, Sydney NSW 2052, Australia
[2]Qatar Environment and Energy Research Institute (QEERI)
HBKU, Qatar Foundation, Doha, Qatar. P.O. Box 5825, Doha, Qatar

ABSTRACT:
The thermal stability of novel hole selective contacts is of key importance when integrating these films into high-efficiency silicon solar cells. In this work, we focus on the thermal stability of atomic layer deposited (ALD) WO_x with and without an a-Si:H interface passivation layer. Hydrogen diffusion from a-Si:H to WO_x layer was detected by both Fourier transform infrared spectroscopy and *in-situ* ultraviolet photoelectron spectroscopy for annealing temperatures above 180 °C. The work function decreased for higher annealing temperatures, but the decrease was slightly less for the samples with the a-Si:H interface passivation layer due to the H doping of WO_x. An annealing temperature higher than 223 °C results in insufficient workfunction for hole selectivity. In terms of surface passivation, it was found that the plasma-enhanced ALD of WO_x introduced some plasma damage on the a-Si:H layer as was detected by injection-dependent effective lifetime measurement. Fortunately, this could be recovered by an annealing treatment at 220 °C, resulting in an implied V_{OC} of 730 mV. The results provide a promising potential of the implementation on ALD WO_x for high-efficiency silicon solar cells.
Keywords: stability, passivation, a-Si:H

1 INTRODUCTION

Carrier selective contacts are currently a hot topic in silicon photovoltaics as they enable to achieve significantly higher efficiencies compared to standard diffused solar cells as demonstrated by the 26.6% all-back-contact heterojunction solar cell from Panasonic [1]. Recently, some transition metal oxides (TMOs) have shown excellent hole-selective properties on c-Si because of their high work function resulting in a desirable band bending in Si. TMOs such as MoO_3, WO_3 and V_2O_5, have been widely used for organic electronic devices as a hole transport material[2]. Moreover, their wide bandgaps allow high light transmission into the solar cell. Thermal evaporated TMOs also appears slightly sub-stoichiometric and the defects (e.g. oxygen vacancies) which are close to the conduction band edge resulting in high work-function with semi-metallic properties[3]. An efficiency of 22.5% was achieved by using thermal evaporated MoO_x as a hole collector on a *n*-type solar cell with an *n*-type amorphous silicon heterojunction contact [4]. Another alternative deposition technique, atomic layer deposition (ALD), provides more controllable film quality and has successfully been introduced in high-volume manufacturing for the deposition of aluminium oxide surface passivation layers [5] However, the thermal stability of such TMO contacts on c-Si has not been comprehensively studied and is of key importance when integrating these films into high-efficiency silicon solar cells. It has been shown that the fill factor (*FF*) of a MoO_x SHJ solar cell degraded after a post-deposition anneal above 130 °C[4]. On the contrary, the hole extraction was enhanced after the annealing at 180 °C using WO_x as a hole-selective layer[6].

In this work, we investigate the thermal stability of WO_x passivated hole-selective contact prepared by plasma enhanced atomic layer deposition (PE-ALD) and the impact of the intervening a-Si:H interface passivation layer. The film properties were studied by Fourier-transform infrared spectroscopy (FTIR) and the evolution of workfunction during annealing is observed by *in-situ* ultraviolet photoelectron spectroscopy (UPS). Finally, we show the excellent surface passivation of the stack on c-Si.

2 EXPERIMENTAL DETAILS

In this study, double-side polished 280 μm boron doped float-zone (FZ) wafer with a resistivity of 1-10 Ω·cm were used as a substrate. After standard Radio Corporation of America (RCA) cleaning process, the wafer was dipped into 1 % hydrofluoric acid to remove the surface oxide. An ~9 nm WO_x layer was grown on the substrate at 200 °C by plasma enhanced atomic layer deposition (PE-ALD) (Fiji G2, Cambridge Nanotech, Veeco/CNT Inc., USA) using bis(tertbutylimino) bis(dimethylamino)tungsten(VI) $[({}^tBuN)_2(Me_2N)_2W]$ (Strem Chemicals Inc., USA) as the W precursor. The precursor was heated to 80 °C and delivered by a boost kit[7]. An ALD cycle consisted of pulsing W precursor followed by O_2 plasma for 20 s (300 W). The reactor chamber was purged for 3.5 and 4 s of 80 sccm argon after the precursor pulse and O_2 plasma exposure, respectively, as depicted in Figure 1. A dual mode plasma enhanced chemical vapour deposition (PECVD) system (Roth&Rau, AK400) was used to deposit the a-Si:H film at 280 °C.

The FTIR (Nicolet 5700, Thermo Fisher) absorption spectra of the WO_x/a-Si:H stack were obtained by comparing the sample's IR transmission before and after the deposition. The effective minority carrier lifetime was measured by quasi-steady-state photoconductance (Sinton, WCT-120) on a symmetrical test structure. The annealing process was performed on a conventional hot plate in air ambience for 20 min. The *in-situ* UPS (ESCALAB 250Xi, Thermo Scientific) was performed in a ultra-high vacuum chamber (< 2×10^{-9} mbar) by a He I source (21.2eV) with a pass energy of 2 eV.

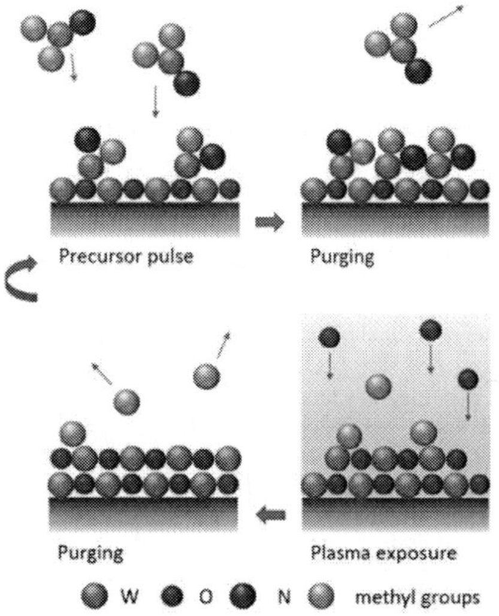

Figure 1 Schematic diagram of a typical PE-ALD cycle.

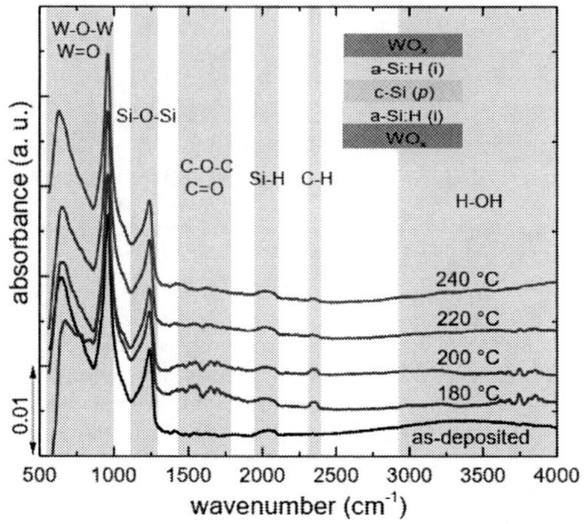

Figure 2 FTIR spectra of WO_x / a-Si:H stack annealed at different temperatures. Shaded regions indicate the principal absorption bands. The spectra were offset vertically for clarity.

Table 1 Absorption bands in the FTIR spectra of WO_x/Si:H stack.

Region (cm^{-1})	Assignment	References
600-1000	W-O-W, W=O	[8]
1100-1300	Si-O-Si	[9]
1350-1800	C=O, O-C-C	[10]–[12]
1950-2150	Si-H	[15]
2300-2400	C-H	[13]
2250-3800	O-H	[16]

3 RESULTS AND DISCUSSION

3.1 FTIR

WO_x (9 nm) / a-Si:H (9 nm) stacks were deposited on both sides of the c-Si substrate. The absorption spectra were taken before and after annealing at different temperatures from 180 to 240 °C, as shown in Figure 2. The main absorption features in the FTIR spectra are summarized in Table 1. The stretching molecular vibrations of WO_x appear in the 600-1000 cm^{-1} region which includes the response of the bridging (W-O-W) and the terminal (W=O) oxygen bonds[8]. A significant Si-O-Si [9] peak was detected in the 1100-1300 cm^{-1} range before annealing, indicating the formation of SiO_x interlayer between WO_x and a-Si:H during the first stages of the PE-ALD process. A higher intensity of Si-O peak was seen in a control sample where WO_x was directly grown on the c-Si surface (not shown). This indicates that the SiO_x layer is thinner when WO_x was grown on a-Si:H surface compared to a H-terminated silicon surface. A group of peaks associated with C=O and C-O-C appeared between 1350 and 1800 cm^{-1} only after annealing at 180 and 200 °C[10]–[12]. The same feature was observed in a control sample without the a-Si:H layer, hence the presence of carbon-oxygen related bonds could be attributed to the decomposition of residual methyl groups in the WO_x film. These features decreased at higher annealing temperatures. After annealing, an apparent double-peak at 2300 to 2400 cm^{-1} related to C-H appeared[13], and it has the same trend of carbon-oxygen related peaks which decreased after annealing at 220 °C. The presence of carbon impurity is due to the nature of ALD deposition, but it has been shown not to affect the level of surface passivation for Al_2O_3[14]. The presence of Si-H responses between 1950 and 2150 cm^{-1} was assigned to the a-Si:H layer[15]. The broadband absorption at 2250-3800 cm^{-1} region of the as-deposited film relates to the hydroxyl (O–H) groups in the film, which disappeared after annealing indicating that it could be related to chemisorbed water from the ambient[16].

To examine the influence of the intervening a-Si:H layer on WO_x during the annealing process, a control sample without a-Si:H was used. Figure 3 indicates the changes of C-H peaks of the samples with and without an a-Si:H passivation layer after annealing at different temperatures. For the sample without an a-Si:H layer, the C-H peaks appeared after annealing at 180 and 200 °C, and the peaks were flattened when the annealing temperature increased to 220 °C. This indicates the release of hydrogen from the ALD WO_x film. The sample with an a-Si:H interlayer showed a similar trend of reduction of the C-H peaks, but the signals could be still detected after 220 and 240 °C annealing. This is possibly due to the hydrogen diffusion from the underlying a-Si:H layer[17]. The influence of additional hydrogen source for the WO_x films will be discussed in the following section.

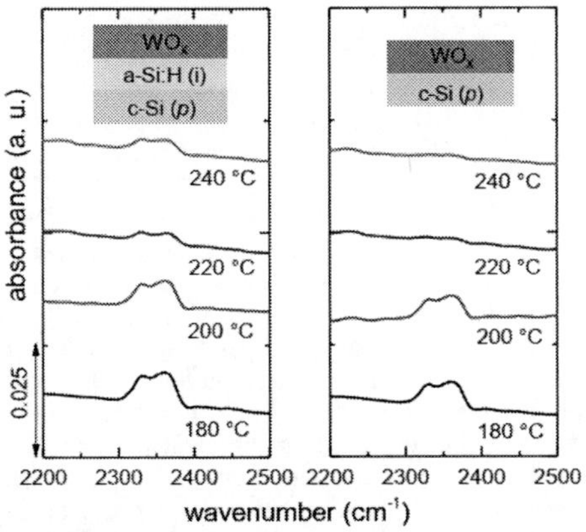

Figure 3 The change of C-H peaks from the samples with (left) and without (right) an a-Si:H passivation layer after annealing from 180 to 240 °C. The spectra were offset vertically for clarity.

3.2 *In-situ* annealing UPS

To understand the effect of annealing on the WO_x film, *in-situ* UPS was used to study the electronic structure and determine the work function of the films during annealing. Figure 4 shows the UPS spectra of the as-deposited and annealed WO_x film on c-Si. In Figure 4 (a), an expanded view of the UPS spectra is shown as well as the region focusing on the secondary electron cut-off and the Fermi edge region presented in Figure 4 (b) and (c), respectively. For the as-deposited film, two main peaks of O 2p and W 5d-O 2p orbitals can be seen at around 6 and 14 eV, respectively[18]. No gap state near the Fermi edge region was observed, implying a stoichiometric as-deposited WO_x film[19]. During the annealing process, an apparent increase of the intensity near the Fermi region and a shift of the spectrum toward lower binding energy can be observed. The shift indicates the change of the electronic structure of the film due to the creation of oxygen vacancy during annealing. The oxygen vacancies in WO_x film result in gap states located at 0.8 and 2.2 eV near the Fermi edge[18]. A high concentration of the gap states could lead to low work function which reduces the band bending for hole collection[20]. Figure 5 gives the comparison of the as-deposited and the annealed (177 °C) WO_x / c-Si as well as the annealed WO_x/a-Si:H/c-Si spectra for comparison. Both of the annealed spectra show the high concentration of gap states near the Fermi region. However, a lower intensity of gap state at around 2 eV for the WO_x/a-Si:H sample could be attributed to the H diffusion into the WO_x film which fills the oxygen vacancies. This hydrogen doping in WO_x is expected to improve the work function of the WO_x film[18].

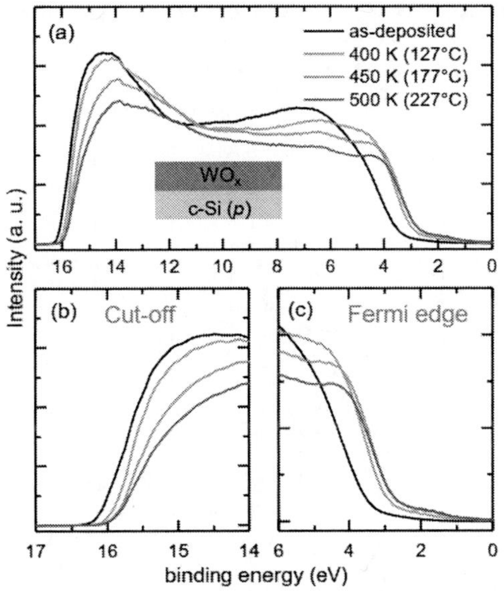

Figure 4 (a) Full range of the *in-situ* UPS spectra of WO_x / c-Si annealed from 127 to 227 °C. Enlarged view of the spectra at the secondary electron cut-off (b) and Fermi-edge (c) region.

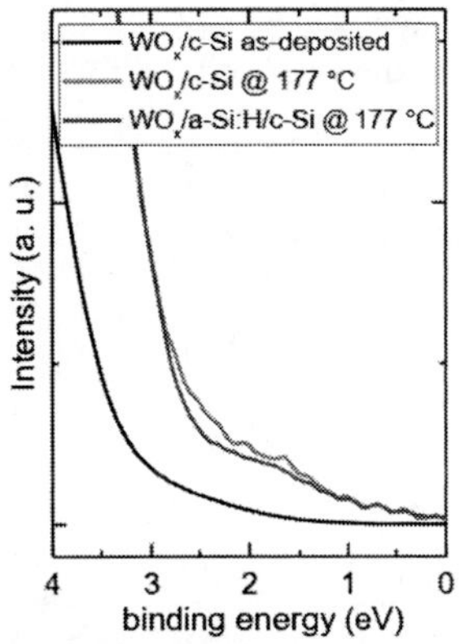

Figure 5 Comparison of the as-deposited WO_x / c-Si and the annealed WO_x/c-Si and WO_x/ a-Si:H/c-Si spectra.

The work function of WO_x and WO_x/a-Si:H under different annealing temperatures were deduced from the UPS spectra, as shown in Figure 6. For the as-deposited films, a work function of above 6.5 eV was detected on WO_x, and a lower value of 6.1 eV was found on the WO_x/a-Si:H sample. The lower work function of WO_x could potentially be attributed to the H diffusion from the a-Si:H layer during the ALD process at 200 °C. Although the H doping lowered the work function of as-deposited WO_x film, additional H source improved the work function when the sample was subjected to a post-deposition annealing. The decrease of work function on the annealing temperature could be seen on both samples. However, at the range from 127 to 327 °C annealing temperature, the WO_x/a-Si:H sample showed a slightly higher work function than that of WO_x sample due to the H doping. Higher annealing temperature results in lower work function, which were observed for both samples due to the increase of oxygen vacancies. This implies the limit of annealing condition for this structure. It's worth to note that the annealing process was performed under vacuum in the UPS chamber without a capping layer.

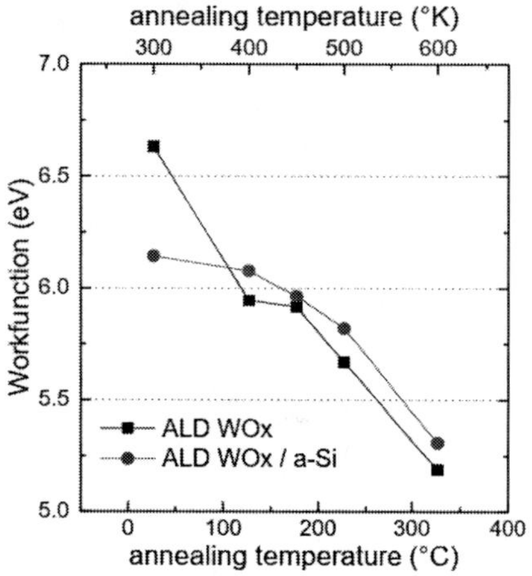

Figure 6 Calculated workfunction as a function of annealing temperature.

3.3 Effective minority carrier lifetime

The potential of ALD WO_x as passivated hole-selective contact on c-Si solar cells can be evaluated by the surface saturation current density. The effective minority carrier lifetime of the WO_x/a-Si:H stack after annealing at 180 to 220 °C and a control sample without a WO_x layer was used for comparison, as shown in Figure 7. The degradation of effective lifetime after WO_x deposition could be seen on the as-deposited stack. The degradation could be attributed to the plasma damage on the a-Si layer during the PE-ALD deposition process[21]. The annealing treatment improved the passivation of the stack structure. The effective lifetime at 1×10^{15} cm^{-3} injection level was increased from 2.1 to 2.6 ms after annealing at 220 °C, resulting in an implied V_{OC} of 730 mV. The result demonstrates that the high quality of surface passivation can be achieved by ALD WO_x/a-SiH stack.

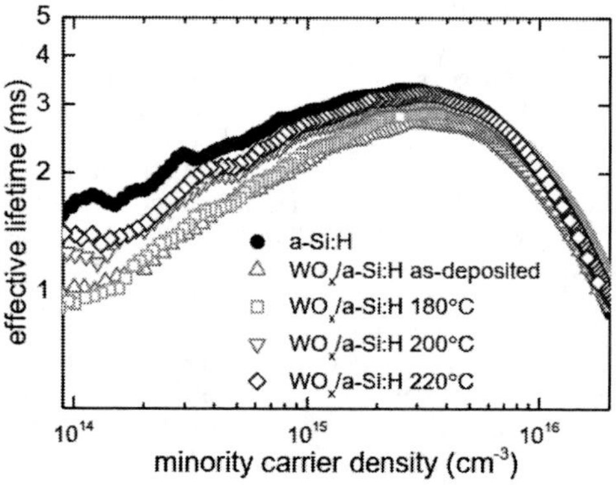

Figure 7 Effective minority carrier lifetime of WO_x/a-Si:H stack after annealing at 180 to 220 °C and a control sample without a WO_x layer (solid dot)

4 CONCLUSION

This work presents the study on the effect of annealing of WO_x/a-Si:H stack. The film property was studied by FTIR measured after annealing at 180 to 240 °C. A significant signal of Si-O and carbon-related bonds were observed. The presence of SiO_x could potentially affect the carrier transportation, but a thinner SiO_x layer could be inferred when the WO_x was deposited on a-Si:H. The influence of impurity on the film properties and carrier mobility should be further studied. Hydrogen diffusion from a-Si:H was detected by both FT-IR and *in-situ* UPS study. The work function of WO_x and WO_x/a-Si:H decreased when the annealing temperature increased, and the additional H source improved the work function of WO_x/a-Si:H stack after annealing. An annealing temperature higher than 223 °C results in insufficient workfunction for hole collection. The plasma damage on the a-Si:H layer after PE-ALD deposition could be seen from the as-deposited WO_x/a-Si:H stack by injection-dependent effective lifetime measurement, and it could be recovered by an annealing process at 220 °C, resulting in an implied V_{OC} of 730 mV. Future work of the implementation will focus on the ability of hole-selectivity by contact resistance measurement and the performance on SHJ solar cells.

5 ACKNOWLEDGEMENT

This publication was made possible by NPRP grant # NPRP9-021-009 from the Qatar National Research Fund (a member of Qatar Foundation). The findings achieved herein are solely the responsibility of the authors.

6 REFERENCE

[1] K.Yoshikawa *et al.*, "Silicon heterojunction solar cell with interdigitated back contacts for a photoconversion efficiency over 26%," vol. 2, p. 17032, Mar.2017.

[2] J.Meyer, S.Hamwi, M.Kroger, W.Kowalsky, T.Riedl, andA.Kahn, "Transition metal oxides for organic electronics: Energetics, device physics and applications," *Adv. Mater.*, vol. 24, no. 40, pp. 5408–5427, 2012.

[3] Y.Guo andJ.Robertson, "Origin of the high work function and high conductivity of MoO3," *Appl. Phys. Lett.*, vol. 105, no. 22, p. 222110, Dec.2014.

[4] J.Geissbühler *et al.*, "22.5% Efficient Silicon Heterojunction Solar Cell With Molybdenum Oxide Hole Collector," *Appl. Phys. Lett.*, vol. 107, no. 8, 2015.

[5] B.Hoex, "Developments in Ultrafast ALD for PV," *Futur. PV*, vol. 7, p. 60, 2012.

[6] M.Bivour, J.Temmler, F.Zahringer, S.Glunz, andM.Hermle, "High work function metal oxides for the hole contact of silicon solar Cells," *Conf. Rec. IEEE Photovolt. Spec. Conf.*, vol. 2016–Novem, pp. 215–220, 2016.

[7] A.Bertuch, B. D.Keller, N.Ferralis, J. C.Grossman, andG.Sundaram, "Plasma enhanced atomic layer deposition of molybdenum carbide and nitride with bis(*tert* -butylimido)bis(dimethylamido) molybdenum," *J. Vac. Sci. Technol. A Vacuum, Surfaces, Film.*, vol. 35, no. 1, p. 01B141, 2017.

[8] M. F.Daniel, B.Desbat, J. C.Lassegues, B.Gerand, andM.Figlarz, "Infrared and Raman study of WO3 tungsten trioxides and WO3, xH2O tungsten trioxide tydrates," *J. Solid State Chem.*, vol. 67, no. 2, pp. 235–247, 1987.

[9] K. B.Koller, W. A.Schmidt, andJ. E.Butler, "In situ infrared reflection absorption spectroscopic characterization of plasma enhanced chemical vapor deposited SiO2 films," *J. Appl. Phys.*, vol. 64, no. 9, pp. 4704–4710, Nov.1988.

[10] D. N.Goldstein, J. A.McCormick, andS. M.George, "Al2O3 Atomic Layer Deposition with Trimethylaluminum and Ozone Studied by in Situ Transmission FTIR Spectroscopy and Quadrupole Mass Spectrometry," *J. Phys. Chem. C*, vol. 112, no. 49, pp. 19530–19539, Dec.2008.

[11] E.Langereis, J.Keijmel, M. C. M.van deSanden, andW. M. M.Kessels, "Surface chemistry of plasma-assisted atomic layer deposition of Al2O3 studied by infrared spectroscopy," *Appl. Phys. Lett.*, vol. 92, no. 23, p. 231904, Jun.2008.

[12] V.Verlaan, L. R. J. G.Van DenElzen, G.Dingemans, M. C. M.Van DeSanden, andW. M. M.Kessels, "Composition and bonding structure of plasma-assisted ALD Al 2O3 films," *Phys. Status Solidi Curr. Top. Solid State Phys.*, vol. 7, no. 3–4, pp. 976–979, 2010.

[13] A. C.Dillon, A. W.Ott, J. D.Way, andS. M.George, "Surface chemistry of Al2O3 deposition using Al(CH3)3 and H2O in a binary reaction sequence," *Surf. Sci.*, vol. 322, no. 1, pp. 230–242, 1995.

[14] H. B.Profijt, S. E.Potts, M. C. M.van deSanden, andW. M. M.Kessels, "Plasma-Assisted Atomic Layer Deposition: Basics, Opportunities, and Challenges," *J. Vac. Sci. Technol. A Vacuum, Surfaces, Film.*, vol. 29, no. 5, p. 50801, 2011.

[15] J. J. H.Gielis, P. J.Van DenOever, B.Hoex, M. C. M.Van DeSanden, andW. M. M.Kessels, "Real-time study of a-Si:H/c-Si heterointerface formation and epitaxial Si growth by spectroscopic ellipsometry, infrared spectroscopy, and second-harmonic generation," *Phys. Rev. B - Condens. Matter Mater. Phys.*, vol. 77, no. 20, pp. 1–16, 2008.

[16] Y.-C.Kim, H.-H.Park, J. S.Chun, andW.-J.Lee, "Compositional and structural analysis of aluminum oxide films prepared by plasma-enhanced chemical vapor deposition," *Thin Solid Films*, vol. 237, no. 1, pp. 57–65, 1994.

[17] W.Beyer andH.Wagner, "The role of hydrogen in a-Si:H - results of evolution and annealing studies," *J. Non. Cryst. Solids*, vol. 59–60, no. PART 1, pp. 161–168, 1983.

[18] M.Vasilopoulou *et al.*, "Hydrogenated under-stoichiometric tungsten oxide anode interlayers for efficient and stable organic photovoltaics," *J. Mater. Chem. A*, vol. 2, no. 6, pp. 1738–1749, 2014.

[19] J.Zhu, M.Vasilopoulou, D.Davazoglou, S.Kennou, A.Chroneos, andU.Schwingenschlögl, "Intrinsic Defects and H Doping in WO3," *Sci. Rep.*, vol. 7, no. December 2016, p. 40882, 2017.

[20] M.Mews, L.Korte, andB.Rech, "Oxygen vacancies in tungsten oxide and their influence on tungsten oxide/silicon heterojunction solar cells," *Sol. Energy Mater. Sol. Cells*, vol. 158, pp. 77–83, 2016.

[21] Y.Kuang *et al.*, "Towards the implementation of atomic layer deposited In2O3:H in silicon heterojunction solar cells," *Sol. Energy Mater. Sol. Cells*, vol. 163, no. August 2016, pp. 43–50, 2017.

HIGH EFFICIENCY LOCALLY LASER DOPED IBC SOLAR CELLS

Marco Ernst[1,^], Evan Franklin[1], Teck Chong[1], Kean Fong[1], Daniel Walter[1], Er-Chien Wang[1], Teng Kho[1], and Andrew Blakers[1]
^Corresponding author: e-mail: marco.ernst@anu.edu.au
[1]Centre for Sustainable Energy Systems, Australian National University, Canberra, ACT 2601, Australia

ABSTRACT: Interdigitated back contact (IBC) solar cells offer one of the most promising routes towards highest device efficiencies and have a high potential for low cost production sequences. In this work, we replace typical patterning steps and high-temperature diffusions in IBC device fabrication by means of laser processing. We report on the fabrication of a 22.8% efficient all-laser-doped IBC solar cell with localized boron and phosphorus doped contacts. By means of three-dimensional numerical device modelling we demonstrate that this cell is limited by recombination at the localized n^+ contacts and determine an efficiency potential of 24.5% with improved laser parameters and surface passivation.
Keywords: c-Si, Characterisation, Devices, High-Efficiency, Laser Processing, n-type, Silicon Solar Cell, Solar Cell Efficiencies

1 INTRODUCTION

One key goal in commercial solar cell fabrication is to improve the energy conversion efficiency. This is reflected in the increasing market share of high-efficiency solar cell concepts [1]. Interdigitated back contact (IBC) solar cells [2-8] are considered as one of the most promising approaches and have recently achieved world highest conversion efficiency of 26.7% in a heterojunction back-contact solar cell concept [2].

Placing both electrical contacts onto one side of the solar cell comes with the advantage of reduced finger shading loss at the front side and thus allowing to optimize optical and resistive losses separately. However, processing of such a device requires more complex steps, for instance patterning the n^+ and p^+ regions and separating of metal contacts.

Dahlinger et al. performed large-area laser doping to minimize the required patterning steps and recently demonstrated a device efficiency of 23.2% with this process [3]. Franklin et al. demonstrated an all-laser-doped IBC cell concept using localized laser doping with a device efficiency of 19.1% while the champion cell with photolithographic patterning steps achieves 24.4% [4].

In this paper, we demonstrate the successful processing of an all laser-doped IBC solar cell with localized doped and contacted areas and an efficiency of 22.8% with non-optimized laser parameters [5]. By means of specifically designed test structures and numerical device simulations we characterise the properties of localized contacts and show that an efficiency of 23.8% is possible by choosing optimized laser-processing parameters. If further improvements to the surface passivation at the front- and rear side are implemented we estimate an efficiency potential of 24.5%.

2 CELL DESIGN AND FABRICATION

All-laser-doped IBC solar cells are fabricated from a 100 Ω cm, n-type float zone wafer. Figure 1 summarizes the process flow for fabricating IBC solar cells by means of localized laser doping.

The localized laser-doped regions are created from spin-coated nanoparticle ink and spin-on dopants by means of pulsed UV laser irradiation with a pulse width of approximately 25 ns.

A second laser process prior metallization is used to locally ablate a silicon nitride film to open the contact regions within the localized doped areas.

Figure 2 shows a schematic of the cell structure with localized p^+ and n^+ contacts.

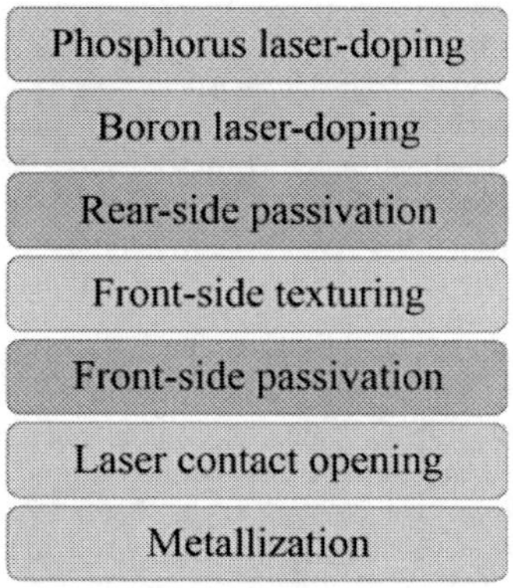

Figure 1. Summarized processing sequence for laser-doped back-contact solar cells.

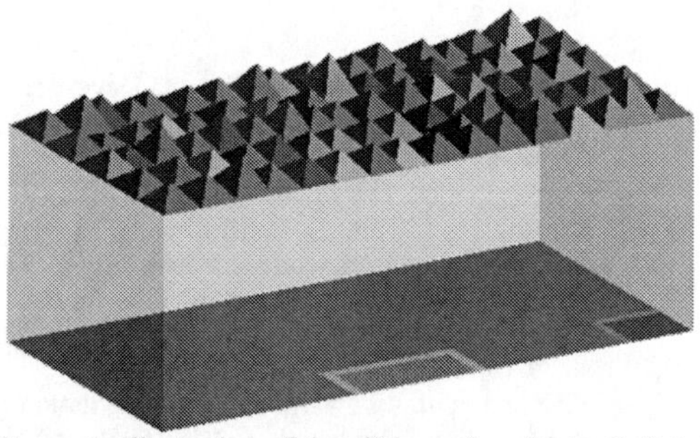

Figure 2. Illustration of the all-laser-doped interdigitated back contact solar cell with localized p^+ (green) and n^+ (red) contacts.

3 CHARACTERISATION OF LOCALIZED CONTACTS

3.1 Recombination of localized contacts

We design and prepare distinct test structures and numerical simulations to accurately determine the recombination parameter J_0 and contact resistivity ρ_c of localized contacts [9].

First, we determine the recombination parameter J_0 from spatially resolved steady-state PL measurements (BT Imaging LIS-R1 PL imaging tool). Figure 3 shows a schematic of the test structure used for determination of the recombination parameter J_0 from PL measurements. Both sides are symmetrically passivated with a SiO_2/Si_3N_4 stack. We create a square array of localized laser-doped areas in regions measuring 12 mm × 12 mm. The metallized local contacts on the rear side are constrained to an area smaller than the laser doped regions (red).

In order to accurately quantify the recombination properties of localized laser-doped regions from PL measurements and numerical simulations we calibrate the simulated PL signal using PCD lifetime measurements in a reference area of the wafer.

Figure 4 shows an example for a calculated J_0-image using this method. Each of the 16 regions has been processed with a variation in laser parameters.

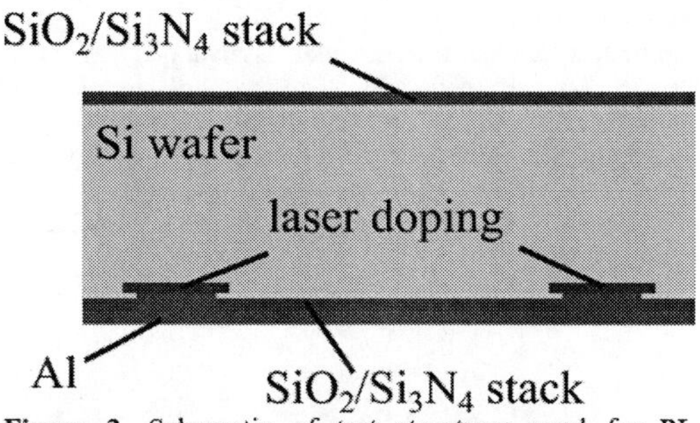

Figure 3. Schematic of test structures used for PL analysis. We refer to the laser-doped as the rear side.

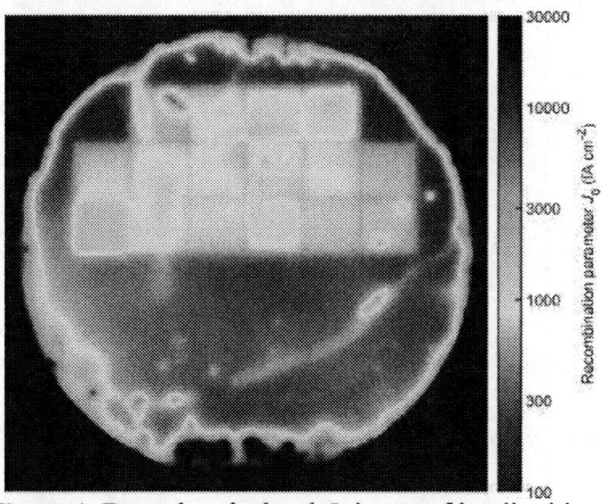

Figure 4. Example calculated J_0-image of localized laser doped contacts.

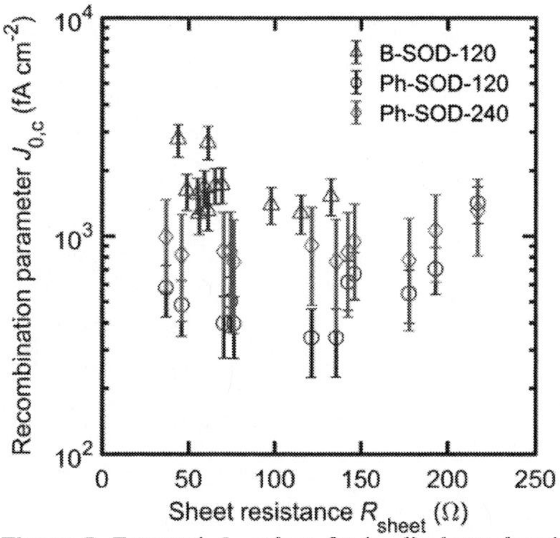

Figure 5. Extracted $J_{0,c}$-values for locally laser doped regions from boron (B-SOD) and phosphorus (P-SOD) spin-on dopant sources.

Figure 5 shows the extracted $J_{0,c}$ parameter as a function of the sheet resistance R_{sheet} for locally boron and phosphorus laser-doped regions. The $J_{0,c}$ value is a lumped parameter for the passivated and metallized regions of the laser-doped areas. For boron spin-on dopant (SOD) we achieve values of 1,000 to 3,000 fA/cm^2 and for phosphorus SOD values are in the range from 500 to 2,000 fA/cm^2. We observe almost no variation of the recombination parameter with the sheet resistance which we explain with a re-distribution of dopants during a high-temperature oxidation when passivating the rear surface.

3.2 Contact resistivity of localized contacts

In order to determine the contact resistivity ρ_c we measure dark *I-V* curves of individual contacts and fit the diode characteristics with three-dimensional device simulations. The final test structure for measuring the dark *I-V* characteristics of the single laser doped localized contacts is illustrated in Figure 6. The rear side has only one single laser-doped contact in the centre of 2 mm × 6 mm-sized rectangular boxes. The front side is a full-area contact with opposite polarity doping.

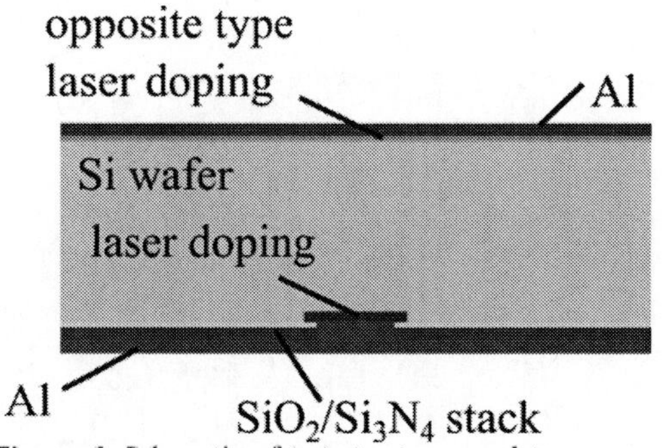

Figure 6. Schematic of test structures used to measure dark *I-V* characteristics of single laser doped localized contacts.

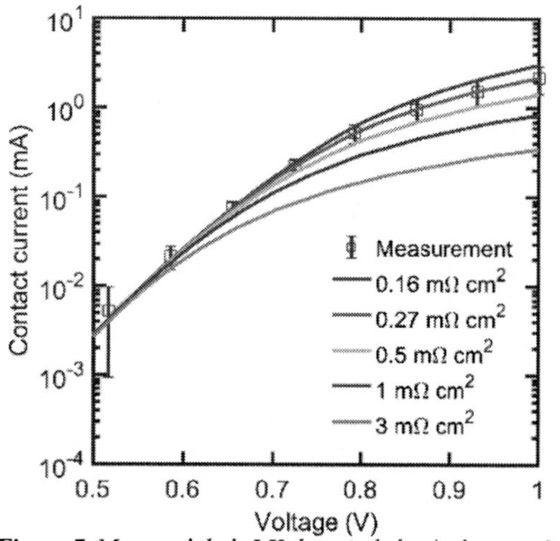

Figure 7. Measured dark *I-V* characteristics (red squares) with 95% confidence intervals from ten measurements (errors bars). The simulated dark *I-V* curves are plotted for J_0 = 1,000 fA/cm^2 for various contact resistivity values ρ_c.

We measure *I-V* curves for each individual diode under forward bias using a four-point probe setup in the voltage range from 0.5 V to 1.0 V and then take the average of each ten measurements for identical laser parameters. We perform three-dimensional numerical simulations of the dark I-V curves using Quokka [10] to extract the contact resistivity of the samples.

Figure 7 shows an example for a measured dark *I-V* curve of single laser doped contact. A recombination parameter J_0 = (1,000 ± 590) fA/cm^2 with 95% confidence interval, as determined by the PL recombination analysis, is used in the simulations. The simulated curves give the best fit to the measurement for a contact resistivity $\rho_c = 0.27^{+0.23}_{-0.11}\ \mathrm{m\Omega\,cm^2}$. Here, we have determined the upper and lower limits of ρ_c using the lower and upper limits for J_0 and the measured contact current, respectively.

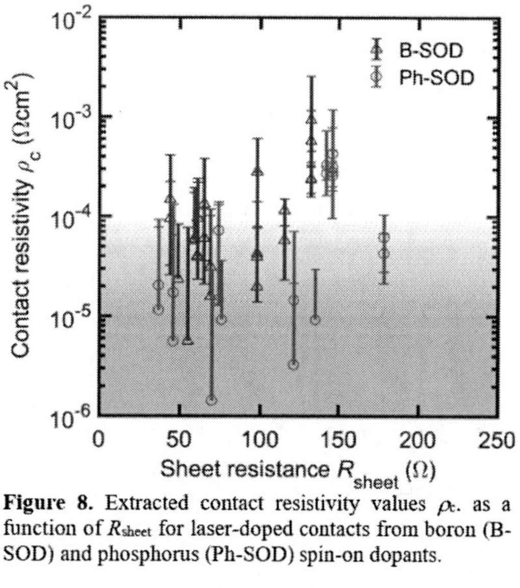

Figure 8. Extracted contact resistivity values ρ_c. as a function of R_{sheet} for laser-doped contacts from boron (B-SOD) and phosphorus (Ph-SOD) spin-on dopants.

Table 1. Parameter of measured and simulated laser-doped IBC solar cell.

Cell parameter	Meas.	Sim.	Potential (Gen2)	Potential (Gen3)
V_{OC} (mV)	679	679	691	713
J_{SC} (mA/cm^2)	41.86	41.87	42.34	42.25
FF	80.2%	80.2%	81.5%	81.5%
η	22.80%	22.80%	23.82%	24.54%

Figure 8 shows the extracted contact resistivity values ρ_c for localized laser-doped contacts from boron and phosphorus spin-on dopant sources. For both dopant sources, we determine values in the range from 10^{-5} to 10^{-3} Ω.cm^2.

4 RESULTS

Table 1 summarizes the results of our first all-laser-doped IBC solar cell with a designated-area of 2 cm × 2 cm. We have achieved an energy conversion efficiency of 22.8%. For these solar cells, we have used a silicon nanoparticle based phosphorus dopant source. We note that the laser processing parameters have not been optimized for the n$^+$ contacts. Figure 9 shows the measured current density *J* and power density *P* as a function of voltage *V* for our first cell.

We perform three-dimensional numerical device simulations using Quokka to estimate the limiting contact properties of the experimental laser-doped IBC cell. Bulk, optical generation profile, geometric and undiffused surface recombination properties are measured values that have been taken from our baseline IBC cell process [4] at the Australian National University. Table 2 lists the input parameters used in our simulations.

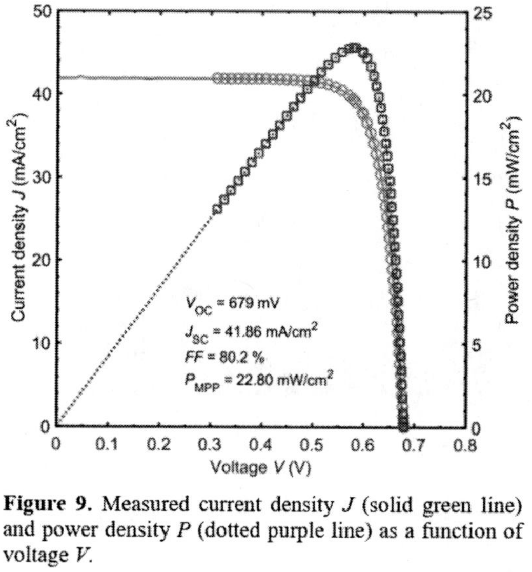

Figure 9. Measured current density *J* (solid green line) and power density *P* (dotted purple line) as a function of voltage *V*.

Table 2. Key properties for 3D numerical device modelling of laser-doped IBC solar cells.

Property	Value	Gen2	Gen3
Thickness W_z (µm)	230	-	-
Bulk resistivity ρ_b (Ω.cm)	100	-	-
Bulk SRH lifetime $\tau_{p0} = \tau_{n0}$ (µs)	5000	-	-
Finger half pitch W_x (µm)	250	-	-
Contact pitch W_y (µm)	110	-	-
Undiffused rear J_0 (fA/cm^2)	21	-	2
Front surface J_0 (fA/cm^2)	5	-	2
Optical generation J_{gen} (mA cm^{-2})	$42.76 \times f$ $f = 0.99$	$f = 1$	-
n$^+$ to p$^+$ contact ratio	1:2	-	-
p$^+$ contact recombination $J_{0,b}$ (fA/cm^2)	1,300	1,200	-
n$^+$ contact recombination $J_{0,ph}$ (fA/cm^2)	2,700	700	-
p$^+$ contact resistivity $\rho_{c,b}$ (mΩ.cm^2)	0.7	0.1	-
n$^+$ contact resistivity $\rho_{c,ph}$ (mΩ.cm^2)	0.7	0.1	-
External series resistance R_s (Ω)	0.05	-	-

The local recombination parameter $J_{0,b}$ = 1,300 fA/cm^2 of the boron laser-doped region and upper limit of contact resistivity $\rho_{c,b} = 7 \times 10^{-4}$ Ω cm^2 have been determined from separately prepared test structures as described above and in Ref. [9]. We found that for the silicon nanoparticle based phosphorus dopant source a laser fluence of 3 J/cm^2 is at the threshold for providing acceptable recombination values for the n$^+$ contacts. Therefore, we fit the recombination parameter $J_{0,ph}$ = 2,700 fA/cm^2 and contact resistivity $\rho_{c,ph} = 7 \times 10^{-4}$ Ω cm^2 to match the open-circuit voltage V_{OC} and fill factor FF of the experimental cell.

The simulation results are shown in Table 1 and agree well with the experimental cell parameters. The green circles and purple squares in Figure 9 show the simulated J-V and power density curve, respectively.

In Ref. [9] we have shown that recombination parameter values $J_{0,b}$ = 1,200 fA/cm^2 and $J_{0,ph}$ = 700 fA/cm^2 for laser-doped boron and phosphorus regions are possible with optimized laser-processing parameters. With these values, we estimate an efficiency potential of 23.8% for laser-doped IBC cells using the same geometric cell design with optimized contact fractions. The varied input parameters for the simulations of the efficiency potential with improved laser-doped contacts are listed in Table 2 under "Gen2". Table 1 lists the simulation results.

If we further improve the surface passivation at the front- and rear side, for instance by implementing our recent developments of ONO passivation stacks [11], we estimate an efficiency potential of 24.5% (Gen3).

ACKNOWLEDGEMENTS

This work has been supported by Trina Solar and the Australian Government through the Australian Renewable Energy Agency. Responsibility for the views, information or advice expressed herein is not accepted by the Australian Government.

REFERENCES

[1] International Technology Roadmap for Photovoltaic (ITRPV.net) Results 2015 (2016).

[2] K. Yoshikawa, H. Kawasaki, W. Yoshida, et al., Nat. Energy **2**, 17032 (2017).

[3] M. Dahlinger, K. Carstens, E. Hoffmann, S. Wansleben, J.R. Köhler, R. Zapf-Gottwick, and J.H. Werner, in 31st European Photovoltaic Solar Energy Conference and Exhibition (2015), pp. 462–465.

[4] E. Franklin, K. Fong, K. McIntosh, A. Fell, A. Blakers, T. Kho, D. Walter, D. Wang, N. Zin, M. Stocks, E.-C. Wang, N. Grant, Y. Wan, Y. Yang, X. Zhang, Z. Feng, and P.J. Verlinden, Prog. Photovolt: Res. Appl **24**, 411 (2016).

[5] M. Ernst, E. Franklin, A. Fell, K. Fong, D. Walter, E.-C. Wang, T. Kho, and A. Blakers, phys. stat. sol. (a) **31**, 1700318 (2017).

[6] R.J. Schwartz and M.D. Lammert, in International Electron Devices Meeting (1975), pp. 350–352.

[7] R.M. Swanson, S.K. Beckwith, R.A. Crane, W.D. Eades, Young Hoon Kwark, R.A. Sinton, and S.E. Swirhun, IEEE T ELECTRON DEV **31**, 661 (1984).

[8] P.J. Verlinden, R.M. Swanson, and R.A. Crane, Prog. Photovolt: Res. Appl **2**, 143 (1994).

[9] M. Ernst, A. Fell, E. Franklin, and K. Weber, IEEE J. Photovoltaics **7** (2), 471 (2017).

[10] A. Fell, "A Free and Fast Three-Dimensional/Two-Dimensional Solar Cell Simulator Featuring Conductive Boundary and Quasi-Neutrality Approximations," IEEE T. Electron. Dev. **60** (2), 733–738 (2013).

[11] T. Kho et al., "ONO for Silicon Solar Cell", 55-C-mon, Silicon PV 2017.

PILOT PRODUCTION OF HIGH EFFICIENT METAL CATALYZED TEXTURED DIAMOND WIRE SAWN MC-SI SOLAR CELLS COMBINED WITH NICKEL-COPPER PLATED FRONT CONTACT PROCESSING

S. Fox[b], X. Jie[b], H. Wu[b], J. Burschik[a], N. Bay[a], K. Crouse[c], H. Kühnlein[a], B. Lee[c], A. Letize[c], M. Passig[a], D. Pysch[a] M. Sieber[a], K. Vosteen[a]

a) RENA Technologies GmbH, Freiburg Solar Technology Center, Hans-Bunte Str. 19, 79108 Freiburg, Germany
b) Jinko Solar Co.,Ltd, No 58 Yuan Xi Road, Haining, Zhejiang, PR China
c) MacDermid Enthone Inc., 245 Freight street, Waterbury, CT06702, USA
eMail damian.pysch@rena.com

ABSTRACT: This paper presents results of combining two promising future technologies: i) metal catalyzed textured (MCT) diamond-wire sawn (DWS) mc-Si wafers and ii) nickel-copper-silver (Ni/Cu/Ag) plated front contact processing. Results of optimizations of laser patterning, annealing and module string soldering are presented. Several 60-cell module with a power output of around 270 W were made, followed by subsequent module reliability testing. All module tests have been passed within the existing boundaries. Pseudo fill-factor (pFF) and open circuit voltage V_{oc} have been identified as the parameters with the highest potential for further improvements. First successful results of plating-emitter development on a MCT-surface are presented.
Keywords: Cu plating; Light-induced plating; Diamond-wire saw, multicrystalline silicon, metal catalyzed texture

1 INTRODUCTION

In the past decade, the PV industry undertook considerable efforts to make Si solar cells cost effective relative to conventional carbon emitting energy generation technologies (coal, gas etc.). This goal has been achieved for several regions in the world [1]. In the past years, due to production overcapacity, the continued and increased pressure to further reduce production costs is driving all promising innovations (both disruptive and evolutionary). One important cost-cutting innovation step in the past years was the replacement of slurry-wire-based wafer sawing technology with diamond-wire-based wafer sawing technology for Cz-wafer material.

Cast multi-crystalline (mc) wafers now try to follow this rout. Amongst other improvements, this is one main step to keep the cost advantage of mc-Si wafers compared to Cz-Si wafers. On one hand, many solar cell producers are switching their production from mc- to Cz-wafers based silicon solar cells, because wafer costs per watt currently favour Cz Si wafers [2]. On the other hand, different approaches are being investigated to cost-effectively texture diamond wire sawn (DWS) mc-Si wafers. The common acidic HNO3/HF-based texture process, optimized for slurry-wire sawn mc-Si wafers, cannot be applied to DWS mc-Si wafers without losses in reflexion and cell-efficiency. One very promising approach is the so-called metal catalyzed texturing (MCT) technology. Its process details, electrical performance and feasibility for combination with Ni/Cu/Ag-plating are discussed in detail in the following sections.

A second very promising approach to significantly reduce production costs is the substitution of cost-intensive screen-printed Ag pastes with Ni/Cu/Ag plating [3, 4]. Furthermore, copper plating technology comes along with a potential to increase the efficiency by 0.1 to $0.3\%_{abs}$ compared to the current screen-printing technology. Due to silver's roles as both a semi-precious investment metal and as an industrial commodity, silver paste prices are highly sensitive to volatile speculative trends in metals exchange markets, which in turn has the undesirable effect of shrinking the already narrow margins of most solar cell manufactures. Replacing the Ag paste with a low-cost price-stable plated Ni/Cu/Ag stack would alleviate this cost pressure. Another important consideration is that due to the limited known reserves of silver, it will certainly become a bottleneck in the scale-up of the PV technology as it seeks to play a more significant role in the global power generation market [5]. Last, but not least, a copper-plated silicon solar cell would fit the broader commercial market's demand of abandoning lead, a toxic component of silver metallization pastes, from the device.

Thus, the two most cost-intensive factors in Si-mc-wafer-based solar cell production (wafer and metallization) are addressed by the work presented.

2 PROCESS AND EQUIPMENT

2.1. Solar cell processing

The following process flow was applied to all solar cells subsequently described herein: After obtaining wafers by diamond-wire sawing of the cast mc-Si ingot, a MCT surface texturing process as depicted in Fig. 2 was applied. The MCT process is discussed in more detail in Section 2.2. Next, a shallow 80-Ω/sq phosphorus-doped emitter was formed via thermal diffusion of POCl3 (see Fig. 5, SOP-emitter profile). After this, PSG is etched from the cell surfaces and the junction is isolated by a single-side silicon etch. Then, a SiNx anti-reflection coating was deposited by tube PECVD on the front side. On the rear side, Al paste (for Al-BSF formation) and silver pads (for cell inter-connection) were printed and fired for all Ni/Cu/Ag contacted solar cells. For all reference cells, a Ag paste was screen-printed on the front side before firing. This comparison of metallization process flows is illustrated in Fig. 1.

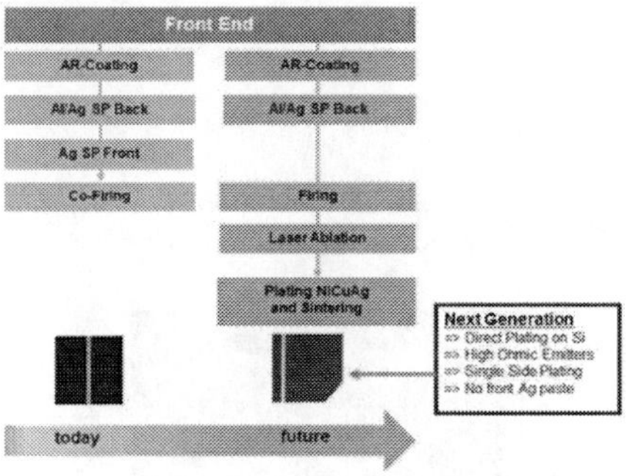

Figure. 1. Process flow overview for conventional solar cells with screen-printed silver front-side contact metallization (left) and solar cells with laser-patterned, plated Ni/Cu/Ag contact metallization, and sintering (right).

2.2. Metal-catalyzed texturing of diamond-wire sawn multi-crystalline Si-wafers

The process flow for metal-catalyzed texturing of diamond-wire sawn multi-crystalline Si wafers is shown in Fig. 2a.

a)

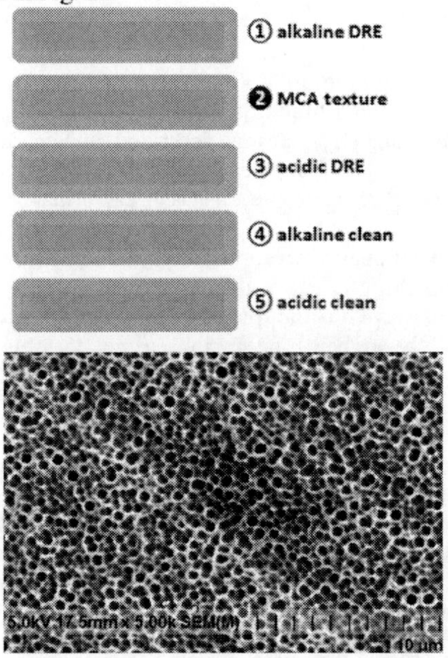

b)

Fig. 2. (a) Overview of the MCT process flow. (b) Representative image of a MCT-textured DWS mc-Si wafer surface.

First, the DWS mc-Si wafers are treated with an alkaline damage-removal etch (DRE), followed by the actual MCT process, which can be sub-divided into a metal deposition step and an acidic etching step [6, 7]. Next, an acidic damage removal step is applied, followed by an alkaline and acidic clean. In-between all process steps, a rinse needs to be conducted. Application of this process chain to DWS mc-Si wafers results in surface morphologies as shown in Fig. 2b.

2.3. Ni/Cu/Ag plating process

In order to deposit Ni/Cu/Ag-plated contacts, the grid pattern needs to be formed by selective UV ps laser ablation (InnoLas Solutions GmbH) of the front PECVD SiN_x:H layer. The subsequent in-line plating relies on single-sided wet chemical processing in order to protect the rear contacts of the solar cells (InCellPlate, RENA Technologies GmbH). The in-line plating process starts with a chemical pre-treatment of the laser-opened areas, to assure plating will start on the exposed emitter surface where intended. Nickel (Ni) is plated by light-induced plating (LIP). A Cu layer is then LIP-plated on top of Ni, to give the contacts sufficient conductivity (see Fig. 3b and 3c). At the end of the plating sequence, single-sided immersion plating of Ag caps the Cu layer. This very thin (100 nm or less) Ag layer facilitates soldering of interconnection ribbons to the metal stack (see Fig. 3a). Intermediate rinsing steps are applied in-between all process baths.

In a last processing step, the complete metal stack is thermally annealed in an inert atmosphere. During this step, the adhesion of the metal stack to Si is improved and the contact resistance and line resistance are reduced via appropriate nickel silicide formation at the nickel-silicon interface [8].

The laser step was performed on an InnoLas Solutions GmbH tool and all plating steps were performed with RENA-Technology GmbH manufactured in-line equipment in conjunction with MacDermid-Enthone-produced plating chemistry.

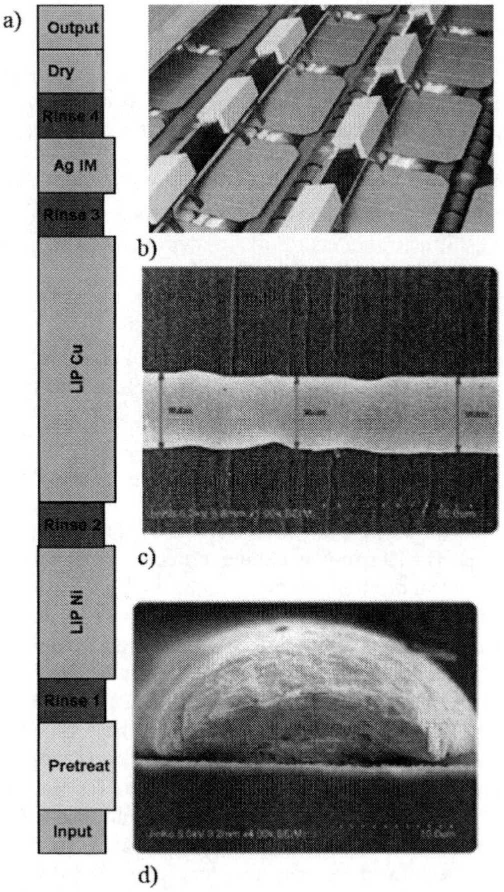

Fig. 3. (a) Sketch of Ni/Cu/Ag plating process flow; (b) picture of light-induced plating LIP Cu process bath. Representative SEM image of plated contact finger on a MCT DWS-Si-wafer: c) Top view of a 22 µm wide finger is shown, d) Cross-section of the same contact finger with a finger height around 10 µm is shown.

3 EXPERIMENTAL WORK

In order to achieve an efficiency increase with plated cells relative to Ag-paste screen printed cells, some process steps have to be individually optimized [9].

3.1. Laser processing and influences on solar cell performance

First, the laser process was optimized to the MCT wafer surface. The goal is to realize a homogenous opening in the SiN_x layer while simultaneously minimizing the creation of a laser-induced damage region (specifically, silicon amorphization) within the emitter's crystal structure.

In Fig. 4a, mean efficiency values (5 cells per group) before and after annealing are shown over varied laser pulse energies. A wide process window, between 0.6 to 0.8 µJ/pulse, results in high efficiencies of above 19%. No significant difference in the investigated dependence can be distinguished before and after the anneal process, aside from an increase in efficiency by approximately 0.2%. In Fig. 4b, a representative surface image after laser patterning is shown for the optimum laser parameter of 0.8 µJ/pulse.

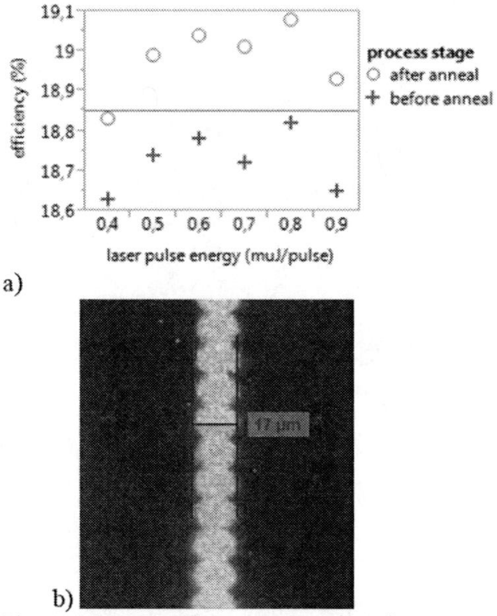

Fig. 4. (a) Mean efficiency values (5 cells per group) before and after annealing over normalized laser power is shown. (b) Representative surface image after laser patterning.

3.2. I-V data comparison of MCT mc-Si solar cells contacted by silver paste screen-printing vs. Ni/Cu/Ag-plating

After the initial process development (process 'base-lining') cycle, as described in Sections 3.1, two groups (10 cells each) of cells are fabricated in order to compare the electrical performance of the plated Ni-Cu-Ag contact technology to conventional screen-printed Ag paste contact technology. The I-V data is shown in Table 1. The efficiency for the plated front contact group is slightly higher. The main reasons for this are a superior short circuit current (I_{sc}) and fill factor FF. The higher FF is mainly explained by a significantly reduced series resistance (R_s) value. Based on the observed R_s difference, an even higher advantage in FF should be observed [11]. Based on this observed difference, it is reasoned that the plated group has a reduced pseudo fill factor. This finding leads to the conclusion that the diode quality of the plated cells is most likely reduced either by a laser-induced emitter damage, a too-shallow junction, Ni-silicide spiking after anneal, or a combination of the above-mentioned influences. Furthermore a 5 mV lower V_{oc} value is observed for the plated contact group.

Table 1. Overview of I-V data (efficiency, V_{oc}, I_{sc}, FF and R_s) for two different front contact metallization: i) Ni/Cu/Ag plating (after annealing), and ii) Ag-paste screen printing. The group size is 10 cells per group.

Group	mean η (%)	V_{oc} (mV)	I_{sc} (A)	FF (%)	R_s (mΩ)
Ni/Cu/Ag plated					
MCT cells	19.04	637	9.11	80.37	1.3
Ag screen printed					
MCT cells	19.00	642	8.99	80.12	2.1

Both, the reduced pFF and V_{oc} value are tackled with an emitter profile adaptation approach [12]. A first approach of a plating emitter created on a MCT-wafer surface is shown in Fig.1. A standard operation procedure (SOP) emitter profile (optimized for Ag-paste technology) is shown as a reference. It is believed that a deeper junction can increase the junction quality by reducing Ni-spiking and/or sensitivity to laser damage.

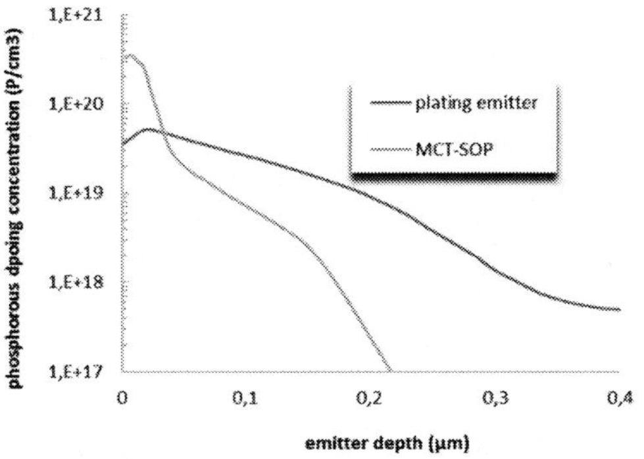

Fig. 5: Phosphorous profile of a typical industrial standard operation process (SOP) emitter (dark green) and first approach for an emitter adapted for plating contacts (light green).

Two groups of MCT mc-Al-BSF solar cells have been processed with the two emitter profiles depicted in Fig. 5. First, the MCT-SOP emitter represents a standard industrial profile. The solar cell group which received the MCT-SOP emitter diffusion have been contacted by optimized Ag-paste screen print technology. Second, the plating-emitter cells have been contacted by the plating technology. The IV-data results are summarized in Tab.2. On the one hand, the gap in V_{oc} could be closed (please compare Tab. 1&2). On the other hand, some difficulties in solar cell processing resulted in inferior V_{oc}, FF, Rs, and η values, compared to the data presented in Tab.1.

Future work will aim for further improvement of the plating emitter in order to increase the pFF-value and close the V_{oc}-gap simultaneously.

Table. 2: IV data measured for MCT mc-Al-BSF solar cells processed with special emitter profiles (show in Fig 5) for plating and Ag-screen print contacts.

group	no.	mean η (%)	V_{oc} (mV)	I_{sc} (A)	FF (%)	R_s (mΩ)
Ni/Cu plated standard emitter	26	18.75 +/- 0.1	637 +/-1	9.11 +/- 0.03	79.35 +/-0.2	2.2 +/-0.2
Ag printed, standard emitter	31	18.78 +/- 0.1	637 +/- 1	9.05 +/- 0.03	80.07 +/- 0.4	1.7 +/- 0.2

4. MODULE RESULTS

Before committing to the fabrication of a pilot production of 1st 50, and 2nd 200 full-sized 60-cell modules, the feasibility of Ni/Cu/Ag-plated MCT mc-Si cells to standard solder technology for interconnecting 12 cells in a string is investigated. In Fig. 6, the observed failure pull strength forces, after standard module-production contact tab soldering, are shown. All values were determined under a pull angle of 60°. The contact annealing temperature and time was varied from 300°C to 450°C and 3 min to 6 min, respectively. Values above 1 N are known to enable a reliable module production process. Longer annealing times perform slightly better compared to the 3 min anneal, with most values well above 2 N. Further, optimization of the soldering process will be part of future experiments.

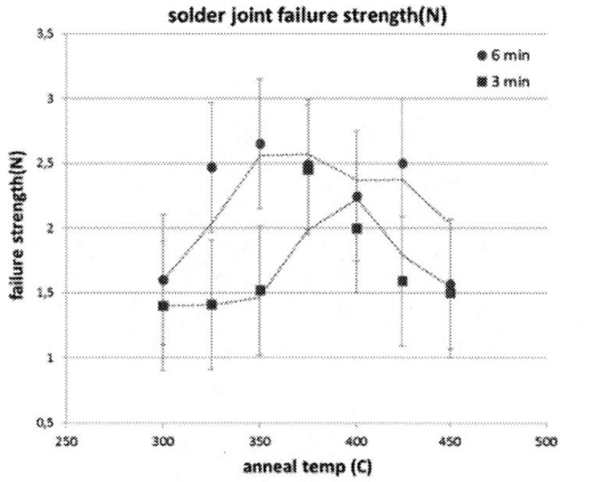

Fig. 6. Measured failure pull strength force after standard contact tab soldering for module production. All values were determined under a pull angle of 60°. The contact annealing temperature and time were varied from 300°C to 450°C and 3 min (red squares) to 6 min (blue circles), respectively.

After obtaining positive results, as discussed in Section 3 and shown in Fig. 6, more than 2000 MCT cells were fabricated and contacted by the plating technology with a mean η of 18.84% (please see I-V data distribution and further statistical information in Fig. 8a). As a reference, more than 10000 MCT cells from the same production-batch were processed and contacted by standard Ag-paste screen print with a mean η of 18.79% (see I-V data distribution and further statistical information in Fig. 8b). Thus, in the presented experiment an efficiency gain of 0.05% could be observed. Typical efficiency gain values range around 0.1% at an efficiency level above 19%.

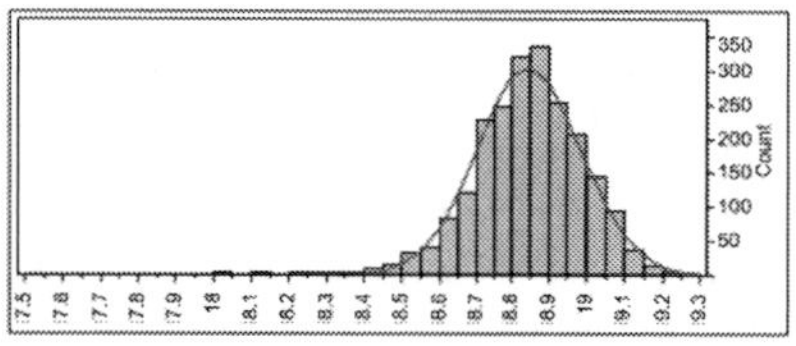

a)

Mean η	18.84%
Std Dev η	0.14%
Std Err Mean η	0.003%
Upper 95% Mean η	18.84%
Lower 95% Mean η	18.83%
No. of MCT cells contacted by plating	2102

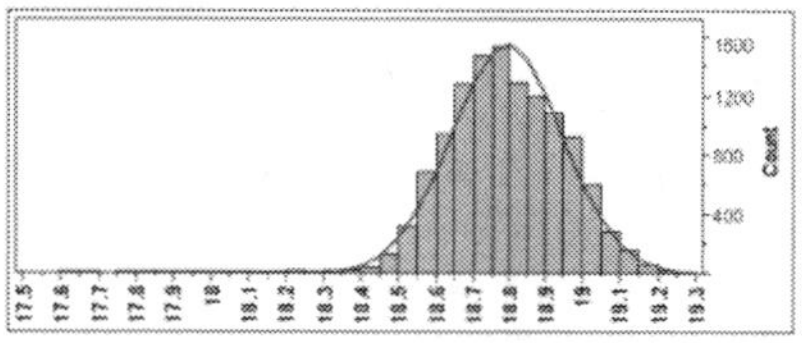

b)

Mean η	18.79%
Std Dev η	0.15%
Std Err Mean η	0.002%
Upper 95% Mean η	18.79%
Lower 95% Mean η	18.78%
No. of MCT cells contacted by screen print	10737

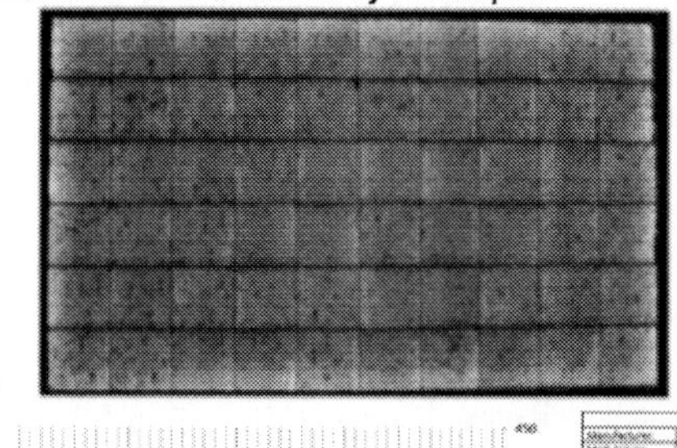

c)

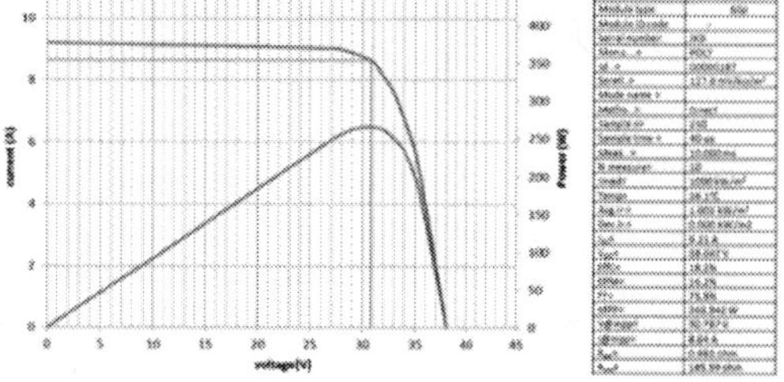

d)

Fig. 7. (a) Efficiency binning distribution of 2102 MCT mc-Si solar cells contacted by a plated Ni/Cu/Ag stack. (b) Efficiency binning distribution of 10737 MCT mc-Si solar cells contacted by Ag-screen print. (c) EL image and (d) I-V data of a representative 60-cell module populated with cells.

Several 60-cell module were subsequently manufactured and tested for long term stability thermal cycling and humidity freeze testing. All module tests have been passed within the existing boundaries. A representative module's electro-luminescence (EL) image is shown in Fig. 8c. Please note that despite no adaptation of the cell interconnection or the module 'lay-up' processes to the presence of plated bus bar pads, neither cell cracks, nor finger interruptions, nor other typical module failure modes are observed through visible inspection of the EL image. The module I-V data is show in Fig. 8d. The power output is 266 W. After completion of further optimization loops, modules in the output range of 280 to 290 W are expected.

5. Cz-PERC SOLAR CELL RESULTS

As a side project, the plating technology has been applied to standard industrial Cz-PERC solar cells. The same efficiency of 21.3% could be observed for the Ni/Cu/Ag plated as well as for the Ag-paste screen printed group. The slightly lower V_{oc}, FF and R_s values for the plated group is compensated by a superior I_{sc} value (see Tab. 3).

Table 3: IV-data of industrial production Cz-PERC solar cells contacted with Cu-plating and Ag-paste screen print.

group	no.	η (%)	max η (%)	V_{oc} (mV)	I_{sc} (A)	FF (%)	R_s (mΩ)
plated Cz PERC	49	21.3 +/- 0.1	21.5	662 +/-1	9.89 +/-0.02	79.5 +/-0.2	2.5 +/-0.1
printed Cz PERC	200	21.3 +/- 0.1	n.a.	665 +/- 1	9.81 +/-0.02	79.6 +/- 0.2	2.4 +/-0.1

6. CONCLUSION AND OUTLOOK

In this paper, we show that combining two emerging technologies - metal catalyzed textured diamond wire sawn mc-Si wafers and Ni/Cu/Ag-plated front contact processing - show very good results, even at the process 'base-lining' stage of development. Results from the initial pilot production and process optimization, including laser patterning, annealing, and soldering are presented.

Several 60-cell module with a power output around 270 W were made, along with subsequent generation of module reliability data. All module tests have been passed within the existing boundaries. Since unit process development and whole process integration are in the early stages, it´s expected that further optimizations will result in rapid improvement of the cell and module performance.

As a primary case-in-point, discussed in Section 3.2, it is noted that the plated cell's I-V response mainly suffers from a reduced pFF and V_{oc}. A very promising solution for this is moving away from an emitter profile which is mainly optimized for silver paste's restrictive sintering parameters (and thus is limiting the contacting properties) and, instead, developing an optimized emitter profile suited for highly-ohmic Ni/Cu/Ag-plated contacts. First successful results of plating-emitter development on a MCT-surface have been presented. Future work will continue to focus on further emitter profile adaptation, evaluation of selective emitter technology and continuous efficiency improvement.

9 REFERENCES

[1] Lazard. Lazard's levelized cost of energy analysis – Version 10.0. 2016. www.lazard.com/media/438038/levelized-cost-of-energy-v100.pdf.

[2] SEMI PV Group Europe 2016, ITRPV Eight Edition 2017, www.itrpv.net/reports/downloads/ (10.4.2017).

[3] Horzel J, Shengzhao Y, Bay N, Passig M, Pysch D, Kühnlein H, Nussbaumer H., Verlinden P. Industrial Si solar cells with Cu-plated contacts. IEEE J. Photovolt. 2015; 5(6):1595-1600.

[4] Shi Z, Wenham S, Ji J. Mass production of the innovative PLUTO solar cell technology. Proc. 34th IEEE Photovoltaic Specialists Conference (PVSC), 2009. DOI: 10.1109/PVSC.2009.5411566.

[5] Meng T. Terawatt Solar Photovoltaics: Roadblocks and Opportunities. Springer Verlag 2014, DOI 10.1007/978-1-4471-5643-7.

[6] Ye X, Zou S, Chen K, Li J, Huang J, Cao F, Wang X, Zhang L, Wang X, Shen M, Su X. 18.45% Efficient Multi-Crystalline Silicon Solar Cells with Novel Nanoscale Pseudo-Pyramid Texture. Adv. Funct. Mater. 2014; 6708–6716.

[7] Wang X, Zou S, Xing G. 19.31% efficient multicrystalline silicon solar cells using MCCE black silicon technology. Photovoltaics International 35th Edition. 02/2017.

[8] Russel R, Tous L, Philipsen H, Horzel J, Cornagliotti E, Ngamo M, Choulat P, Labie R, Bekcers J, Bertens J, Fujii M, John J, Poortmans J, Mertens R. A simple copper metallization process for high cell efficiencies and reliable modules. Proc. 27th EUPVSEC, Germany, 2012.

[9] Bay N. Brand A. Büchler A. Burschik J. Kluska S. Kuehnlein HH. Passig M. Pysch D. Sieber M. Benefits of different process routes for industrial direct front side plating. 7th International Conference on Silicon Photovoltaics. SiliconPV 2017.

[10] Murakami T, Froment B, Ouaknine M, Yoo W. Nickel silicide formation using a stacked hotplate based low temperature annealing system. 203rd Electrochemical Society Meeting. Paris. 2003.

[11] Pysch D, Mette A, Glunz SW. A review and comparison of different methods to determine the series resistance of solar cells. Sol. Energy Mater. Sol. Cells 2007; 91:1698–1706.

[12] D. Pysch, J. Burschik, N. Bay, K. Crouse, H. Kühnlein, B. Lee, A. Letize, M. Passig, M. Sieber, K. Vosteen. Pilot production of highly efficient metal catalyzed textured diamond wire sawn mc-Si solar cells combined with nickel-copper plated front contact processing. Journal: Energy Procedia. EGYPRO32295. PII: S1876-6102(17)34498-3

ACKNOWLEDGMENT

The authors kindly thank all industrial partners for their support.

DEVELOPMENT AND OPTIMIZATION OF A NOVEL INLINE BLACK SILICON TEXTURING PROCESS FOR INCREASED SOLAR CELL PERFORMANCE

W. Jooss, I. Melnyk, J. Jung-König, A. Teppe, T. Werling, O. Voigt, F. Binaie, P. Fath
RCT Solutions GmbH, Line-Eid-Str. 1, 78467 Konstanz, Germany
Tel: (+49) 7531-584 70 17, Email: wolfgang.jooss@rct-solutions.com

B. Hu, Q. Zhang, I. Wang, P. Tian
RCT Automation Equipment (Suzhou) Co., Ltd., No.17 Shiyang Road, New District, Suzhou, Jiangsu, China 215152

JL Huang, JW Yu, JL Qiu, S Zhou, QS Huang, XF Gou
CECEP Solar Energy Technology (Zhenjiang) Co., Ltd., No. 9, Beishan Road, New Area, Zhenjiang, Jiangsu China 212132

ABSTRACT: In this work we introduce a novel texturing approach for an inline equipment based on Metal Catalyst Chemical Etching (MCCE). The texturing process can be varied in a wide parameter range resulting in average reflectivity in the range of 12% up to 22%. An inline tool was developed for the processing of an as-cut diamond wire sawn wafer until final cleaning prior to diffusion with processing time less than five minutes. Al-BSF solar cells were processed on DWS multicrystalline wafers in an industrial production line with average efficiencies of 18.8% and peak efficiencies of 19.1%, demonstrating an efficiency increase of >0.3%abs. as compared to MWSS wafer using acidic isotexturing. The texturing tool in the production line showed consisting etching rates and high throughput, further demonstrating the industrial applicability of this process. A loss analysis revealed that further performance increases can be accomplished by stronger textures in conjunction with emitter optimization. Due to a low Ag consumption in the MCCE process in combination with inline processes, a very competitive CoO could be shown of about 1.5 USDct/wafer, which is only slightly higher than acidic istotexturing, but substantially lower than RIE or MCCE using batch tools and additives. The process in the inline tool is also capable for asymmetric texturing of front and rear surface, thereby enabling a slim process for the manufacturing of PERC solar cells.
Keywords: MCCE, multicrystalline silicon, texturing

1 INTRODUCTION

In recent years, Diamond Wire Sawing (DWS) has been introduced in the manufacturing of crystalline silicon wafer as alternative technology to Multi-Wire Slurry Sawing (MWSS). Until now, DWS has nearly replaced MWSS only for the slicing of monocrystalline wafers. Due to significant cost reductions possible by DWS, there are currently strong efforts undertaken for slicing also multicrystalline wafers. Whereas technological challenges in the slicing process itself have been mainly solved, efforts are currently undertaken to develop a cost-effective and efficient technology for texturing the multicrystalline wafers. This is necessary, since the current standard technology for texturing mc-Si of acidic isotexturing is not as effective due to differences in the (sub)-surface morphology after the slicing process. Another trigger point for the development of a new texturing technology for mc-Si wafer is the need for further efficiency increases, since reflection losses can still be reduced.

The implementation of nano-scale textures have been discussed in recent years as an effective substituting technology, e.g. by Reactive Ion Etching (RIE) [1-2] or Metal Catalyst Chemical Etching (MCCE) [3-4]. Since with those technologies the reflectivity can be reduced well below 10%, the term "black silicon" has been introduced. RIE requires vacuum chamber processes as well as wet chemical pre- and post-treatments. Hence, this approach is currently seen as costly. Additionally, MCCE has been introduced as an alternative approach. In MCCE, metal nanoparticles are deposited. These nanoparticles serve as catalyst and texturing is promoted in the vicinity of those nanoparticles. Depending on the chemicals and process parameters, different surface structures can be accomplished including nanowires, pits or pores. For implementation of MCCE in industrial solar cell manufacturing, only batch-type tools have been considered, mainly using Ag as metal and H_2O_2 as oxidant [3].

The main target of our process development is a process which can be transferred into an inline tool. This requires an overall process time below five minutes from the as-cut to a cleaned wafer to ensure an appropriate tool length. According to our experience inline processes are best suited for texturing, as it is the case for inline acidic isotexturing of MWSS multicrystalline wafers. Stable, identical and reproducible process conditions are more easily accomplished in an inline bath, especially when gases are generated during the texturing process. Other targets of the process development were the avoidance of additives and H_2O_2 as oxidant, since we observed, that the bath stability (meaning reliable and stable concentration for a long period of time in a typical industrial production environment, including long bath lifetime) is a major concern.

3 GENERIC PROCESS FLOW OF MCCE

A generic process flow for Metal Catalyst Chemical Etching (MCCE) of a crystalline silicon wafer is given in Figure 1. It includes all relevant steps necessary to process a wafer from an as-cut state to be ready for emitter diffusions. The saw damage after wafering needs to be removed (about 3 to 5 µm). This can be performed as separate process step or during MCCE. Since MCCE is a process, in which a metal is used as a catalyst to promote etching in the vicinity of the metal, the

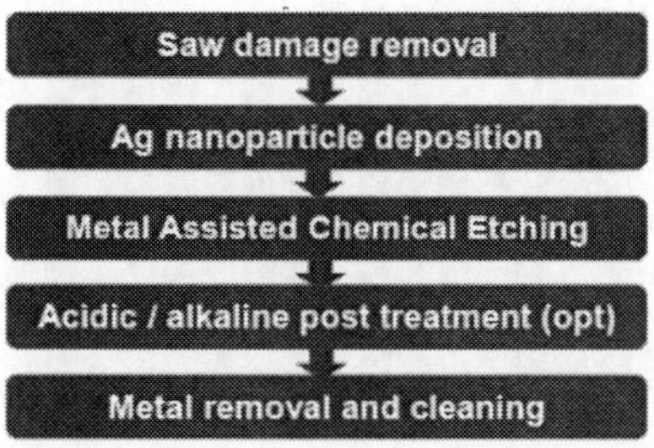

Figure 1: Generic process flow for black silicon based on Metal Assisted Chemical Etching (MCCE).

respective metal precipitates need to be deposited, e.g.Ag nanoparticles. Metal nanoparticle deposition is carried out by dipping the wafer into a wet chemical solution containing metal ions, which is in most cases a solution of $AgNO_3$. Additionally HF needs to be present to initiate deposition. After metal deposition, the texturing process starts, which require an oxidant (e.g. H_2O_2 or HNO_3) and HF. Depending on the bath composition, a porous Si layer can be formed. Also the surface morphologies can vary. As an example, a pit-like structure can be obtained, with typical diameters in the range from 50 to 200 nm and depths in a similar range. Also alternative structures like nanowires or pores can be accomplished. Those structures can lead to a low reflectivity, however the use for solar cells is not ideal (deep and steep structures, with high surface enlargement and challenges for good surface passivation, including low J_{0e}). Hence post-treatments are optionally carried out (either alkaline or acidic) to determine an optimum structure for high solar cell performance. This optimized structure is a trade-off between low reflectivity (necessary for high J_{sc}) and reduced surface enlargement (necessary for high V_{oc} and FF). In addition, the wafer surface needs to be cleaned from any metal and contaminations prior to emitter diffusion. An overview on MCCE can be found in [5].

In general, all steps of the generic process can be carried out individually, but alternatively, two or more of the steps can be combined in one treatment. As an example, Ag nanoparticle deposition and MCCE can be carried out simultaneously, if the bath composition includes $AgNO_3$, HF and oxidant (e.g. H_2O_2 or HNO_3). Also saw damage removal can be included in such a process step, making alkaline saw damage removal prior to texturing obsolete.

4 TEXTURING PROCESS DEVELOPMENT

Based on the generic process flow, we started our process development which had the target to avoid additives and H_2O_2 as oxidant. Also overall process time to fulfil all steps of the generic process needs to be below five minutes in order to be used in an inline production tool. Initial process development was carried out in a lab environment. During process development we have carried out several DoE by mainly varying the bath compositions during Ag nanoparticle deposition / MCCE as well as acidic post treatment. Figure 2 shows the reflectivity curves for four differently MCCE textured DWS wafer, obtained by different process parameters.

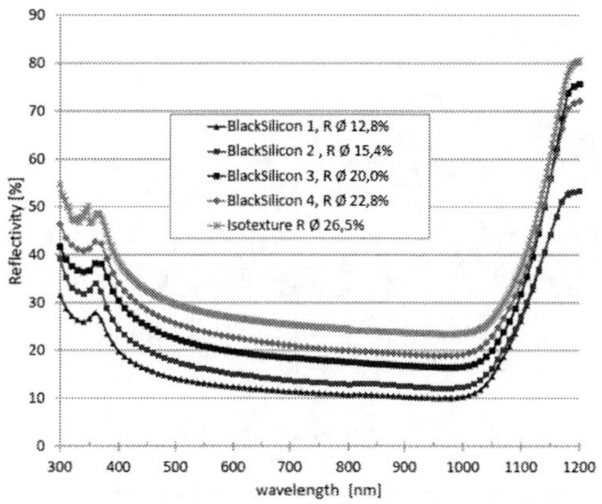

Figure 2: Measured reflectivity curves for four different MCCE textured wafer. The reflectivity of the MCCE process can be tuned in a wide range. For reference, the reflectivity of an acidic isotextured solar cell is given.

Also shown is the reflectivity curve for a typical acidic isotextured wafer. The average reflectivity, calculated as average between 400 nm and 1100 nm, can be varied for MCCE in a wide range between 12.8% and 22.8%, indicating the flexibility of MCCE to obtain different surface structures. The reflectivity is in all cases lower than the acidic isotextured reference wafer. The difference in reflectivity is a consequence of the surface morphology.

Figure 3 shows SEM pictures of different surface morphologies. The SEM picture in the top shows a sample with R_{ave}=12%. The surface is more rough, typical structure sizes are up to 300 nm, with the depth in the similar range. The surface morphology changes

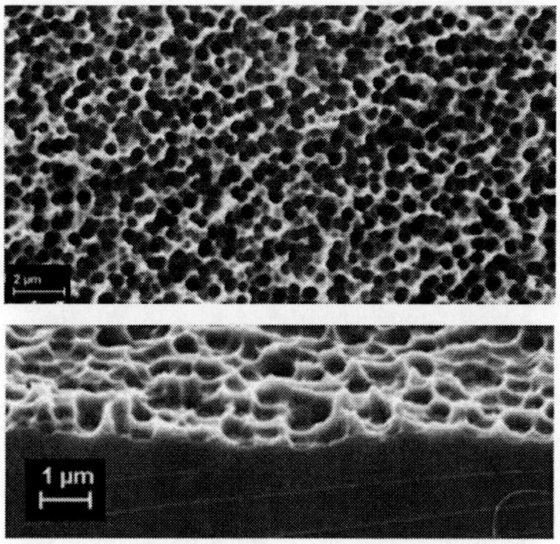

Figure 3: SEM pictures of MCCE textured wafers (top) sample with R_{ave}=12%. The surface is more rough, typical structure sizes are up to 300 nm, (bottom) sample with R_{ave}=22%. The structure sizes are increased and the morphology resembles acidic isotextured wafers.

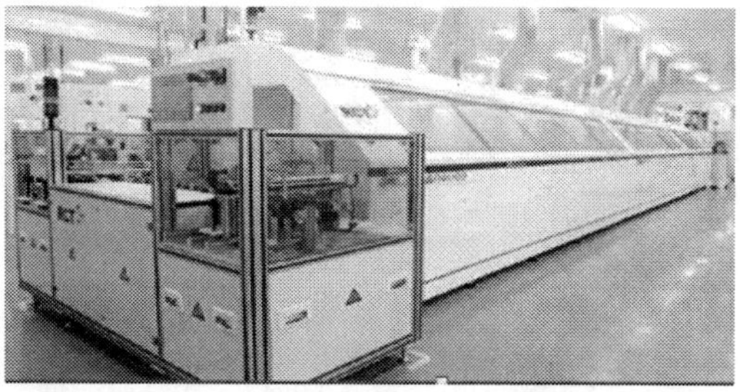

Figure 4: Inline tool for MCCE. The overall length is > 12 m (excluding automation).

significantly for wafer with a higher reflectivity of about 22%. The width of the structures increases beyond 300 nm, whereas the depth decreases. The surface structure resembles the one of acidic isotextured wafers. This structure is easier to be implemented in standard solar cells lines, since adaptations in processes (e.g. diffusions) due to the surface morphology are not required.

5 INLINE EQUIPMENT

Based on the process of section 4, an inline equipment was designed and manufactured. As a key feature of this equipment, the three processes of the generic process (1) saw damage removal (2) Ag nanoparticle deposition and (3) MCCE are carried out in one process bath. This enables the reduction of the process time from loading to unloading below five minutes. The developed process and equipment has been filed as a patent. Also the process uses HNO_3 as oxidant and is free of additives. An image of the inline tool is shown in Figure 4.

The equipment was installed in the manufacturing line of CECEP for ramp-up and implementation in production of Al-BSF solar cells.
Figure 5 shows the etch depth and etch rate as a function of the amount of processed wafer. It can be seen, that consistent and stable etching was achieved. The etch depth was controlled in the range between 3.0 and 3.5 µm/side, which could in principle be reduced below 3,0 µm/side. The lower initial etch rate and etch depth is due to process start-up in the morning and will vanish in continuous production.

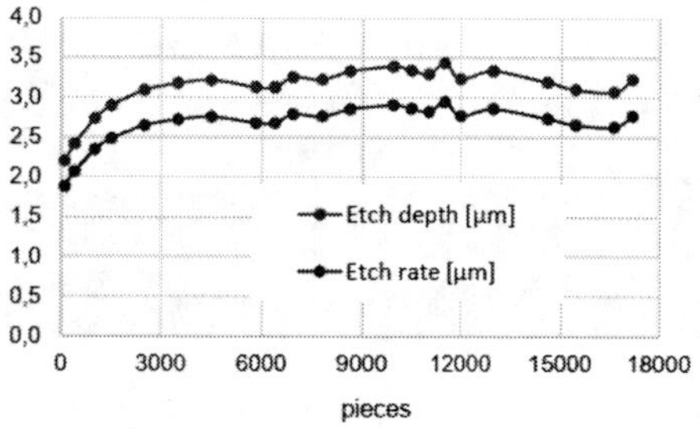

Figure 5: Etch rate and etch depth as a function of processed wafer. The graph shows consistent and stable processing. The lower etch depth for the first wafer is due to process start in the morning and will vanish in continuous production.

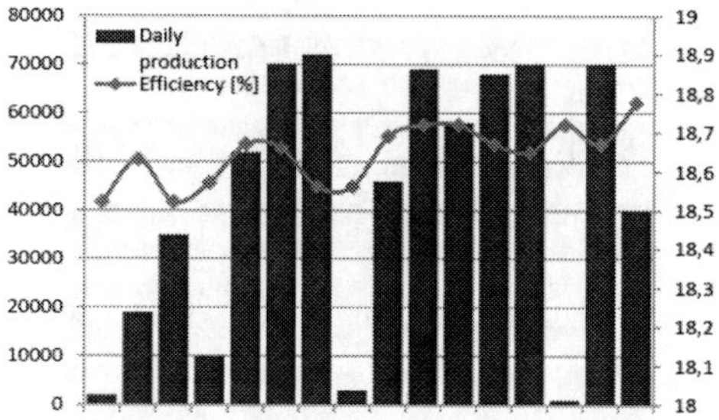

Figure 6: Daily production and average efficiency during the ramp-up phase of the inline equipment at CECEP.

6 SOLAR CELL RESULTS

In the following, details of the solar cell manufacturing on mc-Si DWS wafers at CECEP are given. Standard Al-BSF solar cells have been manufactured with no major adaptations to other process steps (e.g. diffusions). Hence, a surface structure with moderate R_{ave} has been selected, since the surface structure is close to the one of typical acidic isotextured wafer on MWSS. As reference, isotextured solar cells have been processed on MWSS wafer. The illuminated IV-results of typical batches of about 10.000 wafers are shown in Table I. The MCCE textured solar cells show a higher efficiency of 0.38%abs.. V_{oc}, J_{sc} and FF have been increased as compared to the acidic isotextured solar cells on MWSS wafers. The increase in FF of about 0.6%abs. is caused by a lower series resistance R_s and higher pFF. The low R_s indicates a good contact to the MCCE surface, whereas the higher pFF demonstrates, that the surface structure did not lead to higher non-ideal losses, which could be caused by an increased surface area or as a consequences of a rougher surface area. The improved texturing led to lower reflection losses and therefore to an increase in J_{sc} of about 1%rel. V_{oc} was improved by about 3 mV. The increase in V_{oc} is mostly from a lower J_{01}. Further analysis is necessary, whether the lower J_{01} is due to a lower contribution of the emitter J_{01e} or due to a lower contribution from the rear J_{01r} (caused by a planar rear surface and therefore by better BSF formation, see next section). The best solar cell in this batch reached an efficiency of 19.12%, with a very good V_{oc} of 641 mV.

The daily production and average efficiency towards the end of the ramp-up phase are shown in Figure 6. It can be seen, that the average efficiency at the end of the ramp-up was consistently in the range between 18.6% and 18.8%, with daily throughputs exceeding 70.000 solar cells. Until now, more than 1.5 million solar cells have been processed.

7 LOSS ANALYSIS

For the determination of losses in J_{sc}, the spectral response and reflectivity was measured for a typical solar cell. In Figure 7 the Internal Quantum Efficiency (IQE), External Quantum Efficiency (EQE) and Reflectivity (REFL) are shown. Based on those data, a loss analysis was performed to extract optical and electrical losses.

Table I: Average illuminated IV results of differently processed solar cells: acidic isotextured on MWSS wafer and MCCE textured (i-BlackTex) on DWS wafer. Also best solar cells parameters on MCCE textured wafer on DWS are given.

Texture	Wafer	V_{oc} [mV]	J_{sc} [mA/cm^2]	FF [%]	η [%]	R_s [mOhm]	pFF [%]	
isotexture	slurry	634	36.4	79.8	18.42	2.0	82.1	average
i-BlackTex	DWS	637	36.7	80.4	18.80	1.6	82.3	average
i-BlackTex	DWS	641	37.0	80.7	19.12	1.7	82.7	best

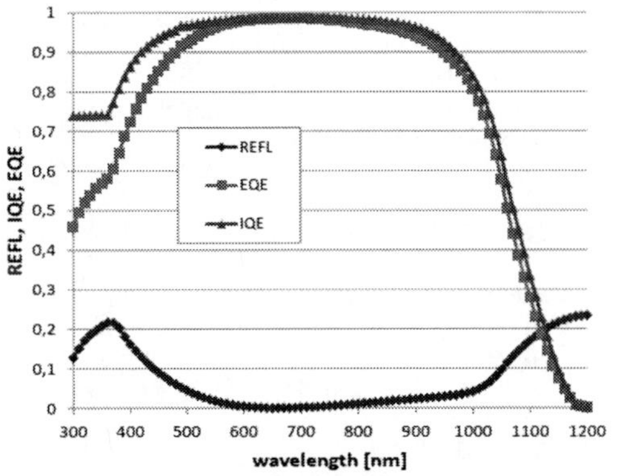

Figure 7: Internal Quantum Efficiency (IQE), External Quantum Efficiency and Reflectivity (REFL) of a typical MCCE textured solar cell with Al-BSF.

Details of the contribution to the losses are shown in pie charts in Figure 8, separately for optical and electrical losses. The major optical loss is due to the shadowing of the contact grid (finger and busbar). Losses due to front surface reflectivity (primary reflectance) amount to 1.21 mA/cm^2. Based on our experience, a reduction below 1 mA/cm^2 is feasible by using stronger textures. However, stronger texture will require optimization of the front emitter to maintain low emitter losses. The major electrical loss is due to emitter recombination, amounting to 0.73 mA/cm^2.

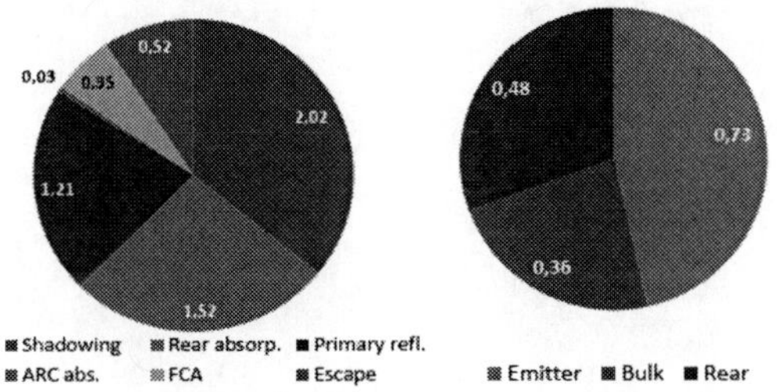

Figure 8: Losses in J_{sc} (in mA/cm^2) as determined from the quantum efficiencies of Figure 7. (left) optical losses (right) electrical losses.

8 COST-OF-OWNERSHIP

Besides performance aspects, manufacturing costs are another important criterion for industrial applicability of a new manufacturing technology. Figure 7 compares the Cost-of-Ownership (CoO) of the new inline approach applying MCCE with the industry standard of acidic isotexturing and an alternative approach of MCCE using batch type processes with additive. In our cost model, the CoO of acidic isotexturing is about 1.2 USDct/wafer. The MCCE on RCT's inline tool leads to slightly higher costs of about 1.5 USDct/wafer. This is caused by a higher equipment cost as well as slightly higher consumable costs for Ag and cleaning. This moderate increase in CoO results from a very low Ag consumption, which is below 0.5 mg/wafer, leading to a cost contribution of less than 10% to the CoO. Ag recycling, which is foreseen to be implemented in the near future, will lead to a further reduction of Ag consumption and therefore CoO. Nonetheless, the costs are substantially lower than batch-type MCCE processes, since the equipment cost and also consumable costs for additives are higher. Hence, MCCE is also attractive from a cost perspective, when compared to alternative approaches like RIE and batch-type MCCE.

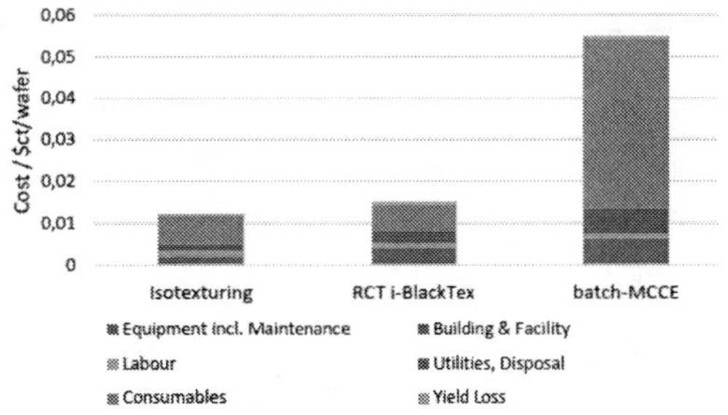

Figure 9: Cost of Ownership analysis of three different texturing processes. (left) acidic isotexturing as applied as standard for texturing of MWSS wafer (middle) MCCE in RCT's inline tool (left) typical MCCE process used in batch-type equipment.

9 ASYMMETRIC TEXTURING

In future, we plan to apply MCCE for the manufacturing of multicrystalline PERC solar cells. Besides an effectively textured front surface, PERC solar cells require, in an ideal case, a planar rear surface. Therefore a different surface treatment on both surfaces is advisable. In general, different measures are currently applied to accomplish such surface structures of textured front and planar rear including masks [Centaurus] or polishing. However, additional process steps or costly acidic polishing are the drawbacks of such approaches. Hence, it is beneficial to obtain such structures without additional measures, e.g. by wet chemical treatments. Whereas in batch type processes, different treatments on front and back side cannot be accomplished due to the specific nature of the process, the situation is different in inline processes. In general, several measures can be undertaken to obtain different conditions on front and rear side of the wafer, e.g. process temperatures or concentrations of the bath components. This can be accomplished by the inline hardware itself or by the process conditions. In one example, if the wafer is not deeply dipped into the bath, the temperature can be different due to the exothermic nature of the reaction. A higher temperature will be obtained on the upper part as compared to the lower part, since there will be only limited exchange of media preventing the formation of identical conditions on both wafer sides. This limited exchange of media can also lead to different concentrations of the bath components. In addition, gas bubbles, which will be generated during the reaction, can leave the wafer surfaces easily on the upper part, whereas they remain for a longer period on the lower part of the wafer in the inline tool. As a consequence, both surfaces are treated differently. By taking benefit of those facts, we have accomplished an average reflectivity of 20% on the front surface (being the lower surface in the inline tool) and 30% on the rear surface, clearly demonstrating the different etching behavior. SEM pictures of the front and rear surface are shown in Figure 10. Whereas the front surface shows the typical structure of MCCE, the rear surface is almost planar, with only some shallow pits visible. For process and hardware of asymmetric texturing, a patent has been filed. A further planarization of the rear surface can be accomplished during wet chemical edge isolation, possibly leading to an almost planar rear surface. This surface can then be well passivated by a dielectric stack for PERC solar cells.

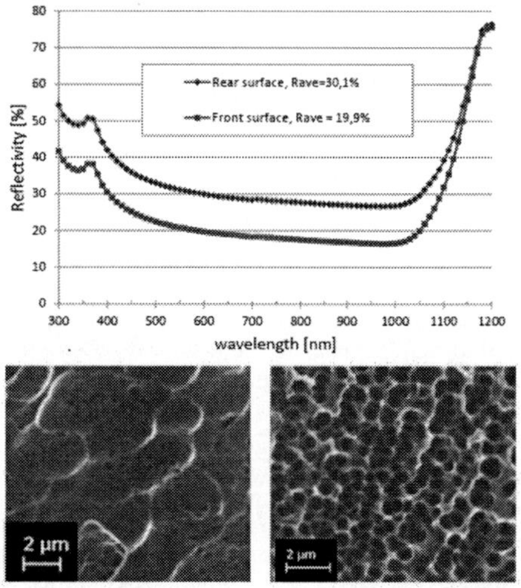

Figure 10: SEM pictures of (left) rear surface and (right) front surface of the same wafer as textured in an inline process. Different process conditions were obtained on upper side (rear) and lower side (front) by hardware and process optimizations. The reflectivity curves are shown in the top.

10 SUMMARY

In this work we have introduced a novel inline approach for MCCE, which has been successfully applied to DWS multicrystalline wafer. The developed MCCE process uses one single step for saw damage etching, Ag nanoparticle deposition and metal catalyst chemical etching. The texturing process can be tuned in a wide range of reflectivity between 12% and 23%. Based on this approach, an inline equipment has been developed and implemented at CECEP for the manufacturing of Al-BSF solar cells. Average efficiencies of 18.8% have been demonstrated, which show an efficiency increase of 0.38%abs. compared to acidic isotextured reference cells on MWSS wafer. Further features of the process and tool include a very competitive CoO and the possibility of asymmetric texturing, meaning different reflectivities on front and rear surface. The latter will enable a lean process for the manufacturing of PERC solar cells. A loss analysis revealed that average efficiencies beyond 19% are feasible in the near future by optimizing the texture in conjunction with emitter diffusion.

11 REFERENCES

[1] Y. Inomata, K. Fukui, K. Shirasawa, "Surface texturing of large area multicrystalline silicon solar cells using reactive ion etching method", Solar Energy Materials and Solar Cells, 48, (1997), 237-242

[2] J. Yoo, G. Yu, J. Yi, "Large-area multicrystalline silicon solar cell fabrication using reactive ion etching (RIE)", Solar Energy Materials and Solar Cells, 95, (2011), 2-6

[3] A. Kumaai, "Texturization using metal catalyst wet chemical etching for multicrystalline diamond wire sawn wafer", Solar Energy Materials & Solar Cells 133 (2015) 216–222

[4] F. Cao, K. Chen, J. Zhang, X. Ye, J. Li, S. Zou, X. Su "Next-generation multi-crystalline silicon solar cells: Diamond-wire sawing, nano-texture and high efficiency", Solar Energy Materials and Solar Cells 141, (2015), 132-138

[5] Z. Huang ,N. Geyer , P. Werner, J. de Boor, U. Gösele, "Metal-Assisted Chemical Etching of Silicon: A Review", Adv. Mater. 23 (2011), 285–308

HIGH THROUGHPUT PRINTING FOR HIGHLY EFFICIENT COST-EFFECTIVE SI SOLAR CELLS

F. Clement[a*], A. Lorenz[a], M. Pospischil[a], D. Biro[a], H. Brocker[b], D. Bangel[b], R. Greutmann[b], M. Lehner[b], T. Ott[c], F. Hage[c], K. Oppelt[c], T. Honold[c], L. Wende[c], M. Ringel[c], A. Senne[d], J. Rohde[e] and R. Preu[a]

a) Fraunhofer Institute for Solar Energy Systems ISE, Heidenhofstraße 2, 79110 Freiburg, Germany
b) Gallus Ferd. Rüesch AG, Harzbüchelstrasse 34, 9016 St. Gallen, Switzerland
c) ASYS Automatisierungssysteme GmbH, Benzstr. 10, 89160 Dornstadt, Germany
d) ContiTech Elastomer-Beschichtungen GmbH, Breslauer Str. 14, 7154 Northeim, Germany
e) Zecher GmbH, Görlitzer Str.2, 33098 Paderborn, Germany

*Corresponding author: Florian Clement | Phone: +49 (0)761 4588 5050 | e-mail: florian.clement@ise.fraunhofer.de

ABSTRACT: One of the main challenges within Si solar cell production is the increase of throughput rates. With throughput rates in the range of 2400 Wafers / hour, metallization lines based on flatbed screen printing are typically the bottleneck of a modern solar cell production. Within this work we present innovative printing technologies allowing for higher throughput rates, e.g. rotational printing technologies as flexographic printing and rotational screen printing and as well multi-nozzle dispensing. In principle, flatbed screen printing processes can be completely replaced by using e.g. rotational screen printing for rear side printing and e.g. flexographic printing or dispensing for front side printing.. Hence, the realization of higher throughput rates and thus lower production costs is addressed. Moreover, new printing technologies allow for Ag paste savings up to 20% and efficiency increases of 1-2% relative by optimizing the finger geometry.
Keywords: Metallization, Printing, Production Technology, Si solar cell

1 MOTIVATION

Flatbed screen printing allows for throughput rates in the range of 2400 wafers / hour. Du to process limitations as for example a restricted squeegee speed and a non-continuous process flow, a further increase is limited. So, there is a big need on the market for alternative approaches allowing for significantly higher throughput rates especially for metallization processes. Innovative printing technologies as rotational printing [1-4] or mulit-nozzle dispensing [5-7] have the potential to increase throughput rates. Hence, this work focusses on the evaluation of innovative printing technologies in order to allow high printing speeds.

Rotational printing technology allows for high printing speeds, an inline process flow, rear and front side metallization processes and a significant increase of throughput rates. Very crucial points on the roadmap to future solar cell production. Furthermore, multi-nozzle dispensing as front side metallization approach is also a promising printing technology allowing high printing speeds and in principle an inline process flow. Moreover, dispensing prevents "meshmarks" and thus increases the contact finger homogeneity.

2 EXPERIMENTAL APPROACH

In this study, latest improvements concerning rotational printing and dispensing are presented. First, rotational screen printing as technology for rear side metallization is discussed. Second, rotational flexographic printing and dispensing are evaluated as printing technology for fine line contacts with contact widths below 30µm on the front side of the solar cell. Third, a future concept for an innovative printing machine allowing a continuous process flow during printing, high throughputs and cost savings is discussed.

3 RESULTS & DISCUSSION

Within this work innovative printing technologies are presented aiming at the realization of much higher throughput rates in comparison to flatbed screen printing. A new machine concept allowing a continuous process flow during printing is discussed. Rotational printing units, e.g. for flexographic or rotational screen printing can be integrated. With this machine concept we see a huge potential to realize the rear and front side metallization of Si solar cells based on high throughput printing technologies leading to significant cost savings.

3.1 Rotary screen printing

Using rotary screen printing (RSP) the realization of a full area rear side Al metallization with high throughput rates is demonstrated on Si solar cells for the first time [4]. With an experimental setup using a roll-to-roll printing machine it was possible to process Si solar cells (see Fig. 1).

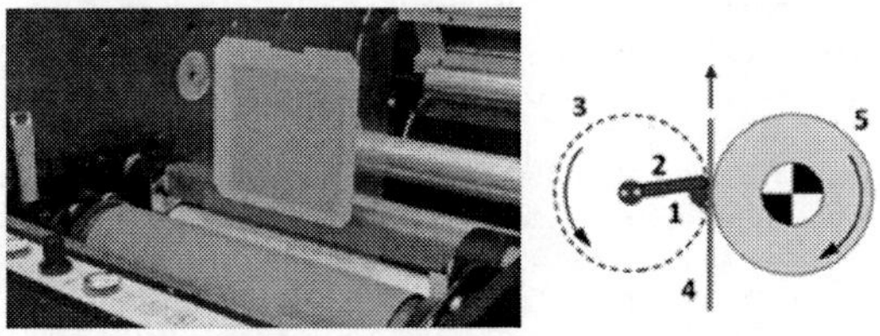

Figure 1: Left: Rotary screen printing unit of the Gallus EM 280 printing machine which is used for the experiment. The silicon solar cell is fixed with tapes on web-based material. Right: Schematic of the RSP setup. Paste (1) is pressed by a fixed squeegee (2) through the openings of the rotating cylinder screen (3). The substrate (4) is guided by an impression cylinder (5).

A first experiment on Al-BSF Cz-Si solar cells shows that the rear side metallization of Al BSF solar cells can be realized by RSP technology without any performance losses. The best group of solar cells with RSP rear side metallization even achieved a slightly better conversion efficiency of 19.4% compared to the flatbed screen printing (FSP) reference group with 19.3% (Fig. 2). A SEM investigation of the best RSP group and the FSP group showed a comparable Al layer thickness of 26μm and depth of the Al BSF of 4-6μm to the FSP reference group (see Fig. 3).

We show in summary that RSP technology is well-suited for rear side metallization of Si solar cells.

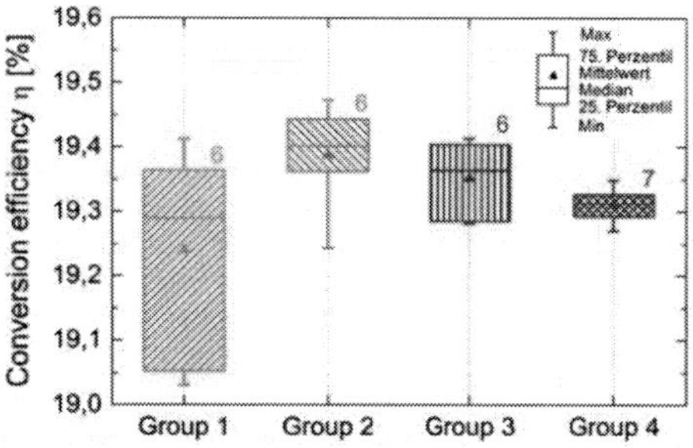

Figure 2: Conversion efficiency of Al-BSF Cz-Si solar cells with RSP rear side metallization (group 1 to 3 represent 3 different rotary screens) and FSP rear side metallization (group 4)

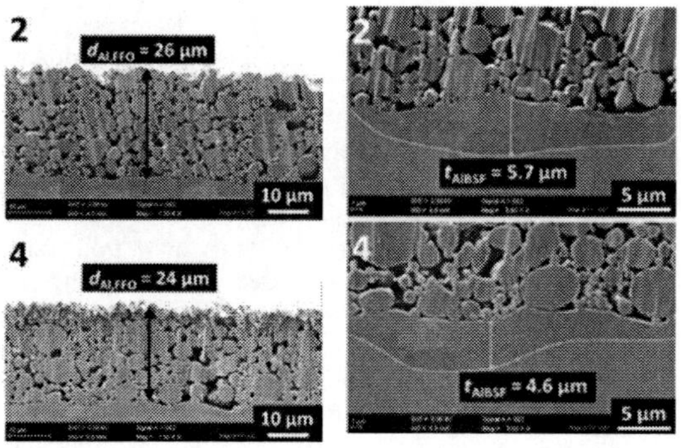

Figure 3: SEM cross sectional images of the rear side metallization after firing realized with a flatbed screen (group 4) and a rotary screen (group 2). The thickness of the printed Al layer (left) and the Al-BSF (right) is comparable.

3.2 Flexographic printing

With rotational flexographic printing (see Fig. 4) fine line printing of contact fingers can be realized. This is shown within an experiment ([3]) using an anilox roller with three different band sections (dip volume and line screening is varied) and laser engraved flexible printing plate with test structures (line width is varied). The experimental setup is shown in Fig. 5. Using an anilox roller with a large ink transfer capacity (band 2) and a highly viscous, in-house developed Ag ink led to contact fingers with a mean width below 30 μm (see Fig. 6) and very low Ag consumptions of ~20mg (calculation with GridMaster). However, realizing such narrow contact fingers with a sufficiently low lateral finger resistance using flexographic technology is still challenging. With mean lateral finger resistances (see Fig. 7) down to ~6 Ω/cm (flatbed screen printing: ~0.4 Ω/cm), flexographic printing is currently well applicable for the front side metallization of busbarless solar cells with multi busbar/wire interconnection technology (i.e. MeyerBurger's SmartWire concept). In principle, depending on printing parameters and paste rheology, the finger height and therefore the lateral finger resistance can be adjusted for different module integration technologies.

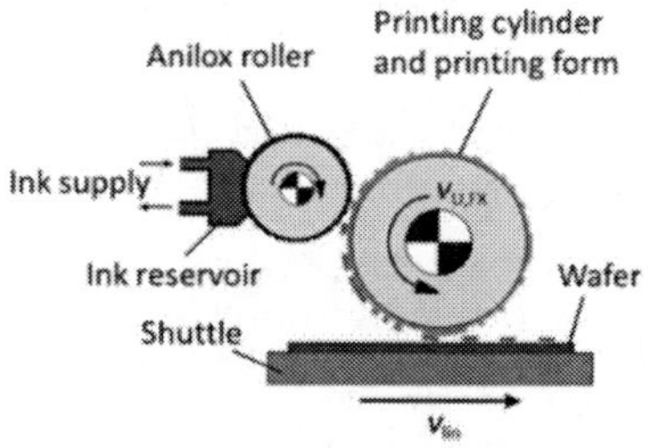

Figure 4: Schematic of a flexographic printing unit for the metallization of Si wafers: The ink is transferred from the annilox roller onto elevated areas of the printing form on the printing cylinder and from there directly onto the wafer.

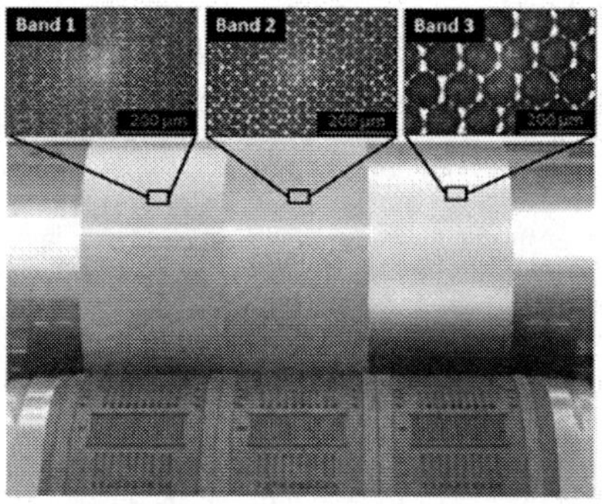

Figure 5: Experimental setup for a flexographic printing test varying different printing parameters. The microscopic images show the annilox roller with three differently engraved band sections. A ContiTech Laserline CSC flexographic printing form is used for the experiment. The test layout contains finger elements with nominal widths between 5 and 50 μm

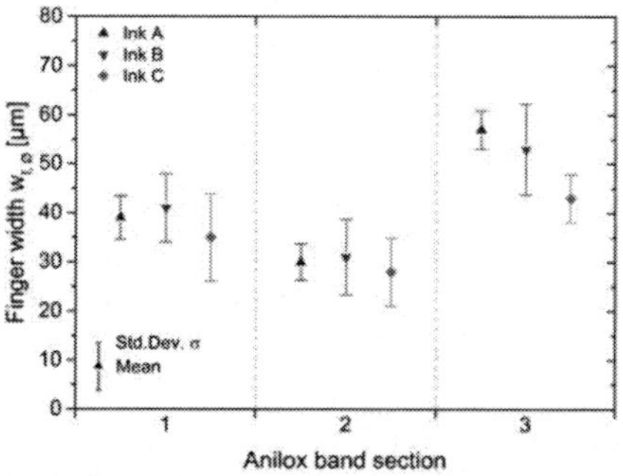

Figure 6: Finger width of flexographic printed contacts using three different in-house developed Ag-inks and three different annilox roller parameters. The nominal finger width on the printing plate was 5μm.

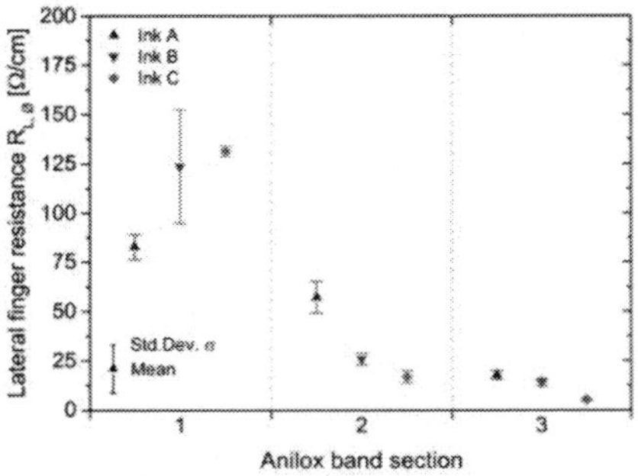

Figure 7: Lateral finger resistance of flexographic printed fingers using three different in-house developed Ag-inks and three different anilox roller parameters. The nominal finger width on the printing plate was 5µm.

We show in summary that flexographic printing technology is well-suited for front side metallization of Si solar cells especially for wire-based or multi-busbar module integration.

3.3 Multi-nozzle Dispensing

Within the experiment, industrial pre-processed PERC solar cells (Cz, p-type, 156x156mm², LCO) were equipped with a screen printed five busbar layout followed by a dispensed contact finger grid (nozzle opening D = 35µm, number of contact fingers: N=100) and compared with single-step screen printed samples (opening width w = 40µm, N = 100). At optimum firing conditions (temperature variation at three temperatures), dispensed samples reached top conversion efficiencies of 21.3% and 21.4% with the two applied industrial screen printing pastes, respectively, best screen printed cells reached 21.1% with both pastes (see Fig. 10). The main reason for this increase was a substantially reduced finger width reaching 33µm and 27µm (see. Fig 9) for the dispensed samples compared to around 45µm reached by screen printing. It was further demonstrated how dispensing benefits from screen printing paste development regarding an improvement of contact geometries: Some pastes already allow the application of 25µm nozzle openings reducing the Ag consumption per cell below 50mg featuring contact geometries close to the simulated optimum regarding aspect ratio and finger width and especially cross section homogeneity. Finally, a dispensing speed of 1m/s demonstrates a high throughput potential.

We show in summary that multi-nozzle dispensing (see Fig. 8) as further solution for high throughput printing of fine line contacts with high aspect-ratio allows for efficiency advantages of ~1.5%rel. (see Fig. 10) and a reduction of Ag paste consumption by ~15% in comparison to flatbed single-step screen printing leading to a significant reduction of production costs.

Figure 8: "Full wafer size" multi-nozzle dispense print head developed by Fraunhofer ISE and integrated in ASYS platform allowing for high printing speeds

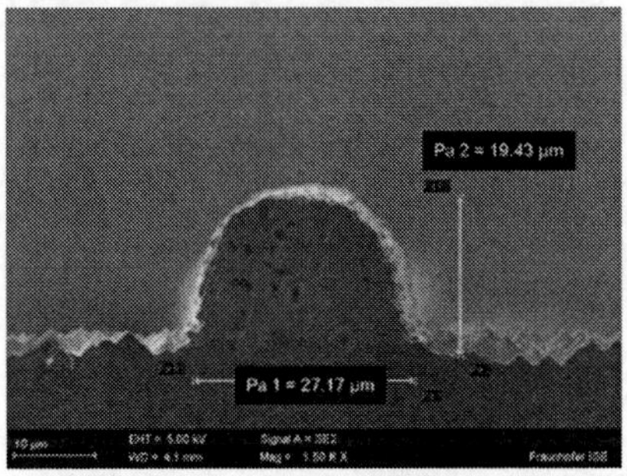

Figure 9: SEM image of a dispensed finger with a finger width below 30µm and finger height of ~20µm. The homogenous finger geometry allows for a more efficient material usage of the applied Ag paste.

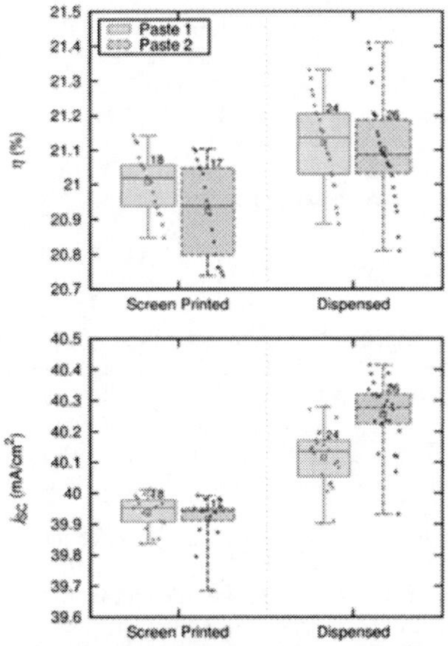

Figure 10: IV parameters (Jsc, eff.) of a PERC cell run based on contact finger metallization by dispensing and single step screen printing applying the same two commercially available screen printing pastes to both technologies. An efficiency increase of around 1.5%rel. is demonstrated for dispensing mainly driven by an increase in the short circuit current due to less shading of the contact fingers.

4 SUMMARY AND OUTLOOK

Within this work, innovative printing technologies are demonstrated aiming at the realization of much higher throughput rates, higher efficiencies and reduced production costs in comparison to flatbed screen printing. The proof of principle for rotary rear side metallization and flexographic as well as multi-nozzle dispensing front side metallization is shown.

At the moment, a new machine concept is under development and construction which addresses a continuous process flow during printing. Rotational printing units for flexographic printing (front side printing) and rotary screen printing (rear side and busbar printing) will be integrated. In principle, the integration of a dispensing unit for high aspect ratio fine line printing is also possible.

Hence, we see a huge potential to realize the rear and front side metallization of Si solar cells based on these high throughput printing technologies leading to higher throughput rates and significantly lower production costs. A first estimation of possible throughput rates shows that values above 6000wph are reachable.

5 ACKNOWLEDGEMENT

The authors would like to thank all co-workers at the Photovoltaic Technology Evaluation Center (PV-TEC) and the mechanical workshop at Fraunhofer ISE for processing of the samples. The authors further thank ASYS Group, Gallus Group, Zecher GmbH and ContiTech Elastomer-Beschichtungen GmbH for the support of this work.

This work was partly supported by the German Federal Ministry of Education and Research (BMBF) within the funding program Photonics Research Germany under the contract number 13N13512 (Rock-Star) and partly by the German Federal Ministry for Economic Affairs and Energy (BMWi) within the research project "GECKO" under contract number 0325404.

6 REFERENCES

[1] H. Kipphan, "Handbuch der Printmedien. Technologien und Produktionsverfahren." (Springer, Heidelberg, 2000)

[2] P. Hahne, "Innovative Druck- und Metallisierungsverfahren für die Solarzellen-technologie.", PhD Thesis, Fernuniversität Hagen (2000)

[3] A. Lorenz, C. Gredy, M. Lehner, R. Greutmann, H. Brocker, J. Rohde et al., "Progress with Rotational Printing for the Front Side Metallization of Silicon Solar Cells." Proc. of the 32nd EUPVSEC, Munich, Germany, 2016, pp. 413–419.

[4] A. Lorenz et al, Silicon PV 2017; DOI: 10.1016/j.egypro.2017.09.343

[5] M. Pospischil et al., Proc. of the 32nd EUPVSEC, Munich, Germany, 2016,.

[6] E. Lohmüller et al., IEEE Electron Device Letters 32, 1719 (2011).

[7] J. Specht et al., in Proceedings of the 25th European Photovoltaic Solar Energy Conference and ExhibitionValencia, Spain, 2010), pp. 1867.

PRODUCTION-COMPATIBLE REGENERATION OF BORON-DOPED CZOCHRALSKI-SILICON IN A COMBINED FAST-FIRING AND REGENERATION BELT-LINE FURNACE

Dominic C. Walter,[1] Verena Steckenreiter,[1] Lailah Helmich,[1] Thomas Pernau[2] and Jan Schmidt[1,3]

[1]Institute for Solar Energy Research Hamelin (ISFH),
Am Ohrberg 1, D-31860 Emmerthal, Germany
[2]centrotherm photovoltaics,
Johannes-Schmid-Straße 8, D-89143 Blaubeuren, Germany
[3]Department of Solar Energy, Institute of Solid-State Physics, Leibniz Universität Hannover,
Appelstraße 2, D-30607 Hannover, Germany

ABSTRACT: Boron-doped Czochralski-grown silicon (Cz-Si), as used in the industrial production of solar cells today, is suffering from a light-induced degradation (LID) of the carrier lifetime during illumination. It has been known since a decade that under lab conditions, this degradation can be permanently cured by illumination at increased temperature, which has been known as 'regeneration' treatment. In this contribution, we examine the permanent regeneration of the carrier lifetime in standard boron-doped Cz-Si using an industrial combined fast-firing and regeneration furnace - the *c.FIRE REG* of centrotherm photovoltaics. Our study reveals that for an optimized firing and regeneration process, very high implied open-circuit voltages $V_{oc,impl}$ exceeding 740 mV are achieved on standard 1-2 Ωcm *p*-type Cz-Si wafers. These very high measured $V_{oc,impl}$ values clearly indicate the suitability of the process for highly efficient PERC cell production. Additionally, as the *c.FIRE REG* furnace is realized in a belt-line configuration, which exposes each wafer for less than 70 s to the firing and regeneration conditions, our experiments clearly demonstrate the excellent suitability of the examined tool for the implementation into industrial solar cell production lines. We also demonstrate the successful application to industrial PERC solar cells produced by two different manufacturers.
Keywords: Czochralski-grown silicon, LID, boron-oxygen defect, lifetime, regeneration

1 INTRODUCTION

For more than 10 years it is known that the lifetime in boron-doped Czocharalski-grown silicon (Cz-Si), which is without further treatment limited by the boron-oxygen (BO) defect centre [1], can be permanently regenerated under illumination at elevated temperatures [2]. The process time for the complete regeneration of the lifetime is influenced by (a) the illumination intensity [3], (b) the temperature [2], and (c) the previous thermal treatments [4, 5] of the sample. In this contribution, we focus on transferring our lab-type regeneration process, which results in very high lifetimes > 1 ms after short process periods [5], into an industrial belt-line furnace with integrated fast-firing and regeneration zones - the *c.FIRE REG* from centrotherm photovoltaics. By adapting the temperature profile optimized at ISFH to the profile in the *c.FIRE REG* furnace, we achieve record-high stable lifetimes of $\tau(\Delta n = 0.1{\times}p_0) = (1460 \pm 80)$ µs on 1.0 Ωcm boron-doped Cz-Si, which is equivalent to an implied open-circuit voltage of $V_{oc,impl}$=745 mV. Additionally, we examine the response of commercially available *p*-type Cz-Si wafers from two different suppliers to our optimized regeneration process. For both suppliers we achieve lifetimes well above $\tau(\Delta n = 0.1{\times}p_0) > 2$ ms on ~1.8 Ωcm boron-doped Cz-Si.

These results clearly demonstrate that the new firing and regeneration furnace is suitable for the production of highly efficient solar cells on industrial scale using conventional boron-doped Cz-Si. We verify our lifetime results on industrially produced PERC solar cells from two different manufacturers.

2 EXPERIMENTAL DETAILS

As mentioned above, we processed lifetime samples from conventional 1.0 Ωcm boron-doped Cz-Si. The samples feature non-textured surfaces and the processing sequence is as follows:

We start with the removal of the surface damage which is followed by a phosphorus diffusion (850 °C for 1:09 hours), resulting in an n^+-region on both surfaces with a sheet resistance of 100 Ω/□. After the phosphorus diffusion the samples are split into two groups. In group 1, the n^+-region is removed and we passivate the surfaces with an Al_2O_3/SiN_x stack, deposited via plasma-assisted atomic layer deposition (plasma-ALD) and plasma-enhanced chemical vapour deposition (PECVD), respectively. On the *p*-type silicon surface, a 10 nm thick passivation layer of Al_2O_3 and a 70 nm thick SiN_x capping layer are deposited (n=2.05). In group 2, the n^+-region remains on the sample and the surfaces are directly passivated with a 70 nm thick SiN_x layer.

Afterwards, the samples of each group are further subdivided for processing, either at ISFH in a conventional belt-line firing furnace (DO-FF-8.600–300, centrotherm) or at centrotherm photovoltaics in the new *c.FIRE REG* furnace. A photograph of this furnace is shown in Fig. 1.

For firing at ISFH we choose our optimum firing conditions with a belt speed of 7.2 m/min and 850 °C peak set temperature [6]. The temperature profiles in both furnaces were recorded using a DATAPAQ Insight Oven Tracker with 0.5 mm sheathed thermocouples attached to

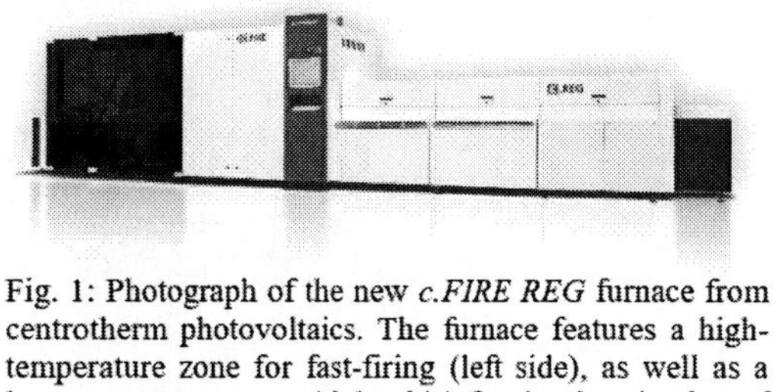

Fig. 1: Photograph of the new *c.FIRE REG* furnace from centrotherm photovoltaics. The furnace features a high-temperature zone for fast-firing (left side), as well as a low-temperature zone (right side) for the deactivation of the BO defect (regeneration of the lifetime).

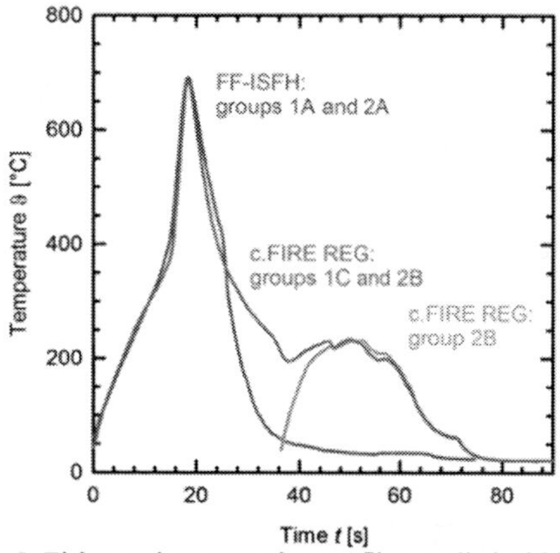

Fig. 2: Firing and regeneration profiles applied within this study. The blue curve shows the temperature profile of the conventional belt-line firing furnace at ISFH, whereas the green curve represents the adapted profile in the *c.FIRE REG* furnace, with integrated firing and regeneration. The orange curve shows the temperature profile of the regeneration-only process in the *c.FIRE REG* furnace.

the samples. Please note that due to the high belt speed, the actual sample temperature is always lower than the set peak temperature. As the temperature treatment prior to the regeneration treatment is of major importance to achieve a fast regeneration to high lifetimes [5], we adapted the temperature profile of our optimized firing process at ISFH [6] to the profile in the *c.FIRE REG* furnace.

For the regeneration treatment in the *c.FIRE REG* furnace, we choose an illumination intensity of 85 % of the maximum intensity and a maximum regeneration temperature (measured) of 234 °C. The three different temperature profiles applied within this study are shown in Fig. 2.

Comparison of the temperature profiles in the firing zone of the conventional firing furnace and the new furnace shows that both temperature profiles agree very well for the first 25 seconds. The firing profile in the *c.FIRE REG* furnace features a slightly faster cooling ramp from the maximum temperature compared to the traditional firing furnace, which is expected [6] to promote shorter regeneration process times. After about 30 to 35 seconds, the samples are transferred into the regeneration unit, where the samples remain for about 40 seconds. Hence, in total the firing and regeneration process takes about 70 seconds.

Table I: Summary of the five different sample groups investigated within this contribution. While group 1 consists of lifetime samples without diffused n^+-regions, group 2 consists of samples with diffused n^+-regions.

Group	**1 w/o n^+-region**		
	A	B	C
Fast-Firing	ISFH	ISFH	*c.FIRE REG*
Regeneration	Lab-type	*c.FIRE REG*	
Group	**2 w/ n^+-region**		
	A		B
Fast-Firing	ISFH		*c.FIRE REG*
Regeneration	Lab-type		

To be able to compare the results obtained using the new *c.FIRE REG* furnace with our laboratory results, we separate the processed samples in five different groups: group 1 in three subtypes A, B and C, and group 2 in two subtypes A and B. An overview of the different groups is given in Table I.

For our lab-type regeneration process, we use a hotplate at 185 °C and a halogen lamp with an adjusted illumination intensity equivalent to 1 sun.

Lifetime measurements are performed using the WCT-120 lifetime tester from Sinton Instruments. If not stated otherwise, all reported lifetimes were extracted at a fixed injection density of $\Delta n = 0.1 \times p_0$, with Δn being the excess carrier density and p_0 the hole concentration measured in darkness. In our case, p_0 equals the doping concentration N_{dop} as only non-compensated boron-doped silicon was examined in this study.

3 RESULTS AND DISCUSSION

In Fig. 3, the measured lifetimes after regeneration of each group are summarized. The blue symbols correspond to the Cz-Si lifetime samples, full symbols without n^+-region, open symbols with n^+-region. Orange symbols correspond to boron-doped float-zone silicon (FZ-Si) reference samples with a resistivity of 1.2 Ωcm, processed in parallel to monitor the surface passivation quality. Please note that both sample types have a final thickness of about 160 µm. As can be seen in Fig. 3, the FZ-Si samples in group 1 feature lifetimes above 2 ms, indicating very low surface recombination velocities. For the Cz-Si lifetime samples we obtain very high lifetimes in all three sample groups: Group 1A features an average lifetime of τ=(926±158) µs, group 1B a lifetime of τ=(870±360) µs and group 1C a lifetime of τ=(1460±90) µs. Hence, we observe the highest lifetime with the lowest scatter in group 1C, which has been fired and regenerated completely in the industrial *c.FIRE REG* furnace. Figure 4 shows the injection-dependent lifetimes of two different samples, one after firing and regeneration using the *c.FIRE REG* furnace (group 1C) and one after complete light-induced

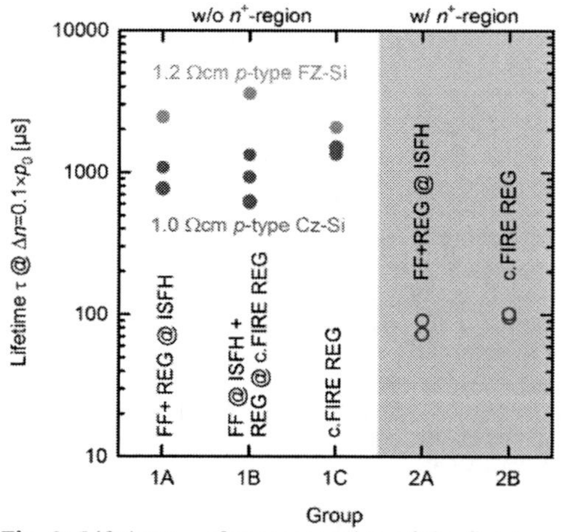

Fig. 3: Lifetimes τ after regeneration of the five sample groups in this investigation. The highest lifetimes with the lowest scatter are obtained in sample group 1C, fired and regenerated completely in the *c.FIRE REG* furnace.

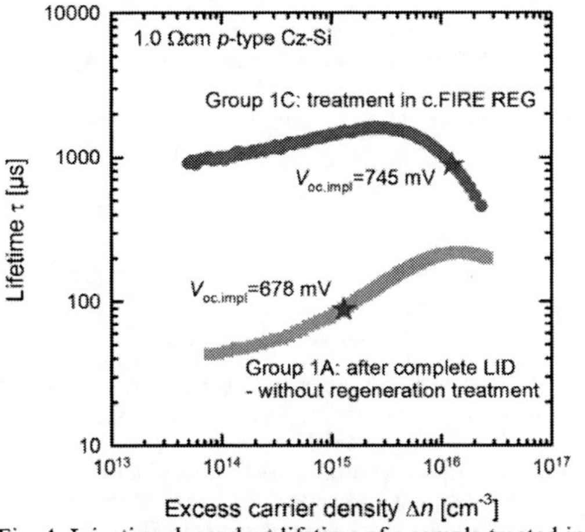

Fig. 4: Injection-dependent lifetime of a sample treated in c.FIRE REG (blue symbols) and a sample of group 1A (orange symbols) without regeneration treatment.

degradation (LID) prior to the regeneration treatment (group 1A). The treatment increases the implied 1 sun-V_{oc}-value from 678 mV after LID to 745 mV after regeneration.

In Fig. 3 on the right-hand side, the results of samples with n^+-region are displayed. The lifetime in these samples is much lower, as it is limited by recombination in the diffused regions. Comparing group 2A and 2B, we conclude that the regeneration of the lifetime works equally well in low-lifetime samples. Please note that with the lower lifetimes also the excess carrier concentration during the regeneration process is much lower as compared to the samples of group 1 and is therefore comparable to the carrier concentration in silicon solar cells. Hence, these results suggest that the regeneration of silicon solar cells can be performed alike. The high lifetimes obtained in group 1C are in excellent agreement with the best lifetime values obtained on the same material using the ISFH lab regeneration process. A comparison of the measured injection-dependent lifetimes

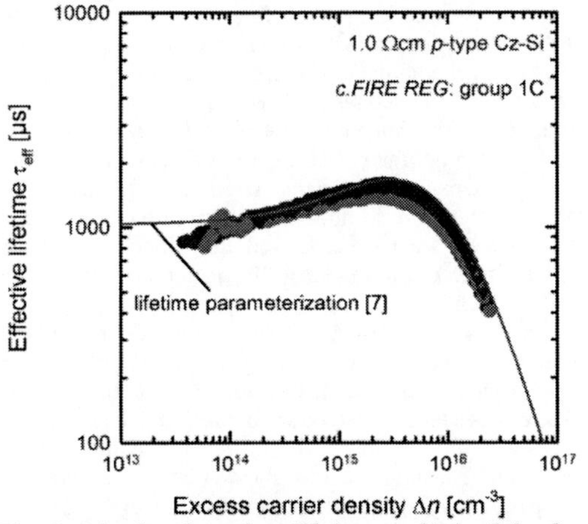

Fig. 5: Injection-dependent lifetime $\tau_{eff}(\Delta n)$ of the four samples of group 1C (colored symbols). The red solid line is the result of our lifetime parameterization after regeneration as presented previously [7].

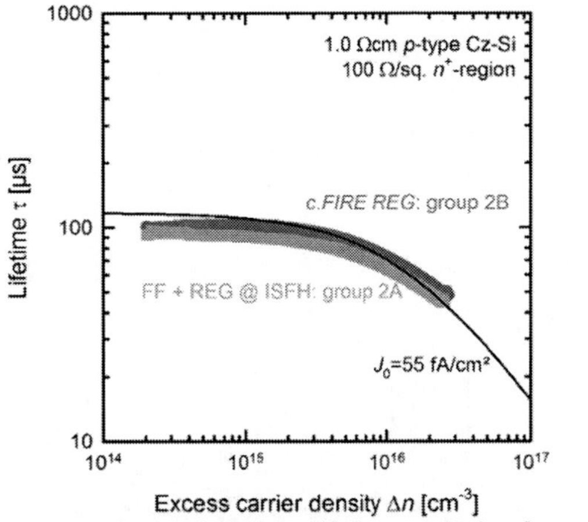

Fig. 6: Injection-dependent lifetime $\tau_{eff}(\Delta n)$ of two samples from two groups: group 2A (orange symbols) and group 2B (blue symbols). The solid black line is the calculated lifetime limit for a sample with a recombination current of J_0=55 fA/cm² at both n^+-diffused and SiN_x-passivated surfaces.

of group 1C with our recent lifetime parameterization after regeneration [7] is shown in Fig. 5. Excellent agreement is observed.

For the samples of group 2 with n^+-region, we measure much lower lifetimes compared to the results of group 1. For group 2A we measure an average lifetime of $\tau = (82 \pm 9)$ μs and for group 2B we measure $\tau = (98 \pm 3)$ μs. Again, the lifetime of the sample group 2B, treated completely in the *c.FIRE REG* furnace, features the higher lifetime with the lower scatter. In Figure 6, the injection-dependent lifetimes of two samples, one from group 2A and one from group 2B, are displayed. The very flat injection dependence towards low injection levels indicates that the lifetimes in these samples are not limited by the BO defect center. In these cases, the lifetimes are obviously limited by recombination in the diffused regions. The black solid line in Fig. 6 is the calculated lifetime limit according to:

$$\tau(\Delta n) = \frac{q n_i^2 W}{2\times J_0(p_0+\Delta n)},$$

with q being the elementary charge, n_i the intrinsic carrier concentration of silicon, W the sample thickness, p_0 the hole concentration in darkness and Δn the excess carrier concentration. Here, a recombination current of J_0=55 fA/cm² is chosen, which is a good value for the applied phosphorus diffusion and SiN_x surface passivation [9]. The measured lifetime curves agree well with this calculated curve. Hence, we conclude that the bulk lifetimes of these samples with n^+-regions have been completely regenerated. However, due to the lower lifetimes and, hence, the lower excess carrier concentrations during regeneration, this sample group features an excess carrier concentration comparable to solar cells treated with the same process. It is known that after fast sample cooling, the regeneration rate constant, i.e. the process time for regeneration, is mainly a function of the excess carrier concentration during regeneration [3, 8, 10] (and temperature [2]). Therefore, we expect that solar cells can be regenerated equally well using the same parameters applied within this

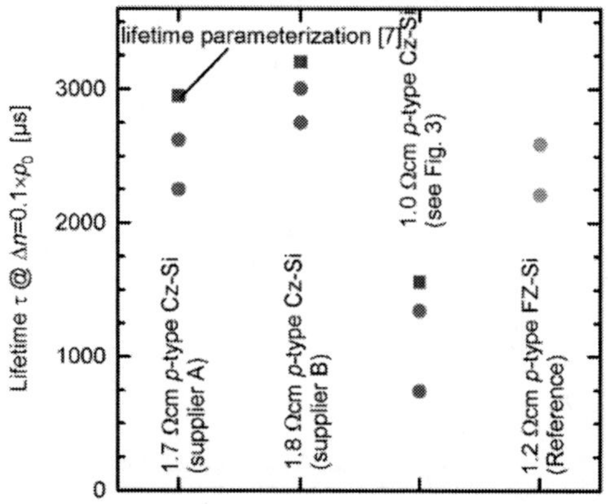

Fig. 7: Lifetime after regeneration in the *c.FIRE REG* furnace of the second experiment, testing standard solar cell materials from two different wafer suppliers.

contribution. In a second experiment, we have tested the response of different commercially available Cz-Si wafers from two different suppliers to the regeneration process in the *c.FIRE REG* furnace described above. The materials included in this experiment are 1.7 Ωcm and 1.8 Ωcm *p*-type Cz-Si wafers from two different suppliers, 1.0 Ωcm *p*-type Cz-Si wafers as used in the previous experiment and 1.2 Ωcm *p*-type FZ-Si wafers as references to monitor the surface passivation quality. Sample processing was performed as described for the previous experiment and the temperature profile in the *c.FIRE REG* furnace was adapted to the profile shown in Fig. 2. The result of this second experiment is shown in Fig. 7. On one wafer of supplier B, we obtained an exceptionally high lifetime of τ=3 ms ($V_{oc,impl}$=744 mV) after the regeneration process. In general, it can be seen that the regenerated lifetimes are for all three materials under investigation close to the calculated lifetime using our recent parameterization [7]. A comparison to the lifetimes obtained on FZ-Si (orange symbols in Fig. 7) reveals that the lifetimes measured on the Cz-Si wafers of the two different suppliers are even higher than the lifetime on FZ-Si. Assuming that the lifetime in FZ-Si material is limited by the surface recombination, we determine according to Sproul [11] a surface recombination velocity S between S=2 cm/s and S=2.6 cm/s. Taking these values as well as the lower doping concentration into account, the lifetime of the Cz-Si material of the two different suppliers can be explained with the recombination at the surfaces. This finding underlines even further the suitability of the process to reach high lifetimes on an industrial level.

Finally, the regeneration was tested on **P**assivated **E**mitter and **R**ear **C**ells (PERC) [12] from two different manufacturers in the industrial inline regenerator *c.REG*. We applied very similar furnace settings: For manufacturer A, we applied an illumination intensity of 90 % and for the PERC cells of manufacturer B, an intensity of 80 % was applied. It should be noted, however, that the same settings lead to different temperatures on the PERC cells compared to the un-metallized lifetime samples. While we measured a maximum temperature of ~235 °C on the lifetime samples, we measured an elevated temperature of ~270 °C on the solar cells during regeneration in the *c.REG*.

Table II: Relative changes in the open-circuit voltage V_{oc} and the conversion efficiency η referenced to the values $V_{oc(0)}$ and η_0 measured before LID prior to the regeneration process. Shown are mean values over 45 cells for manufacturer A and 7 cells of manufacturer B.

	After regeneration in the *c.REG*		After stability test (24 h @ 0.08 suns)	
PERC Manu-facturer	$\frac{\Delta V_{oc}}{V_{oc(0)}}$	$\frac{\Delta\eta}{\eta_0}$	$\frac{\Delta V_{oc}}{V_{oc(0)}}$	$\frac{\Delta\eta}{\eta_0}$
A	0.32 %	0.16 %	0.13 %	-1.40 %
B	-0.10 %	-0.07 %	-0.51 %	-0.97 %

In total, 45 cells of manufacturer A and 7 cells of manufacturer B were investigated. As can be seen from the data presented in Table II, the furnace process does not negatively affect the surface passivation of the cells, as the changes after the regeneration process are very small: $\Delta\eta/\eta_0$=-0.07 % for manufacturer B and even $\Delta\eta/\eta_0$=+0.16 % for manufacturer A, referenced to the conversion efficiency η_0 being the efficiency before light-induced degradation prior to the regeneration process. After the regeneration process, we checked the solar cells for stability under an illumination equivalent to 0.08 suns at 40 °C for ≥24 h. As can be seen from Table II, the efficiency slightly decreases between 0.97 % and 1.4 % relative, which is, however, much less than the typical BO-related degradation of about ~3-5 % for industrial PERC cells. Also, for the cells of manufacturer A, the V_{oc} value after the stability test is even higher than prior to the regeneration process. This could be an indication that the observed slight degradation in efficiency is not related to the typical BO-related degradation. Hence, we conclude that industrial PERC solar cells can be permanently regenerated in the industrial *c.REG* furnace applying our optimized regeneration process. However, please note that the applied regeneration conditions here are at the lower boundary for temperature and light intensity of typical industrial regeneration processes.

4 SUMMARY

In this contribution, we have investigated the permanent regeneration of the lifetime in a production-compatible industrial belt-line furnace, built by centrotherm photovoltaics, which combines the fast-firing and the regeneration step. We have performed a lifetime study to optimize the regeneration process, while the fast-firing profile was adapted to the optimized firing-profile at ISFH. In the optimized regeneration process, the lifetime samples are exposed to the regeneration conditions for less than 40 s with a maximum temperature of 234 °C. Using this process, we have achieved very high lifetimes of $\tau(\Delta n = 0.1{\times}p_0) = (1460 \pm 80)$ µs on 1.0 Ωcm *p*-type Cz-Si, which is equivalent to an implied 1-sun open-circuit voltage of $V_{oc,impl}$=745 mV. Additionally, we showed that these high lifetimes are in very good agreement with our most recent lifetime parameterization for boron-doped Cz-Si after permanent deactivation of the BO defect.

On lifetime samples with a phosphorus-diffused n^+-region on both sides, we have shown that the process leads to a permanent regeneration of the lifetime even when the excess carrier concentration is much lower than in lifetime samples without diffused regions and comparable to the concentration in silicon solar cells.

In a second experiment, we have examined the

performance of standard commercially available boron-doped Cz-Si from two different suppliers with base resistivities of 1.7 and 1.8 Ωcm. For both materials we have achieved lifetimes significantly exceeding 2 ms and implied open-circuit voltages $V_{oc,impl}$≥743 mV.

Finally, we have applied the optimized regeneration conditions to PERC solar cells from two different manufacturers in the *c.REG* regeneration furnace. Stability tests performed on these cells clearly showed that PERC cells can be effectively regenerated using the industrial *c.REG* furnace.

All these results clearly demonstrate that the new industrial furnace is well suitable for the production of highly efficient solar cells on industrial scale using conventional boron-doped Cz-Si.

ACKNOWLEDGEMENT

The authors would like to thank Olaf Romer of centrotherm photovoltaics for the conduction of experiments at the *c.FIRE REG* furnace and Cornelia Marquardt of ISFH for sample processing. This work was funded by the German State of Lower Saxony and the German Federal Ministry of Economics and Energy (BMWi) within the research project "Upgrade Si-PV" (contract no. 0325877B). The content is the responsibility of the authors.

5 REFERENCES

[1] J. Schmidt, A. G. Aberle and R. Hezel, Proceedings 26th IEEE Photovoltaics Specialists Conference, Anaheim, CA, USA, p. 13 (1997)
[2] A. Herguth, G. Schuber, M. Kaes, and G. Hahn, Proceedings 21st European Photovoltaic Solar Energy Conference, p. 530 (2006)
[3] A. Herguth, G. Schuber, M. Kaes, and G. Hahn, Proceedings 22nd European Photovoltaic Solar Energy Conference, p. 893 (2007)
[4] S. Wilking, S. Ebert, A. Herguth, and G. Hahn Proceedings 28th European Photovoltaic Solar Energy Conference, p. 34 (2013)
[5] D. C. Walter, B. Lim, K. Bothe, V. V. Voronkov, R. Falster, and J. Schmidt Applied Physics Letters **104**(4) 042111 (2014)
[6] D. C. Walter, B. Lim, K. Bothe, R. Falster, V. V. Voronkov, and J. Schmidt, Solar Energy Materials and Solar Cells **131** 51 (2014)
[7] D. C. Walter, B. Lim, and J. Schmidt Progress in Photovoltaics: Research and Application **24**(7) 920 (2016)
[8] S. Wilking, S. Ebert, C. Beckh. A. Herguth, and G. Hahn, Proc. 32nd European Photovoltaic Solar Energy Conference, p. 487 (2016)
[9] H. Hannebauer, M. Sommerfeld, J. Müller, T. Dullweber, and R. Brendel Proceedings 27th European Photovoltaic Solar Energy Conference, p. 1360 (2012)
[10] V. Steckenreiter, D. C. Walter, and J. Schmidt AIP Advances **7** 035305 (2017)
[11] A. B. Sproul, Journal of Applied Physics **76** 2851 (1994)
[12] A. W. Blakers, A. Wang, A. M. Milne, J. Zhao, and M. A. Green, Applied Physics Letters **55** 1363 (1989)

ULTRAFAST IN-LINE CAPABLE REGENERATION PROCESS FOR PREVENTING LIGHT INDUCED DEGRADATION OF BORON-DOPED P-TYPE CZ-SILICON PERC SOLAR CELLS

A. A. Brand[1], K. Krauss[1], P. Wild[2], S. Schörner[2], S. Gutscher[1], S. Roder[1], S.Rein[1], J. Nekarda[1]
[1]Fraunhofer Institute for Solar Energy Systems ISE, Heidenhofstr. 2, 79110 Freiburg, Germany
[2]Rehm Thermal Systems GmbH, Leinenstrasse 7, 89143 Blaubeuren, Germany
Phone: +49 761 - 4588 5053; e-mail: andreas.brand@ise.fraunhofer.de

ABSTRACT: Solar cells based on boron-doped p-type Czochralski-grown silicon (Cz-Si) substrates suffer from light-induced degradation (LID) which is related to a metastable boron-oxygen defect and limits the achievable cell efficiency significantly, but can be eliminated by carrier injection at elevated temperatures within a so-called regeneration process. We have developed an ultrafast regeneration process (UFR) to prevent up to 98% of the LID effect in less than 4 seconds process-time. The technology behind this process has been implemented in an in-line prototype tool with belt speeds of up to 8000 mm/min. Compared to competing techniques the technology introduced here enables tools with smaller footprint and broader process window, offering irradiation intensities up to 1 MW/m² and temporal and spatial modulation of the irradiation intensity with up to 100 Hz. Undistorted in-situ temperature measurements can be acquired via commercially available thermographic imaging equipment. Further, we introduce a new parameter called regeneration completeness in order to allow benchmarking for BO-regeneration processes. It quantifies the effectiveness of BO-regeneration processes independent of the overall efficiency in order to allow a direct comparison of the quality of different regeneration processes.
Keywords: LID, BO-defect, regeneration, stabilization, annealing

1 INTRODUCTION

Boron-doped Czochralski-grown silicon (Cz-Si) is widely used as substrate in the photovoltaic industry. However, this material type suffers from a degradation of bulk lifetime within the first hours of light exposure what is known as light-induced degradation (LID) [1]–[8]. The responsible defect is widely known as the boron-oxygen (BO) defect [3].

The efficiency of passivated emitter and rear cells (PERC) [10] can benefit significantly from a higher base doping concentration which reduces the resistive losses in the bulk, allows a reduction of base contact area and, hence, rear-side recombination. However, an increased boron base doping concentration implies a stronger LID effect and, thus, leads to a stronger limitation of bulk lifetime and efficiency, which overcompensates possible gains [4], [11]. Therefore, avoidance or reversal of BO-related LID of is one of the most important singular enhancements for state of the art high-efficiency PERC solar cells on boron-doped Cz-Si substrates. As the LID-related efficiency losses typically amount up to 1-2%$_{abs}$, the avoidance or reversal of the LID directly leads to a respective efficiency advantage without any significant changes in technology [12].

Thus, prevention of the BO-related LID is a key enabler for high-efficiency solar cells manufactured from industry-standard boron-doped Cz-Si wafers. Ways to avoid BO-related LID are the replacement of boron with gallium during crystallization [13] or techniques to reduce oxygen concentration in the crystal [3]. Another way to strongly prevent BO-related LID is applying a so-called *regeneration* [14]. When first reported regeneration required several minutes to hours, but recently approaches for fast regeneration processes with industrially feasible process time in the order of seconds have been reported [16], [27], [17]. Within this paper we present the integration for a fast regeneration process and report on its implementation in an industrial tool. With a large number of solar cell manufacturers adopting the PERC solar cell concept [18], the demand for such a regeneration process is bound to increase.

2 BO-DEFECT

2.1 BO-defect states

An important characteristic of the BO-defect is that according of a model introduced by Herguth et al. [14] it exhibits three different defect states. In the *annealed state* which is metastable upon illumination or carrier-injection, the BO-defect is inactive and bulk lifetime only limited by other defects. This is the state of the BO-defect after fast firing which is typically the last process in the solar cell production. In the *degraded state,* the bulk lifetime is significantly reduced by BO-defect recombination [4], [19]. This state is stable under light irradiation at room temperature and typically limits the performance of boron-doped Cz-Si solar cells [20]. The third state is often called the *regenerated, passivated* or *stabilized state.* In this state the BO-defect is deactivated as in the annealed state but in addition stable upon illumination and carrier injection. Note, that incomplete transitions between the states are possible and lifetime can be limited by a mixture of the states and their recombination. The regenerated state is preferred for solar cells since it offers higher bulk lifetime and thus higher solar cell efficiency over a long period of operation.

2.2 BO-defect state transitions

The BO-defect states are metastable and can be converted from one state to another by combinations of charge carrier injection and temperatures. The transition rates between the states depend on both, carrier injection and temperature, see Ref. [Niewelt] and references therein [1], [14]. Different combinations of both change the ratio of these transition rates, so over time an equilibrium of all three defect states forms. Although there will be a combination of states in most cases, we simplify our naming to the dominant state from now on.

Carrier injection can be achieved by applying a forward current to the solar cell or by light illumination. High-intensity light illumination in the range of tens or hundreds of kilowatts per square meter can reach much higher injection levels than forward current through a solar cell, which would be limited by the series resistance and thus damage or destroy the electrical contacts. Thus, high-intensity illumination is more practical for injecting high charge carrier densities. From here on, we will restrict ourselves to light illumination with photon energies above the silicon bandgap ($E_{\mathrm{ph}} \geq 1.17$ eV) as carrier injection source. Furthermore, we discuss only the open-circuit case when applying carrier injection via light illumination.

The annealed state converts to the degraded state during light illumination at temperatures below 45°C [1], [2], [21]. The full transition from the annealed to the degraded state is in the order of 24 h when light intensities of 1 kW/m² are applied [14].

The degraded state can be converted back to the annealed state by annealing in the dark for few minutes at temperatures > 200°C [1].

The degraded state can also be converted to the regenerated state by light illumination at elevated temperatures > 65°C. This process is often called *BO-regeneration* or just *regeneration* [8], [14]. It is known that illumination with higher light intensities increases the transition rate towards the regenerated state [15], [16] but also causes higher process temperatures which in turn increase the transition rate towards the annealed state [2].

3 REGENERATION COMPLETENESS

3.1 Complete approach

In order to be able to benchmark and compare different regeneration processes on different solar cells, it is important to have a robust and technology-independent measure for the amount of defects which are transferred to the regenerated state upon the applied treatment. An absolute efficiency gain or loss is not a suitable indicator to assess the regeneration result quantitatively since it depends on the solar cell manufacturing process and absolute efficiency.

In order to ensure comparability of the regeneration results for different regeneration processes, we introduce an easy to understand measure of process effectiveness, called *regeneration completeness RC*, which is defined as

$$RC = \frac{\eta_{\mathrm{regenerated}} - \eta_{\mathrm{degraded}}}{\eta_{\mathrm{annealed}} - \eta_{\mathrm{degraded}}}.$$

The *RC* value represents the amount of prevented degradation as a fraction of the total efficiency loss a single solar cell exhibit without any BO-regeneration. The *RC* values should be between 0 and 1 and are higher the the more defects have been transferred to the regenerated state. However, if other properties of the solar cell are altered by the regeneration process, the RC values could extend beyond the range of 0-1.

Proper *RC* calculations are often hindered by transitions between the metastable defect states of iron, i.e., the formation of interstitial iron (Fe_i) upon illumination and the formation of iron-boron pairs (FeB) upon dark storage [22]. Thus, the presence of iron associated with born or as an interstitial defect after each treatment need to be taken into consideration when taking *IV*-measurements at different BO-states [3], [22]–[24]. In figure 1 the left column illustrates the experimental plan of temperature and illumination treatments which is required for the formation and elimination of different defect types as basis for a precise calculation of *RC*. The expected BO and FeB/Fe_i defect states after each process step are listed in the right columns. The procedure is similar to the characterization performed in Ref. [9].

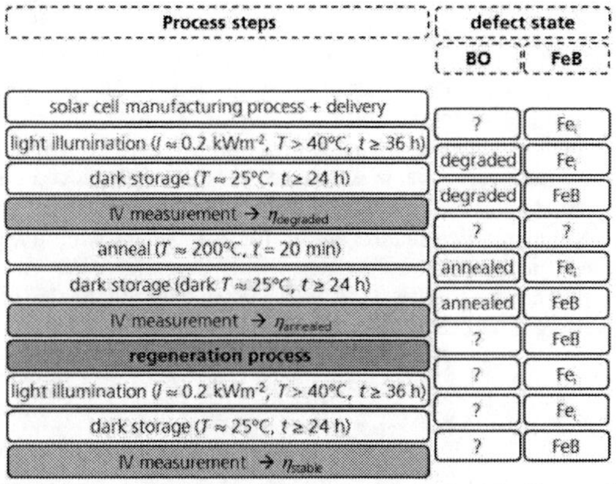

Figure 1: Experimental plan for determination of regeneration completeness *RC*, left column. On the right the according defect states of BO and FeB are listed.

First, we degrade the solar cells by a light illumination step using standard fluorescent tube lamps with a light intensity of roughly 200 W/m². During illumination the wafers reach a temperature slightly above 40°C. This light illumination is maintained for at least 36 h in order to ensure complete degradation. After light illumination possible iron defects are fully present as interstitial iron (Fe_i) as FeB pairs have dissociated.. Dark storage of the solar cells at approximately 25°C for at least 24 h causes the iron to pair (FeB), the BO-defect being still fully active. Subsequently we conduct the first *IV*-measurement which gives η_{degraded}, the efficiency at full BO-related LID but without the influence of Fe_i.

Next, we apply a thermal anneal at 200°C in the dark for 20 minutes using a hotplate in order to convert the BO-defect to its annealed state. It is necessary to do this treatment without any light illumination or other carrier injection to avoid transitions to the regenerated state. This is unwanted at this point since we try to investigate the effectiveness of a different regeneration process. Another dark storage is conducted in order to ensure iron is in the FeB state. We then conduct a second *IV*-measurement of the solar cells which gives η_{annealed}, the efficiency of the solar cells without any BO-recombination. This state is comparable to the one we expect directly after fast firing, the last processing step of the typical PERC solar cell production.

Now, we apply the regeneration process. After regeneration, the entirety of the BO-defects is in an unknown state, which likely is a mixture of the annealed, degraded and regenerated defect state. Due to the light illumination and the high temperatures upon regeneration, iron is in the Fe_i state. We conduct another light illumination step at room temperature to convert all remaining annealed BO defects to the degraded state and then apply another dark storage step to convert iron into the FeB state (for both steps the same parameters as before are used). Afterwards the solar cells are *IV*-measured a third time which gives $\eta_{\mathrm{regenerated}}$.

This concludes the experiment. Inserting the values for all three input values for η_{degraded}, η_{annealed} and

$\eta_{regenerated}$ gives the *RC* for each solar cell.

3.2 Practical approach for time efficient evaluation

The determination of the actual regeneration completeness of each solar cell as described above requires a time consuming series of dark storage, light illumination, annealing and *IV*-measurements. Therefore, we also define a more time efficient way for evaluation of regeneration results of large sample The output parameter of the simplified procedure is called *practical regeneration completeness* (*pRC*) to differentiate it from the precise *RC* value extracted from the full procedure (see Figure 1). Figure 2 illustrates the procedure that is applied for determining *pRC*. Because of its simplicity and increasing robustness with sample size the determining *pRC* is very well suited to evaluate the of different regeneration processes on an industrial scale.

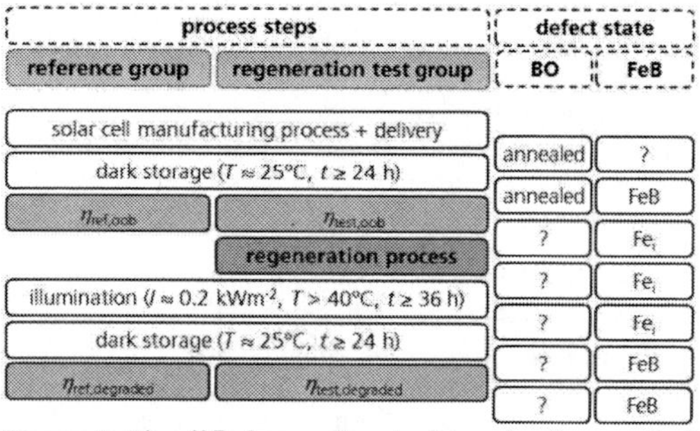

Figure 2: Simplified procedure to determine the practical regeneration completeness *pRC*.

After solar cell manufacturing it is assumed that for all the solar cells almost all the BO-defects are in the annealed state which is usually the case if the solar cells have not been exposed to light or forward bias current after fast firing.

Like in section 3.1, a dark storage step is applied before each *IV*-measurement to ensure that possible iron defects are always in the FeB state upon measurement and are thus fully subtracted out in the RC/pRC calculation. The first *IV*-measurement represents the out of box (oob) efficiency of each solar cell η_{oob}. Randomly picking some of the solar cells as a reference group is not very effective as it could generate a non-representative subset. The out of box *IV*-measurements are evaluated in order to generate a small subset of solar cells for the reference group which have a comparable distribution of the *IV* values as the regeneration group.

The reference group is degraded by applying a light illumination step. This is followed by another dark storage step. Finally, the solar cells of the reference group are *IV*-measured again resulting in $\eta_{ref,degraded}$.

The regeneration test group is treated with the regeneration process. It then undergoes the same light illumination and dark storage steps as the reference group to degrade all BO-defects which remain in the annealed state. Finally, *IV*-measurements of the test group are conducted resulting in $\eta_{test,regenerated}$. The *pRC* can now be calculated for each solar cell as follows:

$$pRC = 1 - \frac{\eta_{\text{test,oob}} - \eta_{\text{test,regenerated}}}{\Delta\overline{\eta}_{\text{ref}}}$$

With $\Delta\overline{\eta}_{ref}$ representing the average efficiency difference for all *n* solar cells of the reference group between out of box and degraded *IV*-measurements.

4 ULTRAFAST REGENERATION TECHNOLOGY

There are some industrial tools available today that use high-intensity lamps or LEDs for BO-regeneration. Laser radiation can deliver much higher intensities. It has been shown by the authors of this paper [17] and other groups [15], [16], [25] that using laser irradiation intensities of several tens to hundreds of kilowatt per square meter can speed up the full regeneration of Cz-Si wafers to as little as 0.2 seconds. However, we are unaware of anyone having developed a laser-based production tool for regeneration at this point. Because of the very fast regeneration that is being enabled by this drastic intensity increase we refer to this regeneration technology ultrafast regeneration (UFR).

Our regeneration technology offers several advantages for inline BO-regeneration systems compared to other technologies used in industrially implemented tools. Some notable advantages are:

- laser irradiation intensities of up to 1 MW/m²
- small footprint tooling
- maximum carrier injection per generated heat
- simple in-situ temperature measurement
- durable illumination components

We use laser radiation sources which offer significantly higher irradiation intensity up to 1 MW/m² and higher compared to competing light sources like lamps and LEDs. Figure 3 illustrates intensities for different irradiation sources in a regeneration tooling setup.

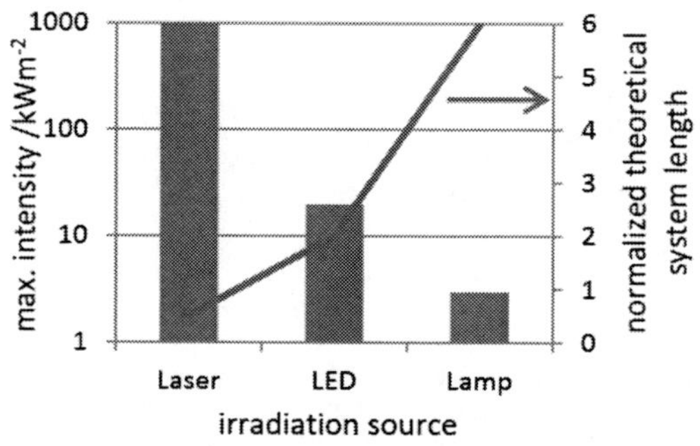

Figure 3: Comparison of the irradiation intensity for different light sources and the theoretical lengths for inline regeneration systems depending on the intensity. An in-line belt speed of 6000 mm/min is assumed.

Because of the higher available irradiation intensity, an UFR in-line production tool can be of much smaller footprint while maintaining the same or higher wafer throughput compared to in-line tools based on other light sources. Figure 3 also displays the estimated system lengths achieving the same degree of regeneration completeness for different irradiation sources. The system lengths for laser is estimated taking into account experimental values for regeneration times in dependence of irradiation intensities < 1 MW/m² and under the assumption of 6000 mm/min belt transportation speed. The system lengths for the LED and Lamp as irradiation source are estimated taking into account publically available information about processing speeds from manufacturers of regeneration tools. The estimation reveals that an UFR regeneration system can probably be built with less than half the length of a system based on high power LEDs.

One of the key problems to adopt high-intensity light sources of any kind for the BO-regeneration process is the heat accumulation which can cause the wafers to become hotter than reasonable for the optimal process window. The heat generated in the solar cells by photon absorption correlates with the intensity. Therefore, it is our goal to maximize the minority charge carrier density in the solar cell while minimizing the temperature. UFR uses photon energies of 1.27 eV which add only 0.1 eV of excessive energy per excited electron since the bandgap of silicon is 1.17 eV. With silicon being an indirect semiconductor using irradiation with 1.17 eV photon energy would not be ideal since a major part of the photonic flux would be transmitted to the rear side of the solar cell where the photons would only add to the heat but not the carrier injection in the silicon.

Controlling the solar cell temperature during the regeneration process is of great importance to the regeneration effectiveness, as mentioned above [14], [26]. Figure 4 shows an example of an in situ temperature measurement during the UFR process. The image was acquired using a standard M-IR camera in a spectral range of 7.5 –13 µm. The emissivity is calibrated for the area in between busbars.

Such thermographic imaging is easily implemented since the laser radiation used for the UFR process has a bandwidth of less than ± 10 nm and no background emissions in the spectrum of the camera. This is a key advantage compared to competing light sources.

The laser components we use for UFR are proven to be highly reliable and maintenance free under 24/7 production scenarios. A continuous test by the manufacturer at 100% output power over > 19000 h operation did result in a less than 2% deviation in irradiation intensity.

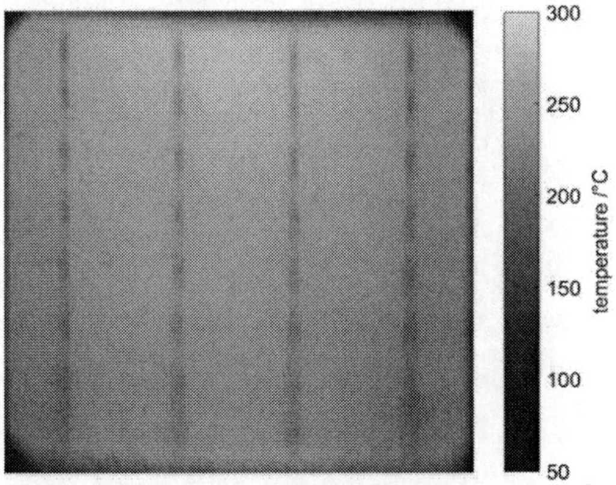

Figure 4: In-line measurement of the temperature of a solar cell along the laser irradiation path. This image is recorded by a commercial infrared camera.

5 LABORATORY SCALE PROCESS

The development of UFR technology was based on a lab-scale tool before integration into an industrial production prototype.
We use industry standard p-type PERC solar cells based on Cz-Si wafers which we obtained from a renowned solar cell manufacturer. The rear side passivation of the solar cells is passivated with an AlO_x/SiN_x PECVD passivation stack. The rear contacts are realized using laser contact opening (LCO) of the rear side passivation and screen printing of aluminium paste. The front contacts consist of screen-printed silver paste. The contact formation is realized by a fast firing process.

Process variations have been performed on a UFR prototype in order to optimize the achieved *pRC*, which is determined as described in figure 2. The best UFR process achieves 98% *pRC* with an irradiation length of less than 4 seconds.

In order to test the robustness of the UFR process we apply the optimized process parameters from the above mentioned experiment to PERC solar cells from four additional solar cell manufacturers, all using boron-doped Cz-Si. The solar cells from these manufacturers may exhibit differences in boron and oxygen concentration. The different manufacturers are also likely to use different production technologies and processing routes all of which are known to have a strong influence on the results of a regeneration process.

The results are illustrated in figure 5. Manufacturer #1 represents the same kind of solar cells which were used for the process optimization with *pRC* = 98%.

The regeneration completeness for manufacturers #2-5 show an average *pRC* of 82.1%. This result illustrates that the UFR process can be very effective at regenerating standard mass produced Cz-Si PERC solar cells even without manufacturer-specific optimization. However, optimization seems obvious when integrating any regeneration processes into a production line and, hence, the need for a tool which can offer this flexibility.

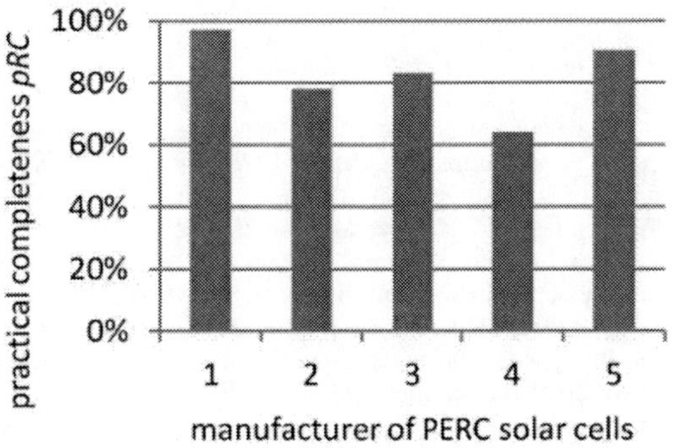

Figure 5: Example for regeneration completeness of an UFR process for solar cells of different manufacturers. These results were achieved using a non-inline laboratory setup. The process parameters applied to all groups are the ones optimized for manufacturer #1.

Since our UFR technology offers a wide parameter space, we have observed that simple process optimization for the other manufacturers yields similarly good results as the *pRC* = 98% from manufacturer #1.

6 IN-LINE TOOL

We have developed a prototype UFR system for solar cell production. Figure 6 shows the completed UFR system with an overall footprint of less than 4 m². Due to the compactness, this tool can be easily integrated into new as well as existing production sites.

This system demonstrates the production-readiness of the UFR technology including an integrated inline temperature control of the solar cells during laser irradiation.

In a first UFR experiment using the recently completed inline tool we achieved a preliminary result of $pRC = 86\%$ at a belt speed of 6 m/min. This value was achieved on solar cells of manufacturer #1, described in section 5.1. The result demonstrates the successful implementation of our UFR technology into a production tool. Other regeneration tools on the market claim to achieve similar levels of LID-prevention although it is unknown to the authors of this article how these are determined.

Figure 6: Prototype of single track inline regeneration tool based on UFR technology. It offers a belt speed of up to 8000 mm/min and a footprint of less than 4 m².

7 CONCLUSION & OUTLOOK

We have introduced a new measurement parameter called *regeneration completeness* (*RC*) which quantifies the effectiveness of BO-regeneration processes independently of overall solar cell efficiency. Furthermore, a more time efficient procedure for large sample sets which results in the *practical regeneration completeness* (*pRC*) has been introduced. This allows for the first time to easily compare the process effectiveness of different regeneration processes on an industrial scale instead of comparing absolute efficiency differences on a very limited number of solar cells.

A technology for ultrafast regeneration to reduce boron-oxygen related light-induced degradation of p-type Czochralski-grown silicon solar cells has been developed. The UFR process has been demonstrated to achieve up to 98% regeneration completeness in < 4 s irradiation time. The ability to monitor and control intensity and temperature distribution during the process allows accurate control of processing parameters and, thus, enables more options for process optimizations to accommodate for differences in the solar cell manufacturing process.

We have implemented this technology in an in-line production tool with industrial belt speeds up to 10 m/min and we have achieved up to 86% practical regeneration completeness in our preliminary tests. To our knowledge, this is the fastest in-line LID prevention process available for manufacturing.

8 ACKNOWLEDGEMENTS

The authors would like to thank their colleagues at the Fraunhofer institute for solar energy systems (ISE) for their help with regeneration experiments and solar cell characterization. This work was funded by the German Federal Ministry for Economic Affairs and Energy within the research project "UFO" under contract number 0324080.

9 REFERENCES

[1] H. Fischer and W. Pschunder, "Investigation of photon and thermal induced changes in silicon solar cells," in *Photovoltaic Specialists Conference, 10 th, Palo Alto, Calif*, 1974, pp. 404–411.

[2] S. Glunz, S. Rein, W. Warta, J. Knobloch, and W. Wettling, "On the degradation of Cz-silicon solar cells," in *Proc. 2nd WC PVSEC*, 1998.

[3] S. Rein, S. Diez, R. Falster, and S. Glunz, "Quantitative correlation of the metastable defect in Cz-silicon with different impurities," in *Photovoltaic Energy Conversion, 2003. Proceedings of 3rd World Conference on*, 2003, vol. 2, pp. 1048–1052.

[4] K. Bothe, R. Sinton, and J. Schmidt, "Fundamental boron-oxygen-related carrier lifetime limit in mono- and multicrystalline silicon," *Prog. Photovolt. Res. Appl.*, vol. 13, no. 4, pp. 287–296, Jun. 2005.

[5] B. Lim, "Boron-oxygen-related defects in crystalline silicon: Formation, recovery kinetics, and impact of compensation," Südwestdeutscher Verlag für Hochschulschriften, Saarbrücken, 2012.

[6] T. U. Naerland, "Characterization of light induced degradation in crystalline silicon," Univ. Sci. Technol., Trondheim, Norway, Saarbrücken, 2013.

[7] J. Lindroos and H. Savin, "Review of light-induced degradation in crystalline silicon solar cells," *Sol. Energy Mater. Sol. Cells*, vol. 147, pp. 115–126, Apr. 2016.

[8] T. Niewelt, J. Schon, W. Warta, S. W. Glunz, and M. C. Schubert, "Degradation of Crystalline Silicon Due to Boron–Oxygen Defects," *IEEE J. Photovolt.*, vol. 7, no. 1, pp. 383–398, Jan. 2017.

[9] F. Fertig, K. Krauß, and S. Rein, "Light-induced degradation of PECVD aluminium oxide passivated silicon solar cells: Light-induced degradation of PECVD aluminium oxide passivated silicon solar cells," *Phys. Status Solidi RRL - Rapid Res. Lett.*, vol. 9, no. 1, pp. 41–46, Jan. 2015.

[10] A. W. Blakers, A. Wang, A. M. Milne, J. Zhao, and M. A. Green, "22.8% efficient silicon solar

cell," *Appl. Phys. Lett.*, vol. 55, no. 13, pp. 1363–1365, Sep. 1989.
[11] S. Rein, W. Warta, and S. Glunz, "Investigation of carrier lifetime in p-type Cz-silicon: specific limitations and realistic prediction of cell performance," in *Photovoltaic Specialists Conference, 2000. Conference Record of the Twenty-Eighth IEEE*, 2000, pp. 57–60.
[12] S. Wasmer, A. A. Brand, and J. Greulich, "Metamodeling of numerical device simulations to rapidly create efficiency optimization roadmaps of monocrystalline silicon PERC cells," *Energy Procedia*, 2017.
[13] S. Glunz, S. Rein, J. Knobloch, W. Wettling, and T. Abe, "Comparison of boron- and gallium-doped p-type Czochralski silicon for photovoltaic application," *Prog. Photovolt. Res. Appl.*, vol. 7, no. 6, pp. 463–469, 1999.
[14] A. Herguth, G. Schubert, M. Kaes, and G. Hahn, "Avoiding boron-oxygen related degradation in highly boron doped Cz silicon," in *Proceedings of the 21st European Photovoltaic Solar Energy Conference*, 2006, pp. 530–537.
[15] P. Hamer, B. Hallam, M. Abbott, and S. Wenham, "Accelerated formation of the boron-oxygen complex in p-type Czochralski silicon: Accelerated formation of the boron-oxygen complex in p-type Czochralski silicon," *Phys. Status Solidi RRL - Rapid Res. Lett.*, vol. 9, no. 5, pp. 297–300, May 2015.
[16] S. Wilking, J. Engelhardt, S. Ebert, A. Herguth, and G. Hahn, "High speed regeneration of BO-defects: Improving long-term solar cell performance within seconds," presented at the EUPVSEC, Amsterdam, 2014.
[17] A. A. Brand, S. Wasmer, M. Graf, J.-F. Nekarda, and R. Preu, "Laser based evolution of the PERC process chain," presented at the PVCellTech, Kuala Lumpur, 2016.
[18] "International Technology Roadmap for Photovoltaic (ITRPV) 2017," *International Technology Roadmap for Photovoltaic (ITRPV)*. [Online]. Available: http://www.itrpv.net/. [Accessed: 17-Sep-2017].
[19] S. Rein, W. Warta, and S. W. Glunz, "Investigation of carrier lifetime in p-type Cz-silicon: specific limitations and realistic prediction of cell performance," Proceedings of the 28th IEEE Photovoltaic Specialists Conference, Anchorage, USA, pp. 57–60, 2000..
[20] J. Schmidt *et al.*, "Impurity-related limitations of next-generation industrial silicon solar cells," *IEEE J Photovolt*, vol. 3, pp. 114–118, 2013.
[21] J. Schmidt, A. G. Aberle, and R. Hezel, "Investigation of carrier lifetime instabilities in Cz-grown silicon," in *Conference Record of the Twenty Sixth IEEE Photovoltaic Specialists Conference - 1997*, 1997, pp. 13–18.
[22] L. J. Geerligs and D. Macdonald, "Dynamics of light-induced FeB pair dissociation in crystalline silicon," *Appl. Phys. Lett.*, vol. 85, no. 22, pp. 5227–5229, 2004.
[23] J. Schmidt and K. Bothe, "Structure and transformation of the metastable boron-and oxygen-related defect center in crystalline silicon," *Phys. Rev. B*, vol. 69, no. 2, p. 024107, 2004.
[24] S. Rein and S. W. Glunz, "Electronic properties of interstitial iron and iron-boron pairs determined by means of advanced lifetime spectroscopy," *J. Appl. Phys.*, vol. 98, no. 11, p. 113711, 2005.
[25] L. Song *et al.*, "Laser Enhanced Hydrogen Passivation of Silicon Wafers," *Int. J. Photoenergy*, vol. 501, p. 193892, 2015.
[26] S. Wilking, C. Beckh, S. Ebert, A. Herguth, and G. Hahn, "Influence of bound hydrogen states on BO-regeneration kinetics and consequences for high-speed regeneration processes," *Sol. Energy Mater. Sol. Cells*, vol. 131, pp. 2–8, Dec. 2014.
[27] B. J. Hallam et al., "Advanced hydrogenation of dislocation clusters and boron-oxygen defects in silicon solar cells," Energy Procedia, vol. 77, pp. 799–809, 2015.

CONSTRUCTING SUBMICRON-TEXTURE ON DIAMOND-WIRE-SAWN MULTI-CRYSTALLINE SILICON SOLAR CELLS VIA COPPER CATALYZED CHEMICAL ETCHING

Xiaodong Su*, Jiawei Zha, Ting wang, Chengfeng Pan, Kexun Chen, Fenqin Hu, Shuai Zou
College of Physics, Optoelectronics and Energy, Collaborative Innovation Center of Suzhou Nano Science and Technology, and Jiangsu Key Laboratory of Thin Films, Soochow University
1 Shizi street, Suzhou 215006, China. Tel: 0086 512 69157039. Email: xdsu@suda.edu.cn

ABSTRACT: Mass production of diamond-wire-sawn multi-crystalline silicon (DWS mc-Si) solar cells reach a tipping point with the maturement of metal catalyzed chemical etching (MCCE) technique. However, nano-textured DWS mc-Si solar cells based on Ag-MCCE still suffer from the certain saw-mark remains and the color differences among grains, and also slightly decrease of open-circuit voltage (V_{oc}). Unoriented etching of Cu-MCCE not only depress the saw marks and the color differences, but also result in random shallow pits, which acted as artificial defects to be easily converted into a submicron-texture by using a conventional HNO3/HF etching. For the first time, we demonstrate that the efficiency of DWS mc-Si solar cells with Cu-MCCE process is over 19%, especially with improved V_{oc} due to better surface passivation. Thereby, cost-effective Cu-MCCE is of great potential in PV industry.
Keywords: Diamond-wire-sawn mc-Si solar cell, metal catalyzed chemical etching, submicron texture, power conversion efficiency

1 INTRODUCTION

Nowadays, several hundred thousand tons of solar-grade silicon ingots are sliced into more than ten-billion wafers by either multi wire slurry sawn (MWSS) or diamond wire sawn (DWS) technique each year. The cost of whole photovoltaic system can be reduced largely by popularizing DWS technique from single-crystalline silicon to the multi-crystalline silicon (mc-Si), due to high slicing speed and more output of wafers per ingot. More important, the efficiency of DWS mc-Si solar cells can reach same level of MWSS ones by using of nano-texture technique. [1,2,3] In general, the nanostructured silicon is usually called as black silicon due to excellent light trapping in wide spectrum (300-2000nm). [4] However, in an actual solar cell, the nanostructure layer overlapped with the phosphorus-diffused n^+ zone will introduce more electron-hole (*e-h*) recombination centers and nonuniform doping, thus degrading the cell performance. [5] As a trade-off, the thickness of nanostructure layer was usually controlled below 200nm to retain both optical and electric gains.[6]

Cost-effective MCCE is a bottom-up approach to produce nano-texture on the silicon surface. Among several metals such as Pt, Au, Ag and Cu, Ag is most investigated in MCCE process due to both excellent catalytic activity and the relative cost-effective. Such highly nanostructured black silicon must be modified by post-etching via either alkali or acid etching for better passivation and less photo-generated electron-hole recombination on the surface. In past few years, we have developed an Ag-MCCE process to make black mc-Si (Bmc-Si) solar cell in production line.[1,2,3] However, with more and more diamond wire sawn (DWS) mc-Si wafers with visible saw-marks came to the market, some limitations of Ag-MCCE emerge, such as the slight decrease of V_{oc}, the certain remains of saw-marks and the color differences among grains. Essentially, Ag-MCCE is an oriented etching because the digging direction of Ag nanoparticles is along <100> directions.[7] As a result, the Bmc-Si cells usually show color differences among different grains after SiN_x coating via conventional PECVD.

In our opinion, an unoriented etching method can eliminate the color differences among grains, while it's able to produce a submicron-texture (more than 300nm) for depressing saw-marks and benefiting the passivation for better V_{oc}. It was reported that the Cu nanoparticles do not have any preferential etching direction and result in shallow pits on the silicon surface due to its much lower redox potential relative to those of silver and gold particles.[8] In this work, we demonstration that the Cu-MCCE plus HNO_3/HF post etching has great potential to produce submicron-textures from 300-600nm, and the efficiency of submicron-texture is over 19%.

2 EXPERIMENTAL

Multi-crystalline Si wafers (15.6 × 15.6 × 0.18 cm^3, p-type, specific resistance ρ = 1-3 Ω.cm, GCL company, China) has been used for this work. The Cu-MCCE process is detailed as following. First, the DWS mc-Si wafer was applied a damage removal etch (DRE) in KOH solution at 80°C for 100s. Second, Cu nanoparticles were deposited on the surface of DRE *mc*-Si wafer by dipping the wafer in $Cu(NO_3)_2$/HF aqueous solution. Third, the samples were etched in HF/H_2O_2 aqueous solution at room temperature to produce nano-pits on the surface. Fourth, the wafer was dipped in a diluted HNO_3 solution in order to remove Cu. Fifth, the wafer was etched in HNO_3/HF aqueous solution to convert above nano-pits into submicron-texture. After whole Cu-MCCE, a standard solar cell's process was applied, i.e., the phosphorus diffusion to form *p-n* junction emitter, the PECVD to deposit the SiN_x antireflection/passivation layer, and the screen printing to metalize top/bottom electrodes.

3 RESULTS AND DISCUSSION

It was reported that Cu are directly adsorbed by taking an electron from the Si due to higher electronegativity than Si.[9] Our systematic experimental confirmed that the presence of fluorine promotes the intermixing of Si and Cu by dissolving the silicon oxide with HF, and the growth rate of Cu particles increases with the diluted HF concentration.[10] Due to the different redox potentials of Ag^+/Ag ($E_{Ag+/Ag}$=0.8V) and Cu^{2+}/Cu ($E_{Cu2+/Cu}$=0.34V) concerning the Si conduction and valence band edge energies (E_c=-0.45V and E_v=0.67V), the reduction behaviors of Ag and Cu are different. Ag nano-particles can be reduced by injecting holes from Ag^+ into the valance band due to much positive potential than E_v of silicon. In contrast, Cu nano-particles may be reduced by drawing the electrons drifted from the conduction band since its redox potential is positioned in the bandgap. When copper ions diffuse near the surface of bare silicon, they take electrons from the underlying silicon and the Cu is deposited there. In same moles of metal reduction, one Cu atom will take two electrons but Ag takes one electron from Si. In the case of Cu reduction, the silicon will be oxidized stronger than that of Ag by losing two electrons or being injecting holes. The oxide layer will be removed by HF to maintain the growth of Cu nano-particles by drawing electrons from Si substrate, meanwhile a pitlike shadow will be remained.[11] It was reported that the silicon nanopores could be produced on silicon surface by using either one-step or two-step Cu-MCCE. [12, 13] However, to meet the requirements of high efficiency and mass production, the both processes must optimize the nanostructure to reduce surface recombination, and shorten whole process time (i.e., one-step process takes more than 2 h in ref.12). In Table I, we propose an alternative Cu-MCCE process, in which the Cu depositing solution consisting of only $Cu(NO_3)_2$ and HF while each step takes maximum 200s.

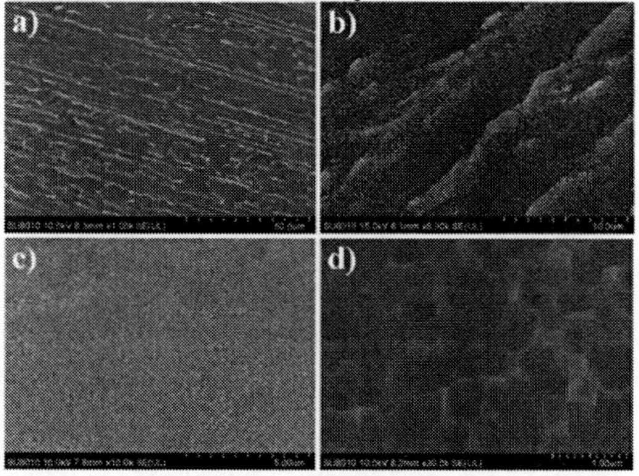

Figure 1: Surface cross-section morphologies of mc-Si wafer after a) DRE, b) Cu-coating, c) HF/H_2O_2 etch, and d) HF/HNO_3 etch.

As shown in **Figure 1**, the saw marks still visible after DRE in KOH, and the Cu nano-particles are much denser above the saw marks than other places, may attribute to the more defects remained here. Therefore, the saw marks will be etched faster in HF/H_2O_2 solution, and resulting in a uniform but rough surface, where the nano-pits are in similar size of Cu nanoparticles (~100nm), as shown in Fig. 2c. Note that, the average size of Cu nanoparticles is large than Ag nanoparticles (~100nm vs. 20-50nm).[1] Therefore, the visible saw marks are removed mostly. After HNO_3/HF etching, the nano-pits will be converted into a submicron-texture of bowl-like shape, as shown in Fig. 2d. Generally, HNO_3/HF solution is used to etch the as-cut mc-Si wafer, where the heavily damaged regions are preferentially etched to form worm-like texture.[14] In the case of DWS mc-Si wafer, it's difficult to form sufficient depth of texture due to shallow sawn-damaged layer. In this work, the shallow nano-pits intentionally introduced by Cu-MCCE can act as random artificial defects for further formation of submicron texture.

Figure 2 presents the reflectivity (R) and external quantum efficiency (EQE) of DWS mc-Si wafers/cells dependences on wavelengths. The inserted images indicate two typical morphologies of submicron-textured wafers after 120s and 150s etching in HNO_3/HF solution. Here, HNO_3 is a strong oxidant to form thin layer of SiO_x on the surface of silicon wafer while HF is etchant to dissolute the SiO_x. The initial shallow nano-pits are now transferred into the bowl-like textures of hundreds nanometers in both length and depth. The opening diameter (d) and the height (h) of the bowl-like submicron-texture increase with the increase of etching time simultaneously. However, the aspect ratio $r = h/d$ varied with the etching time, and a large r means a better light trapping due to more gradient refraction index. Although 150s etched wafer (r=0.83 and R=18.21%) has weaker light trapping ability than that of the 120s etched wafer (r=0.95, and R=15.59%), but its cell EQE values below 400nm are superior than that of 120s etched one. It's well-accepted that the number of recombination centers which is proportional to the surface area.[15] The results imply that the large size of submicron-texture can facilitate the surface passivation by SiN_x coating, and could lead to the better cell performance.[1] Certainly, with the increase of the feature sizes of micron-textures, the light trapping ability will be decrease, and mainly result in a less gain in the short circuit current. Therefore, a tradeoff between optical gain and electric loss must be considered to achieve high efficiency.

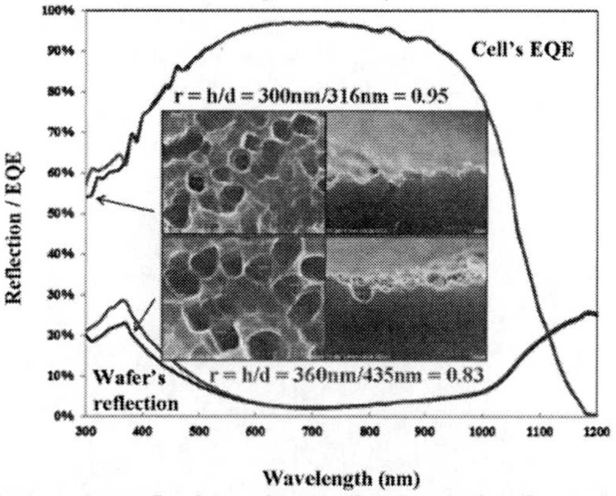

Figure 2: Reflection and EQE of QSC mc-Si wafers and cells with different submicron textures. Insert: Surface and cross-sectional SEM images of submicron textures with different aspect ratios.

Main characteristics of two typical DWS mc-Si solar cells with different submicron-textures are detailed in **Table I**. The measurement area is 15.6×15.6cm² under AM 1.5G illumination. We noted that the size of submicron-texture has influence to the cell's performance. The efficiency of our best submicron-textured DWS mc-Si solar cell with aspect ratio r=0.95 and r=0.83 reaches 18.75% and 19.06%, respectively. With increase of feature size in the later, the V_{oc}, serial resistance (R_s), the shunt resistance (R_{sh}) and filling factor (FF) are improved, and I_{sc} decreases slightly due to higher light reflection. To our knowledge, 19.06% is the highest reported efficiency of DWS mc-Si solar cell based on Cu-MCCE technique.

Table I: Main characteristics of Cu-MCCE DWS *mc*-Si solar cells

r	R (%)	V_{oc} (mV)	I_{sc} (A)	R_s (mΩ)	R_{sh} (Ω)	FF (%)	Eff (%)
0.95	15.6	637.2	9.122	2.4	116	78.5	18.75
0.83	18.2	639.3	9.112	1.9	183	79.6	19.06

Figure 3 compares the etching mechanism of Ag- and Cu-MCCE. In HF/H_2O_2 etching step, Ag nanoparticles can dig along <100> directions of silicon very deep up to several hundred nanometers, thus result in deep nano-hole arrays in the wafer.[16] The nano-hole arrays in the adjacent grains of different orientations may have large separation angle, will certainly bring the color difference to the wafer as well as to the final cell. Here, we like to emphasis that the color difference usually will not influence the efficiency of final cell, but may be not acceptable by some of customs. However, it's still of interests to clarify this issue. In case of Cu, no deep nano-holes but shallow nano-pits formed on the surface, although the surface was roughened to some extent. Moreover, no Cu particles were observed due to the dissolution of Cu particles during the etching process. Therefore, we speculate that there are two factors taking responsibility for the huge difference between Ag- and Cu-MCCE. First, the dissolution of Cu by H_2O_2 needs two electrons, meaning the Si underneath Cu particles will be oxidized stronger than that underneath Ag particles; the former pitlike shadows became deeper after HF etching. Second, Cu particles, about double sizes of Ag ones, are difficult to dig deeply, thus produce shallow nano-pits on the surface of wafer. Such nano-pits can facilitate the further formation of submicron-textures.

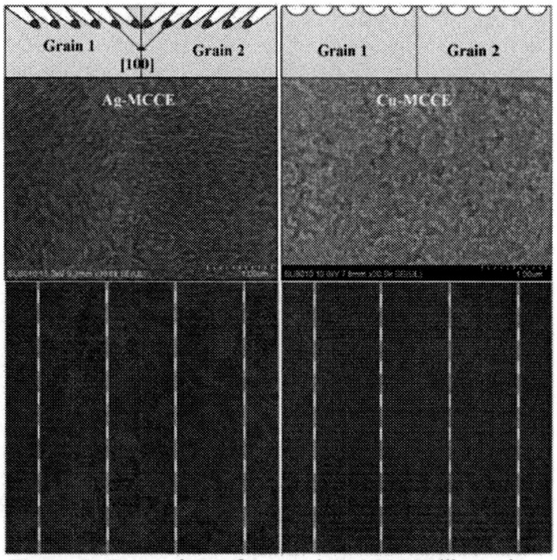

Figure 3: Comparison of Ag- and Cu-MCCE: illustration of etch mechanism (upper), SEM images of nanostructured wafer (middle), and photos of cells (lower). The measurement area is 15.6×15.6cm^2, and the corresponding current density of Pmc-Si is 36.7 mA/cm^2 under AM 1.5G illumination. The efficiencies are measured with IV measurement system (Berger PSL-SCD, Germany) calibrated with the standard cells of Fraunhofer Institute for Solar Energy Systems, Germany.

4 SUMMARY

In summary, 19.06%-efficient DWS mc-Si solar cells with submicron-texture have been fabricated by using Cu-MCCE process on an industrial production line, and this value is comparable to that of Ag-MCCE. Besides, the submicron-texture can depress saw-marks and color differences of the DWS mc-Si cell, also leads to higher V_{oc} due to better surface passivation. Moreover, cost-effective Cu-MCCE is unique and capable to produce submicron-textures for various silicon wafers, i.e. sc-Si, quasi-single crystalline silicon and direct silicon, thereby is of great potential in PV industry.

ACKNOWLEDGMENT: This work was supported by the Key University Science Research Project of Jiangsu Province under Grant No. 16KJA140002, and by the Priority Academic Program Development of Jiangsu Higher Education Institutions (PAPD).

REFERENCES

[1]X. Y. Ye, S. Zou, K. X. Chen, J. Li, J. Huang, F. Cao, X. S. Wang, L. J. Zhang, X. F. Wang, M. R. Shen, and X. D. Su, Adv. Funct. Mater. **24**, 6708 (2014).

[2]F. Cao, K. X. Chen, J. J. Zhang, X. Y. Ye, J. J. Li, S. Zou, and X. D. Su, S Sol. Energy Mater. Sol. Cells 141, 132-138 (2015).

[3]F. Q. Hu, Y. Sun, J.W. Zha, K. X. Chen, S. Zou, L. Fang, and X. D. Su, Sol. Energy Mater. Sol. Cells **159**, 121-127 (2017).

[4]M. Halbwax, T. Sarnet, Ph. Delaporte, M. Sentis, H. Etienne, F. Torregrosa, V. Vervisch, I. Perichaud, and S. Martinuzzi, Thin Solid Films **516**, 6791 (2008).

[5]M. D. Kelzenberg, S. W. Boettcher, J. A. Petykiewicz, D. B. Turner-Evans, M. C. Putnam, E. L. Warren, J. M. Spurgeon, R. M. Briggs, N. S. Lewis, and H. A. Atwater, Nature Materials **9**, 239 (2010).

[6]F. Toor, H. M. Branz, M. R. Page, K. M. Jones, and H. C. Yuan, Appl. Phys. Lett. **99**, 103501 (2011).

[7]U. Gangopadhyay, S. K. Dhungel, K. Kim, U. Manna, P. K. Basu, H. J. Kim, B. Karunagaran, K. S. Lee, J. S. Yoo, and J. Yi, Semicond. Sci. Technol. **20**, 938 (2005).

[8]J. P. Lee, S. Choi, and S. J. Park, Langmuir **27**, 809 (2011).

[9]T. Ohmi, T. Imaoka, I. Sugiyama, and T. Kezuka, J. Electrochem. Soc. **139**, 3317 (1992).

[10]M. K. Lee, H. D. Wand, and J. J. Wang, Solid-State Electronics **41**, 695 (1997).

[11]H. Morinaga, M. Suyama, and T. Ohmi, J. Electrochem. Soc. **141**, 2834 (1994).

[12]Y. T. Lu, and A. R. Barron, J. Mater. Chem. A **2**, 12043 (2014).

[13]F. Toor, J. H. Oh, and H. M. Branz, Prog. Photovolt: Res. Appl. **23**, 1375 (2015).

[14]J. Acker, T. Koschwitz, B. Meinel, R. Heinemann, and C. Blocks, Energy Procedia **38**, 223 (2013).

[15]M. Algasinger, J. Paye, F. Werner, J. Schmidt, M. S. Brandt, M. Stutzmann, and S. Koynov, Advanced Energy Materials **3**, 1068 (2013).

[16]K. Tsujino, and M. Matsumura, Adv. Mater. **17**, 1045 (2005).

EARLY EFFICIENCY PREDICTION OF SILICON HETEROJUNCTION CELLS PROCESSED ON THERMAL DONOR-RICH CZOCHRALSKI WAFERS

J. Veirman, R. Varache, A. Danel, M. Albaric, E. Letty, B. Martel, C. Roux
Univ. Grenoble Alpes, INES, F-73375 Le Bourget du Lac, France
CEA, LITEN, Department of Solar Technologies, F-73375 Le Bourget du Lac, France

ABSTRACT: The crystalline/amorphous silicon heterojunction solar cell technology (SHJ) features among the highest efficiency potentials of all cell architectures, but also the most stringent specifications regarding the substrate quality. In this regard, multi-millisecond carrier lifetime wafers are mandatory to leverage the SHJ architecture benefits. As a consequence, the content of recombination-active defects must be carefully controlled in the silicon substrate, as strong efficiency reductions may otherwise arise. In this context, the presence of defects such as thermal donors can severely limit the final device efficiency, and their presence in excessive concentrations should be detected as early as possible in the cell process in order to better manage the cell production/understand the performance variations. This paper introduces a novel technique, based on OxyMap measurements on as-cut wafers, which allows an upstream prediction of the final device performance assuming that thermal donors are limiting the carrier lifetime.

Keywords: heterojunction solar cell, silicon, thermal donors, oxygen defects

1 INTRODUCTION

Owing to the very efficient passivation provided by hydrogenated intrinsic amorphous silicon (a-Si:H) layers, amorphous/crystalline Si heterojunction (SHJ) can achieve particularly high open-circuit voltages (V_{oc} around 750 mV [1]), allowing efficiencies (η) over 25% on double-side contacted large area devices [2]. In order to leverage this η potential, the substrate – mostly obtained from Czochralski (Cz) monocrystalline ingots - must however feature bulk minority carrier lifetime (τ_{bulk}) well in excess of 1 ms.

Some of the most prominent defects that prevent such lifetimes to be achieved are oxygen (O)-related thermal donors (TD), formed in the course of crystal cooling. While TD are destroyed during industrial homojunction processes, these recombination-inducing defects survive the SHJ process (<250°C) and can reduce by up to 4% absolute the η of the finished device [3-4]. It is therefore of paramount importance to detect wafers featuring excessive TD concentrations ([TD]) and to predict the expected η, in order to better manage the production of a SHJ manufacturing line. This prediction should also ideally be performed at the earliest in the process flow, i.e. at the as-cut stage. This would in turn allow to redirect in time a given wafer quality to a particular production line (targeting a particular application or market for instance), or understand performance variations in retrospect (by decoupling substrate and process related issues).

Despite considerable effort from Cz manufacturers, it is estimated that 10-20% of commercial Cz wafers are affected by O-related defects, such as TD [5]. To the best of our knowledge, no characterization techniques have demonstrated so far the possibility, at the as-cut state, to reliably identify such Cz wafers and to predict the resulting η reduction. As cell η are constantly increasing, the τ_{bulk} requirements become more stringent, and the need for such technique rises.

In this context, the purpose of this work is to demonstrate the possibility to predict at the as-cut stage the η reduction induced by TD present in the wafer bulk. This prediction is a three-step process generated with the OxyMap technique [6] involving:

1) the measurement of the phosphorus ([P]) and TD concentrations,
2) the computation from these values of the resulting bulk lifetime τ_{bulk} = f([TD], [P]) and its dependence with injection level (Δn) using the recombination parameters from M. Tomassini et al. [4],
3) the prediction of η as a function of [TD] and [P] by feeding the computed τ_{bulk}=f(Δn) into a SHJ cell model generated with the Silvaco software.

The different steps are detailed hereinafter and the predictions are validated by means of comparison with experimental details gathered from the SHJ pilot line at CEA-INES.

2 STEP 1: MEASUREMENT OF [P] AND [TD]

The procedure was tested on wafers from a complete commercial n-type Cz ingot labelled as high quality, i.e. with τ_{bulk} >1 ms (as specified by the provider). This ingot was delivered in the form of 45 boxes of a hundred 180 µm thick wafers each, with no box-to-box traceability. The as-cut wafer from the top (i.e. closest to the ingot seed end) of each box was sampled. All wafers were then stacked and placed in an automatic OxyMap tool from AET Technologies, featuring a 100-wafer batch capability (runtime around 4 hours). The OxyMap technique relies on the change in wafer resistivity before/after some specific thermal treatments.

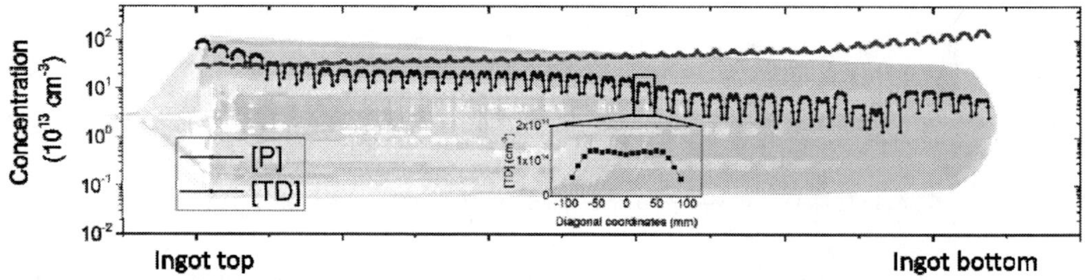

Figure 1: Variations in phosphorus ([P]) and thermal donor ([TD] concentration along the studied high quality ingot

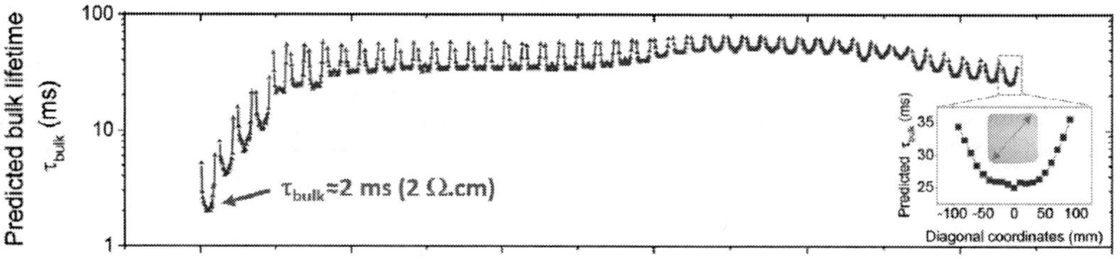

Figure 2: Variations in bulk carrier lifetime (τ_{bulk}) at an injection level of 10^{15} cm^{-3} along the studied high quality ingot, computed from data in Fig. 1.

It yields a wealth of valuable data in the form of a linescan along the wafer diagonal, among which [P] and [TD] (see extensive details in [6]).

The measurement results are shown in Fig. 1, where the linescans for all 45 wafers are stitched together (an arbitrary linescan is given as onset as an illustration). Due to the absence of traceability, it was a prerequisite to determine the relative order of boxes along the ingot. This reconstruction was performed using the [P] data. Indeed, as P experiences segregation during ingot pulling, [P] steadily increases from ingot seed to tail, and can thus be used to re-establish traceability *a posteriori*. Re-ordering based on the as-received ρ alone would have been erroneous since ρ may not reflect the P segregation (and thus the correct order), particularly when [TD] is close to [P], usually at the ingot top.

It is interesting to notice that large radial variations in [TD] are revealed, with [TD] being significantly higher at the wafer center. This is the direct consequence of the higher interstitial oxygen concentration ([O_i]) at the wafer center, typical of Cz ingots. In addition, one can observe that while [TD] is relatively low in most of the ingot, it rapidly increases towards the ingot seed end. This well-known fact stems from the large interstitial oxygen concentration [O_i] and comparatively slow cooling, specific to the seed end. OxyMap results indeed revealed that [O_i] varies from 18 ppma (seed end) down to 13 ppma (tail end) (OxyMap values in compliance with the industry-custom conversion factor DIN of 2.45×10^{17}.cm^{-2}). In parallel, the thermal history index (THI) computed from [O_i] and [TD] data [6], which gives an indication of the cooling velocity, reveals a 6-fold slower cooling at the top than at the tail. Interestingly for our purpose, the measured [TD]~10^{15} cm^{-3} at the seed end exceeds the "safe" threshold value of 5×10^{14} cm^{-3} reported in [4], meaning that an η reduction is to be expected on SHJ cells processed on the concerned wafers.

3 STEP 2: CALCULATION OF THE CARRIER LIFETIME LIMITED BY THERMAL DONORS

While the exact origin of their recombination activity still requires complementary studies, the presence of TD is associated with a non-negligible recombination activity. In this work, the TD-limited SRH lifetime ($\tau_{SRH,TD}$) was modeled using the simplified SRH statistics with the effective recombination parameters determined in [4] as input. These parameters were established in [4] from 5 wafers sweeping a wide range of [TD], where only 3 were representative of what can be met in commercial ingots (up to 2×10^{15} cm^{-3} based on our experience). In addition, the desired [TD] were achieved by the authors using deliberate 450°C anneals, in contrast to a practical case where TD are generated during ingot cool down in the range 350-550°C. Nevertheless, the authors compared their measurements with experimental τ_{eff} values obtained on as-grown ingots, and the comparison revealed a very good correlation, supporting the fact that $\tau_{SRH,TD}$ for a given [TD] is not primarily dependent on how TD were generated (i.e. on the thermal history). Therefore, we decided to trust this model further and use it for practical cases, where TD originate from the ingot thermal history and not deliberate anneals.

τ_{bulk} was subsequently computed using the most up-to-date expression for intrinsic recombination channels (τ_{int}) from the work of Richter et al. [7], combining τ_{SRH} and τ_{int} using the usual inverse law ($\tau_{bulk}^{-1} = \tau_{int}^{-1} + \tau_{SRH,TD}^{-1}$). The equilibrium free electron concentration (n_0), involved in the calculations of both τ_{int} and $\tau_{SRH,TD}$, is set to be 2×[TD]+[P], which takes into account the double donor character of TD in standard resistivity Cz-Si (>1 Ω.cm [8]). Fig. 2 depicts the results of the calculations for all positions across the studied ingot ($\Delta n=10^{15}$ cm^{-3}). Due to the high [TD] at the ingot top, the predicted τ_{bulk} drops to about 2 ms. Such values are in good agreement with τ_{eff} measurements performed after the a-Si:H deposition step (Sinton WCT-120), where τ_{eff} around 1.5 ms were measured (yielding τ_{bulk} = 1.8 ms assuming an effective surface recombination velocity of S_{eff}=0.5 cm.s^{-1}). Note that the radial variations in [TD] discussed in the previous section leads to τ_{bulk} being lower at the wafer center. Also, the predicted values above 10 ms should be regarded with care owing to the the difficulty in measuring high bulk carrier lifetime at the wafer level, and to the lack of data points at low [TD] in [4].

4 STEP 3: MODELLING THE INFLUENCE OF TD ON THE SHJ CELL PERFORMANCE

The 160 µm thick n-type SHJ rear emitter structure was modelled using the 2D Silvaco Atlas package. In order to model a-Si:H and ITO layers, measurable quantities (thicknesses, ITO sheet resistance, a-Si:H activation energy, etc) are used. Ray tracing was activated, and a modified incoming light spectrum was calculated separately and used as input to account for reflections. While the intrinsic a-Si:H was effectively assumed to be undoped (mid-gap Fermi level (E_F)), the activation energy (E_a) was set to E_a=E_c-E_F=0.2 eV for (n)a-Si:H and E_a=E_F-E_V=0.3 eV for (p)a-Si:H. The interfacial defect density (D_{it}) used until now was slightly fine-tuned to better match the η values achieved in practice on the cell production day. To model the busbar-less metallization scheme chosen for the cell process (see section 5), a suitable external series resistance was added to model resistive losses in the metal grid (an extended details about the simulations shall be given in a forthcoming separate paper). Simulations using this model yielded a maximum η of 22.67% under standard test conditions.

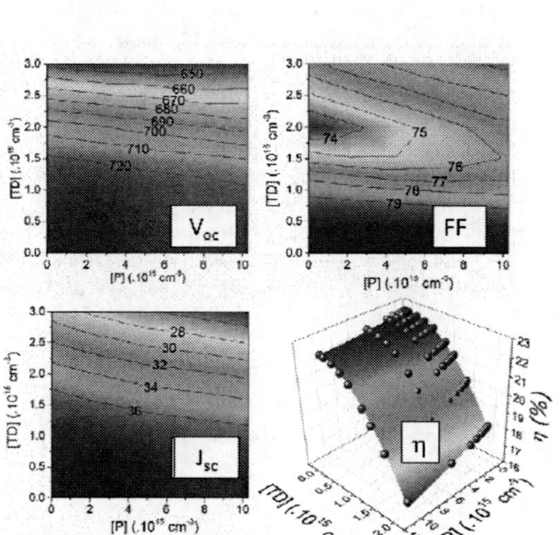

Figure 3: Calculated cell current-voltage parameters as a function of [TD] and [P]. On the efficiency graph, the color surface is the surface polynomial parametrization (see text) derived from the fitting of the simulated data (red balls). Maximum values obtained for [P]=[TD]=0 cm^{-3} equal 733 mV, 80.2, 38.5 mA.cm^2, 22.67%.

To quantify the effect of TD on the cell performance, τ_{bulk} was calculated using the procedure described in section 3 for several usual values of [P] and [TD]. Again, n_0 was set to be 2×[TD] + [P]. As a doubly ionized species, TD are known to scatter free charge carriers 4 times more effectively than singly ionized impurities such as P, thereby particularly hindering carrier mobilities [9]. This enhanced scattering cross section was not taken into so far in our models (only a 2-fold effect was set). Yet, we calculated that this approximation only leads to an overestimation of the minority hole mobility by 8% at [TD]=3×10^{15} cm^{-3}, which is unlikely to have any significant effect on the extracted current-voltage parameters.

Fig. 3 shows the resulting calculated variations in the main cell parameters as a function of [P] and [TD]: short-circuit current density (J_{sc}), fill factor (FF), V_{oc} and η. It calls for several comments. First, it is observed that the reduction in τ_{bulk} upon increasing [TD] (and to a lesser extent [P]) leads to a reduction in V_{oc} and J_{sc}. For V_{oc} (V_{oc}=kT/q×ln(Δn|$_{Voc}$×(Δn|$_{Voc}$+n_0)/n_i^2), the increase in n_0 (i.e. in either [P] and/or [TD]) cannot counterbalance the reduction in $\tau_{eff}|_{Voc}$ and thus in Δn|$_{Voc}$ (=G× $\tau_{eff}|_{Voc}$). The relatively strong dependence of J_{sc} on [TD] is due to a combination of two factors: 1) the large sensitivity of the rear emitter simulated structure to the diffusion length. This structure indeed requires that photogenerated carriers (mostly at the front side) cross the entire substrate thickness and thus must have a higher τ_{bulk} than in front emitter structures 2) the large asymmetry of the TD-related capture cross sections (σ_p>>σ_n) that leads upon increasing [TD] to a fast reduction of $\tau_{SRH,TD}$ at low Δn (where the cell operates at J_{sc}) thereby affecting the collection efficiency faster than other defects with more symmetric σ.

Regarding the FF, an a priori surprising behavior is revealed: for a given [P], the FF is first reduced down to a minimum value upon increasing [TD], and then unexpectedly rises again upon further increase in [TD]. Such a behavior due to the presence of SRH centers with asymmetric capture cross sections has already been experimentally observed and explained in the past, for instance in metallic-rich silicon wafers [10] or boron-doped samples after activation/formation of boron-oxygen complexes [11]. It was shown by the authors of these studies to be inherent to the bulk Si (no influence of the cell process), more precisely to the strong injection-dependent τ_{SRH}. To check if the same effect was at play, we calculated analytically the influence of increasing [TD] on the implied current-voltage curves of an ideal solar cell (no series/shunt/internal resistances, no surface recombination) following a similar formalism as in the WCT-120 spreadsheet from Sinton Instruments. The doping was set to be n_0=5×10^{14} cm^{-3} and the thickness to 160µm. The obtained curves are shown in Fig. 4 and are commented below.

Roughly speaking, from a material point of view, the ideal situation for the FF is achieved when τ_{eff} is highest at Δn corresponding to the maximum power point (Δn|$_{MPP}$), and drops as rapidly as possible towards Δn|$_{Voc}$. This leads to the sharpest possible drop after MMP on the current-voltage characteristics, which occurs in a material where τ_{eff} is limited by intrinsic recombination channels (i.e. no SRH centers, green curve in Fig. 4, iFF=88.3%). Upon gradually introducing TD, τ_{eff} decreases and the Δn range swept along the current-voltage characteristics shifts consequently to lower values. As a result, the difference between τ_{eff} at MPP and V_{oc} operating points first decreases, before the situation becomes reversed (τ_{eff} is lower @ Δn|$_{MPP}$ than at Δn|$_{Voc}$, red curve, iFF=75.9%), after what an intermediate situation is eventually reached, where τ_{eff} is approximately constant for all operating points along the current-voltage characteristic (blue curve, iFF=82.1%). The effect on the corresponding current-voltage curves is also shown as an inset, and reveals a "shoulder" for [TD]=2×10^{15} cm^{-3}, responsible for the poor calculated iFF. Note that the amplitude of these variations in iFF closely match that of FF from the Silvaco simulation, supporting that the effect observed in Fig. 3 is indeed caused by the substrate.

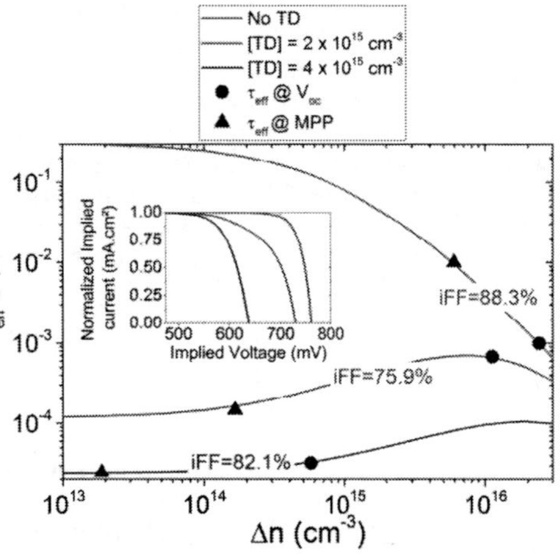

Figure 4: Calculated operating points at MPP and V_{oc} (superimposed to the τ_{eff} curves) for 3 different [TD]. The resulting implied current-voltage curves are plotted as an inset. See text for more detail.

The predicted η as a function of [TD] and [P] from Fig. 3 was subsequently parametrized with an empirical polynomial expression, as no straightforward analytical expression could be hoped for. It was found that the simulated points (red balls in Fig. 3) could not be adjusted by simple empirical expressions, and the following polynomial form was used, with the parameters being listed in Table 1:

$$\eta = z_0+a\times[P]+b\times[TD]+c\times[P]^2+d\times[TD]^2+f\times[P]\times[TD] \quad (1)$$

Table 1: List of parameters used in Eq. (1).

Parameter	Value
z_0	22.67
a	-3.907×10^{-17}
b	3.481×10^{-16}
c	3.041×10^{-34}
d	-1.244×10^{-30}
f	-8.270×10^{-32}

This parametrization was then used to predict η for the 45 wafers sampled along the ingot (Fig. 5). As apparent from Fig. 5, a significant 0.9% absolute η reduction is predicted locally at the wafer center of the first boxes due to the presence of TD. Conversely, the rest of the ingot should not suffer from the presence of TD.

5 VALIDATION OF THE PREDICTIONS

In order to put the predictions to the test, around 1600 busbar-less SHJ solar cells were fabricated on the CEA-INES pilot line from wafers in the boxes depicted by orange stars in Fig. 5. Details regarding the process flow (texturization, a-Si:H layers and transparent conductive oxide deposition) can be found in [12]. A solar simulator from Pasan was used to measure the device current-voltage characteristics. The experimental η results are plotted in Fig. 6, together with the corresponding predictions. The predictions were calculated assuming that the central τ_{bulk} value drives the overall cell performance. Also, the displayed uncertainties take into account the cumulative uncertainties in OxyMap [P] and [TD] measurements, in the surface fitting in section 4, as well as in the $\tau_{SRH,TD}$ model in the [TD] range 0 to 2×10^{15} cm^{-3}. It can be observed that in the TD-rich region, the η predictions are in very good agreement with the experimental data, where the predicted η drop of ≈0.9% absolute is observed in practice. It is worth recalling here that the wafers responsible for this drop still feature τ_{bulk} of around 2 ms, well in excess of current wafer specifications from top-tier manufacturers (mostly 1 ms). This emphasizes how demanding and responsive to variations in bulk quality the concerned SHJ cell architecture is.

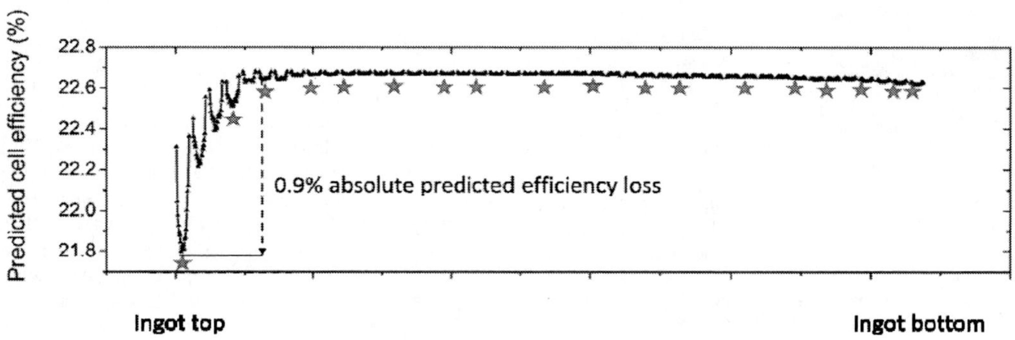

Figure 5: Predicted variations in efficiency (η) along the 45 boxes of the studied high quality ingot. The stars indicate the positions used for the SHJ cell processing.

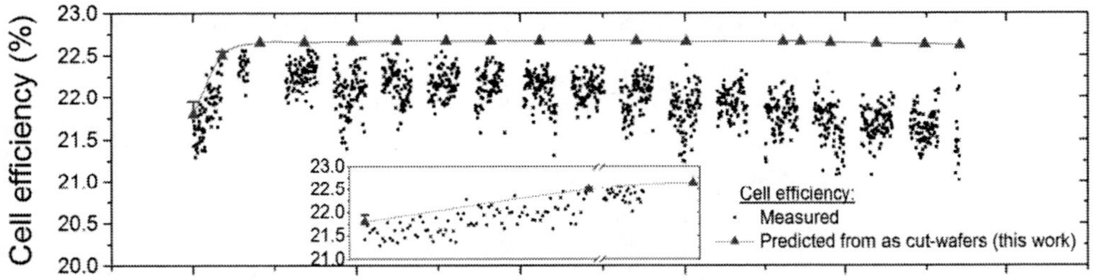

Figure 6: Predicted vs. experimental efficiency (η) variations for the 16 studied boxes. The inset is a zoom on the seed end fraction of the ingot.

On the other hand, the slow and progressive η reduction observed along the ingot is not reproduced by the predictions. This suggests that other defects, unconsidered in our τ_{bulk} model, limit η (possibly involving intrinsic defects and/or metallic impurities).

6 CONCLUSIONS AND PERSPECTIVES

A novel procedure to detect wafers affected by thermal donors and predict the effect of thermal donors on the silicon heterojunction solar cell efficiency was presented. The procedure was validated by means of comparison with experimental cell efficiencies along a complete commercial "high quality" ingot. We demonstrated the effectiveness of the developed procedure in predicting the thermal donor-related efficiency reductions. As the predictions are established from measurements on as-cut wafers (OxyMap), these results highlight the possibility to implement valuable pre-sorting strategies well ahead of production. We hope that this work lays the first stone towards more complete characterization techniques that allow to predict the final device efficiency from upstream measurements on as-cut Cz wafers. The need for such techniques was stressed, as those wafers affected by defects other than thermal donors remain today tricky and costly to detect at the as-cut stage.

The authors would like to thank Meyer Burger for the common developments made in the frame of the CHIC project.

7 REFERENCES

[1] Taguchi, M., Yano, A., Tahoda, S., Matsuyama K., Nakamura, Y., Nishiwaki, T., Fujita, K., Maruyama, E., IEEE J. of Photovoltaics, Vol. 4, No. 1, 2014.
[2] D. Adachi et al., Appl. Phys. Lett. 107, 233506 (2015)
[3] A. Danel et al., in: 31st European Photovoltaic Solar Energy Conference, Hamburg, Germany (2015), pp. 279-283.
[4] M. Tomassini, J. Veirman, R. Varache, E. Letty, S. Dubois, Y. Hu, and Ø. Nielsen, J. of Appl. Phys. 119, 084508 (2016)
[5] Bronsveld, P.C.P. et al. 2015, *Proc. 31st EU PVSEC*, Hamburg, Germany
[6] J. Veirman et al., in Photovoltaics International Vol. 33, September 2016.
[7] A. Richter, S.W. Glunz, F. Werner, J. Schmidt, A. Cuevas, Phys. Rev. B. 86 (2012)
[8] J. Veirman et al, Proc. 6th CssC Workshop, Aix-Les-Bains, France (2012).
[9] M. Claybourn and R. C. Newman, Appl. Phys. Lett 52 (25), 2139-2141 (1988)
[10] D. Macdonald and A. Cuevas, Prog. Photovolt. Res. App. 8, 363-375 (2000)
[11] J. Schmidt et al. Prog.Photovol. Res. Appl. 9, 249-255 (2001)
[12] A. Danel et al., in: 33rd European Photovoltaic Solar Energy Conference, Amsterdam, Netherlands (2017)

RAPID VAPOR-PHASE DIRECT DIFFUSED EMITTER FOR SOLAR CELL APPLICATIONS

Saskia Kühnhold-Pospischil[1], Elke Gust[1], Diana Amiri[1], Armin Richter[1], Bernd Steinhauser[1], Marion Driessen[1], Bernhard Michl[3], Johannes Greulich[1], Jan Benick[1], Stefan Lindekugel[2], Stefan Janz[1]

[1] Fraunhofer Institute for Solar Energy Systems, Freiburg, Germany
[2] now with Sick AG, Waldkirch, Germany
[3] now with, Lechwerke AG, Augsburg, Germany

ABSTRACT: Emitter diffusion using $POCl_3$ or BBr_3 as precursor is a time consuming process and requires chemical treatments to remove silicate glasses after the diffusion process. Moreover, no standard emitter diffusion process using BBr_3 is established so far. A so called *rapid vapor direct diffused emitter* (RVD) might be an alternative to these $POCl_3$ and BBr_3 diffusion processes. Here, we present several RVD emitter profiles in n-type wafers achieved using the *rapid thermal chemical vapor deposition* (RTCVD 160) tool by varying diborane (B_2H_6) concentrations, peak temperatures and durations. Further, temperature induced degradation of carrier lifetimes is observed.

Keywords: Boron, diffusion, rapid thermal processing, doping, n-type

1 INTRODUCTION

In the fabrication chain of Si based solar cells, the emitter diffusion process is time consuming and thus costly. The standard industrial emitter is made by solid phase diffusion of phosphorus or boron from doped silicate glass. The silicate glass is deposited by using $POCl_3$ or BBr_3 and O_2 as precursor gases in a tube furnace. This type of emitter formation process takes more than 45 min and requires additional wet-chemical cleaning steps to remove the silicate glass. Further, even though a p/n-junction is usually only needed on one wafer face, both wafer faces are exposed to the standard tube furnace emitter diffusion. Thus, additional chemical treatments are needed to remove the unwanted emitter on one of the wafer faces (emitter etch back step). Compared to that, the *rapid vapor-phase direct diffused* (RVD) emitter comprises several advantages. During the diffusion process, the hydrogen prevents the formation of a silicate glass layer and thus reduces the time of exposure. In addition, due to the oven design used for the RVD emitter in this contribution, the emitter diffusion is only formed on one wafer surface. Further, as demonstrated by Lindekugel *et al.*, the RVD emitter process can deliver highest solar cell efficiencies [1].

In this contribution, temperatures (980°C to 1100°C) and doping gas concentrations (a mixture of diborane (B_2H_6) and hydrogen) were varied and the resulting RVD emitter profiles in n-type Si wafers were analyzed. To evaluate the influence of temperature induced degradation effects and in order to determine the quality of the emitters, *Photoluminescence* (PL) and *Quasi-Steady State Photo Conductance Decay* (QSSPC) measurements were conducted.

2 APPROACH

The experimental design is shown in Figure 1. FZ, n-type Si wafers (1 Ω cm) were used for the experiments. In addition to that some p-type Si wafers (1 Ω cm) were used for temperature degradation experiments. After standard chemical cleaning steps the wafers were exposed to high temperature treatments within the *rapid thermal chemical vapor deposition* (RTCVD 160) tool [2,3]. For *Group 1*, two temperatures (980°C and 1100°C for 15 min) were applied without any doping gases. After the temperature treatment these wafers were passivated on both sides with Al_2O_3 using atomic layer deposition (ALD) and were subsequently annealed on a hotplate at 450°C for 5 min. Carrier lifetimes were studied by calibrated PL measurements [4].

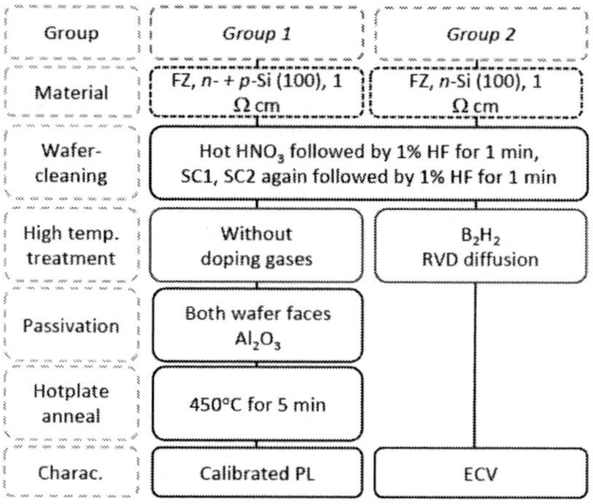

Figure 1: Experimental design for RVD samples.

For *Group 2*, RVD diffusion processes were carried out. To this end, a gas mixture of B_2H_6 diluted in hydrogen (0.25% B_2H_6) and pure hydrogen was used leading to a total B_2H_6:H_2 ratio of around 15:1. Peak temperatures (between 980°C to 1100°C), peak temperature durations (15 min and 30 min) and B_2H_6 gas fluxes (0.3 slm and 1.5 slm) were varied in order to obtain different doping profiles (see Figure 2). The heating ramp of the optically heated furnace was chosen to be 150 K/min. Dopant gas flow started immediately after reaching the peak temperature and stopped after cooling down to 450°C in order to prevent out-diffusion. Doping profiles were studied using the *electrochemical capacitance-voltage profiling* (ECV) method [5].

In order to determine the recombination factor J_{0e} for the diffused boron emitter, textured lifetime samples were processed in the RTCVD reactor. Due to the single-sided nature of the process, an asymmetrical sample layout was chosen with only one side having a random pyramids texture and the diffused emitter passivated by a stack of aluminum oxide and silicon nitride, while the second side was planar and undiffused, but passivated by the same layer stack.

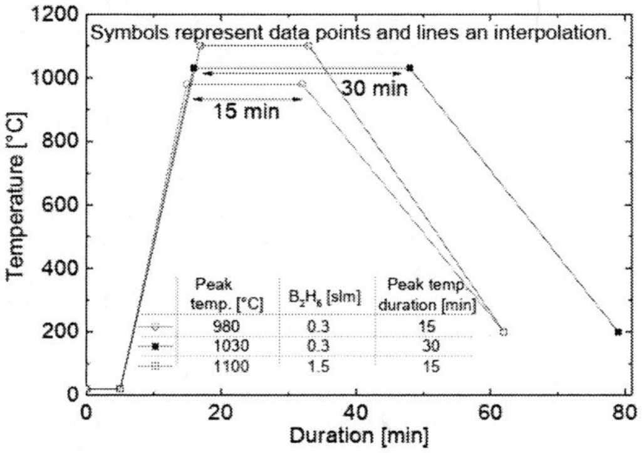

Figure 2: Temperature profiles of *Group 2* samples are given exemplarily. Dopant gas flow starts at peak temperature and stops at 450°C.

3 RESULTS AND DISCUSSION

3.1 RVD emitter profiles

Three different RVD emitter profiles measured by ECV are shown in Figure 3. The three doping profiles correspond to three different temperatures (980°C, 1030°C and 1100°C), temper durations of 15 to 30 min and three different B_2H_2 gas fluxes (0.3 slm and 1.5 slm) as indicated in the small table in the top right corner of the graph. The corresponding temperature profiles to these three doping profiles are given in Figure 2.

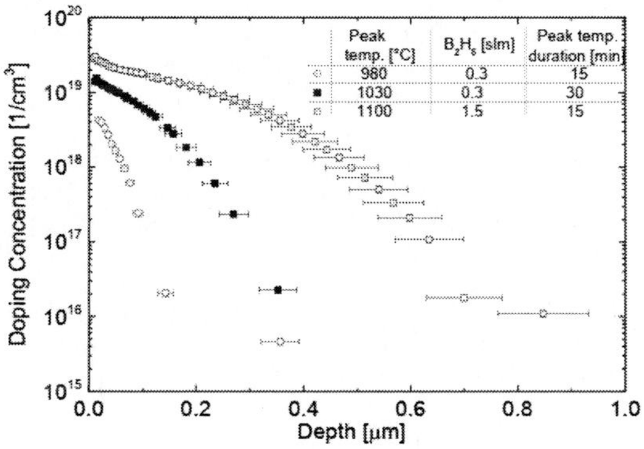

Figure 3: Three different RVD doping profiles obtained by temperature and B_2H_2 doping gas variations (*Group 2*).

Along with these temperature profiles and B_2H_2 variations, surface doping concentrations around 4×10^{18} cm^{-3} to 2×10^{19} cm^{-3} and doping depths of around 0.2 µm to 0.8 µm were obtained. The sheet resistances were calculated from the dopant profiles.

At the lowest investigated peak temperature of 980°C, the resulting emitter profile was too shallow and too lowly doped for application as a homogeneous emitter (sheet resistance of 3000 Ω/sq.). However, it could still be used in combination with a selective emitter structure. At 1030°C peak temperature, the surface concentration would be high enough to result in a low specific contact resistance when contacted by titanium or nickel, but the profile is still very shallow and thus could be prone to metal spiking (sheet resistance of 611 Ω/sq.). The most promising profile shape was determined for 1100°C peak temperature, with a sheet resistance of 185 Ω/sq.. This emitter should allow for a low J_{0e} in the passivated area and at the same time should provide good contact ability and low recombination at the contacts.

At this time, only lifetime results for the blue profile with the highest doping concentration were available. However, as will be discussed in the next section, at this temperature there is already a strong influence of degradation on the determined lifetime and hence we are only able to give an upper limit of the J_{0e} of 100 fA/cm^2 for the passivated surface. This is most likely strongly limited by the degraded bulk as J_{0e} should be in the range of 20 fA/cm^2. Due to the lower doping concentration, J_{0e} for the other two processes would be expected to be 10 fA/cm^2 in case of the black profile (1030°C) or even below in case of the red profile (980°C). However, these values are currently only estimates and remain to be verified by experimental data.

3.2 Temperature induced degradation

To estimate the influence of the temperature treatment in the RTCVD reactor on the bulk lifetime, carrier lifetime maps for samples processed at a peak temperature of 980°C and at 1100°C for 15 min were measured *(Group 1)*. In this case, no doping gas but only hydrogen was used. The calibrated, spatially resolved lifetimes for the *n*-type samples are shown in Figure 4. The mappings show that carrier lifetime is distributed quite inhomogeneously. Striking marks are visible in the PL-maps. It is not clear so far if the different recombination areas are dominated by contaminations or crystal defects (parallel faults). Either the carriers of the reactor made of quartz or inhomogeneous heating are reasonable explanations for both effects.

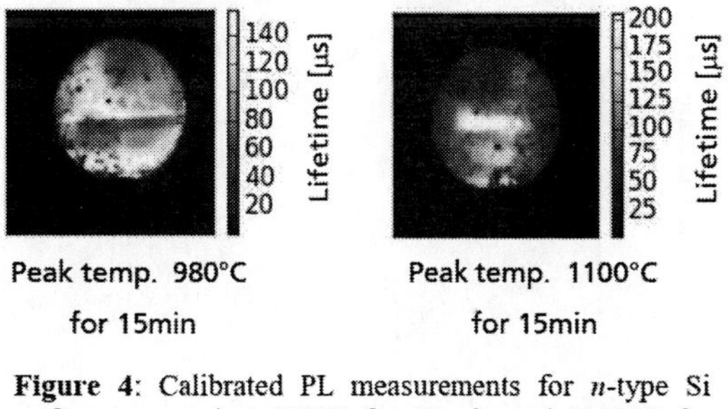

Figure 4: Calibrated PL measurements for *n*-type Si wafers tempered at 980°C for 15 min and 1100°C for 15 min within the RTCVD oven. No doping gases were used.

As shown in Figure 5, similarly processed *p*-type samples revealed even lower lifetimes in the range of 10 µs or below. As *p*-type material is more prone to impurities like iron these results suggest that one degradation mechanism could be due to impurities. Some of the observed structures in the photoluminescence maps could correspond to a variation in the impurity concentration. Unfortunately due to the low signal emitted by the *p*-type samples, quantitatively evaluable iron imaging was not feasible. In addition to the lifetime degradation by impurities, at least at the very high temperature processes at 1000°C and above, it has to be taken into account that intrinsic bulk defects (e.g.) could be introduced by the temperature treatment and thus even in a clean environment, lifetime degradation might be present. Hence, it is important to control the temperature ramps very well as well as providing a clean wafer treatment to prevent such degradation.

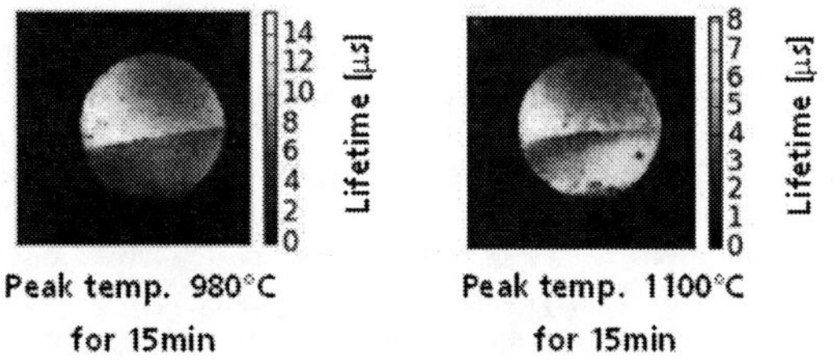

Figure 5: Calibrated PL measurements for *p*-type Si wafers tempered at 980°C for 15 min and 1100°C for 15 min within the RTCVD oven. No doping gases were used.

3.2 Auto-doping

After more than 20 high temperature processes including B_2H_2 gas fluxes within the RTCVD, further variation of doping gas fluxes had no influence on the resulting doping profiles. The observed phenomena correspond to so-called *auto-doping* and have their origin in parasitic depositions inside the reactor. Three doping profiles resulting from this effect are shown exemplarily in Figure 6. For one the plateau concentration is much higher and the emitter depth is also increased compared to the same process in a freshly cleaned reactor (see also blue line in Figure 3). Another observation is that with doubling the diffusion time at 1100°C the dopant depth can be increased by more than 30%.

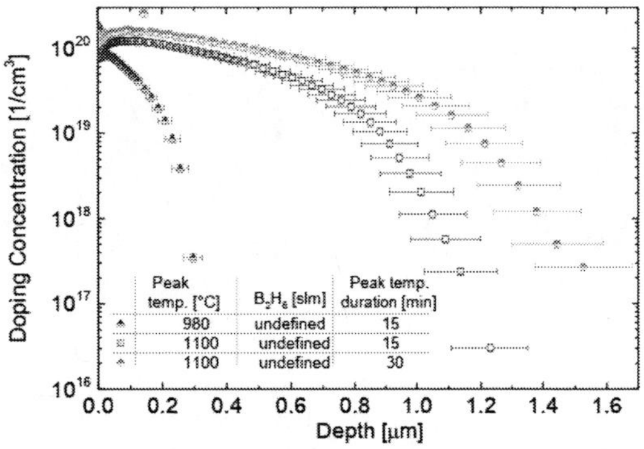

Figure 6: Three different RVD doping profiles obtained by temperature and B_2H_2 doping gas variations after several RVD emitter runs without cleaning the chamber (*auto-doping*).

In our case dummy wafers inside the reactor could be identified as the source of additional doping and a simple exchange of the wafers led again to the original dopant profiles as shown in Figure 3.

4 SUMMARY

In this contribution, different doping profiles in *n*-type wafers were obtained using the RTCVD tool. In order to vary the doping profiles, B_2H_2 concentrations, peak temperatures and peak-temperature durations were varied. At temperatures between 980°C and 1100°C and peak temperature durations of only 15 min to 30 min, doping profiles adequate for highly efficient solar cells with sheet resistance of 185 Ω/sq. could be achieved.

The temperature induced degradation observed seems to be connected to induced crystallographic defects, recombinative impurities or both but needs to be studied further. Furthermore, it needs to be taken into account that auto-doping is present using the RTCVD tool and regular exchange of dummy wafers is needed.

5 ACKNOWLEDGEMENTS

The authors would like to thank Philipp Barth, Astrid Seiler and Karin Zimmermann for wet chemical cleaning processes and passivation layer deposition. Parts of this work were funded by the European Commission (EC GA No. 609788).

6 REFERENCES

[1] S. Lindekugel *et al.* "Emitters grown by rapid vapour-phase direct doping for high efficiency solar cells" 31th EUPVSEC, Hamburg, Germany, (2015).

[2] F. R. Faller *et al.*, "High-temperature CVD for crystalline-silicon thin-film solar cells", IEEE Transactions on electron devices, 1999

[3] S. Reber *et al.*, "The RTCVD160-a new lab-type silicon CVD processor for silicon deposition on large area substrates", Photovoltaic Energy Conversion, 2003

[4] J. A. Giesecke *et al.*, "Minority carrier lifetime imaging of silicon wafers calibrated by quasi-steady-state photoluminescence", Solar Energy Materials and Solar Cells, 2011

[5] E. Peiner *et al.*, "Doping Profile Analysis in Si by Electrochemical Capacitance-Voltage Measurements", Journal of the Electrochemical Society, 1995

CHARGE-CONTROLLABLE MG-DOPED ALOX PASSIVATION LAYERS FOR P- AND N-TYPE SILICON

Hyunju Lee[1], Fumiya Nishimura[2], Haruhiko Yoshida[2], Takefumi Kamioka[1], Naotaka Iwata[1] and Yoshio Ohshita[1]
[1]Toyota Technological Institute, 2-12-1 Hisakata, Tempaku, Nagoya, 468-8511, Japan
+81 52 809 1877, leehyunju@toyota-ti.ac.jp
[2]Graduate School of Engineering, University of Hyogo, Himeji, Hyogo 671-2280, Japan

ABSTRACT: A new class of passivation materials, Mg-doped AlO_x (MAO) was developed and its surface passivation and interface properties on crystalline silicon (c-Si) were investigated in this study. Super-cycle plasma-enhanced atomic layer deposition (PEALD) was employed to fabricate MAO passivation. Capacitance-voltage (C-V) and effective minority carrier lifetime (τ_{eff}) measurements clearly show that Mg doping by super-cycle PEALD can effectively engineer effective fixed charge density (Q_{eff}) and polarity, i.e. field effect passivation of MAO passivation on c-Si without significant degradation of chemical passivation. For instance, Q_{eff} of the MAO passivation layers could be precisely varied from about -5.2×10^{12} to $+1.0\times10^{12}$ cm^{-2} with the increase of Mg dopant concentration from about 4.7 to 51 at.%. Interface trap density (D_{it}) of the passivation layers, however, slightly increased from about 2×10^{10} to 1.5×10^{11} $cm^{-2}eV^{-1}$ with the increase of Mg dopant concentration. In addition, our study also shows that about 4.7 at.% Mg-doped MAO layers can provide superior surface passivation as compared to pure AlO_x layers, and although about 33 at.% Mg-doped MAO layers demonstrated very low Q_{eff} of 1.7×10^{11} cm^{-2}, they achieved considerably low upper-limit effective surface recombination velocity ($S_{eff,max}$) of about 25 and 14 cm/s on p- and n-type c-Si, respectively due to their high chemical passivation.
Keywords: Mg doping, AlO_x passivation, fixed charge control, super-cycle, PEALD

1 INTRODUCTION

Currently, in the field of crystalline silicon (c-Si) photovoltaics, trends toward thinner substrates and higher substrate lifetime have placed an increased importance on surface passivation [1,2]. Recently, AlO_x passivation has been intensively studied and enhanced solar cell efficiencies largely by providing excellent field effect passivation induced by high negative fixed charge density (Q_f) of AlO_x passivation layers and chemical passivation induced by hydrogen diffused from AlO_x passivation layers on p, p^+, and n doped Si surfaces [3]. However, the AlO_x passivation is compromised on lightly doped n and heavily doped n^+ Si surfaces because the minority carrier concentration at the surface is increased by the significant negative fixed charges of the AlO_x passivation, which can lead to parasitic shunting [4] and near surface recombination [5] on n and n^+ doped Si surfaces.

Meanwhile, AlO_x has been investigated as a passivated contact to conventional diffused junction solar cells [6] and a carrier selective contact which extracts one-type of carrier from the absorber for solar cells [7]. However, although AlO_x could provide appreciable passivation to diffused p^+ and n^+ Si surfaces and a relatively low contact resistivity to n^+ Si surfaces, the thermal stability during metal contact annealing and the contact resistivity on p and p^+ doped Si surfaces still need to be intensively optimized for solar cell application [6,7]. To address the problems mentioned above, versatile passivation materials based on AlO_x, which have tunable fixed charges (i.e., tunable field effect passivation), largely enhanced chemical passivation, high thermal stability, and engineered defects at a specific energy level to promote charge transport through the passivating oxide, need to be developed.

In this study, a ternary oxide, Mg-doped AlO_x (MAO) is developed and its passivation performance on Si surfaces is investigated because MgO_x has not only excellent properties such as high permittivity (κ~10), large band gap (~7.8 eV), and high breakdown field (~12 MV/cm) [8,9] but also positive Q_f on Si surfaces [10,11]. In particular, we demonstrate that the fixed charge density and polarity, i.e. field effect passivation of the MAO passivation layers can be well controlled by adjusting the Mg dopant concentration in the passivation layers without significant degradation of chemical passivation, which is achieved by employing a super-cycle plasma-enhanced atomic layer deposition (PEALD) technique. We also demonstrate the effects of the controlled fixed charge density and polarity on passivation performance of MAO nanolayers on p and n doped c-Si surfaces.

2 EXPERIMENTAL

2.1 Fabrication of MAO passivation layers

All experiments were carried out on mirror-polished single-crystalline float-zone (Fz) and Czochralski (Cz) (100) Si wafers. The boron doped p-type and phosphorus doped n-type Si wafers had a resistivity of 1–5 Ωcm. MAO passivation layers were fabricated on the Si wafers by using super-cycle PEALD at 200 °C. Bis(ethylcyclopentadienyl)magnesium [$Mg(EtCp)_2$], trimethyl aluminum [$Al(CH_3)_3$], and O_2 plasma were used for the deposition. As shown in Fig. 1 the Mg dopant concentration was controlled by the Mg and Al precursor cycle ratio R_{pc} (= $[n_{Mg}']/[n_{Mg}'+n_{Al}']$, where n_{Mg}' and n_{Al}' are the numbers of total sub-cycles of Mg and Al precursors, respectively). R_{pc} was varied from 1/2 to 1/20. MgO_x and AlO_x passivation layers were also fabricated for comparison.

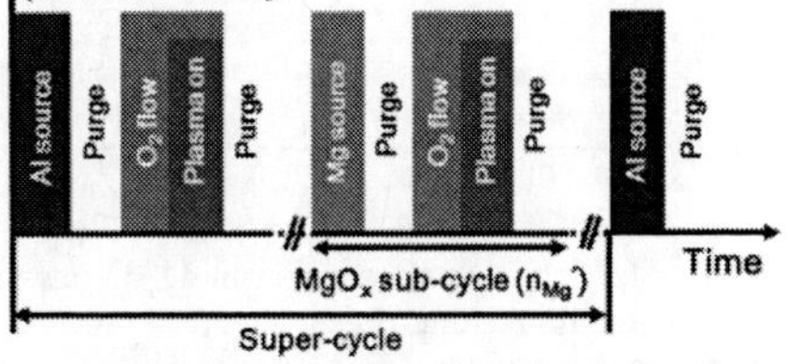

Figure 1: Schematic diagram of the super-cycle PEALD process for fabricating MAO passivation layers.

All samples received post deposition annealing (PDA) at

500 °C in an industrial tube furnace under a N_2 gas atmosphere.

2.2 Characterization of MAO passivation layers

The thickness and refractive index (n) of the fabricated passivation layers were determined by a single-wavelength ellipsometer at the wavelength of 632.8 nm. The composition of the passivation layers was determined by inductively coupled plasma-atomic emission spectroscopy (ICP-AES) analysis. The effective minority carrier lifetime τ_{eff} of the samples was measured by the photo-conductance decay (PCD) technique using a Sinton WCT-120 lifetime tester. 20-nm-thick passivation layers were deposited on both sides of Si wafers for the lifetime measurements. Assuming an infinite lifetime in the bulk, the upper-limit effective surface recombination velocity $S_{eff,max}$ was calculated by

$$S_{eff,max} = W/(2\tau_{eff}), \quad (1)$$

where W is the wafer thickness and τ_{eff} is the effective minority carrier lifetime [12]. The interface properties of the fabricated MAO passivation layer, i.e. effective fixed charge density Q_{eff} and interface trap density D_{it}, were characterized by capacitance-voltage (C-V) measure-ments.

3 RESULTS AND DISCUSSION

3.1 Control of Mg dopant concentration in MAO layers

The Mg doping into MAO passivation layers was realized by inserting single MgO layer (one ALD cycle) discretely into an AlO_x matrix by substituting one TMA-O_2 plasma cycle with one $Mg(EtCp)_2$-O_2 plasma cycle in a single super-cycle as shown in Fig. 1. As shown in Fig. 2, Mg dopant concentration measured by ICP-AES analysis as R_{pc} varies from 1/20 to 1/2 varies from about 4.7 to 51 at.%. Therefore, the Mg dopant concentration calculated from the R_{pc} and ALD growth rate is well matched to the Mg dopant concentration measured by ICP-AES analysis. It should be noted that the Mg concentration was uniform along the direction normal to the surface even though the Mg had been fed discretely. This was confirmed by Rutherford backscattering spectrometry (RBS) and X-ray photoelectron spectro-scopy (XPS) (data not shown here).

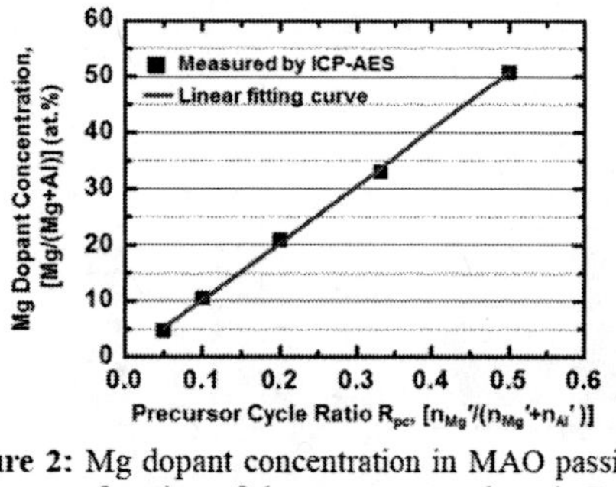

Figure 2: Mg dopant concentration in MAO passivation layers as a function of the precursor cycle ratio R_{pc}. The values were measured by ICP-AES analysis.

3.2 Impact of Mg-doping on the surface passivation

For the investigation of Mg doping effect on field effect and chemical passivation, Q_{eff} and D_{it} of the passivation layers were extracted by C-V measurements. As shown in Fig. 3, the interface properties of AMO passivation layers can be clearly affected by varying the R_{pc} (i.e., Mg dopant concentration). For instance, increasing Mg dopant concentration resulted in the highest negative Q_{eff} of 5.2×10^{12} cm^{-2} at the R_{pc} of 1/20 (i.e., about 4.7 at.% of Mg in the passivation layer), and moreover further increase in Mg dopant concentration induced not only the lowest negative Q_{eff} of 1.7×10^{11} cm^{-2} at the R_{pc} of 1/3 (i.e., about 33 at.% of Mg in the passivation layer) but also positive Q_{eff} of 1.0×10^{12} cm^{-2} at the R_{pc} of 1/2 (i.e., about 51 at.% of Mg in the passivation layer). In particular, high positive Q_{eff} of MAO passivation layers with the R_{pc} of 1/2 is an unexpected result because pure MgO passivation layers showed small positive Q_{eff} of 1.7×10^{11} cm^{-2}. In addition, it should be noted that the measured D_{it} values were below or around 1×10^{11} $cm^{-2}eV^{-1}$ even though Mg dopant concentration increased largely. Therefore, we can conclude that fixed charge density and polarity (i.e., field effect passivation) of MAO passivation layers can be systemically controlled by Mg doping using super-cycle PEALD without significant increase of D_{it} (i.e., significant degradation of chemical passivation) in this study.

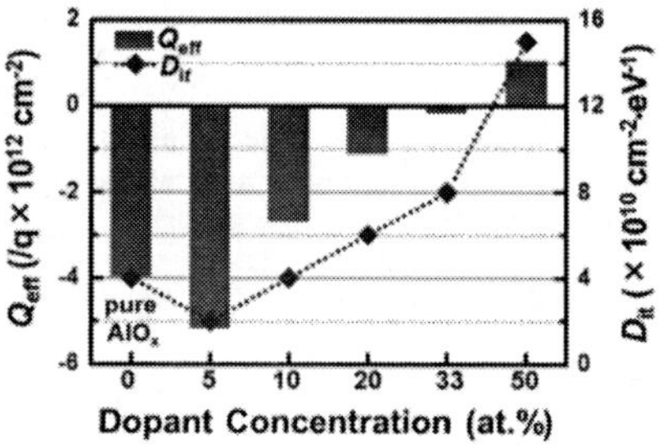

Figure 3: Measured Q_{eff} and D_{it} of 20-nm-thick passi-vation layers deposited on p-type c-Si wafers after PDA at 500 °C in N_2 for 15 min as a function of Mg dopant concentration.

Next, Mg doping effect on $S_{eff,max}$ of the pure AlO_x and MAO passivated p and n doped Si samples was investigated. As shown in Fig. 4, the $S_{eff,max}$ of AMO passivated Si samples is clearly affected by varying the R_{pc}. For example, although pure AlO_x passivation layers showed excellent passivation performance, MAO passivation layers with the R_{pc} of 1/20 showed the lowest $S_{eff,max}$ of 14 and 6 cm/s for p and n doped Si samples, respectively. Further increased Mg dopant concentration led to a continuous increase of $S_{eff,max}$ and resulted in the highest $S_{eff,max}$ of about 25 and 14 cm/s at the R_{pc} of 1/3 for p and n doped Si samples, respectively. However, MAO passivation layers with the R_{pc} of 1/2 showed the lower $S_{eff,max}$ of about 20 and 10 cm/s for p and n doped Si samples, respectively. Therefore, the trend in measured $S_{eff,max}$ values shows good match with the trend in measured Q_{eff} values.

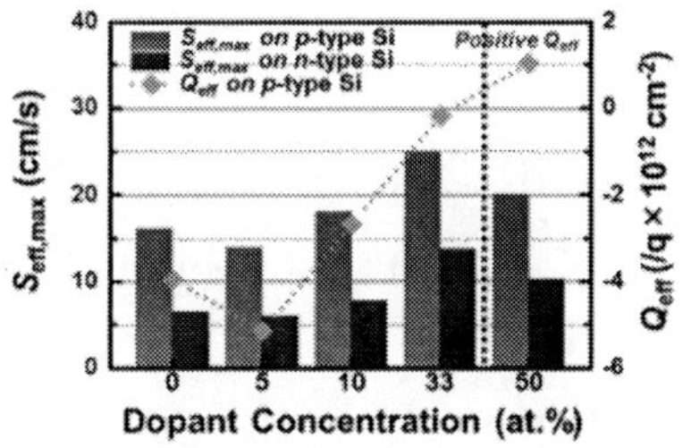

Figure 4: Measured $S_{eff,max}$ of 20-nm-thick layer-passivated p- and n-type c-Si samples and measured Q_{eff} on p-type c-Si wafers after PDA at 500 °C in N_2 for 15 min as a function of dopant concentration.

Finally, it has been well known that SiO_2 passivation layers show poor long-term stability in some cases [13,14] and require a very high temperature process which is not suitable for low-cost industrial processes [15]. It also has been reported that pure AlO_x passivation layers show parasitic shunting and near surface recombination on n and n^+ doped Si surfaces as mentioned before. Therefore, considering the experimental results in the study, MAO passivation layers with small or positive Q_{eff} and considerably low D_{it} could be applied for the passivation of p^+, n, and n^+ Si surfaces instead of SiO_2 and pure AlO_x passivation layers.

4 CONCLUSION

We have developed Mg-doped AlO_x passivation and its surface passivation and interface properties on p and n doped crystalline silicon surfaces were investigated in this study. Capacitance-voltage and effective minority carrier lifetime measurements clearly show that Mg doping can effectively engineer effective fixed charge density and polarity, i.e. field effect passivation of Mg-doped AlO_x passivation on crystalline Si without significant degradation of chemical passivation. Our study also shows that Mg-doped AlO_x layers can provide superior surface passivation as compared to pure AlO_x layers, and show considerably low upper-limit effective surface recombination velocity even without high effective fixed charge density due to their high chemical passivation. Therefore, considering the experimental results, Mg-doped AlO_x layers could be applied for the passivation of p^+, n, and n^+ Si surfaces instead of SiO_2 and pure AlO_x passivation layers.

5 ACKNOWLEDGEMENTS

The authors thank the staffs of Collaborative Clean Room at Toyota Technological Institute for experimental support on sample fabrication. This study is supported by New Energy and Industrial Technology Development Organization (NEDO).

6 REFERENCES

[1] International Technology Roadmap for Photovoltaics (ITRPV) Results 2015. Available from: http://www.itrpv.net/.cm4all/iproc.php/ITRPV%20Seventh%20Edition%20Vers%202.pdf?cdp=a.
[2] J. Schmidt, A. Merkle, R. Brendel, Progress in Photovoltaics: Research and Applications 16 (2008) 461.
[3] G. Dingemans, W. M. M. Kessels, Journal of Vacuum Science & Technology A: Vacuum, Surfaces, and Films 30 (2012) 040802.
[4] S. Dauwe, L. Mittelstädt, A. Metz, R. Hezel, Progress in Photovoltaics: Research and Applications 10(2002) 271.
[5] I. Dirnstorfer, D. K. Simon, P. M. Jordan, T. Mikolajick, Journal of Applied Physics 116 (2014) 044112.
[6] J. Bullock, D. Yan, A. Cuevas, Physica status solidi-Rapid Research Letters 7 (2013) 946.
[7] S. Smit, D. Garcia-Alonso, S. Bordihn, M. S. Hanssen, W. M. M. Kessels, Solar Energy Materials and Solar Cells 120 (2014) 376.
[8] O. Madelung, U. Rössler, M. Schulz, *II-VI and I-VII Compounds; Semimagnetic Compounds, New Series, Group III, vol. 41B*, Heidelberg, Springer, 1999.
[9] A. Posadas, F. J. Walker, C. H. Ahn, T. L. Goodrich, Z. Cai, K. S. Ziemer, Journal of Applied Physics 92 (2008) 233511.
[10] A. Kohn, A. Kovács, T. Uhrmann, T. Dimopoulos, H. Brückl, Applied Physics Letters 95 (2009) 042506.
[11] C.-Y. Su, M. Frederickx, M. Menghini, L. Dillemans, R. Lieten, T. Smets, J. W. Seo, J.-P. Locquet, Thin Solid Films, 520 (2012) 4508.
[12] A. B. Sproul, Journal of Applied Physics 76 (1994) 2851.
[13] A. F. Thomson, K. R. McIntosh, Applied Physics Letters 95 (2009) 052101.
[14] G. Dingemans, C. A. A. van Helvoirt, D. Pierreux, W. Keuning, W. M. M. Kessels, Journal of The Electrochemical Society 159 (2012) H277.
[15] A. G. Aberle, Progress in Photovoltaics: Research and Applications 8 (2000) 473.

LASER-TRANSFERRED NIV-SEED FOR THE METALLIZATION OF SILICON HETEROJUNCTION SOLAR CELLS BY CU-PLATING

A. Rodofili, W. Wolke, L. Kroely, M. Bivour, G. Cimiotti, J. Bartsch, M. Glatthaar, J. Nekarda
Fraunhofer Institute for Solar Energy Systems ISE
Heidenhofstr. 2, 79110 Freiburg, Germany

ABSTRACT: We present results about a laser-based method for the metallization of silicon heterojunction solar cells by Cu-plating. The method consists of first depositing a dielectric layer as plating mask onto the transparent conductive oxide (TCO) and then depositing a NiV seed layer onto the plating mask by laser induced forward transfer (LIFT). Afterwards, the seed layer is fired through the plating mask in a second laser step in order to form a contact to the TCO. By dividing the process into laser transfer and firing (LTF) each step can be optimized separately. The final metallization is produced by Cu-plating. A pulse plating process is applied to further reduce parasitic plating. Different dielectric layers are tested as plating masks for their resistance against parasitic plating. The combination of a 15 nm thick Al_2O_3 layer as plating mask in conjunction with pulse plating is completely free of parasitic plating. Finally an efficiency of 22.2% is reached outperforming the screen printed reference cells by $0.5\%_{abs}$.

Keywords: Laser Processing, Heterojunction, Metallization, Plating

1. INTRODUCTION

Cu-plating as a low-temperature process is well suited for the metallization of silicon heterojunction (SHJ) solar cells, due to their temperature-sensitive passivation. Kaneka has demonstrated an efficiency of 25.1% with a plated SHJ solar cell [1], which represents the efficiency record for both-side contacted SHJ solar cells and is higher than any reported result of solar cells with screen printed contacts.

Screen printing is the most used method for the metallization of silicon homojunction solar cells. After decades of continuous development it has proved to be a robust method with a high performance. But due to the firing temperatures > 700°C it is not directly applicable to SHJ solar cells. Special low-temperature pastes have to be used, which feature a much higher resistivity of the cured fingers [2]. The resistivity of Cu-plating, on the other hand, is close to that of bulk Cu. Furthermore, with plating it is possible to produce thinner fingers compared to screen printed fingers.

With respect to adhesion, low-temperature pastes did not yet undergo the same development as the high-temperature pastes and in fact in many cases the requested peel force of 1 N/mm with soldered ribbons is not reached [3]. Also the development of Cu-plating on transparent conductive oxides (TCO) is still at the beginning, but with an appropriate intermediate layer between the TCO and the plated Cu high peel forces between 3 and 5 N/mm have already been measured [4, 5].

Plating a metal grid on the TCO of SHJ solar cells requires a negative plating mask that defines the shape of the grid, because otherwise the whole TCO area would be plated. Many approaches use patterned organic masks for this purpose [4, 6, 7, 7–11]. These masks, however, have to be removed after the plating process. But the coating materials and the waste water management for chemically removing the mask are quite expensive.

An approach based on a SiO_x plating mask was presented by Adachi et al. [12] and patented by Kaneka [13]. The advantage of this sequence is that the SiO_x layer remains on the cell. The SiO_x layer, however, cannot be deposited in the same tool as the TCO, because one part of the metallization is applied before the SiO_x layer. And since screen printing and plating are combined the lowest achievable finger width is defined by the screen printing finger width.

The use of laser processes for structuring a dielectric plating mask, as it is done for homojunction solar cells [14], would be a preferable choice. But since the dielectric layer and the TCO are both transparent for laser wavelengths in the suitable spectrum, such processes are

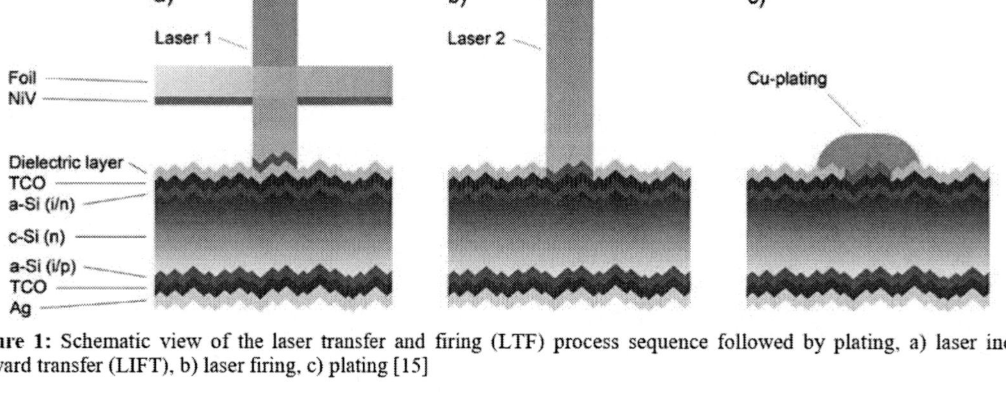

Figure 1: Schematic view of the laser transfer and firing (LTF) process sequence followed by plating, a) laser induced forward transfer (LIFT), b) laser firing, c) plating [15]

not easily applicable for SHJ solar cells. A recently presented laser structuring approach uses an amorphous silicon (a-Si) layer on the dielectric plating mask, which acts as absorption layer for the laser irradiation [16]. The absorption layer, however, has to be removed after plating, which makes the process rather expensive.

In this paper, we present results about a method for plating on SHJ solar cells with a laser structured dielectric plating mask, in which the plating mask can in principle be deposited in the same tool as the TCO and all deposited layers can remain on the finished solar cell. The method is based on laser induced forward transfer (LIFT) [17] and is described in detail in [15].

Different dielectric layers were tested for their resistance against parasitic plating. Furthermore we produced (156x156) mm² solar cells with the LTF and plating approach and compared them to screen printed solar cells.

2. APPROACH

A scheme of the process sequence can be seen in figure 1. A dielectric layer is deposited on the TCO as plating mask. A NiV seed layer grid is deposited by LIFT onto the dielectric layer using inexpensive plastic foil as LIFT carrier substrate. After LIFT the foil is removed and in a second laser step the seed layer is fired to form a contact to the TCO. The metallization is then formed by plating on the seed layer grid, while the dielectric plating mask also protects the TCO from the acidic electrolyte.

The separation of the laser process into laser transfer and firing (LTF) allows for a separate optimization of the two steps. As it was shown in [15] the two steps do not need to be precisely aligned to each other.

We used a reverse pulse plating process to enhance the plating selectivity and thereby reduce parasitic plating [18]. Instead of direct current (DC) we alternatingly applied forward and reverse pulses. During the forward pulses the plated metal grows faster on the seed layer than on the plating mask, because on the plating mask the ions need more time to diffuse to the rare defect sites, where they are eventually deposited. During the reverse pulses, on the other hand, the small nuclei that have formed during the previous forward pulse on the plating mask are completely dissolved, while on the seed layer the metal that has been deposited during the previous forward pulse is only partially dissolved.

3. EXPERIMENTAL

The carrier foils for the LIFT process were 100 µm thick biaxially-oriented polyethylene terephthalate (BoPET) foils. 100 nm NiV (with 7 wt% vanadium) were sputtered on one side of the foils. For laser transfer and laser firing we used a laser with 1064nm wavelength and 15 ns pulse length. The Gaussian laser beam had a diameter of 20 µm in the focal plane. The laser beams for transfer and firing were accurately aligned to each other.

We used non-metallized industrial (156x156) mm² Cz n-type SHJ precursors with indium tin oxide (ITO) as TCO. The front side structure was a-Si(i)/a-Si(n)/ITO and the back side structure a-Si(i)/a-Si(p)/ITO. Both wafer sides were textured.

We deposited different dielectric layers on the front side ITO as plating masks. The tested materials were Al_2O_3 deposited by atomic layer deposition (ALD) and SiN_x deposited by sputtering and plasma enhanced chemical vapor deposition (PECVD). The effect of the deposition on the passivation was measured by a comparison of photoluminscence (PL) images before and after the deposition from which the change of the implied open circuit voltage (V_{OC}) was calculated [15]. The back side metallization was produced by evaporating 1 µm of Ag. The front was metallized by the LTF sequence followed by Cu-plating in a RENA inline-tool, in which only the front side is wetted by the copper sulfate based plating bath and the cell is illuminated from the front side and contacted on the rear side. After the Cu plating a thin Ag capping was plated onto the copper. We also produced reference solar cells with a screen printed low-temperature paste metallization [19], which was printed on the front side ITO of the precursors and cured for 9 min at 200°C. The back side of the reference cells was also fully metallized with 1 µm evaporated Ag.

4. RESULTS AND DISCUSSION

Different dielectric layers were deposited on the ITO of the precursors as plating masks to test their resistance against parasitic plating and to measure the V_{OC} change caused by the deposition. The parasitic plating after Cu-plating was qualitatively categorized (low, mid, high). Even though with standard DC-plating more parasitic plating was expected, we used it for this test instead of pulse plating, because the aim was just to compare the different dielectric layers.

Table I: Dielectric layers tested for their resistance against parasitic plating with DC plating

Layer	Deposition temperature (°C)	Thickness (nm)	Parasitic plating	ΔVoc (mV)
Al_2O_3 (ALD)	180	20	low	0
Al_2O_3 (ALD)	180	15	low	0
Al_2O_3 (ALD)	180	11	low	0
Al_2O_3 (ALD)	180	7	mid	0
Al_2O_3 (ALD)	180	3	mid	0
SiN_x (PECVD)	300	15	mid	-10
SiN_x (sputtered)	150	15	high	1

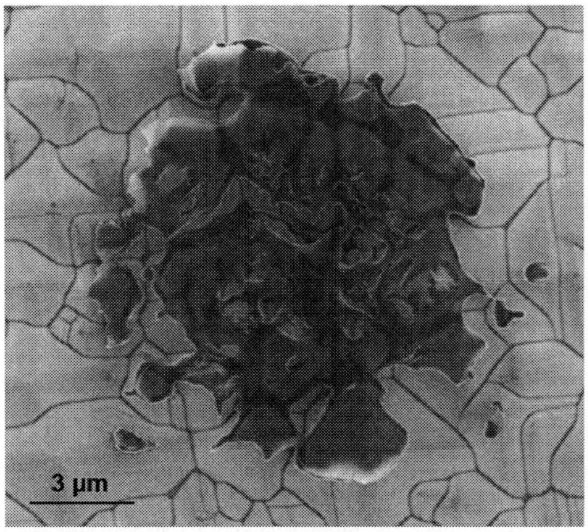

Figure 2: SEM image of NiV LIFT spot on a SHJ precursor coated with 15nm Al_2O_3 produced with a LIFT fluence of 0.25 J/cm^2.

The results are listed in table 1. For all layers except the PECVD layer only a negligible implied V_{OC} change was measured. This can be explained with the deposition temperatures which are also listed in the table. The highest temperature was used for PECVD.

For Al_2O_3 layers with a thickness of 11 nm or higher only weak parasitic plating was observed. Therefore in the further experiments 15 nm Al_2O_3 layers were used as plating mask. The sputtered SiN_x layer, on the other hand, was very prone to parasitic plating. With pulse plating, however, a better performance is expected for all layers. Sputtered layers would be favorable, since they can be deposited in the same tool as the TCO.

Figure 2 shows a SEM image of a NiV spot produced by LIFT. The transferred NiV is very well confined within a small area. It is advantageous to produce a well confined NiV transfer, since the size of the seed layer defines the final width of the contact fingers after plating.

The solar cells metallized by LTF and plating were produced with 15 nm ALD-Al_2O_3 as plating mask. Table 2 shows the IV measurement results of these cells compared to the results of the screen printed reference cells. The J_{SC} and FF are higher for the plated cells, which is clearly the result of the lower bulk resistivity of the plated copper, the lower finger width of 30 µm compared to the 55-60 µm wide screen printed fingers and the lower specific contact resistance of the LTF contacts. For the LTF contacts, we measured a specific contact resistance of 0.7 mΩcm², whereas the screen printed contacts reached 4 mΩcm² [19]. These differences are also reflected by the lower series resistance (R_{SER}) of the plated cells, which is also displayed in the table. An efficiency (η) of 22.2% was reached with the plated LTF cells, which corresponds to a gain of 0.5%$_{abs}$ compared to the screen printing reference. However, the number of cells produced in this first batch is low. More cells need to be produced to support the results.

A cross section of a plated 30 µm wide finger is visible in figure 3. As can be seen in the photograph of a plated grid in figure 4, in contrast to the results with DC-plating, there is no parasitic plating if pulse plating is used with the 15 nm Al_2O_3 layer as plating mask. As for adhesion, the plated fingers passed a tape adhesion test. However, a quantitative peel force measurement needs to be performed.

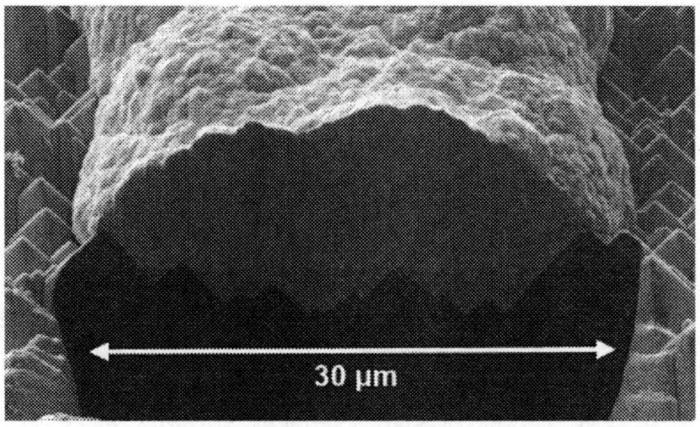

Figure 3: SEM image of a cross section of a finger produced by LTF and plating.

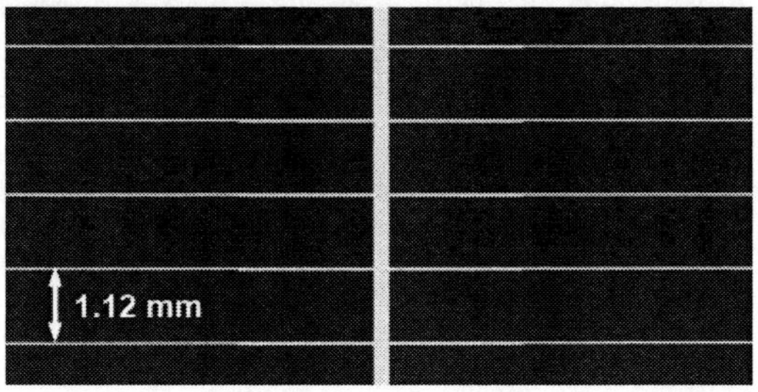

Figure 4: Photograph of a grid produced by LTF and plating.

Table II: IV measurement results of two solar cells metallized by screen printing and two solar cells metallized by LTF and Cu-plating.

Metallization	Area (cm²)	V_{OC} (mV)	J_{SC} (mA/cm²)	FF (%)	R_{SER} (Ωcm²)	η (%)
Screen printing	239.0	726	37.7	79.0	0.80	21.6
Screen printing	239.0	727	37.8	79.1	0.76	21.7
LTF + Cu-plating	239.0	727	38.2	79.7	0.56	22.1
LTF + Cu-plating	239.0	728	38.0	80.1	0.47	22.2

5. CONCLUSION

With the LTF and plating approach for SHJ solar cells a maximum efficiency of 22.2% was reached, which corresponds to a gain of $0.5\%_{abs}$ compared to screen printed reference cells. Moreover, the approach is cost-effective. The laser process employs low-cost plastic foil as LIFT carrier and inexpensive ns IR lasers. The two laser steps are easily combined, because a precise alignment is not necessary. And since the dielectric plating mask remains on the cell the number of process steps is reduced. With DC-plating the ALD-Al_2O_3 plating mask showed the least parasitic plating, while the sputtered SiN_x layer was very prone to parasitic plating. With 15 nm Al_2O_3 and pulse plating no parasitic plating at all was observed. Further experiments will show whether sputtered layers, if necessary with increased thickness, will perform as well against parasitic plating, if pulse plating is used. A sputtered layer could be deposited in the same tool as the ITO, which would further simplify the production.

Even if the LTF solar cells already outperformed the screen printed reference cells in these first experiments, the process is still very likely to improve, because many parameters have not been optimized, yet. Furthermore, adhesion and module tests will show whether the produced cells can be interconnected by standard industrial soldering processes and whether the additional dielectric layer improves the module lifetime by protecting the solar cells from moisture.

6. ACKNOWLEDGEMENTS

The authors would like to thank Jutta Zielonka and Christian Schmiga for the preparation of the SEM and EDX measurements. This work was funded by the German federal ministry for economic affairs and energy (support code: 0325574).

7. REFERENCES

[1] D. Adachi, J. L. Hernández, and K. Yamamoto, Appl. Phys. Lett. **107**, 233506 (2015).

[2] S. D. Wolf, A. Descoeudres, Z. C. Holman, and C. Ballif, green **2** (2012).

[3] A. De Rose, D. Erath, T. Geipel, A. Kraft, and U. Eitner, in: 33rd EU PVSEC, Amsterdam, The Netherlands, 2017 (WIP, München, 2017).

[4] A. Aguilar, S. Y. Herasimenka, J. Karas, H. Jain, J. Lee, K. Munoz, L. Michaelson, T. Tyson, W. J. Dauksher, and S. Bowden, in: IEEE 43rd Photovoltaic Specialists Conference (PVSC), Portland, USA, 2016 (IEEE, Piscataway, NJ, 2016), pp. 1972–1975.

[5] J.-B. Heng, J. Fu, B. Kong, Y. Chae, W. Wang, Z. Xie, A. Reddy, K. Lam, C. Beitel, and Z. Xu, in: 28th EU PVSEC, Paris, France, 2013 (WIP, München, 2013), pp. 952–955.

[6] J. Geissbuhler, S. D. Wolf, A. Faes, N. Badel, Q. Jeangros, A. Tomasi, L. Barraud, A. Descoeudres, M. Despeisse, and C. Ballif, IEEE J. Photovoltaics **4**, 1055 (2014).

[7] J. B. Heng, J. Fu, B. Kong, Y. Chae, W. Wang, Z. Xie, A. Reddy, K. Lam, C. Beitel, C. Liao, C. Erben, Z. Huang, and Z. Xu, IEEE J. Photovoltaics **5**, 82 (2015).

[8] A. Khanna, K.-U. Ritzau, M. Kamp, A. Filipovic, C. Schmiga, M. Glatthaar, A. G. Aberle, and T. Mueller, Applied Surface Science **349**, 880 (2015).

[9] Z. Li, P.-C. Hsiao, W. Zhang, R. Chen, Y. Yao, P. Papet, and A. Lennon, Energy Procedia **67**, 76 (2015).

[10] P. Papet, J. Hermans, T. Söderström, M. Cucinelli, L. Andreetta, D. Bätzner, W. Frammelsberger, D. Lachenal, J. Meixenberger, B. Legradic, B. Strahm, G. Wahli, W. Brok, J. Geissbühler, A. Tomasi, C. Ballif, E. Vetter, and S. Leu, in: 28th EU PVSEC, Paris, France, 2013 (WIP, München, 2013), pp. 1976–1979.

[11] J. Yu, J. Bian, W. Duan, Y. Liu, J. Shi, F. Meng, and Z. Liu, Solar Energy Materials and Solar Cells **144**, 359 (2016).

[12] D. Adachi, T. Terashita, T. Uto, J. L. Hernández, and K. Yamamoto, Solar Energy Materials and Solar Cells **163**, 204 (2017).

[13] D. Adachi, K. Yamamoto, J. L. Hernandez, and N. Valckx, US9246026 B2, (2016).

[14] S. Kluska, C. Fleischmann, A. Büchler, W. Hördt, C. Geisler, S. Hopman, and M. Glatthaar, Solar Energy Materials and Solar Cells **120, Part A**, 323 (2014).

[15] A. Rodofili, W. Wolke, L. Kroely, M. Bivour, G. Cimiotti, J. Bartsch, M. Glatthaar, and J. Nekarda, Sol. RRL **1** (2017).

[16] A. Dabirian, A. Lachowicz, J.-W. Schüttauf, B. Paviet-Salomon, M. Morales-Masis, A. Hessler-Wyser, M. Despeisse, and C. Ballif, Solar Energy Materials and Solar Cells **159**, 243 (2017).

[17] J. Bohandy, B. F. Kim, and F. J. Adrian, Journal of Applied Physics **60**, 1538 (1986).

[18] M. Glatthaar, R. Rohit, A. Rodofili, Y. J. Snow, J. Nekarda, and J. Bartsch, IEEE J. Photovoltaics, DOI 10.1109/JPHOTOV.2017.2748999 (2017).

[19] D. Erath, M. Pospischil, R. Keding, M. Jahn, I. L. Lontchi, A. Lorenz, and F. Clement, in: Energy Procedia, 7th International Conference on Silicon Photovoltaics, SiliconPV (2017).

KEY ASPECTS FOR FABRICATION OF P-TYPE CZ-SI PERC SOLAR CELLS EXCEEDING 22% CONVERSION EFFICIENCY

S. Werner, E. Lohmüller, P. Saint-Cast, J.M. Greulich, J. Weber, S. Schmidt, A. Moldovan, A.A. Brand, T. Dannenberg, S. Mack, S. Wasmer, M. Demant, M. Linse, R. Ackermann, A. Wolf, R. Preu
Fraunhofer Institute for Solar Energy Systems ISE, Heidenhofstr. 2, 79110 Freiburg, Germany
Phone: +49 761 - 4588 5049; e-mail: sabrina.werner@ise.fraunhofer.de

ABSTRACT: This paper gives a close-up insight into recent and future developments that are performed with industrial focus at Fraunhofer ISE's PV-TEC pilot-line to increase the energy conversion efficiency of 6-inch p-type Czochralski-grown silicon (Cz-Si) passivated emitter and rear cells (PERC) to 22% and above. First, the current status of PERC solar cell fabrication allowing for conversion efficiencies up to 21.5% is discussed. Then, we examine four key aspects in detail that need to be considered for optimizing the cells' front side to boost the cell efficiency to the 22% regime. We demonstrate selective emitter laser doping out of the phosphosilicate glass layer, which is formed by a gas phase phosphorus oxychloride diffusion process. After diffusion and wet-chemical emitter etch back, the field emitter features a very low saturation current density of only 31 fA/cm² (textured, SiN_x-passivated). Specific contact resistances of 1 mΩcm² confirm the low-resistance contacting of the laser-doped surfaces using a commercially available silver screen printing paste. Apart from developing an accurate alignment procedure to match laser-structured and screen-printed layouts, we have also optimized our single-step screen-printing process for finger widths of 38 µm at 16 µm height. Based on simulations we find that efficiencies up to 22.5% are possible when the optimized process routes are integrated into PERC solar cells.
Keywords: p-type silicon solar cells, PERC, monocrystalline, selective emitter, passivation

1 INTRODUCTION

The passivated emitter and rear cell (PERC) [1] technology on p-type silicon is of very high interest for both research institutes and industry. It has entered mass production during the last years and it is forecast to gain more and more market share [2]. Thus, improving the energy conversion efficiency of PERC solar cells is of major interest for the photovoltaic community. The improvement of the cell's front side is identified as next step to boost the efficiency to 22% and above [3–7]. Lots of research on this subject is performed, but often details regarding the fabrication process or the implementation of novel/adapted processes are not discussed.

The aim of this work is to give a close-up insight into recent and future p-type silicon PERC technology that is being developed with focus on industrial application at Fraunhofer ISE's PV-TEC pilot-line [8]. We discuss in detail the key aspects that need to be considered for a successful integration of necessary process developments into the baseline sequence to target the 22% cell efficiency regime. Therefore, the integration of a selective emitter using laser doping out of the phosphosilicate glass (PSG) layer and several related key aspects to be tackled are discussed, which are expected to have a high impact on future cell development.

2 APPROACH

This paper discusses the improvement of the p-type silicon PERC baseline fabrication process at Fraunhofer ISE's PV-TEC pilot-line [8]. Our continuous process development comprises a two-step approach. First, we characterize and optimize the processes of interest using test samples like, e.g., charge carrier lifetime samples, aluminum back surface field (Al-BSF) solar cells, etc. In a second step, the most promising processes are integrated into the PERC baseline process. Therefore, we examine variations of different processes on the basis of monthly cell runs with 200 wafers each for continuous optimization. The p-type wafer material used is mainly magnetically-cast Czochralski silicon (mCz-Si), as it offers stable high-quality material properties and also features a significantly less pronounced impact of boron-oxygen-related light-induced degradation in comparison with conventional Cz-Si. However, conventional Cz-Si wafers are also used from time to time to benchmark our baseline process.

In 2016, we mainly focused on rear side and base material optimizations while ensuring a stable and reproducible baseline process. As the front side currently limits the efficiency of our PERC solar cells [4], we carried out intensive investigations regarding possible front side optimizations. We identified the following four key aspects to be tackled for a successful fabrication of p-type Cz-Si PERC solar cells with 22% efficiency and above:

(i) The dark saturation current density j_{0e} of the passivated emitter needs to be decreased. Our approach is to decrease the surface doping concentration of the emitter by optimized diffusion processes as well as by adding a wet-chemical emitter etch back process. Additionally, the surface passivation needs to be adapted for such lowly-doped emitters. Also, the dark saturation current density $j_{0e,met}$ of the front screen-printed and fired silver contacts needs to be decreased, while a low specific contact resistance ρ_C must be ensured. Therefore, a selective emitter approach with laser doping out of the PSG layer [9,10] is pursued.

(ii) The process integration of the selective emitter approach including the emitter etch back process into the baseline sequence needs to be realized. For this purpose, we investigate two different process sequences.

(iii) The accurate alignment of the screen-printed grid onto the laser-structured selective emitter needs to be ensured. Therefore, we develop an advanced alignment procedure.

(iv) The front metallization needs to be improved. Different screens and silver pastes are tested to realize smaller finger widths and maintaining high aspect ratios.

All the experiments performed with respect to the mentioned key aspects provide experimentally derived input parameters for simulation-based estimations of the achievable energy conversion efficiencies when the results are transferred into the PERC baseline.

3 PROCESS SEQUENCES

3.1 PERC solar cell fabrication

Fig. 1 shows the recent PERC baseline process at Fraunhofer ISE. The industrial-oriented solar cell fabrication is carried out in the PV-TEC pilot-line [8]. Pseudo-square p-type Cz-Si wafers with an edge length of 156 mm serve as starting material for PERC solar cell fabrication. After alkaline texturing, a tube diffusion using an industrial furnace with phosphorus oxychloride ($POCl_3$) as liquid dopant precursor forms the phosphorus-doped homogeneous emitter. The used $POCl_3$ diffusion is an industrial process that features an in-situ oxidation step and enables full-boat loadings with 200 wafers with sufficient doping uniformity over the single wafers as well as over the entire boat [11]. Subsequently, an inline wet-chemical etching process removes the rear emitter and the front PSG layer.

A wet-chemical cleaning step precedes the surface passivation. The passivation of the rear surface is ensured by a 6 nm-thin aluminum oxide (Al_2O_3) layer deposited by fast atomic layer deposition (fast-ALD), followed by an outgassing step in a tube furnace with N_2 atmosphere and a temperature plateau at 550°C for 10 minutes. Instead of depositing the Al_2O_3 layer by a fast-ALD process, we also have the possibility to deposit this layer by plasma-enhanced chemical vapor deposition (PECVD) using the MAiA system that is wide-spread in industry. In both cases, a minimum 100 nm-thick silicon nitride (SiN_x) layer, deposited by PECVD, serves as capping layer on top of the Al_2O_3 layer. On the front side, a 75 nm-thick non-graded PECVD SiN_x layer serves as anti-reflection coating and surface passivation.

An infrared laser process locally ablates the rear passivation layer stack in order to form the local contact openings (LCO). The front and rear metallization is applied using screen printing processes with commercially available metal pastes. For the front metallization, we perform a double printing step with printing the same layout a second time after paste drying. The front silver grid features five busbars and 100 fingers. The five busbars are tapered with an effective busbar width $w_{busbar,eff} = 400$ µm over the total busbar length. The finger width w_{finger} after contact firing is measured to be $w_{finger} \approx 55$ µm. The rear electrode consists of full-area aluminum and optional silver solder pads. Finally, contact firing with varying the peak temperature is performed in an industrial conveyor belt furnace.

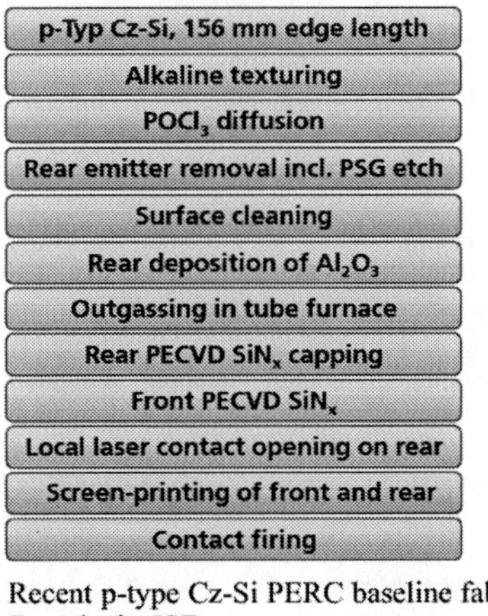

Figure 1: Recent p-type Cz-Si PERC baseline fabrication process at Fraunhofer ISE.

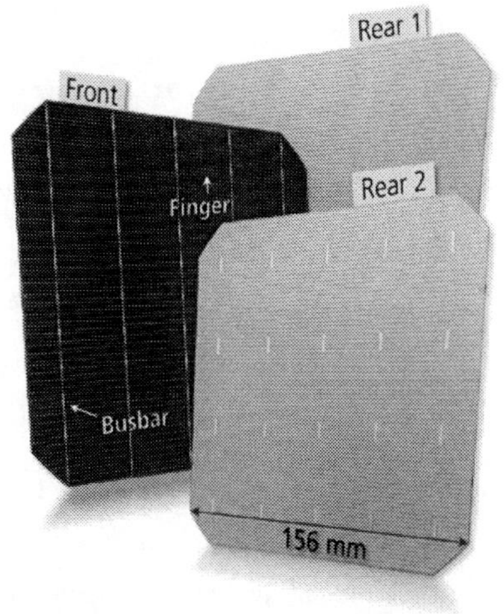

Figure 2: Photograph of a p-type Cz-Si PERC solar cell from the front and the rear side fabricated at Fraunhofer ISE's PV-TEC pilot-line (Rear 1: w/o solder pads, Rear 2: with solder pads).

Fig. 2 shows the front and rear side cell design of our PERC solar cells with line-shaped rear side contacts. The rear side "Rear 2" shows the silver solder pads, which allow the integration of the PERC solar cells into modules.

3.2 Test samples and characterization methods

All j_{0e} values given in this work are determined from quasi-steady-state photoconductance (QSSPC) measurements on symmetric n-type Cz-Si lifetime samples with high base resistivity. Their surfaces are alkaline textured. Although these samples do not feature an Al_2O_3 passivation layer, an outgassing step is performed to resemble the cell fabrication process (see Fig. 1). Then, the front and rear surfaces are passivated by a conventional SiN_x layer. A firing step activates the passivation properties. For evaluation of the QSSPC measurements, the procedure from Ref. [12] is used.

Charge carrier concentration profiles after PSG removal are obtained using the electrochemical capacitance-voltage (ECV) technique. The surface roughness needs to be taken into account as the samples are initially alkaline textured and partly treated by a laser process. The measured charge carrier concentration profiles are corrected to match the sheet resistance R_{sh} by applying the procedure described in Ref. [13]. For this end, four point probe measurements determine the emitter sheet resistance R_{sh} in the region of interest.

To determine specific contact resistances ρ_C, measurements according to the transfer length method (TLM) [14] are performed. The investigated samples undergo alkaline texturing, $POCl_3$ diffusion, optional laser doping, PSG etching, PECVD SiN_x passivation, screen-printing, and finally, contact firing. The screen-printing layout is the same as that used for the solar cells.

4 EFFICIENCY IMPROVEMENTS IN 2016

The conversion efficiency η improvements for PERC solar cells during the year 2016 are depicted in Fig. 3. Starting with a mean value of $\eta_{mean} = 20.8\%$ in January 2016, the efficiency increases to $\eta_{mean} = 21.3\%$ ensuring a stable and reproducible baseline process. The champion

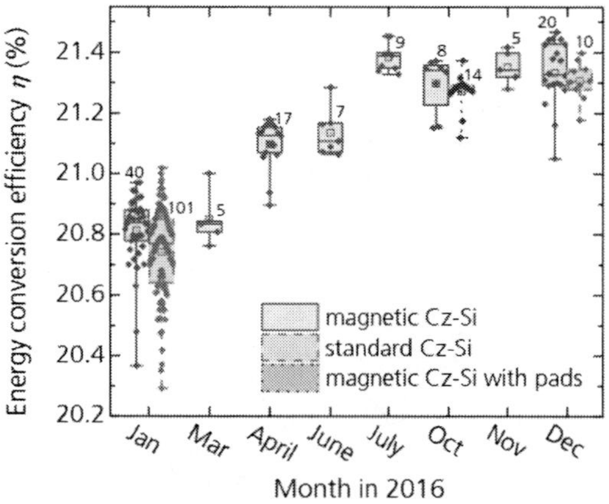

Figure 3: Improvement in conversion efficiency η of PERC solar cells fabricated at Fraunhofer ISE's PV-TEC pilot-line in 2016 for different materials, without or with solder pads. Per experiment, only the most efficient parameter set is shown. All measurements are performed with an industrial cell tester in the "as processed" state. The numbers given above the box plots state the total number of PERC solar cells fabricated within the respective process group.

cell shows $\eta = 21.5\%$ with open-circuit voltage $V_{OC} = 665$ mV, short-circuit current $j_{SC} = 39.7$ mA/cm², and fill factor $FF = 81.2\%$ in the as-processed state (in-house measurement). Our baseline is also compatible with standard Cz-Si material, which performs similar to the mCz-Si wafers (see Fig. 3: January [15] and December). Another experiment (see October) shows the successful integration of silver solder pads with similar $\eta_{mean} = 21.3\%$ for PERC cells either with or without solder pads. In adhesion tests of the front grid and the rear pads, cell breakage confirms sufficient adhesion and thus the module-readiness of our PERC cells.

The following process adaptions/optimizations are the main changes for the achieved efficiency improvements in Fig. 3:

- January to March: change in base resistivity and LCO line pitch
- March to April: evaluation of new aluminum pastes
- April to June: change in LCO line pitch
- June to July: change in base resistivity
- July to October: change in LCO line pitch and implementation of silver solder pads
- October to November: adaption of the rear LCO geometry from lines to dots by increasing the laser pulse distance and adapting the aluminum paste
- November to December: process simplification, i.e., the omission of a wet-chemical rear side polishing step that, until then, has preceded $POCl_3$ diffusion. Simultaneously, an alkaline texturing process resulting in smaller pyramids is implemented.

In addition to these main adjustments, numerous smaller optimizations—that are difficult to quantify—are also incorporated in the continuously evolving fabrication process.

5 KEY ASPECTS FOR INCREASING THE EFFICIENCY OF PERC SOLAR CELLS TO 22% AND ABOVE

In the PERC cell runs in 2016, we mainly focused on rear side and base material optimizations. In parallel experiments that are based on test samples or Al-BSF solar cells, intensive investigations have been carried out separately in order to optimize the front side processes.

5.1 Key aspect one: Lower emitter dark saturation current density

The currently used homogeneous phosphorus emitter with $R_{sh} \approx 90$ Ω/sq allows for $j_{0e} \approx 85$ fA/cm² on alkaline textured surface passivated by a non-graded PECVD SiN_x layer [11]. An approach to reduce j_{0e} of the passivated diffused regions (i.e. to increase the V_{OC}) is the reduction of the phosphorus doping concentration at the silicon surface N_{surf} [16]. We showed the successful low-resistive electrical contacting of very lightly phosphorus-doped surfaces with $N_{surf} \approx 3 \cdot 10^{19}$ cm^{-3} with low specific contact resistances $\rho_C \approx 5$ mΩcm² for a commercial state-of-the-art silver screen-printing paste [13]. On symmetric lifetime samples with alkaline textured surfaces and PECVD SiN_x passivation layer, we achieve very low $j_{0e} \approx 40$ fA/cm² for these emitters with low N_{surf} if the samples undergo the outgassing step before SiN_x layer deposition. We assume that despite the inert atmosphere, a thin passivating silicon dioxide (SiO_2) layer grows on the emitter surface during the outgassing step and improves the surface properties.

Based on these promising findings, we implemented homogenous emitters with low N_{surf} in Al-BSF solar cells. However, the solar cells with these lowly-doped emitters do not show a performance advantage compared to the solar cells with the reference emitter featuring $N_{surf} \approx 1.2 \cdot 10^{20}$ cm^{-3}. This is caused by a lower V_{OC} due to significant larger $j_{0e,met}$ below the silver contacts for the lowly-doped emitter groups [13].

Thus, the novel emitters are very promising in terms of ρ_C and j_{0e}, but still challenging in terms of contact recombination $j_{0e,met}$. Whenever new generations of silver screen printing pastes allow for lower $j_{0e,met}$, these emitters are very interesting for further development using homogenous emitter diffusion.

Until then, a selective emitter is advantageous to decrease $j_{0e,met}$ for the front screen-printed and fired silver contacts, while ensuring low ρ_C, as higher doping underneath the metal contacts significantly reduces $j_{0e,met}$ [17]. Our approach to form a selective emitter in the areas of the metal contacts is local laser doping out of the PSG layer [9,10], as it is highly industrial relevant.

The PSG layer grown during $POCl_3$ diffusion must provide a sufficient amount of phosphorus atoms for local laser doping. On the other hand, N_{surf} in the photoactive area must not be too high to allow for low j_{0e}. Our approach for a $POCl_3$ diffusion process that takes both requirements into account is the use of a moderate deposition at the beginning to allow for low N_{surf} and the addition of a second deposition step with active nitrogen (N_2) flow through the $POCl_3$ bubbler after the drive-in step to provide a PSG layer with high phosphorus concentration [18,19]. The second deposition step is intended to not change the doping profile any further [18,19].

Fig. 4 shows the resulting as-diffused doping profile on alkaline textured surface after PSG removal for this $POCl_3$ diffusion ($R_{sh} \approx 116$ Ω/sq, $N_{surf} \approx 5.7 \cdot 10^{19}$ cm^{-3}).

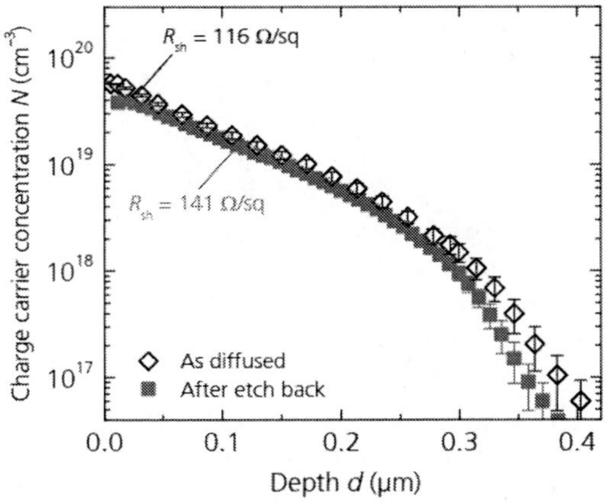

Figure 4: Charge carrier concentration profiles measured by ECV technique after PSG etching (as diffused) and after emitter etch back on alkaline textured surface. The sheet resistances R_{sh} are given.

For symmetric alkaline textured lifetime samples featuring this emitter, j_{0e} is found to be $j_{0e} = (42 \pm 2)$ fA/cm².

To further reduce the emitter recombination in the photoactive area, the emitter is partly etched back in solutions of either hydrofluoric acid (HF) and persulfate [20] or ozone [21], which leads to an increase in R_{sh}. We target a value of 140 Ω/sq $\leq R_{sh} \leq$ 150 Ω/sq such that the lateral conductivity is not decreased too much for our current grid layout. Fig. 4 illustrates the doping profile of the etched-back emitter with $R_{sh} \approx 141$ Ω/sq and $N_{surf} \approx 3.8 \cdot 10^{19}$ cm^{-3}. By comparison with the as-diffused profile, the profile after etch back is shifted by approx. 20 nm, which corresponds to the silicon removal at the emitter surface. We find $j_{0e} = (31 \pm 4)$ fA/cm² for this etched-back emitter.

For the laser doping experiments, we use a pulsed ultraviolet laser with a wavelength $\lambda = 355$ nm [10]. As previously, also symmetrically alkaline textured and diffused carrier lifetime samples are utilized for the examinations. To determine R_{sh}, the saturation current density $j_{0e,laser}$, and ρ_C for the laser-structured areas, test fields are formed either on one or both wafer sides by applying different laser parameters A-D (i.e., different pulse energies, pulse pitches and line pitches).

Fig. 5 shows an excerpt of the obtained results for the laser-doped test fields in terms of $j_{0e,laser}$ and ρ_C in dependence of R_{sh}. For comparison, the result of the etched-back emitter without laser doping is also plotted ("Ref"). The laser processed fields exhibit moderate recombination current densities 125 fA/cm² $< j_{0e,laser} <$ 280 fA/cm² with 80 Ω/sq $> R_{sh} >$ 40 Ω/sq for the different applied laser parameters. Specific contact resistances 1 mΩcm² $\leq \rho_C <$ 4 mΩcm² prove the low-resistance electrical contacting of these laser-doped regions compared to $\rho_C \approx$ 650 mΩcm² when laser-doping is not applied.

Fig. 6 shows the corresponding charge carrier concentration profiles after laser doping and PSG removal. Obviously, the laser processing yields significant higher charge carrier concentrations in the course of the profile compared to the reference profile after etch back. The surface concentration increases significantly from initially $N_{surf} \approx 3.8 \cdot 10^{19}$ cm^{-3} after etch back to a surface-near maximum carrier concentration (A) $N_{max} \approx 5.0 \cdot 10^{19}$ cm^{-3}, (B) $N_{max} \approx 6.0 \cdot 10^{19}$ cm^{-3}, (C) $N_{max} \approx 8.0 \cdot 10^{19}$ cm^{-3}, and (D) $N_{max} \approx 1.1 \cdot 10^{20}$ cm^{-3}.

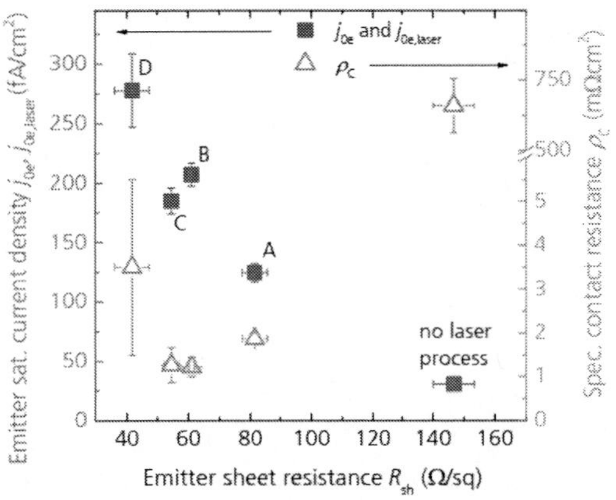

Figure 5: Emitter saturation current densities j_{0e} and $j_{0e,laser}$ for the photoactive and laser-doped areas, respectively (initially textured, SiN_x-passivated, fired), and specific contact resistance ρ_C in dependence on the emitter sheet resistance R_{sh} for different laser doping processes A-D (variation of power, pulse pitch, and line pitch). The reference "Ref" without laser doping is also shown. All samples have been diffused in the same $POCl_3$ diffusion process and etched back in HF and persulfate (compare Fig. 4).

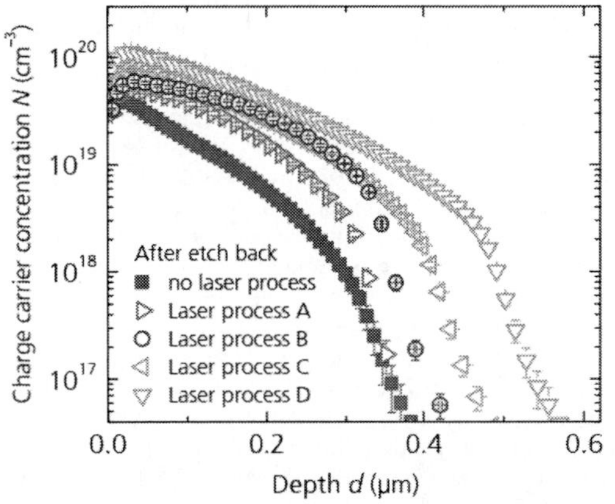

Figure 6: Charge carrier concentration profiles measured by the ECV technique after laser processes A-D and wet-chemical etch back. The profile after etch back from Fig. 4 is shown as reference.

5.2 Key aspect two: Process integration of the emitter etch back step into the PERC baseline

The process integration of the selective emitter approach with respect to the etch back process is another key aspect. For the emitter etch back, there are two options for its implementation into the fabrication sequence.

First, it might be integrated after wet-chemical removal of the rear emitter before surface passivation. At this stage, the doping-type on front and rear side of the wafers is different. In this case we found that R_{sh} does not change significantly during etching. The cause probably lies in the so-called "cathodic protection": the n-doped front is protected from the etching solution and prevents the removal of a significant amount of silicon at the emitter surface.

The second option is to implement the etch back before rear emitter removal. In doing so, the PSG layers on front and rear side are also etched. This is important because these PSG layers usually ensure high wettability of the front side of the wafers with the water film sprayed onto it before rear emitter removal. The water film protects the front emitter from possible wrap-around of the etching solution and gas phase etching. Hence, when no emitter etch back is performed, the water sticks to the front side as the PSG layer is hydrophilic. The PSG layer of the front side is then etched after rear emitter removal in HF within the same inline wet-chemistry tool.

In our case, however, the emitter etch back precedes the rear emitter removal and thus, no PSG layer is present on the front side during the latter. Hence, the front side is hydrophobic and not able to hold the protective water film. To solve this, we grow a thin wet-chemical SiO_2 layer in an ozone-based solution at the end of the emitter etch back. This SiO_2 layer provides a hydrophilic front surface during rear emitter removal.

We tested this approach on PERC cells with homogenous emitter without noticing occurring wrap-around as a variation in the experiment from November 2016 (see Fig. 3). These PERC cells are fabricated according to an adapted process sequence compared to Fig. 1, including the PSG layer removal and the SiO_2 layer growth in ozone-based solution after $POCl_3$ diffusion before rear emitter removal. The measured mean energy conversion efficiency $\eta_{mean} = 21.3\%$ for these cells is similar to the cells of the reference group, which are fabricated according to the standard sequence for homogeneous emitter cells.

5.3 Key aspect three: Accurate alignment of the screen-printed grid onto the laser-structured selective emitter

A further key aspect is a high alignment precision of the screen-printed grid on top of the laser-structured selective emitter. If misalignment occurs, as illustrated in Fig. 7, this will cause significant losses in *FF* as no low-resistive electrical contact can be formed to the adjacent etched-back emitter. Additional losses in V_{OC} will occur due to high recombination in these not well-shielded, metallized areas. Losses in short-circuit current j_{SC} will arise as the laser-doped area that is not covered by the metal fingers yields a higher reflectivity and reduced short-wavelength short-circuit current density. Hence, it is very important to align screen printing and laser processes very precisely.

We developed an alignment procedure that considers inaccuracies of laser and screen printing processes such as screen warping. The result is an adapted laser grid that resembles the actual screen-printed grid. This procedure has already been successfully tested on so-called "pPassDop" solar cells in Ref. [22]. We demonstrated an accurate alignment of 65 µm-wide fingers on 37 µm-wide laser-processed lines over the whole wafer for the rear side metallization of these "pPassDop" solar cells.

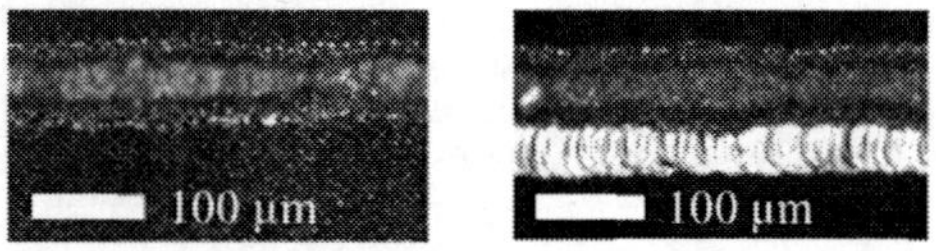

(a) (b)

Figure 7: Light-micrographs of a silver finger on a wafer for which the alignment procedure (a) was applied or (b) was not. The bright laser-doped line is only visible in (b) since it is completely covered by the silver finger in (a). The images are taken from Ref. [22].

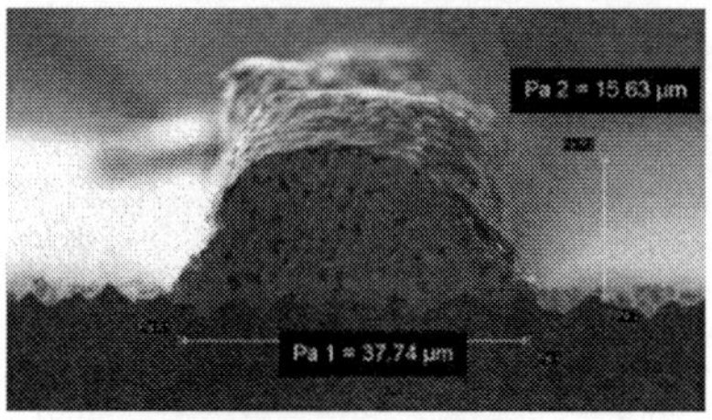

Figure 8: Scanning electron micrograph image of a cross section of a screen-printed and fired finger using a commercially available silver paste.

5.4 Key aspect four: Smaller front side finger widths

The last key aspect discussed in this paper is the front metallization. From an actual measured finger width after contact firing $w_{finger} \approx 55$ µm, we aim to significantly reduce w_{finger}. Therefore, we examined screens and silver pastes from different suppliers. We found a promising new combination of screen mesh and commercially available silver paste that allows for printing shallower fingers. Fig. 8 shows a scanning electron micrograph of a screen-printed and fired silver finger with significantly reduced width of $w_{finger} \approx 38$ µm and a finger height of $h_{finger} \approx 16$ µm resulting from a single printing step. Due to the improved aspect ratio by using a new generation commercially-available silver screen-printing paste, the grid resistance (60 Ω/m per finger) does not change compared to our current 55 µm-wide finger contacts that are formed by double-printing. With the lower finger width and the increased R_{sh} between the contacts when using the selective emitter from section 5.2, the finger number needs to be increased. This increase avoids *FF* losses due to the reduced lateral conductivity in this emitter. An estimation using the software tool GridMaster [23] reveals that the amount of fingers needs to be increased to a number between 110 and 120.

6 SIMULATION RESULTS AND NEXT STEPS

6.1 Simulation of conversion efficiency

All findings concerning the four key aspects discussed in the sections 5.1 to 5.4 are important for increasing the efficiency of our PERC solar cells. With fast metamodeling of numerical device simulations [24], we can easily estimate the cell efficiency potential when implementing the discussed improvements in the form of experimentally derived input parameters.

We integrate a selective emitter to the simulation of the latest PERC solar cells with homogeneous emitter from the experiment in December in Fig. 3. As $j_{0e,met}$ is not determined experimentally, we vary $j_{0e,met}$ for the simulation with values of $j_{0e,met} = \{1000$ fA/cm², 500 fA/cm², 200 fA/cm²$\}$; see also Table 1. The finger number is optimized accordingly in terms of leveling

Table 1: Simulated η for different $j_{0e,met}$ and optimized finger number. The total saturation current density $j_{0,front}$ of the front side is also stated.

$j_{0e,met}$ (fA/cm²)	Finger number	$j_{0,front}$ (fA/cm²)	η (%)
1000	112	73	22.2
500	116	52	22.4
200	120	39	22.5

shading and contact recombination with series resistance losses.

As result of the simulations, the total saturation current density $j_{0,front}$ of the front side is found to be between 39 fA/cm² $\leq j_{0,front} \leq$ 73 fA/cm², see also Table I. This is a significant improvement compared to our current PERC solar cells with $j_{0,front}$ = 152 fA/cm². The conversion efficiencies are computed to values in a range of 22.2% $\leq \eta \leq$ 22.5% for the given $j_{0,front}$. Thus, the simulations clearly show the potential of our performed optimizations and forecast PERC solar cells with efficiencies exceeding 22% after integration of the discussed adaptions from the sections 5.2 to 5.4 into the "new" baseline sequence for selective emitter solar cells.

6.2 PERC solar cells in 2017

After all the preliminary work performed in 2016 as discussed above, we have been ready to implement the developed processes for the selective emitter into a full cell fabrication sequence. This was started in the beginning of 2017. Unfortunately, the running experiment has been lost due to a fire in our PV-TEC laboratory on February 24th, 2017 [25]. Not only have the respective samples been destroyed, but also all other industrial tools for PERC solar cell fabrication that have been located in this laboratory. Thus, the implementation of the selective emitter into fully fabricated PERC solar cells has not been performed yet.

As we have access to other labs and tools at Fraunhofer ISE that are capable to process large-area PERC solar cells, we have been able to re-establish a PERC baseline process in the meantime. We transferred our standard processes from the PV-TEC tools to these interim machines as far as possible. To date, we are again able to fabricate PERC solar cells according to the fabrication sequence shown in Fig. 1 with current peak energy conversion efficiencies of η = 20.9%.

6.3 Outlook to 2018

At the end of the first quarter of 2018, we plan to launch the operation of the PV-TEC laboratory again. New and up-to-date industrial equipment for the frontend processes until passivation will be located in our PV-TEC cleanroom laboratory for industrial equipment [26], while the also new and up-to-date industrial backend equipment will be located in the rebuilt PV-TEC. After the hook-up of the new machines, we will again pick up the above discussed developments and aim at their implementation into PERC solar cells with selective emitter.

7 SUMMARY

Our industrial-oriented PERC baseline process showed stable and reproducible results for 6-inch p-type Cz-Si PERC solar cells achieving mean energy conversion efficiencies η_{mean} = 21.3% and η = 21.5% for the champion cell before the fire in February 2017. Currently, after transferring processes, our baseline achieves 20.9% peak efficiencies. Prior to the fire accident, the cells' front side was identified to be the limiting factor on the way to 22% cell efficiency to date. To improve the cell's front side, we investigated the approach to incorporate a selective emitter using laser doping from the PSG layer into our baseline process. This implementation of a selective emitter and the 22%-efficiency target induces four key aspects:

(i) We show the reduction of the surface doping concentration of the emitter in the photoactive region by an adapted $POCl_3$ diffusion process and a subsequent wet-chemical emitter etch back step. In doing so, we are able to decrease the emitter dark saturation current density to $j_{0e} \approx (31 \pm 4)$ fA/cm² on alkaline textured surface passivated with a standard, non-graded PECVD SiN_x layer.

A high phosphorus concentration within the PSG layer for laser doping is ensured by the above mentioned $POCl_3$ diffusion process that features a second deposition step with active N_2 flow through the $POCl_3$ bubbler after drive-in. For the laser-doped region, we find the emitter dark saturation current density to be $j_{0e,laser} \approx$ 180 fA/cm² (SiN_x passivated, fired) with a sheet resistance of $R_{sh} \approx$ 54 Ω/sq. Specific contact resistances of $\rho_C \approx$ 1 mΩcm² confirm low-resistive electrical contacting after firing using a commercially available silver screen printing paste.

(ii) Concerning process integration, we identify that the emitter etch back needs to be performed before rear emitter removal.

(iii) An accurate alignment of the screen-printed grid on top of the laser-structured selective emitter is crucial for avoiding losses in η. We developed a procedure that considers inaccuracies of laser and screen-printing processes such as screen warping.

(iv) We optimized our screen-printing processes now allowing for a finger width $w_{finger} \approx$ 37 µm at a finger height $h_{finger} \approx$ 16 µm after single printing.

Simulations that are based on the experimentally derived input parameters from this work imply efficiencies up to 22.5% for 6-inch p-type Cz-Si PERC solar cells when the mentioned optimized processes are integrated into the "new" baseline PERC process for selective emitter solar cells at Fraunhofer ISE.

ACKNOWLEDGEMENTS

The authors would like to thank all colleagues at the Fraunhofer ISE PV-TEC for their teamwork and J. Horzel for his support and the fruitful discussions. This work was funded by the German Federal Ministry for Economic Affairs and Energy mainly within the research projects "CUT A" (contract no. 0325823), "POLDI" (contract no. 0324079D), "HELENE" (contract no. 0325777D), "Groschen" (contract no. 0324012B), "AdmMo" (contract no. 0325775E), "CUT B" (contract no. 0325910A).

REFERENCES

[1] A. W. Blakers, A. Wang, A. M. Milne et al., "22.8% efficient silicon solar cell", *Appl. Phys. Lett.*, vol. 55, no. 13, pp. 1363–1365, 1989.

[2] ITRPV, 8th edition, Sept. 2017, "International Technology Roadmap for Photovoltaic (ITRPV) - 2016 Results including maturity report", available onl.: http://www.itrpv.net/Reports/Downloads/.

[3] B. Min, H. Wagner, M. Müller et al., "Incremental efficiency improvements of mass-produced PERC cells up to 24%, predicted solely with continuous development of existing technologies and wafer materials", *Proc. 31st EU PVSEC*, Hamburg, Germany, 2015, pp. 473–476.

[4] P. Saint-Cast, S. Werner, J. Greulich et al., "Analysis of the losses of industrial-type PERC solar cells", *Phys. Status Solidi A*, 2016.

[5] R. Brendel, T. Dullweber, R. Peibst et al., "Breakdown of the efficiency gap to 29% based on experimental input data and modeling", *Prog. Photovolt: Res. Appl.*, vol. 24, no. 12, pp. 1475–1486, 2016.
[6] W. Deng, F. Ye, R. Liu et al., "22.61% efficient fully screen printed PERC solar cell", *Proc. 44th IEEE PVSC*, Washington D.C., USA, 2017.
[7] M. Müller, B. Bitnar, G. Fischer et al., "Loss analysis of 22% efficient industrial PERC solar cells", *Energy Proced.*, in press, 2017.
[8] D. Biro, R. Preu, S. W. Glunz et al., "PV-TEC: Photovoltaic technology evaluation center - design and implementation of a production research unit", *Proc. 21st EU PVSEC*, Dresden, Germany, 2006, pp. 621–624.
[9] L. Ventura, A. Slaoui, J. C. Muller, "Realization of selective emitters by rapid thermal and laser assisted techniques", *Proc. 13th EU PVSEC*, Nice, France, 1995, pp. 1578–1581.
[10] U. Jäger, S. Mack, C. Wufka et al., "Benefit of selective emitters for p-type silicon solar cells with passivated surfaces", *IEEE J. Photovoltaics*, vol. 3, no. 2, pp. 621–627, 2013.
[11] S. Werner, E. Lohmüller, S. Maier et al., "Process optimization for the front side of p-type silicon solar cells", *Proc. 29th EU PVSEC*, Amsterdam, The Netherlands, 2014, pp. 1342–1347.
[12] A. Kimmerle, J. Greulich, A. Wolf, "Carrier-diffusion corrected J_0-analysis of charge carrier lifetime measurements for increased consistency", *Sol. Energy Mater. Sol. Cells*, vol. 142, pp. 116–122, 2015.
[13] S. Werner, E. Lohmüller, S. Maier et al., "Challenges for lowly-doped phosphorus emitters in silicon solar cells with screen-printed silver contacts", *Energy Proced.*, in press, 2017.
[14] H. H. Berger, "Contact resistance and contact resistivity", *J. Electrochem. Soc.*, vol. 119, no. 4, pp. 507–514, 1972.
[15] P. Saint-Cast, I. Reis, J. Greulich et al., "Cz silicon benchmark for p-type PERC solar cells", *Proc. 32nd EU PVSEC*, Munich, Germany, 2016, pp. 1038–1043.
[16] A. Wolf, A. Kimmerle, S. Werner et al., "Status and perspective of emitter formation by $POCl_3$-diffusion", *Proc. 31st EU PVSEC*, Hamburg, Germany, 2015, pp. 414–419.
[17] T. Fellmeth, A. Born, A. Kimmerle et al., "Recombination at metal-emitter interfaces of front contact technologies for highly efficient silicon solar cells", *Energy Proced.*, vol. 8, pp. 115–121, 2011.
[18] S. Werner, S. Mourad, W. Hasan et al., "Structure and composition of phosphosilicate glass systems formed by $POCl_3$ diffusion", *Energy Proced.*, in press, 2017.
[19] S. Werner, E. Lohmüller, J. Weber et al., "Suitability of low recombinative $POCl_3$ diffusion processes with in-situ oxidation for forming laser-doped selective emitters", *Proc. 33rd EU PVSEC*, Amsterdam, The Netherlands, 2017.
[20] A. Lachowicz, K. Ramspeck, P. Roth et al., "Nox-free solution for emitter etch-back", *Proc. 27th EU PVSEC*, Frankfurt, Germany, 2012, pp. 1846–1850.
[21] A. Moldovan, K. Birmann, J. Rentsch et al., "Combined ozone/HF/HCl based cleaning and adjusted emitter etch-back for silicon solar cells", *Solid State Phenomena*, vol. 195, pp. 305–309, 2013.
[22] E. Lohmüller, S. Werner, M. H. Norouzi et al., "Bifacial p-type PERL solar cells with screen-printed pure Ag metallization and 89% bifaciality", *Proc. 33rd EU PVSEC*, Amsterdam, The Netherlands, 2017.
[23] T. Fellmeth, F. Clement, D. Biro, "Analytical modeling of industrial-related silicon solar cells", *IEEE J. Photovoltaics*, vol. 4, no. 1, pp. 504–513, 2014.
[24] S. Wasmer, A. A. Brand, J. M. Greulich, "Metamodeling of numerical device simulations to rapidly create efficiency optimization roadmaps of monocrystalline silicon PERC cells", *Energy Proced.*, in press, 2017.
[25] Fraunhofer ISE (press release), , "Fraunhofer ISE Laborbrand", available online: https://www.ise.fraunhofer.de/de/presse-und-medien/news/2017/brand-in-labor-im-solar-info-center.html, accessed on: 09-21-2017.
[26] J. Rentsch, A. Moldovan, M. Bivour et al., "A new pilot research facility for HJT and selective contact solar cells – PV-TEC SELECT", *Proc. 33rd EU PVSEC*, Amsterdam, The Netherlands, 2017.

LIGHT INDUCED DEGRADATION – DEFECT GETTERING AT GRAIN BOUNDARIES

Tabea Luka*[1,2], Marko Turek[1], Stephan Großer[1], Christian Hagendorf[1]
[1]Fraunhofer Center for Silicon-Photovoltaics CSP, 06120 Halle (Saale), Germany
[2]Anhalt University of Applied Sciences, 06366 Köthen (Anhalt), Germany

ABSTRACT: Under illumination, multi-crystalline PERC solar cells may suffer from a severe efficiency loss up to 20 %rel induced by carrier injection at elevated temperatures which is known as LeTID. Our experiments reveal that LeTID degradation and regeneration occurs rather homogeneously throughout the cell. However, structural defects as grain boundaries are less affected by LeTID during the entire degradation and regeneration process. Based on these results, a gettering of impurities, which are involved in LeTID, during crystallization or cell processing is most likely. It is shown that this effect cannot be explained by a slower degradation at grain boundaries due a to locally reduced charge carrier density or a laterally inhomogeneous regeneration through defect gettering at grain boundaries during the LeTID. The getter range was determined to be around 200-250 µm based on conventional LBIC, supplemented by high-resolution µLBIC at an enhanced lateral resolution of 2-3µm. The maximum getter range of different impurities during the cooling after the firing step was estimated. The calculations indicated that if the LeTID impurity gettering occurs during the firing step then a fast diffuser in silicon (Cu, H, Ni, Co) has to be involved.
Keywords: Light induced degradation, PERC, solar cells, impurity gettering

1 INTRODUCTION

The PERC cell concept is on the rise referring to market penetration. In 2016, already 15 % of all produced silicon solar cells were PERC cells. The technology roadmap predicts a further increase in the overall market share of the PERC/PERT technology of up to almost 60 % in 2027 [1]. Degradation due to light exposure at elevated temperature (short LeTID [2, 3]) can lead to a severe power loss of PERC solar cells up to 20 %. This was first detected at multi-crystalline silicon solar cells [2, 3]. However, recent investigations showed that mono-crystalline cells are also affected by LeTID [4]. A profound understanding of the mechanism which causes light induced degradation on PERC cells [2] and reliable mitigation strategies are required in order to establish the PERC/PERT technology as the leading solar cell architecture.

It has been reported that the solar cell production process can be adapted to reduce LeTID. Mitigation strategies are either (1) reducing the firing temperature, (2) slowing down the cooling rate of the firing step, (3) performing an additional annealing step, or (4) adding a final regeneration step at the end of the cell processing [5-7]. However, these measures imply either a narrowing of the process window leaving fewer options for the optimization of cell properties or an extension of processing time. Thus, a deeper understanding of the defect is favored allowing maximal process flexibility for an optimal efficiency output.

The root cause of LeTID is still unknown. However, it is widely believed that some metallic impurity causes the degradation [5, 8]. Among other defect complexes Ti, W, Mo and Cu are under discussion as degradation sources. Also, hydrogen is discussed being involved in the LeTID degradation [8, 9].

In previous papers, it was shown that the degradation is rather homogenous throughout the entire solar cell. Thus, the LeTID defect is caused by a wide spread and homogenously distributed impurity in the Si bulk [10]. However, significant deviations in the degradation of structural defects were detected. It has been shown that grain boundaries with a high defect density as well as rear contacts show a reduced degradation in comparison to intra grain regions [10, 11]. This indicates a certain analogy to gettering effects at grain boundaries which have been detected on metal rich wafers [12]. However, a detailed investigation of the grain boundaries during LeTID proving that a gettering effect occurred has been missing.

In this paper, we show that other phenomena than the impurity gettering during the crystallization or cell processing cannot be responsible for the reduced LeTID effect at grain boundaries at maximal degradation. In particular, we investigate whether a slower degradation at grain boundaries can be due to locally reduced charge carrier density or a laterally inhomogeneous regeneration which starts at the grain boundaries. At maximal degradation, the samples are investigated using µLBIC with a microscopic spatial resolution of 2-3 µm in order to determine the getter range and the presence of gettering precipitates at grain boundaries with high resolution. Finally, we discuss which elements may be involved in the LeTID grain boundary gettering based on their characteristic diffusion length at higher cell processing temperatures.

2 EXPERIMENTALS

2.1 Experimental set-up

The investigated samples are industrially produced multi-crystalline Si PERC solar cells. The degradation has been carried out using a LID-test setup based on LEDs [patent pending at European patent office, EP15200094.9]. On this setup the intensity is variable adjustable between 0.1 and 2 suns. This allows experiments at constant charge carrier density by continuously increasing the illumination during degradation, so that the open circuit voltage V_{oc} stays constant. The investigated samples were treated at elevated temperatures of 75 °C (with 1 sun illumination) and between 100 °C and 130 °C (at constant carrier injection of approximately $1.6 \cdot 10^{14}$ cm^{-3}) leading to accelerated LeTID degradation. Integral current-voltage measurements as well as spatially resolved light beam induced current (LBIC) measurements are performed under standard test conditions in the solar cell analysis system LOANA.

Degraded samples were cut in smaller pieces and their Al rear-contact was removed chemically. After this

* Corresponding author: e-mail tabea.luka@csp.fraunhofer.de, Phone: + 49 345 5589-5130, Fax: + 49 345 5589-101

preparation, µLBIC measurements were performed at an in-house developed LBIC system consisting of laser scanning microscope LSM700 from ZEISS extended with a LBIC acquisition system developed by point electronic GmbH. Excitation wavelengths of 555 nm and 639 nm have been used.

3 RESULTS

3.1 Temperature dependent LeTID kinetics of integral cell efficiency

To exclude the influence of other (non LeTID related) light induced degradation mechanisms as BO-LID, FeB-LID or Sponge-LID, a pretreatment at 25 °C under 1 sun illumination was performed for 24 h. The observed efficiency change was less than 1 $\%_{rel}$ during this pretreatment. Thus, LeTID is the major light induced degradation mechanism in these samples as the degradation is much more pronounced at elevated temperature. At these elevated temperatures, the samples showed a pronounced efficiency degradation of 12 to 20 $\%_{rel}$, see Fig. 1.

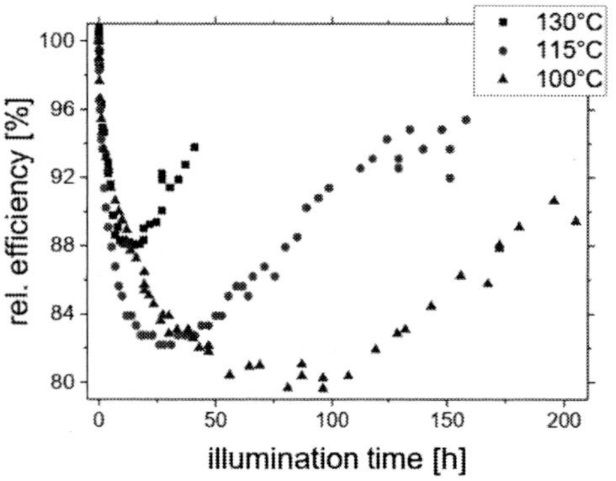

Figure 1: Relative efficiency change during light soaking at elevated temperature at constant carrier injection ~$1.6*10^{14}$ cm^{-3}. The degradation and regeneration rate can significantly be increased using higher temperatures.

Figure 1 shows that the degradation rate can be significantly increased when going to higher temperatures. Also the time scale until maximal degradation agrees with previous investigations on LeTID [3]. Assuming a single exponential fit, the degradation rates were determined to be 0.3 h^{-1} at 130 °C, 0.16 h^{-1} at 115 °C and 0.05 h^{-1} at 100 °C.

3.2 Lateral and time resolved LBIC investigation during degradation and regeneration

Throughout degradation and regeneration LBIC measurements were performed. The following data was determined on a sample, which has been treated at 130 °C and a constant carrier injection of approximately $1.6*10^{14}$ cm^{-3}. A similar behavior was also observed at the other samples.

At the degraded sample, the typical LeTID appearance is detected: grain boundaries and rear contacts show a relatively high lifetime in comparison to intra-grain regions [10], see Fig. 2 (middle). After 41 h of light treatment, the sample is partly recovered from 12 % to 6 % relative efficiency loss. Qualitatively, the same contrast as in the initial state is detected laterally, while the absolute EQE value is at a lower average value, see Fig. 2 (right).

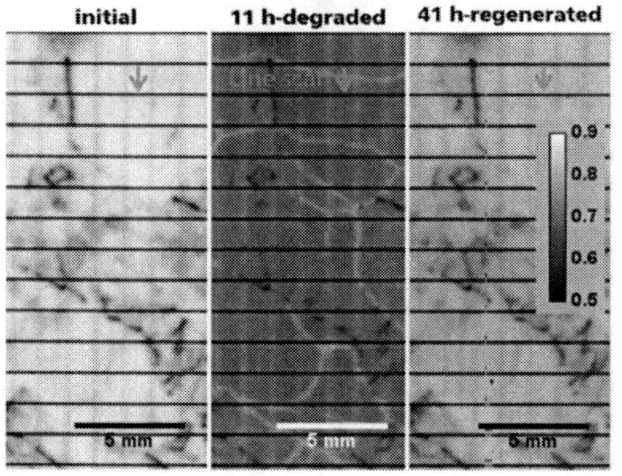

Figure 2: LBIC (EQE 960 nm) images before (left), after 11 h of light treatment (middle) and after 41 h of light treatment (right). The typical LeTID appearance of less affected grain boundaries and rear contacts is observed. The blue arrow indicates where the line scan in Figure was made.

The laterally different degradation extent at grain boundaries, intra-grain regions and rear contacts was investigated locally and time resolved, see Fig. 3. For this analysis, regions with low initial EQE values were neglected, since no significant change during degradation was detected there. As "grain boundaries" only those with an enhanced EQE value compared to the surrounding after 11 h of degradation were chosen. This excludes some grain boundaries, i.e. twin boundaries which behave rather like intra-grain regions as was shown in [13].

In Fig. 3, we see that rear contacts and grain boundaries start at a slightly lower value than intra-grain regions (marked by label "1."). However, after a few minutes of degradation this contrast changes. Throughout the degradation and regeneration process, a higher EQE value and thus a higher lifetime is observed at grain boundaries in comparison to intra grain regions, see Fig. 3 light blue arrows labelled by "2.". Intra-grains and grain boundaries show a similar kinetic as the point of maximal degradation is reached after the same time, see label "3." in Fig. 3. Thus, the maximal LeTID extent at grain boundaries is significantly reduced compared to intra-grain regions. Also at rear contacts, the degradation extent of intra-grains is not reached. This might be the same effect as for grain boundaries. However, the regeneration at rear contacts also appears to be slowed down. Thus, it cannot be excluded that a faster degradation in the Si bulk above the rear contacts interferes with a slower degradation closer to the rear side of the cell. This would also lead to a lower maximal degradation due to the subsequent regeneration.

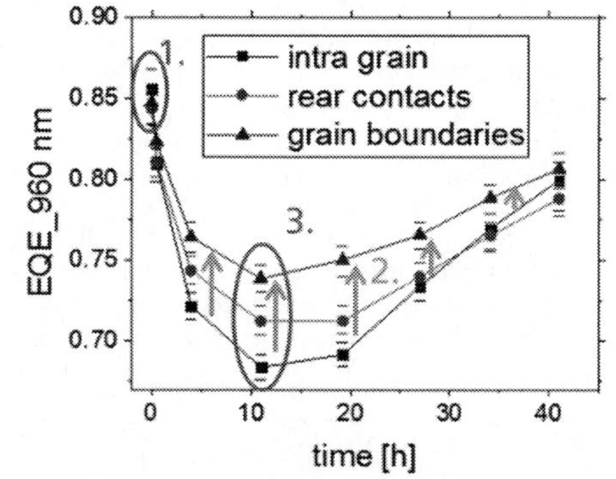

Figure 3: Locally evaluated LBIC (EQE 960 nm) signal over LeTID treatment time. Grain boundaries and intra-grain regions show a similar kinetic behavior.

These results already indicate that the regeneration starts homogenously throughout the cell. This is further investigated by a quantitative analysis of the surroundings of the grain boundaries. It has been proposed that a long scale diffusion process (~100 µm) from the bulk to crystal defects, such as the wafer surface or grain boundaries, causes regeneration [5]. This would imply that the width of the less LeTID-affected region around grain boundaries increases during the light exposure. Line scans throughout degradation and regeneration across a grain boundary at different times are compared in Fig. 4. The EQE intensity profile around the grain boundary can be described by a gaussian distribution. Based on the fit, the full width at half maximum (FWHM) was calculated. Throughout the LeTID treatment the FWHM is around ~260 µm and remains constant during the entire degradation and regeneration. Thus, a wide range diffusion process (~100 µm) from the grain boundary into the grain or conversely during degradation and regeneration at 130 °C can be excluded. Hence, also at lower temperatures no diffusion processes during degradation or regeneration are expected.

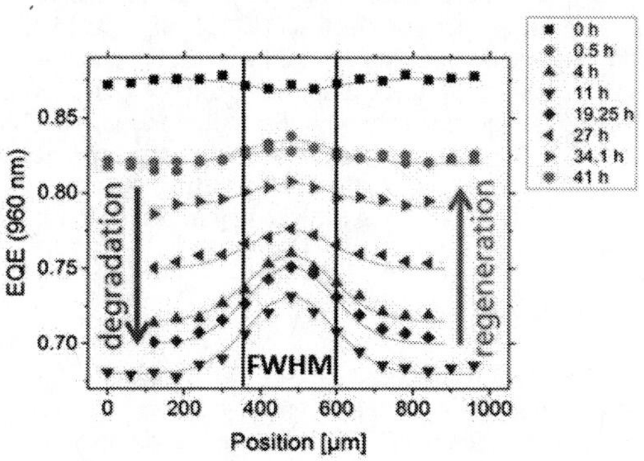

Figure 4: Line scan across grain boundary (position indicated in Figure 2). The symbols are measured data. The lines are fits assuming a normally distributed EQE de-/increase around the GB. During degradation, the EQE values continuously decrease towards the minimum at around 11h (red symbols) The FWHM does not change throughout degradation and regeneration.

The EQE intensity profile with a FWHM of about ~260 µm determined with the LBIC system is in the range of the conventional LBIC resolution ~100 µm. Therefore, this measurement was supplemented by higher resolution µLBIC system on degraded samples for a more exact determination of the getter range and detailed investigations of the grain boundary. The spatial resolution of the µLBIC system is around 2-3 µm and thus goes beyond the conventional LBIC system.

3.3 High resolution µLBIC investigations of defect gettering regions at grain boundaries originating from high temperature processes

The sample investigated in this section has been degraded at 75 °C and 1 sun illumination. After significant efficiency degradation of ~12 $\%_{rel}$, the treatment was interrupted for high resolution lateral investigations. The Al rear contact was chemically removed, which allows lateral investigations from the front and from the rear to get a more profound understanding of lateral variations in bulk lifetime and rear passivation quality. LBIC measurements from both sides show qualitatively the same appearance, see Fig. 5. Decorated recombination active grain boundaries (marked with an A) as well as grain boundaries showing reduced LID extent (marked with a B) are observed.

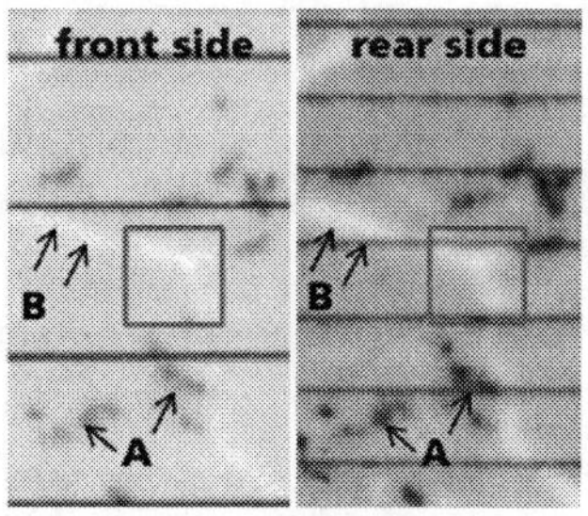

Figure 5: LBIC of degraded PERC cell from the front side (left) and the rear side (right). Decorated grain boundaries (A) and by LeTID less effected grain boundaries (B) are observed. The blue square marks the region where the µLBIC image presented in Figure 6 was taken.

Qualitatively, the same results as with conventional LBIC are detected with µLBIC, since both methods show grain boundaries of type A and B, see Figure 5 and Figure 6. The µLBIC measurement was performed from the rear side of the solar cell. Thus, it contains information about the rear side passivation quality as well as the bulk lifetime. After chemical Al-removal, the rear side passivation is intact on the investigated area. At grain boundaries of type B, a bright area with an enhanced LBIC signal distribution surrounding the GB is detected (indicated by the dashed line in Fig. 6). The LBIC signal intensity decreases from the grain boundary into the intra-grain regions over a distance of around ~220 µm continuously. This is consistent with a lateral gradient of the LeTID defect concentration within the extension of the maximum getter range determined from the conventional LBIC data in section 3.2. Referring to the LBIC results shown above, it has to be noted, that this gettering gradient seems to be frozen in at the typical LeTID degradation and regeneration temperatures between 75-130°C.

Furthermore, it can be observed in Fig. 6 that a locally enhanced lifetime is detected directly at grain boundaries. This may indicate a reduction of recombination activity due to a deactivation of recombination active defect complexes at the very neighborhood of structural defects at grain boundaries.

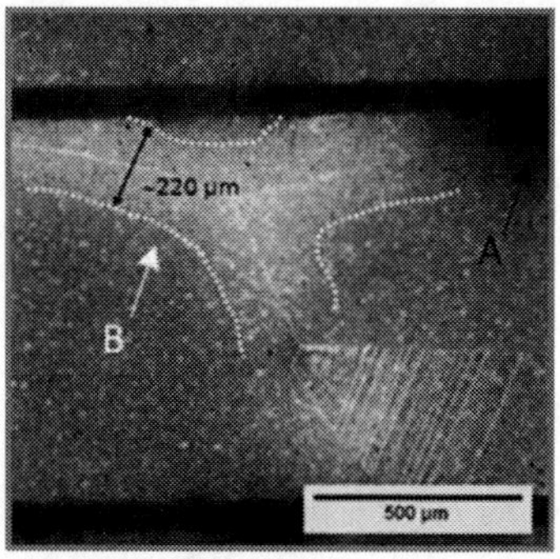

Figure 6: µLBIC image of region marked in Figure 5. Getter range around grain boundary (type B) is around 220 µm.

4 DISCUSSION

The typical appearance of LeTID affected solar cells with grain boundaries showing a reduced LeTID has been under discussion in previous works. Possible explanations that have been proposed were:

(1) a slower degradation at grain boundaries, i.e. caused by a locally lower charge carrier density,

(2) a regeneration starting at the grain boundaries i.e. through impurity gettering at slightly enhanced temperatures during the LeTID light soaking process,

(3) a defect gettering at high temperatures at the grain boundary during crystallization or cell processing.

The first part of our results showed that similar degradation and regeneration kinetics are observed at grain boundaries compared to intra-grain regions. This contradicts the hypothesis (1) that LID is just slowed down at grain boundaries in contrast to a LID reduction.

Secondly, it was shown that the getter range around the grain boundary does not significantly change upon slightly elevated temperatures of 75 - 200 °C present during degradation and regeneration, which contradicts the hypothesis (2) that LeTID impurities diffuse long distances ~100 µm to the grain boundaries. Based on the high-resolution µLBIC results, an impurity gettering profile is present around the grain boundaries at a length scale between 200-250 µm. This defect distribution is frozen in during LeTID cycles between 75 – 130 °C. Accordingly, it is assumed that the observed getter range of 200-250 µm indicates the grain boundary gettering of a LeTID-related impurity during higher temperatures that appear during crystallization or cell processing.

The firing step is one critical process step, which determines the LeTID extend [5-7]. As this is also the last high temperature step in cell production, the impurities might be gettered during this step. Under the assumptions that the LeTID defects are homogeneously distributed before cooling down and that the grain boundary acts as a sink, the getter range of different impurities can be estimated. Input parameters are the cooling rate of the oven, the diffusiveness of the impurities in silicon and the precipitation behavior of the impurity. We estimated the getter range during cool down in a fast firing oven (FFO) with a peak temperature of 800 °C as described in [7]. Diffusion constants are taken from [14]. The precipitation behavior of the impurity at a certain temperature significantly affects the estimated gettering region. The precipitation behavior is relevant to determine the temperature range at which the gettering occurs from a maximal temperature T_{max} to a minimal temperature T_{min}. The maximal temperature T_{max} is the temperature at which metal precipitates or H-Si pairs at a grain boundary are stable. The minimal temperature T_{min} is the temperature until which a significant amount of mobile interstitial ions or atoms are present in the silicon, while at lower temperatures less mobile compounds are formed in the grains (like H_2 or Cu_3Si).

Impurity	estimated FWHM	T_{max} [°C]	T_{min} [°C]
Cu	~380 µm	600	280
H	~320 µm	>800	600
Ni	~100 µm	600	330
Co	~150 µm	>800	620
Li	~60 µm	>800	-
Fe	~40 µm	>800	-
Cr	~16 µm	>800	-
Ti	~0.04 µm	>800	-

Table 1: Estimated FWHM of the potential getter range during cooling in an FFO.

Slow diffusers and medium-fast diffusers can be excluded causing the getter range during firing. Since even without reducing the relevant temperature range getter ranges < 60 µm are determined, see Table 1. Thus, fast diffusers (Cu, H, Ni, Co) have to be considered. For these impurities, T_{min} was chosen to be the temperature were the solubility drops below 10^{10} cm^{-3}. For Cu and Ni T_{max} is chosen to be 600 °C, as even at this temperature Cu and Ni precipitates are unstable [15].

The results of the estimated getter range during the cooling of a firing step in an FFO oven are presented in Table 1. It can be concluded that if the gettering occurs during firing only a fast diffuser (Cu, H, Ni, Co) can cause the reduced degradation around the grain boundaries. The getter ranges of these impurities are in order of the determined getter range on our samples and around 200-250 µm. However, one the basis of our experimental data it cannot be excluded that the gettering occurs during crystallization involving slower impurities which might have been gettered.

4 SUMMARY

Industrially produced mc-Si PERC solar cells showing strong sensitivity to light and elevated temperature induced degradation (LeTID) resulting in a power loss of 12-20 % were investigated. Our experiments have revealed that LeTID degradation and regeneration occurs rather homogeneously throughout the cell. However, structural defects, such as grain boundaries, are less affected by LeTID during the entire degradation and regeneration process. It was shown that this effect can neither be explained by a slower degradation due to reduced carrier injection at grain boundaries nor by a lateral inhomogeneous regeneration through defect gettering at grain boundaries during a typical LeTID light soaking with elevated temperatures below 130 °C. Based on our results, a gettering of LeTID-related impurities at higher temperatures during crystallization or cell processing is most likely. The getter range around the grain boundaries was determined to be approximately 200-250 µm. The typical getter range of different impurities during the cooling after the firing step was estimated. The calculations showed that if the LeTID impurity gettering occurs during the firing step then a fast diffuser in silicon (Cu, H, Ni, Co) has to be gettered. However, it cannot be excluded that the gettering happens during crystallization, where less mobile impurities could cause a similar gettering range.

REFERENCES

[1] International Technology Roadmap for Photovoltaic Results 2017.

[2] K. Ramspeck, et al., *Proceedings of the 27th EU-PVSEC*, pp. 861-865, 2012.

[3] F. Kersten, et al., *Solar Energy Materials and Solar Cells*, 142, pp. 83-86, 2015.

[4] F. Fertig et al., *Energy Procedia 124*, 338-345, 2017.

[5] Bredemeier et al., *AIP ADVANCES*, 6, 035119, 2016.

[6] C. E. Chan, et al., *IEEE journal of Photovoltaics*, 6, pp. 1473-1479, 2016.

[7] R. Eberle et al., *Physica status solidi - rapid research letters* 10, p. 861-865, 2016.

[8] K. Nakayashiki et al., *IEEE journal of Photovoltaics*, 6 (4), pp. 860-868, 2016.

[9] M.A. Jensen et al, *IEEE Journal of Photovoltaics*, 7 (4), 980-987, 2017.

[10] T. Luka et al., *Solar Energy Materials and Solar Cells*, 158, pp. 43-49, 2016.

[11] M. Selinger et al., *Energy Procedia*, 92, pp. 867-872, 2016.

[12] D. Macdonald, J. Tan, and T. Trupke, *Journal of Applied Physics* 103, 073710, 2008.

[13] T. Luka et al., *Energy Procedia 124*, 756-766, 2017.

[14] D.J. Fisher, *Diffusion in Silicon - 10 Years of Research*, 1998.

[15] M.A. Jensen et. al., *Proceedings of the 44th IEEE PVSC*, (to be published 2017).

Acknowledgment

We thank our colleagues Martina Werner, Stefan Eiternick, Steffi Göller, Andreas Graff and Susanne Richter for sample preparation and support by experiments and Dominik Lausch and Florian Schindler for valuable discussions.

Furthermore, we thank the German Federal Ministry for Commerce and Energy and the industry partners within the research cluster "SolarLIFE" (contract no. 0325763 D) for the financial support. Tabea Luka thanks the Federal Ministry of Education and Research for the financial support within the joined research graduate school "StrukturSolar II" (03SF0417 A).

BIFACIAL P-TYPE SILICON PERL SOLAR CELLS WITH SCREEN-PRINTED PURE SILVER METALLIZATION AND 89% BIFACIALITY

E. Lohmüller[1], S. Werner[1], M. H. Norouzi[1], S. Mack[1], M. Demant[1], S. Gutscher[1], P. Saint-Cast[1], S. Wasmer[1], N. Wöhrle[1], B. Bitnar[2], S. Steckemetz[2], P. Palinginis[2], H. Neuhaus[2], M. König[3], A. Wolf[1]
[1]Fraunhofer Institute for Solar Energy Systems ISE, Heidenhofstr. 2, 79110 Freiburg, Germany
[2]SolarWorld Industries GmbH, Berthelsdorfer Straße 111A, 09599 Freiberg, Germany
[3]Heraeus Precious Metals GmbH & Co. KG, Heraeusstr. 12-14, 63450 Hanau, Germany
Phone: +49 761 - 4588 5701; e-mail: elmar.lohmueller@ise.fraunhofer.de

ABSTRACT: We present 6-inch bifacial p-type Czochralski-grown silicon passivated emitter and rear locally-diffused (PERL) solar cells with "pPassDop" layer stack on the cell's rear side. The "pPassDop" layer stack consists of an aluminum oxide and a boron-doped silicon nitride layer serving as both surface passivation and doping source. Local laser processing introduces boron and aluminum atoms from the "pPassDop" layer stack into the silicon. The electrical contacting of the formed line-shaped p-doped back surface field is realized by screen-printed and fired pure silver contacts (i.e. without aluminum). The fabricated PERL solar cells reach high bifaciality of up to 89%. The monofacial peak front side energy conversion efficiency, measured with contact bars on both sides on a black non-conducting chuck, is given by 19.8%. Fill factors of more than 79% and specific contact resistances in the single-digit mΩcm^2-range prove the successful low-resistance contacting on both sides. It is important to stress that the same commercial state-of-the-art firing-through pure silver screen-printing paste is used for the cell's front and rear side metallization. A special alignment procedure ensures that the rear silver grid with finger widths of about 65 µm is placed over the whole wafer on top of the about 37 µm-wide laser-doped and opened structures.
Keywords: p-type silicon solar cell, bifacial, screen-printing, silver contact, boron-doped silicon nitride

1 INTRODUCTION

Bifacial solar cells [1] fabricated from p-type silicon wafers are currently of great interest to the photovoltaic community [2–7] and have experienced a considerable hype since SolarWorld's announcement in 2014 [8] to commercialize a bifacial passivated emitter and rear cell (PERC) variant. The bifacial solar cells allow for increased power generation due to additional rear side illumination. Furthermore, cell and module manufacturing need only minor adjustments with respect to their monofacial device counterparts.

However, bifacial p-type silicon PERC or passivated emitter and rear totally diffused (PERT) solar cells typically feature rather wide aluminum contact fingers and busbars on the rear side (finger widths 100 µm < $w_{finger} \leq$ 255 µm) [2,7,9], as the electrical conductivity of aluminum is rather poor. The associated high aluminum coverage fraction on the rear side negatively affects the cells' energy conversion efficiency η_{rear} when illuminated from the rear side. The bifaciality $\beta = \eta_{rear} / \eta_{front}$ of such p-type silicon PERC/PERT devices is reported to be $\beta \leq 80\%$ [2,3,7]. On the other hand, for bifacial p-type silicon PERT solar cells with silver grid on the rear side allowing for contacts with much smaller finger width, $\beta = 87\%$ is reported [4]. For bifacial n-type silicon PERT solar cells, for which the rear side metallization with silver contacts is state-of-the-art, β exceeds values of 90% [10–14].

The purpose of this work is to realize industrial bifacial p-type Czochralski-grown silicon (Cz-Si) solar cells that feature similar high bifaciality β as bifacial PERT solar cells. It is important that the process complexity compared to industrial bifacial p-type Cz-Si PERC solar cells with aluminum rear side grid is not increased.

2 APPROACH

To address the challenge concerning high bifaciality β, we substitute the rather wide aluminum contact fingers and silver busbars on the rear side as known for conventional bifacial p-type PERC solar cells [2,7,9] by thinner screen-printed and fired silver fingers and busbars. Thereby, the additional printing step for the rear side silver contact pads—necessary for external interconnection in conventional PERC solar cells—is eliminated. We choose the passivated emitter and rear locally diffused (PERL) solar cell structure with local back surface field (BSF) that is shown in Fig. 1. Key for this approach is the so-called "pPassDop" layer stack [15,16] applied on the rear side of the cell. This layer stack consists of a boron-doped silicon nitride (SiN_x:B) layer on top of a thin aluminum oxide (Al_2O_3) layer. It serves as both surface passivation and dopant source during laser doping to locally form highly p-doped line-shaped BSF areas.

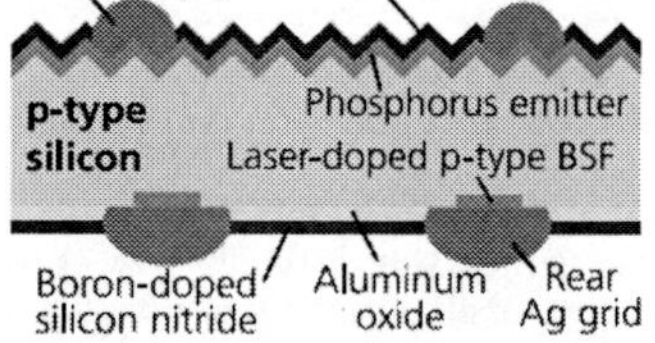

Figure 1: Schematic cross section of the bifacial p-type Cz-Si PERL solar cell structure investigated in this work. The front and rear side are metallized by the same firing-through pure silver (Ag) screen-printing paste. The width of the rear finger contacts is larger than the laser-doped BSF.

3 SOLAR CELL FABRICATION

3.1 Process sequence

For cell fabrication, see Fig. 2, conventional pseudo-square p-type Cz-Si wafers with an edge length L_{edge} = 156.75 mm, a base resistivity $\rho_{base} \approx 2$ Ωcm and an initial wafer thickness t = 190 µm are used that are alkaline textured at SolarWorld Industries. At Fraunhofer ISE, the wafers are then processed into final solar cells according

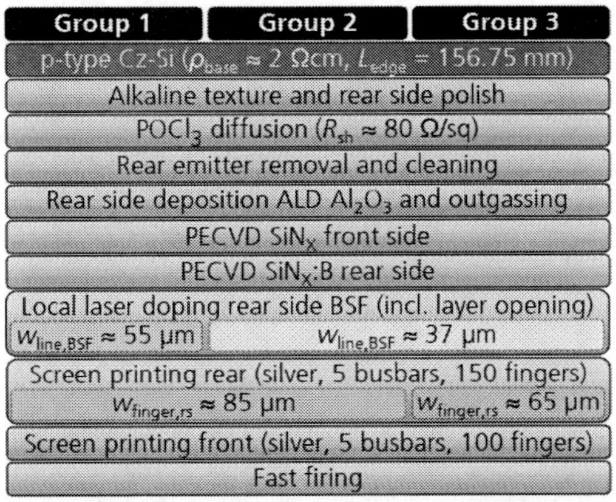

Figure 2: Schematic process sequence for fabrication of the bifacial p-type Cz-Si PERL solar cells with "pPassDop" layer stack on the rear side as illustrated in Fig. 1.

to an industry-oriented PERC baseline process [17] in the PV-TEC pilot-line [18]. A wet-chemical rear side polishing step precedes the tube furnace phosphorus oxychloride ($POCl_3$) diffusion with in-situ oxidation [19]. The resulting emitter features a sheet resistance $R_{sh} \approx 80$ Ω/sq and a dark saturation current density $j_{0e} \approx 85$ fA/cm² determined on alkaline textured and SiN_x-passivated surface according to the procedure described in Ref. [20]. After wet-chemical rear emitter removal and surface cleaning (including the removal of the front side phosphosilicate glass), a 6 nm-thick Al_2O_3 layer is formed on the rear side by fast atomic layer deposition (ALD) followed by an outgassing step in nitrogen ambient in a tube furnace. Plasma-enhanced chemical vapor deposition (PECVD) forms the front side SiN_x passivation and antireflection layer as well as the rear side SiN_x:B layer.

Three groups of solar cells with a variation in the width of the line-shaped p-doped BSF $w_{line,BSF}$ and the width of the rear finger metal contacts $w_{finger,rs}$ are examined. A pulsed infrared diode-pumped solid-state disk laser from Jenoptik with a wavelength $\lambda = 1030$ nm and a repetition rate $f = 30$ kHz is used for both laser doping out of the "pPassDop" layer stack and its simultaneous removal. The distance of the BSF lines is 1.04 mm with two different line widths $w_{line,BSF} \approx 55$ µm (group 1) and $w_{line,BSF} \approx 37$ µm (group 2,3). The line width variation results from two different laser powers. The area of the rear busbars is also doped and opened.

The rear side contact grid with five busbars and 150 fingers (1.04 mm spacing) is applied by screen-printing the commercial state-of-the-art firing-through pure silver paste SOL9621M (w/o aluminum). Two different screen-openings are utilized resulting in rear side finger widths $w_{finger,rs} \approx 85$ µm (group 1,2) and $w_{finger,rs} \approx 65$ µm (group 3) measured after contact firing. The same silver paste is used to apply the front side contact grid, which also features five busbars but 100 fingers (1.56 mm spacing) with width $w_{finger,fs} \approx 64$ µm. Finally, fast firing with variation in the peak temperature for each cell group completes the fabrication sequence.

3.2 Differences to conventional bifacial p-type Cz-Si PERC solar cells with rear side aluminum fingers

Only two minor adaptions are needed to implement the bifacial p-type Cz-Si "pPassDop" PERL approach into an existing PERC production line: the rear side SiN_x capping layer needs to be replaced by a SiN_x:B layer and the laser contact opening process of the rear side passivation layer needs to be adapted to also function as doping process.

As a silver contact grid forms the rear side metallization of the PERL solar cells and soldering onto screen-printed silver is state of the art, no additional rear side solder pads for external interconnection are necessary. This reduces the total number of process steps by one compared to conventional bifacial PERC technology.

3.3 Doping profiles of the rear side BSF

To determine the doping profiles resulting from the two laser powers applied to the rear layer stack, 2x2 cm²-large test fields are lasered with a constant laser line pitch of 25 µm. The samples used for this purpose are processed according to Fig. 2 until passivation on both sides. Fig. 3 shows the charge carrier concentration profiles of the corresponding p-dopings determined by electrochemical capacitance-voltage measurements (ECV) [21]. The profiles are scaled to match the sheet resistances R_{sh}, which are locally determined by 4-point-probe measurements around the ECV spots [22,23]. This adaption requires an area scaling factor of about 1.1. As the initially alkaline textured surface is flattened by the rear side polishing step and the laser processes, the obtained area scaling factor is reasonable. As it has been shown in Ref. [16], the p-doping can consist of both boron from the SiN_x:B layer and aluminum from the Al_2O_3 layer.

Both doping profiles feature the same quite low surface dopant concentration $N_{surf} \approx 3 \cdot 10^{19}$ cm^{-3}. The profile depth d_{prof} determined at a dopant concentration $N = 10^{16}$ cm^{-3} is larger for group 1 with $d_{prof} \approx 5$ µm in comparison to $d_{prof} \approx 3.5$ µm for group 2 and 3. Thus, the higher laser power of group 1 yields both, a larger line width and a deeper doping profile.

3.4 Alignment procedure for the rear side structures

As the rear side finger widths $w_{finger,rs}$ are only about 30 µm (group 1,3) or about 50 µm (group 2) wider than the laser-doped and opened BSF lines $w_{line,BSF}$, an accurate alignment is required between screen printing and laser

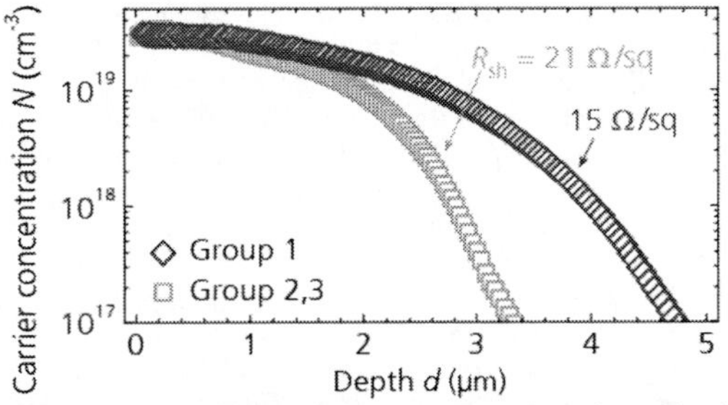

Figure 3: P-type charge carrier concentrations determined by ECV measurements for laser-doped 2x2 cm² large test fields. The sheet resistances R_{sh} are given. The uncertainty in the surface dopant concentration is $\Delta N_{surf} = 10^{19}$ cm^{-3}.

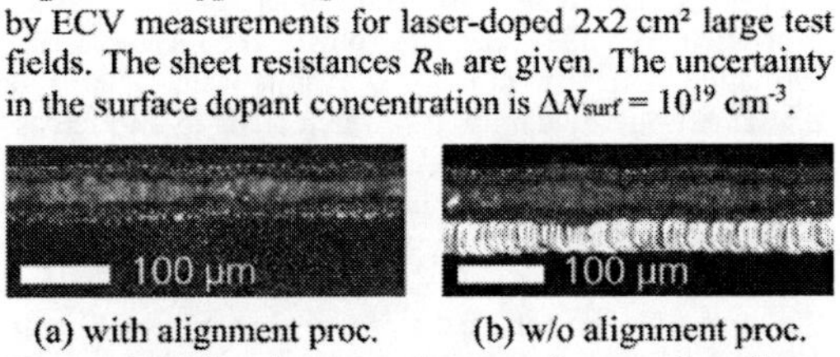

(a) with alignment proc. (b) w/o alignment proc.

Figure 4: Light-microscope images of a rear silver finger on a wafer from (a) group 3 and (b) on a test wafer for which the alignment procedure has not been applied. The finger width in both images is $w_{finger,rs} \approx 65$ µm. Both images are taken at the same position—at the second finger counted from the wafer edge next to the wafers' corner.

process. The transfer of the grid structure from the screen onto the wafer surface is affected by several effects that usually lead to slight deformation. It might thus be necessary to account for deformations by an adapted laser layout.

A special alignment procedure that is based on image recording of both lasered and screen-printed structures with subsequent computer-based data processing ensures an optimal placement of the rear finger grid on top of the laser-doped BSF lines over the whole wafer. Fig. 4(a) illustrates the high accuracy when the alignment procedure is applied. This exemplary image is valid for the entire wafer. The laser-doped BSF line with a width $w_{line,BSF} \approx 37$ µm is not visible as it is completely covered by the screen-printed and fired finger contact, highlighting the successful implementation of our in-house developed alignment procedure.

In contrast, Fig. 4(b) shows an area found on wafers when the alignment procedure has not been applied. It is an example of an extreme case of non-optimal alignment between screen-printed finger and laser-doped BSF line: the bright laser-doped line is clearly visible. However, there are also areas on the wafer for which the alignment is very accurate. Hence, the misalignment as exemplarily shown in Fig. 4(b) is caused by the above mentioned screen deformation and cannot be resolved by offsets in the x- or y-direction or by a rotation of the screen during the printing process.

4 SOLAR CELL RESULTS

Prior to cell metallization, some fully passivated cell precursors are fired and their implied open-circuit voltages iV_{OC} are measured by the quasi-steady-state photoconductance (QSSPC) technique [24]. These QSSPC measurements reveal the voltage potential of the cell precursors prior to laser doping to be $iV_{OC} \approx 670$ mV. As we have been faced with some issues during PECVD layer deposition on front and rear side, the iV_{OC} is somewhat lower than expected.

4.1 Current-voltage measurements

Monofacial current-voltage (*IV*) measurements from front and rear side are both performed with contact bars on both sides on a black non-conductive chuck.

Fig. 5 shows the front side *IV* data of the three solar cell groups for the best firing temperature. The mean front side energy conversion efficiencies $\eta_{front,mean}$ are given by $\eta_{front,mean} = 19.5\%$ (group 1), $\eta_{front,mean} = 19.4\%$ (group 2), and $\eta_{front,mean} = 19.7\%$ (group 3). The slightly higher $\eta_{front,mean}$ by $0.1\%_{abs}$ for group 1 compared to that of group 2 results from both somewhat larger short-circuit current density j_{SC} and open-circuit voltage V_{OC}. The solar cells from group 3 show an efficiency advantage of $0.3\%_{abs}$ compared with the cells of group 2, which are processed with the same laser power for the local BSF formation. This gain in $\eta_{front,mean}$ results from the smaller rear side finger width $w_{finger,rs}$ which leads to an increase in j_{SC} and V_{OC} by 0.2 mA/cm² and 9 mV, respectively. However, the mean fill factor FF_{mean} of group 3 is $0.3\%_{abs}$ lower than that of group 2.

The *IV* data for front and rear side measurements of the most efficient solar cells of the three groups is summarized in Table I. It is seen that the overall most efficient solar cell from group 3 with $\eta_{front} = 19.8\%$ also features the highest rear side efficiency $\eta_{rear} = 17.6\%$, yielding an excellent bifaciality $\beta = 89\%$. The large advantage in η_{rear} opposite to group 1 and 2 results from the smaller rear side finger width $w_{finger,rs} \approx 65$ µm.

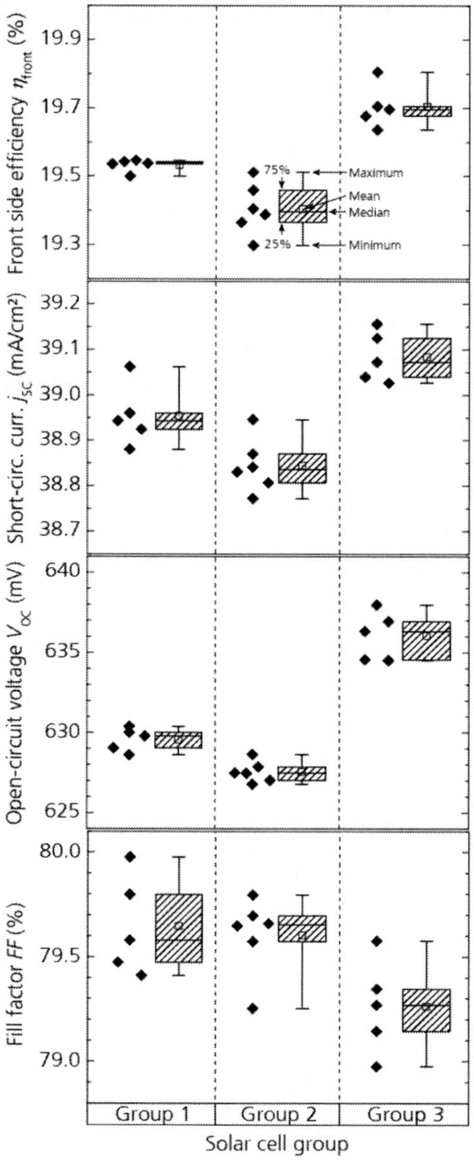

Figure 5: Front side *IV* data of the three solar cell groups for the best firing temperature. The cells are measured with contact bars on both sides on a black non-conducting chuck within an industrial cell tester (cell area $A_{cell} \approx 242.5$ cm²). The points to the left of the box plots refer to the individual cell measurements. The meaning of the box plots is explained exemplary.

Table I: *IV* parameter of the most efficient solar cell per group. The measurements with contact bars on both sides on a black non-conducting chuck are performed by illuminating the solar cells from their (a) front side or (b) rear side.

Group	(a) Front				(b) Rear	
	η_{front} (%)	j_{SC} (mA/cm²)	V_{OC} (mV)	*FF* (%)	η_{rear} (%)	β (%)
1	19.6	38.9	629	80.0	16.1	82
2	19.5	39.0	629	79.7	16.2	83
3	19.8	39.1	638	79.4	17.6	89

4.2 Fill factor analysis and specific contact resistances

The following fill factor analysis refers to the mean IV data shown in Fig. 5 (and others that are not shown). The series resistances r_S are given by mean values of $r_{S,mean} = 0.62\ \Omega cm^2$ (group 1,2) and $r_{S,mean} = 0.68\ \Omega cm^2$ (group 3). As the mean pseudo fill factors $pFF_{mean} \approx 82.7\%$ are similar for all three groups, the $0.3\%_{abs}$ lower FF_{mean} for group 3, see Fig. 5, results from a higher $r_{S,mean}$. This higher r_S can be caused, on the one hand, by a higher rear grid resistance $R_{grid,rear}$ or, on the other hand, by a higher specific contact resistance ρ_C on the rear side.

The mean rear grid resistance for group 3 cells with $R_{grid,rear,mean} = 10.8\ m\Omega$ (measured between adjacent busbars) is twice as high as for the cells in group 1 or 2 with $R_{grid,rear,mean} = 5.5\ m\Omega$ due to the lower $w_{finger,rs}$. An evaluation using GridMaster [25] shows that this difference in $R_{grid,rear}$ leads to an about $0.006\ \Omega cm^2$ higher global r_S for group 3, which corresponds to an about $0.03\%_{abs}$ lower FF.

To examine the quality of the electrical contacting of the screen-printed and fired silver grid to the front side emitter as well as to the rear side laser-doped BSF, ρ_C is determined according to the transfer length method (TLM) [26]. For this purpose, four 10 mm-wide strips are cut out between the five busbars of each of two selected solar cells per group with median efficiencies. On each strip, five TLM measurements are performed on the rear side at various points resulting in total 40 TLM measurements per structure of interest. As the front side is equally processed for all groups, only two exemplary cells are measured on the front side.

For calculation of ρ_C according to the TLM measurements, the contact widths need to be considered. For the measurements on the front side, the contact width is considered to be the screen-printed finger width after firing $w_{finger,fs} = 64\ \mu m$. For the calculation of the rear side ρ_C, two different contact widths are considered. First, the printed finger widths $w_{finger,rs}$ and second, the width of the laser-doped BSF lines $w_{line,BSF}$. As the parts of the metal contacts that overlap the silicon base (i.e. the areas without BSF) do not contribute to the current transport between base and contact grid—as ρ_C is orders of magnitude larger in these areas—, both evaluation approaches are of interest. Note that due to the sample structure, the determined ρ_C values represent upper limits [27].

Fig. 6 shows the correspondingly determined ρ_C values for the front side and the rear side of the solar cells. The measured ρ_C for the front side is $\rho_C = (4.6 \pm 2.1)\ m\Omega cm^2$. The mean values for the rear side contacts range between

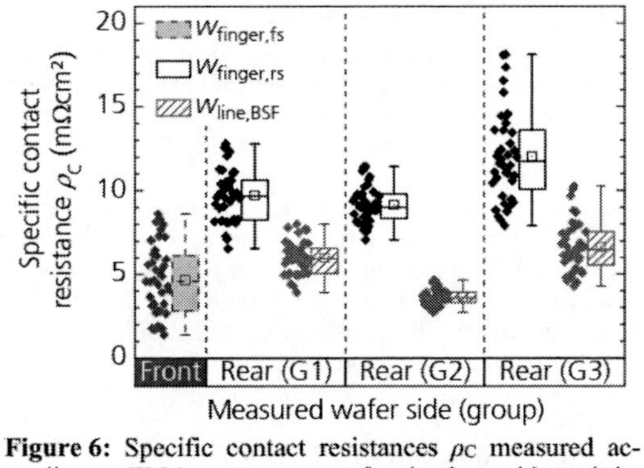

Figure 6: Specific contact resistances ρ_C measured according to TLM measurements for the front side and the rear side of the three different solar cell groups. The contact widths $w_{finger,fs}$, $w_{finger,rs}$, and $w_{line,BSF}$ are used to determine the corresponding ρ_C values.

$3.6\ m\Omega cm^2 \leq \rho_{C,mean} \leq 6.6\ m\Omega cm^2$ (evaluation with $w_{line,BSF}$) or between $9.2\ m\Omega cm^2 \leq \rho_{C,mean} \leq 12.0\ m\Omega cm^2$ (evaluation with $w_{finger,rs}$). The $\rho_{C,mean}$ for group 3 are slightly higher by about $3\ m\Omega cm^2$ compared to those of group 2. Comparing $\rho_{C,mean}$ of group 3 with those of group 1, the difference is about $3\ m\Omega cm^2$ for the evaluation with $w_{finger,rs}$ and about $0.6\ m\Omega cm^2$ for $w_{line,BSF}$. However, the applied commercial state-of-the-art pure silver paste is capable of contacting the laser-processed and p-doped BSF areas despite low surface dopant concentration $N_{surf} \approx 3 \cdot 10^{19}\ cm^{-3}$ (see Fig. 3) with reasonable low ρ_C. In accordance with related investigations published in Ref. [28], the results impressively demonstrate the capability of the applied commercial firing-through pure silver screen printing paste to contact both the cell's front side and rear side with reasonable low electrical resistance.

Again, the evaluation using GridMaster [25] shows that the difference in ρ_C leads to an about $0.07\ \Omega cm^2$ higher global r_S for group 3, which corresponds to an about $0.4\%_{abs}$ lower FF. This is in very good agreement with the measured FF-data in Figure 5 and Table I.

To summarize, the higher r_S and thus, the lower FF for the solar cells from group 3 originate mainly from the higher rear side ρ_C.

4.3 Open-circuit voltage analysis

The mean V_{OC} of the three different groups, see Fig. 5, are given by $V_{OC,mean} = 629$ mV (group 1), $V_{OC,mean} = 627$ mV (group 2), and $V_{OC,mean} = 636$ mV (group 3). In the following two sections, we will analyze these differences in more detail.

4.3.1 Estimation of dark saturation current densities

The dark saturation current densities of the laser-doped and opened BSF lines $j_{0,laser,BSF}$ are determined according to the procedure described in Refs. [15,16]. Based on test fields with varying coverage fraction of the BSF lines, $j_{0,laser,BSF} \approx 1900$ fA/cm² (group 1) and $j_{0,laser,BSF} \approx 2400$ fA/cm² (group 2,3) is extracted.

Due to the two different BSF line widths $w_{line,BSF}$, compare also with Fig. 7, the total coverage fraction f_{BSF} is different for group 1 compared to group 2 and 3, see Table II. The area-weighted total contribution of the BSF regions $j_{0,laser,BSF,tot}$ to the overall j_0 of the entire cell calculates to $j_{0,laser,BSF,tot} \approx 125$ fA/cm² (group 1) and $j_{0,laser,BSF,tot} \approx$

Table II: Summary of dark saturation current densities j_0 and coverage fractions f for the three solar cell groups as discussed in the text. The five busbars with an effective width of 400 μm each are taken into account.

Group	$j_{0,laser,BSF}$ (fA/cm²)	f_{BSF} (%)	$j_{0,laser,BSF,tot}$ (fA/cm²)	$f_{overlap}$ (%)
1	1900	6.6	125	3.0
2	2400	4.8	115	4.8
3	2400	4.8	115	2.8

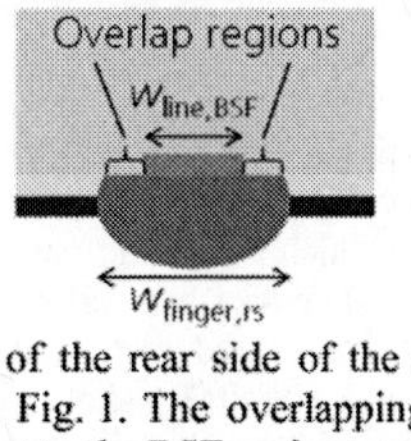

Figure 7: Detail of the rear side of the schematic cross section shown in Fig. 1. The overlapping regions of the metal contacts next to the BSF regions are marked.

115 fA/cm² (group 2,3). These values are estimated under the assumption that the metal contacts (as present in the final cells) have no additional impact on the recombination properties within the BSF regions. As $j_{0,\text{laser,BSF,tot}}$ is lower for group 2 than for group 1, the laser-doped and metallized BSF regions cannot explain the slightly lower V_{OC} of the group 2 cells.

On the other hand, the overlap of the metal contacts over non highly-doped base area, Fig. 7, increases for group 2 compared to group 1 as $w_{\text{line,BSF}}$ is smaller at the same $w_{\text{finger,rs}}$. Table II shows the estimated coverage fractions of the metal contacts that overlap the base area f_{overlap}. These values suggest that the lower $j_{0,\text{laser,BSF,tot}}$ of group 2 cells is overcompensated by the charge carrier recombination, which is related to the overlapping parts of the metal contacts. In turn this means that the surface recombination at the part of the metal contacts that overlaps the silicon base (recombination velocity S_{overlap}) must be larger than the recombination at the part of the metal contacts that lies on top of the BSF.

Regarding the V_{OC} difference between the solar cells from group 2 and 3, the 9 mV higher V_{OC} for the latter results from the lower f_{overlap}; see Table II. The next section deals with the quantification of S_{overlap}.

4.3.2 Estimation of surface recombination velocities at the base regions with metal overlap

In order to assess the recombination properties at the areas with metal overlapping the base region, Fig. 7, we conduct numerical 3D device simulations with the software tool Quokka3 [29] for the most efficient solar cell from group 2 and 3 (they differ only by f_{overlap}). As the front and rear side contact distances have a small least common multiple, we are able to use a symmetry element that represents the full solar cell containing half the region between two adjacent busbars as well as two and three fingers on the front and rear side, respectively.

The metallized region overlapping the base on the rear side is modelled with an effective surface recombination velocity S_{overlap}. Concerning the base, we assume a bulk lifetime of $\tau_{\text{bulk}} = 3$ ms according to the parameterization in Ref. [30] for 2 Ωcm p-type Cz-Si. We account for non-ideal recombination behavior by including an external second diode with ideality factor $n = 2$ and a saturation current density $j_{02} = 14$ nA/cm² (j_{02} is based on two-diode model fits to the dark IV curves of the finished solar cells). For the specific contact resistances of the rear grid, we apply $\rho_C = 3.6$ mΩcm² and $\rho_C = 6.6$ Ωcm² for group 2 and 3, respectively. These are the mean ρ_C values measured on the corresponding solar cells per group as shown in Fig. 6. All other input parameters of the front and rear side rely on the measured values given in the sections 3.1, 3.2, 4.2, and 4.3.1. Regarding further electrical properties of the front side, we refer to the data published in Ref. [31].

Within our simulation results, we find a strong influence of the surface recombination velocity S_{overlap} at the interface between the rear metal contacts and the lowly-doped base accounting for significant V_{OC} losses, see Fig. 8. As S_{overlap} is the only unknown parameter, we can clearly assign the losses in V_{OC} to high S_{overlap} values. It is seen that the group 2 cell exhibits a stronger decrease in V_{OC} with increasing S_{overlap}, which is a strong indication for the effect causing the V_{OC} difference observed in the experiment.

To match the measured and simulated IV data, see Table III, $S_{\text{overlap}} = 3\cdot10^4$ cm/s and $S_{\text{overlap}} = 1\cdot10^4$ cm/s for group 2 and 3 are considered, respectively.

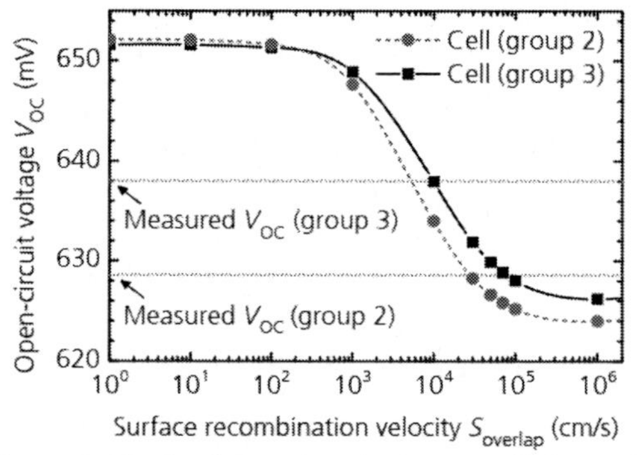

Figure 8: Simulated V_{OC} values for the most efficient cells from group 2 and 3 depending on the recombination velocity S_{overlap} at the interface between the rear metal contacts and the non-highly-doped base. Clearly visible is the huge V_{OC} drop for $S_{\text{overlap}} > 1000$ cm/s, confirming the detrimental impact of the overlapping metal regions.

Table III: Measured and simulated IV parameters of the most efficient solar cell for group 2 and 3, respectively.

Group	Type	S_{overlap} (cm/s)	η_{front} (%)	j_{SC} (mA/cm²)	V_{OC} (mV)	FF (%)	pFF (%)
2	Meas.	-	19.51	38.95	628.6	79.70	82.85
	Simul.	*3·10⁴*	*19.55*	*38.92*	*628.3*	*79.97*	*82.72*
3	Meas.	-	19.81	39.13	638.0	79.35	82.75
	Simul.	*1·10⁴*	*19.94*	*39.18*	*637.9*	*79.79*	*82.74*

With the conducted and validated simulations, the overlap region is clearly identified as being very harmful and thus, the overlap region needs to be reduced.

5 OUTLOOK

From further experiments, we acquired knowledge about recombination and electrical contacting properties for the "pPassDop" PERL approach with respect to a wider range of laser parameters. We already developed a laser doping process that yields significantly lower $j_{0,\text{laser,BSF}} \approx$ 1100 fA/cm². As shown in Ref. [28], the related specific contact resistance ρ_C is found to be $\rho_C \approx 2$ mΩcm² with the same firing-through silver paste as used within this paper. However, this laser process has so far not been tested on cell level. In combination with further decreasing the share of metal overlapping the silicon base f_{overlap}, which is possible by applying our alignment procedure, we pursue the short-term goal to increase the cell performance to a front side energy conversion efficiency $\eta_{\text{front}} \geq$ 21% and a bifaciality $\beta \geq 90\%$.

On the other hand, non-firing-through silver pastes provide an additional efficiency enhancement potential and are less challenging with respect to the overlap between silver fingers and the non-highly-doped BSF. However, corresponding pastes are still limited with respect to ρ_C with values exceeding 40 mΩcm².

6 SUMMARY AND CONCLUSION

The most efficient bifacial "pPassDop" PERL solar cell currently achieves a bifaciality of 89% at a front side energy conversion efficiency of 19.8% (measured with contact bars on both sides on a black non-conducting chuck). With fill factors larger than 79%, we demonstrate the low-ohmic electrical contacting of both the phosphorus-doped front and the laser-p-doped rear surface with the same commercial pure silver screen-printing paste. We show that the firing-through silver paste introduces significant losses in open-circuit voltage due to high recombination in the non-laser processed overlap region. The larger the overlap of the silver contacts over the non-laser-doped areas, the higher is the total charge carrier recombination. Our innovative alignment procedure allows for matching the rear side structuring processes despite low line width of both screen-printed fingers (65 µm) and laser-doped back surface field regions (37 µm).

Compared to the fabrication of industrial bifacial PERC solar cells, the examined bifacial PERL solar cells require even one process step less and thus, implementation in existing PERC production lines is very attractive.

ACKNOWLEDGEMENTS

The authors would like to thank all colleagues at the Fraunhofer ISE PV-TEC; especially H. Brinckheger, C. Harmel, M. Linse, R. Neubauer, M. Retzlaff, and D. Trogus. This work was funded by the German Federal Ministry for Economic Affairs and Energy within the research projects “HELENE” (contract no. 0325777D) and “PV-BAT400” (contract no. 0324145).

REFERENCES

[1] A. Cuevas, "The early history of bifacial solar cells", *Proc. 20th EU PVSEC*, Barcelona, Spain, 2005, pp. 801–805.

[2] T. Dullweber, C. Kranz, R. Peibst et al., "PERC+: industrial PERC solar cells with rear Al grid enabling bifaciality and reduced Al paste consumption", *Prog. Photovolt.: Res. Appl.*, vol. 24, no. 12, pp. 1487–1498, 2016.

[3] S.-Y. Chen, H.-C. Liu, H.-Y. Chang et al., "Bifacial solar cells fabricated by PERC process for mass production", *Proc. 32nd EU PVSEC*, Munich, Germany, 2016, pp. 772–774.

[4] A. Teppe, E. Wang, W. Guo et al., "Pilot production of bifacial multicrystalline PERCT cells achieving 18.5% efficiency and singlefacial more than 19%", *Proc. 32nd EU PVSEC*, Munich, Germany, 2016, pp. 610–614.

[5] ITRPV, 2017, "International Technology Roadmap for Photovoltaic (ITRPV) - Results 2016", available online: http://www.itrpv.net/Reports/ Downloads/2017/.

[6] Taiyang News, 2017, "Bifacial solar module technology - It's time to produce solar power on both module sides", available online: http://www.taiyangnews.info/reports/ taiyangnews-bifacial-solar-report-2017/.

[7] T. Fellmeth, S. Meier, F. Clement et al., "Co-diffused bifacial PERT solar cells", *Energy Proced.*, in press, 2017.

[8] Handelsblatt, 2014, "Asbeck versucht Neustart: Ex-Sonnenkönig kündigt neue Modultechnik an", available online: http://www.genios.de/presse-archiv/artikel/HB/20141215/ asbeck-versucht-neustart/6E27CF9D-B436-48EF-9CEA-EFC3080CD8F5.html.

[9] K. Krauß, F. Fertig, J. Greulich et al., "biPERC silicon solar cells enabling bifacial applications for industrial solar cells with passivated rear sides", *Phys. Status Solidi A*, vol. 213, no. 1, pp. 68–71, 2016.

[10] J. C. Jimeno, G. Bueno, R. Lago et al., "Low cost using ultra-thin bifacial cells", *Proc. 22nd EU PVSEC*, Milan, Italy, 2007, pp. 875–878.

[11] S. Gonsui, S. Goda, K. Sugibuchi et al., "n-type high efficiency bifacial silicon solar cell with the extremely high bifaciality of 96% in average fabricated by using conventional diffusion method", *Proc. 28th EU PVSEC*, Paris, France, 2013, pp. 718–721.

[12] K. Sugibuchi, N. Ishikawa, S. Obara, "Bifacial-PV power output gain in the field test using "EarthON" high bifaciality solar cells", *Proc. 28th EU PVSEC*, Paris, France, 2013, pp. 4312–4317.

[13] K. Nakamura, N. Takeda, T. Soga et al., "N-type bifacial cell using simplified ion doping system", *Proc. 40th IEEE PVSC*, Denver, USA, 2014, pp. 3649–3653.

[14] H. Schulte-Huxel, F. Kiefer, S. Blankemeyer et al., "Flip-flop cell interconnection enabled by an extremely high bifacial factor of screen-printed ion implanted n-PERT Si solar cells", *Proc. 32nd EU PVSEC*, Munich, Germany, 2016, pp. 407–412.

[15] M. H. Norouzi, P. Saint-Cast, E. Lohmüller et al., "Development and characterization of multifunctional PassDop layers for local p+-laser doping", *Energy Proced.*, in press, 2017.

[16] M. H. Norouzi, P. Saint-Cast, U. Jäger et al., "Development and characterization of AlO_x/SiN_x:B layer systems for surface passivation and local laser doping", *IEEE J. Photovoltaics*, vol. 7, no. 5, pp. 1244–1253, 2017.

[17] S. Werner, E. Lohmüller, P. Saint-Cast et al., "Key aspects for fabrication of p-type Cz-Si PERC solar cells exceeding 22% conversion efficiency", *Proc. 33rd EU PVSEC*, Amsterdam, The Netherlands, 2017.

[18] D. Biro, R. Preu, S. W. Glunz et al., "PV-TEC: Photovoltaic technology evaluation center - design and implementation of a production research unit", *Proc. 21st EU PVSEC*, Dresden, Germany, 2006, pp. 621–624.

[19] S. Werner, E. Lohmüller, S. Maier et al., "Process optimization for the front side of p-type silicon solar cells", *Proc. 29th EU PVSEC*, Amsterdam, The Netherlands, 2014, pp. 1342–1347.

[20] A. Kimmerle, J. Greulich, A. Wolf, "Carrier-diffusion corrected J_0-analysis of charge carrier lifetime measurements for increased consistency", *Sol. Energy Mater. Sol. Cells*, vol. 142, pp. 116–122, 2015.

[21] E. Peiner, A. Schlachetzki, D. Krüger, "Doping profile analysis in Si by electrochemical capacitance-voltage measurements", *J. Electrochem. Soc.*, vol. 142, no. 2, pp. 576–580, 1995.

[22] R. Bock, P. P. Altermatt, J. Schmidt, "Accurate extraction of doping profiles from electrochemical capacitance voltage measurements", *Proc. 23rd EU PVSEC*, Valencia, Spain, 2008, pp. 1510–1513.

[23] S. Werner, E. Lohmüller, S. Maier et al., "Challenges for lowly-doped phosphorus emitters in silicon solar cells with screen-printed silver contacts", *Energy Proced.*, in press, 2017.

[24] R. A. Sinton, A. Cuevas, "Contactless determination of current–voltage characteristics and minority-carrier lifetimes in semiconductors from quasi-steady-state photoconductance data", *Appl. Phys. Lett.*, vol. 69, no. 17, pp. 2510–2512, 1996.

[25] T. Fellmeth, F. Clement, D. Biro, "Analytical modeling of industrial-related silicon solar cells", *IEEE J. Photovoltaics*, vol. 4, no. 1, pp. 504–513, 2014.

[26] H. H. Berger, "Contact resistance and contact resistivity", *J. Electrochem. Soc.*, vol. 119, no. 4, pp. 507–514, 1972.

[27] S. Eidelloth, R. Brendel, "Analytical theory for extracting specific contact resistances of thick samples from the transmission line method", *IEEE Electron Device Lett.*, vol. 35, no. 1, pp. 9–11, 2014.

[28] E. Lohmüller, S. Werner, M. H. Norouzi et al., "Low-ohmic contacting of laser-doped p-type silicon surfaces with pure Ag screen-printed and fired contacts", *Phys. Status Solidi A*, 2017, submitted for publication.

[29] A. Fell, J. Schön, M. C. Schubert et al., "The concept of skins for silicon solar cell modeling", *Solar Energy Materials and Solar Cells*, in press, 2017.

[30] D. C. Walter, B. Lim, J. Schmidt, "Realistic efficiency potential of next-generation industrial Czochralski-grown silicon solar cells after deactivation of the boron-oxygen-related defect center", *Prog. Photovolt.: Res. Appl.*, vol. 24, no. 7, pp. 920–928, 2016.

[31] S. Wasmer, A. A. Brand, J. M. Greulich, "Metamodeling of numerical device simulations to rapidly create efficiency optimization roadmaps of monocrystalline silicon PERC cells", *Energy Proced.*, in press, 2017.

RESEARCH OF INDUSTRIAL HIGH EFFICIENCY N-TYPE SOLAR CELL WITH SELECTIVE BACK SURFACE FIELD PROCESS

Dawei Liu[1,2], Ziqian Wang[1,2], Jinye Zhai[1,2], Feng Li[1,2], Jinchao Shi[1,2], Dengyuan Song[1,2]
[1] Yingli Green Energy Co., Ltd, 3399 North Chaoyang Street, 071051, Baoding, China,
[2] State Key Laboratory of Photovoltaic Materials and Technology, Yingli solar, 071051 China

ABSTRACT: The interest in N-type solar cells has increased recently because of high efficiency and low light-induced degradation(LID). The back side of Bifacial N-type Si solar cells has to be highly doped phosphorous to form back surface field (BSF), this results in low contact resistance and good fill factor. However, highly doped back surface field increases surface recombination and then reduces the efficiency of cell. In this paper, N-type bifacial Si solar cells with selective back surface field structure (SBSF) is presented. We developed an industrially feasible etch back process to create a selective back surface field. the areas for back side metallization were shielded for etching by an acid resistant etch barrier to achieve a low contact resistivity. The highly doped BSF was etched back by a wet chemical process. The influences of several process parameters on cell efficiency were discussed. The performance was improved in terms of both short-circuit current density (0.65mA/cm^2)and open-circuit voltage(8mV), which gives rise to 0.43% absolute efficiency gain compare to the cells with homogeneous BSF. It was shown that improvement of cell performance was because of the better spectral response in long wavelength and lower back side recombination.
Keywords: N-type solar cell, Bifacial, selective back surface field, etch back

1 INTRODUCTION

Enhancing the conversion efficiency of solar cells is a fundamental approach to lower the cost of photovoltaic power. In order to improve efficiency, double sides contacted solar cells on Czochralski (Cz) silicon based material attract more and more attention recently. Solar cells from n-type Si wafers have potentially higher efficiency thanks to advantages of n-type materials, The higher tolerance of minority carrier lifetime to common impurities and the performance not suffering B-O complex related light-induced degradation allow such ultra-high efficiency. In 2010, commercial n-type solar cells named PANDA were introduced by Yingli Solar. PANDA cells are bifacial N-type Si solar cells with front-side junction and H-Pattern grids on both sides [1]

The back side of Bifacial N-type Si solar cells has to be heavily doped phosphorous to form back surface field (BSF), which results in low contact resistance and high fill factor. However, surface recombination velocity will be increased on the heavily doped back surface field giving raise to cell efficiency reduction. The motivation to form selective back surface field on n-type cells is from the selective emitter known for p-type cells[2-5]. Therefore, a selective etch-back BSF is to be established in N-type solar cell aiming at lower BSF recombination. We developed an industrially feasible etch back process to create a selective back surface field(SBSF) solar cell, back surface field properties such as surface doping profiles and saturation current densities are investigated and the improvement in cell performance is also demonstrated.

2 SOLAR CELL STRUCTURE AND PROCESSING

2.1 Cell structure

The structures of n-type cell and n-type cell with SBSF structure are illustrated in Figure 1. The rear side of n-type cell have an N+ doping in order to facilitate the formation of an Ohmic contact. The rear side metallization has an open structure. Boron emitter and phosphorous BSF are passivated by stacks of dielectric layers. The different of SBSF structure cell is heavily doped n++ layer underneath the contact fingers to decrease the contact resistance, higher sheet resistance n+ region between the areas intended for metallization to reduce the losses that step from Auger-recombination with the phosphorous heavily doped region.

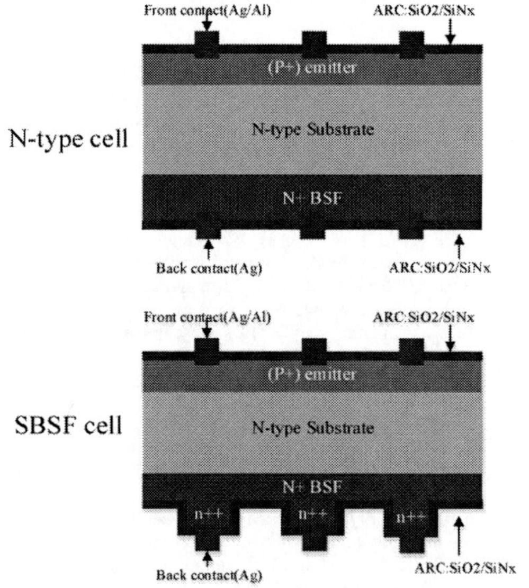

Figure 1:Cross section of N-type cell (Top) and N-type cell with SBSF structure (Bottom)

2.2 Cell process

Figure 2 shows the manufacturing process of N-type cell and N-type solar cells with SBSF structure created by Etch back process. The additional process steps are indicated. The cell was made of 156 × 156mm and 190um thickness n-type Cz-Si wafer. The wafer was textured with both sides using alkaline etching, then phosphorous BSF and the boron emitter were formed by co-diffusion. Surface passivation was prepared using wet chemical oxidation process. SiNx layer was deposited on both sides by PECVD as an anti-reflection coating. The front and rear metallization were screen printed and co-firing. In etch back process, the mask was printed in similar pattern of subsequent rear fingers, and cured at 180℃. Etch back process can be carried out by means of an inline single side chemical etching process, based on a HF/HNO3 mixture. The masked area was protected during acidic etching while the rest of the BSF was etched back. Therefore Rsheet of BSF in the etched areas depends on the etching time. Mask removal was done in the same machine with alkaline bath.

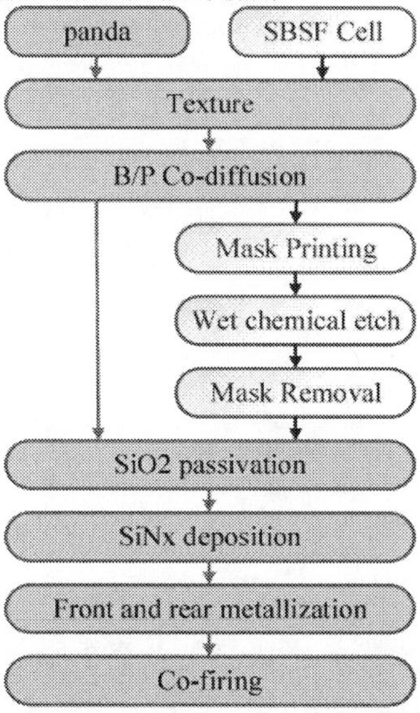

Figure 2: Process sequence of panda cell and SBSF cell

The relationship between the etching time and the sheet resistance after the etch back process was measured. Surface doping concentration profiles were measured by ECV method. The BSF recombination parameter was measured using QSSPC method at the injection level of $5 \times 10^{15}cm^{-3}$. Internal quantum efficiency to determine whether and where had been improved. One-sun I-V measurements of all the cells were done under standard testing conditions.

3 RESULTS AND DISCUSSION

3.1 Diffusion and etch back process

The wet chemical etch back process for SBSF was developed. An acidic solution composed of HF, HNO3 and H_2O provides higher etching homogeneity and better process control. Wafers were treated in etching solution for different time and Rsh was measured by four point probe on surface.

For the experiments two BSF doping profile were used, a shallow BSF doping with 40Ω/sq and a deep BSF doping profile with 16Ω/sq. Figure 3a shows BSF doping profile with two different BSF sheet resistance(16 Ω/sq and 40Ω/sq) were realized by POCl3 diffusion. It shows different surface concentration and depth.

Figure 3b shows the relationship between Rsh and etch time. The etching rate with 40Ω/sq was observed to be faster than 16Ω/sq and the homogeneity is poor. The average BSF sheet resistance was determined to 60Ω/sq, for the heavily doped original 16Ω/sq had a homogeneity with stand deviation of 7 Ω/sq. However lightly doped original 40 Ω /sq with a stand deviation 12 Ω /sq. Inhomogeneity is very important for etch back process.

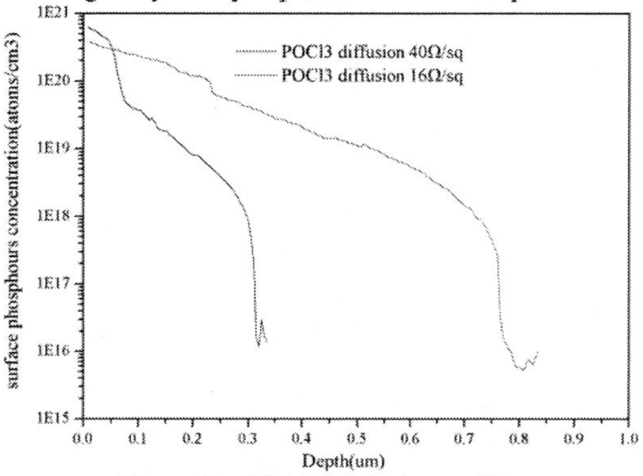

Figure 3a: BSF ECV doping profile

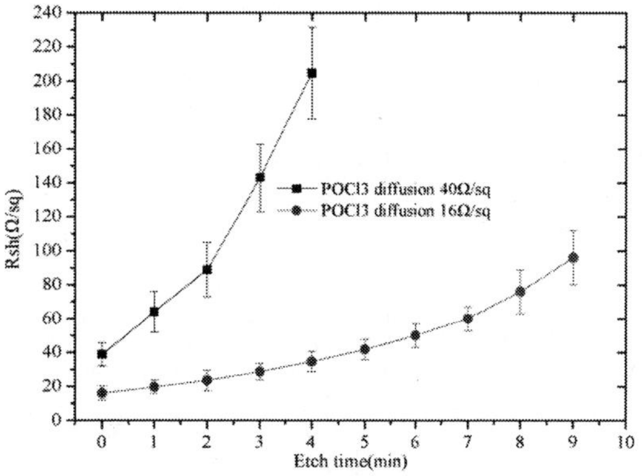

Figure 3b: Sheet resistance versus etch time

3.2 Sheet resistance and Jo investigation

For insert of SBSF process into N-type solar cell, a 16 Ω /sq POCl3 diffused starting BSF with different etching time which results from different profile depths and surface doping concentrations. Four BSF target sheet resistances of 33, 45,61, and 73Ω/sq were carried out by the time of etch. Fig. 4 shows the doping depth was etched resulting the decrease in surface doping concentration and doping depth, significant improvement of back surface J0 was expect.

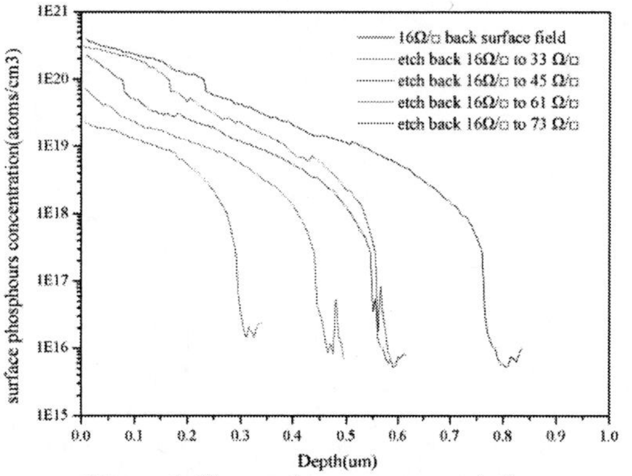

Figure 4: Sheet resistance versus etch time

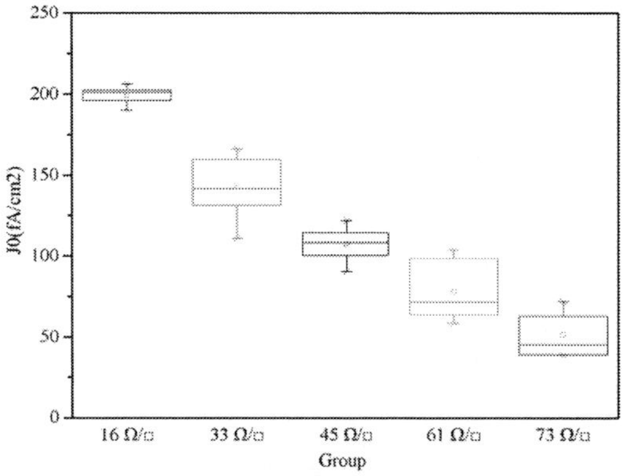

Figure 5: Jo of the n+layer for different sheet resistance

For the measurement of the BSF saturation current density, N-type wafers with bulk reisitivity of 5Ωcm CZ substrate and thickness 180um were used. Double-side texturing, starting from an initially heavily doped BSF(16 Ω/sq) the samples were subsequently etch back to 33, 45, 61 and 73 Ω /sq. All samples were passivatied with PECVD SiNx anti reflection layer. J0 measurement were conducted after firing. It can be seen that along with the increased sheet resistance during the etch back, Jo is reduced drastically. This can be explained less Auger recombination and reduced electrical inactive doping surface concentration by etch back process. Especially for Rsheet=73Ω/sq, Jo has been reduced 50fA/cm2.

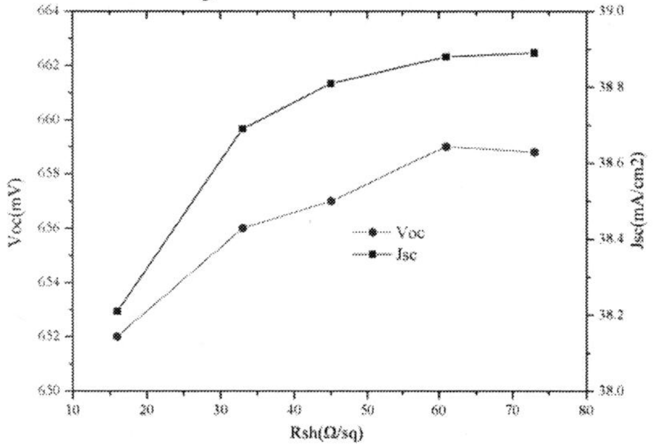

Figure 6: Voc and Jsc as function of Rsh

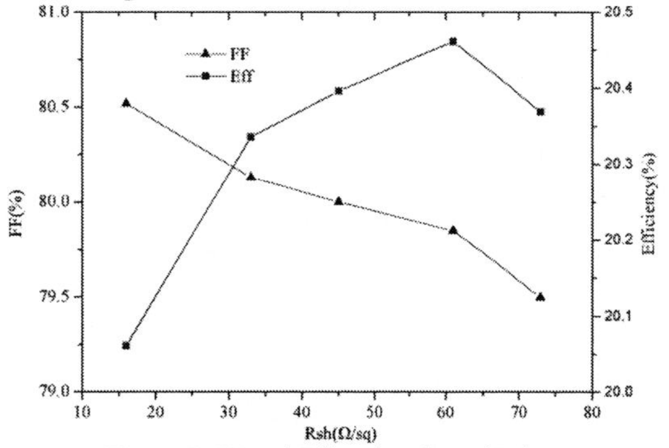

Figure 7: FF and Eff as function of Rsh

The corresponding electrical parameters of different etch Rsh(33Ω/sq, 45Ω/sq, 61Ω/sq, 73Ω/sq) solar cell were also investigated, With increasingly etch depth and higher final Rsh, the Voc and Jsc are improved due to higher long wave response and lower surface concentration in Figure 6 and Figure 7.While, the FF will be decreased due to increased lateral resistance. Lower FF can be compensated by the increasment of Voc and Jsc when increasing the Rsh till 60 Ω /sq. The final efficiency will be suffered from the poor FF,Therefor the etching back to 61Ω/sq is optimum.

3.3 Mask width optimization

Four batches of 400pcs Cz-wafers were fabricated using process flow described in Fig 2. The parameters that varied on the 4 batches are the linewidth of the mask with 300um, 250um, 200um and 150um respectively. The finished cells were measured by inline IV-tester and average value are presented in Table 1.

Table 1: Results of 4 batches with vary mask line width

Mask line width	Voc [mV]	Jsc [mA/cm2]	FF [%]	Eta [%]
300um	657	39.17	79.93	20.57
250um	659	39.21	79.86	20.64
200um	661	39.26	79.85	20.71
150um	660	39.30	79.82	20.68

As shown in table 1, Voc and Jsc are increased with decreasing of mask linewidth due to lower recombination on smaller area where are heavily doped. However, the improvement of efficiency is offset when bellowing 200 um mask linewidth. It is because of the misalignment between fingers and heavily doped region.

3.4 Result analysis for the SBSF solar cell

Two batches of 200pcs cells were fabricated and were processed in parallel in Yingli's product line. We make a comparison between N-panda, the homogeneous BSF structure solar cell, and the SBSF structure solar cell that fabricated through Etch back process. The IV measurement results are shown in table2.

Table 2: Results of 2 batches different BSF process

Process technology	Voc [mV]	Jsc [mA/cm2]	FF [%]	Eta [%]
N-panda	653	38.79	80.93	20.50
Selective BSF	661	39.44	80.23	20.93

Comparing to the N-panda cells, SBSF cell get 0.43% efficiency gain. Voc (+8 mV) and Jsc (+0.65 mA/cm2) were increased due to lower surface recombination, as was proved by the IQE shown in Fig 7. The solar cell with SBSF structure get an obvious improvement in IQE for wavelength above 950 nm.

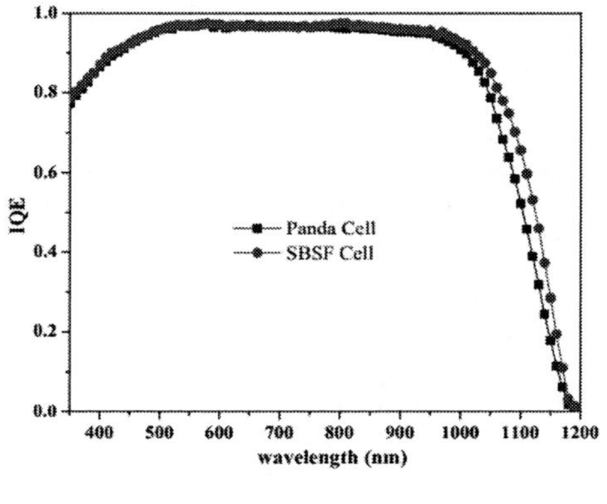

Figure 8: Comparison of internal quantum efficiency (IQE) of Panda cell and SBSF cell.

4 CONCLUSIONS

N-type bifacial Si solar cells with selective back surface field (SBSF) structure are presented. We developed an industrially feasible etch back process to prepare the SBSF cell. Several process parameters on cell efficiency were optimized. Cell performances were improved in terms of both short-circuit current (+0.65 A/cm^2) and open-circuit voltage (+8 mV). The cell efficiency gain was higher than 0.4% absolutely compared to the cell with homogeneous BSF.

References

[1] A.R. Burgers et al. 19% efficient N-type si solar cells made in plot production, 25th EUPVSEC,2010, Valencia, Spain

[2] Haverkamp, e t al. Minimizing the Electrical Losses on the Front Side: Development of a Selective Emitter Process from a Single Diffusion. In Proceedings of the 33rd IEEE Photovoltaic Specialists Conference, 2008, San Diego, CA., USA.

[3] D. Rudolph et al. etch back selective emitter process with single pocl3 diffusion , 26th EUPVSEC,2011, Hamburg, Germany

[4] Mohammad Ziaur Rahman, Status of Selective emitter for P-type C-Si solar cells, Optics and photonics journal, 2012, 2 , 129-134 .

[5]B. Paviet-Salomon et al. 19.3% efficient selective emitter silicon solar cells by industrial process, 26th EUPVSEC, 2011, Hamburg, Germany

[6] Dawei liu et al, Pilot line production of industrial high-efficient bifacial n-type silicon solar cells with selective back surface field process, 30th EUPVSEC, 2015, Hamburg, Germany

LARGE-AREA (6 INCH) SCREEN-PRINTED IBC SOLAR CELLS WITH EFFICIENCY ABOVE 24.1% WITHOUT PASSIVATED CONTACTS

Guanchao Xu, Yang Yang *, Xueling Zhang, Shu Chen, Wei Liu, Yan Chen, Yifeng Chen, Pietro P. Altermatt, Pierre J. Verlinden and Zhiqiang Feng
State Key Laboratory of PV Science & Technology
Trina Solar, Changzhou, 213031, China
* Corresponding author: email: yang.yang02@trinasolar.com

ABSTRACT: A maximum open-circuit voltage of 702 mV and a full area efficiency of 24.13% has been achieved on 6 inch square n-type CZ wafer with a Interdigitated Back Contact (IBC) cell structure, without utilizing the technology of passivated contacts. Conventional industrially feasible and low cost processes are used for this cell fabrication. This result is achieved by applying a selective emitter structure, improved surface passivation and re-designed rear pattern with the assistance of numerical simulation. The contact resistivity on both emitter and BSF region is reduced to less than 0.5 mΩcm^2. The power loss analysis shows that the biggest recombination loss is within the bulk, summing up to 45% of the total recombination current at P_{mpp}.

Keywords: IBC cells, low cost, numerical simulation

1 INTRODUCTION

Although IBC cells have been successfully commercialized for decades, the mainstream of high efficiency silicon solar cells nowadays is still the passivated emitter and rear contact (PERC) cell, due to the lower manufacturing cost. However, in recent two years more and more research activities on IBC cells and efficiency higher than 25% have been reported by several groups. Smith et al. [1] reported 25.2% efficiency of the SunPower X-Series architecture cell with a total wafer area of 153.5 cm^2 with passivated contacts technology, which was developed years ago in the lab [2]. LG solar announced their 365 W NeON® IBC module made of 60 cells with an average efficiency assumed to be higher than 24.5% [3]. Last year, a new record efficiency of 26.3% was reported by Kaneka [4], by integrating their advanced hetero-junction technology with an IBC structure, and in 2017 this number went up to 26.7% [5]. These top results indicate that on the roadmap towards the theoretical efficiency limit of 29% for single junction crystalline solar cells [6], both the IBC structure and passivated contacts are undoubtedly playing important roles. On the other hand, in terms of the levelized cost of electricity (LCOE), an intricate balance has to be made between higher manufacturing cost and better performance, which arises another interesting question: What is the efficiency limit of IBC cells based on conventional processes like tube diffusion and screen printing, without adding additional tools for passivated contacts nor hetero-junction formation?

This is a difficult question to answer because to a certain extent cell efficiency depends greatly on the minimum feature size we can achieve, and this feature size is becoming smaller and smaller. For example, with advanced screen design, the width of screen printing fingers has been narrowed to less than 30 µm, which could have only been achieved by plating a few years ago. Nevertheless, we have been focusing on low cost IBC development based on conventional processes and trying to answer this question by improving efficiency step by step.

In 2016, the IBC cells produced in a pilot line of Trina solar demonstrated an independently confirmed maximum efficiency of 23.5% on 6" substrates with a homogeneous emitter structure design [7]. To achieve this result, we reduced the contact size and reduced the contact resistivity and recombination loss at the same time. With the help of 3D numerical modelling, we also re-designed the diffusion and contact pattern. In the resulting 23.5% IBC cell last year, one of the biggest recombination losses was the loss in BSF region and has meanwhile been significantly reduced, without introducing more resistive loss. Compared to last year, V_{oc} increased by nearly 10 mV, and FF gained 0.5% (abs). Thus, the dominating diffusion recombination moved to the emitter area due to its large area ratio, and the dominating resistive loss moved to the emitter contact, which indicates a selective emitter design should be beneficial to further improve efficiency. So a lightly doped emitter can be used to reduce the rear side J_0, and a small area heavily doped region under the contact to reduce the contact resistive loss. Following this thinking, we re-designed the cell rear pattern and optimized the process flow. As a result, a total diffusion J_0 of 45 fA/cm^2 and 24.13% efficiency has been achieved on a large area (156 x 156 mm square) IBC solar cell. In the following, the characterization of the improved contacts, the diffusion recombination and bulk quality and the power loss analysis, will be described in detail.

Table I: IV parameters of 24.13% champion IBC cell as independently measured at Japan Electrical Safety & Environment Technology Laboratories (JET).

Area [cm^2]	V_{oc} [mV]	I_{sc} [A]	V_{mpp} [mV]	I_{mpp} [A]	FF [%]	Eff [%]
244.9	702.7	10.25	607.9	9.66	81.47	24.13

2 CHARACTERIZATION OF CONTACT AREAS

The measurement of the $J_{0,cont}$ of contacts is critical but difficult as the perimeter region around contacts may introduce artificial measurement effects. It was evaluated with numerical modelling that utilizing the J_0 measured on a large unpassivated area cannot give a good prediction of cell performance. The J_0 under contacts should be measured on structures with the same contact dimensions as in the finished cell. To do that, we prepare samples with the same diffusion process and the same passivation quality as in the cell; and then a series of dots is opened with various pitches to form different ratios between contact and passivated area. A 156 x 156 mm wafer is divided into four 78 x 78 mm areas with different contact area ratios. The $J_{0.cont}$ was then extracted by extrapolating the curve in Figure 1 to 100% contact ratio.

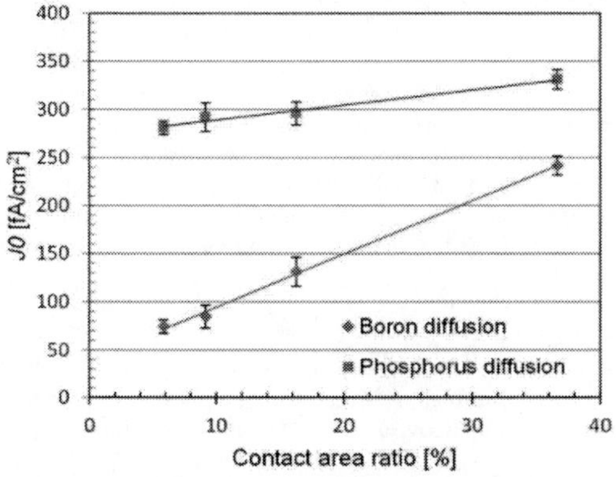

Figure 1: Measured J_0 on boron and phosphorus diffusion samples, respectively, as a function of contact ratios.

The measurement result using this method shows that the $J_{0.cont}$ on the heavily doped emitter reduced from 1425 fA/cm^2 (homogeneous emitter) to 1250 fA/cm^2. The sheet resistivity of the heavily doped emitter decreased from more than 90 ohm/sq (homogeneous emitter) to around 45 Ω/sq. The $J_{0.cont}$ on the BSF area is in the range of 550-600 fA/cm^2.

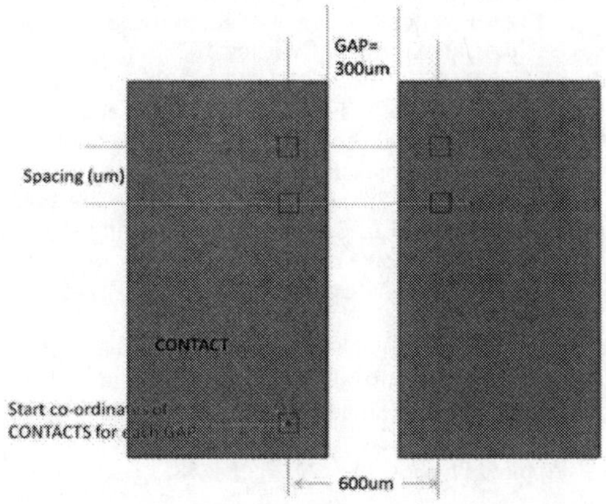

Figure 2: The structure of samples for contact resistivity extraction.

The contact resistivity measurement method used in this paper is based on a fitting of the total measured resistance and simulations on particular samples [8]. The samples are prepared as shown in Figure 2. The spacing between contacts is varied to get a series of total resistance values. By fitting the curve of the measured total resistance as a function of specific contact resistivity with the simulated result, the contact resistivity can be extracted. This method is proved to be more reliable than the TLM method for small dot contacts, especially when the contact resistivity is lower than 5 mΩcm^2. The extracted contact resistivity for the heavily doped emitter is lower than 0.5 mΩcm^2, while on the homogeneous emitter it was around 2 mΩcm^2. The contact resistivity on BSF area is around 0.2 mΩcm^2.

3 DIFFUSION RECOMBINATION AND BULK QUALITY

For a detailed analysis of recombination losses in the cell, monitoring diffusion wafers have been prepared during the cell processing. All the monitoring wafers have a symmetrical structure with the same diffusions and passivation layers on both sides, namely the FSF for the front surface field, the BSF for the back surface field, the L-EMT for the lightly diffused emitter, the H-EMT for heavily diffused emitter, the N-cont for the contact region under the BSF and the P-cont for the contact region under the emitter. These samples also have the same thermal history as the real cell. The area weighted J_0 for the different regions is listed in Table II.

Table II: J_0 values in different regions of the IBC cell (unit: fA/cm^2)

FSF	BSF	L-EMT	H-EMT	N-cont	P-cont	total
4.2	12.7	9.6	3.1	6.1	12.4	48.1

As shown in Table II, although the area of the BSF has been reduced, the biggest recombination loss is still in the BSF, followed by the contact region under the emitter and the lightly diffused emitter. To confirm this test result, Suns-V_{oc} measurements are performed to extract the total diffusion J_0 of the finished cell [9].

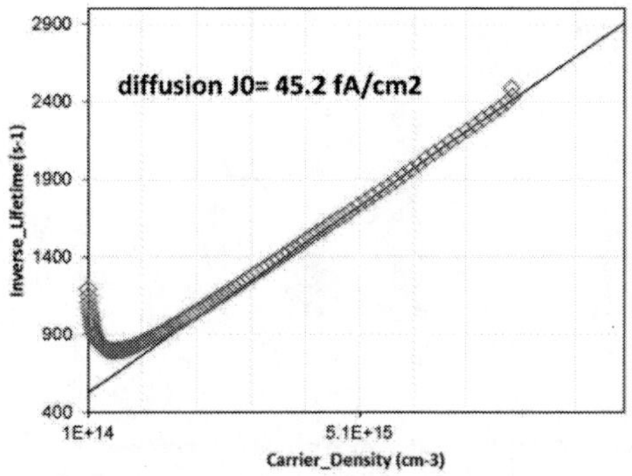

Figure 3: Measured inverse lifetime as a function of carrier density. The diffusion J_0 is extracted at 6×10^{15} cm^{-3}.

As shown in Figure 3, the extracted total diffusion J_0 by the Suns-V_{oc} method is 45.2 fA/cm^2, very close to 48.1 fA/cm^2, the total area weighted J_0 from monitoring wafers. The small difference is already within the uncertainty in determining J_0.

So far, the bulk properties have not been discussed, but actually they are very important and cannot be omitted, especially after high temperature treatment. To complete this analysis, effective lifetime curves, extracted from Suns-V_{oc} measurements, are fitted with the SRH model by assuming two bulk defects: one with its energy level in the midgap and another as shallow defect to represent thermal donors [10]. The τ_n and τ_p parameters of these two defects are ramped until the theory fits the experimental lifetime curves well in the injection range from 10^{14} cm^{-3} to 10^{16} cm^{-3}. The measured J_0 = 45.2 fA/cm^2 is used to calculate the bulk lifetime from the measured effective lifetime. Then, these bulk parameters, together with the diffusion profiles, measured by the ECV method, and the extracted surface recombination velocities are put in a Sentaurus numerical simulation model. As shown in Table III, the simulated IV parameters agree well with experimental data.

Table III: Simulated and measured IV parameters of the IBC cell

	V_{oc} [mV]	J_{sc} [mA/cm^2]	V_{mpp} [mV]	J_{mpp} [mA/cm^2]	FF [%]	Eff [%]
Sim.	706	42.0	602.7	40.0	81.4	24.1
Mes.	703	42.1	607.9	39.7	81.5	24.1

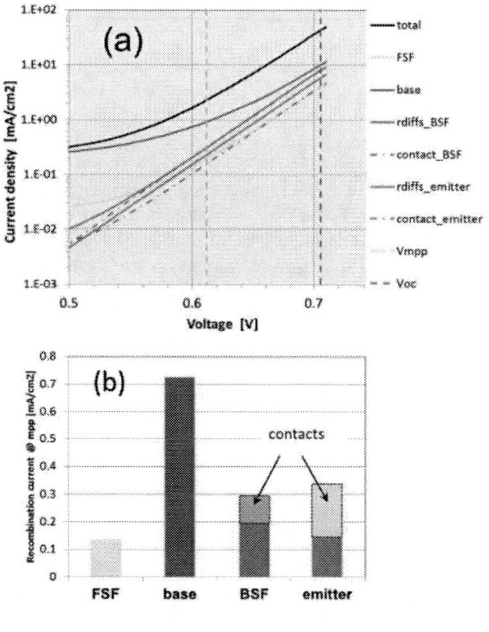

Figure 4: (a) Simulated recombination currents in different regions of the IBC cell as a function of applied voltage; (b) recombination currents at V_{mpp}.

Figure 4 shows the simulated recombination currents integrated over different regions in the cell. The dominating loss is from the bulk, up to 45% of the total recombination loss. The other recombination losses in the diffusion layers are consistent with the result from the monitoring wafers shown in Table II. Because the initial wafer without high temperature process has a much higher lifetime (not shown here), the degradation of bulk lifetime is probably caused by metal contamination and/or oxygen related defects during cell processing. The simulation shows that by improving the bulk lifetime, cell efficiency can potentially increase to 24.5% without other improvements in the process.

4 CONCLUSIONS

In this paper, a 24.13% total-area efficiency IBC cell is achieved on a 6" square n-type Cz wafer, without using a hetero-junction or passivated contact technology. By improving the contact quality and utilizing a selective emitter structure on the rear side, we obtain a good contact resistivity as low as 0.5 mΩcm^2 and reduce the recombination losses without compromising the resistive losses. The biggest loss is from the bulk, up to 45% of total. Further improvement in the bulk lifetime in finished cells might lead to an efficiency of 24.5%, indicating there is still room for efficiency improvements and has a good potential for low-cost manufacturing.

5 REFERENCES

[1] D.D. Smith et al. "Silicon solar cells with total area efficiency above 25." 43th IEEE PV Specialists Conference (2016) 3351-3355.
[2] P.J. Cousins et al. "Generation 3: Improved performance at lower cost." 35th IEEE PV Specialists Conference (2010) 275-278.
[3] http://www.lg.com/global/business/solar/neon-r
[4] K. Yoshikawa et al. "Silicon heterojunction solar cell with interdigitated back contacts for a photo conversion efficiency over 26%", Nature Energy 2 (2017) 17032.
[5] to appear in the proceedings of 2017 IEEE PVSC conference
[6] R.M. Swanson. "Approaching the 29% limit efficiency of silicon solar cells" 31 IEEE PV Specialists Conference (2005) 889-894.
[7]Z. Li et al. "Pilot Production of 6inch IBC Solar Cells Yielding an Average Efficiency of 23% with a Low-Cost Industrial Process", 31st EU-PVSEC (2016) 571.
[8] A. Fell, S. Surve, E. Franklin, and K. J. Weber, "Characterization of Laser-Doped Localized p-n Junctions for High Efficiency Silicon Solar Cells," IEEE Trans. on Electron Dev. 61 (2014) 1943–1949.
[9] R. A. Sinton and A. Cuevas, "A quasi-steady-state open-circuit voltage method for solar cell characterization," 16th EU-PVSEC (2000) 1152 – 1155.
[10] M. Tomassini et al. "Recombination activity associated with thermal donor generation in monocrystalline silicon and effect on the conversion efficiency of heterojunction solar cells". Journal of Applied Physics. 119 (2016) 084508.

SOLAR CELLS ON < 50 μm THICK EPITAXIAL FOILS CONDUCTIVELY BONDED TO LOW-COST SI CARRIER

Hariharsudan Sivaramakrishnan Radhakrishnan[1,*], Nikola Bednar[2], Twan Bearda[1], Robert Roozeman[3], Jarkko Heikkinen[3], Nadja Adamovic[2], Alexander Ulyashin[4], Martin Syvertsen[4], Ivan Gordon[1]
[1] IMEC, Kapeldreef 75, B-3001 Leuven, Belgium
[2] Technische Universität Wien, Institute of Sensor and Actuator Systems, 1040 Vienna, Austria
[3] INKRON, Espoo, Finland
[4] SINTEF, Material and Chemistry, Oslo, Norway
* Corresponding author: phone +32 16 28 1173, email sivarama@imec.be

ABSTRACT: As PV production ramps up and numerous installed solar panels reach their end of life, potentially huge waste sources of silicon become available. Such silicon can be recovered and recycled back into the PV industry, for making solar cells directly on the recycled Si. In cases where the recycled Si is of very low purity and cannot be used directly as active material, they can instead be used to make low-cost Si supporting carriers that are highly-conductive. These supporting substrates have been made using low-cost techniques such as by sintering of low-cost Si powders. These conductive substrates can act as mechanical carriers of thin silicon (< 50 μm) and as rear-contacts in what is called a wafer-equivalent concept. In this work, thin epitaxial silicon foils from the porous silicon-based lift-off process are used. For conductive bonding, low-temperature Ag paste from INKRON (IPC-114) has been preferably used due to its robust bonding properties (low temperature curing, high conductivity and minimal volume change). As a first demonstrative step, thin (<50 μm) epifoil-based and thinned (<50 μm) Cz-based cells conductively bonded onto highly-doped p^+ Cz silicon substrates have been fabricated. We chose the heterojunction technology for cell fabrication, leading to 15.7% efficiency for the best epifoil cell, partially processed after conductive bonding. It was observed that the cell performance of bonded cells are comparable to that of non-bonded epifoil cells that were processed freestanding. Furthermore, in order to reduce costs of the conductive bonding step, extensive simulations have been performed to understand the best configuration and pattern geometry for the bonding layer. This work is done in framework of H2020 EU project CABRISS.
Keywords: Thin silicon, conductive bonding, epitaxial silicon lift-off

1 INTRODUCTION

Efficient silicon utilisation is one of the corner stones of the silicon PV industry. As a result of this drive, there is a keen interest in kerfless technologies [1]–[4] which enable the production of very thin silicon (< 50 μm) with minimal kerf loss, using disruptive methods such as the porous silicon-based lift-off of epitaxial silicon[5]–[11]. In addition, there is also an increased interest in the recycling of critical raw materials such as silicon from silicon waste sources such as kerf, broken wafers in the cell manufacturing plants as well as end-of-life modules. This forms the basis of the H2020 EU project CABRISS, which aims to develop a circular economy mainly for the photovoltaic industry, but also the electronics and glass industries.

One of the ideas pursued in CABRISS is that silicon that is recovered from various silicon waste sources can eventually be transformed into low-cost Si substrates such as by sintering or spraying of low-cost Si powders. Such substrates can be used as inexpensive supporting carriers for the high quality epitaxial foils produced by the porous silicon-based lift-off process, in a wafer-equivalent concept, since the processing of thin silicon with thickness below 50 μm is very challenging as they are highly flexible, fragile and prone to breakage [12]. In addition, thinner silicon cells are also more susceptible to breakage during cell interconnection due to stresses arising from the soldering of ribbons in the tabber-stringers, and to failure during operation through propagation of micro-cracks formed during cell interconnection, eventually leading to reduced reliability and operational lifetime of solar modules [13]. By introducing the wafer-equivalent approach and bonding such high quality epitaxial foils onto low-cost silicon substrates, a mechanical yield and reliability similar to regular wafers may be attainable. This wafer-equivalent concept is schematically depicted in Fig. 1 (a). In this way, recovered and recycled silicon from low-purity sources that cannot be used directly as the active material can have another life in the form of low-cost Si carriers.

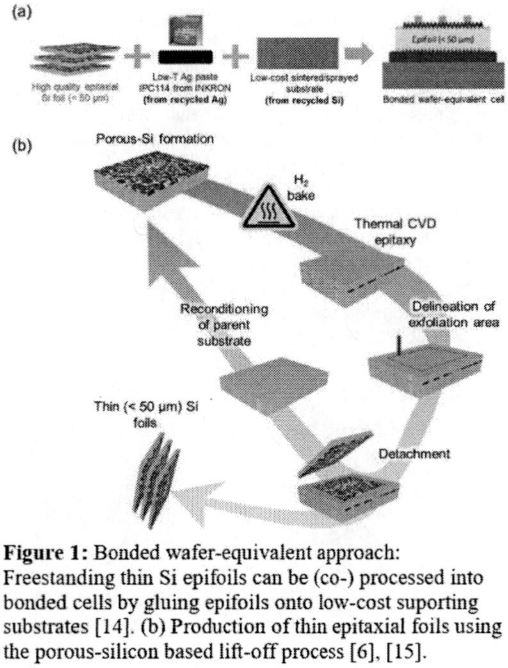

Figure 1: Bonded wafer-equivalent approach: Freestanding thin Si epifoils can be (co-) processed into bonded cells by gluing epifoils onto low-cost suporting substrates [14]. (b) Production of thin epitaxial foils using the porous-silicon based lift-off process [6], [15].

The process leading to the creation of ultra-thin

silicon foils with thicknesses (< 50 μm) that are a factor of 3 less than the current wafer thickness (180 μm), is described schematically in Fig 1 (b). Starting with highly-doped p^+ wafers, a double layer of porous silicon is etched electrochemically, and subsequently reorganised at a high temperature, before epitaxial growth of high quality silicon. This epitaxial silicon layer is then exfoliated from the substrate at the porous layer to produce freestanding thin Si epitaxial foils [5]. The parent substrate is re-used in the next cycle of layer transfer.

For the bonded wafer-equivalent approach, the two main developments needed for the realisation of this concept are: (1) conductive bonding development which enables the wafer-equivalent approach, and (2) development of a cell technology which allows high efficiency solar cells on high-quality epitaxial foils without contamination risks from the low-purity Si carrier or the conductive glue. Previously, we have reported on the development of conductive bonding of thin heterojunction epifoil cells onto highly-doped Si carriers [14]. In this work, the progress in this topic is reported.

2 SINTERING OF LOW-COST SILICON POWDER-BASED INGOTS AND WAFERING

Sintering of low-cost Si powder based ingots, with *in situ* doping, has been done by hot pressing or spark plasma sintering (SPS) of ingots (Fig. 2) using different types of Si powders and was followed by multi-wire sawing to fabricate thin (~170 μm) highly-conductive supporting substrates [16], [17]. Different Si powder feedstocks, including recycled Si powders of 10 and 80 μm particle size distributions have been used. Si powders have been mixed with boron powder to provide *in situ* doping upon hot pressing of ingots. Hot pressing involves subjecting the silicon powder to a high pressure at an elevated temperature. The powder was placed in a cylindrical graphite die with graphite punches entering the cylinder from the top and bottom and transferring the pressure from the rams. The thermal cycle for the hot pressing was as follows: (i) heat from room temperature to 1350°C at a ramping rate of 10°C/min, (ii) apply 30 MPa pressure, while holding at 1350°C for 1 h; (iii) after one hour release the pressure; (iv) ramp down to room temperature at 20°C/min. Resistivity of processed Si powder based wafers in all cases were in the mOhm range, which is sufficient to use them as a supporting part and at the same time as a back side electrode.

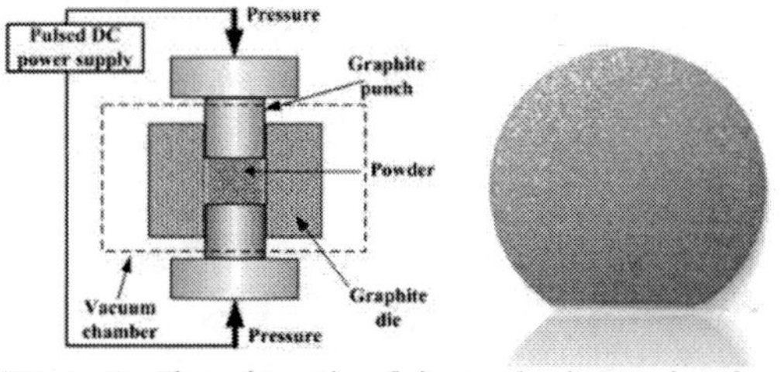

Figure 2: The schematic of the spark plasma sintering (SPS)/hot pressing process/tool, used to produce sintered silicon ingots, and a photograph of a sintered supporting silicon carrier.

3 CONDUCTIVE BONDING DEVELOPMENT

Development of conductive bonding is a key enabler for the wafer-equivalent approach described above. The key criteria for conductive bonding development are the following: (1) excellent mechanical bond strength; (2) excellent electrical contact with minimal fill factor degradation through the achievement of low series resistance and contact resistances for the conductively bonded stack and high shunt resistance by minimising shunt risks; (3) minimal contamination of the high quality epifoil during the bonding process, (4) the bonding method used must be industrially-applicable i.e. compatible with large-area application.

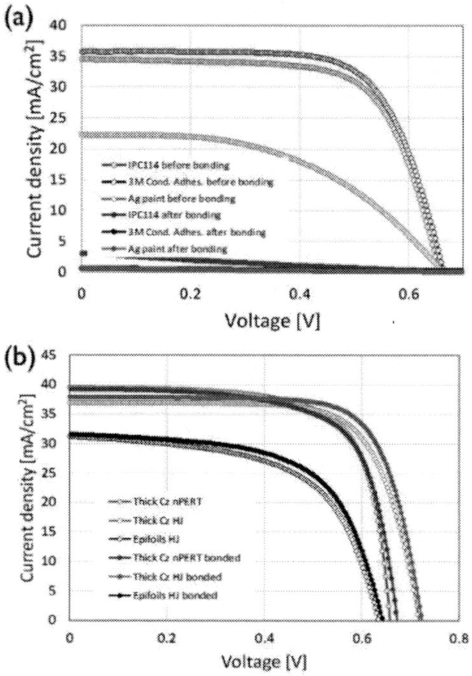

Figure 3: (a) J-V characteristics of three thin epifoil HJ cells before and after bonding using different conducting glues, namely IPC114 low-T Ag paste from INKRON, 3M conductive adhesive tape 9709 and SPI Supplies Ag paint. (b) J-V characteristics of three different types of solar cells (thick Cz nPERT cell, thick Cz HJ cell and thin epifoil HJ cell) before and after bonding to an Al-coated Si carrier using IPC114.

Based on these criteria, several low temperature bonding options, namely, 3M conductive adhesive tape 9709, INKRON IPC-114 low-temperature Ag paste and SPI Supplies Ag paint, were tested, as already reported previously [14]. The experimental procedure used for conductive bonding development is to use the conductive glues listed above to bond different types of functional finished cells to highly-doped Cz substrates (as a model case for highly-doped sintered substrates). The success of the bonding is evaluated by checking the bond strength (qualitatively), bonding-induced cracks (using photoluminescence), electrical performance (through fill factor changes of the cells before and after bonding).

As reported earlier, all three conductive glues exhibited good mechanical bond. However, the electrical performance of the bond for all three types of glues was found to be poor during the preliminary tests [14]. Fig. 3 (a) shows the current-voltage characteristics of 3 thin heterojunction (HJ) epifoil cells bonded using the 3 different bonding agents, before and after bonding. After bonding, the bonded wafer-equivalent solar cells exhibit extremely high series resistance.

Since the nature of the problem is similar for all three types of conductive glues and it is known that all glues have high conductivity, the high electrical resistance can be attributed to the contact resistances between the conductive glue and the highly-doped Si substrate and/or the rear-side metallisation of the cell. Thus, for the subsequent investigations, focus was placed only on one type of conductive glue, namely the low-temperature Ag paste, IPC114, from INKRON. This special Ag paste is well-suited as a conductive glue because it can be cured at a low temperature of 150°C, which is suitable for the heterojunction cell technology that is applied. It also has high electrical conductivity ($3.5x10^{-5}$ Ω.cm), which will ensure minimal resistive loss. It also shows minimal mass loss upon curing (~1%), which means the volume change during curing will be minimal and there will not be formation of large bubbles after curing due to outgassed volatiles. This minimizes the risk of micro-cracks or bubbles. Finally, this Ag paste can be easily screen-printed which is a well-known and cost-effective process in the PV industry, compatible with large-area application.

In order to address the contact resistance issue, the Si carriers were coated with a thin layer (200-1000 nm) of Al on both sides before screenprinting of the Ag paste on the surface of the carrier. For this investigation, three different types of cells (180 µm thick Cz n-PERT cell, 180 µm thick Cz HJ cell and 45 µm thick epifoil HJ cell) were bonded to the Al-coated Si carriers, and current-voltage measurements were performed before and after bonding.

Table I: Summary of cell results before and after bonding for 3 representative cells of each type (180 µm thick Cz n-PERT cell, 180 µm thick Cz HJ cell and 45 µm thick epifoil HJ cell).

Cell type		J_{SC} [mA/cm^2]	V_{OC} [mV]	FF [%]	Eta [%]	R_s [Ω.cm^2]
Thick Cz nPERT	Bef	39.4	658	71.1	18.5	0.15
	Aft	40.0	661	70.3	18.6	0.41
Thick Cz HJ	Bef	36.9	719	71.7	19.0	1.7
	Aft	37.8	724	73.6	20.1	1.4
Epifoils HJ	Bef	31.1	635	58.6	11.6	2.0
	Aft	31.5	643	61.0	12.4	2.1

The addition of an Al inter-layer between the Ag paste and the Si carrier, drastically improves the cell resistance, as evidenced in Fig. 3 (b) and Table I. For the nPERT cell, the FF drops slightly (~1% absolute) after bonding, which is reflected as an increase in the series resistance (extracted using the method that employs two illumination levels during I-V measurements) of ~0.25-0.4 Ω.cm^2. For the thick and thin HJ cells, the FF increases and consequently the efficiency increases after bonding. This is because the changes in series resistance is dominated by the annealing effect of the curing process on the ITO damage, which results in an improvement of both the FF and the Voc. Overall, it can be concluded that any negative effect on the FF due to the bonding of HJ cells to a Si carrier is negligible. With this, we have successfully developed a conductive bonding procedure for finished cells onto highly-doped Si carriers.

4 FABRICATION OF CONDUCTIVELY-BONDED SILICON SOLAR CELLS

For cell development on thin fragile epitaxial foils, a simple, low temperature, heterojunction cell technology based on hydrogenated amorphous silicon (a-Si:H) was chosen. Since the conductive glue that is used is a Ag paste and the Si carrier in the final concept is a low-purity Si carrier, there are major sources of contamination and care must be taken to avoid risk of lifetime degradation due to contamination. Therefore, the cell fabrication process was chosen such that the passivation of both silicon surfaces are completed before bonding.

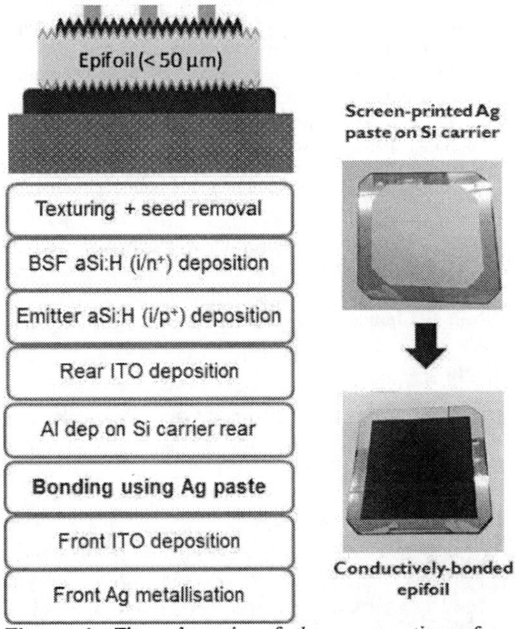

Figure 4: The schematic of the cross-section of a conductively bonded cell, the process flow used for its fabrication, and photographs showing the bonding process.

The process flow leading to conductively bonded HJ cells on thin silicon (thin epifoils and thinned Cz wafers of 50 µm starting thickness) is depicted in Fig. 4, and starts with a double-side texturing which simultaneously removes the porous silicon seed layer. Following a surface cleaning procedure, both surfaces are passivated with a stack of intrinsic and n-doped a-Si (i/n$^+$) BSF and a stack of intrinsic and p-doped a-Si (i/p$^+$) emitter. Subsequently, indium tin oxide (ITO) is deposited on the rear-side of the epifoil. After this, a low-temperature Ag paste (IPC114 from INKRON) as the conductive glue is screen-printed on a p$^+$ silicon Cz carrier wafer that is coated on both sides with a layer of sputtered Al. Real sintered substrates were not yet used for this work. After screen-printing, an epifoil is laid gradually from one edge to the other over the wet Ag paste. A slight pressure is applied. Curing is done in a belt furnace at 150°C.

Subsequently, ITO is also deposited on the front-side of the wafer-equivalent stack for the lateral conduction of charges, where it also acts as an anti-reflection coating (ARC). The front metallisation is completed by Ag screen-printing of the front contacts using a low-temperature Ag paste.

The light IV measurement results of the bonded cells produced as described above on both epifoils and thinned Cz wafers are given in Table II. For comparison, cell results of freestanding cells processed without bonding [18] are also given. It is clear that the V_{OC} of the bonded epifoil cells are much lower compared to that of the bonded thinned Cz cells. This difference is due to the large difference in the bulk quality of the two types of silicon. Effective lifetimes on passivated thin Cz wafers are in the range of 1-2 ms while those of passivated epifoils are in the range of 50-120 µs [19]. This large difference in V_{OC} also results in a lower efficiency for epifoil cells (~2% lower). The best conductively bonded epifoil cell is 15.7% efficient while the best bonded thinned Cz cell is 17.9% efficient. The best freestanding cells are indeed much better than their bonded counterparts in terms of both FF and efficiency. However, when comparing the averages values, it is clear that the cell performances are similar. Thus, it can be concluded that the conductive bonding process does not have a significant negative impact on the cell performance.

Table I: Summary of cell results of conductively bonded epifoil and thinned Cz cells. For comparison, cell results of freestanding cells processed without bonding are given. B = bonded. FS = freestanding.

Split			J_{SC} [mA/cm^2]	V_{OC} [mV]	FF [%]	Eta [%]	Ref.
Epifoil	B	Best	33.2	681	69.2	15.7	This work
		Avg.	32.8	676	69.1	15.3	
Thin Cz	FS	Best	33.2	738	73.0	17.9	
		Avg.	32.7	739	72.1	17.3	
Epifoil	B	Best	33.5	693	73.2	17.0	[18]
		Avg.	32.9	656	72.1	15.6	
Thin Cz	FS	Best	34.0	744	74.4	18.8	
		Avg.	33.9	723	71.6	17.6	

5 SIMULATIONS OF THE PATTERNED PRINTING OF THE BONDING GLUE

While there is a working procedure for conductive bonding, the cost of the adhesive layer is expected to be a significant part of the total cost of a HJ epifoil solar cell on sintered/sprayed silicon substrate, since 80-90% of the low-temperature Ag paste is made of Ag. In order to reduce the costs, use of the adhesive material can be limited by printing it in a patterned style, as opposed to the all-covering blanket screen-printing, reported above.

We investigated how the different patterns (lines, squares, etc.) of printed adhesive layers affect the final solar cell parameters. The patterned adhesive layer causes an additional serial resistance that reduce the fill factor (FF) of the cell and thus the efficiency. Simulations were carried out in order to assess the price/efficiency trade-off.

The model was built using finite element method (FEM) based software package COMSOL Multiphysics 5.2a, first in 1D geometry and subsequently in 2D and 3D geometries. As input data, the conductivities of used materials (cell electrodes, adhesive and substrate) were taken, as well as an I-V curve of an non-bonded cell (FF = 70.8%, η = 18.3%). Significant input data are the contact resistivities between the adhesive and epifoil bottom electrode, as well as between the adhesive and the Al interlayer deposited on the Si carrier, obtained by experimental measurements.

The results of the first simulation step (1D) show the deterioration of the fill factor, and consequently of the efficiency, as the area covered with the adhesive layer is lowered (Fig. 5 (a)). In this step, lateral current flows through the bonded cell are not considered, but only the change of adhesive resistance and contact resistance. The compromise was made here to investigate in more details the case of 60% adhesive coverage ratio, since at this level efficiency drops for 0.3% with saving of 40% of adhesive material.

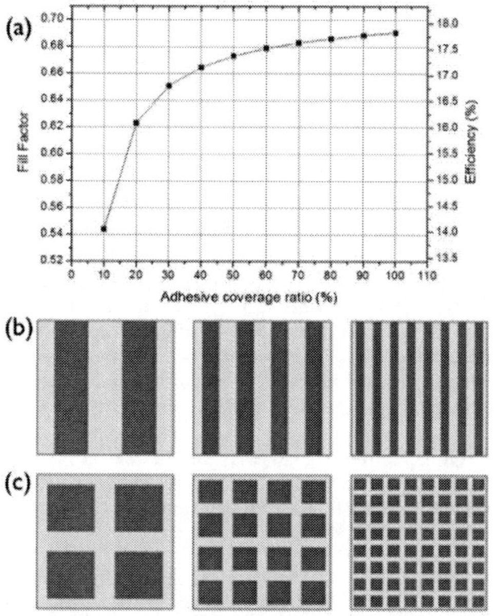

Figure 5: (a) Fill factor and efficiency of bonded cells as a function of the adhesive coverage ratio. Patterns of adhesive layer: (b) 2D – lines, (c) 3D – squares. All patterns have the same area coverage ratio.

The subsequent 2D and 3D simulation steps were carried out in order to determine the impact of pattern shape (lines (Fig. 5 (b), squares (Fig. 5 (c)) and size, while keeping the pattern coverage ratio at 60%. The resulting I-V curves are presented in Fig. 6, and corresponding calculated fill factor and efficiency in Fig. 7. It can be seen that the fill factor and efficiency gradually decrease with an increase in the gap in the patterns, which is caused by inefficient photo-current conduction from the areas uncovered by the adhesive layer. The apparent rise of the fill factor, that is noticeable for the large square patterns (Fig. 7 (a)), is the consequence of the drop of J_{SC} (Fig. 6 (b)). The efficiency in this case is still the lowest. As expected, it can be concluded that smaller patterns give better results (with gaps up to 2 mm). Moreover, linear shapes are preferred to squares, due to better current collection possibilities.

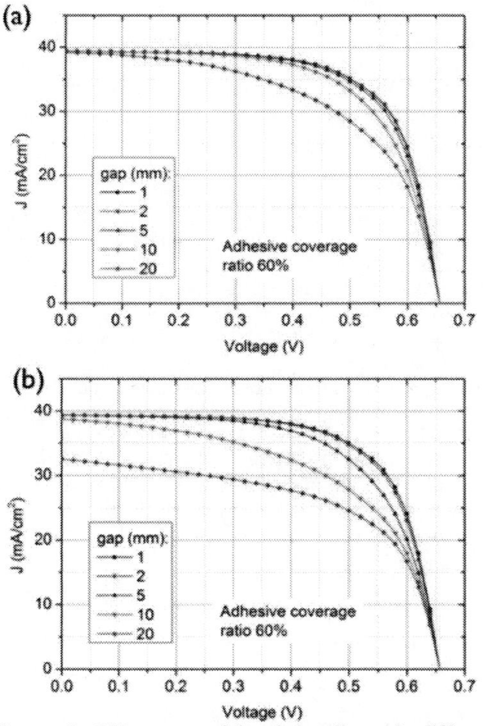

Figure 6: J-V curves of bonded cells with different adhesive pattern shape and size, while maintaining the adhesive coverage ratio of 60%: (a) 2D- lines, (b) 3D - squares.

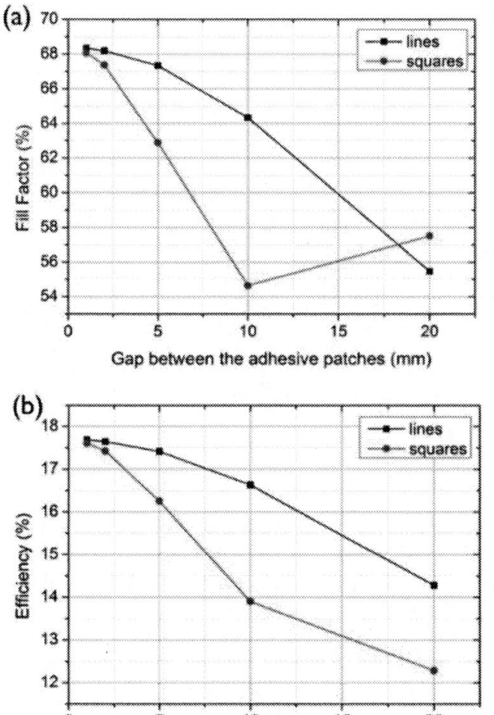

Figure 7: Fill factor and efficiency of bonded cell presented in Fig 4.

6 CONCLUSIONS

The ongoing development of a wafer-equivalent concept is described. In this work, high quality epifoils from the porous silicon-based lift-off process are mechanically supported on a highly doped p^+ Cz substrate. In the final embodiment a low-cost Si carrier, from recovered and recycled Si from waste sources, produced using low-cost techniques such as sintering or thermal spraying is envisaged. The Si carrier is also highly conductive and acts as the rear contact. Thus, the glue used for bonding epifoils onto the Si carriers must also be highly conductive. A working conductive bonding solution based on INKRON's low-temperature Ag paste, IPC-114, has been developed, whereby an Al interlayer between the conductive glue and the Si carrier is necessary for maintaining the cell performance after bonding. A high efficiency heterojunction cell process for thin epitaxial silicon foils, which incorporates conductive bonding in the process flow, has been established. The results shows that the bonding process does not have a significant negative impact on the cell performance with the best bonded epifoil cell reaching 15.7% and the best bonded thinned Cz cell reaching 17.9%. The efficiency of the bonded epifoil cells are limited by the bulk lifetime. The efficiency of the cell technology can also be improved by improving the series resistance and FF. Extensive simulations have been carried to determine the optimal geometry for achieving a patterned print of the conductive glue, which minimises the Ag paste usage with negligible loss of cell performance.

7 ACKNOWLEDGEMENTS

This project has received funding from the European Union's Horizon 2020 CABRISS project (Grant no. 641972).

6 REFERENCES

[1] R. Brendel, "Review of Layer Transfer Processes for Crystalline Thin-Film Silicon Solar Cells," *Jpn. J. Appl. Phys.*, vol. 40, no. 7, pp. 4431–4439, 2001.
[2] F. J. Henley, "Kerf-free wafering: Technology overview and challenges for thin PV manufacturing," *Conf. Rec. IEEE Photovolt. Spec. Conf.*, pp. 1184–1192, 2010.
[3] H. Sivaramakrishnan Radhakrishnan, V. Depauw, I. Gordon, and J. Poortmans, "Thin crystalline silicon: overview of lift-off technologies and module fabrication approaches," *(in Prep.*, 2017.
[4] H. Sivaramakrishnan Radhakrishnan and V. Depauw, "Lift-off approaches for thin crystalline silicon," in *Thin Film Solar Cells Fabrication, Characterization and Applications*, J. Poortmans, Ed. John Wiley & Sons, Ltd, 2016.
[5] K. Van Nieuwenhuysen, V. Depauw, R. Martini, J. Govaerts, M. Debucquoy, H. Sivaramakrishnan Radhakrishnan, I. Gordon, T. Bearda, K. Baert, and J. Poortmans, "High Quality Epitaxial Foils, Obtained by a Porous Silicon Based Layer Transfer Process, for Integration in Back Contacted Solar Cells," in *27th European Photovoltaic Solar Energy Conference and Exhibition*, 2012, pp. 2471–2474.
[6] H. Sivaramakrishnan Radhakrishnan, R. Martini, V.

Depauw, K. Van Nieuwenhuysen, M. Debucquoy, J. Govaerts, I. Gordon, R. Mertens, and J. Poortmans, "Improving the Quality of Epitaxial Foils Produced Using a Porous Silicon-based Layer Transfer Process for High-Efficiency Thin-Film Crystalline Silicon Solar Cells," *IEEE J. Photovoltaics*, vol. 4, no. 1, pp. 70–77, Jan. 2014.

[7] J. H. Petermann, D. Zielke, J. Schmidt, F. Haase, E. G. Rojas, and R. Brendel, "19%-efficient and 43 µm-thick crystalline Si solar cell from layer transfer using porous silicon," *Prog. Photovoltaics Res. Appl.*, vol. 20, no. 1, pp. 1–5, Jan. 2012.

[8] C. Gemmel, J. Hensen, S. Kajari-Schroder, and R. Brendel, "4.5 ms Effective Carrier Lifetime in Kerfless Epitaxial Silicon Wafers from the Porous Silicon Process," *IEEE J. Photovoltaics*, vol. 7, no. 2, pp. 430–436, 2017.

[9] N. Milenkovic, M. Driesen, B. Steinhauser, J. Benick, S. Lindekugel, M. Hermle, S. Janz, and S. Reber, "Epitaxial N-type silicon solar cells with 20% efficiency," in *Proceedings of the 43rd IEEE Photovoltaic Specialists Conference (PVSC)*, 2016, pp. 58–61.

[10] S. Janz, N. Milenkovic, M. Drießen, and S. Reber, "n-Type and p-Type Silicon Foils Fabricated in a Quasi-Inline Epi Reactor with Bulk Lifetimes Exceeding 500 µs," in *Proceedings of the 31st European Photovoltaic Solar Energy Conference and Exhibition*, 2015.

[11] E. Kobayashi, Y. Watabe, R. Hao, and T. S. Ravi, "High efficiency heterojunction solar cells on n-type kerfless mono crystalline silicon wafers by epitaxial growth," *Appl. Phys. Lett.*, vol. 106, no. 22, p. 223504, Jun. 2015.

[12] D. Kray, H. Kampwerth, E. Schneiderlöchner, A. Grohe, F. Kamerewerd, A. Leimenstoll, D. Osswald, E. Schäffer, S. Seitz, S. Wassie, S. Glunz, and G. Willeke, "Comprehensive experimental study on the performance of very thin laser-fired high-efficiency solar cells," in *Proceedings of the 19th European Photovoltaic Solar Energy Conference*, 2004, pp. 608–611.

[13] S. Pingel, Y. Zemen, and O. Frank, "Mechanical stability of solar cells within solar panels," in *Proceedings of the 24th European Photovoltaic Solar Energy Conference*, 2009, pp. 3459–3463.

[14] H. Sivaramakrishnan Radhakrishnan, K. Van Nieuwenhuysen, J. Govaerts, V. Depauw, T. Bearda, M. Debucquoy, R. Roozeman, J. Heikkinen, M. Schumann, R. Buchwald, H. J. Möller, A. Ciftja, G. Stokkan, E. Øvrelid, A. Stonkus, P. Dubravskij, J. Ul, and A. Ulyashin, "Solar cells and mini-modules based on 40 µm-thick epitaxial Si foils: towards conductive bonding onto low-cost Si powder sintered supporting substrates," in *Proceedings of the 32nd European Photovoltaic Conference and Exhibition*, 2016, pp. 1268–1271.

[15] H. Sivaramakrishnan Radhakrishnan, R. Martini, V. Depauw, K. Van Nieuwenhuysen, T. Bearda, I. Gordon, J. Szlufcik, and J. Poortmans, "Kerfless layer-transfer of thin epitaxial silicon foils using novel multiple layer porous silicon stacks with near 100% detachment yield and large minority carrier diffusion lengths," *Sol. Energy Mater. Sol. Cells*, vol. 135, pp. 113–123, Apr. 2015.

[16] R. Buchwald, S. Würzner, H. J. Möller, A. Ciftja, G. Stokkan, E. Øvrelid, and A. Ulyashin, "Microstructural characterization of Si wafers processed by multi-wire sawing of hot pressed silicon powder based ingots," *Phys. status solidi*, vol. 212, no. 1, pp. 25–29, Jan. 2015.

[17] G. Stokkan, A. Johanson, E. Øvrelid, A. Ciftja, J. Hennicke, and A. Ulyashin, "SINTERED LOW RESISTIVITY SUBSTRATES FOR HYBRID SOLAR CELLS," in *Proceedings of 28th European Photovoltaic Solar Energy Conference*, 2013, pp. 427–430.

[18] T. Bearda, A. Umer, S. Jambaldinni, M. Filipič, K. Van Nieuwenhuysen, H. Sivaramakrishnan Radhakrishnan, V. Depauw, I. Gordon, M. Debucquoy, Y. Abdulraheem, J. Szlufcik, and J. Poortmans, "Fabrication of silicon heterojunction cells on 50µm epitaxial substrates," in *Proceedings of the 33rd European Photovoltaic Solar Energy Conference (EUPVSEC)*, 2017.

[19] H. Sivaramakrishnan Radhakrishnan, M. Xu, T. Bearda, M. Filipič, K. Van Nieuwenhuysen, V. Depauw, I. Gordon, M. Debucquoy, J. Szlufcik, and J. Poortmans, "Heterojunction IBC solar cells on thin (< 50 µm) epitaxial Si foils produced from kerfless layer transfer process," in *Proceedings of the 33rd European Photovoltaic Solar Energy Conference (EUPVSEC)*, 2017.

FROM ADVANCED THIN-FILMS MODULES TO HIGH EFFICIENCY SILICON HETEROJUNCTION TECHNOLOGY AT 3SUN

W. Favre[2], G. Condorelli[1], A.-S. Ozanne[2], D. Muñoz[2], A. Canino[1], P. Rotoli[1], A. Battaglia[1], J.-F. Lerat[2], A. Moustafa[2], A. Valla[2], J. Stendera[2], F. Medlege[2], M. Fernandes[2], V. Barth[2], A. Ragonesi[1], P.-J. Ribeyron[2], M. Guercio[1], C. Gerardi[3]

[1]3SUN, Contrada Blocco Torrazze Zona Industriale 95121, Catania, Italy
[2]Univ. Grenoble Alpes, CEA, LITEN, INES, LHET, F-73375 Le Bourget du Lac, France
[3]Enel Green Power, Contrada Blocco Torrazze Zona Industriale 95121, Catania, Italy

ABSTRACT: We report on the first results obtained within the HETNA project: collaboration between 3SUN thin-film modules producer and CEA-INES for the conversion of 3SUN fab into a high efficiency silicon heterojunction (SHJ) solar cells & modules factory. Improvements of SHJ samples transportation between the two sites and samples management at 3SUN site enabled absolute efficiency increase up to 4.5% for cells with hydrogenated amorphous silicon (a-Si:H) layers performed using 3SUN's TEL Plasma Enhanced Chemical Vapor Deposition (PECVD) reactors. It was possible to reach implied open-circuit voltage (iV_{oc}) > 744mV and implied fill factor (iFF) > 84% for 20nm thick undoped a-Si:H layers deposited on textured 156 PSQ silicon wafers. Finally, we present results obtained on 72 cells bifacial modules designed specifically for 3SUN SHJ project made using high efficiency SHJ cells. The glass-glass module architecture, exempt of any frame, demonstrated a nominal power of 380.7W under flash test measurement with dark background.
Keywords: Heterojunction solar cell, surface defectiveness, PL imaging, industrial line, bifacial module

1 INTRODUCTION

In Europe, some companies and institutes have been working for several years on the development of SHJ at pilot line scale such as Meyer Burger and CEA-INES, with cells efficiency up to 22% [1–5]. Recently, actors that were previously mainly focused on the production of thin-films modules found in SHJ a way to increase their existing production capacity with potential limited CAPEX [6, 7]. The European company 3SUN (Enel Green Power/EGP group), based in Catania and well known for its 200MW production capacity of a-Si:H/µc-Si tandem cells is one of them [8, 9]. In this paper we present the first results of the collaboration between 3SUN and CEA-INES for the thin-film fab conversion into a SHJ cells & modules fab. In this scope, SHJ solar cells have been processed with a-Si:H layers prepared in an industrial 13.56MHz multi-chamber PECVD system used in 3SUN for thin film modules fabrication, the other steps being performed in the SHJ pilot line at CEA-INES.

2 EXPERIMENTAL DETAILS

2.1 Samples preparation

All the experiments have been led on as cut 156mm pseudo-square wafers (PSQ). After texturation and cleaning steps at CEA-INES the wafers have been shipped to 3SUN for HF-last dip followed by a-Si:H layer deposition using an industrial 13.56MHz multi-chamber plasma-enhanced chemical vapor deposition (PECVD) system, used in 3Sun for thin film modules manufacturing [8, 9]. After the PECVD deposition, cells were shipped back to CEA-INES for SHJ cell finalization with ITO layers and grid metallization with a 4 busbars configuration on both sides. More experimental details can be found elsewhere [10].

Finally, several 72 SHJ cells bifacial (glass-glass) modules with no frame and no visual defects were realized at CEA-INES with high efficiency SHJ cells.

2.2 Characterization techniques

Passivation performances obtained on cells precursors (no metallization) have been measured using a WTC-120 Quasi-Steady-State Photoconductance (QSSPC) lifetime tester in the transient mode [11, 12].

Photoluminescence (PL) full area images have been performed using a BT imaging LIS-R2 analysis tool. In the images presented, the signal intensity is directly related to the radiative recombination activity: defective regions appear darker than the other ones. Defectiveness level was determined using the grey-level based technique reported in ref. [13].

Scanning Electronic Microscope (SEM) coupled with EDX enabled advanced local defects analysis from a Nova NanoSem 630 FEI tool.

The cells electrical parameters were determined under AM1.5 conditions using a solar simulator with dark background.

At the module level, electroluminescence (EL) images were obtained from a LumiSolarProfessional GreatEyes tool and flash tests from a PASAN tool using multiflashes method to account for the modules' capacitive effects with dark background conditions.

3 SHJ CELLS RESULTS AND PROCESS IMPROVEMENTS AT 3SUN

The first SHJ cells made using 3SUN PECVD reactors demonstrated an efficiency of 15% under STC. It was possible to determine that the cells suffered from high level of defectiveness which was attributed to three main sources: (1) sample transportation between the two sites, (2) cleanliness of working area at 3SUN (3) 3SUN

PECVD tray design not optimized for silicon wafers.

On Figure 1 it is possible to point out that the PL images of cells precursors with 3SUN a-Si:H layers is covered by a large number of dark spots. These dark spots are not dead regions but related to lower values for the local minority carriers' effective lifetime, limited by high recombination rate. From such pictures, it is however impossible to determine properly the type of defect introduced locally (bulk, front surface or rear surface) since the PL image is an integration of the radiative emission generated by the sample on its whole thickness. For further analysis purpose, a specific region of a sample was cut and introduced into a SEM. It was possible to link the low PL signal to the presence of several tens of microns size particles (size estimated to ~40µm for the particle shown on Figure 1. The origin of these particles, mainly composed of carbon and oxygen, was difficult to establish but they could come from both transportation and/or 3SUN working area, prior to the PECVD deposition step. It is worth to note that these particles affected the PL emission on a large range, up to several hundreds of microns around and thus limiting the local cell efficiency.

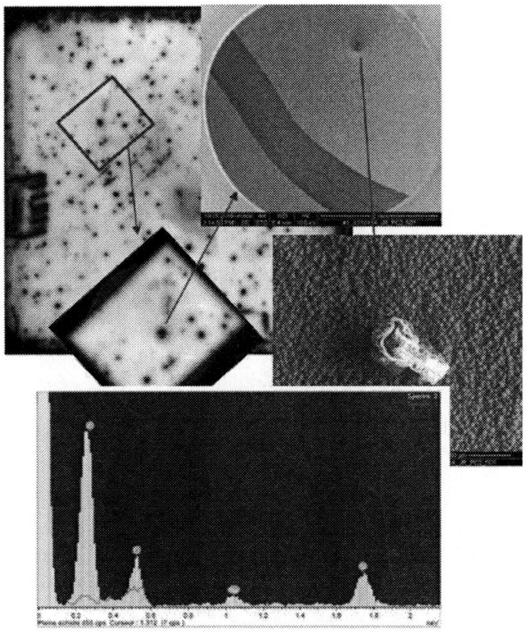

Figure 1: Identification of defectiveness source using PL, SEM and EDX techniques.

On Figure 2, we present typical results obtained from PL imaging of SHJ cells processed with 3SUN PECVD layers (a) to (c) with 1st generation PECVD tray ((a) 1st run, (b) after several runs) and 2nd generation PECVD tray and are compared to a CEA-INES reference cell. We can see a nice improvement after tray optimization that highly reduced defectiveness at the cells border, enabling efficiency increase of 3% and 4.5% absolute from condition (a) to (c) respectively. It is important to note that all these tests have been performed without any optimization of the PECVD layers properties. The best efficiency obtained in these condition was up to 19.5%, whereas standard mean efficiency for CEA-INES SHJ reference in similar 4 busbars configuration is around 21.5%. More details on the effect of cleanliness, samples transportation and PECVD tray design can be found in ref. [10].

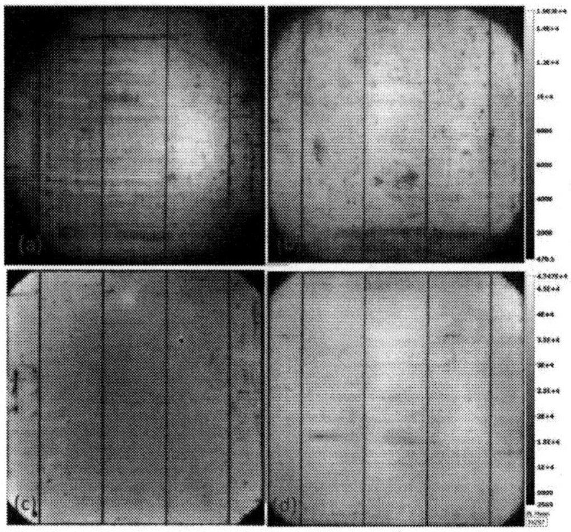

Figure 2: PL imaging of SHJ cells (a) 1st generation tray, (b) 1st generation tray but several runs, (c) 2nd generation tray and (d) CEA-INES reference.

In order to go a step further in the device losses analysis and quantify the possible efficiency losses due to local defectiveness, we used the method based on PL imaging described in ref. [13]. This technique, enabling an automatic rating of the cells local defectiveness has already demonstrated good correlation with Fill Factor (FF) losses and J_{02} when applied to SHJ devices. In this method, we define a so-called "Grey Level" value G_d' which is directly linked to the PL image pixels affected by low signal. On Figure 3, the FF determined under STC is plotted against G_d' obtained from PL images local defectiveness analysis for several batches of cells. We find a clear correlation between the two parameters, for each process conditions. The first processing condition, represented by dark squares on the figure, concerns batches of cells produced on the CEA-INES pilot line in 2014. The FF was affected by Gd' increase, with a possible upper value for FF at this time around 77.5% in case of absence of local defects (G_d' = 0). The second processing condition concerns batches of cells produced with 3SUN PECVD layers. The results are fully aligned on the dark spot and again clearly correlated to FF losses, as high as 10% absolute in case of the cells with the highest defectiveness. We find that cells without any local defects could be obtained with a FF up to 77% for this processing conditions. It is important to note that the good agreement for 1st and 2nd processing conditions is believed to be coincidence since this method must be calibrated for a given production line. Finally, the last processing condition shown corresponds to cells generated on the CEA-INES pilot line in 2016. These cells show a FF increase up to 3% absolute compared to the other processing conditions with values as high as 80%. This improvement is not related to local defectiveness decrease since same G_d' value as for other batches are obtained), but to processing conditions optimization.

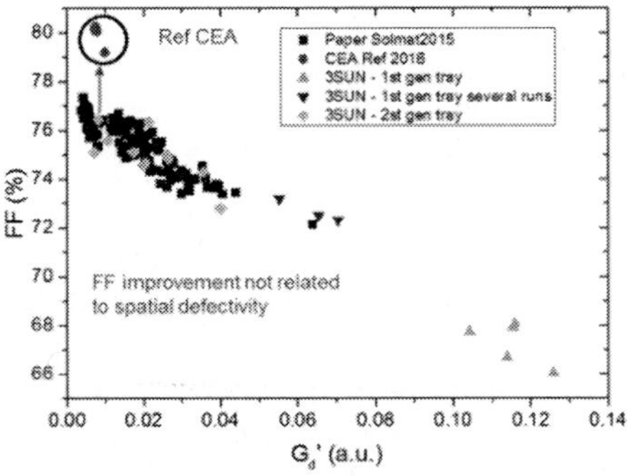

Figure 3: Fill Factor versus Grey Level rating for several set of SHJ cells.

Finally, we estimated that in the case of our various studies, the efficiency losses (after working area cleanliness control) attributed to transportation, waiting between processing steps (often referred as q-time in literature) are respectively in the range of 0.3% and 0.4% absolute. This means that it may be possible to reach efficiency in the range of 20.2% for samples with 3SUN PECVD layers (without any optimization) in case of full processing at the same place. For further cells efficiency improvement, there is a need to revise the fabrication recipes.

4 72 SHJ CELLS GLASS-GLASS MODULES RESULTS AND BIFACIALITY EFFECTS

SHJ cells are compatible with high bifaciality ratios that can benefit from possible rear illumination in outdoor conditions and improve the module energy yield. For this reason, a 72 SHJ cells bifacial glass-glass module was designed. In this part, we report on the best modules results produced at CEA-INES with this design in 2016 and 2017 respectively.

The modules pictures and EL images are presented on Figure 4 and show no apparent defects (no crack, no dark spots). The IV parameters are given for the two sides of these modules in Table 1. We underline that these measurements have been performed using a dark background (no illumination at the non-illuminated side). The peak power for these modules have been determined at 357.2W and 380.7W respectively, with 4 busbars cells configuration. From 2016 to 2017, cells size was increased from so-called M0 to M2, their efficiency was improved and the cells-to-module (CTM) losses was reduced. As a result, the module power was improved by more than 23W.

Module	Side	P_{max} (W_c)	V_{oc} (V)	I_{sc} (A)	FF (%)
Best module 2016	Front side	357.2	52.6	8.82	77.0
	Back side	313.0	52.4	7.66	78.0
Best module 2017	Front side	380.7	53.4	9.15	77.9
	Back side	331.5	53.3	7.95	78.2

Table 1: IV parameters obtained from flash test measurement of the best 72 SHJ cells glass-glass modules produced in 2016 and 2017 respectively at CEA-INES.

Figure 4: (Left) Picture (Right) EL image of high efficiency 72 SHJ cells glass-glass modules produced in 2016 (top) and 2017 (bottom).

In order to take into account the possible module performance benefit from bifacial configuration, a new stardard project IEC 60904-1-2 has been recently proposed [14]. A methodology, called G_E method, has been defined to determine the necessary front irradiance that leads to the same results as a bifacial irradiance based on the following equation:

$$G_{E_i} = 1000\,W.m^{-2} + \varphi.G_{R_i} \qquad (1)$$

where φ is a so-called bifaciality ratio given by:

$$\varphi = Min(\varphi_{I_{SC}}, \varphi_{P_{max}}) \qquad (2)$$

From Table 1, one can easily calculate the modules bifaciality ratios being, for these modules, the ratio between I_{sc} with back side (BS) illumination and I_{sc} with front side (FS) illumination equal to 86.8% and 86.9% respectively. For the best module produced in 2017 it was thus possible to perform flash tests measurements as proposed by the G_E method, considering the three following irradiance levels:

- 1087W/m² (eq 1000W/m² FS + 100W/m² BS)
- 1135W/m² (eq 1000W/m² FS + 157W/m² BS)
- 1173/m² (eq 1000W/m² FS + 200W/m² BS)

The results are shown on Figure 5, on which it is possible to observe that with an equivalent illumination contribution of 100W/m² at the module backside, it is possible to reach P_{max} up to 410W.

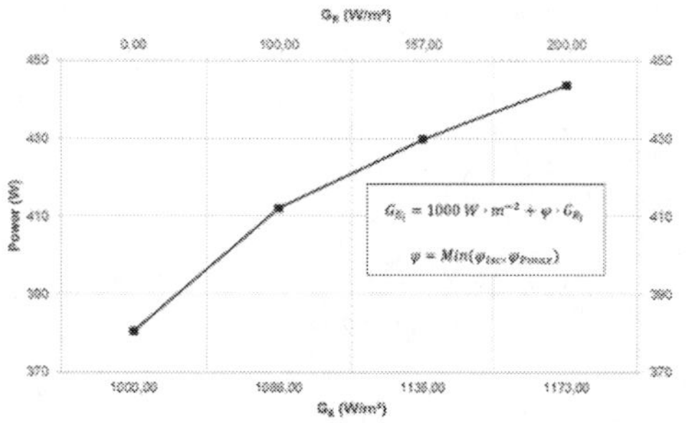

Figure 5: Module power (W) as a function of the irradiance level on the backside (G_E method)

5 CONCLUSION

In this paper we have reported experimental results at the cell and module levels obtained within a strong collaboration between 3SUN and CEA-INES on the deployment of silicon heterojunction technology at large production scale. We demonstrated continuous efficiency improvement for the cells made with the 3SUN a-Si:H layers with a best cell around 19.5% in STC, and estimated a potential higher than 20% without any recipe change. Finally, results of 72 SHJ cells bifacial modules have been presented. The nominal powers of the best module under monofacial illumination conditions and considering an equivalent rear illumination of 100mW/m² are respectively equal to 380.7W_p and 410W_p. giving clear insight of the potential for this PV technology.

6 ACKNOWLEDGEMENTS

The authors would like to thank 3SUN technicians D. Nicotra and A. Di Matteo and the CEA-INES LABFAB, Module and CEA-Leti teams for their technical and practical support. We would also like to thank M. Sciuto for his contribution.

This project has received funding from the European Union's Horizon2020 Programme for research, technological development and demonstration under Grant Agreement N° 745601

7 REFERENCES

[1] P. Papet, T. Söderström, M. Gragert, S. Beyer, B. Bonnet-Eymard, Y. Yao, et al. Investigations on Half Cells for Heterojunction Modules. 32nd Eur Photovolt Sol Energy Conf Exhib. 2016 Jul 25;1–1.

[2] P.-J. Ribeyron, E. Gerritsen, D. Heslinga, D. Muñoz, F. Ozanne, S. Bancel, et al. Recent Progress on the CEA-INES Heterojunction Solar Cell Pilot Line. 31st Eur Photovolt Sol Energy Conf Exhib. 2015 Nov 3;279–83.

[3] L. Korte, F. Korsós, M. Tallián, B. de Gier, E. Macron, J. Dupuis, et al. Status of the EU FP7 HERCULES Project: What Is the Potential of n-Type Silicon Solar Cells in Europe? 32nd Eur Photovolt Sol Energy Conf Exhib. 2016 Jul 25;331–4.

[4] A.Danel, R. Varache, S. Harrison, M. Vandenbossche, N. Rey, P. Lefillastre, J. Gaume, J. Veirman, A. Bettinelli, C. Roux, Versatile pilot line to support the heterojunction solar cell industrial development: busbar and busbar-less configurations, 2DO1.5 paper to be published in the Proc. of the 33rd European Photovoltaic Solar Energy Conference, Amsterdam, The Netherlands (2017)

[5] J. Veirman et al., "Early Efficiency Prediction of Silicon Heterojunction Cells Processed on Thermal Donors-Rich Czochralski Wafers" 2CO.10.2 paper to be published in the Proc. of the 33rd European Photovoltaic Solar Energy Conference, Amsterdam, The Netherlands (2017)

[6] G. Kekelidze, G. Bubnov, N.J. Matsapey, Y. Trouillot R. Turchet, F. Quesnel, et al. A Solar Module Prototype Assembled from Silicon Heterojunction Solar Cells Manufactured in Gen5 KAI PECVD Reactors. 32nd Eur Photovolt Sol Energy Conf Exhib. 2016 Jul 25;1–1.

[7] L. Li et al. Mass production of high efficiency silicon heterojunction solar cells: a low-cost approach by upgrading GEN8.5 thin film solar line. 32nd Eur Photovolt Sol Energy Conf Exhib. 2016;(2BO.9).

[8] C. Gerardi, M. Foti, D. Rapisarda, A. Canino, A. Battaglia, G. Condorelli. Industrial Scale Optimization of SiOx Bottom n-Layer in Tandem Solar Cell. 32nd Eur Photovolt Sol Energy Conf Exhib. 2016 Jul 25;1260–3.

[9] C. Gerardi, A. Battaglia, G. Condorelli, A. Canino. External Quantum Efficiency as Function of Applied Voltage of Multi-Junction Hydrogenated Amorphous Si Based Cell: Performance Optimization after Stabilization. 32nd Eur Photovolt Sol Energy Conf Exhib. 2016 Jul 25;1097–9.

[10] G. Condorelli, P. Rotoli, A. Canino; A. Battaglia, W. Favre, A.-S. Ozanne et al., "Contamination control challenges on SHJ solar cell processing", to be published in the proceedings of the IEEE PV Specialist Conference, 2017.

[11] Nagel H, Berge C, Aberle AG, "Generalized analysis of quasi-steady-state and quasi-transient measurements of carrier lifetimes in semiconductors", J Appl. Phys lett, vol 86, pp 6218-6221, 1999.

[12] R. Sinton, A. Cuevas, "Contactless determination of current–voltage characteristics and minority-carrier lifetimes in semiconductors from quasi-steady-state photoconductance data" Appl Phys Lett. vol 69, pp 2510-2512, 1996.

[13] O. Nos, W. Favre, F. Jay, F. Ozanne et al., "Quality control method based on photoluminescence imaging for the performance prediction of c-Si/a-Si:H heterojunction solar cells in industrial production lines" Sol. Energy Mater. Sol. Cells, 144, pp210-220, 2016.

[14] Power PVTECH, "Two-faced, why solar's next success story has two sides", Vol. 12, p. 20, Sept. 2017

A NEW PILOT RESEARCH FACILITY FOR HJT AND SELECTIVE CONTACT SOLAR CELLS – PV-TEC SELECT

Jochen Rentsch, Martin Bivour, Denis Erath, Frank Feldmann, , Andreas Fischer, Sebastian Mack, Anamaria Moldovan,
Jan Temmler, Martin Hermle, Stefan W. Glunz, Ralf Preu
Fraunhofer Institute for Solar Energy Systems ISE
Heidenhofstraße 2, 79110 Freiburg, Germany
e-mail: jochen.rentsch@ise.fraunhofer.de

ABSTRACT: After 10 years of operation of its Photovoltaic Technology Evaluation Centre PV-TEC Fraunhofer ISE recently significantly expanded its research capabilities with the opening of a new pilot research facility for selective contact solar cells – PV-TEC SELECT. PV-TEC SELECT aims to address major challenges on the way to industrialize high efficiency solar cell concepts like Heterojunction (HJT) or TOPCon technology. In this context, the paper introduces the new research facility with its unique equipment setup and process environment (Clean room class 1000 on more than 800 m²). The development of a robust base line process allows Fraunhofer ISE to expand its research offerings for PERC solar cells to next generation high efficiency cell concepts like HJT and TOPCon.

Keywords: Pilot production, Selective Contacts, Heterojunction, Throughput, Inline processing

1 INTRODUCTION

The race for higher efficient PV modules generates an increasing level of complexity in the processes involving the manufacturing of more sophisticated products and hence the increase in related investment & fabrication costs. Given that the market still requires a reduction of the technology cost, a difficult equation has to be solved: to produce high efficiency industrial devices with a low industrial processing costs. Among the existing solar cell technologies, selective contacts, especially the Heterojunction Technology (HJT), have demonstrated a large potential to address these challenges. Furthermore, the HJT solar module is specifically suited for a bi-facial architecture, further enhancing the electricity production capacity by 15-20%, and opening up many new market segments. The HJT technology will remain during the next decade one of the premium technologies with 1-2% higher efficiency than conventional technologies. In order to support the development of this technology and 10 years of successful operation of our existing Pilot Research platform PV-TEC (Photovoltaic Technology Evaluation Centre) [1], Fraunhofer ISE decided to expand its current technical infrastructure to set up a new pilot facility for solar cells based on selective contacts – PV-TEC-SELECT. The new centre was officially inaugurated in spring 2016 and has been step-wise expanded since then. It is expected to reach full operation at the end of Q1/2018. Within this new facility, companies along the entire production chain to produce highest efficient solar cells can find an optimum framework to test their latest process and equipment innovations on an industrial relevant scale.

The main goal of PV-TEC SELECT is to capitalize on previous lab-based developments on the hetero-junction and selective contact technology and bring the key selected technologies to industrial-scale production level leveraging the potential of a high level of automation to further reduce production costs. We present the design of the service centre that has the goals to offer service at high volume production rates, to be able to compare new and existing production technologies for selective contacts and that is able to be flexibly extended in the future and is able to evaluate upcoming new solar cell processing sequences for highest efficient solar cell structures on an industrial scale outside the industrial production facilities.

2 CHALLENGES IN MANUFACTURING AND AMBITION

PV-TEC SELECT addresses services and R&D topics on relevant existing bottlenecks for industrial scale production of HJT/selective contact solar cells:

- Quality assurance: Incoming inspection of source wafer material becomes mandatory in order to keep overall production performance distributions tight. Focussed parameters to be determined include base resistivity and minority carrier lifetime distributions.
- Defectivity and interface sensitivity: One of the major limitations and cost drivers in current high efficiency production of HJT cells are all necessary efforts to keep wafers clean in between the critical surface sensitive processing steps while keeping wafer throughputs high. R&D efforts concentrate on suitable automation solutions and process ambient specifications as well as alternative process sequences to reduce vacuum breakages and automation needs.
- Throughput: Optimizing productivity is essential to stay cost competitive. Increasing the throughput of the equipment in order to achieve maximum output is therefore a suitable way to reduce tool-related costs per cell. In order to optimize the throughput in a HJT cell production line, both, front-end (chemical and deposition processes) and back-end (metallization and classification) processes should have equal capacity. Current ITRPV expects wafer throughputs per tool of more than 5000 wafer/h from 2019 onwards [2]
- OPEX and CAPEX cost: Efforts to reduce cost sensitive materials like wet chemical solutions or silver pastes for front (and rear) metallisation by decreasing finger width, alternative and/or simplified process sequences showing similar or even superior performance need to be investigated.

For the formation of selective contacts, two basic approaches are followed and further developed at Fraunhofer ISE. They can be separated by the maximum temperature allowed during processing.

The low temperature approach ($T_{max} < 250°C$) represents the classical Heterojunction solar cell process comprising of the deposition of intrinsic amorphous silicon for surface passivation.
Within the high temperature approach ($T_{max} > 250°C$), Fraunhofer ISE further develops its proprietary and patented TOPCon technology. For TOPCon, surface passivation is achieved by tunneling oxide deposition followed by the formation of doped micro- or poly-crystalline silicon layers on both sides.
In both cases, metallization includes TCO deposition and screen-printed, PVD or plated front and rear metallization techniques, in the high temperature case much wider process windows offer advantages over the rather limited process conditions for classical HJT.

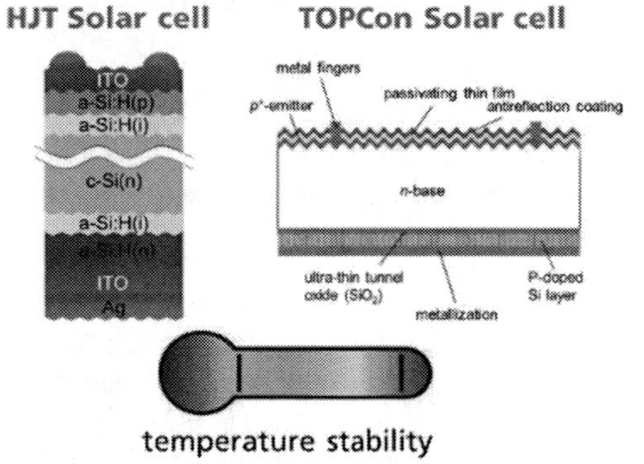

Figure 1: Main device structures for development within PV-TEC SELECT.

3 FACILITY LAYOUT AND SERVICE OFFERINGS

PV-TEC SELECT host the full front end processing chain for HJT and selective contact solar cell structure comprising of inline and batch wet chemistry tools, inline as well as cluster tool based PECVD deposition for intrinsic and doped amorphous and microcrystalline silicon layers as well as PVD deposition of TCO, metal as well as metal oxide layers (Figure 3). The corresponding processing equipment has been selected to be able to operate at higher throughputs (500-1000 Wafer/h) in order to allow design of experiments on statistically relevant data base.
The laboratory environment has been designed to fulfil Cleanroom Class 1000 conditions which allow us to easily switch process sequences without sacrificing the cleanliness of wafer surfaces in-between the different processing steps.
The front end processing within PV-TEC SELECT is well connected with the backend technologies in our already existing PV-TEC pilot line exhibiting different printing based metallisation techniques (screen print, dispensing, inkjet, flexographic print) [3-5] as well as plating technologies. Additionally, the activities are strongly linked to our Module Technology Centre, where cell connection and encapsulation of the new device structures and metallization approaches are tested and qualified. The design of the new evaluation centre is based on the assumption that a solid baseline process for Heterojunction solar cells is needed as reference process and technology set. To be able to further develop the solar cell processes and to implement future concepts like the TOPCon technology [6], the baseline equipment is supplemented with innovative components.

In PV-TEC SELECT the machines are organised as (mostly) automated islands. The major processing tools are:

- Batch wet etching tool for texturing and cleaning
- in-line wet etching tool for cleaning and conditioning
- Reference DC parallel plate PECVD system for intrinsic and doped amorphous silicon layer deposition
- Inductively coupled plasma (ICP)-PECVD system for intrinsic and doped amorphous silicon as well as PECVD based TCO layers
- Inline linear-rf based multi-chamber PECVD system for intrinsic and doped amorphous silicon layers
- Tube LPCVD system for the deposition of doped poly-Si layers
- Tube PECVD system for the deposition of doped poly-Si layers
- PVD cluster system for DC and rf biased sputtering of TCOs, metals and metal oxide layers
- Inline PVD systems comprising DC and rf biased sputtering of TCOs as well as metal oxide layers

The following main service areas are addressed and offered:

- Testing of new equipment in PV-TEC SELECT by bypassing the baseline.
- Solar cell producers should be able to bypass their line and execute interesting processes.
- Equipment suppliers that do not have a complete production line should be able to order partly processed wafers for testing their equipment which can remain at the supplier's site. Finalizing can be done at ISE.
- Qualifying materials e.g wafers, pastes, chemicals.
- Training of foreign staff

4 BASELINE PROCESS

Figure 2, Table 1 and 2 show recent achievements in back end processing to establish a baseline printing based process on well-known industrial-like equipment, enabling for short process times and a high-throughput production capability [3].

Figure 2: SEM images of contact lines screen-printed with a low-temperature silver paste: double printed line with $w_{f,Ø} = 34$ µm and $h_{fmax,Ø} = 20$ µm

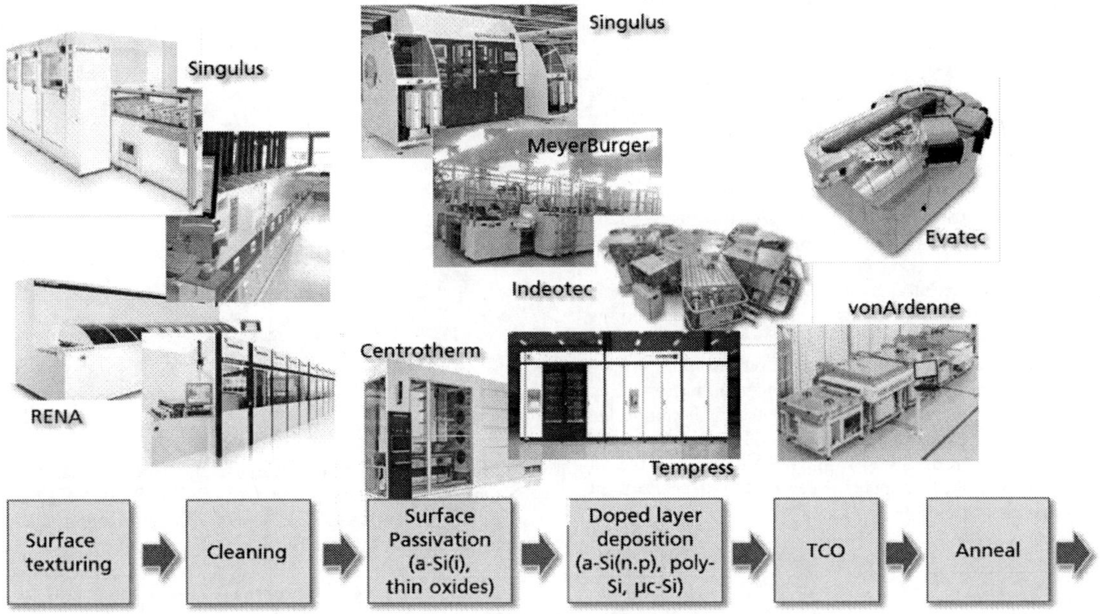

Figure 3: Equipment setup of new Pilot Research Facility PV-TEC SELECT

Screen printing as state-of-the-art metallization technique has shown its ability to produce fine contact fingers on SHJ solar cells. Dispensing is a promising alternative, as it also can be used for the conventional module integration based on busbars, but offers higher aspect ratios even within a single printing step (Table 1). Heterojunction solar cells with a conversion efficiency of up to 21.9 % and FF > 80 % were produced by single screen printing on bothcell sides. Although, highest Fill Factor values of 80.9 % were reached with the dispensing method, increased shading caused by higher printed line width as well as a different busbar geometry led to decreased values of J_{sc}, still enabling an efficiency of 21.6% (Table 2). The high level of V_{oc} proves that the passivation quality is not significantly affected by both metallization techniques.

Table I: Comparison of achieved printed line geometry for all investigated printing methods

	$w_{f,\varnothing}$ (µm)	$h_{fmax,\varnothing}$ (µm)	AR (u.a.)
Screen-printing (single print, 50 µm)	56	13	0.23
Screen-printing (double print, 30 µm)	34	20	0.59
Dispensing	68	30	0.44

Table II: I-V-data of the best screen-printed compared to the best dispensed solar cell. Decreased J_{sc} of the dispensed solar cell is partly due to a different busbar geometry causing a loss of ~0.5 mA/cm²

	V_{oc} (mV)	J_{sc} (mA/cm²)	FF (%)	η (%)
Screen-printing (single print)	727	37.6	80.1	21.9
Dispensing	726	36.9	80.9	21.6

4 SELECTED RESEARCH RESULTS

4.1 Quality assurance throughout the process chain

Important quality requirement for wafers is needed to assure solar cell conversion efficiency improvements and process robustness for the pilot line activities, therefore incoming inspection of source wafer material becomes mandatory in order to keep overall production performance distributions tight. Within PV-TEC SELECT, wafer fine inspection of large area n-type silicon source wafers can be performed within a fully automated inline inspection tool determining all basic geometrical, electronic as well as optical parameters of each individual wafer.

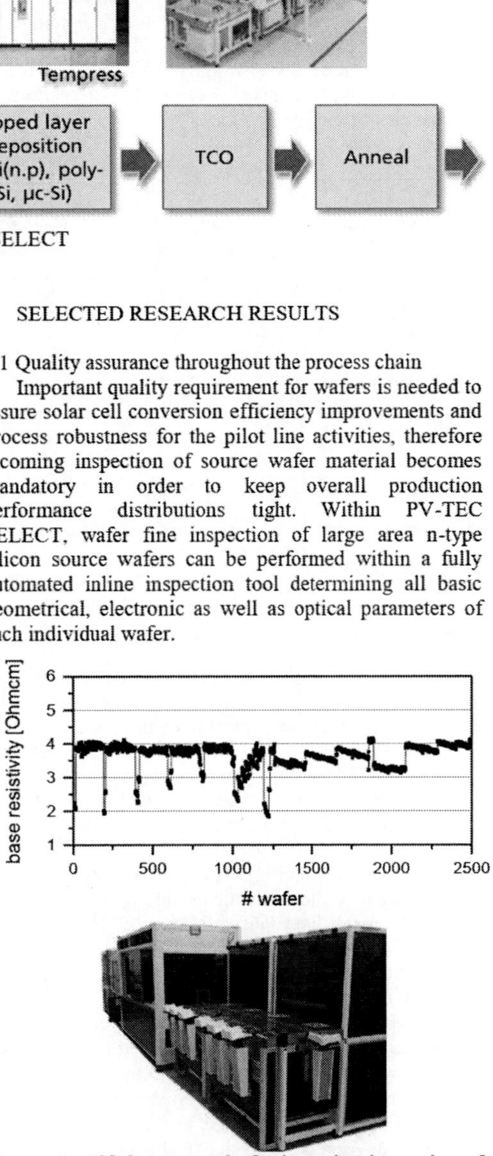

Figure 4: Lifetime example for incoming inspection of large batch of n-type silicon source wafers measured by fully automated inline inspection system: distribution of base resistivity

4.2 Wafer handling of highly sensitive wafer surfaces

One of the major challenges for the industrialization of high efficiency solar cell manufacturing processes represents the automation and handling of the wafers with their highly sensitive surfaces. As typical wafer throughputs of industrial process tools surpass 3600 wafers/h, wafer handling cycles (loading/unloading from cassettes or carriers) need to be performed typically in the range of only a few seconds. Common gripper solutions like Bernoulli, suction or area grippers all show an impact on the resulting electrical properties at critical process interfaces e.g. for wafer handling in between the final surface clean from a batch wet bench and the intrinsic amorphous silicon deposition within a typical PECVD furnace. Earlier studies showed an impact of up to 10 % lifetime reduction in the related areas due to the handling step [7], which in case of HJT solar cells could result in significant losses in the open circuit voltage. In case of Bernoulli grippers, mainly the deposition of particles is responsible.

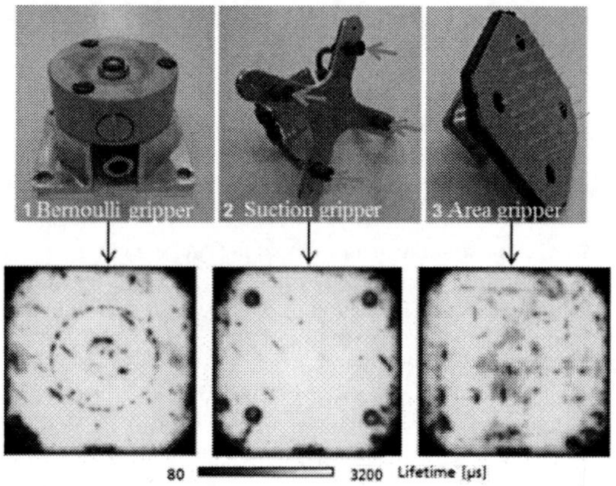

Figure 5: Investigations on the impact of wafer grippers on electrical properties of alkaline textured and a-Si passivated surfaces [7]

The "fingerprint" reflects the distribution of the Bernoulli stream nozzles that blows the air from the inside to the outside to create a lower pressure of the Bernoulli Effect [8]. Future investigations will further concentrate on optimization of Bernoulli based wafer gripping as well as investigations on other touchless wafer transport and handling methods like ultrasonic end-effectors.

4.3 Technologies with high throughput potential

Aiming to address the higher throughput demand of future processing lines (and to reduce the cost of ownership (COO)) processes for the deposition of hydrogenated amorphous silicon (a-Si:H) which is used as passivation and selective layer for silicon heterojunction (SHJ) solar cells are developed on a state-of-the-art inline PECVD deposition system MAiA..The developed processes have to fulfill two important basic requirements:

- excellent passivation of the c-Si absorber
- high selectivity with respect to the charge carriers to obtain the best possible electrical contact.

The a-Si:H(i) process supplies excellent stability concerning passivation even after chamber cleaning by an etching process and although to the fact that other processes carried out within the same deposition chamber. In Figure 6 the minority carrier lifetime τ_{eff} is shown over a larger period of tool operation. One can see a fluctuation of the τ_{eff}, but an overall level of above 3 ms (corresponding to an iV_{OC} >730 mV)..

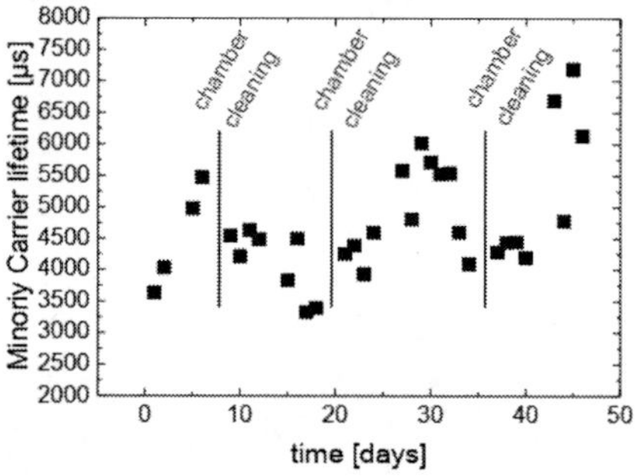

Figure 6: Minority carrier lifetime as a function of the time to proove the reproducibility/stability of the a-Si:H(i) process; red lines show chamber cleaning cycles by an etching process [9].

Fig. 7 depicts the result for the iV_{OC} as a function of the wafer position on the deposition tray. It can be seen that with an optimized process iV_{OC} values over 730 mV were obtained for all positions whereby a slight degradation in transport direction (from top to bottom) can be observed.

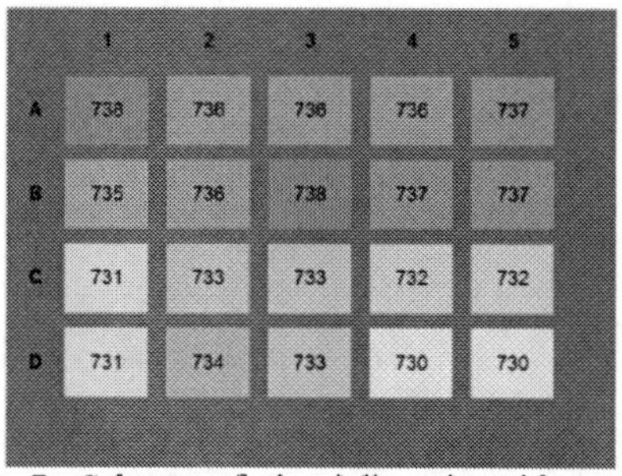

Figure 7: Scheme of the inline deposition tray (5 columns, 4 rows) with iV_{OC} values in mV measured by QSSPC as a function of the tray position [9].

Further results also for n- and p-doped amorphous silicon layers by inline PECVD deposition are presented elsewhere [9]

4.4 New process technologies for cost reduction

In Figure 8, results of a new, simple, cost effective and production-feasible ozone-based surface conditioning process for the controlled rounding and smoothing of textured surfaces is presented [10]. For the rounding of the textured surface a diluted solution consisting of HF and HCl in ozonized DI-H_2O was utilized which additionally offers an improved cleaning performance. The process was carried out in an industrial batch tool. The ozone concentration is kept constant during the process by continuous recirculation through an ozone generator which is coupled with the process bath. Within this experiment the degree of rounding was adjusted by varying the exposure time (2, 4 and 8 min). Considering the impact of the rounding reaction on the size of the pyramids, n-type Cz samples with small size (1-3 µm) and medium size pyramids (4-6 µm) were prepared in potassium hydroxide (KOH) solution in addition with a commercial available organic two component additive. For all c-Si surfaces already the 2 min O_3 treatment resulted in significantly higher efficiencies compared to the standard SC1/SC2 clean.

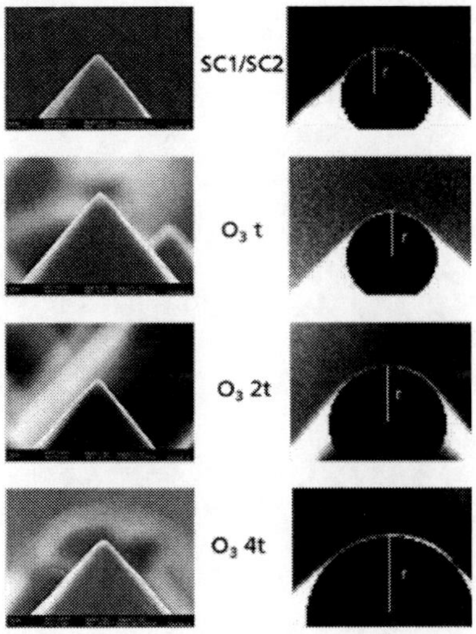

Figure 8: Ozone based surface treatments after texturing (two different pyramid sizes) in comparison to a standard SC1/SC2 clean. The resulting rounding of the pyramid tips was analyzed by high resolution SEM (35 K optical zoom) with manual fitting of a circle in the rounded structure of the pyramid tip in order to estimate the „Rounding radius".

The O_3-based conditioning leads to a cell efficiency improvement of up to 0.4% absolute for the group with small size pyramids and up to 0.8% to maximum values of 22.7% for the medium size pyramids. The results demonstrate the importance of an adequate c-Si surface treatment, which can be obtained with industrially feasible approaches. The results are an important step towards lowering the cost of advanced cell concepts while at the same time increasing the cell efficiency.

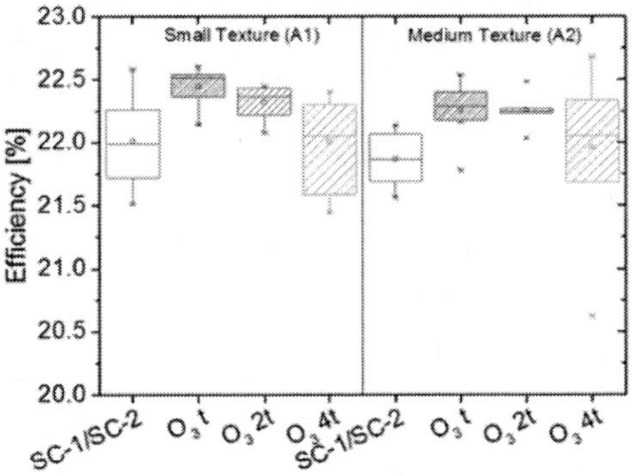

Figure 9: Solar cell efficiency comparison of large area HJT cells with different Ozone based surface treatments after texturing.

One major drawback of state-of-the-art screen printed based metallization approaches for HJT solar cells represents the low temperature anneal requirement which typically leads to rather poor line conductivities of the fine front metallization grid resulting in higher series resistance losses. At Fraunhofer ISE, a new metallization has been developed based on Laser induced forward transfer (LIFT) of metals to form a fine line seed layer for subsequent selective copper and silver plating to form the contact [11]. An additional thin insulating layer needs to be deposited on top of the TCO in order to protect the TCO during plating. Excellent finger geometries and electrical properties have been achieved using this metallization approach (Figure 10). First application of the process sequence for large area (156 x 156 mm²) HJT solar cells revealed an absolute increase in efficiency of 0.4 %mainly due to an improvement in fill-factor as well as J_{sc}.

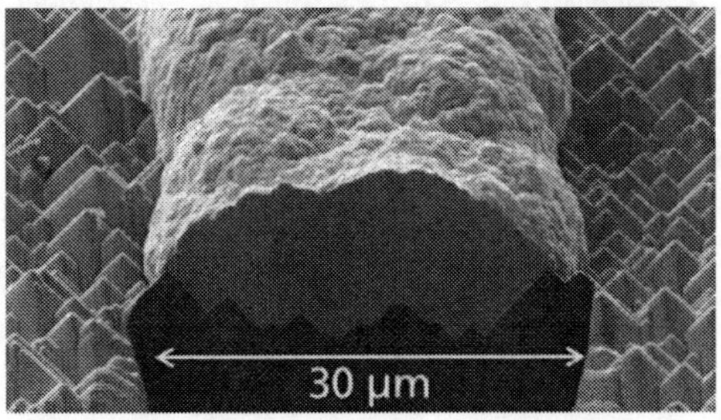

Figure 10: SEM image of cross section of Laser transferred seed-layer and Cu-plated contact finger.

Table III: I-V-data of the best screen-printed compared to the best LIFT process

Metallization	Area	V_{OC}	J_{SC}	FF	η
	[cm²]	[mV]	[mA/cm²]	[%]	[%]
Screen printing	239,0	726	37,7	79,0	21,6
Screen printing	239,0	727	37,8	79,1	21,7
LIFT + Cu-plating	239,0	727	38,2	79,7	22,1
LIFT + Cu-plating	239,0	728	38,0	80,1	22,2

5 CONCLUSION

We have set up a complete pilot line for the front end of solar cells with selective contacts. On an area of more than 900 m² we reach a clean-room quality of 1000, this unique set-up enables to us to perform high efficiency solar cell processing with very high standards for surface cleanliness while still applying individual industrial type systems for the deposition of highly passivating layers. The individual tool set up allows performing research for different potential device structures like HJT or TOPCon in order to find optimum industrial applicable process sequences.

The main systems include: inline and wet etching tools for texturing and cleaning, an ICP PECVD as well as a inline linear rf PECVD tool for intrinsic and doped amorphous silicon layers as well as a PVD tool for DC and rf biased deposition of TCOs.

In the different relevant research areas (wet chemistry for surface texture and cleaning, deposition of amorphous silicon, TCO layers) we have achieved excellent resulzs enabling cell efficiencies for HJT well above 22 %.

We combine "this" with the state-of-the-art back-end processing in our main PV-TEC facility which has approved to process heterojunction solar cells at an efficiency exceeding 22% on high-throughput equipment in a rather pilot lab ambient

6 REFERENCES

[1] D. Biro et.al., "PV-TEC: Photovoltaic Technology Evaluation Center - Design and implementation of a production research unit", Proceedings of the 21st EUPVSEC, September 4-8, 2006, Dresden
[2] ITRPV roadmap, edition 2017.
[3] D. Erath et.al., "Comparison of Innovative Metallization Approaches for Silicon Heterojunction Solar Cells", submitted for publication Silicon PV 2017, Freiburg
[4] A. Lorenz et.al., "Progress with Rotational printing For The front side metallization of Silicon solar cell", Proceedings of the 32nd EUPVSEC, 2016, Munich, Germany
[5] M. Pospischl et.al., "High speed dispensing – A high throughput metallization technology for > 21% PERC type solar cells", Proceedings of the 32nd EUPVSEC, 2016, Munich, Germany
[6] F. Feldmann et.al.," Efficient carrier-selective p- and n-contacts for Si solar cells", Solar energy materials and solar cells 131 (2014), pp.100-104
[7] S. Nold et.al. "Investigations on the impact of wafer grippers on optical and electrical properties of alkaline textured and a-Si passivated surfaces", Proceedings of the EUPVSEC 2011, Frankfurt, Germany
[8] H. Samadi et.al., "Particle Free Handling of Substrates", IEEE Transactions on Semiconductor Manfacturing, VOL. 29, NO. 4, November 2016
[9] J. Temmler et.al.,"Inline PECVD deposition of a-Si layers for heterojunction solar cells on an industrial scale", this conference
[10] A. Moldovan et al, Solid State Phenomena, vol. 195, pp. 305–309, 2012
[11] A. Rodofili, this conference

Partly funded by the German Federal Ministry for Economic Affairs and Energy by the projects „HERA" Contract 0324086A „INNOHET" Contract 0325634B, „SELEKTIV B" Contract 0324141B and PV BAT 400" Contract 0324125.

VERSATILE PILOT LINE TO SUPPORT THE HETEROJUNCTION SOLAR CELL INDUSTRIAL DEVELOPMENT: BUSBAR AND BUSBAR-LESS CONFIGURATIONS

A.Danel, R.Varache, S.Harrison, M.Vandenbossche, N.Rey, P.Lefillastre, J.Gaume, J.Veirman, A.Bettinelli, C.Roux

CEA, LITEN, Department of Solar Technologies, F-73375 Le Bourget du Lac, France

ABSTRACT: In this paper we present the 2016-2017 progress made on the CEA silicon heterojunction solar cell (SHJ) pilot line called 'LabFab'. The pilot line has entered its second generation phase with the installation of a new sputtering tool for Transparent Conductive Oxide (TCO) deposition allowing edge masking for junction opening, as well as the development of busbar-less (BB0) back-end process and an increase of the nominal capacity at 2400 wafers per hour. Following the wafer market, the pilot line tools were also upgraded to handle both M0 and M2 156x156 mm^2 sizes, as well as lower wafer thicknesses (down to 90µm), thereby increasing the LabFab overall versatility. The line is thus capable of producing and sorting two main different industrial products: 4 busbar (BB4) and BB0 rear emitter bifacial cells, with a focus on modules and systems output and reliability. Improving the 20.5% efficiency BB4 baseline end of 2015 with the use of higher quality substrates (high Si bulk lifetime / resistivity ratio) coupled to texturization, PECVD, PVD and screen printing process optimization, the LabFab production flow has recently reached a lot to lot average cell efficiency of 21.5% and 22.0% for BB4 and BB0, respectively. The best 244.3cm^2 cells with industrial bifacial processing show 22.3%, 22.1% and 23.3% efficiency for BB4-160µm, BB4-90µm and BB0-160µm respectively. Modules realized by ribbon conductive gluing on selected BB4 cells show a power cell-to-module ratio of 101+/-1% in the glass-backsheet (GBS) configuration and 98+/-1% with a glass-glass bifacial design (GG). 324Wp and 313Wp 60-cell GBS modules were acheived with 160µm and 93µm cells respectively. A 72-cell GG record module was manufactured at 381Wp (flashed with dark backside) and measured at 444Wp under the standardized GE norm (1000W/m^2 front side + 200W/m^2 rear side illumination). Our first modules based on BBO cells and the multi-wire interconnection show similar power as achieved with BB4 cells (386Wp 72-cell, 318Wp 60-cell, GBS, learning stage) with a high reliability and consistent field data. This allows a short term perspective to equal or slightly increase the power output for BB0 systems compared to the BB4 reference, taking into account an amount of silver per cell divided by about a factor 3 to 4.

Keywords: silicon solar cells, heterojunction, industrial pilot line

1 INTRODUCTION

As now widely recognized, the amorphous / cristalline silicon heterojunction (a-Si:H/c-Si, SHJ) is one of the most attractive solar cell architectures combining high performance and industrial compatibility. Simple, low temperature, high throughput and cost effective processes, bifaciality, thin wafers (<100µm) and possible combination with rear-contact cell design are the SHJ cell main advantages, contributing to its competitiveness with regard to other architectures [1-3]. With such advantages, a SHJ market is currently emerging with 4 to 5 GW production capacity forecast before 2020 [4] considering existing actors expansions, new players, equipment suppliers [5] and thin films reorientation [6, 7].

In this context, the CEA-INES has started to operate by the end of 2011 an industrial heterojunction cell pilot line located at the French National Solar Energy Institute (INES) and operated by the French Alternative Energies and Atomic Energy Commission (CEA). This 'LabFab' was conceived to bridge R&D labscale developments and mass production performance requirements. It offers a <8h turnaround time from as-cut wafers to electrical cell testing and sorting, together with flexible R&D activities for each individual process step [8]. This paper showcases the main 2016-2017 progress of the LabFab which has entered its second generation phase with:

- The installation of a new sputtering tool for Transparent Conductive Oxide (TCO) deposition allowing edge masking for junction opening.
- The development of busbar-less process.
- The upgrade of the nominal capacity to 2400 wafers per hour (excluding screen printing till at about 1000 wph).
- The handling of M0 or M2 156x156 mm^2 wafers.
- The optimization of automation for thin wafers, down to 90µm [9].

The line is thus capable of producing and sorting two different industrial products: 4 busbar (BB4) and busbar-less (BB0) cells. Specific cell efficiency optimization for either BB4 or BB0 configuration is presented in the next sections. The BB4 baseline was improved from 20.5% (before edge exclusion PVD, [8]) to 21.5% Q3 2017, with best runs at 22.0%. The BB0 baseline was established at 22.0%, with best runs at 22.6%. Then, record cells and corresponding modules are discussed.

2 EXPERIMENTAL DETAILS

Commercial n-doped c-Si Czochralski (CZ) 156x156 mm^2 wafers are used. After texturing as-cut wafers in KOH-based solutions and O_3-HF cleaning terminated by an HF-last step, a-Si:H stacks and Indium Tin Oxide (ITO) are deposited on both sides of the cell by Plasma Enhanced Chemical Vapor Deposition (PECVD) and sputtering (PVD), respectively. Then, metallization is performed by low temperature silver paste screen-printing on both front and rear sides of the cell in a BB4 or BB0 rear emitter configuration, insuring the bifaciality of the finished cell.

The current-voltage (I-V) cell parameters are measured under standard test conditions (1000W/m^2, AM

1.5g spectrum, 25°C) using a double comb or a double wire array testers for BB4 and BB0 cells, respectively ('black chuck' conditions). Both testers are daily calibrated thanks to sister cells previously measured by FhG ISE CalLab.

The LabFab runs in production mode several days a week, with several thousands of wafers and typically 5 to 15 batches per day. In 2016 and 2017, some hundred thousand wafers per year were fed into the pilot line to answer our industrial partners' requests and to develop the CEA SHJ know-how. The production baseline performance of first semester 2017 can be seen in Fig. 1 where only reference batches are plotted (weekly recurrence of the same cell elaboration recipes). For BB4 cells, the average efficiency was 21.2% with best runs at 21.5% (56-cell tray) and a best cell at 22.0%. For BB0 cells, due to lower sensitivity to metal finger resistivity and less shadowing (no busbar), the efficiency is 21.9% in average with best runs at 22.2% and a best cell at 22.8%. The intra-batch and day-to-day variability of these control runs is mainly attributed to the incoming wafer quality.

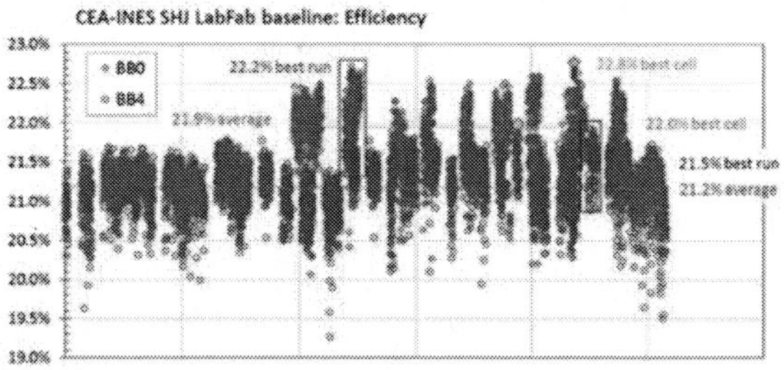

Figure 1: S1 2017 LabFab baseline with new edge exclusion PVD tool. BB4 and BB0 technologies are reported with 23 batches – 6100 cells and 9 batches – 2400 cells, respectively.

3 CELLS RESULTS AND DISCUSSION

3.1 Screen printing of BB0 cells

Despite the BB4 product shows a quite constant efficiency level owing to its long-standing production at CEA-INES (Fig. 1), an important limitation on performances is the metal grid contribution on serial resistance. The front grid process is a balance between low resistance (higher fill factor – FF) and shadowing (lower short-circuit current density – Jsc). Fig. 2 illustrates the finger resistance (R_{line}) requirement for a BB4 grid design optimized with a 1.8mm finger pitch (mainly driven by the ITO lateral conductivity), using a given set of silver paste, screen design (aperture, mesh, emulsion) and printing parameters, and considering a given wafer surface morphology post texturization (size of random pyramids). The dynamic of this test is solely attributed to the print since all data come from a set of sister cells with good and similar pseudo-FF. For our present BB4 cells, the front grid resistance is no longer a first order limiting factor for fingers with R_{line} <0.8 Ohm/cm. Note that reducing further metal resistance is possible, but necessitates higher amount of Ag paste, generally leading to larger lines and so Jsc decay.

On the contrary, due to the specific multi-wire interconnection (SmartWire™), the strong compromise between Jsc and FF for different finger widths almost disappears for BB0. Indeed, considering the 18 wires used to interconnect cells in module, and assuming that the rear grid and the Ag/TCO contact resistance are non-limiting, for the same 1.8mm pitch, the R_{line} specification for BB0 front grid can be released by a factor 18/4, i.e. about <3.6 Ohm/cm.

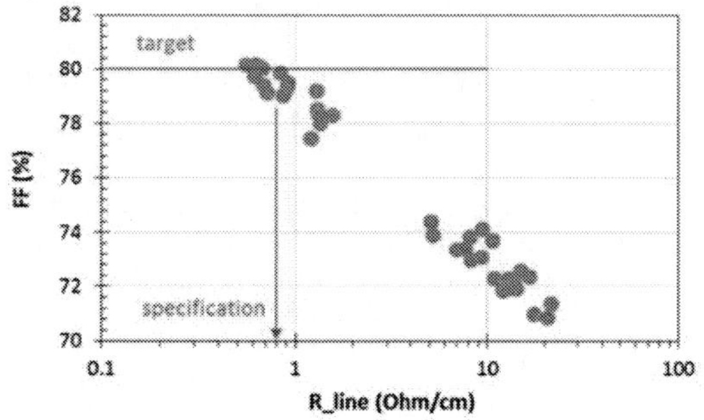

Figure 2: FF against R_{line} for BB4 cells.

Accordingly, the BB0 screen printing optimization can go to very thin lines (and smaller amount of Ag paste). Thus, thorough studies on the interplay between wafer surface morphology post texturization (size of random pyramids, cleaning with or without rounding) and screen-printing properties were conducted. Considering a given set of screen and silver paste type, the printing of narrow fingers is easier using smaller pyramids since the paste spreads out less on smaller pyramid flanks, as illustrated in Fig. 3b (same Ag paste used for BB4 and BB0 cells, double and simple print front side, respectively).

Fig. 3a and in a less extend 3b illustrate the strong interplay between surface morphology and screen printing. Insets in Fig. 3b are typical example of front side fingers printed on '2µm, 4µm and 8µm' texturization (average length of pyramid base calculated from a density measurement is considered here). Open circle points for a few BB4 batches show that the rounding of pyramids performed during our baseline HF/O_3 –based wet clean does not modify the screen printing outputs. The average pyramid size is the key parameter.

Using small pyramids thanks to KOH-additives texturization recipes, shadowing losses are reduced while keeping series resistance unchanged with R_{line} below the specification. As shown in Fig. 3, for the BB4 cell configuration, a good balance between the FF (low line resistivity) and the Jsc (low finger shadowing) has been obtained for a large range of average pyramid sizes. With <5µm pyramids, our baseline data (average) are '1' Ag paste /cell (normalized total mass deposited), 80µm width and 0.65 Ohm/cm for front fingers. For BB0 very small pyramids are mandatory to print lines as thin as possible: so far we have not reached the upper limit for R_{line} specification discussed earlier. Optimizing the screen and printer parameters for BB0 cells, 50µm, 2 Ohm/cm lines and '0.25' total Ag has been achieved in a routine for the startup phase, but these values could be pushed further without losing FF.

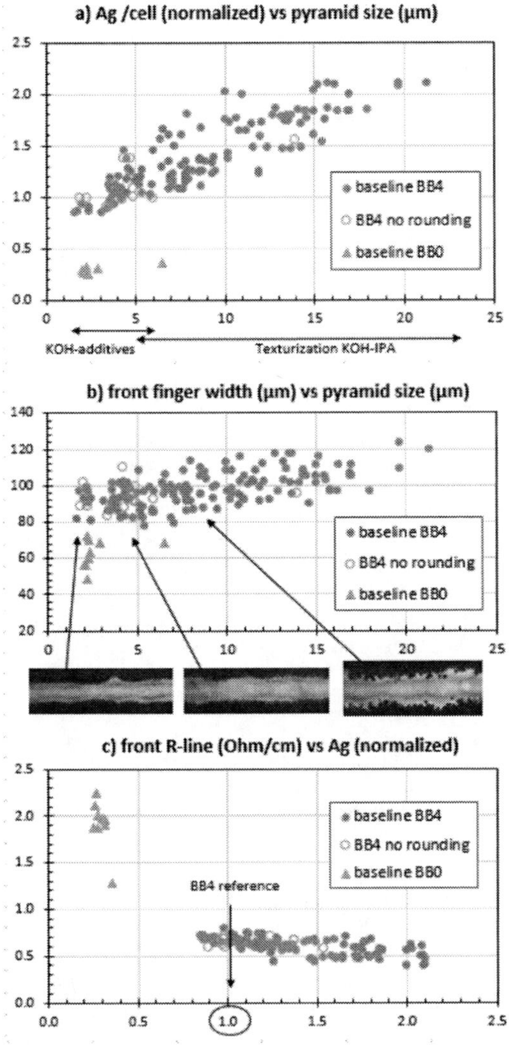

Figure 3: Illustration of wafer surface morphology and silver screen printing interplay – batch to batch average data.

3.2 Substrates quality

The baseline presented in Fig 1 indicates a day-to-day variability, which can be mainly attributed to the different incoming wafer materials. Indeed, it is well-known that wafer quality (bulk effective carrier lifetime and resistivity) strongly impacts the SHJ cells efficiency [10, 11]. In addition to the high quality requirement, the ability to use all the produced material with low cell efficiency variation is critical for a cost-effective SHJ production. Therefore, investigations and understanding of the impact of the main wafer properties (resistivity, interstitial oxygen content, bulk lifetime) is a major topic. In order to study these effects, the cell production from wafers sampled along whole ingots has become a routine on the pilot line. In Fig. 4 two suppliers were benchmarked. The results show that supplier 1 material homogeneity is quite good over the whole ingot length. However cells made from supplier 2 wafers show a big variation on the top (seed part) and bottom (tail). This is a well-known behavior: the top of the ingot contains a high concentration of interstitial oxygen forming recombination-active thermal donors (TDs), whereas the bottom of the ingot is contaminated with metallic impurities. We collected and analyzed results from a large amount of specific runs, shedding light on the relationships between these parameters and the resulting cell efficiency. Based on such qualification results, the CEA aims at developing tools and models to be able to predict already at the as-cut stage the cell efficiency depending on the wafers properties (TD concentration, resistivity for example), in order to estimate the industrial suitability of incoming material [12].

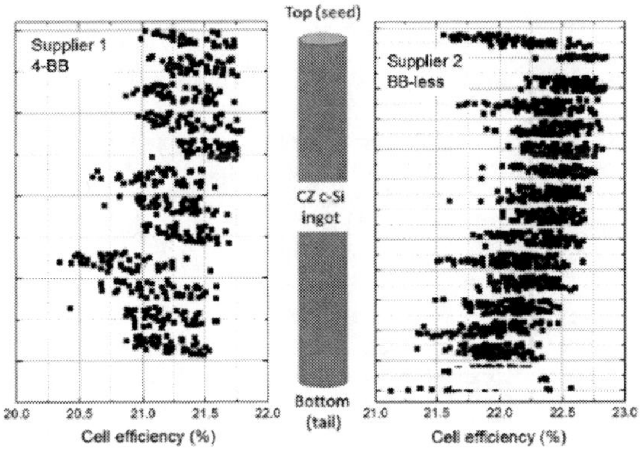

Figure 4: Cells produced along the whole length of two ingots from different suppliers – impact of material quality on cell performance.

3.3 Baseline improvement

Starting from the 20.5% BB4 baseline in 2015, the use of the new PVD with edge exclusion insured a jump to 20.9% as expected from demo tests [8] (Fig. 5, blue points). Then, the average efficiency climbed up to 21.1% after first optimizations (Fig. 5, orange points): substrates = high quality commercial M2 Cz wafers with a bulk lifetime to resistivity ratio >2 ms/Ohm.cm ; wet = repeatable texturing with 2 to 3µm average pyramids using different KOH-additives process ; PVD = new tray design to optimize edge exclusion on M2 wafers ; print = new low resistivity Ag paste offering an improved height to width ratio and allowing a single print on the front side with dedicated screens and printer parameters. Then, in Q3 2017, new aSi:H buffer layers were developed to reach the current 21.5% baseline (Fig. 5, grey points). BB0 design also benefits from these optimizations as shown in table I.

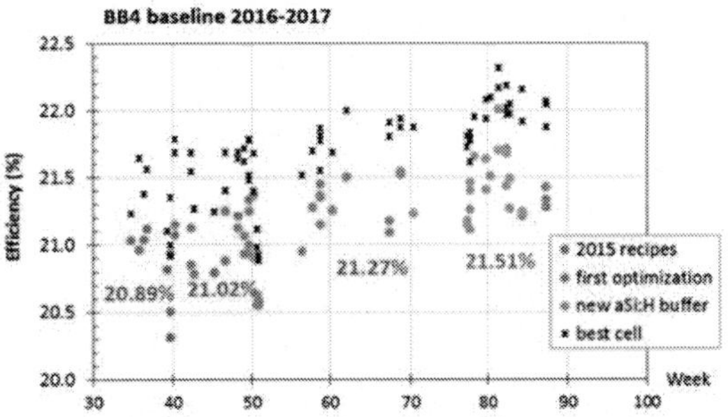

Figure 5: 2016 & 2017 LabFab BB4 baseline improvement with new edge exclusion PVD tool.

Table I: Summary of actual LabFab cell efficiency.

		Efficiency (%)	Voc (mV)	Jsc (mA/cm2)	FF (%)
BB4	**Baseline**	**21.51**	**734.4**	**36.91**	**79.33**
	Best batch	22.01	738.9	36.78	80.98
	Best cell 160μm	22.32	739.7	37.08	81.37
	Best cell 89μm	22.13	747.1	36.45	81.25
BB0	**Baseline**	**21.97**	**736.3**	**37.55**	**79.46**
	Best batch	22.56	738.5	38.00	80.38
	Best cell 160μm	23.27	744.6	38.17	81.86

4 MODULES

Modules made with ribbons connected thanks to conductive gluing on a selection of BB4 cells show typical cell-to-module power ratio (CTM) of 101+/-1% in the GBS configuration and 98+/-1% with a glass-glass bifacial design (GG). A 324Wp 60-cell GBS record was acheived and a 72-cell GG record was characterized at 381Wp (flashed with dark backside) and measured at 444Wp under standardized GE norm (1000W/m^2 front side + 200W/m^2 rear side illumination).

Our first modules with BB0 cells and a multi-wire interconnection show similar powers as acheived with BB4 cells (318 Wp 60-cell GBS and 386Wp 72-cell GBS, learning stage) with a high reliability and consistent field data. This allows a short term perspective to equal or slightly increase the power outputs for BB0 systems compared to the BB4 reference, including a reduction of the silver consomption by a factor of 3 to 4.

Since SHJ cell production lines operate at low temperature and with symmetric deposition and metallization steps, thin wafers can be used without risks of excessive bowing. In addition, the use of thin wafers constitutes a route to enhance the silicon heterojunction technology competitiveness by reducing the material costs [13]. Last year we produced the first 60-cell glass-GBS module using 93μm BB4 M0 cells. 282Wp was reached with a 96% CTM and without breakage. Then, a second 60-cell trial with our best 93μm M2 cells and module optimization was manufactured to reach 313Wp with a CTM of 99.1%. This corresponds to 0.98Wp /g Si in the finished module.

5 CONCLUSIONS AND PERSPECTIVES

With the objective to make the SHJ technology even more competitive, CEA optimizes the process flow for different busbar and busbar-less designs, considering the whole PV chain, taking into account final module and system energy production data ($/kWh considered). At the cell level, this includes the wafer selection, surface morphology adapted to screen-printing and iterative process optimization for a short term target of 23% for the best busbar design cells. Also, a new PECVD cluster is being started-up, making aSi:H deposition available both in parallel plate reactors and "box-in-the-box" configuration. At the module level, the cell bifaciality will be optimized to enhance the GG system energy output. Moreover, the moduling continuous optimization, mainly for BB0 cells, should allows the CEA teams to approach a 400Wp 72-cell module very soon. Finally the standard activity on <100μm cells will continue in order to overtake the symbolic 1Wp /1 g Si target at module level.

ACKNOWLEDGEMENTS

Part of the work on thin substrates has received funding from the European Union's Seventh Program for research, technological development and demonstration under grant agreement No 609788 in the frame of the CHEETAH project, [14].

The authors kindly thank CEA and Meyer Burger teams for the work done throughout the year.

REFERENCES

[1] M. Taguchi et al, "24.7% Record Efficiency HIT Solar Cell on Thin Silicon Wafer", *IEEE J. of Photovoltaics*, Vol. 4-1 (2014), pp. 96-99.

[2] S. De Wolf et al, "High-efficiency Silicon Heterojunction Solar Cells: A Review", *Green*, Vol. 2 (2012), pp. 7-24.

[3] K. Yoshikawa et al., "Silicon heterojunction solar cell with interdigitated back contacts for a photoconversion efficiency over 26%", *Nature Energy*, Vol. 2-5 (2017), pp. 17032.

[4] ITRPV, eighth edition march 2017, www.itrpv.net

[5] J. Zhao et al., "Silicon Heterojunction Solar Cells in Meyer Burger's Demo Line: Results of Pilot Production on Mass Production Tools", 2BO.9.1 paper in the proceedings of the 32nd European Photovoltaic Solar Energy Conference, Munich, Germany (2016).

[6] D. Andronikov et al., "A Solar Module Prototype Assembled from Silicon Heterojunction Solar Cells Manufactured in Gen5 KAI PECVD Reactors", 2AV.2.10 poster in the proceedings of the 32nd European Photovoltaic Solar Energy Conference, Munich, Germany (2016).

[7] W. Favre et al., "from Advanced Thin-Films Modules to High Efficiency Silicon Heterojunction Technology at 3SUN", 2DO.1.2 paper to be published in the Proc. of the 33rd European Photovoltaic Solar Energy Conference, Amsterdam, The Netherlands (2017).

[8] A. Danel et al., "Recent Progress on the CEA-INES Heterojunction Solar Cell Pilot Line" in: 31st European Photovoltaic Solar Energy Conference, Hamburg, Germany (2015), pp. 279-283.

[9] S. Harrison et al., "How to Deal with Thin Wafers in a Heterojunction Solar Cell Industrial Pilot Line: First Analysis of the Integration of Cells Down to 70μm in Production Mode" in: 32nd European Photovoltaic Solar Energy Conference, Munich, Germany (2016), pp. 358-362.

[10] B. Martel et al., "Influence of the wafer position within a n-type monocrystalline Si ingot on the conversion efficiency: a comparative study between different cells architectures", nPV Workshop, Konstanz, Germany (2015).

[11] D. L. Batzner et al., "Silicon Specification for High Efficiency HJT" in: 31st European Photovoltaic Solar Energy Conference, Hamburg, Germany (2015), pp. 1042-1045.

[12] J. Veirman et al., "Early Efficiency Prediction of Silicon Heterojunction Cells Processed on Thermal Donors-Rich Czochralski Wafers" 2CO.10.2 paper to be published in the Proc. of the 33rd European Photovoltaic Solar Energy Conference, Amsterdam, The Netherlands (2017).

[13] A. Louwen et al., "A cost roadmap for silicon heterojunction solar cells", Solar Energy Materials & Solar cells, 147 (2016), pp. 295-314.

[14] J.M. Kroon, "Highlights from the FP7 project on photovoltaics CHEETAH: more power with less material", 7EO.3.5 paper to be published in the Proc. of the 33rd European Photovoltaic Solar Energy Conference, Amsterdam, The Netherlands (2017).

INDUSTRIAL, SCREEN-PRINTED DOUBLE-SIDE CONTACTED POLO CELLS

R. Peibst[1,2], Y. Larionova[1], S. Reiter[1], N. Orlowski[1], S. Schäfer[1], M. Turcu[1], B. Min[1], R. Brendel[1,3], D. Tetzlaff[2], J. Krügener[2], T. Wietler[2], U. Höhne[4], J.-D. Kähler[4], H. Mehlich[5], S. Frigge[5]

[1]Institute for Solar Energy Research Hamelin (ISFH), Am Ohrberg 1, 31860 Emmerthal, Germany
[2]Institute for Electronic Materials and Devices, Leibniz Universität Hannover, Schneiderberg 32, 30167 Hannover, Germany
[3]Institute for Solid State Physics, Leibniz Universität Hannover, Appelstrasse 2, 30167 Hannover, Germany
[4]Centrotherm Photovoltaics AG, Vahrenwalder Str. 269A, D-30179 Hannover, Germany
[5]Meyer Burger (Germany) AG, An der Baumschule 6-8, D-09337 Hohenstein-Ernstthal, Germany

Abstract: We demonstrate an industrial double-side-contacted, screen-printed large area cell with POLO junctions on both sides and an in-house-measured energy conversion efficiency of 22.3 % (A = 244.15 cm^2, V_{oc} =714 mV, FF= 81.1 %, J_{sc} =38.5 mA/cm^2). Most remarkably, this cell shows a very low series resistance < 0.05 Ωcm^2, which supports previous observations of low specific junction resistance for the POLO scheme. The current limitations of the cell are (*i*) the low short-circuit current due to parasitic absorption in the rather thick poly-Si (30 nm), as well as in the ITO, (*ii*) deterioration of the recombination behavior upon TCO sputtering, and (*iii*) shunts near the edge due to non-adapted TCO edge-exclusion.

Keywords: Carrier selective contacts, Polycrystalline Silicon, Passivation, Transparent Conductive Oxide, Silicon Solar Cell

1 INTRODUCTION

Long time after the first reports in the bipolar transistor [1-3] and photovoltaic [4-7] community, poly-silicon on oxide (POLO) –junctions and related approaches with additional components (O, C, ...) in the doped Si layers (SIPOS-like, Si:C, TopCon, ...) have been reconsidered [8-11] and are since then attracting significant research interest. Despite recent progress in the understanding of the working principle [12-14] and in the optimization of the junction quality for both polarities (n^+ POLO and p^+ POLO) [15-16], it seems still not clear how to integrate these junctions best in solar cells, and what is the most viable cell concept for this purpose. Different approaches have been proposed or demonstrated, either based on the integration of "POLO-like" junctions for one polarity (mostly the electron-collecting layer [9,17-19], less often the hole-collecting layer [10,20]), or based on the integration of both polarities. The latter concept has the advantage of strongest minimization of recombination losses. A possible cell structure is an interdigitated back-contacted (IBC) cells with POLO junctions for both polarities on the rear [21-24], or a double-side contacted cell with n^+ POLO on one and p^+ POLO on the other side. The latter cell structure was already suggested by Yablonovitch *et al.* in 1985 [6], and demonstrated on lab-type scale by different groups in the last years [25-26].

ISFH is evaluating both concepts – IBC cells and double-side contacted cells with POLO junctions for both polarities. While we already demonstrated the high efficiency potential of our POLO junctions with a 25 %-efficient IBC cell [23], the transfer to industrialization appears more straightforward for double-side contacted cells since no patterning is required and since the development of the back-end processes is largely supported by the experience made when industrializing *a*-Si:H-based hetero-junction cells.

In this work, we focus on industrial processing of double-side contacted cells with POLO junctions for both polarities (full area 6"Cz Si wafers, no lab-type processes, lean process flow) [16]. We report on our first cell results.

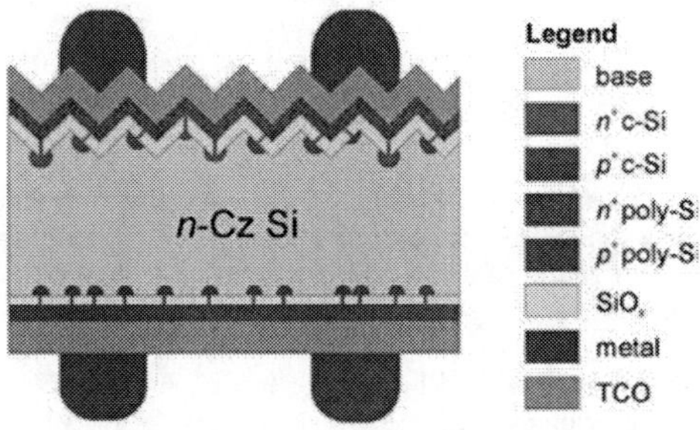

Figure 1: Schematic drawing of final cell double-side contacted POLO cell with TCO (in this case ITO) and low-temperature Ag screen-print metallization on both sides.

2. EXPERIMENTAL DETAILS ON CELL PROCESSING

As substrate material, we use 156.75 mm × 156.75 mm (M2) *n*-type Cz Si wafers from NorSun (4 Ωcm, (100), as-cut thickness 170 µm). The decent passivation quality of p^+ POLO on textured surfaces [27,28], which seems to be a fundamental issue implied by the decent passivation of SiO_x on Si <111> surfaces [28], does require the fabrication of single-side textured wafers to locate the p^+ POLO junction on the planar rear side. For this purpose, a SiN_x protection layer is deposited on the rear-side by Plasma-Enhanced Chemical Vapour Deposition (PE-CVD) after saw damage removal in KOH and subsequent cleaning. Alkaline texturization of the front-side is performed, and the SiN_x protection layer is removed from the rear side by HF. One should note that the preparation of single-side textured wafers could be simplified (the SiN_x deposition/removal could be omitted) by performing a combination of double-side texturization and subsequent wet-chemical polishing of the rear-side. The interfacial oxide is grown wet-chemically in a solution of de-ionized water and ozone. The thickness of the interfacial oxide, as determined by ellipsometry on shiny-etched reference wafers, is 1.6 nm. Intrinsic Si is deposited on both sides of the wafers by Low-Pressure Chemical Vapour Deposition (LP-CVD) in a E2000 horizontal furnace from Centrotherm. The Si layer thickness

	V_{oc} ($V_{oc,impl}$) [mV]	J_{sc} [mA/cm^2]	FF ($pFF_{,impl.}$) [%]	η [%]	J_{01} [fA/cm^2]	J_{02} [nA/cm^2]	R_{shunt} [Ωcm^2]	R_s [Ωcm^2]
QSSPC after frontend / 2-D-fit	(721)	-	(85.6)		25 (surf.+bulk)	0.3		
IV-meas. final cell	714	38.5	81.1	**22.3**				
2-D-Fit *IV* data	714	38.5	81.3	22.3	26	**7.6**	**1695**	< 0.05

Table 1: Summary of measured IV and implied IV quantities. The four columns on the right show the parameters of the 2-diodes model, extracted by fitting experimental curves.

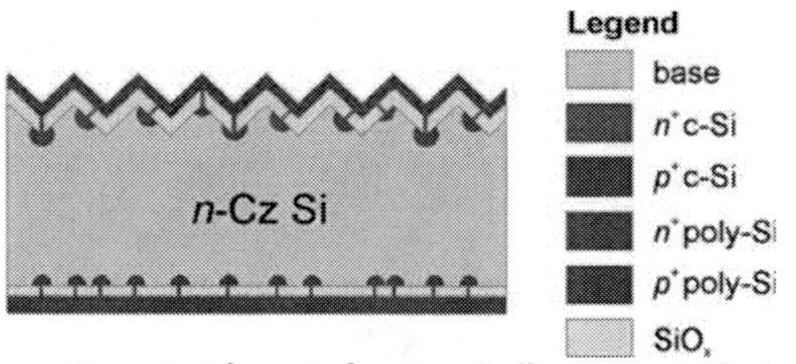

Figure 2: Schematic drawing of cell precursor with n⁺ POLO on textured front-side and p⁺ POLO on planar rear side.

in this experiment is 30 nm. Doping is performed ex-situ by ion implantation of Phosphorus (front) and Boron difluoride (rear). Formation of the POLO junctions for both polarities is performed in one high-temperature step at 870 °C. As a side remark, we did verify the existence of pinholes even for this rather low temperature budget by using our selective wet-chemical etching method [9].

While these front-end processes are performed at ISFH, MBE and Centrotherm, back-end processing is performed by Meyer-Burger Germany using their demo line for high-throughput fabrication of *a*-Si:H/*c*-Si-heterojunction cells . Indium-Tin-Oxide (ITO) is sputtered on both sides in a HELiA PVD tool. Busbarless low-temperature Ag screen-print is applied to both sides of the cell. Figure 1 shows a schematic cross section of the final double-side contacted POLO cell.

The full-area *IV*-measurement is performed in-house at MBG using a grid-touch measurement system from Pasan. The calibration of this system was realized previously by an extensive cooperation between MBG and F-ISE CalLab.

3. CELL RESULTS

Figure 2 depicts a schematic cross section of the cell precursors after front-end processing (i.e., after POLO junction formation). At this stage, we monitor the implied recombination characteristic of the precursors by photo-conductance decay measurements. The implied open circuit voltage $V_{oc,impl.}$ of the cell precursors is determined to be 721 mV, the implied pseudo fill factor is 85.6 %. Applying the Kane and Swanson method, the total $J_{01,suf.}$ value of the doped surfaces (front and rear) is determined to be 20 fA/cm^2. One should note that the passivation quality of the POLO junctions can be further improved ($J_{01,surf.}$ of the surfaces can be halved) by supplying atomic hydrogen, e.g. by the deposition of a H-rich SiN_x layer. For process simplicity reasons, this step is not performed in this experiment. We analyze the suns-$V_{oc,impl.}$ curve with the 2-diodes model. A good fit is obtained for J_{01} = 25 fA/cm^2 and J_{02} = 0.3 nA/cm^2. Thus the contribution of the bulk to J_{01} is around 5 fA/cm^2. Table 1 summarizes all of the measured quantities, as well as the corresponding parameters from the analysis by the 2-diodes model.

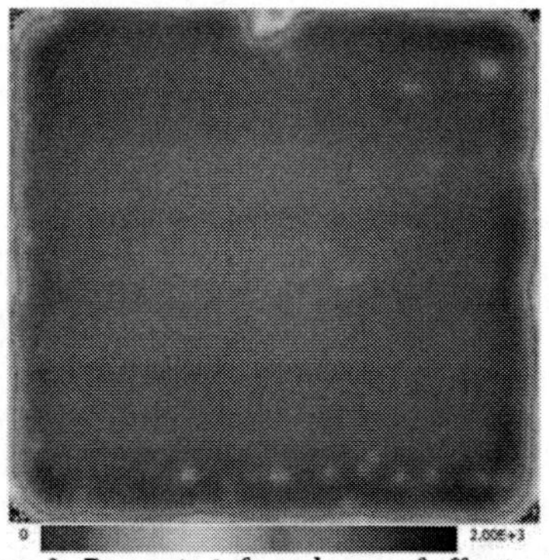

Figure 3: Dynamic infrared map of effective lifetime, measured on a 6" cell precursor at an illumination level of 0.3 suns.

The homogeneity of the passivation quality at precursor level is monitored by Dynamic Infrared Lifetime Mapping (Dy-ILM). Figure 3 shows the resulting map of the effective lifetime at an illumination level of 0.3 suns. Expect for a region attributable to tweezer imprints, the lifetime is homogeneous across the wafer.

The *I-V* curve of our best cell is shown in Fig. 4. The parameters of the *I-V* curve are listed in row 2 of Tab. 1. The energy conversion efficiency η is 22.3 %. The open circuit voltage V_{oc} is 714 mV, i.e., 7 mV lower than $V_{oc,impl.}$ measured on precursor level. This difference can be attributed to sputter damage induced during TCO deposition. The short-circuit current density is 38.5 mA/cm^2. This decent value is a consequence of parasitic absorption in the rather thick (30 nm) poly-Si layers and in the ITO. The fill factor of 81.1 % is limited by a degraded pseudo fill factor (compared to the precursor level) rather than by series resistance effects. Most remarkably, the comparison with the suns-$V_{oc,impl.}$-curve reveals an identical slope of both curves in the vicinity of V_{oc}, where series resistance effects should be visible. This implies that there is no significant series resistance resolvable. This might be a consequence of the short finger length (the Pasan tester measures with 30 wires), as well as a consequence of the low specific junction resistivity of our POLO junctions [22].

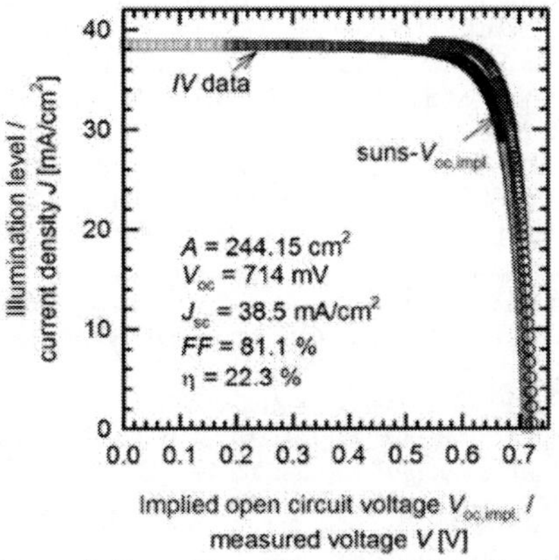

Figure 4: Comparison of measured IV and implied IV-curve (the latter was determined by QSSPC)

The fit of the *I-V* curve by the 2 diodes model actually yields the best agreement for an apparent R_s value of 0 Ωcm². However, one should take into account that the 2-diodes model does not describe Auger recombination in high -evel injection (ideality factor 2/3), as the lowest modelled ideality factor is 1. Since Auger recombination in the bulk (excess carrier concentration comparable to doping concentration) is significant in our cells for voltages close to V_{oc}, the steep slope of the experimental *IV*-curve in this region can hardly be reproduced by the 2-diodes model even for an apparent R_s value of 0 Ωcm². For this reason, the 2-diodes model possibly underestimates R_s. By further analysis (1-diode model analysis with n fixed to 2/3), we determined 0.05 Ωcm² as an upper limit for R_s in our cells.

The 2-diodes model further quantifies the present limitations for the cell efficiency (see Tab. 1): The degradation of the pseudo fill factor is mainly caused by a deterioration of the recombination behavior (J_{02} increased from 0.3 to 7.6 nA/cm²), as well as by a shunting issue (R_{shunt} 1675 Ωcm²). The J_{01} value on cell level (26 fA/cm²) is almost identical to that on precursor level.

4. ANALYISIS OF THE RESULTS

In order to understand the current limitations of our double-side contacted POLO cell – shunting, deterioration of recombination behavior upon TCO sputtering, parasitic absorption in the poly-Si/TCO stack on the front side – we experimentally investigate all these building blocks separately.

First, we perform dark lock-in thermography measurements on our cell. Most prominently, the heat generation can be resolved near the wafer edge. It is known from previous works [15,21] that touching n^+p^+ doped regions in the poly-Si imply high recombination due to the low lifetime in the poly-Si. Since we do not include any edge exclusion in the doping process, it's quite likely that there is a direct contact between the n^+ doped poly-Si on the front and the p^+ on the rear at the wafer edge. However, illumination-based lock-in thermography measurements at precursor level reveal that these edge shunts are not present prior to TCO deposition. Possibly, the recombination at the wafer edge is limited by the transport in the poly-Si at this stage, while the TCO supplies additional conductivity and thus resolves the transport limitation. Of course, there is an edge exclusion included in the TCO deposition for the HELiA PVD tool. However, the width of the non-covered region near the wafer edge was optimized for *a*-Si:H layers. The poly-Si layers are more conductive, and therefore require an enlargement of the width of the TCO-free region.

Second, we investigate the pseudo fill factor deterioration upon TCO deposition. For this purpose, we analyze suns-$V_{oc,impl.}$ curves prior and after TCO deposition, and determined the corresponding implied pseudo fill facto . ITO sputtering (in a tool located at ISFH) on 30 nm poly-Si layers yields a $pFF_{impl.}$ loss up to 3 $\%_{abs}$. For thicker poly-Si layers, this degradation is less pronounced. We therefore conclude that the Si/SiO$_x$ interface between the *c*-Si substrate and the interfacial oxide is quite sensitive to sputter damage. In contrast to *a*-Si:H/*c*-Si heterojunction structures, a post deposition anneal at ~200 °C is not sufficient to repair the damage. It might be possible to further optimize the sputter process in order to reduce the damage. An alternative is to develop other gentle deposition methods such as Atomic Layer Deposition (ALD).

Third, we also considered the issue of low J_{sc} due to parasitic absorption in the poly-Si and the TCO. A further improvement of ~0.6 mA/cm² in J_{sc} is possible when the poly-Si thickness is reduced from 30 nm to 10 nm [30]. Also, the incorporation of oxygen in the poly-Si layers (SIPOS-semi insulating polycrystalline silicon) [11,31] could improve its transparency. Fine-line Ag-print would also reduce optical shading. Combining these measures with a more transparent TCO such as ZnO:Al [32], we think that a J_{sc} improvement of 2 mA/cm² is achievable.

6 CONCLUSIONS

We demonstrate an industrial double-side contacted, screen-printed large area cell with POLO junctions on both sides and an energy conversion efficiency of 22.3 %. One should note that no adaptation of the back-end processes on the requirements of our POLO cells was performed to achieve this result, and only two such runs were been performed until today.

Most remarkably, this cell shows a very low series resistance < 0.05 Ωcm², which supports previous observations of low specific junction resistance for the POLO scheme. The present limitations of the cell are (*i*) the low short-circuit current due to parasitic absorption in the rather thick poly-Si (30 nm), as well as in the ITO, (*ii*) deterioration of the recombination behavior upon TCO sputtering, and (*iii*) shunts near the edge due to non-adapted TCO edge-exclusion. All of these limitations can be addressed, i.e., none of them seems to be a fundamental road blocker.

We would like to remark that we consider our concept not to be in competition with *a*-Si:H/*c*-Si heterojunction, since it uses, expect the quartz tube annealing furnace, in principle similar tools. Rather, we think that this type of POLO cell can be a variant or an upgrade of the heterojunction concept.

7 ACKNOWLEDGEMENTS

We thank B. Gehring, R. Winter, G. Glowatzki, B. Koch and R. Zieseniß for their help with sample processing, and B. Lim, R. Reineke-Koch and M. Rienäcker for valuable discussions. Furthermore, we thank the German Ministry for Economic Affairs and Energy for funding under grant and 0325702 (POLO).

6 REFERENCES

[1] J. Graul, A. Glasl, H. Murrmann, in Proc. of the Electron Devices Meeting, 1975 International, 1975, p. 450-454.
[2] J. Graul, A. Glasl, H. Murrmann, IEEE Journal of Solid-State Circuits **11**, 491-495 (1976).
[3] I.R. Post, P. Ashburn, G.R. Wolstenholme, IEEE Trans. Electron Devices **39**, 1717-1731 (1992).
[4] J. Fossum, M. Shibib, in Proc. of the Electron Devices Meeting, 1980 International, 1980, p. 280-283.
[5] F. Lindholm, A. Neugroschel, M. Arienzo, P. Iles, IEEE Electron Device Lett. **6**, 363-365 (1985).
[6] E. Yablonovitch, T. Gmitter, R.M. Swanson, Y.H. Kwark, App. Phys. Lett. **47**, 1211-1213 (1985).
[7] N.G. Tarr, IEEE Electron Device Lett. **6**, 655-658 (1985).
[8] R. Brendel, T. Dullweber, R. Gogolin, H. Hannebauer, N. Harder, J. Hensen, S. Kajari-Schroeder, R. Peibst, J. Petermann, U. Römer, in Proc. of the 28th European Photovoltaic Solar Energy Conference and Exhibition, 2013, p. 676-691.
[9] F. Feldmann, M. Bivour, C. Reichel, M. Hermle, S.W. Glunz, in Proc. of the 28th European PV solar energy conference and exhibition, 2013.
[10] G. Nogay, J. Stuckelberger, P. Wyss, E. Rucavado, C. Allebé, T. Koida, M. Morales-Masis, M. Despeisse, F.-J. Haug, P. Löper, Sol. Energ. Mat. Sol. (2017).
[11] J. Stuckelberger, G. Nogay, P. Wyss, Q. Jeangros, C. Allebé, F. Debrot, X. Niquille, M. Ledinsky, A. Fejfar, M. Despeisse, F.J. Haug, P. Löper, C. Ballif, Sol. Energ. Mat. Sol. **158**, 2-10 (2016).
[12] R. Peibst, U. Römer, Y. Larionova, M. Rienäcker, A. Merkle, N. Folchert, S. Reiter, M. Turcu, B. Min, J. Krügener, D. Tetzlaff, E. Bugiel, T. Wietler, R. Brendel, Sol. Energ. Mat. Sol. **158**, 60-67 (2016).
[13] T.F. Wietler, D. Tetzlaff, J. Krügener, M. Rienäcker, F. Haase, Y. Larionova, R. Brendel, R. Peibst, App. Phys. Lett. **110**, 253902 (2017).
[14] R. Brendel and R. Peibst, IEEE Journal of Photovoltaics **6**, 1413-1420 (2016).
[15] U. Römer, R. Peibst, T. Ohrdes, B. Lim, J. Krügener, T. Wietler, R. Brendel, IEEE J. Photovolt. **5**, 507-514 (2015).
[16] R. Peibst, Y. Larionova, S. Reiter, M. Turcu, R. Brendel, D. Tetzlaff, J. Krügener, T. Wietler, U. Höhne, J. Kähler, in Proc. of the 32nd European Photovoltaic Solar Energy Conference and Exhibition, 2016, p. 323-327.
[17] A. Richter, J. Benick, F. Feldmann, A. Fell, M. Hermle, S.W. Glunz, Sol. Energ. Mat. Sol. (2017).
[18] Y. Tao, V. Upadhyaya, C.W. Chen, A. Payne, E.L. Chang, A. Upadhyaya, A. Rohatgi, Prog. Photovoltaics **24**, 830-835 (2016).
[19] M. Stodolny, M. Lenes, Y. Wu, G. Janssen, I. Romijn, J. Luchies, L. Geerligs, Sol. Energ. Mat. Sol. **158**, 24-28 (2016).
[20] F. Feldmann, M. Bivour, C. Reichel, M. Hermle, S.W. Glunz, Sol. Energ. Mat. Sol. **120**, 270-274 (2014).
[21] M. Rienäcker, A. Merkle, U. Römer, H. Kohlenberg, J. Krügener, R. Brendel, R. Peibst, Energy Procedia **92**, 412-418 (2016).
[22] M. Rienäcker, M. Bossmeyer, A. Merkle, U. Römer, F. Haase, J. Krügener, R. Brendel, R. Peibst, IEEE J. Photovolt. **7**, 11-18 (2017).
[23] F. Haase, F. Kiefer, S. Schäfer, C. Kruse, J. Krügener, R. Brendel, R. Peibst, Japanese Journal of Applied Physics **56**, 08MB15 (2017).
[24] G. Yang, A. Ingenito, N. van Hameren, O. Isabella, M. Zeman, App. Phys. Lett. **108**, 033903 (2016).
[25] U. Römer, R. Peibst, T. Ohrdes, B. Lim, J. Krügener, E. Bugiel, T. Wietler, R. Brendel, Sol. Energ. Mat. Sol. **131**, 85-91 (2014).
[26] F. Feldmann, C. Reichel, R. Müller, M. Hermle, Sol. Energ. Mat. Sol. **159**, 265-271 (2017).
[27] Y. Larionova, R. Peibst, M. Turcu, S. Reiter, R. Brendel, D. Tetzlaff, J. Krügener, T. Wietler, U. Höhne, J.-D. Kähler, in Proc. of the 32nd European Photovoltaic Solar Energy Conference and Exhibition, Munich, Germany, 2016, p. 452 - 455.
[28] Y. Larionova, M. Turcu, S. Reiter, R. Brendel, D. Tetzlaff, J. Krügener, T. Wietler, U. Höhne, J.D. Kähler, R. Peibst, physica status solidi (a) (2017).
[29] D. Tetzlaff, J. Krügener, Y. Larionova, S. Reiter, M. Turcu, F. Haase, R. Brendel, R. Peibst, U. Höhne, J.D. Kähler, T.F. Wietler, Sol. Energ. Mat. Sol. **173**, 106-110 (2017).
[30] S. Reiter, N. Koper, R. Reineke-Koch, Y. Larionova, M. Turcu, J. Krügener, D. Tetzlaff, T. Wietler, U. Höhne, J.-D. Kähler, Energy Procedia **92**, 199-204 (2016).
[31] Y. Kwark, R. Swanson, J. Electrochem. Soc. **129**, 197-201 (1982).
[32] T.F. Wietler, S. Reiter, Y. Larionova, R..-R. Koch, F. Heinemeyer, R. Brendel, J. Krügener, D. Tetzlaff, R. Peibst, submitted to IEEE J. Photovoltaics (2017)

INTERDIGITATED BACK CONTACT SILICON SOLAR CELLS FEATURING ION-IMPLANTED POLY-SI/SIOX PASSIVATING CONTACTS

Christian Reichel[1], Ralph Müller[1,2], Frank Feldmann[1,2], Armin Richter[1], Martin Hermle[1], Stefan W. Glunz[1,2]
[1]Fraunhofer Institute for Solar Energy Systems (ISE), Heidenhofstrasse 2, 79110 Freiburg, Germany
[2]Department of Sustainable Systems Engineering (INATECH), Albert Ludwigs University Freiburg, Georges-Köhler-Allee 103, 79110 Freiburg, Germany

ABSTRACT: Interdigitated back contact (IBC) silicon solar cells featuring passivating contacts based on tunneling oxides (SiO_x) and n- and p-type polycrystalline silicon (poly-Si) thin films were fabricated with different rear side configurations to show the impact of the recombination in the transition region between p-type and n-type poly-Si on the performance of the solar cells. On the one hand a) p^+ and n^+ poly-Si regions were in direct contact to each other ("no gap") using a local overcompensation (counterdoping) as a self-aligning process and on the other hand b) undoped (intrinsic) poly-Si remained between the p^+ and n^+ poly-Si regions ("gap"). These configurations were investigated in terms of recombination characteristics by illumination- and injection-dependent quasi-steady state photoluminescence (suns-PL) and were compared to solar cells that featured c) etched trenches separating the p^+ and n^+ poly-Si regions ("trench"). While the latter configuration allowed for open-circuit voltages (V_{oc}) of 720 mV, fill factor (FF) of 79.6%, and short-circuit current (J_{sc}) of 41.3 mA/cm^2, resulting in conversion efficiencies (η) of 23.7%, solar cells without a trench showed a poor performance due to non-ideal recombination in the space charge regions with high local ideality factors as well as recombination in shunted regions. Therefore, V_{oc} of only 593 mV and FF of only 61.3% were achieved for the "no gap" configuration whereas for the "gap" configuration higher V_{oc} of 680 mV but also low FF of 65.6% were obtained.

Keywords: amorphous silicon, annealing, back contact, deposition, heterojunction, lifetime, passivation, polycrystalline, recombination, solar cells

1 INTRODUCTION

Interdigitated back contact (IBC) silicon solar cells featuring passivating contacts have pathed the way to efficiencies of more than 25% as presented by SunPower [1], Sharp [2] and Panasonic [3]. Kaneka pushed the limits even further and achieved a world record of 26.7% utilizing intrinsic and doped amorphous silicon (a-Si) thin films as passivating contacts with excellent passivation and excellent transport properties [4,5]. These excellent properties are currently challenged by another passivating contact based on semicrystalline or polycrystalline silicon (poly-Si) and silicon oxide (SiO_x) thin films, named tunnel oxide passivated contacts (TOPCon) or polycrystalline silicon on oxide (POLO) contacts which gained a lot of interest in the past years and re-emerged as a competitive and alternative technology to a-Si thin films [6,7]. The application of poly-Si thin films resulted in efficiencies of up to 25.7%, realized on n-type c-Si with a p-type boron-diffused emitter on the front and a n-type poly-Si thin film as a passivating contact. In fact, this represents the highest efficiency reported for both-sides contacted c-Si solar cells, allowing open-circuit-voltages (V_{oc}) of 725 mV and fill factors (FF) of 83.3%[6].

The integration of these poly-Si thin films into single-side contacted c-Si solar cells however poses a challenge due to high recombination in the transition from p- to n-type poly-Si [8–12]. This was first shown by SunPower in 2009 [8] and later encountered by several other groups [9–12]. In order to show the effects of the recombination on the performance of the solar cells in more detail, the transition region between p-type and n-type poly-Si was created in different configurations. On the one hand a) p^+ and n^+ poly-Si regions were in direct contact to each other ("no gap") using a local overcompensation (counterdoping) as a self-aligning process [9,10] and on the other hand b) undoped (intrinsic) poly-Si remained between the p^+ and n^+ poly-Si regions ("gap") [13]. These configurations were compared to solar cells that featured c) etched trenches separating the p^+ and n^+ poly-Si regions ("trench") [7,12,14], see Figure 1. With illumination- and injection-dependent quasi-steady state photoluminescence (suns-PL) measurements the charge carrier lifetime and open-circuit voltage characteristic of the solar cells were determined. In this respect, the open-circuit voltage characteristics of the solar cells were modeled to reveal the dominating recombination mechanism that is influencing the performance of the solar cells.

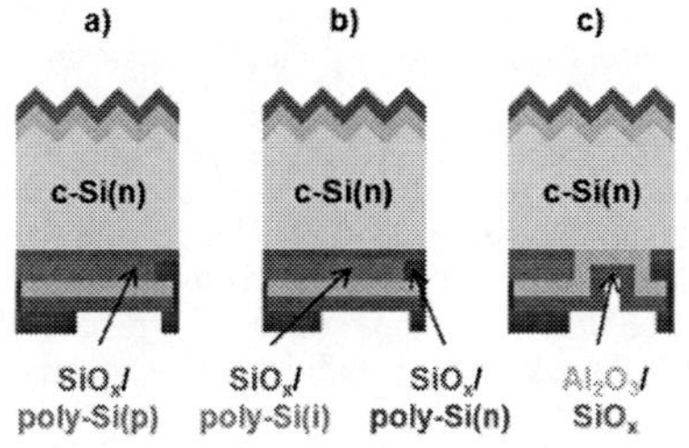

Figure 1: Schematic illustration of the solar cells with different configurations: a) p^+ and n^+ poly-Si regions in direct contact to each other ("no gap"), b) undoped (intrinsic) poly-Si remains between p^+ and n^+ poly-Si regions ("gap") and c) trenches separate p^+ and n^+ poly-Si regions.

2 EXPERIMENTAL

IBC solar cells with an area of 2 cm × 2 cm were fabricated on shiny-etched <100>-oriented phosphorus (P)-doped n-type float-zone silicon wafers with a specific resistivity of 1 Ωcm and a thickness of 200 µm. The front side was textured with random pyramids. Then an ultra-thin tunneling silicon oxide (SiO_x), about 1.5 nm, was wet-chemically grown [15]. Subsequently, about 35 nm of intrinsic amorphous silicon (a-Si) was deposited on the wafers by low-pressure chemical vapor deposition (LPCVD) [16]. Boron monofluoride (BF) ions were then introduced into the intrinsic a-Si by ion implantation at an ion dose of 5×10^{14} cm^{-2} either on the entire rear side for the "no gap" configuration or locally via a SiO_x mask for the "gap" configuration. P ions were then implanted via another SiO_x mask either into the intrinsic a-Si at an ion dose of 1.5×10^{15} cm^{-2} for the "gap" configuration or into the boron-doped a-Si at an ion dose of 2.3×10^{15} cm^{-2} for the "no gap" configuration as a counterdoping process [17], applying an ion energy of 2 keV [18]. The pitch distance of the IBC solar cells was 500 µm, whereas for the "no gap" ("gap") configuration the emitter was 400 µm (360 µm) and the BSF was 100 µm wide. In the "gap" configuration, a 20 µm wide undoped (intrinsic) a-Si remained between the doped a-Si. For the fabrication of the solar cells with a trench, some of the solar cells with the undoped (intrinsic) a-Si, the a-Si was dry-chemically etched via a sacrificial SiO_x mask, creating a 20 µm wide trench. Afterwards, a high-temperature anneal was conducted, transforming the a-Si into poly-Si. BF- and P-implantations resulted in sheet resistances of about 1000 Ω/sq as measured for the whole stack with constant doping densities of about 3.0×10^{19} cm^{-3} in the poly-Si and a shallow doping tail extending to about 50 nm into c-Si. Then a 10 nm aluminum oxide (Al_2O_3) was deposited on both sides, succeeded by a 60 nm silicon nitride (SiN_x) antireflection coating deposition [19] on the front side and a 100 nm SiO_x deposition on the rear side. Afterwards, a hydrogen passivation was applied in a remote-plasma hydrogen passivation (RPHP) system [20]. Prior to the thermal evaporation and the lift-off of several 10 µm wide and in total 100 nm thick titanium (Ti) and palladium (Pd) lines, local openings were etched in Al_2O_3 and SiO_x on the rear. Finally, 5 µm thick aluminum (Al) was thermally evaporated and locally to realize the finger grid and the busbars in an interdigitated pattern. Photolithography was used for the several patterning processes and the lift-off. Additionally, asymmetric control samples with the same front side as the solar cells, but different rear sides (poly-Si(p), poly-Si(n) and poly-Si(i) as well as Al_2O_3/SiO_x) were fabricated using the same processes as described above

The current-voltage (IV) parameters of the solar cells were measured under standard testing conditions (STC). and, furthermore, in forward and in reverse direction without illumination. In addition, the illumination- and injection-dependent charge carrier lifetime characteristics were determined for the solar cells using quasi-steady state photoluminescence (suns-PL) measurements [21,22] and for the asymmetric control samples transient photoconductance decay measurements were performed [23,24]. In this way, effective minority charge carrier lifetimes (τ_{eff}) and implied open-circuit voltages (iV_{oc}), equivalent to the quasi-Fermi level splitting in the silicon volume at an injection level corresponding to open-circuit (OC) conditions, and implied fill factors (iFF) at an injection level corresponding to maximum power point (MPP) conditions were obtained. For the asymmetric control samples, iV_{oc} was used to determine the surface saturation current density (J_{0s}) of the different rear sides. Furthermore, illumination- and injection-dependent quasi-steady state open-circuit voltage (suns-V_{oc}) curves of the solar cells were measured to determine the series resistance (R_s) of the solar cells, calculating R_s from the difference in pFF and FF [25].

3 RESULTS

Table 1 presents the results of the different solar cell configurations. The solar cells without a trench showed a poor performance due to detrimental recombination in the transition region between p-type and n-type poly-Si. This is in good accordance with the observations for lateral poly-Si pn- and pin-diodes where defects at grain boundaries greatly enhanced the recombination [26–28]. The detrimental recombination was even more pronounced for the "no gap" configuration, allowing V_{oc} and FF of only 593 mV and 61.3%, respectively. For the "gap" configuration, V_{oc} of up to 680 mV and FF of 65.6% were obtained. In this case, R_s was twice as high as for the "no gap" configuration. For both configurations additional junction and/or shunt leakage currents were observed when measuring the non-illuminated IV curves which limit the V_{oc} and FF of the solar cells. In contrast, solar cells with trenches allowed for V_{oc} of 720 mV, FF of 79.6% and J_{sc} of 41.3 mA/cm², resulting in η of 23.7%.

Table I: IV and suns-PL parameters of the IBC solar cells with different configurations, showing the best values measured under standard testing conditions (AM1.5G, 100 mW/cm², 25 °C) for a designated area of 4.0 cm².

V_{oc} (mV)	J_{sc} (mA/cm²)	FF (%)	iV_{oc} (mV)	iFF (%)	R_s (Ωcm²)	η (%)
a) "no gap" configuration						
593	41.5	61.3	597	62.7	0.38	15.1
b) "gap" configuration						
680	41.2	65.6	675	68.8	0.67	18.4
c) "trench" configuration						
720*	41.3*	79.6*	724	81.2	0.34	23.7*

* independently confirmed by Fraunhofer ISE CalLab

The V_{oc}-limit of the solar cells was calculated, based on an ideal saturation current density (J_{01}) as a sum of the area-weighted surface and bulk saturation current density (J_{0s}) and (J_{0b}), respectively, of the asymmetrical control samples. J_{0b} was around 8 fA/cm² and J_{0s} was around 3 fA/cm² for the Al_2O_3/SiN_x on the textured front side of the wafer. J_{0s} of the poly-Si(p) and poly-Si(n) was 11 fA/cm² and 5 fA/cm², respectively, whereas Al_2O_3/SiO_x on the rear side of the wafer allowed for 2 fA/cm². The intrinsic the poly-Si(i) thin film has a much higher recombination, resulting in J_{01} of 78 fA/cm². J_{0b} was calculated using the experimentally determined charge carrier lifetime of 7.2 ms for symmetrical lifetime samples [29] with an intrinsic carrier concentration n_i of 8.79×10^9 cm^{-3} at 25°C.

Using these values for the "trench" configuration, the V_{oc}-limit was 724 mV (J_{01} of 21 fA/cm²). For the "no gap" and the "gap" configuration, the V_{oc}-limit was 723 mV (J_{01} of 22 fA/cm²) and 718 mV (J_{01} of 28 fA/cm²), respectively. In the latter case, the V_{oc}-limit is much higher than the actual measured V_{oc} of the solar cells. Thus, the transition region which is not accounted for in this simple evaluation , has a strong impact on the recombination behavior of the cell. .

In order to evaluate the impact of the recombination in the transition region on the performance of the solar cells, iV_{oc} and iFF on finished solar cells were additionally determined by suns-PL measurements. In this case, iV_{oc} is a measure of the splitting of the quasi-Fermi levels of the minority charge carriers which is affected by recombination in the silicon volume and at all surfaces or interfaces, and can be calculated by measuring the average excess carrier density (Δn_{av}) of the solar cells [23,24]. In order to account for diffusion limited lifetime carriers [30], the injection-dependent charge carrier lifetime characteristics of the solar cells were additionally determined under short-circuit (SC) conditions performing suns-PL measurements with the same generation [31]. Subtracting Δn_{av} under SC conditions from Δn_{av} under OC conditions leads to Δn_{av} at the junction of the solar cells so that iV_{oc} and iFF can be extracted from the corrected suns-PL measurements, see Figure 2. The illumination-dependent iV_{oc} curves of the solar cells were then modeled to analyze the origin of the recombination in the transition region. In this case, the injection-dependent charge carrier lifetime characteristics was calculated, using radiative, Auger [32], bulk and surface recombination [33] as well as recombination in space-charge regions and in shunted regions [34], and converted into illumination-dependent iV_{oc} curves. The resulting illumination-dependent iV_{oc} curves were then interpreted in terms of the dominating recombination mechanisms [35].

For the "no gap" configuration, the illumination-dependent iV_{oc} curves from the corrected suns-PL measurements are marked by low iV_{oc} values due to detrimental recombination. Since high-quality silicon wafers were used and since the front surfaces as well as the p^+ and n^+ poly-Si rear surfaces enable high minority charge carrier lifetimes of more than several milliseconds, most of the recombination originates from the recombination in the transition region from the p^+ to the n^+ poly-Si. In this respect, J_{01} of 21 fA/cm^2 was used for the bulk and surface recombination, see Figure 2a (green curve), but additional recombination terms were required to obtain a good fit at moderate and low illumination intensities. At these illumination intensities, recombination in space-charge regions and in shunted regions can occur. Therefore, a saturation current density (J_{scr}) of 17000 nA/cm^2 was used with a local ideality factor of 3.0, for the space-charge regions and a R_{sh} of 0.5 kΩcm^2 was used for the shunted regions, see Figure 2a (orange and brown curve). The high local ideality factors indicate that non-ideal recombination in the space charge regions of these solar cells exists [10,12,13] which limit the iV_{oc} to 597 mV and the iFF to 62.7%. Local ideality factors of 3.0 were also reported for lifetime samples with a direct contact of p^+ and n^+ poly Si regions [10]. Hence, it can be concluded that for the "no gap" configuration non-ideal recombination in the space charge regions as well as recombination in shunted regions are the limiting factors allowing V_{oc} to only 593 mV and FF of only 61.3% for these solar cells.

In comparison, the illumination-dependent iV_{oc} curves of the "gap" configuration shows a different behavior. Here, higher iV_{oc} values were obtained. J_{01} of 28 fA/cm^2 was used for bulk and surface recombination, see Figure 2b (green curve). For the space-charge region recombination, J_{scr} of 3600 nA/cm^2 was used with a local ideality factor of 2.9, whereas for recombination in the shunted region an R_{sh} of 0.57 kΩcm^2 was used to obtain a good fit at moderate and low illumination intensities, see Figure 2b (orange and brown curve).

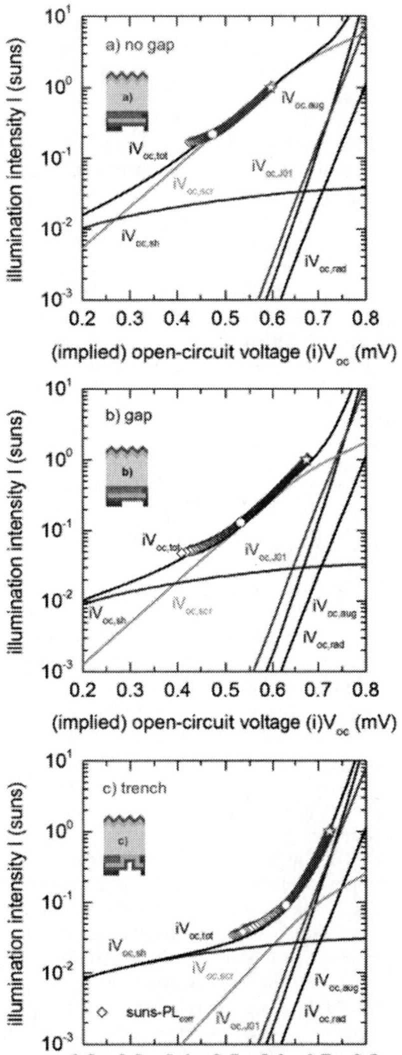

Figure 2: Illumination intensity (I) vs. (implied) open-circuit voltage $(i)V_{oc}$ from corrected suns-PL measurements under open-circuit (OC) conditions of the different solar cell configurations: a) no gap, b) gap and c) trench (schematic illustration, insets). Stars and circles denote the position corresponding to OC and to maximum power point (MPP) conditions, respectively. Radiative (black), Auger (blue), bulk and surface recombination (green) as well as recombination in space-charge regions (orange) and shunted regions (brown) are modeled.

In this case, the local ideality factors are almost in the

same range as compared to the "no gap" configuration and also indicate that non-ideal recombination in the space charge regions of these solar cells exists which limit the iV_{oc} to 680 mV and the iFF to 68.7%.

In contrast to solar cells without a trench, the illumination-dependent iV_{oc} curves of the "trench" configuration shows a completely different behavior. At high illumination intensities Auger and rear surface recombination dominate, whereas at moderate and low illumination intensities recombination in the space-charge region and in shunted regions dominate. For the rear surface recombination, J_{01} was 21 fA/cm^2, see Figure 2c (green curve), whereas for the space-charge region recombination, J_{scr} was 40 nA/cm^2 when a local ideality factor of 2.3 was used and for recombination in the shunted region, R_{sh} was 0.62 kΩcm^2, see Figure 2c (orange and brown curve). The local ideality factors are much lower as compared to solar cells without a trench, indicating a more ideal recombination behavior for the "trench" configuration. Therefore, high iV_{oc} of 724 mV and high iFF of 81.2% were achieved. However, even here additional junction and/or shunt leakage currents were observed but to a much lesser extent, limiting the iFF and thus the FF of these solar cells. In this respect, the influence of shunts due to locally etched poly-Si (defects in the sacrificial SiO_x mask) and/or of junction leakage currents due to the inversion layer in the trench (presence of Al_2O_3 passivation) have to be examined in the future.

4 SUMMARY AND CONCLUSIONS

In summary, the recombination within the highly-defective poly-Si is detrimental for the performance of interdigitated back contact (IBC) silicon solar cells where the p^+ and n^+ poly-Si regions are in direct contact to each other and where undoped (intrinsic) poly-Si remains between the p^+ and n^+ poly-Si regions. Illumination- and injection-dependent quasi-steady state photo-luminescence (suns-PL) measurements revealed that non-ideal recombination in the space charge regions, characterized by high local ideality factors, as well as recombination in shunted regions is dominating the recombination behavior. While for the "pn-junction" configuration, the highest junction leakage currents and lowest shunts resistances were determined, it was observed that intrinsic poly-Si between the p^+ and n^+ poly-Si regions increased the shunt resistance slightly but could not prevent the detrimental recombination in the space charge region. In contrast, the separation of the p^+ and n^+ poly-Si regions by a trench enabled highly efficient solar cells (V_{oc} of 720 mV, FF of 79.6%, J_{sc} of 41.3 mA/cm^2, η of 23.7%), showing that a lowly conducting and highly passivating intermediate layer is mandatory between the p^+ and n^+ poly-Si regions of IBC solar cells with poly-Si passivating contacts.

5 ACKNOWLEDGEMENTS

The authors would like to thank A. Leimenstoll, F. Schätzle, S. Seitz, A. Seiler, A. Lösel, E. Schäffer, and F. Martin for preparation and measurements of the solar cells. The authors also acknowledge M. Bauer from the Department of Microsystems Engineering (IMTEK) at the Albert-Ludwigs-University of Freiburg for the deposition of the polysilicon in the Cleanroom Service Center (RSC). F. Haase and R. Peibst from the Institute for Solar Energy Research Hamelin (ISFH) and J. Krügener from the Institute of Electronic Materials and Devices (MBE) at the University of Hannover are recognized for fruitful discussions within the joint project which was funded by the German Ministry of Economic Affairs and Energy under grant No. 0325827B "26+".

REFERENCES

[1] P. Cousins, Solar Technology Efficiency: More Breakthroughs are Coming: Solar Technology Efficiency: More Breakthroughs are Coming, 2016.

[2] J. Nakamura, N. Asano, T. Hieda, C. Okamoto, H. Katayama, K. Nakamura, Development of Heterojunction Back Contact Si Solar Cells, IEEE J. Photovoltaics 4 (6) (2014) 1491–1495.

[3] K. Masuko, M. Shigematsu, T. Hashiguchi, D. Fujishima, M. Kai, N. Yoshimura, T. Yamaguchi, Y. Ichihashi, T. Mishima, N. Matsubara, T. Yamanishi, T. Takahama, M. Taguchi, E. Maruyama, S. Okamoto, Achievement of More Than 25% Conversion Efficiency With Crystalline Silicon Heterojunction Solar Cell, IEEE J. Photovoltaics 4 (6) (2014) 1433–1435.

[4] K. Yoshikawa, H. Kawasaki, W. Yoshida, T. Irie, K. Konishi, K. Nakano, T. Uto, D. Adachi, M. Kanematsu, H. Uzu, K. Yamamoto, Silicon heterojunction solar cell with interdigitated back contacts for a photoconversion efficiency over 26%, Nat. Energy 2 (2017) 17032.

[5] M.A. Green, Y. Hishikawa, W. Warta, E.D. Dunlop, D.H. Levi, J. Hohl-Ebinger, A.W. Ho-Baillie, Solar cell efficiency tables (version 50), Prog. Photovolt: Res. Appl. 25 (7) (2017) 668–676.

[6] A. Richter, J. Benick, F. Feldmann, A. Fell, M. Hermle, S.W. Glunz, n-Type Si solar cells with passivating electron contact: Identifying sources for efficiency limitations by wafer thickness and resistivity variation, Solar Energy Materials and Solar Cells (2017).

[7] F. Haase, F. Kiefer, S. Schäfer, C. Kruse, J. Krügener, R. Brendel, R. Peibst, Interdigitated back contact solar cells with polycrystalline silicon on oxide passivating contacts for both polarities, Jpn. J. Appl. Phys. 56 (8S2) (2017) 08MB15.

[8] D. De Ceuster, P. J. Cousins, D. D. Smith Trench process and structure for backside contact solar cells with polysilicon doped regions, 2009.

[9] C. Reichel, F. Feldmann, R. Müller, A. Moldovan, M. Hermle, S.W. Glunz, Interdigitated Back Contact Silicon Solar Cells with Tunnel Oxide Passivated Contacts Formed by Ion Implantation, in: Proceedings of the 29th European Photovoltaic Solar Energy Conference and Exhibition, Amsterdam, The Netherlands, 22-26 September 2014, pp. 487–491.

[10] U. Römer, R. Peibst, T. Ohrdes, B. Lim, J. Krugener, T. Wietler, R. Brendel, Ion Implantation for Poly-Si Passivated Back-Junction Back-Contacted Solar Cells, IEEE J. Photovoltaics 5 (2) (2015) 507–514.

[11] D.L. Young, W. Nemeth, V. LaSalvia, R. Reedy, S. Essig, N. Bateman, P. Stradins, Interdigitated Back Passivated Contact (IBPC) Solar Cells

Formed by Ion Implantation, IEEE J. Photovoltaics 6 (1) (2016) 41–47.
[12] M. Rienäcker, A. Merkle, U. Römer, H. Kohlenberg, J. Krügener, R. Brendel, R. Peibst, Recombination Behavior of Photolithography-free Back Junction Back Contact Solar Cells with Carrier-selective Polysilicon on Oxide Junctions for Both Polarities, Energy Procedia 92 (2016) 412–418.
[13] U. Römer, Dissertation, Gottfried Wilhelm Leibniz Universität Hannover: Polycrystalline silicon / monocrystalline silicon junctions and their application as passivated contacts for Si solar cells. Dissertation, Hannover, 2015.
[14] G. Yang, A. Ingenito, O. Isabella, M. Zeman, IBC c-Si solar cells based on ion-implanted poly-silicon passivating contacts, Solar Energy Materials and Solar Cells 158 (2016) 84–90.
[15] H. Kobayashi Asuha, O. Maida, M. Takahashi, H. Iwasa, Nitric acid oxidation of Si to form ultrathin silicon dioxide layers with a low leakage current density, Journal of Applied Physics 94 (11) (2003) 7328–7335.
[16] F. Feldmann, C. Reichel, R. Müller, M. Hermle, The application of poly-Si/SiOx contacts as passivated top/rear contacts in Si solar cells, Solar Energy Materials and Solar Cells 159 (2017) 265–271.
[17] C. Reichel, F. Feldmann, R. Müller, R.C. Reedy, B.G. Lee, D.L. Young, P. Stradins, M. Hermle, S.W. Glunz, Tunnel oxide passivated contacts formed by ion implantation for applications in silicon solar cells, Journal of Applied Physics 118 (20) (2015) 205701.
[18] F. Feldmann, R. Müller, C. Reichel, M. Hermle, Ion implantation into amorphous Si layers to form carrier-selective contacts for Si solar cells, Phys. Status Solidi RRL 08 (09) (2014) 767–770.
[19] D. Schuldis, A. Richter, J. Benick, P. Saint-Cast, M. Hermle, S.W. Glunz, Properties of the c-Si/Al 2 O 3 interface of ultrathin atomic layer deposited Al 2 O 3 layers capped by SiN x for c-Si surface passivation, Appl. Phys. Lett. 105 (23) (2014) 231601.
[20] S. Lindekugel, H. Lautenschlager, T. Ruof, S. Reber, Plasma hydrogen passivation for crystalline silicon thin-films, in: Proceedings of the 23rd European Photovoltaic Solar Energy Conference (EUPVSEC), Valencia, Spain, 1-5 September 2008, pp. 2232–2235.
[21] J.A. Giesecke, B. Michl, F. Schindler, M.C. Schubert, W. Warta, Minority carrier lifetime of silicon solar cells from quasi-steady-state photoluminescence, Solar Energy Materials and Solar Cells 95 (7) (2011) 1979–1982.
[22] T. Trupke, R.A. Bardos, M.D. Abbott, J.E. Cotter, Suns-photoluminescence: Contactless determination of current-voltage characteristics of silicon wafers, Appl. Phys. Lett. 87 (9) (2005) 93503.
[23] R. A. Sinton, A. Cuevas, M. Stuckings, Quasi-Steady-State Photoconductance, A New Method for Solar Cell Material and Device Characterizatio - Photovoltaic Specialists Conference, 1996., Conference Record of the Twenty Fifth IEEE, in: Proceedings of the 25th IEEE Photovoltaic Specialists Conference (PVSC), Washington D. C., USA, 13-17 May 1996, pp. 457–460.
[24] R.A. Sinton, A. Cuevas, Contactless determination of current–voltage characteristics and minority-carrier lifetimes in semiconductors from quasi-steady-state photoconductance data, Appl. Phys. Lett. 69 (17) (1996) 2510–2512.
[25] D. Pysch, A. Mette, S.W. Glunz, A review and comparison of different methods to determine the series resistance of solar cells, Solar Energy Materials and Solar Cells 91 (18) (2007) 1698–1706.
[26] J. Manoliu, T. I. Kamins, p-n junctions in polycrystalline-silicon films, Solid-State Elecronics (15) (1972) 1103–1106.
[27] M. Dutoit and F. Sollberger, Lateral Polysilicon p-n Diodes, J. Electrochem. Soc. Solid-State Science and Technology 125 (10) (1978) 1648–1651.
[28] N.M. Johnson, D.K. Biegelsen, M.D. Moyer, Grain boundaries in p - n junction diodes fabricated in laser-recrystallized silicon thin films, Appl. Phys. Lett. 38 (11) (1981) 900–902.
[29] R.S. Bonilla, C. Reichel, M. Hermle, P.R. Wilshaw, Extremely low surface recombination in 1 Ω cm n-type monocrystalline silicon, Phys. Status Solidi RRL 11 (1) (2017) 1600307.
[30] M.D. Abbott, R.A. Bardos, T. Trupke, K.C. Fisher, E. Pink, The effect of diffusion-limited lifetime on implied current voltage curves based on photoluminescence data, Journal of Applied Physics 102 (4) (2007) 44502.
[31] T. Trupke, E. Pink, R.A. Bardos, M.D. Abbott, Spatially resolved series resistance of silicon solar cells obtained from luminescence imaging, Appl. Phys. Lett. 90 (9) (2007) 93506.
[32] A. Richter, S.W. Glunz, F. Werner, J. Schmidt, A. Cuevas, Improved quantitative description of Auger recombination in crystalline silicon, Phys. Rev. B 86 (16) (2012) 187.
[33] A. Cuevas, R.A. Sinton, Prediction of the open-circuit voltage of solar cells from the steady-state photoconductance, Prog. Photovolt: Res. Appl. 5 (2) (1997) 79–90.
[34] S.W. Glunz, D. Biro, S. Rein, W. Warta, Field-effect passivation of the SiO2Si interface, Journal of Applied Physics 86 (1) (1999) 683–691.
[35] C. Reichel, R. Müller, F. Feldmann, A. Richter, M. Hermle, S. W. Glunz, submitted for publication (2017).

APPROACHING 22% EFFICIENCY WITH MULTICRYSTALLINE N-TYPE SILICON SOLAR CELLS

J. Benick[1], R. Müller[1,2], F. Schindler[1], A. Richter[1], H. Hauser[1], F. Feldmann[1,2], P. Krenckel[1], S. Riepe[1], M. C. Schubert[1], M. Hermle[1], S. W. Glunz[1,2]
[1]Fraunhofer Institute for Solar Energy Systems (ISE), Heidenhofstrasse 2, 79110 Freiburg, Germany
[2]Department of Sustainable Systems Engineering (INATECH), Albert Ludwigs University Freiburg, Georges-Köhler-Allee 103, 79110 Freiburg, Germany

ABSTRACT: The transition to seed assisted growth (high-performance multicrystalline silicon) significantly enhanced the quality of multicrystalline silicon. Combining the new growth technique with the inherent benefits of n-type doping results in a multicrystalline material which should be well suited for the fabrication of high-efficiency solar cells. In this work high-efficiency solar cells with passivating rear contact were fabricated on n-type high-performance multicrystalline silicon, crystallized at Fraunhofer ISE. The material features a high diffusion length >800 µm after application of all high temperature process steps needed for cell fabrication. Applying a black-silicon texture at the font resulted in a weighted reflectance of ~1%, maintaining a good emitter passivation with $J_0 \leq 60$ fA/cm^2 for a 90 Ω/sq emitter. For the resulting n-type multicrystalline silicon solar cells conversion efficiencies up to 22.3% were reached, representing the current record for multicrystalline silicon solar cells.
Keywords: multicrystalline silicon, n-type silicon, high-efficiency solar cells, black silicon

1 INTRODUCTION

Increasing the solar cell conversion efficiency is a powerful lever for a further cost reduction in photovoltaics. To date for industrial screen-printed p-type solar cells on monocrystalline silicon, efficiencies up to 22.6% [1] have been reported, whereas for a comparably processed p-type solar cell on multicrystalline (mc) silicon, the highest certified efficiencies reported to date is 21.3% [2]. Nevertheless, multicrystalline p-type silicon accounts for ~70% of the global solar cell production [3]. Though mc silicon suffers from a higher carrier recombination caused by structural defects and a higher concentration of impurities, the simpler crystallization process of the mc silicon shows a cost advantage potential over monocrystalline silicon. In contrast to marked share, the highest efficiencies for silicon solar cells so far were reported for n-type silicon where for an interdigitated back contact solar cell, a record efficiency of 26.7% was reported [4]. The high material quality of n-type silicon is mainly due to its relative tolerance to common impurities (e.g. Fe) resulting in higher minority carrier diffusion lengths compared to p-type substrates with a similar impurity concentration [5]. With progress in crystallization techniques, such as seed-assisted growth for the fabrication of high-performance multicrystalline silicon (HP mc) [6], the material quality of mc-Si wafers has significantly increased in recent years, mainly due to a reduced density of recombination-active dislocation clusters.

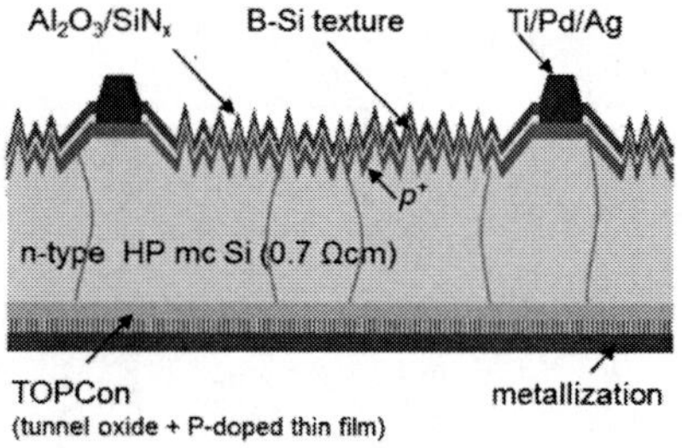

Figure 1: Schematic of the n-type HP mc TOPCon solar cell.

The HP mc process combined with the mentioned benefits of n-type silicon thus might offer opportunities for a low-cost, high-efficiency silicon material which has the potential to reduce the efficiency gap to monocrystalline silicon. In this paper, we apply a high-efficiency solar cell fabrication process with a passivating rear side contact (TOPCon [7]) on high-quality n-type HP mc silicon.

2 n-TYPE SILICON MATERIAL

2.1 Material fabrication

The n-type HP mc silicon applied in this work has been developed at Fraunhofer ISE. The size of the research ingot was G2 which is equivalent to 75 kg of pure polysilicon feedstock. The silicon was crystallized by the directional solidification method with seeded growth. High-purity silicon granules from a Fluidized Bed Reactor (FBR) process were used as seed material and were placed in the bottom of a high-purity fused-silica crucible. As dopant material silicon wafers enriched with phosphorus (1085 ppmw) were used. The crystallization resulted in a resistivity profile from 1.5 Ω cm above the remaining seed material to 0.5 Ω cm at the ingot top. A brick with 156 mm side length was cut from the ingot center. After cropping of 15 mm at bottom and top, the brick was processed into wafers with a thickness of 195 µm by multiwire sawing with SiC slurry and structured wire.

2.2 Material quality

Lifetime samples were fabricated on wafers from the upper ingot part (bulk resistivity of about 0.7 Ω cm) to investigate the electrical quality of the n-type HP mc silicon. The lifetime samples received the same high-temperature processes which are used in the solar cell process sequence, i.e. boron diffusion and annealing of the TOPCon layer. After cleaning (HNO_3, followed by a HF-dip) the lifetime samples received a boron diffusion at 890°C (tube furnace) with one side of the wafer being masked by a PECVD SiO_x. After the diffusion the borosilicate glass and the SiO_x mask were removed in HF and the high temperature step necessary for the annealing of the TOPCon layer was applied (N_2, 800°C). As a last step the boron emitter was removed in a wet-chemical process and the wafers were passivated by SiN_x.

To characterize the quality of the base material (minority carrier lifetime), photoluminescence imaging calibrated by harmonically modulated photoluminescence [8] was performed. Exemplarily, the lifetime image of the middle part of one n-type HP mc silicon wafer (50×50 mm^2, sister wafer of the 21.9% cell) is shown in Figure 2. The dashed square marks the area where the best solar cell was located on a sister wafer. The measurement was performed at constant illumination of 0.05 suns which is representative for the maximum power point of the fabricated solar cells. The best grains show effective lifetimes >1.5 ms. However, near the grain boundaries, at the boundaries itself and at some defective areas the lifetime is significantly lower. Nevertheless, the average lifetime of the 50×50 mm^2 wafer section shown in Figure 2 (square root harmonic average) is still close to 600 µs. This corresponds to a minority carrier diffusion length of more than 800 µm which is more than 4 times the cell thickness (195 µm). Applying the efficiency limiting bulk recombination analysis (ELBA [9]) on this lifetime sample reveals an efficiency potential of 22.7% in the marked area for the cell structure shown in Fig1. Thus, the applied material (n-type HP mc) should be well suited for the processing of high-efficiency solar cells.

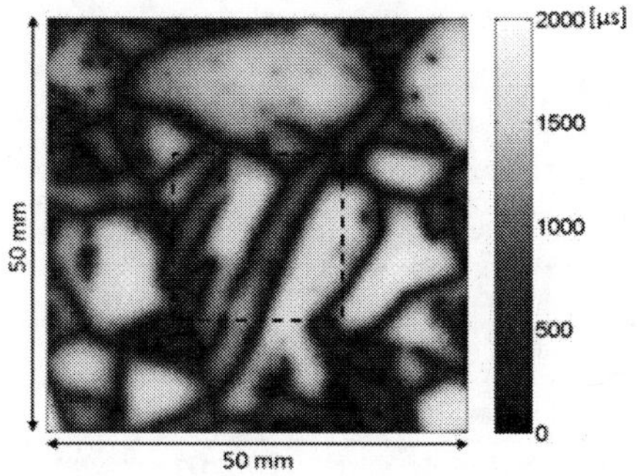

Figure 2: Image of the minority carrier lifetime of an n-type HP mc silicon lifetime sample which received the same high-temperature processes as the final solar cells. The measurement was done at an illumination of 0.05 suns. The marked square in the middle of the shown wafer section indicates the area where the 21.9% cell has been processed on a sister wafer.

3 SURFACE TEXTURE

3.1 Technology

In this work a black silicon texture was realized on the front side of the HP mc silicon solar cells as this surface allows for an almost ideal coupling of the incident light. Fortunately, the previous issue with this texture, the passivation of the nanostructured surface, was solved by an ALD deposition of Al_2O_3 [10–13]. For the realization of the black silicon surface texture an inductively coupled plasma (ICP) process with a very low bias voltage was applied in order to keep the damage as low as possible [14]. This process was performed with an Oxford ICP133 tool using SF_6 and O_2. Note that a separate damage etch back before passivation of the surface is not necessary. A SEM micrograph of the resulting black-silicon texture applied for the solar cells fabricated in this study is shown in Figure 3. As can be seen the ICP texture does not feature perfectly steep needles as known for a "classical" black-silicon texture. The aspect ratio of this texture is in the range of 2.

Figure 3: SEM micrograph of the black-silicon front side texture of a HP mc silicon wafer.

3.2 Surface reflection

The reflectance measurement of a wafer with black-silicon texture on the front side and planar back side is shown in Figure 4. Without any antireflection coating the surface reflection weighted with the AM1.5G solar spectrum in the range from 280 to 1000 nm (R_w) is about 2.5%. To further reduce the surface reflection an antireflection coating (ALD Al_2O_3/PECVD SiN_x) was deposited. The ALD process leads to a uniform coating of 10 nm Al_2O_3, whereas the thickness of the SiN_x layer varies between 10 and ~100 nm due to the non-conformality of the PECVD process on the textured surface. Applying the antireflection coating reduces the weighted reflectance to ~1%. Thus, a very effective coupling of the incident light can be reached with the applied surface texture.

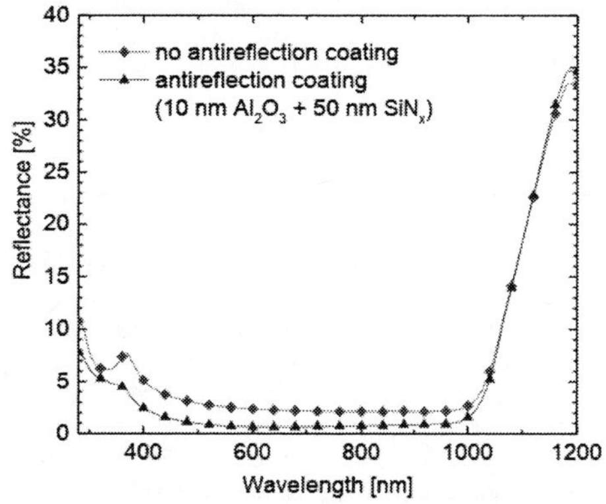

Figure 4: Measured reflectance of the ICP texture (black silicon) without and with antireflection coating.

3.3 Emitter diffusion

As the black-silicon texture has proven to feature very good optical properties, the compatibility of this texture with emitter diffusion and surface passivation has to be verified, too. For this reason lifetime test samples (*n*-type, FZ, 1 or 10 Ω cm) were fabricated. The samples are either planar, feature a both-sides random-pyramids texture or a single side black-silicon texture. All samples received a BBr_3 tube furnace boron diffusion at 890°C (R_{sheet}: 90 Ω/sq, N_{surf}: 8×10^{19} cm^{-3}, depth: 0.3 µm) and were then passivated by PA-ALD deposited Al_2O_3 (10 nm). The saturation current density to quantify the surface recombination was extracted from injection-dependent effective lifetime data measured using QSSPC

[15] according to the high-injection method proposed by [16] and applying the correction routines introduced by [17].

The extracted saturation current densities (for a single surface) are summarized in Table 1. For the planar reference samples a saturation current density in the range of 25 ± 5 fA/cm² was measured. For both, random pyramids and black silicon texture the saturation current density is increased to 50 ± 10 fA/cm² which still allows for high open-circuit voltages > 700 mV (assuming no additional recombination). Applying no surface passivation (e.g. underneath the metal contacts) the saturation current density of the applied boron emitter (calculated by EDNA [18], 25°C) is approximately ~1100 fA/cm².

Table 1: Measured saturation current densities of single surfaces (planar or textured with random pyramids or black silicon). The samples feature the 90 Ω/sq emitter as is also applied for the fabrication of the solar cells.

Surface	J_{0e} [fA/cm²]
Planar	25 ± 5
Random pyramids	50 ± 10
Black silicon	50 ± 10

4 SOLAR CELLS

4.1 Fabrication

High-efficiency solar cells with passivating rear side contacts (TOPCon [7]) were processed on *n*-type HP mc silicon (195 µm, 0.7 Ω cm, fabricated at Fraunhofer ISE) to investigate the excellent quality of the *n*-type HP mc silicon at the device level. The final cell structure is shown in Figure 1. As references solar cells on float-zone silicon were fabricated as well (*n*-type, 1 Ω cm, 200 µm).

The processing of these *n*-type HP mc TOPCon solar cells starts with the definition and texturing of the active cell area (7 solar cells on each wafer, cell size 20×20 mm²). The front-side black-silicon texture applied for the HP mc solar cells is realized by a plasma etching step (inductively coupled plasma ICP as introduced earlier). The FZ reference solar cells were textured in a KOH based solution resulting in a random-pyramids surface. To define the cells on every wafer, emitter windows were opened in an oxide mask (PECVD SiO_x) and the uniform boron emitter was realized by the BBr_3 tube furnace diffusion at 890°C (R_{sheet} = 90 Ω/sq). After the removal of the borosilicate glass in HF the passivating rear side contact (TOPCon) was deposited and activated by an annealing step at 800°C. Then the front side was passivated by an Al_2O_3/SiN_x layer stack deposited via PA-ALD and PECVD, respectively. The front side contacts (area fraction ~1.7%) were realized by photolithography and evaporation of a Ti/Pd/Ag layer stack. To improve the electrical contact and completely activate the passivation of front and rear side, the samples received an annealing step in atomic hydrogen (remote plasma hydrogen passivation, 425°C). For the rear side contact, a 1 µm thick Ag layer was evaporated.

4.2 Cell results

The measured IV data of the best cells are summarized in Table 2. As can be seen, for the FZ reference cells high conversion efficiencies up to 23.3% could be reached with a V_{oc} of 684 mV, a J_{sc} of 41.5 mA/cm² and a *FF* of 82.2%. The comparison of the measured IV data with 3D total area Quokka [19] device simulations (688.2 mV, 41.6 mA/cm², 82.5%) using experimentally-determined input parameters (e.g. $J_{0e,total}$ = 70 fA/cm², $J_{0,TOPCon}$ = 7 fA/cm², measured reflection, including cell perimeter) reveal that potential of the applied cell structure was almost fully exploited and no significant process-related issues occurred during cell fabrication.

For the *n*-type HP mc silicon solar cells of the first batch also very high conversion efficiencies up to 21.9% were measured with a V_{oc} of 673 mV, a J_{sc} of 40.8 mA/cm² and a *FF* of 79.7%. However, there is still a significant gap to the monocrystalline reference solar cells as well as to the efficiency of 22.7% predicted by the analysis of the bulk lifetime (ELBA) of a lifetime sample from a sister wafer to the best solar cell. Thus in the next section a short analysis of the main losses of the first solar cell run will be given. The full loss analysis is still in preparation and will be presented elsewhere.

In a second cell batch the fabrication of the n-type HP mc solar cells with black silicon texture was repeated and a new efficiency record of 22.3% could be achieved. The gain in efficiency is mainly due to small improvements in cell processing like a further reduced surface reflection as well as a lower series resistance.

Table 2: IV measurements (AM1.5g, 100 mW/cm², 25°C) of the best cells of each group.

V_{oc} [mV]	J_{sc} [mA/cm²]	*FF* [%]	*pFF* [%]	η [%]
n-type FZ, random pyramids				
683.9	41.5	82.2	84.1	23.3
n-type HP mc, planar surface				
676.1	37.3	79.5	81.7	20.1
n-type HP mc, black silicon texture				
672.6	40.8	79.7	81.6	21.9*
n-type HP mc, black silicon texture, 2nd run				
674.2	41.1	80.5	81.6	22.3*

* Fraunhofer ISE CalLab (4 cm², aperture area)

4.3 Cell analysis (run 1)

The performance of the FZ reference solar cells defines the limit of the cell fabrication process which was used as input for the ELBA analysis. In this case, the cell fabrication process explicitly enables a V_{oc} of 685 mV. According to ELBA, the recombination in the mc-Si bulk material reduces the V_{oc} limit to 679 mV. As ELBA takes the recombination active structural crystal defects into account, these cannot explain the gap between the V_{oc} of the best mc cell (673 mV) and the ELBA limit (679 mV). Thus, the V_{oc}-potential of the material was not fully exploited in this batch - probably due to technological issues - and there is still room for about 6 mV improvement.

The J_{sc} of the mc cells with black-silicon texture is about 0.8 mA/cm² lower compared to the FZ reference cells with random pyramids. To investigate this difference, reflection (R) and internal quantum efficiency

(IQE) was measured (see Figure 4). The reflection curve proves that the black silicon texture performs similarly well as the random pyramids. Even the path-length enhancement in the silicon bulk due to the front texture seems to be identical as the reflection from the back side (wavelength range 1000 - 1200 nm) is identical for the black silicon and random pyramids texture. In the low wavelength range, the black silicon shows even superior absorption. So the reason for the lower J_{sc} of the mc cells is not caused by optics but a reduced diffusion length of charge carriers due to the higher recombination rate in the bulk. Holes generated at the back side of the mc cell have a lower chance to reach the emitter at the front side. This is visible by the drop of the internal quantum efficiency above 800 nm (compared to the FZ reference) leading to a reduction in J_{sc}.

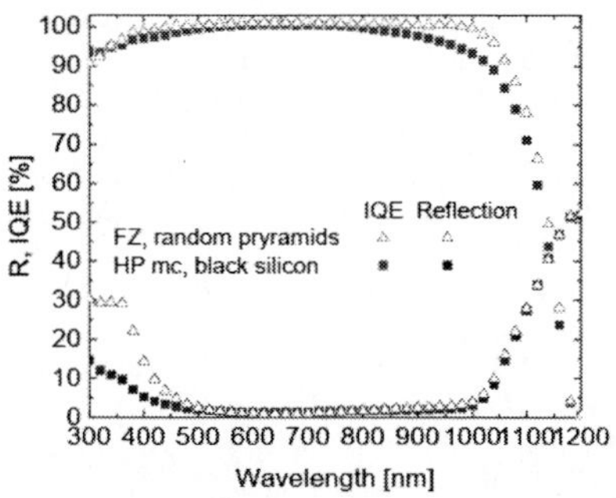

Figure 5: Quantum efficiency and reflection curves of the best *n*-type HP mc solar cell with a black-silicon surface texture as well as of the *n*-type FZ reference with a random-pyramids texture.

The fill factor (*FF*) is often influenced by the series resistance (R_s). In order to separate this effect, SunsV_{oc} measurements were performed to determine the pseudo fill factor (*pFF*) which is not disturbed by R_s. The *pFF* shows that the values of the best cells reach ~84% for the FZ cells and ~82% for the mc cells. No correlation between the *pFF* and the area fraction of RASCD (recombination active structural crystal defects) was found which supports the assumption of a different *pFF* limitation independent of the crystal structure quality.

Whereas the shunt resistance did not lead to a significant loss in *pFF* for the best multicrystalline solar cell, the J_{02} recombination shows a larger influence on the *FF* (~32% of the total recombination at maximum-power point in dark IV measurements). Regarding all multicrystalline cells of the first batch it can be observed that all cells, including the planar ones, exhibit a significantly higher J_{02} than the FZ references. Consequently, the high J_{02} is not an issue of the black-silicon texture, but rather related to the mc-Si material. It can also be observed that the J_{02} does not correlate with the area fraction of recombination active structural crystal defects. Both facts indicate that the high J_{02} might be a more general problem.

Figure 6 shows Suns-V_{oc} curves of the best finished cells together with Suns-$V_{implied}$ curves of the according bulk reference lifetime samples (which received the same high temperature steps as the final solar cells, all diffused areas have been etched) measured with QSSPC. The mc-Si material bulk recombination shows an ideality factor very close to 1 down to 550 mV. However, Suns-V_{oc} measurements of the finished cells reveal that a non-ideal recombination (with ideality factor >1) appears leading to an increased recombination at the maximum-power point (located at 566 mV). From the graphs it becomes obvious, that the mc-Si material is more affected than the FZ material. A possible explanation would be the accumulation of dissolved impurities in (or around) the space charge region during the emitter diffusion. This might lead to the observed non-ideal homogenously distributed recombination that translates into an increased J_{02}.

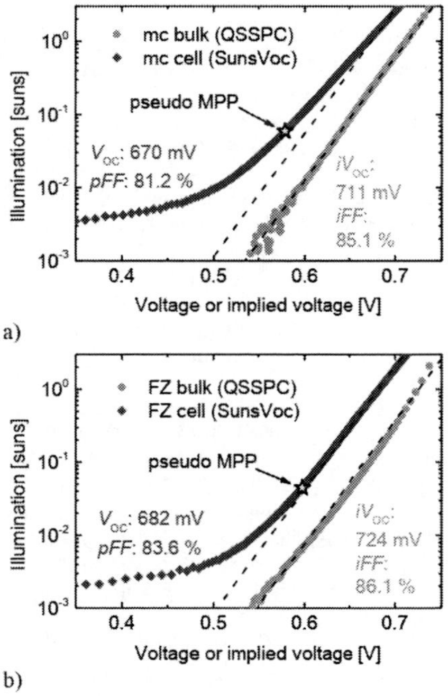

Figure 6: Comparison of the pseudo IV curves of a finished solar cell (Suns-V_{oc}) and a bulk reference lifetime sample, which received the same high temperature processes as the final solar cell (Suns-$V_{implied}$ from QSSPC). Dashed lines indicate the slope of recombination channels with an ideality factor of 1. Measurements are shown for both, the best FZ reference (b) as well as best n-type HP mc silicon solar cell (a).

4 SUMMARY

In this work we could show that n-type HP mc silicon is well suited for the fabrication of high-efficiency solar cells. Applying an adapted TOPCon cell fabrication process with a black-silicon front surface texture and a 90 Ω/sq Al_2O_3-passivated boron emitter a record conversion efficiency of 22.3% could be achieved on the n-type HP mc silicon.

The analysis of the material revealed, that only a part of the efficiency gap to the FZ reference solar cells can be related to recombination active structural crystal defects. A significant portion of the efficiency gap is due to non-ideal recombination, leading to an increased J_{02}. Most probably this might be caused by an accumulation of impurities in the space charge region during emitter diffusion.

ACKNOWLEDGEMENTS
This work was supported by the German Federal Ministry for Economic Affairs and Energy under contract number 0324034 (multiTOP).
The authors would like to thank S. Seitz, A. Seiler, A. Leimenstoll, F. Schätzle, C. Harmel, R. van der Vossen and E. Schäffer for their support with cell processing and measurements.
The authors also would like to thank Wacker Polysilicon for silicon materials and fruitful discussions.

REFERENCES

[1] *Press Releases | Trina Solar.* [Online] Available: http://phx.corporate-ir.net/phoenix.zhtml?c=206405&p=irol-newsArticle&ID=2230468. Accessed on: Feb. 24 2017.
[2] W. Deng *et al.*, "Development of high-efficiency industrial p-type multi-crystalline PERC solar cells with efficiency greater than 21%," *Energy Procedia*, vol. 92, pp. 721–729, 2016.
[3] *Photovoltaics Report - Fraunhofer ISE.* [Online] Available: https://www.ise.fraunhofer.de/de/daten-zu-erneuerbaren-energien.html. Accessed on: Mar. 14 2017.
[4] K. Yoshikawa *et al.*, "Silicon heterojunction solar cell with interdigitated back contacts for a photoconversion efficiency over 26%," *Nat. Energy*, vol. 2, p. 17032, 2017.
[5] D. Macdonald and L. J. Geerligs, "Recombination activity of interstitial iron and other transition metal point defects in p-and n-type crystalline silicon," *Applied Physics Letters*, vol. 85, no. 18, pp. 4061–4063, 2004.
[6] Y. M. Yang *et al.*, "Development of high-performance multicrystalline silicon for photovoltaic industry," *Progress in photovoltaics: research and applications*, vol. 23, no. 3, pp. 340–351, 2015.
[7] F. Feldmann, M. Bivour, C. Reichel, M. Hermle, and S. W. Glunz, "Passivated rear contacts for high-efficiency n-type Si solar cells providing high interface passivation quality and excellent transport characteristics," *Solar energy materials and solar cells*, vol. 120, pp. 270–274, 2014.
[8] J. A. Giesecke, M. C. Schubert, B. Michl, F. Schindler, and W. Warta, "Minority carrier lifetime imaging of silicon wafers calibrated by quasi-steady-state photoluminescence," *Solar energy materials and solar cells*, vol. 95, no. 3, pp. 1011–1018, 2011.
[9] B. Michl *et al.*, "Efficiency limiting bulk recombination in multicrystalline silicon solar cells," *Solar energy materials and solar cells*, vol. 98, pp. 441–447, 2012.
[10] M. Otto *et al.*, "Extremely low surface recombination velocities in black silicon passivated by atomic layer deposition," *Appl. Phys. Lett.*, vol. 100, no. 19, p. 191603, 2012.
[11] P. Repo *et al.*, "Passivation of black silicon boron emitters with atomic layer deposited aluminum oxide," *Phys. Status Solidi RRL*, vol. 7, no. 11, pp. 950–954, 2013.
[12] P. Repo *et al.*, "Effective passivation of black silicon surfaces by atomic layer deposition," *IEEE Journal of Photovoltaics*, vol. 3, no. 1, pp. 90–94, 2013.
[13] M. Algasinger *et al.*, "Improved black silicon for photovoltaic applications," *Advanced Energy Materials*, vol. 3, no. 8, pp. 1068–1074, 2013.
[14] J. Hirsch, M. Gaudig, N. Bernhard, and D. Lausch, "Optoelectronic properties of Black-Silicon generated through inductively coupled plasma (ICP) processing for crystalline silicon solar cells," *Applied Surface Science*, vol. 374, pp. 252–256, 2016.
[15] R. A. Sinton and A. Cuevas, "Contactless determination of current–voltage characteristics and minority-carrier lifetimes in semiconductors from quasi-steady-state photoconductance data," *Applied Physics Letters*, vol. 69, no. 17, pp. 2510–2512, 1996.
[16] D. E. Kane and R. M. Swanson, "Measurement of the emitter saturation current by a contactless photoconductivity decay method," in *18th IEEE photovoltaic specialists conference 1985*.
[17] A. Kimmerle, J. Greulich, and A. Wolf, "Carrier-diffusion corrected J0-analysis of charge carrier lifetime measurements for increased consistency," *Solar energy materials and solar cells*, vol. 142, pp. 116–122, 2015.
[18] K. R. McIntosh and Pietro P. Altermatt, "A freeware 1D emitter model for silicon solar cells," in *35th IEEE Photovoltaic Specialists Conference (PVSC), 2010*.
[19] A. Fell, "A Free and Fast Three-Dimensional/Two-Dimensional Solar Cell Simulator Featuring Conductive Boundary and Quasi-Neutrality Approximations," *IEEE Trans. Electron Devices*, vol. 60, no. 2, pp. 733–738, 2013.

EVALUATION OF TOPCON TECHNOLOGY ON LARGE AREA SOLAR CELLS

F. Feldmann[1,2,*], B. Steinhauser[1], V. Arya[1], A. Büchler[1], A. A. Brand[1], S. Kluska[1], M. Hermle[1], S. W. Glunz[1,2]

[1]Fraunhofer Institute for Solar Energy Systems (ISE), Heidenhofstrasse 2, 79110 Freiburg, Germany
[2]Department of Sustainable Systems Engineering (INATECH), Albert-Ludwigs University Freiburg, Georges-Köhler-Allee 103, 79110 Freiburg, Germany
*Corresponding author: frank.feldmann@ise.fraunhofer.de, +49 (0) 761 4588 5287

ABSTRACT: Passivating contacts, like TOPCon or poly-Si/SiO_x, have pushed the efficiency of crystalline silicon (c-Si) solar cells notably. Currently, the world-record efficiency of 25.7% for a both-sided contact solar cell has been achieved by an n-type solar cell featuring a selective boron front emitter and a passivating rear contact (TOPCon). Like the 25%-efficient PERL cell, this record cell was realized on a small area of 2x2 cm² using lab processes, like photolithography and evaporation of metal contacts. This paper addresses the first steps towards the realization of a large area TOPCon cell by industry-relevant fabrication steps. The scope of this paper lies on the application of laser ablation and NiCu plating to a 1.5µm deep, 140 Ω/sq boron-diffused emitter. The 10x10 cm² cell yielded an efficiency of 22.9%.

Keywords: n-type, Silicon Solar Cell, Passivation, PECVD

1 INTRODUCTION

Passivating and carrier-selective contacts which consist of an ultrathin SiO_x layer and a heavily doped silicon film (e.g. TOPCon [1] or poly-Si [2, 3]) are an appealing technology to further push the efficiency of silicon solar cells. The replacement of a partial rear contact (PRC) scheme by TOPCon as a full-area rear contact has so far resulted in efficiencies up to 25.7% [4]. The potential of such an approach has been shown only on an area of 2x2 cm². The first solar cells of practical size (≥100 cm²), which featured a passivating rear contact and a diffused front side, however, achieved efficiencies of only 20.7% [5] and 21.2% [6]. Both solar cells were -amongst other things- strongly limited by recombination at the front side and, therefore, did not benefit from the excellent surface passivation enabled by the passivating rear contact.

In this paper, we demonstrate 100 cm² n-type solar cells featuring a homogeneous boron emitter and a passivating rear contact. The scope of this paper lies on the formation of the front grid using laser contact opening (LCO) and NiCuAg plating.

2 EXPERIMENTAL DETAILS

2.1 Fabrication of large area solar cells

Solar cells of significant area (10×10 cm²) were fabricated on shiny-etched, 1 Ωcm n-type float-zone silicon. The cell layout is shown in Fig. 1. On the front random pyramids were formed by alkaline texturing and a 140 Ω/sq emitter (N_{surf} = $1x10^{18}$ cm^{-3}, depth = 1.5 µm) was realized by tube furnace diffusion and subsequent drive-in oxidation. At the rear, the TOPCon structure was applied as described in Ref. [1]. The front surface was passivated by a stack of Al_2O_3 and SiN_x. The front contact openings were realized by laser ablation using a UV/ps laser source. Thereafter, the front and rear surface passivation were activated using an atomic hydrogen treatment (RPHP) [7] at 425 °C. The rear contact was formed by thermal evaporation of Ag and the front grid was realized by forward bias plating of Ni, Cu, and Ag. Finally, a second anti reflection layer was applied at the front by evaporation of MgF_2.

2.2 Characterization and analysis

The minority carrier recombination at the laser-formed contact openings ($J_{0,LCO}$) was investigated on a symmetric passivated J_{0e} sample using lifetime calibrated photoluminescence imaging (PL).

The light and dark I-V characteristics of the solar cells were measured using the LOANA setup from pv-tools. To further analyse the performance of the solar cells, 3D device simulations using Quokka v3 were performed. The input parameters, e.g. J_{0e}, $J_{0,LCO}$, were determined from J_{0e} samples.

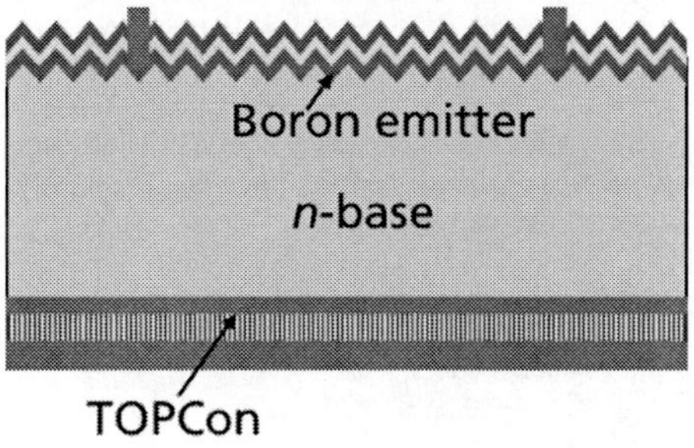

Figure 1: Solar cell structure featuring a homogeneous boron emitter and n-TOPCon as passivating rear contact.

3 RESULTS

3.1 Laser contact opening

The influence of the laser contact opening (LCO) process was determined as a function of laser power on a symmetric boron-diffused and passivated J_{0e} sample. The corresponding PL image after single-sided LCO is depicted in Fig. 2. The sample's total J_0 (including the bulk) before laser averaged to 46.2 fA/cm² which resulted in J_{0e} = 18.1 fA/cm² for either side. After LCO the total J_0 was significantly increased, especially for the highest used laser power J_0 increased by a factor of more than 3. This corresponds to a $J_{0,LCO}$ > 10000 fA/cm² (calculated with ~0.4% contact opening). On the other hand, for the two lowest laser powers the total J_0 increased only to 59 fA/cm² and 62 fA/cm², respectively. The corresponding $J_{0,LCO}$ were 2700 fA/cm² and 3300 fA/cm², respectively.

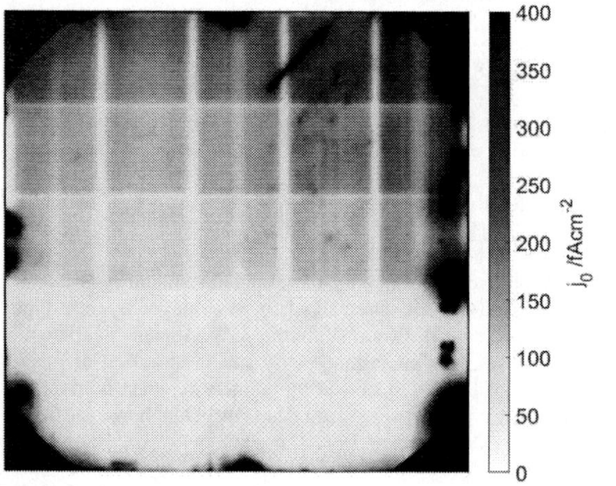

Figure 2: PL image of a J_{0e} sample after LCO. The upper row was subjected to the highest laser power, while the lower row was subjected to a tenth of the maximum laser power.

These values were used to calculate the V_{oc} as a function of the metallized area fraction. The dashed line refers to the calculation with $J_{0,LCO}$ = 3300 fA/cm² and the solid lines refer to the ideal case, wet-chemical contact opening which does not introduce damage to the crystal ($J_{0,WCO}$). Using 1D device simulations [8] $J_{0,CO}$ = 1800 fA/cm² was determined. Since $J_{0,LCO}$ is higher than $J_{0,WCO}$ the V_{oc} is expected to be lower when using LCO instead of wet-chemical contact opening. With $A_{metal} \approx$ 1.2%, a V_{oc} of about 695 mV is expected.

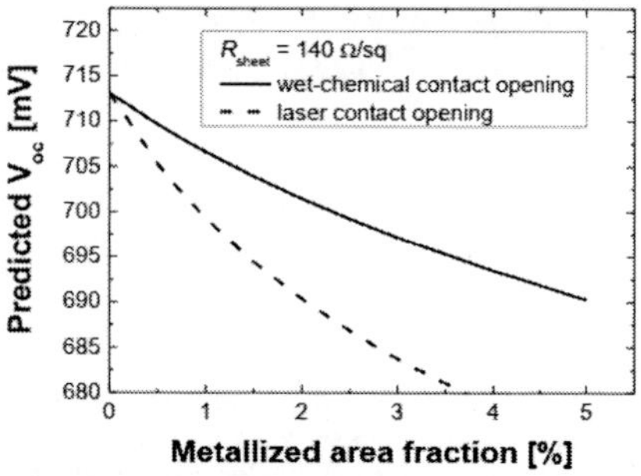

Figure 3: V_{oc} as a function of metallized area fraction for the 140 Ω/sq emitter. The solid line refers to wet-chemical contact openings, which do not damage the crystal and the dashed line refers to LCO, which induces certain crystal damage.

3.2 Solar cell results and analysis

The solar cell results for the best cell are shown in Table I. The V_{oc} of 694 mV matched very well the calculation shown in Fig. 3 and showed that the influence of the laser contact opening was well estimated. The FF and pFF values were 81.0% and 83.9%, respectively. The pFF was slightly lower than the ideal FF_0 (84.6%) as the laser-induced damage probably affected the cell at maximum power point conditions, too. From the difference between pFF and FF the series resistance was calculated to R_S = 0.58 mΩcm². In order to measure the contact resistivity of the fingers by the TLM method, multiple solar cells were diced into stripes. The contact resistivity of the metal fingers averaged to ρ_c = 0.53±0.19 mΩcm² and, therefore, did not affect the FF significantly. Instead, a FF loss of 1%$_{abs}$ can be attributed to the grid resistance. This is supported by the measurement of the grid resistance (busbar to busbar) which revealed a resistance of 80 mΩ, amounting to approx. one third of R_S.

The J_{sc} was 40.8 mA/cm². Due to a metal coverage of ~2%, the main J_{sc} loss can be attributed to shading by the front metal grid. The influence of ghost plating was present but is yet to be quantified.

In total an efficiency of 22.9% was measured. Fig. 4 shows the free energy loss analysis (FELA) obtained from the numerical device simulation. It can be clearly seen that there are two dominant loss mechanisms. The resistive loss incurred by the grid fingers has a share of ~32% of the total loss. The aggregate loss due to recombination in the emitter (including contacted regions) takes a similar share of ~34% of the total energy loss. Hence, the simulation reveals that both the LCO and plating process have the highest potential for improvement.

Table II: Light *I-V* results of 10x10cm² n-type solar cell.

Cell	V_{oc}	J_{sc}	FF	pFF	η
	[mV]	[mA/cm²]	[%]	[%]	[%]
Best	694	40.8	81.0	83.9	22.9

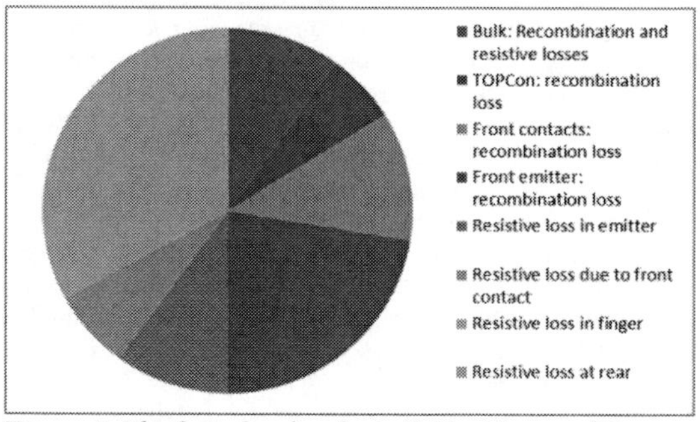

Figure 4: Pie chart showing the individual losses of the FELA.

4 SUMMARY

10x10 cm² solar cells with homogeneous boron emitter and passivating rear contact were realized. Firstly, the impact of laser ablation on the J_0 was studied and it was found that a low laser power has to be used in order to contain crystal damage. Still, the Si crystal was damaged during LCO, which resulted in a significant increase of the recombination current (J_0) at the metal front contacts. Secondly, solar cells featuring NiCu plated front metallization were realized and analyzed. An efficiency of 22.9% was measured. The FELA revealed that the dominant losses were the limited finger conductivity and the minority carrier recombination at the front side. From literature it is known that laser damage can be cured by forming gas annealing or firing [9]. Alternatively, the threshold energy for ps laser ablation can be reduced by using a nanosecond pre-pulse thereby mitigating the impact of laser damage significantly [10]) Hence, future work will focus on a "damage-free" laser process and an optimized front metallization with respect to shading and conductivity.

5 ACKNOWLEDGMENTS

The authors are deeply indebted to Fabian Meyer who set up and improved the laser system. The authors would like to thank A. Leimenstoll, F. Schätzle, A. Seiler, and S. Seitz for processing of the solar cells in the cleanroom and G. Cimiotti is thanked for plating the solar cells. Furthermore, the authors acknowledge F. Martin and E. Schäffer for measuring the *I-V* characteristics of the solar cell.

This work was funded by the German Federal Ministry for Economic Affairs and Energy under contract number 03225877D (PROJECT PEPPER).

6 REFERENCES

[1] F. Feldmann, M. Bivour, C. Reichel, M. Hermle, and S. W. Glunz, Solar Energy Materials and Solar Cells 120 (2014) 270.

[2] U. Römer, R. Peibst, T. Ohrdes, B. Lim, J. Krügener, E. Bugiel, T. Wietler, and R. Brendel, Solar Energy Materials and Solar Cells 131 (2014) 85.

[3] D. Yan, A. Cuevas, J. Bullock, Y. M. Wan, and C. Samundsett, Solar Energy Materials and Solar Cells 142 (2015) 75.

[4] A. Richter, J. Benick, F. Feldmann, A. Fell, M. Hermle, and S. W. Glunz Solar Energy Materials & Solar Cells to be published (2017)

[5] M. K. Stodolny, M. Lenes, Y. Wu, G. J. M. Janssen, I. G. Romijn, J. R. M. Luchies, and L. J. Geerligs, Solar Energy Materials and Solar Cells 158 (2016) 24.

[6] Y. G. Tao, V. Upadhyaya, C. W. Chen, A. Payne, E. L. Chang, A. Upadhyaya, and A. Rohatgi, Progress in Photovoltaics 24 (2016) 830.

[7] S. Lindekugel, H. Lautenschlager, T. Ruof, and S. Reber, in Proceedings of the 23rd European Photovoltaic Solar Energy Conference, Valencia, Spain, 2008, p. 2232.

[8] K. R. McIntosh and P. P. Altermatt, 35th Ieee Photovoltaic Specialists Conference (2010) 2188.

[9] A. Uruena, M. Aleman, E. Cornagliotti, A. Sharma, M. Haslinger, L. Tous, R. Russell, J. John, F. Duerinckx, and J. Szlufcik, Progress in Photovoltaics 24 (2016) 1149.

[10] A. A. Brand, F. Meyer, J. F. Nekarda, and R. Preu, Applied Physics a-Materials Science & Processing 117 (2014) 237.

SCREEN-PRINTED METALLIZATION FOR P-TYPE POLY-SI PASSIVATED CONTACTS FORMED BY LPCVD

S. Mack[1], T. Fellmeth[1], J. Schube[1], F. Feldmann[1], M. Lenes[2], J. M. Luchies[3]
[1]Fraunhofer Institute for Solar Energy Systems ISE, Heidenhofstraße 2, 79110 Freiburg, Germany
[2]Tempress Systems, Radeweg 31, 8171 MD Vaassen, The Netherlands
[3]Amtech Systems Inc., Radeweg 31, 8171 MD Vaassen, The Netherlands
Phone: +49 761 - 4588 5048; e-mail: Sebastian.mack@ise.fraunhofer.de

ABSTRACT: We report on the metallization on passivating contacts by screen-printed metal pastes. The passivating contact consists of a thermal SiO_x layer of 1 to 2 nm thickness, in combination with an in-situ boron doped low pressure chemical vapor deposited polysilicon layer of 300 nm thickness. After thermal annealing, SiN_x deposition and contact firing, we find extremely low dark saturation current densities down to 1 fA/cm^2. Using a commercially available high temperature fire through Ag paste, which is applied by conventional screen printing technology, we achieve a specific contact resistance $\rho_c = 2$ mΩcm^2. This makes the application interesting as rear side contact of industrial type high-efficiency p-type solar cells.
Keywords: LPCVD, passivated contact, screen printing, polysilicon, passivation, recombination

1 INTRODUCTION

Recently, passivating contacts received increased attention, as a method to decrease minority carrier recombination at the metal contacts. Passivating contacts may consist of a tunnel oxide layer in combination with a highly doped polysilicon [1–3] or SiC_x (TOPCon) layer [4]. These passivating contacts do reduce metal induced carrier recombination significantly, for metals being placed on top of the highly doped silicon layer. The use of full area passivating rear contacts in combination with full area evaporated Ag metal contacts has led to several efficiency records earlier this year [5, 6].

Apart from Ag, also other metals have been investigated as a contact material for passivating contacts, such as physical vapor deposited Aluminum [7, 8] or silver [9]. Alternatively, also screen printing technology has been applied, i.e. screen printed Ag pastes for contacting n^+ polysilicon layers [10–12], or AgAl pastes in case of p^+ polysilicon layers [10], and the papers report increased recombination due to interaction of the paste with the poly-Si layer.

In this paper, we fabricate passivating contacts by low pressure chemical vapor deposition (LPCVD) of boron doped amorphous silicon, which is subsequently transformed into polysilicon by a thermal anneal. We use commercially available fire through silver pastes, which are regularly screen printed for contacting phosphorous doped emitters by high temperature contact firing, to directly contact these layers. Moreover, we investigate the carrier recombination in the metallized areas and correlate that with a microstructural analysis of the contacts.

2 EXPERIMENTAL

2.1 Sample preparation

Fig. 1 shows the experiment layout. Phosphorous doped silicon wafers with 156 mm edge length and of resistivity $\rho = 6.1$ Ωcm (determined after thermal donor killing) serve as starting material. After saw damage removal in alkaline solution and wet chemical cleaning, the samples are inserted into a Tempress industrial LPCVD horizontal tube furnace (capable of loading 1000 wafers), for growing the thin tunnel oxide and deposition of a boron-doped polysilicon layer with around 300 nm thickness on both sides. Next, in another tube, a thermal anneal for 30 min at 900 °C activates the dopant and crystallizes the Si layers. The resulting sheet resistance R_{sh} of the polysilicon layer is 200 Ω/sq., and the doping is more or less constrained to the polysilicon layer and there is only a marginal diffusion through the tunnel oxide into the wafer itself.

Next, a SiN_x layer is formed on both sides of the wafer by two direct plasma enhanced chemical vapor deposition (PECVD) processes using a Tempress PECVD system (Spectre series), with a thickness of 80 nm. Four different Ag pastes are printed at the same side on one wafer, one in each corner, applying a screen layout with a Berger structure. After each printing step the metal paste is dried in a conveyor belt furnace. The center of the wafer is left non metallized. A firing variation step in a conveyor belt furnace running at 6200 mm/min with peak set temperatures between 780 °C and 900 °C completes processing.

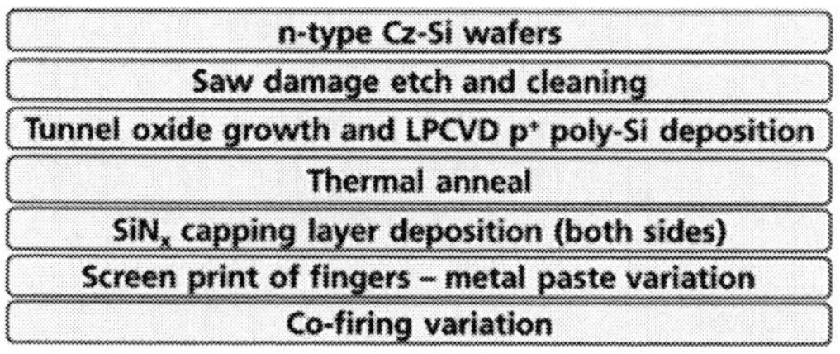

Figure 1: Process flow of the experiment. In total four Ag pastes have been tested.

The resulting sample structures are depicted in Fig. 2, although both on the same wafer, but on different sites.

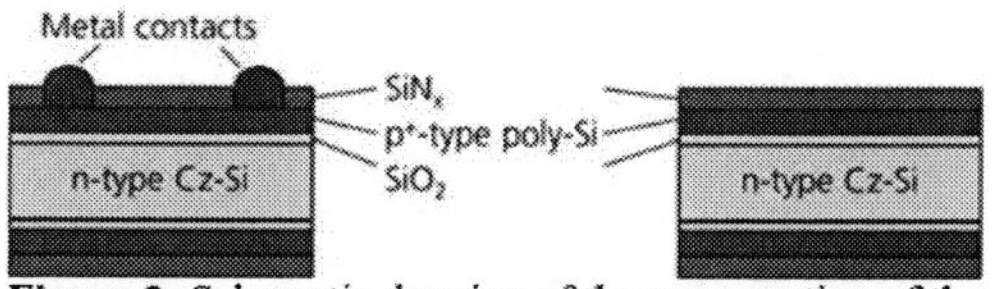

Figure 2: Schematic drawing of the cross section of the resulting metallized (left) and non metallized (right) test structures

2.2 Characterization

We apply the quasi steady-state photoconductance decay (QSSPC) technique for determination of the minority carrier lifetime. Additionally, we perform calibrated photoluminescence (PL) imaging to retrieve information about local distribution of the passivation quality. Following metallization, we use a chip saw to prepare stripes of 1 cm width perpendicular to the finger orientation, which enables us to determine the specific contact resistance by the transfer length method (TLM). For specific samples, we selectively remove the metallization by wet-chemical etching and analyse the samples by scanning electron microscopy (SEM).

3 RESULTS

3.1 Passivation quality

Fig. 3 shows the results of the QSSPC measurements, which have been performed in the non metallized center of the samples. The resulting minority carrier lifetime has been calculated into the dark saturation current density J_0 for one side of the sample, applying newest models [13]. Before contact firing, the passivation quality of the p^+ polysilicon/SiN_x stack is already on a very high level, with $J_0 = 13$ fA/cm² only. This value decreases even further after contact firing, and in general, higher firing set temperatures lead to lower J_0 values, probably due to enhanced hydrogenation. For 780 °C, we determine a J_0 of 5 fA/cm² and for the highest investigated temperature of 900 °C J_0 reaches extremely low values down to 1 fA/cm² (locally). To our knowledge, these are the lowest values ever reported for boron doped polysilicon layers. Including the n-type wafers, this corresponds to an implied open circuit voltage of 732 mV for the symmetric structure, see Fig. 2 (right).

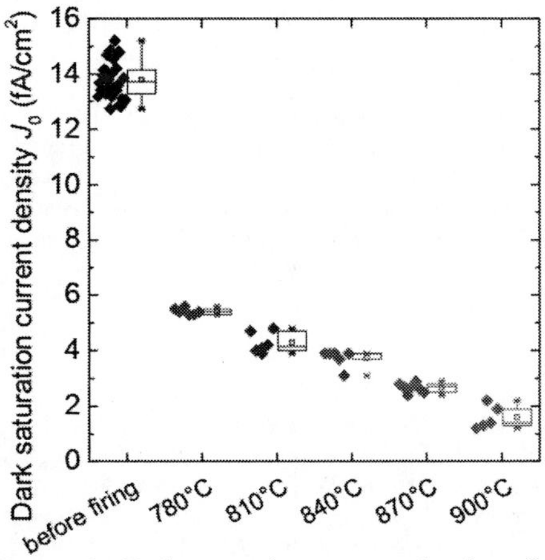

Figure 3: Dark saturation current density of the p^+ polysilicon/SiN_x stack after contact firing at different set temperatures.

3.2 Contact resistance

As pointed out in the experimental section, we apply screen printing technology and four different commercially available Ag pastes for metallization of the polysilicon layer, as screen printing is the state of the art metallization technology today. The results for the specific contact resistance ρ_c, which has been determined from TLM measurements, conducted either in manual or automated measurement modus, are shown in Fig. 4. On the first view it is apparent that higher firing set temperatures lead to lower ρ_c. On the other hand, large differences are visible between the 4 pastes. Results for Ag paste 4 are not shown, as this is a pad paste and, as expected, we could not measure ρ_c as the paste should not contact silicon through the SiN_x layer. From the remaining 3 Ag pastes, Ag paste 2 yields the lowest ρ_c. At 840 and 870 °C firing set temperature, we measure $\rho_c = 2$ mΩcm², which is similar to the values that are achieved when applying this paste for contacting phosphorous-doped emitters with a sheet resistance of around 90 Ω/sq. on a textured surface, for which this paste was developed [14]. The differences between the automated and manual measurements might be related to the measurement setup such as measurement times and subject of further investigations.

The low ρ_c highlights the enormous potential for metallization of boron doped polysilicon/SiN_x stacks by screen-printed Ag pastes.

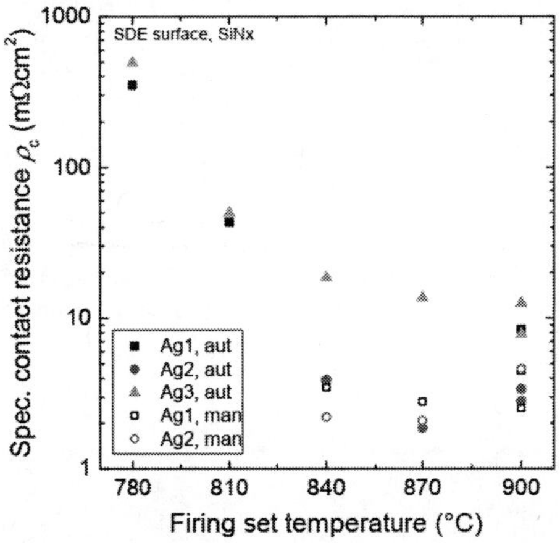

Figure 4: Specific contact resistance ρ_c of three commercially available fire through Ag pastes on p^+ polysilicon/SiN_x stack plotted vs. the firing set temperature.

3.3 Luminescence imaging

To receive information about homogeneity and interaction between paste and passivation, we perform photoluminescence (PL) imaging from the non-metallized side of the wafer using a short pass filter. The results are shown in Fig. 5. For samples fired at 780 °C, the signal in the non metallized regions is on a very high level, as expected from the high iV_{oc} values. The four defects visible result from manual scribing next to the screen printed metal pastes. The Berger structure itself is not visible, which indicates that the passivation quality is preserved under the metal contacts when the sample is fired at 780 °C. On the other hand, this correlates with very high specific contact resistance at this set temperature, see Fig. 4.

For higher firing set temperatures, the grid pattern becomes more and more visible in the PL image, with the exception of Ag pad paste 4, which as pointed out above is supposed not to etch through the SiN_x layer. Thus, for the fire through Ag pastes 1 to 3, the low ρ_c comes at the cost of locally increased recombination at the metal contacts. A detailed investigation of the recombination at the metal contacts is performed in Ref [15].

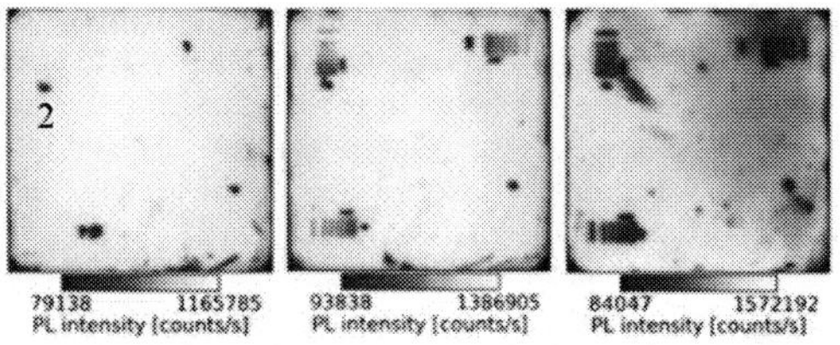

Figure 5: Photoluminescence images taken from non-metallize side of sample fired at 780 °C (left), 840 °C (centre), and 900 °C (right) set temperature, where the grid structure from 3 out of 4 Ag pastes are clearly visible. Ag paste 1 is located in the lower left corner, Ag pastes 2 to 4 follow clockwise.

3.3 SEM images

To further investigate the microscopic origin of the reduced signal at the metallized area, scanning electron microscopy is performed. Prior to this, the metallization as well as the glass layers have been etched by wet chemical processing. On other samples, we have identified that the etching sequence itself removes the SiN_x layer, but leaves the polysilicon layer itself intact. In addition to the experiment plan in Figure 1, which only included samples with saw-damage etched surfaces, here we also included some samples with alkaline textured surfaces. The results are shown in Figure 6, with images taken on samples with saw damage etched surfaces in the left, and samples with textured surface in the right column. Also given is the contact firing set temperature, at which the samples were processed.

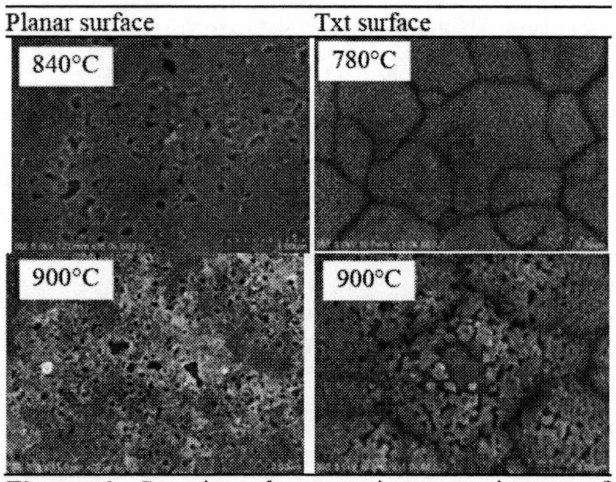

Figure 6: Scanning electron microscopy images of samples with either planar or alkaline textured surfaces, fired at different set temperatures. All samples were metallized with Ag paste 2.

It is apparent, that screen printed Ag pastes interact with the polysilicon surface. On alkaline textured surfaces, this interaction is mainly restricted to the pyramid tips, which are the most protruding structures. Similar behaviour has been observed by several authors when applying such fire through Ag pastes for contacting phosphorous emitters [16, 17]. Here, the metal paste locally completely removes the polysilicon, an effect, which is more pronounced at higher firing set temperatures. Also voids in the polysilicon layer at the facets of the pyramids occur with increasing set temperature.

On planar samples, the polysilicon layer shows similar voids. The size of these voids varies locally, an effect that might be caused by an inhomogeneous particle size distribution within the metal paste before screen printing. Here again, the size of voids increases for higher firing set temperatures, due to the higher reaction kinetics. Apart from the larger voids, with a diameter of around 200 nm, also smaller voids in the double digit nm are visible.

Thus, the partial removal of the polysilicon layer by the Ag paste is the origin of the reduced PL signal at the metal fingers observed in Fig. 5.

4 SUMMARY

In this work, we have fabricated passivating contacts by tunnel oxide formation by thermal oxidation and subsequent deposition of a boron doped polysilicon layer by LPCVD, both of which in the same process. After thermal activation, SiN_x deposition and subsequent contact firing, the layer stack yields dark saturation current densities down to 1 fA/cm^2 (locally) and 1,5 fA/cm^2 (average) on planar samples. In general, there is a strong dependency of J_0 on the firing set temperature, and J_0 decreases for increasing temperatures, which might be due to enhanced hydrogenation. We have tested several screen printed silver pastes for contacting of this stack. Surprisingly, conventional fire through Ag pastes, typically used for contacting phosphorous emitter in p-type solar cells, work quite well. For Ag paste 2 and planar samples, we find a low specific contact resistance $\rho_c = 2$ mΩcm^2 at a firing set temperature of 840 °C, which is compatible with front side firing. However, PL images show a reduced signal in the metallized regions, a sign of increased minority carrier recombination, and this effect becomes more severe with increasing firing temperature. The reason for this is presumably a partial etching of the polysilicon by the applied pastes.

For our samples, there seems to be a process window at 840 or 870 °C, where we find both low ρ_c and low J_0, at a temperature, which should be compatible with front side firing. The investigated structure might not represent a passivated contact in original sense, nevertheless screen printed Ag metallization of boron-doped polysilicon layers is quite promising for application on the rear of p-type solar cells.

ACKNOWLEDGEMENTS

The authors would like to thank their colleagues at Fraunhofer ISE for processing and characterisation. This work was supported by the German Federal Ministry of Economic Affairs and Energy within the research project "HIPPO" under contract 0324086A.

REFERENCES

[1] U. Römer et al., "Recombination behavior and contact resistance of n+ and p+ poly-crystalline Si/mono-crystalline Si junctions," Solar Energy Materials and Solar Cells, vol. 131, pp. 85–91, 2014.

[2] Y. Kwark, R. A. Sinton, and R. M. Swanson, "Low J0 contact structures using sipos and polysilicon films," in 20th IEEE Photovoltaic Specialists Conference Las Vegas, Las Vegas, USA, 1988, pp. 787–792.

[3] J. Y. Gan and R. M. Swanson, "Polysilicon emitters for silicon concentrator solar cells," in 21st IEEE Photovoltaic Specialists Conference Kissimmee, Kissimmee, 1990, pp. 245–250.
[4] F. Feldmann et al., "Efficient carrier-selective p- and n-contacts for Si solar cells," Solar Energy Materials and Solar Cells, vol. 131, pp. 100–104, 2014.
[5] A. Richter et al., "n-Type Si solar cells with passivating electron contact: Identifying sources for efficiency limitations by wafer thickness and resistivity variation," Solar Energy Materials and Solar Cells, 2017.
[6] F. Schindler et al., "Optimized multicrystalline silicon for solar cells enabling conversion efficiencies of 22%," Solar Energy Materials and Solar Cells, vol. 171, pp. 180–186, 2017.
[7] Di Yan, A. Cuevas, Y. Wan, and J. Bullock, "Passivating contacts for silicon solar cells based on boron-diffused recrystallized amorphous silicon and thin dielectric interlayers," Solar Energy Materials and Solar Cells, vol. 152, pp. 73–79, 2016.
[8] U. Römer, Polycrystalline silicon / monocrystalline silicon junctions and their application as passivated contacts for Si solar cells, 2015.
[9] Y. Tao et al., "Large area tunnel oxide passivated rear contact n -type Si solar cells with 21.2% efficiency," Prog. Photovolt: Res. Appl., vol. 24, no. 6, pp. 830–835, 2016.
[10] H. E. Çiftpınara et al., "Study of Screen Printed Metallization for Polysilicon Based Passivating Contacts," in 7th International Conference on Silicon Photovoltaics, Freiburg, Germany, 2017.
[11] M. K. Stodolny et al., "Material properties of LPCVD Processed n-type Polysilicon Passivating Contacts and its Application in PERPoly Industrial Bifacial Solar Cells," in 7th International Conference on Silicon Photovoltaics, Freiburg, Germany, 2017.
[12] R. Naber et al., "n-PERT Solar Cells with Passivated Contact Technology Based on LPCVD Polysilicon and Fire-through Contact Metallization," in 32nd EU PVSEC, München, Germany, 2016, pp. 430–433.
[13] A. Kimmerle, J. Greulich, and A. Wolf, "Carrier-diffusion corrected J0-analysis of charge carrier lifetime measurements for increased consistency," Solar Energy Materials and Solar Cells, vol. 142, pp. 116–122, 2015.
[14] S. Werner, E. Lohmüller, S. Maier, S. Mourad, and A. Wolf, "Challenges for lowly-doped phosphorus emitters in silicon solar cells with screen-printed silver contacts," Energy Procedia, no. N/A, N/A, 2017.
[15] S. Mack et al., "Metallisation of boron-doped polysilicon layers by screen printed silver pastes," Physica Status Solidi Rapid Research Letters, no. submitted, 2017.
[16] P. Rothhardt, S. Meier, R. Hoenig, A. Wolf, and D. Biro, "Influence of surface near doping concentration on contact formation of silver thick film contacts," Sol. Energy Mater. Sol. Cells, vol. 153, pp. 25–30, 2016.
[17] E. Lohmüller et al., "The HIP-MWT+ solar cell concept on n-type silicon and metallization-induced voltage losses," in 29th EU PVSEC, Amsterdam, The Netherlands, 2014, pp. 635–641.

BBr_3 EMITTER PASSIVATION BY ULTRA-THIN BORON DOPED LPCVD POLYSILICON LAYERS

R.C.G. Naber[1]*, M. Lenes[1], J.R.M. Luchies[1], J. Zhai[2], D. Liu[2], Z. Wang[2], F. Li[2]
1. Tempress Systems BV, Radeweg 31, 8171 MD Vaassen, The Netherlands
2. Yingli Energy Co., Ltd., 3399 North Chaoyang Street, Baoding 071051, China
* E-mail: rnaber@tempress.nl; Tel. +31 578 699 200

ABSTRACT: We aim to demonstrate the use of ultra-thin Boron doped LPCVD polysilicon layers for the front-side BBr_3 emitter passivation of n-PERT silicon solar cells. The high passivation performance and the relatively low light absorption of such ultra-thin layers can enable a simple solar cell design with full-area polysilicon passivation on both the front- and the rear-side. To make the cells we apply a single deposition step of undoped polysilicon on both sides, followed by an etch step to make the front-side layer ultra-thin. The doping of the front-side polysilicon layer is achieved by an auto-doping process which uses the emitter as a source of Boron dopant while the rear-side polysilicon doping is obtained by Phosphorus implantation. Solar cell half-fabricates made in this way demonstrated an excellent passivation performance with lifetime values of up to 1.6 ms and a V_{OC} of 685 mV after fire-through contact metallization on the front- and rear-side (100 Ohm/sq emitter). In our preliminary experiments to fully minimize the front-side polysilicon thickness we have so far obtained a best cell efficiency of 20.6% with a V_{OC} of 677 mV (60 Ohm/sq emitter). By retaining a thicker polysilicon layer in the contact-areas during the etch step it may be feasible to obtain a passivated front-side contact using the same process steps.

Keywords: passivation; n-type; silicon solar cell

1 INTRODUCTION

In recent works it has been shown that fire-through metallization can be used to contact doped polysilicon layers that are deposited on the rear-side of n-PERT solar cells in order to reduce the voltage losses that are due to the recombination losses near the metal contacts [1,2]. An interesting prospect is to apply the same technique to the front-side Boron emitter in order to reduce the voltage losses even further [3-5]. Several techniques have been developed for the passivation of Boron emitters to enable high-efficiency n-type solar cell manufacturing processes. Well-known examples are the use of thermal oxide, wet chemical oxide and aluminium oxide layers. In order to combine one of the existing emitter passivation techniques with the polysilicon-based passivated contact technology on the front-side one will need to apply the passivation layer in the non-contact areas and manufacture a Boron-doped polysilicon stack in the contact areas, which is then contacted by the metallization. Such a combination therefore requires structuring of both the oxide passivation layer and the polysilicon stack and the application of Boron doping only to the structured polysilicon stack. It is clear to see that these requirements can result in a relatively complicated manufacturing process.

In our work we tried to simplify the solar cell design and the resulting process flow in order to enable an n-PERT technology with integrated passivated contact technology on the front- and rear-side. To avoid the use of two different passivation schemes on the front-side we implement a polysilicon stack as a full-area passivation layer so that the same LPCVD process can be used to passivate the contact and non-contact areas. In the non-contact areas the layer then has to be ultra-thin in order to minimize absorption losses inside the polysilicon. To avoid a separate doping process for the polysilicon layer we use an auto-doping process that uses the emitter as a source of Boron dopant. We have found that such a process can yield a high passivation quality despite the fact that the Boron is diffused past the tunnel oxide into the polysilicon, which might have been expected to have an adverse effect on the passivation characteristics. With these process and cell design choices we hope to enable a high cell efficiency potential and a relatively simple manufacturing process.

2 EXPERIMENTAL

The figure below presents the structure of the solar cells that were prepared for this study. On the rear-side a stack consisting of an LPCVD in-situ grown oxide tunnelling layer and an n-type doped polysilicon layer is used as a full-area passivation layer and as a contacting layer for the fire-through contact metallization, which has previously been shown to reduce the recombination losses brought by the metallization. On the front-side there is a BBr_3 emitter that is passivated by an LPCVD in-situ grown oxide tunnelling layer and an ultra-thin Boron doped polysilicon layer. Due to the low thickness of the polysilicon layer it is to be expected that the fire-through contact metallization penetrates beyond the polysilicon layer and in this way impairs the ability of the polysilicon layer to reduce the recombination losses brought by the metallization, but it is feasible that such an effect can be enabled by retaining a thicker polysilicon layer in the contact areas. The used process flow readily allows for such a thickness modification, as explained below, but it is not yet implemented in our experiments.

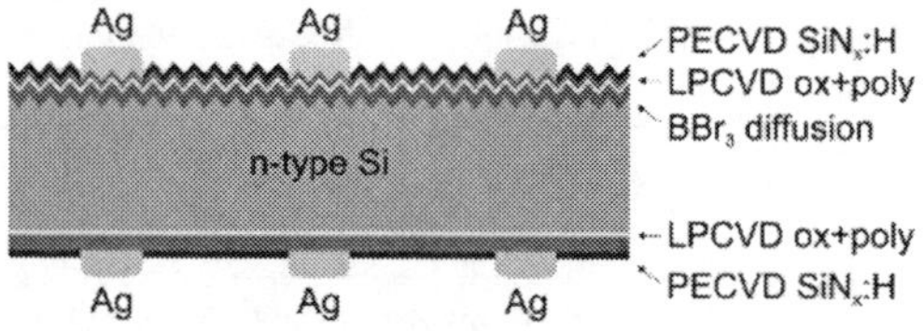

Figure 1. The structure of the developed n-PERT silicon solar cells with polysilicon-based passivation technology on the front- and rear-side.

The process flow presented in figure 2 is similar to the one that is used in the Yingli production line for PANDA bifacial cells and modules except for the omission of the polysilicon deposition step. N-type Czochralski wafers were sourced from this production line with a front-side BBr_3 emitter and rear-side emitter removal and processed to solar cells using the facilities at ECN. An LPCVD process with single-slot loading at 2.4 mm pitch was used to grow a tunnel oxide layer and deposit a ~100 nm thick undoped polysilicon layer on both sides of the wafers.

Doping of the front-side layer was done with an auto-doping process during the thermal oxidation step in which the Boron dopant diffuses from the emitter through the tunnel oxide and into the polysilicon layer. During the same process step the implanted Phosphorus is activated in order to dope the rear-side polysilicon layer. One of the advantages of the auto-doping method is that it only occurs in areas that are adjacent to the emitter and since the emitter is already electrically isolated along the wafer edges from the rear-side doped areas with this production process, the front- and rear-side also remain isolated after the auto-doping process. The presented process flow therefore does not require any additional steps to re-establish the isolation.

To reduce the front-side polysilicon layer thickness we apply a simple etch process which is combined with a normal cleaning step, so it does not increase the number of process steps compared to the normal production process. By masking the front-side contact areas during the etch it is possible to retain a higher polysilicon layer thickness in those areas, for example by ink-jet printing a suitable resist layer. In this way it may be possible to reduce the recombination losses brought by the fire-through contact metallization using a relatively simple process flow modification.

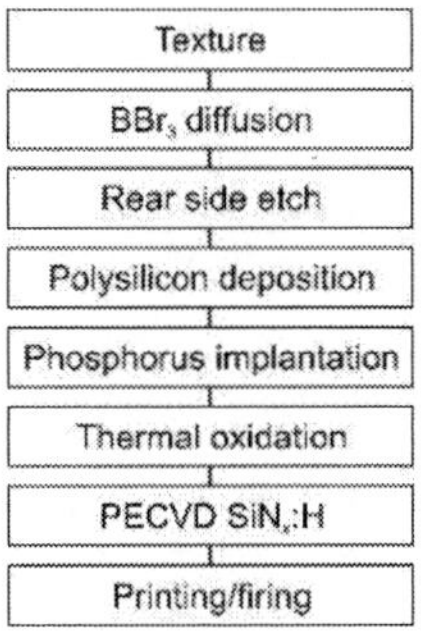

Figure 2. The process flow that was used to make the presented n-PERT solar cells. Standard cleaning steps and a combined cleaning/polysilicon etch step were omitted.

3 RESULTS AND DISCUSSION

3.1 Polysilicon layer characteristics

As part of the experiments that were done so far we have established a reference process that yields an ultra-thin polysilicon layer thickness of ~20 nm on the textured front-side. The reference process has been tested most frequently and gives a high passivation performance in a reproducible way and so for that reason it is used to compare new results to. Figure 3 presents the Boron and Phosphorus doping levels inside the polysilicon layers made with the reference process, measured using ECV profiling (electrochemical capacitance-voltage).

Figure 3a depicts three zones from left to right: (1) a ~20 nm thick Boron doped polysilicon; (2) a transition zone near the tunnel oxide; (3) the top part of the BBr_3 emitter. The dopant level inside the Boron doped polysilicon layer and the BBr_3 emitter is the same at around 5E19 cm^{-3}. With an auto-doping process that uses the emitter as a source of Boron dopant it is to be expected that the concentration in the polysilicon and the emitter would be about the same, so the data is consistent with that expectation. The transition zone (2) shows quite a bit of data scatter, which we believe is due to the ECV measurement technique.

The ECV profile of the rear-side in figure 3b shows that the polysilicon layer has a doping level of 3E20 cm^{-3} and a thickness of around 110 nm. The data also indicates that the layer thickness on the front-side was originally 110 nm after the LPCVD deposition, which later became ultra-thin after the application of an etch step. The dopant tail past the tunnelling layer shows that there was some light diffusion through the oxide, but the large transition in concentration clearly shows that the in-situ grown LPCVD oxide functions as a strong diffusion barrier to the implanted Phosphorus.

In a previous paper we showed that a layer thickness of ~70 nm on the rear-side can yield a better cell efficiency than thicker layers due to a higher I_{SC} [1]. For the reference process we chose to use a thicker layer despite the lower cell efficiency to lower the dependency on the choice in fire-through contacting paste in the cell experiments, which becomes more critical for thinner layers.

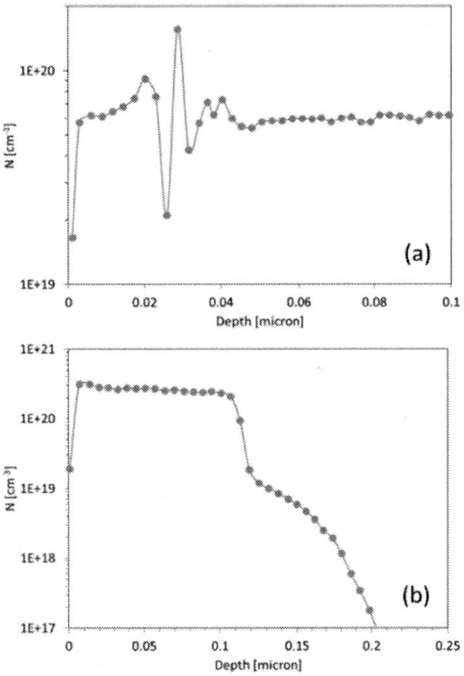

Figure 3. ECV profile measurements. (a) Sample with an ultra-thin Boron doped polysilicon layer on top of a BBr_3 emitter. (b) Sample with a Phosphorus doped polysilicon layer and a dopant tail inside the substrate.

3.2 Passivation performance

Table I presents data obtained on solar cell half-fabricates that were processed using the reference process up until the PECVD SiN_x:H deposition step. For comparison we include one group with a thermal oxide layer as the emitter passivation layer, similar to that reported in an earlier paper [1]. All three groups were processed simultaneously using the same process flow, except that the thermox group was processed with front-to-front loading in the LPCVD polysilicon deposition step. The results with ~20 nm thick polysilicon layers show a high passivation performance, especially when combined with a high-ohmic emitter, which indicates that the potential gain in V_{OC} with this passivation technique is significant. The results can perhaps be improved further by optimizing the emitter dopant profile or by smoothening the alkaline texture in order to reduce the total surface area.

Compared to the thermox passivation the lifetime is almost twice as high for the same 60 Ohm/sq emitter, but before firing the contrast is even larger, as the increase with polysilicon passivation after simulated firing is relatively small, especially for the high-ohmic emitter. An explanation for this observation may be that the passivation quality of the Boron auto-doped polysilicon is less dependent on the atomic hydrogen that is released from the nitride layers during the simulated firing. Such behaviour is unusual because Boron emitter passivation techniques other than thermox, such as wet chemical oxides, are also known to require a firing step to activate the passivation fully [6].

Front passivation	Emitter [Ohm/sq]	τ [ms] (BF)	τ [ms] (AF)	$J_{0,total}$ [fA·cm^{-2}] (AF)
Thermox	60	0.1	0.48	97
Polysilicon	60	0.5	0.98	59
Polysilicon	100	1.3	1.5	34

Table I. QSSPC measurement results on solar cell half-fabricates before (BF) and after (AF) simulated firing. Each group consisted of two samples, the results of which were averaged. Lifetime and $J_{0,total}$ values were extracted at 1E15 and 1E16 cm^{-3} injection level, respectively.

A printing experiment was performed on the sample groups in order to assess the passivation performance on cell level, as presented in Table II. The results show the expected gain in V_{OC} for the polysilicon passivation groups, with a peak value of 685 mV. The cell efficiency of the groups was suboptimal due to a low doping level on the rear-side (caused by a procedural issue) and the relatively high thickness of the front-side polysilicon, but the 100 Ohm/sq group did reach up to 20.5%.

Front passivation	Emitter [Ohm/sq]	No. of cells	V_{OC} [mV]
Thermox	60	7	662
Polysilicon	60	6	678
Polysilicon	100	7	684

Table II. Averaged IV measurement results on solar cell samples from the same groups as in Table I.

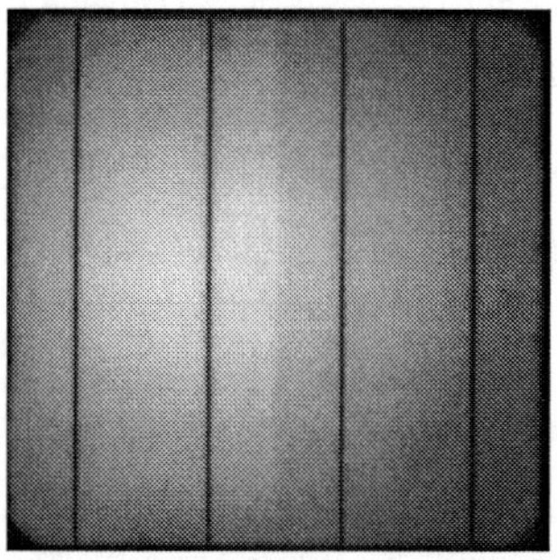

Figure 4. Averaged photoluminescence image for the cells in Table II with a 100 Ohm/sq emitter. The left-right brightness asymmetry is caused by the illumination.

The PL measurements presented in figure 4 show that the passivation is homogeneous across the whole wafer area. This confirms that the Boron auto-doping process is suitable for large-area applications.

3.4 n-PERT solar cell efficiency potential

Having established a reference process with a ~20 nm polysilicon layer thickness with a high passivation performance, we have conducted a few small tests to try to reduce the thickness to less than 10 nm in order to minimize the absorption losses. There are a number of possible approaches towards that goal and we recently did preliminary tests on an approach which now appears to be the most promising and consists of a simple back-etch procedure. Cell results are not yet available however. In Table III we present a preliminary result using another approach to minimize the polysilicon thickness, which enables a ~25% thickness reduction (estimated from the used process settings). The 20.6% peak efficiency combined with a 677 mV V_{OC} is the best cell result so far, which we will use to estimate the solar cell efficiency potential.

Front passivation	V_{OC} [mV]	I_{SC} [A]	FF [%]	η [%]
Best cell	677	9.38	79.0	20.58
Average	676	9.41	78.8	20.45

Table III. IV measurement results on M2-size solar cells with a 60 Ohm/sq emitter and poly-Si front- and rear-side passivation.

The I_{SC} in Table III is a bit lower than what has been achieved in other works on n-PERT solar cells, but with further optimization it should be possible to obtain a considerably higher value using the following adjustments:

- 5 nm front polysilicon instead of ~15 nm
- 50-80 nm rear polysilicon instead of ~110 nm
- High-ohmic emitter (80-100 Ohm/sq)
- Graded PECVD SiN_x:H with lower absorption

The high-ohmic emitter should also increase the average V_{OC} to more than 680 mV, as presented in Table II. Exactly how much I_{SC} gain there will be is difficult to quantify but a reasonable estimate would appear to be 0.3-0.4 A based on our experiment results that encompassed some of the proposed adjustments and the size of the polysilicon thickness change from ~15 to 5 nm. Combined with a fill factor response of 80% that is normally achieved for this cell type with four-busbar screen-printed metallization, the cell efficiency should be able to go beyond 21.5%. This efficiency value then still suffers from a significant V_{OC} loss brought by the front-side metallization. The implied V_{OC} for the 100 Ohm/sq group in Table I is estimated to be ~712 mV and yields an average V_{OC} of 684 mV in Table II, which indicates that the front metallization V_{OC} loss is 25-30 mV. If we overcome these losses by retaining a thicker Boron auto-doped polysilicon layer in the front-side contact areas then a V_{OC} of >700 mV may be feasible, which then enhances the cell efficiency potential to beyond 22.5%. More advanced metallization schemes such as multi-busbar technology that can reduce the resistance losses inside the contact fingers should be able to enhance the cell efficiency even further.

In collaboration with Fraunhofer ISE we have conducted some preliminary tests with ink-jet printed resist layers to try to retain a higher polysilicon in the front-side contact areas. Figure 5 presents a microscope image of such a patterned polysilicon layer. The polysilicon was deposited on top of a PECVD SiN_x:H layer for color perception purposes. A pattern was created using ink-jet printing of a resist and then the etching process was applied that is also used for the previously presented solar cells. The pink color of the stripe pattern indicates that it has the as-deposited polysilicon thickness while the surrounding areas have a blue color due to a reduced polysilicon thickness, which should be ~20 nm based on the used process settings. Work is ongoing towards making a first solar cell with Boron auto-doped polysilicon passivation that has a height pattern made with ink-jet printing.

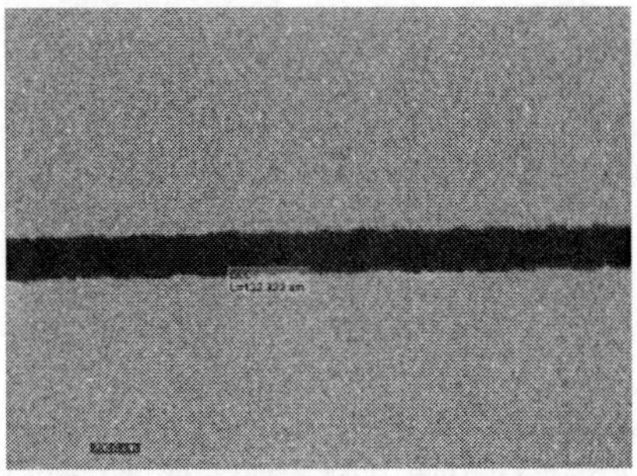

Figure 5. Microscope image of a patterned polysilicon layer deposited on top of a PECVD SiN_x:H layer for colour perception purposes.

The stated cell efficiency gains should also enhance the rear-side cell efficiency to some extent, but it remains to be seen by how much. We previously obtained a bifaciality of 84% with a 70 nm rear-side polysilicon layer [1]. The cell technology should remain suitable for bifacial module applications that can yield 10-30% more power output. The gains in V_{OC} should also help lower the module power temperature coefficient.

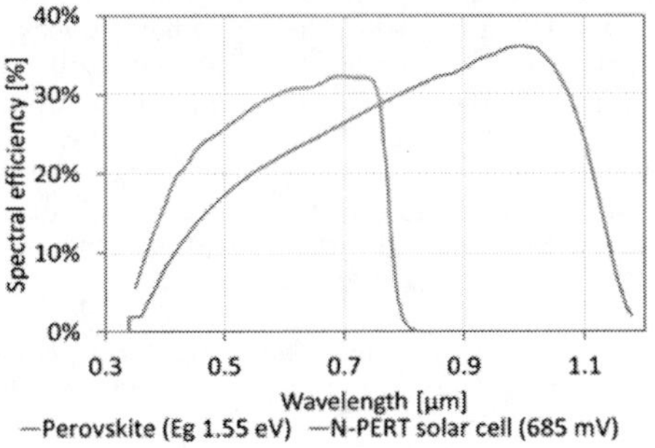

Figure 6. Spectral efficiency for a perovskite solar cell and an n-PERT solar cell with front- and rear-side polysilicon passivation.

Another approach towards higher cell efficiencies is to make tandem solar cells where the n-PERT solar cell serves as a bottom cell that mainly harvests light in the red part of the spectrum. The front-side polysilicon layer could be made thicker in this case, since the absorption in the blue part of the spectrum is done by the top cell. The most simple tandem cell configuration is four-terminal in which the cells are simply stacked and contacted individually. Due to the presence of a conductive Boron-doped passivation layer it may also be feasible to make tandem cells in a two-terminal fashion, where the top cell is prepared directly on top of the polysilicon layer. Figure 6 presents spectral efficiency data for a perovskite solar cell and one of the cells from Table II with a 100 Ohm/sq emitter [7]. It demonstrates that the two solar cells can enhance their mutual performance by providing a higher conversion efficiency in a wider range of the solar spectrum.

4 CONCLUSIONS

The front-side Boron emitter passivation technology using ultra-thin Boron auto-doped polysilicon that has been developed was shown to bring high V_{OC} values of up to 685 mV in combination with fire-through contact metallization. Compared to other Boron emitter passivation techniques it brings some unique advantages in that it can simplify the process flow and cell design and potentially enable passivated front-side contact technology in a way that only requires one additional ink-jet printing step. In this way it may help to increase the industrial adoption of n-type silicon solar cell technology.

5 ACKNOWLEDGEMENTS

We thank Mike Jahn, Roman Keding, Andreas Wolf and Martin Zimmer from Fraunhofer ISE for the preliminary work on patterning polysilicon layers with ink-jet printed resist layers presented in figure 5 and we thank Stefan Luxembourg, Dong Zhang and Bart Geerligs from ECN Solar Energy for the spectral efficiency data presented in figure 6.

6 REFERENCES

[1] R.C.G. Naber *et al.*, "N-PERT solar cells with passivated contact technology based on LPCVD polysilicon and fire-through contact metallization", Proc. 32nd EU-PVSEC, Munich, Germany, 2016.

[2] M.K. Stodolny *et al.*, "N-type polysilicon passivating contacts for industrial bifacial N-PERT cells", Proc. 32nd EU-PVSEC, Munich, Germany, 2016.

[3] R. Peibst *et al.*, "Implementation of n^+ and p^+ POLO junctions on front and rear side of double-side contacted industrial silicon solar cells", Proc. 32nd EU-PVSEC, Munich, Germany, 2016.

[4] F. Feldmann *et al.*, "The application of poly-Si/SiO_x contacts as passivated top/rear contacts in Si solar cells", Sol. Energy Mater Sol. Cells, vol. 159, pp. 265, 2017.

[5] A. Ingenito *et al.*, "Silicon solar cell architecture with front selective and rear full area ion-implanted passivating contact", Sol. RRL, vol. 1, pp. 1700040, 2017.

[6] V.D. Mihailetchi *et al.*, "Nitric acid pretreatment for the passivation of Boron emitters for n-type base silicon solar cells", Appl. Phys. Lett., vol. 92, pp. 063510, 2008.

[7] S.L. Luxembourg *et al.*, "Investigation of industrial crystalline silicon cell architectures as bottom cell in perovskite/c-Si hybrid tandems", Proc. 33rd EU-PVSEC, Amsterdam, The Netherlands, 2017.

BORON REMOVAL FROM SILICON BY MOISTURIZED GASES

Jafar Safarian, Claudio Sanna, Gabriella Tranell
Department of Materials Science and Engineering, Norwegian University of Science and Technology (NTNU), No-7491 Trondheim, Norway

ABSTRACT: The photovoltaic (PV) industry is in rapid growth and a large supply of PV feedstock materials must be provided to maintain this growth. Since silicon is still the dominant material for the fabrication of solar cells, low-cost solar-grade silicon (SoG-Si) feedstock is demanded by the PV industry and a continuous growth for it is expected. The most cost-effective and direct approach for producing SoG-Si is to purify and upgrade metallurgical-grade silicon produced from raw materials. The majority of impurities existing in MG-Si are effectively removed through directional solidification of molten silicon, which is a key process in silicon solar cells production. However, the removal of boron and phosphorus by this method is difficult and expensive due to the relatively large distribution coefficients of these elements between solid and liquid silicon during solidification. Therefore, the elimination of these elements to the levels required for SoG-Si feedstock requires other processes. In the present work, the boron removal by top-gas blowing of H_2, H_2-H_2O, Ar-H_2, Ar-H_2-H_2O, He-H_2O and He-H_2-H_2O over silicon melt is studied. It is shown that the type of gas mixture and the concentration of water vapor in the gas affect the rate of boron removal. Moreover, it is indicated that the kinetics of boron removal from silicon is depending on the mass transport of the gaseous species in the system and also the chemical reactions at the melt surface.

Keywords: Silicon, Solar, Refining, Phosphorous, Boron, Solidification

1 INTRODUCTION

The production of high purity silicon for PV industry through metallurgical processes have been important in recent years with even commercial production, however the most common route to produce SoG-Si from Metallurgical Grade Silicon (MG-Si) is still based around so-called the Siemens technology which includes gasification, distillation and re-deposition of silicon via trichlorosilane, a very energy intensive method which also generates significant amounts of chemical waste. [1] More cost effective metallurgical processes for making solar grade silicon – with less environmental impact – have been studied and developed in the recent years due to the rapid PV market growth. Boron (B) and phosphorus (P) are the two crucial impurity elements that are most difficult to be removed from silicon, since they cannot be effectively separated by directional solidification, a key process step for removing many impurities such as Fe, Ti, Al, Ca, Mg, *etc.*. The removal of B and P requires the dedicated refining processes and has been the motivation of many purification process studies. In addition to boron slag refining technology, which is currently applied industrially by ELKEM Solar AS, the removal of B can be performed through the treatment of liquid silicon by oxidative plasma or other gas refining techniques.

In the oxidative gas refining processes, the dissolved B in molten silicon is gasified through the formation of volatile B-containing compounds.[1] The removal of boron by different reactive gases, such as O_2[2,3], H_2[4,5,6,7,8], CO_2[3], and H_2O[9,10,11,12,13,14,15], has been studied through several research groups. It has been observed that the application of the H_2-O_2 gas mixtures in plasma refining is more effective in terms of the kinetics of B removal compared to the single H_2, O_2, CO_2 gases[16,17]. In our recent publication[6] the removal of boron from silicon by top blowing of humidified hydrogen has been studied in the present work through experimental work, thermodynamic calculations, computational fluid dynamic modeling, and quantum chemistry calculations. The effect of process parameters; temperature, lance diameter, lance distance from the melt surface, gas flow rate, and crucible material on the kinetics of boron removal were studied. It has been shown that the rate of boron removal is decreased with increasing temperature due to the competitive reactions between silicon and oxygen as well as boron and oxygen, which can be confirmed with the increases of p_{SiO}/p_{HBO} in the system. The rate of boron removal is increased with increasing the gas flow rate due mainly to the better supply and transport of the gas over the melt surface, as confirmed by the CFD modeling. Moreover, the rate of boron removal in alumina crucible is the highest followed by that in quartz and graphite crucibles, respectively. Faster B removal in quartz crucible than that in graphite crucible can be attributed to more oxygen dissolves in silicon melts. The fastest boron removal in alumina crucible is attributed to the additional boron gasification through aluminum borate ($AlBO_2$) formation on the melt surface. Thermodynamic properties of the $AlBO_2$ species have thus been revised by quantum chemistry calculations, which were more accurate to describe the formation of gaseous $AlBO_2$ than those in the JANAF Thermochemical Tables.

Based on the previous studies, the main chemical reactions for boron gasification from silicon melts have been proposed as:

In graphite, quartz and alumina crucible:

$$\underline{B} + \underline{H} + \underline{O} = HBO\,(g) \qquad (1)$$

In alumina crucible:

$$\underline{Al} + \underline{B} + \underline{O} = AlBO_2\,(g) \qquad (2)$$

Based on the obtained results, it has been proposed that boron removal from silicon melt by humidified hydrogen is controlled both by the chemical reaction for boron gasification and mass transport in the adjacent gas phase.[6] In the present study the use of top-gas blowing of silicon by H_2, H_2-H_2O, Ar-H_2, Ar-H_2-H_2O, He-H_2O and He-H_2-H_2O over silicon melt is studied through experimental work in order to discover the effect of the gas phase on kinetics of boron removal.

2 EXPERIMENTAL PROCEDURE

2.1 Materials

In the present study high purity polysilicon from Siemens process was melted and alloyed by high purity boron (99.9%). A Si-B master alloy containing 500 ppmw boron was first prepared, and then it was used to dope again polysilicon by boron. The concentration of B in silicon was around 10±1 ppmw for the refining experiments. The refining experiments were done on 400g silicon melt in high quality-high purity graphite crucibles proper for silicon melting.

2.2 Refining experiments

The refining experiments were done in a vacuum induction furnace with controlled atmosphere. The sample (crucible+boron doped silicon) were put in the furnace and it was evacuated and Ar flashed for several times. Then it was heated up under Ar flow and after complete melting the refining gas was purged over the melt with 5 NL/min at 1550°C. The refining was done for different duarations using H_2, H_2-H_2O, Ar-H_2, Ar-H_2-H_2O, He-H_2O and He-H_2-H_2O gases, where the steam was introduced to the dry gases by a humidifier. The amount of H_2O in H_2-H_2O gas mixtures was 2, 4 and 6 vol%. and for the other experiments it was fixed at 4 vol%.

During refining samples from the melt were taken in different refining durations and silicon samples were analyzed by high-resolution Inductively Coupled Plasma-Mass Spectroscopy technique. In this case two parallel samples were analyzed and the average measured boron concentration was considered as the reliable result.

3 RESULTS AND DISCUSSION

The obtained experimental results are presented and discussed as follows.

3.1 Refining by H_2 and H_2-H_2O gases

The measured concentrations of B for different refining times (C_t) were normalized with regard to the initial concentration of B (C_0). The obtained results are shown in Figure 1. As we see in Figure *1-a*, the rate of B removal is significantly affected by the introduction of steam into hydrogen. More over, for the applied refining conditions the highest rate is obtained for the case of 4 vol% H_2O. This may be due to the oxygen potential in the system and also competition between Si and B elements on the melt surface toward oxidation by SiO and HBO gases formation, respectively. Obviously this result may indicate that the rate of B removal is dependent on the chemical reactions at the surface and rate of HBO formation.

3.2 Effect of inert gas introduction

The effect of introduction of inert gases into the refining gas is seen in Figures 1-b and 1-c when Ar and He are introduced into H_2 and H_2-H_2O gases, respectively. It is observed that when Ar and He are introduced into hydrogen with around 50%, the slow rate of B removal becomes a little better. Under these conditions, the removal of B may be mainly through the formation of BH gas as there is not oxygen available in the system. Moreover, the comparison of results for 50%Ar and 50%He gases with hydrogen shows that the rate and extent of B removal by Hydrogen is almost two times higher when He is used. This may indicate that the mass transport of gaseous species such as BH in the gas phase is important and can be a rate limiting step.

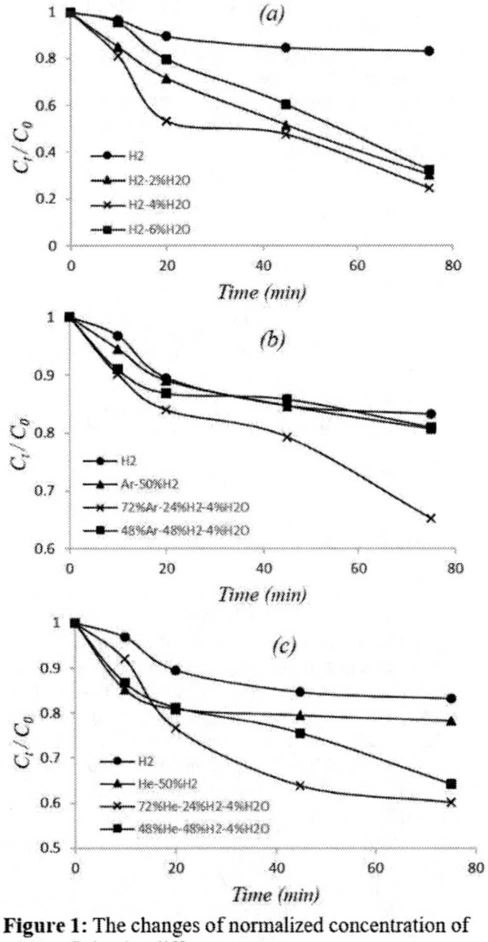

Figure 1: The changes of normalized concentration of B during refining by different gases.

The presented results in Figures 1-b and 1-c indicate also that for the both Ar and He containing gases, the rate and extent of B removal is better when steam is introduced into Ar-H_2 and He-H_2 gases. Obviously, this is due to the changing of the mechanism of B removal, in which B is removed through chemical reaction (2) by HBO formation. It is worth mentioning that HBO is much more volatile than B_xH_y species, and when oxygen is available HBO formation is more favorable. Figure 1 also shows that for the refining by these three-commponent gases, the best refining rate is observed for the case in which less hydrogen is used, which is here 24%H_2. This may indicate that the gas phase composition plays again an important role.

3.3 Mass transfer coefficient

Based on many research works in the area of gas refining of silicon, it is generally accepted that the B removal in this technique is through a first order reaction. Therefore, we may write:

$$ln\left(\frac{C_0}{C_t}\right) = k_B \frac{A}{V} t \quad (3)$$

Where k_B is the overall mass transfer coefficient, A is the melt surface, and V is the volume of the melt. Considering the experimental conditions and the results in Figure 1, the relationship between the left hand side of equation (3) and the $(A/V)t$ may give a line, and k_B can be determined. This is shown for selected experimental results (Figure 1-a) as illustrated in Figure 2.

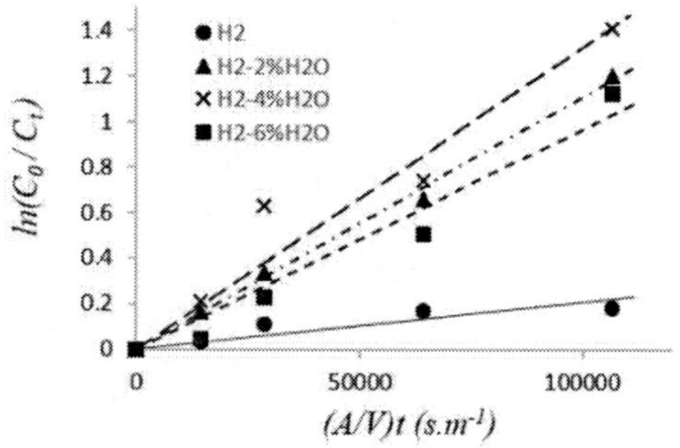

Figure 2: Graphical applied method for calculating the mass transfer coefficients for selected refining trials.

Regarding the determined mass transfer coefficients, it is concluded that the highest rate of B removal for H_2 and H_2-H_2O refining gases is obtained for around H_2-4%H_2O gas as illustrated in Figure 3. The results of our former works in relatively similar process conditions, however different temperatures are also shown in this Figure and they may show that an optimum steam concentration is close to 4% H_2O, and application of high steam concentrations does not increase the process rate significantly.

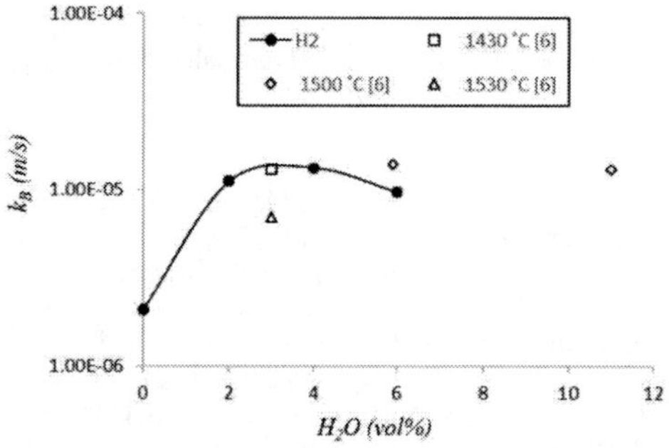

Figure 3: The relationship between the mass transfer coefficient of B and steam concentration for H_2-H_2O refining gas mixtures.

As observed above the rate of B removal for Ar- H_2-H_2O and He- H_2-H_2O is dependent on the type of the inert gas and also the amount of hydrogen, when the steam concentration is fixed. Figure 4 shows the determined k_B values for the related experiments and it indicates that there is higher rate of B removal at the lower H_2 concentration and also when He is used instead of Ar. The dependence of k_B on the type of inert gas must be attributed to the mass transport of the B-containing gaseous product (HBO) in the gas phase.

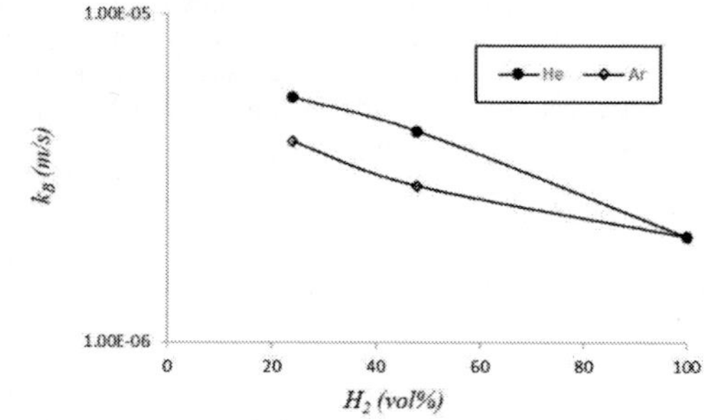

Figure 4: The relationship between the mass transfer coefficient of B and steam concentration for H_2-H_2O refining gas mixtures.

4 CONCLUSIONS

The removal of B from silicon was studied using different gas compositions by top gas blowing technique and using dry and moisturized gases. It was found that rate and extent of B removal is the best when H_2-H_2O gas mixture with around 4vol% H_2O is used. In addition, the mixing of different inert gases with dry and moisturized hydrogen and studing the kinetics of B removal indicated that the mass transport of B-containing gaseous products is affected by the inret gas type and concentration, which is indicating that the process rate is depending on the mass transfer of BH and HBO species in the gas phase.

5 ACKNOWLEDGEMENT

The authors would like to appreciate the financial support provided by the Research Domain 3-Recycling and Refining in in SFI-metal production[18], a center for research-based innovation in Norway.

6 REFERENCES

[1] J. Safarian, G. Tranell, M. Tangstad, Energy Procedia, 20 (2012), 88.
[2] S. Tsao and S.S. Lian: Mater. Sci. Forum, vols. 475–479 (2005), 2595.
[3] K. Suzuki, K. Sakaguchi, T. Nakagiri, and N. Sano: J. Jpn. Inst. Met., 54 (1990), 161.
[4] E. Nordstrand and M. Tangstad: Metall. Mat.. Trans. B, 43B (2012), 814.
[5] S. Rousseau, M. Benmansour, D. Morvan, and J. Amouroux: Sol. Energy Mater. Sol. Cells, 91 (2007), 1906.
[6] J. Safarian, K. Tang, K. Hildal, G. Tranell: Metall.Mat. Trans. E, 1E (2014), 41.
[7] J. Safarian, K. Tang, J.E. Olsen, S. Andersson, G. Tranell, K. Hildal: MMTB, 47 (2016), 1063.
[8] C. Sanna, Master thesis: Boron removal from molten silicon by different gas mixtures, NTNU, 2016.
[9] T. Ikeda and M. Maeda: Mater. Trans., 37 (1996), 983.
[10] Khattak C.P., Joyce D.B., and Schmid F.: Report NREL/SR-520-27593, National Renewable Energy Laboratory, 1999.
[11] C.P. Khattak, D.B. Joyce, and F. Schmid: Sol. Energy Mater. Sol. Cells, 74 (2002), 77.
[12] N. Yuge, M. Abe, K. Hanazawa, H. Baba, N. Nakamura, Y. Kato, Y. Sakaguchi, S. Hiwasa, and F. Aratani: Prog. Photovolt.Res. Appl., 9 (2001), 203.
[13] N. Nakamura, H. Baba, Y. Sakaguchi, and Y. Kato: Mater. Trans., 45 (2004), 858.
[14] J.J. Wu, W.H. Ma, Y.N. Dai, and K. Morita: Trans. Nonferr. Met. Soc. China, 19 (2009), 463.
[15] J. Safarian, K. Tang, K. Hildal, G. Tranell: Metall. Mater. Trans. E, 1E (2014), 41.
[16] C. Alemany, C. Trassy, B. Pateyron, K.-I. Li, and Y. Delannoy: Sol. Energy Mater. Sol. Cells, 72 (2002), 41.
[17] E. Fourmond, C. Ndzogha, D. Pelletier, Y. Delannoy, C. Trassy, Y. Caratini, Y. Baluais, and R. Einhaus: 19th European Photovoltaic Solar Energy Conference, Paris, France, 7–11 June, 2004.

[18] SFI Norwegian acronym for Centre for Research driven Innovation (CRI)

HYDROMETALLURGICAL PURIFICATION OF MAGNESIUM-DOPED SILICON BY DIFFERENT ACIDS

Jafar Safarian, Stine Espelien
Department of Materials Science and Engineering, Norwegian University of Science and Technology (NTNU), No-7491 Trondheim, Norway

ABSTRACT:
The purification of commercial metallurgical grade silicon through alloying and then acid leaching is studied. Magnesium is added into molten silicon and then acid leaching of the solidified Si-Mg alloy by HCl, H_2SO_4 and HNO_3 acids is done. The main parameters that are studied are acid concentration on leaching of particles with a specific size range. Chemical analysis of the samples before and dafter leaching are determined by high resolution Inductively Coupled Plasma-Mass Spectroscopy (ICP-MS). It is shown that the impurity removal from silicon is depending on the type of acid. Moreover, the applied method is efficient for the removal of particular impurities such as P, Fe, Ti, Al, Ca, ... In addition, it is indicated that HCl is more effective than the other types of acids for given process conditions. Moreover, the best silicon purification, and in particular P removal is obtained when HCl with 10% and 15% concentrations is used.
Keywords: Silicon, Solar, Refining, Magnesium, Phosphorous, Acid, Leaching

1 INTRODUCTION

The silicon feedstock for solar cells is mainly produced through the chemical Siemens process. Although this process and the more recent developed Fluidized Bed Reactor (FBR) technology make silicon of high purity i.e. 6 N up to 11N, but these chemical processes have high energy consumptions per unit mass of final product. Considering the fast growing demand for silicon to the photovoltaic industry, significant amount of solar grade silicon feedstock is required in near future. This has lead to the development of metallurgical production routes, which are in principle less energy consuming, environmental friendly and more cost efficient. Hydrometallurgical treatment of silicon has been shown to be an effective method for removal of metallic impurities in the production of solar grade silicon feedstock. [1] Boron (B) and phosphorus (P) are the two crucial impurity elements that are most difficult to be removed from silicon, since they cannot be effectively separated by directional solidification, a key process step for removing many impurities such as Fe, Ti, Al, Ca, Mg, and so on. The removal of B and P requires the dedicated refining processes and it has been the motivation of many purification process studies. In addition to boron slag refining technology, which is currently applied industrially by ELKEM Solar AS, the removal of B can be performed through the treatment of liquid silicon by oxidative plasma or other gas refining techniques. Phosphorus can be removed by different processes such as vacuum refining and solvent refining techniques [1, 2]. Hydrometallurgical refining of silicon in which acid leaching technique is employed to remove the segregated impurities on the silicon grain boundaries is a process that can be applied to remove many impurities and even P largely.

During solidification of molten silicon, a major portion of the impurities will precipitate in the form of silicide at grain boundaries and positions between the silicon grains or a secondary phase such as $CaSi_2$ or Mg_2Si, which is produced through eutectic solidification of Ca or Mg-containing silicon. [2-5] These impurities can be dissolved into acid solutions, while the silicon phase is not attacked by any aqueous acids, except for hydrofluoric acid in combination with an oxidation agent. Hence, leaching by acid will dissolve and remove the impurities located at the grain boundaries, while the silicon matrix is not dissolved, and lumps of silicon are used, they are disintegrated to much smaller particles. Effects of hydrometallurgical conditions on silicon refining has been studied in several studies [6-11] and is currently one of the refining steps in the Elkem Solar Silicon process. One of the earliest published results in this field was done by Dietl (1983), and he used MG-Si as raw material and did leaching by a combination of HCl and HF as acids. He achieved removal of all metallic impurities, except Al, down to less than 10 ppma. The result was therefore very promising for using acid leaching as a refiner method for SoG-Si production, but the other studies have illustrated later that this method is not providing good results for B and P removal because of the high segregation coefficients of these elements. For improving the removal of impurities by leaching, the alloying of silicon with elements that can lower the segregation coefficients of B and P, and enhance the removal of other impurities is interested. The most studied refiner element in literature is Ca, and our recent research works [3-5] has shown that Mg is also a very good candidate to be added as an intermediate refiner into silicon. In the present study, the effect of acid type and concentration on the removal of impurities from Mg-doped silicon are studied.

2 EXPERIMENTAL PROCEDURE

In the present study a commercial metallurgical silicon was doped by Mg and then it was leached by acids. The details of the experimental procedure is described as follows.

2.1 Materials

A commercial metallurgical grade silicon (MG-Si) was mixed with magnesium silicide (Mg_2Si) with purity above 99% and to obtain around 2 wt% Mg. The mixture was then heated and smelted in high quality graphite crucibles at 1500°C in an induction furnace. The obtained Si-Mg alloy was then slowly cooled down and solidified with average of 10°C/min.

Table I: Chemical composition of Mg-doped silicon.

Impurity	Concentration
Mg	1.94 wt%
Fe	1619 ppmw
Al	259.8 ppmw
Ca	252.2 ppmw
Ti	138.4 ppmw
Mn	34.6 ppmw
B	44.9 ppmw
P	12.2 ppmw
Zr	5.5

2.2 Acid leaching

The solidified silicon samples were crushed and particles with 0.315-1.0 mm particle size were separated by sieving and then similar masses were leached by HCl, H_2SO_4 and HNO_3 solutions with 5, 10, 15 and 20% concentrations. The leaching was performed at 60°C and for 3 hours.

Silicon samples before and after leaching were analyzed by high-resolution Inductively Coupled Plasma-Mass Spectroscopy technique. In this case two parallels were analyzed and the average of measured concnetrations were considered as the concentration after leaching. The ICP-MS measurements were accurately done at NTNU and they were supported with testing blank and reference samples analysis with the silicon samples.

3 RESULTS AND DISCUSSION

The obtained experimental results are presented and discussed as follows.

3.1 Effect of acids concentration

In order to study the elimination of impurities form silicon, a removal degree parameter is calculated based on the ICP-MS analysis results as:

$$Removal\ degree = \frac{C_i - C_f}{C_i} \times 100$$

Where C_i and C_f are the concentrations of the impurity before and after leaching, respectively.

The results for leaching with different concentrations of HCl are shown in Figure 1. It is observed that the removal degrees for all impurities are largest for 10% and 15% HCl, the concentration of 10% is slightly better than 15%. A concentration of 20% is shown to be less effective than both 10 and 15% for almost all the elements removal. The low acid concentration of 5% HCl gives significantly the smaller removal degree of all impurities. Boron is removed insignificantly to maximum 9%, while HCl in the process is more effective for P removal and more than 60% of phosphorus is removed in all acid concentrations.

The effect of concentration of HNO_3 on the removal of impurities is shown in Figure 1 and as seen the application of this solvent is effective. The best concentration of acid is 10% or 15% and these two concentrations are giving close results. For more elements, while the higher concentration of 20% is probably effective on the removal of F and Ti. Similar to HCl, the low concentration of 5% is not effective on the impurity removal and the lowest removal degrees for all elements are observed for this concentration. It is worth noting that the results for B removal were not reliable and are not shown in the figure.

The results of leaching of silicon particles by H_2SO_4 are shown in Figure 1 and it is observed that all the impurities are partially removed by different concentrations of H_2SO_4. Moreover, it is observed that all acid concentrations with 5 to 20% concentrations are effective, and the lower B removal degrees are seen for 15% H_2SO_4.

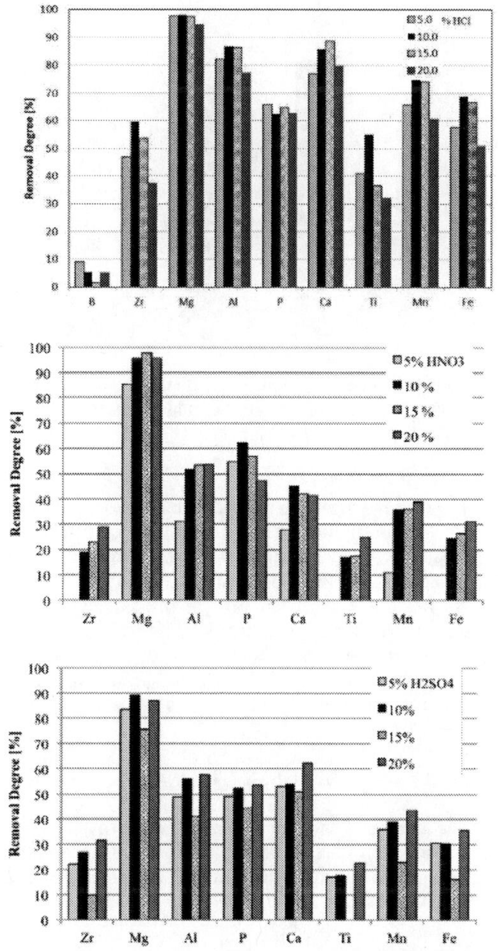

Figure 1: The removal degree of impurities form Mg-doped silicon by different acids with different concentrations.

3.2 Effect of type of acid

The presented results in Figure 1 for different types of leaching agents show that for a given concentration of acid, the removal degree is dependent on the type of acid. For instance, for acids with 10% concentration, we see Fe removal of 68%, 30% and 25 % for HCl, H_2SO_4 and HNO_3, respectively. It is also observed in this figure that the removal of P is in general better by HCl than HNO_3 and that is better than H_2SO_4. We observe more than 60%

P removal by HCl, while we see lower removal degrees for the other two types of acids.

The overall purification degree of silicon in leaching by the three different acids were calculated for all the studied concentrations and the results for the best conditions are shown in Figure 2. As seen the purification degree is better with considering the removal of Mg, which was added in the alloying step. This is due to the easy dissolution of Mg_2Si phase in the acids, which causes also the removal of other impurities. Figure 2 shows also that HCl is more effective than the other two acids and H_2SO_4 shows slightly better results than HNO_3.

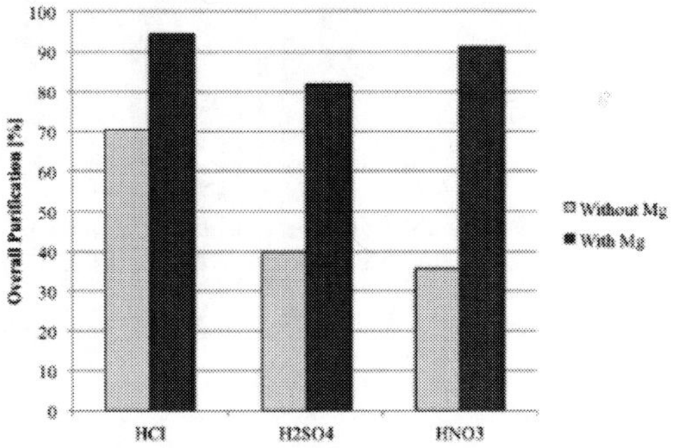

Figure 2: The overall purification degree of impurities calculated with considering the added Mg into silicon and without it.

4 CONCLUSIONS

The refining of silicon was performed through its alloying by Mg in molten state, controlled solidification and employing acid leaching. Different acids with different concentrations were used and it was observed that Si purification is dependent on type of acid, and in addition the concentration of acid. It is found that HCl is a better agent for the process compared to HCl than HNO_3. Moreover, 10% and 15%HCl concentrations are more effective than lower and higher concentrations.

5 ACKNOWLEDGEMENT

The authors would like to appreciate the financial support provided by the Research Domain 3-Recycling and Refining in SFI-metal production[12], a center for research-based innovation in Norway.

References

[1] J. Safarian, G. Tranell, M. Tangstad, Energy Procedia, 20 (2012), 88.
[2] J. Safarian, M. Tangstad, Metall. Mater. Trans. B, 43B (2012), 1427-1445.
[3] S. Espelien, J. Safarian, AIMS Energy, 5(4) (2012), 636-651.
[4] J. Safarian, G. Tranell, 32nd European Photovoltaic Solar Energy Conference and Exhibition, Munich, Germany (2016), 1011–1014.
[5] S. Espelien, G. Tranell, J. Safarian, *Energy Technology 2017*, *Carbon Dioxide Management and Other Technologies*, Springer International Publishing, 355–365.
[6]J. Dietl, *Solar cells* 10 (1983), 145–154.
[7] I.C. Santos, A.P. Gonçalves, C.S. Santos, et al., *Hydrometallurgy* 23 (1990), 237–246.
[8] T. Shimpo, T. Yoshikawa, K. Morita, *Metall Mater Trans B* 35 (2004), 277–284.
[9] Z.L. Yu, W.H. Ma, Y.N. Dai, et al. *Trans Nonferrous Met Soc China* 17 (2007), 1030–1033.
[10] Z. Jian, T. Li, X. Ma, et al. *J Semiconductors* 30 (2009), 22–27.
[11] J.M. Juneja, T.K. Mukherjee, *Hydrometallurgy* 16 (1986), 69–75.
[12] SFI Norwegian acronym for Centre for Research driven Innovation (CRI)

ON THE FABRICATION OF SOLAR CELLS BASED ON NEWLY PRODUCED RECYCLED SILICON FEEDSTOCKS FROM CABRISS – A COMPARATIVE STUDY BETWEEN MATERIAL PROPERTIES AND SOLAR CELLS PERFORMANCES

A. DERBOUZ DRAOUA(1), B. MARTEL(1), S. DUBOIS(1), C. AUDOIN(1), M. SERASSET(1), A. FAUVEAU(1), H. S. RADHAKRISHNAN(2), J. DENAFAS(3), L. PETRENIENE(3), N. SEVERINO(4), N. BEDNAR(4), A.G. ULYASHIN(5), M. SYVERTSEN(5).

(1) Univ. Grenoble Alpes, CEA, LITEN, DTS, Labo, F-38000 Grenoble, France
(2) IMEC, Leuven B-3001, Belgium
(3) SOLITEK R&D, LT-08412, Lithuania
(4) TU Vienna, Vienna A-1040, Austria
(5) SINTEF Materials and Chemistry, No-0314, Oslo.

ABSTRACT: Second life silicon solar cells based on 100% recycled silicon have been recently produced at CEA-INES in the frame of the CABRISS Project. A maximum efficiency of 18.1% has been reached (Al-BSF solar cells), with Mono-like silicon substrate grown from recycled feedstock. During this first period of the project, several Silicon ingots were produced using different conventional methods: Multi-Si standard casting method, Standard Cz-Si pulling and also mono-Si seed-assisted casting method (Mono-like technology). To assess the photovoltaic potential of recycled materials, we developed an overall material qualification protocol to evaluate the effect of the different solar cell processing steps on the material quality. This study shows that all solar cells performances are in good agreement with the data extracted from this protocol, especially with the bulk carrier lifetime determined after each technological step. In addition this study brings new insights into the compatibility between novel Recycled Silicon feedstocks (Re-Si) and the fabrication of standard solar cells. This is definitely an important step to improve the definition of the future Re-Si feedstocks specifications and enhance the development of efficient recycling processes.
Keywords: Recycling, Silicon, Solar Cell.

1 INTRODUCTION

The tremendous growth of the photovoltaic (PV) market and production capacities during the past years encourages the development of recycling processes of end of life modules and PV industry wastes. To explore and develop industrial solutions for such recycling issues, the European project CABRISS was launched. The goal of this project is essentially focused on the salvage of valuables metals such as silver (Ag) and indium (In) but also the silicon (Si) coming from PV wastes. In parallel, the development of these recycling processes will help feeding advanced reflections on how to create a circular economy integrating PV wastes recovery back into the market. During the previous PVSEC conference, some works were presented on the recycling of broken solar cells (or at earlier stages of the solar cell fabrication) and the production of recycled Si (Re-Si) feedstocks [1]. Several p-type Re-Si ingots with different crystallinities (multi-crystalline (mc) and mono-like (ml)) were then successfully grown. Recently, these ingots were sliced into 156 mm² wafers for further material studies and solar cells fabrication.

2 STUDIED INGOTS

Among the different ingots already grown in CABRISS project [2], the wafers used for these material and solar cell studies are depicted in table 1 and presented below:

Monolike ingot A wafers (G2, 4 bricks) originates from ml(mono-like)-Si ingot crystallized with 100% recycled broken solar cells feedstock.

Multi-crystalline ingot B wafers (G1, one brick) was crystallized with a blending between another recycled broken solar cells feedstock and electronic-grade Si (EG-Si).

Multi-crystalline ingot C wafers (G1, one brick) are coming from a mc (multi-crystallne)-Si ingot crystallized with as-cut broken p-type wafers. This case will be considered as the *best case* scenario, since the silicon to be recycled was initially rather pure (no metals, no emitters or silicon nitride layers to be removed before casting).

To obtain reference wafers, a *Multi-crystalline ingot D (G1, single brick)* was grown with EG-Si intentionally doped with boron.

Table I: Re-Si and reference wafers obtained in the frame of the CABRISS project and studied here – Nature of silicon feedstock, crystallization process and resistivity/conductivity type.

Source	*Ingot Casting Process*	*Feedstock*
A G2 (~ 55 Kg)	Mono-like-Si p-type ~ 0.6 Ω.cm	100 % recycled broken Solar Cells feedstock
B G1 (~ 10 Kg)	Multi-Si p-type ~ 2 Ω.cm	24% recycled broken solar cells feedstock 76% Electronic grade Si
C G1 (~ 10 Kg)	Multi-Si p-type ~ 1 Ω.cm	100 % As-cut broken wafers
D G1 (~ 10 Kg)	Multi-Si p-type ~ 2 Ω.cm	Electronic Grade Silicon Reference

All the sources were regularly sampled along the ingot height. They were first characterized in terms of resistivity and thickness. All wafers were scanned through their diagonal with a double side eddy current probe coupled with a capacitive thickness probe. 19 points (1 every cm) were collected for each wafer.

2 RESULTS AND DISCUSSSION

2.1 Materials Evaluation: focus on Ingot A

For this paper, we essentially focused on the interesting A source as the Re-Si is coming from 100% purified solar cells scraps. Even if the preliminary feedstocks analyses predicted standard post-crystallization resistivity targets (~ 0.8 Ω.cm) [1], the resistivity measurements (after crystallization) depicted lower values (between 0.4 and 0.6 Ω.cm), which reveals an unexpected contamination of this Re-Si by dopant species.

In order to evaluate the contamination level of the source A wafers, GDMS measurements were done at three different ingot positions (bottom, middle and top parts). results are presented in Table 2.

The GDMS dopants analyses reveal a high boron concentration which mainly explains the measured low resistivity. The phosphorus concentration remains low, except at the top part (due to segregation effects). The aluminum concentration (Al acts as a dopant but also as a virulent recombination center when paired with O atoms) remains under the detection limit. The same is true for all main metal impurities, their concentrations are under the GDMS detection limit (<10 ppb wt), which is an encouraging result. However this shall not indicate that ingot A is totally free from metal contamination and we should keep in mind that lower concentration levels (even below detection limits) could significantly affect the electrical performances [2].

Table II: Dopants and metallic impurities concentration measured by GDMS on ingot A.

Elements	Bottom (ppm wt)	Middle (ppm wt)	Top (ppm wt)
B	0.18	0.25	0.21
P	0.06	0.07	0.16
Al	<0.01	<0.01	<0.01
Ag	<0.01	<0.01	<0.01
Cu	<0.01	<0.01	<0.01
Fe	<0.05	<0.05	<0.05
Pb	<0.005	<0.005	<0.005
Zn	<0.01	<0.01	<0.01
Ti	<0.005	<0.005	<0.005
Ni	<0.01	<0.01	<0.01
Cr	<0.005	<0.005	<0.005

In parallel, FTIR (Fourier Transformed Infra-Red Spectroscopy) measurements for extracting the interstitial oxygen concentrations [O_i] were done on thick slices at the bottom and top parts of the ingots (close to the first and last wafers positions). They revealed relatively high [O_i] at the bottom part (close to 3×10^{17} cm^{-3}), and much lower [O_i] at the top (~1×10^{16} cm^{-3}), as shown in figure 1. The combination of high boron concentration and relatively high [O_i] at the bottom part might lead to pronounced Boron-Oxygen LID effects (which will be evaluated on the final solar cells).

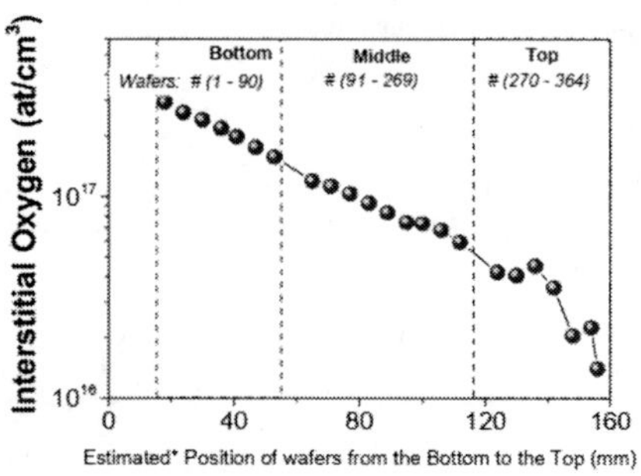

Figure 1: Interstitial Oxygen concentration extracted from FTIR analyses on ingot A.

In order to evaluate ingot A wafers electrical quality, the effective carrier lifetime (τ_{eff}) on as-cut wafers and after phosphorus diffusion was measured by IC-QssPC with a SINTON instrument. All wafers (as-cut and after phosphorus diffusion) underwent a chemical surface polishing and cleaning followed by a double side Aluminum oxide/Silicon nitride layers passivation. Finally a firing step was conducted for all wafers with an infrared belt furnace as shown in figure 2.

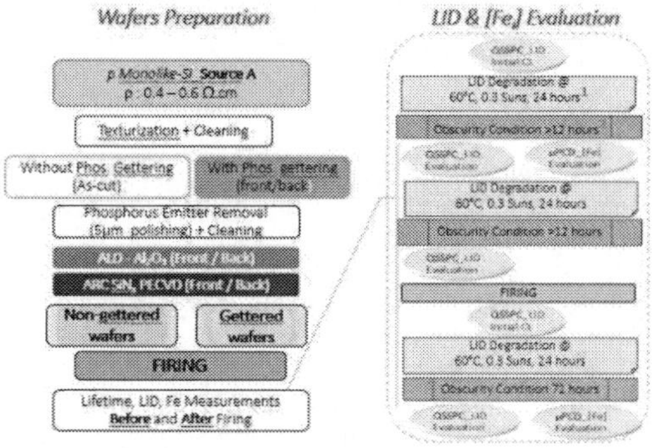

Figure 2: Experimental protocol to assess the recombination properties of there-Si material experiencing the Al-BSF fabrication process.

One can observe from figure 3 a very low as-cut τ_{eff} on the whole ingot (maximum 15 µs). The external gettering effect developed by the phosphorus diffusion is very pronounced as we improve from factors 2 to 20 the initial τ_{eff}. As the Phosphorus-Diffusion-Gettering (PDG) step is particularly efficient for extracting fast diffusing metal impurities from the Si bulk (e.g, Fe, Cr, Cu), this τ_{eff} improvement means that the initial τ_{eff} was essentially dominated by such elements. As a general observation, we can notice a "bell shape" τ_{eff} behavior for as-cut wafers, after phosphorus gettering and after firing treatments. The lower τ_{eff} at the bottom and top parts of the ingot are probably due to the presence of residual impurities decorating subgrains defects as shown by the µW-PCD (MicroWave PhotoConductivity Decay) effective carrier lifetime mappings presented in figure 4.

After subsequent firing of the gettered wafers, we observe an improvement of τ_{eff} for all wafers from ingot A. We have to notice that the firing step is usually necessary to activate the bulk hydrogenation effect (H diffusion into the Si material from the H-rich silicon nitride layer). Therefore this τ_{eff} improvement is likely to be due to the electric passivation of the main defects in the Si material, by hydrogen.

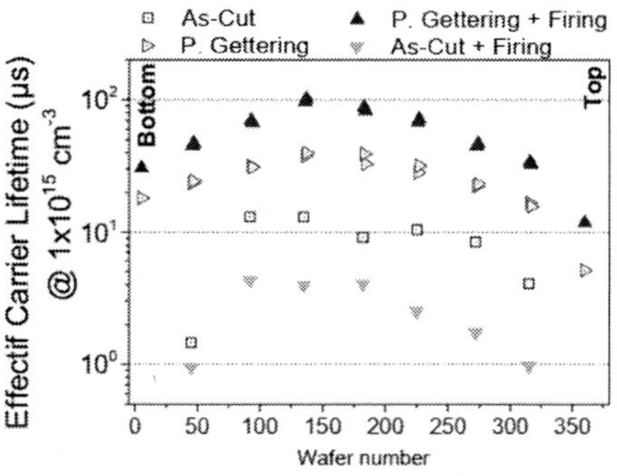

Figure 3: τ_{eff} variation along the ingot height: as-cut wafers, after P-gettering and after firing. SiNx:H passivated ingot A wafers.

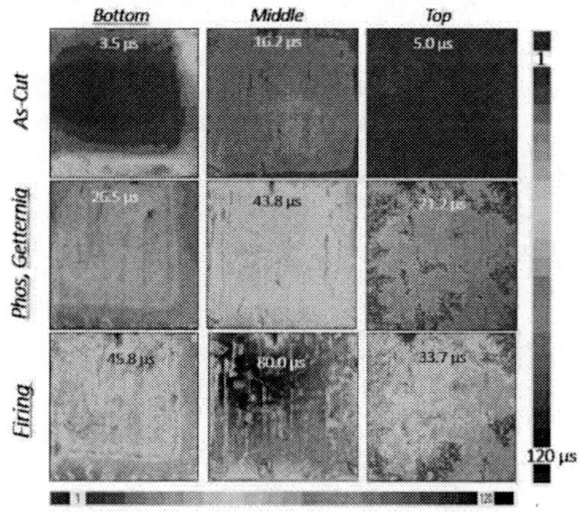

Figure 4: µWPCD carrier lifetime mappings of as-cut, after P-Gettering and after firing of SiNx:H passivated source A wafers. The values in blank are average values.

The Light-Induced Degradation (LID) test protocol consisted in placing the passivated wafers under 0.5 Sun illumination at 50°C for 24 hours. This illumination protocol should be long enough to stabilize the LID associated with the boron-oxygen (BO) complexes. However, even if these mechanisms are still poorly documented in the literature, metal related LID would require several weeks of light exposure to be stabilized. τ_{eff} measurements were done using WCT120 instrument (IC-QssPC analyses) before and after illumination. The carrier lifetime only limited by the LID defects (τ_{LID}) was calculated at an excess carrier density (Δn) $\Delta n=10^{15}cm^{-3}$ using the equation (1) below:

$$\frac{1}{\tau_{LID}} = \frac{1}{\tau_{eff(After\ LID)}} - \frac{1}{\tau_{eff(Before\ LID)}} \quad (eq.\ 1)$$

The wafers (bottom, middle and top) from ingot A that were used in previous studies (as-cut measurements and gettering tests) were used for LID tests. The results in terms of τ_{LID} on ingot A wafers are presented in the Figure 5 below:

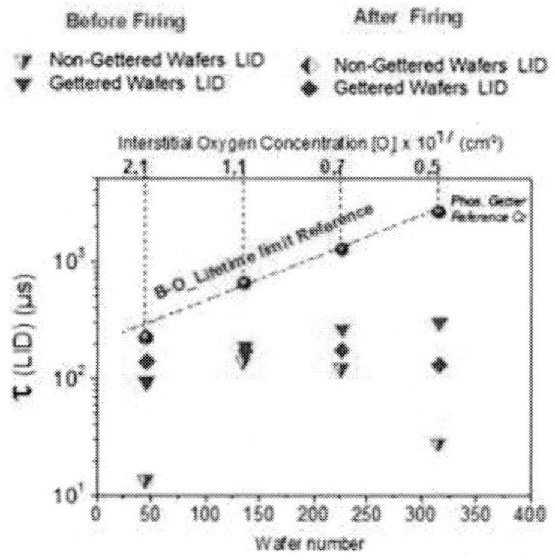

Figure 5: τ_{LID} values for the as-cut wafers, after gettering and after firing treatments. Ingot A. Comparison with the computed values of the carrier lifetime limited by the B-O related complexes [5].

First, we can notice that all τ_{LID} values of Ingot A wafers lie well below the boron-oxygen related carrier lifetime limit for B-doped Czochralski (Cz) Si [5], particularly for the middle and top part of the ingot. This indicates, at least for the middle and top part of the ingot, that the boron oxygen defects do not explain the observed LID effects. It has been shown that LID effects in mc-Si are not only due to BO complexes formation, but could be linked to residual metal impurities such as copper for instance [4]. Indeed, one can observe a significant LID at the top part of mc-Si ingots, where the metal concentrations are usually higher (due to the segregation of metal impurities during crystallization). We can also notice a large reduction of the LID effect after gettering (effect that decreases the metal impurities concentration. After firing step the top parts feature a slight increase of LID effect (as shown in figure 6), this step being known for the dissolution of metal precipitates. Therefore, all these results, indicate that the LID effects of source A wafers from the middle and top parts of the ingot, would essentially be due to the presence of residual metallic impurities, even if the occurence of other LID mechanisms cannot totally be rulled out.

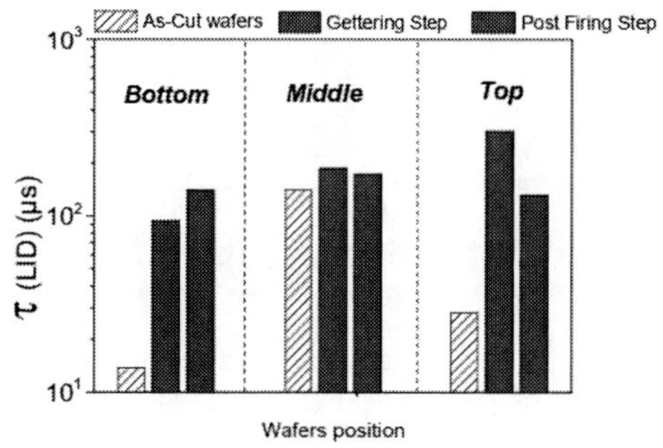

Figure 6: τ_{LID} calculation for the as-received wafers, after gettering and after firing treatments for ingot A.

Interstitial iron concentration [Fe_i] was evaluated on the same passivated wafers after the LID study using the "iron-boron pair dissociation under illumination" method [6]. This method consists in measuring the carrier lifetime increase (at high injection levels) due to the optical dissociation of the electrically active Fe_iB pairs. In order to avoid any interferences between this phenomenon and the LID effect, these measurements were done after the degradation test mentioned above, and after the storage of the wafers in the dark during about 24 hours (time long enough for a complete reassociation of the FeB pairs). The carrier lifetime measurements were done with a Semilab WT2000 instrument, locally, near the edge of the wafers (the same coordinates were used from one wafer to another). We used repeated carrier lifetime measurements (the laser used for the τ_{eff} measurements simultaneously dissociated the FeB pairs) and monitored the change in τ_{eff}. As expected, Figure 7(a) shows an increase of τ_{eff} for the as-cut (with passivated surfaces) wafers after illumination due to the dissociation of FeB pairs. It is not the case of the gettered wafers, where the lifetime is unchanged (figure 7(b)). This is explained by a low [Fe_i] due to the efficient P-gettering.

It should always be kept in mind that there are numerous possible explanations for an observed change in carrier lifetime after exposure to illumination. According to the literature, the characteristic fingerprint of Fe contamination corresponds to a particular excess carrier density (the so-called *crossover point*) at which the carrier lifetime remains unchanged whatever the respective Fe_i and FeB concentrations [6]. In our case, we monitored the changes in τ_{eff} as the samples relaxe to equilibrium conditions with QssPC analyses (Figure 8). We confirmed the existence of a crossover point associated with an excess carrier density around $3x10^{14}$ cm^{-3}. This value is in fair agreement with the value reported by Macdonald et al, for the same boron concentration [7]. Therefore, this means that dissolved Fe is well responsible of the highlighted changes in τ_{eff}.

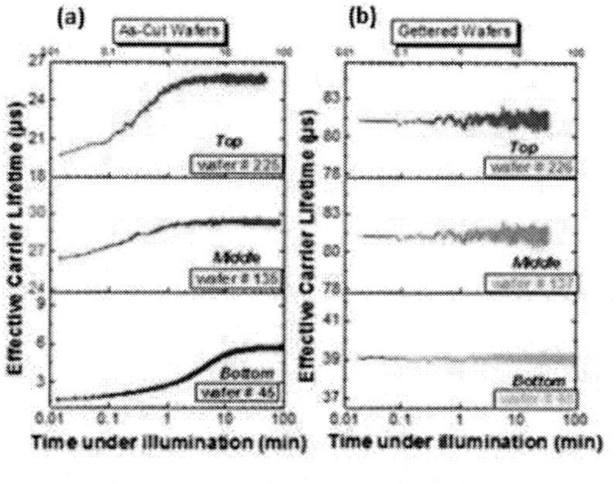

Figure 7: τ_{eff} evolution (under repeated µWPCD measurements) for as-cut and gettered source A wafers.

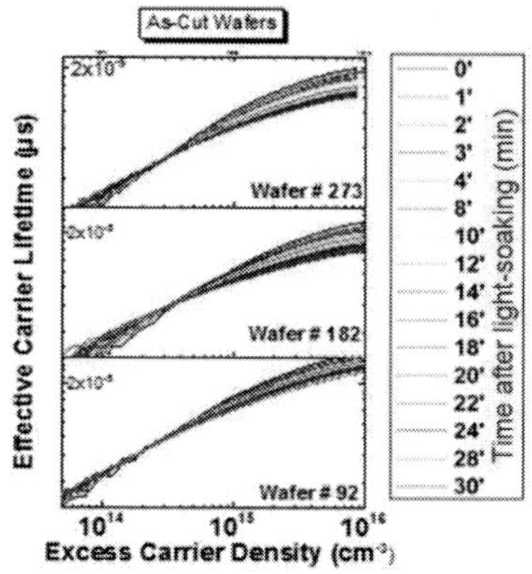

Figure 8: τ_{eff}(QSSPC) as a function of time following the illumination used for dissociating the FeB pairs. As-cut Ingot A wafers.

Figure 9 presents the results of the [Fe_i] determinations for source A wafers. First, one can notice that [Fe_i] is very high on the whole height of ingot A ($\geq 10^{12}$ cm^{-3}), compared to around 5 x 10^{11} cm^{-3} for standard mc-Si [6]. These high [Fe_i] could be explained by the use of 100% of recycled broken solar cells feedstock. The [Fe_i] is higher at the bottom than at the top part. This could be explained by the proximity of the silicon nitride coated silica crucible, source of parasitic contaminations. The effect of phosphorus gettering leads to a significant reduction of [Fe_i]. Most of the [Fe_i] data after phosphorus gettering are below the detection limits of this technique (~10^{10} atoms/cm^3) and reveal the ability of the P diffusion step for extracting fast diffusing impurities. Notice that after the firing step, [Fe_i] are also below the detection limit.

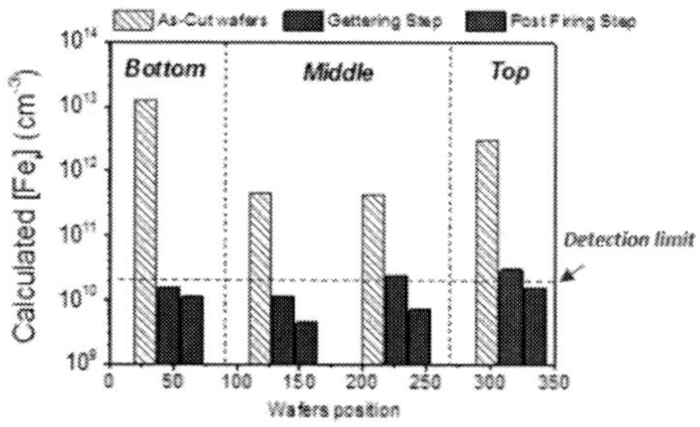

Figure 9: [Fe_i] calculations for the as received wafers, after gettering and firing treatments, for source A materials.

2.2 Solar Cells performances

Sources A wafers (batches of 4 solar cells at different ingot positions) underwent an Al-BSF solar cell fabrication process, mainly composed of a KOH texturation, a double side phosphorus diffusion known to develop good gettering effects, a back side chemical polishing, an H-rich silicon nitride layer deposition, and finally a front and rear metallization (screen-printing) followed by a firing step. The resulting solar cells performances were assessed by I-V characterization under AM1.5 illumination. They are presented in figure 10 and compared with the carrier lifetime values.

One can see from figure 10, that the solar cells efficiencies distribution along the ingot height follows well the preliminary carrier lifetime studies. A best efficiency of 18.1% was reached, compared to 18.5% for our reference (*but not state of the art*) ml-Si material. Upper part of the ingot was not shown here, mainly due to metallization issues interfering with the material evaluation.

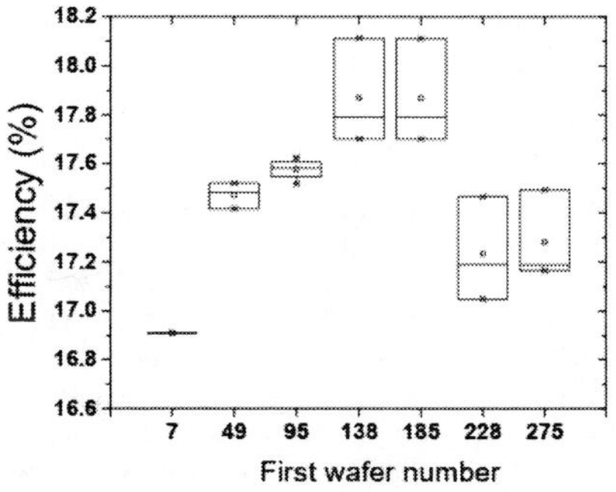

Figure 10: PV conversion efficiency of solar cells from source A wafers.

Similar Al-BSF process was applied to the sources B, C and D after acidic texturisation. The results are illustrated in figure 11. As expected the results show a clear difference in their performances due to the origin of recycled feedstock. Indeed, the performances of the source B cells, which come from a mc-Si ingot crystallized with a blending of 24% re-Si, do not exceed 16.6% in conversion efficiency (the values are even below 16% for the cells fabricated with wafers from the ingot bottom part). We have to notice that the source B present a red zone (region with a low carrier lifetime, likely to be due to a parasitic contamination from the silicon nitride coated silica crucible) at the edges of the wafers, which could be due to the G1 size of the ingot. We have increased the efficiency of some solar cells by 1% absolute after laser-cutting of 1 cm from the edge of solar cell (results not shown here). Concerning source C we reached conversion efficiency values around 17% in the whole of the ingot. These results, indicates that broken as-cut wafers can be successfully reused into the PV chain. One can see that solar cells results for all ingots are in good agreement with lifetime study. The lower the carrier lifetime/ the higher the [Fe$_i$], the lower the solar cell performance in term of efficiency, Fill factor (FF), short circuit density (Jsc), open circuit voltage (Voc).

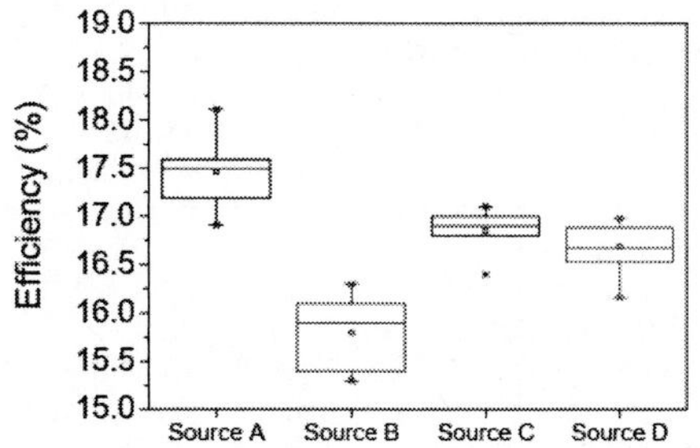

Figure 11: PV conversion efficieny results of A, B and C Re-Si sources compared to a reference mc-Si ingot D.

CONCLUSION

The main goal of this work was to develop an overall recycled material qualification protocol to evaluate the effect of different solar cell fabrication processing steps on the material properties. We showed that all solar cells performances are in good agreement with the data extracted from the protocol (especially the bulk carrier lifetime and [Fe$_i$] extracted after the main solar cell processing steps).

Up to now, three different sources of Re-Si wafers among five available in the frame of the CABRISS project were studied and processed into Al-BSF solar cells. The first one is source A – G2 mono-like wafers coming from recycled solar cells scraps. The lower but still good quality of the resulting Re-Si wafers (as determined by the wafer qualification protocol) led to a best 18.1% efficiency Al-BSF solar cell (compared to 18.5% for the references). Notice that LID effects a priori due to metallic impurities were highlighted via the "material study". Therefore it will be important to study these effects at the solar cell level. The second one is source B – G1 mc-Si wafers coming from a blending between electronic grade Si and recycled broken solar cells feedstock. The efficiency of the associated solar cells did not exceed 16.6 % probably due to the simultaneous presence of high density of metallic impurities and extended crystallographic defects (grain boundaries, dislocations) which make difficult to enhance material quality with phosphorus gettering and firing treatments. In addition, we have to notice that the source B wafers present a red zone at the edges (as shown by the μWPCD lifetime mapping) due to the G1 size. It will be interesting to test the same feedstock with G2 size because when some source B solar cells are submitted to a laser-cut at 1 cm from the edge, the efficiency increases by 1% absolute (results not shown in this study). The third source is source C - G1 multi-crystalline wafers coming from recycled broken as-cut p-type wafers. The good quality (as determined by the wafer qualification protocol) of this Re-Si led to promising Al-BSF solar cells performances (Re-Si solar cells average efficiency of 17.3% compared to 17.0% for the references.

Further works are in progress on :

- The production of Re-Si module based on 100% recycled silicon,
- The production of G6 size ingot using recycled silicon,
- The fabrication of recycled n-PERT and p-PERC solar cells.

REFERENCES

[1] M. Syvertsen et al., Proceeding of the 32nd EUPVSEC conference, Munich (2016)

[2] https://www.spire2030.eu/cabriss

[3] G. Coletti, Progress in Photovoltaics Research and Applications 2013, 21:1163-1170; DOI: 10.1002/pip.2195

[4] A. Inglese et al, Applied Physics Letters 107, 052101 (2015)

[5] K. Bothe et al, Progess in Photovoltaic, Volume 13, Issue 4, June 2005, Pages 287–296

[6] D. H. McDonald et al, Journal of Applied Physics 95, 1021 (2004)

[7] D. H. McDonald et al, Journal of Applied Physics Letters 89, 142107 (2006)

[8] J. Tan et al, Journal of Applied Physics Letters 91, 043505 (2007).

SILICON KERF AS RAW MATERIAL FOR HIGH-CAPACITY LI-ION BATTERY ANODES

Trygve Tveiterås Mongstad[1†], Samson Yuxiu Lai[2], Sean Erik Foss[1], Eivind Johannes Øvrelid[3]

[1] Department of Solar Energy, Institute for Energy Technology (IFE), Kjeller, Norway
[2] Department of Physics, Institute for Energy Technology (IFE), Kjeller, Norway
[3] SINTEF Materials and Chemistry, Trondheim, Norway

Corresponding author: samson.lai@ife.no, Tel: +47 480 09 583
[†]Present address: FUSen AS, Oslo, Norway

ABSTRACT: Of the polysilicon that goes in to manufacturing of silicon wafers for solar cells, more than 50% is lost as saw dust in the wire-sawing process. This waste byproduct, called silicon kerf, currently has low value to the silicon industry and is difficult to recycle cost-effectively. However, because silicon can store up to 10 times more Li ions than current commercial anode materials, kerf could be used as a raw material for Li-ion battery anodes. The performance of silicon kerf as a Li-ion battery anode is modestly comparable to that of Alfa Aesar Si nanoparticles with an average particle size of < 100 nm. For kerf half cells, a reversible discharge capacity of about 1,750 mAh/g_{Si} for 40 or 50 cycles was observed for a loading of 0.59 g_{Si}/cm^2. Surface morphology and particle size distributions were also compared. Cycle life and specific power remain challenges but approaches like limited capacity cycling, passive filtration for particle size control, and industrially facile coatings can be explored while maintaining the economic edge of upcycling kerf.

Keywords: *kerf, silicon, lithium-ion, battery, anode*

1 INTRODUCTION

Cutting of silicon wafers results in more than 50% Si saw dust, usually referred to as kerf. Waste sludge from traditional silicon wafer cutting consists of ethylene glycol, silicon and silicon carbide, as well as particles and metal impurities from the cutting wire. The change in industry from slurry-based cutting to diamond-wire cutting for single crystalline wafers makes it easier to recycle the kerf and results in kerf with high purity in the bulk. However, the surface may be oxidized with and contaminated with metals and organics from the wire, saw, and the coolant.

The properties of the silicon kerf particles are not ideal, especially the size distribution. The best results with silicon anodes have been achieved with silicon particles with dimensions below 100 nm [1], while silicon kerf particles are normally larger. Using the right methods, these challenges can be circumvented. Bao et al. reported a combination of silicon recovery methods, ball milling, and graphene coating to produce high capacity anodes [2]. Kim et al. prepared silicon kerf and graphene nano-composites by spray pyrolysis [3], showing promising results with high capacity and stability. However, implementing these relatively costly techniques into the practical production of kerf-based silicon anodes may erase the advantages from upcycling a waste product.

In this work we have used a more industrially relevant approach, avoiding the high cost of ball milling and graphene coating, but focusing on the electrode preparation designed specifically for high silicon content anodes. We combine a method for separation and purification of silicon kerf with current state-of-the-art electrode preparation methods for silicon-based anodes for Li-ion batteries. In this way we outline how low-value waste material from Si solar cell production can be converted to high-value materials for future high capacity Li-ion batteries.

2 EXPERIMENTAL METHODS

The kerf used in this experimental study contained, in addition to Si, metal levels below 1000 ppm and less than 3% C and 3.5 % O. Slurries were prepared using, by weight, 15% sodium carbomethoxyl carbonate (Aldrich) as binder, 10% graphite (TIMCAL KS6 L), 15% carbon black (TIMCAL Super C65), and balance of as-received silicon kerf. Slurries were homogenized with an aqueous buffer solution using a centrifugal mixer (ARE-250CE, Thinky Co.). The slurry was cast onto a dendritic copper foil (ETP-grade, Schlenk) with doctor blade thicknesses of 12.7 or 25.4 microns. After drying overnight at room temperature, cast slurries were dried in a vacuum oven at 120 °C for 2 hours. The average active material mass loadings were approximately 1.1 and 1.5 mg per electrode, respectively. Electrodes were punched as 15 mm disks from cast slurries. The actual as-prepared electrode thicknesses after discounting the Cu current collector were 4.8 and 9.1 µm, respectively.

Coin cells were assembled using CR2032 stainless steel casings in an Ar filled glovebox (MBraun) with less than 0.2 ppm of oxygen and water vapor. The electrolyte used was 35 µL of 1.0 M $LiPF_6$ in an ethylene carbonate/propylene carbonate/dimethyl carbonate solution in a 1:1:3 weight ratio, respectively, with 5 wt % of fluoroethylene carbonate and 1 wt % of vinylene carbonate as additives. Celgard K2045 was used as a separator (18 mm diameter). The counter and reference electrode was a punched disk of Li foil (99.99%, LinYi Gelon LIB Co.).

Electrochemical performance was measured using a multi-channel Arbin battery cycler, with a schedule of 3 formation cycles at C/20 followed by 100 cycles at C/10 assuming the theoretical discharge capacity of $Li_{15}Si_4$. Galvanostatic cycling was done from 0.05 V to 1.0 V. Rate capability tests were also performed. As-received silicon kerf was examined using conventional scanning and low voltage (30 kV) scanning transmission electron microscopy (LV-STEM, Hitachi S-4800 with transmitted electron detector) with energy dispersive X-ray spectroscopy (EDS) attachment. Transmission electron grids were prepared by loading with kerf and shaking off excess particles.

3 RESULTS AND DISCUSSION

Fig. 1 shows LV-STEM images of the kerf material used in this work. The material consists of small particles and flakes with sizes in the range of 0.1 to 5 µm. Flakes appear to have a plate-like aspect ratio and have sharp edges and interfaces. Particle sizes also appear to vary significantly. In contrast, Fig. 2 shows the particle size and shape of commercial Si nanopowders produced by Alfa Aesar.

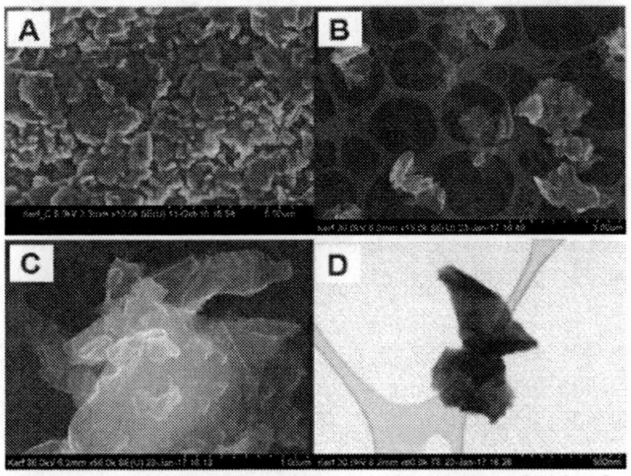

Figure 1: A) Scanning electron microscopy of Si kerf particles (carbon coated for SEM). 5 µm scale bar. B, C) LV-STEM of Si kerf particles (scanning mode). 3 and 1 µm scale bars, respectively. D) LV-STEM of Si kerf particles (transmission mode). 500 nm scale bar.

Figure 2: Commercial Si nanopowders from Alfa Aesar with an average particle size of A) <50 nm and B) <100 nm. 5 µm scale bars.

Occasionally, foreign particles were observed in LV-STEM, shown in Fig. 3. Based on the spiral-like shape and the EDS indicating primarily Si content, these objects are believed to Si turnings from the wire-sawing process. Their influence on battery performance is unknown, but their appearance is relatively infrequent.

The size distributions by volume percent of the kerf and other Si nanoparticles were measured by laser scattering measurements as seen in Fig. 4. The kerf particle size distribution is bimodal with mean particle sizes at about 600 nm and 5.5 µm. The larger may be representative of agglomerates. The similar particle size distribution suggests that kerf may close to an optimal size for effective use in a Li-ion battery anode.

Electrochemical cycling of Li cells made with kerf retained relatively higher discharge capacity compared to

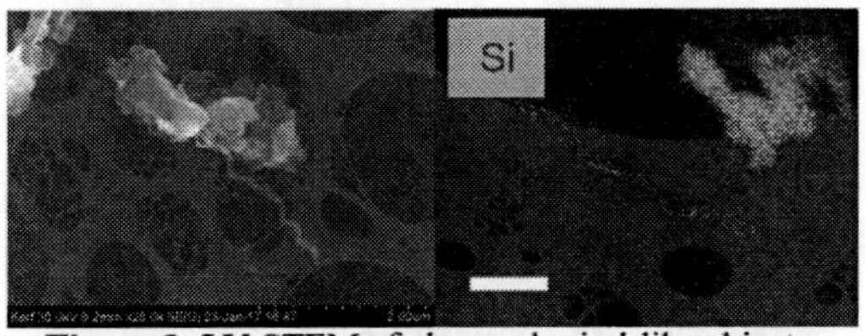

Figure 3: LV-STEM of observed spiral-like objects, believed to be turnings from the wire-sawing process (2 µm scale bar). EDS overlay reveals that turnings contain mostly Si (1 µm scale bar).

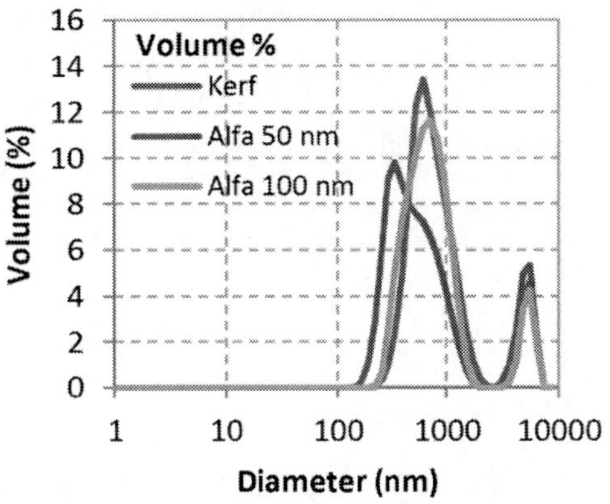

Figure 4: Particle size distributions of silicon kerf relative to commercial Si nanoparticle powders.

nanoparticles and with in-house silicon nanoparticles synthesized from a silane-pyrolyzing free space reactor, as shown in Fig. 5. Cells were cycles at C/10 (approximately equivalent to 0.36 A/g_{Si}), assuming a Si theoretical energy capacity of 3,579 mAh/g_{Si}. The active material loadings are as follows: IFE FSR Si (0.32 mg/cm^2), 50 nm Alfa (0.34 mg/cm^2), 100 nm Alfa (0.45 mg/cm^2), kerf (blue, 0.59 mg/cm^2), kerf (green, 0.86 mg/cm^2). The kerf with lower loading (blue) is a result of the thinner casting and thinner electrodes perform better than thicker electrodes because of lower electronic resistance and shorter ionic diffusion pathways.

The overall discharge capacity of about 1750 mAh/g_{Si} from kerf is modestly comparable to that of 100 nm Alfa Si. It should be noted that both the 50 nm and 100 nm Si half cells experienced testing glitches that lowered the charge/discharge current density, resulting in a small but sudden increase in the discharge capacity midway through testing. While the kerf's performance is somewhat stable for 40 to 50 cycles, the discharge capacity decreases significantly afterward, due to the electrolyte additives being fully consumed in new SEI formation and Si volume expansion-induced degradation. For reference, commercial graphite-based anodes have discharge capacities of 372 mAh/g_C that are stable for at least 300 cycles. Based on the performance of IFE FSR Si and 50 nm Alfa Aesar Si, it is clear that the optimal particle size is still closer to 50 nm, although a lower bound exists where too much surface area is detrimental to battery performance because of excessive solid electrolyte interface (SEI) layer formation.

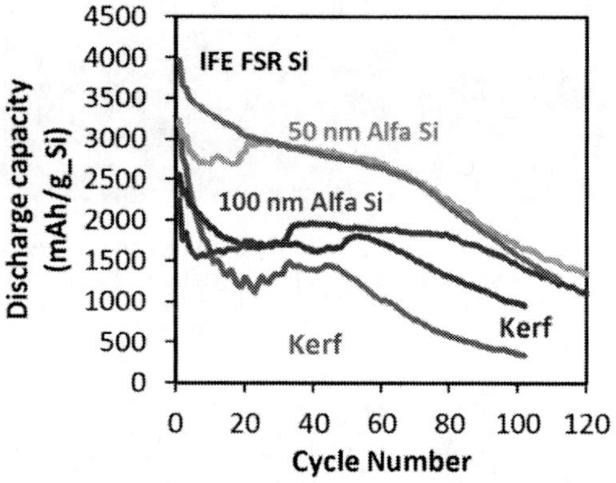

Figure 5: Discharge capacity cycling performance of various half cells. Active material loadings: IFE FSR Si (0.32 mg/cm^2), 50 nm Alfa (0.34 mg/cm^2), 100 nm Alfa (0.45 mg/cm^2), kerf (blue, 0.59 mg/cm^2), kerf (green, 0.86 mg/cm^2).

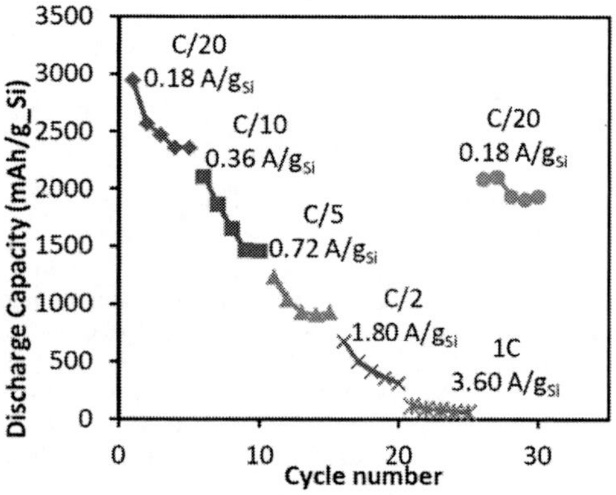

Figure 6: Rate capability of kerf half cell with 0.86 mg/cm^2 loading.

The rate capability test results for the kerf half cell are shown in Fig. 6. Current density values are displayed underneath each of the C-rates. Kerf exhibits poor performance at high specific power. Additionally, as revealed in the capacity cycling data, the kerf decreases in capacity for the first 20 cycles, which skews the rate capability test results. Without additional electrode modifications to enhance conductivity, it seems that the large particle sizes of kerf impede fast redox reactions.

4 CONCLUSIONS

We have shown that silicon kerf as a waste from silicon solar cell wafer production can be utilized to produce high capacity silicon anodes. The capacity deteriorates quickly, providing a relatively stable cycle life for 40 to 50 cycles. This gives promise for kerf as a feedstock material for Li-ion batteries, but the materials and methods need improvement. Reducing the particle size is a tempting option but other approaches such as using limited capacity cycling or industrially facile coatings to enhance electronic conductivity and specific power may be more economically practical. Particle size control through filtration may also be a less energy-intensive approach than ball-milling.

5 ACKNOWLEDGEMENTS

Funding support was provided by the Research Council of Norway through the ENERGIX Project No. 255116.

6 REFERENCES

[1] H. Kim, M. Seo, M.-H. Park, and J. Cho, "A Critical Size of Silicon Nano-Anodes for Lithium Rechargeable Batteries," *Angewandte Chemie International Edition,* vol. 49, pp. 2146-2149, 2010.

[2] Q. Bao, Y.-H. Huang, C.-K. Lan, B.-H. Chen, and J.-G. Duh, "Scalable Upcycling Silicon from Waste Slicing Sludge for High-performance Lithium-ion Battery Anodes," *Electrochimica Acta,* vol. 173, pp. 82-90, 8/10/ 2015.

[3] S. K. Kim, H. Kim, H. Chang, B.-G. Cho, J. Huang, H. Yoo, *et al.*, "One-Step Formation of Silicon-Graphene Composites from Silicon Sludge Waste and Graphene Oxide via Aerosol Process for Lithium Ion Batteries," vol. 6, p. 33688, 09/20/online 2016.

IN-SITU MEASUREMENT OF THE SOLID-LIQUID-INTERFACE DURING THE GROWTH OF SILICON INGOTS BY THE ULTRASONIC SOUND METHOD

M. Trempa[1], M. Hinderer[3], I. Kupka[2], C. Reimann[1,2], J. Friedrich[1,2], P. Czurratis[3]

[1]Fraunhofer IISB, Schottkystrasse 10, 91058 Erlangen, Germany
[2]Fraunhofer THM, Am St.-Niclas-Schacht 13, 09599 Freiberg, Germany
[3]PVA TePla Analytical Systems GmbH, Deutschordenstraße 38, 73463 Westhausen, Germany

Corresponding author´s e-mail: matthias.trempa@iisb.fraunhofer.de

ABSTRACT: For the industrial growth of silicon ingots (multi-crystalline, high-performance multi-crystalline or also mono-like) by directional solidification it is important to know as many as possible details about the growth process in order to achieve a high yield/wafer quality and to reduce the costs. For example, the end of melting phase, the position of the seeding level, the start of solidification or the growth velocity are terms of large interest. For determination of these quantities the so-called dipping method is typically used where a rod made from a ceramic or glassy material is dipped into the melt until it touches the solid-liquid-interface. However, this method has some drawbacks which make it to a non-usable tool for every growth run in industrial production. Within this work an alternative, contactless technique, namely the ultrasonic sound method, was evaluated at a laboratory scale crystallization furnace. The results show that this method could be a suitable tool for in-situ determination of all the relevant growth steps during every solidification process without disturbing the growth process itself.

Keywords: directional solidification, crystalline silicon, measurement tools, ultrasonic sound

1 INTRODUCTION & MOTIVATION

One important issue for the production of high performance multi-crystalline (HP mc) or mono-like silicon ingots with a high wafer yield is to control the silicon seeding level along the complete crucible bottom. A locally melted silicon seed layer, e.g. at the crucible corners lead to unfavorable seeding conditions followed by a reduced wafer quality and therefore will reduce the wafer yield. Additionally, the speed of the moving solid-liquid interface and the shape of the complete phase boundary have to be controlled to obtain a sufficient wafer quality with high yield. An already realized industrial solution to control the seeding level locally and the growth speed uses a rod dipping procedure. It determines the axial position of the feedstock or the ingot height level by an iterative mechanical dipping of a rod inside the melt followed by touching the solid phase, respectively. However, this non-contactless method leads to disturbances at the solid-liquid interface during growth and therefore negatively influences the material quality. Further, some contamination may happen in dependence of the used rod material (e.g. aluminum oxide) or the rod may brake due to mechanical forces during dipping. In consequence this method cannot be used as an in-situ control measurement tool for every growth run in industrial ingot production.

An alternative approach to overcome these issues could be the ultrasonic sound method. Hereby, ultrasonic waves are sent by a transducer, which is located outside of the crystallization furnace, through a wave guide into the silicon melt. After encountering the crucible bottom, the seeding layer or the solid crystal surface the waves are (partly) reflected and are received again by the transducer. The signal can be evaluated to determine in-situ the position of the seed level and the growth speed of the crystal. The wave guide which can be made of a highly pure silica rod has to be dipped into the melt, however will not actively touch the solid level and thereby does not influence the growth front.

Within this work a former used ultrasonic sound measurement setup [1] was further developed and fit with additional hardware components. The new setup was tested on a small crystallization furnace in our laboratory. The results show that this method could be a promising tool for process development and process control in industrial production of silicon ingots by directional solidification.

2 EXPERIMENTAL

2.1 Ultrasonic sound method

In this application, the pulse reflection method is used. A special transducer transmits short ultrasound pulses and receives the echoes of the previously transmitted pulse. To transmit an ultrasonic pulse, the transducer converts high voltage signals into high frequency electromechanical vibrations, which are propagated as a plane parallel wave field through a coupling medium (water) into the wave guide (silica rod).

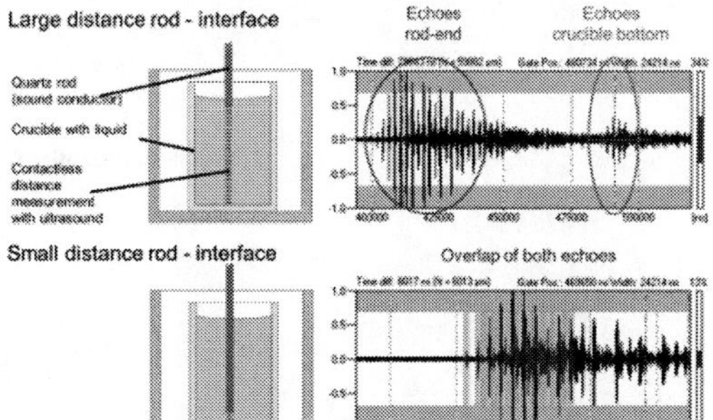

Figure 1: Ultrasonic sound signal for different distances between end of the waveguide and the crucible bottom

To receive ultrasonic signals, the transducer is able to convert reflected waves back to very small, high frequency electrical signals. These signals are amplified, recorded and analyzed. To implement a distance measurement, the timespan between the transmitted pulse and the received echo needs to be evaluated. This timespan represents the transit time of the ultrasonic wave (forward and backward). If the speed of sound for the material, in which the ultrasonic wave propagates, is known, the distance can easily be calculated. In this setup, a pulse repetition rate of 500Hz is used, giving a 500 distance points per second.

In Fig. 1 two examples of the ultrasonic pulse echoes are shown. In the upper case the distance between the end of the waveguide (silica rod) and the crucible bottom is relatively large in the range of several cm. Here the main echoes are clearly separated on the axis of transit time due to the large difference of the way the ultrasonic weaves have to overcome. In the lower case the distance is much smaller in the range of several mm. This results in an overlap of the echoes of rod end and crucible bottom which makes it necessary for the analysis software to extract the echoes from each other. To separate the echoes, the software is capable of using different methods like triggering on the signal and gating the signal. Also, advanced signal processing functions like Fourier, Hilbert and wavelet transformations are available.

2.2 Measurement setup

The ultrasonic measurement setup is schematically shown in Fig.2. The transducer which sends and receives the ultrasonic sound pulses is placed in a small chamber outside the crystallization furnace. It is in direct contact with the waveguide, a highly pure silica rod, which transports the ultrasonic sound pulses into the silicon melt and back to the transducer. The small chamber is filled with water to cool the transducer and to act as couple medium between the transducer and the wave guide. For positioning the waveguide exactly perpendicular to the crucible bottom an x-y-adjusting unit was installed before the waveguide passes the vessel cover plate. Additionally, during the further development a motorized translation unit was implemented, which enables a self automatically translation of the wave guide in axial direction for keeping the distance between rod end and solid-liquid interface on a constant level.

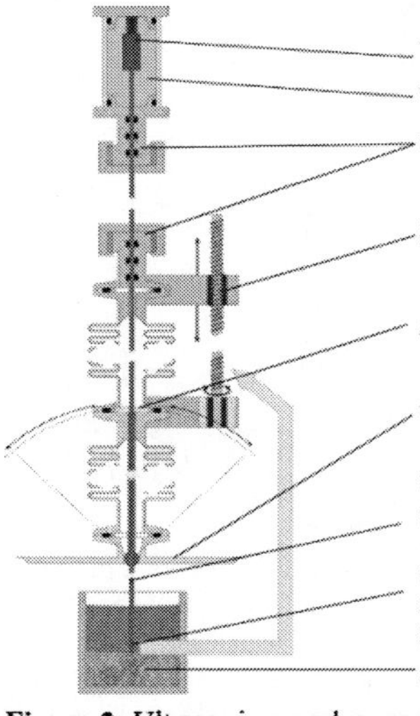

Figure 2: Ultrasonic sound measurement-setup

This provides several advantages concerning the measurement accuracy, the evaluation of the complex ultrasonic sound signal during the process and finally the in-situ observation of the solid-liquid interface which is correlated to the actual motor position.

The setup was tested in a lab-scale furnace containing 4 inch silicon ingots with a height of 60 mm. The silicon material is heated by one induction coil surrounding a graphite susceptor and the silica crucible containing 1100 g silicon chunks, respectively. A standard crystallization process for multi-crystalline silicon (complete melting of the feedstock) was carried out including a 10 mm down movement of the susceptor/crucible (translation speed: 0.1 mm/min) out of the hot-zone in order to initialize the crystallization.

Two experiments were carried out. The first one was done with a constant position of the waveguide in the upper melt region to make a principal test of the performance of the ultrasonic sound measurement equipment. Further, in a second experiment the automatic translation unit was implemented to evaluate the potential of this method for the use in an industrial environment.

3 RESULTS

3.1 1st experiment with fixed waveguide position

In this experiment, after full melting of the silicon feedstock (equals a melt height of ~62 mm), the initial distance between the end of the wave guide (silica rod) and the crucible bottom was set to 34 mm. The sequence of the measurement/crystallization process was as follows:

1. start 10 mm lift down of crucible
2. start crystallization
3. crucible lift down stopped
4. rod end is overgrown by silicon crystal

In Fig. 3 the time frame of the ultrasonic signal from this time slot is shown. One vertical line in this diagram corresponds to one echo response at a given process time as shown in Fig.1, while the color represents the signal intensity (red indicating a strong signal, blue a weak signal). The vertical direction correlates to the transit time in nano-seconds (ns) of the acoustic sound signal. Transit time means the time is needed for the ultrasonic pulse to overcome the distance from the transducer through the waveguide and further through the silicon to the crucible bottom / s-l-interface and back to the transducer. The horizontal direction correlates to the real process time in a hh:mm:ss format. On the outer left hand side the echoes of the rod end (marked with red dots) and the crucible bottom (marked with green dots) are visible as data points of high intensity accumulating to lines with ongoing process time.

At (1) the crucible shift down has started, while the silica rod remains at a constant position. In consequence the transit time increases with increasing distance of the rod end to the crucible bottom resulting in downwards inclined lines of the crucible bottom echoes in Fig. 3.

At (2) a spreading of the original crucible bottom echo can be observed indicating the start of crystal growth. Thereby, the new appearing lines are related to the echoes of the new formed s-l-interface. Hence the growing interface reduces its distance to the rod end over time the lines in Fig.3 are inclined in upwards direction. On the other hand the crucible bottom lines go further downwards until the crucible shift down has stopped (3).

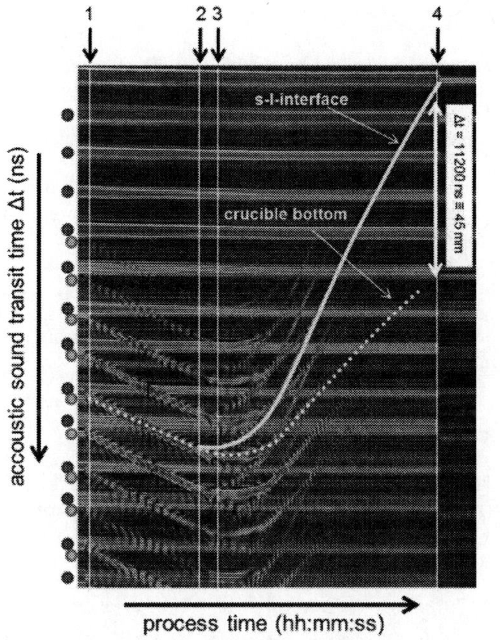

Figure 3: Time frame of the ultrasonic signal during the relevant process steps in the 1st experiment

During further growth the transit time between the fixed rod end and the moving s-l-interface becomes smaller. Also the transit time towards the crucible bottom decreases instead of its constant position. This is due to the higher sound velocity in solid silicon (8000 m/s) in comparison to the liquid (~3900 m/s [1]) and the increasing solid/liquid ratio over process time.

At the end (4) the rod end is overgrown by the silicon crystal and no further sound signal can be detected. By the use of the transit time difference Δt between the lines for the s-l-interface and the crucible bottom (white arrow in Fig. 3), the grown silicon ingot length l can be calculated by

$$l\,(t) = \frac{\Delta t\,(t)}{2} \times v = \frac{\Delta t\,(t)}{2} \times 8000\frac{m}{s} \qquad \text{(eq. 1)}$$

whereas v is the sound velocity in solid silicon. So at (4) the total grown ingot height below the rod end can be calculated to be 45 mm. A measurement on the vertical cut of the grown ingot exactly confirms this value.

3.2 2nd experiment with automatic translation

In a second experiment, the automatic translation unit was implemented in the measurement setup. After full melting of the silicon feedstock (equals a melt height of ~62 mm), the initial distance between the end of the wave guide (silica rod) and the crucible bottom was set to 19 mm. The sequence of the measurement/crystallization process was as follows:

1. start 10 mm lift down of crucible and automatic translation of the rod downwards (8:19h)
2. start crystallization and automatic translation of the rod upwards (9:44 h)
3. 26 mm ingot height are grown (12:21 h)
4. 4 mm manual translation in downwards direction (rod remained in silicon melt) (14:15 h)
5. further 4mm manual translation downwards (14:25 h)
6. automatic rod-translation stopped (14:44 h)
7. rod becomes ingrown by silicon crystal (15:09 h)

In Fig. 4 again the time frame of the ultrasonic signal from this second experiment is shown. At (1) the crucible lift down has started. At the same time the automatic translation unit moves the waveguide also moves downwards to keep the distance of the rod end to the crucible bottom on a constant level of 19 mm resulting in the same transit time as before.

At (2) the silicon crystal growth has started indicated by the splitting of the echoes in one which is still induced by the crucible bottom and in some weaker echoes which are induced by the new formed s-l-interface. In this case the transit time for the pulses reflected by the s-l-interface remains almost constant due to the constant distance to the rod end (nearly horizontal line). Otherwise, the transit time for the crucible bottom increases over time due to the increasing distance of it to the rod end. Again by the use of the transit time difference Δt between the lines for the s-l-interface and the crucible bottom (white arrows in Fig. 4), the grown silicon ingot height h can be calculated for every time step. E.g. at (3) the transit time difference Δt = 6516 ns can be transferred by equation (1) into a grown ingot length of 26 mm.

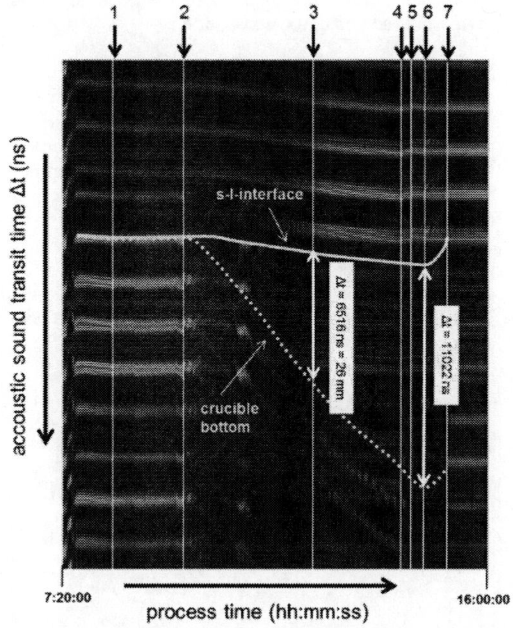

Figure 4: Time frame of the ultrasonic signal during the relevant process steps in the 2nd experiment

After 41 mm growth at (4) the waveguide was shifted downwards by 4 mm to prevent that the waveguide leaves the silicon melt in consequence of the automatic upwards translation. This was repeated short time later at (5) resulting in a final distance of 11 mm of the waveguide to the s-l-interface. At (6) the automatic translation was stopped in order to remain the waveguide inside the residual silicon melt until it was overgrown by the crystal at (7). Consequently, the transit time of both the s-l-interface and the crucible bottom decrease resulting in upwards declined echo lines. With the transit time difference Δt at (6) = 11022 ns the grown ingot height of 44 mm was calculated. If the actual distance between the s-l-interface and the waveguide (11 mm) is added, the final grown ingot height under the rod end should be 55 mm.

If we now compare this calculation with the experimental result measured at a vertical cut of the crystal after growth process (see Fig. 5) it turns out that both values exactly match each other. This means that the accuracy of the ultrasonic sound measurement is quite good to determine the grown ingot height "on the fly".

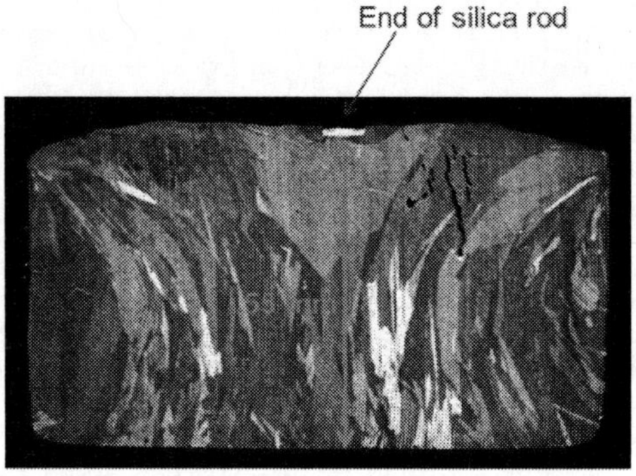

Figure 5: Vertical cut of the grown silicon ingot with ingrowing rod end of the waveguide

In Fig. 6 such an "on-the-fly" curve of the ingot growth is shown which is generated in-situ is by the used software on the basis of the motor position. On the y-axis the axial position of the waveguide (motor position), the crucible and the s-l-interface are plotted; on the x-axis the process time. Additionally, all above mentioned time points (1) to (7) are marked with red dots. Now, by observing this curve, the operator is able to control when the crucible moves downwards, the crystal growth starts, when it ends and how fast it is over time. In this case a mean growth velocity of ~ 20mm/h over the complete ingot height appears.

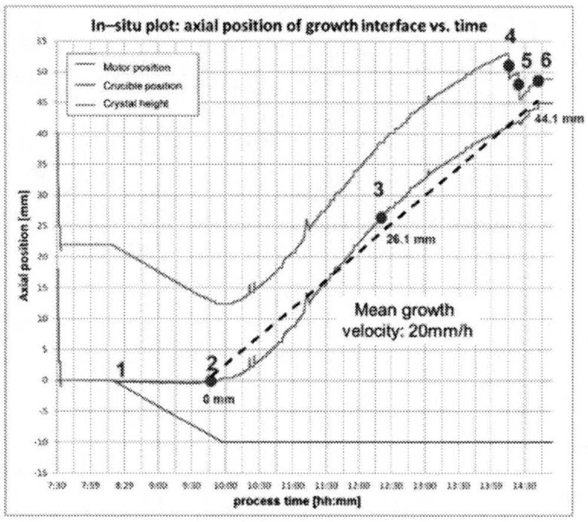

Figure 6: Axial position of the waveguide (rod end) and motorized translation unit vs. process time

4 CONCLUSIONS

In summary it could be shown, that the ultrasonic sound method could be a suitable tool in industry for determination the growth behavior of silicon ingots grown by directional solidification without disturbing the grain growth. The operator receives in-situ information about the starting and end point of growth and also the growth velocity which makes it possible to react on process fluctuations quickly.

For the actual trend in industry, the growth of high performance multi-crystalline silicon ingots by seeding on a remaining feedstock layer, it should be possible to detect the upper surface of this layer as initial s-l-interface by this method, too.

RERERNCES

[1] M. Azizi, E. Meissner, J. Friedrich, Proceedings 25th European Photovoltaic Solar Energy Conference (2010), p. 1520.

COST EFFECTIVE GROWTH OF SILICON MONO INGOTS BY THE APPLICATION OF INCREASED PULL SPEED IN CZ-PULLER

F. Mosel[1], A.V. Denisov[1], B. Klipp[1], R. Kunert[2], P. Dold[2]
[1] PVA Crystal Growing Sytems GmbH, Im Westpark 10-12, 35435 Wettenberg, Germany
[2] Fraunhofer-Center für Silizium-Photovoltaik CSP. Otto-Eißfeldt-Str.12, 06120 Halle (Saale), Germany
frank.mosel@pvatepla.com, phone: +49 64168690-125, fax: +49 64168690-822

ABSTRACT: The permanent demand on the improvement of solar cell efficiency and reduction of production costs has placed much focus on the multiple Cz-pulling techniques from a single crucible. PVA Crystal Growing Systems GmbH has developed a mobile recharge system for Czochralski pullers, the SiCharger MRS. This system and its potential was presented in detail at the 33rd EU PVSEC [1]. Another way to save costs while increasing productivity is a higher pull rate. In order to enable a higher pull speed at a stable crystal growth, the hot zone was investigated and optimized. The entire crystal growth arrangement was simulated with the commercial program package CGSim [2]. As a result, the pull rate could be significantly increased resulting in an appreciable saving of energy and process gas. In the next approach, an active crystal cooling system was implemented. This device consists of a water cooled container, which is placed a few centimeters above the phase boundary of the growing crystal, in order to increase the temperature gradient in the growing crystal close to its solidification interface. With this arrangement the heat transport through the crystal is improved. Crystals with 8 inch diameter were grown with a pull rate of 1.7 mm/min.

Keywords: Crystal Growth Rate, Simulation, Czochralski

1 INTRODUCTION

To meet the demand for an increase of the solar cell conversion efficiency and a reduction of the cost of the high quality monocrystalline substrate material, it is inevitable to optimize the whole value chain of crystalline silicon solar cells. A cost-intensive element in this chain is the conversion from the polycrystalline feedstock to the monocrystals by the Czochralski method. Economic analysis of the Czochralski technique identifies the crucible, furnace parts and the consumption of energy as the main cost drivers. To lower the production costs the application of multiple ingot growth from a single crucible is a powerful tool, as it was demonstrated by the authors [1]. Besides the cost savings an increase in productivity is also necessary. An increase of the pull rate is an effective tool to shorten the process time. Besides the increase of the pull speed and the cost efficiency, the material quality remains the key factor for the success in the PV market. That means that all economical improvements in the process conditions have to be justified by means of the quality of the grown ingots.

2 THEORY

2.1 Limitation of the growth rate in Cz-configuration

The physical rate limiting parameter for the growth of ingots in a Czochralski configuration is the dissipation of the latent heat of fusion of silicon at the interface crystal/melt by heat conduction through the growing ingot. The heat received from a crystal in contact with its melt can be determined as the heat of solidification plus the heat flux from the melt into the ingot. This situation can be written in a simple one dimensional model [3]:

(1) $$\lambda_s \cdot A \cdot \left(\frac{dT}{dx}\right)_S = \lambda_l \cdot A \cdot \left(\frac{dT}{dx}\right)_l + \rho \cdot A \cdot L \cdot v_p$$

In which λ is the thermal conductivity, A the area of the ingot, (dT/dx) is the axial temperature gradient, v_p is the constant growth velocity, ρ is the crystal density, L the latent heat of fusion and s,l are the subscripts for the solid and liquid phase. Assuming an isothermal melt i.e. $(dT/dx)_l = 0$ the maximum pull rate v_{pmax} can be estimated and related to the crystal radius r:

(2) $$v_{p\max} = -\frac{\lambda_m}{\rho \cdot L}\left(\frac{dT}{dx}\right)_{x=0} = \frac{1}{\rho \cdot L}\sqrt{\left(\frac{2\sigma\varepsilon\lambda_m}{3 \cdot r} \cdot T_m^{\,5}\right)}$$

Here σ is the Stefan-Boltzmannn constant, ε is the surface emissivity, m is the subscript for melting temperature. For a detailed derivation of this equation see E. Billig [4].

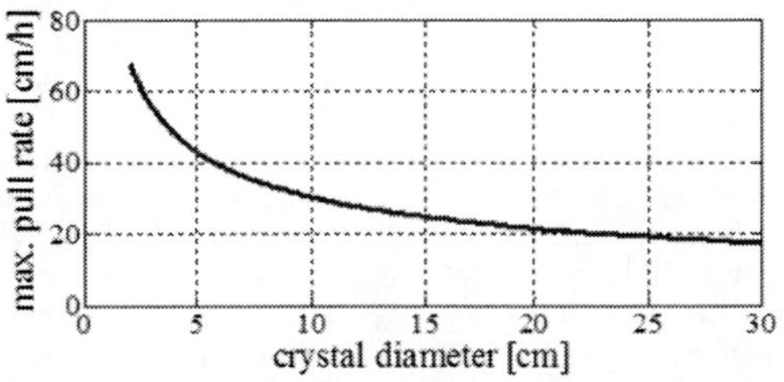

Figure 1: Max. crystal pull speed for silicon versus crystal diameter estimated by the model of E. Billig [4].

The model shows the inverse diameter square root dependency of the pull rate. It assumes the absence of thermal convection and a radiation from the crystal surface to the environment at zero Kelvin. The geometry of the hot zone especially in the hot growth region has a very complex geometry with different material parameters. S.N. Rea [5] has developed a thermal model which takes this situation to a certain degree into account. The calculated maximum pull rates show the same dependency of the crystal diameter as the model of Billig but with significant lower magnitudes. The curve of Rea is shown in Fig 2 together with the predicted maximum growth by models of Ciszek and Wilcox [6].

The interfacial temperature gradient is influenced by the thermal conditions at the crystal surface and not by its length. This holds for a crystal length which is twice its diameter and a Biot number H > 0.15 [7]. This is the case for our growth conditions. The Biot number H for cylindrical surface (3) gives the ratio of the thermal

$$(3)\quad H = \frac{hr}{\lambda}$$

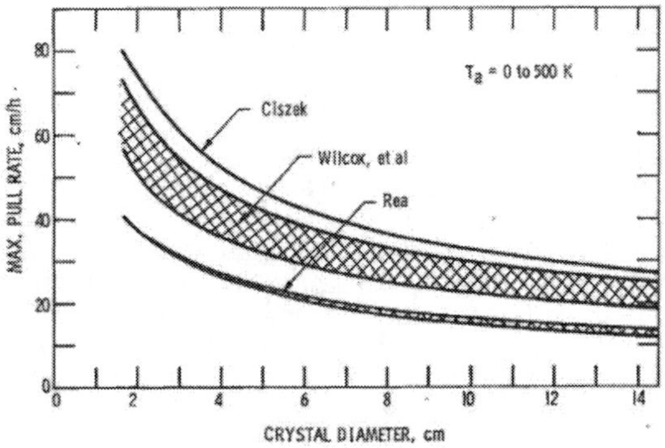

Figure 2: Silicon pull rates predicted by various models. The figure is reproduced from [6].

resistance within and at the surface of the ingot with r is the radius of the crystal, h is the convection heat transfer coefficient an λ is the thermal conductivity. In order to optimize all growth parameters especially the design of the hot zone including the inner heat shield, the application of numerical modeling of the Czochralski process is indispensable. The objective of the numerical simulation is to develop the growth conditions for a high pull rate which permits a stable crystal growth without the risk of losing the shape and crystallinity of the ingot.

2.2 Numerical simulation

The calculations were performed for 8 inch crystals in a 24 inch hot zone in a SC 24/26 crystal puller of PVA Crystal Growing Systems GmbH. In our development work we apply the CGSim software package from STR [2], which is especially developed for the Czochralski configuration. The following strategy was used for the simulation calculations. The standard hot zone was subjected to an extended analysis. The most important evaluation criteria are the deflection of the phase boundary (concavity), the radial temperature distribution on the melt surface and the axial temperature gradients in the center of the crystal and on its surface. The simulation calculations were carried out at a grown crystal length of 500 mm, since from this length no further change of the phase boundary shape is to be expected due to its Biot number. The deflection of the phase boundary should be as low as possible in order to ensure stable growth conditions. The knowledge of the curvature is also of importance for the distribution of intrinsic point defects, which is described by the well-known v/G criterion [8]. The radial temperature gradient in the melt at the crystal edge should be as high as possible and no supercooled areas with temperatures lower than the melt/crystal interface located close to the crystal should occur [9]. The axial temperature gradient in the crystal at the solid/liquid interface should be steep in order to enhance the dissipation of the latent heat in the crystal. The standard hot zone analyzed in this work is shown in Fig.3 on the left hand side. On the right hand side of Fig.3 the optimized design is shown. From the viewpoint of power reduction, the upper insulation was made more efficient. The inner radiation shield was completely reworked. Some of the important parameters of the geometry are indicated by arrows in fig.3.

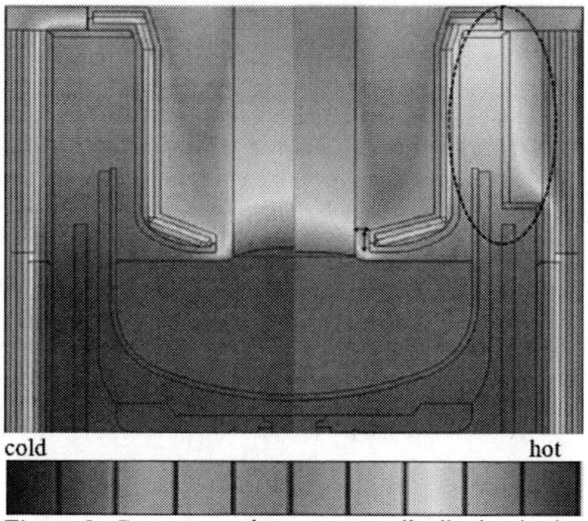

Figure 3: Geometry and temperature distribution in the standard (left hand side) and optimized (right hand side) hot zone.

Fig.5 shows a diagram which is based on the numerical simulations and the experience based on real crystal growth experiments. The interface curvature H is plotted versus the average pull speed in the body phase at a crystal length of 500 mm. 3 different regions are shown.in the diagram. In the stable growth region (green) the system is insensitive to changes in the average pull speed. This region is robust and suitable for industrial production. In the metastable growth region (yellow) all growth parameters have to be well tuned. Small changes can lead to unstable growth with loss of shape, i.e. spiral growth. In the unstable growth region (red) no regular crystal growth is possible.
In the graph the points connected by the lines represent individual results of the simulation calculations for the examined crystal growth arrangements (V1-V4). The open symbols in the graph represent experimental results. In cases V1 and V2 the deflections of the phase boundaries were determined by the LPS technique (see Fig.4).

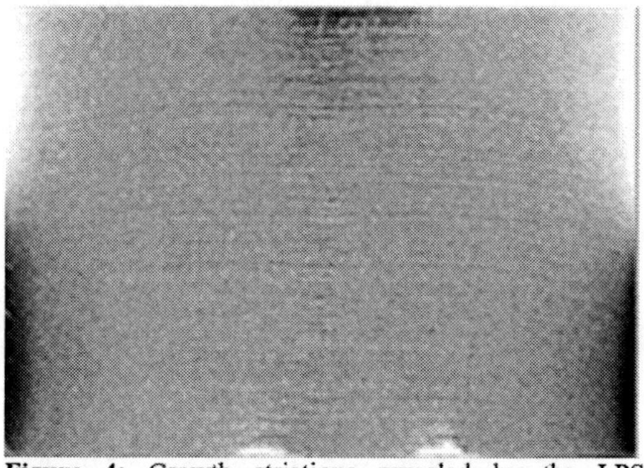

Figure 4: Growth striations revealed by the LPS technique [11] on a vertically sliced sample (205 mm x 250 mm), indicating the deflection of the growth interface.

The growth regions of the crystal growth configurations V3 and V4 have been determined experimentally by variation of the pull speed. The graph does not represent an analytical model but rather an empirically determined state diagram. In our crystal growth configuration the theoretical limit of the pull rate seems to lie in the range of 1.3 mm/min which can be achieved in principle by optimizing the inner heat shield and its integration in the standard hot zone (growth configuration V2).

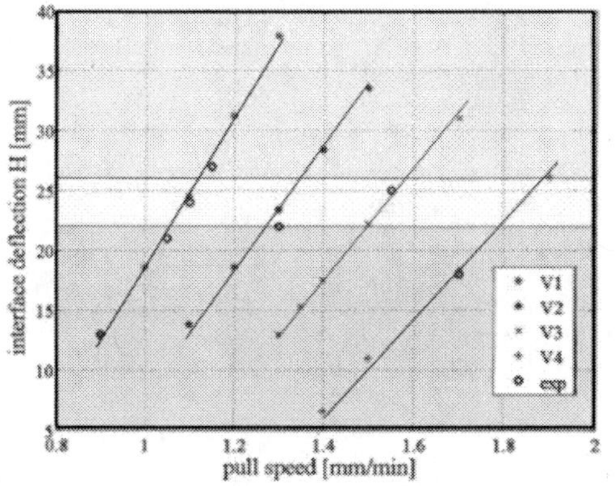

Figure 5: State diagram of the stable (green), metastable (yellow) and unstable (red) growth conditions. The main parameter deflection H of the interface shape is plotted versus the pull speed.

In order to achieve a further increase in the pull speed, an active crystal cooling system has been specifically integrated into this optimized hot zone. This active crystal cooling is a water-cooled container in form of a coil placed between the inner heat shield and the growing crystal. The crystal cooling coil configuration was first published by S.N. Rea in 1977 [6]. The principle of the arrangement is shown in Fig.6.

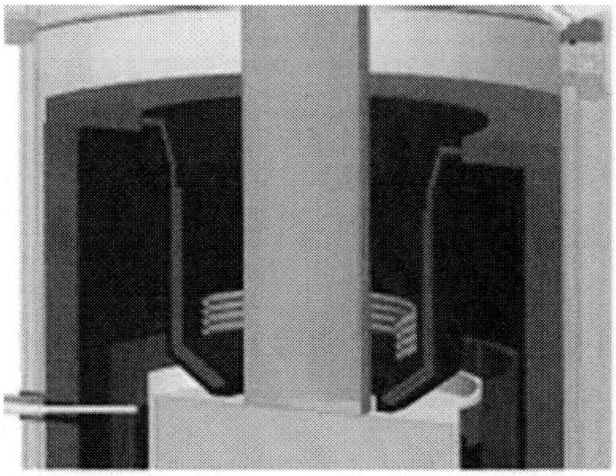

Figure 6: Arrangement of the active crystal cooling device in the hot zone.

The cooling coil is positioned in a certain distance to the melt. In this region the axial temperature gradient in the crystal at the phase boundary can be influenced effectively by the thermal conditions at the crystal surface. The geometry and its surface and especially the position of the cooling container with respect to the melt surface influence effectively the absorption capacity of the cooling device. The results of the numerical calculations are shown in Fig.7.

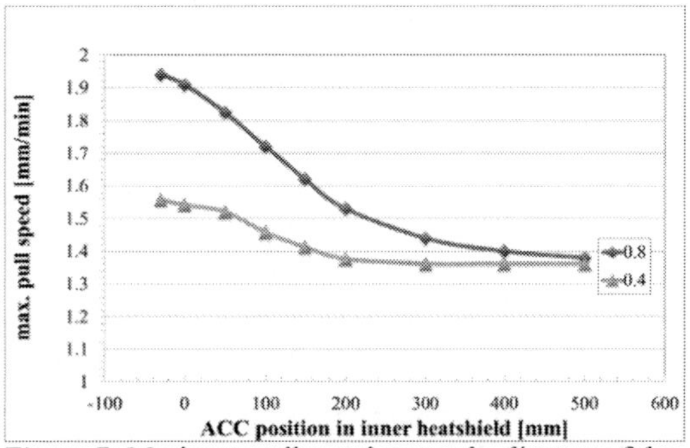

Figure 7: Maximum pull speed versus the distance of the cooling device to the melt surface for the emissivities 0.4 and 0.8. Position 0 represents the applied position of the lower edge of the cooling device in the growth configurations with a distance of 100 mm to the melt surface (-100).

Different geometries with different surface characteristics of the cooling devices have been tested and are still under investigation. Fig.8 shows two examples.

Figure 8: Cooling container with a spiral shape and a cylindrical shape.

It is well known that the temperature gradient at the crystallization interface, which is dependent on the crystal radius and the pull rate influence the distribution of intrinsic point defects. There is a critical ratio of the pull rate v_p to the temperature gradient G(r) of $v_p/G(r) = 0.134\ mm^2/K\ min$ [10]. It was found that for a ratio more than the critical value v/G_{crit} vacancy aggregates develop and for a ratio lower than the critical value self-interstitial defects are predominant. At the critical ratio interstitials and vacancies annihilate each other and the crystal is free of dislocation loops and voids. Fig 9 shows the calculated values for the different crystal growth conditions. All applied growth conditions show a ratio v/G, which cause an intrinsic point defect distribution dominated by vacancies.

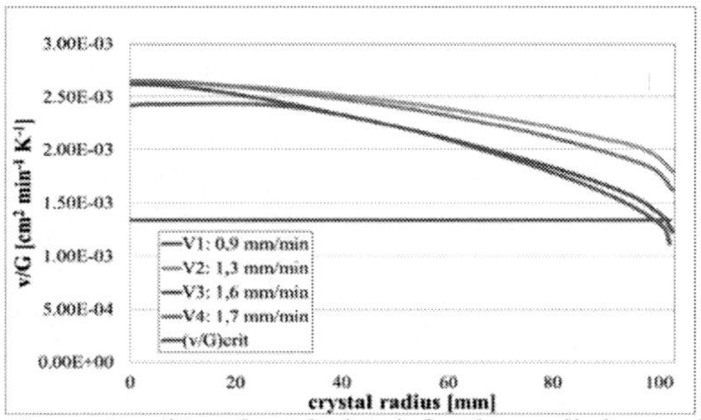

Figure 9: v/G ratio calculated for the applied crystal growth conditions (Tab.I).

3 CRYSTAL GROWTH

Several crystals have been grown in 4 different growth configurations V1 – V4, summarized in table 1. In growth configuration V1 and V2 no active cooling devices were applied, whereas in the growth configurations V3 and V4 two different cooling devices were tested. All growth configurations and their potential regarding the maximum pull speed are shown in the stability diagram in Fig.5.

growth configuration	hotzone	cooling device	emissivty	position of cooling device
V1	standard	no	-	-
V2	optimized	no	-	-
V3	optimized	cylinder	0.4	0
V4	optimized	spiral	0.6	0

Table I: Growth configurations with main parameters

3.1 Crystal Growth without active cooling

Several crystals have been grown under growth conditions without an active cooling device. The reference crystal in the standard hot zone (V1) was pulled with an average growth rate of 0.9 mm/min. The maximum average pull speed at which stable growth can be achieved in this configuration is 1.0 mm / min. At a pull speed of 1.15 mm/min spiral growth during the body phase was observed. From the reference crystal (V1) a vertical cut was prepared and the deflection of the solid/liquid interface was examined by the LPS-technique. The specimen was cut at a crystal length of 500 mm to make a comparison with numerical modeling possible. The concavity of the phase boundary of 13 mm, which was estimated from the photo is in good agreement with the calculated value of 12.5 mm. In the optimized hot zone (V2) a crystal was pulled with an average pull rate of 1.3 mm/min. It turned out that this pull speed already lies in the metastable range of the diagram (Fig.5). In a second growth run stable growth conditions were achieved by adjusting the distance of the inner heat shield to the melt surface. For these growth conditions the measured deflection of the interface was 23 mm compared to the calculated value of 22 mm. The good agreement confirms the results of the numerical simulation calculations.

3.2 Crystal Growth with active cooling

In order to achieve a higher pull speed, the axial gradient in the crystal must be increased according to equation 1. In the optimized hot zone of V2, which is the basis for the growth configuration V3 andV4, this was achieved by applying a water-cooled heat exchanger, hereinafter referred to as active crystal cooling. Such an arrangement was first described and experimentally investigated by S.N. Rea (6). The crystal cooling of Rea, however, led to an unstable growth process and thus this arrangement was not pursued further. Based on the work of S.N. Rea, the PVA Crystal Growing Systems GmbH developed an active crystal cooling system with different cooling devices. Examples are shown in Fig.8. The cooling device often referred to as cooling jacket in the literature is positioned between the inner heat shield and the growing crystal. The cooling element absorbs the radiant heat of the crystal and at the same time shields the crystal from the emitted heat radiation of the graphite inner heat shield. The development of the active crystal cooling system was carried out under strict safety considerations. The safety concept is discussed in detail in chapter 5. With the crystal growth configuration V3 a mean pull speed of 1.6 mm/min was achieved in the body phase from a length of approx. 100 mm after the transition from crown to body. With the crystal growth configuration of V4 a mean pull speed of 1.7 mm/min was possible. In this growth configuration too, the transition from the crown to the body lasted for some time. The process parameters in this area have to be optimized so that the transition to the fast pull speed is as short as possible.

3.3 Characterization of the grown ingots

The grown crystals V1, V2 and V4 were prepared for characterization as shown in Fig.10.

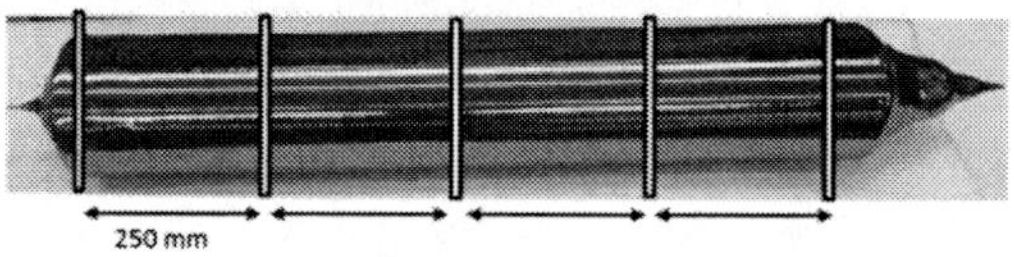

Figure 10: Position of the samples for characterization

5 slices were cut from the crystal at equal intervals and the contents of interstitial oxygen O_I and substitutional carbon C_S were investigated using FTIR. The conversion factor is $3.14x10^{17}$ cm^{-2} for O_I and $1.0x10^{17}cm^{-2}$ for C_S. The measurements on the wafer were performed 7 mm from the edge, half the radius and in the center. The measurement results are shown in Tab II. The samples are designated 1 to 5 from top to tail.

V1	sample 1	sample 2	sample 3	sample 4	sample 5	
center	10.8	9.1	8.2	7.5	6.5	x 10^{17} cm^{-3}
half radius	10.4	8.9	8.1	7.7	6.4	x 10^{17} cm^{-3}
edge	6.3	5.7	4.6	5.6	5.8	x 10^{17} cm^{-3}

V2	sample 1	sample 2	sample 3	sample 4	sample 5	
center	9.5	6.6	5.8	4.6	4.7	x 10^{17} cm^{-3}
half radius	9.2	6.6	5.6	4	4.3	x 10^{17} cm^{-3}
edge	3.6	3.2	2.3	1.6	3.6	x 10^{17} cm^{-3}

V4	sample 1	sample 2	sample 3	sample 4	sample 5	
center	10.8	6.1	6.1	8.6	9.7	x 10^{17} cm^{-3}
half radius	10.2	6	5.9	8.7	9.3	x 10^{17} cm^{-3}
edge	4.1	3.9	3.5	5.9	6.1	x 10^{17} cm^{-3}

Table II: O_I-concentration in the crystals V1, V2, V4. The C_S-concentration in 95% of all samples were below detection limit (<10^{16} cm^{-3})

The measurement results do not have any statistical certainty. However, it can be concluded that the high pull speed in the configurations V2 and V4 do not have a negative effect on the distribution of oxygen and carbon content.

4 ECONOMIC ANALYSIS OF THE HIGH PULL SPEED

The cost savings potentials of the different growth configurations are summarized in Tab.3. All data are based on the body phase only since this growth phase has defined boundary conditions (diameter, length). The other crystal growth phases are often different and not taken into account. The consumption of the heater energy is calculated from the effective power in the body phase. The consumptions of cooling water and gas correlate with the pull rate in the body phase in the case of crystal growth configuration V1 and V2. In the case of crystal growth configuration V3 and V4 the consumption of the cooling water is enhanced due to the different cooling devices.

growth config.	mean pull speed. [mm/min]	time body phase [h]	Δ time body phase [%]
V1	0.9	20.5	0
V2	1.3	14.6	29
V3	1.6	12.05	41
V4	1.7	11.5	44

growth config.	Δ energy [%]	Δ argon [%]	Δ cooling water [%]
V1	0	0	0
V2	32	29	29
V3	41	41	39
V4	43	44	37

Table III: Measures of the productivity of the examined crystal growth configurations in the body phase. The Δ-data represent savings potentials referred to V1.

5 SAFETY CONSIDERATONS OF THE ACTIVE COOLING SYSTEM

5.1 Safety concept

The safety components in the control system of the puller detect faults in the cooling water circuit of the active cooling system and then switch off the water supply and the heater immediately. The water flow through the cooling circuit, the cooling water temperature at the outlet and the pressure inside the growth chamber are used as recognition and trigger criteria. The most important components are duplicated (redundant), so that if a sensor or actuator fails, the second one can take over and reset the machine to a safe state.

The greatest danger to man and machine is the uncontrolled discharge of water in the process chamber, which cannot be prevented, but the quantity of cooling water that may be released by the cooling device is kept low. Two series-connected, pneumatically actuated shut-off valves are located in the cooling water flow, two check valves connected in series in the return. This ensures that, in the worst case, only the quantity of water corresponding to the internal volume of the cooling device can enter into the process chamber, which is about 600 ml. When water enters the process chamber, a spontaneous pressure increase occurs. Two independent monitoring devices are used for overpressure detection. As soon as one of the two monitoring devices responds, the cooling water supply of the cooling device is stopped and the heating device is switched off. The flow rates and temperature of the cooling water of the crystal cooling are also monitored. If the temperature exceeds a fixed limit value or falls below the preset flow rate, the heating device is switched off. A short-term cooling water failure or pressure fluctuations are intercepted by a separate system for about 30 seconds. Before the start of each process the safety devices are automatically checked for perfect functionality.

The functionality of the shutdown due to a pressure increase in the process chamber was tested and confirmed under real crystal growth conditions.

5.2 Risk analysis

A risk analysis was carried out for the case where the cooling device, which is located a certain distance above the melt, has a water leak. Several reactions may occur:

A physical reaction where the transfer of water to vapour creates a steam explosion

A chemical reaction, where water reacts with the silicon metal and form hydrogen may lead in some situation to an oxyhydrogen reaction.

Physical reaction

When the water enters the process chamber, it is heated to 100° C. and evaporated. The evaporation enthalpy required for evaporation is provided by the melt surface and the surface of the internal heat shield. The following consideration was made for a water quantity of 1 liter. This corresponds to 55.6 mol of H_2O. In the puller with a free volume of about 1 m^3 this quantity would lead to a pressure of approx. 8 bar, if the mean gas temperature corresponds to the silicon melting temperature. Since the amount of water from the cooling device cannot be released instantaneously and a significantly lower gas temperature will adjust, a considerably lower pressure in the machine will be set for a short moment.

For the possibility of water entry with consequent steam formation and an increase of the internal pressure (steam explosion) several safety devices are installed in the puller, which react in succession at different pressures. If the pressure inside the growth chamber exceeds a value of 1.023 mbar the safety valve releases the overpressure.

Chemical reaction

(4) $Si + 2H_2O\ (gas) \rightarrow SiO_2 + 2H_2\ (gas)$

The water is poured on the surface of the silicon melt. The water is heated to 100°C and gas is formed. The water reacts with the melt at the surface and H_2 and SiO_2 are formed on the surface. The silicon oxide forms a continuous layer and a small amount of hydrogen is in the gas above the melt. The reaction according to equation (4) stops, while the water gas expands its volume.

The risk analysis was performed by the PVA Crystal Growing Systems GmbH together with SINTEF Materialer og kjemi, Norway.

5.3 Certificates

The set-up of the safety devices described in chapter 5.1 were checked and confirmed according to the guidelines EN ISO 13849-1:2008-12 by the German Technical Supervisory Board TÜV Hessen GmbH.

6 CONCLUSIONS

With an optimized design of the hot zone significantly higher pull speeds can be achieved. In this report an average growth rate of 1.3 mm/min in the body phase is shown. Further optimizations were investigated with numerical simulation calculations. The limit appears to be at 1.4 mm/min. With the use of an active crystal cooling system, an average pull speed of 1.7 mm/min was achieved. A further increase seems quite possible. The process parameters for such high growth rates still have to be optimized. The crystals grown with a high pull speed show good material characteristics with respect to the oxygen- and carbon concentrations compared to conventional pull rates. Statistical certainty is not possible within the scope of this work. The use of active crystal cooling promises a significant increase in productivity combined with a high cost saving potential.

7 ACKNOWLEDGEMENTS

This work was supported by the German Ministry of Economy and Energy under contract number 0325883A.

8 REFERENCES

[1] F. Mosel, A.V. Denisov, B. Klipp, R. Sharma, R. Kunert, P. Dold, Proceedings 33rd European Photovoltaic Solar Energy Conference, München, Germany, (2016), 1064-1068

[2] CGSim package, STR Group,Ltd.

[3] J.C. Brice, J. Crystal Growth 2 (1968) 395-401

[4] E. Billig, Proc. Roy. Soc. (London) 229 (1955) 346-363

[5] S.N. Rea, J. Crystal Growth 54 (1981) 267-274

[6] S.N. Rea, Final Report, ERDA/JPL 954475 (April 1977)

[7] V.H. Kio, W. R. Wilcox, J. Crystal Growth 12 (1971) 191-194

[8] E. Dornberger, W.von Ammon, J. Virbulis, B. Hanna,. T. Sinno, J. Crystal Growth 230 (2001) 291-299

[9] V. Kalaev, A. Sattler, L. Kadinski, J. Crystal Growth 413 (2015) 12-16

[10] Fraunhofer-Center für Silizium-Photovoltaik CSP, Otto-Eißfeldt-Straße 12, 06120 Halle (Saale), Germany

[11] H.J. Schulze, A. Lüdge, H. Riemann, Electrochem. Soc. **143**, 4105 (1998)

STUDY OF IMPURITIES DIFFUSION IN SILICON LIQUID PHASE IN CONDITIONS OF HIGH TURBULENCE OF MELT

Sergey M. Karabanov[1], Dmitry V. Suvorov[1], Oleg A. Belyakov[2], Andrey S. Karabanov[2], Dmitriy Y. Tarabrin[1], Evgeny V. Slivkin[1], Valery L. Dshkhunyan[3]
1- Ryazan State Radio Engineering University, 59/1 Gagarina St., Ryazan 390005, Russia
2- Helios Resource Ltd., 126 Proletarskaya St., Saransk 430001, Mordodia
3 - Solar Consult Ltd., 51B Novaya St., Ryazan 390027, Russia

ABSTRACT: The paper presents the research results of impurity diffusion to silicon melt surface under different conditions to determine optimal purification conditions of silicon for solar cells. The dependences of the impurity flow (B, P, Al, Fe) across the silicon melt boundary on the average radial flow rate on the melt surface are determined. It has been established that, beginning with the average radial flow rate of 0.1 m/s on the melt surface, the intensity of the impurity flow from the melt surface increases sharply. Obviously, it is connected with the transfer to turbulent melt flow conditions. The research is carried out by mathematical modeling using COMSOL Multiphysics and Matlab software.
Keywords: Si, impurity, purification

1 INTRODUCTION

Currently a large number of studies have been devoted to the search for a technology of direct metallurgical-grade silicon purification for solar-grade silicon production. The most well-known technologies are as follows: direct crystallization (solidification), Elkem Solar Silicon Production Process which includes fire refining by addition of compounds containing calcium [1]; Silicon Process - recrystallization in the Al-Si system [2]; direct carbothermic reduction – this method is used in SOLSILC and SPURT projects [3, 17]; Electron-beam melting [4]; Photosil process using plasma processing of silicon melt [5]; NEDO melt-purification process. NEDO MG-Si melt purification is a two-stage process using plasma and electron beam treatments [6].

The results of studies on the use of the plasma-chemical method for metallurgical-grade silicon purification to the solar-grade silicon level are reported in [7-15]. In these papers, the possibility of using this method for silicon purification is shown in principle.

The paper presents the research results of impurity diffusion to the silicon melt surface.

2 MATHEMATICAL MODEL, RESULTS AND DISCUSSION

2.1 Mathematical model

The mathematical model of impurities diffusion is based on approximation of axially symmetrical geometry of a crucible. In [15], detailed description of the model part describing induced mass transfer within silicon melt is presented.

Removal of impurities from silicon melt occurs on the liquid-gas boundary. The impurity concentration in the gas phase above silicon melt is determined by the saturated vapor pressure of the impurity chemical compond or impurity at the temperature equal to the minimum temperature of the reactor cold parts. i.e. the temperature of impurity condensation zone.

Since the temperature of the condensation zone is much lower than the melt surface temperature, the approximation of the calculation model assumes complete removal of impurities on the melt surface, that corresponds to zero impurity concentration on the melt surface:

$$n_i(z=0)=0$$

where n_i is i-impurity concentration, z – an axial axis coordinate.

Due to the plasma flow to the melt surface, laminar and turbulent flows are provided. A detailed mathematical model and a procedure for calculating silicon flows induced by plasma flow is described in our previous paper [15]. The flow direction on the surface of the cylindrical crucible is radial, having a direction from the edges to the center [15]. It was shown that the flow rate on the melt surface is irregular. The average radial rate value determines the transit time Δt of the distance from the crucible edge to the center by the melting point:

$$\Delta t=\frac{r}{avg(v(r))},$$

where r is the crucible radius, avg(v(r)) – average radial rate.

In the model approximation, it is considered that the purification process occurs during alternating time intervals with Δt duration. The initial condition for each of the time intervals is zero impurity concentration on the melt surface:

$$n_i(z=0)=0$$

and uniform concentration distribution along the axial axis.

$$n_i(L\geq z>0)=n,$$

where L is the cylindric crucible length.

Time Δt determines the duration of purification and, accordingly, the effective purification depth of silicon melt from impurities.

The total impurity flow F_1 passing through the melt surface is calculated as the product of the impurity flow Fc_i in from the near-surface layer during time Δt multiplied by the "refresh rate" of the near-surface layers 1/Δt:

$$F_i=\frac{1}{\Delta t}Fc_i(t=\Delta t).$$

Impurity diffusion to the melt surface was calculated using a two-dimensional approximation in a cylindrical coordinate system. The equation describing the mass transfer inside the crucible melt was the equation of continuity:

$$\frac{\partial n_i}{\partial t} + \nabla \cdot (-D_i \nabla n_i) + u \cdot \nabla n_i = R_i ,$$

which includes both diffusion and convective (induced) components.

The flow equation due to diffusion and convection, separately for each impurity type, has the form:

$$F_i = -D_i \nabla n_i + u n_i ,$$

where D_i is the impurity diffusion coefficient in silicon melt. D_i coefficient depends on the temperature and an impurity type.

The problem solution was carried out by mathematical modeling using COMSOL Multiphysics and Matlab software.

Fig. 1 shows axially symmetric geometry of the numerical model and computational grid used for calculation.

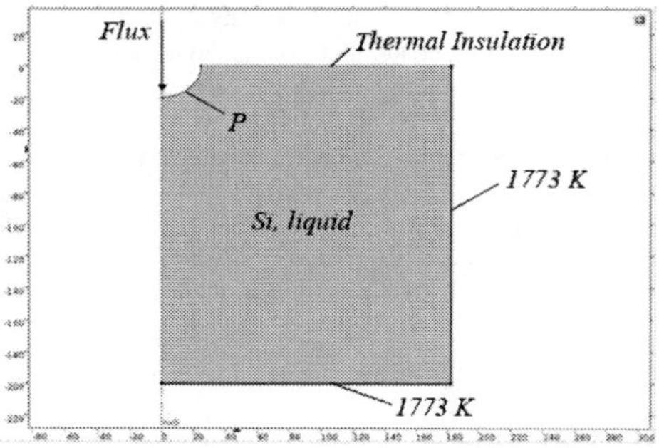

Figure 1: Axially symmetrical geometry of the numerical model

2.2 Numerical calculation conditions

The area of the plasma arc local effect is presented in the form of a hemisphere with the diameter of 40 mm in which the total input power is concentrated. The flow velocity vector of gas dynamic jet has the direction to the crucible bottom. Using a numerical model, calculation for a crucible with the height of 200 mm and the diameter of 360 mm was made under the condition of the constant crucible wall temperature of 1773 K for different rates of induced near-surface silicon flow of 10^{-4} – 0.1 m/s in the area of plasma effect.

The purpose of the performed modeling is to investigate the influence of the average rate of melt radial movement on the purification efficiency. The diffusion coefficient at increasing temperature varies slightly, so the calculations were carried out at constant temperature of 1773 K.

The diffusion of the following impurity types has been studied: boron (B), phosphorus (P), aluminum (Al), iron (Fe). The data on the diffusion coefficient in liquid silicon for these types of impurities are taken from [16]. The diffusion coefficients were assumed to be equal to: $5.62 \cdot 10^{-4}$ cm²/s for P, $1.25 \cdot 10^{-4}$ cm²/s for B, $6.31 \cdot 10^{-4}$ cm²/s for Al and $2.95 \cdot 10^{-5}$ cm²/s for Fe. The initial concentration for P is equal to $9.677 \cdot 10^{-7}$[mol/cm³], i.e. $5.83 \cdot 10^{17}$ cm^{-3}. This concentration corresponds approximately to the experimentally measured one.

2.3 Numerical calculation results

Figure 2 shows the example of distribution of the mass transfer rate inside silicon melt at different flow rates. It can be seen that the initiating flow inside silicon melt is located under the area of plasma jet influence and is directed towards the crucible bottom. Starting from the flow rate of 0.1 m/s, the initiating plasma jet causes turbulent flows inside silicon melt and on the surface melt [15]. The analysis of the melt thermal field shows that the highest temperature area is located along the crucible axis and corresponds to the area of the main mass transfer.

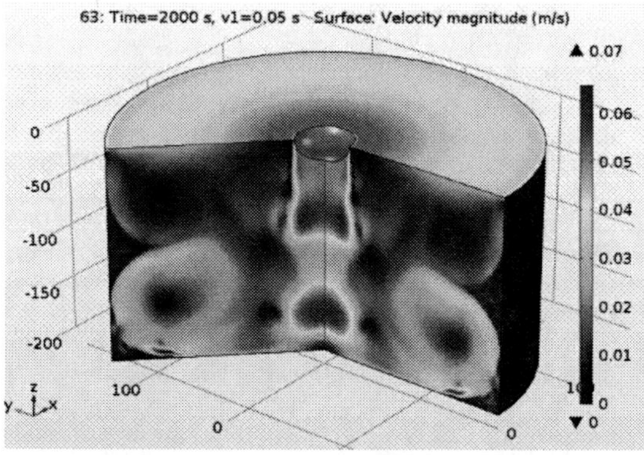

Figure 2: Distribution of the flow rate of liquid silicon in steady mode at the flow rate on the plasma spot surface of 0.05 m/s

Figure 3 shows the dependence of the average radial flow rate on the silicon melt surface on the silicon rate on the surface in the plasma spot area. It can be seen that the dependence is practically linear in the entire range of the inducing flow rates.

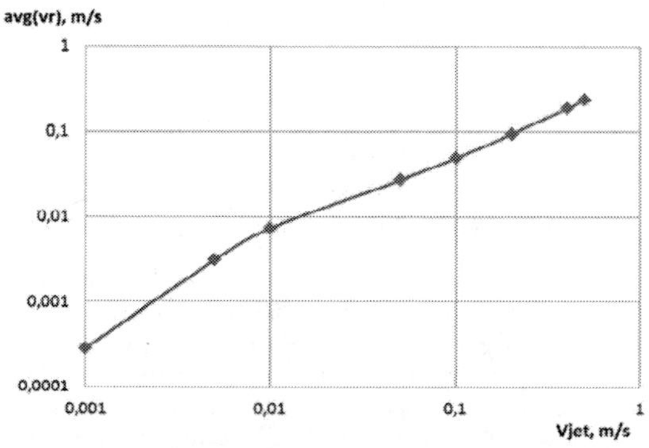

Figure 3: The dependence of the average radial flow rate on the silicon melt surface on the silicon rate on the surface in the plasma spot area

The calculation of the purification depth on which the silicon near-surface layer is purified during transit time of flow from the crucible center to the edge shows that the impurities are effectively removed from the silicon melt surface at the depth of about 0.05 – 1 mm.

Figure 4 shows the dependence of the impurity flow (B, P, Al, Fe) through the silicon melt boundary on the average radial flow rate on melt surface.

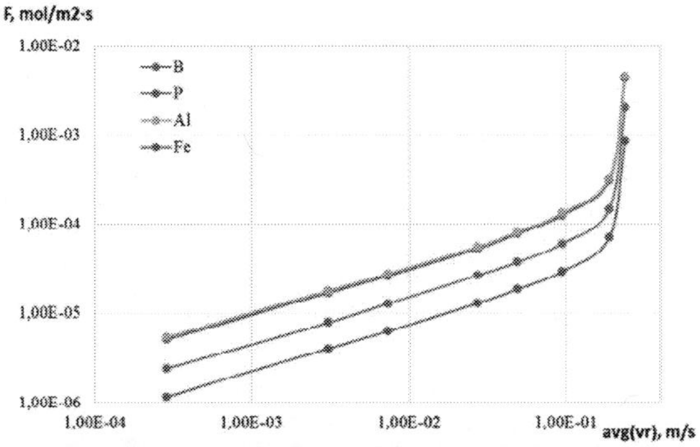

Figure 3: The dependence of the impurity flow (B, P, Al, Fe) through the silicon melt boundary on the average radial flow rate on melt surface

It can be seen that in the area of flow rate of $5 \cdot 10^{-4}$ - $1 \cdot 10^{-1}$ m/s the flow intensity increases linearly and monotonely with the increase of the average radial flow rate on the melt surface. Starting from the rate of $1 \cdot 10^{-1}$ m/s to $2.3 \cdot 10^{-1}$ m/s, fast dependence growth is observed. Obviously, it is connected with a high purification rate of the layer with the thickness not exceeding 100 μm, and also with transition to significant mass transfer turbulence in the crucible. Among the considered impurity types (B, P, Al, Fe), the highest values of flow through the melt surface correspond to P and Al due to their high diffusion coefficients in liquid silicon.

The results of the performed mathematical modeling explain the data of our earlier experimental studies [14].

4 CONCUSION

As a result of the calculations, the following regularities of plasma-chemical purification of silicon melt under conditions of its high turbulence are established:

- - starting from the average radial flow rate on the melt surface of 0.1 m/s, the intensity of the impurity flow from the melt surface rises sharply;
- impurity diffusion from silicon melt takes place from near-surface layers with the thickness of 0.05-1 mm;
- - it is shown that among the considered impurity types (B, P, Al, Fe), the highest values of flow through the melt surface correspond to P and Al.

The performed modelling shows that the realization of a mode providing intensive mass transfer of silicon melt with an average radial rate is greater than 0.1 m/s results in significant intensification of plasma-chemical purification rate.

The research results are planned to be used for the development of the silicon plasma-chemical purification technology and for design of the system for this process realization.

7 REFERENCES

[1] R. Tronstad, A sustainable product for the solar silicon market, Proceedings of the Silicon for the Chemical and Solar Industry XI, Bergen-Ulvik, Norway, June 25–29, 2012, pp. 197–206.

[2] T. Jester, Solar-grade silicon: Viable option to 'EG', Renewable Energy Focus, March/April 2014, pp. 34-35.

[3] L.J. Geerligs, et al., Solar-grade silicon by a direct route based on carbothermic reduction of silica: requirements and production technology, ECN Report, The Netherlands, 2002

[4] A.F.B. Braga, S.P. Moreira, P.R. Zampieri, J.M.G. Bacchin, P.R. Mei, New processes for the production of solar-grade polycrystalline silicon: a review, Sol Energy Mater Sol Cells 2008; 92:418–24.

[5] F. Cocco, D. Grosset-Bourbange, P. Rivat, G. Quost, J. Degoulange, R. Einhaus, M. Forster, H. Colin, Photosil UMG silicon: Industrial evaluation by multi-c p-type ingots and solar cells, Proceedings 28th European Photovoltaic Solar Energy Conference, pp. 1435–1438

[6] N. Yuge, M. Abe, K. Hanazawa, H. Baba, N. Nakamura, Y. Kato, et al., Purification of metallurgical-grade silicon up to solar grade, Prog Photovolt: Res Appl 2001; 9:203–9.

[7] A.S. Skryabin, Patent 2434683 RU, MPK B03B 7/00 (2008.01), Bulletin of inventions, 2011, No. 33.

[8] B.M. Moon, G. Kim, Boron refining and continuous casting of solar-grade polysilicon using steam-hydrogen plasma with electromagnetic induction melting, Poster Presentation, 28th European Photovoltaic Solar Energy Conference.

[9] S.M. Karabanov, V.L. Dshkhunyan, V.I. Yasevich, M. Hoshino, Method for manufacturing ingots of polycrystalline silicon, Patent 2465201 RU, December 13, 2010.

[10] S.M. Karabanov, V.L. Dshkhunyan, V.I. Yasevich, Method for purification of metallurgical-grade silicon by moisturized AC plasma in vacuum, Patent 2465202 RU, November 17, 2010.

[11] S.M. Karabanov, V.L. Dshkhunyan, V.I. Yasevich, M. Hoshino, Method for refinement of metallurgical-grade silicon, Patent 2465200 RU, February 14, 2011.

[12] S.M. Karabanov, V.L. Dshkhunyan, V.I. Yasevich, M. Hoshino, Method for refinement of metallurgical-grade silicon with dry argon plasma with water injection on the surface of melted silicon with subsequent directional crystallization, Patent 2465199 RU, November 17, 2010.

[13] S.M. Karabanov, V.L. Dshkhunyan, V.I. Yasevich and M. Hoshino, The study of the method of plasma-chemical purification of metallurgical-grade silicon, Proceedings of 31st European Photovoltaic Solar Energy Conference, 14-18 September, 2015, Hamburg, Germany, pp. 555 – 558.

[14] S.M. Karabanov, V.I. Yasevich, D.V. Suvorov. A.S. Karabanov, Mathematical modeling and experimental research of the method of plasma chemical purification of metallurgical-grade silicon, 16th International Conference on Environment and Electrical Engineering (EEEIC), pp. 1-5.

[15] S.M. Karabanov, V.I. Yasevich, D.V Suvorov, E.V Slivkin, A.S Karabanov, D.Yu. Tarabrin, Mathematical modeling of metallurgical-grade silicon plasma-chemical purification process,

Proceedings 32nd European Photovoltaic Solar Energy Conference, pp. 1001-1004, ISBN: 3-936338-41-8, Paper DOI: 10.4229/EUPVSEC20162016-2DV.3.4.

[16] K. Tang, E.J. Øvrelid, G. Tranell, M. Tangstad, Thermochemical and kinetic databases for the solar cell silicon materials, Proceedings 12th International Ferroalloys Congress: Sustainable Future, pp.619-630.

[17] S.M. Karabanov, New technologies for solar-grade silicon production, MRS Proceedings, Vol. 716 (2002), pp. 495-499.

THERMOMECHANICAL STRESS MODELLING DURING MELTING AND SOLIDIFICATION OF A MONOLIKE INGOT PROCESS.

A. Lantreibecq[1], M. Legros[2], D. Pelletier[1], J.P. Monchoux[2], E. Pihan[1]
1. Univ. Grenoble Alpes, INES, F-73375 Le Bourget du Lac, France
CEA, LITEN, Department of Solar Technologies, F-73375 Le Bourget du Lac, France
2. CEMES-CNRS, UPR 8011, 29 rue Jeanne Marvig, 31055 Toulouse, Cedex 4, France

ABSTRACT: We investigate the history of the thermomechanical stresses in the seed area during the melting and solidification steps of monolike ingot processing. This is specifically relevant to understand dislocation densities and microstructures at the beginning of growth of such ingots. We found that Von Mises (VM) stresses in seeds are highest at the onset of melting and sufficient to generate dislocation multiplication in seeds before growth begins. As large stresses before growth start might explain the high experimental local dislocations densities in seeds, we also focused on several mechanical boundary conditions applied to the seed for the VM calculations during the heating stage. Results show that locally fixed boundary and arbitrary locally loaded seed conditions used in simulations are significant compared to pure thermally induced stresses.
Keywords: Modelling, Directional solidification system, Seeded growth, Thermomechanical stresses, monolike silicon.

1 INTRODUCTION

Monolike silicon is a promising material for silicon based solar cells. Subgrain boundaries (SGBs) were found to be critical defects [1] in term of electrical efficiency, and further improvements of material quality require a better understanding of their origin. The dislocations that compose these SGBs can have various origins, such as thermomechanical stresses [2], [3] or other identified local sources [4]. It has been specifically identified that SGBs can be generated by growth on defective seeds [4]. Most thermomechanical stress calculations in the literature focus on the growth and cooling stages. As a consequence, we know only little on the values of stresses experienced by the seeds during the heating and melting stage. In this study, we investigate thermomechanical stresses during melting and compare them to stresses during solidification and cooling. In order to do so, we first compare time-dependent calculated VM stresses using a fixed geometry. Then, we explore the influence of the silicon load and thermal conductivity as well as the influence of boundary conditions on these stress calculations. From this, we expect to evaluate possible mechanisms for defect generation in the initially dislocation-free seeds during heating.

2 METHODOLOGY

In order to estimate VM stresses in solid silicon during a seeded growth process, solid silicon thermal field is first extracted from 3D time dependent thermal simulation of a silicon ingot using FLUENT software. Second, Von Mises stresses are evaluated from solid expansion due to thermal fields and various boundary conditions using COMSOL solid mechanics module.

2.1 Thermal models description

The simulated furnace geometry is based on a laboratory scale G2 (68 kg ingots) directional solidification furnace with top and bottom resistive heaters and a radiative heat exchanger. The 3D dynamic thermal field is calculated using methodology described in [5]. Main phenomena taken into account in this model are:

- Heat transfers by radiation, convection and conduction between components
- Natural convection in the silicon melt
- Release of latent heat at the solid-liquid interface during crystallization

In this work, 3 different silicon properties (table 1) and geometry (figure 1) are used to investigate either time dependent thermally induced stress trends either stress in seeds during the heating stage.

In model 1, the feedstock, which will melt during the second step of the ingot elaboration, is considered as fully dense (density = 2.33 g/cm^3). Thermal conductivity, and emissivity have liquid silicon values. With this model, heating ramp, directional melting and solidification are simulated as well as the cooling stage. Silicon properties fit well for the end of melting and solidification but are a poor approximation of the material properties during heating, start of melting (feedstocks is an assembly of silicon chunks) and cooling (solid silicon).

In model 2, the significant changes are the addition of a silicon seed and silicon surface emissivity modification. Surface emissivity pass from liquid silicon value (0.2) to solid silicon one (0.7). This simulation is done on half an ingot and without latent heat. Those hypothesis reduce consequently computation time of thermal simulation without introducing errors. Indeed, simulated furnace is nearly symmetrical and thermal maps after maximum temperature exceeding silicon melting temperature are discard. A solid silicon seed is added changing the ingot height (figure 1) and weight.

In model 3, as real silicon loads are performed with dense monocrystalline silicon seeds and a more or less random assembly of silicon chunks, the feedstock density is therefore lower (1.4 g/cm^3). To maintain total ingot weight, silicon feedstock height is modified. Thermal conductivity is also changed to correspond to an uncompact feedstock of 50 mm chunks thermal conductivity [6].

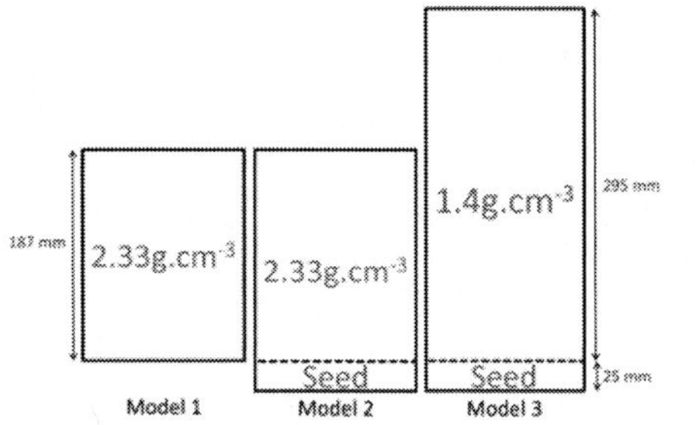

Figure 1: Cut view of quarter ingot for model 2 (left) and model 3 (right)

Table 1: Parameter change in different models

Parameter		unit	Model 1	Model 1	Model 2
Solid height		cm	18.7	20.2	32
Feedstock density		g.cm-3	2,33	2.33	1.4
Seed density		g.cm-3	None	2.33	2,33
Silicon surface Emissivity			0.2	0.7	0.7
Part of process			Full	Only heating	Only heating
Thermal conductivity	1000K	W.m-1.K-1	31.6	31.6	10
	1400K		23.6	23.6	16

2.2 Von Mises stress (VMS) calculations

Von Misses stresses are evaluated from solid expansion due to thermal fields and for various boundary conditions. We have used similar methodology as in [7] for VM calculation except that it is adapted to a 3D square geometry. First, a solid geometry is generated; then, deformed depending on the temperature field. Finally, the material reacts elastically to the deformation, generating stresses following the Hooke's law. This process is the same for the 3 thermal models.

Depending on the thermal stage, ingot geometry used for stress calculation change. For heating and melting steps, only the seed is considered and is modelled as a monoseed ($38 \times 38 \times 2.5$ cm^3). In model 1, as seed is not separated from the ingot, temperature field in a volume corresponding to the seed is extracted. For solidification step, the solid-liquid interface height (h_sl) is found with a matlab routine, in order to calculate stress only in the solid silicon. The matlab routine finds at each x,y coordinate the minimum height with a liquid fraction equal to 1. The geometry is $38 \times 38 \times$ h_sl cm^3. Finally, for the cooling full ingot geometry is used in VMS calculation.

From thermal model 1, we obtained VMS during a full process (heating, melting, solidification and cooling). Results are presented in section §3.1. Then, model 2&3 are used to discuss effects of feedstock parameters (§3.2) and mechanical boundary conditions on seeds (§3.3).

3 RESULTS AND DISCUSSIONS

3.1 Dependence in time

Figure 2 summarizes the time dependent calculated Von Mises Stresses in Model 1. The value considered is the maximum stress on each 3 axis corresponding to center, corner and face (Figure 2a). During heating and melting, stress is probed only inside the seed, and during solidification and cooling, in seed and solidified part of ingot.

During melting, the maximum Von Mises stress is 6.5 MPa, and is located on seed sides, whereas minimum values are found in the corners and center. After a period at the start of growth where stress diminishes, the VM stresses increase and reach ≈ 15 MPa at the end of solidification. At the same time, radial distribution of stress stay maximal on sides and minimal in the corners.

A first question that arises from those results is to determine if those stresses are sufficient to generate or multiply dislocations. From [4] for example, a large number of generation sources have been already identified in the monolike process. As a consequence, in this process, one might not count on dislocation generation to be a limiting step for defect generation in the seed. Concerning dislocation multiplication, we need an estimated value of critical VMS to be able to discuss our time dependent results. A critical threshold for VMS in the range of 1 - 3 MPa can be estimated knowing that 1 MPa value is critical resolved shear stress (CRSS) extrapolated value at melting temperature [8],. As a consequence, Figure 2 demonstrates that stresses are sufficient to multiply dislocations in the solid Si from end of heating stage to the end of solidification. These dislocations may then grow inside the ingot by epitaxy.

The fact that dislocations can move and multiply, might not be sufficient to explain some observed features in monolike ingots. Krause [4] found a critical dislocation density of 2×10^5 cm^{-2} in the seed. If this density is exceeded, Krause found that subgrains were generated above during growth. From [9], an estimated Von Mises stress around 20 MPa could generate such a dislocation density. As our VMS results are far below this estimate, we might consider that our time dependent VMS results imply that stress is not large enough before growth starts to generate subgrain precursors in seeds, even if sufficient to multiply dislocation.

Another feature of our results is that stress in seeds before growth do not exhibit high and localized peak values (see §3.2). As a consequence, in the next sections, we are looking for thermal phenomenon or mechanical boundary conditions that could generate higher VMS values at the end of heating stage. We choose the end heating stage for those case study, because this is the time when simulated stresses are maximum in the seed before growth starts.

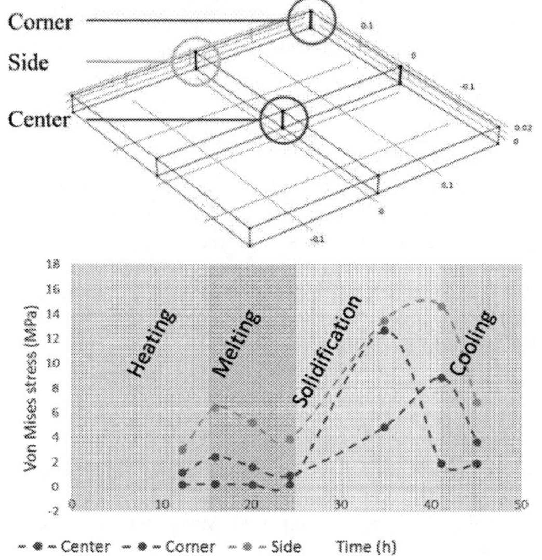

Figure 2: Comparison of time dependent simulated of maximum Von Mises stress calculated from model 1 for center, side and corner positions with free boundary

conditions, and schematic of the 3 axis in seed.

3.2 Influence of feedstock properties

In this part, the influence of feedstock density and conductivity on VMS in seeds at the end of heating stage (process time ≈15h) is investigated. Here we calculate VMS in free boundary conditions applied to the seed. In such case, thermal gradients and geometry in the seed alone are taken into account for stress evaluation. A half ingot horizontal view of stress distribution at seed half height is presented in figure 3a (respectively Figure 3b) for model 2 (respectively Model 3). Spatial distribution of stresses is almost unchanged, with minimum in corners and maximum at sides. Maximum VMS values are smaller in model 3 (6.3 MPa) compared to that obtained in model 2 (8.5 MPa). From this result, silicon load height and thermal conductivity play effectively a role on VMS in seeds in free boundary conditions. Nevertheless, both spatial distribution and maximum VMS values stay of same the order of magnitude. Thus the thermal field changes generated by the silicon load properties in this furnace cannot explain alone high and localized stress values. In the next section we then consider the mechanical possible origins of such stresses through the variation of boundary conditions applied to the seed in the case of model 3.

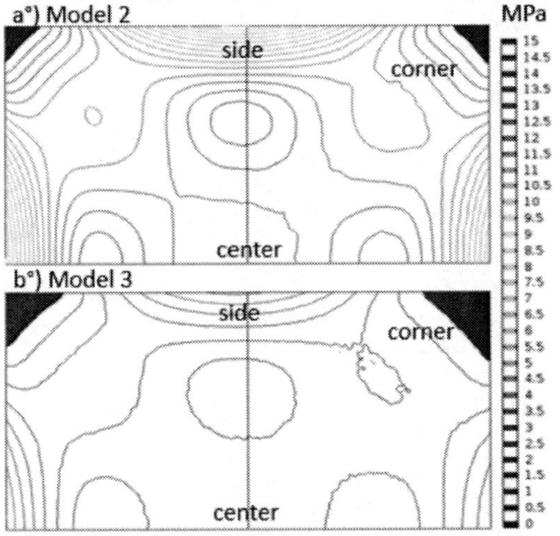

Figure 3: Half seed view of calculated VMS isolines in the horizontal mid height plan of the seed for dense load (Model 2) a) and chunk like load (Model 3) b). Full Blue areas are below 0.5 MPa.

3.3 Seed with mechanical boundary conditions

Different mechanical boundary conditions have been studied to estimate how interaction between seed and its environment is critical. Two practical cases could happen during heating and melting stages:

(i) the mechanical junction between the seed and the load due to a gravity fallen of the silicon melted droplets and its re-solidification at the contact of the seed,

(ii) the localized indentation from a silicon chunk trapped between the seed and the crucible.

The choice of a geometry for the mechanical junction case simulation is not straight forward. The fact that a silicon droplet could fall to the seed could happen from the end of heating stage to the end of melting stage, corresponding to a wide variety of geometries. We choose the extreme hypothesis that both the load has mechanical properties of silicon (Model 3) and the junction is effective on the whole top seed surface with the silicon load.

The choice of a force or pressure for a localized indentation from a silicon chunk is neither easy to do. Trempa [10] investigated top seed surface indentation by feedstock for various contact surface and silicon weight. Here we propose to investigate the case where seed expansion is limited by the presence of a crucible and an intermediate small silicon chunk with a localized contact.

We choose to simulate this boundary condition as a 3 points indentation with an arbitrarily load of 600 N toward the seed ($\vec{F}.\vec{n} = -600$ N). Each indent is localized at a different arbitrary seed height. Another way to simulate this boundary condition could be to fully block the four side surfaces of the seed ($u = v = w = 0$). But this condition seems to be too strict, because feedstock rocks might be reorganized during the seed expansion and because both silicon (seed and chunk) and crucible can partially plastically deform.

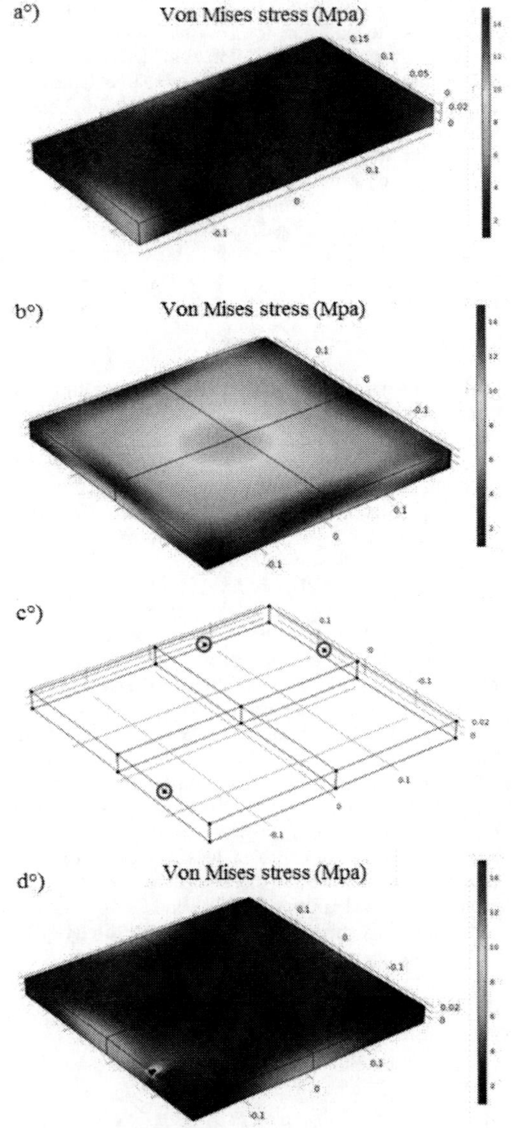

Figure 4: 3D surface view of calculated VM stress for model 3 with different boundary conditions: free boundary reference a), linked to feedstock b) and 3 loaded points position c°) and corresponding VM stress map d).

All simulation are performed at the end of the heating stage. The reference case is the free boundary condition applied on seed surfaces as already presented in §3.2. The 3D surface view of calculated VM stress map for the half seed of the reference is presented in figure 4a), with its maximum on sides around 6 MPa.

Similar VMS simulation results for the case where seed is linked to feedstock are presented in figure 4b). The stress distribution is drastically changed with maximum positioned in the center (in opposition to side in reference) and minimum positioned in bottom corners and top sides (in opposition to corners only in the reference case). Nevertheless the VMS values in the center stays limited to around 10 MPa. Thus, we observe a moderate increase of VMS in seed induced by deformation of the whole silicon block (seed and feedstock) from thermal field alone but a drastic distribution change.

VMS simulation results in the case of 3 indentations points is presented in figure 4c). Locally, the stress increases significantly (here over 100 MPa), but does not change VMS distribution in the rest of the seed. Nevertheless, the local VMS peak at contact area can affect the neighboring area overs some centimeters. Such high localized VMS on sides could be sufficient for both generation and multiplication of dislocations in their vicinity.

From this study, we can conclude that a direct seed/crucible interaction needs to be avoided and that a good feedstock distribution can prevent localized generation of dislocations on seed sides.

4 CONCLUSIONS

Simulations of stress evolution during the melting and solidification of laboratory-scale monolike Si ingots provided new information on thermomechanical stress history:

- Seeds see a peak stress at the end of the heating stage when solid height is the highest.
- The maximum of Von Mises stress (VMS) is attained during solidification and can be twice the values during heating.
- The VMS maxima are located on the sides of the seeds

One important consequence of those findings is that thermomechanical stresses are sufficient at the end of the heating stage to multiply dislocation in the seeds. These dislocations may then grow inside the ingot by epitaxy.

The feedstock simulation parameters change significantly the thermal distribution and VM stresses: better feedstock modelling (containing argon and thus having a lower conductivity) decreases the VM stress in the seeds. This works also provides useful radial stress information on the influence of locally locked boundary conditions. Those may be further compared to experimental post-growth characterization in ingots and seeds.

Additional investigations were performed to clarify the role of boundary conditions on seeds. Results show that locally fixed boundary conditions can modify either the distribution or the local intensities of VM stresses in the seed.

ACKNOWLEDGEMENT
This project has received support from the State Program "Investment for the Future" bearing the reference (ANR -10-IEED -0003).

REFERENCE

[1] F. Jay *et al.*, "Advanced process for n-type mono-like silicon a-Si:H/c-Si heterojunction solar cells with 21.5% efficiency," *Sol. Energy Mater. Sol. Cells*, vol. 130, pp. 690–695, Nov. 2014.

[2] Y. Miyamura *et al.*, "Crystal growth of 50cm square mono-like Si by directional solidification and its characterization," *J. Cryst. Growth*, vol. 401, pp. 133–136, Sep. 2014.

[3] B. Gao and K. Kakimoto, "Numerical investigation of the influence of cooling flux on the generation of dislocations in cylindrical mono-like silicon growth," *J. Cryst. Growth*, vol. 384, pp. 13–20, Dec. 2013.

[4] A. Krause, L. Sylla, and D. Oriwol, "Plastic Deformation as an Origin of Dislocations in Cast Mono," *Energy Procedia*, vol. 92, pp. 833–838, Aug. 2016.

[5] Albaric, M., Pelletier, D., Pihan, E., Plassat, N., and Jouini, A., "Silicon Crystallization Process: Comparison between 3D Transient Simulations and Experimental Results," 2012.

[6] X. Liu, S. Nakano, and K. Kakimoto, "Effect of the packing structure of silicon chunks on the melting process and carbon reduction in Czochralski silicon crystal growth," *J. Cryst. Growth*, vol. 468, no. Supplement C, pp. 595–600, Jun. 2017.

[7] B. Gallien, M. Albaric, T. Duffar, K. Kakimoto, and M. M'Hamdi, "Study on the usage of a commercial software (Comsol-Multiphysics®) for dislocation multiplication model," *J. Cryst. Growth*, vol. 457, pp. 60–64, Jan. 2017.

[8] N. Miyazaki, H. Uchida, T. Munakata, K. Fujioka, and Y. Sugino, "Thermal stress analysis of silicon bulk single crystal during Czochralski growth," *J. Cryst. Growth*, vol. 125, no. 1, pp. 102–111, Nov. 1992.

[9] X. Chen, S. Nakano, and K. Kakimoto, "Three-dimensional global analysis of thermal stress and dislocations in a silicon ingot during a unidirectional solidification process with a square crucible," *J. Cryst. Growth*, vol. 312, no. 22, pp. 3261–3266, Nov. 2010.

[10] M. Trempa *et al.*, "Dislocation formation in seed crystals induced by feedstock indentation during growth of quasimono crystalline silicon ingots," *J. Cryst. Growth*, vol. 454, pp. 6–14, Nov. 2016.

SI WAFER MANUFACTURING BY THERMAL SPRAY OF RECYCLED SI POWDERS

M. Vardavoulias*[1], A. S. Azar[2], T. Halvorsen[3] , M. Moen[3], K. Mork[3], P. A. Carvalho[2], Ø. Dahl[4], A. Ulyashin[2]

[1] Pyrogenesis, Lavrion, Greece
[2] SINTEF, Oslo, Norway
[3] RESITEC, Kristiansand, Norway
[4] SINTEF, Trondheim, Norway
*Corresponding author: mvardavoulias@pyrogenesis-sa.gr

ABSTRACT: The aim of the present work is to demonstrate processing of polycrystalline free-standing Si wafers and thin Si layers on various substrates using thermal spray of relevant Si powders, including recycled ones. Silicon powders were fabricated by milling from two different Si feedstocks: (i) Si kerf produced during sawing of Si wafers from the ingots, and (ii) Si powder from Si scrap (broken wafers). Two different Si powder based structures were produced by thermal spray: (i) free-standing Si-based wafers with thicknesses of about 300-400 micrometres and sizes up to 6 inches, which can be used as supporting parts for "Si wafer equivalent" structures, and, (ii) Si-based layers with thicknesses of about 50-100 micrometres deposited on different supporting substrates, such as aluminium, glass and highly conductive Si powder based sintered substrates. Raman spectroscopy, resistivity, and SEM analyses have been used for characterization of the Si based structures fabricated by thermal spray. It is observed that such layers are poly-crystalline with a low-fraction of the amorphous phase and can be considered as a low-cost alternative for Si PV, which are based on utilization of Si wafers and thin Si layers deposited by conventional methods such as CVD, PECVD, e-beam, etc.

Keywords: Thermal Spray, Silicon, c-Si, Si-Films, Thin Film, Substrates, Grain

1 INTRODUCTION

Thermal spraying technology has been employed for many years to synthesize high performance coatings of a wide range of materials, including Si [1-6]. In general, the process of thermal spraying involves feeding a powder based feedstock into a high-energy plasma or flame, in which powder particles are heated, partly or fully melted, and simultaneously accelerated, towards the target surface of a substrate. Silicon particles cool and then solidify onto the substrate, forming a re-crystallized silicon layer. Two commonly used processes for various applications are plasma spray (PS) and flame spray (FS) [7]. It has been shown earlier, that in plasma sprayed silicon conversion efficiencies of thin Si based solar cells of 4.3% can be achieved [4]. However, it should be noted that in flame spraying, relatively low temperatures can be reached in the flame (~3000 °C) compared to common plasma spraying ones, which are quite high (~15000 °C). Therefore, FS has some advantages compared to plasma spray and can be considered as an appropriate method to produce Si coatings on various substrates. In particular, as soon as Si particles are heated to lower temperatures upon FS compared to PS, cooling of such heated/melted particles can be done much faster in case of FS process and a wider range of the supporting substrates for the deposition of Si layers can be used. The aim of the present work is to show the viability of flame spraying for the fabrication of free-standing Si wafers and thin layers on a wide range of supporting substrates, which can be of interest for the low-cost Si based PV.

2 EXPERIMENTAL PROCEDURE

2.1 Si powders and relevant structures processed by thermal spray

Silicon powders with particle sizes in the range of about 10-50 μm were fabricated by conventional ball milling from two different Si feedstocks: (i) Si kerf produced during sawing of Si wafers from the ingots, and (ii) Si powder from Si scrap (broken wafers). Two different Si powder based structures were produced by the flame based thermal spray: (i) free-standing Si wafers with thicknesses of about 300-400 μm and sizes up to 6 inches, and, (ii) Si-based layers with thicknesses of about 50-100 μm deposited on different supporting substrates, such as aluminium, glass, ceramics and highly conductive sintered Si powder substrates.

2.1 Characterization

Raman spectroscopy, resistivity and SEM analyses have been used for the characterization of Si powder based free-standing wafers and thin Si layers deposited on different substrates by flame based spray.

3 SILICON POWDER BASED WAFERS AND THIN LAYERS DEPOSITED BY FLAME SPRAYING

3.1 Free standing wafers

Optimization of thermal spray parameters resulted in reproducible fabrication of mechanically stable Si wafers with dimensions 50x65 mm^2 (**Figure 1**), 156x156mm^2 (**Figure 2**) and thicknesses between 300 and 500 microns.

Figure 1: *Free standing 50x65 mm^2 Si powder based wafers*

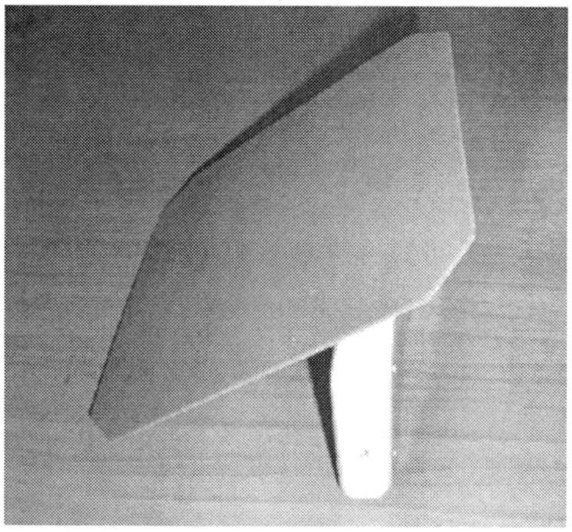

Figure 2: *Free standing 156x156 mm² Si powder based wafer*

SEM images taken at different magnifications of the surface of a thermal sprayed wafer are shown in **Figure**

3.

Figure 3: *SEM images of the surface of a free standing thermal sprayed wafer*

From **Figure 3** it can be concluded that surface of Si powder based wafers consists of grains with different shapes, which are assembled in a "naturally" textured Si based structure.

Figure 4 shows results of Si line Raman mapping for the surface area (100x100 μm²) of a free-standing Si powder based substrate.

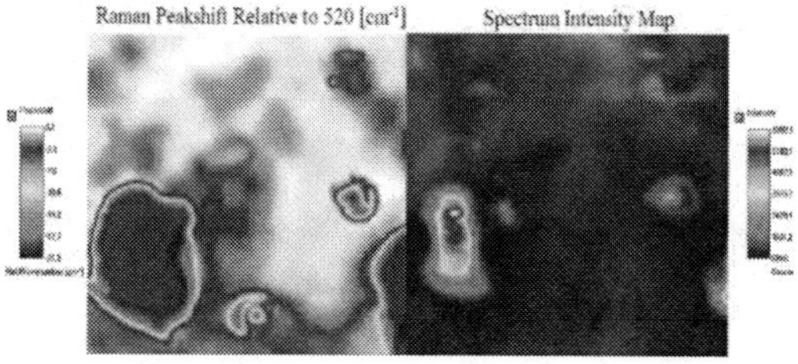

Figure 4: *Si line Raman mapping: free standing Si powder substrate, surface area 100x100 μm².*

From **Figure 4** it can be concluded that Si wafer, sintered by thermal spray, consists of mainly poly-Si grains with small portion of the amorphous Si phase inclusions.
As an option, resistivity of such wafers can be regulated by the mixing of Al and/or B powders prior to thermal spray. **Table 1** shows results of the resistivity measurements for Si wafers processed by thermal spray of a mix of Al/Si powders with different proportions of Al and Si.

Table 1: *Resistivity of free standing*

Al content in Si powder	Resistivity, Ωcm
0%	10
1%	0.6
3%	0.47
5%	0.024
10%	0.018
15%	0.002

From **Table 1** it can be seen that highly conductive Si based wafers can be sintered by thermal spray of Al/Si powders mixtures.

3.2 Si powder based layers deposited on Al substrates by thermal spraying

Figure 5 shows SEM images taken at different magnifications of the surface of a thermal sprayed pure Si powder based layer on an Al substrate and **Figure 6** shows results of the Si line Raman mapping for the surface area (100x100 μm²) for the same structure.
From **Figures 5** and **6** it can be concluded that similar to the case of free-standing Si wafers, surface of Si powder based wafers consists of grains with different shapes and different degree of crystallinity. Mainly poly-Si phase is dominating, with a small portion of the a-Si inclusions. It should be noted that grains in this case are quite smooth and grain boundaries are not as pronounced as in case of free-standing Si powder based wafers processed by thermal spraying. It can be speculated that presence of an Al substrate provides specific conditions for the

crystallization of partly melted Si particles on the Al surface. More detailed analysis is required to clarify the origin of the observed effect.

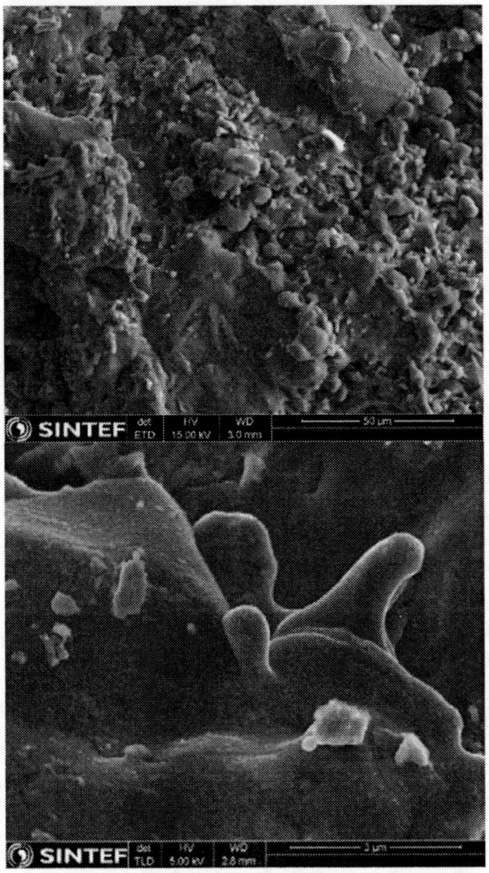

Figure 5: SEM images of the surface of Si powder based thin layer deposited on an Al substrate.

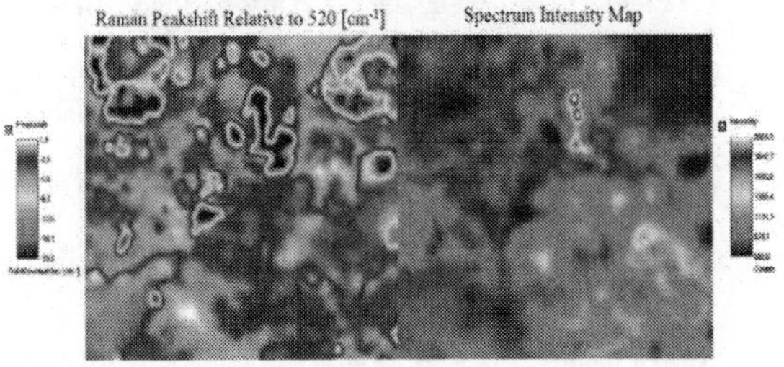

Figure 6: Si line Raman mapping: Si powder based layer (~100 μm thick) deposited on an Al substrate, surface area 100x100 μm².

3.3 Si powder based layers deposited on glass substrates by thermal spray

Figure 7 shows surface and cross section SEM images of a thin Si layer deposited on a glass substrate by thermal spray. Quite dense Si layer with a "naturally" textured Si surface can be formed, as can be seen from **Figure 7**. Several thermal spray scanning velocities were tested in this case and crystallinity of Si layers obtained at different conditions were monitored by micro-Raman. Upon Raman measurements, laser beam was focused on a central part of the largest grains, observed in an optical microscope.

Figure 7: *Surface and cross section SEM images of a thin Si layer deposited on a glass substrate by thermal spray.*

Figure 8 shows results of Raman measurements for the optimized thermal spray process (scanning velocity 5). Raman spectra were taken from the centre of 2 different grains and are indicated in **Figure 8** as Si peak (1) and Si peak (2).

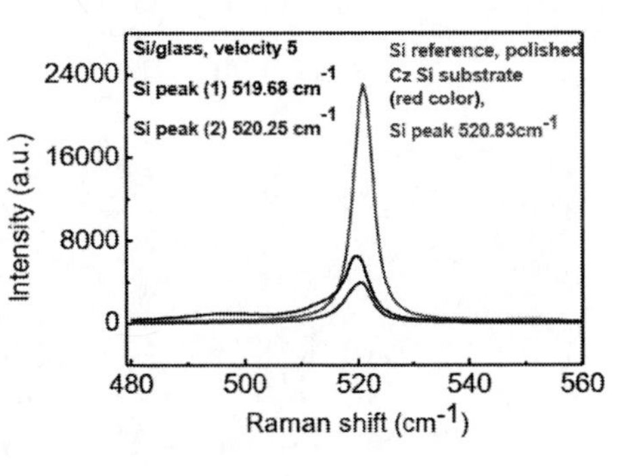

Figure 8: *Raman spectra of Si lines taken from 2 different grains of Si powder based layer deposited on a glass substrate with an optimized scanning velocity.*

From **Figure 8,** it can be seen that Si grains are crystalline, in contrast to similar Raman measurements for Si powder layers deposited on a glass substrate with the not optimized scanning velocity (velocity 3) upon the thermal spray process (**Figure 9**).

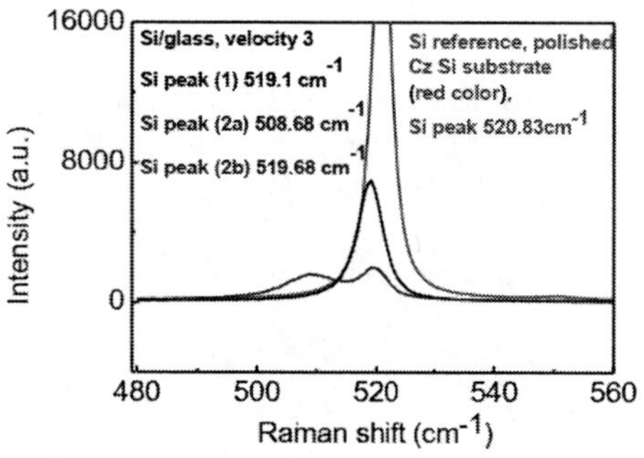

Figure 9: *Raman spectra of Si lines taken from 2 different grains of Si powder based layer deposited on a glass substrate with the non-optimized scanning velocity*

In this case, Si peak (1) (519.1cm^{-1}) is shifted from the standard peak position of crystalline Si (520 cm^{-1}), and Si peak (2) consists of 2 well pronounced lines – 508.68 cm^{-1}, which can be attributed to a poly-Si and 519.68 cm-1 line, which can be attributed to a stressed crystalline Si grain. From **Figures 8** and **9** it can be concluded that scanning velocity in Si powder thermal spray process is an important parameter, which should be taken into account for the process optimization. In an optimized case, substrate is heated by flame to the most suitable temperatures, which can provide proper crystallization of Si powders and relevant Si layers.

3.4 Si powder based layers deposited on Si powder sintered substrates by thermal spray

It has been established, that sintering of Si powder based substrates, including highly conductive ones, can dramatically reduce the thermal budget required [8-11]. Such sintered wafers can be used as the low cost supporting substrates, on which high quality Si layers (solar cell base) can be deposited to fabricate "Si wafer equivalent" [12]. **Figure 10** shows results of the Si line Raman mapping for the surface area (100x100 μm^2) of a thin Si layer deposited on top of the hot pressed and sintered Si powder based substrate [9, 11].

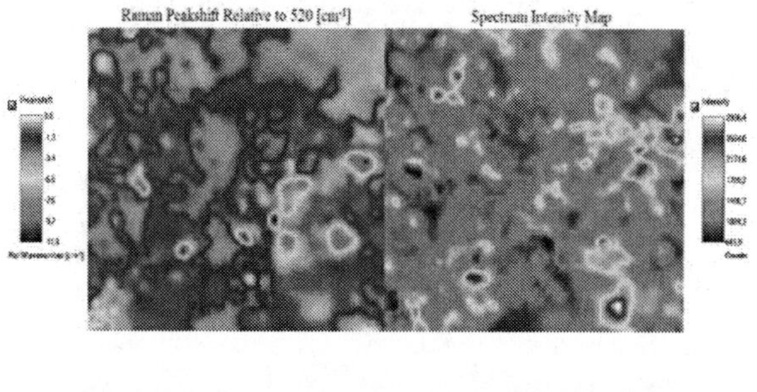

Figure 10*: Si line Raman mapping: Si powder based layer (~100 μm thick) deposited on a sintered Si powder based substrate.*

From **Figure 10** it can be seen, that only poly-Si and crystalline Si phases are present in this case, without any traces of amorphous silicon, which was observed in previous cases (free-standing wafers, Al and glass substrates). It can be concluded that the origin of a supporting substrate plays an important role for the resulting crystallinity of Si layers deposited on supporting substrates by thermal spray of Si powders.

5 CONCLUSION AND OUTLOOK

It was demonstrated that Si powder based structures – free-standing Si wafers or thin Si films can be processed by the flame based thermal spray. Important to note that thin Si layers can be deposited on several types of low-cost substrates (Al, glass, Si) with the low temperature melting points. It is shown, that thermal spray processing conditions and the origin of the supporting substrates play an important role for the crystallinity of the thermal spray deposited Si layers.

It is established that fabricated at appropriate thermal spray processing conditions, such layers are poly-crystalline with a low fraction of the amorphous phase and can be considered as a low-cost alternative for Si PV, which is based on utilization of Si wafers and thin Si layers deposited by conventional methods such as CVD, PECVD, e-beam, etc.

ACKNOWLEDGEMENT

This project has received funding from the European Union's Horizon 2020 research and innovation programme under grant agreement No. 641972 (CABRISS).

REFERENCES

[1] R. Suryanarayanan, G. Brun, M. Akani, Thin Solid Films 119 (1984) 67–73.
[2] B. Kayali, M. Rodot, R. Suryanarayanan, L. Quangnam, R. Mghaieth, R. Gauthier, P. Pinard, Materials Science and Engineering B 5 (1989) 51–56.
[3] M. Akani, R. Suryanarayanan, G. Brun, Thin Solid Films 151 (1987) 343–353.
[4] F. Tamura, Y. Okayasu, K. Kumagai, Solar Energy Materials and Solar Cells 34 (1994) 263–270.
[5] B.D. Kharas, G. Wei, S. Sampath, H. Zhang, Surface and Coatings Technology 201 (2006) 1454–1463.
[6] D.J. Varacalle, H. Herman Jr., G.A. Bancke, T.D. Burchell, G.R. Romanoski, Surface and Coatings Technology 49 (1991) 24–30.
[7] A. Nistal, E. García, C. García-Diego, M.I. Osendi, P.Miranzo. Surface & Coatings Technology 270 (2015) 8–15.
[8] Peter, K., et al., Solar Energy Materials and Solar Cells 74 (2002) 219-223
[9] G. Stokkan, A. Johanson, E. Øvrelid, A. Ciftja, J. Hennicke, A. Ulyashin Proc. EUPVSEC 28, Paris, 2013, pp. 427-430.
[10] Bellanger, P., et al., Journal of Crystal Growth 359 (2012) 92-98.
[11] R. Buchwald, S. Würzner, H. J. Möller, A. Ciftja, G. Stokkan, E. Øvrelid, A. Ulyashin Phys. Status Solidi A 212, (2015) 25–29.
[12] Reber, S., et al., Solar Energy Materials and Solar Cells 65 (2001) 409-416

SILICON POWDER BASED INGOTS AND SUBSTRATES, PROCESSED BY SPARK PLASMA SINTERING

T. Kaden[a], A.S. Azar[b], M. Syvertsen[b], H.J. Möller[a], N. Abrosimov[c], M. Fleissner Sunding[b], J. Hennicke[d], A. Ulyashin[b]

[a]Fraunhofer Technology Center for Semiconductor Materials THM, Am St.-Niclas-Schacht 13, 09599 Freiberg, Germany
[b]SINTEF Materials and Chemistry, Oslo, Norway, [c]Leibniz-Institut für Kristallzüchtung IKZ, Berlin, Germany,
[d]FCT Systeme GmbH, Rauenstein, Germany

*Corresponding author: thomas.kaden@thm.fraunhofer.de

ABSTRACT: This work presents results concerning sintering of silicon powder based wafers and ingots produced the by Spark Plasma Sintering (SPS) method. It is shown that wafering of such ingots can be done by conventional wire sawing. Light microscopy, scanning electron microscopy, energy dispersive spectroscopy, Raman spectroscopy, as well as minority carrier life time measurements were used for the analysis of silicon wafers fabricated by SPS followed by wire sawing. It is found that during the SPS process, material inhomogeneity of sintered ingots and wafers occurs. Nevertheless, the inhomogeneity can be smoothed by proper optimized sintered conditions for the large scale (up to 8 inch) ingots and wafers. It is found that the SPS process, being applied to Si powder, provides formation of Si material (wafers or ingots), which has structural properties similar to those for multicrystalline silicon (mc-Si). In case if silicon powder has solar grade quality, minority carrier lifetime values are comparable with those for mc-Si substrates. It is concluded that SPS silicon structures can be potentially considered as a base for low-cost Si based photovoltaics.

Keywords: Silicon, wafering, diamond wire, multicrystalline silicon

1 INTRODUCTION

Highly conductive sintered silicon is a promising material, which can be used as supporting substrates for high efficiency devices deposited on or bonded to the surface of such substrates. The sintered substrates will function in this case as mechanical support as well as a conductive cell backside. While a thickness of ~50 µm of high quality crystalline silicon is sufficient to achieve high efficiency solar cells, such thin silicon sheets are very fragile and require mechanical support. The ideal material for constructing the mechanical support is silicon of a lower quality than required for the actual device, because this material will experience the same reaction to thermal treatments during cell processing and will be less prone to generation of interface defects between the support and the device. Several routes have been proposed and investigated for using silicon as substrates [1-2].

If a regular multi/c-Si wafer is used as a substrate for epitaxial growth, a similar crystalline structure as the substrate will result, which is beneficial in terms of limiting recombination active defects such as dislocations and grain boundaries. However, there are also drawbacks connected to these substrates, particularly the high cost and energy budget required to melt, crystallize and wafering. All these costs are the same for Si feedstock with any quality and cannot be reduced essentially. Furthermore, if a high degree of doping is desired in order to use the substrate as a conductive cell backside, restrictions are placed on the crystal growth rate in order to avoid constitutional supercooling, making the process less economically attractive [3].

Sintering of highly conductive silicon powder completely avoids the process of melting and crystallizing the feedstock. The high temperature step required for sintering is in the range of ~1 h, which dramatically reduces the thermal budget required to form sintered ingots. While sintering may in principle be conducted to produce wafers directly [4], sintering of large ingots followed by conventional wafer sawing may in practice be the most promising alternative [5]. The sintered material has a certain degree of porosity, which may enable faster and more economical wafer sawing than of mono- or multicrystalline bricks [3].

In this paper we report results of several experiments connected to sintering of silicon powder using SPS method. To verify a feasibility of an up-scaling for the SPS process, sintering of round wafers at a large scale (up to 200 mm in diameter) was performed. To verify a feasibility to make Si wafers out of sintered Si, mini-ingots with diameter 40 mm have been fabricated by SPS and sliced using conventional wire saw method.

2 EXPERIMENTAL PROCEDURE

2.1 Spark Plasma Sintering (SPS)

The scheme of the SPS process is shown in Figure 1. Sintering of Si powder ingots was done using SPS system at ~1200-1300°C and pressures up to 30 MPa. Wafers with a thickness of about 300-500 µm and diameters from 40 mm up to 200 mm, as well as cylindrical mini ingots with diameters of 40 mm lengths about 2-3 cm were sintered. Si powders with diameters in the range of 1-60 µm processed by milling of a solar grade quality Si feedstock have been used.

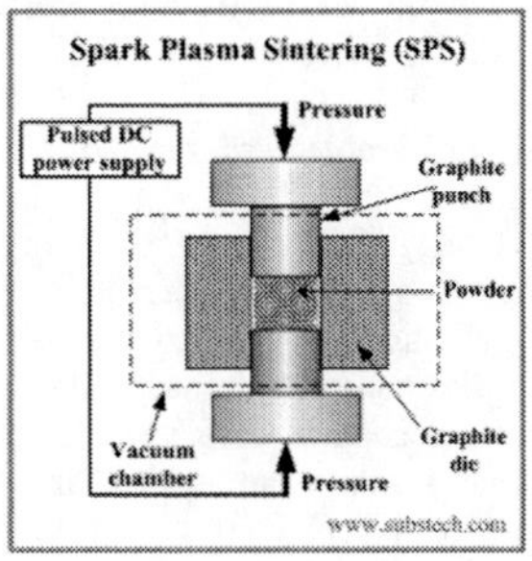

Figure 1: Scheme of Spark Plasma Sintering (SPS).

It should be noted that the SPS process is in principle similar to a well-known hot pressing approach, but instead of a resistance heater heating the pressing tool from the outside, SPS heating up to 1300 °C is done "in-situ" by the Joule heat of a pulsed DC current flowing directly through the pressing tool and/or the sample. To demonstrate that the SPS process can be realized for Si powder at a large scale suitable for PV application, individual wafers with diameters up to 200 mm were sintered at different temperatures. As an example, Figure 2 shows an image of a 200 mm round Si powder based wafer, sintered at 1200 °C, which has a thickness of about 500 µm.

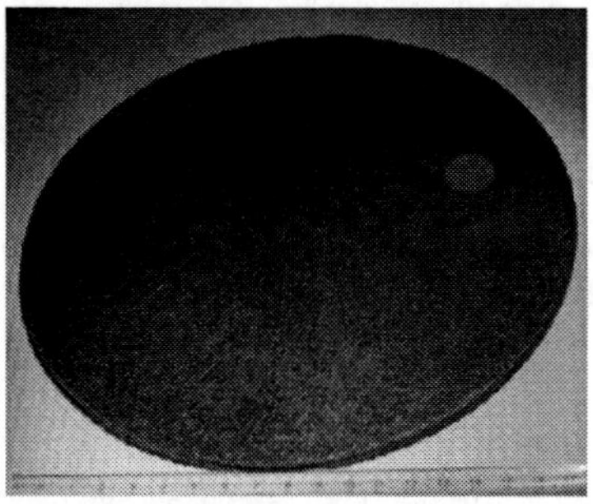

Figure 2: 200 mm wafer produced by SPS of Si powder.

However, sintering of individual wafers requires post-sintering removal of the subsurface layer, which is contaminated upon the sintering process, and to avoid such problem, sintering of mini-ingots followed by conventional wire saw wafering was tested.

Within this work, three SPS mini-ingots with a diameter of 40 mm were used for a detailed analysis. Different qualities of silicon powder feedstock as well as 2 different sintering temperatures were used, as summarized in Table 1.

Table 1: Overview of SPS ingots.

Ingot #	Powder feedstock/Sintering temperature
1	Solar grade/ 1200 °C
2	Solar grade/ 1250 °C
3	Highly boron doped silicon/ 1200 °C

2.2 Wafering

Mini-ingots with 40 mm in diameter were sliced into 180 µm thin samples using multi-wire sawing processes. A slurry-based process using polyethylene glycol (PEG) with SiC abrasive particles was compared to a diamond wire sawing (DWS) process. For both technologies, a wire diameter of 120 µm was used. In the slurry process, the SiC abrasive had the size F600. For the DWS process, a diamond grit size of 8-16 µm was used. The ingots were sliced with comparable feed rates of 0.5 mm/min with both technologies.

Figure 3 (left side) shows an optical image of the surface of an as-cut SPS Si powder based substrate, sintered at 1200 °C (ingot #1). It can be seen that the distribution of grains is quite homogeneous over the whole surface area of such a wafer. For a comparison, Figure 3 (right side) shows an optical image of the surface for a similar wafer, fabricated from ingot #2, sintered at a bit higher temperature of 1250 °C (the same positions in both ingots were selected to extract these wafers). It can be seen that in the central part of the sintered wafer, a quite extended zone with large grains is formed. It can be assumed, that because of the non-homogeneous current distribution in the sintering ingot upon the SPS process, the central part of such ingot has higher temperature compared to that for the rest of the ingot. Therefore, in combination with a high pressure, the central part is partly melted or Si powder in this part is at the conditions above the plasticity limit of Si, which provides remarkable growth of rather large grains.

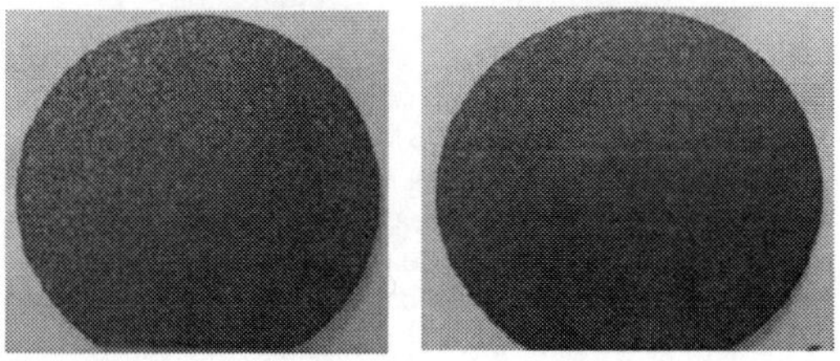

Figure 3: Wafer sliced from SPS ingot #1 sintered at 1200 °C (left) and ingot #2 sintered at 1250 °C (right).

3 CHARACTERIZATION OF WAFERS FROM SPS SILICON INGOTS

3.1 Crystal structure

As it shown above, wafers from SPS ingot sintered at 1250 °C with a diameter of 40 mm exhibit a specific crystal structure. While an outer ring with a microcrystalline structure and higher porosity exists, the central part of the wafer shows a crystal structure comparable to multicrystalline material with very low porosity. A close-up of this multicrystalline structure is shown in the SEM image in Figure 4. Large grains can be recognized in this case.

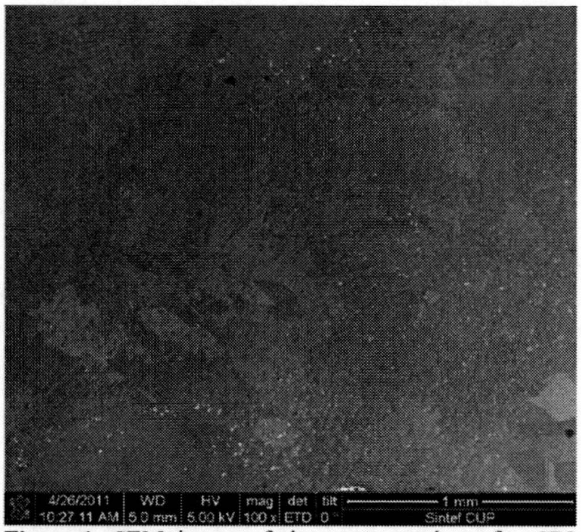

Figure 4: SEM image of the center region of a SPS silicon wafer from the ingot sintered at 1250 °C.

3.2 Structural defects

Although the center region exhibits a densely sintered silicon material, there are still structural defects present in the material. Figure 5 depicts such inclusion as they can be recognized on an etched wafer surface. A further close-up shows that these inclusions mainly contain oxygen and aluminum, which might be gettered during the sintering process, see Figure 6.

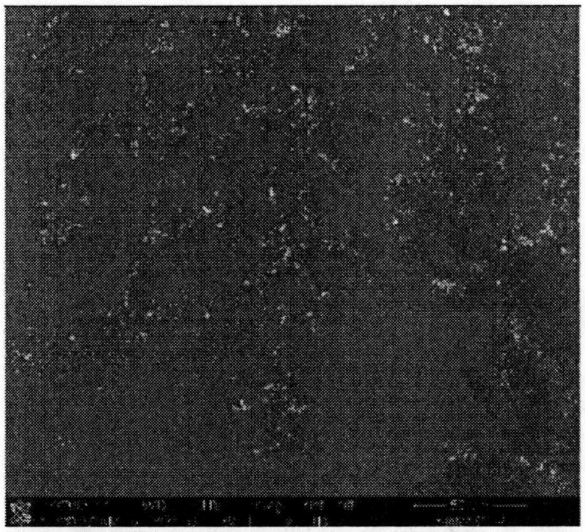

Figure 5: SEM image of an inclusion-rich part of a SPS silicon ingot.

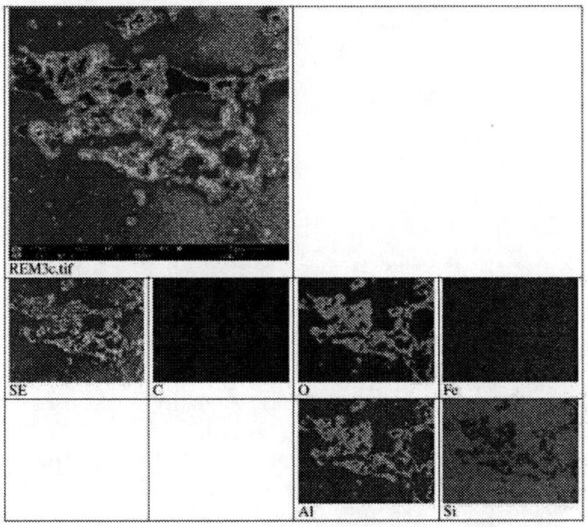

Figure 6: SEM and EDX close-up images showing high oxygen and aluminum concentrations on structural defects.

3.3 Wafer quality

SPS ingots were cut into wafers with a thickness of 180 µm, which is the standard thickness of current silicon based solar cells. In the presence of larger volumes with certain porosity, the SPS silicon shows a lower hardness compared to multicrystalline silicon, so slicing could be done at a higher feed rate. For the evaluation of the sawing processes, basic wafer parameters such as thickness and total thickness variation (TTV) were determined.

In general, both multi-wire sawing techniques are feasible for the production of wafers from SPS ingots. When the sintered silicon is denser or contains a higher concentration of inclusions, then the DWS process is beneficial compared to the slurry process. This is demonstrated by the TTV values shown in Figure 7. The wafers obtained in the slurry process exhibit a significantly higher TTV than wafers from the DWS process. However, the TTV values are mainly out of the wafer specification used in photovoltaics, setting a maximum TTV of 30 µm. This can be improved by optimization of both the SPS and the wafering process.

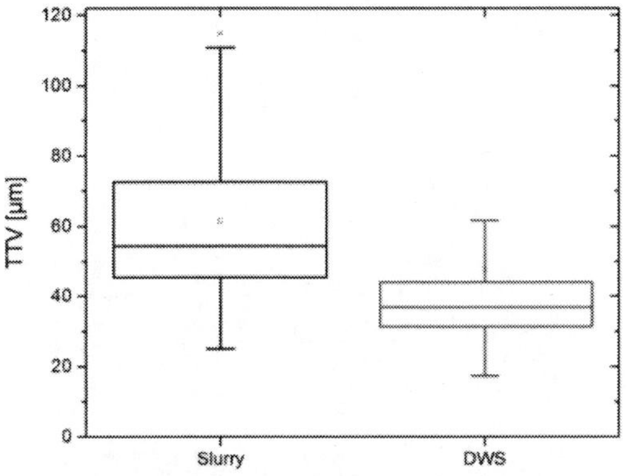

Figure 7: Total thickness variation (TTV) of SPS wafers produced by slurry and diamond wire sawing (DWS) process with comparable feed rates.

3.4 Electrical quality

The electrical quality of the SPS wafers is evaluated in terms of resistivity and minority carrier lifetime. Resistivity was measured by a four-point probe, the carrier lifetime was determined using Microwave-Detected Photoconductance (MDP) at an excess carrier density of $2*10^{16}$ cm^{-3}. The mean values for both measurements are summarized in Table 2.

Resistivity was determined on 5 points in the center of the sample. Significant differences were detected among the three ingots. The high resistivity of ingot #1 may not be reliable since it the measurement is strongly influenced by the numerous grain boundaries within this material. The values for ingot #2 are assumed to be realistic, since there are not many grain boundaries in the center region of the sample. Ingot #3 turns out to have a very low resistivity of 0.32 mΩ cm. Here, the measurement is not influenced by the grain boundaries. The low resistivity explains that there is no lifetime detectable in ingot #3. However, ingot #1 and #2 exhibit lifetimes, measured on the as-cut and non-passivated wafer, which are comparable to multicrystalline silicon.

Table 2: Mean values of resistivity and carrier lifetime for three SPS ingots with diameter of 40 mm.

Ingot #	Resistivity [Ω cm]	Lifetime [µs]
1	287	22.6
2	11	20.4
3	$3.2*10^{-4}$	Not detectable

The particular grain structure of ingot #2 has also an effect on the lifetime distribution. As shown in Figure 8, the multicrystalline center region exhibits a region of very low lifetime in the center surrounded by a ring of the highest lifetimes. The microcrystalline outer ring has again a lower electrical quality. In case of wafer from ingot #1, more homogeneous distribution of the minority carrier lifetime occurs (see Figure 9).

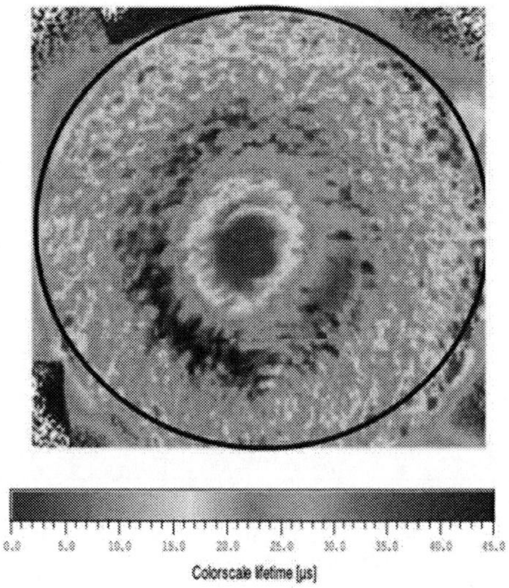

Figure 8: Carrier lifetime mapping on a wafer from ingot #2.

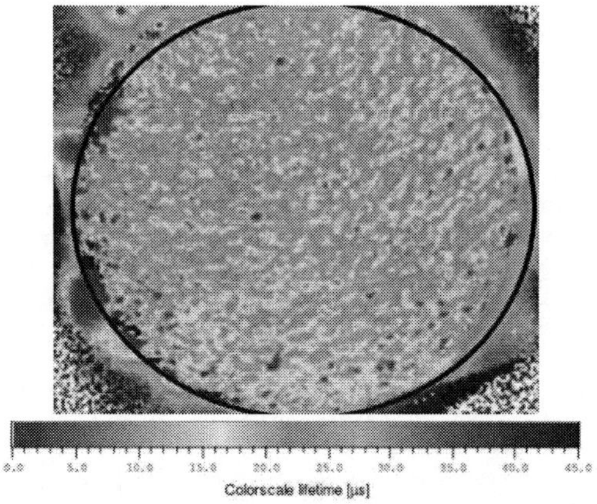

Figure 9: Carrier lifetime mapping on a wafer from ingot #1.

4 ETCHING OF SPS WAFERS

In order to prepare SPS wafers for further solar cell processing steps, the saw damage needs to be removed from the wafer surface after wire sawing. Using a standard alkaline mixture, the etch rate turned out to be too high to get an appropriate etching process. This is due to the porosity of the material that offers an increased surface, which is exposed to the etching solution.

The saw damage removal was better controllable with an acidic etching mixture. Figures 10 and 11 show a comparison of the as-cut diamond-wire sawn surface and the surface after 1 min etching. On the as-cut surface the mostly ductile material removal in the diamond wire sawing process is visible, see Figure 10. Additionally, a high number of inclusions is present. After acidic etching, the surface appears porous with a significantly decreased reflectivity (Figure 11).

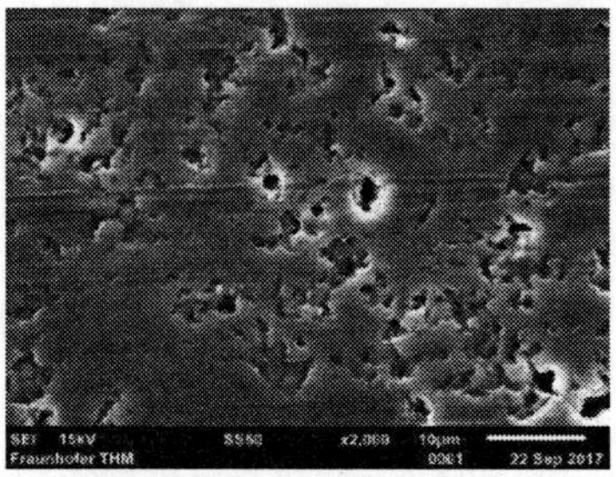

Figure 10: SEM image of as-cut diamond wire sawn wafer surface.

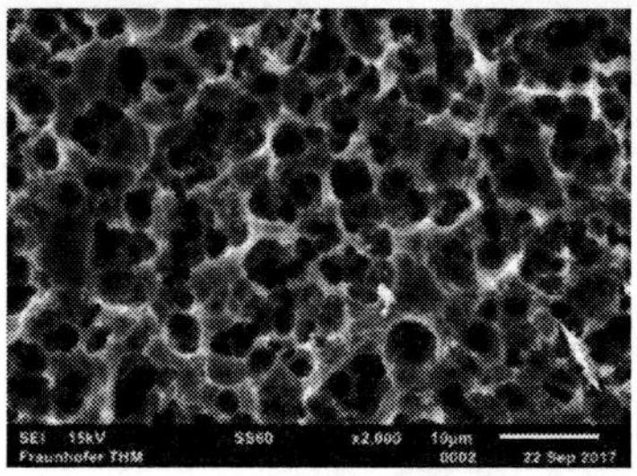

Figure 11: SEM image of wafer surface after 1 min acidic etching.

5 CONCLUSION AND OUTLOOK

It was demonstrated that silicon ingots and wafers can be produced by Spark Plasma Sintering (SPS) of silicon powder. When solar-grade silicon powder as feedstock material, the electrical quality is comparable to standard multicrystalline silicon. For the wafering process it could be concluded that the diamond wire sawing process yields a higher wafer quality if inclusions are present in the SPS silicon ingots.

In the future, a higher material quality can be achieved by a further optimization of the sintering process. Wafer quality can be improved by an adapted wafering process.

ACKNOWLEDGEMENT
This project has received funding from the European Union's Horizon 2020 research and innovation programme under grant agreement No. 641972 (CABRISS).

REFERENCES

[1] Peter, K., et al., Solar Energy Materials and Solar Cells 74 (2002) 219-223
[2] Reber, S., et al., Solar Energy Materials and Solar Cells 65 (2001) 409-416
[3] G. Stokkan, A. Johanson, E. Øvrelid, A. Ciftja, J. Hennicke, A. Ulyashin Proc. EUPVSEC 28, Paris, 2013, pp. 427-430.
[4] Bellanger, P., et al., Journal of Crystal Growth 359 (2012) 92-98.
[5] R. Buchwald, S. Würzner, H. J. Möller, A. Ciftja, G. Stokkan, E. Øvrelid, A. Ulyashin Phys. Status Solidi A 212, (2015) 25–29.

A COMPREHENSIVE DYNAMIC MODEL OF THE DIAMOND WIRE SAWING PROCESS

Daniel Treyer *,1, Sebastian Gaulocher[1], Stefan Niederberger[1], Helge Rafael[2], Alfons Ams[3]
[1]University of Applied Sciences Northwestern Switzerland, Institute of Automation, Windisch, Switzerland
[2]Meyer Burger AG, Gwatt (Thun), Switzerland
[3]TU Bergakademie Freiberg, Institute of Mechanics and Fluid Dynamics, Freiberg, Germany
*Corresponding author: daniel.treyer@fhnw.ch

ABSTRACT: The diamond wire sawing process is a well-established method for the production of monocrystalline silicon wafers. The productivity of the sawing process and the wafer quality are divergent targets that depend on the recipe of wire saw. The correlations between productivity, quality and recipe are difficult to model due to the complex interaction between the diamond wire and the silicon brick and due to the vast amount of process parameters. This work presents a comprehensive dynamic model of the diamond wire sawing process in order to develop a profound understanding based on the analysis of process variables and recipe parameters. Both over process time and wire web coordinate, the model provides insights into wire performance, wire bow and cutting forces, making it a valuable tool for process development.
Keywords: crystalline silicon, wafering, diamond wire sawing, multi-wire sawing, modelling

1 INTRODUCTION

In the photovoltaic industry, the wafering process is used for the production of crystalline silicon wafers. The multi-wire sawing setup consists of a thin abrasive wire helically wound around two wire guide rollers to build a wire web. The rotation of the wire guide rollers induces a transversal wire motion, while a silicon brick is fed through the wire web. The contact forces and the relative motion between brick and diamond wire result in material removal. The removal process alters the contour of the brick inside the kerf, eventually producing thin wafers.

The main objective is to improve productivity and wafer quality of the diamond wire sawing process. The approach chosen was to develop a comprehensive model of the wire sawing process, which includes the recipe of the wire saw and process variables that are relevant to productivity and wafer quality, such as wire performance, wire bow and cutting forces.

Developing a detailed and comprehensive model is justified by the fact that carrying out wire sawing experiments in order to collect measurement data is expensive and time-consuming. Due to the harsh process environment, the use of sensors is possible to a limited extent only. The typical pilgrim motion [1], [2] of the wire causes a great variation of the process variables over the wire web and the cutting progress, which further increases the complexity of the process. By describing the conditions in the entire wire web, this model allows one to gain insights into the correlations between recipe, productivity and wafer quality, caused by the underlying process variables. Beyond process analysis, a comprehensive model is useful for condition monitoring and model-based optimization.

In the literature, several topics concerning the modeling of the wire sawing process have already been investigated in detail.

The silicon material removal process is either described based on microscopic models as in [3] or on macroscopic models as in [4]. Microscopic models are for example simulated using discrete element methods, while macroscopic models on the other hand aim at describing the removal process based on measurable machining parameters.

The interaction between moving wire and silicon brick has been treated in [5] and the loss of wire tension in the wire web has been examined in [6] for the slurry process.

In [7] and [8], the impact of sawing parameters on wafer properties has been examined. Specifically, the effects of diamond wear on wafer quality have been analyzed in [9].

The wear of the diamond wire has been investigated in [1] regarding the machining process and in [2] regarding the silicon contact length. Various diamond wires and their behavior regarding wafer quality have been evaluated in [10]. The aging of diamond wire due to spool winding, damage during first silicon contact and through different beam materials is studied in [11].

Finally, the cutting forces and the cutting ability of diamond wire have been analyzed in [12], specifically for the wire break-in period. In [13], the productivity of the sawing process has been improved by adaptive force control. An analytical force model for diamond wire sawing is presented in [14], although the wire performance has not been taken into account.

In this work, an approach for force modeling is presented, considering wire performance and the recipe parameters of the wire saw. The model is formulated to cover the conditions along the entire wire web.

The remainder of this paper is organized as follows: In Section 2, a comprehensive model of the wire sawing process is presented. In Section 3, parameter identification and model validation are described. In Section 4, results regarding a number of simulation runs are discussed, whereas the last section concludes the paper and gives an outlook on further work.

2 MODEL OF THE WIRE SAWING PROCESS

The aim of this model is to describe relevant process variables such as cutting forces or wire performance over wire web and process time. The model is formulated as a system of first-order differential equations (nonlinear state-space model). Two coordinate systems are introduced; a machine-based one as depicted in Figure 1 and a wire-based one as in Figure 3. Further, the wire web, consisting of a multitude of wire loops, is partitioned into segments, see Figure 2. Each wire web segment consolidates a number of real wires to one single simulated wire. Regarding the process variables, similar conditions are assumed for all wires of a particular segment. This reduces the computational burden, while the model still represents the conditions along the entire wire web.

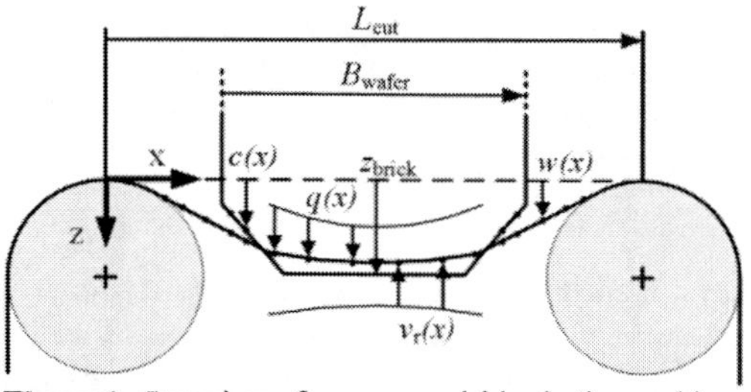

Figure 1: Overview of process variables in the machine-based coordinate system

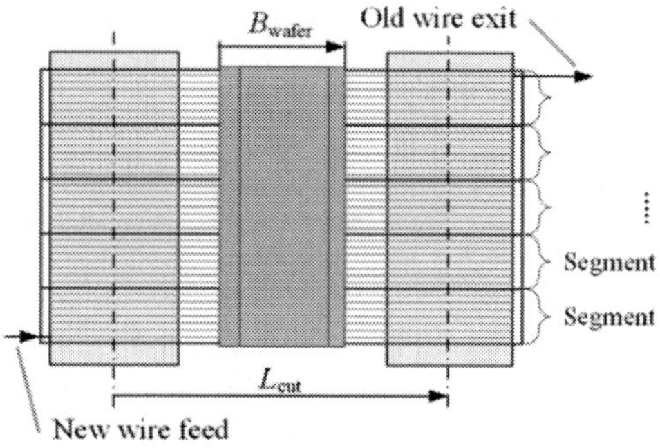

Figure 2: Wire web segments

The effects of the cutting fluid and thermal influences are not considered, and the wire management outside the wire web is not modelled, either.

The wire sawing model consists of several submodels, which will be introduced in the following subsections. The contact model describes the interaction between the deflected wire and the silicon brick. The material removal model characterizes the silicon removal rate, depending on wire speed, cutting force, and wire performance. Finally, the wire performance models implement the transformation between the machine-based and the wire-based coordinate systems.

2.1 Contact model

A model for a single deflected wire as presented in [15] is established first. Figure 1 gives an overview of the process variables employed. It is assumed that the wire does not slip on the wire guide rollers. The Bernoulli beam model (1) describes the relation between the wire bow $w(x,t)$ and the line load $q(x,t)$, neglecting the shear strain. The bending stiffness is denoted by EI_z and the mass distribution by μ. In this particular application scenario, the high pre-tensioning of the wires cannot be neglected. Therefore, the vibrating string model (2) is employed, which describes a continuum without bending stiffness under pretension F_0. Due to the pilgrim motion, loss of wire tension as described in [6] is considered to be less of a problem in the diamond wire process, and therefore the wire tension F_0 is assumed to be constant in the present work. Since Eqs. (1) and (2) are linear, they are combined by superposition. Furthermore, string vibrations $w_{,tt}(x,t)$ are neglected, and the static wire model is formulated as in Eq. (3).

$$\mu w_{,tt}(x,t) + EI_z w_{,xxxx}(x,t) = q(x,t) \quad (1)$$

$$\mu w_{,tt}(x,t) - F_0 w_{,xx}(x,t) = q(x,t) \quad (2)$$

$$EI_z w_{,xxxx}(x) - F_0 w_{,xx}(x,t) = q(x) \quad (3)$$

The contour $c(x)$ of the silicon brick inside the kerf imposes a constraint on the wire deflection. In the static case, the wire is assumed to occupy the state of minimal potential energy. The wire bow $w(x)$ and the line load $q(x)$ can therefore be computed by minimizing the potential energy of the wire V, see Eq. (4). Minimizing the potential energy is a constrained optimization problem, and the constraints are as follows:

i. the line load is restricted to positive values only: $-q(x) \leq 0$,
ii. there is no wire deflection at the support points, which means the boundary conditions are $w(0) = w(L) = 0$, and
iii. the wire does not penetrate the contour of the brick: $w(x) > c(x)$.

The single wire between the support points is discretized along the x axis, with discretization lengths x_h. The discretization applies for the process variables $w(x)$, $q(x)$, $c(x)$ and the material removal rate $v_r(x)$. Based on this discretization, the partial derivatives are approximated by finite differences, where the matrices M_{xx} and M_{xxxx} are the respective differential operators. Solving Eq. (3) for $w(x)$ and complementing with constraint (ii) is represented by $w(x) = M_w \cdot q(x)$. Replacing the integral in Eq. (4) by a discrete sum and substituting $w(x)$ by $M_w \cdot q(x)$ results in a constrained quadratic program (QP) of general form given in Eqs. (5) and (6). Further details on the contact model and on solving the underlying optimization problem are given in [15].

$$\min_{w(x)} V = \min_{w(x)} \frac{1}{2} \int_0^{L_{cut}} \left[F_0 w_{,x}^2(x) + EI_z w_{,xx}^2(x) \right] dx \quad (4)$$

$$\min_{q(x)} V = \min_{q(x)} \frac{1}{2} q^T(x) M_w^T (F_0 M_x^T X_h M_x + EI_z M_{xx}^T X_h M_{xx}) M_w q(x) \quad (5)$$

$$\begin{bmatrix} -I_{N\times N} \\ -M_w \end{bmatrix} \cdot q(x) \leq \begin{bmatrix} 0_{N\times 1} \\ 0_{N\times 1} \end{bmatrix} + \begin{bmatrix} 0_{N\times N} \\ -I_{N\times N} \end{bmatrix} \cdot c(x) \quad (6)$$

By introducing an independent contact model for every wire web segment, the model is extended to represent the conditions along the entire wire web. This leads to a pseudo-2D representation of the brick contour inside the kerf $c(x,y)$, where y is discretized by the number of wire web segments. This extends to the wire bow $w(x)$, the line load $q(x)$ and the material removal rate $v_r(x)$.

2.2 Material removal model

The contact model describes the line load $q(x)$ as a function of the brick contour $c(x)$. To gain insights into the temporal development of the contact forces, the time derivative $\dot{c}(x)$ is described as a function of material removal rate $v_r(x)$ and the feed speed v_f: $\dot{c}(x) = -v_r(x) + v_f$. The material removal inside the kerf, which occurs as a result of the interaction between moving diamond wire, cutting fluid and silicon brick, is a highly complex process on a microscopic level. In this work, a macroscopic material removal model is employed. A widely used removal model is given by the law of Preston, see e.g. [5]:

$$v_r(x) = \gamma p(x) v_{rel} \quad (7)$$

Here, the material removal rate $v_r(x)$ is a function of the Preston coefficient γ, the lapping pressure $p(x)$ and the relative velocity v_{rel}. Since the silicon brick is assumed to be at rest, the relative velocity is replaced by the wire speed v_c. The lapping pressure is substituted by dividing the line load $q(x)$ by the wire diameter D. In order to employ dimensionless variables, the reference values $v_{c,ref}$, $v_{r,ref}$ and p_{ref} are introduced, see Eq. (8).

To account for the conditions specific to the diamond wire sawing process and to allow the model to fit the experimental data better, two further parameters are introduced, with α and β describing the progression of the impact of the wire speed and line load on the material removal rate.

A characteristic specific to diamond wire sawing is that the abrasive particles are fixed to the moving wire, meaning that they enter the wire web multiple times when the wire web is operated in pilgrim mode. With increasing aging of the abrasive particles, their sharpness decreases, having a lessening impact on the material removal rate. A measure to determine wire aging is the wire performance u, given in sawn area of silicon per unit length of the wire [1]. Other measures such as the contact length [2] are also possible. To account for the impact of the wire performance on the material removal rate, the term $f(u_{seg})$, termed the relative wire sharpness, is introduced in Eq. (9). Here, u_{seg} refers to the current wire performance prevailing in the segments of the wire web. The parameter δ describes the gradient of the wire sharpness decrease and k gives a residual wire sharpness for highly aged wire.

These considerations lead to the parametric material removal model given in Eqs. (8) and (9).

$$v_r(x) = \gamma v_{r,ref}\left(\frac{|v_c|}{v_{c,ref}}\right)^{\beta}\left(\frac{q(x)}{Dp_{ref}}\right)^{\alpha} f(u_{seg}) \tag{8}$$

$$f(u_{seg}) = \left(\frac{1+\frac{k}{1-k}\delta u_{seg}}{1+\frac{1}{1-k}\delta u_{seg}}\right)^{\alpha} \tag{9}$$

Using a macroscopic model means that the parameters are process-specific, being for example valid for one specific diamond wire only. A model in this form also implicates that the parameters need to be identified using experimental data.

2.3 Wire performance models

When operating the wire saw with continuous new wire feed [1], a specific length of new wire is taken from the new wire spool in every pilgrim cycle. This results in a wire web with a high variation of the wire performance between the new wire side and the old wire side of the wire web, see Figure 3 for example. To account for these conditions, the mean wire performance in the wire web segments u_{seg} is calculated at every time step of the simulation.

$$u_{seg} = f(u(x_{wire}),x_{web}) \tag{10}$$

Here, $u(x_{wire})$ is the total wire performance in the wire-based coordinate system. For implementation purposes, $u(x_{wire})$ is discretized along the wire direction x_{Wire}, but in contrast to the discretization along the wire in the machine-based coordinate system, a much coarser grid is chosen. The variable x_{web} describes the position of the wire web in the wire-based coordinate system, see Figure 3.

Material removal causes an increasing total wire performance $u(x_{wire})$. For the single-wire model with material removal rate $v_r(x)$, the time derivative of the wire performance is given in Eq. (11) and in Eq. (12) in discretized form.

$$\dot{u}_{seg} = \frac{1}{L_{loop}}\int_0^{L_{cut}} v_r(x)dx \tag{11}$$

$$\dot{u}_{seg} = \frac{1}{L_{loop}}\sum_n x_h(n)v_r(n) \tag{12}$$

Here, L_{loop} is the length of one wire winding. Partitioning the wire web in segments results in a specific material removal rate $v_r(x)$ and therefore in one wire performance derivative $\dot{u}_{seg}$ for every wire web segment.

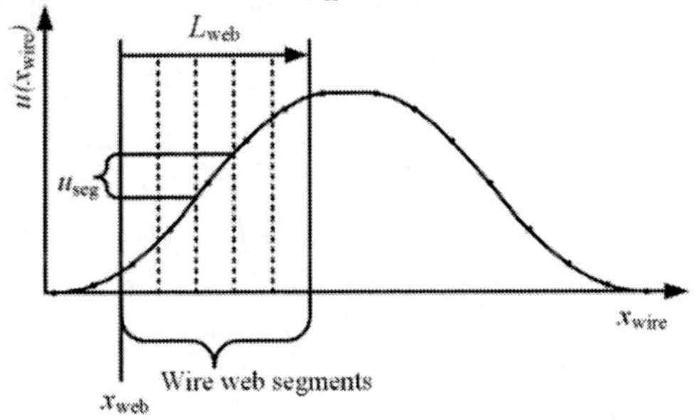

Figure 3: Overview of process variables in the wire-based coordinate system

The time derivative of the total wire performance $\dot{u}$ is computed by allocating the variation of the wire performance in the wire web segments $\dot{u}_{seg}$ from the machine-based to the wire-based coordinate system:

$$\dot{u}(x_{wire}) = f(\dot{u}_{seg}, x_{web}) \tag{13}$$

2.4 Implementation

The recipe of the wire saw is usually generated in a man-machine-interface by defining profiles for wire speed and feed speed over the entire recipe. Therefore, the inputs of the model are given by the wire speed $v_c(t)$ and the feed speed of the brick $v_f(t)$ along the z axis. The wire web position in the wire-based coordinate system is given in Eq. (14), and the feed position of the silicon brick is given in Eq. (15).

$$\dot{x}_{web} = v_c(t) \tag{14}$$

$$\dot{z}_{brick} = v_f(t) \tag{15}$$

To sum up, the state vector of this model is given in Eq. (16) and the input vector is given in Eq. (17).

$$w = [c \quad z_{brick} \quad u \quad x_{web}]^T \tag{16}$$

$$v = [v_c \quad v_f]^T \tag{17}$$

In this way, the wire-sawing model is implemented in a simulation environment and simulated with standard solvers for ordinary differential equations. While the input variables v_c and v_f and the state variables z and x_{web} are scalars, there is a brick contour c, discretized along the single wire, for every wire web segment. Also, the total wire performance u is a vector discretized along the wire coordinate x_{wire}.

3 PARAMETER IDENTIFICATION AND MODEL VALIDATION

The parameters of the material removal model are identified using experimental data. This section outlines the identification and model validation procedure.

3.1 Experimental setup

For a number of sawing experiments with different

sawing recipes, a piezo-electric sensor has been attached between the holding fixture of the wire saw and the beam that the silicon brick is glued to, see Figure 4.

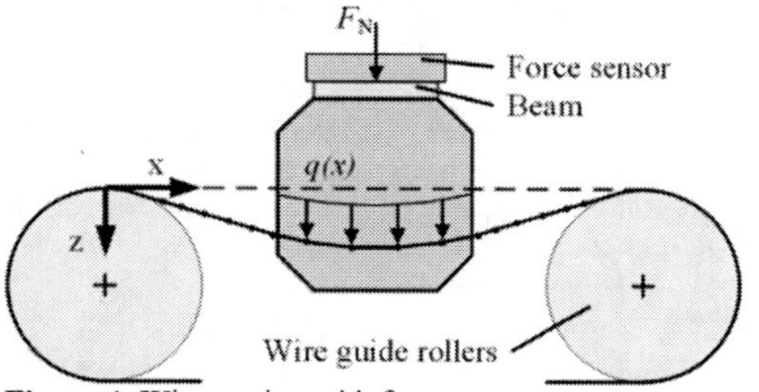

Figure 4: Wire sawing with force sensor

The total measured and simulated normal forces along the z axis are used for the identification of the parameters. The simulated normal force $F_{N,sim}$ is computed by integrating the line load $q(x)$ over one single wire in every wire web segment and summing up this quantity over the entire wire web.

The sawing experiments were executed with a wire speed of $v_c = 25\ m/s$, pilgrim lengths of $600\ m/599\ m$ in forward and backward direction, a wire tension of $F_0 = 19\ N$, a wire core diameter of $D = 100\ \mu m$, a wafer width of $B_{wafer} = 125\ mm$, a total of 348 wire loops in operation and a wire web length of $L_{web} = 710\ m$. Eight experiments with three different feed speeds v_f have been carried out, four of them in a cut series to achieve a high wire performance, hence the wire web is rewound to the initial position before the start of the following experiment (rewind mode). Table I gives an overview on the wire sawing experiments. The data has been partitioned in two sets for identification and validation, I denotes an experiment used for parameter identification and V denotes experiments used for model validation. The index $_{s,n}$ denotes the nth experiment of the cut series.

Table I: Wire sawing experiments (mono c-Si)

$v_f\ [\frac{mm}{min}]$	New wire	Rewind 1	Rewind 2	Rewind 3
0.9	I			
1.4	$I_{s,1}$ V	$I_{s,2}$	$I_{s,3}$	$I_{s,4}$
2.1	I V			

3.2 Parameter identification

Identifying the parameters is an optimization problem, where the root mean square value of the prediction error $e = F_{N,sim} - F_{N,meas}$ is minimized over a number of N_{exp} experiments with an arbitrary number of samples N_s each:

$$\min_{\alpha,\beta,\gamma,\delta,k} J \tag{18}$$

$$J = \sqrt{\frac{1}{N_{exp}} \sum_{n=1}^{N_{exp}} j(n)^2} \tag{19}$$

$$j(n) = \sqrt{\frac{1}{N_s} \sum_{m=1}^{N_s} (F_{N,sim}(n,m) - F_{N,meas}(n,m))^2} \tag{20}$$

This parameter identification problem is solved with standard gradient-based optimization algorithms. During optimization, the simulation runs for computing $F_{N,sim}$ are executed with 10 wire web segments.

3.3 Model validation

The parameters of the material removal model resulting from solving the optimization problem presented above with the identification data set are given in Eq. (21).

$$\begin{aligned} \alpha &= 1.259 \\ \beta &= 1.042 \\ \gamma &= 1.598 \\ \delta &= 164.9 \\ k &= 0.126 \end{aligned} \tag{21}$$

For the identification data set, the in-sample-error is $J_I = 0.0485\ N$, and for the validation data set, the out-of-sample-error is $J_V = 0.0466\ N$.

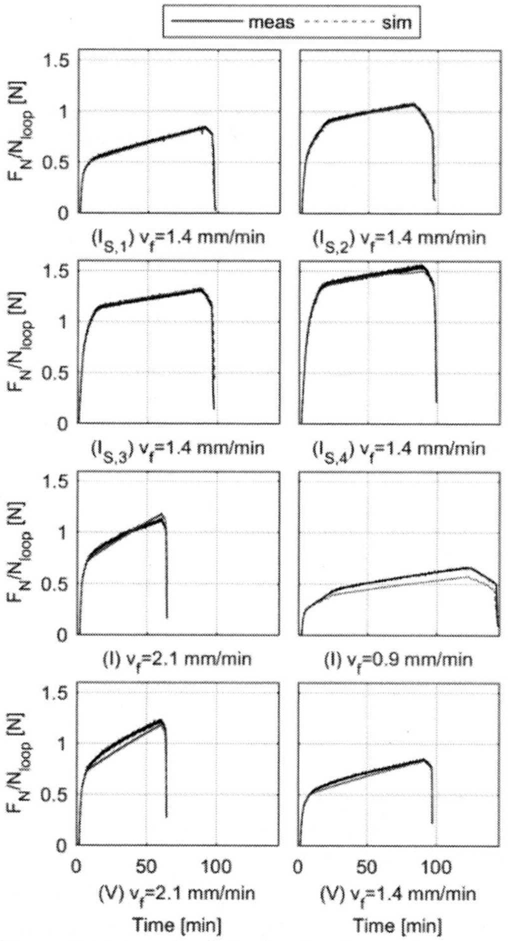

Figure 5: Measured and simulated total normal forces per wire loop in wire sawing experiments

In Figure 5, the total simulated and measured normal forces per wire loop are plotted over process time for all experiments. The impact of the wire performance leads to an increasing normal force over all single cuts and the cut series. The wire performance is represented in the material removal model by means of the relative wire sharpness. For the experiment with $v_f = 0.9\ mm/min$, the measured normal force is underestimated in the simulation. For feed speeds as low as in this experiment, effects not modelled may lead to a faster decrease of relative wire sharpness. For the rest of the experiments, a satisfying fit of the measured global normal force is achieved. At the start and end of each experiment, sawing through the wafer chamfer leads to a smaller contact length and therefore to a decreased normal force.

By using data from experiments executed at different feed speeds, the model is validated at different operating points. The model is capable of estimating the global normal force based on local conditions in the wire web. Consequently, it is now possible to analyze local process variables over wire web and process time.

4 RESULTS

In the following section, some simulation results regarding the variation of the process variables over the wire web are discussed. For a more straightforward interpretation of the plots, these simulation runs were executed with 5 wire web segments, all other simulation parameters correspond to the model validation case in the previous section.

Figure 6 presents the maximal wire bow of a single wire for every wire web segment over process time in the last experiment of the cut series. The corresponding total wire performance in the wire-based coordinate system at the end of each experiment in the cut series is shown in Figure 8. In the rewind mode, the wire web moves in the negative wire direction during an experiment and is reset to the initial position $x_{web} = 0$ after each experiment. Resetting the position changes wire performance in the segments u_{seg} between experiments and causes the distribution of the bow over the wire web to change during the experiments.

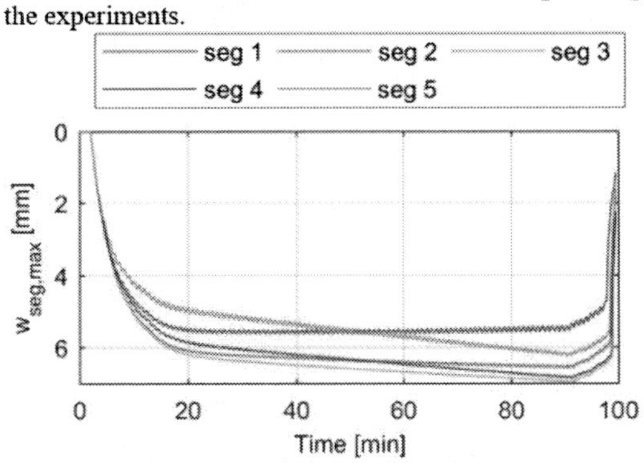

Figure 6: Maximal wire bow in wire web segments, rewind mode, Experiment $I_{5,4}$

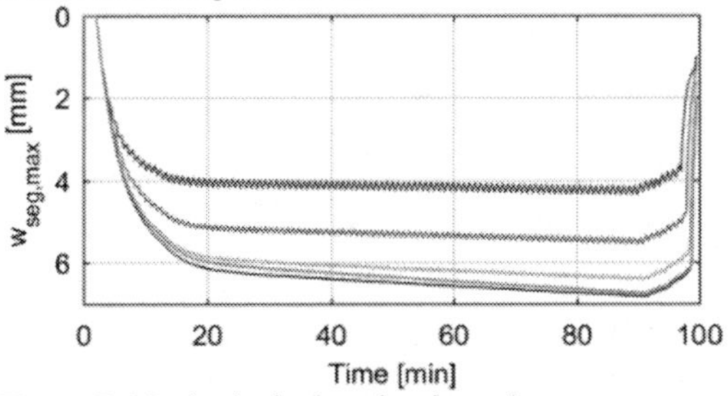

Figure 7: Maximal wire bow in wire web segments, reload mode, Experiment 4 of cut series

To examine the impact of different recipe modes on wire bow, the cut series has been simulated in reload mode, where the position of the wire web remains unchanged between experiments. Figure 7 depicts the maximal wire bow of a single wire for every wire web segment in the last experiment of the reload cut series. The corresponding total wire performance at the end of each experiment is shown in Figure 9. Reload mode causes a more steady distribution of the wire bow over process time, since wire performance u_{seg} is not changed between experiments.

By examining the wire bow at the end of the experiments, predictions about the saw-out characteristics of a certain recipe type can be made. Abruptly decreasing wire bow marks the point where a wire has completely sawn into the beam, where a wire experiences much less resistance.

An effect of the reload mode is that the saw-out characteristic is not as favorable as for the rewind mode for the scenario presented here. Since the wire web in reload mode moves constantly in the negative wire direction, segment 1 on the new wire side moves incessantly through fresh wire with a high relative wire sharpness, leading to a small bow and an early saw-out. The opposite is true for the wire web segments on the old wire side. As a result, the saw-out time increases.

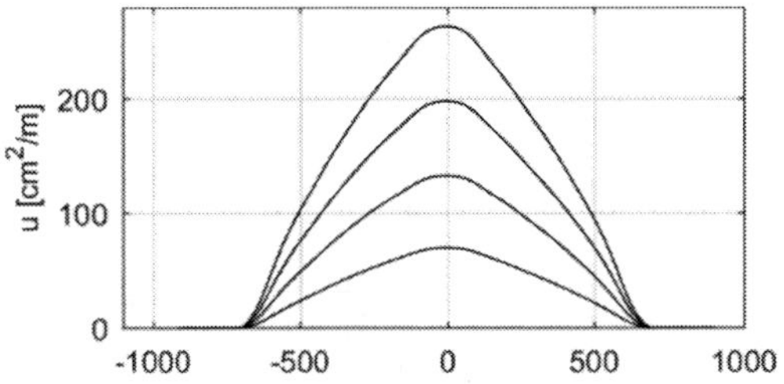

Figure 8: Total wire performance u in the wire-based coordinate system after each experiment of cut series, rewind mode

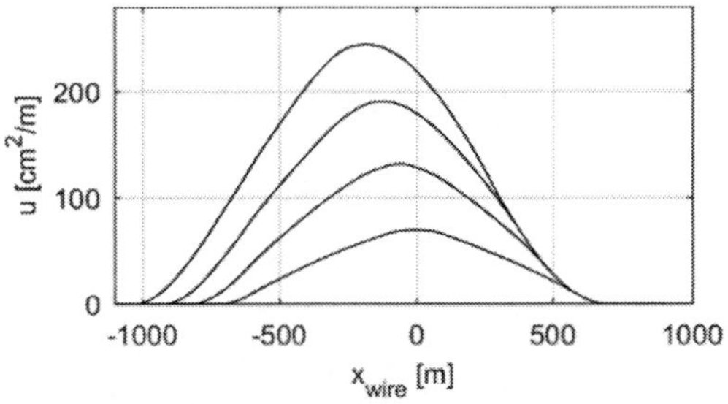

Figure 9: Total wire performance u in the wire-based coordinate system after each experiment of cut series, reload mode

5 CONCLUSIONS

This paper presents a comprehensive model of the wire sawing process, considering aspects such as cutting forces and wire bow, wire performance and the recipe of the wire saw. Based on this model, it is possible to gain insights into the variation of the process variables over wire web and process time, making it a valuable tool for process development.

Concerning the validation of the model, measured and simulated total forces in feed direction have been used to identify the parameters of the material removal model. In addition, it would be possible to use measurement data such as motor torque or wire bow for validation purposes and to further increase confidence in the model.

Other opportunities to extend the model capabilities are the inclusion of effects such as the sawing through the sacrificial beam, which potentially influences the relative wire sharpness. In addition, modeling of reconditioning procedures utilized to increase relative wire sharpness between experiments is possible.

Further work includes the use of the model for process analysis, monitoring and optimization. For analysis purposes, the model may be used to establish a relation between process variables and wafer quality regarding the variations in the wire web. Furthermore, the identified parameters of the material model may reveal characteristics of different diamond wires.

ACKNOWLEDGEMENT

This work was partly supported by the Swiss Commission for Technology and Innovation (CTI). The authors would like to acknowledge Meyer Burger AG for providing the experimental data and for sharing insights into the wire sawing process.

REFERENCES

[1] J. Peguiron, C. Heiniger, S. Habegger, Proceedings 29th European Photovoltaic Solar Energy Conference (2014) 386.
[2] K. Sunder, H. Uhle, S. Knöppel, O. Anspach, Proceedings of 28th European Photovoltaic Solar Energy Conference (2013) 1491.
[3] B. Nassauer, A. Hess, M. Kuna, Int. Journal of Solids and Structures 51 (2014) 2656.
[4] T. Liedke, M. Kuna, Int. Journal of Machine Tools & Manufacture 51 (2011) 711.
[5] M. Lorenz, Dissertation, TU Bergakademie Freiberg (2013).
[6] D. Meissner et al., Solar Energy Materials & Solar Cells 120 (2014) 346.
[7] R. Koepge, C. Klute, F. Kaule, K. Sunder, O. Anspach, S. Schoenfelder, Proceedings 31th European Photovoltaic Solar Energy Conference (2015) poster.
[8] B. Sopori et al., Proceedings IEEE 42nd Photovoltaic Specialist Conference (PVSC) (2015).
[9] A. Kumar, S. Kaminski, S. N. Melkote, C. Arcona, Wear 364-365 (2016) 163.
[10] K. Sunder, O. Anspach, Proceedings of 29th European Photovoltaic Solar Energy Conference (2014) 588.
[11] K. Sunder, O. Anspach, Proceedings of 32nd European Photovoltaic Solar Energy Conference (2016) 300.
[12] S. Lee, H. Kim, D. Kim, C. Park, Int. Journal Advanced Manufacturing Technologies 87 (2016).
[13] S. Li, S. Du, A. Tang, R. G. Landers, Y. Zhang, Journal of Manufacturing Science and Engineering 137 (2015) 061003.
[14] S. Li, A. Tang, Y. Liu, J. Wang, D. Cui, R. G. Landers, Journal of Manufacturing Science and Engineering 139 (2017) 041003.
[15] D. Treyer, S. Gaulocher, S. Niederberger, H. Rafael, A. Ams, Proceedings in Applied Mathematics and Mechanics (2017).

MECHANICAL VIABILITY OF METALLURGICAL SILICON SUBSTRATES FOR THE USE IN ULTRATHIN DEVICES

M. E. Ortiz de Zárate[1], C. Domergue[1], C. Alarcón[1] and J. Barredo[2]
Author for correspondence: Josu Barredo
[1]Technical University of Madrid, c/ José Gutiérrez Abascal 2, 28006 Madrid, Spain.
[2]Center for Modeling in Mechanical Engineering (CEMIM-F2I2), c/ José Gutiérrez Abascal 2, 28006 Madrid, Spain
email: jbarredo@etsii.upm.es

ABSTRACT: The cost reduction of the solar cells devices becomes fundamental in the current situation of the photovoltaic industry. A new device based on very thin upgraded metallurgical silicon (UMG-Si) is being analyzed. The success of these cells requires a good behavior of the silicon substrates from a mechanical point of view. It's well known that the UMG-Si presents higher number of impurities compared with solar grade silicon. Once the cracks originated by the wafering process are removed by chemical baths, these impurities are the origin of the wafer failure. This paper analyzes the strength of these substrates in comparison with the solar grade silicon substrates in order to establish if it's necessary to handle or not with special care these new substrates.
Keywords: Cost reduction, Characterization, Impurities, Metallurgical Silicon

1 INTRODUCTION

The cost reduction of the solar cells devices becomes fundamental in the current situation of the photovoltaic industry. In this context, a high efficiency solar cell on low cost silicon substrates appears as an interesting alternative to be analyzed. This device is based on very thin silicon substrates combined Silicon-Germanium solar cells to form a monolithically integrated tandem structure. The goal of this tandem cell structure is to achieve efficiencies in the range of 19%-20% with a potential to reach 22%-23%. To reduce the cost, upgraded metallurgical silicon (UMG-Si) is going to be used in the device with very low thickness.

The success of these cells requires low breakage rate during the cell production process. Therefore, a good behavior of the UMG-Si substrates from a mechanical point of view is necessary.

The UMG-Si crystal presents a higher number of impurities compared with solar grade silicon [1] becoming cheaper and interesting in a reduction cost process context. These impurities may lead to the failure of the wafer once the cracks originated by the wafering process are removed. Taken into account the very low thickness of the silicon substrates (below 50 μm), the mechanical viability appears as a key since a high breakage rate will lead to an increase of the final cost.

2 BACKGROUND

In previous studies [2], [3] the mechanical strength of the solar grade silicon crystal was obtained. The objective of the study was to determine the depth of cracks generated by the multi wire sawing process. To this end, five sets of samples with different chemical baths duration were analyzed. It was demonstrated that longer duration of the chemical bath results in higher mechanical strength until a constant value is achieved. This means that the mechanical strength is strongly influenced by the cracks originated by the wafering process and a clear way to increase the strength becomes removing the damaged layers. When a constant value of strength is achieved, the cracks coming from the wafering have been eliminated obtaining the strength of the silicon crystal.

After the study, it was concluded that the multi wire sawing process generates cracks with a depth between 13 μm and 20 μm which agreed with other results [4].

The strength of each set of wafers is defined through a probabilistic approach based on the Weibull model (which will be explained in point 5). In table I, the Weibull parameters defining the strength of each set are presented.

Table I: Weibull parameters defining the strength of the five sets of wafers

Bath duration	Thickness removed [μm/face]	Weibull parameters m [-]	σ_u [MPa]	σ_0 [MPa mm^2/m]
3 min	13	2,35	226,0	595,7
6 min	20	2,43	328,9	999,5
9 min	26	2,40	277,0	883,1
12 min	34	2,65	288,4	984,4
15 min	41	2,66	330,6	1093,6

The representation of these Weibull curves (figure 1) shows clearly that set 1 (chemical bath of 3 minutes) is influenced by the cracks coming from the wafering whereas for sets 2 to 5, all the curves are quite similar meaning that the cracks have been removed and this band represents the strength of the solar grade silicon crystal. This band of curves is taken as the reference to compare with the results obtained for the UMG-Si.

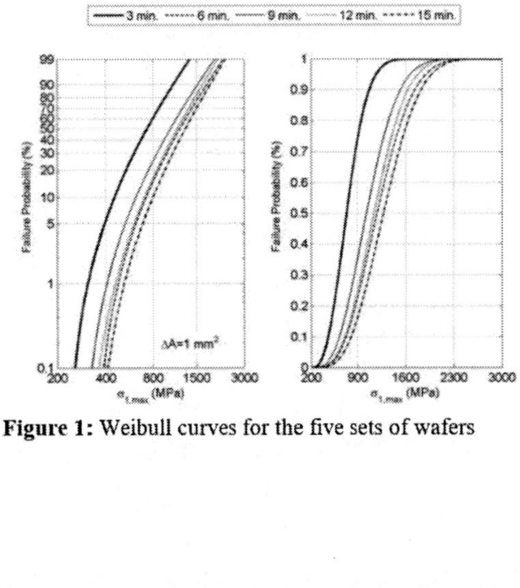

Figure 1: Weibull curves for the five sets of wafers

3 SAMPLES PREPARATION

For this study, a set of UMG-Si samples has been prepared. Mono-crystalline UMG-Si Czochralski wafers of 156 mm x 156 mm and 180 µm of thickness are cut into 52 mm x 52 mm samples. In order to analyze the mechanical strength of the UMG-Si crystal, after the cutting process, samples are chemically bathed to remove the cracks originated by the wafering process.

First, samples are introduced in an alkaline thermostat bath at 80°C. This bath is composed by deionized water, KOH and isopropyl alcohol. After that, samples are washed in water, dried and placed in a dilute hydrochloric acid bath for some minutes. Finally, they are washed in water and dried again. Figures 2 shows some pictures of the process.

Figure 2: Pictures of the bathing process

The bath duration is enough to reduce about 25 µm per face ensuring that these cracks have been eliminated. Therefore, a set of 42 samples of UMG-Si with a mean thickness of 130 µm is ready for the study.

4 RING-ON-RING TEST

The Ring-on-Ring test is employed in the analysis. Figure 3 shows a photograph of one test. In this type of test, stresses inside the lower ring are much higher than in the outer part. Hence, the influence of border cracks that might exist can be neglected.

Figure 3: A photograph of one ring-on-ring test

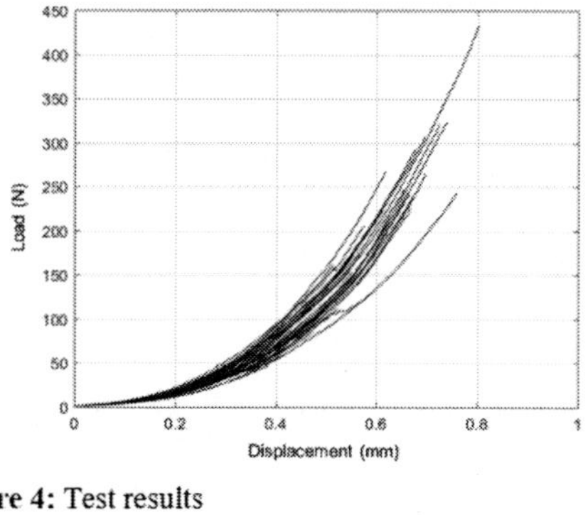

Figure 4: Test results

After the test campaign, load-displacement curves until the failure moment of each sample are collected (figure 4).

4 FINITE ELEMENT MODELS

The test gives information about the behavior of the samples and the maximum load and displacement reached before failure. However, it's necessary to determine the stress state of each sample in the moment of failure. The Finite Element (FE) Method is used for the simulation because both the anisotropy of mono-crystalline silicon crystal and the non-linear behavior of the samples during the test can be taken into account. The non-linearities present in the test are due to large displacements and contact between the wafer and the rings. The anisotropy is considered using the following elastic constants:

c_{11}=165.6 GPa
c_{12}=63.9 GPa
c_{44}=79.5 GPa

In figure 5 the model is presented and in figure 6 the good fitting between the model and the experimental result is shown for one test.

Figure 5: FE model simulating the test

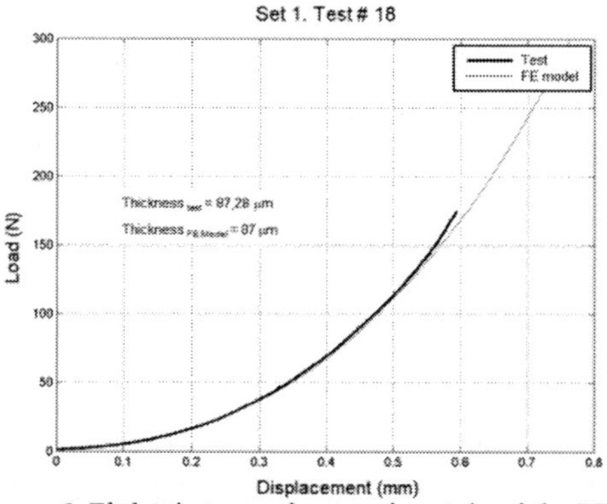

Figure 6: Fitting between the experimental and the FE model for one test

After the simulations, the stress distribution for all samples in the moment of failure is obtained. This result is necessary for the probabilistic model employed to characterize the mechanical strength (as presented in the next point).

5 WEIBULL ANALYSIS

The high scatter observed in the strength characterization of brittle materials is represented through statistical distributions getting a more accurate failure prediction. Based on the weakest link theory, Weibull [5] introduced a probabilistic approach in 1939 that is widely used in the strength characterization of brittle materials. This probabilistic model takes into account the so called size effect i.e. the influence of the size of the stressed area. The most general expression of the failure probability according to the Weibull model is given in (1) which corresponds to the failure probability of a sample subjected to a uniform stress.

$$P_f(\sigma) = 1 - exp\left[-\int_A \left(\frac{\sigma-\sigma_u}{\sigma_0}\right)^m dA\right] \quad (1)$$

In (1), σ_u [MPa] is the location parameter representing a threshold stress which differentiates stresses levels which may lead to failure from those that may not; σ_0 [MPa (mm^2) (1/m)] is the scale parameter giving information about the strength of the material. Finally, m [dimensionless] is the shape parameter (it's also called the Weibull modulus) and it gives information about the scattering of the results. These 3 parameters define the strength of the material according to the Weibull probabilistic model.

The expression (1) is valid for an area subjected to a uni-axial and uniform tensile stress. This assumption is generally not true so it's necessary to obtain a valid expression for the test conditions in order to fit a Weibull cumulative distribution function (cdf) to the test results. In this case, the Ring-on-Ring test results in a non-uniform biaxial stress distribution.

To take into account the multi-axial stress state, the Principle of Independent Action (PIA) has been employed. PIA establishes that each principal stress acts independently. Therefore, the failure probability for a multiaxial stress distribution is obtained through the surveilling probability as it's shown in (2)

$$P_s(\sigma) = P_s(\sigma_1) \cdot P_s(\sigma_2) \cdot P_s(\sigma_3) \quad (2)$$

The non-uniform stress distribution is considered through the equivalent area. By definition, a specimen having the equivalent area A_{eq} subjected in a tensile test to the maximum stress observed in the ring-on-ring test, results in the same probability of failure. Therefore, the $A_{eq,i}$ of the sample i for the three-parameter Weibull distribution can be calculated as follows:

$$A_{eq,i} = \int_A \left(\frac{\sigma-\sigma_u}{\sigma_{max}-\sigma_0}\right)^m dA \quad (3)$$

However, the converse problem exists: the parameters defining the strength must be calculated from the experimentally obtained data. An iterative method to fit the test result to a Weibull cdf, based on the one proposed by Przybilla [6], is applied. The results show that the Weibull parameters of the UMG-Si set are quite similar to the ones obtained for solar grade silicon crystal without cracks from the wafering process:

$m = 2{,}01$
$\sigma_u = 396{,}09$ MPa
$\sigma_0 = 806{,}00$ MPa mm^2/m

Figure 7 shows the representation of the cdf for the UMG-Si with the previous results obtained for solar grade silicon. It can be seen that the UMG-Si curve is inside the reference band of solar grade silicon in the lowest extreme. Therefore, it seems that the lower quality of the crystal doesn't lead to a lower mechanical strength being minimum the influence of the higher number of impurities in the crystal from a mechanical point of view.

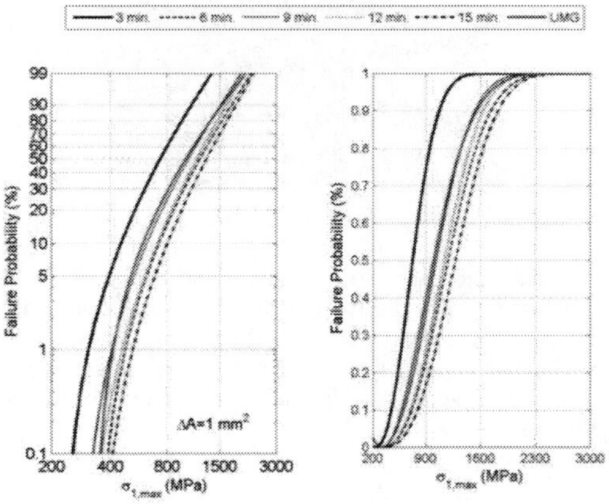

Figure 7: Weibull curves for the five sets of wafers of solar grade silicon and the set of UMG-Si

6 CONCLUSIONS

The comparison of the Weibull parameters defining the strength of UMG-Silicon with solar grade silicon results in similar values. The representation of the Weibull cdf of all sets shows that the UMG-Si set is in the lower extreme of the band of reference valid for solar grade silicon. Therefore, it seems that the bigger number of impurities in the UMG silicon crystal doesn't affect significantly the strength of the material.

From a mechanical point of view, it can be concluded that the UMG-silicon crystal presents similar strength that the solar grade silicon crystal.

7 DISCUSSION

Another set of UMG-Si samples was prepared. The main characteristic of this second set was the low thickness of the samples (mean thickness of 42 μm). The objective of the analysis of this set was to validate the results for ultrathin devices as the ones that are going to be used in the tandem structure.

The study of this second test has become quite difficult and none conclusive result can be achieved.

First of all, due to the high flexibility of the samples, some of them were broken during the manipulation to test them. A more careful handling is necessary.

During the test, small fractures were observed at the beginning. This may be due to the lack of accuracy of the testing tools. A more accurate tolls may be necessary for this analysis.

An excessive distortion was observe during the test (figure 8). This is called the buckling effects and it's due to the very low thickness of the samples and the stiffness of the test. Another types of test are being designed to carry out the study.

Figure 8: Buckling effects in ultrathin samples testing

Finally, in the simulations of the successful test by the FE Method, convergence problems appear. It's though that this is also due to the buckling effects.

Further investigations are being carried out in order to get a correct analysis of the mechanical strength of these ultrathin devices.

8 ACKNOWLEDGEMENTS

This work has been made under the sponsorship of the University State Secretary, belonging to Spanish Ministry of Economy and Competitiveness, in its contract ENE2014-56069-C4-2-R.

The presented results have been obtained in cooperation with Instituto de Energía Solar (IES, www.ies.upm.es).

9 REFERENCES

[1] E. Forniés, L. Méndez and M. Tojeiro, *"Polysilicon vs. upgraded metallurgical-grade silicon (UMG-Si): Technology, quality and costs"* Photovoltaics International - The technology resource for PV professionals, Vol. 31, pp. 29-38, 2016.

[2] J. Barredo *"Estudio de las propiedades mecánicas de obleas de silicio"* PhD. Thesis, Universidad Politécnica de Madrid, 2013

[3] J. Barredo, L. Hermanns, A. Fraile, J. C. Jimeno and E. Alarcón *"The influence of different surface treatments on the mechanical strength of silicon wafers"* Proceedings of the 22nd European Photovoltaic Solar Energy Conference, 1:1164-1167, 2007

[4] Park et al. *"Correlation between wafer fracture and saw damage introduced during cast silicon cutting"*. 15th Workshop on Crystalline Silicon Solar Cells and Modules: Materials and Processes, Colorado, USA, 2005

[5] W. Weibull. *"A statistical theory of the strength of materials"* Technical report, Royal Swedish Institute of Engineering Research, 1939.

[6] C. Przybilla, A. Fernández-Canteli and E. Castillo *"Deriving the primary cumulative distribution function of fracture stress for brittle materials from 3- and 4-point bending tests"* Journal of the European Ceramic Society, 31(4):451 - 460, 2011.

THE STUDY OF WATER-BASED SLURRY FOR WAFER SLICING AND THE TOTALLY RECYCLING OF MATERIAL IN SLICING PROCESS

Teng-Yu Wang[1]*, Cheng-Yao Cheng[3], Pei-Shan Huang[3], and Shen Tsao[1]
[1]Material and Chemical Research Labs, Industrial Technology Research Institute, Hsinchu, Taiwan, R.O.C.
[2]Green Energy Technology Inc., Taoyuan, Taiwan, R.O.C.

ABSTRACT: The wire-sawing is the standard procedure in silicon wafer producing. The stainless steel wire with silicon carbine (SiC) particles are widely used in the conventional wafer cutting process. The cutting slurry are composed of SiC and polyethylene glycol (PEG). After cutting, the silicon particle are added into the waste slurry. The use of SiC accounts for a large part of the wafer production cost. On the other hand, the treatment of waste slurry also requires additional cost. In this study, two methods were used to reduce the wafer cost. First, water-based slurry slicing are used to replace the PEG based slurry. The cutting slurry with 40% water and 60% PEG was used as the lubricating oil. An additive was added into the cutting slurry to increase the dispersion of SiC cutting particles. The cost could be reduced by reducing the amount of PEG used. Second, the SiC particle are recycled from the waste slurry. After separation and cleaning, the reused SiC mixed with new SiC were used on slicing process again. The yield of cutting process with reused SiC and water-based slurry was 90.9%, which was comparable with the standard cutting process (91.0%). Recycling of SiC can provide additional cost reduction. In this study, the use of water-based slurry and the recycling of SiC caused a cost reduction of 4.87 cent/pc and 4.66 cent/pc, respectively.
Keywords: Multicrystalline Silicon; Wire Saw; Recycle.

1 INTRODUCTION

The ingot growth and wafer slicing is the main process for silicon wafer production. In PV industry, all of the wafers are sliced by wire sawn process. Traditionally, silicon wafers were sliced with steel wire and silicon carbine (SiC) particles, which was called slurry-sawn system. The diamond wire were used for single-crystalline silicon wafer slicing recently because of its highly cutting velocity [1]. The mass production of multi-crystalline silicon wafers were still sliced with slurry-sawn system so far. The diamond wire-saw using on multi-crystalline wafers production is under study because the surface texturing will be a problem [2]. The meal-assisted etching (MAE) or reactive-ion etching (RIE) were needing to re-treating the surface of wafer [3]. However, these two approaches will increase the production cost. The slurry-sawn system is still the mainstream for multi-crystalline silicon wafers producing. The used slurry are mixing with the cutting fluid and abrasive SiC particles, as well as the worn down metals from the saw wire. The price of SiC is not cheap, which is usually separated from the slurry for recycling. The technique of recycling of silicon from used slurry was noticed only while the price of poly-silicon is expansive [4]. It is too cheap to recycle now. The production speed of cutting waste is surprisingly fast. In Taiwan, the production of cutting waste is estimated to more than 2,500 metric tons per month. Therefore, the cutting waste recycling is currently an important technology, whether in environmental protection or economic considerations.

2 EXPERIMENTAL

The multi-crystalline silicon was grown by casting method. The wafers were sliced by slurry-sawn system. The water is mixed with PEG in a different composition as the cutting fluid. The water content was varies from 10% to 40%. An additive was added into the cutting slurry to increase the dispersion of SiC cutting particles.

Figure 1 shows the recycling and reused processing in this study. After wafer slicing, SiC with bigger particle size (>7 μm) will recycled and mixed with new SiC particle. The reused of SiC had the same cutting yield and could reduce the production cost. The cutting oil could be recycled and reused again. When the SiC particle became to smaller than 7 μm, the SiC was filtrated and separated from the waste slurry. After drying, the SiC are reused as the blasting sand in surface treatment process for diamond wire cut wafer. All of the SiC was effectively utilized.

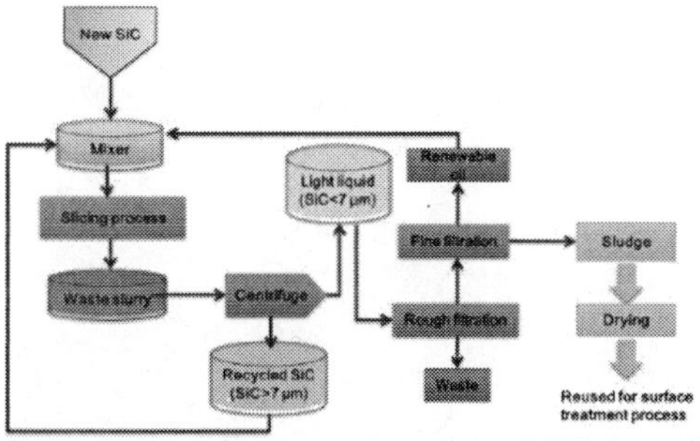

Figure 1: The recycling and reusing of SiC in wafer slicing process.

3 DISCUSSIONS

In commercial process, the silicon wafers are sliced by slurry-sawn process. The cutting fluid is composed of SiC and PEG. This is because the SiC can be suspended in PEG. After wafer slicing, the silicon particles will go into the slurry and mixed with SiC particles. This caused it difficult of separation and recovery of cutting slurry [5]. Besides, the treatment of PEG will increase the cost of recycling process. If the PEG can be replaced by water, the composition of cutting fluid will become simple and easy to recycle. The most important is the cost could be reduced. The viscosity of PEG is 100 times of water, the dispersion of SiC will become poor if the cutting fluid is water-based. Therefore, an additive was added into the cutting fluid to help the dispersion of SiC particle. The additive is a sol-gel material, which will catch the SiC particle and dispersed in the water. Therefore, the SiC particle will not accumulate and precipitates at the bottom of the system. Figure 2 shows how the additive help the dispersion of SiC particle in the cutting fluid.

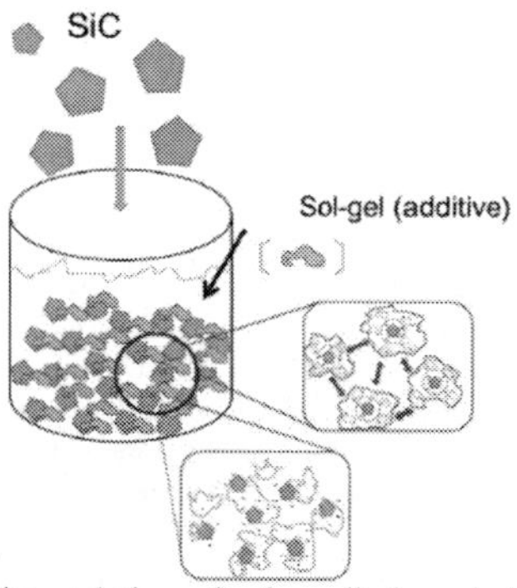

Figure 2: The variation of solar cell characteristics with different grinding speed

Fig. 3 and Fig. 4 are the comparison of cutting results with different cutting fluid composition. The conventional cutting slurry is 100% PEG with SiC particles. When the water was added into the cutting fluid to replace some parts of the PEG, the cutting results were not change much. In this study, the water content in cutting fluid was varied from 10% to 40%. The quality of obtained wafers were closed, but the cost were different.

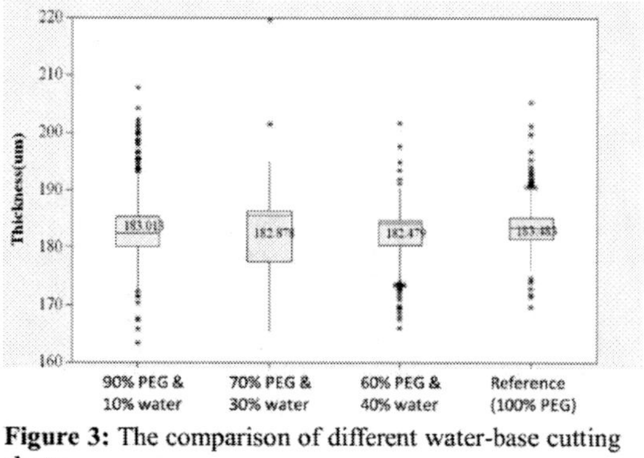

Figure 3: The comparison of different water-base cutting slurry

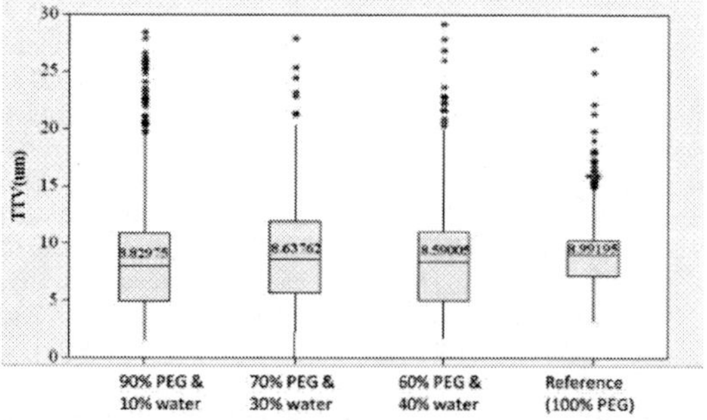

Figure 4: The variation of solar cell characteristics with different sandblasting times.

When the cutting fluid is 100% PEG, the average cutting time is 12 hours. When the water was added into the cutting fluid, the cutting ability will reduced and the cutting process will spend more time. The cutting yield were 91.4% and 90.9% as the cutting fluid is 30% and 40% water, respectively. The more water was added, the slower the cutting speed became. The cutting yield was reduced slightly, too. The only advantage was that the cost was reduced as the PEG was replaced by water. Fig. 5 is the cost analysis of the water-based slicing process. The cutting speed, cutting yield, and material price were considered together and the cost was calculated. As the water content was 40%, the most cost was saved. The cost can be reduced by 1.48 cent per wafer as the 40% water was added into the cutting fluid.

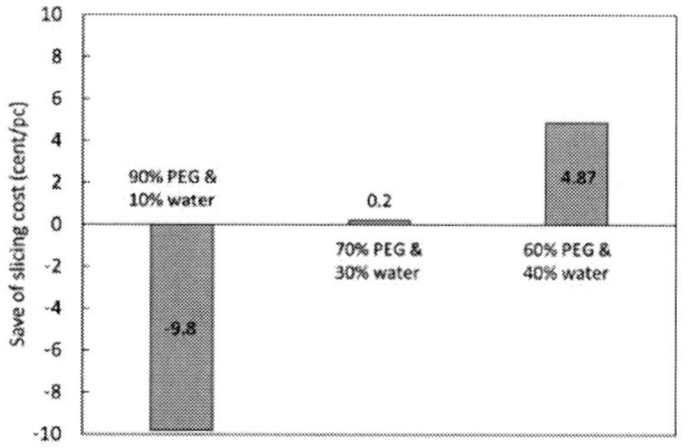

Figure 5: The cost change of different composition of cutting fluid.

The SiC is the leading role in wafer slicing. The surface of SiC particle is very sharp before slicing, the circularity is in the range between 0.87 and 0.89. The circularity is define as:

$$circularity = \frac{circle\ circumference}{perimeter\ of\ the\ particle} = \sqrt{\frac{4\pi A}{p^2}}$$

A and p represent the surface area and perimeter of the SiC particle, respectively. When the circularity closer to 1, on behalf of the particles closer to the circle. The ability of cutting will become worse. In this study, the SiC particle with diameter larger than 7 μm was recycled and reused in wafer slicing process. The circularity of recycled SiC was in the range between 0.9 and 0.92. Although the cutting ability is poor, but it still can be used for wafer slicing. Fig. 6 shows the comparison of surface morphology of SiC particles before and after wafer slicing. The recovered SiC was mixed with unused SiC at a ratio of 3:7 and re-used in the wafer slicing process. Fig. 7 is the cost analysis of the slicing process using recovery SiC. As the 30% of the SiC is from recycled, the cost can be reduced by 4.66 cent per wafer.

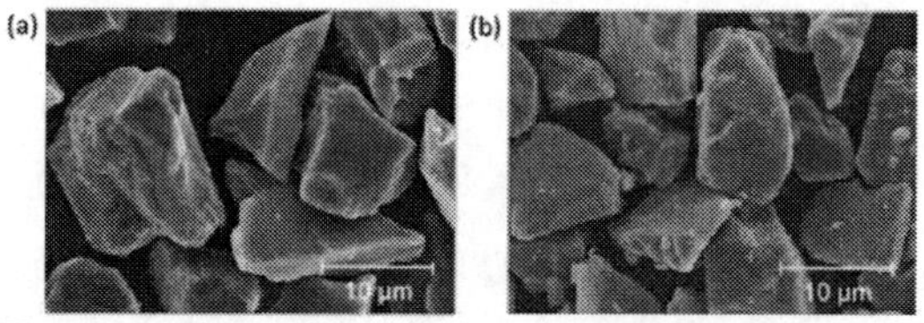

Figure 6: The surface morphology of SiC particles (a) before slicing; (b) After slicing.

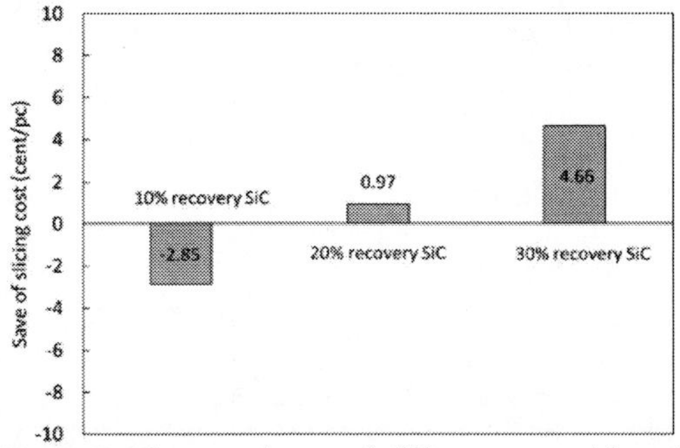

Figure 7: The cost change of different composition of SiC particles.

The PEG could be recycled and reused, too. After separating the SiC particle, the renewable PEG was obtained from filtration. The recovered PEG was mixed with unused PEG at a ratio of 7:3 and re-used in the wafer slicing process. The cutting yield can be maintained higher than 91%. The process cost could be reduced, too. Fig. 8 is the cost analysis of wafer slicing process.

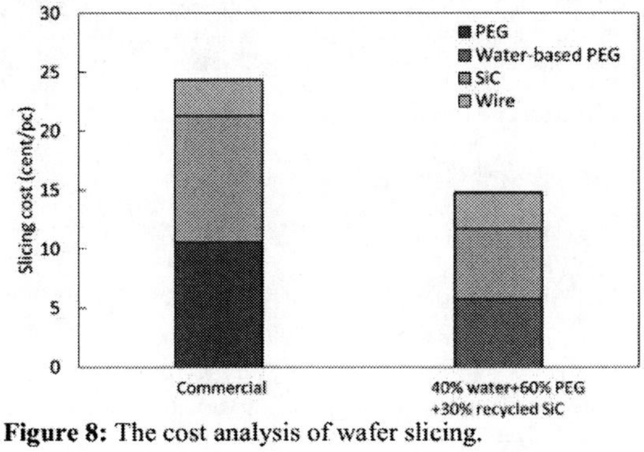

Figure 8: The cost analysis of wafer slicing.

4 CONCLUSIONS

The water-based slurry were used in slicing process to reduce the production cost. An additive was added into the cutting slurry to increase the dispersion of SiC cutting particles. The cost was reduced by reducing the amount of PEG used. The cutting yield is comparable with the totally PEG based slicing. After wafer slicing, the SiC particle are recycled from the waste slurry. After separation and cleaning, the reused SiC mixed with new SiC were used on slicing process again. The yield of cutting process with reused SiC and water-based slurry was 90.9%, which was comparable with the standard cutting process. Recycling of SiC can provide additional cost reduction. When the SiC become smaller, the SiC was separated and used for other application. The PEG in the cutting slurry were also recycled and reused. All the materials used in the slicing process were recycled and no waste was produced.

ACKNOWLEDGEMENT

The financial support provided by Bureau of Energy is gratefully acknowledged.

REFERENCE

[1] H. Wu, Precision Engineering 43 (2016) pp.1-9.

[2] T. Y. Wang, W. J. Lih, C. Y. Cheng, J. Y. Liu, and W. H. Lin, 32rd European Photovoltaic Solar Energy Conference and Exhibition, pp. 307-309, June. 20-24, Munich, Germany. (2016).

[3] A. Kumagai, Solar Energy Materials & Solar Cells 133 (2015) pp.216-222.

[4] J. Yoo, G. Yu, and J. Yi, Solar Energy Materials & Solar Cells 95 (2011) pp.2-6.

[5] T. Y. Wang, Y. C. Lin, C. Y. Tai, C. C. Fei, M. Y. Tseng, and C. W. Lan, Progress in Photovoltaics: Research and Applications 17 (2009) pp.155-163.

A NOVEL APPROACH TO DETERMINE THE DIAMOND OCCUPANCY OF DIAMOND WIRES FOR OPTIMIZED CUTTING PROCESSES FOR CRYSTALLINE SILICON

Lydia Lottspeich[*1], Laura Theophil[*2], Marcel Fuchs[*3], Thomas Kaden[*4]
[*]Fraunhofer Technology Center for Semiconductor Materials,
Am St.-Niclas-Schacht 13, 09599 Freiberg, Germany
[1]Tel.: 0049-3731-2033169, Fax: 0049-3731-2033199, lydia.lottspeich@thm.fraunhofer.de;
[2]laura.theophil@thm.fraunhofer.de; [3]marcel.fuchs@thm.fraunhofer.de; [4]thomas.kaden@thm.fraunhofer.de

ABSTRACT: The production of wafers from crystalline silicon for solar cells is increasingly realized by diamond wire sawing. A steel core wire is used on which diamonds are fixed in a layer of nickel or resin. The cutting efficiency of the wire is determined by the interplay of the material properties of the crystal that is cut, the cooling fluid and the characteristics of the diamond grit on the wire. Hereby, the density of diamonds with respect to the wire length, the shape of the diamonds and the grain size distribution of the diamonds are important. A new approach is presented to remove diamonds from nickel bonded diamond wires and subsequently analyze their density and grain size distribution. The gravimetrically determined mass of diamonds on the wire shows the same trend as the number of diamonds counted manually on microscope images. Thus, a quality control of diamond wires prior to the wafering process as well as a characterization of the wire wear during the cutting process becomes possible with higher accuracy.
Keywords: Wafering, diamond wire, wire wear, silicon, characterization

1 INTRODUCTION

The wafering of silicon for photovoltaics is currently converted from slurry sawing to diamond wire sawing (DWS). While this technology transition is well advanced for monocrystalline silicon, the share of the DWS technology for the wafering of multicrystalline silicon is currently lower, which is partly due to the material properties of mc-Si. The cutting efficiency in DWS is determined by the properties of the material that is cut [1], the coolant that is used and in particular the wire, which has a distinct diamond occupancy.

A standard diamond wire for wafering consists of the wire core, the bonding layer and the embedded diamonds. The wire core is made from steel, core diameters in silicon wafering range from 60 to 120 µm. It is surrounded by the bonding layer with a thickness of 2 to 9 µm that is usually composed of nickel or resin with the abrasive particles, i.e. the diamonds, embedded. The number, the size and the distribution of the embedded diamonds can vary. Thereby the number of diamonds per meter wire length is called occupancy and has a major influence on the cutting speed and the resulting wafer surface quality, which is a topic of current research [2].

The size of the diamonds has a direct influence on the cutting efficiency. With up to 20% of the total cost the diamond wire also represents one of the main cost driving factors in the wafering process. Therefore it is important to get the diamond wire wear lifespan as close to the maximum as possible. Thus, it is important to know how variations of the coolant or material that is cut influence the diamond wire wear.

Therefore, a benchmarking of new diamond wires needs to include a detailed analysis of the diamond particle occupancy. Usually this is done by manually counting the number of diamonds on a microscopic image (see FIG. 1). Since this method uses only a limited wire length, it implies a high inaccuracy when predicting the diamond occupancy of the whole diamond wire with a length of typically 60 km on one wire spool.

Furthermore, it is not possible to accurately measure the size of the diamonds, since these are embedded in nickel (or resin) layers of varying thickness for different wire manufacturers. The SEM/EDX-images of wire cross sections in FIG. 2 illustrate this actuality.

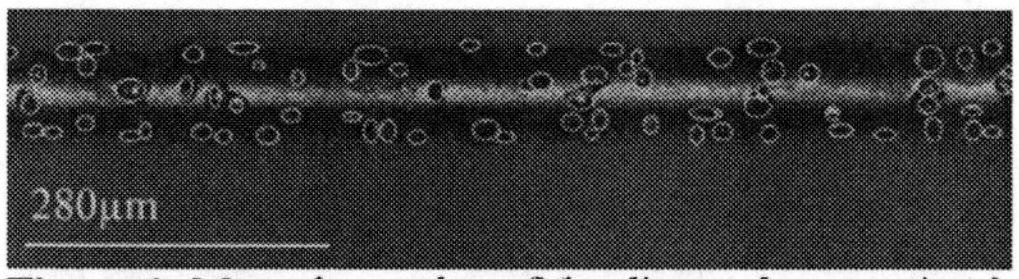

Figure 1: Manual counting of the diamond occupation in a microscope image on a wire length of 1 mm.

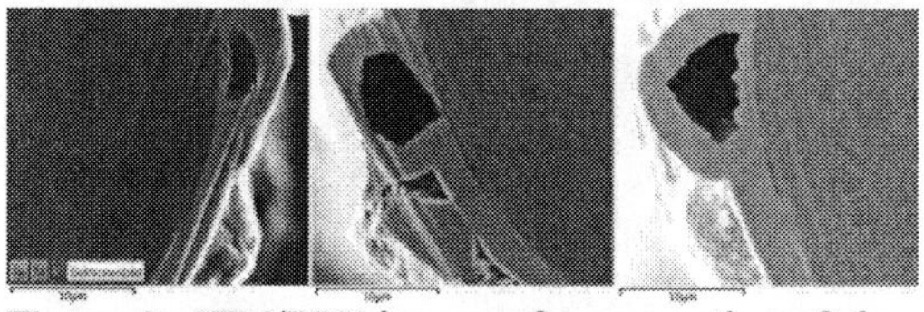

Figure 2: SEM/EDX-images of cross section of three different diamond wires. The varying thickness of the nickel layer (green) is demonstrated.

As can be seen in FIG. 3 the diamonds of wire #4 are located very close to each other appearing as conglomerates. However, the actual number of diamonds beneath the diamond wire coating can only be estimated and therefore an additional error in the inherently inaccurate counting method is bound to occur.

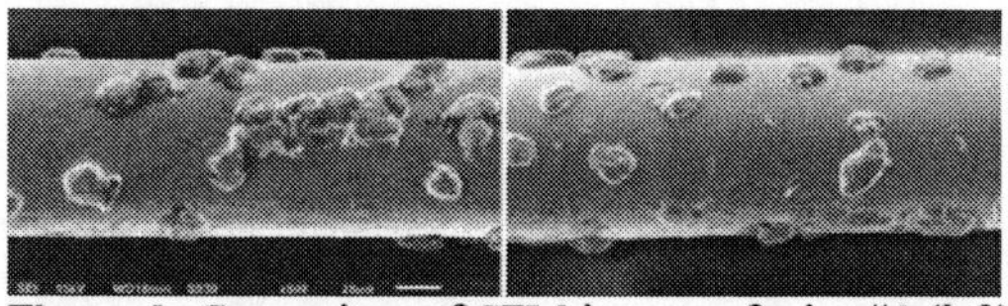

Figure 3: Comparison of SEM images of wire #4 (left) with apparent diamond conglomerates and wire #6 (right) with mostly separate diamonds.

In order to increase the accuracy of the diamond occupancy determination and thus enable for a prediction of the wire wear, a novel approach that offers a higher accuracy is shown in this work.

2 PROCEDURE

The new approach is to first remove the diamonds chemically from the diamond wire and to analyze them separately. First, the bonding nickel layer has to be dissolved from the steel core wire with a single etching step after an appropriate cleaning. It is important to dissolve the nickel layer without affecting the steel core. A mixture of HNO_3 (65%) and H_2O_2 (30%) has been proven to be successful. Due to the H_2O_2 the reaction rate is highly increased compared to when only using HNO_3.

FIG. 4 shows SEM and EDX images of an intact diamond wire before the etching, a diamond wire during the etching process, and a blank and intact diamond wire core after etching. After the etching process the resulting blank and intact core wire is extracted and discarded. By filtration of the thinned etching solution through a polycarbonate membrane with a low pore width (0.2 µm) the diamonds are caught on the filter membrane. The mass of the diamonds per meter of wire is then determined gravimetrically. If required the geometry of the diamonds on the membrane can be determined optically, e.g. with a laser confocal microscope.

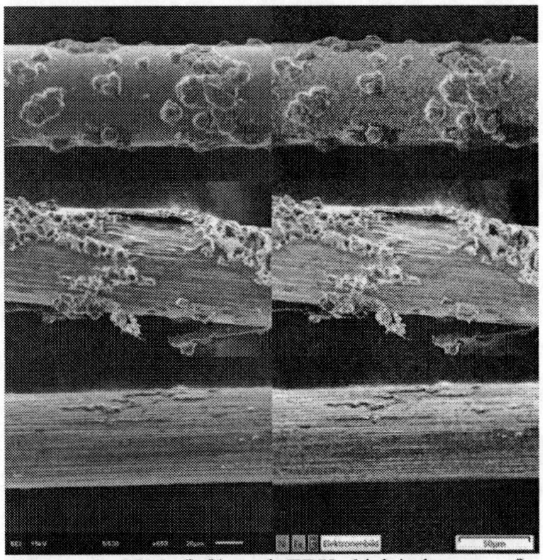

Figure 4: SEM (left) and EDX (right) images of a diamond wire before (top), during (middle) and after (bottom) the etching process.

Furthermore the grain size distribution of the diamonds on the wire can be determined by flushing the diamonds from the filter membrane. This is done by depositing the filter membrane in a centrifuge tube with ethanol and then applying a short ultrasonic treatment.

After the vaporization of the ethanol the diamonds are collected with a few drops of clean water and the grain size distribution is determined by laser diffractometry. By using the grain size distribution data and the gravimetrical data the total number of diamonds per meter wire length is calculated.

To be able to do the gravimetrical determination and the laser diffractometry even on used diamond wires from sawing processes or wires with a low diamond concentration it is recommended to use a diamond wire length of at least 3 - 10m.

3 RESULTS

3.1 Gravimetry and grain size distribution

Six diamond wires with a core diameter of 70 µm from different manufacturers were investigated. The etching solutions were filtered as described above and the masses of the diamonds were determined gravimetrically. Subsequently the grain size distribution was determined by laser diffractometry as shown for all six wire samples in FIG. 5.

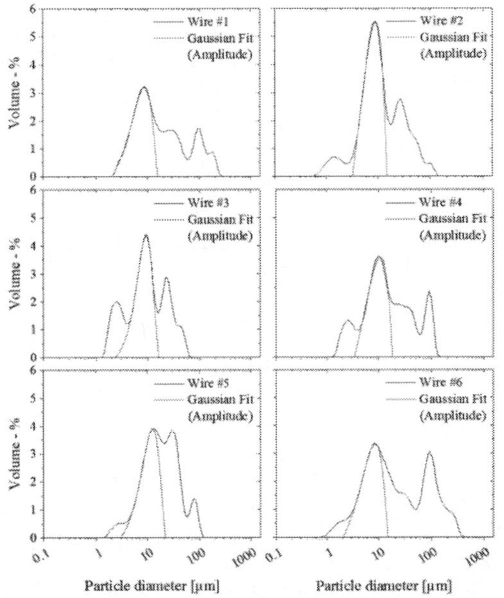

Figure 5: Grain size distribution of dissolute diamonds from six different wires. The peak related to the diamond particles is fitted with a Gaussian Amplitude function.

For comparison, the grain size distribution as measured with the same laser diffraction setup on a pure diamond powder is shown in FIG. 6.

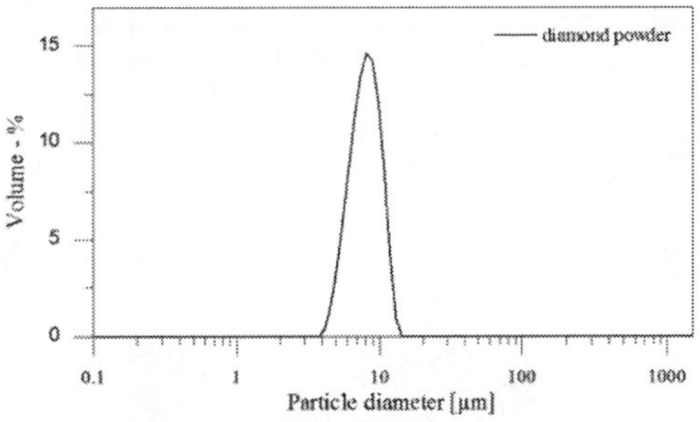

Figure 6: Grain size distribution of a pure diamond powder measured by laser diffractometry with a d50 value of 8.5 µm (specification: d50 = 8-9 µm).

The measured distribution shows only one sharp peak with a d50 value of 8.5 µm. This finding proves that the laser diffraction measurement is a reliable method to determine grain size distributions. In FIG. 5 it is obvious that further peaks next to the respective main peaks in the grain size distribution are present. Those peaks are supposed to be caused by metal residuals that are left in the etching solution and those should be dependent on the composition of the diamond wire coating. In addition the peaks at high particle diameters can be due to measurement artifacts, e.g. caused by bubbles in the measured solution.

The peak near the expected d50 value with respect to the manufacturers' specification was fitted with a Gaussian amplitude function. The d50 value of this peak is extracted and compared with the manufacturer specification in TAB. 1. The comparison shows that the measured and fitted d50 values are in good agreement with the given specifications of the different diamond wire manufacturers. The d50 value from the laser diffraction measurement is in all cases near the average of the distribution specified for the respective wire.

Table I: Measured d50 diamond diameter values compared to the specifications of the different wire manufacturers.

Diamond Wire Number	Manufacturer specifications [µm]	Measured d50 values [µm]
#1	6-12	8.9
#2	6-12	9.0
#3	6-12	9.4
#4	8-12	10.2
#5	8-16	13.3
#6	6-12	8.4

With the gravimetrically determined masses of the diamonds and the d50 values of the fits the number of diamonds per 10 m diamond wire length can be calculated assuming that the diamonds have a spherical shape. In FIG. 7 the so obtained values are compared to the values obtained by the original determination counting method. The depicted results of the different methods show a similar trend. However, the number of diamonds is systematically higher for the etching method. This should be due to the inherent inaccuracy of the manual counting method where only a side view of the diamond wire on a very short wire length is analyzed and then extrapolated. However with the above demonstrated new etching method the accuracy of the calculated number of diamonds per wire length is significantly increased by analyzing a total diamond wire length of 3-10 m.

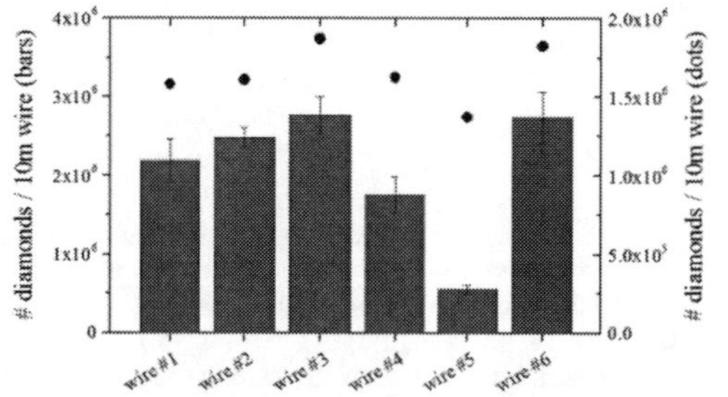

Figure 7: Number of diamonds on the different wires determined with the new etching method (bars) and the manual counting method (dots).

3.2 Analysis of diamond grain size distribution in microscope images

As a next step, the grain size distribution of diamonds can directly be measured in microscope images of the diamonds on the filter membrane. An exemplary image is shown in FIG. 8 for wire #1. The manual analysis of the diamonds present in the microscope image yields the grain size distribution in the right graph of FIG. 8 with a d50 value of 12.1 µm. This value is higher as measured by the laser diffraction method which may be due to the low number of diamonds that were counted. In the future a higher number of diamonds will be incorporated in the analysis by an automated counting of the diamonds in a set of microscope images.

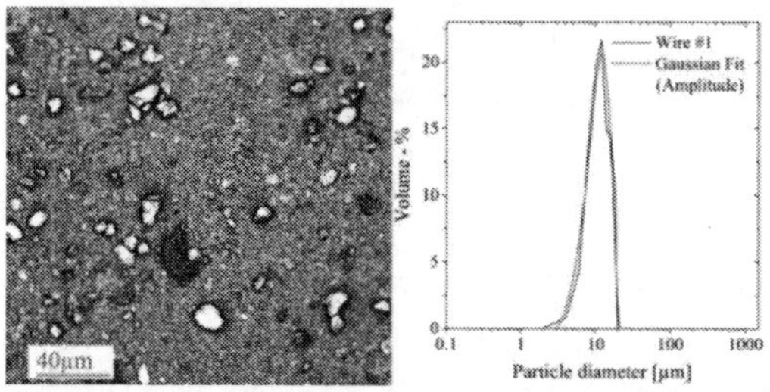

Figure 8: Laser confocal microscope image of diamond particles on a filter membrane extracted by dedicated etching of diamond wire #1 (left), the grain size distribution (right) shows a d50 of 12.1 µm.

3.3 Shape of diamond grains

As additional information the diamond laden filter membranes can be analyzed by laser confocal microscopy before the rinsing step with ethanol. Thus the shape of single diamond grains can be evaluated (see FIG. 9) and it could e.g. be analyzed if the shape of the diamond grains has an influence on the cutting efficiency and the resulting geometry and quality parameters of the wafer.

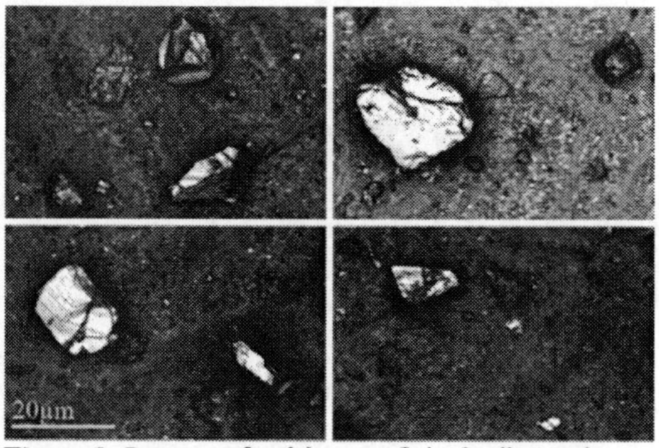

Figure 9: Laser confocal image of single diamonds on a polycarbonate membrane.

4 CONCLUSION

In this contribution, a new method was presented to extract diamonds from nickel coated steel wires used in multi wire sawing of crystalline silicon. The method was applied to state-of-the-art diamond wires with core diameters of 70 µm and diamond grit sizes of 6-12 µm up to 8-16 µm. The d50 values determined by laser diffraction are in good agreement with the mean values of the diamond grit sizes according to the manufacturers' specifications. It was also found that the gravimetrically determined mass of diamonds on the wire shows the same trend as the number of diamonds counted manually on microscope images but offers higher accuracy. A still existing difference of the methods will be investigated further by developing an automated counting of the extracted diamonds on the filter membrane and taking into account the shape of the diamonds.

REFERENCES

[1] T. Kaden et al., "The influence of material properties on the wire sawing process of multicrystalline silicon", in Proc. 33rd Eur. Photovolt. Sol. Energy Conf., 2017.
[2] R. Vedantham, Design and Development of Super Fine Engineered Diamond Wire, PV Days Halle (2016)

ACKNOWLEDGEMENT

This work was funded by the German Ministry for Economic Affairs and Energy (BMWi) with in the project "DIANA" (Contract No. 0324087B), which is gratefully acknowledged.

The Impact of Diamond Wire Quality on The Mechanical Strength of Thin Silicon Wafers for PV Cells

[1]Tetsuo Fukuda, [2]Halubai Sekhar, [3]Katsuto Tanahashi, [4]Katsuhiko Shirasawa,
[5]Hidetaka Takato, [6]Kazuya Ohkubo, and [7]Naoki Toge

[1-5] Renewable Energy Research Center, National Institute of Advanced Industrial Science and Technology (AIST),
2-2-9 Machiike-dai, Koriyama, Fukushima 963-0215, Japan
[1]Phone/Fax/E-mail: +81-29-849-1547/+81-24-963-0824/tetsuo-fukuda@aist.go.jp

[6,7] Noritake Company Limited Co.
210 Takeno, Tanushimaru-machi, Kurume, Fukuoka 839-1215, Japan

ABSTRACT: We have investigated the mechanical strength of monocrystalline silicon wafers to improve the yield of photovoltaic (PV) cell fabrication process. In slicing monocrystalline silicon bricks, diamond abrasives fixed on a wire surface scratch front and backside surfaces of wafers more seriously than wafer edges. Therefore, damages and defects not at edges but on both surfaces cause a serious concern for lowering the mechanical strength. We prepared 200 and 120 μm thick wafers by cutting silicon bricks with a multi-wire saw, in which two kinds of diamond wires, the one with the abrasive-size ranging from 8 to 16 μm in diameter and the other from 6 to 12 μm were equipped. On the wire surface, the dispersion of 6 to 12 μm abrasives was improved to be larger compared with that of 8 to 16 μm abrasives. We report that the improved wire is essential to have wafers with higher mechanical strength and the reason why the wire improvement results in the strength increase.
Keywords: Silicon, Silicon Solar Cell, Crystalline

1 INTRODUCTION

In the production of monocrystalline silicon photovoltaic (PV) cells, the important cost-cutting is achieved by thinning wafers[1, 2]. However, thin wafers are very fragile and often break in handling processes. This is due to damages caused in the process of cutting silicon bricks.

The damages are divided into two categories such as those at wafer edges and those of front and backside surfaces of the wafer. It has been reported that in multi-crystalline silicon wafers, the impact of edge damages on the mechanical strength is far more serious than that of front and backside damages[3]. In slicing a monocrystalline silicon brick, diamond abrasives fixed on a wire scratch not only wafer edges but also both front and backside surfaces. Therefore, damages and defects on both surfaces cause a serious concern for lowering the mechanical strength[4].

We investigate the impact of wire quality (abrasives' size and their dispersion) on the mechanical strength of as-sawn wafers and using Raman spectroscopy, we study cutting damages of wafer surfaces. We report the dependences of the fracture stress on the surface damages.

2 EXPERIMENTAL PROCEDURE

We prepared pseudo square silicon wafers of 156 mm in size and of 200 and 120 μm in thickness by cutting monocrystalline silicon bricks with a diamond multi-wire saw. We employed two kinds of diamond wires on which the sizes of fixed diamond abrasives were in the ranges from 6 - 12 and 8 – 16 μm, respectively. The diameters of the core steel wires were the same (100 μm). From now on, we describe the wire specifications as 100d-M6/12 and 100d-M8/16. The cutting conditions such as wire movement velocity, brick feed velocity and so on were the same and the number of wafers from one brick was on a preproduction scale (600 - 650 wafers for 200 μm thickness and 800 - 850 wafers for 120 μm thickness). After cutting, sample wafers were cleaned in an ultrasonic bath filled with hot water.

Using a test machine (SHIMADZU, AUTGRAPH AG-X plus), we measured fracture strength of pseudo square sample wafers in the way shown in Figure 1. All sample wafers were bent along the direction of saw marks with the cross-head velocity of 3 mm/min. Using the test machine, we calculated the facture stress σ_{frac} in accordance with the equation[5]

$$\sigma_{frac} = (3P_{frac}L) / (2bh^2),$$

where P_{frac} is the fracture load, L the distance between two supporting bars (80 mm), b the width of a sample wafer (156 mm), and h the thickness of a sample wafer (200 and 120 μm), respectively.

Using Raman microscopy (RENISHAW inVia Raman Microscope), we measured surface damages in the center of each wafer. In our Raman measurement, the laser wavelength was 532 nm whose penetration depth was nearly 880 nm and deconvolution technique was employed to quantify the fractions of mono- (Raman shift peak centered at 520 cm^{-1}), multi- (at 510 cm^{-1}), amorphous (at 480 cm^{-1})[6], and other defected silicon, as shown in Figure 2.

3 RESULTS AND DISCUSSIONS

Figure 3 shows various distributions of fracture stress of wafer thickness for 200 μm (a) and for 120 μm (b), respectively. These two kinds of wafers were cut with the same diamond wire of 100d-M8/16. For thick wafers, almost no difference in the distribution of fracture stress was found between fresh and worn-out side wafers. On the contrary, for thin wafers, the distribution in worn-out side wafers was narrower than that in fresh side wafers, that is to say, the average fracture stress and its standard deviation of worn-out side wafers are better than those of fresh side wafers. This is exactly shown in Table 1. These results reveal that diamond wires should be improved for cutting thinner wafers.

Figure 4 shows fracture stress of thin wafers cut with diamond wires of 100d-M8/16 (a) and 100d-M6/12 (b). Surprisingly, the distributions of the fracture stress of both fresh and worn-out wire side wafers were dramatically improved. Since monocrystalline structure of the wafer surface was more seriously damaged by back and forth motion of a wire with larger diamond abrasives, it is obvious that making diamond abrasives smaller is an important method.

It has been pointed out that on the surface of a wire, a few abrasives sometimes agglomerate into larger one. Since this means that a large abrasive particle is fixed on the wire, the suppression of the agglomeration is strictly required. We developed our smaller-abrasive wire (100d-M6/12) to have higher dispersion than larger-abrasive wire (100d-M8/16). We believe that the improved strength shown in Figure 4 (b) was due to both effects.

Figure 5 shows the dependence of the fracture stress on crystallinity factor[6] Φ_c defined by

$$\Phi_c = (I_{520} + I_{510}) / (I_{520} + I_{510} + I_{480}),$$

where I_{520} , I_{510}, and I_{480} are intensities of mono-, multi-, and amorphous silicon Raman shift peaks. Since the factor indicates the crystalline volume fraction, Figure 5 shows that wafer strength becomes lower as the monocrystalline structure is destroyed. Figures 4 and 5 show that the cutting by the improved wire (100d-M6/12) results in less defective surfaces of wafers.

4 CONCLUSIONS

The slicing by a wire with larger size abrasives results in lower strength wafers, particularly, in fresh wire side region. We conclude that the slicing technique with smaller fixed abrasives and their improved dispersion has the high potential to achieve wafers with higher strength. This is due to less defective surfaces caused by the improved-wire cutting. Raman measurement revealed the dependency of the wafer strength on the defective surfaces.

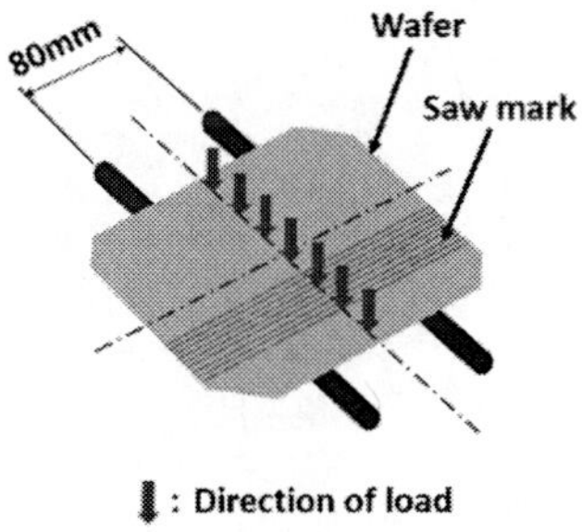

Figure 1. Schematic drawing of our bending test.

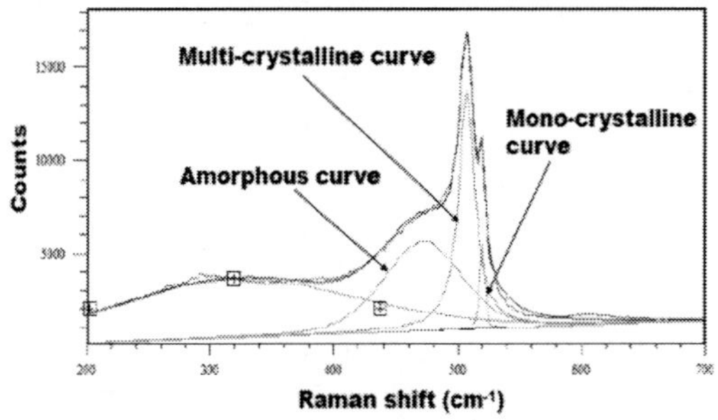

Figure 2. Deconvolution into sevaral Raman shift peaks.

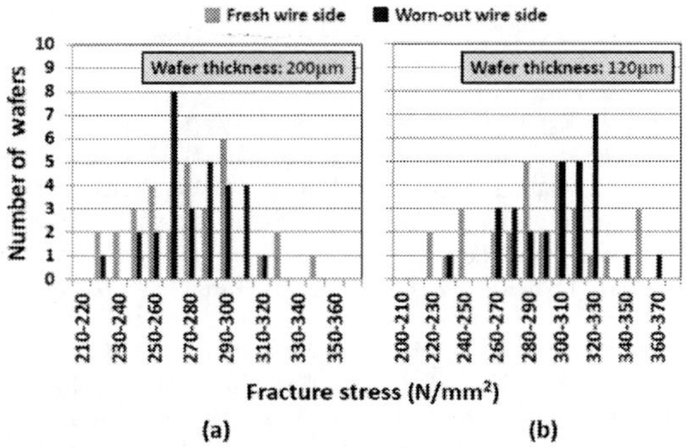

Figure 3. Fracture stress distributions of 200 μm (a) and 120 μm (b) wafers.

Table 1. Fracture stress in various conditions expressed in average ± standard deviation.

Wafer thickness (μm)	Fracture stress (N/mm²)	
	Fresh wire side	Worn-out wire side
200	277±29	267±22
120	269±36	287±20

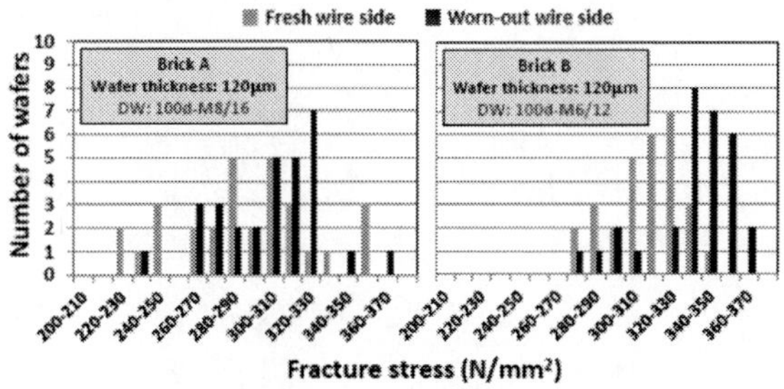

Figure 4. Fracture stress of thin wafers cut by larger abrasives (8 to 16 μm) (a) and by smaller abrasives (6 to 12 μm) (b). In smaller-abrasives wire, also their dispersion was improved to be higher.

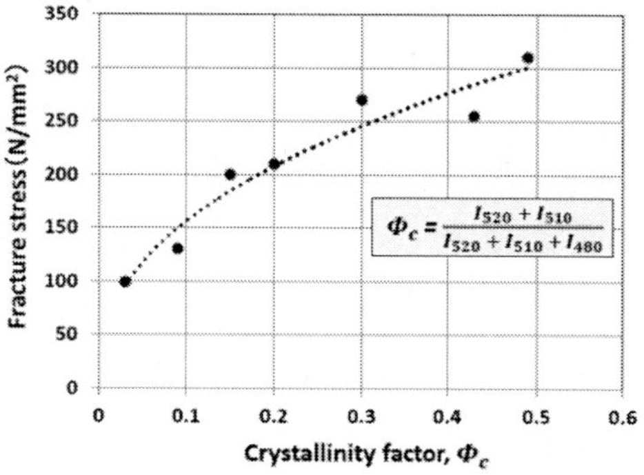

Figure 5. The dependence of fracture stress on crystallinity factor.

References

[1] T. Kinoshita, D. Fujishima, A. Yano, A. Ogane, S. Tohoda, K. Matsuyama, Y. Nakamura, N. Tokuoka, H. Kanno, H. Sakata, M. Taguchi and E. Maruyama, 26th European Photovoltaic Solar Energy Conference and Exhibition, p. 871.

[2] B. Terheiden, T. Ballmann, R. Horbelt, Y. Schiele, S. Seren, J. Ebser, G. Hahn, V. Mertens, M. B. Koentopp, M. Scherff, J. W. Muller, Z. C. Holman, A Descoeudres, S. D. Wolf, S. M. Nicolas, J. Geissbuehler, C. Ballif, B. Weber, P. S.-Cast, M. Rauer, C. Schmiga, S. W. Glunz, D. J. Morrison, S. Devenport, D. Antonelli, C. Busto, F. Grasso, F. Ferrazz, E. Tonelli, and W. Oswald, Phys. Status Solidi A 212, No. 1, p. 13 (2015).

[3] A. Grun, A. Lawerenz, R. Porytskyy and O. Anspach, 26th European PV Solar Energy Conference and Exhibition, p. 961 (2011).

[4] M. Shi, K. Youssef and G. Rozgonyi, 22nd Workshop on Crystal. Si Solar Cells & Modules, p. 161 (2012).

[5] SEMI Standard G86-0303.

[6] C. Droz, E. V.-Sauvain, J. Bailat, L. Feitknecht, J. Meier, and A. Shah, Solar Energy Materials and Solar Cells **81**, p. 61 (2004).

THE INFLUENCE OF MATERIAL PROPERTIES ON THE WIRE SAWING PROCESS OF MULTICRYSTALLINE SILICON

Thomas Kaden*, Elena Ershova, Lydia Lottspeich, Marcel Fuchs
Fraunhofer Technology Center for Semiconductor Materials, Am St.-Niclas-Schacht 13, 09599 Freiberg, Germany
*Corresponding author: thomas.kaden@thm.fraunhofer.de

ABSTRACT: The influence of material properties on the cutting efficiency of multicrystalline silicon was investigated for bricks taken from different positions of industrial-scale G5 multicrystalline silicon ingots in a diamond wire sawing process. It is shown, that bricks from diverse ingot positions reveal significant differences regarding the forces in feed direction when exactly the same process conditions are applied. Higher forces in feed direction imply a stronger bowing of the wire web developing during the process and therefore indicate a lower cutting efficiency. The size of silicon nitride and silicon carbide precipitates in the material are identified as the main factor for the cutting efficiency. The findings show the potential to optimize industrial wafering processes regarding process time or wire consumption.
Keywords: Wafering, diamond wire, multicrystalline silicon, characterization

1 INTRODUCTION

For the production of solar cells from multicrystalline silicon (mc-Si) the so-called High Performance (HP) mc-Si is the state-of-the-art base material. Using n-type HP mc-Si, record conversion efficiencies were recently demonstrated [1]. Even though this material shows a superior electrical quality compared to standard mc-Si, there are still crystal defects and impurities present, which might influence the cutting efficiency in multi wire sawing. Especially for the technology transition in the field of wire sawing of crystalline silicon from slurry to diamond wire sawing (DWS), the material properties of mc-Si are a critical factor. At the time, DWS processes of mc-Si require about 40% more process time compared to Cz-Si.

While many phenomena in the DWS abrasion process are already understood [2-4], the material properties of mc-Si with respect to the cutting efficiency were not investigated in detail yet. In this contribution major material influences in bricks of HP mc-Si were investigated and compared with Czochralski silicon.

2 PRE-CHARACTERIZATION OF BRICKS

2.1 Brick position

The investigations were performed on bricks with a height of 295 mm and a base area of 156 mm x 156 mm taken from different positions of a G5 HP mc-Si ingot. The bricks are named according to Fig. 1. A corner brick is defined by two faces having contact with the crystallization crucible, an edge brick has one such face and a center brick has no contact with the crucible. There are nine such center bricks in a G5 mc-Si ingot. A monocrystalline Czochralski ingot (Cz-Si) was used as reference. All bricks were boron doped.

2.2 Identification and electrical characterization

Since the investigated bricks originate from the same ingot, the doping concentration should be equal in all bricks. This was confirmed by four-point probe measurements of the resistivity. As shown in Fig. 2, the resistivity is in the range of 4 – 10 Ω cm corresponding to a boron doping concentration of $(1.4 - 3.5)\cdot10^{15}$ cm^{-3}. The resistivity of the reference Cz-Si brick is 1.3 – 1.7 Ω cm ($8.6\cdot10^{15} - 1.1\cdot10^{16}$ cm^{-3}).

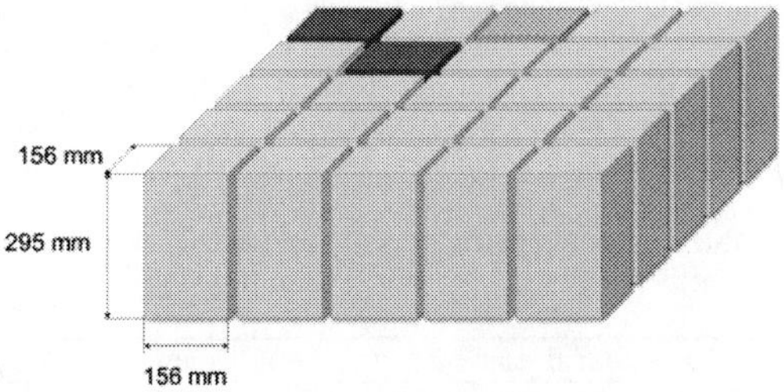

Figure 1: Scheme of brick positions in G5 mc-Si ingots (center – grey, edge – orange, corner – red).

The position of the bricks in the ingot was unambiguously determined by minority carrier lifetime measurements on bottom and top slabs of the bricks. The topographic measurements clearly show areas of reduced lifetime near the crucible (the so-called "red zone") which indicates the position in the ingot.

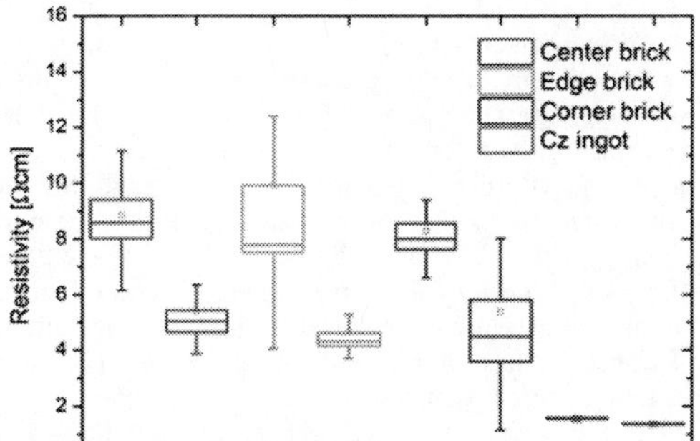

Figure 2: Resistivity of the investigated bricks, measured at the bottom and top, respectively.

2.3 Brick inspection by IR transmission

The HP mc-Si bricks were inspected prior to diamond wire sawing by IR transmission. The used method offers a lateral solution of 200 µm, i.e. inclusion with dimension larger than 200 µm can be found. As it can be seen from Fig. 3, there are no inclusions present in the three bricks, which allows for the application of an industrial wafering process.

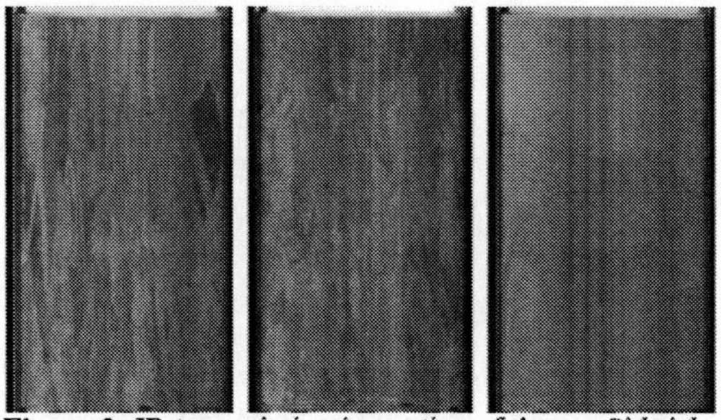

Figure 3: IR transmission inspection of the mc-Si bricks, corner (left), edge (middle) and center (right) position.

3 DIAMOND WIRE SAWING

3.1 Process parameters

The DWS processes were conducted on a Meyer Burger DS264 industrial wire saw in oscillating mode. In order to compare the cutting efficiency for the bricks from different ingot positions, the same process parameters were applied for cutting bricks of 295 mm, respectively. The parameters are summarized in Tab. 1.

Table 1: Process parameters of the diamond wire sawing experiments.

Process parameter	Value
Wire core diameter	120 µm
Diamond grit size	8-16 µm
Pitch	330 µm
Feed rate	0.55 mm/min
Wire speed	18 m/s
Wire tension	28 N
Wire consumption	3 m/wafer
Cooling medium	water
Additive concentration	2%

3.2 Cutting efficiency

During the sawing process, the forces in three directions (parallel and perpendicular to the wire direction, in feed direction) are permanently recorded. Since equal process parameters were used for all investigated bricks, the forces measured during the wire sawing process can be used to evaluate the cutting efficiency. Especially the force in feed direction F_z is an appropriate measure for the evaluation. When the cutting efficiency is low, a higher bowing of the wire web beneath the brick occurs, which results in a higher F_z.

In Fig. 4, the F_z values recorded over the whole process time are displayed for all bricks. It is found that the center brick reveals the highest F_z during cutting, while F_z for the edge and corner bricks is significantly lower. The highest cutting efficiency (lowest F_z) is found for the Cz-Si reference, as it would be expected. However, F_z for the edge and corner bricks is only 11-18 % higher compared to the Cz-Si reference. Thus, DWS processes using almost the same feed rate as for Cz-Si would be possible for these bricks. It should be emphasized that two-thirds of a G5-ingot are edge and corner bricks.

3.3 Wafer geometry and quality

The DWS processes are further evaluated by the quality of the produced wafers. In Tab. 2, the average wafer thickness and the total thickness variation (TTV) of the wafers are shown.

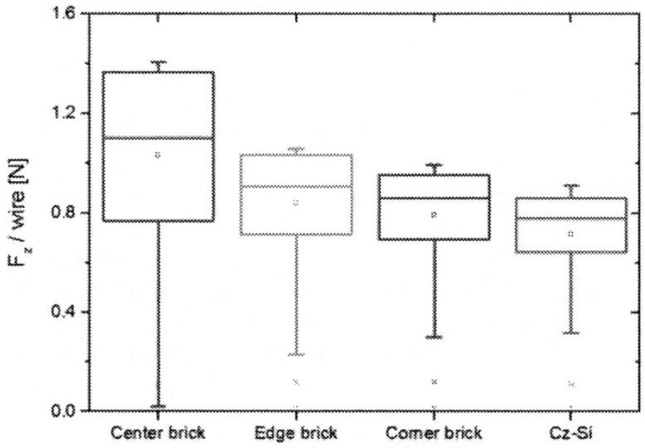

Figure 4: Force in feed direction (F_z) recorded during the entire DWS processes for the different bricks.

The lowest wafer thickness was measured for the reference Cz-Si, while the wafers from the mc-Si center brick yield the highest thickness. The thickness of wafers from the edge and corner bricks is in between the values of center brick and Cz-Si. This is in good agreement with the measured F_z values. The higher F_z is throughout the process, the higher is the wire wear, i.e. the effective wire diameter decreases by an enhanced loss of diamond volume. This leads to a reduction of the sawing channel width resulting in thicker wafers.

The TTV shows a similar trend, the higher the mean value of F_z, the higher is the TTV. It should be mentioned that the TTV for the Cz-Si reference is increased in spite of the given trend. This can be explained by the pseudo-square geometry of the Cz-SI ingot, for which the sawing process was not optimized.

Table 2: Mean values of F_z during the DWS process, wafer thickness and total thickness variation (TTV).

Brick position	Mean F_z/wire [N]	Mean wafer thickness [µm]	TTV [µm]
Center	1.03	192.2	17.7
Edge	0.84	190.1	15.2
Corner	0.79	188.9	14.8
Cz-Si (ref.)	0.71	188.6	16.8

3.4 Mechanical stability of the wafers

It was tested if the differences in the cutting efficiency are also reflected in the mechanical stability of the wafers. The four-line bending test was used to test as-sawn and damage etched wafers. There was no difference detectable for both wafer states.

4 MATERIAL PROPERTIES AND RELATION TO THE CUTTING EFFICIENCY

In order to identify the major material parameter that dominates the cutting efficiency, different material properties were investigated and are discussed in the following.

4.1 Carbon and oxygen concentration

The concentration of dissolved substitutional carbon (C_s) and interstitial oxygen (O_i) was determined by Fourier Transform Infrared Spectroscopy (FTIR) in the bottom and top slabs of the mc-Si bricks and in the top and tail slaps of the Cz-Si brick, respectively. As depicted in Fig. 5, there are no significant differences in the O_i-concentration of the mc-Si bricks. The same holds for the Cs-concentration, see Fig. 6.

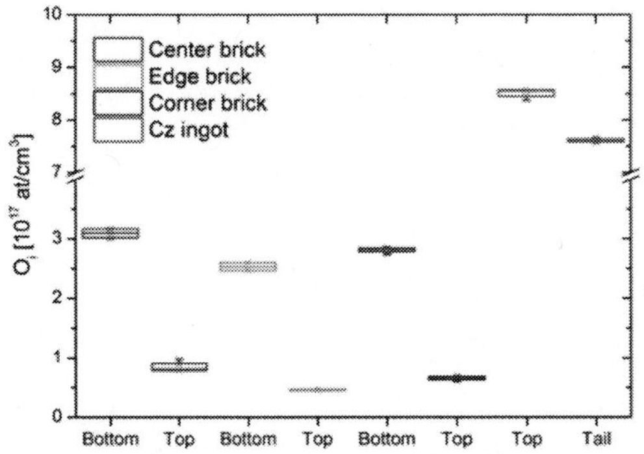

Figure 5: Concentration of interstitial oxygen (O_i) in the different HP mc-Si bricks and the Cz-Si reference.

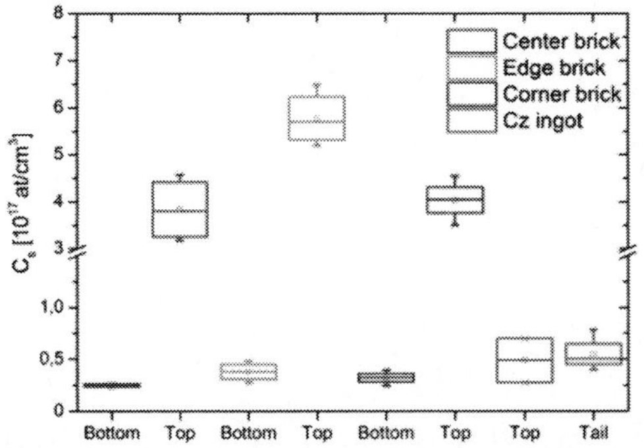

Figure 6: Concentration of substitutional carbon (C_s) in the different HP mc-Si bricks and the Cz-Si reference.

Thus it can be concluded that the cutting efficiency of mc-Si does not depend on the O_i- or C_s-concentration in the concentrations commonly present in the material.

4.2 Grain structure

The grain structure was characterized using X-ray Diffraction (XRD) with respect to the number of grains on a wafer, the occurring crystal orientations and the type of grain boundaries. Thy typical characteristic of HP mc-Si is found, in the bottom of the bricks there is a high number of grains with a small area. The grains are growing in area towards the top of the ingot, the number of grains decreases consequently.

The smallest grains are found in the center brick, where the most pronounced growth in grain size is observed. Since the mean value of the grain size is almost equal for each brick, no influence of the crystal structure on the sawing performance is expected within a single ingot.

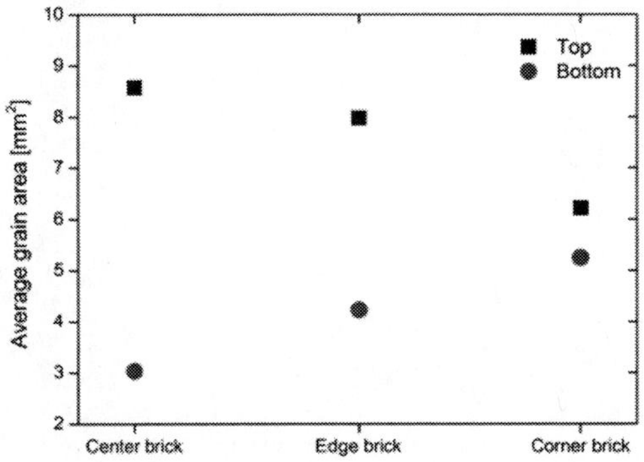

Figure 7: Mean grain area measured on the bottom and top slab of each ingot.

4.3 Precipitates

Precipitates were detected by IR transmission microscopy on the bottom and top slabs of the bricks. In Fig. 8, examples of silicon nitride precipitates (Si_3N_4) occurring as needles are shown. These precipitates are randomly oriented in the silicon crystal as Fig. 8 illustrates. Thus one can display them along the needle or as cross section of the needle. Furthermore, silicon carbide precipitates were found in less extent.

In order to evaluate the influence of the precipitate size on the cutting efficiency, the number of precipitates was automatically counted on slabs of each brick in an area of 10 x 10 cm²; the area of each precipitate was measured. A significant difference of the mean precipitate area between the center and the corner and edge bricks was found.

The values of the mean force F_z in the sawing process and the related mean area of precipitates are summarized in Tab. 3. Although there is a clear difference between the center brick and the bricks near the crucible wall, the correlation is still poor and needs to be improved with a higher number of experiments. The convolution of the precipitate density and the area could also lead to a better correlation and will be conducted in future investigations.

However, it is reasonable to consider the existent Si_3N_4- and SiC-precipitates as the main factor reducing the cutting efficiency in mc-Si since both materials have a higher hardness and fracture toughness compared to silicon, respectively.

It should be mentioned that the situation might be different in the case of other crystal producers. If the precipitates in the edge regions of the ingot would be larger, depending on the melt convection during the crystallization process, the cutting efficiency of edge and corner bricks will be lower compared to center bricks.

Table 3: Mean values of F_z during the DWS process, wafer thickness and total thickness variation (TTV).

Brick position	Mean F_z/wire [N]	Mean area of precipitates [µm²]
Center	1.03	95.7
Edge	0.84	44.3
Corner	0.79	53.2

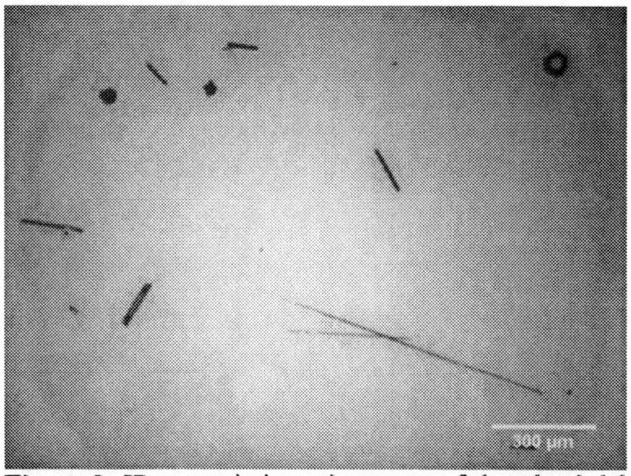

Figure 8: IR transmission microscopy of the edge brick showing silicon nitride needles in longitudinal sense and as cross section.

5 CONCLUSION

The inspection and characterization of multicrystalline silicon is important for the optimization of the diamond wire sawing (DWS) process. It turned out that the standard IR transmission inspection with 200 µm spatial resolution is not sufficient to detect all impurities, which have a relevant influence on the cutting efficiency of the DWS process.

By comparing bricks from different ingot positions, significant differences in the cutting efficiency were found, as demonstrated by the force in feed direction recorded over the entire process time. For a specific crystal manufacturer within this work, the bricks from the edges and corners of an HPM-Si ingot can be easier cut compared to bricks from the ingot center. This was mainly attributed to the presence of silicon nitride precipitates of different size within these ingots as characterized by IR transmission microscopy.

The findings presented here offer the potential for process optimization with a cost reduction potential. Consequently, a presorting of bricks prior to the wafering could be established, which distinguishes center and edge/corner bricks, given that the precipitate distribution is known. After presorting, adapted processes regarding feed rate or wire consumption can be applied for the two groups of bricks.

The relevant precipitate distribution can be monitored with a high resolution IR transmission measurement of the bricks, offering a spatial resolution of 18 µm.

In general, a further reduction of the size of precipitates, especially of silicon nitride needles and SiC clusters is desired to further improve the cutting efficiency of multicrystalline silicon bricks in diamond wire sawing and thereby narrow the gap with respect to DWS processes on solar Cz-Si.

REFERENCES

[1] J. Beneck et al., "High-efficiency n-Type HP mc Silicon Solar Cells", IEEE Journal of Photovoltaics, Vol. 7, No. 5, pp. 1171-1175 (2017).
[2] H.J. Möller, "Basic Mechanism and Models of Multi-Wire Sawing", Advanced Engineering Materials, Vol. 6, No. 7, pp. 501-513 (2004).
[3] R. Buchwald et al., "Analysis of the sub-surface damage of mc- and Cz-Si wafers sawn with diamond-plated wire", Energy Procedia, Vol. 38, pp. 901-909 (2013).
[4] S. Würzner et al., "Characterization of the Diamond Wire Sawing Process for Monocrystalline Silicon by Raman Spectroscopy and SIREX Polarimetry", Energies, Vol. 10, 414 (2017).

ACKNOWLEDGEMENT

Part of this work was funded by the German Ministry for Economic Affairs and Energy (BMWi) within the project "DIANA" (Contract No. 0324087B), which is gratefully acknowledged.

SIMPLE MODEL FOR THE CALCULATION OF WIRE TENSION FORCES IN THE MULTI WIRE SAWING PROCESS

K. Sunder, R. Rataj and O. Anspach*
PV Crystalox Solar Group, Gustav-Tauschek-Str. 2, D-99099 Erfurt, Germany
*Phone: +49 361 60085301, Fax: +49 361 60085109, E-mail: oliver.anspach@pvcrystalox.com

ABSTRACT: The wire tension force is fundamental for the multi wire sawing process with both slurry and diamond wire. It impacts the force between cutting particles and work piece and therewith the wire bow, which again affects the wafer quality. The wire tension force can be altered by all parameters that affect the wire's length and diameter. These are for instance fluctuations of the wire guiding roller groove depth or radii, missing parallelism of the wire guiding rollers, a change of wire bow, plastic wire elongation and wire wear. The possibilities are manifold and tend to change during a cut with position and time. As a consequence an iterative calculation is introduced to follow the changes of wire tension force along the wire web and to enable therewith an understanding which of the mentioned factors is relevant, how they affect each other and how they develop with time.
Keywords: wire tension force, diamond wire process, slurry process, crystalline silicon wafer, multi-wire sawing

1 INTRODUCTION

For the production of silicon wafers via the multi wire sawing process a thin steel wire is wound several thousand times around wire guiding rollers. The guidance of the wire is ensured by grooves that are included in the wire guiding roller covering. The resulting wire loops form a web with a length of several kilometers. During the sawing process this wire web is moved with high speed while pressing silicon blocks into it. As a consequence each wire loop below a silicon block forms a bow until the vertical force suffices for a material removal that equals the feed speed. The material removal takes place via indentation of loose silicon carbide particles and via scratching with fixed diamond particles for the slurry and the diamond wire process, respectively. For both processes the wire bow is known to have a direct impact on wafer quality [1].

During its lifetime each segment of the wire passes the whole wire web and therewith all wire guiding roller grooves, which can slightly vary in distance and groove depth. Also the wire bows vary with time and wire web position. Furthermore for the slurry process the wire's diameter decreases during the cut due to wire wear, whereas for the diamond wire process the wire is wound several times on and off the wire spools without any change of core diameter. All these incidents have an impact on the wire tension force in the wire web, which is fundamental for the multi wire sawing process as it affects the wire bow [2] and therewith the wafer quality.

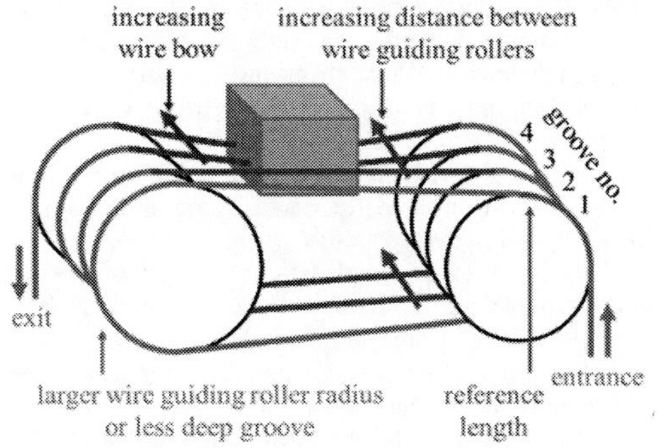

Figure 1: Factors that may increase the stretched wire length l_s along the wire web compared to a reference length given at the wire entrance side.

The aim of this work is to introduce a simple model that enables the calculation of wire tension forces as a function of all mentioned time- and wire web position-dependent factors.

2 THEORY

2.1 Factors that impact the wire tension force

To calculate the wire tension force F of any wire segment at any time and position of the sawing process one has to look deeper on the factors that define it. These are for a steel wire the elastic modulus E of the steel, the wire's cross-section area A and the stretched and relaxed wire lengths l_s and l_r, respectively [3]:

$$F = E \cdot A \cdot \frac{l_s - l_r}{l_r} \qquad (1)$$

If the wire could freely slip on the wire guiding rollers the calculation would be easy as F would be equal for every position and time. Due to the softness of the wire guiding roller covering and the high wire tension forces in the order of several Newton, slippage can only be expected around the vertexes of the wire guiding rollers. As a consequence every wire loop in the wire web could hold a different wire tension force than its neighbor. And at the wire guiding rollers even neighboring segments could hold different wire tension forces. Thus the wire tension force F^i of each wire segment i has to be determined separately by the factors given in Equation (1).

Here the elastic modulus E is approximately constant, while the wire's cross-section area A decreases with wire wear for the slurry process [4] and stays constant for the diamond wire process [5]. The stretched wire length l_s depends on different geometric factors that are illustrated schematically in Figure 1. If l_s equals a reference length (green) at the wire entrance side it will be stretched further and relaxed again as it is moved along the illustrated wire web.

To understand the following detailed explanations it is important to keep in mind that both wire guiding rollers and therewith all wire guiding roller grooves are moved simultaneously. This means if the reference length reaches the first left wire guiding roller groove, which has in Figure 1 an increased radius and therewith holds an increased wire length (orange), it is stretched. After leaving this wire guiding roller groove it relaxes again and stays constant until it leaves the second groove of the right wire guiding roller. As the wire guiding rollers in this example are not perfectly parallel, the distance of the wire guiding rollers increases with increasing groove number. This may have an impact on the stretching of the reference length when it leaves a wire guiding roller groove and enters the lower or upper wire web of groove number two, three and four (violet). In the third and fourth groove in Figure 1 a silicon block is cut. As a consequence a wire bow is formed as a function of the sawing profile [2]. The resulting increase of wire lengths in the upper wire web may cause a further stretching of the reference length as soon as it leaves the third and fourth groove of the right wire guiding roller (red).

The last factor in Equation (1) that has to be discussed is the relaxed wire length l_r. It resembles the wire length the stretched wire length l_s would decrease to if it was cut out of the wire web and reached a wire tension force of 0 N. l_r is known at the edges of the wire web, where the wire enters with a wire tension force set by machine regulation:

$$l_r^0 = \frac{E \cdot A^0 \cdot l_s^0}{F^0 + E \cdot A^0} \quad \text{at wire entrance side} \tag{2}$$

l_r of a wire segment stays constant as long as it is fixed to the wire guiding rollers. But as soon as it leaves this non-slippage area it will balance with the l_r of other free segments. As a consequence l_r is not defined by simple functions or geometric factors, but develops in dependence of the surrounding free wire segments. Thus l_r and therewith F for all wire web positions and times can only be calculated by iteration.

2.2 Method of iteration

The approach for the iteration to calculate the time- and position-dependent relaxed wire lengths l_r is to subdivide the total wire movement in partial movements. After each partial movement the accompanied changes of stretched wire lengths l_s are followed and the corresponding l_r are balanced with those neighboring l_r that are not fixed to the wire guiding rollers. These l_r are called "free" in the following.

Neighboring free l_r are defined in this work as l_r in the same groove of the upper or lower wire web plus the first l_r on the wire guiding roller in wire movement direction. Through this definition a possible slippage at the vertexes of the wire guiding rollers is considered. At the same time the calculation effort is reduced by not taking into account the first l_r on the wire guiding roller against wire movement direction. The error caused by this simplification should be small.

As described by means of the example in 2.1 changes of l_s only take place while entering or leaving the wire guiding rollers. Consequently there are three wire segments that are relevant for the balancing calculation: the wire segment that enters the wire web, the wire segment that leaves the wire web and the wire segment in between. The stretched length of the latter depends on the respective wire guiding roller distance and wire bow. The stretched lengths of the entering and the leaving wire segment are necessarily equal and are determined by a reference length.

This reference length is arbitrarily defined as a fraction of the circumference of the first wire guiding roller groove at the wire entrance side in whole numbers n. If the wire is moved only by reference lengths this

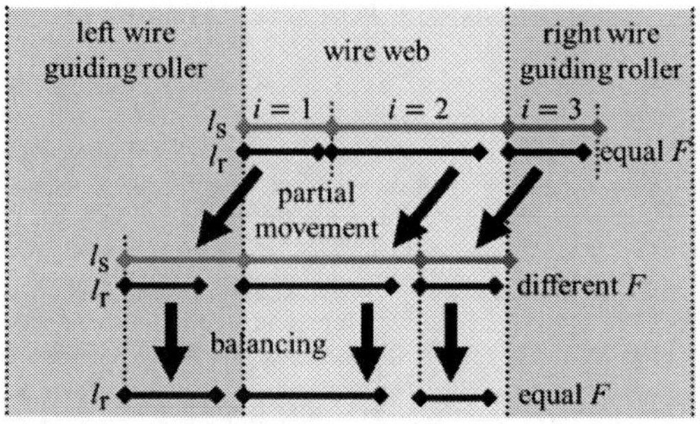

Figure 2: Illustration of the balancing of relaxed wire lengths l_r after partial wire movement (see Equation (3)) by the example of the upper free wire length in the first groove in Figure 1.

definition results in partial wire guiding roller movements by $1/n$ wire guiding roller rotations. n is aimed to be as large as necessary to obtain relevant results. Only for illustration purpose it is set quite small as for example in Figure 1 where $n = 4$.

In practice the balancing of relaxed wire lengths works as follows: after each partial wire guiding roller rotation the three relevant neighboring stretched wire lengths l_s per upper or lower wire web and the corresponding free relaxed wire lengths l_r are identified and new l_r are calculated as follows:

$$l_r^i = \frac{\sum_{k=1}^{3} l_r^k}{\sum_{k=1}^{3} l_s^k} \cdot l_s^i \tag{3}$$

This equation becomes clearer on basis of the example given in Figure 2. In the upper part of the figure the l_s and l_r of the upper first groove of Figure 1 are illustrated. The l_r are set to result in equal wire tension forces for all three wire segments (see Equation (1)). After forward movement by 1/4 wire guiding roller rotation the first l_s is stretched further due to the increased radius of the left wire guiding roller illustrated in Figure 1 (orange). The other l_s and all l_r are not affected by the movement as illustrated in the middle of Figure 2. As a consequence the first segment holds for the moment an increased wire tension force. To establish equal wire tension forces for all three segments the first l_r has to gain length on cost of the second and third l_r as shown in the lower part of Figure 2.

To clarify how different wire tension forces can evolve along the wire web another example is illustrated in Figure 3. The figure shows a wire saw with two wire guiding rollers and a small wire web of three wire loops. For simplicity the setting has no geometrical anomalies, preventing any further stretching of stretched wire lengths l_s along the wire web. For a better visualization the number of stretched wire lengths per circumference of the first right wire guiding groove is with $n = 8$ again quite small. Consequently partial wire guiding roller rotations by 1/8 are possible.

The focus of the explanations will be on the free upper wire length of the first loop. This comprises the first l_s at the left wire guiding roller and the l_s between the wire guiding rollers. These stretched lengths are given additionally at the right hand side of all illustrations in Figure 3 together with the corresponding relaxed wire lengths l_r. The colors of the wire lengths illustrate the present wire tension force that increases from 0 N (black) to maximum (red).

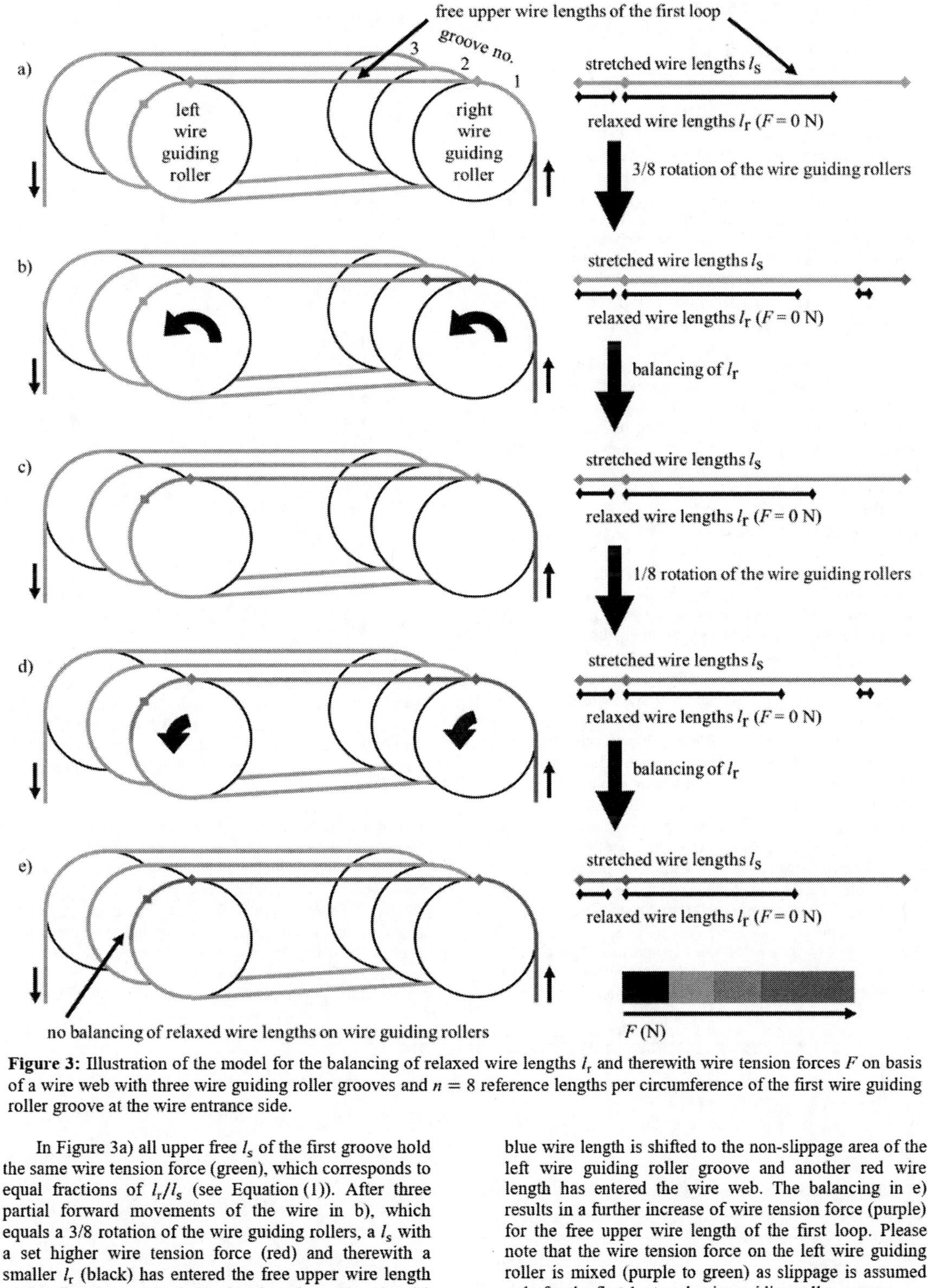

Figure 3: Illustration of the model for the balancing of relaxed wire lengths l_r and therewith wire tension forces F on basis of a wire web with three wire guiding roller grooves and $n = 8$ reference lengths per circumference of the first wire guiding roller groove at the wire entrance side.

In Figure 3a) all upper free l_s of the first groove hold the same wire tension force (green), which corresponds to equal fractions of l_r/l_s (see Equation (1)). After three partial forward movements of the wire in b), which equals a 3/8 rotation of the wire guiding rollers, a l_s with a set higher wire tension force (red) and therewith a smaller l_r (black) has entered the free upper wire length of the first loop. The balancing based on Equation (3) results in equal l_r/l_s for the first upper free wire length, which corresponds to a medium wire tension force (blue) as shown in c). In d) the wire is moved forward by a 1/8 rotation of the wire guiding rollers. As a consequence one blue wire length is shifted to the non-slippage area of the left wire guiding roller groove and another red wire length has entered the wire web. The balancing in e) results in a further increase of wire tension force (purple) for the free upper wire length of the first loop. Please note that the wire tension force on the left wire guiding roller is mixed (purple to green) as slippage is assumed only for the first l_s at each wire guiding roller groove.

3 CALCULATION RESULTS

3.1 Accuracy versus calculation time

The calculation model was realized using GNU Octave. As a first test a simple setup with a constant distance of the two wire guiding rollers of 600 mm, constant wire guiding roller radii of 175 mm, a wire guiding roller length of 1000 mm and constant groove depths of 250 µm was implemented. These values were taken from the machine layout of a Meyer Burger DS271 and were used for all calculations in this work if not noted elsewise. The starting wire web was set to a homogeneous wire tension force of 20 N and fresh wire was added unidirectional with an increased wire tension force of 22 N. The calculation was performed for various partial rotations $1/n$ of the wire guiding rollers with $n = 2$ to 180.

Figure 4a) gives the calculated wire tension forces as a function of wire guiding roller groove number for the upper wire web after wire movement by 4 km for seven different n. For clarity only wire tension forces around groove number 1744 (dotted line) are shown. This is the groove that the first wire segment that entered the wire web with the increased wire tension force has reached. The calculation results match the observation during daily operation of wire saws at PV Crystalox Solar Silicon that the wire tension force changes via a single refreshment of the wire web. Furthermore the resemblance of all curves given in Figure 4a) illustrates that the order of magnitude of partial wire guiding roller rotation $1/n$ has only a minor effect on the results.

Under the assumption that the calculation with the smallest partial rotation ($n = 180$) is most reliable, the accuracy $a(n)$ of the calculation was estimated as the maximal deviation from $F(180)$ as shown exemplarily for $n = 2$ in Figure 4a). The resulting accuracies are given in Figure 4b) together with the calculation time as a function of n. The figure reveals on the one hand that the calculation is even for $n = 2$ with 0.161 N quite accurate and that the accuracy improves with decreasing partial wire guiding roller rotation $1/n$. On the other hand the calculation time increases with a square dependence on n. As a compromise $n = 6$ was used for the following calculations. This corresponds to an accuracy of 0.062 N.

3.2 Change of wire tension force by machine regulation

Remarkable about the calculation results given in Figure 4a) is that the change of wire tension force is not

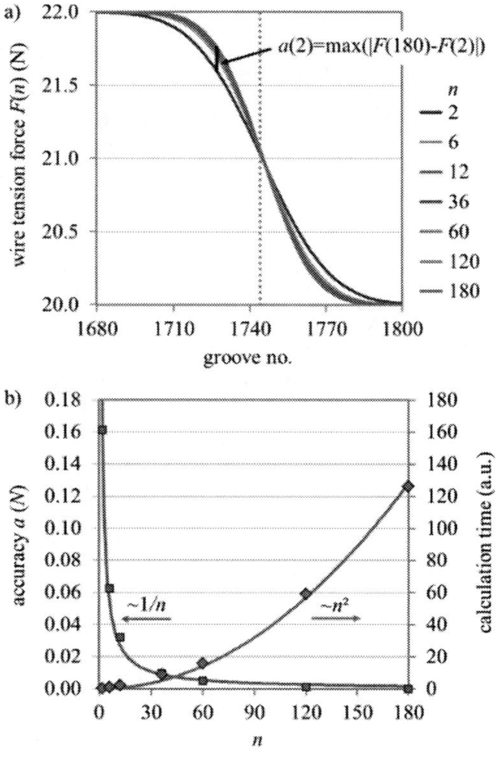

Figure 4: a) Calculated wire tension force F in the upper wire web after forward movement of 4 km of fresh wire with an increased wire tension force for different partial wire guiding roller rotations $1/n$. b) The accuracy a of the calculation improves proportional to the partial wire guiding roller rotation $1/n$, whereas the calculation time increases with a square dependence on n.

abrupt, but smooth. To estimate the significance of this smooth change of wire tension force for multi wire sawing Figure 5 illustrates the wire tension force as a function of wire guiding roller groove number after various wire movements. The dotted lines give again the groove number that the first wire segment that entered the wire web with the increased wire tension force has reached. The solid lines are the results for an unidirectional wire movement like in the slurry process.

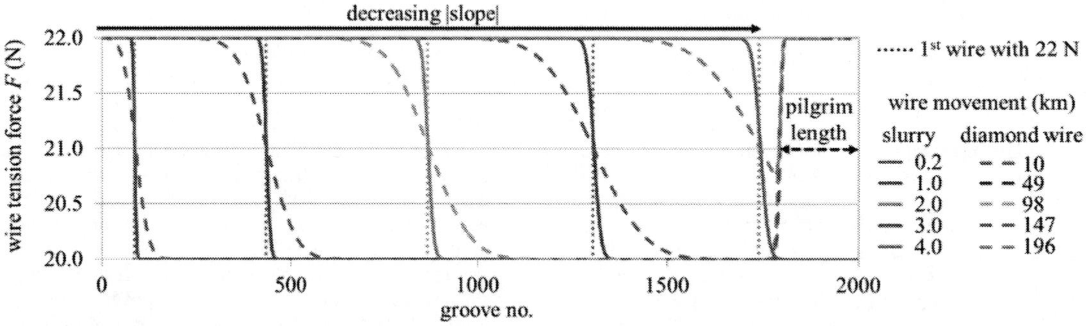

Figure 5: Calculated changes of wire tension force in the upper wire web as a function of wire guiding roller groove number after various wire movements when wire with an increased wire tension force of 22 N is fed to a wire web with a starting wire tension force of 20 N. The dotted lines mark the respective positions of the first wire segment that entered the wire web with the increased wire tension force. The solid and dashed lines were calculated for the unidirectional slurry process and the bidirectional diamond wire process, respectively.

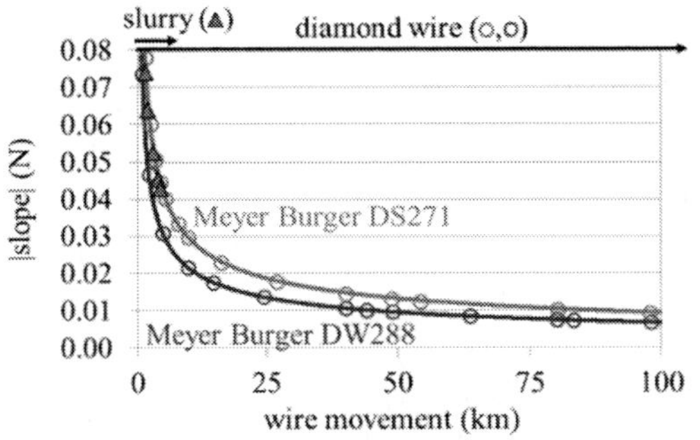

Figure 6: Absolut value of the slopes of the wire tension force curves in parts given in Figure 5 and Figure 7 as a function of wire movement. With decreasing |slope| the change of wire tension force along the wire web gets smoother. This effect is more relevant for the diamond wire process (open circles) and for special designed diamond wire saws like the Meyer Burger DS288 (red).

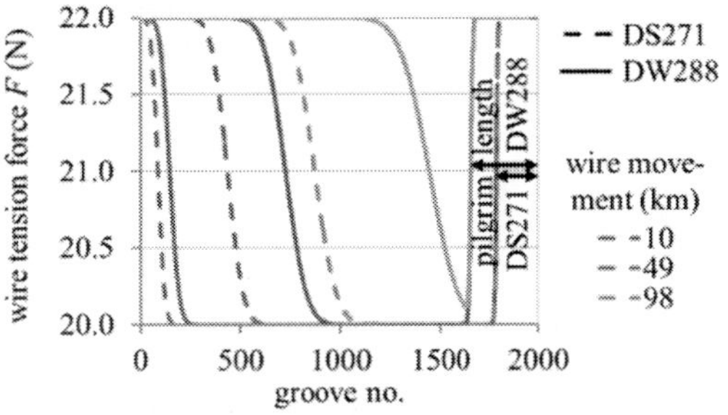

Figure 7: Calculated change of wire tension force in the upper wire web as a function of wire guiding roller groove number after various wire movements when wire with an increased wire tension force of 22 N is fed to a wire web with a starting wire tension force of 20 N. The dashed and solid lines are calculated for the bidirectional diamond wire process on a Meyer Burger DS271 and DW288, respectively.

Here only some kilometers of wire movement are required to refresh the whole wire web.

By contrast, in the diamond wire process, where the wire is moved back-and-forth with a small fresh wire supply per cycle, the wire stays in the wire web for several cuts. To reach the same wire web positions as in the slurry process the diamond wire has to be moved several tenth of kilometers. The results for a bidirectional diamond wire movement with pilgrim lengths of 500 m/480 m are given in Figure 5 by dashed lines.

Figure 5 makes evident that for both the slurry and the diamond wire process the smooth change of wire tension force with groove number becomes more pronounced the farer the wire is moved. This observation can be numbered by the absolute values of the slopes of the wire tension force curves. These |slopes| are given in Figure 6 as a function of wire movement by filled blue triangles and open blue circles for the slurry and the diamond wire process, respectively. The |slopes| for both processes decrease with the same reciprocal square root dependence on wire movement. This indicates that the smooth change of wire tension force along the wire web does not depend on the movement direction, but only on the movement length.

3.3 Machine geometry

Multi wire saws designed for the diamond wire process show principal layout differences compared to slurry multi wire saws. Generally the wire guiding roller radii and their distance are smaller to reduce the wire bow and enable higher feed speeds. Furthermore the wire guiding roller length is usually reduced, which results even for smaller pitches in a reduction of groove number.

To estimate the impact of diamond wire saw geometry on the distribution of wire tension force along the wire web the addition of fresh wire with an increased wire tension force of 22 N to a wire web with 20 N was calculated for the geometry of a Meyer Burger DW288 diamond wire saw [6]. The wire was moved bidirectional with the same pilgrim lengths as in 3.2 of 500 m/480 m. To facilitate the comparison with the slurry saw DS271 the same number over wire guiding roller grooves was implemented, even though the wire guiding roller lengths differ significantly.

Figure 7 illustrates the calculated wire tension force as a function of wire guiding roller groove number after different wire movements for a wire saw with the geometry of a Meyer Burger DS271 (dashed lines) and a Meyer Burger DW288 (solid lines), respectively. The increased wire tension force has proceeded farer into the wire web for the DW288. Furthermore the DW288 results in a smoother change of wire tension force. This becomes more evident in Figure 6, where the absolute values of the slopes of the wire tension force curves are given as a function of wire movement. The |slopes| for the DW288 (red cycles) are smaller than for the DS271 (blue cycles).

3.4 Wire guiding roller radii

The impact of wire guiding roller radii on wire tension force was calculated on basis of a wire saw with linearly increasing wire guiding roller radii. The starting wire tension force in the wire web was set to 20 N and the whole wire web was refreshed unidirectional with the same wire tension force.

Figure 9a) shows the set wire guiding roller radii as a function of wire guiding roller groove number (blue) together with the calculated wire tension force (red). The linear increase of wire guiding roller radii by 0.2 mm within the wire guiding roller length of 1000 mm results in a linear increase of the wire tension force by 2.2 N.

3.5 Wire guiding roller groove depth

To test the theoretical impact of wire guiding roller groove depth on wire tension force a strong steplike variation of the groove depth was implemented. The starting wire tension force in the wire web was set to 20 N and the whole wire web was refreshed unidirectional with the same wire tension force.

Figure 9b) gives the set groove depth as a function of groove number (blue) together with the resulting wire tension force (red). The steplike decrease (increase) of groove depth by 50 µm results in a steplike increase (decrease) of the wire tension force by 0.55 N.

3.6 Free wire lengths

The theoretical impact of free wire lengths on the wire tension force was tested by implementing a cut that started with a steplike increased wire bow of 5 mm between groove number 350 and 550. This resulted in a steplike in- and decrease of the wire tension force in the upper wire web by 0.39 N compared to the remaining

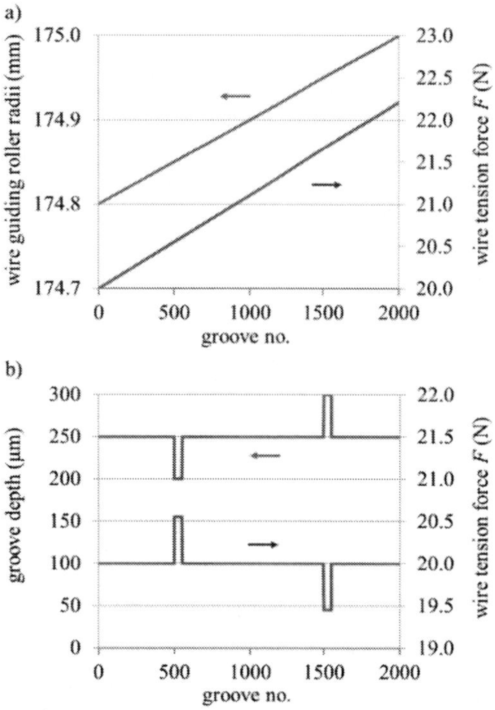

Figure 9: Impact of variations of a) the wire guiding roller radii and b) the wire guiding roller groove depth on the calculated wire tension force in the upper wire web as a function of wire guiding roller groove number after unidirectional refreshment of the whole wire web.

wire web that held 20 N. During two calculations the wire web was refreshed unidirectional as in the slurry process and bidirectional as in the diamond wire process, respectively.

Figure 8 illustrates the calculated wire tension forces in the upper wire web as a function of groove number after various wire movements for the slurry (solid lines) and the diamond wire process (dashed lines). The starting wire tension force distribution is given in black.

The figure reveals that for both the slurry and the diamond wire process the wire loops with the increased wire tension force leave the wire web at the right hand side. Consequently changes of wire tension force due to increased free wire lengths are equalized as soon as the wire web is refreshed. The differences for the slurry and the diamond wire process lay on the one hand in the required wire movement that comprises some kilometers and several tenth of kilometers for the slurry and the diamond wire process, respectively. On the other hand the phenomenon illustrated in Figure 5 results in a smoother change and a reduction of wire tension forces for the diamond wire process before the wire leaves the wire web.

4 DISCUSSION

4.1 Accuracy versus calculation time

The iteration was originally planned to work with as small partial rotations of the wire guiding rollers $1/n$ as possible to get the most precise results. Thanks to the variety of factors that impact the wire tension force and

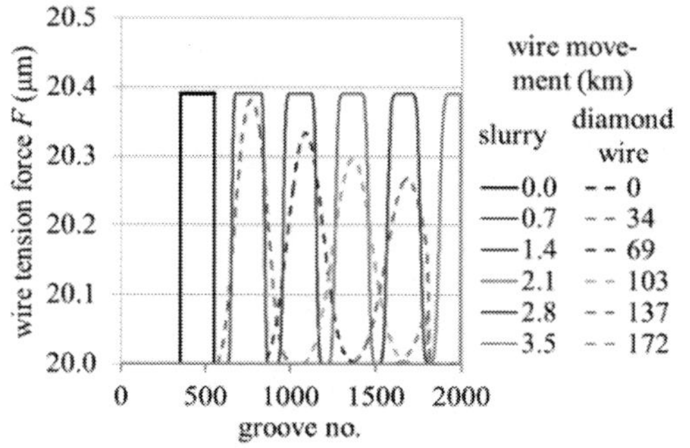

Figure 8: Calculated change of wire tension force in the upper wire web as a function of wire guiding roller groove number after various unidirectional (slurry) and bidirectional (diamond wire) wire movements when the cut starts with an increased wire tension force between groove number 350 and 550 (black) due to 5 mm wire bow.

that partly change with time it became soon obvious that a compromise between precision and calculation time has to be found. The high accuracies that can be reached even with comparable large partial rotations (see Figure 4) were surprising. They demonstrate that the extent of the non-slippage area does not matter. Instead it only seems important that the wire guiding rollers hold the wire tension force at all.

4.2 Plastic wire elongation

For both the slurry [4] and the diamond wire sawing process [7] it was concluded that the applied steel wire suffers a plastic elongation when it starts cutting the silicon material. This results in a decrease of wire tension force depending on wire radius. For instance a decrease due to plastic elongation by around 2 N was evident for the slurry process with a straight wire's diameter of 120 µm.

For the slurry process the wire's diameter and therewith the wire tension force decrease with increasing wire wear along the wire web. Nevertheless the impact of wire elongation and wire wear can be followed comparably easy as a steady state establishes within minutes. As discussed in literature [4] the plastic elongation results in long-term in a steep loss of wire tension force at the wire web entrance, whereas the loss of wire diameter accounts for a smooth decrease of wire tension force along the wire web.

By contrast for the diamond wire process the wire's steel core diameter and therewith the wire tension force are not affected by wire wear [5]. Things become complicated anyway as parts of the plastic elongated wire are wound back to the wire spool and enter the wire web with a freshly set wire tension force. This was simulated in 3.2 and 3.3 by adding fresh wire with an increased wire tension force of 22 N to a wire web with 20 N. The results are shown in Figure 5 for the bidirectional diamond wire process in comparison with the unidirectional slurry process. For the latter it is only a gedankenexperiment, meanwhile for the diamond wire process it actually has relevance.

Depending on wire saw geometry and cutting profile it can take several cuts until the whole wire web holds wire whose tension force was refreshed by machine regulation. Figure 7 illustrates the results for two different wire saw layouts. All calculations are summarized in Figure 6 where the absolute values of the slopes of the wire tension force curves are shown as a function of wire movement. The smaller the |slope| the smoother the change of wire tension force with groove number.

Figure 7 makes clear that a special designed diamond wire saw like the Meyer Burger DW288 results, due to its smaller wire loops, in a faster refreshment of the wire web. Besides are the shorter wire guiding roller lengths of a diamond wire saw an additional advantage for wire tension force equalization. And even for the first cuts, where changes of the wire tension force along the wire web cannot be avoided, the diamond wire saw geometry results at least in smoother changes as can be followed in Figure 6 by smaller values of |slope| for the DW288 diamond wire saw than for the DS271 slurry saw.

4.3 Wire guiding roller radii

The radii of the wire guiding rollers may vary due to manufacturing inaccuracies, but can also be changed deliberately. One way or the other Figure 9a) makes clear that the wire guiding roller radii have a strong impact on the wire tension force. For this it is on the one hand important to have a quality assuring look at the wire guiding roller radii. On the other hand a deliberate increase of wire guiding roller radii with increasing wire wear could counteract the known loss of wire tension force in the slurry process [4].

4.4 Wire guiding roller groove depth

The depth of the wire guiding roller grooves can vary by some tenth of microns due to manufacturing inaccuracies or irregular usage of the wire guiding rollers. If for instance many consecutive cuts with equal block lengths and gaps between the blocks were performed, the groove depth would develop differently.

Figure 9b) illustrates how changes in groove depth theoretically affect the wire tension force. A decreased (increased) groove depth results effectively in increased (decreased) wire guiding roller radii and therewith in increased (decreased) wire tension forces for the wire in the respective groove. Consequently the wire guiding roller groove depth has the same quality assuring relevance and the same potential to change the wire tension force deliberately like the wire guiding roller radii.

4.5 Free wire lengths

There are several factors that may change the free wire lengths permanently or temporarily. If for instance the wire guiding roller distance would increase with increasing groove number as a consequence of inaccurate adjustment the free wire lengths of the lower and upper wire web would increase permanently with groove number as illustrated schematically in Figure 1. By contrast the wire bow only affects the free wire lengths in the upper wire web. It establishes at positions where silicon is cut and it is formed as a function of sawing profile and wire wear. Consequently the wire bow and therewith the free upper wire lengths are both position- and time-dependent. Finally wire jumps between two blocks in the diamond wire process result in a permanent small increase of the respective free wire length.

Figure 8 illustrates how an increased upper free wire length as a consequence of a steplike wire bow affects the wire tension force distribution in the upper wire web for

Table 1: Factors that could impact the wire tension force in the multi wire sawing process and their relevance for the slurry and diamond wire sawing.

wire tension force relevant factor	process relevance slurry	diamond wire
wire diameter	yes	-
plastic wire elongation	minor	yes
wire bow	minor	yes
rewinding on the wire spools	-	yes
wire jump	-	no
wire guiding roller distance	no	no
wire guiding roller radii	yes	yes
wire guiding roller groove depth	yes	yes

the slurry (solid lines) and the diamond wire process (dashed lines). For both processes the wire tension force in the upper wire web is increased at the beginning (black lines) and equalizes as soon as the respective wire segments leave the wire web on the right hand side. This indicates that an increase of the free wire length does not affect the wire tension force as long as the wire web is refreshed. Consequently permanent changes of the free wire length do not affect the wire tension force. Meanwhile for the time-dependent wire bow formation one has to differentiate between the sawing processes. For the slurry process the impact on the wire tension force distribution is only momentarily as the wire web is refreshed within minutes. For the diamond wire process the effect of wire bow formation on wire tension force is relevant as the wire stays for several cuts in the wire web.

5 CONCLUSION

In this work various time- and position-dependent factors that could impact the wire tension force in the multi wire sawing process were discussed. Their relevance for the unidirectional slurry and the bidirectional diamond wire process was estimated on basis of a simple model that enables the calculation of wire tension forces along the wire web. Table 1 gives a brief summary of the results. Here factors that can result in long-lasting wire tension force irregularities along the wire web are rated as relevant for the sawing process.

Some factors like the wire guiding roller radii or groove depth are important for both the diamond wire and the slurry sawing process as only small changes can result in measurable changes of the wire tension force. Contrary variations of the wire guiding roller distance were found to have no impact on the wire tension force at all.

Furthermore some factors have relevance only for one of the processes. For example the loss of wire diameter due to wire wear only occurs for the slurry process. Whereas other factors like the wire bow or the plastic wire elongation take place for both processes, but are only relevant for the diamond wire process. The cause for this difference is that for the slurry process the wire is refreshed within minutes, but stays for several cuts in the wire web for the diamond wire process.

The insights given in this work can help to understand and improve wire tension force irregularities in the multi wire sawing process that may result in poor wafer quality for the affected wafers. To validate the presented calculative observations comparative wire tension force measurements along the wire web are planned.

6 REFERENCES

[1] K. Sunder and O. Anspach, Proceedings of the 29th EU PVSEC (2014) p. 588.
[2] B. Hurka et al., Proceedings of the 25th EU PVSEC (2010) p. 1591.
[3] G. Gottstein, Physical Foundation of Materials Science, Springer, 2004.
[4] D. Meißner et al., Solar Energy Materials and Solar Cells (2014) p. 346.
[5] J.-D. Penot et. al., Freiberger Sawing Symposium (2013) visual presentation.
[6] Meyer Burger, Technical specification wire saw DW288P6 series 3 (2016).
[7] K. Sunder and O. Anspach, Proceedings of the 32nd EU PVSEC (2016) p. 300.

IMPACT OF RESIDUAL ALUMINIUM CONTAMINATION ON THE DETERMINATION OF BORON AND PHOSPHORUS DENSITIES USING HALL EFFECT IN A SOLAR GRADE SILICON INGOT – A COMPARISON TO OTHER CHARACTERIZATION TECHNIQUES

A.Fauveau[1], B.Martel[1], J. Veirman[1], B. Drevet[1], H.Lignier[1], A. Kaminski-Cachopo[2], F.Ducroquet[2]
[1]. Univ. Grenoble Alpes, CEA, LITEN, DTS, Silicon material and PV Cells Division, F38000 Grenoble, France
[2]. Univ. Grenoble Alpes, IMEP-LAHC, Minatec, Grenoble INP, 3 parvis Louis Néel, CS 50 257, Grenoble, France

ABSTRACT:
Up to now, dopant quantification in compensated silicon is routinely performed with physical and chemical characterization techniques based on mass spectroscopy. As an alternative to these techniques, a complementary method based on Hall Effect measurements has been studied. This approach has already been proven to be fast, easy to use and efficient in determining boron and phosphorus densities in materials for which other residual dopant elements were in negligible concentrations. The present paper evaluates the influence of a third dopant on the traditionally used two dopant based Hall effect method, via the study of an Upgraded Metallurgical Grade (UMG) ingot that contains a significant residual Aluminium content. Some possible improvements of the Hall effect-based technique are discussed to improve its ability to simultaneously quantify more than two doping species, a task that remains challenging up to now.

Keywords: compensated silicon, characterization, aluminium, Hall effect,

1 INTRODUCTION

Nowadays, even if intentionally doped electronic-grade silicon (EG-Si) is commonly used in the photovoltaic (PV) industry, several alternative Si sources arouse growing interest. Among them, some recent and significant efforts were made to find alternative Si purification processes in order to recycle Kerf losses [1] and also solar cells scraps coming from production line wastes or even from end-of-life modules [2]. Metallurgical purification routes which have already been developed in the past [3] are intended to be used to purify these new feedstocks in order to reach PV requirements. The resulting solar grade (SoG) Si contains both donor and acceptor impurities, mainly boron (B) and phosphorus (P) species, which makes it compensated. Moreover, residual concentrations of other doping impurities are sometimes found in such materials. The control of dopant concentrations in this kind of Si is of paramount importance, especially because they govern the electrical properties (conductivity type, resistivity (ρ), mobility (μ)) of the bulk material [4], and ultimately the behavior of the finished device [5]. Up to now, this quantification is routinely done by physical and chemical characterization techniques based on mass spectroscopy (MS) [6]. For the past years, a complementary and alternative technique based on Hall Effect measurements has been studied [7]. This method relies on the fitting of experimental values of the charge carrier density (p) as a function of temperature (p(T)) by the Fermi-Dirac statistics [8]. This approach has already been proven to be fast, easy to use and efficient to determine B and P densities ([B] and [P]) in materials for which the residual dopant concentrations were negligible [10]. Theoretically, it can be used to characterize more than two dopants. However, for such Si, the precision of this method is strongly reduced due to the uncertainties on the experimental data. The aim of this work is then to study the impact of a third dopant contamination in SoG-Si materials on the determination of [B] and [P] with the Hall Effect-based method mentioned above. Indeed, this unwanted species, if present in a sufficient amount, could influence the net doping of the Si and thus complicate the procedure. To fulfill our goal, we have characterized a complete SoG-Si ingot which was known to contain a residual Aluminium (Al) contamination, an element often present in appreciable concentrations in SoG-Si. Both routine (mass spectroscopy techniques) and Hall Effect-based characterizations of dopant densities (the latter under the assumption that only B and P are present) were compared at different ingot heights. In parallel, the influence of a third doping impurity on the Hall Effect-based dopant densities determination was assessed from theoretical considerations supported by Fermi-Dirac statistics.

2 EXPERIMENTAL APPROACH

2.1 Materials

The studied generation 5 (400 kg) mc-SoG-Si ingot contains both [B] and [P] in the part per million weight (ppm wt) range. Considering the efficiency of the segregation step to remove metallic impurities from the melt, only traces of Al are expected, at least at the bottom and middle parts of the ingot (the top part could be more contaminated due to the Al segregation). A vertical 1 cm-thick and 4 cm-large rod was cut along the ingot center (wire sawing technique).

Five positions (from A to E) have been selected through the rod height (relative height: $0<H<1$). Most of the samples are chosen in the upper half of the ingot ($H\geq0.5$), as the segregation mechanisms lead to wider variations of compensation ratios and dopant concentrations in this part, and that the aim is to test the techniques over ranges as large as possible of these quantities. Smaller samples (1×1 or 2×2 cm^2 and 3 mm thick) were then extracted from the rod by wire sawing (as close as possible to each other, as depicted in Fig. 1). Routine techniques such as Second Ion Mass Spectroscopy (SIMS), Glow Discharge Mass Spectroscopy (GDMS), Inductively Coupled Plasma Mass Spectroscopy (ICPMS) were performed on these samples in parallel of Hall effect measurements of p(T), performed in the [50K-290K] temperature range.

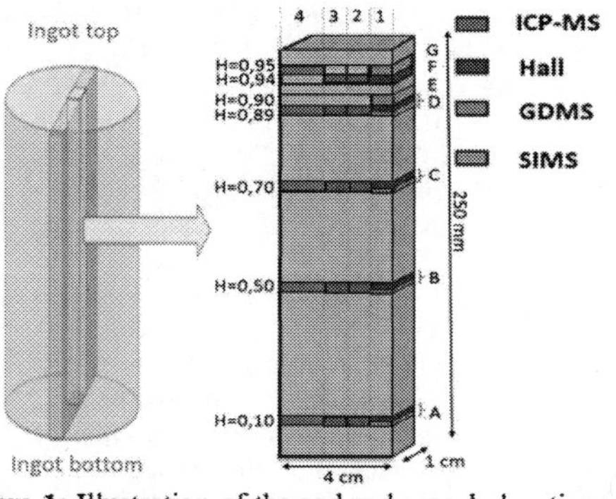

Figure 1: Illustration of the rod and sample locations for the dopant analyses

In addition, a conductivity type transition from p to n has been detected with a Semilab p/n tester around H=0.9. Note that both SIMS and Hall Effect measurements have been performed on the same samples, except at the top part (due to samples preparation issues, we had to cut twice, noticed E- and E+ on Fig. 1-left). The p(T) data were corrected using the appropriate Hall factors (Szmulowicz for p-type samples [9], and Fauveau et al. for n-type samples [10]). Then, they were fitted with the theoretical equation for the variation of charge carrier density with temperature (Fig. 2), assuming only two main doping species are present, leading to two unknown parameters, respectively [B] and [P] [10].

The main underlying theoretical basis required for the calculation of p(T) is recalled here. In p-type compensated Si, it can be expressed from the electro-neutrality equation (1), expressed here for dopants releasing one charge, such as B, Al or P, combined with the Fermi-Dirac statistics [8]. Note that the n-type case can be treated in the same way by switching donors (N_D) as majority dopants and acceptors (N_A) as minority ones.

$$p(T) = \sum_i N_{Ai}^-(T) - \sum_i N_{Di}^+(T) \quad (1)$$

In compensated Si, minority dopants are known to be all ionized whatever the temperature [8]. We can thus assume $N_{Di}+ = N_{Di}$ in (1). In addition, the total majority ionized dopant density (ΣN_{Ai}^-) can be expressed as a function of the total acceptor density (ΣN_{Ai}), their ionization energy (E_i) and the temperature (2).

$$\sum_i N_{Ai}^-(T) = \sum_i N_{Ai}\left[1 + g^{-1}.\exp\left(\frac{E_i - E_F(T)}{kT}\right)\right]^{-1} \quad (2)$$

where g is the degeneracy factor, E_F the Fermi level and k the Boltzmann constant. Acceptor dopants are here considered to all have a degeneracy factor of 4.

Combining (1) and (2) and the Boltzmann statistics to describe the level occupancy in the majority band, one can obtain a mathematical description of p(T) as a function of both N_{Ai}, E_i and N_{Di} [8].

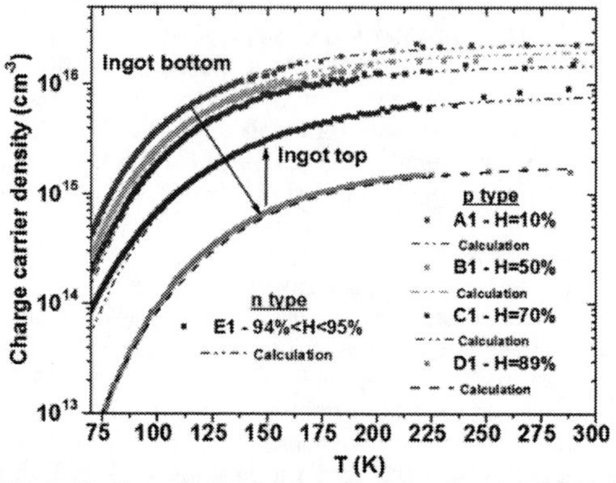

Figure 2: Experimental data of n(T) corrected by Hall factor (Szmulowicz [9] for p-type and Fauveau [10] for n-type) and fitted with Fermi-Dirac statistics using the method presented in [10]. The energy level used for B is 44.6 meV.

2.2 Results

The measured dopant densities are depicted for all the tested techniques in Fig. 3.

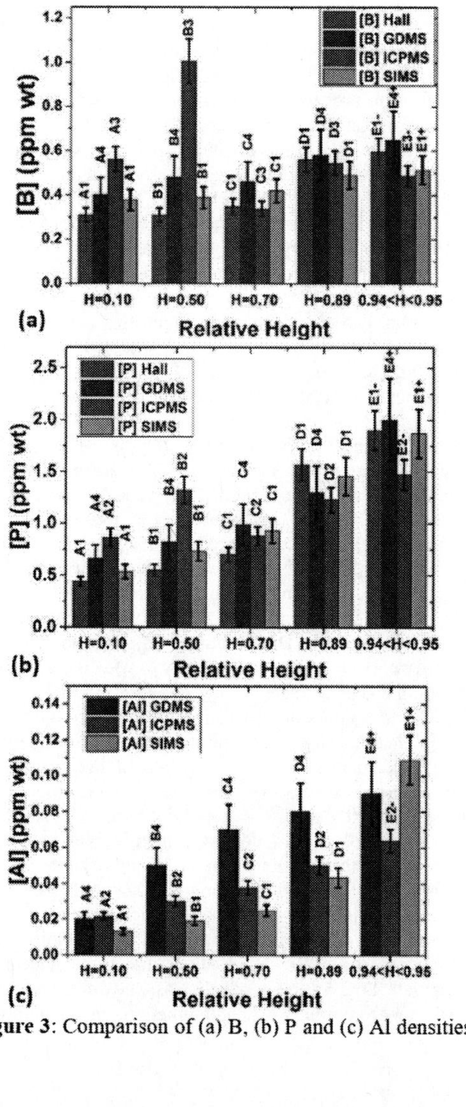

Figure 3: Comparison of (a) B, (b) P and (c) Al densities

determined by Hall effect, GDMS, ICPMS and SIMS for 5 selected positions in the ingot. Al densities are not measured by Hall effect technique, as it is a 2-dopant based method.

First, as a general remark, dopant densities increase along the ingot height. This is in qualitative agreement with segregation laws for impurities featuring lower-than-unity segregation coefficients, such as B, P, Al. Also, it can be observed that [B] profiles are flatter than [P] counterparts, again in good agreement with the relatively higher equilibrium segregation coefficient of B (k_B=0.80) as compared to that of [P] (k_P=0.35). The [B] and [P] determined by each technique are fairly comparable: the residual differences are mainly in the order of magnitude of the uncertainties of each method (evaluated between 10 and 20%, either internally or according to the subcontractor).

Most of the SIMS and Hall Effect measurements have been performed on the same samples. Observed differences in the concentrations extracted are thus consequences of the measurement in itself: they are not due to a difference resulting from possible radial segregation effects due to a curved solidification front, as can be observed in practice. In some cases, dopant concentrations measured by HE are lower than those measured by SIMS or GDMS. Among the reasons to explain this difference, is the fact that HE measures only electrically active dopants whereas the other two techniques reveal all dopant atoms whatever their electrical configuration. It would suggest that part of the dopants are not active (and thus, that the thermal history of the ingot favored aggregates or inactive dopant precipitates). Another significant difference between the Hall effect based technique, and GDMS and SIMS is that the first one is a volume measurement whereas the other two are surface measurements. However, due to the low sample thickness (less than 1 mm), the difference in dopant concentration (due to the segregation process) between the two sides of the samples is too low (between 0.1% and 2% for [B] depending on its position in the ingot) to explain the observed experimental difference between those two techniques.

Samples located at H=89% and H=94% are highly compensated as they are close to the conductivity type transition. For those two samples, and contrary to the other ones, dopant densities obtained by HE are not lower than those measured by the other techniques. As mentioned earlier, the ingot features a residual Aluminum content ([Al]), especially in the last crystalized part (again, due to segregation). These Al concentrations, which reach a non-negligible percentage (≈5%) close to the transition in conductivity type, have not been taken into account in the calculation of [B] and [P] presented in Fig. 3 (hypothesis : N_A=[B] et N_D=[P], where N_A and N_D are respectively the acceptor and donor concentrations). Therefore, it is reasonable to suppose that Al concentrations may influence the precision of the extracted [B] and [P]. In the next section we will evaluate if the presence of Al could explain our findings. Fig. 3-c compares the Al densities measured with mass spectroscopy techniques. More discrepancies are noticed on [Al], mainly explained by the fact that we are approaching the detection limits of the employed techniques. To assess the impact of [Al] on the Hall Effect-based [B] and [P] determinations, we focus on three sample positions. The p-type sample at position A was selected because it contains only traces of Al, position D since it is also p-type but contains higher [Al], and finally position E which also contains a higher [Al] but features a n-type conductivity.

3 CALCULATIVE APPROACH

By a calculative approach, this part evaluates the influence of Al on the dopant densities extracted with the two dopant Hall Effect-based method. The analytical expression of p(T) previously described in part 2 was used to fit the experimental p(T) data. In our practical case, we first neglected the presence of Al and assumed that $\Sigma N_{Ai}^-(T)$ = $[B^-](T)$, where $[B^-](T)$ is the ionized B density at the temperature T, E_B - E_V = 45 meV (with E_V the valence band energy) and ΣN_{Di} = [P]. However, as the extra Al dopants are taken into account, we shall assume that ΣN_{Ai}^- (T) = $[B^-]$ (T)+$[Al^-]$ (T), E_B - E_V = 44.6 meV, E_{Al} - E_V = 69 meV and ΣN_{Di} = [P]. In order to understand the impact of the presence of Al on the p(T) behavior, we have calculated p(T) for each A, D and E positions (with appropriate adaptations for the n-type case [8]) integrating or not the [Al] contribution. Note that the dopant densities used for this calculative approach were taken from Mass spectrometry experimental data. The results are depicted in Fig. 4.

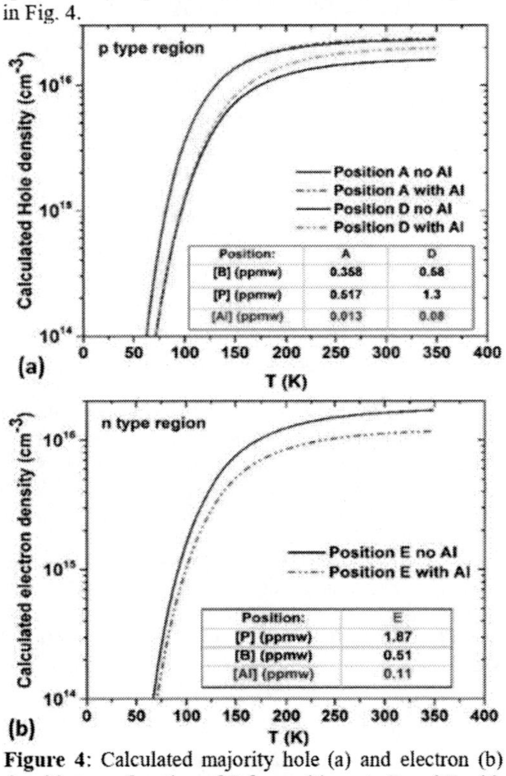

Figure 4: Calculated majority hole (a) and electron (b) densities as a function of T for positions A, D and E with and without Al contaminations.

For position A, one can see from Fig. 4-a that traces of Al (only 0.01 ppm wt) do not affect noticeably the p(T) data. However, moving to position D, the difference becomes more pronounced, especially upon increasing T. This can be explained from the difference of E_i between Al and B atoms. Indeed, because Al dopants introduce a deeper E_i within the Si bandgap, they will become neutral faster than the B atoms when cooling down the sample (i.e. from right to left). Consequently we can observe for position D (Fig. 4-a) a difference in the freeze-out velocity in the p(T) curves whether there is Al or not. Position E in n-type region is also affected by the presence of Al (Fig. 4-b). However, contrary to the position D, both B and Al are fully ionized whatever the temperature (in n-type Si, B and Al being minority dopants). Thus, one can observe no difference on the freeze-out velocity of p(T) curves, with or without Al.

If we now fit the calculated p(T) curves containing the [Al] contribution, without taking into account the presence of Al in the mathematical expression of p(T), we can assess the error made on the determination of [B] and [P] for an Al-containing sample using the standard Hall effect analysis (2 dopants). This was done for the three positions of interest, and the results are summarized in Table 1.

Table 1 : Comparison between the [B] and [P] values expected from the mass spectroscopy measurements and those extracted by fitting the calculated p(T) curves (with Al) without taking into account the Al contribution. [Al] are respectively 0.01, 0.08 and 0.11 ppm wt for A, D and E.

Position	Expected [B] (ppm wt)	Fitted [B] (ppm wt)	Deviation (%)	Expected [P] (ppm wt)	Fitted [P] (ppm wt)	Deviation (%)
A	0.358	0.367	2.45	0.517	0.530	2.45
D	0.58	0.72	19.44	1.3	1.61	19.25
E	0.51	0.53	3.77	1.87	1.87	0

As expected from Fig. 4, traces of Al (0.01 ppm wt) in position A lead to a minimal error on [B] and [P]. This behavior is hard to confirm on our experimental data (Fig. 2) as other sources of error, intrinsic to each characterization technique, may mask it. However, a higher [Al] (0.08 ppm wt) leads to a significant overestimation of [B] and [P]. This stems from a stronger consequence of an unjustified use of the 2 dopant based p(T) model, linked with the higher Al content.

This last result correlates well with our experimental data. Indeed, looking closer at the Hall Effect experimental results of position D (Fig. 3), one can see that the [P] (and to a lesser extent [B]) is slightly overestimated compared to the other methods. In addition, we can also observe a strong increase of the [P] compared to sample C1, which does not obey the expected segregation laws (not shown here). This last point holds also true for the [B]. Finally, we can notice (Table 1) that the Al contamination of an n-type sample (position E) only leads to an overestimation of the [B]. This is in good agreement with the fact that all minority dopants are ionized, thus we expect to estimate the [B]+[Al] dopant densities instead of the [B] only in that case.

4 CONCLUSIONS

It was shown by a calculative approach that traces of Al (0.01 ppm wt) have a poor impact (<3%) on [B] and [P] determinations by a traditional 2-dopant Hall Effect analysis ([B] and [P] in the range 0.5 -1.2 ppm wt). However, a substantial overestimation (up to 20%) of [B] and [P] has been revealed with higher Al densities ([Al]) (0.08 ppm wt). The latter result was explained by the difference of ionization energy between B and Al atoms, which strongly modifies the carrier freeze-out dynamics. This effect is confirmed experimentally at the top part of the studied ingot where the [Al] is higher. Including higher temperatures (beyond 350 K) to the Hall effect temperature sweep could bring complementary p(T) data points that could potentially help determine the concentration of the residual doping impurity, in addition to [B] and [P].

5 ACKNOWLEDGMENTS

This project has received support from the State Program "Investment for the Future" bearing the reference (ANR-10-ITE-0003).

7 REFERENCES

[1] E. Provent et al., proceeding of the 26th European Photovoltaic Solar Energy Conference, Hamburg (2011)
[2] M. Syvertsen et al., CABRISS project, proceeding of the 31st EUPVSEC conference (2016)
[3] F. Cocco et al., Proceedings of the 28th European Photovoltaic Solar Energy Conference, Paris (2013).
[4] J. Veirman PhD Thesis, from INSA Lyon (2011)
[5] S. Dubois et al., proceeding of the NREL Workshop (2011)
[6] Sergio Pizzini, R. S. Hockett, June 2012, DOI: 10.1002/9781118312193.ch6
[7] B. Martel et al., Silicon Material Workshop, Roma (2013)
[8] E.F. Schubert, Physical foundations of solid-state devices, Rensselaer Polytechnic Institute, Troy, New York, (2007)
[9] F. Szmulowicz, Calculation of the mobility and the Hall factor for doped p-type silicon, Phys. Rev. B, 34, 6 (1986), p4031.
[10] A. Fauveau et al., Energy Procedia, Volume 96, (2016), pages 691-696

Effect of high temperature firing on the bond configuration and composition of a-SiN:H passivation layer deposited by PECVD

S. Jafari[1,2,*], M. Gläser[3], N. Bernhard[2], D. Lausch[1]
[1]Fraunhofer Institute CSP, Center for Silicon Photovoltaics, Halle (Saale), Germany
[2]Anhalt University of Applied Sciences, Köthen, Germany
[3]Martin Luther University, Karl-Freiherr-von-Fritsch Staße 3, Halle (Saale), Germany
[*]sahar.jafari@csp.fraunhofer.de, +49 345 5589 5124

Bond configuration and microstructure of hydrogenated amorphous silicon nitride films (SiNx:H) deposited by PECVD have been investigated after deposition of SiN as well as after the firing process. Deposition parameters were varied in a statistical full factorial design (DoE) and their effect on the film properties was studied in both steps, after the deposition and after the firing. Result revealed that the change in gas flow ratio of $NH_3/(SiH_4+NH_3)$ and the temperature of the deposition are the two significant factors for the composition and the bond structure of SiN. The low [N]/[Si] ratio deduced by XPS in Si-rich SiN layers results better surface passivation with high minority carrier lifetime after deposition. The density of Si-N bonds was low in these layers and the densification occurs mostly due to the Si-H breaks after firing. The improvement of minority carrier lifetime has been observed in N-rich SiN layers deposited by identical gas flow of ammonia and silane which shows a high density of Si-N after deposition. It has been seen that the atomic density of SiN plays a key role on the thermal stability of SiN at high temperatures.

Keywords: SiN:H , PECVD, hydrogen passivation, firing treatment

1 INTRODUCTION

The efficiency of silicon solar cells is influenced by the presence of defects which induce recombination active centers in the band gap of the semiconductor. Using the amorphous hydrogenated SiN on the surface of silicon shows a significant enhancement in the recovery of efficiency. There are several reasons for that, it is firstly because that SiN acts as an anti-reflective coating layer and increases the light absorption in solar cells. It also can passivate the dangling bonds and impurities on the surface of silicon substrates which called surface passivation. The other important reason is that the plasma deposited SiN contains up to 40% hydrogen [1] which contributes to the bulk passivation [2] and deactivates the recombination centers at high temperature in silicon bulk. The co-firing process which takes place after the deposition of SiN supplies the sufficient energy for hydrogen diffusion into the bulk and passivation of the defect centers and impurities.

Microstructure and bond configuration of SiN have significant impact on the diffusion mechanisms of hydrogen [3] and it is important to explore the appropriate layer composition to improve the bulk passivation. The other main factor is the thermal stability of SiN at the high temperature which avoids the deterioration of the passivation quality during the firing process.

The impact of the deposition parameters on the hydrogen bond structure and composition of the film have been studied in this work. The SiN films have been deposited by PECVD technique with various deposition parameters designed by DoE and then they got fired at the temperature of 850 °C in a standard industrial belt furnace for few seconds. Main object of this work was to find out the optimized deposition recipe which can result in a high amount of hydrogen, meanwhile, this recipe also provides a film with sufficient thermal stability.

The result shows that the deposition temperature as well as the gas flow ratio of $NH_3/(NH_3+SiH_4)$ influence significantly the film composition and the passivation quality of SiN. The large increase of bond configuration has been observed for Si-N bonds during the firing process. Densification of SiN supposed to be accompanied with the release of hydrogen. Using identical gas flow of ammonia and silane seems to provide a SiN film with high density of Si-N bonds.

2 EXPERIMENTAL

The statistical full factorial design has been applied to deposit the SiN with various deposition parameters. Both sides of p-type FZ silicon wafers have been deposited by PECVD techniques. Table I shows the applied values.

Table I: the various deposition parameters has been used for deposition of SiN by PECVD

Parameter	Low	Medium	high
$NH_3/(NH_3+SiH_4)$	0.1	0.5	0.9
Temperature (°C)	200	300	400
Power (W)	50	100	150
Pressure (mTorr)	500	750	950
Total gas flow (mTorr)	500	1000	1500

The gas mixture of NH_3 and SiH_4 has been diluted by nitrogen in the deposition chamber. The medium values were the daily center point of experiments and they provide the reliable statistical analysis and the evidence for the reproducibility of the deposition process. To determine the effect of firing on the film properties and the composition, samples were located in an industrial belt furnace for few seconds at 850 °C after the deposition. The atomic structure and bond configuration of the films were monitored before and after firing by FTIR. The absorbance peaks of Si-N, Si-H and N-H bond have been analyzed and the relative bond concentrations were calculated by the model developed by Lanford et al[4]. The absorption cross section of three mentioned bonds is respectively equal to 2.4×10^{-17}, 7.4×10^{-18}, 5.3×10^{-18}[5]. The effect of deposition parameters on the

[N]/[Si] ratio in SiN has also been investigated by XPS spectroscopy. The optical properties such as the thickness and the refractive index of the films were measured by using optical spectroscopic ellipsometry after deposition and after firing. Moreover, the minority carrier lifetime has been monitored by QSSPC in both steps.

3 RESULT AND DISCUSSION

3.1 Bond configuration after deposition

Figure 1 shows the contour diagrams of temperature-gas flow ratio dependence of the hydrogen bond configuration (Si-H and N-H) and the Si-N bond density measured by FTIR spectroscopy after deposition. Shifting from dark blue to dark green color shows the enhancement of the bond density in the diagrams. It can be seen in Figure 1a and 1b that the variety in the hydrogen bonds, Si-H and N-H is proportional to the gas flow ratio of NH_3/ (NH_3+SiH_4). As expected, the high ammonia gas flow increases the N-H bond concentration and similarly the high silane gas flow grows a layer with more Si-H bonds. Relative density of Si-N bond is reflecting the atomic density of SiN film and it shows that the films density increases by applying identical mixture of ammonia and silane during deposition.

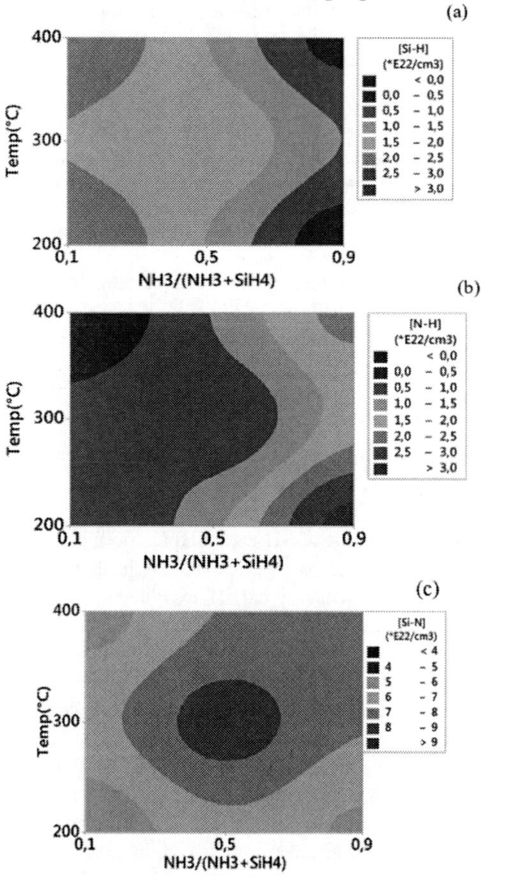

Figure 1: relative bond concentration of a) [Si-H], b) [N-H] and c) [Si-N] in SiN after the deposition as a function of gas flow ratio and deposition temperature deduced by FTIR

Figure 2 demonstrates the correlation of [N]/[Si] ratio deduced from XPS with the refractive index of the film. Seven samples with different refractive indexes and various bond configurations have been selected for XPS measurement. The result in Figure 2 shows that the higher is the N concentration in the SiN, the lower is the refractive index which shows the properties of a stoichiometric film but the Si-rich layers have mostly high refractive index up to 3.3.

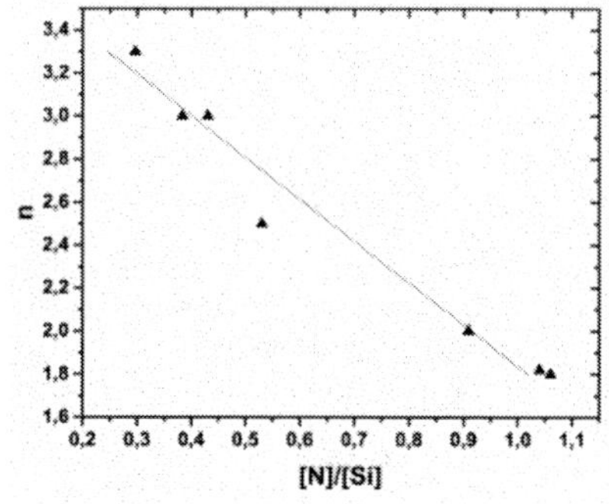

Figure 2: refractive index of SiN for different ratio of [N]/[Si]

3.2 Bond configuration after firing

In figure 3, the contour diagram indicates the relative hydrogen bond concentration and Si-N density present in SiN films after the firing process. Figure 3a and 3b show the bond concentration of Si-H and N-H in SiN as a function of temperature and gas flow ratio of NH_3/(NH_3+SiH_4) respectively. The change in Si-H is more sensible than in N-H bond concentration due to the firing process compared the two first diagrams in Figure 1 and 3. It can be attributed to the fact that binding energy of Si-H (3.1 ev) is lower than N-H (4.3 ev) [6]. Reduction of Si-H, means the release of hydrogen, happens mostly in the Si-rich layers which deposited with low gas flow ratios in the left side of the diagram shown in Figure 3a. The change in N-H bond concentration was not considerable after firing and it stays relatively unchanged. It is reported [3] that N-rich SiN has more density of Si-N bond that Si-rich layers. The presence of Si-Si bonds is likely to be high in Si-rich compared to Si-N bonds layers considering the binding energies. The increase in the Si-N density is observed after firing for most of the samples shown in Figure 3c compared to 1c. It is clearly seen that the layers deposited by high temperature and high gas flow of ammonia turn to be denser after firing. The influence of densification on the passivation quality is shown in Figure 4.

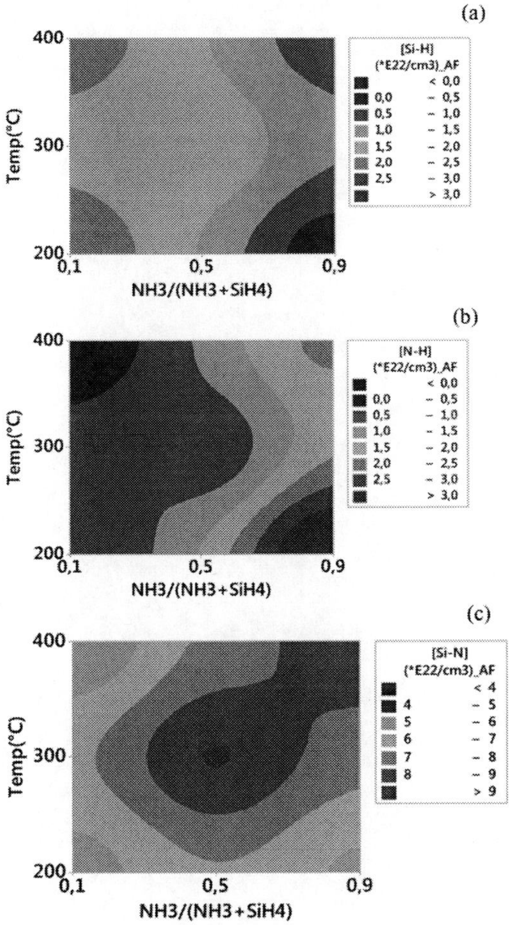

Figure 3: Clear line drawings are essential

3.3 Minority carrier lifetime

The increase of Si-N bond concentration is correlated to the increase of carrier lifetime by comparing the Figures 3c and 4b. It can be attributed to the hydrogen release during the formation of Si-N out of broken Si-H and N-H at high temperature which influences the surface and bulk passivation of underlying silicon [3]. Whether the release is diffusion into the underlying bulk silicon or an effusion into the ambient air it is unclear so far and will be clarified in an upcoming publication. Following equation describes the possible reaction during firing causes the densification in SiN.

$$[N-H]+[Si-H]\xrightarrow{firing}2H+[Si-N]$$

Dekker et al [7] showed that in layers with high mass density no molecular hydrogen was observed and the atomic hydrogen diffuses through those layers. It can explain the increase of lifetime in high density layers.
Result shows the more densification due to the firing is also accompanied with the release of hydrogen and as it is shown in Figure 4b it improved the passivation quality. However the Si-rich layers show high lifetime after deposition in Figure 4a, but the increase of lifetime after firing happens in N-rich layers by the medium deposition values. These layers deposited with identical amount of ammonia and silane and they show the reflective index around 2 close to stoichiometric Si_3N_4. It is confirmed the fact that Si-N density plays a key role on the thermal stability and improvement of passivation quality at high temperature.

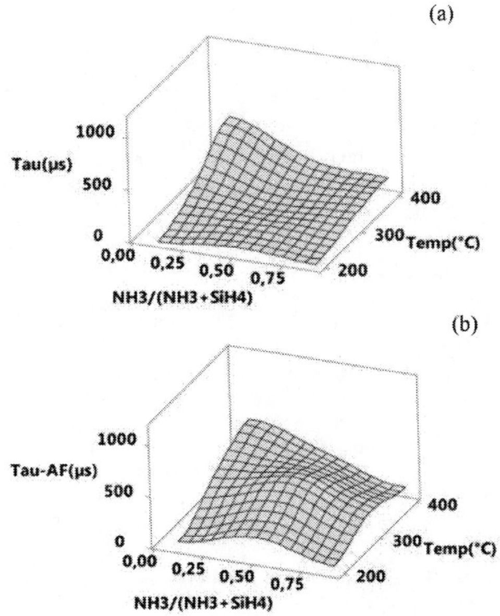

Figure 4: Minority carrier lifetime measured by QSSPC as a function of gas flow ratio and deposition temperature a) after deposition and b) after firing

4 CONCLUSION AND OUTLOOK

Statistical full factorial design has been carried out. The results show that the gas flow ratio of $NH_3/(NH_3+SiH_4)$ and the temperature are the most significant deposition parameters for the film composition and bond configuration. Linear correlation between the refractive index and the [N]/[Si] ratio in SiN has been shown. SiN layers with high N concentration show dielectric properties and refractive indexes close to 2. The bond concentration of hydrogen is highly influenced by the amount of ammonia or silane gas flow during the deposition, high ammonia gas flow causes more N-H, thus Si-H increased by high silane gas flow. Since the Si-H bond transition after firing was more considerable than N-H bonds, the hydrogen release is more likely in Si-rich SiN. The increase in Si-N density has been observed which shows the densification due to the firing. The passivation quality and minority carrier lifetime after deposition of SiN was high for Si-rich layers, however the improvement in lifetime has been observed in the samples with high N concentration deposited by identical gas flow ratio of ammonia and silane and medium temperature. The densification during firing is accompanied with release of hydrogen and the lifetime improvement can be attributed to the hydrogen passivation. Since the relative change in hydrogen bond concentration does not provide sufficient evidences for lifetime improvement and hydrogen diffusion and effusion, the absolute value of hydrogen in different depths is desired and it has been planned to be measured by elastic recoil detection analysis ERDA in future.

5 AKNOWLEDGEMENT

This work was carried out within the joint research graduate school "StrukturSolar II" of Anhalt University of Applied Sciences and Martin-Luther-University Halle-Wittenberg which is funded by the German Ministry of Education and Research under identification code 03SF0417 A.

5 REFERENCE

[1] A. Aberle, Solar Energy Materials and Solar Cells, 65 (2001) 239.
[2] K.A. Münzer, Proceedings 24th European Photovoltaic Solar Energy Conference, (2009), 1558.
[3] J. Hong, W. M. M. Kessels, W. J. Soppe, Journal of Vacuum Science & Technology, B 21 (2003) 2123.
[4] W. A. Lanford, M. J. Rand, Journal of Applied Physics, 49 (1978) 2473.
[5] Z. Yin, F. W. Smith, Physical Review B, Vol. 42 (1990) 3666.
[6] J. Robertson, Philosophical Magazine B, Vol.69, No. 2 (1994) 307-326.
[7] H. F. W. Dekkers, G. Beaucarne, Applied Physics Letters 89 (2006) 211914.

REGENERATION OF BORON-OXYGEN RELATED DEGRADATION IN Cz-Si PERC-TYPE SOLAR CELLS AT HIGH TEMPERATURES

Axel Herguth, Christian Derricks and Giso Hahn
University of Konstanz, Department of Physics, 78457 Konstanz, Germany
Author for correspondence: axel.herguth@uni-konstanz.de

ABSTRACT: Boron-oxygen related light induced degradation (BO-LID) may be eliminated by applying a regeneration process which converts the recombination active 'degraded' defects responsible for BO-LID into an inactive 'regenerated' state. As industrial mass production demands for fast, inline-capable processes and as the regeneration reaction accelerates with increasing temperature, it seems straightforward to increase treatment temperature. However, it is found that a reverse reaction destabilizes the 'regenerated' state at high temperatures and thus the 'regenerated' state is not necessarily the preferred defect state at high temperatures. Within this contribution it is investigated up to which temperature the regeneration process can be applied for boron-doped Cz-Si PERC solar cells. Experiments are accompanied and compared to theoretical predictions. In addition, it is found that the samples under investigation suffer from a second LID effect. Whether this can be attributed to LeTID, degradation of surface passivation quality or a different mechanism remains unclear.
Keywords: Degradation, BO-LID, LeTID, Cz-Si, PERC, Simulation

1 INTRODUCTION

The efficiency potential of p-type PERC (passivated emitter and rear) solar cells surpasses that of full area Al alloyed cells due to its better rear side passivation. As the additional process complexity comes down to higher costs per cell, this advantage only pays off from an economic point of view, if the excess charge carrier lifetime in the bulk τ_b is sufficiently high so that rear side passivation becomes more and more the limiting factor.

However, it is known that rather high qualitative boron-doped monocrystalline Cz-Si suffers from boron-oxygen related light induced degradation (BO-LID) that can, if the worst comes to the worst, push τ_b down into the low two-digit µs region [1] – too low for the PERC advanced rear passivation to pay off.

As of 2006 it is known that BO-LID is curable by a treatment called 'regeneration' consisting of excess carrier injection (often done by illumination) while the sample is kept at an elevated temperature [2,3]. Within the regeneration treatment the harmful defect converts into an inactive 'regenerated' state which is, in contrast to the initial state prior to defect activation, insensitive to illumination under cell/panel working conditions.

Of course, the regeneration treatment can only suit industrial mass production, if it can be applied either in inline mode (time frame of dozens of seconds) or batch mode (time frame of minutes). Both modes require ways to accelerate the regeneration treatment.

It was found that higher temperatures accelerate the regeneration reaction, as well as injection does [2,3]. However, from lifetime studies it was also found that a counter reaction (Redegradation or Destabilization) exists that renders the regenerated defect state more and more instable for temperatures above ~200°C [2,3,4]. Hence, the regeneration treatment at temperatures too high does not result in a majority of defects in the regenerated state. In other words, stabilizing lifetime and thus efficiency via a regeneration treatment does not work anymore.

Within this contribution, regeneration of BO-LID in Cz-Si PERC solar cells in the high temperature regime >200°C is investigated both from a theoretical point of view and in experiments in order to verify the theoretical construct that can be used to predict the impact of the regeneration process even on unknown samples.

2 THE THEORY BEHIND BO-LID (IN PARTS)

The defect responsible for the BO-LID is known to feature at least three different states (see Fig. 1, [2,3]): the recombination inactive 'annealed' state A which is instable under illumination at typical cell/panel working conditions; the recombination active 'degraded' state B; the recombination inactive 'regenerated' state C which is stable under typical cell/panel working conditions. The defect may convert between these states as shown in Fig. 1 where also typical temperature and illumination conditions are summarized under which the respective reactions are known to occur. From experiment it is clear that there exists a reaction path converting the defect from C to A, however, it is hard to decide whether it is a direct reaction connecting C and A or whether it is a two-step reaction C to B to A. In the following this reaction will be treated as redegradation reaction from C to B. State conversion may be described by rate equations of the kind

$$\dot{n}_B = -\left(\kappa_{BA} + \kappa_{BC}\right)\cdot n_B + \kappa_{AB}\cdot n_A + \kappa_{CB}\cdot n_C \quad (1)$$

where n_j ($\Sigma_j n_j = 1$) is the occupation of the j state in the defect pool and κ_{ij} represent the conversion rates. The whole reaction scheme including especially the complete rate equation system used within this contribution is described in [3] in more detail.

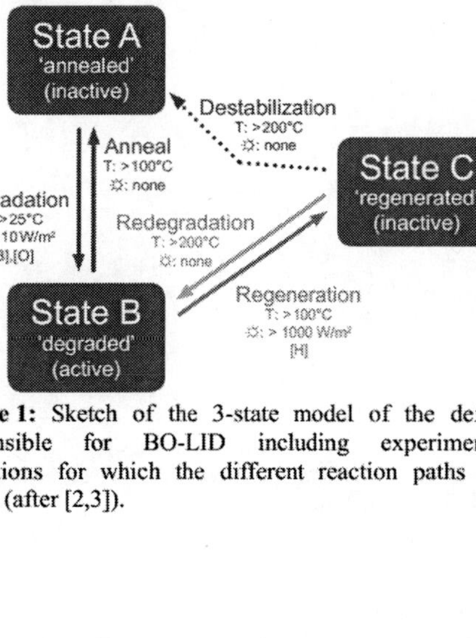

Figure 1: Sketch of the 3-state model of the defect responsible for BO-LID including experimental conditions for which the different reaction paths are active (after [2,3]).

All reactions seem to be describable by an Arrhenius-type temperature dependence (thermally activated) following the formula

$$\kappa_{ij}(T, p_0, \Delta n) = \kappa^*_{ij}(p_0, \Delta n) \cdot \exp\left(-\frac{E_{ij}}{k_B T}\right) \quad (2)$$

with κ^*_{ij} being the specific pre-exponential factor and E_{ij} the specific activation energy of the respective reactions between state i and j (i,j = A,B,C).

However, one should keep in mind that the kinetic parameters of the different reactions at work were typically not determined in the whole temperature range used in this contribution.

3 THEORETICAL PREDICTIONS

The activation energies of degradation $E_{AB} = 0.475$ eV and anneal $E_{BA} = 1.32$ eV were taken from [5], while those of regeneration $E_{BC} = 1.0$ eV and redegradation $E_{CB} = 1.25$ eV originate from [4]. The pre-exponential factors of anneal $\kappa^*_{BA} = 7.4\times10^{12}\,s^{-1}$ [5] and redegradation $\kappa^*_{CB} = 5\times10^{9}\,s^{-1}$ [4] are treated as constant. The factor of degradation was adapted from $\kappa^*_{AB}(p_0 = 1\times10^{16}\,cm^{-3}) = 4\times10^{3}\,s^{-1}$ [5] to the doping of the used PERC cells using $\kappa^*_{AB} \sim p_0^{2}$ [5]. An enhancement of degradation due to injection level as proposed by e.g. Hamer [6] was neglected, as injection level Δn under the experimental conditions applied in this investigation was considered small compared to doping p_0.

The reaction rate of regeneration depends on various parameters like the thermal history (especially the fast firing process) [7], therefore the reaction rate of regeneration was 'calibrated' by an in-situ monitored regeneration treatment at reasonably low temperature [8] yielding $\kappa^*_{CB} = 3\times10^{9}\,s^{-1}$.

The resulting reaction rates in dependence of temperature are shown in Fig. 2. As can be seen, anneal rate κ_{BA} and regeneration rate κ_{BC} are almost comparable, but for higher temperatures, anneal rate surpasses regeneration rate. In addition, both rates exceed the rate of degradation κ_{AB} for high temperatures by far.

The colored regions in Fig. 2 show the worst case estimation of uncertainty of the respective rates taking a 1% uncertainty in E_{ij} and κ_{ij} into account in order to provide a feeling for the uncertainty encountered in the simulation. The uncertainty of all above mentioned values in literature exceeds 1% thus meaning that simulation probably depicts correctly the general trend, but fails to deliver exact results.

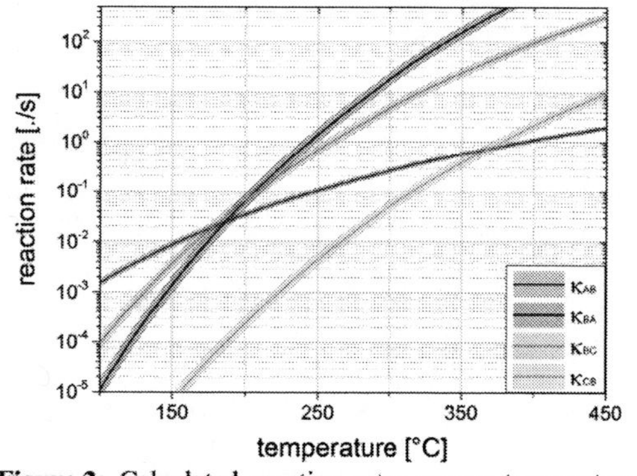

Figure 2: Calculated reaction rates versus temperature. Colored areas depict the uncertainty.

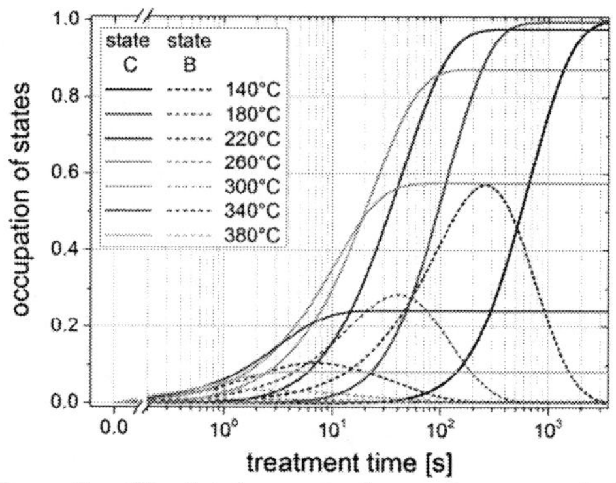

Figure 3: Simulated occupation of states during treatment at different temperatures.

Fig. 3 shows the simulated occupation of states during the treatment at the set temperatures used later on in the experiment. Curves were calculated using the reaction system described above (and in more detail in [3]) implemented in CASSANDRA [9].

Five details should be noted: (I) Long-term equilibrium of state C (solid lines) (and of states A and B as well) is reached faster for increasing temperature. (II) Maximal occupation of state B (dashed lines) occurs faster, but less pronounced with increasing temperature. Above 260°C almost no occupation of state B occurs anymore. (III) Long-term equilibrium value of state C decreases with increasing temperature. As long-term occupation of state B is always negligible, a decreasing occupation of state C implies an increasing occupation of state A. (IV) As both states A and C feature negligible recombination activity, only the occupation of the recombination active state B has an observable influence. Therefore, monitoring the occupation of states during the treatment, gets harder with increasing temperature. (V) Even though the occupation of state A and C may not be distinguished during the treatment, defects in state A and C will react differently when exposed to degradation conditions (low temperature, low intensity illumination): Defects in state A will convert to state B observable as BO-LID while defects in state C will remain stable.

Long-term equilibrium occupation of states versus temperature is depicted in Fig. 4 showing the continuous shift of preference within the defect pool. The colored areas represent the worst case estimation of uncertainty resulting from the uncertainty of rates in Fig. 2.

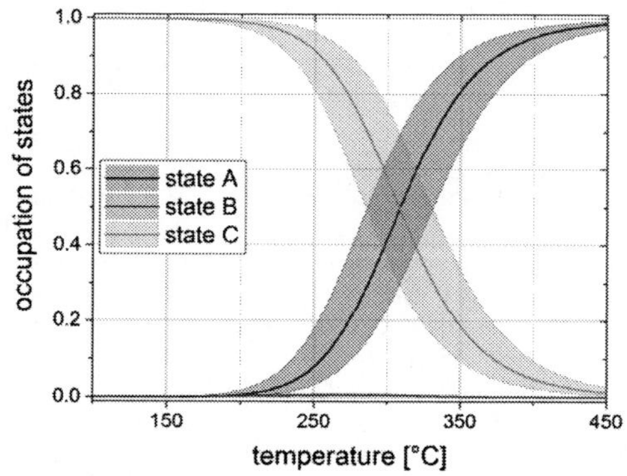

Figure 4: Simulated long-term equilibrium occupation of states versus temperature using the rates shown in Fig. 2. Colored areas depict the uncertainty.

4 EXPERIMENT

4.1 Setup

State-of-the-art boron-doped (~1.2 Ωcm) Cz-Si PERC-type solar cells are used in the experiments. The solar cells were neither annealed, nor degraded prior to the experiments.

Two cells were treated separately: The first was used to quantify the extent of BO-LID by low temperature, low intensity degradation (gray data in Fig. 7). The second was used to obtain the regeneration rate at ~150°C (Fig. 5) used as 'calibration' for the simulation.

The rest of the cells was subjected to illumination on a hotplate in the temperature range of 140 to 380°C. It should be noted that the actual sample temperature may deviate from these set values. Halogen incandescent lamps were used to illuminate the samples with ~2 suns (current equivalent according to [10]).

In order to monitor the progress of the treatment, samples were removed from the hotplate and V_{OC} was measured according to STC.

After the treatment, the cells were exposed to degradation conditions (~35°C, ~0.15 suns, current equivalent according to [10]) to convert defects in state A to state B.

4.2 Preliminary test

As the experiment aims at the long-term equilibrium occupation of states, monitoring the progress might seem irrelevant, however, a first test revealed that monitoring is essential. Fig. 5 shows the evolution of V_{OC} of a cell treated at 150°C and 1 sun intensity.

As expected, V_{OC} drops and recovers due to boron-oxygen related degradation and regeneration in the first 10^4 s (~2.5 h). Then, however, V_{OC} drops and (partly) recovers for a second time – an effect not expectable for BO-LID. It remains unclear, whether this second drop is due to LeTID [11] also occurring in Cz-Si as suggested in [12,13], a degradation of surface passivation quality as shown in [14] or a completely different effect.

In consequence, long-term equilibrium concerning BO-LID is virtually impossible to determine. The best estimation seems to be the maximum value enforced by BO regeneration and the second drop. Unfortunately, this means that the occupation of state C is systematically underestimated giving rise to the asymmetric uncertainty bars in V_{OC} shown in Fig. 8 and that monitoring the progress is mandatory, even if this induces other uncertainties as well.

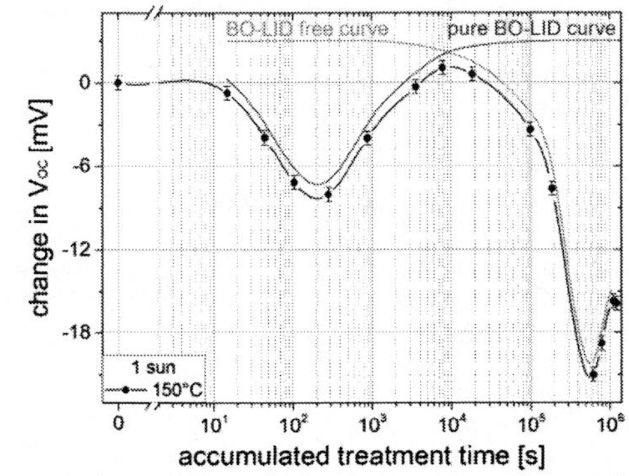

Figure 5: Evolution of V_{OC} during treatment at 150°C and 1 sun illumination intensity. The green and red line depict hypothetical progressions.

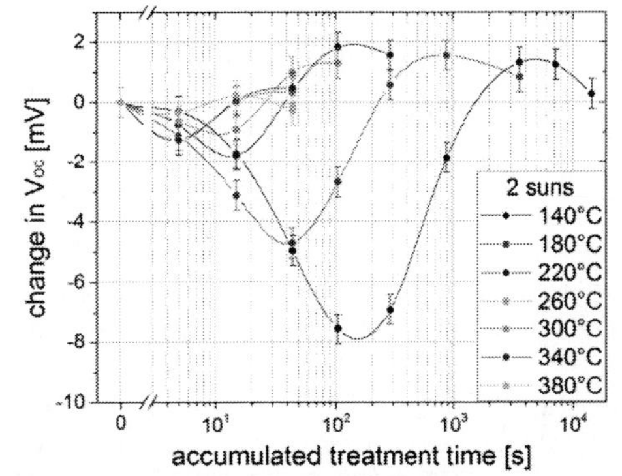

Figure 6: Evolution of V_{OC} during the treatment at various temperatures and 2 suns illumination. Spline lines are only a guide to the eye.

4.3 High temperature treatment

Fig. 6 shows the monitoring results during the treatment at different temperatures. As can be seen, the coldest data set (140°C, black) features the slowest degradation-regeneration-cycle. With rising temperature (180°C, red; 220°C, blue), dynamics are accelerated and the maximal loss decreases, just like simulation has predicted. For even higher temperatures, amplitude decreases further and becomes more and more insignificant, again just like simulation has predicted.

The slight decrease of V_{OC} in the end of every curve shows the beginning of the second drop as depicted in Fig. 5 and marks the point at which treatment was aborted. Nevertheless, a slight loss already occurred.

4.4 Subsequent BO-LID

Fig. 7 shows the results of the subsequent BO-LID treatment at 35°C and 0.15 suns illumination used to convert defects in state A into state B making them distinguishable from defects in state C.

Almost no loss is observable for the lowest treatment temperature (140°C, black) thus meaning almost all defects were -as intended- in state C and not A. Increasing treatment temperature leads to more pronounced losses, indicating a gradual shift of equilibrium from state C to A, again just as simulation has predicted. For even higher temperatures losses would probably converge to the full degradation drop of an untreated sample (gray).

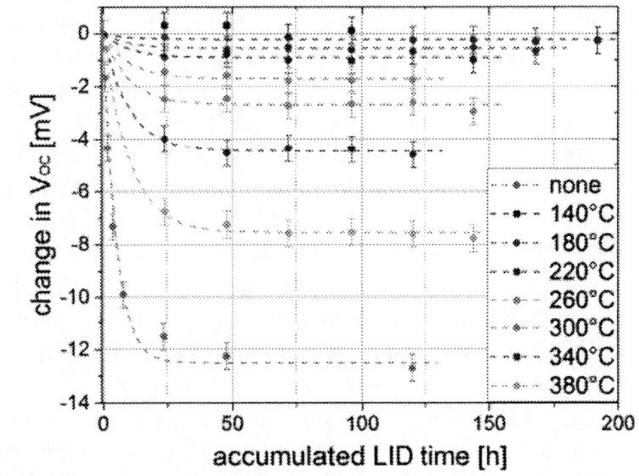

Figure 7: Evolution of V_{OC} during the subsequent BO-LID treatment at ~35°C and 0.15 suns illumination. Dashed lines represent exponential decay curves yielding the long-term drop in V_{OC}.

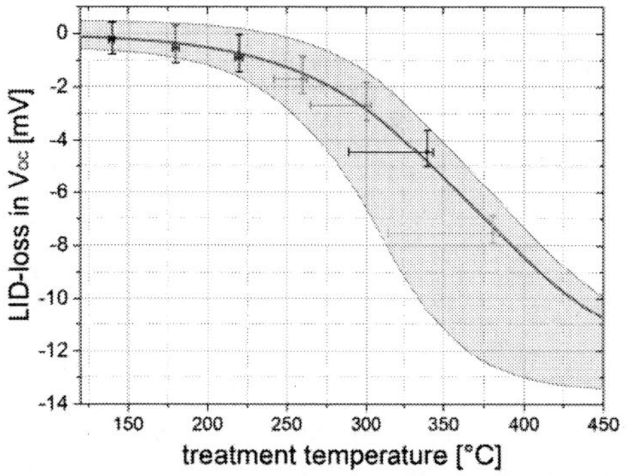

Figure 8: Loss in V_{OC} after BO-LID (Fig. 7) versus treatment temperature. The gray line represents a fit to the data, the gray area takes the experimental uncertainty into account.

4.4 Discussion

The long-term BO-LID loss in V_{OC} from Fig. 7 versus treatment temperature is plotted in Fig. 8. At first glance, Fig. 8 seems to resemble the predicted 'S'-shaped curve in Fig. 4, however, with regard to slope the graphs differ. One could object that Fig. 4 and Fig. 8 do depict substantially different entities and that the 'S'-shape in occupation (Fig. 4) does not necessarily translate to V_{OC} (Fig. 8) without distortion. Therefore, simulation was not only done for the occupation of states but also for the corresponding V_{OC} curve showing that this is not the reason for the discrepancy of Fig. 4 and Fig. 8.

Uncertainty in the input parameters of the simulation, of course, lead to uncertainty in the prediction of occupation as shown in Fig. 4. This may indeed lead to a slight change in slope, but the major effect is a shift on the temperature axis.

The major point of uncertainty is probably found in the experimental data at high temperatures. Firstly, BO related long-term equilibrium was probably not reached as the treatment was broken off too early taking the second LID drop into account. This implies an overestimation of V_{OC} losses and thus introduces a systematic asymmetric uncertainty in V_{OC} loss. Secondly, removing the sample multiple times for monitoring also means that the sample has to heat up again delaying dynamics. Especially for short treatment durations, 'true' temperature is applied shorter than intended and dynamics will have evolved less than expectable from Fig. 3 (calculated with constant temperature). The third point concerns temperature itself. At high temperatures, solar cells tend to bow due to bimetallic stress between the printed Al contact (large coefficient of thermal expansion) and the silicon wafer (small coefficient of thermal expansion). Thus direct thermal contact is hardly present if the sample is not pressed/sucked to the hotplate – which was not the case in the experiment at least for temperatures exceeding 300°C. In consequence, sample temperature may deviate strongly from hotplate temperature and may show lateral inhomogeneity.

The uncertainty in temperature in Fig. 8 was estimated from radiative heat transfer (Stefan-Boltzmann law) resulting in a rather large uncertainty at high temperatures (>300°C) and comparably low uncertainty at low temperatures (<300°C). Taking these uncertainties into account, the slopes of Fig. 4 and Fig. 8 match better.

5 CONCLUSIONS

The 3-state model of BO-LID can be successfully used to describe at least the trend at high temperatures for state-of-the-art PERC-type solar cells: Occupation of the regenerated state decreases with increasing temperature above a certain threshold temperature.

However, uncertainty in the input parameters of the simulation yields, of course, uncertainty in prediction complicating exact predictions. In addition, experimental uncertainty in temperature encountered in the presented results hampers a clear assessment of the threshold temperature up to which BO-LID is limited to a certain degree.

Furthermore, it was found that the PERC-type solar cells under investigation suffer from a second LID effect. Whether this can be attributed to LeTID also occurring in Cz-Si, a degradation of surface passivation quality or a completely different effect, remains unclear.

ACKNOWLEDGEMENT

Part of this work was supported by the German Federal Ministry for Economic Affairs and Energy under contract numbers 0325877C and 0324001. The content is the responsibility of the authors.

REFERENCES

[1] K. Bothe, R. Sinton, J. Schmidt, Progress in Photovoltaics **13** (2005) 287
[2] A. Herguth, G. Schubert, M. Kaes, G. Hahn, Proceedings 21st European Photovoltaic Solar Energy Conference (2006), 530
[3] A. Herguth, G. Hahn, Journal of Applied Physics **108** (2010) 114509
[4] S. Wilking, C. Beckh, S. Ebert, A. Herguth, G. Hahn, Solar Energy Materials & Solar Cells **131** (2014) 2
[5] K. Bothe, J. Schmidt, Journal of Applied Physics **99** (2006) 013701
[6] P. Hamer, B. Hallam, M. Abbott, S. Wenham, Physica Status Solidi RRL **9**(5) (2015), 297
[7] S. Wilking, S. Ebert, A. Herguth, G. Hahn, Journal of Applied Physics **114** (2013) 194512
[8] S. Wilking, M. Foster, A. Herguth, G. Hahn, Solar Energy Materials & Solar Cells **142** (2015) 87
[9] A. Herguth, S. Wilking, Energy Procedia **124** (2017) 60
[10] A. Herguth, Energy Procedia **124** (2017) 53
[11] F. Kersten et al., Solar Energy Materials & Solar Cells 142 (2015) 83
[12] F. Fertig et al., Energy Procedia **124** (2017) 338
[13] D. Chen et al., Solar Energy Materials & Solar Cells **172** (2017) 293
[14] D. Sperber, A. Graf, D. Skorka, A. Herguth, G. Hahn, IEEE Journal of Photovoltaics, accepted, doi: 10.1109/JPHOTOV.2017.2755072

INFLUENCE OF SILICON NITRIDE AND ITS HYDROGEN CONTENT ON CARRIER-INDUCED DEGRADATION IN MULTICRYSTALLINE SILICON

Carlos Vargas[1], Kyung Kim[1], Gianluca Coletti[1,2], David Payne[1], Catherine Chan[1], Stuart Wenham[1] and Ziv Hameiri[1]
[1] School of Photovoltaic and Renewable Energy Engineering, University of New South Wales, Sydney, Australia
[2] Energy Research Centre of the Netherlands, the Netherlands

ABSTRACT: Multicrystalline silicon (mc-Si) wafers show degradation due to carrier injection. This carrier-induced degradation particularly limits the performance of passivated emitter rear cell (PERC) made with mc-Si and is, therefore, gaining increasing attention from the photovoltaic community, as it is expected that PERC architecture will become dominant in the near future. In this study, we investigated the degradation behaviour of mc-Si wafers with six different silicon nitride passivation layers which were then separately fired at six different peak temperatures (a matrix of 36 conditions). We found that wafers fired at a low peak temperature (< 650 °C) had similar degradation rates independent of the silicon nitride deposition conditions, whilst the degradation rate of samples fired at higher temperature showed a clear dependence on these conditions. At high firing temperatures normally used for contact formation, it appears that there is a correlation between the degradation rate and the amount of hydrogen released from the dielectric during firing. Furthermore, we verified that no degradation of the surface passivation quality occurs, indicating that the degradation is primarily associated with a bulk-related defect.

Keywords: Multicrystalline silicon, Degradation, Silicon-Nitride, Hydrogen

1 INTRODUCTION

Light and elevated temperature induced degradation (LeTID) in multicrystalline silicon (mc-Si) wafers were first reported by Ramspeck *et al.* in 2012 [1]. Kersten *et al.* [2] found that this degradation can be induced by current injection and is faster under open-circuit than under short-circuit conditions; therefore, it is also known as carrier-induced degradation (CID) [2], [3]. The formation of a boron-oxygen (B-O) complex has been discarded as a possible cause: (i) as mc-Si doped with gallium (Ga) also presents CID [1], (ii) mc-Si is characterized by low interstitial oxygen (O_i) concentrations in comparison with Czochralski (Cz) silicon, and (iii) the time scales of CID formation defect and regeneration rates are much slower than those of B-O [2].

Fertig *et al.* [4] reported more pronounced degradation in mc-Si cells with an aluminium oxide (AlO_x) passivation layer than in full-area aluminium back surface field mc-Si cells. Therefore, they concluded that the AlO_x surface passivation could play a role in CID formation. Recently, Kersten *et al.* [5] reported different CID rates for mc-Si lifetime samples passivated with AlO_x, silicon nitride (SiN_x) and AlO_x/SiN_x stack layers. Particularly, they used two SiN_x layers deposited by two different industrial systems and reported differences in the degradation rates. This was the first indication that the SiN_x properties can influence CID rate. However, the deposition conditions and the SiN_x properties were not reported.

It is also known that CID has a strong dependence on the firing conditions [6], [7]. Higher peak firing temperatures lead to more pronounced degradation. CID is less severe in samples fired at low temperature (< 600 °C) and it is not present in non-fired samples [6].

In this study, we vary the deposition parameters to investigate their impact on the CID extent and its formation and regeneration rates.

2 EXPERIMENTAL METHOD

Neighbouring six inch conventional *p*-type mc-Si wafers with a thickness of ~ 190 μm and resistivity of 1.9 Ω.cm were used to fabricate symmetrical lifetime test structures. Processing included: RCA (Radio Corporation of America) cleaning followed by a light phosphorous diffusion (both sides; carried out in a Tempress $POCl_3$ tube) to form a doped layer with a sheet resistance of 80 Ω/□. Subsequently, the wafers received a hydrofluoric acid (HF) dip followed by a SiN_x deposition [8]. These test structures were divided into six different groups, where each group was symmetrically deposited with a SiN_x layer on both sides using a Meyer Burger MAiA plasma-enhanced chemical vapour deposition (PECVD) system (see Table I for the deposition conditions and properties of each layer). In parallel, six Cz double-side polished wafers were prepared using the same RCA cleaning process. Each of them was deposited with SiN_x only on one side. These polished wafers were used to characterize the optical properties of the layer [thickness and refraction index (RI)] and for Fourier transform infrared spectroscopy (FTIR) measurements to estimate their hydrogen content (see Table I).

Table I: Deposition conditions and layer properties

Layer label	Temperature (°C)	Gas ratio (NH_3/SiH_4)	Pressure (mBar)	Thickness (nm)	RI (at 633 nm)	H fraction as deposit
SiN_x-1	442	2.25	0.17	121.9	2.13	9.92 %
SiN_x-2	442	4.00	0.33	107.7	2.14	10.34 %
SiN_x-3	400	2.25	0.17	120.5	2.14	11.13 %
SiN_x-4	400	4.00	0.33	109.6	2.14	11.36 %
SiN_x-5	350	2.25	0.17	121.8	2.12	12.15 %
SiN_x-6	350	4.00	0.33	113.0	2.12	12.41 %

The six-inch mc-Si wafers were laser cleaved into 3.9 cm × 3.9 cm tokens and underwent firing at six different peak temperatures (589 °C, 645 °C, 676 °C, 748 °C, 798 °C and 858 °C - actual temperature measured on wafer). Subsequently, the wafers were degraded using a laser at 938 nm with an intensity of 46 kW/m^2 and a sample temperature of 140 °C, following the laser-accelerated process proposed by Payne *et al.* [3]. This process was applied for different time-steps, starting with a short time (2 s) in progressive steps until partial lifetime recovery was reached (600 s), allowing for clear identification of any differences in the time dependence of CID at the most degraded point for each wafer. At this stage, the Cz wafers were fired at 748 °C and 795 °C, and the change in the hydrogen fraction of the films was measured using FTIR.

After each laser treatment step, the changes in the effective lifetime of the sample were measured using a photoconductance (PC) lifetime tester (WTC-120, Sinton instruments) along with photoluminescence (PL) imaging (BTI LIS-R1).

3 RESULTS AND DISCUSSION

3.1 *Degradation differences between the different SiN_x layers*:

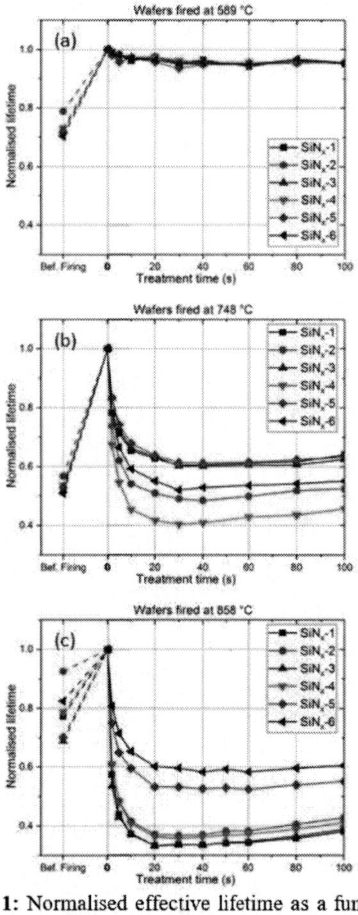

Figure 1: Normalised effective lifetime as a function of the laser treatment time for different SiN_x layers fired at: (a) 589 °C, (b) 748 °C and (c) 858 °C. The firing point corresponds to the zero point in degradation time.

Despite using samples from neighbouring wafers, variations within 10% of the effective lifetime were found between samples before firing; after firing these differences were within 15%, depending on the firing peak temperature [6]. Therefore, in order to identify the changes in the lifetime during the degradation process, independently of the sample effective lifetime after firing (τ_{fire}), the normalised lifetime ($\tau_{norm} = \tau(t) / \tau_{fire}$, where $\tau(t)$ is the effective lifetime after each laser treatment step, including the value before firing) was used. The effective lifetimes were taken at an excess carrier density Δn = 1×10^{15} cm^{-3}.

Figures 1 (a)-(c) presents τ_{norm} of the tokens fired at peak temperatures of 589 °C, 748 °C and 858 °C as a function of laser illumination time. At low firing temperature [see Figure 1 (a)], the degradation was weak and slow, as expected [6], and no differences in the CID extent between the different SiN_x layers could be observed. Higher firing temperatures increased the degradation effect [see Figures 1 (b) and (c)] and differences in degradation characteristics between the different SiN_x layers become apparent.

Figures 2 (a) to (d) present PL images of two samples (passivated with SiN_x-1 and SiN_x-4 layers) fired at the same temperature of 748 °C, after firing and in the most degraded state. It is possible to observe how these neighbouring samples degraded differently due to the different SiN_x layers.

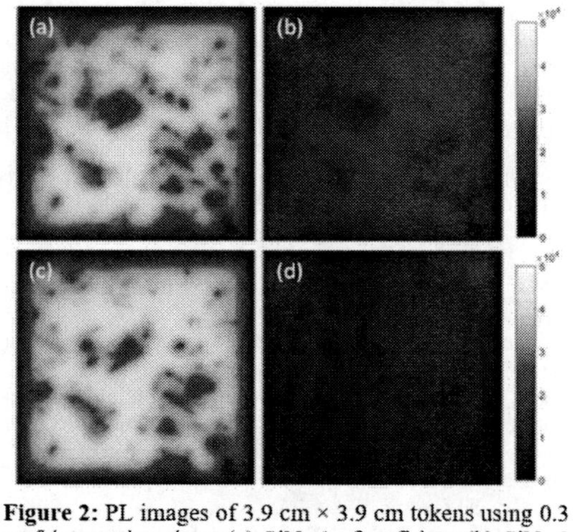

Figure 2: PL images of 3.9 cm × 3.9 cm tokens using 0.3 s of integration time: (a) SiN_x-1 after firing, (b) SiN_x-1 after 30 s of laser treatment, (c) SiN_x-4 after firing, and (d) SiN_x-4 after 30 s of laser degradation.

Figure 3 presents the relative change in the effective lifetime [defined as $(\tau_{deg}-\tau_{fire})/\tau_{fire}\times 100\%$, where τ_{deg} is the effective lifetime in the most degraded state] due to the degradation. At low firing temperatures (< 650 °C) the degradation is similar for all the samples, independently of the SiN_x layer. However, at higher firing temperatures, a significant difference can be noticed between the different layers. This difference is relevant as most screen-printing pastes require peak firing temperatures of around 800 °C. Similar results were found for other wafer sets and for other firing conditions (not shown here; they will be presented in a future publication).

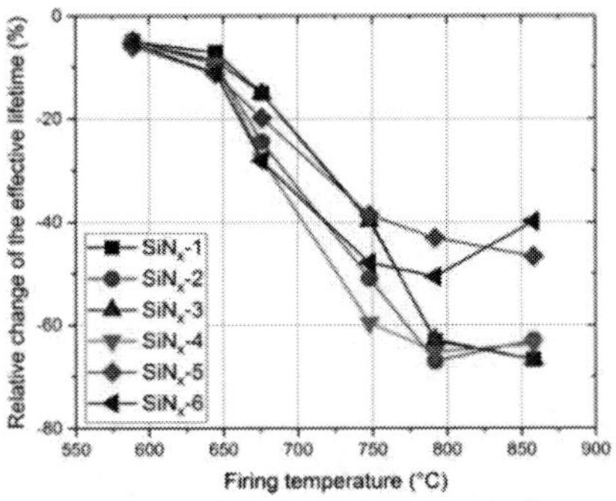

Figure 3: Relative lifetime change due to CID for different firing temperatures and SiN_x passivation layers.

3.2 *Impact of the deposition conditions and hydrogen released during the firing on CID*:

To estimate the amount of hydrogen released from the dielectric layer during firing, we compared the hydrogen content in the films before and after firing. Despite the fact that only a fraction of the released hydrogen would diffuse. Despite the fact that only a fraction of the released hydrogen would diffuse into the bulk, we can infer that greater hydrogen release would correlate to a greater amount of hydrogen penetrating into the wafer. Figure 4 presents the CID extent (defined as the absolute value of the relative change in the effective lifetime) as a function of the fraction of hydrogen released during the firing. The degradation was more pronounced for wafers with SiN_x layers that released a higher hydrogen fraction. Similar results were obtained for other firing temperatures.

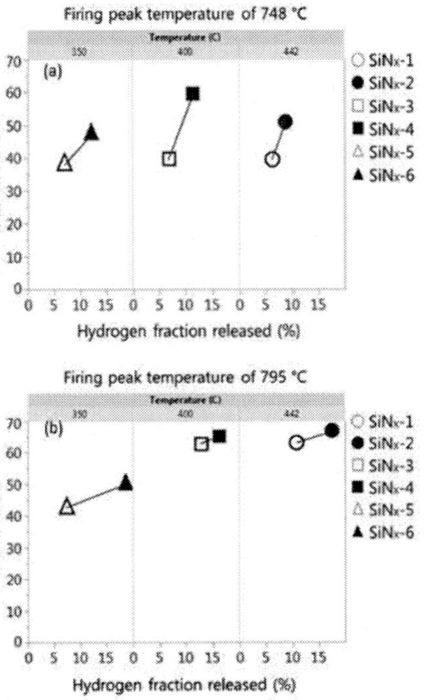

Figure 4: Comparison between the lifetime changes for the wafers fired at 748 °C (a) and 798 °C (b) in terms of the hydrogen fraction released in the firing step and the temperature and pressure of the passivation layer deposition.

This is a first indication that CID may be related to the amount of hydrogen released during the firing step. This indication supports Chan *et al.* [9] and Jensen *et al.* [10] recent suggestion that hydrogen has a role in the formation and passivation of the CID defect. The secondary dependence of the CID extent with the deposition temperature of the SiN_x could be associated with different layer density [8] or differences in the thermal history of the samples [9].

3.3 *Surface recombination changes due to CID*:

Figures 5 (a) and (b) plots the surface recombination saturation current (*Jos*) and the high-injection bulk lifetime (τ_{bulk}; using the Kane and Swanson method [11] as implemented in the Sinton Instruments lifetime tester) extracted at Δn = 2×10^{16} cm^{-3}, before and after the degradation for the tokens fired at 748 °C. No significant change of the surface passivation quality can be noticed (in some cases an improvement can be observed), whilst τ_{bulk} changes considerably. Similar results were found for the wafers fired at other peak temperatures. Therefore, it can be concluded that CID is not associated with changes in the surface recombination and is most likely related to a bulk-defect.

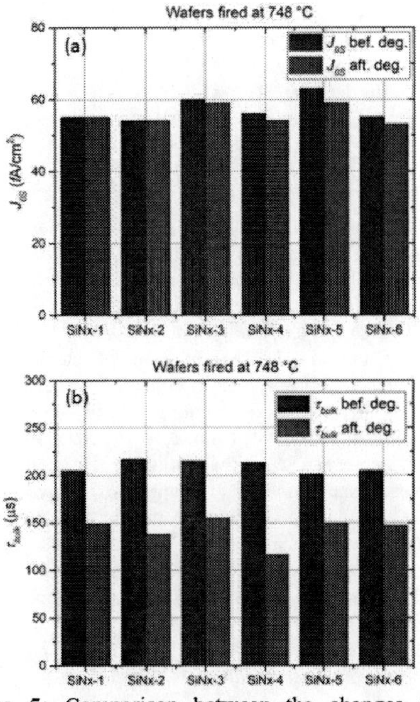

Figure 5: Comparison between the changes in the surface saturation current density (a) and the high-injection bulk lifetime (b) before and after the degradation of wafers fired at 748 °C.

4 CONCLUSIONS

In this contribution, we demonstrated that the SiN_x deposition conditions influence the extent of CID when high firing temperatures are used. A correlation between the hydrogen released during the firing and the degradation was found, indicating a possible link between them. Moreover, we do not detect significant changes in the surface recombination and therefore we concluded that the CID effect is primarily associated with a bulk defect.

REFERENCES

[1] K. Ramspeck *et al.*, "Light induced degradation of rear passivated mc-Si solar cells," *27th Eur. Photovolt. Sol. Energy Conf. Exhib.*, pp. 861–865, Oct. 2012.

[2] F. Kersten *et al.*, "Degradation of multicrystalline silicon solar cells and modules after illumination at elevated temperature," *Sol. Energy Mater. Sol. Cells*, vol. 142, pp. 83–86, Nov. 2015.

[3] D. N. R. Payne *et al.*, "Acceleration and mitigation of carrier-induced degradation in *p*-type multi-crystalline silicon," *Phys Status Solidi–Rapid Res Lett*, vol. 10, no. 3, pp. 237–241, Mar. 2016.

[4] F. Fertig, K. Krauß, and S. Rein, "Light-induced degradation of PECVD aluminium oxide passivated silicon solar cells," *Phys. Status Solidi - Rapid Res. Lett.*, vol. 9, no. 1, pp. 41–46, Jan. 2015.

[5] F. Kersten, J. Heitmann, and J. W. Müller, "Influence of Al_2O_3 and SiN_x Passivation Layers on LeTID," *Energy Procedia*, vol. 92, pp. 828–832, Aug. 2016.

[6] C. E. Chan *et al.*, "Rapid stabilization of high-performance multicrystalline *p*-type silicon PERC cells," *IEEE J. Photovolt.*, vol. 6, no. 6, pp. 1473–1479, Nov. 2016.

[7] D. Bredemeier, D. Walter, S. Herlufsen, and J. Schmidt, "Lifetime degradation and regeneration in multicrystalline silicon under illumination at elevated temperature," *AIP Adv.*, vol. 6, no. 3, p. 035119, Mar. 2016.

[8] Z. Hameiri, N. Borojevic, L. Mai, N. Nandakumar, K. Kim, and S. Winderbaum, "Low-absorbing and thermally stable industrial silicon nitride films with very low surface recombination," *IEEE J. Photovolt.*, vol. 7, no. 4, pp. 996–1003, Jul. 2017.

[9] C. Chan *et al.*, "Modulation of carrier-induced defect kinetics in multi-crystalline silicon PERC cells through dark annealing," *Phys Status Solidi–Rapid Res Lett*, vol. 1, no. 2, p. 1600028, Feb. 2017.

[10] M. A. Jensen, A. E. Morishige, J. Hofstetter, D. B. Needleman, and T. Buonassisi, "Evolution of LeTID defects in *p*-type multicrystalline silicon during degradation and regeneration," *IEEE J. Photovolt.*, vol. 7, no. 4, pp. 980–987, Jul. 2017.

[11] D. Kane and R. Swason, "Measurement of the emitter saturation current by a contactless photoconductivity decay method," *18th IEEE Photovolt. Spec. Conf.*, vol. 18, pp. 578–583, 1985.

INVESTIGATING POSSIBLE CAUSES OF LIGHT INDUCED DEGRADATION IN BORON-DOPED FLOAT-ZONE SILICON

D. Sperber, A. Herguth, G. Hahn
University of Konstanz, Department of Physics, 78457 Konstanz, Germany

ABSTRACT: The use of different silicon nitride deposition tools is found to change the degree of light induced degradation (LID) of B-doped float-zone silicon after a fast firing step. In addition, a thermally grown SiO_2 interlayer further suppresses LID after firing. Possible mechanisms and a potential link to Light and elevated Temperature Induced Degradation (LeTID) are discussed. Furthermore, it is shown that LID is not related to an earlier described class of thermally activated defects in float-zone silicon and that phosphorous gettering does not influence the occurrence of LID in B-doped float-zone silicon significantly.
Keywords: Light induced degradation, LeTID, Stability, Degradation, Silicon nitride, Minority carrier lifetime

1 MOTIVATION

Light and Elevated Temperature Induced Degradation (LeTID) is a subtype of light induced degradation (LID) and poses a significant challenge as it may severely limit bulk minority charge carrier lifetime τ_b in multicrystalline silicon (mc-Si) [1-3]. Highly purified B-doped float-zone silicon (FZ-Si), on the other hand, has been shown to suffer from LID, too, after a fast firing step [4-6]. There are a number of similarities between LID in mc-Si and FZ-Si, leading to the question whether LID in FZ-Si is connected to LeTID.

Only recently, it has been supposed that LeTID also affects monocrystalline Czochralski-grown silicon (Cz-Si) [7,8], giving rise to the idea that LeTID affects crystalline silicon in general. If LeTID also affects B-doped FZ-Si, this would offer the possibility to study LeTID in a well controlled and highly purified material. Hence, the aim of this contribution is to further investigate possible root causes of LID in FZ-Si and compare it to LeTID.

2 EXPERIMENTAL

All shown samples are made of B-doped FZ-Si wafers of thickness 250 µm and specific resistivity ~1 Ωcm, equaling a doping density N_d of ~$1.5\cdot10^{16}$ cm^{-3}. The wafers already received a saw damage etch and chemical clean by the manufacturer and were shipped with a thin chemically grown oxide on the wafer surface. Part of the wafers did not receive further chemical pre-treatment while others were subjected to a dip in hydrofluoric acid (HF) to remove the thin oxide. Selected samples received a clean in a solution of H_2O_2 and H_2SO_4 at 80°C, followed by another dip in HF (Piranha clean).

Two different plasma enhanced chemical vapor deposition (PECVD) tools were used for two-sided deposition of SiN_x:H. A type 1 SiN_x:H was grown in a lab-tool using a direct plasma at a frequency of 13.56 MHz. The deposition temperature was 400°C and a two-sided deposition took ~20 min. Type 2 SiN_x:H was grown in an industrial direct plasma tool with a frequency of 40 kHz. In this case, the deposition temperature was 450°C and a two-sided deposition took ~1.5 h. Accordingly, samples with different SiN_x:H deposition have seen a different temperature load during processing. Both SiN_x:H depositions resulted in a refractive index ~2.0 (600 nm) and thickness ~75 nm. As precursor gases, NH_3 and SiH_4 were used in both PECVD tools, resulting in a hydrogen-rich SiN_x:H layer in both depositions.

Afterwards, most samples were laser-cut into square samples of edge length 5 cm. Most samples were then fired at a measured peak sample temperature ~800°C in a fast firing belt furnace. Two samples were, however, fired as part of a whole wafer and the temperature measurement might have underestimated the real wafer temperature. In this case, firing temperatures are indicated as ≥800°C. After firing, samples were stored in darkness until LID treatments started.

LID treatments were carried out on illuminated hotplates at 80°C and ~1 sun equivalent illumination intensity, achieved by matching the short circuit current of a solar cell to that under a solar spectrum simulator [9]. During treatment, effective minority carrier lifetime τ_{eff} was repeatedly determined by photoconductance decay, using a Sinton Instruments lifetime tester (WCT-120). One sample was additionally characterized with photoluminescence (PL) imaging at different points of sample treatment.

3 RESULTS

3.1 Influence of different SiN_x:H on LID in FZ-Si

Two samples which received no further chemical pre-treatment were coated with different types of SiN_x:H before firing at ~800°C. The results of subsequent LID treatment at 80°C and ~1 sun are shown in Fig. 1. As can be seen, τ_{eff} reaches a minimum, denoted with the roman number I, after ~1 h of treatment in both samples. Feature I has previously been shown to be caused by degradation of τ_b and a subsequent regeneration of τ_b then leads to a maximum II of τ_{eff} [4]. The subsequent long-term decline of τ_{eff} has been shown to be caused by a decrease of surface passivation quality [4,10] and will not be the focus of this paper.

As can be seen, feature I is significantly stronger in the sample which received a type 2 SiN_x:H. This is also visible in the change of effective defect density ΔN_{eff} (Fig. 2) which was calculated with reference to the first measurement point after treatment started to exclude initial effects of illuminating a sample (*e.g.* changing the state of FeB complexes):

$$\Delta N_{eff}(t) = 1/\tau_{eff}(t) - 1/\tau_{eff}(t_1)$$

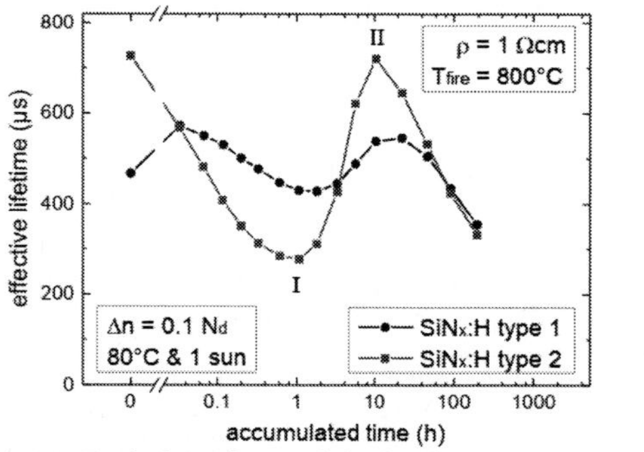

Figure 1: Evolution of τ_{eff} at injection level $\Delta n = 0.1\ N_d$ of two samples that differed in SiN_x:H deposition only and were treated simultaneously at 80°C and ~1 sun.

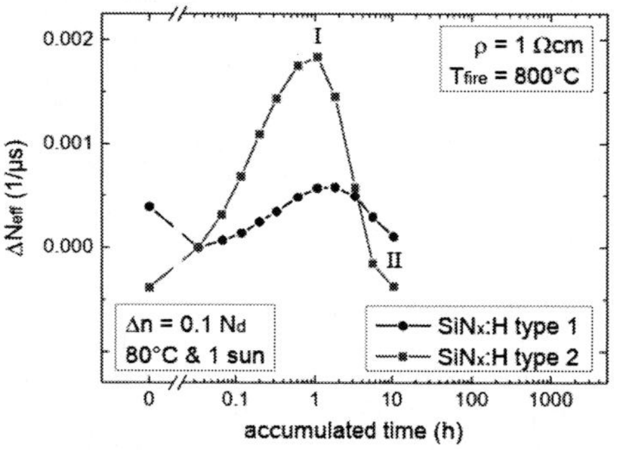

Figure 2: Evolution of ΔN_{eff} ($\Delta n = 0.1\ N_d$) of the samples shown in Figure 1.

So far, measurement data were only evaluated at $\Delta n = 0.1\ N_d$. In [11], an injection dependent visualization has been introduced to show a wide injection range of measurement data while still providing good temporal resolution as well. The injection resolved plots of the samples discussed before are shown in Figs. 3 and 4 and provide additional information: It can be seen that the biggest difference of degradation at feature I can be found at lower injection (blue) in accordance with stronger Shockley-Read-Hall recombination in the FZ bulk. The values at higher injection (red) and a calculation of the surface saturation current density J_{0s} additionally reveal that the sample with type 1 deposition suffers from more surface related recombination, explaining the lower τ_{eff} before sample treatment and at feature II.

Because both samples did not receive a chemical pre-treatment and differ in the PECVD step only, this step should have caused the different degradation behavior. One possible difference lies in a different hydrogenation of the samples. Hydrogen has been speculated to be involved in LID of FZ-Si [4,6], and it is known that different PECVD deposition temperatures result in a different degree of hydrogenation of the silicon bulk in a subsequent high temperature step [12]. Additionally, another substance besides hydrogen may affect LID via in- or out-diffusion during the firing step.

Aside from diffusion related differences, the samples have also seen a different temperature load during processing due to different deposition temperature and duration. It is therefore possible as well that a different degradation behavior arises due to different temperature load and future experiments should aim on varying temperature load independent of SiN_x:H deposition type to further investigate a possible influence of temperature load on LID behavior.

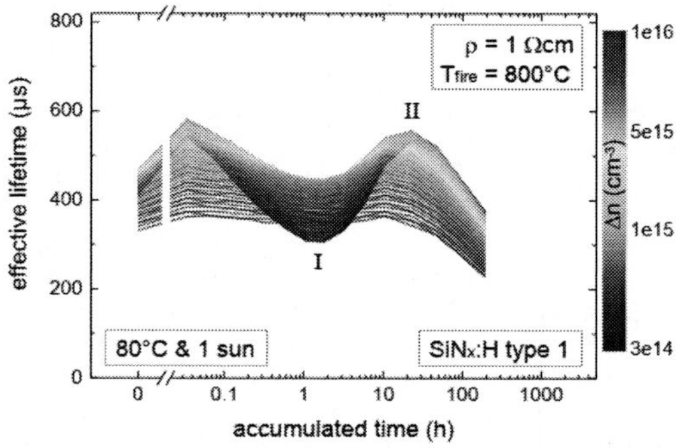

Figure 3: Injection resolved evolution of τ_{eff} of the sample with type 1 SiN_x:H. Δn is color coded, ranging from lower injection (blue) to higher injection (red).

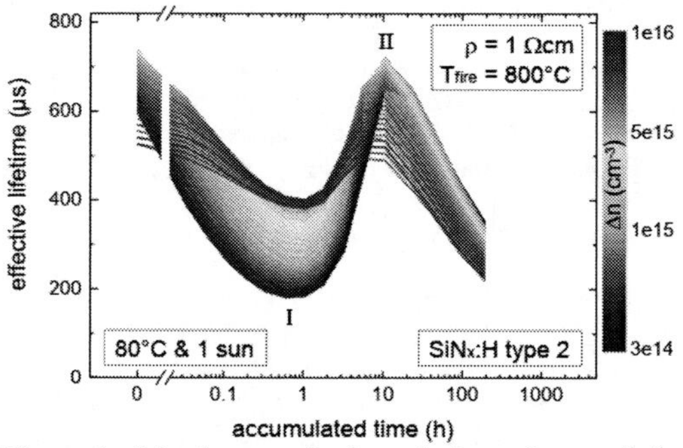

Figure 4: Injection resolved evolution of τ_{eff} of the sample with type 2 SiN_x:H.

3.2 Influence of 1000°C anneal and $POCl_3$ gettering

In [13] it has been shown that both B-doped and P-doped FZ-Si may suffer from thermally activated defects after being treated in a temperature range between ~400°C and ~800°C. However, this type of defects can be permanently annihilated by annealing a sample at a temperature ≥1000°C [13]. To investigate a possible link between these thermally activated defects and LID in FZ-Si, the sample shown in Fig. 5 first received a 1000°C anneal in oxygen ambient for 30 min before receiving an HF dip and a Piranha clean. After a type 2 deposition and firing, the sample still shows significant LID.

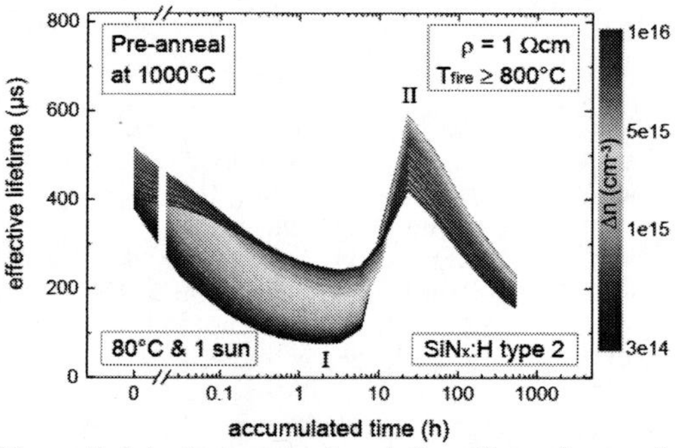

Figure 5: Injection resolved evolution of τ_{eff} of a sample passivated with type 2 SiN_x:H. The sample received an anneal at 1000°C for 30 min before sample processing.

Degradation is even stronger as in the samples discussed before which is probably related to a slightly higher firing temperature of the sample. Consequently, it can be stated that LID is not related to thermally activated defects as discussed in [13]. A similar conclusion has been drawn in [6] concerning AlO_x:H/SiN_x:H passivated samples.

Another sample first received a $POCl_3$ diffusion (840°C for ~45 min) which may getter impurities. After etching off the resulting emitter, the sample received a Piranha clean and a type 2 SiN_x:H before firing. The sample shows a very similar degradation behavior compared to samples without gettering step (data not shown), leading to the conclusion that the defects responsible for LID in FZ-Si are not significantly influenced by a $POCl_3$ gettering step.

3.3 Influence of SiO_2 interlayer

Another sample was passivated with a layer stack consisting of 10 nm SiO_2 (thermally grown at 840°C for 20 min) capped with type 2 SiN_x:H. As can be seen in Fig. 6, this sample shows no significant bulk related degradation (no dip in blue). The sample received a Piranha clean before thermal oxidation. However, other Piranha cleaned samples have shown strong bulk related degradation before [4,5,10]. Therefore, it appears that the thermal oxidation step caused the difference in degradation behavior. Because a phosphorous gettering step at the same temperature (840°C for ~45 min) does not significantly influence LID, as discussed in the previous section, the changed degradation behavior of the SiO_2/SiN_x:H sample is probably not related to a different temperature load. Accordingly, different hydrogenation or, more generally, diffusion-related differences are likely causes for the suppression of LID in this sample. Similar to the sample discussed here, another SiO_2/SiN_x:H sample with even better surface passivation (due to a higher oxidation temperature of 900°C) shows very weak bulk related degradation [10].

3.4 Inhomogeneous degradation behavior

One sample passivated with type 2 SiN_x:H was measured with PL imaging at different points of sample treatment. As can be seen in Fig. 7, the sample shows a rather homogeneous surface before treatment and after regeneration. Only the lower left sample corner has suffered significantly in the regenerated state (II) due to sample handling.

However, in the degraded state (I), a ring-like pattern can be observed. As the sample was cut out of the upper right part of a wafer, the ring-like pattern appears to be

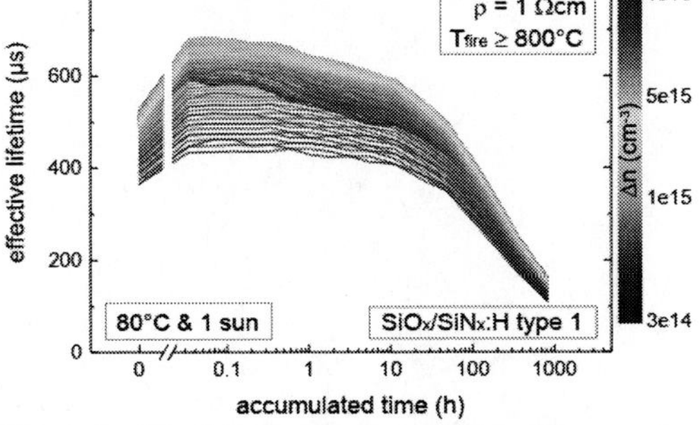

Figure 6: Injection resolved evolution of τ_{eff} of a SiO_2/SiN_x:H passivated sample (type 2 SiN_x:H).

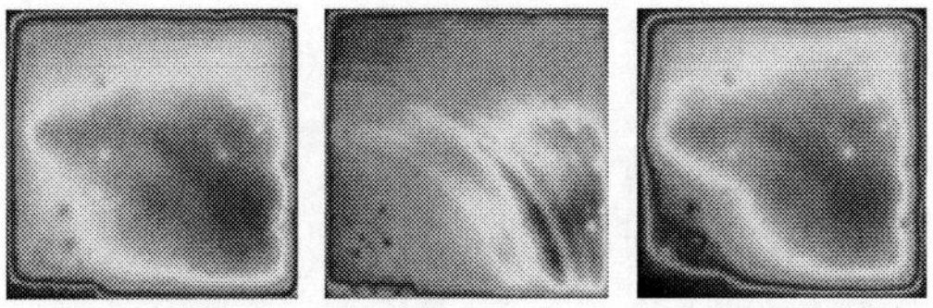

(a) before treatment (b) degraded (I) (c) regenerated (II)

Figure 7: PL images of a sample passivated with type 2 SiN_x:H, fired at ~835°C and treated at 80°C and ~1 sun (a) before treatment (b) after bulk degradation (I) and (c) after bulk regeneration (II).

concentric to the wafer center. A similar observation was made in [6] where it has been concluded that the pattern cannot occur only due to spatial inhomogeneous doping (which would cause slight variations in PL intensity) but probably reflects an inhomogeneous distribution of the defect causing LID. It remains, however, possible that slight doping variations influence the distribution of defects in the degraded state: As LID does not occur in P-doped FZ-Si [6,11], it is possible that an acceptor dopant (*e.g.* boron) takes part in defect formation. Slight variations in dopant density may then result in an inhomogeneous defect distribution.

3.5 Comparison of LID in FZ-Si with LeTID

As shown before, LID in FZ-Si may be related to the degree of hydrogenation of the silicon bulk. This possible role of hydrogen may also link LID in FZ-Si to LeTID in mc-Si: Both LID phenomena increase in strength with higher peak firing temperature [14,15] and depend on the presence of hydrogen rich layers during firing [6,16]. Accordingly, both degradation mechanisms might require an in-diffusion of hydrogen from dielectric layers into the silicon bulk during firing at elevated temperatures and possibly, LID in FZ-Si and LeTID in mc-Si share a common root cause.

If a similar mechanism is causing LID in FZ-Si and LeTID in mc-Si, further studies on FZ-Si may allow to investigate the underlying degradation mechanism in a purified and well controlled material. By optimizing dielectric deposition conditions, both temperature load and hydrogenation could be influenced to lessen the impact of LID on crystalline silicon.

4 SUMMARY

It has been shown that B-doped FZ-Si samples coated with different SiN_x:H deposition tools express a different degree of LID after a fast firing step. A SiO_2/SiN_x:H passivated sample showed no significant LID. Potential reasons for these observations include different hydrogenation of the silicon bulk and different temperature load during sample processing. Neither annihilation of thermally activated defects as described in [13] nor a gettering step influenced LID significantly. In the degraded state, an inhomogeneous distribution of the underlying defect could be observed. Finally, a potential link of LID in FZ-Si to LeTID has been discussed.

ACKNOWLEDGEMENTS

The authors would like to thank A. Heilemann, L. Mahlstaedt, J. Rinder, S. Joos, B. Rettenmaier, F. Mutter and J. Engelhardt for technical support. Part of this work was supported by the German Federal Ministry for Economic Affairs and Energy under contract numbers 0325763B and 0324001. The content is the responsibility of the authors.

REFERENCES

[1] K. Ramspeck, S. Zimmermann, H. Nagel, A. Metz, Y. Gassenbauer, B. Birkmann, A. Seidel, Proceedings 27th European Photovoltaic Solar Energy Conference (2012) 861.
[2] F. Fertig, K. Krauß, S. Rein, Physica Status Solidi RRL 9 (2015) 41.
[3] F. Kersten, P. Engelhart, H.C. Ploigt, A. Stekolnikov, T. Lindner, F. Stenzel, J.W. Müller, Solar Energy Materials and Solar Cells 142 (2015) 83.
[4] D. Sperber, A. Heilemann, A. Herguth, G. Hahn, IEEE Journal of Photovoltaics 11 (2017) 463.
[5] D. Sperber, A. Herguth, G. Hahn, Physica Status Solidi RRL 11 (2017) 1600408.
[6] T. Niewelt, M. Selinger, N.E. Grant, W. Kwapil, J.D. Murphy, M.C. Schubert, Journal of Applied Physics 121 (2017) 185702.
[7] Fertig et al., Energy Procedia 124 (2017) 338.
[8] D. Chen, M. Kim, B.V. Stefani, B.J. Hallam, M.D. Abbott, C.E. Chan, R. Chen, D.N.R. Payne, N. Nampalli, A. Ciesla, T.H. Fung, K. Kim, S.R. Wenham, Solar Energy Materials and Solar Cells 172 (2017) 293.
[9] A. Herguth, Energy Procedia 124 (2017), 53.
[10] D. Sperber, A. Graf, D. Skorka, A. Herguth, G. Hahn, IEEE Journal of Photovoltaics, accepted, doi: 10.1109/JPHOTOV.2017.2755072.
[11] D. Sperber, A. Graf, A. Heilemann, A. Herguth, G. Hahn, Energy Procedia 124 (2017), 794.
[12] H.F.W. Dekkers, G. Beaucarne, Applied Physics Letters 89 (2006) 211914.
[13] N.E. Grant, V.P. Markevich, J. Mullins, A.R. Peaker, F. Rougieux, D. Macdonald, J.D. Murphy, Physica Status Solidi A 11 (2016) 2844.
[14] D. Sperber, A. Herguth, G. Hahn, Energy Procedia 92 (2016) 211.
[15] D. Bredemeier. D. Walter, S. Herlufsen, J. Schmidt, AIP Advances 6 (2016), 035119.
[16] F. Kersten, J. Heitmann, J.W. Müller, Energy Procedia 92 (2016) 828.

IMPACT OF TEMPERATURE AND DOPING ON LETID AND REGENERATION IN MC-SI

Jakob Fritz, Annika Zuschlag, Daniel Skorka, Andreas Schmid, Giso Hahn
University of Konstanz, Department of Physics, 78457 Konstanz, Germany
jakob.fritz@uni-konstanz.de, Tel: +49 7531 885296, Fax: +49 7531 883895

ABSTRACT: Light and elevated temperature induced degradation can be observed in multicrystalline Si PERC solar cells and results in efficiency losses of up to 12%$_{rel}$. This work investigates the impact of temperature as well as doping on the degradation and regeneration behavior of lifetime samples. While ungettered Ga-doped samples did not show any regeneration effect within the duration of the experiment, increasing the treatment temperature sped up the degradation and regeneration where it could be observed. Ga-doped samples showed a slower degradation compared to B-doped samples. For higher temperatures (≥200°C) the lifetimes during regeneration exceeded the initial lifetimes before starting the illumination treatment for the gettered samples. After regeneration at higher temperature, the stability of lifetime at 75°C was tested. When tested after treatment at 150°C and before the lifetime exceeded the initial value, it stayed stable. After treatment at 250°C until the lifetime exceeded the initial lifetime, it decreased again during the stability test. However, the kinetics of this decrease was much slower than during the initial degradation.
Keywords: Degradation, Lifetime, Multicrystalline Silicon

1. INTRODUCTION

LeTID is a recently discovered effect on mc-Si PERC (passivated emitter and rear cell) solar cells [1-3]. This effect can cause efficiency losses up to ~10%$_{rel}$ under illumination [3] and cannot be attributed to standard BO complex formation or FeB pair dissociation [1]. Therefore, it can be a showstopper for applying PERC technology to mc-Si. Also a regeneration process after degradation has been observed in [3]. The kinetics of degradation and regeneration is influenced by illumination and temperature, while the underlying mechanisms causing the degradation and regeneration of mc-Si PERC solar cells are still unknown.

This work investigates the influence of doping and temperature on the dynamics of LeTID and regeneration by varying the treatment temperature for differently processed sister wafers for lifetime studies. The motivation for the increase of temperature is to study the kinetics of degradation and regeneration and check the stability of bulk lifetime after regeneration. The outcome is of high importance for implementation of regeneration schemes for mc-PERC solar cells in industrial production where short treatment times and stability of lifetime after regeneration are essential.

2. EXPERIMENTAL

Four sets (A-D) of sister wafers (5x5cm^2, 1.3-1.4 Ωcm, comparable grain and defect structure) are used to compare the influence of different degradation conditions on B- (thickness 200 µm) and Ga-doped (150 µm) mc-Si materials. The neighbouring sample sets A and B are B-doped, sets C and D Ga-doped. Wafers originate from the middle of the ingot height and are treated equally within one sample set. All wafers are chemically etched to remove saw damage. Sample sets A (B-doped) and C (Ga-doped) receive only a surface passivation by firing of a direct plasma-enhanced chemical vapour deposition (PECVD) SiN_x:H layer. Sample sets B (B-doped) and D (Ga-doped) are gettered using an open tube $POCl_3$ diffusion process (55 Ω/□). The emitters are removed and the same surface passivation as in case of sample sets A and C was applied. This process sequence is shown in Fig. 1.

Different degradation conditions are applied to the samples. While illumination with halogen lamps (0.9±0.05 suns) is held constant, the applied temperature is varied from 25°C to 250°C. Effective minority charge carrier lifetime (τ_{eff}) is measured repetitively by the fast and self-calibrated time resolved photoluminescence imaging (TR-PLI) [4] method at room temperature, resulting in a series of spatially resolved lifetime maps for each sample over degradation time.

To determine whether the observed maxima after regeneration are stable at 75°C, samples are put on a hotplate after they were degraded and regenerated at higher temperatures such as 150°C or 250°C. Samples at these higher temperatures show a development that includes two decreases and increases in lifetime. The samples treated at 250°C are tested on lifetime stability in the final maximum whereas the stability of the first maximum was tested with samples treated at 150°C (see Fig. 2 and 3).

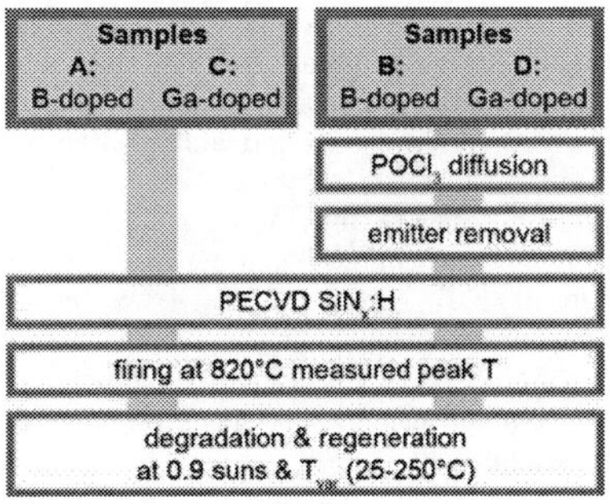

Figure 1: Process sequence of the investigated lifetime samples.

3. RESULTS AND DISCUSSION

3.1 Harmonic average lifetime values

The behaviour of lifetime samples degraded at 75°C and 1 sun was examined in multiple studies and was published before (*e.g.* [3, 5]) Degradation at this temperature is often used in literature, therefore the choice of this temperature is helpful for comparison of results from different experiments, and was proposed as standard for degradation experiments [3]. Nevertheless, it is tempting to speed up the process of degradation and regeneration by increasing the temperature.

This can be seen in Fig. 2, where the development of τ_{eff} over time is shown for B-doped P-gettered samples. As for all plots, the first 0.02 h (slightly over 1 min) are plotted linearly to show the behavior at the start of the treatment. After the break in the x-axis the plot is shown in logarithmic scale to better represent the whole timespan of the examination.

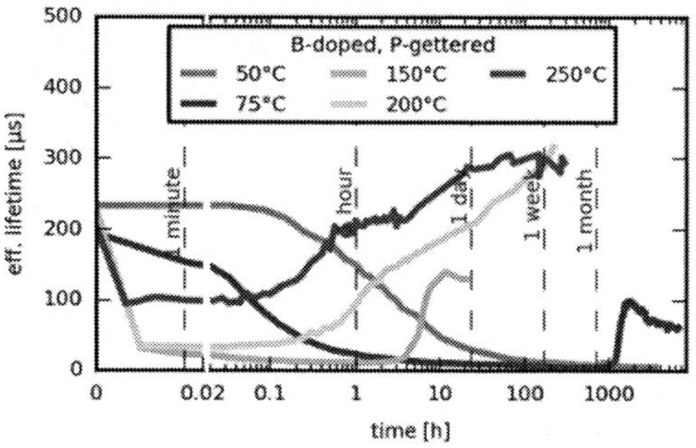

Figure 2: Overview of the harmonic averages of τ_{eff} for each point in time for different degradation temperatures. The samples are B-doped and P-gettered.

As expected, degradation is faster and regeneration sets in earlier with increasing temperature. This also means that the minimum in τ_{eff} is reached faster with increasing temperature. As the regeneration reaction sets in earlier with increasing temperature, it starts to overlap with the degradation leading to less pronounced minima for higher temperatures. As a result, the minimum of τ_{eff} at 200°C and 250°C is less pronounced compared to the minimum for 75°C and 150°C. These minima at 75°C and 150°C have a comparable τ_{eff}, which can be explained by the assumption that regeneration does not overlap too much with degradation. At lower temperature, the degradation reaction is almost completed before significant regeneration kicks in.

To quantify the increasing dynamics of degradation and regeneration, the time of treatment after which the regeneration visibly sets in can be used.

For the B-doped sample treated at 200°C (Fig. 2), the developing of a slight shoulder can be seen at approximately 2 h. Afterwards, τ_{eff} increases further to values even above the initial τ_{eff} before the treatment. The same can be observed for the sample treated at 250°C. Here the shoulder can be seen at ~1 h before the initial lifetime is exceeded.

The harmonic averages of τ_{eff} over time for P-gettered Ga-doped samples can be seen in Fig. 3. Degradation and regeneration of the samples treated at 75°C and 150°C are similar but slower than for the B-doped samples shown in Fig. 2. For the sample degraded at 200°C, the maximum visible at around 1 h is reached much faster than for the 150°C sample. After that a second minimum can be observed at ~10 h. Following this minimum, τ_{eff} exceeds the initial lifetimes, as for the B-doped samples. So the shoulders seen in the Ga-doped sample at 250°C and in the B-doped samples at 200°C and 250°C are probably this first maximum which overlaps with a second decrease and increase. This is why the maximum can only be detected as a shoulder. Again, all these kinetics are accelerated by the increase in temperature.

If the temperature on the other hand is too low, the second maximum does not occur within the examined time of, *e.g.*, ~5,000 h at 75°C.

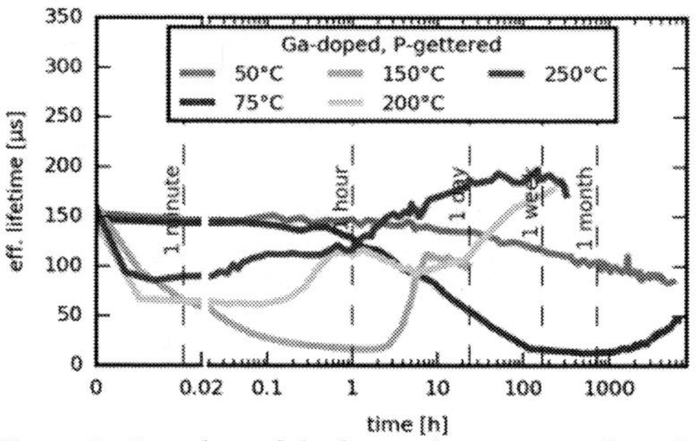

Figure 3: Overview of the harmonic averages of τ_{eff} for each point in time for different degradation temperatures. Tests for stability are not shown here. The samples are Ga-doped and P-gettered.

3.2 Spatially resolved lifetime values

Often only averages of τ_{eff} are used to discuss the effects on a mc-Si sample. This has the disadvantage that information on local lifetime distribution is not taken into account. Therefore, further analyses of the degradation and regeneration behavior will be discussed based on rainbow plots like shown in Fig. 4. Each single line in Fig. 4 represents one of 2,500 small areas on the lifetime sample. The applied color coding, based on the initial τ_{eff} values of the sample at the beginning of degradation shows that for the relative τ_{eff} distribution is kept within the sample during the experiment. So the areas with highest initial τ_{eff} remain the areas with highest τ_{eff} throughout the experiment. This was also observed before (*e.g.* [5, 6]).

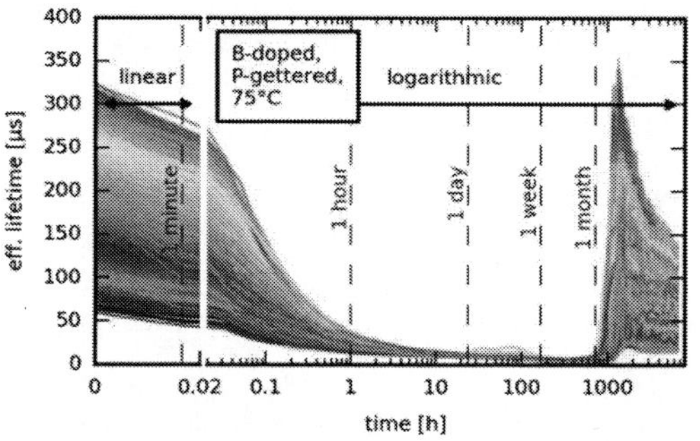

Figure 4: Development of τ_{eff} during treatment at 75°C and 1 sun for a P-gettered, B-doped sample.

To better compare the results for increased temperatures, first the τ_{eff} data of a P-gettered B-doped sample degraded at 75°C is shown in Fig. 4. The ungettered, B-doped samples show a similar development but on a lower τ_{eff} level.

The ungettered Ga-doped samples show a different behaviour. They degrade but do not regenerate within the investigated time span of roughly 5,000 h (at 75°C). But they also do not regenerate within the investigated time span when treated at higher temperatures (*e.g.* within 3,000 h at 225°C).When treating the samples at 75°C, the maximum of τ_{eff} after degradation and regeneration is reached after roughly 1,000 h (~42 d). Due to these long timescales at 75°C, an increase in temperature is quite tempting as it speeds up the kinetics.

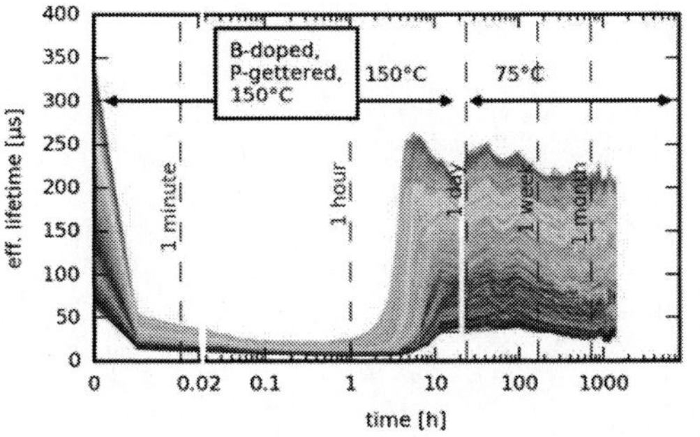

Figure 5: A sister sample (gettered and B-doped) of the one shown in Fig. 4 during treatment at 150°C and 1 sun and a following test for stability at 75°C.

By using equally processed sister samples, the influence of temperature during treatment is the dominating parameter when comparing the results. As can be seen in Fig. 5 (150°C), the whole process of degradation and regeneration accelerates. The lifetimes measured at maximum degradation are nevertheless comparable to the lifetimes at maximum degradation for the sample degraded at 75°C (*cf.* Fig. 2). The following test for stability at 75°C and 1 sun shows that the sample stays approximately at this τ_{eff} level for the rest of the examined time span of over 1,300 h.

When trying to accelerate the process even further by increasing temperature to 250°C (Fig. 6), samples reach the (less deep) minimum in τ_{eff} and a first maximum (or shoulder, respectively) earlier, but they also start increasing towards a second maximum directly afterwards. This is why the first maximum at around 1 h cannot be seen clearly when examining data from samples treated at 250°C (*cf.* Fig. 2). Nevertheless, the first maximum can still be seen as a shoulder in the increasing lifetime towards a second maximum that is reached after roughly 200 h.

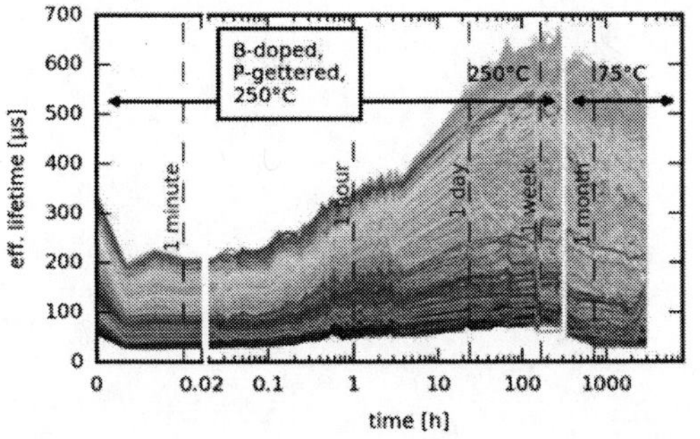

Figure 6: Another sister sample (gettered and B-doped) of the ones shown in Fig. 4 and 5 during treatment at 250°C and 1 sun and a following test for stability at 75°C (starting at ~300 h).

At the time this maximum is reached, the measured lifetimes are at least equal but mostly considerably higher than the initial τ_{eff} values.

The samples were then also tested for stability at 75°C to get further insight in the underlying mechanisms. This second maximum seems not to be stable. The B-doped (P-gettered and ungettered) samples show a slight decrease in τ_{eff} directly after lowering the temperature and afterwards a slight degradation. The P-gettered Ga-doped sample shows a steady decline in lifetime when checking for stability of the second maximum but no immediate decrease in lifetime after changing the temperature.

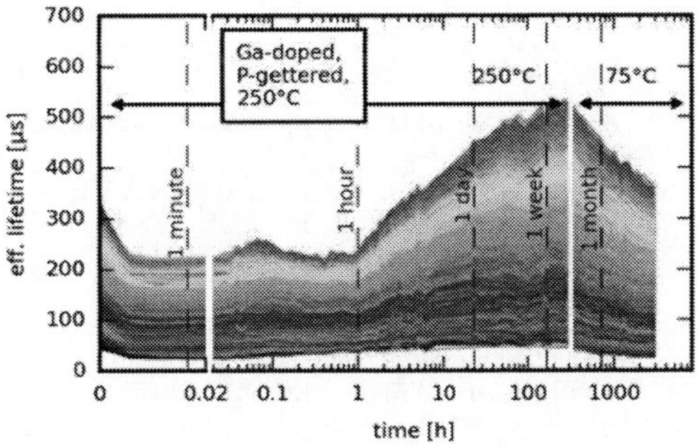

Figure 7: Gettered Ga-doped sample during treatment at 250°C and 1 sun followed by a stability test at 75°C (starting at ~300 h).

This decline can be seen in Fig. 7 starting at around 300 h for the gettered Ga-doped sample. Another aspect that can be seen in the data of the Ga-doped sample is that the increase after the first minimum and the decline after the first maximum do not overlap as much as with comparable B-doped samples. Therefore, the first maximum is more clearly visible compared to the B-doped sample (Fig. 6), where it can only be seen as shoulder (around 1 h).

One question arising from the fact that at higher temperatures two declines and increases are observed is to determine whether those effects originate from the bulk or the surface of the sample. According to investigations presented in [5, 7], the first minimum is due to bulk effects, whereas the second one arises from surface effects (temporarily degrading surface passivation overlapping with beginning regeneration). These results were gained examining Cz and FZ samples, but they match very well the timescales and behaviour of the mc-Si samples shown here.

4 SUMMARY

The influence of different temperatures on LeTID and regeneration could be shown using spatially resolved data from TR-PLI.

The higher the used temperature, the faster the samples degrade and regenerate. This roughly follows the law of Arrhenius for the examined temperatures of 75°C, 150°C, 200°C and 250°C, as described in [6].

For samples treated at 150°C, a similar minimum lifetime compared to 75°C sister samples could be determined. For higher temperatures, the minimum is less pronounced due to competing regeneration and/or annealing effects.

A second increase could be seen when treating samples with temperatures of ~200°C or higher. This second maximum reaches lifetimes higher than the initial lifetimes of the sample but did not seem to be stable at a following treatment at 75°C 1 sun.

The first maximum was also tested for stability using samples that were degraded and regenerated at 150°C. This first maximum is more stable for the examined time of ~1,300 h.

5 REFERENCES

[1] K. Ramspeck et al., Proc. 27th EUPVSEC (2012) 861.
[2] F. Fertig et al., Physica Status Solidi RRL 9(1) (2014) 41.
[3] F. Kersten et al., Solar Energy Materials And Solar Cells 142 (2015) 83.
[4] D. Kiliani et al., Journal of Applied Physics 110 (2011) 054508.
[5] A. Zuschlag et al., Progress in Photovoltaics: Research and Applications 25(7) (2017) 545.
[6] J. Fritz, A. Zuschlag, D. Skorka, A. Schmid, G. Hahn,, Energy Procedia 124 (2017) 718.
[7] D. Sperber et al., IEEE Journal of Photovoltaics (2017), accepted, doi: 10.1109/JPHOTOV.2017.2755072.

INVESTIGATION ON THE PHOSPHORUS DIFFUSION GETTERING MECHANISM OF CHROMIUM IN MULTI-CRYSTALLINE SILICON

N. Khelifati[1,2*], D. Bouhafs[1], S-E-H. Abaidia[2] and Y. Kouhlane[1]
[1] Centre de Recherche en Technologie des Semi-conducteurs pour l'Energétique (CRTSE), Division de Développement des Dispositifs de Conversion à Semiconducteurs (DDCS), Bd. 2 Frantz Fanon, les sept merveilles B.P.140, Algiers, Algeria
[2] Université M'hamed Bougara (UMBB), Faculté des Sciences, Route de la Gare Ferroviaire, 35000–Boumerdès, Algeria
[*] e-mail : khelifatinabil@crtse.dz

ABSTRACT: In this work, we combine experimental results and their simulation to investigate the phosphorus diffusion gettering (PDG) of chromium in multi-crystalline silicon (mc-Si), in order to get an insight on the dominating gettering mechanism. We first performed experiments using sets of neighboring p-type mc-Si wafers underwent a surface cleaning followed by phosphorous diffusion to form n^+ emitter. The diffusion process was carried out in a tube furnace by using liquid $POCl_3$ as the phosphorus source in the temperature range 850-950°C. The samples were characterized mainly by using Secondary Ion Mass Spectroscopy (SIMS), where the phosphorus and chromium profiles were determined. We then used software developed in our center to simulate the SIMS data and extract active phosphorus (P_s) and phosphorus-vacancy defects (P_4V) profiles, as well as Cr-Ps and Cr-P_4V concentrations. Based on a quantitative comparison between simulation and experimental findings, we adequately explained the dominating mechanism of Cr gettering in $POCl_3$ emitter.
Keywords: Silicon, Chromium impurity, Phosphorus diffusion, Gettering, SIMS, Simulation.

1 INTRODUCTION

Phosphorus diffusion gettering (PDG) of metal impurities, such as Cr, Fe, Ni and Cu, is well known phenomenon that can be occurred during the silicon-based solar cells fabrication (e.g., emitter formation of p-type solar cell). It results in metals segregation to phosphorus diffused region, due to the increasing of their solubility compared to that in material bulk. Despite the extensive studies, the physical mechanism of metals segregation in heavily phosphorus doped layer still remains unclear and under investigation in the photovoltaic community, especially the segregation related to certain metals like chromium. An important simulation work done by Chen et *al.* [1] combining *ab initio* and continuum approaches, demonstrates that during phosphorus diffusion gettering the main trapping sites responsible for metals gettering are P_4V defects situated near the surface which is rich in inactive phosphorus (case of a heavy emitter). In the case of iron, a comparison between their simulation results and those obtained experimentally by Talvitie et *al.* [2] showed a close agreement between calculations and experimental aspects. In further investigation, Schön et al. [3] discussed deeply the impact of active and clustered phosphorus, oxygen as well as phosphor-silicate glass on the gettering efficiency of iron in diffused phosphorus emitter. Their findings indicate the dominance of two different PDG mechanisms; (i) segregation gettering through active phosphorus, and (ii) immobile oxygen.

In this work, we investigate the phosphorus diffusion gettering mechanism of chromium in multi-crystalline silicon by combining experimental results and their simulations.

2 EXPERIMENT AND SIMULATION

In our experiment, mc-Si boron doped p-type adjacent wafers with resistivity of ~1 Ω.cm were used. Firstly, the wafers were chemically treated in NaOH:H_2O and HCl:H_2O solutions to remove the saw damage defects from the surface and neutralize it, respectively. Later on, the wafers were divided into three sets; A, B and C, and underwent a standard PDG process for 90 minutes at 850 °C, 900 °C and 950 °C, respectively. The PDG setup was carried out in a tube furnace using a phosphorus oxychloride ($POCl_3$) liquid as the phosphorus source. After gettering experiments, the PSG were removed using diluted HF. Phosphorus and chromium profiles were measured by secondary ion mass spectroscopy (SIMS) technique.

The investigation of SIMS profiles has been done via software so-called "*GettProg*" developed in our group for the simulation of phosphorus in-diffusion, phosphorus-vacancy clusters formation and metallic impurities gettering. All clustered phosphorus is assumed to be in the form P_4V. More details about the theoretical basis used in the simulation of phosphorus diffusion and metals gettering can be found in [5] and [1, 3, 4], respectively.

3 RESULTS AND DISCUSSION

3.1 SIMS profiles of phosphorus and chromium

Measured depth profiles of phosphorus and chromium obtained for sets A, B and C are shown in figure 1. In the same figure we presented the simulation profiles of total and active phosphorus (solid lines).

Firstly, we observe that SIMS profiles of the three sets differ significantly. Indeed, the overall profiles depths increase with process temperature to be around 0.6 µm for set A, 1 µm for set B and more than 1.5 µm for set C. We also remark an increment of surface phosphorus concentration from 5 x 10^{19} cm^{-3} for set A to 1.5 x 10^{20} cm^{-3} and 2 x 10^{20} cm^{-3} for sets B and C, respectively.

The simulation data of the phosphorus agree fairly well with the measured findings, as shown clearly in figure 1. However, the simulated phosphorus concentration for the set A in the deeper region and for the set C near the surface are slightly overestimated as compared to SIMS data. It can be due to the error in simulation and/or in SIMS analysis.

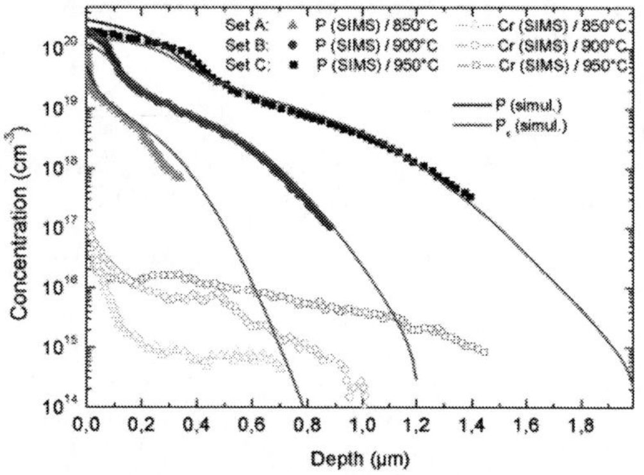

Figure 1: Phosphorus and chromium SIMS profiles after gettering during 90 min at 850°C, 900°C and 950°C compared to simulation data (total and active phosphorus).

Regarding the chromium profiles, all the samples of three sets showed maximum concentrations near the surface. They also exhibit a steep decrease within the first 100-200 nm. After that, the Cr concentration decreases slightly until the detection limit of SIMS is reached (~7 x 10^{14} cm^{-3}). The gettering process at 900°C leads to higher Cr concentrations in the depth of ~150 nm, while the Cr accumulation in the deeper region is higher for the samples gettered at 950°C.

Within the limit of the SIMS measurement accuracy, we note that the dominant gettering takes place in the heavily phosphorus doped region as well as in the moderately phosphorus doped region. The accumulated Cr in both regions indicates the presence of two possible gettering mechanisms through the reaction of interstitial Cr with substitutional phosphorus (P_s) and phosphorus-vacancy defects (P_4V). The last one dominates typically near the surface, which is inactive phosphorus rich region.

3.2 Gettering mechanism of chromium

In the following, we investigate the PDG mechanisms of Cr, through a quantitative comparison between SIMS and simulation data. The model used is based on the reaction between metal impurity (Cr) and a given gettering site. A detailed description of this model can be found in the references [1, 4].

According to this model, the reaction rate is defined as:

$$\frac{dC_{Cr-GS}}{dt} = 4\pi a \, . D_{Cr} \left(C_{Cr_i} \, C_{GS} - K_{Cr/GS} \, . C_{Cr-GS} \right) \quad (1)$$

where GS indicates gettering site (i.e. P_s or P_4V), D_{Cr} is the diffusivity of Cr in Si, *a* is the capture cross section of the reacting species with a typical value of 0.5 nm. $K_{Cr/GS}$ is the equilibrium constants related to the formation energy of $Cr\text{-}P_4V$ clusters or $Cr\text{-}P_s$. C_{Cri} is the interstitial chromium concentration. In this study, the initial concentration of interstitial chromium was taken 2.5 x 10^{12} cm^{-3} for all samples.

As a first investigation on the dominant gettering mechanisms (see figure 2), one can clearly observe that the simulated concentration of $Cr\text{-}P_4V$ complexes decreases differently according to the gettering temperature. Furthermore, the depth of these complexes increases proportionally as a function of temperature. The calculated $Cr\text{-}P_4V$ concentration was found to be higher than experimental profile except the case of set A in the near surface region (~25 nm).

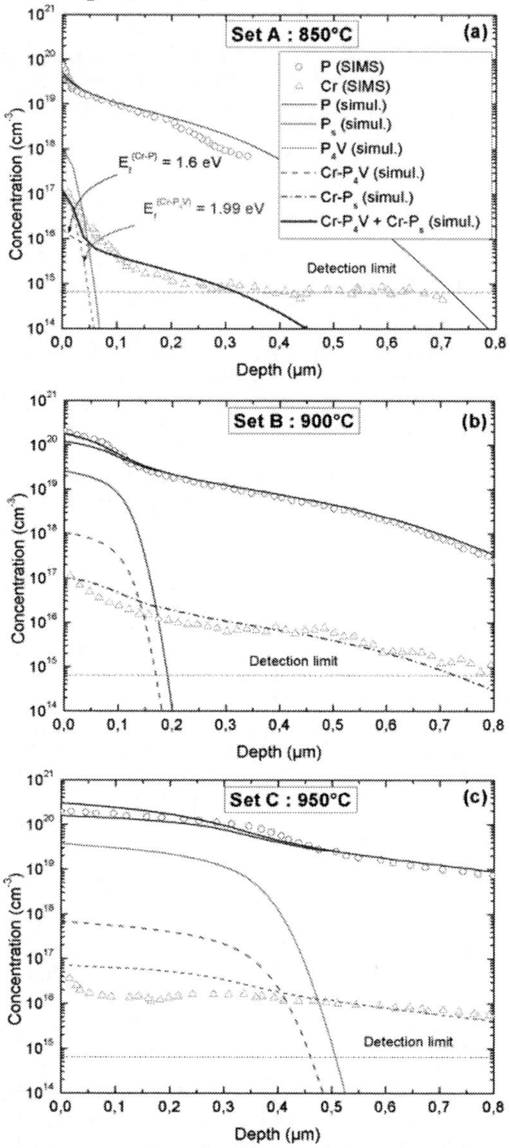

Figure 2: Phosphorus and chromium SIMS profiles, and simulated total phosphorus, substitutional phosphorus (P_s), P_4V clusters, $Cr\text{-}P_4V$ and $Cr\text{-}P_s$ profiles for the three samples sets A, B and C. The initial concentration of interstitial chromium was taken 2.5 x 10^{12} cm^{-3}.

For the set A gettered at 850°C, we remark that the simulated $Cr\text{-}P_s$ profile don't completely fit the SIMS data of Cr. Indeed, the simulation provides an acceptable fitting of the Cr profile only beyond 100 nm depth, but in the surface region (i.e. 0-100 nm) the agreement between $Cr\text{-}P_s$ and SIMS data disappears. However, the calculation of Cr trapped by P_4V clusters showed that Cr- P_4V profile match well with the SIMS one. So, the steep decrease of the Cr profile near the surface correlates adequately with the decrease of the $Cr\text{-}P_4V$ concentration, indicating the dominance of Cr gettering thought its trapping in P_4V clusters. On the other hand, the smooth decrease of Cr concentration beyond ~100 nm is in agreement with simulation findings and can be explained by the dominance of gettering mechanisms through P_s. The total concentration of trapped Cr calculated by combination mechanisms shows an appropriate fitting of SIMS profile. More care in the simulation can improve its agreement with the experimental. We have to note that the Cr-Ps profiles were generated by using a formation energy $E_f^{Cr\text{-}Ps}$ = 1.6 eV, which is significantly higher than the calculated value; 0.22 eV reported by Chen et al. [1]. This energy formation was not used in our simulation because it has not been validated by experimental data on the one hand and it leads in our case to underestimated concentrations of Cr-P_s in the range 10^{10}-10^{12} cm^{-3} even for very high initial Cr concentration, on the other hand.

Figures 2 (b) and (c) illustrate the results obtained for set B and C gettered, respectively, at 900°C and 950°C, and their simulation with the combined model including both segregation mechanisms; Cr-P_s pairing and Cr-P_4V complexes formation. We observe that Cr concentration in the heavily phosphorus doped region in set B is higher than that mesured in set D ($1x10^{17}$ cm^{-3} vs. $4x10^{16}$ cm^{-3}). However, the total amount of Cr for set C, in the deeper regions of emitter, is higher than both sets B and A.

In the near surface region dominated by P_4V, the calculated Cr- P_4V profiles of sets B and C don't show any similarity with the measured SIMS data, either in the order of magnitude of the concentration or in the shape of the profiles. On the contrary, the shapes and the slopes of the simulated Cr-P_s showed a good agreement with SIMS profiles in the whole phosphorus diffusion region. These results provide evidence that at relatively high temperatures (900-950°C), the segregation gettering through the substitutional phosphorus P_s is dominant.

4 CONCLUSION

Conventionally, metal impurity gettering is characterized by measuring the remaining metal concentration in the wafer bulk which indeed is the measure of the actual gettering efficiency. However, in order to get a deeper insight on the actual gettering mechanisms, we have extended our study to characterize the chromium profiles also in the heavily doped region after three phosphorus diffusion processes. Our findings demonstrate a possible domination of segregation gettering mechanism at 850°C by immobile phosphorus complex (P_4V) near the surface. Nevertheless, at this temperature the gettering mechanism in deeper doped regions can be attributed to substitutional phosphorus (P_s). Then, the gettering of Cr at 850°C can be probably done through both mechanisms (formation of Cr-P_4V and Cr-P_s). At higher temperature, i.e. 900°C and 950°C, the dominated mechanism shift towards that associated to P_s.

ACKNOWLEDGMENTS

The authors would like to acknowledge the funding from Directorate General for Scientific Research and Technological Development (DGRSDT) / Algeria.

REFERENCES

[1] R. Chen, B. Trzynadlowski, and S. T. Dunham, J. Appl. Phys. 115, (2014) 054906.
[2] H. Talvitie, V. Vahanissi, A. Haarahiltunen, M. Yli-Koski, and H. Savin, J. Appl. Phys. 109, (2011) 093505.
[3] J. Schöon, V. Väahäanissi, A. Haarahiltunen, M. C. Schubert, W. Warta, and H. Savin, J. Appl. Phys. 116, (2014) 244503.
[4] A. Yazdani, R Chen, and S. T. Dunham, J. Appl. Phys. 121, (2017) 095702.
[5] A. Bentzena and A. Holt, J. Appl. Phys. 99, (2006) 064502.

HOW TO PROPERLY ASSESS BORON-OXYGEN RELATED DEGRADATION IN CRYSTALLINE SILICON

Axel Herguth
University of Konstanz, Department of Physics, 78457 Konstanz, Germany
Author for correspondence: axel.herguth@uni-konstanz.de

ABSTRACT: Boron-Oxygen related light induced degradation (BO-LID) may seriously limit the efficiency of silicon solar cells. It is therefore essential to assess its impact correctly. Within this contribution it is shown by means of simulations how temperature and doping level as well as injection level influence the degradation behavior and on which time scale degradation occurs. It is furthermore shown how regeneration may influence BO-LID results leading to spoiled interpretations. A suggestion for appropriate degradation conditions is given.
Keywords: Degradation, BO-LID, Simulation

1 INTRODUCTION

The efficiency potential of silicon solar cells depends among other things on the lifetime of excess charge carriers in the base of the cell. It is known for quite a while that this lifetime is subject to light induced degradation in boron-doped silicon contaminated with oxygen [1-4] leading to noticeably reduced lifetimes [5]. As typical single silicon crystals grown by Czochralski's (Cz) method contain more oxygen than typical multicrystalline (mc) ingots, this effect mainly affects Cz-Si but is detectable in mc-Si as well.

Even though the general impact of the BO-LID effect is widely known, there seems to exist confusion on the duration required for complete degradation under certain degradation conditions. Furthermore, if degradation is done at elevated temperatures, be it intentionally or not, e.g. in order to accelerate the process, it may happen that the sample not only degrades, but also begins to regenerate as well.

Within this contribution, it will be shown, based on theoretical calculations, under which combination of light intensity, temperature and duration degradation should be performed and what happens, if these entities are varied.

2 THE THEORY BEHIND BO-LID (IN PARTS)

To the current state of knowledge, the BO related defect may appear in at least three different defect states: the recombination inactive but instable annealed state A,

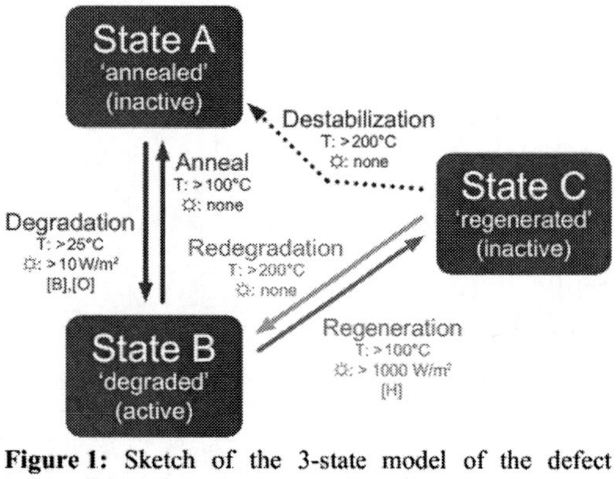

Figure 1: Sketch of the 3-state model of the defect responsible for BO-LID including experimental conditions for which the different reaction paths are active (after [2,3]).

the recombination active metastable degraded state B, and the recombination inactive stable regenerated state C [6,7] which is hardly distinguishable from the annealed state A. The different defect states may convert into each other via reaction paths, e.g. from A to B via the degradation reaction A→B. The mathematical approach presented in [7] is used within the CASSANDRA framework [8] in the following to describe and simulate the defect pool.

2.1 Reaction rates

All reactions seem to be describable by an Arrhenius-type temperature dependence (thermally activated) following the formula

$$\kappa_{ij}(T, p_0, \Delta n) = \kappa^*_{ij}(p_0, \Delta n) \cdot \exp\left(-\frac{E_{ij}}{k_B T}\right) \quad (1)$$

with κ^*_{ij} being the specific pre-exponential factor and E_{ij} the specific activation energy of the respective reactions between state i and j ($i,j = \mathrm{A,B,C}$).

To the current state of knowledge, the anneal reaction rate κ_{BA} is believed to only depend on temperature with activation energy $E_{BA} = 1.32\,\mathrm{eV}$ and trial frequency $\kappa^*_{BA} = 7.4\times10^{12}\,\mathrm{s}^{-1}$ [4] determining the reaction rate.

The degradation reaction rate κ_{AB} (of the slow forming defect) features more complex dependencies. On the one hand it is believed to follow an Arrhenius-type behavior with activation energy between 0.46 and 0.48 eV [4].

In earlier investigations from Bothe [6] and others the trial frequency was found to depend quadratically on the boron content and not to react to illumination intensities above ~1/100 suns [9]. However, later investigations with compensated material revealed that it only correlates with the boron content as the hole concentration equaled the boron content, but hole concentration is the actual driving force. Hence, as Kim et al. [10] pointed out, degradation does indeed depend on light intensity, but it becomes only observable if the generated excess carrier density becomes high compared to the doping level.

Therefore, the question is, under which circumstances excess hole concentration matters. Under the assumption of excess electron carrier lifetime being that high that diffusion length is bigger than cell thickness, excess carrier concentration may be estimated according to the equation

$$n \cdot p = (n_0 + \Delta n)\cdot(p_0 + \Delta p) = n_i^2 \cdot \exp\left(\frac{qV}{k_B T}\right) \quad (2)$$

with p_0 and $n_0 = n_i^2/p_0$ being the carrier concentrations in dark thermal equilibrium, Δn and Δp the excess carrier density or injection level (neutrality demands for $\Delta n = \Delta p$), n_i the intrinsic carrier concentration, V the voltage and q/k_BT the thermal equivalent voltage. The degrading cell's voltage is typically left floating, meaning $V = V_{OC}$, however the equation remains valid for any voltage. Here the situation becomes complicated, because V_{OC} depends not only on degradation conditions (intensity and temperature) and doping level p_0 but also on other parameters like emitter saturation current and rear surface recombination velocity.

Table I shows an exemplary comparison of a 1 and 3 Ωcm (high performance PERC-type) solar cell applying strong and weak illumination. As can be seen, strong illumination should have a strong influence especially on weakly doped material. However, as will be discussed later on, applying weak illumination around 0.1 suns yields no significant acceleration especially taking into account that BO-LID (and its dynamic) is most relevant on higher doped material such as 1 Ωcm anyway.

Table I: Comparison of different resistivity base material applying strong and weak illumination

ρ [Ωcm]	1		3	
p_0 [10^{15} cm^{-3}]	15		4.7	
ϕ [suns]	1	$^1/_{10}$	1	$^1/_{10}$
V_{OC} [mV]	670	611	665	606
Δp [10^{15} cm^{-3}]	1.3	0.14	2.4	0.34
p/p_0	+9%	+1%	+50%	+7%
$\kappa_{AB}(p)/\kappa_{AB}(p_0)$	+18%	+2%	+130%	+15%

2.2 Temporal dynamics in the 2-state model

Only taking the annealed state and the degraded state of the defect in Fig. 1 into account and ignoring the presence of the fast forming defect (or stage) or in other words assuming that only the slow forming defect (or stage [11]) is relevant in the long run, the defect kinetic may be described by a system of coupled differential rate equations as described in [7].

$$\frac{d}{dt}\begin{pmatrix} N_A \\ N_B \end{pmatrix} = \begin{pmatrix} -\kappa_{AB} & +\kappa_{BA} \\ +\kappa_{AB} & -\kappa_{BA} \end{pmatrix}\begin{pmatrix} N_A \\ N_B \end{pmatrix} \qquad (3)$$

Here (N_A, N_B) with $N_A + N_B = 1$ represents the normalized defect density or occupation of the individual defect states in the defect pool. (κ_{AB}, κ_{BA}) are the reaction rates converting the defect states into each other. If the reaction rates are treated as constants, the system has the following solution for the degraded state N_B

$$N_B(t) = \frac{\kappa_{AB}}{\kappa_{AB} + \kappa_{BA}} \cdot \left[1 - \left(N_{A0} - \frac{\kappa_{BA}}{\kappa_{AB}} N_{B0}\right) \cdot e^{-\lambda t}\right] \qquad (4)$$

with (N_{A0}, N_{B0}) being the occupation at $t = 0$. The system evolves towards equilibrium (degradation is nothing else than a relaxation of a non-equilibrated system) in an exponentially saturating way with the effective rate constant $\lambda = \kappa_{AB} + \kappa_{BA}$. The rates are shown in Fig. 2 for 1 Ωcm and 3 Ωcm under the assumption of $p = p_0$, meaning low intensity illumination. As can be seen, κ_{AB} exceeds κ_{BA} by far below 100°C and thus κ_{BA} does not contribute significantly to the effective rate constant λ. Note that stronger illumination will enhance κ_{AB}(3 Ωcm) more than κ_{AB}(1 Ωcm) narrowing the gap between red and green.

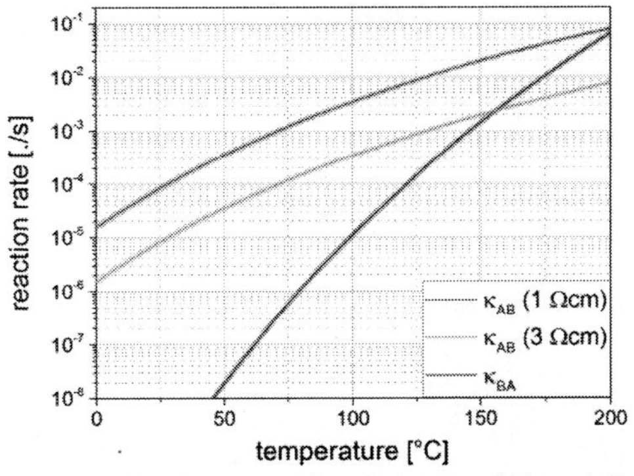

Figure 2: Reaction rates of anneal κ_{BA} and degradation κ_{AB} for 1 Ωcm and 3 Ωcm under the assumption $p = p_0$ (low intensity) including uncertainty (colored area).

3. BO-LID IN THE 2-STATE MODEL

3.1 Long-term equilibrium versus temperature

Long-term occupation of states is given by the pre-factor in Eq. 3 and depends on the reaction rates. Again under the assumption of $p = p_0$, meaning low intensity illumination, the occupation of states for 1 and 3 Ωcm is shown in Fig. 3. The temperature at which κ_{AB} equals κ_{BA} corresponds to the crossover of the occupation of degraded and annealed state. In general, the degraded state is the favored state up to a certain 'threshold' temperature setting an upper temperature limit within a degradation treatment. Threshold temperature depends on the extent of BO-LID that should still be achievable in the treatment. E.g., temperature cannot exceed ~107°C if at least 95% degradation in 3 Ωcm material is desired.

However, two effects were ignored in the above calculations: (a) $p = p_0 + \Delta p$ exceeds p_0 and thus $\kappa_{AB}(p)$ exceeds $\kappa_{AB}(p_0)$ especially for high resistivity material and higher illumination intensity. In consequence, the degraded state remains favored even at higher temperatures and the upper limit of applied temperature increases. (b) The assumption of constant reaction rates does not hold if voltage drops during the degradation treatment. For low resistivity material, where Δp is even under 1 sun illumination almost negligible to p_0, the degradation rate κ_{AB} is almost constant. For high resistivity material, where Δp is comparable to p_0, the degradation rate κ_{AB} varies noticeably as Δp drops with voltage. For long term equilibrium, the degraded voltage is relevant.

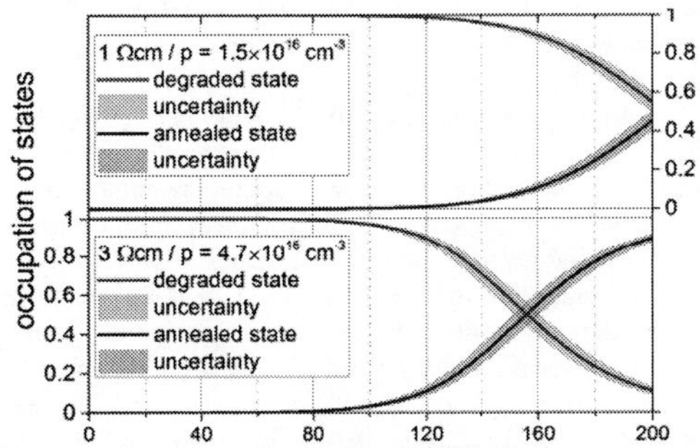

Figure 3: Long-term equilibrium occupation of states resulting from the reaction rates shown in Fig. 2 including uncertainty.

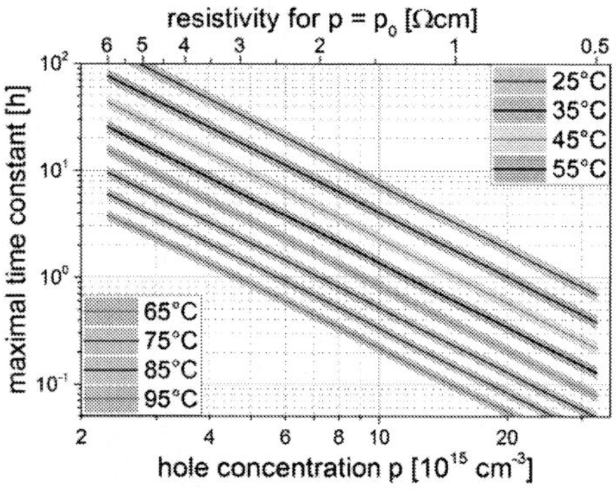

Figure 4: Degradation time constant in dependence of hole concentration p and applied temperature. For low intensity illumination, hole concentration p can be approximated by p_0 and thus resistivity. However, p always exceeds p_0 and thus the time constant t_0 given here is in respect thereof only an upper limit.

3.2 Needed duration for BO-LID

The expectable duration required for complete degradation till equilibrium scales with effective rate constant λ (from Eq. 4) defining a time constant $t_0=\lambda^{-1}$ which is depicted in Fig. 4. However, it should be noted that interrupting the degradation treatment at t_0 will not result in complete degradation as the exponential function then has only decayed to $e^{-1}\approx 37\%$. Some durations for characteristic completeness are summarized in Table II.

For example, 1.5 Ωcm ($p_0\approx 10^{16}$ cm^{-3}) materials degrade at 25°C with $t_0\approx 7.3$ h and thus degradation requires ~22 h ($3\cdot t_0$) for $N_B(t)=95\%$ and ~34 h ($4.6\cdot t_0$) for $N_B(t)=99\%$. Increasing temperature to 45°C reduces t_0 to ~2.3 h thus degradation would take ~7 h for $N_B(t)=95\%$ and ~11 h for $N_B(t)=99\%$.

Table II: Durations needed for a specific degree of completeness of BO-LID

$N_B(t)$	63%	90%	95%	99%
t/t_0	1.0	2.3	3.0	4.6

4. BO-LID IN THE 3-STATE MODEL

So far, degrading properly was only a matter of keeping $\kappa_{AB}>\kappa_{BA}$ and waiting long enough. And with respect to Fig. 3 and Fig. 4, the duration of BO-LID should be reducible by raising temperature to, e.g., 100°C even for 3 Ωcm material without compromising the achievable extent of degradation in long-term equilibrium too much.

However, there exists third, 'regenerated' defect state (see Fig. 1) which is accessible from the degraded state via the regeneration reaction. The regenerated state is recombination inactive and therefore not directly distinguishable from the annealed state.

Unfortunately, the regenerated state is the preferred defect state under degradation conditions; therefore Fig. 3 is, strictly speaking, only valid for vanishing regeneration reaction rate κ_{BC}. The regeneration rate κ_{BC} does not only depend on temperature [6,7], but also on injection [6,7,12], providing the possibility to intentionally manipulate this reaction. Herguth [6,7] and Wilking [12] have shown that regeneration rate κ_{BC} seems to scale more than linear with injection level thus changing illumination intensity by a factor of 10 will change the regeneration rate κ_{BC} by at least a factor of 10. In contrast, degradation rate κ_{AB} is for higher doped material (like 1 Ωcm) virtually constant with respect to this strong change in illumination intensity (see Table I) and weakly reacting for lower doped materials.

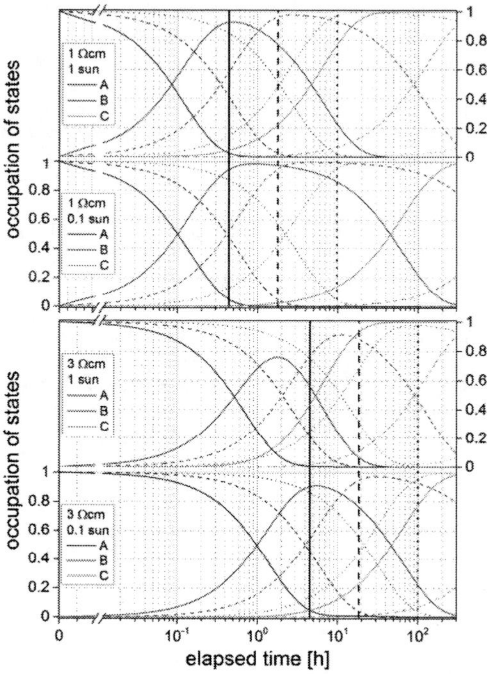

Figure 5: Time resolved evolution of the occupation of states in the defect pool including the regeneration reaction for 1 Ωcm (top) and 3 Ωcm (bottom) material. Each graph includes strong (1 sun) and weak (0.1 suns) intensity conditions. Different line styles represent different temperatures: full 85°C, dashed 55°C, dotted 25°C. The black lines mark $3\cdot t_0$ where 95% completeness would be expected (see Table II).

Fig. 5 depicts the temporal evolution of the defect pool according to the complete mathematical approach in [7] (computed with CASSANDRA [8]) for 1 and 3 Ωcm material for relatively strong (1 sun) and weak (0.1 suns) illumination conditions as well as for different treatment temperatures (25, 55, 85°C) assuming a rather high regeneration rate κ_{BC} (as found for well hydrogenated lifetime samples and PERC-type cells) and $\kappa_{BC}(1\ \text{sun})=10\cdot\kappa_{BC}(0.1\ \text{suns})$. First thing to note is that the degraded state B (red) starts to fill and the annealed state A (blue) starts to empty in the beginning; the sample degrades. The second thing to note is that the degraded state does not reach unity any more, but features a more or less sharp maximum. In the long run the defect pool's preference turns to the regenerated state C (green). However, if the degraded state does no longer reach unity after a certain waiting time, interpretation becomes harder. This shall be illustrated in a virtual experiment.

A first experimenter degrades 3 Ωcm (Fig. 5 bottom) at 25°C and 1 sun (dotted red line) and waits $3\cdot t_0$ (3·33 h, dotted black vertical line) according to Fig. 4. He then reports 98% of full degradation amplitude.

Another experimenter with the same sample wants to accelerate the degradation, chooses therefore 85°C and 1 sun (full red line) and a waiting time of $3\cdot t_0$ (3·1.5 h, full black vertical line) ignoring an injection level enhanced degradation rate κ_{AB}. He reports then only 56% of full degradation amplitude. Even if he would have monitored the temporal evolution, and would report the maximal degradation after 1.8 h, he would only report 76% of full degradation amplitude.

If the second experimenter would have used 0.1 suns illumination instead still waiting $3 \cdot t_0$ (3·1.5 h, full black vertical line), he would have reported 90% of full degradation amplitude – a value at least closer to the result of the first experimenter.

Hence the same material seems to degrade differently strong in the range of 56% to 98% only because the experimenters used different degradation conditions and did not realize (by in-situ monitoring) that a complex dynamic passes off in the background.

5. CONCLUSIONS

Regarding the 2-state model of Boron-Oxygen related light induced degradation (BO-LID), degradation conditions are fairly tolerant to variations of illumination and temperature. BO-LID takes a certain duration mainly depending on applied temperature and doping level, but always results in the long-term in a complete degradation, at least below a threshold temperature, which mainly depends on doping level. Up to this threshold, BO-LID may be sped up by increasing temperature.

However, in reality the 3-state model describes the dynamics better. Unfortunately, not the degraded but the regenerated state is the preferred state and therefore maximum occupation of the degraded state is no longer mandatorily achieved in the long-term. Maximal observable occupation of the degraded state and thus maximal extent of degradation depends on applied temperature and illumination intensity (injection level).

If dynamics are not sufficiently well (in-situ) monitored, the results may severely differ for different experimental conditions. The best results, meaning a degradation as complete as possible, are obtained for low temperature, low intensity degradation conditions. Especially increasing temperature and intensity simultaneously can result in spoiled interpretations.

ACKNOWLEDGEMENT

Part of this work was supported by the German Federal Ministry for Economic Affairs and Energy under contract number 0324001. The content is the responsibility of the author.

REFERENCES

[1] H. Fischer, W. Pschunder, Proceedings 10th IEEE Photovoltaic Specialists Conference (1973) 404.
[2] J. Schmidt, A.G. Aberle, R. Hezel, Proceedings 26th IEEE Photovoltaic Specialists Conference (1997) 13.
[3] S.W. Glunz, S. Rein, W. Warta, J. Knobloch W. Wettling, Solar Energy Materials & Solar Cells **65** (2001) 219.
[4] K. Bothe, J. Schmidt, Journal of Applied Physics **99** (2006) 013701.
[5] K. Bothe, R. Sinton, J. Schmidt, Progress in Photovoltaics **13** (2005) 287.
[6] A. Herguth, G. Schubert, M. Kaes, G. Hahn, Proceedings 21st European Photovoltaic Solar Energy Conference (2006), 530.
[7] A. Herguth, G. Hahn, Journal of Applied Physics **108** (2010) 114509.
[8] A. Herguth, S. Wilking, Energy Procedia **124** (2017) 60.
[9] J. Schmidt, K. Bothe, R. Hezel, Proceedings 3rd World Conference on Photovoltaic Energy Conversion (2003) 919.
[10] B. Hallam et al., Energy Procedia **77** (2015) 799.
[11] M. Kim, M. Abbott, N. Nampalli, S. Wenham, B. Stefani, B. Hallam, Journal of Applied Physics **121** (2017) 053106.
[12] S. Wilking, S. Ebert, C. Beckh, A. Herguth, G. Hahn, Proceedings of the 32nd European Photovoltaic Solar Energy Conference (2016) 487.

INFRARED IMAGE PROCESSING ALGORITHM FOR SOLAR CELL DEFECT ASSESSMENT

A. Hovhannisyan, A. Petrosyan
National Polytechnic University of Armenia (Polytechnic), Teryan str. 105, Yerevan, Armenia
E-mail: anna.l.hovhannisyan@gmail.com

ABSTRACT: The detection and assessment of solar cell defects are highly important during the development, production and usage of photovoltaic modules. Solar cell defects can be detected by different methods including electroluminescence, photoluminescence and thermography. In this paper, the algorithm for solar cell defect assessment by infrared image processing is proposed. The algorithm is based on comparison of infrared images of testing cell and the sample of high quality solar cell. For the algorithm realization, the NI LabVIEW and NI Vision Assistant applications are used. The suggested method of defects assessment allows to realize the solar cell infrared image processing automatically with the high definition.
Keywords: Defects, Electroluminescence, Solar Cell, Qualification and Testing, Digital Image Processing

1 INTRODUCTION

Solar energy is one of the most promising energy resource for the near future. As a solar energy source solar panels are used. So, the detection of their defects is crucial. Nowadays there are different methods of solar cell investigation. Module defects, such as cracks, soldering and interconnection faults, potential induced degradation, and hot spots can often be characterized by imaging techniques including photoluminescence (PL), electroluminescence (EL), and thermography. These methods are the contactless methods of the solar cells defects detection. Infrared imaging allows anomalies to be seen clearly and can be used to scan installed solar panels during normal operation. The aim of this paper is to find solar cell defects automatically by doing infrared image processing. For that it is developed an algorithm, which uses some pre-processed filters and other digital image processing methods (like particle analysis) to find the patterns for the solar cells with the defects.

2 THE ELECTROLUMINESCENCE METHOD

In this paper the electroluminescence method is used for solar cell investigation. Electroluminescence is a contactless and fast method of solar cell characterization. The EL method is used to determine values of the local effective diffusion length Leff [1], the carrier collection length Lc [2], as well as the local junction voltage [3] and the series resistance [4, 5] of a solar cell.
In the figure 1 it is presented the hardware setup for the infrared image acquisition by elecroluminescence method.

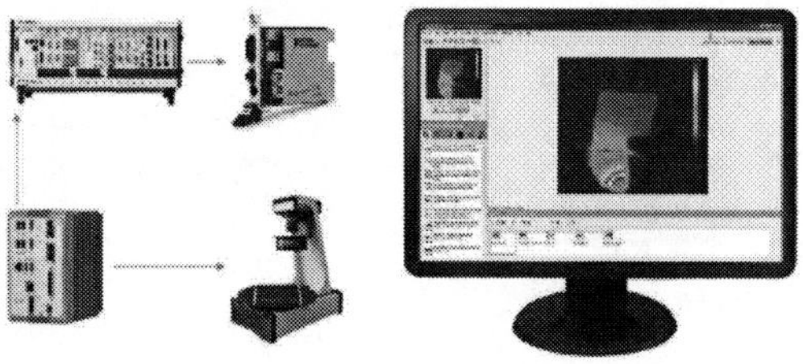

Figure 1: Hardware Setup for IR Image Acquisition

As a power source, it is used NI SMU 4130, which is a precise programmable power source unit. To change the voltage programmatically and do the measurements it is written a code using LabVIEW programming environment.

Infrared images of solar cell are captured from the IR camera. Then the images are transferring to NI IC-3173, which is an industrial controller with reconfigurable I/O. The whole image processing algorithm is implemented on the controller, which has reconfigurable FPGA thanks to which the algorithm speed is high.

3 THE ALGORITHM FOR SOLAR CELL DEFECT ASSESSMENT AND EXPERIMENTAL RESULTS

The new optimized algorithm is developed for the IR image digital processing.

By using color detection technics, as well as filtering and analyzing methods of image processing, it is possible to implement an algorithm to detect solar cell defects, such as cracks, hot spots, etc. automatically.

In the figure 2 it is presented an algorithm implemented on a FPGA.

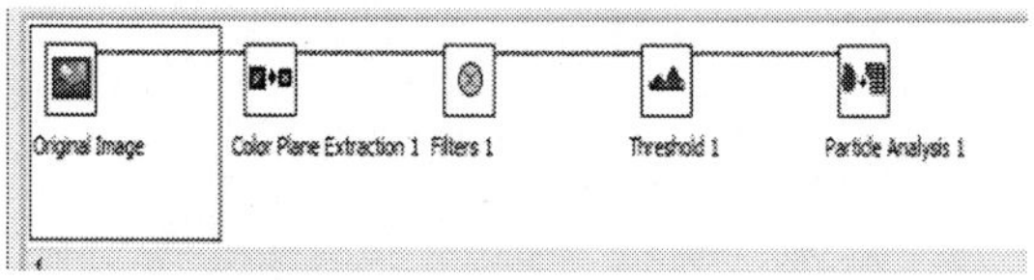

Figure 2: The Algorithm Implemented on an Industrial Controller IC-3173

In the first step the original IR image is captured (Fig. 3). It contains thermal information about the surface in camera field of view. Hot parts are colored by red color, warmer parts with yellow, cool parts with green and cold parts with blue color. So, the treated parts are those, which colored red.

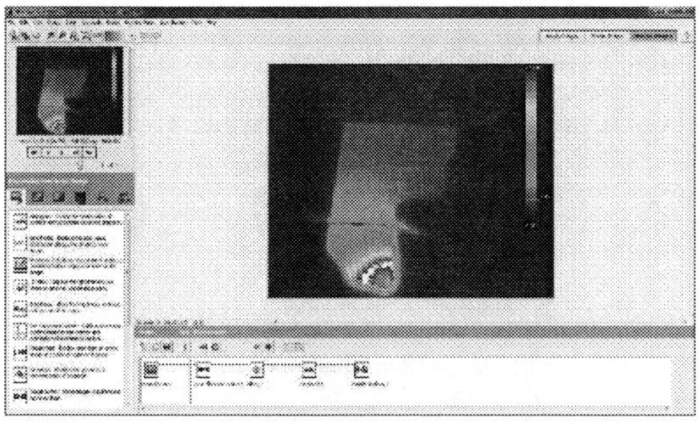

Figure 3: Original Infrared Image

Then, color segmentation is used to detect regions of specific colors within an image and separate these from regions containing unknown colors. Using this method, it is possible to separate objects of different colors that appear in an inspection image to determine whether all expected colors are present or which proportions of each color are present, separate colored objects or regions from background colors, also, check whether an undesired object or feature characterized by a different color is present.
For some of these use cases, color segmentation may be required to obtain regions of interest (ROIs) representing particular colors prior to running further analysis (for example, the region location, size, color, and other shape characteristics).

Color segmentation techniques apply to a range of applications involving quality and process control [6]. This block extracts the three-color planes (RGB, HSV, or HSL) from an image. In the proposed case the treated (defected) part of image is colored by red, so the R plane from RGB is taken for processing.

In the Fig. 4 it is presented the result of Color Plane Extraction (Red color is taken).

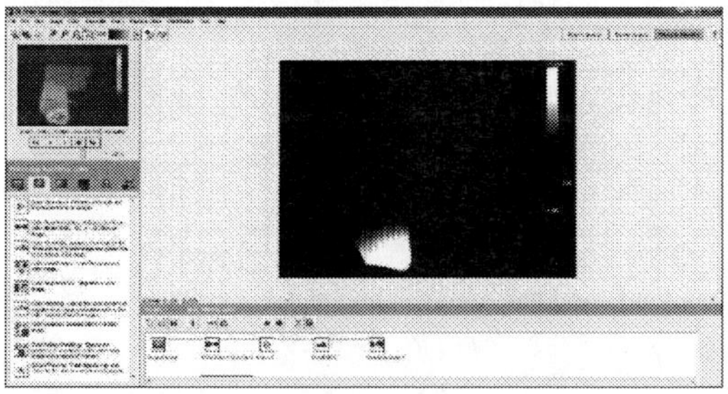

Figure 4: Result of Color Plane Extraction (Red color is taken)

As a filter, it is used Median filter. The median filter is a lowpass filter. It assigns to each pixel the median value of its neighborhood, effectively removing isolated pixels and reducing detail. For each pixel $P\ (i, j)$ in an image where i and j represent the coordinates of the pixel, the convolution kernel is centered on $P\ (i, j)$. Common cases for median filter are 3x3, 5x5, 7x7 windows.

$P\ (i, j)$ = median value of the series $[P\ (n, m)]$.

In the Fig. 5 filtering result is presented.

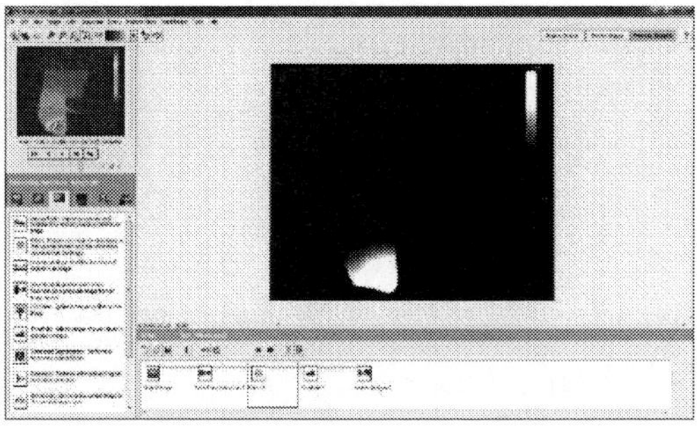

Figure 5: Filtering Result

Thresholding segments an image into a particle region, which contains the objects under inspection, and a background region based on the pixel intensities within the image. The resulting image is a binary image.

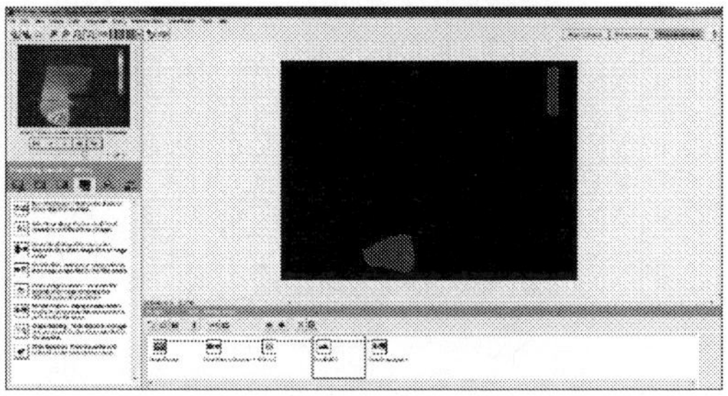

Figure 6: Thresholding Result

Particle analysis consists of a series of processing operations and analysis functions that produce information about particles in an image. Using particle analysis, we can detect and analyze any 2D shape in an image. A typical particle analysis process scans through an entire image, detects all the particles in the image, and builds a detailed report on each particle. We can use multiple parameters such as perimeter, angle, area, and center of mass to identify and classify these particles. Using multiple parameters can be faster and more effective than pattern matching in many applications [7, 8, 9].

In the fig. 7 particle analysis's implementation is presented, and in the fig. 8 it is shown the main information about the defected area of the solar cell.

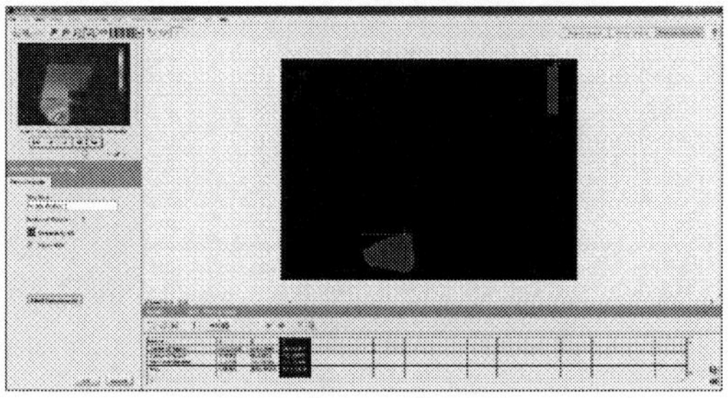

Figure 7: Particle Analysis Implementation

Results ...	1	2	3
Center of Mass X	623.25000	609.02946	247.08380
Center of Mass Y	7.00000	68.21551	432.38877
Max Feret Diameter	3.16228	111.30588	116.76044
Area	4.00000	2682.00000	6945.00000

Figure 8: Particle Analysis Results

The suggested algorithm for the image processing of a solar cell was implemented on the industrial controller with the FPGA. This allows to do a solar cell defect assessment faster that an implementation on a host computer. In the figures 9 and 10 it is shown the comparison analysis of the implementation the same algorithm both on the host machine, and the FPGA target. An estimated of the time required to implement the algorithm on a host machine is 19 ms, while an estimated of the time for the implementation the same algorithm on a FPGA target is 8.73 ms.

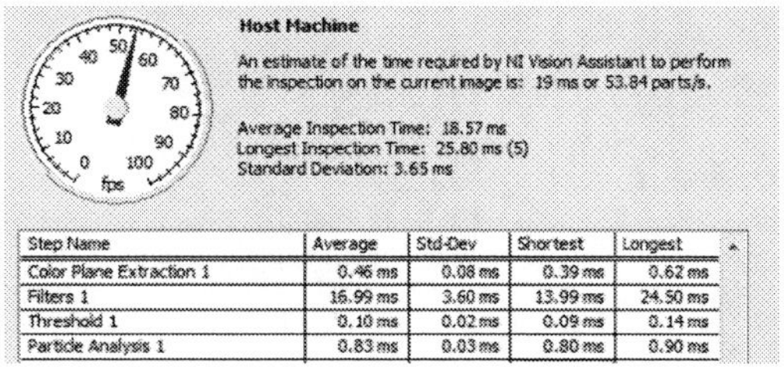

Figure 9: Algorithm Performance Meter on a Host Machine

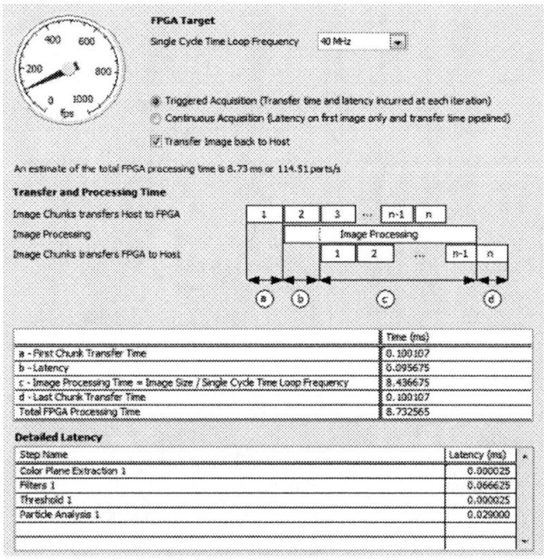

Figure 10: Algorithm Performance Meter on a FPGA Target

4 CONCLUSION

The methodology presented in the paper allows to define solar cell defects using IR image digital processing methods.

The algorithm, implemented in an FPGA is faster than the implementation in a real time.

The algorithm can detect solar cell defects with high precision. The red part of the original image shows that there are cracks in the corner of the solar cell, which is detected by the suggested algorithm.

The suggested algorithm can be used at the manufacturing process to detect none usable parts of the product without completing production process. This will make production more cost-effective.

5 ACKNOWLEDGEMENTS

We would like to thank State Committee of Science of Armenia, National Instruments Yerevan, Armenian National Engineering Laboratories (ANEL) for supporting this work.

6 REFERENCES

[1] P. Wurfel, T. Trupke, T. Puzzer, E. Schaffer, W. Warta, S. Glunz, Journal of Applied Physics 101 (2007) 123110.
[2] K. Bothe, D. Hinken, K. Ramspeck, B. Fischer, R. Brendel, Proceedings of 22nd European Photovoltaic Solar Energy Conference, Italy (WIP, Milan, 2007) 1673.
[3] K. Bothe, P. Pohl, J. Schmidt, T. Weber, P. Altermatt, B. Fischer, R. Brendel, Proceedings of 21st European Photovoltaic Solar Energy Conference, Germany (WIP, Dresden, 2006) 597.
[4] D. Hinken, K. Rampspeck, K. Bothe, B. Fischer, R. Brendel, Applied Physics Letters 91 (2007) 182104.
[5] J. Haunschild, M. Glatthaar, M. Kasemann, S. Rein, E. R. Weber, Phys. Status Solidi RRL 1-3 (2009) DOI 10.1002.
[6] ni.com
[7] Ya-Hui Tsai, Du-Ming Tsai, Wei-Chen Li and Shih-Chieh Wu, Defect Detection of Solar Cells Using EL Imaging and Fourier Image Reconstruction, Proceedings of the Institute of Industrial Engineers Asian Conference 2013, 53-62 (2013).
[8] E.R. Davies, Computer and Machine Vision, Fourth Edition, Elseiver (2012).
[9] National Instruments Corporation, NI Vision Concepts Manual (2008).

STUDY OF OXYGEN IN CZOCHRALSKI SILICON BY HYPERSPECTRAL PHOTOLUMINESCENCE

E. Olsen[1]*. L. Kvalbein[1], M. Helander[1], T. Mehl[1], E. Letty[2], W. Favre[2], J. Veirman[2], I. Burud[1]

[1]Norwegian University of Life Sciences, N-1432 Ås, Norway
[2]CEA, LITEN, INES-Department of solar technologies, F-73375 Le Bourget du Lac, France
* Email: Espen.Olsen@nmbu.no

ABSTRACT: The possibility of using hyperspectral imaging of photoluminescence to detect oxygen in n-type Cz-Si wafer was investigated. Wafers from the seed and tail end of an ingot were studied. The method OxyMap was used on twin wafers for quantification of the oxygen content. PL signals previously associated with radiative recombination of charge carriers through traps introduced by interstitial oxygen, Oi, were detected in the same area as was associated with a high oxygen content (seed end). The P-line and what may be phonon replicas were seen as pronounced signals in the seed end in cooled samples (90 K). For room temperature a broad defect related luminescence signal with center peak at 0,72 eV was recorded in addition to luminescence from the band to band transition at 1,1 eV.

1 INTRODUCTION

Thermal donors (TD) are considered a major problem in the silicon industry using Czochralski (Cz) grown monocrystalline wafers in general as well as in the photovoltaics (PV) branch. Thermal donors lead to increased levels of thermally excited electrons altering the expected resistivity. Further, they introduce defects in the crystal structure and trap levels in the otherwise forbidden band gap of silicon. TD are reported to be caused by interstitial oxygen and are mainly formed at temperatures around 450 °C [1]. The seed-end in a Cz ingot has a naturally higher oxygen concentration, and during pulling this part of the ingot experiences a slow cooling which results in longer time at the temperature range for the formation of TDs as compared to the tail end [2].

Hyperspectral imaging is a method where images are recorded for a multitude of spectral bands. The most advanced hyperspectral cameras have spectral resolution of a few nm for the whole sensitivity range of the detector. The detectors are frequently semiconductor based charge-coupled devices (CCDs) and the sensitivity range is determined by the material used. Silicon based detectors are used in the range from UV (400 nm) to NIR (1000 nm), while InGaAs and HgCdTe (MCT) detectors are used in the NIR and SWIR range (1000-2500 nm).

Spectrally resolved photoluminescence (SPL) has been used for studies of multicrystalline and mono-like wafers for solar cell purposes in a number of previous studies [3-6]. In this work SPL is used for studies of defect related emissions due to the presence of interstitial oxygen in n-type Cz silicon wafers. Photoluminescence due to radiative recombination of photoexcited charge carriers from the conduction- to the valence band of Si - referred to as band-to-band PL (BB-PL), is an inherent material property in the material giving rise to emissions with energy matching the band gap (1,1 eV, 1150nm). The intensity is directly related to the number of excited charge carriers and may so be used for studies of the material quality and carrier lifetime. This is routinely performed and via proper calibration routines give highly resolved images of wafers with the associated carrier lifetime. In silicon, defects may introduce states in the forbidden band gap. These can act as recombination active sites through the Shockley-Read-Hall (SRH) mechanism and is the direct cause of low carrier lifetimes at defect rich sites in wafers. Recombination through this mechanism may be radiative - emitting a photon, or non-radiative, keeping energy conservation through interaction with one or more phonons. Radiative emission caused by states in the band gap will have lower energy (longer wavelengths) than the BB-PL and is referred to as defect related luminescence (DRL) [7]. Recently, new detector materials have been introduced in hyperspectral cameras (MCT) which are now able to record images in the range 1000 nm - 2500 nm range (1.3 eV - 0.49 eV) enabling imaging of DRL from a multitude of SRH-related mechanisms as well as the BB-PL.

In this work we present images and spectra obtained from SPL on n-type Cz silicon wafers from the top and bottom end of an ingot at room temperature (295 K) and cooled with liquid nitrogen (90 K). Images of low-energy DRL emissions related to the oxygen content are presented.

2 EXPERIMENTAL

A Specim SWIR hyperspectral camera equipped with a MCT detector giving a spatial and spectral resolution of 300 µm and 6 nm, respectively, was employed in the studies. A Lasiris Magnum II laser with wavelength 808 nm and irradiated power density of 2 W/cm^2 was used for excitation of charge carriers. Two n-type Cz-Si samples with nominal resistivity of 10 Ωcm were studied. The samples were cut in half and then texture etched and passivated. The sample was positioned on a polished aluminum holder cooled with liquid nitrogen (LN_2), for the recordings at 295 K, the sample holder did not contain LN_2. Si BB-PL - as well as inter-band gap luminescence was studied at 295 K and 90 K. The oxygen content was measured by the OxyMap technique on neighbouring wafers [8]. The oxygen content in the seed end was measured to be $12 \cdot 10^{17}$ cm^{-3} while the content in the tail end wafer was measured to be $4 \cdot 10^{17}$ cm^{-3} in the center part of the whole wafers. Visualization of the data sets were performed by image manipulation with the software packages MATLAB and GIMP. The experimental setup is depicted in Figure 1.

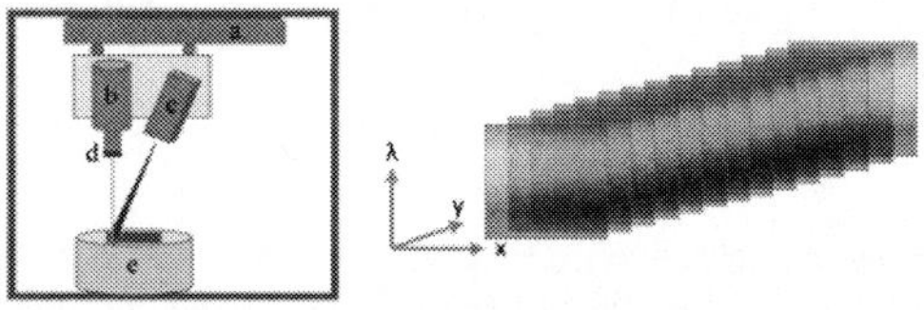

Figure 1: Experimental set up. a) translational stage, b) hyperspectral camera, c) excitation source, d) 1000 nm low-pass filter, e) cryogenic sample holder.

3 RESULTS AND DISCUSSION

Hyperspectral recordings were first performed at room temperature (295 K). The total photoluminescence (integrated) signal as function of photon energy for the whole specimen was calculated from the data set. The spectra emitted from the seed- and tail wafers are shown in Fig. 2. The band-to-band radiative recombination leads to a strong BB-PL signal with energy reflecting the silicon band gap at 1.1 eV. The intensity of this signal from the seed- and tail ends are close to identical indicating similar material properties with respect to lifetime of excited charge carriers. For the seed end wafer, a separate, broad signal with center energy around 0.72 eV was found. Such signals associated with lower energy than the band gap are well studied and believed to be caused by radiative recombination through allowed states - or traps, via the SRH-mechanism introduced by defects in the silicon lattice. Such emissions are frequently referred to as defect related luminescence. The tail wafer did not exhibit this characteristic. A similar emission has previously been reported on cooled specimens and related to thermal donors formed from original interstitial oxygen (O_i) [1].

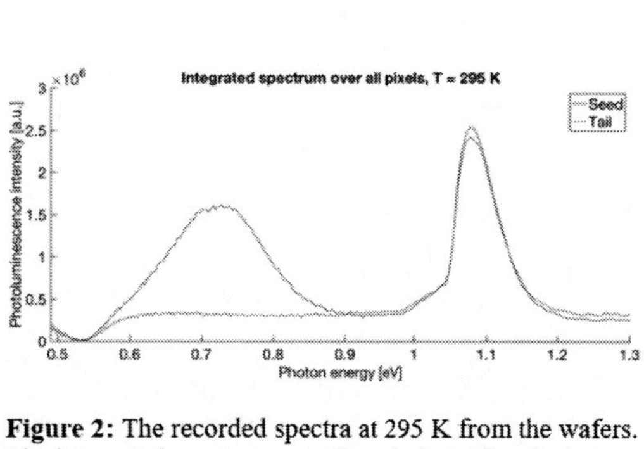

Figure 2: The recorded spectra at 295 K from the wafers. The integrated spectrum over the whole surface is shown

The samples were then cooled by liquid nitrogen (LN_2) to 90 K and similar recordings were made. The two associated spectra from the wafers are depicted in Fig. 3. The BB-PL signal is stronger and with significantly lower half width (sharper) for both wafers. This is as expected due to a sharper distribution of phonons being present for indirect band-to-band recombination leading to a significantly sharper peak for the BB-PL signal. The unresolved, broad shoulder on the low energy side of the BB-PL peak seen for the samples measured at 295 K is now experienced as a separate signal with energy 1.05 eV. This is a well studied emission known to be associated with multi-phonon recombination over the band gap and called a phonon replica of the band-to-band emission [9]. The DRL signal, on the other hand, now show some strikingly interesting features. The broad emission with a center energy of around 0.72 eV recorded at 295 K is now resolved into several individual signals. 4 peaks are clearly identifiable, while two more may be extracted from the low-energy shoulder. Two mechanisms may be associated with this behaviour. As the BB-PL signal has its lower energy phonon replicas experienced as a broadening shoulder on the low energy side, this may also be the case for the mechanisms leading to the broad 0.72 eV signal seen in Fig. 2. In this case, the primary recombination mechanism is associated with the resolved signal with highest energy in this cluster (0.77 eV). The lower energy resolved signals can then be interpreted as phonon replicas of this. Another explanation may be that different thermal donor clusters associated with slightly different energy states in the band gap have formed during ingot cooling. The suite of signals then represents recombination through each of these. The sharp emission with highest energy has previously been studied in cooled wafers and reported as the "P"-line [1]. This study, however, had a low-energy detection limit of 0.75 eV and the equipment were so not able to record possible emissions with lower energy and it is not yet possible to conclude on this subject. The same study first reporting the P line also identifies a second DRL emission called the H line at 0.91 eV. This can be barely distinguished in our spectra.

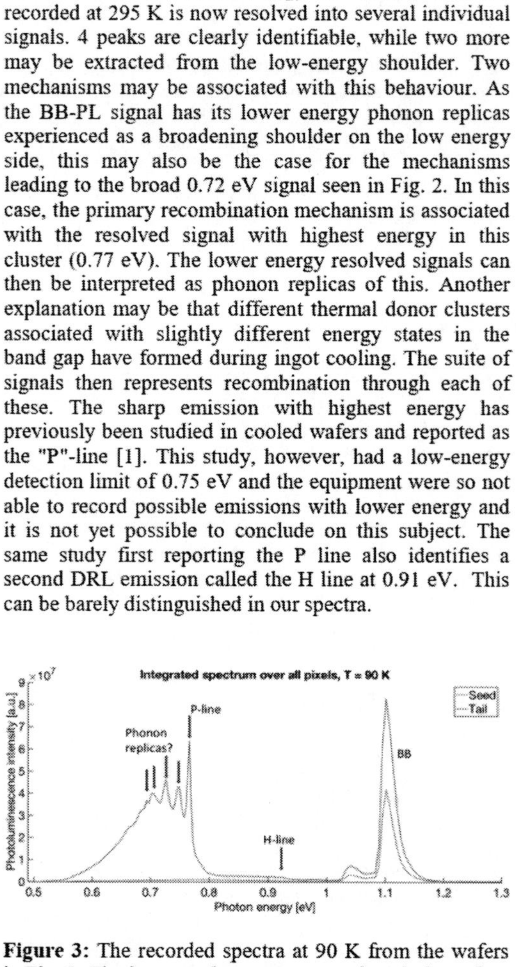

Figure 3: The recorded spectra at 90 K from the wafers in Fig. 2. The integrated spectrum over the whole surface is shown. Possible phonon replicas of the P-line (0.77 eV) are indicated.

Images of the DRL signal as function of position in the cooled wafers are shown in Fig. 4. The images represent the total defect emission in the energy interval 0.66-0.77 eV. The right image represents the emission in the tail end wafer of the ingot, while the left image represents the seed end wafer. It can be seen that the intensity of DRL is significantly lower in the tail end of the ingot. The blue line (low signal) in the middle part of the image is likely due to differences in surface etching due to the handling procedure. In this wafer, the most DRL is found in the outer, circumferential part. The DRL signal in the seed end wafer shows a clear, characteristic, circular pattern frequently attributed to inclusion of oxygen due to the circulating interface between molten Si and the ingot interface during the crystal pulling process. DRL is most pronounced in the inner part of the ingot, consistent with a higher oxygen content.

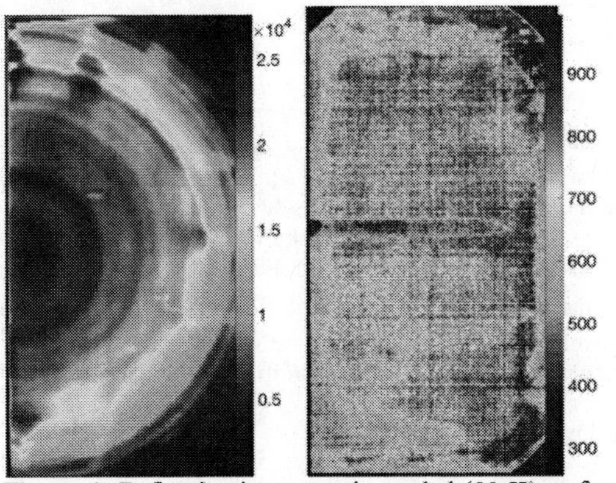

Figure 4: Defect luminescence in cooled (90 K) wafers from the seed- (left) and tail end (right) of a n-type Cz-Si ingot. The emission in the band 0.66 - 0.77 eV is depicted. Signal strength is given in arbitrary units.

4 CONCLUSIONS

Hyperspectral imaging has been applied for the study of oxygen in the seed and tail end of a laboratory Cz-Si ingot. The spectrum after heat treatment show increased intensity for all energy levels compared to before heat treatment, especially in the energy range 0.6 eV - 0.8 eV corresponding to defect related sub band gap luminescence. In general, the intensity of the seed-end was higher than for the tail-end in this energy range, both before and after heat treatment. The signal in the energy range 0.6 eV - 0.8 eV appeared as several intensity maxima when cooled with LN_2 (90 K). At room temperature (295 K), there was only one broad, unresolved signal in this energy range.

7 REFERENCES

1 N.S. Minaev and A.V. Mudryi, Phys. Stat. Sol. **68** (1981).

2 J. Veirman, R. Varache, A. Danel, M. Albaric, E. Letty, B. Martel, C. Roux, "“Early Efficiency Prediction of Silicon Heterojunction Cells Processed on Thermal Donors-Rich Czochralski Wafers”, Proc. of the 33rd European Photovoltaic Solar Energy Conference, Amsterdam, The Netherlands (2017).

3 E. Olsen and A.S. Flø, Appl. Phys. Lett., **99**, 011903 (2011).

4 A.S. Flø, I. Burud, K. Kvaal, R. Søndenå, E. Olsen, AIP Advances, **3** (2013).

5 I. Burud, A.S.Flø and E. Olsen, AIP Advances, **2**, (2012).

6 T. Mehl, I. Burud, E. Letty, E. Olsen, Energy Procedia, **124**, (2017).

7 D. Lausch, T. Mehl, K. Petter, A. S. Flø, I. Burud and E. Olsen, J. Appl. Phys. **119**, (2016).

8 J. Veirman, B. Martel, N. Enjalbert, S. Dubois, Photovolt. Int. **33**, (2016).

9 M. Grundmann, "The Physics of Semiconductors, an Introduction including Nanophysics and Applications". Springer Verlag Berlin, Heidelberg, Germany, 2010, pp 326.

22% EFFICIENT N-TYPE REAR JUNCTION PERT SOLAR CELL WITH 100µM-THIN INDUSTRIAL MONOCRYSTALLINE 6 INCH SILICON WAFERS

Yoonseok CHOI, Jongchul LEE, Jieun LEE,
Taejun KIM*, Myungick HWANG, and Seungwon LEE
Hyundai Heavy Industries Green Energy Co., Ltd., Yongin 446-912, South Korea

ABSTRACT: We have developed 22%-efficient n-Type Rear Junction PERT (Passivated Emitter Rear Totally-diffused) Cells that have the wafer thickness around 100µm since 2014. In the course of this development, we optimized gas composition for boron emitter by Atmospheric Pressure Chemical Vapor Deposition (APCVD) in 2015 and laser opening process conditions, such as power and pulse duration were optimized in 2016. Finally, we have achieved 22% efficiency by optimizing the surface concentrations of boron and phosphorus doped layers. IV parameters of the champion cell are V_{oc} of 685.3mV, I_{sc} of 9.49A, FF of 80.95% and Efficiency of 22.03%. (Calibrated by SERIS, Singapore) The final wafer thickness of the champion cell is 106µm. Efficiency have been improved mainly by V_{oc} increment, which is about 15mV compared to the previous cell IV results obtained at 2015. This result has been obtained mainly due to the reduction of surface recombination by minimized surface doping concentrations.
Keywords: n-type, recombination, boron doping, phosphorus doping

1 INTRODUCTION

In PV industry various types of recombination reduction technologies have been developed for the higher conversion efficiency of solar cells. With this development flow, a PERC (Passivated Emitter Rear Cell) structure has been developed to passivate the back side of substrate to reduce rear recombination, and many companies are now mass producing or preparing [1]. Most of PERC cells that are mass produced use p-Type substrate. However, in order to fundamentally solve the problems such as the light induced degradation (LID) and the low minority carrier life time of the p-Type substrate, many studies have been made on the n-Type PERT cells and some companies are mass producing [2-6].

Our group has also developed n-Type PERT cells. Especially we focus on rear junction PERT cells with 100 µm-thin industrial n-Type mono-crystalline silicon wafers because rear junction structure has similar production process to industrial p-Type PERC cells and less sensitive to metal-induced recombination. And we chose 100µm-thin wafer because the thinner the wafer, the more economically advantageous. Our previous works were on boron emitter optimization by controlling gas composition in atmospheric pressure chemical vapor deposition (APCVD) process [7] and optimization of laser contact opening (LCO) process conditions such as laser power and laser pulse duration time [8]. In this paper we modified the annealing conditions to further optimize the boron emitter and optimize the phosphorus doping layer to reduce recombination current in the front side of the cells.

2 DEVICE FABRICATION PROCESS

Figure1 shows our final cell structure. 130µm thin n-Type Czochralski silicon wafers (6", 239cm^2) were used but final wafer thickness of cell level became under 110µm after saw damage etching (SDE) process. The wafer thickness of our champion cell is only 106µm. To fabricate n-Type rear-junction PERL cells, SDE process was carried out in first. And then we deposited boro-silicate glass (BSG) layer on one side of wafer by using the APCVD system. Thermal drive-in process was performed in tube furnace at above 900°C after BSG deposition. Residual BSG layer was etched in HF bath and boron rich layer (BRL) was removed by additional acid etching process. To make front side of cells, boron doped back side was capped with silicon oxy nitride layer by PECVD and texturing process was performed. Phosphorus atoms were doped on textured silicon wafer by $POCl_3$ diffusion tube furnace. Phospho-silicate glass (PSG) layer and silicon oxy nitride layer were simultaneously removed by wet cleaning process. Thermally grown silicon oxide formation process was followed by front dielectric layer deposition by PECVD. Aluminum oxide layer deposited after short cleaning process and backside dielectric layer was deposited by PECVD. LCO process was carried out for local contact with metal electrode on the back side of the cells. And selective phosphorus laser doping process which is using phosphoric acid was performed on the front electrode area. Rear metal electrodes were screen-printed and fired in belt furnace. Finally plated metal electrodes were formed on the front side of cells.

Electrochemical Capacitance Voltage measurement (ECV) was used to observe the boron and phosphorus doping profiles according to thermal drive-in conditions.

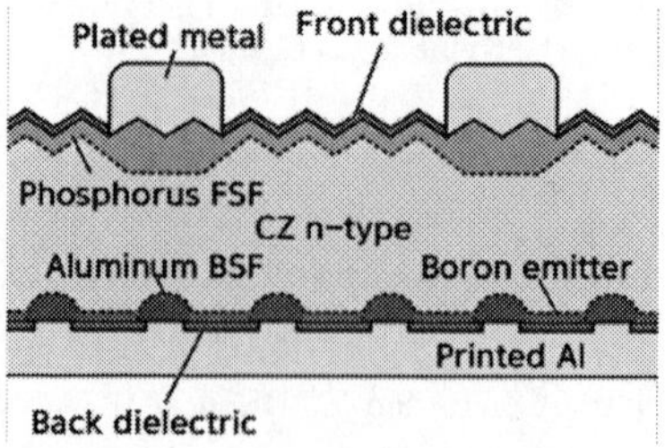

Figure 1: Schematic diagram of n-Type RJ-PERT cell

Suns-Voc measurement was followed for the analysis of recombination characteristics. Cell performance was measured by IV measurement system.

3 OPIMIZATION OF BORON DOPING

Reducing the boron surface concentration is one of the most effective ways to minimize surface recombination. However, as the boron surface concentration decreases, the sheet resistance may increase, which may lead to the problem of increasing the electrical resistance in the emitter region of n-type PERT cells. Therefore, the method of lowering the boron surface concentration while keeping the sheet resistance at a similar level is needed. We have chosen a way to keep the diffusion temperature constant and increase the diffusion time to realize the above method. This is because, in general, lowering the diffusion temperature to lower the boron surface concentration tends to cause an increase in the sheet resistance due to a decrease in solid solubility of boron.

In order to investigate boron doping characteristics such as doping profile and sheet resistance according to the diffusion time, the APCVD deposition was performed on the saw damage etched n-type wafers under the same deposition conditions. Thermal diffusion process was performed under three different time conditions. Figure2 shows the boron doping profiles according to diffusion time conditions. As diffusion time increased, the surface boron concentration decreased and doping depth increased while sheet resistance remained constant value of about 180ohm per square. n-Type rear junction PERT cells were fabricated using the above three diffusion time conditions. I-V characteristics and Suns Voc measurement results are presented in figure 3 and Table I respectively. V_{oc} was improved due to decrease of surface concentration which leads to J01 decrease and I_{sc} was not changed despite of doping depth increase. It can also be seen that J02, saturation current density which is due to recombination in the depletion region of the p-n junction, decreases as the emitter depth becomes deeper. This decrease of J02 is the cause of increase of pFF and FF of real cells.

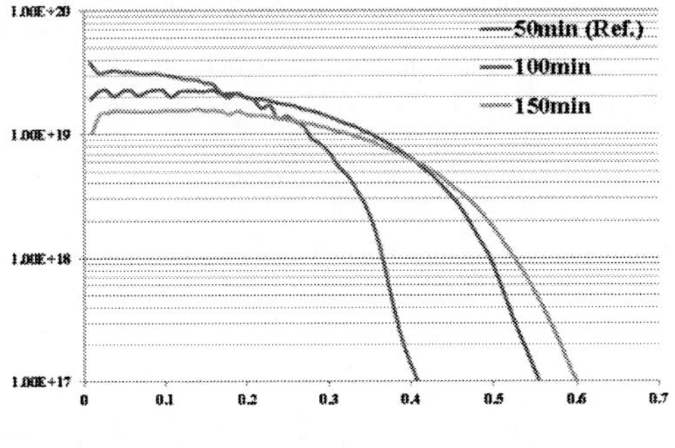

Figure 2: Electrochemical Capacitance Voltage (ECV) measurement results according to diffusion time.

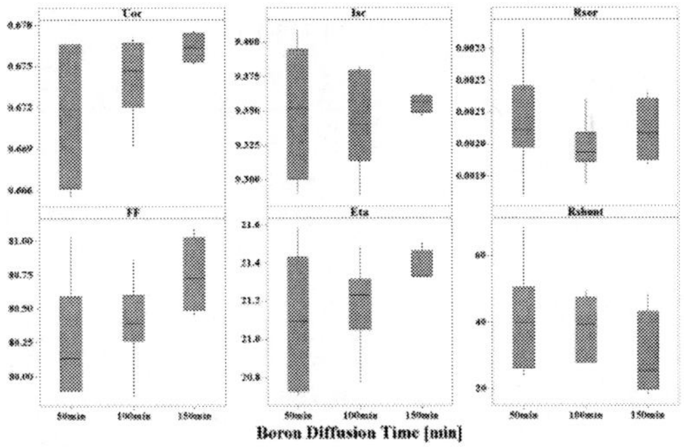

Figure 3: I-V measurement results according to diffusion time.

Table I: Suns-Voc measurement results measured by Sinton Instrument.

Diffusion Time (min)	pFF [%]	J01 [pA/cm^2]	J02 [nA/cm^2]
50	82.3	0.182	11.9
100	82.5	0.174	11.3
150	83.1	0.147	10.7

3 OPIMIZATION OF FRONT SURFACE FIELD

The opposite side of boron emitter, phosphorus doped front surface field (FSF), is also a major part of surface recombination. To reduce the recombination at the front side of n-Type rear junction PERT cells, experiments varying phosphorus doping gas concentration and thermal drive-in time were performed. Diffusion temperature was fixed at 815℃ in order to test the solid solubility at a constant level. Phosphorus doping gas concentration was controlled by adjusting the flow rate of nitrogen gas that carries $POCl_3$ atoms into the diffusion tube furnace. Phosphorus doping profile changes according to experiment conditions as shown in figure 4. As the flow rate of nitrogen which is carrying $POCl_3$ atoms is decreased, the surface concentration of phosphorus gradually decreases and it decreases rapidly below a certain level. And as the diffusion time increase, the surface concentration of phosphorus does not change significantly but the diffusion depth increases by about 0.5μm. I-V results of n-Type rear junction PERT cells according to experiment conditions which are mentioned above are shown in Figure 5. Boron diffusion time of 150min was used in the cell fabrication process. The reduction of the surface phosphorus concentration leads to the improvement of the V_{oc} and the increase of the diffusion depth causes the I_{sc} decrease in the cell level. We could achieve the champion cell efficiency of 22.03% at the experiment condition of 700sccm and 2min. I-V results of champion cell measured by SERIS calibration center are shown in Figure 6 and Table II.

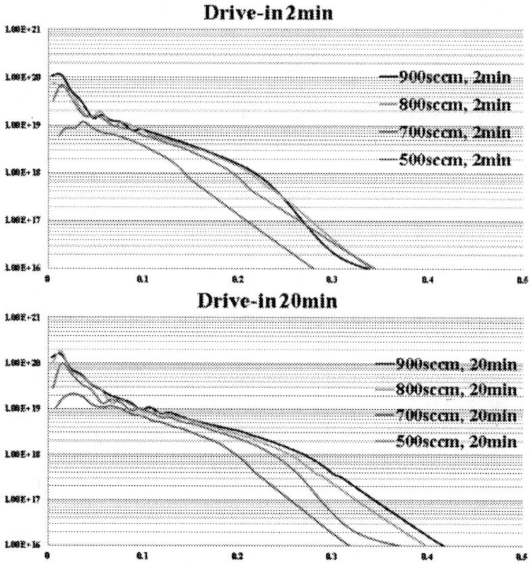

Figure 4: Doping profiles according to diffusion time and carrier N_2 flow rate measured by ECV.

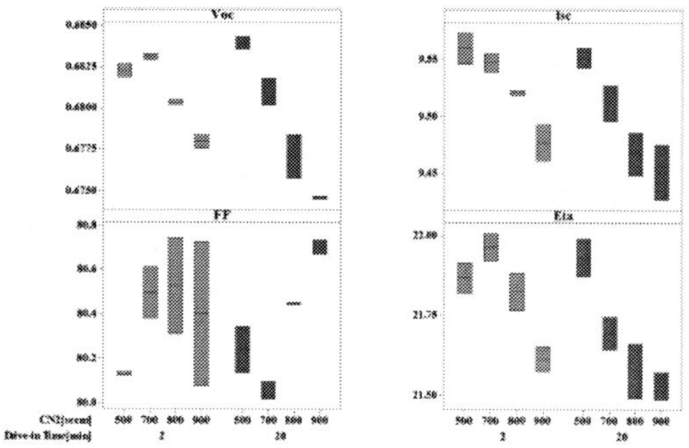

Figure 5: Reflectance of sample with different recipe of capping layer.

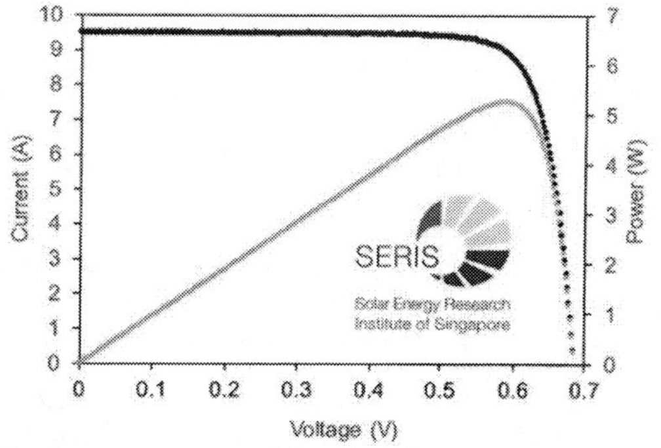

Figure 6: I-V curve of the champion n-Type RJ-PERT cell measured by SERIS calibration center

Table II: I-V parameters of n-Type RJ-PERT cell measured by SERIS calibration center

	Thickness [μm]	V_{oc} [mV]	I_{sc} [A]	FF [%]	η [%]
N-PERT cell with diffusion conditions of 700sccm, 2min	106	685.3	9.49	80.95	22.03

4 SUMMARY

We have investigated the effect of surface doping concentrations of boron and phosphorus on the n-Type rear junction PERT solar cells because high surface doping concentration is known to increase surface recombination current and also deteriorate open-circuit voltage of silicon solar cells. It could be clearly seen that V_{oc} increases with decreasing surface doping concentrations of boron and phosphorus which is due to reduction of surface recombination current. Especially, it was found that I_{sc} also increase with decrease of front phosphorus doping depth, which is due to the increase of quantum efficiency (QE) in the short wavelength region while the boron doping depth of back side of cell does not affect I_{sc}. These front and back surface doping concentration reductions resulted in the cell efficiency over 22% even with the thin wafer thickness around 100μm.

REFERENCES

[1] A.W. Blakers, A. Wang, A.M. Milne, J. Zhao, and M.A. Green, Appl. Phys. Lett. 55, 1363 (1989)

[2] Y.-W. Ok, Solar Energy Material Solar Cells 123, 92–96 (2014)

[3] P. Rothhardt, Energy Procedia 55, 287 – 294 (2014)

[4] F. Duerinckx, Proceeding of 29th European Photovoltaic Solar Energy Conference and Exhibition (2014)

[5] V. Mertens, 28th European Photovoltaic Solar Enrgy Conference and Exhibition (2013)

[6] Jia Chen, Prog. Photovolt: Res. Appl. (2015)

[7] J. Lee et al, Boron Emitter optimization of n-Type Back Junction PERT Cells with Atmospheric Pressure Chemical Vapor Deposition, European Photovoltaic Solar Energy Conference (2015)

[8] J. Lee et al, The Optimization of Laser Contact Opening Process for n-Type Rear Junction Printing PERT Solar Cells, European Photovoltaic Solar Energy Conference (2016)

OPTIMIZATION AND APPLICATION OF A SINGLE-STAGE CO-DIFFUSION PROCESS FOR INDUSTRIAL N-TYPE SILICON SOLAR CELLS

Nadine Wehmeier[1*], Anja Nowack[1], Silke Dorn[1], Fabian Kiefer[1], Till Brendemühl[1], and Sarah Kajari-Schröder[1]

[1] Institute for Solar Energy Research Hamelin (ISFH), Am Ohrberg 1, 31860 Emmerthal, Germany

ABSTRACT: Industrial *n*-type silicon solar cells with high efficiencies usually require a complex fabrication process with two separate furnace diffusions. In this work, we aim for an optimized single-stage co-diffusion process for the simultaneous formation of the p^+-emitter and the n^+-front surface field (FSF) of industrial *n*-type rear emitter cells. A stack of boron silicate glass (BSG) and silicon nitride (SiN_z) deposited by means of plasma-enhanced chemical vapor deposition (PECVD) acts as boron diffusion source. The co-diffusion is performed in a $POCl_3$ quartz tube. At a single temperature plateau of only 850 °C, which is the lowest co-diffusion temperature reported so far, the n^+-region is diffused from $POCl_3$ gas while simultaneously boron is driven-in from the BSG and forms the emitter. For tailoring the phosphorus doping profile, we integrate two in-situ oxidation steps into the co-diffusion process and investigate the influence of the oxidation times to achieve adequate phosphorus sheet resistances. With a first in-situ oxidation for 15 min and a second 60 min oxidation, we achieve phosphorus sheet resistances of $R_{sheet,P}$=117 Ω/□ and saturation current densities as low as $J_{0,P}$=47 fA/cm² on symmetrically diffused and passivated test structures. A contact resistivity of 3 mΩ cm² can be reached with a commercially available Ag paste. In addition, we vary the parameters of the PECVD BSG diffusion source controlling the boron content. Thus, we can adjust the sheet resistance to values between about 110 Ω/□ and 300 Ω/□ and reach emitter saturation current densities as low as 9 fA/cm² and implied open-circuit voltages up to 718 mV. We perform cell simulations in order to identify optimum sheet resistance values for our selected single-stage co-diffusion and contacting geometry and identify optimum parameters of $R_{sheet,B}$=150-200 Ω/□. With a lean process flow, we fabricate co-diffused *n*-type passivated emitter and rear totally diffused (PERT) back junction (BJ) cells with a high efficiency of 20.7%.

Keywords: *N*-type silicon solar cell, PECVD boron silicate glass, Single-stage co-diffusion

1 INTRODUCTION

N-type solar cells allow for high and stable efficiencies. In particular, for *n*-type passivated emitter and rear totally diffused (PERT) back junction (BJ) cells, efficiencies of up to 22.5% have been demonstrated [1]. The industrial screen-printed *p*-type passivated emitter and rear cells (PERC) that are now being introduced into production [2] yield efficiencies of up to 22.6% [3], but the efficiencies might decrease about 0.5%$_{abs.}$ [4] due to light-induced degradation (LID) [5,6].

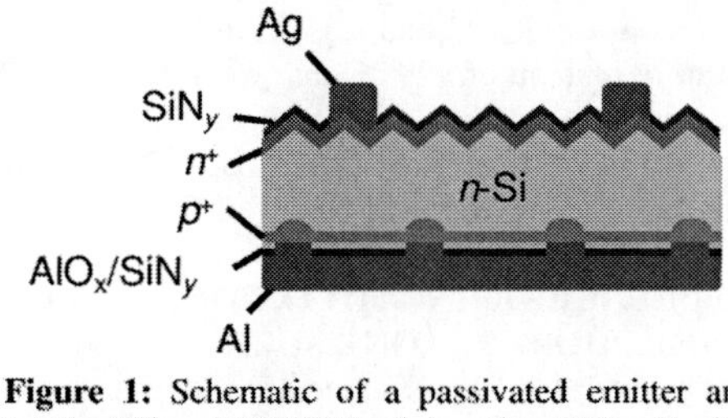

Figure 1: Schematic of a passivated emitter and rear totally diffused (PERT) back junction (BJ) cell with an *n*-type silicon substrate.

The *n*-PERT BJ cell as shown in Fig. 1 can for the most part be processed in a very similar way as PERC. However, the formation of the full-area boron diffusion at the rear of the cell that acts as the emitter adds process complexity. In laboratory scale processing of *n*-PERT BJ cells, the emitter and the FSF are commonly formed during separate BBr_3 and $POCl_3$ furnace diffusion. Thus, an additional high-temperature boron diffusion and the deposition of a dielectric layer, which protects the emitter from parasitic phosphorus indiffusion during the subsequent formation of the front surface field (FSF), are required. We aim for simplified *n*-PERT BJ solar cells with a process complexity comparable to that of *p*-PERC.

We develop boron silicate glass (BSG) layers capped with silicon nitride (SiN_z) as boron diffusion sources. The BSG/SiN_z stack is deposited by means of plasma-enhanced chemical vapor deposition (PECVD). We combine a $POCl_3$ diffusion with a boron drive-in from the solid boron diffusion source. Thus, in a single high-temperature process, the p^+-emitter and the n^+-FSF are formed simultaneously.

In former work, by using a two-stage co-diffusion process with two temperature plateaus for the drive-in of B and P [7], we fabricated *n*-PERT BJ cells with 21.0% efficiency [8,9]. With a CVD BSG diffusion source and a fast co-diffusion process at a temperature of 950 °C, 20.1% efficient bifacial *n*-type solar cells were demonstrated by Meier et al. [10].

Here we present a single-stage co-diffusion with a single temperature plateau of only 850 °C in order to reduce process time and costs. Because of the different diffusion mechanisms of boron and phosphorus, it is challenging to optimize both dopant profiles in a co-diffusion with a single diffusion temperature. The boron diffusion from the PECVD BSG source needs to be sufficiently strong despite the low diffusion temperature in order to provide sufficient lateral conductivity. As a drawback, for high B concentrations at the BSG-Si interface exceeding the solubility of B in Si, a boron-rich layer (BRL) is formed that leads to higher recombination losses [11]. Thus, the PECVD parameters of the BSG need to be adapted to allow a sufficient indiffusion of B into the silicon while at the same time avoiding BRL formation. The phosphorus surface concentration needs to be carefully tuned to allow both effective contacting with a screen-printed Ag paste and minimization of recombination losses. We use in-situ oxidation steps for tailoring the phosphorus doping profile and especially the surface concentration to allow for both low contact resistances and low recombination in the FSF. In-situ oxidation steps before and after the P drive-in have already been applied to $POCl_3$ furnace diffusions [12,13]. Based on the $POCl_3$ diffusion in [12], we integrate two in-situ oxidation steps into our co-diffusion process and investigate the influence of the oxidation durations on the resulting P-FSF.

2 BORON DIFFUSION SOURCE DEPOSITION AND SINGLE-STAGE CO-DIFFUSION

As a solid boron diffusion source, we deposit PECVD BSG layers single-sided in a cluster system (CS 400 P, von Ardenne) using an inductively coupled plasma (ICP) source. The precursor gases silane SiH_4, nitrous oxide N_2O and diborane B_2H_6 (3% diluted in 97% H_2) are used. We vary the gas flow ratio SiH_4/B_2H_6 and the total gas flow in order to control the boron content in the BSG diffusion source and the resulting sheet resistance, as investigated in [7]. In addition, we vary the BSG layer thickness between 30 nm and 40 nm, based on previous studies that showed that the saturation current density decreased for layer thicknesses ≤ 40 nm [8].

The BSG layers are capped by a 120 nm-thick PECVD silicon nitride SiN_z layer deposited in a SiNA tool (Meyer Burger). The SiN_z capping layer acts as phosphorus diffusion barrier during co-diffusion and as etching barrier that prevents the removal of the BSG during chemical cleaning and etching steps, as they would be applied during *n*-PERT BJ cell fabrication (see section 6).

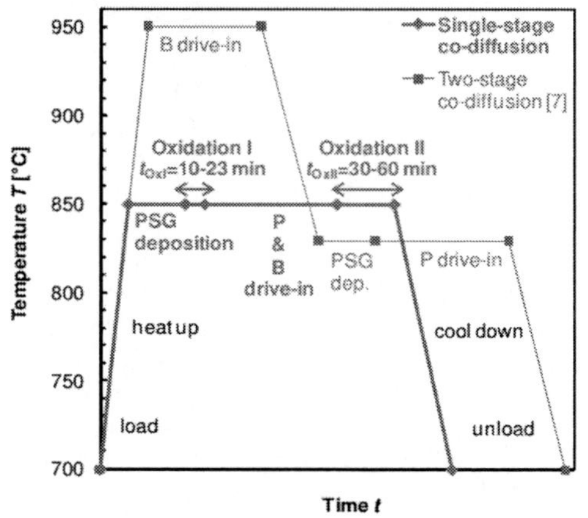

Figure 2: Single-stage co-diffusion (blue) with two in-situ oxidation steps (Oxidation I and Oxidation II). At a single temperature plateau of 850 °C, phosphorus is diffused from the PSG at the same time as boron is driven-in from a PECVD BSG/SiN_z diffusion source. The resulting phosphorus doping profile and the surface concentration can be controlled by the in-situ oxidation steps. For comparison, the previously applied two-stage co-diffusion [7] with two temperature plateaus for the B and P drive-in is shown (red).

The temperature profile of the single-stage co-diffusion is shown in Fig. 2 (blue). After heating up to 850 °C, a phosphosilicate glass (PSG) is deposited from $POCl_3$ gas. A first in-situ oxidation (Oxidation I) step with a duration of t_{OxI}=10-23 min follows that forms a diffusion barrier and slows down the P diffusion. During the drive-in step, P is diffused from the PSG and B is driven-in from the pre-deposited PECVD BSG/SiN_z diffusion source. The succeeding second in-situ oxidation (Oxidation II) with a duration of t_{OxII}=30-60 min oxidizes the P-diffused silicon surface, so that the P surface concentration peak, which is then contained in the silicon oxide, can be removed in a following etching step. In addition, P from the doped silicon surface and B from the BSG are further driven in. In Fig. 2, also the two-stage co-diffusion with a B drive-in step at 950 °C and a P drive-in at a lower temperature plateau of 829 °C that we applied in [7] is shown (red). The single-stage co-diffusion process allows for a lower thermal budget and reduces the process time and cost significantly.

3 INFLUENCE OF IN-SITU OXIDATIONS ON THE PHOSPHORUS FRONT SURFACE FIELD

We investigate the influence of different in-situ oxidation times during single-stage co-diffusion (see Fig. 2) on the phosphorus doping profile. Therefore, on planar *p*-type Cz-Si samples after diffusion and PSG removal, electro-chemical CV (ECV) measurements and four point probe (FPP) measurements are performed.

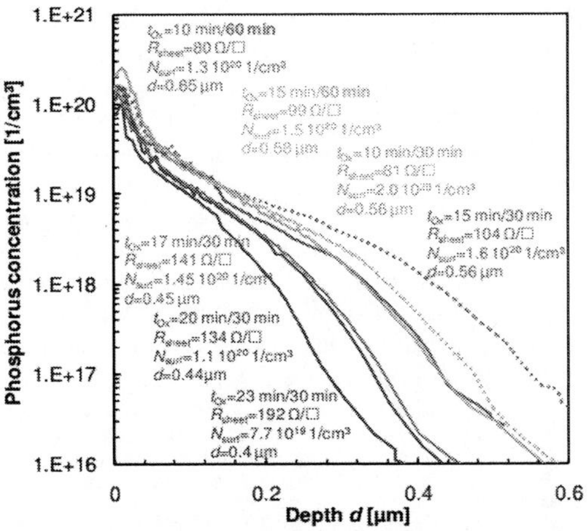

Figure 3: Phosphorus doping profiles for increasing duration of Oxidation I and for two different durations of Oxidation II (30 min and 60 min). The phosphorus concentration is measured by means of ECV on planar 156×156 mm² samples after diffusion and PSG removal.

Figure 3 shows the P doping profiles obtained from co-diffusions with durations of Oxidation I varying from 10 min to 23 min and for a duration of Oxidation II of either 30 min or 60 min. The profile depth corresponds to the measured or extrapolated depth at a concentration of 1×10^{16} 1/cm³.

For an increasing time of Oxidation I (see solid lines from green to violet in Fig. 3), the surface concentration N_{surf} and the profile depth d decrease resulting in increasing sheet resistances R_{sheet} from 81 Ω/□ to 192 Ω/□. When the duration of Oxidation II is doubled from 30 min to 60 min, while keeping t_{OxI} constant (compare solid green and dotted orange lines in Fig. 3), N_{surf} is lowered and d is increased while R_{sheet} reaches a similar value of about 80 Ω/□ (t_{OxI}=10 min) and 100 Ω/□ (t_{OxI}=15 min), respectively.

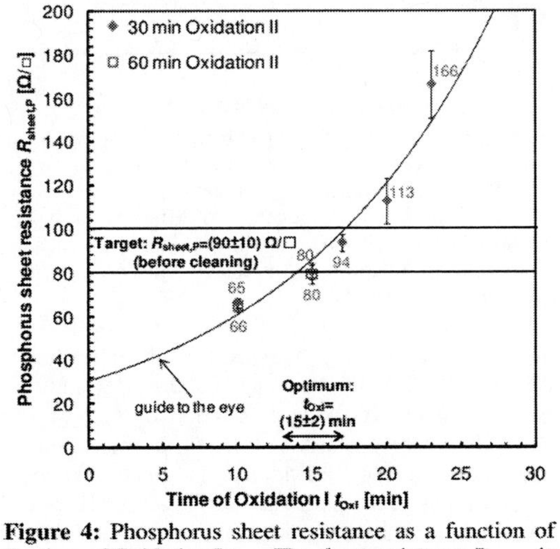

Figure 4: Phosphorus sheet resistance as a function of the time of Oxidation I t_{OxI}. The sheet resistance $R_{sheet,P}$ is measured by means of FPP on planar 156×156 mm² samples after diffusion and PSG removal before cleaning.

Figure 4 shows the mean sheet resistance $R_{sheet,P}$ that is determined from 121 measurement points on a wafer area of 156×156 mm² as a function of the duration of Oxidation I. The error bars correspond to the standard deviation on one wafer.

With increasing time of Oxidation I t_{OxI}, $R_{sheet,P}$ increases. For t_{OxI}=10 min and 15 min, the duration of Oxidation II is doubled from 30 min to 60 min. The measured sheet resistances agree within the measurement uncertainty, an impact of the duration of Oxidation II on $R_{sheet,P}$ is not observed.

In order to fabricate highly efficient n-type cells with a P-FSF that allows for both low contact resistances and low recombination, we consider $R_{sheet,P}$=155 Ω/□ as a maximum target value for the sheet resistance on the cells. With this $R_{sheet,P}$ value and a corresponding $J_{0,P}$=46 fA/cm², Lim et al. demonstrated 21.2%-efficient n-PERT BJ cells [14] with an open-circuit voltage V_{oc}=674 mV. For the P-FSF as applied in [14], it was observed that $R_{sheet,P}$ increases for a textured silicon surface that is RCA-cleaned after diffusion due to a reduction of the P surface concentration.

Thus, we aim for a sheet resistance of (90±10) Ω/□ on planar samples before RCA-cleaning and find an optimal duration of Oxidation I t_{oxI}=(15±2) min.

In the following, we focus on two single-stage co-diffusions with different oxidation times t_{OxI}/ t_{OxII} of 17 min/30 min and 15 min/60 min and characterize further test structures before processing solar cells (see section 6). On textured n-type Cz-Si wafers that are symmetrically diffused, RCA-cleaned, SiN_x-passivated and fired, we measure the photoconductance decay (PCD) with a photoconductance lifetime tester (WCT-120, Sinton Instruments) using an intrinsic carrier concentration of Si of n_i = 8.6×10^9 1/cm³. The measured data is evaluated by the Kane & Swanson method to extract the saturation current density J_0 [15]. Simultaneously, we measure R_{sheet} inductively and extract the implied open-circuit voltage $V_{oc,impl}$. For the PCD measurements, the estimated measurement uncertainty is 10% for R_{sheet} and J_0 and 1% for $V_{oc,impl}$, respectively. We also perform transfer lengths measurements (TLM) [16] in order to determine the contact resistivity ρ_c to a screen-printed Ag-metallization. The results are summarized in Tab. 1.

t_{OxI}/ t_{OxII} [min]	$R_{sheet,P}$ [Ω/□] (planar, before RCA)	$R_{sheet,P}$ [Ω/□] (textured, after RCA)	$J_{0,P}$ [fA/cm²]	$V_{oc,impl}$ [mV]	ρ_c [mΩ cm²]
17/ 30	94 ±4	150 ±10%	56 ±10%	687 ±1%	6.0 ±0.7
15/ 60	80 ±5	117 ±10%	47 ±10%	688 ±1%	3.0 ±0.8

Table 1: Measurement results for phosphorus diffusions on n-type test structures obtained from single-stage co-diffusions with different in-situ oxidation times.

With our single-stage co-diffusion at 850 °C with t_{OxI}=17 min, we obtain a P-FSF with a $J_{0,P}$ of 56 fA/cm² and an implied voltage $V_{oc,impl}$ of 687 mV at $R_{sheet,P}$=150 Ω/□. Due the high P sheet resistance, a contact resistance ρ_c of (6.0 ± 0.7) mΩ cm² to a commercially available Ag paste is measured. By choosing a shorter duration of Oxidation I of t_{OxI}=15 min, a stronger phosphorus diffusion with $R_{sheet,P}$=117 Ω/□ is obtained. This results in a lower contact resistance of only (3.0 ± 0.8) mΩ cm². The low saturation current density of 47 fA/cm² indicates low recombination losses which might result from the P surface concentration that is lowered during the second oxidation step of 60 min (compare dotted orange doping profile in Fig. 3). Due to these promising results on test structures, we select the co-diffusion process with t_{OxI}=15 min and t_{OxII}=60 min for the fabrication of highly efficient cells (see section 6).

4 BORON EMITTER DIFFUSION FROM PECVD BSG/SIN$_Z$

We evaluate the boron diffusions obtained from PECVD BSG/SiN$_z$ diffusion sources at a temperature of 850 °C during the single-stage co-diffusion with varying oxidation times.

Boron doping profiles are measured by means of ECV on planar n-type Cz-Si samples after diffusion. PCD measurements on symmetrically diffused, ALD-AlO$_x$/PECVD-SiN$_y$ passivated and fired planar samples are performed and the boron sheet resistances $R_{sheet,B}$, saturation current densities $J_{0,B}$ and implied open-circuit voltages $V_{oc,impl}$ are extracted.

In order to find a boron diffusion suitable for the p^+-emitter of a high-efficiency solar cell we vary the BSG layer thickness and the gas flow ratios during PECVD which influence the boron content in the boron diffusion source and thus the resulting doping profile.

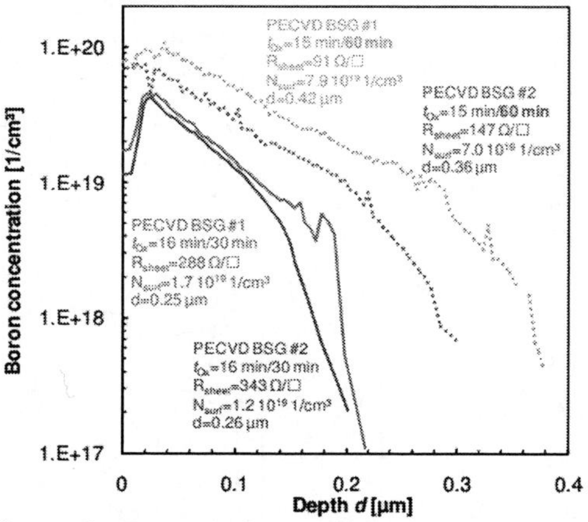

Figure 5: Boron doping profiles for two different PECVD BSG/SiN$_z$ diffusion sources and for durations of Oxidation II of 30 min or 60 min. The B concentration is measured by ECV on planar 156x156 mm² samples after diffusion and BSG/SiN$_z$ removal.

Figure 5 shows the B doping profiles for two PECVD BSG/SiN$_z$ diffusion sources #1 and #2. The boron drive-in is performed during single-stage co-diffusions with two different durations of Oxidation II t_{OxII}=30 min (blue solid lines) or 60 min (green dotted lines) and with t_{OxI}=15 min. For the same BSG diffusion source, the 30 min Oxidation II results in shallow emitter diffusions with a depth of about 0.25 µm and sheet resistances $R_{sheet,B}$>280 Ω/□. Doubling the time of Oxidation II results in deeper emitter diffusion with increased surface concentrations and $R_{sheet,B}$<150 Ω/□.

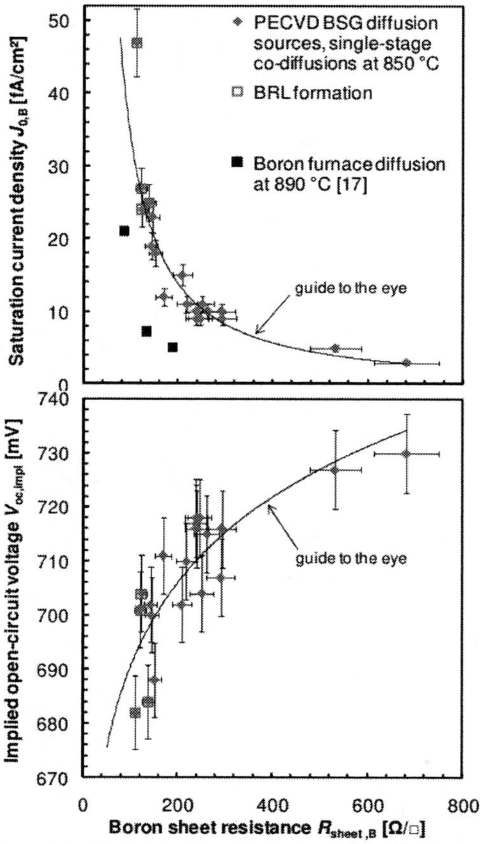

Figure 6: Saturation current density and implied open-circuit voltage as a function of $R_{sheet,B}$ for different PECVD BSG/SiN$_z$ diffusion sources. The boron drive-in is performed during single-stage co-diffusions at 850 °C with different in-situ oxidation times. The $J_{0,B}$ results are compared to literature best values for boron furnace diffusions at 890 °C [17].

The $J_{0,B}$ and $V_{oc,impl}$ results for all investigated PECVD BSG diffusion sources and oxidation times are shown in Fig. 6 as a function of $R_{sheet,B}$. The error bars correspond to the PCD measurement uncertainties. We find adjustable sheet resistances from about 110 Ω/□ to 300 Ω/□. The sheet resistance values >600 Ω/□ are not relevant for solar cell fabrication.

The saturation current density decreases with increasing $R_{sheet,B}$ to values below 10 fA/cm². The

$J_{0,B}$ results that we achieve with single-stage co-diffusions from PECVD BSG/SiN_z are higher than literature best values for boron furnace diffusions at 890 °C [17]. This might be explained by our low boron diffusion temperature of 850 °C leading to shallow boron diffusions with pronounced peak concentrations. With the BSG boron diffusion source, high $V_{oc,impl}$ values far above 700 mV are reached. For the strongest boron diffusions with sheet resistances <140 Ω/□, the formation of a boron rich layer (BRL) is observed. The presence of a BRL increases surface recombination and thus results in high $J_{0,B}$ and low $V_{oc,impl}$ results. Thus, for cell fabrication, we only consider the PECVD BSG parameters that avoid BRL formation but allow for sufficient emitter diffusion to provide the lateral conductivity.

5 CELL SIMULATIONS

We perform cell simulations to determine the efficiency η and the open-circuit voltage V_{oc} of *n*-PERT BJ cells using the CoBo model [18] as implemented in the Quokka software [19]. The CoBo model uses experimental input parameters such as $R_{sheet,P}$, $J_{0,P}$ and ρ_c from Tab. 1 and $R_{sheet,B}$ and $J_{0,B}$ from Fig. 6. All other parameters are taken from [14]. The influence of $R_{sheet,B}$ on the cell parameters is investigated, as shown in Fig. 7. For comparison, also the simulated results of published *n*-PERT BJ cells obtained with either two-stage co-diffusion [9] or separate $POCl_3$ and BBr_3 diffusions [14] are plotted (large symbols).

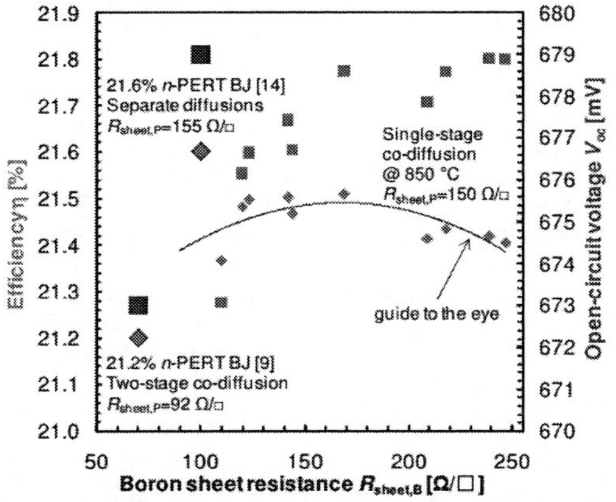

Figure 7: Simulated parameters of *n*-PERT BJ cells applying the single-stage co-diffusion yielding $R_{sheet,P}$=150 Ω/□. The efficiency (blue symbols) and open-circuit voltage (red symbols) depend on $R_{sheet,B}$. For comparison, the simulated results of published *n*-PERT BJ cells [9,14] are plotted (large symbols).

The P-FSF has a major influence on the cell parameters. When, in previous studies, we increased $R_{sheet,P}$ from 92 Ω/□ [9] to 155 Ω/□ [14], the efficiency increased by 0.4%$_{abs.}$ (large blue diamonds). For our single stage co-diffusion with t_{OxI}=17 min and t_{OxII}=30 min resulting in $R_{sheet,P}$=150 Ω/□ (see Tab. 1), V_{oc} can be increased up to 679 mV (red squares) by increasing $R_{sheet,B}$ and thus decreasing $J_{0,B}$. However, due to increasing series resistances and decreasing fill factors for $R_{sheet,B}$>200 Ω/□, η reaches a maximum value (blue diamonds). For our single-stage co-diffusion at 850 °C and the applied screen-print contact layout, we find a maximum simulated efficiency of 21.5% for a boron diffusion source yielding $R_{sheet,B}$=120-200 Ω/□. Taking the observed BRL formation into account, we conclude that, for our cell fabrication process, an optimal boron emitter allowing for high-efficiency co-diffused cells should have a sheet resistance of 150-200 Ω/□.

6 CO-DIFFUSED SOLAR CELLS

For the fabrication of co-diffused *n*-PERT BJ solar cells, as shown in Fig. 1, 156×156 mm² *n*-type Cz silicon wafers with a resistivity of 4.7 Ω cm are saw damage etched and RCA-cleaned. The BSG and SiN_z capping layer depositions are performed by means of PECVD on the rear side. PECVD BSG parameters are chosen to allow for $J_{0,B}$=10 fA/cm² without BRL-formation. During the following alkaline texture, the multifunctional BSG/SiN_z layer stack acts as an etching barrier so that only the front side is textured. We apply our single stage co-diffusion with T=850 °C and t_{OxI}=15 min, t_{OxII}=60 min to simultaneously form the n^+-FSF on the cell front side and the p^+-emitter on the rear side in only one high-temperature process. After PSG and BSG removal, a stack of atomic layer deposited (ALD) aluminum oxide (AlO_x) and silicon nitride (SiN_y) is applied as a rear side passivation, while a SiN_y-layer acts as front side passivation and anti-reflection coating. The metal contacts are formed by screen-printing of commercial Ag paste on the front side with a five busbar layout and a full-area Al screen-printing on the rear side after laser contact opening. Co-firing finalizes the contact formation. Further processing details and a comparison to PERT BJ and PERC cell processes can be found in [8].

Co-diffusion	η [%]	J_{sc} [mA/cm²]	V_{oc} [mV]	*FF* [%]	R_s [Ohm cm²]
Single-stage T=850 °C t_{OxI}/ t_{OxII}= 15/60 min	20.7 ±0.3 *	39.1 ±0.5	663 ±2	80.0 ±0.6	0.47
Two-stage *T*=950/ 829 °C	21.0 ±0.4 ** [8,9]	39.1 ±0.7	666 ±2	80.5 ±0.5	0.54

Table 2: IV-parameters of the best co-diffused *n*-PERT BJ cells for a single-stage and a two-stage co-diffusion process [8,9]. The cell results are measured on a cell area of 244 cm² and 239 cm², respectively, with screen-printed contacts applying a five busbar layout.
* Calibrated measurement by ISFH CalTec.
** Independently confirmed by Fraunhofer ISE CalLab.

Table 2 summarizes the *n*-PERT BJ cell results of our best co-diffused cells. By applying a single-stage co-diffusion instead of a two-stage co-diffusion with two different temperature plateaus for the drive-in of B and P (see Fig. 2), we could lower the thermal budget and thus the processing cost and time. Despite the relatively low boron diffusion temperature of 850 °C, our best cell reaches 20.7% efficiency. Thus, compared to the 21.0% efficient co-diffused *n*-PERT BJ cell that we fabricated with a two-stage co-diffusion [8,9], we could both further simplify and improve the co-diffusion process and optimize the PECVD BSG/SiN_z doping source.

7 SUMMARY AND CONCLUSION

In order to simplify the fabrication of high-efficiency industrial *n*-type silicon solar cells, we combine a pre-deposited boron diffusion source with a co-diffusion process. A PECVD BSG/SiN_z acts as a single-sided, solid boron diffusion source. By $R_{sheet,B}$ and $J_{0,B}$ measurements and cell simulations, we identify optimal boron sheet resistances of $R_{sheet,B}$=150-200 Ω/□. The selected BSG/SiN_z stacks allow for a BRL-free formation of p^+-emitters with both low recombination ($J_{0,B}$<10 fA/cm²) and sufficient lateral conductivity. The n^+-FSF is formed simultaneously during the 850 °C drive-in step of our single-stage co-diffusion process. We apply two in-situ oxidation steps for tailoring the phosphorus doping profile. We observe a significant influence of the Oxidation I duration on $R_{sheet,P}$ and for t_{OxI}=15 min we reach a FSF with low recombination ($J_{0,P}$=47 fA/cm²) and a low contact resistance of ρ_c=3.0 mΩ cm² to screen-printed Ag fingers. We apply our optimized co-diffusion for the fabrication of *n*-PERT BJ cells and reach an efficiency of 20.7% with a lean process flow. The PECVD BSG/SiN_z combined with our co-diffusion replaces the separate BBr_3 and $POCl_3$ furnace diffusions that are required for standard PERT BJ cells. The single-stage co-diffusion with a diffusion temperature as low as 850 °C allows to further reduce the thermal budget and the process time and cost compared to a two-stage co-diffusion. Thus, the fabrication of industrial *n*-type cells with high and stable efficiencies is possible with a process complexity comparable to *p*-type PERC cells. The single-stage co-diffusion and the pre-deposited boron diffusion source might enable PERC cell manufacturers to migrate to *n*-PERT cell fabrication with only small changes in the production line.

8 ACKNOWLEDGEMENTS

We thank Miriam Berger and Anja Christ for cell processing, Sarah Spätlich for diffusion optimization, David Sylla and Tobias Neubert for laser processes and Tobias Gandy for cell measurement. This work was funded by the German State of Lower Saxony and the German Federal Ministry for Economic Affairs and Energy under Grant 0325880A (PERC 2.0).

9 REFERENCES

[1] A. Urueña, M. Aleman, E. Cornagliotti, A. Sharma, J. Deckers, M. Haslinger, L. Tous, R. Russell, J. John, Y. Yao, T. Söderström, F. Duerinckx, J. Szlufcik, Proc. 31th EUPVSEC, pp. 410-413 (2015).

[2] T. Dullweber, S. Gatz, H. Hannebauer, T. Falcon, R. Hesse, J. Schmidt and R. Brendel, Prog. Photovolt: Res. Appl. 20, 630-638 (2012).

[3]http://ir.trinasolar.com/phoenix.zhtml?c=206405&p=irol-newsArticle&ID=2230468, press release, 19.12.2016.

[4] T. Dullweber, C. Kranz, U. Baumann, R. Hesse, D. Walter, J. Schmidt, P. Altermatt, R. Brendel. Proc. 39th IEEE Photovolt. Specialist Conf., pp. 3074–3078 (2013).

[5] Y. Gassenbauer, K. Ramspeck, B. Bethmann, K. Dressler, J.D. Moschner, M. Fiedler, E. Brouwer, R. Drössler, N. Lenck, F. Heyer, M. Feldhaus, A. Seidl, M. Müller, A. Metz, IEEE J. Photovolt. 3, 125-130 (2013).

[6] J. Schmidt, Solid State Phenom. 95-96, pp. 187-196 (2004).

[7] N. Wehmeier, B. Lim, A. Merkle, A. Tempez, S. Legendre, H. Wagner, A. Nowack, T. Dullweber, and Pietro P. Altermatt, IEEE J. Photovolt., Vol. 6, No. 1, pp. 119-125 (2016).

[8] N. Wehmeier, B. Lim, A. Nowack, J. Schmidt, T. Dullweber, and R. Brendel, Phys. Status Solidi RRL 10, No. 2, 148–152 (2016).

[9] N. Wehmeier, A. Nowack, B. Lim, T. Brendemühl, S. Kajari-Schröder, J. Schmidt, R.. Brendel, and T. Dullweber, Solar Energy Materials & Solar Cells, 158, pp. 50-54 (2016).

[10] S. Meier, S. Maier, C. Demberger, A. Wolf, D. Biro, and S. W. Glunz, Sol. RRL, Vol. 1, 1600005 (2017).

[11] M. A. Kessler, T. Ohrdes, B. Wolpensinger, R. Bock, and N.-P. Harder, Proc. 34th IEEE Photovolt. Specialists Conf., pp.1556-1561 (2009).

[12] T. Dullweber, H. Hannebauer, S. Dorn, S. Schimanke, A. Merkle, C. Hampe, and R. Brendel, Prog. Photovolt: Res. Appl., Vol. 25, pp. 509-514 (2017).

[13] A. Wolf, A. Kimmerle, S. Werner, S. Maier, U. Belledin, S. Meier, D. Biro, Proc. 31st European Photovoltaic Solar Energy Conference and Exhibition, pp. 414-419 (2015).

[14] B. Lim, T. Brendemühl, T. Dullweber, R. Brendel, IEEE J. Photovolt., Vol. 6, No. 2, pp. 447-453 (2016).

[15] D. E. Kane, R. M. Swanson, Proc. 18th IEEE Photovolt. Specialists Conf., pp. 578-583 (1985).

[16] D. K. Schroder und D. L. Meier, IEEE Transactions on Electron Devices, Vol. 31, No. 5, pp. 637–647 (1984).

[17] A. Richter, J. Benick, and M. Hermle, IEEE J. Photovolt., Vol. 3, No. 1, pp. 236-245 (2013).

[18] R. Brendel, Prog. Photovolt: Res. Appl., Vol. 20, pp-31-43 (2012).

[19] A. Fell, IEEE Trans. Electron Devices 60 (2), pp. 733-738 (2012).

* corresponding author, e-mail: wehmeier@isfh.de, phone: +49-5151-999-325

N-TYPE MONOLIKE SILICON BIFACIAL SOLAR CELL: AN ALTERNATIVE WAY OF HIGH EFFICIENCY AND LOW COST

Chia-Lung Lin*, Yu-Ta Cheng, Yu-Hsiang Huang, Chia-Ching Wang, Jung-Wu Chien
lin.cl@inventecsolar.com
Inventec Solar Energy Corp.
No. 349, Sec. 2, Renhe Rd., Daxi Dist., Taoyuan City 33547, Taiwan

China "Front Runner Program" leads to huge demand of high efficient solar cells started from 2016. Every company is enthusiastic to develop their own high efficient solar cell technologies and products to keep the competence in this field. sc-Si and mc-Si PERC cells are the most popular among high efficient solar cells, however, the shortage of p-type sc-Si wafer, Al2O3 equipment and LIR tools hinder the cost and development of sc-Si PERC. Besides, extra black silicon texturing and severe LID are still the issues for mc-Si PERC. To solve the difficulties suffered in sc-Si and mc-Si PERC cells, n-type monolike silicon bifacial solar cell is developed and presented in this article. The front cell efficiency up to 20.8% in average was realized by industrial-available process and facilities. Zero LID because of the intrinsic characteristic of n-type silicon wafers was proven without LIR process. Module power of 299Wp was achieved with 60 pcs cells and white backsheet. With the development and commercialization of double glass module, it is expected that low cost and high efficient bifacial solar cells will get more and more attention.

Keywords: n-type，monolike，bifacial

1 INTRODUCTION

Recently, the rapid fall in photovoltaic prices drives researchers to continue developing multifarious technology about efficiency improvement and cost reduction in solar cell manufacture process. sc-Si and mc-Si PERC cells (Passivated Emitter and Rear Cell) are the most popular among high efficient solar cells. However, it is interesting that there are around 15GW capacity of PERC cells in worldwide 2016, but only less than 4GW PERC cells were actually produced in 2016. This phenomenon reveals the key for the PERC cells is the mass production stability, not the capacity. Besides, the p-type sc-Si wafers shortage, there are many other factors hinder the cost and development of sc-Si PERC cells, such as the Al2O3 equipment stability and the additional LIR (light induced regeneration) tool required for LID (light induced degradation) reduction. As for the mc-Si PERC cells, the mc-Si wafers cost is not as cheap as previous compared with sc-Si wafers, also, a complicate black silicon texturing process is needed for the novel diamond wire saw technology. And unfortunately, the LID issue on mc-Si PERC cells cannot be solved even using LIR process.

To reduce solar energy cost is to elaborate solar cells reaching higher efficiencies by developing cost-effective processes [1, 2]. At the present day, the highest efficiencies obtained so far on industrial silicon solar devices involve technologies are implemented on n-type substrates. And monolike silicon is considered as a promising material as it could allow the fabrication of high efficiency solar cells at lower production cost. However, the most studies are generally carried out on p-type monolike silicon. As a consequence, the potential of n-type monolike silicon is seldom investigated. One potential limiting factor to n-type monolike silicon is uncertain considering the impact of cell process flow on the bulk lifetime [3].

Recent years, the PERT (passivated emitter, rear totally diffused) solar cells become popular due to its special application and high power output which generates from both of front and rear sides [4]. The n-type PERT technology could also be well suited to achieve high efficiency on a wide range of resistivity and quality material. Recent results published show the potential of n-type monolike silicon using PERT cell technology [5].

In this work, the cell architecture considered here is a bifacial n-type PERT structure developed by Inventec Solar Energy Co. [6]. Cell scheme is given in FIG 1. A new boron diffusion recipe was developed by Inventec to avoid thermal degradation by controlling boron diffusion process time to less than 90 minutes. The sunny side was doped by BCl3 or BBr3 gaseous diffusion. The rear side was doped by PH3 ion-implant or POCl3 gaseous diffusion. Subsequently, SiNx film deposition and metal printing were processed. The high efficiency n-type monolike silicon bifacial solar cell with front cell efficiency up to 20.8% in average was realized by industrial-available process and facilities. Zero LID because of the intrinsic characteristic of n-type silicon wafers was proven without LIR process. Module power of 299Wp was achieved with 60 pcs cells and white backsheet. With the development and commercialization of double glass module, it is expected that low cost and high efficient bifacial solar cells will get more and more attention.

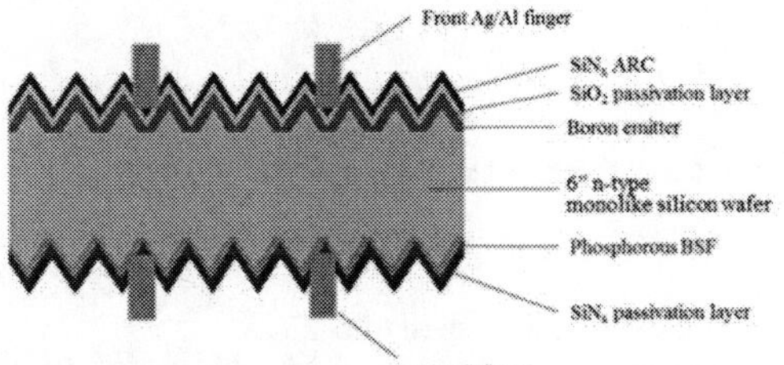

Figure 1. Scheme of n-type bifacial silicon solar cell

2 EXPERIMENT

There are three topics involved in this article. The first one focus on anneal and passivation conditions; the next discuss the influence of raw wafer resistivity and forming gas annealing (FGA) treatment. Finally, the advanced Gen2 n-PERT solar cell technology developed by Inventec [6] was adopted on such n-type moonlike silicon wafers.

FIG 2 shows our advance process for n-type monolike bifacial cell fabrication. First of all, in order to find out the best process condition for monolike wafer, anneal and passivation process after ion implantation was fine-tuned by adjusting the anneal temperature and the oxygen portion. Therefore three processes were designed; called process I, II and III, on monolike wafers with similar resistivity and grain appearance to verify the difference of passivation effect by various anneal and passivation process.

Furthermore, n-type monolike wafers were investigated with two kinds of resistivity: 1~3 Ω-cm and 3~6 Ω-cm. For confirming if hydrogen can offer an extra passivation effect or not, FGA was adopted.

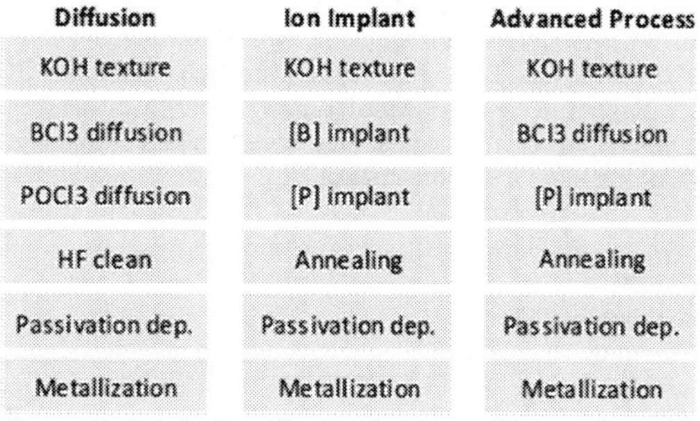

Figure 2. Fabrication flows of n-type silicon solar cells with diffusion, ion-implantation and advanced process flow (right).

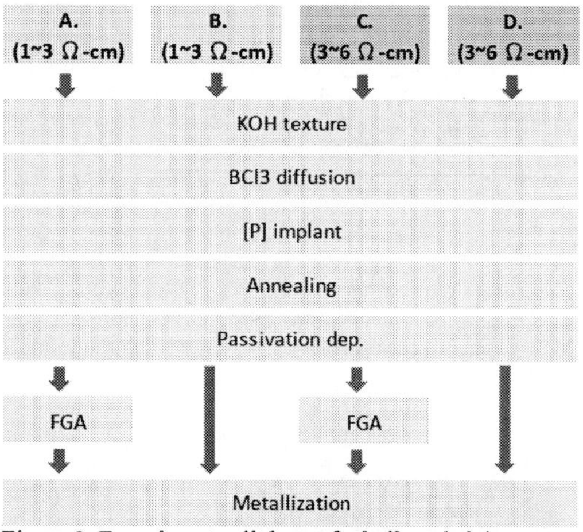

Figure 3. Experiment split by wafer bulk resistivity.

3 RESULTS AND DISCUSSION

3.1 Anneal and Passivation Conditions

The PL image measured on monolike wafers treated by process I, II and III are shown in FIG 4. In process I and III, two different passivation processes were treated on monolike wafers with similar resistivity and grain appearance. In previous experiment, process I and process III are sure to have comparable passivation on Cz-crystalline wafers. However, it reveals a different result on monolike wafers. Between PL image on process I and III shown in FIG 4, comparable PL intensity can only be observed on large size grain but not on small grain and grain boundaries. It can be noted the Voc from process I is 3 mV higher than from process III. This enhancement is mainly caused by the passivation effect on small grain and grain boundaries. Both PL image and Voc indicates that the major area need to be take care (passivated) on moonlike wafers is small grain, grain boundary and dislocation. On monolike wafers, a special passivation process, like process I in this work, is actually needed. Based on the most suitable passivation process I, the anneal conditions in process II was developed to obtain an extra enhancement on wafer quality. In process II, PL intensity exhibit obviously enhancement not only on large size grains, but also the small size grain boundaries. Furthermore, Voc from process II is higher than the other two processes.

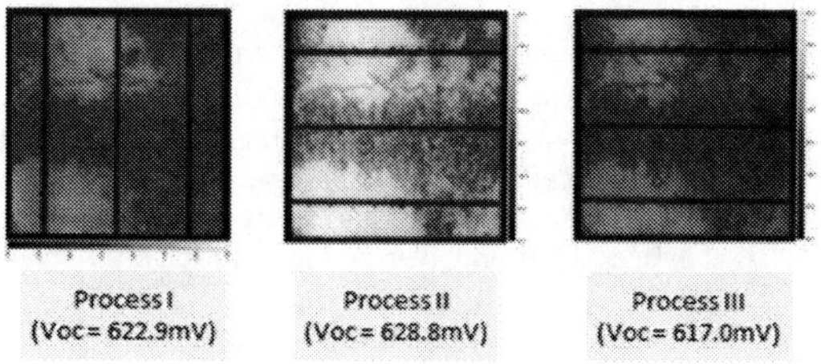

Figure 4. PL image measured on monolike wafers treated by process I, II and III.

3.2 Influence of Resistivity and FGA

In the next, the influence of different wafer resistivity and FGA process was studied. FIG 5 reveals PL images on two kinds of wafer resistivity by process step sequence. It can be seen the PL intensity on small grains increases along with process step sequence. In passivation process, it is well-known that hydrogen could cause passivation effect on dangling bounds. After that, FGA step could further offer a similar effect on these wafers.

With different wafer resistivity, PL intensity is comparable before FGA step. After FGA, wafer with higher resistivity reveals a brighter image than wafer with lower resistivity. Hence, hydrogen passivation effect is seems more effective on high resistivity wafers.

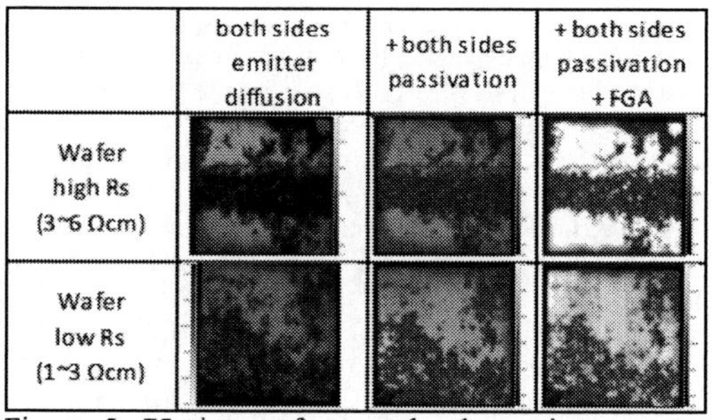

Figure 5. PL image for samples by various process conditions.

3.3 Light Current-Voltage

FIG 6~8 show the cell performances of experiment splits in FIG 3. It only reveals unobvious difference on Voc as shown in FIG 6. Alternatively, samples with high resistivity exhibit higher Isc than low resistivity samples. FIG 7 shows relative FF trend for various samples. The higher resistivity samples exhibit lower FF due to higher Rs, even though Rsh is also higher at the same time. To summarize total effects on these samples, the trend figure of conversion efficiency vs. Isc is shown in FIG 8. It is observed an obviously linear relationship between efficiency and Isc. Hence, the higher wafer resistivity samples with higher Isc and can be passivated easily enhances better cell performance. It is suspected that the wafers higher resistivity moonlike wafers localized in the bottom of ingot position with less dislocation can be easily passivated.

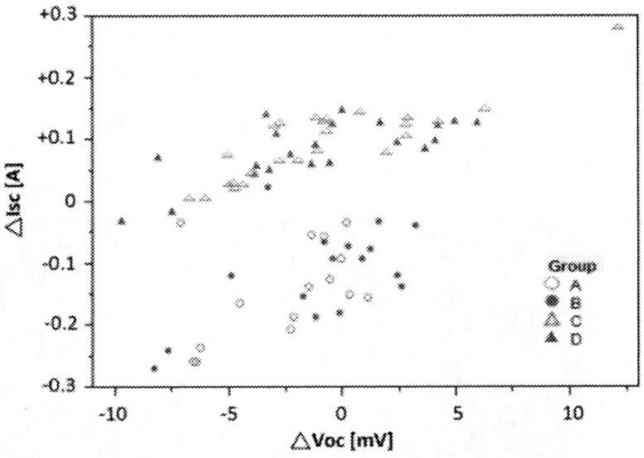

Figure 6. Isc and Voc results on various samples.

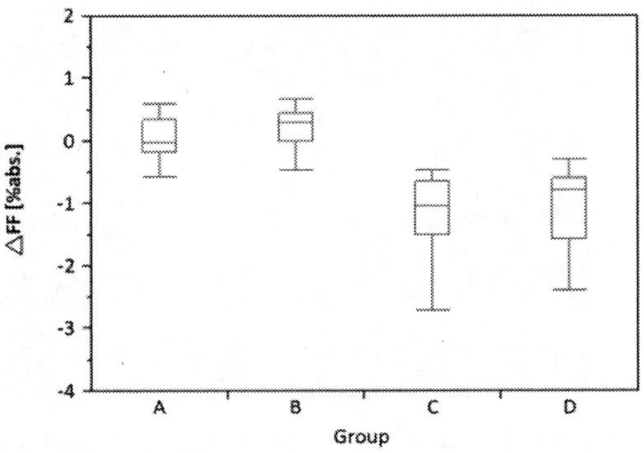

Figure 7. FF trend for various samples.

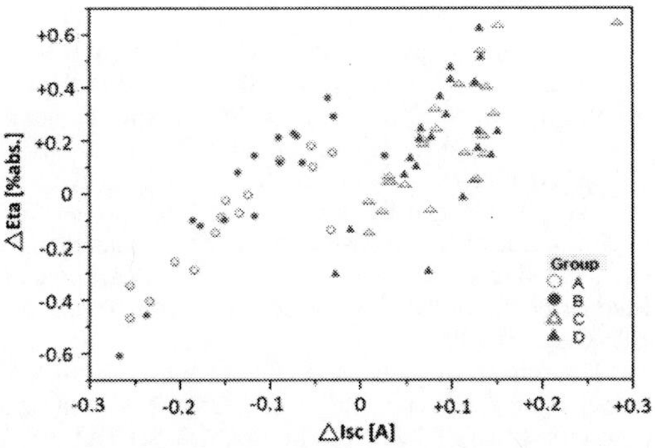

Figure 8. Trend figure of conversion efficiency vs. Isc

3.4 n-type Moonlike Bifacial Cell with High Efficiency and Low Cost

One way to reduce solar cell cost is to maintain cell efficiency using cheaper silicon wafers with a quality satisfying PV conversion demands. Thus, the ISEC Gen2 n-type moonlike bifacial cell has been developed with the combination of ISEC Gen2 n-PERT technology [6] and implemented on the high quality n-type moonlike wafers [5].

With ISEC Gen2 n-PERT technology, not only 4 busbar design, other parameters such as emitter, BSF profile, front and rear side electrodes were also further optimized. Cell performance was studied on the wafers along two different side bricks C2 and E3. The wafer bulk resistivity and the cell performance compared with n-type Cz silicon wafers are shown in FIG 9 and FIG 10 respectively.

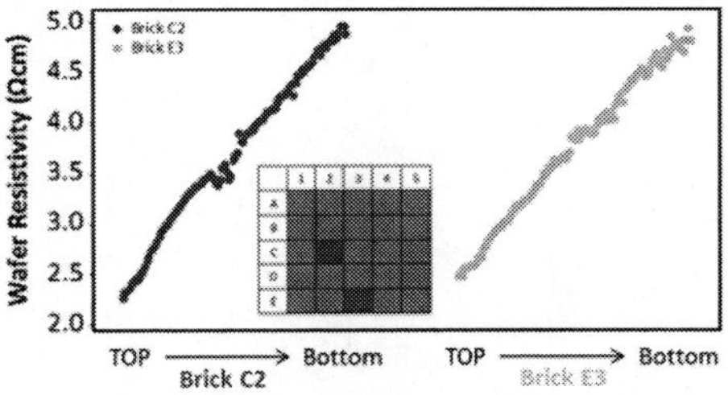

Figure 9. Wafer bulk resistivity along two different side bricks C2 and E3.

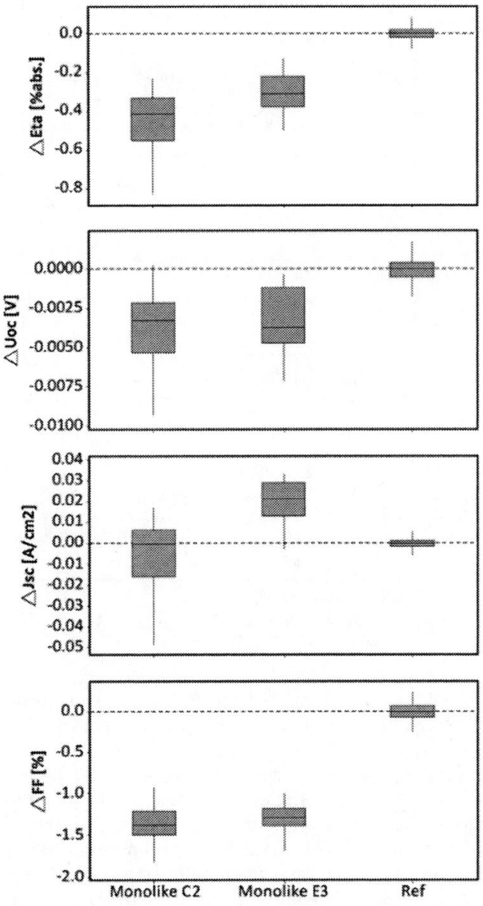

Figure 10. Cell performance compared with n-type Cz silicon wafers

From the comparison, the cells performance on moonlike wafers is 0.3~0.4%abs. lower than on reference Cz wafers, which is mainly dominated by Voc and FF. Monolike cells showed reduced Voc values compared to reference Cz indicates that this monolike material has a lower quality even with high quality casting technology.

The lower FF might be resulted from the relative higher bulk resistivity compared to reference Cz. If FF of monolike wafer is the same as reference Cz, the efficiency gap should be less than 0.2%. From Isc, the low reflectivity could be achieved due to such almost full <100> orientation material same as Cz wafers.

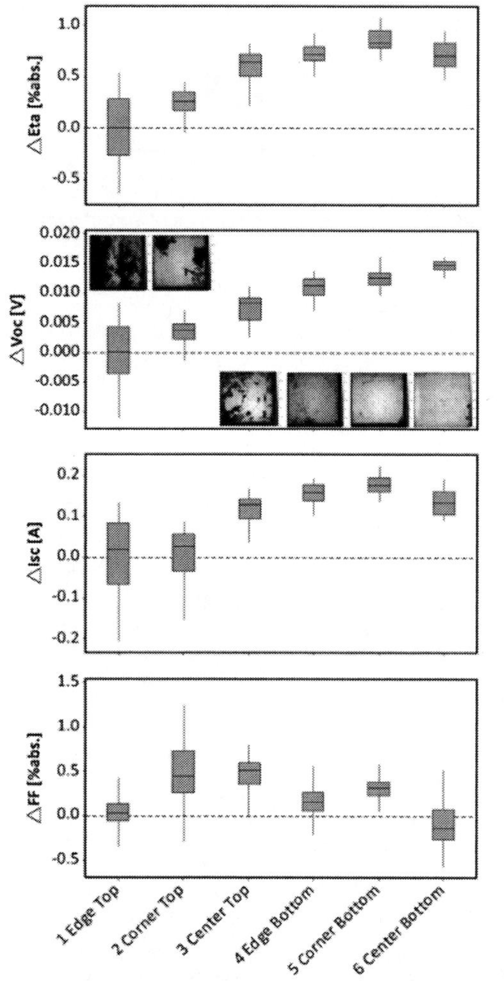

Figure 11. Cell performance and PL images comparison with different wafers at different moonlike ingot position.

From the results shown in FIG 11, it can be noticed that moonlike cell efficiency decreases linearly from bottom to top of the brick because of both Voc and Isc tailing. Also, the wafers at different ingot position play an important role. The same finding has been published in previous [4,7]. Within monolike casting ingot, dislocations multiplication is observed from bottom to top of the ingot and the phenomenon looks to be accentuated in side and corner bricks. On the wafers from bottom portion of brick, the cell process is well-suited to reach high efficiency with a narrow distribution. On the wafers from top portion of brick, it shows much pronounced variation in the efficiency which is clearly correlated to lifetime variation. As discussed in section 3.2 and FIG 5, an extra-step as hydrogenation might be a method required to reduce the efficiency range in this area, although the effect of FGA treatment is not clear by IV test discussed in section 3.3.

Table 1. Electrical data of n-PERT modules by using ISEC Gen2 n-type PERT solar cells with sc-Si and moonlike wafers.

	Power (W)	Voc (V)	Isc (A)	FF (%)	Vmp (V)	Imp (A)
sc-Si	305.73	38.78	10.09	78.2	32.24	9.48
monolike	299.07	38.51	10.11	76.8	31.70	9.44

(a)

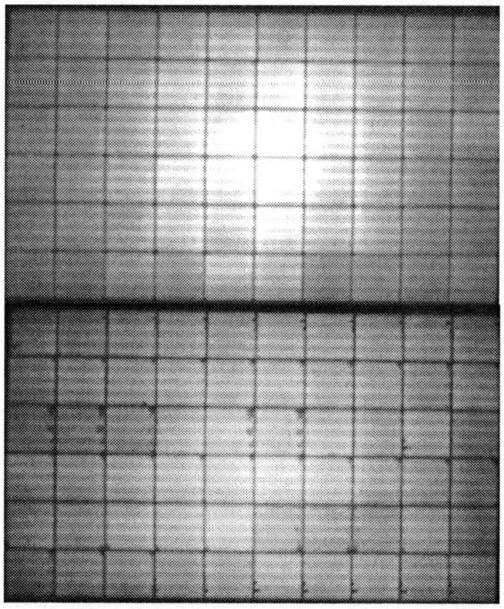

(b)

Figure 12. EL images of n-PERT modules by using ISEC Gen2 n-type PERT solar cells with (a) sc-Si and (b) moonlike wafers.

Finally, this study was completed at the module level. The ISEC Gen2 n-type moonlike bifacial cell was developed on selected high quality n-type moonlike wafers. The high efficiency n-type moonlike silicon bifacial solar cell with front cell efficiency up to 20.8% in average was realized by industrial-available process and facilities. Zero LID because of the intrinsic characteristic of n-type silicon wafers was proven without LIR process. Module power of 299Wp was achieved with 60 pcs cells and white backsheet.

The electrical data and EL images of n-PERT modules are shown in Table 1 and FIG 12. The module power is up to 305.7W (CTM loss 1.26%) and 299W (CTM loss 0.5%) by using sc-Si and moonlike wafers respectively.

4 SUMMARY

N-PERT got more and more attention due to high module power than traditional p-type solar cells. In this article, the high efficiency and low cost ISEC Gen2 n-type moonlike bifacial cell was developed to solve the difficulties of sc-Si and mc-Si PERC cells. Through fine-tuning anneal and passivation process, wafers in the bottom portion of brick, with higher resistivity and less dislocation, could obtain better electrical characteristics due to higher Isc and Voc. Regarding the influence of FGA treatment, although we can observe the difference on PL images, it reveals unobvious enhancement by LIV test. The high efficiency n-type moonlike silicon bifacial solar cell with front cell efficiency up to 20.8% in average was realized by industrial-available process and facilities. Zero LID because of the intrinsic characteristic of n-type silicon wafers was proven without LIR process. Module power of 299Wp was achieved with 60 pcs cells and white backsheet.

5 ACKNOWLEDGEMENT

The authors would like to acknowledge ECM Green Tech. for providing high quality n-type moonlike and Win Win Precision Technology Corporation (WINAICO) for supporting the advanced module technologies.

6 REFERENCE

[1] Mihailetchi VD, Jourdan J, Edler A, Kopecek R, Harney R, Stichtenoth D et al. Screen printed n-type silicon solar cells for industrial application. Proc. 25th European Photovoltaic Solar Energy Conference, Valencia, Spain; 2010, pp.1146-1148

[2] Burgers AR, Geerligs LJ, Carr AJ, Gutjahr A, D.S.Saynoval DS, Xiong J et al. 19.5% efficient n-type Si solar cells made in production. Proc. 26th European Photovoltaic Solar Energy Conference, Hamburg, Germany; 2011, pp.1144-1147.

[3] Y. Veschetti, S. Manuel, V. Sanzone, R. Monna, G. Fortin, E. Pihan, P. Lefillastre, B. Novel, A. Jouini. Potential of n-type monolike silicon using PERT cell technology. Proc. 28th European Photovoltaic Solar Energy Conference, Paris, France; 2013, pp.727-730.

[4] Sugibuchi K, Ishikawa N, Obara S., "Bifacial-PV power output gain in the field test using "EarthON" high bifaciality solar cells", Proc. 28th European Photovoltaic Solar Energy Conference, Paris, France, 2013. P.4312-4317.

[5] Raphaël C, Nicolas E, Etienne P, Nelly P, Gautier F, Lotfi B, Sébastien D. CRYSTALMAX silicon for high efficiency/ low-cost solar cells. Proc. 32nd European Photovoltaic Solar Energy Conference, Munich, Germany; 2016, pp.285-288.

[6] Chia-Ching Wang, Chia-Lung Lin, Yu-Ta Cheng, Yu-Hsiang Huang, Chin-Pao Tsao, Jung-Wu Chien, I-Liang Chen, Chin-Huang Chen, Cheng-Lien Wang. High efficient n-type and p-type PERT solar cells by industrially feasible process. Proc. 32nd European Photovoltaic Solar Energy Conference, Munich, Germany; 2016, pp.602-605.

[7] R.Caba, B.Grange, L.Bounaas, R.Monna, N.Plassat, E.Pihan, Y.Veschetti. 20% PERT technology adapted to n-type moonlike silicon: simplified process and narrowed cell efficiency distribution. Proc. 29th European Photovoltaic Solar Energy Conference, Amsterdam, Netherland; 2014, pp.648-652.

INVESTIGATION OF IN-SITU ANNEALING DURING PHYSICAL VAPOR DEPOSITION OF AL REAR CONTACTS ON *N*-PERT REAR JUNCTION CRYSTALLINE SILICON SOLAR CELLS

Zih-Wei Peng[1], Thomas Buck[1], Manfred Dörr[2], Achim Hain[2], Henning Nagel[3], Philip Hartmann[3], Peter Wohlfart[2], Radovan Kopecek[1]
[1]International Solar Energy Research Center (ISC) Konstanz, Rudolf-Diesel-Str. 15, 78467 Konstanz, Germany
[2]SINGULUS TECHNOLOGIES AG, Hanauer Landstrasse. 103, 63796 Kahl am Main, Germany
[3]Fraunhofer Institute for Solar Energy System (ISE), Heidenhofstr. 2, 79110 Freiburg,Germany
Corresponding author: TEL: +49 7531 3618362; Fax: +49 7531 3618311; E-mail: zih-wei.peng@isc-konstanz.de

ABSTRACT: We introduce a new thermal evaporation system featuring good contact formation without a post deposition annealing process developed by SINGULUS. Compared to an in-line high-rate thermal evaporation system, the new metallization technology showed no losses in efficiency on *n*-PERT rear junction solar cells. In this work, we analyze the electrical properties of the processed solar cell devices to understand effects induced by the new metallization technique. It is shown that the pre-heating unit for substrates of the new evaporation system helps to avoid a post deposition annealing step. The saved process time can increase throughput and helps to reduce production costs. Therefore this technique eliminates another factor separating evaporated metallization from transfer to industrial mass production so far.
Keywords: *n*-PERT, Rear Junction, J_{0e}, Al-evaporation

1 INTRODUCTION

Physical vapor deposition (PVD) of thin or thick metal layers is used in various fields with a wide range of applications (e.g. for optical, mechanical or electrical coatings). Today, for advanced solar cell concepts on lab scale, we see approaches replacing screen printing paste metallization by PVD-metal layer deposition. The PVD technique causes no direct mechanical stress while processing. It offers therefore the advantage of reduced wafer breakage and opens the option to process thinner wafers. The lower thermal budget compared to fast firing thick film metallization gives more freedom to combine the method with many advanced passivation technologies (e.g. tunneling oxide). So far, the crucial point to be addressed is how throughput can be improved to reduce the cost/W_P in the end for opening the access of PVD metallization for relevant mass production applications. Although there are many PVD deposition methods, they are basically categorized in two general groups: sputtering (e.g. Direct Current (DC), Radio Frequency (RF) sputtering) and evaporation deposition (e.g. thermal or electron-beam evaporation). For both deposition methods the atoms of the vaporized material are controllably transferred in high vacuum from a source to the substrate where the layer is grown atoms by atoms. In this work we focus on a thermal evaporation type which was introduced and developed by SINGULUS [1]. It was shown that the technology establishes a good rear contact to our n-type Passivated Emitter and Rear Totally diffused (*n*-PERT) and rear junction (RJ) solar cell concept without a post deposition annealing step and achieves the same solar cell performance on these substrates as layers processed with an in-line high-rate thermal evaporation system [2].

2 EXPERIMENTAL

2.1 *n*-PERT-RJ cell with rear Al-evaporated contact

The cell concept we used for the evaporation tests in this work is the *n*-type Passivated Emitter and Rear Totally diffused (*n*-PERT) solar cell structure with a rear boron emitter or so-called rear junction (RJ) [3]. We start with 6-inch *n*-PERT-RJ solar cell precursors, print and sinter the front side thick film silver paste fingers first. At the rear un-texturized boron side which is coated by SiN_X, we use a Rofin StarFemto FX laser system for local ablation of the dielectric to open the contact area for metallization. A stretcher /compressor unit was used to generate laser pulse durations of 250 fs with 514 nm wavelength. Finally, we evaporate the Al layer on the fully rear side with two different evaporation systems. The reference process was made with the in-line tool ATON-500 fabricated by Applied Materials, while the other category was coated with a new prototype developed by SINGULUS. The process flow and cell structure are shown in Fig. 1 and 2.

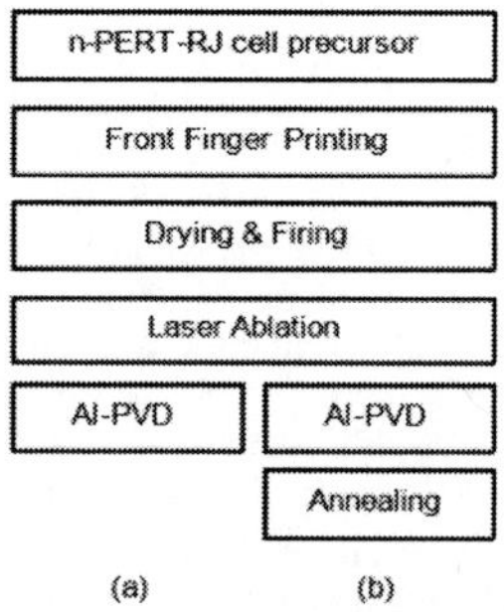

Figure 1: Process flow of the cell samples with different evaporation systems. (a) New evaporation system without post deposition annealing. (b) In-line high-rate thermal evaporation system with post deposition annealing step.

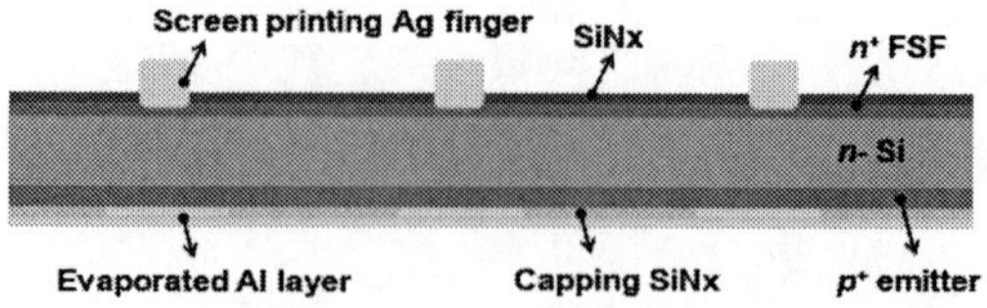

Figure 2: The solar cell structure of the *n*-PERT-RJ with full area Al-evaporation on the locally opened passivation area.

2.2 The evaporation system

The investigated Al films were deposited with a system developed by SINGULUS [1]. The construction has two major qualities that differ from the reference system: deposition is made in a diffused way and the vacuum chamber makes use of a specific treatment to prevent the Al vapor from condensing at the vacuum chamber walls. This setup is well suited to deposit metal layers of uniform thickness. A cross section of the design of the chamber is depicted in Fig. 3(a). In the top-down deposition geometry, the wafer is placed at the bottom on a transport table in an opening connected to a lock. The ceramic made evaporation boat will be heated up over 1200 °C to emit the Al vapor with approximately 10^{-2} mbar towards the top cover of the chamber. The particles are reflected by the pre-coating anti-condensing agent and then stick uniformly distributed to the target. The processing concept can be seen in Fig. 3(b). In contrast to the in-line high-rate thermal evaporation system, the substrates are pre-heated to about 400 °C before Al evaporation is started. This procedure will be called consecutively the pre-heating step. The step can be merged easily into the process flow of the evaporation. With the pre-heating step, we don't need an extra annealing process after deposition as we directly observe similar solar cell efficiency (Table I) as found for the reference process of route (b).

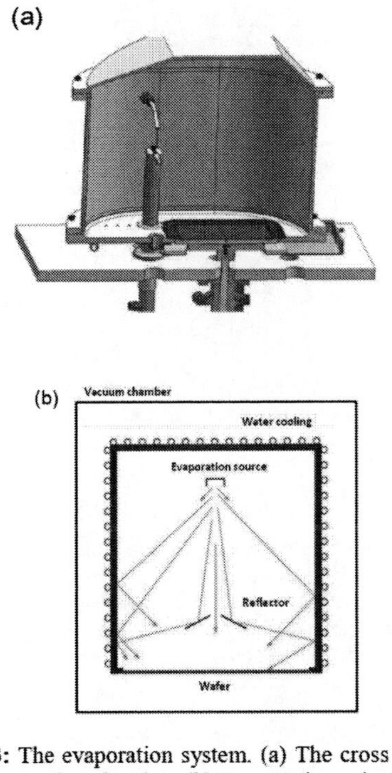

Figure 3: The evaporation system. (a) The cross section of the evaporation chamber. (b) Evaporation principle.

Table I. Electrical properties of the *n*-PERT-RJ solar cells with Al-evaporation rear contact deposited by different evaporation systems. G1: evaporated by SINGULUS system; G2/G2$_S$ (before/after a post deposition annealing step): evaporated by the in-line system.

	J_{SC} (mA/cm^2)	V_{OC} (mV)	FF (%)	Eta (%)	FF-pFF	pFF-FF0
G1	38.0	643	78.7	19.2	-4	-0.8
G2	37.8	642	53.9	13.1	-29	-0.8
G2$_S$	38.6	642	77.6	19.2	-4	-2.3

3 RESULTS AND DISCUSSION

3.1 Evaporation process comparison

Table I illustrates the electrical properties, of the three different investigated groups. Each group represents six cell samples. G1 was Al coated with the SINGULUS system; G2/G2$_S$ are deposited by the in-line system showing the I-V parameters before and after an annealing step respectively. Tempering was done in a rapid thermal processing (RTP) furnace at 400 °C for two minutes. The solar cell efficiency of G1 is comparable with the performance found for G2$_S$. The data states that for the in-line tool the devices need an additional temperature treatment to establish a reasonable contact and solar cell performance. We suppose that the contact formation at the Si-Al interface has the demand of a higher thermal budget to get established properly. This can be proven by the difference between fill factor (FF) and pseudo fill factor (pFF) [4].The solar cells in G2 exhibit a much higher FF-pFF difference than G1 and G2$_S$. As the pFF, represents the FF without series resistance effect the difference of both reveals how performance is affected by series resistance (R_S). This value is directly related to the contact quality, if we assume that all other contributions in series resistance are constant. After the annealing process the contact quality of the Al-Si interface is improved, the FF-pFF value is reduced. Under the thermal budget (400°C, 2min), the Si starts to dissolve by a solid state diffusion process into Al and to saturate at a 0.24 at% composition as can be learned from the phase diagram [5].The solubility of silicon in aluminum rises as the temperature increases. For that reason the annealing process helps to form good contact quality and to improve the FF and the current density (J_{SC}) in G2$_S$.

The difference between ideal fill factor (FF_0) [6] and pFF quantifies the FF-losses due to the recombination in the space charge region (SCR). This value is increased after the post deposition annealing process. It is because as a matter of the Si diffusion process voids occur at the interface that open the potential to be filled by Al [7]. The so called spiking makes shunting more likely for the rear p^+ emitter and increases the recombination in SCR. The post deposition annealing is a key process to "activate" the PVD metal contact deposited by the in-line evaporation system. For the new evaporation system developed by SINGULUS annealing after our Al deposition is not needed for contact formation on our devices. We achieve comparable solar cell performance of high uniformity without this treatment and thus establish already a good contact at the interface between Si and Al (Fig. 4(a)).

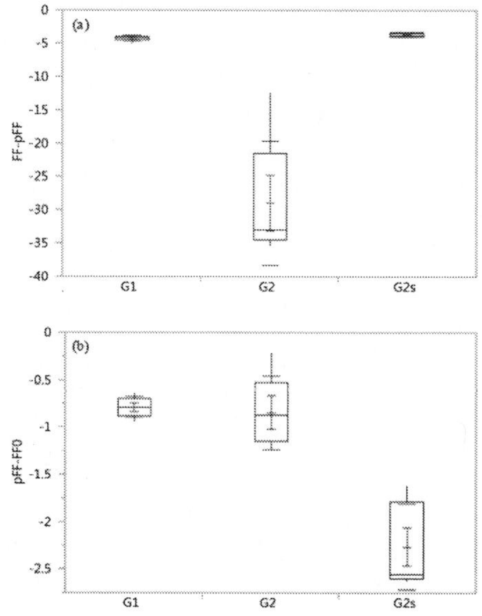

Figure 4: Electrical properties of devices processed with different evaporation tools and processes. (a) pFF-FF value, reveals series resistance effect. (b) pFF-FF0 values, reveals shunt resistance effect.

Data from spectral response measurements (Fig. 5) show that G1 has a lower performing IQE for long wavelengths compared to $G2_S$, and suggest optical matters to explain the lower current density of G1. Experiments are on the way to investigate the micro structure of the interface to improve and clarify the observed effects.

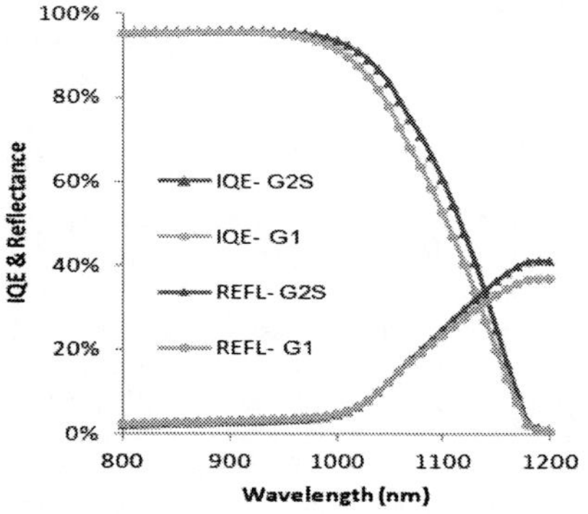

Figure 5: Internal quantum efficiency and reflectance measured on G1 and $G2_S$ solar cells.

3.2 Edge exclusive mask application

Due to the changed deposition geometry the Al particles can hit the target under a wide spectrum of angles. The fact is especially important if we consider the edges of our devices. Appropriate screening is needed to avoid metallization on the edge of the solar cells and to avoid parasitic shunting (Fig. 6). After assembling an adapted 1mm edge exclusive mask in the evaporation chamber, we improved the shunt resistance from 350 Ohm·cm^2 to 12000 Ohm·cm^2 in average and eliminated leakage current on the edge of the solar cells as depicted in Fig. 7.

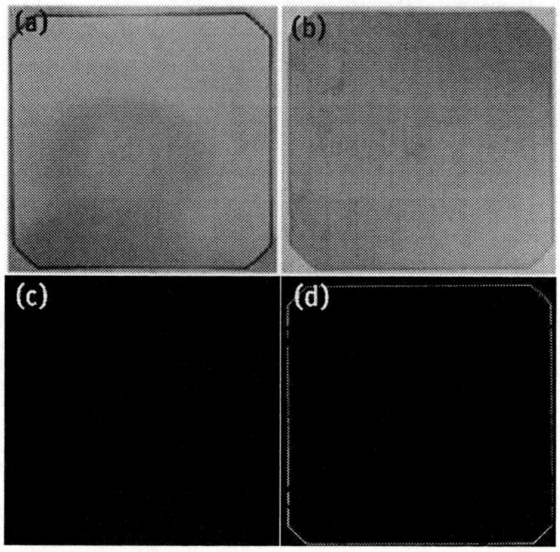

Figure 6: (a) Al-evaporation with 1 mm edge exclusive mask. (b) Without edge exclusive mask. Reverse-bias EL images of samples with 1 mm edge exclusive mask (c), without 1 mm edge exclusive mask (d).

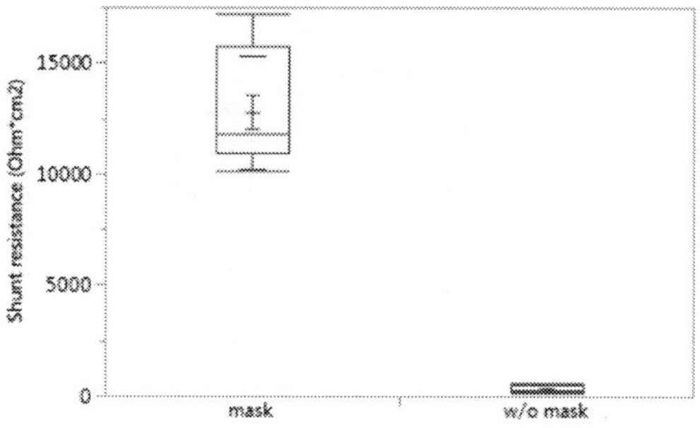

Figure 7: Shunt resistance: Al-evaporation with 1mm edge exclusive mask (left). Without edge exclusive mask (right).

4 SUMMARY

Al films, evaporated by the setup developed of SINGULUS are suited for the rear side metallization of *n*-PERT rear junction solar cells and lead to efficiencies that are comparable to devices processed with an in-line high-rate thermal evaporation system. It was found that a post deposition annealing step is disposable for the new technique opening the option of increased throughput and cost reduction. After process optimization -considering mainly the laser power and contact pattern design- we achieved with improved cell precursors 20.8% efficiency on 156 mm-large industrial Cz wafers. (Table II). An improvement of current density is addressed by further experiments.

Table II: I-V of *n*-PERT-RJ champion cell with evaporated-Al rear contact

	Jsc (mA/cm^2)	Voc (V)	FF (%)	Eta (%)
mean	38.7	0.674	79.3	20.71
best cell	38.7	0.675	79.6	20.82

5 ACKNOWLEDGEMENT

This work was funded by the BMWi within the IdeAl project (Industrielles PVD-Al für hocheffiziente kristalline Si-Solarzellen und Module) under project number 02E2-41V7142.

6 REFERENCES

[1] M.R. Huber, M. Dörr, P. Wohlfart, "New Evaporation Technology for Rear Side Metallization of High Efficiency Solar Cells," 27th EUPVSEC, pp. 1185-1187, 2012.

[2] J. Nekarda, D. Reinwand, A. Grohe, P. Hartmann, R. Preu, R. Trassl, S. Wieder, "Industrial PVD Metallization for high efficiency crystalline silicon solar cells," in 34th IEEE Photovoltaic Specialists Conference (PVSC), pp. 892-896, 2009.

[3] T. Dullweber, S. Gatz, H. Hannebauer, T. Falcon, R. Hesse, J. Schmidt, R. Brendel, "19.4% -Efficient Large Area Rear-Passivated Screen-Printed Silicon Solar Cells," 26th EUPVSEC, pp. 811-816, 2011.

[4] J. Greulich, M. Glatthaar, S. Rein, "Fill factor analysis of solar cells' current-voltage curves," Progress in Photovoltaics 18(7), pp. 511-515, 2010.

[5] J.R. Black: "Electromigration of Al-Si Alloy Films," 16th Ann. Proc. Reliab. Phys., pp. 233-240, 1978.

[6] M. A. Green Solar Cells: Operating Principles, Technology and System Applications. UNSW: Kensington, 1986.

[7] K. Katkhouda: "Aluminum-based PVD rear-side metallization for front-junction nPERT silicon solar cells," Ph.D thesis, 2015.

SELECTIVE EPITAXY AS CONTACT PASSIVATION APPROACH IN BIFACIAL N-TYPE PERT SOLAR CELLS

M. Recamán, I. Kuzma, Y. Li, S. Singh, E. Cornagliotti, S. Jambaldinni, R. Russell, J. John, F. Duerinckx, J. Szlufcik and J. Poortmans[2]
imec, Kapeldreef 75, B-3001 Leuven, BE
[2] imec, Kapeldreef 75, B-3001 Leuven, BE; KU Leuven, B-3001 Leuven, BE and UHasselt, B-3590 Diepenbeek, BE
Corresponding author: phone: +32 1628 1478, e-mail: Maria.RecamanPayo@imec.be

ABSTRACT: This works focuses on the introduction of selective epitaxy as a contact passivation approach for bifacial n-type PERT solar cells. Because of the use of a metal grid on their rear side, bifacial cells provide a higher energy yield through rear illumination but have larger recombination losses than their monofacial counterparts with point contacts. A breakdown analysis of the recombination losses in bifacial cells fabricated with a homogeneous emitter, and for which efficiency values above 22.5 % were already achieved, reveals that there is still an important contribution of the emitter $J_{0,metal}$ (3000-4000 fA/cm^2 for 1.2 % metal contact fraction) and that the emitter $J_{0,pass}$ can also be further improved to increase efficiency. This works aims at reducing the recombination losses in these devices by introducing in one single step an epitaxially grown highly doped layer at the p-type regions to be contacted. The structures here fabricated up to the step just prior to metallization (including patterning of both emitter and BSF) demonstrate the potential of this approach to significantly reduce the recombination losses compared to the reference without a selective epitaxial contact passivation scheme: (a) homogeneous emitter with SiO_2 passivation: $J_{0,total}$=90 fA/cm^2, iV_{oc}=681 mV, (b) homogeneous emitter with Al_2O_3 passivation: $J_{0,total}$=70 fA/cm^2, iV_{oc}=688 mV, and (c) selective emitter via selective epitaxy with SiO_2 passivation: $J_{0,total}$=55 fA/cm^2, iV_{oc}=695 mV. Keeping in mind that this approach also brings the possibility to independently optimize the emitter in the passivated regions to reduce Auger recombination, further gain in passivation may still be possible, which increases the potential of this structure to achieve higher efficiency values at cell level.
Keywords: bifacial PERT, contact, passivation, epitaxy

1 INTRODUCTION

This works focuses on the introduction of selective epitaxy as contact passivation approach in bifacial n-type PERT solar cells. The main feature of bifacial cells is that they collect sunlight both on the front and the rear side, as they also capture light reflected from the surface beneath the module and the environment [1-2]. Thanks to the implementation of a metal grid on their backside, so that the light can also reach the active region from the rear, bifacial cells provide a higher energy yield through rear illumination but also larger recombination losses than their monofacial counterparts with a fully metallized backside and point contacts. Bifacial PERT solar cells with the layout presented in Figure 1, and fabricated using laser doping for the n^{++} BSF contact, have already reached efficiency values above 22.5 % with a homogenous p-type emitter [3]. Nonetheless, a breakdown analysis of the recombination losses reveals that there is still an important contribution of the $J_{0,metal}$ on the emitter side, in the range of 3000-4000 fA/cm^2 (for 1.2 % metal contact fraction), which is approximately two times larger than the $J_{0,metal}$ measured for the laser doped n^{++} BSF contact. On top of that, the $J_{0,pass}$ for the emitter can also be further improved. We should keep in mind that the use of a homogeneous doping profile results in a trade-off between resistive and recombination losses.

In an attempt to achieve higher efficiency values, this work aims at reducing the recombination losses of these devices by implementing heavy doping at the contact regions of the p-type emitter. Overall the final goal would be the independent optimization of the doping profile in the passivated and the metallized areas of the p-type emitter region. In our approach, selective epitaxy will be the doping technology introduced to locally grow that highly doped p-type layer on the emitter contact regions.

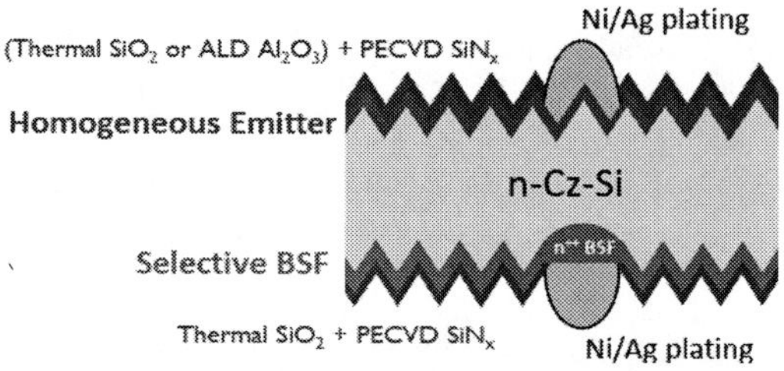

Figure 1: Layout for the reference bifacial PERT solar cells of this work with co-plated n-type and p-type contacts, homogeneous emitter and selective BSF (by n-type laser doping).

Although selective epitaxy has proved its potential for high efficiency and process simplification in the fabrication of n-type PERT and IBC solar cells [4-12], the integration of this technology for the specific case of the local contact in bifacial PERT devices is not that evident. There are two main challenges to be tackled: (1) the surface damage after laser ablation, and (2) the impact of the epitaxial thermal budget on the passivating properties of both front and rear dielectrics. Selective epitaxy has been successfully applied at cell level on SDR and textured surfaces, but no proof of concept has been demonstrated yet on laser ablated surfaces. Laser ablation of dielectrics to open the contact regions not only introduces damage, and therefore the presence of additional defects [13], but also gives rise to a very rough surface morphology which is not ideal for the growth of a high-quality epitaxial layer. Besides, dielectrics such as ALD Al_2O_3 and PECVD SiN_x are sensitive to a high thermal budget which can degrade their passivation properties as a consequence of an increase in D_{it}, hydrogen effusion from the films, crystallization or other forms of chemical, physical and structural damage [14-17].

Taking the previous remarks into consideration, the first part of this work addresses two main aspects: (1) the surface preparation after laser ablation and thus prior to epitaxy, and (2) the passivation performance of the dielectrics after the thermal budget induced by the selective growth of the epitaxial layer at the emitter contact regions. After the development of a selective epitaxial process with the required conditions, the second part of this work focused on the implementation of that step in the fabrication sequence of bifacial PERT solar cells.

2 EXPERIMENTAL

Semi-device structures mimicking a bifacial PERT solar cell were fabricated to study the impact on passivation of the laser ablation process, the epitaxial thermal budget and the co-plating process [18]. Those structures were fabricated on standard n-type monocrystalline Cz silicon textured wafers following the "masking" sequence depicted in Figure 2 (*top*). This approach allows to preserve the fully textured surface on both sides of the final device reaching bifaciality values close to 99 %. Initially, two alternatives for the passivation of the BSF and the emitter were analyzed. The first option, hereinafter named as "SiO_2 route", consisted of using the stack thermal SiO_2/PECVD SiN_x for the passivation of both BSF and emitter. In the second approach, herein so called "Al_2O_3 route", the stack ALD Al_2O_3/PECVD SiN_x was alternatively used for the passivation of the emitter.

The "emitter removal" flow represented in Figure 2 (*bottom*) was more specifically applied when investigating the implementation of the epitaxial step in the fabrication sequence of bifacial PERT cells. Compared to the "masking" sequence, the main advantages of this flow where the rear surface is not fully textured anymore, are a better optical performance at module level [19], and a lower J_0 for the passivated emitter regions. The lower J_0 values for the passivated emitter regions are the result of two main effects: 1) the smoother rear surface with a much lower contribution of the (111) silicon orientation [6, 10] and, 2) the additional drive-in of the diffused emitter that takes place in this sequence during the BSF formation and which is not happening in the "masking" sequence (see Figure 2).

Passivation was assessed by PL-QSSPC on structures with blanket dielectric stacks, and on samples where the grid pattern for the emitter and the BSF contacts was created by laser ablation and laser doping, respectively.

3 RESULTS AND DISCUSSION

3.1 Dielectrics passivation after epitaxial thermal budget

In this experiment, structures with blanket dielectric stacks have gone through the epitaxial step with only thermal budget, that is, without actual epitaxial growth occurring. The goal was to assess the maximum epitaxial temperature and (deposition) time that the dielectrics could withstand without losing their passivation properties. The results, in terms of $J_{0,total}$ for the device as a function of epitaxial temperature and (deposition) time at that temperature, are presented in Figure 3 for the structures of both "SiO_2 route" and "Al_2O_3 route". The passivation was measured after epitaxial thermal budget, and compared to the values obtained for the reference case with only firing (and no epitaxy).

For the "SiO_2 route", the results confirm that there is a remarkable loss in passivation after an epitaxial thermal budget above 750°C (≥30 min). A similar study performed on symmetric BSF and emitter structures reveal that this degradation happens simultaneously on both doped regions (see Table I).

Nevertheless, there is a promising processing window at 700°C where, for a broad range of (deposition) time, the passivation after epitaxial thermal budget can even outperform the reference values with only firing (and no epitaxy). This condition is also related to the contribution of hydrogen passivation from PECVD SiN_x, which is known to be released around that temperature [20-21].

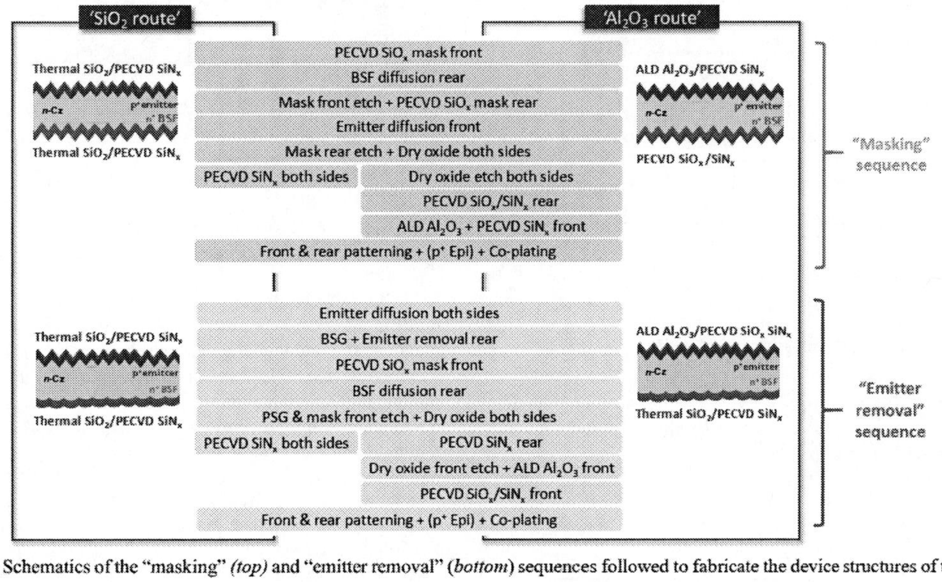

Figure 2: Schematics of the "masking" *(top)* and "emitter removal" (*bottom*) sequences followed to fabricate the device structures of this work.

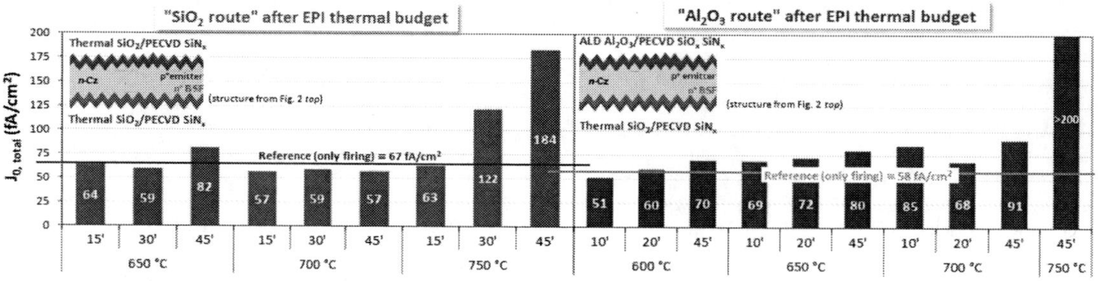

Figure 3: $J_{0,total}$ (fA/cm²) extracted at $10^{16}cm^{-3}$ after epitaxial thermal budget: *(left)* "SiO_2 route", and *(right)* "Al_2O_3 route".

The same investigation on semi-devices of the "Al_2O_3 route" confirms that selective epitaxy would introduce a thermal budget which is too large to keep the excellent passivation obtained after a firing step. Only if epitaxy took place at a temperature below 600°C we could not only keep but also improve the performance compared to the reference case with firing. At higher deposition temperature the passivation collapses because of different forms of damage (increase in D_{it}, hydrogen desorption from defects, stress or crystallization of the films) [14-17].

Table I: J_0 (fA/cm²) extracted at $10^{16}cm^{-3}$ for symmetric structures of the "SiO_2 route" after epitaxial thermal budget.

Passivation structure (Fig. 2 *top*)	EPI thermal budget	$J_{0,BSF}$ or $J_{0,emitter}$ (fA/cm²)
Thermal SiO_2/PECVD SiN_x / n-Cz, n⁺ BSF, n⁺ BSF / Thermal SiO_2/PECVD SiN_x	700°C, 15'	10
	750°C, 45'	61
Thermal SiO_2/PECVD SiN_x / n-Cz, p⁺emitter, p⁺emitter / Thermal SiO_2/PECVD SiN_x	700°C, 15'	47
	750°C, 45'	123

Table II: J_0 (fA/cm²) extracted at $10^{16}cm^{-3}$ for symmetric structures of the "Al_2O_3 route" after epitaxial thermal budget.

Passivation structure (Fig. 2 *top*)	EPI thermal budget	$J_{0,BSF}$ or $J_{0,emitter}$ (fA/cm₂)
PECVD SiO_x/SiN_x / n-Cz, n⁺ BSF, n⁺ BSF / PECVD SiO_x/SiN_x	600°C, 10'	24
	750°C, 45'	63
ALD Al_2O_3/PECVD SiN_x / n-Cz, p⁺emitte, p⁺emitte / ALD Al_2O_3/PECVD SiN_x	600°C, 10'	27
	750°C, 45'	>>200

When the experiment was realized on symmetric BSF and emitter structures, the results, for which an example can be found in Table II, demonstrate that this passivation loss is mainly dominated by the quicker degradation of the stack ALD Al_2O_3/PECVD SiN_x which is less thermally stable than, for example, the stack PECVD SiO_x/SiN_x.

In view of the previous results, the implementation of an epitaxially grown p-type layer in the contact regions of bifacial PERT cells focused on the "SiO_2 route". An additional test where these structures have gone through the epitaxial step at 700°C for (deposition) times between 10 and 90 min, revealed that above 15 min the stack thermal SiO_2/PEVCD SiN_x outperforms its reference with only firing (and no epitaxy), and it is also at the same level than the "Al_2O_3 route" reference (see Table III).

Table III: Passivation results for the structures of this work.

Passivation structure (Fig. 2 *top*)	Thermal Budget	$J_{0,total}$ (fA/cm²)	iV_{oc} (mV)	τ_{eff} (µs)
ALD Al_2O_3/PECVD SiN_x / n-Cz, p⁺ emitter, n⁺ BSF / PECVD SiO_x/SiN_x	FIRING (reference)	44	700	1152
Thermal SiO_2/PECVD SiN_x / n-Cz, p⁺emitter, n⁺ BSF / Thermal SiO_2/PECVD SiN_x	FIRING (reference)	81	680	506
	EPI@700°C (*best*)	33	710	2043

3.2 Plating of epi layers grown on laser ablated surfaces

To locally grow the p-type highly doped layer at the emitter contact regions of bifacial PERT solar cells, a selective epitaxial process at 700°C was developed on laser ablated textured surfaces.

As for any doping process, a cleaning step must be introduced between the laser ablation process used to pattern the dielectrics in the emitter side and the selective epitaxial step. The cleaning is required to remove any impurity or native oxide present on the open silicon surface, which can prevent a selective deposition or unnecessarily degrade the quality of the grown film and therefore its electronic properties. The standard iPV cleaning we normally use in cell processing, and which consists of a 2-step sequence based on DIW/O_3/HCl and DIW/HF/HCl was also applied here prior to epitaxy. Any additional silicon etching step introduced to remove laser damage or smoothen the surface prior to epitaxy did not bring any advantage from the passivation point of view. The solutions with an etching behavior investigated in that specific study were TMAH (5 % at 80 °C), HF/HNO_3 (1:240 at room temperature), KOH (2.5 % at 50 °C) and the so called i²PV, under development in the author's group, and whose chemistry simultaneously incorporates oxidation, and silicon/oxide etching processes in one single step.

The SEM images of Figure 4 show the morphology of a half a micron thick p-type epitaxial layer selectively grown on a laser ablated textured surface after the standard iPV cleaning. That would be the type of layer to be contacted in the cell processing.

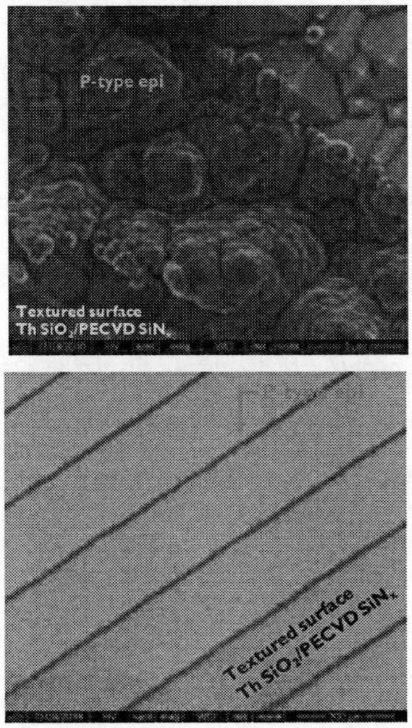

Figure 4: SEM images of a p-type epitaxial layer grown on laser ablated (LA) textured surfaces.

The co-plating process used for the metallization of the bifacial n-type PERT cells in this work simultaneously plates both polarities [18]. In this particular case, the full plating sequence was successfully applied to the epitaxial layers grown on laser ablated surfaces (see Figure 5), achieving as good adhesion properties as for the reference case without epitaxy. With respect to the recombination losses, Table IV shows that a p-type highly doped epitaxial layer in the range of 500-750 nm incorporated at the emitter contact regions of bifacial cells can already provide a better performance than the reference.

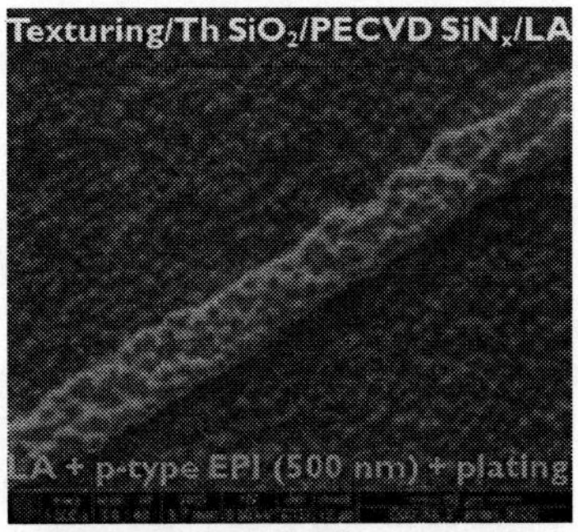

Figure 5: SEM image of a plated p-type epitaxial layer grown on laser ablated (LA) textured surfaces.

Table IV: Passivation results ($J_{0,total}$ extracted at 10^{16}, τ_{eff} extracted at 10^{15}, iV_{oc} at 1 Sun and *i*FF) and contact resistance (ρ_c) after plating.

	LA + firing + plating (reference - no epi)	LA + EPI (500 nm) + plating
Structure: *Fig. 2 (top)*	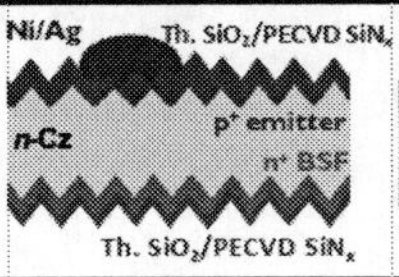	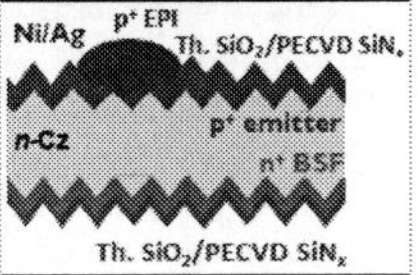
$J_{0,total}$ (fA/cm²)	96	56
τ_{eff} (µs)	657	1279
iV_{oc} (mV)	677	696
*i*FF (%)	84.1	84.5
ρ_c (mΩ.cm²)	1.6	1.6

3.3 Selective epitaxy in the solar cell flow

When introducing the p-type highly doped epitaxial layer at the emitter contact regions of these bifacial devices, the epitaxial step could be performed before or after the n^+ laser doping step (LD) at the BSF side. When laser ablation and p-type selective epitaxy (LA+EPI) are carried out before LD, the main advantage is that the dielectrics at the BSF side, which are not patterned yet, prevent any parasitic deposition during the selective growth of the p^+ epitaxial layer at the emitter contact regions [4-12]. In this approach, a firing step after LD would still be required to heal any laser damage introduced during that step. If LD is applied before LA+EPI, a few nanometer thick dielectric mask, for example PECVD SiO_x, should be deposited on the BSF side after LD and prior to LA+EPI. This mask would be required to prevent any parasitic deposition on the open silicon areas of the BSF during the selective growth of the p^+ epitaxial layer at the emitter contact regions. The main advantage here is that an annealing of the damage introduced during LD can take place during epitaxy and the firing step can be omitted.

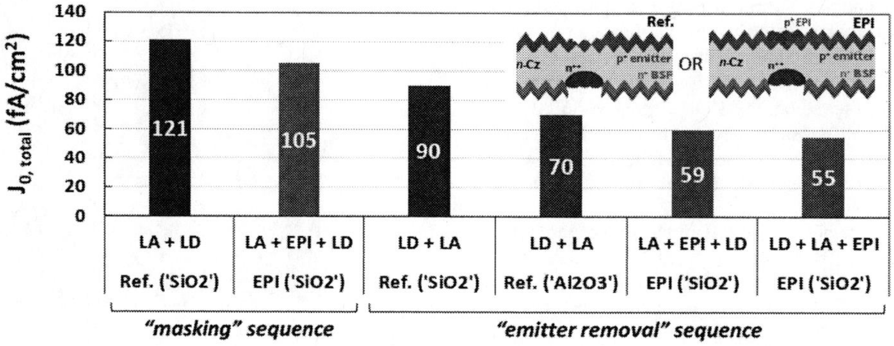

Figure 6: $J_{0,total}$ (fA/cm^2) extracted at 10^{16}cm^{-3} after patterning by laser ablation (LA) and laser doping (LD) for reference structures with only firing (Ref.) and structures incorporating a p-type highly doped epitaxial layer at the contact regions (EPI). Device fabrication was realized using the "SiO_2 route" or "Al_2O_3 route" as well as the "masking" sequence or "emitter removal" sequence.

In order to define which is the most promising process sequence in terms of recombination losses, semi-device structures were fabricated using the "masking" sequence and the "emitter removal" sequence (from Figure 2), and measured by PL-QSSPC. In the epitaxial flow (EPI), the patterning for these structures was accomplished doing LD before or after LA+EPI. Reference (Ref.) samples (with only firing and no epitaxy) were also included in order to benchmark the performance of the epitaxial flow. The reference samples were fabricated following both the "SiO_2 route" and the "Al_2O_3 route". Figure 6, which summarizes the $J_{0,total}$ for the different structures investigated, points out that the "emitter removal" sequence used to fabricated these bifacial cells outperforms in terms of passivation the "masking" sequence. The results also show that the sequence where LD is applied prior to LA+EPI is slightly better than LA+EPI before LD. Table V collects the main passivation results comparing the best structures for both the reference and the epitaxial flows. This data confirms that the structures introducing an epitaxial layer provide lower recombination losses than both references using thermal SiO_2 or ALD Al_2O_3 passivation.

Table V: Passivation results ($J_{0,total}$ extracted at 10^{16}, τ_{eff} extracted at 10^{15}, iV_{oc} at 1 Sun and iFF) after full patterning (prior to metallization) for best reference and epitaxial structures.

	LD + LA + firing (Ref.: "Al_2O_3 route")	LD + LA + EPI (750 nm) ("SiO_2 route")
Structure: *Fig. 2 (right)*	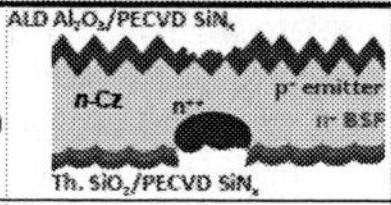	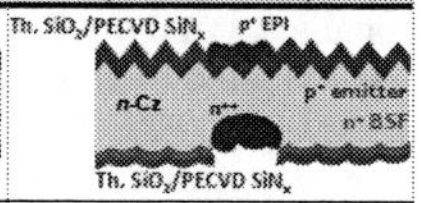
$J_{0,total}$ (fA/cm^2)	70	55
τ_{eff} (µs)	965	1248
iV_{oc} (mV)	688	695
iFF (%)	84.0	84.2

Overall introducing a p-type highly doped epitaxial layer at the emitter contact regions of these cells is not only reducing the recombination losses, but also opens new opportunities to further gain in cell performance. Regarding this last point, we have to consider that this approach brings in the possibility for an independent design of the emitter in the passivated areas, which can be fully optimized to reduce Auger recombination [7], increasing the potential of this technology to increase cell efficiency.

4 CONCLUSIONS

The main goal of this research work is to reduce the recombination losses at the emitter contact regions of bifacial n-type PERT solar cells. With that purpose, selective epitaxy is introduced to create in one step a very highly doped layer at the p-type contact regions of these devices.

The passivation results on structures fabricated till the step just prior to metallization (including both the emitter and BSF patterning) demonstrate that the technology can outperform the reference structures which do not include a selective epitaxial contact scheme: (a) homogeneous emitter with SiO_2 passivation: $J_{0,total}$=90 fA/cm^2, iV_{oc}=681 mV, (b) homogeneous emitter with Al_2O_3 passivation: $J_{0,total}$=70 fA/cm^2, iV_{oc}=688 mV, and (c) selective epitaxial emitter with SiO_2 passivation: $J_{0,total}$=55 fA/cm^2, iV_{oc}=695 mV.

Keeping in mind that this approach also brings the possibility to independently optimize the emitter in the passivated regions to minimize Auger recombination, further gain in passivation may still be possible, which increases the potential of the approach to achieve higher efficiency values at cell level.

5. REFERENCES

[1] Cuevas, A. et al. 1982 . *50 per cent more output power from an albedo-collecting flat panel using bifacial solar cells*. Solar Energy, 29 (5), pp. 419-420.

[2] Sala, G. et al. 1983. *Albedo collecting photovoltaic bifacial panels*. Proceedings of the 5th International Photovoltaic Solar Energy Conference, Athens.

[3] Imec news release, 18 April 2017. Available at https://www.imec-int.com/en/articles/imec-develops-bifacial-n-pert-solar-cell-with-a-record-22-8-percent-front-side-efficiency [accessed on 14 September 2017].

[4] Recamán Payo, M. et al. *Method for forming patterns of differently doped regions*. Patent WO 2013/020868 A1.

[5] Recamán Payo, M. et al. 2013. *Boron-doped selective silicon eitaxy: High efficiency and process simplification in interdigitated back contact cells*. Prog. Photovoltaics Res. Appl., 22 (4), pp. 711-725.

[6] Recamán Payo, M. et al. 2014. *Opportunities for*

silicon epitaxy in bulk crystalline silicon photovoltaics. Proceedings of the 29th EU PVSEC, Amsterdam, pp. 497-502.
[7] Recamán Payo, M. et al. 2015. *Advanced doping profiles by selective epitaxy in n-type PERT cells*. Proceedings of the 30th EU PVSEC, Hamburg, pp. 433-439.
[8] Li, Y. et al. 2015. *Process simplification and improvement of rear side of IBC cells by means of PECVD SiO_x and epitaxy*. Proceedings of the 30th EU PVSEC, Hamburg, pp. 466-469.
[9] Kuzma-Filipek, I. et al. 2015. *Integration processes for nPERT Si solar cells using single-side emitter epitaxy and front side laser doping*. Energy Procedia, 77, pp. 527-533.
[10] Recamán Payo, M. et al. 2015. *The use of silicon epitaxy in advanced n-type PERT and IBC silicon solar cell designs*. Photovoltaics International, 28, pp. 45-56.
[11] Recamán Payo, M. et al. *Method for fabricating crystalline photovoltaic cells*. Patent US 2016/072001 A1.
[12] Li, Y. et al. Submitted on 1 September 2017. *Epitaxial selective n-type doping and its application in crystalline silicon IBC solar cells*. Progress in Photovoltaic: Research and Applications, *under Review*.
[13] Dang, Ch. et al. 2014. *Investigation of laser ablation induced defects in crystalline silicon solar cells*. Energy Procedia, 55, pp. 649-655.
[14] Barbos, C. et al. *Characterization of Al_2O_3 thin films prepared by thermal ALD*. Energy Procedia, 55, pp. 558-564.
[15] Veith, B. et al. 2011. *Comparison of the thermal stability of single Al_2O_3 layers and Al_2O_3/SiN_x stacks for the surface passivation of silicon*. Energy Procedia, 8, pp. 307-312.
[16] Dingemans, G. et al. 2012. *Status and prospects of Al_2O_3-based surface passivation schemes for silicon solar cells*. J. Vac. Sci. Technol. A, 30 (4), pp. 040802-1 - 040802-27.
[17] Liu, Y. et al. 2011. *Thermally induced damages of PECVD SiN_x thin films*. J. Mater. Res., 26 (19), pp. 2552-2557.
[18] Russell, R. et al. 2017. *Simultaneous fabrication of n & p contacts for bifacial cells by a novel co-plating process*. Proceedings of *this* conference.
[19] Duerinckx, F. et al. 2017. *Optical simulation of bifacial silicon solar cells at module level*. Proceedings of *this* conference.
[20] Duerinckx, F. 1999. *Bulk and surface passivation of screen printed multicrystalline silicon solar cells based on plasma enhanced CVD of silicon nitride*. Doctoral dissertation, KUL and imec, ISBN:90-5682-210-1.
[21] Yelundur, V. et al. 2000. *PECVD SiN_x induced hydrogen passivation in string ribbon silicon*. Conference Record of the 28th IEEE Photovoltaic Specialists Conference, Alaska, pp. 91-94.

Acknowledgements
The authors gratefully acknowledge the financial support of Imec's Industrial Affiliation Program for Photovoltaics (IIAP-PV).

GETTERING EFFICACY OF APCVD PSG AND BSG LAYERS IN MC-SI

Christian Fischer, Johannes Fichtner, Annika Zuschlag, Giso Hahn
University of Konstanz, Department of Physics, 78457 Konstanz, Germany
christian.fischer@uni-konstanz.de, Tel: +49 7531 885296, Fax: +49 7531 883895

ABSTRACT: APCVD (atmospheric pressure chemical vapor deposition) offers promising improvements for the PERT (passivated emitter rear totally diffused) solar cell concept. One-sided diffusion of P as well as B allows reduced processing steps. In this work the gettering efficacy of P- and B-doped layers deposited by APCVD on mc (multicrystalline) Si (silicon) material is evaluated by minority charge carrier lifetime measurements.
Comparison with the influence of thermal load showed that P-gettering resulted in significant lifetime increase, whereas temperature treatment without PSG (phosphor silicate glass) induced the opposed effect. Close to the crucible wall, the local defect structure in the wafer leads in ungettered samples to negative effects that can be strongly reduced by P-gettering. In addition, the APCVD PSG layers showed at least comparable gettering efficacy to standard $POCl_3$-gettering via diffusion in a quartz tube furnace. Finally, co-diffusion of APCVD PSG and BSG layers was tested for a PERT concept.
Keywords: Gettering, Multicrystalline Silicon, Lifetime, APCVD

1 INTRODUCTION

Most industrial crystalline Si solar cell concepts use open tube diffusions for fabrication of doped layers. The most prominent example is P-diffusion, which normally is carried out in an open quartz tube furnace using, *e.g.*, $POCl_3$ as precursor. For emitter fabrication in n- or p-type PERT solar cells, normally a BBr_3-based open tube diffusion is used. Due to the slower diffusion of B in Si, B-diffusion is usually applied at higher temperatures compared to P-diffusion.

For the PERT solar cell concept, a one-sided diffusion of P as well as B would be beneficial to reduce processing steps (masking or etching). Therefore, CVD(chemical vapor deposition)-based doping layers are of high interest for industrial processing of PERT solar cells, as they can be deposited on one side only and allow for a following single co-diffusion step for both doping types. Compared to PE(plasma-enhanced)CVD, AP(atmospheric pressure) CVD [1] could be even more beneficial as it does not involve vacuum processing and allows for a high throughput. In addition, the temperature load on the material is different to standard diffusion processes. When applying the PERT solar cell concept to mc-Si, care has to be taken that effective minority charge carrier lifetime τ_{eff} is not negatively influenced by the higher temperatures normally needed for B-diffusion.

Standard P-diffusion is known to have a gettering effect on lifetime limiting impurities, especially in defect-rich mc-Si. But also B-diffusion can have a positive gettering effect. Gettering efficacy of $POCl_3$- and BBr_3-based open quartz tube diffusions have been studied for mc-Si (*e.g.* [2]) and dependence of gettering efficacy on the material quality has been shown (*e.g.* [3]). However, to our knowledge the gettering efficacy of doped APCVD SiO_2 layers (phosphor-silicate glass (PSG) and boro-silicate glass (BSG)) on mc-Si has not been studied so far.

Therefore, in this contribution the gettering efficacy of P- and B-doped APCVD layers in standard mc-Si is investigated. It is compared to wafers with same temperature load, but without doping glasses. In addition, the gettering efficacy is studied depending on wafer position in the ingot because distance to crucible walls is a critical parameter for contaminations and defects. The gettering efficacy of a standard $POCl_3$-based layer is compared to the APCVD-based processes.

2 EXPERIMENTAL

156x156 mm^2 B-doped industrial standard mc-Si wafers (~1 Ωcm) from one standard ingot were obtained with known position and defined orientation of each wafer in the block. Hence gettering efficacy can be studied in dependence of vertical and horizontal distance to the crucible walls. If not specified, gettering efficacy on samples from mid ingot height with about 10 cm in horizontal distance to the crucible wall are discussed. Sister wafers shown here origin from a relatively small range in height resulting in a very similar defect structure.

5x5 cm^2 samples were laser cut out of the 156x156 mm^2 wafers. After saw damage etch, wafer surfaces were cleaned. A reference (A) wafer was not further processed and only a surface passivation was realized by firing of a remote PECVD SiN_x:H layer. For samples B, C, D, E, and F different gettering sequences were applied. The doping glasses for those samples are deposited by APCVD. APCVD allows depositing BSG- and PSG-layers on one side of the wafer without capping the other side. An overview of the experiment with the applied APCVD process sequences is given in Fig. 1.

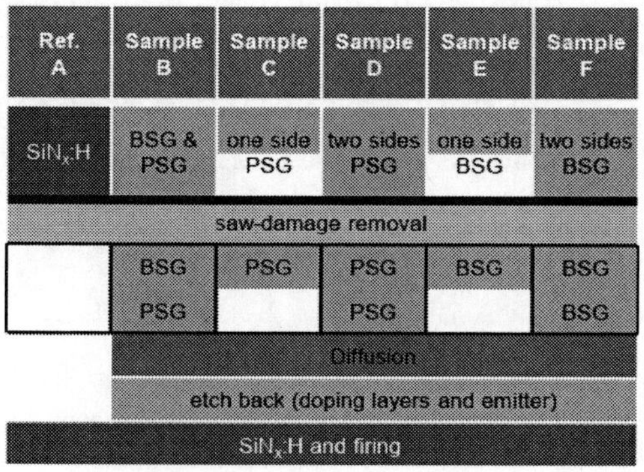

Figure 1: Overview of process steps and APCVD layers.

Sample C received a one-sided and sample D a two-sided PSG. Sample B received PSG on one side and BSG on the other. Sample E features a one-sided BSG and sample F a two-sided BSG. To separate temperature induced effects from P-gettering, for sample C and D sister wafers received the same thermal load but without any deposition (process steps not shown in Fig. 1). After deposition, respectively temperature treatment, the dopants were driven into wafers in a standard open quartz tube diffusion furnace at same temperature and duration for all samples in the same process. This resulted in a 60 Ω/□ P-diffusion and a 110 Ω/□ B-diffusion. Afterwards, PSG and BSG and the diffused layers were removed by wet chemical etching and PECVD SiN_x:H was deposited on both sides and fired in a belt furnace at 800°C peak firing wafer temperature.

To determine gettering efficacy in areas close to the crucible wall, two sister wafers with one side showing the effect of the nearby crucible wall were additionally processed analogous to the reference sample A and sample D. For comparison of two-sided APCVD P-gettering with $POCl_3$-gettering, two sister wafers from top ingot height were used as wafers from top height show elevated concentrations of contaminations. $POCl_3$ diffusion was performed in an open quartz tube furnace with parameters in accordance to achieve similar sheet resistance (R_{sh}) as the APCVD samples. Due to process requirements, the $POCl_3$ sample received a different SiN_x:H from a direct plasma process. Therefore, SiN_x:H was removed after firing from the sister wafers with $POCl_3$- and APCVD P-gettering and a wet chemical surface passivation was applied for comparison.

Afterwards, τ_{eff} of the samples was measured by self-calibrated time resolved photoluminescence imaging (TR-PLI [4]) and the harmonic mean of τ_{eff} is calculated.

3 RESULTS AND DISCUSSION

3.1 APCVD gettering efficacy

Fig. 2 shows selected spatially resolved τ_{eff} data for the samples described above. Comparing the reference sample A (Fig. 2A, harmonic mean: 66 µs) with the single-sided P-gettered one (sample C, Fig. 2C, 111 µs), a pronounced P-gettering effect can be seen. This effect is further improved for the two-sided P-gettered sample (sample D, Fig. 2D, 129 µs). From literature it is known that having P-gettering layers present on both wafer sides results in a more pronounced gettering effect in mc-Si (*e.g.* [5]). This seems to be true for APCVD-based PSG layers as well.

On the other hand, a sample with the same temperature load as the one-sided P-gettered sample C shows a decrease in τ_{eff} in most sample areas (Fig. 3a, 44 µs). Therefore, we conclude that for the pronounced positive gettering effect the presence of a PSG layer and/or the diffused P-profile are needed. The sample that received the same thermal load as the two-sided PSG sample D shows an even more decreased τ_{eff} (Fig. 3b, 35 µs). This could be caused by the second run through the APCVD reactor and the additional temperature load involved therein, or by additional in-diffusion of impurities, as no masking layer is present for the second run in contrast to the two-sided PSG sample D where the other side is protected by the PSG deposited in the first run through the machine.

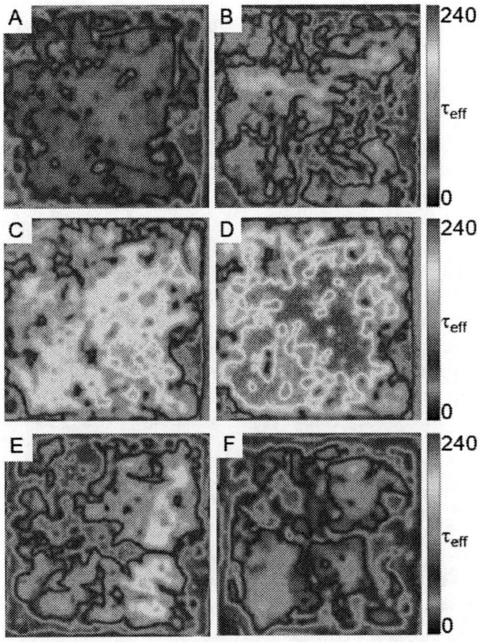

Figure 2: τ_{eff} maps of the 5x5 cm² samples processed according to Fig. 1 from mid ingot position. A) reference sample, B) co-gettered with BSG and PSG, C) one-side P-gettered, D) two-sides P-gettered, E) one-side B-gettered, F) two-sides B-gettered sample.

The one-sided B-gettered sample E (Fig. 2E, 58 µs) shows in some regions a positive gettering effect, too, but significantly less effective than the one-sided P-gettering (Fig. 2C). Standard B-gettering is usually less effective than P-gettering, but the efficacy seems to depend more critically on the applied gettering parameters [2]. Especially as the two-sided B-gettered sample F shows an overall decrease of τ_{eff} (Fig. 2F, 47 µs). Nevertheless, it demonstrates that also a B-diffusion from an APCVD-sourced BSG layer can have a positive gettering effect.

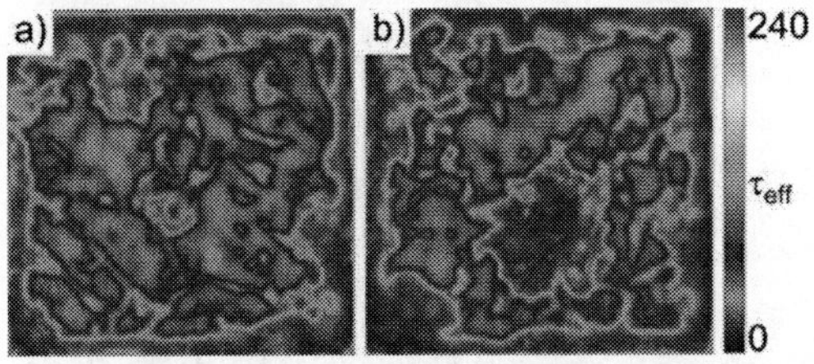

Figure 3: τ_{eff} maps of the 5x5 cm² samples from mid ingot position. a) same temperature load as one-side P-gettered sample C, b) same temperature load as two-sides P-gettered sample D.

Sample B with APCVD PSG on one side and APCVD BSG on the other (Fig. 2B, 63 µs) shows a mixed behavior. In some areas a significant improvement can be detected, while other areas do not respond well to the gettering layers or show even decreased τ_{eff} values. The reason for this behavior is still unclear. But it is expected that the local defect structure in combination with the applied temperatures may play a crucial role. In addition, unintentional contamination during processing might play a role and should be avoided.

3.2 Influence of crucible wall

All samples discussed so far resulted from a column of mid ingot position far away from the crucible walls. Samples from a position close to the crucible wall were also included in this study. As the gettering efficacy should be strongly influenced by the local concentration of impurities as well as their interaction with extended defects, a more complete picture is expected when investigating samples with varying concentration of impurities. It can be stated that in samples close to the crucible wall, an area of strong decrease in τ_{eff} is visible in the reference sample (red zone effect, Fig. 4a, 18 µs), which can be almost completely removed after APCVD-PSG-based P-gettering (Fig. 4b, 108 µs).

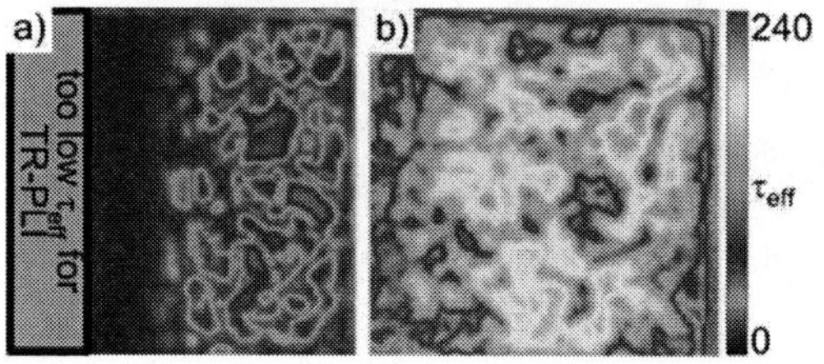

Figure 4: τ_{eff} maps of the 5x5 cm^2 samples from mid ingot position close to the crucible wall. a) sample processed according to reference sample A, b) two-sides P-gettered sample according to sample D.

3.3 Comparison of APCVD PSG and $POCl_3$ gettering

In addition to the described APCVD-based gettering schemes, also $POCl_3$-based standard gettering steps are carried out and provide reference gettering scenarios with similar R_{sh}. Due to process requirements, the $POCl_3$-gettered sample received a different SiN_x:H passivation. Hence for comparing both samples with a surface passivation of same quality, SiN_x:H layers were removed after firing from both samples and surface passivation was realized with wet chemical passivation. With comparable surface passivation the APCVD P-gettered sample (Fig. 5b, 329 µs) shows slightly higher τ_{eff} than the $POCl_3$-gettered one (Fig. 5a, 277 µs). In addition, both samples show similar τ_{eff} in low quality areas.

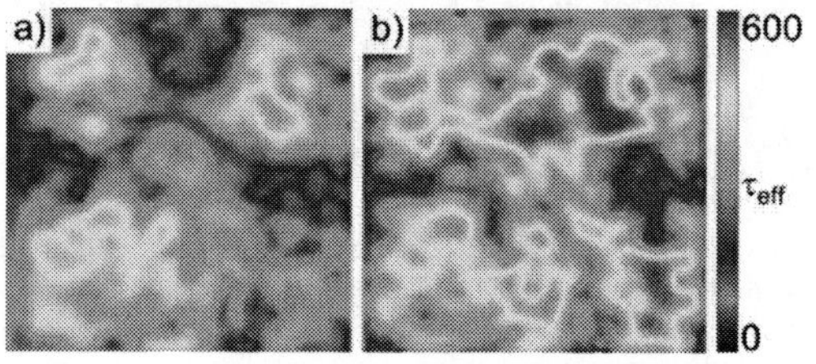

Figure 5: τ_{eff} maps of the 5x5 cm^2 samples from top ingot position with wet chemical surface passivation. a) $POCl_3$-gettered, b) two-sides APCVD P-gettered sample.

4 SUMMARY AND OUTLOOK

Deposition of PSG and BSG layers using APCVD is a very promising alternative to $POCl_3$- and BBr_3-based standard diffusion as it offers the possibility of higher throughput, less process complexity and lower cost. It offers a very attractive processing scheme especially for PERT-type solar cell architecture as PSG and BSG can be deposited on each side of the wafer, and emitter as well as back surface field can be formed in only one co-diffusion step.

Especially for mc-Si, gettering efficacy plays an important role and is a key point for reaching high efficiencies. Therefore, the gettering efficacy of APCVD PSG and BSG layers has been investigated. Results on samples from mid ingot position point out that one-side P-gettering is less effective than two-sides gettering, and B-gettering is less effective than P-gettering at the same diffusion temperature. Spatially resolved measurement of τ_{eff} allows for investigation of the influence of local defect structure on gettering efficacy.

As expected, gettering efficacy is strongly affected by the local defect structure. Especially wafers from a position close to the crucible wall with higher density of extended defects and higher transition metal concentration benefit from P-gettering steps. Regarding the contamination level this holds true also for wafers of the top ingot height.

Comparison of $POCl_3$-based diffusion gettering as the state-of-the-art reference gettering scheme in today's industrial production with APCVD P-gettering has shown comparable τ_{eff}.

5 ACKNOWLEDGEMENTS

Part of this work was funded by the German BMWi MuSkAD project (034041A) and project no. 0324001. The content of the publication is the responsibility of the authors. The authors would like to thank M. Fleck, J. Lindroos, L. Mahlstaedt, F. Mutter, J. Rinder and D. Skorka for their assistance with this study.

6 REFERENCES

[1] K.O. Davis, PhD thesis, University of Central Florida (2015).
[2] S.P. Phang, D. Macdonald, J. Appl. Phys. 109 (2016) 073521.
[3] S. Gindner, P. Karzel, B. Herzog, G. Hahn, IEEE J. Photovolt. 4(4) (2014) 1063.
[4] D. Kiliani, G. Micard, B. Steuer, B. Raabe, A. Herguth, G. Hahn, J. Appl. Phys. 110 (2011) 054508.
[5] A. Schneider, R. Kopecek, G. Hahn, S. Noël, P. Fath, Proc. 31st IEEE PVSC, Lake Buena Vista 2005, 1051.

ENHANCING PERFORMANCE OF UPGRADED METALLURGICAL GRADE SILICON SOLAR CELLS NANO-TEXTURED BY USING METAL CATALYZED CHEMICAL ETCHING

Volker Hoffmann[1,*], Shuai Zou[2], José Manuel Míguez[1], Xiaodong Su[2]

[1] Silicio Ferrosolar (Ferroglobe group), Polígono Industrial de Sabón, Spain
[2] College of Physics, Optoelectronics and Energy, Collaborative Innovation Center of Suzhou Nano Science and Technology, and Jiangsu Key Laboratory of Thin Films, Soochow University, 1 Shizi street, Suzhou 215006, China
[*] Phone: +49 173 3506051, E-mail: volker.hoffmann@ferroglobe.com

ABSTRACT: Photovoltaic electricity generation has reached a significant share on total electricity production in several countries, already, due to a continuous decrease in production cost and increase in solar cell efficiency. The present study combines both approaches. Upgraded metallurgical grade silicon (UMG-Si) has been developed as an alternative to the standard polysilicon (poly-Si) feedstock being cheaper in production and investment cost. With a significantly lower carbon footprint. Metal catalyzed chemical etching (MCCE) forms a nano-texture on the wafer surface shown to increase the efficiency in the solar cells. The MCCE process has been applied to wafers made of UMG-Si produced by Silicio Ferrosolar (CS-Silicon™). In the case of Ag-MCCE process, the etching behavior of multicrystalline UMG-Si wafers was almost the same as that of the mc-Si wafers made of poly-Si. MCCE-textured UMG-Si solar cells showed an efficiency gain of 0.37%-0.48% absolute over the standard acidic textured ones, and the champion cell reaches an efficiency of 19.2%. The efficiency increase in nano-textured cells is related to an improved light absorption, a flatter emitter and a better contact formation in. Both, efficiency gain through MCCE texture and absolute cell efficiency are the same for UMG-Si cells and poly-Si reference cells within the statistical relevance.
Keywords: Nano-texture, metal catalyzed chemical etching, Solar Grade Silicon, UMG-Si

1 INTRODUCTION

By far most of the new PV installations worldwide are still based on crystalline silicon solar cells. The production of the high purity Si feedstock (poly-Si) is highly energy intensive with a high investment cost and long lead time involved. The cost reduction of photovoltaic electricity generation stays in the focus of research and development in order to further increase its global share. It can be achieved by both, decrease in production cost and increase in solar cell efficiency. The present study combines both approaches.

In the last years several companies like Elkem Solar and Silicio Ferrosolar have developed an alternative Si feedstock with significantly lower energy consumption by upgrading silicon metal (mg-Si) through the metallurgical route [1,2]. The so called upgraded metallurgical grade Si (UMG-Si) is also known as solar grade Si (SoG-Si). Beside the relatively low energy consumption with its much lower carbon footprint, the significantly reduced investment cost and the scalability of production capacity are inherent competitive advantages compared to the standard poly-Si production.

Metal catalyzed chemical etching (MCCE) is a method developed at Soochow University that forms a nano-texture on the wafer surface. In addition to the increased light absorption in nano-textured cells, also the front contact and emitter formation, as well as the surface passivation, are strongly influenced by the surface texture [3,4]. Therefore, the texture has a significant influence on the cell efficiency. The MCCE nano-texturing process is not as sensible to the wafer surface condition as the standard acidic texture, which needs to be adjusted when switching cell production between slurry based wire sawn wafers and diamond wire sawn wafers. This is a big advantage in terms of cell production flexibility. Therefore, the MCCE has the potential to become the standard texture process in multicrystalline-Si solar cell production [5].

The aim of this work is to study the performance of the MCCE process on multicrystalline wafers made of UMG-Si with respect of the formation of the nano-texture as well as the efficiency gain compared to the acidic texture, and compared to cells made of poly-Si. UMG-Si is a compensated feedstock, containing both dopants, B and P, which needs to be considered when using it. In order to lever the cost advantage of UMG-Si up to the generated PV electricity, the efficiency of solar cells made of UMG-Si must not be lower than of cells made of poly-Si. A standard cell process is more forgiving with respect to impurities than an advanced cell process. Therefore, it is important to study the performance of UMG-Si wafers with advanced cell processes, as the nano-texturing in the present case.

The MCCE nano-texture is applied to UMG-Si wafers to learn if there is an influence of the increased dopant concentration in UMG-Si wafers on the etching performance. Both, etching performance and final cell efficiency of MCCE nano-textured UMG-Si cells are studied in this work in order to evaluate the capability of CS Silicon™ to be used in combination with the MCCE nano-texture, and to compare it with the performance of standard poly-Si cells.

2 EXPERIMENTAL

2.1 CS Silicon™ purification process

The SoG-Si production of Silicio Ferrosolar is integrated in the silicon metal (mg-Si) production facility of Ferroatlántica S.A. in Sabón, Spain. The purification process starts with mg-Si of a better than standard quality which is produced in special production campaigns. This special mg-Si is refined by a slagging process in induction furnaces (see Fig. 1).

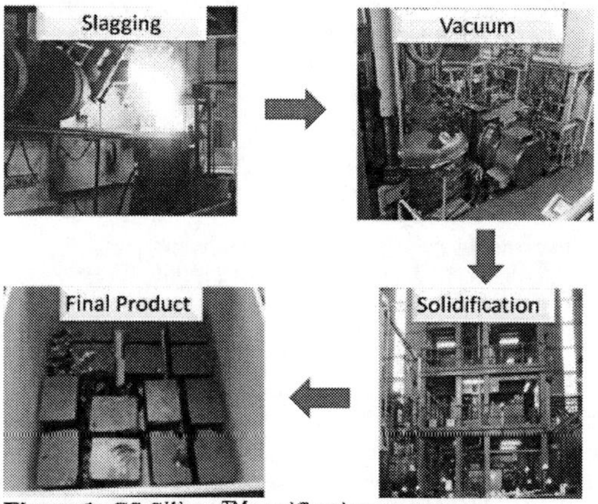

Figure 1: CS Silicon™ purification process

The slagging process is repeated as many times as necessary to obtain the target B concentration for this step. Next, there is a vacuum purification step to reduce the P concentration down to the target value. The last step is a directional solidification process to get out metals and inclusions by segregation. Then, the resulting ingot is cut into bricks which undergo a final inspection and cleaning treatment. The target range of dopant concentrations for the feedstock is 0,1 – 0,2 ppmw B and 0,1 – 0,6 ppmw P.

2.2 UMG-Si ingot and wafers

An industrial multicrystalline ingot made of compensated CS Silicon™ feedstock from Silicio Ferrosolar has been grown in a directional solidification furnace. The dopant concentrations in the feedstock were 0,16 ppmw B and 0,34 ppmw P. With Ga added to the crucible charge the resistivity of the resulting p-type ingot was in the range 1,2–2,0 Ωcm over the usable brick lengths, and the total brick yield was as of ingots made with poly-Si feedstock. The bricks of this ingot were sliced into wafers with 180 μm thickness.

2.3 Ag-MCCE nano-texture

The metal catalyzed chemical etching MCCE process was developed at the Soochow University to nano-texture mono – and multicrystalline Si-wafers. Figure 2 shows the process sequence for the MCCE texturing.

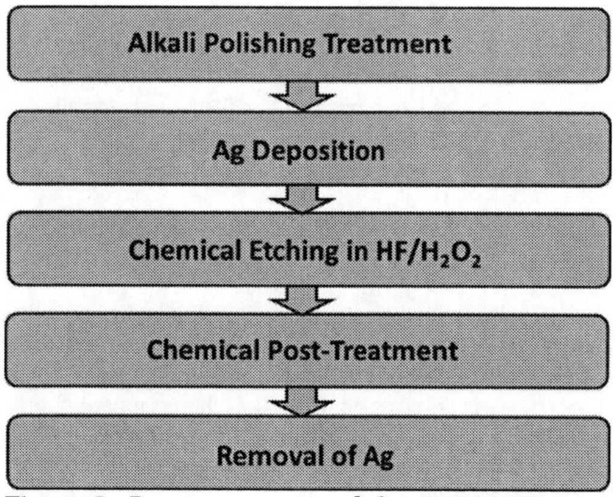

Figure 2: Process sequence of the Ag-MCCE nano-texture

After an alkali polishing treatment Ag particles are deposited by immersing the Si wafers in a solution containing 10^{-4}M $AgNO_3$ and 10^{-3}M HF for 3min. In the next step, the Ag-loaded wafers were etched in a mixed solution of 49% HF and 30% H_2O_2 (2:1) for 5-7mins to form a microstructure of a nano-hole array on the surface. Then, a chemical post-treatment was applied to convert nano-holes into a nano-texture with a larger opening than that of nano-holes. Finally, remained Ag particles were removed by using a home-made cleaning process.

4 batches of UMG-Si wafers from different bricks of the same ingot were divided into 2 equivalent groups, each, forming 2 sets of similar wafers. One set was nano-textured at Soochow University by their Ag-MCCE process, and the other set was textured with the standard acidic etching. Another 2 sets of standard multicrystalline poly-Si wafers, but from different ingots, were processed in the same way as a reference.

2.3 Solar cells

The solar cells were produced at the cell production line of Canadian Solar Inc. with a back surface field BSF process. The process sequence of the solar cell production is shown in Figure 3.

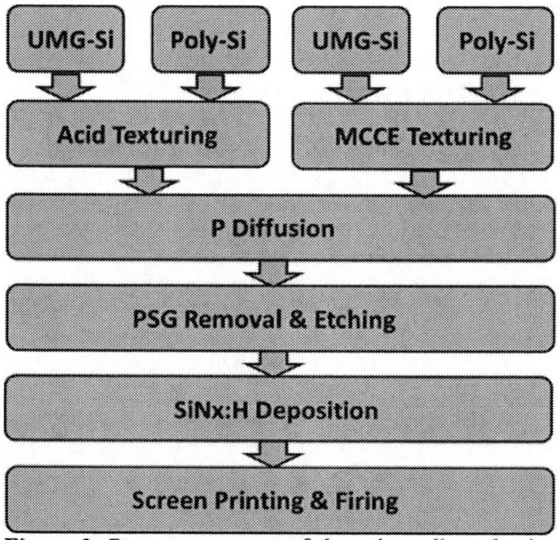

Figure 3: Process sequence of the solar cell production with UMG-Si and poly-Si wafers

Each set of wafers from UMG-Si and poly-Si was processed into cells, batch by batch, in sequence in order to decrease the influence of process variations.

3 RESULTS AND DICUSSION

Figure 4a) is a typical surface SEM image of UMG-Si wafers after a standard acid texturing process, showing a micron-texture with homogeneous oval pits, which have an opening size about 3-5 μm. In the case of Ag-MCCE, figure 4b) displays a desired nano-texture consisting of nearly round nano pits. Specially, both depth and width of the nano-texture were manipulated to obtain a range about 600–800 nm, which can ensure a low surface recombination while keeping low reflectance. The surface morphology of the acidic textured UMG-Si wafers as well as of the MCCE nano-textured UMG-Si wafers showed no significant difference to the poly-Si reference wafers.

Thus, no influence of the increased dopant concentration in UMG-Si wafers on the etching behavior could be noticed.

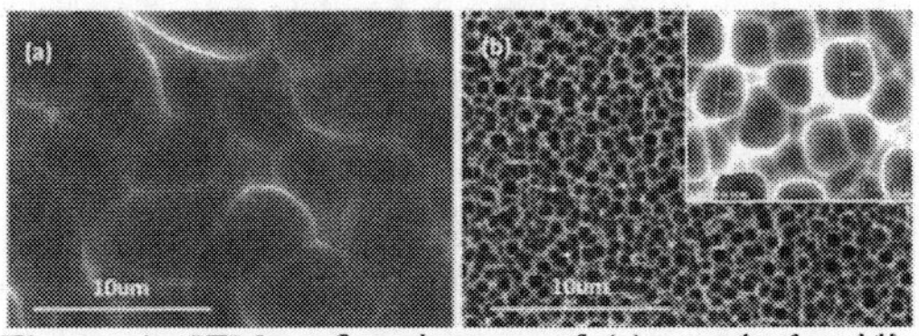

Figure 4: SEM surface images of (a) standard acidic texture, and (b) MCCE nano-texture

The mean parameters of MCCE textured multicrystalline UMG-Si and poly-Si solar cells are shown in table I. The average cell efficiency of the 4 groups of UMG-Si wafers which were sliced from the same ingot, reached 18.9% - 19.1% with MCCE texture. Moreover, the champion UMG-Si solar cell reached an efficiency of 19.2%. The variation between the 4 batches in each set was rather small, resulting in a good confidence level of this result. With an average of 19,0% efficiency, the UMG-Si cells resulted in even 0,1% higher efficiency than the poly-Si reference cells with 18,9% in average.

MCCE	Isc [A]	Uoc [mV]	FF [%]	Efficiency
UMG-223	8,97	639	81,1	19,1%
UMG-233	8,94	636	80,7	18,9%
UMG-234	8,96	637	80,9	19,0%
UMG-443	8,95	637	81,0	19,0%
UMG avg.	8,96	637	80,9	19,0%
best UMG cell	**8,99**	**640**	**81,2**	**19,2%**
Poly-Si	8,98	637	80,3	18,9%

Table I: Main characteristics for MCCE textured multicrystalline UMG-Si and poly-Si solar cells

The results of the reference cells with acidic texture are shown in table II. Also in this group, the variation between the 4 different batches of UMG-Si cells is small. The efficiency of acidic textured UMG-Si cells is 18,4% - 18,7%, and 18,5% in acidic textured poly-Si reference cells. The efficiency gain of MCCE textured UMG-Si cells compared to acidic textured cells is 0,37% - 0,48% absolute. Also in the acidic textured cells, the UMG-Si cells resulted in 0,1% higher efficiency than the poly-Si reference cells, with 18,6% in average compared to 18,5% in poly-Si cells.

Acidic	Isc [A]	Uoc [mV]	FF [%]	Efficiency
UMG-223	8,88	638	80,5	18,7%
UMG-233	8,85	634	79,8	18,4%
UMG-234	8,86	636	80,0	18,5%
UMG-443	8,86	636	80,3	18,6%
UMG avg.	8,86	636	80,1	18,6%
Poly-Si	8,91	634	79,7	18,5%

Table II: Main characteristics for acidic textured multicrystalline UMG-Si and poly-Si solar cells

As expected, the short-circuit current (I_{SC}) in nano-textured solar cells had a significant increase up to 100 mA in average in UMG-Si cells. But also the fill factor (FF) showed a significant increase of 0,77% absolute, and the Voc was increased by 1,4 mV in average in nano-textured UMG-Si cells, resulting in an overall efficiency increase of 0,42% absolute in average.

In poly-Si cells I_{SC} increased by 70 mA, the FF by 0,57%, and V_{OC} by 2,5 mV in average with MCCE texture. The efficiency gain at the cell level was 0,35% in average, which is slightly lower than for UMG-Si cells in this production run. This can be due to a slightly different quality between the 2 sets of poly-Si wafers, which unfortunately were not taken from the same ingot. Nevertheless, the cell parameter improvement in MCCE textured cells was found for both types, UMG-Si and poly-Si cells to be similar.

The relative difference of the cell parameters of all groups are shown in figure 5.

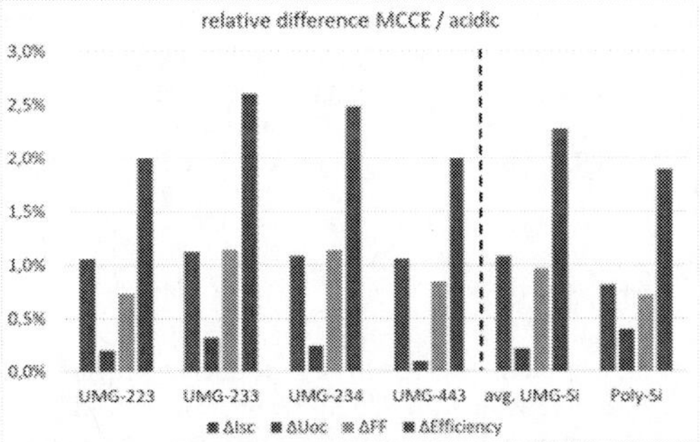

Figure 5: Relative difference of cell parameters for MCCE texture over acidic texture of multicrystalline UMG-Si and poly-Si solar cells

The relative efficiency increase in MCCE textured UMG-Si cells was +2,3%, and in poly-Si cells +1,9%. The main contribution to the efficiency gain came from an increased I_{SC}, with +1,08% in UMG-Si and +0,81% in poly-Si cells. This indicates on a better light absorption, and also on a better surface passivation in MCCE nano-textured cells. The fill factor increase in MCCE textured cells was also significant. In UMG-Si cells it was 0,96% relative in average, and in poly-Si cells 0,72%. This correlates with a reduced series resistance R_S and increased shunt resistance R_{SH}. These results indicate, that the nano-texture is favorable to improve the contact formation and the emitter formation. The V_{OC} increased, too, in MCCE textured cells, although not as much as I_{SC} and FF. But still, the gain is 0,21% relative in UMG-Si and 0,39% in poly-Si cells. Again, the different relative gains in poly-Si compared to UMG-Si cells could be due to the fact that the 2 sets of poly-Si wafers were taken from 2 different ingots, possibly with slightly different resistivity.

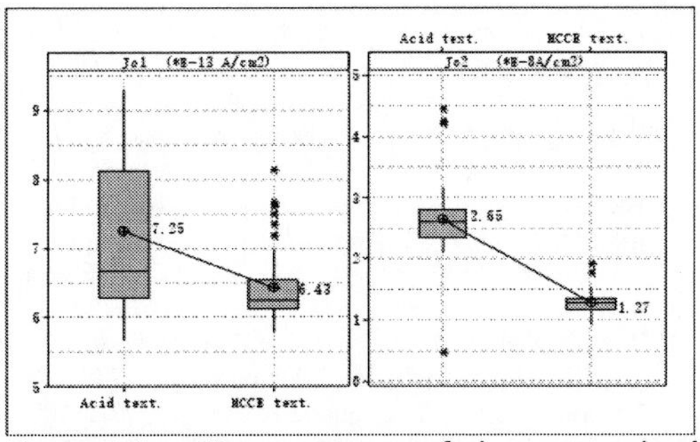

Figure 6: Suns-Voc measurement of micron-textured and nano-textured UMG-Si solar cells.

The Suns-V_{OC} measurement of UMG-Si solar cells resulted in a J_{01} of 725 fA/cm² and a J_{02} of 26,5 nA/cm² in acidic textured cells, and 643 fA/cm² and 12,7 nA/cm² in MCCE textured cells, respectively. Both, J_{01} and J_{02} were lower in MCCE textured UMG-Si cells compared to acidic textured cells as shown in figure 6.

Based on a two-diode model, J_{01} is due to recombination of electron-hole pairs in the base and emitter, and J_{02} is related to the injection dependent Shockley-Read-Hall recombination in the space charge region [6,7]. With both textures, J_{02} is more than four orders of magnitude larger than J_{01}. Therefore, J_{02} dominates the dark current in the cells, and it can be concluded, that the nano-texture is well passivated by the SiN_x coating. J_{02} is twice as high in acidic textured cells, than in MCCE textured ones. This indicates on a flatter pn-junction in nano-textured UMG-Si solar cells with a lower injection recombination, whereas a the rougher surface of acidic textured cells may result in a distorted pn-junction with a higher recombination rate in the space charge region [4,5].

4 CONCLUSIONS

In this study the Ag-MCCE texturing process was successfully applied to UMG-Si wafers, achieving an average cell efficiency of 19,0%, which was even slightly higher than in the reference poly-Si cells. The morphology of nano-textured UMG-Si wafers was the same as of nano-textured poly-Si reference wafers, showing no influence of the higher dopant concentration in UMG-Si. Also the significant efficiency increase of 0,42% absolute in UMG-Si and 0,35% in poly-Si cells compared to the standard acidic texture is very similar for both materials. The improved performance of cells with Ag-MCCE nano-texture could be attributed to a better light absorption, improved front surface passivation, a flatter emitter, and better contact formation.

With the same performance of nano-textured UMG-Si and poly-Si solar cells, the cost advantage and the lower carbon footprint of UMG-Si, together with the efficiency increase by MCCE nano-texture, can result in a significant further cost reduction of PV electricity generation.

5 ACKNOWLEDGEMENTS

The authors would like to thank Joaquín Diéguez from Silicio Ferrosolar for taking care about the material flow between the companies involved in this work.

6 REFERENCES

[1] V. Hoffmann, K. Petter, J. Djordjevic-Reiss, E. Enebakk, J.T. Håkedal, R. Tronstad, T. Vlasenko, I. Buchovskaja, S. Beringov, M. Bauer, First results on industrialization of Elkem Solar Silicon at Pillar JSC and Q-Cells, Proc. of the 23rd EU PVSEC, Valencia (2008), 1117.

[2] V. Hoffmann, J.M. Míguez, J. Bullón, I. Buchowska, T. Vlasenko, S. Beringov, J. Denafas, K. Meškerevičienė, L. Petrėnienė, R. Ordás, Effect of total dopant concentration on the efficiency of solar cells made of CS-Silicon, Proc. of the 31st EU PVSEC, Hamburg (2015), 559.

[3] X.Y. Ye, S. Zou, K.X. Chen, J.J. Li, J. Huang, F. Cao, X.S. Wang, L.J. Zhang, X.F. Wang, M.R. Shen, and X.D. Su. 18.45%-efficient multi-crystalline silicon solar cell with nano-scale pseudo-pyramid texture. Adv. Funct. Mater. 24, 6708–6716 (2014).

[4] F. Hu, Y. Sun, J. Zha, K. Chen, S. Zou, L. Fang, X.D. Su, Pre-texturing multi-crystalline silicon wafer via a two-step alkali etching method to achieve efficient nanostructured solar cells, Solar Energy Materials & Solar Cells 159, 121–127 (2017).

[5] F. Cao, K.X. Chen, J.J. Zhang, X.Y. Ye, J.J. Li, S. Zou, X.D. Su, Next-generation multi-crystalline silicon solar cells: Diamond-wire sawing, nano-texture and high efficiency, Solar Energy Materials & Solar Cells 141, 132–138 (2015).

[6] R. Hussein, D. Borchert, G. Grabosch, W.R. Fahrner, Dark I-V-T measurements and characteristics of (n) a-Si/(p) c-Si heterojunction solar cells, Sol. Energy Mater. Sol. Cells 69, 123-129 (2001).

[7] J. Greulich, M. Glatthaar, A. Krieg, G. Emanuel, S. Rein, JV characteristics of industrial silicon solar cells: influence of distributed series resistance and Shockley Read Hall recombination, Proceedings of the 24th EU PVSEC, Hamburg (2009).

IMPACT OF GLASS CHEMISTRY ON CONTACT FORMATION FOR SILVER METALLIZATION PASTES

Lindsey Karpowich, Ryan Mayberry, and Matthias Hörteis
Heraeus Precious Metals North America Conshohocken LLC
West Conshohocken, PA 19428 USA
Tel. +1-610-825-6050
Fax +1-610-825-7061
lindsey.karpowich@heraeus.com

ABSTRACT: The glass constituent of silver metallization paste plays an important role in the electrical contact formation for crystalline silicon solar cells. Glass chemistry is critical for controlling the complex redox reactions that occur during rapid thermal firing. In this work, we show how modifying ratios of lead and bismuth in the glass can affect the etching of the silicon nitride anti-reflective coating (ARC) and the subsequent crystallization of silver onto the silicon wafer. Reducing lead content while increasing bismuth content results in a glass composition that can sufficiently fire through the silicon nitride layer. However, the formation of silver crystallites and colloids is suppressed, and the final glass microstructure at the interface is also impacted. This results in a much higher series resistance due to poor contact between bulk silver and silicon emitter. By systematically changing the concentrations of lead and bismuth in the glass, we can better understand the fundamental reactions that govern contact formation and electrical properties of silicon solar cells.
Keywords: contact, metallization, glass frit, silicon solar cell

1 INTRODUCTION

The most common method for the metallization of crystalline silicon (Si) solar cells is screen-printed silver (Ag) thick film technology [1]. Screen printing has many advantages, including the maturity of the technology and much lower cost than other metallization techniques. Silver metallization pastes can significantly improve the efficiency of a solar cell by reducing contact resistance between the emitter and conductive finger line. However, the paste must also etch away the protective silicon nitride passivation layer in order to contact the emitter. The chemistry of the metallization paste must balance the low contact resistance necessary to achieve higher fill factor (FF) and maintain sufficient passivation for higher open-circuit voltage (Voc).

The redox reactions that govern contact formation have been studied and modeled in various works ([2-6] and many others). These reactions occur during the rapid thermal firing process during which the printed Ag paste is co-fired with the silicon wafer at temperatures up to 850°C. It is commonly understood that the reactive component in the Ag metallization paste is the glass frit, and in particular lead oxide (PbO) and/or bismuth oxide (Bi_2O_3) incorporated into the glass. The first reaction generally agreed to occur above 500°C [7] involves the etching and oxidation of the silicon nitride (SiNx) passivation layer:

$$4PbO + 2SiN_x \rightarrow 4Pb + 2SiO_2 + xN_2 (\uparrow) \quad (1)$$

Bismuth oxide will behave in a similar way according to reaction 2, although at a slightly higher temperature [6]:

$$2Bi_2O_3 + 3SiN_x \rightarrow 4Bi + 3SiO_2 + \frac{3x}{2} N_2 (\uparrow) \quad (2)$$

The reactions above effectively "open" the ARC layer, exposing the silicon to the molten glass frit during firing. In order to form electrical contact between the emitter and bulk Ag finger line, authors generally agree that the Ag must first oxidize and dissolve into the glass frit above 650°C via reaction 3:

$$2Ag + \tfrac{1}{2} O_2 \rightarrow Ag_2O \quad (3)$$

The dissolved Ag ions can then diffuse to the glass-silicon interface and oxidize the Si, resulting in metallic Ag being deposited onto the Si surface via reaction 4:

$$2Ag_2O + Si \rightarrow SiO_2 + 4Ag \quad (4)$$

Any dissolved Ag that is not reduced at the Si surface during firing can precipitate out of the glass as the Ag solubility decreases upon cooling.

Most authors agree that both Ag crystallites and colloids in the interfacial glass layer contribute to electrical conductivity between the emitter and bulk Ag. However, the exact mechanisms and kinetics of Ag crystallite and colloid formation are not yet completely understood. In order to crystallize at the silicon-glass interface, dissolved Ag ions must be able to diffuse through the glass during the firing process. In order to precipitate as colloids during cooling, the Ag solubility and stability within the glass must also be critical. Thus the glass properties are expected to impact the Ag ion mobility and ability to form Ag crystallites and colloids. In this work, we aim to understand how the concentrations of reactive species PbO and Bi_2O_3 can impact the glass properties and hence contact formation.

2 EXPERIMENTAL APPROACH

Thick film Ag metallization paste formulations that are common to the industry typically contain 85-90 wt.% Ag powder, 5-15 wt.% solvent-based organic vehicle, and 1-5 wt.% glass or inorganic. In this study, a standard lead-silicate glass formulation was chosen as a baseline composition. The molar percentage of lead (Pb) was then systematically decreased while the molar percentage of bismuth (Bi) was increased by the same amount. The ratios of Pb and Bi in the glasses are found in Table I, normalized to relative molar percentages.

Oxide powder precursors of glass components were mixed together and placed into ceramic crucibles. The powder blends were then heated at 1000°C for 30 minutes to form a glass melt. The melts were quenched into DI water at room temperature and subsequently milled using various planetary and ball mills to form a fine ~1 µm powder. The thermal properties of each glass were characterized via differential scanning calorimetry (DSC) as well as hot-stage microscopy (HSM), using a heating rate of 20°C/minute.

Table I: Molar ratios of Pb and Bi oxides in glasses.

Oxide	A	B	C	D	E
PbO	1.00	0.92	0.85	0.77	0.69
$BiO_{1.5}$	-	0.08	0.15	0.23	0.31

After the glasses were prepared, silver metallization pastes were manufactured using industry-standard thick film processes. The ratio of silver powder to glass was approximately 44:1, with the remaining balance being comprised of a solvent-based organic vehicle containing resins and binders. The mixtures were then homogenized using a 3-roll mill. Once milled, the pastes were screen printed onto P-type multicrystalline silicon wafers with a sheet resistance of ~90 Ohm/sq. The printed wafers were dried at 150°C for 10 minutes in a box oven. After drying, the wafers were co-fired with aluminum paste (previously printed and dried on the rear side of the wafer) using an IR belt furnace at a peak temperature of 800-850°C.

The solar cells were characterized using h.a.l.m. cetisPV cell tester to measure the current-voltage (I-V) response. In addition, fired wafers were etched with nitric acid HNO_3 (to remove the Ag) and hydrofluoric acid HF (to remove the glass layer) in order to observe microstructural interactions with the SiNx layer and wafer using scanning electron microscopy (SEM).

3 RESULTS & DISCUSSION

3.1 Glass thermal properties

The influence of Pb:Bi ratio on glass thermal properties can be seen in Figure 1. Based on the DSC heat flow response, it is clear that increasing the amount of Bi in the glass leads to a larger crystallization event between 500-600°C (exotherm).

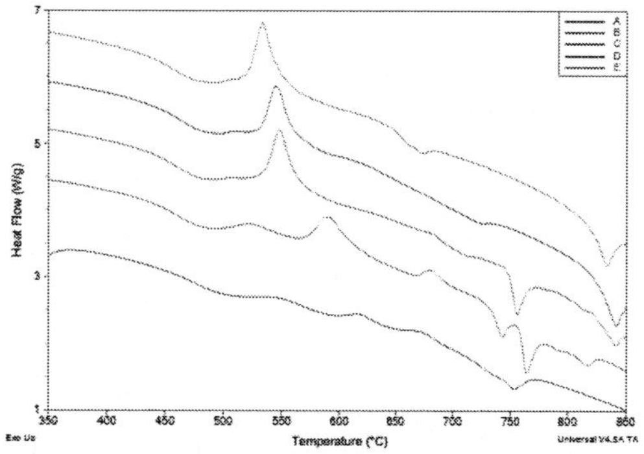

Fig. 1: DSC heat flow response of glass frits.

The onset of the crystallization also shifts to lower temperatures as Bi% increases. In general, a crystallizing or de-vitrifying glass can exhibit higher viscosity due to the blocking effect of the crystals [8]. It is possible that the high-Bi glasses with a more pronounced crystallization may have a higher viscosity and less flow during rapid thermal processing and contact formation. This is also supported by the melting endotherm shifting to higher temperatures. Thus as the Bi content is increased, the glass exhibits stronger crystallization beginning at lower temperatures and even higher melting temperatures.

In order to further investigate the glass during firing, HSM analysis was performed to observe physical behavior. Figures 2 and 3 show an exemplary data set illustrating the change in cross-sectional area of a glass sample as a function of temperature. Specific transition points with their correlating physical form are also defined.

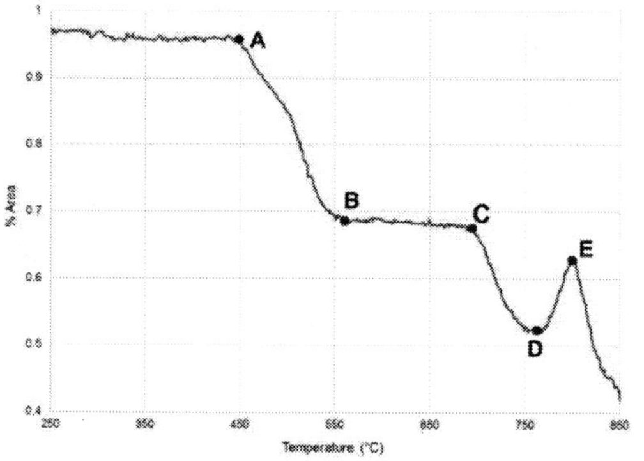

Fig. 2: Typical HSM curve showing glass behavior during heating. A) sintering onset; B) crystallization onset; C) crystallization end; D) deformation; and E) sphere.

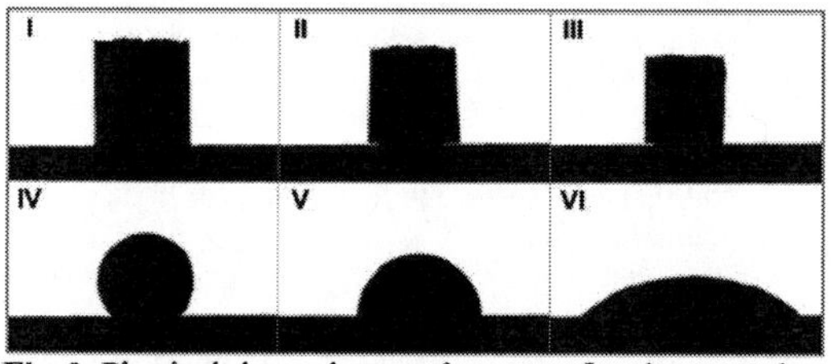

Fig. 3: Physical shape changes that occur for glass samples in HSM, including: I) initial; II) sintering onset; III) deformation; IV) sphere; V) hemisphere; and VI) flow.

The HSM data in Table II correlate well with the DSC results. As the Bi content increases in the glass, crystallization onset shifts to lower temperatures. The crystallization "window" (temperature between onset and end) also increases with Bi content. The final flow temperature is increased as well. All of the results indicate that the thermomechanical properties of the glass are greatly influenced by the Pb:Bi ratio, especially the crystallization and flow behaviors.

Table II: HSM transition points.

	A	B	C	D	E
Sinter Onset (°C)	449	452	440	440	435
Cryst. Start (°C)	631	558	513	485	472
Cryst. End (°C)	735	695	651	639	626
Flow (°C)	783	816	830	825	860

3.2 I-V performance

The electrical performance of silver metallization pastes containing the Pb-Bi glasses is shown in Figure 4.

As the molar percentage of Bi increases, the series resistance increases almost exponentially. This results in a corresponding decrease in fill factor (FF) and overall cell efficiency. The series resistance of a solar cell is the sum of numerous individual resistive contributions: gridline resistance, emitter sheet resistance, bulk resistance, contact resistance, and others. In this case, due to all other resistive components being relatively equal, it is most likely that the contact resistance is increasing as the Bi content in the glass is increased.

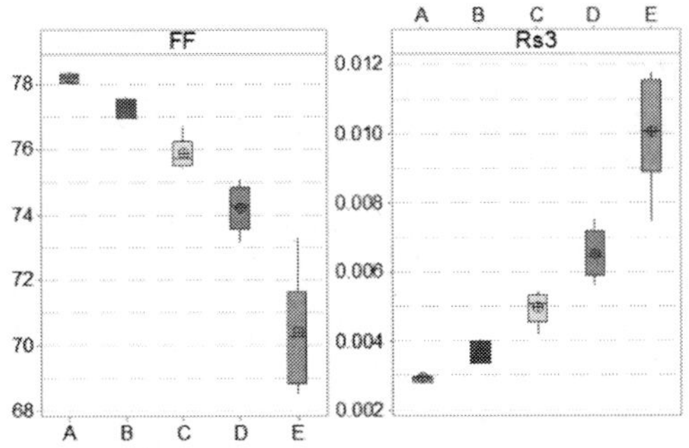

Fig. 4: Fill factor and series resistance for metallization made with different glasses.

As previously described, contact formation is believed to be a two-step process involving high temperature redox reactions to etch the SiNx layer and form silver precipitates and colloids to facilitate current transport. The initial reaction to oxidize the SiNx layer can be accomplished by the reduction of either PbO or Bi_2O_3 from the glassy state. The second reaction involves the dissolution of Ag into the glass, Ag ion transport to the interface, and subsequent reduction back to metallic Ag in the form of crystallites on the silicon surface or colloids suspended in the interfacial glass. This reaction is more complex and can be impacted by solubility as well as diffusion of Ag ions within the glass.

3.3 Glass and interface microstructure

In order to understand which reaction during contact formation is being impacted by Pb and Bi content in the glass, SEM analysis was performed. Metallized cells were etched with HNO_3 to remove the sintered Ag and observe the underlying interfacial glass layer, as well as subsequent etching with HF to remove the glass and observe the microstructure of the wafer surface. Figure 5 shows the interfacial glass of both high-Pb glass A (control sample) and glass E (highest Bi content). The images were created using backscattered electrons, which scatter more effectively from elements with higher molecular weight (MW). It is clear that the sample with high-Pb glass has more areas with high-MW composition in the glass, shown by the brighter regions. In contrast, the sample with Glass E (high-Bi) has fewer regions with high-MW composition in the glass. This could be due to a deficiency in the higher MW components of the glass system at the interface (Pb, Bi, Ag). Fewer Ag colloids in the glass could indicate that Ag is less soluble in the Bi-rich glass. Figure 5E could also suggest that there is less glass at the interface, which would correlate with the DSC and HSM data indicating that the high-Bi glass is less vitreous due to crystallization.

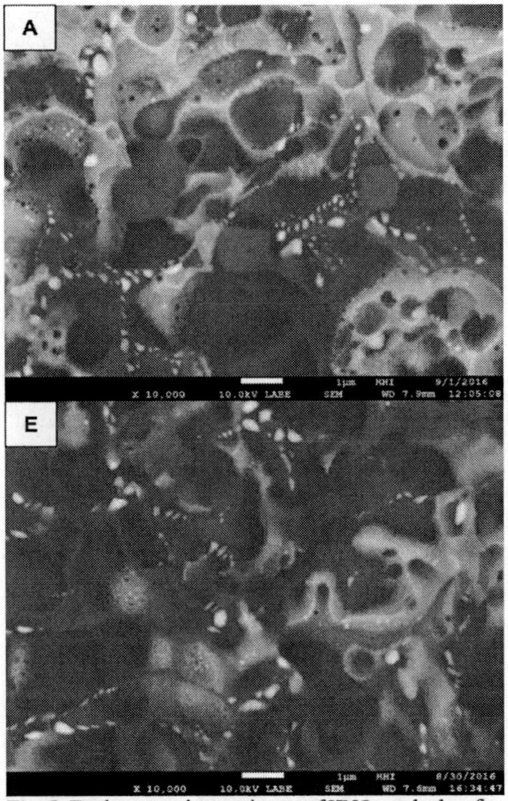

Fig. 5: Backscatter electron image of HNO_3-etched wafers for high-Pb glass (A), and high-Bi glass (E).

Figure 6 shows the surface of the metallized solar cell after both the Ag and glass layers have been removed via etching. Secondary electrons (SE) are emitted by atoms near the surface of a sample, thus the image is more related to the topography and has greater resolution. The SE images show that for both glass A and glass E, the SiNx has been sufficiently etched away to expose the silicon emitter (darker regions on the surface). It also appears that fewer Ag crystallites have grown into the silicon for glass E, suggesting that the Pb:Bi ratio in the glass plays a role in the second redox reaction necessary for contact formation. There could be a difference in the potential for Pb and Bi to help reduce Ag_2O in the glass to form Ag in crystallite or colloid form. However, as it is primarily the oxidation of Si to SiO_2 that facilitates the formation of Ag crystallites on the surface [7], it is unlikely that the Pb:Bi ratio would impact this reaction. Thus it is more likely that the change in chemical composition of glass led to changes in physical behavior impacting glass flow and potentially Ag diffusion to the interface. This could limit Ag crystallite formation and increase the contact resistance at the interface, leading to the drop in FF seen in Figure 4. It is also possible that there is sufficient Ag crystallite density to contact the emitter, but the source of high contact resistance is due to the impediment of other current transport mechanisms. This is supported by the difference in glass morphology evidenced in Figure 5.

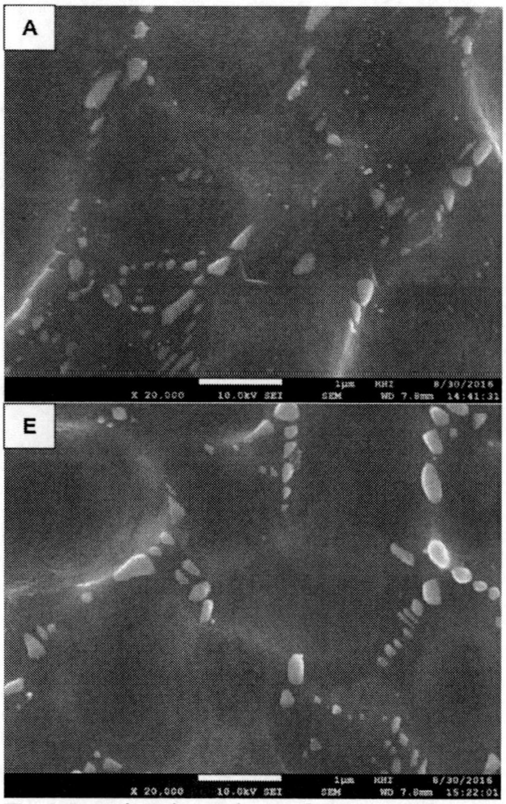

Fig. 6: Secondary electron image of HNO_3- and HF-etched wafers for high-Pb glass (A), and high-Bi glass (E).

4 CONCLUSIONS

Contact formation for crystalline silicon solar cells is a complex process involving redox reactions between a glass frit in the metallization paste and the SiNx/Si surface. While there is a general understanding of the nature of these reactions, it can be difficult to draw conclusions about the glass composition and contact formation. Changing the main conditional glass former (in this case Pb and Bi) can strongly influence both the thermomechanical properties of the glass as well as the contact resistance of the metallization paste. By decreasing Pb content and increasing Bi content, the glass exhibits a stronger tendency to crystallize at lower temperatures, indicated by DSC heat flow behavior. While lower-Pb glasses were still able to sufficiently etch the SiNx ARC layer, they resulted in higher contact resistance despite the fact that Ag crystallites were observed on the emitter surface. The relationship between glass composition and contact formation is critical for optimizing the electrical performance and efficiency of silicon solar cells with screen-printed metallization pastes. Although SEM analysis can help identify the potential root cause of poor contact formation, it is clear that the reaction is complex and there is room for further investigation.

5 REFERENCES

[1] International Technology Roadmap for Photovoltaic (ITRPV) Eight Edition, September 2017.

[2] Schubert, G., Thick film metallisation of crystalline silicon solar cells 2006, Universität Konstanz: Konstanz.

[3] Schubert, G., et al., "Physical understanding of printed thick-film front contacts of crystalline Si solar cells - review of existing models and recent developments," *Sol. Energy Mater Sol. Cells*, **90** (2006).

[4] Hilali, M.M., et al., "Effect of glass frit chemistry on the physical and electrical properties of thick-film Ag contacts for silicon solar cells," *J. Electron. Mater.*, **35** (2006).

[5] Hong, K.-K., et al., "Role of PbO-based glass frit in Ag thick-film contact formation for crystalline Si solar cells," *Met. Mater. Int.*, **15-2** (2009).

[6] Hörteis, M., Fine-line printed contacts on crystalline silicon solar cells 2009, Universität Konstanz: Konstanz.

[7] Fields, J. D. et al., "The formation mechanism for printed silver-contacts for silicon solar cells," *Nat. Commun.*, **7**:11143 (2016).

[8] Shelby, J. *Introduction to Glass Science and Technology*, Cambridge: The Royal Society of Chemistry (2005).

INDUSTRIALLY MCCE TEXTURED MULTI-CRYSTALLINE PERC WITH 20% EFFICIENCY

Zhuo Xu[1], Hongfang Wang[1], Yingchao Wang[1], Feng Li[2], Jinchao Shi[1], Dengyuan Song[2], Dawei Liu[1]

[1] Yingli Energy (China) Co. Ltd,.
[2] State Key Laboratory of Photovoltaic Materials and Technology,
3399 North Chaoyang Street, Baoding, China

ABSTRACT: In this paper, Metal catalyzed chemical etching (MCCE) was introduced into texture of diamond wire saw (DWS) multi-crystalline silicon wafers, which shows better optical effect. Different relectance samples were tested, and opimum reflectance between 15-17% was found. Furthermore, MCCE was combined with PERC structure, and 0.82% efficiency gain was found, 20.04% efficiency was achieved. It is found that the PERC structure performs better with MCCE surface texture than with the conventional acid texture.There is still room for improving, and can be expected for future mass production of multi-crystalline Silicon solar cells.
Keywords: MCCE, texture, PERC

1 INTRODUCTION

In today's industrial cell production, it becomes more and more critical to control the cost of ownership. For this purpose, there are normally two different ways, one is to reduce the material consumption involved in production, and the other is to get a higher efficiency for a lower cost per watt. The cost of wafers cut by diamond wire saw (DWS) can be decreased very much, because more wafers can be produced from a same ingot. The DWS wafers can not be textured by conventional acid texture systme due to less damaged layer. There are 3 texture ways: RIE, MCCE and conventional acid system plus addative. Another way to get better cost of ownership is to improve the cell performence. It is reported that the aluminum oxide film possesses a strong surface passivation effect on boron doped silicon, which corresponds to the rear surface of the conventional p-type solar cell [1-5]. Combined with PERC structure, great efficiency gain can be acheaved[6--8].

In this work, we introduce MCCE technology to make texture of DWS wafers. To futher improve the efficiency of multi-crystalline Si solar, PERC is made on the MCCE textured surface.

2 EXPERIMENTAL DETAILS

2.1 MCCE optizization

The substrate used was 190um-thick DWS multi-crystalline wafers (156.75mm × 156.75mm). The wafses were divided into 5 groups. The etching time was increasing from group 1 to group 5, which resulted in different surface morphology and different reflectivity. Around 95 Ω/sq emitter was formed from POCl3 source by diffusion. After edge isolation and phosphorous silicate glass (PSG) removal, 80nm SiNx film was coated by plasma enhanced chemical vapor deposition (PECVD). Then the metallization was applied and IV data of the cell was tested.

The reflectivity was tested by in integrating sphere reflector. The morphology was measured by SEM.

2.2 PERC structure

Figure 1 shows the PERC cell structure studied in this work. Homogeneous phosphorous emitter is on front side of the boron doped p-type silicon substrate, passivated front SiNx. AlOx is on the rear surface of the substrate to passivate the back side with SiNx capped on for protection. After opening of the rear stack layer with laser, aluminum paste is printed and fired to form local BSF. And front metallization is made for contact.

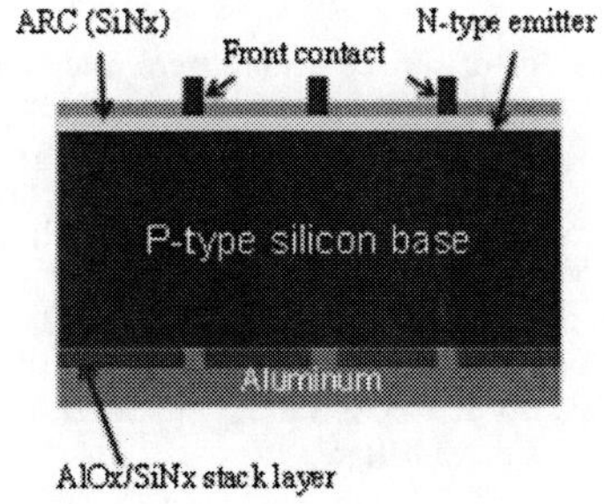

Figure 1: PERC cell structure

2.3 PERC processing

Figure 2 shows the corresponding process flow diagram for the PERC solar cells fabricated in this work. The wafer surface was textured with acid system for slurry cut wafers, and MCCE for DWS wafers. Thermal diffusion of POCl3 was used to form the homogeneous emitter. And the emitter isolation and PSG removal was carried out by wet chemical process. Around 5nm aluminum oxide film was formed by atomic layer deposition (ALD) on the rear side of PERC groups. Then an around 80nm thick silicon nitride layer was deposited on front side as an anti-reflective coating by PECVD. To ensure the passivation effect would not be influenced by aluminum paste, aluminum oxide layer was capped by a thicker silicon nitride film, which formed a AlOx:H/SiNx:H stack layer on rear side. The stack layer then received a line pattern of local contact opening with laser ablation. The wavelength of the laser is 532nm. Afterwords, rear side Al printing and top side Ag printing was applied and these finished PERC cells were fired.

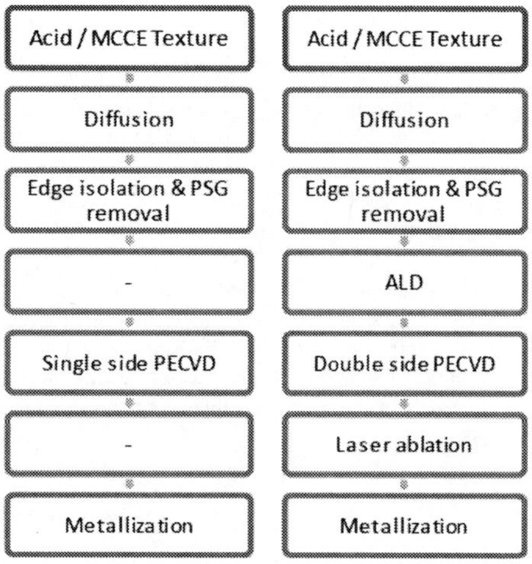

Figure 2: Process flow of conventional cell (left) and PERC (right)

3 RESULTS AND DISCUSSION

3.1 Lifetime

Two kinds of surface texture were made by different MCCE recipes, seen in figure 3. The size of etching pit is around 500nm and 1100nm, respectively. With Silicon Nitride film coated on double side and firing, the samples were measured by Sinton PCT-120. The result is shown in figure 4. The lifetime of larger etching pit is much higher than the smaller one. It can be explained as it is more difficult for PECVD to form SiNx film in small size pitch, which leads to poor passivation. For larger size pit, the passivation is much better.

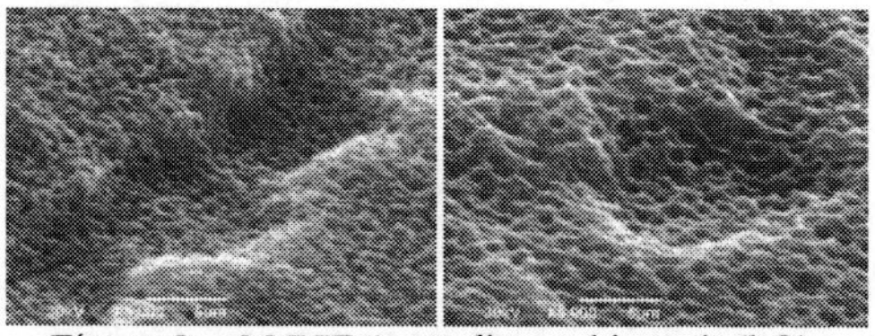

Figure 3: MCCE-1, smaller etching pit (left) and MCCE-2, larger etching pit (right)

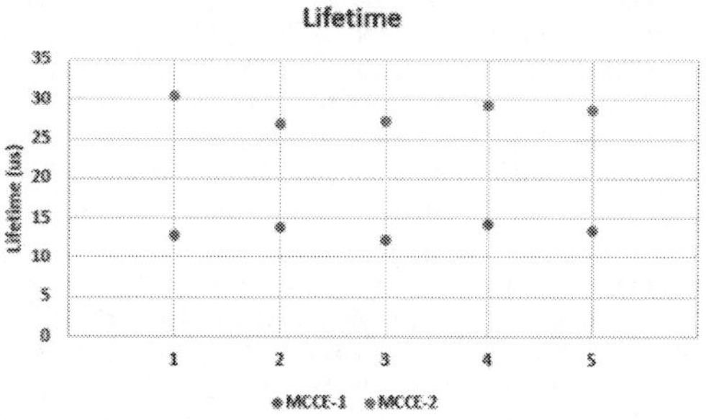

Figure 4: Lifetime of different surface

3.2 Optimization

Figure 5 shows the SEM images of different wafers with different reflectivity. With the reflectivity increasing, the size of etch pits enlarges. It is probably due to the more time etched by the MCCE chemicals.

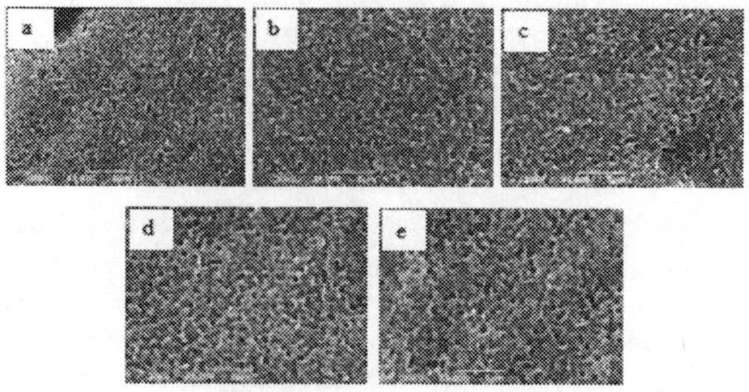

Figure 5: SEM images of MCCE textured wafers with different reflectivity (a)11%, (b) 13%, (c) 15%, (d) 17%, (e) 19%.

Table 1 shows the cell results of different MCCE process. There is a clear increase of Voc from group 11% to group 19%. It is mainly because the size of etch pit is enlarged, which leads to a better passivation. Espetailly, when the size of pit is below 500nm, the passivation become difficult, and Voc drops quickly. The Isc increase from 11% to 15%, but decrease from 17% to 19%. The increasing of Isc is mainly because the passivation gets better, even the reflectance increases. The drop from 17% to 19% is due to lees light trapping. Therefore, there is a balance between the light and passivation. And a optimum reflectance of 15-17% is get.

Table I: IV results of different reflectivity

	Voc (mV)	Isc(A)	FF(%)	Eff(%)
11%	627	8.84	81.16	18.32
13%	635	8.98	81.29	18.86
15%	636	9.11	81.31	19.17
17%	636	9.10	81.19	19.13
19%	637	8.98	81.01	18.88

3.3 PERC combined with MCCE texture

Table 2 shows the result of PERC cells with different surface textures. The efficiency gain from acid textured PERC to acid textured Al-BSF cells is 0.67%, while the gain fromf MCCE textured PERC to MCCE textured Al-BSF cells is 0.82%.

Table 2: IV results of different reflectivity

	Voc (mV)	Isc(A)	FF(%)	Eff(%)
Acid-BSF	632	9.01	80.99	18.75
Acid-PERC	648	9.30	79.21	19.42
MCCE-BSF	638	9.12	81.15	19.21
MCCE-PERC	655	9.38	80.16	20.04

Figure 6 shows both the blue and the red response of MCCE textured PERC cells are better than the acid textured conventional cells. The increase of the blue response is due to much the lower reflectivity in between 350nm and 500nm. More light trapped in the surface leads to a better external quatum efficiency (EQE) in the corresponding wavelength. The gain of red response is from PERC structure. Aluminum Oxide offers an excllent passivation on back surface of the cell. And AlOx/SiNx stack layer performs as a reflector for the long wavelength light between 1050-1200nm, which can be seen in figure 6.

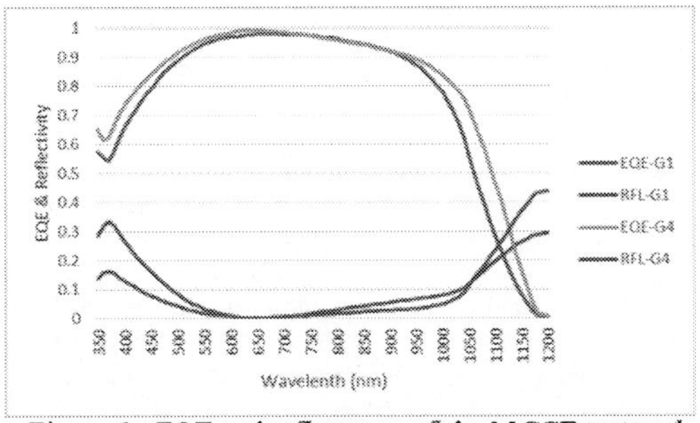

Figure 6: EQE and reflectance of the MCCE textured PERC and acid textured conventional cell

4 CONCLUSION

Different relectance samples were tested, and opimum reflectance between 15-17% of MCCE texture was found. Furthermore, MCCE was combined with PERC structure, and 0.82% efficiency gain was found, 20.04% efficiency was achieved. It is found that the PERC structure perfoms better with MCCE surface texture than with the conventional acid texture.There is still room for improving, and can be expected for future mass production of multi-crystalline Silicon solar cells.

REFERENCES:

[1] E. Cornagliotti, L. Tous, A. Uruena de Castro, 28th European Photovoltaic Solar Energy Conference and Exhibition, (2013) 976.

[2] O. Hegr, P. Cech, M. Kusko, 28th European Photovoltaic Solar Energy Conference and Exhibition, (2013) 2047

[3] Gijs Dingemans, Erwin Kessels, American Vacuum Society. vol. 30,(2012) 040802-1.

[4] X. Gay, F.M.M. Souren, B. Dielissen, 28th European Photovoltaic Solar Energy Conference and Exhibition, (2013) 1220.

[5] P. G. Vermont, V. Kuznetsov, and E.H.A. Granneman, 26th European Photovoltaic Solar Energy Conference and Exhibition, (2011),1644.

[6] A. Li, W.-H. Chen, Y.-M. Lin, 28th European Photovoltaic Solar Energy Conference and Exhibition, (2013), 1867.

[7] B. Vermang, H. GoverdeH, 26th European Photovoltaic Solar Energy Conference and Exhibition, (2011) 1739..

[8] Kapila Wijekoon, Himal Khatri, 27th European Photovoltaic Solar Energy Conference and Exhibition, (2012) 608..

LASER ABLATION INDUCED RECOMBINATION LOSSES OF N-PERT-RJ SOLAR CELLS WITH REAR CONTACTS FORMED BY AL EVAPORATION

Zih-Wei Peng[1], Thomas Buck[1], Jens Theobald[1], Valentin D. Mihailetchi[1], Radovan Kopecek[1]
[1]International Solar Energy Research Center (ISC) Konstanz, Rudolf-Diesel-Str. 15, 78467 Konstanz, Germany
Corresponding author: TEL: +49 7531 3618362; Fax: +49 7531 3618311; Email: zih-wei.peng@isc-konstanz.de

ABSTRACT: In this work, we use photoluminescence (PL) imaging to extract V_{oc} and J_{01} as a function of laser ablation of the rear side dielectric layer of *n*-PERT-RJ solar cell precursors. The linear fit of the experimental data yields different values for the emitter recombination current density according to the ablation patterns and varied laser settings. We quantified the damage on the laser treated surface by extraction of the laser ablation induced emitter recombination current density ($J_{0e\text{-}ablation}$) and found a decrease from 2200 fA/cm² to 780 fA/cm² in our experimental matrix. We also monitored the impact on J_{0e} induced by cleaning and metallization after the laser process. With the quantification of the emitter surface recombination and Optical surface inspection, we improved our *n*-PERT-RJ solar cells with Al-evaporated rear contact to an efficiency of 20.8%.
Keywords: *n*-PERT, Rear Junction, Laser ablation, J_{0e}, Al-evaporation

1 INTRODUCTION

During recent years the ablation of dielectric layers by laser became more and more common, for the production of advanced solar cell structures, (e.g., PERC, PERT, IBC) [1-3]. How to reduce laser ablation induced damage on the Laser Contact Opening (LCO) surface will be a key issue to have a good cell performance. In this work, we focus on *n*-type Si wafers and the concept of Passivated Emitter Rear Totally Diffused (PERT) solar cells, which have the boron emitter at the back side to form a rear junction (RJ). The metallization of the rear side is done by an evaporated PVD-Al layer while front contacts are established by thick film screen printing technique. We adjust different laser settings (e.g., ablation fluence, pulse duration) and patterns (dot and lines) to perform the ablation on the *n*-PERT-RJ's rear capping SiN_X. Evaporated Al metallization disagrees in its needs and capabilities to the traditional screen printing Al paste contact formation. Due to a lower thermal budget and less Al metal-Si interaction [4], the evaporation forms no Si-Al alloy and hence misses the ability to eliminate ablation induced defects as found for Al paste metallization with significant inter-diffusion between Si and Al during the high temperature firing process. For this reason, a conservative laser ablation with a minimum of induced defects at the surface and in the bulk is so important for the considered PVD metallization of the advanced solar cell concepts. We present an experimental approach to extract the recombination current density for n^+-n-p^+ test devices after laser ablation ($J_{0e\text{-}ablation}$), a post ablation clean ($J_{0e\text{-}clean}$) and finally for samples with evaporated Al contacts on top of the opened and cleaned rear side ($J_{0e\text{-}evp}$).

2 EXPERIMENTAL

The *n*-type Passivated Emitter and Rear Totally diffused (*n*-PERT) and rear junction (RJ) cell concept [5] precursors are used as substrates (as shown in Fig. 1) for our investigations.

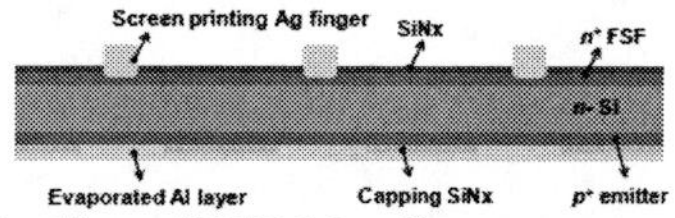

Figure 1: The *n*-PERT-RJ cell precursors as ablation substrate.

The presented experiments have been carried out on a Rofin StarFemto FX laser system with VIS (514 nm) wavelength. A stretcher /compressor unit was used to generate pulse durations of 250 fs and 2 ps. We performed different laser settings to ablate the 150nm thick SiN_X dielectric layer on top of the untexturized boron emitter with 85 Ohm/sq sheet resistance with five different ablation fractions ($F_{ablation}$) in one test sample (Fig. 2). After dielectric film ablation, the images are measured by photoluminescence (PL) technique and are transcoded first to spatial resolved and subsequently to integrate implied open-circuit voltage (V_{oc}) images [6]. With the V_{oc} values, we can extract J_{01} from the one-diode model given by

$$V_{OC} = V_T \cdot \ln (J_{sc} / J_{01}) \qquad (1)$$

to quantify all recombination processes in the bulk and in the emitter volume [7].

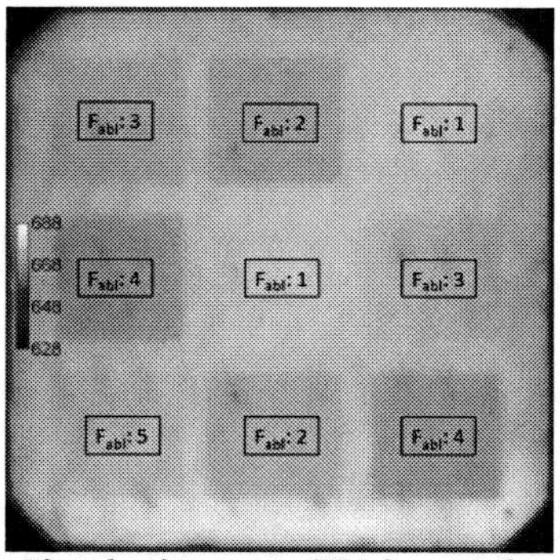

Figure 2: Photoluminescence (PL) image of test sample with different laser ablation fractions.

According to the one-diode model, V_T represents the thermal voltage at 25 °C and J_{sc} is the short circuit current density. Equation (1) is only valid, if losses related to the second diode with an ideality factor of n = 2 are negligible at V_{OC}. The dark saturation current density J_{01} (eq. 2) of the ablated device is increased when $F_{ablation}$ is raised.

$$J_{01} = J_{0b} + J_{0e} \cdot (1 - F_{ablation}) + J_{0e\text{-}ablation} \cdot (F_{ablation}) \qquad (2)$$

By fitting of equation (2) to the J_{01} - $F_{ablation}$ graphs, the recombination current density of the ablated boron emitter (J_{0e}) can be extracted from the slope that is directly correlated with the damage induced by different laser ablation settings and patterns. The so called $J_{0e\text{-}ablation}$ then gives us direct information about the damage induced by the laser process.

The data was extracted at 3 different process stages from the samples. Apart from measurements directly after laser ablation an additional examination was done after a post ablation cleaning given by a 60 seconds 2% HF/HCl dip treatment to judge the recombination current density induced by the cleaning step ($J_{0e\text{-}clean}$). To validate the desired removal of the dielectric film, microscopic images are taken by Laser Scanning Microscopy (LSM) for optical inspection after both steps. A final PL imaging procedure was taken after Al-evaporation on the ablated area and we applied the same method to extract again the change in recombination current density of the samples which induced by the metallization at the emitter–interface ($J_{0e\text{-}evp}$)."

3 RESULTS AND DISCUSSION

3.1 Laser ablation induced recombination current density

Table I recaps the laser ablation induced emitter recombination current density ($J_{0e\text{-}ablation}$) for different laser settings. The data is extracted from the slope of the J_{01} vs. laser ablation area fraction in Fig. 3. We found a

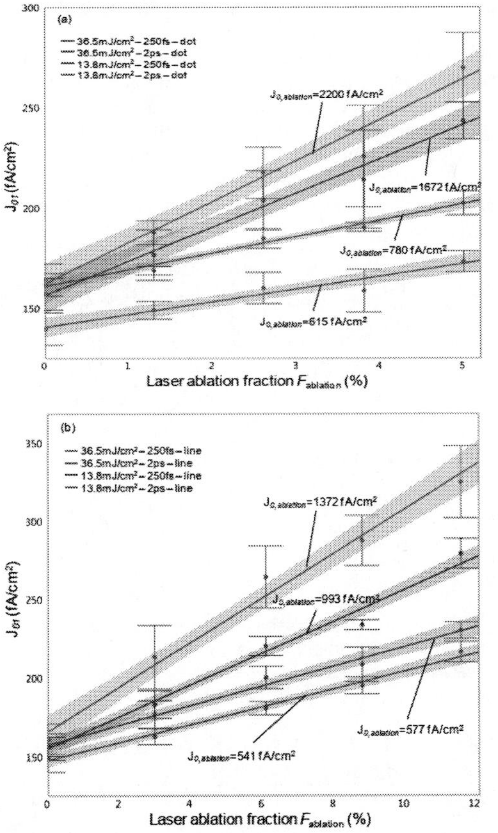

Figure 3: J_{01} as a function of laser ablation area fraction with different laser settings. (a) dot ablation, (b) line ablation.

minimum for $J_{0e\text{-}ablation}$ representing less ablation induced recombination at the lower ablation fluence. Optical observations and qualitative comparison of various LSM images as depicted in Fig. 4 support this result. The application of higher ablation fluence to the devices (Fig. 4a and 4b), results in dark color regions in the center of the ablation area and indicates the formation of residuals [8]. The residuals were attributed to remaining SiN_X which is modified in its structure but not removed due to transition from one- to two-photon absorption and consequential too low fluence for direct evaporation [8]. Residuals occur only for laser fluencies clearly above the ablation threshold and are independent of the pulse duration in the tested range from 250 fs to 2ps [9]. The residual coverage is elevated for increasing pulse energy.

Table I: Extraction of the rear emitter recombination current density with different laser settings and after different process steps.

ablation shape	dot				line			
ablation fluence (mJ/cm²)	36.5		18.3		36.5		18.3	
pulse duration	250 (fs)	2 (ps)	250 (fs)	2 (ps)	250 (fs)	2 (ps)	250 (fs)	2 (ps)
$J_{0e\text{-}ablation}$ **(fA/cm²)**	2200	1672	780	615	1372	993	577	541
$J_{0e\text{-}clean}$ **(fA/cm²)**	168	195	196	176	186	167	190	168
$J_{0e\text{-}evp}$ **(fA/cm²)**	118	142	145	126	135	105	155	122

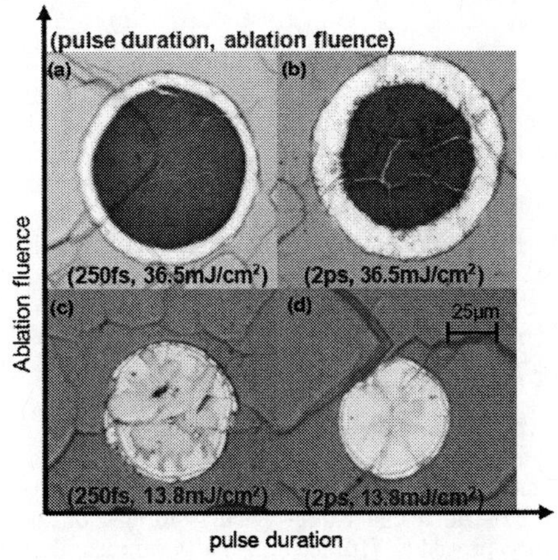

Figure 4: The LSM images of dot ablation with different laser settings.

Without the residual effect, selected ablation fluence (F_{abl}) should be above the ablation threshold fluence (F_{th}) and below the residual threshold fluence (F_{res}) [8]. The laser ablation induced emitter recombination current density ($J_{0e\text{-}ablation}$) can be reduced from 2000 to 780 fA/cm² with dot ablation pattern when changing ablation fluence from 36.5 mJ/cm² to 13.8 mJ/cm². As a consequence the V_{oc} of the samples is increased by 3 to 10 mV (Fig. 5) according to different ablation area fractions.

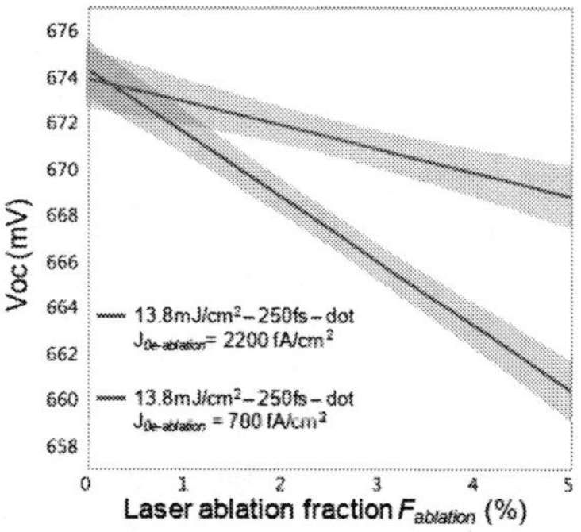

Figure 5: V_{OC} measured by PL with different ablation induced recombination current density.

Line shape ablation features lower $J_{0e\text{-}ablation}$, for all laser settings compared to dot shape opening. Line shape ablation, is formed by overlap of the dot procedure. Less residual are found [9] and the SiN_X is removed more clearly than for dot shape ablation as can be seen by comparison of LSM images in Fig. 4 and Fig. 6.

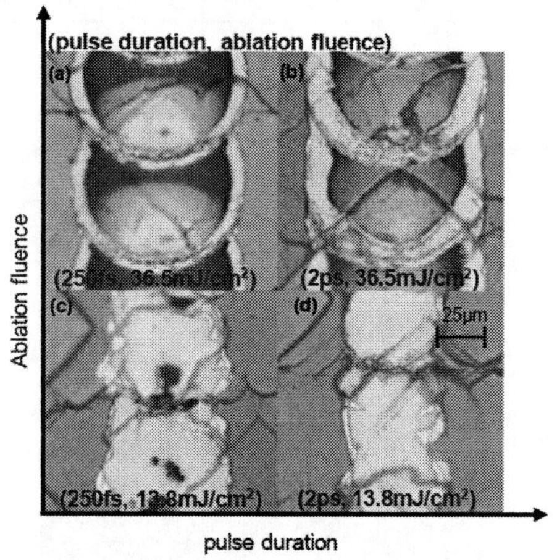

Figure 6: The LSM images of line ablation with different laser settings.

3.2 Post ablation cleaning induced recombination current density

After performing the post ablation clean, it was found that for all groups the J_{0e} was increased around 180 fA/cm^2 in average (Table I). The effect is supposed to occur due to a thin oxide which is formed during the ablation procedure [10] leading to a lower recombination because of its passivation quality. The layer would be removed then during the HF cleaning step and elevates the recombination. We name this increase post ablation cleaning induced recombination current density ($J_{0e\text{-}clean}$).

3.3 Al-evaporation induced recombination current density

We monitored the emitter recombination current density (J_{0e}) continuously. After the post ablation cleaning process a 3µm thick Al layer was evaporated on the opened test structure. With the same extraction approach, it was found a 130 fA/cm^2 constant elevation in average (Table I). The term "$J_{0e\text{-}evp}$" has been introduced to indicate, the recombination current density caused from evaporated Al at the contact area.

3.4 Cell performance with different LCO pattern designs

To understand how the emitter recombination current affects the performance on a solar cell device level, we design two different LCO patterns on *n*-PERT-RJ solar cells. In this experiment the *n*-PERT-RJ solar cells are with front side screen printing Ag finger and rear fully evaporated Al contact. We use VIS (514 nm), 250 fs pulse duration and 13.8 mJ/cm^2 fluence to perform ablation of the dielectric above the rear boron emitter with dot (450 µm pitch) and line (750 µm pitch) LCO pattern designs. Based on the emitter saturation current density extraction approach in this paper, we found $J_{0e\text{-}total}$ for dot and line ablation of 1121 fA/cm^2 and 922fA/cm^2, respectively. The $J_{0e\text{-}total}$ is the summation of $J_{0e\text{-}ablation}$, $J_{0e\text{-}clean}$ and $J_{0e\text{-}evp}$. The V_{OC} difference between these two LCO pattern designs predicted by the recombination current density extraction method is around 8mV. It matches well the V_{OC} difference identified in the I-V measurement (Fig.7 and Table II). Apart from the emitter recombination current density, we also need to consider the effective opened area fraction to determine the best processing conditions for solar cell devices. For dot-LCO we saw a higher recombination current density per unit ablated area, but due the 450µm pitch pattern design on solar cells, the total ablation area fraction for each solar cell is around 0.6%. This percentage is enough to reach a comparable fill factor (FF) as a line-LCO pattern design with an 8% opening of the dielectric (Table II). In total the combination of $J_{0e\text{-}total}$ and F_{abl}, (the ablation area weighting effect), results in a 8mV superior V_{OC} for the dot-LCO pattern on solar cell devices. .

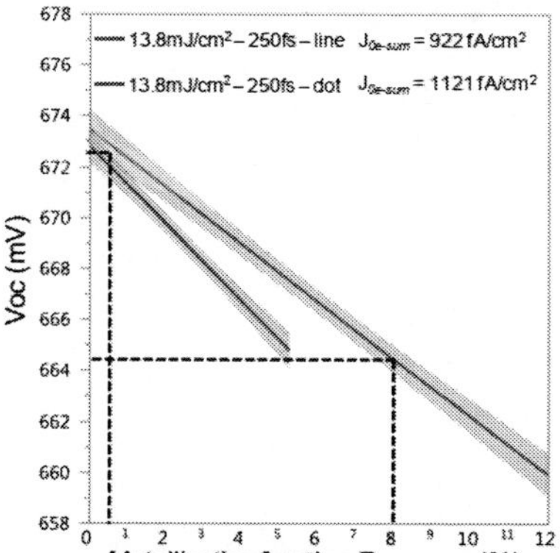

Figure 7: Calculated Voc difference between two different ablation pattern designs.

Table II: I-V of *n*-PERT-RJ cell with evaporate-Al rear contact but different LCO design

F_{abl} (%)	LCO-shape	J_{SC} (mA/cm^2)	Voc (V)	FF (%)	Eta (%)
0.6	dot	38.8	0.661	77.6	19.9
8	line	38.7	0.653	77.7	19.6

4 SUMMARY

The goal of this work was to quantify the emitter recombination current density at the laser ablation surface, and the interface between evaporated Al and emitter. By varying the laser ablation fractions on the PL measurement samples, we can monitor the change of the emitter recombination current density step by step. Three main processes for rear contact formation (laser ablation, post ablation clean and Al evaporation) are focused, and after extraction from the J_{01} - $F_{ablation}$ graphs, we can quantify the process induced recombination current density. Laser ablation induced defects dominate the recombination current density (Fig. 8); the $J_{0e\text{-}ablation}$ decreased from 2200 to 780 fA/cm^2 and is highly correlated to the ablation fluence settings (from 36.5mJ/cm^2 decreased to 13.8mJ/cm^2).

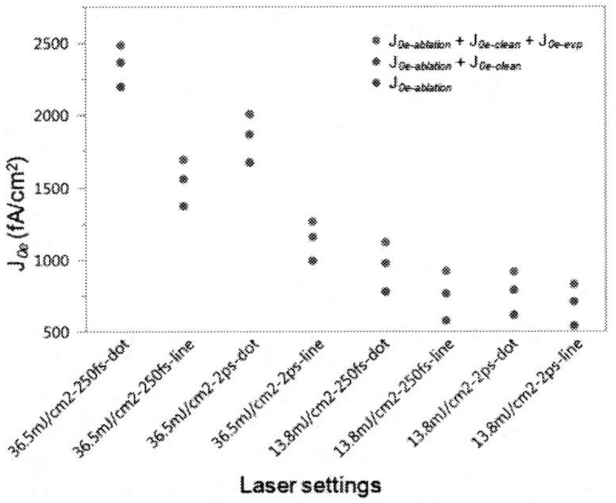

Figure 8: Summary of rear emitter recombination current with different laser settings and after ablation, cleanin, or evaporation processes.

The additional $J_{0,e}$ induced by post ablation clean and Al-evaporation are 180 and 130 fA/cm^2 respectively. We have been working on the laser ablation process development for *n*-PERT-RJ cells with rear fully Al evaporated contact. After modifications of the ablation fluence (13.8 mJ/cm^2) and contact pattern design (1% point ablation fraction) we achieved with new *n*-PERT-RJ solar cell precursors so far 20.8% efficiency on 156mm-large industrial Cz substrates (Table III).

Table III: I-V of *n*-PERT-RJ solar cells with full area Al-evaporation on the locally opened passivation area & front Ag screen printing.

	Jsc (mA/cm^2)	Voc (V)	FF (%)	Eta (%)
mean	38.7	0.674	79.3	20.71
best cell	38.7	0.675	79.6	20.82

5 ACKNOWLEDGEMENT

This work was funded by the BMWi within the IdeAl project (Industrielles PVD-Al für hocheffiziente kristalline Si-Solarzellen und Module) under project number 02E2-41V7142.

6 REFERENCES

[1] A. Uruena De Castro, "Advanced Rear-Side Contact Schemes on i-PERC (industrial Passivated Emitter and Rear Cell) solar cells," Ph.D thesis, 2013.

[2] Y. Tao et al., "20.7% efficient ion-implanted large area n-type front junction silicon solar cells with rear point contacts formed by laser opening and physical vapor deposition," Prog. Photovolt: Res. Appl., 22, pp. 1030–1039, 2014.

[3] M. Rienäcker et al., "Recombination Behavior of Photolithography-free Back Junction Back Contact Solar Cells with Carrier-selective Polysilicon on Oxide Junctions for Both Polarities," Energy Procedia 92, pp. 412-418, 2016.

[4] J.O. McCaldin, et al., "Diffusivity and Solubility of Si in the Al Metallization of Integrated Circuits," Appl. Phys. Lett. 19, pp. 524-527, 1971.

[5] Z. Peng, et al., "Investigation of laser contact opening pitch affecting the void formation on n-Type rear emitter silicon solar cells," Energy Procedia, 124, pp. 665–670, 2017.

[6] T. Trupke, et al., "Spatially resolved series resistance of silicon solar cells obtained from luminescence imaging," Appl. Phys. Lett., 90, 093506, 2007.

[7] W. Shockley, "The theory of p–n junctions in semiconductors and p–n junction transistors," Bell Syst. Tec. J. 28, pp. 435–489, 1949.

[8] T. Wütherich et al., "investigation of laserablation of different dielectric layers with ultra short pulses", Energy Procedia, 27, pp. 537-542, 2012.

[9] J. Theobald, et al., "Study on surface residual phenomena for sub-picosecond laser ablation on dielectrics for perc solar cells," 29th EUPVSEC, 2014.

[10] Z. Du, et al., "Impact of Laser-Induced Oxidation on Silicon Wafer Solar Cells' Performance," IEEE Journal of Photovoltaics, 6(3), pp. 617-623, 2016.

Novel wet chemical cleaning concepts for high efficiency silicon solar cells

M. Haslinger[1*], M. Soha[3], S. Jambaldinni[1], A. Hajjiah[2], J. Szlufcik[1], J. Poortmans[1,4,5] and J. John[1]
[1]IMEC, Leuven, Belgium, [4]KU Leuven, Leuven, Belgium, [5]UHasselt, Diepenbeek, Belgium
[2]Kuwait University, College of Engineering and Petroleum, Electrical Engineering Department, Safat, Kuwait
[3]University of Debrecen, Debrecen, Hungary
*Corresponding author: Phone: +32-16-28 1256; Fax: +32-16-28 1501; e-mail: michael.haslinger@imec.be

ABSTRACT: We report on the progress of wet chemical process development for silicon surface cleaning of n-type passivated emitter with rear totally diffused (n-PERT) solar cells. An Ozone based wet chemical cleaning process sequence has been developed which provides both, improved effective minority carrier lifetime and reduction in cost of ownership.

The proposed cleaning solution (i^{2}PV: imec industrial **P**hoto**V**oltaic Clean), is a mixture of ozonated DI water (DIW/O3) and diluted Hydrochloride and Hydrofluoric acid (HCl/HF). The silicon surface clean is performed in a single tank and provides a hydrophobic or hydrophilic silicon surface after the clean depending on the ozone concentration. In order to quantify the cleaning efficiency of the solution we have determined surface contamination concentration of prominent transition metals before and after wet chemical cleaning. We also determined the impact on solar cell performance by measuring effective minority carrier lifetime in large area 156x156 mm^2 commercial n-type cz-Silicon, 3-5 Ohm.cm, 180um thickness, after a high temperature process and passivation with thermal SiO_x. We benchmarked the cleaning efficiency of the solution by determining the spatial distribution of the effective minority carrier lifetime using photoluminescence measurements. The impact of pre-treatment of the silicon surface has been investigated.

The low cost of ownership is mainly determined by the low concentration of the involved chemicals (HCl and HF less than 2%), the ozone that is switched on and off before and after the cleaning process, the application at room temperature and the short processing time of only 2-5min. Despite the low chemical concentrations the solution shows a highly effective impurity removal quality. Metal removal efficiency is determined and effective minority carrier lifetime is extracted on passivated wafers. As a novel parameter a measure for the spatial distribution of the effective minority carrier lifetime is introduced on large area Cz-Si.

The uniformity of the lifetime distribution has been defined by measuring the color distribution in the PL spatial image using the histogram data for red in the RGB bitmap. This provides us a scale from 50 to 75. In this case 75 represents a very good uniformity and 50 a rather poor uniformity. The minority carrier lifetime distribution uniformity is an important parameter for potential solar cell efficiency losses. A uniformity vs effective minority carrier lifetime graph is depicted resulting from a parameter sensitivity screening design of experiment (DoE). There is a clear trend towards hydrophobic pre-treatment in both, minority carrier lifetime and uniformity, respectively.

The large process window, the low chemical consumption and the high metal impurity removal efficiency makes this solution an interesting choice for silicon surface cleaning in solar cell production.

Keywords: wet chemical cleaning, ozone, hydrofluoric acid, effective minority carrier lifetime, lifetime uniformity distribution, Metal removal efficiency

1 INTRODUCTION

Silicon solar cells based on crystalline silicon (c-Si) are the dominating technology in photovoltaics (PV) industry and have a market share of more than 85% of the modules produced for roof top installations [1]. The PV Industry continues to demand even lower cost per watt solar cells so that PV electricity can ultimately compete with conventional sources of energy. One viable approach is to increase the conversion efficiency and simultaneously reduce the manufacturing cost of the device [2]. The need for cheaper levelized cost of PV-originated electricity (LCOE), i.e. cost per KWh, translates in a need for cheaper solar cells with an increase in conversion efficiency [3]. In order to reach this goal, novel cell concepts and novel fabrication methods [4] have to be developed for implementation in industrial process lines [5]. In this context, wet chemical cleaning processes should also evolve towards easy to implement, well controlled, and cost effective steps [6-7].

Process development using wet chemical processing is triggered by two main cost driven objectives: first, reducing the amount of chemicals to the minimum needed to reach the specified performance target and second, reducing the processing time to increase the throughput required in mass fabrication. On the performance side the requirements are, first, a high surface metal removal efficiency and, second, a homogeneous silicon surface cleaning and conditioning. In order to reach the first cost objective, low cost chemicals with increased chemical bath lifetimes[8] and reduced chemical consumption by using highly diluted solutions were applied. Moreover, in order to reach the second cost objective, it is essential to reduce the process time for the sequential wet chemical steps by means of optimizing the wet chemical processes themselves. The performance of a wet chemical cleaning can be benchmarked by two figures of merit. The capacity to remove metal impurities effectively from the silicon surface (surface metal removal efficiency) and the effective minority carrier lifetime which is a measure for the remaining metal contamination on the surface and in the bulk of the silicon material.

2 EXPERIMENTAL APPROACH

2.1 Clean/Etch mechanism

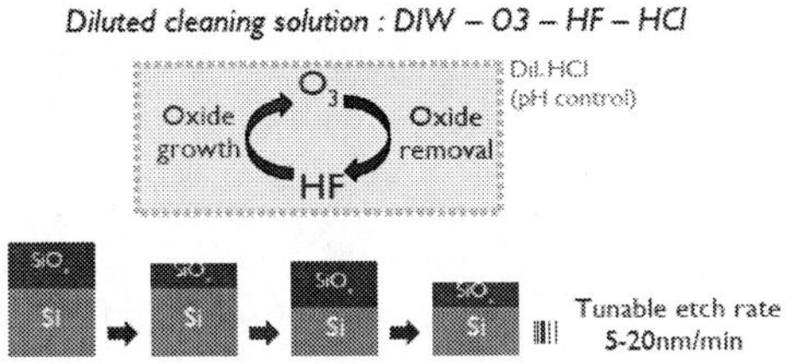

Fig.1:Schematic drawing of the clean/etch mechanism of the i²PV solution on silicon.

In fig.1 the clean/etch mechanism of the i²PV clean is schematically depicted. A simultaneous process of oxidation of the silicon surface by ozone and the removal of the oxide by HF is proposed. Silicon material is consumed and the silicon surface is etched with a tunable etchrate between 5 to 20nm/min.

2.2 Effective minority carrier lifetime

In order to determine the potential impact on solar cell efficiency the effective minority carrier lifetime of the solar cell wafers is extracted using PL measurements. The process flow for this procedure is depicted in figure 2. After removing the saw damage the wafer surface is pre-conditioned either hydrophilic with DIW/O3 or hydrophobic with HCl/HF. Then the i²PV clean is applied using different process parameters shown in table 1. The wafers are passivated with SiOx using a wet oxidation process at 950C to mimic the worst case scenario that surface impurities diffusing into the silicon bulk when applying a high temperature process.

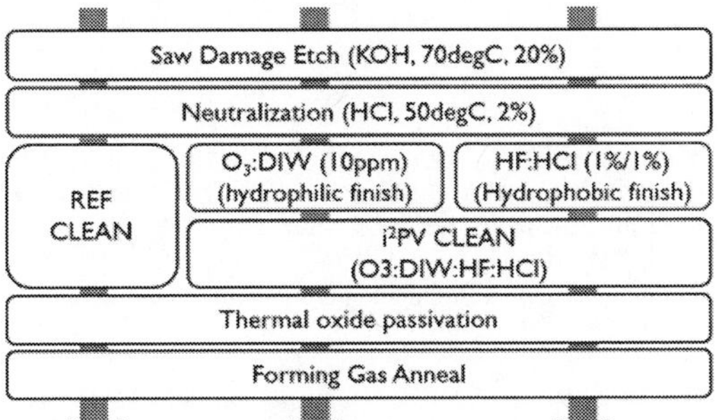

Fig.2:Process flow for the evaluation of the impact of the pre-treatment on the i²PV cleaning results.

All the results are benchmarked against the reference process which is a two tank clean of subsequent DIW/O3 and HCl/HF.

3 RESULTS AND DISCUSSION

3.1 Metal removal efficiency

The surface metal removal efficiency of the cleaning solution is tested using the controlled contamination procedure. Silicon wafers are exposed to well defined concentrations of different transition metals in an acidic solution with the aim that the metals are outplated from the solution to the silicon surface. The surface concentration of the outplated metals is determined by TXRF measurements. Fig.3 depicts the metal surface concentration of prominent transition metals before and after cleaning.

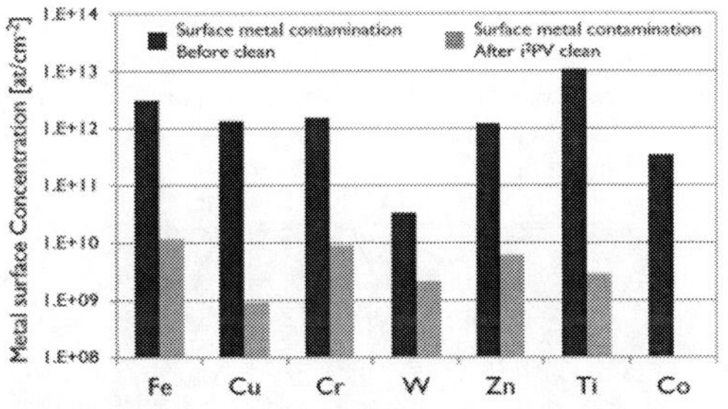

Fig.3: Metal removal efficiency of i²PV for different transition metals. The initial silicon surface concentration is given in black. The silicon surface concentration after the cleaning with i²PV is depicted in grey. Note: for Co the surface concentration after the clean was below the detection limit of 1e8/cm².

3.2 Impact of the pre-treatment

Fig.4 shows the impact of the pre-treatment on the effective minority carrier lifetime. Hydrophilic wafer showing a very narrow distribution and values around 2ms. The distribution for the hydrophobic case is larger but lifetime values up to 3.5ms have been measured with an average of 2.5ms (5 samples per group). The reference clean, sequential DIW/O_3 and HCL/HF was measured at $\tau_{eff} < 1$ms (3 samples).

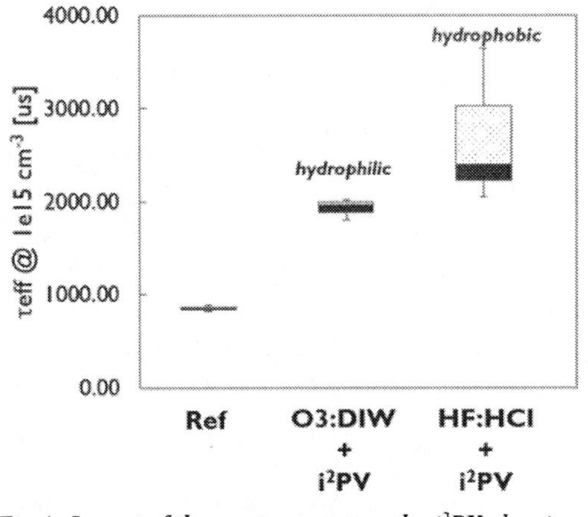

Fig.4: Impact of the pre-treatment on the i²PV cleaning. Hydrophobic pre-treatment reaches lifetimes up to 5ms.

The difference in effective minority carrier lifetime between the two different cleaning concepts iPV and i²PV is explained by the sub-surface etching behaviour of the i²PV clean. Metal impurities located shallow under the surface are removed by the i²PV clean.

3.3 Spatial distribution uniformity of the effective minority carrier lifetime

To extract the uniformity data the wafers have been processed as described in the process flow in fig.2 The uniformity of the lifetime distribution has been defined by measuring the color distribution in the PL spatial image extracting the histogram pixel data for the red color (high lifetime data) in the 256 RGB bitmap normalized to the full RGB spectrum. This provides us a scale from 50 to 75 in arbitrary units (a.u.). In this case 75 represents a very good uniformity and 50 a rather poor uniformity as depicted in fig 5.

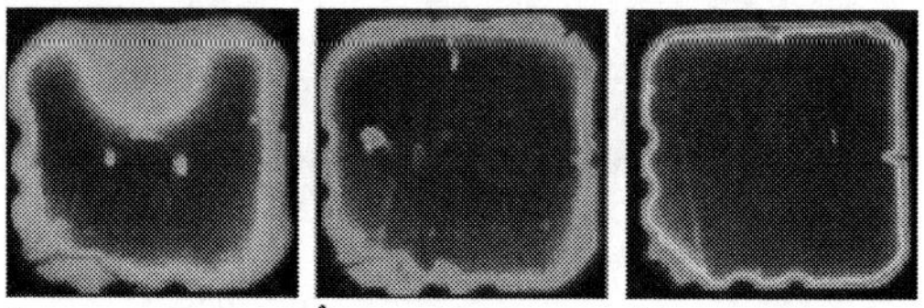

Fig.5: Effective minority carrier lifetime uniformity of the high lifetime part (red)
a. 50, avg(τ_{eff})=2.5ms @1e15/cm^3
b. 65, avg(τ_{eff})=3ms @1e15/cm^3
c. 70, avg(τ_{eff})=4ms @1e15/cm^3

3.4 Cleaning process window (parameter sensitivity screening)

In a design of experiment (DoE) the sensitivity of the novel clean on the applied process parameters has been investigated (parameter sensitivity screening). The screened process parameters are indicated in table 1.

Chemicals	DIW:O_3:HF:HCl
Process time	2-5min
HF conc.	0.5% – 2%
pH	0.5 – 3
O3 conc.	5 – 20ppm
Temp.	RT – 35degC

Table 1.: Process parameters for i²PV clean. Note: the HCl concentration determines the pH of the solution.

The result of the sensitivity screening design of experiment is depicted in figure 6. Per group 4 wafers have been processed, the marker in the graph shows the average value per group. We can conclude that

- All groups show average effective minority carrier lifetime of >2ms
- Best average effective minority carrier lifetime up to 5ms have been measured
- The spatial lifetime distribution uniformity is higher than 60 out of 75 (except of one group)
- There is a trend that hydrophobic pre-treated wafers have a higher lifetime and a higher uniformity (blue circle)

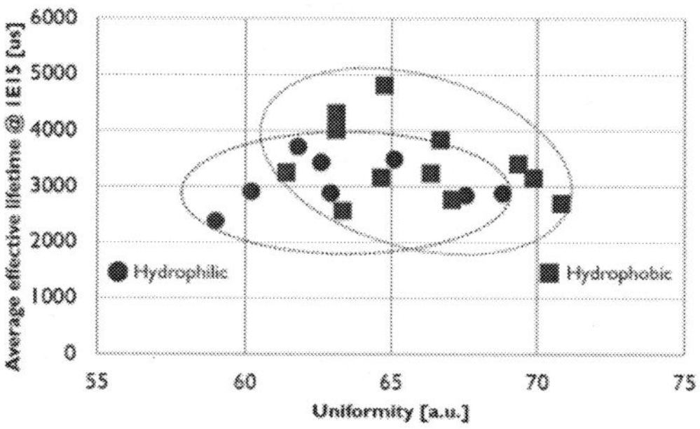

Fig.6: Effective minority carrier lifetime vs uniformity of the lifetime distribution measured on 156x156mm², n-type Cz-Si, 3-5 Ohm.cm wafer. The different marks correspond to different process conditions. The blue color represents hydrophobic pre-treatment of the silicon surface, the red color represents hydrophilic pre-treatment.

4 CONCLUSION AND OUTLOOK

We have investigated a wet chemical cleaning solution with low cost of ownership potential that can provide state of the art figure of merits when applied on commercial available n-type cz-silicon material.

The metal removal efficiency of prominent transition metals has been shown to exceed 99%.

Effective minority carrier lifetime up to 5ms@1e15/cm^3 injection level (1-3 Ohm.cm, cz-silicon, 156x156mm^2) has been measured and a measure for spatial lifetime distribution uniformity has been introduced.

A parameter sensitivity screening DoE has been performed and a large process window for the involved process parameters could be concluded.

This, together with the potentially low cost of ownership, due to extended tank lifetimes and low concentration of process chemicals, makes the proposed i²PV clean an interesting candidate for application in silicon solar cell production lines.

5 ACKNOWLEDGEMENT

The authors like to acknowledge the funding agencies and the industrial partners of the imec Industrial Affiliation Program in SiPV, and KFAS under project code: P115-15EE-01

6 REFERENCES

[1] Jäger-Waldau, A. JRC Scientific and Policy Reports- PV Status Report 2012 (2012).
[2] C. T. M. International Technology Roadmap for Photovoltaics- Results 2012. ITRPEV. net 2 (2012).
[3] Verlinden, P. et al., presentation for the 29th EUPVSEC (2013)
[4] Recamán Payo, M. et al., ,Proceedings of the 30th EUPVSEC (2014)
[5] John, J. et al., EPJ Photovolt., 3 (2012), p. 35005
[6] Rentsch, J. et al., Proceedings of the 25th EUPVSEC (2009), p1113
[7] Schweckendiek, J. et al., Proceedings of the 11th UCPSS (2012), p283
[8] Haslinger, M. et al., Proceedings of the 13th UCPSS (2016), p344.

SUITABILITY OF $POCL_3$ DIFFUSION PROCESSES WITH IN-SITU OXIDATION FOR FORMING LASER-DOPED SELECTIVE EMITTERS WITH LOW CARRIER RECOMBINATION

S. Werner, E. Lohmüller, J. Weber, A. Wolf
Fraunhofer Institute for Solar Energy Systems ISE, Heidenhofstr. 2, 79110 Freiburg, Germany
Phone: +49 761 - 4588 5049; e-mail: sabrina.werner@ise.fraunhofer.de

ABSTRACT: Laser-doped selective emitters feature the advantage of more effective shielding of minority charge carriers from the metal contacts while allowing for low emitter saturation current density j_{0e} in the photoactive area. The formation of emitters by diffusion processes using phosphorus oxychloride ($POCl_3$) with incorporated in-situ oxidation gains more and more attention as it allows for low j_{0e} below 90 fA/cm² on textured surface with silicon nitride passivation in industrial cycle times. Hence, the combination of both—$POCl_3$ diffusion with in-situ oxidation and laser doping—is very interesting. We examine four different $POCl_3$ diffusions with in-situ oxidation and one reference $POCl_3$ diffusion without in-situ oxidation in terms of their suitability for selective emitter laser doping. Detailed characterizations of the grown layers on the silicon surface are performed after diffusion with respect to the individual layer thicknesses of the phosphosilicate glass (PSG) and the intermediate silicon dioxide (SiO_2) layer as well as the stacks' total phosphorus doses. The as-diffused depth-dependent charge carrier concentration profiles show that a second PSG deposition step attached after drive-in to the diffusion processes hardly impacts their course. We find that $POCl_3$ diffusions with in-situ oxidation—especially those with second deposition step—allow for effective laser doping. Thereby, the intermediate SiO_2 layer thickness plays a key role: the thicker the layer is the less phosphorus can be incorporated additionally from the PSG layer into the silicon.
Keywords: in-situ oxidation, $POCl_3$ diffusion, selective emitter, laser doping, phosphosilicate glass

1 INTRODUCTION

Tube furnace diffusion processes using phosphorus oxychloride ($POCl_3$) as liquid dopant precursor are the dominating emitter formation technology for p-type silicon solar cells [1]. Improving $POCl_3$ diffusion processes and the resulting emitter doping profiles is essential for further increasing the energy conversion efficiency of silicon solar cells. The reduction of emitter recombination, enabled by a fast progress in front contact screen printing paste properties, has been of major interest in the past few years.

An approach to realize less emitter recombination is the reduction of the surface doping concentration by implementing an in-situ oxidation into the $POCl_3$ diffusion process [2–9]. When applied to passivated emitter and rear solar cells (PERC) [10], energy conversion efficiencies beyond 21% are reported [6,11,12]. As the charge carrier recombination at the metal contacts is high for that kind of doping profiles with low surface doping concentration [13], one option for its decrease is the selective emitter approach.

Recently, selective emitters came back into focus for PERC solar cells targeting the 22% efficiency regime and above [6,11,12,14,15]. The selective emitter can be formed by, e.g., local laser doping out of the phosphosilicate glass (PSG) layer [16]. Correctly, $POCl_3$ diffusions form a stack layer consisting of a PSG layer and a silicon dioxide (SiO_2) layer: the SiO_2 layer separates the PSG layer from the silicon surface [9,8,17,18]. Hereby, the SiO_2 layer features a much lower phosphorus concentration than the phosphorus-rich PSG layer.

For laser doping, sufficiently high phosphorus content within this PSG/SiO_2 stack layer is necessary. Hence, this paper investigates the combination of tube furnace $POCl_3$ diffusion with in-situ oxidation and the formation of selective emitters by laser doping out of the grown PSG/SiO_2 stack layer.

2 APPROACH

The approach pursued in this paper to ensure sufficient phosphorus content in the PSG/SiO_2 stack layer for effective laser doping is the implementation of a second deposition step with active nitrogen (N_2) gas flow through the $POCl_3$ bubbler (i.e. active N_2-$POCl_3$ flow) at the end of the diffusion process.

We investigate two pairs of $POCl_3$ diffusion processes that feature an in-situ oxidation and that are performed with and without a second deposition step, respectively. $POCl_3$ diffusion without in-situ oxidation but second deposition step serves as reference. Table I summarizes the examined $POCl_3$ diffusion processes. Process details and the labeling of the processes are discussed in detail in section 2.1.

The schematic experiment plan in Fig. 1 depicts the three groups of test samples fabricated and the different characterization approaches applied in this work. Full-square p-type Czochralski-grown silicon (Cz-Si) wafers with an edge length of 156 mm serve as starting material. After either saw damage etching or alkaline texturing, the five different tube furnace $POCl_3$ diffusion processes form the emitter and the PSG/SiO_2 stack layer.

The different $POCl_3$ diffusion processes, the applied

Table I: Summary of the five investigated $POCl_3$ diffusion processes. The emitter dark saturation current densities j_{0e}, determined on alkaline textured surface after passivation by a conventional SiN_x layer and firing, are taken from Refs. [4,11,13]. (n.d.: not determined)

$POCl_3$ process	1st dep. (N_2-$POCl_3$)	drive-in (O_2 share)	2nd dep.	j_{0e} (fA/cm²)
Ref 2nd dep	med	< 4%	yes	42
in-situ thin	low	100%	-	73
in-situ thin 2nd dep	low	100%	yes	n.d.
in-situ thick	high	100%	-	85
in-situ thick 2nd dep	high	100%	yes	n.d.

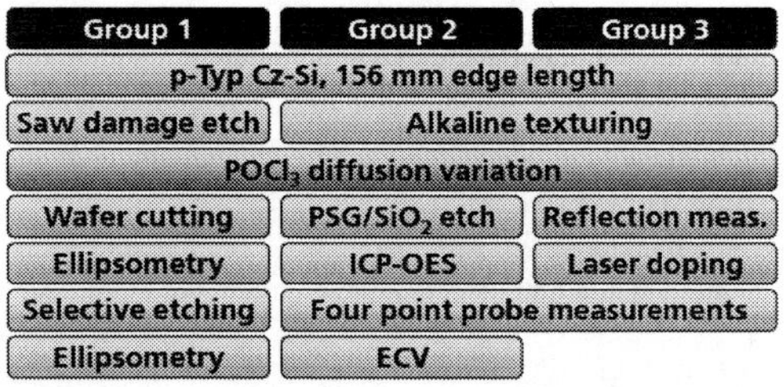

Figure 1: Experiment plan for investigating laser doping out of the SiO_2/PSG stack layer formed by different $POCl_3$ diffusion processes. (ECV: electrochemical capacitance-voltage measurements; ICP-OES: inductively coupled plasma optical emission spectrometry).

further process steps, and the utilized characterization methods are being discussed in the following sections.

2.1 $POCl_3$ diffusion processes

We target at low content of inactive phosphorus at the surface in the as-diffused emitter and thus, low charge carrier recombination in the homogeneously-doped area. For this purpose, we select the five $POCl_3$ diffusion processes summarized in Table I. For three of the five processes, we previously characterized their emitter dark saturation current density j_{0e} in Refs. [4,11,13]. The j_{0e} values on alkaline textured surface passivated by a conventional silicon nitride (SiN_x) passivation layer that is formed by plasma-enhanced chemical vapor deposition are found to be between 42 fA/cm² $\leq j_{0e} \leq$ 85 fA/cm² after simulated contact firing (see also Table I, rightmost column).

Table I also contains characteristic process parameters for the examined $POCl_3$ diffusion processes. We label the processes with respect to the used in-situ oxidation and the thickness of the resulting total PSG/SiO_2 stack layer. The reference process "Ref 2nd dep" has no in-situ oxidation as the oxygen (O_2) share in the gas atmosphere during drive-in is below 4% but features a second PSG deposition with active N_2-$POCl_3$ flow after the drive-in. For the other four processes, the drive-in with 100% O_2 share is similar. Corresponding to the name convention, the processes "in-situ thin" and "in-situ thick" feature either a thin or a thick PSG/SiO_2 stack layer after diffusion, respectively (see section 3.1). The N_2-$POCl_3$ and O_2 gas flows during the first deposition are a factor two to three lower for the process "in-situ thin" than for the process "in-situ thick". The suffix "2nd dep" indicates that a second PSG deposition step with active N_2-$POCl_3$ flow follows the drive-in step. The two process variations with the same second deposition step as used for the reference process are accordingly named "in-situ thin 2nd dep" and "in-situ thick 2nd dep".

2.2 Determination of layer thicknesses by means of selective etching

In order to extract the individual thicknesses of the PSG and SiO_2 layers formed during the five $POCl_3$ diffusions, the wafers with saw damage etched surface of group 1 in Fig. 1 are cut into 36 pieces each after diffusion. Ellipsometry measurements are performed before and after stepwise etching the stack layers in 0.1wt% hydrofluoric acid (HF) solution [9]. As the etching rate is dependent on the phosphorus concentration within the respective layer [8,9], the thickness of each layer can be extracted. The detailed procedure is described in Ref. [9].

Applying this procedure, we examine the individual thicknesses of the PSG and SiO_2 layers of the five diffusion processes and the impact of a second deposition step on these layer thicknesses for diffusion processes with in-situ oxidation.

2.3 Determination of phosphorus doses and charge carrier concentration profiles

We use the inductively coupled plasma optical emission spectrometry (ICP-OES) [19] to determine the phosphorus dose within the total PSG/SiO_2 stack layer on alkaline textured surfaces. Therefore, the stack layers of the wafers of group 2 in Fig. 1 are etched in diluted HF and the solution is analyzed with respect to its phosphorus concentration.

After the PSG/SiO_2 stack layer has been removed, the samples are further characterized with respect to their doping properties. Electrochemical capacitance-voltage measurements (ECV) [20] yield the charge carrier concentration profiles. The profiles are scaled to match the sheet resistances R_{sh} [13,21], which are locally determined by four point probe (4pp) measurements around the ECV spots.

2.4 Suitability for laser doping

To examine the suitability of the $POCl_3$ diffusion processes with respect to laser doping, we use a pulsed ultraviolet (UV) ns-laser with a wavelength λ_{laser} = 355 nm [22].

Subsequent to reflection measurements of the PSG/SiO_2 stack layer on textured surface for the wafers of group 3 in Fig. 1, we prepare 2 x 2 cm²-large test fields as shown in Fig. 2 applying four different pulse energies $E_{p,set}$ = {62 µJ; 74 µJ; 90 µJ; 120 µJ} at constant pulse-to-pulse and line-to-line distances d_{pulse} = 15 µm and d_{line} = 25 µm, respectively. Then, the evolution of R_{sh} for the applied laser processes for the different $POCl_3$ diffusions is determined by 4pp measurements.

3 RESULTS

3.1 PSG/SiO_2 layer thicknesses after $POCl_3$ diffusion

Fig. 3 shows the individual thicknesses of the PSG and SiO_2 layers d_{PSG} and d_{SiO2}, respectively, for the five diffusion processes investigated in this work.

For the reference process "Ref 2nd dep", the PSG layer with d_{PSG} = (24 ± 2) nm is about four times thicker than the SiO_2 layer with d_{SiO2} = (6 ± 1) nm. Chen et al. [18] reported a nearly constant SiO_2 layer thickness $d_{SiO2} \approx$ 6 nm for different deposition times and $POCl_3$ flow rates for common diffusion processes without in-situ oxidation, while they found the total stack layer thicknesses d

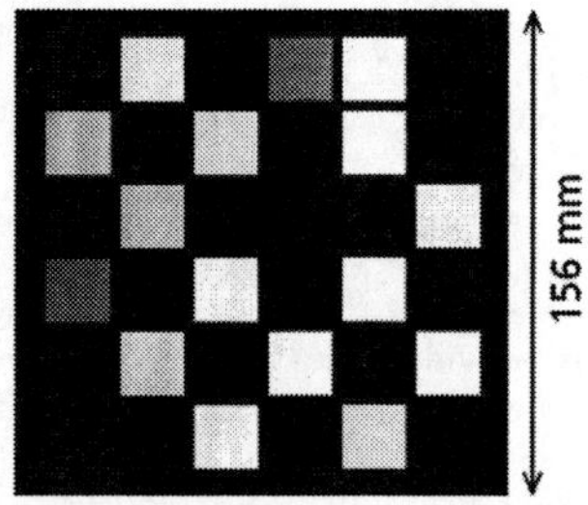

Figure 2: Image scan of an alkaline textured Cz-Si wafer with PSG/SiO_2 stack layer after laser doping. The visible square fields with a size of 2 x 2 cm² were processed with different laser parameters.

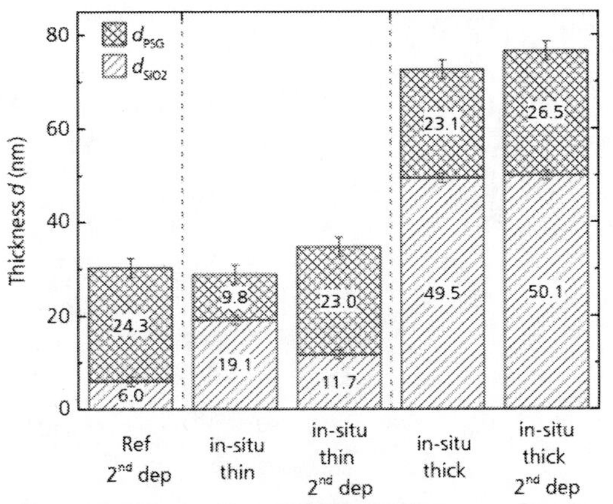

Figure 3: Individual layer thicknesses d_{PSG} and d_{SiO2} of the PSG and the SiO_2 layer, respectively, formed on the silicon surface during the five $POCl_3$ tube furnace diffusions from Table I. The individual thicknesses are extracted by the selective etching procedure described in Ref. [9].

to be between 10 nm < d < 40 nm. Hence, the layer thicknesses found for our reference process without in-situ oxidation "Ref 2nd dep" are in accordance with those found by Chen et al., although Chen et al. investigated processes without second PSG deposition.

In contrast, d_{SiO2} for the process "in-situ thin" in Fig. 3 is significantly larger with $d_{SiO2} = (19 \pm 2)$ nm. This thicker intermediate SiO_2 layer is formed by the in-situ oxidation [9]. During in-situ oxidation, the oxygen from the atmosphere diffuses through the PSG and SiO_2 layers and reacts with the silicon at the surface to form SiO_2.

Attaching a second PSG deposition for process "in-situ thin 2nd dep", the second deposition clearly leads to an increase in d_{PSG} to $d_{PSG} = (23 \pm 1)$ nm, while d_{SiO2} is reduced to $d_{SiO2} = (12 \pm 2)$ nm. This result (i.e. the increase in d_{PSG} while d_{SiO2} is decreased) is in accordance with results we found in Ref. [9], where we investigated the influence of a second PSG deposition step on the PSG/SiO_2 stack layer for a diffusion process without in-situ oxidation. We propose that the reduction in d_{SiO2} results from the diffusion of phosphorus from the PSG layer to the PSG/SiO_2 interface, where they react with SiO_2 to form additional PSG [18]. The PSG layer thickness for process "in-situ thin 2nd dep" is comparable to that of the reference process.

Processes "in-situ thick" and "in-situ thick 2nd dep" feature significantly higher total layer thicknesses d than the other processes as the N_2-$POCl_3$ and O_2 gas flows during first PSG deposition are two to three times higher; see Table I. The strong first PSG deposition for "in-situ thick" results in a total thickness of $d \approx 70$ nm, while the weak first PSG deposition for the reference process and both "in-situ thin"-processes results in $d \approx 30$ nm. For diffusion "in-situ thick", $d_{SiO2} = (50 \pm 2)$ nm is more than twice as large as $d_{PSG} = (23 \pm 2)$ nm. The attached second PSG deposition for process "in-situ thick 2nd dep" hardly affects these layer thicknesses: $d_{SiO2} = (50 \pm 2)$ nm and $d_{PSG} = (27 \pm 2)$ nm. Only d_{PSG} is about 3 nm larger.

In contrast to the process "in-situ thin 2nd dep", d_{SiO2} is not decreased for process "in-situ thick 2nd dep" by adding the second PSG deposition step. As the intermediate SiO_2 layer is more than twice as thick for "in-situ

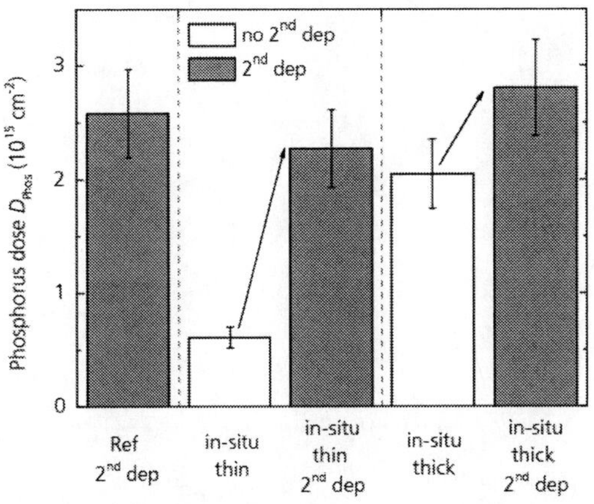

Figure 4: Total phosphorus dose within the PSG/SiO_2 stack layer for $POCl_3$ diffusion processes with and without second deposition step, determined by ICP-OES.

thick", it seems that there is either a critical SiO_2 layer thickness, which represses the increase of d_{PSG} by a second deposition step, or a critical PSG layer thickness, from which point the PSG layer growth occurs slower.

Also remarkable is the fact that for all processes that feature the (same) second deposition step, similar d_{PSG} values between 23 nm $\leq d_{PSG} <$ 27 nm are found, irrespective on the gas flows used during first deposition and drive-in. On the other hand, d_{SiO2} varies significantly between 6 nm $\leq d_{SiO2} <$ 50 nm.

3.2 Phosphorus dose of the PSG/SiO_2 stack layers

The phosphorus dose within the total PSG/SiO_2 stack layers on textured surface is determined by ICP-OES. The ICP-OES measurement gives the phosphorus weight within the examined solution in µg/l. This quantitiy is converted to the amount of phosphorus atoms and then divided by the sample area. As the diffusion occurs on both wafer surfaces, that value is further divided by two to obtain the phosphorus dose D_{Phos} for one wafer side.

Fig. 4 shows the obtained phosphorus doses D_{Phos} for the five diffusion processes. It is evident that the second PSG deposition step leads to significantly higher D_{Phos} that ranges between $(2.3 \pm 0.3) \cdot 10^{15}$ cm$^{-2} \leq D_{Phos} \leq (2.8 \pm 0.4) \cdot 10^{15}$ cm^{-2}. For process "in-situ thin", the second deposition leads to an increase in D_{Phos} from $D_{Phos} \approx (0.61 \pm 0.09) \cdot 10^{15}$ cm^{-2} to $D_{Phos} \approx (2.3 \pm 0.3) \cdot 10^{15}$ cm^{-2}, which corresponds to a factor of four. Process "in-situ thick" with very thick intermediate SiO_2 layer, see Fig. 3, shows a slighter increase in D_{Phos} due to the second deposition step. This is consistent with the assumption that the majority of the phosphorus is located in the PSG layer [9,18] and hence a slight increase in d_{PSG} correlates with a slightly higher D_{Phos}.

3.3 Phosphorus doping profiles after $POCl_3$ diffusion

Fig. 5(a) shows the as-diffused charge carrier concentration profiles of the reference process and the two diffusion processes "in-situ thin" and "in-situ thin 2nd dep", determined by ECV measurements on textured surface. The profile depth d_{prof} at a dopant concentration $N = 10^{17}$ cm^{-3} is between 320 nm < d_{prof} < 400 nm. It is clear that the surface charge carrier concentration $N_{surf} = (6.5 \pm 0.7) \cdot 10^{19}$ cm^{-3} for the reference process "Ref 2nd dep" is lower than for the "in-situ thin"-process with $N_{surf} = (2.1 \pm 0.2) \cdot 10^{20}$ cm^{-3}. On the other hand, the course of the profiles for processes

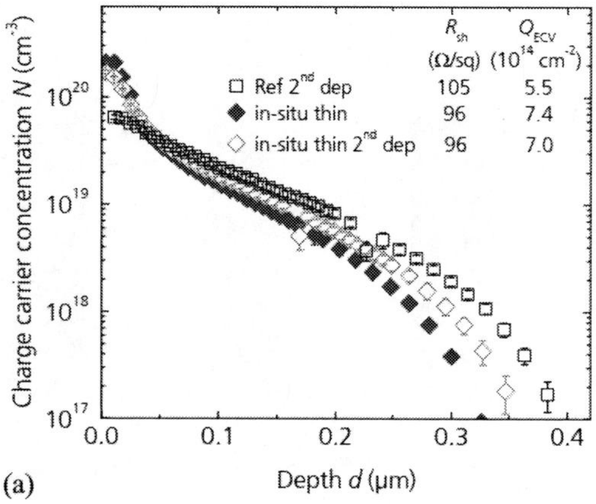

(a)

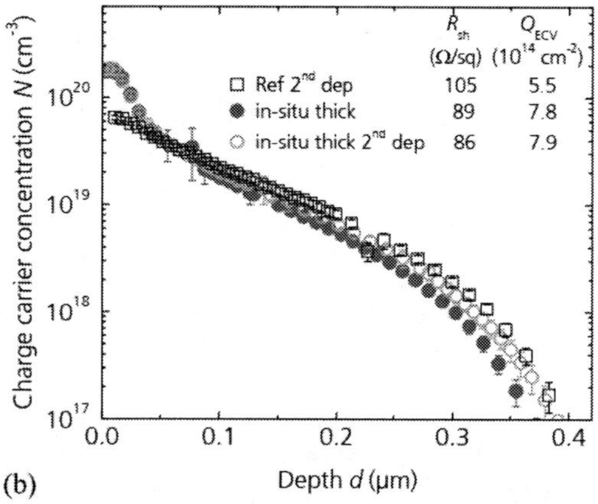

(b)

Figure 5: Charge carrier concentration profiles obtained by the ECV technique after PSG etching on textured surface for the five diffusion processes stated in (a) and (b). The emitter sheet resistances R_{sh} as well as the integrated total charge carrier dose Q_{ECV} are stated.

"in-situ thin" and "in-situ thin 2nd dep" is very similar. Hence, the performed second deposition step hardly affects the resulting doping profile. This is confirmed by the integrated total charge carrier dose Q_{ECV} given in Fig. 5(a), which is similar for both processes within the measurement uncertainty.

Although the PSG/SiO_2 layer thicknesses are very similar for the processes "Ref 2nd dep" and "in-situ thin 2nd dep" (Fig. 3), the courses of the profiles are different.

Fig. 5(b) shows the charge carrier concentration profiles of, the reference process and the two diffusion processes "in-situ thick" and "in-situ thick 2nd dep". Here, d_{prof} is very similar for all processes with $d_{prof} \approx 400$ nm. Both "in-situ thick"-processes with $N_{surf} = (1.8 \pm 0.2)\cdot 10^{20}$ cm^{-3} result in very similar charge carrier concentration profiles as obtained for the two "in-situ thin"-processes. Again, it is clear that the performed second deposition step hardly affects the resulting doping profile: the charge carrier concentrations are very similar either without or with second deposition. This is also confirmed by the integrated total charge carrier dose Q_{ECV} given in Fig. 5(b), which is similar for both processes.

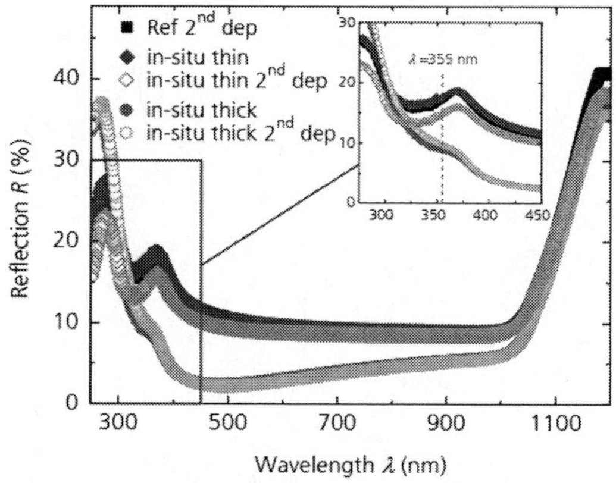

Figure 6: Reflection measurements on textured surface with the PSG/SiO_2 stack layers from the indicated $POCl_3$ diffusion processes.

3.4 Impact of laser doping on sheet resistance

As the laser doping process is at least partly dependent on the coupling of the laser radiation into the silicon material, the absorption A of the samples is of interest. In order to obtain A, we measure the reflection R which is dependent on the surface morphology and the total layer thickness of the PSG/SiO_2 stack. Commonly, lasers with wavelengths between 355 nm $\leq \lambda_{laser} \leq$ 1030 nm are used for laser doping approaches [23]. Within this work, we use a UV ns-laser at $\lambda_{laser} = 355$ nm.

Fig. 6 shows R as a function of wavelength λ. R is significantly lower at $\lambda_{laser} = 355$ nm for both "in-situ thick" diffusion processes compared to the other processes as they feature a thicker PSG/SiO_2 stack layer. As the transmission is zero for λ_{laser}, the absorption is $A = 1 - R$. For the results plotted in Fig. 7, we multiplied the applied pulse energy $E_{P,set}$ with the corresponding A, in order to account for the differences in reflection. However, note that the reflectance changes dramatically during the application of each laser pulse due to local heating and melting of the wafer surface.

Fig. 7 summarizes the obtained sheet resistances $R_{sh,4pp}$, determined by 4pp measurements after laser doping. All three images (a) to (c) have in common that $R_{sh,4pp}$ decreases continuously with increasing E_P. For the reference $POCl_3$ diffusion process with second deposition "Ref 2nd" (no in-situ oxidation) in Fig. 7(a), $R_{sh,4pp}$ is reduced from $R_{sh,4pp} \approx 100$ Ω/sq to $R_{sh,4pp} \approx 20$ Ω/sq if E_P is increased to $E_P \approx 100$ µJ. For both diffusion processes with in-situ oxidation and second deposition in Fig. 7(a), also a significant reduction in $R_{sh,4pp}$ to $R_{sh,4pp} \approx 25$ Ω/sq and $R_{sh,4pp} \approx 30$ Ω/sq is seen for the processes "in-situ thin 2nd dep" and "in-situ-thick 2nd dep", respectively. However, the reductions in $R_{sh,4pp}$ are somewhat weaker pronounced than for the reference.

Referring to Figs. 3 and 4, it is clear that the different thicknesses of the intermediate SiO_2 layer is the dominating property which causes the differences in the obtained $R_{sh,4pp}$ values after laser doping in Fig. 7(a), as all three $POCl_3$ diffusion processes feature comparable d_{PSG} and D_{Phos}. The thicker the intermediate SiO_2 layer is, the less is the achievable reduction in $R_{sh,4pp}$. The presence of a thick intermediate SiO_2 layer ($d_{SiO2} \approx 50$ nm) thus retards the in-corporation of phosphorus from the PSG layer into the silicon surface considerably compared to the thinner SiO_2 layer ($d_{SiO2} \approx 12$ nm) present for the "in-situ thin

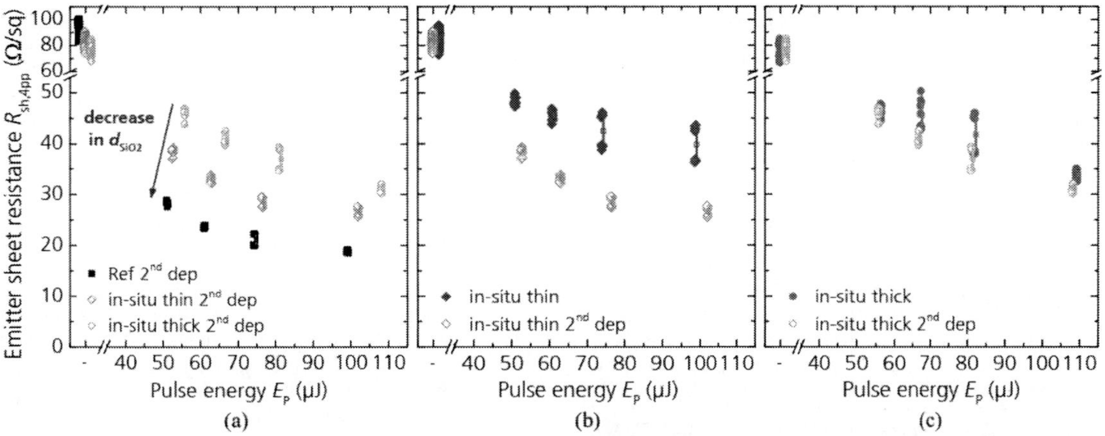

(a) (b) (c)

Figure 7: Emitter sheet resistances $R_{sh,4pp}$ after laser doping for the pulse energies E_P—which correspond to the applied pulse energies $E_{P,set}$ times the absorption A—of the respectively indicated diffusion processes in (a) to (c). The pulse-to-pulse as well as the line-to-line distance is constant for all E_P. One hundred 4pp measurements are performed for all diffusion processes without laser doping (-), while ten 4pp measurements are performed per E_P and per diffusion process.

2nd dep" diffusion process. The even thinner intermediate SiO_2 layer ($d_{SiO2} \approx 6$ nm) of the reference process allows for an even stronger laser doping.

Fig. 7(b) shows the impact of the second deposition step during $POCl_3$ diffusion for the "in-situ thin" processes. Obviously, the second deposition and thus, the higher D_{Phos} as shown in Fig. 4, allows for significantly higher doping resulting in lower $R_{sh,4pp}$ values. Remember that the second deposition step not only increases D_{Phos} but also leads to the decrease in the intermediate SiO_2 layer thickness; see Fig. 3. A thinner SiO_2 layer also allows for higher laser doping, compare Fig. 7(a).

For the "in-situ thick" processes in Fig. 7(c), $R_{sh,4pp}$ decreases with increasing E_P, but only slightly lower values are seen for the $POCl_3$ diffusion process with second deposition at each E_P. This can be explained by the only slight increase of the phosphorous dose due to the second deposition (see Fig. 4). The thick intermediate SiO_2 layer (see Fig. 3) seems to prevent the in-diffusion of phosphorus atoms from the PSG into the silicon for both "in-situ thick" processes, although slightly more phosphorus is provided within the PSG/SiO_2 stack layer for "in-situ thick 2nd dep" (see Fig. 4).

5 SUMMARY AND CONCLUSION

Five different diffusion processes utilizing phosphorus oxychloride ($POCl_3$) as liquid dopant precursor are examined with respect to their as-diffused properties and their suitability with laser doping from phosphosilicate glass (PSG) to form a selective emitter structure. Therefore, variations in deposition parameters, in-situ oxidation (e.g. high oxygen flow during drive-in), and the influence of a second PSG deposition step (e.g. active N_2 flow through the $POCl_3$ bubbler after drive-in) are investigated.

The $POCl_3$ diffusions with in-situ oxidation result in grown stack layers consisting of silicon dioxide (SiO_2) and PSG, whose intermediate SiO_2 layers with a thickness $d_{SiO2} = (19 \pm 2)$ nm or $d_{SiO2} = (50 \pm 2)$ nm are twice as thick as the PSG layers with thicknesses $d_{PSG} = (10 \pm 1)$ nm and $d_{PSG} = (23 \pm 1)$ nm, respectively. The total PSG/SiO_2 stack layer thickness depends on the chosen gas flows during deposition. The higher $POCl_3$ and O_2 gas flows result in thicker stack layers.

The implementation of a second PSG deposition step for the diffusion process with in-situ oxidation results in increased PSG layer thicknesses, while the charge carrier concentration profiles hardly change. The SiO_2 layer thickness decreases if the intermediate SiO_2 layer before second deposition does not exceed a critical thickness. Or, there is a critical PSG layer thickness from which the PSG layer growth occurs slower. The reference $POCl_3$ diffusion without in-situ oxidation but with second deposition features a thin SiO_2 with $d_{SiO2} = (6 \pm 2)$ nm, while the PSG layer is about four times thicker.

We find that the $POCl_3$ diffusion processes with second PSG deposition feature a higher total phosphorus dose D_{Phos} within the PSG/SiO_2 stack layers. D_{Phos} ranges between $(2.3 \pm 0.3)\cdot 10^{15}$ cm$^{-2} \leq D_{Phos} \leq (2.8 \pm 0.4)\cdot 10^{15}$ cm^{-2} for all processes with second deposition.

Laser doping applied to the PSG/SiO_2 stack layers reduces the emitter sheet resistance R_{sh} with increasing laser pulse energy. It is found that for the $POCl_3$ diffusion processes with second PSG deposition, R_{sh} can be significantly more reduced compared with the $POCl_3$ diffusion processes without second deposition as long as the intermediate SiO_2 layer thickness is a few 10 nm or less. From $R_{sh} \approx 90$ Ω/sq after diffusion, R_{sh} can be reduced to 15 Ω/sq $< R_{sh} <$ 35 Ω/sq by laser doping. This result suggests that the additionally provided phosphorus from the second deposition step is advantageous for laser doping. For $POCl_3$ diffusion processes with in-situ oxidation, the reduction in R_{sh} by the laser process is weaker for thicker intermediate SiO_2 layers. For SiO_2 layer thicknesses above about 50 nm, laser doping results in similar R_{sh} values measured on samples either diffused with or without second PSG deposition step.

In summary, our results show that $POCl_3$ diffusion processes with in-situ oxidation are suitable for the formation of selective emitters by laser doping.

ACKNOWLEDGEMENTS

The authors would like to thank all colleagues at the Fraunhofer ISE PV-TEC; especially C. Harmel for laser processing, A. Fischer and S. Maus for selective etching, A. Moldovan for ICP-OES measurements, S. Schmidt and S. Dholipet Nagendra Kumar for ECV measurements and M. Berlin for 4pp measurements. This work was funded by the German Federal Ministry for Economic Affairs and Energy within the research project "POLDI" under contract number 0324079D.

REFERENCES

[1] ITRPV, 2017, "International Technology Roadmap for Photovoltaic (ITRPV) - Results 2016", available onl.: http://www.itrpv.net/Reports/ Downloads/2017/.

[2] A. Cuevas, "A good recipe to make silicon solar cells", *Proc. 22nd IEEE PVSC*, Las Vegas, USA, 1991, pp. 466–470.

[3] A. Kimmerle, R. Woehl, A. Wolf et al., "Simplified front surface field formation for back contacted silicon solar cells", *Energy Proced.*, vol. 38, pp. 278–282, 2013.

[4] S. Werner, E. Lohmüller, S. Maier et al., "Process optimization for the front side of p-type silicon solar cells", *Proc. 29th EU PVSEC*, Amsterdam, The Netherlands, 2014, pp. 1342–1347.

[5] S. Werner, E. Lohmüller, A. Wolf et al., "Extending the limits of screen-printed metallization of phosphorus- and boron-doped surfaces", *Sol. Energy Mater. Sol. Cells*, vol. 158, pp. 37–42, 2016.

[6] T. Dullweber, H. Hannebauer, S. Dorn et al., "Emitter saturation currents of 22 fA/cm^2 applied to industrial PERC cells approaching 22% conversion efficiency", *Proc. 32nd EU PVSEC*, Munich, Germany, 2016, pp. 335–339.

[7] A. Wolf, A. Kimmerle, S. Werner et al., "Status and perspective of emitter formation by $POCl_3$-diffusion", *Proc. 31st EU PVSEC*, Hamburg, Germany, 2015, pp. 414–419.

[8] P. A. Basore, J. M. Gee, M. E. Buck et al., "Simplified high-efficiency silicon cell processing", *Proc. 7th Int. PVSEC*, Nagoya, Japan, 1993, pp. 61–64.

[9] S. Werner, S. Mourad, W. Hasan et al., "Structure and composition of phosphosilicate glass systems formed by $POCl_3$ diffusion", *Energy Proced.*, in press, 2017.

[10] A. W. Blakers, A. Wang, A. M. Milne et al., "22.8% efficient silicon solar cell", *Appl. Phys. Lett.*, vol. 55, no. 13, pp. 1363–1365, 1989.

[11] S. Werner, E. Lohmüller, P. Saint-Cast et al., "Key aspects for fabrication of p-type Cz-Si PERC solar cells exceeding 22% conversion efficiency", *Proc. 33rd EU PVSEC*, Amsterdam, The Netherlands, 2017.

[12] P. Saint-Cast, S. Werner, J. Greulich et al., "Analysis of the losses of industrial-type PERC solar cells", *Phys. Status Solidi A*, 2016.

[13] S. Werner, E. Lohmüller, S. Maier et al., "Challenges for lowly-doped phosphorus emitters in silicon solar cells with screen-printed silver contacts", *Energy Proced.*, in press, 2017.

[14] G. Fischer, M. Müller, S. Steckemetz et al., "Model based continuous improvement of industrial p-type PERC technology beyond 21% efficiency", *Energy Proced.*, vol. 77, pp. 515–519, 2015.

[15] M. Müller, B. Bitnar, G. Fischer et al., "Loss analysis of 22% efficient industrial PERC solar cells", *Energy Proced.*, in press, 2017.

[16] L. Ventura, A. Slaoui, J. C. Muller, "Realization of selective emitters by rapid thermal and laser assisted techniques", *Proc. 13th EU PVSEC*, Nice, France, 1995, pp. 1578–1581.

[17] A. Cuevas, P. A. Basore, G. Giroult-Matlakowski et al., "Surface recombination velocity of highly doped n-type silicon", *J. Appl. Phys.*, vol. 80, no. 6, pp. 3370–3375, 1996.

[18] R. Chen, H. Wagner, A. Dastgheib-Shirazi et al., "A model for phosphosilicate glass deposition via $POCl_3$ for control of phosphorus dose in Si", *J. Appl. Phys.*, vol. 112, no. 12, p. 124912, 2012.

[19] Nölte, Joachim, "ICP - Emissionsspektrometrie für PraktikerWiley, 2002.

[20] E. Peiner, A. Schlachetzki, D. Krüger, "Doping profile analysis in Si by electrochemical capacitance-voltage measurements", *J. Electrochem. Soc.*, vol. 142, no. 2, pp. 576–580, 1995.

[21] R. Bock, P. P. Altermatt, J. Schmidt, "Accurate extraction of doping profiles from electrochemical capacitance voltage measurements", *Proc. 23rd EU PVSEC*, Valencia, Spain, 2008, pp. 1510–1513.

[22] U. Jäger, S. Mack, C. Wufka et al., "Benefit of selective emitters for p-type silicon solar cells with passivated surfaces", *IEEE J. Photovoltaics*, vol. 3, no. 2, pp. 621–627, 2013.

[23] U. Jäger, M. Okanovic, M. Hörteis et al., "Selective emitter by laser doping from phosphosilicate glass", *Proc. 24th EU PVSEC*, Hamburg, Germany, 2009, pp. 1740–1743.

FULLY ION IMPLANTED *N*-TYPE SILICON BIFACIAL SOLAR CELL WITH 20.1% EFFICIENCY

Katsuto Tanahashi, Masaaki Moriya, Shalamujiang Simayi, Yasuhiro Kida, Tetsuo Fukuda,
Satoshi Utsunomiya, Katsuhiko Shirasawa, and Hidetaka Takato
Renewable Energy Research Center, National Institute of Advanced Industrial Science and Technology (AIST)
2-2-9, Machiikedai, Koriyama, Fukushima 963-0298, JAPAN

ABSTRACT: We fabricated the fully ion-implanted *n*-type passivated emitter, rear totally-diffused (PERT) silicon solar cells with elemental boron-implanted emitters and phosphorus-implanted back surface field (BSF). The commercial base *n*-type (phosphorus doping) magnetic field-applied Czochralski (MCZ) silicon wafers of 156 mm × 156 mm in size were used in this study. A maximum front side conversion efficiency η of 20.4% was obtained. The bifacial factor η_{rear}/η_{front} of the best-performing solar cell was 93.1%. Photoluminescence (PL) images revealed the ring-like dark contrasts of the solar cells with low conversion efficiencies. The conversion efficiencies were degraded due to oxygen precipitation during the fabrication of the solar cells.

1 INTRODUCTION

N-type silicon solar cells have several advantages over the *p*-type cells, because they have a much higher bulk lifetime [1] and no light-induced degradation associated with the formation of boron-oxygen complexes [2]. There have already been many studies on the *n*-type passivated emitter, rear totally-diffused (PERT) cell. In spite of these advantages, there are many challenges to realizing the *n*-type PERT silicon solar cell with a high efficiency and low cost, because it requires many fabrication steps and a boron-doped emitter on one side and phosphorus-doped back surface field (BSF) on the other side. Furnace diffusions with boron tribromide (BBr_3) and phosphorus oxychloride ($POCl_3$) are popular as a method to form a boron-doped emitter and a phosphorus-doped BSF, respectively.

Ion implantation has been developed as an alternative to the furnace diffusion process [3], [4], [5], [6]. *N*-type PERT silicon solar cells with boron-implanted emitter and phosphorus-implanted BSF were reported. Implantation of elemental boron requires high temperature annealing, typically above 1050°C, to eliminate the residual crystalline defects and activate boron atoms [7], [8]. It is well known that high temperature annealing causes the degradation of bulk lifetime due to oxygen precipitation in the silicon wafer [9]. So far, there are reports on the oxygen precipitation in the silicon wafer during the thermal treatment of solar cell fabrication [10], [11], [12]. However, the effects of oxygen precipitation on the electrical characterisctics of fully ion-implanted *n*-tpye PERT silicon solar cells are not known well.

In this paper, We fabricated the fully ion-implanted *n*-type PERT silicon solar cells with elemental boron-implanted emitters and phosphorus-implanted BSF. The aim of this paper is to investigate the impact of oxygen precipitation in the silicon wafer on the electrical characteristics of the solar cells. We discuss the effect of oxygen precipitation on the conversion efficiency of the solar cells by using the photoluminescence (PL) characterization.

2 EXPERIMENTAL

A commercially available magnetic field-applied Czochralski (MCZ) grown silicon was used, which is usually used for the fabrication of ultra-large scale integrated circuits. Oxygen concentration determined by a Fourier transform infrared (FTIR) spectrometer was 6.1 × 10^{17} atoms/cm^3. The conversion factor of oxygen was 4.81 (ASTM '79). Carbon concentration determined by a secondary ion mass spectroscopy (SIMS) was 3 × 10^{15} atoms/cm^3.

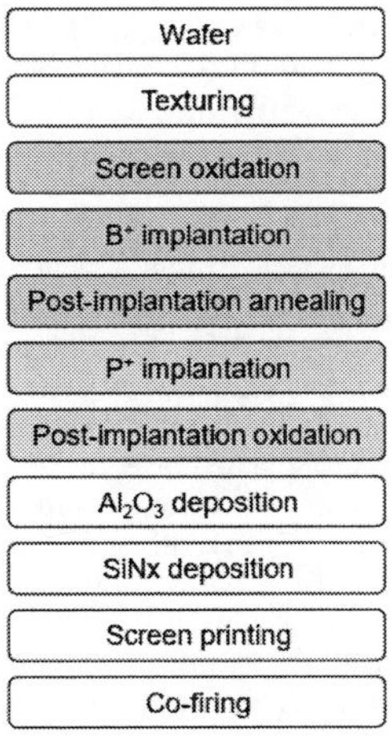

Fig. 1. Process flow for the fabrication of the fully ion-implanted *n*-type PERT silicon solar cells with elemental boron-implanted emitters and elemental phosphorus-implanted BSF

Fig. 1 shows the fabrication process of a solar cell. PERT solar cells were fabricated on an *n*-type (phosphorus doping) MCZ silicon wafer with a size of 156 mm × 156 mm, a resistivity of 3 Ω·cm, and a thickness of 180 μm. Following the wafer texture, a *p*-type emitter was fabricated on the front side of the wafer using the beamline implanter equipped with mass analyzer. A screen oxide layer was grown on the textured surface. Implantation of the elemental boron was implemented with an acceleration voltage of 10 keV and a dose of 3 × 10^{15} atoms/cm^2. Step rotational implantation of the elemental boron was implemented with a tilt angle of 35° and a rotation angle of 15° [13]. Isothermal annealing at 1050 °C for 30 min was performed as a post-implantation annealing in a nitrogen atmosphere, to recover the crystalline damage caused by implantation and activate the boron atoms. The sheet resistances of the emitter were 94.2–99.5 Ω/sq. for the 11 solar cells. After the removal of the screen oxide by dipping in hydrofluoric acid, the BSF was fabricated on the rear side. Implantation of elemental phosphorus was implemented with an acceleration voltage of 5 keV and a dose of 4 × 10^{15} atoms/cm^2. Step rotational implantation of the elemental phosphorus was also implemented with the same conditions as the implantation of boron. Recently, we found that the post-implantation oxidation improved cell characteristics in comparison with the post-implantation annealing in a nitrogen atmosphere for a phosphorus-implanted emitter solar cell with Al-BSF [13]. In this study, the post-implantation oxidation was also performed at 900 °C for 10 min. The oxide layers on both sides of the wafer were removed by dipping in hydrofluoric acid. Subsequently, the wafers were passivated with a stack of AlO/SiN on the front side and with SiN on the rear side. Al/Ag and Ag were metalized on the front and rear sides, respectively, by screen-printing, and co-firing was performed.

3 RESULTS AND DISCUSSION

3.1 Characterization of boron-implanted emitter in n-type PERT silicon solar cell

First, we characterize the boron-implanted emitter in the *n*-type PERT silicon solar cell. Fig. 2 shows the depth profile of boron concentration with SIMS measurement of a boron-implanted mirror polished wafer. Before implantation of boron, a screen oxide was grown on the surface. Implantation of boron was performed with an acceleration voltage of 10 keV and a dose of 3 × 10^{15} atoms/cm^2. The wafer was subjected to isothermal annealing at 1050 °C for 30 min in a nitrogen atmosphere and wet oxidation at 900 °C for 10 min, which correspond to the thermal treatments of the post-implantation annealing of the boron-implanted emitter and the post-implantation oxidation of phosphorus-implanted BSF, respectively. Thermal treatments corresponding to the Al_2O_3 deposition, SiN_x deposition and firing were omitted, since they have insignificant effects on the the diffusion of boron in comparison with the above. Prior to SIMS measurement, the oxide layer of the wafer was removed by dipping in hydrofluoric acid. Segregation of boron (i.e. boron rich layer and boron clustering) is not observed in the sub-surface of the depth profiles shown in Fig. 2. The sub-surface profile of boron in oxidized silicon tends to drop off due to the segregation factor of SiO_2/Si [14]. An extreme decrease in boron concentration is not observed in Fig. 2. This

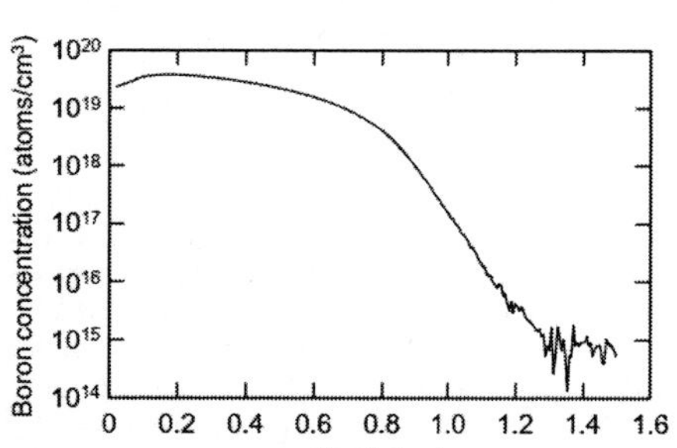

Fig. 2. Depth profile of boron concentration with SIMS measurement of a boron-implanted mirror polished wafer, with an acceleration voltage of 10 keV and dose of 3 × 10^{15} atoms/cm^2. Isothermal annealing at 1050 °C for 30 min in a nitrogen atmosphere, and wet oxidation at 900 °C for 10 min, were performed.

means the impact of post-implantation oxidation at 900 °C for 10 min on the sub-surface profile is small. The tail of the boron profile reaches 1.3 μm in depth. This is mainly due to the isothermal annealing at 1050 °C for 30 min.

3.2 Electrical characterization of *n*-type PERT silicon solar cells

Table I shows the *I-V* parameters of the fully ion-implanted *n*-type PERT silicon solar cells with the maximum front side conversion efficiency (in-house measurement under standard testing conditions; AM 1.5G, 100 mW/cm2, 25°C). The short circuit current J_{sc}, open circuit voltage V_{oc}, fill factor *FF*, and conversion efficiency η, are listed in Table I. Fig. 3 shows the *I-V* curve of the solar cell listed in Table I, which is measured for the front side of the cell. The best-performing solar cell achieved 20.4% front side conversion efficiency with J_{sc} of 39.52 mA/cm^2, V_{oc} of 646.7 mV, and *FF* of 79.7%. The bifacial factor η_{rear}/η_{front} of the best-performing solar cell was 93.1%.

Fig. 4 shows the front side conversion efficiency distribution of *n*-type silicon solar cells. The maximum front side conversion efficiency reaches 20.4%, but the minimum is 19.0%. In spite of the fabrication of *n*-type PERT silicon solar cells under the same process conditions, the electrical parameters therefore have great dispersion.

TABLE I. I-V PARAMETERS OF THE FULLY ION-IMPLANTED N-TYPE PERT SILICON SOLAR CELLS. THE RESULT OF THE BEST FRONT CONVERSION EFFICIENCY SOLAR CELL IS LISTED IN THE TABLE.

	Front	Rear
J_{sc} (mA/cm^2)	39.52	36.63
V_{oc} (mV)	647.8	646.6
FF (%)	79.7	80.1
η (%)	20.4	19.0
η_{front}/η_{rear} (%)	93.1	

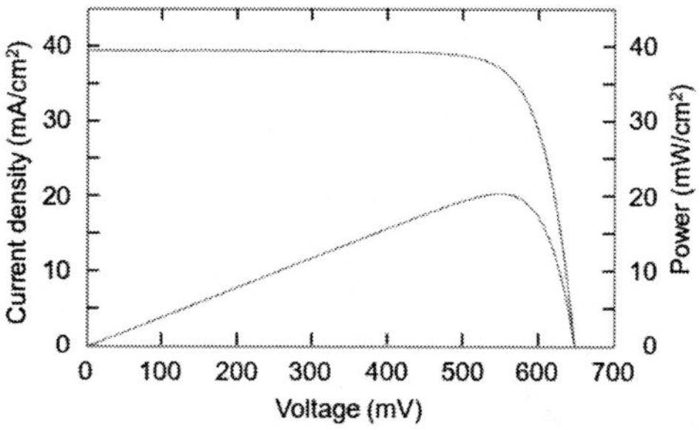

Fig. 3. *I-V* curve of the maximum front side conversion efficiency solar cell.

3.3 Impact of oxygen precipitation in silicon substrates on electrical characteristics of fully implanted *n*-type PERT silicon solar cells

Fig. 5 shows the PL images of the fully ion-

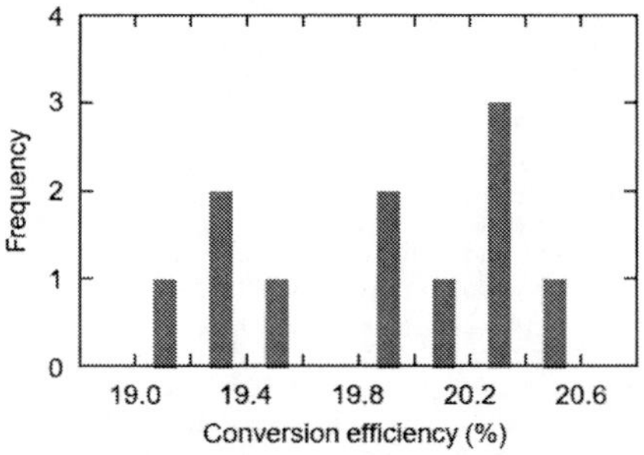

Fig. 4 Front side conversion efficiency distribution of *n*-type PERT silicon solar cells.

implanted *n*-type silicon solar cells at room temperature. We observed PL images of the solar cells without de-processing of AlO/SiN and SiN layers. The laser diode of wavelength at 850 nm was used as an excitation light. Penetration depth of excitation light was 20 μm in depth (1/e). PL from the solar cell was detected by a Charge Coupled Device (CCD) image sensor. The intensities of the defect-related deep-level emissions from silicon at room temperature, such as dislocation and oxygen precipitate in the silicon, are several times weaker than those of band-edge emission of silicon [15]. Therefore, the PL images in Fig. 5 are mainly due to the band-edge emission of silicon. Uniform contrast PL image was observed over the entire wafer in the solar cell with the front side conversion efficiency of 20.4%. However, the ring-like dark contrasts were observed in the solar cells with the front side conversion efficiencies of 19.0% and 20.1%. The maximum PL intensity of the solar cell with conversion efficiency of 20.4% is larger than that of 19.0% and 20.1%. In Fig. 5, we adjusted the contrast of each PL image depending on the intensity of the individual solar cell. The band-edge emission of silicon is proportional to the minority carrier lifetime [16]. This suggests that the degradation of bulk lifetime causes the decrease of the conversion efficiencies of the *n*-type silicon solar cells.

N-type silicon solar cells used in the PL observation were subjected to thermal treatments accoding to the process flow in Fig. 1, typically, annealing at 1050 °C for 30 min and oxidation at 900 °C for 10 min. Therefore, the ring-like contrast is obviously due to oxygen precipitation during the fabrication process of fully ion-implanted *n*-type PERT silicon solar cells. Here, it should be noted that the behavior of oxygen precipitation differs from wafer to wafer: the solar cells with conversion efficiencies of 19.0% and 20.1% show the ring-like contrasts of the PL images, but that of 20.4% shows the uniform contrast over the entire wafer. Moreover, the radial distributions of the ring-like contrast of a solar cell with conversion efficiency of 19.0% is differnt from that of 20.1%. According to the classical nucleation theory, homogeneous nucleation and growth of oxygen precipitates at a given temperature depend on the oxygen concentration [17]. Carbon in silicon has a role of heterogeneous nucleus of oxygen precipitate [18]. However, oxygen and carbon concentrations of silicon wafer with conversion efficiencies of 19.0%, 20.1%, 20.4% are almost the same value, which were 6.1×10^{17} atoms/cm^3 and 3×10^{15} atoms/cm^3, respectively. Therefore, another factor apart from the oxygen and carbon concentration obviously has an effect on the oxygen precipitation during the fabrication process of *n*-type PERT silicon solar cells: point defects such as vacancies and self-interstitials. Recent semiconductor-grade MCZ silicon crystal growth controls the radial and axial distribution of point defects, suppressing the grown-in void defects [19]. It is well known that vacancies enhance the oxygen precipitation and self-interstitials, and vice versa [20]. It is speculated that the ring-like contrast of PL images, i.e. the ring-like distribution of oxygen precipitates, are due to the point defects incorporated at the melt/crystal interface during crystal growth. The control of point defects is sensitive to the growth conditions of MCZ silicon. A large amount of vacancies were incorporated at the ring-like dark contrast areas of the PL images in Fig. 5. This enhanced the oxygen precipitation during the fabrication processes of the solar cells.

Thus, we found the effect of oxygen precipitation on the conversion efficiency of the fully ion-implanted *n*-type PERT silicon solar cells. The suppression of oxygen precipitation is one of the key technologies needed to achieve a high conversion efficiency of *n*-type PERT silicon solar cells.

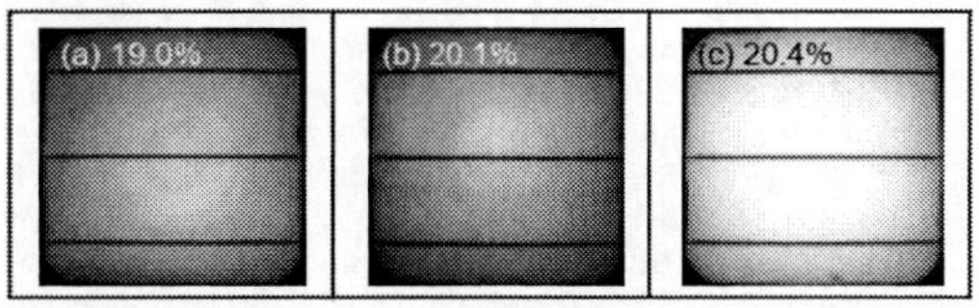

Fig. 5. PL images of the fully ion-implanted *n*-type solar cells with conversion efficiencies of (a) 19.0%, (b) 20.1%, and (c) 20.4%.

4 CONCLUSIONS

Fully ion-implanted *n*-type PERT silicon solar cells with elemental boron-implanted emitters and phosphorus-implanted BSF were fabricated on a commercial *n*-type phosphorus CZ silicon wafer of size 156 mm × 156 mm. Implantation of elemental boron was implemented with an acceleration voltage of 10 keV and a dose of 3×10^{15} atoms/cm^2. Post-implantation annealing was performed at 1050 °C for 30 min. Implantation of elemental phosphorus was implemented with an acceleration voltage of 5 keV and a dose of 4×10^{15} atoms/cm^2. Post-implantation oxidation was also performed at 900 °C for 10 min.

The best-performing of 11 solar cells achieved 20.4% front side conversion efficiency, while the worst achieved 19.0%. The electrical parameters have a very wide dispersion.

PL images revealed the ring-like contrast of the solar cells with low conversion efficiencies. This ring-like contrast is due to oxygen precipitation during the fabrication processes of the fully ion-implanted *n*-type silicon solar cells.

ACKNOWLEDGEMENT

The authors would like to thank Dr. Toshimitsu Mochiduki for useful discussions. This work was supported by New Energy and Industrial Technology Development Organization.

REFERENCES

[1] D. Macdonald and L. J. Geerligs, "Recombination activity of interstitial iron and other transition metal point defects in *p*- and n-type crystalline silicon," *Appl. Phys. Lett.*, vol. 85, no. 18, pp. 4061–4063, 2004.

[2] S. W. Glunz, S. Rein, J. Y. Lee, and W. Warta, "Minority carrier lifetime degradation in boron-doped Czochralski silicon," *J. Appl. Phys.*, vol. 90, no. 5, pp. 2397–2404, 2001.

[3] Y. Tao, Y.-W. Ok, F. Zimbardi, A. D. Upadhyaya, J.-H. Lai, S. Ning, V. D. Upadhyaya, and A. Rohatgi, "Fully ion-implanted and screen-printed 20.2% efficient front junction silicon cells on 239 cm^2 n-type CZ substrate," *IEEE J. Photovolt.*, vol. 4, no. 1, pp. 58–63, 2014.

[4] F. Kiefer, J. Krügener, F. Heinemeyer, M. Jestremski, H. J. Osten, R. Brendel, and R. Peibst, "Bifacial, fully screen-printed n-PERT solar cells with BF_2 and B implanted emitters," *Sol. Energy Mater. Sol. Cells*, vol. 157, pp. 326–330, 2016.

[5] T. S. Böscke, D. Kania, C. Schöllhorn, D. Stichtenoth, A. Helbig, P. Sadler, M. Braun, M. Dupke, M. Weiβ, A. Grohe, J. Lossen, and H.-J. Krokoszinski, "Fully Ion Implanted and Coactivated Industrial n-Type Cells With 20.5% Efficiency," *IEEE J. Photovolt.*, vol. 4, no. 1, pp. 48–51, 2014.

[6] A. Lanterne, J. Perchec, S. Gall, S. Manuel, M. Coig, A. Tauzin, and Y. Verschetti, "Understanding of the annealing temperature impact on ion implanted bifacial n-type solar cells to reach 20.3% efficiency," *Prog. Photovolt.*, vol. 23, no. 11, pp. 1458–1465, 2015.

[7] J. Krügener, R. Peibst, F. Alexander Wolf, E. Bugiel, T. Ohrdes, F. Kiefer, C. Schöllhorn, A. Grohe, R. Brendel, and J. Osten, "Electrical and Structural Analysis of Crystal Defects After High-Temperature Rapid Thermal Annealing of Highly Boron Ion-Implanted Emitters," *IEEE J. Photovolt.*, vol. 5, no. 1, pp. 166–173, 2015.

[8] R. Müller, J. Benick, N. Bateman, J. Schön, C. Reichel, A. Richter, M. Hermle, and S. W. Glunz, "Evaluation of implantation annealing for highly-doped selective boron emitters suitable for screen-printed contacts" *Sol. Energy Mater. Sol. Cells*, vol. 120, pp. 431–435, 2014.

[9] D. C. Walter, B. Lim, R. Falster, J. Binns, and J. Schmidt, "UNDERSTANDING LIFETIME DEGRADATION IN CZOCHRALSKI-GROWN *N*-TYPE SILICON AFTER HIGH TEMPRATURE PROCESSING," in *Proc. 28th European Photovoltaic Specialist Conf. and Exhibition (EuPVSC)* 2012, pp. 699–702.

[10] L. Chen, X. Yu, P. Chen, P. Wang, X. Gu, J. Lu, and D. Yang, "Effect of oxygen precipitation on the performance of Czochralski silicon solar cells," *Sol. Energy Mater. Sol. Cells*, vol. 95, pp. 3148–3151, 2011.

[11] P. J. Cousins, D. D. Smith, H-C Luan, J. Manning, T. D. Dennis, A. Waldhauer, K. E. Wilson, G. Harley, and W. P. Mulligan, "GENERATION 3: IMPROVED PERFORMANCE AT LOWER COST," in *Proc. 35th IEEE Photovoltaic Specialist Conf. (PVSC)* 2010, pp. 275–278.

[12] G. Galbiati, V. D. Mihailetchi, A. Halm, H. Chu, and R. Kopecek, "LARGE AREA ZEBRA SOLAR CELLS IN PILOT PRODUCTION: THE RESULTS OF EP7 HERCULES PROJECT," in *Proc. 32th European Photovoltaic Specialist Conf. and Exhibition (EuPVSC)* 2016, pp. 980–982.

[13] K. Tanahashi, M. Moriya, Y. Kida, S. Utsunomiya, T. Fukuda, K. Shirasawa, and H. Takato, "Impact of Post-implantation Annealing Conditions on Electrical Characteristics of a Phosphorus-implanted Emitter Crystalline Silicon Solar Cell," *IEEE J. Photovolt.*, vol. 7, no. 3, pp. 741–746, 2017.

[14] A. S. Grove, O. Leistiko, Jr., and C. T. Sah, "Redistribution of Acceptor and Donor Impurities during Thermal Oxidation of Silicon," *J. Appl. Phys.*, vol. 35, no. 9, pp. 2695–2701, 1964.

[15] M. Tajima, Y. Iwata, F. Okayama, H. Toyota, H. Onodera, and T. Sekiguchi, "Deep-level photoluminescence due to dislocations and oxygen precipitates in multicrystalline Si," *J. Appl. Phys.*, vol. 111, no. 11, pp. 113523-1–113523-6, 2012.

[16] M. Tajima, Z. Li, S. Sumie, H. Hashizume, and A. Ogura, "Comparison of Silicon-on-Insulator Wafer Mapping between Photoluminescence Intensity and Microwave Photoconductivity Decay Lifetime," *Jpn. J. Appl. Phys.*, vol. 43, no. 2, pp. 432–438, 2004.

[17] K. F. Kelton, R. Falster, D. Gambaro, M. Olmo, M. Cornara, and P. F. Wei, "Oxygen precipitation in silicon: Experimental studies and theoretical investigations within the classical theory of nucleation," *J. Appl. Phys.*, vol. 85, no. 12, pp. 8097–8111, 1999.

[18] F. Shimura, "Carbon enhancement effect on oxygen precipitation in Czochralski silicon," *J. Appl. Phys.*, vol. 59, no. 9, pp. 3251-3254, 1986.

[19] V. V. Voronkov and R. Falster, "intrinsic Point Defects and Impurities in Silicon Crystal Growth," *J. Electrochemical Soc.*, vol. 149, no. 3, pp. G167–G174, 2002.

[20] H-S Kim, K-S Lee, B-Y Lee, H-D Yoo, S-H Pyi, C-G Koh, B-S Hong, and Y-W Kim, "Observation of Micro-Oxygen Precipitates in the Vicinity of the Oxidation-Induced Stacking Fault Ring and Their Effects on Thin Gate Oxide Breakdown," *Jpn. J. Appl. Phys.*, vol. 40, part 2, no. 12A, pp. L1286–L1289, 2001.

REAR-SURFACE LASER CONTACT OPENING DESIGN OPTIMIZATION FOR PERC SOLAR CELLS

E.Picard, M.Pirot*, R.Monna and S.Dubois
INES, CEA, LITEN, Department of Solar Technologies
Address F-73375 Le Bourget du Lac, France
*Corresponding author: email: marc.pirot@cea.fr Phone: +33 479 792 9 34

ABSTRACT: This paper reports on the analysis of the photovoltaic (PV) potential of PERC structures and focuses on the carrier recombination and series resistance related losses caused by the back-side Al-Si local contacts. More precisely, we propose a *3-steps* procedure to define a process window for the back-side LCO (Laser Contact Opening) design of a PERC solar cell. This method takes into account the 3 main PV performance limitations coming from the LCO design: void occurrence limitation, lateral resistance limitation and carrier recombination limitation. The proposed procedure starts with a simplified determination of the substrate main bulk properties followed by an evaluation of surfaces recombination parameters via symmetrical and asymmetrical passivated structures based on high quality p-type Cz wafers. Then, the *void occurrence* is addressed with a specific approach combining variations of the opening width and changes in the firing temperature, supported by LBIC (Light Beam Induced Current) analyses (for voids density calculation) and illuminated IV measurements. Eventually, an experimental matrix modulates the LCO fraction (for an opening width and firing conditions limiting void occurrence) in order to define the good trade-off between low lateral resistances and reduced carrier recombination at the Al-BSF/Si interface. By applying these optimization procedures, PERC cells with V_{oc} above 665 mV were fabricated with performances essentially limited by front-surface and bulk recombinations.

Keywords: PERC Solar cells; Back Contact, Laser Contact Opening, Recombination

1 INTRODUCTION

Passivated emitter and rear contact (PERC) solar cells could quickly become the industrial high efficiency mainstream technology due to significant improvements in process maturity. Industry is regularly reporting above 21% average performances in mass production. These results are mostly driven by high open circuit voltage (V_{OC}) spawned by higher quality p-type silicon (Si) substrates and a specific passivation multilayer between the silicon and the rear metallization. This passivation multilayer enhances the rear light reflection and strongly reduces the free carrier recombination rate at the back surface. Al_2O_3/SiN_x stacks are known to well fulfill these requirements with surface recombination velocity falling far below 5 cm/s and implied V_{OC} above 730 mV for Al_2O_3/SiN_x coated symmetrical Cz p-type structures.

One of the main challenges of PERC developments is then to maintain these high implied Voc allowed by the rear passivation, at the level of the complete solar cell. Consequently this challenge includes on the one hand the use of high bulk carrier lifetime (Tau bulk) materials. On the other hand it requires the decrease of the recombination currents on the front side of the cell (Joe for the non-metallized n+ emitter and Jo.met for the metal-n+(Si) interface) and on the back side of the cell (Surface recombination velocity Spass for the Si-dielectrics interface and Smet for the local Si-Al interface). Additionally, this development cannot be disconnected from resistive losses optimizations and has to take into account a specific PERC issue known as 'voids effect' occurring during local BSF (Back Surface Field) formation, usually done by laser ablation of the dielectric layers before the aluminum (Al) layer deposition by screen-printing and firing. Therefore, dedicated PERC process optimization procedures are necessary to drive process improvements further. In this paper, we propose a procedure (method) to optimize the back-side Laser Contact Opening (LCO) design of a PERC solar cell.

Prior to any metallization steps, in order to evaluate the influence of the back-side local contacting step (parallel lines, cf Figure 3), 2 experimental matrixes of back-side LCO design were prepared (with samples that experienced the ASYMPERC fabrication process).

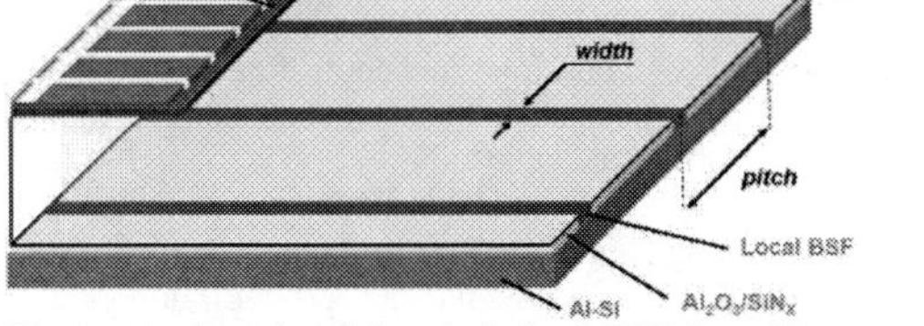

Figure 1: Sketch of the optimized PERC architecture, focussing on the LCO design (parallel lines) of the rear surface.

2 EXPERIMENTAL APPROACH

In this paper, 156×156 mm² p-type Cz KOH-textured wafers of 1 to 3 Ω.cm resistivity were used. As shown on Figure 1, wafers were preliminary characterized using Oxymap tool to determine their thickness, resistivity, the interstitial oxygen density ([Oi]), the thermal donor concentration (in the as-received samples), and their Thermal History Index (THI) parameter [1]. Then, two families of symmetrical structures were prepared in order to identify the recombination parameters of the non-metallized front and rear surfaces. For the assessment of the front surface properties, the wafers were double side Phosphorus doped and SiNx:H passivated (SYMPHOS). An equipment from SEMCO TECHNOLOGIES was used to conduct the P-diffusion step (LYDOP process). For the assessment of the rear surface properties, the symmetric wafers consisted in double side polished and Al2O3/SiNX passivated wafers (SYMALOX). In addition, wafers with an asymmetrical structure were made (ASYMPERC) for cells purpose (SiN_x:H passivated n^+ emitter on the textured front surface, and Al2O3/SiNx stack on the polished rear surface). Notice that the Al2O3 layers were deposited by thermal ALD (Atomic Layer Deposition) and the SiNx:H layers by PECVD (Plasma-Enhanced Chemical Vapor Deposition). WCT-120 tool was used to characterize the related charge carrier bulk lifetimes, the surface recombination parameters and the implied Voc on each structure.

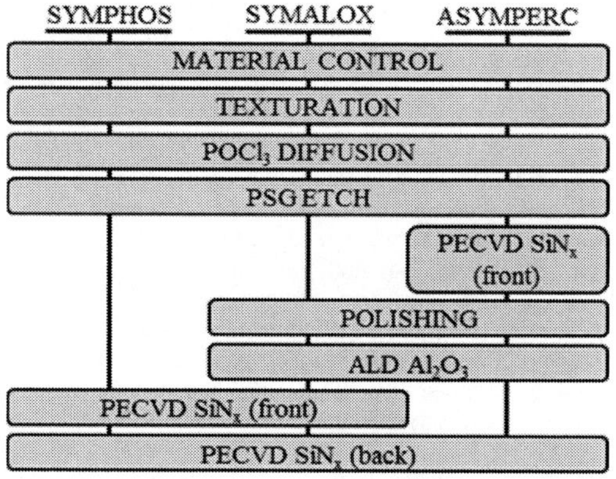

Figure 2: Process flow of the 3 structures used to determine the carrier recombination parameters. SYMPHOS & SYMALOX are symmetrical structures and ASYMPERC is a non-metallized PERC precursor (also used for cells manufacturing).

The aim of the first experimental matrix (Figure 2) is to evaluate both the effects of the opening width and the firing T on voids occurrence, in order to define a process window (opening width, firing-T), for avoiding voids on the rear surface. Done at a fixed contact opening fraction, it also allows to evaluate the influence of voids density on the PV performance losses.

The second experimental (Figure 3) matrix modulates the LCO fraction (for a given opening width) in order to optimize the back-side contact and lateral resistances. Done at a fixed firing temperature and with a design which prevent any void formation, we can then estimate the related recombination current generated by the local BSF.

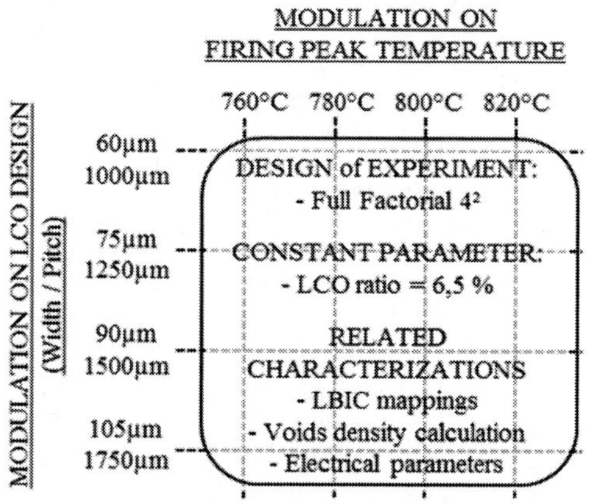

Figure 3: First experimental matrix on voids occurrence and its related charcaterizations

This experimental procedure done in three steps allows extracting all necessary information to optimize the back-side of a PERC solar cell design (summarized in Figure 3). Indeed, the combination of recombination currents determinations with resistive losses analyses and voids densities within one common empirical model allows the determination of the main influent factors of the LCO design on the PERC PV performances.

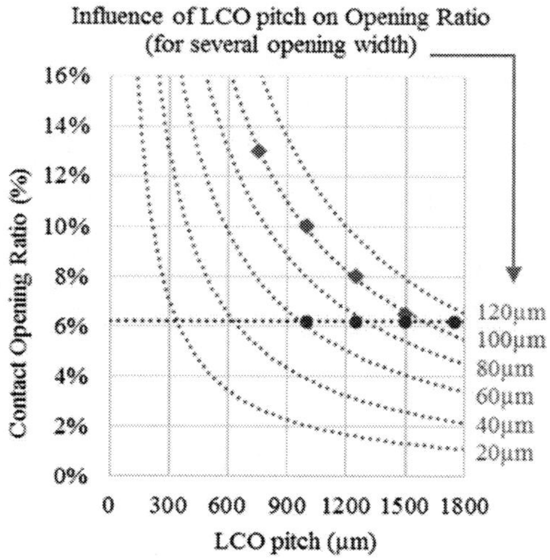

Figure 3: First (red •) and second (blue ♦) experimental matrix represented as fonction of back-side design main parameters (LCO width and pitch).

3 RESULTS

3.1 Voids versus LCO width and firing

In this paper, we call void any lack of local back surface field which weaken the local contact enough to avoid any current collection [2]. The first experimental matrix supported by 40×40mm² LBIC characterization @946nm allows to visualize the void formation. As shown on Figure 4, the void occurrence is enhanced by increases in the firing peak-T, and decreases in the LCO width.

This qualitative information can be extended to a quantitative parameter f_{voids} in order to determine the void occurrence fraction:

$$f_{voids} = \frac{A_{voids}}{A_{LCO}}$$

Where A_{voids} can be described as the area affected by strongly weakened local back surface field and A_{LCO} is the total area opened by laser.

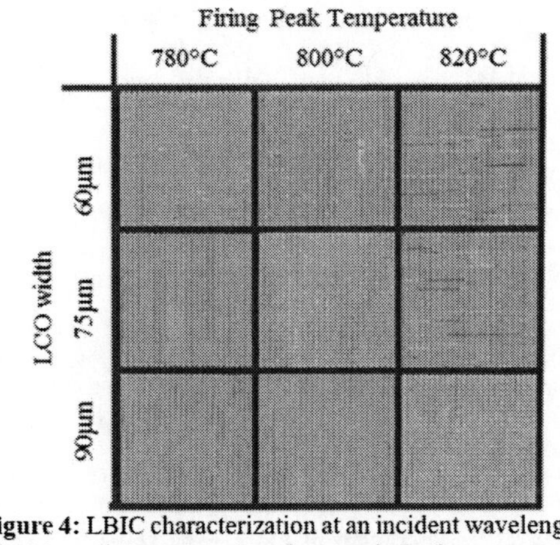

Figure 4: LBIC characterization at an incident wavelength of 946 nm (40 x 40mm² surface area). Red zones reveal lack of current collection corresponding to voids and/or weakened local bsf.

As represented on Figure 5, an empiric model can be built to link the void occurrence f_{voids} with LCO width and firing peak temperature.

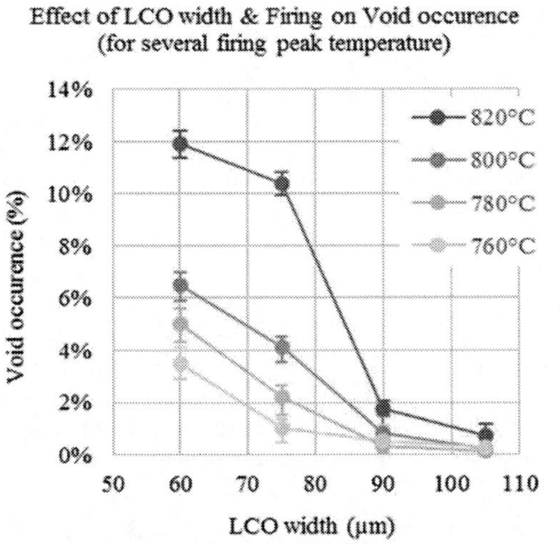

Figure 5: Quantitative extraction of void occurrence fraction f_{voids} from LBIC mappings. f_{voids} as a function of the LCO width.

Moreover, by linking the f_{voids} to illuminated-I(V) parameters we observe a consistent correlation between f_{voids} and FF such as 1% increase in f_{voids} lowers the FF by 0.13%.

Based on the analysis of above graph and related data, a threshold can then be defined on LCO minimum width and maximum firing peak temperature to ensure a limited void formation in PERC cell processing. This limit will have also to take into account other process aspects such as front-side optimum firing peak temperature and back-side aluminum paste quality (which can both significantly influence the void formation threshold).

Therefore, to prevent any influence from voids on cell performance in the 2nd experimental matrix, a LCO width of 100µm has been chosen.

3.2 Lateral resistance versus LCO pitch

By fixing the LCO width to a value hindering voiding effects, the next aspect of the design optimization is related to the contribution of lateral resistance losses due to the LCO pitch. Indeed, series resistances of classical PERC structures made without any additional Boron BSF, are influenced by the LCO pitch, the substrate resistivity ($\rho_{substrate}$) and the substrate thickness ($t_{substrate}$), as summarized by Figure 6 with r_{pitch} defined by [3]:

$$r_{pitch} = \frac{1}{3} \cdot \left(\frac{pitch}{2}\right)^2 \cdot \frac{\rho_{substrate}}{t_{substrate}}$$

The weighty influence of substrate resistivity on associated resistive losses can be then reduced by adjusting the pitch to a value that will lowly affect the FF. This value has been empirically defined on 2 Ω.cm substrates during second experimental matrix. Related results are reported in Figure 6 which are confirming the influence of the pitch on FF loss.

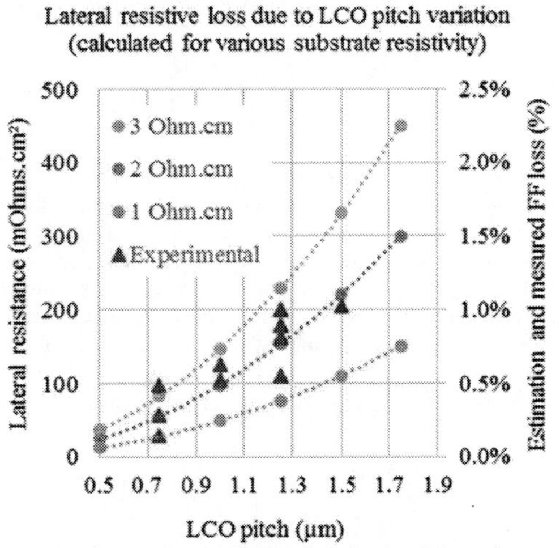

Figure 6: Lateral resistive loss influenced by substrate resistivity and LCO pitch. Red triangles are representing experimental data from 2nd experimental matrix on 2 Ω.cm substrates.

3.3 Recombination versus LCO fraction

As described on Figure 1, several symmetrical and asymmetrical structures were prepared to determine main recombination parameters based on QSSPC carrier lifetime measurements. Therefore, the bulk lifetime τ_{bulk}, the emitter saturation current J_{oe} and the surface recombination velocity S_{pass} have been quantified through mentioned structures.

	SYMPHOS	SYMALOX	ASYMPERC
τbulk			333 µs
Spass		1,9 cm/s	
Joe	52 fA/cm²		
impliedVoc	675 mV	735 mV	690 mV

Figure 7: Summary of non-metallized symmetrical and asymmetrical structures and related main parameters measured after firing. The bulk lifetime is extracted from asymmetrical strucrture as it corresponds to manufactured cells

In order to implement metal/Si recombination parameters for front-side and back-side in the total cell recombination, the approach introduced by Fischer [4] has been used. For the back side calculations, the effective surface recombination velocity has then been determined as follow:

$$S_{eff}=\left(\frac{l_P}{2\sqrt{\Pi f_{LCO}}}\cdot arctan\left(\frac{2}{l_P}\cdot\sqrt{\frac{\Pi}{f_{LCO}}}\right)-exp\left(\frac{1}{l_P}\right)+\frac{1}{f_{LCO}\cdot S_{met}}\right)^{-1}+\frac{S_{pass}}{1-f_{LCO}}$$

With the normalized period, $l_P = pitch/t_{substrate}$ and the LCO fraction, $f_{LCO} = A_{LCO}/A_{CELL}$. By modulating the LCO fraction in the 2nd experimental matrix, we are then able to extract the S_{met} parameter by quantifying the V_{OC} loss caused by an increasing LCO fraction such as:

$$\frac{\Delta(V_{OC})}{\Delta(f_{LCO})}=-0,72mV\ /\ \%$$

Moreover, as expected, a small but consistent reduction of J_{SC} is observed while increasing LCO fraction such as:

$$\frac{\Delta(J_{SC})}{\Delta(f_{LCO})}=-0,05\cdot\left(\frac{mA}{cm^2}\right)/\ \%$$

The complete analysis can then apply for an optimum LCO design, meaning with limited voiding effect, reduced lateral resistive contribution and optimized local BSF recombination. In the summary shown on Figure 8, an optimum process window has been used, and we clearly notice a substantial limitation of the PV potential of the studied PERC by the relatively low bulk carrier lifetime.

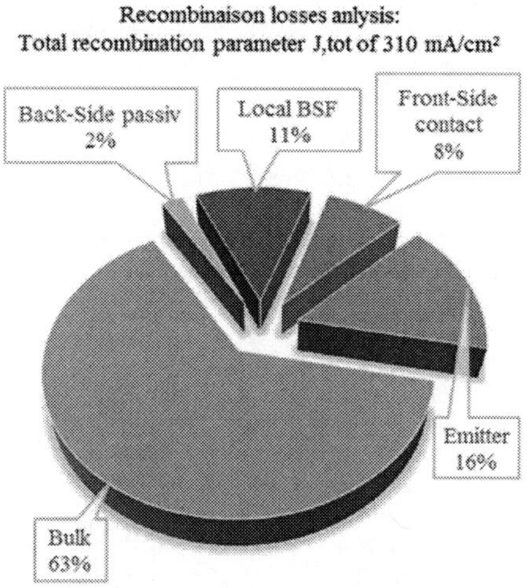

Figure 8: Summary of relative recombination losses of an optimized PERC cell with V_{OC} of 665mV.

4 CONCLUSION - OPTIMUM WORKING POINT AND PERSPECTIVES

We have here presented a 3 steps method to define a process window for the back-side LCO design of a PERC solar cell. This method takes into account the 3 main cell performance limitations coming from the back-side LCO design as shown on Figure 9 (void occurrence limitation, lateral resistance limitation and carrier recombination limitation).

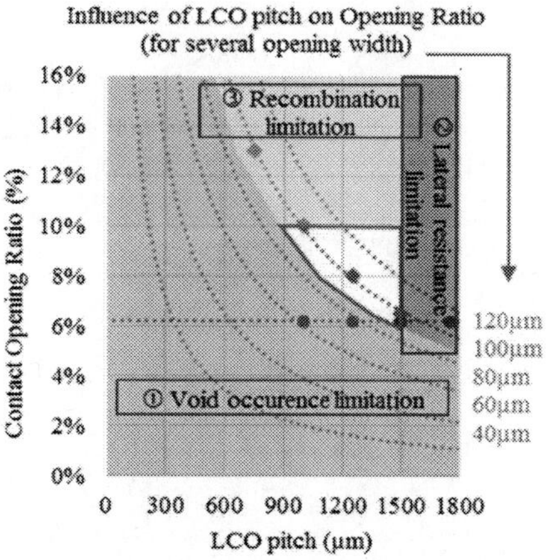

Figure 9: Description of the 3 main cell performance limitation defining the process windows of LCO back-side design.

Via a stabilized LCO process windows and related recombination analyses, improvements have been made on cell level to reduce contributions of bulk and emitter on recombinations. A cell V_{OC} of 668 mV have thus been obtained and could be significantly optimized by improving p-type Cz wafer quality with reduced [O_i] and THI.

Moreover, regarding the mentioned limitations, the back side Boron doping as additional homogeneous BSF would be a significant technological breakthrough to release the limitation related to the LCO pitch. Eventually, the void formation limit can be strongly reduced by optimizing the Al paste selection and the firing process as already reported by [5].

Also, further optimizations of the rear passivation will be driven on the PECVD TWYN developed by SEMCO. Recently, promising results have been reached with SYMALOX structures (AlOx / SiNx PECVD layers): iVoc equivalent to the values obtained with ALD layers, combined with excellent passivation homogeneity.

Figure 10: SEMCO PECVD TWYN – SiN_x, SiO_x, SiO_xN_y, SiC_x, SiC_xN_y, a:Si, PSG, BSG, Al_2O_3, more …

5 REFERENCES

[1] J. Veirman et al, "Oxygen-defect characterization for improving R&D relevance and Cz-Si solar cell efficiency", Photovoltaic I., Sept. 2016.

[2] R. Horbelt, "Void formation on PERC solar cells and their impact on the electrical cell parameters verified by luminescence and scanning acoustic microscope measurements", Energy Procedia 84 (2015) 47-55.

[3] D.L. Meier, "Contact resistance: Its measurement and Relative importance to Power Loss in a Solar Cell", IEEE Transactions on Electron Devices vol.31 (1984) p.647.
[4] B. Fischer, "Loss Analysis of Crystalline Silicon Solar Cells Using Photoconductance and Quantum Efficiency Measurements", Thesis University of Konstanz, 2003.
[5] D. Chen, "Preventing the Formation of Voids in the Rear Local Contact Areas for Industrial-Type PERC Solar Cells", 28th EU-PVSEC (2013).

Double side Cu-plated technology on front junction n-PERT solar cells

Kuang-Chieh Lai[*], *Yueh-Lin Lee, Ming-Shiou Lin,*
Chia-Chih Chuang, Chi-Chun Li
Advance Technology Development Department,
Motech Industries, Inc, No.96, Dashun 7th Rd., Xinshi Dist., Tainan City 74145, Taiwan
* Corresponding author; E-Mail: Jay_Lai@motech.com.tw

ABSTRACT: In this paper, we evaluated n-type passivated emitter and rear totally diffused (n-PERT) solar cells featuring nickel/copper (Ni/Cu) plated double side contacts. The cell features a rear surface field formed by fully diffused phosphorous and fully diffused BBr_3 front emitter with thermal oxide passivation. Laser process conditions are demonstrated to from the metal contact open by ultraviolet (UV) wavelength laser system. The UV nanosecond laser (ns-laser) is used to ablate the passivation layer on the double-side of the solar cells. In this study, our process combines laser ablation for the patterning of the dielectrics on front-side with a front metallization based on Ni/Cu forward bias plating and Ni electro-less plating (ELP) for the rear contacting electrode. The large-area (238.9 cm^2) silicon solar cells manufactured on 170 μm thick, 3Ω-cm, n-type Cz wafers using this technology, reach confirmed efficiencies up to 18.93%; with Voc of 641 mV, a jsc=38.5 mA/cm^2 and FF of 76.7 %.

Keywords: n-type solar cells, electro-less plating, forward bias plating.

1 INTRODUCTION

The replacement of Ag screen printing with Ni/Cu plating is a promising approach to reduce the material costs for metallization processes in various solar cell designs. The successful application of plated Ni/Cu contacts was recently demonstrated for various kinds of p-type [1,2] and n-type [3] silicon solar cells. Motech developed n-type bifacial cells based on PERT (passivated emitter, rear totally diffused) structure, and previously reported >20.6% efficiency for n-PERT cells made with screen-printed electrodes on both sides of the cell [4] and >21% efficiency for n-PERT cells made with plated Cu electrodes on front sides of the cell [5].
However, the conventional metallization approach is relatively expensive for bifacial cells. Compared with p-type cells, the usage of Ag pastes on bifacial n-type cells roughly doubles. Therefore, a low cost alternative to Ag paste metallization, such as Cu plating studied in this work, is particularly important for n-type cells. In addition to the material cost savings, the electrical performance of the solar cell also benefits from Cu plating, because the bulk conductivity of plated Cu is higher than that of Ag pastes, and the contact to Si of plated Ni/Cu is superior to sintered Ag contact. The Cu fingers defined by laser contact opening (LCO) can also be made thinner than screen-printed Ag fingers, further reducing the shading loss for the former.

We recently demonstrated the fully electro-less plating (ELP) Ni and electrolytic plated Cu rear electrodes on n-PERT cells. The rear-side plated Ni/Cu electrode can be a cathode for front-side forward bias plating (FBP) process [6].

2 EXPERIMENTS

A schematic diagram of solar cells is shown in Figure 1. We used 6'' n-type CZ wafers with thickness of 170um and resistivity of 3~5 Ω-cm as starting materials to fabricate the solar cells. The rear side was polished with random pyramids and phosphorous thermal diffusion on the rear–side surface formed a rear surface field to enhance passivation. Amorphous SiNx/SiO_2 thin films (~80 nm) were deposited on the rear side of cells in a PECVD reactor to passivate the doped silicon surface. The front side was textured with random pyramids and Boron thermal diffusion on the front–side surface formed a front surface field to enhance passivation. Amorphous SiNx/SiO_2 thin films (~80 nm) were deposited on the front side of cells in a PECVD reactor to passivate the doped silicon surface. The Anti-Reflective Coating (ARC) is then ablated with a UV ns-laser to form the rear and front pattern. The double side metallization is performed in cluster equipment from Motech, with a sequence as follows: the rear electro-less plated Ni seed metal layer with plated Cu layer is used on rear-side electrode and the front forward bias Ni+Cu plating is used on front-side electrode. Finally, a rapid thermal anneal (RTA) is performed in a belt furnace from Despatch to improve the front contact resistance and form a nickel silicide.

The process flow for this front junction cell can be seen on Figure 2. The cells are manufactured on ~170 μm thick, n-type Cz-Si wafers. After texturing, the wafers are cleaned and a phosphorus implant are applied to form a n+ BSF with an Rsheet= 50 ± 5Ω/sq. The boron diffusion the BSG is removed and the dopants are driven-in through a wet-oxidation step. The resulting boron emitter features a sheet resistance of 75 ± 5 Ω/sq. A SiNx coating is deposited by PECVD on the front side as an ARC. A SiNx layer is also deposited on the rear side (on top of the thermal oxide).

3 RESULTS AND DISCUSSIONS

Figure 3(a) shows the front-side metal finger cross-section by SEM analysis. The metal contact was opened by UV-ns laser patterning on passivated dielectric films. The opening line contact is around 30μm width. The plated front-side finger on laser open area is about 46μm width and 12μm height. The seed Ni layer thickness was around 1um.

Figure 3(b) shows the rear-side electrode, including fully plated ELP Ni and electrolytic Cu. The thickness of ELP Ni is around 0.3~3μm. The locally shunting of Nickel silicide formation could be occurred from thick Ni seed layer. The uniformity of Ni film is needed to improve on rear n+ surface next work.

The cell performance of printed Ag and plated Ni/Cu rear electrodes is shown in table 1. The cell's efficiency for rear plated electrodes show low current density (Jsc) and open-circuit voltage (Voc). The Voc decrease for rear

ELP Ni case might be attributed to rear-side laser contact opening damage. The rear-side laser condition and the open ratio needed to be optimized next work, too. The Jsc drop might be attributed to electrical and to optical losses. The laser damage on opening contacts or non-optimal open ratio both will affect to result Jsc losses. On the other hand, the absorption of evanescent waves at rear-side metal surface may be due to optical losses issue [7].

Figure 4 shows external quantum efficiency (EQE) to compare rear printed Ag and plated Ni/Cu cases. The poor internal reflection of light at the rear side, as can be seen from the low reflectance in the range of λ>1050 nm, can be explained by the absorption of evanescent waves at the metal surface.

4 CONCLUSION

In this study, the electro-less plating and the forward bias plating Ni/Cu with UV ns-laser contact opening processes were employed to the rear- and front-side of n-PERT silicon solar cells. We used the low cost ELP Ni processes rear electrodes to replace PVD metal or printing metal paste. The double side Cu-plated cell efficiency needs still to be improved on n-type PERT solar cells. However, the rear laser processes optimal, dielectric layers scheme and ELP Ni film quality need to be motified in future work.

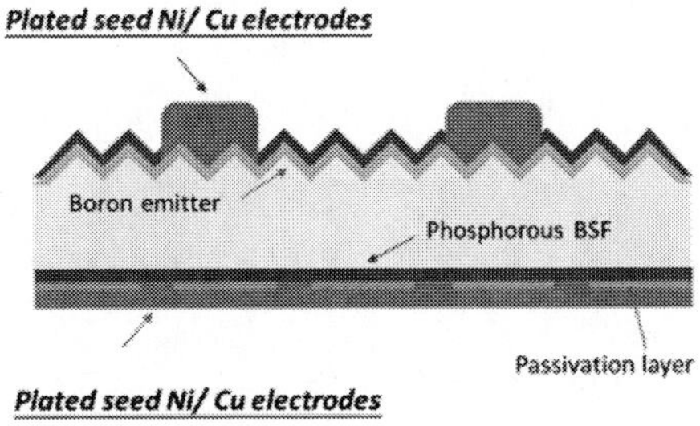

Figure 1 A schematic cross-section of n-type silicon solar cells

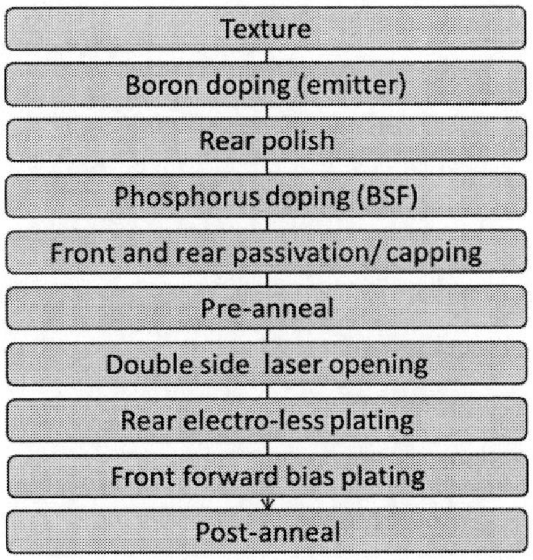

Figure 2 The flow chart with the processing steps

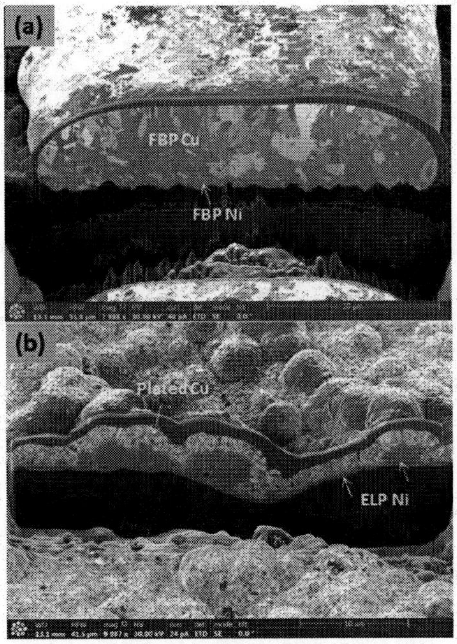

Figure 3 Metalization electrodes by SEM analysis: (a) front-side forward bais plated Ni/Cu finger; (b) rear-side electro-less Ni/ plated Cu

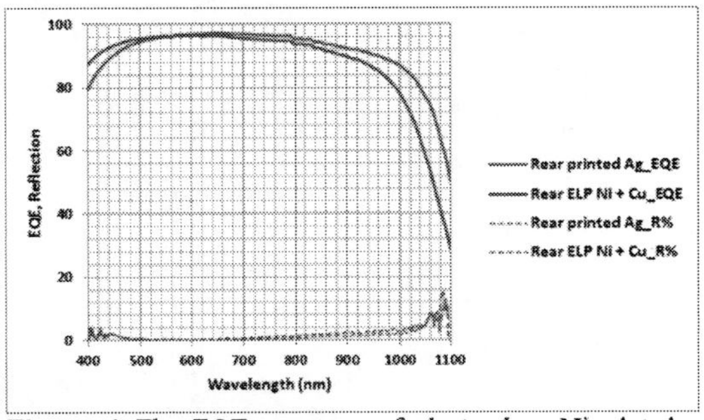

Figure 4 The EQE response of electro-less Ni plated rear-side electrodes for n-PERT solar cells

Table 1 Normalized cell efficiency with different metallization rear electrodes

Conditions	Eff. (%)	FF	Voc (mV)	Jsc (mA/cm^2)
Rear printed Ag	1	1	1	1
Rear ELP Ni+Cu	0.955	1.023	0.982	0.951

5 REFERENCES

[1] Metz, A. Et al. 2014, "Industrial high performance crystalline silicon solar cells and modules based on rear surface passivation technology", *Solar Energy Materials and Solar Cells,* 120:417–25

[2] Horzel, J. Et al. 2014, "Low Cost Metallisation Based On Ni/Cu Plating Enabling High Efficiency Industrial Solar Cells", *Proc. 29th EU-PVSEC* 507–12.

[3] Bartsch, J. Et al. 2014, "21.8% Efficient n-type Solar Cells with Industrially Feasible Plated Metallization", *Energy Procedia* 2014; 55:400–9

[4] Yu, S. et al. 2014, "20.63% nPERT Cells and 20% PR Gain Bifacial Module", *40th IEEE PVSC*, Denver, Colorado, USA

[5] Kuang-Chieh, L. et al. 2016, "Cu-plated electrodes with green nano-laser opening metal contact on n-type silicon solar cells", *Proceedings of 32th EU PVSEC*, Munich, Germany

[6] Kamp, M. 2016, "Forward bias plating for p-contacts", *Electrochemical Processes For Metallization of Novel Silicon Solar Cells*, pp. 61-64

[7] Feldmann, F. Et al. 2014, "Passivated Rear Contacts for High-Efficiency n-type Si Solar Cells Providing High Interface Passivation Quality and Excellent Transport Characteristics", *Solar Energy Materials & Solar Cells*, pp. 270-274

BIFACIAL PERC+ SOLAR CELLS AND MODULES: AN OVERVIEW

T. Dullweber[1], H. Schulte-Huxel[1], C. Kranz[1], S. Blankemeyer[1], U. Baumann[1], R. Witteck[1], R. Peibst[1,2], M. Köntges[1], R. Brendel[1,3], Y. Yao[4]

[1]Institute for Solar Energy Research Hamelin (ISFH), Am Ohrberg 1, D-31860 Emmerthal, Germany
[2]Institute for Electronic Materials and Devices, Leibniz Universität Hannover, Schneiderberg 32, D-30167 Hanover, Germany
[3]Department of Solar Energy, Institute of Solid-State Physics, Leibniz University of Hanover, Appelstrasse 2, D-30167 Hanover, Germany
[4]Meyer Burger Technology AG, Schorenstrasse 39, CH-3645 Gwatt (Thun), Switzerland

ABSTRACT: Since its first publication in 2015, the PERC+ cell concept has been rapidly adopted by several solar cell manufacturers worldwide. The fast industrial implementation is facilitated by the very similar process technology of bifacial PERC+ cells and main stream monofacial PERC cells. Conversion efficiencies of industrial PERC+ solar cells up to 21.6% with front side illumination and 17.3% with rear side illumination were reported. Meanwhile, four companies are offering commercial bifacial PERC+ modules with maximum rating power around 300 W_p when illuminated from the front side, only. These modules apply 60 PERC+ cells with 4 or 5 busbars, which are interconnected by conventional stringing and tabbing technology. First small scale outdoor installations verify an increase of the energy yield relative to monofacial PERC modules between 13% and 22%. Two large scale out door installations with 2 MW_p and 20 MW_p are under construction in Taiwan and China, respectively. We report for the first time in detail on a novel bifacial PERC+ prototype module by applying the Smart Wire Connection Technology. We interconnect 18 halved PERC+ solar cells by soldering 18 wires directly to the Ag front and Al rear fingers. The resulting prototype module exhibits independently confirmed front and rear side efficiencies of 19.8% and 16.4%, respectively. Additionally, Meyer Burger certified a full-size PERC+ SWCT module according to the IEC 61215 norm thereby demonstrating the long term reliability of this novel module technology.

Keywords: silicon solar cells, PERC, PERC+, bifacial, Smart Wire Connection Technology

1 INTRODUCTION

The photovoltaic (PV) industry is currently converting their production capacity from silicon solar cells with full-area aluminum rear contact to Passivated Emitter and Rear Cells (PERC) [1,2]. Accordingly, the photovoltaic technology roadmap ITRPV forecasts an increased market share of PERC solar cells from presently 15% towards 50% until 2026 [3]. These industrial PERC cells apply p-type wafers and a full-area screen-printed aluminum (Al) rear layer which only locally contacts the silicon wafer at areas where the rear passivation has been removed by laser contact opening (LCO) [2]. The full-area aluminum layer prevents any transmission of sunlight from the rear side into the silicon wafer and hence any bifacial applications of these industrial PERC cells.

On the other hand, bifacial solar cell concepts receive increasing interest for applications e.g. in PV power plants where the produced electricity can be increased by up to 20% using bifacial instead of monofacial solar modules [4,5]. Accordingly, the photovoltaic technology roadmap ITRPV predicts a market share of bifacial solar modules of 30% until 2026 [3]. Presently, industrial bifacial solar cell concepts are mainly applying n-type wafers such as Passivated Emitter and Rear Totally diffused (PERT) solar cells [6,7,8,9]. However, a challenge of these bifacial PERT cells is that they typically require screen-printed silver (Ag) finger grids on both wafer sides and hence consume a high amount of expensive Ag paste. Also, the n-PERT cells use single sided boron and phosphorus doping which requires additional or other process steps compared to p-type PERC cells and hence may increase manufacturing costs.

At ISFH and in parallel at SolarWorld, in 2015 we introduced a bifacial PERC solar cell named PERC+ [10] which applies a screen-printed Al finger grid at the rear side enabling front side efficiencies up to 21.5% and rear side efficiencies up to 16.7% [11]. Depending on the detailed installation conditions in a PV power plant, PERC+ cells can increase the energy yield by up to 25% due to their bifacial nature when applied to glass-glass modules [10, 12]. Another potential application is building-integrated PV where the bifacial PERC+ cells may provide higher aesthetic appearance compared to solar cells with full-area Al layer [13]. SolarWorld pioneered mass production of bifacial PERC+ cells [14]. Meanwhile, several additional solar cell manufacturers introduced bifacial PERC+ into large volume production and are now offering commercial bifacial glass/glass modules applying PERC+ solar cells.

This paper provides an overview of the status of the implementation of the PERC+ solar cell concept into industrial solar cell and bifacial module production. Section 2 of this paper explains the process technology of PERC+ solar cells, summarizes published front and rear side efficiencies of industrial PERC+ solar cells, and highlights the implications of the Al finger print on the Al-Si allying process and Al contact formation. The third section outlines the status of commercial bifacial PERC+ glass/glass modules applying conventional tabbing and stringing interconnection technology, which are since recently available by several solar cell manufacturers. In addition, section 3 shows examples of first outdoor field installations of bifacial PERC+ modules on roof top installations or PV power plants and their bifacial gain compared to monofacial PERC reference modules. In section 4, we demonstrate a novel prototype module where we apply the Smart Wire Connection Technology

(SWCT) [15] from Meyer Burger AG, Switzerland, to busbar less PERC+ solar cells by soldering 18 wires directly to the front Ag fingers and rear Al fingers without the use of Ag pads. Hence, the Ag paste consumption is reduced to 55 mg per PERC+ cell to print the front side Ag fingers. The resulting PERC+ SWCT prototype module exhibits independently confirmed conversion efficiencies of 19.8% when illuminated from the front and 16.7% when illuminated from the rear. We have shown first results of this PERC+ SWCT prototype module in an oral presentation at the PVSEC-26 in 2016 [16] and now publish this new module concept for the first time.

2 INDUSTRIAL PERC+ SOLAR CELLS

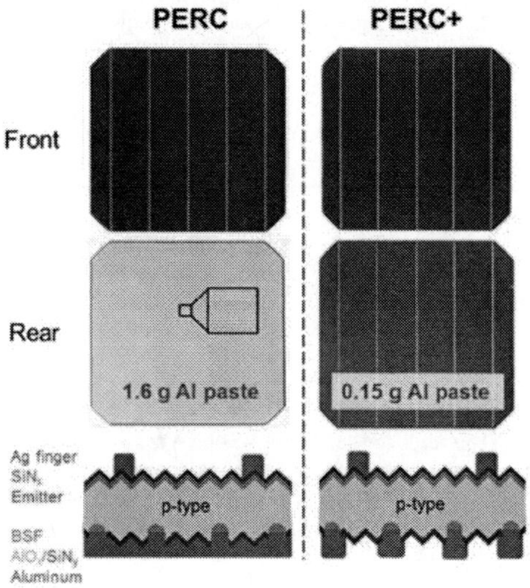

Figure 1: Photographs of the front and rear side as well as schematic cross section drawings of industrial PERC and PERC+ solar cells. PERC+ cells enable bifacial applications and reduce the Al paste consumption while applying the same processing sequence as industrial PERC solar cells.

ISFH and SolarWorld first independently and later jointly started the development of a bifacial PERC solar cell design in 2015 by applying a screen-printed rear Al finger grid instead of the conventional full-area aluminum (Al) rear layer while using the same PERC manufacturing sequence with only minimal recipe modifications for rear passivation, laser contact opening and Al screen printing [10], see Figure 1. Hence, a monofacial PERC cell production line can be switched to produce bifacial PERC solar cells without requiring any invest into new or different production tools. We named this novel cell concept PERC+ and demonstrated several advantages in the initial publications [10,11]. In particular, the Al finger grid enabled bifacial applications of the PERC+ cells with front side efficiencies up to 21.2% and rear side efficiencies up to 16.7% measured with a black chuck [11]. The corresponding bifaciality was up to 80.0%. When measured with a reflective brass chuck, PERC+ cells demonstrated front side efficiencies up to 21.5% compared to conventional PERC cells with 21.1% efficiency [11]. The Al paste consumption of the PERC+ cells was drastically reduced to 0.15 g instead of 1.6 g for the conventional PERC cells [10]. Hence, PERC+ solar cells are attractive for both, bifacial and monofacial module applications [10] which is why we choose the naming convention PERC+ rather than e.g. biPERC or bifiPERC.

In 2015, two additional publications addressed the concept of bifacial PERC+ cells. Trina Solar published bifacial glass glass modules applying bifacial PERC+ solar cells designed for aesthetic optical appearance in building integrated photovoltaic applications [13]. The Fraunhofer ISE assessed the concept of bifacial PERC+ cells mainly by numerical simulations of the potential front and rear conversion efficiencies and corresponding bifacial gains [12]. Since then, several solar cell manufacturers have introduced PERC+ solar cells into pilot production or mass production. Table I summarizes published PERC+ conversion efficiencies when illuminated from the front or rear side. Whereas ISFH set the benchmark in 2015 for high PERC+ front and rear side efficiencies as outline above, first published PERC+ conversion efficiencies of solar cell manufacturers were 20.3% [13] in 2015 and 20.7% [17] in 2016 which then continuously improved to 21.6% in 2017 as reported by LONGi Solar [20] and by ISFH in this paper. The 21.6% efficiency obtained by ISFH are independently confirmed by ISFH CalTeC with V_{oc} = 667 mV, J_{sc} = 39.2 mA/cm^2, FF = 82.4%. Benefiting from continuous improvement of the industrial PERC cell technology with a current record efficiency of 22.6% [21], very likely PERC+ solar cells with front side efficiencies above 22% will be demonstrated soon. At the same time, the conversion efficiency when illuminated from the rear side of PERC+ cells produced by cell manufacturers improved from 13.9% [17] in 2016 towards 17.3% [20] in 2017. For the 21.6% efficient PERC+ cell fabricated by ISFH in 2017 we did not measure a rear side efficiency as the rear Al finger grid has been optimized for monofacial applications and not for high bifaciality. Unfortunately, several leading solar cell manufacturers such as SolarWorld and Trina Solar which are producing bifacial PERC+ cells and modules (see Table II) did not publish any or no recent PERC+ cell efficiencies and hence do not appear at all or only with initial results in Table I.

The limited Al volume of the Al fingers changes the alloying process with the silicon wafer during furnace firing resulting in deeper Al-BSFs [10,22] compared to PERC cells with full-area Al layer. Figure 2a) shows measured Al-BSF depths demonstrating that PERC+ cells obtain up to 8 µm deep Al-BSFs compared to PERC with up to 6 µm. The black line represents a fit using a published model to the PERC data. The dashed red line shows a fit to the PERC+ data using an extended model that takes the limited Al volume of the Al fingers into account [22]. The limited Al volume of the Al fingers leads to a higher silicon concentration in the screen-printed aluminium during furnace firing causing thicker Al-BSFs during the epitaxial re-growth in the cool-down phase [22]. This effect becomes more pronounced for narrow LCO widths around 50 µm which are industrially preferred in order to increase the throughput of the LCO tool. Figure 2 uses different symbol styles where filled circles indicate a filled local Al contact, half-filled circles a partially voided Al contact and open circles a fully voided Al contact as determined by scanning electron microscopy. Whereas for PERC we find rather many voided contacts in particular for narrow LCOs as displayed in Fig. 2, for PERC+ cells we find no fully voided contact. Further analysis reveals that voids in particular occur for Al contacts where the Al-Si eutectic extends more than 20 µm deep into the Si wafer [23]. To explain this finding, we proposed an analytical model that calculates the surface energies of the liquid Al-Si melt, the Si wafer surface and the screen-printed Al particle surface [23]. According to this model, voids form for deep contacts since then during furnace firing a sufficient amount of Al-Si melt is available in order to wet the large surface area of Al particles rather than the small Si wafer surface area. The Al fingers reduce the Al contact depth by about 7 µm which is the reason why PERC+ cells do not exhibit voids [23]. The increased Al-BSF thickness and the reduced amount of voids of PERC+ cells compared to conventional PERC cells result in up to 3 mV higher open circuit voltages V_{oc} due to reduced rear contact recombination [10,23].

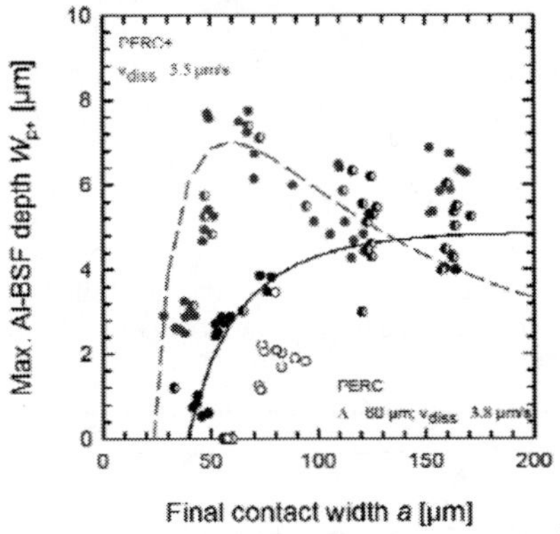

Figure 2: Maximum Al-BSF depth of PERC (black) and PERC+ (red) solar cells for different contact line widths. The black and red line represent fits applying analytic models [22]. PERC+ cells obtain deeper Al-BSFs compared to PERC cells since the Al fingers confine the Si diffusion from the wafer into the Al paste during furnace firing.

3 COMMERCIAL BIFACIAL PERC+ MODULES AND FIELD INSTALLATIONS

SolarWorld pioneered mass production of bifacial PERC+ solar cells and fabrication of novel PERC+ glass/glass bifacial modules named Bisun which have been launched at the Intersolar 2015 [14], see Figure 3. Since then, Neo Solar Power, Trina Solar and LONGi Solar followed this technology route and are now offering commercial bifacial glass/glass modules applying PERC+ solar cells as well [25-27] as summarized in Table II. All manufacturers apply 4 or 5 busbar designs and obtain maximum rating powers between 290 W_p and 305 W_p with 60 PERC+ cells per module [24-27]. The rating power is stated for front side illumination, only. When applying additional rear side illumination, the output power increases accordingly. For example, with 10% additional rear side illumination the output power will increase by approx. 8% towards 330 W_p total output power.

A higher number of busbars such as 4 or 5 is preferred for PERC+ modules, since the specific resistivity of Al fingers is 6 times higher compared to Ag front fingers [10]. A higher number of busbars reduces the Al finger length thereby reducing the Al finger line resistance and related resistive power losses. Since the first publication of a PERC cell applying a 5 busbar design [28], the PV industry is migrating from 3 busbars to 4 or 5 busbars which is beneficial to the industrial adoption of PERC+ cells as explained above. The interconnection of PERC+ cells to strings is accomplished using conventional tabbing-stringing technology where the Cu ribbons are soldered to the Ag front busbars and to Ag pads on the PERC+ rear side.

Figure 3: Photographs of the front and rear side of a commercial Bisun module from SolarWorld applying PERC+ solar cells. Image taken from [10]. Whereas this photograph still shows a 3 busbar design, more recent Bisun modules and other bifacial PERC+ modules apply a 4 or 5 busbar designs as summarized in Table II.

a)

b)

Figure 4: Photographs of field installations of bifacial modules applying PERC+ solar cells. a) 13 kWp Bisun modules from SolarWorld mounted on a 1-axis tracker [29.]. b) 2 MWp roof top installation currently under construction by Neo Solar Power applying their GloryBifi modules [31].

Table I: Published efficiencies of industrial PERC+ solar cells when illuminated from the front or from the rear side. Several leading solar cell manufacturers such as SolarWorld and Trina Solar are producing bifacial PERC+ cells and modules (see Table II) but did not publish any or recent PERC+ cell efficiencies.

Year	Efficiency [%] front / rear	Organization	Source / comments
2015	21.5 / 16.7	ISFH	[11]; industrial process flow. No rear Ag pads
2015	20.3 / n.p.	Trina Solar	[13]; optimized for optical appearance in BIPV
2016	20.7 / 13.9	Big Sun Energy Technology Inc	[17]
2017	21.1* / 15.8*	JinkoSolar	[18]
2017	21.4 / n.p.	Neo Solar Power	[19]
2017	21.6 / 17.3	LONGi Solar	[20]
2017	21.6* / n.p.	ISFH	This paper; rear side optimized for monofacial applications

n.p.: not published; *independently confirmed

Table II: Commercially available bifacial modules applying PERC+ solar cells. Maximum rating power stated for front side illumination under standard test conditions (STC, AM1.5, 25°C) and without rear side illumination.

Company / Product	Maximum rating power [Wp]	Comments	Source
SolarWorld / Bisun	290	5 BB, 60 cells	[24]
Neo Solar Power / Glory Bifi	300	4 BB, 60 cells	[25]
Trina Solar / DUOMAX	300	5 BB, 60 cells	[26]
LONGi Solar / LR6 -60PD	305	4 BB, 60 cells	[27]

Table III: Examples of field installations of bifacial modules applying PERC+ solar cells. As far as available, simulated and measured out door performance values are provided. Increase in energy yield refers to the additional energy yield generated by bifacial PERC+ modules relative to monofacial reference PERC modules.

Company	Installation	Increased energy yield	Source
SolarWorld, ISFH, ISE	Theoretical calculation	Up to 25% (80% albedo, 0.5 m mounting height)	[10]
SolarWorld	3.2 kW, Bisun modules, roof installation, Germany	13% measured, 13.3% calculated (74% albedo, 0.28 m mounting height)	[29]
SolarWorld	13 kW, Bisun modules, 1-axis tracker, Germany	22% measured (17% albedo, 0.9 m mounting height)	[29]
Trina Solar	20 MWp, DUOMAX modules, solar power plant, China	Under construction on sandy ground with high diffuse reflection	[30]
Neo Solar Power	2 MWp, Glory Bifi modules, roof installation, Taiwan	Under construction on a Taiwan government building	[31]

When applying bifacial PERC+ modules to outdoor field installations, in particular the albedo of the ground as well as the mounting height of the module are important parameters to maximize the additional diffuse illumination of the modules rear side and hence the additional energy yield of bifacial PERC+ modules compared to monofacial PERC modules. Numerical simulations predict that the energy yield of PERC+ modules can increase by up to 25% when mounted 0.5 m above ground with an albedo of 80% [10]. SolarWorld took the effort to measure the energy yield of Bisun modules in two different small scale out door test installations as summarized in Table III. A 3.2 kWp installation on top of a flat roof with a high albedo of 74% due to white ballast stones demonstrated an increased energy yield of 13% compared to a monofacial PERC reference module which compared well to the calculated energy yield of 13.3% [29]. When mounted to a 1-axis tracker system, the Bisun modules obtained 22% additional energy yield despite a relative low albedo of 17% of the sandy ground [29]. Presently, two large scale bifacial PERC+ field installations are under construction as listed in Table III. Trina Solar is supplying its bifacial DUOMAX modules to a 20 MW_p power plant installation in China on sandy ground with high albedo [30]. At the same time, Neo Solar Power is constructing a 2 MW_p roof top installation on Taiwan government building using its bifacial Glory Bifi modules [31]. Both installations will be important to verify the predicted energy yield increase of bifacial PERC+ modules in real large-scale outdoor field installations.

5 NOVEL PERC+ PROTOTYPE MODULE APPLYING THE SMART WIRE CONNECTION TECHNOLOGY (SWCT)

One step further in terms of reducing resistive losses caused by PERC+ rear Al fingers is to move from 4 or 5 busbar designs to PERC+ solar cells without any busbars. In this case the module interconnections is accomplished by 18 wires which are soldered directly to the Ag front and Al rear fingers applying the Smart Wire Connection Technology (SWCT) of Meyer Burger [15]. We have shown first results of a PERC+ SWCT prototype module in an oral presentation at the PVSEC-26 in 2016 [16] and now report on this new module concept for the first time in detail.

The PERC+ cells of the PERC+ SWCT prototype module have been fabricated at ISFH with a process sequence which is described in detail in Ref. 10. In the reference split group 1, we apply a 5 busbar Ag front grid design. These conventional 5 busbar PERC+ cells are used later on to calibrate the IV tester which measures the busbar less PERC+ cells. The PERC+ cells of split group 2 which are later used for the SWCT module receive no Ag busbars on the front. Instead, for these cells just the Ag fingers are printed and hence the resulting PERC+ cells of split group 2 are busbar less. For the Al screen print, the 5 busbar reference PERC+ cells of group 1 are printed using an Al screen design with 5 busbar H-pattern whereas the PERC+ cells of split group 2 are printed with another screen without Al busbars but which prints only the Al fingers. Both aluminum screens have a finger opening width of 100 µm and a pitch p which is identical to the LCO pitch. The Al fingers are printed aligned to the laser contact openings. Since the busbar less PERC cells receive only an Ag finger print but no Ag front busbars and no rear Ag pads, the Ag paste consumption is reduced to 55 mg per busbar less PERC+ cell. The Ag front and Al rear contacts are fired in a conventional belt furnace during which the Al paste locally alloys with the silicon wafer at areas where the rear passivation has been removed by laser ablation. A schematic drawing of the resulting bifacial PERC+ solar cells is shown in Fig. 1. Finally, 18 busbar less PERC+ cells are laser-cut into half cells which are later on used for the PERC+ SWCT module fabrication.

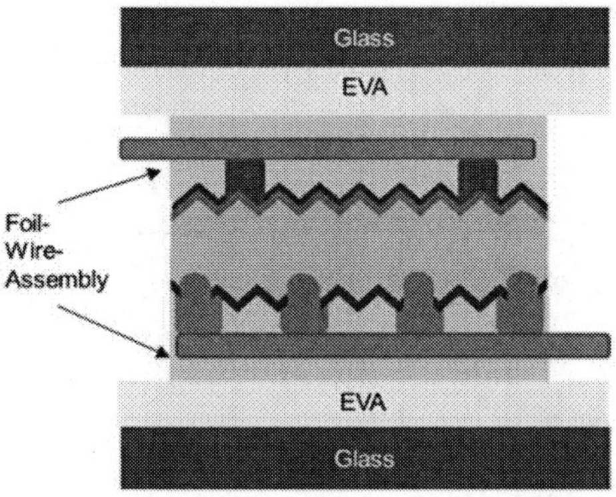

Figure 5: Schematic drawing of the interconnection of PERC+ cells by applying the Smart Wire Connection Technology. 18 InSn-coated wires are soldered directly to the Ag front and Al rear fingers during the lamination process.

We measure the IV characteristics of the 5 busbar reference PERC+ cells using a conventional IV tester which contacts the rear side full-area with a brass chuck and the front side busbars with contact bars. The busbar less PERC+ cells are measured using a grid touch IV tester from PASAN which contacts the front Ag fingers with 30 wires and the rear Al fingers with a brass chuck. We calibrate the PASAN IV tester with a 5 busbar reference PERC+ cell which has been measured with the conventional IV tester.

We use 18 halved busbar less PERC+ solar cells to manufacture a bifacial prototype module applying the Smart Wire Connection Technology (SWCT) of Meyer Burger which is installed at the ISFH SolarTeC technology center. 18 wires with a diameter of 200 µm coated with InSn as low-temperature solder are embedded in a transparent foil using the foil-wire-assembly (FWA) tool from Meyer Burger. Afterwards, the PERC+ SWCT prototype module is assembled as shown schematically in Figure 5. The wires contact the Ag front and Al rear fingers directly without using Ag busbars or Ag pads. The PERC+ cells including the FWA are encapsulated by ethylene-vinyl acetate (EVA) foils with enhanced UV-Transmission [32] on both sides in a glass/glass module configuration. Both glasses are coated with an anti-reflection layer in order to reduce optical losses. In-between neighboring PERC+ cells we place light-reflective films (LRF) produced by the company 3M, USA, which guide the incident light towards the PERC+ cells thereby increasing the light absorption in the module [33]. The module is laminated using a conventional lamination tool. During lamination, the InSn-coated wires are soldered to the Ag front fingers and Al rear fingers. Photographs of the front and rear side of the resulting PERC+ SWCT prototype module are shown in Figure 6 a) and b), respectively. The IV parameters of the PERC+ SWCT prototype module are measured independently by TÜV Rheinland, Germany, for both, front and rear side illumination conditions.

a)

b)

Figure 6: Photographs of the a) front side and b) rear side of the PERC+ SWCT prototype module using 18 halved busbar less PERC+ cells.

When illuminated from the front, 5 busbar (BB) reference PERC+ cells exhibit efficiencies η_{front} up to 21.1% with a very narrow distribution, whereas the front side efficiency of the busbar less PERC+ cells ranges between 20.0% and 20.8%. This decrease in efficiency is caused by a decrease of the Fill Factor (FF) from 80.8% for the 5 BB PERC+ cells to between 76% and 78% for the busbar less PERC+ cells. One part of this FF decline is caused by Ag finger interruptions which occurred due to a non-optimized screen print process for the busbar less PERC+ cells. Another contribution could be due to a not yet optimized calibration procedure of the PASAN IV tester at ISFH. The open circuit voltage V_{oc} ranges from 657 mV to 663 mV for both, 5 BB and no BB PERC+ cells. The no BB PERC+ cells obtain the highest J_{sc} values up to 40.4 mA/cm^2 due to the missing busbar shadowing, whereas the 5 BB PERC+ cells exhibit values up to 37.7 mA/cm^2. Assuming that the busbar less PERC+ cells should exhibit at least as high FF values as the 5 BB reference cells when the issues with screen printing and possibly IV testing are fixed, their short term efficiency potential is $\eta_{front} = 21.5\%$ due to the high J_{sc} values.

When illuminated from the rear, the busbar less PERC+ cells exhibit conversion efficiencies η_{rear} between 15.8% and 16.5%. This relatively low rear side efficiency is primarily caused by a low J_{sc} of around 32 mA/cm^2. As analyzed and explained in detail in Ref. 10, the low J_{sc} values are caused by a high reflectance of the PERC+ rear side due to the relatively wide Al fingers accounting for approximately 10% metallized area as well as due to non-ideal anti-reflection properties of the AlO_x/SiN_y rear passivation in-between the Al fingers. The bifaciality of the busbar less PERC+ cells is defined as η_{rear} divided by η_{front} and reaches values up to 79%. As a figure of merit of bifaciality, we define the equivalent bifacial efficiency as

$$\eta_{eq,0.1} = \eta_{front} + 0.1\times \eta_{rear} \qquad (1)$$

The factor 0.1 describes the additional stray light intensity irradiating to the PERC+ cells rear side relative to the AM1.5g front side illumination. In practice, this factor can vary between 0 and 0.25 depending e.g. on the albedo of the ground and on the detailed mounting geometries of bifacial modules [10]. Taking the measured front and rear side efficiencies above, the busbar less PERC+ cells exhibit an equivalent bifacial efficiency $h_{eq,0.1}$ up to 22.4% with a short term improvement potential up to 23.1% when solving the Ag finger interruptions and possibly the IV tester contacting issues.

The IV parameters of the PERC+ SWCT prototype module have been independently measured by TÜV Rheinland, Germany, and are summarized in Table IV. When illuminated from the front side, the module exhibits an aperture conversion efficiency η_{front} of 19.8%, a V_{oc} of 11.8 V, a short circuit current I_{sc} of 4.8 A, and a *FF* of 78.7%. The module V_{oc} corresponds well to the sum of the V_{oc} values of 11.9 V of the 18 busbar less PERC+ solar cells as displayed in line 3 in Table I. Also, the average I_{sc} of 4.91 A of the 18 PERC+ cells corresponds well to the module I_{sc} taking into account optical losses e.g. due to the higher reflectance of the module glass compared to the PERC+ cell. However, the module *FF* is 1.4% higher compared to the average PERC+ cell *FF* which is an indication that the PASAN IV tester may underestimate the PERC+ cell *FF* e.g. due to a not yet optimized calibration procedure. When illuminated from the rear side, the module I_{sc} decreases to 3.94 A resulting in a rear side module efficiency η_{rear} of 16.4% which corresponds well to the respective PERC+ cell rear side efficiencies. Accordingly, the module bifaciality η_{rear} / η_{front} is 83%. Applying equation 1 results in an equivalent bifacial efficiency $\eta_{eq,0.1}$ of 21.4% of the PERC+ SWCT prototype module. For comparison, the highest monofacial PERC module efficiency reported so far is 20.2% [33] which demonstrates the additional energy yield to be expected by the novel bifacial PERC+ SWCT module when applied in suitable out door field installations.

In addition to high conversion efficiencies, another important criterion for a new module technology is the reliability of the technology in order to allow 20 or 25 years of module operation without significant degradation of the nominal power out. For the novel PERC+ SWCT module technology, in particular the reliability of the interconnection between the Al fingers and the wires has to be demonstrated as well as other reliability tests. Hence Meyer Burger recently fabricated a PERC+ SWCT module applying 60 full-size PERC+ solar cells which were interconnected with the SWCT module technology using 18 wires per PERC+ cell [34]. The wires have a diameter of 300 µm and are coated with an In-free low-temperature solder material. The PERC+ SWCT module has been externally certified [34] at Certisolis, France, according to the norm IEC 61215, which tests all reliability relevant issues such as module performance under temperature cycling and mechanical load. In addition, the PERC+ SWCT module passed [34] the norm IEC 61730 which tests the product safety, e.g. in terms of fire protection standards.

Table IV: The IV parameters of the PERC+ SWCT prototype module are measured independently by TÜV Rheinland for both, front and rear side illumination conditions. Additionally, for the 18 PERC+ cells which have been used for the module fabrication the sum of the V_{oc} values as well as the average I_{sc}, FF and η values are provided.

	V_{oc} [V]	I_{sc} [A]	FF [%]	η [%]
PERC+ SWCT module, front	11.8	4.80	78.7	19.8*
PERC+ SWCT module, back	11.8	3.94	78.8	16.4*
18 busbar less PERC+ cells, front	11.9$^{\Sigma}$	4.91^{ø}	77.3^{ø}	20.5^{ø}

*independently confirmed by TÜV Rheinland; $^{\Sigma}$ = sum; ^{ø} = average

6 CONCLUSIONS

Since its first publication by ISFH and SolarWorld in 2015, the PERC+ cell concept has been rapidly adopted by several solar cell manufacturers worldwide. The fast industrial implementation is facilitated by the very similar process technology of bifacial PERC+ cells and monofacial PERC cells, which are becoming main stream in the PV industry. Published conversion efficiencies of industrial PERC+ solar cells increased in the past two years up to 21.6% when illuminated from the front side and up to 17.3% when illuminated from the rear. Further efficiency increases can be expected in the near future since PERC+ cells benefit from the continuous technology improvements of monofacial PERC cells. Additionally, the limited volume of the Al fingers change the alloying properties during furnace firing resulting in deeper Al-BSFs and hence higher open circuit voltages of PERC+ cells compared to PERC cells. That is why PERC+ cells are attractive for both, monofacial and bifacial module applications.
Meanwhile, four companies are offering commercial bifacial PERC+ modules with maximum rating power around 300 W_p when illuminated from the front side, only. These modules apply 60 PERC+ cells with 4 or 5 busbars, which are interconnected by conventional stringing and tabbing technology. First small scale outdoor installations by SolarWorld of bifacial PERC+ modules measured an increase of the energy yield relative to monofacial PERC modules between 13% and 22% depending on the installation conditions. Two large scale out door installations with 2 MW_p and 20 MW_p of bifacial PERC+ modules are under construction in Taiwan and China, respectively.
In this paper, we published for the first time a novel bifacial PERC+ prototype module by applying the Smart Wire Connection technology (SWCT). We interconnect 18 halved PERC+ solar cells by soldering 18 wires directly to the Ag front and Al rear fingers without using Ag busbars or Ag pads. The resulting prototype module exhibits independently confirmed front and rear side efficiencies of 19.8% and 16.4%, respectively. These values correspond to an equivalent bifacial efficiency of 21.4% which exceeds the world record monofacial PERC module efficiency by more than 1% absolute. Additionally, Meyer Burger certified a full-size PERC+ SWCT module according to the IEC 61215 norm thereby demonstrating the long term reliability of this novel module technology.

ACKNOWLEDGEMENTS

We thank the German Federal Ministry for Economic Affairs and Energy for funding part of this work under contract no. 032577C (HELENE) as well as SolarWorld for the fruitful collaboration within the HELENE project. In addition, we thank Hua Li, LONGi Solar, for contributing their latest PERC+ cell results to this paper.

REFERENCES

[1] A. W. Blakers, A. Wang, M. Adele, J. Zhao, M. A. Green, Appl. Phys. Letters **55**, 1363 (1989).
[2] T. Dullweber and J. Schmidt, IEEE J. Photovoltaics **6** (5), 1366 (2016).
[3] International Technology Roadmap for Photovoltaic (ITRPV.net), 2015 Results, 7th edition, October 2016. Available at http://www.itrpv.net/Reports/Downloads/
[4] C. Comparotto, M. Noebels, L. Popescu, A. Edler, J. Ranzmeyer, T. Klaus, V. D. Mihailetchi, R. Harney, J. Lossen, T. S. Böscke, D. Schär, H. Nussbaumer, T. Baumann, F. P. Baumgartner, Proc. 29th European Photovoltaic Solar Energy Conference, 2014, p. 3248.
[5] G. J. M. Janssen, B. B. Van Aken, A. J. Carr, A. A. Mewe, Energy Procedia **77**, 364 (2015).
[6] I. G. Romijn, B. B. Van Aken, J. Anker, P. Barton, A. Gutjahr, Y. Komatsu, M. Koppes, E. J. Kossen, M. W. P. E. Lamers, D. S. Saynova, K. C. J. J. Tool, Y. Zhang-Steenwinkel, P. R. Venema, A. H. G. Vlooswijk, C. Schmitt, H. Kühnlein, N. Bay, M. König, A. Stassen, Proc. 28th European Photovoltaic Solar Energy Conference, 2013, p. 736.
[7] D. Song, J. Xiong, Z. Hu, G. Li, H. Wang, H. An, B. Yu, B. Grenko, K. Borden, K. Sauer, T. Roessler, J. Cui, H. Wang, J. Bultman, A. H. G. Vlooswijk, P. R. Venema, Proc. 38th IEEE Photovoltaic Specialists Conference, 2012, p. 3004.
[8] V. D. Mihailetchi, J. Jourdan, A. Edler, R. Kopecek, R. Harney, D. Stichtenoth, J. Lossen, T. Boescke, H. J. Krokoszinski, Proc. 25th European Photovoltaic Solar Energy Conference, 2010, p. 1446.
[9] T. Dullweber, N. Wehmeier, A. Nowack, T. Brendemuehl, S. Kajari-Schroeder, and R. Brendel, Phys. Status Solidi A **213** (11), 3046 (2016).
[10] T. Dullweber, C. Kranz, R. Peibst, U. Baumann, H. Hannebauer, A. Fülle, S. Steckemetz, T. Weber, M. Kutzer, M. Müller, G. Fischer, P. Palinginis, and H. Neuhaus, Prog. Photovolt: Res. Appl. **24**, 1487 (2016)
[11] T. Dullweber, C. Kranz, R. Peibst, U. Baumann, H. Hannebauer, A. Fülle, S. Steckemetz, T. Weber, M. Kutzer, M. Müller, G. Fischer, P. Palinginis, and H. Neuhaus, Proc. 31st European Photovoltaic Solar Energy Conf., 2015, p. 341.
[12] K. Krauß, F. Fertig, J. Greulich, S. Rein, and R.

Preu, Phys. Status Solidi A **213**, 68 (2016).
[13] B. Liu, Y. Chen, Y. Yang, J. Dong, H. Shen, Z. Feng, P.J. Verlinden, Proc. 31st European Photovoltaic Solar Energy Conf., 2015, p. 659
[14] SolarWorld press release, 2015. Available at http://www.pv-tech.org/news/intersolar_europe_solarworld_to_launch_glass_glass_bifacial_modules
[15] A. Faes, M. Despeisse, J. Levrat, J. Champliaud, N. Badel, M. Kiaee, T. Söderström, Y. Yao, R. Grischke, M. Gragert, J. Ufheil, P. Papet, B. Strahm, B. Cattaneo, J. Cattin, Y. Baumgartner, A. Hessler-Wyser, C. Ballif, Proc. 29th European Photovoltaic Solar Energy Conference, 2014, p. 2555.
[16] T. Dullweber, H. Schulte-Huxel, C. Kranz, S. Blankemeyer, U. Baumann, R. Witteck, M. Köntges, R. Brendel, oral presentation at the 26th Photovoltaic Science and Engineering Conference, Singapore, 2016
[17] S.-Y. Chen, H.-C. Liu, H.-Y. Chang, Y.-Y. Chiu, Y.-H. Wang, Y.-H. Lin, S.-H. Yu, W.-J. Lih, C.-H. Du, Proc. 32nd European Photovoltaic Solar Energy Conference (2016), p. 772
[18] Communication with JinkoSolar
[19] https://www.nsp.com/nspsolarcells?lang=en
[20] Communication with H. Li, LONGi Solar
[21] W. W. Deng, F. Y. Ye, R. Liu, Y. Li, H. Chen, Z. Xiong , Y. Yang, Y. Chen, Y. Wang, P. P. Altermatt, Z. Feng, P. J. Verlinden, 44th IEEE Photovoltaic Specialists Conference, 2017, in press
[22] C. Kranz, B. Wolpensinger, R. Brendel, T. Dullweber., IEEE J. Photovoltaics, **6** (4), 830 (2016)
[23] C. Kranz, U. Baumann, B. Wolpensinger, F. Lottspeich, M. Müller, P. Palinginis, R. Brendel, T. Dullweber, Solar Energy Materials and Solar Cells, **158**, 11 (2016).
[24] https://www.solarworld.de/en/products/sunmodule-bisun-protect/
[25] https://www.nsp.com/nspsolarmodules?lang=en
[26]http://www.trinasolar.com/de/product/duomax4060/duomax-twin-deg5c07ii
[27]http://en.longi-solar.com/Home/Products/module/id/12_.html
[28] H. Hannebauer, T. Dullweber, U. Baumann, T. Falcon, R. Brendel, Physica Status Solidi RRL **8**, 675 (2014)
[29] H. Neuhaus (SolarWorld), presented at PV CellTech Conference, Malaysia (2017)
[30] http://www.trinasolar.com/en-uk/resources/newsroom/mon-05012017-1500
[31] https://www.pv-tech.org/news/nsp-to-construct-first-commercial-rooftop-system-using-its-bifacial-solar-m
[32] M. R. Vogt, H. Holst, H. Schulte-Huxel, S. Blankemeyer, R. Witteck, D. Hinken, M. Winter, B. Min, C. Schinke, I. Ahrens, M. Köntges, K. Bothe, R. Brendel, Energy Procedia **92**, 523 (2016).
[33] H. Schulte-Huxel, R. Witteck, S. Blankemeyer, S. Spätlich, M. R. Vogt, D. Hinken, H. Holst, U. Sonntag, T. Brendemühl, I. Ahrens, T. Neubert, H. Hannebauer, U. Baumann, T. Dullweber, M. Köntges, K. Bothe, R. Brendel, IEEE J. Photovoltaics **7** (1), p. 25 (2016)
[34] Y. Yao et al., oral presentation at the SNEC 11th International Photovoltaic Power Generation and Smart Energy Conference & Exhibition, Shanghai (2017)

EFFECTS OF TELLURIUM OXIDE IN SILVER PASTE ON THE ELECTRICAL LOSSES IN SILICON SOLAR CELLS

*Takayuki Aoyama [1,2], Mari Aoki [3], Isao Sumita [3], Yasushi Yoshino [1], Yoshio Ohshita [3], and Atsushi Ogura [2]
[1] Noritake Co., Limited, 300, Higashiyama, Miyoshi-Coh, Miyoshi, Aichi 470-0293, Japan
[2] Meiji University, 1-1-1, Higashimita, Tama-ku, Kawasaki, Kanagawa 214-8571, Japan
[3] Toyota Technological Institute, 2-12-1, Hisakata, Tempaku-ku, Nagoya, Aichi 468-8511, Japan

*Corresponding author
Phone: +81-561-34-6217, Fax: +81-561-34-4997, E-mail: takayuki_aoyama@n.noritake.co.jp

ABSTRACT: Silver paste contacting to emitter of silicon solar cells creates shunting of p-n junction and induces recombination in the cells, due to metal silver crystallites on the emitter. Constituents of glass frit in the silver paste strongly affect creating the crystallites. However, effects of the glass frit constituents, the most well-known of which are silicon oxide and lead oxide, on the electrical losses are not understood enough, but recently tellurium oxide is highly concerned as a constituent of the frit to achieve high-efficiency of conventional p-type solar cells. In this study, the effects of the tellurium oxide in the silver paste on the electrical losses of the conventional solar cells are investigated by using the effectual "floating contact method". Our study based on the floating contact method will demonstrate that the tellurium oxide can reduce the shunting and the recombination caused by the silver paste in the conventional solar cells. It can be concluded that the tellurium oxide is an effective constituent of the frit to achieve high-efficiency solar cells in terms of its effects on the electrical losses.

Keywords: silicon solar cell, metallization, screen printing, shunts, recombination

1 INTRODUCTION

Silver paste contacts for emitter of silicon solar cells have been pointed out to cause shunting of p-n junction and to induce carrier recombination in the cells [1-3]. Constituents of glass frit in the silver paste strongly affect these electrical losses. Lead oxide and silicon oxide have been typically used as the main constituents. Meanwhile, recently, tellurium oxide is proposed as a constituent of the glass frit in the silver paste for n^+ emitter of conventional p-type solar cells to achieve high efficiency [4], and widely used for commercial products. However, the detail effects of the tellurium oxide on the silver paste have not been clear yet, especially for the electrical losses. In this study, the effects of tellurium oxide on the electrical losses in the conventional p-type solar cells are investigated by using the effectual "floating contact method." Our study will clearly show that the tellurium oxide mitigates the electrical losses caused by the silver paste in the conventional p-type solar cells.

2 EXPERIMENTAL

2.1 Floating contact method

The floating contact method can effectively evaluate effects of paste constituents on the cell parameters, because the method can evaluate the cell parameters under eliminating the paste effects on measuring the parameters: for instance the paste effects on the series resistance of the cells. The contact configuration of the floating contact method shown in Figure 1 consists of two kinds of contacts: the one is "grid-contact" of standard H-pattern, and the other is the "floating contacts". The floating contacts are placed between the gridlines of the grid-contact, being electrically and geometrically isolated [5]. The grid-contact can only play a role of an electrode to measure the *I-V* characteristics of the cells, regardless of whether the cell has the floating contacts or not. On the other hand, the floating contacts are so isolated from the grid-contact that they can hardly affect the grid-contact to measure the cell characteristics, but contrarily affect carrier behavior: for instance shunting or recombination in the emitter, just beneath the floating contacts depending on the paste used for them. Then, the effects of the floating-contacts on the cell characteristics can be measured through the grid-contact by comparing the characteristic of the cell having the floating contacts with the one having nothing. In addition, the area "fraction" of the floating contacts, which is defined as the ratio of the area covered by the floating contacts to the full area of a solar cell, was changed in the range of 0% to about 12% by varying the number of the floating contacts to investigate the effects of the fraction on the electrical losses. Thus, this method can evaluate the electrical losses caused by the paste or the paste constituents used for the floating contacts with changing their fraction.

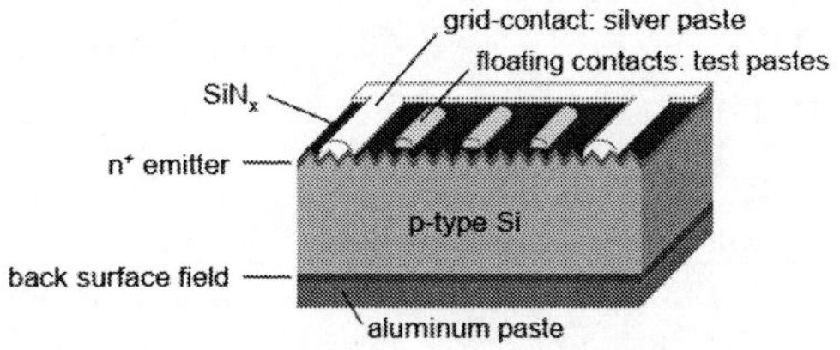

Figure 1: Configuration of the floating contacts on conventional p-type solar cells. The floating contacts are electrically and geometrically isolated from the grid-contact.

2.2 Fabrication of conventional p-type solar cells with the floating contacts

The conventional p-type solar cells shown in Figure 1 were made, using p-type Cz-Si wafers (156 mm × 156 mm), on both sides of which the surfaces were textured in a random pyramid structure. The sheet resistance of the phosphorus-doped n^+ emitter, which was passivated by a SiN_x layer, was approximately 80 Ω/square. The metallization on the n^+-contact was fabricated by screen-printing with a commercially available silver paste as the grid-contact, whereas the metallization for the back surface field (BSF) were done with a commercially available aluminum paste by screen-printing. The grid-contact consists of 3 busbars and 64 fingers. After screen-printing the grid-contact and drying, the floating contacts were additionally screen-printed with test pastes on the n^+ emitter side. Subsequently, the screen-printed pastes were co-fired with an infrared belt furnace at 925 °C.

In this study, three types of homemade test pastes shown in Table 1 were prepared for the floating contacts. All the test pastes contain at least silver powder and organics in common. Ag/glass paste contains certain glass frit, and Ag/glass/TeO_2 paste contains the same quantity of the frit as that of the Ag/glass paste, and a fewer quantity of tellurium oxide (TeO_2) than the frit. Furthermore, the glass frit mainly consists of lead oxide (PbO), silicon oxide (SiO_2), boron oxide (B_2O_3), and zinc oxide (ZnO), not containing the TeO_2.

Table I: Constituents of test pastes

Test paste	Silver	Glass frit	Tellurium oxide
Only-Ag paste	✓	None	None
Ag/glass paste	✓	✓	None
Ag/glass/TeO_2 paste	✓	✓	✓

2.3 Evaluation of the electrical losses

The dark *I-V* curves were measured with an *I-V* tester (ADC4601) in a dark environment. The analysis on the measured dark *I-V* curves was carried out by a numerical fitting method based on the two-diode model to extract some parameters such as shunt resistance and recombination current density. The illuminated *I-V* curves were measured with the same *I-V* tester and a solar simulator (WACOM WXS-156-10) under an air mass 1.5G spectrum with an intensity of 100 mW/cm^2.

Table 2 shows the electrical properties of the conventional solar cell with only the grid-contact (no floating contacts) under the illumination, which are benchmarks on that of solar cells with the floating contacts.

Table II: Electrical properties of conventional solar cell with only the grid-contact

Open-circuit voltage	Short-circuit current density	Fill factor	Efficiency
0.632 V	35.9 mA/cm^2	77.3%	17.6%

3 RESULTS AND DISCUSSION

3.2 Dark *I-V* curves

Figure 2 shows the dark *I-V* curves of the solar cells with only the grid-contact (no floating contacts) and the cells with the floating contacts of each test paste at the floating contact fraction of about 12%. In Figure 3 (a), the dark *I-V* curve of the solar cell with the only-Ag paste is considerably close to that of the solar cell with only the grid-contact. The only-Ag paste has no effects on the electrical losses, because it cannot fire-through the SiN_x layer to contact to the n^+ emitter directly. In contrast, as shown in Figure 3 (b), the dark current of the Ag/glass paste obviously increases with the voltage more rapidly than that of the only-Ag paste. By contrast, the current increase with the voltage becomes lower by adding the TeO_2 to the Ag/glass paste. The rapid-increasing of the current, which is "hump" current, at the low voltage range typically originates in either shunt current or recombination current, or both [6]. Hence, in the case of the Ag/glass paste, adding the glass frit to the only-Ag paste increases the shunt or the recombination current in the solar cells, but adding the TeO_2 to the Ag/glass paste can decrease the current increase.

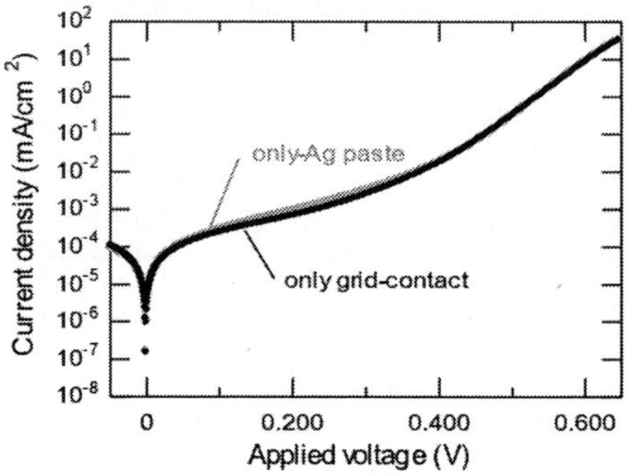

(a)

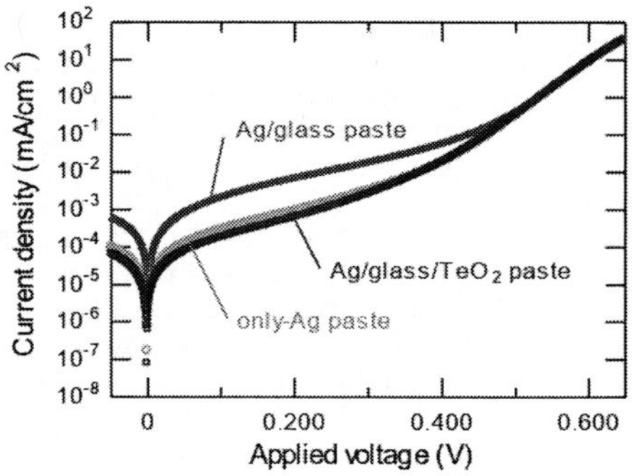

(b)

Figure 2: Dark *I-V* curves of n-type solar cells with only the grid-contact and with the floating contacts of each test paste. (a) Only grid-contact and only-Ag paste at the floating contact fraction of about 12%, and (b) only-Ag paste, Ag/glass paste, and Ag/glass/TeO_2 paste at the floating contact fraction of about 12%. The only-Ag paste has no effects on the electrical losses. On the other hand, the dark current of the Ag/glass paste obviously increases with the voltage more rapidly than that of the only-Ag paste, but adding the TeO_2 to the paste can decrease the current increase as same level with that of the only-Ag paste.

3.2 Shunt resistance and recombination current

The dark *I-V* curves as obtained above were analyzed by applying the two-diode model to extract the shunt resistance R_{sh} and the recombination current density J_{02}. The dependence of the R_{sh} and the J_{02} on the floating contact fraction is shown in Figure 3, on which "zero" percent corresponds to the R_{sh} or the J_{02} of only the grid-contact. The floating contacts of the only-Ag paste and the Ag/glass/TeO_2 paste keep the R_{sh} nearly constant at the same level as only the grid-contact does. By contrast, the R_{sh} of the Ag/glass paste decreases with increasing the fraction, being lower than that of the only-Ag paste and the Ag/glass/TeO_2 paste. Hence, adding the TeO_2 to the Ag/glass paste can mitigate the R_{sh} decrease.

On the other hand, a similar tendency can be observed in the dependence of the J_{02} on the test pastes. The J_{02} of the only-Ag paste and the Ag/glass/TeO_2 paste stays almost constant through the fraction, being comparable with that of only the grid-contact, whereas the J_{02} of the Ag/glass paste significantly increases with increasing the fraction. Thus, adding the TeO_2 can mitigate the J_{02} increase due to the Ag/glass paste. Summarizing the above results, the TeO_2 can mitigate the R_{sh} decrease and the significant J_{02} increase caused by the silver paste with the glass frit.

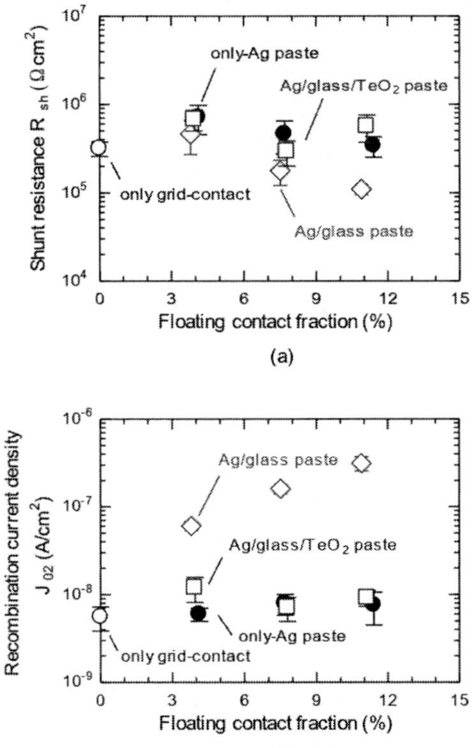

Figure 3: Dependence of (a) shunt resistance R_{sh} and (b) recombination current density J_{02} on the floating contact fraction. Adding the TeO_2 to the Ag/glass paste can reduce the R_{sh} decrease and the J_{02} increase caused by the paste.

3.3 Open-circuit voltage

Figure 4 shows dependence of the open-circuit voltage V_{oc} on the floating contact fraction. The V_{oc} of the solar cells with the only-Ag paste slightly decreases with increasing the fraction, whereas the V_{oc} of the other pastes significantly decreases with same tendency on the fraction. Even though normally V_{oc} decreases due to shadowing [7], but the V_{oc} has decreased monotonically with the fraction, closely relating to shunting of p-n junction [8], [9] and also relating to increasing diffusion or recombination current [9], [10]. The shadowing effect due to the floating contacts of the test pastes on the V_{oc} has to be almost same among the test pastes; therefore, the difference of the V_{oc} between the only-Ag paste and the other pastes is caused by the effects of the glass frit and/or the tellurium oxide in the paste.

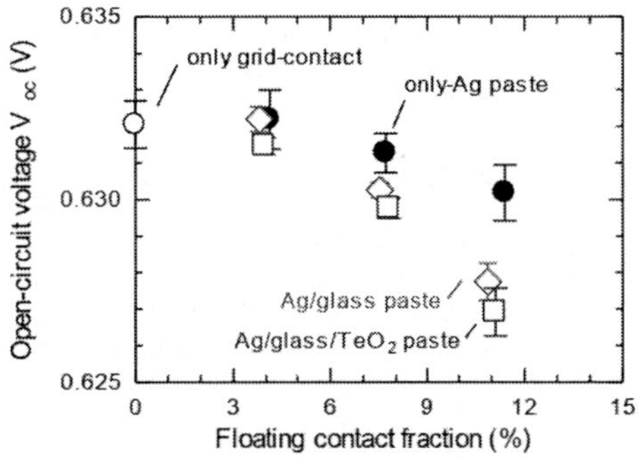

Figure 4: Dependence of open-circuit voltage V_{oc} on the floating contact fraction. The V_{oc} of the solar cells with the only-Ag paste slightly decreases with increasing the fraction, whereas the V_{oc} of the other pastes significantly decreases with same tendency on the fraction.

4 Summary

Lead oxide and silicon oxide have been used as constituents of glass frit in typical silver paste for emitter of silicon solar cells. However, tellurium oxide is proposed as the constituent and widely used for the frit in commercially available silver paste.

In this study, effects of the tellurium oxide on the silver paste are investigated by using the effectual floating contact method. The tellurium oxide in the silver paste can decrease the electrical losses in the conventional p-type solar cells. Hence, it can be concluded that the tellurium oxide is an effective constituent of the frit for reducing the electrical losses to achieve high-efficiency solar cells.

References

[1] B. Ross, "Screenable silver and base metal solar cell contacts," in Proc. 14th IEEE Photovolt. Spec. Conf., San Diego, CA, USA, Jan. 1980, pp. 787–792.

[2] K. Emery et al., "Electro-optical properties of screen printed solar cells," in Proc. 16th IEEE Photovolt. Spec. Conf., San Diego, CA, USA, Sep.1982, pp. 947–951.

[3] M. M. Hilali, A. Rohatgi, and B. To, "A review and understanding of screen-printed contacts and selective-emitter formation," in Proc. 14th Workshop Crystalline Silicon Sol. Cells Modules, Winter Park, CO, USA, Aug. 2004, pp. 109–116.

[4] A. F. Carroll, K. W. Hang, B. J. Laughlin, K. R. Mikeska, C. Torardi, P. D. Vernooy, "THICK-FILM PASTES CONTAINING LEAD-AND TELLURIUM-OXIDES, AND THEIR USE IN THE MANUFACTURE OF SEMICONDUCTOR DEVICES," U.S. Patent 8,497,420 B2, Jul. 30, 2013.

[5] R. Hoenig, "Evaluation and microstructure analysis of thick film contacts for industrial silicon solar cells," Ph.D. dissertation, Fraunhofer Inst. Sol. Energy Syst., Freiburg, Germany, 2014.

[6] J. Beier and B. Voss, "Humps in dark I-V-curves-

analysis and explanation," in Proc. 23rd IEEE Photovolt. Spec. Conf., Louisville, KY, USA, May 1993, pp. 321–326.
[7] M. B. Prince, "Silicon solar energy converters," J. Appl. Phys., vol. 26, pp. 534–540, 1955.
[8] P. A. Payne and E. L. Ralph, "Low temperature and low solar intensity characteristics of silicon solar cells," in Proc. 8th IEEE Photovolt. Spec. Conf., Seattle, WA, USA, Aug. 1970, pp. 135–141.
[9] B. Sopori, "Silicon solar-cell processing for minimizing the influence of impurities and defects," J. Electron. Mater., vol. 31, pp. 972–980, 2002.
[10] R. J. Stirn, "Junction characteristics of silicon solar cells," in Proc. 9th IEEE Photovolt. Spec. Conf., Silver Spring, MD, USA, May 1972, pp. 72–82.

OPTIMIZED BACKSIDE REFLECTANCE FOR ADVANCED METALLIZATION P-TYPE BIFACIAL PERC SOLAR CELLS

Sung-Yu Chen, Yu-Hsuan Lin, Jhen-Fu Huang, Chorng-Jye Huang, Li-Yu Li, Shih-Peng Hsu, Chen-Hsun Du
Green Energy & Environment Research Lab., Industrial Technology Research Institute
No.195, Sec. 4, Chung Hsing Rd., Chutung, Hsinchu, Taiwan 31040, R.O.C

ABSTRACT: In this paper, we proposed the highly potential structure p-type bifacial passivated emitter and rear cells (bi-PERC or PERC+) by optimizing the thickness of rear silicon nitride (SiNx) layer and applying the advanced metallization instead of conventional screen printing on the front surface. The reflectance from the rear side shows obviously vibration phenomenon by varying the thickness of rear SiNx layer, and therefore the effect between the amplitude and period of Al_2O_3/SiNx stack layers were investigated. When the thickness of rear SiNx layer was increased from 90 nm to 190 nm, the front efficiency keeps the value of 20.3% ~ 20.5% but the rear efficiency dramatically drops from 13.36% to 10.9%. Therefore, the appropriate thickness of rear SiNx layer is around 90 nm to 115 nm. For the front side grid metallization we evaluate 4 different split groups which differ in the process, including standard screen printing, dual screen printing, copper electroplating and knotless screen printing for improving cell efficiency. Comparison of the standard and dual screen printing, the cell efficiencies were almost the same. It shows the efficiency gain for copper metallization and knotless printing are 0.07% and 0.27% respectively on mono-crystalline silicon with an area of 244 cm^2.
Keywords: Bifacial、Silicon Solar Cell、Metallization、High-Efficiency、Knotless screen printing

1 INTRODUCTION

The low-cost and high-efficiency passivated emitter and rear cells (PERC) have been successfully transfer to industrially mass production from 2014 and it was predicted that the market share will be more than 50% in 2027 due to the wonderful cell performance [1-2].

In the past two years, p-type bifacial PERC solar cells (bi-EPRC or PERC+) were gradually attached great importance according to the mostly mature technology of PERC [3-5]. Bi-PERC can not only reduce the rear electrode material consumption but also generate double-sided power, therefore it is inevitable tendency in the future.

In this paper, we investigated the influences of different thickness of silicon nitride (SiNx) layer on the rear surface for Bi-PERC by analyzing the both-sided optical and photovoltaic characteristics

In addition, we evaluate 3 different metallization processes, dual screen printing, copper electroplating and knotless screen printing instead of conventional screen printing and estimate the potential of advanced metallization to improve the front efficiency of Bi-PERC by studying the electrode profile and cell efficiencies

2 EXPERIMENTAL

Figure 1 shows the process flow and the cross-section structure for p-type bi-PERC which is based on the PERC fabrication. In this paper, 156.75 mm × 156.75 mm p-type single-crystalline silicon CZ wafer with 200 μm thickness was adopted and the resistivity is about 1~1.5 Ω-cm. The size of wafer is M2, which means that the diameter is 210 mm and the area is 244 cm^2.

There are contaminations and native oxide on the surface of bare wafer therefore a wafer clean procedure is necessary before the facial texturization by using alkali solution. Emitter formation was carried out with sheet resistance of 95 ohm/sq. on the both side by POCl3 diffusion and then phosphor-silicate glass (PSG) was removed by dipping hydrofluoric (HF) acid solution.

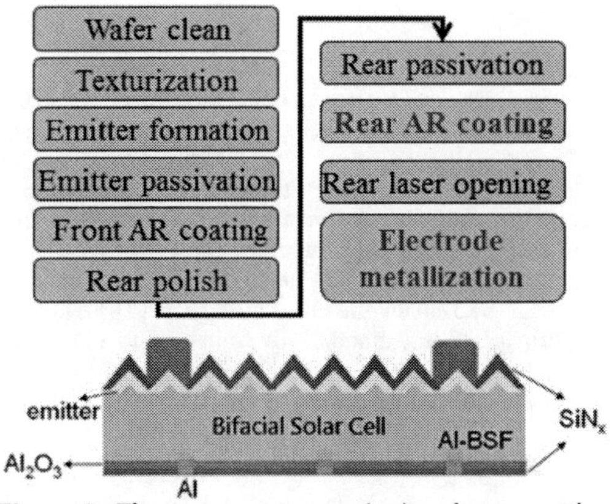

Figure 1: The process sequence (top) and cross-section structure (bottom) for p-type bifacial PERC solar cells.

After that, The following step is thermal oxidation process at 800℃ for 20 min in order to passivated emitter and optimized the doping depth at the same time. The SiNx anti-reflective film with the thickness of 90 nm was coated on the front surface by plasma-enhanced chemical vapor deposition (PECVD). Rear emitter removal and surface polish were completed immediately by single-step etching in alkali solution. In the following, An ultra-thin oxidation layer with less than 2 nm was formed at 550℃ and then the passivated films of aluminum oxide (Al_2O_3) was deposited on the rear flat surface by atomic layer deposition (ALD). After annealing, the anti-reflective layer with different thickness of SiNx was deposited onto Al_2O_3 for forming an Al_2O_3/SiNx stack layers in order to investigate the optical properties for both surface. Passivated stack layers was locally opened by a laser ablating. Finally, 4 different grid metallization methods were used to the fabrication of solar cells [6-8].

3 DISCUSSIONS

The optical reflectance with different thickness of

rear SiNx was measured from the front and rear sides respectively for semi-finished cells before the step of rear laser opening (as shown in Figure 2). For the front side, the optical reflectance is almost the same for all conditions. For rear surface, there is obviously vibration for all cases due to the flatness of the rear surface. Figure 2 (b) shows the standing wave phenomena and there is a red shift as increasing the rear SiNx thickness. We focus on the reflectance within from 600 nm to 900 nm, the reflectance was increasing with the thickness from 90 nm to 190 nm. In other words, the optical absorption was decreased as increasing the thickness of SiNx. The mainly range of absorption for silicon bulk is within 600 nm to 900 nm, and therefore the total reflectance within this range would influence the cell performance of rear side.

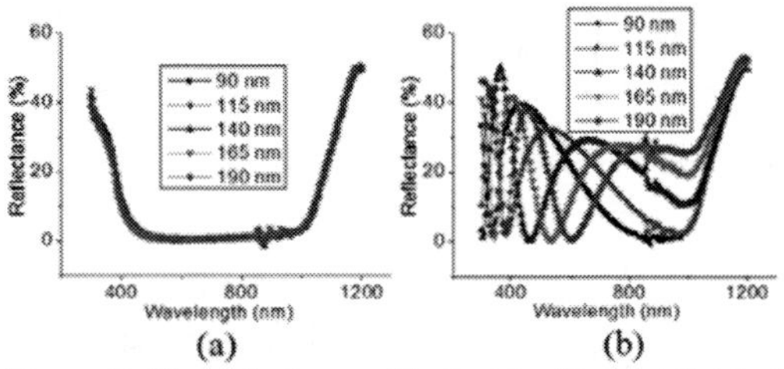

Figure 2: The reflectance of both sides for semi-finished cells: (a) front surface and (b) rear surface

Figure 3 shows the front and rear efficiency variations with different rear SiNx thickness. The front side keeps the cell performance of 20.3% to 20.5% and the difference is small and negligible therefore the front side can be remain the high efficiency. The rear cell performance decay dramatically from 13% to 11% as the rear SiNx thickness increasing from 90 nm to 190 nm and that means it is very sensitive to the thickness.

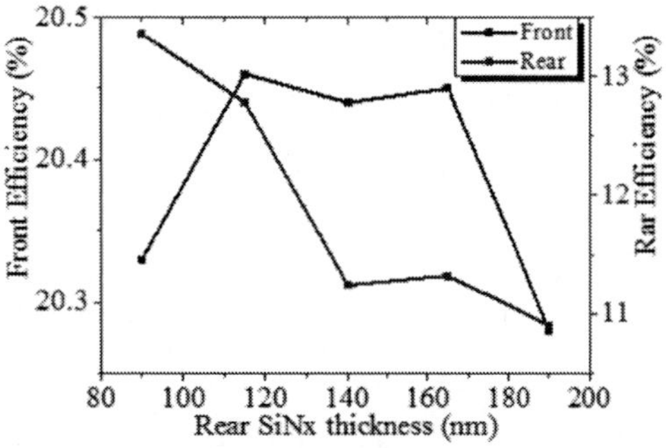

Figure 3: The efficiencies for bifacial PERC solar cells by varying rear silicon nitride thickness

Table I shows the photovoltaic characteristics for both sides of Bi-PERC with different thickness of rear SiNx. For the front side, the efficient difference by varying the thickness within 90 nm to 190 nm is 0.18% absolute. The only lower fill factor (F.F.) occurs on the cells with 190 nm SiNx and this is because of inappropriate sintering condition. Compare to the front efficiencies, the value in rear side is low due to the less open circuit voltage (Voc) and short circuit current density (Jsc) which is related to the flatness of rear surface. The efficient of rear side which differ in varying the thickness within 90 nm to 190 nm is 2.46% absolute which was directly contributed from Jsc obviously.

Table I: The photovoltaic characteristics for front side of bifacial PERC solar cells with different thickness of rear SiNx

	SiNx [nm]	Voc [mV]	Jsc [mA/cm^2]	F.F. [%]	Eff. [%]
Front	90	658	39.11	79.00	20.33
	115	659	39.10	79.40	20.46
	140	660	39.16	79.09	20.44
	165	661	39.07	79.19	20.45
	190	660	39.08	78.63	20.28
Rear	90	644	26.12	79.42	13.36
	115	643	24.94	79.69	12.78
	140	642	22.32	78.44	11.24
	165	642	22.34	78.93	11.32
	190	642	21.85	77.70	10.90

In this paper, three advanced metallization technologies were estimated: first is dual screen printing which including two step printings of finger and standard pattern; second is copper electroplating with opened anti-reflection layer by laser ablation; the third is knotless screen printing with fine line finger. The cross section images of finger by different metallization methods were compared by SEM measurement (as shown in figure 4). Figure 4 (a) shows the finger shape of conventional screen printing which is 41.4 μm width with 16.5 μm height and the aspect ratio is 0.4. The following result of dual printing is 33.9 μm width with 17.9 μm height and the aspect ratio is 0.53. The finger of copper electroplating is 28 μm with 9 μm height and the poor aspect ratio of 0.32. Finally, both the finger width and height are almost near 23 μm by hiknotless printing and thus the excellent aspect ratio of 0.98 [9-10].

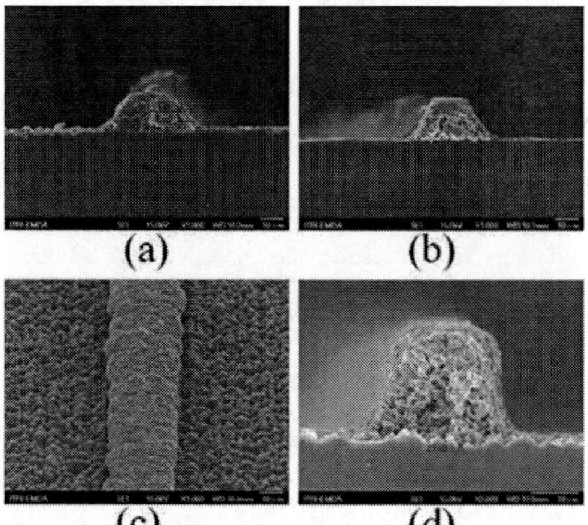

Figure 4: SEM images for different advanced metallization: (a) conventional screen printing (b) dual screen printing (c) copper electroplating (d) knotless screen printing

There I-V properties between standard and dual screen printing are almost the same due to the latter has wider finger more than 70 μm in the corners due to different screen board.

Comparing to conventional metallization, the Jsc and F.F. were increased by copper electroplating due to fewer metal coverage and semiconductor/metal contact ability improvement, shown in Table Ⅱ. Therefore, the efficiency gain of 0.07% absolute was achieved on the front side.

Table II: IV characteristics for front side of bifacial PERC solar cells with different thickness of rear SiNx

Metallization	Voc [mV]	Jsc [mA/cm^2]	F.F. [%]	Eff. [%]
Conventional	663	38.72	79.75	20.46
Cu plating	661	38.84	79.98	20.53

Comparison of conventional and Knotless metallization, the metal coverage of knotless screen printing was reduced almost by half and the height of finger was increased near 1.4 times. Thus the efficiency gain of 0.27% contributed by Jsc was achieved, shown in Table Ⅲ. It shows the knotless printing will be one of more competitive technologies for mass production.

Table III: IV characteristics for front side of bifacial PERC solar cells with different thickness of rear SiNx

Metallization	Voc [mV]	Jsc [mA/cm^2]	F.F. [%]	Eff. [%]
Conventional	664	39.21	78.96	20.54
Knotless	664	39.89	78.56	20.81

4 CONCLUSIONS

The effects of Bi-PERC solar cells with different thicknesses of rear SiNx layer were investigated. The front side keeps the cell performance of 20.3% to 20.5% but the rear efficiency decayed dramatically from 13% to 11% as the thickness increasing from 90 nm to 190 nm. Choosing appropriate thickness of backside SiNx thickness and refractive material without affecting the front side efficiency is important. Moreover, compare to conventional screen printing, the cell grid metallized by knotless screen with an efficiency gain of 0.27% absolute and the best efficiency of 20.81% were demonstrated. Knotless screen printing shows a highly potential metallization technology for mass production in the future.

ACKNOWLEDGMENT

The financial support provided by Bureau of Energy, Ministry of Economic Affairs in Taiwan is gratefully acknowledged.

REFERENCES

[1] International Technology Roadmap for Photovoltaic (2017) 35.
[2] Martin A. Green, Solar Energy Materials & Solar Cells 143 (2015) 190.
[3] T. Dullweber et al., 31st European Photovoltaic Solar Energy Conference, (2015) 341.
[4] Haibing Huang et al., Solar Energy Materials & Solar Cells 161 (2017) 14.
[5] Binhui Liu et al., 31st European Photovoltaic Solar Energy Conference (2015) 659.
[6] Marco Galiazzo et al., Energy Procedia 67 (2015) 116.
[7] Atteq ur Rehman1, Soo Hong Lee1, Solar Cells - New Approaches and Reviews (2015) 279.
[8] Hearous, https://www.pv-tech.org/products/heraeus-launches-front-side-silver-pastes-for-knotless-screen-printing (2017).
[9] T. Dullweber et al., 32nd European Photovoltaic Solar Energy Conference, (2016) 335.
[10] Feng Ye et al., 43rd IEEE photovoltaic specialists conference, (2016) 3360.

PASTE DEVELOPMENT FOR ELECTROCHEMICAL SCREEN PRINTING TO STRUCTURE METAL LAYERS OF BACK CONTACT SOLAR CELLS

Gensowski Katharina, Kamp Mathias, Efinger Raphael, Klawitter Markus, Pospischil Maximilian, Eckert Jonas, Bartsch Jonas
Fraunhofer Institute for Solar Energy Systems ISE, Heidenhofstraße 2, 79110 Freiburg, Germany
Phone: +49 (0) 761/45 88-2083
katharina.gensowski@ise.fraunhofer.de

ABSTRACT: An electrochemical etching process is developed to realize the contact pattern of back contact solar cells. It combines ECM technology (Electrochemical Machining) with screen printing to allow local removal of metallic layers for maskless formation of electrically isolated conductive paths. The present work focused on water-based paste development for this application. A plain sodium nitrate solution is used as electrolyte which allows electrochemical etching. Thickener and rheological additives are added to form a screen-printable paste. Additionally different additives are varied to improve printing paste behavior and printing results. The homogeneity of the printing paste is a major factor in the quality of etched structures. The multi-level homogenizing process developed in the present work reduces the particle size by about 35%. Using the optimized $NaNO_3$-based paste, etched lines are formed in 100 nm aluminum layers by the Electrochemical Screen Printing (ESP) process. The smallest lines show widths of 80 µm and the adjacent aluminum regions are electrically separated up to a resistance of 1.6 MΩ. The process time is less than 2 seconds for 156x156 mm^2 samples. Further process modifications ensure another particle size reduction of 33%. First IBC test structures are etched with line widths of 160 µm by using the novel process.
Keywords: printing paste, screen printing, electrochemical etching, IBC solar cell

1 INTRODUCTION

The highest efficiencies of silicon solar cells are reached with a back contact cell design. A heterojunction back contact solar cell fabricated by Kaneka achieves an efficiency of 26.7% [1]. Still, the process complexity, e.g. the metallization, and the high production costs limit back contact solar cells to relatively low production volumes. Possibilities to lower the production costs of high-efficiency back contact solar cells are described by NREL [2]. To use low-cost and simple metallization techniques to produce back contact solar cells is one promising option. Fig. 1 illustrates the difference in process steps between the patented SunPower back end processing route (left) and an alternative route based on electrochemical etching which replaces three costly process steps. Local electrochemical etching is a simple and low-priced method to structure metal layers. Within the present work, electrochemical etching is realized by the Electrochemical Screen Printing (ESP) approach.

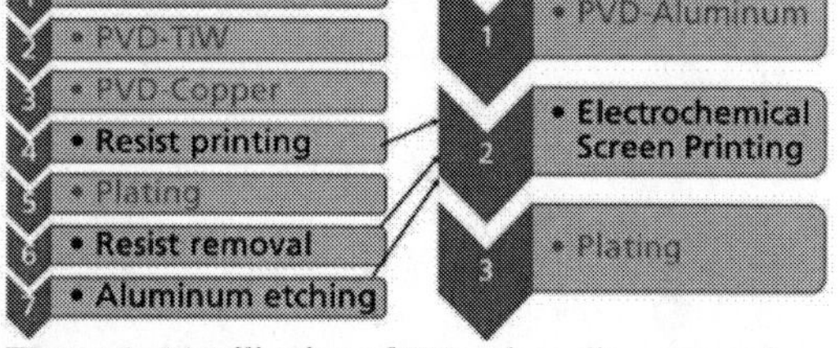

Figure 1: Metallization of IBC solar cells – comparison of the process steps between patented SunPower back end processing route [3] and alternative processing route (ESP) [3–5]

The principle of the Electrochemical Screen Printing is explained in [5, 6]. ESP allows to structure metal layers and generates conductive path structures in these layers e.g. for metallization of solar cells, of circuit boards or electrical sensors. It is a simple, one-step-process. Short process durations and excellent etching control by applied current are its advantages. Optimized commercial printing pastes are not available for the ESP process until now, but the printing paste has a big influence on the process result. Therefore, an optimal printing paste is needed for highest quality of metal structuring. Some requirements and challenges for optimized electrochemical printing pastes for the ESP resemble those of ink systems for other applications described in literature, such as those of Stüwe et al. [7] for inkjet technology to etch metal layers. Still, the processes differ importantly in many aspects. In this work the development and optimization of printing pastes for the ESP are presented.

2 APPROACH AND THEORY

2.1 Electrochemical Screen Printing

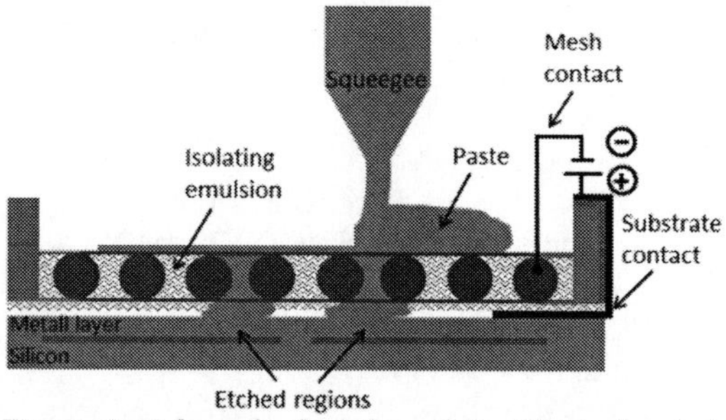

Figure 2: Schematic depiction of the Electrochemical Screen Printing (ESP) process during printing and etching [8]

The ESP, which was developed at Fraunhofer ISE, combines screen printing and electrochemical etching to structure thin metal layers like aluminum, copper and nickel. Fig. 2 depicts the ESP process while printing. The screen, a stainless steel mesh, is coated with an insulating emulsion and functions as cathode. During the printing process, the screen emulsion touches the silicon-wafer coated with the PVD metal layer, which represents the anode. The squeegee transfers the conductive printing paste onto the metal surface. Thus, the current circuit gets closed and the metal layer will be etched at the printed positions [5, 6].

2.2 Electrochemical printing pastes

Printing pastes for electrochemical etching should possess high etching speed (current proportional to etching rate at similar current efficiency) and satisfactory printing properties (depending on printing techniques), which include rheological properties, paste homogeneity and wettability. Other requirements are high electrical conductivity and aging behavior. A simple and non-toxic paste should be developed to achieve the requirements of conductive printing pastes. The electrochemical printing fluids are water-based systems which need to feature at least an electrolyte of such as sodium nitrate, sodium chloride or sulfuric acid and a rheology modifier. In addition, additives such as wetting agents and foam suppressors can be added to improve the performance.

It is well-known that $NaNO_3$ solution is able to etch metals electrochemically, e.g. as electrolyte for the ECM technology [9–11]. To perform the etching locally, it is necessary to restrict paste spreading by increasing the viscosity and yield point. In terms of sodium nitrate, sodium chloride or sulfuric acid electrolyte, the associated reactions may differ between dissolution of the metal and transformation of metal. $NaNO_3$ and NaCl will dissolve aluminum while H_2SO_4 is especially suited for aluminum treatment, transforming it into anodic aluminum oxide (AAO). AAO is non-conductive and will electrically isolate metal regions from each other [5, 12]. The present work focused on $NaNO_3$ to increase the conductivity of water which functions as solvent. The main reactions which take place at the cathode and at the anode during ESP are the following:

Cathodic: $2\ H_2O + 2e^- \rightarrow H_2\uparrow + 2\ OH^-$

$NO_3^- + H_2O + 2e^- \rightarrow NO_2^- + 2\ OH^-$

Anodic: $2\ Al \rightarrow 2\ Al^{3+} + 6\ e^-$

Overall: $2\ Al + 6\ H_2O \rightarrow 2\ Al^{3+} + 6\ OH^- + 6\ H_2\uparrow$

3 EXPERIMENTAL DETAILS

3.1 Production, homogenizing and characterization of electrochemical printing pastes

The electrochemical printing pastes are prepared in a beaker by using a stirrer to solve the solid components. As is commonly known in paste manufacturing, homogenization of paste is a crucial aspect, and different processing options such as ultrasonic agitation, three roll milling and filtration can be employed. Often, combinations of the mentioned techniques are needed for an optimal result. After stirring the $NaNO_3$-based fluids, they are homogenized by utilizing a multistage process which consists of ultrasonic treatments, a three-level roller mill processing and a two-stage filter processing. The aim is to generate a homogenous paste with small particles, without agglomerates and without air inclusions.

To define the optimal concentration of the paste ingredients a designed experiment was applied. The following four factors were analyzed:

1. Different concentrations of $NaNO_3$ solution
2. Soluble thickeners – thickener 1 (ST1) und thickener 2 (ST2) with a constant concentration
3. Different concentrations of insoluble, particle-based thickener (IPBT)
4. Different concentrations of liquid rheology modifier for thixotropic paste properties (LRM)

Different temperatures of the $NaNO_3$ solution are necessary to solve the thickener completely and more easily – 75°C for soluble thickener ST1, 40°C for the soluble thickener ST2. 48 pastes variations result from the designed experiment.

Moreover additives like humectants and foam suppressors were added to printing pastes. A humectant should improve the duration till the pastes dry out in the screen meshes and the storage life. A foam suppressor should reduce the air bubbles in the paste, so the paste layer shows no interruptions.

The printing pastes are characterized concerning rheological properties, wettability, homogeneity, printability and storage life. The rheological properties, yield point and viscosity, are determined by the rotational rheometer Anton Paar MCR 502 with a two plate rotational geometry. LS13 320 Laser Diffraction Particle Size Analyzer from Beckman Coulter is applied to measure the particle size and distribution of the printing pastes.

To reduce the particle size even more, additional processes are tested. One possibility is to add a dispersant to the printing pastes. Another option is to change the roller distances of the three-level roller mill or to use thinner membranes during the filtration process (see chapter 4.1.2).

3.2 Electrochemical Screen Printing

The ESP process is demonstrated by using monocrystalline silicon wafers with a PVD aluminum layer on top. A stack of silicon-oxide and silicon nitride separates substrate material and metal layer, enabling the measurement of electrical separation. 100 nm und 250 nm aluminum layers are used. For the ESP process a commercial ASYS EKRA screen printing machine S5 STA, which is modified for electrochemical printing, is used.

Depending on the used screens, the mesh widths differ. In the present work, pastes with particle sizes of below 5 µm should obtain process reliable printing results. Also the paste layers, which can be reached, depend on the utilized screen.

4 RESULTS AND DISCUSION

4.1 Electrochemical printing pastes and characterization

To create the first generation paste several rheology modifiers were explored. Thickener ST1 and thickener ST2 resulted in each case in an elastic paste, while the thickener IPBT leads to a waxy consistency with solid agglomerates. The printing fluids which were thickened with only one thickener, did not exhibit the right consistency to use for ESP. Thickener ST2 forms smaller networks than thickener ST1. A combination of two thickeners, thickener ST1 and thickener IPBT or thickener ST2 and thickener IPBT, improved the rheological properties of the medium. Additionally the printability of the paste is optimized with another additive, the modifier LRM. This rheology modifier changes the thixotropic flow behavior of the medium and builds a three-dimensional molecular network configuration, thus increasing the stability under load.

4.1.1 Results of designed experiment

In the next step the paste mixture is optimized by using a designed experiment and by developing a multilevel homogenizing process. 20% $NaNO_3$ solution thickened with thickener ST2 works well. Higher concentrated $NaNO_3$ solution and thickener ST2 result in inhomogeneous pastes. By using thickener ST1 in 20% to 40% $NaNO_3$ solution, the viscosity of the medium increased without any flocculation. In addition, the multi-stage homogeneity process worked well with thickener ST1-based paste.

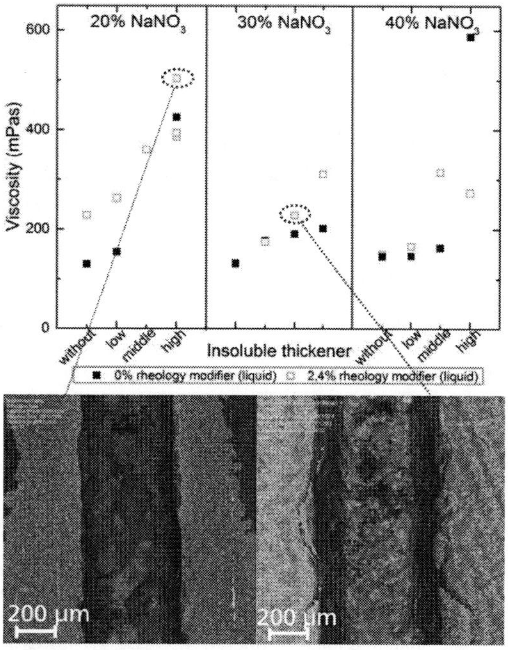

Figure 3: Diagram: viscosity values at shear rate of $100\ s^{-1}$ of thickener ST1 thickened printing fluids at different concentration levels; microscope images: two printing results for two different printing pastes (screen opening: 300 µm)

The viscosity of printing fluids depends on the concentration and combination of the additives (Fig. 3). Thickener ST1 combined with the three different concentrated $NaNO_3$ solutions gives very similar viscosities; about 130 mPas at shear rate of $100\ s^{-1}$. Thickener IPBT generates a viscosity increase depending on the concentration. If high level of thickener IPBT is added to the thickener ST1 based 20% $NaNO_3$ solution, the viscosity is 426 mPas at $100\ s^{-1}$. Furthermore, the viscosity increases also with the modifier LRM. The effect of the rheology additives are lower in a higher concentrated $NaNO_3$ medium compared with the 20% $NaNO_3$ solution.

The two images of etching lines in Fig. 3 visualize that different rheological properties of fluids resulted in different etching widths. The etched line on the left is narrower (428 µm) than the structure on the right (616 µm). The reason for the various line widths is that the used paste for the thinner structure is more viscous (503 mPas) than the utilized fluid on the right (229 mPas), but also the contact angles show differences. As expected the paste spreading is less with a more viscous electrochemical printing fluid.

Most of the printing pastes have got a particle size between 40 µm and 90 µm (d_{90}-value) (Fig. 4). The particles are too large, thus the huge particles clog the screen meshes during the ESP process. The metal surface won't be covered with fluid and no electrochemical reaction will take place. To dispense a paste on an aluminum layer, the measured contact angle is between 75° and 85°. The contact angle should ideally be about 90°, so the paste does not spread and narrow structures can be etched.

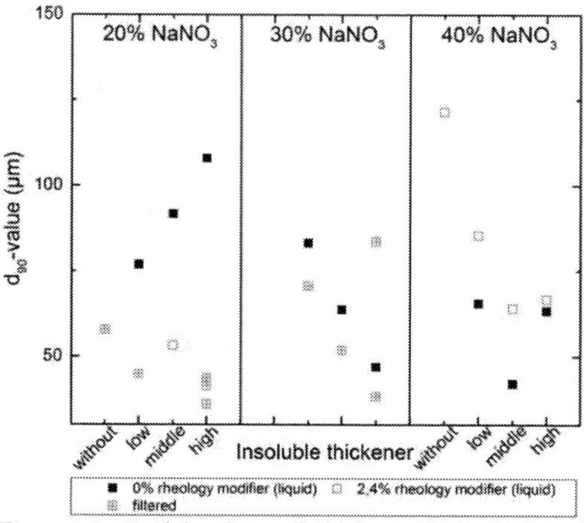

Figure 4: Particle sizes of thickener ST1 thickened printing fluids (different concentration levels of additives)

After characterizing the printing pastes from the designed experiment a standard recipe could be identified. The 20% $NaNO_3$ solution is thickened with low level of thickener ST1. High level of thickener IPBT and high level of modifier LRM are also added. This printing paste has got a viscosity of 503 mPas (shear rate $100\ s^{-1}$) and a contact angle on aluminum layer of 84°. The particle size, d_{90}-value, is 36 µm. It is possible to store the paste for 30 days without any properties changing.

4.1.2 Further optimization of standard paste

As already mentioned the particle size of printing fluids could be minimized from 47 µm to 30 µm (d_{90}-values), which corresponds to a reduction of about 35% (first generation printing pastes). It is expected that the breakup of agglomerates is one reason for reducing the d_{90}-values. The agglomerates consist of the thickener IPBT, because this additive tends to form agglomerates. The particles of 30 µm or larger are still too big for fine screen meshes. The aim is to create particle size smaller than 5 µm. This is why further optimization of the standard paste from the designed experiment was made.

The first possibility to minimize even more the particle size is to add a dispersing agent to the standard recipe from the designed experiment. The concentration of the dispersant was varied. Fig. 5 depicts the particle distribution of these printing fluids. All printing pastes with dispersant have got significantly smaller particle sizes than the best paste from the designed experiment (d_{90} = 36 µm). The printing fluid with high level of dispersing agent shows the most improved particle distribution because of the border areas. This printing fluid contains very small particles and only few agglomerates in the range of 100 µm to 200 µm. The precise particle values are presented in Table I.

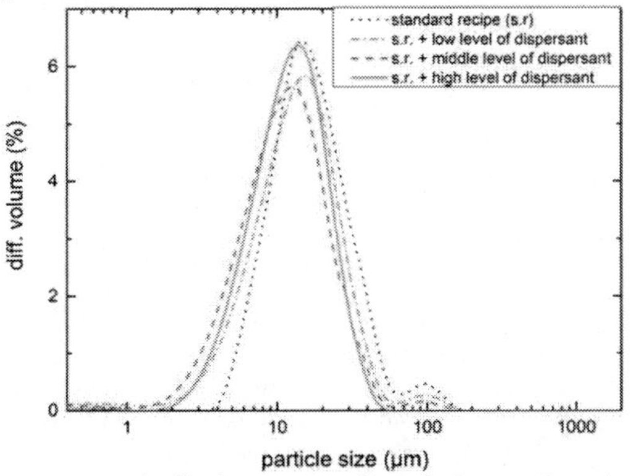

Figure 5: Distribution curves of standard paste from the designed experiment (s.r.) and different concentration of a dispersing agent

Table I: Particle values of standard paste from the designed experiment and different concentration of dispersant

	d_{10} (µm)	d_{50} (µm)	d_{90} (µm)
Standard recipe (s.r.)	8	16	36
S.r. + low level of dispersant	5	13	29
S.r. + middle level of disp.	4	11	25
S.r. + high level of disp.	5	12	24

To influence the particle size the use of the contact mode at the three roll mill is also possible. The mode is controlled by a force value. The distance of the rolls is 0 µm, so the agglomerates are even more broken up. The risk to form flakes during the homogenization process does not exist in this case, because no metal particles are used which might be deformed permanently.

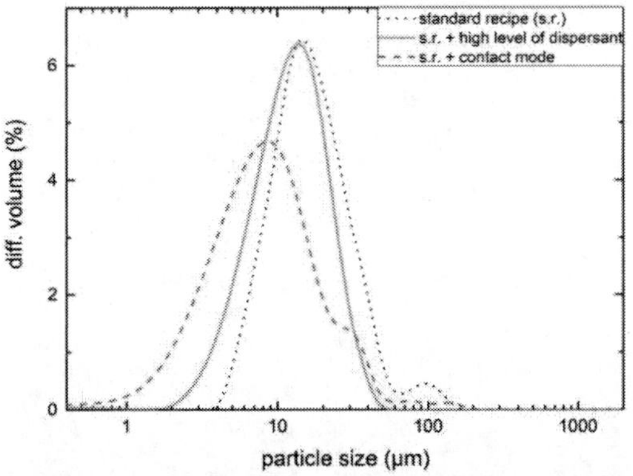

Figure 6: Distribution curves of three different printing pastes – standard printing fluid from designed experiment, printing paste with high level of dispersing agent and printing paste which is homogenized with contact mode

Table II: Particle values of further optimized printing pastes

	d_{10} (µm)	d_{50} (µm)	d_{90} (µm)
Standard recipe	8	16	36
Standard recipe + dispersant	5	12	24
Standard recipe + contact mode	3	8	24

Table II shows the different particle size values of three fluids. One is the best fluid from the designed experiment, one paste contained a dispersant in addition to the standard recipe, in one case the standard paste was run over the three roll mill in contact mode. In comparison to d_{90}-values of the standard recipe, the d_{90}-values of the two improved pastes concerning particle size is significantly smaller. The d_{90}-value is reduced again by about 33%. Both methods to minimize the particle size work well. Fig. 6 depicts the difference in the particle distribution curves of the two homogenization processes. The green, solid curve shows a narrower curve shape that means that the particle size distribution is smaller. The paste is more homogeneous than the printing fluid which associated to the magenta, dashed curve. Additionally, the printing paste with dispersant (green, solid graph) does only exhibit very few agglomerates in the scale of 100 µm to 200 µm. The advantage of the particle distribution of the magenta, dashed curve is that the printing paste has got more really small particles (> 2 µm) than the paste with dispersant.

A third possibility to reduce the particle size is to apply smaller membranes during the filtration. This method lowers the large agglomerates. However, if there are too many large agglomerates left, the membranes are frequently clogged and much paste is lost. In the future these methods of reducing the particle size will be combined.

4.2 Electrochemical Screen Printing

Using the first generation of developed paste, printing processes through screen openings of 50 µm to 100 µm could be realized on PVD-aluminum layers successfully, but with an optimization potential (Fig. 7). The widths of the etched lines were twice as large as the screen openings. Each structure features a length of 28 mm. Another difficulty is also shown in Fig. 7 (red marks). The etched structures exhibit some interruptions, consequently the aluminum regions are not electrically separated. Air bubbles, agglomerates, too thin paste layers and dried out pastes in screen meshes might be reasons. The printing paste has to optimized to achieve improved etched structures. Also, process conditions such as current/voltage characteristics could be adjusted.

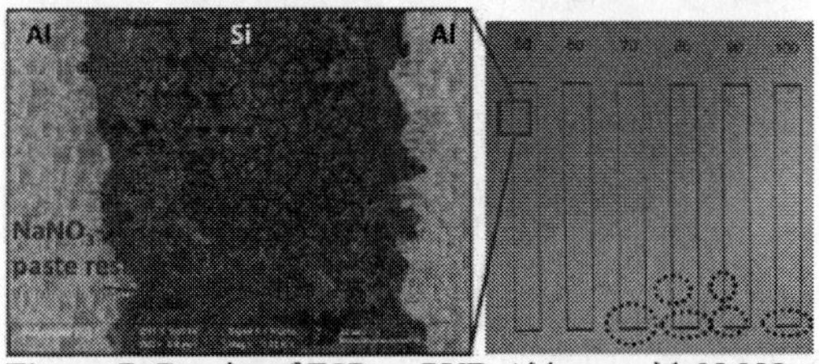

Figure 7: Results of ESP on PVD-Al layer with $NaNO_3$-based paste; Photograph of etched structures with interruptions by using 50 µm to 100 µm screen openings (right); SEM image of etched aluminum and paste residuals by using 50 µm screen opening (left)

A paste from the designed experiment was used to structure a 100 nm aluminum layer (Fig. 8). The printing speed which corresponds to the etching time is 150 mm/s. The aluminum areas are electrically separated (1.6 MΩ) and the line width is around 80 µm for a screen opening of 50 µm. Thus, rheological properties and paste homogeneity of the printing fluids could be optimized, whereby the reached printing results are satisfactory.

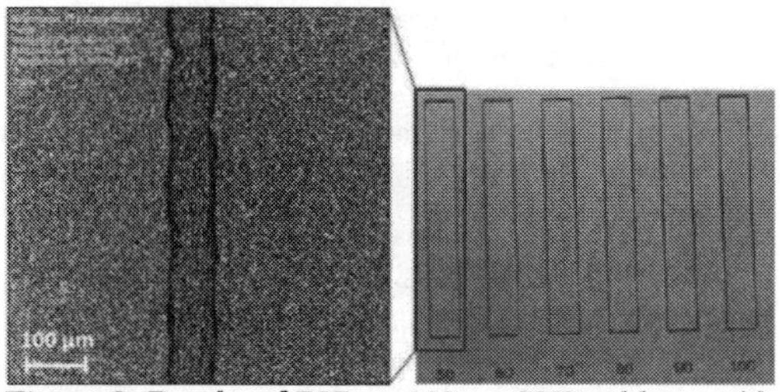

Figure 8: Results of ESP on 100 nm PVD-Al layer with $NaNO_3$-based paste; Photograph of etched structures by using 50 µm to 100 µm screen openings (right); microscope image of etched structure with a line width of 80 µm aluminum achieved with 50 µm screen opening (left)

A paste from the designed experiment was used to structure a 100 nm aluminum layer (Fig. 8). The printing speed which corresponds to the etching time is 150 mm/s. The aluminum areas are electrically separated (1.6 MΩ) and the line width is around 80 µm for a screen opening of 50 µm. Thus, rheological properties and paste homogeneity of the printing fluids could be optimized, whereby the reached printing results are satisfactory.

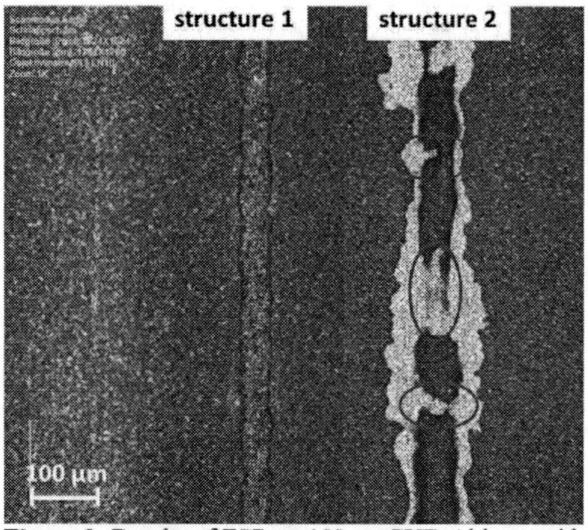

Figure 9: Results of ESP on 100 nm PVD-Al layer with $NaNO_3$-based paste; microscope images of etched structures by using 50 µm screen opening (left: first print experiment; right: second print experiment)

Fig. 9 depicts two different etched lines on aluminum layers. On the left side it is the same structure as shown in Fig. 8. Structure 2 (right) was the next print after structure 1 (left). The results exhibit a big difference, e.g. concerning contour precision, interruptions and therefore electrical separation. The challenge for the future is to create a constant quality of the etched structures between consecutive printing experiments, as in an industrial production several thousand prints have to be performed without any degradation of results. The reason for this behavior has not been fully understood yet. The most likely causes are fast drying of the paste within the screen and clogging of screen meshes by too large particles. The optimization potential is again concerning the printing paste, mainly particle size, drying behavior and paste layer. The particle size should be preferably smaller than 5 µm to reliably pass the typical mesh openings of about 40 µm. In addition, the printing step can be adjusted to flood the screen immediately after printing, thus keeping the paste wet.

The electrochemical printing fluid is an important influencing parameter on the ESP process. In addition, some process parameters affect the quality of etched structures. Fig. 10 depicts several etched lines on aluminum layers created with different printing speeds. The paste spreading can be reduced by using higher printing speeds and more viscous printing pastes. A disadvantage of this might be a shorter reaction time to etch the aluminum completely.

The thicker the aluminum layer, the more challenging is the dissolution of the aluminum. Due to the fact that the paste is consumed by the process it is necessary to transfer enough paste to the metal surface to etch the aluminum completely.

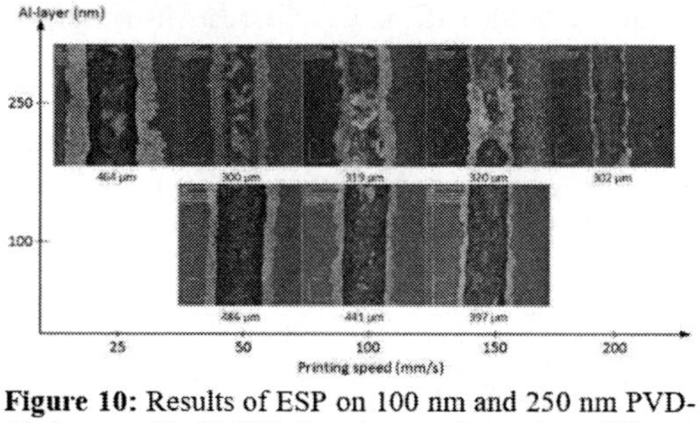

Figure 10: Results of ESP on 100 nm and 250 nm PVD-Al layer with $NaNO_3$-based paste by using different printing speeds

After the ESP process and removal the paste residuals, the etched aluminum layer is plated with a Ni-Cu stack. The processing sequence and SEM images of the plated structures will be published in [8].

Our latest results show the satisfactory structuring of thin aluminum and copper with the ESP process. Further paste improvements are aspired, where a special focus is set on bringing the particle size below 5 µm. Also, the paste transfer behavior will be optimized, to allow structuring of thicker metal layers

Adjusted pastes will also allow higher flexibility towards other processing schemes such as electrochemical dispensing, where the particle size is even more critical.

4.3 First electrochemical screen printed IBC test structure

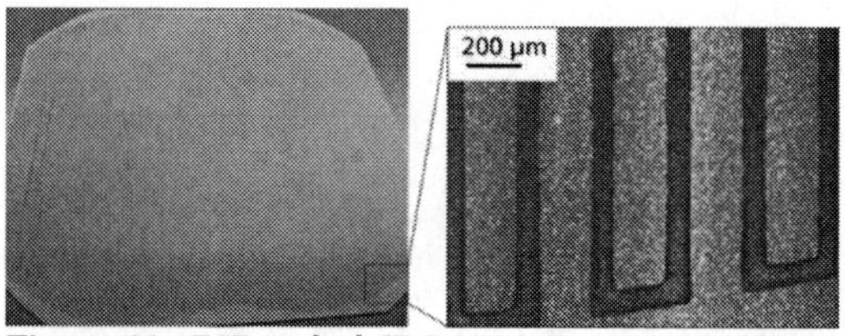

Figure 11: ESP etched IBC test structure on 100 µm PVD-Al layer by using 150 µm screen openings (etched line widths 160 µm)

Fig. 11 illustrates one of the first electrochemical screen printed IBC test structures. The printing duration for the length of 125 mm was 2.5 s. To transfer the process from small test structures to a 125x125 mm² wafer is a challenging task concerning the printing homogeneity, the current distribution and thus the etching regions. This very first result is considered satisfactory, as it already shows a precise resolution between the different lines and corners.

As already mentioned, the paste has to be improved, but also the current distribution for structuring larger areas has to be optimized. Several approaches to achieve this are currently under investigation. With an improved paste and printing setup, the creating of functional IBC solar cells will be aspired in the future.

5 CONCLUSION

To structure metal layers, a paste for a new etching process (ESP) is developed, with an aspired application in a simplified back contact solar cell back-end process. The simple, first generation $NaNO_3$-based paste consists of thickener ST1, thickener IPBT and rheology modifier LRM. The ESP process was successfully demonstrated on PVD-aluminum layers – line widths of 80 µm are achieved and electrical separation up to 1.6 MΩ is demonstrated. The quality of printing results is not yet similar for repeated printing of the same structure. The aim is to improve the whole etching process (process parameters) and to make the process suitable for high-throughput production.

The first paste mixtures are optimized concerning particle size, rheological properties and printability (etching results) during this work. Also a manufacturing process and a multi-stage homogenization process are developed for printing pastes. For the future there is still a high potential of optimization concerning the homogeneity and composition of printing pastes. First experiments concerning etching IBC test structures are demonstrated. Printing on large area substrates in an industrially relevant speed will be improved.

6 ACKNOWLEDGEMENT

This work was funded by the German Federal Ministry for Economic Affairs and Energy within the project "KAluS50", grant No. 0324009A.

7 REFERENCES

[1] M. A. Green *et al.*, "Solar cell efficiency tables (version 50)," *Prog. Photovolt: Res. Appl.*, vol. 25, no. 7, pp. 668–676, 2017.

[2] National Renewable Energy Laboratory, *Silicon Materials and Devices R&D | Photovoltaic Research | NREL.* [Online] Available: https://www.nrel.gov/pv/silicon-materials-devices-rd.html. Accessed on: Aug. 09 2017.

[3] W. P. Mulligan, "Solar cell and method of manufacture," US 7897867 B1.

[4] M. Kamp *et al.*, "Zincate processes for silicon solar cell metallization," *Solar Energy Materials and Solar Cells*, vol. 120, pp. 332–338, 2014.

[5] M. Kamp *et al.*, "Electrochemical Contact Separation for PVD Aluminum Back Contact Solar Cells," *Energy Procedia*, vol. 67, pp. 70–75, 2015.

[6] M. Kamp, J. Bartsch, and M. Glatthaar, "Method for structuring layers of oxidizable materials by means of oxidation and substrate having a structured coating," US20160204699 A1, Aug 18, 2016.

[7] D. Stüwe, D. Mager, D. Biro, and J. G. Korvink, "Inkjet technology for crystalline silicon photovoltaics," (eng), *Advanced materials (Deerfield Beach, Fla.)*, vol. 27, no. 4, pp. 599–626, 2015.

[8] M. Kamp *et al.*, "Structuring of metal layers by electrochemical screen printing for back contact solar cells: IEEE PVSC, under review,"

[9] M. M. Lohrengel, K. P. Rataj, and T. Münninghoff, "Electrochemical Machining—mechanisms of anodic dissolution," *Electrochimica Acta*, vol. 201, pp. 348–353, 2016.

[10] K. P. Rajurkar, M. M. Sundaram, and A. P. Malshe, "Review of Electrochemical and Electrodischarge Machining," *Procedia CIRP*, vol. 6, pp. 13–26, 2013.

[11] S. Wang, Y. Zeng, Y. Liu, and D. Zhu, "Micro wire electrochemical machining with an axial electrolyte flow," *Int J Adv Manuf Technol*, vol. 63, no. 1-4, pp. 25–32, 2012.

[12] K. Schwirn, "Harte Anodisation von Aluminium mit verdünnter Schwefelsäure: Dissertation," Zentrum für Ingenieurwissenschaften der Martin-Luther-Universität Halle-Wittenberg, 2008.

nPERT solar cells with a high bifaciality > 93%

Ping-Kuan Chang[1]*, Liang-Tang Wang[1], Shih-Wei Chiu[1], Yuan-Jun Lin[1], Wen-Tai Chung[1], Chun Kuo[1], Chi-Chun Li[1]

[1]Motech Industries, Inc. Science Park Branch, 74145 Tainan, Taiwan

*Corresponding author: Ping-Kuan Chang, Phone: +886-6505-0789 ext 0684, Email: ken_chang@motech.com.tw

Abstract **— Bifaciality is a highly emphasized factor while developing n-type passivated emitter rear totally diffused (nPERT) bifacial solar cells. It is important to increase the rear-side efficiency of rear side of the bifacial solar cell for higher bifaciality. In this study, nPERT bifacial solar cell made of 156 x 156 mm n-type Cz-wafer with high bifaciality of 93% was presented. Various etch depths of rear side were provided to investigate the optical and electrical properties of the nPERT solar cells. The results showed that a suitable etching depth of the rear side leads the lower surface reflectivity and higher bifacility was obtained with 5.6% enhancement in short-circuit current density (J_{sc}).**

Index Terms **— nPERT solar cell, Reflectivity, Bifaciality**

1 INTRODUCTION

Recently, solar cells manufactured with crystalline silicon accounted for more than 85% market share [1]. Among those, more than 90% were made with p-type substrates and the remaining with n-type substrate. However, the p-type substrate suffers from light-induced degradation (LID) effect, and is more sensitive to metallic impurities contamination [2]. The n-type wafers' advantages include longer minority carrier diffusion length and less sensitivity to metallic impurities contamination [3]. Therefore, n-type silicon has an enormous potential for wide scale applications in the photovoltaics industry.

N-type silicon solar cells are often made into bifacial structures, which can produce even higher output power, with contribution from the rear side. Bifacial photovoltaic (PV) systems can be used for specific applications and areas such as white painted rooftops, vertical noise barriers along highways or railways, desert, and snow area. More importantly, bifacial n-type cells produced by industrial processes are cost competitive.

This study proposes an optimized rear-side texturization in order to improve bifaciality of the nPERT solar cells. We further improved the efficiency of rear-side of the cell by the optimizing etch depth. The influences of various etch depths on the optical and electrical properties of the cells are presented.

2 EXPERIMENTAL

Figure 1 and Figure 2 show the schematic cross section and process flow of nPERT bifacial cells, respectively. N-type Cz-silicon (100) wafers with a thickness of 200 μm, 156 x 156 mm^2, and resistivity of 1-6 Ω cm are used. Random pyramid textures on Si surface are achieved by wet-chemical etching. Boron emitter was made by using BBr_3 precursor in industrial tube furnaces. The rear-side light trapping and efficiency were developed and optimized by wet chemical process [4-5]. Rear-side BSF was formed by phosphorous implantation. Specific passivation layers developed in our previous study are deposited on top of the diffusion layers. Anti-reflection layers of SiN_x are deposited by plasma enhanced chemical vapor deposition (PECVD). Standard screen printers in production line were used with commercial Ag/Al and Ag pastes. Firing profiles were carefully designed to match up the various printing conditions. The reflectivity (R%) of rear-side texture was measured from 380 nm to 1100 nm with the UV-VIS-NIR spectrophotometer. The surface root-mean-squared (rms) of the rear side of the cells is studied using three dimension optical microscopy measurement. The current-voltage (I-V) characteristics of the nPERT solar cells are measured under 1000 W/m^2, AM1.5G illumination from Class A solar simulator.

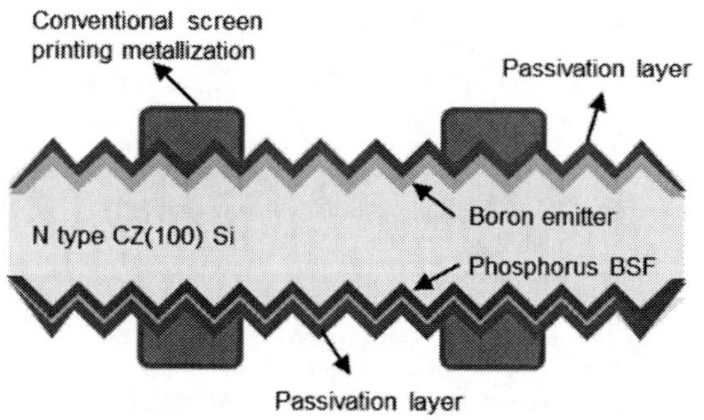

Fig. 1 Schematic cross-section of n-type bifacial solar cells.

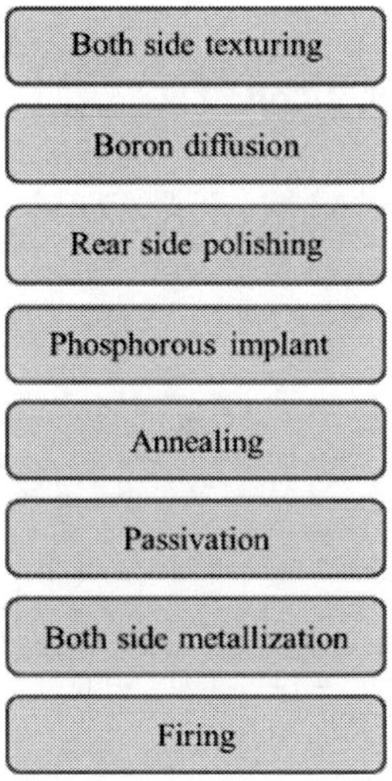

Fig. 2 Manufacturing process of nPERT bifacial cells.

3 RESULTS AND DISCUSSION

Figure 3 shows the reflectivity of rear-side textured surface and rear-side J_{sc} ratio against standard cell as a function of etch depth. Less etch depth shows a better optical result with lower reflectivity and higher J_{sc} ratio can be obtained. It is clearly seen that relative 5.6% enhancement in J_{sc} of rear side of the cells with less etch depth. The increased J_{sc} is attributed to reduce reflections from silicon solar cells and improve light absorption by light trapping.

Among the discussion above, it is obvious that the rear-side efficiency should also be improved since the J_{sc} was increased by the contribution of the lower reflectivity at the rear surface. Figure 4 demonstrates the comparison of the rear-side efficiency ratio against standard cell, higher cell efficiency is obtained by less etch depth. The result is positive related with the improved rear side J_{sc} ratio shown in Fig. 3. It also can be understood that the bifaciality of nPERT solar cell was improved over 5%, with the raised rear side cell efficiency. The bifaciality with such optimized rear side process can be achieved to 93%.

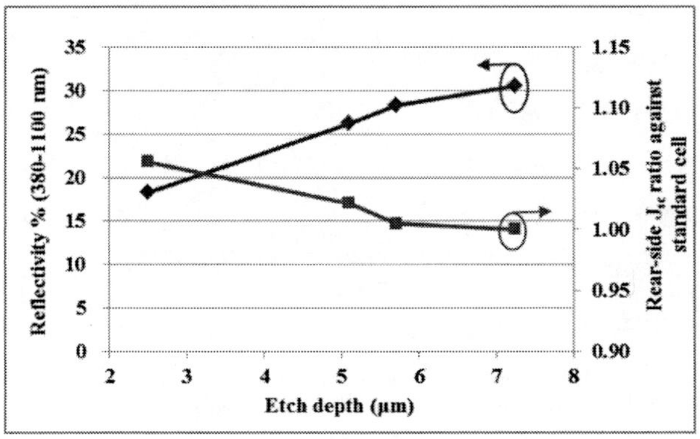

Fig. 3 Reflectivity and rear-side J_{sc} ratio against standard cell for various etch depths.

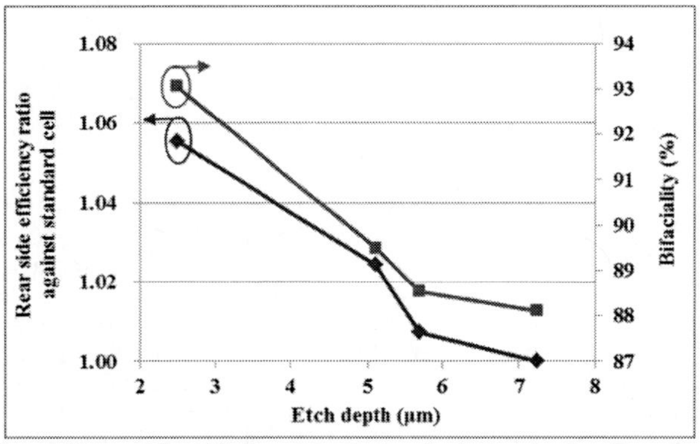

Fig. 4 Rear-side efficiency ratio against standard cell and cell bifaciality for various etch depths.

4 CONCLUSIONS

This study investigated the rear side etching process optimization of nPERT bifacial solar cells and the related cell characteristics. Among the experiment, the etch depth fine-tune and morphology of rear-side to improve the bifaciality of nPERT solar cell were discussed. The less etch depth is believed to decrease the reflectivity of rear-side, will have a good J_{sc} and efficiency. As a result, a nPERT bifacial cell process in mass-production level with high bifaciality of 93% was demonstrated.

5 REFERENCES

[1] Photon International March 2009, p170.

[2] J. Schmidt, A.G. Aberle, R. Hezel, Proceedings of the 26th IEEE PVSC (1997).

[3] D. Macdonald, L. J. Geerligs, Appl. Phys. Lett., 85 (2004) 4061-4063.

[4] G. Kastner, M. Zimmer, K. Birmann, F. Souren, J. Rentsch, R. Preu, Proceedings of the 25th EUPVSEC (2010).

[5] E. Cornagliotti, A Uruena, J. Horzel, J. John, L. Tous, D. Hendrickx, V. Prajapati, S. Singh, R. Hoyer, F. Delahaye, K. Weise, D. Queisser, H. Nussbaumer, J. Poortmans, Proceedings of the 27th EUPVSEC (2012).

A SIMPLE B/P THERMAL DIFFUSION APPROACH FOR N-TYPE PERT SOLAR CELL

Shude Zhang[1]*, Sanyang Zhang[1]*, Xiaorui Liu[1], Chenyang Wu[1], Shuanglong Yu[2], Qingzhu Wei[1,2], Zhichun Ni[1,2]
[1] Suzhou Talesun Solar Technologies Co., Ltd., Changshu, Jiangsu, 215542, China
[2] Nanjing University of Aeronautics & Astronautics, Nanjing, Jiangsu, 210016, China
* These authors contributed equally to this work. E-mail: shude.zhang@talesun.com

ABSTRACT: A simple boron/phosphorus thermal diffusion approach is proposed, which can considerably simply the fabrication procedure of n-type PERT (Passivated Emitter Rear Totally-diffused) crystalline silicon solar cells with even higher efficiencies. The boron-doped emitter and phosphorus-doped back surface field were fabricated in one co-diffusion process, because the boron paste coated and dense BSG (boron silicate glass) layer formed on the surface of the front side could act as a barrier from the diffusion of phosphorus. With this B/P co-diffusion method, the maximum efficiency of 20.41% on the front side of PERT solar cell was achieved, and 18.78% on the rear side achieved. The differences between co-diffusion method and traditional two-step diffusion method were compared and discussed in detail. These results indicate that B/P co-diffusion might be a promising alternative method for the mass production of n-type PERT solar cells.

Keywords: n-type, Silicon Solar Cell, Fabrication

1 INTRODUCTION

In recent years, the photovoltaic industry has developed rapidly, but the conventional p-type crystalline silicon solar cells still dominate the market. P-type solar cells have high temperature coefficient, serious light-induced degradation (LID) [1,2] and severe potential-induced degradation (PID), all of which will affect the power output. In contrast, n-type crystalline silicon solar cells have low temperature coefficient, no LID effect and slight PID effect. Besides, n-type solar cells can attain higher conversion efficiency since the minority carrier lifetime is long and less affected by impurities and defects [3,4]. There are several kinds of n-type solar cell technologies, such as HIT [5], IBC [6], TOPCORN [7] and PERT. Compared with the three formers, the fourth one is the most suitable technology for mass production at present, because of the relatively simple process and moderate cost. In addition, the bifacial structure of PERT can take advantage of the diffuse reflection light on the backside, which will increase the power output significantly. However, there are still some problems with PERT technology. Generally, boron diffusion [8] is adopted to prepare p-type emitter at first, then phosphorus implantation or diffusion adopted to prepare n-type back surface field. However, the cost of ion implantation facility is huge, and the two-step diffusion procedure including multiple annealing treatments will greatly increase the process risk. In this work, the front emitter and back surface field fabrication processes were integrated into one step through co-diffusion method. During the co-diffusion process, the boron paste coated and dense BSG layer formed on the surface of the front side could prevent the diffusion of phosphorus [9,10]. This one-step fabrication method of n-type PERT solar cell provides an excellent route to cost reduction and productivity improvement.

2 EXPERIMENTS

In this work, complete solar cells and corresponding symmetrical cells were fabricated to investigate the co-diffusion method and compare the differences between it and two-step diffusion method. All samples were fabricated with n-type monocrystal silicon wafers with side length of 156.75 mm, thickness of 180 μm and resistivity of 1–5 Ω/cm. The preparation of solar cells and symmetrical cells with different methods is shown below.

1) Preparation of n-type PERT solar cells with co-diffusion method (Figure 1):

Boron paste was spin-coated on one side of the wafer after texturing. Subsequently, boron and phosphorus were diffused in one process with boron paste as boron source and $POCl_3$ as phosphorus source. After etching and removing BSG, the sample was thermally oxidized to form SiO_2 and deposited with SiN_x on both sides. Finally, metallization was applied to both sides with screen printing.

2) Preparation of n-type PERT solar cells with two-step method (Figure 1):

After boron paste was spin-coated, one diffusion process was executed to form the emitter. Then, SiN_x mask was deposited to protect the boron emitter from phosphorus diffusion, which was another diffusion process to form the back surface field with $POCl_3$ as phosphorus source. It should be noted that SiO_2 on the backside was removed before phosphorus diffusion. After etching and removing mask & BSG, the sample was thermally oxidized to form SiO_2 and deposited with SiN_x on both sides. Finally, metallization was applied to both sides with screen printing.

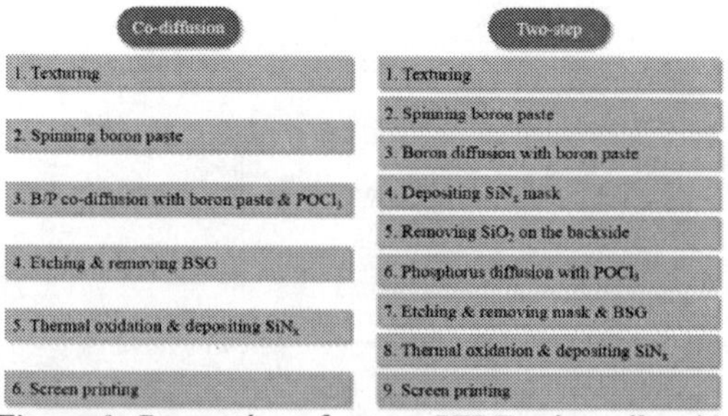

Figure 1: Preparation of n-type PERT solar cells with different methods.

3) Preparation of symmetrical cells (Figure 2):

Boron paste was spin-coated on both sides of the textured wafer and sandwiched between two paste-coated wafers to imitate the actual diffusion environment during the diffusion process. According to the preparation procedures of co-diffusion and two-step methods, two kinds of symmetrical cells were fabricated. In order to find out the optimal diffusion temperature, different temperatures varying between 940 °C and 955 °C were adopted to diffuse boron.

Figure 2: Preparation of symmetrical cells.

200 n-type PERT solar cells were fabricated with two different methods respectively. Subsequently, a series of characterization techniques were applied to analyze the cell performance, such as IV test, internal quantum efficiency (IQE) measurement and electrochemical capacity voltage (ECV). In addition, quasi steady state photo conductance (QSSPC) technique was applied to the symmetrical cells.

3 RESULTS AND DISCUSSION

The quality of the emitter was evaluated with symmetrical cells. By using QSSPC technique, the iV_{oc} and J0e of symmetrical cells fabricated with two kinds of methods and different temperatures varying between 940 °C and 955 °C are shown in Figure 3. It can be found that iV_{oc} of the co-diffusion sample is higher and J_{0e} lower than the two-step sample, which means slower recombination rate, higher open circuit voltage and higher quality of emitter [11]. Therefore, it could be concluded that the boron paste coated and dense BSG layer formed on the surface of the front side could act effectively as a barrier from the diffusion of phosphorus during the co-diffusion process. What's more, the quality of the emitter prepared with two-step diffusion method appears to be influenced by the complicated procedure containing multiple high-temperature treatments. At 950 °C, the highest iV_{oc} of 680 mV and lowest J_{0e} of 41.7 fA/cm² are obtained for both kinds of samples, which means that 950 °C is the optimal temperature for boron diffusion. Therefore, the ECV curve of 950 °C sample with co-diffusion method was measured and is shown in Figure 4. The surface concentration of boron is 8.98×10^{19} cm^{-3} and the junction depth is 0.52 μm.

Table I and Figure 5 shows the electrical properties and efficiency distribution of 200 cells prepared with co-diffusion method. It can be found that most of the efficiencies on the front side are distributed between 19.9% and 20.2%. The maximum efficiency of 20.41% on the front side was achieved. Considering that these are only preliminary experimental results, it can be predicted that the cell efficiency on the front side should break through 20.8% easily by the optimization of each process such as thermal oxidation and metallization. On the rear side, the average and maximum efficiencies of 18.32% and 18.78% were achieved respectively.

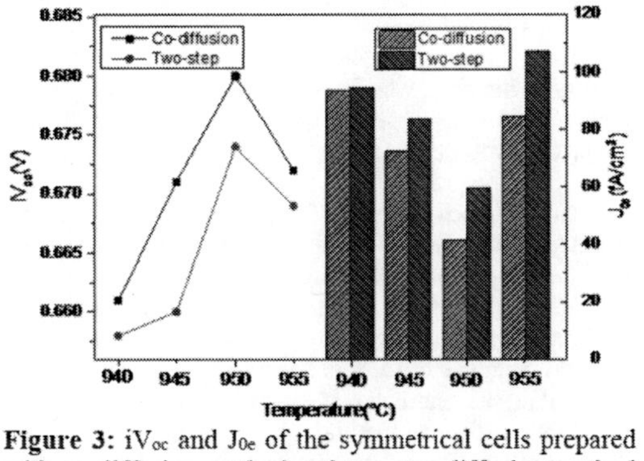

Figure 3: iV_{oc} and J_{0e} of the symmetrical cells prepared with co-diffusion method and two-step diffusion method respectively.

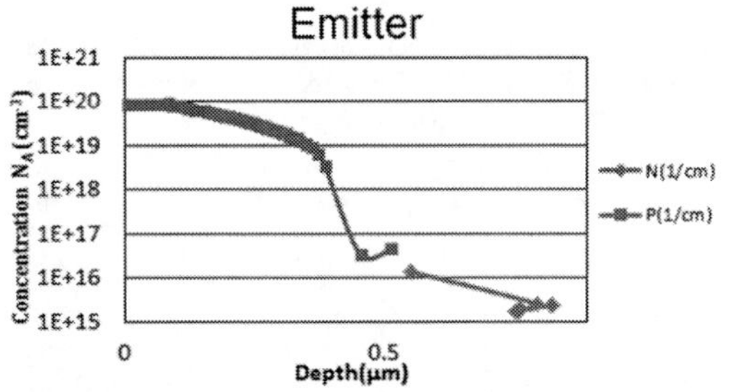

Figure 4: ECV curve of the emitter of the sample prepared with co-diffusion method.

Table I: Electrical properties of 200 cells prepared with co-diffusion method.

		V_{oc} (mV)	J_{sc} (mA/cm²)	FF (%)	Eff (%)
Front Side	AVG	647.6	38.10	79.26	20.00
	MAX	648.7	38.41	80.08	20.41
Rear Side	AVG	644.5	34.29	81.13	18.32
	MAX	645.8	34.90	81.54	18.78

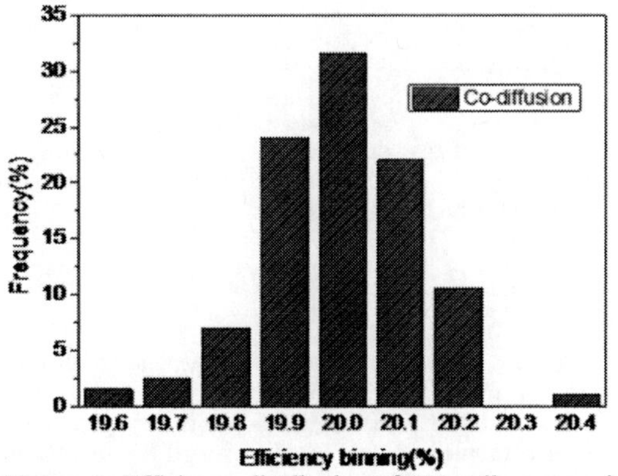

Figure 5: Efficiency distribution of 200 cells prepared with co-diffusion.

Table II compares the average electrical properties of 200 cells prepared with two different methods respectively. It can be found that the average electrical properties of cells prepared with co-diffusion method on the front side are higher than those with two-step diffusion method, which is consistent with the above results regarding iV_{oc} and J_{0e}. However, the average electrical properties on the rear side of cells prepared with co-diffusion are lower than those with two-step diffusion, which is probably because SiO_2 on the backside was removed before phosphorus diffusion in the two-step method. According to the internal quantum efficiency curves of solar cells prepared with two different methods respectively shown in Figure 6, the spectral response of solar cell prepared with co-diffusion method at the short wavelength regions is better than two-step diffusion method, but worse at the long wavelength regions. This spectral response result is in line with the comparison result of solar cell electrical properties with two different fabrication methods.

Table II: Average electrical properties of 200 cells prepared with two different methods respectively. (Co is short for co-diffusion, and two for two-step.)

		V_{oc} (mV)	J_{sc} (mA/cm^2)	FF (%)	Eff (%)
Front Side	Co	647.6	38.10	79.26	20.00
	Two	645.8	38.01	78.69	19.75
Rear Side	Co	644.5	34.29	81.13	18.32
	Two	645.1	34.54	81.25	18.49

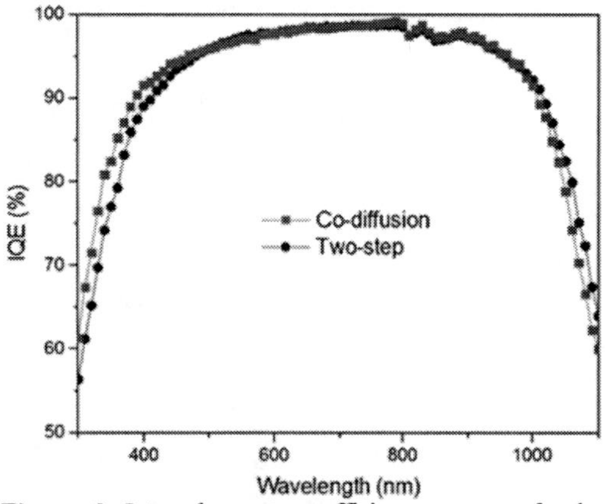

Figure 6: Internal quantum efficiency curves of solar cells prepared with co-diffusion method and two-step diffusion method respectively.

4 CONCLUSION

In this work, a simple boron/phosphorus thermal diffusion approach called co-diffusion is proposed, which is much simpler than the traditional two-step diffusion method adopted to fabricate n-type PERT crystalline silicon solar cells. The boron paste coated and dense BSG layer formed on the surface of the front side could act effectively as a barrier from the diffusion of phosphorus during the co-diffusion process. Without multiple high-temperature treatments, the quality of the emitter prepared with co-diffusion method seems to be better than the two-step diffusion method, which results in higher open circuit voltage and efficiency on the front side. With this B/P co-diffusion method, the maximum efficiency of 20.41% on the front side of PERT solar cell was achieved, and 18.78% on the rear side achieved. These results indicate that B/P co-diffusion might be a promising alternative method for the mass production of n-type PERT solar cells.

5 ACKNOWLEDGEMENT

The authors would like to thank Toray Industries, Inc. for the support of boron paste and experiments.

6 REFERENCES

[1] F. Wolnya, T. Webera, M. Müllera, G. Fischera, "Light Induced Degradation and Regeneration of High Efficiency Cz PERC Cells with Varying Base Resistivity", Energy Procedia 38 (2013) 523.

[2] S.W. Glunz, S. Rein, J.Y. Lee, W. Warta, "Minority Carrier Lifetime Degradation in Boron-Doped Czochralski Silicon", Journal of Applied Physics 90 (2001) 2397.

[3] D. Macdonald, L.J. Geerligs, "Recombination Activity of Interstitial Iron and Other Transition Metal Point Defects in p- and n-type Crystalline Silicon", Applied Physics Letters 85 (2004) 4061.

[4] J.E. Cotter, J.H. Guo, P.J. Cousins, M.D. Abbott, F.W. Chen, K.C. Fisher, "P-type Versus n-type Silicon Wafers: Prospects for High-Efficiency Commercial Silicon Solar Cells", IEEE Transactions on Electron Devices 53 (2006) 1893.

[5] D. Fujishima, H. Inoue, Y. Tsunomura, T. Asaumi, S. Taira, T. Kinoshita, M. Taguchi, H. Sakata, E. Maruyama, "High-Performance HIT Solar Cells for Thinner Silicon Wafers", Photovoltaic Specialists Conference (PVSC) 35th IEEE (2010) 3137.

[6] P.J. Cousins, D.D. Smith, H.C. Luan, J. Manning, T.D. Dennis, A. Waldhauer, K.E. Wilson, G. Harley, W.P. Mulligan, "Generation 3: Improved Performance at Lower Cost", Photovoltaic Specialists Conference (PVSC) 35th IEEE (2010) 275.

[7] F. Feldmannn, M. Bivour, C. Reichel, H. Steinkemper, M. Hermle, S.W. Glunz, "Tunnel Oxide Passivated Contacts as an Alternative to Partial Rear Contacts", Solar Energy Materials & Solar Cells 131 (2014) 46.

[8] Y. Schiele, S. Fahr, S. Joos, G. Hahn, B. Terheiden, "Study on Boron Emitter Formation by BBr_3 Diffusion for n-type Si Solar Cell Applications", European Photovoltaic Solar Energy Conference and Exhibition 28th (2013) 1242.

[9] C. Duran, "Bifacial Solar Cells: High Efficiency Design, Characterization, Modules and Applications", (2012) 12.

[10] Y.J. Cho, H.S. Chang, "Enhanced Boron Gettering Effect of N-Type Solar Grade Si Wafers by in situ Oxidation", Metals and Materials International 19 (2013) 1377.

[11] A. Kimmerle, P. Rothhardt, A. Wolf, R.A. Sinton, "Increased Reliability for J_0-analysis by QSSPC", Energy Procedia 55 (2014) 101.

NEW CHEMICAL ATTACK OF AG-CATALYZED ON SI IN HF-H2O2-AGNO3 MEDIUM. APPLICATION TO SI SOLAR CELLS TREATMENT

Waly Bodian[1,2], Diouma Kobor[1], Jean-Marc Joubert[2], Stéphane Bastide[2], Ulf Blieske[3]
[1]Laboratoire de Chimie et de Physique des Matériaux (LCPM) Université Assane Seck de Ziguinchor, BP 523, Sénégal
[2]Institut de Chimie et des Matériaux Paris-Est, 2-8 rue H. Dunant, 94320 Thiais, France
[3]Cologne Institute for Renewable Energy, TH Köln, Germany
email: dkobor@univ-zig.sn

ABSTRACT: The paper presents the investigation of a new three component (HF, $AgNO_3$ and H_2O_2) metal assisted etching treatment on Si. Future possible applications are in particular the texturization of multicrystalline Si substrates. Currenttreatments of these substrates are known to be complex and moderately effective, especially when diamond wire saws are used for wafer production leaving less surface defects.
A first study of the chemical attack of (100) silicon catalyzed by a metal using the above mentioned components is presented. The presence of $AgNO_3$ leads to the deposition of Ag and it is shown that H_2O_2 re-oxidizes a part of this latter, resulting in different surface structure depending on its concentration. At low H_2O_2 concentration, the formation of Si nanowires obtained is inhibited in favor of a surface roughness, which makes possible the reduction of the reflectivity. At high concentration of H_2O_2, Si electro-polishing is obtained with the formation of isolated and well-defined size Ag nanoparticles. These results open new fields of investigation both to the dissolution mechanisms involved and to the potential applications to surface treatments of Si solar cells.
Keywords: Silicon, silver assisted, metal assisted etching, chemical etching, texturization, surface texture, optical properties

1 INTRODUCTION

Scientific research and industrial development had opened the miniaturization leading to the appearance of nanosciences and nanotechnology disciplines.

Silicon based nanostructures have attracted wide attentions due to their potential applications in photovoltaic, lithium battery, biomedicine, and optoelectronic devices [1–5]. Porous silicon and silicon nanowires have been particularly concerned and studied over the past decades because of their outstanding properties of antireflection and photoluminescence, respectively [6–12]. Surface texturing is a key process in silicon solar cell fabrication. It reduces the optical reflection losses to well below 10 %, compared to a polished silicon surface, which reflects about 40 % of the incident light in the visible region. Several methods of elaboration of nanomaterials have developed among which we can mention the conventional methods such as PVD (Physical Vapor Deposition) and CVD (Chemical Vapor Deposition), but these techniques require sophisticated and expensive equipment requiring high vacuum and/or high temperature. Recently, a new method based on metal-assisted chemical etching (MACE) has attracted increasing attentions for their simplified and low-cost fabrication process [13,14]. MACE is essentially a wet etching method,producing anisotropic high aspect ratio semiconductor micro and nanostructures without incurring lattice damage. These high aspect ratio structures can potentially be used for the formation of periodic nanostructures. Ag is one of the most frequently used noble metals.

For this purpose, there are known two-component systems, the most traditional one being HF-$AgNO_3$, which leads to the deposition of Ag dendrites network and concomitantly to the formation of Si nanowires [15] and HF-H_2O_2, which makes possible to form mesopores from previously deposited Ag nanoparticles [16, 17]. From a fundamental point of view, it may be asked how these two systems could be combined, knowing that in the presence of the three components an additional reaction occurs, the oxidation of deposited Ag by H_2O_2 ($E°_{Ag+/Ag}$ = 0.80 V/ENH vs. H_2O_2/H_2O = 1.78 V/ENH): Is the amount of the deposited Ag as a function of [H_2O_2]? What are the shape and the size limit of deposits using high concentration of H_2O_2? Is there Si dissolution and according to what regime? From a practical point of view, does this 3-component system make possible to texturize the surface of multicrystalline Si substrates, which current treatments are known to be complex and moderately effective?
In this paper, beyond the fundamental aspects, the focus will be on the potential applications of this new type of Si-3-component attack in the field photovoltaic materials such as poly – Si texturization.

2 EXPERIMENTAL PROCEDURE

2.1 Nanowires fabrication and new attack method

A boron doped P-type (100) silicon wafer of 600 – 650 µm of thickness with resistivity of 1–5 Ω.cm of the polished face was purchased from BT Electronics. For nanowires fabrication MACE method was used following the method described elsewhere [18]. Some samples were pretreated with Piranha solution and the others are free Piranha pretreatment with different etching time. For the new attack method, 100 ml of 0.3 M HF solution was prepared. One considers a variable f, which is the molar fraction of H_2O_2 according to the following relation:

$$f_{H_2O_2} = [H_2O_2]/([H_2O_2] + [HF]) \qquad \text{(eq. 1)}$$

Its values vary between 0, when [H_2O_2] = 0, and 1, when [HF] = 0. The concentration and the volume of H_2O_2 were calculated for different values of f, according to:

$$[H_2O_2] = 0.3\, f_{H2O2}/(1 - f_{H2O2}) \qquad \text{(eq. 2)}$$

Table 1 below shows the different concentration and volume values of H_2O_2 that have been tested. Concretely, we realized these compositions for each solution by taking the adequate volume V_{H2O2} 30 %wt, adding 1.32 ml of 40 %wt HF. The solution is then completed to 100 ml with ultrapure water. Then, 20 mg of $AgNO_3$ are weighed and introduced into the solution to finally obtain a solution of HF - H_2O_2 - $AgNO_3$ in a Teflon beaker in which samples are immersed for different etching time.

Table 1: f_{H2O2}, H_2O_2 concentration and HF and H_2O_2 volumes used for the different solutions

f_{H2O2}	V_{HF} (mL)	$[H_2O_2]$ (mol.L^{-1})	V_{H2O2} (mL)
0.95	1.32	5.700	57.980
0.90	1.32	2.700	27.460
0.80	1.32	1.120	12.200
0.70	1.32	0.700	7.120
0.60	1.32	0.450	4.570
0.50	1.32	0.300	3.050
0.40	1.32	0.200	2.030
0.30	1.32	0.120	1.300
0.20	1.32	0.075	0.075
0.10	1.32	0.033	0.330
0.00	1.32	0.000	0.000

Finally, the samples were rinsed in deionized water and subsequently immersed in a HNO_3 solution (0.1 M) for 3 min to remove the etchant.

2.2 Surface characterization

Surface morphology of the samples was observed using a MERLIN FEG of Zeiss scanning electron microscopy model at ICMPE (Paris-Est University). SEM images treatment using ImageJ software permitted to plot the numbers of nanoparticles histograms according to their sizes. The total reflectance for normal incidence was measured using a spectrophotometer Hitachi UV-VIS-NIR 4001 with an integrating sphere model at ICMPE (Paris-Est University) in the wavelengths range between 350 and 1200 nm.

3 RESULTS AND DISCUSSION

3.1 Nanowires fabrication: importance of piranha pretreatment

Figure 1 shows the mechanism of silver assisted etching. For the initial electroless deposition of silver from HF/$AgNO_3$ solution on silicon, simultaneous electrochemical processes including cathodic (silver-deposition reaction) and anodic reactions (silicon oxidation and dissolution) occur on the silicon surface. First, Ag+ ions in the vicinity of the silicon surface capture electrons from the Valence Band (VB) of silicon and are deposited in the form of metallic silver nuclei. With the deposition and growth of silver nuclei, SiO_2 is formed simultaneously underneath the silver nanoparticles which growth to form dendrites (Fig. 1b). As the SiO_2 underneath the silver nanoparticles is etched away as by HF, then pits are formed immediately at the same location.

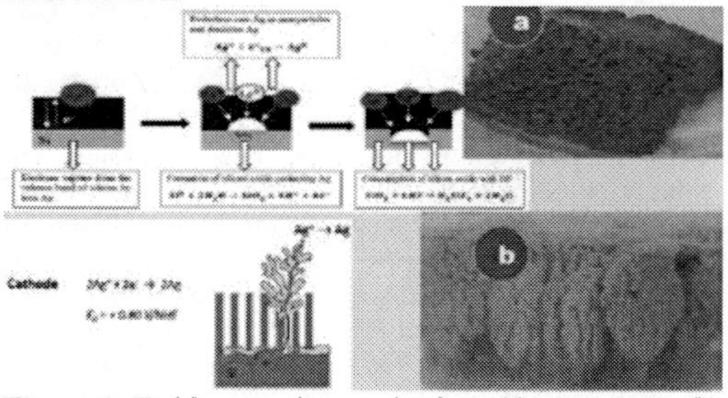

Figure 1: Etching reaction mechanism, (a) macropores (b) silver dendrites foam.

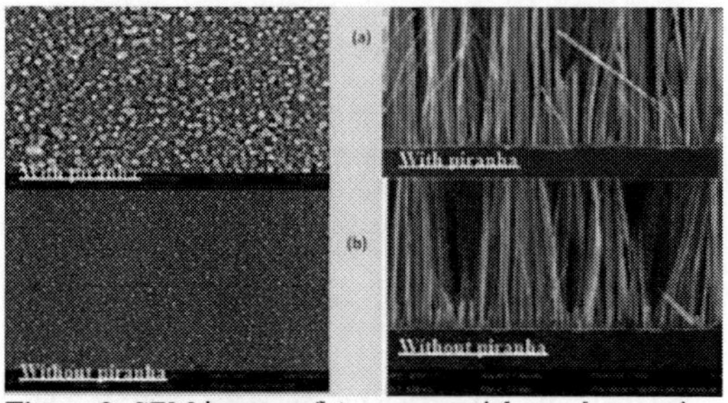

Figure 2: SEM images of Ag nanoparticles and nanowires (a) using piranha pretreatment for Si sample and (b) without piranha pretreatment.

So the silver nanoparticles sink into the pits as they are formed [19]. Pores are then created after the migration of the silver nanoparticles trough these pits. The foam volume depends to the etching time and so the nanowires length (Fig. 1b). The following equations explain the chemical reactions that occur in the process:

Anode (Si)

$$Si + 6HF \rightarrow H_2SiF_6 + 2H^+ + H_2 + 2e^-$$

Cathode

$$H_2O_2 + 2H^+ + 2e^- \rightarrow 2H_2O \qquad E_0 = +1.76\ V/NHE$$

Cathode

$$2Ag^+ + 2e^- \rightarrow 2Ag \qquad E_0 = +0.80\ V/NHE$$

At first, we asked ourselves if a preprocessing of the cleaning of the samples by piranha method before the formation of the nanowires was necessary.

Figure 2 presents SEM images of Ag nanoparticles formed after cathode reaction and Si nanowires after etching with and without piranha pretreatment. We find that deposit silver nanoparticles is quite different after 5 min in HF-$AgNO_3$ medium with and without piranha. The nanowires grow slower but are denser with the piranha treatment. We also observed that the results were more reproducible with this treatment. This shows that even a simple modification of the treatment protocol can have important consequences on the results.

3.2 Surface characterization

We first investigated the effect of an increasing of H_2O_2 amount from the HF-$AgNO_3$-H_2O_2 system at the very beginning of the deposition. The SEM images of Fig. 3 show the deposition of Ag nanoparticles, after 5 min of treatment, obtained when the fraction of H_2O_2 relative to HF is increased. In Fig. 3, we observe that the silver nanoparticles deposition is well influenced by the presence of H_2O_2. We see that for f = 0, many Ag nanoparticles are present on the silicon surface after only 5 min of treatment. For f = 0.1 ($[H_2O_2]$ = 0.033 mol/L) there is less Ag nanoparticles present on the Si surface. As the volume of H_2O_2 increased, the number of Ag nanoparticles deposited on the silicon surface, as well as their size, decreases.

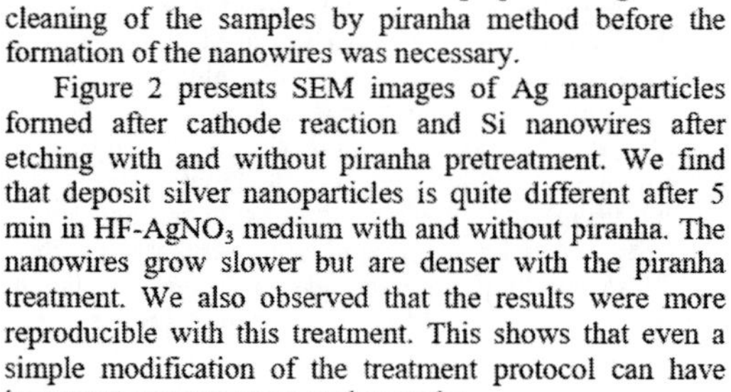

Figure 3: Influence of H_2O_2 concentration on Ag nanoparticles deposition for 5 min etching time

It is therefore seen that the presence of H_2O_2 influences the deposition of the Ag nanoparticles from the beginning of the etching. We have done the same experiments by increasing the deposit time to 1h45min. Such a long time makes it possible to suppose that a kind of equilibrium has been attained between the Ag deposition and its dissolution by H_2O_2. The SEM images of Fig. 4 show the deposition of Ag nanoparticles when the fraction of H_2O_2 increased after 1h45 min of treatment. As observed above, the deposition is also very dependent to H_2O_2 fraction.

For the system without H_2O_2 (f = 0), the particles are mainly dendrites with variable length, up to 15 µm. The section of these dendrites varies between 1 and 1.5 µm. There are also spherical nanoparticles but hidden between the nanowires, so it is difficult to estimate their size. For a H_2O_2 fraction of f = 0.1, the particles are predominantly dendrites but clearly smaller, with variable length up to 1 µm. When H_2O_2 amount is increased beyond f = 0.2, the dendrites can no longer be formed. Instead, there are only Ag nanoparticles (Fig. 4a). From f = 0.2 to f = 0.9, one observes that the nanoparticles become less important and smaller. It is the first time for the observation of such behavior. The section size of these dendrites varies from 200 to 300 nm. There are also spherical nanoparticles with a diameter less than or equal to 100 nm. Ag nanoparticles were dissolved in nitric acid after 1h45min treatment and SEM images are presented in Fig. 4b. There is a very clear evolution of the surface nanostructuring as a function of the H_2O_2 fraction. For f = 0, obtained nanowires are very small in size. From f = 0.05 to f = 0.1, no nanowires were fabricated but rather craters or mesopores, which size varies between 100 and 200 nm in diameter. The surface of these samples appears rather dark. For f ≥ 0.2, surfaces remain polished. The formation of Ag nanoparticles is accompanied by a surface electro-polishing, which, therefore, remains flat.

We wanted to determine the characteristics of these nanoparticles as a function of f. Silver nanoparticles size distribution was thus established by processing the SEM images using the ImageJ software. Figure 5 gives two examples for f = 0.6 and 0.9. For f = 0.6, the nanoparticles are spherical, with variable sizes and an average diameter equal to 25 nm. The Ag nanoparticles coverage rate is 72 % (only 28 % of the silicon surface is not covered). For f = 0.9, the mean diameter of the nanoparticles (spherical) is ≤ 100 nm, with an average size of 22 nm, and a coverage rate of 43 %, slightly the half surface area. From these results, it is seen that it is possible to adjust the size of the nanoparticles and the coverage rate of the substrate by modulating the H_2O_2 fraction.

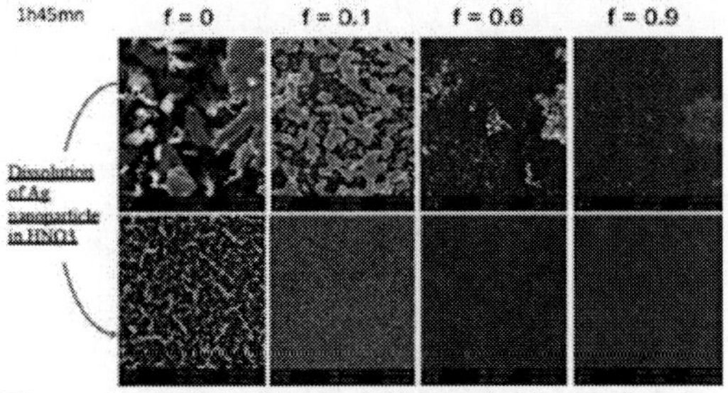

Figure 4: Ag deposition and Si texturization as function of H_2O_2 fraction for 1h45min etching time.

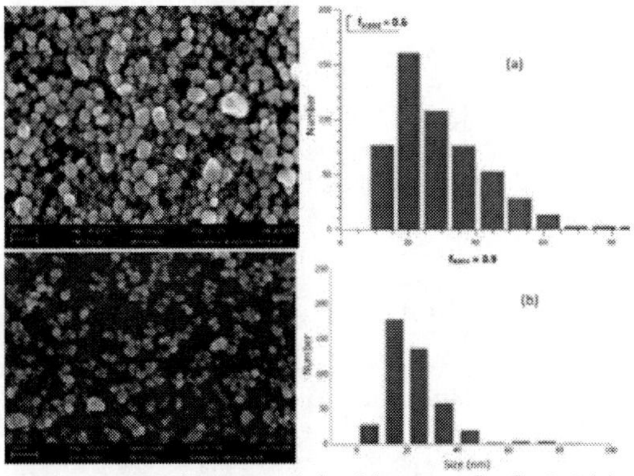

Figure 5: SEM images and size histogram for f = 0.6 (a) and 0.9 (b) deduced by image processing.

It is therefore a relatively easy control mean, which can be used for practical applications of silicon metallization. Figure 6 shows the reflectivity spectra of the samples after 1h45min of treatment with HF-$AgNO_3$-H_2O_2 system for different f values. It is found that for $f_{H2O2} > 0.2$, the reflectivity is close to that of the mirror polished silicon before treatment, which is in good agreement with the surface SEM images (Fig. 4b). On the other hand, for f = 0.1 and especially f = 0.05, the reflectivity is relatively very low, almost as low as that with nanowires obtained without H_2O_2. For f = 0, the average reflectivity is about 10 %, which is slightly lower than the reflectivity obtained by the standard monocrystalline silicon KOH etching industry solar cell, which is 12 %. Thus, these results are very interesting for p-type silicon cells that require (n+) rather thick emitters (0.25 µm). Indeed, for this type of solar cells, the presence of nanowires on the surface is not desirable because the diffusion of the n+ emitter would result in an overdoping of the nanowires and their electronic quality would then be bad. It is therefore necessary to avoid forming too long and too fine structures. For f = 0.05 to 0.1, the treatment could be used to texturize the silicon for solar cells because the obtained structures on the surface are small (100 - 200 nm) and nevertheless allow the reflectivity to be reduced to interesting values (< 10 – 15 %).

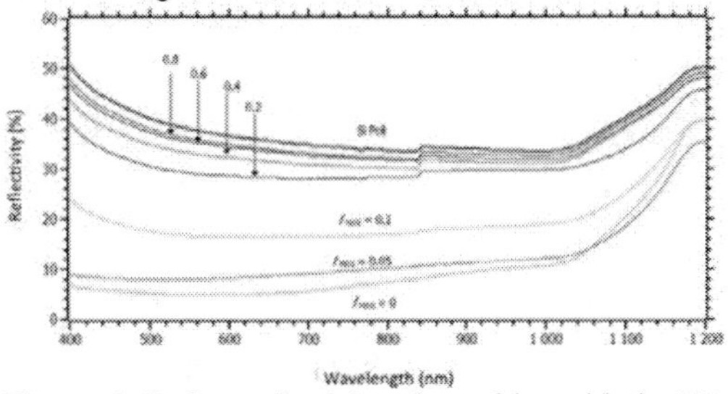

Figure 6: Surface reflectivity after etching with the HF-$AgNO_3$-H_2O_2 system for different f values.

4. CONCLUSION

The treatment of (100) Si by HF - $AgNO_3$ - H_2O_2

Shows the following results: The possibility of the control of Ag nanoparticles deposition by varying the H_2O_2 fraction is demonstrated.

The results show the determination of a H_2O_2 concentration allowing texturisation without formation of nanowires (f_{H2O2} = 0.05 - 0.1)

The new treatment allows the reduction of the surface

reflectivity below 10 % over the entire solar spectrum (for comparison R = 12 % for KOH-textured Si-c). It is a new promising texturization treatment.

In perspectives, investigations will be done on (111) oriented c-Si to investigate the potential of the new texturization treatment for multicrystalline silicon.

References

[1] E. Garnett, P. D. Yang, NanoLett.10 (2010) 1082.

[2] K. Q. Peng, X. Wang, S. T. Lee, Appl. Phys. Lett. 92 (2008) 163103.

[3] S. K. Srivastava, D. Kumar, P. K. Singh, M. Kar, V. Kumar, M. Husain, Sol. Energy Mater. Sol. Cells 94 (2010) 1506.

[4] Y. F. Huang et al, Nat. Nanotechnol. 2 (2007) 770.

[5] C. K. Chan et al, Nat. Nanotechnol. 3 (2008) 31.

[6] A. I. Hochbaum et al, Nature 451 (2008) 163.

[7] D. D. D. Ma, C. S. Lee, F. C. K. Au, S. Y. Tong, Science 299 (2003) 1874.

[8] K. Q. Peng et al, Chem. Eur. J. 12 (2006) 7942.

[9] J. Zhu et al, NanoLett. 9 (2009) 279.

[10] X. Li, P. W. Bohn, Appl. Phys. Lett. 77 (2000) 2572.

[11] M. Lipinski, S. Bastide, P. Panek, C. Levy-Clement, Phys. Status Solidi 197 (2003) 512.

[12] J. Y. Lee, W. K. Han, J. H. Lee, Sol. Energy Mater. Sol. Cells 95 (2011) 77.

[13] H. P. Wang, K. Y. Lai, Y. R. Lin, C. A. Lin, J. H. He, Langmuir 26 (2011) 12855.

[14] W. Chern et al., NanoLett.10 (2010) 1582.

[15] K. Q. Peng, Z. P. Huang, J. Zhu, Adv. Mater. 16 (2004) 73.

[16] K. Tsujino, M. Matsumura, Adv. Mater. 17 (2005) 1045.

[17] C. Chartier, S. Bastide, C. Lévy-Clément, Electrochim. Acta. 53 (2008) 5509.

[18] C. Chartier, S. Bastide, C. Levy-Clement, Electrochimica Acta 53 (2008) 5509.

[19] A. Md Asgar, M. Hasan, F. Md Huq, ZH Mahmood, J Nanomed Nanotechnol. 5 (2014) 221,

NANOTEXTURED SILICON SURFACES USING TAILORED VOLTAGE WAVEFORM PLASMAS: IMPACT OF ION BOMBARDMENT ENERGY ON ETCHING DYNAMICS AND PASSIVATION

Guillaume Fischer[1,2,*], Etienne Drahi[3], Fabien Lebreton[3,2], Pavel Bulkin[2], Gilles Poulain[3], Erik V. Johnson[2]

[1]Institut Photovoltaïque d'Ile-de-France (IPVF) - 8 rue de la Renaissance, 92160 Antony, France
[2]LPICM – CNRS - Ecole polytechnique, Université Paris-Saclay, 91128 Palaiseau, France
[3]Total S.A. Renewables, 2 Place Jean Millier, 92078 Paris La Défense Cedex, France
*Corresponding author: guillaume.fischer@ipvf.fr

ABSTRACT: Front surface nanotexturing of silicon is investigated using a capacitively coupled reactive ion etching SF_6/O_2 discharge at room temperature. The plasma is either powered with a standard RF voltage at 13.56 MHz, or an innovative multi-frequency approach named Tailored Voltage Waveforms (TVWs). It has been shown that by using TVWs, it is possible to significantly modify the ion bombardment energy (IBE) at the substrate while keeping the discharge power constant. Measurements suggest that increasing IBE is beneficial in achieving low reflectance values in a short process time, although the role of other mechanisms at play has still to be clarified. Nanotextured samples have then been coated by thermal ALD with a 7 nm thick Al_2O_3 passivation layer. The simultaneous effects of surface area enhancement and plasma induced damage of the passivation quality are successfully dissociated, and it is shown that increasing the IBE (in the range from 80 to 200 eV) strongly increases the surface recombination velocity (SRV). However, one can achieve both low reflectance (less than 5%) and good passivation quality (front SRV around 40 cm/s), without the need for an additional chemical damage removal etching. Furthermore, the possibility of recovering the passivation through a "dry" damage removal etching is examined.

Keywords: c-Si, Plasma texturing, Reflectance, Passivation

1 INTRODUCTION

Solar light absorption in a crystalline silicon (c-Si) solar cell can drastically be increased through the simultaneous use of front surface texturing and anti-reflective coatings [1]. Microscale texturing using wet processes is now standard at industrial level [2], for instance producing pyramidic morphologies in the case of alkaline etching. On the other hand, nanoscale texturing (or "nanotexturing") has raised much interest for many years in the field of photovoltaics [3], as it produces very low reflectance surfaces (referred to as black silicon) without the need for an anti-reflective coating.

Various methods have been investigated for the nanotexturing of c-Si surfaces. In particular, SF_6/O_2 plasma etching has been explored as a possible solution for solar cells, as the *in-situ* formation of nanostructures had been reported in the field of microelectronics as a means to attain the best conditions for deep plasma etching [4]. Bombardment of the c-Si surface by ions during SF_6/O_2 plasma etching plays a key role in the *in-situ* texturing process. Indeed, this phenomenon is attributed to a complex balance of competing mechanisms: ion assisted etching of c-Si by fluorine radicals, formation of micro-masking species (of type SiF_xO_y) thanks to the presence of oxygen radicals, and sputtering of the micro-mask by ion bombardment [5].

However, the integration of nanotextured c-Si surfaces at the front surface of high efficiency c-Si solar cells still proves to be challenging due to the need for surface passivation. Indeed, the high aspect ratio of such nanostructures prevents the conformal deposition of passivation layers (such as plasma enhanced chemical vapor deposited amorphous silicon nitride). Moreover, nanotextured surfaces have a high surface area enhancement factor (ratio between the real surface area and the projected surface area) which intrinsically increases recombination events at the surface [6]. While the conformality issue can be solved using atomic layer deposition (ALD) of amorphous alumina (Al_2O_3) for instance, the surface area enhancement cannot be avoided.

A key point of interest in the present study is this issue of the passivation of plasma textured c-Si. Indeed, both ultraviolet light and ion bombardment from the plasma can induce defects in the c-Si, such as vacancies or impurities [7,8], which will act as recombination centers. Therefore, a wet damage removal etching (DRE) step is usually performed after plasma texturing in order to remove the damaged layer and achieve good surface passivation level [9]. However, DRE usually leads to a re-increase of the surface reflectance [10,11].

In the present study, a new type of plasma excitation technique, named Tailored Voltage Waveforms (TVW, described in next section) will be used to texture c-Si samples. The latter will then be passivated using an ALD-Al_2O_3 layer in order to assess the surface passivation quality that can be reached depending on the process conditions.

2 METHODS

2.1 Samples and processes

In this study, 180 µm thick n-type doped c-Si CZ (100) samples have been used, with a resistivity of 4.5 Ω.cm. A saw damage removal (SDR) process has first been applied.

The subsequent steps of the process flow are schematically described in figure 1. The samples were HF cleaned, then textured on one side using an SF_6/O_2 plasma. The pressure was set to 30 mTorr, and the gas flow ratio of SF_6/O_2 was 1.3 for all processes. The substrate temperature during processing was comprised between 25 and 40 °C, while the total coupled power and type of voltage excitation were varied. The details of the plasma excitation production and control are described in the next section. Reference samples for passivation did not go through these two first steps. Notice that no wet DRE has been performed after texturing.

Another HF dip is then performed, before the sample is loaded in the ALD reactor (Picosun R-200[1]). Both surfaces of the wafers were passivated by 7 nm of Al_2O_3 synthetized from TMA and H_2O at 150 °C during the same deposition run. The post-deposition annealing was carried out at 350 °C for 30 min, followed by surface illumination at 1 mW/cm² overnight to ensure trap saturation in Al_2O_3 [12].

(a)

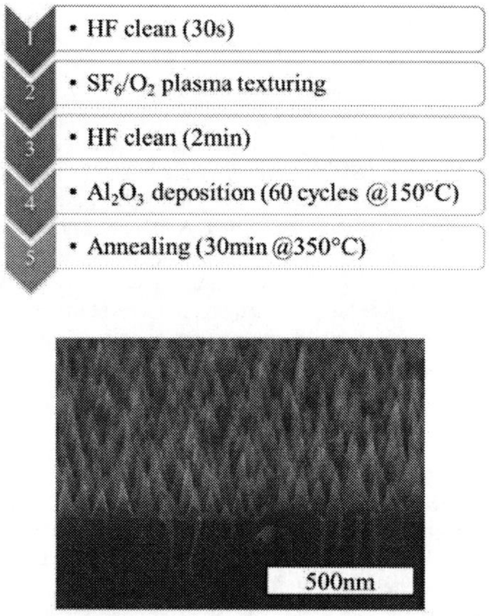

Figure 1: (a) Performed process flow, (b) cross-section view SEM image of a nanotextured c-Si sample (before Al_2O_3 deposition).

2.2 Tailored voltage excitation of SF_6/O_2 plasma

Nanoscale texturing of the c-Si samples has been achieved by reactive ion etching (RIE) with an SF_6/O_2 discharge, using a Nanomaster NRE-3500 reactor. A schematic of the reactor is shown in figure 2(a). This capacitively coupled plasma (CCP) reactor can either be powered with a standard radiofrequency (RF) voltage at a frequency $f = 13.56$ MHz, or with Tailored Voltage Waveforms (TVW). The latter excitation voltage is obtained by adding upper harmonic components to a basis frequency (again at 13.56 MHz) and precisely controlling the relative amplitudes and phase-shifts of the different components, therefore obtaining tailored excitation signals – examples are depicted on figure 2(b). The control of the waveform tailoring is ensured by a system similar as the one described in [13].

In the implementation used for this study, similar to the one proposed in [14], the waveforms can be expressed as:

$$V(t) = V_0 \sum_{k=1}^{n} \frac{n-k+1}{n} \cos(2\pi k f t + \Theta) \qquad (1)$$

where V_0 is a pre-factor tuning the peak-to-peak voltage, n the number of harmonic components, and Θ the angular phase-shift between the components.

Tailored Voltage Waveforms were first envisioned and tested [15,16] as a means to decouple the ion flux from the ion bombardment energy in a CCP discharge. This effect comes from the amplitude asymmetry of the waveforms; tuning the difference in minimum and maximum values of voltage in the waveforms induces variations on the voltage drops among the sheaths of the plasma. Another effect has later been discovered, namely the slope asymmetry effect [17], which stems from the difference in the sheath expansion and collapse rates in front of the electrodes, and can lead to significant variations in the ion flux at the substrate electrode.

The magnitude of both the amplitude and the slope asymmetry effect strongly depends on the discharge properties. In the case of an electronegative SF_6/O_2 discharge, both effects have been observed, and their preponderance depends in particular on the discharge pressure, on the gas flux ratio SF_6:O_2, and on the coupled power [18]. Figure 2(c) shows the evolution of IBE_{max} and the total ion flux with the applied waveform in an SF_6/O_2 discharge (same gas composition and pressure as for texturing), at a coupled power of 25 W. These measurements were obtained using a retarding field energy analyzer. The graph shows that variations around 40 % in both parameters can be obtained.

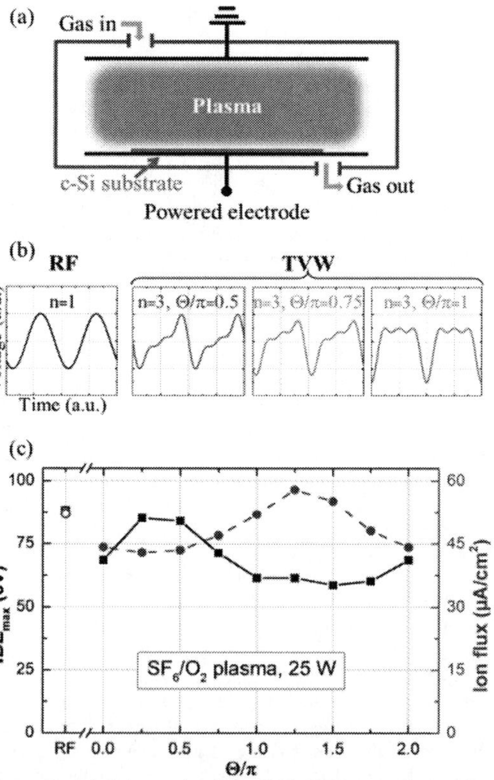

Figure 2: (a) Schematic of the RIE-CCP reactor used for nanotexturing. (b) RF signal compared to examples of TVW signals used in the study (n and Θ values in equation 1). (c) Maximal ion bombardment energy (squares) and total ion flux (bullets) at the substrate electrode as a function of the applied waveform.

As it has been discussed previously, the maximal ion bombardment energy at the substrate, IBE_{max}, is of fundamental interest when studying the etching and the

[1] Certain commercial equipments are identified in this paper in order to specify the experimental procedure adequately. Such identification is not intended to imply recommendation or endorsement by the authors' organizations, nor is it intended to imply that the materials or equipment identified are necessarily the best available for the purpose.

creation of defects in c-Si. This parameter is estimated by monitoring the voltage waveform $V(t)$ at the powered substrate electrode, and given by:

$$IBE_{max} = e\left[\left(\frac{-V_{min}}{V_{PP}}\right)\left(\left(\frac{V_{max}}{V_{PP}}\right)V_{PP} + V_{DC}\right) - V_{DC}\right] \quad (2)$$

where V_{PP} is the peak-to-peak voltage, V_{DC} is the DC self-bias, while V_{min}=min($V(t)$)-V_{DC} and V_{max}=max($V(t)$)-V_{DC} respectively represent the minimum and maximum values of voltage with respect to the DC self-bias. One may note that the first term on the right hand side of equation 2 corresponds to the average plasma potential, and as the sheaths are collisional, only represents the maximum possible IBE value (and not the mean value, nor the most probably value).

2.3 Characterization

The total reflectance of the samples has been measured using a Perkin-Elmer Lambda 900 UV-Vis spectrophotometer equipped with an integrating sphere, acquired for a wavelength range λ from 250 to 1100 nm and the effective reflectance R_{eff} has been computed as:

$$R_{eff} = \frac{\int_{280nm}^{1100nm} R(\lambda) S_{AM1.5g}(\lambda) d\lambda}{\int_{280nm}^{1100nm} S_{AM1.5g}(\lambda) d\lambda} \quad (3)$$

where R is the spectral reflectance and $S_{AM1.5g}$ the solar spectral irradiance as defined by the AM1.5g standard [19].

The effective lifetime, τ_{eff}, of the passivated samples has been measured by photo-conductance decay using a WCT-120 from Sinton instruments. Then τ_{eff} was used to compute the front surface recombination velocity [20], S_{front}:

$$S_{front} = \left(\frac{1}{\tau_{eff}} - \frac{1}{\tau_{bulk}}\right) W - S_{back} \quad (4)$$

where τ_{bulk} is the bulk lifetime (computation from [21]), W the sample thickness, and S_{back} is the back surface recombination velocity. For reference (symmetric) samples, we assume S_{front} and S_{back} are identical, while for single-sided textured samples, S_{back} is taken as the average of S_{back} obtained for the ten reference samples, with a value of 3.7 cm/s.

The etching rate was estimated using stylus profilometry, by probing the step created between a completely masked area and an etched area on mirror polished samples processed simultaneously with SDR samples.

Finally, corona-voltage measurements performed on a Semilab PV-2000A tool have been used to extract surface area enhancement factors. The surface area enhancement factor can be estimated by the inverse of the slope of the contact potential measurements under light, as these slopes are inversely proportional to the capacitance of the Al_2O_3 layer [22].

3 RESULTS

In the first subsection, the texturing dynamics and passivation degradation will be investigated for different process conditions using TVW excitation. Subsequently, the particular contribution of surface area enhancement to surface recombination velocity degradation will be assessed and isolated from ion bombardment damaging effects. Finally, design rules for the formation of low reflectance low surface recombination velocity black silicon surfaces using plasma etching are drawn.

3.1 Nanotexturing dynamics and passivation degradation

Figure 3 shows the evolution of the effective reflectance and front surface recombination velocity as a function of process time for different voltage excitations and ion bombardment energies. In TVW mode the different values of IBE have been obtained for the same coupled power of 35 W by changing the waveform, which correspond to the waveforms depicted on figure 2(b).

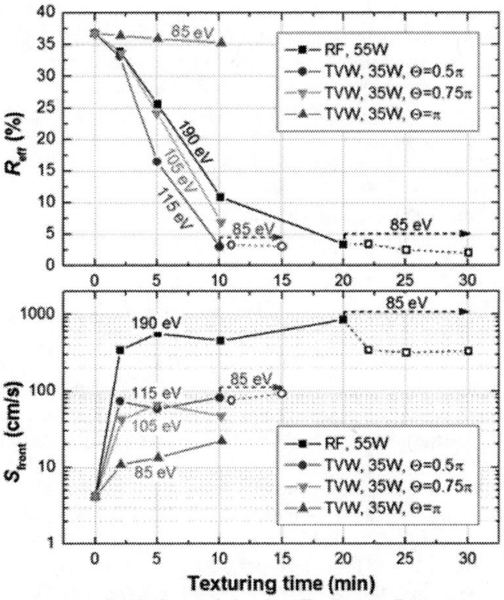

Figure 3: Evolution of (top) effective reflectance and (bottom) front surface recombination velocity with the texturing time for different process conditions. Labels indicate the maximal IBE.

After only 2 min of processing, a small decrease (from 0.3 to 3.5 % absolute) of R_{eff} is observed for all the samples, showing that the surface texturing has effectively begun. Afterwards, R_{eff} continues to decrease with a rate depending on the process conditions. From the samples etched in TVW excitation mode, it is suggested that the higher the ion bombardment, the faster the decrease; after 10 min, the reflectance of the samples is about 35%, 7% and 3 % for IBE_{max} of 85 eV, 105 eV and 115 eV, respectively. However, this hypothesis is challenged by the fact that the samples textured in RF mode (at 55 W, IBE_{max} = 190 eV and R_{eff} = 11 % after 10 min) do not fall within this trend. The hidden parameters favoring a faster decrease of reflectance remain to be understood, with a likely candidate being the ion flux, as strong variations of this property have been observed as previously highlighted.

Figure 3 also shows that the surface recombination velocity jumps after 2 min of texturing, and that the magnitude of the rise is directly correlated to the ion bombardment energy. At 85 eV, S_{front} increases by a factor 2.6, while at 190 eV, S_{front} rises by almost two orders of magnitude. The magnitude of this effect cannot solely be explained by the inherent surface area enhancement occurring during texturing, and is attributed to the rapid formation of a plasma damaged layer. The dependence on the maximal ion bombardment energy suggests that the depth of the damaged layer and/or the density of electrically active defects mostly depend on the energy of incoming ions. A contribution from ultraviolet light to defect generation is however not discarded.

As the process continues, S_{front} remains fairly stable relative to the observed initial jump. These observations prove that the damaged layer is formed in less than 2 min, after which the depth of this damaged layer and the density of electrically active defects are not considerably affected.

Finally, a few attempts were made to attempt to recover a lower surface recombination velocity (open symbols data points in figure 3). These samples have been obtained by a two-step texturing process; they were first textured either at IBE_{max} = 190 eV during 20 min, or at IBE_{max} = 115 eV during 10 min, chosen so that they show the same effective reflectance after this first step (R_{eff} = 4 %). The second texturing step was performed by changing the excitation voltage to achieve a lower value of ion bombardment energy (IBE_{max} = 85 eV) for a shorter period (between 30 s and 10 min).

When performing a second process step at lower IBE_{max}, two hypotheses can be formulated: (i) either the initial damaged layer on top of the nanostructures can be totally removed and replaced by a thinner damaged layer (or one with fewer defects), (ii) some parts of the initial damaged layer (for instance at the top or at the bottom of the nanostructures) cannot be removed and only a partial recovery can occur. Obviously, the validity of either these hypotheses depends on the nanotexturing mechanisms and on the morphology of the damaged layer (for instance, if the damaged layer has a uniform thickness below the nanostructures).

For the samples initially etched at IBE_{max} = 190 eV (black squares in figure 3), it is first observed that the effective reflectance continues to decrease during the second texturing step: R_{eff} = 2 % after a total process time of 30 min. Moreover, a recovery is indeed observed: after 2 min at 85 eV, S_{front} has dropped by a factor of 2.5, and then remains stable. The etch rate at IBE_{max} = 85 eV has been estimated to be around 24 nm/min, suggesting that the plasma damaged layer (containing electrically active recombination centers) is in this case less than 50 nm deep. For the samples initially etched at IBE_{max} = 115 eV (red bullets in figure 3), both the effective reflectance and the surface recombination velocity remain fairly stable during the second process step, and no recovery is evident.

Further tests will be necessary to definitively assess the possibility of recovery for any initial process condition, as well as the magnitude of this effect. This possibility would also open up the option of a dry DRE step after nanotexturing and, most importantly, using the same process tool and plasma recipe.

3.2 Contribution of surface area enhancement

Figure 4 shows the evolution of S_{front} as a function of IBE_{max} for 10 min textured samples. The graph simultaneously shows the local and effective S_{front} values for each sample, where the local S_{front} corresponds to the ratio between the effective S_{front} (computed using equation 4) and the surface area enhancement factor A_{text}/A_{ref} of the corresponding sample. This factor has by definition a value of unity for reference (i.e. un-textured) samples. The dichotomy between local and effective S_{front} helps to separate two effects that occur simultaneously during nanotexturing: surface area enhancement and plasma induced damage of the c-Si. It is clearly observed in figure 4 that both local and effective S_{front} drastically increase with the maximal ion bombardment energy experienced by the substrate during the process.

The increased surface area is therefore not the sole factor behind the greater S_{front} observed: increasing ion bombardment energy also leads to higher local densities of electrically active defects and/or deeper defects. It should be highlighted that while the computation of a local S_{front} helps one to understand the phenomena at play, the effective S_{front} is the relevant parameter for devices, as the surface area enhancement is unavoidable when reducing reflectance through texturing. This surface area enhancement should however remain as low as possible.

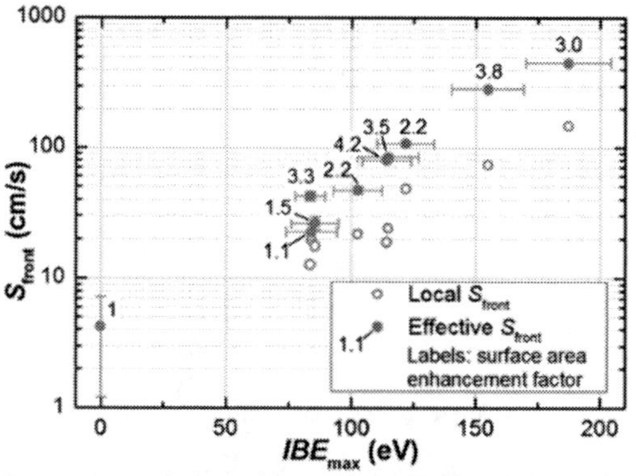

Figure 4: Evolution of local and effective front surface recombination velocity with maximal ion bombardment energy for 10 min textured samples. Labels indicate the surface area enhancement factor. Error bars are only shown for effective S_{front} to ensure readability.

3.3 Design rules

Figure 5 shows a synthesis of the major results in terms of effective front surface recombination velocity and effective reflectance values, obtained for all the studied samples. One should be reminded that the final goal is to simultaneously obtain a low S_{front} and a low R_{eff}. The best compromise obtained during this study has the following characteristics: R_{eff} = 4.5 % and S_{front} = 42.1 cm/s (corresponding to a measured effective lifetime of τ_{eff} = 364 µs).

This graph finally allows us to point out two major "design rules" for SF_6/O_2 plasma nanotexturing of c-Si surfaces for applications in solar cells. First, it is observed that in order to reach surface recombination velocities below 100 cm/s, the ion bombardment energy has to be maintained below 120 eV, either as a single process step or in case of dry DRE, as proposed previously.

Furthermore, figure 5 shows that effective reflectance values below a threshold of 5 % have been obtained for samples with surface area enhancement factors ranging roughly from 2.3 to 4.2. These values are only relevant for the type of nanostructures obtained with the studied experimental conditions. However, it gives a necessary condition to reach fairly low reflectance. More importantly, the lowest reflectance values have not been obtained for the samples having the highest surface area enhancement factors; the latter can therefore also be tuned to further decrease the effective surface recombination velocity.

However, and in addition to the ion bombardment energy, the influence of both UV light and ion flux on nanotexturing and defect creation remain to be understood.

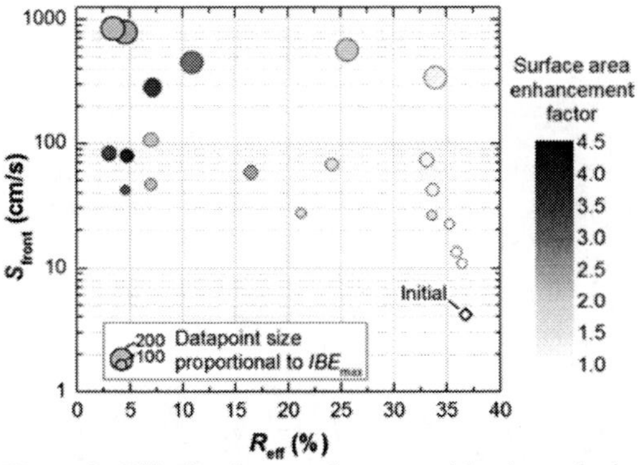

Figure 5: Effective front surface recombination velocity as a function of effective reflectance for all the obtained samples.

4 CONCLUSION

Silicon surface nanotexturing has been obtained through SF_6/O_2 plasma etching in a capacitively coupled discharge. Using either Tailored Voltage Waveforms or standard RF excitation, it has been possible to vary the maximal ion bombardment energy experienced by the surface without changing the discharge power.

It has been shown that increasing ion bombardment energy is beneficial in decreasing the effective reflectance and rapidly obtaining black silicon surfaces, although the bombardment energy is not the only parameter at play. However, increasing ion bombardment energy in the range from 80 to 200 eV drastically increases the surface recombination velocity, as measured after Al_2O_3 passivation. This effect is attributed to the formation of a plasma damaged layer, while the inherent surface area enhancement induced by nanotexturing only further worsens this passivation loss.

Design rules for the process have been extracted from the obtained results: in order to simultaneously reach front surface recombination velocities below 100 cm/s and effective reflectance below 5 %, the ion bombardment energy has to be kept below approximately 120 eV, while the surface area enhancement factor has to be superior to 2. Finally, results also suggest that the constraint on the maximal admissible ion bombardment energy can be partly released using a "dry" damage removal etching, i.e. a second process step using the same equipment and recipe but a lower ion bombardment energy that will partly remove the previously formed damaged layer.

ACKNOWLEDGMENTS

This project has been supported by the French Government in the frame of the program of investment for the future (Programme d'Investissement d'Avenir - ANR-IEED-002-01).

REFERENCES

[1] C.S. Solanski and H.K. Singh, *Anti-reflection and Light Trapping in c-Si Solar Cells*, Springer (2017).

[2] "International Technology Roadmap for Photovoltaic Results 2016," ITRPV (2017) 8th Edition.

[3] M. Otto, M. Algasinger, H. Branz, B. Gesemann, T. Gimpel, K. Füchsel, T. Käsebier, S. Kontermann, S. Koynov, X. Li, V. Naumann, J. Oh, A.N. Sprafke, J. Ziegler, M. Zilk and R.B. Wehrspohn, *Advanced Optical Materials* 3 (2015) 2.

[4] H. Jansen, M. de Boer, R. Legtenber and M. Elwenspoek, *Journal of Micromechanics and. Microengineering* 5 (1995) 2.

[5] M. Schnell, R. Ludemann and S. Schaefer, *Proceedings of the 28th IEEE Photovoltaic Specialists conference* (2000).

[6] J. Oh, H.-C. Yuan and H.M. Branz, *Nature Nanotechnology* 7 (2012) 11.

[7] S. Samukawa, B. Jinnai, F. Oda and Y. Morimoto, *Japanese Journal of Applied Physics* 46 (2007) 3.

[8] G.S. Oehrlein, *Materials Science and Engineering: B* 4 (1989) 1.

[9] G. Kumaravelu, M.M. Alkaisi, A. Bittar, D. Macdonald and J. Zhao, *Current Applied Physics* 4 (2004) 2.

[10] J. Yoo, G. Yu and J. Yi, *Materials Science and Engineering: B* 159–160 (2009).

[11] G. Xiao, B. Liu, J. Liu, and Z. Xu, *Materials Science in Semiconductor Processing* 22 (2014).

[12] F. Lebreton, R. Lachaume, P. Bulkin, F. Silva, S. Filonovitch, E.V. Johnson and P. Roca i Cabarrocas, *Energy Procedia* 124 (2017).

[13] E. V. Johnson, T. Verbeke, J.-C. Vanel and J.-P. Booth, *Journal of Physics D: Applied Physics* 43 (2010) 41.

[14] B. Bruneau, T. Novikova, T. Lafleur, J.-P. Booth and E.V. Johnson, *Plasma Sources Science and Technology* 23 (2014) 6.

[15] B.G. Heil, U. Czarnetzki, R.P. Brinkmann and T. Mussenbrock, *Journal of Physics D: Applied Physics* 41 (2008) 16.

[16] J. Schulze, E. Schüngel and U. Czarnetzki, *Journal of Physics D: Applied Physics* 42 (2009) 9.

[17] B. Bruneau, T. Gans, D. O'Connell, A. Greb, E.V. Johnson and J.-P. Booth, *Physical Review Letters* 114 (2015) 12.

[18] G. Fischer, E. Drahi, G. Poulain, B. Bruneau and E.V. Johnson, oral presentation at the 69th Gaseous Electronics Conference (2016).

[19] ASTM International, standard G173-03(2012).

[20] S.-Y. Lien, C.-H. Yang, K.-C. Wu and C.-Y. Kung, *Nanoscale Research Letters* 10 (2015) 1.

[21] A. Richter, S. W. Glunz, F. Werner, J. Schmidt and A. Cuevas, *Physical Review B* 86 (2012) 16.

[22] M. Wilson, Z. Hameiri, N. Nandakumar and S. Duttagupta, *Proceedings of the 40th Photovoltaic Specialist Conference* (2014).

ELECTROLESS-PLATED METALLIZATION FOR N-TYPE SILICON SOLAR CELLS

Yueh-Lin Lee*, Ming-Shiou Lin, Kuang-Chieh Lai, Chia-Chih Chuang, and Chi-Chun Li
Motech Industries, Xinshi Dist., Tainan City, 74145, Taiwan
(*Corresponding author: phone/fax/e-mail:+886-6505-0789-0608/+886-6505-8446/yuehlin_lee@motech.com.tw)

ABSTRACT: In this work, both front- and rear-side metallization based on electrochemical plating contacts is proposed for n-type crystalline silicon solar cells. The influences of laser open fraction of rear-side surface on cell performances are investigated. Electroless-plated Ni is used as seed layer on rear-side surface, which is combined with rear-side electroplated Cu and front-side electroplated Ni/Cu metallization. The results of Suns-Voc measurement indicate the surface damage and recombination issues, resulting from laser open fraction of rear-side surface of cells. A difference of conversion efficiency between 3.9 and 9.5% open fraction is about 0.74% for Ni/Cu metallization cells.

Keywords: Metallization, Silicon solar cell, Performance

1 INTRODUCTION

Recently, the majority of crystalline silicon solar cells manufactured in the industry are mono-facial cells with an aluminum based metallization on the rear side. Conventional Ag/Al or Ag screen-printed contacts usually have high contact resistance and lead to junction shunting issue during the post firing process [1]. The Ag/Al screen printing can be replaced with Ni/Cu electrochemical plating technologies for front- and rear-side metallization. Ni has been empolyed as seed/ barrier layer to prevent Cu diffusion [2, 3]. Therefore, both front- and rear-side metallization of cells can be achieved via electroless-plated and electroplated technologies without using some costly vacuum equipments such as sputter [3]. However, some challenges still must be overcame for electrochemical plating metallization technique, including the improvements of adhesion, unformity, and contact property. On the other hand, a simple patterned process and advanced metallization plays are interesting for silicon solar cell technology. Low surface recombination loss in silicon solar cells is achieved by passivated surface. A method with local and low-damage opening is needed [4]. Laser process provides a feasible way to achieve pattern design, which is accurate and fast.

2 EXPERIMENTAL

The 156 × 156 mm^2 n-type mono-crystalline silicon wafers with a random pyramid texture, boron-doped emitter, and phosphorus-doped surface field were used as substrates, consisting of the following: A saw-damage etching, alkaline texturing, borosilicate (BSG) removal, and drive-in dopants. The plasma enhanced chemical vapour deposition (PECVD) SiNx based dielectric was used as passivation and anti-reflection coating (ARC). Prior to metallization, H-pattern of four busbars was defined in the dielectric layer using UV laser ablation. The front side of cells was used for laser open fraction of 4%, and the rear side surface was designed with different open fractions. The open fraction of rear-side surface varied between 3.9 - 9.5%.

Electroless-plated Ni (ELP-Ni) metallization was employed for n-type Si solar cells. ELP-Ni was deposited on rear-side surface according to the following chemical reaction [1, 3]:

$$Ni^{+} + 2H_2PO_2^{-} + 2H_2O \rightarrow Ni + H_2 + 4H^{+} + 2HPO_3^{2-}$$

Ni deposition was performed in the alkali electroless Ni bath. The deposition temperature was kept in the range of 60-70℃. The 0.1 μm ELP-Ni layer was annealed under N_2 atmosphere at 400℃ for 10 min. After annealing process, the residual Ni was removed by using the diluted nitric acid. The 1.0 μm ELP-Ni deposition was performed as rear-side seed layer. Subsequently, electroplated Cu (EP-Cu) layer was deposited as rear-side conductive layer.

Electroplated Ni (EP-Ni) deposition was performed for front side contact layer at 50℃ in the acidic Ni bath. The 0.1 μm EP-Ni layer was annealed at 400℃ for 10 min, and then the residual Ni was removed. Finally, the 1.0 μm EP-Ni and EP-Cu layers were deposited as front-side metallization. Each front-side metallization was used by forward bias plating (FBP). The schematic structure of our cells with laser ablated, electroless plated, and electroplated metallization is shown in Fig 1.

3 RESULTS AND DISCUSSION

Fig 2(a) shows the scanning electron microscope (SEM) image of rear-side ELP-Ni layer. ELP-Ni layer is deposited on the full area of rear side surface, including both laser open and non-open regions. Fig 2(b) shows the energy dispersive spectrometer (EDS) analysis of rear-side ELP-Ni layer. The nickel and phosphorus signals are detected for ELP-Ni film. An amount of below 7 wt.% of phosphorus is attributed to the reactions of the reducing agent.

Fig 3 shows the cross-section SEM image of rear-side metallization, which was created by focus ion beam (FIB) cutting. The thicknesses of rear-side Ni and Cu layers are about 1.0 and 2.7 μm, respectively. It can be seen that ELP-Ni film grows to follow the rear-side surface structure. From FIB cross-section image, the challenges of ELP-Ni good adhesion and forming a barrier layer for Cu diffusion are obvious.

Fig 4 shows the SEM images of front-side metallization. Prior to front-side electroplated Ni/Cu, the finger width of UV laser ablation is about 13.0 μm. The finger width of electroplated Ni/Cu is 40.5 μm, and the height is 13.8 μm. So, electroplated Cu deposition occurs in both vertical and lateral direction simultaneously. The narrow Ni/Cu finger width is due to the narrow laser open width, resulting in the low shading loss.

Fig 5 shows V_{oc} and J_{01} measured by Suns-V_{oc} for different laser open fraction of rear-side cells. V_{oc} value reduces and J_{01} value increases with increasing the laser open fraction. The loss in V_{oc} and J_{01} value is attributed to more surface damages and higher recombination at the contacts.

Table 1 shows the performances of Ni/Cu plated cells with different laser open fraction of rear-side surface. The open-circuit voltage (V_{oc}) value decreases with increasing laser open fraction, resulting from the more damages of rear-side surface. The improvements of fill factor (FF) and series resistance (R_s) are due to an increase fraction of rear-side electrode contact. The increase of contact area will reduce R_s and enhance FF values. Due to metal/Si contact fraction increases with laser open fraction. The loss in short-circuit current density (J_{sc}) could be attributed to the decraeses of reflectivity and internal quantum efficiency in the long wavelength region [5]. Finally, the difference of conversion efficiency (η) between 3.9 and 9.5% open fraction is about 0.74%.

4 CONCLUSIONS

We have presented electroless-plated Ni coating is combined with electroplated Ni/Cu metallization for n-type silicon solar cell. Electroless-plated Ni has been deposited as seed layer on the full area of rear side surface, which includes 6.8 wt.% of phosphorus. Selective laser ablation and Ni/Cu electroplating are key processes for both front- and rear-side metallization. The finger width and height of front-side Ni/Cu with UV laser ablation process are about 40.5 μm and 13.8 μm, respectively. According to Suns-V_{oc} results, the contact damages and recombination current are related to laser open fraction of rear-side cells. Therefore, the differences of Ni/Cu plated cell performances such as open-circuit voltage (V_{oc}), short-circuit current density (J_{sc}), fill factor (FF), and conversion efficiency (η) between 3.9 and 9.5% open fraction are 0.03 mV, 0.88 mA/cm^2, 0.02%, and 0.74%, respectively.

5 ADDITIONAL COMPONENTS

5.1 Illustrations

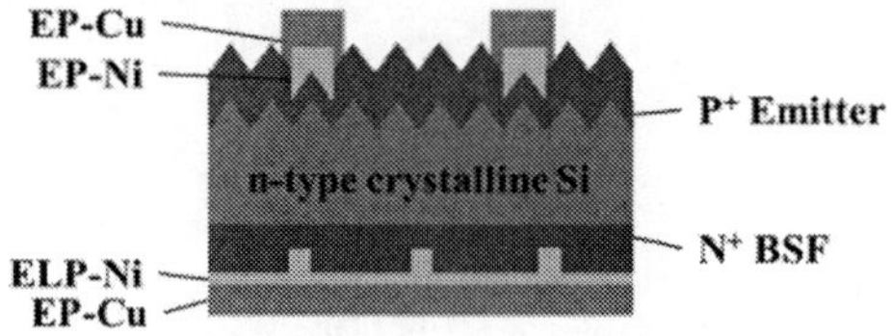

Figure 1: Schematic structure of Ni/Cu metallization cell.

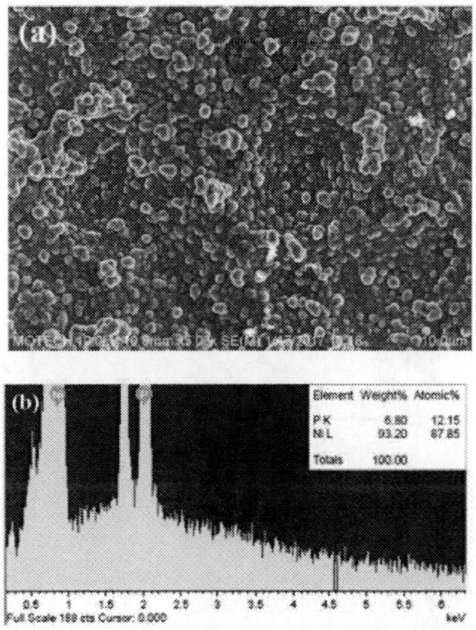

Figure 2: (a) SEM top-view image and (b) EDS spectrum of rear-side electroless plated Ni (ELP-Ni) layer.

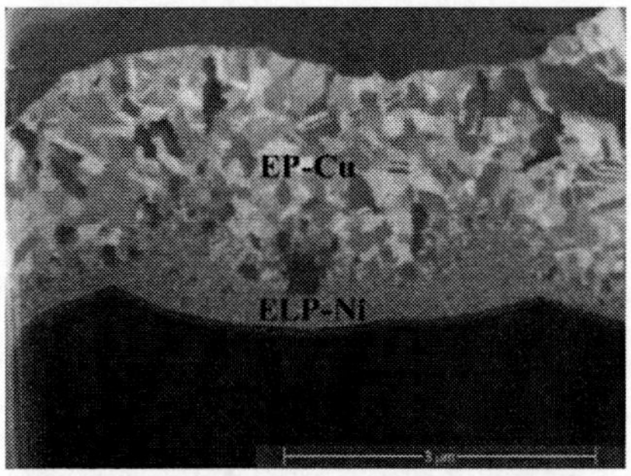

Figure 3: FIB cross-section image of rear-side Ni/Cu metallization.

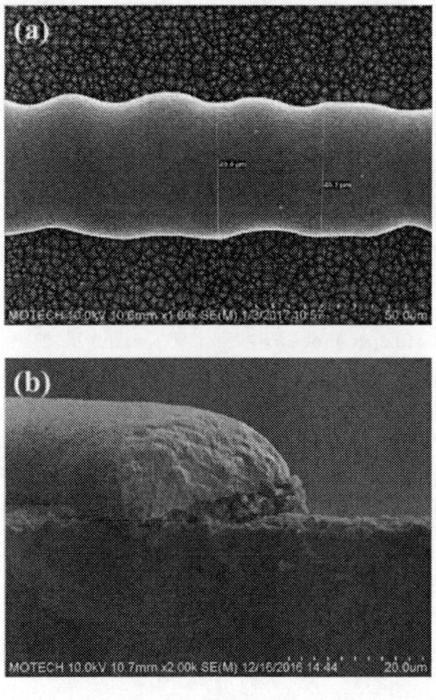

Figure 4: (a) SEM top-view and (b) cross-section images of front-side Ni/Cu plated finger.

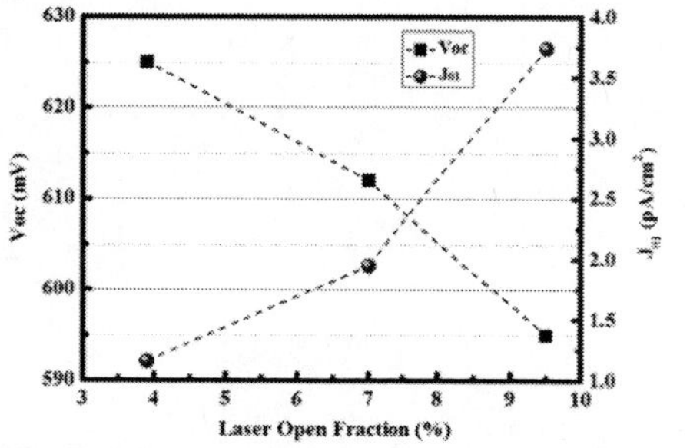

Figure 5: V_{oc} and J_{01} as a function of laser open fraction of rear-side cells.

5.2 Table

Table 1: Performances of Ni/Cu metallization cells with different laser open fraction of rear-side surface.

Rear open ratio of laser ablation	η (%)	FF (%)	Jsc (mA/cm²)	Voc (mV)	Rs (mΩ)
3.9%	Δη	ΔFF	ΔJsc	ΔVoc	ΔRs
7.0%	Δη-0.17	ΔFF+0.01	ΔJsc-0.03	ΔVoc-0.01	ΔRs-0.25
9.5%	Δη-0.74	ΔFF+0.02	ΔJsc-0.88	ΔVoc-0.03	ΔRs-1.85

5.3 References

[1] A. ur Rehman and S. H. Lee, "Crystalline Silicon Solar Cells with Nickel/Copper Contacts", published in Solar Cells - New Approaches and Reviews, (2015).

[2] A. K. Singh, V. K. Bajpai, and C. S. Solanki, "Effect of Light on Electroless Nickel Deposition for Solar Cell Applications", Energy Procedia 54, pp. 763 – 770, (2014).

[3] M. C. Raval and C. S. Solanki, "Review of Ni-Cu Based Front Side Metallization for c-Si Solar Cells", Journal of Solar Energy, Vol. 2013, pp. 1-20, (2013).

[4] P. Engelhart, S. Hermann, T. Neubert, H. Plagwitz, R. Grischke1, R. Meyer, U. Klug, A. Schoonderbeek, U. Stute, and R. Brendel, "Laser Ablation of SiO_2 for Locally Contacted Si Solar Cells With Ultra-short Pulses", Prog. Photovolt. Res. Appl. 15, pp. 521–527, (2007).

[5] S. Gatza, J. Müller, T. Dullweber, and R. Brendel, "Analysis and optimization of the bulk and rear recombination of screen-printed PERC solar cells", Energy Procedia 27, pp. 95 – 102, (2012).

FASHIONING "BLACK" SILICON BY NICKEL-FILM ASSISTED CHEMICAL ETCHING

Marius Treideris, Alfonsas Rėza, Mindaugas Kamarauskas, Vladimir Agafonov, Arūnas Šetkus
Center for Physical Sciences and Technology
Saulėtekio av. 3, LT-10257 Vilnius, Lithuania.
tel.: +370-5-2626737, fax: +370-5-2627123, e-mail: marius.treideris@ftmc.lt

ABSTRACT: In this work, we demonstrate a promising and cost competitive technology for "black" silicon formation on c-Si wafers. For this we proposed a method for wafer size surface texturing of silicon substrate based on a large area surface coating by a nanometer-thin nickel film and subsequent thin nickel film assisted chemical etching (TNFACE). We investigated a relationship between the technological conditions, the surface morphology and optical reflection. We demonstrate that a large area of crystalline Si wafer can be covered homogeneously by the "black" silicon with average reflectivity less than about 2% in the visible light range after etching for 5 min.
Keywords: texturisation, optical properties

1 INTRODUCTION

Silicon based solar cells are the most commercialised type of the cells that are available to various users on the market. However, unlocking complete potential of the silicon solar cells still depends on the progress in balancing between an increase of efficiency and decrease of manufacturing costs. One of the ways to increase the silicon solar cell efficiency is to decrease the total reflectivity of the surface. For this, the surface texturing with micrometer size structures is used in combination with antireflection layers [1]. New and most promising approach is known to be so called "black" silicon obtained by formation of nanostructures on the silicon surface and solar cells with efficiencies 22,1% were demonstrated [2]. However, until now, this type of surface modification is frequently obtained by using two main technological procedures: etching silicon surface with RIE that actually is an expensive and small area process [2] or using two step metal assisted etching with noble metal as catalyst [3]. However metal assisted etching still isn't cost friendly because of the noble metals used during the process and has difficulties on the deposition of the metal catalyst on the surface due to small metal salt concentrations in the metal catalyst deposition baths. Therefore, there is still great demand for a cheap and simple technology of the "black" silicon on the large area with simple instrumentation, fast process and easy control of the structure parameters by process parameters.

This study presents an original technology which is cost effective and promising way for the future "black" silicon formation over the large area with homogeneous coverage and easy control over formation conditions.

2 MATERIALS AND METHODS

Commercially available substrates of p-type CZ-grown crystalline Si with (100) orientation and the resistivity 1-5 Ω cm obtained from SIEGERT WAFER GmbH were carefully prepared by the standard procedures before etching. Simple 3 step thin nickel film assisted chemical etching (TNFACE) (Fig.1) was carried out by etching samples with magnetron sputtered thin nickel layers (20-80nm) in $HF:H_2O_2:H_2O$ etchant with different ratios and different etching time (1-10 min.). After etching nickel film residues are removed by immersion in HNO_3. The scanning electron

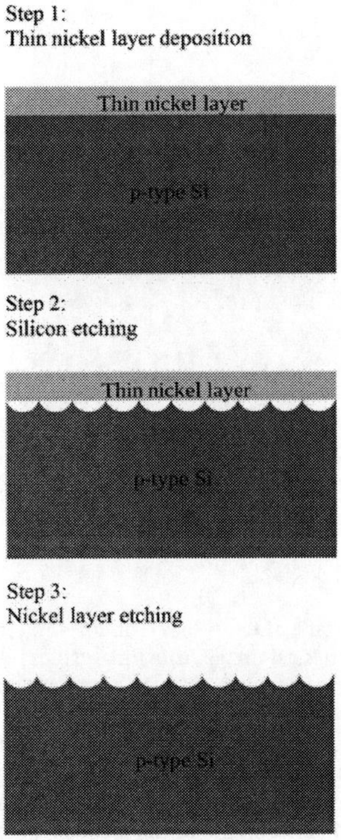

Figure 1: Schematic diagram of simple 3 step TNFACE

microscope (SEM) EVO-50 (Carl Zeiss SMT) with EDS spectrometer INCA (Oxford Instruments) was used for the Energy-Dispersive X-ray Spectroscopy (EDS). The surface topography was obtained by the scanning probe microscope SPM-Dimension 3100/Nanoscope IVa from Veeco Metrology. Analysis of the surface images was performed using software package for nano- and microscale image processing SPIP (Image Metrology).The optical reflectance was measured using SHIMADZU UV-3600 spectrometer with an integrating sphere.

3 RESULTS AND DISCUSSIONS

Samples after etching look homogeneous and black in ambient light at any tilt angle as it is illustrated by image in Fig.2a. From the AFM topography it is clear that the cone-like structure was formed by etching on the surface Fig.2b.

a)

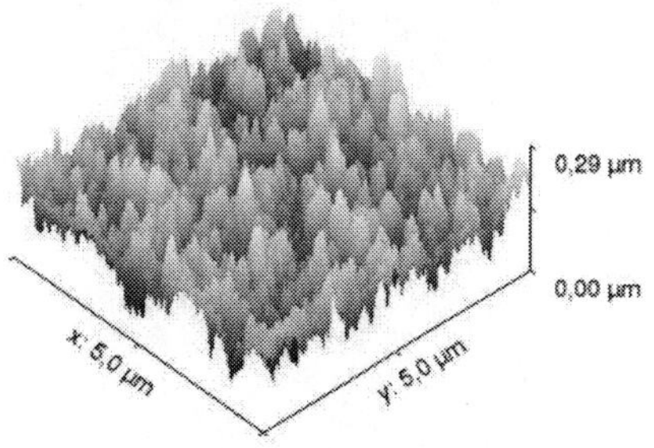

b)

Figure 2: Surface of the "black" silicon formed by TNFACE: a) optical image, b) AFM surface topography (80nm nickel film, H_2O:H_2O_2:HF=10:8:4, t=5min.)

Homogeneity of the surface was analyzed by topography images in several selected areas, namely in the center and in the corner of the piece of wafer shown in Fig.2a. The results proved that cone structures are distributed uniformly in all places on the wafer with slight statistical differences due to random

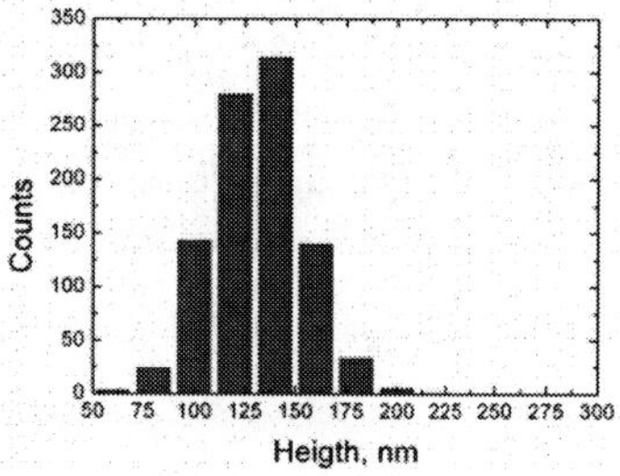

a)

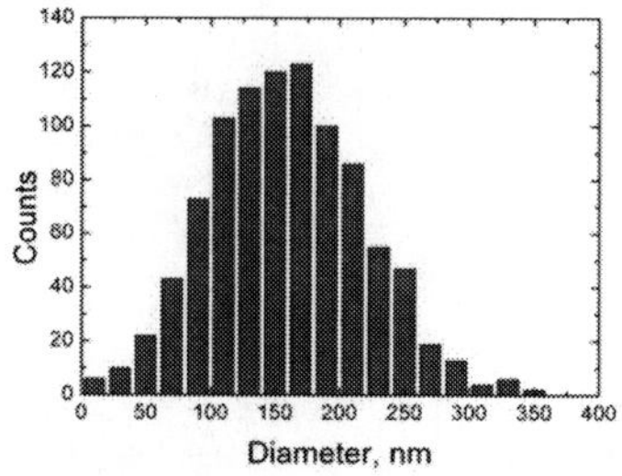

b)

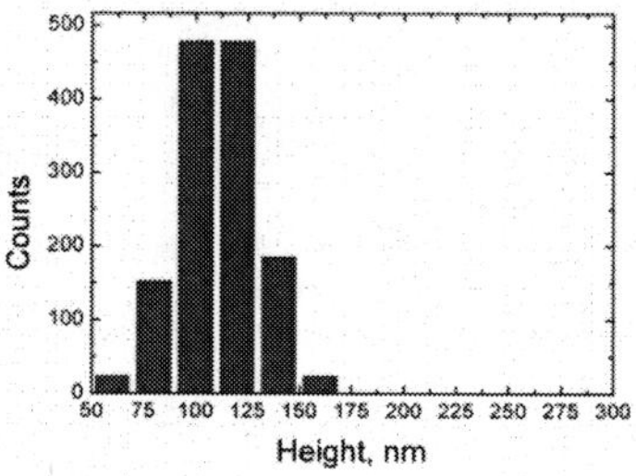

c)

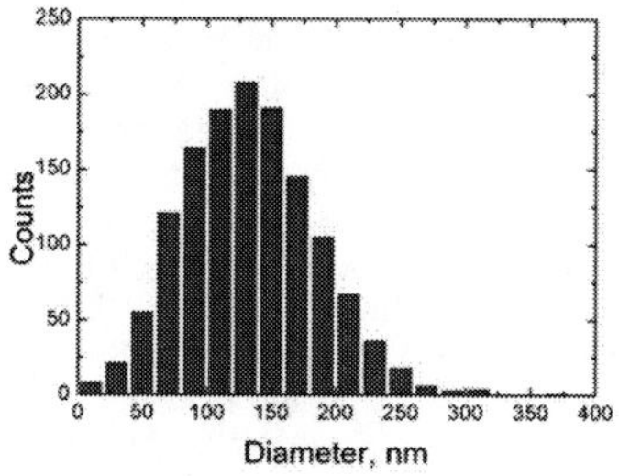

d)

Figure 3: Cones like structure: a)-height and b)-diameter distribution in the sample center and c)-height and d)-diameter distribution in the sample corner (80nm nickel film, H_2O:H_2O_2:HF=10:8:4, t=5min.)

nanostructure etching with average height 100-150 nm and diameter 120-180 nm for the cone structures etched for 5 min. Fig.3.

With the increase in the etching time the dimensions of a cone like structures increases in comparison to etching for 5 min. After 10 min. etching, the cone like structures were measured being of height equal to 150-220nm and diameter about 150-220nm as it follows from the results in Fig.4.

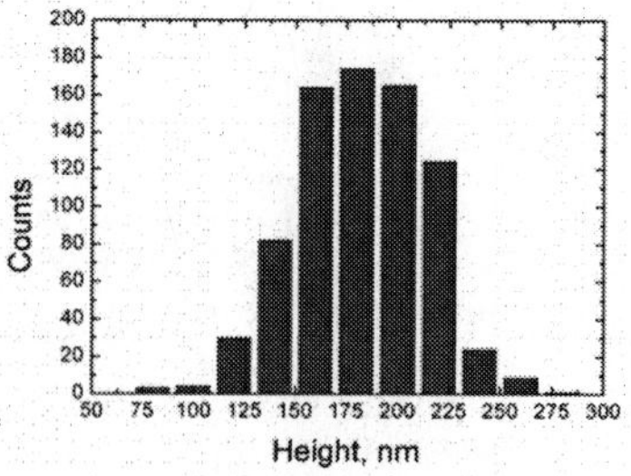

a)

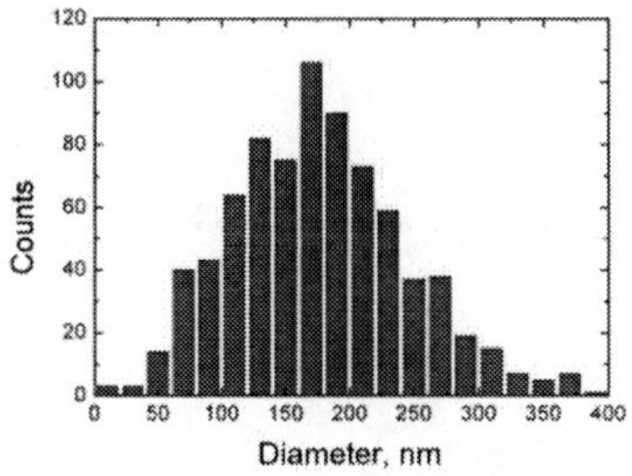

b)

Figure 4: Cones like structure: a)-height and b)-diameter distribution (80nm nickel film, $H_2O:H_2O_2:HF=10:8:4$, t=10min.)

Based on the experiment results we prove that the etching time and the related change in the cone-like structure dimensions has a significant influence on the optical reflectance. Typical results of the reflectance are illustrated in Fig. 5.

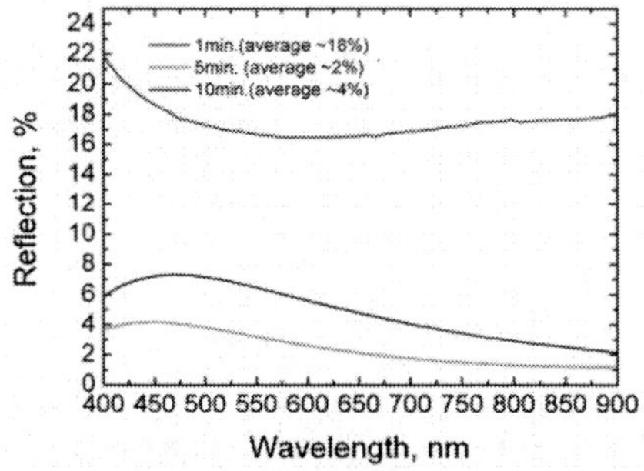

Figure 5: Optical reflectance spectra for the samples etched for different time (1,5 and 10 min.) (80nm nickel film, $H_2O:H_2O_2:HF=10:8:4$)

The reflectance average was about 18 % for the samples etched for 1 minute. For samples etched for 5min, the average reflectance was about ~2 % with the minimum reflectance of less than 1% at about 900nm and for samples etched for 10 min. average reflectance ~4% with the minimum reflectance of less than 2% at about 900nm.

EDS measurements after Ni residues removal in HNO_3 shows no trace of nickel on the formed "black" silicon surfaces.

4 CONCLUSIONS

In conclusion cone like structures on monocrystalline silicon was formed by using a large area surface coating by a nanometer-thin nickel film and subsequent thin nickel film assisted chemical etching (TNFACE). It was demonstrated that structures on the surface are homogeneous and the structure size and surface reflectance depends on the etching conditions. In this development state "black" silicon with average reflectance ~2% in the 400-900 nm range was demonstrated.

5 ACKNOWLEDGMENT

This research was funded by a grant (No. MIP-70/2015) from the Research Council of Lithuania.

References:

[1] Huang, Bohr-Ran, Ying-Kan Yang, and Wen-Luh Yang, Applied Surface Science 266 (2013) 245-249.
[2] Savin, H., Repo, P., Von Gastrow, G., Ortega, P., Calle, E., Garín, M., & Alcubilla, R, Nature nanotechnology, 10(7) (2015) 624-628.
[3] Toor, F., Miller, J. B., Davidson, L. M., Nichols, L., Duan, W., Jura, M. P., ... & Black, M. R., (Nanotechnology 27(41) (2016) 412003.

C-SI SURFACE PASSIVATION OPTIMIZATION OF ALD ALOX DEPOSITED LAYERS

R.Monna*[1], C.Denis[1], A.Veau[1], L.Bounaas[2] and S.Dubois[1]
[1] Univ. Grenoble Alpes, CEA, LITEN, DTS, LHMJ, F-38000 Grenoble, France
[2] ECM Greentech, 109 rue Hilaire de Chardonnet, F-38100 Grenoble, France
*Corresponding author: Email: remi.monna@cea.fr, Phone: +33 479 792 931

ABSTRACT: This study focuses on the optimization of the passivation properties of AlOx layers deposited either by ALD. Different thicknesses of AlOx layers have been deposited by ALD on p-type Cz silicon wafers and annealed at different temperatures. In the absence of "SiNx capping", the increase of the forming gas annealing temperature improves the passivation regardless to the ALD deposited AlOx layer thickness. Very high passivation performances were obtained before SiNx capping (i-V_{oc} around 720 mV) with AlOx ALD deposited layers annealed at 550°C. The deposition of a silicon nitride capping layer remarkably improves the surface passivation (i-V_{oc} gain of 30 to 40 mV) on low temperature annealed AlOx layers but has no influence on the passivation quality of high T annealed AlOx layers. After SiNx capping, i-V_{oc} around 720 mV are obtained regardless to the ALD deposited AlOx thickness and annealing temperature. The same passivation quality is obtained when an AlOx layer is directly covered with a SiNx layer (the AlOx was not annealed before).
Keywords: Passivation, Annealing, PECVD

1 INTRODUCTION

Aluminum Oxide (AlOx) is widely used for the backside passivation of PERC solar cells. A very high negative fixed charges density up to 10^{13} cm^{-2} located near the AlOx-silicon interface enables to provide an excellent field-effect passivation on p-type substrates [1]. AlOx also acts as an effective hydrogen reservoir, supplying hydrogen to passivate the dangling bonds on the silicon surface during a subsequent annealing step. Furthermore, the possibility to deposit very thin AlOx films enables flexibility in designing passivation schemes by stacking the material with other films.

Aside from the issue of costs and quality, aluminum oxide films for rear surface passivation must be sealed, in most case with a silicon nitride layer (SiNx). While PECVD (Plasma-Enhanced Chemical Vapor deposition) is a well-established technology for the deposition of silicon nitride antireflection coating, it can also be employed for depositing a capping layer on the rear side. In this work, we have studied the influence of the post deposition annealing step of the AlOx layer on the surface passivation of SiNx capped AlOx layers.

Several tests have been carried out in order to study the influence of the *charge activation* annealing temperature after the deposition of AlOx layers of variable thicknesses. After annealing, a SiNx capping layer was deposited by PECVD to assess its influence on the passivation quality. When AlOx/SiNx surface passivation films are implemented in screen printed PERC cells, the multilayer thermal stability during a high temperature firing process is critical. The impact of this firing step on the passivation properties has also been studied and evaluated.

2 EXPERIMENTAL APPROACH

Symmetrical test structures were processed for the determination of the surface passivation properties. Dielectric layers were deposited on both sides of 1.5 to 4 Ω.cm KOH polished p-type Cz silicon wafers. The wafers were first cleaned in a H_2SO_4 / H_2O_2 ultrasonic bath to remove the organic contaminants and in an HCl / H_2O_2 bath to remove metallic contaminants. The cleaning sequence ultimate step enables the formation of a thin silicon oxide (~ 1 nm). Precautions regarding storage time of the substrates after cleaning were necessary, as a passivation degradation was observed on some samples when they were not rapidly (less than a few days) capped by a dielectric layer. Indeed, increasing the storage time promotes the probability of surface contamination which can lead to a poor chemical passivation. In the same way, the storage time between the AlOx deposition and the activation annealing steps needed to be minimized in order to avoid natural oxidation and contamination. The passivation performances of the structures were however permanently stabilized after the capping of the AlOx layers with SiNx.

For the AlOx deposition, a gas mixture comprising Ozone (O_3) and trimethylaluminum (TMA) was used in an ALD equipment from BENEQ. The silicon nitride capping layer was deposited using a PECVD direct RF plasma reactor combining Ammonia (NH_3) and Silane (SiH_4) gases. The test samples were annealed in a quartz tube furnace with forming gas and a conventional IR belt furnace is used to simulate the thermal profile undergone by the solar cells during the metallization step. Carrier lifetime measurements were carried out on a Sinton WCT-120 from which the 1 SUN implied-Open-Circuit-Voltage (i-V_{oc}) were extracted.

3 RESULTS AND DISCUSSIONS

3.1 Effect of the AlOx layer thickness, of the activation temperature and of the "capping" with a SiNx layer

AlOx layers of different thicknesses (5 nm, 10 nm and 20 nm) were deposited by ALD and activated at different temperatures for 30 minutes. As shown in figure 1, the i-Voc measurements revealed no improvement of the passivation quality when increasing the AlOx layer thickness. In the absence of "SiNx capping", the increase of the annealing temperature improves the passivation regardless to the AlOx thickness and very high passivation performances (around 720 mV) were obtained with an AlOx layer activated annealed at 550°C. This phenomenon (monotonic increase in i-Voc with annealing Temperature) contrasts a little with the literature, where it was seen that the annealing temperature of 425°C is optimal even for small AlOx thicknesses [2]. On the one hand, increasing the annealing temperature usually improves the charge activation of the AlOx layer. On the other hand, for high temperature annealed substrates, the passivation quality should be lower due to loss of hydrogen. However, the structure of the AlOx and SiOx interfacial layers is also changing during the annealing step. Therefore the passivation quality on high Temperature annealed wafers could depend on the initial density and features of the interface defects that have to be saturated, which could be different for our samples in comparison with works reported in the literature.

The deposition of a silicon nitride capping layer remarkably improves the surface passivation (i-Voc gain of 30 to 40 mV) for low AlOx annealing temperatures. At this low annealing temperature, the field effect passivation is limited due to low negative fixed charges density generated at the Si/AlOx interface. The thermal budget duringthe deposition of the SiNx capping layer enhances the charge activation of the AlOx layer and thus improves the surface passivation. For the alumina layers annealed at 450°C, a gain of approximately 10 mV was observed after capping. This improvement can be compared with the results obtained by Veith et al. [3]. In this case, the hydrogen supply coming from the SiNx layer would essentially explain the increase in performance.

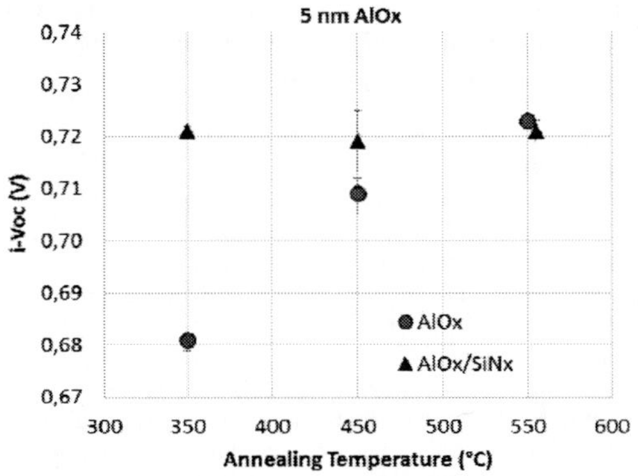

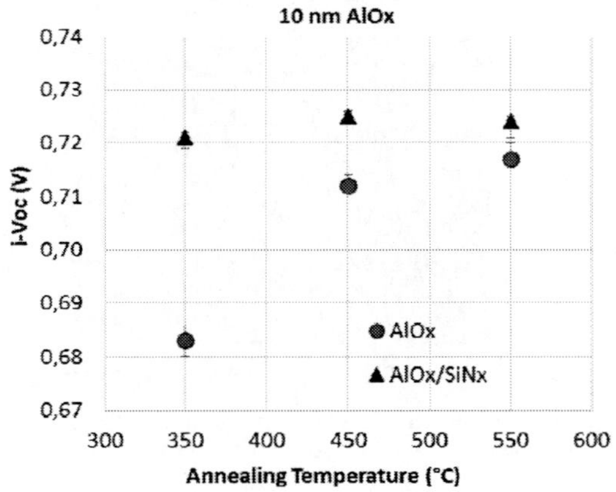

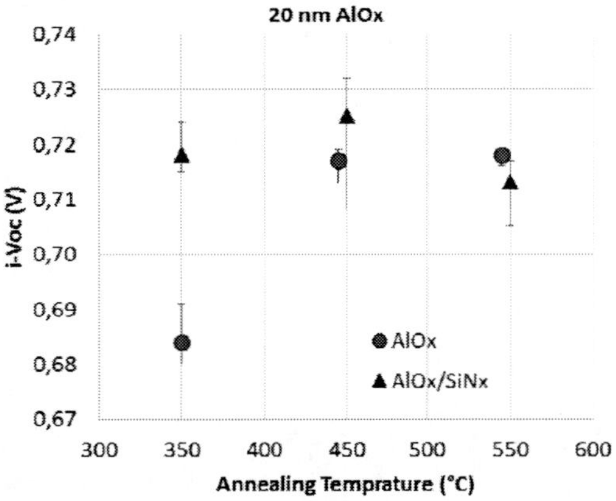

Figure 1: i-Voc for different AlOx thicknesses deposited by thermal ALD, as a function of different annealing temperatures. The measurements were carried out directly after the annealing of the AlOx layers, and after capping with 70 nm of SiNx.

Finally, we can observe that after SiNx capping, i-Voc around 720 mV is obtained regardless to the AlOx thickness and annealing temperature.

3.2 Effect of the firing step on AlOx layers deposited by ALD and capped with a SiNx layer

The measurements carried out on these stacks after the firing step are displayed in Figure 2. The firing does not significantly influence the i-Voc of the coated wafers. A relatively good thermal stability of the stacks is observed with an excellent surface passivation (i-Voc > 720 mV).

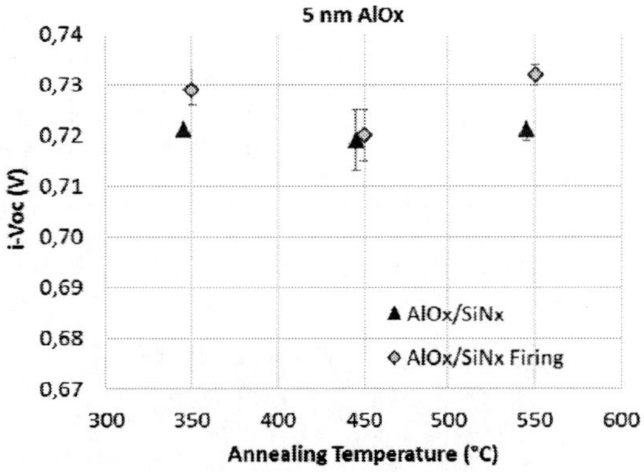

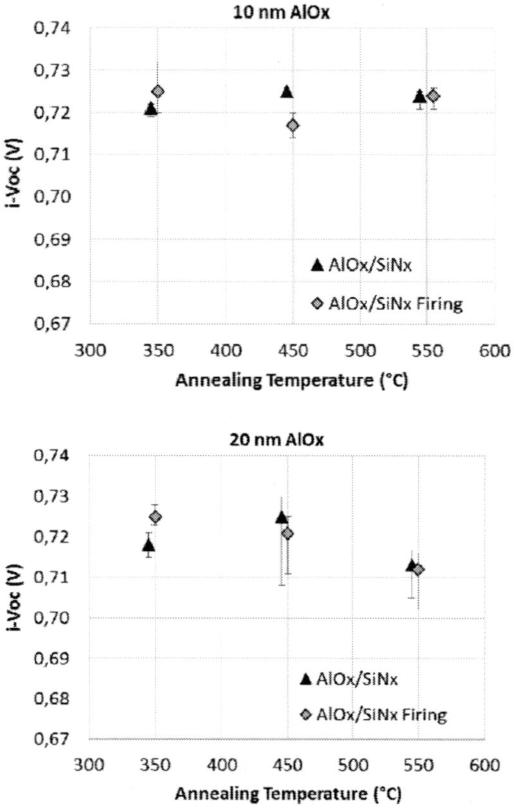

Figure 2: Influence of the firing step on the passivation quality of AlOx / SiNx layer as a function of the AlOx annealing temperature and thickness.

A slight decrease in passivation, which results in a loss of i-Voc of a few mV, is observed after deposition of the SiNx layer on a 20 nm AlOx thick layer followed by a firing step. On these wafers, the presence of "blisters" with diameters between 0.1 and 0.2 mm was observed (see figure 3). This phenomenon is generally attributed to the large amount of hydrogen present in the AlOx layer and may also be responsible for some of the passivation performance losses.

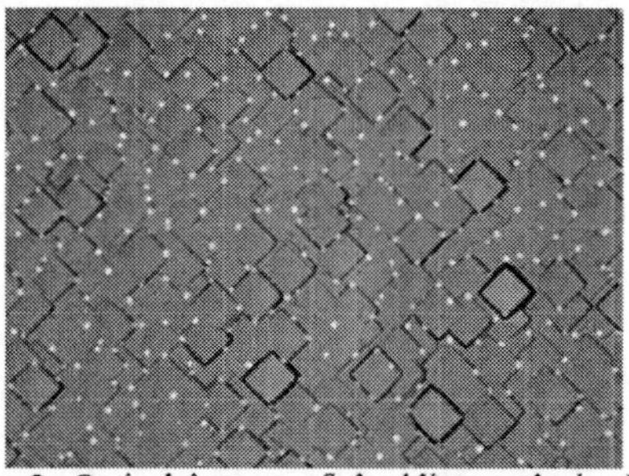

Figure 3: Optical images of the blisters obtained after firing of a thick AlOx (> 20nm) layer covered with SiNx

3.3 Passivation of AlOx/SiNx layers without AlOx activation

We have finally removed the activation step in order to check if the thermal budget of the "SiNx capping" is sufficient to activate the passivation of the AlOx layer, as suggested in the literature. High i-Voc values obtained after capping and after firing have revealed very good surface passivation properties of the stack, as can be seen in figure 4.

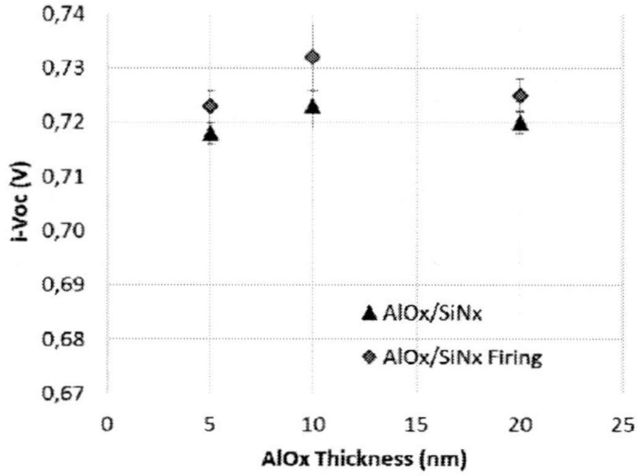

Figure 4: Influence of the firing step on the passivation quality of AlOx / SiNx stack as a function of the AlOx thickness (the AlOx layer has not been annealed before the SiNx deposition).

Indeed, after the deposition of the "capping layer", it can be observed that for all AlOx thicknesses, i-Voc values are in the range of 720 mV. This is close (within a few mV) to the best values obtained when the AlOx layer undergoes an activation annealing before "SiNx capping". In addition, it is noted that the firing step improves the passivation quality for the three studied AlOx thicknesses with a i-Voc gain of 5 to 9 mV. This observation is in accordance with the results obtained by Veith et al. [3].

Figure 5 (image of the photoluminescent signal) shows that a good homogeneity of the surface passivation is obtained with these passivation stacks.

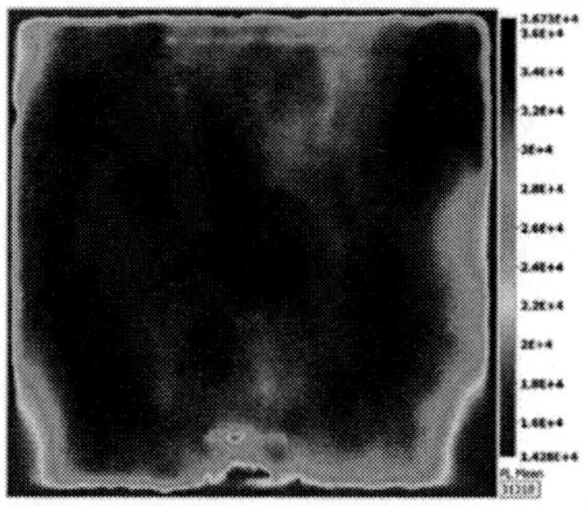

Figure 5: Photoluminescence image of the AlOx / SiNx layers deposited by ALD; 5 nm AlOx layer

4 CONCLUSION

The industry has been relying on the successful screen printing Al-BSF technology for the last few decades, but it is expected that PERC solar cells featuring dielectric rear side passivation will become the predominant technology in the next few years. The AlOx layer is an excellent candidate for the rear side passivation of p-type cells due to high surface passivation properties and the absence of an inversion layer.

In this study, AlOx layers have been deposited using an ALD lab-scale equipment from BENEQ. Different thicknesses of AlOx layers have been deposited and annealed at different temperatures. The i-Voc measurements revealed no improvement of the surface passivation quality when increasing the thickness of the ALD deposited AlOx layer. In the absence of "SiN capping", the increase of the annealing temperature improves the surface passivation regardless to the alumina thickness. Very high passivation performances (around 720 mV) were obtained with AlOx layers activated at 550°C. The deposition of a silicon nitride capping layer remarkably improves the surface passivation (gain of 30 to 40 mV) for low AlOx annealing temperatures but has no influence on the passivation quality of high T annealed AlOx layers. After capping with a SiNx layer and firing, i-Voc over 720 mV are obtained regardless to the AlOx thickness and annealing temperature. We have obtained the same passivation quality when an AlOx layer is directly covered with a SiNx layer (the AlOx was not annealed before the SiNx capping).

5 REFERENCES

[1] G.Dingemans et al., Status and prospects of Al_2O_3-based surface passivation schemes for silicon solar cells, J.Vac.Sci.Technol.A 30(4), Jul/Aug 2012.

[2] B. Kafle and J. Rentsch, Impact of thermal treatment on PECVD Al2O3 passivation layers, Energy Procedia 27, P273-279, 2012.

[3] B. Veith, F. Werner et al., Comparison of the Thermal Stability of Single Al_2O_3 layers and Al2O3/SiNx Stacks for the Surface Passivation of Silicon, Energy Procedia 8, P307-312, 2011.

OPTIMIZED PERC SILVER PASTE FOR HIGH EFFICIENCY EMITTERS

Giuseppe Scardera[1], Charlie Torardi[2], Paul D. VerNooy[2], QiJie Guo[2], Roman Petres[1], Shannon Dugan[1], and Brian J. Laughlin[2]

[1]DuPont Silicon Valley Technology Center, DuPont Photovoltaic and Advanced Materials, 965 E Arques Ave, Sunnyvale, CA 94085, USA
[2]DuPont Experimental Station, DuPont Photovoltaic and Advanced Materials, 200 Powder Mill Rd, Wilmington, DE 19880, USA

ABSTRACT: A new generation PERC Ag paste, DuPont™ Solamet® PV20x, designed for lower recombination emitters is presented. Improved contact performance is demonstrated on (i) lower surface concentration emitters, (ii) high sheet resistance shallow emitters, and (iii) post-diffusion low temperature annealed emitters. An average specific contact resistivity of 1.27 mOhm.cm^2 is demonstrated on a 1.3e20 cm^{-3} peak concentration emitter. Improved contact is achieved while maintaining high open circuit voltage. PERC cells made with a 1.3e20 cm^{-3} peak concentration emitter yield a 7 mV gain over those made with a 2.9e20 cm^{-3} peak concentration emitter. The combined emitter and Ag paste optimization yields a +0.2%$_{abs}$ efficiency gain. A +0.1%$_{abs}$ efficiency gain is also demonstrated on a 2.1e20 cm^{-3} peak concentration emitter treated with a post-diffusion low temperature anneal.

Keywords: phosphorus emitter, emitter saturation current, silver paste, specific contact resistivity.

1 INTRODUCTION

The Passivated Emitter and Rear Cell (PERC) has progressed from a lab R&D solar cell in the early 1980's to the next generation architecture of choice for manufacturers [1]. PERC is projected to be the dominant cell technology by 2024 [2]. The higher efficiency ceiling provided by the PERC architecture over conventional Al-BSF cells requires cell makers to revisit emitter optimization as well as metal contact performance.

Figure 1 shows the recombination losses at maximum power point determined from a Quokka [3] simulation of an industrial PERC with a 20.7% efficiency along with the corresponding losses for an Al-BSF cell with the same front side emitter and metal. The input parameters were chosen to best match the efficiency of current state of the art industrial cells [2].

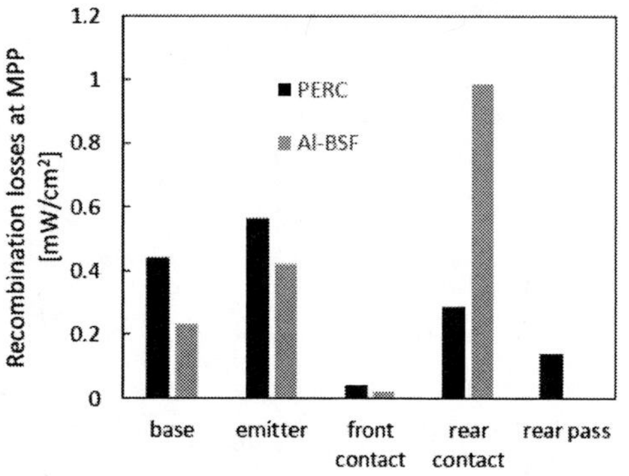

Figure 1: Recombination losses at maximum power point determined from a Quokka simulation of an industrial PERC solar cell with a 20.7% efficiency. The recombination losses for an Al-BSF cell with the same front side configuration are also shown for reference.

The simulation results indicate that emitter recombination (i.e. passivated, non-metal region) constitutes the largest portion of recombination losses in a PERC, while metal recombination at the emitter represents a marginal portion of losses. These findings are similar to other PERC simulations [4]. The simulation results also indicate that the efficiency entitlement from improving emitter recombination is greater for PERC than for Al-BSF cells. With the growing adoption of PERC, revisiting phosphorus emitter optimization is critical to capture significant efficiency gains.

Significant improvements to emitter recombination were made possible in the past for industrial Al-BSF cells with the introduction of DuPont™ Solamet® PV17x Ag paste. Conventional industrial emitter profiles at the time had peak concentrations in the 4e20 cm^{-3} to 9e20 cm^{-3} range. Advances in the Ag paste formulation led to dramatic improvements in contact capability for lower surface concentration emitters [5], enabling industrial emitters to move surface concentrations down closer to ~3e20 cm^{-3}, thus capturing efficiency entitlement from lower recombination emitters [6]. The combination of optimized emitter and Ag paste led to dramatic efficiency gains of +0.3% absolute for Al-BSF cells [6]. A similar jump in the contact capability of Ag paste, targeting even lower peak concentration emitters, is again required, this time for PERC.

The aim of emitter optimization is to minimize both resistive losses and recombination losses. Resistive losses include lateral losses between metal fingers associated with sheet resistance and losses at the metal-semiconductor contact, typically characterized by its specific contact resistivity, ρ_c. Emitter recombination losses occur in both the passivated and metal regions and are typically characterized by their emitter dark saturation current densities J_{oe} and J_{om}, respectively. In the passivated emitter regions, J_{oe} is affected by multiple recombination mechanisms, namely (i) Auger recombination, which is dependent on the amount of active dopant, (ii) emitter Shockley-Read-Hall (SRH) recombination associated with recombination sites in the bulk of the emitter, and (iii) surface SRH recombination which is characterized by the surface recombination velocity (SRV) and is dependent on the dopant surface concentration [7]. Emitter SRH recombination has been attributed to SiP precipitates which form in the presence of excess inactive phosphorus [8-10]. Lowering the surface concentration of industrial emitters has been shown to reduce the excess of inactive phosphorus and to significantly reduce emitter SRH recombination [11].

For a homogeneous emitter solar cell, the open circuit voltage (V_{oc}) directly depends on finding the best compromise between J_{oe} and J_{om} since minimizing these two parameters have opposing emitter profile requirements [12]. Lower J_{oe} can be achieved by reducing the emitter profile surface concentration and/or profile depth [13] while J_{om} is minimized by increasing surface concentration and/or emitter depth [14]. The emitter requirements for J_{om} are essentially the same as for ρ_c, which improves with more active dopant [15].

For the passivated regions, emitter optimization can also be viewed in terms of impact on V_{oc} and short circuit current density (J_{sc}). Increasing emitter sheet resistance reduces photo-generated current density lost due to recombination in the emitter (J_{lost}) [13] leading to higher J_{sc}. This can be easily achieved with a shallower emitter, however shallow emitters require significantly higher sheet resistance to achieve the same J_{oe} as a lower surface concentration and deeper emitter [6].

In this study, emitters with peak concentrations below 3e20 cm^{-3} are studied, targeting lower J_{oe} by minimizing emitter SRH recombination as well as SRV. In order to capture the efficiency entitlement for PERC associated with a lower recombination emitter, Ag paste must be able to make contact and maintain low metal-induced recombination. A new generation of silver paste is tested with three emitter optimization approaches available to cell makers namely, (i) deeper emitters with lower surface concentration achieved with increased thermal budget, (ii) shallow emitters with higher sheet resistance, and (iii) annealed emitters employing an additional post-diffusion low thermal budget process.

2 EXPERIMENTAL

PERC cells were fabricated using 156 mm (M2) p-type Cz-Si wafers with 2.2 Ohm-cm resistivity. Wafers were prepared with an alkaline random pyramid texture. $POCl_3$ diffusions were performed in an MRL Industries atmospheric pressure quartz furnace. The rear side diffused layer was removed with a wet chemistry etch. Rear side passivation consists of a PECVD AlOx/SiNx stack deposited with a Meyer Berger AK1000 and a Roth & Rau SiNA. Front side silicon nitride passivation was performed with a Roth & Rau SiNA. Rear side laser openings were created using a 3D MicroMac employing a line pattern with 45um opening and 1mm pitch. A 4BB design with 110 fingers at a 34um screen opening was used for the front side metal pattern. DuPont™ Solamet® PV19L and PV20x silver pastes were used for the front side metallization. A blanket Al print of DuPont™ Solamet® PV36S was applied on the rear side of the cells. Both front and rear side metals were deposited using a Baccini screen printer. Cells were co-fired in a six zone Despatch inline firing furnace. IV was measured using a Halm flash tester calibrated with a certified PERC calibration cell. Specific contact resistivity, ρ_c, was determined using the transmission line method (TLM) measured on three full length strips laser cut from different regions of the cells.

Emitter profiles were measured using dynamic secondary ion mass spectroscopy (SIMS) on polished wafers that accompanied textured wafers in the diffusion furnace. Carrier lifetime was measured on symmetric test structures (double sided n+ phosphorus emitter and silicon nitride passivation fired in a conventional inline belt furnace) with a Sinton WCT-120 tool using the quasi-steady state photoconductance (qssPC) method. The J_{oe} values were extracted at an injection level of $1x10^{16}$ cm^{-3} using the method described in ref [16].

Lower peak concentration emitters were created using a two temperature plateau approach with separate $POCl_3$ deposition and drive-in using both inert and oxidizing ambient. Emitter profiles with both decreasing peak concentration and increasing depth were chosen in order to maintain a relatively narrow sheet resistance range, thus allowing for direct cell performance comparison using the same front metal grid design. The shallow emitter was prepared with a single temperature plateau process employing reduced $POCl_3$ concentration. The annealed emitter was achieved by taking a two-temperature-plateau-processed emitter, followed by an HF-based clean to remove phosphosilicate glass (PSG) from the wafer surfaces, and finally loaded into another quartz furnace for a short thermal anneal at a temperature lower than that used for the $POCl_3$ deposition. Fourier Transform Infra-red (FTIR) spectroscopy was used to verify oxide formation during the anneal process. Double-side polished 10,000 Ohm.cm Si wafers were used as test substrates. A 60,000 scan measurement was performed in order to maximize the signal-to-noise ratio. A wafer from the same batch was used for background subtraction with its native oxide left on the wafer in order to verify formation of additional oxide.

3 RESULTS AND DISCUSSION

3.1 Lower peak concentration emitters

Figure 2 shows the SIMS profiles for three phosphorus emitters with simultaneously decreasing peak concentration and increasing depth. The emitters shown have peak concentrations of 2.9e20 cm^{-3}, 2.4e20 cm^{-3}, and 1.3e20 cm^{-3}, with sheet resistance values of 75 Ohm/sq, 68 Ohm/sq, and 66 Ohm/sq, respectively.

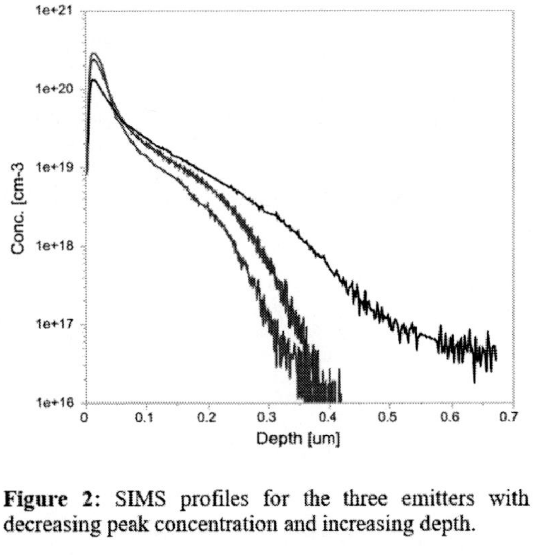

Figure 2: SIMS profiles for the three emitters with decreasing peak concentration and increasing depth.

Figure 3 shows the extracted J_{oe} values for the emitters shown in Figure 2, with profiles identified by their peak concentration. The green diamonds correspond to the 95% confidence intervals. As expected, J_{oe} is reduced with decreasing peak concentration. The J_{oe} values for the 2.9e20 cm^{-3}, 2.4e20 cm^{-3}, and 1.3e20 cm^{-3} conditions are 123.9 fA/cm^2, 107.8 fA/cm^2, and 68.1 fA/cm^2, respectively.

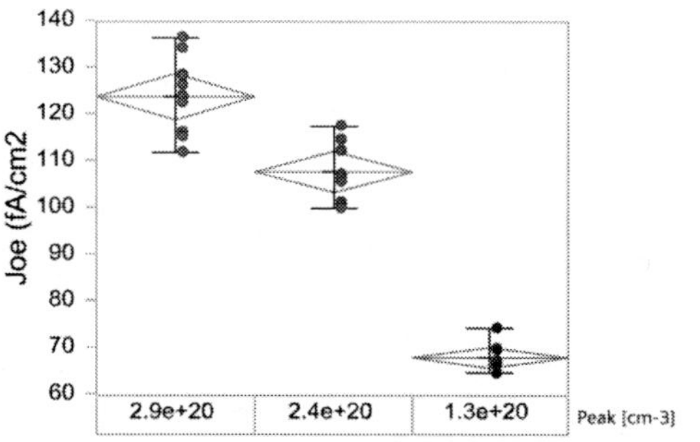

Figure 3: Emitter saturation currents for the profiles shown in Figure 2 identified by their peak phosphorus concentrations.

The corresponding IV results for cells made with both PV19L and PV20x Ag paste are shown in Figure 4 (15-17 cells per condition). The green diamonds correspond to the 95% confidence intervals. Clear V_{oc} gains are demonstrated for decreasing emitter peak concentration, consistent with the J_{oe} trend. A 7 mV gain is observed from 2.9e20 cm^{-3} to 1.3e20 cm^{-3}. PV20x Ag paste exhibits significantly improved FF performance over PV19L in all cases, with a very large offset for the most aggressive (1.3e20 cm^{-3}) emitter, where the previous generation paste (PV19L) starts to drop off dramatically. PV20x Ag shows only a slight drop in FF for the 1.3e20 cm^{-3} emitter. The observed J_{sc} drop for the 1.3e20 cm^{-3} emitter may likely be caused by increased J_{lost} associated with a deeper profile.

The superior contact performance of PV20x is clearly demonstrated with the specific contact resistivity results shown in Figure 5. PV20x yields a ρ_c of 1.27 mOhm-cm^2 for the 1.3e20 cm^{-3} emitter compared to 2.62 mOhm-cm^2 achieved with PV19L. Furthermore, PV20x is able to improve contact without sacrificing metal recombination as indicated by the V_{oc} remaining unchanged versus PV19L for all the emitters.

The average IV and ρ_c values are shown in Table I along with IV maximum values. Two decimal places are shown wherever the 95% confidence intervals in the data set indicate statistical significance, thus warranting such precision.

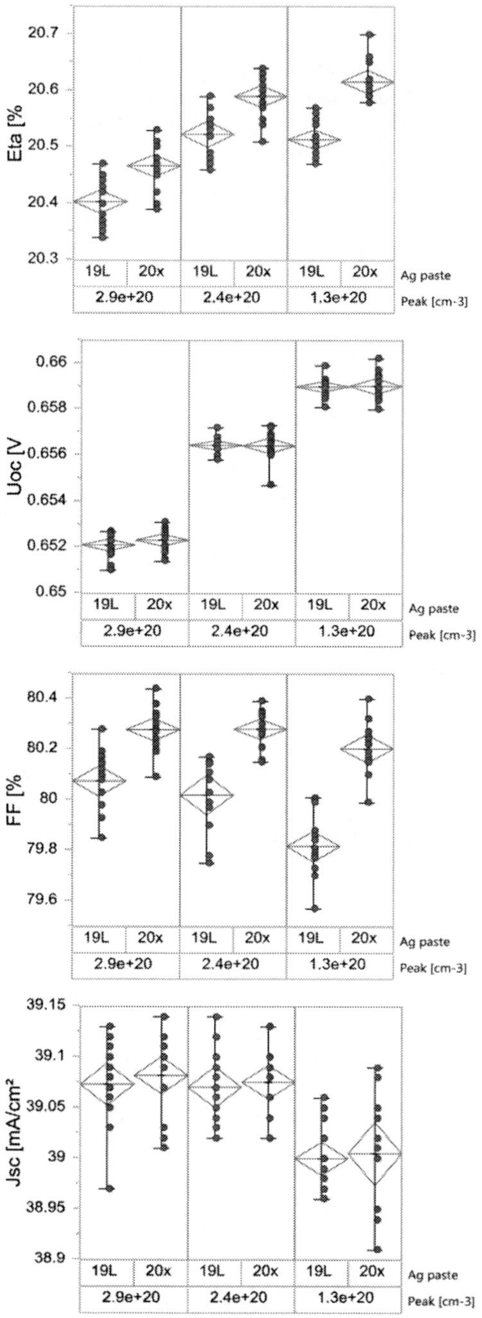

Figure 4: IV results for PERC cells made with the three emitters shown in Figure 2 (identified by their peak phosphorus concentrations) and metallized with both current and next generation Ag paste.

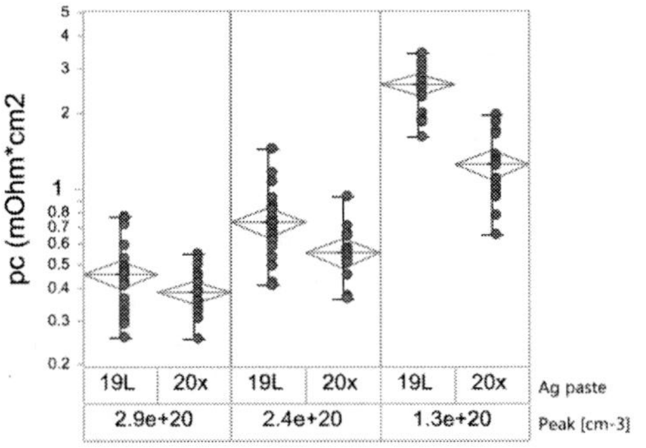

Figure 5: Specific contact resistivity values for the cells shown in Figure 4.

Table I: Average IV and ρ_c values for PERC cells shown in Figures 4 and 5 with peak emitter concentration, N_s, indicated. *Maximum IV values shown in italics.*

N_s (cm^{-3})	Ag	*Eff* *(%)*	V_{oc} *(mV)*	J_{sc} *(mA/cm^2)*	*FF* *(%)*	ρ_c *(mΩ.cm^2)*
2.9e20	**19L**	**20.40**	**652**	**39.07**	**80.1**	**0.46**
		20.47	*653*	*39.13*	*80.3*	
2.9e20	**20x**	**20.47**	**652**	**39.08**	**80.3**	**0.39**
		20.53	*653*	*39.14*	*80.4*	
2.4e20	**19L**	**20.52**	**656**	**39.07**	**80.0**	**0.75**
		20.59	*657*	*39.14*	*80.2*	
2.4e20	**20x**	**20.59**	**656**	**39.1**	**80.3**	**0.56**
		20.64	*657*	*39.13*	*80.4*	
1.3e20	**19L**	**20.51**	**659**	**39.00**	**79.8**	**2.62**
		20.57	*660*	*39.06*	*80.0*	
1.3e20	**20x**	**20.62**	**659**	**39.00**	**80.2**	**1.27**
		20.70	*660*	*39.09*	*80.4*	

3.2 Shallow emitters

Solar cell manufacturers may opt for a relatively shallow emitter profile, as it requires less thermal budget and allows for higher throughput as well as still reducing J_{oe} and enabling higher J_{sc} via reduced J_{lost}. Figure 6 shows the SIMS profile for a 100 Ohm/sq shallow emitter along with a deeper profile with the same peak concentration (and thus lower sheet resistance) for reference. The inset of Figure 6 shows the corresponding J_{oe} values for these profiles. As expected, for the same peak concentration, the shallower emitter yields a lower J_{oe} value, however these values are still higher than those for 70 Ohm/sq profile with 1.3e20 cm^{-3} peak shown in Figure 3.

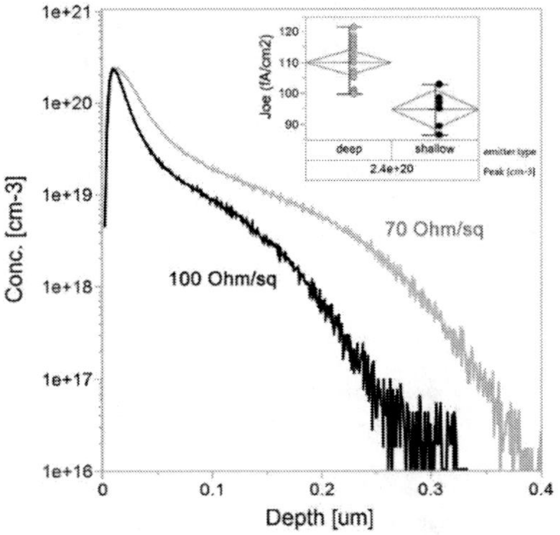

Figure 6: SIMS profiles for two emitters with the same peak concentration (2.4e20 cm^{-3}) and different depths, also identified by their sheet resistance. Inset shows the corresponding J_{oe} values.

It should be noted that doping uniformity (both across-the-wafer and across-the-furnace) can be a challenge when employing higher sheet resistance values. A four-point probe sheet resistance map of a shallow emitter process with a high standard deviation is shown in Figure 7. Such a wide variation in doping uniformity especially for the highest sheet resistance values poses a challenge for consistent contact quality. Both PV19L and PV20x were printed and fired on the emitter represented in Figure 7. The ρ_c results from TLM are shown in Figure 8. PV20x yields significantly better contact performance with an average value of 1.49 mOhm.cm^2 compared to 3.98 mOhm.cm^2 for PV19L, demonstrating better robustness to wide emitter sheet resistance variation.

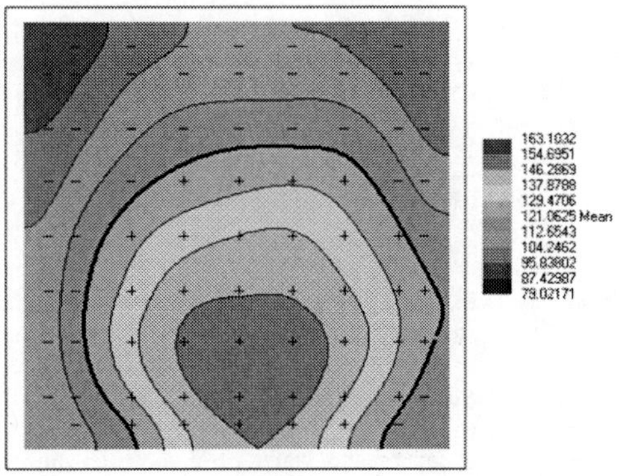

Figure 7: Sheet resistance map (156mm wafer) of a shallow emitter with poor uniformity (standard deviation of 14.4%).

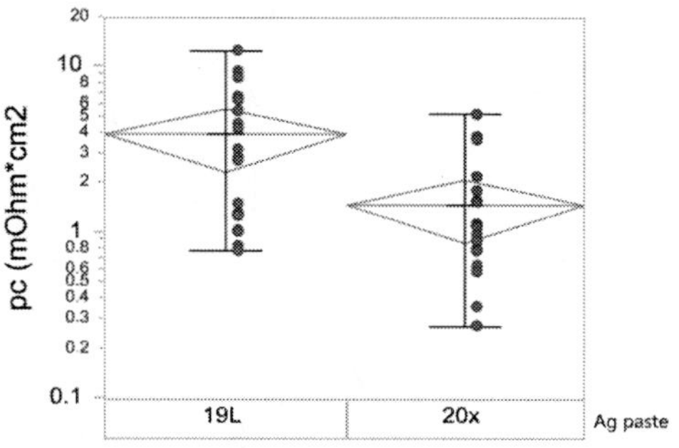

Figure 8: Specific contact resistivity, ρ_c, values measured for PV19L and PV20x Ag pastes on the emitter shown in Figure 7.

3.3 Annealed emitters

Another approach to optimizing emitter recombination performance is to employ a separate post-diffusion annealing process. That is, after the diffusion process is completed, the resulting phosphosilicate glass (PSG) is removed with an HF-based clean and the wafers are loaded into a quartz furnace for a second time and subjected to a low thermal budget anneal employing a temperature lower than that used for the $POCl_3$ process. The profile depth is not expected to change due to the limited thermal budget as observed elsewhere with low temperature processing [17]. Also, the post-anneal sheet resistance increased by less than 5 Ohm/sq. The FTIR absorbance spectrum shown in Figure 9, with a broad peak in the 1000-1100 cm^{-1} range associated with the Si-O stretching peak cm^{-1} [18], confirms the formation of a thin surface oxide.

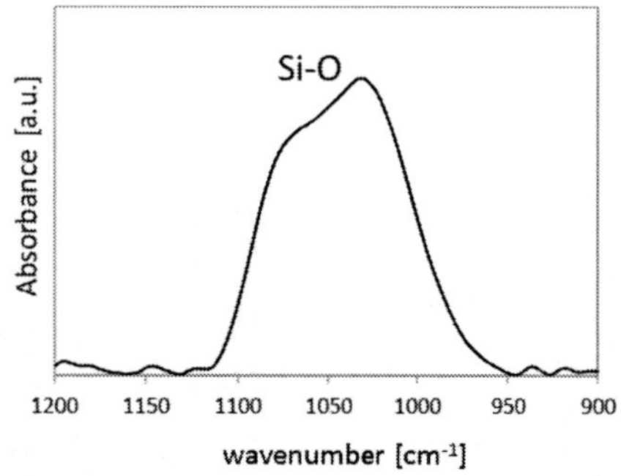

Figure 9: FTIR spectra of thin oxide layer formed during low temperature anneal process.

The oxide layer is expected to slightly etch back the surface concentration as well as further passivate the surface as it is left on the wafer prior to silicon nitride deposition. Silicon oxide is known to provide lower SRV than silicon nitride for decreasing emitter surface concentration [7].

Figure 10 shows the J_{oe} values for a 2.1e20 cm^{-3} peak emitter processed with and without a post-PSG-removal low temperature anneal. A significant reduction in J_{oe} from 75.1 fA/cm^2 to 64.4 fA/cm^2 is observed with the anneal process. In fact, the J_{oe} values for the annealed 2.1e20 cm^{-3} peak concentration emitter is lower than that of the non-annealed 1.3e20 cm^{-3} emitter previously shown in Figure 3.

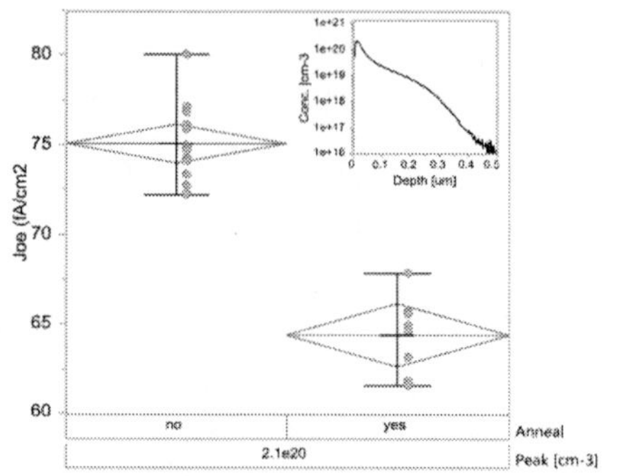

Figure 10: J_{oe} values for an emitter with a 2.1e20 cm^{-3} peak concentration (inset) with and without a post-PSG-removal low temperature annealing process.

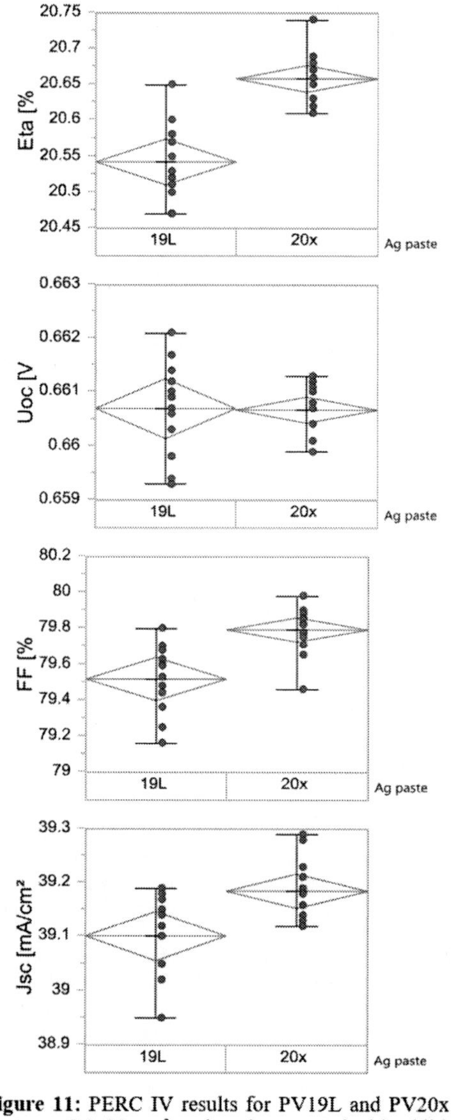

Figure 11: PERC IV results for PV19L and PV20x Ag pastes on 2.1e20 cm^{-3} peak emitter treated with additional low temperature anneal.

Figure 11 shows the IV results for PERCs made with both PV19L and PV20x Ag paste on 2.1e20 cm-3 peak concentration emitters treated with an additional low temperature anneal. A significant FF gain is observed for PV20x. This gain is associated with the improved contact performance demonstrated with the ρ_c results shown in Figure 12. Again, the improved contact is achieved without compromising metal recombination, as indicated by the V_{oc} being maintained. The average IV and ρ_c values are shown in Table II along with maximum IV values.

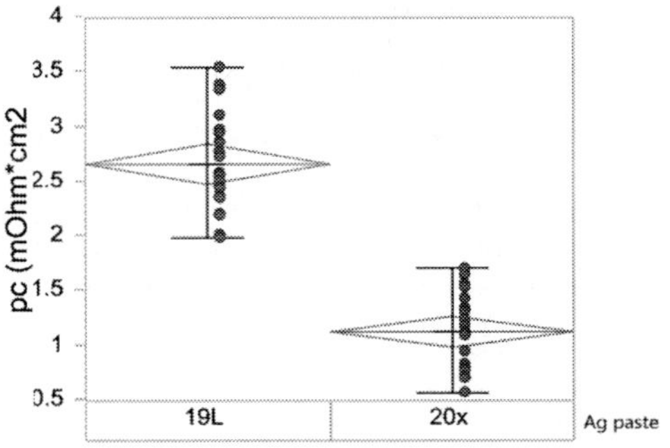

Figure 12: Specific contact resistivity values for the cells shown in Figure 11.

Table II: Average IV and ρ_c values for PERC cells shown in Figures 11 and 12. *Maximum IV values shown in italics.*

Ag	*Eff (%)*	V_{oc} *(mV)*	J_{sc} *(mA/cm²)*	*FF (%)*	ρ_c *(mΩ.cm²)*
19L	**20.54**	**660.7**	**39.10**	**79.5**	**2.67**
	20.65	*662.1*	*39.19*	*79.8*	
20x	**20.66**	**660.7**	**39.19**	**79.8**	**1.13**
	20.74	*661.3*	*39.29*	*79.9*	

4 CONCLUSIONS

PERC solar cells have greater efficiency entitlement, available through reducing emitter recombination, compared to Al-BSF cells. With the growing adoption of PERC in manufacturing, cell makers must revisit emitter optimization to capture significant efficiency gains. Such optimizations will lead to lower peak concentration emitters and will require Ag paste to make excellent contact and maintain low metal recombination.

A new generation screen printable Ag paste, DuPont™ Solamet® PV20x, designed for PERC solar cells with lower recombination emitters is presented. PV20x provides superior contact to lower surface concentration emitters with an average specific contact resistivity of 1.27 mOhm.cm² demonstrated on a 1.3e20 cm^{-3} peak concentration emitter. Improved contact is achieved while maintaining high V_{oc}. PERC cells made with a 1.3e20 cm^{-3} peak concentration emitter yield a 7 mV gain over those made with a 2.9e20 cm^{-3} peak concentration emitter. The combined emitter and Ag paste optimization yields a +0.2%$_{abs}$ efficiency gain. PV20x is also shown to provide more robust contact performance on high sheet resistance, shallow emitters with poor doping uniformity. Improved contact and cell performance is also demonstrated on a post-diffusion low temperature annealed emitter.

5 REFERENCES

[1] M.A. Green, Solar Energy Materials & Solar Cells 143 (2015) 190.
[2] International Technology Roadmap for Photovoltaic (ITRPV), Eighth Edition, March 2017 (www.itrpv.net)
[3] A. Fell, IEEE Transactions on Electron Devices, Vol 60 (2), (2012) 733.
[4] B. Min, H. Wagner, M. Muller, H. Neuhaus, R. Brendel, and P.P. Altermatt, Proceedings of 31st European Photovoltaic Solar Energy Conference (2015) 473.
[5] K.R. Mikeska, A.F. Carroll, L.G. Hanna, M.E. Lewittes, Z. Lee, L. Liang, A. Meisel, G. Scardera, S. Subramoney, P.D. VerNooy, Proceedings of the European Photovoltaic Solar Energy Conference (2012) 1717.
[6] G. Scardera, A. Meisel, K.R. Mikeska, A.F. Carroll, M.Z. Burrows, D. Poplavskyy, M. Abbott, F. Lemmi, H. Antoniadis, Proceedings of European Photovoltaic Solar Energy Conference (2012) 923.
[7] M.J. Kerr, PhD Dissertation, Australian National University (2002)
[8] D. Nobili, A. Armigliato, M. Finnetti, and S. Solmi, Journal of Applied Physics 53(3) (1982) 1484.
[9] P. Ostoja, S. Guerri, P. Negrini, and S. Solmi, Solar Cells 11(1) (1984) 1.
[10] B. Min, H. Wagner, A. Dastgheib-Shirazi, A. Kimmerle, H. Kurtz, and P.P. Altermatt, Physica Status Solidi Rapid Research Letters 8(8) (2014) 680.
[11] M. Abbott, G. Scardera, K.R. McIntosh, A. Meisel, Proceedings of 38th IEEE Photovoltaic Specialist Conference (2013) 1383.
[12] A. Cuevas, P.A. Basore, G. Giroult-Matlakowski, C. Dubois, Journal of Applied Physics 80 (1996) 3370.
[13] R.R. King, A. Sinton, R.M Swanson, IEEE Transactions on Electronic Devices 37 (1990) 365.
[14] U. Jager, S. Mack, A. Kimmerle, A. Wolfe, and R. Preu, Proceedings of 35th IEEE Photovoltaic Specialist Conference (2010) 3185.
[15] S.M. Sze, Physics of Semiconductor Devices 2nd Edition (1981) 304.
[16] A. Blum, J. Swirhun, R. Sinton, and A. Kimmerle, Proceedings of the European Photovoltaic Solar Energy Conference (2013) 1521.
[17] H. Li, M. Abbott, B. Hallam, and S. Wenham, Energy Procedia 92 (2016) 466.
[18] D.V. Tsu, G. Lucovsky, M.J. Mantini, Physical Review B. 33 (1986) 7069.

Point Contact Formation Using Silicon Nanoparticle Dispersed SiO_2

Hiroshi Nagayoshi[1], Hirotomo Demura[1], and Alexander Ulyashin[2]

1 Tokyo National College of Technology
1120-2, Kunugida-machi, Hachoji, Tokyo, Japan
2 SINTEF Materials and Chemistry, Metallurgy Department, Forskningsveien 1,
2Blindern, NO-0314 Oslo, Norway

A nanometer size point contact property using P^+Si nanoparticle(P^+SiNP) embedded SiO_2 has been investigated. The SiO_2 was formed from the Perhydropolysilazane (PHPS), which is an inorganic liquid source of SiO_2, showing good passivation property on Si with the combination of the hot steam annealing. The P^+SiNP embedded SiO_2 layer was coated on an n-type silicon substrate by spin coat, followed by the hot steam annealing at 600°C. The Au/(P^+SiNPs/SiO_2)/N-silicon hetero junction diode property was evaluated. The diode showed clear rectifying property with very low reverse current. A forward bias current increased by illumination, showing no memory effect between photo/dark conditions. This property can be explained by a combination of nano-size point contact and the field effect of MOS structure surrounding the contact. Electrons accumulate at the silicon surface under forward bias, forming an n^+p^+ tunnel point contact. The whole surface area becomes depression state under the reverse bias condition, showing very low reverse current.

Keywords: Silicon nanoparticle, Passivation, PHPS, SiO_2, c-Si, Selective electrode

1. INTRODUCTION

Nanocomposites, which include the phase of nanometer range size materials have been attracting attentions in many fields. There have been investigated wide variety of combinations since nanocomposite materials have a potential to be suitable alternatives to overcome limitations of conventional materials. Semiconductor nano particles, which show unique properties including quantum effect, have been intensively investigated to create novel electronic, optical, and opt electric applications. In semiconducter device applications, interface peroperties between composite materials and semiconductor surface are crusible. Liquid source dielectrics such as the films made by sol-gel method show poor interface property compare to the filmes of dry processing technology. On the otherhand, preparing nanocomposite materials such as nanosilicon embeded SiO_2 requires high temperature and long processing time. From the point of practical application in solar cell devices, liquid souce dielectic material with keeping good interface propety is attractive. We have been investigating the passivation effect of liquid source SiO_2 prepared from the PHPS, showing good passivation property for silicon surface when it combined with the hot steam annealing. That means the SiO_2 made from PHPS has an ideal property to create new composite materials, having unique function with keeping good interface property. The SiO_2 film made from the PHPS can be prepared less than 450°C, showing good passivation property. Compare to the ordinally organic binder materials, it is quite stable against ultra violet light and heat, which is suitable for solar cell device applications. In addition, the SiO_2 coating using PHPS have been used in automotive industries and architecture fields. The combination of silicon nanoparticles with liquid source SiO_2 is an interesting option as a practical point of view. Nanoparticle embedded SiO_2 using the PHPS has a suitable property to create new approaches in photovoltaic application such as a new type selective electrode. The purpose of this work is to create a new type thin contact structure for Si solar cells with low process temperature as well as low cost. This paper describes a junction property of P^+ silicon nanoparticles(P^+SiNPs) embedded SiO_2/N-Si. [1-3]

2. EXPRIMENTAL

A non catalyst type PHPS was used as a binder material to avoid a possibility of deterioration of the interface property by the catalyst. The PHPS is an organosilicate polymer consisting of cyclic $(SiH_2\text{-}NH)_n$ units, which changes into pure SiO_2 by heat treatment lower than 450°C. The PHPS was coated by spin coater. The film thickness was controlled by rotation speed, time and concentration of the PHPS. A spray coat is also applicable. To obtain thin layer, density of the PHPS was controlled by adding xylene solvent. After the spin coat, the samples were dried at room temperature. A SiO_2 formation process using the PHPS is shown as follows.

$$(\text{-}SiH_2NH\text{-}) + 2\,H_2O \rightarrow (\text{-}SiO_2\text{-}) + NH_3 + 2H_2 \qquad (1)$$

This reaction does not contain organic component like sol-gel source materials, so residual carbon in the film is avoidable. The PHPS release H_2 and NH_3 during the process, which contribute to improve the SiO_2/Si interface peoperty. The pure PHPS needs around 450°C of reaction temperature to complete the reaction. By using catalyst such as amine or vanadium, SiO_2 films can be obtained even at room tempreature. In addition, thick pure SiO_2 layer of more than 10 μm could be easily obtained by low temperature process, that means the SiO_2 made from the PHPS has a variety of potential applications like MEMS devices, composite materials with the combinations of nano particles, etc. [1-3]

A SiO_2 passivation layer thickness was controlled by the concentration of PHPS in xylene solvent. After dried at room temperature, the hot steam annealing was carried out at 600 °C, 1 hour. This is an optimized condition to obtain good passivation effect.[1-2] P^+SiNPs are added

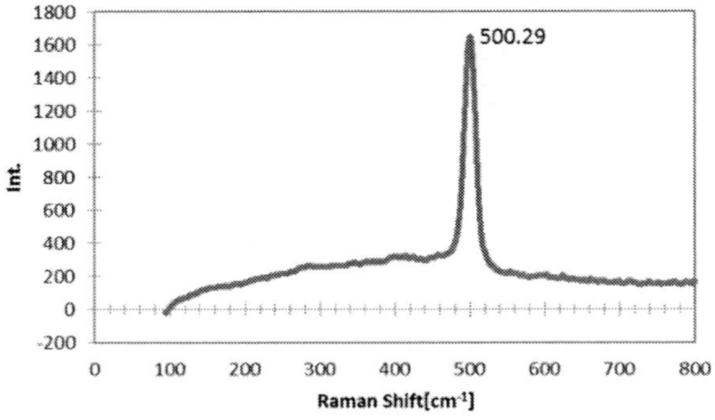

Fig.1 Raman spectra of nano silicon particle.

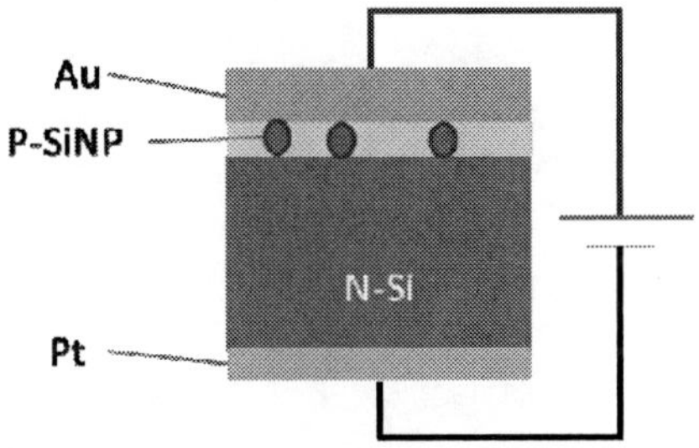

Fig.2 Schematic of the Au/(P^+NPs/SiO_2)/N-silicon hetero junction diode.

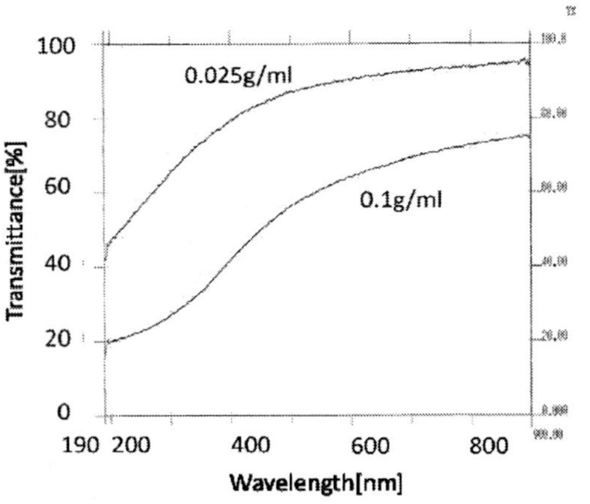

Fig.3 Example of the transmittance spectra of SiP^+NPs/SiO_2 composite film.

into the PHPS. The average size of P^+ silicon nanoparticle is 90nm. Raman spectra suggested that most are crystalline. The peak position is lower than that of bulk silicon caused by a phonon confinment effect or residual stress. Figure 2 shows the schematic of the sample. Semi transparent Au electrode, 3mm in diameter was prepared by evaporation. A CZ N-type <100> 1 Ω·cm wafer was used as a substrate. Figure 2 shows the schematic of the Au/(P^+NPs/SiO_2)/N-silicon hetero junction diode. The film thickness is 90nm, which is the same size of the particles. A semi transparent thin gold electrode was preapred as a front electrode to test the photo response. The back surface of the sample has platinum electrode, making ohmic contact. Figure 3 shows the transmittance spectra of the film. The transmittance of the 0.025g/mL decreased at the short wave length region caused by the photo absorption of silicon nano particles. The spectra shifted to lower level when the film has higher silicon nanoparticle concentration in the source liquid. Figure 4 shows the IV characteristics of the Au/(P^+NPs/SiO_2)/N-silicon hetero junction diode. The sample showed clear rectifying property. However, series resistance is high. The particle size has distribution, so not all particles in the film contribute to the forward current. The contact between nanoparticles and the silicon surface has a mere physical contact, not a junction. The particle concentration was not optimized yet. In spite of high defect density at the interface of N-Si surface/p^+SiNPs, the sample showed very low reverse current. The contact area between the substrate and the SiNP is nm size, so the contact property must be affected by the MOS field effect surrounding the contact point. The forward current increased under illumination (AM1.5, 100mW/cm^2). In the previous paper, p+nano silicon powder composite showed memory effect of photo current, however in the result using the present nano particle indicated the change of photo/dark current was reversible.4 The size distribution of nanopowder is quite smaller than the previous powder, including less floating nanopowders in the SiO_2 layer. The threshold voltage is lower than the ordinally silicon pn junction diode, suggesting a different mechanism exists. The most part of electrode area is forming MOS structure, having an accumulation mode under the forward bias. The total contact area between P^+SiNP and the substrate is quite small and contact property must be easily affected by the field near the contact. In the forward bias, silicon surface is essentially N^+, making P^+N^+ tunnel junction with P^+ nanopowder. It needs pulse response test to clear wheather the tunnel current or ordinally minority carrier injection. Figure 5 shows the image of band diagram under forward bias. The P^+ nanoparticles drain electrons from the accumulated surface. The SiO_2 layer made from the PHPS tend to have positive charge in the film after the hot steam annealing.

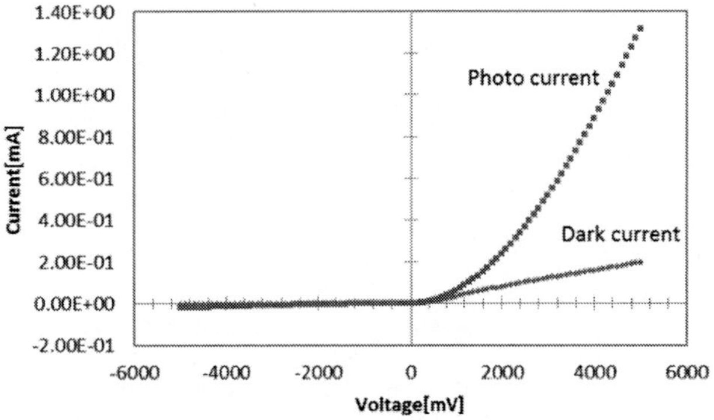

Fig.4 I-V characteristics of the Au/(P^+NPs/SiO_2)/N-silicon hetero junction diode.

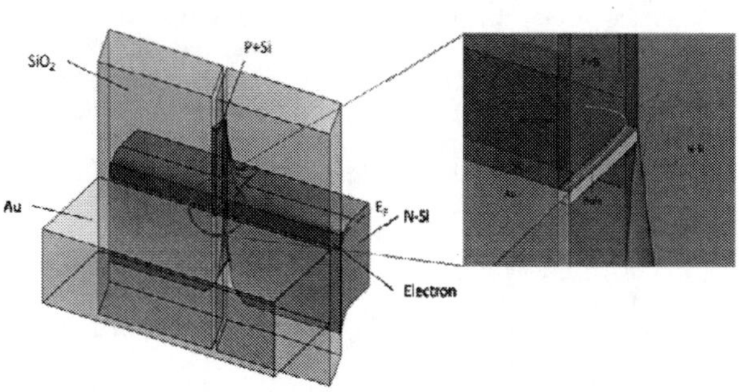

Fig. 5 Band diagram under forward bias.

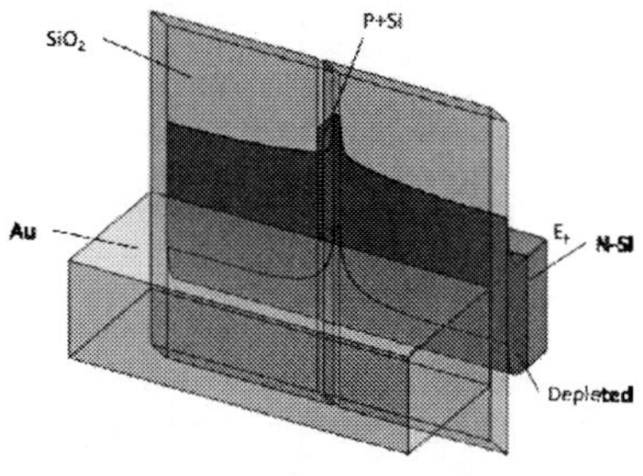

Fig. 6 Band diagram under forward bias.

Those charge might contribute to lower the series resistance. In the reverse bias, the surface of silicon including point contact area become depression mode, showing very low reverse current even when the point contact has defects. Figure 6 shows the image of band diagram under the reverse bias condition. When the silicon surface has accumulation mode caused by a positive charge in the film, this composite film works as a selective electrode of electron.

SUMMARY

We have investigated the Au/(P^+SiNPs/SiO_2)/N-silicon hetero junction diode property. The sample showed clear rectifying property, showing very low reverse current and low threshold voltage under forward bias. This property can be explained by the combination of p^+n^+ tunneling contact and the MOS field effect surrounding the contact point between p+ silicon nanoparticle and the substrate. The series resistance is still high and further optimization is needed.

REFERENCES

[1] H. Nagayoshi and C. Hagiwara, Proceedings on 30th European Photovoltaic Solar Energy Conference and Exhibition, (2014)1000-1002.

[2] Hiroshi Nagayoshi, and Chihiro Hagiwara , Energy Procedia , Volume 77, August 2015, Pages 822-826.

[3] Hiroshi Nagayoshi, and Takuya Murooka, Energy Procedia, Volume 77, August 2015, Pages 242-247

[4] H. Nagayoshi, H. Demura, and A. Ulyashin, Proceedings on 32nd European Photovoltaic Solar Energy Conference and Exhibition, (2016)242-244.

Effect of Laser Ablation Process on High Efficiency Silicon Solar Cells

Ming Shiou Lin*,Yueh Lin Lee, Kuang Chieh Lai ,
Chia Chih Chung, and Chi-Chun Li
Motech Industries, Xinshi Dist., Tainan City, 74145, Taiwan
(*Correspondingauthor:phone/fax/e-mail:+886-6505-0789-0617/+886-6505-8446/Joshua_lin@motech.com.tw)

Abstract—Laser techniques have been applied in mass production of silicon based crystalline solar cell, such as PERC and n-PERT. The detail laser work plays an important role in the fabrication of high efficiency solar cells. On the other hand, the conventional Ag screen-printed contacts usually have high contact resistance and lead to junction shunting issue during the post firing process. The Ag screen printing can be replaced with Ni/Cu plating to reduce the production cost of solar cells. Before Ni/Cu plating, the patterned removal of SiNx is operated by 532nm green laser and 355nm UV laser, respectively.

However, there are still drawbacks discovered during laser process. It is observed that if the output energy and focus are not suitable, there is silicon nitride(SiNx) left on the surface or damage of wafer discovered by scanning electron microscope (SEM). By the way, the characterization of SiNx residue is identified by Energy Dispersive Spectrum(EDS). Based on the measurements, it is also believed that the open-circuit voltage(Voc) and high cell efficiency(η) both depend on the Laser ablation strongly. The filling factor(FF) and cell efficiency both indeed correspond to variation of the opening area and metallization. According to our investigation, the optimization of laser ablation process is available and the removal of SiNx residue is under control effectively.

Key words: Laser ablation, Solar cell, Plating

I.Introduction

Today, photovoltaic power is one of the clean energy techniques for the global environmental concerns. n-PERT solar cells are one of the advanced solar cells and have a high potential to achieve excellent performance with simple process. It is essential that the most industrial n-PERT solar cells are metallized by screen printing and PVD[1]. However, the plating process is fabricated below 100℃, which avoids the thermal effect of initial high level quality of the crystalline silicon substrate[2]. The electrode patterns are obtained by laser ablation before plating process. Besides, the open-circuit voltage(V_{oc}) and cell efficiency(η) may be resulted from wafer damage due to laser working[3]. In this study, the laser ablation influence and cell performance are discussed.

II. Experiments

The 156 × 156 mm^2 n-type mono-crystalline silicon wafers with a random pyramid texture, boron-doped emitter, and phosphorus-doped surface field were used as substrates, consisting of the following: A saw-damage etching, alkaline texturing, borosilicate (BSG) removal, and drive-in dopants. The plasma enhanced chemical vapour deposition (PECVD) SiNx based dielectric was used as passivation and anti-reflection coating (ARC). Prior to metallization, H-pattern of four busbars was defined in the dielectric layer using the 532 nm and 355 nm wavelength of nanosecond pulse laser. Different laser fluences were used to ablate the contact opening regions. Conventional screen printing of silver paste on rear side as well as firing was performed. Ni/Cu plating metallization was employed as front side contact for n-type Si solar cells. The plated Ni contact layer and plated Cu conductive layer were applied to n+np+ front emitter solar cell structure. Each Ni/Cu metallization was used by forward bias plating (FBP).

The performance of cell efficiency was measured by Burger(Lichttechnik). SunsV_{oc} measurements were available by Sinton. The micro-structures were observed by SEM(Hitachi,S-4800) and the elements at the interface were identified by energy dispersive X-ray spectroscopy (EDS, HORIBA).

III. Results and Discussion

In this study, it is supposed that the difference of open-circuit voltage(V_{oc}) and cell efficiency(η) between the samples is affected by laser ablation. Therefore, an experiment is designed to ensure and clarify the appearance. The energy and focus are both key parameters for the experimental design. The control of energy depends on the output intensity. The cells depend on different laser ablation are defined as A, B, C, and D. Figure 1 is the curves of cells of open-circuit voltage (Voc) with top view SEM migraphs of various laser ablated areas on the SiNx surface of cells. There are 4 cells under the experimental design and they are (A),(B),(C),and(D), respectively. In our experiments, the widths ablated by green and UV laser are 20 μm and 15 μm, respectively. Obviously, the open-circuit voltage (Voc) corresponds to the laser and output power strongly.

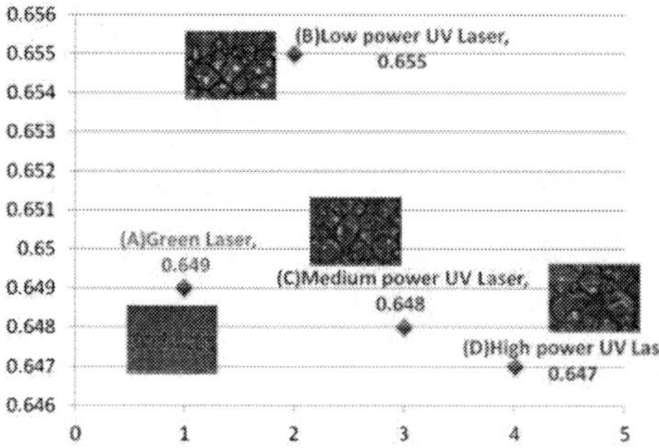

Figure 1. Cells of open-circuit voltage (Voc) with top view SEM migraphs of various laser ablated areas on the SiNx surface of ce. There are 4 cells under the experimental design and they are (A),(B),(C),and(D), respectively.

Table 1 is the performances of Ni/Cu plated cells of the Laser-ablated cells. The performance of cell (B) is superior to the others because of the less damage during laser ablation. Cell (B) exhibits a 0.54 mA/cm^2 increase of Jsc, 0.77% increase of FF, and 2.47% increase of η.

Table 1 Performances of Ni/Cu plated cells of the Laser-ablated cells

Cell	Eff.(%)	FF	Jsc(mA/cm²)
(A)Green Laser	△η+0.7	△FF+0.22	△Jsc+0.20
(B)Low power UV Laser	△η+2.47	△FF+0.77	△Jsc+0.54
(C)Medium power UV Laser	△η+0.48	△FF+0.29	△Jsc
(D)High power UV Laser	△η	△FF	△Jsc

Table 2 shows the related parameters measured by Suns-Voc of the Laser-ablated cells. The improvements of Voc and PFF value of Cell (B) is attributed to the less surface damages and lower recombination at the contacts.

Table 2 The Suns-V_{oc} measurements of the Laser-ablated cells

Cell	Voc(mV)	Pseudo-Eff.(%)	PFF
(A)Green Laser	△Voc	△η+1.26	△FF+0.050
(B)Low power UV Laser	△Voc+15	△η+2.35	△FF+0.079
(C)Medium power UV Laser	△Voc+6	△η+0.85	△FF+0.031
(D)High power UV Laser	△Voc+10	△η	△FF

In general, the filling factor(FF) and cell performance depend on the metal contact behavior strongly. It is believed that the suitable laser ablation leads to the close metal contact for the lower specific contact resistance. In this study, the performance of cell (B) is superior to the others. In summary, the optimization of laser ablation process is available and the removal of SiNx residue is under control effectively. According to the data, the improvement of the filling factor(FF) and the benefit of cell performance are attributed to the optimization of laser irradiation working.

IV. Conclusions

In a word, the effect of laser ablation on the metal contacts of Ni/Cu plating at the interfaces between metallization layer and opening area for n-type solar cell affect the cell performance obviously. On the other hand, the lower open-circuit voltage(V_{oc}) corresponding the promotion of cell efficiency(η), which results in the enhancement of cell performance. In our study, the optimization of laser ablation process is available to form close metal contact and the metallization is also under control effectively.

References

[1]K. Katkhouda, A. Limia, L. Bornschein, R. Koseva, T. Geppert, A. Grohe, H. Krokoszinski,and P. Schaaf, Aluminmm-Based Rear-Side PVD Metalliztion for nPERT Silicon Solar Cells, IEEE JPV, **4**(1), 160-167 (2014)

[2]A. Kraft, C. Wolf, A. Lorenz, J. Bartsch, M. Glatthaar, S. W. Glunz, Long Term Stability Analysis of Copper front Side Metallization for Silicon Solar Cells, Energy Procedia 55, 478-485 (2014)

[3]G. Heinrich, M. Bahr, K. Stolberg, T. Wutherich, M. Leonhardt, A. Lawerenz, Investigation of Ablation Mechanisms for Selective Laser Ablation of Silicon Nitride Layers, Energy Procedia, 8, 592-597 (2011)

INVESTIGATION ON DIFFERENT SURFACE MODIFICATIONS USING LASER TEXTURING

Behrad Radfar [1,2,*], Firat Es [1], Rasit Turan [1,2,3]
[1]The Center for Solar Energy Research and Applications (GÜNAM), Middle East Technical University
[2]Micro and Nanotechnology Graduate Program, Middle East Technical University
[3]Department of Physics, Middle East Technical University
Dumlupinar Blvd. 1, 06800 Ankara – Turkey

Email: esfirat@metu.edu.tr, Tel: +90 312 210 5090
Email: turanr@metu.edu.tr, Tel: +90 312 210 5069
*Corresponding Author, behrad.radfar@metu.edu.tr, Tel: +90 312 210 4303

ABSTRACT: One of the important factors that determines the performance of a crystalline silicon solar cell is the surface structuring. Surface not only affects the electrical properties with recombination but also the optical properties via light trapping. To minimize reflection from the flat surface and improve light trapping, the crystalline silicon wafers must be textured. For monocrystalline Si cells, anisotropic alkaline etchants are commonly used to create pyramids on the surface. However, this method is not suitable for multi-crystalline Silicon due to grains of random crystallographic orientation. The aim of this study is to employ laser texturing which is an anisotropic texturing process as an alternative method for texturing multi-c solar cells. The approach utilizes a laser to create pits on the surface of multi-crystalline silicon wafers. The laser processing parameters are justified by performing series of experiments for different values of processing parameters. In order to determine the effectiveness of the process, the reflectance and Scanning Electron Microscopy (SEM) images of produced textures are taken into account. As a result, proper parameters for laser and laser motor is founded. The SEM images show that textures are deep enough to trap light inside and the reflection measurements indicate that samples have a low reflection.
Keywords: Multi-Crystalline Silicon, Solar Cell, Surface Treatment, Laser Texturing.

1 INTRODUCTION

The growing needs for energy and the depletion of conventional sources of energy like oil and coal, developing a new source of energy has become important. One of the most important alternatives for conventional energies is solar energy. The direct conversion of solar energy into electricity is performed by solar cells [1].

Most of the solar cells are created from the crystalline silicon wafer. Form the cost point of view, multi-crystalline silicon is more favourable than mono-crystalline silicon [2][3]. However, solar cells based on multi-crystalline silicon are commonly less efficient than mono-crystalline silicon solar cells. One of the reasons is the difficulty of surface texturing [4]. Surface texturing is a technique which improves the efficiency of light trapping in solar cells. It can be formed on a monocrystalline wafer by anisotropic wet etching [2][3]. For mono-crystalline silicon with a (100) surface orientation, the industrial solution for anisotropic etching is KOH and IPA. The alkaline chemical etches (100) planes and stops on (111) planes, hence creates pyramids on the surface. These pyramids are typically 5-10 um in height and randomly located across the surface [5].

Since incoming light rays experience multiple bounces from the pyramids of the surface texture, the light has more chances to absorb into the solar cell (Figure 1). This effect decreases reflectance and hence increases the absorption. On average each beam bounces a little more than twice, reducing the minimum reflectance to ~10% (before anti-reflective coating) at ~900 nm, in comparison to a value of ~35% for a bare polished silicon wafer. Multiple bouncing depends on the shape of texture unit, and the pattern density [5][4].

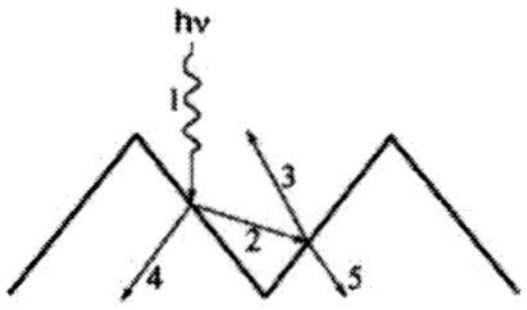

Fig 1 The influence of surface texture on light reflection of the textured surface. (1) Incident light, (2 and 3) reflected light and (4 and 5) absorbed light [6].

Unfortunately, anisotropic etching is not possible for multi-crystalline silicon due to the variety of crystal orientations at the surface. One of the ways for isotexturing is laser texturing [7][8]. This method of texturing generates an array of pits ablated individually [9]. At the beginning, individual pits are arranged in the form of square arrays. In order to reduce reflectance, it is important to maximize the area of the texture pits or wells. However, one of the disadvantages of this technique is related with removing residual slag [9]. Also, the shape, depth and spacing of the laser-ablated pits play an important role in their effectiveness. The shallow or widely spaced pits reflect more light and have lower optical performance. The deeper pits reduce the reflection but often contain residual ablation slag that reduce the electrical performance [9].

2 EXPERIMENTS

For this study, a p-type multi-crystal 15.6cmx15.6cm Silicon wafer with 200µm thickness has been used. During the process, nanosecond Infrared laser with 1064 nm wavelength, irradiated on the poly c-Si wafer. In the Fig 2, the laser texturing process is shown schematically. The laser moves in line to create pits with or without overlap. Then, laser starts another line.

First, each wafer has been divided into different regions in order to investigate the uniformity across the wafer while texturing with the laser.

Then, different patterns with different laser motor parameters and spacing between laser lines have been investigated. Also, laser parameters such as laser pulse frequency and power have been varied to see their effect on surface morphology and wafer reflectance.

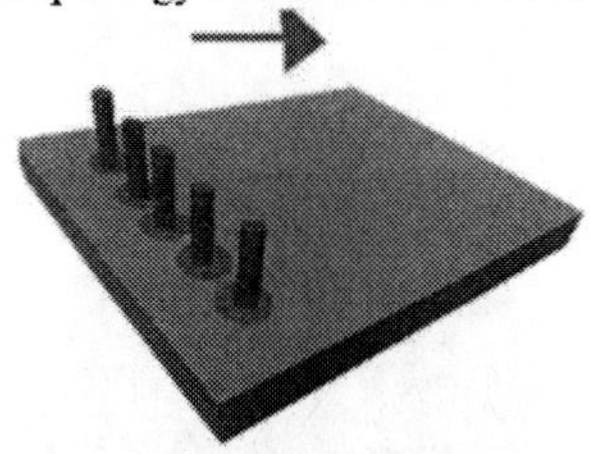

Fig 2. Laser texturing scheme

3 RESULTS AND DISCUSSIONS

3.1 Uniformity across the wafer

A full wafer has been divided into 64 regions with dimension 14mmx14mm each. Each region has 140 vertical and 140 horizontal lines with 100μm space. This wafer was textured with laser pulse frequency 55kHz (number of pulses per second) and 80% of total power of equipment. The Optical image from the sample is shown in Fig 3 and the SEM images of selected regions are shown in Fig 4.

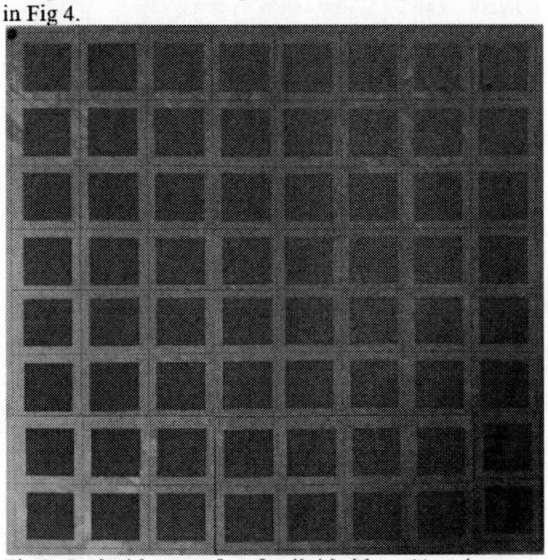

Fig 3. Optical image of wafer divided into 64 regions

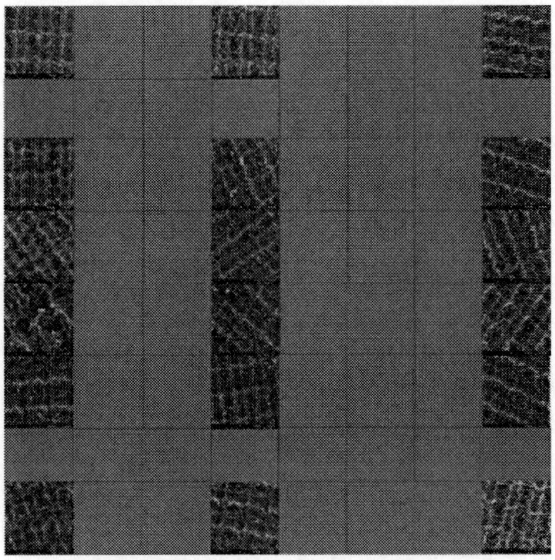

Fig 4. SEM images of selected regions

As it is seen in the optical images, the textures are quite uniform across the wafer and the SEM images of selected regions prove uniformity of the shapes. Hence, it's possible to texture full-size wafer.

3.2 Morphology dependence to laser pulse frequency, number of scans, laser motor's parameters

In this experiment, the dependence of the morphology of the surface on the laser parameters such as laser pulse frequency, scanning motor's draw step, and step period have been studied. Laser motor's draw step is a unit step of scanning motor when the laser is on. When draw step is large, the marking time will be short. The step period is a unit time of mark step. Table 1 shows different parameters used for this experiment. Also, the depth of grooves can be controlled by changing a number of scan with the same laser parameters.

Table 1. Parameters to investigate surface morphology

	Freq. (kHz)	**Pattern**	**Number of scan**	**Speed (mm/s)**	**Overlap (%)**
1st Row	30	Line	5	3605	-100
2nd Row	55	Criss-cross	3	572	87
3rd Row	55	Line	5	3605	18

The optical images of the sample are provided in Fig 5. The top view SEM images for this set of samples have been taken to inspect the morphology of surface as shown in Fig 6. The cross-section SEM indicates the depth of grooves and is shown in Fig 7.

Fig 5. Optical image of samples with different overlap between spots

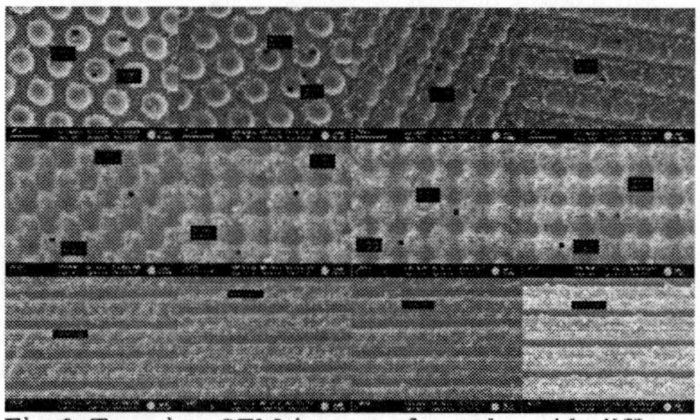

Fig 6. Top view SEM images of samples with different overlap between spots

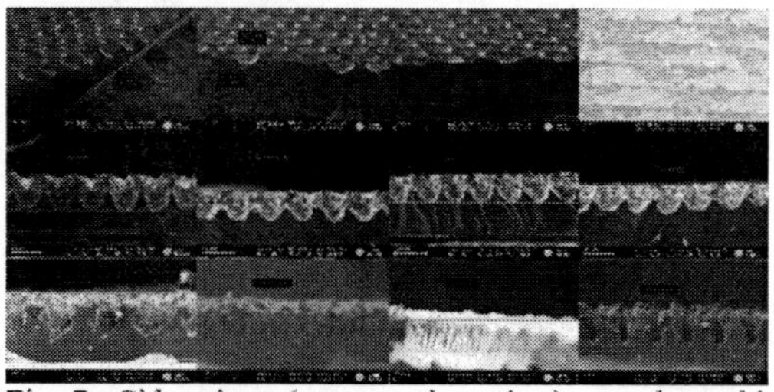

Fig 7. Side view (cross-section view) samples with different overlap between spots

The uniformity along each row is acceptable. In addition, it's possible to have various surface morphology. As it's seen in the optical image, two rows are almost dark black which indicate their low reflection. The spot size for all cases is around 70 μm and the depth varies from 50 to 100 μm.

3.3 Reflection dependence to number of scans, laser motor's parameters

The optical reflectance from the laser textured samples displayed in Table 2 is shown in Figure 9.

Table 2. Parameters used for investigation of reflectance

	Frequency (kHz)	Number of Scan	Speed (mm/s)	Overlap (%)
1st Row	30	1,2,3,5	3605	-100
2nd Row	30	1,2,3,5	572	76
3rd Row	55	1,2,3,5	572	87

The SEM images of these samples are shown in Fig 8. These images show the surface morphology versus different iterations and parameters.

Fig 8. SEM images of samples with different process parameters displayed in Table 2. Images have the same order as this table.

The weighted reflection shown in Fig 9 indicates that reflection can be as low as 6% in some cases.

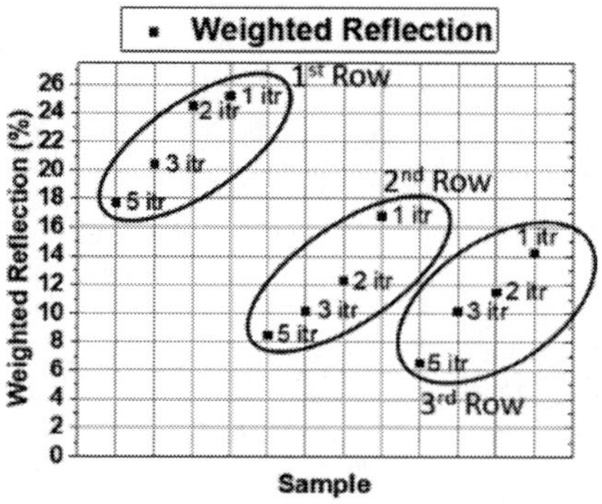

Fig 9. Weighted reflection of samples with different morphology and iterations.

The SEM images show extreme depth dependency to a number of scans which increased from right to left in each row. The weighted reflection also indicates that deeper structures have a lower reflection. The high reflection of the first row (separated spots structurer) is because of many untextured areas among spots. Moreover, the similarity of reflection of second and third row comes from their similar morphology and depth. In addition, the difference between them indicates that higher pulse frequency increases the depth which slightly decreases weighted reflection.

For comparison, a reference sample with KOH texturing has been prepared. The weighted reflection of the KOH textured sample was 11.87% which is showing the potential of the laser texturing approach.

4 CONCLUSION

Texturing multi-crystalline Si wafer with a laser as a

function of laser parameters has been studied in this work. We have shown that a variety of surface structures can be formed on the surface by adjusting the laser and process parameters. Very low optical reflections can be reached by optimizing these parameters.

5 REFERENCE

[1] L. A. Dobrzański, A. Drygała, K. Gołombek, P. Panek, E. Bielańska, and P. Zięba, "Laser surface treatment of multicrystalline silicon for enhancing optical properties," J. Mater. Process. Technol., vol. 201, no. 1–3, pp. 291–296, May 2008.
[2] L. A. Dobrzański and A. Drygała, "Laser processing of multicrystalline silicon for texturization of solar cells," J. Mater. Process. Technol., vol. 191, no. 1–3, pp. 228–231, 2007.
[3] L. a Dobrzański and a Drygała, "Laser texturization in technology of multicrystalline silicon solar cells," J. Achiev. Mater. Manuf. Eng., vol. 29, no. 1, pp. 7–14, 2008.
[4] K.-R. Kim, T.-H. Kim, H. Park, S. Kim, S. Cho, J. Yi, and B. Choi, "UV laser direct texturing for high efficiency multicrystalline silicon solar cell," Appl. Surf. Sci., vol. 264, pp. 404–409, Jan. 2013.
[5] B. C. Vineis, M. Levy-finklshtein, J. Carey, G. Knight, E. Wefringhaus, and R. Harney, "Ultrafast Laser Texturing for Enhanced Solar Cell Performance and Lower Cost," 2011.
[6] A. Goetzberger and V. U. Hoffmann, Photovoltaic Solar Energy Generation, vol. 112. Berlin/Heidelberg: Springer-Verlag, 2005.
[7] L. A. Dobrzański, A. Drygała, P. Panek, M. Lipińki, and P. Zieba, "Development of the laser method of multicrystalline silicon surface texturization," Arch. Mater. Sci. Eng., vol. 38, no. 1, pp. 5–11, 2009.
[8] J. C. Zolper, S. Narayanan, S. R. Wenham, and M. A. Green, "16.7% efficient, laser textured, buried contact polycrystalline silicon solar cell," Appl. Phys. Lett., vol. 55, no. 22, p. 2363, 1989.
[9] M. Abbott and J. Cotter, "Optical and electrical properties of laser texturing for high-efficiency solar cells," Prog. Photovoltaics Res. Appl., vol. 14, no. 3, pp. 225–235, May 2006.

LOW-TEMPERATURE SOLDERING FOR THE INTERCONNECTION OF SILICON HETEROJUNCTION SOLAR CELLS

Angela De Rose*, Denis Erath, Torsten Geipel, Achim Kraft, and Ulrich Eitner
Fraunhofer Institute for Solar Energy Systems ISE, Heidenhofstraße 2, 79110 Freiburg, Germany
*Phone: +49 (0)761 4588 5856, angela.de.rose@ise.fraunhofer.de

ABSTRACT: Interconnecting silicon heterojunction (SHJ) solar cells by low-temperature ribbon soldering allows the use of standard stringing equipment and might therefore be the cheapest and most straightforward implementation in existing fabrication lines. However, solder joints on low-temperature metallization pastes of SHJ cells are known for a weak adhesion to the cell surface. This work is dedicated to a better understanding of the interaction between solder and low-temperature metallization on SHJ solar cells. We evaluate soldering during a lamination process with standard copper ribbons featuring six different low-temperature solder alloys. The mechanical adhesion of the solder joints and microstructural changes in the metallization pastes during the soldering process are analyzed. Our study includes three metallization pastes, two of them show poor adhesion on the wafer surface after soldering (up to 0.03 N/mm) and one paste performs slightly better (up to 0.28 N/mm). We find this difference to be caused by an interaction of the pastes and the liquid solder, neither depending on the composition of the solder alloy nor on the soldering time. A fine grain structure of the Ag-particles of such pastes correlates with a higher penetration of solder components (Sn and Bi) resulting in dissolution of the metallization and detachment from the cell surface.
Keywords: Interconnection, Heterojunction, Soldering, Low-temperature

1 INTRODUCTION

Crystalline silicon solar cells based on heterojunction technology are one of the most promising concepts to obtain very high energy conversion efficiencies. The world record efficiencies reached within the last years underline the competitiveness of this technology [1–3]. Recently, Yoshikawa *et al.* presented a record cell efficiency above 26% combining heterojunction and back-contact technology [4]. A lean production chain with low-temperature processes and carrier selective hetero-structure for excellent passivation are main advantages of silicon heterojunction (SHJ) solar cells. Due to their higher open circuit voltage and their lower temperature coefficient of power, they offer a superior module performance compared to homojunction cells.

Consequently, an industrially feasible interconnection process in photovoltaic module fabrication becomes important. Recent activities address the use of alternative and non-standard interconnection approaches like electrically conductive adhesives [5] or wire interconnection [6]. To ensure an easy and cost-efficient implementation with conventional stringers, our work focuses on an interconnection by ribbon soldering with lead-free low-temperature solder alloys.

Since the passivation by the amorphous silicon layers of SHJ cells cannot withstand temperatures above 250 °C [7, 8], low-temperature soldering is considered as a suitable technology. The main challenge is to overcome the known weak adhesion between metallization paste and wafer surface, observed after soldering on SHJ solar cells [9].

Bismuth-based solder alloys possess lower melting points than common tin-lead-based solders and can therefore be processed at temperatures below 200°C. The approach followed in this work is to interconnect three commercially available low-temperature metallization pastes for SHJ solar cells with copper ribbons featuring six different solder coatings by soldering and examine the interaction between the materials. We analyze the differences and show relations between mechanical, optical and microstructural properties of the three pastes under investigation.

2 EXPERIMENTAL

2.1 Sample fabrication

The experimental procedure including sample fabrication (green) and characterization techniques (orange) is depicted in Figure 1. Our experiments are performed with bifacial monocrystalline SHJ solar cells (156×156 mm²) from Meyer Burger Technology AG. The cells are industrially pre-processed on n-type Cz-Si 6'' wafers with a-Si:H(i)/a-Si:H(n+)/TCO on the front side and a-Si:H(i)/a-Si:H(p+)/TCO on the rear side [10]. Based on previous investigations focusing on screen printing quality [11], the busbars at the front and rear side are prepared in a five-busbar design with three different metallization pastes "A" to "C" (Ag content between 85 - 95%; different amount of solids). The pastes are cured for t_{cur} = 4.5 min at T_{cur} = 200 °C in an inline infrared/convection furnace after each screen printing step on the front and rear side.

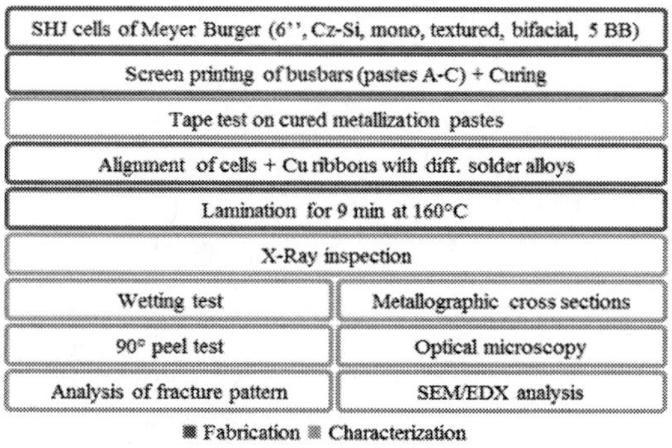

Figure 1: Schematic process sequence for fabrication and characterization of low-temperature soldered joints on SHJ solar cells.

A set of samples is fabricated for mechanical peel tests and for preparation of cross sections, respectively. Our analysis includes bare and solder-coated copper ribbons (1.5 mm × 0.18 mm) with six bismuth-based solders and a tin-lead solder as reference, see in Table I.

Table I: Overview on composition, melting point and delivery of the processed solder alloys.

	Composition (wt. %)	Melting Point (°C)	Form
SnBi	43/57	139	coating
	60/40	139 - 150	coating
	42/58	139 - 140	paste
SnBiAg	41/57/2	138 - 142	coating
	60/38/2	153 - 156	coating
PbSnBi	42/37/21	118 - 122	paste
SnPb	60/40	183 - 187	coating

The solder pastes are manually stencil printed onto the cell and covered with bare copper ribbons, whereas the ribbons with solder coating are attached to the busbars without additional solder paste. The melting point of the bismuth alloys allows the use of a lamination process step ($t_{lam} \approx 9$ min, $T_{lam} = 160$ °C) for interconnection and therefore a process simplification. As reference, SnBi 43/57 and SnPb 60/40 coated ribbons are soldered manually onto the metallization of SHJ cells for comparison to shorter soldering times ($t_{man} \approx 1$ - 2 s, $T_{man} = 160$ - 220 °C).

2.2 Characterization of solder joints

The fabricated samples are characterized by X-ray inspection, wettability analysis, peel tests and microscope as well as SEM/EDX images at prepared cross sections. Prior to the solder joint analysis, tape tests on the metallization pastes (after curing) are performed to qualify the initial adhesion (before soldering) on the wafer. An adhesive tape is pulled off in an angle of 90° from the pastes whereas the force is detected.

X-ray inspection as a non-destructive method is well suited to provide an insight into the quality of the solder joints [12]. Voids, caused by air trapped within the joint, reduce the contact area and thus result in a higher electrical resistance and lower adhesion strength. They can be detected by a difference in contrast within the X-ray image.

Prior to soldering, the wettability of the screen printed and cured low temperature metallization with the liquid solder alloys is analyzed. The cells are placed on a hotplate and pre-heated to 200 - 220 °C so that the solder melts as soon as it comes in contact with the metallization. A flat shiny appearance of the solder (low contact angle) indicates good wettability whereas solder drops (high contact angle) are referred to de-wetting.

The mechanical strength of the interconnection is characterized by peel tests according to DIN EN 50461, requiring forces larger than 1 N/mm [13]. The adhesion force of the soldered ribbon is recorded against the position along the busbar. The ribbons are peeled off in an angle of 90° to the cell surface with a speed of 50 mm/min. The resulting fracture pattern gives information about the weakest interface. We distinguish between an adhesive fracture where the metallization is separated from the solar cell, and a cohesive fracture within the metallization paste.

In order to inspect the cross section of the solder joints, a part of the soldered busbar is cut by a precision saw, embedded into a nickel-containing epoxy, ground and polished. Optical microscopy is used to inspect the formation of the solder joints.

Scanning electron microscopy (SEM, Carl Zeiss Microscopy Auriga 60) is performed on a crossbeam workstation in combination with energy diffractive X-ray (EDX, Bruker Quantax XFlash 6|60 detector) spectroscopy. Both methods reveal information on the cross section of the solder joint, especially on intermetallic phase formation [14].

3 RESULTS AND DISCUSSION

3.1 Wetting and mechanical adhesion

Based on the knowledge of a weak adhesion between metallization and solar cell after curing but before soldering [9], tape tests along the metallization of all three pastes are carried out in pretests. The measured average normalized peel forces exceed values above 0.9 N/mm for all pastes, limited by the adhesion of the tape on the paste (see Figure 2 (a)). These values are considered high enough for a successful interconnection.

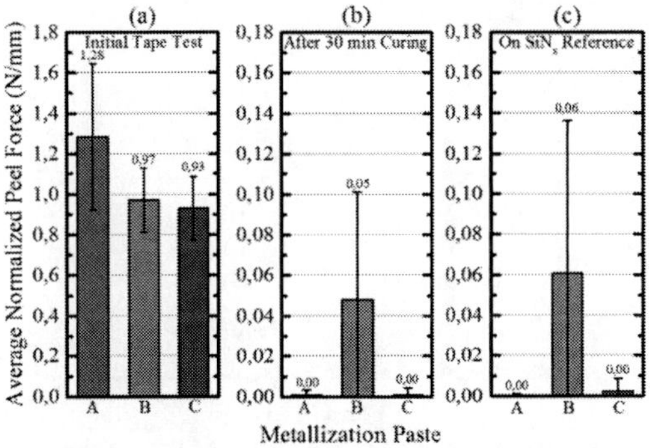

Figure 2: Experimental results of 90° tape test on the initial pastes after curing for 4.5 min (a) and of 90° peel test after soldering with SnBiAg 41/57/2 (b, c) as a function of metallization paste (A - C). Each box contains data of two peel tests, showing average and standard deviation. Note the different scaling of (b) and (c) compared to (a).

A set of the aforementioned SHJ solar cells is employed to test the wetting behavior of the metallization pastes and to investigate the mechanical adhesion after soldering (see left column (orange) of process sequence in Figure 1). All three pastes show good wettability for the used solder alloys which is the requirement for a successful solder process. The results of the mechanical peel tests, including each the data for one ribbon peeled off, are shown in Figure 3. Average and standard deviation of the normalized peel force are plotted as a function of metallization paste (A - C) and solder alloy (green: ribbon coatings, blue: solder pastes, grey: ribbon coatings soldered manually). In agreement to previous investigations by Gierth *et al.* [9], all combinations show poor adhesion. Paste B yields considerably higher values with 0.21 N/mm and 0.28 N/mm when using PbSnBi 42/37/21 and SnBi 43/57, respectively. We exclude the contact area of the solder joint to be the reason for this prominent deviation between pastes A - C due to homogenous solder joint formation proven by X-ray imaging. Exemplarily, the X-ray images for SnBiAg 60/38/2 on paste A, B and C are given in the inset of Figure 3, showing homogeneous solder joints along the busbar. The formation of those solder joints without any voids seems to be supported by the vacuum (~ 1 mbar) during the lamination process.

Figure 4 presents the force-path diagram for SnBi 43/57 on paste A (orange), on paste B (green) and for SnPb 60/40 soldered manually onto paste B (grey). The adhesion on paste B is higher along the whole length of the metallization. The fracture patterns (top-view microscopy images of busbars after peel test), also given in Figure 4, reveal a mostly cohesive fracture within the metallization for paste B. For paste A and paste C, the fracture is identified to be an adhesion failure between metallization paste and cell surface (TCO), which correlates with the measured adhesion. This indicates that the interface of paste A and C to the TCO is the weakest interface in this material combination, independent from the used solder alloys.

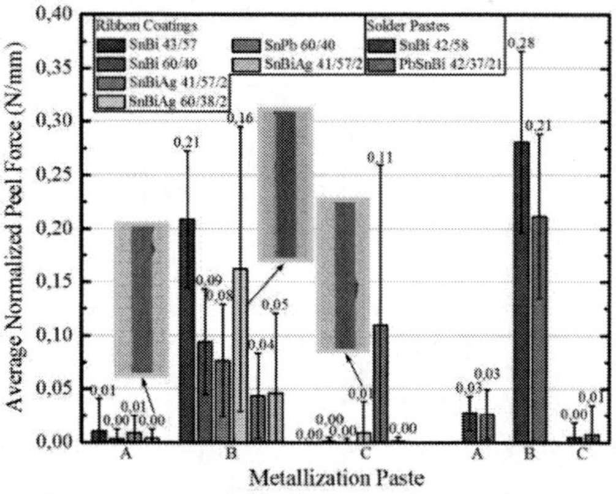

Figure 3: Experimental results of 90° peel test as a function of metallization paste (A - C) and solder alloy (color code). Four of the alloys are used as ribbons (green) and two are applied as solder pastes (blue), both processed during lamination for 9 min at 160 °C. As reference, two additional alloys are soldered manually (grey). Each data set gives average and standard deviation over one ribbon peeled off. In the inset, three X-ray images for SnBiAg 60/38/2 are shown exemplarily for a void-free solder joint formation.

In contrast to low-temperature pastes, the mechanical adhesion to the wafer for common high-temperature screen printing pastes is achieved by a firing step performed at temperatures up to 850°C. The contact formation is related to Ag-crystallites grown into the emitter thus ensuring electrical current path and mechanical stability [15]. Our pastes are dried within a short curing step for $t_{cur} = 4.5$ min at $T_{cur} = 220$ °C which leads to evaporation of solvents and to sintering of the Ag-particles. These effects differ fundamentally from the implications of the firing step. To improve the mechanical adhesion of paste A and C within our possibilities, a variation of the curing time up to 30 min is done [11], yielding no effect on wetting behavior or mechanical adhesion after soldering. Figure 2 (b) shows the corresponding peel test data with better results for paste B. As a reference measurement, homojunction solar cells with an anti-reflection coating (SiN_x) on the front side are processed and screen printed with the investigated metallization pastes A - C. Soldering and peel tests are performed in the described procedure. Most of the fracture patterns are also adhesive between paste and wafer resulting in zero adhesion (see Figure 2 (c)). Consistently to the peel tests on TCO, paste B performs best with 0.06 N/mm, which suggests the TCO on the SHJ cells not to be the reason for the poor adhesion.

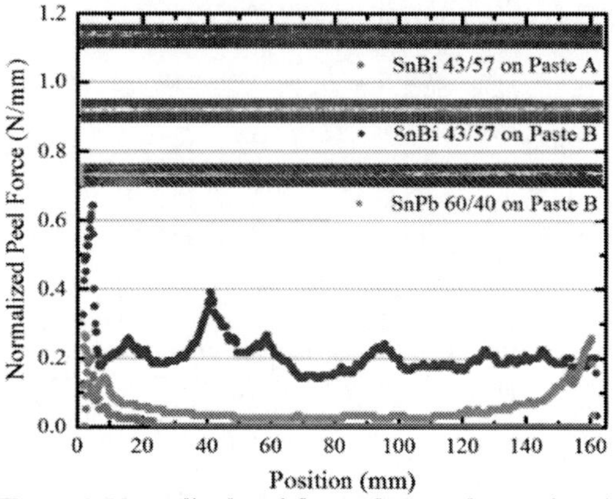

Figure 4: Normalized peel force of 90° peel tests plotted against the position along the busbar metallization. The graph includes the data for a ribbon coated with SnBi 43/57 on paste A (orange) and paste B (green) and as reference SnPb 60/40 on paste B (grey, soldered manually). The corresponding fracture patterns are shown above.

The fact that the initial adhesion is much higher than after soldering (see Figure 2 (a)) leads to the assumption that the loss in adhesion is induced by the liquid solder. This is confirmed by the performed wetting tests, exemplarily shown for a Bi- and a Pb-based solder alloy given in Figure 5. The application of the liquid solders onto metallization paste A leads to spalling from the cell surface in both cases, whereas paste B remains visually unaffected. Due to the fact that the TCO is not wettable, the liquid solder deforms. We conclude a fast reaction of paste A with the liquid solder which is supported by results from manually soldered solar cells where short soldering times of 1 – 2 s are used. The Pb as well as the Bi references (grey in Figure 3 and Figure 4) show similar results which underlines that the loss of adhesion is caused within seconds after the contact with liquid solder, not depending on the composition of the solder alloy. The root cause of better adhesion of paste B may be found in the microstructure, wherefore cross sections of the solder joints are of great interest.

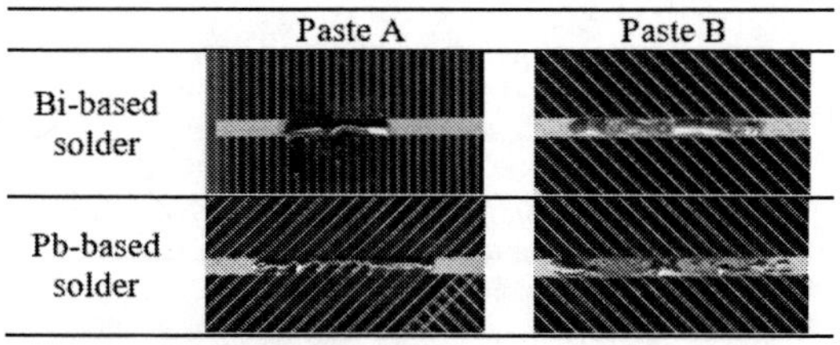

Figure 5: Wetting test with Bi- and Pb-based solder alloys on metallization paste A and B. Directly after contacting paste A with the liquid solder, the paste is spalling from the cell surface in contrast to paste B.

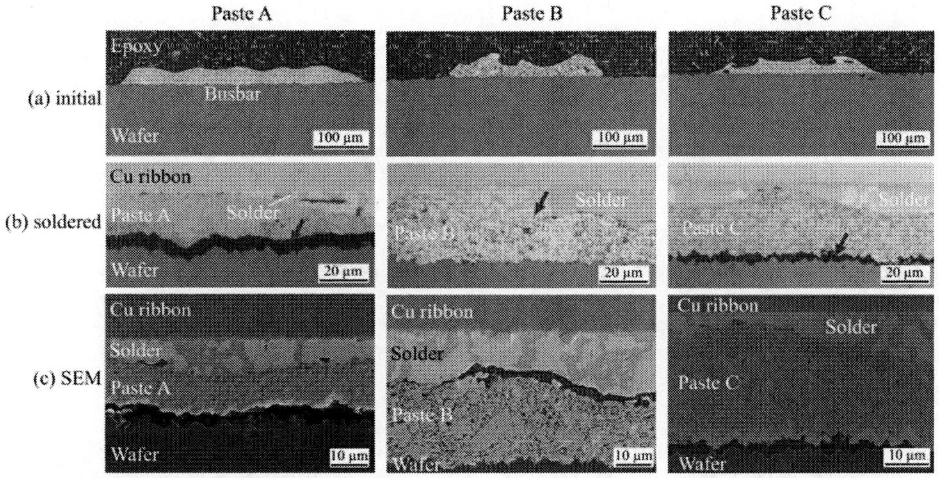

Figure 6: Optical microscopy images of metallographic cross sections of an unsoldered SHJ cell (a) and contacted with copper ribbons coated with SnBi 43/57 (b) on paste A, B and C. The red arrows mark the position of the gap after soldering. (c) Detailed SEM images of the cross sections shown above.

3.2 Microstructural characterization

Metallographic cross sections reveal a very small section of the entire soldered busbar and should therefore be interpreted carefully. Figure 6 (a) shows optical microscopy images with cross sections of bifacial SHJ cells in the uncontacted state. All pastes are initially in direct contact with the textured wafer surface which is in agreement with the good initial adhesion. The width of the busbars differs because of a tapered busbar design.

The metallization on the rear side is contacted with SnBi 43/57 coated copper ribbons and soldered during lamination (see Figure 6 (b)). The cross sections clearly reveal a gap between paste A and C and the wafer surface correlating with zero adhesion measured by the peel test. In case of paste B, the solder is in direct contact with the paste, as well as the paste with the wafer surface. Additionally, paste B features a coarser grain structure compared to paste A and C which is also confirmed by top view microscopy images (not shown here).

To focus in detail on the microstructure at those three interconnections, SEM is performed with higher magnification at the relevant interfaces (Figure 6 (c)). The difference in grain structure of Paste A and C compared to Paste B is clearly visible. Paste A and C appear finely grained and homogeneous whereas Paste B is relatively coarsely structured. The interfaces between paste A and C and solder do not show any cracking.

In contrast, paste B features larger and non-uniform grains and partially a gap between solder and paste. This SEM image shows a different position than given in Figure 6 (b) for paste B. Microscopy images with minor magnification reveal that the gap between paste B and solder appears only in parts. These differences in grain structure and the existence of the gap are suspected to influence the mechanical adhesion.

This idea is supported by energy dispersive X-ray (EDX) spectroscopy performed on the above shown cross sections. The results for paste B and C are given in Figure 7. EDX-mapping of the solder joints reveals a penetration of Sn (green) into the busbar which consists mainly of Ag (red). The Sn-penetration is more homogeneous and distinct for paste C (Figure 7 (b)), as well for paste A (not shown here). The finer grain structure of paste A and C seems to assist infiltration of liquid solder into the metallization paste because of an accompanying higher penetrability. An EDX element scan supports the assumption of an Ag_3Sn-phase formation at the solder-metallization-interface penetrating into the entire metallization. The growth of intermetallic compounds in solder joints on high-temperature metallization has already been studied in literature [14, 16] and depends strongly on temperature and exposure time. We exclude the time to be the relevant parameter due to similar results for the manually

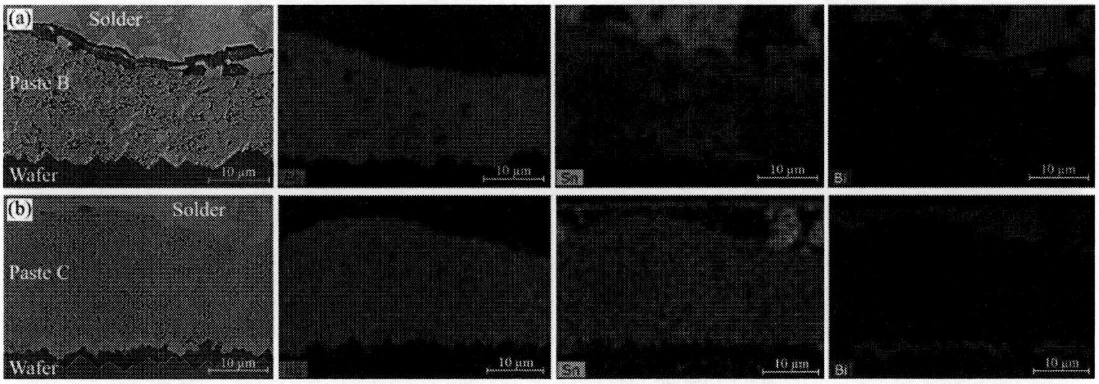

Figure 7: EDX element mapping of SnBi 43/57 solder joints on (a) paste B and (b) paste C.

soldered joints, as already assumed above. For paste B, the interaction with the liquid solder is found to be different. Beside the comparably low Sn-penetration, Bi (blue in Figure 7) seems to form clusters within the Ag-metallization. The occurring structure of Bi-phases is completely different for paste C. Bi agglomerates at the interface to the wafer and forms a solid Bi-layer. Compared to Sn or Pb, Bi is brittle [17] and shows a negative thermal expansion when melted and expands when solidified [14, 17–19]. The latter characteristic supports the ablation of the metallization from the cell surface. This finding is consolidated by the fracture pattern which is adhesive for paste C and cohesive when Bi is enclosed within the metallization (paste B).

The interaction of liquid solder and metallization paste strongly changes the structure of the metallization. Short soldering times are sufficient to induce an almost complete dissolution of the metallization and detachment from the cell surface. Intermetallic phase formation as well as Bi agglomeration at the wafer interface is found, which contribute to the embrittlement of the joint and ablation of the metallization.

4 CONCLUSION AND OUTLOOK

Peel tests on SHJ cells with six solder alloys on three different low-temperature metallization pastes reveal weak adhesion forces between cell surface and metallization after low-temperature soldering. Reference measurements using tape tests on unsoldered busbars prove that the loss in adhesion is caused by the solder diffusion. We show a relation between the mechanical adhesion of the pastes on the wafer after soldering, the grain structure of the pastes and penetration of Sn and Bi into the metallization. Paste B, which shows slightly higher peel forces, features a coarser grain structure which seems to impede the Bi penetration towards the wafer surface. Here, the fracture pattern after peel test shows a cohesive fracture within the paste, whereas an adhesive fracture is observed for two pastes with nearly zero adhesion. An interaction between liquid solder and paste occurs within the first seconds during soldering and leads to dissolving of the paste.

To realize the successful interconnection of SHJ solar cells with low-temperature soldering, an adaptation of the metallization pastes is indispensable. The penetration of liquid solder into the busbar should be reduced. Our results indicate that for instance a coarser grain structure may be beneficial.

5 ACKNOWLEDGEMENT

This project was funded by the German Ministry for Economic Affairs and Energy within the research project "HERA" under contract number 0325835B.

6 REFERENCES

[1] K. Masuko *et al.*, "Achievement of More Than 25% Conversion Efficiency With Crystalline Silicon Heterojunction Solar Cell," *IEEE J. Photovoltaics*, vol. 4, no. 6, pp. 1433–1435, 2014.

[2] M. A. Green *et al.*, "Solar cell efficiency tables (version 48)," *Prog. Photovolt: Res. Appl.*, vol. 24, no. 7, pp. 905–913, 2016.

[3] D. Adachi, J. L. Hernández, and K. Yamamoto, "Impact of carrier recombination on fill factor for large area heterojunction crystalline silicon solar cell with 25.1% efficiency," *Appl. Phys. Lett.*, vol. 107, no. 23, p. 233506, 2015.

[4] K. Yoshikawa *et al.*, "Silicon heterojunction solar cell with interdigitated back contacts for a photoconversion efficiency over 26%," *Nat. Energy*, vol. 2, no. 5, p. 17032, 2017.

[5] M. Späth *et al.*, "Hetero-junction module technology," in *Proceedings of the 26th European Photovoltaic Solar Energy Conference and Exhibition*, 2011, pp. 3121–3124.

[6] P. Papet *et al.*, "New Cell Metallization Patterns for Heterojunction Solar Cells Interconnected by the Smart Wire Connection Technology," *Proceedings of the Fifth Workshop on Metallization for Crystalline Silicon Solar Cells*, vol. 67, pp. 203–209, 2015.

[7] B. A. Korevaar *et al.*, "Influence of annealing on performance for hetero-junction a-Si/c-Si devices," in *Proceedings of the 23rd European Photovoltaic Solar Energy Conference and Exhibition*, 2008.

[8] S. de Wolf and M. Kondo, "Nature of doped a-Si: H/c-Si interface recombination," *Journal of Applied Physics*, vol. 105, no. 10, p. 103707, 2009.

[9] P. Gierth *et al.*, "Comparison of NiV and Polymer Paste Metallization as Low Temperature Interconnection of High Efficiency Heterojunction Solar Cells," in *Proceedings of the 28th European Photovoltaic Solar Energy Conference and Exhibition*, 2013, pp. 464–467.

[10] B. Strahm, "Heterojunction Solar Cells," Talk at nPV Workshop, Meyer Burger Research AG, 2017.

[11] D. Erath *et al.*, "Comparison of innovative metallization approaches for silicon heterojunction solar cells," *Silicon PV*, 2017, tbp.

[12] B. G. Batchelor, *Machine Vision Handbook*, 2nd ed.: Springer London Dordrecht Heidelberg New York, 2012.

[13] Solar cells - Datasheet information and product data for crystalline silicon solar cells, EN 50461: 2007-03, 2007.

[14] T. Geipel *et al.*, "Intermetallic compounds in solar cell interconnections: Microstructure and growth kinetics," *Solar Energy Materials and Solar Cells*, vol. 159, pp. 370–388, 2017.

[15] M. Hoerteis, "Fine-line printed contacts on crystalline silicon solar cells," PhD thesis, University of Konstanz, Germany, 2009.

[16] T. Geipel *et al.*, "Comprehensive study of intermetallic compounds in solar cell interconnections including lead-free, low melting point solders," in *Proceedings of the 32nd European Photovoltaic Solar Energy Conference and Exhibition*, 2016, pp. 479–486.

[17] H. H. Manko, *Solders and Soldering*, 3rd ed. New York, NY: S&P Global, 1992.

[18] M. J. Burek *et al.*, "Grain boundary effects on the mechanical properties of bismuth nanostructures," *Acta Materialia*, vol. 59, no. 11, pp. 4709–4718, 2011.

[19] Z. Mei and J. Morris, "Characterization of eutectic Sn-Bi solder joints," *JEM*, vol. 21, no. 6, pp. 599–607, 1992.

Nanocrystalline vs. amorphous n-type silicon front surface field layers in silicon heterojunction solar cells: Role of thickness and oxygen content.

A.B. Morales-Vilches, L. Mazzarella, M. Hendrichs, L. Korte, R. Schlatmann, B. Stannowski

Helmholtz-Zentrum Berlin für Materialien und Energie GmbH, Schwarzschildstr. 3, 12489 Berlin, Germany.

***Abstract* – We report on a systematical investigation of the influence of layer thickness and oxygen content of different silicon-based n-doped layers deposited by Plasma Enhanced Chemical Vapor Deposition (PECVD) on silicon heterojunction (SHJ) solar cell performance. N-type amorphous (a-), nanocrystalline (nc-) hydrogenated silicon (Si:H) and silicon oxide (SiO_x:H) films were used as front surface field (FSF) layers on rear emitter (RE) SHJ solar cells. The benefit, both in electrical and optical properties, of using nc-silicon layers, is demonstrated. The best cell in the oxygen variation experiment was obtained with an (n)nc-SiO_x:H FSF having a refractive index n of 2.72 at 633 nm, which leads to an increase in short circuit current density, j_{sc}, of more than 1 mA/cm^2 as compared to a reference cell with an a-Si:H FSF. This higher j_{sc} obtained in our best nc-SiO_x:H prototypes leads to a conversion efficiency, η, of 21.4%. It is also remarkable our best cell result, certified by ISFH CalTeC, with an improved nc-Si:H FSF that enabled an improved j_{sc} of 38.6 mA/cm^2 and an η of 22.5%.**

I. INTRODUCTION

Developing FSF layers with reduced parasitic absorption which allow an increase in j_{sc} while maintaining or improving V_{oc} and FF is highly interesting for further enhancement of the conversion efficiency of rear emitter silicon heterojunction (RE SHJ) solar cells. A-Si:H is an extensively used material in silicon photovoltaics applications because of its excellent passivation properties [1] and its ability to collect efficiently the photogenerated charge from the c-Si bulk [2]. However, it has some drawbacks due to its high refractive index (n) and its high parasitic absorption [3]. In this work we report on three optimization studies developed at PVcomB-HZB aiming at a reduction of both n and parasitic absorption and, consequently, an improved j_{sc}. Initially, a remarkably conductive and transparent nc-Si:H layer was developed. Due to a high nanocrystalline fraction with a low absorption coefficient, this layer is already more transparent than the a-Si:H layer [4]. The growth of nc-Si:H layers at low temperatures on top of passivating a-Si:H surfaces in SHJ cells is challenging, as discussed elsewhere [5, 6]. To improve the nucleation of the nc- phase, we introduced a buffer layer between the i-a-Si:H and the (n) nc-Si:H FSF layer. In parallel we also introduced a highly conductive contact layer between the nc-Si:H FSF and the TCO layer in order to reduce the contact resistance at this interface. Finally, oxygen was introduced into both a-SiO_x:H and nc-SiO_x:H in order to increase the transparency of the respective layers [7–9]. From this set of optimized materials, some were selected and RE SHJ cells with different FSF layer thicknesses were fabricated in order to demonstrate the j_{sc} gain in finished cells.

II. EXPERIMENTAL DETAILS

N-type a-Si:H and nc-Si:H layers, with and without oxygen, were deposited in an Applied Materials (AKT1600) PECVD cluster tool operating at 13.56MHz. More details regarding the reactor are described elsewhere [10, 11]. All films were deposited on glass substrates in order to characterize them electrically (lateral conductivity, σ) and optically (refractive index n, k). Additionally, RE SHJ cells grown on 5-inch n-type Cz-Si textured wafers (figure 1 (b)), were manufactured employing selected a-Si:H and nc-Si:H FSF layers, varying their thickness as well as oxygen content, by increasing the CO_2/SiH_4 flow ratio in the PECVD process from 0 to 2.8. Concerning the nanocrystalline material growth we have reported elsewhere [9] an optimization of the deposition conditions in order to obtain the desired crystalline structure, minimizing the thicknesses of the incubation and nucleation phase. When reducing the thickness of the nc-Si:H FSF layer on final SHJ devices an alternative structure was introduced (Fig. 1 (c)) where an intrinsic seed layer was successfully implemented as a nucleation layer to increase the passivation and the crystallization of the nc-Si bulk structure. A contact layer was also introduced to improve the contact between the nc-Si FSF and the front TCO layers. The silane dilution in hydrogen was also varied in the non-oxidic FSF layers in order to increase the layer transparency.

ITO was used as front and rear TCO, sputtered in a Leybold Optics DC sputtering tool. On the rear side of the devices a full area silver layer was also deposited by sputtering. All

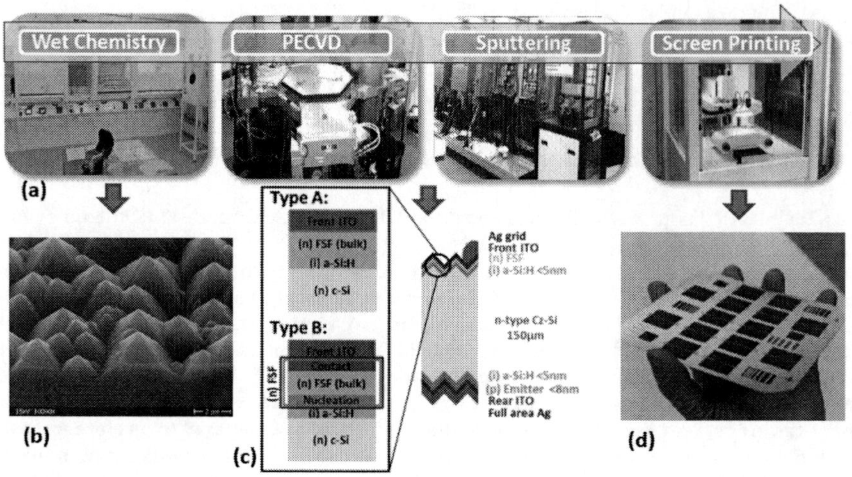

Fig. 1: (a) SHJ process sequence at HZB, (b) SEM image of a textured surface, (c) Schematic cross-section of the rear emitter (RE) silicon heterojunction (SHJ) solar cell structure fabricated in this work with a magnification on the two different FSF structures used, type A (only bulk) and type (nucleation/contact approach), (d) picture of the 5'' CZ-wafers with the 14 cells layout used in the devices of this study.

the sputtering depositions were done through a shadow mask which defines 14 identical cells per wafer, each with 4 cm^2 active area (figure 1 (d)). A low-temperature-curing silver grid was screen printed with two different grid patterns; an initial cell design with a total busbar+grid shading of 7-8% was used in the experiments from Fig 3 and 4. In the thickness variation experiment on non-oxidic nc-Si:H FSFs a busbar-less silver grid was used in order to reduce the metallic shading area down to <4% to obtain the highest j_{sc} possible without compromising the FF. Fig. 1 shows the process sequence of the SHJ cells developed at HZB (a), together with a SEM image of our textured wafer surfaces (b), the schematic cross-section of the RE cell structure (c) and an image of the 14 cells layout on 5 inch wafers (d). The current-voltage (J-V) characteristics were measured on finished cells under standard test conditions in an AAA+ dual-source sun simulator.

III. RESULTS

a) Influence of oxygen dilution on amorphous and nanocrystalline n-doped FSF layers

Fig. 2 shows the lateral conductivity as function of the refractive index of different a-Si(O_x):H and nc-Si(O_x):H layers deposited on glass substrates, where the oxygen to silane flow ratio CO_2/SiH_4 was varied from 0 to 2.8. The crystalline fraction of the nc-Si:H films led to a higher conductivity of those samples in comparison with the amorphous ones. From this data we can also observe that the oxygen incorporation in the material decreases its conductivity while reducing n, as expected from literature [7].

Selected a-Si:H and nc-Si:H layers were used as FSF layers in RE SHJ solar cells. J-V parameters of the best cell in each variation are depicted in Fig. 3 where we can observe the comparison between a-Si:H (circles) and nc-Si:H (triangles), the influence of the oxygen content (open vs. full symbols) and a thickness variation for selected FSF layers (lines).

We start by discussing the cells with nanocrystalline FSF without additional oxygen. They reveal 1 to 2% (abs.) higher efficiencies than the amorphous FSF cells for all thicknesses and CO_2/SiH_4 ratios. Together with the lower n shown in the nc-Si:H FSF layers that improve j_{sc}, the higher conductivity of the nc-Si:H layers and their carefully selected power, pressure, H_2 and CO_2/SiH_4 dilution during

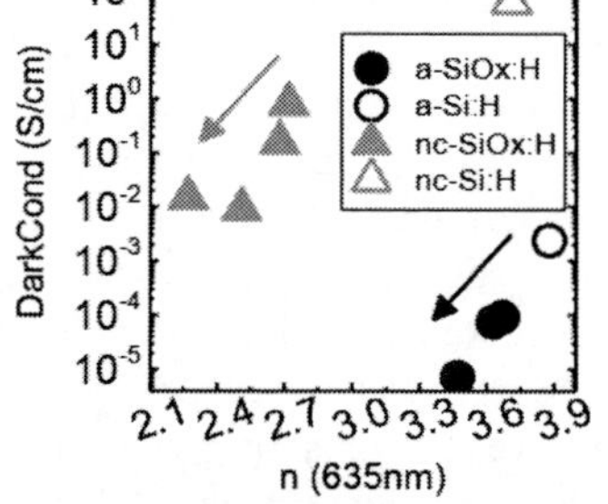

Fig. 2: Dark lateral conductivity (σ) versus refractive index (n) for n-type a-Si(Ox):H (circles) and nc-Si(Ox):H (triangles) with increasing CO_2/SiH_4 ratios (arrows indicate the direction of increasing oxygen content). Open symbols are references without oxygen.

the PECVD depositions yield both enhanced V_{oc}s and FFs of the cells.

Increasing the oxygen content in the FSF layer allows to obtain more transparent layers, therefore in the cells with SiO_x:H-based FSFs we obtained cells with higher j_{sc}. Using both nc- and a-SiO_x:H materials the j_{sc} on cells increased by ~1 mA/cm^2 when increasing the oxygen content. In spite of that improvement in j_{sc}, the a-SiO_x:H samples experienced an η degradation of almost 2% (abs.). This decrease is related to the low V_{oc} and FF that is created by the poor contact to the TCO and/or to the i-a-Si:H that the a-SiO_x:H layers provide.

b) Thickness reduction of FSF layers with and without oxygen

It is important to note, that when introducing these nc-Si:H layers in SHJ solar cells we have to consider the possible damage on the surface passivation due to the higher power and hydrogen dilution conditions required by the plasma when growing nanocrystalline films on top of the amorphous passivating layer. As seen in Fig. 4, V_{oc} and FF are strongly affected when decreasing the thickness of the n-doped layer of the FSF. In order to prevent this, an intrinsic nanocrystalline buffer layer was introduced between the a- and the doped nc-Si layers. This buffer acts as a nucleation layer and improves the crystallinity in the

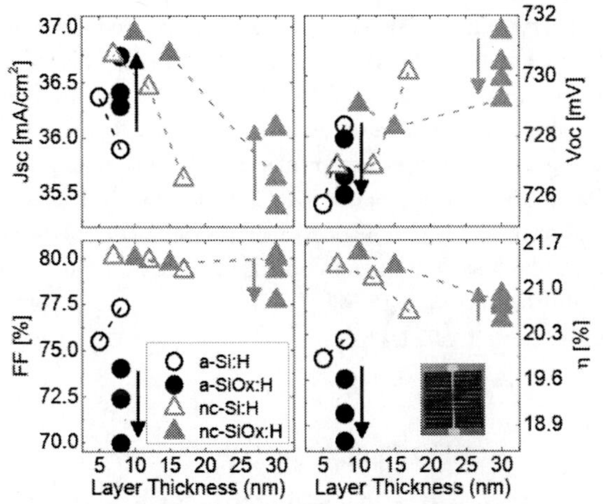

Fig. 3: N-type Cz-Si RE SHJ solar cell (4 cm^2 active area) characteristics of the best cell per group with different a-Si:H (circles) and nc-Si:H (triangles) FSF layers and different n-doped layer thicknesses (estimated from film thicknesses measured on polished c-Si using a correction factor of 1.7 to account for the textured surface). Lines are guides to the eye. Open symbols represent FSF layers without oxygen; full symbols correspond to layers varying the oxygen content (arrows indicate the direction of increasing oxygen content). Cell parameters for nc-SiO_x:H was taken from Ref. [9] Inset: front grid layout used in the cells shown in this experiment (7% metallization area).

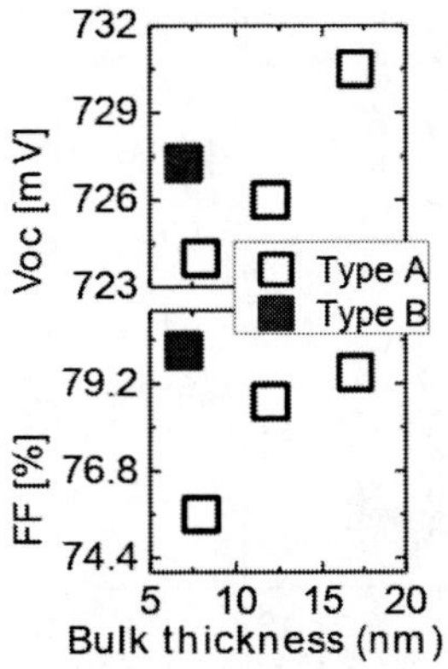

Fig. 4: a) N-type Cz-Si RE SHJ solar cell (4 cm^2 active area, 7% front grid metallization area) open circuit voltage (V_{oc}) and fill factor (FF) characteristics of the best cell per group reducing the nc-Si:H FSF layers thickness. Symbols in blue show the characteristics of one FSF sample with nucleation and contact layer approach (Type B).

thinner nc-Si:H FSF bulk layer. This structure (Type B, figure 1 (c)) was successfully implemented in finished SHJ cells and yielded good FF and V_{oc} while reducing the FSF n-doped bulk layer thickness down to 7 nm, demonstrating the necessity to use Type B FSF.

Regarding thickness comparison between the a-Si:H and the nc-Si:H (without oxygen) FSF layers, its thickness was decreased, from 8 to 5 nm and 17 to 7 nm respectively, calculated with a 1.7 factor to account the larger effective surface area of the textured surface in the a- and nc- growth. Additionally, selected nc-SiO_x:H samples were compared (n = 2.7, and CO_2/SiH_4 ratio = 1.2) reducing its FSF thickness to comparable values with the non-oxidic samples, from 30 to 10 nm thickness. It is clearly observable from the data (Fig. 3), that the 30 nm thick nc-SiO_x:H FSFs absorb too much light, leading to losses in j_{sc}. Reducing the FSF thickness to 15 nm a 1 mA/cm^2 gain in j_{sc} is obtained. Even the j_{sc} in the non-oxidic samples is enhanced by 0.6 - 1 mA/cm^2 (for the a- and the nc-Si:H, respectively) when reducing their corresponding FSF thickness. Still, we have to take into account the FF and V_{oc} effect on the final η values. In both nc-Si:H and nc-SiO_x:H samples we were able to maintain the FF values when reducing the FSF layer thickness, which leads to the idea that the contact among layers and the crystallinity are well preserved even for thinner nanocrystalline FSF layers. On the contrary, FF and V_{oc} on thinner a-Si:H FSF samples dropped dramatically, leading to a reduction in η.

c) Silane dilution and FSF thickness optimization

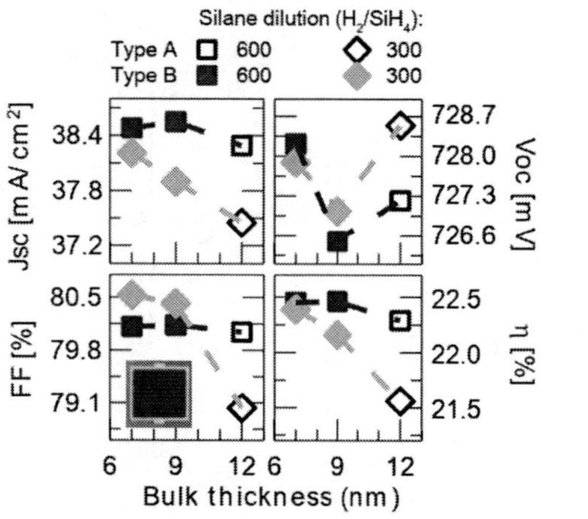

Fig. 5: N-type Cz-Si RE SHJ solar cell (4 cm² designated area, busbarless layout) characteristics of the best cell per group reducing the nc-Si:H FSF layers using two different silane dilution. Open symbols are samples with Type A FSF. Closed symbols are samples with Type B FSF. Inset: front grid layout of the cells used in this experiment (<4% metallization area).

Further FSF thickness variations in non-oxidic FSF SHJ solar cells were done varying additionally the silane dilution of the FSF layers. Both types of cell structures were used, Type A with only a doped FSF bulk film, and Type B with nucleation and contact layers. A busbarless screen printed silver contact layout was used in these cells, providing a reduced metallization area of <4%. As seen from the cell results shown in Fig. 5, in cell structure Type A the FSF layers grown with higher silane dilution are more transparent providing j_{sc} up to 1 mA/cm² higher than the lower silane diluted FSF samples. When reducing the thickness of both silane diluted bulk FSF samples, using Type B structure, FF and V_{oc} are successfully maintained allowing the efficiency to follow the j_{sc} trends. In the samples with higher silane dilution the j_{sc} increases up to 0.5 mA/cm² while for the cells with lower silane dilution FSF the j_{sc} improvement is >1 mA/cm², reducing the difference between different dilution samples in the thinner FSF cells. The best cell result in this variation was obtained with a higher silane dilution and Type B FSF with a bulk thickness of 7 nm, providing an efficiency η of 22.4% and a short circuit current density j_{sc} of 38.4 mA/cm².

As seen in Fig. 6 the best cell with an optimized FSF, using a bifacial layout and an improved screen printed silver metallic contacts, was certified by ISFH CalTeC with a final conversion efficiency, η, of 22.5%.

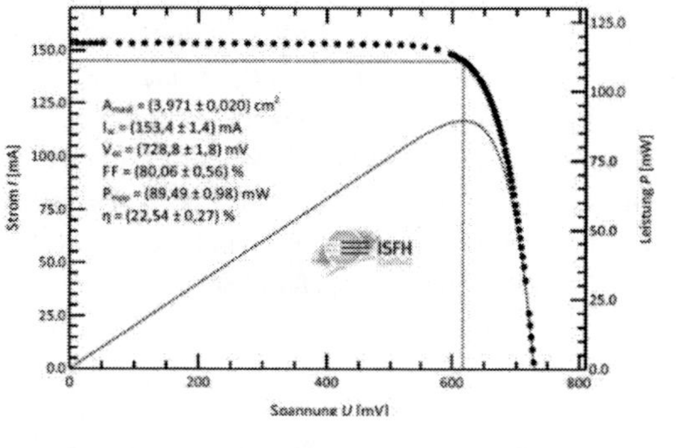

Fig. 6: Certification by ISFH CalTeC of the best non-oxidic FSF cell.

CONCLUSIONS

To summarize, in this work we studied effect of varying thickness and oxygen incorporation both in a-Si:H and nc-Si:H FSF layers in rear emitter silicon heterojunction (RE SHJ) cells. Introducing oxygen to the FSF layers we gain in j_{sc} due to the lower parasitic absorption and lower n values, but only in the nc-SiO_x:H FSF cells this gain in j_{sc} leads to a gain in η. Decreasing the thickness in the non-oxidic samples we further improve the j_{sc}, however we have to find a compromise in order to maintain the FF and V_{oc} to further transfer the gain into an improved efficiency. In the experiments with oxidic FSFs we obtained our best result using nc-SiO_x:H FSF layers: in 4 cm² aperture area cells with a front silver grid of 7% metallization area, a best cell with a conversion efficiency, η, of 21.4%, a V_{oc} of 728.3 mV and a FF of 79.8% could be demonstrated. Our best result on non-oxidic samples was obtained with a very thin nanocrystalline FSF layer (7 nm on textured wafers, as calculated from film thicknesses measured on polished c-Si references with a 1.7 factor to account for the larger effective surface area on textured wafers) without oxygen, using a busbarless screen printed silver front contact with a low metallization area (<4%). This improved FSF structure and front grid layout leads to a short circuit current density, j_{sc}, as high as 38.2 mA/cm², a fill factor, FF, of 80.5 %, an open circuit voltage of 727.8 mV and a conversion efficiency, η, of 22.4%.

ACKNOWLEDGMENTS

The authors would like to thank M. Hartig, K. Mayer-Stillrich, M. Zelt, H. Rhein, K. Jacob, A. Cruz Bournazou, G. Farias, D. Debrassine, K. Bhatti and M. Werth for assistance in processing and characterization of the devices analyzed here. Furthermore, we gratefully acknowledge the German Federal Ministry of Economic Affairs and Energy for financial support within the HERA project (reference number: 03258251). Parts of this work were supported by the European Union's Horizon 2020 Programme for research, technological development and demonstration under grant agreement no. 727523 (project NextBase).

REFERENCES

[1] J. I. Pankove and M. L. Tarng, "Amorphous silicon as a passivant for crystalline silicon," *Appl. Phys. Lett.*, vol. 34, no. 2, pp. 156–157, Jan. 1979.

[2] K. Matsuura, H.; Okuno, T. ; Okushi, H.; Tanaka, "Electrical properties of an-amorphous/p-crystalline silicon heterojunctions," *J. Appl. Phys.*, vol. 55, pp. 1012–1019, 1984.

[3] Z. C. Holman *et al.*, "Current Losses at the Front of Silicon Heterojunction Solar Cells," *IEEE J. Photovoltaics*, vol. 2, no. 1, 2012.

[4] P. Roca i Cabarrocas, N. Layadi, T. Heitz, B. Drévillon, and I. Solomon, "Substrate selectivity in the formation of microcrystalline silicon: Mechanisms and technological consequences," *Appl. Phys. Lett.*, vol. 66, no. 26, pp. 3609–3611, Jun. 1995.

[5] J. P. Seif *et al.*, "Strategies for Doped Nanocrystalline Silicon Integration in Silicon Heterojunction Solar Cells," *IEEE J. Photovoltaics*, vol. 6, no. 5, pp. 1132–1140, 2016.

[6] G. Nogay *et al.*, "Microcrystalline Silicon Carrier Collectors for Silicon Heterojunction Solar Cells and Impact on Low-Temperature Device Characteristics," *IEEE J. Photovoltaics*, vol. 6, no. 6, pp. 1654–1662, 2016.

[7] P. Cuony *et al.*, "Silicon Filaments in Silicon Oxide for Next-Generation Photovoltaics," *Adv. Mater.*, vol. 24, no. 9, pp. 1182–1186, Mar. 2012.

[8] L. Mazzarella *et al.*, "Nanocrystalline Silicon Oxide Emitters for Silicon Hetero Junction Solar Cells," *Energy Procedia*, vol. 77, pp. 304–310, 2015.

[9] L. Mazzarella *et al.*, "N-type ncanocrystalline silicon oxide front contacts for silicon heterojunction solar cells: photocurrent enhancement on planar and textured wafers," *submited*, 2017.

[10] O. Gabriel, S. Kirner, M. Klick, B. Stannowski, and R. Schlatmann, "Plasma monitoring and PECVD process control in thin film silicon-based solar cell manufacturing," *EPJ Photovoltaics*, vol. 5, p. 55202, Feb. 2014.

[11] S. Kirner, O. Gabriel, B. Stannowski, B. Rech, and R. Schlatmann, "The growth of microcrystalline silicon oxide thin films studied by in situ plasma diagnostics," *Appl. Phys. Lett.*, vol. 102, no. 5, p. 51906, Feb. 2013.

INFLUENCE OF THE TRANSPARENT ELECTRODE SPUTTERING PROCESS ON THE INTERFACE PASSIVATION QUALITY OF SILICON HETEROJUNCTION SOLAR CELLS

Leonard Tutsch, Martin Bivour, Winfried Wolke, Martin Hermle, Jochen Rentsch
Fraunhofer Institute for Solar Energy Systems ISE
Heidenhofstraße 2, 79110 Freiburg, Germany
e-mail: leonard.tutsch@ise.fraunhofer.de

ABSTRACT: The efficiency of silicon heterojunction solar cells is limited by the parasitic absorption within the intrinsic and doped hydrogenated amorphous silicon (a-Si:H) layers, hence minimizing their thickness by simultaneously preserving the high-level passivation quality is crucial. The a-Si:H must withstand subsequent processing steps, in particular the deposition of a transparent electrode. Here sputter deposition of transparent conducting oxides (TCOs) is commonly the method of choice, since this well established technique can yield uniform TCO layers with suitable electrical and optical properties. This paper aims at the design of a soft sputter deposition process, which allows utilizing well established DC magnetron sputtering on top of ultrathin a-Si:H layers, without a significant degradation of passivation quality. For this purpose, the influence of the a-Si:H thickness, the chamber pressure and the applied power are reviewed for In_2O_3:Sn (ITO) deposition. As a simple approach to combine excellent surface passivation, acceptable electrical TCO bulk properties and high throughput, an ITO double layer stack, featuring a few nanometers thin softly deposited intermediate ITO layer below a high rate deposited thicker ITO layer, is presented. Hereby the implied open-circuit voltage of test structures could be improved by almost 30 mV.

Keywords: TCO transparent conducting oxides, sputtering, heterojunction

1 INTRODUCTION

Silicon heterojunction (SHJ) solar cells receive more and more attention in recent years due to their high performance potential, owing to the effective absorber surface passivation capability of hydrogenated amorphous silicon (a-Si:H). The SHJ design, consisting of a monocrystalline silicon (c-Si) wafer (most commonly n-type) sandwiched between few nanometers thick intrinsic a-Si:H followed by a doped a-Si:H layer on each side, forming the emitter (p-type) and the back-surface field (n-type), enables record open-circuit voltages V_{oc} without the need for complex patterning steps [1]. Unfortunately, the intrinsic (*i*) and doped (*p,n*) a-Si:H layers absorb light parasitically, meaning that the charge carriers generated here contribute only marginally to the short-circuit current (J_{sc}) [2]. In [3] optimum *i* and *p*-layer thicknesses of 4 and 3 nm are reported, respectively, because no gain in the resulting V_{oc} and fill factor *FF* is achievable for thicker layers and the J_{sc} decreases gradually.
As the conductivity within the amorphous layers is low, transparent electrodes are deposited on top to facilitate lateral charge transport to the local metal contacts. For this purpose tin doped indium oxide In_2O_3:Sn (ITO) is a widespread transparent conducting oxide (TCO) and the highest quality films have routinely been produced via the magnetron sputtering technique [4]. Magnetron sputtering is also an appealing approach for the deposition of a high work function oxide of transition metals like Molybdenum or Tungsten [5]. A highly transparent film thereof can be used as a replacement for the p-type a-Si:H emitter layer and will be deposited directly on the a-Si:H(i) passivation layer. In general, during such sputtering processes the substrates are exposed to high-energy particles (sputtered atoms, reflected neutrals, ions and electrons with kinetic energies of up to the target voltage multiplied by the elementary charge) as well as to plasma radiation, which modify the nanometer thin layers and interfaces below and can lead to a strong increase in recombination and thus to a loss in V_{oc} and potentially in *FF*.
Monte Carlo simulations via the Stopping and Range of Ions in Matter (SRIM) software [6] (Fig. 1) estimated that oxygen ions O^-, which are expected to be the majority species of negatively charged ions incident on the substrate [7], can penetrate up to 7 nm deep into the a-Si:H film when accelerated by voltages of the order of the target – substrate voltage (e.g. 252 V for the 40 W process considered later). The penetration depth of argon atoms of the same energy, coming from the target after reflection and neutralization, is slightly smaller, due to their larger atomic diameter. It can be assumed that the sputtering processes can lead to significant damages to a depth of roughly 10 nm within the passivation layer and the a-Si:H / c-Si interface region.

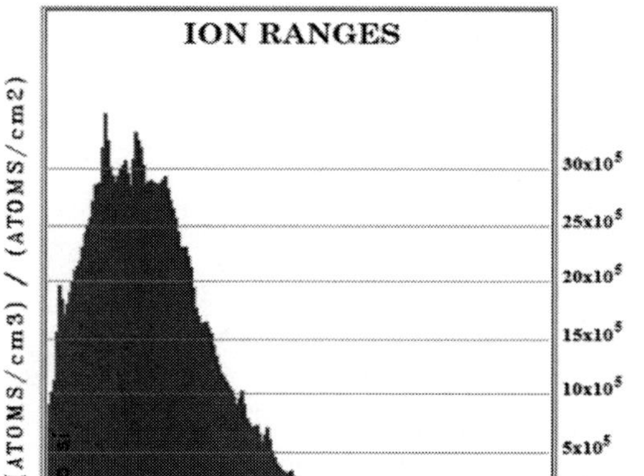

Figure 1: SRIM Monte Carlo simulation for the penetration depth of oxygen atoms with an initial kinetic energy of 252 eV in a-Si:H. The amount of oxygen atoms within a cubic centimeter of a-Si:H, normed by the number of impacting particles per area, is shown for a depth of up to 100 angstrom below the surface.

According to the simulations those damages do not only consist of broken atomic bonds, but also of atomic defects, which are hardly or not at all remediable by the low temperature annealing procedure typically applied for SHJ cells.
Varying the applied power and the chamber pressure enables to control the flux and energy of the involved particles, which significantly influences the growth and thus the properties of the resulting films, as well as the hereby caused passivation degradation. This paper aims at investigating the influence of these two externally controllable parameters on the sputter-induced a-Si:H / c-Si recombination losses.
Depositing the first nanometers of ITO in a relatively gentle and slow sputter process, followed by a thicker layer at a high growth rate, turned out to be a reasonable compromise between passivation degradation, electrical TCO bulk properties (sheet resistance, charge carrier mobility and density) and process duration.

2 EXPERIMENTAL

Planar n-type float-zone c-Si wafers with a thickness of 200 µm and a resistivity of 1 Ωcm were used to quantify the degradation of the passivation quality during ITO sputter deposition. The wafers were cleaned according to the RCA cleaning procedure. After the final HF dip, the a-Si:H(i) was deposited by means of plasma-enhanced chemical vapor deposition (PECVD). After activating the c-Si / a-Si:H(i) interface passivation in a 200 °C step for 10 minutes on a hotplate at ambient air, all samples had minority charge carrier (mc) lifetimes of a few milliseconds at an injection level of 10^{15} cm^{-3}. For faster processing, some of the rear sides of the samples were coated by well-passivating silicon nitride (SiN_x). Lifetimes were always determined via photo conductance measured by the Sinton WCT-120 tool operated in generalized mode [8].
For higher sensitivity and simplification of the process sequence, doped a-Si:H layers, which are usually applied to form junctions vital for charge carrier separation in solar cells, are omitted in these lifetime samples. Thus the about 75 nm thick ITO is expected to induce a band bending within the absorber and the resulting field effect is likely to influence the charge carrier recombination at the interface to some extent. Nevertheless, since trends in the chemical passivation capability as a function of the preceding sputter processes are observable and in order to be closer to the actual cell, ITO was not etched after its deposition.
The DC magnetron sputtering was performed by an Oxford Instruments cluster tool using a circular ITO target with a composition of In_2O_3/SnO_2: 90/10 weight percent, a diameter of 25.4 cm and a purity of 99.99 %. The gap between target and the sample is about 10 cm. Besides the argon and oxygen flows, the total pressure within the process chamber can be set by means of restricting the pump capacity by a butterfly valve. Hereby a relatively large pressure of up to 85 mTorr can be reached. The substrate holder is continuously heated to 100°C.
Providing good access to quantify the sputter-induced degradation of the interface passivation, the mc lifetime within the passivated c-Si wafer at a fixed minority charge carrier density (mcd = 10^{15} cm^{-3}, if not stated otherwise), was consistently measured before ITO deposition, afterwards and subsequent to an annealing step on a hotplate. The mc lifetime generally shows similar trends as the implied V_{oc} (iV_{oc}) at 1 sun illumination, which is a key performance indicator of a potential solar cell. Every data point represents the average of two samples taken from different wafers and the error bars indicate the standard deviation.

3 RESULTS AND DISCUSSION

3.1 Impact of the a-Si:H(i) layer thickness

In order to evaluate the passivation capability, the standard baseline ITO DC magnetron sputtering process (referred to as ITO std) was applied to double-side passivated planar c-Si wafers, whereby the a-Si:H(i) thickness on the side to be ITO coated was varied. The ITO std recipe, whose main parameters are a sputtering power of 200 W, 2 mTorr chamber pressure, a growth rate of 0.44 nm/s and argon and oxygen flows of 30 and 0.7 sccm respectively, is selected due to moderate damage for above 10 nm thick a-Si:H layer stacks and good electrical bulk properties of the resulting ITO film. Those are a charge carrier mobility μ of above 40 cm²/Vs at a charge carrier density of around 10^{20} cm^{-3}, leading to a sheet resistance of less than 300 Ω/□ for a 75 nm thick film.
Fig. 2 shows that the initial mc lifetimes and the degree of sputter-induced degradation are highly dependent on the a-Si:H(i) layer thickness. The level of surface passivation before sputtering saturates for about 7 nm thick films on a solid level of around 3 ms. While the sputter-induced deterioration of the passivation gradually decreases with a-Si:H(i) layer thickness, low temperature annealing allows to partly cure this damage for films > 7 nm and a similar level is obtained for 12, 18 and 23 nm.

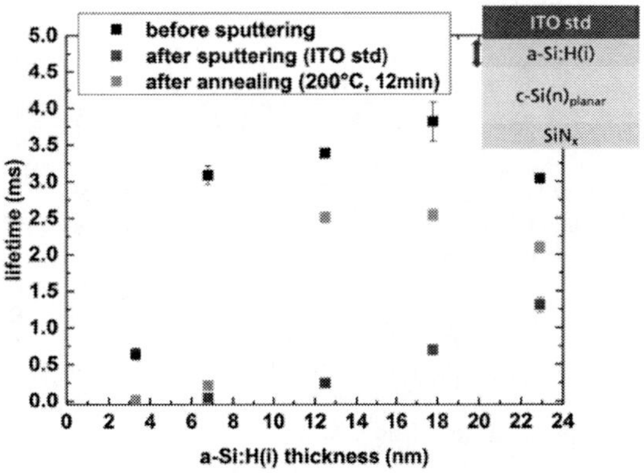

Figure 2: Minority charge carrier lifetimes for various a-Si:H(i) layer thicknesses before (black squares) and after (red squares) ITO deposition and a subsequent annealing (green squares).

In a-Si:H the penetration depth of energetic particles (here primary sputtered atoms, reflected neutrals and ions at energies of up to the target voltage multiplied by the elementary charge) can be estimated to be smaller than 10 nm (Fig. 1). Hence for the thicker a-Si:H(i) layers, only a negligible amount of harmful particles is expected to arrive at the vicinity of the a-Si / c-Si interface. Furthermore, the absorption length of ultra violet

radiation of 350 nm wavelength amounts about 10 nm in amorphous silicon [9]. This photon energy corresponds to the strength of the passivating Si-H bonds at the c-Si surface and is thus able to degrade the passivation quality.
However, to conclude, the ITO std sputtering process is unsuited to preserve the excellent surface passivation of 7 nm a-Si:H(i) films, which calls for an adjustment of the procedure.

3.2 Influence of the applied sputtering power

To isolate the impact of the applied power, ITO was deposited on 8 nm thick a-Si:H(i) layers under otherwise equal process conditions. The power was varied from 200 W (std) to 40 W and 500 W (Fig. 3), coming along with a change in the corresponding growth rates and cathode voltages, depicted in Tab. 1. Decreasing the power to 40 W softens the process clearly, visible in the lifetimes measured after sputtering and more pronounced in the values taken after the annealing steps. Annealing the samples at 210 °C showed a slight enhancement of the mc lifetime compared to the preceding 180 °C step whereas annealing at further elevated temperatures (not shown) was detrimental.
The difference in ITO deposition-induced damage between the 200 W and the 500 W processes is present, but far less significant.

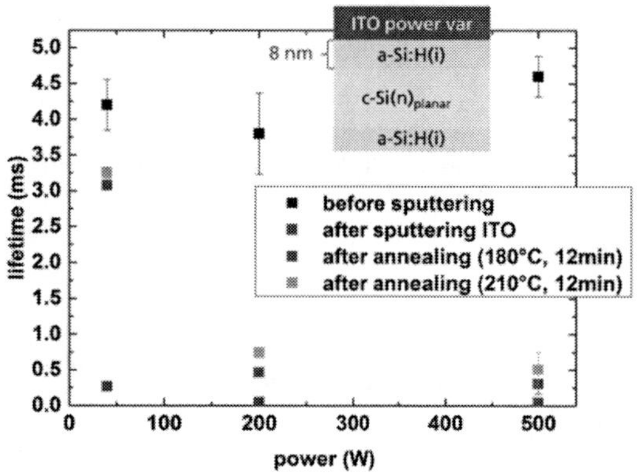

Figure 3: Lifetimes before and after sputtering at different power and after consecutive annealing steps at 180 and 210 °C.

During low power processes, a lower amount of harmful particles hit the substrate per time unit, indicated by the lower deposition rate (Tab. 1) and the sputter current (power divided by cathode voltage) and also the plasma radiation is less dense.

Table I: Deposition rate and DC target voltage for various power and pressure settings

Power [W]	Pressure [mTorr]	Rate [nm/s]	DC Voltage [V]
40	2	0.072	252
200	2	0.44	283
500	2	1.16	313
200	5	0.38	295
200	10	0.277	305
200	20	0.18	315
200	40	0.068	324
200	85	0.018	330

While this is likely compensated by the extended deposition duration to get the same thickness of about 75 nm, the applied power also determines the maximum kinetic energy of the sputtered atoms and reflected neutrals, since it correlates with the target voltage. This limitation of the impact energy and the consequential penetration depth could be a reason for the less severe passivation damage, but probably not the complete explanation, since the reduction in target voltage between the 200 W and the 40 W process amounts only 11 %. Here a further and more detailed investigation is necessary, e.g. by means of plasma analysis.
However, in terms of throughput, reducing the growth rate for the ITO single layer is not a suitable solution.

3.3 Influence of the total chamber pressure

In a next step the total chamber pressure was varied at a constant power of 200 W (Fig. 4). To improve comparability, the measured minority carrier lifetimes are normalized by the related pre-sputtering values, which were on the same level for all samples. A clear tendency of lower degradation for higher process pressures is observable, which is more pronounced after annealing. Although the DC voltage increases with pressure (Tab. 1), particles arrive at the substrate surface with quite low velocities, due to collisions within the denser gas. At the high pressure of 85 mTorr and for the geometry of our sputter system, the plasma between target and substrate is dense enough to thermalize almost all energetic particles coming from the cathode [10]. For these high pressure processes the remaining damage, likely primary induced by plasma radiation, could be cured virtually completely in a 200 °C step.
Unfortunately, films grown under these conditions show relatively poor electronic bulk properties as discussed in the next subsection. Since also process durations are increased by a multiple, depositing the whole ITO layer at a high pressure will neither lead to a better solar cell performance nor be an economically attractive process.

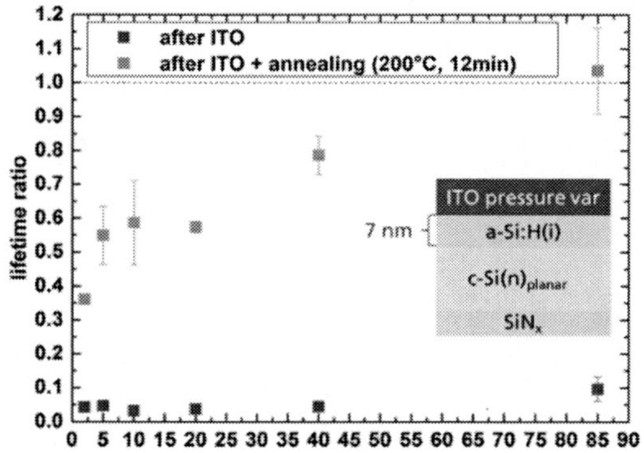

Figure 4: Minority carrier lifetimes after ITO sputter processes at various chamber pressures and subsequent to annealing. The values are normalized by the individual initial pre-sputter mc lifetime (at a level of 1.5 ms) to simplify comparison.

3.4 Inserting a softly deposited intermediate ITO layer

Taking advantage of the softer sputter process at either low power or high pressure, these layers were combined with the standard ITO procedure in a 75 nm thick layer stack, with the aim of enhancing electrical bulk properties of the film and to shorten deposition durations. Fig. 5 shows a comparison with regard to sputter-induced degradation of five recipes, namely (i) a single layer of the standard ITO (sl_std), (ii) a single layer with a reduced power of 40 W (sl_low_power), (iii) a single layer at an enhanced pressure of 85 mTorr (sl_high_pressure) and two layer stacks consisting of a 20 nm layer deposited at (iv) either 40 W (stack_power) or (v) 85 mTorr (stack_pressure) followed by 55 nm ITO std, summarized in Tab. 2.

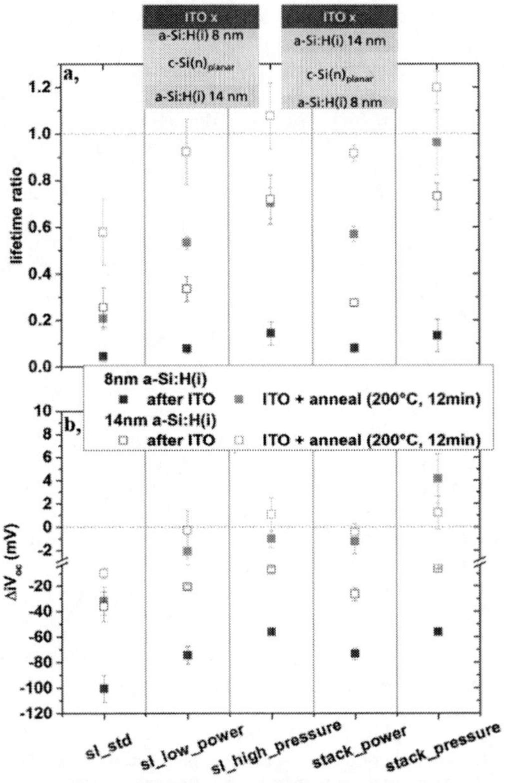

Figure 5: Normalized mc lifetimes (a) and change in iV_{oc} (b) during sputtering and annealing for five ITO deposition recipes (listed in Tab. 2) applied on a-Si:H(i) passivation layers of 8 nm (filled squares) and 14 nm (empty squares) thickness.

Table 2: ITO sputter recipes evaluated in Figs. 5 and 6

Name	Power [W]	Pressure [mTorr]	Duration [min:sec]
sl_std	200	2	03:10
sl_low_power	40	2	16:12
sl_high_pressure	200	85	64:49
stack_power	40	2	04:38
	+ 200	2	02:16
stack_pressure	200	85	18:31
	+ 200	2	02:16

The processes were applied for a single-sided ITO coating of lifetime samples with asymmetrically thick (8 and 14 nm) a-Si:H(i) layers on both sides. Before sputtering, this resulted in initial mc lifetimes of around 2 ms and an iV_{oc} level of 726 ± 3 mV. By this means the sputter impact can be evaluated for two passivation layer thicknesses by depositing on top of one of the two opposing sample sides. The difference in iV_{oc} before and after sputtering ΔiV_{oc} (Fig. 5b) shows the same trends as the post-sputter mc lifetime normalized by the initial values (Fig. 5a). Even after annealing, the reference ITO procedure (sl_std) causes a loss of 30 mV for a coating of the 8 nm thick a-Si:H(i) layer. For the single ITO layers deposited under soft process conditions as well as for the ITO double layer stacks, the degradation is less than 2 mV after annealing.

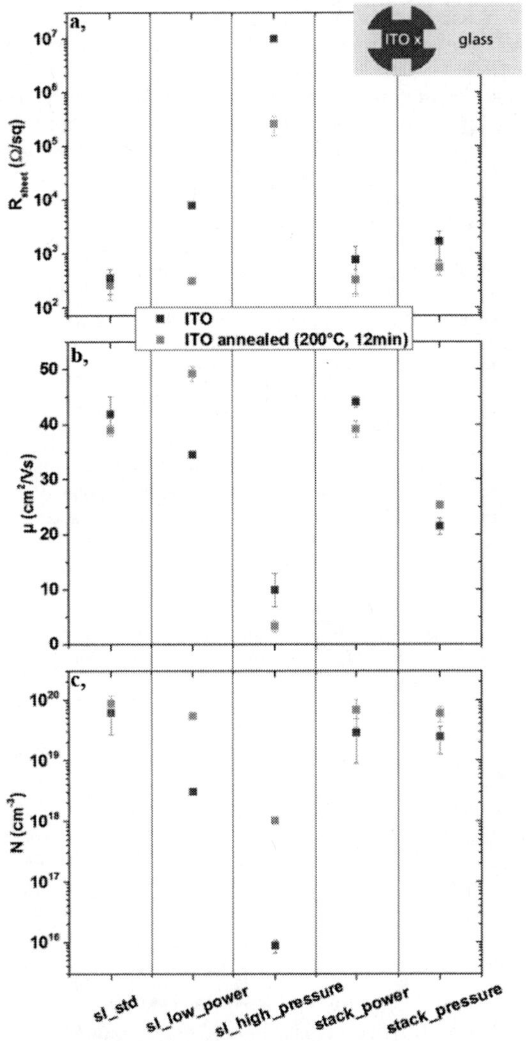

Figure 6: Electrical film properties sheet resistance (a), charge carrier mobility (b) and concentration (c) of 75 nm thick ITO deposited on glass according to the recipes in Tab. 2 and subsequent to an annealing step at 200 °C, measured via the van der Pauw method.

A similar effect is observable for the 14 nm thick a-Si:H(i) layer, whereby this layer is less sensitive to sputter damage, visible in the smaller absolute value of ΔiV_{oc} for every process.

The main message from this comparison is that it suffices to sputter a nanometer thin protective layer gently before the residual ITO follows by harsh process conditions optimized for bulk properties and a high growth rate.

In Fig. 6 the corresponding electrical bulk properties of the ITO films deposited on glass are shown. The sheet resistance and the mobility and density of the charge carriers are determined via the van der Pauw method using the Hall Effect. Although the films grown on glass substrates do not utterly resemble the ones within the SHJ cell, it can be concluded that the high pressure deposited ITO cannot match the good electrical properties provided by the standard process. The values for sheet resistance and charge carrier mobility of the annealed high / low pressure layer stack (stack_pressure) ($R_{sheet} = 550\ \Omega/\square$, $\mu = 25\ cm^2/Vs$) are in between the high pressure single layer (sl_high_pressure) ($R_{sheet} = 2.5x10^5\ \Omega/\square$, $\mu = 3\ cm^2/Vs$) and the standard ITO film (sl_std), but slightly worse than one might expect by treating the layer stack as a parallel circuit of the two ITO films. This suggests that the adverse growth of the 20 nm thick low-rate deposited protective layer continues for a while in the subsequent standard processed film.
For the film sputtered at low power and also for the related layer stack, the electrical properties are at a similar level as for the ITO std process. Remarkably, the charge carrier mobility within the annealed film deposited at 40 W reaches the value of 50 cm^2/Vs.
However, the prolonged process durations of the gently deposited films and the poor electrical properties of the high pressure layers motivate to reduce their thickness to a minimum necessary to function as an effective protection layer. For this purpose the sputter-induced degradation is quantified in Fig. 7 as a function of the thickness of this intermediate layer deposited at 85 mTorr for 7 nm thick a-Si:H(i) films with initial mc lifetimes of around 1.5 ms.

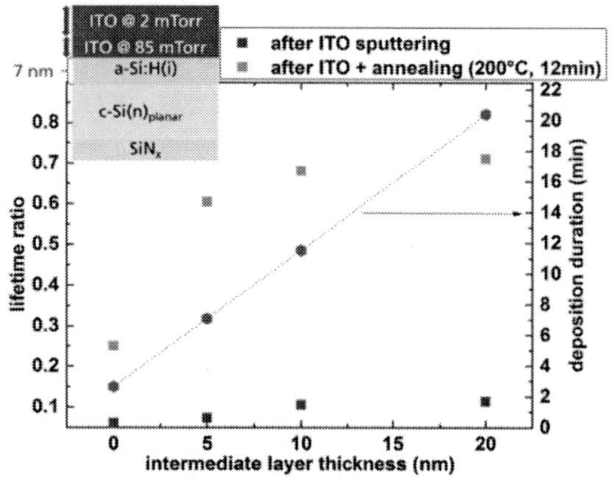

Figure 7: Influence of the thickness of the gently deposited protective ITO layer on the sputter-induced degradation and the process duration of the entire ITO film. The soft layer was deposited at 85 mTorr and 200 W, before the remaining ITO to build a 75 nm thick film followed under standard conditions (2 mTorr, 200 W).

Already a 5 nm thick ITO shielding is sufficient to increase the normalized mc lifetimes after sputtering and annealing at 200 °C by a factor of 2.4, compared to the deposition without intermediate layer. Thicker protective layers (10 nm and 20 nm) prevent sputter-induced degradation slightly further. In contrast to the previous experiment (Fig. 5), this time annealing could not boost the mc lifetime to the full initial value for the 20 nm protective ITO layer, which can be due to a slight variation in the process chain or in the annealing procedure. Nevertheless, when a gently deposited layer is present, the sputter-induced loss in iV_{oc} was merely between 2 and 4 mV (at an initial level of 715 ± 2 mV). Hence, surface passivation can be much better preserved during the sputter deposition, which here resulted in an iV_{oc} improvement of 26 mV in comparison to the standard ITO single layer.

4 CONCLUSION

As was estimated by SRIM simulations, the highest-energy atoms present during ITO sputter coating are able to modify the underlying a-Si:H film and the interface region to the c-Si absorber up to a depth of 10 nm below the substrate surface. Since the optimum thickness of the intrinsic / doped charge carrier selective a-Si:H layer stack is of the same order, the excellent interface passivation quality is highly sensitive to the sputtering process conditions, in particular the applied power and the chamber pressure. Both a decrease in power down to 40 W and an enhanced pressure of 85 mTorr, compared to the (200 W, 2 mTorr) standard process, showed a significant decrease in sputter-induced damage and the resulting degradation in iV_{oc} was almost completely remediable during low temperature curing. While the low power process also yielded an ITO film with good electrical properties, the high pressure turned out to be detrimental for the layer growth, resulting in a high resistivity. It has to be mentioned that these processes were not optimized yet and reducing the oxygen flow during the high pressure ITO deposition will likely enhance the charge carrier concentration within the TCO layer and thereby lower its sheet resistance.
Applying these gently deposited layers as a shielding in a stack previous to the standard ITO process is an effective way to alleviate sputter-induced degradation by a multiple. At least for the low / high power ITO stack, a good compromise between preserving the absorber passivation, good electrical bulk properties of the TCO film and a reasonable process duration was achieved.

5 REFERENCES

[1] S. De Wolf et al., *Green*, **2** (2012) 7.
[2] Z. Holman et al., *IEEE J. Photovolt.*, **2** (2012) 7.
[3] H. Fujiwara et al., *J. Appl. Phys.*, **101** (2007) 054516.
[4] D. Ginley, *Handbook of Transparent Conductors*, Springer (2010).
[5] M. Bivour et al., *Sol. Energ. Mat. Sol. Cells.*, **14** (2015) 34.
[6] J. Ziegler et al., *Nucl. Instr. Meth. Phys. Res. B*, **268** (2010) 1818.
[7] T. Welzel et al., *ViP*, **25** (2013) 52.
[8] H. Nagel et al., *J. Appl. Phys.*, **86** (1999) 6218.
[9] F. Lebreton et al., *Appl. Phys. Lett.*, **108** (2016) 051603.
[10] R. Somekh, *J. Vac. Sci. Technol. A*, **2** (1984) 1285.

"Partly funded by the European funded project DISC under the European Union's Horizon 2020 research and innovation program under grant agreement N°727529"

INLINE PECVD DEPOSITION OF A-SI LAYERS FOR HETEROJUNCTION SOLAR CELLS ON AN INDUSTRIAL SCALE

J. Temmler, A. Moldovan, D. Putra, M. Bivour, J. Rentsch
Fraunhofer Institute for Solar Energy Systems ISE
Heidenhofstrasse 2, 79110 Freiburg, Germany
Phone/Fax: +49 761 4588 2137 / +49 761 4588 9250
E-mail: jan.temmler@ise.fraunhofer.de

ABSTRACT: In this work the recent results concerning inline deposition of amorphous silicon obtained with a PECVD tool from Meyer Burger Germany (MBG) are presented. The depositions are done with an adapted MAiA tool which for this matter contains radio frequency linear plasma sources for the intrinsic as well as the doped amorphous silicon layers. In order to investigate the properties of these layers/layer stacks, especially concerning passivation quality and selectivity, we prepared various sample structures to determine minority carrier lifetime/implied open-circuit voltage, band bending and electrical properties. Minority carrier lifetimes of up to 7 ms and average implied open-circuit voltages of above 730 mV are verified for intrinsic amorphous silicon. We could proof a sufficient band bending for both n- and p-doped and demonstrate an n-doped inline process which leads to external open-circuit voltages above 700 mV and pseudo fill factor values above 80 %. It is also shown for the layer stack consisting of intrinsic and n-doped amorphous silicon that we can reach a sufficient net doping to obtain required selectivity of the contact for the extraction of excess charge carriers.
Keywords: silicon heterojunction, amorphous silicon, inline PECVD, passivation, selective contacts

1 INTRODUCTION/MOTIVATION

Silicon Heterojunction (SHJ) solar cells represent a key technology to approach very high conversion efficiencies close to the theoretical limit of silicon solar cells. The main advantages of SHJ solar cells are a lean production chain with low temperature processes and carrier selective hetero-structures for excellent passivation. Both–side contacted SHJ solar cells featuring amorphous thin-film layer stacks enable high conversion efficiencies up to 25.1 % [1]. To achieve such high efficiencies different features as the front grid electrode, the transparent conductive oxide (TCO), the quality of the hydrogenated amorphous silicon (a-Si:H) layers and the c-Si surface structure must be carefully optimized. Aiming to reduce the cost of owner-ship (COO) an increased throughput is required which can be realized within an inline PECVD deposition process. Hydrogenated amorphous silicon (a-Si:H) which is used as passivation and selective layer for silicon heterojunction (SHJ) solar cells has to fulfill two important basic requirements:

- excellent passivation of the c-Si absorber
- high selectivity with respect to the charge carriers to obtain the best possible electrical contact.

In this work results concerning passivation quality and selectivity regarding photo-generated excess charge carriers of the layer stack consisting of intrinsic and doped a-Si:H obtained with an adapted MAiA inline PECVD tool from Meyer Burger Germany (MBG) are presented.

2 PASSIVATION QUALITY

First step is the development/optimization of an intrinsic a-Si:H layer for passivation. For this purpose first experiments were carried out to screen the whole deposition parameter space (process pressure, process temperature, plasma power and gas flows (SiH_4, H_2). Having determined the temperature as a key parameter for excellent passivation, we conducted an experiment with a variation of only the process temperature aiming to determine the optimal process range (chapter 2.1). Next we examined the homogeneity over one wafer, over the whole deposition tray and also over time (chapter 2.2).

To obtain the passivation properties of the a-Si:H(i) layers lifetime samples (textured n-type Cz c-Si, 180 µm, 3.5 Ω*cm with 12 nm a-Si:H(i)) on both sides) were prepared and analyzed by Quasi Steady-State Photoconductance measurements (QSSPC) in order to determine the minority carrier lifetime (τ_{eff}) and implied V_{OC} (iV_{OC}) [2]. We reached very good passivation quality with iVoc > 730 mV (see Fig. 4).

2.1 Process Temperature

Fig. 1 and 2 depict the lifetime τ_{eff} and iV_{OC} values achieved for different process temperatures of the inline PECVD process and after annealing for 15 min at 180°C. The process temperature was varied in the range between 200°C and 300°C.

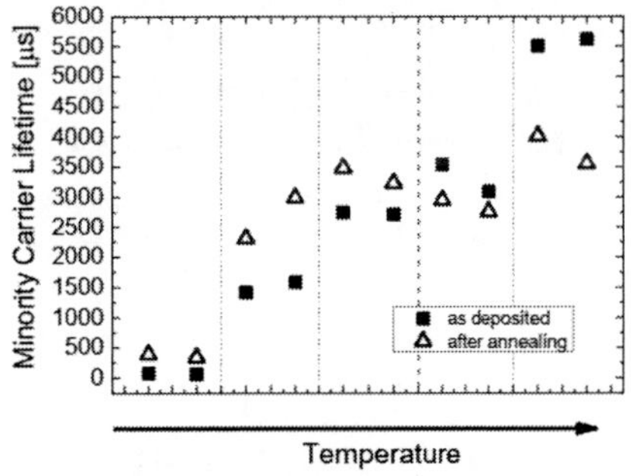

Figure 1: Minority carrier lifetime τ_{eff} in µs as a function of the process temperature; as deposited and after annealing (15min, 180°C).

One can observe an overall rise of both parameters τ_{eff} /iV_{OC} with increasing process temperature. For the lower process temperature range there is a rise of τ_{eff} /iV_{OC} after annealing, whereas to the higher range there is a decrease of τ_{eff}/iV_{OC} after annealing. We assume that for higher temperatures an in-situ annealing and also partly epitaxial growth of the layers might occur, which lowers the lifetime after annealing and hence does not lead to improvement by annealing.

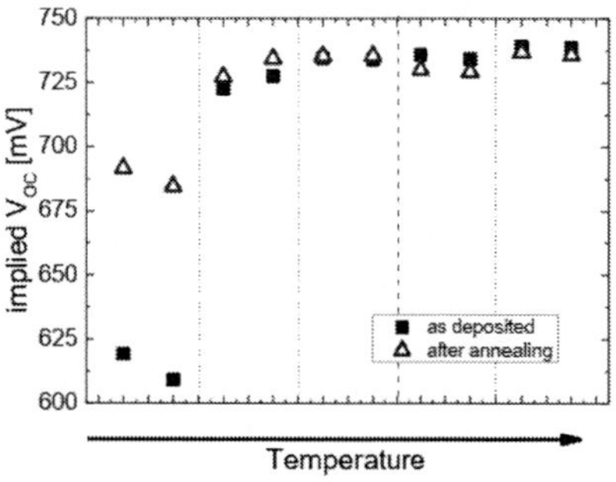

Figure 2: Implied V_{OC} in mV as a function of the process temperature; as deposited and after annealing (15min, 180°C).

2.2 Process Homogeneity

In the next step the homogeneity of the passivation concerning wafer area, position on deposition tray (consisting of 5 columns and 4 rows) and also over time was analyzed. Fig 3. shows a photoluminescence (PL) image of one wafer deposited with 12nm a-Si:H(i) on both sides measured immediately after deposition. The defects at the edge or in the middle of the wafer can be traced back to tweezers and wafer markings.

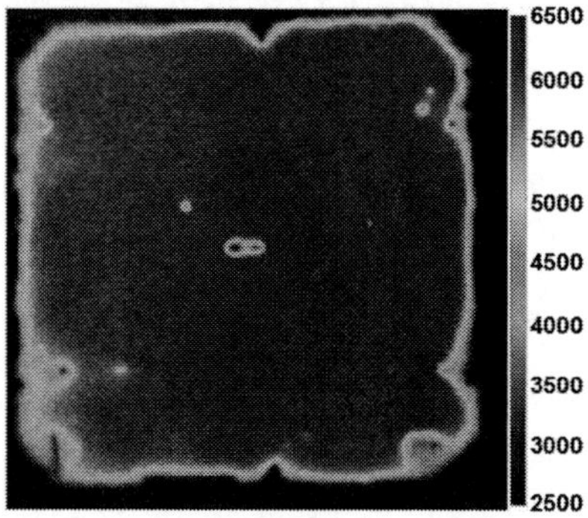

Figure 3: PL image of the minority carrier lifetime τ_{eff} in µs after deposition with 12nm a-Si:H(i) layer on both sides; scale from 2500 to 6500µs.

Fig. 4 depicts the result for the iV_{OC} as a function of the wafer position on the deposition tray. It can be seen that with an optimized process iV_{OC} values over 730 mV were obtained for all positions whereby a slight degradation in transport direction (from top to bottom) can be observed. In Fig. 5 the lifetime τ_{eff} is shown as a function of time. One can see a fluctuation of the τ_{eff}, but an overall level of above 3 ms (corresponding to an iV_{OC} >730 mV). The a-Si:H(i) process supplies excellent stability concerning passivation even after chamber cleaning by an etching process and although to the fact that other processes carried out within the same deposition chamber.

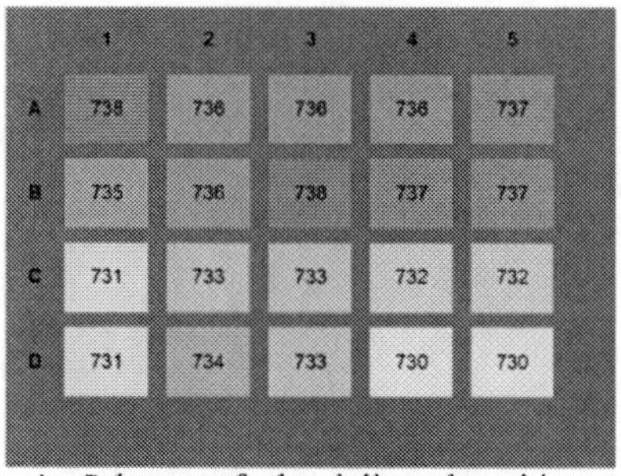

Figure 4: Scheme of the inline deposition tray (5 columns, 4 rows) with iV_{OC} values in mV measured by QSSPC as a function of the tray position.

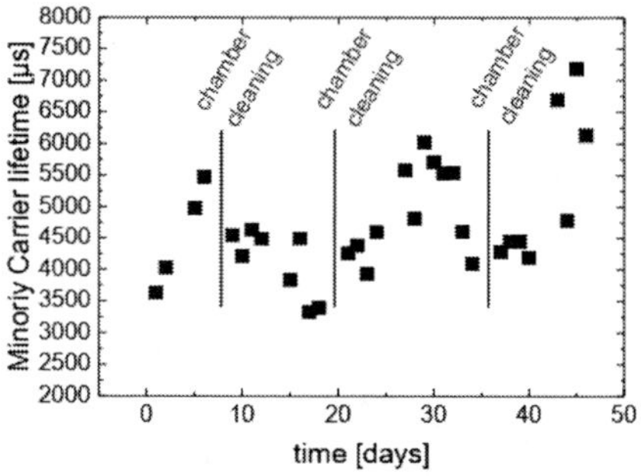

Figure 5: Minority carrier lifetime τ_{eff} in µs as a function of the time to outline the reproducibility and/or stability of the a-Si:H(i) process; black lines show chamber cleaning by an etching process.

3 ELECTRICAL PROPERTIES (SELECTIVITY)

The doped a-Si:H layers have to provide a sufficient net doping to induce a strong band bending and therefore a high built-in potential within the absorber. This causes the formation of the high/low junction and the selective charge carrier extraction from the absorber to the external contacts. For investigation of the electrical properties two different sample structures were prepared (Fig. 6)

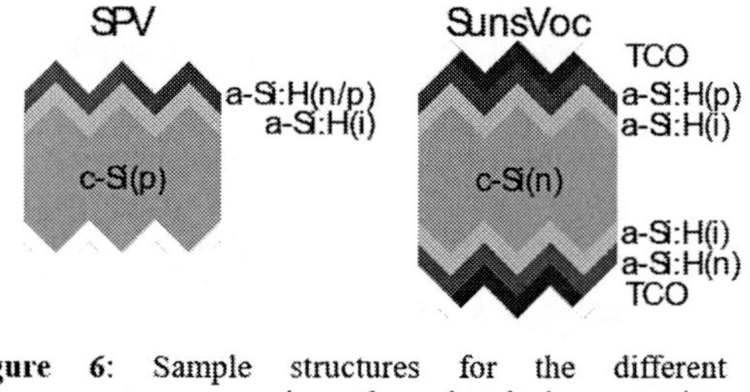

Figure 6: Sample structures for the different measurements concerning the electrical properties; samples with only a-Si:H(i) and the layer under investigation (either a-Si:H(n) or a-Si:H(p), left) for SPV measurements and samples with layer stacks of intrinsic and n-doped layer and ITO on top for SunsVoc measurements (right) to investigate the a-Si(i/n) layer stack.

For the investigations concerning band bending of a-Si:H(n) we used textured FZ c-Si, p-type, 250 µm,

0.5 Ω*cm whereas for the band bending of a-Si:H(n) and for the SunsVoc samples textured FZ c-Si, n-type, 200 µm, 1.0 Ω*cm were utilized. For both experiments the buffer layer (intrinsic a-Si:H) was 12 nm on textured surfaces and was deposited with the MAiA tool. The doped a-Si:H layers have a thickness of 10 nm and the ITO layers were 70 nm thick. Doping is realized by adding B_2H_6 (a-Si:H(p)) respectively PH_3 (a-Si:H(n)) to the deposition process.

3.1 Band Bending (SPV)

Fig. 7 shows the band bending $V_{bb,dark}$ which is induced by the doped a-Si:H layer into the c-Si absorber. It can be seen that in the case of n-doped a-Si:H layers a saturation of the band bending for a gas phase doping above 1000 ppm is reached whereas for the p-doped a-Si:H is a slight rise (below 2000 ppm) to be observed. Above that value no clear trend for both doping types reveals.

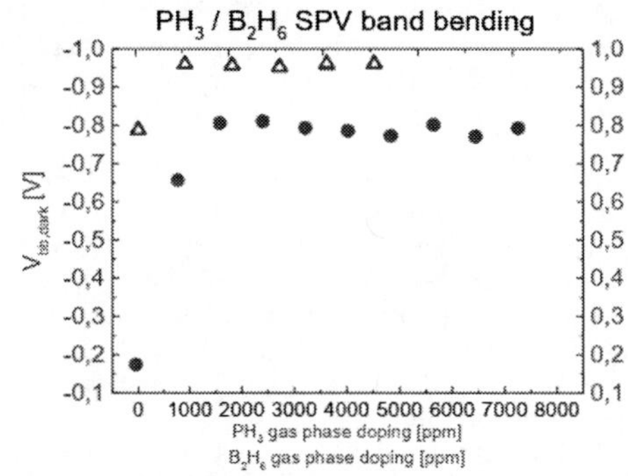

Figure 7: Band Bending $V_{bb,dark}$ in V as a function of the gas phase doping in ppm for either p-doped a-Si:H (B_2H_6) or n-doped a-Si:H (PH_3) layers with 12nm intrinsic buffer layer between c-Si and doped a-Si:H.

3.2 Electrical parameters (SunsVoc)

SunsVoc samples are a simple and quick way to further analyze and evaluate various electrical properties (e.g. of the doped a-Si:H layers) on preliminary test structures without having to prepare complete solar cells. The first experiment deals with the investigation of the n-doped a-Si:H (PH_3) layer more specifically the layer stack consisting of intrinsic and n-doped a-Si:H. The layer stack and also the intrinsic buffer layer on the other side of the wafer were deposited with the MAiA tool. All other layers (p-doped a-Si:H, ITO) come from a reference PECVD tool.

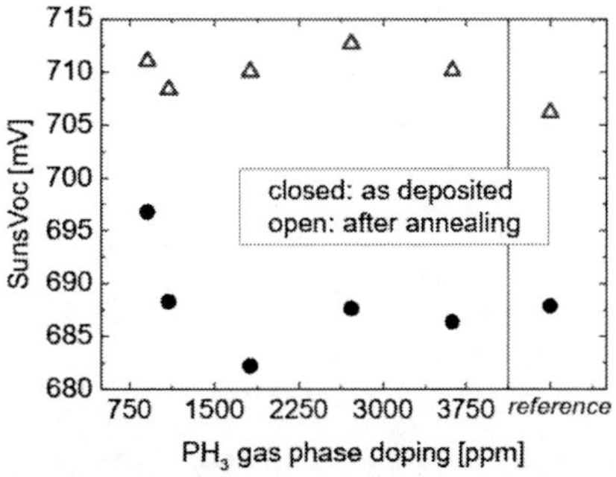

Figure 8: External V_{OC} in mV as a function of the PH_3 gas phase doping; values as deposited (closed circles) and after annealing (open triangles).

Fig. 8 depicts the values of V_{OC} as a function of the PH_3 doping gas flow. After annealing values above 700 mV were reached. Just like for the SPV investigation, there is no clear trend. The slight increase for the regime of low gas phase doping (starting from 1000 ppm to 2700 ppm) could be due to the fact that more doping material leads to an improved extraction of the photo-generated excess charge carriers whereas the decrease with higher gas phase doping (above 2700 ppm) could be explained by incorporation of more doping material within the doped layers which creates more defects and therefore increased recombination.

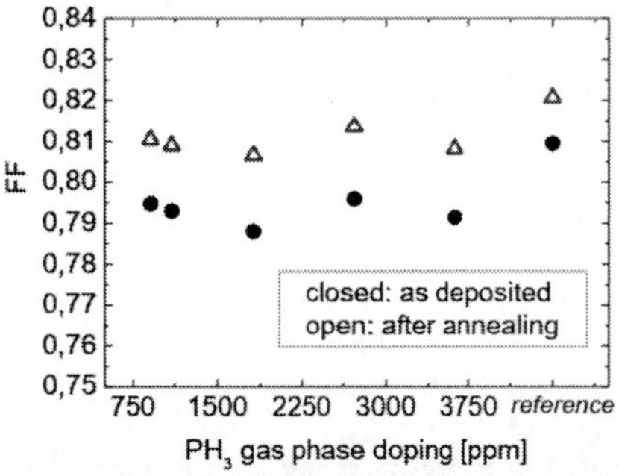

Figure 9: Pseudo fill factor as a function of the PH_3 gas phase doping; values as deposited (closed circles) and after annealing (open triangles).

In Fig. 9 one can see the Pseudo fill factor (pFF) as a function of the PH_3 gas flow. For all variations we get values around 0.8. There is a slight increase of pFF with PH_3 which could be due to improved selectivity of the doped layer and therefore more efficient extraction of photo-generated excess charge carrier. For more information on the net doping SHJ test structures are analyzed by means of Suns-V_{OC} measurements for high illumination intensities [6] (Fig. 10,11).

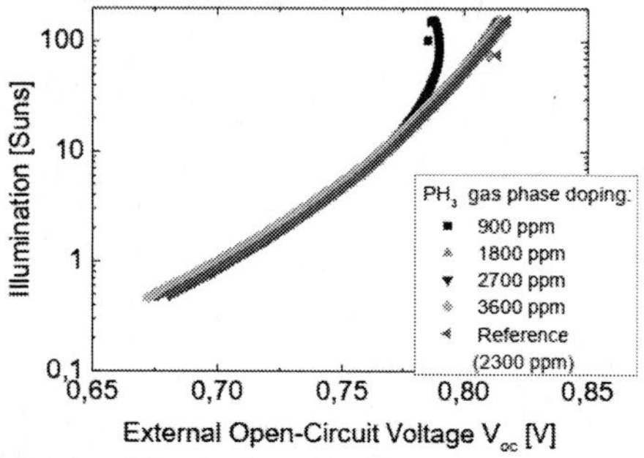

Figure 10: Illumination (Suns) as a function of the external V_{OC} for the n-doped a-Si:H (PH_3) layers. Insufficient selectivity is characterized by a kinking of the curve towards lower V_{OC} values.

Under these conditions the contact is "pushed" towards flat band and thus high-injection conditions with increasing illumination intensity which is reflected in a characteristic "back bending" as shown in Fig. 10 for the lowest doping (black curve). To get a qualitative interpretation the slope of this external V_{OC} at 100 suns (ideality factor $n_{100suns}$) is analyzed. Fig. 11 shows this slope as a function of the PH_3 gas phase doping. Towards higher doping $n_{100suns}$ increases and especially for the highest doping approaches the value of the reference. That indicates a sufficient selectivity and therewith associated the absence of a pronounced ITO / a-Si:H(n) Schottky contact.

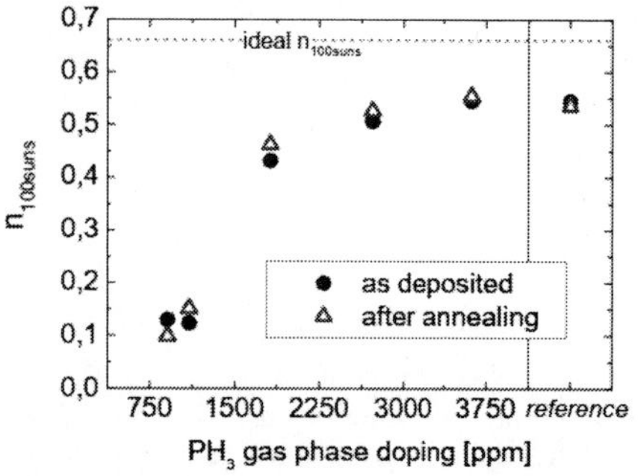

Figure 11: Ideality factor $n_{100suns}$ as a function of the PH_3 gas phase doping; as deposited (closed circles) and after annealing (open triangles).

4 SUMMARY/CONCLUSION

Within development/optimization of inline deposition of a-Si:H(i) as passivation layer for SHJ the temperature turned out to be a key parameter for excellent passivation quality. We could show lifetimes τ_{eff} of up to 7 ms and on average iV_{OC} values above 730 mV. An excellent homogeneity over the wafer area and a very good homogeneity over the whole deposition tray and for the process over time was verified (especially also after chamber cleaning by a plasma etching).
Concerning inline deposition of doped a-Si:H SPV measurements revealed a sufficient band bending for both n- and p-doped a-Si:H. Further experiments regarding a-Si:H(n) led to V_{OC} values above 700 mV after annealing and pFF above 80 %. $SunsV_{OC}$ measurements with high illumination intensities showed the necessity of the highest gas phase doping to reach a sufficient net doping to obtain the required selectivity respectively absence of a pronounced ITO / a-Si:H(n) Schottky contact.

ACKNOWLEDGEMENTS

The authors would like to thank all colleagues at the Fraunhofer ISE for their support without which the experiment wouldn't have been possible. This work was funded by the German Federal Ministry for Economic Affairs and Energy within the research project "HERA" (contract no. 0325825B).

REFERENCES

[1] D. Adachi, J. L. Hernández, and K. Yamamoto, Appl. Phys. Lett, vol. 107, no. 23, p. 233506, 2015.

[2] R. A. Sinton and A. Cuevas, "Contactless determination of current–voltage characteristics and minority-carrier lifetimes in semiconductors from quasi-steady-state photoconductance data", Appl. Phys. Lett., 69(17), 2510–2512 (1996).

[3] D. Decker, H. Schlemm, J. Mai, P. Wolf, H. Mehlich; 27th European Photovoltaic Solar Energy Conference and Exhibition; p. 1624 - 1629; 2012.

[4] D. Decker, P. Wolf, S. Pieper, H.-P. Sperlich, H. Schlemm, J. Mai; 28th European Photovoltaic Solar Energy Conference and Exhibition, p. 1919 - 1927, 2013.

[5] A. Moldovan et al.; Energy Procedia 2016, accepted for publication

[6] M. Bivour, M. Reusch, S. Schroer, F. Feldmann, J. Temmler, H. Steinkemper, and M. Hermle, IEEE Journal of Photovoltaics (2014).

LOW-COST FABRICATION OF PATTERNED ELECTRODES IN HETERO-JUNCTION BACK-CONTACT SILICON SOLAR CELLS BY PLASMA ION IMPLANTATION

Koichi Koyama, Keisuke Ohdaira, Hideki Matsumura
Japan Advanced Institute of Science and Technology (JAIST)
1-1 Asahidai, Nomi, Ishikawa 923-1292, Japan

ABSTRACT: An interdigitated back contacted silicon heterojunction (IBC-SHJ) solar cell is one of the most promising structures for high-efficiency solar cells. However, the patterning of back-contact electrodes causes the increase in process steps and cost. Thus, to simplify the fabrication of IBC-SHJ solar cells, we attempted to form interdigitated p-a-Si and n-a-Si regions by plasma ion implantation. This method aims at simplification by converting p-a-Si to n-type. We previously reported that the conduction type of a-Si can be converted by the plasma ion implantation of phosphorus (P) atoms into p-a-Si / i-a-Si / c-Si regions, and also that the quality of passivation can be kept sufficiently high, the same as that before ion implantation, when the samples are annealed at around 250 °C and also when the energy and dose of ion implantation are appropriately chosen for fitting to a-Si layer thickness and bulk c-Si carrier density. This paper is to present the first successful results of conventional SHJ solar cells in which all n-a-Si layers are formed by conversion from p-a-Si by plasma ion-implantation of P atoms.
Keywords: Heterojunction, Passivation, Contact, Ion implantation, Cat-CVD

1 INTRODUCTION

The world top three efficiencies of crystalline-silicon (c-Si) solar cells, 26.6%, 26.3% and 25.6%, are all achieved by IBC-SHJ structure [1–3]. Although superiority of the structure is apparent, the cost for fabrication of IBC-SHJ is not so low, since the back electrodes are usually formed by complicated patterning process such as photolithography. Thus, it is strongly desired to develop a new low-cost technology fabricating the patterned back-electrodes. For achieving the purpose, the patterning of electrodes using ion implantation with a hard-mask process [4] appears one of the most expectable ways. Among various methods of ion implantation, plasma ion implantation appears the most attractive method owing to the low system cost and short process time for high dose ion implantation. For instance, the ion implantation of P atoms with a dose of 10^{16} cm^{-2} can be completed within a few second [4]. We have previously reported that phosphorus (P) plasma ion implantation can be achieved with no deterioration of a-Si/c-Si interface quality by annealing at only 250 °C [5]. Thus, here, we attempted to apply plasma ion implantation to make patterned p-a-Si/i-a-Si/c-Si and n-a-Si/i-a-Si/c-Si electrodes by converting the conduction type of some patterned areas of p-a-Si/i-a-Si/c-Si to n-a-Si/(i-a-Si)/c-Si. We also measured the contact resistance of patterned electrodes made by conversion from p-a-Si to n-a-Si, and found that it was almost equivalent to that of initially formed n-a-Si electrodes. In addition, we revealed that the performance of SHJ solar cells using converted n-a-Si from p-a-Si was comparable with that of the conventional SHJ solar cells using deposited n-a-Si.

2 EXPERIMENTAL METHODS

The passivation quality of the a-Si/c-Si heterojunction was evaluated from the carrier lifetime (τ_{eff}) measured by the microwave photoconductive decay method (μ-PCD) using a KOBELCO LTA-1510EP apparatus [6–8]. The n-type mirror-polished 300-μm-thick (100) Float-zone (FZ) c-Si (resistivity of 1–5 Ω cm) was used as substrates. Photo-carriers were excited by the exposure of laser light with a wavelength of 904 nm, and τ_{eff} was measured at a photon density of 5 × 10^{13} cm^{-2}.

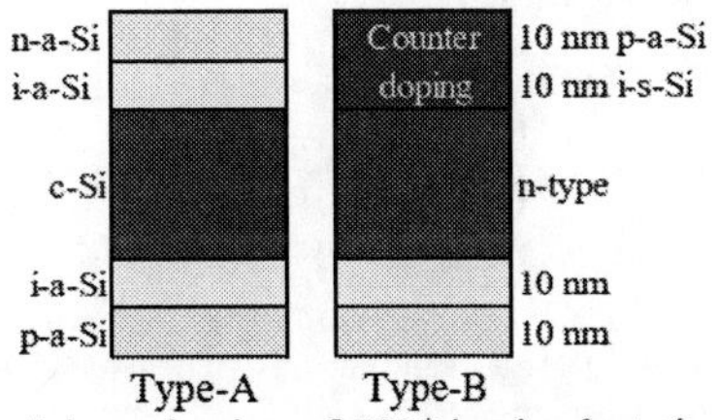

Figure 1: Schematic view of PH_3^+ ion implantation into the p-a-Si/i-a-Si sample (Type-B).

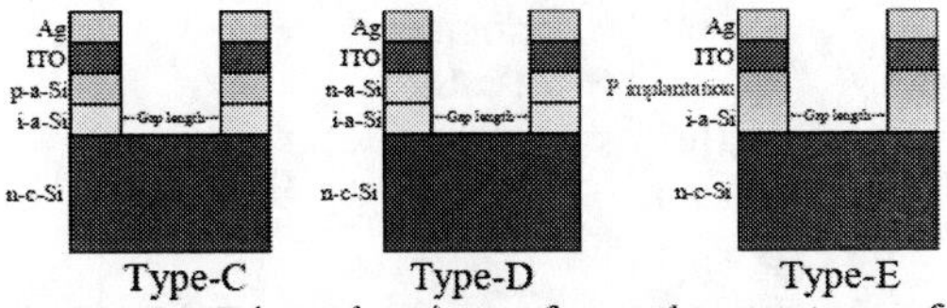

Figure 2: Schematic view of sample structure of investigation for contact resistance.

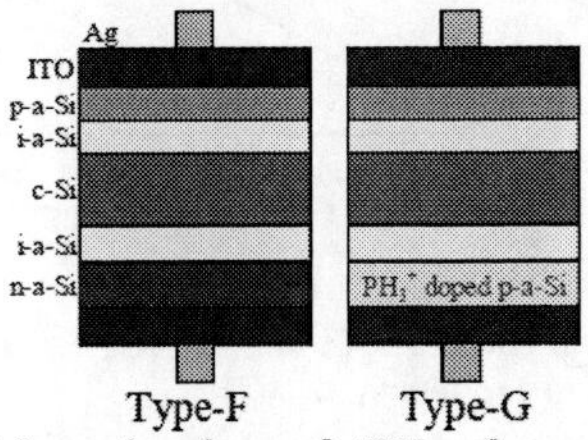

Figure 3: Schematic view of SHJ solar cells using counter doping (Type-G).

The c-Si wafers were cleaned using 5% diluted hydro fluoric acid solution to remove native oxide. Fig. 1 shows the structure of ion implanted samples to evaluate SHJ interface quality. p-a-Si/i-a-Si, or n-a-Si/i-a-Si layers were deposited on the c-Si substrates by catalytic chemical vapor deposition (Cat-CVD or hot-wire CVD). P atoms were implanted onto them with 10^{16} cm^{-2} doses and 5 keV ion energy. Ion implantation was carried out with ULVAC PVI-3000 equipped without mass separator.

The contact resistance of ion implanted a-Si/indium tin oxide (ITO)/Ag was measured by forming coplanar electrodes, as schematically shown in Fig. 2. ITO layers were inserted between a-Si and metal to prevent their inter-mixing. All the samples were also patterned to make coplanar electrodes. The gap of two coplanar electrodes was varied to evaluate the contact resistance from the relationship between measured resistance and the electrodes-distance. The ITO layers were formed by the sputtering method. The samples were annealed at 200 °C for 30 minutes after forming electrode structures. The similar process was performed for p-a-Si/i-a-Si/c-Si without implanted samples for comparison. Type-F and Type-G of Fig. 3 shows the schematics of SHJ structures. The area of the cells was ~2 cm^2. Ag finger grids were formed by screen printing and successive annealing at 200 °C for 30 min.

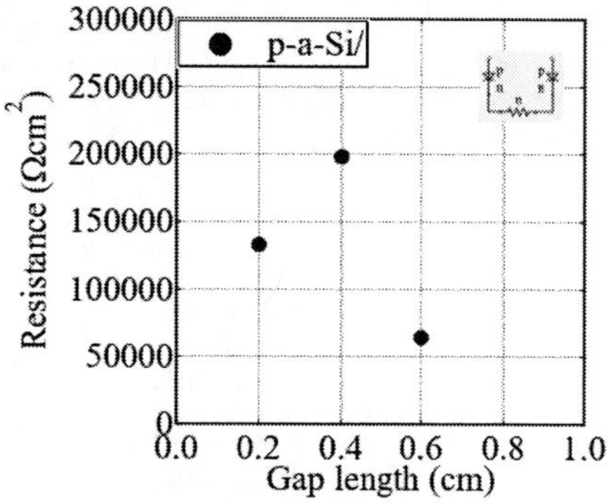

Type-C

Figure 4: Type-C samples resistance as a function of electrode gap length.

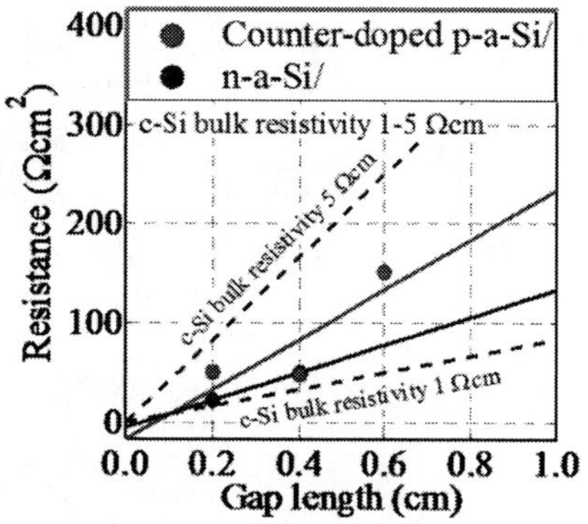

Type-D,E

Figure 5: Type-D and E samples resistance as a function of electrode gap length.

3 RESULTS AND DISCUSSION

The τ_{eff} of ion implanted sample (Type-B) is about only 10 μs. However the τ_{eff} is recovered to 217 μs by annealing at 250 °C for 100 min. This recovered τ_{eff} is approximately equal to 228 μs which is τ_{eff} of the sample with deposited n-a-Si (Type-A). This recovering may be owing to the termination of defects created by ion implantation [5]. This defect termination at low temperatures probably results from the existence of hydrogen atoms implanted together with P atoms in non-mass-separated plasma ion implantation. Since the interface quality is not changed by ion implantation after annealing, plasma ion implantation can be used for SHJ solar cells.

Figs. 4 and 5 show the resistance for three types of samples: (Type-C) p-a-Si/c-Si/p-a-Si, (Type-D) n-a-Si/c-Si/n-a-Si, and (Type-E) P implanted p-a-Si c-Si/P implanted p-a-Si as a function of electrode gap length. Since the Type-C includes two diodes in the current path, as shown in the inset of Fig. 4, it becomes highly resistive. On the other hand, Type-D and -E demonstrate that these contact resistance is almost zero. We can confirm that, as the result of P implantation, p-a-Si is converted to n-a-Si. In the case of the conventional ion implantation with mass-separator into c-Si, the annealing at temperatures over 800 °C is always required to eliminate defects created by ion-implantation. The plasma ion implantation into a-Si/c-Si hetero-structures is so different from the conventional ion implantation into c-Si, and annealing at temperatures as low as 250 °C is sufficient for defect elimination. This appears applicable to the fabrication of back-electrodes of IBC-SHJ solar cells in which the high temperature annealing is prohibited to keep quality of a-Si. The contact resistance evaluated above appears low enough and comparable with that of initially formed n-a-Si layers.

Fig. 6 shows *J–V* curves of two types of SHJ solar cells: the cells with deposited n-a-Si (black dash line), and those with converted n-a-Si from p-a-Si (red line). These characteristics are not sufficiently high as that of the recently reported high efficiency SHJ solar cells, which may be owing to flat cell surface inducing large optical reflectance and unoptimized ITO layers. However, the performance of SHJ solar cells with counter doped a-Si is not largely different from that of SHJ solar cells with deposited n-a-Si. We demonstrate that counter doping can be applied to SHJ solar cells.

From the results and discussions above, it is concluded that low-cost fabrication of patterned back-electrodes with low contact resistance in IBC-SHJ solar cells is possible by using plasma ion implantation and successive low temperature annealing.

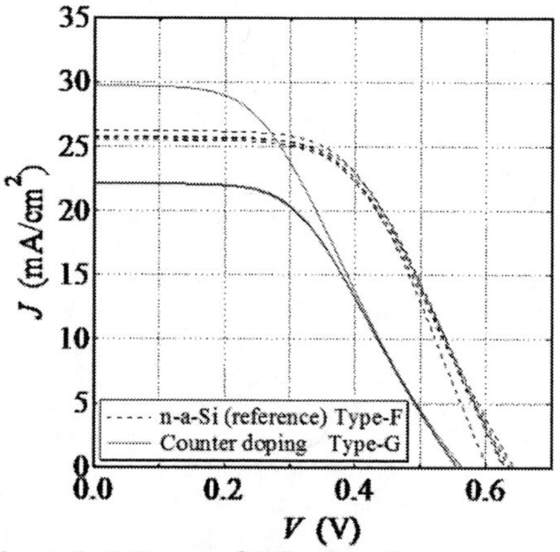

Figure 6: *J–V* curve of SHJ solar cells.

4 CONCLUSIONS

By plasma ion implantation of P atoms for only a few second and successive annealing at as low as 250 °C, heterojunction p-a-Si/i-a-Si/c-Si layers can be converted to n-a-Si/(i-a-Si)/c-Si layers. The contact resistance after converting p-a-Si to n-a-Si becomes almost zero. We also demonstrate that converting p-a-Si to n-a-Si can be applied to SHJ solar cells. The results appear to show the promising feasibility for simplifying the fabrication process of IBC-SHJ solar cells or developing low cost fabrication processes by using plasma ion implantation.

ACKNOWLEDGEMENT

This work is supported by The New Energy and Industrial Technology Development Organization (NEDO). We would also like to thank ULVAC, Inc. for help of the ion implantation.

5 References

[1] K. Masuko, M. Shigematsu, T. Hashiguchi, D. Fujishima, M. Kai, N. Yoshimura, T. Yamaguchi, Y. Ichihashi, T. Mishima, N. Matsubara, T. Yamanishi, T. Takahama, M. Taguchi, E. Maruyama, and S. Okamoto, IEEE J. Photovoltaics 4, 1433 (2014).
[2] H J. Nakamura, N. Asano, T. Hieda, C. Okamoto, H. Katayama, and K. Nakamura, IEEE J. Photovoltaics 4, 1491 (2014).
[3] http://www.kaneka.co.jp/kaneka-e/images/topics/1473811995/1473811995_101.pdf
[4] https://www.ulvac.co.jp/wiki/en/crystalline-sillicon-solar-cells/
[5] K. Koyama, N. Yamaguchi, D. Hironiwa, H. Suzuki, K. Ohdaira, and H. Matsumura, Jpn. J. Appl. Phys. 56, 08MB21-1–4 (2017)
[6] D. T. Stevenson and R. J. Keyes, J. Appl. Phys. 26, 190 (1955).
[7] R. L. Watters and G. W. Ludwing, J. Appl. Phys. 27, 489 (1956).
[8] M. Kunst and G. Beck, J. Appl. Phys. 60, 3558 (1986).

AUTHOR INDEX

AUTHOR INDEX

AUTHOR INDEX

AUTHOR INDEX

AUTHOR INDEX

AUTHOR INDEX

AUTHOR INDEX

AUTHOR INDEX

AUTHOR INDEX

AUTHOR INDEX

AUTHOR INDEX

AUTHOR INDEX

AUTHOR INDEX

AUTHOR INDEX

AUTHOR INDEX

AUTHOR INDEX

AUTHOR INDEX

AUTHOR INDEX

AUTHOR INDEX

AUTHOR INDEX

AUTHOR INDEX

AUTHOR INDEX

AUTHOR INDEX

33rd European Photovoltaic Solar Energy Conference and Exhibition (EU PVSEC 2017)

Amsterdam, The Netherlands
25-29 September 2017

Volume 2 of 4

Editors:

A. Smets
P. Helm
N. Taylor

ISBN: 978-1-5108-5397-3

Printed from e-media with permission by:

Curran Associates, Inc.
57 Morehouse Lane
Red Hook, NY 12571

Some format issues inherent in the e-media version may also appear in this print version.

Printed by Curran Associates, Inc. (2018)

For permission requests, please contact WIP – Renewable Energies
at the address below.

WIP – Renewable Energies
Sylvensteinstr. 2
81369 Munchen
Germany

Phone: +49 89 72012735
Fax: +49 89 72012791

wip@wip-munich.de

Additional copies of this publication are available from:

Curran Associates, Inc.
57 Morehouse Lane
Red Hook, NY 12571 USA
Phone: 845-758-0400
Fax: 845-758-2633
Email: curran@proceedings.com
Web: www.proceedings.com

33rd European Photovoltaic Solar Energy Conference and Exhibition

Proceedings of the International Conference held in Amsterdam, The Netherlands

25 September – 29 September 2017

Edited by:

A. SMETS
Delft University of Technology
Professor Solar Energy
Delft, The Netherlands

N. TAYLOR
European Commission Joint Research Centre
Ispra, Italy

P. HELM
WIP Renewable Energies
Munich, Germany

Coordination of the Technical Programme:
European Commission Joint Research Centre
Via E. Fermi 1
21020 Ispra (VA)
Italy

Institutional Support:
European Commission
UNESCO – United Nations Educational, Scientific and Cultural Organization, Natural Sciences Sector
WCRE – World Council for Renewable Energy

Supporting Organisations:
ARE – Alliance for Rural Electrification
EREF – European Renewable Energies Federation
ESA – European Space Agency
EUFORES –The European Forum for Renewable Energy Sources
EUREC – The Association of European Renewable Energy Research Centres
European Technology & Innovation Platform PV
IEA PVPS – International Energy Agency, Photovoltaic Power Systems Programme
REN21 – Renewable Energy Policy Network for the 21st Century
TKI Urban Energy - Topsector Energie

Institutional PV Industry Cooperation:
SolarPower Europe
SOLARUNITED

Supporting Associations:
AIE – European Association of Electrical Contractors
AREA – African Renewable Energy Alliance
CABA – Continental Automated Buildings Association
EPIC – European Photonics Industry Consortium
GENSED - Turkish Solar Energy Industry Association
MPIA - Malaysian Photovoltaic Industry Association
SASIA - Saudi Arabia Solar Industry Association

EU PVSEC realised by:
WIP
Sylvensteinstr. 2, 81369 München, Germany
Tel: +49 89 720 12 735, Fax: +49 89 720 12 791
Email: pv.conference@wip-munich.de
www.photovoltaic-conference.com
www.wip-munich.de

Proceedings produced and published by:
WIP
Sylvensteinstr. 2, 81369 München, Germany
Tel: +49 89 720 12 735, Fax: +49 89 720 12 791
Email: pv.conference@wip-munich.de
www.photovoltaic-conference.com
www.wip-munich.de

ISSN 2196-0992

33rd EUROPEAN PHOTOVOLTAIC SOLAR ENERGY CONFERENCE AND EXHIBITION
25 SEPTEMBER – 29 SEPTEMBER 2017
AMSTERDAM, THE NETHERLANDS

EU PVSEC COMMITTEES

INTERNATIONAL SCIENTIFIC ADVISORY COMMITTEE (ISAC)

Chair
P. Szymanski, European Commission, JRC, Director of Energy, Transport and Climate, Petten, The Netherlands

Secretary
N. Taylor, European Commission, JRC, Institute for Energy and Transport, Ispra, Italy

Committee Members
P. Frankl, Head of the Renewable Energy Division, International Energy Agency, France
M. Getsiou, European Commission, DG RTD, Brussels, Belgium
S.W. Glunz, Head of Division "Solar Cells – Development and Characterization", Fraunhofer ISE, Freiburg, Germany
D. Habermann, Chief Innovation Officer, Meyer Burger Technology, Gwatt (Thun), Switzerland
P. Malbranche, General Director, CEA INES, France
P. Menna, European Commission, DG ENER, Brussels, Belgium
S. Nowak, Managing Director of NET Nowak Energy & Technology, St. Ursen, Switzerland
W.C. Sinke, Program Development Manager, ECN Solar Energy, The Netherlands
A.H.M. Smets, Professor Solar Energy at Delft University of Technology, The Netherlands
M. Topič, Chairman ETIP PV, Head of Laboratory of Photovoltaics and Optoelectronics of the University of Ljubljana, Slovenia
P. Verlinden, Chief Scientific and Vice Chair, State Key Lab. TRINA Solar, China
A. Grassi, Managing Director, ETA-Florence Renewable Energies, Florence, Italy
P. Helm, Scientific & Managing Director, WIP - Renewable Energies, Munich, Germany

CONFERENCE EXECUTIVE COMMITTEE

Conference General Chairman
A.H.M. Smets, Professor Solar Energy at Delft University of Technology, The Netherlands

Technical Programme Chairman
N. Taylor, European Commission, JRC, Institute for Energy and Transport, Ispra, Italy

Committee Members
H. Ossenbrink, Former European Commission, Italy
P. Helm, Scientific & Managing Director, WIP - Renewable Energies, Munich, Germany
W.C. Sinke, Program Development Manager, ECN Solar Energy, The Netherlands
S. Nowak, Managing Director of NET Nowak Energy & Technology, St. Ursen, Switzerland
A. Grassi, Managing Director, ETA-Florence Renewable Energies, Florence, Italy

2017 SCIENTIFIC COMMITTEE
Topic Organisers and Paper Review Experts

Topic 1 – New Materials and Concepts for Photovoltaic Devices

N.J. Ekins-Daukes, Imperial College London, United Kingdom
A. Martí Vega, UPM, Madrid, Spain
J. Poortmans, imec, Leuven, Belgium
J.A.M. van Roosmalen, ECN, Petten, The Netherlands

V. Bukauskas, Center for Physical Sciences and Technology, Vilnius, Lithuania
J. Govaerts, imec, Leuven, Belgium
J.F. Guillemoles, CNRS, Chatou Cedex, France
M.A. Hernandez-Fenollosa, UPV, Valencia, Spain
J.C. Hummelen, University of Groningen, The Netherlands
D. König, UNSW Australia, Sydney, Australia
I. Konovalov, University of Applied Sciences Jena, Germany
C. Lévy-Clément, CNRS, Thiais, France
K.R. McIntosh, PV Lighthouse, Coledale, Australia
Y. Okada, University of Tokyo, Japan
B. Rech, HZB, Berlin, Germany
F. Roca, ENEA, Portici, Italy
M. Rusu, HZB, Berlin, Germany
V. Saly, Slovak University of Technology, Bratislava, Slovakia
R.E.I. Schropp, Utrecht University, Driebergen, The Netherlands
H. Shirai, Saitama University, Saitama-City, Japan
W. Warta, Fraunhofer ISE, Freiburg, Germany

Topic 2 – Silicon Photovoltaics

O. Anspach, PV Crystalox Solar, Erfurt, Germany
D.L. Bätzner, Meyer Burger Research, Hauterive, Switzerland
R. Brendel, ISFH, Emmerthal, Germany
J. Cárabe, CIEMAT, Madrid, Spain
F. Ferrazza, eni spa, San Donato Milanese, Italy
S.W. Glunz, Fraunhofer ISE, Freiburg, Germany
G. Hahn, University of Konstanz, Germany
D. Muñoz, CEA, Le Bourget du Lac, France
S. Riepe, Fraunhofer ISE, Freiburg, Germany
M. Topic, University of Ljubljana, Slovenia
A.W. Weeber, ECN, Petten, The Netherlands
P. Wohlfart, Singulus Technologies, Kahl am Main, Germany

A.G. Aberle, SERIS, Singapore
N. Azkona, UPV/EHU, Zamudio, Spain
J. Bagdahn, Anhalt University of Applied Sciences, Köthen, Germany
R. Barinka, Fill Factory, Rožnov pod Radhoštm, Czech Republic

P.A. Basore, PV Specialist Services, Solana Beach, USA
S. Bowden, Arizona State University, Tempe, USA
T.M. Bruton, Progress in Photovoltaics, Woking, United Kingdom
A. Danel, CEA, Le Bourget du Lac, France
B. Dauksher, Arizona State University, Tempe, USA
M. Debucquoy, imec, Leuven, Belgium
P. Delli Veneri, ENEA, Portici, Italy
T. Desrues, CEA, Le Bourget du Lac, France
D. Dimova-Malinovska, Bulgarian Academy of Sciences, Sofia, Bulgaria
S. Dubois, CEA, Le Bourget du Lac, France
T. Dullweber, ISFH, Emmerthal, Germany
P. Fath, RCT-Solutions, Konstanz, Germany
S. Gall, HZB, Berlin, Germany
I. Gordon, imec, Leuven, Belgium
J. Haschke, EPFL, Neuchâtel, Switzerland
K. Hesse, Wacker Chemie, Burghausen, Germany
J. Horzel, CSEM, Neuchâtel, Switzerland
J.-R. Huang, Motech Industries, Taoyuan, Taiwan
J. John, imec, Leuven, Belgium
D.-W. Kang, Cheongju University, Republic of Korea
S.D. Kim, San Jose, USA
A. Meeder, Geo-En Energy Technologies, Berlin, Germany
J. Meier, Meier Technologies, Berlingen, Switzerland
A. Metz, Alzenau, Germany
C. Molpeceres, UPM, Madrid, Spain
J.W. Müller, Hanwha Q CELLS, Bitterfeld-Wolfen, Germany
Y. Ohshita, Toyota Technological Institute, Nagoya, Japan
S. Peters, Hanwha Q CELLS, Bitterfeld-Wolfen, Germany
K. Petter, Hanwha Q CELLS, Bitterfeld-Wolfen, Germany
R. Preu, Fraunhofer ISE, Freiburg, Germany
J. Rentsch, Fraunhofer ISE, Freiburg, Germany
E. Sauar, Brighterlite, Oslo, Norway
H. Schlemm, Meyer Burger, Hohenstein-Ernstthal, Germany
J. Schmidt, ISFH, Emmerthal, Germany
M.C. Schubert, Fraunhofer ISE, Freiburg, Germany
B. Terheiden, University of Konstanz, Germany
M. Vetter, IPHT, Jena, Germany
G.P. Willeke, Fraunhofer ISE, Freiburg, Germany
D. Wood, Hemlock Semiconductor, Hemlock, USA
A. Yamada, Tokyo Institute of Technology, Japan
D.L. Young, NREL, Golden, USA
M. Zeman, Delft University of Technology, The Netherlands
J. Zhao, Meyer Burger, Hohenstein-Ernstthal, Germany

Topic 3 – Thin Film Photovoltaics

C.J. Brabec, University of Erlangen-Nuremberg, Germany
A.N. Tiwari, EMPA, Dübendorf, Switzerland
S.C. Veenstra, ECN, Eindhoven, The Netherlands
W. Wischmann, ZSW, Stuttgart, Germany

G. Bardizza, European Commission JRC, Ispra, Italy
R. Birkmire, University of Delaware, Newark, USA
D. Cheyns, imec, Leuven, Belgium
S. Cros, CEA, Le Bourget du Lac, France
T. Dalibor, AVANCIS, Torgau, Germany
B. Dimmler, Manz CIGS Technology, Schwäbisch Hall, Germany
K. Durose, Stephenson Institute for Renewable Energy, Liverpool, United Kingdom
S. Huang, UNSW Australia, Sydney, Australia
S.J.C. Irvine, Swansea University, St. Asaph, United Kingdom
C.-F. Lin, NTU, Taipei, Taiwan
M. Lira-Cantú, CIN2, Barcelona, Spain
M.Ch. Lux-Steiner, HZB, Berlin, Germany
M. Meuris, imec, Leuven, Belgium
J. Parisi, University of Oldenburg, Germany
R. Po, eni spa, Novara, Italy
M. Powalla, ZSW, Stuttgart, Germany
A. Redinger, University of Luxembourg, Luxembourg
A. Romeo, University of Verona, Italy
V. Sittinger, Fraunhofer IST, Braunschweig, Germany
A. Virtuani, EPFL, Neuchâtel, Switzerland

Topic 4 – Concentrator and Space Photovoltaics

G. Flamand, imec, Leuven, Belgium
C. Signorini, European Space Agency, Noordwijk, The Netherlands

I. Antón, UPM, Madrid, Spain
A.W. Bett, Fraunhofer ISE, Freiburg, Germany
R. Campesato, CESI, Milan, Italy
V. Khorenko, Azur Space, Heilbronn, Germany
R. King, Arizona State University, Tempe, USA
F. Rubio Berenguel, Soitec Solar, Freiburg, Germany
G. Siefer, Fraunhofer ISE, Freiburg, Germany
G. Timò, RSE, Piacenza, Italy

Topic 5 – Performance, Reliability and Sustainability of Photovoltaic Modules and Balance of System Components

M. de Wild-Scholten, SmartGreenScans, Groet, The Netherlands
R. Gottschalg, Loughborough University, United Kingdom
N.M. Pearsall, Northumbria University, Newcastle upon Tyne, United Kingdom
M. Perrin, CEA, Le Bourget du Lac, France
T. Sample, European Commission JRC, Ispra, Italy
K. Wambach, Wambach-Consulting, Aindling, Germany

K. Baert, KU Leuven, Heverlee, Belgium
G. Beaucarne, Dow Corning, Seneffe, Belgium
R. Einhaus, Apollon Solar, Lyon, France
G. Friesen, SUPSI, Canobbio, Switzerland
V. Fthenakis, Columbia University, New York, USA
W.J. Gambogi, DuPont, Wilmington, USA
L. Gomes de Freitas, Federal University of Uberlândia, Brazil
G. Graditi, ENEA, Portici, Italy
Y. Hishikawa, AIST, Tsukuba, Japan
U. Jahn, TÜV Rheinland Energy, Cologne, Germany
S. Krauter, University of Paderborn, Germany
A.R. Lagunas, CENER, Sarriguren-Navarra, Spain
C. Monokroussos, TÜV Rheinland, Shanghai, China
M.A. Muñoz-García, UPM, Madrid, Spain
H. Nagel, Fraunhofer ISE, Freiburg, Germany
H. Nussbaumer, ZHAW, Winterthur, Switzerland
S. Oktik, Sisecam, Istanbul, Turkey
G. Oreski, PCCL, Leoben, Austria
W. Palitzsch, Loser Chemie, Zwickau, Germany
J.N. Roy, IIT Kharagpur, India
R. Sinton, Sinton Instruments, Boulder, USA
M. Van Iseghem, EDF R&D, Moret-sur-Loing, France
E. Voroshazi, imec, Leuven, Belgium
A. Wade, First Solar, Mainz, Germany

Topic 6 – PV System Performance and Integration

F.P. Baumgartner, ZHAW, Winterthur, Switzerland
F. Bonemazzi, ENEL, Rome, Italy
F. Frontini, SUPSI, Canobbio, Switzerland

P. Lechner, ZSW, Stuttgart, Germany
O. Mayer, GE Global Research, Garching, Germany
F. Montanari, ENEL Green Power, Rome, Italy
C. Protogeropoulos, EEPS, Athens, Greece
A. Scognamiglio, ENEA, Portici, Italy
W.G.J.H.M. van Sark, Utrecht University, The Netherlands
I. Weiss, WIP - Renewable Energies, Munich, Germany

C. Alonso-Tristán, UBU, Burgos, Spain
J. Binder, ZSW, Stuttgart, Germany
P. Blanc, MINES ParisTech, Paris, France
P. Bonomo, SUPSI, Canobbio, Switzerland
F. Bourry, CEA, Le Bourget du Lac, France
I. Da Silva, Strathmore University, Nairobi, Kenya
B. Gaiddon, Hespul, Lyon, France
M. Grottke, WIP - Renewable Energies, Munich, Germany
I.B. Hagemann, Architekturbüro Hagemann, Aachen, Germany
C. Helmke, Clake Energy, Erfurt, Germany
T. Huld, European Commission JRC, Ispra, Italy
E. Kaijuka-Okwenje, Rural Electrification Agency, Kampala, Uganda
S. Koopman, CSIR, Pretoria, South Africa
P.A. Leicester, Loughborough University, United Kingdom
P. Malbranche, CEA, Le Bourget du Lac, France
G. Mütter, Alternative Energy Solutions, Vienna, Austria
C. Nyman, Soleco, Porvoo, Finland
K. Peter, ISC Konstanz, Germany
T. Reijenga, BEAR-iD, Gouda, The Netherlands
J. Remund, Meteotest, Bern, Switzerland
K.T. Roro, CSIR, Pretoria, South Africa
M. Schroedter-Homscheidt, German Aerospace Center, Wessling, Germany
M. Sengupta, NREL, Golden, USA
D. Stellbogen, ZSW, Stuttgart, Germany
J. Sutterlueti, Gantner Instruments, Schruns, Austria
S. Tselepis, Pikermi, Athens, Greece
X. Vallvé, Trama TecnoAmbiental, Barcelona, Spain
M. Zehner, Rosenheim University of Applied Sciences, Germany

Topic 7 – PV Economics, Markets and Policies

C. Breyer, Lappeenranta University of Technology, Finland
M. Getsiou, European Commission DG RTD, Brussels, Belgium
G. Masson, Becquerel Institute, Brussels, Belgium
T. Nordmann, TNC Consulting, Feldmeilen, Switzerland
S. Nowak, NET Nowak Energy & Technology, St. Ursen, Switzerland
E. Perezagua, Consultores de Energía Fotovoltaica, Madrid, Spain

S. Caneva, WIP - Renewable Energies, Munich, Germany
I. Kaizuka, RTS, Tokyo, Japan
D. Mayer, MINES ParisTech, Paris, France
P. Mazzucchelli, EUREC, Brussels, Belgium
S. Szabó, European Commission JRC, Ispra, Italy

SUBJECT INDEX

New Materials and Concepts for Photovoltaic Devices

Sessions 1AP.1, 1AO.1, 1AO.2, 1AO.3, 1CO.1, 1CO.2, 1CV.3

Silicon Photovoltaics

Sessions 2BP.1, 2AO.4, 2AO.5, 2AO.6, 2BO.1, 2BO.2, 2BO.3, 2BO.4, 2CO.9, 2CO.10, 2CO.11, 2CO.12, 2DO.1, 2DO.2, 2DO.3, 2AV.1, 2AV.2, 2AV.3, 2CV.2

Thin Film Photovoltaics

Sessions 3CP.1, 3AO.7, 3AO.8, 3AO.9, 3BO.9, 3BO.10, 3BO.11, 3BO.12, 3CO.3, 3CO.4, 3DO.7, 3CV.1, 3DV.2

Concentrator and Space Photovoltaics

Sessions 4DO.4, 4CV.4

Performance, Reliability and Sustainability of Photovoltaic Modules and Balance of System Components

Sessions 5DP.1, 5BO.5, 5BO.6, 5CO.5, 5CO.6, 5CO.7, 5CO.8, 5DO.5, 5EO.1, 5BV.4, 5DV.3

PV System Performance and Integration

Sessions 6DP.2, 6BO.7, 6BO.8, 6CO.13, 6CO.14, 6CO.15, 6CO.16, 6DO.6, 6DO.10, 6DO.11, 6DO.12, 6EO.2, 6BV.1, 6BV.2, 6BV.3

PV Economics, Markets and Policies

Sessions 7EP.1, 7DO.8, 7DO.9, 7EO.3, 7DV.1

Topic Code

1 New Materials and Concepts for Photovoltaic Devices
2 Silicon Photovoltaics
3 Thin Film Photovoltaics
4 Concentrator and Space Photovoltaics
5 Performance, Reliability and Sustainability of Photovoltaic Modules and Balance of System Components
6 PV System Performance and Integration
7 PV Economics, Markets and Policies

Session Type

P = Plenary Session
O = Oral Session
V = Visual Session

Day Codes

A = Monday, 25 September 2017
B = Tuesday, 26 September 2017
C = Wednesday, 27 September 2017
D = Thursday, 28 September 2017
E = Friday, 29 September 2017

e.g. 3CP.1 ⇒ 3=Thin Film Photovoltaics, C=Wednesday, P=Plenary Session, 1=Session 1

TABLE OF CONTENT
of EU PVSEC 2017 Proceedings Papers

Silicon Photovoltaics

Visual PRESENTATIONS 2AV.3 Heterojunction Solar Cells

Thin Film Photovoltaics

Plenary SESSION 3CP.1 Thin Film PV: Pushing the Limits with Breakthroughs in Industry and Research

Oral PRESENTATIONS 3AO.7 Optical Losses and TCO`s

Oral PRESENTATIONS 3AO.8 Module Stability and Characterisation

Oral PRESENTATIONS 3BO.9 Manufacturing and Performance Improvements

Visual PRESENTATIONS 3DV.2 CI(G)S, CdTe and Related Thin Film Solar Cells and Modules (II) / Perovskite, Organic and Dye-Sensitised Devices

Concentrator and Space Photovoltaics

Oral PRESENTATIONS 4DO.4 III-V-Based Devices for Terrestrial and Space Applications

Visual PRESENTATIONS 4CV.4 III-V-Based Devices for Terrestrial and Space Applications

Performance, Reliability and Sustainability of Photovoltaic Modules and Balance of System Components

Plenary SESSION 5DP.1 Performance, Reliability and Sustainability of Photovoltaic Modules and Balance of System Components

PV System Performance and Integration

Plenary SESSION 6DP.2 PV System Performance and Integration

Oral PRESENTATIONS 6BO.7 Advanced Field Performance Estimation

Oral PRESENTATIONS 6BO.8 Failure Modes and Degradation

Visual PRESENTATIONS 6BV.2 Design and Operation of PV Systems (II)

PV Economics, Markets and Policies

Oral PRESENTATIONS 7DO.8 Global PV Economics and Market Trends

Oral PRESENTATIONS 7DO.9 Innovative National PV Market Economics Business Cases

Oral PRESENTATIONS 7EO.3 PV-Related Policies, Strategies and Societal Issues

Visual PRESENTATIONS 7DV.1 PV Economics and Markets / PV-Related Policies, Strategies and Societal Issues

FOREWORD

The EU PVSEC 2017 in Amsterdam concluded with the strong message that the photovoltaic revolution is just getting started. Research and innovation across the sector, together with fast deployment and numerous application options, are paving the way to putting PV at the centre-stage of sustainable energy systems. This year 2,516 participants from 89 countries took part in this most inspiring platform. It provided a vital forum for information exchange in the field of PV solar energy research, demonstration and applications. In the exhibition 64 companies from all parts of the world welcomed visitors and presented their products and services.

The EU PVSEC 2017 Proceedings give a comprehensive overview of the photovoltaic solar sector, its current status and future prospects in science, research, innovation, development and deployment on more than 4,500 pages. Selection for inclusion in the conference was made by paper review experts and topic organisers (see the listing on pages III-V), to whom we express our sincere gratitude for their comprehensive review work and overall contribution to the success of the conference. In addition to the 619 submitted papers, the proceedings include 70 presentations (slides) shown during the plenary and oral presentations as well as 232 poster files of the visual presentations. In total this amounts to 921 publications.

The 2017 EU PVSEC Proceedings maintain our commitment to providing quick and open access to high quality scientific results. They constitute a powerful tool for targeted and quick information search and retrieval, enabling you to search by keyword, paper title and authors. We are sure that these features will help to simplify the use and exploit the full potential of this extensive source of information.

The Conference Proceedings are published as downloadable DVD file and are also fully accessible online. A DOI code (Digital Object Identifier) has been assigned to each paper. This ensures unequivocal and permanent identification and full citability. The papers can be viewed and downloaded in a full free open access from the EU PVSEC Proceedings website www.eupvsec-proceedings.com.

We are confident that these Proceedings will play an important role in providing a comprehensive overview of the current actors and activities in the global PV sector and that they will disseminate information on the state-of-the-art of technologies and applications. This can generate further research, add momentum to innovation and promote interest in PV worldwide.

We would like to cordially thank all authors and participants of the EU PVSEC 2017 for their contributions and look forward to welcoming you in 2018 at the 35th EU PVSEC 2018 in Brussels, Belgium.

The Editors

EU PVSEC 2017

25 - 29 September 2017

33rd European PV Solar Energy Conference and Exhibition

RAI Convention & Exhibition Centre • Amsterdam • The Netherlands

Selected Highlights

Nigel Taylor, Technical Programme Chair

.... with help from the session chairs and others

European Commission

Silicon

European Commission

2nd Fabrication Run of n-Type HP mc Solar Cells
IV Results

Fraunhofer ISE

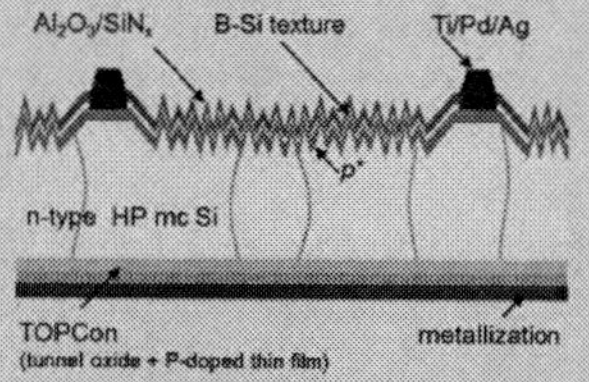

V_{oc} [mV]	J_{sc} [mA/cm^2]	*FF* [%]	*pFF* [%]	η [%]
674.2	41.1	80.5	81.6	22.3*

Cell area 2×2 cm², aperture area (*) certified by Fraunhofer ISE CalLab

- 2nd cell batch
- Applying of „lessons learned" in cell fabrication
- Same material as previous solar cells

→ New record efficiency of 22.3% for a mc silicon solar cell

2DO.3.1 Approaching 22% Efficiency with Multicrystalline n-Type Silicon Solar Cells
J. Benick et al, Fraunhofer ISE, Freiburg, Germany

European Commission

Q.ANTUM technology on p-Cz substrates in mass production

Hanwha Q CELLS

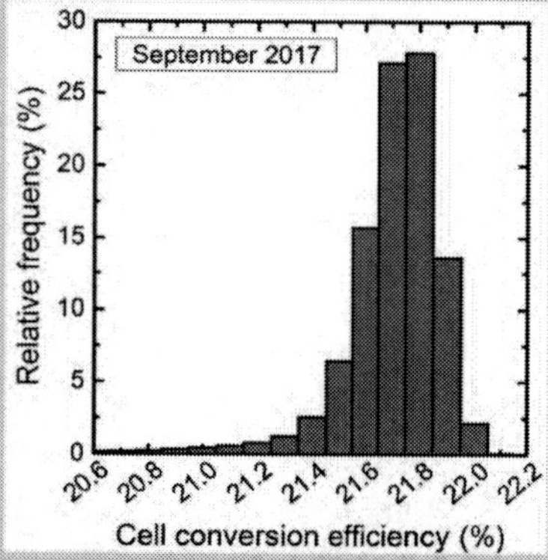

Platform change to Q.ANTUM[1] secures:

- Efficiency increase of +0.6%abs. per year
- Evolutionary development process[2] and upgrade of existing tool set
- Silver consumption: -50%
- Throughput increase
- **Avg. cell efficiency of 21.7% in production (max. >22%)**

2CO.9.6 Mass Production of Q.ANTUM Solar Cells and Modules on p-Type Cz Silicon Substrates
F. Fertig et al, Hanwha Q CELLS, Bitterfeld-Wolfen, Germany

European Commission

24.13% efficiency IBC cell

Trinasolar

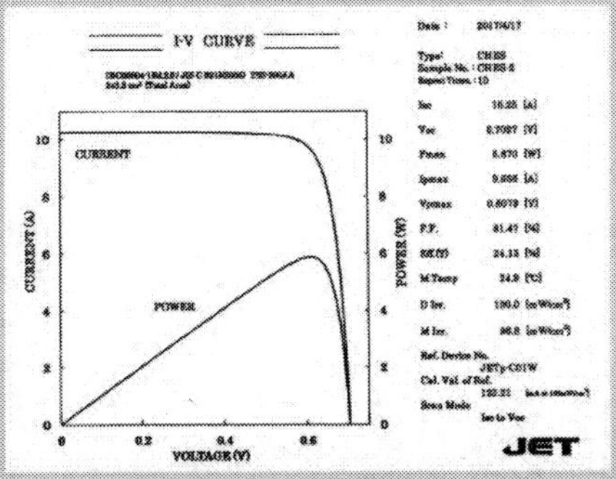

Compared to 23.5% cell (2016):

- 10 mV gain in V_{oc} by using selective emitter structure;
- J_{sc} is also improved by using less absorption in SiN_x;
- Higher FF is due to the reduced R_s, mainly from lower ρ_c on emitter.

Area	V_{oc}	I_{sc}	V_{mpp}	I_{mpp}	FF	Eff
[cm²]	[mV]	[A]	[mV]	[A]	[%]	[%]
244.9	702.7	10.25	607.9	9.66	81.47	24.13

* Independently measured at Japan Electrical Safety & Environment Technology Laboratories (JET)

2CO.11.5 Large-Area (6 Inch) Screen-Printed IBC Solar Cells with Efficiency Approaching 24% without Passivated Contacts

G. Xu et al, Trina Solar Energy, Changzhou, China

European Commission

Advantages of a-Si/c-Si heterojunction (HJ) back contact (BC) Si solar cell

a-Si/c-Si HJ : High V_{oc} and FF by excellent a-Si/c-Si interface

Back contact : High J_{sc} and FF by small optical and resistive loss

Si wafer-based: Reasonable production cost

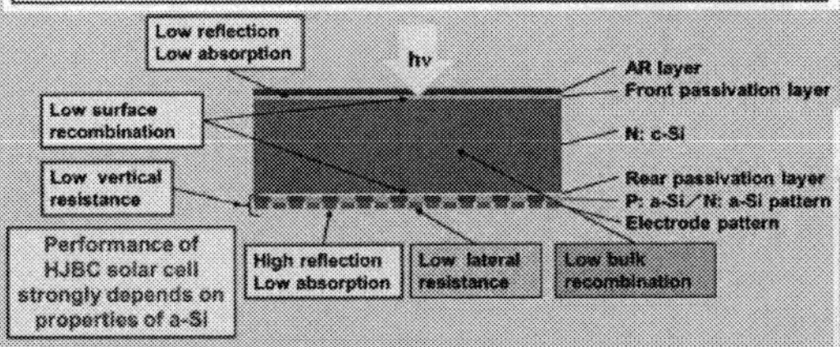

Kaneka

High efficiency c-Si solar cell; PV parameters

*da: designated area *ta: total area *ap: aperture area

	V_{OC} (V)	J_{SC} (mA/cm²)	FF (%)	Eff (%)	Area* (cm²)	Cell type
Panasonic	0.740	41.8	82.7	25.6	143.7 (da)	HJBC [1]
SunPower	0.737	41.3	82.7	25.2	153.5 (ta)	IBC [2]
Fraunhofer ISE	0.725	42.5	83.3	25.7	4.017(da)	TOPCon [3]
Kaneka	0.738	40.8	83.5	25.1	151.9 (ap)	HJ [4]
Kaneka	0.744	42.3	83.8	26.3	180.4 (da)	HJBC [5]
Kaneka	0.740	42.5	84.7	26.6	179.7 (da)	HJBC [6]
Kaneka	0.738	42.7	84.9	26.7	79.0 (da)	HJBC
Theoretical limit	0.753	43.6	89.2	29.3		100µm-thick wafer [7]
Ratio (26.7%/limit)	0.980	0.979	0.952	0.911		

- High V_{oc}, J_{sc} and FF obtained by excellent PECVD passivation and structure design
- Challenge for improvement of FF is a key to achieve >27% efficiency

[1] K. Masuko et al., IEEE Journal of Photovoltaics, 4, 1433 (2014). [2] "SunPower 2015 Analyst Day" (2015). [3] A. Richter et al., Silicon PV proceedings (2017). [4] D. Adachi et al., Applied Physics Letters, 107, 233506 (2015). [5] K. Yoshikawa et al., Nat. Energy 2, 17032 (2017) [6] K. Yoshikawa et al., Silicon PV proceedings (2017) [7] A. Richter et al., IEEE J. Photovolt. 3, 1184 (2013).

2BP1.1 Record-Breaking Efficiency Back-Contact Heterojunction Crystalline Si Solar Cell and Module

K. Yamamoto et al, Kaneka, Osaka, Japan

J.L. Hernández, Kaneka, Westerlo-Oevel, Belgium

European Commission

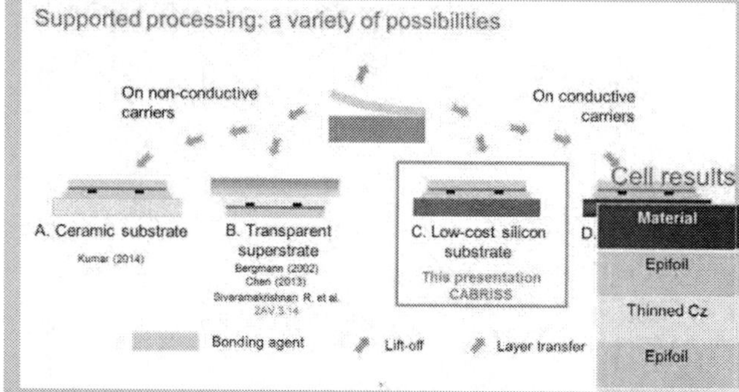

imec

TU WIEN · SINTEF · PyroGenesis · CABRISS · INKRON

Cell results

Material	Configuration		J_{sc} [mA/cm²]	V_{oc} [mV]	FF [%]	Eff. [%]	
Epifoil	Bonded to p⁺ Cz substrate	Best Avg.	33.2 32.8	681 676	69.2 69.1	15.7 15.3	*This work*
Thinned Cz	Bonded to p⁺ Cz substrate	Best Avg.	33.2 32.7	738 739	73.0 72.1	17.9 17.3	*This work*
Epifoil	Freestanding	Best Avg.	33.5 32.9	693 656	73.2 72.1	17.0 15.6	*Bearda et al. [2AV.3.26]*
Thinned Cz	Freestanding	Best	34.0 33.9	744 723	74.4 71.6	18.8 17.6	*Bearda et al. [2AV.3.26]*
Comparison with literature (cells bonded to conductive substrates)							
Epifoil (18 µm)	Bonded to steel	Best	34.5	632	77.2	16.8	*Amberwave/UNSW*
Slim-cut (25 µm)	Bonded to steel	Best	33.6	580	76.7	14.9	*Astrowatt*

2CO.12.2 Solar Cells on < 50µm Thick Epitaxial Foils Conductively Bonded to Low-Cost Si Carrier

H. Sivaramakrishnan Radhakrishnan et al, imec, Leuven, Belgium
N. Bednar & N. Adamovic, Vienna University of Technology, Austria
R. Roozeman & J. Heikkinen, INKRON, Espoo, Finland
A. Ulyashin & M. Syvertsen, SINTEF, Oslo, Norway

European Commission

Thin Film

European Commission

Key techniques implemented into the record sub(mini)modules

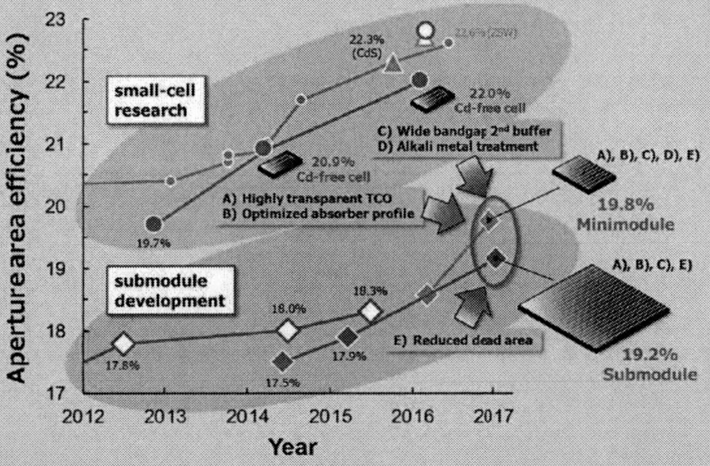

SOLAR FRONTIER

3CP.1.1 Present Status of Solar Frontier Cu(In,Ga)(Se,S)2 Record Efficiencies and Overall Progress

V. Bermudez et al, Solar Frontier, Atsugi, Japan

European Commission

HIGH EFFICIENCY MODULES

- Current in-house 30×30cm² champion module with 18.2% efficiency
- high Ga/high S absorber, wide bandgap buffer - highly Na doped In_xS_y:Na/i-$Zn_{0.85}Mg_{0.15}O$

AVANCIS CNBM

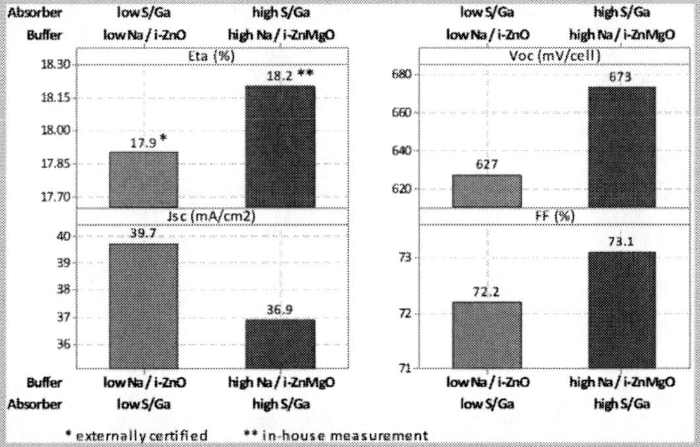

3BO.9.1 Wide Bandgap Sequential Absorber with Tunable Buffer Bandgap for CIGSSe Solar Modules at 18% Efficiency

R. Lechner et al, Avancis, Munich, Germany

European Commission

SOLIBRO Hi-Tech
A Hanergy Company

FULL SIZE: CHAMPION MODULE GEN 3

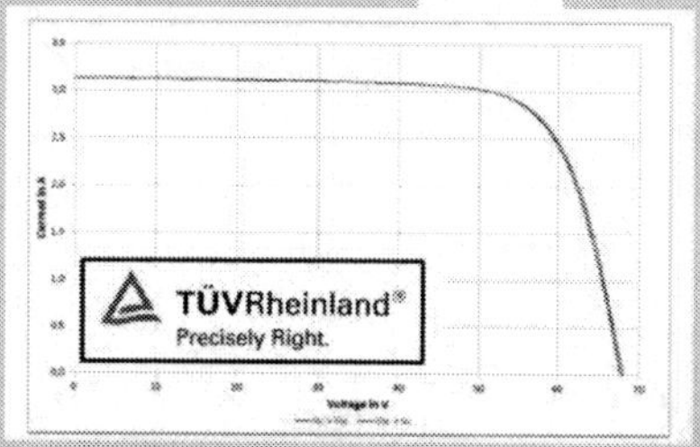

P_{mpp} [W]	159.4
V_{oc} [V]	67.9
I_{sc} [A]	3.12
FF [%]	75.2
η	17.0 %

- 17.9% Aperture Area Efficiency
- Externally verified
- Processed in SL2 production
- Module with future target process GEN3

17.0% @ 0.94m²
totale area efficiency
New world record

CELL AND MODULE DESIGN

Cell design

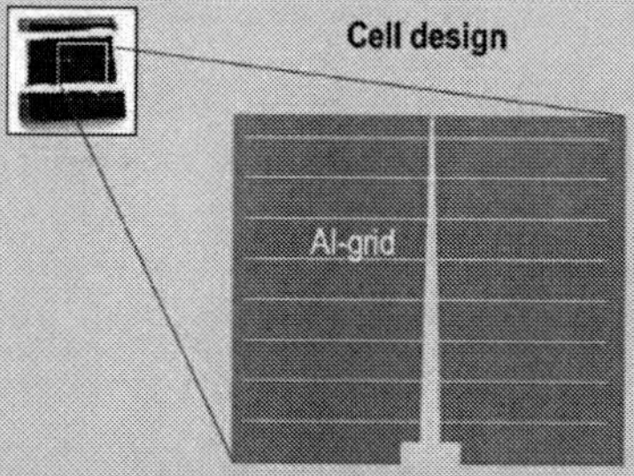

- Sheet resistance: 70 Ohms/sq
- Cell area: 10x10 mm²
- Dead area loss: ~ 3%
- Encapsulation: MgF_2

3CP.1.2 **17% Total Area Efficiency at Commercial Size CIGS Module**

P. Kratzert et al, Solibro Research, Uppsala, Sweden

Double Vision

Light Trapping

IV Results of III-V//Si Solar Cells

- Increase in J_{sc}
  - 11.6 → 12.7 mA/cm²
- Efficiency: 33.3%
- Record efficiency for 2T multi-junction solar cell

V_{oc} [V]	J_{sc} [mA/cm²]	FF [%]	η [%]
Gen 1			
3.046	11.7	87.5	31.3
Gen 2			
3.125	11.6	86.5	31.4
Gen 3			
3.127	**12.7**	83.8	**33.3**

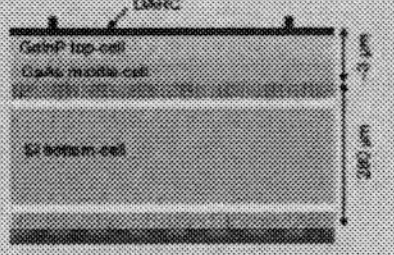

1AP.1.3 Monolithic III-V//Si Multi-Junction Solar Cell Exceeding an Efficiency of 31%

Fraunhofer ISE, Freiburg, Germany

Surplus collection of carriers at 3rd terminal injected from top cell

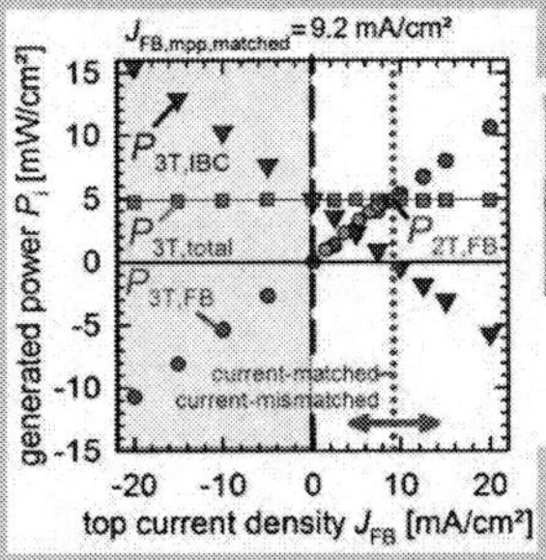

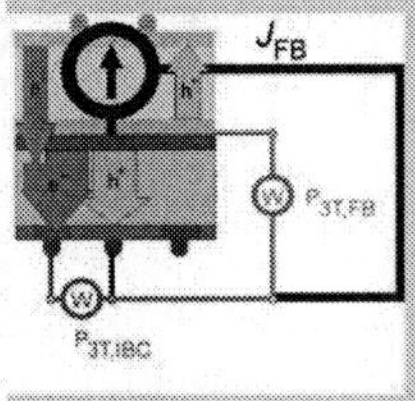

1AP.1.2 : Maximum Power Extraction Enabled by Monolithic Tandems Using Interdigitated Back Contact Bottom Cells with Three Terminals

ISFH and NREL

European Commission

22.7% Perovskite-CIGS thin film tandem solar cells

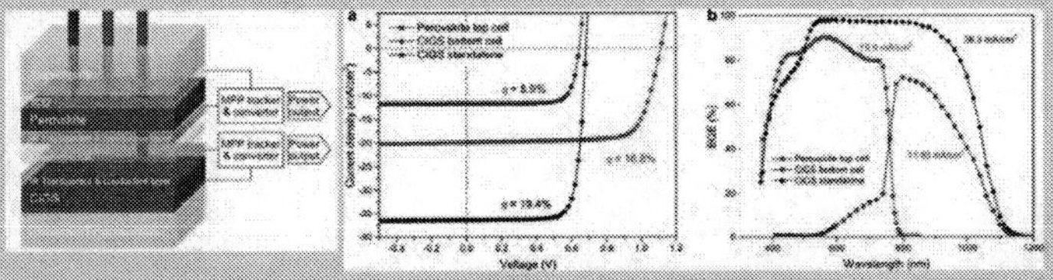

Solar Cells	V_{OC} (V)	J_{SC} (mA/cm²)	FF (%)	η (%)	MPP (%)	Cell area (cm²)	Efficiency gain
CIGS (stand alone)	0.684	36.4	78.2	19.4	19.4	0.213	
CIGS bottom cell	0.645	11.6	78.1	5.9	5.9	0.213	
Perovskite top cell	1.116	19.9	75.7	16.8	16.8	0.273	
Perovskite-CIGS 4-terminal tandem					22.7	0.213	3.3

Further improvement in Perovskite-CIGS thin film tandem solar cells via bandgap engineering, optical coupling etc.

Laboratory for Thin Films and Photovoltaics

3DO.7.4 Efficient and Stable NIR-Transparent Perovskite Solar Cells Prepared by Partial Ion Ex-Change Method for All-Thin-Film Tandem Applications

F. Fu, S. Pisoni, T. Feurer, A. Wäckerlin, S. Nishiwaki, A.N. Tiwari & S. Buecheler

EMPA, Dübendorf, Switzerland

European Commission

Performance

15

European Commission

~~MWp~~ MWh/MWp

Energy rating uncertainty

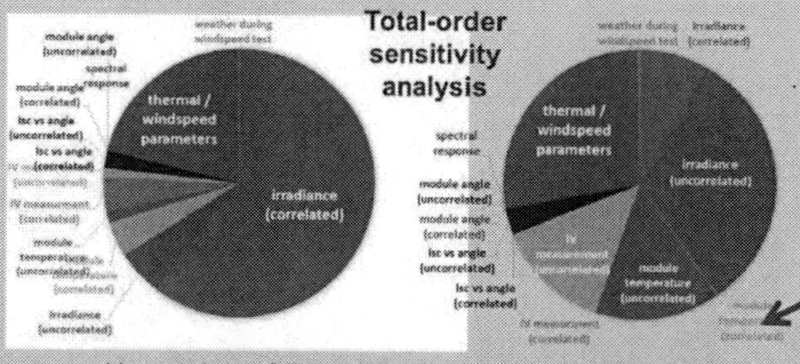

Uncertainty of $E_{mod,year}$
2.16 % (95% conf.)

Uncertainty of ***CSER***
1.84 % (95% conf.)

Simulation of 1000 modules, mixed types.
Uncertainty model based on survey of accredited test labs.

IEC 61853 Energy rating standard

Part 1 – Measure performance matrix

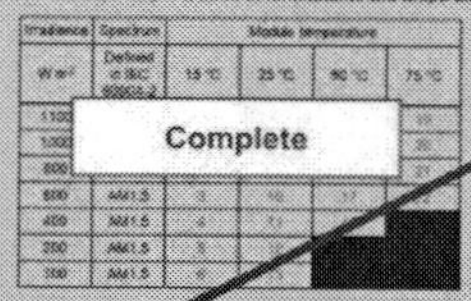

Complete

Part 2 – Other measurements

Complete

- Ang...
- Spe...
- Thermal & wind speed parameters

Part 3 – Standardised model

Final stages

***CSER* = Climate specific energy ratio**
Performance relative to STC as %

Part 4 – Reference climates
Several (currently 6) profiles
Hourly values for:

- Ambi...
- Wind ...
- Sun e...
- Sun incidence angle
- Global horizontal irradiance
- Direct horizontal irradiance
- Global in-plane irradiance
- Direct in-plane irradiance
- Spectral irradiance

Negotiating which climates to include

5CO.7.4 Progress in Energy Rating Standards: Accuracy and Optimisation
J.C. Blakesley, NPL, United Kingdom
T. Huld & H. Müllejans, European Commission JRC, Ispra, Italy

PhotoClass PV-Enerate

16

European Commission

Bifacial impact in the field

Exceptional service in the national interest — Sandia National Laboratories

Bifacial Single Axis Tracker (NM)

- Daily Potential Bifacial Energy Gains were estimated from front and back irradiance data using reference cells.
- Potential gains increase when tracker is off-track, but yield decreases

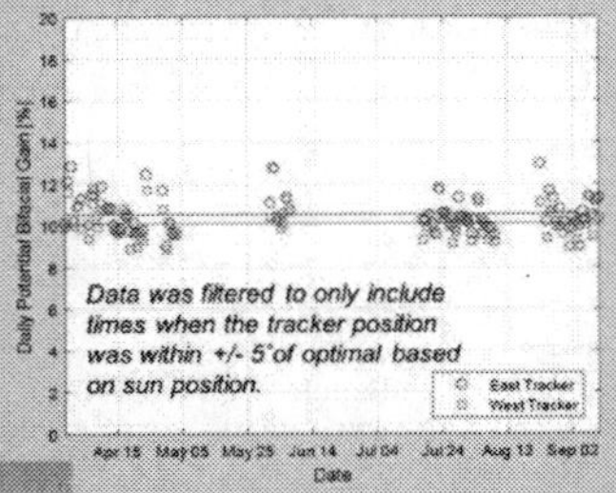

16

6CO.13.3 Outdoor Field Performance from Bifacial Photovoltaic Modules and Systems

J.S. Stein et al, Sandia National Laboratories, Albuquerque, United States
C. Deline, NREL, Golden, United States
F. Toor, University of Iowa, United States

European Commission

Light and elevated Temperature Induced Degradation

Piercing the Puzzle Together

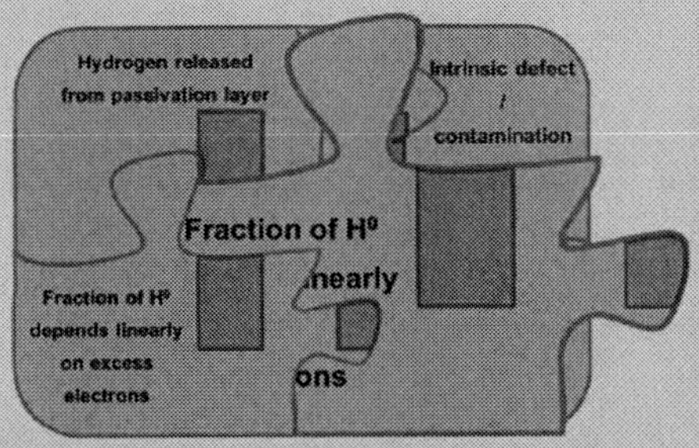

Model
Degradation

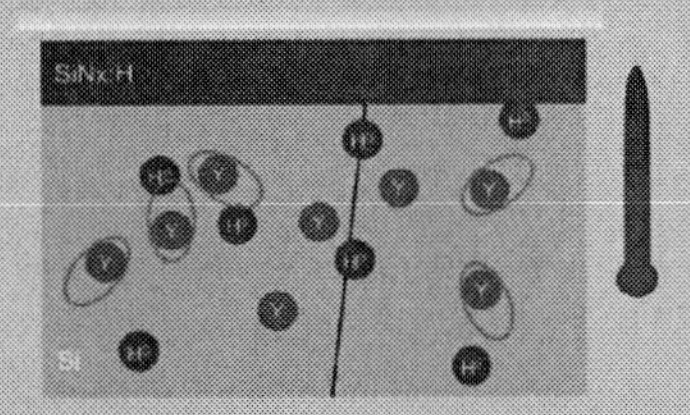

- LeTID observed in all p-type Si materials
- Defect activation only possible after fast firing of SiN_x:H-layer
- LeTID defect involves one fast-diffusing species
- Defect activated by inducing a change in charge state

→ Key to controlling LeTID: Manipulation of hydrogen

- Promising route for investigations: Float-zone silicon

2BP.1.4 Understanding Light-Induced Degradation in Multicrystalline Silicon: Possible Complex Formation Mechanisms

F. Schindler et al, Fraunhofer ISE, Freiburg, Germany

European Commission

Big data can help (1)

More than one million PV systems < 30 kWp have been installed in Germany.

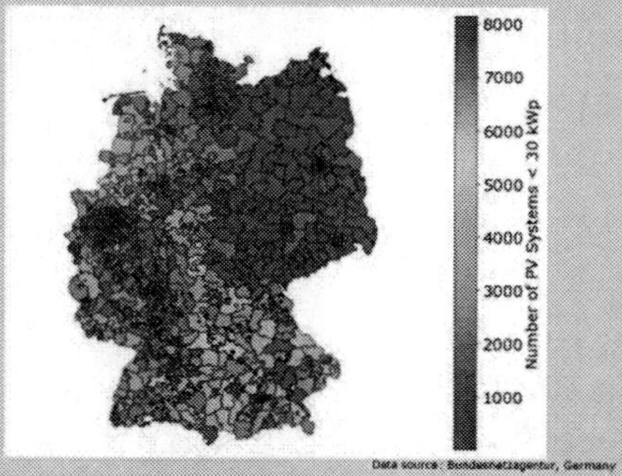

There is an annual loss of €450m to €500m by residential roof-top PV systems in Germany.

Year	Annual electric energy production in TWh	Annual yield loss in TWh	Monetary loss in million EUR
2014	36.1	1.46	€460m
2015	38.7	1.59	€489m
2016	37.6	1.58	€460m

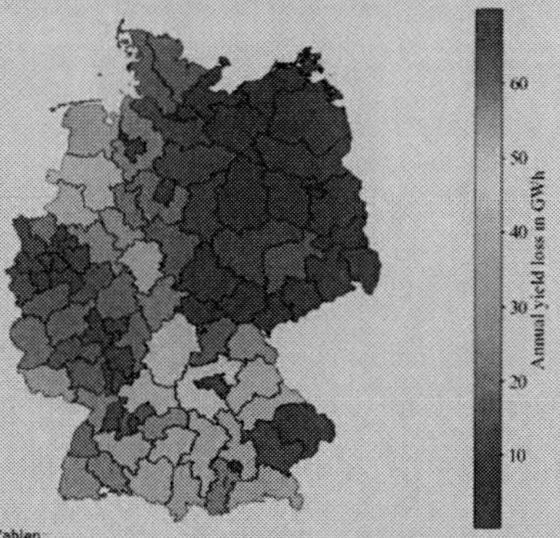

Source for mean Feed-In Tariff: BMWI, EEG in Zahlen: Vergütungen, Differenzkosten und EEG-Umlage 2000 bis 2017.

6BO.7.1 Survey on Yield of PV Systems in Germany 2014 to 2016

H. te Heesen & M. Rumpler
Trier University of Applied Science, Neubrücke, Germany
V. Herbort Ulm University of Applied Sciences, Germany

European Commission

Big data can help (2)

Model development: Model Architecture

i-EM

Two parallel approaches

- **Generic Fault Prediction (Supervision-Diagnosis Model):** *predict inverter deviations from nominal behavior*
- **Specific Fault Prediction (Short-Term Fault Prediction Model):** *predict a specific fault class*

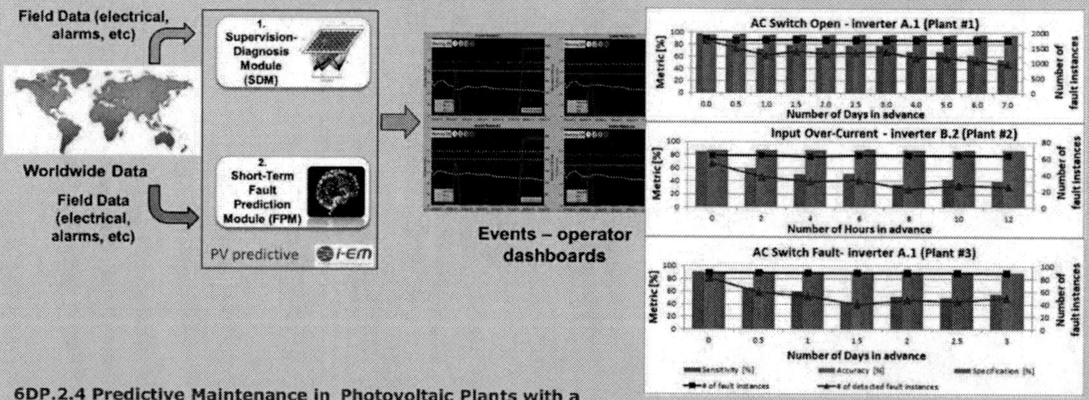

6DP.2.4 Predictive Maintenance in Photovoltaic Plants with a Big Data Approach

Alessandro Betti et al, i-EM (Livorno, ITALY), Enel Green Power (Rome, Italy)

European Commission

Non-destructive testing increasingly quantitative

Indication for PID

PID
- Homogeneous T increase
- Located at one string end
- Trend of PID-ratio increase towards the end
- < 50% of modules affected
- Incubation time
- PID-cells close the frame

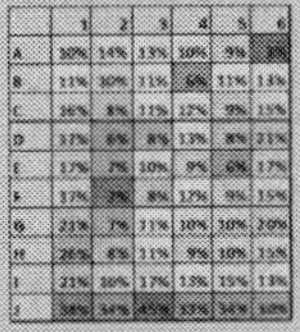

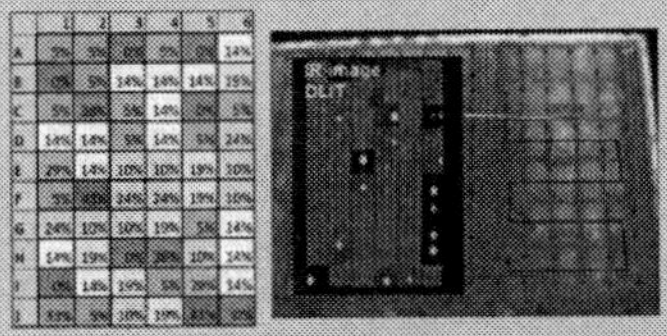

- Homogeneous T increase
- Arbitrarily distributed
- many modules may be affected
- Close to commissioning
- Affected cells everywhere

6BO.8.2 Quantitative Study of Potential Induced Degradation of a Roof-Top PV-Installation with IR-Imaging

C. Buerhop-Lutz et al, ZAE Bayern, Erlangen, Germany
C. Zetzmann Rauschert, Pressig, Germany
C. J. Brabec University of Erlangen-Nuremberg, Germany

European Commission

Which module is an insurance case?

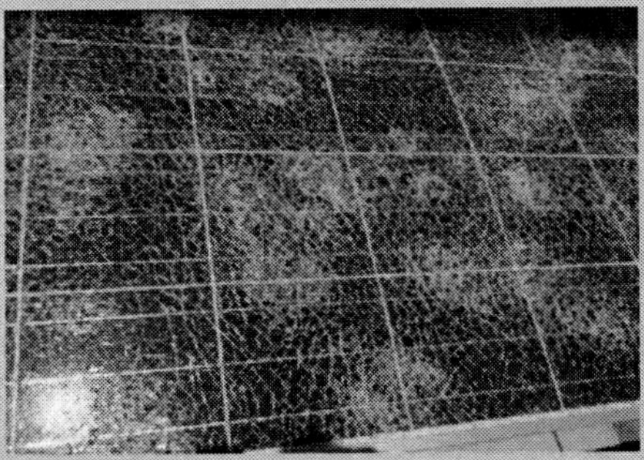

6BO.8.3 Scientific Investigation of a PV Generator After Hail

W. Mühleisen et al, CTR, Villach, Austria
M. Spielberger PVSV, Guttaring, Austria
H. Sonnleitner ENcome, Klagenfurt, Austria
Y. Voronko OFI, Vienna, Austria

European Commission

Product Environmental Footprint Category Rules (PEFCR) for PV

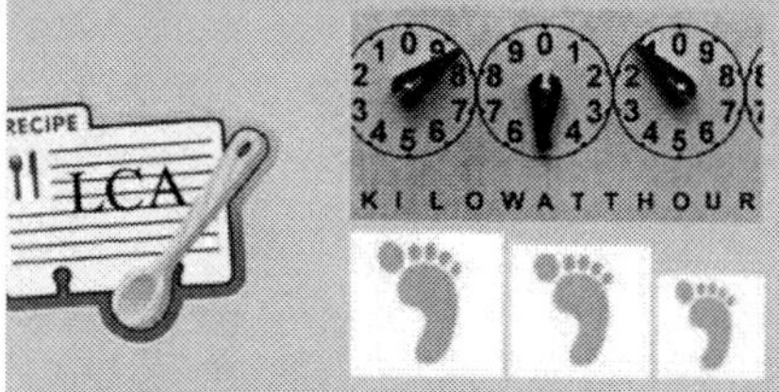

The PEFCR provides a harmonized set of rules to evaluate the environmental impacts of electricity produced from photovoltaic modules installed in the region of the EU28 and EFTA countries.

Circular Economy

Is it feasible to aggregate the complex results of a Product Environmental Footprint evaluation into a single score?

Communication Vehicles (CVs)

Performance label	Loyalty schemes	Campaigns targeting user behaviour
Performance label + QR code	Leaflets, catalogues, etc.	Marketing campaigns/ advertising
Performance improvement label	Instruction manuals	Declaration/ product passport
Barcode	Websites (producer, vendor)	On invoices
Pictogram, POS product advertisement	Websites (3rd party)	PEF external communication report
Consumer receipt – basket comparison	App based on performance	Performance tracking report

Communication vehicles tested during the PEF Pilot Phase for the product category of PV electricity production

5DP.1.4 The Product Environmental Footprint (PEF) of Photovoltaic Modules – Lessons Learned from the Environmental Footprint Pilot Phase on the Way to a Single Market for Green Products in the European Union

A. Wade, First Solar, Mainz, Germany
P. Stolz & R. Frischknecht Treeze, Uster, Switzerland
G. Heath, NREL, Golden, United States
P. Sinha, First Solar, Tempe, United States

European Commission

Product Innovation

European Commission

BIPV / VIBV / IIPV

Solar module for automotive applications

TOYOTA

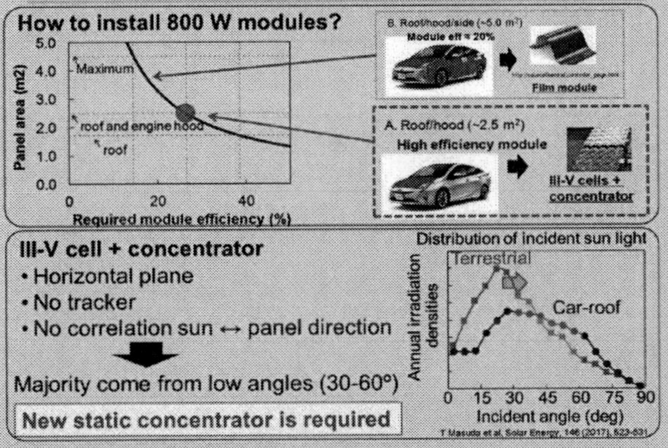

- New static low concentrator was proposed for the automotive application.

- Prototype module was fabricated by using cutting tool for lens processing.

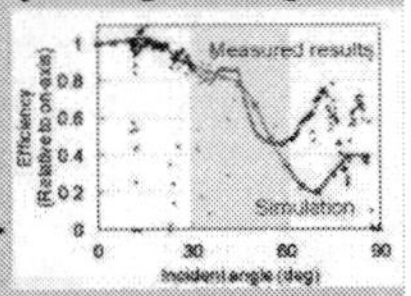

- Characteristics of lens is in good agreement with calculated results.

4DO.4.6 Solar Powered Vehicles with Static Concentrator Photovoltaics

T. Masuda et al, Toyota, Susono, Japan

K. Araki & M. Yamaguchi, TTI, Nagoya, Japan

European Commission

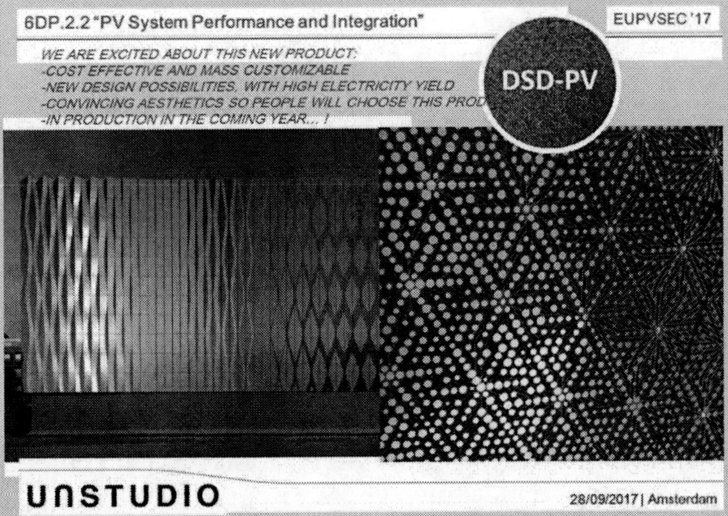

6DP.2.2 Aesthetics and Performance of PV

T. Minderhoud, UNStudio, Amsterdam, Netherlands

European Commission

Terrawatt

European Commission

Different Phases of VRE Integration – Towards Sector Coupling

Phase	Description	Country Examples
1	VRE capacity is not relevant at the all-system level	Most countries, incl. Mexico, Indonesia, South Africa
2	VRE capacity becomes noticeable to the system operator	Brazil, China, India, Sweden, Texas
3	Flexibility becomes relevant with greater swings in the supply/demand balance	Italy, Germany, Portugal, Spain, UK
4	Stability becomes relevant. VRE capacity covers nearly 100% of demand at certain times	Ireland, South Australia, Denmark
5	Structural surpluses emerge; electrification of other sectors becomes relevant	
6	Bridging seasonal deficit periods and supplying non-electricity applications; seasonal storage & synthetic fuels	

Electricity only accounts for around one fifth of total final energy demar
renewables will require multiplying their uses in buildings, indus

Acceptance
Landscapes
Clean Energy for All

EUPVSEC 2017, Opening, Paolo Frankl, IEA

European Commission

Open access proceedings

European Commission

Thanks!

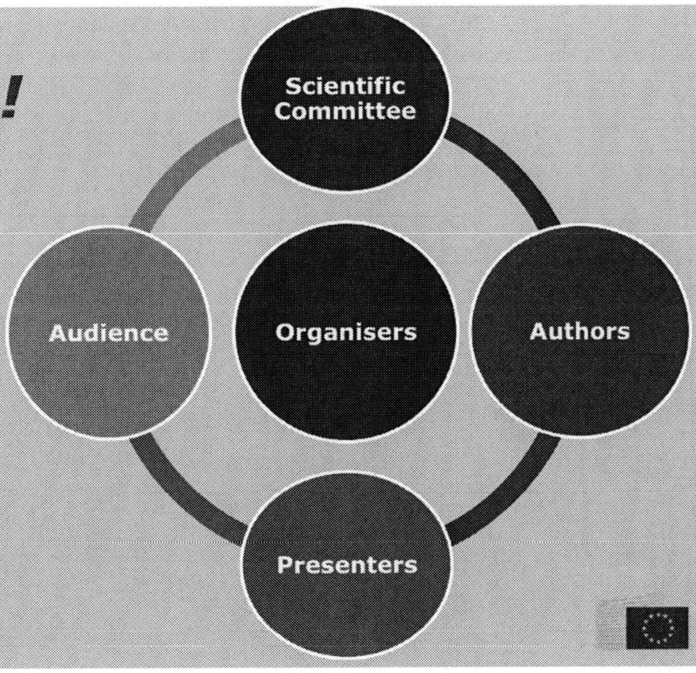

European Commission

LIST OF EXHIBITORS
(in alphabetical order)

3D-Micromac	DE
Abet Technologies	US
Advanced Silicon Group	US
ASYS	DE
Bentham Instruments	UK
BERGER Lichttechnik	DE
Berlin-Brandenburg Energy Network	DE
CSEM SA	CH
Delft University of Technology	NL
Dutch Energetics Management	NL
ECN	NL
Engineered Materials Systems	US
ETA Florence Renewable Energies	IT
EUREC	BE
European Commission, DG JRC	BE
FME	NL
greateyes	DE
h.a.l.m. elektronik	DE
Heraeus Photovoltaics	DE
Holland Solar	NL
HydroPV Technologies	NL
IEA PVPS	CH
InnoLas Solutions	DE
Ionx Cleaning Facilities	NL
ISC Konstanz	DE
IZOVAC Technologies	BY
Jonas & Redmann	DE
JRT Metallization Lines	DE
Kipp & Zonen	NL
KOPEL / KYOSHIN ELECTRIC	JP
Luvata Pori Oy	FI
Newport Spectra-Physics	DE
pv magazine group	DE
pv-tools	DE
Quantum Zurich	CH
RENA Technologies	DE
Semilab	HU
SINGULUS TECHNOLOGIES	DE
Sinton Instruments	US
Sisecam Flat Glass	TR
Solar Swiss Connect	CH
SOLARC Innovative Solarprodukte	DE
SOLARUNITED	DE
Solaxess SA	CH
Solibro	DE
Solliance	NL
Stäubli Electrical Connectors	CH
Sunprojects	NL
SUPSI ISAAC	CH
SVCS	CZ
Tempress Systems	NL
TFSC-Instrument	FR
TKI Urban Energy	NL
University of Ljubljana, LPVO	SI
Valentin Software	DE
Van der Valk Solar Systems	NL
Vela Solaris	CH
VITRONIC	DE
VON ARDENNE	DE
Wiley	US
WIP Renewable Energies	DE
ZSW Baden-Württemberg	DE

We thank the EU PVSEC 2017 Sponsors

BERGER Lichttechnik

LONGi Solar

RENA

SINGULUS

A SUCCESSFUL CONVERSION OF SILICON THIN-FILM SOLAR MODULE PRODUCTION TO HIGH EFFICIENCY HETEROJUNCTION TECHNOLOGY

D. Andronikov[1], A. Abramov[1], S. Abolmasov[1], K. Emtsev[1], G. Ivanov[1], A. Kukin[1], I. Nyapshaev[1], D. Orekhov[1], A. Semenov[1], G. Shelopin[1], E. Terukova[1], E. Terukov[1]
N. Belkova[2], A. Dubrovskiy[2], P. Ishmuratov[2], A. Ivanov[2], D. Saykin[2], I. Shakhray[2], A. Smirnov[2], V. Tarasov[2], V. Timakov[2], A. Tomchinskiy[2]
G. Kekelidze[3]

[1] R&D Center of Thin Film Technologies in Energetics, Polytechnicheskaya str. 28, 194064 St. Petersburg, Russia
[2]Hevel LLC, Shorshelskiy 12, 429950, Novocheboksarsk, Russia
[3]Moscow Technological Institute, Leninskii Avenue 38A, off. 104, 119334, Moscow, Russia

ABSTRACT: In this work we report first results of successful conversion of Hevel LLC micromorph production line to Si HJ technology with production kicked-off in Q1 2017. Implementation of in-house Si HJ cell technology developed by R&D Center TFTE using Gen5 (1.4 m^2) KAI PECVD reactors with rear emitter bifacial cell design, O3 based IPA free texturing technology from Singulus, PVD tools and SmartWire module assembly line from Meyer Burger allowed to reach average cell efficiency of 21,7% with respective average 60 cell module power of 300W at production level and cell efficiencies as high as 22,77% at R&D. Achieved average module power after conversion from TF to Si HJ technology has almost doubled the annual line capacity from 97 to 160 MW$_p$.

Keywords: Gen5 KAI PECVD reactors, silicon heterojunction, HJT modules mass production

1 INTRODUCTION

It is clear, that despite of significant progress in silicon thin film (Si TF) PV technology in last two decades, recent drop of silicon price makes this technology noncompetitive in most of PV application for coming years. This situation has been confirmed by escape of main suppliers of equipment for Si TF PV technology from the market and resulted in severe problems with return of investments for companies entering this field. Fortunately, excellent crystalline silicon (c-Si) surface passivation by hydrogenated amorphous silicon (a-Si:H) gives a chance for employing the most expensive part of the Si TF production lines, i.e. PECVD systems, for silicon heterojunction (Si-HJ) technology. Simple structure of Si-HJ solar cells in combination with their high efficiency and low temperature processing make them very attractive for the PV industry. The record R&D Si-HJ cell with standard cell architecture developed by Panasonic [1] has an efficiency of 24.7%, while record IBC HJT cell by Kaneka shows an efficiency as high as 26,3% [2]. In light of these events a number of PV players started developing high-efficiency, crystalline silicon solar cells with a-Si:H films as surface passivation layers [3].

The results of application of industrial KAI-MT PECVD system from TEL Solar for manufacturing of Si HJ cells [4] and their subsequent implementation in modules [5] have been presented recently. It has been shown that Gen5 KAI PECVD reactors provide excellent uniformity of optical and electrical properties of hydrogenated amorphous silicon layers across entire surface of a 110×130 cm^2 wafer carrier and allow producing Si-HJ solar cells with efficiencies above 21%. Here we report first results of successful conversion of Hevel LLC's micromorph production line to Si HJ technology (HJT) by using an in-house solar cell technology developed by R&D Center TFTE – Hevel LLC's R&D division, implemented with support of equipment suppliers, and SmartWire module assembly line from Meyer Burger.

2 TECHNOLOGY DEVELOPMENT AT R&D LEVEL

It is well known, that PECVD systems and their infrastructure are the most expensive part of HJT production line. Therefore, the main objective of our project was to adapt KAI-MT systems and their automation, designed by TEL Solar (former Oerlikon Solar) for micromorph modules production, to the requirements of heterojunction technology.

Main attention of R&D Center at the first stage of the project was focused on development of Si-HJ process compatible with PECVD systems of production line because it became clear from the very beginning of our project that this task is one of the main challenges of the project. Therefore, even if some, especially preliminary, steps of PECVD process development have been performed using Oxford Instrument cluster tool, which includes several PECVD chambers, the main part of this work was done on KAI1-1200 reactors of R&D pilot line, which have exactly the same design as reactors of KAI-MT systems of production line.

Since both internal and external KAI-MT handling systems are designed for operation with 1.1x1.3 m glasses used in micromorph technology as a superstrate for silicon thin films deposition, application of KAI systems for heterojunction technology required design of carriers for loading of c-Si wafers into PECVD reactors. These carriers with 4 mm thickness made of special alloy compatible with requirements of PECVD deposition of high quality amorphous silicon layers needed for HJ technology and fluorine based cleaning process were designed by R&D Center.

HJT process development has been started at R&D Center TFTE in March 2014 [4]. At the beginning, our process capabilities were limited by the set of equipment intended for Si thin film technology, so we focused our efforts on small lab-scale cells with maximal size of 4x4 cm^2 with subsequent transfer of the achieved results to 100x100 cm^2 and full-size 6" solar cells (see Figure 1). Besides the PECVD process, special attention has been

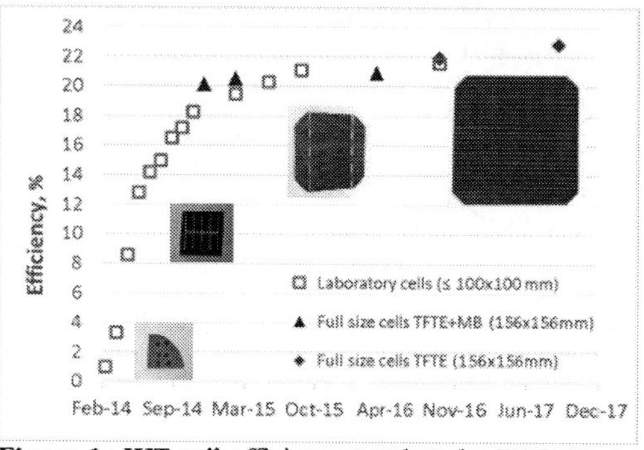

Figure 1: HJT cell efficiency trend at the R&D Center TFTE

paid to the development of the TCO deposition process on front and rear side of the cells. Main results of the study focused on the effect of target composition and oxygen flow during DC sputtering of front and rear indium tin oxide (ITO) layers on the performance of 6" bifacial Si-HJ are presented in [6]. Wet chemical treatment of wafers has been developed based on own know-how and support from wet chemistry equipment supplier Singulus. Special care has been taken to the proper choice of raw materials (wafers, TCO target material, Ag metallization paste etc.) for development of high efficiency HJT cells, which was supported by Meyer-Burger. These efforts allowed us to achieve 20,4% efficiency on full-size 6" cells in the end of 2014 with further efficiency improvement up to 22,77% in mid 2017 (See Figure 2). During this development the cell configuration has undergone a transformation from front-emitter mono-facial to rear-emitter bifacial design in accord with world trends for Si HJ technology.

3 PROCESS TRANSFER TO PRODUCTION LINE

One of the major challenges for implementing a new process in a mass production is its upscaling from laboratory to production level. Identical design of reactors in R&D and production PECVD machines allowed us to provide a smooth process transfer to Hevel production line.

Process transfer from R&D to production has been started in November 2016 before installation and application works on new cell production equipment have been finished. Therefore, at this stage wafers textured at

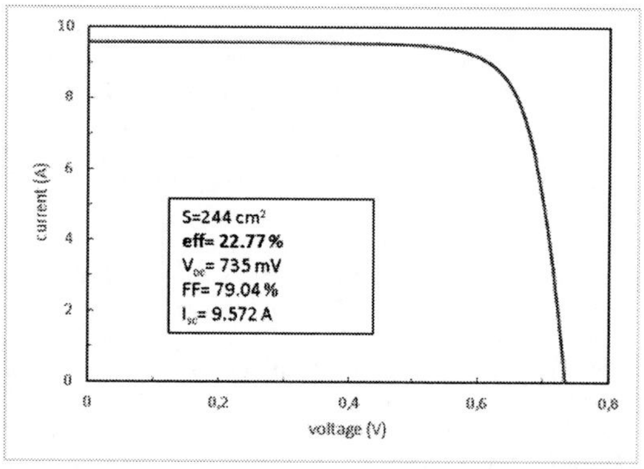

Figure 2: Best R&D HJT cell IV curve and characteristics.

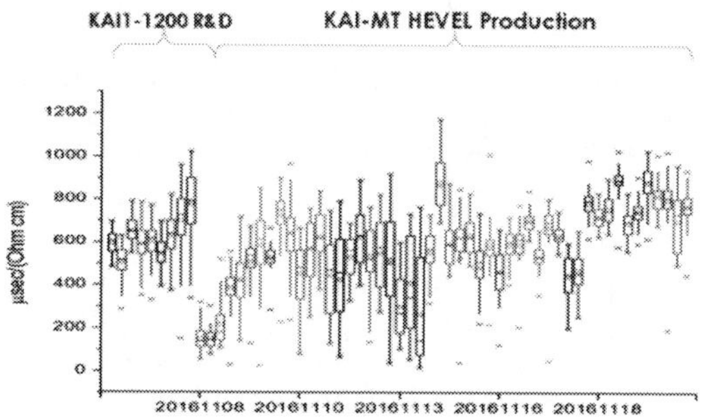

Figure 3: MCLT/resistivity ratio for R&D PECVD tool and first runs on production PECVD machines

the R&D center with an in-house developed process were used for processing in the production PECVD machines. Since minority carrier lifetime (MCLT)/resistivity ratio is a major criteria of passivation quality a Sinton WCT-120 tester was used to estimate the quality of PECVD process on KAI-MT production tools. In parallel, a specially designed wafer carrier housing 48 wafers, instead of the glass superstrate previously used in silicon thin-film ("micromorph") technology, was tested for compatibility with the KAI-MT internal and external transport in respect to its mechanical properties and temperature behavior. Special care was taken to design the carrier in such manner that no excessive deformation inside or outside the PECVD chambers was allowed which could influence the transport of the carriers inside and between the PECVD machines and loading/unloading stations. A new process flow for internal and external transport of wafer carriers was also designed. Figure 3 clearly illustrates that the PECVD process transfer has been successfully accomplished in less than two weeks. Optimizing wet chemistry treatment and wafer loading procedure has allowed increasing and even surpassing the MCLT/resistivity ratio values achieved previously at R&D KAI1-1200 pilot line tool and minimizing their process distribution. The distribution of average MCLT values between different KAI-MT pairs and the reactors inside is presented in Figure 4. One can see from the figure, that the average MCLT value over all reactors of production line of about 2,5 ms with deviation in the range of ±10 % has been reached.

To complete the HJT cell production line the following equipment has been acquired from major

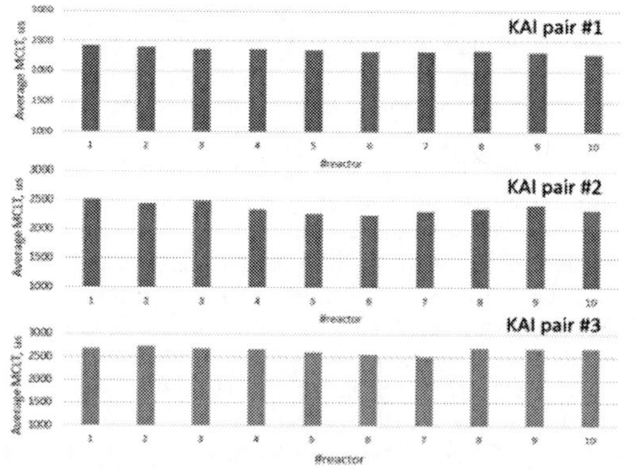

Figure 4: MCLT distribution between different KAI-MT pairs and reactors inside.

suppliers: two Hennecke Wafer Inspection Systems (throughput 3600 wafers/hour), a Silex 5200 Wet chemistry tool from Singulus (throughput 5200 wafers/hour), a newly designed wafer carrier loading/unloading and wafer flipping unit by Jonas&Redmann company (throughput >6000 wafers/hour), an Asys print line (throughput 4800 wafers/hour), two Helia PVD tools from Meyer Burger (throughput 2400 wafers/hour each), two Hennecke Cell Inspection Systems with flashing units from Pasan (throughput 2400 wafers/hour).

For module assembly with glass/backsheet design, a Meyer Burger SWCT (SmartWire Cell Technology) line was supplied. A special lamination recipe for Buerkle laminators remaining from the "micromorph" line has been tested and implemented for producing SWCT modules.

Efficient process support from equipment suppliers have ensured swift implementation of new technology and ramp-up of production capacity and yield (see Figure 5) with cell efficiencies above 21%.

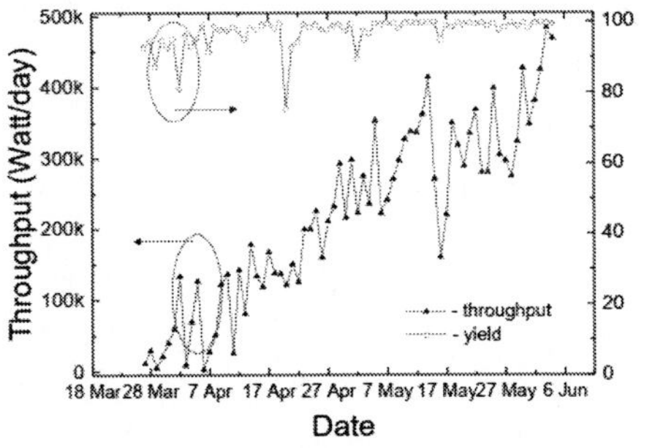

Figure 5: Progress in line throughput and cell yield during ramp-up of the line

The overall project schedule is presented in Figure 6 – the transition to Si HJ technology from shutting down the "micromorph" line and end of ramp-up of the new cell and module line has been accomplished in just 8 months.

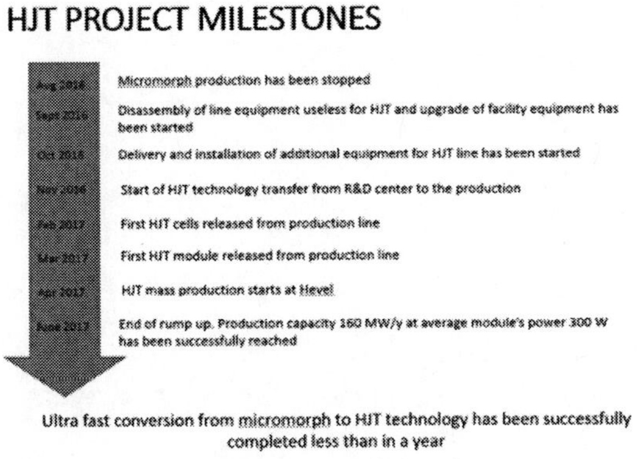

Figure 6: Total project time schedule

4. TECHNOLOGY FINE TUNING

Further increase of cell efficiency and manufacturing stability during full-scale production is only possible by implementing statistical analysis of process parameters and cell characteristics largely due to natural production related distribution and to relatively small achievable efficiency gains possible (on the orders of 0,1-0,2 abs %). This process was facilitated by introduction of cell line integrated manufacturing execution system (MES).

During 6 months of production we have optimized PECVD process parameters in terms of intrinsic and doped layer thicknesses, doping levels, wafer carrier internal transport in terms of line throughput and stability, ITO deposition process parameters in terms of Ar/O_2 ratio, screen printing process in terms of Ag paste printability and increase of the line throughput. As a result an average production cell efficiency of 21,7% has been achieved with best result of 22,6%. A typical production trend is presented in Figure 7.

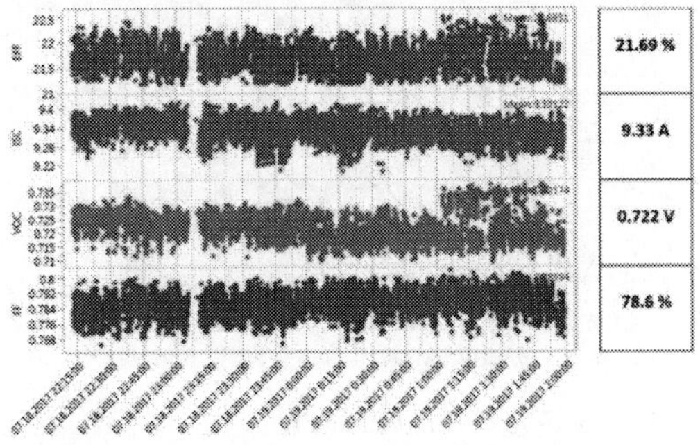

Figure 7: A typical production trend of cell characteristics

A progress in cell efficiency has allowed achieving an average production 60-cell module power of 300W, which has been independently confirmed by Fraunhofer ISE (Figure 8). The Hevel annual line production capacity has almost doubled from 97 to 160 MW_p

Manufacturer / Type — HEVEL SOLAR — Fraunhofer ISE
Hersteller / Typ

IV-curve parameter under standard test conditions (STC) — CalLab PV Modules
IU-Kennlinienparameter unter Standardtestbedingungen (STC)

Measurand Messgröße		Measured value* Messwert	Rel. uncertainty Rel. Unsicherheit
Short circuit current Kurzschlußstrom	I_{SC} / A	9.132	15 x 10^{-3}
Open circuit voltage Leerlaufspannung	V_{OC} / V	43.737	6 x 10^{-3}
Current at maximum power Strom bei maximaler Leistung	I_{MPP} / A	8.510	18 x 10^{-3}
Voltage at maximum power Spannung bei maximaler Leistung	V_{MPP} / V	35.346	19 x 10^{-3}
Maximum power Maximale Leistung	P_{MPP} / W	300.792	28 x 10^{-3}
Fill factor Füllfaktor	FF / %	75.312	26 x 10^{-3}
Efficiency Wirkungsgrad	η / %	17.965	29 x 10^{-3}

Figure 8: An extract from Fraunhofer ISE module measurement report

As a result, on September 19th, only in one year after closing the "micromorph" production, a 20 MW solar power plant constructed from HJT modules produced at Hevel was put in operation in Altay region in Russia (Figure 9).

Figure 9: A 20 MW solar power plant in Altay region in Russia constructed from Hevel HJT modules

Further progress in cell and module power as well as production cost reduction lie in continuous enhancement of raw material quality, reduction of wafer thickness, and subsequent optimization of all process steps based on analysis of production trends as well as advanced module assembly solutions.

4 CONCLUSIONS

In record-breaking schedule Hevel thin-film "micromorph" production line has been converted to high-efficient silicon heterojunction technology by implementing an in-house cell production process developed by its daughter company R&D Center TFTE. In less than a year after shutting-down the thin-film line an average cell efficiency of 21,7% with respective average 60 cell module power of 300W at production level and cell efficiencies as high as 22,8% at R&D have been reached. Achieved average module power after conversion from TF to Si HJ technology has almost doubled the annual line capacity from 97 to 160 MW_p

The authors wish to thank Meyer Burger, Singulus, Jonas&Redmann Company, ASYS for efficient support in implementing Si-HJ technology to Hevel production line, Izovac for their support at the early stages of the project, and Skolkovo Foundation for granted access to laboratory equipment.

REFERENCES

[1] M. Taguchi, A. Yano, S. Tohoda, K. Matsuyama, Y. Nakamura, T. Nishiwaki, K. Fujita, E. Maruyama, IEEE Journal of Photovoltaics 4 (2014) 96.
[2] K. Yoshikawa et el. Nature Energy, 2017;
[3] S.D. Wolf, A. Descoeudres, Z.C. Holman, C. Ballif, Green 2 (2012) 7.
[4] S. Abolmasov et al., Proc. of the 31st EUPVSEC, pp. 1046 – 1049, 2015
[5] D. Andronikov et. al, Proc. of the 32nd EUPVSEC 2AV.2.10, 2016
[6] S. Abolmasov, Proc. of the 33rd EUPVSEC, 2AV.3.39, 2017

IMPROVEMENT OF SILICON HETEROJUNCTION SOLAR CELLS WITH ARGON PLASMA TREATMENT

O. Sergeev, A. Neumüller, M. Vehse, C. Agert
DLR Institute of Networked Energy Systems
Carl-von-Ossietzky-Str. 15, 26129 Oldenburg, Germany

ABSTRACT: The quality of interfaces in silicon heterojunction (SHJ) solar cells is highly important for the solar cell conversion efficiency. This work studies the interface properties of hydrogenated amorphous silicon thin films on crystalline silicon wafers after argon plasma treatment (APT) compared to hydrogen plasma treated (HPT) and non-treated films. The investigation extends our previous study by examining the structural changes resulting from the post-deposition plasma treatment and corresponding influence on silicon thin film solar cells. We analyzed the infrared spectra of our samples to gain a deeper understanding of the fundamental plasma treatment effects. By using post-deposition APT and HPT, we were able to improve the structure of these layers. Our results show that APT yields a more compact material structure which provides better chemical passivation of silicon wafer surfaces. We propose to introduce APT as a post-deposition process step in the fabrication of silicon heterojunction solar cells and show preliminary results of improved performance of bifacial SHJ solar cell.
Keywords: Silicon Heterojunction Solar Cell, Argon, Hydrogen, Post-deposition Plasma Treatment, Microstructure

1 INTRODUCTION

In wafer based silicon heterojunction (SHJ) solar cells the interface or junction region between the absorber, passivation layer and p- or n- doped layers is very critical. In the case of SHJ solar cells back diffusion of photogenerated minority carriers is very likely and recombination occurs mainly on the defect states at the interface. To suppress this effect and to reduce recombination losses, hydrogen plasma treatment (HPT) is usually applied [1, 2]. This work studies the interface properties of hydrogenated amorphous silicon thin films on crystalline silicon wafers after post-deposition HPT and after argon plasma treatment (APT) as highly promising alternative approach. Single intrinsic hydrogenated amorphous silicon (i-aSi:H) layers followed bypost-deposition APT or HPT were deeply analyzed. As this layer forms the heterointerface to the n-doped crystalline silicon (c-Si) absorber which is the most important for the performance of SHJ solar cells.

To investigate single layers and to gain a deeper understanding of the fundamental plasma treatment effects the surface photovoltage (SPV), X-ray photoelectron spectroscopy (XPS), spectroscopic ellipsometry, infrared spectroscopy and minority carrier lifetime measurements were conducted.

Besides single layer characterization the origin of enhanced interface properties in SHJ after post-deposition APT solar cells was studied.

2 EXPERIMENTAL

Within this study we fabricated different sets of samples as shown on Fig. 1. Double side polished (100) float zone (FZ) n-doped silicon wafers were used for fabrication of single i-aSi:H layers (Fig. 1a). Textured (111) n-type Czochralski (CZ) silicon (c-Si) 6-inch wafers with an average resistivity of 5 Ω·cm and a thickness of 130 μm were used to fabricate the test layer heterostructure (Fig. 1b) and bifacial SHJ solar cells (Fig. 1c).

Prior to the deposition, the silicon wafers were immersed into a 1% HF solution for 3 minutes. Then i-aSi:H layers were deposited on the wafer surface from one side (Fig. 1a) or consequently in separated processes from both sides to form symmetrical heterostructure (Fig.1 b,c). After the deposition, the i-aSi:H layers were exposed to an argon or hydrogen plasma which led to ion bombardment on the surface [1,2,3].

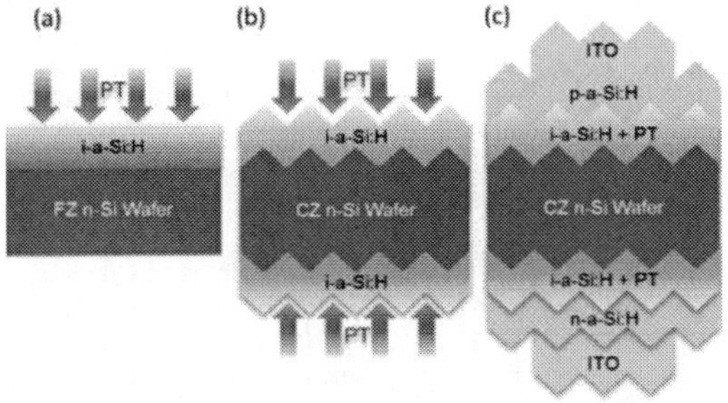

Figure 1: Sample structures developed in this work for investigation of post-deposition plasma treatment (PT) effects on single i-aSi:H layers (a), on the wafer passivation quality (b), on SHJ solar cell efficiency (c)

Fabrication and post-deposition plasma treatment of all amorphous hydrogenated silicon layers have been done with 13.56 MHz PECVD at Von Ardenne cluster tool CS 400 PS.

First, to evaluate the effect of APT or HPT onto a-Si:H thin films we produced the series of samples with i-aSi:H thickness variation from 10 nm to 140 nm, labeled as TS (Fig 1a). The samples were deposited at 19 mW/cm² power density (RF-P), pressure of 1 mbar, the distance between electrodes of 20 mm, the heater temperature of 200 °C and hydrogen dilution $R = \varphi(H_2)/\varphi(SiH_4)$ of 2.5.

Further, two sets of samples, labelled LP for low-power (19 mW/cm²) and HP for high-power (38 mW/cm²) deposition of a-Si:H were fabricated each as a symmetrical heterostructure (Fig 1b). The hydrogen dilution, R was set to 2.5 at constant silane flow of 40 sccm, while other process parameters were the same as in series TS. The process time was correspondingly adjusted to fabricate layers of the same thickness.

Post-deposition plasma treatment was performed under the same conditions for all series, i.e. at 200 sccm gas flow (argon or hydrogen), power of 19 mW/cm², electrode distance of 25 mm, pressure of 1 mbar and at 200 °C heater temperature. Treatment time was 60 s and

120 s for APT and HPT, correspondingly. To reduce the dangling bonds effect the samples were annealed for 90 minutes at 190 °C in air [4].

Structural modifications in i-aSi:H thin films were analyzed with Fourier-transform infrared spectroscopy (FTIR) Perkin Ellmer Spectrum 400. Surface potential changes (SPV) were measured by scanning Kelvin probe system (SKP5050) equipped with 2 mm gold-covered vibrating tip and under pulsed laser irradiation (670 nm, 20 mW/cm^2 intensity) [5]. Further, XPS measurements were performed with an Omicron XPS system under normal emission geometry, and employing Al K_α radiation [6]. The effective minority carrier lifetime (τ_{eff}) was determined using Sinton instruments WCT-120 equipment at a carrier injection level of $\Delta n = 1x10^{15}$ cm^{-3} and 1 sun.

Finally, we fabricated a bifacial solar cell as shown on the Figure 1c. Symmetrical 5-6 nm thick i-aSi:H layers were deposited at a power density of 38 mW/cm^2 with a hydrogen dilution of R = 2.5. After that either APT or HPT was applied to both i-aSi:H layers prior to the subsequent deposition of the n- and p-doped amorphous silicon layers. Finalizing the cell structure, 75 nm thick ITO front and back electrodes were deposited through the shadow mask with 2x2 cm^2 square openings with DC-sputtering in the PVD-chamber of the VAAT cluster tool. The value of short circuit current density J_{sc} was obtained from external quantum efficiency measurement. The pseudo fill factor (pFF), the pseudo open-circuit voltage (pV_{oc}) and the pseudo solar conversion efficiency ($p\eta$) was determined using a Suns-V_{oc} set-up (Sinton WCT-120).

All measurement errors were determined by means of multiple measurements, either in different points or on different samples. Error bars of individual measurements were determined from the standard deviation. Otherwise, an instrumental error of 5% was assumed, as specified by the tool suppliers.

3 RESULTS AND DISCUSSIONS

Several publications show that the microstructure factor of a-Si:H thin films correlates with the defect density [7,8]. Therefore, the ms factor is an ideal indicator to get a deeper understanding of general a-Si:H material properties after plasma treatment (PT).

Infra-red spectra were recorded in order to define a microstructure and hydrogen content in analyzed samples. The difference in IR-spectra of non-treated and plasma treated 10 nm thick i-aSi:H layer from series TS is shown in Figure 2.

The microstructure factor (ms) was calculated as the ratio of the integrated intensities of the absorption peak around 2100 cm^{-1} corresponding to polyhydride bonds and all the absorption peaks related to monohydride stretching modes around 2000 cm^{-1} [8].

Fig. 3 reveals a strong dependence of the ms value on the plasma treatment type and on the thickness of i-aSi:H layer. For APT samples (black points), the ms values is lower than that for HPT and non-treated samples for any i-aSi:H thickness. At the same time, the HPT, commonly used in SHJ solar cell fabrication, obviously leads to increase of the microstructure factor.

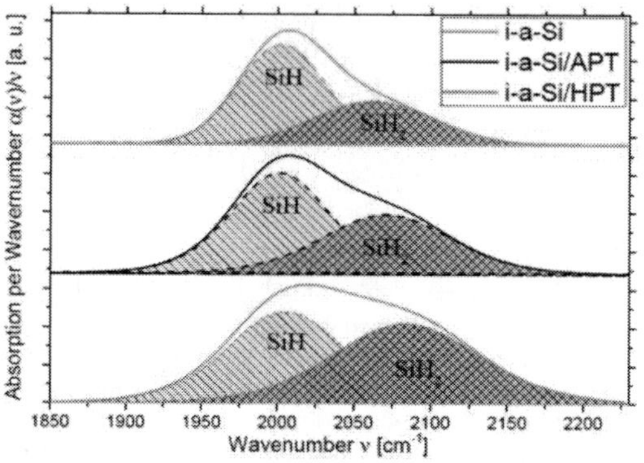

Figure 2: IR spectra of i-aSi:H layers: initial non-treated and subjected to argon- and hydrogen plasma treatment

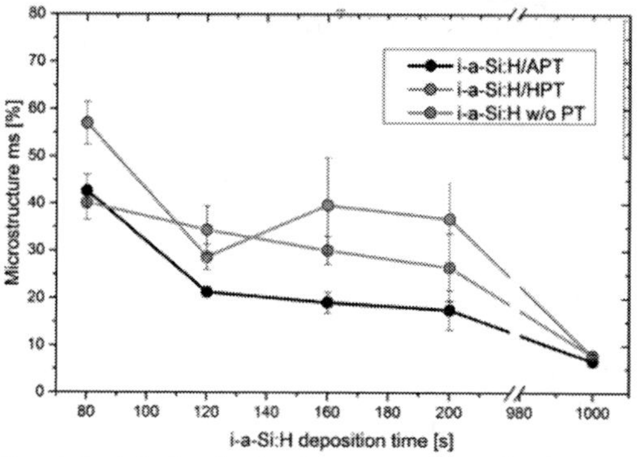

Figure 3: Calculated microstructure results of series TS

This effect was already described as loss of a-Si:H film compactness and reorganization of H bonding in favor of multi-vacancy-related bonding configuration which is accompanying the HPT [9]. The hydrogen concentration increases after HPT owing to enrichment of the amorphous silicon layer with hydrogen atoms and their diffusion to the i-aSi:H/c-Si-interface.

The same effect can be seen in Figure 4, where the ratio between hydrogen concentrations corresponding to main silicon–hydrogen vibrational modes is plotted for five different i-aSi:H film thicknesses depending on the post-deposition plasma treatment method. The change of the hydrogen concentration which contributes to the different silicon-hydride bonding is calculated as described by Langford et al. [10]

We observed, that the total hydrogen content in i-aSi:H layers (blue spots in Fig.4) changes with the film thickness in divers ways for different kind of samples. It increases for initial non-treated samples from 21 % at 10 nm thickness to the maximum value of 29 % at 20 nm thickness and then stabilizes at values around 20-21% to thicker films. Important to notice, that the fraction of defect-free isolated monohydride bonding (C_{H2000}) in total hydrogen content steadily increases from 38 % to 83 %. At the same time HPT samples show intensive rise of total hydrogen content from 23 % to 33 % with increasing the thickness from 10 nm to 20 nm and then slow recession to 20% for the thickest layer. According to these results the HPT has only an impact to very thin layers. However, it is clear seen, in comparison to initial samples the higher hydrogen content in HPT samples is related to the increase of the week clustered polyhydride fraction (C_{H2001}), which is attributed to the voids formation.

Our observations indicate that APT does not change the total hydrogen concentration with film thickness, but the structural properties improve. The week polyhydride fraction after the APT is significantly smaller compared to the initial and HPT samples and also steadily reduces from 57% to 17% with thickness. This effect is beneficial for fabrication of high-quality amorphous silicon layers [11].

The effects of HPT and APT are marginal for thicker films. In our case 140 nm thick i-aSi:H layer shows the same hydrogen concentration as well as the same ratio of bonding fractions for initial and plasma-treated samples.

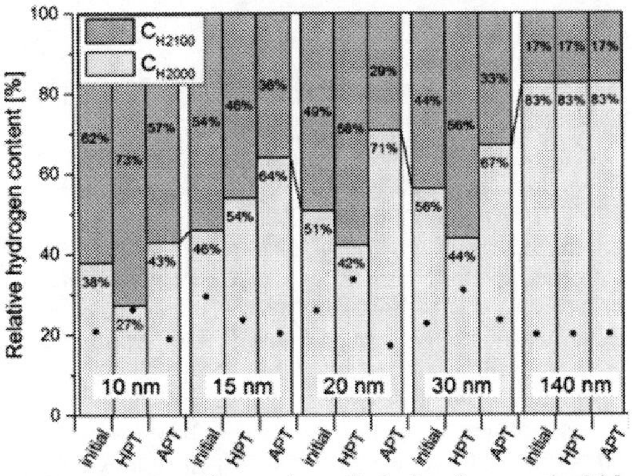

Figure 4: The ratio of isolated monohydride concentration (C_{H2000}) to the clustered polyhydride concentration (C_{H2001}) in initial and plasma-treated i-aSi:H layers of different thickness. Blue spots reveal total hydrogen concentration in the film.

In order to estimate the influence of improved compact structure on the electrical properties of plasma treated i-aSi:H material the surface photovoltage and Kelvin Probe measurement techniques were applied (Fig. 5).

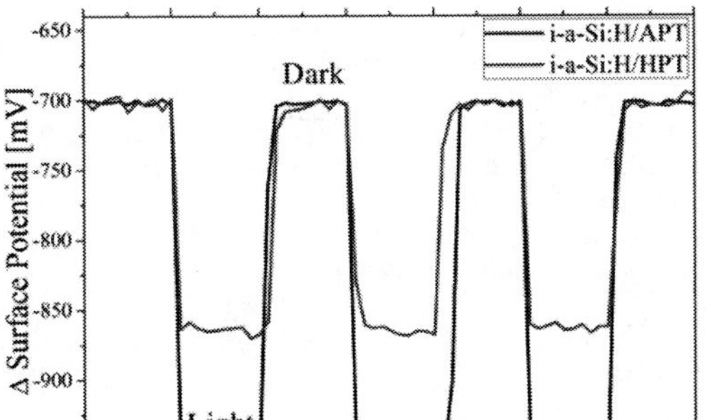

Figure 5: Several surface photovoltage (illuminated) and Kelvin probe (dark) measurements after APT or HPT.

In dark, both samples with HPT (red) and with APT (black) show the same relative surface potential. Under the illumination the differences between two samples is evident. The APT sample shows 29 mV difference in the surface potential under dark and illuminated conditions which is nearly two times better as for HPT sample.

This fact leads us to the assumption that the minority carrier diffusion length increases and the interface or surface defect states reduces significantly after the APT compared to the HPT. This effect can be explained again with neutralized material stress at the surface [12] and above mentioned conversion of hydrogen from a higher silicon-hybrid to a monohybrid state at the interface.

According to Palade et al. a heavy ion trap within a silicon network, in our case argon, would produce lattice defects which act as trapping centers and local fields of strains [13]. That would have a detrimental effect on the i-aSi:H/c-Si interface. Therefore, if the argon plasma treatment leads to argon implantation or diffusion, as reported by Neitzert et al. for hydrogen and helium [14] the defect density would increase. In our XPS measurements (Fig. 6), neither an Ar 2p nor an Ar 2s photoelectron signals was detected. From this measurement, it can be concluded that no argon is implanted in the near-surface region of these layers. Further, no possible doping is provoked due to the absence of argon.

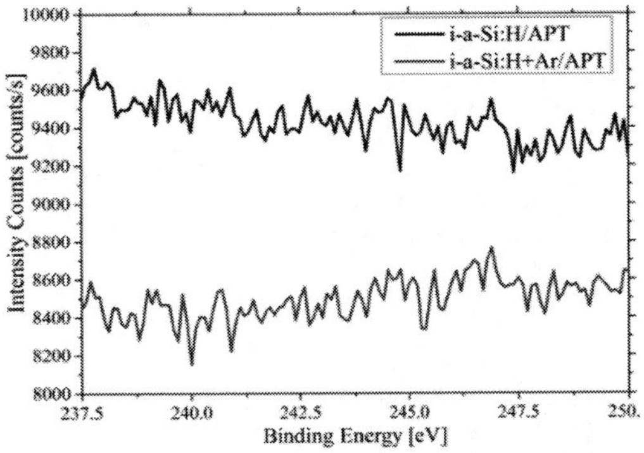

Figure 6: XPS Signal of a post deposition APT sample and the XPS signal of an i-aSi:H layer diluted with argon during the deposition.

As APT efficiently improves very thin i-aSi:H layers (< 20 nm), it is very promising to apply this kind of treatment for the enhanced passivation in SHJ solar cells. Two sets of textured wafers were passivated symmetrically with 9 nm thick i-aSi:H films (set LP for low-power and set HP for high-power deposition). One part of them was subjected both sides to the HPT, second part to the APT.

We studied the recombination at the i-aSi:H/c-Si heterointerface as suggested by Olibet et al. taking the amphoteric nature of silicon dangling bonds into account [15]. The effective lifetime (τ_{eff}) was measured depending on the minority carrier density (Δn) and depicted on Figure 7. As expected, the effective lifetime for all APT samples is significantly higher as for HPT samples by all injection levels. At minority carrier density of $1 \cdot 10^{15}$ cm^{-3} lifetime was 915 µs and 453 µs for APT and HPT samples of high power series and 665 µs and 411 µs for APT and HPT samples of low power series correspondingly.

Basically, with the same type of plasma treatment the samples fabricated at high power provide better lifetime as low-power deposited samples (as shown in Fig.7).

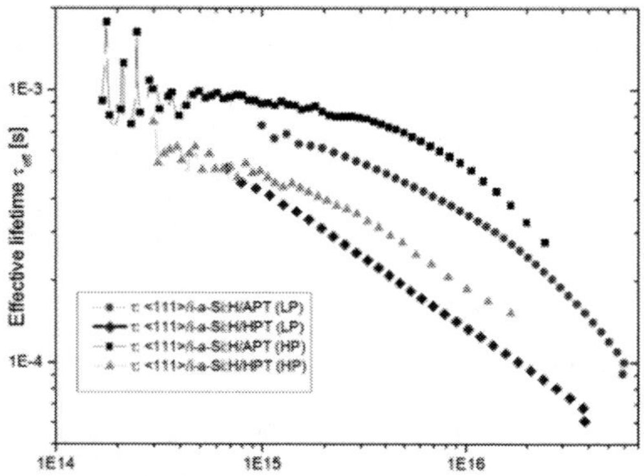

Figure 7: Injection level dependent effective lifetime for samples from each series LP and series HP after APT and HPT

After introducing argon plasma treatment on thin i-aSi:H layers and demonstrating beneficial effects especially for the minority carrier lifetime after APT, we implemented interface passivation steps in SHJ solar cells. The average short current density (j_{sc}) obtained by external quantum efficiency measurements was 35 mA/cm^2, and the Suns-Voc results are summarized in Tab. 1. The standard deviation is obtained from measurements on 16 individual SHJ solar cells of 2 x 2 cm^2 lab size each one.

Table I: Suns-Voc results

	pVoc, mV	pFF, %	pη, %
Initial	657.9 ± 36.8	75.1 ± 10.6	17.4 ± 3.0
HPT	689.6 ± 6.8	80.5 ± 1.0	19.3 ± 0.3
APT	708.0 ± 11.6	79.3 ± 1.0	19.6 ± 0.4

The values of pseudo solar conversion efficiencies pη and pV_{oc} increase significantly on average with APT compared to HPT and initial reference samples. APT and HPT samples have a similar pFF which is again higher as for non-treated sample. It can be also seen that with APT and HPT the variance of pη, pV_{oc} and pFF is lower across the sample. Obviously, non-uniformities of thickness and of layer quality are reduced with improved compact structure after post-deposition plasma treatment.

4 CONCLUSION

We demonstrate that our approach of post-deposition plasma treatment leads to the improvement of the i-aSi:H/c-Si interface by restructuring the i-aSi:H layer itself. It is discussed that argon or hydrogen plasma treatment applied to i-aSi:H/c-Si structures can also improve chemical passivation. By using either post-deposition APT or HPT, we were able to reduce the material stress and improve the structure of these layers. However, our results show that APT yields even more compact material and therefore the passivation quality improves after APT. This was shown with the help of effective lifetime and surface photovoltage measurements.

In addition, SHJ solar cells were fabricated to demonstrate the difference between APT and HPT on the open-circuit voltage and the influence on the solar conversion efficiency. As a result, a significant increase in the open-circuit voltage was observed after APT.

ACKNOWLDEGMENT

The authors thank Dr. Anamaria Moldovan (Fraunhofer ISE) for the Si-wafer pre-treatment and Ivan Shutsko, Martin Kellermann, and Maike Ahrlich for general assistance. Financial support from the German Federal Ministry of Economics and Technology (BMWi, reference number 0325825H, HERA) is gratefully acknowledged.

REFERENCES

[1] A. Descoeudres et al. Applied Physics Letters, 99(12), 2011. DOI:10.1063/1.3641899.

[2] C.-H. Lee, B.-J. Kim, and M. Shin. Progress in Photovoltaics: Research and Applications, 22(3):362-370, 2014. DOI:10.1002/pip.2281.

[3] A. Defresne, O. Plantevin, I. Sobkowicz, J. Bourçois, and P. Roca i Cabarrocas. Nuclear Instruments and Methods in Physics Research Section B: Beam Interactions with Materials and Atoms, 2015. DOI:10.1016/j.nimb.2015.04.009.

[4] W. Beyer, D. Lennartz, P. Prunici, and H. Stiebig. In Symposium A: Amorphous and Polycrystalline Thin-Film Silicon Science and Technology, volume 1321 of MRS Proceedings, pages mrss 11-1321-a17-14 (6 pages), 2011. DOI:10.1557/opl.2011.817.

[5] N. Revathi et al. Journal of Vacuum Science & Technology A, 32(6):061506, 2014. DOI: 10.1116/1.4896334.

[6] C. Schulz, T. Schmidt, J. I. Flege, N. Berner, C. Tessarek, D. Hommel, and J. Falta. Physica status solidi (c), 6(S2):S305-S308, 2009. DOI: 10.1002/pssc.200880845.

[7] T. Shimizu, K. Nakazawa, M. Kumeda, and S. Ueda. Physica B+C, 117:926- 928, 1983. DOI: 10.1016/0378-4363(83)90696-4.

[8] A. H. Mahan, P. Raboisson, and R. Tsu. Applied Physics Letters, 50(6):335-337, 1987. DOI: 10.1063/1.98192.

[9] M. Mews, T. F. Schulze, N. Mingirulli, and L. Korte. Applied Physics Letters, 102(12):122106, 2013. DOI: 10.1063/1.4798292.

[10] A.A. Langford, M.L. Fleet, B.P. Nelson, W.A. Lanford, N. Maley. Phys. Rev. B 45 (1992) 13367–13377, DOI: 10.1103/PhysRevB.45.13367.

[11] M.H. Brodsky, M. Cardona, J.J. Cuomo. Phys. Rev.B16, 1977,3556–3571, DOI: 10.1103/ PhysRevB.16.3556.

[12] S. De Wolf, S. Olibet, and C. Ballif. Applied Physics Letters, 93(3):032101, 2008. DOI: DOI: 10.1063/1.2956668.

[13] C. Palade, S. Lazanu, and M. L. Ciurea. In Semiconductor Conference, 2014 International, p.125-128, 2014. DOI: 10.1109/ SMICND.2014.6966411.

[14] H. C. Neitzert, N. Layadi, P. Roca i Cabarrocas, R. Vanderhaghen, and M. KunstJournal of Applied Physics, 78(3):1438-1445, 1995. DOI: 10.1063/1.360302.

[15] S. Olibet, E. Vallat-Sauvain, and C. Ballif. Phys. Rev. B, 76:035326, 2007. DOI: 10.1103/PhysRevB.76.035326.

HETEROJUNCTION IBC SOLAR CELLS ON THIN (< 50 µm) EPITAXIAL SI FOILS PRODUCED FROM KERFLESS LAYER TRANSFER PROCESS

Hariharsudan Sivaramakrishnan Radhakrishnan[1,*], Menglei Xu[1,2], Twan Bearda[1], Miha Filipič[1], Kris Van Nieuwenhuysen[1], Valérie Depauw[1], Ivan Gordon[1], Maarten Debucquoy[1], Jozef Szlufcik[1], Jef Poortmans[1,2,3]
[1] imec, Kapeldreef 75, B-3001 Leuven, Belgium
[2] KU Leuven, Kasteelpark Arenberg 10, B-3001 Leuven, Belgium
[3] Universiteit Hasselt, Martelarenlaan 42, B-3500 Hasselt, Belgium
* Corresponding author: phone +32 16 28 1173, email sivarama@imec.be

ABSTRACT: The i^2-module concept is one of the approaches for processing very thin (< 50 µm) silicon foils into highly-efficient back-contacted devices. In this work, we process kerfless epitaxial foils, that are lifted-off and layer-transferred onto glass, into heterojunction interdigitated back-contacted (HJ-IBC) devices. Ethylene vinyl acetate (EVA) is used as the bonding agent. As reference, thick and thin float zone wafers are used. With a carefully-selected and optimized process sequence that is compatible with EVA-bonded silicon, we obtain extraordinary V_{OC} values of 738 mV and 751 mV for the thick and thin FZ cells, respectively, which demonstrates the high quality rear-side passivation achieved with our process. However, epifoil cells have a much lower V_{OC} compared to the FZ cells and this is attributed to a lower bulk lifetime (30 µs). The best efficiency for an epifoil cell is 16.1% and that for a thick FZ cell bonded using EVA is 21.1%. Improving the epifoil lifetime to a level comparable to that of FZ material could increase the epifoil efficiency to above 20%.
Keywords: heterojunction, back-contact cell technology, epitaxial silicon lift-off

1 INTRODUCTION

Kerfless technologies [1]–[4] are mooted as the eventual replacement for the conventional wafer production route, since they allow thin silicon wafers (< 100 µm) and foils (< 70 µm) to be produced with negligible kerf losses, potentially enabling much higher silicon utilization (<5g/W_p), which has been one of the key drivers in the silicon PV industry [5]. One of extensively investigated routes to produce thin silicon foils is the epitaxial silicon lift-off technology based on porous silicon [6]–[13].

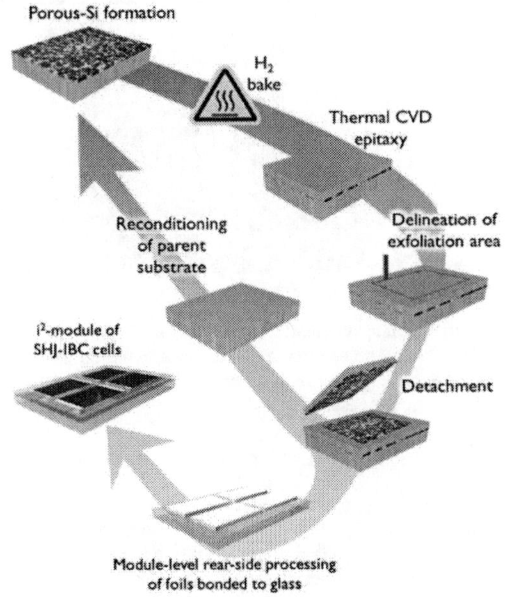

Figure 1: The i^2-module concept: schematics depicting the porous silicon-based lift-off of epitaxial silicon [13], the layer transfer of the epifoils onto glass and subsequent module-level cell processing of these epifoils into a HJ-IBC module.

As thin (< 50 µm) silicon foils (a factor of 3-4 times thinner than today's wafers) start to become available, there is a need for the development of concepts that enable processing of thin silicon into highly-efficient devices. The i^2-module ("integrated interconnect-module") concept [14] is one of the approaches pioneered by imec for processing such thin silicon foils of < 50 µm thickness into high-efficiency solar modules. While this concept is generically valid for thin Si produced by any method, Fig. 1 schematically describes the i^2-module concept for the case of an epitaxial foil ("epifoil") produced from the kerfless porous silicon-based lift-off process [6]. In this sequence, an epitaxial layer is grown on top of a sacrificial porous silicon seed layer using chemical vapor deposition (CVD) at temperatures above 1100 °C. This epitaxial layer is then lifted off from the parent substrate using the porous silicon layer as the detachment layer and layer-transferred onto module glass for further processing into interconnected solar cells.

The salient features of the i^2-module concept are that (1) the entire rear-side cell processing and cell interconnection are carried out at the module level, which has the cost benefits of large-area (m^2) processing, and (2) the thin Si foil is mechanically supported throughout the cell processing and cell interconnection stages, thus enabling high mechanical yield during device processing.

For the processing of epitaxial foils in the i^2-module approach, we adopted the silicon heterojunction (HJ) solar cell technology based on hydrogenated amorphous silicon (a-Si:H), and the interdigitated back-contacted (IBC) solar cell architecture for the following reasons: (1) the presence of bonding agents that glue the epifoils onto the glass constraints the maximum rear-side processing temperatures to 250 °C, which suggests the use of low-temperature heterojunction technology; (2) after bonding epifoils to the module glass, only the rear side is accessible for processing, which necessitates a back-contacted solar cell architecture; and (3) this technology has been applied successfully to achieve high efficiencies, as demonstrated by a series of world record efficiencies in the last few years, with the highest efficiency being 26.7% achieved on HJ-IBC solar cells by Kaneka [15], [16].

Previously, we reported on the ongoing process developments for bonded HJ-IBC solar cells on thick and thin float zone (FZ) wafers . The best reported efficiency for thick (190 µm) cells bonded using silicone was 21.7% [17], [18] and that bonded using EVA was 20.2% [18]. The best efficiency obtained for a thin (56 µm) EVA-bonded cell was 20.1% [18].

In this paper, we report on the implementation of the developed bonded HJ-IBC process flow on thin epitaxial foils from the porous silicon-based lift-off process, outlined in Fig. 1. The device performance of epifoil cells are bench-marked against thick and thin cells based on FZ material, which allows us to compare the influence of the cell thickness as well as the bulk material quality on the cell performance.

2 EXPERIMENTAL DETAILS

The schematic cross-section of the bonded HJ-IBC device is depicted in Fig. 2, together with a flow chart detailing the main steps of the process flow. Monocrystalline n-type (arsenic-doped, 10^{16} cm^{-3}) epitaxial layers with a thickness of 50 µm are grown on parent substrates with a detachable seed layer at 1130 °C in a CVD reactor. The area of the epitaxial layer to be detached is defined using laser ablation, which releases the epifoil from the parent substrate. Large-area epifoils with two different sizes of 10×10 cm^2 and 12.5×12.5 cm^2 are used for this work. For bench-marking, 200 µm thick n-type (3 Ω.cm) float zone (FZ) wafers and thinned down (70 µm) FZ wafers are processed together with the epifoils.

Prior to texturing, one of the two surfaces of the wafers/foils is protected with a dielectric layer in order to achieve single-side texturing. For epifoils, the surface with the seed layer is textured, so that the seed layer is simultaneously removed. A dilute TMAH solution with additives at 80 °C is used for texturing. The textured samples are then subjected to a cleaning sequence of sulphuric-peroxide mixture (SPM) and a dilute hydrofluoric acid (HF) dip. The final thicknesses of the epifoils, thinned FZ wafers and thick FZ wafers are 47 µm, 65 µm and 190 µm, respectively.

The textured side is the front-side of the cell and is passivated using a thin layer of intrinsic a-Si:H at 200 °C. An anti-reflection coating (ARC) of silicon nitride (SiN_x) is then deposited on top using PECVD at 225 °C. Note that the total mechanical yield of all the freestanding processing steps for large-area epifoils with a size of 12.5×12.5 cm^2 was only ~15% and those with a size of 10×10 cm^2 was 50%.

In the next step, the front-side processed epifoils and FZ wafers are bonded using ethylene vinyl acetate (EVA) sheets to module glass at a temperature of 150 °C and out-gassed for 2 hours in vacuum at a temperature of 200 °C. Note that the area of the EVA sheets is larger than the size of the silicon, such that EVA is exposed to process chemicals and ambient process gases during the rear-side processing. This configuration is important to prove the compatibility of the process flow to module-level processing.

The rear-side processing of the bonded samples begins with a wet cleaning sequence that removes surface contaminants on the rear-side. Subsequently, the rear-side is passivated using a stack of intrinsic and n-doped a-Si (i/n^+) at 175 °C using PECVD, which forms the back surface field (BSF) of the cell. A PECVD silicon oxide (SiO_x) layer, which is used as an etch mask and a sacrificial layer for lift-off, is then deposited on top at 225 °C.

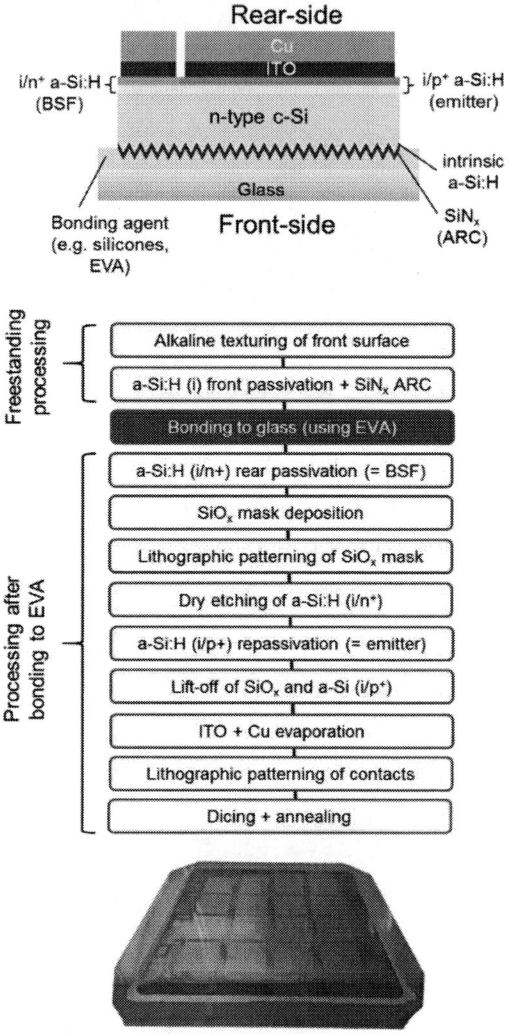

Figure 2: The schematic cross-section of a bonded HJ-IBC device, the process sequence implemented in this work and the photograph of an EVA-bonded epifoil during the process.

Lithography and wet etching is used to pattern the SiO_x etch mask and expose areas of a-Si (i/n^+) BSF, corresponding to the emitter fingers. The opened areas are etched using a dry etch process based on NF_3 plasma at 175 °C. The etched areas are passivated using a stack of intrinsic and p-doped a-Si (i/p^+) at 175 °C, which forms the emitter of the cell.

Lift-off of the remaining SiO_x mask together with the a-Si (i/p^+) on top is then carried out in dilute HF, resulting in an interdigitated pattern of emitter and BSF on the rear-side. Finally, indium tin oxide (ITO) is deposited by sputtering and copper by e-beam evaporation. Lithography and etching are used to pattern the rear contacts. After dicing the bonded wafers into small-area (2.5×2.5 cm^2) cells, the cells are annealed for different times between 150-175 °C.

Illuminated current-voltage (I-V) measurements at different illumination intensities (1 sun and 0.1 sun) are performed with an aperture area of ~2×2 cm^2. Spectral response measurements of the cells were performed to calculate the external quantum efficiency (EQE) of the cells. Suns-V_{OC} measurements were also done to extract the pseudo-fill factor (pFF) of the cells. Note that the front side of the cell is covered by the glass and EVA during all measurements.

3 RESULTS AND DISCUSSION

One of the main challenges of processing HJ-IBC devices on silicon that is bonded to glass using polymeric glues such as EVA or silicone, is that all rear-side processing must be compatible with the entire stack of materials, and in particular the bonding agent (in this case: EVA). Chemicals during wet processing or gases during plasma processing, that attack the bonding agent, could contaminate the surfaces of the bonded wafers with organic volatiles, which can result in poor passivation. This has been the subject of extensive studies in our previous works [17]–[19].

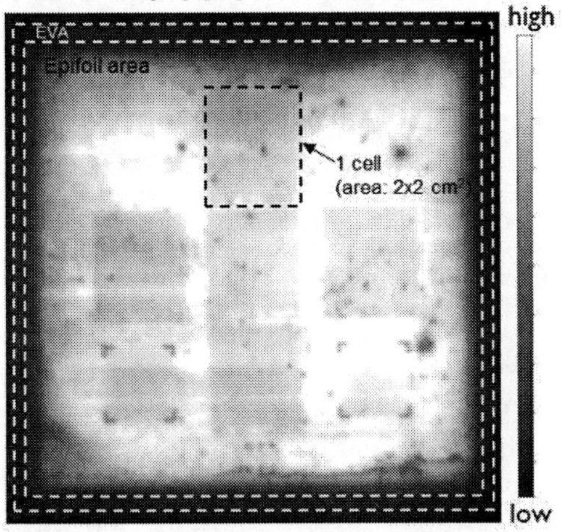

Figure 3: (a) Uncalibrated PL image at the end of the process of a 10 × 10 cm^2 epifoil, processed while bonded to glass with EVA (area: 11 × 11 cm^2). Of the 5 cells (area: 2 × 2 cm^2) fabricated on this epifoil, one of them is indicated with a dotted line.

Figure 3 shows the uncalibrated photoluminescence (PL) image at the end of the cell process of a 10×10 cm^2 epifoil, processing while bonded to glass using EVA. The area of EVA is larger than the epifoil such that it is exposed to all processing chemicals and gases. The process sequence, including wet cleaning and dry etching of a-Si:H, compatible with EVA has been developed and reported previously [18]. In Figure 3, five small-area cells can be discerned (one of them indicated using a dotted line), displaying uniform high quality passivation in both emitter and BSF areas, despite exposure of the bonding agent, EVA, to process chemicals and gases. This unambiguously demonstrates the compatibility of the described process flow for processing EVA-bonded silicon.

The cell performance results obtained from light I-V measurements on thick FZ cells, thin FZ cells and epifoil cells are summarized in Table I. For the best thick and thin FZ cells, remarkably high open-circuit voltage (V_{OC}) values of 738 mV and 751 mV are attained, which are the best values reported on bonded HJ-IBC cells to-date. The higher V_{OC} values for the thin FZ cells can be understood from the fact that in thinner cells, volume recombination is lower compared to thicker cells, which implies an increase in V_{OC} as thickness reduces. Notwithstanding the fact that the results shown in Table I are uncertified, the V_{OC} of 751 mV obtained for the thinned FZ cell is the highest reported so far for crystalline Si solar cells [15]. This demonstrates once again the very high quality passivation obtained on these bonded wafers, following the described process sequence.

Table I: Summary of cell performance results of the 3 splits

Cell	Anneal		J_{SC} [mA/cm^2]	V_{OC} [mV]	FF [%]	η [%]
190 µm thick FZ	150°C, 30 min	**Best**	**40.7**	**738**	**70.0**	**21.1**
		Avg.	*40.5*	*732*	*69.4*	*20.6*
65 µm thin FZ	150°C, 30 min	**Best**	**38.7**	**751**	**67.5**	**19.6**
		Avg.	*38.6*	*742*	*67.8*	*19.4*
47 µm epifoil	175°C, 5 min	**Best**	**35.0**	**673**	**69.0**	**16.1**
		Avg.	*34.3*	*666*	*66.8*	*15.2*

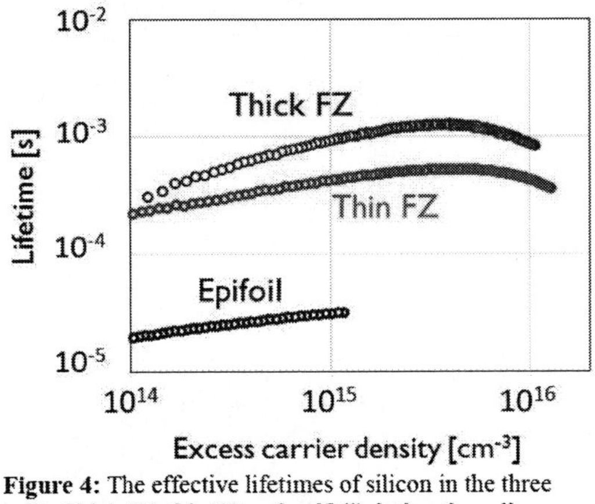

Figure 4: The effective lifetimes of silicon in the three splits (thick FZ, thin FZ and epifoil) during the cell process.

In comparison, the best V_{OC} for our epifoil cell is 673 mV, which is far lower than that of the cells fabricated on FZ material. Since the processing is similar for all cell types, this difference is attributed to the lower bulk lifetimes of epifoils compared to the FZ wafers. Figure 4 compares the effective lifetimes of the three cell types during the cell processing. At the injection level of 10^{15} cm^{-3}, epifoils have a lifetime of only ~30 µs, which is an order of magnitude lower than the FZ counterparts, which show lifetimes of ~400 µs (thin FZ) and ~1 ms (thick FZ). Spectral response measurements also show that the external quantum efficiency (EQE) of the epifoil cell is significantly lower than both the thick and thin FZ cells throughout the entire spectral range (Fig. 5).

The short-circuit current density (J_{SC}) of the different cells show a clear decreasing trend with thickness. In order to understand this trend, an analysis of the J_{SC}

losses, following ref. [20], was performed and the results are shown in Fig. 5. For all cell types, there is a strong parasitic absorption of UV light by EVA, which results in a loss of ~0.74 mA/cm^2 in J_{SC}. In addition, as the cell thickness reduces, the front escape loss of light, that is emitted out of the cells without being absorbed, increases. For the epifoil cell, losses in J_{SC} due to bulk recombination predominate.

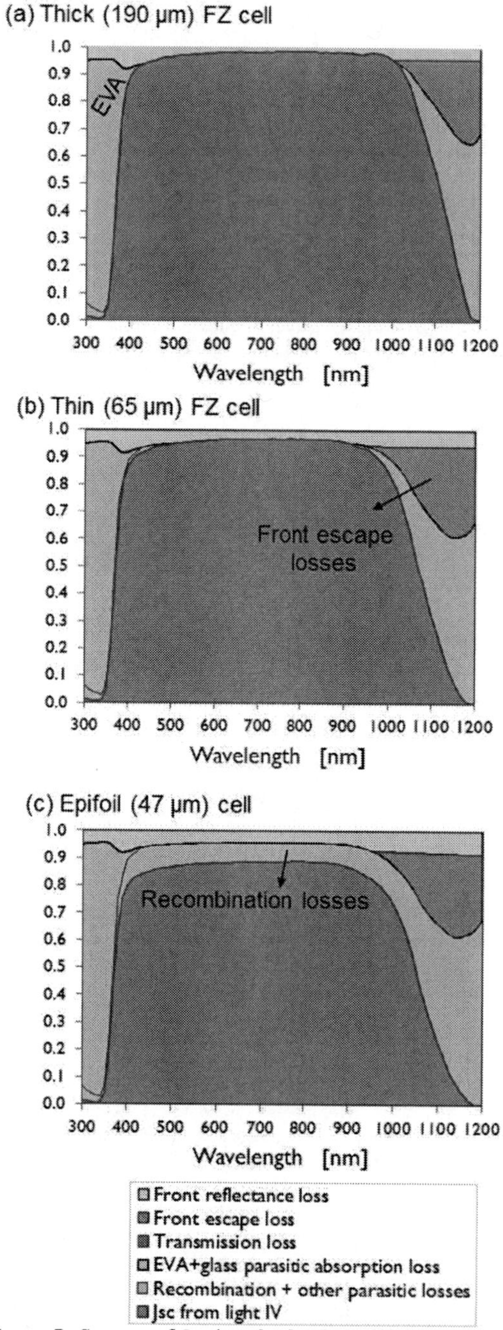

Figure 5: Sources of J_{SC} loss in the three different cell types: thick FZ cell, thin FZ cell and epifoil cell (based on [20]).

For all cell types, a major limitation in the cell performance is the fill factor (FF). A fill factor analysis (not shown here) was performed. It was found that the FF losses are due to series resistance and J_{02}-related recombination. However, annealing the cells in optimised conditions eliminates the J_{02} contribution to FF loss. Thus, the FF is mainly limited by series resistance, arising from high contact resistivity at the a-Si:H/ITO interface.

The best cell efficiency of 21.1% for an EVA-bonded thick cell was obtained, which is ~1% absolute higher than that previously reported for this cell type [18]. The best HJ-IBC epifoil has an efficiency of 16.1%.

Improvements to the cell performance of future epifoil devices are focused on (1) the optimization of the epitaxial Si growth conditions to increase the bulk lifetime [21], which would in turn improve the V_{OC}, (2) the optimization of the contact stack (a-Si, ITO, Cu) to improve the FF, (3) the use of alternative bonding agents to reduce losses in J_{SC} in the UV wavelength range [17], [18] and (4) the implementation of advanced light trapping structures such as inverted nano-pyramids to enhance light absorption in thin silicon cells [22].

4 CONCLUSIONS

In the i^2-module approach, HJ-IBC cells on silicon bonded to glass using EVA were fabricated using a carefully-optimized process flow that is compatible with such a material stack. This resulted in extraordinary V_{OC} values of 738 mV and 751 mV for the thick and thin FZ cells, respectively, demonstrating high quality rear-side passivation. Epifoil cells suffered from much lower V_{OC} and J_{SC} due to much lower bulk lifetimes compared to the FZ material. The best efficiencies of 16.1% for an epifoil cell and 21.1% for a thick FZ cell with EVA as bonding agent were attained. Improvements to the epifoil lifetime, together with cell FF improvements as well as switching from EVA to alternative bonding agents are needed to enable efficiencies above 20% for epifoil-based HJ IBC devices.

5 ACKNOWLEDGEMENTS

This work is financially supported by IMEC's Industrial Affliation Program for Photovoltaics (IIAP-PV). This project has received funding from the European Union's Horizon 2020 research and innovation programme under the Marie Sklodowska-Curie grant agreement No 657270.

6 REFERENCES

[1] R. Brendel, "Review of Layer Transfer Processes for Crystalline Thin-Film Silicon Solar Cells," *Jpn. J. Appl. Phys.*, vol. 40, no. 7, pp. 4431–4439, 2001.
[2] F. J. Henley, "Kerf-free wafering: Technology overview and challenges for thin PV manufacturing," *Conf. Rec. IEEE Photovolt. Spec. Conf.*, pp. 1184–1192, 2010.
[3] H. Sivaramakrishnan Radhakrishnan, V. Depauw, I. Gordon, and J. Poortmans, "Thin crystalline silicon: overview of lift-off technologies and module fabrication approaches," *(in Prep.*, 2017.
[4] H. Sivaramakrishnan Radhakrishnan and V. Depauw,

"Lift-off approaches for thin crystalline silicon," in *Thin Film Solar Cells Fabrication, Characterization and Applications*, J. Poortmans, Ed. John Wiley & Sons, Ltd, 2016.

[5] SEMI, "International Technology Roadmap for Photovoltaic (ITRPV) 2014 Results," 2015.

[6] K. Van Nieuwenhuysen, V. Depauw, R. Martini, J. Govaerts, M. Debucquoy, H. Sivaramakrishnan Radhakrishnan, I. Gordon, T. Bearda, K. Baert, and J. Poortmans, "High Quality Epitaxial Foils, Obtained by a Porous Silicon Based Layer Transfer Process, for Integration in Back Contacted Solar Cells," in *27th European Photovoltaic Solar Energy Conference and Exhibition*, 2012, pp. 2471–2474.

[7] J. Govaerts, C. Trompoukis, H. Sivaramakrishnan Radhakrishnan, S. N. Granata, E. G. Carnemolla, R. Martini, M. Karim, I. Sharlandziev, T. Bearda, V. Depauw, K. Van Nieuwenhuysen, I. Gordon, J. Szlufcik, and J. Poortmans, "Solar cells from epitaxial foils : an epifoil epiphany," *Energy Procedia*, vol. 77, pp. 871–880, 2015.

[8] J. H. Petermann, D. Zielke, J. Schmidt, F. Haase, E. G. Rojas, and R. Brendel, "19%-efficient and 43 µm-thick crystalline Si solar cell from layer transfer using porous silicon," *Prog. Photovoltaics Res. Appl.*, vol. 20, no. 1, pp. 1–5, Jan. 2012.

[9] C. Gemmel, J. Hensen, S. Kajari-Schroder, and R. Brendel, "4.5 ms Effective Carrier Lifetime in Kerfless Epitaxial Silicon Wafers from the Porous Silicon Process," *IEEE J. Photovoltaics*, vol. 7, no. 2, pp. 430–436, 2017.

[10] N. Milenkovic, M. Driesen, B. Steinhauser, J. Benick, S. Lindekugel, M. Hermle, S. Janz, and S. Reber, "Epitaxial N-type silicon solar cells with 20% efficiency," in *Proceedings of the 43rd IEEE Photovoltaic Specialists Conference (PVSC)*, 2016, pp. 58–61.

[11] S. Janz, N. Milenkovic, M. Drießen, and S. Reber, "n-Type and p-Type Silicon Foils Fabricated in a Quasi-Inline Epi Reactor with Bulk Lifetimes Exceeding 500 µs," in *Proceedings of the 31st European Photovoltaic Solar Energy Conference and Exhibition*, 2015.

[12] E. Kobayashi, Y. Watabe, R. Hao, and T. S. Ravi, "High efficiency heterojunction solar cells on n-type kerfless mono crystalline silicon wafers by epitaxial growth," *Appl. Phys. Lett.*, vol. 106, no. 22, p. 223504, Jun. 2015.

[13] H. Sivaramakrishnan Radhakrishnan, R. Martini, V. Depauw, K. Van Nieuwenhuysen, M. Debucquoy, J. Govaerts, I. Gordon, R. Mertens, and J. Poortmans, "Improving the Quality of Epitaxial Foils Produced Using a Porous Silicon-based Layer Transfer Process for High-Efficiency Thin-Film Crystalline Silicon Solar Cells," *IEEE J. Photovoltaics*, vol. 4, no. 1, pp. 70–77, Jan. 2014.

[14] J. Govaerts, C. Boulord, M. Fisher, G. Beaucarne, I. Gordon, J. Poortmans, and K. Baert, "I-module technology: evaluation and evolution," in *Proceedings of the 28th European Photovoltaic Solar Energy Conference and Exhibition*, 2013, pp. 59–62.

[15] M. A. Green, Y. Hishikawa, W. Warta, E. D. Dunlop, D. H. Levi, J. Hohl-Ebinger, and A. W. H. Ho-Baillie, "Solar cell efficiency tables (version 50)," *Prog. Photovoltaics Res. Appl.*, vol. 25, no. 7, pp. 668–676, 2017.

[16] K. Yoshikawa, H. Kawasaki, W. Yoshida, T. Irie, K. Konishi, K. Nakano, T. Uto, D. Adachi, M. Kanematsu, H. Uzu, and K. Yamamoto, "Silicon heterojunction solar cell with interdigitated back contacts for a photoconversion efficiency over 26%," *Nat. energy*, vol. 2, pp. 1–8, 2017.

[17] M. Xu, T. Bearda, H. Sivaramakrishnan, S. Kiran, M. Hasan, S. Malik, M. Filipi, V. Depauw, K. Van, Y. Abdulraheem, M. Debucquoy, I. Gordon, and J. Szlufcik, "Silicon heterojunction interdigitated back-contact solar cells bonded to glass with efficiency >21%," *Sol. Energy Mater. Sol. Cells*, vol. 165, pp. 82–87, 2017.

[18] H. Sivaramakrishnan Radhakrishnan, T. Bearda, M. Xu, S. K. Jonnak, S. Malik, M. Hasan, V. Depauw, M. Filipic, K. Van Nieuwenhuysen, Y. Abdulraheem, M. Debucquoy, I. Gordon, J. Szlufcik, and J. Poortmans, "Module-level cell processing of silicon heterojunction interdigitated back-contacted (SHJ-IBC) solar cells with efficiencies above 22%: Towards all-dry processing," in *Proceedings of the 43rd IEEE Photovoltaic Specialists Conference (PVSC)*, 2016, pp. 1182–1187.

[19] M. Xu, T. Bearda, H. Sivaramakrishnan Radhakrishnan, S. K. Jonnak, V. Depauw, K. Van Nieuwenhuysen, M. Filipic, M. Debucquoy, I. Gordon, J. Poortmans, and J. Szlufcik, "Process Development of HJ-IBC cells on Silicon Bonded to Glass," in *Proceedings of the 33rd European Photovoltaic Solar Energy Conference*, 2016.

[20] B. Paviet-Salomon, A. Tomasi, A. Descoeudres, L. Barraud, S. Nicolay, M. Despeisse, S. D. Wolf, and C. Ballif, "Back-Contacted Silicon Heterojunction Solar Cells: Optical-Loss Analysis and Mitigation," *IEEE J. Photovoltaics*, vol. 5, no. 5, pp. 1293–1303, 2015.

[21] H. Sivaramakrishnan Radhakrishnan, R. Martini, V. Depauw, K. Van Nieuwenhuysen, T. Bearda, I. Gordon, J. Szlufcik, and J. Poortmans, "Kerfless layer-transfer of thin epitaxial silicon foils using novel multiple layer porous silicon stacks with near 100% detachment yield and large minority carrier diffusion lengths," *Sol. Energy Mater. Sol. Cells*, vol. 135, pp. 113–123, Apr. 2015.

[22] C. Trompoukis, O. El Daif, P. Pratim Sharma, H. Sivaramakrishnan Radhakrishnan, M. Debucquoy, V. Depauw, K. Van Nieuwenhuysen, I. Gordon, R. Mertens, and J. Poortmans, "Passivation of photonic nanostructures for crystalline silicon solar cells," *Prog. Photovoltaics Res. Appl.*, vol. 23, no. 6, pp. 734–742, Jun. 2015.

AMORPHOUS SILICON DEPOSITED WITH PLASMA EXCITATION FREQUENCIES LARGER THAN 100 MHZ FOR HETEROJUNCTION SOLAR CELLS

C. Strobel*[1], B. Leszczynska[1], S. Leszczynski[1], M. Albert[1], F. Stahr[2], J. Kuske[2], J. W. Bartha[1]

[1] Dresden University of Technology, Semiconductor and Microsystems Technology Laboratory, 01062 Dresden, Germany, *Phone: +49 351 463 33151, e-mail: carsten.strobel@tu-dresden.de

[2] Forschungs- und Applikationslabor Plasmatechnik GmbH, Gostritzer Str. 67, 01217 Dresden, Germany

ABSTRACT: Intrinsic hydrogenated amorphous silicon layers (a-Si:H) for heterojunction solar cells with intrinsic thin layers (HIT) are deposited by means of radio frequency (RF) and very high frequency (VHF) plasma-enhanced chemical vapor deposition (PECVD). The VHF plasma excitation frequency used here (140 MHz) is much higher than typically used in the VHF range. It is widely known, that the ion energy in the plasma decreases with increasing plasma excitation frequency. Thus, it can be expected that the reduced ion bombardment energies in VHF plasmas results in a better passivation of the wafer surface. The properties and the passivation potential of intrinsic a-Si:H layers fabricated with a plasma excitation frequency of 140 MHz and 13.56 MHz are compared. It is demonstrated, that the effective carrier lifetimes remain higher for VHF a-Si:H layers grown near the onset of epitaxial growth as compared to RF a-Si:H layers. This leads to a higher degree of freedom for the fabrication of passivation layers by VHF-PECVD with high hydrogen dilution ratios in the plasma. High carrier lifetimes up to 13 milliseconds are achieved. The static and dynamic deposition mode of a-Si:H passivation layers is compared. A drawback of the dynamic deposition method could not be observed. Instead, a slightly increased carrier lifetime for the dynamic deposition is demonstrated. High solar cell efficiencies above 20 % are achieved for heterojunction devices with intrinsic thin a-Si:H layers.

Keywords: PECVD, amorphous silicon, heterojunction solar cell, very high frequency

1 INTRODUCTION

Heterojunction solar cells with high quality intrinsic a-Si:H passivation layers are a promising technology for future photovoltaic power generation. The efficiency potential for this kind of solar cells is the highest amongst the crystalline silicon solar cell technologies. Recently, world record efficiencies of 26.3 % [1] were achieved for this technology which is very close to the theoretical efficiency limit of Si solar cells (approx. 29.1 % [2,3]). According to the ITRPV roadmap [4], the market share of this technology will strongly increase in the future.

Up to now, the dominating a-Si:H deposition technology for heterojunction solar cells is RF-PECVD (13.56 MHz). However, an increased plasma excitation frequency has beneficial effects regarding higher deposition rates at constant material properties [5]. This is an important point for the productivity of HIT solar cell mass production. Another aspect, as was demonstrated by Heintze et al. [6], are the reduced ion energies and increased ion fluxes for high plasma excitation frequencies (VHF). These changed deposition conditions might influence the critical c-Si / a-Si:H interface of the heterojunction devices. On the one hand, a reduced damage of the interface due to decreased ion bombardment can be expected for VHF-deposited a-Si:H with low ion energies in the plasma. On the other hand, the initial growth regime of a-Si:H on c-Si is also influenced by the different ion fluxes, which may lead to different passivation properties.

The range of plasma excitation frequencies larger than 100 MHz is rarely investigated so far. For thin film silicon solar cells high efficiencies and high rates have been achieved for microcrystalline silicon solar cells using 140 MHz [7]. Amorphous silicon could be deposited at 140 MHz with increased deposition rates at constant material and thin-film solar cell properties [5] which is not possible for RF-deposition. However, it is a novel research field to apply a plasma excitation frequency of larger than 100 MHz to the crystalline silicon heterojunction solar cell technology. In this study we demonstrate differences in the material and passivation properties between RF- and VHF-deposited i-a-Si:H layers. As the deposition at very high plasma excitation frequencies (> 100 MHz) requires dynamic deposition processes, carrier lifetimes of a-Si:H layers deposited statically and dynamically will be presented as well. A complete HIT solar cell structure is fabricated and analysed concerning the most important solar cell parameters (efficiency, FF, J_{sc}, V_{oc}, Rs, Rsh, n_1, J_{01}).

2 EXPERIMENTAL

A deposition system with linear plasma sources and substrate movement (dynamic deposition) is used to deposit a-Si:H layers with RF and VHF (140 MHz) homogeneously on large areas [8]. Thus, the non-uniformity problem for the deposition at large area electrodes with high plasma excitation frequencies (standing waves) is overcome by this fabrication method. The chosen frequency of 140 MHz is a compromise between increasing technical effort and reduced ion energies at increased plasma excitation frequencies. The deposition rate for both frequencies (13.56 MHz, 140 MHz) is equally low (< 30 nm/min).

Amorphous silicon i-layers were deposited at 180 °C in a low power and low pressure regime with different silane concentration ratios. Thereby, the silane concentration (SC) is the ratio of the silane gas flow (SiH_4) and the total gas flow of silane and hydrogen (SiH_4 + H_2). Trimethylborane (TMB) and phosphine were used as dopant gases for the p- and n-layer deposition. For carrier lifetime measurements double side polished (DSP) n-float zone (FZ) wafers covered with thin amorphous silicon layers on both sides of the wafer were used. For HIT solar cells 4 inch texturized n-type

FZ wafers (1-5 Ohm-cm, 300 µm) were used. Thin intrinsic amorphous silicon layers were used to passivate the wafers on the front- and backside. A a-Si:H p-doped layer on the front side and an a-Si:H n-doped layer at the backside set up the p-n junction and back surface field of the solar cell. Indium tin oxide and a silver grid are used as front contact. The back contact consists of a full area ITO/Ag stack. The active area of the solar cells was 0.81 cm^2.

To characterize the silicon layers and solar cells several measurement procedures were applied. Raman spectroscopy was performed with a 514 nm excitation laser to evaluate the amorphous Raman intensity at 480 cm^{-1}. The layer thicknesses were measured by a Veeco surface profilometer (Dektak 8). The effective carrier lifetime was measured by microwave detected photoconductivity (MDP) (pulse 50.000 µs) in a high injection regime. The system is equipped with a mapping function (here: 2 mm resolution). The solar cell characteristics were determined using a sun simulator developed in house at Technische Universität Dresden. A xenon arc lamp in combination with an infrared-filter ensures a good fit of the emitted light to the AM 1.5 spectrum. The solar simulator is calibrated against a class A simulator. The external quantum efficiency (EQE) was measured with a SPEQUEST system from LOT-QuantumDesign. The system can measure the EQE in a wavelength range from 300 - 1800 nm.

3 RESULTS

In Fig. 1 a) the effective carrier lifetimes of thin intrinsic a-Si:H layers deposited by RF-PECVD (red diamonds) and VHF-PECVD (140 MHz) (green diamonds) on n-type FZ wafers are shown. Thereby, the silane concentration ratio was varied between 10 % and 100 %. For RF-deposited a-Si:H the lifetime decreases abruptly for decreased silane concentration ratios. On the other hand, for VHF-deposited a-Si:H the lifetime remains almost constant up to silane concentration ratios of 50 %. For lower silane concentration ratios the lifetime strongly decreases.

One reason for the low carrier lifetimes at decreased silane concentration ratios is an increased epitaxial growth. In Fig. 1 b) the intensity of the Raman shift at 480 cm^{-1} is shown. The Raman shift at 480 cm^{-1} is attributed to the amorphous phase in the deposited silicon material. For both deposition methods the 480 cm^{-1} Raman intensity decreases at declining silane concentration ratios. The reduced amorphous volume fraction in the material can be due to epitaxial growth on the crystalline silicon substrate. The difference between the a-Si:H materials grown with RF-PECVD and VHF-PEVCD is that the lifetime remains high for a plasma excitation frequency of 140 MHz even if epitaxial growth has probably started. This is not the case for RF-deposited a-Si:H where the lifetime decreases at once in the epitaxial growth regime. One cause for this effect may be attributed to the reduced ion bombardment energies in case of the VHF-deposition. For the soft 140 MHz deposition the epitaxially grown crystallites are well passivated whereas for high energetic ion bombardment (RF-PECVD) the interface crystallites are probably damaged and defect rich. An observation which supports the better passivation of VHF-a-Si:H is the higher porosity of the VHF-deposited amorphous silicon as was deduced from the materials refractive indexes measured by UV-VIS-NIR transmission. An increased porosity correlates with a larger volume fraction of silicon-di-hydride and tri-hydride in the material as was also confirmed by FTIR measurements. It is known, that annealing-induced silicon passivation originates to some extend from the transfer of hydrogen from a higher silicon-hydride state in the a-Si:H layer to a monohydride state on the c-Si surface [9]. As all the samples were annealed at 200 °C/ 10min, the higher hydrogen content of the more porous VHF a-Si:H film can facilitate the passivation process.

However, due to Raman measurement uncertainties it is not entirely clear whether the epitaxial growth has already started for VHF-deposition at medium silane concentration ratios. It is also possible, that the onset of epitaxial growth is shifted towards lower SC values for VHF-deposited a-Si:H. Independent of the microscopic reason for the different passivation properties of VHF and RF a-Si:H, the degree of freedom for the VHF a-Si:H fabrication at increased hydrogen dilution ratios in the plasma is larger compared to RF deposited a-Si:H.

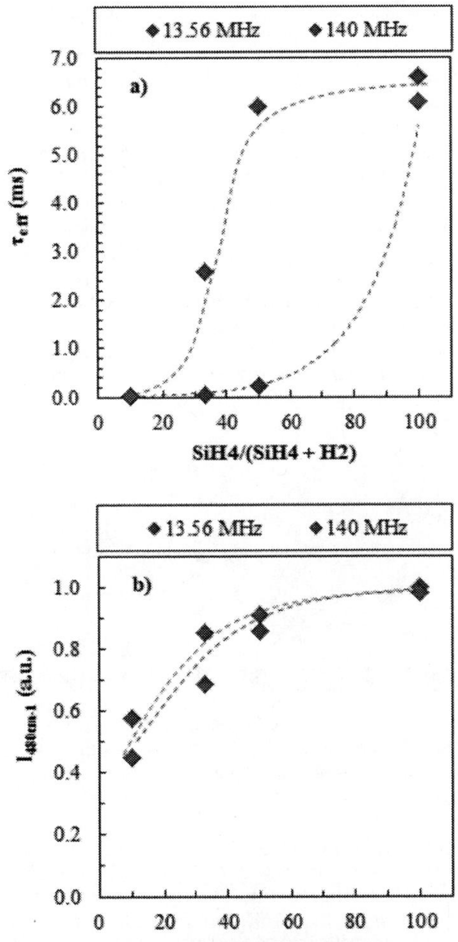

Figure 1: a) Effective carrier lifetime of thin intrinsic a-Si:H layers deposited by RF-PECVD (red diamonds) and VHF-PECVD (green diamonds) on DSP n-type FZ wafers. b) Raman intensity (480 cm^{-1}) of a-Si:H deposited by the different deposition methods.

Further on, the static and dynamic deposition of a-Si:H was compared (see Tab. I). The effective carrier lifetime of dynamically deposited a-Si:H (3.4 ms) exceeds that of the statically deposited material (2.6 ms). Note that all deposition parameters (power, pressure, silane concentration, gas flow, thickness) besides the type of deposition (static / dynamic) were kept constant for this comparison. The absolute values of the lifetime are lower than reported before because of a higher applied plasma power to facilitate plasma ignition during the static deposition. The substrate velocity during the dynamic deposition was 248 mm/min. The possibility to deposit dynamically without a reduction of the passivation quality is an important result for the VHF mass production of HIT solar cells.

Table I: Comparison of effective carrier lifetimes for statically and dynamically deposited a-Si:H with otherwise constant deposition parameters.

deposition	τ_{eff} (ms)	deposition rate
static	2.6	37.5 nm/min
dynamic	3.4	1.58 nm m/min

The best effective carrier lifetime (12.6 ms) achieved in this study for i-a-Si:H / n-c-Si / i-a-Si:H deposited statically by RF-PECVD is shown in Fig. 2. The substrate used for the evaluation of the lifetime was a quarter 4 inch n-FZ (1-5 Ohm-cm) wafer. The best lifetime was obtained in the middle of the substrate, which was not affected by wafer handling. The i-a-Si:H layer thickness was very thin (< 10 nm) as required for good HIT solar cells. To achieve these high lifetimes a double layer structure of i-a-Si:H deposited with different silane concentration ratios was applied. For VHF-deposited a-Si:H comparable effective carrier lifetimes were obtained.

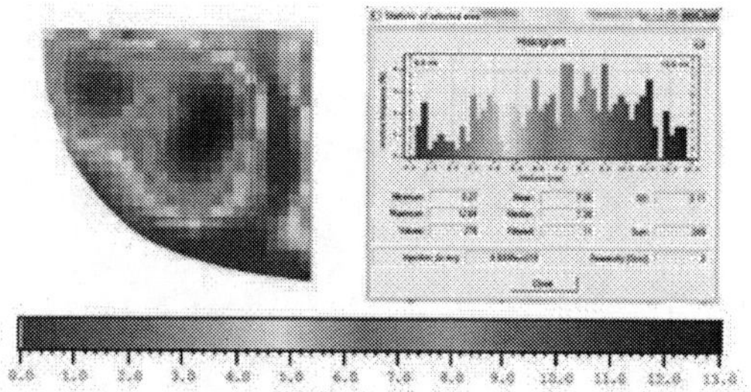

Figure 2: Map of the effective carrier lifetime (i-n-i) for RF-deposited a-Si:H on a quarter 4 inch wafer substrate as measured by MDP.

Finally, complete HIT solar cells on textured substrates were fabricated (see Fig. 3). Therefore a standard test structure with ITO-silver contacts was used. For comparison, the a-Si:H i-layers of the solar cells were deposited by RF- and VHF-PECVD. The most important HIT solar cell parameters are given in Tab. II. Efficiencies of approximately 20 % were achieved for both investigated deposition methods. The high V_{oc}'s demonstrate the excellent passivation quality of the deposited i-a-Si:H layers. Note that these V_{oc}'s were obtained on 300 µm thick wafers. For thinner wafers increased V_{oc} values can be expected.

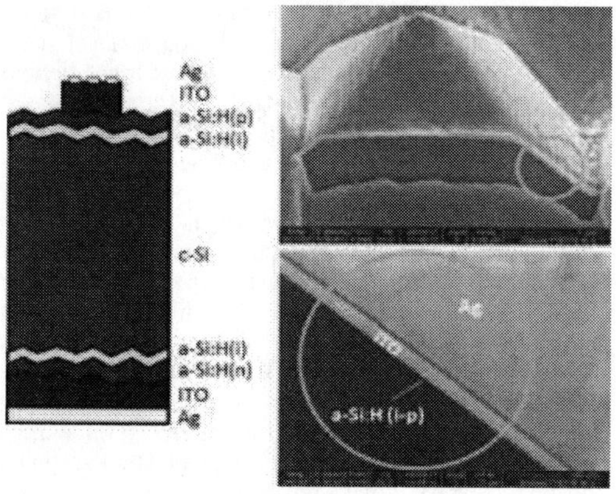

Figure 3: Left: Structure of the HIT solar cell as fabricated in the present study. Right: SEM cross section images of the HIT solar cell front contact.

Table II: Solar cell parameters of HIT devices whose a-Si:H layers were deposited by RF- and VHF-PECVD.

i-a-Si:H	13.56 MHz	140 MHz
η (%)	20.03	19.87
FF (%)	77.0	76.0
V_{oc} (mV)	723	719
J_{sc} (mA/cm²)	35.98	36.41
$J_{0,1}$ (mA/cm²)	8.03×10^{-10}	1.59×10^{-09}
n_1	1.17	1.19
$J_{0,2}$ (mA/cm²)	5.09×10^{-06}	1.99×10^{-05}
n_2	2.0	2.14
R_s (Ohm)	2.24	2.3
R_{sh} (Ohm)	95453	19906

The dark IV-curves of the HIT solar cells were also measured and fitted by a common 2-Diode model,

$$J = J_{0,1}\left(exp\left[\frac{U - IR_s}{n_1 U_T}\right] - 1\right) + J_{0,2}\left(exp\left[\frac{U - IR_s}{n_2 U_T}\right] - 1\right) + \frac{(U - IR_s)}{R_{sh}}$$

with n_1 and n_2 the diode ideality factors of diode one and two, R_s the series resistance, R_{sh} the shunt resistance, $J_{0,1}/J_{0,2}$ the dark saturation current densities of diode one and two. The fit-parameters for both types of solar cells are presented in Tab. II, as well. The slightly lower V_{oc} of the cell with a VHF a-Si:H i-layer is due to a higher $J_{0,1}$, because V_{oc} is indirectly proportional to $J_{0,1}$ according to the following equation [10],

$$V_{oc} = \frac{n_1 kT}{q} ln\left(\frac{J_{phot}}{J_{0,1}} + 1\right)$$

with q the elementary charge and J_{phot} the photo current of the HIT cell. Further on, $J_{0,1}$ can be described by,

$$J_{0,1} \approx \frac{q n_i^2 D_{min}}{L_{eff} N_{D/A}}$$

with n_i the intrinsic carrier density, D_{min} the minimum diffusion constant, L_{eff} the effective diffusion length and $N_{D/A}$ the acceptor or donor density. As $N_{D/A}$, n_i and D_{min} are equal for both investigated HIT devices, the reason for the higher $J_{0,1}$ of the VHF-cell is a lower effective diffusion length. The effective diffusion length itself is related to surface recombination velocities at the a-Si:H / c-Si interface.

The fill factor of the cell with RF a-Si:H i-layers is slightly increased due to a higher shunt resistance R_{sh}. This can be related to the lower porosity of the RF-i-layers which successfully impedes shunts.

4 CONCLUSION

For comparison, amorphous silicon for heterojunction solar cells was deposited by RF- and VHF-PECVD. The applied VHF plasma excitation frequency of 140 MHz is much higher than typical values in the VHF range. Peak lifetimes of close to 13 ms were achieved for RF- and VHF-deposited a-Si:H. For silicon heterojunction solar cells an efficiency of approximately 20 % was obtained for both investigated deposition methods. Especially the high V_{oc} values (> 720 mV at 300 µm wafer thickness) demonstrate the high passivation quality of RF- and VHF-deposited a-Si:H. For the dynamic deposition of a-Si:H slightly higher carrier lifetimes are achieved compared to the static deposition. Amorphous silicon fabricated by 140 MHz plasma excitation can be deposited with higher hydrogen dilution ratios while the effective carrier lifetime remains stable. This results in a higher flexibility during the VHF fabrication of i-a-Si:H passivation layers for heterojunction solar cells. Further on, higher deposition rates for VHF manufactured a-Si:H can be obtained without losing passivation quality. Thus, the VHF deposition of a-Si:H is a promising technology for highly productive and highly flexible mass production of HIT solar cells.

5 ACKNOWLEDGEMENT

This work is supported by the Bundesministerium für Bildung und Forschung (BMBF), Germany, under contract number 03SF0538B.

6 REFERENCES

[1] K. Yoshikawa, H. Kawasaki, W. Yoshida, T. Irie, K. Konishi, K. Nakano, T. Uto, D. Adachi, M. Kanematsu, H. Uzu and K. Yamamoto, Silicon heterojunction solar cell with interdigitated back contacts for a photoconversion efficiency over 26 %, Nature Energy 2, 17032 (2017)

[2] M. J. Kerr, A. Cuevas and P. Campbel, Limiting efficiency of crystalline silicon solar cells due to Coulomb-enhanced Auger recombination. Prog. Photovolt. Res. Appl. 11, 97–104 (2003).

[3] A. Richter, M. Hermle and S. W. Glunz, Reassessment of the limiting efficiency for crystalline silicon solar cells. IEEE J. Photovolt. 3, 1184–1191 (2013)

[4] http://www.itrpv.net/Reports/Downloads/

[5] B. Leszczynska, C. Strobel, M. Albert, J.W. Bartha, U. Stephan, J. Kuske, High rate PECVD deposition of silicon thin films at very high excitation frequencies (81.36-140 MHZ), in: Proceedings of the 27th European Photovoltaic Solar Energy Conference and Exhibition (2012) 2507–2510.

[6] M. Heintze and R. Zedlitz, New diagnostic aspects of high rate a-Si:H deposition in a VHF plasma, J. Non-Cryst. Solids 198-200, 1038-1041 (1996).

[7] C. Strobel, B. Leszczynska, U. Merkel, J. Kuske, D.D. Fischer, M. Albert, J. Holovský, S. Michard, and J.W. Bartha, High efficiency high rate microcrystalline silicon thin-film solar cells deposited at plasma excitation frequencies larger than 100 MHz, Sol. Energy Mater. Sol. Cells 143, 347 (2015).

[8] C. Strobel, T. Zimmermann, M. Albert, J.W. Bartha, J. Kuske, Productivity potential of an inline deposition system for amorphous and microcrystalline silicon solar cells, Sol. Energy Mater. Sol. Cells 93 (2009) 1598–1607.

[9] S. De Wolf, S. Olibet and C. Ballif, Stretched-exponential a-Si:H/c-Si interface recombination decay, Appl. Phys. Lett. 93, 032101 (2008).

[10] T. F. Schulze, L. Korte, E. Conrad, M. Schmidt, and B. Rech, Electrical transport mechanisms in a -Si : H / c -Si heterojunction solar cells, J. Appl. Phys. 107, 023711 (2010)

SiO_x:H PASSIVATION LAYER FABRICATED BY ATOMIC LAYER DEPOSITION FOR HETEROJUNCTION SOLAR CELLS

Mickaël Lozac'h[1], Shota Nunomura[1], Hitoshi Sai[1], Takuya Matsui[1], and Koji Matsubara[1]
[1] Research Center for Photovoltaics (RCPV), Advanced Processing Team, National Institute of Advanced Industrial Science and Technology (AIST), 1-1-1 Umezono, Tsukuba, 305-8568, Japan

ABSTRACT: Hydrogenated silicon oxide (SiO_x:H) has been developed as an alternative passivation layer for heterojunction solar cells and deposited by atomic layer deposition (ALD) technique. Such process allows a precise control of ultra-thin film thickness in the order of 10 to 20 Å, necessary for the carrier transport via tunneling effect. We demonstrate that hydrogen atoms from SiO_x:H films greatly enhance the effective lifetime up to 0.5 ms of symmetric structure p-type amorphous silicon on Fz-Si wafers after a thermal annealing between 250-300°C using only such thin passivation layer. We investigated the effective lifetime and the constitution properties of amorphous silicon layer / hydrogenated silicon oxide layers using QSSPC method, FTIR analysis, and Raman spectroscopy in function of the annealing temperature by thermal process. The effective lifetime was related to Si-H binding, the hydrogen concentration, and the crystallinity of amorphous silicon layers at each annealing temperature.
Keywords: Passivation layer, Atomic Layer Deposition, hydrogenated silicon oxide, heterojunction solar cell, Fourier Transform Infra-Red, hydrogen concentration

1 INTRODUCTION

Hydrogenated amorphous silicon (a-Si:H) layer has impressive properties for the surface passivation of silicon wafers [1]. However due to its relative low bandgap about 1.7 eV, the absorption within the solar spectrum range decreases the optical properties of the silicon heterojunction (SHJ) solar cell devices. An alternative material such as silicon oxide is a candidate of choice that combines both passivation and optical properties suitable for the passivation layer due to its high transparency. Still, the use of silicon oxide is often related to thermal annealing step at temperature above 800°C that may degrade the bulk quality and inflate the fabrication costs [2]. In this study we use the atomic layer deposition (ALD) technique to deposit ultra-thin layer of silicon oxide below 20 Å with a precise control of the film thickness to allow carrier transportation via tunneling effect. The ALD technique also allows the incorporation of hydrogen atoms at each cycle during the film deposition and thus fabricate hydrogenated silicon oxide layer (SiO_x:H), these hydrogen atoms are already localized very close to the interface between the silicon wafer and a-Si:H layer. We found optimized temperature regarding the effective lifetime about 250-300°C for both n- and p-type layers which is a relatively low temperature for such material and suitable for low thermal fabrication process of SHJ solar cells. Using hydrogenated silicon oxide as passivation layer between the silicon wafer and (p) a-Si:H we also underline an enhanced passivation properties with effective lifetime up to 0.5 ms using such thin passivation layer. In this article we will discuss the passivation effect in term of effective lifetime related with the hydrogen concentration, the binding structure, and the film crystallinity depending on the annealing temperature.

2 EXPERIMENTAL DETAILS

2.1 Sample structure and wafer cleaning

We fabricated symmetric structure to characterize the effective carrier lifetime on Fz silicon wafers substrate (n-type doped, 1-5 Ωcm, 280 µm thick, oriented ⟨100⟩, polished double sides), as in Fig. 1. Prior to the deposition of the silicon oxide layers by atomic layer deposition (ALD) technique, the wafers were chemically cleaning by using the IMEC process [3]. The IMEC cleaning process consists of a self-heated piranha solution at 75°C for 15 min (H_2SO_4:H_2O_2 ratio 4:1), an etching step for 2 min (HF:HCl:H_2O ratio 1:1:20), followed by an oxide growth for 15 min (HCl:H_2O_2:H_2O ratio 1:1:5) and removed in a diluted HF for 30 sec (HF:H_2O ratio 1:20) as the last step. Just after this cleaning process a natural oxide growth about 5±1 Å was measured by spectroscopic ellipsometry using Tauc-Lorentz model. The wafers were transported into a vacuum box between each process step in order to reduce the natural oxide growth.

2.2 Deposition processes

ALD system (Oxford Instruments) was used to deposit the passivation layers consisting of hydrogenated silicon oxide (SiO_x:H) layers. The substrate temperature was about 300°C during the full process. The precursor tris(dimethylamino)silane (3DMAS) was used as silicon source. O_2 plasma was used as oxygen source with O_2 pressure of 15 mTorr, an inductively coupled plasma (ICP) with RF power of 250 W and a discharge time of 3 sec. Concerning the incorporation of hydrogen atoms, hydrogen (H) plasma was generated using the following growth condition: H_2 pressure of 38 mTorr with ICP-RF power of 250 W for 20 sec discharge. The thickness were estimated by spectroscopic ellipsometry (SE) measurement in situ using Cauchy model.

The wafers were then placed into plasma enhanced chemical vapor deposition (PECVD) for the deposition of hydrogenated amorphous silicon (a-Si:H). The intrinsic (i) a-Si:H layer was deposited at 180°C using pure silane (SiH_4) gas, the n-type a-Si:H was deposited at 200°C with a mixture of SiH_4, H_2 and PH_3, and the p-type a-Si:H was deposited at 150°C with a mixture of SiH_4, H_2, and B_2H_6. The thicknesses were confirmed by SE using Tauc-Lorentz model at 20±1 nm for both (i) and (n) a-Si:H layers and about 23±1 nm for (p) a-Si:H. An illustration of the final symmetric structure is presented in Fig. 1.

2.3 Characterization of deposited films

The effective lifetime for minority carriers were characterized by quasi-steady-state photoconductance

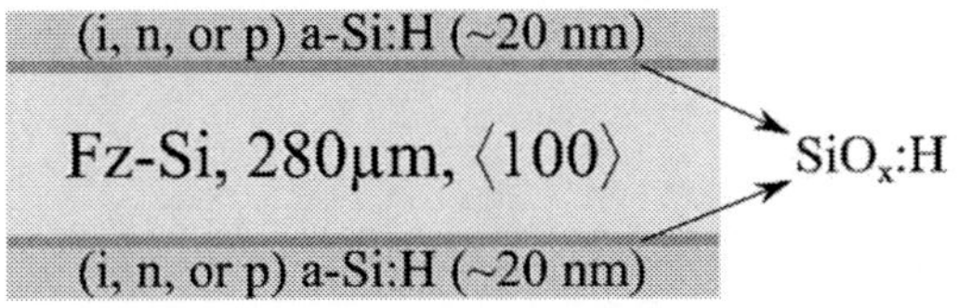

Figure 1: Symmetric structures realized with ultra-thin hydrogenated silicon oxide (SiO_x:H) films on Fz-Si substrate prior to a-Si:H layers.

(QSSPC) using a Sinton WTC-120 instruments at 1×10^{15} cm^{-3} injection level [4]. The implied open-circuit voltage (iV_{oc}) was calculated from the minority of carrier density, the resistivity and the wafer thickness [5].

The thicknesses of the deposited films were characterized by spectroscopic ellipsometry using Cauchy model in situ of ALD facility and thermal oxide model for SiO_x:H films after thermal annealing. For a-Si:H films we used the Tauc-Lorentz model to determine the thickness, roughness and the bandgap.

The hydrogen concentration were determined using Fourier transform infra-red (FTIR) measurements with the integration of the double peaks located at 2000 cm^{-1} and 2100 cm^{-1} related to the Si-H_x stretching mode [6]. The proportionality constants (or inverse cross-section) used were $A_{2000}=9.0\times10^{19}$ cm^{-2}, and $A_{2100}=2.2\times10^{20}$ cm^{-2} [6], the hydrogen concentration (C_H) were then determined by the following equation 1 [6]:

$$C_H = A_{2000} \times I_{2000} + A_{2100} \times I_{2100}$$
$$\text{with } I = \int_{\omega_s} \frac{\alpha(\omega)}{\omega} d\omega \quad \text{(Eq. 1)}$$

where $\alpha(\omega)$ is the absorption coefficient of the film at the wavenumber ω, ω_s stands for the stretching mode around 2000-2100 cm^{-1}, and I is the integrated area around these specific wavenumbers. The low frequency mode at 2000 cm^{-1} is usually associated with monohydrides (SiH), while the high frequency mode at 2100 cm^{-1} is commonly assigned to dihydrides or clustered monohydrides (SiH_x) [7]. The microstructure factor (noted R*) are estimated and deduced from integrated area of Si-H_x stretching mode using the ratio $I_{2100}/(I_{2000}+I_{2100})$ [8,9]. This R* parameter is supposed to relate the porosity of materials; higher values of R* may reflect more porous structure.

The Raman spectroscopy was used to underline the amorphous silicon signature around 480 cm^{-1} beside the silicon substrate at 521 cm^{-1} and judge about the crystallinity of the a-Si:H layers [10].

3 RESULTS AND DISCUSSIONS

3.1 Thickness control

The ALD technique uses a number of cycles to reach a target thickness, and each cycle has a deposition step, a residual purge step, and a pressure stabilization step prior to the new cycle. This process allows a precise control at atomic scale. Figure 2 presents the thicknesses measured in situ using the Cauchy model from SE. A thickness about 20 Å or below is necessary for the transport of photogenerated carrier via tunneling effect [11], which is validated for recipes below 7 cycles for SiO_x:H films.

3.2 Passivation properties

The passivation properties of SiO_x:H ultra-thin films has been investigated in term of effective lifetime for symmetric structures with intrinsic, n-type, and p-type amorphous layers. Figure 3 presents the effective lifetime in function of the annealing temperature with and without SiO_x:H layer of 15±2 Å (ALD 3 cycles process) at the interface. Intrinsic (i) a-Si layer presents high passivation properties after 240°C annealing with an effective lifetime about 3 ms that degrades at higher annealing temperature (Fig. 3 a). Inserting SiO_x:H layer prior to (i) a-Si:H deposition does not improve the effective lifetime, a peak about 200 μs is observed after 400°C annealing (Fig. 3 a). For the case of symmetric structures without SiO_x:H layer between Fz-Si and n-type a-Si:H layer, a lifetime about 1 ms is observed as desposited and degrades after annealing, contrary to symmetric structures using SiO_x:H at interface, which present a lifetime about 0.4 ms after 300°C annealing (Fig. 3 b). However for p-type a-Si:H a clear improvement of the effective lifetime about one order of magnitude is underlined for structure with SiO_x:H films reaching a lifetime about 0.5 ms after 300°C annealing. The implied open-circuit voltage (iV_{oc}) reported is about 656 mV for such structure using only SiO_x:H as passivation layer. Thus, ultra-thin SiO_x:H films as passivation layer can be used at low temperature annealing about 300°C for p-type a-Si:H. We also notice the enhanced lifetime at higher annealing temperature (> 700°C) for both n- and p-type a-Si:H symmetric structure, suggesting a field effect passivation due to the doped amorphous silicon layer.

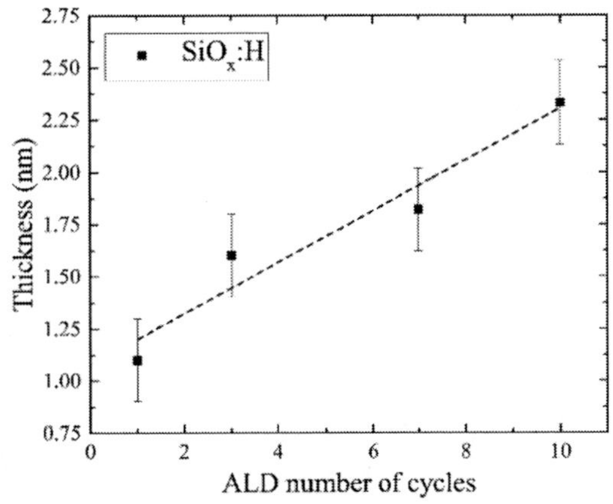

Figure 2: Thicknesses of hydrogenated silicon oxide (SiO_x:H) films in function of the number of cycles from ALD technique.

3.3 Low temperature passivation

FTIR measurements were carried out on p-type a-Si:H symmetric structures after each thermal annealing step to underline the role of hydrogen atoms. To distinguish the Si-H stretching mode signal at 2000-2100 cm^{-1} between (p) a-Si:H and SiO_x:H layers, different SiO_x:H thicknesses with 1, 3 and 10 ALD cycles were fabricated. Figure 4 presents the effective lifetime, the hydrogen concentration and the microstructure factor in function of the annealing temperature. Up to 200°C annealing, the hydrogen concentration remains relatively constant about $7\pm1\times10^{21}$ cm^{-3}, then the hydrogen diffusion start and effuse from the material. This threshold at 200°C is accompanied by an enhancement of the effective lifetime due to the hydrogen diffusion that passivate dangling bonds and possibly

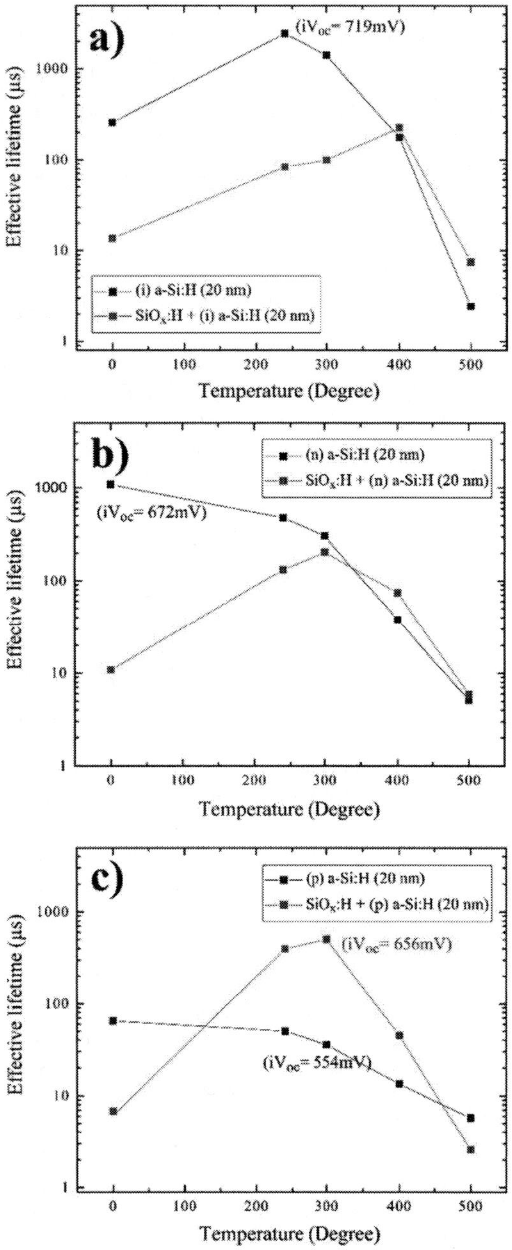

Figure 3: Effective lifetime in function of the thermal annealing for symmetric structure on Fz-Si with SiO_x:H (red) or without (black) at interface with a) intrinsic, b) n-type, or c) p-type a-Si:H of about 20 nm, implied open-circuit voltage (iV_{oc}) values are also reported.

decreases the donor/acceptor Gaussian distribution defects related to the defect-pool model [12] inside p-type a-Si:H bandgap. This hydrogen diffusion is facilitated by reorganization of microstructure that occurs between 200-300°C where R* seems enhanced (Fig. 4) [8,9]. After 200°C annealing, samples that present higher hydrogen concentration also display higher lifetime.

The FTIR signal evolution in function of the annealing temperature is displayed for the symmetric structure SiO_x:H

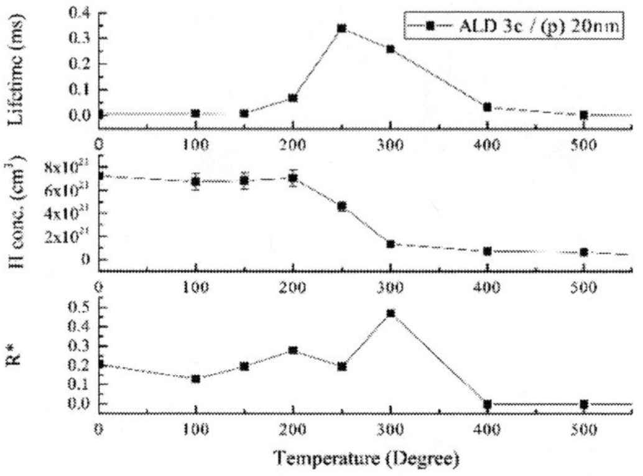

Figure 4: Effective lifetime, hydrogen concentration (H conc.) and microstructure factor (R*) in function of the annealing temperature for symmetric structure using (p) a-Si:H with different thicknesses of SiO_x:H passivation layer.

(ALD 3 cycles) / (p) a-Si:H on Fig. 5. From as deposited to 300°C annealing temperature a double Gaussian fit is observable and necessary for an accurate fitting of the experimental data. The ratio $R^* = I_{2100}/(I_{2000}+I_{2100})$ seems to be peaked around 300°C, which may suggest formation of a larger microvoid (nano porous) structure, facilitating the mobility of hydrogen atoms inside a-Si and at interface. After 400°C annealing only one Gaussian is necessary, no signal at 2100 cm^{-1} is observable and only a weak signal from monohydrides (SiH) is present and slightly shifted toward lower wavenumbers. This indicates dissociation of weak bond dihydrides modes, accompanied with dangling bond defect creation at annealing temperature above 300°C. Because of that, the lifetime is lowered at high temperature annealing.

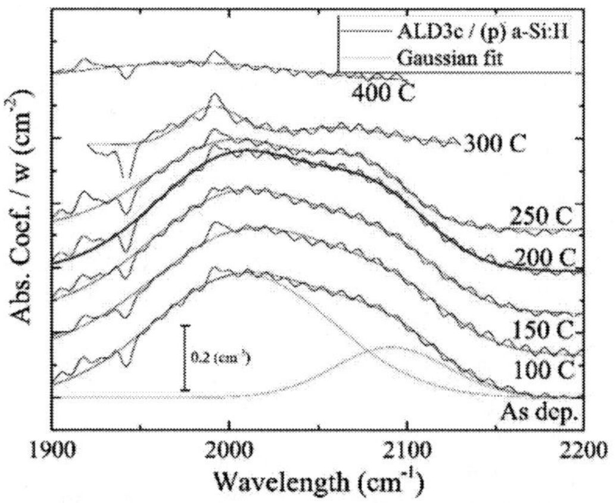

Figure 5: Absorption coefficient divided by the thickness (w) in function of the wavelength evaluated by FTIR measurements from as deposited to 400°C annealing temperature.

The Raman spectroscopy measurements (not presented here) present an amorphous silicon signal at 480 cm^{-1} beside the silicon substrate at 521 cm^{-1} until an annealing temperature of 600°C. Thus, at lower annealing temperature the amorphous silicon layer remains in amorphous state, and the improvement of the effective lifetime at 250-300°C is primarily attributed to the passivation effect of hydrogen atoms.

4 CONCLUSION

In this study we presented hydrogenated silicon oxide ultra-thin films deposited by ALD as alternative passivation layer. An improvement of the effective lifetime of one order of magnitude is underlined using such ultra-thin films on symmetric structure with (p) a-Si:H layer. The effective lifetime of reach 0.5 ms after 300°C annealing. The hydrogen atoms, their concentration and the microstructure factor were underlined in the enhancement of the effective lifetime.

Acknowledgments

This research was supported by the New Energy and Industrial Technology Department Organization (NEDO) Project. The authors would like to thank Dr. Masashi Yamazaki for his expertise on ALD system. This work was partially conducted at the AIST Nano-Processing Facility, supported by "Nanotechnology Platform Program" of the Ministry of Education, Culture, Sports, Science and Technology (MEXT), Japan.

References

[1] K. Yoshikawa, H. Kawasaki, W. Yoshida, T. Irie, K. Konishi, K. Nakano, T. Uto, D. Adachi, M. Kanematsu, H. Uzu, K. Yamamoto, *Nat. Energy* **2017**, *2*, 17032.

[2] R. Tromp, G. W. Rubloff, P. Balk, F. K. Legoues, *Phys. Rev. Lett.* **1985**, *55*, 2332.

[3] S. N. Granata, T. Bearda, F. Dross, I. Gordon, J. Poortmans, R. Mertens, *Energy Procedia* **2012**, *27*, 412.

[4] R. A. Sinton, A. Cuevas, *Appl. Phys. Lett.* **1996**, *69*, 2510.

[5] R. A. Sinton, A. Cuevas, *Appl. Phys. Lett.* **1996**, *69*, 2510.

[6] A. A. Langford, M. L. Fleet, B. P. Nelson, W. A. Lanford, N. Maley, *Phys. Rev. B* **1992**, *45*, 13367.

[7] M. Mews, T. F. Schulze, N. Mingirulli, L. Korte, *Appl. Phys. Lett.* **2013**, *102*.

[8] A. Descoeudres, C. Allebé, N. Badel, L. Barraud, J. Champliaud, G. Christmann, L. Curvat, F. Debrot, A. Faes, J. Geissbühler, J. Horzel, A. Lachowicz, J. Levrat, S. M. De Nicolas, S. Nicolay, L. Senaud, A. Tomasi, C. Ballif, M. Despeisse, *IEEE Conf. PVSC 44th* **2017**.

[9] B. Macco, J. Melskens, N. J. Podraza, K. Arts, C. Pugh, O. Thomas, W. M. M. Kessels, *J. Appl. Phys.* **2017**, *122*.

[10] Q. Li, K. Tao, Y. Sun, R. Jia, S. M. Wang, Z. Jin, X. Y. Liu, *Sol. Energy* **2016**, *135*, 487.

[11] F. Feldmann, M. Bivour, C. Reichel, M. Hermle, S. W. Glunz, *Sol. Energy Mater. Sol. Cells* **2014**, *120*, 270.

[12] M. J. Powell, S. C. Deane, *Phys. Rev. B* **1993**, *48*, 10815.

COPPER PLATING PROCESS FOR BIFACIAL HETEROJUNCTION SOLAR CELLS

A. Lachowicz[1], J. Geissbühler[1], A. Faes[1], J. Champliaud[1], F. Debrot[1], E. Kobayashi[2], J. Horzel[1], C. Ballif[1], M. Despeisse[1]
[1] CSEM PV-center, Jaquet-Droz 1, 2002 Neuchâtel, Switzerland; email: agata.lachowicz@csem.ch
[2] Choshu Industry Co., Ltd., 3740, Shin-yamanoi, Sanyo-Onoda, Yamaguchi 757-8511, Japan

ABSTRACT: We describe a process sequence for bifacial heterojunction (HJT) solar cells with simultaneous plating on both sides from a developed copper electrolyte. The process comprises the deposition of a PVD seed layer, patterning by advanced inkjet printing of a hotmelt-ink mask and copper electroplating, followed by optional deposition of a capping layer such as immersion silver to facilitate interconnection in modules. Subsequently, the mask and the seed layer are chemically removed.
Our copper electrolyte yields layers with very low internal stress when plating at high plating rates above 3 μm/min. The combination of a sputtered seed layer and electroplated copper with low internal stress allows for good adhesion even for very thick plated layers as required for instance for shingled solar cell interconnection or 6" IBC solar cells. An aperture conversion efficiency above 24% has been independently confirmed for a large area solar cell with a four busbar front grid utilizing an industrial heterojunction precursor.

Keywords: copper plating, silicon heterojunction, high efficiency

1 INTRODUCTION

The heterojunction technology allows for highest efficiencies while offering the advantage of simple manufacturing in only a few low temperature processing steps [1]. The highest reported efficiency of 26.6% has been achieved with an interdigitated back contact silicon heterojunction cell [2] and even with a simple contacting scheme on both sides, with copper plated metallization, 25.1% were reached [3].

Different patterning and plating approaches are published and applied in the industry. For example a screen or inkjet printed metal paste may be used as a seed for plating. A dielectric layer like silicon nitride or silicon oxide is then deposited on the entire wafer surface in order to prevent plating on the thin conductive oxide (TCO). Copper is subsequently electroplated where the seed is present since the dielectric layer is not continuous on the metal paste [4, 5]. The contacting for copper plating occurs through the patterned seed (with busbar layout).

Another approach often applied is PVD deposition of metal seed layers and patterning by photolithography or by hotmelt inkjet printing. [6, 7] An overview of different processing routes including also plating of nickel and copper directly on ITO is given in [8]. Metals like Ag, Ni, NiV, Cr and Ti are used as adhesion and barrier layers. Metal layer stacks including a barrier layer and a copper layer as a conductive plating base are described e.g. for (diffused) IBC cells [9].

In this work we have focused on plating of bifacial solar cells and in particular of bifacial busbar-less solar cells for SmartWire interconnection. A conductive seed layer enables the simultaneous plating of both sides of such busbar-less solar cells at high plating rates. Hotmelt inkjet patterning is a mature industrial technology and offers at the same time high flexibility for R&D developments. Printing strategies have been developed for good finger definition and low ink consumption [10].

Besides good adhesion, high finger conductivity and low contact resistance the compatibility of the plated metallization with the entire process sequence and the resulting process complexity need to be considered. In particular the etch-back of the seed layer at the end of the process sequence and the adequate application of a capping layer were challenges that required good process integration.

For narrow finger plating the process as well as the equipment had to be optimized and adapted. The hotmelt-ink mask has a hydrophobic surface. The narrower the openings are the more care had to be taken to ensure good wetting and avoid finger interruptions. Currently, a highly reproducible and reliable integrated processing sequence has been established at CSEM including module interconnection.

2 PROCESS

2.1. Process sequence

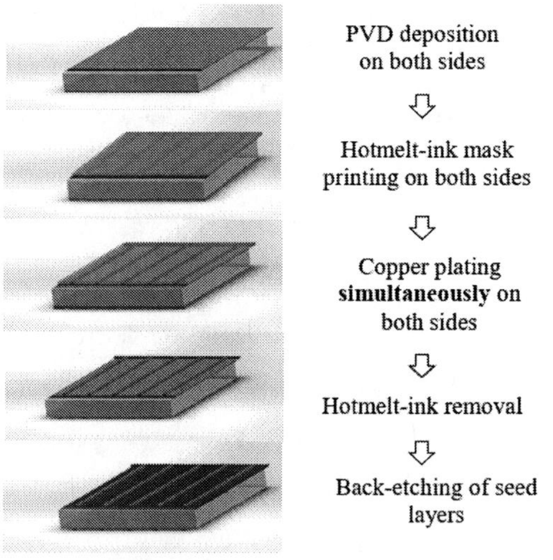

Figure 1: Plated metallization process applicable to solar cells with and without busbars

The process sequence, applicable to cells with and without busbars, is schematically described in figure 1. Industrial size (6" M2) HJT precursors with TCO on both sides are used. PVD seed layers are deposited on both sides followed by hotmelt-inkjet printing. Copper is then co-plated on both sides. The current density (plating rate) can be adjusted for front and rear side individually. This enables to find an overall optimum with respect to the front and rear finger height, the minimum plating time and material consumption. Subsequently, in the same wet chemical sequence the mask is removed and the seed layer etched back.

With an advanced hotmelt inkjet printing process with low ink consumption finger widths below 30 µm are obtained. Details of the process are described in [11]. An example of typical finger dimensions is depicted in fig. 2. In this case an immersion silver layer is deposited on top of the plated copper.

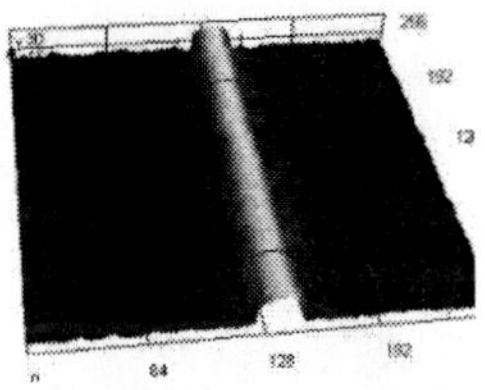

No.	Width [µm]	Height [µm]
1	25.184	17.901
2	24.846	17.600

Figure 2: Confocal microscope picture (3D LEXT microscopy) of a Cu finger plated (left) and the corresponding measurement values of finger width and finger height.

2.2. Interface TCO - metal

A void-free interface between the metallization and the TCO is crucial for good adhesion and low contact resistance (fig. 3). Besides a good adherent seed layer and low-stress copper deposit the etch-back process has be optimized in order to avoid too strong under-etching of copper. Furthermore, this etch-back process should not impact the capping layer.

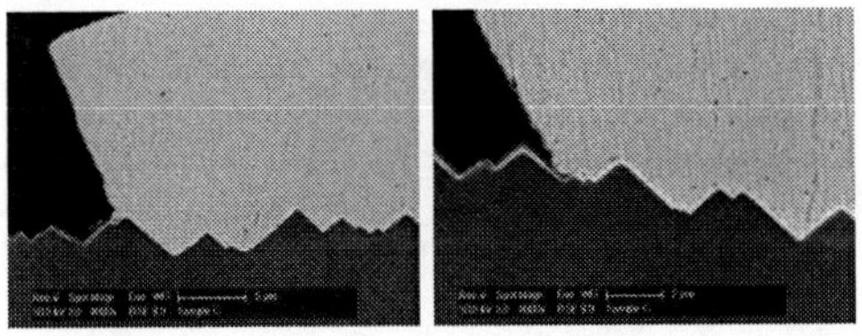

Figure 3: Cross sections showing examples of plated lines with well adherent seed and plated layers

2.3. Contact resistance

We have extracted the TCO-metal contact resistance from dedicated TLM measurements applying TLM patterns to the p-type side of n-type silicon heterojunction solar cells. The obtained contact resistance values are listed in Table I. In our opinion such low contact resistance values do not noticeably limit the FF and efficiency of the plated HJT solar cells – even when busbar-less solar cells with fingers of < 30µm contact width are plated.

Table I: Specific contact resistivity

TCO	Metallization	ρ_c [$\Omega \cdot cm^2$]
IWO	PVD seed layer 1, plated	$1.8 \cdot 10^{-5}$
ITO	PVD seed layer 1, plated	$5.8 \cdot 10^{-5}$
ITO	Screen printing	$1.1 \cdot 10^{-5}$

3 MODIFIED COPPER ELECTROLYTE

3.1 Measurement of internal stress of plated Cu deposits.

We measured internal stress of the plated Cu layers by the stress analyzer method [12, 13], also called split strip method. In this analysis Cu-Fe alloy strips of the type PN 1194, provided by Specialty Testing and Development Inc., were plated at different current densities in our pilot plating line. These initially parallel two strips are coated on opposing surfaces by insulating material so that plating occurs on the respective non-coated sides. The plating time has in all cases been adjusted in order to obtain a final plated thickness of 25 µm for all tested samples. The maximum distance between the tips of both strips was measured directly after deposition. The internal stress was then calculated from these values according to the Stoney formula as described in [15].

For the calculation of the internal stress of the plated Cu layer the copper thickness has in a first order approximation been calculated from the applied current density, the plated area and the plating time. Since some ghost plating occurred on the insulating coating (on one side of the strip) the thickness has been then verified by measurement of cross sections of the strips. In figure 4 corrected values, calculated with the measured thickness are depicted.

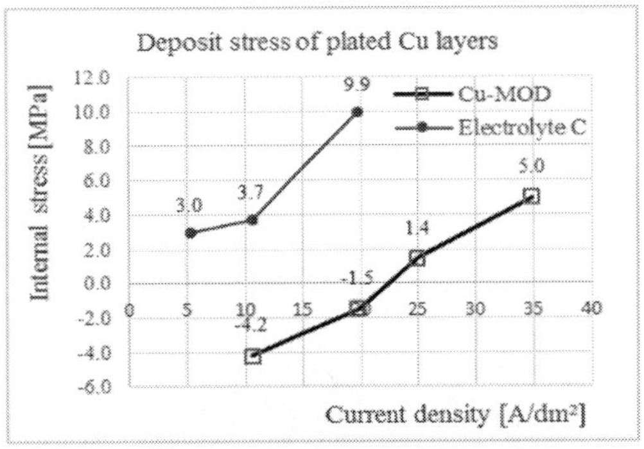

Figure 4: Comparison of deposit stress in dependence on current density. Negative values indicate compressive stress, positive values tensile stress. Cu-MOD is the new electrolyte, Electrolyte C is a commercially available electrolyte.

The developed copper electrolyte shows compressive stress at low current densities changing to low levels of tensile stress for current densities above 20 A/dm². At a current density of 20 A/dm² the internal stress is minimized. This corresponds to a copper plating rate of ~4.5 µm/min. Low internal stress at high plating rates helps to shorten the overall processing time and therefore increases the throughput in plating (cost reduction).

This is especially beneficial for large area IBC solar cells, in order to maintain good adhesion for highly conductive metal fingers while shortening the plating time and minimizing bending of the solar cells.

The above reported results for internal stress have been obtained with direct current plating (DC).

3.2. Line resistance

Figure 5 shows line resistance per unit length as function of plating height for a finger width of 30 µm. The optimum plated copper thickness depends on the grid layout and interconnection scheme and can be chosen accordingly in a wide range of plated Cu thickness values that can be achieved. Thus plating throughput and cost can be co-optimized with performance of the plated solar cells for each grid layout and interconnection scheme individually.

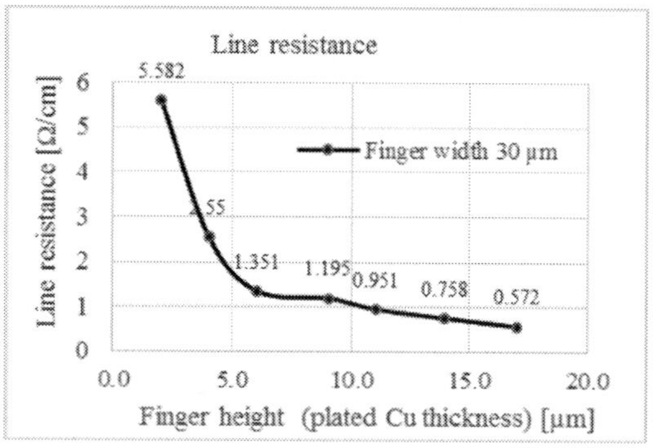

Figure 5: Line resistance per unit length measured with a 4-probe setup on busbar-less cells

The specific resistivity of the plated copper layer as calculated from the measured line resistance and from finger dimensions as evaluated by 3D laser assisted confocal microscopy was determined to be ~ 2.6 µΩ·cm, which is only slightly above the value for pure bulk copper (1.7 µΩ·cm).

4 EXPERIMENT WITH TILTED FINGER SURFACE

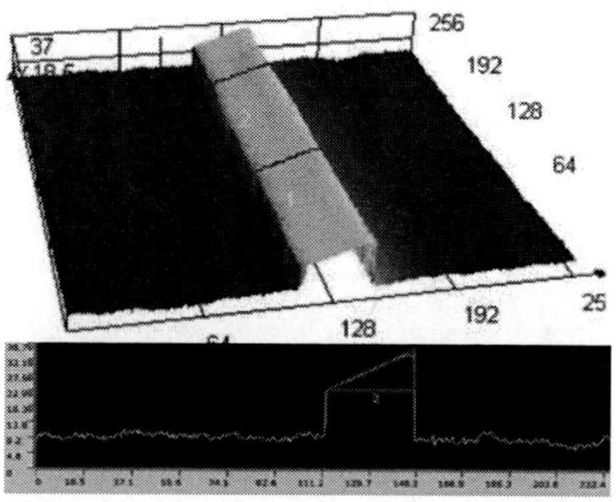

Figure 6: Confocal microscope picture (3D LEXT microscopy) and the corresponding profile of a Cu finger plated at 20 A/dm² for 5 minutes. The average finger width is ~35 µm. The finger height for the tilted part is 11.9 µm. The corresponding tilt angle is close to 20°.

Tilted finger surfaces can be reproducibly achieved with modified chemistry, i.e. with an organic additive and dedicated processing conditions.

The measured tilt angle compared to the precursor surface plane is so far about 20° (fig. 6) and will have to be further increased to take full benefit from improved light trapping in PV modules.

5 CELL AND MODULE RESULTS

5.1 Cells for SmartWire modules

Bifacial busbar-less solar cells for two 60-cells-SmartWire-modules have been manufactured with the process as described above in our pilot plating line. Solar cells were arranged vertically and fully immersed in the wet chemical sequence for Cu co-plating of front and rear contacts (bifacial solar cells). In figure 7 the cell efficiency distribution for a pilot plating run is given as well as a picture of the front side of interconnected cells in a module that has been manufactured from this plating run. The narrow efficiency distribution confirms the good reproducibility of this patterning and plating approach, with median efficiency at 22.3% and best cell at 22.8%.

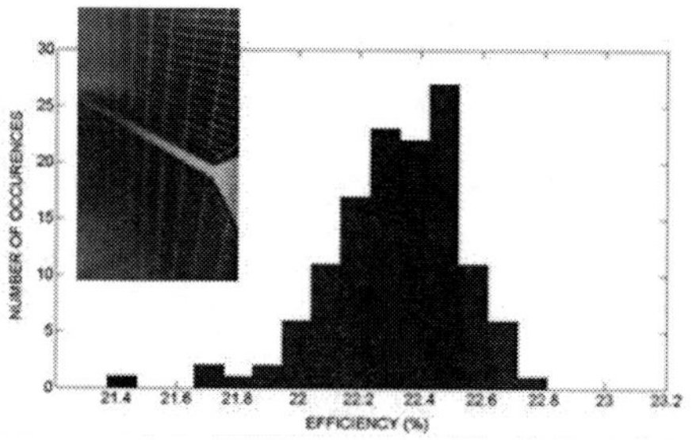

Figure 7: Efficiency distribution of solar cell batches produced for 60-cell modules with CSEM pilot line Cu plating and a photo of a cell interconnected in a glass-glass module from this run.

On these cells no capping layer has been applied after Cu plating. The low melting point alloy of the SmartWire interconnection allows for good connection to these bare copper surfaces. Both, InSn coated wires as well as new generation In-free wires have been applied for the 60-cell modules yielding good performance (Table II). Further details about In-free SmartWire interconnection are described in [15].

Table II: Performance of 60-cell glass-glass modules with copper plated bifacial HJT solar cells (measurement with white background).

Wire type	Voc [V]	FF [%]	Pmax[W]
InSn coated wires	43.59	78.74	307.89
In-free coated wires	43.50	78.01	303.40

5.3. High efficiency Cu plated HJT solar cells

We applied the plating sequence as depicted in fig. 1 to industrial 6" M2 heterojunction precursors. For the solar cell reported in Table III we applied a 4-busbar-layout on both sides and a second antireflection coating on the front side. The cell has been measured on a gold chuck with aperture by Fraunhofer ISE CalLab. An efficiency of 24.1% and a FF of 82.5% have been independently confirmed in this measurement.

Table III: Cell efficiency certified by Fraunhofer CalLab

Cell	Area [cm²]	Jsc [mA/cm²]	Voc [mV]	FF [%]	Eff [%]
4BB	224.8	40.0	731.2	82.5	24.1

The excellent FF for large area metallization achieved for our Cu plating metallization sequence indicates that this approach is compatible with high efficiency processing and not limiting the FF (at this level). It also confirms that the sputtering, patterning, plating and chemical processing including all handling and contacting steps do not impact negatively the solar cell passivation.

6 SUMMARY

Through optimization of single process steps, of the copper electrolyte and of the plating equipment high efficiency solar cells are reproducibly achieved. The application of a seed layer offers the possibility to simultaneously plate at high deposition rates on both sides of bifacial cell, especially of busbar-less solar cells. The hotmelt inkjet patterning is a mature industrial method and advanced printing strategy enable fingers <30 µm in width at low ink consumption, i.e reduced cost.

The discussed metallization process based on plated Cu metallization has demonstrated to be compatible with high efficiency processing where FFs are not limiting the solar cell performance to levels beyond 24%.

7 ACKNOWLEDGEMENTS

This work has been supported by the Swiss Federal Office of Energy with funding of the Swiss Inno HJT project and by the European Union's Horizon 2020 research and innovation program with funding of the DISC project under grant agreement N°727529 and with funding of the AMPERE project under grant agreement N° 745601.
We thank Choshu Industry Co. for providing us excellent quality precursors for plating.

8 REFERENCES

[1] S. De Wolf, A. Descoudres, Z. C. Holman and C. Ballif, "High-efficiency Silicon Heterojunction Solar Cells: A Review", *Green*, vol. 2, 2012

[2] K. Yoshikawa, W. Yoshida, T. Irie, H. Kawasaki, K. Konishi, H. Ishibashi, T. Asatani, D. Adachi, M. Kanematsu, H. Uzu and K. Yamamoto, "Exceeding conversion efficiency of 26% by heterojunction interdigitated back contact solar cell with thin film Si technology", *Solar Energy Materials and Solar Cells*, in press, 2017

[3] D. Adachi, J. L. Hernández and K. Yamamoto, "Impact of carrier recombination on fill factor for large area heterojunction crystalline silicon solar cell with 25.1% efficiency", *Applied Physics Letters* 107 (2015)

[4] European patent number EP 2489076 B1

[5] D. Adachi, T. Terashita, T. Uto, J. L. Hernández and K. Yamamoto, "Effect of SiO_x barrier layer prepared by plasma-enchanced chemical vapor deposition on improvement of long-term reliability and production coast for Cu-plated amorphous Si/crystalline Si heterojunction solar cells", *Solar Energy Materials and Solar Cells* 163 (2017) 204-209

[6] J. L. Hernández, D. Adachi, D. Schroos, N. Valckx, N. Menou, T. Uto, M. Hino, M. Kanematsu, H. Kawasaki, R. Mishima, K. Nakano, H. Uzu, M. Yoshimi and K. Yamamoto, "High Efficiency Copper Electroplated Heterojunction Solar Cells and Modules – The Path Towards 25% Cell Efficiency", *28th European Photovoltaic Solar Energy Conference and Exhibition*, 2013

[7] F. Ishimura, W. Li, E. Kobayashi, K. Hashimoto, S. Sato, Y. Watabe, E. Bende and G. Coletti, "Metal Wrap Through Heterojunction solar Cell with Plated Electrode", *32nd European Photovoltaic Solar Energy Conference and Exhibition*, 2016

[8] A. Aguilar, S. Herasimenka, J. Karas, H. Jain, J. Lee, K. Munoz, L. Michaelson, T. Tyson, W. Dauksher and S. Bowden, "Development of Cu Plating for Silicon Heterojunction Solar Cells", *IEEE 43rd Photovoltaic Specialists Conference*, 2016

[9] S. Singh, B. O'Sullivan, M. Kyuzo, S. Jambaldinni, L. Tous, R. Russel, M. Debucquoy, J. Szlufick and J. Poortmans, "Application of seed and plate metallization to 15.6cm x 15.6cm IBC cells", *Photovoltaics International*, Vol. 29, 2015

[10] J. Hermans, R. van Knippenbert, E. Kamp, W. Brok, P. Papet, B. Legradic and B. Strahm, "Inkjet Printing for Solar Cell Mass Production on the PIXDRO JETX Platform", *28th European Photovoltaic Solar Energy Conference and Exhibition*, 2013

[11] J. Geissbühler, A. Lachowicz, A. Faes, N. Badel, J. Horzel, J. Champliaud, L. Curvat, P. Papet, B. Strahm, J. Hermans, C. Ballif and M. Despeisse, "High efficiency silicon heterojunction solar cells with electrodeposited copper contacts: progress in process development for bifacial cells", *33rd European Photovoltaic Solar Energy Conference and Exhibition*, 2017 (this conference)

[12] ASTM B975-15, Standard Test Method for Measurement of Internal Stress of Metallic Coatings by Split Strip Evaluation (Deposit Stress Analyzer Method)

[13] B. Stein, "A Practical Guide to Understanding, Measuring and Controlling Stress in Electroformed Metals", *AESF Electroforming Symposium*, 1996

[14] F. H. Leaman, "Measurement Methods and Calculations to Determine Deposit Stress", *SURFIN Conference*, 2015

[15] A. Faes, M. Despeisse, A. Lachowicz, N. Badel, J. Champliaud, J. Levrat, J. Geissbühler, B. Paviet-Salomon, J. Horzel, P. Papet, D. Lachenal, B. Strahm, Y. Yao, T. Soderstrom, J. Hermans and C. Ballif, "Indium-free coating and advanced metallization for SmartWire Connection Technology", *33rd European Photovoltaic Solar Energy Conference and Exhibition*, 2017 (this conference)

ITO SPUTTERING DAMAGE TO SILICON HETEROJUNCTION SOLAR CELLS WITH CAT-CVD A-SI FILMS AND ITS RECOVERY

Takeo Konishi, Keisuke Ohdaira
Japan Advanced Institute of Science and Technology
1-1 Asahidai, Nomi, Ishikawa, JAPAN
Phone: +81-76-51-1563 E-mail: takeo.konishi@jaist.ac.jp

ABSTRACT: We investigate the influence of indium tin oxide (ITO) sputtering damage to silicon heterojunction (SHJ) solar cells with catalytic chemical vapor deposited (Cat-CVD) a-Si films. a-Si was deposited on crystalline Si by Cat-CVD, and ITO was sputtered on the a-Si films. Effective minority carrier lifetime (τ_{eff}) of the sample after ITO deposition decreases drastically due to RF plasma damage to a-Si films and resulting degradation of the passivation quality of a-Si films. The samples after ITO deposition and post-annealing show much smaller τ_{eff} recovery, and it was not able to recover to the initial τ_{eff}. On the other hand, the samples after removing ITO films show significant τ_{eff} recovery to the initial τ_{eff}. QSSPC measurement revealed that the lower τ_{eff} of the sample after post-annealing is actually caused by carrier recombination at p-a-Si/ITO interface. Furthermore, we confirmed open-circuit voltage (V_{OC}) improvement of SHJ cells by performing post-annealing. We can thus conclude that sputtering damage to SHJ solar cells during ITO sputtering can be recovered by performing post-annealing.
Keywords: ITO, Lifetime, Heterojunction, Sputtering, Solar Cell, Cat-CVD

1 INTRODUCTION

Silicon heterojunction (SHJ) solar cells have attracted much attention because of their high conversion efficiency and several industrial benefits due to the low temperature process for cell fabrication [1,2]. An SHJ solar cell contains a crystalline silicon (c-Si) wafer and amorphous Si (a-Si), and the termination of dangling bonds by hydrogen atoms contained in the a-Si films results in good a-Si/c-Si interfacial quality. Thanks to the existence of a-Si passivation films, SHJ solar cells generally have high open-circuit voltage (V_{OC}). a-Si films are conventionally formed by plasma-enhanced chemical vapor deposition (PECVD), in which gas molecules are decomposed in a plasma. On the other hand, catalytic CVD (Cat-CVD) is known as a method for the preparation of high-quality a-Si passivation films, since gas molecules are decomposed on a heated catalyzing wire, and thus, there is no plasma damage to substrates and films [3,4]

Due to low conductivity of doped a-Si, SHJ solar cells require transparent conductive oxide such as indium tin oxide (ITO), for sufficient lateral carrier transport [5]. Sputtering method is, in general, one of the conventional processes to form ITO films. However, because of high-energy ion bombardment to a-Si passivation films, the passivation quality of a-Si films degrades if ITO films are prepared by sputtering on the a-Si films [6,7]. The previous investigation of ITO sputtering damage for a-Si formed by PECVD has revealed that sputtering at room temperature (R.T.) and successive post-annealing are effective for the passivation quality of a-Si films [8]. However, the effect of sputtering damage to Cat-CVD a-Si films has not been sufficiently studied so far.

In our previous work, we have investigated the influence of ITO sputtering to the passivation quality of various types of a-Si films deposited by Cat-CVD [9]. It was revealed that the passivation quality of Cat-CVD a-Si films was also degraded by ITO sputtering. The degradation of passivation quality by sputtering to n-a-Si/i-a-Si stacked layers easily recovers after short post-annealing. On the other hand, the degradation of passivation quality by sputtering to p-a-Si/i-a-Si stacked layers does not recover sufficiently by post-annealing, which is caused by carrier recombination at p-a-Si/ITO interface. These experiments were conducted for the samples with stacked a-Si layer, but not for SHJ solar cell structures. In this work, we investigate the influence of ITO sputtering damage to actual SHJ solar cells by measuring minority carrier lifetime and *J*–*V* characteristics.

2 EXPERIMENTAL DETAIL

Fig. 1 shows the schematics of the samples for the experiment. We used 290-μm-thick floating-zone-grown n-type crystalline Si (c-Si) (100) wafers with a resistivity of 1–5 Ω cm and a bulk carrier lifetime (τ_b) of >10 ms. c-Si wafers were cleaved into 2 cm × 2 cm-sized pieces, and dipped in 5% HF for 30 seconds to remove native oxide and then in 4% H_2O_2 for 30 seconds for the formation of ultra-thin oxide layers to avoid epitaxial growth [10]. 10 nm-thick intrinsic a-Si (i-a-Si) layers were deposited on the both sides of c-Si wafers, followed by the deposition of 10 nm-thick n-a-Si and p-a-Si layers on each side of the samples by Cat-CVD. 70 nm-thick ITO films were then deposited in the both sides of the samples by RF magnetron sputtering at R.T.. A Sn-doped (5wt%) In_2O_3 target with a diameter of 5 cm was used for the sputtering. Deposition condition of a-Si and ITO films are summarized in Table 1. Finally, combed-shaped Ag electrodes were evaporated on both sides of ITO films and sample edges were cut out to avoid shunting. The active area of the cells was 1.6 cm × 1.6 cm. Samples were annealed at 200 °C in air condition. The passivation quality of a-Si films was evaluated by measuring effective minority carrier lifetime (τ_{eff}) using microwave photoconductivity decay (μ-PCD) measurement having the function of a pulse laser with a wavelength of 904 nm and a photon density of 5×10^{13} cm^{-2}, and using quasi-steady-state photoconductivity (QSSPC) measurement. The thicknesses of a-Si and ITO layers were measured by spectroscopic ellipsometry through analyses using parametric semiconductor model and Cauchy model, respectively [11]. Solar cell performance was characterized by Suns–V_{OC} measurement to obtain pseudo *J*–*V* characteristics.

Table 1 Deposition conditions for a-Si and ITO films

	T_{sub} [°C]	T_{cat} [°C]	RF power [W]	Duration	Presurre [Pa]	Gas flow rate [sccm]				
						SiH_4	$He:PH_3$	$He:B_2H_6$	H_2	Ar
i-a-Si	125	1800	N/A	45 s	1	10	-	-	-	-
n-a-Si	250	1800	N/A	35 s	2	10	4.4	-	-	-
p-a-Si	250	1800	N/A	80 s	2	10	-	16	50	-
ITO	R.T.	N/A	50	25 min	0.5	-	-	-	-	13.6

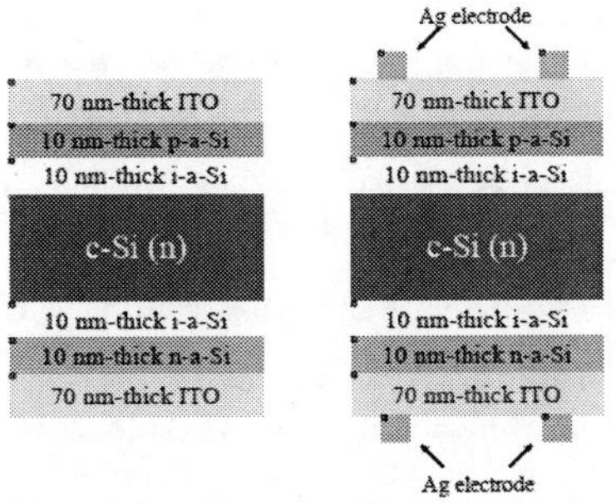

Fig. 1 Schematics of the sample structures for lifetime measurement (left) and *J–V* measurement (right).

3 RESULTS AND DISCUSSSION

3.1 τ_{eff} measurement

Fig. 2 shows the τ_{eff} of the samples measured by μ-PCD before and after ITO sputtering and successive post-annealing for 0, 1, and 5 h, and after the removal of ITO films by dipping in 30wt% of HCl for 2 minutes. Note that Ag electrodes were not formed on the samples in this experiment, and we have, in advance, confirmed that the presence of ITO films on the illuminated side does not affect the measurement. As can be seen in Fig. 2, τ_{eff} was decreased drastically after ITO deposition, and τ_{eff} did not recover even after post-annealing. However, τ_{eff} was recovered significantly after the removal of the ITO film. As already reported in previous papers [6,8], τ_{eff} reduction after ITO deposition is because of the degradation of a-Si passivation quality due to plasma damage. τ_{eff} should thus be recovered by terminating the dangling bonds through post-annealing. τ_{eff} was not recovered despite of post-annealing for 1 and 5 h. However, only when ITO films were removed by HCl etching, τ_{eff} was recovered significantly to the initial τ_{eff}. Note that we separately confirmed that only dipping in HCl does not affect the passivation quality.

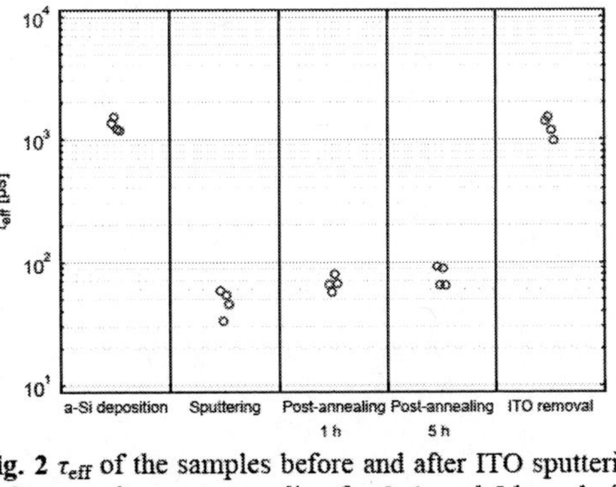

Fig. 2 τ_{eff} of the samples before and after ITO sputtering and successive post-annealing for 0, 1, and 5 h, and after the removal of ITO films.

3.2 Cause of τ_{eff} reduction

Fig. 3 shows the τ_{eff} of the samples measured by QSSPC after a-Si deposition, ITO sputtering, 5-h post-annealing, and the removal of ITO films, plotted as a function of excess carrier density (Δn). The τ_{eff} of all the samples decreases at higher Δn region, which is caused by Auger recombination. After the ITO deposition, τ_{eff} was decreased drastically, and τ_{eff} was not recovered by performing post-annealing 5 h. Then, the τ_{eff} was recovered significantly for the samples after the removal of ITO films. These results are quite similar to Fig. 2. However, it is noteworthy that a reduction in τ_{eff} at lower Δn region is observed for the sample after ITO deposition and 5-h post-annealing, while such a τ_{eff} reduction is not observed for the samples before ITO deposition, and after the removal of ITO films. As already mentioned in a previous paper [12], reduction in τ_{eff} at a lower Δn region can be explained as carrier recombination at the ITO/p-a-Si interface. Due to this phenomenon, the passivation quality of the sample after ITO deposition and 5-h post-annealing seemed not to be recovered; in fact a-Si/c-Si passivation quality was already recovered by performing post-annealing. It is worth mentioning that the sample just after ITO deposition and its removal showed τ_{eff} as low as the sample of after ITO deposition. This clearly indicates that sputtering damage actually affects the sample passivation quality. We therefore conclude that passivation quality degradation by sputtering damage can be recovered by performing post-annealing.

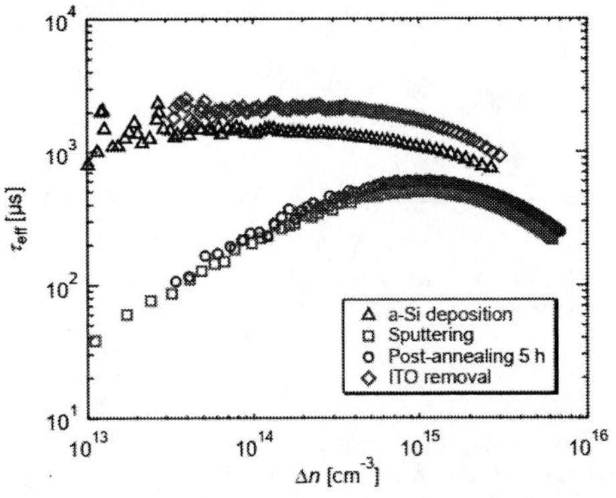

Fig. 3 τ_{eff} of the samples before and after ITO sputtering and successive post-annealing for 5 h, and the removal of ITO films as a function of Δn.

3.3 *J–V* characteristics

Based on the previous results, we investigated the effectiveness of post-annealing on the recovery of the performance of SHJ cells with Cat-CVD a-Si passivation films. Fig. 4 shows the pseudo *J–V* characteristics of an SHJ solar cell before and after post-annealing. Note that J_{SC} values were separately characterized through *J–V* measurement under 1-sun illumination. The pseudo *J–V* curve of the SHJ cell after 1-h post-annealing showed significant improvement in V_{OC} and pseudo fill factor (pFF) compared to the cell before annealing. V_{OC} improvement indicates that a-Si/c-Si interfacial quality was recovered by post-annealing. Improvement in pFF may be an indication of the reduction of recombination current density due to the recovery of a-Si/c-Si interfacial passivation quality of the cell. J_{SC} was also improved by performing post-annealing. To analyze the improvement of

J_{SC}, external quantum efficiency (EQE) measurement was conducted, as shown in Fig. 5. Improvement of EQE after performing 1-h post-annealing was confirmed especially at long wavelength region. This is possibly due to the improvement of a-Si/c-Si interfacial quality at the rear side. From the result, we confirmed that ITO sputtering damage to SHJ cells can also be recovered by post-annealing.

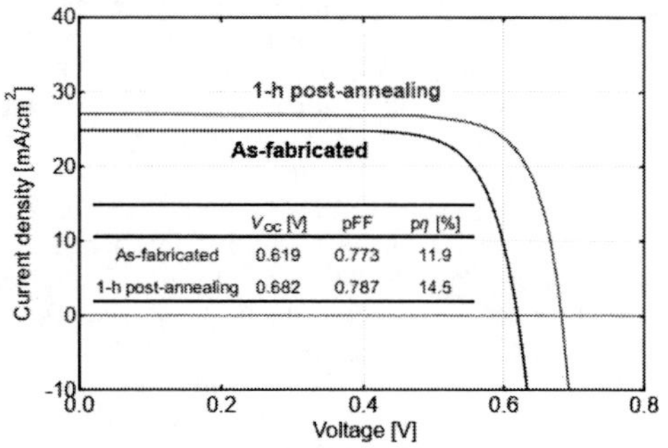

Fig, 4 Pseudo *J–V* characteristics of an SHJ solar cell before and after performing post-annealing.

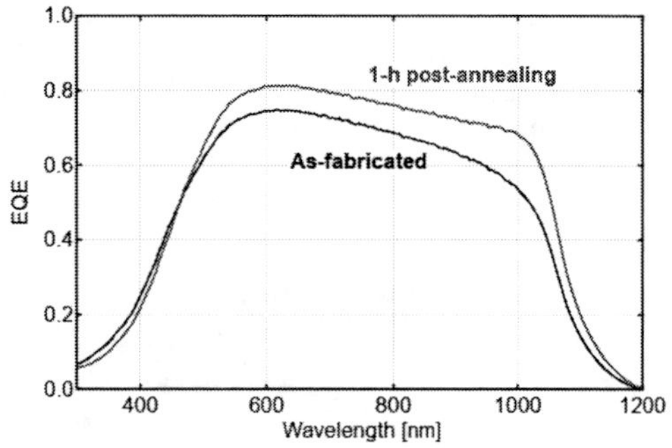

Fig. 5 EQE spectra of an SHJ solar cell before and after performing post-annealing.

4 SUMMARY

The passivation quality of Cat-CVD a-Si films on SHJ solar cells is degraded by ITO sputtering due to plasma damage. The degradation of passivation quality by sputtering to SHJ solar cell seemed not to be recovered by post-annealing since no significant τ_{eff} recovery was observed. The passivation quality of a-Si films in an SHJ cell was, however, already recovered by performing post-annealing. The cause of lower τ_{eff} after post-annealing is considered to be carrier recombination at ITO/p-a-Si interface. Improvement of V_{OC} was confirmed in an SHJ cell after performing post-annealing. We therefore consider that ITO sputtering damage to an SHJ solar cell can be recovered by performing post-annealing.

5 REFERENCE

[1] M. Tanaka, M. Taguch, T. Matsuyama, T. Sawada, S. Tsuda, S. Nakano, H. Hanafusa, and Y. Kuwano, J. Appl. Phys. 31 (1992) 3518.
[2] A. Descoeudres Z. C. Holman, L. Barraud, S. Morel, S. De Wolf, and C. Ballif, IEEE J. Photovolt. 3 (2013) 83–89.
[3] H. Matsumura, K. Higashimine, K. Koyama, and K. Ohdaira, J. Vac. Sci. Technol. B 33 (2015) 031201.
[4] S. Tsuzaki, K. Ohdaira, T. Oikawa, K. Koyama, and H. Matsumura, Jpn. Appl. Phys. 54 (2015) 072301.
[5] M. Filipič, Z. C. Holman, F. Smole, S. De Wolf, C. Ballif, and M. Topič, J. Appl. Phys. 114 (2013) 074504.
[6] A. Morales-Vilches, C. Voz, M. Colina, G. López, I. Martín, P. Ortega, A. Orpella, and R. Alcubilla, Energy Proc. 44 (2014) 3–9.
[7] B. Demaurex, S. De Wolf, A. Descoeudres, Z. Charles Holman, and C. Ballif, Appl. Phys. Lett. 101 (2012) 171604.
[8] D. Zhang, A. Tavakoliyaraki, Y. Wu, R. A. C. M. M. van Swaaij, and M. Zeman, Energy Proc. 8 (2011) 207–213.
[9] T. Konishi, and K. Ohdaira, Thin Solid Films 635 (2017) 73–77.
[10] T. Oikawa, K. Ohdaira, K. Higashimine, and H. Matsumura, Current Appl. Phys. 15 (2015) 1168–1172.
[11] Guide to Using WVASE32TM: Software for Vase and M-44 Ellipsometers, J.A.Woollam Co., Inc., 1989. 159–165.
[12] W. Favre, J. Coignus, N. Nguyan, R. Lachaume, R. Cabal, and D. Muñoz., Appl. Phys, Lett. 102 (2013) 181118.

FABRICATION AND SIMULATION OF ZNS/P-SI HETEROJUCTION SOLAR CELLS

Kaifu Qiu[1], Qi Xie[1], Depeng Qiu[2], Zongcun Liang[1,2], Hui Shen[1,2]
[1] Institute for Solar Energy Systems, Sun Yat-Sen University, 5th Floor, C Block, Building of Engineering, 132 Wai Huan Dong Road, Higher Education Mega Center, Guangzhou, Guangdong, P.R. China, 510006
[2] Shunde-SYSU Institute for Solar Energy, No 1, Sanle Road, Beijiao, Shunde, Guangdong, P.R. China, 528300
Address: Institute for Solar Energy Systems, Sun Yat-Sen University, 5th Floor, C Block, Building of Engineering, 132 Wai Huan Dong Road, Higher Education Mega Center, Guangzhou, Guangdong, P.R. China, 510006
Phone: +86-13512705550, Email: qiukaifu@mail2.sysu.edu.cn

ABSTRACT: Ag/HAZO/n-ZnS/polished p-Si/Ag heterojunction solar cells were fabricated by thermal evaporation method for the first time. The influence of different ZnS thickness on device was investigated. Result shows that thickness of 10 nm is preferable, thickness more or less than 10 nm results in lower fill factor (FF) and short circuit current density (J_{sc}). 10 nm ZnS was then deposited on the textured p-Si and Al was used as back electrode. The device exhibits power conversion efficiency (PCE) of 8.83% with open circuit voltage (V_{oc}) of 0.517 V, J_{sc} of 31.047 mA/cm^2 and FF of 55.0%. The efficiency is the highest value reported so far to our knowledge for n-ZnS/p-Si heterojunction solar cells. Simulation of the device was carried out by wx-AMPS. It shows that the contact barrier of front surface mainly influences the FF, while the contact barrier of back surface influences the V_{oc}. The simulation points the way for efficiency improvement of the device. Experiments according to the simulation confirmed the result.
Keywords: II VI Semiconductor, ZnS thin films, Heterojunction solar cells, Simulation

1 INTRODUCTION

ZnS is an important semiconductor with direct wide bandgaps of 3.72 and 3.77 eV for cubic and hexagonal phases, respectively. It has a little absorption coefficient in the visible range of optical spectrum. Besides, ZnS is a kind of cheap and nontoxic semiconductor material. Also, ZnS thin films are easy to be fabricated by a variety of deposition methods, such as thermal evaporation [1], chemical bath deposition (CBD) [2], RF magnetron sputtering [3], and so on. These characteristics make it qualified for using as windows layer and emitter for Si based heterojunction solar cells [4].

Although ZnS has lots of advantages, the undoped ZnS layers are highly resistive in nature with the electrical resistivity of ~10^7 Ω·cm and carrier concentration of about 10^{11} cm^{-3} [5]. So ZnS thin films with thin enough thickness are essential, otherwise high resistance would limit its application on the optoelectronic device.

There are many reports of ZnS-based heterojunction solar cells by CBD method, such as with CIGS used as the n-type emitter layer [6]. But little work on Si-based heterojunction solar cells applying ZnS as the n-type emitter layer had been reported.

Y. J. Hsiao[7] reported the AZO/n-ZnS/planar p-Si heterojunction solar cells for the first time, whose ZnS of 300 nm thickness were fabricated by CBD method with efficiency of 2.72%. Liang Wen Ji [8] investigated the influence of annealing temperature on the efficiency of the AZO/n-ZnS/textured p-Si solar cells, whose maximum PCE is 3.66%. Both of the ZnS thin films were fabricated by CBD method, the crystallinity is poor. Besides, the thickness of ZnS is so large that more than 100 nm, considering the highly resistive of ZnS, the series resistance of device must be quite large and influence the FF, thus decrease efficiency.

This study aimed to deposit ZnS thin films on p-type silicon wafer using the thermal evaporation method and investigated the influence of thickness of ZnS on the device performance. The thermal evaporated n-ZnS/textured p-Si heterojunction solar cells were fabricated for the first time, and their IV characteristic were analyzed. The simulation of n-ZnS/p-Si heterojunction solar cells by wx-AMPS was also carried out for the first time. Experiments according to the Simulation result in better device performance.

2 EXPERIMENTAL PROCEDURE

ZnS thin films were deposited by thermal evaporation with substrate temperature of 150°C. ZnS particles (99.99% purity) were placed in a molybdenum boat. The chamber was pumped to 6×10^{-6} Torr prior to deposition. The deposition rate and the thickness were measured by digital film thickness monitor using a quartz crystal sensor set at 6 MHz (QI8010, Fil-Tech Inc). P-type Si (100) oriented CZ silicon wafers 200 μm thick with a bulk resistivity of 1 Ω·cm and 1*1 cm^2 area were subjected to planar surface for initial device and textured surface for optimized device. The p-Si substrates were cleaned by standard RCA processing followed by hydrofluoric acid (10%) treatment to remove native oxide and passivate the c-Si surface. The TCO layer and Ag electrode were deposited by RF sputtering and thermal evaporation, respectively. For the initial device, thermal evaporated Ag was used as the back electrode while for optimized device post thermal-treated Al was used as back electrode.

The J-V characteristics were evaluated on a NewPort system under AM 1.5 G sunlight, and recorded by a Keithley model 2400 digital source meter. Transmission electron microscope (TEM, FEI Tecnai G2 F30) was used to study the high resolution cross-sectional graph of the device.

3 RESULTS AND DISCUSSION

Figure 1. (a) shows the structure of the planar ZnS/p-Si heterojunction solar cells, in which n-ZnS layer serves as emitter and 320 nm hydrogen doped AZO (HAZO) acts as TCO layer, respectively. Ag back electrode with thickness of 300 nm was thermal evaporated directly on the Si. Figure 1. (b) shows the side-view SEM image of the HAZO and ZnS coated planar p-Si. The HAZO deposited on the substrate exhibits very good smoothness and density without any pores. It can also be observed that the sharp interface and column HAZO grew vertically on the ZnS layer. There are some cavities between Ag bus bar and HAZO, indicating the not good enough contact of Ag and HAZO and the necessity of post thermal treatment.

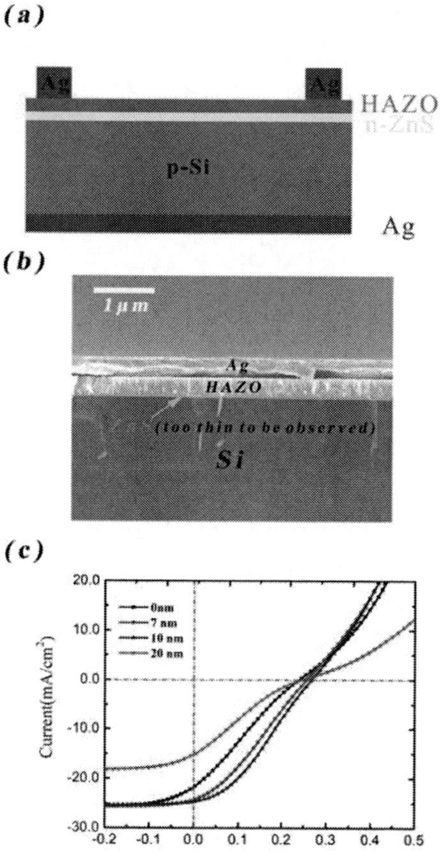

Figure 1: (a) The schematic diagram of the planar ZnS/p-Si heterojunctoin solar cell. (b) The side-view SEM image of the HAZO and ZnS coated planar p-Si. (c) The IV curves of the device with different thickness of ZnS.

Figure 1. (c) shows the illuminated J-V characteristic of the device with different thickness of ZnS. Table 1 shows the influence of thickness of ZnS to the device performance. 0 nm represents that the HAZO layer was deposited directly on the p-Si. ZnS thin film layer can reduce the reflection and increase the J_{sc} subsequently. Besides, ZnS can also decrease the damage during the HAZO sputtering process, which will introduce interface defects and lower the FF. 7 nm ZnS may not be continuous and would introduce the short circuit channel between the HAZO layer and p-Si, which results in the worse FF compared with the device with 10 nm ZnS. 20 nm ZnS is so thick that the series resistance becomes too large and it affects the carriers' transportation, the FF decreases consequently.

Table I: The influence of thickness of ZnS to the devices' performance

Thickness /nm	0	7	10	20
V_{oc}/V	0.237	0.257	0.267	0.246
J_{sc}/mA·cm^2	20.07	24.30	24.96	15.18
FF/%	22.71	29.88	34.01	21.59
PCE/%	1.08	1.87	2.27	0.80
R_{sh}/Ω·cm^2	6.40	5.95	4.80	11.90

Considering the non ohmic contact between Ag and p-Si, which would introduce a Schottky barrier and hinder the carrier transportation, thus decrease V_{oc} and FF. Besides, the planer Si increases the front surface reflection, then decreases the J_{sc}. For optimization, textured Si and ohmic contact between p-Si and back electrode are needed.

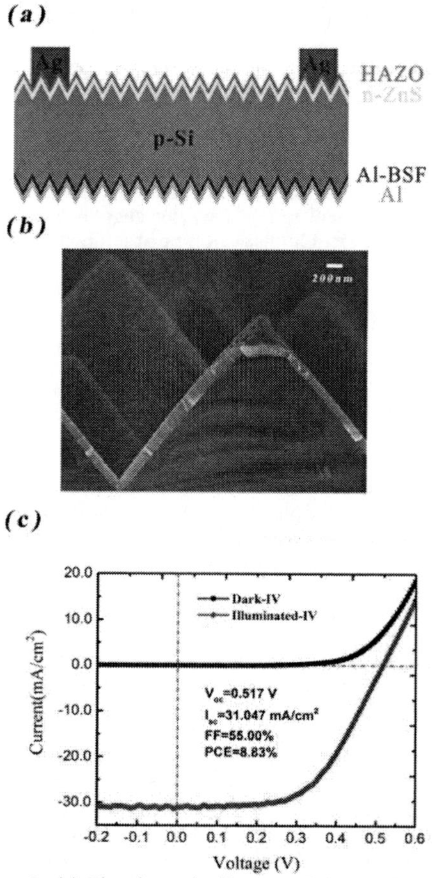

Figure 2: (a) The shematic diagram of the ZnS/textured p-Si heterojunction solar cells. (b) The cross section micrograph of the HAZO/ZnS/textured p-Si layer. (c) The dark and illuminated J-V characteristic of the ZnS/textured p-Si heterojunction solar cells.

Figure 2. (a) shows the schematic diagram of the ZnS/textured p-Si heterojunction solar cells. Al electrode was deposited on the textured Si by thermal evaporation, then followed by furnace annealing at 500℃ for 30mins in N_2 ambient to form the ohmic contact. The thickness of HAZO layer was about 150nm for better antireflection.

Figure 2.(b) shows the side-view SEM image of the HAZO and ZnS coated textured p-Si. The HAZO deposited on the substrate exhibits good coverage and density without any pores. Column HAZO growing vertically on the ZnS laye can also be seen. Figure 2. (c) shows the dark and illuminated J-V characteristic of the device. The electrical parameters increase after textured Si was applied and Al was used as back electrode. There is a Schottky barrier between p-Si and Ag, while Al electrode deposited on the Si and followed by furnace annealing can form the ohmic contact and Al-BSF. These processes increase the V_{oc} and FF. The front surface of textured Si enhances the light trapping and increases the J_{sc}. The efficiency is the published highest value so far to our knowledge for n-ZnS/textured p-Si heterojunction solar cells [9].

Wx-AMPS was used to simulated the ZnS/p-Si heterojunction solar cells. No bulk defect states and interface defects were taken into consideration. Some assumptions were as following: T=300 K, R_{front}=0, R_{back}=1, $S_{r\text{-}front\text{-}e\&h}$=1E8 $cm^{-3}\cdot s^{-1}$, $S_{r\text{-}back\text{-}e\&h}$=1E4 $cm^{-3}\cdot s^{-1}$. R_{front} and R_{back} represent the reflectance of front and back surface. $S_{r\text{-}front\text{-}e\&h}$ and $S_{r\text{-}back\text{-}e\&h}$ represent the recombination rate of electron and hole on the front and back surface.

Table II: The simulation parameters of ZnS/p-Si device

Parameter and units	p^+-Si	p-Si	n-ZnS
Thickness/um	0.2	200	2
Electron affinity (eV)	4.05	4.05	3.9
Relative dielectric constant	11.9	11.9	8.9
Band gap (eV)	1.12	1.12	3.7
N_c (cm^{-3})	2.8E19	2.8E19	2.2E18
N_v (cm^{-3})	1.04E19	1.04E19	1.8E19
m_e ($cm^2\ V^{-1}\ s^{-1}$)	1350	1350	100
m_h ($cm^2\ V^{-1}\ s^{-1}$)	450	450	25
N_a(cm^{-3})	1E20	5E15	1
N_d(cm^{-3})	1	1	1E11-1E16

Where, N_c and N_v represent the conduction and valence band effective state density, m_e and m_h represent the electron and hole mobility, N_a and N_d represent the acceptor and donor concentration.

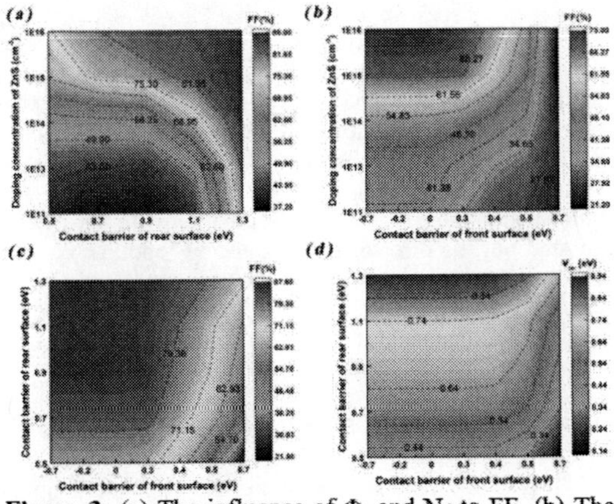

Figure 3: (a) The influence of Φ_r and N_d to FF. (b) The influence of Φ_f and N_d to FF. (c) The influence of Φ_f and Φ_r to FF. (d) The influence of Φ_f and Φ_r to V_{oc}.

Figure 3 shows the simulation result of the impacts of contact barrier of back surface (Φ_r), contact barrier of front surface (Φ_f) and doping concentration of ZnS (N_d) on the device. Figure 3. (a) shows the larger Φ_r and N_d are, the larger of FF will be. However, the influence of N_d is little on the FF when the Φ_r is bigger than 1.2 eV and the influence of Φ_r is not obvious on the FF when the N_d is bigger than 1E15 cm^{-3} or smaller than 1E14 cm^{-3}. Figure 3. (b) shows the Φ_f has negative impact on FF. FF will decrease if Φ_r increases, especially when the N_d is bigger than 1E14 cm^{-3}. However, when the Φ_f is bigger than 0.5 eV, the influence of N_d is negligible on the FF and it will be very bad.

Figure 3. (c), (d) show FF is mainly influenced by Φ_f, while V_{oc} is mainly influenced by Φ_r. Large Φ_f results in low FF and V_{oc}. Simulation results showed that extreme high and low work function materials should be applied on the back surface of p-Si and front surface of ZnS, respectively. The Φ_r is equal to the work function of material deposited on Si minus the electronic affinity of Si (4.05 eV). If Φ_r is required to be more than 1.2 eV, the work function should be larger than 5.25 eV. Extreme high work function materials such as V_2O_5, MoO_x, WO_3 [10, 11] will be best. As for the Φ_f, which is equal to the work function of material deposited on ZnS thin film minus the electronic affinity of ZnS (3.9 eV). If the Φ_f is required to be less than 0 eV, the work function should be smaller than 3.9 eV. Extreme low work function materials such as C_2CO_3, LiF, MgF_2 [12, 13] can meet the requirement. Further simulation shows that if the Φ_f and Φ_r are equal to 1.2 eV and 0 eV, respectively, even if the N_d is equal to 1E15 cm^{-3}, which means that the ZnS thin film is slightly doped, the device represents an efficiency of 23.78% with V_{oc} of 0.83 V, J_{sc} of 39.72mA/cm^2 and FF of 72.37%. The result showed the prospect of ZnS/p-Si heterojunction solar cells.

According to the simulation, we deposited the extreme high work function material WO_3 onto the p-Si. Figure 4 shows the ohmic contact forming mechanism of WO_3 and p-Si from the aspect of energy band. The energy band of p-Si bends up and the holes accumulate in the interface region. Inversion p+ layer formed when the hole concentration reaches a certain value. The very thin p+ region act as back surface field (BSF) layer and carriers pass through the region via tunneling. Thus, ohmic contact is formed.

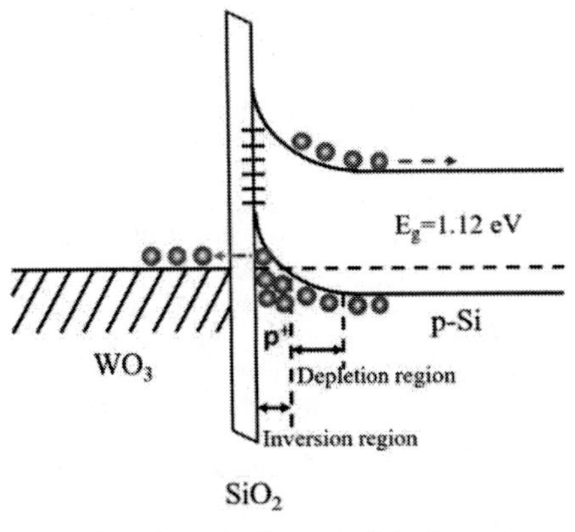

Figure 4: The schematic diagram of ohmic contact forming mechanism of WO_3 and p-Si.

Figure 5 shows the contact resistivity (ρ_c) of WO_3 and p-Si obtained via the TLM method [14]. The insert graph on the upper left and bottom right are the schematic diagram of the structure for TLM testing and fitting result, respectively. It can be seen that the contact exhibits Ohmic I-V behavior allowing accurate extractions of ρ_c. The contact resistivity of 12.7 $m\Omega \cdot cm^2$ was deduced from the fitting result. The Ohmic I-V behavior was resulted from the accumulated layer at the surface of Si as depicted above.

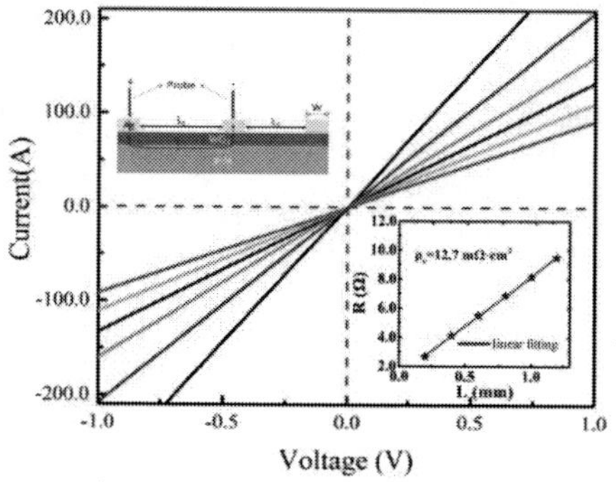

Figure 5: The contact resistivity of WO_3 and Si

Figure 6. (a) shows the schematic diagram of the textured ZnS/p-Si/WO_3 heterojunctoin solar cell. It should be noted that the cell ares is 2*2 cm^2, which is four times than the previous solar cells. WO_3 thin films about 9 nm thick was deposited onto the back surface of Si by thermal evaporation at RT. Ag back electrode was deposited onto the WO_3 thin films directly after the WO_3 deposition accomplished. Figure 6. (b) demonstrated the TEM cross-sectional graph of the p-Si/WO_3/Ag. It can be seen that the WO_3 thin films turned out to be amorphous and the Ag atoms diffuse into WO_3 thin films. The SiO_2 layer was formed by the oxidation of Si after RCA cleaning process and before putting into the vacuum chamber. Figure 6. (c) shows the PCE of the above ZnS/Si heterojunction solar cells. Better anti-reflection result from the optimized thickness of ITO results in higher Jsc. The promotion of V_{oc} and FF are benefit from the insert of WO_3 between Ag and p-Si, which can form the back-surface field on the rear side of Si and is beneficial to the carriers transportation. The result confirmed the part of conclusions obtained from the simulation.

The relative low V_{oc} is due to the small difference of fermi energy (ΔE_F) between undoped ZnS and p-Si. ΔE_F of 0.578 V can be calculated according to the fermi energy deduced from the carrier concentration. Besides, absent of passivation layer also accounting for the poor V_{oc}. The poor FF was mainly caused by the big series resistance (R_s) of 2.14 $\Omega \cdot cm^2$, which may mostly came from the large bulk resistance of undoped ZnS. So, doping of ZnS should be utilized for bigger V_{oc} and FF. Moreover, higher work function materials such as MoO_x even V_2O_5 replacing WO_3 and low work function materials such as MgF_2 and MgO deposited on ZnS would also improve the performance of the device.

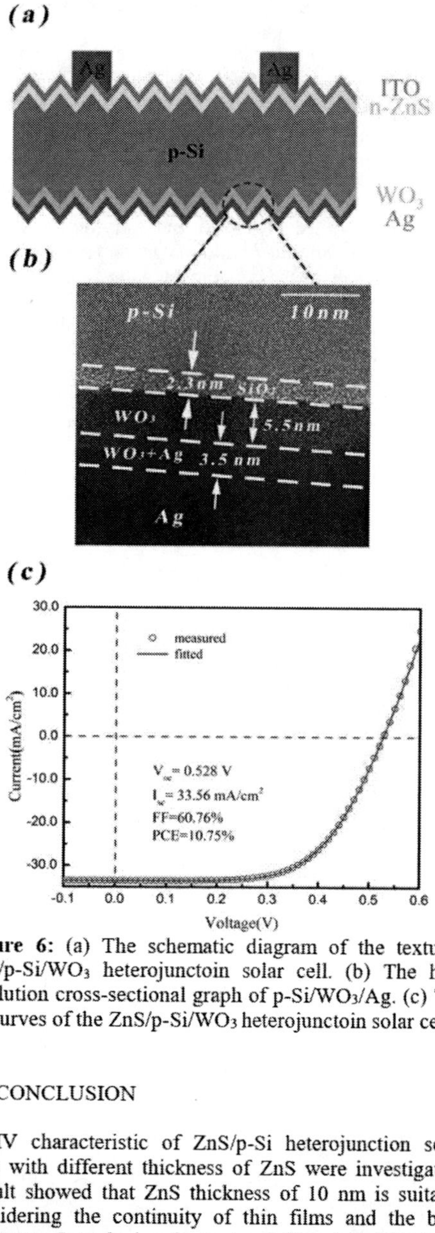

Figure 6: (a) The schematic diagram of the textured ZnS/p-Si/WO_3 heterojunctoin solar cell. (b) The high resolution cross-sectional graph of p-Si/WO_3/Ag. (c) The IV curves of the ZnS/p-Si/WO_3 heterojunctoin solar cell.

4 CONCLUSION

IV characteristic of ZnS/p-Si heterojunction solar cells with different thickness of ZnS were investigated. Result showed that ZnS thickness of 10 nm is suitable considering the continuity of thin films and the bulk resistance. Introducing the textured Si and Al BSF layer increased the J_{sc}, V_{oc} and FF, which obtained the published highest PCE so far. Simulation with wx-AMPS pointed out the direction of V_{oc} and FF optimizing. The later experiment according to the simulation confirmed the result, and results in the best device performance up to now.

ACKNOWLEDGMENT

This work was supported by the National Natural Science Foundation (Grant No. 61774173), the National High-tech R&D Program of China (863 Program; Grant No.2015AA050303 and 2015AA050608), the Guangzhou collaborative innovation Major Project of producing, teaching and researching (Grant No.

201508010011)), and the Jiangsu Collaborative Innovation Center of Photovoltaic Science and Engineering (Grant No. SCZ1405500002).

REFERENCES

[1] R. Vishwakarma, Journal of Theoretical and Applied Physics 9 (2015) 185-192.
[2] G. Liang, P. Fan, C. Chen, J. Luo, J. Zhao, D. Zhang, J Mater Sci: Mater Electron 26 (2015) 2230-2235.
[3] P. Chelvanathan, Y. Yusoff, F. Haque, M. Akhtaruzzaman, M.M. Alam, Z.A. Alothman, M.J. Rashid, K. Sopian, N. Amin, Applied Surface Science 334 (2015) 138-144.
[4] S. Ummartyotin, Y. Infahsaeng, Renewable and Sustainable Energy Reviews 55 (2016) 17-24.
[5] J. Bosco, F. Tajdar, H. Atwater, 38th IEEE Photovoltaic Specialists Conference (2012) 002513-002517.
[6] Y.-J. Hsiao, C.-H. Lu, T.-H. Fang, International Journal of Photoenergy (2013) 5.
[7] Y.J. Hsiao, T.H. Meen, L.W. Ji, J.K. Tsai, Y.S. Wu, C.J. Huang, Journal of Physics and Chemistry of Solids 74 (2013) 1403-1407.
[8] L.-W. Ji, Y.-J. Hsiao, I.-T. Tang, T.-H. Meen, C.-H. Liu, J.-K. Tsai, T.-C. Wu, Y.-S. Wu, Nanoscale Research Letters 8 (2013) 470.
[9] K. Qiu, D. Qiu, L. Cai, S. Li, W. Wu, Z. Liang, H. Shen, Materials Letters 198 (2017) 23-26.
[10] C. Battaglia, S.M.d. Nicolás, S.D. Wolf, X. Yin, M. Zheng, C. Ballif, A. Javey, Applied Physics Letters 104 (2014) 113902.
[11] L.G. Gerling, S. Mahato, A. Morales-Vilches, G. Masmitja, P. Ortega, C. Voz, R. Alcubilla, J. Puigdollers, Solar Energy Materials and Solar Cells 145 (2016) 109-115.
[12] H.-H. Liao, L.-M. Chen, Z. Xu, G. Li, Y. Yang, Applied Physics Letters 92 (2008) 173303.
[13] Y. Wan, C. Samundsett, J. Bullock, T. Allen, M. Hettick, D. Yan, P. Zheng, X. Zhang, J. Cui, J. McKeon, ACS applied materials & interfaces 8 (2016) 14671-14677.
[14] S. Luan, G.W. Neudeck, Journal of Applied Physics 72 (1992) 766-772.

FABRICATION OF SILICON HETEROJUNCTION CELLS ON 50μm EPITAXIAL SUBSTRATES

Twan Bearda[1], Abdullah Umer[1,2], Shruti Jambaldinni[1], Miha Filipič[1], Kris Van Nieuwenhuysen[1], Hariharsudan Sivaramakrishnan Radhakrishnan[1], Valerie Depauw[1], Ivan Gordon[1], Maarten Debucquoy[1], Yaser Abdulraheem[3], Jozef Szlufcik[1], Jef Poortmans[1,4,5]

[1] IMEC, Kapeldreef 75, B-3001 Leuven, Belgium
[2] Ruhr Universität Bochum, Universitätsstraße 150, D-44801, Bochum
[3] Kuwait University, P.O. Box 5969, 13060 Safat, Kuwait
[4] KU Leuven, Kasteelpark Arenberg, 10 B-3001 Heverlee, Belgium
[5] Hasselt University, Hasselt 3500, Belgium

ABSTRACT: We demonstrate solar cell fabrication on large area (125mm × 125mm) epitaxial wafers of 50μm thickness. During handling and transport no mechanical support was used for the wafers, and the process was identical to the process that was used for 50μm and 160μm Czochralski wafers. Due to the wafer preparation procedure, thin epitaxial wafers were found to be mechanically weaker than Czochralski wafers. This resulted in wafer breakage especially during wet processing. The efficiency of epitaxial cells reached 17%, which was lower than the 19% and 20% efficiencies obtained with 50μm and 160μm Czochralski wafers. The cell results were thus found to be a good metric for substrate quality.
Keywords: Thin crystalline silicon, Cost Reduction, Heterojunction, Yield, Substrates

1 INTRODUCTION

In order to reduce the material cost for silicon solar cells, different approaches to make very thin (<100μm) monocrystalline wafers are being investigated [1,2]. Solar cells on thin substrates have also been demonstrated previously [3,4,5,6,7]. However, due to the fragility of the wafers, the procedure often involves small substrate sizes (<100cm^2), small batches and/or bonding of the thin substrate to a carrier substrate. In this contribution we will discuss various adaptations in our standard cell process which allowed us to set up a baseline for batches of large area (156cm^2) thin (50μm) and thick (160μm) substrates without the support of a carrier substrate. This allows a fair comparison of the qualities of different types of substrates.

2 EXPERIMENTAL

2.1 Starting material

In our experiments we target the fabrication of silicon heterojunction (SHJ) cells on epitaxial wafers of 50μm thickness. The process for the fabrication of these epifoils has been extensively described elsewhere [2]. It is relevant to mention here that after detachment of the epifoil from the parent wafer, there is still a porous silicon layer at the side of the cleavage plane. During the cell process, this porous layer will have to be removed.

As a reference in our experiments, we also included Czochralski substrates with thicknesses of 50μm and 160μm. The different thicknesses were obtained by etching in 20% KOH. This allowed to assess the effect of substrate thickness as well as the effect of material quality.

Each batch of wafers consisted of six substrates of each type, with no further splits in the processing. While these numbers do not allow accurate yield predictions for large batches, they did allow us to identify critical process steps. The toolset at imec does not allow epitaxial substrates with a size of 156mm × 156mm. Accordingly, all substrate types were laser diced to 125mm × 125mm (156cm^2).

2.2 Texturing

For efficient light coupling, obviously the front surface of the cell needs to be textured. In addition, in the case of epitaxial wafers, the remainders of the porous silicon layer need to be removed as well. Both can be achieved by means of alkaline etching, resulting in random pyramid formation on both sides of the wafer.

In our reference texturing process [8], the total amount of silicon that is removed is of the order of 12μm. This is a significant fraction of the substrate. For this reason we have developed a new texturing process which results in a total silicon removal of only 6μm with much smaller pyramids (Figure 1). At the same time a low reflection of <10% is achieved. Ref. 9 discusses in more detail the texturing process and the beneficial effect of the smaller pyramid size on the bending strength of the wafers.

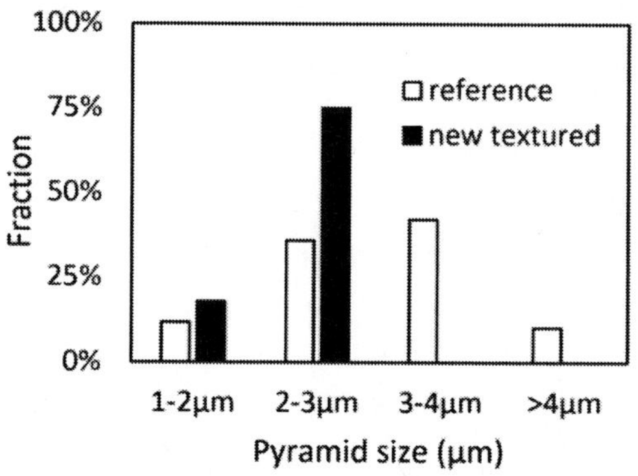

Figure 1. Pyramid size distribution for old and new texturing process.

2.3 Wet processing

All wet processing was done in semi-automatic wetbenches using cassettes. The wet processing has been found to be most challenging in terms of breakage for a number of reasons

- Loading of the wafers in the cassettes was done manually with tweezers. Yield loss by handling was reduced over time as the operators gained more experience.
- Due to the wafer flexibility, capillary forces cause stiction when the wafers are lifted out of the chemical baths. The problem was solved by increasing the pitch between the wafers.
- During cleaning and rinsing, turbulence is used to improve mixing in the baths. However, this also causes breakage of thin wafers. The breakage was prevented by reducing turbulence without affecting the overall cleaning performance.
- An automatic hot water dryer was used to significantly reduce the breakage of the wafers instead of the conventional spin dryer.

2.4 Front and rear passivation

Amorphous silicon was deposited by PECVD to form the front surface field and rear emitter. Processing of thin wafers was safe, although the venting channel was slightly modified to prevent displacement of the substrates during venting.

2.5 ITO deposition

On each wafer four cells of 40mm × 40mm were defined by deposition of Indium Tin Oxide (ITO) through a shadow mask. The wafer and the shadow mask were held in place by a magnetic foil, which provided mechanical stability during deposition on thin wafers.

2.5 Metallization

Low temperature silver paste was screenprinted on both sides of the wafers, followed by 175°C curing. In order to reduce thin wafer bowing, the rear side was not fully covered with metal, but by squares of 9mm × 9mm squares and an area coverage of 80% (Figure 2).

Wafer breakage during screenprinting occurs only if the wafer is already damaged or has cracks due to previous processing. As such it is expected that most yield loss is ultimately related to the wet processing and/or the epitaxial wafer preparation.

A considerable number of thin wafers was processed up to this stage without breakage. After screenprinting, however, some cells on epifoil substrates were partly shunted (Figure 3). It appeared that these shunts coincide with small cracks which were previously unnoticed, and which did not result in wafer breakage. After screenprinting, these cracks cause short-circuiting of the rear emitter by the metal square.

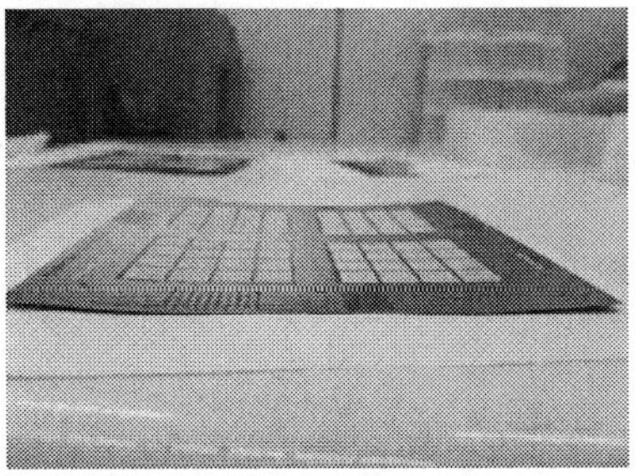

Figure 2. Rear side of a wafer after processing.

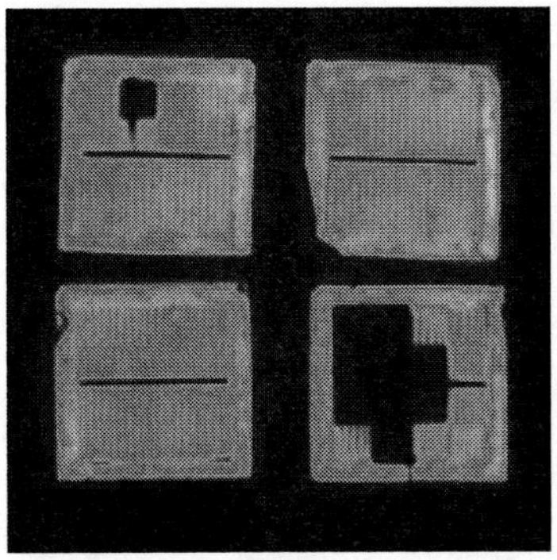

Figure 3. Photoluminescence map of some finished cells after dicing.

3 RESULTS

The process yield loss for the different process steps is shown in Figure 4. Overall the epitaxial wafers suffered higher yield loss than the thinned down Cz wafers. This is due to the current preparation method of the wafers inducing mechanical defects in the substrates . As discussed before, the largest number of wafers was broken during the wet processing.

The device results are reported in Table I. The external quantum efficiency (EQE) is shown in Figure 5. The best thin cell reached a very high V_{oc} of 744mV. Between the Czochralski cells there is considerable variation of FF and V_{oc}. Further experiments have indicated that this is caused by insufficient cleaning after saw damage removal and texturing. Since the epifoils did not receive saw damage removal, the variation in FF was much smaller compared to the Cz wafers.

The short circuit current J_{sc} is quite reproducible for all cells, being 2mA/cm^2 lower in the thin Cz cells compared to the thick Cz cells due to less efficient absorption at long wavelengths. The epitaxial cells have lower J_{sc} and V_{oc} than the thinned-down Cz wafers due to a higher recombination rate in the bulk (see also Figure 5).

Table I. Cell results for different substrate types. Averages and best cells are shown, with the number of measured cells between brackets.

	J_{sc} [mA/cm^2]	V_{oc} [mV]	FF [%]	η [%]
160µm CZ (12)	35.8	722	70.6	18.3
best cell	35.9	730	77.2	20.3
50µm CZ (20)	33.9	723	71.6	17.6
best cell	34.0	744	74.4	18.8
50µm Epifoil (17)	32.9	656	72.1	15.6
best cell	33.5	693	73.2	17.0

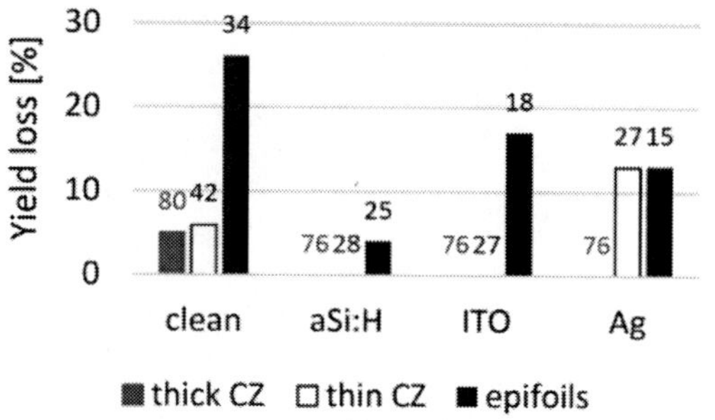

Figure 4. Yield loss per process step as determined over several batches. The total number of wafers per process step is indicated.

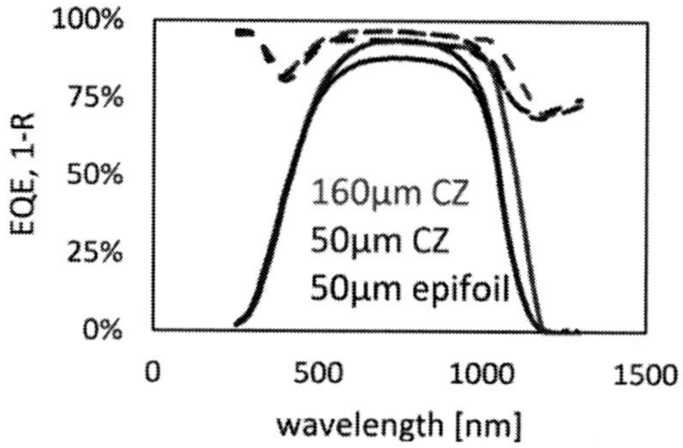

Figure 5. External Quantum Efficiency (EQE, drawn lines) and reflection (1-R, dashed lines) for three different substrate types.

3 CONCLUSIONS

We have demonstrated solar cell fabrication on large area (125mm × 125mm) epitaxial wafers of 50μm thickness. During handling and transport no mechanical support was used for the wafers, and the process was identical to the process that was used for 50μm and 160μm Czochralski wafers. Due to the wafer preparation procedure, thin epitaxial wafers were found to be mechanically weaker than thin Czochralski wafers. This resulted in increased wafer breakage. The efficiency of epitaxial cells reached 17%, and was found to be a good metric for substrate quality.

ACKNOWLEDGEMENTS

This work is financially supported by IMEC's Industrial Affliation Program for Photovoltaics (IIAP-PV). This work was partly funded by H2020 EU project CABRISS under Grant Agreement No. 641972.

REFERENCES

[1] R. Niepelt, J. Mat. Res. 30 (2015), 3227—3240
[2] K. Van Nieuwenhuysen, Proceedings 38th IEEE Photovoltaic Specialists Conference (2012) 1833.
[3] M. Reuter, Solar Energy Materials & Solar Cells 93 (2009) 704—706.
[4] R. Brendel, IEEE J. Photovoltaics 1 (2011) 9—15
[5] C.W. Chen, Proceedings 39th Photovoltaic Specialists Conference (2013) 2179—2182
[6] J. Han, IEEE J. Photovoltaics 6 (2016) 1160—1166
[7] P. Kapur, Proceedings 28th European Photovoltaic Solar Energy Conference (2013) 2228—2231.
[8] A. Hajjiah, Proceedings 31st European Photovoltaic Solar Energy Conference (2015) 956—961
[9] K. Van Nieuwenhuysen, Proceedings 33rd European Photovoltaic Solar Energy Conference (2017)
[10] H. S. Radhakrishnan, Proceedings 33rd European Photovoltaic Solar Energy Conference (2017)

FUNDAMENTAL CONSTRAINTS IMPOSED BY THERMIONIC EMISSION BARRIER AT THE HETERO-INTERFACE AND BY PN JUNCTION DIFFUSION BARRIER ON THE FILL FACTOR AND EFFICIENCY OF SHJ CELLS

Moustafa Y. Ghannam and Yaser Abdulraheem
EE Department, Kuwait University, P.O. Box 5969, Safat 13060, Kuwait
mghannam@aucegypt.edu

ABSTRACT: A model that links the relatively low Fill Factor (FF) in p^+/n amorphous/crystalline silicon (a-Si:H/c-Si) hetero-junction solar cells (SHJ) with perfectly ohmic contacts to hole reflection at the hetero-interface under illumination is proposed. Such a reflection results from the partial emission and transport of holes imposed by the thermionic emission boundary condition at the hetero-interface. The non-emitted holes scatter back and diffuse towards the substrate, enhancing the cell forward current under illumination to levels well above the cell dark current leading to Fill Factor and efficiency degradation. The probability of hole reflection depends on the thermionic emission parameters and also on the diffusion barrier associated with the pn junction adjacent to the hetero-interface. Hole reflection is found to be particularly significant in cells with deficient active doping in a-Si:H. The model is supported by the results of AFORS-HET simulations at different operating temperatures.
Keywords: SHJ cells – Hole reflection – a-Si/c-Si hetero-interface - Fill Factor degradation.

1 INTRODUCTION

The record silicon solar cell conversion efficiency is still detained by the silicon hetero-junction solar cell (SHJ) in which ultrathin double layer (p^+/intrinsic) amorphous silicon a-Si(p^+/i):H contact is deposited on n-type crysalline silicon (c-Si) substrate. A very high open circuit voltage (V_{oc}) reaching 760 mV and an efficiency of 26.3% could be achieved with such cell [1]. The main problem usually encountered in SHJ cells is its Fill Factor (FF) which rarely exceeds 78%. Most of the research related to the FF of the SHJ cell concentrate on reducing the contact resistances, and selecting the proper transparent conductiong oxide (TCO) used at the cell front contact. The relatively low FF, however, persists in SHJ cells even if the cell contacts have a low resistance and are perfectly ohmic. Therefore, it is highly probable and widely accepted that the low FF is linked to a fundamental property associated with the nature of the SHJ structure. The purpose of the present study is to help identifying this fundamental property. Therefore, a rigorous analysis in which the behavior of the physical quantities detrmining the performance of the SHJ cell are looked at very closely. The analysis is supported by AFORS-HET simulations [2] of SHJ cells with different doping levels in a-Si:H. A model will be proposed which consitently interpret the behavior of these quantities. Finally, an optimum SHJ cell design is proposed taking into consideration the trade-off's predicted by the proposed model.

2. DEVICE DESCRIPTION AND PHYSICAL MODELS

The SHJ cell under study sketched in Fig.1 basically consists of a 200 μm thick n-type high quality crystalline silicon (c-Si) substrate on top of which an ultrathin (5 nm) intrinsic a-Si:H(i) layer followed by a (10 nm) p-type doped a-Si:H(p^+) layer are deposited. Finally, a perfectly ohmic front contact is ensured by a 60-80 nm thick (TCO) layer having a proper work function and covering the front surface. The default value of the n-type substrate doping concentration N_D is set to 1.5×10^{16} cm^{-3}, which is not necessarily the optimum value. The

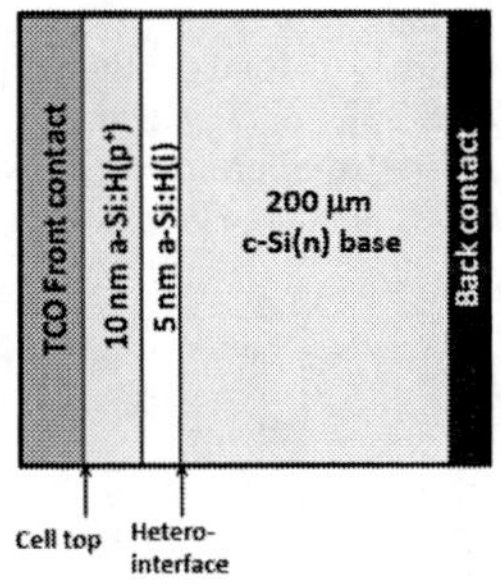

Figure 1: Structure of the Silicon Hetero-Junction (SHJ) solar cell under study

Shockley-Read-Hall (SRH) carrier lifetime in the n-type substrate is fixed at 5 ms.Default values of the a-Si:H band-gap energy E_G and electron affinity χ in a-Si:H are set to 1.72 eV and 3.9 eV [3], respectively, compared to 1.124 eV and 4.05 eV for c-Si.

Four cells with p-type doping concentrations $N_{a\text{-}Si}$ in a-Si:H(p^+) equal to 5×10^{18} cm^{-3} (SHJ_1), 10^{19} cm^{-3} (SHJ_2), 5×10^{19} cm^{-3} (SHJ^3) toand finally 10^{20} cm^{-3} (SHJ_4) are considered. It is well known that the active doping level or equivalently the free hole concentration in a-Si:H(p^+) is only a small fraction (1% to 10%) of the total (chemical) doping level due to trapping of holes by dangling bond defects [4-6]. Default models for the dangling bond defects implemeted in AFORS-HET simulator are maintained with a peak defect density, N_{DB0} ($cm^{-3}eV^{-1}$), set equal to the value of the doping concentration $N_{a\text{-}Si}$ in a-Si:H(p^+) [7]. N_{DB0} is fixed at 1.385×10^{16} $cm^{-3}eV^{-1}$ in a-Si:H(i). The per area peak dangling bond defect density D_{it0} at the a-Si:H(i)/c-Si hetero-interface is fixed at 10^{10} $cm^{-2}eV^{-1}$, ensuring high quality interface passivation [4,7].

AM1.5G solar spectrum is incident on the textured surface and ten internal reflections are considered. Radiative, Auger and SRH recombination are considered, while band-gap narrowing (BGN) is ignored as it should be negligible in the p^+ inversion layer emitter [7]. Carrier transport across the energy barrier at the a-Si:H/c-Si hetero-interface is assumed in the simulations to occur by thermionic emission. The simulations are carried out at 300 K unless stated differently.

3 ANALYSIS AND RESULTS

3.1 Energy band diagrams

The simulated band diagrams **at thermal equilibrium** are displayed in Fig.2 for cells SHJ_1 and SHJ_3. Using the band line-up theory and electron affinity rule [8] the conduction band offset ΔE_C at the hetero-interface is determined from the difference between the electron affinities of c-Si and of a-Si:H. The valence band offset ΔE_V covers the difference ΔE_G - ΔE_C, where ΔE_G is the difference between the band-gap energies of a-Si:H and of c-Si. When default values assigned for these parameters, ΔE_C and ΔE_V at the hetero-interface are estimated to 0.15 eV and 0.446 eV, respectively. Such values are confirmed in Fig.2, which also agree well with reported experimental values [9-11]. Due to a larger work function difference between the a-Si:H and c-Si in contact, the band bending in cell SHJ_3 with larger $N_{a\text{-}Si}$ is larger which ensures: 1) a strong p-type inversion layer emitter and a higher peak hole concentration at the hetero-interface, p_{peak}, crucial for the proper operation of the SHJ cell [7,12,13], and 2) a larger pn junction built-in potential in c-Si at equilibrium, V_{bi}, given by the total valence (or conduction) band bending in c-Si.

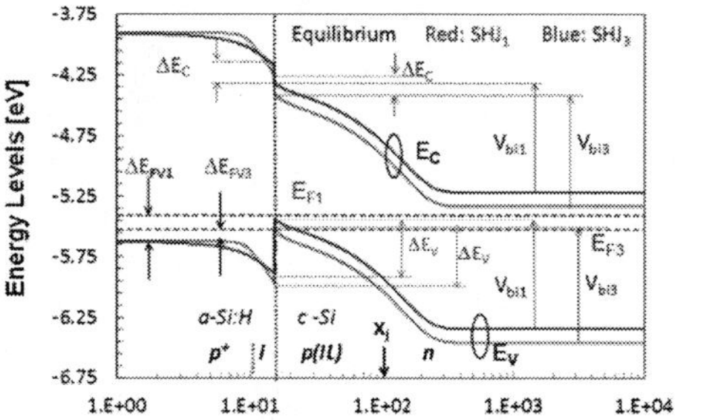

Figure 2: Simulated energy band diagram at equilibrium of default cells SHJ_1 (red) and SHJ_3 (blue). V_{bi} represents the built-in potential in c-Si. All energy levels refer to the vacuum level.

The simulated band diagrams for SHJ_1 and of SHJ_4 **under illumination** and short circuit condition, displayed in Fig.3a and Fg.3b, respectively, indicate a split in the Fermi energy into electron and hole quasi-Fermi energies throughout the pn junction space charge region. Such a separation effectively imposes a forward bias V_{SC} on the pn junction despite the short circuit condition, which amounts to 446 mV in SHJ_1 but decreases to around 100 mV in SHJ_4. Such a split includes though a voltage drop δ of 76 mV to 78 mV associated with the cell series resistance.

The short circuit Fermi energy split disappears if drift-diffusion transport is assumed in the simulations instead of thermionic emission transport across the hetetro-interface, which confirms that V_{SC} is associated with the latter.

3.2 I-V Characteristics under illumination

The simulated I-V characteristics under illumination of the four default cells SHJ_1 to SHJ_4 are plotted and

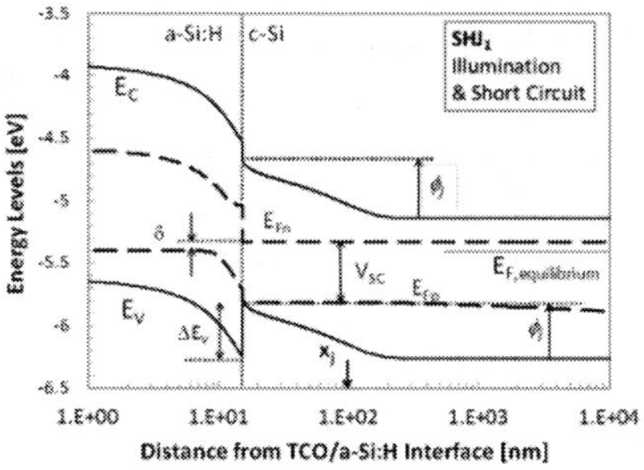

Figure 3a: Simulated energy band diagram for SHJ_1 cell under illumination and short circuit condition. V_{SC} is the pn junction short circuit voltage under illumination, ϕ_j = V_{bi} is the pn junction diffusion barrier, and δ is the voltage drop due to series resistance.

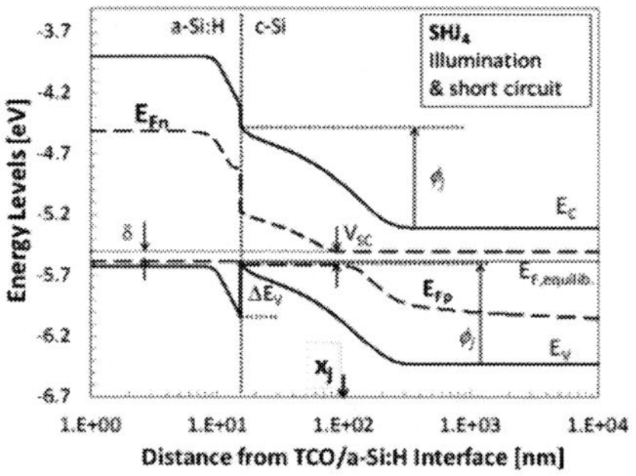

Figure 3b: Simulated energy band diagram for SHJ_4 cell under illumination and short circuit condition. Definitions for V_{SC}, ϕ_j and δ in Fig.3a apply.

compared in Fig.4. The I-V characteristics of SHJ_3 and of SHJ_4 have the expected shape with decent values for V_{OC} and I_{SC} but the FF is limited to 72% and 75.5%, respectively. On the other hand, the I-V characteristics of SHJ_1 and of SHJ_2 suffering from doping deficiency are seriously distorted and exhibit strong degradation of the FF to 40% and 52%, respectively.

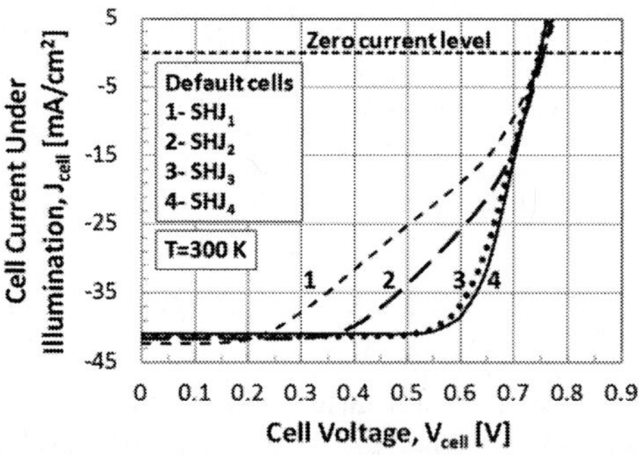

Figure 4: Simulated I-V characteristics of the four default cells SHJ_1 to SHJ_4, under illumination.

4. PROPOSED HOLE REFLECTION MODEL

Under illumination, the photogenerated hole current, J_p, diffuse in the reverse direction in the c-Si substrate, cross the pn junction to reach the a-Si:H/c-Si hetero-interface. The boundary condition at the hetero-interface, and the maximum thermionic emisssion hole current that can be collected are determined by the thermionic emission parameters, namely effective Richardson coefficient (A_h), temperature (T) and thermionic emission barrier (ΔE_V). If the photgenerated hole current J_p is larger than the maximum allowed thermionic emission current some holes will be blocked from crossing the barrier and will scatter back, that is reflected and flow down the diffusion barrier of the pn junction in the forward direction towards the n-type substrate. Such a reflected hole diffusion current forces the pn junction to establish a forward bias, V_{SC}, which explains the Fermi energy split at zero bias that appears at the pn junction in the band diagrams of Fig.3a and Fig.3b. Although the reflected hole current at short circuit condition may be much smaller than the emitted current and hence does not influence significantly the total cell current, the short circuit voltage V_{SC} adds up to the junction voltage at forward bias, which strongly enhances the cell forward current well above its value in the dark, which invalidates the simple one diode superposition circuit model of the cell and leads to a premature decrease in the total cell current and to the distorted I-V characteristics shown in Fig.4.

4.1 Leverage of doping levels

Values of V_{SC} extracted from the simulated band diagrams under illumination and short circuit conditions are plotted versus $N_{a\text{-}Si}$, together with the corresponding values of the cell FF, in Fig.5. The inverse relationship between V_{SC} - FF is confirmed for the cells under study. For low values of $N_{a\text{-}Si}$ V_{SC} is large, hole reflection is substantial leading to serious FF degradation. In addition, in such a case, the inversion layer is weak, p_{peak} is small, the band bending, V_{bi} (or ϕ_j) is small which stimulates hole diffusion and supports hole reflection.

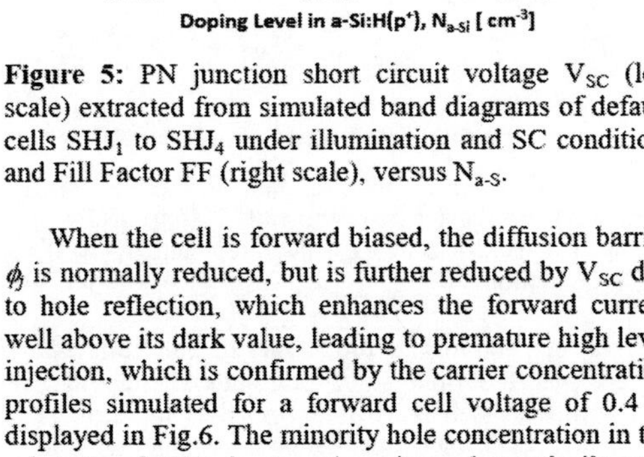

Figure 5: PN junction short circuit voltage V_{SC} (left scale) extracted from simulated band diagrams of default cells SHJ_1 to SHJ_4 under illumination and SC condition, and Fill Factor FF (right scale), versus $N_{a\text{-}S}$.

When the cell is forward biased, the diffusion barrier ϕ_j is normally reduced, but is further reduced by V_{SC} due to hole reflection, which enhances the forward current well above its dark value, leading to premature high level injection, which is confirmed by the carrier concentration profiles simulated for a forward cell voltage of 0.4 V displayed in Fig.6. The minority hole concentration in the substrate of SHJ_1 is strongly enhanced practically reac-

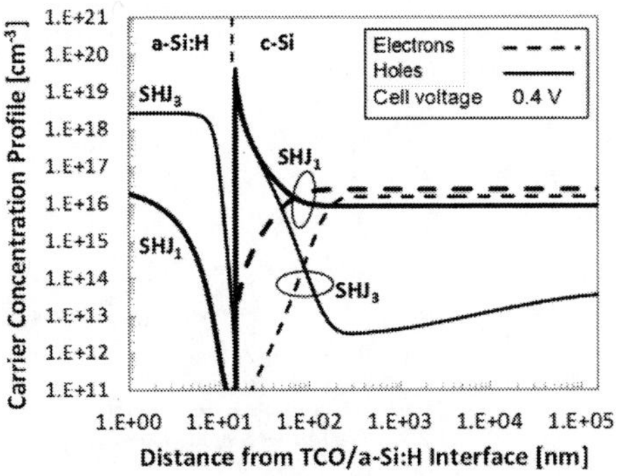

Figue 6: Simulated carrier concentration profiles in default cells SHJ_1 and SHJ_3under illumination for a cell forward bias of 0.4 V.

ing high level injection, which is not the case for SHJ_3.

On the other hand, the built-in potential V_{bi} and ϕ_j also increase with increasing substrate doping concentration N_D, reach a maximum then decrease. This leads V_{SC} to exhibit a minimum and the FF to exhibit a peak when N_D is close to 10^{17} cm^{-3} in SHJ_1 and to reach a peak as high as 80% in SHJ_4 when N_D is equal to 5x10^{17} cm^{-3}. Such improvement, however, is still far from being effective in resolving the drastic degradation in the FF and efficiency of of cells like SHJ_1 suffering from doping deficiency in a-Si:H(p$^+$), as depicted in Fig.7.

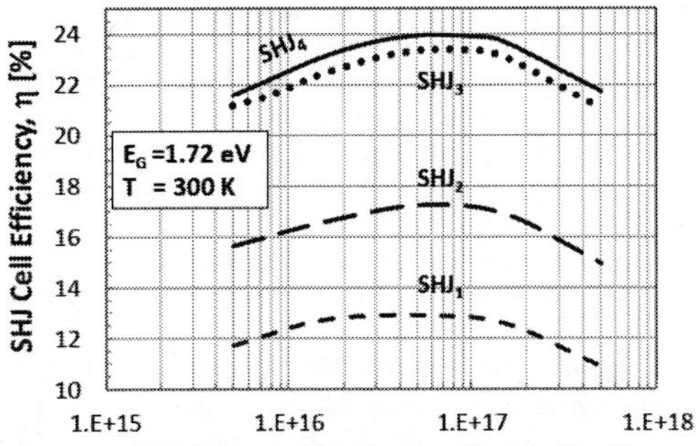

Figure 7: Conversion efficiency of default cells SHJ_1 to SHJ_4 versus the n-type doping concentration, N_D, in the c-Si substrate, as predicted by AFORS-HET simulations.

4.2 Impact of the thermionic emission parameters

Leverage of the effective hole Richardson constant A_h: The default value for the effective Richardson coefficient for holes at the hetero-interface is set in AFORS-HET simulator by the specific thermionic model settings A_h=9.56 A/cm^2/K^2. Such a value is used in all the simulations carried out in the present work, and the results of the simulations are interpreted accordingly.

As depicted in Fig.8, V_{SC} decreases as A_h increases and finally converge in all cells to the minimum level dictated by the series resistance voltage drop δ. As A_h increases the maximum allowed thermionic current increases which reduces the probability of hole reflection. In order to reach the minimumV_{SC} in SHJ_1, the value of A_h should be around 80 A/cm^2/K^2 in SHJ_1, which may not be physically possible.

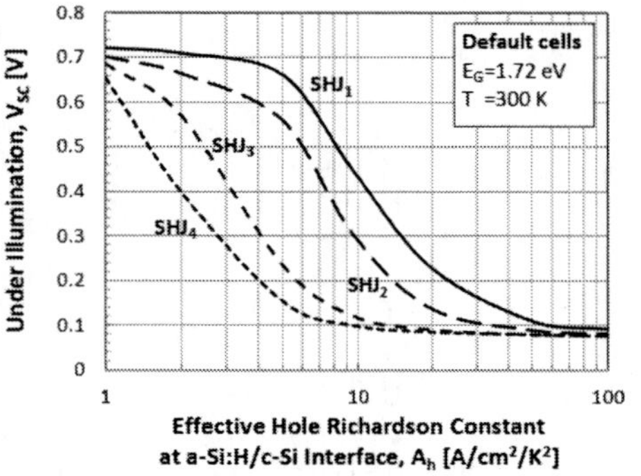

Figure 8: PN junction short circuit voltage (V_{SC}) in SHJ_1 to SHJ_4 extracted from band diagams under illumination, versus the effective hole Richardson constant A_h.

Leverage of the thermionic barrier height ΔE_V: The height of the thermionic energy barrier at the hetero-interface, ΔE_V, has a major influence on the probability of hole thermionic emission and on hole reflection at the hetero-interface. So far in the present analysis, the default value of the band-gap energy of a-Si:H, E_G, has been fixed at 1.72 eV resulting in a thermionic emission barrier height ΔE_V = 0.446 eV at the hetero-interface. Since the conduction band discontinuity (ΔE_C) is determined independently from the difference between $\chi_{c\text{-}Si}$ and of $\chi_{a\text{-}Si}$, the value of ΔE_V can be varied by varying E_G. Reducing E_G leads to a smaller barrier height ΔE_V, hence to a maximum allowed thermionic emisssion current, which decreases the probability of hole reflection at the a-Si:H/c-Si hetero-interface, reduces V_{SC} and finally results in an improved FF. This is confirmed in Fig.9 displaying V_{SC} extracted from band diagrams simulated under illumination at short circuit condition for the four cells SHJ_1 to SHJ_4 as a function of the a-Si:H band-gap energy, E_G, varied in the range 1.6 eV to 1.74 eV. A linear dependence is assumed in the simulations between the electron affinity χ and E_G.

It appears that V_{SC} continuously decreases in all cells with decreasing E_G (or ΔE_V), and reaches the minimum level, when E_G is equal to 1.6 eV, as depicted in Fig.9. While in such a case a moderate rise to 78.2% is achieved in the FF of SHJ_4, the improvement in the performance of SHJ_1 is impressive, since its I-V characteristics is restored, as depicted in Fig.10, and its FF is boosted to 77.2%, as depicted in Fig.11, compared to previous 40% with E_G=1.72 eV. These results confirm the very strong impact of ΔE_V on the hole reflection probability and particularly in cells which initially suffer severe degradation due to a-Si:H doping deficiency. The maximum efficiency for the four cells converge to 24% at E_G=1.6 eV and, like the FF in Fig.11, is maintained in a larger E_G range in cells with a heavily doped a-Si:H layer.

Leverage of the operating temperature, T: The impact of the operating temperature, T, on the performance of the SHJ cell is investigated assuming that the band-gap energies, the electron affinities and A_h do not vary with temperature. The temperature dependence of the intrinsic carrier concentration n_i, however, is taken into consideration.

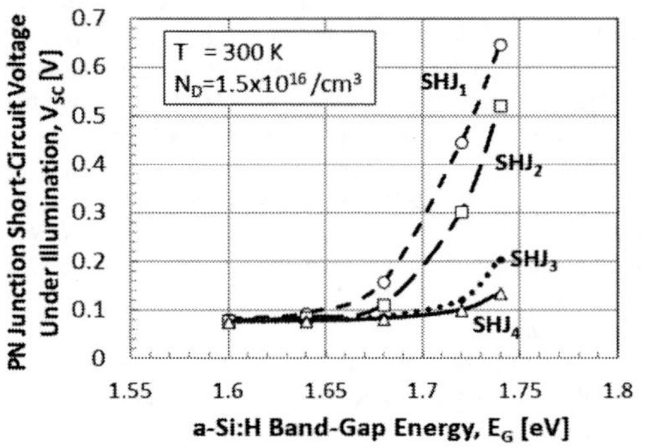

Figure 9: PN junction short circuit voltage, V_{SC}, extracted from band diagrams under illumination and short circuit conditions of default cells SHJ_1 to SHJ_4 versus a-Si:H band-gap energy, E_G.

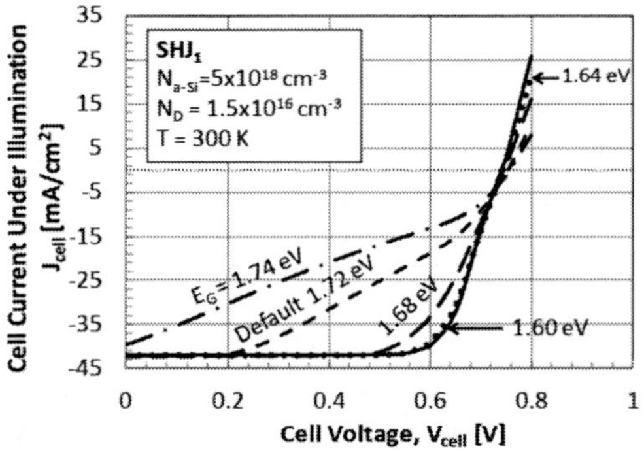

Figure 10: Simulated I-V characteristics under illumination of default cell SHJ_1 for different values of a-Si:H band-gap energy, E_G.

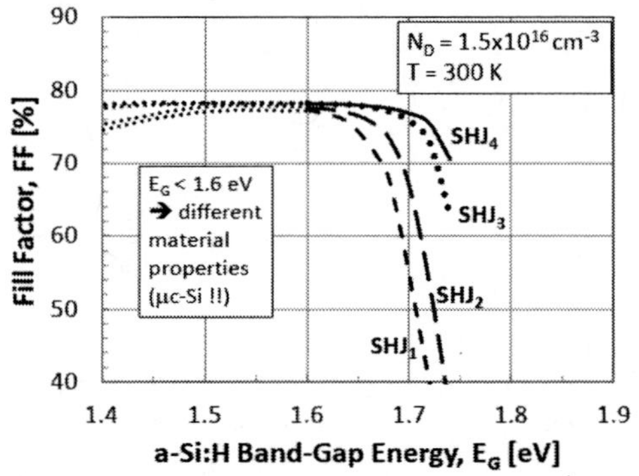

Figure 11: Fill Factor in default cells SHJ_1 to SHJ_4 versus the a-Si:H band-gap energy, E_G, predicted by AFORS-HET simulations.

At low temperature, the probability of hole reflection is expected to be high since hole thermionic emission probability is small. This is confirmed by the increase of V_{SC} with decreasing temperature observed in Fig.12 for

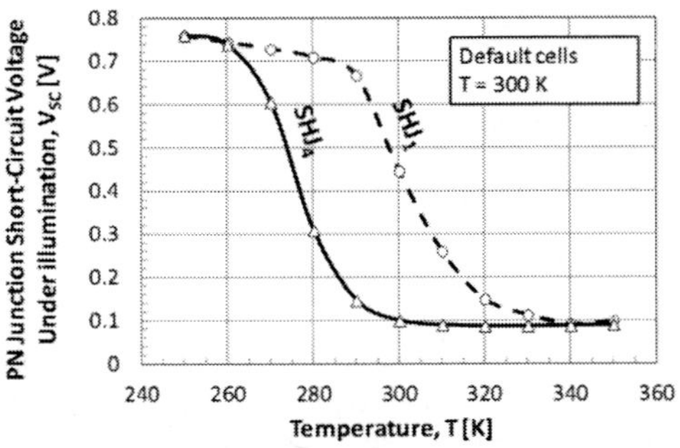

Figure 12: PN junction short circuit voltage, V_{SC}, extracted from band diagrams under illumination and short circuit conditions of default cells SHJ_1 to SHJ_4 versus operating temperature T.

$SHJ_1$and SHJ_4 cells. The useful temperature range of SHJ_4, however, starts at around room temperature, while the degraded performance of SHJ_1 extends up to the 330-340 K range, as confirmed by its I-V characetristics displayed in Fig.13.

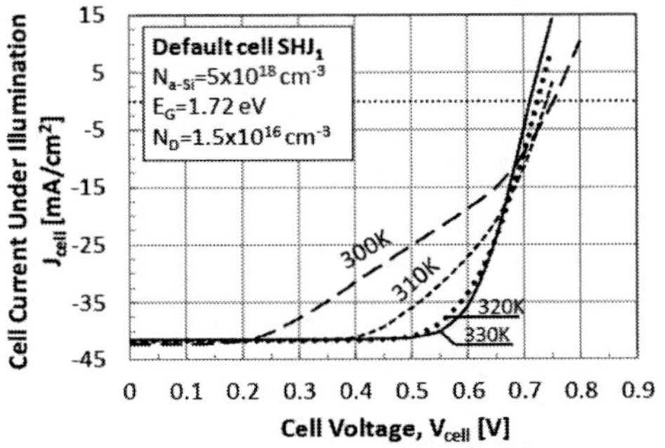

Figure 13: I-V characteristics under illumination of default cell SHJ_1 at different operating temperatures (300K, 310 K, 320 K, and 330 K) as predicted by AFORS-HET simulations.

5 SUMMARY AND CONCLUSION

A model is proposed which relates the FF degradation observed in p^+/n SHJ cells with perfect ohmic contacts to hole reflection at the thermionic emission barrier created by the valence band offset at the a-Si:H/c-Si hetero-interface. The model is based on the fact that the maximum thermionic emission current allowed to cross the a-Si:H/c-Si hetero-interface sets a boundary condition at that interface that forces some of the holes initially travelling within the short circuit current stream to scatter back. The reflected holes form a forward current that establishes an internal forward bias at the pn junction under illumination and short circuit condition. Such a bias adds up at least partially to the forward cell voltage which enhances the cell forward current causing: 1) premature high level injection, 2) I-V characteristics distortion, 3) shift of the maximum power point to lower voltage, and finally 4) FF and cell efficiency degradation. The performance is improved when the thermionic emission paramaters tend to increase the maximum allowed thermionic emission hole current, that is when the hole Richardson constant is relatively high, the thermionic emission barrier height ΔE_V is low, and the operating temparture is high. It is found that decreasing ΔE_V by decreasing the a-Si:H band-gap energy is very effective in improving the cell performance. However, the optimum band-gap energy stops at 1.6 eV as smaller band-gap does not provide a strong p^+ inversion layer capable of sustaining proper cell operation.

ACKNOWLEDGEMENTS

This work is supported by Kuwait University Research Administration project GE01/08 and KFAS/KU-IMEC funded project P115-15EE-01

6 REFERENCES

[1] Kaneca Press News Release:, September 14, 2016. http://www.kaneka.co.jp/kaneka-e/images/topics/1473811995/1473811995_101.pdf
[2] R. Stangl, C. Leendertz, and J. Haschke, R.D. Rugescu (ed.), Solar Energy, Vukovar, Croatia: Intech ISBN 978-953-307-052-0, 2010, pp. 319-352.
[3] M. Tucci, L. Serenelli, S. De Iuliis, M. Izzi, G. de Cesare, and D. Caputo, Physics and technology of amorphous-crystalline heterostructure silicon solar cells, W.J.H.M. Sark, L. Korte, F. Roca (eds.), Springer-Verlag Berlin Heidelberg ISBN: *9783642270079* 2012, p.337.
[4] L. Korte, Physics and technology of amorphous-crystalline heterostructure silicon solar cells, W.J.H.M. Sark, L. Korte, F. Roca (ed), Springer-Verlag Berlin Heidelberg ISBN: *9783642270079*, 2012, chapter 6, pp.160-221.
[5] M. Stutzmann, D. Biegelsen, and R. Street, Phys. Rev. B 35 (1987) 5666-5701.
[6] D. Caputo, G. de Cesare, F. Irrera, A. Nascetti, and F. Palma, J. Non-Cryst. Solids 266-269 Part 1 (2000) 565-568.
[7] M. Ghannam, G. Shehada, Y. Abdulraheem, and J. Poortmans, Sol. Energy Mater. Sol. Cells 132 (2015) 320-328.
[8] J.P. Kleider, Physics and technology of amorphous-crystalline heterostructure silicon solar cells, W.J.H.M. Sark, L. Korte, F. Roca (eds), Springer-Verlag Berlin Heidelberg ISBN: *9783642270079*, 2012, chapter 12, pp. 405-444.
[9] H. Eschrich, J. Bruns, L. Elstner, C. Swiatkowski, J. Non-cryst. Solids 164-166 (1993) 717-720.
[10] M. Schmidt, L. Korte, A. Laades, R. Stangl, C. Schubert, H. Angermann, E. Conrad, K. Von Meydell, Thin Solid Films 515 (2007) 7475-7480.
[11] J-P. Kleider, A. Gudovskikh, and P. Cabarrocas, Appl. Phys. Lett. 92 (2008) 162101 (1-3).
[12] M. Ghannam, Y. Abdulraheem, and G. Shehada, Sol. Energy Mater. Sol. Cells 145 (2016) 423-431.
[13] K. Ghosh, C. Tracy, S. Herasimenka, C. Honsberg, and S. Bowden, Proceedings of the 35th IEEE Photovoltaic Specialists Conference, IEEE Piscataway NJ, 2010, pp. 1383-1386.

COMPARISON BETWEEN A-SIOX:H AND A-SI:H AS PASSIVATION BUFFER LAYER FOR HETEROJUNCTION SOLAR CELLS

L. Martini[1], L. Serenelli[1,2], F. Menchini[2], M. Izzi[2], R. Asquini[1], G. de Cesare[1], D. Caputo[1], M. Tucci[2]
[1] DIET University of Rome "Sapienza", via Eudossiana 18, 00184 Rome – Italy
[2] ENEA Research Centre "Casaccia" DTE-FSN-TEF, via Anguillarese 301, 00123 Rome – Italy

ABSTRACT: In crystalline silicon based heterojunction solar cells the surface passivation quality is fundamental to obtain high efficiency. Intrinsic a-Si:H, as obtained by PECVD process from silane dissociation, is a good candidate for surface passivation, but UV absorption of this material limits the current generation. Moreover, surface passivation quality can be compromised when fabrication steps, following the a-Si:H deposition, exceed the a-Si:H deposition temperature. Both drawbacks can be overcome introducing a source of oxygen in PECVD process, obtaining a-SiOx:H layer that provides at the same time high quality passivation, wider optical bandgap and less susceptible to temperature steps than a-Si:H. In this work we compared two heterojunction solar cells different only for front side passivation layer, consisting of a-Si:H or a-SiOx:H. We monitored lifetime and implied Voc during fabrication steps and evaluated the final I-V characteristics and quantum efficiency. We found that the UV light soaking together with heating during subsequent sputtering process, increased lifetime of cell with a-SiOx:H more than that with a-Si:H. Likewise the cell having a-SiOx:H as passivation layer showed electrical parameters higher than the a-Si:H counterpart. Indeed Jsc and Voc were 1mA/cm2 and 20mV respectively higher in cell with a-SiOx:H than in cell with a-Si:H. In particular cell with a-SiOx:H layer achieved an implied Voc of 751mV.
Keywords: heterojunction, amorphous, oxide

1 INTRODUCTION

The photovoltaic market is dominated by silicon based solar cell, with Silicon Heterojunction (SHJ) technology becoming continuously more important, as currently detains the conversion efficiency record [1]. Heterojunction technology is based on a crystalline silicon wafers (c-Si), usually n-type doped, as light absorber and silicon amorphous layers (a-Si:H) deposited on silicon surface by PECVD technique [2] [3] [4], used to passivate silicon wafers surface and form the junction. There are many advantages in SHJ as simplify and decreasing the energy total amount consumption during manufacturing steps. Two mainly drawbacks of SHJ technology are the amorphous layer filtering effect [5] [6] of UV light, that can limit the efficiency and its low thermal stability that prevent the possibility to use temperatures above 250°C during manufacturing steps [7]. Replacement of a-Si:H with other more transparent and more thermally stable materials can helpful allow further efficiency improvement because of higher current produced and lower contact resistance, actually limited by the low temperature during metallization step by screen printing technique [8]. The hydrogenated amorphous silicon sub-oxide (a-SiOx:H) is a good candidate to substitute a-Si:H thanks to the higher transparency and thermal stability, together with a good chemical compatibility whit silicon surface. Indeed a-SiOx:H allow a high silicon surface passivation level [9]. Another strong key point is the possibility to use the same well established deposition technique to grow a-Si:H, but including in the gas mixture an oxygen source, like CO2 or NO2. In this work we compare two identical solar cells with two different passivation buffer layer on the light exposed side: one made with a-Si:H and another with a-SiOx:H. The superior optical and electrical characteristics of a-SiOx:H in comparison to a-Si:H are demonstrated by complete cells characterization.

2 EXPERIMENTAL

The wafers used for the two heterojunction solar cells manufacturing were p-type FZ 1-5 Ωcm (100) oriented, 250 µm thick, both side polished and characterized by a declared lifetime greater than 1 ms. The fabrication steps of two cells in Figure 1 are schematized. Is important to note that the fabrication steps were the same for two cells except for the front passivation layer that is made with a-SiOx:H in one case (Cell A) and with a-Si:H in other case (Cell B). In this way the comparison between two buffer passivation layers are simplified.

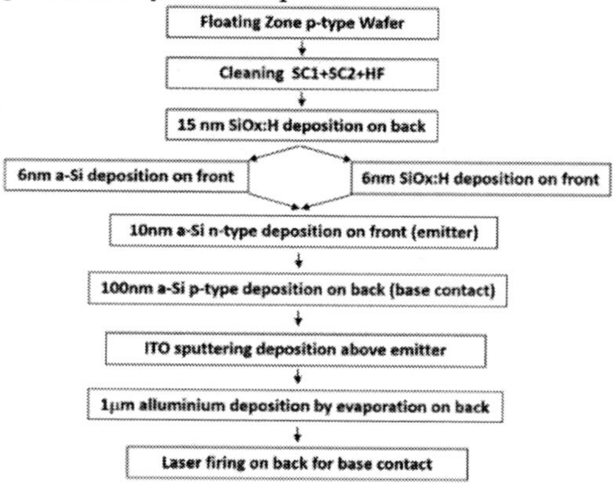

Figure 1: Scheme of fabrication steps of two heterojunction solar cells differing only for the front buffer layer

The wafers were cleaned with standard SC1 + SC2 procedure [10] and dipped 1 minute in hydrofluoric acid at 2% just before entering the PECVD chamber. The first deposition step was the back passivation layers with 15 nm of a-SiOx:H. In PECVD system BenchmarkII by Axic, according to optimized deposition parameter reported elsewhere (epvsec anno scorso?).

On the cells front side a buffer layer of thickness 6 nm was desired to guarantee the surface passivation and minimize the filtering effect of both layers.

While the a-SiOx:H buffer in Cell A was deposited in the same Axic system, the a-Si:H was grown in a three chambers PECVD, where each one is dedicated to different doped amorphous silicon (intrinsic, p-type and n-type). A load-lock chamber connect the three chamber to load and move the sample. After passivation layers depositions, 10 nm of n-type a-Si:H on both cells as emitter layers were grown. On the back of both cells 100 nm of p-type a-Si:H were deposited, to form the base contact. Details of a-Si:H layers deposition in Table I are reported.

Table I: amorphous layers deposition parameters

	SiH_4 sccm	B_2H_6 sccm	PH_3 sccm	P mTorr	T °C	time
i	40	-	-	680	210	28”
n	40	-	10	300	300	58”
p	40	20	-	700	210	4’30”

The process time of a-SiOx:H and a-Si:H layers were calculated using previous test samples; the obtained thicknesses were evaluated by using the fitting by XOP [11] software on the reflectance measured by Perkin Elmer lambda950 UV-VIS-NIR spectrophotometer.

The front side contact was completed by Indium Tin Oxide (ITO) layer obtained by a Kenosystek in-line DC-magnetron sputtering adjusting RF power to 1.2 W/cm^2, at 230°C temperature on sample, Ar flux of 40 sccm and keeping pressure during the glow discharge at $1.1 \cdot 10^{-3}$ mbar. The ITO deposition process parameter were previously optimized to limit damaging on amorphous layers due to the ions bombarding during the discharge. The obtained thickness of 80 nm resulted in a sheet resistance, measured whit four point probe [12], of 30 Ω/sq.

The effective lifetime (τ_{eff}) and implied V_{oc} of each sample was monitored with the aid of QSS.PCD in transient mode by Sinton Consulting WTC200 tester, evaluating lifetime at MCD of 10^{15}.

The front contacts were finally obtained by thermal evaporation of 1 µm thick silver grid, while the base contact on p-type wafer was produce by means of laser firing [13] on 1 µm thick aluminum layer, thermally evaporated on the p-type a-Si:H layers. The laser focus was optimized on the sample surface to promote the migration of both aluminum and boron from p a-Si:H layer toward the silicon surface, minimizing at the same time the surface damages. In this sense the space between the laser spots was 500 µm, that is a compromise to obtain a good Fill Factor preserving the back surface passivation offered by the 15 nm of a-SiOx:H on the back [8].

Finished cells were characterized in terms of dark and AM1.5G illuminated J-V characteristics and Quantum Efficiency.

3 RESULTS AND DISCUSSION

After the deposition of amorphous layer on both cells, reflectance profile were collected and thicknesses evaluated with the help of XOP software. Reflectaces and fits in Figure 2 are reported.

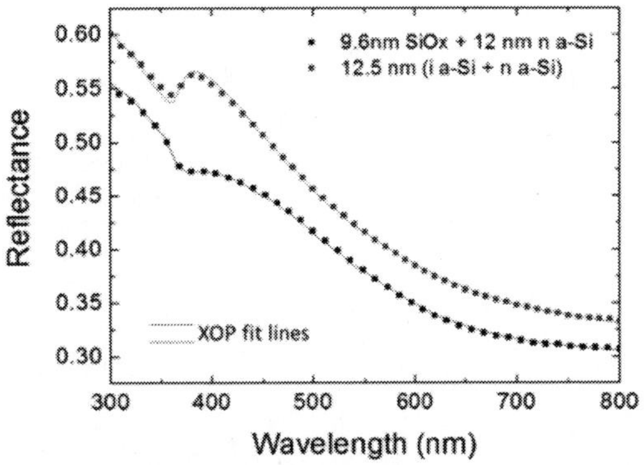

Figure 2: Reflectance measurements (point) and their fits with XOP software (line) of Cell A (black) and Cell B (red) front side.

As observable in Figure 2, On Cell A the obtained thickness of a-SiOx:H was 9.6 nm instead of designed 6 nm, probably due to the different deposition rate on p-type wafer respect to n-type where the rate was calculated. At the same way the thickness of n-type a-Si:H resulted of 12 nm instead 10 nm, this is due to the different substrate on which is grown (a-SiOx:H indeed a-Si:H) besides a variability of deposition process. The overall thickness of intrinsic a-Si:H and n-type a-Si:H on Cell B were 12.5 nm instead 16 nm, having he n-type a-Si:H thickness of about 7.5 nm and the intrinsic layer 5 nm. It is like an unexpected mismatching between impedance of RF generator and those of the chamber reduced the RF power during deposition, decreasing the rate. Nevertheless, despite the different thicknesses between the two cells a comparison of optical and electrical proprieties of two cells was possible: looking at the Figure 2 is already evident the Cell A has an overall reflectance lower than Cell B.

To understand the effect of ITO deposition on passivation surface quality, the lifetime was monitored before and after deposition: for Cell A the lifetime before and after sputtering process was 760µs and 3100µs respectively; for Cell B 990µs and 1060µs respectively. This confirmed the beneficial effect of UV exposition (during glow discharge) together thermal treatment for a-SiOx:H that improved its lifetime four times, while a-Si:H improvement was very slight.

The implied V_{oc} after ITO deposition were 751 mV for Cell A and 730 mV for Cell B. If the deposition was performed at room temperature, the lifetime degrades due to the UV exposure [14] and the damage caused by argon ions but, after a subsequent thermal annealing, the lifetime was recovered. In Figure 3, the effective lifetime as a function of minority carrier density and the related implied V_{oc} values after the sputtering process are reported.

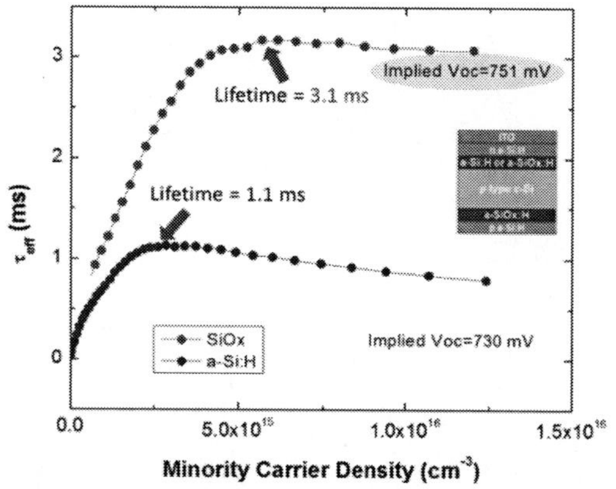

Figure 3: Effective lifetime versus minority carrier density of Cell A (red) and Cell B (blue) and the relative implied V_{oc} values. In the inset the structure of the measured cells is sketched.

After cells completation characterization of electrical and optical proprieties of Cell A and Cell B were performed to make a direct comparison. In Figure 4 the experimental reflectance profiles (focussed on short wavelength range) of two cells are shown as point, black for Cell A and red for Cell B, togheter with the correspective fitting with XOP software illustrated as lines. Cell A show a reflectance lower than Cell B but, due to the difference between cells layers thicknesses, it was difficult compare direclty the reflectance to understand the optical proprieties at the same condition. For this reason we performed a simulation of Cell A reflectance with the aid of XOP using the same thicknesses of all layers involved in Cell B: 80nm for ITO, a-Si:H emitter layer of 7 nm and a-SiOx:H buffer layer of 5 nm. The result of this simulation is again reported in Figure 4 in light blue. Is is immediate to note that both black and light blue lines are lower than red line (referred to Cell B) especially in the region between 350 nm and 450nm, confirming that the cell with a-SiOx:H reflectance is lower, even in the case of higher tichnesses, than cell having buffer layer of a-Si:H. It follows than transmittance of entire front cell is higher.

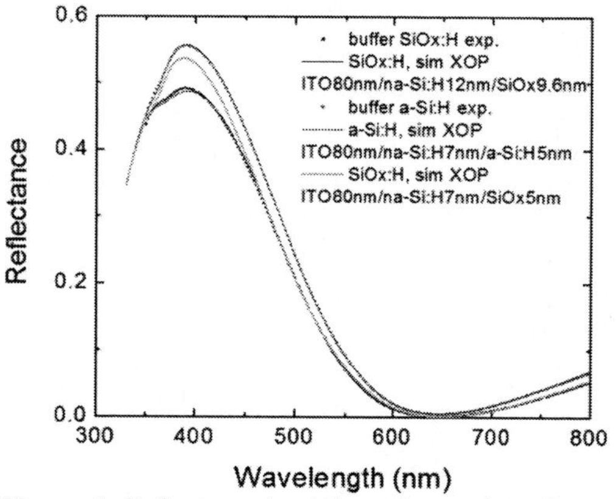

Figure 4: Reflectance (experimental as points, fits as line) of Cell A in black, Cell B in red and simulated reflectance of Cell A with thickness layers equal to Cell B in light blue line.

To compare the External Quantum Efficiency (EQE) shown in Figure 5 we adopted the same methodology used in Figure 4. In this case the simulation take in account the absorption coefficient of varius layers: α_{ITO}, $\alpha_{a\text{-}Si:H}$ and $\alpha_{a\text{-}SiOx:H}$. For this specific connparison EQE are collected in finger-free areas. From these measurements it was possible to calculate the cells current density, finding Cell A characterized by Jsc 0.5 mA/cm^2 higher than Cell B, despite its ticker amorphous layers. The a-SiOx:H advantage is furthermore evidenced when considering the light blue line, obtained by simulation with the same thickness layers of Cell B. In this case, Jsc gain reached 1.0 mA /cm^2

In Figure 6 and Figure 7 the I-V characteristic of Cell A and Cell B are presented, respectively in AM1.5G light and dark condition. As expected the V_{oc} of two cells, reported in Figure 6, were lower than implied V_{oc}: for Cell A implied V_{oc} of 751mV resulted in effective 678mV, while for Cell B, a potential of 730mV dropped to 660mV of V_{oc}. This was mainly due to the laser firing process performed to produce the base contact on p-type silicon, that decrease the passivation on both cells backside and hence reducing the V_{oc}. Indeed if on one hand laser firing technique is able to produce a good ohmic contact, the migration of aluminium and boron on crystalline silicon surface strongly increase the local state of density, because of the surface contamination and dangling bonds production. Anyway, Cell A maintain a V_{oc} higher then Cell B, due to the higher starting lifetime.

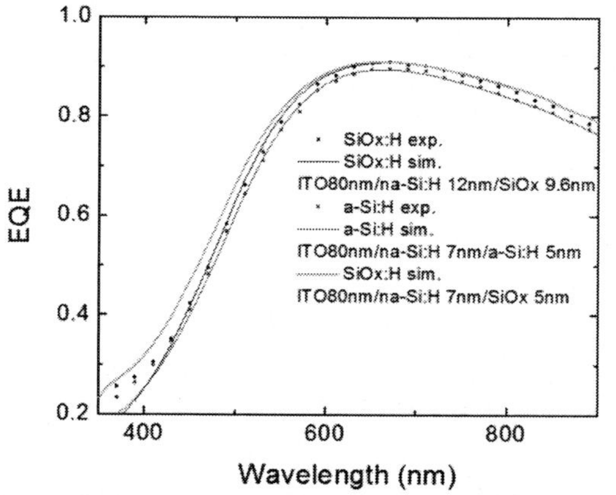

Figure 5: Comparison between EQE (experimental as points, simulation as line) of cell A in black, Cell B in red and simulation of Cell A with thickness layers equal to Cell B in light blue line.

As depicted in Figure 6, Cell A current density was 32.6 mA/cm^2, 0.5 mA/cm^2 higher than Cell B that was 32.1 mA/cm^2. Also the Cell A Fill Factor (FF) was higher than Cell B, due to the higher passivation quality of a-SiOx:H in comparison with a-Si:H. In the Figure 7, from the J-V characteristic in dark, it can be seen the reverse current of Cell B higher than Cell A, confirming the higher surface recombination on Cell B.

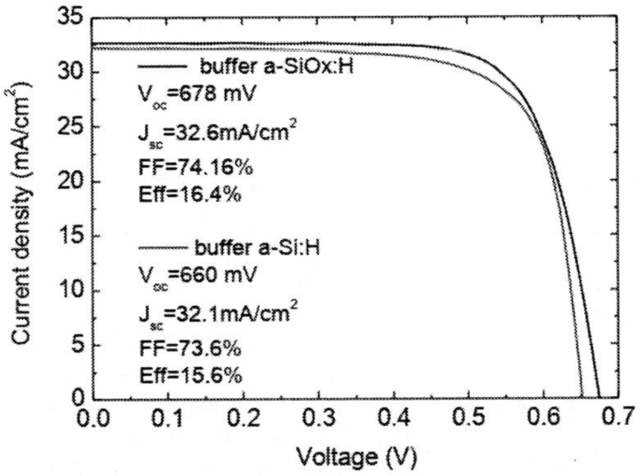

Figure 6: J-V characteristics of Cell A (black line) and Cell B (red line) under AM1.5G.

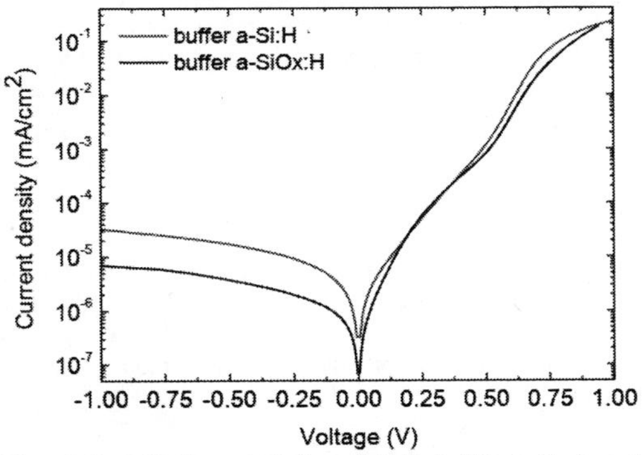

Figure 7: J-V characteristics of Cell A (black line) and Cell B (red line) in dark.

4 CONCLUSIONS

We manufactured two heterojunction solar cells to compare the proprieties of a-SiOx:H as passivation buffer layer in comparison to a-Si:H. The two cells were different only front passivation buffer, while the rear passivation was ensured on both by a-SiOx:H layer.

The better passivation proprieties and the higher transparency of a-SiOx:H were appreciable at cell level: the minority carriers effective lifetime of cell passivated with a-SiOx:H was 3.1ms having an implied Voc of 751mV. In comparison, the cell having buffer contitued by a-Si:H showed a lifetime of 1.1ms and implied Voc of 730mV. The difference between current density due to the higher transparency of a-SiOx:H, calculated by EQE measurement, is 0.5 mA/cm^2 despite an unwanted higher thicknesses of emitter and buffer layers in a-SiOx:H passivated wafer. A simulation was performed to understand the effective current difference if the thickness of emitter and buffer layers of two cells were equal: the Jsc of cell with a-SiOx:H could be up to 1mA/cm^2 higher than the other cell. Also the Fill Factor and the efficiency of cell were higher: 74.16% and 16.4% respectively were obtained for cell with a-SiOx:II buffer, while 73.6% and 15.6% respectively for a-Si:H buffered layer cell.

Further improvements are expected when n-type a-SiOx:H emitter will be involved.

5 AKNOWLEDGEMENT

This work is supported by the Italian Ministry of Economic Development in the framework of the Operating Agreement with ENEA for Research on the Electric System.

6 REFERENCES

[1] K. Yoshikawa, H. Kawasaki, W. Yoshida, T. Irie, et al., Nature Energy, 2017,Vol. 2, 17032.
[2] B. Strahm, Y. Andrault, D. Bätzner, D. Lachenal, et al., Proceedings of the Mater Res Soc Symp, San Francisco CA, USA, 2010, 1245, A01-04.
[3] S. De Wolf, M. Kondo, Appl Phys Lett, 2007, Vol. 90, 042111.
[4] H. Fujiwara, M. Kondo, Appl Phys Lett, 2005, Vol. 86, 032112.
[5] Z. C. Holman, A. Descoeudres, L. Barraud, et al., IEEE Journal of Photovoltaic , 2012, Vol. 2, p. 7-15.
[6] M. Tanaka, M. Taguchi, T. Matsuyama, et al., Jpn J Appl Phys, 1992, Vol. 31, p. 3518-3522.
[7] S. De Wolf, A. Descoeudres, Z. C. Holman, C. Ballif, Green, 2012, Vol. 2, p. 7-24.
[8] L. Serenelli, M. Miliciani, M. Izzi, R. Chierchia, A. Mittiga, M. Tucci, IEEE 40th Photovoltaic Specialist Conference (PVSC), Denver, CO, 2014, p. 2528–2532.
[9] T. Mueller, S. Schwertheim, W. R. Fahrner, Journal of Applied Physics, 2010, Vol. 107, 014504.
[10] W. Kern, J. Electrochem 1990, Vol. 137, n. 6, pp. 1887–1892.
[11] European Synchrotron Radiation Facility (ESRF). http://www.esrf.eu/Instrumentation/software/data-analysis/xop2.4.
[12] F. M. Smits, Bell System Technical Journal, 1958, Vol. 34, pp. 711-718.
[13] M. Tucci, E. Talgorn, L. Serenelli, E. Salza, M. Izzi, P. Mangiapane, Thin Solid Films, 2008, Vol. 516, p. 6767.
[14] B. Demaurex, S. De Wolf, A. Descoeudres, Z.C. Holman, C. Ballif, Appl.phys.Lett. 101, 171604.

A NOVEL BIFACIAL C-SI CELL STRUCTURE AND PROCESS FOR HIGH EFFICIENCY AND LOW COST

Haibin Huang[1],Gangyu Tian[1], Lang Zhou[1]*, Jiren Yuan[1]*, Wolfgang R. Fahrner[1], Wenbin Zhang[2], Xingbing Li[2], Wenhao Chen[3], Renzhong Liu[3]
[1] Xuefu Avenue, No.999, Institute of Photovoltaics, Nanchang University, Nanchang, Jiangxi Province, China
[2] Qigang Rd, NO.738, Yangwang Economic Area, Fengxian District, GCL System Integration Technology Co. Ltd., Shanghai, China
[3]Pingjiang Rd, No. 88, Hareon Solar Co. Ltd., Taicang, Jiangsu Province, China
*Tel. +86 0791 83969552/Fax. +86 0791 83969329/E-mail: lzhou@ncu.edu.cn

ABSTRACT: A novel structure of Ag grid/SiN_x/n^+-c-Si/n-c-Si/i-a-Si:H/p^+-a-Si:H/TCO/Ag grid is designed in order to increase the efficiency of bifacial amorphous silicon/crystalline silicon based solar cells (HJT) and reduce the rear materials consumption and manufacture cost. The simulation result done by AFORS-HET shows that the new structure could get higher efficiency compared with the normal bifacial HIT cell for the increase of the short circuit current, while it could keep most advantages of the HJT, such as: high voltage, low temperature coefficient and good weak light performance. A real cell composed of the novel structure with a size of 75mm*75mm was tried out by a special processing recipe, in which most techniques and machines are normally used in c-Si solar cell manufacture. Without any parameters optimization, the efficiency with light injected from the SiN_x and TCO side reached 19.4% and 18.1% (AM1.5G, 25°C), respectively. The performance and the quantum spectra of the cell are compared with a normal bifacial HIT cell, and the result is constant with the simulation analysis. It could be concluded that the design is a advantageous structure and the processing recipe is worked.
Keywords: heterojunction, silicon solar cell, bifacial

1 INTRODUCTION

The famous HIT structured amorphous silicon /crystalline silicon (a-Si:H/c-Si) have the highest energy conversion efficiency in all the bifacial-structure c-Si solar cells. But the high cost due to the special production equipment and the large amount consumption of ITO prevents its world-wide mass-production. The light absorption loss in TCO layers and amorphous silicon (a-Si:H) layers and the large series resistance of Ag grids made by low temperature fired paste are still barriers to improving the efficiency of the cells[1-3].

It is thought that a structure based on heterojunction of a-Si:H and c-Si (hereafter referred to as HAC cell) and production recipe shown in Fig. 1(a) may be a solution to overcome the above disadvantages of HIT cells. It is expected that, through this solution, first, the light absorption loss on the main light-induced side may be reduced; secondly, the series resistance may be reduced by low temperature solidified Ag grids on the back surface field (BSF) side replaced to high temperature fired Ag grids; thirdly, a half of ITO materials may be saved; fourthly, some manufacturing steps may use cheaper machines which have been used in conventional c-Si solar cell production now.

The part idea of this novel HAC cell structure was mentioned by Fraunhofer Institute for single side light-injection solar cell [4,5], but details of the design is not good enough for mass-production. In this report, the properties of the HAC cell and a referenced HIT cell with dual side light injection are compared by AFORS-HET[6] simulations. Also some samples are made with a designed special recipe, which maybe is suit for mass-production.

2 EXPERIMENTAL

Firstly, comparison of a HAC cell and normal bifacial HIT cell as shown in Fig.1 is done by AFORS-HET. The performance of the cells with light injected from the p+ layer side (thereafter named: Rear side) and n+ layer side (thereafter named: Front side) of the cells are all simulated, separately. In the simulations, the parameters of the wafer and the layers on the rear side are totally same for the two modules in Fig. 1. The reflectance of each surface outside is set to zero. And the parameters are roughly optimized before the comparison simulation.

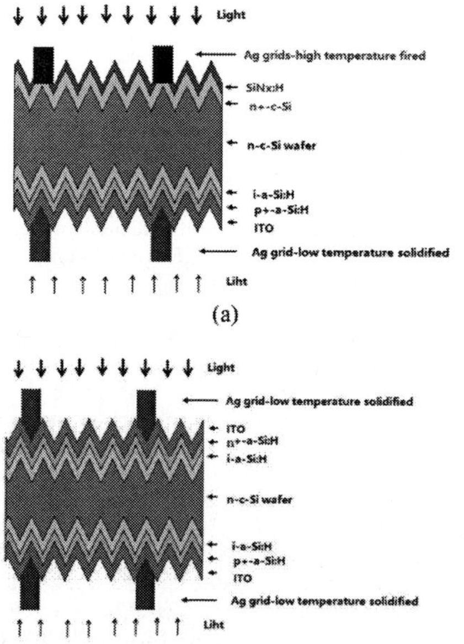

Figure 1: (a) Schematic of the bifacial HAC cell; (b) schematic of a bifacial HIT cell

After the simulation analysis, a HAC cell with a size of 75mm*75mm (Fig.2(b)) was tried out with the processing recipe shown in Fig.2 (a), without any processing parameters optimization. The cell is made with normal n-type solar cell grade single crystalline silicon wafer, (100) oriented, ~180 μm, 2-5 Ω·cm. The wafer is textured by KOH solutions and rounded by HNO3/HF mixed solutions, then cleaned by RCA solutions, and dipped in HF, dried by N2 in the end. About 20 μm Si is etched each side. During the process, the front side is made firstly, and these steps need higher temperatures than those for the other side of the cell. All the steps are done in a normal production line for p-type c-Si solar cells. Which means that the n+ c-Si layer is made by POCl3 diffusion, the SiNx:H layer is made by PECVD, and the Ag grids are made by screen printing and post-fired with high temperature. In the "cleaning III" step, only the naked side of the wafer is cleaned, while the finished side of the wafer is protected. The naked side of the wafer is cleaned by HF, HNO3/HF mixed solution, SC1, SC2, HF, in sequence. The intrinsic and p-type a-Si:H layers are made by PECVD (13.56MHz), and ITO layer is deposited by reactive plasma deposition method. The Ag grids on the ITO layer are made by screen printing method with low temperature Ag paste, and were solidified at ~220 °C. After the cell is made, it is cut into a size of 75mm*75mm by laser, for the reason of uniformity.

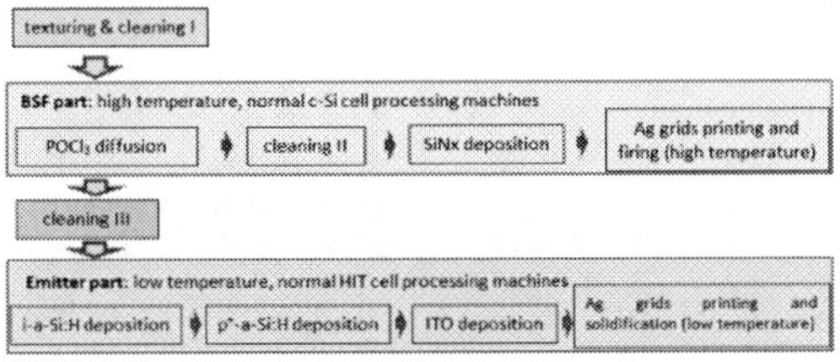

(a)

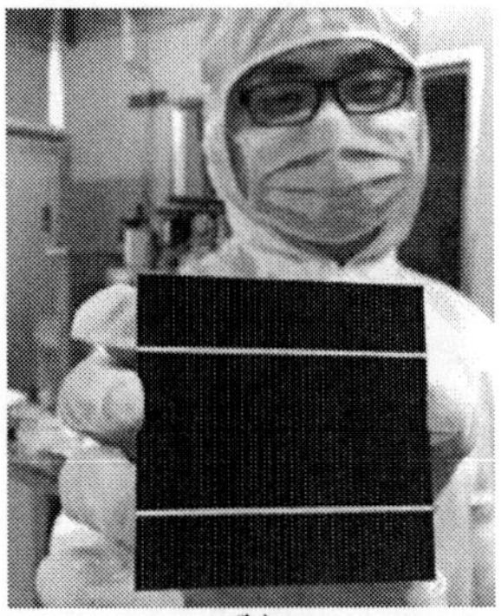

(b)

Figure 2: (a) a processing recipe for the HAC cell; (b) a sample of the novel HAC cell (75mm*75mm)

IV curves of the cells are tested at AM1.5G, 25 °C conditions with light injected from the front side and rear side, separately. The external and internal quantum efficiencies (EQE and IQE) are also tested, and compared with the results of a HJT cell with an efficiency of ~21.5%. The resistances of the fired Ag grids and the solidified Ag grid on the both side of the cell with a same length are tested with an IV tester.

3. RESULT and DISCISSION

Fig.3 shows the simulation results of the cells with a wafer thickness of 150μm. It is found that the HAC cell with the light injected from the front side has the best efficiency, about 0.5% (absolute value) while the HJT cell with the light injected from the front side has the worst performance. From the simulated IV curves of the HAC cell and he HIT cell with light injected from each side, it is clear that the efficiency gain of the HAC compared with the HJT cell comes from the Jsc increase, about 0.7mA/cm2. This meets our expectation that the new design structure on the main light injection side could reduce the parasitic absorption loss of the normal HJT cell. it is clear that the Vocs of the HAC cell has no obvious difference with the light injecting from both side, and they also has no obvious difference with HJT cells. It means that the HAC cell keeps the high Voc feature of the a-Si:H/c-Si structure. For the reverse saturation currents of the HAC and HJT cell are almost the same as shown in Fig. 3(b), it could be concluded that the HAC cell would keep the low temperature coefficient and good weak-light-performance features of the a-Si:H/c-Si structure[7-9], as good as HIT.

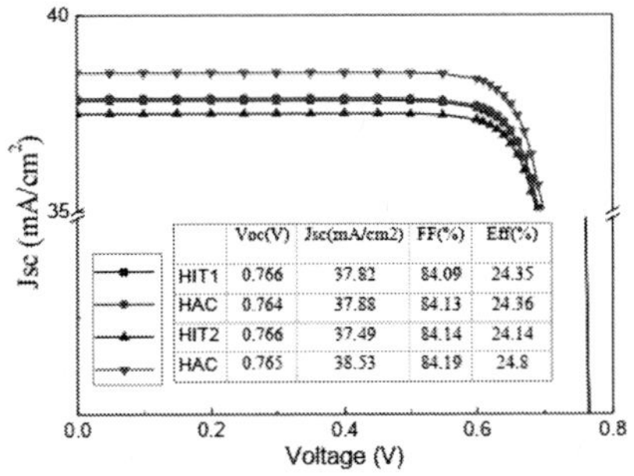

(a)

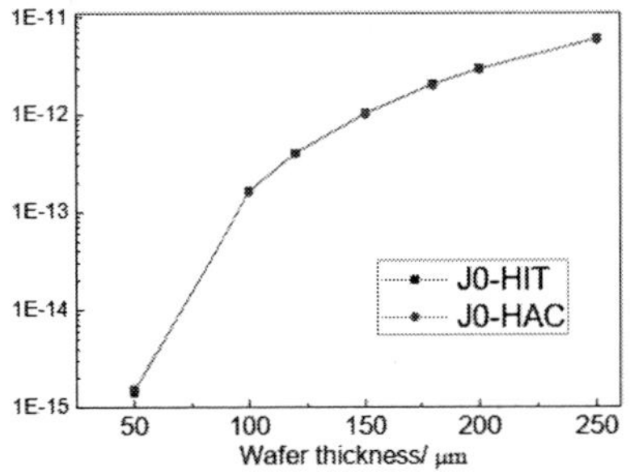

(b)

Figure 3: (a) The simulation result of the HAC cell and bifacial HIT cell with a wafer thickness of 150μm. HAC-1, HIT-1 means light injected from the SiNx side, HAC-2, HIT-2 means light injected from the ITO side. (b) The reverse saturation current J_0 of the two kinds of cells with different wafer thickness.

The performance of the sample made with the recipe shown in Fig.2 without any processing parameters optimization is shown in Table I. The Jsc 40.13 mA/cm2 of the cell with light injected from the front side is higher than 39.5 mA/cm2 of the world-record HJT cell [10], although the efficiency is much lower than that of the record. The Voc of the tests with light injected from either side has no obvious difference. Both phenomena agree with the simulated result. The energy conversion efficiencies with light injected onto the front and the back sides are 19.4% and 18.1% (AM1.5G, 25 °C), respectively. It is believed that the performance of cell could increase by processing parameters optimization. So it is thought that the processing recipe is reasonable.

Table I: Performance of the HAC cell under AM1.5G, 25℃, and the best performance of the HJT cell [8] is as reference

Light incidence side	Voc (mV)	Jsc (mA/cm2)	FF (%)	η (%)
Rear	672.3	**40.13**	72.04	19.4
Front	670.7	39.85	68.24	18.1
unknown	750.0	**39.5**	83.2	24.7

Fig. 4 shows the EQE and IQE of the HAC cell with light injected from the front side and the rear side, in which an EQE spectrum of a HJT cell with the efficiency of ~21.5% is set as a reference. In order to exclude the effect of the Ag grids shadowing, the test is done with a small light spot totally beat on the area without Ag grids. Comparing the EQE results, it is clear that the HAC cell has great response at the short wavelength range with light injected from the front side, much better than that of the normal HJT cell and HAC cell with light injected from the rear side. This is because the SiNx:H film/n+-c-Si structure has much lower parasitic absorption loss than ITO/n+-a-Si:H/i-a-Si:H structure, for the higher bandgap of SiNx:H than that of ITO layer and the higher carrier mobility of the n+-c-Si than that of n+-a-Si:H/i-a-Si:H layers. The phenomenon also agrees with our initial assumption.

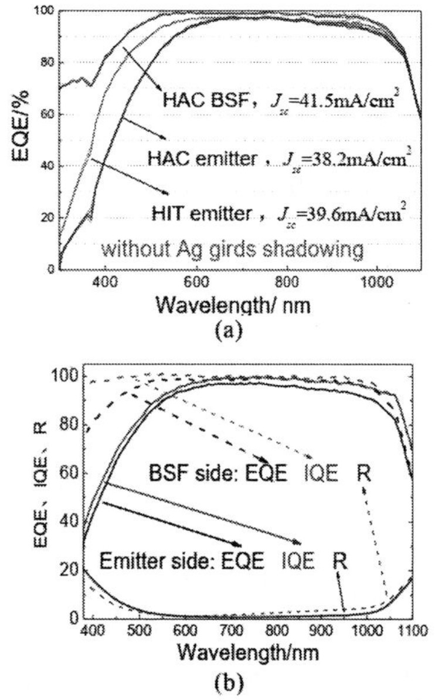

Figure 4. The EQE, IQE spectra of the HAC cell with light injected from the rear or front side, an EQE spectrum of a HJT cell is set as a reference. All the EQE spectra are tested without Ag grid shadowing.

4 CONCLISION

A novel bifacial a-Si:H/c-Si rear junction solar cell is designed, with a structure of Ag grid/SiNx/n+-c-Si/n-c-Si/i-a-Si:H/p+-a-Si:H/TCO/Ag grid in order to increase the efficiency of the a-Si:H/c-Si based solar cells and low down the cost of the cells. According to the simulation result, the new structure could increase the efficiency of the HJT cell compared with the bifacial HIT cell for the increase of Jsc when the light injected from the rear side of the cell, while most of the advantages of the HJT cell still exist. With a special designed processing recipe, a HAC cell with a size of 75mm*75mm is made successfully. Although without any optimization, the efficiency of the cell 19.4%, while Jsc reaches 40.13 mA/cm2, higher than that of the world-record bifacial HIT cell. The results show that the cell structure design has great potential for higher efficiency and it is feasible for mass production.

ACKNOWLEDGEMENT

This work is supported by Jiangxi Provincial Key Research and Development Foundation (Nos. 2016BBH80043) , Open Fund of Jiangsu Key Laboratory of Materials and Technology for Energy Conversion (Grant Nos. NJ20160032) , the National Natural Science Foundation of China (Grant Nos. 61464007, and 51561022).

The authors would like to acknowledge Mr. Hong Wang from Ideal Deposition Equipment and Applications (Shanghai) Ltd. for their help in the sample making.

REFERENCE

[1] K. Maki, D. Fujishima, H. Inoue, Y. Tsunomura, T. Asaumi, S. Taira, T. Kinoshita, M. Taguchi, H. Sakata, H. Kanno, E. Maruyama. 37th EEE Photovoltaic Specialists Conference (2011) 000057.
[2] E. Maruyama, A. Terakawa, M. Taguchi,Y. Yoshimine, D. Ide, T. Baba, M. Shima, H. Sakata, M. Tanaka. 4th IEEE World Conference on Photovoltaic Energy Conference (2006) 1455.
[3] H. Sakata, T. Nakai, T. Baba, M. Taguchi, S. Tsuge, K. Uchihashi, S. Kiyama. 28th IEEE Photovoltaic Specialists Conference (2000) 7.
[4] M. Bivour, M. Rüdiger, C. Reichel, K.U. Ritzau, M. Hermle, S. W. Glunz. Energy Procedia 8 (2011) 185.
[5] M. Bivour; C. Meinhardt; D. Pysch; C. Reichel; K. -U. Ritzau; M. Hermle; S. W. Glunz. 35th IEEE Photovoltaic Specialists Conference (2010) 001304.
[6] A. Froitzheim, R. Stangl, L. Elstner, M. Kriegel, W. Fuhs. 3rd World Conference on Photovoltaic Energy Conversion (2003) 279.
[7] T. Mishima, M. Taguchi, H. Sakata, E. Maruyama. Solar Energy Materials & Solar Cells 95(2011) 1.
[8] M. Tauchi, A. Terakawa, E. Maruyama, M. Tanaka. Progress in Photovoltalcs: Research and Applications. 13 (2005) 481.
[9] N. H. Reich, W. Van Sark, E. A. Alsema, et al. Proceedings of the 20th European Photovoltaic Solar Energy Conference 2005 2120.
[10] M. Taguchi, A. Yano, S. Tohoda, et al. IEEE Journal of Photovoltaics 4 (2013) 96.

A NOVEL PROCEDURE FOR FABRICATING SUB-MICRON TEXTURES ON VARIOUS THICK CRYSTALLINE-SILICON SOLAR CELLS DOWN TO 50 µm WITH LOW-REFLECTIVITY IN WIDE WAVELENGTH

Cong Thanh Nguyen*, Koichi Koyama, Huynh Thi Cam Tu, Shigeki Terashima, Hideki Matsumura
JAIST (Japan Advanced Institute of Science and Technology)
1-1 Asahidai, Nomi, Ishikawa 923-1292, JAPAN
E-mail: nguyen.cong.thanh@jaist.ac.jp

ABSTRACT: Hetero-junction back-contact (HBC) solar cell is the structure realizing the highest efficiency ever obtained. For lowering the cost of crystalline-silicon (c-Si) solar cells, thinning of the original c-Si substrates is also required to reduce the material cost. In such c-Si substrates thinner than 100 µm, the optical reflection of sunlight at the back surface becomes an important key technology, and thus, texturing of the back surface is to be also important. However, in HBC solar cells, the electrode patterning must be performed on such textured back-surface. Then, small size textures should be developed for realizing both high reflectivity at the back-surface and efficient pattern-formation of the back-electrodes. Small size textures are also required for thin c-Si, since several µm size textures fluctuate the c-Si thickness itself and induce unstable performance. In the paper, it is demonstrated that the size of texture can be controlled from 10 µm or more to sub-micron by adding glass microparticles in the conventional alkaline-based texture-forming chemicals, and that such textured structures exhibited the reflectivity lower than 10% in the range of wavelength from 600 nm to 950 nm even for the 50 µm-thick c-Si substrates, before anti-reflection coating.
Keywords: HBC solar cells, thin c-Si solar cells, textured c-Si, sub-micron texture, glass microparticles

1 INTRODUCTION

An a-Si/c-Si (amorphous-silicon/crystalline-silicon) hetero-junction back-contact (HBC) appears to be one of the best structures to obtain high efficiency c-Si solar cells, realizing the world top three efficiencies, 26.6, 26.3 and 25.6 % [1,2]. To reduce the production cost of the solar cells and for even further improvement of their efficiency, extra efforts are still needed. For instance, we are trying to reduce c-Si material cost by using c-Si substrates with a thickness less than 100 µm. In such case, all process must be fitted to the thin substrates. The textured structure may be required even in the back-side of c-Si for the better reflection of sunlight passing through the thin substrate. If we can use the same textured structures on both sides of c-Si, the formation process of the textures becomes simple, since the substrates can be simply dipped into a texture-forming solution. However, if the textures are formed on the back-side, such textures should be small enough for fitting the patterning process of the back electrodes in the HBC solar cells. The reflection at the front surface also should be low enough even for using such small size textures. Prior to deposition of passivation films, the surface of such small-scale textured c-Si substrates should be cleaned enough and passivation films should cover all textured surface with enough step-coverage to realize low surface recombination velocity (SRV) or very long carrier lifetimes. We have also developed a novel chemical cleaning method which is suitable for the textures with sub-µm scale structures [3,4]. The cleaning chemicals should penetrate fine structures. We have discovered that hot conc. sulfuric acid (H_2SO_4) at 140°C has enough low viscosity to penetrate the sub-µm fine structures. To control surface tension of other chemicals, we added methanol (CH_3OH) as a surfactant into them, and that addition of methanol contributes to cleaning up the fine regions of the textures. Therefore, in this paper, only texturing procedures will be focused on for satisfying such requirements of future low cost and high efficiency c-Si solar cells.

At first, we have developed a novel method for quick formation of sub-µm scale textured structures on c-Si. By adding some kinds of materials, glass microparticles, to a commercially available alkaline-based texture-forming chemical, SUN-X600 sold by Wako Pure Chemical Industries Ltd. Texturing time can be shortened to only several mins and also c-Si loss during texture-formation can be small as few µm. Bare textured c-Si (without coating anti-reflection layers) shows reflectivity lower than 10% (8% at the minimum) in the range of wavelength from 600 nm to 1,050 nm. As far as the authors know, these values appear to be the best ones obtained by simple alkaline-based texturing structures and appear to be acceptable for industrial implementation.

2 EXPERIMENTS AND RESULTS

Figure 1 shows two models for texturing of very thin c-Si with an assumed thickness of 50 µm. In Fig. 1, we assumed that (a) 30 µm and (b) 1 µm textured structures

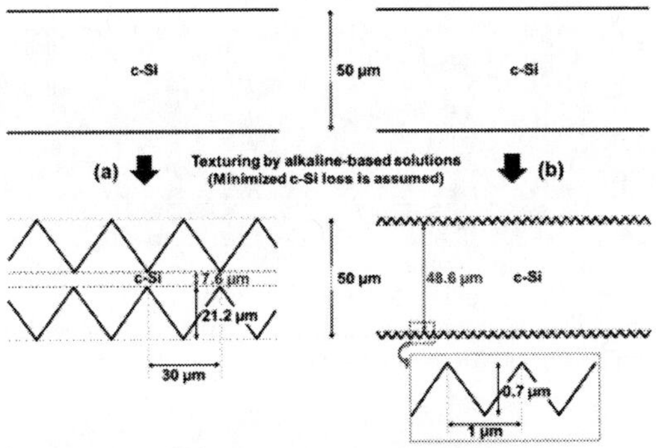

Figure 1. Models for texturing of very thin c-Si substrates with assumed original thickness of 50 µm. Textured c-Si substrates with different pyramid sizes, such as 30 µm (a), and 1 µm (b) formed by the alkaline-based solutions. In these models, c-Si losses by etching are assumed to be negligible.

could be formed by the alkaline-based solutions. Even the c-Si loss is assumed to be negligible, the formation of textured structure in Fig. 1(a) is not realistic at all because of the very thin c-Si body with only 7.6 µm remaining. From this simple consideration, large textured structures cannot be used for very thin c-Si. Development of very small textured structure as shown in Fig. 1(b) with 48.6 µm-thick c-Si body, is inevitable. Naturally, the small textured structure is needed to maintain c-Si volume, reduce etching time, and avoid cracking c-Si, leading to improvement of fabrication throughput and yield.

In the development of the procedures for the formation of the very small textured structure with sub-µm scale size, here, we show the results of two texture-forming chemicals in Fig. 2. One is original texture-forming alkaline-based etchant (a), SUN-X600, and the other is a new chemical produced by mixing an additive (micro-glass particles) with SUN-X600 (b). We discovered that by adding the additive sub-µm textured structures could be realized. The additive is to adjust the surface tension of the chemicals and to control size and number of bubbles generated by texture forming reactions. Both solutions are heated at 80-90°C, and mirror-polished c-Si substrates are dipped into the solution for 50 min in case of the solution (a) and only 4 min in the solution (b).

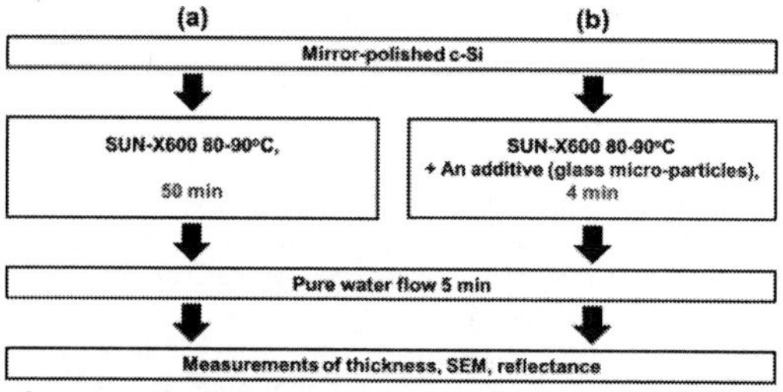

Figure 2. Texturing procedures using pure SUN-X600 without an additive (a), and with an additive (glass microparticles) (b).

Surface morphology of the textured samples observed by scanning electron microscope (SEM) is demonstrated in Fig. 3(a) and (b) for samples made by the solutions (a) and (b), respectively. Size of pyramids are estimated about 10 µm or more are seen in case of the solution (a), and from 0.3 to 2 µm in case of the solution (b).

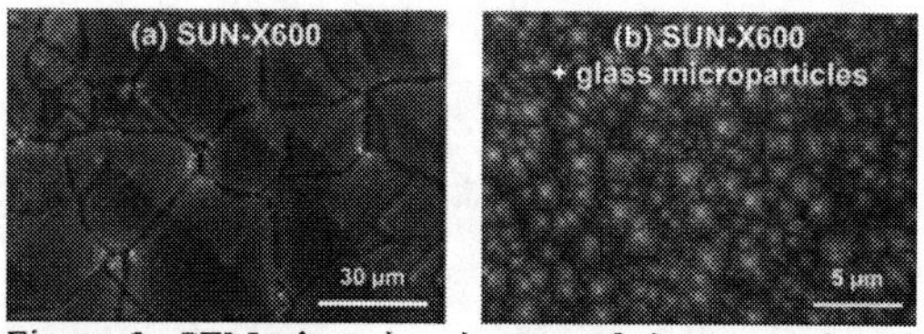

Figure 3: SEM plan view images of the textured c-Si etched by the pure SUN-X600 without the additive (a), and with an additive (glass microparticles) (b), as shown in Fig. 2 (a) and (b), respectively.

We summarized the difference between two procedures (shown in Fig 2(a) and (b)) in Table I to demonstrate the effect of the additive (glass microparticles). It is clearly shown that the additive is effective to reduce (1) the pyramid sizes enormously (from 10-50 µm to < (0.3−2µm)), (2) the c-Si losses (from 40-50 µm to 1-4 µm) and (3) texturing times (from 50 min to 4 min), leading to considerable reduction of the texturing cost.

We also measured the reflectivity of the textured structures as shown in Fig. 4. After adding the additive, the reflectivity of the textured structure reduces about 1.8-2.4% in the range of 400 to 1050 nm wavelength. At about 1000 nm, both textured structures have the minimum

TABLE 1. Comparison of the procedures, with and without using an additive (glass microparticles), as shown in Fig. 2 (a) and (b), respectively.

Procedure	Fig. 2(a)	Fig. 2(b)
Initial c-Si thickness [µm]	283	
Textured c-Si thickness [µm]	235-245	279-282
c-Si loss by etching [µm]	40-50	1-4
Texture / pyramid size [µm]	10-50	< (0.3-2)
Texturing time [min]	50	4

reflectance, 9.9% and 8.0% without any anti-reflection layers for procedures in Fig. 2(a) and (b), respectively. We also compared our textured structures with reported one [5] as shown in Fig. 4 by blue dot-points. They also produced the textured c-Si with a size of 1 µm. Obviously, our textured c-Si made with the additive has lower reflectance than that of any other textures. To know whether our texturing procedure (in Fig. 2(b)) is applicable to very thin c-Si or not, we employed mirror-like c-Si substrates with various thicknesses for the solution (b). c-Si losses are less than 4 µm for all these samples, and their textured structures are similar. Reflectance is plotted by color solid lines. Finally, we can conclude that our texturing procedure is realistic even for c-Si with thickness down to 50 µm.

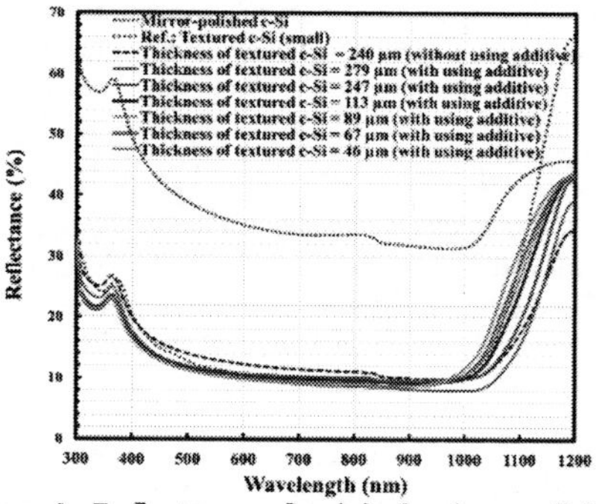

Figure 4. Reflectance of original mirror-polished c-Si (black dot), textured c-Si formed by SUN-X600 without the additive (black dash-line), textured c-Si formed by SUN-X600 with the additive (glass microparticles) with various thicknesses such as 279 µm (pink solid line), 247 µm (brown solid line), 113 µm (blue solid line), 89 µm (purple solid line), 67 µm (green solid line), and 46 µm (orange solid line) as functions of wavelength. A reference textured c-Si 4 is also plotted (blue dot).

3 SUMMARY

We have succeeded to develop a novel procedure for forming sub-micron textured c-S with various thicknesses down to 50 µm and low-reflectivity in wide wavelength used for solar cells. We also discovered the great effect of the additive (glass microparticles which is reusable, very low cost and stable) to adjust the surface tension of the chemicals and to control size and number of bubbles generated by texture forming reactions.

ACKNOWLEDGEMENTS
This work was financially supported by the New Energy and Industrial Technology Development Organization (NEDO), Japan.

REFERENCES

[1] K. Yoshikawa, H. Kawasaki, W. Yoshida, T. Irie, K. Konishi, K. Nakano, T. Uto, D. Adachi, M. Kanematsu, H. Uzu, and K. Yamamoto, Nat. Energy 2, 17032 (2017).
[2] K. Masuko, M. Shigematsu, T. Hashiguchi, D. Fujishima, M. Kai, N. Yoshimura, T. Yamaguchi, Y. Ichihashi, T. Mishima, N. Matsubara, T. Yamanishi, T. Takahama, M. Taguchi, E. Maruyama, and S. Okamoto, IEEE J. Photovoltaics 4, 1433 (2014).
[3] C.T. Nguyen, K. Koyama, S. Terashima, C. Okamoto, S. Sugiyama, K. Ohdaira, and H. Matsumura, in 2016 IEEE 43rd Photovolt. Spec. Conf. (IEEE, 2016), pp. 1169–1172.
[4] C.T. Nguyen, K. Koyama, K. Higashimine, S. Terashima, C. Okamoto, S. Sugiyama, K. Ohdaira, and H. Matsumura, Jpn. J. Appl. Phys. 56, 56502 (2017).
[5] K. Nakamura, M. Aoki, I. Sumita, H. Sato, Y. Kumagai, Y. Kawata, and Y. Ohshita, in 2013 IEEE 39th Photovolt. Spec. Conf. (IEEE, 2013), pp. 2239–2242.

FABRICATION OF HIGH DENSITY NANO-MICRO HIERARCHICAL SUBWAVELENGTH STRUCTURE FOR ENHANCING LIGHT TRAPPING PROPERTIES IN A FEW SECONDS

Harsh A. Chaliyawala, Abhijit Ray, Ranjan Pati, Indrajit Mukhopadhyay
Solar Research and Development Center, School of Solar Energy, Pandit Deendayal Petroleum University, Raysan, Gandhinagar-382007, Gujarat, India.

ABSTRACT: Nanostructures on silicon shows a great potential for use in photovoltaic applications due to its outstanding broadband enhancement in light-harvesting characteristics. Recently, a lot of attention is being paid to combine silicon nanowires (SiNWs) and the Si micro-pyramidal structures (SiMPs) leading to a nano-micro hierarchical structures (nM-Hs) to improve optoelectronic performance of various operational nanoscale devices. Here, we have fabricated hierarchical Si structures consisting of microscale pyramids and nanoscale wires through very facile wet anisotropy technique. The SiNWs above the SiMPs have been fabricated by Ag-induced electroless chemical etching (ECE) in a few seconds. The morphological, optical and electrical performance of the as-prepared hierarchical structure with time of ECE have been characterized by field emission scanning electron microscopy (FESEM), UV-Visible spectroscopy and Hall effect measurements. The uniform distribution of disordered nanowires for various ECE (15, 30 and 60 sec) on 4 μm long micropyramidal structure gives an indication on the formation of the subwavelength nanocavities on the micropyramids to improve light trapping activities. The optimized hierarchical structure for 30 sec ECE, exhibits a maximum photon collection efficiency by reducing the total reflectance R_{tot} to as low as ~5 % in broad spectral range (300-1200 nm). Moreover, the resistivity drastically increases from 4.48 x 10^{-4} to 3.1 x 10^{-3} Ωcm as the ECE time increases. In addition, the bulk concentration (n) shows one order of decrement in the values in contrast to the carrier mobility (μ).

Keywords: Electroless Chemical Etching (ECE), nanowires, nano-micro hierarchical structure (nM-Hs), light trapping.

1. INTRODUCTION

Since 1990's, Si nanowire arrays (SiNWs) are considered to be among the most outstanding semiconductors due to their unique physical properties which make them suitable for many different applications in nanoelectronic [1] or optoelectronic devices [2]. It is recognized now a days due to its strong broadband optical absorption due to the light trapping effect and its quantum confinement effect [3-7]. These extraordinary properties of SiNWs could be further enhanced by implementing it on the micro-structured architectural design to enhance the surface area, and hence improve the potential applications in many different fields of research.

In this paper, the considerable efforts has been made to synthesize the micropyramids on the Si substrate. Hereafter, SiNWs were embedded on the micropyramids by using very facile electroless chemical etching (ECE) process at room temperature. Furthermore, the optimization for the nanowires on micropyramids with the ECE time has been carried to achieve the high broad absorption enhancement in the whole spectral range from 300-1200 nm.

2. EXPERIMENTAL PROCEDURE

2.1 Fabrication methodology

Single crystalline n-type Si (100) substrates having a thickness of T = ~ 350 μm and ρ =2 -10 Ωcm with one side polished surface, were used to fabricate the SiMPs and nM-Hs by electroless chemical etching (ECE) process. The Si wafers were first cut into small pieces with a dimension of 1cm x 1cm and were cleaned sequentially by trichloroethylene, methanol, acetone and de-ionized water for 5 min, respectively, using ultrasonicator. Further, the substrates were immersed in piranha solution with 3:1 mixture of 98% H_2SO_4 and 30% H_2O_2 for 15 min to remove native oxides from the surfaces. Hereafter, the as-cleaned Si substrates were dipped into the optimized 1M KOH/iso-propanol (IPA) aqueous solution for 15 min at 70±5°C to prepare the micropyramids on the Si surface, as schematically presented in Fig. 1.

Figure 1: Schematic representation for the preparation of nano–micro hierarchical design (nM-Hs) by electroless etching (ECE) process.

Hereafter, the as-prepared samples were cleaned under DI water for few minutes and dipped into the $AgNO_3$ (5mM) / (4.8M) HF aqueous solution for electroless deposition of Ag nanoparticles (AgNPs) on the surface at room temperature (~25 °C) for different ECE time (see Fig. 1(c)). After the electroless deposition, as-prepared Si substrates were soaked in aqueous HNO_3 solution (in the ratio of 10:1 using 50%w/v HNO_3) to remove the residual AgNPs, cleaned with de-ionized water and dried at room temperature in natural air.

2.2 Characterization techniques involved

The plane and cross-sectional view of as-prepared micropyramids and nM-Hs design was investigated by field emission scanning electron microscopy FE-SEM (Zeiss, Ultra-55) equipped with EDX (Oxford Instruments). Reflectance (diffuse and total) spectra was recorded with a UV-Vis (UV-2600, Shimadzu), spectrophotometer over a spectral range of 300-1200 nm equipped with an integrating sphere. Spectral measurements were performed for samples on an Edinburgh FLS980-d2d2 spectrometer. The excitation arm of the system was equipped with a Xenon lamp (450W). The emission arm was equipped with a standard visible PMT detector. The electrical characterization was carried out for different ECE variations by using ECOPIA Hall Effect measurement (HMS5000) with the current range of 0-20 mA. Micro Raman measurements has been carried out for different ECE variations using (Renishaw inVia system) with an excitation laser beam focused at 532 nm.

3. RESULT AND DISCUSSION

Fig. 2 (a) - (d), shows the well-controlled surface and cross-sectional view of the as-prepared micropyramidal structure along with the nM-Hs design for 30 sec by the ECE treatment. Firstly, the anisotropic etching were carried out for the optimized time of 15 min to form a micropyramids with the height of ~ 4 µm as shown in Fig. 2 (a) and (c). Subsequently, the highly dense structure consisting of very large surface area and uniform nM-Hs design combining micropyramids (MPs) and nanowires (NWs) were obtained by metal assisted electroless etching technique (ECE) by varying process time, one of them done for 30 sec is shown in Fig. 2 (b) and (d).

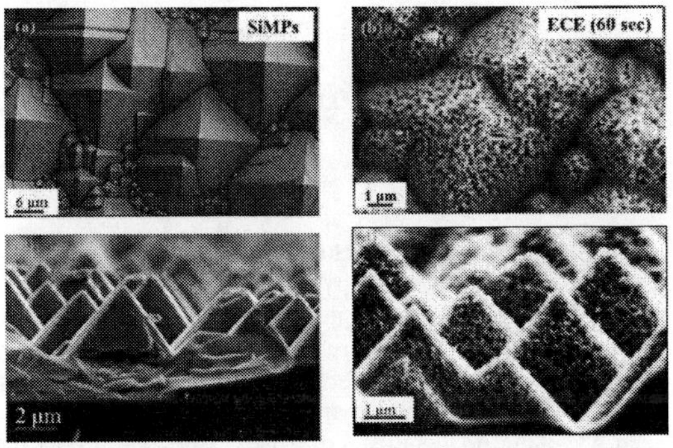

Figure 2: Represents the FESEM surface and cross-sectional images, indicating the evolution of ECE process for the formation of nano - micro hierarchical structure before (a)-(b) and after (c)-(d) ECE etching.

One can see that after the ECE time variations, the micropyramidal structure becomes more porous and very rough with the nanoholes inside the pyramids. This highly dense nanoholes with interspacing yield the formation of random nanowires with very close interspacing between them on the top of the micropyramids due to the existence of Ag nanoparticles [6,8,9]. Notably, this nM-Hs structure can make it possible to suppress the reflectance properties and enhances the light trapping properties as compared to the ideal micro pyramidal structure.

To evaluate the nanowire induced light enhancement properties, the total reflectance R_{tot} and diffusive reflectance R_{diff} was studied over the high broadband range from 300-1100 nm for different ECE variation from 0-60 sec as shown in Fig. 3. The micropyramidal structure exhibits the average R_{tot} value of 11.8% in the range of 350-750 nm. Whereas, when treated for different ECE processes from 15-60 sec, the effective R_{tot} reduces from 11.8% to 6.02% as represented in Fig. 3 (a). In addition, as shown in Fig. 3 (b), the R_{diff} also decreases from 11.8% to as low as 4.94% for ECE 30 sec over the whole visible range (350-750 nm). These light harvesting properties of the nM-Hs design can be attributed to the multiple scattering by the micro-scale pyramidal structure and the nM-Hs induces the subwavelength features of the nanowires on the micropyramids [10].

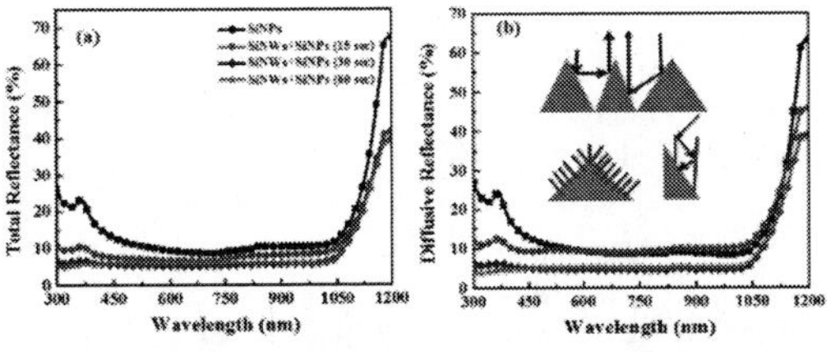

Figure 3: Demonstrates the total reflectance R_{tot} and diffusive reflectance R_{diff} for ECE time variations over the wavelength region from 300-1200 nm. Inset of Fig. 2 (b), shows the schematic structure of the micropyramids interacting with the incoming light.

These nM-Hs on the micropyramids act as an effective medium with the effective refractive index gradually increases from nanowire/air interface to bottom of the nanowire/pyramid interface, thus enhances the probability of captured photons. Furthermore, the absorption curve derived from the transmission (here, T=0) and reflectance data is shown in Fig. 4, which demonstrates that there is a strong residual absorption of ~ 95% for 30 sec ECE process over a broad range of wavelength, below the band gap of the Si. Due to the combined effect of multiple light scattering effect with the gradient index of refraction the nM-Hs design exhibits the maximum photon collection efficiency.

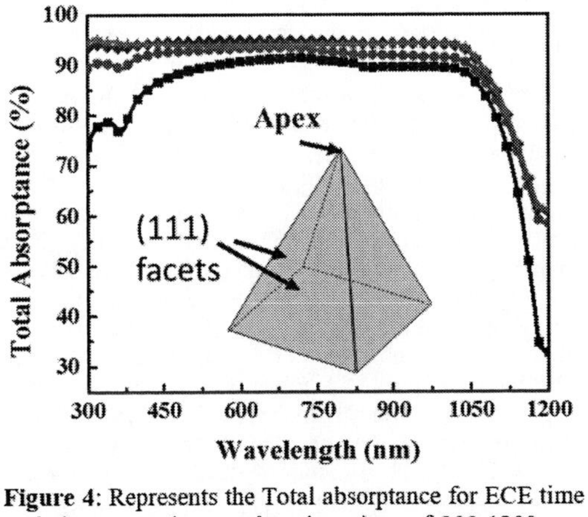

Figure 4: Represents the Total absorptance for ECE time variations over the wavelength regions of 300-1200 nm. Inset of Fig. 2 (b), shows the schematic structure of the micropyramids with (111) facets.

In order to have an insight of the optical confinement effect into the nM-Hs design, photoluminescence (PL) properties has been studied in detail by comparing the ideal micropyramids with the high absorption nM-Hs design done by optimized 30 sec ECE. Fig. 5, shows the room temperature PL spectra with an excitation wavelength of 468 nm. The two PL emission peaks at 694 nm and 780 nm were observed for Si pyramidal and hierarchical structure. It is to be noted that, after ECE process for 30 sec the intensity of the light emission at 780 nm drastically increases as compared to the ideal micropyramidal structure. The existence of the strong intense peak are predominantly due to the quantum confinement effect of free excitons within the Si nanocrystals and/or the SiO_x/Si interface defects states for the indirect band gap Si material [11,12]. Considering the fact that the characteristic size of the Si-nanocrystallites should be less than the Bohr radius of the free exciton of bulk Si, which is around 5 nm, we have calculated the crystallite size for nM-Hs by the following Eq. (1) which comes out to be ~5.8 nm.

$$E = E_0 + \frac{3.73}{d^{1.39}} \quad (1)$$

Where, E_0 = 1.12 eV, RT bandgap for the bulk Si and E is the average nansostructure energy obtained from the PL spectrum. Thus, we believe that the presence of Si nanocrystallites on the walls of nanowires after the ECE process on pyramidal structure, causes a strong quantum confinement and, therefore there is a strong enhancement

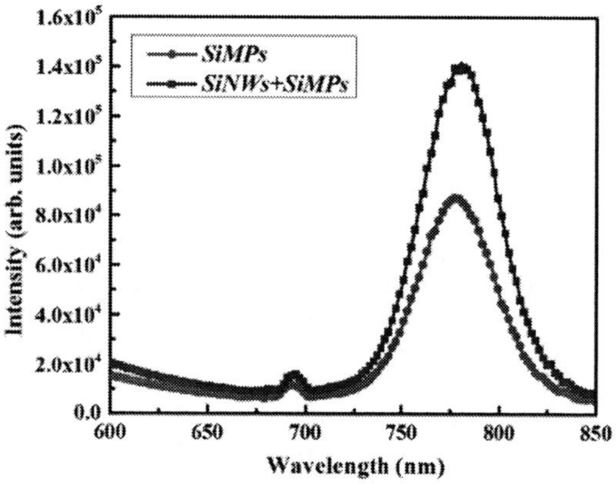

Figure 5: Shows the photoluminescence (PL) spectra of the as-prepared SiMPs and nano-micro hierarchical structure for ECE time for optimized 30 sec.

in the observed broadband PL at room temperature. To evaluate the existance of the quantum confinement effect, raman spectra for different etching time has been carried out. Fig. 5, represents the strong raman signals for nM-Hs design at different ECE time. The strong raman signal at 520.39 cm^{-1}, is observed for micropyramidal structure. These intense peak is corresponding to the first order transverse optical mode (1TO) of c-Si. As the ECE time increases, the nM-Hs design contributes to the shift in the raman peak with an improvement in the intensity distribution. The shift and the improvement in the Raman signals is due to the existense of the quantum confinement effect and the increase in lattice defect states [13].

The electrical properties of the etching variation has been carried out and the results are displayed in table 1. Although the mobility (μ) and bulk concentration (n_c) of all the treated samples shows more or less simialr behavior, but the resistivity (ρ) drastically decreases from untreated to 60 sec ECE variation The increase in the ECE variation thus outperforms the electron path way to conduct the motion, not resisting the flow of the electric current.

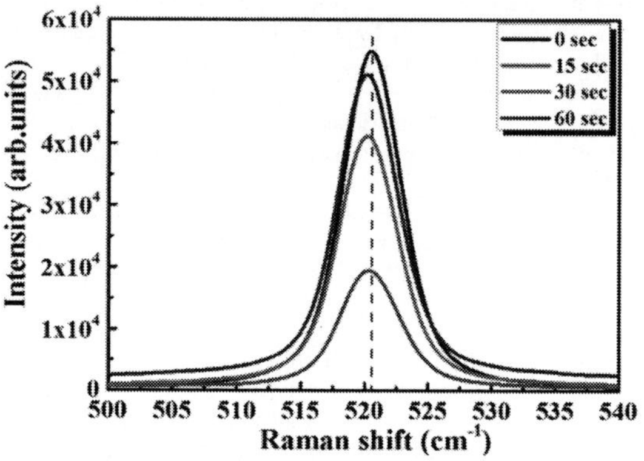

Figure 6: Demonstrates the Raman spectra of nM-Hs design for different etching time.

Table 1: The electrical properties of nM-Hs design for various ECE process

ECE variation (sec)	Resistivity (ρ) x 10^{-3} (Ωcm)	Mobility (μ) (cm^2/V*S)	Bulk concentration (n_c) x 10^{18} (cm^{-1})
0	0.44±0.27	1352.4	-0.103
15	1.25±068	1356.6	-3.69
30	4.25±0.96	1304.6	-1.12
60	3.1±0.54	1310.7	-1.53

4. CONCLUSION

The nano-micro hierarchical structures have been successfully fabricated by using electroless etching process. This nM-Hs design composed of nanowires and micro-pyramids in different scales exhibit favorable broadband light-trapping ability. The multiple scattering by the micro-scale pyramidal structure and the subwavelength features of the nanowires on the micropyramids leads us the reflectance to a very low value of 4.94% with 30 sec ECE, over the whole visible range (350-750 nm). The existence of the strong intense PL peak at 780 nm is due to the quantum confinement effect of free excitons within the Si nanocrystals or the SiO_x/Si interface defects states for the indirect band gap Si material. The improvement in the Raman signals confirms the existence of quantum confinement effect. The shift in the Raman spectra is attributed to the increase in the lattice defects as the etching time increases. The variation in the ECE overtakes and ease to the electron path way not resisting the flow of the electric current. This novel concept of hierarchical structures will open the doors for developing large-area and cost-effective solar energy applications.

5. ACKNOWLEDGEMENT

We thank Dr. Govind Gupta; CSIR-NPL, Delhi, for the photoluminescence (PL) measurements of nano-micro hierarchical structure. Financial support by Pandit Deendayal Petroleum University (PDPU) for carrying out this research, is thankfully acknowledged.

References

[1] V. Schmidt, J. Wittemann, S. Senz and U. Gösele, Adv. Mater., 21 (2009) 2681-2702.
[2] J. Bae, H. Kim, X. Zhang, C. Dang, Y. Zhang, Y. Choi, A. Nurmikko, Z. Wang, Nanotechnology, 21 (2010) 095502.
[3] J. Badán, R. Marotti, E. Dalchiele, D. Ariosa, F. Martín, D. Leinen et al., J. Mat. Res. 30 (2015) 753-760.
[4] N. Anttu, ACS Photonics. 2 (2015) 446-453.
[5] Q. Hong, Y. Cao, J. Xu, H. Lu, J. He, J. Sun, ACS Appl. Mat. & Int. 6 (2014) 20887-20894.
[6] R. Prajesh, A. Agarwal, Bionanoscience, 2 (2012) 218-222.
[7] K. Peng, X. Wang, S. Lee, Appl. Phys. Lett., 95 (2009) 243112.
[8] W. Wei, M. Tsai, S. Ho, S. Tai, C. Ho, S. Tsai, C. Liu, R. Chung, J. He, Nano Lett., 13 (2013) 3658-3663.
[9] Z. Pei, H. Hu, S. Li and C. Ye, Langmuir, 33 (2017) 3569-3575.
[10] R. Ghosh, P. Giri, K. Imakita and M. Fujii, J. of Alloys and Compounds 638 (2015) 419-428.
[11] R. Ghosh, P. Giri, K. Imakita and M. Fujii, Nanotechnology 25 (2014) 045703.
[12] L. Lin, S. Guo, X. Sun, J. Feng, Y. Wang, Nano. Res. Lett., 5 (2010) 1822-1828.
[13] S. A. Razek, M. A. Swillam, N. K. Allam, Appl. Phys. Lett., 115 (2014) 194305.

INFLUENCE OF DC-SPUTTERED ITO LAYERS ON PERFORMANCE OF SILICON HETEROJUNCTION SOLAR CELLS

S. Abolmasov, A. Abramov, D. Andronikov, K. Emtsev, G. Ivanov, I. Nyapshaev, D. Orekhov, A. Semenov, G. Shelopin, E. Terukov

R&D Center of Thin Film Technologies in Energetics, Polytechnicheskaya str. 28, 194064 St. Petersburg, Russia

ABSTRACT: The influence of target composition and oxygen flow during DC sputtering of front and rear indium tin oxide (ITO) layers on the performance of 6" bifacial Si-HJ solar cells was examined. An in-line DC magnetron sputtering system equipped with two rectangular planar cathodes was used to simulate sputtering conditions close to those used in industrial ITO sputter coaters. Sintered ceramic ITO targets having four various SnO_2 concentrations were employed for DC sputtering of ITO thin films at room temperature. A comparison of the electro-optical properties of ITO films on glass substrates and their influence on both front and rear emitter Si-HJ solar cells are presented.

Keywords: silicon heterojunction, bifacial solar cells, ITO thin films, DC sputtering, room temperature

1 INTRODUCTION

Among various transparent conductive oxides (TCO) used in silicon PV industry, only indium tin oxide (ITO) can combine properties of both antireflection coating (ARC) and TCO layer acting as window layer in high-efficiency solar cells such as silicon heterojunction (Si-HJ) solar cells [1, 2]. The properties of both front and rear ITO layers have a great influence on the performance of Si-HJ solar cells [3-5] and depend on many parameters such as type of magnetron discharge (DC, RF or pulsed), target and working gas composition, gas pressure, substrate temperature, target to substrate distance and discharge power.

Unlike many studies on the subject, which mostly focused on the electro-optical properties of ITO thin films deposited on glass [6-10], this work is PV industry oriented and therefore deals with DC sputtering of ITO layers from ceramic targets during manufacturing of 6'' Si-HJ solar cells using n-type crystalline silicon (c-Si) wafers grown by Czochralski (Cz) process.

The focus of this study is on the influence of target composition and oxygen flow during DC sputtering of front and rear ITO layers on the performance of Si-HJ solar cells. A comparison between front (FE) and rear emitter (RE) Si-HJ solar cells is also presented.

2 EXPERIMENTAL DETAILS

The Si-HJ solar cells used in this study had bifacial design. The device process sequence was as follows. First, 6-inch, 180-µm-thick, CZ c-Si wafers were textured and cleaned, then intrinsic and doped amorphous silicon (a-Si:H) thin films were deposited in two Gen5 KAI PECVD reactors of our R&D pilot line [11], using 40.68 MHz (VHF) excitation frequency and 110×130 cm^2 wafer carriers. Before a-Si:H deposition, the wafers were immersed into a 5% HF solution in order to remove the native oxide and produce a hydrogen-terminated surface. To avoid the influence of wafer texturing and PECVD processes on the performance of Si-HJ solar cells, 48 wafers with close values of resistivity have been processed simultaneously in a single chemical sequence followed by a single PECVD sequence. An in-line DC magnetron sputtering system, equipped with two rectangular planar cathodes, was used to obtain good ITO film uniformity ($\leq$ 5 %) across 6'' c-Si wafers. An industrial screen printer and a calibrated Grid^{TOUCH} system were employed for printing the busbar less contact grid using a low-temperature Ag-based paste with subsequent curing and measuring the IV characteristics of produced Si-HJ solar cells, respectively.

Table I: Sputtering conditions.

Substrate	6'' n-type Cz c-Si wafer, textured, 170 µm thick
Two targets	Rectangular ceramic ITO targets with different wt. %
Working gas	$Ar:O_2$
Pressure	4 mTorr
DC power density	2 W/cm^2
Target to substrate distance	7 cm
Substrate temperature	Room temperature
Thickness of ITO layers	~ 100 nm
Wafer carrier speed	150-300 mm/min

The effective minority carrier lifetime (τ) and wafer resistivity were measured prior to ITO deposition by using a Sinton Consulting WCT-120 tester. The lifetime measurements were taken on c-Si wafers with as deposited a-Si:H films. For measuring the optic and electric properties of ITO thin films on glass substrates, a Cary Series UV/VIS/NIR spectrometer (Agilent Technologies) and an Ecopia HMS-5000 Hall effect probe were employed. ITO film thickness measurements were performed using an Alpha Step D-120 stylus profilometer (KLA-Tencor). Base pressure in the in-line DC magnetron sputtering system was below 10^{-6} Torr, whereas the process pressure was about 4 mTorr. DC sputtering conditions are summarized in Table I. Rectangular sintered ceramic ITO targets from Advanced Nano Products (ANP) of various compositions (containing 3, 5 and 10 wt.% SnO_2 in the target) with a density above 99% of theoretical density and high purity (4N) were used, including special composition targets (NCOT).

3 RESULTS AND DISCUSSION

3.1 ITO thin film properties on glass

First, ITO deposition experiments were conducted to evaluate the influence of target composition and oxygen flow (or, in other words, oxygen concentration in Ar/O_2 mixture) on the properties of 100-nm-thick ITO thin films. It is well known that the power density and substrate temperature during magnetron sputtering of ITO thin films as well as post-annealing process greatly affect the properties of ITO films and the performance of Si-HJ solar cells on their basis [12]. In our case, ITO films were sputtered at room temperature on glass substrates and then annealed in ambient air at 180 °C for 20 min prior to taking measurements. Hence, the post-annealing process of ITO thin films on glass mimics curing of Ag paste during manufacturing of Si-HJ solar cells. The power density was fixed at 2 W/cm^2, which is comparable to that used in industrial sputter coaters with rotary targets (~ 10 kW/m) [13]. It should also be stressed that the quality of sputtered ITO thin films also greatly depends on the quality of used ceramic ITO targets (density and purity) as well as on the target surface conditions. In this study, we used ITO targets from one supplier with well-defined characteristics and employed for all targets the same cleaning procedure (DC sputtering in Ar atmosphere for 1h).

Figure 1 shows the effect of oxygen flow on the optical and electrical parameters of ITO layers obtained by DC sputtering of ceramic ITO targets with different SnO_2 concentrations. It can be seen that for all types of ITO targets the curves of optical transmittance and Hall mobility have maximum at the same value of oxygen flow equal to 0.02 l/h while the Hall resistivity (sheet resistance) is close to its minimum. Among three commercially available ITO grades, which currently widely used for mass production of Si-HJ solar cells [13, 14], 97/3 ITO grade result in the highest values of optical transmittance and Hall mobility (93% and 28 cm^2/V·s in our case). Note that these values are similar to those obtained by Umicore group [13]. Nevertheless, we also tested novel ITO targets offered by ANP, so-called NCOT, which over perform standard industrial ceramic ITO targets, as can be seen from Fig. 1. Due to this reason, 97/3 and NCOT targets were chosen in this study for further manufacturing of Si-HJ solar cells.

3.2 Performance of 6" bifacial Si-HJ cells

Figures 2a and 2b show the effect of oxygen flow during DC sputtering of front and rear ITO layers using two 97/3 ITO targets on the conversion efficiency of 6" bifacial Si-HJ solar cells with front and rear emitter design, respectively. Three values of oxygen flow were examined, with 0.01 l/h taken as the reference, i.e. this value was fixed either for front or rear ITO layer while the oxygen flow was varied during deposition of opposite ITO layer. To avoid the influence of wafer texturing and PECVD processes on the performance of Si-HJ solar cells, 48 wafers after PECVD process were divided into six equal groups (= 8 wafers in the group).

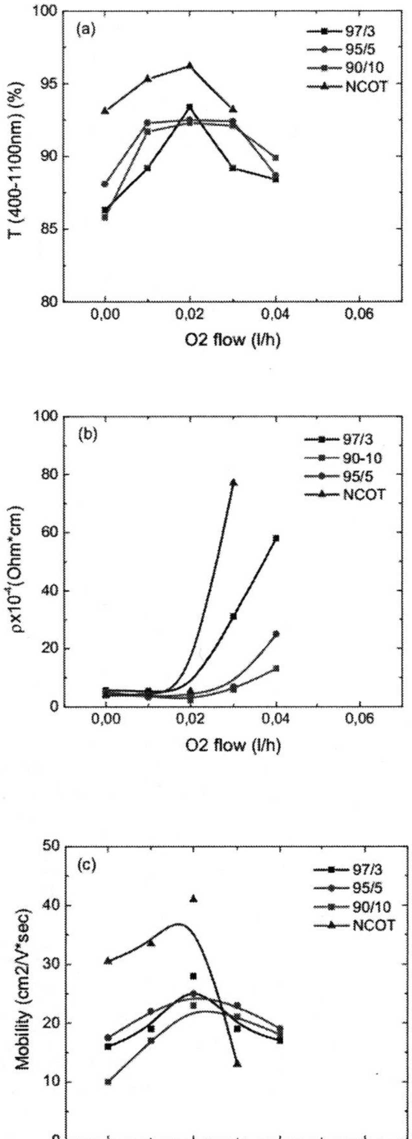

Figure 1: Influence of oxygen flow on parameters of 100-nm-thick ITO layers deposited on glass substrates using ITO targets with different tin oxide content (3, 5, 10 wt.% and NCOT): (a) optical transmittance in the range of 400-1100 nm, (b) Hall resistivity and (c) Hall mobility.

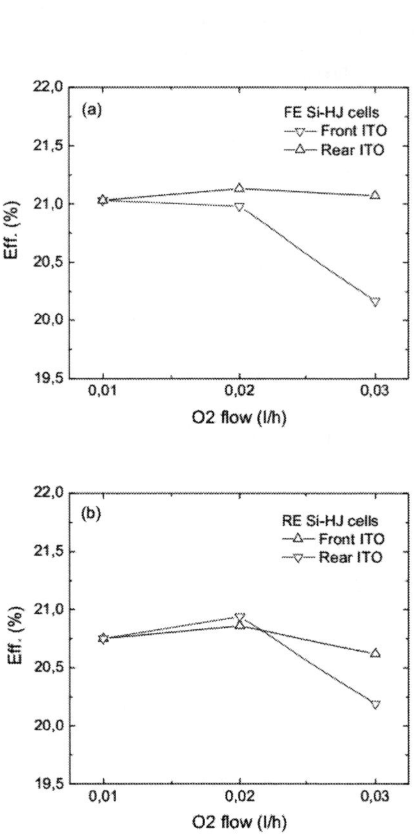

Figure 2: Dependence of average (over 8 cells) value of conversion efficiency of bifacial Si-HJ solar cells with (a) front (FE) and (b) rear emitter (RE) on the oxygen flow during DC sputtering of front and rear ITO layers using 97/3 ITO targets.

Figure 2 illustrates that small variations of oxygen flow in Ar/O_2 working gas mixture around 0.02 l/h (optimum value according to measurements of ITO parameters on glass substrates, see Sec. 3.1) during DC sputtering of ITO layer on n-doped amorphous silicon layer (n-a-Si:H) has small influence on the efficiency of both front and rear emitter bifacial Si-HJ solar cells. On the other hand, the same variations in oxygen flow above the optimum value = 0.02 l/h during DC sputtering of ITO layer on p-doped amorphous silicon layer (p-a-Si:H) result in strong influence on the efficiency of both FE and RE Si-HJ solar cells. The latter is most likely due to barrier formation at p-a-Si:H/n-ITO interface. This effect is more severe for FE Si-HJ solar cells in which most of the solar radiation is absorbed near p-a-Si:H/ITO interface rather than for RE Si-HJ solar cells.

Figures 3, 4 and 5 show a comparison of parameters of bifacial Si-HJ solar cells with rear emmiter manufactured using three types of ITO targets with the same oxygen flow (= 0.01 l/h) during deposition of both front and rear ITO layers. In this case, 48 wafers after PECVD process with close values of resistivity (see upper graphs on Figs. 3-5) were divided into three groups (= 16 wafers in the group). It is easy to see that the use of 97/3 ITO grade results in higher short circuit current (Isc), but lower value of fill factor (FF) in comparison to the use of 90/10 ITO grade. However, gain in Isc is compensated by loss in FF for 97/3 ITO grade and is opposite for 90/10 ITO grade resulting in almost the same value of conversion efficiency for both grades. Nevertheless, as shown above (Fig. 2), the n-a-Si:H/n-ITO interface is less affected by oxygen flow. Therefore, the use of higher oxygen flows during deposition of front ITO layers in RE Si-HJ solar cells may result in an increase of Isc without substantial loss in FF. Further increase in the efficiency of Si-HJ solar cells requires however the use of ITO layers with ultra-high Hall mobility, as demonstrated by the use of NCOT targets (see Figs. 1 and 5). Alternatively, as shown by many research groups, see e.g. [15], the use of hydrogen-doped pure I_2O_3 (IOH) can yield thin films with much higher mobility (> 120 $cm^2/(V\ s)$ [16]). However, un-doped I_2O_3 ceramic sputtering targets are very brittle so that it is unlikely that they can be produced in rotary format with high production yield. According to Umicore group [13], hydrogen addition to the sputter gas atmosphere can bring significant improvement of ITO conductivity by a factor of 2.5 for 97/3 ITO grade and appears to be more promising for industrial use.

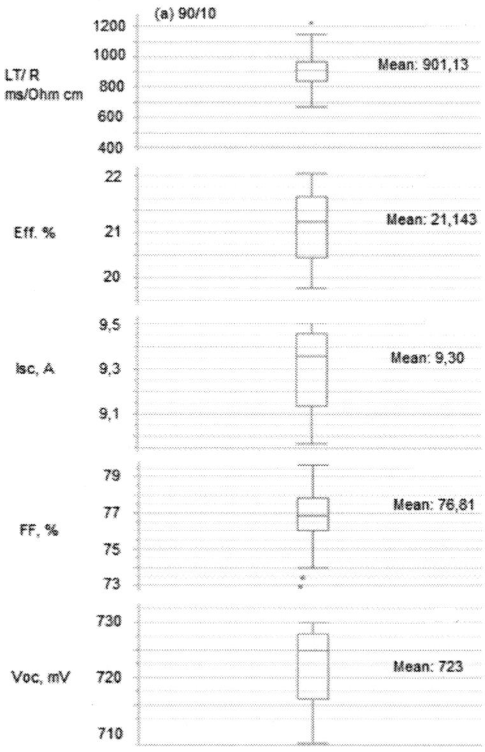

Figure 3: Ratio of effective lifetime to wafer resistivity (LT/R) and parameters of bifacial Si-HJ solar cells with rear emitter manufactured using 90/10 ITO targets and oxygen flow of 0.01 l/h for both front and rear ITO layers. Statistics is shown for 16 cells.

4 CONCLUSIONS

The effects of target composition and oxygen flow during DC sputtering of front and rear ITO layers on the performance of 6" bifacial Si-HJ solar cells with FE and

RE design have been investigated.

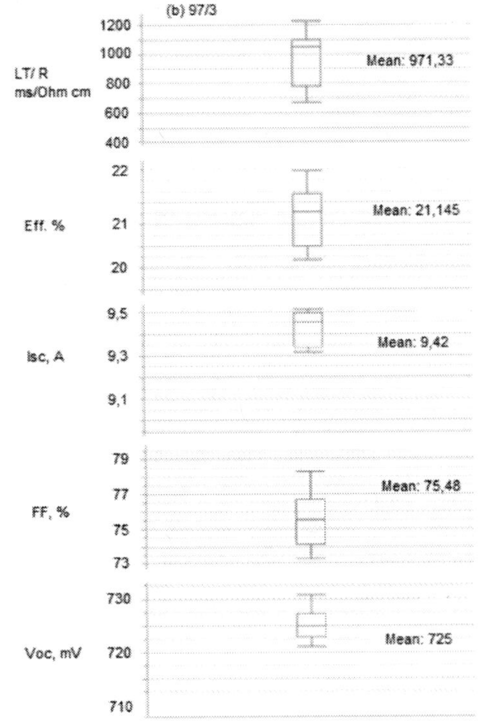

Figure 4: The same as in Fig. 3 with 97/3 ITO targets used.

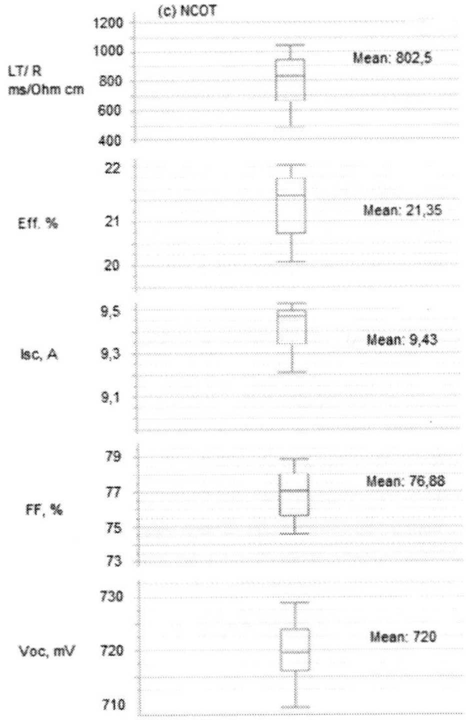

Figure 5: The same as in Fig. 3 with NCOT targets used.

Sintered ceramic ITO targets from one supplier having four various SnO_2 concentrations have been tested. It was found that for all four types of ITO targets the curves of optical transmittance and Hall mobility have maximum at the same value of oxygen flow equal to 0.02 l/h while the Hall resistivity (sheet resistance) is close to its minimum. This value of oxygen flow is also optimal for both FE and RE bifacial Si-HJ solar cells, resulting in the highest efficiency values. Oxygen flow has more influence on p-a-Si:H/ITO interface rather than on n-a-Si:H/ITO interface in both FE and RE Si-HJ solar cells. Among three commercially available ITO grades (90/10, 95/5, 97/3), which currently widely used for mass production of Si-HJ solar cells, 97/3 ITO grade yields the best performance. The use of targets that yield thin films with ultra-high Hall mobility (> 30 $cm^2/(V\ s)$) results in even better performance of Si-HJ solar cells; however the use of such targets in mass production should be validated.

The authors wish to thank Advanced Nano Products for manufacturing and short-term delivery of ITO targets.

REFERENCES

[1] E. Fortunato, D. Ginley, H. Hosono, D.C. Paine, MRS Bulletin 32 (2007) 242.
[2] S. De Wolf, A. Descoeudres, Z.C. Holman, C. Ballif, Green 2 (2012) 7.
[3] M. Bivour, C. Reichel, M. Hermle, S. W. Glunz, Solar Energy Materials and Solar Cells 106 (2012) 11.
[4] Z. C. Holman, M. Filipič, A. Descoeudres, S. De Wolf, F. Smole, M. Topič, C. Ballif, Journal of Applied Physics 113 (2013) 013107.
[5] A. Tomasi, *et al.*, IEEE Journal of Photovoltaics 6 (2016) 17.
[6] W. Deng, T. Ohgi, H. Nejo, D. Fujita, Japanese Journal of Applied Physics 40 (2001) 3364.
[7] H.C. Lee, O.O. Park, Vacuum 75 (2004) 275.
[8] F. Kurdesau, G. Khripunov, A.F. Cunha, M. Kaelin, A.N. Tiwari, Journal of Non-Crystalline Solids 352 (2006) 1466.
[9] O. Tuna, Y. Selamet, G. Aygun, L. Ozyuzer, Journal of Physics D: Applied Physics 43 (2010) 055402.
[10] S. Middya, A. Layek, P.P. Ray, Journal of Applied Science 12 (2012) 1676.
[11] S. Abolmasov, *et al.*, Proceedings 31st European Photovoltaic Solar Energy Conference, Vol. I (2015) 1046-1049.
[12] D. Zhang, A. Tavakoliyaraki, Y. Wu, R.A.C.M.M. van Swaaij, M. Zeman, Energy Procedia 8 (2011) 207.
[13] C. Frei, D. Chiu, P. Lippens, Proceedings 28th European Photovoltaic Solar Energy Conference, Vol. I (2012) 2029-2033.
[14] P. Lippens, D. Chiu, C. Szepesi, Proceedings 27th European Photovoltaic Solar Energy Conference, Vol. I (2011) 2079-2082.
[15] L. Barraud, *et al.*, Solar Energy Materials & Solar Cells 115 (2013) 151.
[16] T. Koida, H. Fujiwara, M. Kondo, Japanese Journal of Applied Physics 46 (2007) L685.

TRANSVERSE-COLLECTION MECHANISMS IN SILICON-HETEROJUNCTION SOLAR CELLS

A. Casado, R. Barrio, J.J. Gandía & J. Cárabe
e-mail: julio.carabe@ciemat.es
CIEMAT. Renewable-Energy Division
Avda. Complutense 40. E-28040 Madrid. SPAIN

ABSTRACT: Most solar-cell models assume, either explicitly or tacitly, that carrier collection proceeds normal to the junction before charge separation and parallel thereto once the junction has been surpassed. In the case of silicon-heterojunction cells, this latter transverse collection process is supposed to take place basically across the transparent conductive oxide deposited on the front side of the device. In order to confirm this, cells with and without a transparent conductive oxide (which we call respectively standard and bare cells) have been prepared and tested. Although the predictions of the so-called normal-plus-parallel collection hypothesis for a bare cell are a very high series resistance and a current density being essentially flat across the cell active-area surface, experimental results are clearly different. Bare cells show series resistances of the order of 1 Ohm and a current collection limited to the cell areas being sufficiently near the front contacts. In consequence, transverse-current collection mechanisms of SHJ solar cells have to be revised. The role of an inversion layer at the interface is discussed. Then, a qualitative model is proposed. This model is based on the fact that, once created in the absorbent, a free minority carrier diffuses not only straight towards the junction, but also in all other directions from the point where the photon has been absorbed. This consideration not only makes the above-mentioned series-resistance estimate invalid, but also limits effective collection to a distance from the metal finger of the order of the minority-carrier diffusion length. Internal-quantum-efficiency and LBIC measurements have yielded results in accordance with the simple model proposed.
Keywords: Silicon Solar Cell, TCO Transparent Conducting Oxides, Heterojunction

1 INTRODUCTION

As it has already been pointed out in previous work [1], current collection mechanisms on the front side of silicon-heterojunction (SHJ) solar cells substantially differ from those of conventional screen-printed cells [2], [3], [4]. Whereas screen-printed devices have a front diffused emitter whose thickness and conductivity allow for the transverse transport of charge carriers to the metal finger or bus without large energy losses (i.e. without boosting cell series resistance), SHJ cells have a very thin (10 to 20 nm) and resistive (10^2 Ω cm or higher) emitter, so that a transparent conductive oxide (typically SnO_2:F, ITO or ZnO:Al; abbreviated TCO) is required on top of it in order to provide a channel for the collection of carriers after charge separation in the rectifying junction.

For some reason, probably just for simplicity, most -if not all- of the models both for screen-printed and for SHJ cells assume more or less explicitly that, after generation, minority carriers travel normally to the front junction until they reach it and then cover the distance to the contact finger or bus by moving parallel thereto. This assumption will be referred to as the *normal-plus-parallel* assumption. In the case of SHJ devices, this transverse charge transport takes place across the said transparent conductive oxide, which is typically an 80-nm thick layer having a sheet resistance of the order of 100 Ω/sq.

In order to evaluate the role of the TCO and the validity of all these considerations, SHJ cells respectively with and without a TCO (respectively referred to as *standard* and *bare* cells) have been prepared and tested. The application of the normal-plus-parallel-collection assumption yields the following estimate of the expected contribution of the emitter to the series resistance of bare cells:

$$R_S \approx \frac{\rho_{emitter} \cdot S}{d_{emitter} \cdot L_F}$$

Eq. 1

$d_{emitter}$ (about 10 to 20 nm) being the emitter thickness, $\rho_{emitter}$ (~ 10^2 Ω cm) its resistivity, and S and L_F respectively the spacing between fingers and the finger length. If S is assumed to be of the order of 1 mm and L_F of the order of 1 cm, Eq. 1 yields $R_S \approx 10^9 \Omega$, well above 1 MΩ for basically whatever reasonable contact dimensions.

Certainly this estimate does not take into account the presence of an inversion layer at the front interface, caused by the heterojunction band offset, as it has been largely reported in the literature [5], [6], [7]. One could assume that once the minority carriers have approached the junction, they do not immediately cross it to reach the emitter, but rather travel parallel thereto through the inversion layer, located in the crystalline side of the interface, before crossing the junction only when they have reached the area under the contact finger. This scenario appreciably modifies the lateral resistance affecting the minority carriers, but is still unable to justify measured series resistances of a few ohms, as well as other features that will be addressed below. Different authors have studied the inversion layer at the front side of an SHJ cell and reported their electrical-resistance values. For instance, Kleider et al. [8] have measured conductances of 6×10^{-4} S for the inversion layer of lab-size (of the order of 1 cm × 1 cm) silicon-heterojunction cells, corresponding to a resistance of 1700 Ω. Maslova et al. [9], in turn, by using AFORS-HET simulation, calculate inversion-layer resistivities varying from 10^{+1} to 10^{-3} Ω cm within 0.2 μm in thickness. A simple calculation allows to estimate the corresponding inversion-layer electrical resistance (for a 1-mm × 1-cm cell) to be not lower than 0.5 to 1 kΩ, in reasonable agreement with Kleider's figures.

In summary, the assumption of carrier transport normal to the junction and then parallel thereto, leads to predicting series resistances of at least 0.5 to 1 kΩ. As it will be illustrated below, experimental results largely differ from these predictions. In addition, the normal-plus-parallel assumption discussed contradicts the well-established model according to which carrier collection in crystalline silicon takes place by diffusion, i.e. isotropically (since the crystalline-silicon absorber is quasi-neutral and there are no preferential minority-carrier gradients within it, with the only exception of the junction interface). The conclusion is that transverse-collection mechanisms in silicon-heterojunction cells must be revised.

In this work, a hypothesis is formulated in order to explain the mechanisms actually governing SHJ solar-cell charge collection. From this hypothesis, it is deduced that, depending on the presence and properties of the TCO, an effective active area, the so-called *collection area*, A_{COL}, can be defined around the metallic contact, as the area such that the probability for carriers absorbed thereon to be collected before being recombined is substantial, in contrast to *dead zones*, where no significant collection takes place. This A_{COL} can be used instead of the A_{ACT}^{NOM} (the nominal active area, defined as the difference between the cell total area and the area shaded by the metallic contact) as the effective active area of the cell, contrarily to what has been usually considered [10].

For a bare cell, A_{COL} can be estimated from the minority-carrier diffusion length in the crystalline-silicon absorber (not to be confounded with the effective diffusion length of the whole device). The figures obtained in this way are consistent with those derived from both internal quantum-efficiency measurements and LBIC maps.

The work opens up the possibility to undertake further studies to develop a numerical model, probably based on finite-element calculations, which would allow to obtain accurate values of A_{COL} and provide a very useful tool to design front-contact grids for silicon-heterojunction cells.

2 EXPERIMENTAL DETAILS

The development of laboratory devices involves the progressive addition of steps one by one in order to understand all the mechanisms underlying the operation of the system under study and identify the individual contributions of the different elements. As part of a research work aimed at studying the collection mechanisms at the front side of silicon-heterojunction solar cells, and more specifically at designing optimum front contacts for them, standard (with a TCO) and bare (without any TCO) SHJ solar cells have been fabricated on 1-Ω·cm-resistive, p-type <100>-oriented polished float-zone (FZ) monocrystalline-silicon wafers. In all cases, a ~12-nm-thick a-Si:H(n) emitter layer has been deposited by 13.56-MHz plasma-enhanced chemical-vapour deposition (PECVD) at 204°C onto a c-Si(p) wafer onto which an aluminium back contact had been previously evaporated [11]. The samples have been thermalised for 1 h in a helium atmosphere in the chamber prior to the PECVD process. The TCO of standard cells is typically an 80-nm aluminium-doped zinc-oxide (AZO) layer [12], deposited by radio-frequency magnetron sputtering. Such a TCO, sputtered at 100 °C has been used in the present study. The front contact is a thermally evaporated grid (18 nm Ti, 130 nm Ag) consisting of three fingers interconnected by one bus bar in the middle.

The geometric characteristics of the front grid have been measured from pictures taken by an optical microscope and analysed by calibrated software. A Sinton-Instruments QSSPC (Quasi-Steady-State PhotoConductance) system has been used to measure the minority-carrier lifetime and diffusion length of the p-type crystalline silicon wafers [13].

The current-voltage characteristics have been measured in darkness and under calibrated 100-mW/cm² AM1.5G irradiance. Internal quantum efficiencies (*IQE* or η_i) have been obtained from absolute spectral responses and cell hemispherical-reflectance spectra. An LBIC system designed in our laboratory has been employed for mapping the electrical response of the cell to an incident He-Ne laser beam (632.8 nm), with a spot diameter of 0.6 mm.

3 RESULTS AND DISCUSSION

3.1 Initial problem and hypothesis

Fig. 1 shows the I-V characteristics of one representative pair of cells: one standard and one bare. The main photovoltaic parameters extracted from these curves are presented on Table I.

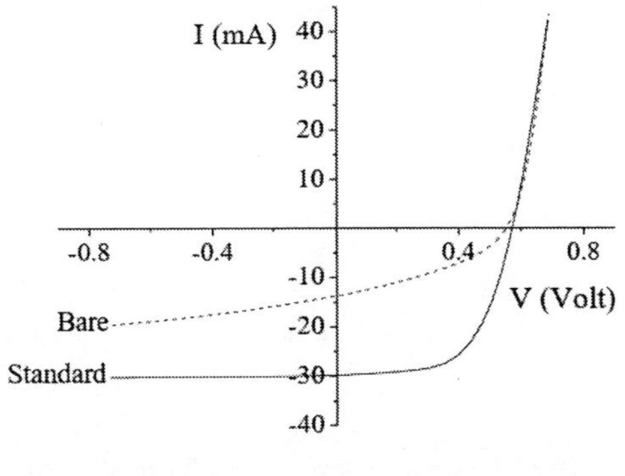

Fig. 1: Current-voltage curves under 100-mW/cm² simulated AM1.5G irradiance corresponding to twin cells differing only in what concerns the presence or absence of a front TCO layer.

The cells have been made with the only purpose to investigate their front-side collection mechanisms, so their back surface had not been previously optimised for low recombination. This explains the modest open-circuit voltages. Apart from this, the most notable feature of these results is that both of the series resistances, that of the bare and that of the standard cell are below 1.5 Ω, i.e., orders of magnitude lower than any value derived from the parallel-plus-transverse hypothesis. A current reduction about 55% and an important fill-factor reduction from 60% to 38% are otherwise not surprising.

Table I: Main photovoltaic parameters of the SHJ solar cells studied: short-circuit current, I_{SC}, open-circuit voltage, V_{OC}, series resistance, R_s and fill-factor, FF.

Cell ID	TCO	I_{sc} (mA)	V_{oc} (mV)	R_s (Ω)	FF (%)
A	Yes	29.8	570	<1.5	60
B	No	13.7	550	<1.5	38

The assumption of minority carriers following the normal path to the p-n junction is thus definitely wrong, even considering the transverse transport mechanism provided by the inversion layer, whose conductance is not high enough to justify the low series resistances found in cells having no front TCO. Instead, another scenario is considered to be more realistic. It is well known that carrier transport in crystalline silicon proceeds by diffusion [2]. When an electron-hole pair is created as a consequence of the absorption of an incoming photon, each of the minority carriers generated (as well as the majority ones) has an equal chance to move away in any direction around its origin [14], until it reaches the junction or is recombined. Since the average path before recombination is determined by the minority-carrier bulk diffusion length, L_n, it can be said that carriers created within a distance of the order of L_n from the region located strictly under the front electrode will contribute to the current as illustrated in Fig. 2, and a corresponding collection area, A_{COL}, can be consistently defined around the front metallic contact.

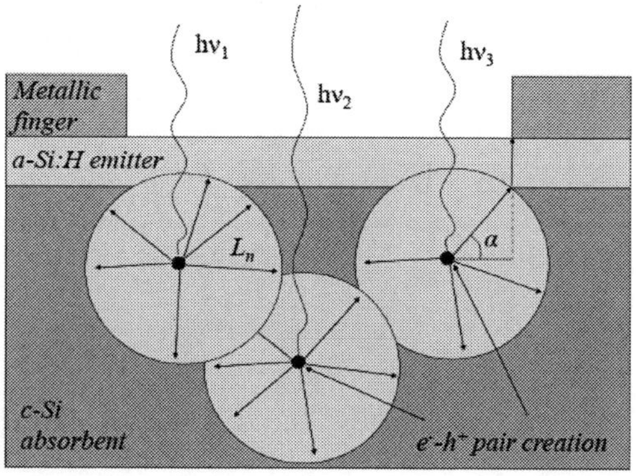

Fig. 2: Schematic model of the internal-collection and transport mechanisms inside the solar cell, from the formation of an electron-hole pair, to the collection of the charge carrier by the front metallic contact after diffusion through the c-Si absorber.

3.2 Geometric calculation of A_{COL}

The collection area, A_{COL}, can be calculated from purely geometric considerations on the basis of the above-described transport mechanisms. A schematic picture of this area surrounding the metallic contact can be seen on Fig. 3.

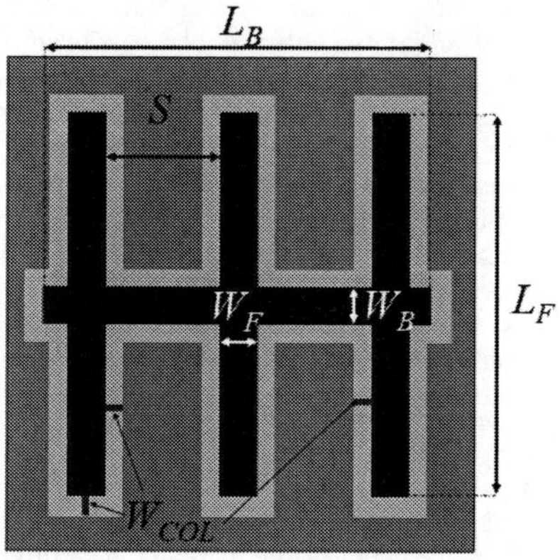

Fig. 3: Basic geometric scheme of the front side of the cells investigated. A limited area (soft grey) around the metallic contact (black) is supposed to correspond to effective carrier collection in bare cells. Its width, W_{COL}, is estimated to be comparable to the diffusion length of minority carriers in the absorbent bulk. L_B, W_B, W_F, L_F are the bus- and finger length and width, respectively.

Table II: Dimensions of the geometric parameters of Fig. 3. A_T is the total area of the cell, and A_{ACT}^{NOM} represents the nominal active area, i.e. the region of the cell surface not covered by the front metallic contact.

S (cm)	L_b=L_f (cm)	W_b=W_f (cm)	A_{ACT}^{NOM} (cm²)	A_T (cm²)
0.37	1.00	0.04	1.06	1.21

Letting A' be the area constituted by the metallic contact area (A_{METAL}) plus the collection area (A_{COL}), then:

$$A_{COL} = A' - A_{METAL}$$

Eq. 2

Since

$$A' = (2W_{COL} + W_B) \times (2W_{COL} + L_B) + \\ +N \times (2W_{COL} + W_F) \times [2W_{COL} + L_F - (2W_{COL} + W_B)]$$

Eq. 3

And

$$A_{METAL} = W_B \times L_B + N \times W_F \times (L_F - W_B)$$

Eq. 4

The collection area can be expressed as:

$$A_{COL} = 4W_{COL}^2 + 2W_{COL} \times [L_B + N \times L_F - (N-1) \times W_B]$$

Eq. 5

Where N is the number of fingers of the front electrode (3 in the present work), and the rest of the parameters are as defined in Fig. 3 and Table II.

Most carriers generated by photons absorbed at a horizontal distance x from the finger must cover a distance equal to, or slightly greater than x to reach it. In addition, the minority-carrier diffusion length is related to the distance covered by carriers before undergoing recombination only in statistical terms, i.e. subject to a substantial dispersion. Therefore it is considered reasonable to assume that collection distance W_{COL} is proportional to the minority-carrier diffusion length, i.e., that it is possible to define a positive real number $n \in (0, 1)$ such that

$$W_{COL} = n \times L_n \qquad \textit{Eq. 6}$$

L_n being the bulk diffusion length of minority carriers in the substrate, which –it must be stressed- should not be confounded with the effective overall diffusion length L_{eff}, which is not only dependent on L_n, but also on surface- and interface recombination. It is then possible to derive A_{COL} from L_n. Table III shows the values obtained for W_{COL} and the corresponding A_{COL} as calculated using Eq. 5 and Eq. 6, for a set of possible values of n. The geometric parameters have been taken from Table II, and the diffusion length has been assumed to be $L_n = 2.4$ mm as derived from the effective minority-carrier lifetime τ_n(1 sun) = 1.8 ms, measured by QSSPC.

Table III: Collection-area, A_{COL}, values calculated geometrically for a 2.4-mm minority-carrier diffusion length and different estimates of parameter n (Eq. 6). The corresponding collection width, W_{COL}, is also indicated.

N	W_{COL} (mm)	A_{COL} (cm^2)
0.20	0.48	0.39
0.25	0.60	0.49
0.30	0.72	0.59
0.35	0.84	0.69
0.40	0.96	0.79
0.45	1.08	0.90
0.50	1.20	1.00

The choice of a given value for parameter n is not possible *a priori* on the only basis of these considerations, but, as it will be demonstrated below, it can be deduced from other measurements.

3.3 Determination of W_{COL} from LBIC maps

So far, a carrier transport mechanism has been proposed and a corresponding collection area has been calculated accordingly, but an evidence of the restriction of carrier collection to a region in the vicinity of the front electrode is still required. In order to confirm or refuse the existence of such a region, LBIC maps of both standard and bare cells have been obtained.

Fig. 4, in fact, evidences the shrinking of the collection area from the whole uncovered region (*nominal active area*, A_{ACT}^{NOM}) in a standard cell, to a zone next to the fingers and bus bar (*collection area*, A_{COL}) in a bare cell, confirming our hypothesis and sharply refuting the parallel-plus-transverse assumption.

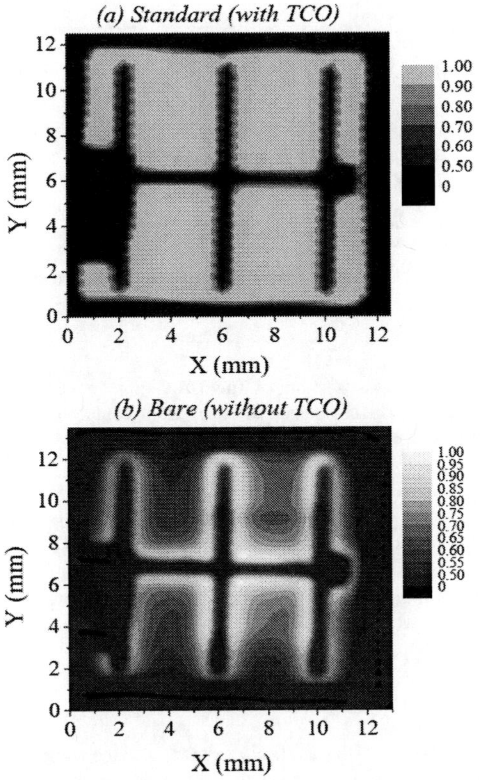

Fig. 4: LBIC images of respectively (a) a standard SHJ cell (having a front TCO) and (b) a bare SHJ cell (having no front TCO). The legend shows the short-circuit current, normalised to a maximum value of 1, obtained when the laser impinges on the cell at short-circuit condition. The collection area can be clearly observed on the cell with no TCO.

Certainly the LBIC maps do not provide a sharp drop of the signal at a given distance from the electrode edge, in part owing to the modest accuracy determined by the laser-spot diameter (0.6 mm, far from negligible with respect to the 3.7-mm finger pitch and even larger than the 0.4-mm finger width), but they can still provide some reasonable estimate of the collection-area width, W_{COL}. With this purpose, a cross section of the LBIC map at Y = 9 mm has been extracted (see please Fig. 5).

In both cells, the signal reduction for X values corresponding to the centre of each of the fingers is limited to 50% of the maximum (instead of dropping to zero), in agreement with the said low geometrical resolution of the laser. For the same reason, the signal reaches a maximum about 1 mm away from finger centre. It remains then constant between fingers in the standard cell, whereas it falls progressively, showing a minimum in the middle of the spacing between fingers in the bare cell. If W_{COL} is estimated, for instance, as the distance from the next finger edge for which the signal has dropped by 2/3, the LBIC data yield $W_{COL} = 0.8 \pm 0.2$ mm. Since this calculation involves a subjective choice, it should not be expected to yield accurate quantitative data but can be used as a reinforcement of the hypothesis formulated.

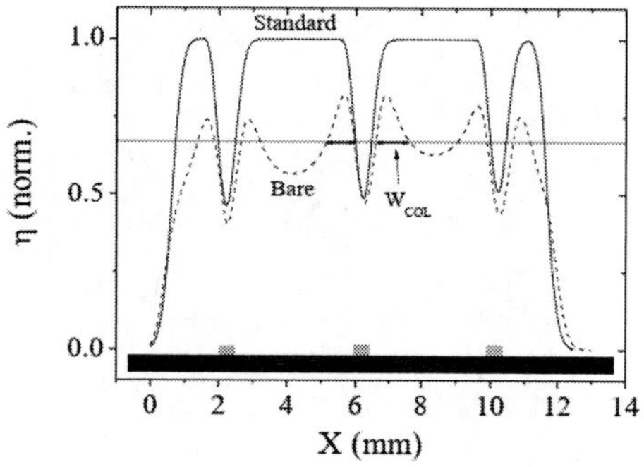

Fig. 5: Cross section of the LBIC map of Fig. 4 at Y=9mm. The cross section of the front-contact fingers is sketched at the bottom.

3.4 Calculation of A_{COL} from internal quantum efficiency

An additional way to calculate the collection area is provided by quantum-efficiency measurements. The internal quantum efficiency of a cell can be expressed as (Markvart et al. (2003)):

$$\eta(\lambda) = \frac{h \times c}{q \times \lambda} \times \frac{\delta I(\lambda)}{[1 - R(\lambda)] \times A \times e(\lambda) \times \delta\lambda}$$

Eq. 7

Where h is the Planck constant, q the elementary charge, c the light speed in vacuum, I the short-circuit current, $R(\lambda)$ the spectral reflectance, λ the wavelength, A the cell area and $e(\lambda)$ is the spectral irradiance.

The application of Eq. 7 respectively to a bare (b) and to a standard (s) cell allows to express the *IQE* ratio as:

$$Q = \frac{\eta_b(\lambda)}{\eta_s(\lambda)} = \frac{\delta I_b}{\delta I_s} \times \frac{1 - R_s(\lambda)}{1 - R_b(\lambda)}$$

Eq. 8

In fact, taking into account that LBIC images are obtained by using monochromatic illumination, the results shown on Fig. 4 are indeed internal-quantum-efficiency maps at laser wavelength. This means that the internal quantum efficiency, *IQE*, of a cell can be interpreted as an average of its position-dependent *IQE* over the whole device. In these conditions, the collection area of the bare cell can be extracted from the condition that both standard and bare cells do have the same IQE measured on their respective collection areas.

With this in mind, the monochromatic short-circuit currents of Eq. 8 can be expressed as:

$$\delta I_i(\lambda) = \frac{q \times \lambda}{h \times c} \times \eta(\lambda) \times [1 - R_i(\lambda)] \times A_{COL}^i \times e(\lambda) \times \delta\lambda$$

Eq. 9

Where i represents either s (standard cell) or b (bare cell). Please note that the *IQE* appearing in Eq. 9 has no subscript, indicating that $\eta(\lambda)$ represents not the overall measured internal quantum efficiency (η_b, η_s of Eq. 8), but the internal quantum efficiency selectively evaluated only on the collection area. This *IQE* is assumed to be equal for all the cells under study since they only differ in what concerns the TCO.

Replacing now Eq. 9 in Eq. 8 and taking into account that the collection area of the standard cell is the nominal active area A_{ACT}^{NOM}, the following expression is found for the *IQE* ratio:

$$Q = \frac{\eta_b(\lambda)}{\eta_s(\lambda)} = \frac{A_{COL}^b}{A_{ACT}^{NOM}}$$

Eq. 10

This means that the bare-cell collection area can be calculated as:

$$A_{COL} = Q \times A_{ACT}^{NOM}$$

Eq. 11

The validity of the above considerations, however, can be sustained only in case the ratio Q is basically a constant with respect to λ, which means that the experimental data yield a very weak dependence of Q on λ. In order to check this, the respective internal quantum efficiencies of the standard and bare cells are shown on Fig. 6.

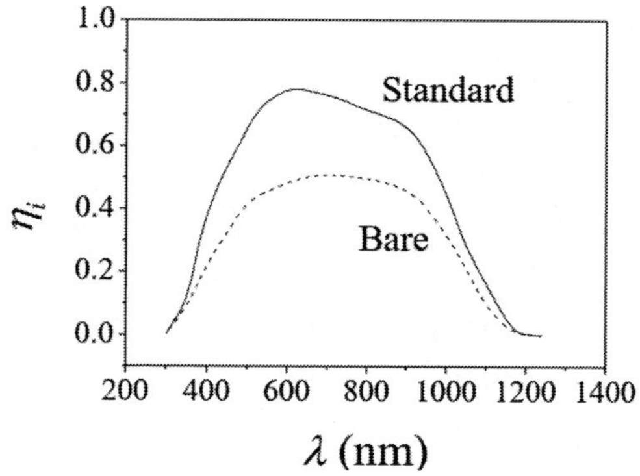

Fig. 6: Internal quantum efficiency of a standard and a bare cell, differing only in the presence of a front TCO in the case of the former.

Such a dependence can be appreciated on Fig. 7, which ultimately confirms the validity of the arguments deployed. Please note that $Q(\lambda)$ is approximately constant for the whole domain of the *IQEs*, with the exception of the *IQE* tails (in the close vicinity of 300 and 1200 nm respectively). This effect is obviously due to experimental errors associated with the very low values of both the *IQEs* in these specific spectral regions.

The value for the bare-cell collection area obtained by replacing in Eq. 11 the figure for Q obtained by the linear fit shown in Fig. 7 and the nominal active area of Table II, is

$$A_{COL} = (0.70 \pm 0.04)\ cm^2$$

Eq. 12

This value confirms the validity of the estimate based on LBIC measurements and allows to make a choice for parameter n in Table III, which turns out to be 0.35 for minority-carrier diffusion length assumed to be 0.24 cm. The corresponding collection width is $W_{COL} = 0.08 \pm 0.02$ cm.

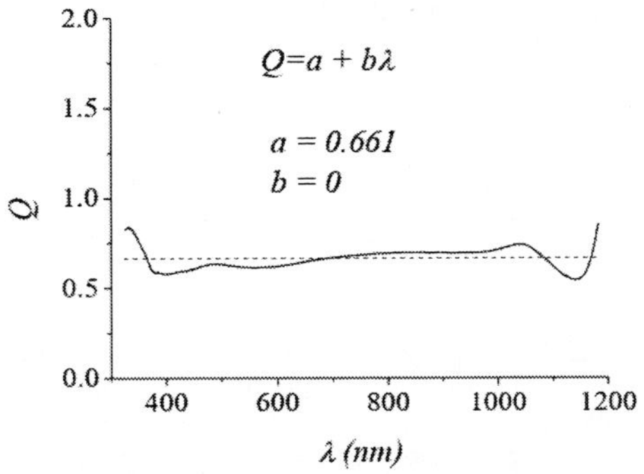

Fig. 7: Ratio Q of the internal quantum efficiencies of the respective standard and bare cells under study. The inset shows the result of a linear fit to $Q=a$ (forcing $b=0$). The constant obtained from this fit, a, is used as the constant Q in the calculation of the collection area of the bare cell.

These evidences are considered to confirm the validity of the model proposed in this work, showing experimentally that even in silicon-heterojunction cells having no front TCO, there is an area around the metallic front contact, being smaller than the nominal active area, where carrier collection is not negligible. The width of this so-called collection area in our bare cells has been found to be about a third (n ~ 0.35) of the minority-carrier bulk diffusion length, L_n. The result is found to be fully consistent with the hypothesis formulated and the 1/3 factor applied to the diffusion length is considered to be in part a consequence of the criterion used to define the collection area, which, being formulated in terms of quantum efficiency, should be determined by the fact that, for a given distance to the contact finger, only carriers diffusing away within a restricted solid angle are in fact collected.

4 CONCLUSIONS

Theoretical estimates of the series resistance of a silicon-heterojunction solar cell and their comparison to experimental values, lead to appreciable contradictions if carriers are assumed to take the shortest (normal) path in their way to the junction.

A model has been proposed for the mechanisms governing the collection of current inside a silicon-heterojunction solar cell. Experimental evidences of the existence of an area around the metallic front contact, where the carriers are collected even if no front transparent conductive oxide is present, have been found. The width of this area has been found to be a significant fraction of the minority-carrier diffusion length, L_n. This validates the assumption proposed, i.e. that instead of the shortest way to the p-n junction, carriers follow all of the multiple possible paths distributed in a sphere centred on the absorption point. Therefore the front transparent conductive oxide determines not so much the series resistance of the cell as the effective collection area, i.e., as the short-circuit current which can be obtained. These considerations are of primary importance for the right design of the front grid of silicon-heterojunction solar cells.

5 ACKNOWLEDGEMENTS

The authors would like to thank colleagues from "Universitat Politècnica de Catalunya" (UPC), especially Dr. Cristóbal Voz, for their cooperation providing wafer passivation and QSSPC measurements and results.

Partial financial support for this work has been provided by the Spanish Research Agency and Ministry of Economy, Industry and Competitiveness under project CHENOC (ENE2016-78933-C4-3-R).

6 REFERENCES

[1] A. Casado, R. Barrio, S. Fernández, J. Cárabe and J. Gandía, in *Proc. 25th European Photovoltaic Solar-Energy Conference*, Valencia, 2010.

[2] S. Fonash, Solar-Cell Device Physics, Elsevier, 2010.

[3] A. Fahrenbruch and R. Bube, Fundamentals of Solar Cells. Photovoltaic Solar-Energy Conversion, London: Academic Press, 1983.

[4] A. Flat, *Solar Energy*, vol. 23, p. 289, 1979.

[5] J. Kleider, A. Gudovskikh y P. Roca i Cabarrocas, *Applied Physics Letters*, vol. 92, p. 162101, 2008.

[6] J. Li, R. Crandall, D. Young, M. Page, E. Iwaniczko y Q. Wang, *Journal of Applied Physics*, vol. 110, p. 114502, 2011.

[7] R. Varache, J. Kleider, W. Favre y L. Korte, *Journal of Applied Physics*, vol. 112, p. 123717, 2012.

[8] J. Kleider, Y. Soro, R. Chouffot, A. Gudovskikh, P. Roca i Cabarrocas, J. Damon-Lacoste, D. Eon y P. Ribeyron, *Journal of Non-Crystalline Solids*, vol. 354, p. 2641, 2008.

[9] O. Maslova, J. Álvarez, E. Gushina, W. Favre, M. Gueunier-Farret, A. Gudovskikh, A. Ankudinov, E. Terukov y J. Kleider, *Applied Physics Letters*, vol. 97, p. 252110, 2010.

[10] A. Burgers, *Progress in Photovoltaics Research and Applications*, vol. 7, p. 457, 1999.

[11] R. Barrio, Tesis: Células Solares de Heterounión de Silicio, Madrid: CIEMAT, 2008.

[12] W. Van Sark, L. Korte and F. Roca, Physics and Technology of Amorphous Crystalline Heterostructure Solar Cells, Springeer, 2012.

[13] R. Sinton and A. Cuevas, *Applied Physics Letters*, vol. 69, p. 2510, 1996.

[14] S. Sze, Physics of Semiconductor Devices, New York: Wiley, 1981.

Optimal surface texturing in highly dilute koh solution: a comparison with plasma texturing for thin silicon solar cells with light trapping

Ali T. Hajjiah*, Olfat T. Hamdan
Kuwait University, College of Engineering and Petroleum, Electrical Engineering Department, P.O. Box 5969, Safat 13060-Kuwait
*Corresponding author: phone: +965 24985829, fax: +965 24817451, e-mail: ali.hajjiah@ku.edu.kw

ABSTRACT: It is shown that alkaline texturing in a process does not involve RCA cleaning consisting of pre-texturing saw damage removal (SDR) for 10 min in 30% KOH followed by texturing in 5%wt KOH for 30min at 85°C results in a minimum average front surface reflectance of 8% without Anti Reflection Coating (ARC), and down to 1% with ARC. Applying such a texturing process results in an improvement of 3 mA/cm2 in the photo-generated current and in an increase of 2% (absolute) in the conversion efficiency of a 25 μm thick epitaxial silicon solar cell implementing a porous silicon Bragg reflector at the back side compared to a similar cell in which the front surface is textured by plasma etching. It is found that with a high quality epitaxial active region and good surface passivation, the efficiency of the cell with an optimally alkaline textured surface exceeds 20% while a maximum efficiency of 18.2% could be attained with a plasma textured surface. Such an advantage of the proposed alkaline texturing process is attributed to a smaller reflectance from the surface and to more efficient light trapping resulting from improved light diffusing properties.

Keywords: Alkaline silicon texturing, plasma texturing, porous silicon back reflector, surface reflectance, thin film silicon solar cells

1 INTRODUCTION

Front surface texturing is applied to silicon solar cells primarily to reduce the optical surface reflection due to the so called "second bounce", which increases the probability of coupling the incident light into silicon to more than 85% through a textured surface, which enhances the cell photocurrent and conversion efficiency [1-5]. In addition, surface texturing helps diffusing the light in the cell which is of great importance to thin cells that require enhanced optical path and light trapping to increase the probability of light absorption, and hence the photocurrent and cell efficiency [6-9]. Surface texturing by plasma etching was found inefficient in providing enough light trapping in thin silicon solar cells [10] and hence should be replaced by other surface texturing techniques.

Different methods have been demonstrated and are commonly used in texturing the surface of wafer-thick silicon solar cells. These methods are mainly based on selective anisotropic etching of the <100> Si surface using alkaline solutions of sodium hydroxide (NaOH) [11,12] or potassium hydroxide (KOH) [13,14] mixed with Iso-propyl alcohol (IPA) which helps stabilizing and controlling the nucleation of the texturing process [15]. Other etchants were also used to texture the silicon surface such as K2CO3 [16,17] and $(CH_3)_4NOH$(TMAH) [18-21]. Dry etching processes, such as plasma etching [22] and reactive ion etching [23] have also been suggested as low cost surface texturing methods applicable to industry but their performance did not reach the level achieved by wet anisotropic etching. Texturing in a mixture of KOH with Isopropyl alcohol (IPA) in water remains by far the simplest and the most cost effective surface texturing method that provides low surface reflectance and meets industrial requirements.

Alkaline anisotropic etching in KOH solution results in randomly distributed upright pyramids covering the silicon surface. The size of the pyramids, surface homogeneity and coverage depend on the etching parameters namely the solution temperature, alkaline and IPA concentrations, etching time, and finally surface preparation prior to texturing. The latter relates to removing the damage and defects created by the wire saw at the wafer surface prior to the texturing. Usually classical RCA standard cleaning, or saw damage removal (SDR) in concentrated alkaline solution are used for pre-texturing surface preparation. However, the effect of either method on the surface morphology is different which results in a distinct influence on texturing nucleation and uniformity. In the present work, the impacts of these two pre-texturing preparation treatments on the surface texturing and hence on the reflectance are compared. In a previous report [24], it was shown that the smallest surface reflectance is obtained after performing a combined pre-texturing surface treatment in which SDR treatment was followed by RCA cleaning prior to texturing in a low %wt KOH/IPA solution.

In the present work, we check the necessity of maintaining the expensive and environmentally harmful RCA pre-texture cleaning, and whether SDR surface preparation in 30% KOH alkaline solution alone prior to texturing process would be enough to ensure minimum surface reflectance. Texturing conditions leading to minimum surface reflectance are investigated and an optimal texturing process is proposed. Finally, the performance of thin silicon solar cells with such an optimally textured surface and implementing a Bragg back reflector is investigated and compared to the performance of similar cells with a plasma textured surface.

2 EXPERIMENTAL: OPTICAL TEXTURING PROCESS

The starting material consists of 4 inch circular 500 μm thick Czochralski (Cz) mirror like silicon wafers, <100> oriented and p-type (Boron doped) with resistivity of 1–5 Ω.cm. The wafers are cut into square samples (3cmx3cm) and divided into two groups. The first group will receive a SDR treatment prior to texturing. The second group will not receive the SDR but will be subject only to a complete RCA cleaning sequence.

2.1 Pre-texturing treatments

Saw damage removal (SDR) carried out on a first group of samples consists of dipping the samples in an aqueous solution of 30%wt KOH heated to 70°C in a capped vessel placed on a hot plate. The measured etch rate is about 0.75 µm/min which is slower than that of other previously reported SDR processes [25]. The samples were very well rinsed in deionized water (DI) having a resistivity of 18 MΩ and dried by blowing nitrogen. The surface morphology is investigated by means of Atomic Force microscopy (AFM) using an Agilent 4520 AFM and Scanning Electron Microscopy (SEM) using a JEOL JSM-700 IF Field Emission microscope. The surface roughness after 5 to 20 min SDR treatment measured by AFM is in the range 1.4-2 nm as by AFM image (Figure1a). On the other hand, SEM indicates that sites for hillocks or pyramidal structure formation are nucleated after SDR (Figure 2a).

A second group of samples received a standard three sequence RCA cleaning in1) piranha solution made of boiling H_2SO_4/H_2O_2 (4:1) at 95°C for 10 minutes, followed by 2) Standard Clean SC-1in a solution made in of NH_4OH/H_2O_2 (4:1) heated at 75°C for 10 minutes, then 3) Standard Clean SC-2 in a solution made of HCL/ H_2O_2 (4:1) heated at 75°C for 10 minutes. The samples were dipped a few seconds in a diluted HF to remove any silicon dioxide (SiO_2) formed during the cleaning. Finally the samples were very well rinsed in DI water and dried. The average surface roughness estimated by AFM amounts to 0.15nm (Figure 1b) which is ten times smaller than that of samples pre-treated by SDR as confirmed by the SEM images (Figure 2a and Figure 2b)

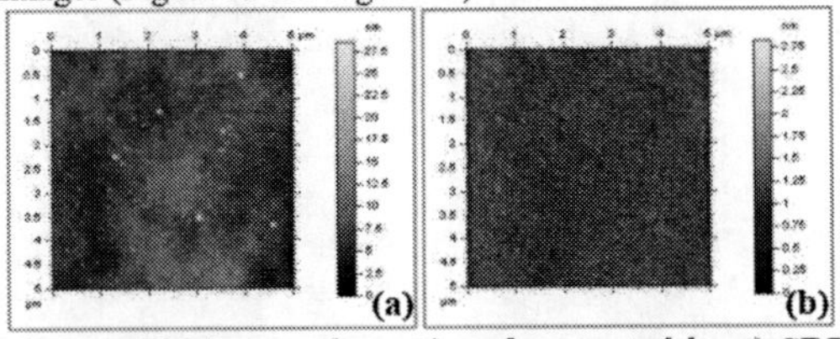

Fig. 1. AFM images for c-Si surface treated by a) SDR and b) ARC cleaning.

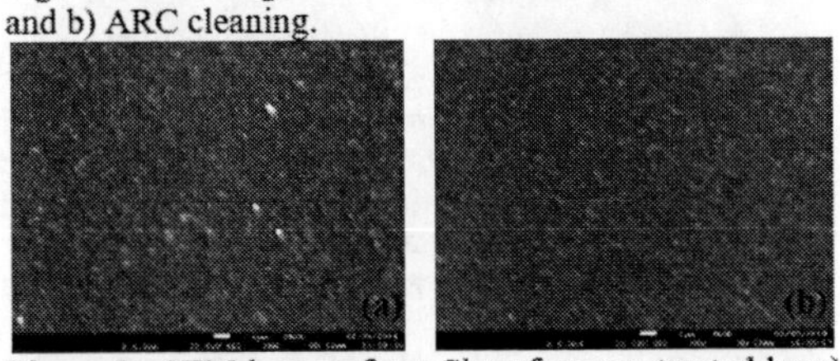

Figure 2. SEM images for c-Si surface pre-treated by a) SDR and b) RCA cleaning.

2.2 Surface texturing

The samples of both SDR and RCA groups are textured in an aqueous solution made of KOH pellets dissolved in heated DI water with the presence IPA. Three KOH concentrations 1%, 3% and 5% by weight are investigated. The texturing solution temperature was varied in the range 70°C to 85°C in a step of 5°C. The solution temperature and the concentration of the KOH solution are maintained constant throughout the complete texturing cycle. The IPA concentration is fixed at 6% of the water volume. The texturing time was varied from 10 min to 40 min in steps of 10 min while maintaining magnetic stirring. After texturing the samples were rinsed in DI water and dried by blowing nitrogen. Some samples went through thermal oxidation in dry oxygen at 1000°C for 150 minutes to grow 115 nm of silicon SiO_2 to be used as ARC.

2.3 Determination of the optimum texturing conditions

The total surface spectral reflectance of the textured samples is measured in the wavelength range 200-1200 nm using an integrated sphere spectrophotometer SolidSpe-3700. The spectral reflectance of a calibrated crystalline silicon polished surface is used as a reference. The optimum texturing process is determined from the combination of parameters that leads to the minimum surface reflectance at a wavelength λ = 700 nm at which the AM1.5G solar spectrum is close to its maximum irradiation. These parameters include KOH concentration, texturing time and solution temperature.

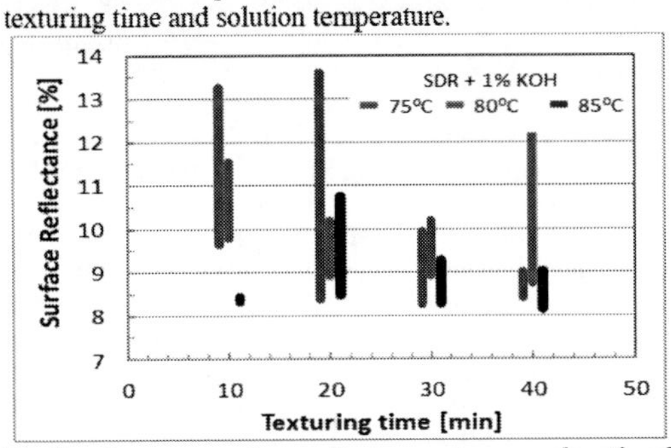

Fig.3a. Average surface reflectance at a wavelength of 700 nm vs. texturing time in 1%wt KOH solution at 75°C, 80°C, 85°C, for samples which received 10 min SDR.

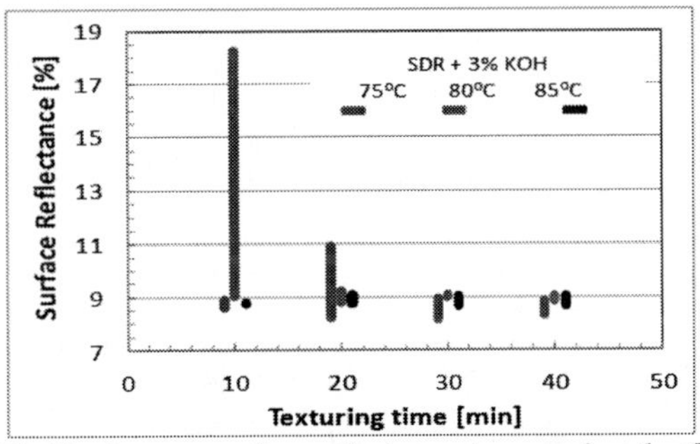

Fig.3b. Average surface reflectance at a wavelength of 700 nm vs. texturing time in 3%wt KOH solution at 75°C, 80°C, and 85°C, for samples which received 10 min SDR

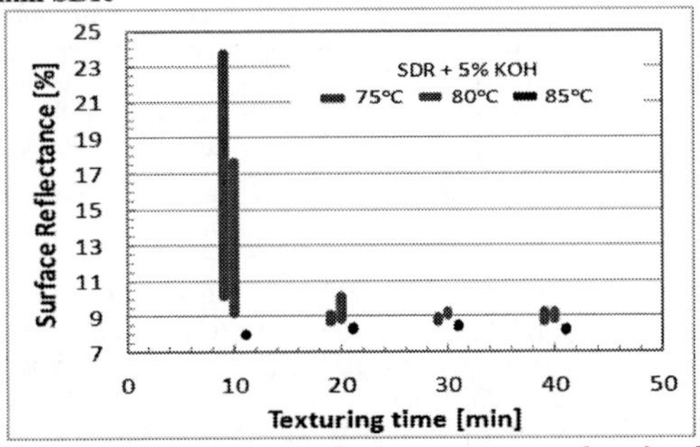

Fig.3c.Average surface reflectance at a wavelength of 700 nm vs. texturing time in 5%wt KOH solution at 75°C, 80°C, and 85°C, for samples which received 10 min SDR.

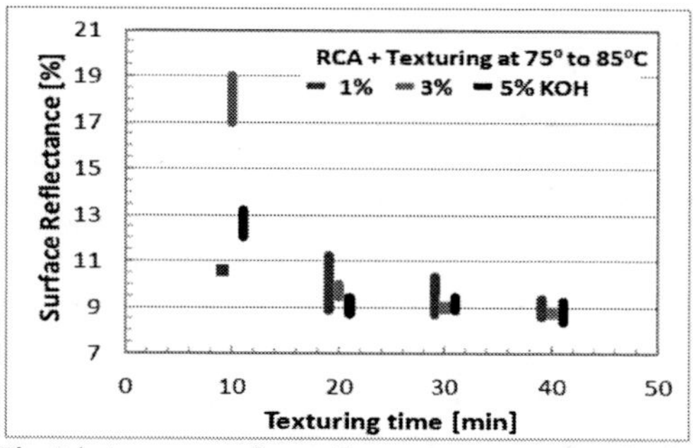

Fig. 3d. Average surface reflectance at a wavelength of 700 nm vs. texturing time in 1%, 3% and 5%wt KOH solution in the temperature range 75°C-85°C, for samples which received an RCA cleaning prior to texturing.

The surface reflectance for different texturing carried out in aqueous solutions with a KOH concentration of 1%, 3% and 5% on samples with SDR pre-texturing treatments show that the minimum reflectance occurs when the texturing temperature and duration lies in the range 75°C-85°C and 20-40 min, respectively (Figure 3a, Figure 3b, Figure 3c). The reflectance after 10 min texturing or at temperature smaller than 75°C is larger which indicates that the texturing is incomplete. Texturing SDR samples in 1% KOH solution results in a large non-uniformity in the textured pattern which leads to a large reflectance, (Figure 3a). Uniformity is established when texturing is carried out in 3% KOH solution at 80°C and 85°C resulting in an average minimum reflectance of approximately 9% (Figure 3b). The optimal texturing procedure is defined for SDR samples textured in 5% KOH at 85°C for 30 min, which results in a minimum reflectance of 8% (Figure 3c). Such a minimum reflectance represents an improvement compared to the 10% minimum reflectance previously reported for other texturing schemes using low KOH/IPA concentrations and potassium silicate additives [26]. It is advised to pursue the texturing for 30 min to ensure that minimum reflectance is reached. Finally, RCA samples textured in 1%, 3% and 5% KOH solution show an average minimum reflectance1-2% larger than the reflectance of SDR samples (Figure 3d), which can be attributed to the difference in starting surface roughness observed earlier (Figure 1 and Figure 2).

As mentioned above, texturing time shorter than 20 min may be sufficient to achieve the minimum surface reflectance (Figure 3c). The size of the pyramids covering the surface, however, grows from 3-5 μm after 10 min texturing to 8-10 μm after 40 min texturing, as confirmed by SEM images (Figure 4). The average minimum reflectance, however, is constant within the range of texturing time 20-40 min (Figure 3c) which means that the pyramid size in the range 3-10 μm has little effect on the surface reflectance as long the pyramids cover the whole surface, which is in agreement with what is reported in [26]. The reflectance starts to increase again after 40 minutes of texturing which could be considered a beginning of deviation from the optimum texturing conditions.

The weighted average reflectance (WAR) for optimally textured samples in the wavelength range 300 nm to 1000 nm with respect to AM1.5G solar spectrum is estimated to 10.8%, which is 1% smaller than recent results reported for similar texturing in aqueous solutions with 1.4 to 2.5% wt KOH with silicate (K_2SiO_2) additives [26]. Finally, the minimum reflectance decreases to less than 1% when a 115 nm thick SiO_2 ARC layer is applied on top of the optimally textured surface (Figure 5).

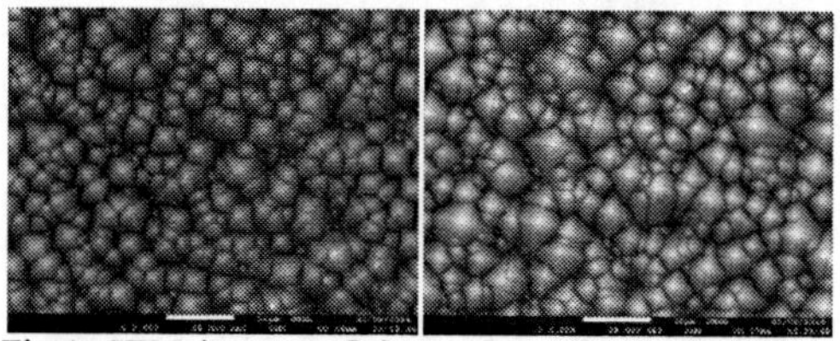

Fig.4. SEM images of the surface of samples textured following the optimal texturing procedure in 3%KOH/IPA at 85°C for a) 10 min ,b) 40 min (Resolution X=2,000).

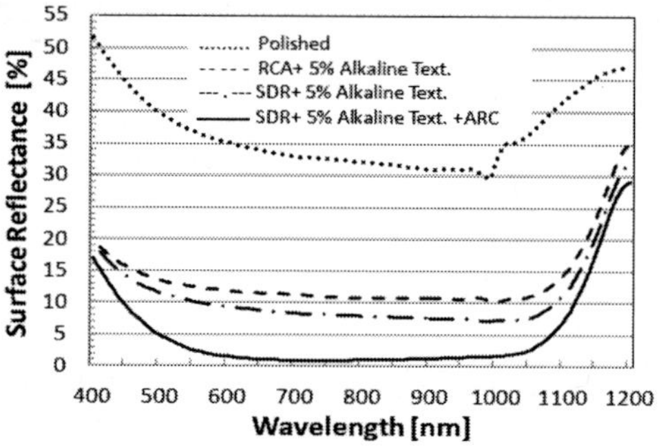

Fig.5. Surface reflectance of 1) Reference crystalline silicon polished surface, 2) Optimal texturing procedure with 10 min SDR, 3) Optimal texturing procedure with RCA cleaning, 4) Optimal texturing procedure with 10 min SDR and 115 nm SiO_2 ARC.

3 IMPACT OF OPTIMAL ALKALINE SURFACE TEXTURING ON THE PERFORMANCE OF THIN SILICON CELLS WITH POROUS SILICON BRAGG REFLECTOR

The main purpose of the present study is to investigate the merits of implementing the relatively cheap and clean texturing process, free of acid waste and with low KOH concentration, proposed here on light trapping in thin n+/p junction c-Si solar cells (tens of microns thick) with Bragg back reflector. The performance of such a cell will be compared to that of a similar cell in which the front surface is textured by plasma dry etching. A relatively recent study [10] revealed that the optical properties of the latter are modest with large surface reflectance and poor light diffusion properties, which results in relatively poor light trapping in the thin active layer and to only 8% boost in the photocurrent. In addition, it is found that enhanced front surface recombination caused by defects generated during plasma texturing is a mjor reason for the degraded cell performance. In the present section, the same ray tracing analysis and simulations proposed in [27] and detailed in [10] will be applied here to investigate the enhancement in light trapping expected when the optimal alakline surface texturing process proposed here is implemented.

3.1 Cell structure

The cell to be studied is proposed in [10] consists of a 25 μm thick p-type c-Si epitaxial layer with a resistivity equal to 0.1 Ω cm grown on a porous silicon back reflector (BR). To investigate the cell performance a diffused emitter region consisting of a shallow n+ Gaussian diffusion with a front surface concentration ND=4 x1019 cm-3 and a depth factor is 0.2 µm resulting in a sheet resistance of 80 Ω/sq is implemented. In such a thin cell, light trapping is relatively ensured by the enhanced optical path due to light diffusion at entry and by back reflection. It has been reported that a maximum equivalent back reflection of 40% can be obtained with a Bragg back reflector made of a stack of porous silicon layers having alternating high/low porosity layers. With high porosity layers 240 nm thick with a porosity of 55%, and with low porosity layers 160 nm thick with a porosity of 22.5%, five layers in the stack would provide 99% of the maximum current that could be reached with this type of back reflectors [10]. Therefore, these conditions are assumed in the device considered here. The performances of both types of cells: 1) with optimal alkaline surface texturing and 2) with plasma surface texturing, are compared.

3.2 Device simulation

The optical performance of the thin silicon cell including the external quantum efficiency, spectral photo-generated current, and finally the integrated photo-generated current Jph over the whole solar spectrum wavelength range, are calculated according to a ray-tracing model formulated in a MATLAB code developed in-house [10,27], and assisted by the PC1D one dimensional device simulations [28]. The ray tracing model treats specular and diffuse reflectance and transmission at the entry and exit through the front surface, as well as reflection at the back interface. The front surface escape reflectance includes two components: 1) direct surface reflectance, extracted experimentally from reference wafers with no back reflection, and 2) light escaping through the front surface and emanating from reflection at the back interface. The spectral reflectance measured at the surface of a reference wafer with optimally alkaline textured surface [this work] is significantly smaller than the measured reflectance reported in [22] for a reference wafer with a plasma textured surface (Figure 6), which confirms the superiority of the optical performance of the former. These data are provided as input data to the simulation software in addition to the parameters of the porous silicon layers and of the active epitaxial silicon layer. All these parameters can be used as independent variables for the sake of optimization. The default values of the electron diffusion length in the epitaxial layer is equal to 30 µm which reflects the quality of the material, the hole diffusion length in the n-type diffused emitter is equal to of 2 µm, and the recombination velocity at the front surface and at the back interface is equal to 104 cm/s and 1000 cm/s, respectively. These values describe the practical collection probability. The ideal cell performance is also estimated at unity collection and compared with the practical situation.

Moreover, the software includes a model that calculates the refractive index of porous silicon for each porosity which leads to an accurate determination of the back reflection at the porous silicon stack. More details of the models used in the MATLAB simulations calculations can be found in [10,27].

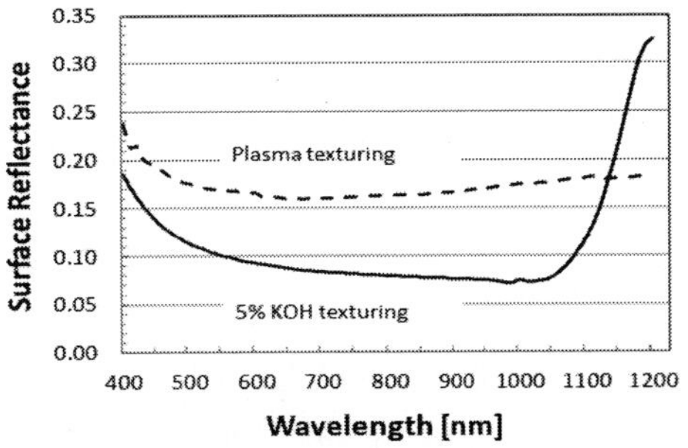

Fig.6. Experimental front surface reflectance as a function of wavelength for alkaline textured c-Si wafer and for plasma textured surface. The curve for plasma texturing is taken from [22].

In parallel with MATLAB simulations essentially carried out to determine the optical performance of the cells, the one dimensional simulator PC1D [28] is used to determine the cell parameters including the photocurrent, open circuit voltage and Fill factor. AM1.5G global solar spectrum in the wavelength range 300-1200 nm and total incident power of 0.1 W/cm2 is considered in the simulations using either program. For proper comparison the number of light bounces at the back surface is adjusted such that the total reflectance calculated by means of PC1D matches that determined using the MATLAB code. It turns out that back reflection provided by the porous silicon stack is equivalent to a lumped back reflectance percentage equal to 40%. The short circuit currents calculated using either program are compared for checking the accuracy of the results. Finally, the simulations are carried out at 300K.

3.3 Results of cell simulations

3.3.1 MATLAB simulations: escape reflectance, external quantum efficiency and spectral photocurrent

The escape front reflectance and External Quantum Efficiency (EQE), and the spectral photogenerated current (Jph(λ)) are determined using MATLAB simulations for the cell described above(Figure 7a and Figure 7b). The EQE and the spectral photogenerated current of the cell with alkaline textured surface are larger than for the cell with plasma textured surface, which is expected and is partially due to the smaller surface reflectance and hence larger light absorption in the former (Figure 6). Furthermore, the calculated escape reflectance is larger than the measured direct reflectance of the reference wafers at long wavelengths due to the light reflected from the back surface bouncing back to the front (Figure 7a and Figure 7b).The enhancement, however, is less significant in the alkaline textured cell than in the plasma textured cell due to better light diffusing properties of the former resulting in longer optical path prior to back reflection.

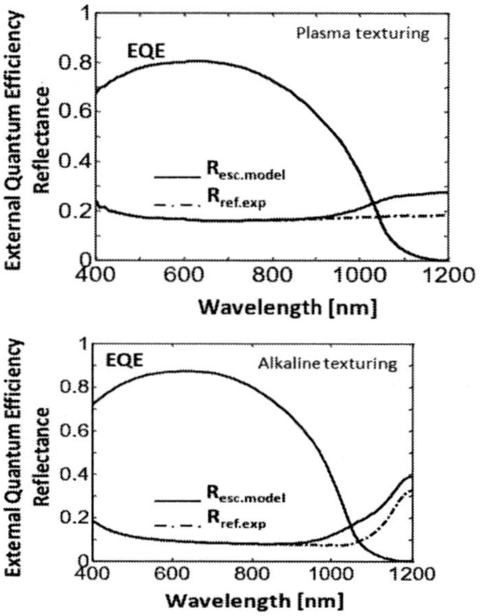

Fig.7a. Escapefront reflectance (Resc.model) and External Quantum Efficiency (EQE) calculated using MATLAB for cells with a porous Si stack BR and with surface textured: upper: by plasma etching, and lower: by optimum 5% alkaline texturing. The experimental Rref.exp is that of a reference sample with no back reflection.

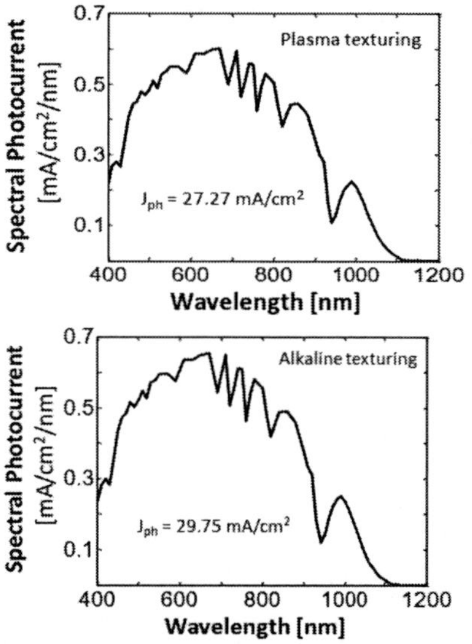

Fig.7b. Photo-generated current calculated using MATLAB for cells with a porous Si stack BR and with surface textured: upper: by plasma etching and lower: by optimum 5% alkaline texturing.

For the practical collection probability defined above, the total photogenerated current J_{ph} (short circuit current J_{sc}) of the alkaline textured cell and of the plasma textured cell with a base width of 25 µm is equal to 29.75 mA/cm2 and 27 mA/cm2, respectively, as listed in Table I. The current in the plasma textured cell increases due to back reflection at the porous silicon stack by 2 mA/cm2 which represents 8%, while the increase in the alkaline textured cell represents only 7% which is due to the longer optical path in the latter confirmed by the escape reflectance results (Figure 7a and Figure 7b). On the other hand, at unity collection probability, defined here by a diffusion length much longer than the layer width (400 µm) and negligible front and back surface recombination velocities Sf and Sb = 1 cm/s, Jph increases to 32.75 mA/cm2 for alkaline textured cell and to 30 mA/cm2 for plasma textured cell.

Assuming practical collection probability, J_{ph} increases with increasing epitaxial layer (active base width) thickness up to xepi ≈ 25 µm, then decreases with further increase in the layer thickness due to enhanced bulk recombination (Figure 8). On the other hand, at unity collection bulk recombination does not limit the collection of the photocurrent, which leads to a continuous increase in the photocurrent with increasing epitaxial layer thickness (Figure 8).

Finally, as expected, the improvement in the cell current is more significant in thin cells, especially with a thickness smaller than 20 µm (Figure 9). On the other hand, as deduced from Table I assuming practical collection probability, the improvement in the photogenerated (Jph) ensured by alkaline texturing with respect to plasma texturing is practically constant (2.7-2.8 mA/cm2), though slightly decreasing as we go from No back reflection (BR) to partial back reflection to ideal back reflection.

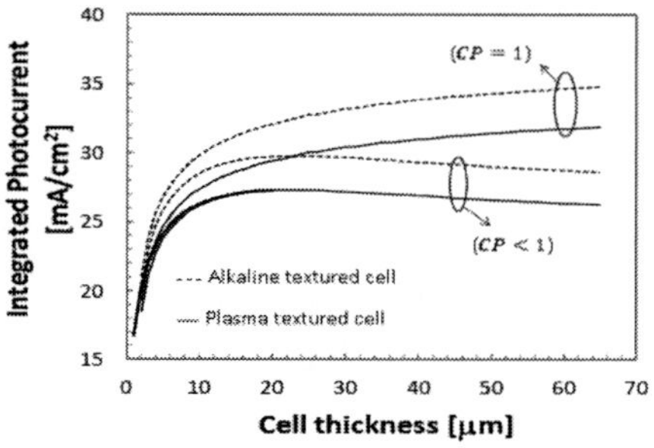

Fig.8. Photo-generated current, Jph, as a function of epitaxial layer thickness, xepi for cells with a plasma textured surface and 5% alkaline textured surface for collection probability (CP): 1) smaller than unity with Ln=30 µm, Sf=104 cm/s and Sb=103 cm/s, and 2) Unity emulated with Ln>400 µm and Sf=Sb=1 cm/s.

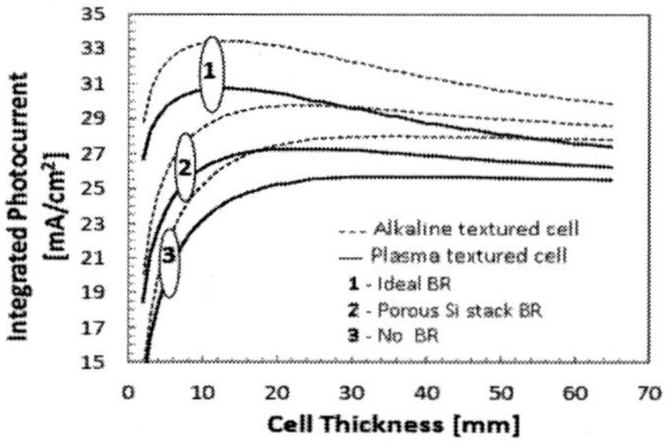

Fig.9. Photocurrent in cells with: 1) Ideal back reflector, 2) porous silicon stack back reflector, 3) No BR.

3.3.2 PC1D simulation: cell efficiency

The values of Jph determined from PC1D simulations are listed in Table II. These are a bit smaller than those estimated from MATLAB calculations due to recombination in the bulk and at the surface of the heavily doped emitter as well as in the pn junction depletion region. These emitter junction recombination terms are ignored in the MATLAB simulations. The results of PC1D simulations, however, still confirm that Jph is 2.1 to 2.6 mA/cm2 larger in the alkaline textured cell than in the plasma textured cell. The cell parameters listed in Table II are determined for textured surface cells with no ARC and assuming relatively severe recombination conditions at the front and back surfaces, which justifies the low efficiencies listed in Table II. With high quality epitaxial layer and improved surface passivation, however, it is possible to achieve a much larger collection probability close to unity. In that case, the PC1D results listed in Table III obtained assuming an equivalent back reflection of 40% indicate that efficiency of the alkaline textured cell can boosted from 15.3% to 18.8%, and hence to 20.1% if ARC is applied, while the efficiency of the plasma textured cell with ARC would raise to 18.2%. Finally, for the hypothetical case of perfect light trapping with 100% back reflection in the 25 µm thick cell, the photo-generated current of the alkaline textured cell would be boosted to 39 mA/cm2 bringing its efficiency close to 22.6%, while the maximum efficiency of the plasma textured cell will be limited to 20%. These results are to be compared with Jph = 42 mA/cm2 and 25% cell efficiency achieved with optimized PERL cell employing high quality material, optimum front and back surface passivation, as well as advanced sophisticated front surface texturing using inverted pyramids [29].

Table I. Results of MATLAB photocurrent simulations for cells with plasma and alkaline textured surfaces (epitaxial layer thickness (x_{epi})=25 µm).

Type of Back Reflection	J_{ph} (mA/cm^2) Plasma texturing	J_{ph} (mA/cm^2) Alkaline texturing	J_{ph} (mA/cm^2) Plasma texturing	J_{ph} (mA/cm^2) Alkaline texturing
	Practical Collection		100% Collection	
No BR	25	27.83	27	30
Porous (40%)	27	29.75	30	32.75
Ideal (100%)	30	32.7	33.4	36.4

Table II. Cell parameters resulting from PC1D simulations for cells with plasma and alkaline textured surfaces (x_{epi} =25 µm). Practical collection probability with S_f =104 cm/s, S_b =103 cm/s and Ln=30 µm is assumed.

Back reflector BR	J_{ph} mA/cm^2	V_{oc}(mV)	FF(%)	η (%)
	Plasma Texturing			
No BR	24.9	640.4	82.8	13.2
Porous stack 40%	26.3	641.6	82.4	13.9
Ideal 100%	28.9	644.1	83.3	15.5
	Alkaline Texturing			
No BR	27	642.3	83	14.4
Porous stack 40%	28.6	643.8	83.1	15.3
Ideal 100%	31.5	646.3	82.5	16.8

Table III. Cell parameters resulting from PC1D simulations for cells with alkaline textured surface (x_{epi} =25 µm) assuming unity collection (No ARC)

	J_{ph} mA/cm^2	Voc(mV)	FF	η (%)
No BR	29.9	696.9	83.5	17.4
Porous stack 40%	32.75	698.6	82.2	18.8
Ideal 100%	36.4	701.5	82.6	21.1

4 CONCLUSION

An optimal silicon surface texturing process is proposed and consists of saw damage removal (SDR) carried out for 5-20 min in 30% KOH directly followed by texturing in 3-5%wt KOH for 30-40 min at a temperature of 80°-85°C. The minimum average reflectance measured using spectrophotometry for a silicon wafer subjected to such a texturing process amounts to 8% without Anti-Reflection-Coating which decreases to 1% when a 110 nm of SiO_2 layer is used as an ARC. This reflectance is smaller than that obtained for a surface that received a standard RCA surface cleaning prior to texturing. Atomic Force Microscopy (AFM) and Scanning Electron Microscopy (SEM) indicate that the reflectance remains higher than the minimum value as long as texturing does not cover the whole sample surface. On the other hand, it is found that pyramid size in the range 3 µm to 10 µm has little impact on the surface reflectance as long the pyramids cover the whole surface. Applying such a texturing scheme to the surface of thin (25 µm) epitaxial layer silicon solar cells in which a porous silicon stack back reflector is implemented results in better light trapping than texturing the surface by plasma etching. The improvement amounts to 3 mA/cm2 in photo-generated current and to 2% in absolute cell efficiency. In addition, if the quality of the epitaxial active silicon layer is high and the surface is well passivated, the efficiency of such a thin cell with optimal alkaline surface texturing exceeds 20% compared to 18.2% for a similar cell with plasma texturing that could be increased to 22.6% and 20%, respectively, when a proper ARC is applied. Such an advantage is attributed to smaller reflectance and to better light diffusing properties of the former.

5 ACKNOWLEDGMENTS

This work is supported by Kuwait Foundation for the advancement of Sciences under project number P115-15EE-01. Experimental work is carried in the semiconductor research facility GE01/08. SEM characterization is done in the nanotechnology research facility GE01/07.

6 REFERENCES

[1] L. Forbes, "Texturing, reflectivity, diffuse scattering and light trapping in silicon solar cells," Solar Energy, vol.86, pp. 319–325, 2012.

[2] M. Green, "Crystalline and thin-film silicon solar cells: state of the art and future potential," Solar Energy, vol.74, pp. 181–192, 2003.

[3] E. Manea, E. Budianu, M. Purica, D. Cristea, I. Cernica, R. Muller and V. M. Poladian, "Optimization of front surface texturing processes for high-efficiency silicon solar cells," Solar Energy Materials and Solar Cells, vol.87, pp. 423–431, 2005.

[4] Z. Q. Xi, D. R. Yang, W. Dan, C. Jun, X. H. Li, and D. L. Que, "Investigation of texturization for monocrystalline silicon solar cells with different

kinds of alkaline," Renewable Energy, vol.29, pp. 2101–2107, 2004.

[5] Q. B. Vu, D. A. Stricker and P. M. Zavracky, "Surface characteristics of (100) silicon anisotropically etched in aqueous KOH," Journal of the Electrochemical Society, vol. 143, pp. 1372-1375, 1996.

[6] R. A. Arndt, J. F. Allison, Jr. A. Meulenberg and J. G. Haynos, "Optical properties of the COMSAT non-reflective cell," Eleventh Photovoltaic Specialists Conference, vol. 1, pp. 40-43, 1975.

[7] A. Goetzberger, "Optical confinement in thin solar cells by diffuse back reflectors," Proceeding of the fifth IEEE Photovoltaics Specialists Conference, vol. 1, pp. 867-870, 198.

[8] E. Yablonovitch, "Statistical ray optics," Journal of the Optical Society of America, vol. 72,pp. 899-907, 1982.

[9] P. Campbell and M. A. Green, "Light trapping properties of pyramidally textured surfaces," Journal of Applied Physics vol. 62, pp. 243, 1987.

[10] M. Y. Ghannam, A. A. Abouelsaood, A. S. Alomar and J. Poortmans, "Analysis of thin-film silicon solar cells with plasma textured front surface and multi-layer porous silicon back reflector," Solar Energy Materials and Solar Cells, vol. 94, pp. 850–856, 2010.

[11] P. K. Singh, R. Kumar, M. Lal, S. N. Singh and B. K. Das, "Effectiveness of anisotropic etching of silicon in aqueous alkaline solutions," Solar Energy Materials and Solar Cells, vol. 70, pp. 103-113, 2001.

[12] S. Tietun, C. Dong and C. Rongqiang, "Surface texturing of crystalline silicon and effective area measurement," Proceeding of SPIE, vol. 4086, pp. 116-119, 2000.

[13] J. W. Müller, A. Metz, and R. Hezel, "A new and simple approach for fabricating inverted pyramids on crystalline silicon solar cells," Proceedings of the 17th European Photovoltaic Solar Energy Conference and Exhibition, pp.347, 2001.

[14] D. L. King and M. E. Buck, "Experimental optimization of an anisotropic etching process for random texturization of silicon solar cells," In Photovoltaic Specialists Conference, Conference Record of the Twenty Second IEEE: 303-308, 1991.

[15] P. Verlinden, O. Evrard, E. Mazy and A. Crahay, "The surface texturization of solar cells: a new method using V-grooves with controllable sidewall angles," Solar Energy Materials and Solar Cells, vol. 26(1), pp. 71-78, 1992.

[16] T. Lauinger, J. Schmidt, A. G. Aberle and R. Hezel, "Record low surface recombination velocities on 1 Ω cm p- silicon using remote plasma silicon nitride passivation," Applied Physics Letters, vol. 68(9), pp. 1232-1234, 1996

[17] R. Chaoui, M. Lachab, S. Chiheub, and N. Seddiki, "Texturization of monocrystalline silicon with potassium carbonate solutions," Proceedings of the 14th European Photovoltaic Solar Energy Conference, Barcelona, pp. 812-814, 1997.

[18] O. Tabata, R. Asahi, H. Funabashi, K. Shimaoka and S. Sugiyama, "Anisotropic etching of silicon in TMAH solutions," Sensors and Actuators A: Physical vol. 34, pp. 51-57, 1992.

[19] L. M. Landsberger, S. Naseh, M. Kahrizi, and M. Paranjape, "On hillocks generated during anisotropic etching of Si in TMAH," Journal of microelectromechanical systems, vol. 5(2), pp. 106-116, 1996.

[20] J. S. You, D. Kim, J. Y. Huh, H. J. Park, J. J. Pak and C. S. Kang, "Experiments on anisotropic etching of Si in TMAH," Solar Energy Materials and Solar Cells, vol. 66(1), pp. 37-44, 2001.

[21] P. Papet, O. Nichiporuk, A. Kaminski, Y. Rozier, J. Kraiem, J. F. Lelievre, A. Chaumartin, A. Fave and M. Lemiti, "Pyramidal texturing of silicon solar cell with TMAH chemical anisotropic etching," Solar Energy Materials and Solar Cells, vol. 90(15), pp. 2319-2328, 2006.

[22] F. Duerinckx, K. Van Nieuwenhuysen, H. J. Kim, I. Kuzma-Filipek, G. Beaucarne, and J. Poortmans, "Light trapping for epitaxial thin film crystalline silicon solar cells," In 20th European Photovoltaic Solar Energy Conference, pp. 1190-1193, 2005.

[23] H. L. Chen, W. Fan, C. C. Cheng, C. H. Lin and K. T. Huang, "Fabrication of texturing antireflection structures in solar cells by using the defocusing exposure in optical lithography," Journal of the Electrochemical Society, vol. 153(9), pp. G802-G806, 2006.

[24] A. T. Hajjiah, S. K. Zachariah and M. Y. Ghannam, "Study of surface morphology and optical characterization of crystalline and multi-crystalline silicon surface textured in highly diluted alkaline solutions," Journal of Engineering Research, vol. 2, 2014.

[25] P. K. Basu, H. Dhasmana, N. Udayakumar, F. Khan and D. K. Thakur, "Regulated low cost pre-treatment step for surface texturization of large area industrial single crystalline silicon solar cell," Solar Energy Materials and Solar Cells, vol. 94(6), pp. 1049-1054, 2010.

[26] P. K. Basu, A. Khanna, Z. Hameiri, "The effect of front pyramid heights on the efficiency of homogeneously textured inline-diffused screen-printed mono-crystalline silicon wafer solar cells," Renewable Energy, vol. 78, pp. 590-598, 2015.

[27] M. Y. Ghannam, A. A. Abouelsaood, I. K. Filipek, F. Duerinckx, J. Poortmans, "Optical modeling of capped multi-layer porous silicon as a back reflector in thin-film solar cells," In Photovoltaic Energy Conversion, Conference Record of the

IEEE 4th World Conference (Hawaii), vol. 2, pp. 1362-1364, 2006.

[28] D. Clugston and P. Basore, "PC1D version 5: 32-bit solar cell modeling on personal computers," Photovoltaic Specialists Conference, Conference Record of the Twenty-Sixth IEEE, pp. 207-210, 1997.

[29] M. Green, K. Emery, Y. Hishikawa, W. Warta and E. Dunlop, "Solar cell efficiency tables," (version39) Progress in Photovoltaics: Research and Applications, vol. 20(1), pp. 12-20, 2012.

THIN IBC C-SI SOLAR CELLS BASED ON CONVENTIONAL TECHNOLOGIES

C. Jin, I. Martín, E. Calle, P. Ortega, G. López, R. Alcubilla
Departament d'Enginyeria Electrònica, Universitat Politècnica de Catalunya, C/Jordi Girona 1-3, Mòdul C4, 08034 Barcelona, Spain. Tel.: +34-93-4054193; fax: +34-93-4016756; E-mail: chen.jin@upc.edu

ABSTRACT: Reducing the wafer thickness for c-Si solar cell fabrication is an effective approach for cost-savings. Interdigitated Back Contacts (IBC) technology is a promising candidate to be applied to thin c-Si substrates due to its potential to facilitate thin device processing: thin c-Si solar cell could be processed attached to a glass with its rear surface, where all the contacts are to be defined, still accessible. The understanding of thin IBC c-Si solar cell performance has great significance in guiding the design of such devices. In this work, we explore the performance of thin IBC c-Si solar cells by both 3D TCAD device simulations and experimental fabrication based on conventional technology developed in our research group. On one hand, an optical model is proposed using 2D ray tracing method whose results are applied to 3D ATLAS TCAD simulator which is used to simulate the electrical performance of the devices. As a result, efficiencies in the range of 20 % are expected for substrates of 10-20 μm without changing our technology. On the other hand, a ~30 μm IBC solar cell is fabricated by thinning down a previously developed IBC c-Si solar cell on conventional thick high quality substrates demonstrating a 12.1% of efficiency. The front surface passivation provided by Al_2O_3 is deduced by comparing with the simulation results revealing a front surface recombination velocity of about 1500 cm/s.
Keywords: IBC; Thin silicon solar cell; etching; simulation

1 INTRODUCTION

Crystalline silicon (c-Si) wafer technology has dominated the PV market for decades, and it will probably do in the near future. This dominant position is based on a constant cost-reduction strategy applied throughout all the value chain. One of the most important trends is the reduction of the substrate thickness, since c-Si wafer accounts for nearly half of today's cell price [1]. Different approaches to thin c-Si solar cells could be found in the literature. For example, efficiencies of 19.1 %, 16.8 % and 7.3 % have been reported for 43, 18 and 5.5 μm-thick double side contacted solar cells, respectively [2][3][4]. As it can be seen, slimming down the substrates reduces the efficiency mainly because of lower photon absorption and thus, lower short-circuit currents. In order to maintain photo-generation high, light trapping techniques must be introduced in the devices resulting very often in lower electrical performances (see for example [5]). The Interdigitated Back Contacted (IBC) structure shows all the contacts at the rear side relaxing the front surface requirements to a good surface passivation and optical properties and allowing a more flexible application of light trapping techniques.

In last years, our research group has developed high efficiency IBC c-Si solar cells with excellent results [6-7]. These devices are based on conventional technologies, namely photolithography, thermal diffusions and silicon oxide passivation, applied to thick (~300 μm) high quality substrates.

In this work, we focus on exploring the performance of IBC c-Si solar cell using this technology but applied to thin c-Si substrates. First, we develop a 3D numerical model using commercial TCAD software (Silvaco-ATLAS, Silvaco Inc., Santa Clara, CA, USA [8]) to simulate the latest p-type IBC c-Si solar cell prototype developed in our group [7]. Special attention is paid to the optical model, since light trapping characteristics play a major role in solar cell efficiency when the c-Si substrate is slimmed down. The objective of these simulations is to determine the maximum potential efficiencies reachable with our current IBC technology on thin c-Si substrates. In the second part of the work, we report on an experimentally fabricated thin IBC c-Si solar cell. The thickness of ~30 μm is reached by Si etching of a prototype cell fabricated on thick substrate through front surface. After thickness reduction, the front surface is processed again to recover surface passivation and optical properties: random pyramids are defined on the front surface and passivated by an Al_2O_3 film. Despite this approach of cell fabrication on thick c-Si wafer followed by a subsequent substrate thickness reduction is not optimal from the point of view of material utility, it allows us to check out in real devices the conclusions deduced from simulations.

2 DEVICE SIMULATION

The simulation reproduces a 2.5×2.5 cm p-type IBC c-Si prototype solar cell, whose main features are stripe-like n+ and p+ diffusions with square-like n++ region under the emitter contacts (see Fig. 2, (b, c)). As it can be observed, the repeating unit of the cell is a cube with anode and cathode located at the edges of one long side on the rear surface, which is extracted as the simulation unit. The length and width of the unit is fixed to 200 μm and 100 μm, while substrate thickness is set as a variable ranging from 280 to 5 μm.

In order to reduce the complexity of the simulations, we perform the optical simulation using the wafer ray tracer provided by PV lighthouse [9]. First of all, in order to validate the calculations of the wafer ray tracer, we simulate the light absorption of a 30 μm-thick bare c-Si wafer and compare it to experimental results. The resulting curves are shown in Fig. 1 with a reasonable concordance. Once we have checked out the validity of the wafer ray tracer, we reproduce the optical properties of our solar cells applying the following procedure to adapt the results to Silvaco ATLAS.

To calculate the light absorption, we define in the ray tracer software front pyramids 3.536 μm high, 5 μm wide and with an angle of 54.78 °. These parameters are provided as default values and they accurately represent the random texturing technology of our cells. In the simulations and for the sake of clarity, this front surface random pyramids texturing is considered in all IBC cells regardless the substrate thickness. Although this technology is not feasible when very thin substrates (< 10 μm) are used, this approach allows us to compare the results underlining the effect of the front surface passivation, which is the main objective of this work. This front surface passivation is provided by a 90 nm-thick Al_2O_3 film that is also included in the optical simulations.

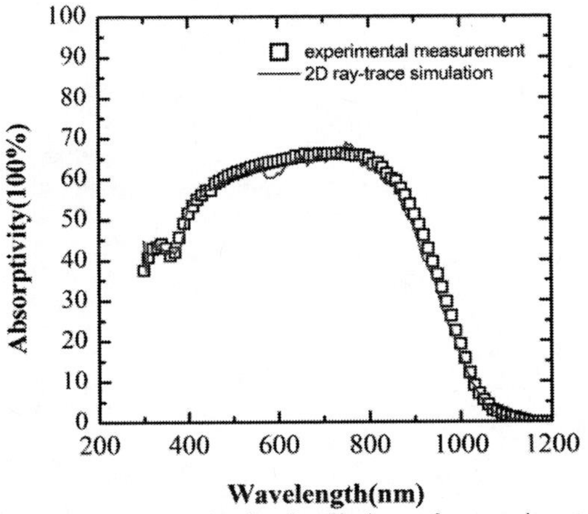

Figure 1. Simulated (line) and experimental (symbols) values of a 30 μm-thick plane c-Si substrates light absorption

Below this front surface, a c-Si substrate of different thicknesses is defined with a flat rear surface with three possible configurations: c-Si/SiO_2/Al, c-Si/Al, and c-Si/SiO_2/air corresponding to regions A1, A2 and A3 in Fig. 2, (b) respectively. The different rear configurations lead to variations in the rear internal reflectance strongly impacting photon absorption for very thin substrates. For every wavelength, the absorption with every rear surface configuration is simulated. Finally, the total absorption ($A_{raytracer}(\lambda)$) is calculated by weighting the absorption of every region by its area coverage (see the Fig. 2 caption where these data are shown).

On the other hand, we carry out similar calculations but in the simplified structure introduced into Silvaco ATLAS. In this case, flat bare c-Si surfaces are considered with Aluminum pads at the rear contacts (see Fig. 2, (a)). Then, we can calculate the total absorption of ATLAS ($A_{ATLAS}(\lambda)$) applying the same procedure explained above. Finally, the spectrum introduced to ATLAS ($S_{ATLAS}(\lambda)$) is calculated using equation (1) reproducing in Silvaco ATLAS the optical absorptance simulated by the raytracer software.

$$S_{ATLAS}(\lambda) = S_{AM1.5}(\lambda)\frac{A_{raytracer}(\lambda)}{A_{ATLAS}(\lambda)} \quad (1)$$

Regarding the electrical modeling, a detailed description of the simulated structure is included in Appendix. The most important features of the structure like doping profiles, surface recombination velocity and geometrical dimensions have been obtained from experimental characterization performed during the device development in our research group. In particular, the recombination velocity on rear surface and series resistance in simulation are validated by comparing External Quantum Efficiency (EQE), open-circuit voltage (V_{oc}) and Fill Factor (FF) with experimental results of a completed 280 μm thick prototype cell. It should be mentioned that surface recombination is modeled by transferring surface recombination velocity to carrier lifetime in a 10 nm thick region beneath every surface following the procedure reported in [10].

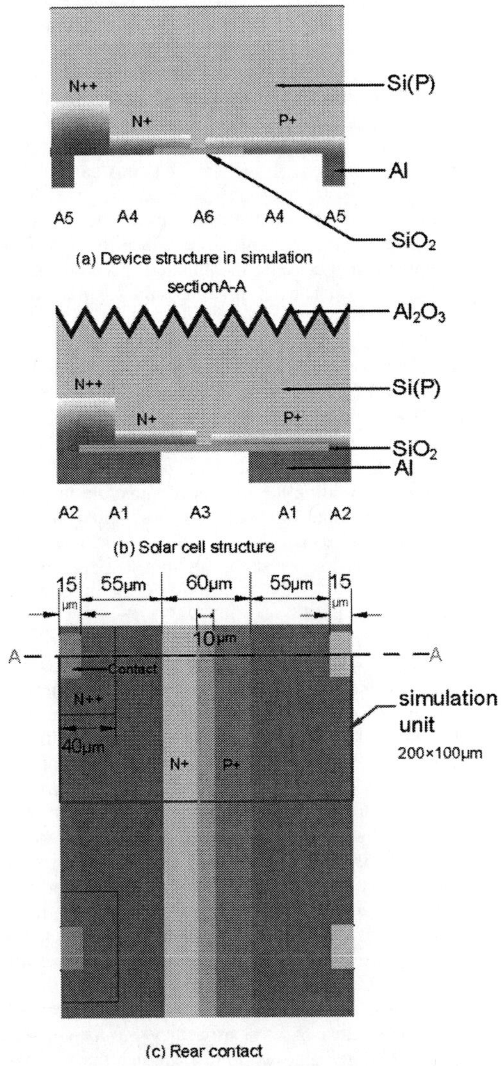

Figure 2: Schematic figure of solar cell structure: (a) Schematic figure of practical solar cell structure: A1-67.75% area [Al_2O_3-90nm/Si-textured/SiO_2-100nm/Al]; A2-2.25% area [Al_2O_3-90nm/Si-textured/Al]; A3-30% area [Al_2O_3-90nm/Si-textured/SiO_2-100nm].
(b) Schematic figure of device structure in ATLAS simulation: A4-67.75% area [Air/Si-flat/Air]; A5-2.25% area [Air/Si-flat/Al]; A6-30% area [Air/Si-flat/SiO_2-100nm]. (c) Unit region for simulation

3 SIMULATION RESULTS AND DISCUSSION

In the simulations reported in this work, we focus on the impact of front surface passivation to the solar cell performance. This parameter is crucial to collect photogenerated carriers in back contacted solar cells and it becomes even more critical when substrate thickness is reduced. Additionally, the front surface is prone to lose passivation quality when sophisticated light trapping techniques are introduced.

In figure 3, we plot the photovoltaic figures obtained from the simulations with substrate thicknesses ranging from 5 to 280 μm. Front surface recombination velocity (S_{front}) is varied while at the rear surface we reproduce the electrical characteristics of our technology. Regarding the short-circuit current (J_{sc}), front surface passivation is strongly impacting it. For well passivated front surfaces J_{sc} is reduced with substrate thickness due to the lower light absorption resulting in 24.1 mA/cm^2 for 5 μm compared to 40.7 mA/cm^2 for 280 μm. On the other hand, for highly recombining front surfaces thinner substrates are more favorable, since the carriers are photogenerated closer to the junction and more efficiently collected leading to 8.1 mA/cm^2 and 3.4 mA/cm^2 for 5 and 280 μm, respectively.

Fill Factor (FF) shows a complex dependence on the front surface passivation. For low S_{front} values, FF is limited due to the relatively high lumped resistance value 0.7 Ωcm^2 related to the metal grid combined to high J_{sc} values. Notice that in this range, higher FF values are obtained for thinner substrates. For $S_{front} \geq 10^3$ cm/s, J_{sc} decreases revealing the impact of substrate thickness on FF. Now higher FF values are obtained for thicker substrates. Interestingly, FF higher than 75 % are obtained even for substrates as thin as 5 μm indicating that the FF loss is not the limiting effect when thin c-Si substrates are used.

Focusing on V_{oc}, a weak dependence of this parameter on substrate thickness is observed resulting in similar values for all substrates except for the case of very high S_{front} values, where lower values are deduced for thinner substrates. This result indicates that, given a reasonable level of front surface passivation, no degradation of V_{oc} is expected when c-Si substrates are reduced. However, this weak dependence of V_{oc} on substrate thickness is unexpected. For example for S_{front}= 1 cm/s J_{sc} is reduced almost 17 mA/cm^2 when substrate thickness is slimmed down from 280 to 5 μm and one would expect that this change in J_{sc} results in a V_{oc} reduction of more than 13 mV, but it does not. Further research is needed to clarify this point and the results will be published elsewhere.

Finally, our simulations demonstrate that remarkable efficiencies close to 17 % are achievable for substrates in the 10-20 μm range without changing the rear surface technology.

4 THIN IBC C-SI SOLAR CELL FABRICATION AND CHARACTERIZATION

In order to experimentally check out the results, we slim down one of our finished IBC prototype solar cells to ~30 μm. The rear surface where the contacts are located is protected by a ~20 μm polyimide film to improve the operability. Then, RIE plasma etching is conducted firstly to reduce the cell thickness to about 100 μm, followed by a further wet TMAH (25% wt.) etching on the front surface. After these steps, random pyramids are defined on the front surface and passivated by a 90 nm Al_2O_3 film. Finally, a 375 °C annealing is performed

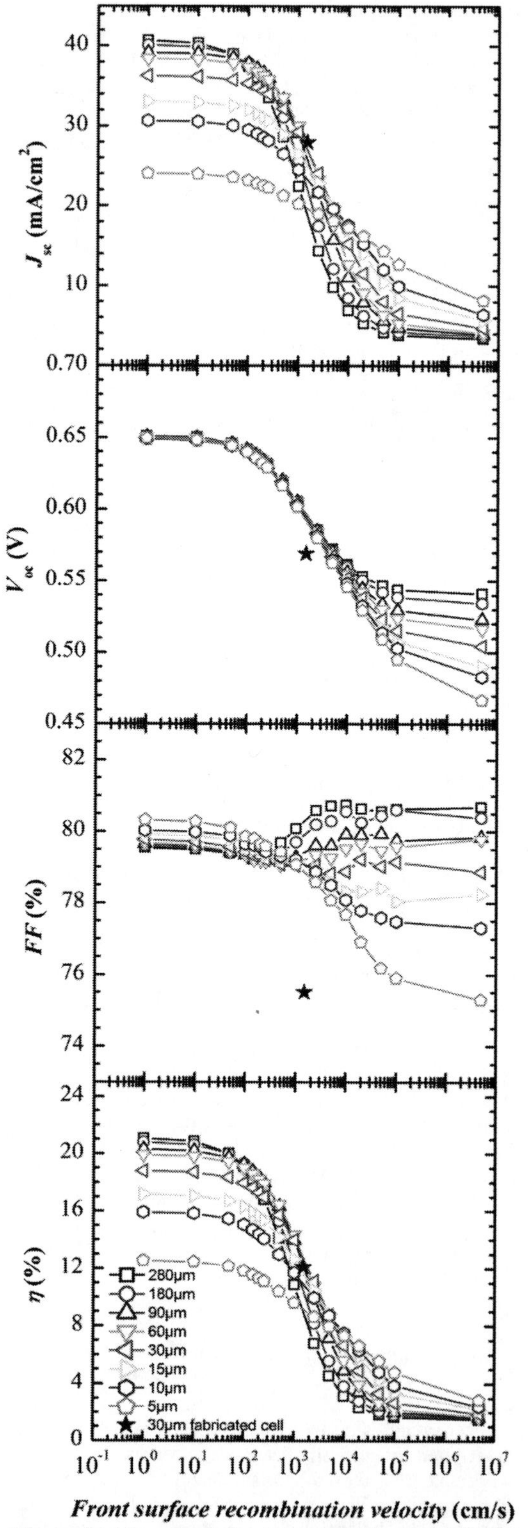

Figure 3. Atlas simulation of c-Si IBC solar cell (0.7 Ω·cm^2 are added in a lumped element to the series resistance) with different bulk Si thickness and front surface S_{front}. PV figures of the fabricated ~30 μm cell is starred

to activate the front surface passivation. A picture of the finished device is shown in Fig. 4.

The obtained photovoltaic figures measured under standard conditions (AM1.5G, 100 mW/cm^2, 25 °C) are summarized in Table 1 and presented in Fig. 3 by starring. The measured current-voltage curve as well as that derived from Suns-V_{oc} measurement are presented in Fig. 5.

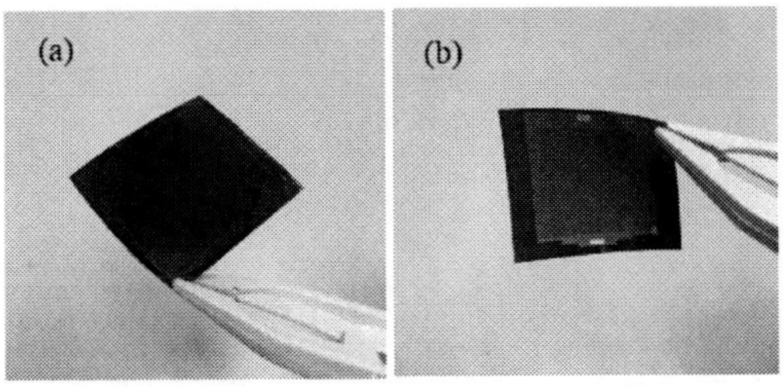

Figure 4. Pictures of a thin c-Si IBC solar cell. (a) front surface; (b) rear surface.

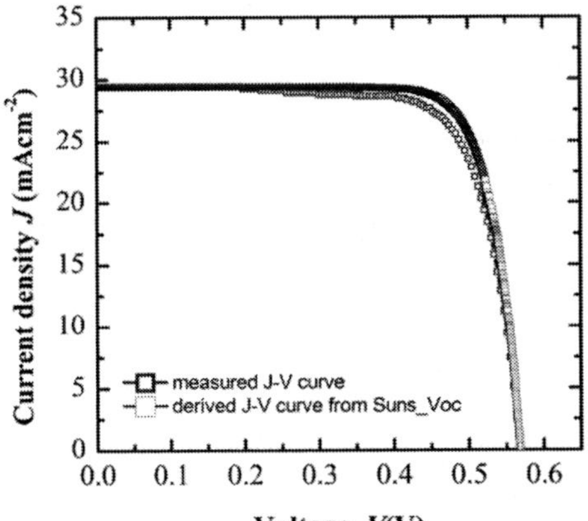

Figure 5. J-V curve of fabricated ~30µm IBC solar cell.

Table 1. Photovoltaic figures of fabricated ~30µm IBC solar cell and S_{front} value deduced from EQE measurements

Jsc (mA/cm2)	Voc (mV)	FF(pFF) (%)	η (%)	Deduced S_{front}
28.1	569	75.5(81.5)	12.1	1500cm/s

In order to compare the PV characteristics of the real cell with the calculated values from simulation, the front surface recombination velocity of the real device should be determined. Since quantum efficiency in short wavelength range is largely determined by S_{front}, the External Quantum Efficiency (EQE) of the fabricated IBC cell is measured for comparing with those from simulation and determine S_{front}. The Internal Quantum Efficiency (IQE), i.e. the electrical loss via recombination, is extracted from ATLAS simulation under different S_{front} values. Then, the optical losses are calculated by the wafer ray tracer software and EQE is obtained by multiplying both results.

Figure 6 shows the best fit of the EQE experimental data with our model resulting in a S_{front} of 1500cm/s. It can be observed that the measured and simulated EQE coincide well in blue/visible part of the spectrum while a small deviation can be seen in the infrared part. This difference is probably attributed to an inaccurate model of the light confinement properties, since these photons are the ones that could reach the rear surface of the device. For example, polymide has not been included in the model and it could have an effect in regions where there is no metal covering at rear surface (regions A3 in figure 2).

Using the S_{front} value determined by EQE, we can see in Fig. 3 that the measured V_{oc} and FF values are lower than the ones predicted by simulation, resulting that the efficiency of the device is nearly 1% lower. Regarding V_{oc}, a possible origin for this deviation could be the RIE plasma etching used for silicon thinning. The plasma etching process may damage the Si crystal that degrades the substrate quality [11]. Focusing on FF, the big difference between its value and the pFF value (see Table 1) indicates a high series resistance which is not included in the simulations. This big difference could be probably related to the difficulties found in the contact between the measurement probes and the metal grid once the device was slimmed down.

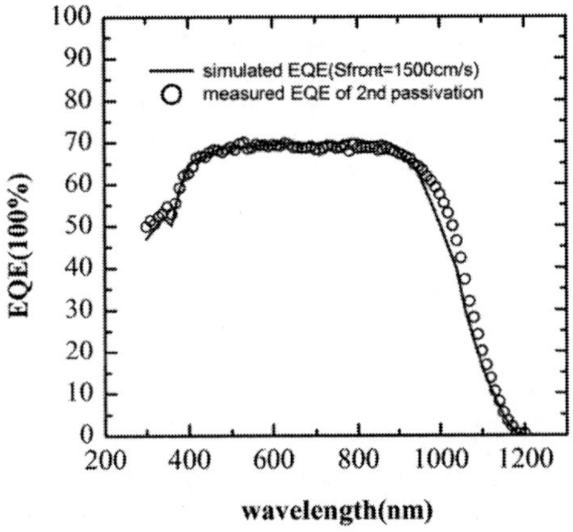

Figure 6. EQE comparison between simulation result and measurement of the fabricated ~30 µm cell

6 CONCLUSION

In this work, we focus on revealing the potential performance of thin c-Si substrates based IBC solar cell applying conventional technology by both simulation and experimental methods. Using 3D TCAD simulation, the performance of IBC c-Si solar cell with substrate thickness ranging from 5 to 280 µm is presented, underlining the front surface passivation which shows a significant impact on thin IBC cell performance. The simulation results demonstrates that, with conventional front surface pyramids texturing, remarkable efficiencies close to 17 % are achievable for substrates in the 10-20 µm range without changing the rear surface technology. On the other hand, a ~30µm-thick IBC cell is fabricated from a finished prototype cell based on thick substrate to check out simulation results. This cell is random pyramids textured and passivated by 90 nm Al_2O_3 layer on front surface to be identical to the simulation counterpart, resulting in an efficiency of 12.1 %. The front surface recombination velocity is deduced to be 1500 cm/s by comparing EQE with simulation results.

ACKNOWLEDGMENTS

This work has been supported by the Spanish Government through the projects TEC2014-59736-R and ENE2016-78933-C4-1-R. Part of the work is also supported by the project REFER COMRDI15-1-0036 funded by ACCIÓ and European Regional Development Fund (FEDER). C. Jin thanks the China Scholarship Council (CSC) for the financial support.

REFERENCES

[1]Technology Roadmap Solar Photovoltaic Energy. 2014 edition from International Energy Agency. https://www.iea.org/publications/freepublications/publication/TechnologyRoadmapSolarPhotovoltaicEnergy_2014edition.pdf
[2] J.H. Petermann et al. "19%-efficient and 43 μm-thick crystalline Si solar cell from layer transfer using porous silicon", Progress in Photovoltaics: Research and Applications 10, 1 (2010).
[3] L. Wang et al. "Development of a 16.8% Efficient 18-μm Silicon Solar Cell on Steel", IEEE Journal of Photovoltaics 4, 1397-1404 (2014).
[4] R. Cariou et al. "Ultrathin PECVD epitaxial Si solar cells on glass via low-temperature transfer process", Progress in Photovoltaics: Research and Applications 24, 1075-1084 (2016).
[5] V. Depauw et al., "Micrometer-Thin Crystalline-Silicon Solar Cells Integrating Numerically Optimized 2-D Photonic Crystals", IEEE Journal of Photovoltaics 4, 215-223 (2014)
[6] H. Savin, H. et al. "Black silicon solar cells with interdigitated back-contacts achieve 22.1% efficiency", Nature Nanotechnology 10, 624–628 (2015).
[7] Calle, E. et al. High efficiency interdigitated back-contact c-Si (p) solar cells. in 2015 10th Spanish Conference on Electron Devices (CDE) 1–4 (IEEE, 2015).
[8]ATLAS(http://www.silvaco.com/products/device_simulation/atlas.html)
[9]PVlighthouse (https://www2.pvlighthouse.com.au/calculators/wafer%20ray%20tracer/wafer%20ray%20tracer.html)
[10] López-González, J. M., Martín, I., Ortega, P., Orpella, A. & Alcubilla. R. "Numerical simulations of rear point-contacted solar cells on 2.2 Ωcm p-type c-Si substrates: Numerical simulations of rear point-contacted solar cells". Progress in Photovoltaics: Research and Applications 23, 69–77 (2015)
[11] Trompoukis, C. et al. "Photonic nanostructures for advanced light trapping in thin crystalline silicon solar cells", Phys. Status Solidi A 212, No. 1, 140–155 (2015)
[12] J.W. Slotboom and H.C. De Graaf, "Measurements of Bandgap Narrowing in Silicon Bipolar Transistors", Solid State Electronics, Vol. 19, p. 857-862, 1976.
[13] D.M. Caughey and R. E. Thomas, "Carrier Mobilities in Silicon Empirically Related to Doping and Field", Proc. IEEE 55, 1967, p. 2192-2193.

APPENDIX

Table I. Summary of the most relevant parameters used in the simulations

Carrier statistics	Fermi-Dirac, complete ionization of impurities
Badgap narrowing	Slotboom and de Graaf model [12]
Carrier mobility	The Analytic Low Field Mobility Model (Concentration and temperature dependent Caughey-Thomas model [13])
Bulk Recombination	
Shockley-Read-Hall (bulk)	Shockley-Read-Hall (SRH) Recombination model. TAUN0=0.5 s TAUP0=0.5 s ETRAP=0
Auger	Injection-dependent Auger coefficients
Radiative	$C = 4.7\times10^{-15}$ cm^3s^{-1}
Surface recombination	
Passivated front surface	Recombination velocity (S_n=S_p) varies from 1 to 5×10^6 cm/s
Passivated rear surface	Recombination velocity (S_n=S_p) of thermal SiO_2 passivated rear surface at each region is determined by fabricated cell as discussed in the text. SRV_{N+}=SRV_{P+}=1.6×10^4 cm/s, SRV_{N++}=5.6×10^4 cm/s, SRV_{SiO2}=80 cm/s.
Doping parameters	
Bulk acceptor density, N_A	5.71×10^{15} cm^{-3}
n^{++} and n^+ regions	Gaussian distribution of Donor, Characteristic length=characteristic length of lateral profile=0.84 Concentration=6.1×10^{19}
	Gaussian distribution of Donor, Characteristic length=characteristic length of lateral profile=0.5483 Concentration=9×10^{18}
p^+ regions	Gaussian distribution of acceptor, Characteristic length=characteristic length of lateral profile=2.094 Concentration=2×10^{18}
Contact parameters	
Anode and cathode contact	Material=aluminum (ohmic contact)
	Finite surface recombination velocities=5×10^6
Lumped series resistance	0.7 Ωcm^2
Optical parameters	
Incident light spectrum	Spectrally enhanced AM 1.5G
Front reflectivity	Front reflections determined by *2D ray tracer* and *3D Luminous*
Back reflectivity	number of reflections traced=2

IN-SITU CHARACTERIZATION OF THE PROTON IRRADIATION INDUCED DEGRADATION OF THIN FILM LIQUID PHASE CRYSTALLIZED SILICON ON GLASS BASED HETEROJUNCTION SOLAR CELLS WITH INTERDIGITATED BACK CONTACTS

Heinz-Christoph Neitzert[1], Carmine Pellegrino[1], Giovanni Landi[1], Juergen Bundesmann[2], Sophie Seidel[2], Andrea Denker[2], Tim Frijnts[2] and Stefan Gall[2]
[1] Dipartimento di Ing. Industriale (DIIN), Università di Salerno, Fisciano, Italy
[2] Helmholtz-Zentrum Berlin für Materialien und Energie GmbH (HZB), Berlin, Germany

ABSTRACT: Heterojunction solar cells with thin film crystalline silicon absorber layer and amorphous silicon emitter have been prepared, using Liquid Phase Crystallized (LPC) Silicon on glass technology, combined with the use of interdigitated back contacts in order to minimize reflection and shadowing losses. The solar cells have been exposed to a proton beam of 68 MeV, that creates a homogeneous depth profile of defects in the solar cells. A first partial irradiation of the solar cell created a very high local defect concentration and after a strong initial decrease of the short circuit current, the degradation almost saturates. A subsequently homogeneous irradiation of the whole solar cell area – however – resulted in a continuous further degradation of the investigated solar cell. The inhomogeneous defect concentration of the solar cell after this 2-step degradation has been evidenced by LBIC imaging.
Keywords: Thin Film Solar Cell, Degradation, Space Cells

1 INTRODUCTION

Solar cells based on crystalline silicon absorber and amorphous silicon emitter with interdigitated back contacts are holding the record efficiency for silicon based solar cells with more than 26% [1]. In this case no expensive Float-zone material, but Czochralski single crystal Silicon wafers have been used. In order to lower the material costs even further, thin film crystalline silicon material has been produced by laser crystallization of amorphous silicon films deposited on glass substrates. For this type of liquid phase crystallized (LPC) silicon solar cells interesting conversion efficiencies of more than 10 % have been achieved [2]. It should be mentioned that the laser crystallization of amorphous silicon is also an important technology for the realization of thin film polycrystalline silicon transistors [3] and for the realization of polycrystalline silicon films on substrates, that do not allow for CVD processes at higher temperatures [4].

For space applications silicon solar cells are still playing an important role, even if high efficiency III-V material based multijunction solar cells [5], CIGS [6], CdTe [7] and nowadays perovskite [8,9] based thin film solar cells have been shown to have a much better radiation tolerance. With decreasing active layer thickness, often a better radiation tolerance can be achieved in electronic devices and it is therefore worth testing, if thin-film crystalline silicon solar cells with absorber layers of < 10 µm can achieve a better radiation tolerance than the classical c-Si solar cells with absorber layer thicknesses of > 100 µm.

Here we report on in-situ monitoring of the short circuit current of LPC type IBC silicon heterojunction solar cells during the exposition to 68 MeV protons. In particular also the influence of inhomogeneous radiation on the photocurrent reduction during irradiation and the resulting light beam induced current map are shown.

2 EXPERIMENTAL

2.1 Investigated solar cells

The back-contact design for the Point Contact Cell is based on the approach, that had been introduced by CSG Solar [10]. The mechanical stability of the point contact cell is provided by a 3.3 mm thick borosilicate glass substrate. High thermal endurance as well as a matched silicon thermal expansion coefficient make this glass an ideal substrate for the technology. Between the glass and the absorber, a thin (~200 nm) SiNx/SiO2/SiOxNy interlayer stack (ILs) is deposited through Plasma-Enhanced Chemical Vapour Deposition (PECVD). The absorber silicon precursor layer is of low electronic quality, being generally amorphous or nano-crystalline, and it is also deposited by means of PECVD, with a thickness of about 10 µm. After the annealing process to remove Hydrogen, the silicon is crystallized directly via the liquid phase with a continuous-wave LIMO diode laser in air at 808 nm, with an energy density of about 2 J mm^2. The sample is then passivated in a hydrogen plasma and etched with potassium hydroxide (KOH) to create a surface texture for light trapping. The hetero-junction is formed by PECVD of a thin intrinsic or lightly doped a-Si:H/nc-Si emitter, followed by a In_2O_3:Sn (ITO) or a ZnO:Al (AZO) transparent conductive oxide (TCO) layer stack. Then a laser scribe, removing c-Si, emitter and TCO, defines and isolates the cell area. Subsequently, a white resin is spin coated on the sample. The white resin is opened locally with an inkjet printer, using a KOH containing ink and this is followed by a TCO and emitter etch, thereby exposing the c-Si absorber for contacting. The sample is then put into solvent atmosphere for several minutes, during which the white resin flows several µm into the opening thereby covering the TCO edge and preventing shunts. Subsequently, another inkjet KOH etch is used to open the resin and expose the TCO.

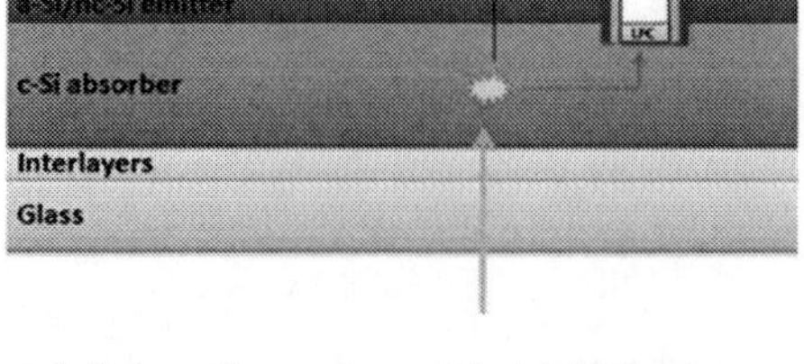

Figure 1: Schematic structure of the LPC-Si point contact solar cell

Then a 20/100 nm Ag/Al (p-type cells) layer stack is deposited and separated into interdigitated contact fingers using an UV laser. Finally, laser firing is used to create a low resistance absorber contact. The final samples realized consist of a 5 cm x 5 cm substrate that contains four 0.6 cm x 1.7 cm (1 cm^2) cells. For more details of the sample preparation the reader is referred to [11,12].

2.2 Proton irradiation

The proton irradiation has been done at the ISL accelerator facility at the HZB with a proton beam of 68 MeV. Taking into count the absorber thickness of the solar cells under test (~10 µm), proton radiation with energies higher than 1 MeV produces a homogeneous defect profile in the samples, according to the SRIM simulation for mono c-Si reported in [13,14]. The ion beam has been focused to a small area of about 3 mm x 3 mm and in case of the need of a homogeneous degradation of the whole solar cell area, scanned over the interesting area.

In the particular case of this work, in a first step the beam has been not scanned during a first irradiation period of about 580s and therefore a rather high dose of about 1.1e13 protons/cm^2 has been delivered to a restricted area of about 9 mm^2. After a short interruption, the beam has been scanned over the whole area for the same irradiation time. The corresponding proton dose in this second irradiation is 1e12 protons/cm^2. This particular irradiation sequence permitted to investigate the influence of a locally high defect concentration on the overall solar cell performance and on the dynamics of the photocurrent decrease under irradiation.

3 RESULTS AND DISCUSSION

3.1 In-situ monitoring of the short circuit current during proton irradiation

The in-situ monitoring of the solar cell short circuit current before, during and after irradiation is shown in Figure 2. Before the start of the first irradiation period, a very stable photocurrent is observed. The positive steps at the beginning and the negative steps at the end of the irradiation period are due to the radiation induced current (RIC). During the irradiation the measured short circuit current of the solar cell decreases monotonically. After the stop of the irradiation, some small recovery of the photocurrent signal, which is probably due to annealing of defects, is observed. In order to compare the dynamics of the short circuit decrease during step 1 and step 2 of the proton induced degradation directly, in Figure 3 the normalized monitored current only during proton irradiation has been plotted together.

Table I: Photocurrent(I_{ph}) and radiation induced current (RIC) before and after the different steps of a 2-step irradiation with a proton beam of 68 MeV

parameter		before	after irradiation
Isc (nA)	1. step	32000	23400
RIC (nA)	1. step	2800	500
Isc (nA	2. Step	23000	11090
RIC (nA)	2. Step	720	420

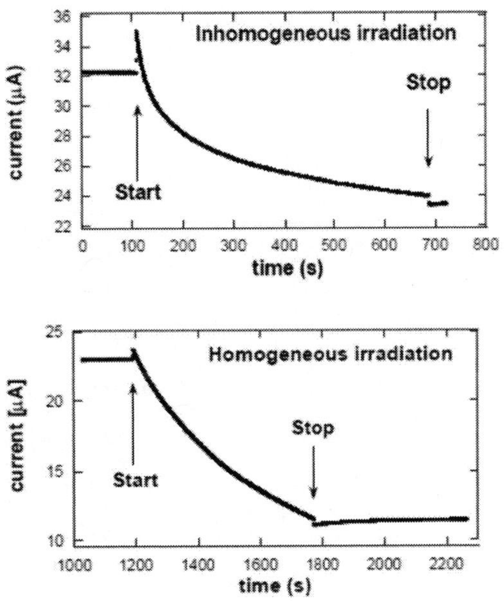

Figure 2: In-situ monitoring of the LPC-Si solar cell short-circuit current with constant illumination during a (upper figure) local irradiation with 68 MeV protons up to a dose of 1e13 protons/cm^2, and a (lower figure) subsequent homogeneous irradiation, with a dose of 1e12 protons/cm^2.

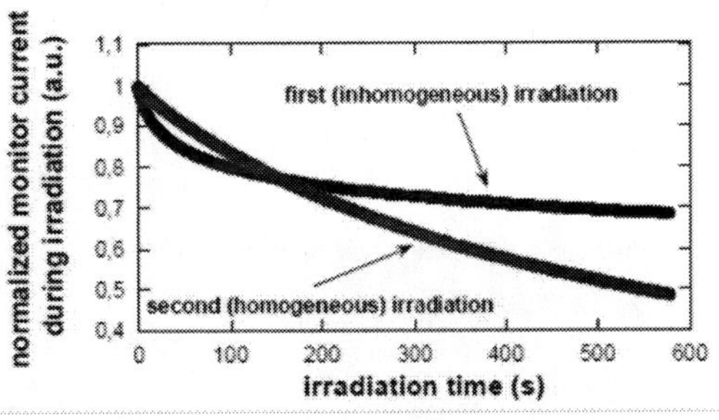

Figure 3: Comparison of the normalized current traces, measured in-situ during the two steps of proton irradiation, shown in Figure 2-

In Table I the changes of the measured photocurrent and of the RIC before and after the respective irradiation steps are listed. It can be observed, that during the first irradiation step with the proton beam focused only onto a small area, the RIC decreased to a value of 18% of the value before degradation, while the photocurrent decreased only to a value of 73% of the value before irradiation. This can be easily explained, because the beam destroys almost completely the area, useful for the RIC, but only 10% of the area, useful for the photocurrent based current generation.

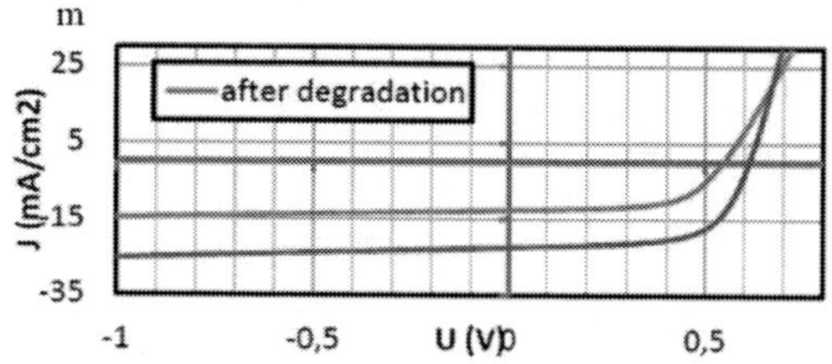

Figure 4: Comparison of the current density - voltage characteristics under AM1.5 illumination of the solar cell before (red line) and after (blue line) a 2-step irradiation with 68 MeV protons.

During the second homogeneous irradiation step, all the area is now contributing as well to the RIC as to the photocurrent. There is now a stronger decrease of the photocurrent of another 52% and a much smaller decrease, as compared to the degradation during the first irradiation step, of about 42% of the RIC.

The overall influence of the 2-step irradiation on the solar cell light current-voltage characteristics can be seen in Figure 4 and the extracted solar cell parameters are summarized in Table II. The initial cell efficiency of 8.82 % has been decreased by the irradiation by about 55 % to a new value of 4.15 %.

Table II: Solar cell parameters before and after 2-step irradiation with a proton beam of 68 MeV

Solar cell parameter	before 2-step irradiation	after 2-step irradiation
Efficiency (%)	8.82	4.15
J_{sc} (mA/cm2)	22.45	12.62
V_{oc} (mV)	611	547
FF (%)	64.2	60.2
R_s (Ohm cm^2)	3.48	7.94
R_p (Ohm cm^2)	341.9	448.4

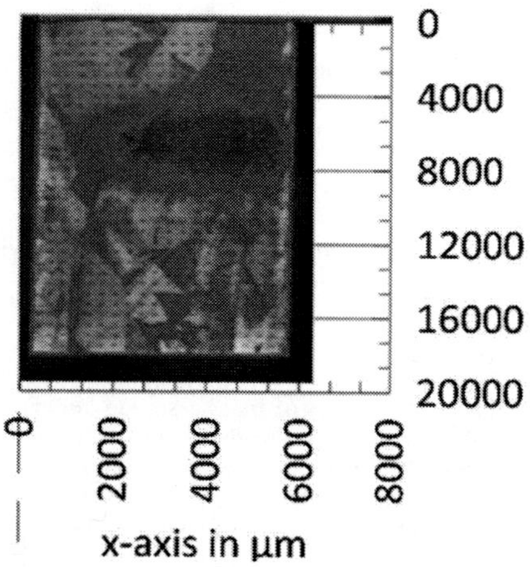

Figure 5: LBIC image of the LPC-Si solar cell after a 2-step (first focussed ion beam without scanning and then with scanning over the whole solar cell area) irradiation with 68 MeV protons.

The LBIC map, seen in Figure 5, shows all principal features of the investigated solar cell after degradation. Clearly visible are the backside point contacts, the polycrystalline character of the absorber layer with relatively large grains and an about 3mm x 3mm large dark spot, showing the highly defective region, where the focused proton beam has been hitting the cell during the first inhomogeneous degradation.

5 CONCLUSIONS

For the first we investigated the radiation stability of liquid-phase crystallized silicon thin-film solar cells on glass featuring a 10 micrometer thick multi-crystalline Si absorber layer. A 2-step degradation process was applied using a 68 MeV proton beam. In a first step the cell was irradiated only in a small area with a high dose. Subsequently the whole area was homogeneously irradiated with a lower dose. The photocurrent was monitored in-situ during irradiation in order to study the dynamics of the degradation. The first degradation step led to a fast initial degradation followed by a saturation region. In contrast the second homogeneous degradation step resulted in a slow continuous degradation without saturation within the investigated time period. The results discussed here give just a first insight into the degradation behavior of LPC-based silicon thin-film solar cells. Further investigations are required to get a deeper understanding of the radiation stability of this type of thin-film solar cell.

References

[1] K. Yoshikawa, H. Kawasaki, W. Yoshida, T. Irie, K. Koishi, K. Nakano, T. Uto, D. Adachi, M. Kanematsu, H. Uzu, K. Yamamoto, Nature Energy 2 (2017) 17032.

[2] M. A. Green, K. Emery, Y. Hishikawa, W. Warta, E. D. Dunlop, Progress in Photovoltaics: Research and Applications 24 (2016) 3.

[3] G. Fortunato, L. Mariucci, R. Carluccio, A. Pecora, V. Foglietti, Applied Surface Science 154-155 (2000) 95.

[4] A. Imparato, C. Minarini, A. Rubino, P. Tassini, F. Villani, D. Della Sala, E. Amendola, M. Kunst, H.C. Neitzert, S. Bellone, Thin Solid Films 487 (2005) 58.

[5] R. D. Harris, M. Imaizumi, R. J. Walters, J. R. Lorentzen, S. R. Messenger, J. G. Tischler, T Ohshima, S. Sato, P. R. Sharps, N. S. Fatemi, IEEE Transactions on Nuclear Science 55 (2008) 3502.

[6] A. Jasenek, U. Rau, K. Weinert, H. W. Schock, J. H. Werner, Proceedings 3rd World Conference on Photovoltaic Energy Conversion (2003) 593. (CIGS)

[7] D. L. Baetzner, A. Romeo, M. Terheggen, M. Doebeli, H. Zogg, A. N. Tiwari, Thin Solid Films 451-452 (2004) 536. (cdte)

[8] F. Lang, N. H. Nickel, J. Bundesmann, S. Seidel, A. Denker, S. Albrecht, V. V. Brus, J. Rappich, B. Rech, G. Landi, H. C Neitzert, Advanced Materials 28 (2016) 8726.

[9] V. V. Brus, F. Lang, J. Bundesmann, S. Seidel, A. Denker, B. Rech, G. Landi, H. C. Neitzert, J. Rappich, N. H. Nickel, Advanced Electronic Materials 2 (2017) 201600438.

[10] E. Schneiderlöchner, R. Preu, R. Lüdemann, S. W. Glunz, Progress in Photovoltaics: Research and Applications 10 (2002) 29.
[11] T. Frijnts, S. Gall, H. Rhein, P. Sonntag, L. Mazzarella, S. Kirner, C. Matarazzo, B. Rech, R. Schlatmann, Proceedings 43rd IEEE Photovoltaic Specialists Conference (2016) 2425.
[12] O. Gabriel, T. Frijnts, N. Preissler, D. Amkreutz, S. Calnan, S. Ring, B. Stannowski, B. Rech, R. Schlatmann, Progress in Photovoltaics: Research and Applications 24 (2016) 1499.
[13] H. C. Neitzert, P. Spinillo, S. Bellone, G. D. Licciardo, M. Tucci, F. Roca, L. Gialanella and M. Romano, Solar Energy Materials & Solar Cells 83 (2004) 435.
[14] H. C. Neitzert, M. Ferrara, M. Kunst, A. Denker, Z. Kertesz, B. Limata, L. Gialanella, M. Romano, Phys. Stat. Sol. (b) 245 (2008), 2877.

LASER-INDUCED CRYSTALLIZATION OF SPUTTERED UNHYDROGENATED SILICON AT LOW TEMPERATURES

E. Saugar[1], J.P. González[1], S. Fernández[1], J.J. Gandía[1], J. Cárabe[1], D. Canteli[2], M. Morales[2], C. Molpeceres[2], F. García-Pérez[3], M.B. Gómez-Mancebo[3].
[1]Energy Department, CIEMAT, Madrid, 28040, Spain
[2]Centro Láser, Universidad Politécnica de Madrid, Madrid, 28038, Spain
[3]Chemistry Division, CIEMAT, Madrid, 28040, Spain

ABSTRACT: 10-µm-thick non-hydrogenated amorphous-silicon (a-Si) films were deposited at relatively high rates (≥ 10 Å/s) by radio-frequency magnetron sputtering (RFMS) on different large-area buffer-layer-coated glass substrates at deposition temperatures ranging from room temperature (RT) to 300°C. These amorphous samples were subsequently crystallized by means of a continuous-wave diode laser, looking for conditions to reach liquid-phase crystallization. The influence of deposition conditions on the quality of the final micro-crystalline silicon films has been studied.

Keywords: amorphous silicon, liquid-phase crystallization, multicrystalline silicon

1 INTRODUCTION

The current photovoltaic scenario is dominated by wafer-based silicon technology, which has led the worldwide market for a long time owing to its good performance-to-cost ratio [1]. Nevertheless, and in spite of the dramatic cost reduction undergone by mono- and multicrystalline silicon wafers in the last decade, the further gain in competitiveness of photovoltaic devices requires intensive search for new approaches in what concerns semiconductor devices [2]. The road to find new solutions and new forms of silicon go through thickness reduction, kerf avoidance and manufacturing-cost lowering. From this point of view, liquid-phase crystallization (LPC) has arisen as a promising and novel method [3] above solid-phase crystallization (SPC), for which a higher thermal budget is needed. Furthermore, cells made from SPC silicon exhibit lower open-circuit voltages [4].

Previous works have demonstrated satisfactory LPC-silicon results in terms of material grain sizes and cell open-circuit voltages well above 600 mV [5]. While it is true that those results are promising, the substrate temperatures involved exceed 600°C, thus limiting the attractiveness of the process[6].

In this work the quality of polycrystalline silicon (poly-Si) obtained from a-Si deposited at relatively low temperatures (300°C) is analyzed in what concerns crystalline fraction and grain size. The effects of deposition conditions, buffer layer, grain size and crystal preferential orientation are discussed.

2 EXPERIMENTAL

10-µm-thick non-hydrogenated a-Si films were deposited onto Corning Eagle XG® glass substrates coated with a 200-nm SiO_2 layer by using an RFMS system. Prior to the deposition, the substrates were cleaned in an ultrasonic bath, subsequently rinsed in ethanol and dried with nitrogen. Two series were deposited respectively at room temperature and 300°C, under a working pressure of 1.07 Pa. The measurement of substrate temperature was performed by means of a type-K reference thermocouple.

The a-Si was treated with a continuous-wave diode laser emitting at 940 nm and forming a spot which is linear along the X axis and Gaussian in the Y axis, and whose dimensions in focus are 16.5 mm × 2.16 mm. During the crystallization process, in order to reduce thermal stress and prevent cracks, the samples were subjected to high temperatures (650°C) for a short time. The samples were moved in a straight line using a constant scanning speed ranging from 400 mm/min to 1000 mm/min and irradiance from 900 J/cm^2·s to 1650 J/cm^2·s. Each sample was crystallized in distinct zones at different laser conditions.

The quality of the micro-crystalline silicon obtained has been evaluated by X-Ray diffraction (XRD) using a PANalytical X'pert Pro diffractometer operating in $\theta - \theta$ configuration, with CuKα radiation (45kV – 40 mA), in the angular range $20° < 2\theta < 80°$. Phase identification was obtained by comparison to the Inorganic Crystal Structure Database (ICDS). In order to study the material modifications caused during the laser process, Raman spectra were measured by using a commercial InVia Renishaw Raman spectrometer employing the 514.5-nm line of an Ar laser.

3 RESULTS AND DISCUSSION

3.1 Buffer layer and stability

As known from literature [7], diffusion barriers play an important role in the laser crystallization process. On the one hand, they are necessary in order to avoid contamination of the silicon due to diffusion of species from the glass substrate [8]. On the other hand, they prevent the formation of droplets due to surface tension during liquid-phase crystallization [3][8]. With this in mind, we have mutually compared the behavior of different samples on glass substrates with and without a 200-nm SiO_2 coating during the laser process.

Table 1: Process conditions and results obtained for two samples deposited at room temperature. A1 on bare glass and B1 on 200-nm SiO_2-coated glass.

ID	Zone	Irradiance (J/cm^2·s)	Scan speed (mm/min)	Status
A1	1	1164 ± 49	600	Damaged
B1	1	1164 ± 49	600	Crystallized

Table I shows the effect of the diffusion barrier on the laser crystallization process. The diffusion barrier not only prevents the formation of droplets [8] and contamination but also favors crystallization, as it is shown on Table 1.

3.2 Structural properties

Once the need for diffusion barriers had been established, the influence of substrate temperature on the crystallization process was investigated. Although very high deposition temperatures, such as those previously cited [6] (about 600 °C) were avoided, samples C1 and D1 were respectively deposited by sputtering at 300 °C and room temperature (RT). Different regions of these samples were subsequently laser-processed in different conditions (see please Table 2).

Figure 1a and 1b show the corresponding Raman spectra of the samples, including a <100> monocrystalline-silicon wafer for reference. The Raman spectra were deconvoluted into three Gaussian peaks respectively at 480 cm^{-1} (corresponding to the amorphous-silicon phase), 510 cm^{-1} (corresponding to a defective crystalline phase) [9] and at 520 cm^{-1} (corresponding to silicon crystallites). From the integration of the deconvoluted Gaussian peaks it is possible to calculate the crystalline fraction (X_C) through the following equation:

$$X_C = \frac{(I_{520} + I_{510})}{(I_{520} + I_{510} + I_{480})}$$

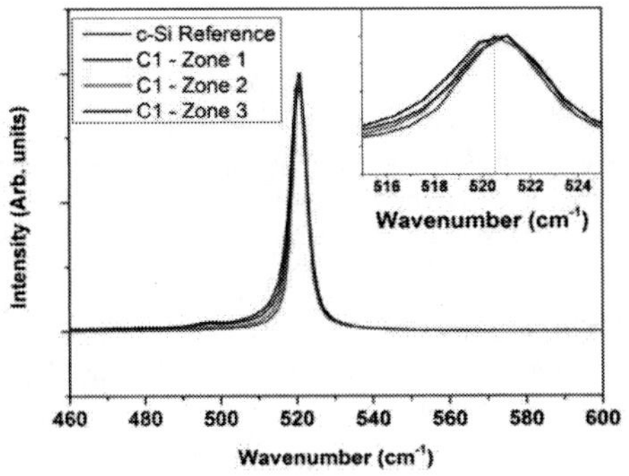

(a)

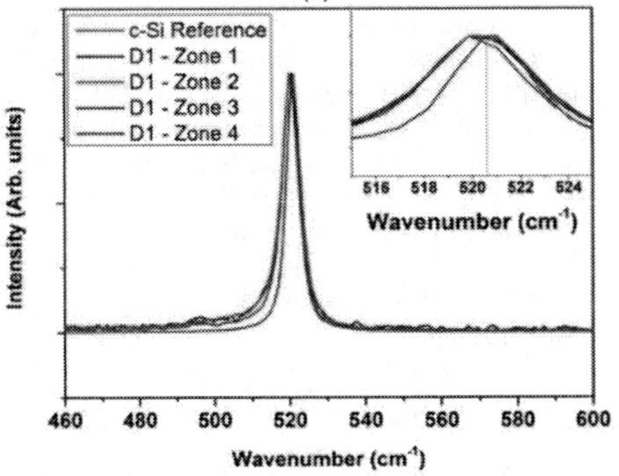

(b)

Figure 1: Raman spectra of 10-µm-thick magnetron-sputtered silicon films deposited at different temperatures: sample C1 at 300°C (a) and D1 at room temperature (RT) (b). The spectrum of a c-Si reference has been added. The graph inset shows a detail of the peak region.

As it can be observed on Figure 1, the spectrum of the c-Si reference (green curves) has its peak at 520 cm^{-1} with a full-width half maximum (FWHM) of 4.5 cm^{-1} approximately. However, the peaks of samples deposited at respectively 300°C and room temperature differ both mutually and with respect to the c-Si reference peak both in position and in FWHM. Furthermore, for sample D1, a slight contribution due to the amorphous phase present in the material can be observed at 490 cm^{-1}.

The sample deposited at relatively high temperature (C1) allows higher irradiances and this means higher crystalline fraction values of up to 20% more than the sample deposited at room temperature.

These samples have also been evaluated concerning their crystal orientation and grain size from X-Ray diffraction (XRD). Figure 2 shows the respective XRD patterns of samples C1 (a) and D1 (b).

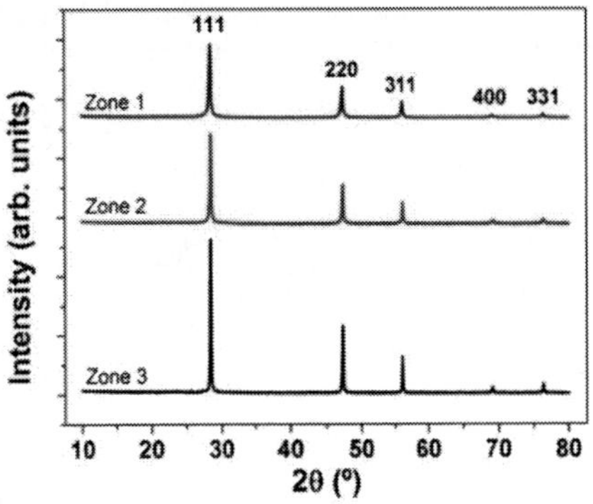

(a)

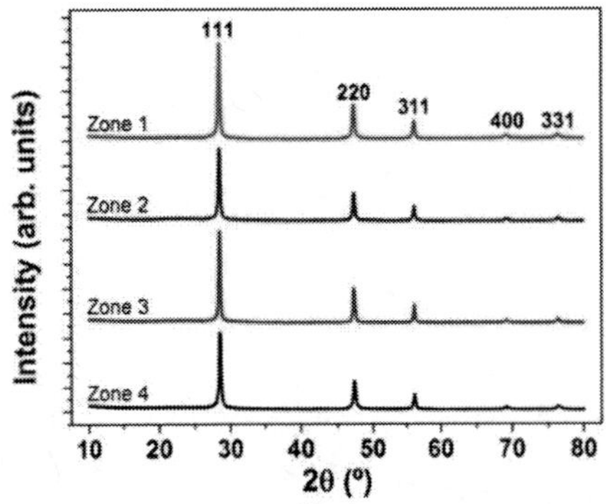

(b)

Figure 2: X-Ray diffraction patterns (XRD) of respectively sample C1, deposited at 300°C (Figure 2a) and sample D1, deposited at RT (Figure 2b). The different patterns in each plot correspond to different zones over the samples.

The XRD results evidence that laser-induced crystallization generates different crystal orientations in different zones, i.e. for different laser irradiances. However, by comparing the relative intensities of the XRD patterns it becomes easy to notice that <111> is the main orientation in both samples. Table 2 shows the crystalline fraction of each zone and the respective grain size as obtained from the Scherrer equation [10].

Table 2 also reveals that grain size increases abruptly, up to 2450 nm, when crystalline fraction reaches high values, beyond X_C = 95%. Furthermore, as suggested by Figure 3, not only irradiance but also sample temperature favors the jump to really large grain sizes. Sample C1 is believed to have undergone LPC.

Table 2: Crystalline fraction derived from the integration of deconvoluted Raman peaks, and grain size obtained from XRD pattern by using Scherrer equation, of two samples deposited respectively at 300°C (C1) and RT (D1).

ID	Zone	Irradiance / Scan speed	X_C	Grain Size (nm)
C1	1	1384 ± 51 W cm^{-2} 1000 mm min^{-1}	91 ± 3	70 ± 2
	2	1538 ± 52 W cm^{-2} 1000 mm min^{-1}	93 ± 3	134 ± 5
	3	1630 ± 53 W cm^{-2} 1000 mm min^{-1}	96 ± 3	2450 ± 400
D1	1	914 ± 47 W cm^{-2} 400 mm min^{-1}	84 ± 3	50 ± 1
	2	963 ± 48 W cm^{-2} 400 mm min^{-1}	81 ± 2	57 ± 1
	3	1164 ± 49 W cm^{-2} 600 mm min^{-1}	75 ± 1	57 ± 1
	4	1222 ± 50 W cm^{-2} 600 mm min^{-1}	85 ± 3	71 ± 2

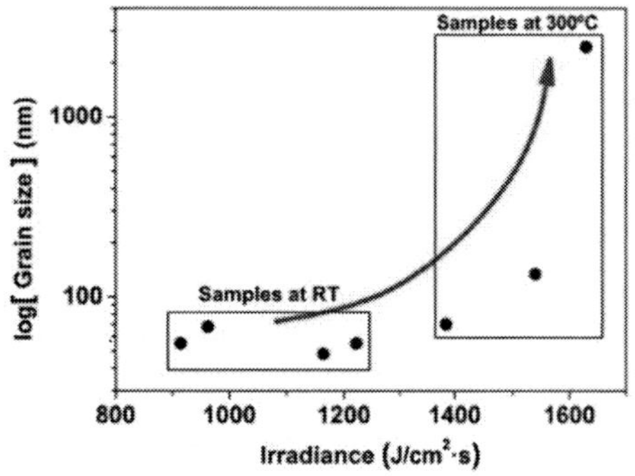

(a)

Figure 3: Grain size in logarithmic scale versus irradiance.

In order to further investigate how substrate temperature affects the samples and their compactness, the refractive index of different samples deposited at different substrate temperatures has been studied. It is well known that refractive index is an important optical parameter related to mass- and void density [11]. From the Clausius-Mossoti equation, mass density can be drawn [12].

$$\rho_{a-Si} = \left(\frac{n_\infty^2 - 1}{n_\infty^2 + 2}\right)\frac{3m_{Si}}{8\pi\alpha_{Si-Si}}$$

where m_{Si} is the mass of the silicon atom, n_∞ is the refractive index at long wavelengths and α_{Si-Si} is the bond polarizability in the amorphous phase. The value $\alpha_{Si-Si} = 1.96 \times 10^{-24}$ cm^3 is taken from [12].

Four a-Si samples have been deposited in the same conditions as for C1 and D1 except for substrate temperature, which has been varied between RT and 300°C.

Figure 4 shows the refractive indexes of four samples deposited at 1.07 Pa at different temperatures ranging from RT to 300°C. In the whole range covered an abrupt refractive-index growth with increasing substrate temperature is appreciated. This result is attributed to a reduction of void density [13].

From the analysis of refractive-index dispersion by means of the Wemple-DiDomenico dispersion model [14], below the interband absorption edge it is possible to obtain important information about atomic structure and mass density. The refractive index values were fitted to the WDD dispersion model according the following equation:

$$n(E)^2 - 1 = \frac{E_d E_0}{(E_d^2 - E^2)}$$

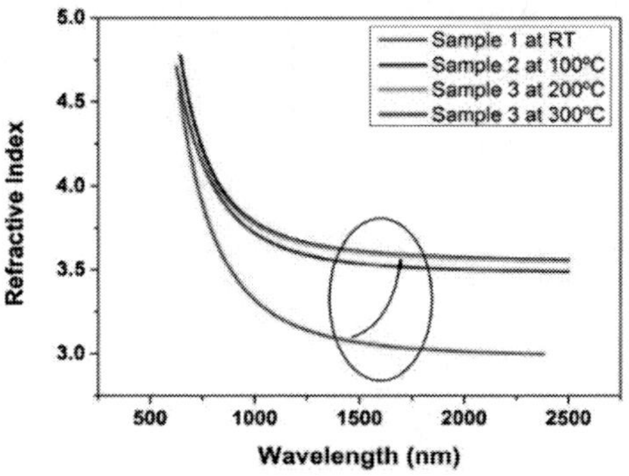

Figure 4: Refractive indexes of four a-Si samples deposited at different substrate temperatures. The circle inset illustrates the refractive-index rise with temperature.

E_d is an important parameter, which is closely related to the structural characteristics of the material and obeys the following relationship:

$$E_d = \beta N_c Z_a N_e,$$

where N_C is the coordination number of the cation nearest neighbor to the anion, and Z_a, N_e, β are constants [14]. The voids and the low packing of the amorphous lattice result in low density and coordination-number values [12]. In this sense, the estimate of the coordination number can provide information about the evolution of void density with substrate temperature. According to Wemple [15], the following expression allows us calculate the coordination number:

$$\frac{E_d^a}{E_d^x} = \left(\frac{\rho^a}{\rho^x}\right)\left(\frac{N_C^a}{N_C^x}\right)$$

where ρ is the density, x and a refer to crystalline- and amorphous phases respectively.

Figure 5 shows the mass density drawn from the Clausius-Mossoti equation and the coordination number derived from the fitting of the Wemple-DiDomenico dispersion model. As it can be observed, mass density and coordination number increase with substrate temperature as a consequence of silicon-atom rearrangement due to substrate-temperature enhancement. At higher temperatures, the diffusion processes are more likely, so greater compactness is induced.

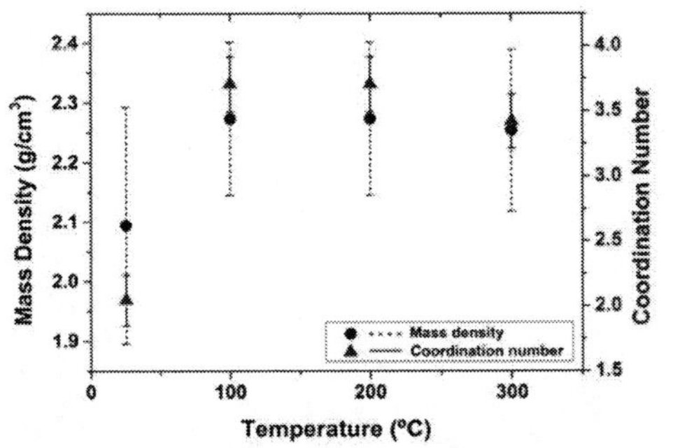

Figure 5: Mass density (black dots with short dash error bars) and coordination number N_c (blue triangles with solid error bars) versus substrate temperature.

The mass-density values obtained increase from 2.1 g/cm^3 for sample deposited at RT to 2.3 g/cm^3 for sample deposited at 300ºC, whereas the coordination number stretches from 2.1 to 3.7 in the same substrate-temperature range. The typical values, according to bibliography [12][16], are ρ = 2.33 g/cm^3 and N_c = 4 for c-Si, ρ = 2.29 g/cm^3 and N_c = 3.7 for a-Si.

In view of the results obtained, the samples deposited at moderately high deposition temperatures show higher compactness than those deposited at RT. Such a result is consistent with the finding that high crystalline fractions and grain sizes of the order of microns are only obtained from amorphous-silicon samples deposited at moderately high temperatures (of the order of 300 ºC, yet well below the high temperatures largely found in the literature). The emergence of this microcrystalline material is attributed to the achievement of liquid-phase crystallisation in the conditions indicated.

4 CONCLUSIONS

In this work several samples have been deposited onto different substrates, either bare or coated with a 200-nm SiO_2 layer, with different substrate temperatures. Firstly, it has been noted that the buffer layer plays a fundamental role to obtain a successful crystallization process by improving wettability and enhancing thermal stability under high laser irradiances.

Secondly, the relevance of substrate deposition temperature has been assessed. It has been observed that the samples deposited at moderately high (although much lower than those typically found in the literature) substrate temperatures show higher compactness and mass density than those deposited at RT. The effect is attributed to the activation of different diffusion processes during deposition, leading to the rearrangement of silicon atoms. Furthermore, these samples offer better performance under high irradiances, resulting in increased crystalline fractions and grain sizes. The results, in summary, suggest that the samples having micron-size crystallites are those for which process conditions trigger liquid-phase crystallization. Such an achievement in spite of the moderate substrate temperatures (not higher than 300ºC) is considered quite relevant.

5 ACKNOWLEDGEMENTS

This research has been partially supported by the Spanish Research Agency (AEI) and Ministry of Economy, Industry and competitiveness through project CHENOC (ENE2016-78933-C4).

6 REFERENCES

[1] International Technology Roadmap for Photovoltaics (ITRPV) Results 2015 (2016) by SEMI PV Group.

[2] G. Bye and B. Ceccaroli, "Solar grade silicon: Technology status and industrial trends," *Sol. Energy Mater. Sol. Cells*, vol. 130, pp. 634–646, Nov. 2014.

[3] J. Haschke, D. Amkreutz, and B. Rech, "Liquid phase crystallized silicon on glass: Technology, material quality and back contacted heterojunction solar cells," *Jpn. J. Appl. Phys.*, vol. 55, no. 4S, p. 04EA04, Apr. 2016.

[4] S. Gall and B. Rech, "Technological status of polycrystalline silicon thin-film solar cells on glass," *Sol. Energy Mater. Sol. Cells*, vol. 119, pp. 306–308, Dec. 2013.

[5] J. Haschke, D. Amkreutz, L. Korte, F. Ruske, and B. Rech, "Towards wafer quality crystalline silicon thin-film solar cells on glass," *Sol. Energy Mater. Sol. Cells*, vol. 128, pp. 190–197, 2014.

[6] D. Amkreutz, J. Haschke, T. Häring, F. Ruske, and B. Rech, "Conversion efficiency and process stability improvement of electron beam crystallized thin film silicon solar cells on glass," *Sol. Energy Mater. Sol. Cells*, vol. 123, pp. 13–16, 2014.

[7] D. Amkreutz, J. Müller, M. Schmidt, T. Hänel, and T. F. Schulze, "Electron-beam crystallized large grained silicon solar cell on glass substrate," *Prog. Photovoltaics Res. Appl.*, vol. 19, no. 8, pp. 937–945, Dec. 2011.

[8] J. Dore, S. Varlamov, and M. A. Green, "Intermediate Layer Development for Laser-Crystallized Thin-Film Silicon Solar Cells on Glass," *IEEE J. Photovoltaics*, vol. 5, no. 1, pp. 9–16, Jan. 2015.

[9] C. Droz, "Thin Film Microcrystalline Silicon Layers and Solar Cells : Microstructure and Electrical Performances," Ph.D. dissertation Université de Neuchâtel, Neuchâtel, 2003.

[10] A. L. Patterson, "The Scherrer formula for X-Ray particle size determination," *Phys. Rev.*, vol. 56, pp. 978–982, 1939.

[11] J. Müllerová, L. Prusáková, M. Netrvalová, V. Vavrunková, and P. Sutta, "A study of optical absorption in amorphous hydrogenated silicon thin films of varied thickness," *Appl. Surf. Sci.*, vol. 256, no. 18, pp. 5667–5671, 2010.

[12] A. H. M. Smets, W. M. M. Kessels, and M. C. M. Van de Sanden, "Vacancies and voids in hydrogenated amorphous silicon," *Appl. Phys. Lett.*, vol. 82, no. 10, pp. 1547–1549, 2003.

[13] H. Chen and W. Z. Shen, "Perspectives in the characteristics and applications of Tauc-Lorentz dielectric function model," *Eur. Phys. J. B*, vol. 43 (4), pp. 503–507, 2005.

[14] S. H. Wemple and M. DiDomenico, "Behavior of the electronic dielectric constant in covalent and ionic materials," *Phys. Rev. B*, vol. 3, no. 4, pp. 1338–1351, 1971.

[15] S. H. Wemple, "Refractive-Index Behavior of

Amorphous Semiconductors and Glasses," *Phys. Rev. B*, vol. 7, no. 8, pp. 3767–3777, 1973.
[16] K. Laaziri, S. Kycia, S. Roorda, M. Chicoine, J. L. Robertson, J. Wang, and S. C. Moss, "High-energy x-ray diffraction study of pure amorphous silicon," *Phys. Rev. B*, vol. 60, no. 19, pp. 13520–13533, Nov. 1999.

GROWTH OF SILICON ON REORGANIZED POROUS SILICON SUBSTRATES BY STEADY-STATE SOLUTION GROWTH FOR PHOTOVOLTAIC APPLICATIONS

Christian Ehlers[1*], Roman Bansen[1], David Uebel[1],
Thomas Teubner[1] and Torsten Boeck[1]

[1]Leibniz Institute for Crystal Growth (IKZ)
Max-Born-Str. 2, 12489 Berlin, Germany

*christian.ehlers@ikz-berlin.de

ABSTRACT: A silicon solar cell requires only several ten micrometers of silicon in order to absorb sunlight efficiently. Therefore, we have developed a process that allows for the continuous deposition of silicon on reorganized porous silicon substrates from a tin solution by steady-state liquid phase epitaxy. The grown epilayer can be detached from the silicon substrate with the intention of processing the epilayer into a solar cell, while reusing the substrate multiple times for a porosification and reorganization process and subsequent epitaxy. We will present cross section scanning electron microscope (SEM) images showing successful growth results of detachable layers.

Keywords: Steady-state liquid phase epitaxy, Silicon, Thin Film

1. Introduction

Common silicon solar cells are produced from bulk silicon crystals by either directional solidification or Czochralski growth. During wafering of the bulk silicon crystal, material is wasted due to kerf loss. Furthermore the produced silicon wafers have a thickness of around 180 µm which is more than is required for efficiently absorbing the sunlight [1].
Therefore, we have developed a process for depositing a silicon layer on reorganized porous silicon (RPS) substrates. We grow the silicon layer by steady-state liquid phase epitaxy (SSLPE) from tin solution. The growth equipment which is described elsewhere [2], allows for the deposition of silicon, on RPS foils with a (100) and (111) orientation [3], as well as on glass substrates with a seed layer [4][5][6][7][8].
The RPS substrates are produced by a lithography process at IMEC, Belgium [9]. The grown silicon layer has a thickness of 40 – 80 µm which is appropriate for an absorber layer of a silicon solar cell [10]. The goal of our research is to grow an epilayer, which can be detached and processed into a solar cell, and then to reuse the original substrate multiple times for porosification, reorganization and epitaxy [11][12].

2. Experimental details

Reorganized porous silicon (RPS) substrates used for these experiments were provided by IMEC, Belgium and are described in detail in [9]. A 4 × 4 cm^2 substrate, with a porosified and reorganized area of 2.5 × 2.35 cm^2 is shown in figure 5a. A RPS substrate cross section is shown exemplary in figure 1. It consist of a thin silicon foil sitting on top of a silicon substrate only held in place by a few remaining Si pillars.

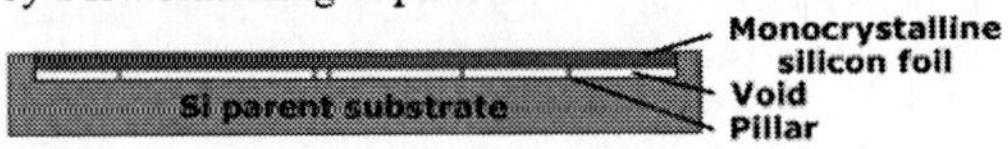

Figure 1: Schematic drawing of a reorganized porous silicon substrate.

These substrates were made from silicon wafers by formation of periodically arranged cylindrical voids by deep-UV lithography, SF_6/O_2 plasma etching and reorganization in a hydrogen atmosphere [9].
The silicon is deposited on the RPS by SSLPE. A cross section of the growth apparatus used for these experiments is shown in Figure 2.
RPS substrates are transferred to the growth chamber via a load-lock. Evacuation of the load-lock chamber containing a RPS substrate, has to proceed slowly to preserve the thin foil from unintentional detachment due to abrupt changes in pressure.
In the center of the growth apparatus, a graphite crucible is located which contains a Si source at the bottom. The crucible is filled with 6N tin. During growth at 700 °C or 800 °C in a hydrogen atmosphere a RPS foil with the Si parent wafer is placed upside down on the liquid tin. Bottom, top and side heaters are controlled using two thermocouples located at different heights in the graphite crucible. A difference of 5 K between the two thermocouples indicates the temperature gradient that causes the Si to migrate by diffusion and convection from the source at the bottom, where it dissolves into the liquid tin, to the RPS foil, where it re-crystallizes eventually.

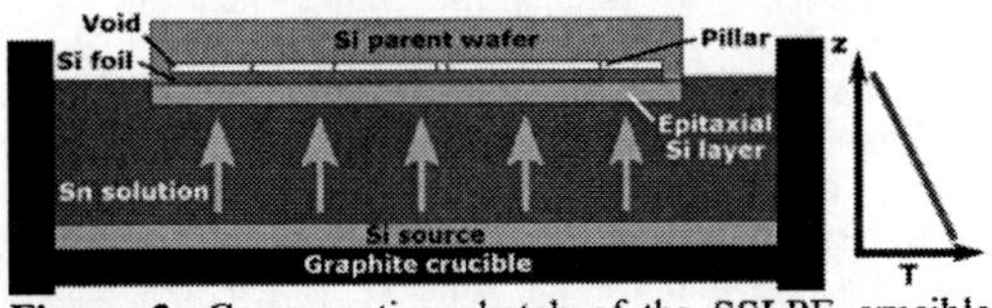

Figure 2: Cross section sketch of the SSLPE crucible during growth.

The samples were analyzed by scanning electron microscopy (SEM) with a NOVA 600 NanoLab™ dual beam SEM/FIB (FEI) equipped with a TEAM™ trident analysis system (EDAX).
Current-voltage (IV) and capacitance-voltage (CV) measurements were made with a Keithley 237 high voltage source measure unit and a Boonton 7200 capacitance meter using Au contacts.
Detachment of the grown layers was done with the assistance of a thermal specimen mounting wax (Crystalbond™).

3. Results and discussion

As the growth from tin solution on RPS foils with a Si (100) substrate leads to the formation of partially interconnected monocrystalline pyramids [3], growth was conducted on RPS foils produced on Si (111) substrates. Successful growth results of a smooth 50 µm thick Si epilayer on RPS foils with a (111) substrate is shown in the SEM image in figure 4.

A previously used approach for growing Si on RPS foils by SSLPE from Sn solution requires a so-called 'melt-back step'[3]. Hereby a reverse temperature gradient is applied for a short moment causing a minor intentional dissolution of the foil. However, since the foil is approximately 1 µm thin, there is a risk that it will dissolve completely in some places. As a result, there is no continuous void left below the epitaxially grown layer. In addition, the Sn solvent can enter the void around these areas. The Sn solvent remains in the void after growth and can prevent detachment of the grown epilayer. This problem is shown in Figure 3, where the existence of Sn inclusions is verified by an energy-dispersive X-ray spectroscopy (EDX) linescan.

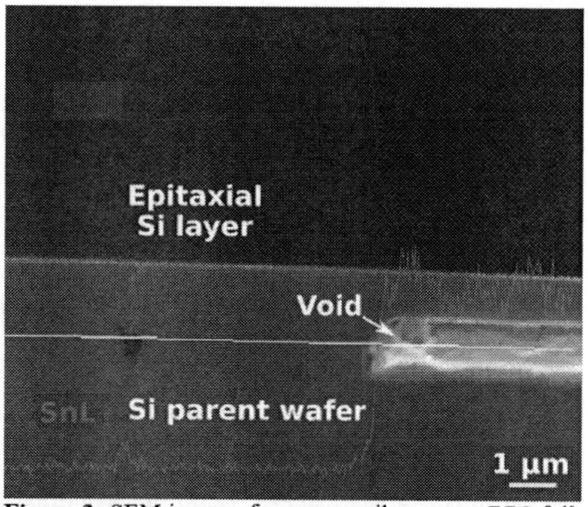

Figure 3: SEM image of a grown epilayer on a RPS foil. Shown in the overlay is an EDX linescan. The epilayer is held in place by Sn which infiltrated the void area.

To avoid the infiltration of the void space with Sn, a different approach for achieving continuous layers that can be detached was developed. Hereby, the substrate with the RPS is slightly undercooled before immersing the substrate into the tin solution. If the temperature difference is too high, thermal stress leads to the breakage of the epifoil, but correctly chosen, growth of detachable layers is possible as shown in figure 4.
In order to detach the grown layer, a specimen mounting wax was used. A similar approach is described for example in [13], where a two-component silicone adhesive is used for such a detachment purpose. The wax is applied on the epilayers in liquid form, after being heated on a hotplate and brought into contact with a flat glass carrier. After a cooldown period, the layer can be detached together with the glass carrier as shown in figure 5b, from which it can then be removed by dissolving the wax in acetone. Such a detached freestanding epitaxial Si layer which still has some pillars attached can be seen in figure 6.

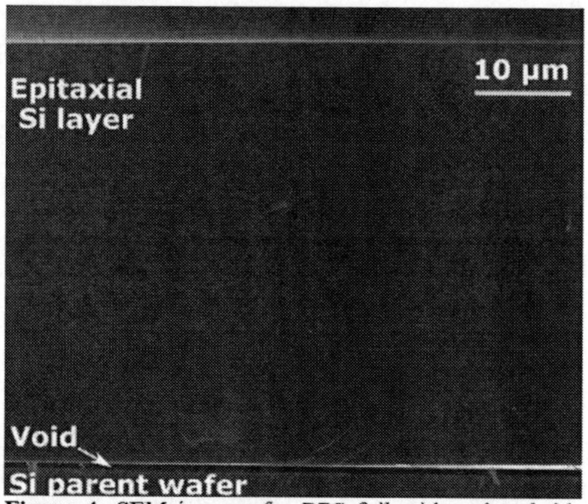

Figure 4: SEM image of a RPS foil with a detachable epilayer still attached to the Si (111) parent wafer produced with the improved growth process.

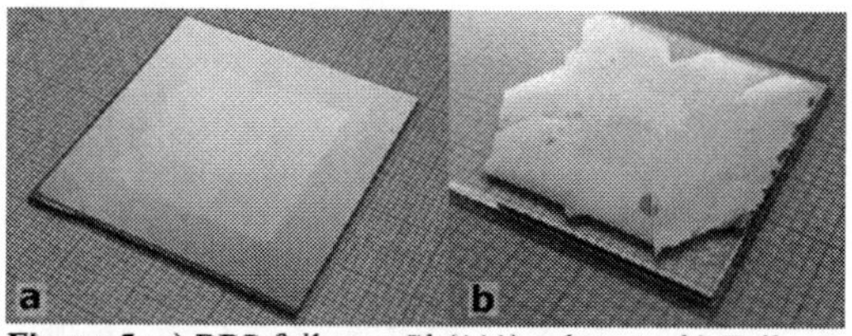

Figure 5: a) RPS foil on a Si (111) substrate. b) Epilayer grown on a RPS foil after detachment using a specimen mounting wax and glass substrate.

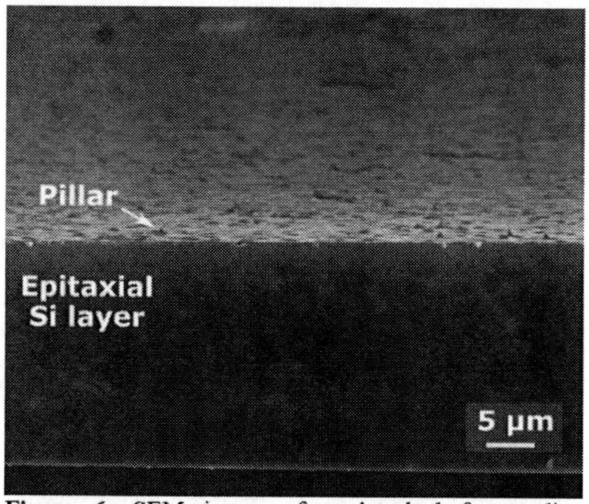

Figure 6: SEM image of a detached freestanding epitaxial Si layer. Also visible are the small pillars which were connected to the Si parent wafer.

The surface of the layers grown at 700 °C with this new approach has a roughness of up to ±10 µm as determined by a Profilometer. However increasing the growth temperature to 800 °C leads to more even surfaces, varying only about ±2 µm, which is in accordance with [14].

IV and CV measurements were conducted with Au Schottky contacts. The n-type Si material has a carrier concentration of 4×10^{15} cm^{-3}, which is sufficient for an absorber layer of a silicon solar cell.

4. Summary

It is possible to grow silicon on reorganized porous silicon substrates with a (111) orientation by steady-state liquid phase epitaxy from tin solution in a continuous fashion. Due to the development of an improved growth process, the grown epilayers can be detached using a specimen mounting wax. The grown material has a low carrier concentration which makes it suitable as an absorber layer for a silicon solar cell.

5. Acknowledgment

We acknowledge K.v. Nieuwenhuysen and V. Depauw from IMEC, Belgium, for providing us with reorganized porous Si substrates.

The research leading to these results has received funding from the European Union Seventh Framework Programme (FP7/2007-2013) under grant agreement number 609788.

6. References

[1] "International Technology Roadmap for Photovoltaic (ITRPV) Eight Edition," 2017.

[2] T. Teubner, R. Heimburger, K. Böttcher, T. Boeck, and R. Fornari, "Equipment for Low Temperature Steady-State Growth of Silicon from Metallic Solutions," *Cryst. Growth Des.*, vol. 8, no. 7, pp. 2484–2488, Jul. 2008.

[3] C. Ehlers, R. Bansen, T. Markurt, D. Uebel, T. Teubner, and T. Boeck, "Solution growth of Si on reorganized porous Si foils and on glass substrates," *J. Cryst. Growth*, vol. 468, pp. 268–271, Jun. 2017.

[4] R. Bansen, C. Ehlers, T. Teubner, T. Markurt, J. Schmidtbauer, and T. Boeck, "Continuous polycrystalline silicon layers on glass grown from tin solutions," *CrystEngComm*, vol. 18, no. 11, pp. 1911–1917, 2016.

[5] R. Bansen, R. Heimburger, J. Schmidtbauer, T. Teubner, T. Markurt, C. Ehlers, and T. Boeck, "Crystalline silicon on glass by steady-state solution growth using indium as solvent," *Appl. Phys. A*, vol. 119, no. 4, pp. 1577–1586, Jun. 2015.

[6] R. Bansen, C. Ehlers, T. Teubner, K. Böttcher, K. Gambaryan, J. Schmidtbauer, and T. Boeck, "Silicon on glass grown from indium and tin solutions," *J. Photonics Energy*, vol. 6, no. 2, p. 25501, May 2016.

[7] R. Bansen, C. Ehlers, T. Teubner, and T. Boeck, "Steady-state solution growth of microcrystalline silicon on nanocrystalline seed layers on glass," *J. Semicond.*, vol. 37, no. 9, p. 93001, Sep. 2016.

[8] T. Boeck, F. Ringleb, and R. Bansen, "Growth of crystalline semiconductor structures on amorphous substrates for photovoltaic applications," *Cryst. Res. Technol.*, vol. 52, no. 1, p. 1600239, 2017.

[9] V. Depauw, I. Gordon, G. Beaucarne, J. Poortmans, R. Mertens, and J. Celis, "Large-area monocrystalline silicon thin films by annealing of macroporous arrays: Understanding and tackling defects in the material," *J. Appl. Phys.*, vol. 106, no. 3, p. 33516, 2009.

[10] M. J. Kerr, P. Campbell, and A. Cuevas, "Lifetime and efficiency limits of crystalline silicon solar cells," in *Conference Record of the Twenty-Ninth IEEE Photovoltaic Specialists Conference, 2002.*, 2002, pp. 438–441.

[11] F. Dross, K. Baert, T. Bearda, J. Deckers, V. Depauw, O. El Daif, I. Gordon, A. Gougam, J. Govaerts, S. Granata, R. Labie, X. Loozen, R. Martini, A. Masolin, B. O'Sullivan, Y. Qiu, J. Vaes, D. Van Gestel, J. Van Hoeymissen, A. Vanleenhove, K. Van Nieuwenhuysen, S. Venkatachalam, M. Meuris, and J. Poortmans, "Crystalline thin-foil silicon solar cells: where crystalline quality meets thin-film processing," *Prog. Photovoltaics Res. Appl.*, vol. 20, no. 6, pp. 770–784, Sep. 2012.

[12] T. Bearda, I. Gordon, H. Sivaramakrishnan Radhakrishnan, V. Depauw, K. Van Nieuwenhuysen, M. Xu, L. Tous, M. Filipič, S. K. Jonnak, A. Hajijafarassar, X. Liu, M. Debucquoy, Y. Abdulraheem, J. Szlufcik, and J. Poortmans, "Thin Epitaxial Silicon Foils Using Porous-Silicon-Based Lift-Off for Photovoltaic Application," *MRS Adv.*, vol. 1, no. 48, pp. 3235–3246, May 2016.

[13] H. Sivaramakrishnan Radhakrishnan, R. Martini, V. Depauw, K. Van Nieuwenhuysen, T. Bearda, I. Gordon, J. Szlufcik, and J. Poortmans, "Kerfless layer-transfer of thin epitaxial silicon foils using novel multiple layer porous silicon stacks with near 100% detachment yield and large minority carrier diffusion lengths," *Sol. Energy Mater. Sol. Cells*, vol. 135, no. May, pp. 113–123, Apr. 2015.

[14] T. Ujihara, E. Kanda, K. Obara, K. Fujiwara, N. Usami, G. Sazaki, A. Alguno, T. Shishido, and K. Nakajima, "Effects of growth temperature on the surface morphology of silicon thin films on (111) silicon monocrystalline substrate by liquid phase epitaxy," *J. Cryst. Growth*, vol. 266, no. 4, pp. 467–474, Jun. 2004.

A LIGHTWEIGHT FLEXIBLE AMORPHOUS SILICON PHOTOVOLTAIC MODULE

Y. Vygranenko[1*], M. Fernandes[1,2], P.Louro[1,2], M. Vieira[1,2]
[1]CTS-UNINOVA, Campus da Caparica, 2829-516 Caparica, Portugal
[2]Departamento de Engenharia Electrónica e Telecomunicações e de Computadores, ISEL, Lisboa, 1950-062, Portugal
*Corresponding author (Tel. +351 21 831 7289, e-mail: yvygranenko@deetc.isel.ipl.pt)

ABSTRACT: This article reports on a monolithic 10 cm × 10 cm area PV module integrating an array of 72 a-Si:H n-i-p cells on a 100 μm thick polyethylene-naphtalate (PEN) substrate. The top metal grid is implemented to reduce the series resistance of the cells with thin (~180 nm) ZnO:Al electrodes and to interconnect them. The metal grid geometry optimization and device performance analysis are performed using a two-dimensional distributed circuit model of the photovoltaic cell. The circuit simulator SPICE is used to calculate current and potential distributions in a network of sub-cell circuits at various loads under AM1.5 illumination conditions, and also to map Joule losses in the front electrode and the metal grid. Comparison of simulated and experimental current-voltage characteristics shows that the shunt leakage is one of the factors reducing the device performance. Using the laser beam induced current (LBIC) and SEM technique, the presence of multiple shunts in the n-i-p cell was detected. A mechanism of micro-shunts formation is proposed and discussed.
Keywords: a-Si:H, flexible substrate, modelling, PV module, shunts.

1 INTRODUCTION

Flexible hydrogenated amorphous silicon (a-Si:H) solar cells on thin plastic substrates are of great interest for a wide variety of engineering applications. Flexible devices can be rolled for transportation, installed on curved surfaces, and they are less likely to be damaged with motions and vibrations. Lightweight photovoltaics can also be used in mobile devices and various electric appliances to cover part of their power demand from ambient illumination using their free-form cases.

Regardless of the intense research efforts directed to the development of flexible a-Si:H solar cells on low-cost plastic substrates, these devices still exhibit considerably lower performance in comparison to that for glass-based equivalents. The major technological challenge is deposition of doped- and undoped a-Si:H layers with required electronic properties at temperatures lower than that for solar cells on glasses or metal foils [1-3]. The increased shunt leakage in Si:H cells on the plastic substrate is another challenging issue [4].

In this contribution, we report on a monolithic a Si:H-based photovoltaic module utilizing the 100 μm thick polyethylene-naphtalate (PEN) substrate. The detailed description of the device design and fabrication steps is the starting point of the presented work. Then, the modeling results are reported and compared to device output characteristics. The impact of the shunt leakage on the device performance and nature of observed micro-shunts in the p-i-n cells are also discussed.

2 DEVICE DESIGN AND FABRICATION

Figure 1 shows a photograph of the developed PV module and cross-sectional view of two cells connected in series. The module of 10 cm × 10 cm area consists of 72 rectangular cells on the PEN substrate. The individual cells are connected in series forming eight rows with connection pads. The novelty in the device design is in the backside encapsulation and front buffer silicon-oxynitride layers incorporated for device integrity. The a-Si:H n-i-p cell integrates the Al/Cr and ZnO:Al layers as the back and top electrodes, respectively. Two 0.3 mm wide Al fingers are symmetrically placed on the ZnO:Al electrode to reduce the emitter resistance. The photosensitive area is 1 × 0.8 cm^2.

In the developed fabrication process, three shadow masks are used in sputtering steps to form the bottom and top electrodes, and top metallization. The first step is sputtering of Al/Cr layers through mask #1. Here, Cr serves as an adhesion layer providing a low contact

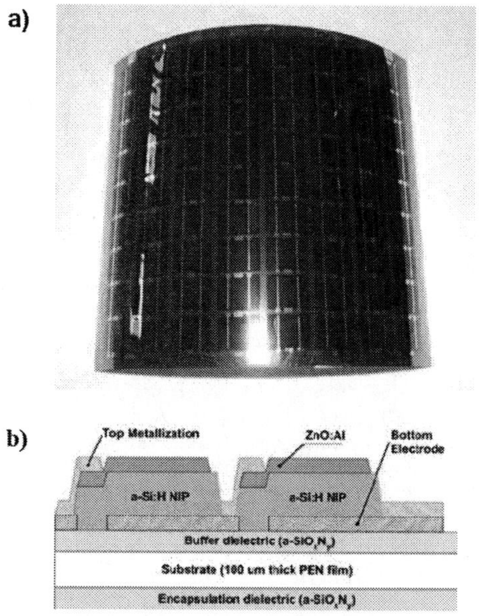

Figure 1: (a) Photograph of the PV module, (b) cross-sectional, and (c) top views of two cells connected in series.

resistance at the metal/n-layer interface. Then, the n-i-p stack is deposited using a 13.56 MHz PECVD system at 150 °C substrate temperature. The deposition conditions are reported elsewhere [5]. The formation of ZnO:Al top electrodes is performed by sputtering through shadow mask #2 at 140 °C substrate temperature. To perform via opening, the n-i-p stack is selectively etched through shadow mask #3 in the reactive ion etching (RIE) system. Finally, the Al fingers and contact pads are sputtered through shadow mask #3.

3 DEVICE MODELLING

The design optimization and device performance analysis are performed using a two-dimensional distributed circuit model of the photovoltaic cell [6]. The circuit simulator SPICE was used to calculate current and potential distributions in a network of sub-cell circuits, while MATLAB was used to create so-called "net list" file, to readout data from the output file, and to visualize

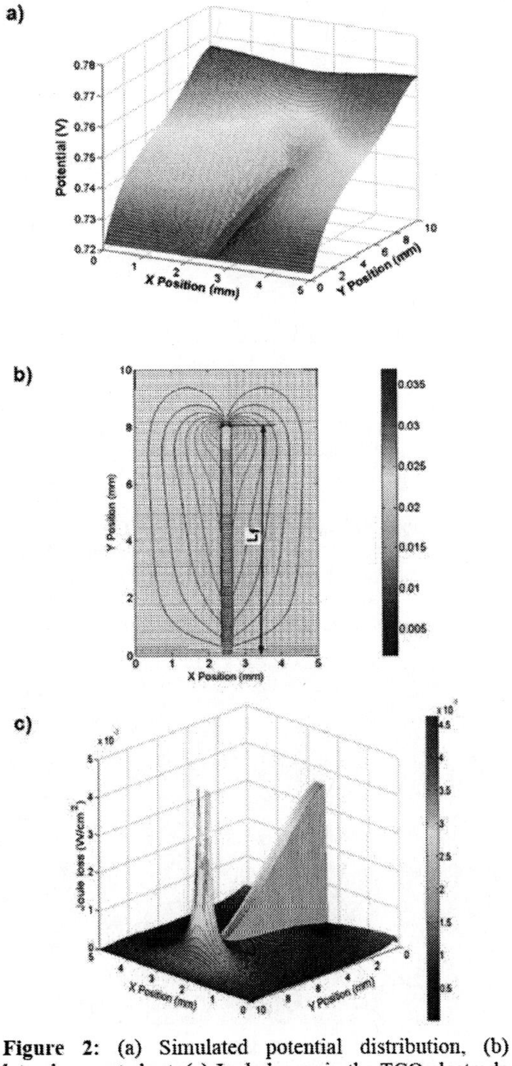

Figure 2: (a) Simulated potential distribution, (b) lateral current chart, (c) Joule losses in the TCO electrode and metal finger under AM1.5 illumination conditions.

the results. Figure 2 shows the simulated potential distribution, lateral current chart, and map of Joule losses in the TCO electrode and one metal finger. The simulation is performed for AM1.5 illumination conditions at a maximum power point (V_{max} = 0.722 V, I_{max} = 4.8 mA). In Fig. 2(a), the voltage drop across the metal finger is 12 mV, while the voltage variation across the transparent electrode is up to 45 mV at 0.2 and 60Ω/sq sheet resistances of metalization and TCO layers, respectively. The significant increase in the current density at the metal finger corners is observed in Fig. 2(b). The 'corner effect' is more prominent in Fig. 2(c), where two sharp peaks indicate the local increase in Joule losses. The Joul loss in the metal finger is also significant.

The grid electrode optimization requires balancing the resistance losses against other related losses such as grid electrode shadowing. We have calculated the power conversion effeiciency (PCE) for various grid electrode geometries. Fig. 3(b) shows the PCE as a function of distance between metal fingers (D_f) at various metal finger lengthes (L_f). The calculations were performed for a metal finger width of 0.3 mm at 60 and 0.2 Ω/sq sheet resistances of TCO and Al top layers, respectively. From the results, the values D_f = 5 mm and L_f = 8 mm are to be optimal for 10 mm × 10 mm cell having two fingers. Fig. 3(b) shows PCE = $f(D_f)$ curves calculated at various sheet resistances of the TCO layer. For cells with 10 and 60 Ω/sq TCO layers the difference in power conversion efficiency is up to 0.23%. In practice, this difference may be even lower due to an increased absorption loss in the thicker TCO layer. Thus, the optimized cell with a thinner transparent electrode yields almost the same power conversion efficiency in comparison to the cell of conventional design with 10 Ω/sq TCO.

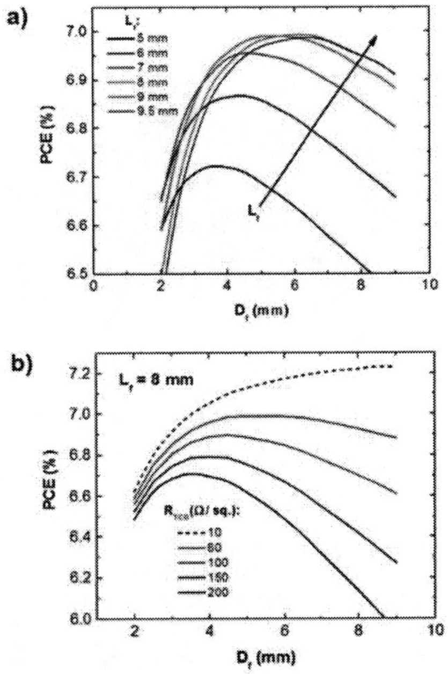

Figure 3: Power conversion efficiency as a function of of distance between metal fingers (D_f) at various (a) metal finger lengthes (L_f) and (b) sheet resistances of the TCO layer.

4 DEVICE CHARACTERIZATION

Fig. 4 shows the typical I-V curve of the module section composed of 18 cells under AM1.5 illumination conditions. The open circuit voltage and fill factor here are lower than what is supposed to be due to existence of resistive shunts in some cells in the module. Current-voltage characteristics of all individual a-Si:H p-i-n cells in the module were measured and analyzed to better understand the issue. It has been found that the shunt current varies linearly with biasing voltage, what can be represented as a shunt resistance (R_{sh}). The statistics on shunt resistances is shown in Fig. 5, where the number of cells with certain ranges of shunt resistance in the module is shown in terms of its probability. Here, the shunt resistance varies in wide range from 10 to 10^5 Ω.

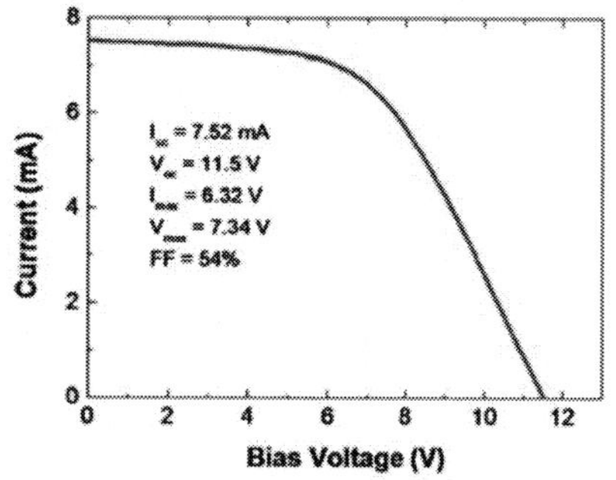

Figure 4: Current-voltage characteristics of the module section (18 individual cells connected in series) under AM1.5 conditions.

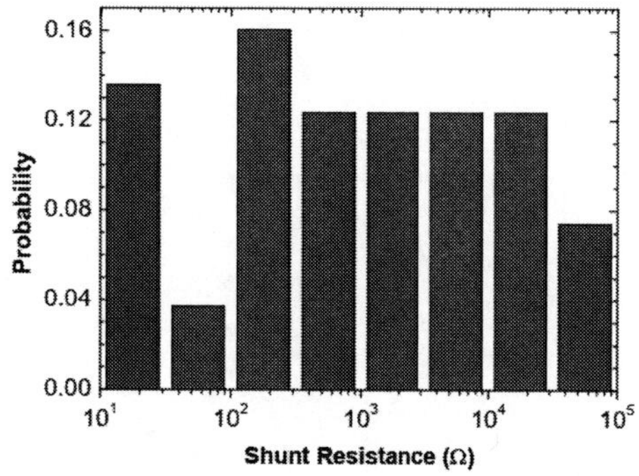

Figure 5: Probability histogram showing distribution of cells with various shunt resistances in a module.

The performed laser beam induced current (LBIC) experiments show the presence of multiple shunts. The characterization was performed in dark and unbiased conditions to reduce the noise and to avoid artificial signal variations. The *ac* component of the output voltage was measured when the device impedance was low (< 1 kΩ). Fig. 6(a) shows a LBIC signal map indicating the shunt position manifested as decreased LBIC signal. In the absence of shunts the signal is constant over the whole area, excluding the metal finger area due to light blockage. Using the developed SPICE model, the LBIC signal was also simulated considering a single shunt at different locations. Fig. 6(b) shows the X-direction scan curves calculated at 1 μA laser-induced photocurrent, and 100 Ω shunt resistance. Modelling shows that the shunting microdefect affects the LBIC signal from an area much larger than its actual size since current from the surrounding area is shunted through the defect. The magnitude of signal variation depends on the sheet resistance of the TCO layer [7]. In the area shielded by the metal finger, the LBIC signal is almost constant that is in good agreement with the experimental LBIC results.

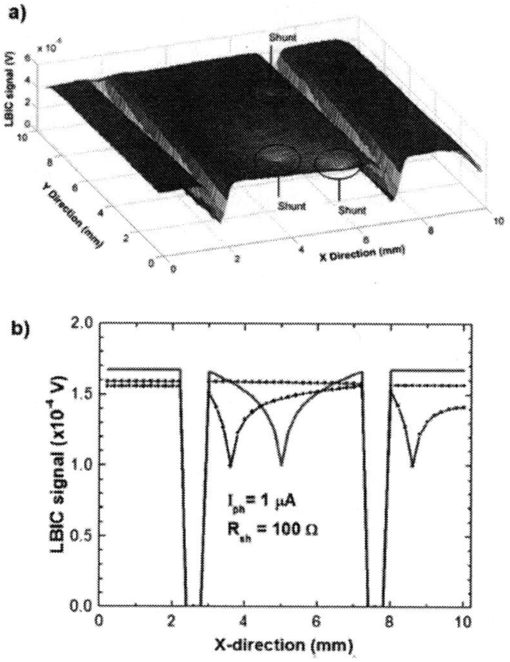

Figure 6: (a) LBIC signal for cell with multiple shunts; (b) Calculated LBIC signals for the cell with 100 Ω shunt at various positions.

The point defects are also obsrved in the SEM image of the cell with shunts in Fig. 7. The density of these defects is much higher than what is expected from external contamination. To understand the nature of electrical shunts, we analyzed the change in the surface

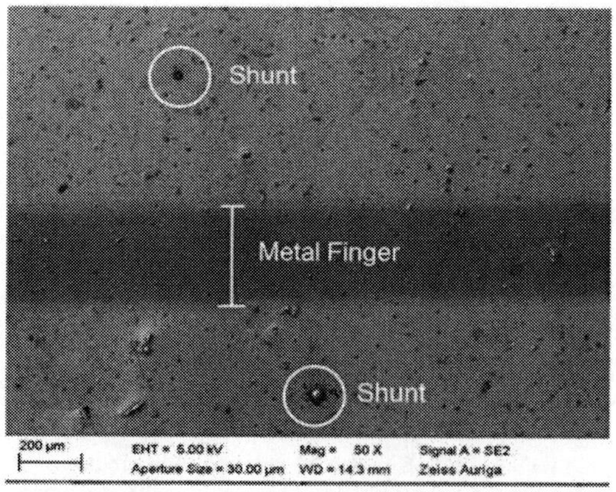

Figure 7: SEM image of the cell with microshunts.

roughness of all layers comprising individual solar cells throughout fabrication process. Multiple experiments have confirmed that the defects appear after deposition of dielectric or semi-conductor layers, or even after sputtering of metal layers on the PEN substrate. The reason is that cyclic oligomers, which are present in PEN, can migrate to the surface forming crystals, if the film is held at temperatures >100°C for tens of minutes [8]. The observed substrate defects may form microscopic pinholes, which are filled with highly conductive top layer material at the final process step, resulting in ohmic shunts.

6 CONCLUSIONS

A monolithic PV module integrating an array of 72 a-Si:H n-i-p cells on a PEN substrate has been fabricated. The metal grid geometry has been optimized using a two-dimensional SPICE model of the photovoltaic cell. The device characterization shows that the shunt leakage is a major factor reducing the output characteristics. The presence of multiple ohmic micro-shunts is confirmed by the SEM and LBIC techniques. The formation of micro-shunts is attributed to surface defects in plastic foils, which are thermally induced during the device fabrication.

ACKNOWLEDGEMENTS

The authors are grateful to the Portuguese Foundation of Science and Technology through fellowship SFRH/BPD/102217/2014 and to Instituto Politécnico de Lisboa (ID&CA project-SolWin) for financial support of this research.

REFERENCES

[1] X. Bories-Azeau, W. A. MacDonald, and J. M. Mace. Proceedings 5th World Conference on Photovoltaic Energy Conversion (2010) 603.

[2] D.D. Fischer et al. Proceedings 29th European Photo-voltaic Solar Energy Conference and Exhibition, Vol. 1 (2010) 1929.

[3] J. K. Rath, M. Brinza, Liu Y., A. Borreman, and R.E.I. Schropp. Sol. Energ. Mater. Sol. Cell. 94 (2010) 1534.

[4] M. Fortes, E. Comesana, J.A. Rodriguez, P. Otero, and A.J. Garcia-Loureiro. Solar Energy 100 (2014) 114.

[5] K. H. Kim, Y. Vygranenko, D. Striakhilev, M. Bedzyk, J.H. Chang, A. Nathan, T.C. Chuang, G. Heiler, and T. Tredwell. J. Non-Crystal. Solids 354 (2008) 2590.

[6] Y. Vygranenko, M. Fernandes, M. Vieira, A. Khosropour, and A. Sazonov. Energy Procedia 60 (2014) 96.

[7] P. Vorasayan, T.R.Betts, R.Gottschalg. Sol. Energ. Mater. Sol. Cell. 95 (2011) 111.

[8] MacDonald, M. K. Looney et al. Journal of the SID 15 (2007) 1075.

INTERDIGITATED LASER-CONTACTED SOLAR CELL ON LIQUID-PHASE CRYSTALLIZED SILICON ON GLASS

Michael Vetter[1)]*, Gema Lopez[2)], Pablo Ortega[2)], Isidro Martin[2)], Gudrun Andrä[1)], David Muñoz[3], Carlos Molpeceres[3)]
[1)] Leibniz-Institute of Photonic Technology (IPHT), Dept. Functional Interfaces, Albert-Einstein-Str. 9, 07745 Jena, Germany
[2)] Micro and Nanotechnologies Group MNT, Departament d'Enginyeria Electrònica, Universitat Politècnica de Catalunya UPC, C/ Jordi Girona 1-3, Modulo C-4, 08034
Barcelona, Spain
[3)] Centro Láser UPM, Universidad Politécnica de Madrid, Ctra. De Valencia Km 7.3, 28031 Madrid, Spain
*Corresponding author: Phone +34 616 368 575
Email: michael.vetter.privat@gmx.net

ABSTRACT: 10 µm thick liquid-phase crystallized silicon (Si) layers on 3.2 mm Borofloat 33 glass (5cm x 5cm) are fabricated by continuous wave line focus laser (808 nm). A sputtered SiO_2/SiON layer stack has been implemented as barrier layer at the glass Si interface. Solar cells with interdigitated back contact are prepared on these multi-crystalline layers by using low temperature (<400°C) Doped-by-Laser (DopLa) fabrication concept. ALD-Al_2O_3/PECVD amorphous intrinsic Si carbide (SiC_y(i)) stack for emitter passivation and a-SiC_y(i)/P-doped a-SiC_z(n)/a-SiC_x(i) stack for base passivation are deposited. With short pulse (ns) UV laser (355 nm) p-type emitter (by Al-diffusion from Al_2O_3) and point contacts with back surface field (by P-diffusion from a-SiC_z(n) layer) are fabricated on n-type mc-Si absorbers layers. Solar cell process steps are monitored by charge carrier lifetime measurements using the quasi steady-state photoconductance method. A strong dependency of charge carrier lifetime on injection level and doping density of absorbers is observed. Higher lifetimes are found for lower absorber doping concentrations. In 0.7 Ωcm LPCSG absorber highest $\tau_{eff} \approx 300$ ns corresponding to $L_{eff} > 20$ µm is found. Laser-doping and contacting using UV laser resulted in small (< 20 mV) loss of V_{oc}(1sun). Crack formation in LPCSG absorbers after laser crystallization process presents technological problems in the preparation of the interdigitated metal contact.

Keywords: multi-crystalline silicon, thin film, laser crystallization, laser doping, carrier lifetime, quasi steady-state photoconductance, IBC

1 INTRODUCTION

Liquid-phase crystallized silicon on glass (LPCSG) is a promising material to fabricate high quality silicon thin films on glass, e.g. for solar cells. A high electronic material quality is achieved by first depositing barrier layers (e.g. SiO_x, SiN_x, SiO_xN_y) and then an amorphous silicon (Si) or nano-crystalline Si layer on the glass followed by crystallization with a line focus laser or electron beam. Solar cells (≈1cm^2) made with LPCSG absorbers have shown open circuit voltage (Voc) of 649 mV (Jsc=27.3 mA/cm^2, FF=68.8%, η=12.1%) [1] demonstrating high material quality approaching highest efficient multi-crystalline wafer solar cells [2]. LPCSG could merge the advantages of crystalline silicon wafer technology with its high efficiency potential and thin film technology with low Si consumption and low cost monolithic integration for module fabrication. Therefore, LPCSG technology could be an approach to overcome the emerging limits for further cost reduction in standard wafer-based module technology. However, a suitable process technology still has to be developed.

Different fabrication concepts for the solar cell emitter, including phosphorus diffusion [3], laser processing [4], or hetero junction formation by amorphous Si layers [5] have been implemented. The fabrication of well passivated absorber contacts has become in the focus recently [6]. In this work, we implement for the first time dielectric layers namely an atomic layer deposited (ALD) aluminum oxide (Al_2O_3)/plasma-enhanced chemical vapor deposited (PECVD) amorphous intrinsic silicon carbide (SiC_x(i)) stack and a-SiC_y(i)/phosphorus (P)-doped a-SiC_z(n)/a-SiC_x(i) stack. In conjunction with short puls (ns) laser (355 nm) processing we fabricate p-type emitter (by Al-diffusion from Al_2O_3) and n-type back surface field (BSF) point contacts (by P-diffusion from a-SiC_z(n) layer) on n-type LPCSG absorbers.

2 EXPERIMENTAL

2.1 LPCSG absorber fabrication

LPCSG absorbers were processed on Borofloat 33 substrates with a sputtered SiO_2/SiON layer stack. Then, an intrinsic a-Si layer (10 µm) was deposited by e-beam evaporation. Doping of absorbers was achieved by deposition of thin (5-50 nm) highly phosphorus doped a-Si films by PECVD on the intrinsic film. Next, crystallization was performed by continuous wave line focus (13 mm long line with 0.1 mm full width at half maximum) laser irradiation at 808 nm at a scan rate of 10 mm/s and a power of 12 kW/cm^2. In this process the Si film is completely molten for some milliseconds and the doping atoms are distributed homogenously in the layer before cooling down and crystallization, forming a multi-crystalline (mc) layer. A hydrogen plasma treatment in a parallel plate reactor was performed at about 500°C for 30 minutes to passivate electronic defects in the mc-Si. The absorbers were fabricated at IPHT in Jena, Germany.

2.2 Charge carrier lifetime measurement

After hydrogen plasma treatment, a surface etch and a RCA cleaning is performed with a final HF (2%)-dip and charge carrier lifetime is measured immediately using the H-termination of the Si surface resulting from the HF-dip as electronic surface passivation. The HF-dip provides a very good surface passivation and this first lifetime measurement gives information on the electronic quality of the absorbers at the initial stage of the solar cell fabrication process. We use injection level dependent lifetime data generated by the quasi steady-state photoconductance (PC) decay method (QSSPC) with a modified (IPHT) WCT-120 PC tool (Sinton Consulting) to characterize LPCSG absorbers [7]. This contactless method allows then a monitoring of the different solar cell process steps [8].

2.3 Solar cell fabrication

We fabricate an interdigitated back contact (IBC) solar cell scheme with 3cm x 3cm cell area on 3.2 mm Borofloat 33 (Schott) substrates going about a factor of 10 beyond the actual size of lab cells for this technology. Figure 1 presents a side view scheme of the solar cell structure (a photo of a processed LPCSG absorber is shown in Fig. 8).

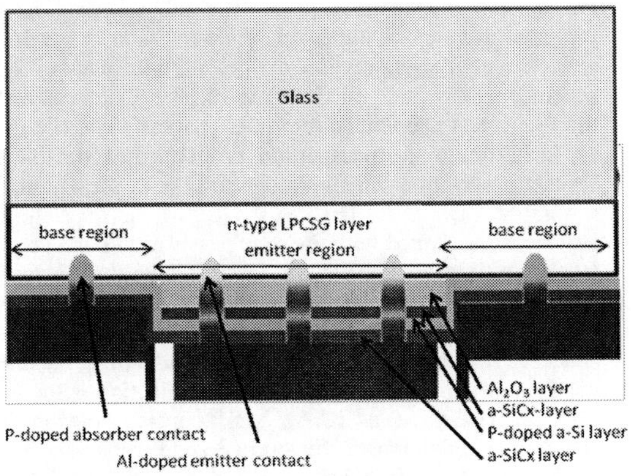

Figure 1: Side view scheme of IBC cell.

The technology is adapted from the so-called (Doped by Laser) DopLa-IBC [9] fabrication concept that has demonstrated efficiencies of 20% in wafer solar cells and is perfectly suited for the processing of solar cells on glass substrates since all process steps are performed at temperatures lower than 400°C [10, 11, 12]. In addition, the whole fabrication process includes only two lithography steps which in the future possibly also could be performed by laser-assisted processing.

3 RESULTS/DISCUSSION

Fig. 2 presents minority charge carrier lifetime curves of four LPCSG absorbers with different specific resistivity. Latter is achieved by depositing P-doped a-Si(n) layers with different thickness on the absorber before the laser crystallization process. Doping density presents strong impact on lifetime of LPCSG absorbers resulting in higher maximum lifetime for lower doping density, resp. higher resistivity. Samples with higher doping density present a stronger decay of lifetime in low injection, while samples with lower doping density present a stronger decay of lifetime in high injection. A plot of the inverse lifetime versus injection level presents a straight line in high injection. From its slope a reverse saturation current density (j_o) has been calculated (0.2 Ωcm: j_o=1.4x10^{-11}A/cm^2, 0.3 Ωcm: j_o=1.7x10^{-11}A/cm^2, 0.7 Ωcm: j_o=1.2x10^{-11}A/cm^2). The surface passivation by HF-dip of the top surface of all samples is the same. It results in a drastic reduction of surface recombination centers at the top surface. This does not create the strong dependency on injection level at high injection. Therefore, we attribute the shape of the lifetime curve at high injection to recombination properties at the glass-silicon interface. Most probably an electric field has been created at the SiON/Si interface after the laser crystallization process. A detailed analysis of the lifetime data is still under work. In any case the maximum of the lifetime of each curve gives an indication of the lower limit of the bulk lifetime and diffusion length in the absorber. The minority carrier lifetime higher than 300 ns of the n-type absorber with 0.7 Ωcm indicates a bulk diffusion length longer than 20 μm in the 10 μm thick absorber.

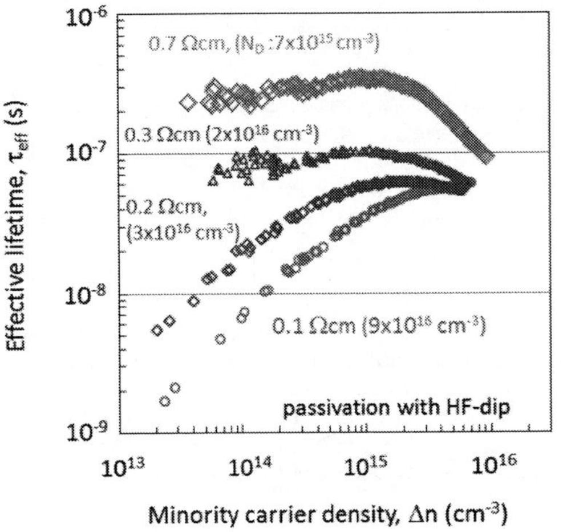

Figure 2: Lifetime curves calculated from photoconductance (PC) data of LPCSG absorbers with different specific resistivity of the mc-Si layer. A strong dependency of lifetime on injection level and doping density is observed. Data at low injection levels is limited by the sensitivity of the WCT-120 PC tool. (top surface passivation by HF-dip.

Fig. 3 shows respective Sun-V_{oc} curves calculated from the PC data of samples of Fig. 2. The slope of the Suns-Voc curve (typically between 60-120 mV/dec) gives information about recombination mechanism in the samples. One sees that the dominant recombination mechanism changes from high to low injection for all samples.

Extrapolation of Suns-V_{oc} curves to 1 sun in low and high injection parts of the curves indicates the implied V_{oc}(1sun) at standard test conditions which can be achieved under the different recombination mechanism. An example is given by the continuous black lines for the blue rhombus data of the 0.7 Ωcm sample. The high injection part indicates a V_{oc}(1sun) of 594 mV, however in low injection a recombination process reduces V_{oc}(1sun) to 552 mV.

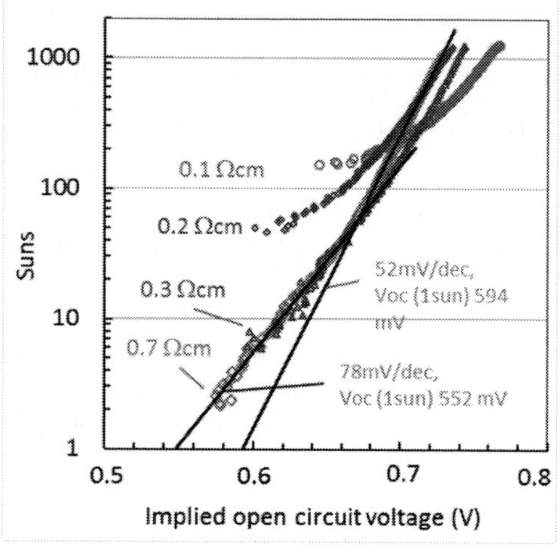

Figure 3: Sun-V_{oc} curves calculated from the PC data of samples in Fig. 2. Continuous black lines are fitting blue rhombus data of the 0.7 Ωcm sample in low resp. high injection. In this way an implied Voc (1sun) at standard test conditions is estimated for the limiting recombination mechanism in low and high injection.

Fig. 4 presents lifetime data of the 0.7 Ωcm precursor of Fig. 2 after different process steps to fabricate the interdigitated contact structure of the solar cell (see photo of a finished solar cell in Fig. 8). These processes were performed at the UPC and all details of this kind of process can be found in references [10-12].

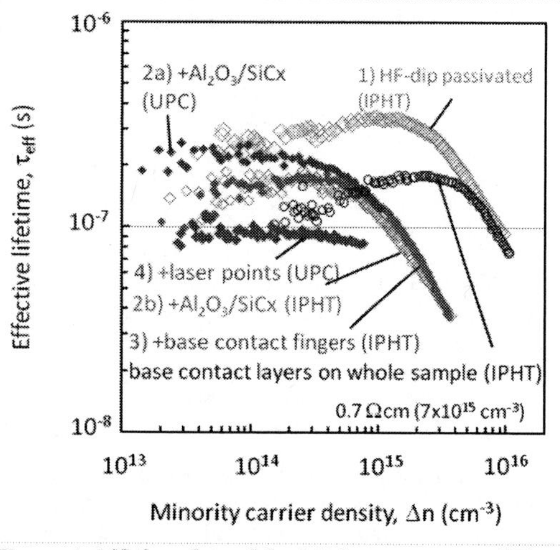

Figure 4: Lifetime data of the 0.7 Ωcm precursor of Fig. 1 after different process steps to fabricate the interdigitated contact solar cell structure.

The solar cell fabrication starts with ALD of Al_2O_3 layer and PECVD of SiC_x(i) layer which presents the basic dielectric passivation scheme of the emitter. After this process step lifetime measurement is performed at the UPC a standard WCT-120 instrument (indicated by violet rhombus and 2a) in Fig. 4) with and also at IPHT (indicated by green rhombus and 2b in Fig. 4), where the modified WCT-120 provides lifetime data at very high injection level. One sees that the data nearly coincides in the range of $\Delta n = 10^{15}$ cm^{-3} while a difference is seen at low injection levels. This might be due to slight degradation of passivation by handling and transportation. Comparing HF-dip lifetime data with the ones obtained after Al_2O_3/SiCx deposition, a clear shift of high injection lifetime to lower values can be seen, indicating that now the recombination at the top surface is dominating here. Fig. 5 presents the respective Suns-Voc curves, here we see in high injection a shift of about 20 mV to lower voltage after depositing the emitter passivation scheme. However, in low injection the change in slope is less resulting in nearly the same V_{oc}(1sun) as before.

Next, a lithography step is performed to open a window for the base passivation scheme in the emitter passivation layers. The width of the emitter fingers is about 500 μm and of the base fingers about 250 μm. Then, the SiC_y(i)/SiC_z(n)/a-SiC_x(i) layer stack for laser doping and passivation of the base contact is deposited by PECVD on the whole surface. The lifetime data after these process steps is indicated by the red rhombus and 3) and present the passivation quality of both passivation schemes side by side on the same surface. One sees that the lifetime is maintained after this process steps. To get an insight into the passivation properties of the back contact passivation scheme Fig. 4 and Fig. 5 present also a sample with a similar resistivity where the whole surface is passivated with the base passivation layer stack (shown by black circles). We see that here that in high injection the carrier lifetime decreases in the same way as with HF-dip passivation, indicating that here also the SiON/Si interface is dominating recombination, but the maximum lifetime and lifetime in low injection is lower. The SiC_y(i)/SiC_z(n)/a-SiC_x(i) back contact scheme is based on a reduction of the surface state density by the Si-rich SiC_y(i) layer and the creation of an accumulation layer by the P-doped SiC_z(n)layer (n/n+ hetero junction BSF).

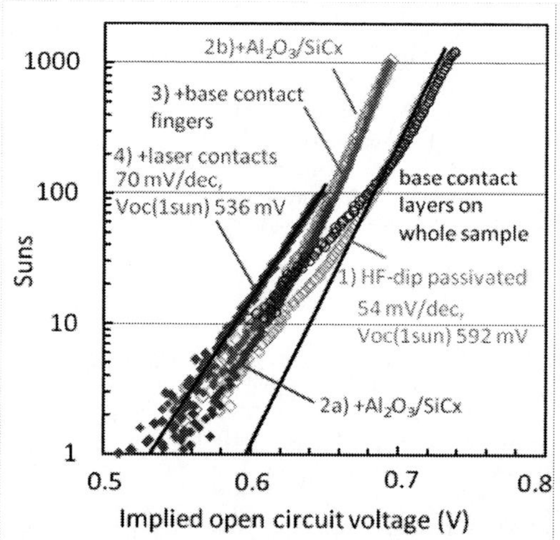

Figure 5: Respective Suns-Voc data which will be discussed in the paper (continuous black lines are fitted to the data of step 3 (red rhombus).

The next step is the fabrication of laser point contacts which has been performed by an UV laser at 355 nm. A typical laser spot is shown in Fig. 6 where the inner 9.5 μm radius presents the electrical contact area, the 14 μm radius is limit of a "dead area" and the 19.4 μm radius is the limit of an "ablation area" where part of passivation stack is partially removed. The arrangement of laser point contacts in the base (dark) and emitter region (bright) is shown in Fig. 7. Details on the UV-laser process parameters will be published elsewhere.

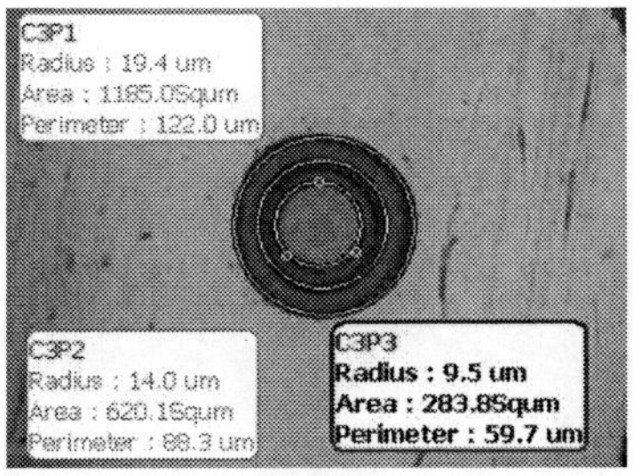

Figure 6: Typical laser spot produced with 355 nm UV laser.

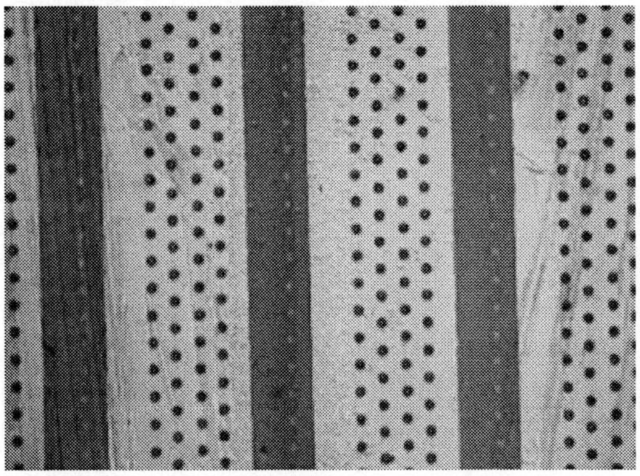

Figure 7: Arrangement of laser point contacts in the base (dark) and emitter region (bright).

The lifetime is measured after the laser process again and data is shown in Fig. 4 and 5 by the dark blue rhombus (step 4). One sees that lifetime in low injection is only reduced by a factor of two and the extrapolated V_{oc}(1sun) is 536 mV not very far from the value extrapolated at low injection of the initial measurement with HF-dip passivated surface (552 mV see Fig. 1). The emitter grid passivation consist finally in a $Al_2O_3/SiC_x(i)//SiC_y(i)/SiC_z(n)/a\text{-}SiC_x(i)$ layer stack with Al-diffused p-type emitter point contacts produced by the laser process. The absorber contact passivation scheme consists in a $SiC_y(i)/SiC_z(n)/a\text{-}SiC_x(i)$ layer stack where the laser forms point contacts with a n/n^+ BSF. Finally, electrical contacts are made by evaporation of Al on the whole top surface of the LPCSG absorber. By lithography and an etch step the emitter and base contacts are separated.

Crack formation in LPCSG absorbers after laser crystallization process presents technological problems in the preparation of the interdigitated metal contact structure. Optimization of the laser crystallization process to prevent glass bending and crack formation is necessary. Nevertheless, partial solar cells present well-rectifying diodes.

4 CONCLUSION

Quasi steady-state photoconductance method is a very helpful method to determine electronic quality of thin (about 10-20 μm) LPCSG absorbers and to monitor solar cell processing. However, sensitivity of WCT-120 photoconductance instrument limits acquisition of data in low injection. Nevertheless, lifetime and Suns-Voc curves give useful information about absorber quality and impact of process steps on recombination. In 0.7 Ωcm LPCSG absorber the $\tau_{eff} \approx 300$ ns corresponds to bulk diffusion length $L_{eff} > 20$ μm larger than the absorber thickness. Extrapolation of Sun-Voc curves in high injection indicates a limit of V_{oc}(1Sun) ≈ 600 mV representing bulk quality, however recombination in low injection reduces V_{oc}(1Sun) to about 550 mV. The similarity of Sun-V_{oc} curve and lifetime curve in high injection of H-passivated LPCSG absorber and absorber with back passivation scheme which is based on accumulation of the Si surface suggests that also at the glass/Si interface an accumulation layer is present. Simulation of charge carrier lifetime could help to better understand the different recombination mechanism and to improve surface passivation. Laser-doping and contacting applying UV laser resulted in small (<20 mV) loss of V_{oc}(1sun). Therefore, UV laser doping presents a very promising method for the fabrication of sophisticated and miniaturized contact structures.

Figure 8: 5cm x 5cm solar cell with 3cm x 3cm interdigitated contact structure and metallization.

5 ACKNOWLEDGEMENT

This project has received funding from the European Union's Horizon 2020 research and innovation programme under the Marie Sklodowska-Curie grant agreement No. 657115 (cSiOnGlass, M.V.). This work has been supported by MINECO through project TEC2014-59736-R and has been funded by ENE2016-78933-C4-1-R and ENE2013-48629-C4-3-R. The authors would like also to thank the Thüringer Aufbaubank (TAB) and the European Commission (ESF) for funding the project Bi-PV (FKZ 2015 FGR 0078).

References

[1] T. Frijnts,S. Kühnapfel, S. Ring, O. Gabriel, S. Calnan, J. Haschke, B. Stannowski, B. Rech, R.Schlatmann, Sol. Energy Mater. Sol. Cells 143, (2015) 457.

[2] M. A. Green, K. Emery, Y. Hishikawa, W. Warta, E. D. Dunlop, Progr. Photovol. Res. Appl. 24 (2016) 905.

[3] J. Dore, D. Ong, S. Varlamov, R. Egan, M.A. Green, IEEE J. Photovolt.4 (1) 33 (2014).
[4] A. Gawlik, I. Höger, J. Bergmann, J. Plentz, T. Schmidt, F. Falk, and G. Andrä, Phys. Status Solidi RRL 9 (7) (2015) 397.
[5] J. Haschke, D. Amkreutz, L. Korte, F. Ruske, B. Rech B, Sol. Energy Mater. Solar Cells 128 (2014)190.
[6] P. Sonntag, J. Haschke, S. Kühnapfel, T. Frijnts, D. Amkreutz, B. Rech, Progr. Photovolt. Res. Appl. 24 (2016) 716.
[7] M. Vetter, A. Gawlik, J. Plentz, G. Andrä, Energy Procedia 92 (2016) 248.
[8] M. Vetter, A. Gawlik, J. Plentz, G. Andrä, Proc. 32nd Europ. Photovolt. Solar Energy Conf. Exh., Munich, Gernany, 2016 (WIP Munich, 2016) pp. 1100.
[9] I. Martin, M. Colina, A. Coll, G. Lopez, P. Ortega, A. Orpella, R. Alcubilla, Energy Procedia 55 (2014) 255.
[10] G. Lopez, P. Ortega, I. Martin, C. Voz, A. Orpella, R. Alcubilla, Energy Procedia 92 (2016) 652.
[11] G. Lopez Rodríguez, Interdigitated Back-contacted (IBC) c-Si solar cells based on laser processed dielectric layers, PhD Thesis, Universitat Politècnica de Catalunya, Barcelona (2016).
[12] P. Ortega, G. López, D. Muñoz, I. Martín, C. Voz, C. Molpeceres, R. Alcubilla. Sol. Energy Mater. Sol. Cells 169 (2017) 107.

UNDERSTANDING CONTACT FORMATION ON N-PERT-RJ SOLAR CELLS

Corrado Comparotto[1], Jens Theobald, Jan Lossen, Valentin D. Mihailetchi
International Solar Energy Research Center (ISC) Konstanz, Rudolf-Diesel-Straße 15, 78467 Konstanz, Germany
[1]Corresponding author. Tel.: +49-7531-361-83-372; E-mail address: corrado.comparotto@isc-konstanz.de

ABSTRACT: In this work we investigated the metallization process of screen-printed n-type passivated emitter rear totally diffused (n-PERT) rear junction (RJ) silicon solar cells. We used the voltage calibrated photoluminescence (PL) method to extract the saturation current density on the laser ablated surface before metallization ($J_{0\text{-}abl}$) and at the metal-silicon interface after metallization ($J_{0\text{-}met}$). Symmetrically diffused and passivated test samples, printed with specific metallization patterns, were fabricated for this purpose. We investigated the influence on $J_{0\text{-}met}$ varying boron (p^+) and phosphorous (n^+) diffusions, cleaning after laser ablation, metal pastes and peak firing temperature. The optimum parameters were then applied to fabricate n-PERT-RJ cells.
Keywords: Silicon Solar Cell, n-type, Recombination, Metallization, Photoluminescence

1 INTRODUCTION

The majority of crystalline silicon solar cells fabricated today is based on boron-doped (p^+) silicon and features a full-area aluminum (Al) contact on the rear side, which results in high rear surface recombination. In order to reduce this loss and therefore allow a high open-circuit voltage (V_{OC}), an increasing number of cell manufacturers are currently bringing the Passivated Emitter and Rear Cell (PERC) concept into production [1]. It features a full-area dielectric rear passivation and only local Al rear contacts. Using B-doped Czochralski-grown silicon (Cz-Si), a new record efficiency of 22.61% was recently reported for industrial PERC solar cells [2]. However, after light-induced degradation (LID), the efficiency of PERC solar cells can decrease up to 1.0%$_{abs.}$ depending on wafer resistivity [3,4,5]. A way to circumvent the detrimental effect of LID is to use n-type Si wafers and the n-PERT-RJ solar cell concept, which features a phosphorus-doped (n^+) front surface field (FSF) at the front and a boron-doped emitter at the rear. It commonly uses a screen-printed contact grid on both sides, as schematically shown in Figure 1. This concept is very similar to p-PERC solar cells in terms of cell architecture and processing sequence and can therefore be easily industrialized.

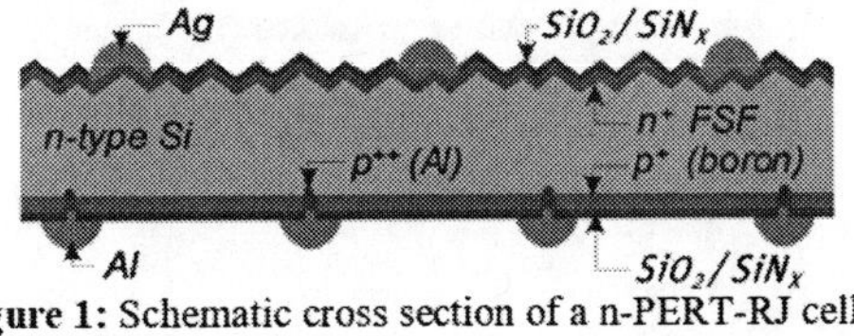

Figure 1: Schematic cross section of a n-PERT-RJ cell

In the first place we compared two different methods for the extraction of $J_{0\text{-}met}$ and chose the most suitable one for this study. Then we focused on investigating and minimizing the metallization-induced recombination losses in order to improve cell performance.

2 COMPARISON BETWEEN *I-V* MEASUREMENTS AND PL IMAGING FOR THE EXTRACTION OF $J_{0\text{-}met}$

We fabricated 156×156 mm² n-PERT-RJ solar cells with varying metal contact fractions (F_M) on the front side. F_M is defined as the ratio of the contact areas to the total area of solar cell. *I-V* measurements and PL imaging were performed on the cells in order to extract $J_{0\text{-}met}$, applying methods proposeded by Hoenig et al. [6] and Shanmugam et al. [7]. The saturation current density (J_{01}) was calculated using the one-diode model and plotted versus F_M in Figure 2. $J_{0\text{-}met}$ was determined from the linear fit of J_{01} vs F_M.

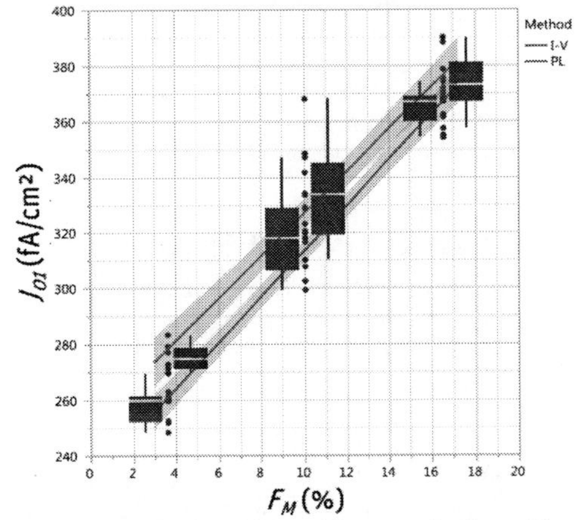

Figure 2: J_{01} of cells with varying F_M on the front side

We observed a good agreement between the *I-V* and PL characterization methods. The extracted $J_{0\text{-}met}$ is found to be of 885 fA/cm² and 823 fA/cm², respectively. The difference in $J_{0\text{-}met}$ between these two values corresponds to a difference in V_{OC} of ≤ 0.7 mV, which lies within the measurement uncertainty.

The PL based method was chosen for the following investigation in this work, as it does not require working solar cells for determining J_{01}. It can therefore be applied at intermediate cell fabrication steps, e.g. after laser ablation and before metallization.

3 EXPERIMENTAL APPROACH

3.1 $J_{0\text{-}met}$ on n^+

In order to determine $J_{0\text{-}met}$ of our n-PERT-RJ cell concept, expressly designed symmetrically textured and P-diffused test samples, with a specific metallization pattern, were fabricated.

156×156 mm² high ohmic Cz-Si n-type pseudo-square wafers were saw damage etched and texturized in an alkaline solution and then cleaned in a piranha solution. Afterwards, the n^+ diffused region was formed in a quartz tube furnace using a $POCl_3$ gas. Subsequently, the passivation and anti-reflection coating (ARC) was deposited on both sides of the sample by plasma-enhanced chemical vapor deposition (PECVD). The metallization was applied by screen printing and firing of Ag metal pastes using a specifically designed grid pattern on one side of the wafer, consisting of nine regions of five different F_M, as shown in Figure 3. Hence, only one screen is needed to print all different regions of interest, as also proposed by Shanmugam et al. [7] and Fellmeth et al. [8]. The schematic cross section of the final test samples is depicted in Figure 4.

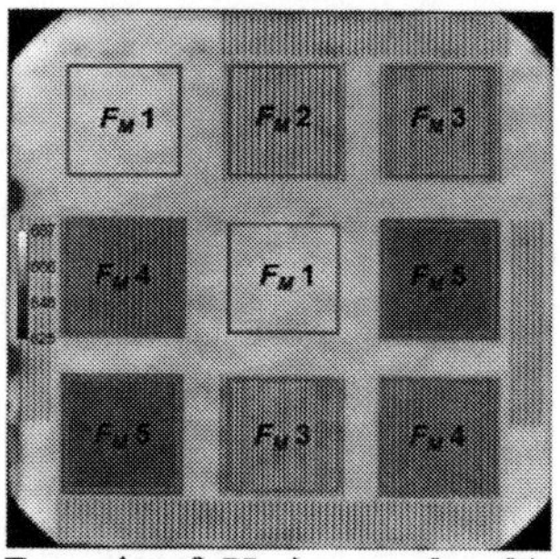

Figure 3: Example of PL image of a fabricated test sample for the determination of $J_{0\text{-}met}$. The color scale shows the absolute V_{OC} values in mV

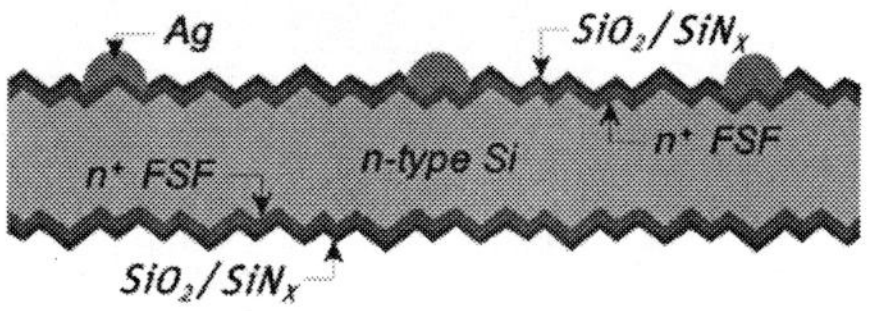

Figure 4: Schematic cross section of the test samples used for the determination $J_{0\text{-}met}$ on n^+

In order to quantitatively image the local solar cell V_{OC}, the PL method was applied on the fabricated samples. It relies on calibrating the local luminescence intensity to the local junction voltage. The measurement was performed from the metalized side, which corresponds to the sunny side of n-PERT-RJ cells. V_{OC} was averaged within each of the nine wafer regions with varying F_M (highlighted blue empty squares in Figure 3) and converted into J_{01} using the one-diode model. The photogenerated current was determined by means of the wafer ray tracer tool from PV Lighthouse [9], simulating the cell structure in Figure 4 and correcting for metallization shading. J_{01} was then plotted versus the correspondent F_M (an example is shown in Figure 5). The slope from the linear fits yielded the metal contact contribution $J_{0\text{-}met}$ of the samples for given sets of parameters (n^+ diffusion, metal paste, peak firing temperature). $J_{0\text{-}pass}$ was determined by QssPC from the metal free regions (F_M 1).

The P diffusion was adjusted in order to yield R_{sheet} of 80 Ω/□, 101 Ω/□, 133 Ω/□ or 140 Ω/□. Three different Ag firing through pastes were also tested (labeled Paste A, Paste B and Paste C). Additionally, the peak firing temperature was varied from 830°C to 870°C with steps of 20°C. The results are plotted in Figure 6.

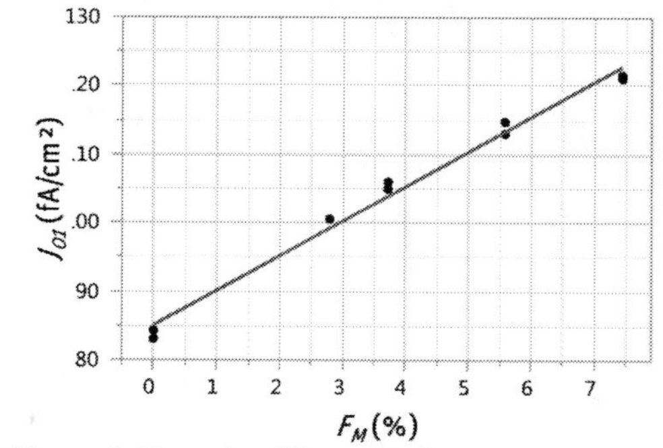

Figure 5: Example of linear fit of J_{01} vs F_M for the test sample in Figure 3

We observed that $J_{0\text{-}met}$ depends on all tested parameters: Al paste, P diffusion and peak firing temperature. Lowest $J_{0\text{-}met}$ is found for lowest R_{sheet}, lowest peak firing temperature and mostly when using Paste B.

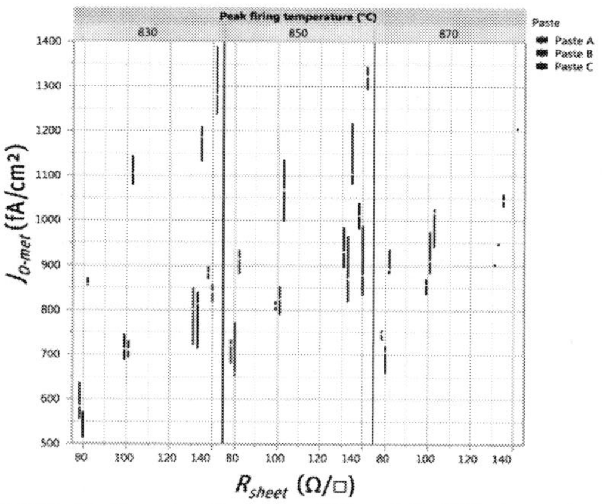

Figure 6: $J_{0\text{-}met}$ vs R_{sheet} for different P diffusions, Ag pastes and peak firing temperatures

On the contrary, $J_{0\text{-}pass}$ is lowest for highest R_{sheet}, as can be seen in Figure 7. This indicates that the best P diffusion in terms of cell performance (for homogeneous diffusions) is the one that minimizes the area weighted combined contributions of $J_{0\text{-}met}$ and $J_{0\text{-}pass}$ together.

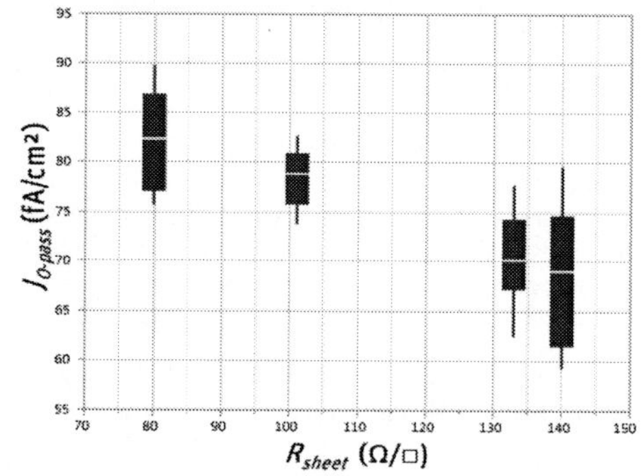

Figure 7: $J_{0\text{-}pass}$ vs R_{sheet} of different B diffusions

3.2 $J_{0\text{-}met}$ on p^+

Similarly to the approach above, specifically

designed test samples were fabricated with the purpose of determining $J_{0\text{-}met}$ on a p^+ diffused region.

The starting wafer material and the initial chemical steps were the same as in the previous case, with the exception of the texturization, which was omitted this time, resulting in symmetrically flat samples, as depicted in Figure 8. The p^+ region was formed in a quartz tube furnace containing a BBr_3 gas. The surface passivation of the p^+ diffused regions was achieved by a stack of in situ grown thermal SiO_2 and a PECVD deposited SiN_X ARC layer [10]. This time the dielectric layers were ablated by laser on one side of the test sample using the lines pattern visible in Figure 3. Afterwards, a short cleaning was performed in a 2% HCl/HF solution. Finally, Al lines were screen printed on top of the laser openings and the test samples were fired. PL imaging was performed this time from the non-metalized side, in order to simulate the working operation of n-PERT-RJ cells.

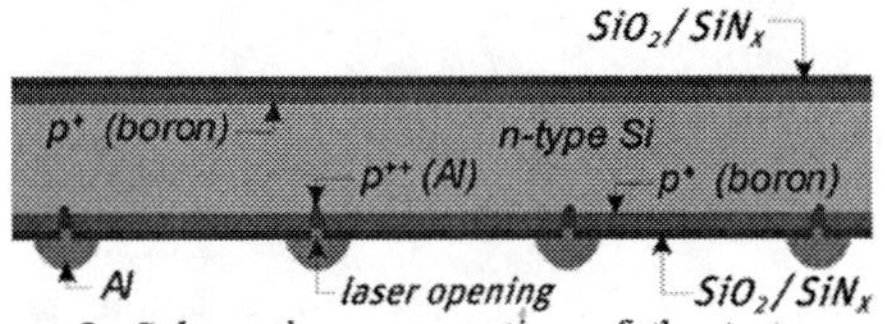

Figure 8: Schematic cross section of the test samples used for determining $J_{0\text{-}met}$ on p^+

In this case, we can identify three differently passivated areas at the sample surface: the well passivated p^+-SiO_2/SiN_X region, the silicon-p^{++} interface (which defines the contact fraction F_M) and the SiO_2/SiN_X-Al area next to the local contact openings. We focused first on the latter in order to understand whether the passivation in this region degrades after the fast firing process. For this purpose, the laser ablation step was omitted for some test samples. The metal lines were printed with different Al pastes directly on top of the passivation stack and fired. An example of PL image from one of such samples is shown in Figure 9. The visible bright stripes indicate a high PL signal coming from the printed Al lines and can be seen for all Al pastes used and for all samples. It follows that there is no passivation loss in the SiO_2/SiN_X-Al area. The fact that the lines are brighter than the rest of the passivated area is caused by the higher internal reflection of the Si/SiO_2/SiN_X/Al stack as compared with only Si/SiO_2/SiN_X. This effect is found to be in the order of few fA/cm² and will be neglected in the rest of the investigation in this sub-section.

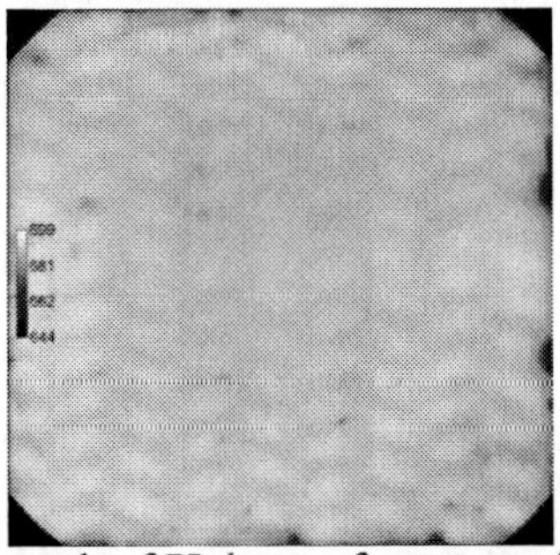

Figure 9: Example of PL image of a test sample where the Al paste was directly screen printed on top of the passivation stack, omitting laser ablation

Employing the test sample structure in Figure 8, two experiments were carried out.

In the first one we investigated the influence on $J_{0\text{-}met}$ varying three different B diffusions, adjusted to yield R_{sheet} of 85 Ω/□, 100 Ω/□ or 140 Ω/□. A reference Al paste and a commercial Al paste produced by Toyal, i.e. TB-06K-B, were tested. The peak firing temperature was also varied, i.e. 830°C, 850°C or 870°C.

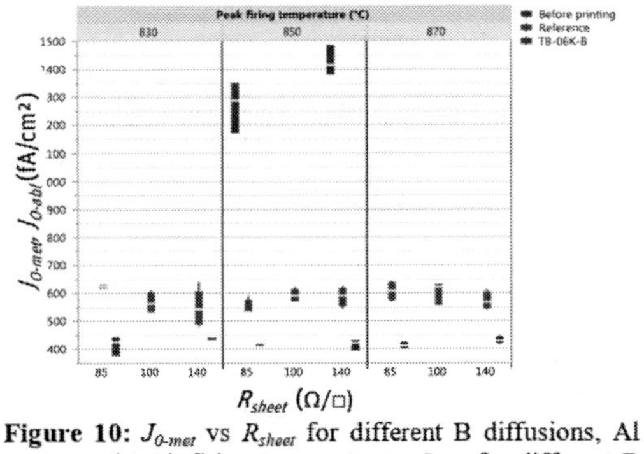

Figure 10: $J_{0\text{-}met}$ vs R_{sheet} for different B diffusions, Al pastes and peak firing temperatures; $J_{0\text{-}abl}$ for different B diffusions

Two main conclusions can be drawn from the results in Figure 10. Firstly, we observe that $J_{0\text{-}abl}$ (before metallization) is much higher than $J_{0\text{-}met}$ after screen printing and firing of either Al paste. This is because the laser ablation process causes a significant surface damage to the p^+ emitter. Secondly, it can be seen that $J_{0\text{-}met}$ is strongly dependent on the Al paste used but is almost independent from R_{sheet} of the p^+ formed emitter and from the peak firing temperature, at least in the tested ranges of 85 Ω/□ to 140 Ω/□ and 830°C to 870°C. This indicates that the Al p^{++} emitter is uniformly well formed and the Al metal is not in direct contact with the p^+ (boron) emitter, which will otherwise result in a significant enhancement of $J_{0\text{-}met}$.

In the second experiment we investigated the influence on $J_{0\text{-}met}$ and contact resistivity (ρ_c) of a cleaning in a 2% HCl/HF solution post laser ablation and the influence of different commercial Al pastes produced by Toyal, i.e. TB-06K-A and TB-06K-B. The Si content by weight in TB-06K-A was either 0% or 5%. The results are plotted in Figure 11.

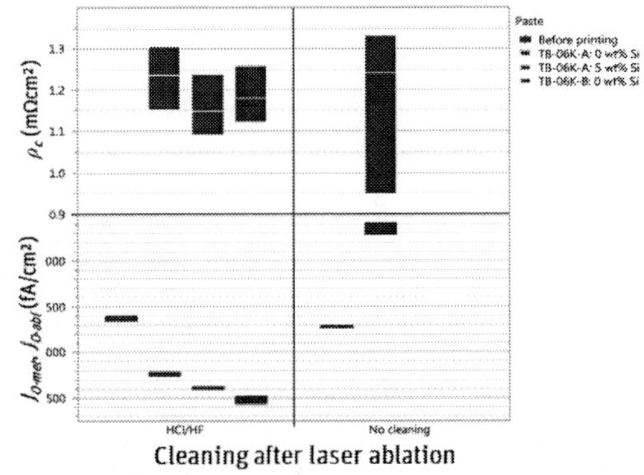

Figure 11: $J_{0\text{-}met}$, $J_{0\text{-}abl}$ and ρ_c vs cleaning after laser ablation for different Al pastes

It can be seen that the cleaning step strongly reduces $J_{0\text{-}met}$, although it has almost no impact on $J_{0\text{-}abl}$ and ρ_c. The Si content in TB-06K-A of 5 wt% reduces both $J_{0\text{-}met}$ and ρ_c, as compared with similar Al paste but without Si (TB-06K-A: 0 wt% Si). However, the lowest $J_{0\text{-}met}$ of around 500 fA/cm² is achieved with a paste of different chemical composition (TB-06K-B: 0 wt% Si).

4 CELL RESULTS

The optimum parameters found in this study were then applied to fabricate n-PERT-RJ cells.

For this purpose, we used 156×156 cm² n-type Cz wafers with base doping of 13.5 Ωcm and an as-cut thickness of 200 µm. Prior to the diffusion processes, all wafers were etched in an alkaline solution and cleaned in a piranha solution. The n^+ and p^+ diffused regions were formed in a quartz tube furnace containing either a BBr_3 or a $POCl_3$ gas, using improved diffusion recipes specifically developed for this cell concept. R_{sheet} of the n^+ diffused region was adjusted to yield 150 Ω/□, whereas the p^+ diffusion was adjusted to 140 Ω/□. On both sides, the in situ grown thermal SiO_2 was capped by PECVD SiN_X for passivation [10]. On the rear side, line pattern contact openings were ablated by laser. Afterwards, the cells were cleaned in a 2% HCl/HF solution. Paste B (Ag) and TB-06K-B from Toyal (Al) were screen printed on the front and rear side, respectively, both using an H-pattern grid design. The optimum co-firing conditions were applied. A schematic cress section of the finished cell is shown in Figure 1. The *I-V* parameters of the champion cell are presented in Table I and the correspondent internal quantum efficiency (IQE) in Figure 12. $J_{0\text{-}tot}$ is calculated from the one-diode model equation.

Table I: *I-V* parameters of the champion cell

J_{SC} (mA/cm²)	V_{OC} (mV)	$J_{0\text{-}tot}$ (fA/cm²)	FF (%)	Efficiency (%)
38.6	**676**	**150**	79.4	20.7

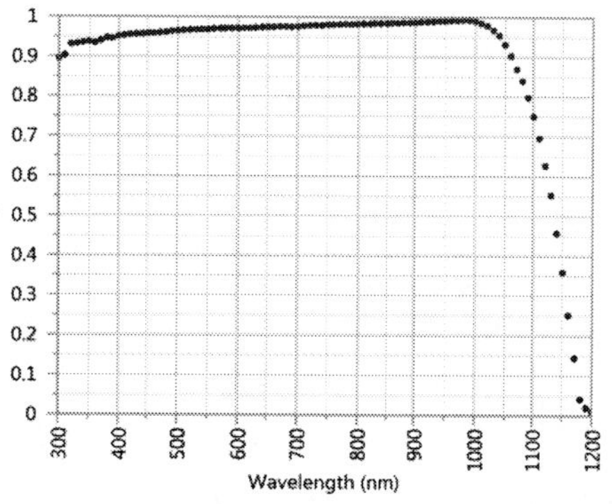

Figure 12: IQE of the champion cell

The champion cell shows high V_{OC} for a screen printed solar cell, only 8 mV lower than the world record PERC cell made by Deng et al. [2]. FF and J_{SC} still need to be improved. A simple way is to reduce F_M on the front side of the cell, e.g., by increasing the number of busbars and at the same time either reducing the number of fingers or narrowing the fingers width, or both.

In Table II the J_0 contributions of each cell region are shown.

Table II: J_0 contributions of the champion cell

Region		J_0 (fA/cm²)	Area (%)	J_0 area weighted (fA/cm²)
Front	Pass	25	96.2	24
	Met	1250	3.8	48
Rear	Pass	30	93.1	28
	Met	500	6.9	34
Bulk				16
Total				150

The largest J_0 contribution comes from the metal-silicon interface on the n^+ region. Therefore, in order to further improve V_{OC}, the optimization of the front side metallization will be targeted first, by reducing F_M and further decreasing $J_{0\text{-}met}$.

5 CONCLUSIONS

We investigated the metallization-induced recombination losses on both sides of our n-PERT-RJ solar cell concept. On the front side, $J_{0\text{-}met}$ is strongly dependent on the Ag paste used and is minimized for low R_{sheet} and low peak firing temperature. On the rear side, $J_{0\text{-}met}$ is strongly dependent on Al pastes and is reduced if a cleaning post laser ablation is performed. Here, we did not observe a significant dependence on R_{sheet} and peak firing temperature. Using these findings, we fabricated n-PERT-RJ solar cells with 20.7% efficiency, a V_{OC} as high as 676 mV and $J_{0\text{-}tot}$ as low as 150 fA/cm².

6 ACKNOWLEDGEMENT

This work was financially supported by the German Federal Ministry for Economic Affairs and Energy within the project KosmoS (contract number: 02E2-41V7204).

7 REFERENCES

[1] S. K. Chunduri, M. Schmela, PERC Solar Cell Technology 2017 Edition, TAIYANGNEWS, 2017.
[2] W. Deng, F. Ye, R. Liu, Y. Li, H. Chen, Z. Xiong, Y. Yang, Y. Chen, Y. Wang, P. P. Altermatt, Z. Feng, P. J. Verlinden, 22.61% Efficient fully Screen Printed PERC Solar Cell, 44th IEEE Photovoltaic Specialists Conference, 2017. To be published.
[3] F. Wolny, T. Weber, M. Müller, G. Fischer, Light induced degradation and regeneration of high efficiency Cz PERC cells with varying base resistivity, Energy Procedia 38, pp. 523 – 530, 2013.
[4] Y. Gassenbauer et al., "Rear-surface passivation technology for crystalline silicon solar cells: A versatile process for mass production," IEEE J. Photovoltaics, vol. 3, no. 1, pp. 125–130, 2013.
[5] T. Dullweber et al., "Silicon wafer material options for highly efficient ptype PERC solar cells," Proc. 39th IEEE Photovoltaic Spec. Conf., Tampa, FL, USA, pp. 3074–3078, 2013.
[6] R. Hoenig, M. Glatthaar, F. Clement, J. Greulich, J.

Wilde, D. Biro, New measurement method for the investigation of space charge region losses induced by the metallization of silicon solar cells, Energy Procedia, Volume 8, pp. 694-699, 2011.

[7] V. Shanmugam, T. Mueller, A. G. Aberle, Johnson Wong, Determination of metal recombination parameters for silicon wafer solar cells by photoluminescence imaging, Solar Energy, 118, pp. 20-27, 2015.

[8] T. Fellmeth, A. Born, A. Kimmerle, F. Clement, D. Biro, R. Preu, Recombination at metal–emitter interfaces of front contact technologies for highly efficient silicon solar cells. Energy Proc. 8, pp. 115–121, 2011.

[9] https://www.pvlighthouse.com.au/.

[10] V. D. Mihailetchi, H. Chu, J. Lossen, R. Kopececk, Surface passivation of boron diffused junctions by borosilicate glass and in situ grown silicon dioxide interface layer, 44th IEEE Photovoltaic Specialists Conference, 2017. To be published.

EVALUATIONS OF PASSIVATED SILICON SURFACES WITH LASER TERAHERTZ EMISSION MICROSCOPE (LTEM) AND CORONA CHARGING

Toshimitsu Mochizuki[1], Akira Ito[2], Hidetoshi Nakanishi[2], Jonathon Mitchel[1], Katsuto Tanahashi[1], Masaaki Moriya[1], Yasuhiro Kida[1], Iwao Kawayama[3], Masayoshi Tonouchi[3], Katsuhiko Shirasawa[1], and Hidetaka Takato[1]
[1]Fukushima Renewable Energy Institute, The National Institute of Advanced Industrial Science and Technology
2-2-9 Machiikedai, Koriyama, Fukushima 963-0298, Japan
[2]SCREEN Holdings, Co., Ltd.
322, Fukukawa-cho, Hazukashi, Fushimi-ku, Kyoto, Kyoto 612-8486, Japan
[3]Institute of Laser Engineering, Osaka University
2-6, Yamadaoka, Suita, Osaka 565-0871, Japan
e-mail : toshimitsu-mochiduki@aist.go.jp

ABSTRACT: Surface band bending evaluation by assessment of Terahertz (THz) emission from silicon wafers using a laser terahertz emission microscope (LTEM) was extended to an evaluation of oxide-passivated n-type silicon surfaces in combination with a corona charger. The THz amplitude of the sample was nearly uniform, indicating electron accumulation at the entire semiconductor-oxide interface. Deposition of negative charge decreased the THz amplitude and further deposition inverted the THz waveform, showing an inversion of surface band bending. We expect that deposited negative charge density equalled to positive oxide charge when the THz amplitude crossed 0 and waveform inverted. A drop of lifetime of the charged sample was also observed at the deposited charge density where we can expect that the Fermi level at the surface was near the midgap and surface recombination rate is maximised. The systematic change in LTEM signal with the deposited charge indicates that LTEM can optically measure and map the surface potential and dielectric charge of passivated surfaces of semiconductors.
Keywords: Silicon, SiO_2, Evaluation, Interfaces

1 INTRODUCTION

Surface passivation is an essential part of all solar cells. Regarding silicon solar cells, various materials such as a-Si:H, SiO_x, SiN_x, and AlO_x are deposited as passivation layers to minimise the surface recombination velocity. Overall quality of the passivation layers can be evaluated by assessing surface recombination velocity *via* I-V characteristics, minority carrier lifetime, or internal quantum efficiency. If the quality of a passivation layer is lower than designed or expected, surface defect density is high and/or surface band bending is inappropriate so that product of density of electrons and holes at the interface are higher than expected. Therefore, evaluation of surface band bending contribute to the optimisation of fabrication processes. The surface band bending is often evaluated by measuring capacitance-voltage (C-V) and voltage-charge (V-Q) relationships, or by electron-based spectroscopy. C-V is poor at detecting local information, V-Q is difficult to apply to passivation layers which are not insulating, and electron-based spectroscopy instruments are bulky and complex. Hence we have proposed laser THz emission microscope (LTEM) as a new, simple tool for evaluating local surface potential[1–4]. Previously, we have reported the systematic relationship between the surfaces potential and laser terahertz emission of oxidised silicon using semi-transparent metal-oxide semiconductor sample.

In this work, we demonstrate the use of a LTEM in combination with a corona charger. THz emission from a Si wafer with SiO layer with various deposited charge density. We have compared the obtained waveforms with the photoluminescence intensity measured using a LED and a cooled Si-CCD camera in the LTEM, and lifetime measured with a QSS-μPCD technique. Deposition of negative charge decreased the THz amplitude and deposition of -3×10^{-11} cm-2 or more negative charge inverted the THz waveform, showing an inversion of surface band bending. A drop of lifetime of the charged sample was also observed. The observed results shows that LTEM in combination with corona gun can measure total dielectric charge of passivation layer on semiconductors.

2 SAMPLE AND EXPERIMENT

Figure 1 shows the structure of the measured sample and experiment. The sample used is an n-doped CZ c-Si wafer coated in a 90 nm wet oxidised layer with a dopant density of 9.8×10^{14} cm^{-2}. The oxidation is done at 900 °C in nitrogen gas and water vapour. After the oxidation, the sample was annealed in 3% H_2/97% N_2 gas at 450 °C for 20 minutes. The thickness of oxidised layer was measured using an ellipsometer. Negative or positive ions are deposited on oxidised surface with a corona charger covering a 12 mm spot size with various charge density, which is uniform over an area 6 mm diameter. Then the sample is transferred to the LTEM for THz time-domain spectroscopy measurement and mapping. While the THz emission was measured in 1 hour after the deposition, the deposited charge was stable so that the measuring the THz emission 24 hour after the deposition did not significantly changed the results.

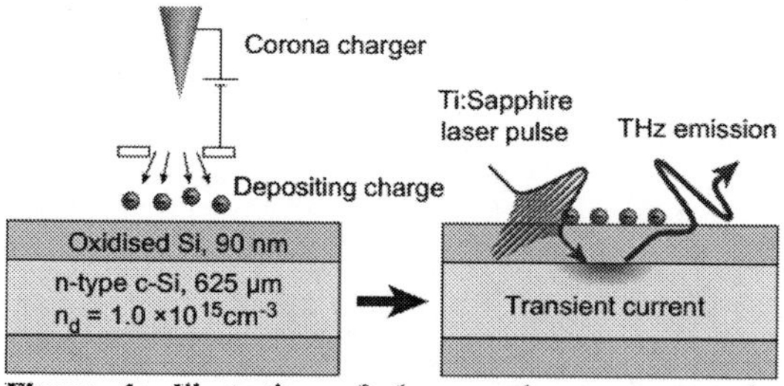

Figure 1: Illustration of the sample structure and experiment.

Figure 2 illustrates the THz time-domain spectroscopy measurements using LTEM. Femtosecond pulses from the Ti:Sapphire laser are focused on the sample and the LT-GaAs based THz detector. The sample is photoexcited at 80 MHz by 800 nm source at an incident angle of 45°, covering a $1/e^2$ diameter of 10 mm. Averaged power for 100 fs pulse width was 100 mW. The THz emission is focused on the detector with the pair of elliptic mirrors. The sample is placed on a 3-axis stage, which enables mapping of the THz waveform. Measurement of the THz waveform is done by sweeping the delay stage and detecting the THz signal with the lock-in amplifier.

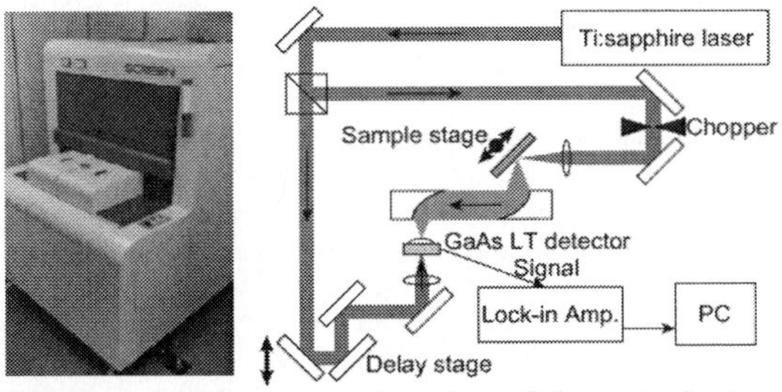

Figure 2: Appearance and optics of laser terahertz emission microscope (LTEM).

Photoluminesence images were taken using a setup in the LTEM, consisting of a LED and an electrically cooled Si-CCD camera. Minority carrier lifetime was measured using Semilab PV-2000A with a QSS-μPCD technique, exciting the sample by 2.5 MHz 904 nm source covering a diameter of 1.1 mm. The photon number per 200 nm pulse was 1.5×10^{13}.

3 RESULTS AND DISCUSSION

3.1 LTEM and PL image of the sample with deposited charge

Figure 3 shows changes of THz amplitude and PL intensity by four charge deposition spots of different charge density, measured and mapped using the LTEM. Fig. 3(a) and (b) shows amplitude of THz emission at 10.5 ps and PL before the charge deposition, respectively. THz amplitude is uniform across the sample and PL intensity is approximately uniform. Fig. 3 (c) shows positions and charge density of the deposition spots. Fig. 3(d) and (e) shows THz amplitude and PL intensity after the deposition. Deposition of negative charge decreased the THz emission and the deposition -4×10^{-11} cm^{-2} or more negative charge changed the sign of the peak amplitude, indicating the transition to the inverted surface. The deposited charge is seen to alter the band bending at the silicon surface. Negative charge deposition also changed the PL intensity. The PL intensity is decreased by deposition up to -4×10^{-11} cm^{-2} and further deposition increased the PL intensity. As discussed in detail later, the change in THz emission and PL intensity is consistent and LTEM is seen to measure and map the change in surface potential induced by charge deposition.

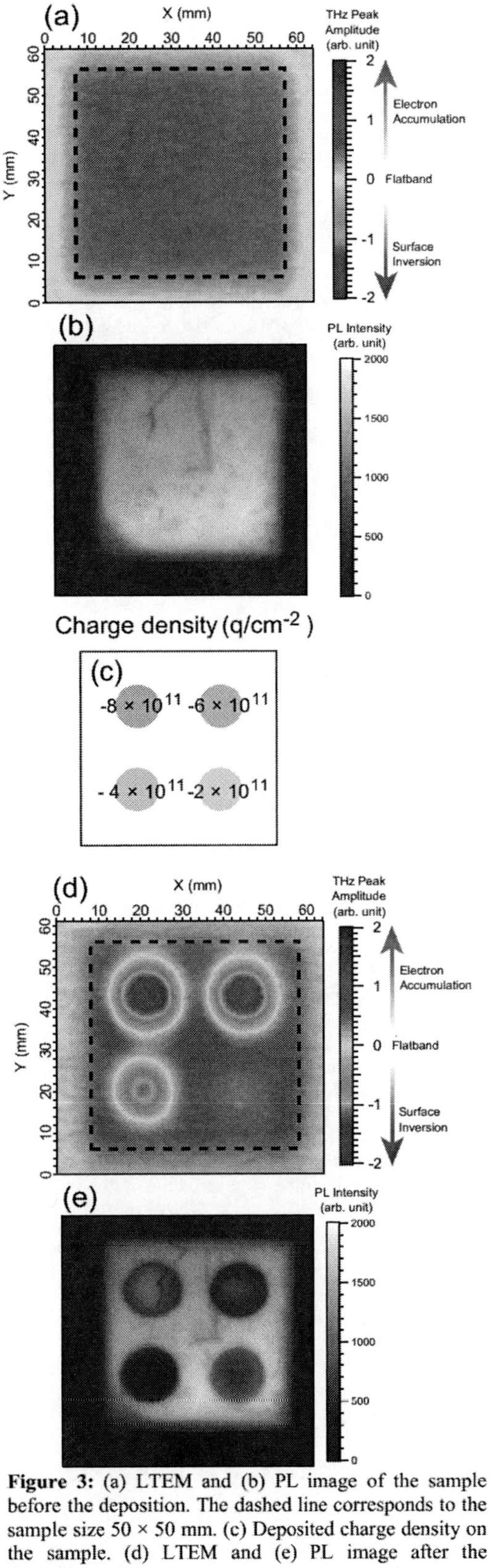

Figure 3: (a) LTEM and (b) PL image of the sample before the deposition. The dashed line corresponds to the sample size 50 × 50 mm. (c) Deposited charge density on the sample. (d) LTEM and (e) PL image after the deposition.

3.2 THz emission from the centre of the sample with various charge density

More systematic relation between THz emission and deposited charge density is studied at the centre of the sample. THz emission was measured using the LTEM and carrier lifetime was also measured at the same time. Figure 4 shows the waveform with various charge density. To change the charge density, sample was rinsed with ultrapure water to remove the deposited charge and then corona charge was deposited at a desired density.

Figure 4 shows waveforms of THz emission from the centre of the sample with various charge density. Deposition of positive charge did not significantly changed the waveform. Deposition of negative charge with density -2 × 10^{-11} cm^{-2} decreased the amplitude but did not inverted the waveform. Waveform is inverted by deposition of -4 × 10^{-11} cm^{-2} or more inverted the waveform and waveform at -6 × 10^{-11} cm^{-2} and -8 × 10^{-11} cm^{-2} is almost identical.

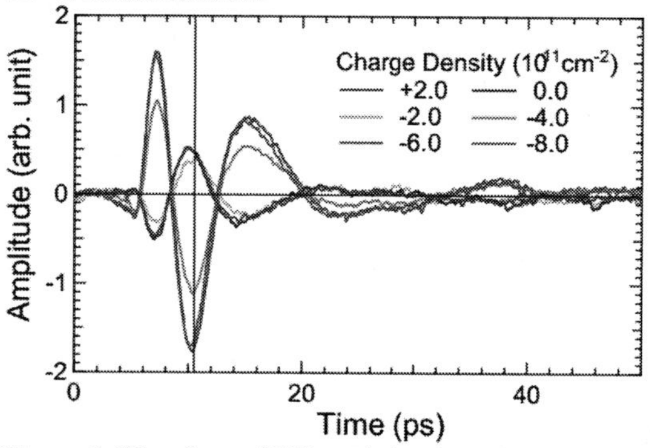

Figure 4: Waveform of THz emission from the centre of the sample with various deposited charge density. Vertical line shows 10.5 ps as peak position in this paper.

Figure 5 shows THz amplitude at 10.5 ps at various deposited charge density by black circles and minority carrier lifetime measured with a QSS-μPCD technique by red open circles. The THz amplitude and lifetime are insensitive to the positive charge deposition. Deposition of negative charge with density -2 × 10^{-11} cm^{-2} or less decreased the intensity and lifetime. Deposition of -3 × 10^{-11} cm^{-2} or more negative charge inverted the waveform and negative amplitude at 10.5 ps is observed. The amplitude changed further down to around -1.7 and becomes insensitive to deposition of -5 × 10^{-11} cm^{-2} or more. The lifetime is minimised at -4 × 10^{-11} cm^{-2}, and further negative charge deposition increased the lifetime.

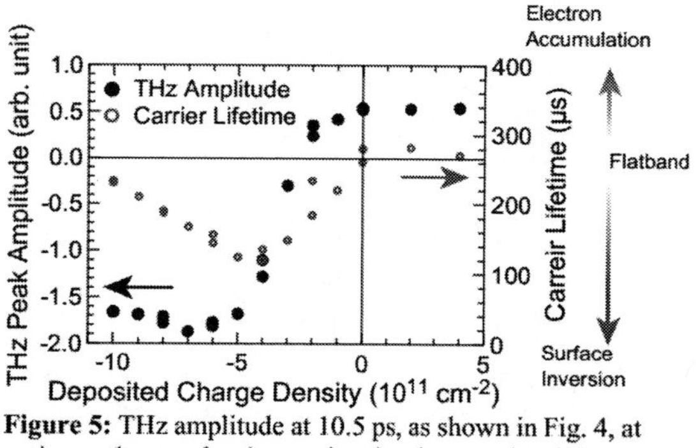

Figure 5: THz amplitude at 10.5 ps, as shown in Fig. 4, at various charge density and minority carrier lifetime measured with a QSS-μPCD technique.

Previously, we have reported that the intensity of THz emission from oxidised surface is very similar to the capacitance-voltage curve. When external voltage is applied using MOS structure, THz amplitude is insensitive as long as the surface is strongly inverted or accumulated, and is sensitive to surface potential near the flatband voltage. The THz amplitude crosses zero close to the flatband voltage.

Based on the above understanding, the observed behaviours of THz amplitude and minority carrier lifetime can be interpreted as follows: the deposited charge acted very similarly to the external voltage and insensitivity of the amplitude and lifetime to the positive charge deposition indicate that electron is accumulated at the oxidised surface in its initial state and positive charge deposition only changed the internal field of the oxidised layer and the densely accumulated layer which is very thin and closest to the surface. THz emission is insensitive to the changes in these layers. The insensitive minority carrier lifetime to the positive charge deposition indicates that the bulk was the dominant factor of minority carrier recombination and possible reduction of surface recombination by positive charge deposition did not significantly changed the carrier lifetime. Decrease and inversion of the peak amplitude by negative charge deposition can be interpreted that deposited charge counteracted the dielectric charge and decreased the surface potential, and then the surface accumulation layer changed to inversion layer. We can estimate that the total dielectric charge of the sample was between 2 × 10^{-11} cm^{-2} and 3 × 10^{-11} cm^{-2}.The dip in minority carrier lifetime at -4 × 10^{-11} cm^{-2} can be interpreted that the surface recombination velocity is maximised because Fermi level at the surface is close to the midgap, where product of electron and hole density at the surface is maximised.[5] This is consistent with the charge density dependence of THz peak amplitude, which indicates that electron is accumulated at the surface with charge density of -2 × 10^{-11} cm^{-2}, and the surface inversion layer is formed at -5 × 10^{-11} cm^{-2}.

4 SUMMARY

In summary, we measured and mapped the laser terahertz emission from oxidised silicon surface with various deposited charge density using laser terahertz emission microscope (LTEM). The waveform inverted by depositing -3 × 10^{-11} cm^{-2} or more negative charge, indicating that the deposited charge counteracted the dielectric charge and electron accumulation layer at the surface transitioned to inversion layer. We also estimated that the total dielectric charge was between 2 × 10^{-11} cm^{-2} and 3 × 10^{-11} cm^{-2}. We also observed a dip of minority carrier lifetime at -3 × 10^{-11} cm^{-2}, where we believe that Fermi level crossed the midgap and consistent to the deposited charge density dependence of the peak THz amplitude. The results strongly indicate that LTEM can be a non-invasive tool to optically evaluate field effect passivation of semiconductor devices. LTEM can also measure and map dielectric charge of passivation layers in combination with corona charging.

REFERENCES

[1] M. Tonouchi, Nat. Photonics **1**, 97 (2007).
[2] Y. Sakai, I. Kawayama, H. Nakanishi, and M. Tonouchi, Sci. Rep. **5**, 13860 (2015).

[3] H. Nakanishi, A. Ito, K. Takayama, I. Kawayama, H. Murakami, and M. Tonouchi, AIP Adv. **5**, 117129 (2015).
[4] T. Mochizuki, A. Ito, J. Mitchell, H. Nakanishi, K. Tanahashi, I. Kawayama, M. Tonouchi, K. Shirasawa, and H. Takato, Appl. Phys. Lett. **110**, (2017).
[5] S.W. Glunz, D. Biro, S. Rein, and W. Warta, J. Appl. Phys. **86**, 683 (1999).

Dry Etch Black Silicon with Low Surface Damage: Effect of Low Capacitively Coupled Plasma Power

Beniamino Iandolo[1], Maksym Plakhotnyuk[1], Maria Gaudig[2], Rasmus S. Davidsen[1], Dominik Lausch[3], Ole Hansen[1]

1 Department of Micro- and Nanotechnology, Technical University of Denmark, Ørsteds Pl., 2800 Kongens Lyngby, Denmark

2 Hochschule Anhalt, Fachbereich 6, Köthen (Anhalt), Germany

3 Fraunhofer Center for Silicon Photovoltaics (CSP), 06120 Halle (Salle), Germany

Abstract — Black silicon fabricated by reactive ion etch (RIE) is promising for integration into silicon solar cells thanks to its excellent light trapping ability. However, intensive ion bombardment during the RIE induces surface damage, which results in enhanced surface recombination velocity. Here, we present a RIE optimization leading to reduced surface damage while retaining excellent light trapping and low reflectivity. In particular, we demonstrate that the reduction of the capacitively coupled power during reactive ion etching preserves a reflectance below 1% and improves the effective minority carrier lifetime thanks to reduced ion energy. Surface passivation using atomic layer deposition of Al_2O_3 improves the effective lifetime to 7.5 ms and 0.8 ms for black silicon n- and p-type wafers, respectively.

1 INTRODUCTION

Black silicon textured surfaces generated using maskless reactive ion etching (RIE) are a potential alternative to wet-chemical textured surfaces in silicon solar cell fabrication [1,2]. Black silicon is an excellent absorber in an extremely broad spectral range, and RIE offers the possibility to texture any Si-based wafer regardless of its crystallinity. The first attempts to integrate black silicon fabricated by RIE into solar cells reached only moderate conversion efficiency [3] due to increased surface recombination caused by ion impact damage [4]. Mitigating the damage induced by RIE is therefore crucial to make black silicon truly competitive. In this work, we demonstrate how the RIE process can be tuned to yield reduced surface damage, and consequently improved effective carrier lifetime. Passivation with Al_2O_3 deposited by atomic layer deposition (ALD) leads to a further increase in effective lifetime on both p- and n-type wafers.

2 MATERIALS AND METHODS

2.1 *Fabrication*. Black Si was obtained on both n- and p-type doped 4'' Si wafers (Czochralski monocrystalline (100), resistivity between 1 and 20 Ωcm, thickness of 350 ± 25 µm) by RIE (Pegasus system, STS). Inductively coupled plasma (ICP) and capacitively coupled plasma (CCP) were used simultaneously in order to fine-tune the RIE process. The following parameters were used: gas flow of 70 sccm and 100 sccm for SF_6 and O_2 respectively (100 sccm); wafer stage temperature of -20 °C; CCP power between 10 and 80 W; ICP power of 3000 W; chamber pressure of 38 mTorr. Following RIE, the samples were RCA cleaned. Subsequently, wafers were coated with 380 cycles of ALD Al_2O_3 synthesized from trimethyl aluminum (TMA) and H_2O at 180 °C. The passivation layers were activated by post-deposition annealing at 500 °C for 5 min in N_2 atmosphere. As reference samples, polished wafers (n- and p-type) were also Al_2O_3 passivated.

2.2 *Characterization*. The thickness of the Al_2O_3 layer was measured on polished samples using ellipsometry (VASE J.A. Woollam Co., Inc.). The reflectivity R was measured with a PerkinElmer Lambda 1050 UV/Vis/NIR spectrophotometer. Morphology analysis was performed using scanning electron microscopy (SEM, Carl Zeiss Supra 40VP) at an acceleration voltage of 5 kV. Transmission electron microscope images were acquired using a TECNAI T20 G^2 microscope (FEI) operated at an accelerating voltage of 200 kV. Single nanostructures were isolated by pressing a Cu TEM grid with holey carbon on top of textured wafers. The effective minority carrier lifetime τ_{eff} was measured using the microwave detected photoconductivity (MDP) method in transient and injection dependent single point modes, using a MDP mapper (Freiburg Instruments). Quasi-steady-state photoconductivity (QSSPC) measurements were performed in injection dependent mode with a Sinton WCT-120 instrument.

3 RESULTS

Fig. 1 (a)-(b) shows scanning electron microscopy characterization (SEM) of an example of the nanostructures. These have the appearance of randomly distributed hillocks with a certain variation in shape, diameter between 200 and 300 nm, and a maximum height of about 550 nm.

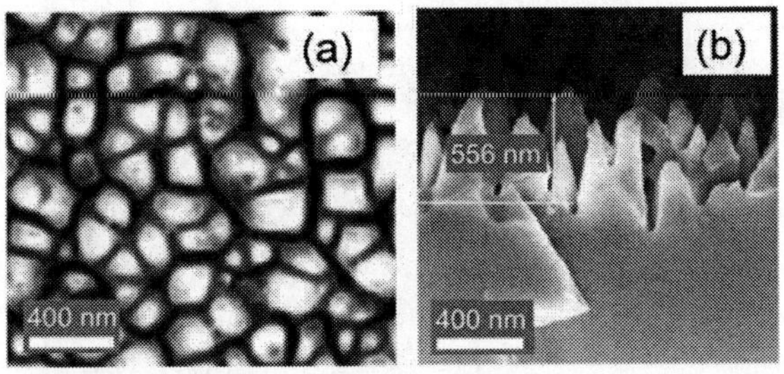

Figure 1. (a) Top-view SEM image of black Si. (b) Cross-section SEM view of the nanostructures.

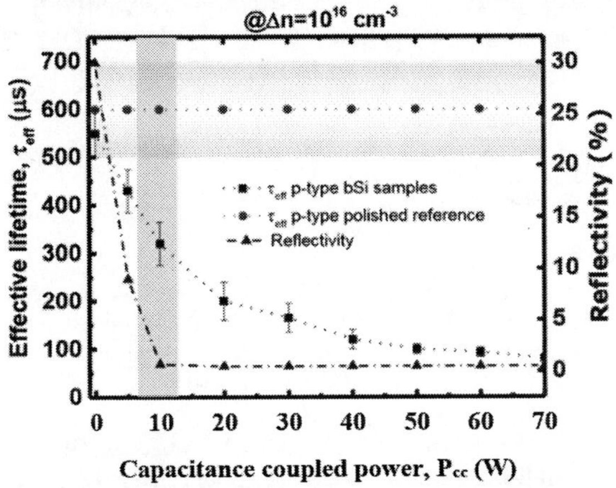

Figure 2. Effective minority carrier lifetime τ_{eff} (left y-axis) and average reflectivity R (right y-axis) as function of CCP power. The graph illustrates the trade-off between high effective lifetime and low reflectivity.

We found that the CCP power is the key parameter to adjust in order to increase the effective lifetime τ_{eff}, thanks to the reduced kinetic energy of ions in the plasma during etch. Fig. 2 presents results in terms of τ_{eff} and average R as function of the CCP power. Clearly, it is possible to decrease the CCP power from 70 W to 10 W without an increase in average R over 1%. On the other hand, the increase in τ_{eff} is of over a factor of 3, as measured by MDP at an injection level of $\Delta n = 10^{16}$ cm^{-3}. All the trends were confirmed with test measurements on quasi-steady state photoconductance (QSSPC, Sinton WCT-120) with an injection level measured at 10^{15} cm^{-3}. For CCP lower than 10 W, R increases abruptly, thus we chose 10 W CCP power as the best compromise between reflectivity and lifetime.

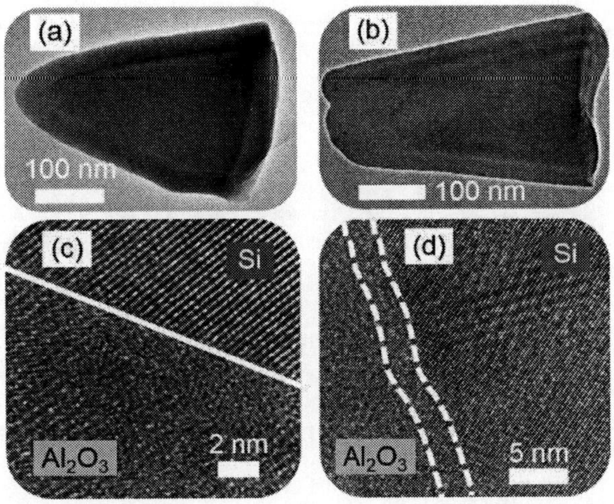

Figure 3. (a) TEM image of a nanostructure fabricated with low CCP 10 W. The Al_2O_3 coating is conformal as expected from the ALD process. (b) TEM image of a nanostructure fabricated with CCP of 60 W. (c) HRTEM image of the Si/Al_2O_3 interface from nanostructure in (a). HRTEM image of the Si/Al_2O_3 interface from nanostructure in (b). A

TEM characterization of nanostructures obtained using different values of CCP corroborate the observed trends in terms of lifetime. Fig.3 (a)-(b) show low magnification TEM images of single nanostructures fabricated using 10 W and 60 W CCP, respectively. In both cases, The Al_2O_3 coating on Si is conformal as expected from the ALD process. HRTEM images, however, reveal clear differences in the quality of the Si/Al_2O_3 interface, which is crucial for surface passivation. In the low CCP case (Fig. 3(c)), it is lattice planes of Si with a distance of 3.13 Å are readily identified, corresponding to the (111) set of planes. The atomic columns of Si extend to the boundary with Al_2O_3 without visible damage, indicating a high quality interface. On the other hand, in the high CCP case (Fig. 3(d)) we observe a region of a few nm in thickness, in which the Si appears to have become amorphous as result of the more intense ion bombardment during RIE.

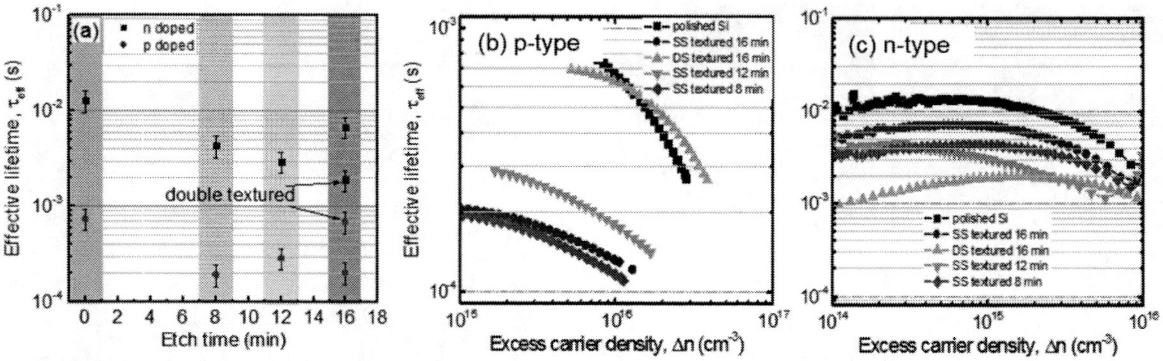

Figure 4. (a) Measured effective lifetime as function of etch time at an injection level of 10^{15} cm^{-3}. (b) Injection dependent effective carrier lifetime for p-type wafers. (c) Injection dependent effective carrier lifetime for n-type wafers.

We further investigated the influence of etching time on the electronic properties of our black Si, for both n- and p-type wafers. It is known that Al_2O_3 act as passivation layers for both types of Si, by virtue of different electrostatic (i.e. physical) passivation mechanism: inversion for n-type Si, and depletion for p-type [5]. The τ_{eff} was affected by the etch time, and results are summarized in Fig.4. For n-type Si, we measured τ_{eff} =12 ms on polished n-type Si, and τ_{eff} = 0.8 ms for p-type Si, at an injection level Δn =10^{15} cm^{-3}. As expected, the nanostructured samples exhibit lower values of a τ_{eff} due to the surface damage. The latter was however mitigated by the milder etch conditions after process optimization. Interestingly, we recorded the longest τ_{eff} on n-type Si for an etching time of 16 min, and on p-type Si for an etching time of 12 min. The measurements for double side (DS) textured samples are also shown. The effective lifetime for a DS textured p-type is higher than for SS textured p-type wafers. We tentatively attribute such surprising trend to the additional RCA cleaning done for the DS-textured sample, which may remove some surface damage. Fig. 3(b) and 3(c) show τ_{eff} as function of excess carrier density. These plots confirm that an etching time of 16 and 12 min is optimal for n- and p-type wafers, respectively.

4 CONCLUSIONS

In this work, we have investigated the effect of RIE parameters on the optoelectronic properties of black Si. We found that reduced CCP power (to ~10W) decreases the surface damage of Si and therefore leads to higher effective minority carrier lifetime, while retaining average optical reflectivity below 1%. Also, we applied a passivation scheme based on ALD Al_2O_3 deposition on native Si oxide interlayer. This allows for remarkably high effective lifetime, i.e., 12 ms and 0.8 ms for polished n- and p-type wafers, respectively; and highest values of 7.5 ms and 0.8 ms for n- and p-type black silicon, respectively. These findings prove that it is possible to achieve high lifetime on both single- and double-sided nanostructured surfaces and therefore to use them in the fabrication of Si solar cells with advanced architectures and efficient light trapping mechanisms without additional AR coatings.

5 REFERENCES

[1] H.M. Branz, V.E. Yost, S. Ward, K.M. Jones, B. To, P. Stradins, Nanostructured black silicon and the optical reflectance of graded-density surfaces, Appl. Phys. Lett. 94 (2009) 231121. doi:10.1063/1.3152244.

[2] S. Koynov, M.S. Brandt, M. Stutzmann, Black nonreflecting silicon surfaces for solar cells, Appl. Phys. Lett. 88 (2006) 203107. doi:10.1063/1.2204573.

[3] R.S. Davidsen, H. Li, A. To, X. Wang, A. Han, J. An, J. Colwell, C. Chan, A. Wenham, M.S. Schmidt, A. Boisen, O. Hansen, S. Wenham, A. Barnett, Black silicon laser-doped selective emitter solar cell with 18.1% efficiency, Sol. Energy Mater. Sol. Cells. 144 (2016) 740–747. doi:10.1016/j.solmat.2015.10.018.

[4] G. Kumaravelu, M. Alkaisi, A. Bittar, D. Macdonald, J. Zhao, Damage studies in dry etched textured silicon surfaces, Curr. Appl. Phys. 4 (2004), 108–110. doi: 10.1016/j.cap.2003.10.008.

[5] J. Hirsch, M. Gaudig, N. Bernhard, D. Lausch, Optoelectronic properties of Black-Silicon generated through inductively coupled plasma (ICP) processing for crystalline silicon solar cells, Appl. Surf. Sci, 374 (2016), 252-256. doi: 10.1016/j.apsusc.2015.11.241

THE SPEER SOLAR CELL – SIMULATION STUDY OF SHINGLED BIFACIAL PERC-TECHNOLOGY-BASED STRIPE CELLS

Nico Wöhrle, Tobias Fellmeth, Elmar Lohmüller, Puzant Baliozian, Andreas Fell, and Ralf Preu
Fraunhofer Institute for Solar Energy Systems ISE, Heidenhofstraße 2, 79110 Freiburg, Germany
Telephone: +49 761 4588 5964, Fax: +49 761 4588 7621, e-mail: nico.woehrle@ise.fraunhofer.de

ABSTRACT: Increasing the output power density of a photovoltaic module is a reliable way of lowering electricity production costs. Besides increasing the solar cells' conversion efficiency, a further option is lowering electrical and optical cell to module losses. The method of shingling singulated monofacial solar cell stripes is known since Dickson Jr.'s patent in 1956. First, it increases the packing density of active cell area in the module. Second, the active cell area is busbar-less reducing shading losses. Third, due to the reduced area of the solar cell stripes, the generated current per cell is less which results in a reduction of the overall series resistance of the cell interconnection within the module. We call our cell concept for this approach – which is based on the p-type silicon bifacial passivated emitter and rear cell (PERC) concept – "shingled passivated edge, emitter, and rear" – or "SPEER". These cells are then to be interconnected by shingling the p-busbar of the first cell onto the n-busbar of the second cell, constituting the first bifacial shingled module of its kind. Each adjacent shingle covers the busbar with active cell area and minimizes spacing losses in the module. This work covers the optimization of the SPEER concept on cell level with the simulation tool Quokka3. The optimized cell provides the basis for full usage of the cell-to-module gains compared to standard modules. Key issues for optimizing the SPEER cells, which will form the module, concern the characteristics and amount of recombination at their cut edges. This directly affects the question of ideal contour-to-area ratio and thus, the width of the SPEER cells. Using latest experience from bifacial PERC cells and literature values for edge recombination as simulation input, we are able to define a region of interest between 15 mm and 25 mm stripe width for building the first SPEER prototypes. We identify the need for edge treatment, be it an emitter window or edge passivation, as crucial for the success of stripe cells against more conventional cell layouts.
Keywords: bifacial shingle solar cell, PERC, SPEER, stripe cell, Quokka3

1 INTRODUCTION

The first appearance of a shingled solar cell interconnection pattern dates back to 1956 with a US patent filed by Dickson Jr. [1], which is only two years after the first publication of a silicon solar cell by Chapin *et al.* [2]. A number of companies have shown approaches for shingled solar modules. A recent overview can be found in Ref. [3]. Hence, the idea of singulated monofacial solar cells interconnected by a shingling design as such is not a new thing. Early publications of shingling module approaches have been mostly motivated by particular design requirements for specific applications [4, 5]. Later, publications started to make use of the potential for higher module power densities of this technique compared to standard-module cell interconnection, see e.g. Ref. [6]. Consequently, a few large module manufacturers [7, 8] seem to rediscover the potential of monofacial shingling technology to reduce cell-to-module (CTM) losses. The recent International Technology Roadmap for Photovoltaic (ITRPV) projects a world market share of 7% for shingled interconnection technology by 2027 [9].

Another string of technological evolution spreading in the photovoltaic industry is the concept of bifacially illuminated solar cells, which has been extensively covered in a recent article by Kopecek and Libal [10].

As the demand for modules with high power density is large, the opportunity is at hand to combine the bifacial cell technology with the shingle cell module technology. Hereby, the bifacial solar cells profit from additional light coming from the rear side. The busbars on front and rear side for the shingle cells are covered by active area from the adjacent cells, leading to a virtually busbar-free cell string. This combination of features is able to create the highest power density on module-level for silicon PERC technology without concentration of illumination. Another aspect is that this cell/module design does not change the basic ideas of PERC process technology and therefore intrinsically profits from improvements on cell process level, e.g. improved emitter diffusions, surface passivation or bulk carrier lifetimes.

While the conceptual idea of shingled module layouts is to profit from lower CTM-losses compared to standard modules, the development has to start on cell level. Since the standard wafer formats dominate the market, most early approaches for stripe cells will rely on the singulation of a various number of cells out of one 6-inch (pseudo)square wafer. Initially, these stripe cells will face the issue of cleaved (i.e. blank) edges and thus, potentially high edge recombination. Hence, the first focus in the development chain towards shingled modules has to be the singulated solar cell. It has to be designed and processed on cell level in a way that it stays as competitive as possible to full area PERC cells rather than outperforming it by a lot. That is the focus of this work. We call this bifacial stripe solar cell "shingled passivated edge, emitter, and rear cell", or "SPEER". The concept and its true bifacial modelling are both a novelty.

2 APPROACH AND MODEL

The SPEER concept uses (pseudo)square format cells separated into stripes along the busbars. In the module, the cells are interconnected by shingling the p-busbar of the first cell onto the n-busbar of the second cell (see Figure 1). This way the busbars operate only as large pads yielding effectively busbarless strings. Each adjacent shingle covers the busbar with active cell area and minimizes spacing losses in the module. The small stripe size reduces resistive power losses on module level by lowering the total current.

These singulated solar cell stripes have a larger contour-to-area ratio than standard (pseudo)square cells,

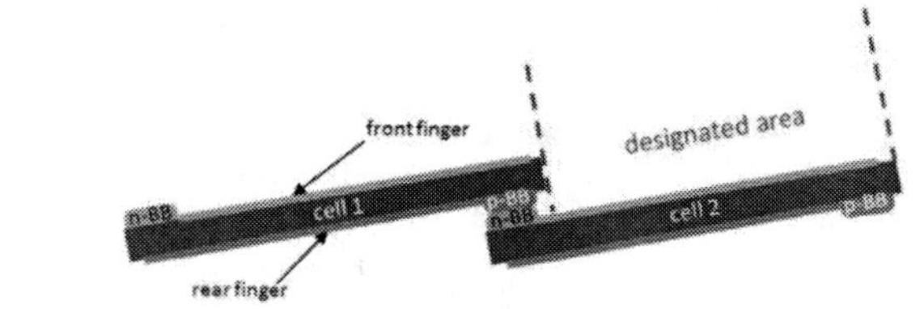

Figure 1: Shingling scheme of two solar cells placing the bottom-busbar of cell 1 onto the top-busbar of cell 2.

which is exemplary shown in Table I for a 2.5 cm- and a 5 cm-wide stripe.
Moreover, (pseudo)square cells undergo passivation processes which also passivate the edges to some extent, while the stripes are singulated after metallization and contact firing leaving the edges initially blank. This poses the question of potential losses by recombination at those edges, which can be divided into three sub-regions for the simulations, sketched in Figure 2:

1. Surfacing bulk region, implicating ideal surface recombination (ideality factor $n = 1$). This can be modeled by using an effective surface recombination velocity between $S_{eff,edge} = 8$ cm/s for excellent passivation (e.g. reported by Saint-Cast *et al.* [11]) and $S_{eff,edge} = 5 \cdot 10^6$ cm/s for an unpassivated surface with high defect density (e.g. reported by Glunz and Dicker [12, 13]).
2. Surfacing highly doped emitter region, implicating ideal surface recombination.
3. Surfacing space-charge-region (SCR), implicating non-ideal surface recombination activity (ideality factor $n \approx 2$). Dicker parameterized this recombination for a single recombining edge using the second diode in the two-diode-model naming it $j_{02,edge}$ [13]. This local recombination current density is determined to be $j_{02,edge} = 13$ nA/cm, scaling with the contour-to-area ratio. An updated value of $j_{02,edge} = 19$ nA/cm has been recently published by Fell *et al.* [14], so that $j_{02,edge,tot}$ = contour/area·19 nA/cm². This yields the $j_{02,edge,tot}$ numbers shown in Table I for exemplary stripe widths.

For the examined cells, Dicker [13] concluded that the recombination of surfacing bulk region and space charge region contribute equally to the total edge recombination while the emitter region surface recombination has negligible influence due to its small extent and is therefore neglected in this study. We will have a closer look into this and distinguish the quality and quantity of recombination influences in the quasi-neutral bulk and the SCR.

The shingling itself (see Figure 1) also has influence on cell optimization. The upper busbar and adjacent area of each cell is covered by active area of the overlying

Table I: Exemplary calculation for contour-to-area ratio and resulting $j_{02,edge,total}$ after the formula of Ref. [13] with $j_{02,edge} = 19$ nA/cm, which shows that a 2.5 cm-wide stripe has almost a fourfold influence on $j_{02,edge,tot}$ recombination compared to a regular squared cell (for a non-passivated edge).

Stripe/cell size	Contour / area (1/cm)	$j_{02,edge,tot}$ (nA/cm²)
2.5 x 15.6 cm²	0.93	**17.7**
5.0 x 15.6 cm²	0.53	**10.1**
15.6 x 15.6 cm²	0.26	**4.9**

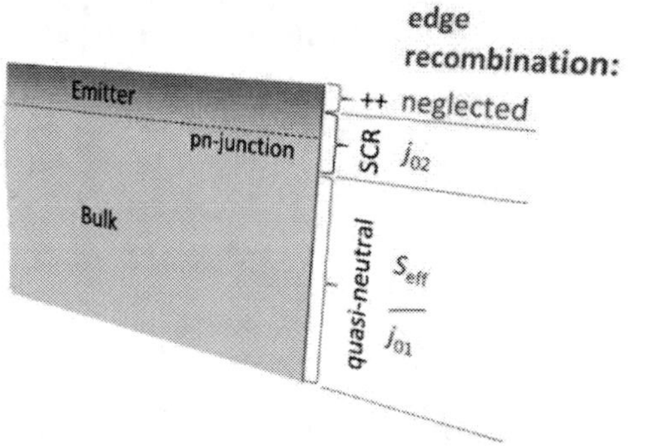

Figure 2: Sketch of the three edge recombination regions used for the simulations with the highly doped area (++), the space-charge-region (SCR), and the quasi-neutral bulk.

cell. Thus, only the marked designated area is relevant for determination of the short-circuit current density j_{SC}. The model has to consider this fact already on cell simulation level; otherwise the stripe width is optimized under inappropriate assumptions.

A simulation tool which can cover the edge recombination as well as the optical features is the recently developed "Quokka3". Due to a lumped skin approach (an expression originally phrased by Cuevas *et al.* [15]) for non-neutral regions, the mesh fineness can be reduced to a minimum. In consequence, this allows the modeling of much larger domains, in our case a whole cell stripe, resulting in a generalized model without spatial simplifications (opposed to the usual unit cell approach) like potential or series resistance distributions. With the latest addition of a vertical resistance and full injection dependence to the skin parameterization [16], the skins can be described by lumped parameters without error compared to, e.g., explicitly accounting for doping profiles. The cell stripe model is depicted in Figure 3.

The input parameters, summarized in Table II, are taken exemplary from PERC processes available at Fraunhofer ISE to have a representative front side conversion efficiency of 21% under 1000 W/m² AM1.5g illumination and a bifaciality of 75% (see Ref. [17]).

Table II: Simulation input parameters.

Emitter	j_{0e} = 45 fA/cm² j_{0met} = 800 fA/cm² R_{sh} = 85 Ω/sq
Rear side	j_{0bsf} = 700 fA/cm² R_{sh} = 60 Ω/sq S_{pass} = 10 cm/s
Bulk minority carrier lifetime τ_b	500 µs
Finger width	front: 40 µm rear: 150 µm (contact opening: 50 µm)
Busbar width	front: 500 µm rear: 1000 µm
Resistivities	Contact front: ρ_c = 2 mΩcm² Ag-finger sheet res.: $R_{sh,f}$ = 3.15 mΩ Contact rear: ρ_c = 5 mΩcm² Al-finger sheet res.: $R_{sh,f}$ = 11.7 mΩ
Optics	random pyramids, 75 nm SiN_x, j_{Ph} (1 sun, no shading) = 41.76 mA/cm²

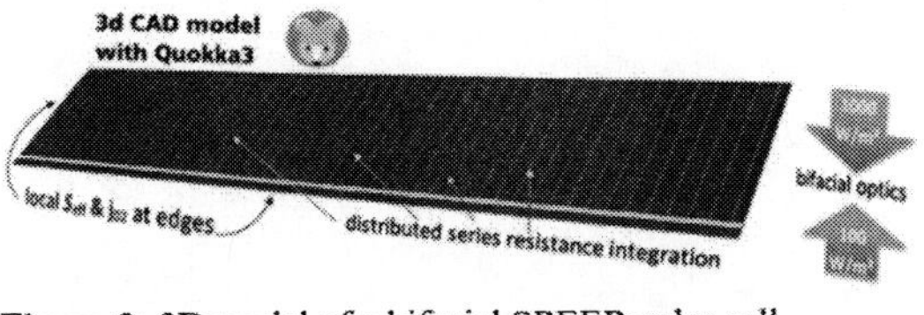

Figure 3: 3D model of a bifacial SPEER solar cell.

Since the publication of the first SPEER simulation results in Ref. [3], the simulation algorithm of Quokka3 has improved: the $j_{02,edge}$ is implemented locally instead of a lumped j_{02}, and the optics model has been updated slightly.

3 EXPERIMENT

We follow the practical process evolution steps of the stripe cell to provide a better understanding of the peculiarities of this concept. The goal of the simulation is to optimize the width of the stripe cell and determine the impact of edge recombination.

The **first simulation** will be the 6-inch pseudo square PERC cell with 205 mm diameter, which serves as reference. The **second simulation** will have a change of layout by virtually slicing the cell right next to the busbar, so that every stripe cell has a "comb-like" metal pattern (Figure 3). The rear grid is 180° rotated, so that the busbar is at the opposite edge than on the front side. Due to the cleaving step, the edges of the stripe cell are potentially highly recombination-active as explained in section 2, thus a negative impact on fill factor *FF* and open-circuit voltage V_{OC} is expected.

In the **third simulation**, the measurement technique is changed to designated area, as explained in section 2 (Figure 1), to optimize the stripe cell in the context of the specific shingle module layout.

The **fourth simulation** is divided in case A and B corresponding to different technical approaches reducing edge recombination:

Case A adds an emitter window, i.e. the diffusion ends 200 µm in front of the edges of the front side, to decouple the recombining edge from the rest of the cell.

Case B, instead, adds edge passivation to reduce the recombination itself. The realization of such an edge passivation technique which is in the best case also suitable for mass production is being examined with high priority in currently ongoing work at Fraunhofer ISE.

As a virtual **side experiment**, located between the third and the fourth simulation, we will separately switch the $j_{01,edge}$- and the $j_{02,edge}$-recombination to work out the separate influence of each recombination region and determine to which degree they add up.

Table III: Applied recombination levels for the cell edges for an unpassivated or a passivated edge.

	$S_{eff,edge}$ (cm/s)	$j_{02,edge}$ (nA/cm)
unpassivated edge	10^6	19
passivated edge	10	0

4 RESULTS

As name convention, we will always use the term output power density p_{out} (mW/cm²) instead of energy conversion efficiency η (%) for measurement data referring to bifacial illumination as it is the less ambiguous unit. The scale is chosen such that with 1000 W/m² monofacial irradiation, the respective numerical value for p_{out} and η are identical.

4.1 Stripe size optimization

The "red star" in Figure 4 represents the **first simulation**, the reference value for the bifacial PERC. It is located at 15.6 mm "stripe width" as this is the equivalent value of the half busbar distance for a five-busbar (5BB) cell and thus has comparable current paths. Under the given illumination of 1000 W/m² front and 100 W/m² rear intensity, p_{out} = 23.4 mW/cm² is obtained.

Cutting this cell intro stripes introduces edge recombination, therefore the **second simulation** features p_{out} = 22.1 mW/cm² for 15.6 mm stripes (orange line). The optimum shifts towards larger stripes of 25 mm width and reaches p_{out} = 22.5 mW/cm². The sliced stripes therefore initially lose 1 mW/cm² towards the full area reference PERC due to *FF* and V_{OC} losses.

Opposed to the reference PERC, the stripe cells will exhibit a shingled layout in the module (Figure 1), and therefore have to be calculated with a designated area excluding the region covered by the overlap of the adjacent cell. This **third simulation** logically increases j_{SC} towards 44 mA/cm² (reminder: 1100 W/m² total irradiance) as the excluded area involves the busbar-shaded region. Moreover it dissolves the dependency of j_{SC} to the stripe width for the same reason, as can be seen in Figure 4b comparing the black and the orange line.

The crucial step to attain the SPEER's competitiveness with the reference is a reduction of edge recombination, performed in the **fourth simulation**.

If this is done by an emitter window of 200 µm inwards from the edge (**Case A**, without edge passivation), V_{OC} increases by roughly 3 mV and *FF* by $1\%_{abs}$ to $2\%_{abs}$ (grey dashed line). The resulting p_{out} = 23.5 mW/cm² for 20 to 25 mm-wide stripes is on par with the reference.

The edge passivated cell (**Case B**, without an emitter window) can exceed the emitter window-cell mainly by a higher *FF*. As a result, the SPEER cell with edge passivation matches the reference cell's *FF* and V_{OC} but exceeds j_{SC} by 1.5 mA/cm² due to larger active cell area (i.e. designated area).

In the end, with high quality edge passivation, the optimal stripe width is located between 15 mm and 20 mm. However the actual preferable stripe size will also depend on handling difficulties in module assembly, which will possibly set a larger lower limit to stripe width.

4.2 Influence of edge recombination types

If the $j_{01,edge}$ and $j_{02,edge}$ recombination are switched separately in the simulation, see Figure 5, one can see the proportional influence of the SCR recombination and base recombination to the total recombination loss of the edges. In Figure 5, both the green dashed line and the black line are identical to the ones in Figure 4.

If $j_{02,edge}$ alone is switched on, the green dashed curve is reduced almost all the way down to the black line. This shows that our SPEER cell is sensitive to the emitter area where it meets the edges. This influence is dominated by the *FF*. On V_{OC}, both $j_{01,edge}$ and $j_{02,edge}$ have similar influence and the changes in j_{SC} are rather small. Additionally it is important to notice that the losses of $j_{01,edge}$ and $j_{02,edge}$ recombination do not add up completely if both are activated.

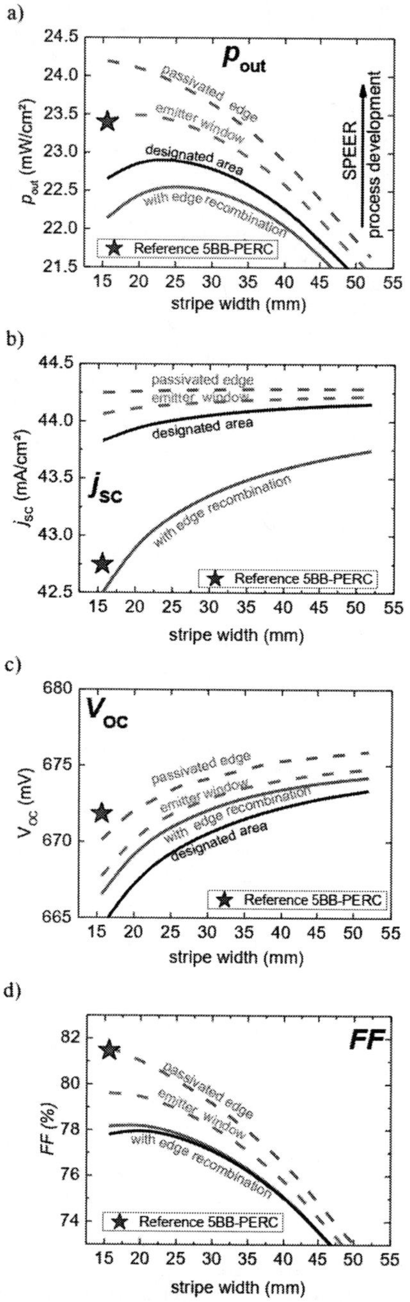

Figure 4: Simulations one to four for the SPEER cell

Current-voltage (*IV*) parameters of the stripe cells at varied cell stripe width for an illumination intensity of 1000 W/m² and 100 W/m² from the front and the rear side, respectively.

First simulation (red star): PERC reference

Second simulation (orange line): $j_{01} + j_{02}$ edge recombination

Third simulation (black line): designated area

Fourth simulation, case A (grey dashes): emitter window

Fourth simulation, case B (green dashes): passivated edge

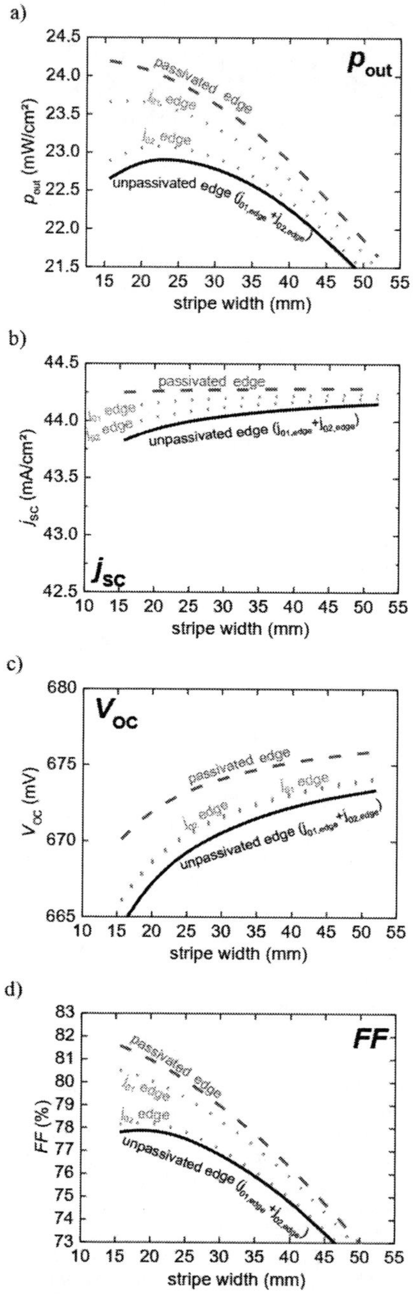

Figure 5: Comparison of influence of edge recombination types

IV parameters of the stripe cells at varied cell stripe width for an illumination intensity of 1000 W/m² and 100 W/m² from the front and the rear side, respectively, with designated area calculation.

Green dashes: passivated edge

Light green dots: j_{01} edge recombination only

Blue dots: j_{02} edge recombination only

Black line: unpassivated edge, $j_{01} + j_{02}$ edge recombination

The explanation is simple: A carrier that has been "consumed" by one recombination type cannot be affected by the second one again. Concerning simulation methodology this means, that both recombination types have to be implemented locally (as Quokka3 does) and not as separate lumped recombination. This would lead to overestimation of the losses.

4 DISCUSSION

The simulation steps 1-4 reflect the process development that has to be made for making the SPEER cell competitive to the full area PERC on cell level.

It is clearly visible that the strip cell layout is initially inferior to a full cell equivalent. On cell level, it accesses its advantages by the projected shingling layout and therefore larger active cell area. This advantage, however, can only be effective if the process-inherent edge recombination losses are minimized. Two suitable approaches are the introduction of an emitter window or an edge passivation.

Reaching this goal ensures the full utilization of the advantages of the shingled module approach which relies on reduced CTM loss towards standard module configurations. Further information about the whole approach of shingled modules can be found in Ref. [3].

5 CONCLUSION

The SPEER solar cell, as cell technology for shingled modules, poses several new challenges in processing, amongst which the most important is the edge recombination. We could show that unpassivated-edge cells have their optimum at 25 mm stripe width, but fall back against the large area PERC reference by 1 mW/cm² or 0.5 mW/cm² (calculated with designated area). The introduction of an emitter window, decoupling the edge recombination from the bulk brings the SPEER on level with the reference. A successful passivation of the edge yields an advantage of 0.7 mW/cm² at 1.1 suns bifacial illumination. The resulting optimal stripe width is located between 15 mm and 20 mm. This value might be influenced by handling considerations in module assembly.

If a suitable edge treatment can be achieved, a shingled module with SPEER cells is put into position to fully access its CTM advantages over a standard module.

6 OUTLOOK

The SPEER solar cell concept is not only a simulation study. The development steps, analogously to the presented simulations, are currently performed at Fraunhofer ISE for PERC as base technology as well as for advanced cell concepts.

ACKNOWLEDGEMENTS

The authors acknowledge the funding of the German Federal Ministry for Economic Affairs and Energy (BMWi) in the frame of the project "PV-BAT400" (contract number 0324145).

REFERENCES

[1] D. C. Dickson, Jr., "Photo-voltaic semiconductor apparatus or the like," US2938938A, USA 2,938,938, May 31, 1960.

[2] D. M. Chapin, C. S. Fuller, and G. L. Pearson, "A new silicon p-n junction photocell for converting solar radiation into electrical power," *J. Appl. Phys.*, vol. 25, no. 5, p. 676, 1954.

[3] N. Wöhrle *et al.*, "Solar cell demand for bifacial and singulated-cell module architectures," *Photovoltaics International*, vol. 36, pp. 48–62, 2017.

[4] J. H. Myer, "Photovoltaic generator," US3369939, US 3,369,939, Feb 20, 1968.

[5] H. Gochermann and J. Soll, "Shingle-type solar cell generator prodn. - allowing formation of curved or domed product," DE3942205C2, Germany 3942205, Feb 1, 1996.

[6] J. Zhao *et al.*, "20000 PERL silicon cells for the "1996 World Solar Challenge" solar car race," *Prog. Photovolt: Res. Appl.*, vol. 5, no. 4, pp. 269–276, 1997.

[7] SunPower Corporation, *SunPower Introduces New Solar Panel: The Performance Series.* [Online] Available: https://us.sunpower.com/blog/2015/11/12/sunpower-introduces-performance-series-solar-panel/. Accessed on: Mar. 21 2017.

[8] B. Yang, P. Nguyen, J. B. Heng, A. Reddy, and Z. Xu, "High Efficiency Solar Panel," US20150090314A1, US 14/563,867, Apr 2, 2015.

[9] ITRPV, "International Technology Roadmap for Photovoltaic: 2016 Results," 2017. [Online] Available: http://www.itrpv.net/Reports/Downloads/.

[10] R. Kopecek and J. Libal, "Quo vadis bifacial PV?," *Photovoltaics International*, vol. 35, 2017.

[11] P. Saint-Cast *et al.*, "High-Efficiency c-Si Solar Cells Passivated With ALD and PECVD Aluminum Oxide," *IEEE Electron Device Lett.*, vol. 31, no. 7, pp. 695–697, 2010.

[12] S. W. Glunz *et al.*, "High-efficiency silicon solar cells for low-illumination applications," in *29th IEEE Photovoltaic Specialists Conference New Orleans*, New Orleans, 2002, pp. 450–453.

[13] J. Dicker, "Analyse und Simulation von hocheffizienten Silizium-Solarzellenstrukturen für industrielle Fertigungstechniken," Dissertation, Fakultät für Physik, Universität Konstanz, Konstanz, 2003.

[14] A. Fell *et al.*, "Modelling edge recombination in silicon solar cells," *IEEE J. Photovoltaics*, submitted, 2017.

[15] A. Cuevas *et al.*, "Skin care for healthy silicon solar cells," in *42nd IEEE Photovoltaic Specialists Conference (PVSC)*, New Orleans, LA, USA, 2015, pp. 1–6.

[16] A. Fell, J. Schön, M. C. Schubert, and S. W. Glunz, "The concept of skins for silicon solar cell modeling," *Solar Energy Materials and Solar Cells*, 2017.

[17] N. Wöhrle *et al.*, "Understanding the Rear-Side Layout of p-Doped Bifacial PERC Solar Cells with Simulation Driven Experiments," *Energy Procedia*, vol. 124C, pp. 225–234, 2017.

AUGER RECOMBINATION IMPACT FOR LIMITING EFFICIENCY OF SILICON SOLAR CELLS

Jongwon Lee, Maxwell K. Cotton, Yongjie Zou, Stanislau Herasimenka and Christiana B. Honsberg
School of Electrical, Computer and Energy Engineering, Arizona State University
7700 South River Parkway, Tempe, AZ, 85284

ABSTRACT: The detailed balance equations including material parameters are important to predict the theoretical efficiency for specific materials. The temperature dependence of recombination processes in silicon solar cell are discussed due to sensitivity of electrical parameters with single junction and multiple exciton generation solar cells (MEGSC). Among these characteristics, the impact of Auger recombination (AR) is significant in high doping or high injection level so that the effects of AR are applied into the detailed balance equations of silicon solar cells for single junction and multiple exciton generation concept. After we parameterize the material properties, the temperature sensitivity of the open circuit voltage shows as the most significant impact to the efficiency. Further, we investigate various Auger coefficient modeling to see the effect of temperature sensitivity. After detailed analyzation, we could also find the potential benefits of using MEGSC in comparison to SJ approaches.

Keywords: detailed balance, silicon, Auger recombination, multiple exciton generation

1 INTRODUCTION

Limiting efficiencies of the silicon solar cell is an important key to see the possible expectations. Under the conventional detailed balance (CDB) theory, the limiting efficiency of silicon solar cell with one sun illumination under the terrestrial spectrum is 33.4% [1]. This calculation is only based on one radiative recombination while excluding other non-radiative recombination (NR) components [1]. After including NR processes with material parameters, the limiting efficiency of a silicon solar cell is 29.8% at 100 μm cell thickness with simular results of other publications [2,3]. The NR process is generally composed of Shockley-Read-Hall (SRH) recombination and Auger recombination (AR). These recombination processes limit the silicon solar cell efficiency. Further, silicon is an indirect bandgap material so that the effect of radiative recombination is relatively small compared to other direct bandgap materials. Therefore, SRH and AR are critical to determine the lifetime of silicon solar cells. Typically, high injection level has a dominant component of AR due to the three carrier interactions. On the other hand, the SRH recombination process can be significant at the low-injection level. Hence, the carrier concentration dependence of silicon solar cell is an actual recombination component to decide the carrier life time and its limiting conversion efficiencies.

While showing the limit of Si solar cells, multiple exciton generation solar cell (MEGSC) has shown an alternative path to surpass the single junction (SJ) limit due to many electron and hole pairs (EHP) generation per one incoming photon by impact ionizaiton [4]. Further, its reverse process is also thermodynamically explained by AR. Its mathematical expression of recombination process is expressed as quantum yield (QY) with chemical potential ($\mu = q \cdot QY \cdot V$, q is the element of charge, V is the operating voltage) [4]. But, with different view point, AR process is also material-specific approaches that it is strongly related to carrier concentration of each material. Therefore, alternative expression for recombination with silicon parameter can be good way to analyze the detailed balance of MEGS without using $\mu = q \cdot QY \cdot V$. We will discuss it in section 2.2

Even using 3rd generation photovoltaic (PV) devices to boost the efficiency, solar cell operation can be strongly influenced by the variation of temperatures. In other words, increasing the operating temperature lowers the bandgaps of the materials, and hence typically lowers the open circuit voltage (=V_{OC}) and raises the absorption. In the meantime, carrier mobility also decreases at elevated temperature, which contributes to a lower fill factor, and possibly a lower short circuit current [5]. This temperature sensitivity can also provide the impact of MEGSC it we can include the material non-idealities into detailed balance equations .

In this paper, we will discuss the impact of temperature sensitivity of NR between Si SJ and MEG phtovoltaic devices. Initially, we will investigate the optimum cell thickness between Si SJ and MEG with regarding material parametrs. After deciding the optimum thickness between two devices, we will simulate the ideal detailed balance equation with excluding the NR to see the effect of temperature variation of radiative recombination itsetls. Then, we will see the impact of temperature variation with including of NR. While comparing these approaches, we will discuss the paths for improving Si solar cells with considering temperature and impacts.

2 THEORY

2.1 Single junction solar cell approach

Due to an indirect-bandgap properties of crystalline Si, the impact of radiative recombination (direct-transition) is relatively small while NR processes becomes important in high doping or high injection level.

The inverse of minority carrier lifetime can be decomposed into SRH, radiative recombination and AR. The relationship is described by Eq. 1. The radiative recombination of Si is too small to impact of carrier life time. Thus, we deal with lifetime based on SHR and AR.

$$\frac{1}{\tau_{eff}} = \frac{1}{\tau_{SRH}} + \frac{1}{\tau_{rad}} + \frac{1}{\tau_{Auger}} \approx \frac{1}{\tau_{SRH}} + \frac{1}{\tau_{Auger}(\propto \frac{1}{n^2 \cdot C_a})} \quad (1)$$

where τ is the lifetime, n is the carrier concentration and C_a is the ambipolar Auger coefficient.

In high doping or high injection level, the effect of AR (C_a) is more important. Further, temperature dependence can change the lifetime since C_a is proportional to T_C^{γ} where γ is a constant [6-9]. First, we selected C_a as $3.88x10^{-31}$ cm^6s^{-1} to verify [2]. Then, we investigate the temperature dependence from [6,9]. These relationships are shown from Eq. 2 to Eq. 6.

$$C_a = 11\times10^{-31}\cdot\left(\frac{T}{300}\right)^{1.8} \quad (2)$$

$$C_n(n,T) = 2.8\times10^{-31}\cdot\left(\frac{T}{300}\right)^{0.6} + \frac{5.3\times10^{-31}\cdot(T/300)^{2.27}}{1+(n/N_{ref})^{\alpha_{Auger}}} \quad (3)$$

$$C_p(p,T) = 0.99\times10^{-31}\cdot\left(\frac{T}{300}\right)^{0.6} + \frac{1.9\times10^{-31}\cdot(T/300)^{2.27}}{1+(p/N_{ref})^{\alpha_{Auger}}} \quad (4)$$

where N_{ref}=$2x10^{18}cm^{-3}$, α_{Auger}=2

$$C_a = C_n + C_p \quad (5)$$

$$C_a = \frac{1.1\times10^{-28}}{T-193} + 2.1\times10^{-31}\cdot T \quad (6)$$

We use these equations to check the temperature sensitivity for electrical parameter variations. Further, we modified the CDB equation for specific materials including absorbance of silicon, absorption coefficients and Auger coefficients [2].

$$\phi(E_1,E_2,T,\mu) = \frac{2\pi}{h^3c^2}\int_{E_1}^{E_2}\frac{E^2}{\exp[(E-\mu)/kT]-1}dE \quad (7)$$

$$J = q\cdot C\cdot\int_{E_i(=0.31eV)}^{E_{max}} a(E)\cdot\Gamma(E)_{AM1.5}dE - q\cdot\phi(E_g,\infty,T_C,\mu) - q\cdot L\cdot C_a\cdot n_i^3\exp\left(\frac{3}{2}\cdot\frac{q\cdot V}{k\cdot T}\right) \quad (8)$$

where ϕ is the particle flux given by Planck's equation at temperature T with a chemical potential μ in the photon energy range between E_1 and E_2, h is Planck's constant, c is the speed of light, while μ (=q·V) is the chemical potential of the SJ. Also, V is the operating voltage, k is the Boltzmann constant, J is the current density of the solar cell, q is the charge of an electron, C is the optical concentration, f_S is the geometry factor (=1/46200), E_g is the bandgap energy, T_C is the solar cell's temperature (=300K), E_{max} (=4.428 eV at 280 nm wavelength) is the highest photon energy of terrestrial spectrum, and $\Gamma(E)_{AM1.5}$ is the converted terrestrial spectrum from ASTMG 173-03. The intrinsic carrier concentration is n_i and L is the cell thickness.

We use fitting approaches given Eq. 9 to calculate the absorbance from [10] due to differences of data points between [10] (121 points) and AM1.5 spectrum (2002 points). Then, it is applied into Eq. 8.

$$a(E) = \frac{\alpha(E)}{\alpha(E)+\frac{1}{4n^2\cdot L}} \approx d + \frac{a-d}{1+\left(\frac{E}{c}\right)^b} \quad (9)$$

where a, b, c and d are the coefficients, $\alpha(E)$ is the absorption coefficient, n is the refractive index and L is the thickness of the solar cells.

Further, we also include the temperature coefficient from [10] to modify the temperature dependent-absorption coefficient and its correspoinding absorbance. It is shown in Eq. 10. We also include the temperature sensitivity of n_i and bandgap (Eq. 11 and 12) and it is applied into Eq. 8.

$$\alpha(E,T)=\alpha(E,300K)\cdot(1+\Delta T\cdot C_\alpha) \quad (10)$$

where C_α is the temperature coefficient for absorption coefficient. ΔT=T_C-300K, T_C is the temperature of cell.

$$E_g(T) = E_{g0} - \frac{A\cdot T^2}{B+T} \quad (11)$$

where A =$4.74x10^{-3}$ [eV/K], B=634 for Si devices and E_{g0}=1.17 eV

$$n_i = 5.29\times10^{19}(T/300)^{2.54}\exp(-6726/T) \quad (12)$$

2.2 Multiple Exciton Generation Approaches

For MEG simulation, first, we use the CDB MEG with temperature variations to see the effect of ideal cases. After including NR, we modified the recombination current like Eq. 8 except for light-generated current is the same as CDB MEG. And, its mathematical expressions are shown in Eq. 13 and 14 that the recombination process of MEGSC is explained by a material-dependent NR. We choose QY as 2 that it generates two EHPs per one incoming photon. Thus, two electrons at the conduction band contribute to do AR. The material parameters are used from section 2.1

$$\text{Ideal } QY(E) = \begin{cases} 0 & 0 < E < E_g \\ m & m\cdot E_g < E < (m+1)\cdot E_g \quad m = 1, 2, 3,\ldots \\ M & E \ge M\cdot E_g \end{cases} \quad (13)$$

where m is the number of multiple electron hole pairs generated, M is the maximum number of electron hole pairs which are generated, E_g is the bandgap, and E is the photon energy.

$$J = q\cdot C\cdot\int_{E_i(=0.31eV)}^{E_{max}} a(E)\cdot QY(E)\cdot\Gamma(E)_{AM1.5}dE - q\cdot\phi(E_g,\infty,T_C,\mu) - q\cdot L\cdot C_a\cdot n_i^3\exp\left(\frac{3}{2}\cdot\frac{q\cdot V}{k\cdot T}\right) \quad (14)$$

After comparing two approaches, we could see the clear temperature effect of MEGSC and we will discuss it in section 3.

3 RESULTS

In this section, we will discuss the effect of temperature sensitivity with AR for both Si SJ and MEG PV devices. First, we will discuss the approaches of [2] to find the optimum thickness for both Si SJ and MEG with considering absorption coefficient of Si and fixed Auger coefficient [2,10]. Then, theoretical efficiencies and electrical parameters between two PV devices were investigated with Auger coefficient variations (increasing or decreasing) to see the impact of temperature sensitivities with ideal CDB cases of SJ and MEG (not including non-radiative recombination).

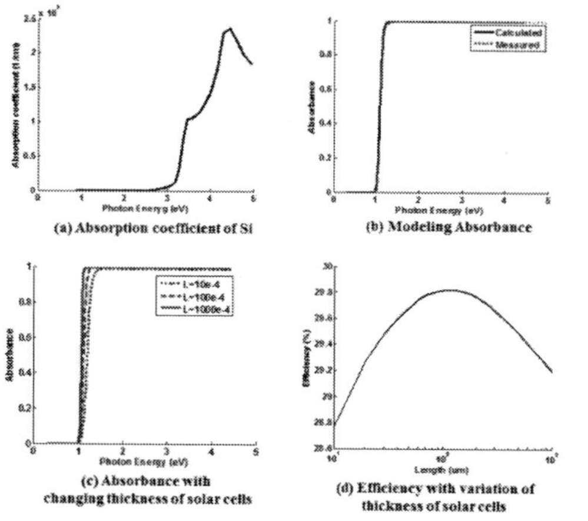

Figure 1: The simulation results of modified balance equations with including absorbances of silicon. The step size of thickness is 10μm

We initially verify DB based on [2]. The simulation results are shown in Fig. 1 (a), (b) and (c) and its corresponding theoretical efficiencies (=Fig.1 (d)) are similar to [2]. For studying of temperature sensitivity with Auger recombination, we initially calculate CDB of SJ case and we disregard n_i effect with temperature. Then, we include the bandgap's temperature sensitivity into Eq. 8 for the next simulation. After verifying Eq. 8, we also compare theoretical efficiencies between SJ and MEG. The inclusion of MEG into DB shows reducing the optimum thickness of MEG from 120μm of SJ to 60μm of MEG. And, its theoretical efficiency increases from 29.8% of SJ to 36.5% of MEG. (see Figure 2 and Table 1). Therefore, MEG of silicon solar cell make it possible to fabricate thin substrate with enhanced theoretical efficiencies.

Table I: The thickness of solar cells and its corresponding efficiency

	Efficiency	Thickness
SJ	29.8 %	120 μm
MEG	36.5 %	60μm

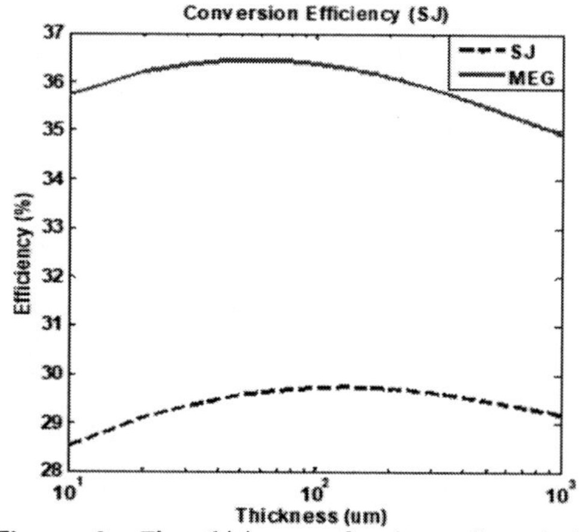

Figure 2: The thickness of solar cell and its corresponding theoretical efficiency at 300K.

And, we use each optimum thickness for SJ (=120 μm) and MEG (=60 μm) respectively to investigate the temperature dependence with other parameters. Typically, we researched the temperature sensitivity of C_a to show the theoretical efficiency for SJ and MEG case [2,6,9] (see Figure. 3 and 4).

n_i and effective bandgaps are strongly influenced by temperature (Eq 10 and 11). Therefore, combining these two effects into DB shows V_{OC} decrease as increasing temperature due to increasing n_i.

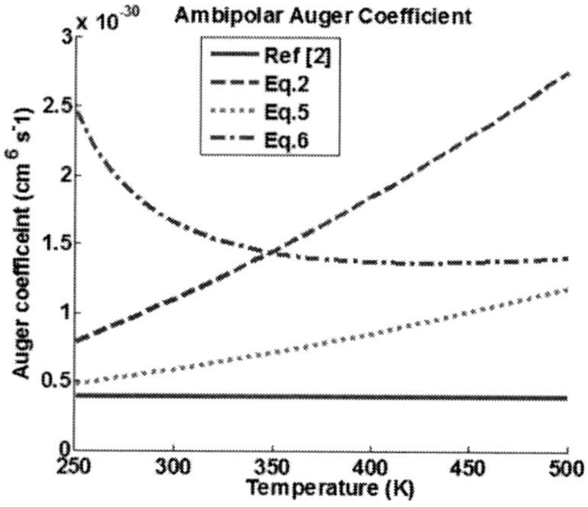

Figure 3: Ambipolar Auger coefficient vs temperature

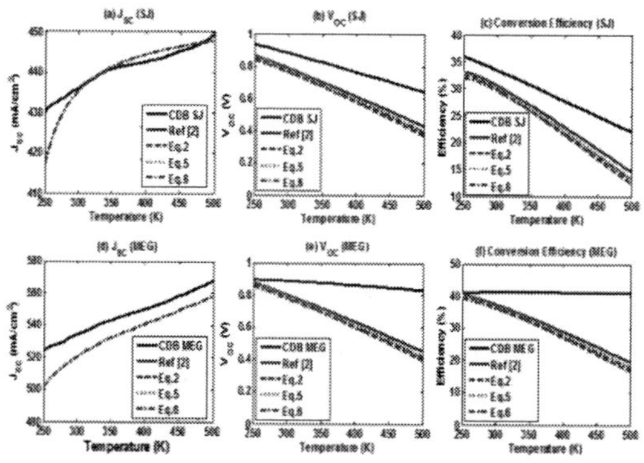

Figure 4: J_{SC}, V_{OC} and theoretical efficiency variations between SJ and MEG with changing temperature from 250K to 500K where CDB is the conventional detailed balance equation of silicon solar cell

C_a also can limit the theoretical conversion efficiency and V_{OC}. But, the differences of both V_{OC} and efficiencies from 4 different Auger coefficients (see Fig.3) are not large if the range of C_a is 10^{-30} to 10^{-31} cm^6s^{-1}. It can state that AR can give the impact to the performance of solar cell but, the variation of C_a does not provide huge impact if the variation of C_a is within a certain range.

Therefore, inclusion of AR effect into DB decrease about 3~4% theoretical efficiency in comparison to CBD of Si SJ due to correlating temperature dependent-parameters (n_i, E_g and C_a). Except for CBD of MEG case (less sensitive to temperature variation), most results even in MEG have the similar tendencies that V_{OC} is more temperature-sensitive to J_{SC}. (see Fig.4)

Table II: The theoretical efficiencies

300K	CDB	Ref2	Eq.2	Eq.5	Eq. 6
SJ	33.4%	30.4%	29.9%	30.3%	29.9%
MEG	41.2%	37.3%	36.4%	36.9%	36.4%

We also found the less temperature sensitive ideal MEG device (see Fig. 4(e)) that V_{OC} of CDB MEG demonstrate a small variation to temperature changes in comparison to other cases (including material parameters, Ref [2], (c), (d) and (e) in Fig. 4(e)). From these simulation results, Si MEG have two main advantages: (1) near-ideal MEG device can make it possible to less temperature sensitive devices due to negligible non-radiative recombination and (2) Si MEG device shows great performances at lower temperature region (from 250K to 300K) with considering less impact of non-radiative recombination. (see Fig. 4 (b) and (e) for V_{OC} comparison, Fig. (c) and (f) for efficiency comparison)

4 CONCLUSION AND FUTURE WORKS

We have researched the detailed balance equations of SJ and MEG for silicon solar cells to see the impact of Auger recombination processes. While including MEG concept into DB equations, thinning Si substrate is available due to many EHP generations. Due to the temperature dependences of electrical parameter, the V_{OC} based on correlations with temperaturedependent-C_a, n_i and E_g are significant factors to determine the entire performances. The overall efficiencies decreases as increasing the temperature while considering n_i, E_g and C_a. Thus, temperature-dependent AR is important non-radiative recombination process. But, from the various modeling of C_a, the impact of the C_a changes is not great due to showing small differences of both efficiency and V_{OC} if the variation in C_a is within 10^{-30} to 10^{-31} cm^6s^{-1}. But, from this research, we could find additional advantages of Si MEG device. (1) MEG approaches can make it possible to develop the thinning devices with near-perfect nanostructures. (2) Si MEG can be less temperature sensitive devices in the assumption of near-perfect materials due to negligible non-radiative recombination. (3) While considering non-radiative recombination, Si MEG shows good results at low temperature region due to small difference of V_{OC} between CDB MEG and other cases.

REFERENCE

[1] W. Shockley and H. J. Queisser, J. Appl. Phys., 32, 3, pp510-519, 1961

[2] T. Tiedje, E. Yablonovitch, G. D. Cody and B. G. Brooks, in *IEEE Transactions on Electron Devices*, vol. 31, no. 5, pp. 711-716, 1984.

[3] A. Richter, M. Hermle and S. W. Glunz, *IEEE J. Photovoltaics*, vol. 3, no. 4, pp. 1184-1191, 2013.

[4] J.H. Werner, R. Brendel, H.J. Queisser, tor solar cells", in Proc. of the IEEE 1st World Conference on Photovoltaic Energy Conversion, Waikoloa, HI, USA, pp. 1742–1745, 1994.

[5] M.A. Green, Prog. Photovolt: Res. Appl.; 11, pp 333-340, 2003;

[6] P. Jonsson, H. Bleichner, M. Isberg, and E. Nordlander, *J.Appl. Phys,* 81, 5, pp. 2256-2262, 1997.

[7] K.G. Svantesson and N.G. Milsson, J. Phys. C: Solid State Phys., 12, pp5111-5120, 1979.

[8] M.J. Kerr and A. Cuevas, J. Appl. Phys., 91, 4, pp 2473-2480. 2002

[9] S. Wang and D. Macdonald, J. Appl. Phys, 112, pp113708 1-4, 2012

[10] M.A. Green, Solar Energy Materials & Solar Cells, 92, 1305– 1310, 2008.

MODELING OF EDGE RECOMBINATION LOSSES IN HALF-CELLS

A. Fell[1], H. Sträter[2], M. Müller[2,3], H. Steinkemper[1], J. Schön[1,4], M. Hermle[1], M. C. Schubert[1], D. H. Neuhaus[2] and S. W. Glunz[1,4]

[1]Fraunhofer Institute for Solar Energy Systems, Freiburg, Germany
[2]Solarworld Innovations, Freiberg, Germany
[3]now with Technical University Bergakademie Freiberg, Institute of Applied Physics, Freiberg, Germany
[4]Department of Sustainable Systems Engineering (INATECH), Albert Ludwigs University Freiburg, Freiburg, Germany

ABSTRACT: A new approach to model the impact of edge recombination within silicon solar cells using a 3D simulation of the entire cell geometry within Quokka3 is presented. The contribution of edge recombination within the space charge region (SCR), which is not directly accounted for in Quokka3's skin concept, is included by an effective property, a localized edge-length specific saturation current density with an ideality factor of 2: $J_{02,edge}$. A worst-case value of 19 nA/cm was found to be largely invariable for varying device properties. The new model is applied to predict the worst-case influence of the cut-edge within PERC half-cells. A moderate efficiency loss of 0.1 $\%_{abs}$ is predicted, for which the SCR and non-SCR edge recombination contribute non-additive. Differentiating those two loss mechanisms reveals a very similar impact on light-JV characteristics dominantly on pseudo-fill-factor, whereas only the SCR mechanism is visible in dark-JV curves as an increased J_{02}. This means that non-SCR edge loss contributions will be missed by the commonly applied dark-J_{02} characterization approach. Dedicated experiments to reveal the different mechanisms by measuring front and rear laser-scribed half-cell designs before cleavage are carried out and compared to the simulations results.
Keywords: simulation, modelling, silicon solar cell, edge loss

1 INTRODUCTION

The use of half-cells for higher module efficiency is becoming an increasingly popular topic both in research [1, 2] and industry [3, 4]. Its main advantage is related to lower ohmic interconnection losses, increasing mainly the fill factor and subsequently the power of the module considerably. However, the efficiency of the individual cells can be adversely affected, due to imperfections at the cut-edge, and an overall higher edge-to-area ratio. Two main loss mechanisms both originating from a high surface recombination at the cell edges are differentiated: i) recombination at the main part of the bulk which can be assumed quasi-neutral (qn), and ii) in the case the pn-junction extends right to the edge, additionally recombination in the space-charge-region (SCR) [5]. A quantitative understanding and differentiation via suitable characterization methods of those loss mechanisms is important when designing and optimizing half-cells.

Previous work on modeling and characterizing losses due to cut edges did commonly use an external diode model, where an increase of the second diode J_{02} was attributed to the entire edge losses, motivated by the SCR recombination theory having an ideality of 2 [5–9]. This paper aims to differentiate the contributions within and outside of the SCR by modeling and experiments.

In a first modeling step, Sentaurus Device [10] simulations of a 2D edge domain are performed. The same domain is subsequently setup in Quokka3 [11], and validated against the Sentaurus simulations. For this, Quokka3 supports an edge-length specific saturation current density with an ideality factor of 2, $J_{02,edge}$, localized at the edges, to represent SCR recombination, see [12] for details. Then Quokka3 is used to simulate the entire half-cell geometry in a single 3D domain, including metal resistance, to straightforward and accurately model the (measurable) influence of the edge region on the cell characteristics. Finally, half-cell layouts are cut with different conditions at Solar World Innovations (SWIN), and IV-results are compared to the simulations.

2 MODELING

The cell design and device properties for all simulations are based on the PERC cell in [13].

2.1 2D edge domain

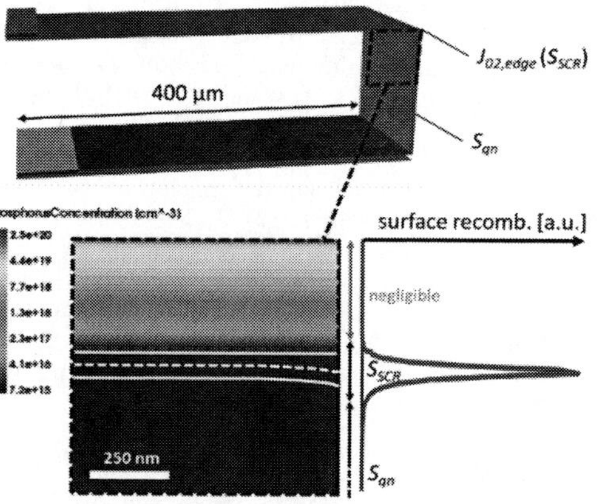

Figure 1: Sketch of the 2D edge solution domain; *upper*: overview as produced by Quokka3; *lower:* corner region as produced by Sentaurus, showing the phosphorus concentration (colorplot), metallurgical junction depth (red dotted line), SCR (white lines, dotted line corresponds to equal carrier densities), and exemplary extracted surface recombination rate along the edge.

In Figure 1 the geometry of the 2D edge domain, representing a small stripe including the cut-edge, is shown. The same domain is setup in Sentaurus Device as well in Quokka3. In Sentaurus, surface recombination in the SCR, S_{scr}, can be set independently of the other surface recombination at the quasi-neutral part of the bulk edge, S_{qn}, by defining a small separate surface region around the pn-junction. Due to the conductive boundary approach in Quokka3, the SCR is not directly modeled and therefore S_{scr} cannot be directly defined. Rather a local edge-length specific $J_{02,edge}$ at a small region at the edge is applied, representing SCR recombination commonly known to show an ideality factor of 2. The absolute value of $J_{02,edge}$ is not known a-priory for a given S_{scr}, but fitted to the Sentaurus simulation results.

2.2 3D half-cell domain

To accurately model the influence of the edge on the half-cell performance, the recent performance boost through Quokka3 is used to discretize and solve the entire cell geometry in 3D, see Figure 2. The domain includes the metal current transport, fully accounting for its distributed nature.

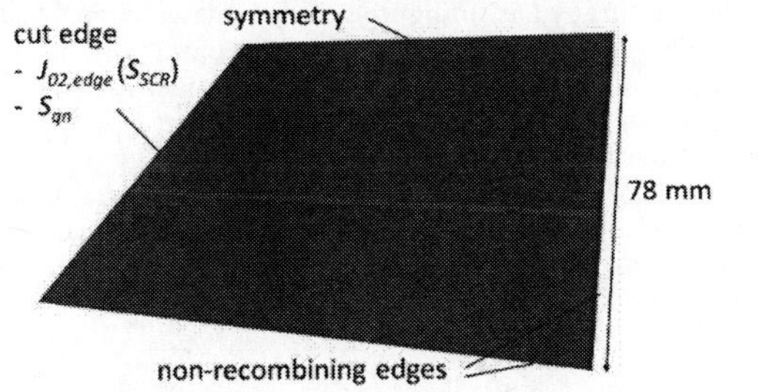

Figure 2: (Half of a) half-cell solution domain as defined in Quokka3. The entire geometry is discretized in 3D (including the bulk thickness) and solved, accurately quantifying the influence of the cut-edge properties on the cell characteristics.

3 EXPERIMENTAL

For validation of the modeled differences in edge recombination multicrystalline Al-BSF solar cells are prepared with a half cell metallization layout comprising neither front nor rear paste laydown in the half cell region where the laser scribing process takes place to prevent molten metallization. The laser scribing process is carried out with an IR disc laser from front and rear side, respectively. After scribing the cells effectively represent half-cells for which in the case of front scribing high additional SCR recombination is to be expected, whereas for rear scribing only additional non-SCR recombination is introduced. The half cells are obtained by mechanical cleavage.

The solar cells are characterized for their light and dark current-voltage characteristics with a standard HALM IV-tester measuring in a hysteresis-free setup. The calibration of short-circuit current I_{SC} for the half cells is carried out by defining I_{SC} of the full cells similar to the sum of I_{SC} of the two corresponding half cells. This approach is supported by the simulation results showing negligible I_{SC} losses due to edge recombination. The calibration is averaged for three full cells and their six half cells. The electrical characterization is carried out before laser scribing, after laser scribing and after mechanical cleavage.

4 RESULTS AND DISCUSSION

For a worst-case scenario, assuming surface recombination velocities at the limit of thermal velocity $S_n = S_p = 10^7$ cm/s, a value of $J_{02,edge}$ = 19 nA/cm within Quokka3 provides good agreement to the Sentaurus simulations. In [12] it was found by a substantial variation of device properties that it in fact states a largely invariable value, thus valid for any pn-junction bordering a highly recombining edge within (typical) silicon solar cells. Having such a fundamentally valid $J_{02,edge}$ value, makes the consideration of SCR edge losses within Quokka3 particularly straightforward, as the Sentaurus fits do not have to be redone for varying device properties.

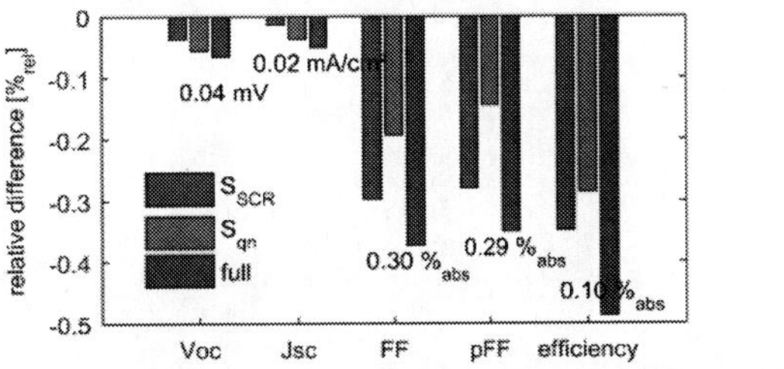

Figure 3: Simulated change of main light-JV parameters of the PERC half-cell due to different edge recombination cases, compared to the reference case without edge recombination; inserted values denote the absolute loss for the case of full edge recombination.

In Figure 3, simulated losses of light-JV parameters of the half-cell for three worst-cases are shown, relative to the case of no edge recombination: a) $S_{scr} = 10^7 cm/s$, b) $S_{qn} = 10^7 cm/s$, c) $S_{qn} = S_{scr} = 10^7 cm/s$. It can be seen that both loss mechanisms mainly influence pFF and show similar absolute impact on efficiency for this worst-case example. They are not additive, which is due to the transport of minority carriers to the neighboring recombination regions being limited by diffusion. Overall, a moderate worst-case efficiency loss due to cut-edge recombination of 0.1 $\%_{abs}$ is predicted.

Looking at the simulated dark-JV's in Figure 4, it can be seen that despite both loss mechanisms introducing similar non-idealities in the light-JV, solely the SCR loss contribution is visible as a non-ideality in the dark-JV. This means that an experimental determination of edge losses via the commonly applied J_{02} extraction from dark JV-curves is not generally sufficient: for the case that non-SCR edge recombination dominates, a possibly significant loss for light-JV performance is missed.

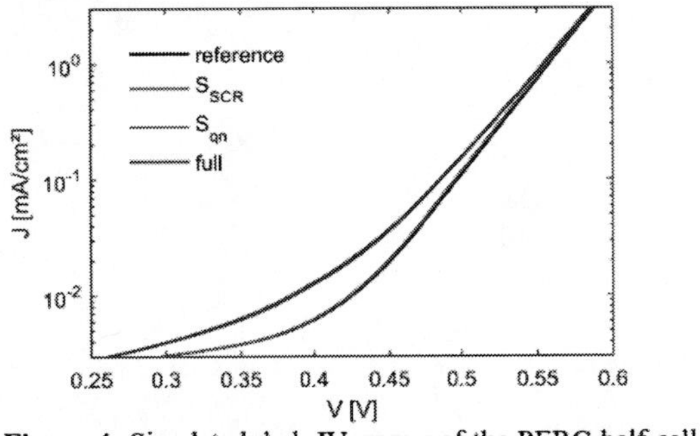

Figure 4: Simulated dark JV-curves of the PERC half-cell for the different edge loss cases; a typical external shunt resistance of 10^5 Ωcm² is assumed

In Figure 5 the experimental results for Voc, FF and (dark-) J_{02} are shown for the two different cut conditions (front and rear side), before cleavage of the cell. As expected in both cases no notable impact on Voc is observed. The loss for front scribing is even larger than predicted by the worst-case simulations, meaning that the laser did likely introduce damage to the SCR extending some distance into the material, or lead to surface area enlargement due to a non-clean cut. The *FF* loss due to rear-scribing is well comparable with the S_{qn} simulations, the apparent increase of J_{02} is however statistically not significant, and does not allow to prove or disprove the dark-JV simulation results. Improved experiments on a PERC design have been started but were not finished by the time of the conference.

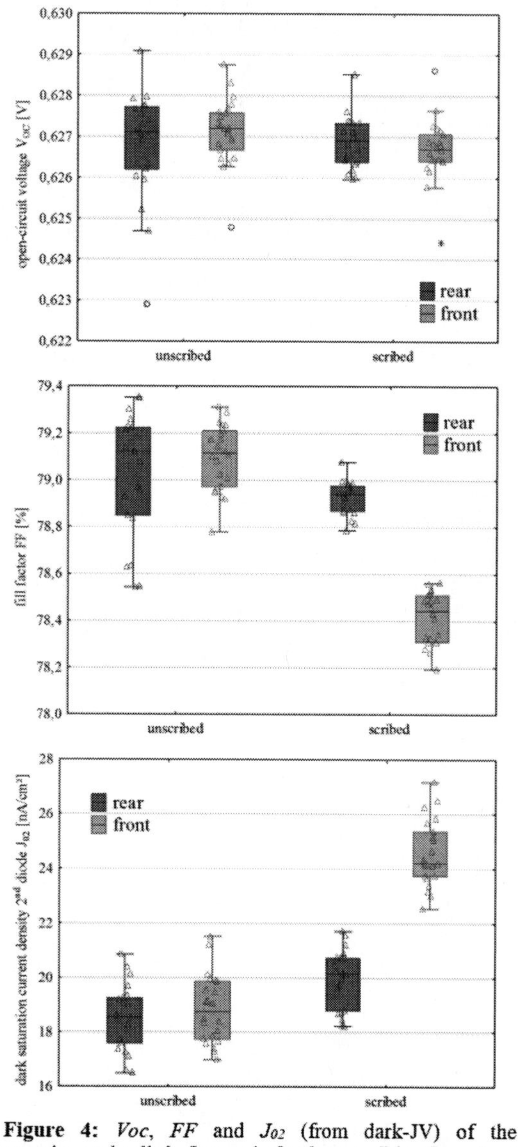

Figure 4: *Voc*, *FF* and J_{02} (from dark-JV) of the experimental cells before and after laser scribing from front and rear side.

5 CONCLUSIONS

A largely invariable localized $J_{02,edge}$ value of 19 nA/cm was found to well represent worst-case SCR edge recombination within Quokka3's skin concept. It is applied to accurately predict the worst-case influence of the cut-edge within PERC half-cells, by simulating the entire cell geometry in 3D. A moderate efficiency loss of 0.1 $\%_{abs}$ is predicted, for which the SCR and non-SCR edge recombination contribute non-additive. Differentiating those two loss mechanisms reveals a very similar impact on light-JV characteristics dominantly on *pFF*, whereas only the SCR mechanism is visible in dark-JV curves as an increased J_{02}. This is important for the experimental characterization of edge losses, which is commonly carried out by determining the increase of dark J_{02}: for a case where non-SCR recombination dominates, e.g. by edge design changes to suppress SCR recombination, this common characterization approach misses potentially significant edge losses.

Dedicated experiments to reveal the different mechanisms by measuring front and rear laser-scribed half-cell designs before cleavage were carried out. The front laser scribing (through the emitter) reveals J_{02} increase beyond the simulated worst-case, indicating increased recombination due to laser damage extending into the material or surface area enlargement. The non-SCR losses show a *FF* loss similar as predicted by the simulations, other trends are however not statistically relevant to prove or disprove the simulations. Refined experiments are under way.

ACKNOWLEDGEMENTS

This work was financially supported by the German Federal Ministry for Economic Affairs and Energy (BMWi) within the research project "PVBat400" under Contract No. 0324145.

The authors thank Nico Wöhrle (Fraunhofer ISE) for thorough beta-testing of the edge loss implementation in Quokka3.

Andreas Fell acknowledges funding by the European Commission through the Marie-Curie fellowship "Quokka maturation".

REFERENCES

[1] S. Eiternick, F. Kaule, H.-U. Zühlke, T. Kießling, M. Grimm, S. Schoenfelder, M. Turek, High Quality Half-cell Processing Using Thermal Laser Separation, Energy Procedia 77 (2015) 340–345.

[2] R. Witteck, D. Hinken, H. Schulte-Huxel, M.R. Vogt, J. Müller, S. Blankemeyer, M. Köntges, K. Bothe, R. Brendel, Optimized Interconnection of Passivated Emitter and Rear Cells by Experimentally Verified Modeling, IEEE Journal of Photovoltaics 6 (2) (2016) 432–439.

[3] C. Arndt, D. Stichtenoth, N. Müller, M. Reinicke, D. Grote, U. Schröder, M. Scherr, M. Weser, F. Lange, P. Voigt (Eds.), Manufacturing of 300Wp Modules by Industrialization of PERC Solar Cell Technology Including Crystal, Cell and Module Development, 2014.

[4] S. Zhang, X. Pan, H. Jiao, W. Deng, J. Xu, Y. Chen, P.P. Altermatt, Z. Feng, P.J. Verlinden, 335-W World-Record p-Type Monocrystalline Module With 20.6% Efficient PERC Solar Cells, IEEE Journal of Photovoltaics 6 (1) (2016) 145–152.

[5] R. Kuhn, P. Fath, E. Bucher (Eds.), Effects of pn-junctions bordering on surfaces investigated by means of 2D-modeling. Conference Record of the Twenty-Eighth IEEE Photovoltaic Specialists Conference - 2000 (Cat. No.00CH37036), 2000.

[6] M. Hermle, J. Dicker, W. Warta, S.W. Glunz, G. Willeke (Eds.), Analysis of edge recombination for

high-efficiency solar cells at low illumination densities, IEEE, 2003.
[7] C. Chan, M. Abbott, B. Hallam, M. Juhl, D. Lin, Z. Li, Y. Li, J. Rodriguez, S. Wenham, Edge isolation of solar cells using laser doping, Solar Energy Materials and Solar Cells 132 (2015) 535–543.
[8] K. Rühle, M.K. Juhl, M.D. Abbott, L.M. Reindl, M. Kasemann, Impact of Edge Recombination in Small-Area Solar Cells with Emitter Windows, IEEE Journal of Photovoltaics 5 (4) (2015) 1067–1073.
[9] D. Bertrand, S. Manuel, M. Pirot, A. Kaminski-Cachopo, Y. Veschetti, Modeling of Edge Losses in Al-BSF Silicon Solar Cells, IEEE Journal of Photovoltaics 7 (1) (2017) 78–84.
[10] Synopsis, Sentaurus Device User Guide, release I-2013.12, Zurich, Switzerland, 2013.
[11] A. Fell, J. Schön, M.C. Schubert, S.W. Glunz, The concept of skins for silicon solar cell modeling, Solar Energy Materials and Solar Cells (2017).
[12] A. Fell, J. Schön, M. Müller, M.C. Schubert, S.W. Glunz, Modelling edge recombination in silicon solar cells, submitted to IEEE Journal of Photovoltaics (2017).
[13] A. Fell, K.R. McIntosh, P.P. Altermatt, G.J.M. Janssen, R. Stangl, A. Ho-Baillie, H. Steinkemper, J. Greulich, M. Muller, M. Byungsul, K.C. Fong, M. Hermle, I.G. Romijn, M.D. Abbott, Input Parameters for the Simulation of Silicon Solar Cells in 2014, IEEE Journal of Photovoltaics 5 (4) (2015) 1250–1263.

2D/3D SIMULATIONS OF BLACK-SILICON INTERDIGITATED BACK-CONTACTED c-Si(n) SOLAR CELLS

E. Calle[1], D.Carrió[1], P.Ortega[1], G. Von Gastrow[2], H. Savin[2], I. Martín[1], R. Alcubilla[1]

[1] Universitat Politècnica de Catalunya, Micro and Nano Technologies Group MNT,.
C/ Jordi Girona 1-3, Mòdul C4, 08034 Barcelona, Spain, pho. +34 934054193, e.mail: eric.calle@upc.edu
[2] Aalto University, Department of Electronics and Nanoengineering Tietotie 3, 02150 Espoo, Finland
e-mail: hele.savin@aalto.fi

ABSTRACT: black silicon (b-Si) reduces drastically light reflectance in the front side of c-Si solar cells to values near zero for the whole absorbed solar spectrum. In this work, we apply 2D and 3D simulations to explore the efficiency limits of interdigitated back-contacted c-Si(n) solar cells with line or point contacts respectively, using ALD Al_2O_3 passivated b-Si in the front surface. Realistic physical and technological parameters involved in a conventional oven-based fabrication process are considered in the simulations, especially those related to surface recombination on the b-Si as well as high doped p+/n+ strip regions. One important issue is the temporal stability of surface passivation on b-Si surfaces. In this work experimental long-term b-Si surface passivation data after two years and its impact on cell performance are studied. Simulations demonstrate initial and final photovoltaic efficiencies over 24.6% and 23.2% respectively for an emitter coverage of 80% independently of the cell contact strategy. A photocurrent loss about 1.3 mA/cm^2 occurs when surface recombination velocity at the b-Si surfaces degrades from 6 cm/s to a final value of 28 cm/s.
Keywords: c-Si, 3D simulations, black-silicon, IBC solar cell, point-like contacts, surface passivation

1 INTRODUCTION

Nano-textured silicon, also referred as black silicon (b-Si), is an excellent choice to drastically reduce crystalline Si solar cell reflectance in the useful solar spectrum [1]. One of the major issues related to b-Si surfaces is the possible increase of surface/bulk recombination due partly to the enlarged surface area of such nanostructures and partly to a possible lattice damage during etching. These drawbacks have been overcome successfully in the last years applying less aggressive etching techniques based on inductively-couple plasma reactive ion etching (ICP-RIE) combined with surface passivation based on atomic layer deposited (ALD) Al_2O_3 films. Thereby, effective surface recombination velocities (S_{eff}) below 10 cm/s on p- and n-type c-Si substrates have been achieved [2-5]. Black silicon surfaces passivated by Al_2O_3 have been used either in front-back-contact or all-back-contact solar cells [6-8]. In particular, Interdigitated Back Contacted (IBC) architecture is very well suited to b-Si. In this cell structure, the front surface passivation and antireflection coating are totally decoupled from the rear electrodes design. Recently IBC cells with b-Si front surface, from now on b-Si IBC cells, have reached promising efficiencies beyond 22% independently of substrate polarity [9, 10]. One important issue related to b-Si is the temporal stability of surface passivation. In a previous work we have reported long-term degradation of S_{eff} parameter on b-Si surfaces both for n- and p-type substrates [5], but its impact in the IBC cell photovoltaic performance is not explored yet.

In this work, we study the long-term efficiency potential of ALD Al_2O_3 passivated b-Si IBC c-Si(n) cells as well as the influence of the cell geometry, i.e. emitter coverage (f_e), in the photovoltaic performance by means of 2D/3D simulations (Synopsys Sentaurus Device). Realistic physical parameters for front and rear surfaces are considered in the simulations based on separate experimental recombination lifetime measurements. Two types of structures are simulated: IBC cells with contact lines (2D) and alternatively using locally point square contacts (3D).

2 CELL DEFINITION AND OPTICAL MODEL

2.1 Unit cell geometry definition

Solar cells are large area devices but symmetrical enough to be studied as a repetition of a simple domain or unit cell, reducing simulation time and hardware computational requirements. The corresponding IBC solar cell is a periodical repetition of the unit cell, once it is flipped respect to the symmetry axis, with a periodicity related to the contact pitch in the x and y directions. In fig. 1 the simplest unit cell for the 3D and 2D IBC c-Si(n) solar cell architectures are shown. Notice that the p+ emitter is partly passivated with ALD Al_2O_3 and the rest with thermal SiO_2. In this way the electrical behavior of the whole structure can be obtained considering that each unit cell is electrically connected in parallel with the others (current scaling up).

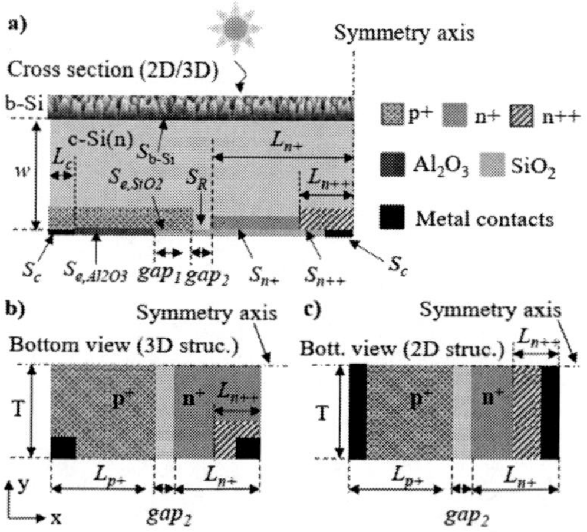

Figure 1: a) 2D cross section sketch of the simplest unit cell, with only one base and one emitter contacts, is shown. b) and c) bottom views for the 3D and 2D unit cells considering point-like and line-like contacts respectively.

Notice that the unit cell contains only the half area of a single emitter and base contacts and strip p+ and n+ doped regions are used in the simulations. Additionally, we include square and strip n++ regions beneath electron contacts in the 3D and 2D simulations respectively.

Table I summarizes the geometrical values as well as the parameters related to doped regions used in the study. Gaussian doping profiles are introduced in the simulations considering the expected surface doping N_s and the sheet resistance R_{sh} for each doped region using our convectional fabrication process. Notice that changing the length of p+ regions (L_{p+}), from 98 to 798 µm and fixing the other length parameters, the emitter coverage f_e ranges from 50 to 89%.

Table I: Geometrical and doping profile parameter values.

Geometrical parameters (µm)							
w	T	gap_1	gap_2	L_{n+}	L_{n++}	L_{p+}	L_c 3D/2D
260	100	98(*)	4	98	50	98-798	25/15

Doping profile parameters					
$N_{s,n++}$ (cm^{-3})	$R_{sh,n++}$ (Ω/sq)	$N_{s,n+}$ (cm^{-3})	$R_{sh,n+}$ (Ω/sq)	$N_{s,p+}$ (cm^{-3})	$R_{sh,p+}$ (Ω/sq)
7×10^{19}	7.0	3.5×10^{18}	110	5×10^{18}	190

(*) For L_{p+}-L_c<100 µm then gap_1=L_{p+} - L_c

2.2 Optical model

Optical modelling of b-Si surfaces is quite complex in general. However, we can use a good approach to simulate the photogeneration inside the structure from experimental reflectance data as follows: 1) a modified AM1.5G solar spectral irradiance is introduced in the simulations (I_{mod}) considering the measured reflectance of b-Si coated with 20 nm thick ALD Al_2O_3 layer ($R_{b\text{-}Si}$) as I_{mod} =$I_{AM1.5}$ (1-$R_{b\text{-}Si}$) (see Fig.2). 2) Total photon transmission (T=100%) and internal front reflectance (R_{int_Front}=100%). 3) An average internal rear reflectance (<R_{int_rear}>) is assumed in the simulations taking into account contacted, metalized and non-metallized non-contacted areas with local reflectance of 75, 96 and 85% respectively. 4) Ray tracing is performed using for absorption and refractive index the default wavelength dependent program values for c-Si.

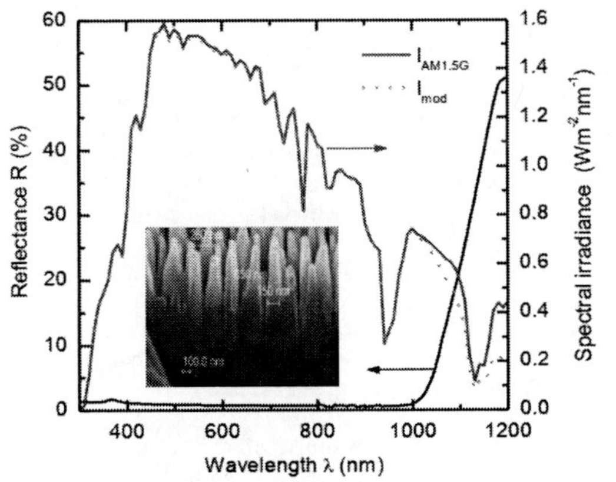

Figure 2: b-Si reflectance (sample coated with 20 nm thick ALD Al_2O_3 layer), AM1.5G and modified spectral irradiances used in the simulations. A SEM/FIB image of the b-Si surface is also shown in the inset.

The simplified simulated structure, from an optical point of view, is shown in Fig. 3 where gap_{metal} is the distance between electron and hole metal electrodes, 80 µm in our devices.

Flat energy bands have been considered at the passivated semiconductor surfaces. Therefore, there are no inversion or accumulation layers below passivated surfaces. Additionally, physical models, e.g. carrier statistics, band gap narrowing, and carrier mobilities are the same as considered in [11].

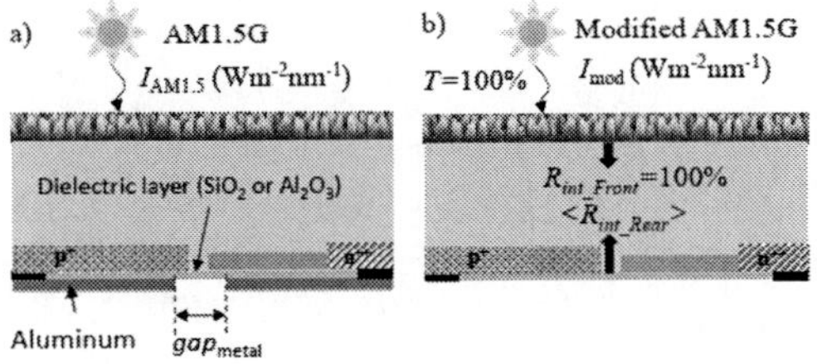

Figure 3: a) Cross section of a b-Si IBC cell and b) its equivalent structure considering the simplified optical model.

3 SURFACE RECOMBINATION

In order to have accurate simulations is important to introduce realistic surface recombination velocities values (S) at the different surface regions of the IBC device. In this section, we determine these parameters evaluated in the b-Si surface as well as in the p+, n+ and n++ doped regions. Experimental symmetrical test samples using high quality c-Si (n/p) FZ <100> substrates (270 µm thick and 2 ± 1 Ωcm) were prepared to perform effective lifetime (τ_{eff}) measurements by the QSS-PC technique.

3.1 Surface recombination at b-Si surfaces

The most critical parameter in the IBC simulations is the surface recombination velocity at the nanostructured surface ($S_{b\text{-}Si}$). In Figure 4 $S_{b\text{-}Si}$ values, once b-Si surfaces are passivated with 20 nm ALD Al_2O_3, are extracted from τ_{eff} data on c-Si(n) substrates considering a time window study of two years. Experimental details are reported in [5]. Surface recombination velocities are calculated using eq. (1) and evaluated at a carrier excess density of Δn=10^{15} cm^{-3}, being w the sample thickness. In this study we consider that bulk lifetime is dominated by intrinsic recombination mechanisms (auger and band-to-band), being τ_{in} the related lifetime, using Richter et al parameterization [12]. b-Si data are directly compared with random pyramids textured surfaces passivated with the same Al_2O_3 layer as a reference.

$$S_{b-Si} \cong \frac{w}{2}\times\left(\frac{1}{\tau_{eff}}-\frac{1}{\tau_{in}}\right) \quad (1)$$

As it can be seen in Fig. 4, excellent initial $S_{b\text{-}Si}$ values extracted about 6 cm/s are achieved in the nanostructures despite the unfavorable large b-Si surface area. However this value increases after two years to 28 cm/s.

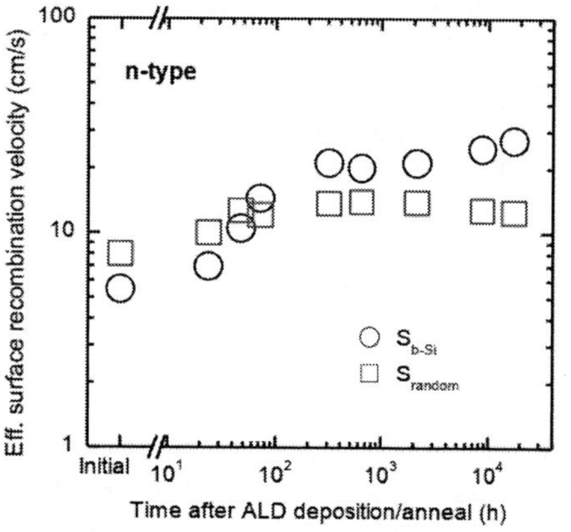

Figure 4: Surface recombination velocities for b-Si ($S_{b\text{-}Si}$) and random pyramids (S_{random}) textured surfaces on n-type substrates.

3.2 Surface recombination at p+ and n+/n++ surfaces

Surface passivation at the doped regions was determined by the extraction of the emitter recombination current density J_{oe} parameter from τ_{eff} measurements using eq. (2). Where n_i, q, and N_{bulk} are the intrisic concentration (8.56×10^9 cm^{-3} at T=25°C), q the elementary charge and N_{bulk} the bulk doping respectively. Notice that n- or p-type substrates were used to form p+ and n+/n++ emitters respectively.

$$\frac{1}{\tau_{eff}} - \frac{1}{\tau_{in}} \cong \frac{2J_{oe}}{qn_i^2 w}\left(N_{bulk} + \Delta n\right) \qquad (2)$$

Samples were doped at different temperatures using solid dopant sources and underwent a drive-in stage at 1080°C during one hour in pure O_2 ambient growing a thermal SiO_2 film of 110 nm as a passivation layer. Final emitter sheet resistance (R_{sh}) covers a range between 60-600 and 6-200 Ω/sq in boron and phosphorous doped emitters respectively. Finally, an annealing in forming gas during 10 min was made in all samples. For the boron emitters, the SiO_2 layer was removed and replaced by an ALD Al_2O_3 film deposited at 200°C (30 nm thick) as alternative passivation layer. In this case, an extra annealing at 400°C 10 min was made to activate surface pasivation.

In Figure 5 J_{oe} data vs. R_{sh} for phosphorous emitters are shown. Filled symbols and empty symbols correspond to passivated annealed and unpassivated (bare) n+ emitters respectively. Lines correspond with PC1D simulations considering the fundamental surface recombination velocity for holes (S_{po}) as a free simulation parameter. In that case doping profiles introduced in the PC1D program were generated using SUPREM program in agreement with the diffusion and drive-in stages followed in our test devices. As can be seen in Fig. 5, SiO_2 passivated n+ emitters follows the expected trend, i.e. the lower R_{sh} the higer J_{oe} value. The estimation of the S_{po} parameter from eq. (3) as a function of the phosphorous doping at the surface N_s allows a close fitting of the J_{oe} values, as it is already reported in the literature [13].

Experimental J_{oe} data meaured in boron doped emitters are shown in Fig. 6. ALD Al_2O_3 films offer a better performance than SiO_2 passivated boron emitters in the whole R_{sh} range, being the fundamental surface recombination velocity for electrons (S_{no}) around 200 cm/s for a 100-200 Ω/sq emitter, much lower than its counterpart using SiO_2 as a passivated layer (1500 cm/s).

$$S_{po} = 100 \times \frac{N_s}{10^{18}} \qquad (3)$$

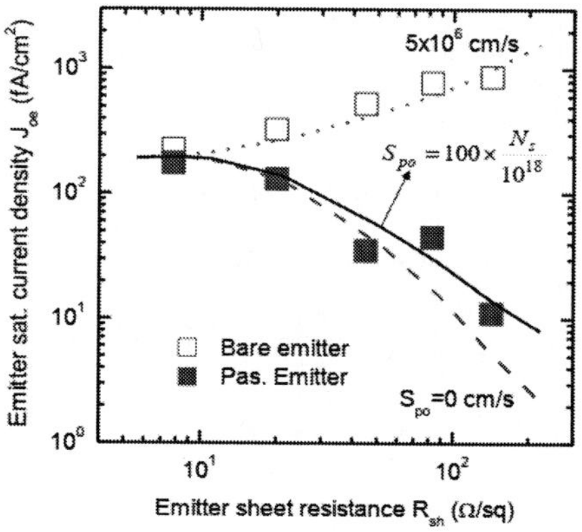

Figure 5: J_{oe} vs. R_{sh} for passivated (fill symbols) and unpassivated bare (empty symbols) phosphorous emitters. Lines correspond with PC1D/SUPREM simulations considering S_{po} as a free simulation parameter.

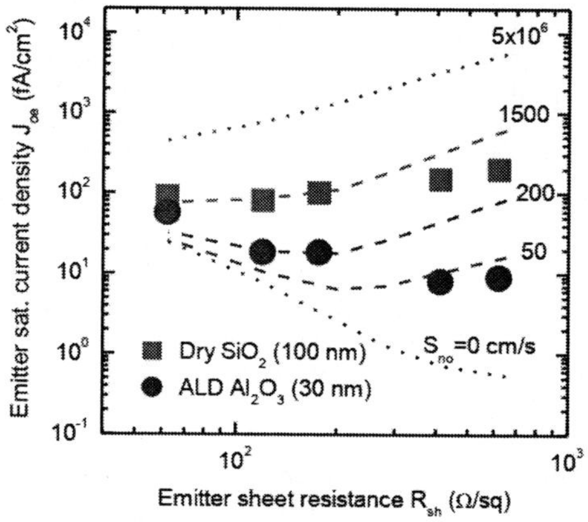

Figure 6: J_{oe} vs. R_{sh} for passivated boron emitters. Square and circles symbols correspond with passivated emitters with thermal SiO_2 and ALD Al_2O_3 respectively. Lines correspond with PC1D/SUPREM simulations considering S_{no} as a free simulation parameter.

Finally, table II summarizes all surface recombination velocities used in the 2D/3D simulations. Notice that surface recombination velocities on n+ and n++ regions are calculated taking into account eq. (3) with the surface doping given in table I.

Table II: Surface recombination velocities at the different device regions (see Fig. 1). Fundamental surface recombination velocities are considered the same in the simulations, i.e. S_{no}=S_{po}=S_i, in all cases.

Surface recombination velocities (S_{no}=S_{po}) (cm/s)					
$S_{e,Al2O3}$	$S_{e,SiO2}$	S_{n+}	S_{n++}	$S_{b\text{-}Si}$	S_R
200	1500	400	7000	6-28	10

4 b-Si IBC SOLAR CELL SIMULATION RESULTS

Simulation results for 2D and 3D b-Si IBC structures under AM1.5G solar spectrum (1 kW/m^2, T=25°C) are shown in Fig. 7. Excellent initial efficiencies (η) up to 24.7% for emitter coverages around 80% independently of structure typology are achieved. The efficiency is limited mostly by the relatively low open circuit voltage (V_{oc}) values < 710 mV reached using our baseline fabrication process compared with the state-of-the-art around 750 mV obtained in silicon heterojunction based IBC c-Si(n) solar cells [14].

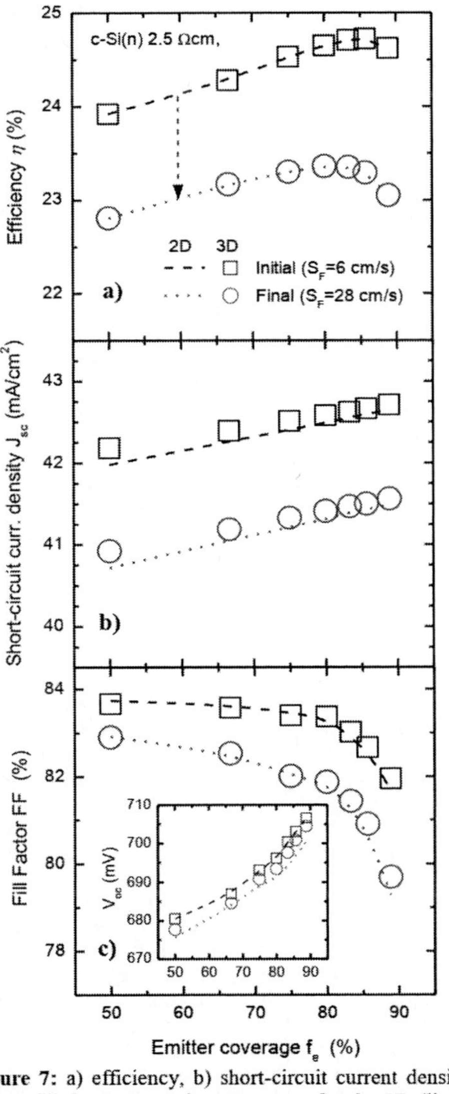

Figure 7: a) efficiency, b) short-circuit current density, and c) fill factor vs. emitter coverage for the 2D (lines) and 3D (symbols) of simulated b-Si IBC solar cells. Expected open circuit voltages are also shown in the inset of plot c).

A drop of 1.5% in absolute efficiency is expected after two years, due in part to a degradation of short-circuit current density (J_{sc}) of ~1.2 mA/cm^2 from outstanding initial and stabilized values of 42.7 and 41.5 mA/cm^2 respectively. The efficiency trend with f_e is the expected behaviour considering a trade-off between recombination and series resistance losses, i.e. higher f_e values leads to lower recombination losses, higher J_{sc} and V_{oc}, but higher ohmic losses, i.e. lower fill factors (*FF*).

5 CONCLUSIONS

In this work, we apply 2D/3D simulations to explore efficiency limits of black silicon IBC solar cells using realistic physical and technological parameters involved in a conventional oven-based fabrication process. One important issue is the temporal stability of surface passivation on b-Si surfaces. In this work long-term b-Si surface passivation and its impact on cell performance is studied. Simulations demonstrates initial and final (after two years) photovoltaic efficiencies over 24.6% and 23.2% respectively for an emitter coverage of 80% independently of the cell contact strategy. A photocurrent loss about 1.3 mA/cm^2 occurs when surface recombination velocity at the b-Si surfaces changes from 6 cm/s to a final stabilized values around 28 cm/s.

ACKNOWLEDGMENTS

This work was partially funded by the Spanish MINECO (PCIN-2014-055) and Finnish TEKES (40329/14) agencies under Solar-Era.Net FP7 European Network. Part of the work was also supported by the project REFER COMRDI15-1-0036 funded by ACCIÓ and European Regional Development Fund (FEDER) as well as the Spanish MINECO projects (ENE2016-78933-C4-1-R and TEC2014-59736-R).

REFERENCES

[1] M. Algasinger et al, *Advanced Energy Materials* 2013, **3**; 1068-1074.
[2] T. Allen et al, *In Proc. of 40th IEEE Photovoltaic Specialist Conference* 2014, Denver Colorado USA, pp. 562-566.
[3] P. Repo et al, *IEEE Journal of Photovoltaics* 2013, **3**(1); 90-94.
[4] G. von Gastrow et al, *Solar Energy Materials & Solar Cells* 2015, **142**; 29-33.
[5] E. Calle et al, *Energy procedia* 2016, **92**; 341-346.
[6] M. Otto et al, *Advanced Optical Materials* 2015, **3**; 147-164.
[7] J. Oh et al, *Nature Nanotechnology* 2012, **7**; 743-748.
[8] S. Jeong et al, *Nature Communications* 2013, **3**, doi:10.1038/ncomms3950.
[9] H. Savin et al, *Nature Nanotechnology* 2015, **10**; 624-628.
[10] P. Ortega et al, *Progress in Photovoltaics: Research and Applications* 2015, **23**(11); 1448-1457.
[11] J.M. López-González et al, *Progress in Photovoltaics: Research and Applications* 2015; **23**(1); 69-77.
[12] A. Richter et al, *Physical Review B* **86**: 165202 (2012).
[13] A. Cuevas et al, *Progress in Photovoltaics: Research and Applications* 2000; **8**(6); 603-616.
[14] K. Masuko et al, *IEEE Journal of Photovoltaics* 2014; **4**(6); 1433-1435.

Simulating the Effect of Partial Rear Contacts on Si Solar Cells by a Finite Element Circuit Simulator

Yong-Han Lin*, Huang-Yu Chen, and Chi-Chun Li
Motech Industries, Inc., Science Park Branch,
No. 96, Dashun 7th Rd., Xinshi Dist., Tainan 74145, Taiwan
*E-mail: yonghanlin@gmail.com

ABSTRACT: TCAD simulation of Si solar cells requires physical parameters as inputs. However, knowing their correct values is not trivial, especially when analytical tools are not available. For optimizing complex contact patterns, such as localized contacts at the rear, defining a 3D TCAD mesh can be more intricate if reducing the cell to a small unit is difficult. Here, we present a work showing that the effect of rear dashed line contacts in PERC-type cells can be simulated with ease by Griddler, a finite element method based simulator developed by SERIS. The cell properties can be described simply with a plausible set of parameter values, obtained first by calibrating the model with a few cells of rear solid line contacts. Although these values are not individually verified and may not be physically correct, the combination does depict the experiments qualitatively well, indicating that this approach can be a quick guide for complex pattern design.
Keywords: simulation, metallization pattern, PERC

1 INTRODUCTION

Numerical simulation of Si solar cells with Technology Computer Aided Design (TCAD) software packages has recently been widely used in the photovoltaics community. It helps in analyzing fabricated cells and predicting cell behaviors when processes or material properties are modified [1]. To achieve high accuracy, values for the input parameters of relevant physical models in TCAD packages need to be carefully considered and examined specifically to the system under study. Unfortunately, obtaining these values is not an easy task and can only be realized under certain theoretical assumptions if measuring tools are not available.

In the case of optimization of an H-grid metallization pattern, which is a stage having an important role in the commercialization of designed cells, implementation of a detailed TCAD simulation of a representative three-dimensional (3D) unit structure could be cumbersome and insufficient as power dissipation due to current transport along the metal bars cannot be properly considered. Therefore, analytical expressions assuming a current flow pattern perpendicular or parallel to the metal bars are usually applied and suffice for power loss calculations.

However, for optimizing contact patterns with complex geometries, such as metal-wrap-through (MWT) patterns or diverse localized contacts at the rear, both the above two means could become ineffective. For example, the assumption of the rectangular zigzag current path that leads to the analytical power loss formulae no longer applies, and defining a 3D mesh-structure using TCAD tools becomes barely achievable if the pattern is too complicated to be reduced down to an amenable unit structure.

In order to tackle this optimization problem in a more suitable and general manner, a finite-element-method based solar cell simulator named Griddler has been developed by SERIS [2, 3]. Griddler breaks down a complete cell (i.e., full-area front and rear contact patterns, along with the wafer) into thousands of nodes. Each node is treated as a tiny solar cell characterized by the two-diode model, and neighboring nodes are connected to each other by a resistor of a value dependent on the locations of the nodes (overlaid or not overlaid with metals), turning a full-size solar cell into a distributed network of diodes and resistors.

In this article, we present a work showing that the effect of rear dashed line (DL) contacts in PERC-type cells can be simulated with ease by Griddler. Interestingly, we find that the cell properties can be described simply with a plausible set of parameter values, obtained first by calibrating the model with a few cells of rear solid line (SL) contacts. Although these values are not individually verified and may not be physically correct, the combination does depict the experimental data qualitatively well, indicating that this approach can be a quick guide for complex pattern design.

2 METHODOLOGY

Griddler imports the front and rear contact patterns from CAD or image files as its input. The macroscopic material properties (i.e., material input parameters) for diodes and different sorts of resistors are then specified separately for both sides of the cell. In this work, the enhancement of the internal rear reflectance in PERC-type cells [4], resulting from the outperformance of the reflectance of the Si/dielectric/metal stack compared to that of the Si/metal one, is modeled through a combination of the rear side metal optical transparency and the illumination conditions. Figure 1 shows the schematic plot of the front and rear contact patterns for the cells and the geometrical variables that we use for the description of the layout of our rear DL contacts.

Four different types of cells (cell types A—D) were designed and fabricated in this work to compare with the simulation results. The corresponding symbols that we use in the following figures to indicate the respective experimental data of these types of cells and the differences between them are listed in Table I. The key point is that the front side designs (contact patterns, passivation layers, etc.) for all these cell types are different from each other, while the rear side designs for all of them are identical in all respects except the geometrical contact patterns. (That is, all the cell types have the same rear side "materials recipe".)

Different cell types were fabricated with different purposes. The experimental data of either cell type A or B is used to extract the values of the parameter set that is

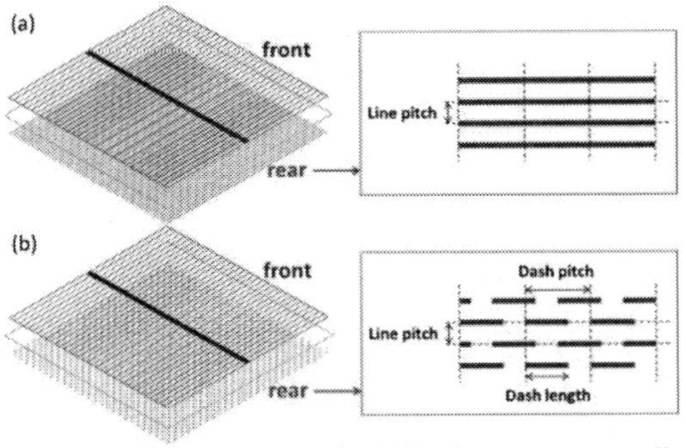

Figure 1: Schematic plots of Griddler input patterns: cell rear contacts for (a) solid lines (SLs) and (b) dashed lines (DLs), respectively, and the three geometrical variables used to characterize dashed lines. In adjacent dashed lines in (b), the dashes are horizontally shifted by half a dash pitch.

Table I: Differences between the four types of the fabricated cells.

Cell type	Symbol	Front side design	Rear side design	
			Contact pattern	Rest
A	□	Recipe A	Solid lines (with line pitch varied)	Identical recipe
B	○	Recipe B	Solid lines (with line pitch varied)	
C	△	Recipe C	Dashed lines (with line pitch varied, while dash pitch and length fixed)	
D	▽	Recipe D	Dashed lines (with all 3 variables varied)	

required for the description of the effect of partial rear contacts for the rear side materials recipe under study (i.e., the rear side model). This process is regarded as the calibration of the model. On the other hand, the data obtained from cell types C and D are utilized to test the applicability of the model.

3 MODEL CALIBRATION AND ITS APPLICABILITY

As mentioned earlier, the simulation input parameters are divided into two different sets: one for the front side, and one for the rear side. We first fit (calibrate) the experimental data of type A by adjusting all these parameters and derive the respective models that account for the front and rear side properties. Figure 2 shows this calibration result, in which the experimental data (open-circuit voltage V_{oc}, short-circuit current density J_{sc}, fill factor (FF), and efficiency η) and the simulation results as a function of rear SL pitch are displayed.

We notice that, to complete a calibration using a specific cell type (either A or B), a total of 4 — 5 experimental groups of varied rear SL pitches should be enough as long as the resulting variation of η with rear SL pitch exhibits a peak value. This is usually can be attained by choosing a wide range of rear SL pitch and having them evenly spaced.

Inspection of Fig. 2 reveals that the experimental data of type B as a function of rear SL pitch behave similarly to that of type A. This is not a surprise since they have the same rear side materials recipe. To examine whether the established rear side model can be applied to type B, we then fit its data by adjusting only front side

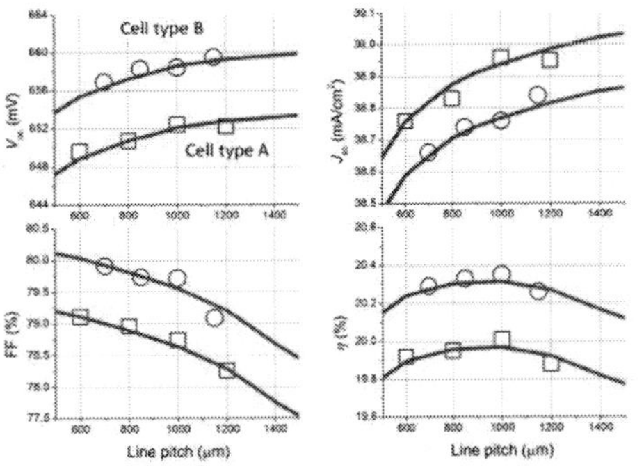

Figure 2: Cell properties as a function of rear SL pitch for types A and B. The symbols are the experimental data and the curves are the corresponding simulation results.

parameters and keeping the rear side parameters totally unchanged. Such tuning, practically speaking, is equal to shifting the simulation curves upwards or downwards. The result is shown in Fig. 2. As expected, the data of type B can also be depicted qualitatively well by the simulation curve using the same set of rear side parameter values.

The established rear side model is further tested by applying it to cell types C and D. More complex rear DL contact patterns are employed in these two types of cells. For type C, with fixing two of the three geometrical variables in the description of the rear DL contacts (i.e., using constant dash pitch and dash length, see Fig. 1), the dependences of cell properties on DL pitch are investigated. The result is displayed in Fig. 3.

For type D, the design is more randomized in which all the three geometrical variables are varied. A total of five experimental groups (patterns) are tested for this type of cell. Each group has its own combination of line pitch, dash pitch, and dash length. The variations of the experimental data with designed pattern are then compared with those of the simulated cell properties. Such a comparison is shown in Fig. 4. Clearly, as can be seen in Figs. 3 and 4, even for a more complex rear DL contact pattern, the established rear side model also delineates the experimental data to an acceptable degree.

We mention in passing that, in the course of rear side metallization of PERC-type cells, voids may form between the Si/metal interface, in company with the formation of back surface field, as a result of imbalance interdiffusion between Si and metal (typically, Al) [5]. In this case, the voids apparently can block the current flow and jeopardize the cell properties. If a significant number of voids form during the metallization process, it should be noted that the presented approach may not be applicable.

4 CONCLUSION

In this work, we show that the effect of partial rear contacts on Si solar cells can be simulated with ease by Griddler, a finite element circuit simulator developed by SERIS. With a suitable set of parameter values, simulation results exhibit a good agreement with experimental data. Despite the fact that these parameter values are not properly verified and may not be physically correct, our results indicate that Griddler,

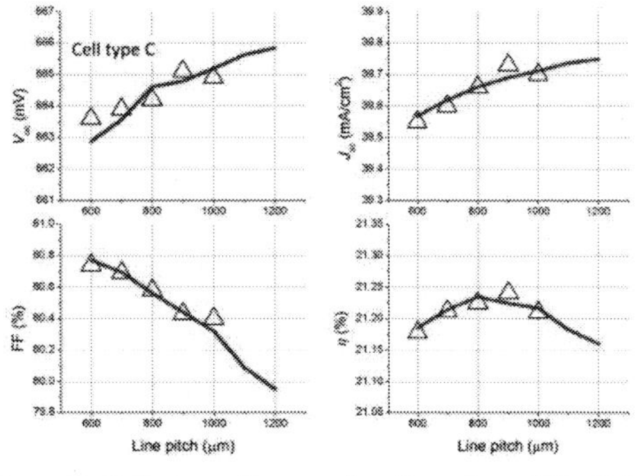

Figure 3: Cell properties as a function of rear DL pitch for type C. The symbols are the experimental data and the curve is the simulation result.

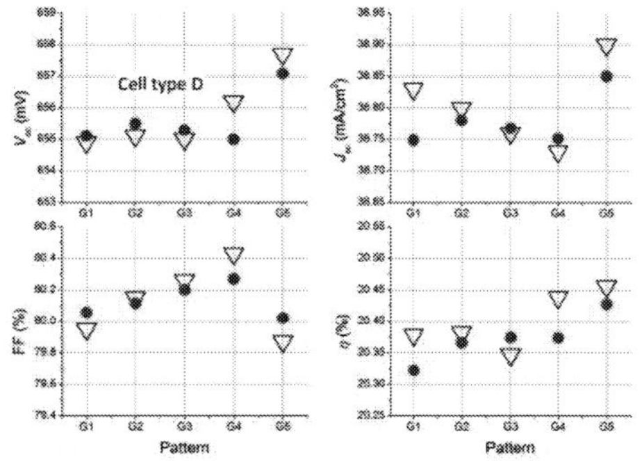

Figure 4: Cell properties varied with different rear DL patterns for type D. The empty triangles are the experimental data and the solid circles are the simulation result.

together with the presented approach, can serve as a quick guide for complex pattern design.

REFERENCES

[1] R. De Rose, K. Van Wichelen, L. Tous, J. Das, F. Dross, C. Fiegna, M. Lanuzza, E. Sangiorgi, A. Uruena De Castro, and M. Zanuccoli, Energy Procedia 27 (2012) 197.

[2] J. Wong, T. Mueller, R. Sridharan, X. Zhang, Y. Yang, Z. Feng, Q. Huang, P. Verlinden, and A. G. Aberle, Proceedings IEEE 38th Photovoltaic Specialists Conference (2012) 000501.

[3] J. Wong, Proceedings IEEE 39th Photovoltaic Specialists Conference (2013) 933.

[4] A. Ristow, M. Hilali, A. Ebong, and A. Rohatgi, Proceedings 17th European Photovoltaic Solar Energy Conference (2001) 1335.

[5] E. Urrejola, K. Peter, H. Plagwitz, and G. Schubert, Applied Physics Letters 98 (2011) 153508.

THE PC1D DIFFUSION MODEL IN THIN-FILM SOLAR CELLS

Luigi Abenante and Massimo Izzi
ENEA, Italian National Agency for New Technologies, Energy, and Sustainable Economic Development
Via Anguillarese 301, 00123, Roma, Italy

ABSTRACT: We show that the optical diffusion model embedded in the numerical simulation program PC1D systematically underestimates light-generated currents in thin textured Si solar cells. The demonstration exploits both the exact analytical solution to minority-carrier transport in uniformly doped regions and the fact that, at the first pass of internal light, optical direction is determined by surface texture. We provide a simple correction procedure to remove the detected systematic error.
Keywords: Light Trapping, Simulation, Solar Cell

1 INTRODUCTION

There is increasing interest in designing ultrathin Si solar cells with active layer thickness of a few micrometers because of their reduced cost of production compared to conventional devices. In such thin films, efficient light absorption is achieved with both broadband antireflection coatings and effective light-trapping structures [1]. In the present work, we deal with light trapping.

Light trapping depends on both internal reflectance and average propagation angle, θ, with respect to the device normal of the light crossing the device. The limit for θ is 60°, which occurs at standard Lambertian optical regimen.

Angle θ may vary at each pass of light across the considered device. Depending on the capability of the front and rear surfaces of diffusing light rays, θ=60° can be reached at the first, second, or third pass of light across the device.

In the present work, we show that the first two cases can be modeled in ultrathin Si solar cells by the numerical simulation program PC1D [2], provided that an optical diffusion model embedded in the program is taken into account. This diffusion model is activated when one or both internal surfaces are imposed to be optically diffusive [3].

We show that the PC1D diffusion model underestimates light-generated currents at low device thickness. A simple correction procedure is though provided, which allows simulating the performance of textured Si solar cells with PC1D safely at any device thickness. If it is used according to the directions given in this work, PC1D can so become a very useful tool in designing ultrathin Si solar cells.

2 STANDARD LAMBERTIAN OPTICS IN PC1D

In the following, θ will be denoted as θ_1 at the first optical pass and θ_n at the second and successive optical passes. θ=60° at all optical passes across a Si sheet can be consequently achieved by simultaneously assigning θ_1=60° and θ_n=60°.

PC1D allows assigning θ_1 through opportune assignment to the facet angle, δ (see Fig. 1). At PC1D default value for refractive index, we have θ_1=60° at δ=75.705°. At θ_1=60°, θ_n=60° is simultaneously assigned in PC1D by simply checking "specular" in both the "rear surface-" and "front surface-" sections of the "Reflectance" dialog box.

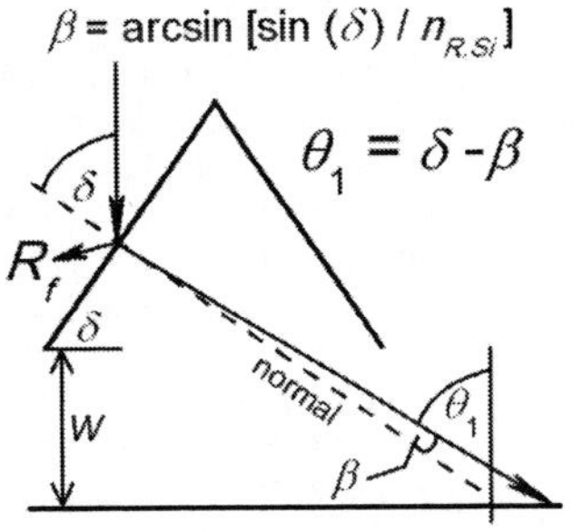

Figure 1: The first pass of a light ray across a Si sheet with pyramidal front-surface texture of facet angle δ. R_f is the front-surface reflectance and W is the sheet thickness. Pyramid dimensions are exaggerated. Relationships among internal angles are indicated, where $n_{R,Si}$ is the refractive index of Si.

In PC1D, on the other hand, θ_n=60° can be also imposed by checking "diffuse" in the aforementioned "rear surface" section of the "Reflectance" dialog box [3]. In this case, θ_n=60° is set regardless of θ_1. In PC1D, however, checking "diffuse" in the "rear surface" section of the "Reflectance" dialog box triggers automatically a built-in optical model called "diffusion model" [3].

Concerning the third and subsequent passes of internal light across a given device, this model is such that, *"because only weakly absorbed photons survive the first round trip of the device thickness, subsequent photogeneration resulting from these trapped photons is assumed to occur uniformly over the device thickness"* [3]. In thin devices, also non-weakly absorbed photons may survive the first round trip. In the case of thin devices, consequently, assuming uniform absorption after the first round trip of the device thickness is likely to underestimate light-generation.

3 CHECK AND CORRECTION PROCEDURE

In Fig. 2, curves of external quantum efficiency (*EQE*) as calculated with PC1D at both the settings mentioned in the previous section are shown together with *EQE*-curves calculated with the exact analytical expressions derived from the analytical solution to minority-carrier transport in uniformly doped regions [5], which are reported in the appendix. The curves are relevant to cells having the structure and transport parameters listed in Table 1 and various values of thickness, W. The internal reflectance of these cells is R_{FI}=0.921 at the front and R_B=0.982 at the back. The cells have R_f=0 (see Fig. 1). Such devices have been already simulated elsewhere [6,7]. As can be seen, analytical exact and PC1D curves only overlap in the case, where, in PC1D, specular internal surfaces are imposed. The PC1D-curves calculated at diffuse rear surface are systematically lower than the remaining ones up to W=100 μm.

We conclude that, as expected, the PC1D diffusion model underestimates light-generation in thin cells.

In Fig. 3, the discrepancy ΔJ_{sc} between the values of short-circuit current density J_{sc}, calculated with PC1D at the two aforementioned settings is reported for W ranging from 0.06 μm to 300 μm. Since, in both settings, the light-generation at the first optical pass is the same, ΔJ_{sc} only depends on the second and successive passes of trapped light. This observation allows making up a correction procedure for simulations performed with PC1D at θ_1<60° and θ_n=60°, which require imposing diffuse rear surface. In these cases, a previous determination of a ΔJ_{sc}-curve (or value) at $\theta_1=\theta_n$=60° is necessary, to be added to the J_{sc}-curve (or value) calculated at θ_1<60°, θ_n=60°, and diffuse rear surface. This removes the systematical underestimation of J_{sc} in thin devices, which is associated to assigning diffuse rear surface in PC1D. In the case of the cells simulated in the present work, the ΔJ_{sc}-curve to be added is shown in Fig. 3.

5 CONCLUSION

A method to model standard Lambertian optics with PC1D in thin and ultrathin Si solar cells has been given. How to determine θ_1 has been explained.

If $\theta_1=\theta_n$=60°, then specular internal surfaces must be imposed. If θ_1<60° and θ_n=60°, and hence, assigning diffuse rear surface is compulsory, then a correction procedure must be applied. An example of correction procedure has been given.

References

[1] A. Ingenito, O. Isabella, M. Zeman, ACS Photonics (2014) 140213140005007. doi:10.1021/ph4001586

[2] P. A. Basore, D. A. Clugston, PC1D version 5.9, University of New South Wales (2003).

[3] P.A. Basore, IEEE Trans. Electron Devices 37 (1990) 337. doi:10.1109/16.46362

[4] M. A. Green, Prog. Photovoltaics Res. Appl. 10 (2002) 235. doi:10.1002/pip.404

[5] L. Abenante, Sol. Energy. 129, (2016) 204. doi:10.1016/j.solener.2016.02.009

[6] L. Abenante, J. Appl. Phys. 117 (2015) 026101. DOI:10.1063/1.4905182

[7] A. Bozzola, P. Kowalczewski, L. C. Andreani, J. Appl. Phys. 117 (2015) 026102. DOI: 10.1063/1.4905183

Table I: Device and transport parameters of the modeled cells.

Region	Thickness (μm)	Doping density (cm^{-3})	Surface Recombination velocity (cm/s)	Diffusion length (μm)
Emitter	0.05	10^{19}	0	20
Base	$W-0.05$	10^{16}	0	200

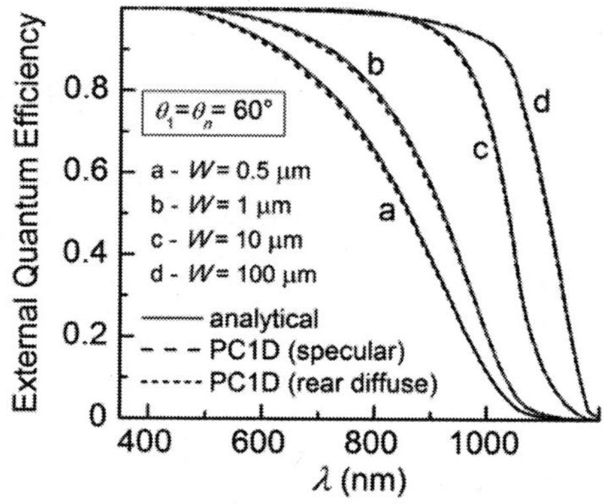

Figure 2: Analytical and numerical *EQE*-curves of solar cells having equal device, transport and optical parameters and different W-values. PC1D-curves are relevant to two settings of internal surfaces.

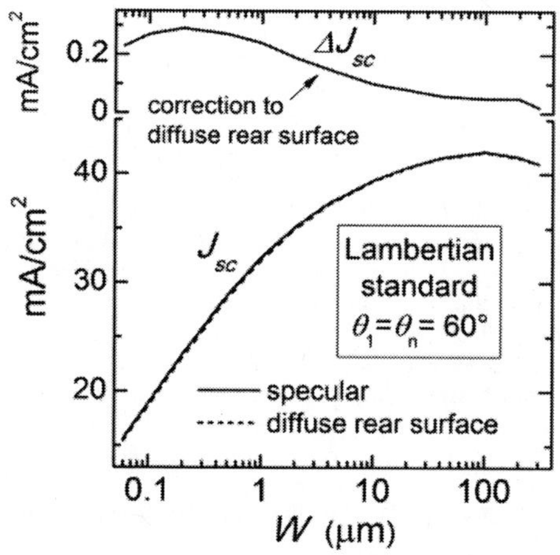

Figure 3: $J_{sc}(W)$-curves of solar cells having same device, transport and optical parameters as the cells in Fig. 2. Curves are calculated with PC1D at two settings for internal surfaces. The discrepancy $\Delta J_{sc}(W)$ between curves is also reported.

Appendix

With the generation rate

$$g(x)=\frac{\alpha(1-R_F)}{\cos\theta}\times\frac{e^{-\frac{x\alpha}{\cos\theta}}+R_BT^2e^{\frac{x\alpha}{\cos\theta}}}{1-R_BR_{FI}T^2},$$

where α is the absorption coefficient and $T=\exp(-\alpha W/\cos\theta)$, and with $g'(x)$ being its first derivative, L the diffusion length, D the diffusivity, and S the surface recombination velocity for the relevant region, under the usual boundary conditions, the exact analytical solution to minority-carrier tranport in uniformly doped regions [5] gives the *EQE* contributions from emitter and base as

$$EQE_{em}=\frac{(1-\alpha^2L^2)^{-1}(1-R_{front})L^2}{\cosh(H_{em}/L)+(LS/D)\sinh(H_{em}/L)}\times\left\{g'(x_{em})\left[\cosh\left(\frac{H_{em}}{L}\right)+\frac{LS}{D}\sinh\left(\frac{H_{em}}{L}\right)\right]+\right.$$

$$\left.-\frac{1}{L}\left[Lg'(x_0)-\frac{LS}{D}g(x_0)+g(x_{em})\times\left(\sinh\left(\frac{H_{em}}{L}\right)+\frac{LS}{D}\cosh\left(\frac{H_{em}}{L}\right)\right)\right]\right\}$$

and

$$EQE_{ba}=\frac{(1-\alpha^2L^2)^{-1}(1-R_{front})L^2}{\cosh(H_{ba}/L)+(LS/D)\sinh(H_{ba}/L)}\times\left\{g'(x_{ba})\left[\cosh\left(\frac{H_{ba}}{L}\right)+\frac{LS}{D}\sinh\left(\frac{H_{ba}}{L}\right)\right]+\right.$$

$$\left.-\frac{1}{L}\left[g(x_{ba})-g(x_{ba})\left(1+\frac{LS}{D}\right)\exp\left(\frac{H_{ba}}{L}\right)-\left(Lg'(W)+\frac{LS}{D}g(W)\right)\right]\right\},$$

respectively, where x_0 is the emitter surface position, x_{em} the position of the interface between emitter and space-charge region (SCR), H_{em} the emitter width, x_{ba} the SCR-base interface position, and H_{ba} the base width.

IMPACT OF DIFFERENT TREATMENT TECHNOLOGY AND HIGHLY ACCELERATED STRESS TEST FOR THE MONO SILICON PERC SOLAR CELLS

Cheng-Wen Kuo, Ta-Ming Kuan, Li-Guo Wu, Chih-Chiang Huang, and Cheng-Yeh Yu
TSEC Corporation, No.85, Gaungfu N. Rd., Hsin-Chu 30351, Taiwan.

ABSTRACT: When the PERC technology introduced, the light induced degradation (LID) effect will become serious. In this work, the different LID treatment technology result and the accelerated tests with light soaking conditions have been investigated. We fabricated 1*1 single-module samples to verify the different treatment technology. The samples exposed with the 3-suns light intensity at the temperature of 70°C after LID treatment, which is known as highly accelerated stress test (HAST). The power decay ratio of the cells with the illuminance exceeds 700 kWh/m^2 and could be resulted from the decrease in short-circuit current. There is no particularly significant dark area in electroluminescent images after HAST LID test. It was demonstrated that the PERC solar cells after the treatment process, LID effect has been significantly improved, while strengthening the test can also maintain LID effect < 3%.
Keywords: LID, Mono, PERC

1 INTRODUCTION

The passivated emitter and rear cell (PERC) structure for p-type wafer is currently able to achieve high effeciency with the reasonable cost of production. However, the bulk lifetime degradation of light-induced degradation (LID) is found to have a larger detrimental effect on high efficiency PERC cells, which the impact of LID increases when a low back surface recombination velocity is achieved [1]. The LID phenomenon of p-type high effeciency solar cells must be overcome. It could be explained that the boron-oxygen complex is the origin of LID of solar cells made from Czochralskigrown (Cz) silicon [2-3]. There are few methods to reduce LID effect which can use low oxygen Magnetic-CZ material, high resistivity base or Ga doping wafers [2,4]. Another ways to reduce the LID effect is regeneration effect of the boron-oxygenr complex if simultaneously annealing the H-rich sample at elevated temperature and carrier injection [5].

In recent years, many equipment was designed to solve the LID phenomenon, it has been demonstrated that the boron-oxygen related defects can be decreased under a light-induced regeneration (LIR) treatment, which combines actions of temperatures and an external illumination source. In this work, there are different LIR equipment to apply for reduced LID effect. We fabricated 1*1 single-module samples to verify the different treatment technology. The samples exposed with the 3-suns light intensity at the temperature of 70°C after LIR is also investigated in this paper.

2 EXPERIMENTAL

The high efficiency 6-inch (157.75 mm x 157.75 mm) PERC solar cells were fabricated on 190 μm thick p-type boron doped Cz-silicon wafers with a resistivity of 1~3 Ω-cm. Chemical wet etching methods were used for saw damage removal and surface texturing, which include both acidic etching for multi-crystalline silicon wafers and alkaline etching for mono-crystalline silicon wafers. The solution composition of acidic etching was nitric acid (HNO3) : hydrofluoric acid (HF) : deionized water (DI) = 1 : 2 : 1.5, and the solution composition of alkaline etching was potassium hydroxide (KOH) : deionized water (DI) = 1 : 50. Then, the rear surface is polished in order to reduce the roughness and its impact on the effective charge carrier lifetime [6]. Conventional phosphorous diffusion was employed by furnace to form the n+ emitters with a sheet resistance of 80 – 100 Ω/. The diffusion process also improved to uniformity for this work.

The phosphosilicate glass (PSG) was then removed by HF-contained solutions, formed edge isolation in the meantime. In order to decrease PID effect and increase efficiency, we optimized the CVD process parameter to improve. An antireflection coating (ARC) layer of silicon nitride (SiN_x: H) and/or SiN_x/SiO_x dual ARC was deposited by plasma enhanced chemical vapor deposition (PECVD) with thickness of 80 nm. The passivation was carried out on rear side also using a PECVD. Afterward, the rear side laser opening ratio is also optimized [7]. All PERC solar cells have the same laser-ablated opening width in the rear-side passivation stacks which was about 45 μm. The front side silver and backside silver and aluminum metallization was formed by screen printing technique and followed by co-firing process, which resulted in formation of front silver grid contact and local back-surface-field. Finally, the current-voltage measurements of solar cells under AM 1.5 illumination were performed. The PERC solar cells would be divided into three groups, using different treatment process. The 60 KW LID result is shown in Table I.

3 RESULTS AND DISCUSSION

Table I summarized the average, max and min LID results of conversion efficiency. It could be found that the LID result of PERC cells without treatment is between 2 % and 4 %. The result was similar to the result of the typical PERC solar cell which is lower 5 %. Besides, the LID results of the treatment process PERC solar cells were lower 2 %. It could be demonstrated the boron-oxygen related defects can be deactivated under a LIR treatment, which combines actions of temperatures and an external illumination source [1].

Table I: The 1 sun LID effect of the PERC solar cells with different treatment process.

	LID (%) (Avg)	LID (%) (Max)	LID (%) (Min)
Ref.	2.77	3.56	2.03
Group A	0.87	1.60	0.32
Group B	1.17	1.73	0.61
Group C	1.35	1.78	1.09

In order to further understand the LIR PERC solar cells under high stress LID test, we fabricated 1*1 single-module samples to expose with the 3-suns light intensity at the temperature of 70°C. Table II, III and IV shown the 3-sun LID result of the Group A, Group B and Group C, respectively.

Table II: The 3 sun LID effect of Group A solar cell with treatment process.

Illuminance (kWh/m^2)	Pmax (W)	Voc (V)	Isc (A)	FF (%)	Vmpp (V)	Impp (A)
0	**4.48**	**0.6520**	**9.80**	**70.12**	**0.5013**	**8.94**
126	**4.48**	**0.6544**	**9.75**	**70.25**	**0.5026**	**8.92**
252	**4.46**	**0.6516**	**9.76**	**70.12**	**0.5022**	**8.88**
378	**4.47**	**0.6491**	**9.79**	**70.37**	**0.4998**	**8.95**
504	**4.54**	**0.6496**	**9.82**	**71.26**	**0.5035**	**9.02**
630	**4.51**	**0.6487**	**9.79**	**70.93**	**0.5018**	**8.98**
756	**4.50**	**0.6485**	**9.76**	**71.05**	**0.4999**	**9.00**

Table III: The 3 sun LID effect of Group B solar cell with treatment process.

Illuminance (kWh/m^2)	Pmax (W)	Voc (V)	Isc (A)	FF (%)	Vmpp (V)	Impp (A)
0	**4.59**	**0.6586**	**9.86**	**70.70**	**0.5062**	**9.07**
126	**4.61**	**0.6585**	**9.86**	**71.00**	**0.5095**	**9.05**
252	**4.60**	**0.6598**	**9.77**	**71.44**	**0.5111**	**9.01**
378	**4.60**	**0.6561**	**9.80**	**71.60**	**0.5086**	**9.05**
504	**4.52**	**0.6550**	**9.74**	**70.92**	**0.5040**	**8.98**
630	**4.55**	**0.6557**	**9.75**	**71.23**	**0.5092**	**8.94**
756	**4.52**	**0.6541**	**9.77**	**70.74**	**0.5035**	**8.98**

Table IV: The 3 sun LID effect of Group C solar cell with treatment process.

Illuminance (kWh/m^2)	Pmax (W)	Voc (V)	Isc (A)	FF (%)	Vmpp (V)	Impp (A)
0	**4.48**	**0.6553**	**9.82**	**69.66**	**0.4975**	**9.01**
126	**4.52**	**0.6545**	**9.84**	**70.27**	**0.4980**	**9.08**
252	**4.41**	**0.6520**	**9.78**	**69.06**	**0.4897**	**9.00**
378	**4.43**	**0.6518**	**9.76**	**69.61**	**0.4915**	**9.01**
504	**4.42**	**0.6502**	**9.77**	**69.60**	**0.4917**	**8.99**
630	**4.39**	**0.6497**	**9.74**	**69.31**	**0.4876**	**8.99**
756	**4.37**	**0.6482**	**9.77**	**69.05**	**0.4867**	**8.98**

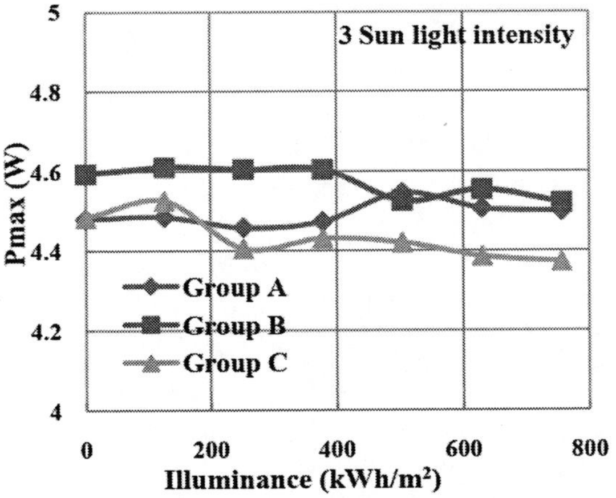

Figure 1: The relationship between the Pmax and the illuminance of group A, and Group B and Group C under 3 sun LID test.

Figure 1 showed relationship between the Pmax and the illuminance of group A, and Group B and Group C under 3 sun LID test. It was observed to the Pmax decreased with the accumulated illuminance increased for Group B and Group C. It may be the treatment condition is not the best parameters. This is because of the modified treatment parameter is based on the one sun LID test. Group B and Group C are off-line equipment, the adjusted treatment parameter is for the keep the PERC solar cells efficiency. The Group A used the in-line equipment to treatment, the advantage of the equipment is the PERC solar cell efficiency can be decreased. However, the treatment parameter influences the LID effect.

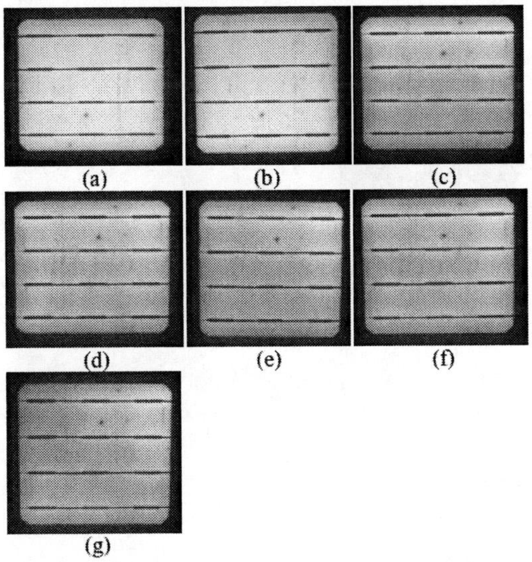

Figure 2: The group A EL images of the 1*1 module at (a) initial and after (b) 126 kWh/m^2, (c) 252 kWh/m^2 , (d)378 kWh/m^2, (e) 504 kWh/m^2, (f) 630 kWh/m^2,and (g) 756 kWh/m^2.

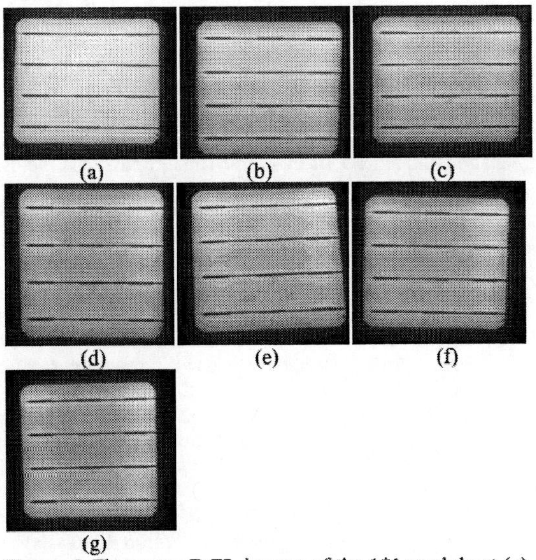

Figure 3: The group B EL images of the 1*1 module at (a) initial and after (b) 126 kWh/m^2, (c) 252 kWh/m^2 , (d)378 kWh/m^2, (e) 504 kWh/m^2, (f) 630 kWh/m^2,and (g) 756 kWh/m^2.

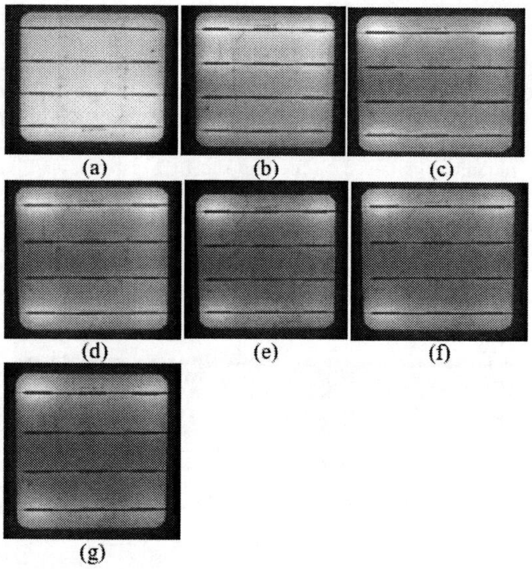

Figure 4: The group C EL images of the 1*1 module at (a) initial and after (b) 126 kWh/m^2, (c) 252 kWh/m^2 , (d)378 kWh/m^2, (e) 504 kWh/m^2, (f) 630 kWh/m^2,and (g) 756 kWh/m^2.

Figure 2, 3, and 4 shown the EL of the 1*1 module at initial and after 3 sun LID test. No EL darkened area was found after 3 sun LID test for the Group A, B and C. However, it can be observed that the EL images show slightly darkened as the accumulated illuminance increases.

4 CONCLUSIONS

In this work, there are three LIR treatment technology and the accelerated tests with light soaking conditions have been investigated. The average cell efficiency of Group A only loses 0.87% compared to reference cells which do not use LIR treatment. It could be demonstrated that the boron-oxygen related defects can be deactivated under the LIR treatments. On the other hand, it was observed that the PERC solar cells after the different treatment process, 3-suns LID effect has also been significantly improved, while strengthening the test can also maintain LID effect < 3%. The Group A has the best LIR result as cumulative illuminance > 378 kWh/m^2.

5 REFERENCES

[1] K.-Y. Yen, J.-R. Huang, Y.-F. Lin, S. Su, S.H.T. Chen, and L.-W. Cheng, "The Development of In-Line Regeneration Tool for the Effective Suppression of Light-Induced-Degradation on P-type Silicon Solar Cells" 32nd EUPVSEC, pp. 495, 2016.

[2] X. Meng, Z. Xia, J. Woo, X.-S. Wang, G. Xing, "Regeneration of B-Doped Solar Cell by Carrier Injection and Extra Energy" 31st EUPVSEC, pp. 407, 2015.

[3] F. Fertig, J. Greulich, J. Broisch, D. Biro, and S. Rein, "Stability of the regeneration of the boron-oxygen complex in silicon solar cells during module integration", Solar Energy Materials and Solar Cells, vol 115, pp189, 2013.

[4] F. Mosel, A. Molchanov, B. Fischer, D. Krischel, A. Seidl, and B. Birkmann, "Growth of High Quality Silicon Mono Ingots by the Application of a Magnetic Cusp Field in Cz-Puller" 27th EUPVSEC, pp, 933, 2012

[5] A. Herguth, G. Schubert, M. Kaes, and G. Hahn., "Investigations on the long time behavior of the metastable boron–oxygen complex in crystalline silicon" Prog. Photovoltaics. Vol. 16, pp. 135, 2007.

[6] E. Urrejola, R.Petres, J. Glatz-Reichenbach, K. Peter, E. Wefringhaus, H. Plagwitz, G. Schubert., "High Efficiency Industrial PERC Solar Cells with all PECVD-Based Rear Surface Passivation" 26th EUPVSEC, pp. 2233, 2011.

[7] C.-W. Kuo, T.-M. Kuan, C.-C. Liu, C.-C. Huang, L.-G. Wu, C.-Y. Yu, "Impact of Laser Opening Ratio for Mass Production High Efficiency Mono-Crystalline Silicon Solar Cells" 29th EUPVSEC, pp. 1392, 2014

REVIEW OF TOOLS AND APPROACHES FOR INLINE QUALITY CONTROL IN HIGH EFFICIENCY SILICON SOLAR CELL PRODUCTION

J. Haunschild, J. Greulich, H. Höffler, S. Wasmer, G. Emanuel, A. Krieg, L. Friedrich, S. Rein
Fraunhofer Institute for Solar Energy Systems ISE, Heidenhofstraße 2, 79110 Freiburg, Germany
Telephone: +49 761 4588 5634, Fax: +49 761 4588 7621, e-mail: Jonas.haunschild@ise.fraunhofer.de

ABSTRACT: Multiple tools and methods for the optical and electrical quality control of high efficiency silicon solar cells during their industrial production are available on the market, even more are discussed in the literature. We summarize the possibilities of these tools along the value chain from wafer to cell for the case of passivated emitter and rear cells (PERC) and discuss some showcases. We recommend a broad inline quality control before solar cell production in order to discard wafers with insufficient quality. Especially PL-Imaging can reveal material defects and process variations and thus acts as a versatile tool for the whole production chain. An economic evaluation shows that a CAPEX spending of 250,000 EUR for metrology equipment is already justified if cell efficiency is increased by only 0.03 %abs from a reference efficiency 20.61 % to 20.64 %. Combining technological and economical aspects, this paper gives an overview over recent inline metrology tools and upcoming challenges.

Keywords: solar cell, quality, metrology, PL-Imaging, economics

1 INTRODUCTION

In the photovoltaic industry, each market player faces extreme competition. Solar cell and module manufacturers aim at reducing their costs and use of consumables and at the same time at improving throughput and uptime, yield, process stability, solar cell reliability and solar cell output power. However, the higher potential of novel devices goes along with a higher sensitivity towards material and process variations. Concerning material quality, a certain kind of incoming test is highly prioritized by cell manufacturers. Especially geometrical properties can be precisely tested at the beginning of the cell production line or at the end of the wafer production line in order to prevent off-spec material to enter the production chain.

To better control variations during the production process and in order to find further potential of process improvements, an intelligent use of inline characterization techniques ideally combines the required investigation of material and device properties with real-time process and production control. In general, the most challenging question for cell manufacturers and metrology suppliers concerning inline metrology is which tools and methods are required and economically advisable. The central question is how to control and improve the yield, the reliability, the mean cell efficiency and the scattering of the efficiency distribution at which cost.

To be able to discuss these questions, different tools and approaches are discussed in this article including an economic assessment. While this article gives an overview, more details are published elsewhere [1].

2 APPROACH

In the Photovoltaic Technology Evaluation Center PVTEC [2] at Fraunhofer ISE, a broad pool of inline and off-line metrology tools is available and used for characterization and quality control during solar cell production. Most experiments in the PVTEC include systematic variations of process parameters which need to be quantified and correlated to the resulting IV parameters of the solar cells. Other experiments include broad variations of material quality or are done without variation in order to analyze process fluctuations. The data from these experiments provide information about the process variations which may be detected by means of the various metrology tools and about their relevance for the output parameters of the final solar cell. From this, the economic benefit of the different tools may be assessed, strongly depending on the cell concept under test.

3 INCOMING CONTROL

The electrical and mechanical quality of wafers can be checked during final inspection of wafer production or during incoming inspection of solar cell production. Poor quality wafers should be identified and discarded at an early stage in the process to avoid unnecessary costs. Inline accessible mechanical wafer properties are wafer thickness and its variation, the size and geometry, saw marks and roughness, chipping, holes and cracks. Inline-accessible electrical properties are base resistivity, effective lifetime and crystal defects [3, 4]. In addition, surface contamination and reflectivity are optical properties which are worth to be investigated. The measurement procedure of most of these properties is covered by recent SEMI standards.

The wafer thickness is typically measured capacitance-based. Tools measuring e.g. three traces with several hundreds of measurement points each allow to detect not only the mean thickness (typically ~180µm), but also the total thickness variation (~20µm) and more details of the wafer's shape (see Figure 1).

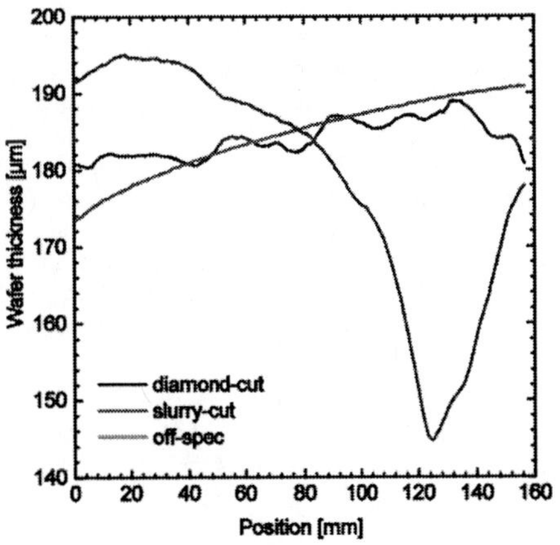

Figure 1. Three examples of thickness profiles. The diamond-cut wafer shows a typical large-scale saw mark structure, the slurry-cut wafers shows a strong gradient, while the off-spec wafer shows a very strong saw mark.

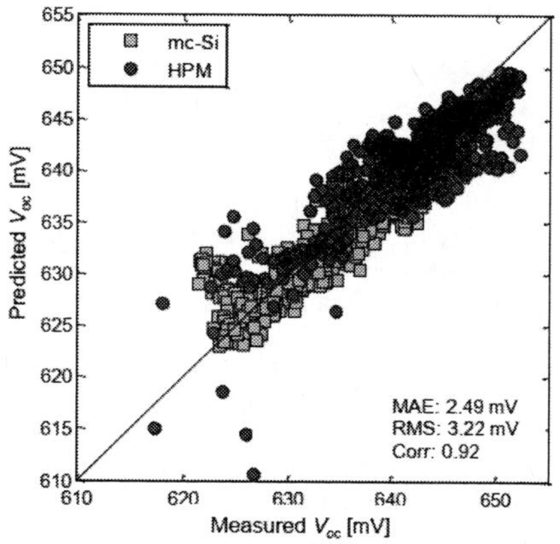

Figure 2. Prediction results of the open-circuit voltage V_{OC} for more than 1300 passivated emitter and rear cells (PERC) obtained with a regularized regression model based on empirical data, adapted from Ref [5].

The bulk lifetime of minority charge carriers is one of the most important parameters to characterize the electrical quality of wafers. Unfortunately, in the as-cut state, only an effective lifetime can be measured which is strongly limited by surface recombination and thus does not allow a correlation to final solar cell parameters. Only wafers with very low lifetime (e.g. due to crucible contaminations) can be identified and sorted out. Nevertheless, lifetime measuring systems based on microwave-detected photoconductivity (MPD, [6]), microwave-detected photoconductance decay (MW-PCD, [7]) and quasi-steady-state photoconductance (QSSPC, [9]) measurements are available. To overcome the limit of surface recombination, these techniques can be used to inspect the lifetime in later process stages (e.g. after diffusion or passivation). Nevertheless, there is a very promising approach for material quality rating in the as-cut stage which has yielded excellent results on multicrystalline wafers [8]. It is based on the analysis of PL images and Figure 2 shows the predicted open circuit voltages plotted against the measured ones on the final cells, showing a very good prediction accuracy [5, 8].

4 PRODUCTION PROCESS

Inline quality control between the incoming test of the as-cut wafers and the final outcoming test of the finished solar cells is not known to be very widespread. It is nevertheless a prerequisite to quickly detect arising problems during production and to constantly achieve high solar cell conversion efficiencies, which gains importance as the efficiency potential of the cell structures increases. Many of the following methods require the combination of several measurements (*e.g.* resistivity and wafer thickness, or resistivity as-cut and after diffusion). Thus, the data need to be assigned to one another wafer-specifically. For this purpose, single-wafer-tracking methods e.g. using data-matrix-codes have been developed [10, 11].

During texturing, the concentration of chemicals in the bath can be controlled continuously using near-infrared spectroscopy. Acidic (HF, HNO3) and alkaline (KOH, organic additives) baths typically used in silicon photovoltaics can be analyzed [12, 13]. After the texturization process, the reflectance and the thickness of the wafer can be measured and used to control the quality of the reflection properties and of the silicon removal, respectively. Figure 3 depicts photoluminescence images that are taken after three different process stages. From these, recombination-active defects that are induced or activated within the different steps can be detected spatially resolved, which often allows to identify their origin [14, 15]. An overview of the inline metrology for emitter diffusion and the following process steps can be found in Ref [1].

4 FINISHED SOLAR CELL

After contact formation on finished solar cells, extensive inline characterization is available. The most important inline characterization for finished solar cells is the measurement of the current-voltage (IV) characteristics under standard test conditions (25°C, 1000 W/m^2 illumination with AM1.5g spectrum), from which the energy conversion efficiency can be deduced and the cells can be sorted into the corresponding bins. Due to mismatch effects [16], the choice of the reference cells, which are calibrated in the CalLab [17], can have a significant impact on the measured values of short-circuit current and conversion efficiency and needs to be closely adjusted to the cell technology under test. Besides the measurement under constant illumination, also the suns-V_{oc} characteristic as well as the forward and reverse IV characteristics in the dark can be measured, which allow a basic analysis of non-ideal recombination, series resistance, ohmic and non-linear shunting, and hot-spot danger [18].

Line or matrix cameras are used for visual inspection of the front side of the cells to detect paste residuals and chipping defects, measure the finger width and the cell dimensions and the color of the anti-reflection coating [19]. By inspecting the rear of the PERC cells with full-area Al print, the darkening of the aluminum at the local contact openings due to silicon alloying during the contact formation can be detected, which may be used to detect inhomogeneous formation of the local back-surface field. For bifacial solar cells, the visual inspection of the cell's rear is in principle the same as that of the cell's front: paste residuals, chipping, finger width and color can be detected.

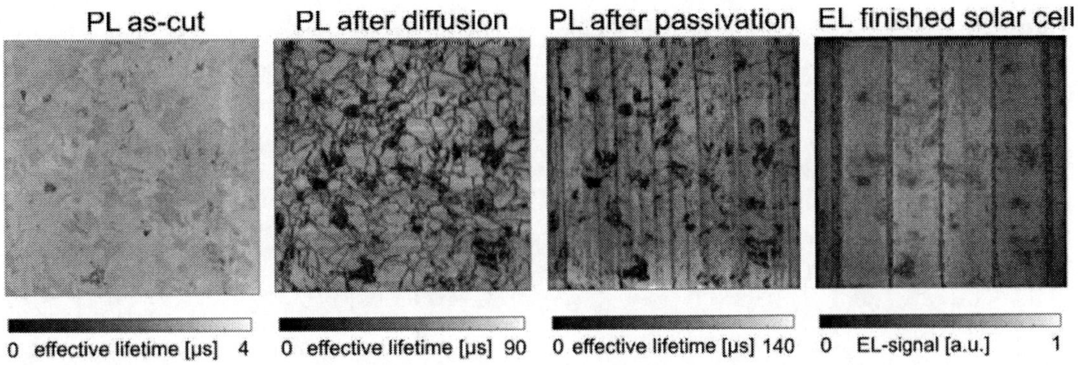

Figure 3: Photoluminescence (PL) and electroluminescence (EL) images of a mc-Si wafer at different stages of the PERC production sequence. In the as-cut state, dislocations and grain boundaries are visible that become even more pronounced after emitter formation. After surface passivation, line-shaped defects possibly induced by saw marks become apparent. In the finished cell, several material- and process-induced defects (possibly saw marks, crystal dislocations, finger interruptions, edge shunts) overlie.

Besides the measurement of the IV characteristics of the solar cells, advanced upgrades for the IV testers are available. Electroluminescence (see Figure 3) and photoluminescence imaging allow to detect cracks, finger interruptions and dark areas [20, 21]. Thermography is used to detect cells with hot spots and to predict their temperature behavior in the finished module [22]. Such a local analysis is preferable compared to an analysis of the global dark reverse current because inhomogeneous reverse current and power dissipation within the cell is the reality, not the exception. This is exemplified in Figure 4, where the thermography image of an mc-Si PERC cell shows a significant temperature increase of 5 K after only 40 ms reverse current load, which is likely to damage the module. Since the cell's reverse current at -12 V is below 2 A, this hot spot danger cannot be predicted from only the IV characteristics [23]. A systematic comparison of inline hot spot detection and evaluation via the above mentioned cell's reverse current, temperature increases in short-term measurements and two more advanced approaches is presented in Ref. [24].

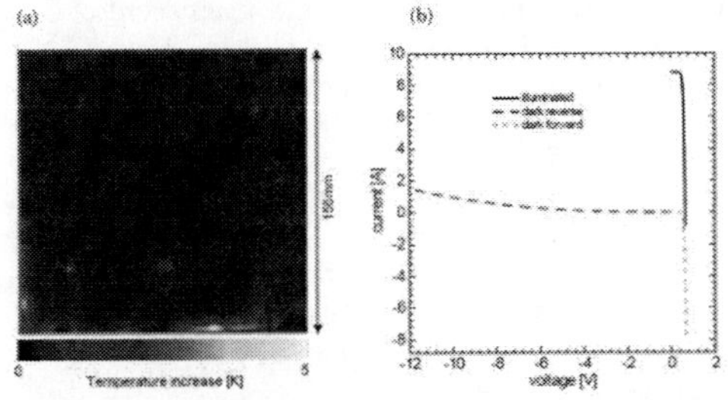

Figure 4. (a) Thermography image of an mc-Si PERC cell with a severe temperature increase of up to 5 K after applying a reverse voltage of -12 V for only 40 ms, indicating hot spot danger for the module. (b) IV characteristics of the same cell showing only a moderate reverse current < 2 A at -12 V reverse voltage.

5 ECONOMIC ASSESSMENT

One of the cell and module manufacturers main concerns is the expected return of investment, when they are faced with decisions to invest in advanced inline characterization techniques. Looking at real production environments, parameters like the uptime or the yield of a production tool are randomly influenced by failures, consumables, wafer quality, and other factors. So, it often seems not to be clear how to distinguish between positive effects of inline quality control and other factors influencing the performance of a production tool. This makes it difficult to appropriately prepare an investment decision for an inline characterization technique.

To better understand the economic impacts of an integration of inline characterization techniques into a production environment, a 500 MWp/year monocrystalline PERC cell production process was simulated and examined with our cost of ownership (COO) calculation tool 'SCost' [25]. We calculated the essential productivity improvement (in terms of cell efficiency gain) for the cell manufacturer with respect to the prospected capital expenditures (CAPEX) on inline characterization in the production line, i.e., how much CAPEX can be spent to break even the expected production performance enhancement.

Figure 5 shows the resulting break-even analysis based on the W_P-cost equivalence before and after the application of the inline characterization techniques for a timeframe of five years. If the cell manufacturer realizes a higher cell efficiency by using the inline characterization technique, the all-in cell costs (all-in module costs) decrease compared to the reference value. This influence of cell efficiency enhancement to the all-in cell costs is analyzed in Figure 6.

This finding shows how small the efficiency increase is to justify inline metrology, but also confirms that a well-founded investment decision requires the metrology-induced efficiency gain to be demonstrated. However, as inline metrology can also improve other production parameters, such as production yield, equipment uptime and consumable utilization, it often may turn to account also without induced efficiency gains. Such effects may be very tool-specific and thus have to be assessed individually.

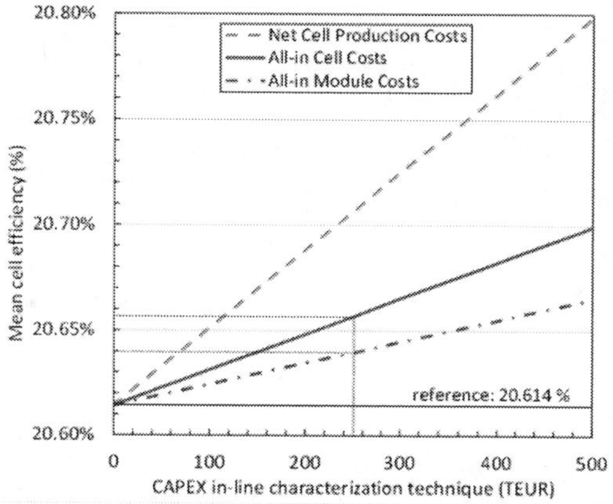

Figure 5. Break-even analysis of an inline characterization technique, integrated into a 500 MWp/year monocrystalline PERC cell production line. The three lines indicate for the 'net cell production costs', the 'all-in cell costs' and the 'all-in module costs' the respective gain in mean cell efficiency to be reached by the characterization technique in order to break even within five years the expended costs for the additional characterization technique.

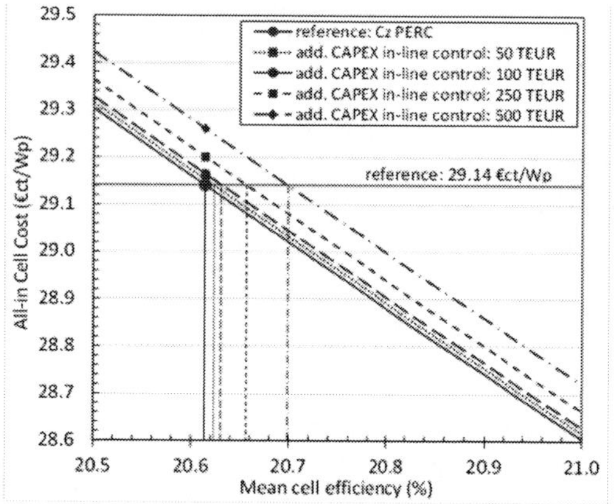

Figure 6. Sensitivity analysis of the mean cell efficiency influence on 'All-in Cell Costs'. The red line indicates the costs of the Cz PERC reference and the red dot the reference cell efficiency of 20,61 %, respectively. The dashed blue lines indicate the costs including additional CAPEX for inline characterization. All figures are calculated for a 500 MWp/year monocrystalline PERC cell production.

6 OUTLOOK

After the fire incident in the PVTEC laboratory at Fraunhofer ISE, new and updated inline characterization tools will be available soon for test and service. The automated inspection system for as-cut wafers will be equipped with 16 different measurement systems, including waferidentification, tools for optical and geometrical inspection, photolumineszence imaging, a new tool for microcrack detection and a tool for grain structure analysis in mc-Si wafers. The system will be able to measure not only standard but also diamond-wire-cut wafers with shiny surface more accurately. Partially processed wafers will be able to be measured in a separate inspection system which is equipped among other techniques with lifetime calibrated photoluminescence imaging, combining high spatial resolution and physical relevance of the data. For finished solar cells, an inline cell tester will be available with IV, spectral response, thermography, electroluminescence, combined electro- and photoluminescence and high-resolution print inspection units which will allow inline in-depth analysis with high spatial resolution. Beyond the high-quality measurement equipment, the system will be able to handle not only standard cells, but also bifacial, busbarless and back-contacted solar cells.

7 CONCLUSION

In this article, we gave an overview over state-of-the-art inline metrology for material and process control. We recommend incoming inspection during solar cell production or outgoing inspection during wafer production as a very important step for quality control, as wafers with insufficient mechanical or electrical properties can be detected and discarded. During solar cell processing, inline metrology helps to quickly identify upcoming problems, and appropriate tools can be applied after each production step. All finished solar cells need to be IV-tested. Different approaches to analyze IV curves are available, as well as additional metrology tools to monitor spatially resolved specific defects such as cracks or hot-spots. While during process optimization extensive characterization is necessary, even a running production can benefit from a comprehensive metrology as the quality of the individual processes has to be known on a statistically relevant basis to be able to reduce variations and increase the standards and thus to improve the overall efficiency gain. Economic and technological challenges are addressed to answer the question how much gain in efficiency can be justified from investments in metrology.

ACKNOWLEDGEMENTS

This work was supported by the German Ministry of Economic Affairs and Energy within the frame of the project "CUT-B" (0325910A).

REFERENCES

[1] J. M. Greulich *et al.*, "In-line quality control in high efficiency silicon solar cell production," *Photovoltaics International*, vol. 35, 2017.

[2] D. Biro *et al.*, "PV-Tec: Retrospection to the three years of operation of a production oriented research platform," in *24th EU PVSEC*, Hamburg, 2009, pp. 1901–1905.

[3] J. Haunschild *et al.*, "Rating and sorting of mc-Si as-cut wafers in solar cell production using PL imaging," *Sol. Energy Mater. Sol. Cells*, vol. 106, pp. 71–75, 2012.

[4] J. Haunschild, I. Reis, J. Geilker, and S. Rein, "Detecting efficiency-limiting defects in Czochralski-grown silicon wafers in solar cell production using photoluminescence imaging," *Phys. Status Solidi RRL*, vol. 5, no. 5-6, pp. 199–201, 2011.

[5] M. Demant *et al.*, "Inline quality rating of multi-crystalline wafers based on photoluminescence images," *Prog. Photovolt: Res. Appl.*, vol. 24, no. 12, pp. 1533–1546, 2016.

[6] K. Dornich, N. Schüler, B. Berger, and J. R. Niklas, "Fast, high resolution, inline contactless electrical semiconductor characterization for

photovoltaic applications by microwave detected photoconductivity," *Materials Science and Engineering: B*, vol. 178, no. 9, pp. 676–681, 2013.
[7] J. Schmidt and A. G. Aberle, "Accurate method for the determination of bulk minority-carrier lifetimes of mono- and multicrystalline silicon wafers," *J. Appl. Phys.*, vol. 81, no. 9, pp. 6186–6199, 1997.
[8] M. Demant, M. Glatthaar, J. Haunschild, and S. Rein, "Analysis of luminescence images applying pattern recognition techniques," in *25th EU PVSEC*, Valencia, 2010, pp. 1078–1082.
[9] R. A. Sinton and A. Cuevas, "Contactless determination of current–voltage characteristics and minority-carrier lifetimes in semiconductors from quasi-steady-state photoconductance data," *Appl. Phys. Lett.*, vol. 69, no. 17, pp. 2510–2512, 1996.
[10] A. Krieg, C. Schmitt, and S. Rein, "Wafer identification using laser marked data matrix codes," in *25th EU PVSEC*, Valencia, 2010, pp. 1337–1341.
[11] P. Waver, U. Vom Bauer, J. Müller, and D. P. Schreite, "Every solar cell is an original: laser marking of silicon solar cells yields new opportunities in quality control," *Photovoltaics International*, no. 10, pp. 16–22, 2010.
[12] T. Dannenberg, M. Zimmer, and J. Rentsch, "Fully automated concentration control of the acidic texturisation process," *Energy Procedia*, vol. 27, pp. 617–623, 2012.
[13] L. Mohr, T. Dannenberg, M. Zimmer, and J. Rentsch, "Alternative inline analysis of acidic etching baths," in *32nd EU PVSEC*, Munich, 2016, pp. 929–934.
[14] J. Haunschild, M. Glatthaar, S. Riepe, and S. Rein, "Quality control using luminescence imaging in production of mcsilicon solar cells from umg feedstock," in *35th IEEE Photovoltaic Specialists Conference Honolulu*, Hawaii, 2010, pp. 812–816.
[15] T. Trupke, J. Nyhus, and J. Haunschild, "Luminescence imaging for inline characterisation in silicon photovoltaics," *Phys. Status Solidi RRL*, vol. 5, no. 4, pp. 131–137, 2011.
[16] J. Hohl-Ebinger and W. Warta, "Uncertainty of the spectral mismatch correction factor in STC measurements on photovoltaic devices," *Progress in Photovoltaics: Research and Applications*, vol. 19, no. 5, pp. 573–579, 2011.
[17] W. Warta *et al.*, "Precise measurement of solar cell performance in production," in *24th EU PVSEC*, Hamburg, 2009, pp. 1382–1386.
[18] A. Krieg *et al.*, "Accuracy of inline IV measurements under industrial conditions," in *22nd EU PVSEC*, Milano, 2007, pp. 368–371.
[19] A. Krieg, J. Greulich, M. Tondorf, and S. Rein, "Anti-reflection-coating thickness measurements on textured silicon surfaces: Evaluation and accuracy of different measurement techniques," in *28th EU PVSEC*, Paris, 2013, pp. 1820–1824.
[20] K. Bothe *et al.*, "Electroluminescence imaging as an in-line characterisation tool for solar cell production," in *21st EU PVSEC*, Dresden, 2006, pp. 597–600.
[21] J. Haunschild, M. Glatthaar, M. Kasemann, S. Rein, and E. R. Weber, "Fast series resistance imaging for silicon solar cells using electroluminescence," *Phys. Status Solidi RRL*, vol. 3, no. 7-8, pp. 227–229, 2009.
[22] K. Ramspeck, S. Schenk, D. Duphorn, A. Metz, and M. Meixner, "In-line thermography for reliable hot spot detection and process control," *Energy Procedia*, vol. 55, pp. 133–140, 2014.
[23] F. Fertig, C. Willibald, I. Geisenmeyer, M. C. Schubert, and S. Rein, "Illumination and temperature dependence of breakdown mechanisms in multi-crystalline silicon solar cells," *Energy Procedia*, vol. 38, no. 0, pp. 32–42, 2013.
[24] S. Wasmer, I. Geisemeyer, D. Pfengler, J. M. Greulich, and S. Rein, "Comparison of inline hot spot detection and evaluation algorithms for crystalline silicon solar cells," in *33rd EU PVSEC*, Amsterdam, The Netherlands, to be published, 2017.
[25] S. Nold *et al.*, "Cost Modeling of Silicon Solar Cell Production Innovation along the PV Value Chain," (eng), pp. 1084–1090, 2012.

COMPARISON OF INLINE HOT SPOT DETECTION AND EVALUATION ALGORITHMS FOR CRYSTALLINE SILICON SOLAR CELLS

S. Wasmer[1], I. Geisemeyer[1], D. Pfengler[2], J.M. Greulich[1], S. Rein[1]
[1]Fraunhofer Institute for Solar Energy Systems ISE, Heidenhofstr. 2, 79110 Freiburg, Germany
[2]InfraTec GmbH Infrarotsensorik & Messtechnik, Gostritzer Str. 61-63, 01217 Dresden, Germany
Phone: +49 761 - 4588 5075; e-mail: sven.wasmer@ise.fraunhofer.de

ABSTRACT: We present a detailed comparison of five inline-feasible hot spot detection and evaluation algorithms that are tested at the hands of inline thermography measurements of 108 industrial monocrystalline silicon solar cells. We evaluate the algorithms concerning their ability to predict the measured equilibrium temperature increases of the reversely-biased solar cells from short-term measurements and compare the requirements of each method. Beyond that, we apply and test for the first time a novel approach using the temporal evolution of the temperature of a hot spot right after turning on a direct voltage, only needing a series of temperature-calibrated thermography images (time series approach). We show that with the most accurate method of iterative convolution of power-calibrated thermography images with temperature response functions, equilibrium temperature increases can be predicted with a mean absolute error of only 3 K and a correlation coefficient of 0.90. Furthermore, relying both on similar physical considerations, we find the new time-series approach and the temperature increase T(100 ms)-T(80 ms) in the short-term measurement to yield almost as accurate results (correlation coefficients of 0.86 and 0.87 respectively and mean absolute error of 3.5 K for the time series approach), while implementation is less challenging for these.
Keywords: Silicon Solar Cell, Characterisation, Hot Spot, Reliability, Thermography

1 INTRODUCTION

If a solar cell in a solar module operating in the field is fully or partially shaded, it can be reverse-biased by the other cells in the same string of the module. This is not harmful for usual voltage ranges unless the shaded cell contains breakdown channels. In this case, localized currents can lead to a significant heating of the cell, also referred to as "hot spot", which can eventually irreversibly damage the module components [1, 2]. According to TUV Rheinland, in a global study for the period between 2011 and 2013, almost 6% of 6200 tested modules were sorted out due to hot spot danger [3]. Being able to reliably detect dangerous cells inline after the cell process or prior to module fabrication, while not sorting out too many harmless cells, would be beneficial concerning the yield and quality of production lines. However, since hot spots are often localized, global criteria such as the current I_{rev} at a fixed reverse testing voltage V_{rev} often fail. In contrast, thermography imaging is capable of detecting the local heat dissipation, but it was shown that local short-term temperature increases only weakly correlate with the equilibrium temperature increases $\Delta T_{eq.}$ [4].

As of today, these issues seem not to be resolved and inline hot spot detection and evaluation is not yet common in solar cell production. Furthermore, publications on inline hot spot evaluation algorithms and their reliability are scarce [5, 6]. In this work, we aim for a reliable and relevant hot spot detection method. For this purpose, we compare five algorithms concerning their ability to predict $\Delta T_{eq.}$ of about 100 industrial silicon aluminium back-surface field solar cells with the help of inline measurements and analysis times in the range of 100 ms.

As pointed out above, still many modules do not pass hot spot tests. Furthermore, with the introduction of novel and more complex cell concepts that are possibly more prone to high breakdown currents as well as the potential future increase of the number of cells per string in a module, raising the potential reverse bias applied to a shaded cell, hot spots and their evaluation are still an important topic in inline cell characterisation.

To the authors' knowledge, this is the first systematic comparison of different methods on a statistical basis. Beyond that, we apply and test for the first time a novel approach using the temporal evolution of the temperature of a hot spot right after turning on a direct voltage, only needing a series of temperature-calibrated thermography images (see Sec. 2.2).

2 APPROACH

2.1 Overview

We make use of an uncooled "VarioCAM® hr head" microbolometer camera installed in a "PV-LIT" inline measurement tool from InfraTec incorporated in the cell tester at the "PV-TEC" lab at Fraunhofer ISE. For each of the cells that are mounted in air during the measurement, we first acquire a series of 5 images with a framerate of 50 Hz ("short-term") and second a series of 50 images with an image acquisition frequency of 1 Hz ("equilibrium"). From the equilibrium measurement, we deduce the equilibrium temperature increase $\Delta T_{eq.}$, which we then try to predict solely from the short-term measurements. The system is synchronized such that the time of application of the constant reverse voltage V_{rev} is equal to the start of the image acquisition which is carried out line-by-line, that is the lines of the first image T_{20ms} are acquired corresponding to points in time of 0 to 20 ms after reverse-biasing. Based on the distinct temporal evolution of the temperature of a hot spot, we make use of a novel approach to extract its physical parameters that can be used to predict $\Delta T_{eq.}$. We then determine the correlations between the measured $\Delta T_{eq.}$ with the following methods:

1) Global current $I_{rev}(V_{rev} = -12\ V)$.
2) Temperature increase $\Delta T_{100\leftrightarrow 20} = T_{100ms} - T_{20ms}$ in the short-term measurement. We use this to represent a temperature difference $\Delta T_{100\leftrightarrow 0}$ with an image T_{0ms} acquired before biasing as with our line-by-line

reading camera, the lines of an image $\Delta T_{100\leftrightarrow 0}$ would correspond to significantly different durations of biasing.

3) Temperature increase $\Delta T_{100\leftrightarrow 80}$ in the short-term measurement.
4) Predicted temperature increase by our time series approach (see Sec. 2.1).
5) Approach based on the iterative convolution of power-calibrated thermography images with temperature response functions, as described in Ref. [7]. We use power-calibrated $\Delta T_{100\leftrightarrow 20}$ images for this end.

We test the methodologies at the hands of measurements of 108 industrial monocrystalline silicon solar cells that apart from hot spots show homogeneous electrical properties. We conduct the measurements using a reverse voltage of $V_{rev} = -12$ V and acquire two time series of thermography images with the uncooled microbolometer camera: firstly with an image acquisition frequency of 50 Hz for 100 ms in order to garner the temporal evolution of the hot spots right after application of the direct voltage and secondly with 1 Hz for 50 s, which proves to be enough to reach equilibrium and determine the resulting temperature increase ΔT_{eq}. We then analyse the correlations between ΔT_{eq} with each of the methods given above. Before illustrating the results of that comparison, we first give in the next section a short review of the methodology of the time series approach.

2.2 Time-series approach

If a solar cell with diode breakdown channels is reversely biased, current flowing through these channels can lead to a significant local heating of the solar cell, also referred to as the term "hot spot". The temporal evolution of this heating in the range of milliseconds to seconds right after applying the reverse bias can reveal important physical properties of the examined hot spot [8]. The simulations are carried out by an analytical model, where a hot spot is assumed to be a cylindrical breakdown channel with radius r_0 and power P_{in}, spanned over the whole cell thickness. The boundary conditions of convection and radiation are considered to be negligible on these short time scales. The temperature increase distribution $\Delta T(r,t)$ depends on time t and distance r from the hot spot origin (in cylindrical coordinates). $\Delta T(r,t)$ can be calculated (see also Ref. [9]) as:

$$\Delta T(r,t;P_{in},r_0) = \frac{4P_{in}}{\pi^3 r_0^3 dk}\int_0^\infty \left(1-e^{-\kappa u^2 t}\right)\cdot \frac{J_0(ur)\cdot J_1(ur_0)}{u^4\left(J_1(r_0u)Y_0(r_0u)-J_0(r_0u)Y_1(r_0u)\right)^2}du \qquad (1)$$

with J_n and Y_n the Bessel functions of the first and second kind of order n, d the cell thickness and k and κ denote the thermal conductivity and diffusivity of silicon.

By applying this result, we first acquire a series of thermography images right after the application of the reverse bias. We then adjust the analytical model to fit the measured temporal evolution of the hottest pixel on the cell and extract its characteristic parameters P_{in} and r_0. These can then be mapped to an expected equilibrium temperature increase ΔT_{eq} with a look-up table, allowing to evaluate the possible danger of a cell prior to module production in the range of 100 ms for the measurement and 10 ms for the analysis, suitable for inline application. This approach using a time series of thermography images is further illustrated and explained in Figure 1, where we depict the measurement and simulation of the time-dependent temperature evolution of a) a hot spot in the cell area and b) a hot spot in the corner of a silicon solar cell after the application of a reverse voltage at time $t = 0$ s. Temperatures are shown for the hottest pixel, measured with the uncooled microbolometer camera with a resolution of 355 µm per pixel and a framerate of 50 Hz.

Given are the best fits of r_0 and P_{in} to match the measurement data until 0.1 s (first 5 frames, dashed line) and until 1 s (all 50 frames, full line). For both hot spot sites, our analytical model can well describe the temporal evolution of the hot spot, with only slight relative deviations (+5% and -11%) in case of extrapolating the

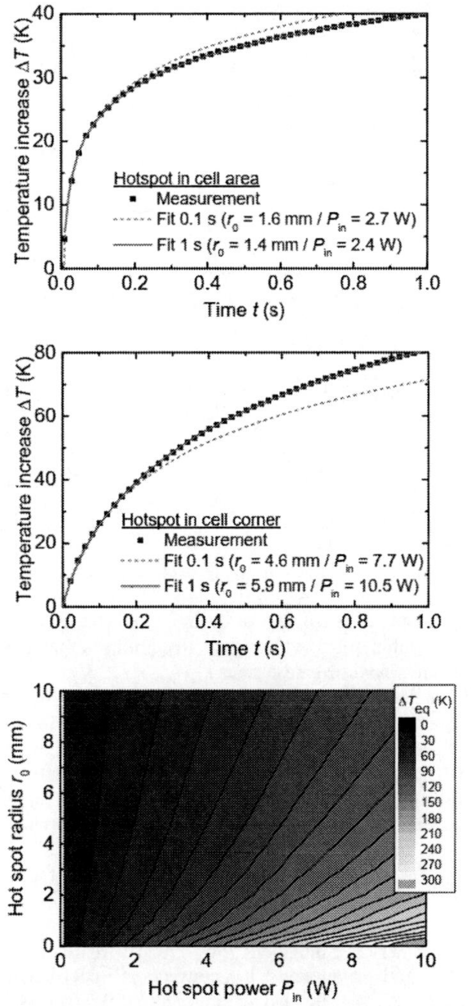

a)

b)

c)

Figure 1: Short-term measurement and simulation of the time-dependent temperature evolution of the hottest pixel of **a)** a hot spot in the cell area and **b)** a hot spot in the corner of a silicon solar cell.

c) Numerically simulated equilibrium temperature increases that the camera would measure if the hot spot origin would lie in the centre of a pixel.

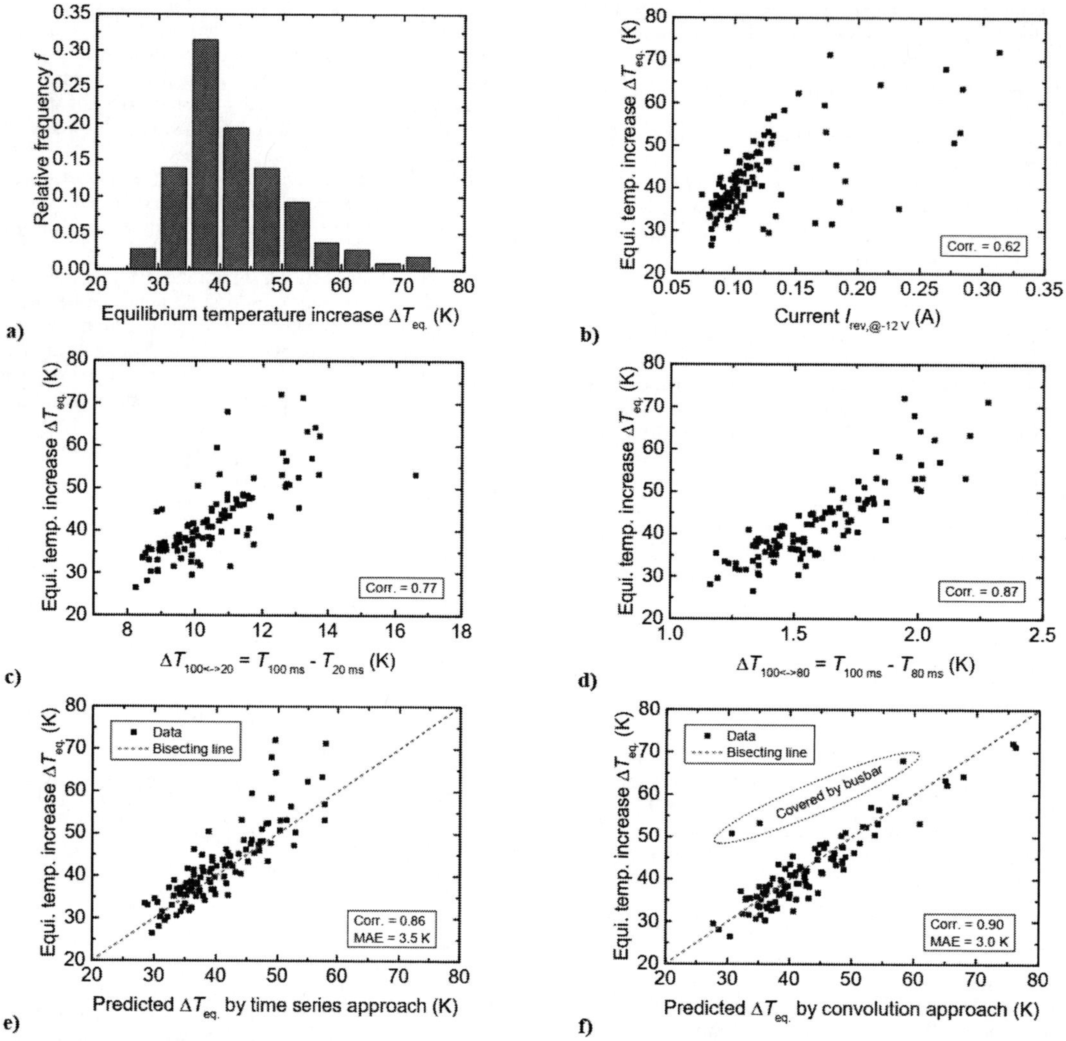

Figure 2: a) Histogram of the measured equilibrium temperature increase of the hottest pixels of the 108 silicon solar cells. Correlation plots of the measured equilibrium temperature increase $\Delta T_{eq.}$ with
b) the global reverse current $I_{rev}(V_{rev} = -12\ V)$,
c) the measured temperature increase $\Delta T_{100\leftrightarrow 20}$ in the short-term measurement,
d) the measured temperature increase $\Delta T_{100\leftrightarrow 80}$ in the short-term measurement,
e) predicted temperature increase by our introduced "time series approach" and
f) the approach based on the iterative convolution of power-calibrated thermography images with temperature response functions [7] (using power-calibrated $\Delta T_{100\leftrightarrow 20}$ images). Parts of the signal of the hot spots of the three marked points were covered by busbars, which introduces a large deviation.

fit from the first 5 to all 50 measurement points, i.e. from 0.1 s to1 s.

In c) the resulting equilibrium temperature increase by numerical simulations for a broad variation of r_0 and P_{in} and assuming a convection coefficient of $h_{conv} = 8\ W/(m^2 \cdot K)$ and an emissivity of 0.9 on both sides is depicted. Given are the temperatures that the camera would measure if the hot spot origin would lie in the centre of a circular cell with radius of 15.6/2 cm that is mapped to the centre of a pixel. These simulations serve as a look-up table for our time series approach, where first r_0 and P_{in} are extracted from the temporal evolution of the hot spot temperature and which then are converted to a prediction of the equilibrium temperature.

As an intermediate result, we find our analytical model to describe the temporal evolution of the temperature of a hot spot right after switching on the reverse voltage to be well suited to fit our measurements. Fundamentally, the signal-to-noise ratio of the thermography images seems to be high enough to analyse the time-dependent temperature development even on the pixel level.

3 RESULTS

The measured equilibrium temperature increases ΔT_{eq} are given in Figure 2(a) and lie within 26 to 72 K with a median of 40 K, and the comparison of the correlations of the five tested hot spot detection and evaluation methods with ΔT_{eq} are given in Figure 2(b-f).

Concerning our approach based on the temporal evolution of hot spot temperatures, we find the hot spot radii to lie within 1.3 to 4.5 mm with a peak at 3.4 mm and hot spot powers of 1.5 to 3.7 W, with most being around 2.3 W. Note that these hot spot powers significantly exceed the global powers $V_{rev} \cdot I_{rev}$, which is due to the fact that almost all hot spots happen to be located at the cells' edges or corners. Within the time series approach, the increased apparent hot spot powers due to the mirroring of heat waves at the cells' edges are inherently correctly included (in the convolution approach, this has to be taken care of explicitly).

Regarding the approach based on the iterative convolution with temperature response functions, we calibrate temperature difference images $\Delta T_{100\leftrightarrow20}$ with a calibration factor of 0.00073 W/K.

Concerning methods 4) and 5), we find a convection coefficient of $h_{conv} = 8$ W/(m²·K) to describe most accurately the cooling conditions in the measurement chamber of our inline cell tester during the measurements.

Finally, we find Pearson correlation coefficients of $\rho = 0.62$ for the global criterion of the current I_{rev} as well as 0.77 and 0.87 for the short-term temperature increases $\Delta T_{100\leftrightarrow20\,ms}$ and $\Delta T_{100\leftrightarrow80}$, respectively. The more advanced methods of time series and convolution approach exhibit $\rho = 0.86$ and 0.90 with mean absolute errors MAE = 3.5 and 3.0 K.

4 DISCUSSION

Concerning the correlation analysis for the measured equilibrium temperatures increases ΔT_{eq}, we expectedly identify the global criterion of the current I_{rev} and the short-term temperature increase $\Delta T_{100\leftrightarrow20}$ to show only weak prediction accuracies.

We obtain significantly more accurate results for the more advanced approaches, namely the newly introduced time series approach and the convolution method. Our results show that both evaluation algorithms can be utilized to predict ΔT_{eq} accurately enough based on inline-feasible measurements that last only 100 ms.

Table I: Tentative comparison of the requirements and prediction accuracies of the five hot spot evaluation algorithms.

	Number of images	Precision of synchronisation	Simplicity of calibration	Complexity of image processing	Prediction accuracy of ΔT_{eq}
1) $I_{rev}(V_{rev} = -12$ V)	0	++	++	++	--
2) $\Delta T_{100\leftrightarrow20}$	2	0	+	+	-
3) $\Delta T_{100\leftrightarrow80}$	2	-	+	+	0
4) Time series	4	--	-	0	+
5) Convolution	2	0	--	--	++

However, they still can fail on certain occasions, as we find maximal deviations of -18.8 K and -20.2 K for the time series and convolution approach respectively. We explain most of the outliers to be due to superposed heat flux of accumulated hot spots in the vicinity to each other, not detectable by the time series approach in the short-term measurement. In case of the convolution approach, we find all three severe outliers to be caused by busbars shading parts of the power signal.

Interestingly, the tested method 3), that is the temperature increase $\Delta T_{100\leftrightarrow80}$ in the short-term measurement, shows a significantly better correlation than that of the temperature increase $\Delta T_{100\leftrightarrow20}$, although exhibiting much lower absolute signals that are more prone to noise. An explanation for this is that according to the findings in Ref. [8] and Eq. (1), the most important characteristic of hot spot power P_{in} is weaker correlated to the absolute temperature increase $T(t_1)$-$T(t_0 \approx 0$ ms) than to the difference between the temperatures at two time points t_0 and t_1, with t_0 considerably greater than zero.

As these results are achieved on cell level, care has to be taken while transferring these results on module level. This could be done by changing the look-up table in Figure 1(c) for the time series approach and the point spread functions of the convolution approach (Ref. [7]) to those of the applied module layout. However, as we believe that there is a strong correlation between the measured $\Delta T_{eq,cell}$ of cells mounted in air with the measured $\Delta T_{eq,module}$ at the backsheet of the finished module, an empirical, monotonous relation $\Delta T_{eq,module}(\Delta T_{eq,cell})$ could be found experimentally.

4.1 Comparison of requirements

In order to enable a better assessment of the requirements of the different methods, we give in Table I an overview of the requirements categorized in 5 classes based on our experiences, ranging from "--" (difficult to implement) to "++", meaning basically no effort has to be put into. Approach 1), that is the global current, has of course the least restrict requirements, however, the use for predicting ΔT_{eq} is rather minimal. Better in this respect are the temperature increases in the short-term measurements 2) and 3), with $\Delta T_{100\leftrightarrow80}$ being more reliable albeit posing more problems concerning the synchronization as two images within a distinct time interval after applying the reverse voltage are needed, contrary to one image before, where the exact timestamp is irrelevant, and one image after biasing. We have to note that here both images $\Delta T_{100\leftrightarrow20}$ and $\Delta T_{100\leftrightarrow80}$ are calculated from images after biasing, while the first is actually used to represent a temperature difference $\Delta T_{100\leftrightarrow0}$ with an image T_{0ms} acquired before biasing, see also Sec. 2.1, bullet point 2), which should be easier to implement.

Better prediction results can be achieved with the methods 4) and 5), referred to as time series and convolution approach, which also are the only ones that give direct predictions of ΔT_{eq} in units of K. By utilizing 4), however, more difficulties arise, as at least 4 temperature-calibrated thermography images perfectly synchronized in equidistant time steps have to be acquired and analysed. Concerning 5), calibration is even more restrained as power-calibrated short-term thermography imaging measurements are needed. Furthermore, the edges of the cell in the image have to be known in order to mirror the power sources before the challenging image processing step of convolution with temperature response functions. Both of the advanced methods furthermore need numerical simulations to calculate the expected equilibrium temperature increases and the temperature response functions respectively as well as require a one-time calibration of the convection coefficient to the cooling conditions of the measurement chamber.

5 SUMMARY AND CONCLUSION

We compared five hot spot detection and evaluation algorithms concerning their ability to predict the equilibrium temperature increase ΔT_{eq} of reversely biased silicon solar cells and as well assessed their hard- and software requirements. The methods were tested at the hands of about 100 industrial silicon solar cells using inline-feasible measurements and analyses.

We found the highest prediction accuracy with a mean absolute error of only 3 K to be possible with the approach based on the convolution of power-calibrated thermography images. While it is the physically soundest approach and especially outperforms the others in case of planar or many local hot spots in the vicinity of each other, inline feasibility is still given.

Our newly introduced algorithm based on the analysis of the temporal evolution of hot spot temperatures was a bit less accurate (mean absolute error 3.5 K), while implementation is somewhat less challenging. Using the same physical background as for this time series approach, we identified the temperature increase $\Delta T_{100\leftrightarrow 80}$ in the short-term measurement to show quite a good correlation of $\rho = 0.87$ with the reference ΔT_{eq}, thus representing a good compromise between prediction accuracy and requirements.

The comparison conducted in this work lays ground for reliable classification algorithms of hot spots, where the next step of transfer to module level can be easily carried out. Then, the critical equilibrium temperature increase $\Delta T_{eq,critical}$ has to be defined depending on the module layout and mounting location as well as different shading scenarios.

Despite that, $\Delta T_{eq,critical}$ can be tuned for a trade-off of quality and yield of the sorting algorithm, for example using its receiver operating characteristic curve.

ACKNOWLEDGEMENT

This work was funded by the German Federal Ministry for Economic Affairs and Energy within the research project “iMage” (contract no. 0324045A).

We thank Klaus Ramspeck from h.a.l.m. elektronik GmbH for valuable discussions.

Sven Wasmer gratefully acknowledges the support by scholarship funds from the State Graduate Funding Program of Baden-Württemberg.

REFERENCES

[1] F. Fertig, S. Rein, M. C. Schubert, and W. Warta, “Impact of Junction Breakdown in Multi-Crystalline Silicon Solar Cells on Hot Spot Formation and Module Performance,” in *26th EUPVSEC Hamburg*, 2011, pp. 1168–1178.

[2] I. Geisemeyer, F. Fertig, W. Warta, S. Rein, and M. C. Schubert, “Prediction of silicon PV module temperature for hot spots and worst case partial shading situations using spatially resolved lock-in thermography,” *Solar Energy Materials and Solar Cells*, vol. 120, pp. 259–269, http://www.sciencedirect.com/science/article/pii/S0927024813004856, 2014.

[3] G. TamizhMani *et al., Eds., Regional influence on module design quality: Qualification testing failure rate results from six regional labs of TUV rheinland around the world*. IEEE 40th Photovoltaic Specialist Conference (PVSC), 2014.

[4] T. Roth, R. Siebert, and K. Meyer, “From Short-term Hotspot Measurements to Long-term Module Reliability,” *Energy Procedia*, vol. 55, pp. 504–508, 2014.

[5] J. Hudson, L. Vasilyev, J. Schmidt, and G. Horner, Eds., *Economic impacts and approaches to address hot-spot defects in photovoltaic devices*. 35th IEEE Photovoltaic Specialists Conference, 2010.

[6] K. Ramspeck, S. Schenk, D. Duphorn, A. Metz, and M. Meixner, “In-line thermography for reliable hot spot detection and process control,” *Energy Procedia*, vol. 55, pp. 133–140, 2014.

[7] I. Geisemeyer, F. Fertig, W. Warta, S. Rein, and M. C. Schubert, “Fast Hot Spot Evaluation,” in *29th EUPVSEC Amsterdam*, Amsterdam, The Netherlands, 2014, pp. 2429–2434.

[8] S. Wasmer et al., “Analytical modelling of the temporal evolution of hot spot temperatures in silicon solar cells,” to be published.

[9] H. S. Carslaw and J. C. Jaeger, *Conduction of heat in solids*, 2nd ed. Oxford [Oxfordshire], New York: Clarendon Press; Oxford University Press, 1986, ©1959.

CELL DESIGN OPTIMIZATION FOR SHINGLED MODULES

Dominik Rudolph[1]*, Jorge Rabanal-Arabach[1], Ingo Ullmann[1], Andreas Halm[1], Andreas Schneider[1], Thomas Fischer[2]

[1]International Solar Energy Research Center – ISC Konstanz, Rudolf Diesel-Str. 15, 78467 Konstanz, Germany
[2]Teamtechnik Maschinen und Anlagen, Planckstraße 40, 71691 Freiberg, Germany
*Corresponding author: Phone: +49 7531 36183-54; Fax: +49 7531 36183-11
Email: dominik.rudolph@isc-konstanz.de

ABSTRACT: In order to identify an optimum cell design in respect to cell width and corresponding front metallization finger grid design for cells used in shingled modules, simulations on the power output for cell stripes with varying width and front metal grid are performed. Based on this information, Al-BSF cells with a different amount of busbars are fabricated and cut into 4 to 6 stripes. For verification of the simulation a comparison between experimental and simulation results is made, which shows that the simulation predicts the correct cell to stripe power loss and hence could be used to find the optimum metallization layout. Further, these stripes are used to assemble different types of shingled mini modules to optimize the stripe overlap.
Keywords: Simulation, Module, Shingle concept

1 INTRODUCTION

Shingled modules were developed previously for space applications due to the space restrictions of such systems [1]. Accordingly further improvements and inventions were achieved by Nielsen and Myer [2, 3]. For terrestrial application shingling was not considered for years until recently due to the existing strong competition in the solar module market and the aim to increase module power and module efficiency without changing the module area. Shingling just became a current research topic of module manufacturers world-wide and is also in the scope of many research institutes [4 to 6].

It is obvious that using cells with a standard metallization design for shingled modules is not the best solution to get maximum power output. Therefore we are optimizing the cell and the module design in this work, first by simulation and afterwards in an experiment to verify the simulated results. In regard to the power output of the single cell stripes and for optimizing the module layout for highest power output we optimized the cell stripe layout and the amount of stripes. Shingling of cells in a module may result in a higher power output compared to a standard module even if using the same module area due to the ultimate exploitation of the available space in a module. In addition less cell interconnectors are needed and one additional opportunity is that the cell mismatch can be reduced by sorting of the cell stripes.

The main challenge for this concept is to reduce the cell breakage rate in the whole process, which is due to laser scribing and following stripe cleaving more critical than for the standard cell process.

2 EXPERIMENTS

2.1 Preparation of cells with different amount of busbars

P-type multicrystalline cells of two different materials A und B, are fabricated with a phosphorous diffusion resulting in a sheet resistivity of around 65 Ohm/sq. The surface is passivated by a PECVD SiNx deposition. After the passivation step the wafers are metallized by screen printing: first the rear side pads with an AgAl paste, then the rear side Aluminum and afterwards the front side Ag grid is screen printed. Figure 1 shows the metallization design, which is adequate to cut the cell into six 26 mm wide stripes or into two stripes of 52 mm width and two stripes of 26 mm width. For the other stripe sizes we used a metallization layout with four or five busbars (BB) respectively. Electrical contacting of the cell and building the Al back surface field (Al-BSF) are achieved in a co-firing step.

All cells were characterized by using an IV tester. Laser grooves were applied to the rear side by an infrared cutting laser and afterwards the cells were measured again in an IV tester to verify the influence of the laser scribing to cells performance. After a second IV measurement the cells were cleaved into four, five and six cell stripes. All cell stripes were supposed to be IV measured again after cutting. Unfortunately until now we were not able to measure the 26 mm wide cell stripes because of technical restrictions. But the other stripe sizes of 31.2 mm and 39 mm were measured by contacting the cell on one side of the stripe from the bottom and on the other side from the top. Table I summarizes the different cell layouts and the corresponding cell stripes.

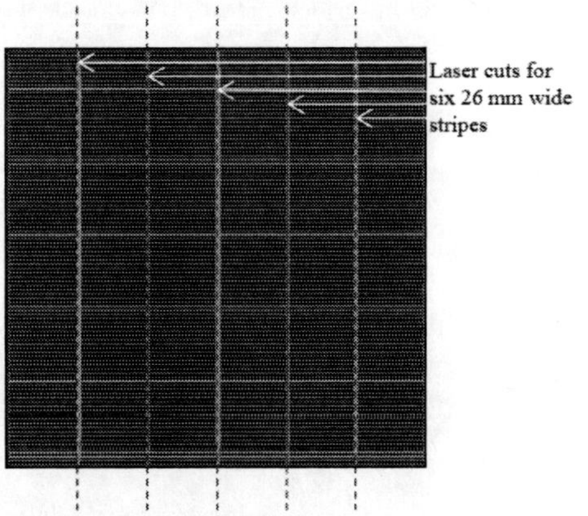

Figure 1: Laser scribing paths for cutting a three busbar cell into six, 26 mm wide stripes

Table I: Cells and cell stripes tested in this work

Concept	BB	Stripe width [mm]	Amount of printed metal lines	IV results
S1/3	3	52	100	X
S1/4	4	39	120	✓
S1/5	5	31.2	105	✓
S1/6	3	26	100	X

2.2 Simulation of cell stripes

For cell simulation we are using Griddler 2.5 developed by Johnson Wong and Ranjani Sridharan [7]. First the full cell IV results are used to simulate the cells and to find out the adequate recombination current values for the passivated area, the metallized area and the edge [8]. The used values are listed in Table II, while k is the periphery to area ratio in cm^{-1}, which varies for each cell stripe type from 0.51 cm^{-1} for the 1/3 stripes up to 0.90 cm^{-1} for the 1/6 stripes. By using these values the different cell stripes were simulated varying the amount of Ag lines on the front. The cell stripes were simulated for the shingle interconnection and not for standard interconnection, which means that it is connected only on one edge on each side of the cell, like it is shown schematically in Figure 2.

Table II: Cell recombination current values of the passivation, the metallization and the edge used for simulation. Starting point was the literature values from Bertrand et al [8].

	Front pass	Front metal	Rear metal	Edge
J_{01} (fA/cm^2)	62	1100	530	582 * k
J_{02} (nA/cm^2)	10	30	14	20.6 * k

2.3 Processing of mini modules

By using 1/4, 1/5 and 1/6 stripes (equivalent to the concept, such as S1/4), different mini modules were fabricated. The interconnection between the stripes is achieved by dispensing of ECA, subsequent curing on a hot plate and in addition with heating by an IR lamp. Figure 2 shows a schematical side view of such shingled string of four cell stripes.

Figure 2: Schematical side view of a shingled interconnection between four cell stripes

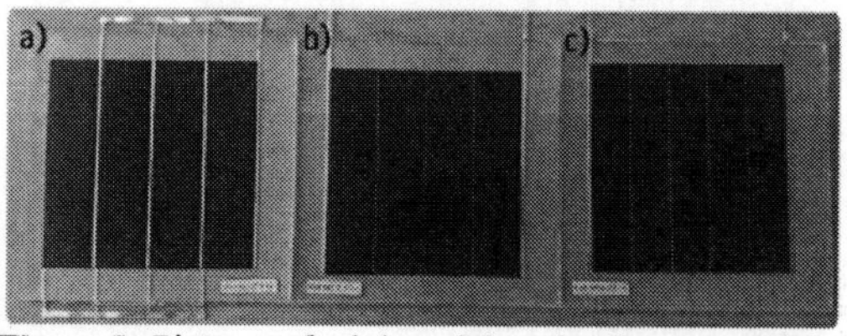

Figure 3: Pictures of mini-modules, a) reference module with four bus bars, b) S1/4 module and c) S1/5 module.

The first and the last stripe are connected to a ribbon interconnector allowing to extract voltage and current for IV measurements. Examples of such interconnected shingle modules are shown in Figure 3 b) and c). The overlap between the cell stripes was varied between 1.5 and 2 mm. Uncut cells were used to fabricate reference modules by contacting each bus bar to an interconnection ribbon, which is shown in Figure 3 a). In order to have comparable results the interconnection was also made by using ECA with subsequent curing. The shingled-modules and the reference modules are manufactured by using 3.2 mm textured glass, EVA and transparent backsheet. Only modules without cracks and without obvious manufacture failure were selected for further investigation.

3 RESULTS AND DISCUSSION

3.1 Experimental results for 1/4 and 1/5 stripes

As reported in literature laser scribing has only a minor influence on the cell performance as long as the as the groove was performed on the rear side to avoid the formation of shunts [8]. This is verified by the pFF determined before and after laser scribing, which is plotted in Figure 4 for both materials A and B. Nevertheless the influence is more detrimental after cleaving the cell into stripes due to higher edge recombination (as pointed out by green points in Figure 4). The edge recombination is dependent of the ratio between periphery and area of the cell stripe. Therefore pFF is reduced by around 0.37 to 0.55$\%_{abs}$ from cell after scribing to cell stripe. The variation of the pFF for the stripes is higher than for the cells due to local cell inhomogeneities, which of course could influence one stripe, while the other stripes of the same cell are not influenced.

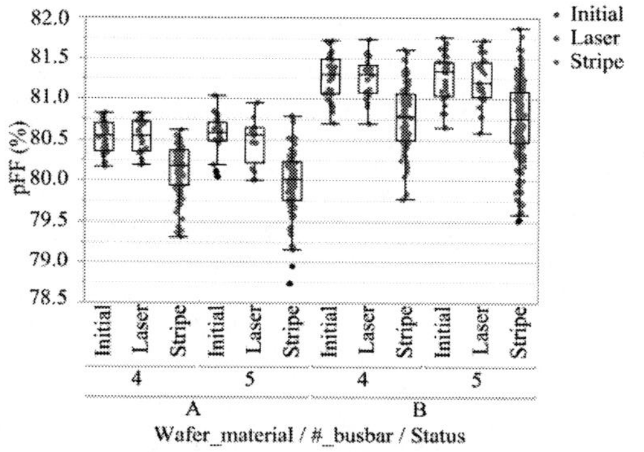

Figure 4: Influence of laser scribing and wafer cleaving on pFF in dependence of the material and the number of busbars

The FF, which is plotted in Figure 5, is strongly reduced for the cell stripes (around 1.1 $\%_{abs}$). One part of this reduction is due to the pFF reduction but the rest is only explainable by an increased series resistance R_S. Therefore the loss of the FF caused by the series resistance of the stripe position is plotted in Figure 6. There is a decreasing trend for larger stripe number for both cell concepts (S1/4 and S1/5) with the highest step between the first and the second stripe. The high R_S of the first stripe can be explained by the metallization layout on the cell front side. The metallization lines are not connected at the end of stripe one, while for the other cell stripes the lines are connected at the end. The difference between the other stripes is due to inhomogeneous front side metallization and surface dependent contact formation. The material has no significant influence on the loss due to R_S. The main part of the series resistance loss is explainable due the long way the electrons have to flow from one side of the cell to the bus bar, which is double the way than for the case of a regular cell.

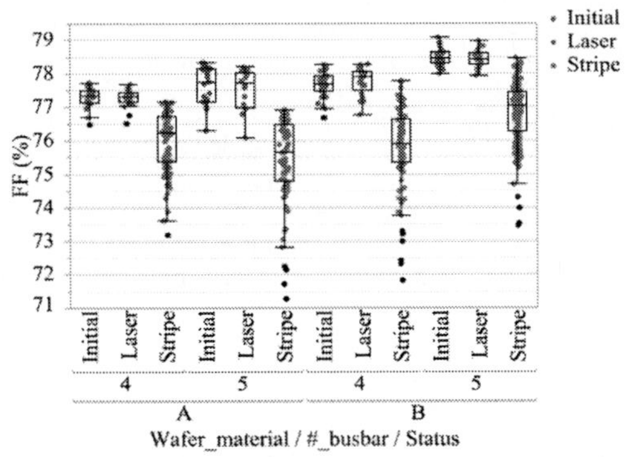

Figure 5: Change in fill factor during wafer processing. FF of the stripes is around 1.1%$_{abs}$ lower than before cleaving

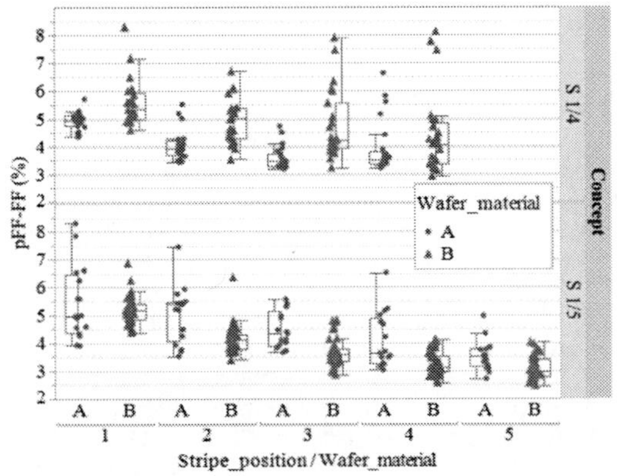

Figure 6: Calculated difference between pFF and FF regarding to the stripe position and shingle concept

The experimental results for the 26 mm wide stripes cannot be reported here so far due to problems with the IV measurement. It was only possible to generate a few data points, which cannot be guaranteed to be correct. But nevertheless we simulated the stripes using the parameters gained from the experimental results of the full cells and the other cell stripes.

3.2 Simulation results

Simulation results for different stripe sizes and amount of screen printed metal lines are carried out by Griddler. The FF dependency is shown in Figure 7. For smaller stripe size the FF increases while the FF is low for larger stripe size. The simulations show that 1/6 cell stripes do not suffer as much from an increased R_S than the 1/3 cell stripes, which is obviously corresponding to the halved way the electrons have to pass from one side of the cell to the bus bar tab. But the 1/6 cell stripes are influenced detrimentally by an increased edge recombination due to a higher periphery to area ratio of $k = 0.90\ cm^{-1}$ in comparison to $k = 0.51\ cm^{-1}$. This directly influences the V_{oc} and therefore the V_{mpp} value of the cell stripe. Hence the maximum stripe's P_{mpp}, is reached when the negative influence of edge recombination and negative influence by R_S are both minimal. Measurement results of 1/4 and 1/5 cell stripes are included in blue squares and red diamonds showing that the mean value fits quite well to the simulated value but due to metallization inhomogeneities exhibits a stronger variation. The cell stripe P_{mpp} is plotted in Figure 8 showing that for each stripe width there is an optimum amount of metallization lines on the front side. Comparing the simulated P_{mpp} results with the measurement results it seems that even a lower amount of metallization lines on the front side would be sufficient to reach the maximum power. Besides, simulation outcome fits to the experimental data for the power output of the stripes. Figure 9 shows the calculated power output for a whole cell by multiplying the power output of one stripe with the number of stripes. This calculation assumes that there is no further loss due to interconnection between the stripes. For the stripe size of 26 mm (1/6) it seems to be not possible to overcome the detrimental influence of the edge recombination on the cell P_{mpp} by varying the metallization layout, so that the optimum stripe width seems to stay in between 26 mm and 52 mm.

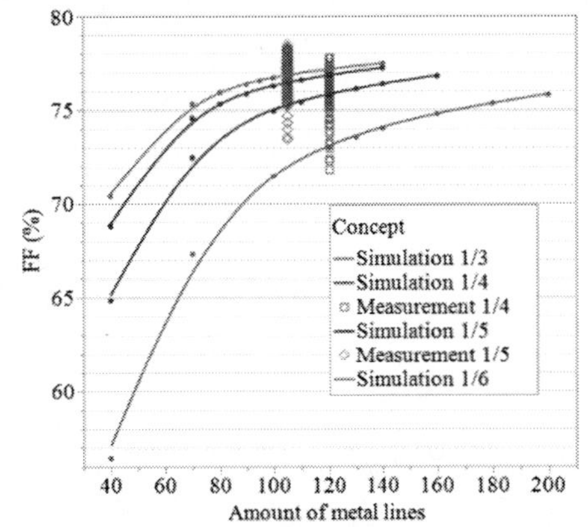

Figure 7: Simulated FF dependency of different cell stripe sizes and amount of metal lines. Blue squares and red diamonds mark the experimental IV results

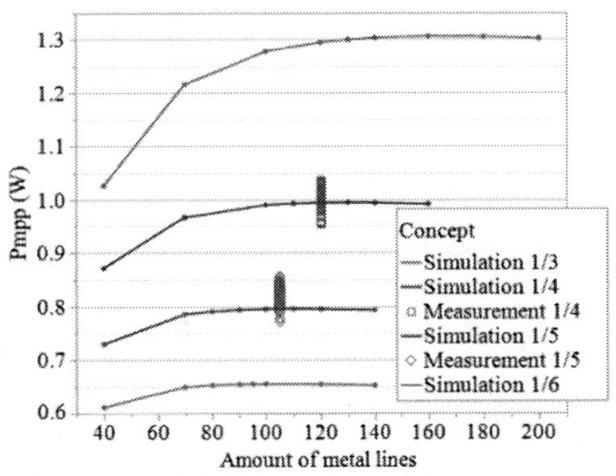

Figure 8: Simulated P_{mpp} of single cell stripes with different amount of metal lines.

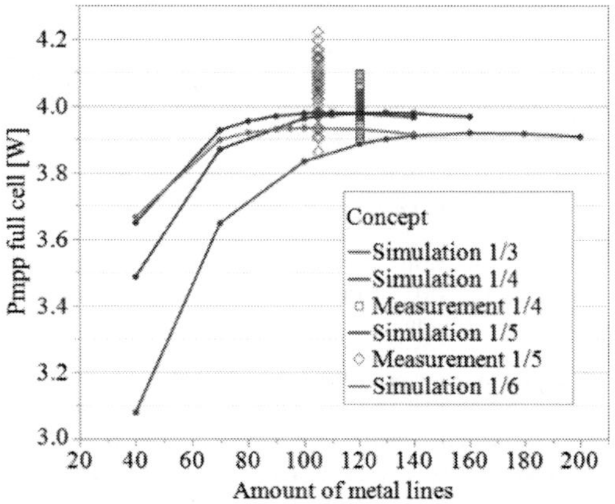

Figure 9: Calculated P_{mpp} of a full cell for varying width of cell stripes assuming no further losses due to interconnection. Experimental results are shown in blue squares and red diamonds

3.3 Mini module results

Figure 10 shows the active area module efficiency of the S1/4 and the S1/5 shingle modules and the corresponding reference modules with four or five busbars respectively. Whatever overlap is chosen the active area module efficiency is higher for the shingled modules than for the reference modules. Regarding these results the overlap between the shingled cell stripes has only a minor influence and the S1/5 concept seems to be the better one. Detailed analysis of the module results is done in the work of Rabanal et al., where also simulations of large modules are discussed [9].

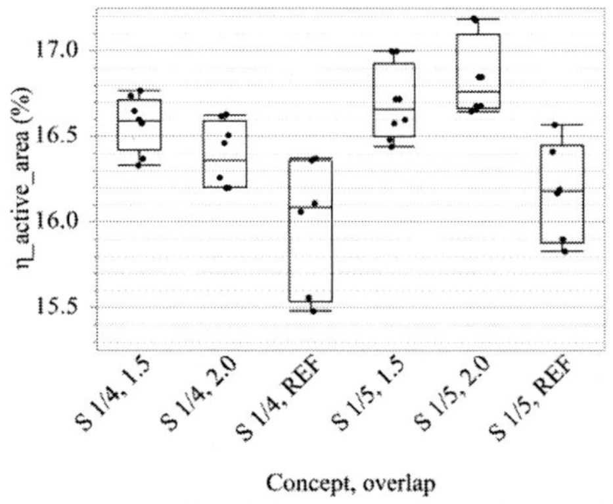

Figure 10: Module efficiency regarding active area for different module concepts

4 CONCLUSION

The simulations have shown that cutting the cell into four or five stripes (e.g. 39 mm or 31.2 mm) results in the highest power output regarding the whole cell, without power losses due to interconnection taking into account. Experimental results with 1/4 and 1/5 stripes prove the results obtained by the simulation in respect to power output and FF but with results for 1/3 and 1/6 stripes are still missing. Voltage and current could not been simulated sufficiently due to material inhomogeneities of the measured cells/stripes and calibration issues regarding the stripe measurement. In a next step stripe results will be incorporated into the simulation (also work for 1/6 and 1/3 stripes) and simulations will be carried out assuming different wafer quality and wafer inhomogeneities. Nevertheless the so far obtained simulation results of the stripes show that the fill factor loss can be reduced by increasing the amount of Ag lines on the front side of the cell. Due to the high cost of silver paste it is not economical to increase the amount of lines used for a standard three bus bar cell significantly. In this case the benefit of the shingled module resulting in less costs due to the reduced amount of interconnectors is reduced or even reversed. Therefore it is absolutely necessary to find a compromise between the amount of fingers and power output of the cell/module.

First results on mini modules show that with the shingled module concept the module efficiency in respect to the active area is higher than for comparable reference mini modules with four or five busbars connected by ribbons.

5 ACKNOWLEDGEMENTS

This work was supported by the German Federal Ministry of Economic Affairs and Energy (BMWi) under contract number FKZ 03241098 (Schindel - Neuartige Verbindungskonzepte für Solarmodule)

6 REFERENCES

[1] Patent US2938938, D.C. Dickson, JR, 1960
[2] Patent US3116171, R.J. Nielsen, 1963
[3] Patent US3369939A, J.H. Myer, 1968
[4] J. Summhammer and Z. Halavani, "High-voltage PV-modules with crystalline silicone solar cells", 28th EU-PVSEC (2013), pp. 3119-3122.
[5] Patent US20150090314A1, B. Yang, 2015
[6] G. Beaucarne, "Materials challenge for shingled cells interconnection", Energy Procedia 98 (2016) 115 – 124
[7] J. Wong, "Griddler: Intelligent computer aided design of complex solar cell metallization patterns," in Proc. IEEE 39th Photovoltaic Spec. Conf., 2013, pp. 0933–0938.
[8] D. Bertrand, S. Manuel, M. Pirot, A. Kaminski-Cachopo, Y. Veschetti, „Modeling of Edge Losses in Al-BSF Silicon Solar Cells" IEEE Journal of Photovoltaics, Vol. 7, No. 1, January 2017
[9] J. Rabanal-Arabach, D. Rudolph, I. Ullmann, A. Halm, A. Schneider, T. Fischer, „Cell-to-Module Conversion Loss Simulation for Shingled-Cell Concept", Proceedings of the 33rd European PV Solar Energy Conference and Exhibition, 2017

OPTICAL SIMULATION OF BIFACIAL SILICON SOLAR CELLS AT MODULE LEVEL

F. Duerinckx, M. Aleman, A. S. H. van der Heide, E. Voroshazi, J. Szlufcik

Imec, Kapeldreef 75, B-3001 Leuven, Belgium
Corresponding author: phone: +32 1628 1162, e-mail: Filip.Duerinckx@imec.be

ABSTRACT: While bifacial modules are receiving increased attention in the PV community, questions remain on the best methods to maximize light trapping in the active device, especially with cells expected to become thinner in the future to further reduce material costs. This work focusses on optimizing the light trapping behavior of such bifacial cells in a module by looking at the impact of the cell's rear surface morphology. Ray tracing simulations using the "Module ray Tracer" from PVLighthouse have been used to quantify the impact of changing the angle of the pyramids on the rear surface of a bifacial cell (from polished to alkaline textured and all conditions in between). The result of these simulations is that neither a polished nor a fully textured rear surface is ideal for light trapping at the module level. A semi-polished rear surface (pyramids with a base angle ~ 25°) results in the highest simulated Jsc for glass/glass encapsulated bifacial cells, even when considering bifacial illumination. This has been confirmed experimentally from reflectance and transmittance measurements for encapsulated bifacial samples with different rear surface morphologies.
Keywords: bifacial, simulation, light trapping

1 INTRODUCTION

Bifacial modules are receiving increased attention in the PV community due to the additional energy generation by the rear surface illumination and the associated positive impact on the Levelized Cost of Electricity (LCoE)[1]. However, questions still remain on the best methods to maximize light trapping in the active device, especially with cells becoming thinner in the future to further reduce material costs [2]. The silicon loss in thinner cells could result in a decrease of the amount of absorbed photons, and therefore a potential Jsc loss. This is particularly critical for bifacial silicon solar cells since they do not feature a rear metal reflector as their monofacial counterparts, resulting in an increased light transmission for the former. Solutions for this loss through light transmission can be found by implementing light trapping features at the module level, for example by using a Light Trapping Film [3]. But complementary solutions can also be found at cell level.

This work focusses on finding ways to optimize the light trapping behavior of silicon bifacial cells (for instance bifacial n-type Passivated Emitter, Rear Totally diffused or nPERT cells), in a module by investigating the impact of the cell's rear surface morphology (from polished to alkaline textured and everything in between).

As simulation software, the "Module Ray Tracer" from PV Lighthouse [4] was used. It is based on Monte Carlo ray tracing in combination with thin film optics calculations at each interface. It outputs the photocurrent generated in the cell as well as the losses due to transmission, reflectance or parasitic absorption in the materials that make up the module. The following section details the overall methodology as well as the input and output parameters used in the simulations. The results are discussed in section 3, while the experimental verification of these simulations results is discussed in section 4.

2 METHODOLOGY

A cross section of the simulated bifacial module structure used in the Module Ray Tracer from PV Lighthouse is shown in Figure 1. It incorporates the bifacial cell and the different components of the bifacial module.

The following components can be distinguished:

- The front and rear glass is a flat, 2 mm thick soda lime glass with low Fe content [5]. A standard AR coating (Enki cleanARC-SA [6]) with a thickness of 115 nm and refractive index of approximately 1.40 is foreseen on the front glass but not on the rear.
- As encapsulant (both sides of the cell), an EVA is chosen with a thickness of 400 μm [7].
- The cell itself has a textured front surface (alkaline texturing with front upright pyramids, 5 μm pyramid width, 3.53 μm pyramid height, pyramid angle: 54.7°) with a 75 nm SiN_x layer as dielectric [8].
- The cell's thickness is varied from 180 μm down to 20 μm.
- On the rear surface, the surface morphology is varied between the same upright pyramids and a perfectly flat surface. The base width of the pyramids is always kept constant (5 μm) but the pyramid angle to the base is varied from 54.7° (alkaline texturing) to 0° (polished surface). The same SiN_x layer is used on this rear surface. Figure 1 shows the textured rear surface as an example.
- No metallization was applied on the cell to simplify the interpretation of simulations, focusing solely on the impact of the rear surface morphology.

Other important parameters used in the simulations are the following:

- The illuminating spectrum was AM1.5g in the wavelength range from 300 to 1200 nm. 500,000 rays were used for each simulation.
- The rays were distributed across the wavelength range weighted by the spectrum. This results in a more accurate result for the generated current.
- Illumination was either on the front surface (perpendicular illumination), or the rear surface (perpendicular or diffuse), or on both sides simultaneously.

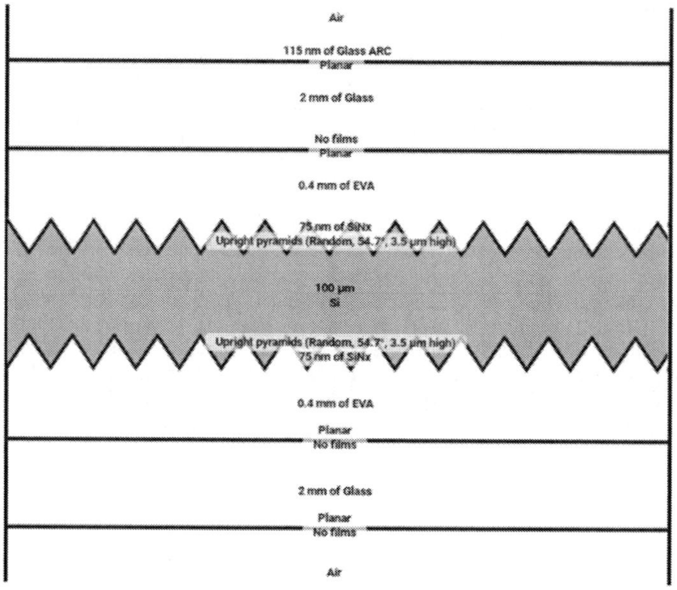

Figure 1: Schematic of simulated bifacial module structure in PV Lighthouse (textured rear surface shown but the simulations describe result of random pyramids with pyramid angles of 0 – 54.7°)

- The output results of the simulations were the following:
  - Losses by transmission (out of the module), by reflection (incl. escape reflection of light being lost by having entered the module but escaping out of the illumination side) and by parasitic absorption (in glass, encapsulant, dielectric).
  - When subtracting these losses from the max available current (i.e. if all photons in the relevant spectrum are generating a carrier with perfect collection (~46.5 mA/cm^2)), the photogenerated current in silicon with perfect carrier collection is obtained. By using the collection efficiency (calculated (see [9]) from the Internal Quantum Efficiency of a typical bifacial nPERT cell), the short-circuit current density is calculated.

The next section describes the results of the simulations focusing on the impact of the cell's rear surface on the different optical loss mechanisms.

3 SIMULATION RESULTS

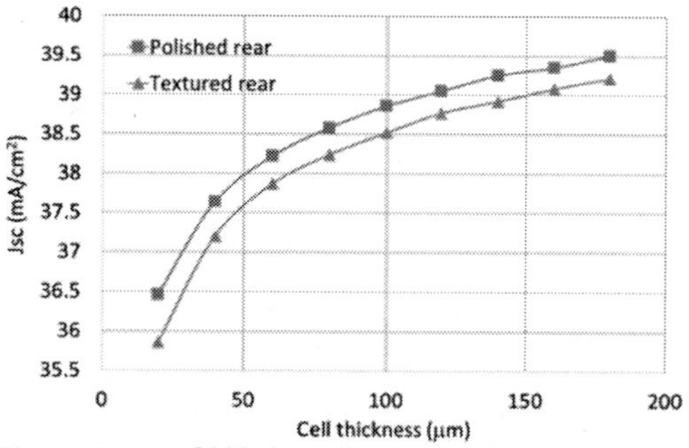

Figure 2: Jsc of bifacial cells in a module with flat and textured rear surface as function of the cell thickness (front illumination only).

Figure 2 shows the resulting Jsc of a bifacial cell in a module as a function of thickness for 2 extreme cases of the rear surface morphology: a polished (pyramid angle 0°) vs a fully textured rear surface with upright pyramids (pyramid angle 54.7°). The illumination for this simulation is 1 sun directed perpendicular onto the front surface of the module so no rear illumination is considered yet. As can be seen for both cases the Jsc drops rapidly towards thinner cells while the cells with a polished rear surface have a small advantage.

More details are available in Figures 3 and 4, showing respectively for the cases of the textured and polished rear surface, the losses in generated photocurrent as function of the cell thickness due to transmission, reflection (sum of direct and escape reflection) and absorption losses (in the glass and the encapsulant) with only front illumination. For the textured rear surface, transmission losses out of the module are the highest while reflection due to a higher escape loss is the major problem for cells with a polished rear surface. This difference can be understood when considering the light path through these devices and how the rear surface morphology impacts this path and the resulting effect it has on the escape angle and Total Internal Reflection (TIR). Naturally, these transmission/reflection losses increase for thinner cells. The absorption losses on the other hand are constant.

Neither a flat nor a textured rear surface is close to ideal. This becomes clear from Figure 5 which shows, as function of the angle of the pyramid on the rear surface (going from a flat rear at the left of the picture to a textured rear on the right), the losses at the front of the cell, the rear and the sum of the two. The front losses are the sum of the reflection losses and the absorption losses in the front glass and encapsulant. The rear losses are similarly determined by transmission losses and the rear absorption losses. The only source of illumination here is, like before, at the front (1 sun, AM1.5g). This simulation is done for bifacial cells with a thickness of 100 µm. Clearly, the front related losses decrease when going from a flat to a textured rear surface while the behavior for the rear losses is opposite. Looking at the sum of the two, the lowest losses are for small pyramids with a pyramid angle between 10° and 25°.

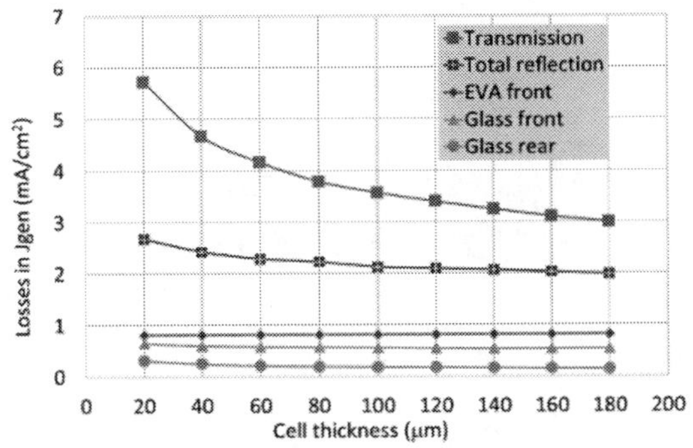

Figure 3: Losses at module level in generated photo-current for bifacial cells with a textured rear surface (front illumination only)

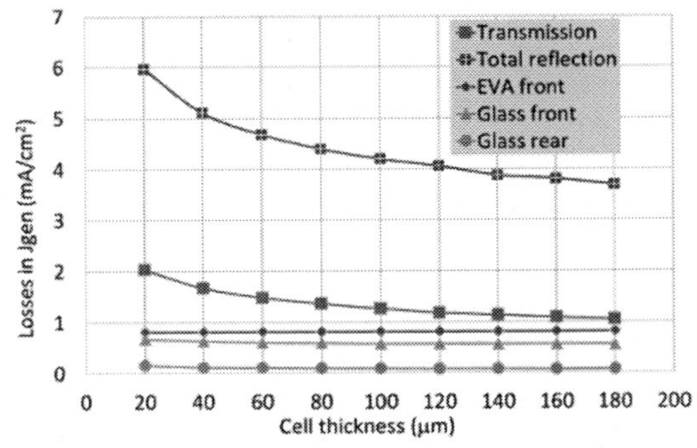

Figure 4: Losses at module level in generated photo-current for bifacial cells with a polished rear surface (front illumination only)

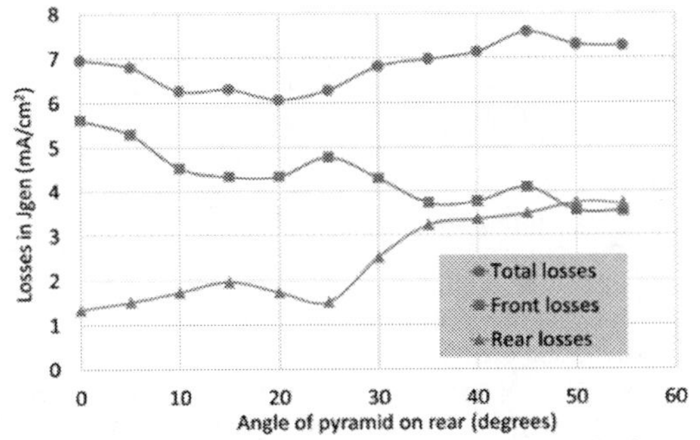

Figure 5: Losses at module level in generated photo-current for 100 μm thick bifacial cells (only 1 sun front illumination) with a variable rear surface morphology

A similar exercise can be performed for rear illumination onto these devices. Figure 6 displays the losses in photogenerated current for the same 100 μm bifacial cell as in Figure 5 but in this case for 1 sun perpendicular rear illumination. While this illumination level is not a likely scenario, it allows direct comparison with Figure 5. The higher total losses can be explained by 2 reasons: (i) the higher rear reflection losses for small pyramid angles (clearly a disadvantage for non-textured surfaces) and (ii) the absence of an ARC of the rear glass as mentioned in section 2.

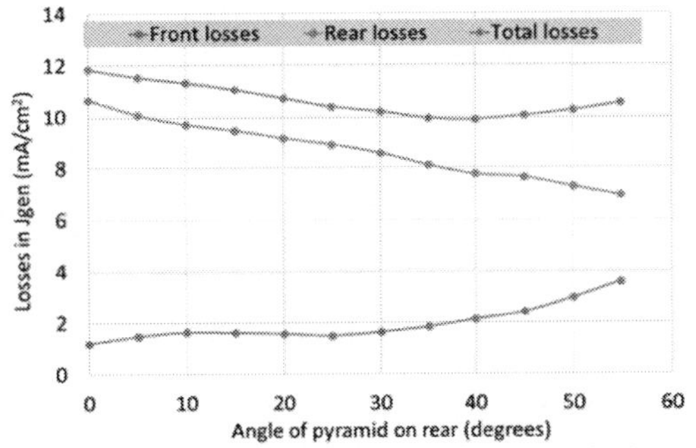

Figure 6: Losses at module level in generated photo-current for 100 μm thick bifacial cells (only 1 sun perpendicular rear illumination) with a variable rear surface morphology

When bringing both illumination sources together the final effect on bifacial cells can be visualized. The resulting Jsc is shown in Figure 7 for 2 illumination conditions:

- 1 sun perpendicular front + 0.2 sun perpendicular rear illumination
- 1 sun perpendicular front + 0.2 sun diffuse rear illumination

The best condition for light trapping occurs for small pyramid angles between 20° and 25° even with the addition of 0.2 sun rear illumination. The results for a textured rear surface are not as good as intuitively could be expected. The reason for this is that, although the reflection with rear illumination is low, the transmission of light through the cell out of the front surface is quite high.

Note that the polished and textured cases give practically the same result. Further simulations have shown an advantage for a polished rear surface as long as the rear light contribution is below 0.2 sun. A textured rear surface is more beneficial for a rear illumination above 0.2 sun.

In conclusion, these simulations predict optimal light trapping for bifacial cells with a rear surface made up of pyramids with angles of 20-25°. The next section will present the results of our experimental verification.

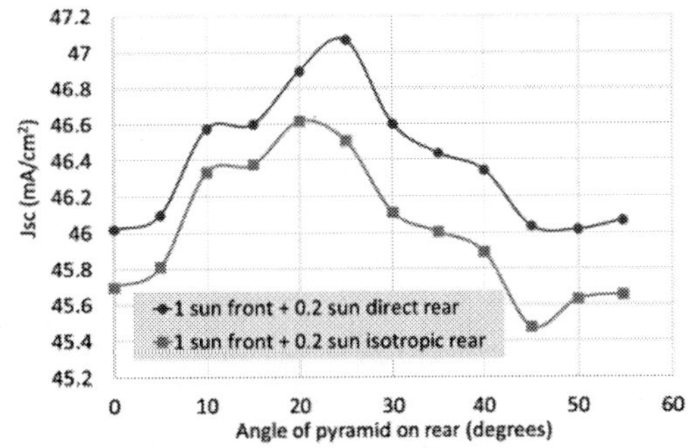

Figure 7: Simulated Jsc for bifacial cell in module with bifacial illumination (1 sun perpendicular illumination on front and 0.2 sun diffuse or perpendicular illumination on rear) as function of the cell's rear surface morphology.

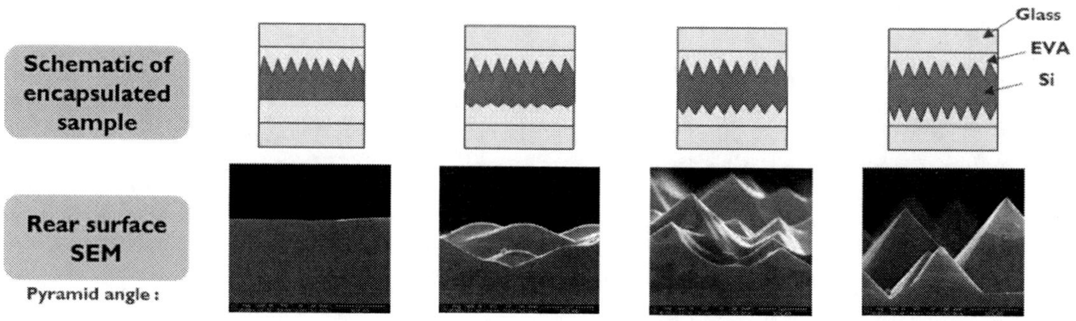

Jsc (mA/cm²) 1 sun front	39.1	39.4	38.9	38.8
Jsc (mA/cm²) 1 sun front + 0.2 sun rear	46.4	46.9	46.8	46.5

Figure 8: Jsc calculated from reflectance + transmission measurements with front-only (1 sun) or bifacial illumination (1 sun front + 0.2 sun rear) for encapsulated silicon samples with different rear surface morphologies. A schematic of the structure as well as a SEM picture of the cell's rear surface and the estimated pyramid angle (extracted from SEM pictures) are also shown.

4 EXPERIMENTAL VERIFICATION

To verify the simulated results, an experiment using encapsulated Si samples was set-up. The Cz-Si wafers (original as-cut thickness was 190 μm) were encapsulated using standard EVA in a glass/glass sandwich as schematically depicted in Figure 8. Standard wafers were used (no cells) to avoid complications due to parasitic absorption in highly doped areas (e.g. Free Carrier Absorption) and other optical effects due to the metallization. Four different sample types were encapsulated:

- A standard silicon wafer, alkaline textured on both sides of the cell. The pyramid angle in this case is ~54.7°.
- A silicon wafer originally textured on both sides but with slight polishing of the rear surface resulting in a pyramid angle of ~36°.
- A similar wafer with a standard polishing process on the rear surface resulting in pyramid angles of ~25°.
- A similar wafer with a stronger polishing process on the rear surface resulting in a macroscopically flat surface.

The polishing in question was performed in a Rena wetbench for inline polishing using a HF/HNO_3 etching solution. The different polishing conditions were achieved by changing the process speed of the wafers in the system (for the slight and standard polishing) or by increasing the number of passes though the tool (for the heavy polishing condition). SEM pictures of each of the different rear surface morphologies are also shown in Figure 8. The wafers received the same dielectrics as real bifacial cells, namely an Al_2O_3/SiN_x stack on the front surface (by respectively Atomic Layer Deposition (ALD) and Plasma Enhanced Chemical Vapor Deposition (PECVD)) and a PECVD SiO_x/SiN_x on the rear surface.

These encapsulated samples were characterized by reflectance and transmission measurements from both sides of the laminate. Based on this data, the resulting absorption in Si was calculated (using a laminate without a Si wafer as reference to determine the glass/EVA absorption). From this, the maximum achievable Jsc in the silicon wafer was determined.

The 2 bottom rows in Figure 8 show the calculated currents for the cases of a front-only illumination (1 sun) and for bifacial illumination (1 sun front + 0.2 sun rear). For the first case, the samples with a standard polishing of the rear surface (pyramid angle ~25°) clearly come out on top. This corresponds well to the simulations results shown in figure 5 where the lowest losses were found to be for pyramid angles in the range of 10-25°. The experimental characterization confirms that this is due to a lower total loss from reflection and transmission compared to the 2 extremes of either a textured or flat rear surface.

When adding a realistic amount of rear illumination (0.2 sun like for the simulations), the experimental data still shows the best Jsc and hence the best light trapping for the rear surface pyramids with an angle of 25°, closely followed by the ones of 36° (only slight polishing). The case of a textured rear surface comes closer but is still clearly worse compared to these 2 cases. This was also predicted by the simulation and can be attributed to too high transmission losses out of the front surface with rear illumination, notwithstanding the advantage this surface morphology has for minimizing rear reflection losses. Note also that the wafers with a textured rear surface are slightly thicker than the other wafers which received a polishing step. Even then the Jsc of those cells is better than for the thicker, textured cell.

Although this work focuses on the optical aspects, we can state that a semi-polished rear surface will also bring electrical advantages in the form of a lower recombination current J_0. This will translate into an improved open-circuit voltage of the bifacial cells.

Therefore, a semi-polished surface in the end can be advantageous both optically and electrically.

5 CONCLUSION

For bifacial cells, a textured rear surface seems at first sight the most logical choice to maximize light trapping with bifacial illumination since it can minimize reflection losses for rear illumination. However, transmission losses for both front and rear illumination are high compared to wafers with a semi-polished rear surface. Both ray tracing simulations at module level and experiments using glass/glass encapsulated Si samples have shown that a semi-polished rear surface (texturing followed by a polishing step resulting in pyramids with a base angle of 25°) lead to better light trapping results. In addition, there will also be an electrical advantage due to a lower rear recombination current.

6 ACKNOWLEDGEMENT

The authors would like to acknowledge the financial support of Imec`s Industrial Affiliation Program for PV (IIAP-PV).

7 REFERENCES

[1] R, Kopecek, I. Shoukry, J. Libal, Photovoltaics InternationalVol. 30, pp. 88-97, 2016
[2] International Technology Roadmap for Photovoltaics (ITRPV), 8th edition, 2017.
[3] M. Mrcarica, C. Panofen, P. Pasmans, A. Schneider, J. Rabanal-Arabach, Bifacial Workshop, Miyazaki, Japan, 2016
[4] www.pvlighthouse.com.au/module-ray-tracer
[5] PV Lighthouse website, refractive index library, low-Fe sodalime glass
[6] PV Lighthouse website, refractive index library, RI data supplied by Enki Technology
[7] H. Nagel, A.G. Aberle, R. Hezel, Progress in Photovoltaics: Research and Applications 7(4), pp. 245–260, 1999
[8] S. Duttagupta, F. Ma, B. Hoex, T. Mueller, A.G. aberle, Energy Procedia 15(0), pp. 78–83, 2012
[9] L. Abenante, IEEE transaction on electron devices, Vol 53, No 12, pp 3047-3053, 2006

A ONE-SUN SPECTRUM-SPLITTING MINIMODULE WITH PRISMATIC ENCAPSULATION: RAY-TRACE SIMULATION USING ZEMAX OPTICSTUDIO

Bruno Concha Ramon*, Mark J. Keevers, Martin A. Green
School of Photovoltaic and Renewable Energy Engineering, University of New South Wales, Sydney, Australia
Contact: b.concharamon@student.unsw.edu.au

ABSTRACT: A computer ray-trace model of a one-sun spectrum-splitting minimodule has been developed using Zemax OpticStudio. The minimodule consists of a fused silica triangular prism, a Si solar cell, a GaInP/GaInAs/Ge triple junction solar cell and a spectrum-splitting dielectric filter. In 2016, such a minimodule set a world record efficiency of 34.5% for the conversion of unconcentrated sunlight to electricity. Here we describe initial ray-trace simulations of the minimodule, assuming a perfect AR coating and idealised AOI-independent pillbox-shaped filter, for either AM1.5 direct or global normal illumination. Our investigation of the AOI response of the minimodule efficiency reveals acceptance angles of 180° and 114° for the TJ-first and Si-first cases, respectively.
Keywords: Encapsulation, Energy Performance, Modelling, Optical Properties, Optical Losses, Performance, Ray Tracing, Simulation

1 INTRODUCTION

Two techniques to improve the performance of solar cells are to concentrate the solar flux and/or to split the solar spectrum into spectral bands. Concentrating the solar flux improves the conversion efficiency while at the same time reducing the required solar cell area [1]. Separating sunlight into different wavelength bands and directing each band to the most efficient convertor mitigates the mismatch between spectrum and bandgap [2]. At UNSW, we are researching spectrum splitting between a GaInP/GaInAs/Ge triple junction (TJ) cell and a Si cell, leading to world record efficiencies [3-5].

The one-sun spectrum-splitting minimodule splits the sunlight using a dielectric band-reflect filter and traps the reflected component inside prismatic encapsulation by total internal reflection (TIR), Fig. 1. The smaller-area cell receives selected wavelengths diverted from the larger-area cell by a dielectric filter, with a small concentration of $\sqrt{3} = 1.73$ for the 30° prism considered.

Ray-trace simulations enable investigation of optical losses, alternative configurations and geometries, sensitivity to angle of incidence (AOI) and spectrum, energy yield, etc. Here we present results of our first ray-trace simulations, which assume AM1.5D or AM1.5G normal illumination, and a minimodule with a perfect AR coating and idealised AOI-independent 'pillbox' filter.

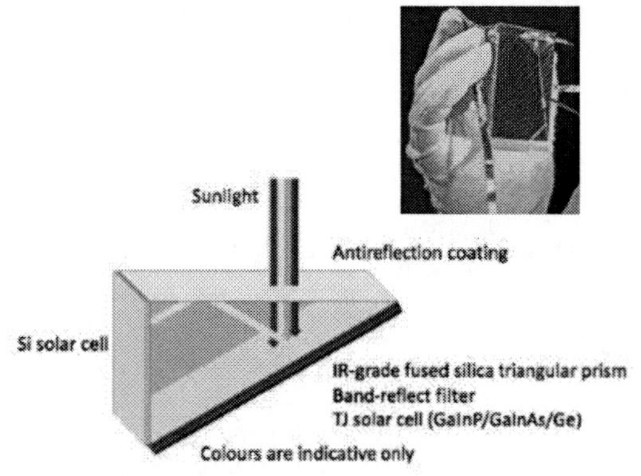

Figure 1: Spectrum-splitting minimodule: schematic and a photo of a fabricated prototype.

2 METHOD

The optical model of the minimodule was developed in the ray-tracing software Zemax OpticStudio [6]. It consists of the following elements:

Light sources: Direct irradiance is modeled by a 'radial source', and diffuse irradiance by a hemispherical 'object source', Fig. 2(a). The hemisphere is rendered as a collection of triangles where each triangle becomes a source emitter, with a cosine angular distribution [7]. With some fine tuning on the size of each triangle, the emitted cosine angular distribution and the number of rays, we were able to generate a uniform flux distribution over the top face of the minimodule, Fig. 2(b). When both sources are active, the minimodule is under the AM1.5G, 1 kW/m^2 standard spectrum. When performing direct irradiance simulations, we turn off the hemisphere source and set the direct source at AM1.5D, 1 kW/m^2.

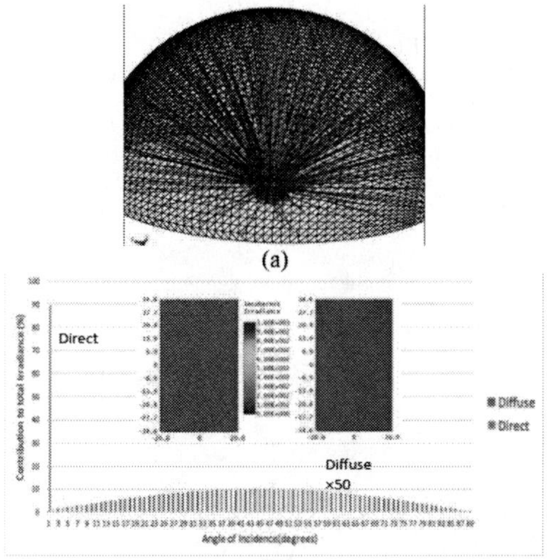

Figure 2: (a) Zemax OpticStudio model of the direct and diffuse light sources (green and blue rays, resp.), and (b) corresponding AOI and flux distribution at the minimodule aperture face (the diffuse contribution is exaggerated ×50 for clarity).

Prism: This is a right-angle triangular prism with apex angle (α) of 30°, Fig. 3. The top (entrance) face is 27.7 cm^2, hypotenuse face 32 cm^2, and side face 16 cm^2. We assume a wavelength dependent refractive index, with Sellmeier dispersion formula [7].

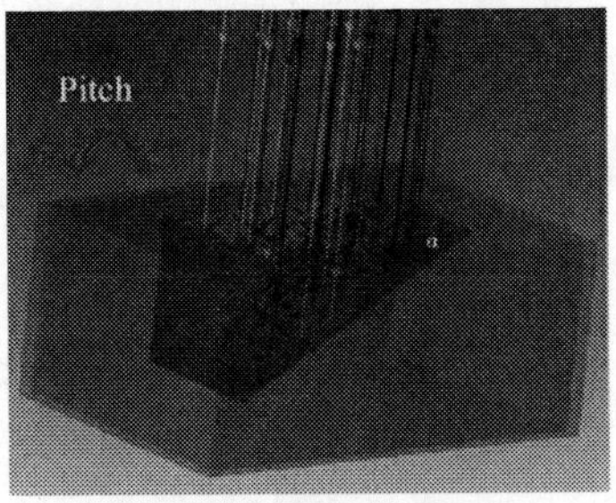

Figure 3: Zemax OpticStudio model of the minimodule, also showing the light-tight aperture mask box (made semi-transparent in this figure) and our definition of 'pitch angle' for varying AOI in the present work.

Aperture mask box: The minimodule is situated in a box that blocks (absorbs) light from all directions except the top face, Fig. 3.

Coatings: The top face has an ideal anti-reflection (AR) coating, and the hypotenuse face has a spectrum splitting filter between it and the adjacent solar cell. We assume the filter has an ideal AOI-independent pillbox wavelength response. For example, for the TJ-first case (meaning that the TJ cell is on the hypotenuse face) and the AM1.5D spectrum, the filter is a band-reflect (BR) filter with 100% reflectance for 890-1025 nm, and 100% transmittance at other wavelengths, Fig. 4. The side faces of the prism do not have any coating.

Solar cells: The Si and TJ cells are represented as perfectly absorbing flux detectors in Zemax OpticStudio, thus ignoring any reflected rays from the prism/encapsulant interface or the cell itself. We also ignore any effects of nonuniform irradiance on cell performance, i.e. each cell is considered a single-pixel detector.

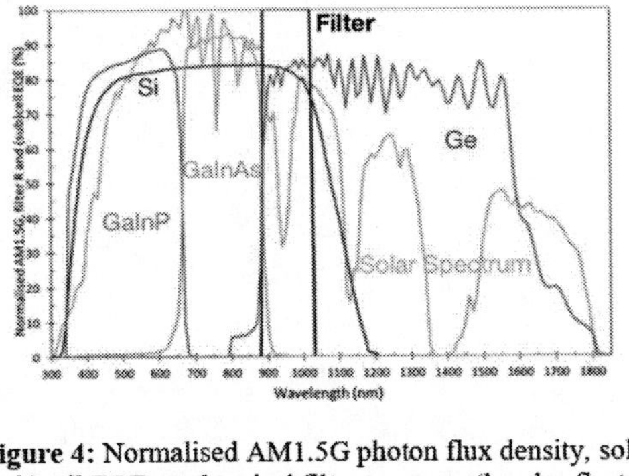

Figure 4: Normalised AM1.5G photon flux density, solar (sub)cell EQEs and typical filter response (band-reflect filter used in TJ-first case).

Zemax OpticStudio calculates the total flux (in W) and flux versus wavelength at each solar cell 'detector', enabling calculation of the spectral irradiance at each cell plane, Fig. 5. This is used as input to Matlab calculations to determine the solar cell light current-voltage (LIV) curves and hence solar cell and minimodule (four-terminal) efficiency. Integration of EQE and cell spectral irradiance is used to calculate current density for each (sub)cell. A simple 1-diode model of each (sub)cell is assumed. The input EQE and baseline LIV parameters are provided by the solar cell manufacturers: Azur Space for the TJ cell [8] and Trina Solar for the Si IBC cell [9] (baseline one-sun AM1.5G efficiencies are 36.0% and 21.5%, resp.). The solar spectrum, EQEs and a typical pillbox filter wavelength response are shown in Fig. 4.

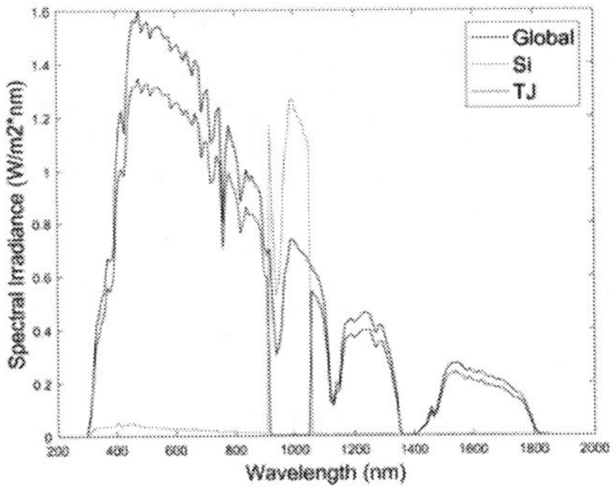

Figure 5: Example for TJ-first case of spectral irradiances determined by Zemax OpticStudio: input AM1.5G spectrum (black), TJ solar cell (blue) and Si solar cell (green). Note that the 30° prism geometry provides concentration factors of √3/2 = 0.87 and √3 = 1.73 for the TJ and Si cell irradiances, respectively.

3 SIMULATION RESULTS AND DISCUSSION

What is the optimum spectrum-splitting filter? The pillbox filter cut-on and cut-off wavelengths were optimised for TJ-first minimodule (four-terminal) efficiency, for both AM1.5 direct and global normal irradiance, resulting in an 890-1025 nm filter for direct and 895-1015 nm for global. The resulting (sub)cell current densities, cell and minimodule efficiencies are presented in Table I for the optimum case and two near-optimum filters. The optimum filter is one that diverts just enough IR photons from the TJ bottom Ge subcell to the Si cell, without detriment to the current-limiting TJ middle GaInAs subcell, Fig. 4 and Table I.

Table I: Calculated (sub)cell current densities, cell and minimodule (four-terminal) efficiencies, for the TJ-first case and either AM1.5 direct or global normal illumination. The optimum filter is topmost in each table.

Direct normal illumination

		Top	Mid	Bot	TJ	Si	Total
BR filter	Cell Eff (%)				34.7	3.4	38.1
890-1025nm	JL(mA/cm²)	12.9	13.0	14.0	12.9	11.5	
BR filter	Cell Eff (%)				35.1	2.5	37.6
915-1020nm	JL(mA/cm²)	12.9	13.4	14.1	12.9	8.6	
BR filter	Cell Eff (%)				35.2	1.8	37.0
935-1015nm	JL(mA/cm²)	12.9	13.4	15.0	12.9	6.1	

Global normal illumination

		Top	Mid	Bot	TJ	Si	Total
BR filter	Cell Eff (%)				34.7	2.8	37.5
895-1015nm	JL(mA/cm²)	13.3	12.9	13.5	12.9	9.5	
BR filter	Cell Eff (%)				34.9	2.3	37.2
920-1015nm	JL(mA/cm²)	13.3	13.1	13.6	13.1	7.8	
BR filter	Cell Eff (%)				34.4	2.6	37.0
925-1030nm	JL(mA/cm²)	13.3	13.1	12.9	12.9	8.7	

The efficiency benefit of introducing spectrum-splitting prismatic encapsulation for the various cases considered: TJ-first and Si-first, AM1.5 direct and global normal illumination, are summarised in Table II. Here we assume that the optimum filter is the same for the Si-first case as determined for the TJ-first case (we will revisit this assumption in future work).

Table II: Calculated cell and minimodule efficiencies for TJ-first and Si-first configurations, AM1.5 direct and global normal illumination.

TJ-first with band-reflect filter

Illumination	Case	TJ	Si	Total
Direct	Baseline (TJ flat plate)	35.7	-	35.7
	Spectrum splitting prism	34.7	3.4	38.1
Global	Baseline (TJ flat plate)	36.0	-	36.0
	Spectrum splitting prism	34.7	2.8	37.5

Si-first with bandpass filter

Illumination	Case	TJ	Si	Total
Direct	Baseline (Si flat plate)	-	21.5	21.5
	Spectrum splitting prism	34.7	3.3	38.0
Global	Baseline (Si flat plate)	-	21.5	21.5
	Spectrum splitting prism	33.6	2.4	36

The TJ-first case shows a significant benefit of prismatic encapsulation spectrum-splitting to a Si cell compared to a flat-plate TJ baseline, at least for our idealised filter model. Not unexpectedly, the benefit is greater for direct illumination (no diffuse component), where the efficiency is increased from 35.7% (baseline TJ flat plate) to 38.1%, a relative improvement of almost 7%. For global illumination the relative improvement is a more modest 4%.

Of course, for the Si-first case, the benefit of introducing prismatic encapsulation spectrum-splitting to a TJ cell is massive compared to a Si flat plate baseline. Supplementing a Si module with a high efficiency TJ or simpler solar cell (e.g. GaAs) via spectrum-splitting prismatic encapsulation may be a viable long term approach to improving Si module efficiency, compared to monolithic or mechanical tandem stacks which suffer from epitaxial growth and/or parasitic absorption issues, respectively.

Next we investigate the AOI response of minimodule performance for rotations about an axis bisecting the long edges of the top (entrance) face of the minimodule. In other words, we vary the 'pitch angle' from -90° (apex angle edge rotated downwards away from light source) to +90° (apex angle edge rotated upwards towards light source), Fig. 3. This is the single axis most relevant to TIR occurring (or not) inside the prism. We consider two performance metrics: optical efficiency, defined as the ratio of flux at each solar cell relative to the flux incident on the minimodule top face; and overall PV efficiency of the solar cells and minimodule. The simulation results for AM1.5 direct illumination are shown in Figures 6 and 7.

In the TJ-first case, Fig. 6, the total = TJ + Si optical efficiency exhibits the expected cosine response until the pitch angle becomes sufficiently negative (-24°) that TIR ceases at the internal top face for rays reflected from the 890-1025 nm band-reflect filter and otherwise destined for the Si cell (see insets in Fig. 6). In the absence of TIR, these IR rays 'leak' out of the top face of the prism. Owing to the fact that in the TJ-first case we use a narrow band-reflect filter, this leakage loss effects only a small portion of the solar spectrum. Hence, while TIR and therefore spectrum splitting switch off at such pitch angles, this has only a relatively small effect on total optical efficiency. The pitch angle response of the overall minimodule PV efficiency is largely a result of this TIR effect, compare Fig. 6 top and bottom plots. Spectrum-splitting prismatic encapsulation thus provides an efficiency boost at small pitch angles and a (pitch) acceptance angle of essentially 180°.

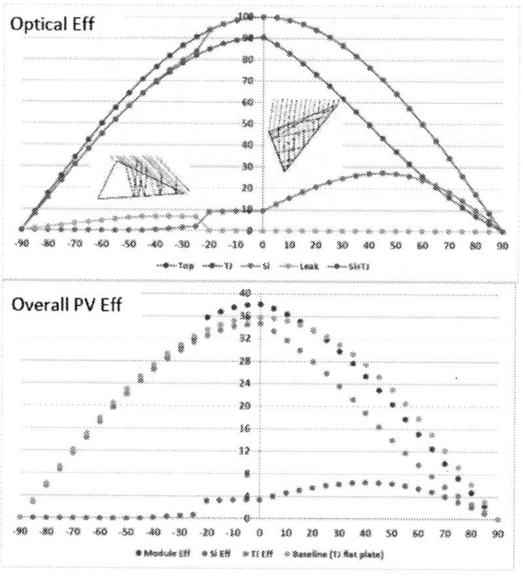

Figure 6: TJ-first direct irradiance pitch angle response of optical efficiency (top) and overall PV efficiency (bottom). Insets are ray diagrams showing the absence (left) or presence (right) of TIR at the top face.

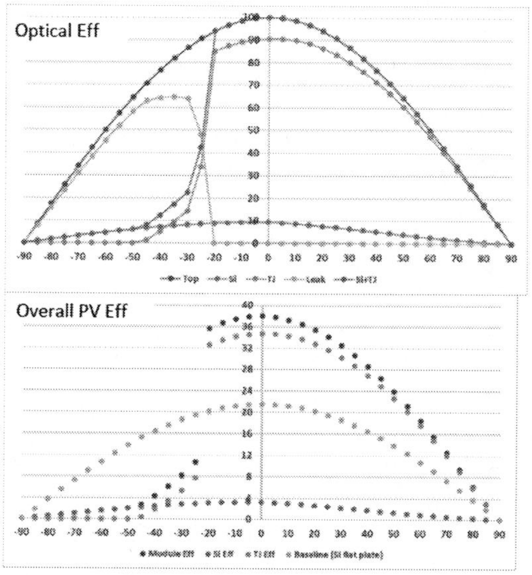

Figure 7: Si-first direct irradiance pitch angle response of optical efficiency (top) and overall PV efficiency (bottom).

A minimodule with swapped cells (Si-first) and a (895-1015 nm) bandpass filter exhibits the same TIR physics but its impact on both optical and overall PV efficiency is now dramatic, Fig. 7. This is because most of the sunlight is reflected by the bandpass filter and most of the output of the minimodule is from the TJ cell, which is lost when TIR and therefore spectrum splitting ceases. Nevertheless, even in this case, the minimodule acceptance angle is considerable at 114° (-24° to +90°), and importantly, within this acceptance angle the minimodule efficiency is greatly improved by the addition of prismatic encapsulation spectrum splitting to a TJ cell. This acceptance angle is the same as that for a 30° apex angle prismatic concentrator [10], as expected. Given the limited acceptance angle of the Si-first approach, its potential increases when any kind of solar tracking is utilised, such as single-axis tracking which is increasingly being used for utility scale PV systems.

When the minimodule is simulated under global illumination (using the combined light source model of Fig. 2, AM1.5G, 1 kW/m^2), the efficiencies are slightly lower, Tables I and II, due to the (approx. 10%) diffuse component of the spectrum. The pitch angle response, however, is similar to that shown in Figures 6 and 7. Further investigation of the diffuse light behaviour is under way.

4 CONCLUSIONS AND FUTURE WORK

A ray-trace model of a spectrum-splitting prismatic minimodule has been established and initial simulation results are presented for a simple case, namely: standard spectra, perfect AR coating, pillbox AOI-independent filter, and neglect of flux nonuniformity. We consider both the TJ-first case (the same configuration as the fabricated minimodule) and the Si-first case, and the AOI response of optical and overall PV efficiency in each case.

The simulations show that the best minimodule performance is achieved in the TJ-first case under direct irradiance. Not only does the TJ-first case show the benefit of spectrum splitting to a Si cell, it has an excellent AOI response, exhibiting a full 180° acceptance angle. The Si-first case shows the massive benefit of spectrum splitting to a TJ cell (assuming the Si cell as baseline), but with a reduced AOI response: the 114° acceptance angle is consistent with that of a prismatic concentrator for a prism with 30° apex angle [10]. These findings can be generalised to other high-efficiency (high-cost) / low-efficiency (low-cost) cell combinations (e.g. Si/perovskite tandems).

With the Zemax OpticStudio optical model and Matlab efficiency calculator established, we will investigate the impact of more realistic assumptions (e.g. AR coating, AOI-dependent filter), alternative prism geometries (e.g. different apex angles, non-right angle triangular prisms, etc.), two-dimensional AOI variations (i.e. pitch and roll), as well as spectrum variations. We will also calculate annualised energy yields. In addition, work is underway to measure the outdoor LIV angle response of the fabricated minimodule, which will help to validate the simulation with experimental results.

5 ACKNOWLEDGEMENTS

This work is funded by the Australian Renewable Energy Agency (ARENA). We are grateful for the supply of solar cells, and corresponding EQE and LIV data, from Azur Space (TJ cells) and Trina Solar (Si IBC cells).

6 REFERENCES

[1] W. Blocker, "High-efficiency solar energy conversion through flux concentration and spectrum splitting", *Proc. IEEE*, vol. 66, no. 1, pp. 104–105, 1978.

[2] A. Mojiri, R. Taylor, and E. Thomsen, "Spectral beam splitting for efficient conversion of solar energy – A review", *Renewable and Sustainable Energy Reviews*, 2013.

[3] M. A. Green, M. J. Keevers, B. Concha Ramon, Y. Jiang, I. Thomas, J. B. Lasich, P. J. Verlinden, Y. Yang, X. Zhang, K. Emery, T. Moriarty, R. R. King, and W. Bensch, "Improvements in sunlight to electricity conversion efficiency: Above 40% for direct sunlight and over 30% for global", 31st EUPVSEC, 2015, pp. 1–5.

[4] https://newsroom.unsw.edu.au/news/science-technology/unsw-researchers-set-world-record-solar-energy-efficiency (accessed 25 January 2015).

[5]http://www.sbs.com.au/topics/science/fundamentals/article/2016/05/19/world-record-solar-cell-efficiency-set-aussie-scientists (accessed 19 May 216).

[6] http://www.zemax.com (accessed 20 September 2017).

[7] Zemax LLC, 2017, OpticStudio 16.5 SP5 Help files.

[8] A.W. Bett, et.al., "Highest efficiency multi-junction solar cell for terrestrial and space applications", 24th EUPVSEC, Hamburg, 2009.

[9] X. Zhang, Y. Yang, W. Liu, K. Zhang, and Y. Chen, "Development of high efficiency interdigitated back contact silicon solar cells and modules with industrial processing technologies", *6th World Conference on Photovoltaic Energy Conversion*, 2014.

[10] D. R. Mills and J. E. Giutronich, "Ideal prism solar concentrators", *Solar Energy*, vol. 21, pp. 423–430, Jan. 1978.

RAPID TESTING OF SPECTRALLY RESOLVED OPTICAL CELL QUALITY USING LED SOLAR SIMULATORS

Kai Sporleder*, Christian Hagendorf, Marko Turek
Fraunhofer Center for Silicon Photovoltaics CSP, Otto-Eißfeld-Straße 12, 06120 Halle (Saale), Germany
*Corresponding author: Tel. +49 (0) 345/5589-5070, kai.sporleder@csp.fraunhofer.de

ABSTRACT: The efficiency of solar cells depends on their electrical and optical properties. The standard in-line electrical characterization yields integral information on key cell parameters. However, since neither spectrally nor spatially resolved quantities are obtained by a standard I-V-measurement, little information is gained about individual cell production process steps. To overcome these limitations, we present an approach for a spatially resolved measurement of the reflectivity of a solar cell which is fully in-line applicable. Based on a LED solar-simulator setup, measurement times of less than 500 milliseconds are achieved. Our method extends existing advanced measurements approaches using LED solar simulators, such as rapid-EQE measurement procedures. As a result, the lateral homogeneity of the anti-reflective coating can be quantified. This optical information together with quantum efficiency data from a rapid-EQE-test could eventually also be used to extract the internal quantum efficiency and apply more advanced loss analysis algorithms.
Keywords: optical inspection, LED solar simulator, solar cell reflectivity

1 INTRODUCTION AND MOTIVATION

LED based solar simulators are powerful cell characterization tools both in production and research environment. Compared to Xenon based solar simulators, they provide high flexibility in illumination spectrum and intensity allowing a cell characterization beyond current-voltage-characteristics at standard test conditions, e.g. testing of external quantum efficiency [1], current and illumination dependent series resistance measurements [2], or the prediction of cell to module losses [3].

Besides the electronic properties of the semiconductor and the cell contacts, the efficiency of a silicon solar cell is determined by its optical properties. Therefore, an anti-reflective coating (ARC) is applied to the front side which minimizes the fraction of reflected light from the cell surface. To this end, the optical properties of the anti-reflective coating, such as its thickness and refractive index are crucial for an optimized cell performance. Furthermore, it is required that the coating process leads to a spatially homogeneous surface coating such that the ARC-layer is equally thick at all positions of the solar cell.

Several approaches to obtain optical data on silicon solar cells have been compared in reference [4]. In our work, two in-line capable set-ups are presented which allow for a reliable optical characterization in significantly less than 1 sec measurement time [5]. To this end, an earlier approach for the rapid testing of the external quantum efficiency using LED solar-simulators [1] is modified by adding a detector measuring the reflected light. In one case, a large-area detector is applied which yields integral information of a larger region of the solar cell without any spatial resolution. In a second case, a conventional silicon-based luminescence camera is adapted such that it measures the optical back-reflection of the illuminated solar cell instead of the luminescence image.

The widely used standard approach using filter wheels or grating mono-chromators are limited by the long measurement times of several seconds up to minutes. In particular, the measurement of large areas, such as complete solar cells or even modules, is very time consuming with conventional optical setups. While our approach yields a reduced spectral information due to the limited number of LED-channels, it is very fast and thus fully in-line capable. Furthermore, the measurement of the reflectivity data can be combined in a single solar simulation tool without any need for major hardware upgrades. Hence, our approach is best suited as an extension to existing in-line tools which makes it relevant for a large range of industrial applications.

2 EXPERIMENTAL SETUP

2.1 Measurement approach

The measurement setup is schematically presented in Figure 1. It consists of a LED-based solar simulator light source (labeled by "2"). It provides a lateral homogeneous light field for up to 21 different wavelengths. Each of these 21 LED-channels can be operated independently of the others allowing for a high flexibility of the spectral properties of the illumination. The switching time between the different channels is in the millisecond range.

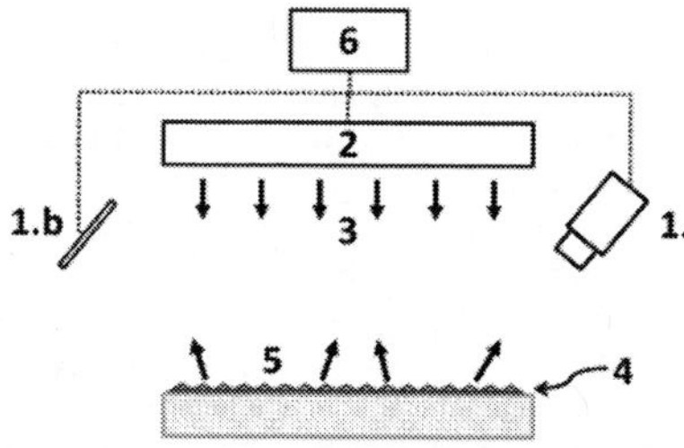

Figure 1: Experimental setup consisting of the LED solar simulator light source (2) emitting mono-chromatic light (3). The sample (4) back- reflects some of the light (5) which is measured by a detector (1). Communication between light source and detector is established over PC (6).

Typical examples of the spectrum of individual channels are shown in Figure 2. It can be observed, that in contrast to a mono-chromator setup, the spectral broadening of the LEDs is much more pronounced. Typical peaks widths are between 20 nm and 100 nm.

This spectral broadening has to be accounted for by a subsequent data analysis in order to achieve reasonable measurement accuracy.

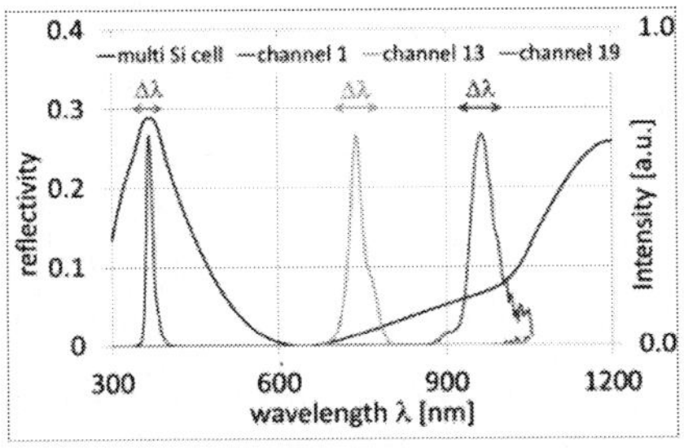

Figure 2: Examples of the spectral broadening of the individual LED channels: blue – low wavelengths, green – intermediate wavelength, red – high wavelength. For comparison, a typical reflection curve of a silicon solar cell is shown (black).

The sample is positioned on the standard I-V-measurement chuck. In case of silicon solar cells, the back-reflected light from the cell surface is highly diffusive. It is measured by a detector leading to a quantified reflection signal. The detector has to be synchronized with the light source. It should provide a high enough sensitivity in order to be capable of measuring the reflected light with a sufficient signal to noise ratio in short times. However, since solar simulator light sources provide a rather high light intensity, this does not impose any serious constriction on the detector.

The detector can either be a camera which then yields a spatially resolved image of the optical properties of the sample or a large area photo-diode or solar cell yielding integral information on the amount of reflected light. In our experiments, the large area detector was realized by positioning a 2 cm x 15.6 cm Si solar cell up-side down in a 45° angle next to the sample. It has been positioned such that the detector's surface is not illuminated by the incident light directly but points to the sample surface.

2.2 Measurement routine

The first step in our measurement approach is the measurement of a reference sample which serves as a calibration of the system. The optical data of the reference sample as well as the test samples has been measured using the LOANA loss analysis system by pv tools. Thus, the reflectivity of each sample is known in a range between 300 nm and 1100 nm in 20 nm steps. Then, the reference cell is measured on the LED solar simulator using our rapid-reflectivity-test approach by subsequently switching the individual channels on and off. In case of the large area detector, its short circuit current is determined to quantify the amount of reflected light. In case of the camera setup, the pixel counts for each of the 21 channels are stored and used as a reference value.

In a second step, the test samples are measured with exactly the same illumination conditions as the reference cell. In order to check the stability of the light source, the data of the internal spectrometer is used. Finally, the measured detector current or pixel counts of the camera are compared to the values of the reference sample yielding quantitative information on the reflectivity for each individual channel. Thus, the spectrally (and spatially) resolved optical properties of the sample are obtained for 21 different wavelengths.

2.3 Data analysis

For a N-channel measurement setup, the data analysis that is applied to all measured data should yield N pairs of wavelength and reflectivity $\{\lambda_{n=1..N}, R_{cell,n=1..N}\}$ which represent the spectral dependence of the reflectivity. In order to find such pairs on a mathematically sound basis, one has to figure out which properties of the broadened illumination peaks is the relevant one, e.g. the position of the peak maximum, the mean wavelength, the medium wavelength or some other quantity. To this end, we consider the measured current $i_{det,n}$ of a detector which depends on the spectral flux density $I_n(\lambda)$ of the n-th channel of the light source, the detector's quantum efficiency $EQE_{det}(\lambda)$ and the spectral dependent reflectivity $R_{cell}(\lambda)$ of the surface of the test sample:

$$\frac{hc}{q} i_{det,n} = \int d\lambda \cdot I_n(\lambda) \cdot \lambda \cdot R_{cell}(\lambda) \cdot EQE_{det}(\lambda) \quad (1)$$

Defining the moments of any wavelength dependent function $f(\lambda)$ for each of the N channels as

$$\langle f(\lambda) \rangle_n \equiv \int d\lambda \cdot I_n(\lambda) \cdot f(\lambda)\,, \quad (2)$$

one can rewrite the measured detector current (1) as

$$\frac{hc}{q} i_{det,n} = \langle \lambda \cdot R_{cell}(\lambda) \cdot EQE_{det}(\lambda) \rangle_n\,. \quad (3)$$

The averaging $<..>_n$ over the broadened peaks $I_n(\lambda)$ does not allow to extract the reflectivity $\{\lambda_{n=1..N}, R_{cell,n=1..N}\}$ directly from this expression (3). Thus, one needs to find two parameters λ' and λ'' such that

$$\frac{hc}{q} i_{det,n} = \lambda' \cdot R_{cell}(\lambda'') \cdot EQE_{det}(\lambda'') \quad (4)$$

is valid as the reflectivity could then be obtained by a simple division. By comparison of Eq. (3) and (4) one finds

$$\lambda' = \langle \lambda \rangle_n \quad \text{and} \quad \lambda'' = \frac{\langle \lambda^2 \rangle_n}{\langle \lambda \rangle_n}\,. \quad (5)$$

Therefore, the correct wavelength λ_n for each spectrally broadened illumination $I_n(\lambda)$ is given by the first and second moment $\lambda_n = <\lambda^2>_n / <\lambda>_n$ of the n-th channel obtained by Eq. (2). The quantum efficiency of the detector $EQE_{det}(\lambda)$ is implicitly determined by performing the reference measurement. This wavelength together with the measured current of the detector yields N pairs $\{\lambda_{n=1..N}, R_{cell,n=1..N}\}$ which represent the spectral dependent reflectivity. Note, that our approach reproduces the results obtain earlier for the rapid EQE testing [1] by assuming a constant reflectivity of $R_{cell}=1$.

3 RESULTS

3.1 Camera based reflection measurement

First, the accuracy of the camera based reflection measurement method was evaluated using a single cell

"1", see Figure 3 (left). In this test, the cell has been used both as reference and sample. The results are represented by the black circles in Figure 4 showing a high degree of coincidence with the reference measurement.

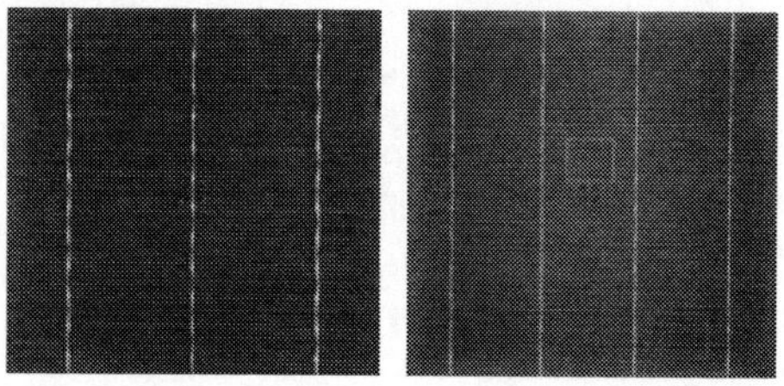

Figure 3: Photographs of multi-crystalline solar cell with homogeneous anti-reflective coating (cell "1", left), and multi-crystalline solar cell with inhomogeneous coating (cell "2": region "blue" and region "green" indicated, right)

In a second step, a lateral inhomogeneous cell "2" has been tested as the measured sample, see Figure 3 (right). Already from a pure optical inspection, it can be observed that the color in the middle of the cell is different from the edge regions. As can be seen in Figure 3, the center region appears green due to thicker silicon nitride layer, while border region of the cell shows higher reflection of blue light. The resulting spectral reflectivity obtained by our rapid testing using the LED solar simulator is shown in Figure 4 by the blue and green circles, respectively. Two major observations can be made from this result: First, the spectral reflectivity of cell "2" can clearly be distinguished from cell "1". In particular, the short wavelength reflectivity is reduced and also the position of the reflectivity minimum at mid-wavelengths is different. Second, the green center region and the blue edge region can be furthermore distinguished from each other. As the blue and the green area are measured within the same data acquisition steps, the spatial inhomogeneity of the anti-reflective coating can be quantified. The differences in the wavelength range from 300 nm to 600 nm are detectable using the LED solar simulator approach while the smaller differences at large wavelengths are reproduced with some uncertainties only.

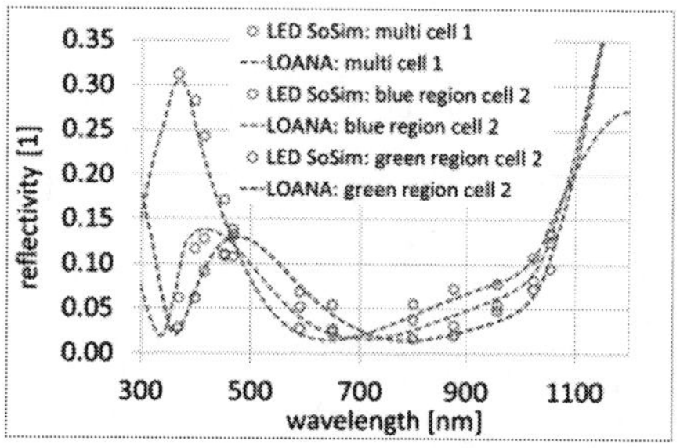

Figure 4: Camera based spectral reflectivity measurement. The lines represent reference values measured with the LOANA system. The circles are obtained by the rapid reflectivity test using the LED solar simulator. The reflectivity measurements of cell "2" were performed at the "green" and "blue" region indicated in Figure 3.

3.2 Detector based reflection measurement

In our second set up, the global cell reflectivity averaged over a larger cell region is measured. In contrast to the previous camera-based setup, a single large-area detector has been used to determine the channel-wise reflectivity. Figure 5 shows the reflectivity of a mono-Si solar cell. The circles represent the data obtained by the rapid reflectivity testing using the LED solar simulator with a multi-Si solar cell used as calibration. It can be observed, that the rapid testing yields reflectivity data in very good agreement with the reference data. Although the detector based method does not give any spatial information it allows for a much faster measurement in less than 100 msec.

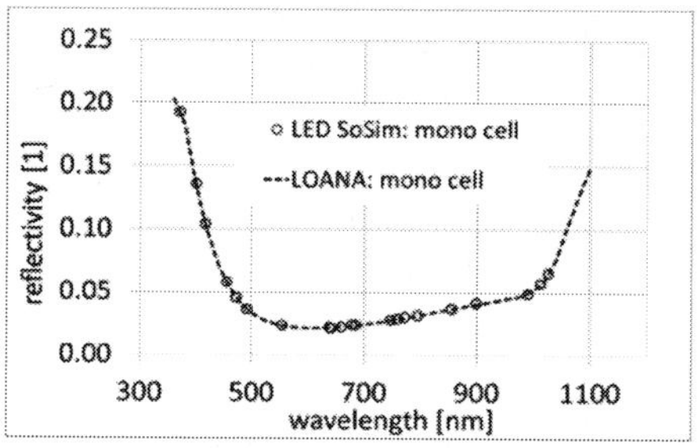

Figure 5: Detector based spectral reflection measurement. Reference values are measured with LOANA system. Circles are measured using LED solar simulator.

4 CONLUSIONS

We have presented a new measurement approach using a LED solar simulator that allows for a rapid determination of spectrally resolved reflectivity of solar cells. With measurement times of 100 milliseconds to 500 milliseconds, our method can be applied in-line to yield quantitative data on optical properties based on a single LED solar simulator as a measurement device. Production relevant quality issues such as the thickness and homogeneity of the passivation layer can rapidly be controlled. Together with a rapid LED solar simulator based EQE-testing, our approach can easily be extended towards a measurement of the internal quantum efficiency. Based on the information of optical and electrical properties, an advanced automated loss analysis can be implemented into a production line to uncover quality issues and provide the opportunity for a more specific cell sorting going beyond the power classification.

REFERENCES

[1] T. Luka et al: "Rapid testing of external quantum efficiency using LED solar simulators"; Energy Procedia 77, pp. 113-118 (2015)

[2] M. Turek; "Current and illumination dependent series resistance of solar cells"; J. Appl. Phys. 115, 144503 (2014)

[3] T. Brammer et al "Modified AM1.5 spectra for inline measurement of cell-to-module effects by using an LED solar simulator"; SNEC 9th, April 27-30th, Shanghai (2015)

[4] A. Krieg at al: "Anti-reflection-coating thickness measurements on textured silicon surfaces: Evaluation and accuracy of different measurement techniques"; 28th EUPVSEC, Sept 30th – Oct 4th 2013, Paris, France
[5] Patent pending: European patent office EP17192864.1

RAPID OPTICAL MODELLING OF PLASMA TEXTURED SILICON

David Payne[1], Malcolm Abbott[2], Alex Claville Lopez[1], Yiyu Zeng[1], Tsun Hang Fung[1], Keith McIntosh[2], Jose Cruz-Campa[3], Rasmus Davidson[4], Maksym Plakhotnyuk[4] and Darren Bagnall[5]
[1]School of Photovoltaic and Renewable Energy Engineering, UNSW Sydney, NSW 2052, AUSTRALIA
[2]PV Lighthouse, Coledale, NSW 2515, AUSTRALIA
[3]1366 Technologies, Bedford, MA 01730, USA
[4]Department of Micro- and Nanotechnology, Technical University of Denmark (DTU), Denmark
[5]Macquarie University, NSW, 2109, Australia

Correspondence: David Payne d.n.payne@unsw.edu.au Ph: +61 2 9385 0165

ABSTRACT: Plasma based texturing is a well-established technique for the fabrication of black silicon and is of broad interest since it can be applied to any type of substrate to achieve excellent optical properties. In many cases these surfaces consist of a combination of nano and micro-scale features, providing both anti-reflection and scattering effects. This type of texturing cannot easily be simulated using the more common approach based on the transfer matrix method combined with raytracing, which has been used to optimize cells, modules and systems with more traditional textures and coatings. Other techniques tend to be slow, computationally expensive, require pre-solving and/or are limited to one dimension. Here we test the optical validity regimes for a broad range of plasma textured black silicon samples and present progress on the development of modelling approaches which are rapid and compatible with three dimensional ray tracing. The work presents the optical characterization and simulation of a variety of plasma textured surfaces from both industrial and academic sources. Modeling approaches based on the effective medium and graded index approaches are combined with more traditional techniques and assessed based on their impact on simulated device performance.

Keywords: Optical properties, Texturing, Light trapping, Simulation, Black silicon, RIE

1 INTRODUCTION

Black silicon has long been studied for its unique optical properties, but recent advances in fabrication techniques and surface passivation have led to a dramatic increase in interest. In the wafering industry there is also a growing need to decrease waste and lower costs, leading to a trend towards diamond wire sawing and kerfless processes. In many cases, traditional chemical texturing approaches which rely on crystallographic orientation or surface damage are ineffective for such wafers, thus further increasing the need for more universally applicable texturing. This growing popularity is driving demand for a fast optical modelling tool that can rapidly and adequately simulate typical black-silicon surfaces, their interaction with AR coating and performance within modules. Development of such a tool is non-trivial as black silicon texturing can consist of nanoscale features that produce both anti-reflection and scattering effects, meaning that conventional fast modelling approaches such as those which rely on standard geometric optics alone are not sufficient. While this problem is relatively new in the crystalline silicon world focused on standard solar cell architectures, it has long been faced by the thin-film community and by groups developing specialized nano-optics or working with tandem structures. This has driven the successful development of solvers that can simulate the scattering parameters of rough interfaces between arbitrary materials [1]. Recently these techniques were expanded to cover applications in which thin layers are applied on top of a wafer absorber, with modelling tools such as GENPRO4 and OPTOS now available [2,3]. In such applications, textured interfaces with nano-scale morphologies can be modeled by generating an angular-dependent scattering matrix using an AFM surface scan as an input. This approach appears well validated and although it does not account for larger 3D structures (such as metal fingers) it would be useful in the study of black-silicon on crystalline silicon wafers. In principle a similar approach could be integrated into a ray tracer that solves the full set of 3D module optics. However, before doing so it is worth exploring if, for the case of a black-silicon textured wafer, a simpler approach combining geometric optics, effective medium and simple lambertian scattering is sufficient.

This paper presents an update from an ongoing project to develop rapid models for nano-textured silicon solar cells and modules. It starts by first developing an extensive dataset of well characterized black silicon samples, investigates the limitations of current rapid modelling before presenting a discussion of potential pathways towards modelling improvements. While there are many different methods for fabricating black silicon, including metal assisted chemical etching and various laser based processes [4], in this study we focus on plasma textured surfaces fabricated in both research level and industrial scale reactive ion etching (RIE) tools.

2 EXPERIMENTAL METHODS

2.1 Sample Fabrication

To achieve a broad data set applicable to industry and research environments, two primary groups of plasma textured silicon surfaces were fabricated using different RIE tools. Sample Set 1 was fabricated using an STS DRIE Pegasus research level tool at the Technical University of Denmark (DTU) and sample Set 2 was fabricated using an industrial scale Wonik IPS Solar RIE tool operated by *1366 Technologies*. Two control samples are also used, (1) a standard upright random pyramid (RP) texture and (2) a double side polished wafer.

Polished silicon wafer substrates (4 inch round) were used for sample Set 1 and the RIE processing time was varied to achieve different surface textures, with samples processed at times of 4, 6, 8 and 16 minutes used within this work. All samples were processed with an O_2 flow rate of 100 sccm and SF6 flow rate of 70 sccm with a coil power of 3 kW. For sample Set 2, industrial grade 6 inch square wafer substrates were used. For this set the coil power was varied, with settings of 14, 16, 18 and 22 kW used to create four different textures. For all four of these samples an O_2 flow rate of 900 sccm was used with a processing time of 150 seconds (2.5 minutes).

2.2 Topographical and Optical Characterization

Surface topography was determined using both scanning electron microscopy (SEM) and atomic force microscopy (AFM) techniques. AFM scans were performed over an area of 10 by 10 µm. The AFM data was analysed using the open-source *'Gwyddion'* software to extract key roughness parameters and various statistical distributions including height, incline and correlation factors. The autocorrelation length (*T*) was determined by fitting the one-dimensional autocorrelation function with the Gaussian function shown in Equation 1.

$$G_x(\tau_x) = \sigma^2 \exp\left(-\tau_x^2 / T^2\right) \tag{1}$$

where σ is the rms of the surface height.

Optical characterization was carried out for all samples using a commercial PerkinElmer Lambda 1050 spectrophotometer to determine the wavelength dependent reflectance and transmittance, including a separation of the specular and diffuse components by use of the integrating sphere based measurement technique, as depicted in Figure 1.

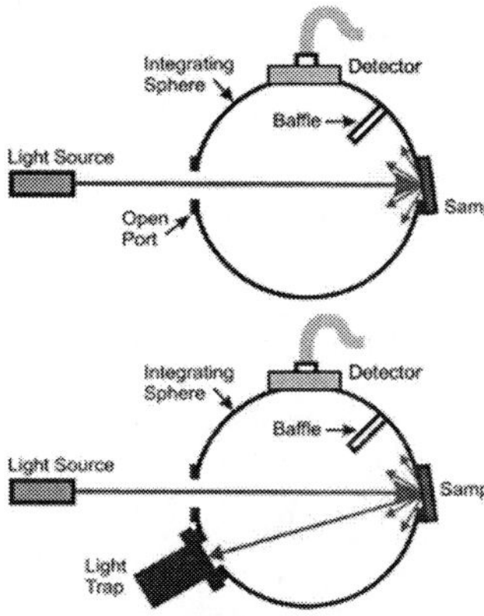

Figure 1 Measurement principle for total (top) and diffuse (bottom) reflectance measurements in the PerkinElmer tool

2.3 Modelling techniques

Modelling of the reflection, absorption, and transmission was performed using the *PV Lighthouse* ray tracer tool which combines monte-carlo based ray tracing with the transfer matrix method (to solve thin optical interfaces) [5]. This tool was designed for use with the more traditional forms of texturing i.e. random pyramids, grooves and isotexture. For the surface topographies in this study the ray tracer's geometrical optics approach becomes increasingly less valid as the incident wavelength increases. The purpose of this work is to (1) investigate the wavelength range where this transition applies for typical plasma textures and (2) to investigate rapid techniques to solve the optical performance in each regime.

To simulate the nano-textures as thin, coherent films (at long wavelengths) we have adopted the effective medium approach. A variety of custom materials data was generated by combining the wavelength dependent n and k values of silicon and air. Several mixing models are described in the literature for this application and numerous methods can be used for converting topography data to an effective medium [6]. In this case we have opted to use the Bruggeman mixing algorithm with various ratios ranging from 9:1 silicon/air to 1:9. These materials were then applied in stacks of thin layers with varying thicknesses in order to create an approximate graded index of an effective medium. Note that this modelling approach deals with the reflection of the surface but does not necessarily describe the scattering of the surface. For samples with the largest features (Sample Set 2), known surface morphology models developed for use with the more traditional surface texture were applied.

3 RESULTS AND DISCUSSION

3.1 Surface Topography

AFM and SEM results confirm that the desired broad range of textures was adequately achieved, as illustrated in the height distribution plot of Figure 2 which shows the average height of features increasing with processing time and power, as well as variations in the shape of the distributions. It also demonstrates that for all samples the distribution of heights was described by a Gaussian function, lending confidence that the AFM scan area was adequate to sample the textures. The one-dimensional auto-correlation functions for each texture (not shown) were also found to be well described by the Gaussian equation. The RP control sample is not shown on the figure, in that case the height distribution started at over 1000 nm.

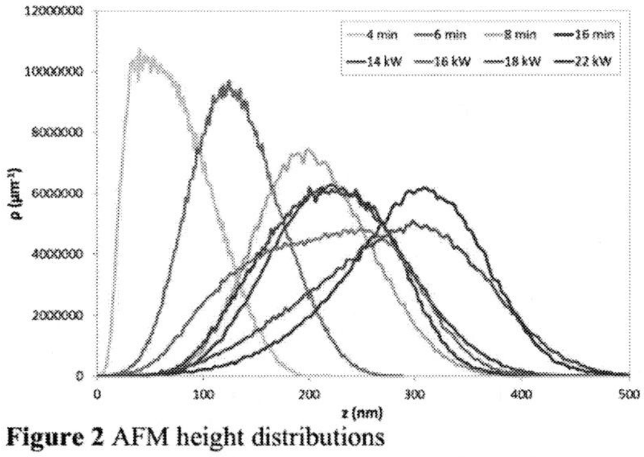

Figure 2 AFM height distributions

The variations in surface features can be further observed from the SEM images, with examples of the range limits for samples set 1 and 2 given in Figure 3 and 4 respectively.

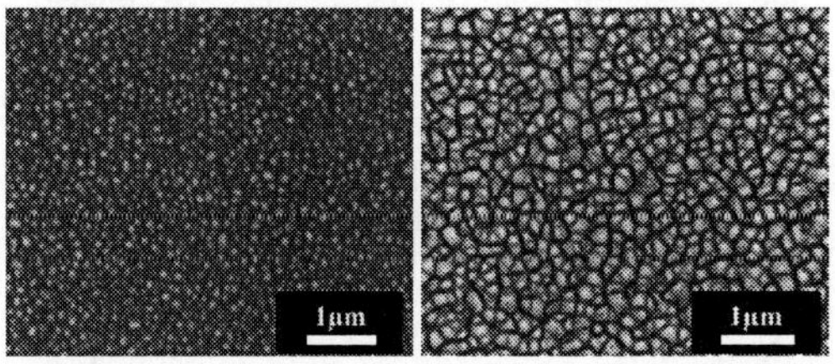

Figure 3 SEM images of 4 minute processed sample (left) and 16 minute processed sample (right)

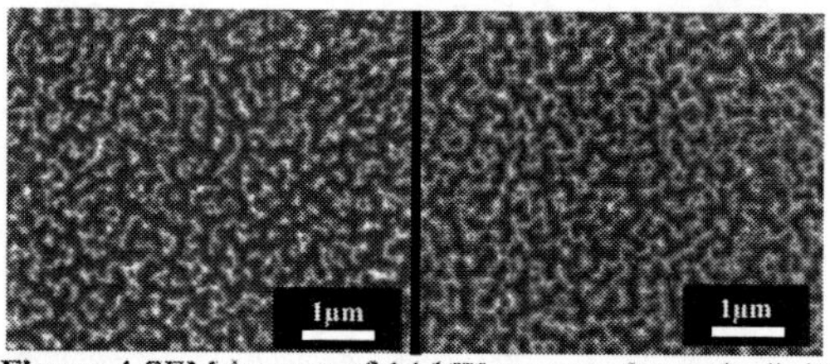

Figure 4 SEM images of 14 kW processed sample (left) and 22 kW processed sample (right)

For a direct quantitative comparison of sample topographies, key roughness parameters of the maximum height, average height, and the calculated RMS roughness and correlation length are presented in Table 1.

Table 1. Roughness parameters for all samples

Sample name	Max height (nm)	Average height (nm)	RMS roughness (nm)	Correlation length (nm)
4 min	222	74	36	45
6 min	290	132	52	55
8 min	517	207	55	90
16 min	938	231	62	125
14 kW	456	217	59	100
16 kW	409	211	72.5	90
18 kW	671	283	80	110
22 kW	703	296	68	115
RP	6169	2866	1042	1750

As expected, increased process time or coil power led to larger features as indicated by the correlating trends found for both maximum and average heights characteristics. In most cases, RMS roughness also increased accordingly, however, there is an exception for the 22 kW sample in which the height of features increased relative to the 18 kW condition, but the RMS roughness decreased. This could be an indicator of a more dramatic change in surface feature shape rather than just an increase in aspect ratio and effective elongation of features. In general, sample Set 1 (process time varied) shows a broader range of topographies whereas sample Set 2 (power varied) shows more subtle variations in texture. This can also be observed in the SEM images of Figures 3 and 4 in which the differences in Figure 3 (Set 1) are clearer than those of Figure 4.

3.2 Model Validity

Based on the statistical roughness parameters of the various textures, it is possible to assess the validity of using geometric optics to simulate the optical performance. Here we present this based on the work of Tang et al. who generated regions of validity determined using a number of nano-textured surfaces and comparing different modelling approaches [7]. Figure 5 reproduces their original diagram and superimposes the 4 min, 16 min, 14 kW, 18 kW and RP samples. These samples were selected as they cover the broadest range or RMS roughness values. In each case the wavelength is varied from 200 nm to 1200 nm with every 100 nm step depicted by the symbols. This assumes an incident angle normal to the surface. Note that creating more acute angles will shift the bars to the left (i.e. will progressively make the geometric optics less valid).

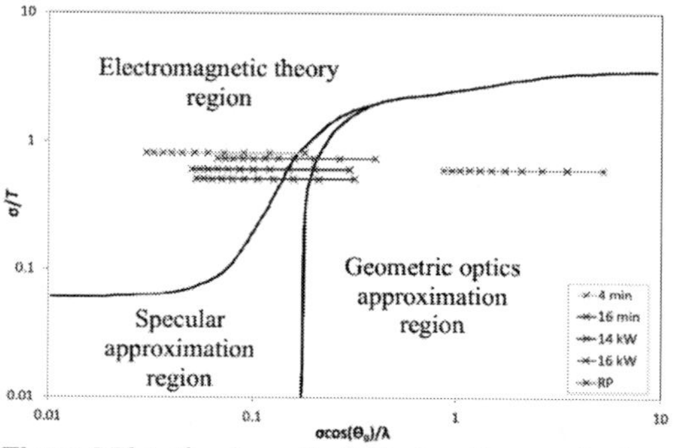

Figure 5 Plot of regions of validity from Tang et al. where E_{SR}=0.2 and E_d=0.2 [7]. Various textures from this study shown from 200 nm to 1200 nm with symbol every 100 nm

From the plot it is clear that in all cases the nano-textures studied are not expected to be well approximated with geometric optics alone. For the shallowest texture (4 min) there are no wavelengths within the range. The 16 kW plasma sample with the largest σ_{rms} is expected to be valid up to 500 nm wavelength. In contrast the control sample with random pyramids is valid under the geometric optics approximation for all wavelengths of interest. In fact it would not become invalid until 7500 nm.

This figure highlights the problem at hand. For the nano-textured surfaces the long wavelength reflection and scattering cannot be easily predicted via the use of geometric optics.

3.3 Optical Results and Model Fitting

Here we first verify the range that both EM and GO can be applied to describe the reflection of the samples with varying levels of texture. Of the two sample sets, the industrially processed samples of Set 2, in which the coil power was varied typically have larger features than the academic sample set. The validity plot of Figure 5 indicates that these samples may be well approximated at short wavelengths using geometric optics. This was tested by modelling the sample textures in the *PV Lighthouse* ray tracer as inverted spherical caps with a 1.5 μm height and the characteristic angle varied to achieve a best fit for each of the different industrial textures. The measured and GO modelled total reflectance characteristics are shown in Figure 6, along with results for planar silicon for comparison.

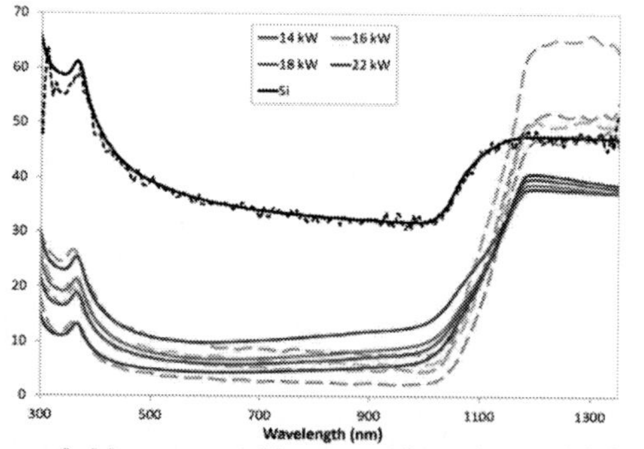

Figure 6: Measurement Vs. geometric optics model for sample Set 2.

The model shows good agreement for the planar silicon example case across the full wavelength range investigated, however, for the textured samples good agreement is only observed at the shorter wavelengths, thus supporting the validity plot of Figure 5. The GO modelling approach underestimates reflectance at longer wavelengths approaching the bandgap and then severely overestimates the reflectance beyond the bandgap. The underestimation is due to the GO model scattering long wavelength light, increasing optical path lengths and consequently absorption below bandgap, whereas in reality these wavelengths cannot distinguish the texture and will therefore not be scattered. The modelled scattering may also explain the overestimation at wavelengths beyond the bandgap due to increased reflectance caused by an effective non-normal angle of incidence at the rear of the sample.

The results of Figure 6 clearly show that GO modelling alone will not suffice for the plasma textured black silicon samples studied here. For the smaller featured sample Set 1, both the GO and EMT approaches were investigated. In this case the EMT agreement was optimized by empirical adjustment of the layer thicknesses and mixing ratios, with total thickness values of the layer stack set to correlate with the height values gathered from the AFM data. The measured and modelled results are shown for comparison in Figure 7.

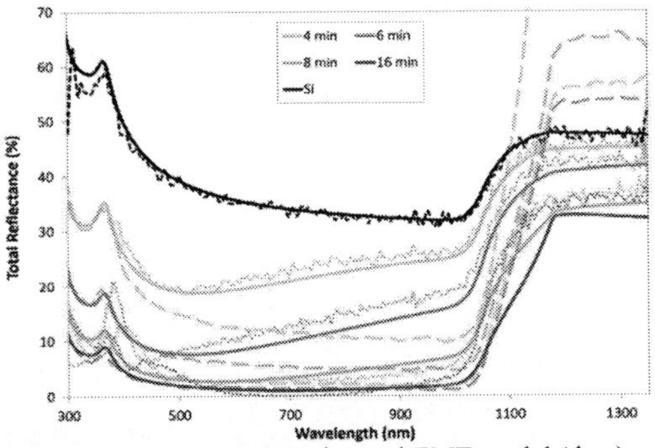

Figure 7: Measurement Vs. layered EMT model (dots) and GO model (dashes) for sample Set 1.

From the curve fitting results it is evident that the reflection (including at the long and short wavelengths limits) of the 4 min sample was reproduced well via the EMT approach. On the validity chart, this was the sample that was entirely within the electromagnetic theory region. For all other samples there is a clear trend of GO modelling fitting best at short wavelengths and EMT providing the best fit over the mid to long wavelength range. It is evident that for the majority of samples investigated here the EMT approach gives the best broadband agreement but neither the GO or EMT method are truly sufficient for accurate simulation across the full wavelength range of interest. These findings are supported by the validity plot and also correlate with the measured reflectance haze characteristics shown in Figure 8, where haze has been calculated as the ratio of scattered to total reflectance.

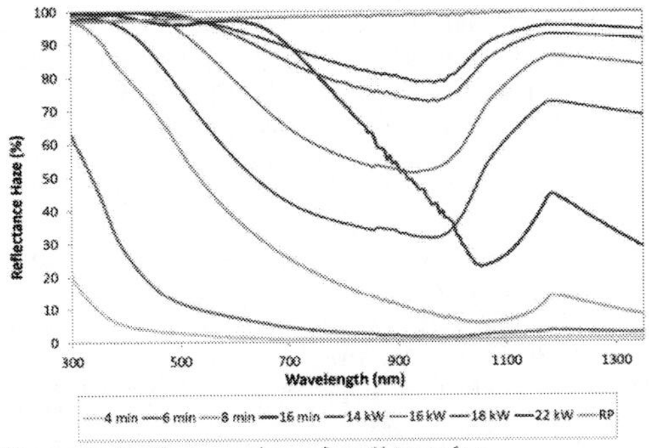

Figure 8: Reflectance haze for all samples

Analysis of the 4 min sample haze reveals that the texture produced very little scatter, particularly at longer wavelengths. All other samples show increased scatter, with those of sample Set 2 showing near 100% haze up to wavelengths of around 600 nm. For all samples, there is a clear correlation between the wavelengths where reflectance haze is high and the accuracy of the GO modelling approach. Reflectance haze could therefore provide a more convenient method for gauging the model validity regime rather than the autocorrelation length used for Figure 5, which requires precise topographical data that is typically difficult and time consuming to obtain.

For nearly all samples, the haze plot shows that there is still partial scattering across the majority of the wavelength range. This indicates that whilst the EMT approach can achieve reasonable broadband agreement for these samples, it does not faithfully simulate the full optical characteristics, as the EMT model results in no scattering. This is further illustrated in the 16 min sample reflectance characteristics which show increased error between measurement and EMT modelling at wavelengths beyond the silicon bandgap (Figure 7). As the Haze plot of Figure 8 shows, this sample is still partially scattering at these wavelengths, leading to a reduction in measured total reflectance that cannot be reproduced using the graded EMT modelling approach. While this does not significantly impact the reflection measured under air, it will create errors once additional layers are added on top (e.g. EVA and glass). It is also interesting to note the different haze characteristics of the two datasets. In general the Set 2 substrates were found to have much more haze than might be expected from the AFM scans. We believe this to be related to larger scale variations in the surface of those substrates caused by their industrial nature (saw damage, etched but not polished). Measurement of a non-plasma textured control sample (not shown) confirmed a significant degree of scatter from the bare substrate. This will be an important consideration going forward as a full model to describe these substrates may require a broadband layer of GO ray tracing combined with treatment of the surface as an additional nano-textured thin layer. It is worth noting that AFM scans alone will not detect these larger scale variations.

For rapid optical modelling of the nano-texture to be accurately achieved, alternatives or extensions to the standard GO or EMT methods must be developed and utilized. Whilst good agreement to total reflectance under air can be achieved using a wavelength dependent combination of GO and EMT, this approach would not faithfully model the scattering properties and therefore the light trapping and absorption. One possible route towards enhancing the EMT approach is the incorporation of scalar scattering theory [1,8]. This would allow for the model to predict the overall scattering intensity for the types of textures studies here, however, it may not accurately reproduce the angular distribution of scatter and therefore the light-trapping. In order for any rapid modelling approach to be truly validated, good agreement must therefore be achieved to not only the conventional reflectance and transmittance characteristics, but also the wavelength and angle resolved scattering. Going forward, these full optical characteristics will be accurately measured followed by investigation of modelling extensions to achieve agreement between all optical characteristics.

4 CONCLUSION

Almost all samples across the broad range of plasma textured black silicon samples from both academic and industrial sources were found to lie across both the geometrical optics and electromagnetic theory validity regions at the wavelengths of interest for silicon photovoltaics. The only exception to this was the 4 min academic sample which had the smallest features and could therefore be well approximated using a layered effective medium approach. For typical plasma textured black silicon, we conclude that the EMT approach is only valid for samples with low roughness (< 40 nm RMS) or a low reflectance haze (< 15%). For plasma textured silicon with more substantial roughness, and therefore higher haze, the GO approximation can only be reliably applied for wavelength ranges over which the haze is approaching 100%. As typical plasma textured black silicon has a reflectance haze between these proposed limits across much of the wavelength range of interest, alternative models or adaptations are required for complete and accurate rapid simulation. Existing models, not investigated within this work, such as scalar scattering theory may provide a solution to this. However, care must be taken to ensure that both the scattering intensity and the angular distribution of scattered light are correctly modelled, as this naturally plays a key role in wavelength dependent path-length enhancement as well as further light-trapping from additional layers such as passivating dielectrics and encapsulants.

ACKNOWLEDGMENTS
The authors would like to acknowledge the Australian Centre for Advanced Photovoltaics (ACAP) and the Australian Renewable Energy Agency (ARENA) for providing funding for this work through ACAP collaboration grant RG123662-O.

REFERENCES

[1] K. Jäger, M. Fischer, R.A.C.M.M. van Swaaij, M. Zeman, A scattering model for nano-textured interfaces and its application in opto-electrical simulations of thin-film silicon solar cells, J. Appl. Phys. 111 (2012) 83108.

[2] R. Santbergen, T. Meguro, T. Suezaki, G. Koizumi, K. Yamamoto, M. Zeman, GenPro4 Optical Model for Solar Cell Simulation and Its Application to Multijunction Solar Cells, IEEE J. Photovoltaics. 7 (2017).

[3] N. Tucher, J. Eisenlohr, P. Kiefel, O. Höhn, H. Hauser, M. Peters, C. Müller, J.C. Goldschmidt, B. Bläsi, 3D optical simulation formalism OPTOS for textured silicon solar cells, Opt. Express. 23 (2015) A1720.

[4] X. Liu, P.R. Coxon, M. Peters, B. Hoex, J.M. Cole, D.J. Fray, Black silicon: fabrication methods, properties and solar energy applications, Energy Environ. Sci. 7 (2014) 3223–3263.

[5] K.R. McIntosh, PV Lighthouse: Module Ray Tracer, PV Light. (2017). https://www.pvlighthouse.com.au (accessed September 28, 2017).

[6] T. Rahman, S. Boden, Optical Modeling of Black Silicon for Solar Cells Using Effective Index Techniques, IEEE J. Photovoltaics. 7 (2017).

[7] K. Tang, R.A. Dimenna, R.O. Buckius, Regions of validity of the geometric optics approximation for angular scattering from very rough surfaces, Int. J. Heat Mass Transf. 1 (1996) 49–59.

[8] C.K. Carniglia, Scalar Scattering Theory for Multilayer Optical Coatings, Opt. Eng. 18 (1979) 182104.

EMISSIVITY CONTROL IN TEXTURED SILICON SOLAR CELLS

D. Alonso-Álvarez[1,*], L. Ferre-Llin[2], Y. Liu[1], A. Mellor[1], A. Riverola[3], D. Chemisana[3], D. J. Paul[2] and N. Ekins-Daukes[1]
[1]Department of Physics, Imperial College London, London SW7 2AZ, United Kingdom
[2]School of Engineering, University of Glasgow, Glasgow G12 8LT, United Kingdom
[3]Environmental Science Department, University of Lleida, 25001, Lleida, Spain
*e-mail: d.alonso-alvarez@imperial.ac.uk

ABSTRACT: Surface textures are a key part on most silicon-based solar cells. Typically, they are optimized for maximum coupling of sunlight into the solar cell, in the wavelength range from 0.3 µm to 1.1 µm. However, recent applications such as radiative cooling and hybrid photovoltaic-thermal (PV-T) systems require to consider also their performance at much longer wavelength, in the mid-infrared (MIR, 5-20 µm). In this work, we study experimentally the emissivity from 0.3 µm to 20 µm of upright and inverted pyramids for a range of pyramid sizes. We conclude that pyramids with sizes in the range of 0.7-2 µm are able to result in high absorptivity in the solar range and low emissivity in the MIR, suitable for PV-T, while the largest pyramids result in high absorptivity and emissivity at all wavelengths, suitable for radiative cooling.
Keywords: Texturisation, Silicon Solar Cell, Optical Properties, Hybrid

1 INTRODUCTION

Surface textures are an intrinsic part of all silicon solar cells. These serve to increase the coupling of sunlight into the material, and increase light absorption due to an enhancement of the path length inside the material. While many types of textures have been used and optimized along the years – ranging from inverted pyramids [1], and diffraction gratings [2] to nanopillars (black silicon) [3], moth-eye structures [4] or nanoplasmonics [5] – the most extended method in industry is based on the preferential etching of crystalline silicon surfaces, resulting in random upright pyramids with typical base sizes between 2 and 4 µm. This is a range of sizes where ray optics can provide a satisfactory description of the coupling of sunlight with energy above the silicon bandgap.

Recent innovations in silicon solar cells require the light coupling problem to be tackled at much longer wavelengths. Optimal radiative cooling of solar cells [6] seeks to suppress the absorption of the un-used solar spectrum (1.1-2.5 µm), and couple the thermal emission in the mid infrared (MIR, 5-12 µm) directly to the cold space beyond the atmosphere. In hybrid photovoltaic-thermal (PV-T) applications [7], [8], the goal is to absorb all incoming sunlight, not just above the silicon bandgap, and suppress the thermal emission. Such applications target a very wide spectral range with a single interface, and aim to deliver contrary absorptive/reflective behavior in different wavelength ranges.

In a recent work, we studied the optical and electrical properties of sputtered indium tin oxide (ITO) and aluminum zinc oxide (AZO), and optimized their use as low emissivity coatings, discussing the trade-off between high transmissivity in the solar range and low emissivity in the MIR, suitable for PV-T applications [9]. In this work, we address the role of textures in a similar context, presenting experimental results of the absorptivity and the emissivity of textured silicon solar cells over the 0.3 – 20 µm wavelength range as a function of the texture properties. We focus on the two most common structures: random upright pyramids and periodic inverted pyramids.

2 EXPERIMENTAL METHODS

2.1 Sample fabrication

All samples were made on high resistivity, double polished, crystalline silicon substrates. The random upright pyramids were created by dipping the substrates in an aqueous solution of KOH for a range of concentrations of KOH and isopropanol, as well as etching time [10]. The etching took place at 70 °C in an ultrasonic bath. For the periodic inverted pyramids, a pattern with the chosen periodicity (1, 2, 5, 10 and 20 µm) was first defined by e-beam lithography on a silicon nitride mask. An aqueous solution of KOH was then used to etch the exposed areas, and finally the remains of the mask are removed.

The back of the samples, textured for the random pyramids and flat for the periodic ones, is covered with aluminum to reflect back to the front any light that is transmitted through the silicon wafer.

2.2 Surface characterization

The surface of the random pyramids was characterized by atomic force microscopy (AFM) in a WITec Alpha 300RS system. The AFM tips had curvature radius <15 nm, heigh >10 µm and half cone angle ~30°, sharp enough to measure the slopes of silicon pyramids (with an ideal slope of 54.7°). Inverted pyramids were measured with a scanning electron microscope (SEM) in order to get information about the typical pyramid size and distribution.

2.3 Emissivity measurements

In both ranges, emissivity is obtained indirectly based on reflectivity measurements of the samples using the fact that there is no transmission due to the aluminum reflector and that absorptivity equal emissivity:

$$\varepsilon \equiv A = 1 - R$$

For the solar range, the samples were attached to one of the ports of a Spectralon integrating sphere. Monochromatic light entering through the opposite port was reflected and scattered by the surface, bouncing multiple times inside the sphere and, ultimately, being collected by a detector. Two detectors, made of silicon and germanium, where used to cover a wider range of wavelengths. A silver mirror was used as reference. The sample had a 10° tilt to prevent specular reflection to be reflected back to the entry port. For the MIR range, a Vertex FTIR system connected to a microscope were used. In this case, the reference was a gold mirror. This configuration presents the limitation that a reduction in the measured reflectivity can be associated to either a reduction in the real reflectivity or a change in the reflected pattern of the light affecting the amount of it that is coupled back in the microscope objective. As we will discuss later, this has the effect of overestimating the measured ε (=A).

3 RESULTS

3.1 AFM and SEM

Fig. 1a shows two examples of the SEM micrographs corresponding to the 1 µm and 5 µm inverted pyramids. The periodic pyramids all show a side length close to the nominal ones. A narrow frame, 200 nm to 800 nm wide in the case of the biggest pyramids, is present around the motifs where no etching took place. These regions could behave as specular reflectors at short wavelengths, reducing the absorptivity in the solar range.

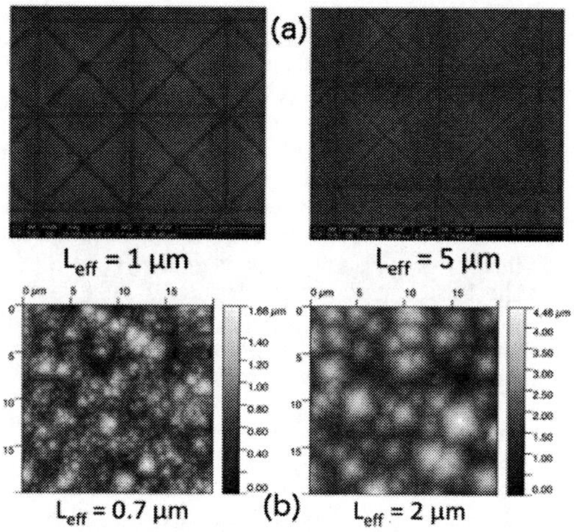

Fig. 1: (a) SEM micrographs of two of the periodic inverted pyramids. (b) AFM image of two of the random upright pyramids. In both cases, the effective pyramid side length L_{eff} is included.

Selected AFM images of the random pyramids are shown in Fig. 1b. The effective pyramid side length L_{eff} in the case of the random samples is estimated by counting the number of pyramids N in each AFM image and assuming they are all identical and covering the full area of the image L_{AFM}^2:

$$NL_{eff}^2 = L_{AFM}^2$$

Values ranging from 0.5 µm to 4.5 µm are obtained in this case. While not measured, it was expected that the back surface had features of similar size. The surface roughness (RMS) in this case is obtained directly from the surface profile, while it is calculated numerically for the periodic pyramids based on the measured geometry, including the frame.

3.2 Emissivity results

Fig. 2a and 2b show the emissivity results for the two sets of samples. In both cases, the absorptivity in the solar range increases significantly with respect the flat silicon counterpart. This increase is highest for the sub-bandgap absorptivity (λ > 1100 nm). The random sample with L_{eff} = 0.53 µm shows the lowest increase. The interpretation is that such small features are not able to maximize the coupling of light into the sample, even at these short wavelengths.

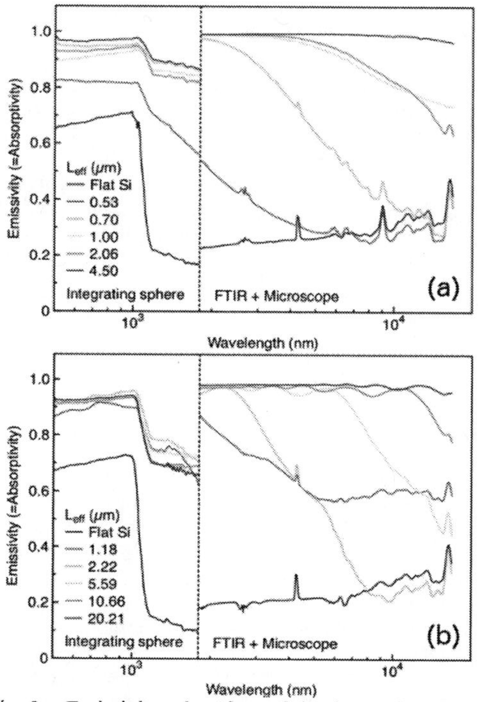

Fig. 2: <Emissivity> (=<Absorptivity>) as a function of wavelength of (a) the random upright pyramids and (b) the periodic inverted pyramids.

At longer wavelengths (λ > 1800 nm), several phenomena can be observed. First, there is a sudden jump in the measured emissivity when changing the experimental technique. This is related with the effect discussed above of the microscope objective not being able to collect all the scattered light when the surface is rough. As it was anticipated, the measured MIR emissivity is overestimated where λ < L_{eff} since there is non-specular reflection for this condition.

The next thing to observe in the MIR range is that smaller pyramids have lower emissivity than the bigger counterparts. That is the case for both, random and periodic pyramids. The explanation is similar to that of the 0.53 nm random sample in the visible: small pyramids compared to the wavelength of the light will not be able to couple the radiation in/out of the material, behaving more as specular surfaces. As the pyramid size becomes larger, radiation will be correctly approximated as rays and geometric optics, situation where it is well known that a rough surface reduces reflection and increases absorption.

For any given pyramid size, the emissivity reduces the longer the wavelength. This decay is monotonous for the random pyramids Fig 2(a) but shows clear oscillations in the periodic ones Fig. 2(b). Those oscillations are related with the optical modes of the periodic structure, affecting both the amount of light coupled into the sample and the amount of reflected light that is collected by the microscope objective.

4 DISCUSSION

The ultimate goal of a surface texture in a silicon solar cell is to reduce reflection and increase the path length of the light inside the material, resulting in an increased absorption. Specifically, for PV-T applications in addition to high absorptivity in the solar range we need low emissivity in the MIR. To assess the performance of each of the textures above in this two aspects, we calculate the average absorptivity over the solar spectrum <Abs> and the average emissivity over the spectrum of a black body at 100 °C, as it was discussed in [9].

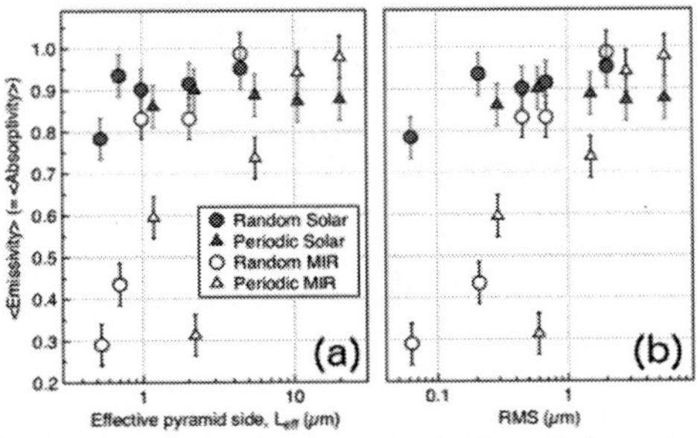

Fig. 3: <Emissivity> (=<Absorptivity>) of all the textures as a function as a function of (a) the effective pyramid side, L_{eff} and (b) the RMS.

Fig. 3 shows the <Abs> and the <Em> as a function of the effective pyramid length and the RMS of the surface. Except for the 0.53 μm random pyramids, all the others have a similar <Abs> in the solar range, of around 0.9 within the experimental error. This absorptivity is typical of commercial solar cells, even considering that this samples have no anti-reflective coating. That dependence is stronger for the <Em> in the MIR, both as a function of L_{eff} and RMS. The general trend is that small L_{eff} is associated with a small RMS – as expected, given that both are related by the pyramids slope of 54.7° – and small <Em>. Out of the samples with highest <Abs> the ones having simultaneously lowest <Em> are the random pyramids 0.7 μm and the periodic pyramids 2.22 μm, with <Em> of 0.43 and 0.32, respectively. It should be noted that the random samples have textures on both sides, which are likely responsible for the generally high <Em> of that series.

The sample with periodic inverted pyramids of 1.18 μm has a <Abs> too low and <Em> too high, breaking the general trend of the series, but it is unclear at this point if this is the real behavior of the sample or an artefact of the measurement as described above.

5 CONCLUSSIONS

To summarize, in this work we have characterized the broadband emissivity (0.3 μm – 20 μm) of textured silicon samples with different types of textures and pyramids sizes. The goal was to evaluate how the texture affects the emissivity and if there is an optimum for its use in PV-T applications.

Our conclusion is that small textures will be generally more suited, being able to deliver the opposing behavior of high absorptivity in the solar range and low emissivity in the MIR range. In particular, out of all the samples analyzed, the best performance is achieved with random upright pyramids of Leff = 0.7 μm and periodic inverted pyramids with Leff = 2 μm.

Further work is underway to evaluate the emissivity theoretical, for which different approximation for solving the Maxwell's equations will need to be used to the different ranges of wavelengths and pyramids sizes.

5 ACKNOWLEDGMENTS

This research was supported by the sUK Engineering and Physical Science Research council (EPSRC) [grant number EP/M025012/1]. A. Mellor was supported by the European Commission through Marie Sklodowska Curie International Fellowship, Grant No. DLV-657359.

REFERENCES

[1] A. W. Blakers, A. Wang, A. M. Milne, J. Zhao, and M. A. Green, "22.8% efficient silicon solar cell," *Appl. Phys. Lett.*, vol. 55, no. 13, pp. 1363–1365, 1989.

[2] R. Dewan and D. Knipp, "Light trapping in thin-film silicon solar cells with integrated diffraction grating," *J. Appl. Phys.*, vol. 106, no. 7, p. 074901, 2009.

[3] J. Oh, H.-C. Yuan, and H. M. Branz, "An 18.2%-efficient black-silicon solar cell achieved through control of carrier recombination in nanostructures," *Nature Nanotech*, vol. 7, no. 11, pp. 743–748, Sep. 2012.

[4] C.-H. Sun, P. Jiang, and Bin Jiang, "Broadband moth-eye antireflection coatings on silicon," *Appl. Phys. Lett.*, vol. 92, no. 6, p. 061112, 2008.

[5] Y. Zhang, N. Stokes, B. Jia, S. Fan, and M. Gu, "Towards ultra-thin plasmonic silicon wafer solar cells with minimized efficiency loss," *Sci. Rep.*, vol. 4, no. 1, pp. 1–6, May 2014.

[6] L. Zhu, A. P. Raman, and S. Fan, "Radiative cooling of solar absorbers using a visibly transparent photonic crystal thermal blackbody," *Proc Natl Acad Sci USA*, vol. 112, no. 40, pp. 12282–12287, Oct. 2015.

[7] M. Lämmle, T. Kroyer, S. Fortuin, M. Wiese, and M. Hermann, "Development and modelling of highly-efficient PVT collectors with low-emissivity coatings," *Solar Energy*, vol. 130, pp. 161–173, 2016.

[8] A. Mellor, D. Alonso-Álvarez, I. Guarracino, A. Ramos Cabal, A. Riverola-Lacasta, L. Ferre Llin, A. Murrel, S. Thoms, D. J. Paul, D. Chemisana, C. N. Markides, and N. J. Ekins-Daukes, "Emissivity control in hybrid photovoltaic thermal generators for improved exergy efficiency," *Energy and Environmental Science.*

[9] D. Alonso-Álvarez, L. Ferre Llin, A. Mellor, D. J. Paul, and N. J. Ekins-Daukes, "ITO and AZO films for low emissivity coatings in hybrid photovoltaic-thermal applications," *Solar Energy*, vol. 155, pp. 82–92, Oct. 2017.

[10] S. Bilgen, F. Es, and R. Turan, "Comparison of mechanical and ultrasonic agitation methods for mono c-Si texturing," *phys. stat. sol. (a)*, vol. 213, no. 10, pp. 2721–2726, Jun. 2016.

LAMBERTIAN OPTICS IN TEXTURED SI SOLAR CELLS WITH NOT-RANDOMIZING FRONT SURFACE

Luigi Abenante and Massimo Izzi
ENEA, Italian National Agency for New Technologies, Energy, and Sustainable Economic Development
Via Anguillarese 301, 00123, Roma, Italy

ABSTRACT: We study a Lambertian light-trapping scheme for textured solar cells, which is not included in the Lambertian models that are considered in the literature, but can be implemented in the numerical simulation program PC1D. We call this scheme Quasi-Lambertian model. In the Quasi-Lambertian model, at the first pass of coupled light, the optical propagation angle is determined by the texture facet angle. Light direction randomizes at the rear surface so that, at the second and remaining optical passes, the optical propagation angle is assigned a mean value equal to 60°. We show that the presented model is practically equivalent to a Lambertian optical model reported in the literature, where front surface randomizes light direction as soon as light enters the device.
Keywords: Light Trapping, Simulation, Solar Cell.

1 INTRODUCTION

Three possible Lambertian light-trapping schemes for textured solar cells have been considered in the literature [1]: one, where both surfaces are optically diffusive (randomizing), another, where front surface is randomizing and rear surface is flat, and a third one, where front-surface is flat and rear surface is randomizing.

For the first two optical models, the propagation angle of trapped light, θ, has been calculated in two ways. In the first way, which we call "standard", θ=60° at all optical passes and wavelengths, λ. In the second way, which we call "Green-Lambertian", $\theta=\theta(\lambda,W)$, where W is the device thickness, at all optical internal passes [1].

In the present paper, we take into account a fourth possible Lambertian scheme. Such scheme is represented in Fig. 1. As can be seen, in this scheme, front surface is textured, but not randomizing, so that, at the first pass of trapped light, the optical propagation angle, θ_1, is determined by the facet angle, δ, which ordinarily is 54.74°. At refractive index 3.58, θ_1=41.6°. Light direction randomizes at the rear surface so that, only at the second and remaining optical passes, the propagation angle has the averaged value θ_n=60°. We call this model "Quasi-Lambertian".

All models can be implemented in the widely used numerical simulation program PC1D. However, while the Green-Lambertian model only requires a suitable assignment to absorption coefficient, the Quasi-Lambertian model needs the correction procedure reported in [2].

Moreover, light-generated currents can be calculated analytically by using the general solution to minority-carrier transport in [3] at $\theta_1=\theta_n$ only. Consequently, the Quasi-Lambertian model, where $\theta_1\neq\theta_n$, cannot be used in an analytical approach.

In the present work, however, we show both theoretically and numerically that the Green- and Quasi-Lambertian models are practically equivalent.

Consequently, the Quasi-standard model can be implemented in both PC1D and the aforementioned analytical solution by using the Green-Lambertian model.

2 THEORY

In analytical Lambertian models, the propagation angle of internal light, θ, is assigned a mean value, θ_m, calculated with

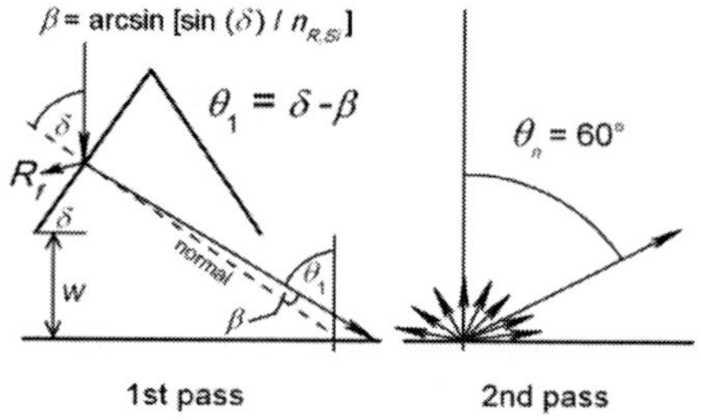

Figure 1: Quasi-Lambertian model.

$$\exp\left(-\frac{\alpha_{Si}W}{\cos\theta_m}\right)=\frac{\int_{0^\circ}^{\theta_{MAX}}\sin\theta\cos\theta\exp\left(-\frac{\alpha_{Si}W}{\cos\theta}\right)d\theta}{\int_{0^\circ}^{\theta_{MAX}}\sin\theta\cos\theta\,d\theta} \quad (1)$$

where α_{Si} is the Si absorption coefficient and θ_{MAX} is the maximum θ-value allowed in the considered Si sheet. Eq. (1) is valid in the weak absorption limit [1]

In the standard approach to solve (1), the exponentials are developed in series obtaining

$$\theta_m=\cos^{-1}\left[\frac{1}{2}\left(\frac{\sin^2\theta_{MAX}}{1-\cos\theta_{MAX}}\right)\right] \quad (2)$$

θ_{MAX}=90° is assigned and θ_m=60° is obtained.

A different way to solve (1) at θ_{MAX}=90° is adopted in [1] resulting in a solution, which is well approximated by an effective θ, θ_{op}, given by

$$\theta_{op}(W)=\cos^{-1}\left[\frac{1+a(\alpha_{Si}W)^b}{2+a(\alpha_{Si}W)^b}\right], \quad (3)$$

where a = 0.935 and b = 0.67 [1].

From (3) it follows that θ_{op} ranges from 0° to 60° at any W-value. A mean value of θ_{op} can be obtained by assigning θ_{MAX}=60° in (2). This gives

$$\theta_m=\cos^{-1}\left[\frac{3}{4}\right] \quad (4)$$

that is θ_m=41.41°. As can be seen, this value is not far from θ=41.6° at the first pass of light across the considered device, which is peculiar to the Quasi-Lambertian model.

At the passes of internal light successive to the first one, it can observed that, in the Green-Lambertian model, the value θ=41.41° corresponds to $\alpha_{Si}W$=3.11, which is relevant to one-pass transmittance, T_a, equal to 0.0446. According to the Basore modeling of textured cells [4] the absorptance, A_R, at the second and successive passes of light across a Si sheet, is

$$A_R = (1-R_f)R_B T_a (1-T_a)\frac{1+R_{FI}T_a}{1-R_{FI}R_B T_a^{\ 2}}. \quad (5)$$

At T_a=0.0446 and same assignments to the remaining parameters as in the device in [1], which are R_f=0, R_{FI}=0.921, and R_B=1 [1], (5) calculates A_R= 0.0444. According to Table 1 in [1] such a value of absorptance corresponds to θ>59.85°.

At T_a=0.0446 and same assignments to the remaining parameters as in the devices in [2], which are R_f =0, R_{FI}=0.921, and R_B=0.982 [2], (5) calculates A_R=0.0436. According to Table 1 in [1] such a value of absorptance corresponds to θ>59.85° as well.

The same occurs for all R_f>0 and all remaining R_B>0. The practical equivalence between the Green-Lambertian and quasi-Lambertian models is therefore general. It is worth remarking that, for surface texture, R_{FI}=1−$(1/n_{R,Si}^{\ 2})$=0.921, where $n_{R,Si}$ is the Si refractive index, is a not-adjustable parameter [1].

2 NUMERICAL CHECK

To verify the aforementioned equivalence, we have calculated light-generated currents, J_{sc}, in the same set of Si solar cells of various thickness, W, and identical remaining parameters as used in [2] by implementing in PC1D the Green-Lambertian model and the Quasi-Lambertian model and comparing the results. Since, to be implemented in PC1D, the quasi-Lambertian model requires that the rear surface is set to diffuse, the J_{sc}-values obtained with this model were corrected according to [2]. The comparison is shown in Fig. 2. As can be seen, the discrepancy, Δ, between the models is practically negligible.

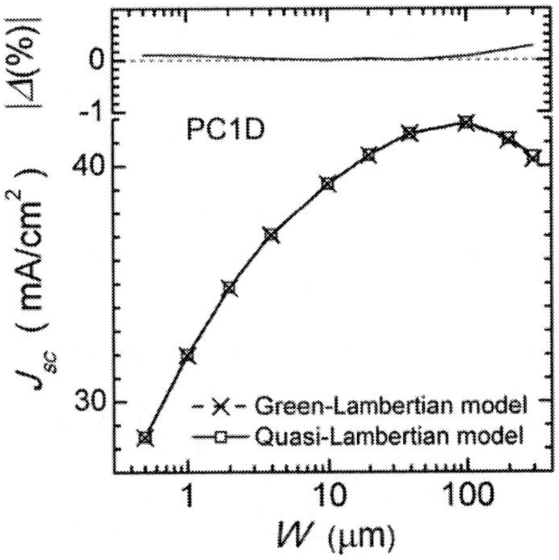

Figure 2: Comparison between numerical simulations performed using the optical models studied in this work.

3 CONCLUSIONS

We have found theoretically that the quasi-Lambertian model in Fig. 1 is practically equivalent to the Lambertian model presented in [1]. The finding has been confirmed by checks on simulated devices.

References
[1] M. A. Green, Prog. Photovoltaics Res. Appl. 10 (2002) 235 doi:10.1002/pip.404.
[2] L. Abenante, "The PC1D diffusion model in thin-film solar cells", 33rd European Photovoltaic Solar Energy Conference (2017).
[3] L. Abenante, Sol. Energy. 129 (2016) 204. doi:10.1016/j.solener.2016.02.009.
[4] P. A. Basore, Twenty Third IEEE PVSC (1993) 147

SCREEN PRINTED MONO-CRYSTALLINE SI SOLAR CELLS: ASSESSING THE MICROSTRUCTURE AND DOPANT CONCENTRATION AT THE FRONT SIDE METALLIZATION INTERFACE BY ELECTRON MICROSCOPY AND NANOSIMS

Praveen Kumar, Michael Pfeffer, Oliver Eibl
Institute of Applied Physics, Eberhard Karls University Tuebingen, 72076 Tuebingen, Germany
Phone: +49 7071 2976015, Email: praveen.kumar@uni-tuebingen.de

ABSTRACT: Screen printed mono-crystalline Si solar cells yield efficiencies that are limited by electrical losses due to the front side metallization. Screen printing affects the emitter and space charge region of the solar cell, particularly with respect to the dopant concentrations. Standard SEM analysis of the metallization interface showed hardly any differences between the optimally and the overfired cells. A set of previously studied cells were investigated: the same front side paste was used, one cell was optimally fired and the other overfired at a too high temperature. NanoSIMS (Secondary Ion Mass Spectroscopy) was used to acquire high-quality Si-, SiB- and P- ion maps of the front side metallization interface of solar cells with a lateral resolution of 80 nm. NanoSIMS was used to image the dopant (P, B) distributions across the Si-metallization interface. The overfired cell showed a dramatically increased of the contact resistance of 131 mΩ.cm^2 and low efficiency 13.6% as compared to the optimally fired cell with contact resistance of 21 mΩ.cm^2 and efficiency of 16.9%. It was found that due to the overfiring, the emitter P dopant profile got completely disrupted yielding a dramatic increase in the contact resistance and thereby decreasing the cell efficiency as compared to the optimally fired cells for which the emitter structure was fully preserved.

Keywords: Silicon Solar Cell, Screen Printing, Metallization, dopants, Electrical Properties

1 INTRODUCTION

In the past 30 years, significant efforts have been applied to understand the contact formation process of silver (Ag) thick film front side metallization in monocrystalline silicon (Si) solar cells [1–5]. The contact formation is realized by screen printing technology, because of its simplicity, high throughput, robust and cost-effectiveness.
Screen printed front side pastes require thermal annealing, which affects the dopant distribution across the Si/Ag interface. Details of the microstructural features appear at the contact interface can be found in Refs. [6–7]. The influence of the firing conditions on the microstructure and its effect on the electrical properties has been documented in the literature [1, 8, 9]. However, no systematic analysis has been carried out, particularly with respect to dopant distribution at the Si/metallization interface. Due to the high-temperature firing process, atoms from the metallization are assumed to diffuse as impurities into the emitter as well as space charge region of the solar cell [9, 10]. This in turn increases the recombination current of the emitter and space charge region of the cell, and degrades the cell efficiency [10].
An important factor that limits the solar cell efficiency is the electrical resistance across the Si-metallization interface [9]. The spatial distribution of dopants across this interface play a key role in determining the overall efficiency. The relatively large dopant concentrations in emitters of Si solar cells make NanoSIMS ideal for studying the dopant distribution in such devices.
Therefore, correlative analyses by SEM/TEM and NanoSIMS of solar cells have been carried out in [9] and here as well. NanoSIMS investigation offers much better detection limits than EDX in the SEM/TEM, EDX has a sensitivity of about 1 at.% which is not sufficient for detecting dopants. Moreover, NanoSIMS combines a focused ion beam with a SIMS yielding a lateral resolution of about 80 nm. This correlative analysis yields a structural characterization of the metallization and the dopant distribution in the Si close to the metallic contact [9]. Such an information is crucial for understanding how a paste interacts with the emitter structure and space charge region.
In this paper we present NanoSIMS and SEM measurements obtained on the optimally and the overfired cell to investigate the role of firing conditions on the electrical losses. By this unique technique (NanoSIMS), new insight about the lateral distribution of the dopant such as (P and B) can be obtained in the emitter and space charge region, yielding device-relevant information not available by other techniques. The resistive losses in mono-crystalline Si solar cells are still under intense debate and with this paper we believe to provide a key study for understanding such losses.

2 EXPERIMENTAL DETAILS

2.1 Solar cells processing and electrical characterization

p-type textured cells were processed on Cz grown <100> Si wafers with P doped emitters. Details of the solar cells processing of p-type cells are described in Refs. [6,7,9]. Cells #1 and #2 with P doped emitters (p-type) were contacted with the same front side metallization paste FSP1. The only difference is the firing temperature, cell #T1 was fired under optimal condition at 900 °C, yielding efficiency of 16.9%. Cell #T2 was fired at a higher set peak temperature of 960 °C and yielded a low efficiency of 13.6%. Details of the cell processing and electrical parameters are given in Table I. Electrical properties were measured at room temperature on large wafers by a standard solar cell tester at Fraunhofer ISE.
Specific contact resistance was measured by TLM measurement on the small cells. Due to overfiring cell #T2 showed a high specific contact resistance of 131 mΩ cm^2 at room temperature, whereas the optimally fired cell yielded a small contact resistance of 21 mΩ cm^2.

Table I. Efficiencies (η), fill factor (FF), open circuit voltage (Voc), series resistance (Rs) and firing condition (T_{FFO}) of p-type cells are given in Table I.

Solar Cells	η [%]	R_S [Ω.cm²]	FF [%]	ρ_C [mΩ.cm²]	T_{FFO} [°C]
#T1	16.9	0.87	77.1	21	900
#T2	13.6	3.91	61.8	131	960

2.2 Secondary Ion Mass Spectrometry with high lateral resolution (NanoSIMS)

SIMS is particularly suited for the study of dopants in silicon such as B and P. In this study, a Cameca NanoSIMS 50 equipped with a double focusing magnetic sector spectrometer was used in imaging mode [11]. The essential features of this instrument are its outstanding lateral resolution together with excellent detection limits enabling the analysis of the dopants in Si. It could be shown that a lateral resolution down to 50 nm can be reached in the negative secondary ion mode under Cs+ primary ion irradiation, for details see [11]. Further, details of the instrument are given in Refs. [9,11]

3 RESULTS AND DISCUSSION

3.1 Microstructure analysis by SEM

Fig. 1a–b show the microstructure of cell #T1 and the overfired cell #T2 in cross section. At the contact interface, pyramidal and lens-shaped Ag nano-crystals are embedded at the Si emitter surface.

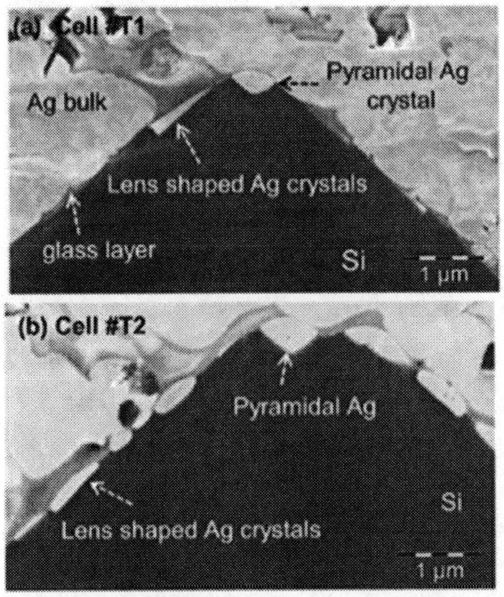

Figure 1: Cross-section SEM images of (a) cell #T1 and (b) the overfired cell #T2. Pyramidal and lens-shaped Ag nano-crystals are embedded at the Si emitter. The thickness of the glass layer increased slightly for the overfired cell #T2 as compared to cell #T1.

These nano-crystals are covered by a glass layer containing a small density of Ag colloids [6, 9]. For both cells #T1 and #T2 the microstructural features are apparently very similar, however, due to overfiring, a slight increased of the Ag nano-crystals size and the glass layer thickness was observed at the contact interface.
The contact resistance of the overfired cell #T2 was significantly larger (more than 6 times) than cell #T1. The glass phase thickness was found to be increased in the overfired cell and is considered one of the reasons for the increased contact resistance [9].

3.2 Chemical mapping by NanoSIMS on the doping level for cells #1 and #2

Correlative nanoanalytics by electron microscopy and NanoSIMS can be carried out to image the contact structure with a 100 nm resolution i.e., emitter and space charge region of the cell. The ion maps (in Fig. 2 and 3a–b) are presented as RGB overlays by assigning different colors: red for Si, green for P and blue for B (detected as SiB^- secondary ions).
Ion maps yield the elemental distribution in the metallization and the emitter with a high lateral resolution. In [8] a clear guideline is given for reading the maps and assigning the different device relevant areas of the cell. For interpretation of the NanoSIMS results we start with P map image as shown in Fig. 2b: the marked areas labelled as: (1) metallization region, (2) highly doped emitter (homogenous layer) best seen in P map, not in the Si map, (3) the emitter region, and (4) similar to (2), however, an inhomogenous layer (Fig. 3c) with varying P concentration, varies from green (high P) to grey (low P) can be seen in the image. Note: (2) is easily visible in P map, (4) is imaged (Fig. 2c) in the superimposed map (Si, P, SiB).
The emitter structure in cell #T1 displays a high-intensity P signal (green line, Fig. 3a) along the interface with the glass phase, seen here as a bright green line (Fig. 3a and c). Such signal are not observed for the overfired cell #T2 (Fig. 3b and d), i.e. the emitter structure got disrupted in the overfired cell. Instead, P appears in small P rich domains in the metallization area but nowhere in the Si emitter area. This indicates that the original emitter structure has not been preserved in the overfired cells. In contrast, the emitter structure is totally preserved in the optimally fired cell #T1. However, slight inhomogeneities were observed at the contact interface yielding areas of P rich lines and areas containing no P rich lines.
A schematic sketch of the Si emitter, metallization and bulk Si for cell #T1 and the overfired cell #T2 has been introduced and is shown in Fig. 3c–d, respectively.

3.3 Dopant (P, B) distributions by NanoSIMS for the overfired cell #2

Different to the SEM analysis, in NanoSIMS large structural differences were observed for the two cells, which explain the large differences in the efficiency values. In addition, the overfired cell #T2 was investigated in cross-sections (Fig. 4a–c) and plan view (Fig. 4d–f).
The overfired cell #T2 yielded a diffusion cloud that extends over 2-5 µm below the metallization at an area that has been assigned to the emitter. B rich pockets are observed at this interface extending over several microns marked as light yellow in Fig. 4b.

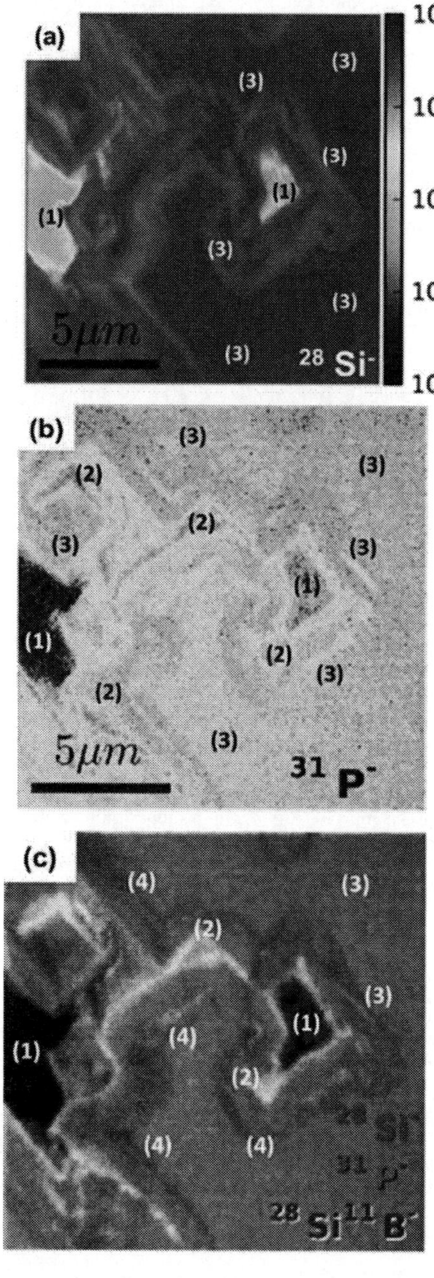

Figure 2: Cell #T1, NanoSIMS results (a) Si map, (b) P map and, (c) RGB overlay ion maps of Si, P, and SiB.

In the overfired cell a diffusion zone is identified below the contact several μm wide with an increased P content best seen in the cross-sectional map (see Fig. 4c). It develops B rich regions and P rich regions, which give rise to the enhanced series and contact resistance [9].

Line profiles for Si, P and B were obtained along the arrow indicated in Fig. 5a. The P counts (100 counts, Fig. 5c) below the metallization is constant for around 2 μm from the emitter surface, and it falls to a few counts (1-2 counts) below that, while B counts remain constant throughout the substrate. Note, the P concentration profile fall off at the edge of the green to blue transition seen in (c) over two orders of magnitude.

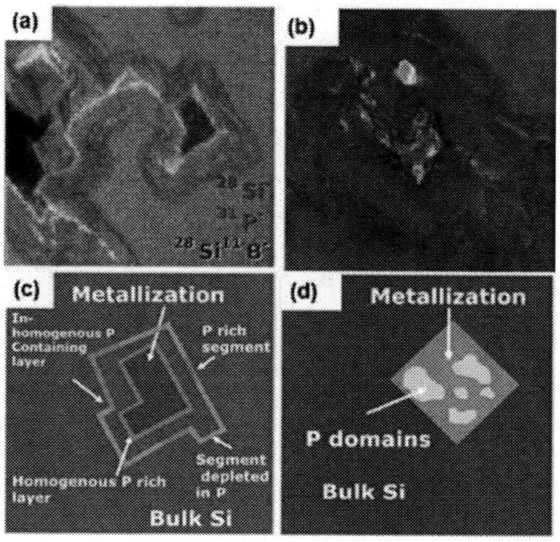

Figure 3: NanoSIMS results obtainmed in plan-view of cell #T1 (a), the overfired cell # T2 (b). RGB overlays are composed of Si (red), P (green) and SiB (blue) signals. Schematic sketch of cell #T1 (Fig. 3c) with preserved emitter structure and the overfired cell #T2 (Fig. 3d) with disrupted emitter structure is seen.

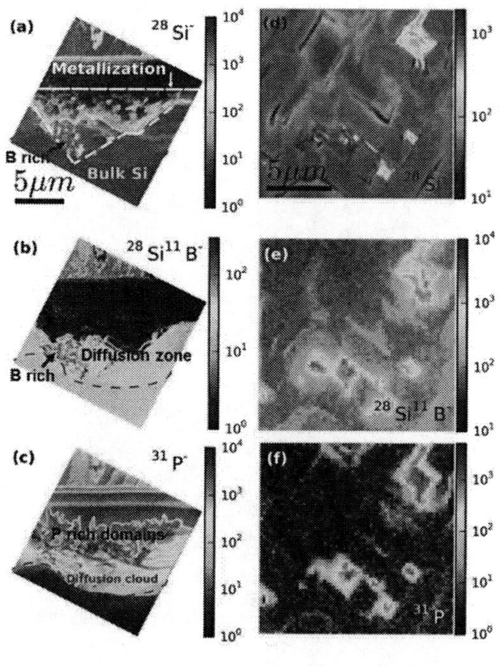

Figure 4: RGB overlays of the overfired cell #T2 (a-c) in cross-sections and (d-f) in plan view.

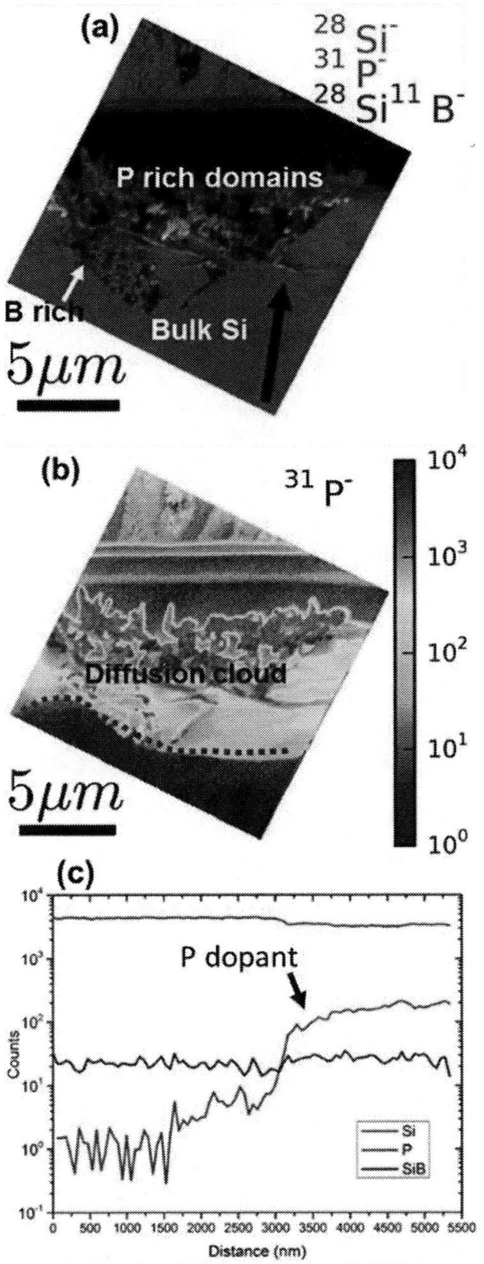

Figure 4: Details of the overfired cell #T2 in cross-section: (a) RGB overlay of Si, P and B (detected as SiB- secondary ions) maps in red, green and blue, respectively in a linear scale (b) P map in the log color scale, and (c) line profiles along the arrow indicated in the image.

In summary, as a major result we can prove quantitatively by NanoSIMS that due to the overfiring the emitter P dopant profile got completely disrupted yielding a dramatic decrease in efficiency. This result could not be obtained without the excellent detection limits of the SIMS method. It yields great possibilities for understanding device properties and effects introduced by the processing of the front side metallization in the emitter and space charge region.

4 CONCLUSIONS

In summary, by electron microscopy and NanoSIMS we have demonstrated a high-resolution and high-sensitivity for understanding the electrical losses of the overfired cells. It is indeed remarkable that the high sensitivity of NanoSIMS enables direct imaging of dopant distributions across the metallization interface in comparison to conventional analytical tools such as EDX.

It was found that the emitter structure was preserved for cell #T1 thereby spatially confining B and the P. While in the overfired cell the diffusion barrier (SiN_x) layer was completely disintegrated leading to barrier-free diffusion of B rich pockets (extending over several μm) into the substrate. Due to the complete disintegration of the emitter structure below the metallization, a diffusion cloud is found yielding a diffuse interface. With such structures an increased contact resistance could be explained.

5 ACKNOWLEDGEMENTS

We acknowledge the group of D. Biro and F. Clement, Fraunhofer Institute for Solar Energy Systems at Freiburg, for providing high-quality p-type solar cells and electrical measurements on the large cells. NanoSIMS data were acquired at Luxembourg Institute of Science and Technology (LIST), Belvaux, Luxembourg.

6 REFERENCES

[1] M.M. Hilali, S. Sridharan, C. Khadilkar, A. Shaikh, A. Rohatgi, S. Kim, J. Electron. Mater. 35 (2006) 2041–2047.

[2] M. Hörteis, T. Gutberlet, A. Reller, S.W. Glunz, Adv. Funct. Mater. 20 (2010) 476–484.

[3] G. Schubert, F. Huster, P. Fath, Sol. Energy Mater. Sol. Cells. 90 (2006) 3399–3406.

[4] C. Ballif, D. M. Huljić, G. Willeke, A. Hessler-Wyser, Appl. Phys. Lett. 82 (2003) 1878–1880.

[5] D. Pysch, A. Mette, A. Filipovic, S. W. Glunz, Prog. Photovoltaics Res. Appl. 17 (2009) 101–114.

[6] P. Kumar, B. Willsch, M. Dürrschnabel, Z. Aabdin, R. Hoenig, N. Peranio, F. Clement, D. Biro, O. Eibl, Energy Procedia. 67 (2015) 31–42.

[7] P. Kumar, M. Pfeffer, B. Willsch, O. Eibl, Sol. Energy Mater. Sol. Cells. 145 (2016) 358-367.

[8] P. Kumar, M. Pfeffer, B. Willsch, O. Eibl, L. Yedra, S. Eswara, J-N. Audinot, T. Wirtz, Sol. Energy Mater. Sol. Cells. 160 (2017) 398–409.

[9] S. Bin Cho, H. S. Kim, J. Y. Huh, Acta Mater. 70 (2014) 1–7.

[10] A. Edler, V.D. Mihailetchi, L.J. Koduvelikulathu, C. Comparotto, R. Kopecek, R. Harney, Prog. Photovoltaics Res. Appl. 23 (2014) 620–627,

[11] C. W. Mueller, P. K. Weber, M. R. Kilburn, C. Hoeschen, M. Kleber, J. Pett-Ridge Adv. Agron. 121 (2013) 1.

MEASUREMENT SETUP FOR IN-SITU QUANTUM YIELD CHARACTERIZATION OF SOLAR CELLS DURING HIGH ENERGY PARTICLE IRRADIATION

Heinz-Christoph Neitzert[1], Vincenzo Carrano[1], Giovanni Landi[1] and Lucio Gialanella[2,+]
[1] Dipartimento di Ing. Industriale (DIIN), Università di Salerno, Fisciano, Italy
[2] INFN Sezione di Napoli, Italy
[+] Dipartimento di Matematica e Fisica, Università della Campania L.Vanvitelli, Caserta, Italy

ABSTRACT: An extremely simple on-line quantum yield measurement tool was developed, that permits to measure the photocurrent spectra of solar cells under harsh measurement conditions without the need of optical windows or moving parts. The small photocurrent, induced by the sequential illumination by a series of LEDs with different emission wavelengths, distributed within the responsivity range of the tested solar cell, is converted to a proportional frequency and then monitored using the audio-card of a personal computer and an appropriate Labview-based Software. Subsequently the quantum yield measurements were performed in-situ during the irradiation of a conventional crystalline Silicon solar cell with 1.7 MeV protons in the spectral range between 400 nm and 880 nm. In this case the complete photocurrent spectrum can been monitored continuously with good temporal resolution.
Keywords: Proton irradiation, Degradation, Space Cells, In-situ characterization

1 INTRODUCTION

Solar cells for space applications have to be characterized and certified regarding their radiation resistance. In-situ characterization of solar cell materials and solar cells during high energy particle irradiation has been frequently reported in literature [1,2,3] but usually just the conductivity [1], photoconductivity [2] or solar cell current-voltage characteristics [3,4] has been monitored during the irradiation. For a better understanding of the degradation mechanisms of the solar cells in space, however, it would be desirable also to obtain spectrally resolved photoelectric properties already during irradiation.

Instead of using bulky monochromators or filter wheels, an alternative in order to generate spectrally resolved light can be the use of a series of LEDs with different wavelengths.. Such a solution for a solar cell simulator has been already proposed in literature [5]. Here we will show a very simple and compact measurement system for quantum yield measurements, which does not require any moving parts and can be easily integrated into a vacuum system without the need of optical windows. Additionally it will be shown that after conversion of the usually small photocurrents into a proportional signal frequency, the data taking can be done simply by using the audio-card of a personal computer. It has already been demonstrated that an audio-card based measurement system can be used for the realization of a sensitive light beam induced current (LBIC) measurement system [6].

2 EXPERIMETAL

2.1 Measurement setup

The in-situ solar cell quantum yield characterization system uses an illumination system with 8 LEDs with different emission spectra, covering point-wise the requested wavelength range. The normalized emission spectra of the 7 monochromatic LEDs, used in the here presented experiment are shown in Figure 1, together with the quantum yield spectrum of the characterized crystalline silicon (c-Si) solar cell. It can be seen, that they cover the wavelength range between 400 nm and 880 nm and hence an important part of the spectral range of crystalline silicon solar cells and photodetectors.

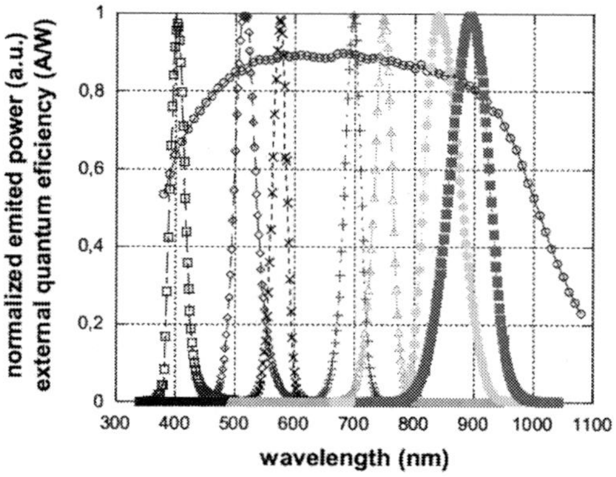

Figure 1: Normalized emission spectra of 7 monochromatic LEDs and the quantum yield spectrum (orange open circles) of a crystalline Silicon solar cell.

Additionally for synchronization purposes a conventional white LED, based on InGaN blue emitting emitter, combined with a phosphor for the longer wavelength emission, has been added. These 8 LEDs are one by one addressed in a sequential way by a control electronics, consisting basically of a clock generator, a shift register and LED drivers with adjustable output currents, which without synchronization in an endless cycle lights up the LEDs. The interface electronics, schematical shown in Figure 2, consists of a transimpedance amplifier, that converts the solar cell photocurrent, induced by the LED illumination, into a proportional voltage, an operational amplifier, that adds to this signal an additional small offset voltage and finally a highly linear voltage-frequency converter and an output buffer. The resulting ac-signal, with a frequency proportional to the input current is then fed to the microphone input of a PC-audiocard. A Labview based program determines the ac-signal frequency with high precision and memorizes the resulting frequency values as a function of time. A detailed description of the receiving part of the measurement system has been given in [6].

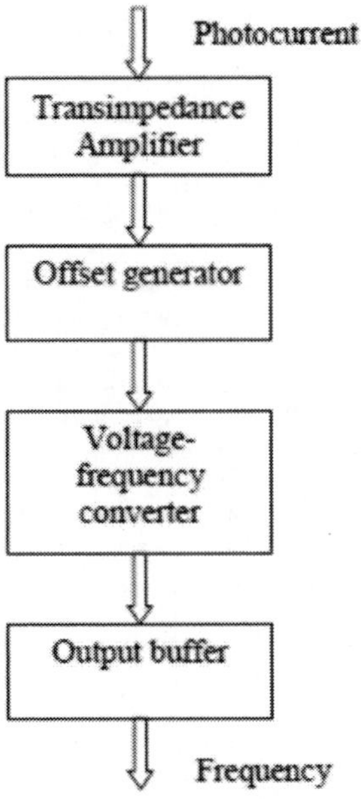

Figure 2: Block diagram of the interface electronics.

The good linearity of the voltage-frequency converter is shown in Figure 3. The whole measurement system with the multiple wavelength LED sources and the above described interface electronics fits into a vacuum tube, as shown in the photo in Figure 4, and only three electrical connections to the external ambient are requested: The two power supply connections and one connection for the output ac- signal, which carries the frequency coded information of the solar cell short circuit current.

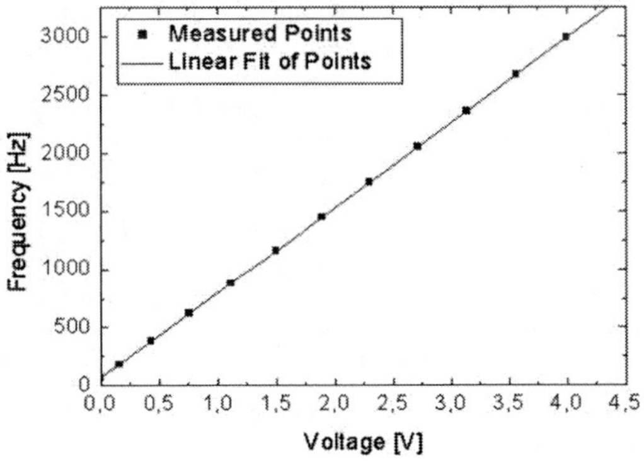

Figure 3: Linearity check of the voltage-frequency converter

2.2 Proton irradiation

The proton irradiation has been done using the TANDEM accelerator of the INFN at Naples university. In the particular case a proton energy of 1.7 MeV has been used. This is the energy value, that for crystalline silicon solar cells with about 300 μm thickness produces the maximum radiation induced damage. As shown by TRIM simulation [7] is the projected range in silicon at this energy about 37 μm, so that the charge carrier diffusion length in the silicon absorber is strongly decreased for irradiation doses exceeding 1e12 protons/cm^2. (see [8,9]). The proton beam has been focused and then scanned across the solar cell area in order to obtain an average homogeneous irradiation over the whole device area.

As device under test a commercial homojunction monocrystalline silicon solar cell has been used and connected to the input of the transimpedance amplifier.

Figure 4: Photo of the LED interface electronics.

3 RESULTS AND DISCUSSION

3.1 In-situ monitoring of the short circuit current without proton beam

A typical measurement sequence for the spectral characterization of the solar cells before the proton irradiation is shown in Fig.4, where a frequency logging plot for 120s is shown, when the different LEDs – first the white LED and then successively the 7 monochromatic LEDs with increasing wavelength from 400 nm up to 880 nm are illuminating the solar cell. It should be mentioned that for every wavelength about 20 measurements have been done. The resulting frequency-logging plot shows very stable and noise-free operation of the measurement system. The LED current values have been adjusted in a way, that the different colors induce distinct photocurrent values and hence clearly distinguishable frequencies at the output of the interface electronics. In order to facilitate the identification of the actual LED color, each cycle starts with a white light illumination, that is biased in a way, that it induces in the investigated solar cell the highest photocurrent.

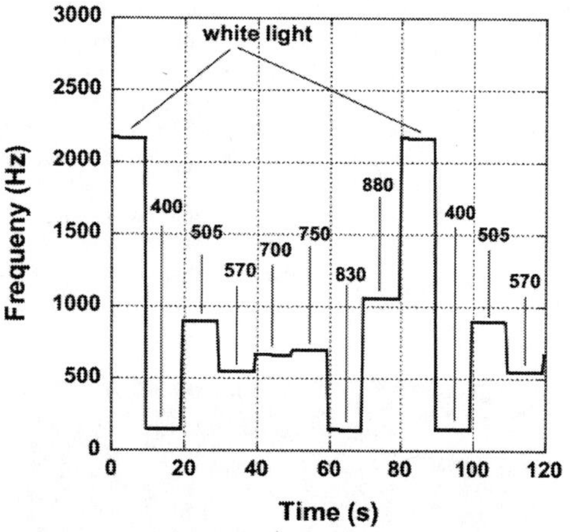

Figure 5: Frequency monitoring before the irradiation of a crystalline silicon solar cell, that is illuminated with a LED sequence with different wavelengths.

3.2 In-situ monitoring of the short circuit current with proton beam

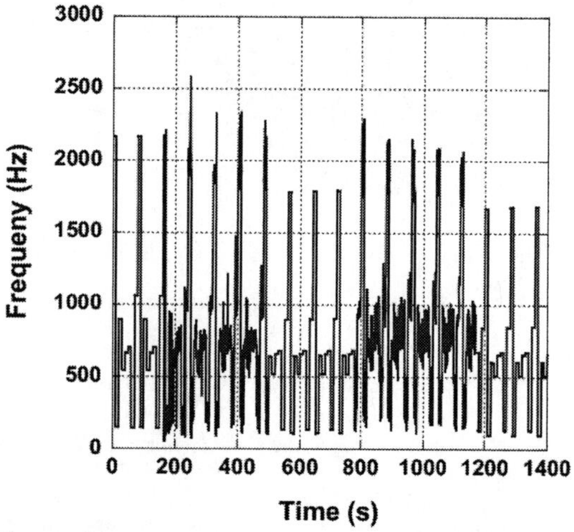

Figure 6:.Frequency monitoring during the irradiation with a proton beam of 1.7 MeV of a crystalline silicon solar cell, that is illuminated with a sequence of LEDs with different wavelengths. The proton irradiation is periodically interrupted.

In Figure 6 a measurement sequence with about 18 cycles is shown, with the proton beam being switched on and off in regular intervals. When the proton beam is switched on after about 180 s, one observes initially an increasing value of the measured frequency, due to the additional contribution of the radiation induced current (RIC) to the total measured solar cell short circuit current. Then the maximum signal amplitudes are continuously decreasing due to the solar cell damaging. It can also be seen, that during proton irradiation the signal trace gets rather noisy. This is due to the fact, that the proton beam is only partially hitting the solar cell. Therefore the RIC contribution to the overall signal is varying chaotically in time. After a time of about 500 s the proton beam has been switched off and the signal amplitudes stabilize again until the proton beam is switched on again at a time of about 800s and switched off again after 1200 s. For a further analysis of the taken frequency logging data, the following procedure is proposed:

The average signal amplitude for the respective LED wavelengths is determined during the periods, when the proton beam has been interrupted and then plotted as a function of the received irradiation time or proton dose. When the quantum yield spectrum of the solar cell under test has been determined before irradiation by an independent ex-situ measurement, the relative change determined in the breaks between the proton irradiation periods, can be translated into a piecewise reconstructed quantum yield spectrum.

4 CONCLUSIONS

A very compact measurement system for the in-situ characterization of the wavelength dependent photoresponse of solar cells and photodetectors has been developed. The LED and PC-audio card based system with a high dynamic range enables the fast measurement of the wavelength dependence of the solar cell photoresponse without the need of optical windows, moving parts or additional electronic instrumentation. The good stability of the measurement system has been verified during the irradiation of a monocrystalline silicon solar cell with high energy protons.

References

[1] A. Battaglia, S. Coffa, F. Priolo, C. Spinella, S. Libertino, Nuclear Instruments and Methods in Physics Research B 96 (1995) 219.

[2] N. Kishimoto, H. Amekura, K. Kono, C. G. Lee, Journal of Non-Crystalline Solids 227-230 (1998) 238.

[3] A. Filevich, C. J. Bruno, J. Fernandez Vasquez, M. Alurralde, I. Prario, M. J. L. Tamasi, M. G. Martinez Bogado, J. C. Pla, J. Duran, J. Schuff, A. Burlon, P. Stoliar, D. Minsky, A. J. Kreiner, R. Mayer, IEEE Trans. Nuclear Science 50 (2003) 2380.

[4] F. Lang, N. H. Nickel, J. Bundesmann, S. Seidel, A. Denker, S. Albrecht, V. V. Brus, J. Rappich, B. Rech, G. Landi, H. C Neitzert, Advanced Materials 28 (2016) 8726.

[5] S. Kohraku, K. Kurokawa, Proceedings of the 3rd World Conference on Photovoltaic Energy Conversion (2003) 1977.

[6] H. C. Neitzert, N. G. Rainone, Proceedings 22nd European Photovoltaic Solar Energy Conference, Conference (2007) 261.

[7] H. C. Neitzert, P. Spinillo, S. Bellone, G. D. Licciardo, M. Tucci, F Roca, L. Gialanella, M. Romano, Solar Energy Materials & Solar Cells 83 (2004) 435.

[8] H. C. Neitzert, M. Ferrara, M. Kunst, A. Denker, Z. Kertesz, B. Limata, L. Gialanella, M. Romano, Phys. Stat. Sol. (b) 245 (2008), 2877.

[9] M. Scherff, R. Drzymalla, R. Goesse, W. R. Fahrner, M. Ferrara and H. C. Neitzert, J. Opitz-Coutereau, A. Denker, R. Stangl, B. Limata, L. Gialanella and M. Romano, J. of the Electrochemical Society 153 (2006) G1117.

Electrical and Optical Characterization of Crystalline Silicon Solar Cells Using Luminescent Down-Shifting of MAPbBr3 Perovskite Nanophosphors Deposited by Spin-on Film Technique

Zong-Xian Lin, Wen-Jeng Ho*, Guan-Yi Li, Bang-Jin You, and Jheng-Jie Liu
Department of Electro-Optical Engineering, National Taipei University of Technology
No. 1, Sec. 3, Zhongxiao E. Rd., Taipei 10608, Taiwan, R.O.C.
*e-mail: wjho@ntut.edu.tw

ABSTRACT:The photovoltaic-performance characterizations of crystalline silicon (C-Si) solar cells based on the combination of plasmonics scattering of silver nanoparticles (Ag NPs) and luminescent downshifting (LDS) of Eu-doped phosphor-particles are experimentally demonstrated in this work. The surface morphologies and plasmonics scattering of Ag NPs of various dimensions are characterized by the measurements of scanning electron microscopy (SEM), optical reflectance and absorbance, and external quantum efficiency (EQE). Furthermore, the LDS characteristics of Eu-doped phosphor-particles silicate layer coated by spin-on film technique over the surface of Ag NPs are revealed on the spectrum of optical reflectance and EQE. The photovoltaic-performance improvements of the C-Si solar cells with the combination of plasmonics scattering and LDS are confirmed by photovoltaic current-voltage (I-V) measurement under one sun AM 1.5G illumination. In this study, the efficiency enhancement of 21.95% for the cell with a antireflective coating (ARC) of SiO_2, 23.62% for the cell with Ag NPs embedded in a pure SiO_2 ARC, and 26.49% for the cell with Ag NPs embedded in a Eu-doped phosphor-SiO_2 ARC were obtained, respectively, compared to the reference bare solar cells.
Keywords: Luminescent down-shifting, $MAPbBr_3$, nanophosphors, perovskite, textured crystalline silicon solar cell

1.Introduction

A higher recombination loss would be exhibited in the surface of photovoltaic devices because the incident photons of higher energy were absorbed within a short distance from the surface. Thus, the reduction in surface recombination loss can be a factor to enhance the conversion efficiency of solar cells. There are two principal approaches to achieve a more efficient utilization of the short wavelength part ($h\nu > E_g$) of the solar spectrum to enhance the conversion efficiency of solar cells. The first is used the multi-junction structure and the second is used the spectral conversion of the luminescent down-shifting (LDS) of the incident spectrum [1]. The LDS materials can absorb high-energy photons and re-emitted lower-energy photons for the applications of solar cells to improve low spectral response of crystalline silicon (C-Si) solar cells at short wavelength band [2-4]. However, it is difficulty to uniformly deposit LDS materials on the textured C-Si solar cells, which affects the antireflection of the pyramid textured-surface.

In this study, the electrical and optical characteristics of the textured C-Si solar cells coated with LDS material of $MAPbBr_3$ perovskite nanophosphors [5-7] by spin-on film technique using two-step spinning rate [8] are demonstrated. The surface morphologies and coverage and the optical properties of the perovskite nanophosphors of various $MAPbBr_3$ concentrations are characterized by the measurements of scanning electron microscopy (SEM) and optical reflectance. Photoluminescence (PL) is used to examine the emitted wavelength of MAPbBr3 nanophosphor. The reflective and external quantum efficiency (EQE) spectra are used to reveal the LDS characteristics of $MAPbBr_3$ nanophosphor. The photovoltaic performances of the cells depending on the concentration of $MAPbBr_3$ nanophosphor are examined by using the photovoltaic current-voltage (I-V) measurement. The effectiveness of the LDS effects of $MAPbBr_3$ nanophosphor on textured C-Si solar cells is demonstrated for the cells having much high efficiency with the proper spinning and concentration conditions.

2. Experiments

Before to characterize electrical and optical properties of textured C-Si solar cells coated with $MAPbBr_3$ nanophosphors material, the textured C-Si solar cells are fabricated firstly as the reference solar cells by following processes. (1) The C-Si wafer was cut into C-Si sample in size of 1×1 cm^2. (2) The C-Si sample with saw damaged surface was removed by dipping the sample in a H_2O/KOH solution. The surface of the sample was then etched by dipping in a solution of H_2O/KOH/IPA, which pyramid textured-surface was formed. (3) After standard cleaning, an n^+-Si emitter layer with a sheet resistance of 80 Ω/sq was applied to the textured samples using a $POCl_3$ diffusion process in a tube diffusion chamber at 850 °C for 3 min. (4) After phosphorus diffusion, the oxide-layer of phosphorus glass on the surface of the samples was removed using hydrogen fluoride (HF) solution. (5) A SiNx layer of one-quarter wavelength thick, as an antireflective coating (ARC), was deposited by PECVD on the emitter surface. (6) Al paste was printed on the rear side and dried at 200 °C. Then the grid pattern of Ag was screen-printed on the front side. The samples with Ag and Al contacts were co-fired at a peak temperature of 750 °C to form a good Ohmic-contact with p-Si and n^+-Si semiconductor. After thermally annealing, the bare textured C-Si solar cells with SiNx ARC (which called the reference cells) were obtained. Finally, the 5, 10, 40 mg/ml of $MAPbBr_3$ nanophosphors mixing with toluene solutions were coated on the bare textured C-Si solar cells by spin-on film processing. The schematic diagram of the fabricated textured silicon solar cell coated with a $MAPbBr_3$ nanophosphors layer is shown in Fig. 1.

The surface morphologies and coverage of the perovskite nanophosphors of various $MAPbBr_3$ concentrations are characterized by SEM. The fluorescence emission spectrum of $MAPbBr_3$ nanophosphors was characterized by PL measurement at room temperature. The optical reflectance and EQE response measurements are used to reveal the LDS characteristics of $MAPbBr_3$ nanophosphor. Furthermore, the photovoltaic current density-voltage (J–V) characteristics of the cells coated with perovskite material of various $MAPbBr_3$ nanophosphor concentrations and without coated perovskite material of $MAPbBr_3$ nanophosphor were also measured and compared to confirm the conversion efficiency improved by inducing of the LDS effects of $MAPbBr_3$

nanophosphors.

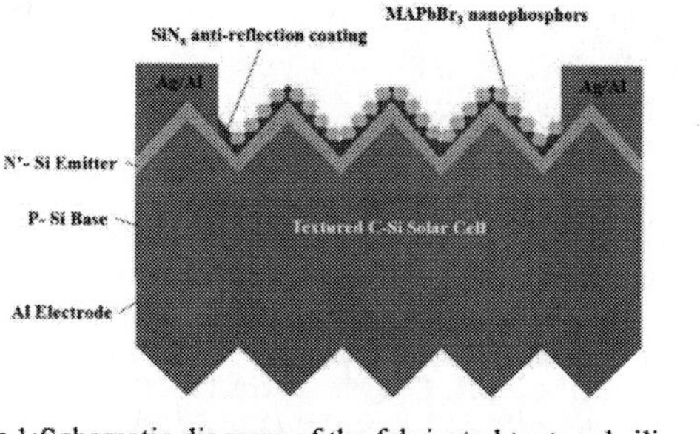

Fig.1:Schematic diagram of the fabricated textured silicon solar cell with a $MAPbBr_3$ nanophosphors layer.

3. Results and Discussion

Figure 2 shows the top-view SEM images of $MAPbBr_3$ nanophosphors deposited on the textured silicon solar cell with concentration of (a) 5 mg/ml, (b) 10 mg/ml, and (c) 40 mg/ml. The images results show that the surface coverage of phosphor particles increases with the deposited concentration of $MAPbBr_3$ nanophosphor. The dimension of single nanophosphor-particles was approximately 20-50 nm. Especially, MAPbBr3 nanophosphor particles are uniformly distributed over the textured surface of C-Si cell with the concentration of 10 mg/ml. however, the nanophosphor particles accumulated on the trough of pyramidal textured-surface of the cell with the concentration of 40 mg/ml can be found which with the same spinning condition. A higher coverage value was obtained for a high deposited concentration. Fig. 3 displays the PL fluorescence spectrum of the used $MAPbBr_3$ nanophosphors measured at room temperature. The PL emission peak is presented approximately at 530 nm with a full width at half maximum of about 25 nm. Fig. 4 shows the optical reflective spectrum of the reference cell without phosphor particles and the cells with the perovskite nanophosphors of various $MAPbBr_3$ concentrations. The reflectance of the cell with phosphor particles is significantly lower than that of the cell without phosphor particles below the wavelength of 475 nm which is attributed to the absorption of high energy incident photons by phosphor particles. A lower reflectance for the cell with a higher concentration of $MAPbBr_3$ nanophosphor was found because of the high-energy photons absorbing by much amount of $MAPbBr_3$ nanophosphor particles. It is noticeable that the reflectance beyond the wavelength of 500 nm which a higher reflectance exhibited for the cell with concentration of 40 mg/ml one due to the $MAPbBr_3$ nanophosphor particles accumulated on the trough of pyramidal textured-surface that affects the antireflection of pyramidal textured-surface. Fig. 5 shows the EQE response of the reference cell without phosphor particles and the cells with the perovskite nanophosphors of various $MAPbBr_3$ concentrations. The obtained EQE results show that the EQE response is almost agreed to the reflective spectrum of the proposed cells. In addition, the EQE values at the wavelength range of 325-450 nm for the cells with the perovskite nanophosphors of various $MAPbBr_3$ concentrations were higher than that of the reference cell without of MAPbBr3 nanophosphor particles. Therefore, the obtained optical reflectance and EQE results are well showed the LDS characteristics of $MAPbBr_3$ nanophosphor. Fig. 6 displays the photovoltaic current density-voltage (J-V) characteristics of the reference cell without phosphor particles and the cells with the perovskite nanophosphors of various $MAPbBr_3$ concentrations. The photovoltaic performances of these cells are also summarized in Table I. In general, the levels of the quantum efficiency of a solar cell can indicate the amount of photo-generated current. Integrating the quantum efficiency of a cell over the entire solar energy spectrum would make it possible to evaluate the amount of the short-circuit current density (J_{SC}) that the cell could produce when exposed to sunlight. Thus, the J_{SC} of the photovoltaic devices is proportional to the EQE response. In present work, the J_{SC} of 37.62 mA/cm^2 and conversion efficiency (η) of 15.70% for a textured C-Si solar cell coated with a $MAPbBr_3$ nanophosphor concentration of 10 mg/ml were better than that of J_{SC} = 35.72 mA/cm^2 (η=14.95%) for the cell with a concentration of 40 mg/ml and J_{SC} = 37.45 mA/cm^2 (η=15.25%) for the cell with a concentration of 5 mg/ml. Thus, the LDS effect of $MAPbBr_3$ nanophosphor on the textured C-Si solar cells is depended on the proper spinning and concentration conditions. For the textured C-Si solar cell to reach a high efficiency, the optimal spinning and concentration conditions are undergoing study in our laboratory.

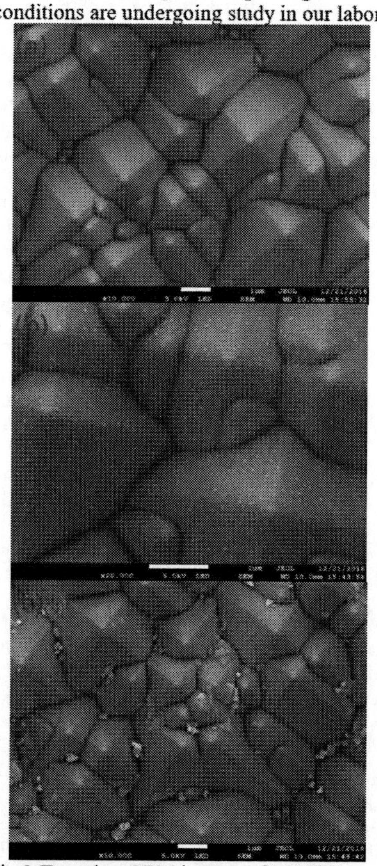

Fig.2:Top-view SEM images of $MAPbBr_3$ nanophosphors deposited on the textured silicon solar cell with concentration of (a) 5 mg/ml, (b) 10 mg/ml, and (c) 40 mg/ml.

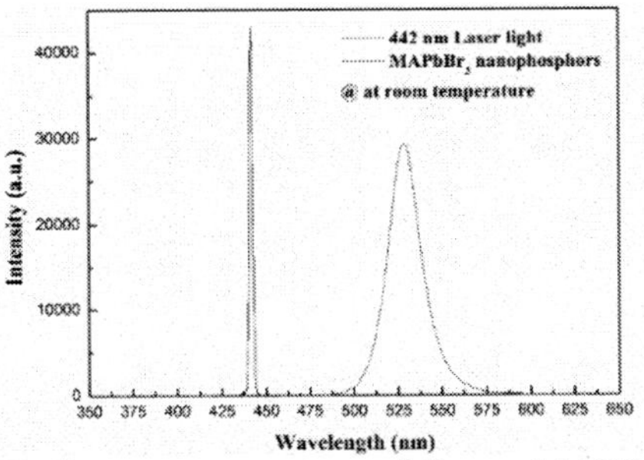

Fig.3:PL fluorescence spectrum of the used $MAPbBr_3$ nanophosphors measured at room temperature.

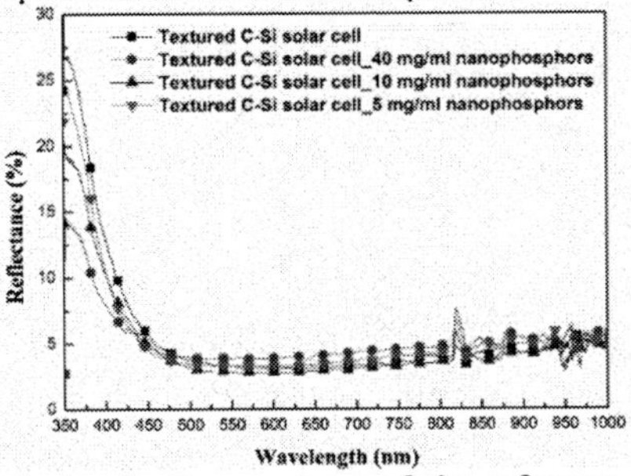

Fig.4:Optical reflective spectrum of the reference cell without phosphor particles and the cells with the perovskite nanophosphors of various $MAPbBr_3$ concentrations.

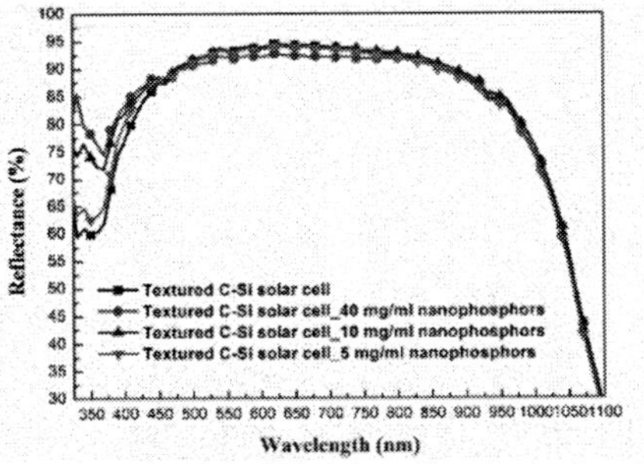

Fig.5:EQE response of the reference cell without phosphor particles and the cells with the perovskite nanophosphors of various $MAPbBr_3$ concentrations.

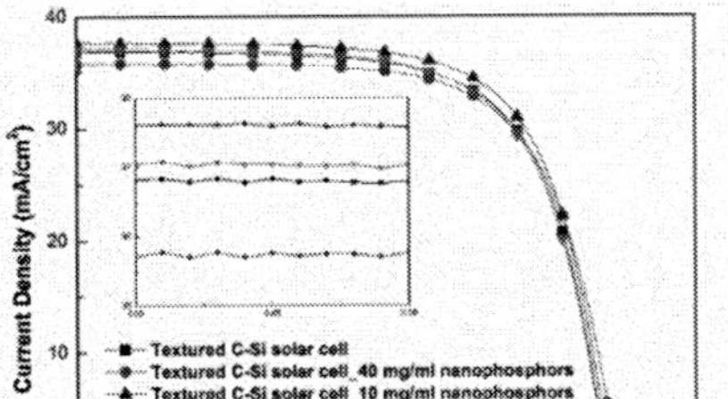

Fig.6:Photovoltaic current density-voltage (J-V) characteristics of the reference cell without phosphor particles and the cells with the perovskite nanophosphors of various $MAPbBr_3$ concentrations.

	V_{oc} (mV)	J_{sc} (mA/cm^2)	FF (%)	η (%)	ΔJ_{sc} (%)	Δη (%)
Textured C-Si SC_A	610.4	36.36	68.48	15.19	-	-
Textured C-Si SC_A (40 mg/ml nanophosphors)	610.4	35.72	68.59	14.95	-1.76	-1.58
Textured C-Si SC_B	598.2	36.83	68.82	15.16	-	-
Textured C-Si SC_B (10 mg/ml nanophosphors)	602.5	37.62	69.3	15.70	2.14	3.56
Textured C-Si SC_C	598.3	37.2	67.96	15.12	-	-
Textured C-Si SC_C (5 mg/ml nanophosphors)	598.6	37.45	68.06	15.25	0.67	0.86

Table I:The photovoltaic performances of the reference cell without phosphor particles and the cells with the perovskite nanophosphors of various $MAPbBr_3$ concentrations.

4.Conclusion

The electrical and optical characteristics of the textured C-Si solar cells coated with LDS material of $MAPbBr_3$ perovskite nanophosphors by spin-on film technique have been demonstrated. The optical reflectance and EQE response measurements are revealed the LDS characteristics of $MAPbBr_3$ nanophosphor with various concentrations. A textured C-Si solar cell coated with a $MAPbBr_3$ nanophosphor concentration of 10 mg/ml under the same spinning condition exhibited a higher photovoltaic performance. The spinning condition and the concentration of $MAPbBr_3$ perovskite nanophosphors are still needed to optimize for the textured C-Si solar cells applications.

5. Acknowledgement

The authors would like to thank the Ministry of Science and Technology of the Republic of China for financial support under Grant MOST 103-2221-E-027-049-MY36.

References

[1] X. Y. Huang, et al., ***J. Appl. Phys. 109***, 113526 (2011).

[2] W. B. Hung, et. al., ***J. of Ceramic Processing Research. 15(3)***, 157 (2014).

[3] Guojian Shao, et. al., ***J. Lumin. 157***, 344 (2015).

[4] S. D. Hodgson, et. al., ***Prog. Photovolt. Res. Appl. 23***, 150 (2015).

[5] L. C. Schmidt, et. al., ***J. Am. Chem. Soc. 136***, 850 (2014).

[6] M. Cha, et al., ***J. Am. Chem. Soc. 138***, 8581 (2016).

[7] S. S. Mali, et al., ***NPG Asia Materials 7***, e208 (2015).

[8] X. Pi, et al., ***Sol. Energy Matter. Sol. Cells 95,*** 2941 (2011).

VALIDATION OF ANALYTIC MODELLING OF LOCAL REAR CONTACTS IN PERC/PERL SOLAR CELLS

Pierre Saint-Cast, Nico Wöhrle, Johannes Greulich
Fraunhofer Institute for Solar Energy Systems ISE
Heidenhofstraße 2, 79110 Freiburg, Germany

ABSTRACT: This work focuses on the comparison of analytical modelling of local rear contacts with two- and three-dimensional (2D and 3D) numerical simulation. The analytical models were developed to calculate the effective surface recombination velocity (S_{eff}) of the rear surface and the bulk contribution to the series resistance ($Rs_{\_rear}$). These parameters represent an effective rear surface and when combined to a one-dimensional (1D) simulation can be used to model the current-voltage characteristics of the solar cell. The first achievement of this paper is a solid comparison of the analytical model and the numerical simulation. The deviation is found to be lower than 20% for more than 95% of the data reported. However the model of S_{eff} is found to be limited to open circuit (OC) condition. Therefore a correction, which extended the validity range of the S_{eff} calculation to maximum power point (MPP) condition, is proposed without loss of accuracy. The last achievement is the comparison of the efficiencies obtained using the 1D simulation with the ones obtained using the 2D and 3D numerical simulation. This method ensures with 95% confidence that the deviation in efficiency will not exceed ± 0.18%$_{abs}$ for point contact and ± 0.28%$_{abs}$ for line contacts.

Keywords: Analytical modelling; Simulation; PERC

1 INTRODUCTION

"Passivated Emitter and Rear Cell" (PERC, [1]) technology is quickly spreading out in crystalline silicon photovoltaics, becoming the industry standard for the next decades. The structuring of the rear of solar cells poses the problem of trade-offs between high rear surface recombination and low series resistance with its mutual variable being contact distance. So far, the theoretical models proposed have been focusing on finding the optimal contact geometry for one given contact technology [2-10]. Most of the simple models [2-4, 9] are based on or derived from the model of Fischer [2]. This allows for an accurate calculation of the effective surface recombination velocity in open-circuit conditions, which might not be accurate at maximum power point. Other models are based on numerical simulations or on calculation spreadsheets [5-8]. They have the advantages to be more accurate than the analytical solutions, but this higher degree of accuracy is paid for by longer computing time, license costs, potentially reduced understanding and the precious time of the best experts. In contrast, suitable, simple, widely applicable and accurate analytical expressions on the other hand would embrace all possible solutions within a few formulae.

The goal of many of the analytical models [2-4] of the PERC solar cell is to replace the structured rear surface by an effective rear surface that is characterized by two parameters: the effective surface recombination velocity (S_{eff}) for the rear surface, and an external series resistance ($Rs_{\_rear}$) which accounts for the resistance in the bulk. The first goal of this paper is the comparison of the values of $Rs_{\_rear}$ and S_{eff} resulting from an earlier published analytical model [4] with its counterpart obtained from three-dimensional (3D) and two-dimensional (2D) numerical simulations.

The $Rs_{\_rear}$ and S_{eff} obtained need to be implemented in a separate one-dimensional (1D) model (for example Pitchmaster [10], PC1Dmod [12]), in order to model a solar cell. It is often assumed implicitly that this multidimensional problem (2D or 3D) is equivalent to a one dimensional problem with an effective rear; however it has never been verified to be valid. The second goal of this paper is to verify that this method works. Therefore S_{eff} and $Rs_{\_rear}$ are use as parameters of a 1D numerical simulation to model the solar cell. The efficiency obtained is then compared to the one directly obtained from the 3D and 2D simulations.

2 MODEL AND SIMULATION

2.1 Approach

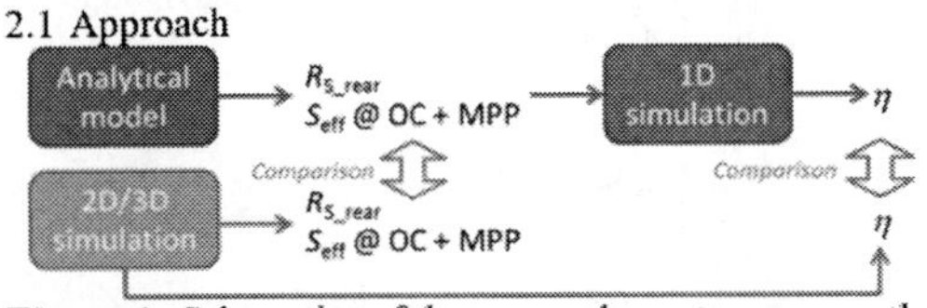

Figure 1: Schematics of the approach use to compare the analytical models and the numerical simulations.

In the schematics Figure 1, the steps followed in this paper are represented visually. First, 2D and 3D numerical simulations are carried out in a range of parameters that is technologically relevant. The two parameters representing the rear surface $Rs_{\_rear}$ and S_{eff} are extracted from these simulations by analysing the 2D and 3D information about Joule heating due to local currents and the rear surface recombination rate, respectively. The same simulations were also used to obtain the energy conversion efficiency (η) of these virtual solar cells.

Then, $Rs_{\_rear}$ and S_{eff} are calculated using the analytical models. In order to obtain the efficiency, a 1D simulation was carried out using the $Rs_{\_rear}$ and S_{eff} values obtained previously. Finally, the analytically and numerically calculated values of $Rs_{\_rear}$, S_{eff} and η are compared.

2.2 Analytical model

The model of Saint-Cast [4] is used in order to analytically calculate the spreading resistance ($Rs_{\_rear}$). This model solves the Poisson Equation for point- and line-shaped contacts assuming a constant current on the front side of the solar cell. This boundary condition should be more suitable for solar cells than a constant voltage on the front considering the homogenous light generation close to the front surface. For contact fractions lower than 10% a simplified formula obtained by parametrisation of the analytical model [9] is available. For line contacts the formula is the following:

$$R_{S\_rear} = \frac{a}{\sigma}\left\{\begin{aligned}&\left(37f - 2 - \frac{0.3}{f}\right) + \\ &\frac{2.82w^{0.88}\times f^{0.64}}{\tanh\left(2.82w^{0.88}\times f^{0.64}\right)}\frac{1}{3w}\left(\frac{1}{f}-1\right)^2 + w\end{aligned}\right\}$$

and for point contacts:

$$R_{S\_rear} = \frac{a}{\sigma}\left\{\begin{aligned}&\left(2.6 - \frac{0.9}{\sqrt{f}} + \frac{0.24}{f}\right) + \\ &\frac{f^{0.25}}{\tanh\left(2w\times f^{0.25}\right)}\frac{-f^2+4f-3-2\ln(f)}{4\left(1-\sqrt{f}\right)f} + w\end{aligned}\right\}$$

where a is the contact half width or radius, f is the contact fraction, w is the wafer thickness divided by a and σ is the wafer conductivity.

For the analytical modelling of the recombination at the local rear contacts, the model published in Ref. 9 is chosen here. This model is close to the model by Fischer [2]. For the cases of this study only a relatively small difference with Fischer's model has been noticed. The model used here is more general than the one by Fischer and gives a higher accuracy in the case where the contact recombination is not strongly dominating the overall rear recombination [4]. The model is valid for open circuit (OC) condition as no majority current is considered in this model [9] (as for Fischer's model [2]).

The formula used for the calculation is the following [9]:

$$S_{eff} = \frac{f\times S_{cont} + (1-f)\times S_{pass} + f(1-f)S_{cont}S_{pass}qR_{diff}}{1 + f(1-f)qR_{diff}\left(f\times S_{pass} + (1-f)\times S_{cont}\right)}$$

where S_{cont} and S_{pass} are the recombination velocity at the contact and the passivation, f is the contact fraction, q is the electron charge and R_{diff} is the diffusion resistance of the minority carriers between the contact and the passivation. An analytical calculation for R_{diff} in the case of line and point contacts can be found in Ref. 9.

2.3 Simulation

For this study the numerical simulations are carried out using a state-of-the-art finite-element software, Sentaurus Device [11]. A large set of parameters is varied in order to obtain a sample which is relevant for applicable technologies. These parameters are listed in Table I. Apart from the parameters reported in this table all the other parameters setting up the simulation are the same, the value of S_{pass} was 10 cm/s for all simulations. The semiconductor model, the optical model, the emitter, the external resistance are the same for all the simulations regardless whether it is a 1D, 2D or 3D simulation. This insures a consistency within the whole simulation data set under study. The main possible inconstancy which might remain is related to meshing issues, which might affect the results when changing the contact geometry.

Table I: Parameters varied in the numerical simulation. For the evaluation of the series resistance, only those setups with a metal fraction below 10% are considered. SRV stands for surface recombination velocity.

Parameter	Range	# steps
Point contact (3D simulation)		
Contact distance	0.2-0.6 mm	7
Contact diameter	20-80 µm	3
Contact SRV	10^2-10^4 cm/s	5
Bulk resistivity	1-3 Ω cm	3
Total		315
Line contact (2D simulation)		
Contact distance	0.4-2 mm	5
Contact width	30-100 µm	3
Contact SRV	10^2-10^4 cm/s	5
Bulk resistivity	1-3 Ω cm	3
Total		180
1D simulation		
External resistance	0.5-2 Ω cm²	16
Effective rear SRV	10-10^4 cm/s	74
Bulk resistivity	1-3 Ω cm	3
Total		3552

As explained in the last section, the purpose of the 1D simulations is different from the one of the multidimensional simulations. The 1D simulations are used to obtain the solar cell simulation based on the effective rear parameters R_{S_rear} and S_{eff}. However, instead of running a new simulation for each parameter calculated in this study, a map of efficiency is created (see figure 2) with more and 1000 points for each bulk resistivity. The efficiency for a specific R_{S_rear} and S_{eff} parameters couple is determined by a geometrical mean of the four nearest points.

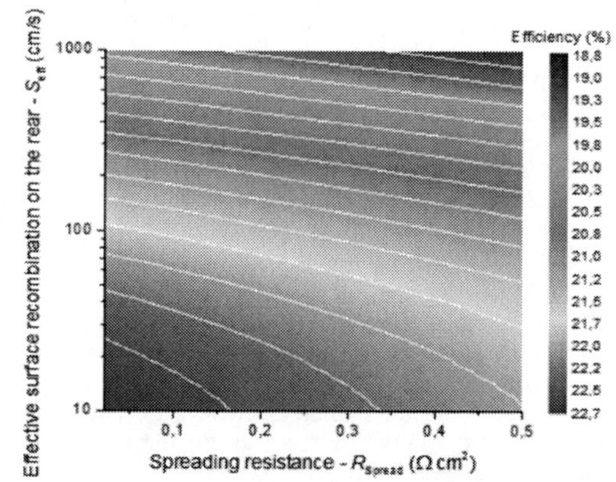

Figure 2: Efficiency density plot as a function R_{S_rear} and S_{eff} parameters. The efficiency values were obtained from 1D simulation for a bulk resistivity of 1 Ω cm.

3 RESULTS AND DISCUTION

3.1 Comparison of $Rs_{\_rear}$ and S_{eff}

In Figure 3, the relative difference between the spreading resistance ($Rs_{\_rear}$) obtained by the numerical simulation and the analytical model is presented. As the numerically simulated results are considered to be the reference, this difference between the simulation and the analytical model, will be in the rest of the paper referred as "the deviation" (the deviation of the analytical model from the simulation). The deviation is expressed as the relative deviation. A positive deviation means that the analytical $Rs_{\_rear}$ value is higher than its simulated counterpart. The analytical model seems to be in a very good agreement with the simulation. More than 95% of the data shows an deviation below ± 10%.

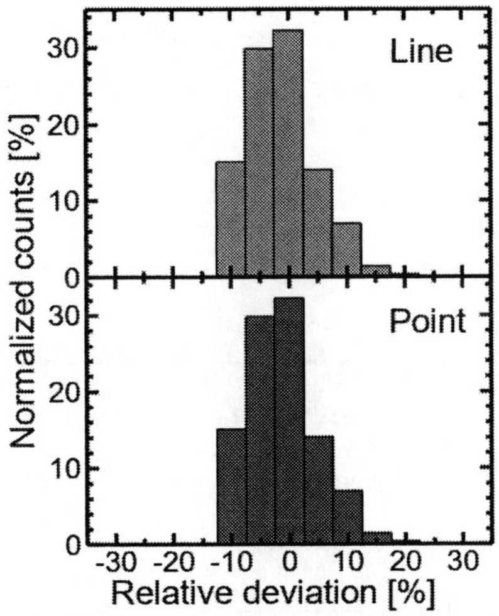

Figure 3: Histogram of the deviation for the spreading resistance model for line contacts and point contacts. As mentioned in section 2.2 the $Rs_{\_rear}$ model is only valid for a metal fraction below 10%, therefore only the result in this range are compared here

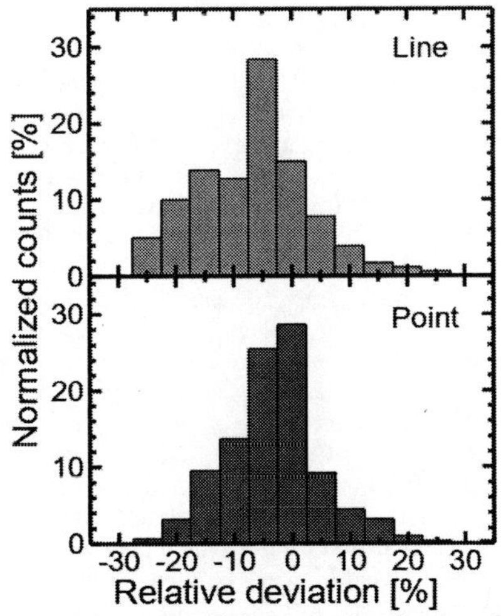

Figure 4: Histogram of the deviation for the effective rear surface recombination velocity (S_{eff}) for line contacts and point contacts in open-circuit conditions.

In Figure 4, the relative deviation of the effective rear surface recombination velocity (S_{eff}) obtained by the analytical model is presented. For this comparison we have chosen to use the values simulated in open-circuit conditions, as in fact the analytical model considers no current flow for its calculation. The analytical model seems to be in a fairly good agreement with the simulation. 94% of the calculated points show an deviation below ± 20% for the line shape contact and below ± 15% for the point shape contacts. An deviation of 20% might look important, but please consider that the recombination rate impacts the voltage only logarimically. Therefore even in the case where the cells' recombination would be largely dominated by the rear surface recombination, the deviation on the open-circuit (OC) voltage would be only about 5 mV.

3.2 S_{eff} at the maximum power point (MPP)

The comparison of the S_{eff} provided in open-circuit condition would be sufficient only if it can be shown that the S_{eff} is independent from the working point of the solar cell. However this is not the case. In fact, the S_{eff} obtained by simulation is different under OC as the one under MPP condition. When comparing the S_{eff} obtained analytically with the S_{eff} obtained by simulation in MPP condition only 50% of the points show an deviation less than ± 20%, with a clear tendency to be overestimated (Figure 5).

This change in S_{eff} over the current-voltage characteristics of the cell can be attributed to the majority carrier current flow. In order to understand this more in detail, we need to start from the definition of S_{eff}. As defined in Ref. 4, S_{eff} is the average of the local surface recombination velocity (SRV) weighted by the local minority carrier density. Therefore in a defined structure (with a geometry and SRV) the S_{eff} can only change due to a change in the minority carrier distribution. In the analytical models used so far [4,9], it is supposed that the minority carrier distribution only depends on the structure's geometry and the local SRV value; the majority carrier flow is not considered in these models. The majority carrier current is subject to series resistance. This induces an increase of the potential at the *pn* junction, and therefore an increase of the minority carrier density. This increase mainly occurs far from the rear contact, as current flowing laterally over longer distances encounters more series resistance. Therefore it is expected that the series resistance in the bulk or $Rs_{\_rear}$ induces an increase of minority carrier density above the passivated area. Following our definition of S_{eff}, the weight factor of the SRV on the passivation (S_{pass}) needs to be multiplied by the factor $\exp(Rs_{\_rear}\, j / V_{th})$, where j is the majority current and V_{th} is the thermal voltage. This leads to an S_{eff} depending on j, which can be expressed as follows,

$$S_{eff}(j) = \frac{S_{eff}(0) - S_{pass}}{e^{jR_{S\_rear}/V_{th}}} + S_{pass}.$$

This correction is applied to the value obtained analytically for OC condition (which became $S_{eff}(0)$) in order to obtain the maximum power point values. In Figure 5 the corrected and uncorrected S_{eff} are compared to S_{eff} obtained from the numerical simulation at the maximum power point (MPP). After correction a fair agreement is obtained with 95% of the calculated points showing an deviation less than ± 20%.

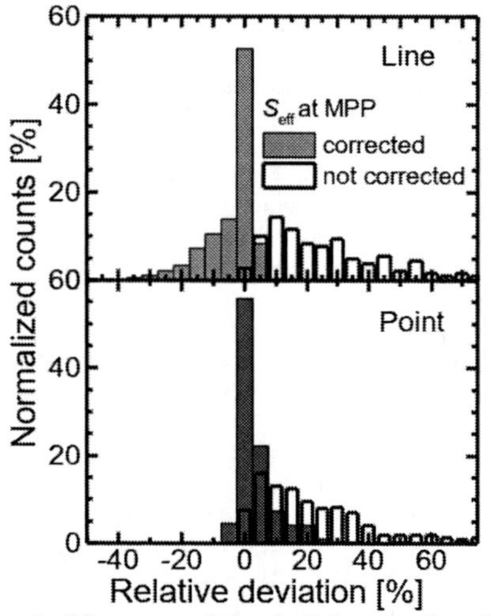

Figure 5: Histogram of the deviation for the effective rear surface recombination velocity (S_{eff}) for line contacts and point contacts in MPP conditions. The empty bars represent the deviation for the uncorrected S_{eff} and the colored histograms represent the deviation for the corrected S_{eff}.

3.3 Comparison of the efficiencies

The efficiency reference values are the ones obtained from the 2D and 3D simulations (for this part the full set of data was used including f> 10%). They are compared to the efficiencies obtained from the 1D simulations using $Rs_{\_rear}$ and S_{eff} as inputs parameters. Several $Rs_{\_rear}$ and S_{eff} as inputs parameters might be used, we choose to focus on the following scenarios I, II and III:

I. The analytical $Rs_{\_rear}$ value and the analytical S_{eff} corrected for MPP condition.
II. The analytical $Rs_{\_rear}$ value and the analytical S_{eff} for OC condition, in order to evaluate the gain of accuracy obtained by the correction.
III. The $Rs_{\_rear}$ and S_{eff} values obtained by the simulation, which will allow to check the accuracy of the method as we are using or reference values for $Rs_{\_rear}$ and S_{eff}.

In Figure 6, the histogram or the deviation of the efficiency is presented. In this case the deviation is express directly as the deviation in efficiency in percent absolute.

For the point contacts, we observe a clear gain in accuracy thanks to the correction at MPP condition (scenario I), where more than 96% of the points present an deviation in efficiency of less than ± 0.15%$_{abs}$. The efficiencies obtained using $Rs_{\_rear}$ and S_{eff} obtained by the 2D/3D numerical simulation (scenario III) are hardly more precise and present a very similar deviation distribution. This suggests that in this case the main source of deviation does not emanate from the analytical model but rather from the method used or the computation accuracy.

Concerning the line shaped contacts, the deviation distribution is much narrower for the values corrected for MPP conditions (scenario I). However, a shift toward higher efficiency is observed in this case. Is seems that the obtained efficiencies are biased by about + 0.15%. A similar bias is observed for the efficiencies obtained using the $Rs_{\_rear}$ and S_{eff} values obtained by the 2D/3D numerical simulation (scenario III). This suggests again that the main source of deviation comes from the method used or the computation accuracy. This method might be more suited for the point contact geometry than for the line geometry, but this reason in not understood yet. Another possible explanation might be that the structure of the mesh which varied between 2D and 3D simulation has a unwanted influence.

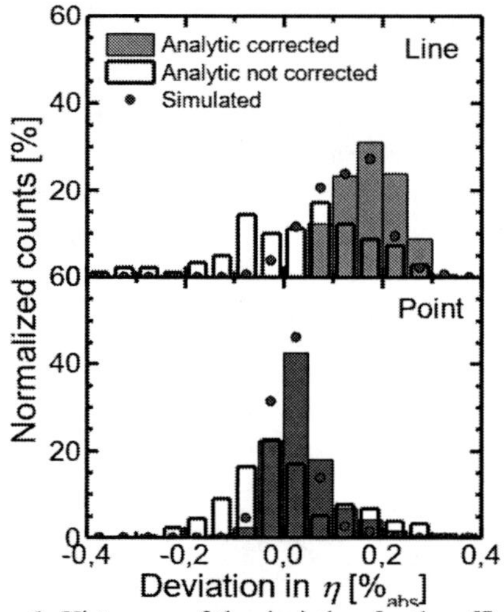

Figure 6: Histogram of the deviation for the efficiency by the 1D simulation. The coloured columns are for the analytical $Rs_{\_rear}$ values and the analytical S_{eff} value corrected for MPP (scenario I). The empty columns are for the analytical $Rs_{\_rear}$ value and the analytical S_{eff} value for OC condition (scenario II). The points are for $Rs_{\_rear}$ and S_{eff} values determined from the numerical 2D/3D simulations (scenario III).

4 CONCLUSION

This work investigated the accuracy of analytical models for local rear contacts of PERC/PERL solar cells, by confronting them to numerical simulation. The deviation made in the calculation of the series resistance $Rs_{\_rear}$ and the effective surface recombination velocity S_{eff} is small enough for most of the practical applications if the parameter range of the model's validity is not left. There is a great confidant that the deviation for $Rs_{\_rear}$ is smaller than ± 10% and that the deviation for S_{eff} is smaller than ± 20% for open-circuit conditions. The study of the simulation results shows that S_{eff} varies significantly with the solar cell working point. A correction has been proposed for S_{eff}. This correction maintains the same level for accuracy in maximum power point (MPP) conditions.

The correction of S_{eff} does not only increase the accuracy in MPP condition but also the accuracy of the efficiency obtained using 1D simulations to model the solar cell. This method insures with a confidence of 95% that the deviation in efficiency will not exceed ± 0.18%$_{abs}$ for point contact and ± 0.28%$_{abs}$ for line contacts. The difference between the accuracies for line contact compared to point contact stills needs to be investigated.

5 REFERENCES

[1] A. W. Blakers, A. Wang, A. M. Milne, J. Zhao and M. A. Green, Applied Physics Letters, 55(13), 1363–1365 (1989).

[2] B. Fischer, PhD Thesis, University of Konstanz, 2003.

[3] H. Plagwitz, R. Brendel, Progress in Photovoltaics: Research and Applications 2006; 14: 1–12.

[4] P. Saint-Cast, *et al.* Journal of Applied Physics 2010; 108: 013705-013705-013707.

[5] A. Cuevas, IEEE Journal of Photovoltaics 2012; 2: 485–493.

[6] A. Cuevas, JH. Petermann, IEEE Journal of Photovoltaics 2013; 3: 587–589.

[7] A. Cuevas, Journal of Applied Physics 2013; 113: 164502.

[8] A. Cuevas, D. Yan, F. Haasse, J. Petermann, R. Brendel. Energy Procedia 2013; 38: 13–21.

[9] P. Saint-Cast, PhD Thesis, University of Konstanz, 2012. Chapter 2, p.15-70

[10] A. Wolf, D. Biro, J. Nekarda, S. Stumpp, A. Kimmerle, S. Mack, and R. Preu, J. Appl. Phys. 108, 124510 (2010)

[11] Sentaurus Device User Guide, Version H-2013-03. Synopsys, Inc.: Mountain View, CA, 2013.

[12] H. Haug, J. Greulich. "PC1Dmod 6.2–Improved simulation of c-Si devices with updates on device physics and user interface." Energy Procedia 92 (2016): 60-68.

FRONT SIDE METALLIZATION OF P- AND N-TYPE SI SOLAR CELLS: MICROSTRUCTURE OF THE GLASS LAYER

Praveen Kumar, Michael Pfeffer, Oliver Eibl
Institute of Applied Physics, Eberhard Karls University Tuebingen,
Auf der Morgenstelle 10, D-72076 Tuebingen, Germany
Phone: +49 7071 2976015, Email: praveen.kumar@uni-tuebingen.de

ABSTRACT: High-efficiency cells (~18.0% for p-type and ~20 % for n-type cells) show similar microstructure and chemical composition of the glass layer and a similar temperature dependence of the series resistance. Details of the glass layer with respect to its quantitative chemical composition, microstructure, and electrical conductivity have not been reported in the literature. However, we consider it as an important microstructural feature affecting the physical properties of the cell: (i) the glass layer reflects the chemical composition of the applied paste, (ii) it contributes to the current path. Therefore, a detailed analysis of glass layers is provided here by applying a combined SEM and TEM-EDX spectroscopy. The chemical composition of the investigated glass layers was $(SiO_x)Pb$, with Pb >10 at.%. Other elements like Ag, Zn, Al, and Ti were detected in the glass layer as minor constitutes. In high-efficiency n- and p-type cells, Al and Zn mole fractions in the of range 1-2 at.% were detected. The chemical composition of the glass layer serves as a chemical bar code of the applied paste used for the screen printing process and affects the contact resistance at the Si/metallization interface. Cells processed with an Al-free paste and a Zn-rich paste were investigated and yielded Al-free and Zn-rich glass layers, respectively. The role of additives in the glass layer, like Al and Zn, were found to be important for the reduction of the contact resistance. Glass layers of high-efficiency cells contain a large density of precipitates and Ag colloids and are considered as dirty semiconductors rather than insulators. The percolation model is, therefore, appropriate to describe the current path in such layers.

Keywords: Silicon Solar Cell, Screen Printing, Metallization, Characterization, Electrical Properties

1 INTRODUCTION

For screen printing front side metallization technology, commercially available pastes are used for contacting phosphorous (P) and boron (B) doped emitters [1–5]. The pastes typically consist of glass frits, silver (Ag) powder, organic binders and a small amount of dopant/additive elements such as zinc (Zn), aluminum (Al), titanium (Ti), B and P [1–5]. Due to the given chemical complexity, the exact composition of pastes, particularly the dopant contents and its effect on the contact quality is largely unknown.
The microstructure at the Si/metallization contact interface depends on the orientation of the Si surface [6, 7], the chemical composition of the paste and the processing conditions (firing temperature and time) and affects the overall series resistance of the cell. Previous studies [1–5] showed that the front side Si/metallization interface consists of the Si emitter embedded with Ag nanocrystals [1–5], a glass layer typically <1 μm thick which contained a high density of nano-Ag colloids and on top of it a bulk metallization finger [6–11].
The presence and significance of the glass layer was identified long time ago, however, hardly any quantitative chemical analysis of the glass layer has been reported in the literature. The glass phase contained Si, Pb and O as main constituents wherein Zn, Ti, Ag and other elements were dispersed in the glass layer [6, 8–10].
In the literature, there are few experimental studies that show the glass layer itself contributes to the current conduction: forming gas annealing (FGA) treatment as reported by a number of groups had significantly improved both the contact resistance and conductivity of the glass layer [12, 13].
The chemical composition of a paste is not known and it is not disclosed by the supplier. Therefore, the microstructural and chemical analyses of the glass layer are of relevance for understanding paste specific differences and its correlation to the electrical properties of the cells. Another important point for the investigation of the glass layer is related to the relevant current path across the Si/metallization interface. For high-efficiency cells a percolative charge transport is assumed through the glass layer [8, 10, 11].
In the present study, we take a correlative approach by applying both the SEM-EDX and analytical TEM-EDX spectroscopy in order to quantify the chemistry of the glass layer. By this approach, we obtained unprecedented insights about the role of the glass layer and established a cross-correlation to the electrical losses across the contact interface. For high-efficiency p- and n-type cells we can prove that the chemical composition of the glass layer affects the density of Ag colloids.
Moreover, the quantitative chemical analysis of the glass layer at the nanoscale yield new possibilities for understanding and optimizing the chemical composition of the glass layer and hence the overall efficiency of the solar cell.

2 EXPERIMENTAL DETAILS

2.1 Solar cells processing and electrical characterization

In this study, we investigated both planar p-type and textured p- and n-type cells. Planar cells with two different orientation <111> and <100> Si surfaces were processed on FZ grown wafer and contacted with different front side metallization pastes referred to as FSP1 and FSP3. Details can be found in Refs. [6, 8].
p-type textured cells were processed on Cz grown <100> Si wafers with P doped emitters. Details of the solar cells processing of p-type cells are described in Refs. [6, 8]. n-type PERT (Passivated Emitter Rear Totally Diffused) cells with B doped emitters were fabricated on Cz grown <100> Si grown wafers with base doping of 3 Ω-cm. For

details see Ref. [9]. Cells #1 and #2 with P doped emitters (p-type) were contacted with two different front-side metallization paste FSP1 and FSP2 respectively, while cell #3 with B doped emitter (n-type) was processed with an AgAl paste. Pastes FSP2 and FSP3 have a similar chemical composition with paste FSP3 being a newer batch of paste FSP2. Details of the cell processing and electrical parameters are given in Table I.
The electrical measurements were performed on small cut pieces of the solar cells ($1\times1.5cm^2$) in an optical cryostat from 80K to room temperature (RT). In a next step, I-V curves of the cells were evaluated with respect to the series resistance as reported in [14].

Table I: Process parameters and electrical properties of the planar and textured (p- and n-type) cells measured on the large wafer (15.6×15.6 cm^2) i.e., firing temperature T_{FFO}, efficiency η are summarized in Table I. Specific contact resistance ρ_c was measured on a small-sized cell by the TLM method.

Cell	Solar cell	Paste	T_{FFO} [°C]	η [%]	ρ_C [mΩ.cm^2]
		Planar cells			
#P1	<100> p-type Si	FSP1	930	7.8	33
#P2	<111> p-type Si	FSP3	930	17.1	4.9
		Textured cells			
#T1	p-type Si	FSP1	900	16.9	21
#T2	p-type Si	FSP2	900	17.8	4.7
#T3	n-type Si	AgAl	830	19.96	< 4

2.2 SEM/TEM specimen preparation and strategy for the quantitative chemical analysis

Microstructural analysis of the Si/metallization requires a careful sample preparation both in plan view and cross-sections. The essential steps are: mechanical grinding, polishing, followed by ion etching to get artifact free high-quality imaging and chemical analysis of the glass layer. High-resolution secondary electron (SE) images were acquired at 50,000 magnification to image nanoparticles in the glass layer. For quantitative EDX in the SEM, a small primary voltage was applied to keep the lateral resolution as small as possible.

The quantitative chemical analysis in the TEM of the glass layer was carried out using a Cliff-Lorimer k-factor method. EDX spectra were acquired with a spot size of 32 nm and an acquisition time of 300 s with a stray aperture to avoid stray radiation and to improve the detection limit for the dopant elements. Furthermore, a quantitative correlation of SEM-EDX mole fractions with TEM-EDX mole fractions could be established.

3 RESULTS AND DISCUSSION

3.1 Effects of Si surface orientation on the microstructural and electrical properties of planar cells

Fig. 1a–b show high-resolution cross-sectional SEM images of planar cells obtained at the Si/metallization contacts interface. Cell #P1 with <100> oriented Si surfaces (Fig. 1b) revealed pyramidal shaped Ag nanocrystals at the Si surface, whereas <111> oriented Si (Fig. 1a) yielded lens-shaped Ag nanocrystals and a high density of Ag colloid in the glass layer.
Fig. 1c shows the dark current measurements of the two planar cells one with <100> oriented Si (red color), and the other being <111> oriented Si (blue color). Dark current measurements of the two planar cells showed significant differences in the I-V curve, especially for <111> oriented Si a large current of about 0.4A was observed at small voltages (close to 0.8V). On the other hand, for <100> oriented Si, the magnitude of the current at 0.8V was 0.03A, which is one order of magnitude lower than <111> oriented Si. For <100> oriented Si we observed a maximum current of 0.1A at higher voltages (higher than 1V).
In summary, planar <100> oriented Si yielded a high contact resistance and pyramidal Ag nanocrystals, whereas <111> oriented cells yielded a low contact resistance but no pyramidal shaped Ag nanocrystals. Therefore, it was concluded that the pyramidal shaped Ag nanocrystals are not necessary for a low contact resistance [6, 8]. Moreover, the shape of the Ag nanocrystals depends strongly on the orientation of the Si surfaces. Also, for n-type high-efficiency cells we did not observe pyramidal Ag nanocrystals at the Si emitter [9].

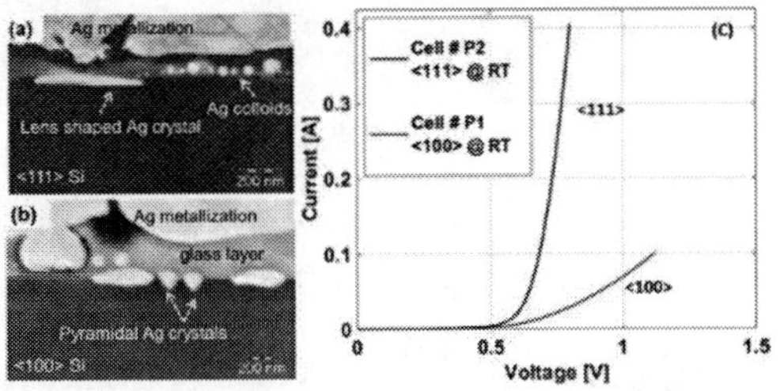

Figure 1: (a–b) Cross-sectional SE images of planar <111> and <100> and oriented Si surfaces obtained at the Si/metallization interface, (c) Dark I-V curves of the <111> and <100> oriented Si planar cells measured at room temperature [1].

3.2 Structure-property correlation of high-efficiency p- (#T2) and n-type (#T3) textured cells

Electrical data of large cells, microstructure analyses and the temperature dependent series resistance measurements of small cells yielded a detailed characterization of high-efficiency p- and n-type cells, which is necessary to establish a structure-property correlation. High-efficiency p- (#T2) and n-type (#T3) cells yielded similar microstructure of the glass layer i.e. a matrix with a high density of Ag colloids covering the emitter and the Ag bulk, similar chemical composition of the glass layer and the similar contact and series resistances (see Fig. 2a-f). Besides this the temperature dependent series resistance (Fig. 2a and b) is similar for n- and p-type cells (#T2 and #T3), both in absolute values and with respect to the temperature dependence [8, 9]. Therefore, it can be concluded that the microstructure of the glass layer determines the series and contact resistance of the front side metallization.

The formation of Ag nanocrystals at the Si emitter is different: in p-type cells Ag nanocrystals are formed, in n-type cells these nanocrystals do not appear [9]. Since the series and contact resistance of high efficiency n-type and p-type cells are comparable, therefore, it is concluded that the pyramidal shaped Ag nanocrystals are not necessary for yielding a low contact resistance.

This conclusion contradicts Refs. [4, 15, 16], in which Ag nanocrystals are suggested to be relevant for the low contact resistance and authors suggested that they contribute significantly to the current path. However, our experimental results do not support this argument, but indicated that the glass layer is essential for the current path and that the relevant understanding is a percolation transport model [10].

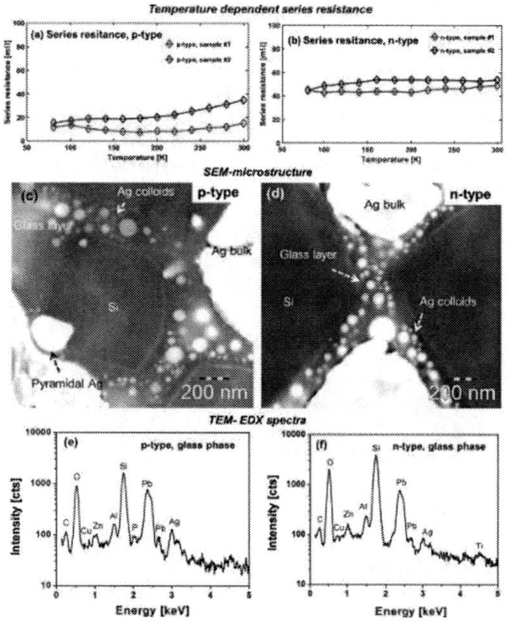

Figure 2: High-efficiency p- (#T2) and n-type (#T3) cells: (a–b) temperature dependent series resistance, (c–d) microstructure at the Si/metallization interface by SEM and (e–f) TEM-EDX spectra of the glass layer.

3.3 Quantitative chemical analysis of the glass layer by SEM and TEM-EDX spectroscopy

The glass layer was investigated by SEM-EDX spectroscopy and spectra corresponding to each cell are shown in Fig. 3a–d. SEM EDX analysis was carried out at low primary energy of 5 keV to improve the lateral resolution. The quantitative chemical analysis of the glass layer was also investigated by TEM-EDX and the average values are summarized in Table II.

The chemical composition of the glass layer is correlated with the chemical composition of the paste applied for the front side metallization. For a more detailed understanding of the EDX spectra of such glass layers arrows point to the relevant peaks, Zn and Al, in Fig.3. :

- Fig. 3a is a spectrum of a glass layer of cell #T1 (p-type) processed with a Zn rich paste, clearly a Zn peak is enhanced in the spectra. Note that in the spectrum no Al peak is visible, but B peak could be detected in the glass layer.
- Fig 3b is a spectrum of a glass layer of cell #T2 (p-type) with the highest efficiency, a reduced Zn peak was observed as compared to cell #T1, see Fig. 3a. An Al peak was also detected in the glass layer.
- Fig. 3c is the spectrum of a glass layer of cell #T3 (n-type) with the highest efficiency, in addition to a Zn peak, the paste contains Al in the glass layer.
- In Fig. 3d another paste of an n-type cell which was Al free, no Al peak was observed in the EDX spectrum of the glass layer [9].

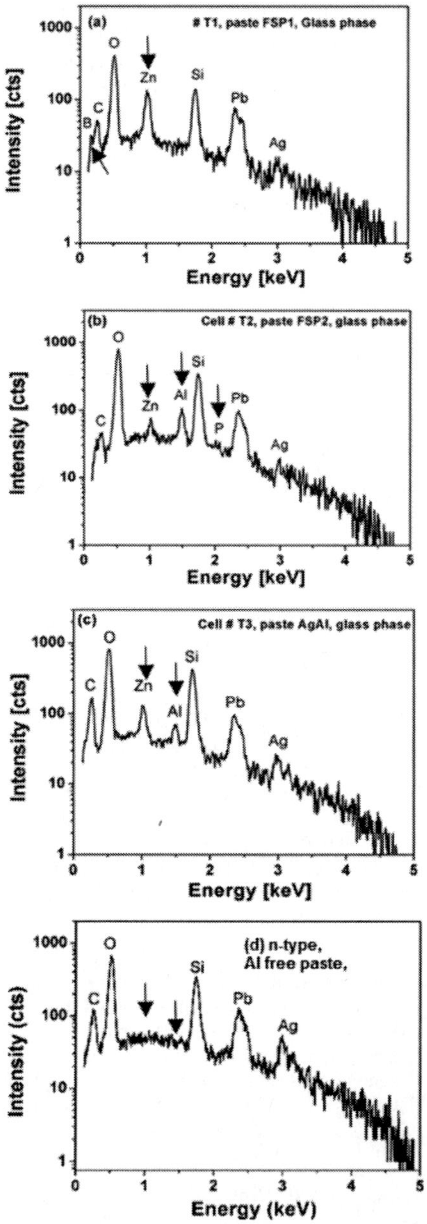

Figure 3: (a–c) SEM-EDX spectra of the glass layer of

cell # T1, # T2 (p-type) and # T3 (n-type), respectively plotted on a logarithmic scale in order to highlight the dopant elements. (d) EDX spectrum of the glass layer of an n-type cell processed with an Al-free paste [9].

The glass layer can be understood as a (SiO_x) Pb matrix, elements such as Ag Zn, Al, and Ti, were detected in the glass layer. For p-type cells, Si/Pb mole fraction ratio was 5.15 for cell #1 and slightly higher 5.28 for cell #2. In case of n-type cell #3, the Si/Pb mole fraction ratio was higher than 7.

Table II: TEM-EDX quantitative chemical analysis of the glass layer of cell # T1, # T2 and # T3. For quantification oxygen was excluded from the list of elements.

Glass layer	Si [at.%]	Pb [at.%]	Ag [at.%]	Zn [at.%]	Al [at.%]	Ti [at.%]	P [at.%]
k-factor	1.00	3.04	1.76	1.54	1.03	1.09	0.99
#T1	***69.83***	13.56	4.26	***9.84***	-	***2.51***	-
#T2	***76.15***	14.42	5.6	***1.08***	***1.76***	***0.17***	***0.82***
#T3	***82.88***	11.82	2.23	***0.87***	***1.96***	***0.24***	-

3.4 Comparison of SEM-EDX mole fractions versus TEM-EDX spectroscopy

Fig. 4 shows a correlation diagram of SEM-EDX mole fractions versus TEM-EDX mole fractions of the glass layer plotted on a log-log scale of cells #T1, #T2 and # T3.

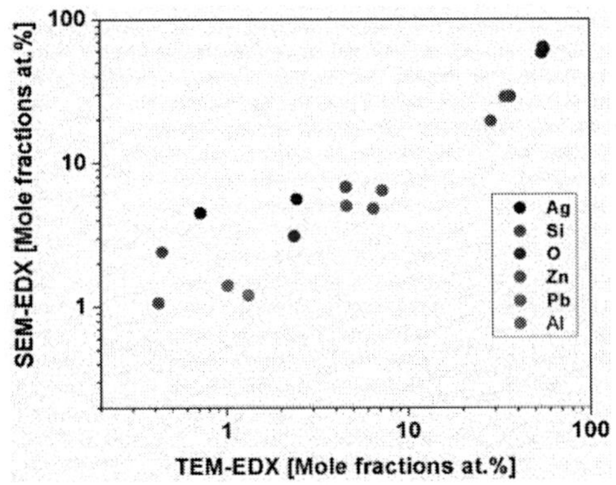

Figure 4: A log-log plot of correlation diagram of SEM-EDX mole fractions against TEM-EDX mole fractions measured in the glass layer of cell #T1 (p-type), and #T2 (p-type) and cell #T3 (n-type). The quantitative data of six different elements (Ag, Zn, O, Si, Al and Pb) are shown in the plot.

The mole fractions values are very close in both measurements, i.e. TEM results follow SEM results. These results could be achieved if 5 keV primary beam was applied in the SEM-EDX. For Ag and Zn mole fractions we observed significant differences: SEM results show a slightly higher mole fractions as compared to TEM-EDX. This result is not trivial since the acquisition conditions are very different. Since analysis in the SEM is much less tedious than in the TEM and therefore these results are of great practical value. In case of additives/impurity elements such as Ag, Zn and Al SEM results showed a slightly higher mole fractions as compared to the TEM results. This is because of the poor resolution of the SEM. On the other hand TEM yielded more accurate mole fractions of the additives/impurity elements with a lateral resolution of 30 nm.

Furthermore, a large number of bright-field, dark-field and energy filtered TEM (EF-) images (see Fig. 5a-c) were acquired of the glass layer of different cells.

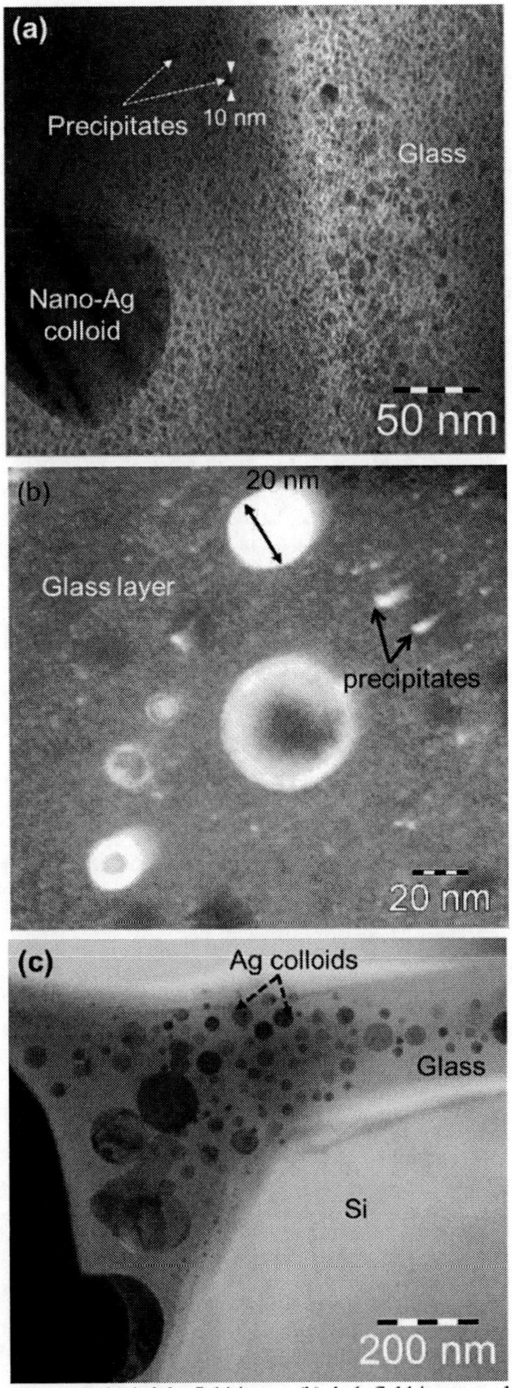

Figure 5: (a) bright-field image (b) dark-field image and (c) EF-TEM image of the glass layer of high-efficiency p-type cell #T2.

They prove the presence of metallic precipitates and Ag colloids in the glass layer and suggest that the glass layer of such cells is a dirty semiconductor rather than an insulator. Therefore, its electrical conductivity generates current paths as discussed in the percolation model.

4 CONCLUSIONS

In summary, the role of different pastes, in particular the microstructure and quantitative chemical composition of the glass layer has been discussed in detail. The chemical composition of the glass layer serves as a chemical barcode of the applied paste used for the screen printing process and affects the contact microstructure and hence the overall efficiency of the cell. These findings are of relevance for the paste optimization and charge transport across the Si /metallization interface.

High-efficiency cells #T2 (~18.0% for p-type) and #T3 (~20% for n-type cells) show similar features: microstructure and chemical composition of the glass layer (a more pronounced wetting and the higher density of nano-Ag colloids), contact resistance less than 5mΩ.cm^2 and the similar temperature dependence of the series resistance.

Given the large density of Ag colloids and precipitates as observed in the glass layers of high-efficiency p- and n-type cells, a percolative charge transport across the Si/metallization interface has been proposed. Details of the percolation model are given in the companion paper [11].

5 ACKNOWLEDGEMENTS

We acknowledge the group of D. Biro and F. Clement at the Fraunhofer Institute for Solar Energy Systems in Freiburg for providing high-quality p-type Si solar cells and L.J. Koduvelikulathu from ISC-Konstanz for providing high- efficiency n-type Si solar cells.

6 REFERENCES

[1] M.M. Hilali, S. Sridharan, C. Khadilkar, A. Shaikh, A. Rohatgi, S. Kim, J. Electron. Mater. 35 (2006) 2041–2047.

[2] M. Hörteis, T. Gutberlet, A. Reller, S.W. Glunz, Adv. Funct. Mater. 20 (2010) 476–484.

[3] G. Schubert, F. Huster, P. Fath, Sol. Energy Mater. Sol. Cells. 90 (2006) 3399–3406.

[4] C. Ballif, D. M. Huljić, G. Willeke, A. Hessler-Wyser, Appl. Phys. Lett. 82 (2003) 1878–1880.

[5] D. Pysch, A. Mette, A. Filipovic, S. W. Glunz, Prog. Photovoltaics Res. Appl. 17 (2009) 101–114.

[6] P. Kumar, B. Willsch, M. Dürrschnabel, Z. Aabdin, R. Hoenig, N. Peranio, F. Clement, D. Biro, O. Eibl, Energy Procedia. 67 (2015) 31–42.

[7] R. Hoenig, M. Duerrschnabel, W. Van Mierlo, Z. Aabdin, J. Bernhard, J. Biskupek et al., Energy Procedia. 43 (2013) 27–36.

[8] P. Kumar, M. Pfeffer, B. Willsch, O. Eibl, Sol. Energy Mater. Sol. Cells. 145 (2016) 358-367.

[9] P. Kumar, M. Pfeffer, B. Willsch, O. Eibl, L. Koduvelikulathu, V. Mihailetchi R. Kopecek, Sol. Energy Mater. Sol. Cells. 157 (2016) 200-208.

[10] M. Pfeffer, P. Kumar, O. Eibl, Journal of Electronic Materials 45 (2016) 5764-5772.

[11] M. Pfeffer, P. Kumar, O. Eibl, submitted to EU PVSEC Amsterdam, 2017.

[12] S. Bin Cho, H. S. Kim, J. Y. Huh, Acta Mater. 70 (2014) 1–7.

[13] M.M. Hilali, M.M. Al-Jassim, B. To, H. Moutinho, A. Rohatgi, S. Asher, J. Electrochem. Soc. 152 (2005) G742.

[14] B. Willsch, P. Kumar, O. Eibl, Journal of Electronic Materials 45 (2016) 2656.

[15] S. Kontermann, G. Willeke, J. Bauer, Appl. Phys. Lett. 97 (2010) 10–13.

[16] E. Cabrera, S. Olibet, J. Glatz-Reichenbach, R. Kopecek, D. Reinke, G. Schubert, J. Appl. Phys. 110 (2011) 114511–5.

INCREASING THE EFFICIENCY OF MULTICRYSTALLINE SILICON PERC SOLAR CELLS FROM CURRENTLY 19% TO 20%

J.M. Greulich, E. Lohmüller, P. Saint-Cast, S. Werner, S. Wasmer, A.J.C. van der Horst, R. Preu
Fraunhofer Institute for Solar Energy Systems
Heidenhofstr. 2, 79110 Freiburg, Germany

ABSTRACT: We discuss a detailed analysis of the potential efficiency improvements of state-of-the-art industrial p-type multicrystalline silicon solar cells with passivated emitter and rear with an average energy conversion efficiency of 19.0%. The aim is to identify the limitations of cells currently fabricated in industry, not of record laboratory cells, and to demonstrate approaches for research institutes and industry to further optimise the devices using industrially applicable processes. According to the analysis, front surface reflection of the active cell area bears the largest potential for increasing the efficiency by replacing the standard acidic texture by a honeycomb texture, which is estimated to improve the efficiency from 19% to 20%. Reducing shading and recombination at the front metal is the second largest potential efficiency increase. We expect that reducing the metal finger width from 50 μm to 40 μm or below is feasible and increases the efficiency further to 20.2%.
Keywords: multicrystalline silicon, characterisation, optical losses, recombination, simulation

1 INTRODUCTION

Passivated emitter and rear solar cells (PERC) [1] based on p-type multicrystalline silicon (mc-Si) are relevant for industry and research for two main reasons: First, mc-Si is the workhorse of photovoltaics with a market share of more than 60% [2] and is expected to keep cost-competitive as simple solutions for texturing diamond-wire cut mc-Si wafers are available [3, 4]. Second, PERC technology is being introduced into mass production in the past few years and is rapidly gaining market share also due to the recent record cell [5, 6] and module efficiencies [7]. To stay cost-competitive, research and industry aim at increasing the energy conversion efficiency without increasing the manufacturing costs. The key to improve the solar cell efficiencies is to know of the bottlenecks of the solar cells in terms of lost power in its different regions.

The free energy loss analysis introduced by Brendel *et al.* [8] and extended to optical losses by Greulich *et al.* [9] quantifies the power losses at each mesh node of the numerical simulations due to electrical and optical effects, while integrating over certain device regions yields their contribution to the total power losses. These losses are directly comparable. However, elaborate numerical device simulations are needed.

Aberle *et al.* [10] published a method which compares the single, experimentally determined loss mechanisms in units of mW/cm^2, but only took into account the overall recombination occurring in the solar cell and does not differentiate the contributions by the different solar cell regions.

The experimental loss analysis conducted by Wong *et al.* [11] and Saint-Cast *et al.* [12] divides the examinations into a breakdown of the loss mechanisms impacting the short-circuit current density j_{sc}, the open-circuit voltage V_{oc} and the fill factor *FF*. This has the drawback that the single losses cannot be compared, as the performance parameters j_{sc}, V_{oc} and *FF* are investigated separately.

Further approaches for optimising the efficiency [13, 14] allow very deep insights into the physics of the device, but require elaborate numerical device simulations and do not directly rely on measured data.

In the present work, we characterise and disclose the losses and potential efficiency improvements of state-of-the-art industrial p-type mc-Si PERC cells. An advanced analysis that does not only derive the losses and potential improvements for maximum power point (mpp) conditions from measurements at short and open circuit, but directly measures recombination currents at mpp is applied. Besides analysing the losses, also the realistic efficiency potential is examined using PC1Dmod [15] simulations.

2 THEORY

The basis of the advanced efficiency potential analysis is formed by three components: (i) optical losses, (ii) recombination losses and (iii) transport losses. It is based on measured data and includes only analytical calculations. All losses are expressed as power density. The details of the approach are discussed in Ref. [16], here only the basics are summarised.

The optical losses comprise reflection by the front metal grid, reflection by the active cell area, escape light and parasitically absorbed light. For their quantification, the external quantum efficiency (EQE) and the spectral reflectance (R) are measured. By weighting with the photon flux of the AM1.5g spectrum, the internal quantum efficiency at maximum power point, the loss currents $j_{opt,x}$ corresponding to the individual losses "*x*" (e.g. reflectance at front metal or active cell area) are determined. By multiplying this quantity with the voltage V_{mpp} at maximum power point, the corresponding power loss or potential power gain is calculated.

The recombination losses comprise recombination in the emitter, in the base, at the rear surface and at the front and rear metal-semiconductor contacts. Corresponding lifetime samples (samples to determine the emitter dark saturation current density j_{0e}; samples to determine the implied V_{oc} (iV_{oc}) with and without front / rear metal contacts) are required and are characterised by means of photoluminescence calibrated to the (implied) V_{oc}. The excitation conditions are carefully chosen to ensure similar injection conditions in the lifetime samples as in the finished solar cells at mpp (i.e., the laser intensity is tuned to generate an implied voltage of the lifetime samples equal to the junction voltage at mpp). From these measurements, the recombination current at mpp, corresponding to the above losses, is derived. Again, by multiplying this quantity with the voltage V_{mpp} at mpp, the corresponding power loss or potential power gain is calculated.

Finally, we consider losses due to series resistance as follows. From each of the single, measurable resistances of emitter sheet resistance R_{sh}, front contact resistivity ρ_C, front grid resistance R_{grid} and rear aluminium resistance R_{Al}, we calculate the single contributions $R_{s,x}$ to the total series resistance R_s via analytical formulae. As we want to rely on measurements and analytical modelling only, we estimate the base resistance by subtracting the sum of the aforementioned contributions from the total R_s of the solar cells. In order to obtain loss power densities, the single series resistance contributions $R_{s,x}$ are simply multiplied by the squared current density flowing at mpp j_{mpp}.

This approach gives a rough orientation to identify which losses are dominant and shall be improved first, and which can be neglected to first order.

The efficiency potential of selected process improvements, i.e., an advanced front surface texture and an improved front metal grid, is estimated by applying measured data to a PC1Dmod [15] simulation model.

3 EXPERIMENTAL

Solar cells and lifetime samples are fabricated in the PV-TEC pilot line [17] at Fraunhofer ISE. Industrial tools are used for each of the solar cell fabrication processes. A group of six cells, which is part of a larger experiment, is analysed here.

The solar cell fabrication process is presented in Figure 1. Regular sized wafers with an edge length of 156 mm are used. The p-type base material is high-performance mc-Si with a base resistivity ranging between 1.4 Ωcm and 1.8 Ωcm. After saw-damage removal, the wafers undergo acidic texturing using hydrofluoric (HF) and nitric acid (HNO_3). The emitter is formed by phosphorus diffusion in a tube furnace utilizing phosphorus oxychloride ($POCl_3$). After diffusion, the emitter is removed at the rear using a single-sided wet-chemical etching process. Prior to passivation, the samples are cleaned using HF and HNO_3. The passivation layer stack on the rear surface is formed by fast atomic layer deposition of aluminium oxide (Al_2O_3) covered by a silicon nitride (SiN_x) capping using plasma-end chemical vapour deposition (PECVD). The front surface is passivated by a PECVD SiN_y layer. A double-layer antireflection coating (ARC) is also applied on a separate group. The rear passivation is opened using

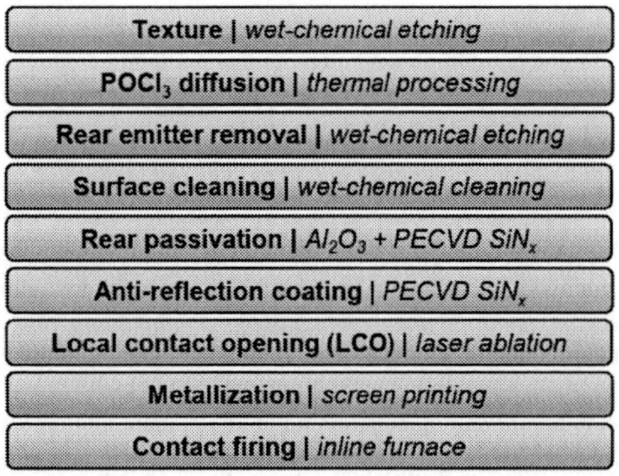

Figure 1: Process flow for the fabrication of p-type mc-Si PERC solar cells applied in this work.

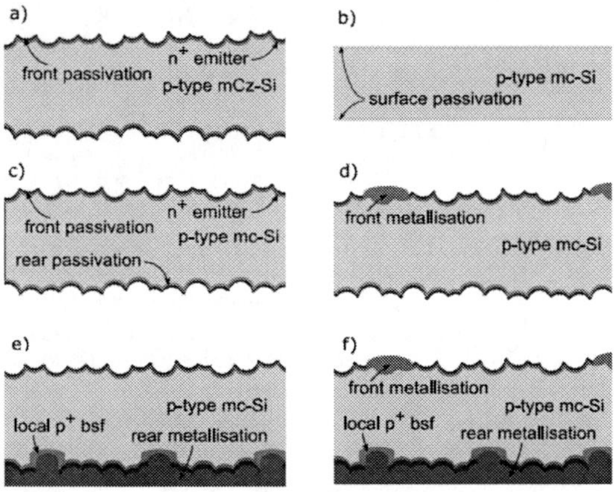

Figure 2: a) The symmetric j_{0e}-samples are based on monocrystalline p-type mCz-Si wafers with acidic texture, emitter and SiN_y passivation. b) The bulk-lifetime (τ_b) samples consist of etched-back iV_{oc}-samples with a smoothened surface passivated with PECVD Al_2O_3. c) The iV_{oc}-samples are based on mc-Si, have the same front as the j_{0e}-samples, but the rear is passivated with Al_2O_3 and SiN_x. d) iV_{oc}-samples including front metallisation. e) iV_{oc}-samples including rear local contact opening and rear metallisation. f) Finished solar cells. All samples are fired in a conveyor belt oven prior to characterisation.

a laser ablation process. The front and rear metallisation are obtained by screen printing, where a five busbar design is applied. The finger width after firing is $w_f \approx 50$ µm. After contact firing, the solar cells are characterised.

The cross sections of the lifetime samples and the cells are sketched in Figure 2. The emitter dark saturation current density $j_{0e} \approx 80$ fA/cm² is measured on symmetrical lifetime samples (Figure 2a) based on monocrystalline p-type mCz-Si wafers applying a refinement of the Kane-Swanson method [18]. The iV_{oc} of the corresponding lifetime samples in Figure 2c is determined using lifetime-calibrated photoluminescence imaging (τ-PLI). Its value is determined to $iV_{oc} = 665$ mV with a standard deviation within each sample of approximately 10 mV.

The current-voltage (IV) characteristics, the external quantum efficiency (EQE) and the spectral reflectance R of the cells are measured. The cells have a mean $j_{sc} = 37.1$ mA/cm², a $V_{oc} = 649$ mV and an average conversion efficiency $\eta = 19.0\%$. The peak efficiency of this batch is 19.4%, which is confirmed by Fraunhofer ISE CalLab PV Cells. When applying a double-layer anti-reflection coating, an efficiency of 19.6% is independently measured by Fraunhofer ISE CalLab PV Cells.

4 RESULTS

We find that first optical, then recombination and finally resistive losses or potential efficiency improvements play subsequent roles in the overall analysis, as shown in Figure 3. Above all, reducing the reflection at the active front side could improve the efficiency by $1.6\%_{abs}$, e.g. by replacing the common

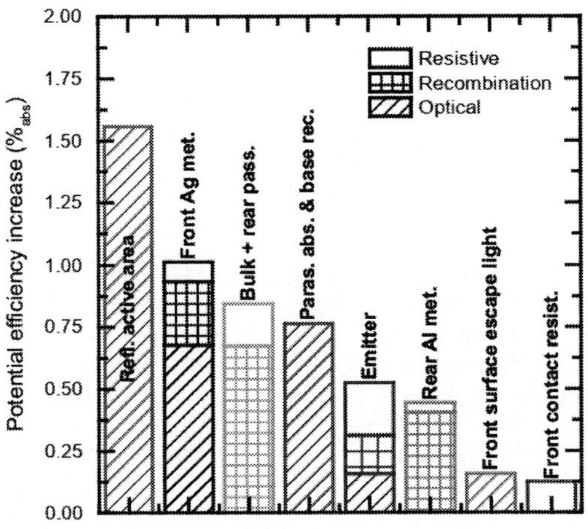

Figure 3: Potential efficiency improvements subdivided in optical, recombination and resistive contributions for different cell regions.

acidic iso-texturing process used for mc-Si wafers. Furthermore, the mc-Si cell should then be optimized concerning the bulk material quality with a potential efficiency gain of 0.7%$_{abs}$, as here the recombination at the front side plays only a minor role (0.4%$_{abs}$). For “Paras. abs. & base rec.”, PC1Dmod simulations showed that most is due to parasitic absorption and mainly occurring at the rear side in the aluminium metallization. Therefore, 0.8%$_{abs}$ could be gained next by mitigating the parasitic absorption, as. Technologically, this could be addressed by optimizing the rear side passivation stack for optical ends [19]. The potential efficiency gains due to resistive influences are not dominant as these sum up to only 0.6%$_{abs.}$

This analysis shows the potential efficiency improvements of individual loss mechanisms if that loss mechanism could be reduced to zero. A more realistic estimation of the potential efficiency is done using a PC1Dmod simulation model. First, it is calibrated using the measured data (reflectance, EQE, *IV* data). The above analysis suggests to first focus on reducing the front reflection. Hence, the first optimisation step is an advanced surface texture, *i.e.* a honeycomb texture. Integrating this process will reduce the reflection and potentially increase the recombination at the emitter as shown by Volk *et al.*in Ref. [20]. We use the measured reflectance of honeycomb textured test samples with anti-reflection coating as input of the modified PC1Dmod model. Thus, the weighted reflectance of the active area

Table I: Important input and output of the PC1Dmod simulation model for the experimentally realised baseline and the two simulated optimisation scenarios.

	baseline	improved texture	improved grid
R_{wa} [%]	6.8	2.3	2.3
$j_{0e,eff}$ [fA/cm²]	100	130	125
w_f[µm]	50	50	40
$R_{s,ext}$ [Ωcm²]	0.39	0.39	0.41
j_{sc} [mA/cm²]	37.2	39.2	39.5
V_{oc} [mV]	646.0	645.6	645.8
η [%]	19.0	20.0	20.2

R_{wa} is decreased, and the effective emitter saturation current density $j_{0e,eff}$, which contains contributions of the metallised and the passivated emitter, is increased by applying an advanced front surface texture (see Table I). The simulation then predicts an efficiency increase from 19.0% to 20.0%. In an additional second optimisation step, the front grid finger width w_f is reduced by assuming an advanced screen-printing process, which also slightly reduces the recombination at the metal-emitter interface, which is reflected in the PC1Dmod model by a lower $j_{0e,eff}$. Improving the front metal shading by reducing the finger width from 50 µm to 40 µm would decrease the shading losses and recombination at the metal-emitter interface, but it would also slightly increase the series resistance. This results in a simulated efficiency of 20.2%.

5 SUMMARY AND CONCLUSIONS

In this work, we applied a method to conduct a thorough efficiency gain analysis to 19%-efficient p-type multicrystalline silicon (mc-Si) PERC solar cells fabricated in an industrial manner. The method relies exclusively on experimental input without the need of simulations. For that end, we determined the contribution of every loss mechanism under maximum power point conditions of the solar cell.

The gain analysis revealed that the reflectance of the active front side area bears the largest potential for optimisation, and the front metallisation the second largest potential. Based on these findings, we conducted PC1Dmod simulations to estimate the achievable efficiency under realistic assumptions. When applying a honeycomb texture, the efficiency is predicted to increase to 20.0%. Reducing the front grid finger width from currently 50 µm to 40 µm is estimated to further increase the efficiency to 20.2%.

The present work presents a systematic method for research institutes to guide research activities and allows cell manufacturers to identify, understand and minimise losses of their cells by comparing their internal results with ours. These results can be used to steer future research activities.

6 ACKNOWLEDGMENT

This work was funded by the German Federal Ministry for Economic Affairs and Energy within the projects CUT A (contract number 0325823) and CUT B (contract number 0325910A).

7 REFERENCES

[1] A. W. Blakers, A. Wang, A. M. Milne, J. Zhao, and M. A. Green, “22.8% efficient silicon solar cell,” *Appl. Phys. Lett.*, vol. 55, no. 13, pp. 1363–1365, 1989.

[2] ITRPV, “International Technology Roadmap for Photovoltaic: 2016 Results,” 2017. [Online] Available: http://www.itrpv.net/Reports/Downloads/.

[3] SCHMID Group | Gebr. SCHMID GmbH, *SCHMID develops new process for the texturing of diamond-wire-cut multi-wafers: Press release, Feb 2, 2017.*

[4] RENA Technologies GmbH, *Advanced texturing solutions and large orders from major Asian customers.* [Online] Available: https://www.rena.com/de/unternehmen/news/news/advanced-texturing-solutions-and-large-orders-from-major-asian-customers/. Accessed on: Aug. 31 2017.
[5] Trina Solar, *Trina Solar Announces New Efficiency Record of 21.25% Efficiency for Multi-crystalline Silicon Solar Cell: Press release, Nov 09, 2015.*
[6] Fraunhofer Institute for Solar Energy Systems (ISE), *Multicrystalline Silicon Solar Cell with 21.9 Percent Efficiency: Fraunhofer ISE Again Holds World Record*, 2017.
[7] Hanwha Q CELLS, *HANWHA Q CELLS Achieves New World Record in Multicrystalline Solar Module Efficiency: Press release, Jun 20, 2016.*
[8] R. Brendel, S. Dreissigacker, N.-P. Harder, and P. P. Altermatt, "Theory of analyzing free energy losses in solar cells," *Appl. Phys. Lett.*, vol. 93, no. 17, p. 173503, 2008.
[9] J. Greulich, H. Höffler, U. Würfel, and S. Rein, "Numerical power balance and free energy loss analysis for solar cells including optical, thermodynamic, and electrical aspects," *J. Appl. Phys.*, vol. 114, no. 20, p. 204504, 2013.
[10] A. G. Aberle, W. Zhang, and B. Hoex, "Advanced loss analysis method for silicon wafer solar cells," *Energy Procedia*, vol. 8, pp. 244–249, 2011.
[11] J. Wong, S. Duttagupta, R. Stangl, B. Hoex, and A. G. Aberle, "A Systematic Loss Analysis Method for Rear-Passivated Silicon Solar Cells," *IEEE J. Photovoltaics*, vol. 5, no. 2, pp. 619–626, 2015.
[12] P. Saint-Cast *et al.*, "Analysis of the losses of industrial-type PERC solar cells," *Phys. Status Solidi A*, vol. 214, no. 3, p. 1600708, 2017.
[13] B. Min *et al.*, "Incremental efficiency improvements of mass-produced PERC cells up to 24%, predicted solely with continuous development of existing technologies and wafer materials," in *31st EU PVSEC*, Hamburg, Germany, 2015, pp. 473–476.
[14] Sven Wasmer, Andreas A. Brand, Johannes M. Greulich, "Metamodeling of numerical device simulations to rapidly create efficiency optimization roadmaps of monocrystalline silicon PERC cells," in *7th International Conference on Silicon Photovoltaics*.
[15] H. Haug and J. Greulich, "PC1Dmod 6.2 – Improved Simulation of c-Si Devices with Updates on Device Physics and User Interface," *Energy Procedia*, vol. 92, pp. 60–68, 2016.
[16] Sven Wasmer, Abraham van der Horst, Pierre Saint-Cast, Johannes Greulich, "Experimental Efficiency Gain Analysis of Passivated Emitter and Rear Silicon Solar Cells, to be published," *IEEE J. Photovoltaics*, 2017.
[17] D. Biro *et al.*, "PV-TEC: Retrospection to the three years of operation of a production oriented research platform," in *24th EU PVSEC*, Hamburg, 2009, pp. 1901–1905.
[18] A. Kimmerle, J. Greulich, and A. Wolf, "Carrier-diffusion corrected J0-analysis of charge carrier lifetime measurements for increased consistency," *Sol. Energy Mater. Sol. Cells*, vol. 142, pp. 116–122, 2015.
[19] Z. C. Holman *et al.*, "Parasitic absorption in the rear reflector of a silicon solar cell: Simulation and measurement of the sub-bandgap reflectance for common dielectric/metal reflectors," *Solar Energy Materials and Solar Cells*, vol. 120, pp. 426–430, 2014.
[20] A. Volk *et al.*, "Honeycomb structure on multi-crystalline silicon Al-BSF solar cell with 17.8% efficiency," *IEEE Journal of Photovoltaics*, vol. 5, no. 4, pp. 1027–1033, 2015.

NUMERICAL ANALYSIS OF SILICON HETEROJUNCTION SOLAR CELL BASED ON MOLYBDENUM OXIDE AS A BACK SURFACE FIELD (BSF)

Haris Mehmood[1*], Hisham Nasser[2], Tauseef Tauqeer[3], Rasit Turan[2]
[1] National University of Sciences and Technology (NUST), Islamabad, Pakistan
[2] Centre for Solar Energy Research and Applications (GÜNAM),
Middle East Technical University (METU), Ankara, Turkey
[3] Information Technology University, Lahore, Pakistan
* Correspondence author: haris.mehmood@seecs.edu.pk

ABSTRACT: Solar cell technology based on carrier-selective contact layers has been envisaged as a prominent photovoltaic (PV) design for enhancing the efficiency of silicon heterostructure devices. Because of its higher work function, the carrier-selective layer such as Molybdenum oxide (MoO_x) assists in pinning the Fermi level at the interface between crystalline silicon (c-Si) and MoO_x. This enables easier extraction of charge carriers from the interface for electrical conduction, thus improving the device performance. In this paper, MoO_x has been employed as a back surface field layer (BSF) of p-n junction solar cell for efficient separation of photogenerated charge carriers. The correlation of various parameters, such as MoO_x work function, its thickness, and c-Si wafer doping concentration to the PV performance of the device has been demonstrated by employing Silvaco TCAD simulation. Comparative analysis has also been drawn to the reference structure based on p-a-Si:H BSF film. The simulated results exhibited an efficiency of 14.7% for MoO_x-based device as compared to 13.6% for the reference solar cell. The obtained efficiency of MoO_x-based devices can be boosted to ~19% by the inclusion of surface passivation layers.
Keywords: Back-Surface-Field, Carrier-selective, Heterostructure, Molybdenum oxide, Simulation, Silicon.

1 INTRODUCTION

Photovoltaic (PV) community is continuously striving to design and develop cost-effective and efficient solar cell technology that is viable for its widespread implementation. The current record efficiency for crystalline silicon (c-Si) solar cell is 26.6% which was achieved from the interdigitated back-contact design technology [1]. However, commercial production of such device is a daunting task from an industrial aspect; mainly because of the requirement for complex patterning and layer optimization at the rear. The heterojunction with intrinsic thin layer technology (HIT) alleviates the lithography issues in back-contact solar cell as both sides are capped with electrode layers and is rather simple to fabricate. Such technology makes use of doped amorphous silicon (a-Si:H) for separating the photogenerated charge carriers. However, these doped layers restrict the device performance due to the technical challenges associated with a-Si:H such as narrow band gap of ~1.7 eV, higher defects state density, and doping related issues [2].

As part of assuaging such issues originating due to doped a-Si:H layer, solar cell structures based on carrier-selective hetetrocontact configuration have attracted significant research interests because of its ability to facilitate easy collection of charge carriers [3]–[6]. Such carrier-selective contacts have been demonstrably integrated with c-Si based solar cells with maximum efficiency of 22.5% experimentally reported so far in the literature [6]. Molybdenum oxide (MoO_x) is one of the carrier-selective contacts offering higher work function that results in Fermi level pinning close to the valence band edge of the c-Si. Subsequently, this band bending of c-Si results in the formation of an energetic conduction band offset at the c-Si/MoO_x interface that prevents the recombination of electrons at an anode terminal. Simultaneously, the energy barrier for the hole charge carriers reduces at the interface, thus facilitating enhanced collection efficiency of the device.

So far, in the literature, device based on carrier-selective contacts has been reported to act as a junctionless device with n-type c-Si wafer [7]. When high work function material is applied to the c-Si, accumulation of holes occurs near the interface that results in a reduced hetetrocontact resistivity and an efficient separation of the photogenerated charge carriers. Although the device has been investigated for n-type c-Si in our previously reported work [8], [9], the effect of MoO_x on p-n junction is yet to be discussed in the literature. In addition to block the minority charge carriers as a result of favourable band offset, employing MoO_x in conjunction with p-n layers would make it emulate the Back-Surface-Field (BSF) layer. Motivated by this, we report on assessing the PV performance of p-type c-Si/MoO_x by varying different design parameters such as MoO_x work function, its thickness, and c-Si doping concentration in this work. An industrial simulator Silvaco TCAD has been used as a mathematical tool for numerical designing of the proposed structure. The reference structure is based on p-n junction configuration where p-a-Si:H has been utilized as a BSF layer. The doping concentration of p-type c-Si absorber and p-type a-Si:H BSF layer have been varied during the simulation and the results are compared with the proposed structure based on MoO_x BSF layer.

2 MODELING OF THE SOLAR CELL

The silicon heterostructure solar cell with the following configuration has been simulated in Silvaco TCAD simulator: ITO (70 nm)/n^+-c-Si (1 μm)/p-c-Si (150 μm)/MoO_x/Ag. The proposed structure is shown in Fig. 1(a) along with the reference structure in Fig. 1(b). The c-Si base and p-type a-Si:H BSF doping concentrations were also varied in order to understand the effect of doping concentrations on the PV performance. The basic material parameters for various layers used in the simulation have been tabulated in Table I.

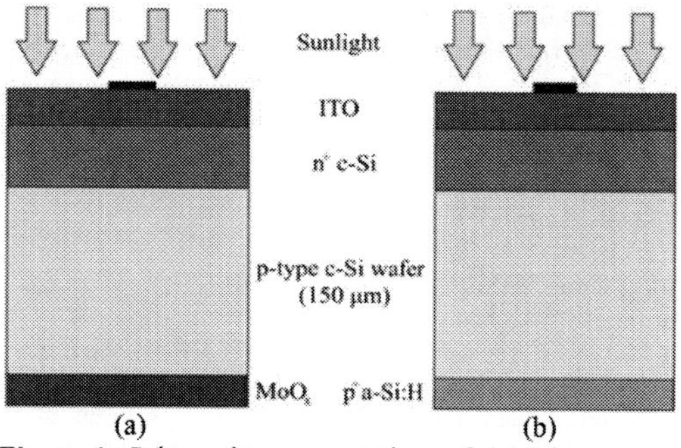

Figure 1: Schematic representations of the (a) proposed and (b) reference devices simulated in Silvaco TCAD

Optical dispersion data of MoO_x have been extracted from our work [10]. The defects parameters for p-a-Si:H were extracted from the literature [11]–[13]. The following assumptions were also considered during the simulation (i) MoO_x layer is treated as a contact layer [7]; (ii) interface recombination velocity of 10^5 cm/s at the top and bottom of the c-Si wafer; and (iii) 20% reflection of incident light from the front planar surface. Various mathematical models pertaining to the fundamental semiconductor equations such as Fermi-Dirac Statistics (FERMI), Auger (AUGER) and SRH recombination (SRH), concentration-dependent SRH (CONSRH), interface thermionic emission (THERMIONIC), band gap narrowing (BGN), have been invoked with the MODELS statement in the simulator. These equations have been numerically solved for iteration and convergence using newton (NEWTON) method. All measurements were performed at a simulated temperature of 25^0C and the Air Mass (AM) 1.5 has been used as a power source.

3 RESULTS AND DISCUSSION

The work function of MoO_x plays crucial role in determining the charge transport properties of the proposed solar cell. It has been well-established that the atmospheric exposure of MoO_x degrades the work function of more than 1 eV [14]. Fig. 2 illustrates the transport mechanism involved with the low and high work function MoO_x at the p-c-Si/MoO_x interface of the proposed device in Fig. 1(a). MoO_x basically acts as a

Table I: Basic material parameters

Parameters	c-Si	intrinsic a-Si:H	p-type a-Si:H
Electron affinity χ (eV)	4.17	3.90	3.90
Energy band gap E_g (eV)	1.12	1.72	1.72
Electron mobility μ_n (cm^2/V-s)	1300	10	10
Hole mobility μ_p (cm^2/V-s)	450	2	2
SRH Carrier lifetime τ_n, τ_p (sec)	10^{-3}	10^{-6}	10^{-6}
Electron density of states N_C (cm^{-3})	2.8×10^{19}	2.5×10^{20}	2.5×10^{20}
Hole density of states N_V (cm^{-3})	1.0×10^{19}	2.5×10^{20}	2.5×10^{20}

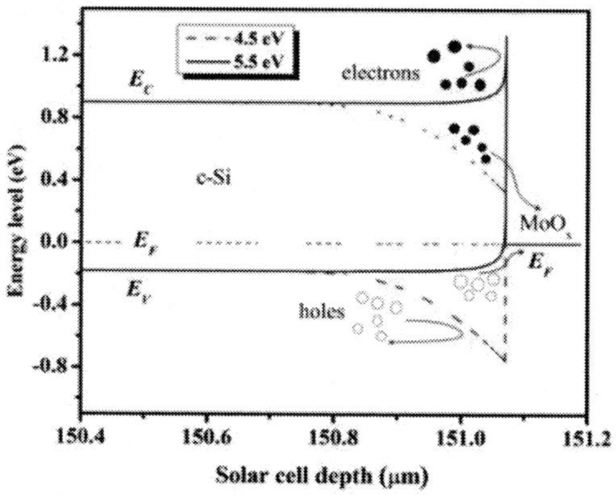

Figure 2: Energy band diagram of the proposed structure under thermal equilibrium. The flow of charge carriers at c-Si/MoO_x interface is represented by arrows

hole extraction layer by virtue of its higher work function. It can be observed that the high work function of MoO_x results in Fermi level of c-Si absorber pinning close to the valence band edge at the c-Si/MoO_x interface. Consequently, the energetic conduction band offset for electrons and low Schottky barrier height for holes are formed. The energetic barrier at the conduction band edge prevents the electrons from recombining with holes at the interface and subsequently electrons are reverted back into the absorber region. The contact recombination is thereby reduced. When MoO_x is annexed with the p-n junction as a BSF layer, the reduced barrier height at the valence band edge in connivance with the high electric field stimulate holes to be transported towards anode terminal with minimum resistance. Thus, much higher open circuit voltage (V_{oc}) of 560.4 mV and fill factor (FF) of 76.3% have been observed with MoO_x work function of 6.9 eV, as shown in Fig. 3.

With the reduction in MoO_x work function, the Schottky barrier height for holes increases and the blocking potential for electrons decreases at the c-Si/MoO_x interface. The electric field decreases from 2.56×10^7 V/cm at work function of 6.9 eV to 3.13×10^4 V/cm at 4.5 eV stamping the BSF properties of high-work function MoO_x. Subsequently, ineffective separation of photogenerated charge carriers at the c-Si/MoOx interface decreases the hole-selective capacity of MoO_x and deteriorates the PV performance of the proposed device as evident from Fig. 3. For the cell with the lowest simulated MoO_x work function level of 4.5 eV, the power conversion efficiency (η) was found to be 6.9% as compared to 14.7% for the work function of 6.9 eV. Similarly, V_{oc} and FF were measured to be 342.5 mV and 65.8%, respectively, at the lowest simulated work function, which were further improved with higher MoO_x work function values. The short-circuit current density (J_{sc}) also increased gradually with the work function and J_{sc} of 34.4 mA/cm^2 was observed for 6.9 eV work function. This is attributed to the feasible collection of charge carriers. From these simulated results, it is believed that strictly monitoring the growth of MoO_x is pertinent to obtain higher work function values which in turn can enhance the overall PV performance of the device.

Fig. 4 illustrates the effect of MoO_x thickness on the FF and η of the proposed device. For a smaller thickness of MoO_x (less than 15 nm), the efficiency and FF remain

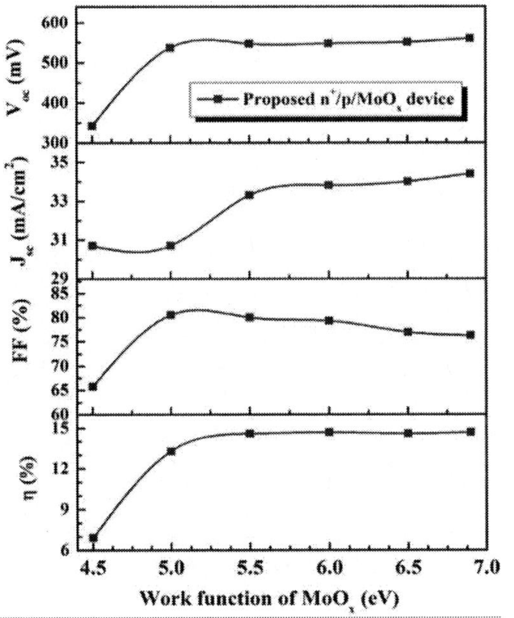

Figure 3: Photovoltaic charcteristics of the proposed device as a function of MoO_x work function levels

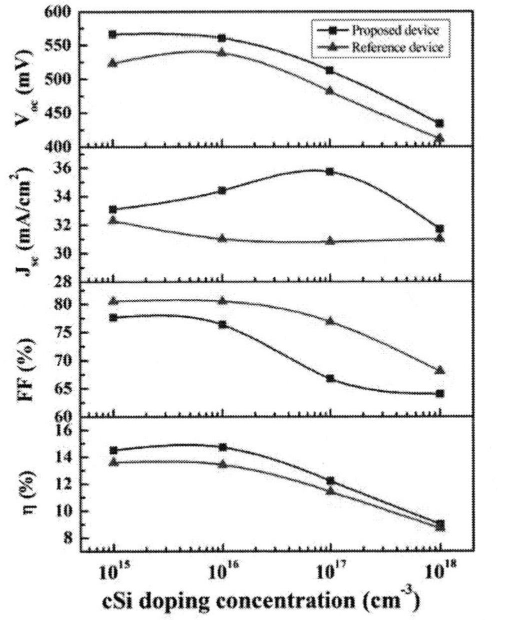

Figure 5: Effect of wafer doping concentration on the output charcteristics of the proposed and the reference solar cells

larger than 14 and 70%, respectively. However, further increasing the thickness lead to a substantial reduction in FF due to the increase in the sheet resistance. Moreover, considerable portion of sunlight reflected from the rear side gets parasitically absorbed within the thicker MoO_x layer. Correspondingly, J_{sc} decreases by 20% as MoO_x thickness is increase from 10 nm to 40 nm (not shown here). From the simulated results, the optimized thickness of MoO_x has been found to be 10 nm with a maximum simulated η of 14.7% achieved from the proposed device.

The effect of wafer doping concentration on the PV characteristic of the proposed device is shown in Fig. 5. The simulated results of the reference structure shown in Fig. 1(b) are also shown and compared with the proposed device in Fig. 5. With the MoO_x thickness of 10 nm, it can be seen that the efficiency of the proposed device slightly increases from 14.5 to 14.7% by increasing the doping concentration to 1×10^{16} cm^{-3}. It then plunges to 9% at elevated doping value of 10^{18} cm^{-3}. Comparatively,

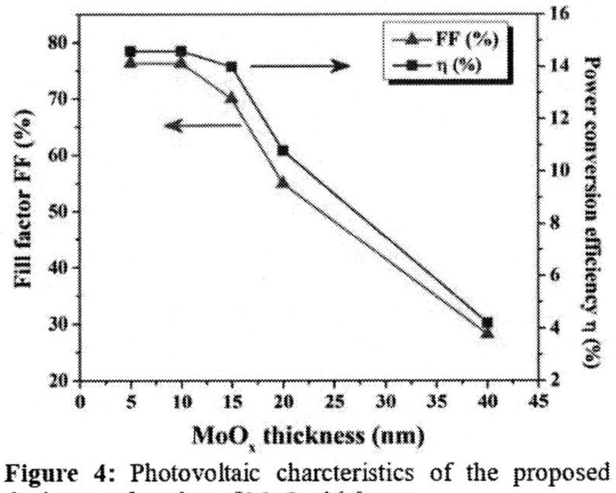

Figure 4: Photovoltaic charcteristics of the proposed device as a function of MoO_x thickness

the reference device demonstrated η of 13.6% at doping concentration of 1×10^{15} cm^{-3} which degraded to less than 9% for high p-type doping concentration. The J_{sc} for the proposed device remains almost the same for a simulated range of doping concentration. A significant reduction in V_{oc} has been observed for both devices at elevated levels of doping concentrations. Due to the higher work function of MoO_x, V_{oc} of 560.4 mV has been observed as compared to 538.4 mV for the reference device at 1×10^{16} cm^{-3}. However, V_{oc} decreases to 433.5 mV and 411.3 mV for the proposed and reference devices, respectively, at doping concentration of 1×10^{18} cm^{-3}. Such reduction in V_{oc} with a corresponding increase in the doping concentration can be related to the band gap narrowing effects. Moreover, the FF of the proposed device drops from 77.6% at 1×10^{15} cm^{-3} to less than 65% at 1×10^{18} cm^{-3} that can be attributed to the tunnelling of electrons at c-Si/MoO_x interface. Such tunnelling phenomenon arises due to the narrowing of conduction band edges that subsequently increases the contact recombination. From Fig. 5, the comparatively low performance of the reference device is due to the small band gap and the presence of defects density of states in a-Si:H. Photogenerated charge carriers get trapped and recombined at defects states reducing both J_{sc} and V_{oc}. Moreover, MoO_x band gap (~3.3 eV) and work function (~6.9 eV) is higher than p-a-Si:H which can be another reason for elevated levels of V_{oc} for MoO_x-based device. The PV performance of the solar cell structure based on n^+-c-Si/n-c-Si/MoO_x layers was recently reported in our work [9]. With the experimental and TCAD models, simulated η of 10.9% and 11.2%, respectively, the reported device exhibited low J_{sc} of 29.7 mA/cm^2 and FF of 65% due to an increase in Schottky barrier height for holes. The results of the proposed structure in this work demonstrated that utilising p-type c-Si absorber with MoO_x results in the better charge transport properties than its n-type counterpart as MoO_x will act as a BSF

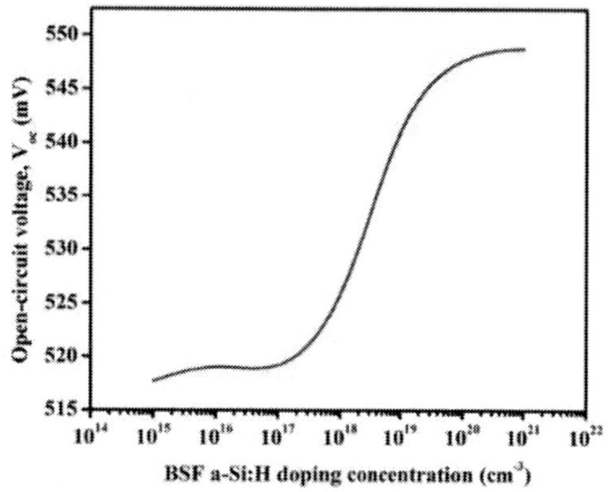

(a)

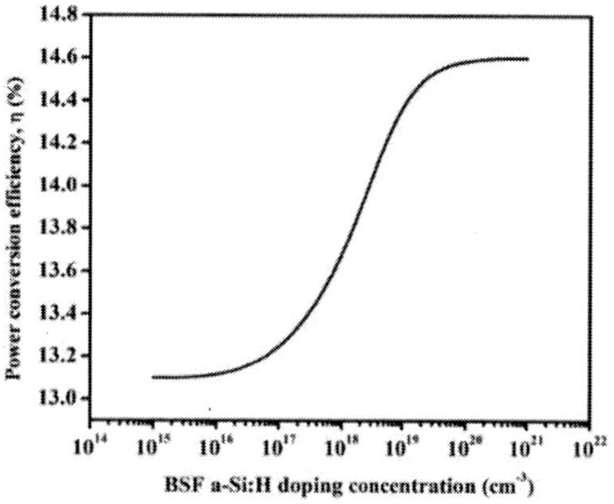

(b)

Figure 6: Effect of p-a-Si:H BSF doping concentration on (a) V_{oc} and (b) η of the reference solar cell.

layer to shore up the electric field at the c-Si/MoO_x interface. Much reduced barrier height for holes is observed at the valence band edges with p-type wafer when compared to the n-type wafer.

The effect of BSF layer doping concentration on the PV parameters of the reference device has also been analysed and elucidated in Fig. 6. Efficiency of 13.6% has been measured with the feasible doping concentration of 1×10^{18} cm^{-3}. With the increase in doping, the electric field becomes strengthened at the rear interface, effectively separating the photogenerated charge carriers. The maximum saturated η of 14.6% has been observed for the reference device with higher doping levels, which is still less than 14.7% for the proposed device. The V_{oc} of the proposed device is considerably low because of the absence of the passivation layers. It is expected that the including passivated intrinsic a-Si:H will lead to an enhancement in V_{oc} due to the reduction in the recombination velocity. Consequently, a final optimized device based on the proposed structure was designed on Silvaco TCAD by inserting i-a-Si:H at wafer's both sides. Much higher V_{oc} of 724.8 mV has been numerically achieved along with the following characterisation parameters: J_{sc} of 33.25 mA/cm^2, FF of 78.3%, and η of 18.9%. These simulated results indicate that MoO_x is an integral part of the PV structure requires for the development of highly-efficient and cost-effective carrier-selective contact based solar cells.

4 CONCLUSION

Silvaco TCAD simulation of the carrier-selective contact-based solar cell has been undertaken in order to ascertain hetetrocontact formation between the MoO_x hole-selective layer and c-Si. The results have clearly shown that employing p-type wafer allows MoO_x to act as a BSF layer that enhances the electric field at the interface. Higher work function of MoO_x induces reduction in holes Schottky barrier height, thereby, enhancing the hole-selectivity of MoO_x. Increase in c-Si wafer doping concentration implies reduction in PV performance that can be attributed to the narrowing of conduction band edges at the rear interface, thus allowing electrons to recombine with holes. The simulated PV characteristics for the proposed device measured the V_{oc} of 560.4 mV, J_{sc} of 34.4 mA/cm^2, FF of 76.3% and η of 14.7% as compared to η of 13.6% for reference device. By including the passivation effects, the efficiency of the device boosted to 18.9% with 1.5 Ω.cm p-type Si wafer. It can be convincingly stated that MoO_x is deem to be an integral part for the development of highly-efficient and cost-effective carrier-selective contact based solar cells.

5 ACKNOWLEDGMENT

This research was supported by Scientific and Technological Research Council of Turkey (Türkiye Bilimsel ve Teknolojik Araştırma Kurumu, TÜBİTAK), Turkey under 2216 - Research Fellowship Programme for International Researchers vide Award No. 21514107-115.02-56312.

6 REFERENCES

[1] K. Yoshikawa, H. Kawasaki, W. Yoshida, T. Irie, K. Konishi, K. Nakano, T. Uto, D. Adachi, M. Kanematsu, H. Uzu, K. Yamamoto, Nature Energy 2 (2017) 17032.

[2] K. J. Bullock, M. Hettick, J. Geissbühler, A. J. Ong, T. Allen, C. M. Sutter-Fella, T. Chen, H. Ota, E. W. Schaler, S. De Wolf, C. Ballif, A. Cuevas, A. Javey, Nature Energy 1 (2016) 15031.

[3] K. Habashy, V. Steenhoff, M. Vehse, M. Becker, M. Schiek, M. Wark, C. Agert, Prceedings 32nd European Photovoltaic Solar Energy Conference and Exhibition, (2016) 1310 – 1315.

[4] L. G. Gerling, S. Mahato, A. Morales-Vilches, G. Masmitja, P. Ortega, C. Voz, R. Alcubilla, J. Puigdollers, Solar Energy Materials and Solar Cells 145 (2016) 109–115.

[5] D. Ahiboz, H. Nasser, R. Turan, Proceedings 4th International Renewable and Sustainable Energy Conference (IRSEC), IEEE Xplore (2016) 144–151.

[6] J. Geissbühler, J. Werner, S. Martin de Nicolas, L. Barraud, A. Hessler-Wyser, M. Despeisse, S. Nicolay, A. Tomasi, B. Niesen, S. De Wolf, C. Ballif, Applied Physics Letter **107** 8 (2015) 81601.

[7] S. Chuang, C. Battaglia, A. Azcatl, S. McDonnell, J. S. Kang, X. Yin, M. Tosun, R. Kapadia, H. Fang, R. M. Wallace, and A. Javey, Nano Letters 14 3 (2014) 1337–1342.

[8] H. Mehmood, H. Nasser, E. Özkol, T. Tauqeer, S. Hussain, R. Turan, Proceedings International Conference on Engineering and Technology

(ICET2017), (2017) IEEE Xplore - *In Press*.
[9] H. Nasser, G. Kökbudak, H. Mehmood, R. Turan, Proceedings 7th International Conf. on Silicon Photovoltaics (SiPV), Energy Procedia 124C (2017) 418–424 In Press.
[10] H. Mehmood, H. Nasser, E. Özkol, T. Tauqeer, S. Hussain, R. Turan, International Journal of Energy Research (2017) In Press.
[11] B. S. N. F. M. Suria, S. Hussain, H. Mehmood, G. Ali, Sains Malaysiana, **43** 6 (2014) 895–898.
[12] A. V. Sachenko, V. P. Kostylyov, M. R. Kulish, I. O. Sokolovskyi, A. Shkrebtii, F. Gaspari, Proceedings 29th European Photovoltaic Solar Energy Conference and Exhibition, (2014) 1116 – 1119.
[13] H. Noge, K. Saito, A. Sato, T. Kaneko, M. Kondo, Japanses Journal of Applied Physics 54 (2014) 08KD17.
[14] J. Meyer, S. Hamwi, M. Kröger, W. Kowalsky, T. Riedl, A. Kahn, Advanced Materials **24** 40 (2012) 5408–5427.

A COST-DRIVEN RESEARCH STRATEGY TOWARDS INDUSTRIALLY FEASIBLE HIGH-EFFICIENCY BACK-CONTACT BACK-JUNCTION SILICON SOLAR CELLS

J. D. Huyeng[1], A. Spribille[1], S. Nold[1], R. Efinger[1], R. Keding[1], O. Doll[2], F. Clement[1]

[1]Fraunhofer Institute for Solar Energy Systems (ISE), Heidenhofstrasse 2, D-79110 Freiburg, Germany
[2]Merck KGaA, Frankfurter Str. 250, D-64293 Darmstadt, Germany

ABSTRACT: The development and improvement of silicon solar cells is often based on a top-down approach, achieving highest conversion efficiencies and then translating it to industrial equipment and fabrication routes. This work presents an alternative option to derive research and fabrication strategies focused on cost of ownership calculations. Given the necessary tools and insight into feasible production routes, this attempt can be very helpful and rewarding for (small) research facilities, existing solar cell fabrication plants or interested investors in the solar industry. A research strategy is exemplarily derived for the goal of industrially feasible back-contact back-junction solar cells, but many aspects can be generalized for different purposes. Several different fabrication routes are shown and compared to a PERC fabrication route. We find that a co-diffusion approach can offer significant cost reduction potential on cell level, from +21 $\%_{rel}$ down to -2$\%_{rel}$.
Keywords: Co-diffusion, Cost calculation, Interdigitated Back-Contact

1 INTRODUCTION

After a period of drought, the growth in photovoltaic industry has continued its formerly impressive run. With new capacities build up, more and more companies are also looking for novel and improved technology ("high-efficiency concepts"). One way to obtain highest efficiencies has been long known as the "interdigitated back-contact" (IBC) concept [1], where all electrodes are placed on the non-illuminated rear side, avoiding shading losses. However, proven industrial feasibility is pending.

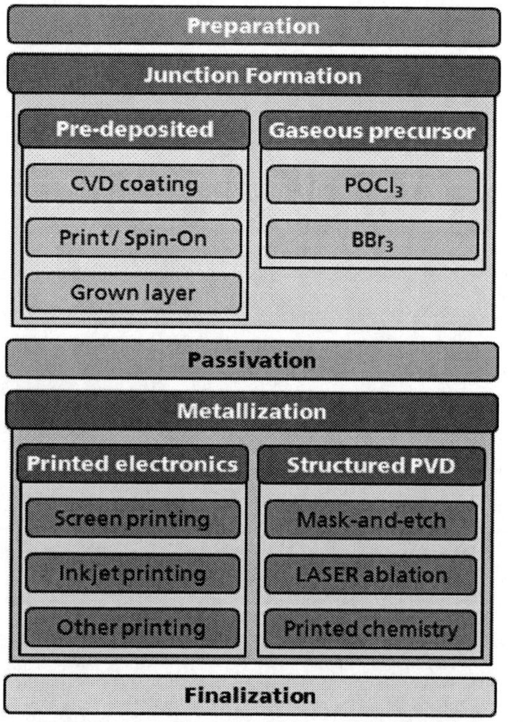

Figure 1: Simplified solar cell fabrication route focusing on different options for "Junction Formation" and "Metallization" (without claim for completeness). The presented studies focus on certain combinations of those, to build an IBC solar cell fabrication plant.

In earlier work by our institute, the performance of screen-printed boron-doping paste in the fabrication of high-efficiency IBC solar cells was investigated [2]. Back-contact back-junction silicon (BCBJ) solar cells with a mean efficiency of η = 20.9 % ± 0.3 % were fabricated and analyzed to reveal the possibilities and limitations of the chosen geometry and process details [3]. All cells were fabricated using the semi-industrial research facilities at Fraunhofer ISE PV-TEC [4], to allow an easy transfer of the evaluated process route into existing PV plants.

For a matured industry like photovoltaics, the cost-competitiveness of a fabrication sequence plays a major role. This work illustrates a method to quantitatively derive research and/or business strategies, focusing primarily on this aspect. Based on the results of these evaluations, cell experiments are currently running to demonstrate lean fabrication of high-efficiency IBC solar cells with minimal costs.

2 METHOD

In the following, an alternative research strategy is developed with a clear emphasis on cost of ownership calculations (COO). Remarkably, this attempt can be implemented with minimal resources, as long as a suited cost-calculation tool is available. Here, the bottom-up "SCost" tool [5] is utilized, designed after SEMI standards E35-0312 [6] and E10-0814E [7].

The strategy is developed in three steps: First, a previously published (research focused) fabrication route is translated into SCost, with the setup of a COO calculation for each process step and relating process equipment. Second, different alternative interlinked process routes are build, altering a small number of process steps each, with different equipment. Third, these alternatives are implemented as well and a COO calculation for each process route is carried out. The generated data is then analyzed in a generalized framework, as described below, for a simplified but still insightful discussion about the suggested alternatives.

2.1 Generalized Framework

It has earlier been described how the competitiveness of different cell concepts can be compared along the PV

value chain by their COO normalized to their output power ("Cost-per-Wattpeak") [5]. As process routes can vary significantly, a simplified scheme for the depiction of the fabrication routes is chosen, consisting of five generic steps:
"Preparation" such as cleaning and surface texturing, "Junction Formation" with different dopant sources and high-temperature steps, "Passivation" of wafer surfaces, "Metallization" of *p*- and *n*-type dopings and "Finalization" of the production cycle such as I-V or other quality tests. This scheme and a number of different options for "Junction Formation" and "Metallization" (without claim for completeness) are shown in Fig. 1.

For a given fabrication route, industrial tools are chosen within SCost for COO calculations. The function of these tools is accredited to one of the five generalized steps and their cost contribution is summed up, for comparison. For all routes a most likely fabrication scheme has been assumed.

2.2 Mathematical Description

The cost of ownership is calculated as a sum of all utilized tools and fabrication steps, including factors like yield loss and capacity utilization within SCost [8]. The cost for any given fabrication route i is then

$$C_i = \sum_j c_j,$$
$$i \in \{\text{PERC}, \text{IBC}_1, \text{IBC}_2, \dots\},$$
$$j \in [1, \dots, n_i],$$

with different numbers of process steps n_i.

Each fabrication route implies a different mean output power under standard test conditions of the finished devices P_i^{out}. We calculate a value based cost of ownership, given as the ratio of C_i and P_i^{out}, which is often referred to as "Cost-per-Wattpeak"

$$\xi_i = C_i / P_i^{out},$$
$$[\xi_i] = 1\ \text{€ct}/W_p.$$

To derive a cost-driven research strategy, we need to be able to compare two different fabrication routes. Here, we take ξ_i as the sole figure of merit for different routes and translate the order to rate the routes:

$$\xi_i < \xi_j \Rightarrow i \prec j,$$
$$i, j \in \{\text{PERC}, \text{IBC}_1, \text{IBC}_2, \dots\}.$$

In reality, this certainly does not hold true for all business decisions. For example, both values should not be to close, and/or the relation should hold true for some time. Also, if, for example, an existing plant has to be transitioned to achieve an alternate route, the difference needs to include a significant margin, covering additional costs like necessary trainings, downtime during migration etc., not included in this COO calculation. With uncertainties in the scalability between routes, high variance in the production outcomes or other factors that cannot directly be accounted for in this quantity, the derivation of true relation between process routes becomes even more involved.

Neglecting these limitations in the following, we shape a first order statement, to separate the two a priori unknown quantities C_i and P_i^{out} ($\equiv P_i$ in the following). Assuming parity between two different routes we have

$$\xi_i = \xi_j \Leftrightarrow \frac{C_i}{P_i} = \frac{C_j}{P_j} \Leftrightarrow \frac{C_i}{C_j} = \frac{P_i}{P_j}$$

Under standard test conditions, one derives P_i from the input Power P_0, with the photon conversion efficiency η according to

$$P_i = \eta \cdot P_0,$$

allowing for the identity

$$\frac{C_i}{C_j} = \frac{P_i}{P_j} = \frac{\eta_i \cdot P_0}{\eta_j \cdot P_0} = \frac{\eta_i}{\eta_j}.$$

For comparison, the costs for different routes C_i are normalized to an industrial *p*-type LCO PERC process sequence [9] C_{PERC} in the following. This also eliminates some potential systematic errors in the calculations.

Finally, one can relate the differences in costs and efficiency by rearranging

$$\Delta C = C_i - C_j \Leftrightarrow C_i = C_j + \Delta C,$$

which leads to

$$\frac{C_i}{C_j} = \frac{C_j + \Delta C}{C_j} = 1 + \frac{\Delta C}{C_j}$$

and analogously for η. Plugging these into the former relation equals

$$\frac{C_i}{C_j} = \frac{\eta_i}{\eta_j} \Leftrightarrow 1 + \frac{\Delta C}{C_j} = 1 + \frac{\Delta \eta}{\eta_j} \Leftrightarrow \frac{\Delta C}{C_j} = \frac{\Delta \eta}{\eta_j}.$$

This means, a cost difference normalized to the reference route cost has to equal the efficiency difference normalized to the reference efficiency (including sign) for both routes assumed to be equal. All these relations are of course only valid for one chosen reference point at a time. One might as well do the same with an inequality between two routes (e.g. for an improvement), carefully checking for signs of quantities.

The relation between costs and efficiency can also be graphically displayed by "cost-efficiency indifference curves" as shown in [10]. In this work, we avoid the assumption of absolute efficiencies (*cf.* Section 4).

If one finds a relative cost difference between two routes $\Delta C/C_{Ref}$ one can then rearrange the last equation to get a first order estimate for a suitable efficiency difference $\Delta\eta$ between these two routes

$$\Delta\eta = \eta_{Ref} \cdot \frac{\Delta C}{C_{Ref}},$$

with the same limitations for validity as mentioned during this section.

3 COST OF OWNERSHIP CALCULATION

In the following, first the COO calculations of different routes are discussed. As mentioned above, we utilize SCost as the tool of choice. It has been developed and updated at Fraunhofer ISE since the first predecessors from as early as 1998 [11], to represent an accurate and up-to-date database. This was only possible due to helpful insight from projects with our partners and various sources from within the industry.

For comparability, all of the following calculations are based on a European PV plant placed on a green field, scaled to an output of 12,000 cells/hour[1], which amounts to about 500 MW depending on the actual cell efficiency.

The COO can be calculated for all stages up to system cost and even LCOE within SCost. For simplicity, the following arguments are only based on cell level production costs (€/Cell), neglecting overhead costs like Selling, General and Administrative Expenses (SG&A), Research and Development Expenses (R&D), capital cost, and wafer costs. This of course further restricts the applicability of the results, but avoids inclusion of uncertain or varying aspects like module integration or system installation costs. Nold *et al.* [5] have shown that at all stages of the PV value chain, area proportional costs are related to the cell output power, favoring high efficiency concepts even more the further downstream the PV value chain the economic assessment is executed. These positive attributes are also neglected in the following.

With the neglection of wafer cost, also the difference between *p*-type doped substrates and *n*-type doped substrates is excluded here. As IBC solar cells are usually fabricated on the latter, more expensive, wafers, this can lead to diametral cost effects avoided at this point.

3.1 Costs for Proof-of-Principle IBC Route

As a starting point, the proof-of-principle investigations regarding screen-printed boron-paste are chosen, as described in earlier work [2]. The fabrication of high-efficiency IBC cells was demonstrated on large-size wafers with active cell areas of 4 cm² (designated area, shaded busbars) and 243 cm² [3]. More details on properties of screen-printed boron-paste or process details can be found in the given references.

For "Junction Formation" and "Metallization", full area deposition technologies (PECVD, PVD) have been used in combination with wet-chemical based mask-and-etch processes. These involve multiple steps, a number of tools and therefore result in a high process complexity. When translated into a virtual solar cell plant via SCost calculations, this also shows high fabrication costs.

To set the goals for further research, different options are possible. One could increase the mean efficiency of finished devices, such that they compensate the additional costs, for example with variations of the employed geometries. We will show, why this is not always an option, when considering a viable reference technology. Another way is to alternate the fabrication sequence such that efficiencies are increased or at least stable, while fabrication costs are reduced.

Both ways can be pursued with a wealth of options. We show in the following, how a cost-driven research strategy can help to identify those who can compete on a level of industrial feasibility.

3.2 PERC Reference Process

We implement a PERC fabrication sequence simplified in the same fashion for SCost as above, based on published work from our institute [9] for accessibility. We chose this cell type as a reference technology,

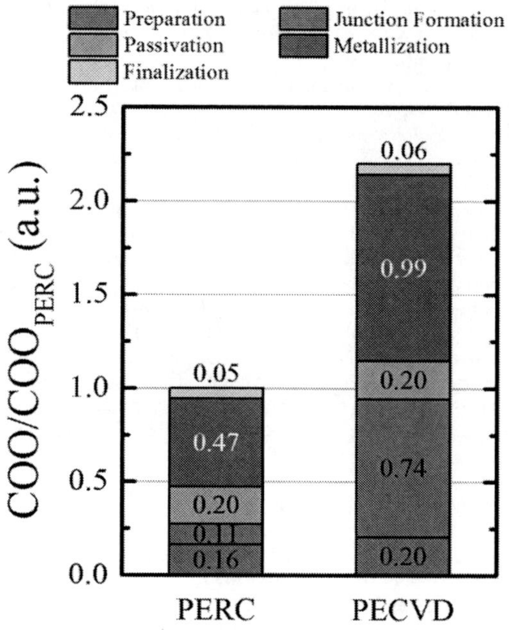

Figure 2: A normalized cost distribution is given for a *p*-type Cz PERC LCO reference process and the research oriented approach, as presented in last year's paper [2], translated into an industrial fabrication plant. Major increases are found for "Junction Formation" and "Metallization" due to the high number of mask-and-etch steps. Different strategies to reduce these contributions are given in the text.

although the dominant fabrication route is still the Al-BSF solar cell. Nevertheless, more and more manufacturers are changing to or installing new capacity for PERC. The comparison between PERC and IBC routes might therefore be more relevant, but as mentioned above arbitrarily limits the applicability of the results.

When comparing both fabrication routes, we group fabrication steps into the generalized framework. We normalize the costs per piece to the PERC sequence, as explained above. Again, this does not include wafer costs or a scaling to the output power of the device. The result is shown in Fig. 2.

We develop six different IBC routes as shown in Fig. 3, with a focus on different process steps for "Junction Formation" and "Metallization". The details are given in the following section.

3.3 Alternative Process Steps for IBC Fabrication Routes

Other research groups have shown IBC process sequences based on sequential diffusion, where areas are also modified with mask-and-etch processes between several high-temperature steps [12]. As full process sequences are rarely disclosed, we compare the publications of different groups and make some educated guesses to derive a fabrication sequence for implementation in SCost. Such routes have shown very high conversion efficiencies of up to 24.4 % for a silicon homojunction [13, 14] and are therefore highly justified especially in a research context. Considering a cost calculation with current prices however, an industrial

1 As mentioned before, this way the results are not directly transferable to (highly specific) transition processes for existing plants, but are more general.

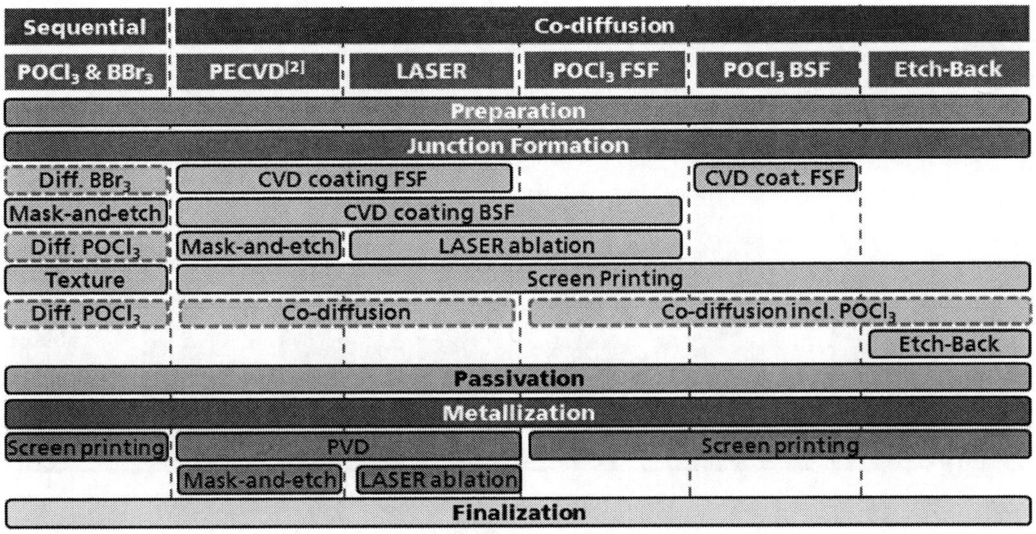

Figure 3: Different combinations of process options for IBC cell fabrication are shown. Orange (dashed) boxes indicate high-temperature steps. Details for processes are given in the text, as well as the calculated COO.

implementation is questionable, as shown below. This of course changes, if new tools or materials are introduced and scaled for mass market fabrication, which cannot be accounted for in this paper.

The sequential diffusion might offer some benefits, like improved gettering of defects, but requires multiple high-temperature steps, which are cost-intensive and therefore increase the "Junction Formation" costs. To effectively avoid multiple high-temperature processes, simultaneous "co-diffusion" of several dopants has been introduced in other work. This is enabled with a combination of different doping sources, *e.g.* solid CVD coatings (structured or full-area), gaseous precursors ($POCl_3$) or printed sol-gels. All the other process sequences incorporate such a diffusion approach, to reduce the increased "Junction Formation" cost in relation to the PERC process.

A way to avoid mask-and-etch steps, the other factor severely increasing the cost in the proof-of-principle process sequence, is direct structuring, *e.g.* with LASER ablation or printed chemistry [15]. The combination with co-diffusion drastically reduces the fabrication costs and is shown below to be the key advantage over sequential diffusion. To reduce the costs of "Metallization", structured screen printing of metal pastes is a suitable choice. For "Junction Formation", the formerly established screen-printable boron-doping paste offers the unique possibility of structured application and minimal process steps in combination with gaseous precursors.

3.4 Alternative IBC Routes

In the following, we describe the six IBC routes addressed in this work and their process details. The results of the SCost implementation can be found in Fig. 4, sorted by their (normalized) COO. They will be discussed in this order in the following. In contrast to Fig. 2, where CVD coatings were applied via PECVD, we additionally show a similar process there, which only substitutes PECVD for APCVD. While APCVD is a cheaper process than PECVD, this does only lead to a minor cost reduction for the "Junction Formation" as its effect is diminished due to the many other involved steps at this point. The following variations are nevertheless based upon APCVD, if coatings are implemented.

The implementation of a sequential diffusion shows a significant cost contribution of "Junction Formation" due to several high-temperature steps. As texturing of the front side is done later, within the process sequence (see Fig. 3), a part of the cost is shifted from "Preparation" to "Junction Formation" in our description.

The "LASER" termed variation demonstrates alternative structuring of CVD and PVD layers with LASER ablation. Given that cell performance is not constrained, this displays an effective way to reduce the costs down to a mere 20 % premium compared to the PERC process. The effect is larger in the "Metallization" costs, as less steps are involved overall (*cf.* PECVD vs APCVD above). If metallization is realized by one screen printing step instead (not shown), some additional costs can be saved (~15 % premium), while also reducing the number of process steps. For this, we assume a single Ag metallization paste for both polarities. All other process routes have been calculated with screen-printed metallization (see Fig. 3).

With the introduction of a gaseous precursor ($POCl_3$) as dopant source for front side doping instead of a CVD source ("$POCl_3$ FSF"), additional costs can be saved without increasing complexity and reasonably expecting better passivation properties, as shown in earlier work [16]. This change is in principle, decoupled from other process parameters, which has also been shown. It is therefore a very suitable and promising approach for cost-effective IBC fabrication and one of the objectives for the derived research strategy.

If the gaseous precursor is instead used for the back surface field doping ("$POCl_3$ BSF") the process is cheaper, as a screen-printed boron-paste can be additively applied on a substrate as a pattern, saving one additional step compared to the "$POCl_3$ FSF" approach. From a technological point of view, the process route becomes more complex, as some interdependencies between the rear side dopings exist in such a setup. This is not covered in the COO calculations, but has to be kept in

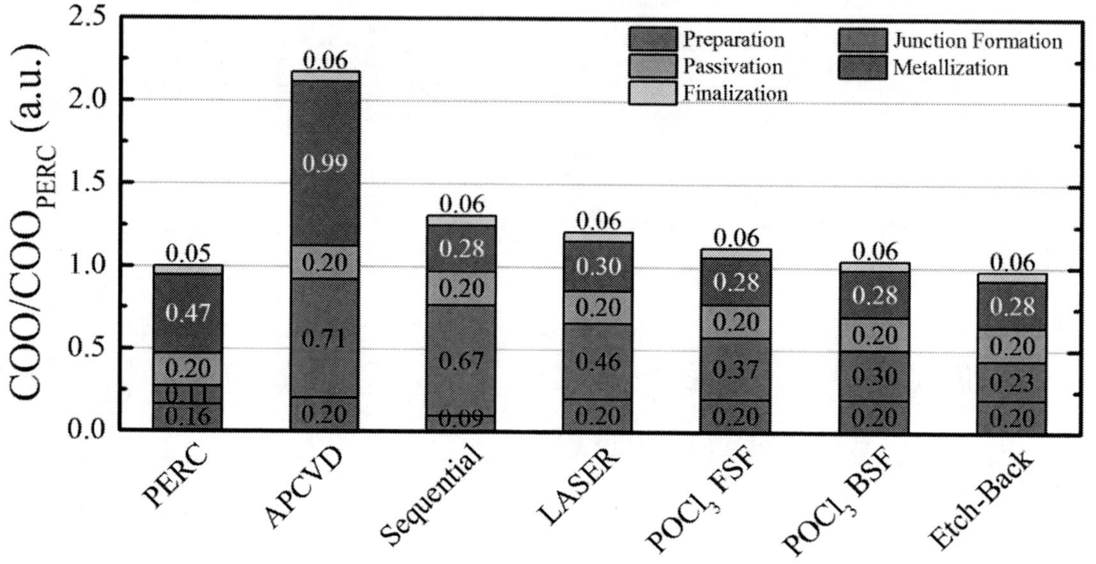

Figure 4: A number of different variations in the process chain are shown with the resulting costs per piece (no wafer costs, no scaling to output power) normalized to the PERC reference route. The main differences are realized in "Junction Formation" and "Metallization", while the other contributions remain similar for all routes.

mind for further decisions.

If no CVD dopant source is used at all, the front and rear side are similarly doped by a gaseous precursor. In such a case, a front side etch-back should be introduced to satisfy the different demands for front- and rear side Phosphorus dopings ("Etch-Back"). This does not change the total number of process steps, but shifts the cost contribution to a wet-chemical process instead, which may be even cheaper with a suitable setup according to our calculations.

As can be seen from Fig. 4, the costs for the last three fabrication routes are close to unity, when normalized to the PERC reference process. We will discuss the conclusions and the thereof derived research strategy in the next section.

4 RESULTS AND DISCUSSIONS

With the normalized COO values, one can now estimate, simulate or test the cell efficiencies of the displayed fabrication routes for a complete comparison. Experimental results of large sample sizes offer the most reliable reasoning for practical business decisions. The use of simulations is highly dependent on the quality, reliability and applicability of input parameters. Where possible this should be done with supporting experiments as closely to the final solar cells as possible. A pure estimation of achievable efficiency should only be used for a first impression and inherently offers large uncertainties, rendering quantitative comparisons impossible.

Lacking experimental results, we choose a different way for the interpretation of the obtained figures. As derived earlier, we relate the change in normalized COO to the solar cell efficiency to estimate a suitable efficiency difference. This number is then used to deduce a decision, which fabrication routes should be taken into consideration for a cost-driven research strategy. The results are shown in Fig. 5. As the efficiency for the reference technology has a quite significant impact, we show three different assumptions from 21 % to 22 %.

Unsurprisingly, the "APCVD" route has to be considered not-feasible for industrial fabrication as it has been derived from a proof-of-principle investigation.

We find that a sequential diffusion needs to surpass a PERC efficiency by more than 6 $\%_{abs}$, based on our calculations and current market prices. Such an advantage has so far never been shown. As mentioned earlier, these calculations do not include wafer prices or scaling of area proportional costs in later fabrication stages. Current *n*-type Cz wafer prices exceed the *p*-type Cz wafer price remarkably and we do not expect the scaling reductions to cover those. Summarizing all these

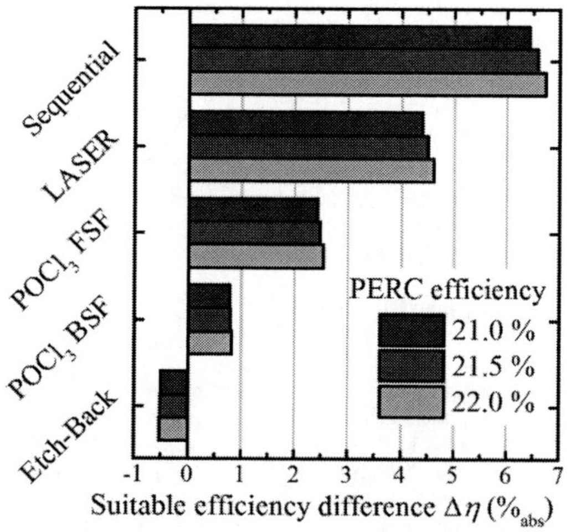

Figure 5: Suitable efficiency difference with respect to different PERC efficiencies based on COO calculation.

aspects makes the sequential diffusion very unlikely for industrially feasible fabrication.

The "LASER" process route, based on thermal co-diffusion, but laser structuring of full area APCVD and PVD depositions, is significantly cheaper than the sequential diffusion, but still requires a huge efficiency increase for parity with a PERC fabrication route. Although one could argue about scaling effects and other benefits, we discard this approach for now, given the high uncertainties and the significant amount of needed improvement. We instead address the co-diffusion setups, with a gaseous dopant source.

The most expensive process flow here ("$POCl_3$ FSF") needs an efficiency improvement of some percent relative to the PERC solar cells, to allow industrially feasible fabrication. This is done via co-diffusion from three different dopant sources, simultaneously driven into the wafer. These decoupled sources make for a high tunability of the fabrication process and should therefore be addressed in future research.

The two more sophisticated process routes of this kind are even closer in their costs to the PERC reference process. It should hence be aimed at trying to solve the inherent challenges of these routes, to maximize the gap between these routes and the PERC process in the value based cost of ownership, to cover for the (volatile, but momentarily) increased wafer price. Any additional efficiency gain is still desirable for additional area-related scaling cost effects.

5 CONCLUSIONS AND OUTLOOK

The herein derived calculations imply that in the near term, only the introduction of co-diffusion offers substantial cost reduction potential towards industrially feasible routes for IBC cell fabrication. Additionally, the use of screen-printed boron-doping paste is shown to close the gap towards the reference technology, due to its unique properties, when used as a simultaneous dopant source and barrier in a co-diffusion setup.

Consequently, we focus our ongoing research on the establishment of lean co-diffusion processes utilizing screen-printed boron-doping paste and the involved challenges of gaseous and solid dopant sources. Based on the experimental results, the derived research strategy aims at the most promising fabrication route for industrially feasible solar cell production.

6 ACKNOWLEDGEMENTS

The authors would like to thank anyone involved in the setup, maintenance and expansion of SCost, the research teams at Fraunhofer ISE PV-TEC and Merck KGaA and all colleagues for valuable discussions.

J.D.H acknowledges funding by the center of renewable energy of the University of Freiburg, within the interdisciplinary graduate school "DENE".

This work was supported by the German Federal Ministry for Economic Affairs and Energy within the research project "SiBoKo" under contract number 0324075B.

7 REFERENCES

[1] R. J. Schwartz and M. D. Lammert, "Silicon solar cells for high concentration applications," in *1975 International Electron Devices Meeting*, pp. 350–352.

[2] J. D. Huyeng *et al.*, "Co-Diffused Back-Contact Back-Junction Silicon Solar Cells with a Novel Screen-Printed Boron-Doping Paste," in *32nd EUPVSEC 2016*, pp. 575–579.

[3] J. D. Huyeng *et al.*, "Analysis of full-wafer size co-diffused BCBJ silicon solar cells with a novel screen printed boron-doping paste," in *SiliconPV 2017*, tbp.

[4] D. Biro *et al.*, "PV-Tec: Retrospection to the three years of operation of a production oriented research platform," in *24th EU PVSEC*, Hamburg, 2009, pp. 1901–1905.

[5] S. Nold *et al.*, "Cost Modeling of Silicon Solar Cell Production Innovation along the PV Value Chain," in *27th EU PVSEC*, Frankfurt, 2012, pp. 1084–1090.

[6] *Guide to Calculate Cost of Ownership (COO) Metrics for Semiconductor Manufacturing Equipment*, SEMI-E35, 2012.

[7] Semiconductor Equipment and Materials International (SEMI), *Specification for Definition and Measurement of Equipment Reliability, Availability, and Maintainability (RAM)*. San Jose, CA, USA: Semiconductor Equipment and Materials International (SEMI).

[8] *Technologiebewertung - Fraunhofer ISE*. [Online] Available: https://www.ise.fraunhofer.de/de/geschaeftsfelder/photovoltaik/silicium-photovoltaik/technologiebewertung.html.

[9] P. Saint-Cast *et al.*, "Cz silicon benchmark for p-type PERC solar cells," in *32nd EU PVSEC*, Munich, 2016, pp. 1038–1043.

[10] J. Jaus, J. Rentsch, and R. Preu, "Determining trade-offs between costs of production and cell/module-efficiency with the concept of indifference curves," in Barcelona, Spain, 2005, pp. 3178–3181.

[11] R. Preu *et al.*, "Innovative production technologies for solar cells - SOLPRO," in *16th EU PVSEC*, Glasgow, 2000, pp. 1451–1454.

[12] Z. Li *et al.*, "Pilot Production of 6" IBC Solar Cells Yielding a Median Efficiency of 23% with a Low-Cost Industrial Process," 2016.

[13] E. Franklin *et al.*, "Design, fabrication and characterisation of a 24.4% efficient interdigitated back contact solar cell," *Prog. Photovolt: Res. Appl.*, vol. 24, no. 4, pp. 411–427, 2014.

[14] G. Xu *et al.*, "6 inch IBC cells with efficiency of 23.5% fabricated with low-cost industrial technologies," in *2016 IEEE 43rd Photovoltaic Specialists Conference (PVSC)*, Portland, OR, USA, 2016, pp. 3356–3359.

[15] A. Spribille *et al.*, "MWT solar cell processing by use of isishape® solaretch® SiD for rear contact and edge isolation," in *27th EU PVSEC*, Frankfurt, 2012, pp. 2001–2006.

[16] N. B. Tanvir *et al.*, "Codiffusion Sources and Barriers for the Assembly of Back-Contact Back-Junction Solar Cells," *IEEE J. Photovoltaics*, vol. 5, no. 6, pp. 1813–1820, 2015.

p-Si BASED BIFACIAL SOLAR CELL WITH IMPROVED PERT STRUCTURE

Y. Eisenberg[1], J. Arumughan[2], L. Kreinin[1], N. Bordin[1], N. Eisenberg[1]
1-SolAround Ltd., Lev Academic Center, Havaad Haleumi Str. 21, Jerusalem 9611001, Israel; ygal@sol-around.com
2-International Solar Energy Research Center - ISC Konstanz, Rudolf-Diesel Str. 15, D-78467 Konstanz, Germany

ABSTRACT: Photovoltaic and recombination characteristics of new generation bifacial solar cells are demonstrated. The cells were fabricated using 6" pseudo square wafers of 5 to 6 Ω.cm single crystalline Cz p-Si. Their front efficiency is exceeding 20 % with back to front short circuit current ratio 89 - 92%. The design is characterized by high bulk minority carrier lifetime above 0.5 ms, not degraded during the fabrication process. Rear p^+ layer was prepared by controllable doping process using preliminary deposited thin B contained solid layer. Effective back surface recombination is below 10 cm/s. Measured implied V_{oc} values of the PERT structure exceeding 700 mV are evidencing the intrinsic potential of the structure to provide the cell efficiency far exceeding 22 %. Textured front surface in combination with smoothly etched back provides effective light trapping. The equivalent efficiency of a bifacial solar cell, which characterizes its energy generation capability, will achieve values in the range 23 – 27 % as a function of use conditions.
Keywords: Silicon Solar Cell, Bifacial Cell, PERT structure, p-type, Boron, High-Efficiency.

1. INTRODUCTION

Solar cell structure n^+-p-p^+ or p^+-n-n^+ with passivated emitter and rear totally diffused (PERT) is very promising for achieving high conversion efficiency and extraordinary bifaciality (ratio of short circuit currents at back and front illumination). High photovoltaic parameters were demonstrated for laboratory and industrially produced n-Si based cells [1], [2]. Some laboratory experiments validated as well chance to prepare p-Si based PERT cells [3]. Co-diffusion of boron and phosphorous with CVD deposited BSG and $POCl_3$ as respective doping sources was used. The parameters of I-V characteristics of the PERT cells were lower than PERC (local BSF cells), and the best efficiency was measured for structure with complete back Al contact. Authors of [4] published outstanding efficiency (24.5 %) of p-Si based PERT cell. However, not regular, but magnetically confined Czochralski grown silicon was used as starting material.

The main factors, which determine success or failure in achieving high cell efficiency and bifaciality, are low bulk and rear surface recombination losses. The importance of high lifetime for achieving front and back high efficiencies have been shown theoretically [5] and experimentally (for example, [6] and [7]).

The Cz p-Si based PERT cells dominating in PV industry are fabricated using aluminum alloying on the rear side, and therefore not optimal for achieving uppermost efficiency due to relatively high back surface recombination. Moreover, this kind of cell obviously does not have back light response and does not have bifacial properties.

Boron diffusion in the rear cell side should provide effective p-p^+ barrier needed for the best front efficiency and bifacial properties of the cell. The main problem of applying boron doping process to p-Si wafers is the degradation of bulk lifetime τ_b during the high temperature treatment (in contrary to n-Si, which is more resistant to thermally induced defects). Degradation of τ_b during the cell fabrication process or use of low τ_b starting Si leads to decreased solar cell efficiencies [8].

Authors of [9] proved the possibility of τ_b retaining during industrially fabrication of p-Si based bifacial PERT cells with B doped p^+ back layer. No degradation of high starting τ_b (in the 0.5 - 1 ms range) during the cell processing was found. Bifaciality of described cells was limited in the range 74 – 79 % mainly due to not controllable heavy boron doping of the rear p^+-layer. Effective back surface recombination, S_{eff}, was of one order of magnitude lower than in Al alloyed back cell structure, and was kept at values of 55 – 95 cm/s. The optical cell design, providing effective light trapping, was used in the mentioned research: front cell surface was texturized and rear one was chemically planar etched.

To avoid the limitations of over-doping at p^+ layer formation, as it happened in [9], a fine controllable B doping procedure was developed. Higher quality of p^+ layer prepared without negative effect on the bulk lifetime was achieved. Some results of such processing focused on improvement of recombination parameters of bifacial solar cell with p-PERT structure are presented in the paper.

2 EXPERIMENTAL

2.1 Solar cell structure and fabrication.

The bifacial n^+-p-p^+ solar cells under discussion were fabricated using 6" pseudo square wafers, with resistivity 3 to 6 Ω.cm, single crystalline Cz Si as starting material. The wafer front was textured, and the back was smoothly etched. $POCl_3$ gas phase diffusion process was used for uniform n^+ layer doping by P. The rear p^+ layers with different sheet resistance, R_{sh}, in the range 50 -150 Ω/□ were doped by high temperature diffusion using preliminary deposited B containing thin layer. Combined thermally grown or CVD deposited thin SiO_2 layers and CVD deposited SiN films were applied to both sides as passivating/antireflective coatings. Contact grids with floating four bus bars were screen printed on both sides.

2.2 Measurements.

I-V curves at one sun illumination (irradiance 1000 W/m^2, light spectrum equivalent to solar spectrum under air mass AM 1.5) and in the dark were measured using h.a.l.m. tracker PV-CTL1 of h.a.l.m. electronics. During the measurements under illumination, the receiving of scattered or reflected light by opposite cell side was prevented.

LOANA IQE-SCAN (pv-tools) was used for front and back spectral response determination. Influence of chuck reflectance on spectral response measurement was evaluated experimentally. Comparative measurements using reflective golden and black chucks demonstrated some difference in long wavelength spectral response (above ~980 nm).

The lifetime measurements by Quasi Steady State Photoconductivity Decay (QSSPCD) technique were undertaken for determination of light injected carriers in plain wafers before and after cell fabrication processing. The measurement tool was WCT-120 Silicon Wafer Lifetime Tester (Sinton Consulting Inc.). The same technique was used for evaluation of recombination in complete structure (before metallic contacts deposition). Implied open circuit voltage at one sun illumination, iV_{oc}, and reverse saturation current density, J_o, are distinguishing the recombination quality of the semiconductor structure of a cell.

Properties of doped layers were evaluated using sheet resistance, R_{sh}, determined by four-probe measurement tool. At least nine points per wafer were checked for characterizing the doping uniformity over the wafer area. Depth distribution of doping impurity was analyzed based on the data of Secondary Ion Mass Spectroscopy, SIMS, and the data of Electrochemical Capacitance-Voltage, ECV, characterization using wafer profiler CVP21 (WEP). Witness n-Si samples were used for more correct determination of B doped layer parameters.

3 MEASUREMENT RESULTS

3.1 Structural properties of bifacial cells.

Examples of the B doping profile are shown in Fig. 1. The profiles are characterizing the p^+ layers with sheet resistances 65 and 140 Ω/□ indicating practically whole doping range studied. Maximal B concentrations are relatively low, and are below B solubility limit in Si. The doping is properly deep, mainly when sheet resistance is going down.

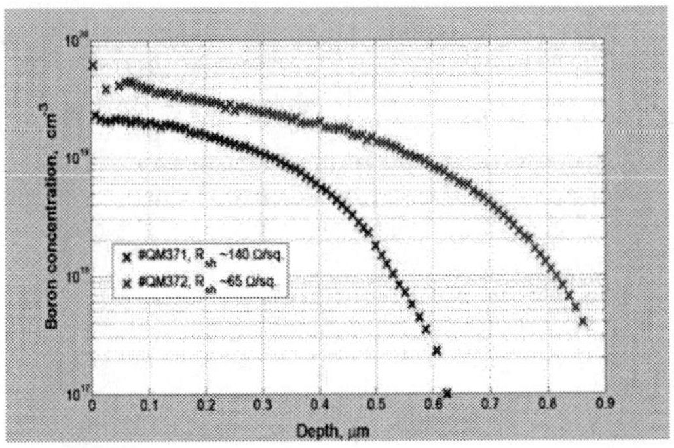

Figure 1: Boron distribution profiles showing the range of doping levels used in the experiments

Important data describing the recombination quality of the fabricated cell diode structures is given in Fig. 2. The implied V_{oc} as a function of the p^+ layer doping level (characterized by R_{sh}) was measured for the structures with R_{sh} of the front n^+ layer ~ 120 Ω/□. Over 10 structures of each shown nominal R_{sh} value was measured. Even with some decrease of iV_{oc} with higher doping values the measurements indicates high cell open circuit voltage expectation.

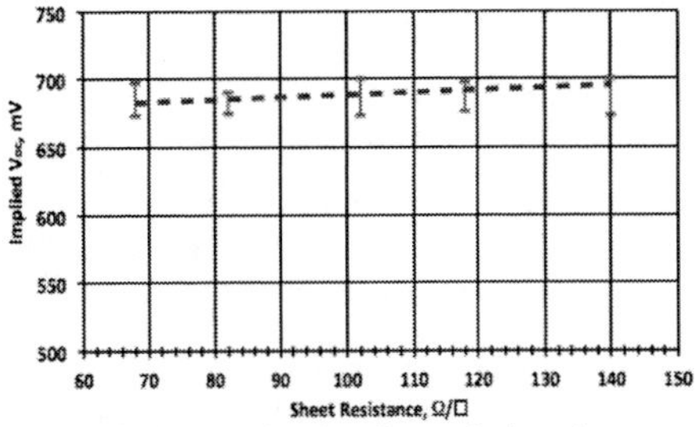

Figure 2: Implied open circuit voltage of n^+-p-p^+ structures vs. doping level of a p^+ layer.

The other illustration of recombination quality of the structure (excluding the edge losses) is the minimal reverse saturation current measured for the structures with each given back R_{sh}. It can be seen in the Table 1.

Table 1. Minimal reverse saturation current densities, J_o, corresponding to iV_{oc} of the n^+-p-p^+ cell structures with different R_{sh} of p^+ layer.

p^+ layer sheet resistance, R_{sh}, Ω/□	68	82	102	118	140
Reverse saturation current density, J_o, fA/cm^2	42	58	36	42	34

3.2 Solar cell internal quantum efficiences.

Spectral response and IQE data are effective means for illustration and analysis of the physical properties of solar cells. Fig. 3 represents the front IQE of a bifacial cell. The measurements were made using reflective golden chuck (with long wavelength reflectivity ~0.9) and "black" low reflecting chuck (with reflectivity ≤ 0.06). Front spectral reflectance of the cell installed on each chuck is shown in the same Figure 3. Some difference of the two IQE curves due to chuck reflection can be seen at the wavelength longer than ~1000 nm. The impact of the chuck reflectance on the correct determination of energetic cell parameters at solar illumination can be evaluated by integration of the cell spectral response over standard solar radiation spectrum and comparison of the results for the cases of using reflective and black chucks.

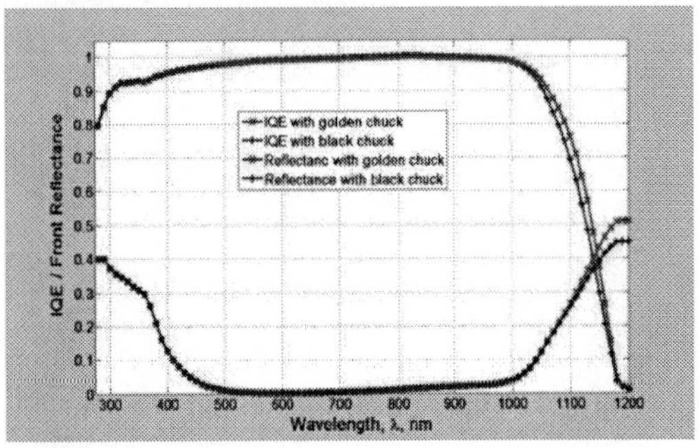

Figure 3: Front internal quantum efficiency, IQE, of the bifacial cell placed on reflective and black chucks.

Such kind of calculations based on the spectral response data for the cell, IQE-s of which are shown in Fig. 3, results in short circuit current densities, J_{sc}, of 39.47 and 39.28 mA/cm^2 for the cases of reflective and black chucks use. Therefore, the relative difference in determination of solar cell current when measured using two type of a chuck is ~ 0.5 %. The relatively small effect of the chuck reflection is the result of effective trapping of the light beams entering through the cell front surface [9]. The same small difference will be resulted in the determination of cell efficiency values.

Fairly high cell spectral response can be seen in the short wavelength range, proving the high n$^+$ emitter quality.

IQE data for the back illuminated solar cell are presented in Fig. 4. Short wavelength IQE at back illumination is only slightly lower than the front IQE. The depth and profile of doping are explaining this difference. Like in previous figure, spectral reflectance of the cell is shown. Since the rear illuminated surface is not textured the reflectance of the cell is remarkably higher. A little lower short wavelength IQE, higher reflectance and ~ 1 % greater contact shading result in lower back short circuit current, $J_{sc\ b}$. The $J_{sc\ b}$ values calculated as above by integration of spectral response over standard solar spectrum for reflective and black chuck data are: 36.60 and 36.41 mA/cm^2, respectively. These values are ~92.7 % of the respective front J_{sc}. It means, that bifacial factor for the given cell is above 92 %.

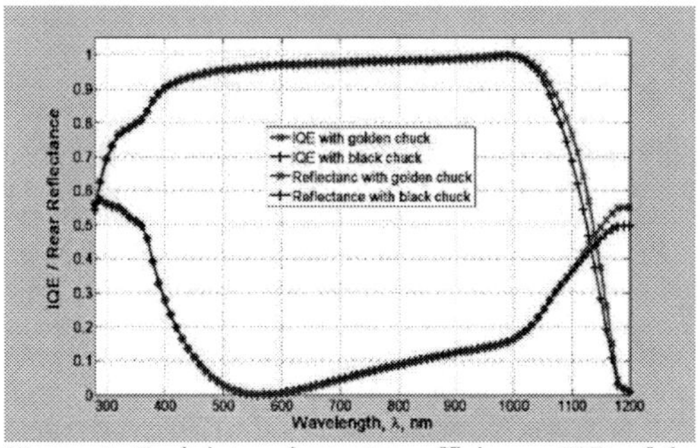

Figure 4: Back internal quantum efficiency, IQE, of the bifacial cell placed on reflective and black chucks.

3. 3 The I-V curves.

The front and back I-V curves of a bifacial cell of ~20.2 % front efficiency at simulated 1 sun illumination are shown in Fig 5. The V_{oc} value of the cell is significantly lower than iV_{oc} measured for the plain structures (see Fig. 1). The obvious explanation is the influence of the contacts.

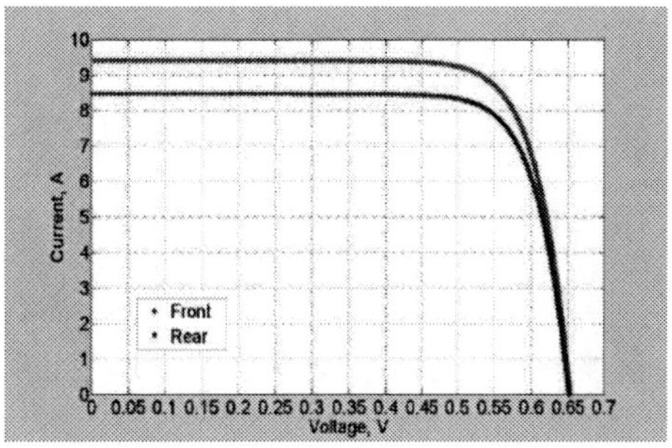

Figure 5: The I-V characteristics of a bifacial cell at front and back 1 sun illumination

A more comprehensive analysis of fabricated bifacial cells can be done based on Table 2. This Table contains the approximate statistical data for the batch (excluding breakage and samples made using mistake operations).

Table 2: Approximate statistical data for a ~ 40 bifacial cell lot

Parameter	Range	Expected
I_{sc}, mA/cm^2	39.2 – 39.6	>40
V_{oc}, V	650 - 658	>660
FF, %	78 - 79	>80
Eff. (front), %	19.9 - 20.3	>21
(I_{scb}/I_{scf})x100, %	89 - 92	>92

Despite a relatively wide range of cell parameters, due to some technological variations in the fabrication process, efficiencies of the cells are quite high and bifacial symmetry exceeding 89 % is good. The data demonstrate an intrinsic superiority of the bifacial n$^+$-p-p$^+$ design and fabrication technology used not only in comparison with regular p-PERT cells with Al alloyed back, but as well with p-PERT cells of the first generation [9]. The shown in Table 2 expected parameters of bifacial solar cells are evaluated using the values, which can be improved without any significant change in fabrication processing.

4 DISCUSSION

The described experimental fabrication, quite close to industrial one, demonstrates that the fabrication technology of bifacial n$^+$-p-p$^+$ Si solar cells provides the main requirements for high efficiency cells.

The most imperative factor for attaining the highest cell efficiency is a high τ_b. The fabrication technology retains very high τ_b in Cz wafers. The measured τ_b values are in the range 0.5 – 1.2 ms, i.e. in the range typical for the tested starting wafers before thermal processing.

Properties of high-low back barrier are of the first order of significance in influence on the front and back cell efficiencies. Controllable B doping of the p$^+$ layer forms a very effective BSF resulting after proper passivation in low S_{eff}. The S_{eff} values below 10 cm/s were measured on symmetrically B doped passivated n-Si wafers (as it was described in [10]). Both above factors in combination with high quality emitter with sheet resistance of ~ 120 Ω/□ are responsible for low recombination losses of solar cell structure and due to this for high implied voltages, as can be seen in Fig. 1. iV_{oc} is varying in the range 680 - 702 mV depending on B doping level. Large iV_{oc} values and low reverse saturation currents (Table 1) are proving the intrinsic potential of the structure to provide cell efficiency far exceeding 22 %.

Effective light trapping due to high back internal reflectance for the light rays refracted on the front textured surface ($R_{in\ b} \approx 0.76$) [9] leads to a significant contribution in I_{sc} improvement compared to a regular mono-facial cell with an Al alloyed back.

Low S_{eff} in combination with large τ_b are responsible for good back spectral response. Small losses in short wavelength response are explained by relatively deep B doping. High back spectral response results in a high bifaciality factor. The back IQE-s similar to curve shown in Fig. 3 in combination with optimal SiN antireflective layer result in $I_{sc\,b}/I_{sc\,f}$ exceeding 92 %.

The above data on high cell efficiency and on retaining during the fabrication procedure of high carrier lifetime leads to conclusion that Cz p-Si used for fabrication of n^+-p-p^+ bifacial cells can be competitive with n-Si as a material for high efficiency cell production.

The quality of bifacial cells cannot be evaluated only by front efficiency as far as regular mono facial cell. The adequate quality criteria may be given by evaluation of energy generated by a solar cell. This can be described by "equivalent efficiency". (See Table 3). Equivalent efficiency is determined as the efficiency of a mono-facial cell needed for generating the same energy as a bifacial cell under the given conditions. The data in the Table are based on the outdoor monitoring of bifacial modules under different working conditions [11], [12], [13]. The table demonstrates how high the values of equivalent efficiency are for a bifacial cell described here with front efficiency of 21 %. It can be seen that the equivalent efficiency of these bifacial cells is always higher than 23 % and for most typical applications exceeds 25 – 27 %.

Table 3: Equivalent efficiency as quality criteria of bifacial cell.

Front efficiency, %	Back contribution (energy gain), %	Equivalent efficiency, %
21	10 (field installation with low albedo)	23.1
21	20 (roof installation with intermediate albedo)	25.2
21	30 (roof installation with high albedo)	27.3

5 CONCLUSIONS

- n^+-p-p^+ with uniformly B doped p^+-layer is a promising structure for industrially produced terrestrial bifacial cells.
- B diffusion using preliminary deposited B containing solid layer allows controllable doping of p^+ layer.
- Retained high bulk lifetime, low back S_{eff} and high quality emitter provide implied V_{oc} of the PERT structure exceeding 700 mV.
- Back to front short circuit current ratio (bifaciality factor) is in the range **89 - 92 %.**
- Equivalent efficiency as quality criteria of PERT bifacial cell reaches **23 - 27 %** in the most typical applications.
- High quality starting Cz p-Si can compete with n-Si as a material for cell production with front efficiency above **22 %.**

Acknowledgments

We are grateful to our colleague Asher Karsenti for decisive help in the paper preparation.

References

[1] Valentin D.Mihailetchi, Haifeng Chu, Giuseppe Galbiati, Corrado Comparroto, Andreas Halm, Radovan Kopecek "A Comparison Study of *n*-type PERT and IBC Cell Concepts with Screen Printed Contacts", *Energy Procedia v. 77, August 2015, Pp. 534-539*

[2] Mark Osborne "Imec pushes commercially ready n-PERT solar cell to record 22% conversion efficiency", *PV-Tech, Jan 14, 2015*

[3] S. Meier, S. Wiesnet, S. Mack, S. Werner, S. Maier, S. Unmussig, C. Demberger, H. Knauss, D. Biro, A. Wolf *"Co-diffusion for p-Type PERT Solar Cells Using BSG Layers as Boron Doping Source", The 32nd Europeaan PV Solar Energy Conference, 20-24 June 2016, Munich, Germany, pp. 565 - 570*

[4] Jianhua Zhao, Aihua Wang, Martin A Green,"24.5% efficiency PERT silicon solar cells on SEH MCZ substrates and cell performance on other SEH CZ and FZ substrates", *Solar Energy Materials and Solar* ***Cells*** · *2001, February, v. 66, pp. 27 – 36*

[5] N. M. Bordina, T. M. Golovner, V. V. Zadde, A. K. Zaitseva, A. P. Landsman, V. I. Strel'tsova, "Operation of thin silicon photoconverter under Illumination on both sides", Applied *Solar Energy,* 6, 81-86, 1975.

[6] N. Bordin, L. Kreinin, J. Broder, N. Eisenberg, G. Grigorieva, K. Zvyagina, M. Kagan, "Recombination Parameters of Solar Cells with Back Surface Field Formed by Ion Implantation", *Proceedings 4th World Conference on Photovoltaic Solar Energy Conversion", 2006.*

[7] C. Duran, T. Buck, R. Kopecek, J. Libal, F. Traverso, "Bifacial Solar Cells with Boron Back Surface Field", *Proceedings 5th World Conference on Photovoltaic Energy Conversion*, 2010.

[8] A. C. Pan, C. del Canizo and A. Luque, " Effect of Thickness on Bifacial Silicon Solar Cells", *6th Spanish Conference on Electronic Devices,* 30 Jan – 2 Feb 2007.

[9] Lev Kreinin, Ninel Bordin, Naftali Eisenberg, Peter Grabitz and Gerhard Wahl, "Industrially Fabricated Bifacial Si Solar Cells with n^+-p-p^+ Structure", *The 28th European Photovoltaic Solar Energy Conference and Exhibition, Paris, France, September 30 - October 4, 2013*

[10] Y. Eisenberg, L. Kreinin, N. Bordin, N. Eisenberg, "Effective surface recombination of p^+ layers doped using ion implantation or surface deposited B sources", *6th International Conference on Silicon Photovoltaics, Silicon PV 2016,* Energy Procedia, Volume 92, August 2016, pp. 16-23.

[11] Lev Kreinin, Ninel Bordin, Asher Karsenty, Avishai Drori, Naftali Eisenberg, "Outdoor valuation of Power Output Improvement of the Bifacial Module" *37th IEEE PVSC,* 2011.

[12] Lev Kreinin, Ninel Bordin, Asher Karsenty, Avishai Drori, Naftali Eisenberg "Experimental Analysis of the Increases in Energy Generation of Bifacial over Monofacial PV Modules", *Proc. of the 26th European Photovoltaic Solar Energy Conf., 2011, 4CO.7.1*

[13] L. Kreinin, A. Karsenty, D. Grobgeld, N. Eisenberg, „PV Systems Based on Bifacial Modules: Performance Simulation vs. Design Factors", *43rd IEEE Photovoltaic Specialists Conference, Portland, Oregon, 5- 10 June 2016.*

AMPERE: A new Project for Innovative Heterojunction Manufacturing Solutions to Improve Competitiveness of the European PV Manufacturing Industry.

Battaglia A. (1); Gerardi C.(2); Scalari S.(2); Bizzarri F.(2); Strahm B.(3); Soederstroem T.(4); Munoz D.(5); Ribeyron P-J. (5); Izzi M.(6); Tucci M.(6); Delli Veneri P.(6); Despeisse M.(7); Perret-Aebi L.(7); Ballif C.(7); Nielsen O.(8); Hartlin B.(9); Aquino C (9); Zink O.(10); Melzer B.(10); Tallian M.(11); Lombardo S. (12); Balucani M. (13); Rentsch. J. (14)

1. *3SUN, Contrada Blocco Torrazze Zona Industriale 95121, Catania, Italy*
2. *EGP-Enel Green Power Viale Regina Margherita 125, 00198 Roma, Italy*
3. *MBR-Meyer Burger Research , Rouges Terres 61, 2068 Hauterive Switzerland*
4. *MBS-Meyer Burger AG,Schorenstrasse 39, CH-3645 Gwatt (Thun) Switzerland*
5. *CEA-LITEN, Institut National de l'Energie Solaire (INES), Savoie Technolac, F-73375 Le Bourget du Lac, France*
6. *ENEA Research Center, Via Anguillarese 301, 00123 Roma;*
7. *CSEM, PV-center, Jaquet Droz-1 , 2000 Neuchâtel, Switzerland*
8. *Norway NorSun AS,Karenslyst Allé 9C,NO-0278 OSLO, Norway*
9. *ERM , Saint Mary Axe 33, EC3A 8AA London, United Kingdom*
10. *J&R-Jonas & Redmann Automationstechnik GmbH, Kaiserin-Augusta-Allee 113, 10553 Berlin, Germany*
11. *Semilab Co. Ltd.,Prielle Kornélia u.2.,1117 Budapest, Hungary*
12. *CNR, Istituto per la Microelettronica e Microsistemi (IMM) Zona Industriale, Ottava Strada, 5, 95121, Catania, Italy*
13. *RISE Technology, 18 Via Monte Bianco, San Martino di Lupari – PD, 35018 Italy*
14. *F-ISE, Fraunhofer Institute for Solar Energy Systems (ISE), Heidenhofstrasse 2, 79110 Freiburg, Germany*

Corresponding Author: *Anna Battaglia,* Anna.Battaglia@3sun.com; *+390956364276*

Abstract

The European PV manufacturing industry has faced strong foreign competition in the last years, which has led to a dramatic reduction of its production capacity. It has completely shifted from a massive leadership at the beginning of the new millennium, with five EU companies in the top ten manufacturers, to a situation of survival where few of them are struggling in an actual extremely competitive market. Europe still has solid industrial assets consisting of silicon & material suppliers, and high quality equipment suppliers that constitute the core of a value chain to be rebuilt.

AMPERE, acronym for **A**utomated photovoltaic cell and **M**odule industrial **P**roduction to regain and secure **E**uropean **R**enewable **E**nergy market, was born for these purposes. AMPERE project aims to the setting-up of an innovative 100 MWp/y full-scale automated pilot line at the 3SUN module PV production line delivering a power module of 380 Wp (bifacial G/G 72-cells). Soon after the end of the project a rapid scale up to 240 MWp/y is planned and a route paving the way to the GWp/y scale will deliver the roadmap to reach production capacity stated. The project will operate with the support of the technological platforms for solar cells at CEA-INES and modules at MBR, the platform for advanced module technologies at MBS along with the support of high level research institutes (CSEM, EPFL, F-ISE, ENEA, CNR) as well as key EU companies covering the whole PV value chain (Norsun, Rise technology, Jonas & Redmann, ERM). Ampere has EGP as internal end-user and with the support of the project it will target an LCOE reduction of at least 15% compared to conventional PV mc-Si technology.

Keywords: production-line, heterojunction.

1 INTRODUCTION

A race towards lower costs and higher competitiveness in PV sector has led to a readily decrease of module prices over time as the historical PV Learning Curve teaches; Asian manufacturers seem to have regained margins and their overcapacities still represent a risk for price decrease and price stabilization cannot reasonably be expected in the coming years [1]. The EU value chain has not entirely disappeared yet, as European equipment suppliers have temporarily benefited from the strong Asian demand. But on long term basis, due to its large domination in PV manufacturing, Asia will attract and create a large part of the forthcoming PV material and equipment suppliers, thus reducing even more the PV European system. European research institutes and equipment manufacturers are still global technology leaders. But the loss of their domestic European customer base seriously jeopardizes this position and their financing. If Europe wants to play a role in the future of photovoltaics, one of the main pillars of future energy supply, it needs to maintain a vital ecosystem along the whole value chain including R&D labs, equipment manufacturers and manufacturing companies. The Ampere project aims to assure the full value chain sustainability in Europe. The sustainability assessment will consider how extended life and improved efficiency will improve known upstream and downstream issues and the results can inform future PV design aspects and criteria. Since the global market for very high efficiency cells is still small but rapidly expanding, upgrading and expanding capacities for high efficiency technology and building up new capacities in Europe with a technology based on high innovation such as Heterojunction, might be achieved rapidly without jeopardizing the efforts of existing European manufacturers to compete on global markets with incremental improvements of mature technologies.

AMPERE will try to follows this goal, aiming at establishing a next generation PV technology with cell efficiencies of 22.5% at production level in the 100MWp/y (and to improve efficiency on the subsequent development of 250 MWp/y lines after the project, to achieve an average efficiency of 23.5% and more), very high quality, high innovation rates, large scale manufacturing and a certain extent of vertical integration for reaching and maintaining a globally competitive position. Moreover, industrial deployment of PV manufacturing and market presence requires considerable size and capital to be competitive at a global level: Having large players in Europe mastering subsequent technology generations is a precondition for maintaining this ecosystem, investing in new technology generations, re-creating employment in the PV sector, re-conquering a strong position on global markets and ensuring a lasting European capacity in this key technology for the transition to sustainable energy supply systems. For the purpose it will focus on the industrialization and automation of innovative HJT technologies, based on a well-developed, assessed and demonstrated HJT owning to the main partners of the consortium [2], [3], [4]. The project involves a strong industrial and academic partners with a wide value chain represented by European leading industries, from technology providers/developers (NORSUN and MEYER BURGER) to manufacturer (3SUN) and finally to end-user (EGP). The main objective of AMPERE is to develop and secure the European supply chain of PV manufacturing by implementing a large scale cell and module industrial pilot lines. The implementation of innovative technologies into these industrial manufacturing lines will demonstrate that a larger production of PV products at a competitive COO and subsequent LCOE compared to non-European manufacturer is possible.

AMPERE main technology is focused on the demonstration of feasibility of bi-facial cells and modules based on heterojunction technology (HJT) at a 100 MWp/y industrial pilot line. Soon after the project it should lead to a short-scale upgrade at 250 MWp/y and then show the path to the GWp/y-scale; the project will deliver the roadmap to reach the production capacity stated. The project will be focused on the improved PV module power, reliability and durability to reduce the cost needed to fix and maintain PV plant and decrease its operational risks. The metric will be applied on the existing PV module industrial production site of 3SUN in Catania, Italy[4] (Fig.1).

Figure1: 3SUN PV module industrial production site, Catania, Italy.

2 APPROACH

During the recent past, it has been proven that silicon heterojunction solar cells can achieve efficiencies more than 26 % as Kaneka, with the support of IMEC, has showed recently [5]. However, most of these record cells have been achieved at the R&D scale and only few companies are producing Si-HJT cells and modules on a mass production level. Though SHJ technology has long shown great potential in achieving high cell & module efficiency, its production capacity has been low and further expansion is slow, mainly due to difficulties in developing high efficiency solar cell technology, as well as high capital cost to build SHJ production line [6].

Ampere basic approach starts through modification and upgrading of the pre-existing 3-SUN thin film silicon PV. So far, 3SUN has produced more than 6 million of thin film silicon PV modules, with annual capacity of 200MWp/y. The new investment plan has the scope of renewing the technology and reusing at least part of the existing expensive equipment (such as the PECVD Cluster tools for a-Si deposition) and facilities, which today are used for thin film PV manufacturing. HJT can leverage on higher annual energy generation owing to bifacial characteristics and to a low thermal coefficient.

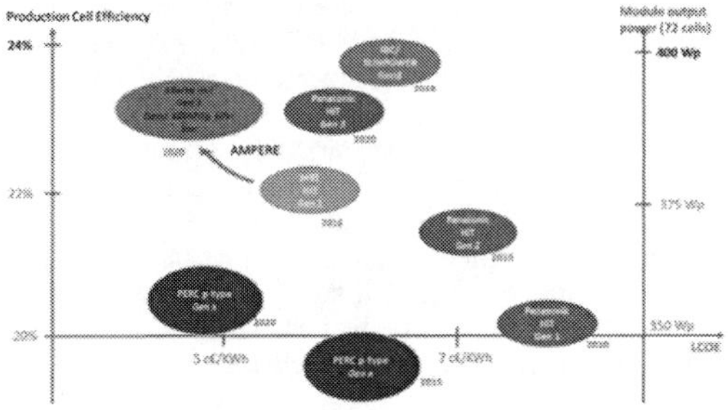

Figure 2: Ampere technology expected positioning in terms of LCOE compared with premium PV market.

Moreover, the glass-glass technology, which 3SUN has utilized for several years in its manufacturing process, will be fundamental for improving the reliability and hence the lifetime of the 3SUN modules installed in EGP power plants, allowing the energy company to realize plants with minimized energy generation costs (LCOE). Key advantages of this approach includes the cell efficiency improvement from 10% to >22% and lower capital cost. The major items inherent to the line conversion are: the modification of equipment PECVD, transfer and material handling systems, integration of thin-film and c-Si production equipment.

AMPERE will start from existing technologies that have been demonstrated their effectiveness during previous European successful projects. Cells Technological Platform of CEA-INES (TRL6) [2] and Module Technological Platform of MBR (TRL6)[3] will be the first two bricks of the project where a technology selection process will enable the implementation of the best materials and instruments into the existing industrial production line of 3SUN (TRL 7-8) [4], which will work as demonstration bed. By establishing a cell and module industrial production line of 100 MWp/y representative of the large scale production, AMPERE will trigger new activities in the European PV value chain to revitalize the sector. It is expected that with HJT bifacial modules (380 Wp bifacial G/G 72-cells) the LCOE will be reduced of at least 15% compared to the main technologies on the market in most of the geographical areas of interest [5]. The recent trend shows a continuous decrease of module prices which have reached <0,5 €/Wp for premium market modules and < 0.36 €/Wp for mainstream [1]. AMPERE project promise to be competitive with the Premium Market consisting of PERC, IBC and current HJT; AMPERE performance assessment in terms of LCOE is presented in figure 2.

Key aspects for European PV market, to compete with mainstream providers, pass through fundamental steps that embrace technological and marketing aspects that Ampere will purchase, such as:

- Improve performances without affecting the process complexity to maintain the CAPEX at a limited level
- Improve reliability towards 40 years lifetime for better LCOE
- Use of abundant, non-toxic materials
- Recyclable and low footprint technology
- Increase market need for the technology

Among the existing solar cells technologies, HJT has demonstrated a large potential to address these challenges. Furthermore, the HJT solar module can be fabricated in a bi-facial architecture, further enhancing the electricity production capacity by at least 20% and opening up many market segments[1]. A first breakdown of costs and a draft of industrial plan has been analyzed and settled. It comes especially from the high level of automation and process control inside the innovation transfer that allow a global cost reduction around 30%. Table 1 summarize cost reduction efforts within AMPERE in the entire PV value chain.

Table I: Ampere Cost Reduction Efforts

	Ambition within AMPERE	Cost reduction reached at the end of the project (2019)
At wafer level	Thinner wafers	~25%
At cell level	· Lower Silver content · Cheaper TCO · Use of Cu for metallization	~40%
At module level	· Replace Indium in wires · Develop 72 cells · Develop half cells	~30%
At process level	Lower CAPEX for production (PVD, PECVD) of cells and SWCT (module)	~20%
Global reduction	-	~30%

3 THE NEW AMPERE SHJ PRODUCTION LINE AT 3-SUN.

Ampere project route will start by modifying and upgrading the pre-existing 3-SUN Thin Film silicon PV production line in such a way to reduce CAPEX investment. The new investment plan has the scope of renewing the technology and reusing at least part of the existing expensive equipment (such as the PECVD Cluster tools for a-Si deposition) and facilities, which today are used for thin film PV manufacturing. HJT can leverage on higher annual energy generation owing to bifacial characteristics and to a low thermal coefficient. Moreover, the glass-glass technology, which 3SUN has utilized for several years in its manufacturing process, will be fundamental for improving the reliability and hence the lifetime of the 3SUN modules installed in EGP power plants, allowing the energy company to realize plants with minimized energy generation costs (LCOE). The project will follows two complementary routes during its three years of project. They includes the acquisition and installation of new equipment for the manufacturing of HJT bifacial modules and the reuse of part of actual Thin Film production line . The line will be set up in three steps in the next three years. A first step includes, by mid-2018, the realization of an assembly line with a capacity of 100MWp/y. The line will be capable to start a limited production of PV module, by assembling state-of-the art bifacial solar cells (either PERT or HJT) available on the market. Such an activity will enable EGP to start with "proof-of-concept" power plants based on bifacial technology. A 100MWp/y HJT cell manufacturing line will be then installed by mid-2019. Both module and cell lines will operate with state-of-the art equipment with limited automation. Part of the removed thin film equipment will be reused in the new HJT line being upgraded within the activities of AMPERE Project (Fig.3). By the end of 2021 the HJT line will be therefore extended to a maximum capacity of 250MWp/y with a full investment in terms of CAPEX in the range of 0,36€/Wp.

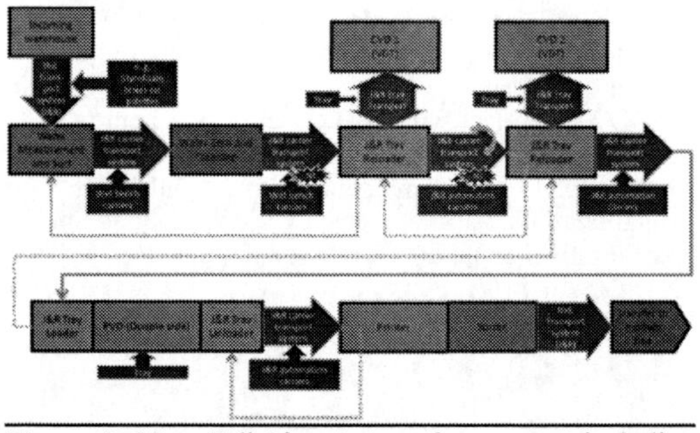

Figure 3: 3Sun Cell Line Automation concept including carrier return loops by using existing CVD tools.

4 AMPERE IMPLEMENTATION.

The outcomes of AMPERE Project will be used by 3SUN to implement extended automation for higher volume manufacturing, 250MWp/y, allowing to significantly reduce the final manufacturing costs, thus incrementing the factory competitiveness (Fig. 6). A fully automated 250MWp/y manufacturing factory will enable EGP to break into the utility scale market with an innovative and robust technology PV solution. Though the expected costs of HJT module will be higher than low cost mainstream technologies (mc-Si), their better performances (bifaciality, efficiency, higher durability), will lead to reduce LCOE of at least 10%.

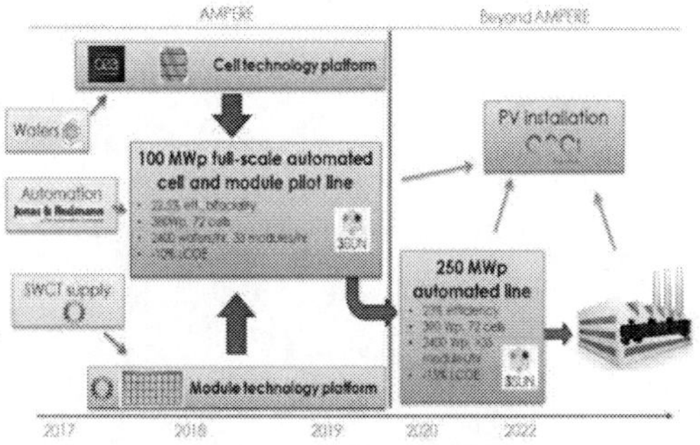

Figure6. Ampere yearly targets and its implementation.

For this reason, in the segment of the utility power plants market, a scale of 250MWp/y can be enough to allow EGP to compete with the conventional technologies (mostly mc-Si and CdTe) as reported in Table II.

Table II: Cost of goods (COGS) of 3SUN expected 250 MWp industrial line

250MWp Manufacturing Line	2020	2022	2024
HJT Costs (€/Wp)	0,42	0,38	0,34
Estimated reduction in LCOE with respect to mc-Si (in the same geographical area)	10%	12%	15%

From 3SUN viewpoint the factory will operate with a positive margin of about 9% in order to allow an adequate return on investment for 3SUN and to maximize the benefits in terms of LCOE for the mother company (EGP). The technology roadmap for EGP/3SUN with the main expected improvements related to new technology bricks implementation developed in the AMPERE projects is reported in Table 3 for the period 2020 – 2024.

Table III. Timeline of expected improvements for the implementation of new technology bricks

Year	2020	2022	2024
Cell Eff (%)	22,5	23,5	24
Module Power (Wp)	380	390	400

Ampere project will deliver the roadmap to reach 1 GW production capacity. Increasing to 1GWp scale will reduce the costs. 3SUN has estimated a budgetary cost under 0.3€/Wp for a gigawatt scale manufacturing in the 3SUN factory, in which they could partially leverage on the already existing facilities.

The benefits arising from the economy of scale are reported in Table IV, and they will position the HJT technology in a leading position not only for the utility scale market but also for the roof top sector where the technology will be surely considered as premium one and will compete with the high performance technologies allowing even larger margins due to the lower manufacturing costs.

Table IV. Cost of goods (COGS) of 3SUN expected 1 GWp industrial line

1GWp/y Manufacturing Line	2022	2024
HJT Costs (€/Wp)	0.28	0.25
Estimated reduction in LCOE with respect to mc-Si (in the same geographical area)	15%	20%

5. CONCLUSION.

A premium technology with high efficiency, high reliability, low temperature coefficient and consequently low LCOE will be industrialized in Europe, through the Ampere project. Premium segment is a very attractive option for utility scale PV market segments as it translates directly into very low LCOE. For such market segments, bifacial solar cells increase the annual generation of energy. The advantages in terms of energy gain can be even improved by using HJT technology, due to lower thermal budget and expected lower degradation rate[8]. The aim of the AMPERE project is to demonstrate EU innovative and sustainable manufacturing of bifacial modules in a pilot line capable of manufacturing up to 100MW/y of modules. These expected outcomes will be assessed and evaluated as a solid milestone for the viability of a 250 MWp/y production scale factory in 2020.

The manufactured products are based on silicon hetero-junction technology PV cells with mean efficiency of 22.5%. These will be industrialised applying innovative production processes that require fewer steps than conventional crystalline-silicon PV cell technologies. The route to implement to 250 MWp/y the production line has been settled together with the efficiency targeted at 23.5% after the end of project. The project will deliver the roadmap to reach the GW/y-scale production capacity with a corresponding LCOE reduction of 15% respect to the mainstream mc-Si technology and will provide investment pathways for the exploitation of the results to support the GW facilities of the next 5-10 years.

6. ACKNOWLEDGE.

This project has received funding from the European Union's Horizon2020 Programme for research, technological development and demonstration under Grant Agreement N° 745601.

7 References

[1] International Technology Roadmap for Photovoltaic (ITRPV).Seventh edition: march 2017, 54
[2] http://liten.cea.fr/cea-tech/ liten/en/Pages/ Liten%20text%20(general%20corporate%20text)/Platforms/PhotovoltaicSolarPlatform.aspx
[3] http://www.meyerburger.com/en/products-systems/ industries/photovoltaics/
[4] http://www.3sun.com/
[5] K. Yoshikawa, H. Kawasaki, W. Yoshida, T. Irie, et al., Nature Energy, 2017,Vol. 2, 17032.

and 2BP.1.1 this conference
[6] Global Market Outlook , Solar Power / 2015 – 2019
[7] A Louwen, W Van Sark, R Schropp, A Faaij - Solar Energy Materials and Solar Cells 147(2016)295.
[8] S. De Wolf , A. Descoeudres, Z. C. Holman, C. Ballif, High-efficiency Silicon Heterojunction Solar Cells: A Review, Green, Volume 2, Issue 1, 2012, 7,

A MICRO-DROPLET ETCHING APPROACH FOR TEXTURIZATION OF DIAMOND WIRE SAWN MULTI-CRYSTALLINE SILICON WAFERS

Lang Zhou, Zhigang Xiao, Zhihao Yue, Haibin Huang, Wenhao Chen, Xinfang Ye
Institute of Photovoltaics, Nanchang University, 999 Xuefu Ave., Nanchang, China

Shanming Jin, Longfei Gong
GCL Suzhou Photovoltaic Technology, Suzhou, China

Hong Chen, Zhongdong Xin
ZhangJiaGang Ultrasonic Electric Equipment Ltd, Wuxi, China

(Corresponding author: Lang Zhou, +86 791 83969552, lzhou@ncu.edu.cn)

ABSTRACT: Development of diamond wire sawing of multicrystalline silicon solar cell wafers has been inhibited by its incompetence with conventional acidic wet etching texturization process. A novel texturization method based on localized etching by micro-droplet of $HF-HNO_3-H_2O$ etchant on wafers has been developed. The micro-droplet is generated by controlled condensation of thermal vapor from a $HF-HNO_3-H_2O$ bath. Micron scale etch-pit type surface textures are made by applying the micro-droplet etching (MDE) process to diamond wire sawn mc-Si wafers. The light reflectivity is reduced to ~18%, and the sawmarks are removed. A few tens of full size DWS mc-Si wafer samples were manually MDE textured, and send to a solar cell fabrication line along with ordinary slurry sawn wafers, without any adjustment of processing parameters for the MDE textured samples. Encouraging I-V characteristics of the MDE textured samples have been obtained. Both J_{sc} and V_{oc} are found to increase as compared to the reference alkaline textured samples, and up to 0.56% absolute increase of average energy conversion efficiency is obtained.
Keywords: Multicrystalline silicon, Texturization, Solar cell, Diamond wire saw

1 INTRODUCTION

Diamond wire sawing (DWS) of silicon wafers for solar cells is well known to be superior to conventional slurry wire sawing in production rate, wafer quality, capability of thin wafering, cleanness, and most importantly now, in cost. Mono-crystalline silicon wafers are now mass produced with the DWS. However, production of multi-crystalline silicon wafers has not yet been benefited from the DWS, due to the problem that the DWS wafers are not compatible to the conventional low-cost wet acid texturization process. First, the anti-reflection effect is not sufficient, and secondly, the saw marks, an unacceptable appearance to the market, kept visible.

The origin of and the solutions to this problem have been intensively studied. However, none of the current approaches is satisfactory in cost, stability and environmental concern, as compared to the conventional wet texturization for the slurry sawn multi-crystalline silicon wafers. The present paper reports a novel approach we developed for a better solution to this problem. Some fundamental studies of the process has been reported previousely [1,2]. It still bases on $HF-HNO_3-H_2O$ bath. But instead of wet etching in the solution, localized etching by micro-droplet of the solution takes place. The micro-droplet is generated in-situ on the wafer surface by condense of thermal vapor out of a $HF-HNO_3-H_2O$ bath. The micro-droplet etching (MDE) texturization of full size DWS multi-crystalline silicon wafers for solar cells have been experimented. The textured wafers were then sent to a conventional production line of multi-crystalline silicon solar cells for performance tests, following their routine recipe but skipping off their texturization parts. The topography, light reflectivity of the MDE textured wafers, and the I-V characteristics of the solar cells fabricated with them were examined. The final results are encouraging. First, the saw marks are removed; secondly, the light reflectivity is lowered to 18~22%, a level better than conventional wet etched slurry sawn multi-crystalline wafers; thirdly, energy conversion efficiency of 18.65% has already been achieved in first trial without any adjustment to the existing cell process recipes, which were optimized for conventional wet etched slurry sawn multi-crystalline silicon wafers.

An additional example of the MDE's capability is also reported. The multi-crystalline ribbon silicon wafers, which are free-of any surficial mechanical damages, have been found to be even more difficult to texturize with the conventional wet acid etching. Their texturization with the present MDE approach was attempted. The result is promising. A textured surface is formed in less than one minute with the MDE, while for five minutes no surface rouhening can be observed with the wet acid etching.

Most of the thermal vapor from the HF-HNO3-H2O bath, which does not react with the silicon, may condense on the wall of the chamber and drop back to the bath. This recycling can make both the cost of and environmental load of the MDE texturization process significantly lower than the conventional wet etching texturization of multi-crystalline silicon wafers.

It should be noted that similar method has been attempted very earlier, well before the emergence of DWS technology, though their purpose and the outcome are totally different from the current work, neither the MDE concept was identified [3,4]. They were aiming for generating secondary nanometer scale texture on wet acid textured slurry sawn wafers. No later report of continued study or commercialization of this work was found.

2 EXPERIMENTALS

Diamond wire sawn (DWS) multicrystalline silicon wafer sample of 156 x 156 mm^2 size and 1-3 $\Omega\cdot cm$

resistance were provided by GCL. The samples were then etched with a 20% NaOH solution. Micro-droplet etching texturization was conducted on the alkaline etched wafers with a single wafer MDE device. Some wafers were left without MDE texturing as reference. An illustration of the MDE concept is given in Fig.1.

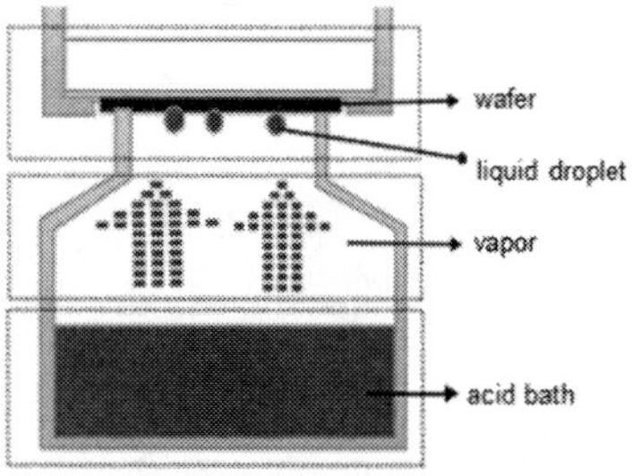

Figure 1: An illustration of the MDE method

A few tens of MDE textured DWS mc-Si wafers have been manually prepared. These wafers and the left reference wafers were sent to a conventional production line of multi-crystalline silicon solar cells in Jinko Ltd for performance tests, following their routine recipe but skipping off their texturization parts, without any adjustment of process parameters for the MDE textured wafers.

Some multicrystalline ribbon silicon wafer samples, provided by 1366 Ltd, namely Direct Wafer, were also experimented with MDE for texturization.

3 RESULTS AND DISCUSSION

3.1 Morphology of the textured wafers

Multiple types of texture can be obtained by the MDE method, depending on the processing parameters including bath composition, bath temperature, wafer-bath distance and wafer temperature. From "black silicon"-like nanometer-scale structured texture with light reflectivity down to 11% (averaged over the wave length range of 400~900 nm, here after the same for other light reflectivity), to micron-scale round etch-pit structure with 22% light reflectivity. To balance reduction of light reflection and compatibility with existing cell process, a texture type with relatively small and deep micron-scale etch-pit was adopted. Figure 2 shows a micrograph of such textured surface.

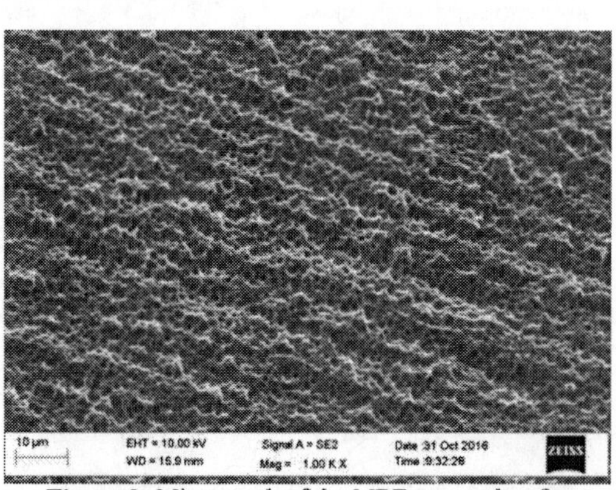

Figure 2: Micrograph of the MDE textured wafer

Figure 3 shows photographs of the full wafers, the MDE etched and the reference. No cutting marks are left visible with naked eyes, though in Fig.2 we can identify their traces in microsclae.

Figure 3: Photograph of the MDE textured wafer (left) and the reference wafer (right) in full size

3.2 Photovoltaic performance of the textured wafers

The solar cells made with the MDE textured DWS mc-Si wafers and the reference wafers were electrically characterized in line. The energy conversion efficiencies measured are plotted in Fig.4. Detailed average photovoltaic parameters are given in Table I. As can be seen, compared with the references, the MDE texturization raised both short circuit current and open circuit voltage of the cells, and the fill factors are increased, too, leading to a significant absolute increase of energy conversion efficiency, 0.8% maximum, and 0.39% in average. The most encouraging aspect of the results is that the open circuit voltages of the MDE textured cells can be higher than the alkaline etched reference cells. The latter are smooth on micro-scale, and are supposed to have relatively high open circuit voltage, while the MDE textured cells' open circuit voltage is even higher than it.

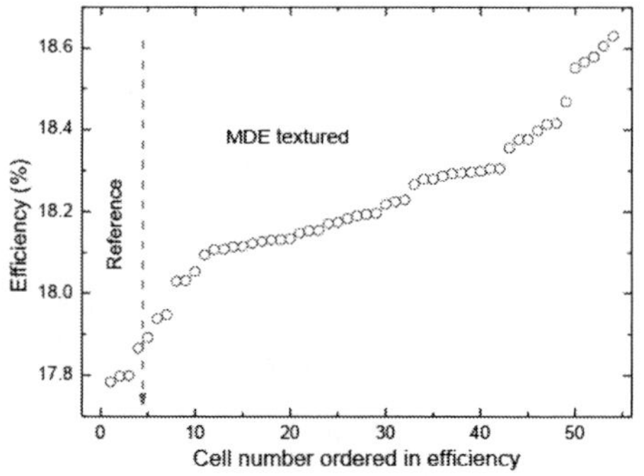

Figure 4: Energy conversion efficiency of the solar cells made with the MDE textured and the reference DWS mc-Si wafers

Table I: Average PV characteristics of the experimental solar cells

	Uoc	Isc	FF	Eta	count
MDE textured	0.6299	8.763	80.395	18.235	50
Reference	0.6267	8.697	79.702	17.849	4

3.3 MDE of multicrystalline ribbon silicon

Figure 5 presents the surface micrograph of the MDE textured ribbon silicon wafer sample. A round etch-pit structured texture is obtained. This is very encouraging considering the fact that the etching time here is only one minute, while in a conventional wet acid bath for texturing of mc-Si wafers, the sample can remain shiny after five minutes. Ribbon silicon is known to be free from any mechanical damages on its surface. The results confirm that first, conventional wet acid texturization strongly relies on presence of mechanical damages on the surfaces, and secondly, MDE texturization is free from such reliance.

Figure 5: Surface micrograph of the MDE textured ribbon silicon wafer

4 CONCLUDING REMARKS

Effectiveness of the micro-droplet etching (MDE) for texturization of diamond wire sawn multicrystalline silicon wafers has been demonstrated. Surfaces with micron-scale etch-pit, with light reflectivity of ~18%, and without visible saw marks, have been made on the MDE textured DWS mc-Si wafers. Increase in both short circuit current and open circuit voltage of the solar cells has been achieved, leading to a 0.39% absolute average increase of cell efficiency, over the alkaline etching textured reference cells. Extended experiment on a ribbon silicon (Direct Wafer) shows that MDE is a promising method for texturization of this type of kerf-free silicon wafers. The fact that MDE texturization of silicon does not rely on presence of surface mechanical damages on silicon is further confirmed.

REFERENCES

[1] W.Chen, X. Liu, M.Li, C.Yin, L.Zhou, Materials Science in Semiconductor Processing, 27(2014) 220.

[2] Z.Xiao, G.Geng, X.Wei, Z.Yue, L.Zhou, Materials Science for Semiconductor Processing 53(2016)8

[3] Z.Xiao, G.Geng, X.Wei, Z.Yue, L.Zhou, Semicond. Sci. Technol., 31(2016)115018.

[4] M. Ju, M. Gunasekaran, K. Kim, K. Han, I. Moon, K. Lee, S. Han, T. Kwon, D. Kyung, J. Yi, Mater. Sci. Eng. B, 153 (2008) 66.

[5] M. Ju, N. Balaji, Y. Lee, C. Park, K. Song, J. Choi, J. Yi, Solar Energy Materials and Solar Cells 107 (2012) 366.

ACKOWLEDGEMENT

Support of the present work by the National Natural Science Foundation of China (Grant 61574072) is gratefully acknowledged.

FAST ALKALINE TEXTURING PROCESS FOR HIGH THROUGHPUT PERC SOLAR CELL MANUFACTURING

Frank Strinitz[a], Florian Schoerg[a], Martina Fuerst[a], Ahmed El Jaouhari[a,*], Holger Kuehnlein[b]
[a]RENA Technologies GmbH, Am Röthenbühl 7, 92348 Berg, Germany
[b]RENA Technologies GmbH, Hans-Bunte-Strasse 19, 79108 Freiburg im Breisgau, Germany
*Author for correspondence: ahmed.eljaouhari@rena.com

ABSTRACT: In the last decade, a lot of effort was dedicated to decrease the texturing time of monocrystalline silicon wafers in order to increase the throughput of the manufacturing tools and competitiveness with multicrystalline inline processes. At the beginning of additive-based texturing, replacement of the volatile isopropyl alcohol (IPA) was a sufficient achievement. Nowadays, surface quality and reduced process times are in the focus of industrial manufacturing. In this work, a fast alkaline texturing process, using a new generation of additive that allows the manufacturing of tools with minimized footprint, is reported. With the additive monoTEX-F®, wafers from all major suppliers can be textured within a timeframe of 5 to 7 minutes, while easily achieving triple digit bath lifetimes and keeping the concentration of organics below 0.02 %. Along with the decrease in etching time comes a higher process stability, a decrease in KOH consumption and a decrease in average reflectivity (400 – 1100 nm) to 10.7 – 11.3 % under continuous production. Furthermore, changes in pre-treatment allow influencing the pyramid size distribution to optimize the surface for subsequent deposition steps.

Keywords: Etching, Silicon, Texturisation

1 INTRODUCTION

Wet chemical etching processes are a reliable starting point for crystalline silicon solar cell production. The main tasks are the removal of parasitic saw damage, reduction of the reflectivity, removal of undesired organic and inorganic impurities and conditioning of the final interface.[1] Lately, the switch from Al-BSF cells to PERC or related cells is underway at most large-scale manufacturers and the wet-chemical treatments have been adapted accordingly. A market share for PERC-/PERL-/PERT-based cells of approximately 20 % has been predicted for 2018 with a steady increase thereafter.[2] Back contact and silicon heterojunction cells are regarded as further upcoming key technologies. The solution to the different demands in surface conditioning is well-established on a laboratory scale and will be briefly reviewed. The adaption for high throughput tools, evolution of additive-based texturing and results at major manufacturers will be described in this study.

2 RESULTS AND DISCUSSION

2.1 Evolution and state-of-the-art of alkaline texturing

Recently, M. F. Abdullah *et al.* have given a review on the importance of texturisation processes for silicon solar cells with a focus on different physical and chemical methods as well as post-treatments and alternative processes.[3] Their focus for the chemical processes were the etchants KOH, NaOH, TMAH and Na_2CO_3, without further investigating the role of IPA and additives in the generation of random pyramidal structures.

Within this chapter, a short look into the evolution of modern alkaline texturing for monocrystalline silicon wafers will be presented, focusing on the impact on current PERC cells (Figure 1). Due to the limited access to exact dates all reported timeframes are rough estimates, especially regarding the introduction in mass production. The use of IPA-based texturing dates back to before 2007 and was discarded due to a series of disadvantages. These were ranging from unstable concentrations due to evaporation of IPA, high costs combined with high IPA consumption and health hazards including toxicity and explosiveness.

The trend for industrial production were mixtures based on IPA and polyethylene glycol and appeared first around the year 2002 with the endeavor to reduce the expensive IPA consumption, shorten texturing times and increase process stability.[4] Parallel developments were based on the use of alcohols with a higher boiling point like butanol isomers.[5]

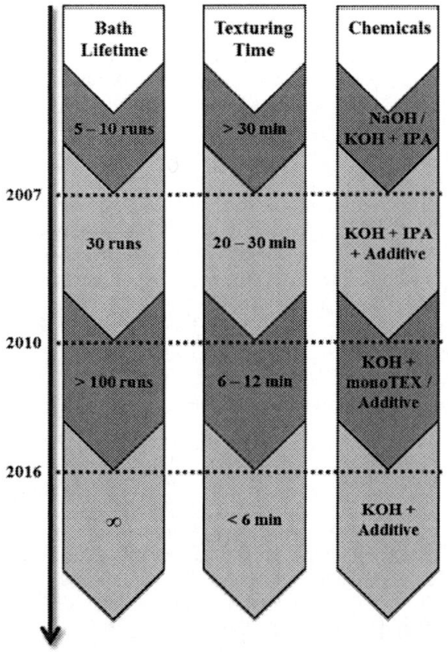

Figure 1: Basic timeline of alkaline texturing processes used on industrial scale (batch size of 200 or 400 wafers per run).

While these systems were introduced in industrial production, the first research groups focused on surfactant-based processes eliminating all volatile compounds like Di Franca *et al.* reported in 2007. They used a dish detergent and managed to obtain reflectivities as low as 12 % with the texturing time reduced to 15 minutes.[6] Two years later, E. Wefringhaus *et al.* presented studies leading to a 10 minute process achieving mean reflectivities (400 – 1100 nm) around 11.1 %.[7]

In the same timeframe, industrial solutions for combinations of low amounts of IPA with added surfactants were presented in 2009 and introduced into mass production the year after.[8] The year thereafter, solutions without IPA were published and established in industrial production.[9] Since then, the transition to next generation additive-based texturing processes has been completed for almost every manufacturer and the necessity of IPA has been overcome. The current benchmark for industrial production is set by the throughput values reported in the ITRPV and commonly achieved with texturing times below 12 minutes down to 6 minutes.

2.2 Parameters influencing texturing quality

After coping with the fundamental problems of IPA-based alkaline texturing, the focus was placed on a more detailed approach towards surface geometry. In the initial IPA-based processes, the removal of the saw damage and the following formation of random pyramidal structures was performed in two independent process tanks with etch depths well above 10 µm/side. After the introduction of additive-based texturing processes, several manufacturers switched to a combined SDR and texturing approach and performed the entire alkaline etching in the texturing solution reducing the etch depth below 10 µm/side. Currently, both approaches towards alkaline texturing are in industrial use. The first one kept the approach with a combined SDR and texturisation step and is mainly employed for texturing additives with slower kinetics and texturing times above 12 minutes. The second one is the conventional use of an independent saw damage removal and texturisation step. This allows a huge increase in throughput for fast texturing additives, without effects on the tool footprint. For the current generation of monoTEX-F® a texturing duration below 7 minutes is established in mass production and in the following section research results on A) the general process window, B) the effects of the additive concentration on surface geometry, C) the influence of temperature changes and finally D) the influence of the surface after the saw damage removal on the texturisation results.

For the initial assessment of the available process window of monoTEX-F®, several assumptions were made. The texturing time should on a lab scale be limited to 6 minutes, process temperatures above 80 °C are disfavored in an industrial scale and as etchant, KOH at a concentration of 1.5 % was selected as baseline for all reported experiments (For further information see chapter "EXPERIMENTAL").

In a first step, the texturing process was evaluated for monocrystalline wafers at the aforementioned KOH concentration at three different temperatures (72 °C, 76 °C and 80 °C). While 80 °C are a common standard temperature in industrial production, lower temperatures, without extended process times, could potentially decrease tool downtimes and power consumption. The additive concentration was varied in the range of 1 mL/L and 12 mL/L, i.e. from 1 mL additive per liter of etchant solution up to the twelvefold amount. The achieved reflectivity allows a quick assessment of the quality of the generated texture and is depicted in Figure 2. Plotted is the weighted reflectivity between 400 and 1100 nm, with our empirical value for a complete homogeneous texture being ≤11.5 % for this measurement range.

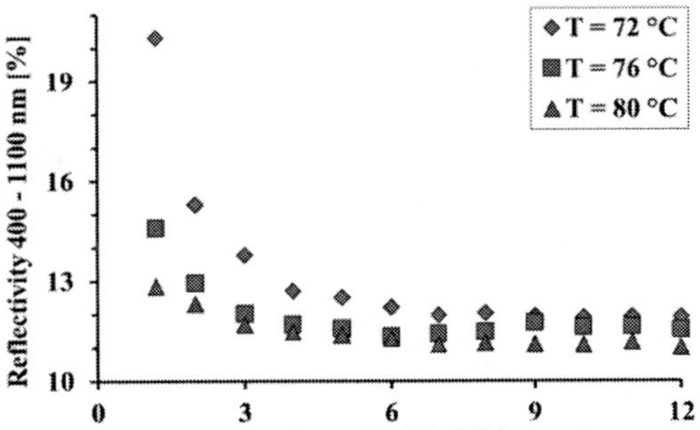

Figure 2: Reflectivities after texturing for 6 minutes with different additive concentrations at 72, 76 and 80 °C.

It is obvious that at high additive concentrations above 6 mL/L the effect of the temperature differences diminishes, however, at lower concentrations the texturisation results at 72 °C are poor and would need a longer process time to compensate the lower etch rate. The comparison between 76 and 80 °C reveals that at both temperatures, a good process window for a 6 minute process can be found, however, at 80 °C this ranges from 3 – 12 mL/L, while at 76 °C the window is smaller and ranges approximately from 4 – 8 mL/L. The lower additive concentration level is mainly determined by a too low selectivity of the etchant on the (111) surface, the higher limit is coming from a decrease in etch rate due to the deceleration of the additive (Table I).

Table I: Influence of temperature and additive concentration on etch depth (total etch depth including SDR step depicted).

	2 mL/L	7 mL/L	12 mL/L
Etch depth (72 °C) [µm/side]	9.11	8.04	7.55
Etch depth (76 °C) [µm/side]	9.40	8.79	7.95
Etch depth (80 °C) [µm/side]	11.44	9.48	8.90

After this initial assessment of the process window of monoTEX-F®, the focus was on a more in-depth analysis of the process parameters on the final surface geometry. While it is quite common for texturing additives to form smaller pyramids compared to IPA-based texturing solutions, the exact size varies significantly. To quantify the pyramid size and the corresponding size distribution, confocal laser scanning microscope (LSM) images were recorded and analyzed by the ISC Konstanz.[10] As comparative figure, the pyramid height Z and its distribution will be used in this work. P. K. Basu *et al.* reported a positive correlation for smaller pyramid heights on the corresponding fill factors for inline-diffused screen-printed cells, with no significant variation in all other cell parameters, except efficiency.[11] For better clarity, the pyramid height at the distribution curve maximum ("maximum of pyramid height distribution", MPHD) is reported. Commonly, a value around ± 2 μm can be found as the distribution width around the maximum. The first investigation was the dependency of the pyramid height distribution on the additive concentration. The correlation between etch depth and additive concentration was already mentioned in the previous section. This decrease in etch rate at higher additive concentration favors the formation of smaller pyramids as can be observed from Table II. As can be taken from the table, the increase of additive concentration from 3 mL/L to 7 mL/L decreases the etch rate approximately 0.7 μm/side and leads to a decrease in pyramid height of 0.9 μm. (The resolution of the pyramid height distribution is 0.3 μm.) In this case, the behavior at 85 °C and a texturing time of 6 minutes is depicted, because the differences are more obvious compared to the same results at 80 °C.

Table II: Influence of additive concentration on geometry of wafer surfaces at 6 minutes process time (80 °C).

	3 mL/L	5 mL/L	7 mL/L
Etch depth [μm/side]	3.98	3.72	3.29
Reflectivity [%]	11.03	11.01	10.77
MPHD [μm]	4.6	4.3	3.7

The next investigation dealt with the correlation between process temperature and additive concentration. In this case, the texturing process in a 5 minute timeframe was tested at three different temperatures i.e. 80, 85 and 90 °C (Table III). The trend that an increase in additive concentration leads to smaller pyramids, can again be observed for the shorter process times and at different process temperatures. In addition, an increase in temperature leads to an increase in pyramid height as expected. However, at higher additive concentrations this correlation is not as pronounced as for the low additive concentration.

Table III: Influence of process temperature on the maximum of the pyramid height distribution at 5 minutes process time and varying additive concentrations.

	3 mL/L	5 mL/L	7 mL/L
MPHD (80 °C) [μm]	3.7	4.0	*
MPHD (85 °C) [μm]	5.2	4.9	4.0
MPHD (90 °C) [μm]	4.9	4.6	4.6

*In this case, the distribution could not be resolved.

The final comparison dealt with different pretreatments of the silicon wafers. As previously mentioned, an SDR step is common in industrial production to decrease the texturing process time and increase bath stability and uniformity. Currently, two different approaches are viable in industrial production. The first is the well-established use of highly concentrated KOH solutions with concentrations between 10 and 45 % and an etch depth up to 10 μm/side can be observed. The advantage for this approach is the high uniformity of the obtained surface and the guaranteed removal of the entire saw damage. The second approach is the use of highly diluted KOH ("dKOH") solutions in the range of 1 to 2 %. This low concentrated etchant allows a much lower etch rate and etch depths in the range of 2 to 3 μm/side are common. This reduces the silicon loss and the consumption of KOH, however, the surface after the dKOH step shows more inhomogeneity and the texturing process has to overcome this disadvantage. For the polished wafers ("SDR wafers") it can be observed that at a temperature of 80 °C and a process time of 6 minutes, no distinctive variation in pyramid height can be observed with additive concentration (Table IV, Figure 3a). This can be contributed to the already small pyramid size at lower additive concentrations and shows that the homogeneously polished surface allows the fast generation of well-formed pyramids.

Table IV: Influence of the SDR pretreatment on the geometry of wafer surfaces at 6 minutes process time (80 °C).

	3 mL/L	5 mL/L	7 mL/L
Etch depth [μm/side]	3.30	2.92	2.66
Reflectivity [%]	11.63	11.35	11.02
MPHD [μm]	4.0	3.7	4.0

A different behavior can be observed for the wafers that have seen a pretreatment with dilute KOH ("dKOH wafers"). The inhomogeneity after the dKOH step leads to an unfinished texture at a concentration of 3 mL/L monoTEX-F® (Table V, Figure 3b). Even at higher additive concentrations, the pyramid size remains decisively larger with 2.4 μm more at 5 mL/L and 0.9 μm more at 7 mL/L.

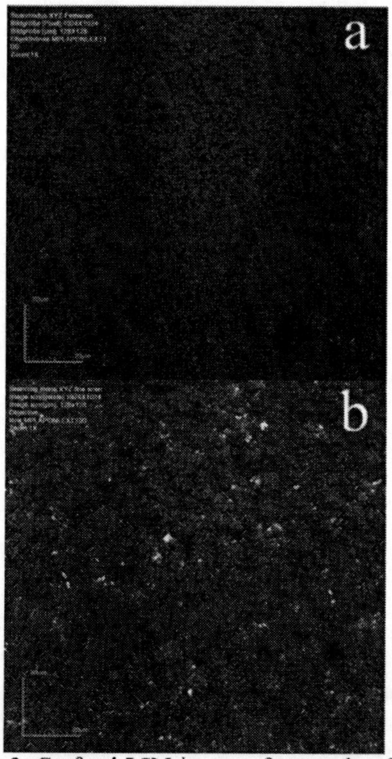

Figure 3: Confocal LSM images of textured wafers, a) SDR pretreatment, 6 minute texture at 80 °C and 3 mL/L additive, b) dKOH pretreatment, 6 minute texture at 80 °C and 3 mL/L additive.

Table V: Influence of the dKOH pretreatment on the geometry of wafer surfaces at 6 minutes process time (80 °C).

	3 mL/L	5 mL/L	7 mL/L
Etch depth [µm/side]	3.58	3.15	2.86
Reflectivity [%]	11.93	11.17	10.93
MPHD [µm]	*	6.1	4.9

*In this case the distribution could not be resolved.

2.3 Application in solar cell manufacturing

In this final part, the influence of these state-of-the-art texturing recipes on the throughput and process stability for PERC cell lines is investigated. The topic that was neglected in the previous parts is the bath lifetime which becomes more important when considering industrial production lines. Currently downtimes of one to three hours are, depending on the facility and the tool, necessary for a full bath change routine. For a common wet-chemical tool (with a capacity of 400 wafers per batch) with one SDR and two texturing tanks and adequate cleaning and rinsing baths, the downtime percentage has been calculated. Especially, limiting the amount of texturing tanks to two per tool decreases the footprint a lot. Further assumptions were that the 6 minute texturing process is the throughput limiting step, the necessary handling times have been included and a 1 hour bath change is achievable regarding the tool setup. Figure 4 shows the influence of the bath lifetime of the texturing bath on the uptime and downtime neglecting scheduled maintenance times and unscheduled failures. It is clearly visible that double digit run numbers force high downtimes and starting at 100 runs per texturing bath lifetime the downtime percentage stabilizes on acceptable level of 8.7 %. Further increases to 250 runs per texturing bath decreases the downtime further to 3.5 %.

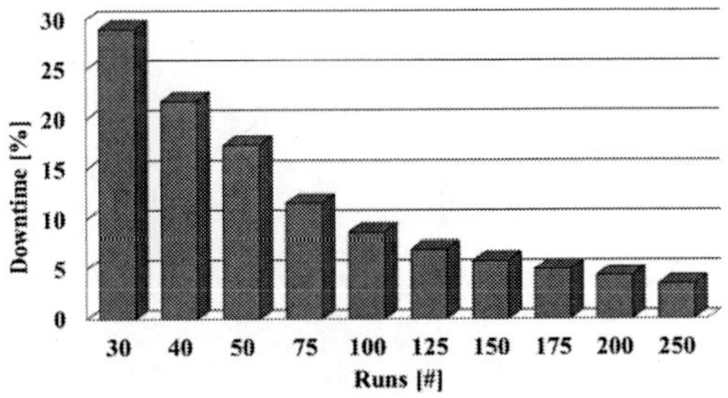

Figure 4: Correlation between downtime and bath lifetime of the texturing bath for a 6 minute texturing process and 1 hour bath change.

To show these numbers descriptively, 250 runs per texturing bath under full utilization means a production of 200 000 wafers in a timeframe of 29.7 hours, including the bath change. We previously reported the high bath lifetime exceeding 150 runs with monoTEX-F® in mass production and depicted a stable reflectivity below 11.3 % as well as an etch amount between 0.65 and 0.75 grams.[12]

To further confirm the stability of this industrial process, for several samples the pyramid height distribution was determined during continuous production. The process used in this production situation resembles best the sequence described in Table V with a dKOH pretreatment step and a texturing process of 6 minutes at a temperature of 80 °C, using an additive concentration around 6 mL/L. The height distribution maxima are in the range of 3.4 – 4.0 µm and are considerably lower than our experimental values that have been reported in the previous section. This is in agreement with our previous experiences that the ageing of the additive under process conditions, the introduction of higher silicate concentrations and the upscaling of the process bath volume leads to smaller pyramid heights. With the current trend to smaller pyramid sizes this simplifies an adaption of laboratory procedures to production tools.

3 SUMMARY

The process time for alkaline texturing as well as the quality of the obtained pyramidal surface has seen major improvements in the last decade. For monoTEX-F®-based texturisation, process times between 5 and 7 minutes have been extensively studied and validated in a laboratory setting as well as in mass production. This new wetting and moderating agent allows the increase of tool throughput while maintaining control over pyramid size distribution via pretreatment and texturing bath composition. Further investigations to explore sub-5 minutes process windows are underway.

4 EXPERIMENTAL

All experiments were conducted with monocrystalline DWS wafers from representative suppliers. For the experiments on the parameters influencing the texturing quality, beaker glasses with a volume of 4.5 L equipped with a stirring bar and a closed lid were used. The typical process sequence was SDR/dKOH, followed by a pSC1 cleaning step, texturing with monoTEX-F®, a second pSC1 cleaning step and after a HF/HCl metal clean, the wafers were dried in a warm air dryer. In between all process steps, the wafers were rinsed in DI water.

All wafers (except the ones where "dKOH" was mentioned) were treated for 3 minutes at 80 °C in a SDR bath at a concentration of 20 % KOH leading to approximately 8 µm/side etch removal and a polished surface.

For the dKOH process, the wafers were treated for 3 minutes at 80 °C in a dKOH bath at a concentration of 1.5 % KOH leading to approximately 4 µm/side etch removal and an uneven surface with the appearance of smaller pyramidal structures.

The KOH concentration for all texturing baths was set to 1.5 %. The monoTEX-F® concentration as well as the process time and temperature are given in the respective experiments. All texturing baths were conditioned by etching one wafer in front of the relevant wafer.

5 REFERENCES

[1] D.-H. Neuhaus, A. Münzer, Advances in OptoElectronics (2007) 1 – 15.
[2] International Technology Roadmap for Photovoltaics (ITRPV) Results 2016 (2017) by SEMI PV Group.
[3] M. F. Abdullah, M. A. Alghoul, H. Naser, N. Asim, S. Ahmadi, B. Yatim, K. Sopian, Renewable and Sustainable Energy Reviews 66 (2016) 380.

[4] K. Holdermann, U.S. Patent, 6,451,218, 2002.
[5] H. Park, J. S. Lee, H. J. Lim, D. Kim, S. Kwon, S. Yoon, Journal of the Korean Physical Society, 55 (2009) 1767.
[6] M. Di Franca, N. Gawehns, R. Auer, Proceedings 22nd European Photovoltaic Solar Energy Conference, Vol. I (2007), 1683.
[7] E. Wefringhaus, A. Helfricht, 24th European Photovoltaic Solar Energy Conference, Vol. I (2009) 1860.
[8] J. Krümberg, I. Melnyk, M. Schmidt, M. Michel, T. Fidler, M. Kagerer, P. Fath, L. Heiliger, H. Nussbaumer, Proceedings 24th European Photovoltaic Solar Energy Conference, Vol. I (2009) 1748.
[9] M. Moynihan, C. O'Connor, B. Barr, S. Tiffany, W. Braun, G. Allardyce, J. Rentsch, K. Birmann, Proceedings 25th European Photovoltaic Solar Energy Conference, Vol. I (2010) 1332.
[10] E. Wefringhaus, C. Kesnar, M. Löhmann, Energy Procedia 8 (2011) 135.
[11] P. K. Basu, A. Khanna, Z. Hameiri, Renewable Energy 78 (2015) 590.
[12] F. Strinitz, A. El Jaouhari, F. Schoerg, M. Fuerst, M. Plettig, H. Kuehnlein, Energy Procedia (2017), accepted, in press.

HIGH QUALITY INDUSTRIAL PHOSPHORUS EMITTER DOPING OBTAINED WITH INNOVATIVE PLASMA IMMERSION ION IMPLANTATION (PIII) PROCESSES

T. Desrues*[1], T. Michel[3], J.-F. Lerat[1], A. Veau[1], A. Lanterne[1], M. Coig[2], F. Milesi[2], F. Mazen[2], F. Torregrosa[3], L. Roux[3], S. Dubois[1], R. Monna[1]

[1] Univ. Grenoble Alpes, CEA, LITEN, DTS, LHMJ, F-38000 Grenoble, France
[2] Univ. Grenoble Alpes, CEA, LETI, MINATEC Campus, F-38000 Grenoble, France
[3] IBS, Rue Gaston Imbert Prolongée, Z.I. Rousset-Peynier, 13790 Peynier, France
*Corresponding author: Email: thibaut.desrues@cea.fr, Phone: +33 479 792 877

ABSTRACT:

We present an innovative approach for reducing the cost of industrial solar cells using Plasma Immersion Ion Implantation (PIII). High quality PIII Phosphorus emitters were obtained on the PULSION® Solar equipment from IBS (Ion Beam Services), capable of simultaneous and homogeneous implantation of 21 wafers (156*156mm^2) in a very short time (≈30 s). Using Phosphine (PH_3) as gas precursor, various doping profiles were obtained, reaching J_{0e} values of 40 fA/cm^2 for R_{Sheet} at 110 Ω/sq. The throughput and reproducibility of the equipment have been tested with the fabrication of ≈500 Al-BSF cells showing a very low dispersion and maximum performances above 19.3%. A 60-cells module with an output power of 268 W has been fabricated with these solar cells, indicating the high quality of the devices. Using PULSION® Solar, we show that PIII could be one promising solution to provide an efficient doping technology compared to conventional approaches.
Keywords: c-Si, doping, solar cells

1 INTRODUCTION

Industrial crystalline silicon (c-Si) solar cells technologies need doping processes to form n^+ and/or p^+ type regions at the surfaces of c-Si wafers. Conventional techniques involve the use of diffusion furnaces where the dopant concentration may be difficult to tune precisely. This could be a limitation especially for high efficiency devices where thin and homogeneous doped layers are needed. We investigate in this work a new approach of doping by using the plasma-immersion ion implantation (PIII) technique. With PIII, c-Si wafers are immersed into a plasma of ion species with low energy while a bias voltage is applied. Relatively simple and easy to operate and maintain, the PIII system requires both low capital investment and running costs. Previous results have shown the capability to obtain very good doping processes for industrial solar cells, either for Phosphorus [1,2] or Boron [3] atoms, in a standard PIII equipment designed for microelectronics purposes (single-wafer chambers). This technique can even be applied on thin Si layers to form passivated contacts [4].

We present in this study experimental results obtained using a PULSION® Solar equipment from the IBS (Ion Beam Services) company. This innovative tool is optimized to fulfill the requirements of the PV industry in terms of cost and productivity. It allows the simultaneous and homogeneous implantation of 21 wafers (156*156mm^2) in a very short time (≈30 s) through the use of two plasma sources. Upgrading this tool with a larger chamber size could therefore allow the (single side) implantation of more than 3600 wafers per hour. The PIII process step is followed by a dopant-activation anneal (eg. in a tube furnace) which can also be used to grow a thin passivating oxide on the c-Si surface. This work report on the development of Phosphorus (P) doping by PIII in the PULSION® Solar equipment and industrial Al-BSF solar cells fabrication using this technique.

2 EXPERIMENTAL APPROACH

After texturing and cleaning, 156 PSQ p-type Cz c-Si wafers (2-3 Ω.cm, 239 cm^2) were implanted on the front side by PIII to form the homogeneous Phosphorus emitter. All PIII processes were performed at CEA-INES in a PULSION® Solar equipment (Figure 1).

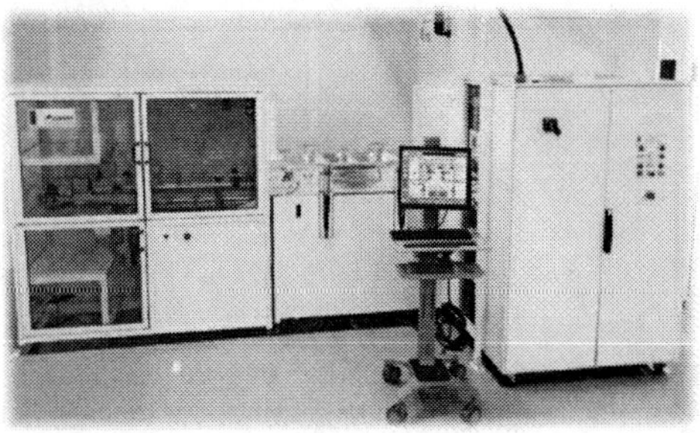

Figure 1: Image of a PULSION® Solar equipment from the IBS Company installed at CEA-LITEN/INES facility

First, we evaluated the impact of PIII parameters by measuring the emitter sheet resistance (R_{sheet}) values after furnace anneal. Homogeneous Phosphorus doping is achieved after PIII process optimization, showing the excellent ion energy distribution across the wafer during PIII, and the great combination with optimized annealing conditions. Identical annealing steps were successively performed in a tube furnace (O_2 ambient) for several minutes to activate the dopants and oxidize the c-Si surface to form a SiO_x layer of few nanometers. An anti-reflective coating (SiN_x:H) was deposited at the front side by PECVD. Industrial Al-BSF solar cells were fabricated - using the simple process described in Figure 2 - on wafers having a "PIII-emitter" and screen printed contacts (4 busbars, Ag contact pads at the rear). The fingers pitch used in this study was fixed (1.8 mm), as well as the fingers width (60 µm). Finally we fabricated a 60 cells module (corresponding to 3 runs of PIII process) with "PIII-emitter" Al-BSF solar cells to assess their capability to withstand the interconnection process as well as ageing tests.

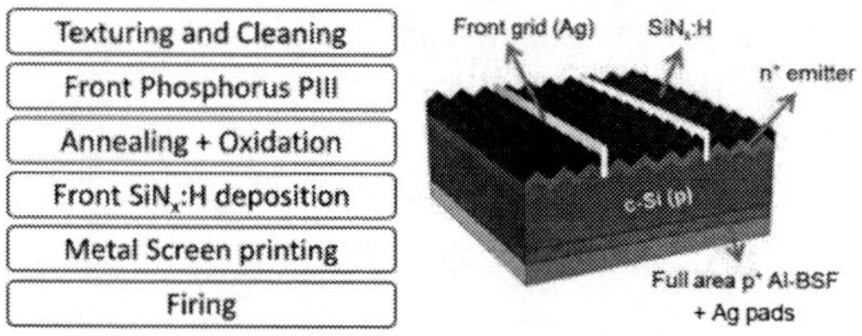

Figure 2: Al-BSF cells manufacturing process and sketch of the cells structure of this work.

3 RESULTS AND DISCUSSIONS

3.1 PIII doping process

The R_{sheet} values targeted for homogeneous industrial emitters depend on the contact technology, and are usually expected to be between 70 and 100 Ω/sq. As shown in Figure 3, using a PIII doping process the R_{sheet} values are easily tunable by simply changing the ion dose used during the implantation step. In this case, the main difference between the doping profiles for different PIII doses is the peak doping concentration at the c-Si surface.

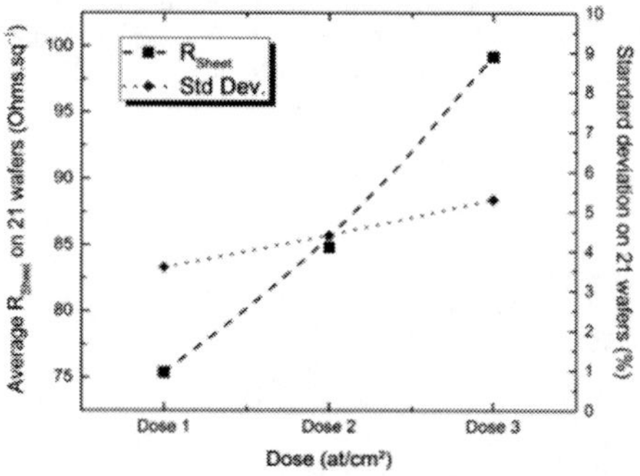

Figure 3: Example of different average R_{Sheet} values obtained on 21 wafers after one PIII step and annealing process depending on the implanted ion dose

The differences observed for the average R_{sheet} values measured on the 21 wafers appear to be quite stable (standard deviation about 5%) even for highly resistive emitters. Optimizations of the PIII process parameters are ongoing to further improve the doping homogeneity on each wafer and between the 21 samples implanted in one run. 10 Al-BSF cells have been fabricated for each dose (out of 21) and the corresponding I-V parameters are shown in Figure 4.

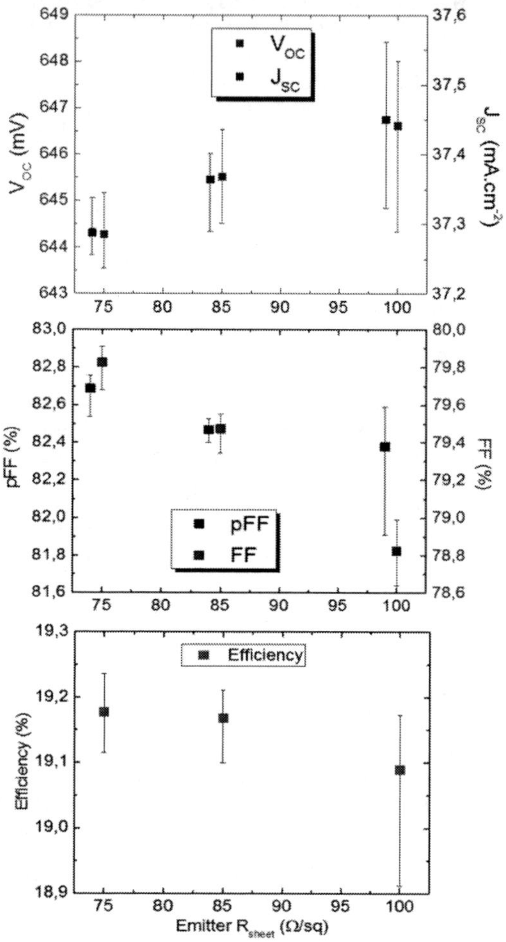

Figure 4: I-V parameters of the Al-BSF cells with different PIII emitter R_{Sheet} values

Higher emitter R_{Sheet} values allow better J_{SC} and V_{OC} due to lower J_{0e}, unfortunately the front metallization scheme (with 1.8mm pitch) limits the FF (series resistance) and therefore the cell efficiency. Using a smaller (1.5mm) pitch and 40µm-wide fingers, efficiencies above 19.7% are targeted due to improved FF and J_{SC} values.

3.2 Fabrication of 500 Al-BSF solar cells featuring PIII emitter and a 60-cells module

For the first "large" batch of ≈500 Al-BSF cells, we used a standard 75 Ω/sq emitter as a starting point, with the same metallization scheme as in the previous section. 24 runs of PIII in PULSION® Solar were necessary to perform the doping process of these cells. As can be seen on Figure 5, an average efficiency of 19.1% (with a very low dispersion) and maximum performances above 19.3% have been reached. These performances have been obtained with a standard cell process which could be enhanced using optimized metallization (smaller pitch and reduced finger width).

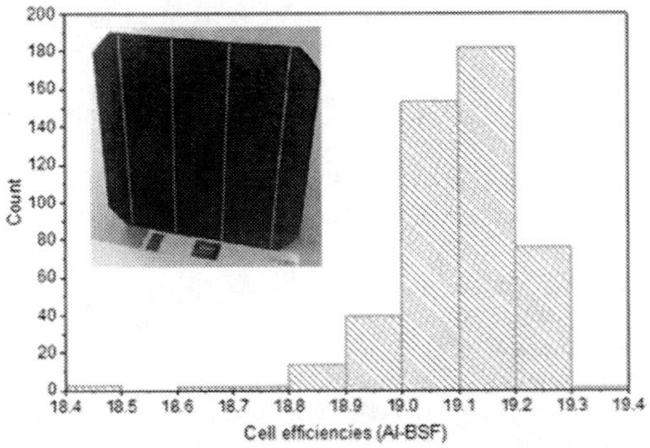

Figure 5: Performances of the ≈500 Al-BSF cells fabricated with the PULSION® Solar equipment for the Phosphorus emitter doping process (75 Ohms/sq)

A 60-cells module has been fabricated with "PIII-emitter" Al-BSF solar cells, using an automated interconnection system. An output power of 268 W has been obtained through flash lamp testing. A homogeneous electroluminescence image of this module is shown in FIG. 6, indicating the high quality of the fabricated cells.

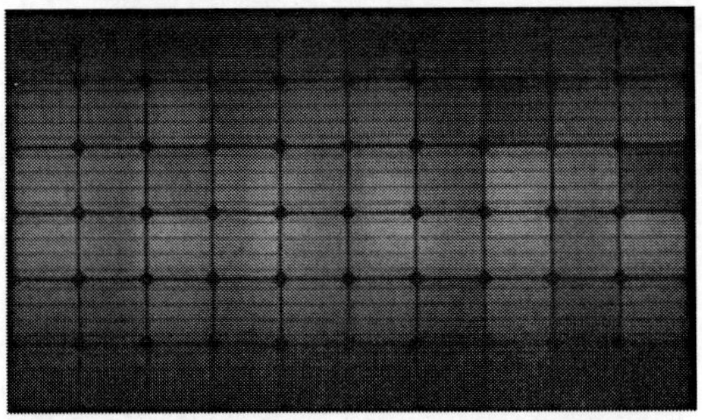

Figure 6: Electroluminescence image of a 268W module made from 60 Al-BSF cells fabricated with a "PIII-Emitter" (75 Ohms/sq)

4 CONCLUSION

This study shows that PIII doping processes are relevant for the fabrication of efficient industrial solar cells at lower cost. High quality n^+ emitters have been obtained with Pulsion Solar® PIII equipement, with J_{0e} values of 40 fA/cm² for R_{Sheet} at 110 Ω/sq.

Industrial Al-BSF solar cells with PIII emitter have been fabricated and an average efficiency of 19.1% has been obtained with a very low dispersion and maximum performances above 19.3%. These performances have been obtained with a standard - industrial - cell process. A 60-cells module has been fabricated with "PIII-emitter" Al-BSF solar cells and an output power of 268 W has been obtained.

Experiments are ongoing using more resistive PIII emitters at the cell level and optimized metallization processes. This doping technique is now also being tested with promising results for PERC and PERT [5] structures, as well as for selective doping with in situ masking.

5 REFERENCES

[1] J. Le Perchec et al, "19.3% efficiency on p-type silicon solar cells by Pulsion® plasma immersion implantation", Energy Procedia, Vol. 33, pp. 18-23, 2013

[2] T. Michel et al, « Phosphorus emitter engineering by plasma-immersion ion implantation for c-Si solar cells", Solar Energy Materials and Solar Cells, Vol. 133, pp. 194–200, 2015

[3] J.-F. Lerat et al, "Boron emitter formation by plasma immersion ion implantation in n-type PERT silicon solar cells", Energy Procedia, Vol. 92, pp. 697–701, 2016

[4] D. L. Young et al, "Low-cost plasma immersion ion implantation doping for Interdigitated back passivated contact (IBPC) solar cells", Solar Energy Materials and Solar Cells, Vol. 158, Part 1, pp. 68–76, 2016

[5] A. Lanterne et al, "Plasma Immersion Ion Implantation for Emitter and BSF Doping in N-type PERT Solar Cells with a Single Activation Anneal", Proceedings of the 44th IEEE PVSEC, Washington DC, 2017

LOW RECOMBINATION EMITTER PROFILE WITH IN-SITU OXIDE PASSIVATION FOR MULTI-CRYSTALLINE SOLAR CELLS

F. Buchholz[1] (1), P. Preis[1] (2), S. Eisert[1] (3), C. Peter[1] (4), E. Wefringhaus[1](5), J. Denafas[2] (6), V. Čyras[2,3] (7), M.P. Bellmann[4] (8)
[1]International Solar Energy Research Center Konstanz, Rudolf-Diesel-Str. 15, D-78462 Konstanz, Germany,
(1) florian.buchholz@isc-konstanz.de, Phone: +49 7531 36 183 553 fax: +49 7531 36 183 11
(2) pirmin.preis@isc-konstanz.de
(3) stephan.eisert@isc-konstanz
(4) christoph.peter@isc-konstanz.de
(5) eckard.wefringhaus@isc-konstanz.de
[2]UAB Soli Tek R&D, Vilnius, Lithuania
[3]Center For Physical Science And Technology (FTMC), Vilnius, LITHUANIA
(6) julius.denafas@solitek.eu
(7) valius.cyras@solitek.eu
[4]SINTEF Materials and Chemistry, Trondheim, Norway
(8) martin.bellmann@sintef.no

ABSTRACT: In situ oxidation during phosphorus diffusion recently has been shown to enable superior surface passivation without adding additional process steps. On the contrary, the PSG removal in diluted HF can be left out. We have optimized such a diffusion process especially for multi crystalline wafers, taking the material requirements in terms of thermal budget and cooling speeds into account, yielding emitter saturation current values below 60 fAcm^{-2}. One of the reasons, why such process recipes have not yet gained widespread acceptance is that the performance of the accordingly diffused emitters, especially on multicrystalline wafers, may vary. To our knowledge the reasons for this behavior has not been explained, yet. We deliver an explanation by studying in depth the passivation behavior taking the front surface and bulk passivation properties into account.
Keywords: phosphorus diffusion, in situ oxidation, multi crystalline silicon, bulk passivation.

1 INTRODUCTION

In situ oxidation during thermal emitter or back-surface field diffusions has been reported in the literature for the use in manufacturing of high efficiency solar cells [1,2]. This kind of diffusion processes rely on diffusion from a limited dopant source. First, a dopant containing thin layer is deposited as silicate glass at relatively low temperatures. Subsequently, the dopant gas flow is switched off and the temperature in the furnace is increased for the drive-in in oxygen containing atmosphere. During this step a high quality oxide is grown and a low surface concentration emitter profile is generated that features a very high quality oxide surface passivation [1].

A scheme of a temperature-time profile of such a phosphorus diffusion recipe is plotted in **Figure 1**. Another short $POCl_3$ step may be added before cooling, when contacting of the low surface concentration turns out to problematic [2].

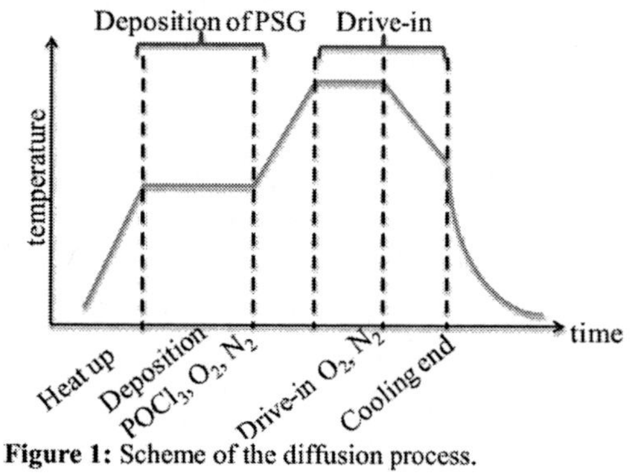

Figure 1: Scheme of the diffusion process.

In contrast to typical diffusion recipes the in situ grown PSG layer does not need to be removed and may be kept as surface passivation layer.

2 EXPERIMENTS AND RESULTS

P-type multicrystalline sister wafers from commercial suppliers and from an experimental ingot as delivered by Sintef were used in the following. The commercial material was medium to low quality standard mc material. The experimental ingot wafers were from a high performance mc Gen1 ingot, crystallized in a reusable crucible made from silicon nitride (4th run) [3]. In addition, for emitter saturation current measurements and as reference in the optimization p-type Cz wafers were used.

Symmetrical lifetime samples and full Al-BSF solar cells were fabricated in the ISC Konstanz pilot line using standard industrial equipment. The wafers were first textured in HF/HNO_3 then cleaned in piranha solution to avoid the interference of potential surface contamination. After cleaning the wafers were diffused using a commercial Centrotherm tube diffusion furnace. Solar cells and lifetime samples were then either PSG etched in diluted HF or not and passivated using PECVD SiNx. Lifetime samples were accordingly treated on both sides. Solar cells were obtained after screen printing metallization (five busbar silver grid on front side, full aluminum on rear side). The last step for all samples was fast firing in a Centrotherm conveyor belt furnace. IV curves were measured by an H.A.L.M. flasher setup. Lifetime samples and cells were measured by QSSPC, IV tester and calibrated photoluminescence. IQE curves were measured by a PVtools spectro-response device.

2.1 Emitter optimization

The starting point for the emitter optimization was a diffusion recipe used for Cz solar cell (referred to as A1) following the scheme in **Figure 1**.This diffusion had been tuned to 90 Ω /sq. The J_{0E} values obtained with and without the generated oxide are shown in **Figure 2**. The J_{0E} values obtained, by the diffusion recipe optimized for the use for mc wafers (MC01) is also shown. Note that this diffusion had been tuned to 75 Ω /sq. The low J_{0E} values of the oxide passivated surface increase, when removing the oxide and replacing by silicon nitride passivation, indicating the superior surface passivation properties.

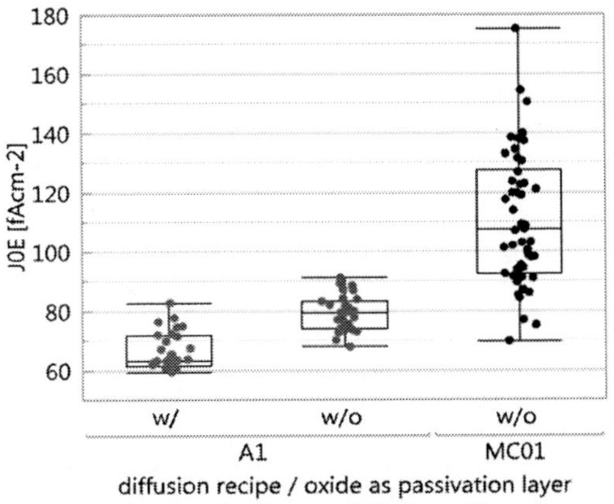

Figure 2: J_{0E} values for the new diffusion with and without the oxide as passivation layer. Note that the sheet resistance differs (A1: 95 Ω/sq vs. MC01: 75 Ω/sq).

Figure 3 shows the homogeneity of the oxide thickness across the diffusion boat and the wafer (five measuring points on one wafer are shown). With a variation in the range of 1 nm a homogeneous front side outcome for the whole diffusion boat can be guaranteed.

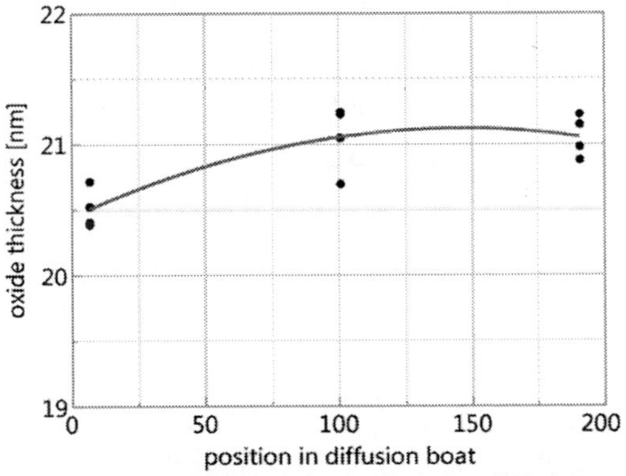

Figure 3: Oxide thickness resulting from A1 diffusion across the diffusion boat (as measured by ellipsometry).

Multicrystalline wafer material is known to contain multiple defects and impurities that are sensitive to the thermal budget and in particular to cooling ramps [4,5,6]. In order to study the impact of peak temperature and cooling ramp on the lifetime of mc wafers, five recipe variations based on diffusion A1 were created (see **Table I**). The peak temperature in three steps was lowered from 920°C to 880°C (group A1-3). The lower temperature is compensated by increasing the plateau time during drive-in/oxidation. By letting the boat cool down to a lower "cooling end temperature" inside the tube a gentler cooling ramp was realized (group B1 and B3). In addition, the speed of the diffusion boat leaving the furnace was slowed. The resulting diffusion profiles and the reference standard industrial MC01 diffusion profile are plotted in **Figure 4**.

Table I: Optimization experiment for the diffusion of mc-wafers.

Diffusion	A1	A2	A3	B1	B3
Peak temperature [°C]	920	900	880	920	880
Plateau time [sec]	240	360	720	240	720
Cooling end [°C]	840	840	840	650	650
Speed of boat leaving the tube	fast	fast	fast	slow	slow

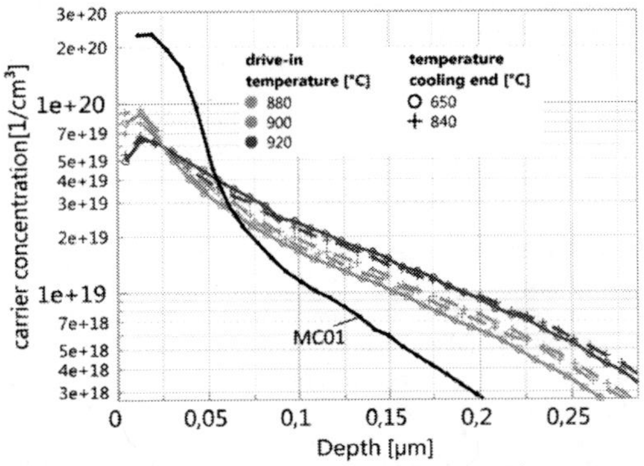

Figure 4: Diffusion profiles (carrier concentration as measured by ECV) of the recipes summarized in **Table 1**. MC01 is a "typical" diffusion for mc wafers and used as a reference throughout this work.

In comparison to the MC01 reference diffusion profile, the much lower surface concentration and the deeper diffusion profile become obvious. Lowering the drive-in temperature from 920°C to 880°C and simultaneously increasing the plateau time from 240 to 720 sec slightly increases the surface concentration and reduces the diffusion depth (as measured by ECV). The measured sheet resistances and standard deviations are summarized in **Table II**.

Table II: Resulting sheet resistance for the 5 diffusions recipes.

Diffusion	A1	A2	A3	B1	B3
Sheet resistance [Ω/sq]	95	90	100	104	104
Std. deviation [Ω/sq]	6.7	3.0	2.5	6.4	5.1

Symmetrical lifetime samples were fabricated from Cz, "good" and "bad" mc material. The *implied* V_{oc} values obtained by these diffusion recipes can be found in **Figure 5**.

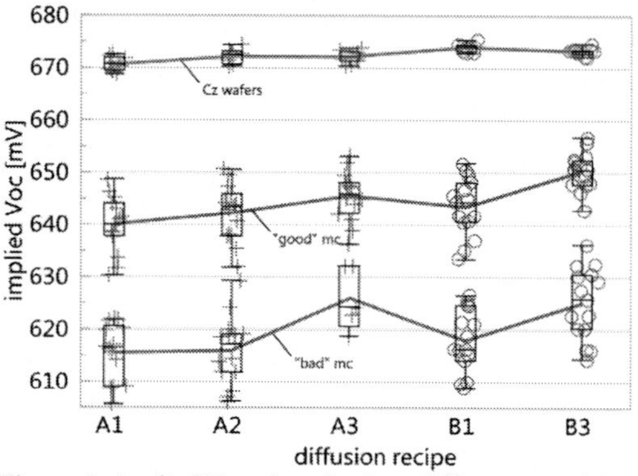

Figure 5: *Implied* V_{oc} values for three different materials

As expected, the additional gettering steps and the low peak temperature have little effect on the Cz material. The "good" mc wafers show a small increase with low peak temperature (the slightly higher sheet resistance also plays a role). A significant increase is measured, when applying the gettering step. The biggest effect is observed on "bad" mc material, showing a similar trend. About 10 mV could be gained in the latter case. As expected, less harsh temperature conditions are beneficial for lower quality mc material.

2.2 Solar cell processing

In order to test the new, optimized emitter diffusion recipe (B3) and compare it with the reference diffusion recipe (MC01), lifetime samples and solar cells were fabricated. The experimental high performance mc wafers and wafers from a commercial supplier were used. From PL-pictures after acidic texturization, it was found that the commercial material had relatively large grains and several severe dislocation clusters. In contrast to this the high performance material provided by Sintef had evenly distributed small grains with little clustering of dislocations.

The resulting solar cell efficiencies reached about 18%. The values, however, would be misleading here, as the thickness of the passivation stack for the oxide had not been fully optimized. Thus only the V_{oc} values are compared as indicator for the emitter performance.

In **Figure 6** the V_{oc} values of the solar cells and the *implied* V_{oc} values as extracted from lifetime samples are shown for the respective four groups. Furthermore, the differences between the two values are displayed.

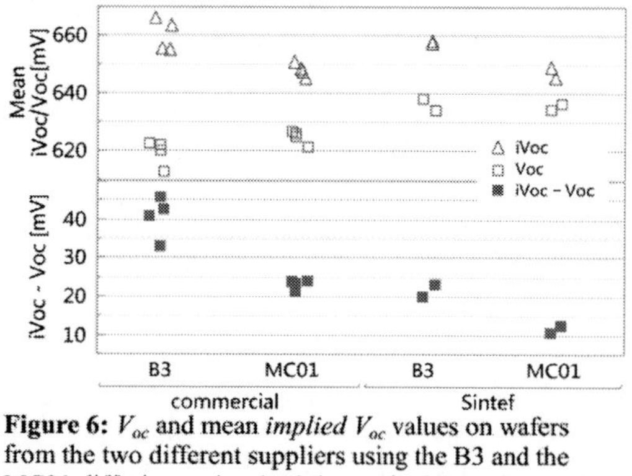

Figure 6: V_{oc} and mean *implied* V_{oc} values on wafers from the two different suppliers using the B3 and the MC01 diffusion recipe (and the *implied* V_{oc} to V_{oc} loss).

For both materials the B3 diffusion recipe shows the highest *implied* V_{oc} values, above 660 mV on the commercial material and slightly lower values on the Sintef material. The MC01 diffusion recipe on the other hand yields similarly low values in the range of 645 to 650 mV on both types of wafers.

However, when having a closer look at the losses from *implied* V_{oc} to V_{oc} after metallization, the losses are the biggest for the group that shows the highest *implied* V_{oc} in the first place. The drop is as large as 40 to 50 mV, yielding very poor V_{oc} values of around 620 mV. A much lower loss of around 25 mV is measured for the reference MC01 recipe, yielding slightly higher final V_{oc} values on this material.

There is a gap between the losses from before to after metallization caused by the two diffusion recipes also for the Sintef material of about 10 vs. 20 mV. Yet, in this case the B3 diffusion leads to similar V_{oc} values in the range of 635 to 640 mV for both types of material. From this it can be concluded, the new diffusion does work on some materials and not on others, however not making full use of the potential as indicated by the implied V_{oc}.

The questions that remain from these results are the following. Why are the *implied* V_{oc} to V_{oc} losses so tremendous for the "commercial" material? Why can the potential indicated by the good *implied* V_{oc} values not be transformed to significantly higher cell V_{oc} values, even in the experimental Sintef material?

In order to shed light on these two questions, IQE curves were measured of cells from the four groups. The resulting mean curves (mean values from three spots per wafer) are displayed in **Figure 7**. When zooming in on the front side, the superior performance as emitter of the B3 diffusion becomes obvious (solid lines in the graph).

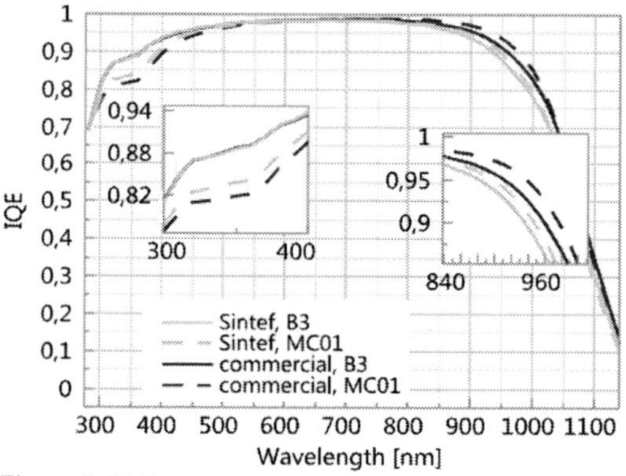

Figure 7: IQE measurements (mean curves from three measurement spots per wafer for each experimental group).

The bulk part (long wavelengths) of the IQE indicates higher bulk lifetimes of the commercial material and slightly higher lifetimes for the reference diffusion MC01. These results mirror the performance of the lifetime sample; however, they apparently contradict the observations from the solar cell data.

A hint towards the answer to the questions is given by looking at PL images of lifetime samples and solar cells (as it is done **Figure 8**).

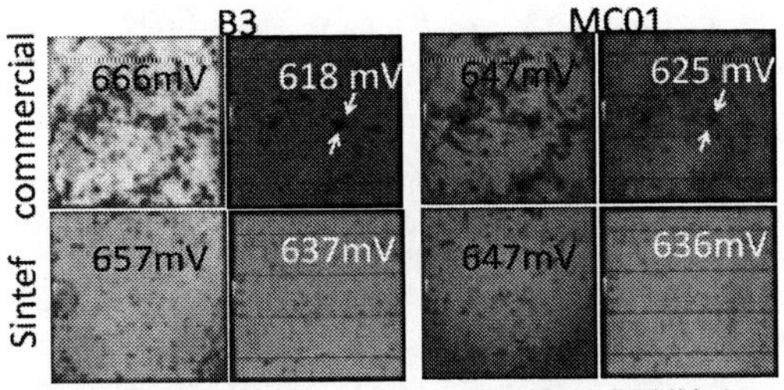

Figure 8: Calibrated PL pictures of symmetrical lifetime samples and solar cells. Mean iV_{oc} and V_{oc} values are added. The white arrows indicate the different size of a dislocation cluster.

In the PL images the dislocation clusters of the commercial material become apparent. The B3 diffusion leads to locally higher lifetimes in the lifetime samples, especially for the large grains in the commercial material. Overall, lifetimes are lower in the case of the MC01 diffusion, yet the losses when metalizing the samples are small, even yielding higher cell V_{oc} values on the commercial material.

After metallization, the dislocation clusters in the case of the commercial material show up in the PL images to a much stronger extend (as indicated by the arrows) when using the B3 diffusion recipe. The lateral conductivity of the solar cells is much higher than in the lifetime sample due to the metallization. In consequence the dislocation clusters as recombination source impact the overall lifetime under open circuit conditions to a much stronger extend, explaining the particularly low solar cell V_{oc} values. Figuratively speaking, they act as recombination sink. As the IQE curves are extracted locally under short circuit conditions, only local information on lifetime is obtained – hiding the effect of "equilibrium" lifetimes under open circuit conditions.

To conclude, apparently the diffusion recipe that produces an in situ oxide affects the recombination activity of dislocation clusters, and, to lesser degree, the bulk lifetime. The activity of the dislocation clusters completely overcompensates for gains from front side performance (surface passivation and low dopant concentration) and potential gains in high lifetime grains. While the material that contains small grains of lower lifetime without dislocation clusters seems to perform neutrally towards the diffusion – gains of surface passivation quality are neutralized by losses in bulk lifetime. Both facts indicate that the bulk recombination is increased, when using the oxide passivation, however, for different reasons: dislocation clusters in the case of the dislocation rich material and grain lifetime in the case of the small grain size material.

It is known from the literature [7,8] that in particular dislocation clusters are commonly deactivated by hydrogenation. Hence, a possible reason explaining the findings is that the thin in situ oxide may act as a barrier against hydrogen in-diffusion thus yielding very poorly passivated dislocation clusters.

3 SUMMARY AND OUTLOOK

A diffusion recipe with in situ oxide growth was introduced to a multicrystalline wafer based solar cell process. The recipe allows keeping the "PSG" and using it as highly efficient surface passivation layer. Based on symmetrical lifetime samples the process was optimized for mc wafers showing improved performance when going to lower drive-in temperatures and less steep cooling ramps. During the optimization it was discovered that mc wafers with high dislocation densities, despite good *implied* V_{oc} values, perform very poorly, when full-Al-BSF solar cells are fabricated. After studying IQE and PL measurements the poor performance could be related to highly recombination active dislocation clusters. We speculate that this can be explained by the function of the oxide as hydrogen diffusion barrier yielding poor defect passivation. Further work is ongoing to check this hypothesis, including an experiment, in which the in situ oxide is removed before SiN_x passivation to check whether bulk lifetimes indeed increase as predicted.

4 ACKNOWLEDGEMENTS

This work has received funding from the Europeans Union's Horizon 2020 research and innovation programme under grant agreement No. 679692.

5 REFERENCES

[1] Mihailetchi VD, et al. Surface passivation of boron diffused junctions by borosilicate glass and in situ grown silicon dioxide interface layer, Proceedings of the 44th IEEE PVSC, Washington DC, 2017, to be published.
[2] Werner S, et al. Structure and composition of phosphosilicate glass systems formed by POCl3 diffusion. Energy Procedia. 2017;124:455-463..
[3] Bellmann, MP, et al. Eco-Solar Factory: Multicrystalline Silicon Ingot Crystallisation from Reusable Silicon Nitride Crucibles, Proceedings of the 33rd EUPVSEC, Amsterdam, 2017, to be published.
[4] Bentzen A, et al. Gettering of Transition Metal Impurities During Phosphorus Emitter Diffusion in Multicrystalline Silicon Solar Cell Processing. Journal of Applied Physics. 2006;99:093509.
[5] Buonassisi T, et al. Engineering Metal-Impurity Nanodefects for Low-Cost Solar Cells. Nature Material. 2005;4:676-679.
[6] Istratov AA, et al. Control of Metal Impurities in "Dirty" Multicrystalline Silicon for Solar Cells. Materials Science and Engineering: B. 2006;134(2-3):282-286.
[7] Martinuzzi S, et al. Hydrogen passivation of defects in multicrystalline silicon solar cells. Solar energy materials and solar cells. 2003;80(3):343-353.
[8] Dekkers HFW, et al. Carrier trap passivation in multicrystalline Si solar cells by hydrogen from SiNx:H layers. Applied Physics Letters. 2006;89(1):013508.

UPGRADE OF AN INDUSTRIAL Al-BSF SOLAR CELL LINE INTO PERC USING 3600 WAFERS/HOUR ALD Al_2O_3+SiN_X SOLUTION RAMP-UP

Floor Souren[1], Bas Dielissen[1], Xavier Gay[1], Roger Görtzen[1], Peter Venema[2], Marten Renes[2], Jan-Marc Luchies[3]
[1]SoLayTec, Dillenburgstraat 9G, 5652 AM, Eindhoven, The Netherlands
[2]Tempress Systems, Radeweg 31, 8171 MD, Vaassen, The Netherlands
[3]Amtech Systems, Radeweg 31, 8171 MD, Vaassen, The Netherlands
e-mail address: Floor.Souren@solaytec.com
Telephone number: +31 40 2380508

ABSTRACT: In this paper, we present the results of the upgrade of a standard monocrystalline Aluminum Back Surface Field (Al-BSF) solar cell line to a passivated emitter and rear cell (PERC) solar cell line and ramp-up. Three additional process steps are added: spatial ALD Al_2O_3 using SoLayTec's InPassion ALD, rear side SiN_x capping layer deposited by a direct plasma system, using e.g. Tempress' SPECTRUM, and a laser to open the rear side dielectric layers, using a conventional laser system. The Al_2O_3 post-deposition anneal is integrated into the SiN_x capping layer deposition process. Besides a good PERC solar cell efficiency, also the wafer appearance of these solar cells is essential. The wafer appearance is a combination of scratches, marks and hazing. In this work, the wafer appearance is improved by optimization of the TMA consumption and the temperature of the integrated-Anneal SiN_x (iA-SiN_x) rear side capping recipe while maintaining the solar cell efficiency. Further adjustments are made in order to improve the throughput and reliability of the equipment. This results in a PERC process with a high solar cell efficiency, good wafer appearance and a throughput above 3600 wafers/hour.
Keywords: PERC, ALD, Al_2O_3, PECVD SiN_x, Manufacturing and Processing

1 INTRODUCTION

Within the last five years, the passivated emitter and rear cell (PERC) solar cell concept has been ramped-up in PV production by many manufacturers. Currently, the PERC solar cell concept has the largest market share of all advanced solar cell concepts of >20% of the total c-Si solar cell production, according to the ITRPV roadmap [1]. The upgrade to PERC leads to a solar cell efficiency increase of 0.5-0.8$\%_{abs}$ and 0.6-1.0$\%_{abs}$ with respect to the standard Al-BSF solar cell based on p-type multi-crystalline and p-type monocrystalline Si, respectively [2,3,4].

For an upgrade of the current Al-BSF solar cell line into PERC technology, the following three systems are added: spatial ALD Al_2O_3 using SoLayTec's InPassion ALD, rear side SiN_x capping layer deposited by a direct plasma system, using e.g. Tempress' SPECTRUM, and a laser to open the rear side dielectric layers, using a conventional laser system.

After ALD Al_2O_3 deposition, a post-deposition annealing step is required to activate the chemical and field-effect passivation. In addition, this annealing process drives out the gaseous H-species built in the ALD Al_2O_3 layer to prevent blistering of the ALD Al_2O_3 film, referred to as 'out-gassing' [5]. The annealing process is integrated into the rear side SiN_x capping process, which is only possible for direct plasma PECVD systems. This combined process is called integrated Anneal-SiN_x capping process (iA-SiN_x), as previously presented by Gay et al. [6]. Besides solar cell efficiency gain for PERC with respect to Al-BSF solar cells, also the wafer appearance is important to customers. The wafer appearance is judged by the customer and is a combination of scratches, marks and hazing. In this work, after achievement of the solar cell efficiency gain for PERC with respect to the Al-BSF solar cell as reference, the wafer appearance of the PERC solar cell is improved. Finally, ramp-up of the PERC solar cell line is started by increasing the wafer numbers and improving the reliability of the equipment of SoLayTec and Tempress.

2 EXPERIMENTAL SETUP

The process sequences for the standard Al-BSF and for the PERC solar cell manufacturing are described in Figure 1. It should be noted that the edge isolation and SiN_x anti reflection coating processes are identical for the Al-BSF and PERC process flows. The metallization and firing processes are optimized for Al-BSF and PERC solar cells, respectively.

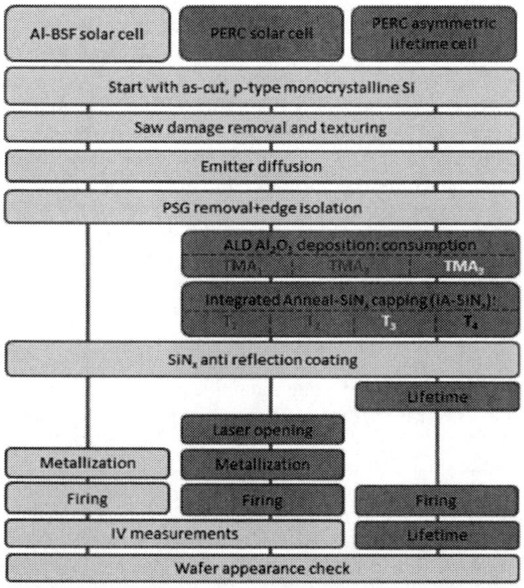

Figure 1: Process sequences for Al-BSF and PERC solar cells. For a faster investigation, asymmetric samples are used. These structures are equivalent to solar cells without laser and without metallization.

In this work the parameters TMA consumption and temperature of iA-SiN_x rear side capping process are both used to get a good wafer appearance while maintaining solar cell efficiency. In addition to the processing flow for complete Al-BSF solar cells and PERC solar cells, asymmetric lifetime cells are made. For these asymmetric lifetime samples different Trimethylaluminum (TMA) consumptions, which is precursor for the Al_2O_3 deposition, are tested as well as different temperatures of the iA-SiN_x rear side capping recipe. Three different settings of the TMA consumption are used: $TMA_1>TMA_2>TMA_3$. The temperature of the iA-SiN_x process is also varied and four different temperatures are tested with the following conditions: $T_1>T_2>T_3>T_4$. Successively, the process conditions for the asymmetric lifetime samples are tested on complete PERC solar cells.

3 RESULTS

In a first phase of the ramp-up, the PERC process is optimized. The focus is put more on the solar cell efficiency improvement. This results in a baseline process sequence with an ALD Al_2O_3 recipe corresponding to the TMA consumption TMA_1 and with an iA-SiN_x recipe using temperature T_1. The obtained solar cell IV-parameters are presented in Table I.

Table I: IV-parameters and appearance for the Al-BSF and PERC solar cells. TMA consumption TMA_1 and iA-SiN_x temperature T_1 are used.

	V_{oc} (mV)	J_{sc} (mA/cm²)	FF (%)	Efficiency (Eff.) (%)	Wafer appearance
Al-BSF	647±1	37.8±0.1	81.1±0.2	19.9±0.1	Good
PERC: TMA_1 and T_1	660±1	39.0±0.1	80.2±0.1	20.7±0.1	Poor

The PERC solar cells obtained, show an efficiency gain of 0.8%$_{abs}$ with respect to Al-BSF solar cells. However, the employed process results in a poor solar cell appearance. In order to improve the solar cell appearance, first asymmetrical lifetime samples are used, see Figure 1 for the process flow. The implied V_{oc} of the samples is measured before and after firing and their appearance is checked in terms of scratches, marks and hazing. The obtained results are listed in Table II.

Table II: Implied V_{oc} results before and after firing and wafer appearance qualification for the investigated samples ($TMA_1>TMA_2>TMA_3$ and $T_1>T_2>T_3>T_4$).

TMA consumption	iA-SiN_x temperature	Implied V_{oc} before firing (mV)	Implied V_{oc} after firing (mV)	Wafer appearance
TMA_1	T_1	667±1	681±2	Poor
TMA_1-repeat	T_1-repeat	670±2	684±2	Poor
TMA_1	T_3	666±2	681±2	Poor
TMA_1	T_4	666±2	681±2	Poor
TMA_2	T_2	668±4	680±4	Poor
TMA_2	T_3	668±3	680±3	Good
TMA_3	T_3	662±2	677±3	Good
TMA_3	T_4	662±4	682±2	Good

The results in Table II show that for good wafer appearance, it is essential to lower the TMA consumption as well as to lower the temperature of the iA-SiN_x capping process. For good passivation, before and after firing, it is essential to use the higher TMA consumptions, TMA_1 or TMA_2. Lowest TMA consumption, TMA_3, leads to lower passivation, even though good wafer appearance. Good passivation as well as a good wafer appearance is obtained for the combination of TMA consumption TMA_2 and iA-SiN_x temperature T_3. TMA consumption TMA_2 in combination with T_2 leads to a poor wafer appearance caused by the too high temperature of the iA-SiN_x capping process.

In order to confirm these results on solar cell level, a batch of solar cells is prepared. Again, the TMA consumption as well as the temperature of the iA-SiN_x capping process are varied. The results are presented in Table III.

Table III: IV-parameters and wafer appearance for PERC solar cells at different TMA consumptions and iA-SiN_x temperatures.

TMA	iA-SiN_x temperature	V_{oc} (mV)	J_{sc} (mA/cm²)	FF (%)	Eff. (%)	Wafer appea-rance
TMA_1	T_1	664±2	39.1±0.1	79.7±0.5	20.7±0.2	Poor
TMA_1-repeat	T_1-repeat	664±1	39.4±0.1	79.7±0.3	20.8±0.2	Poor
TMA_2	T_2	664±1	39.4±0.1	79.3±0.2	20.8±0.1	Poor
TMA_2	T_3	663±1	39.5±0.1	79.3±0.3	20.8±0.1	Good
TMA_3	T_3	660±2	39.1±0.1	79.8±0.4	20.5±0.1	Good
TMA_3	T_4	658±2	39.1±0.1	79.6±0.4	20.5±0.2	Good

The first two groups in Table III have the same conditions and show very similar IV-results. This shows the reproducibility of the results. For good wafer appearance it is essential to lower the TMA consumption as well as to lower the iA-SiN_x temperature, which is in agreement with the lifetime results. For good solar cell efficiency, it is essential to use TMA consumption TMA_1 or TMA_2. As in the case of the lifetime samples, the combination of the intermediate TMA consumption (TMA_2) with temperature T_3 of the iA-SiN_x deposition shows optimal results. For these settings, an efficiency on the same level as the reference process is achieved together with a good solar cell appearance. Note that TMA consumption TMA_2 combined with temperature T_2 of iA-SiN_x capping process has a good solar cell efficiency, but still leads to poor wafer appearance caused by the too high iA-SiN_x temperature T_2. Further reducing the TMA consumption to value TMA_3 results in a lower solar cell efficiency, despite the good wafer appearance. The lower solar cell efficiency for TMA consumption TMA_3 is mostly due to a reduced open circuit voltage caused by a reduced passivation quality. Hence, the lifetime sample results are confirmed on solar cell level. Settings of TMA consumption TMA_2 and iA-SiN_x temperature T_3 are found that deliver good solar cell efficiency as well as a good wafer appearance.

4 MACHINE PERFORMANCES

In a second phase of the ramp-up, the focus is casted on the improvement of the machine performances for the cluster SoLayTec's InPassion ALD and Tempress' SPECTRUM PECVD system. The throughput and rework rate of both systems are optimized.

4.1 InPassion ALD

Figure 2 shows the number of processed wafers during system ramp-up. Over a timespan of one month, the number of wafers processed per day was doubled from 30.000 up to 60.000. During the same time period, the InPassion ALD rework rate was reduced from 1.5% down to <0.25%. The main reasons for the reduced rework rate are standard adjustments and optimizations of the wafer scheduling. The breakage rate is not shown, because it is below 0.025% using the optimized so-called 'DU2.0' design of the InPassion ALD.

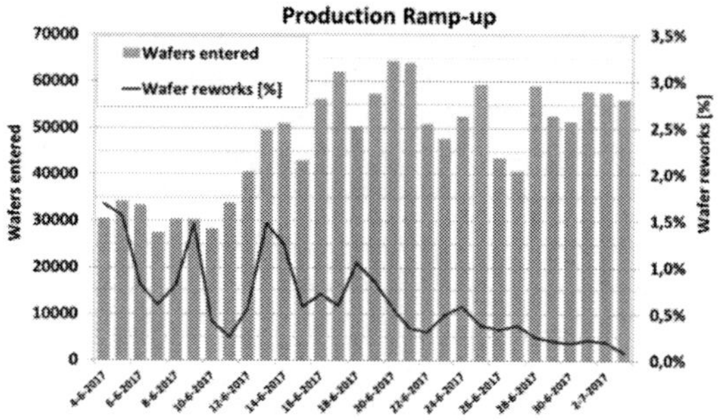

Figure 2: Production increase and rework rate reduction during the initial ramp-up.

4.2 SPECTRUM PECVD

In the process sequence used for the manufacturing of PERC solar cells, the rear side SiN_x capping layer is deposited prior to the front side SiN_x anti reflection coating. Prior to the rear side PECVD SiN_x deposition, the wafers are loaded automatically by a handling robot into a graphite boat with the front side of the solar cell in contact with the graphite plates. If the handling is not accurate, the front side surface of the wafers might be scratched during loading or unloading and the visual appearance of the final solar cells will suffer from that. Due to variations in the assembly of the graphite plates (referred to as 'boat'), which can be a result of manufacturing tolerances, thermal expansion, wear during use and cleaning, scratching of at least part of the cells seemed inevitable. However, by making adjustments to the boat and the handling robot system, which is also part of the so-called 'Tempress-Baumann scratch-free solution', it is demonstrated that complete suppression of the rework rate can be reached. This can be seen in Figure 3.

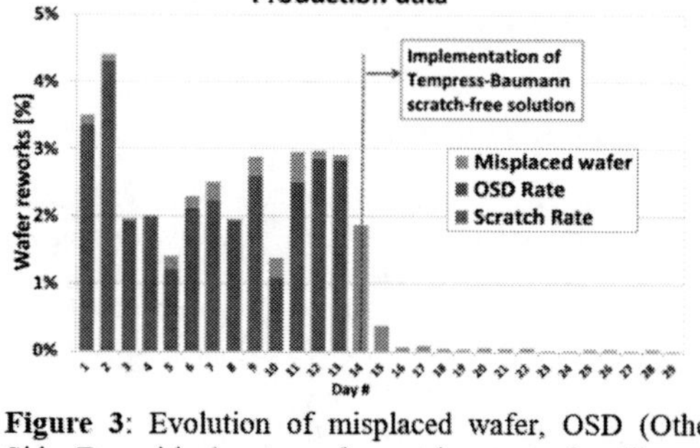

Figure 3: Evolution of misplaced wafer, OSD (Other Side Deposition) rate and scratch rate at handling of PECVD system of Tempress.

5 CONCLUSION

In this paper, we have shown the results of the successful upgrade of a standard Al-BSF monocrystalline solar cell line into a PERC solar cell line and its ramp-up. The addition of the InPassion ALD of SoLayTec, a direct PECVD system such as SPECTRUM PECVD of Tempress for rear side SiN_x capping and a laser tool allows for an efficiency gain of at least $0.8\%_{abs}$ absolute for PERC. The post-deposition anneal of the Al_2O_3 is integrated in the rear side SiN_x process. By tuning the TMA consumption of the ALD Al_2O_3 deposition recipe and the temperature of the iA-SiN_x, the appearance of the solar cells is improved while maintaining solar cell efficiency. Optimum is TMA consumption TMA_2 and iA-SiN_x temperature T_3. For the final ramp-up, the rework rate of the InPassion ALD is improved to <0.25%. In addition, for the SPECTRUM PECVD, the rework rate can be completely reduced by adjustments to the boat and the handling robot system, which is also part of the so-called 'Tempress-Baumann scratch-free' solution. This demonstrates the attractive potential of the cluster of InPassion ALD of SoLayTec for ALD Al_2O_3+SPECTRUM PECVD of Tempress for iA-SiN_x PECVD for PERC upgrades for >3600 wafers/hour.

6 REFERENCES

[1] International Technology Roadmap for Photovoltaic (ITRPV) 2016, 7th edition, March 2017.
[2] R. Sastrawan et al., Implementation of a multicrystalline ALD-Al_2O_3-PERC technology into an industrial pilot production, 28th European Photovoltaic Solar Energy Conference and Exhibition, Paris, France.
[3] D. Chen et al., 21.40% Efficient Large Area Screen Printed Industrial PERC Solar Cells, 31st European Photovoltaic Solar Energy Conference and Exhibition, Hamburg, Germany.
[4] F. Souren et al., Upgrade of an industrial Al-BSF solar cell line into PERC using spatial ALD Al_2O_3, 32nd European Photovoltaic Solar Energy Conference and Exhibition, Munich, Germany.
[5] G. Dingemans, W.M.M. Kessels, Status and prospects of Al_2O_3-based surface passivation schemes for silicon solar cells, J. Vac. Sci. Technol. A, 30(4), 040802, 2012.
[6] Xavier Gay et al., Post-Deposition Thermal Treatment of Ultrafast Spatial ALD Al_2O_3 for the Rear Side Passivation of p-type PERC Solar Cells, 28th European Photovoltaic Solar Energy Conference and Exhibition, Paris, France.

EASY PLATING – STUDY ON CONTACT INTERFACE PROPERTIES OF PARASITIC PLATING-FREE NI/CU SOLAR CELLS

Benjamin Grübel, Andreas Büchler, Sven Kluska, Jonas Bartsch, Gisela Cimiotti, Andreas A. Brand Markus Glatthaar
Fraunhofer Institute for Solar Energy Systems ISE, Heidenhofstr. 2, 79110 Freiburg, Germany
Phone: +49 (0) 761/45 88-5019
Benjamin.gruebel@ise.fraunhofer.de

ABSTRACT: The Easy Plating process sequence is an elegant way to produce Al-BSF or PERC solar cells with laser-structured Ni/Cu-plated contacts. By eliminating the HF-treatment before metal deposition one can avoid parasitic plating at defects of the anti-reflection coating by utilizing native oxide growth to passivate these defects. Recent work demonstrated a significant improvement in terms of metal recombination, optical shading and aesthetical appearance compared to the standard plating approach. However, the Easy Plating sequence is sensitive to laser-induced and native oxide formation at the Si/Ni interface in the laser defined local contact openings.
This work investigates the influence of interface oxides between the Si/Ni layer and the beneficial impact of the thermal anneal for the contact resistance. Therefore, the contact resistance is studied by transmission line measurement (TLM) and SEM is applied for visualization of silicides. Our findings are important for advanced process optimization and an advanced understanding of the processes at the interface.
Keywords: c-Si solar cell, Metallization, Annealing

1 INTRODUCTION

Metal plating of a stack of Ni/Cu/Ag is an industrially feasible alternative to the state-of-the-art Ag screen printing for the front side metallization of silicon solar cells [1, 2]. The substitution of silver by copper leads to material cost reduction. Nickel, as primary layer of the contact, acts as seed layer and diffusion barrier. Furthermore, a low contact resistivity even to lightly doped emitters is feasible. A thin silver capping (< 1 µm) prevents the corrosion of the copper layer.

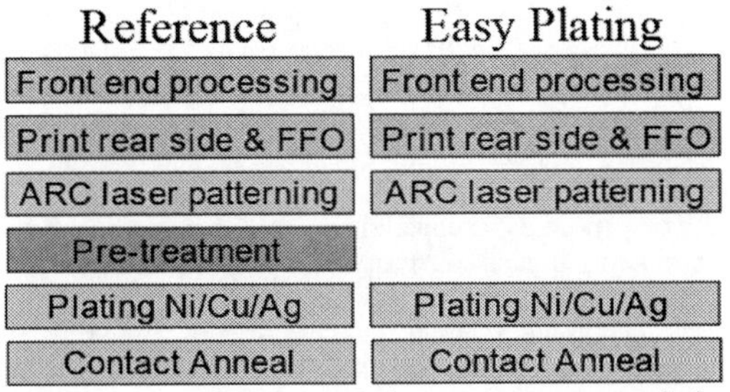

Figure 1: Comparison of the process steps of the *Reference* and *Easy Plating* process for Al-BSF solar cells

At the Fraunhofer ISE the state-of-the-art plating process sequence (*Reference*) is applied as shown in Figure 1 left column [3]: After local removal of the passivation layer (antireflective coating ARC) by laser ablation, oxide layer on bare silicon surfaces are removed by a wet-chemical pre-treatment (e.g. HF) followed by an immediate deposition/plating of metal layers. The silicide formation due to a thermal anneal, can improve the contact system and lead to a reduction of the contact resistance [2]. Unintended metal deposition in non-patterned areas of the ARC is commonly named parasitic plating (PP) or ghost plating. PP is promoted by ARC defects and harms the aesthetics and cell performances by shading and surface recombination [4]. Native oxide can electrically insulate these defects and can prevent the unintended metal deposition. Unfortunately, the HF pre-treatment removes the passivation of ARC defects and provokes an increase of parasitic plating. The *Easy Plating* process sequence, shown in the right column of Figure 1, makes the pre-treatment obsolete, in order to use the native oxides as electrical insulation on the ARC defects. Though, the oxide layers in the laser ablated area are not removed. This requires minimizing process induced oxide formation. While laser-induced oxidation can be minimized by the right choice of laser parameters [5], the delay time between laser ablation and metal deposition is crucial for the native oxide growth [6, 7]. Therefore, within the experiment we tried to reveal the dependency of the native oxide growth from the delay time between laser ablation and plating. The following thermal anneal is assumed to promote the formation of nickel-silicide to improve the contact resistance. While authors stated that the silicide formation was required to reduce the contact resistance [8], no clarified results have shown the direct correlation for contacts plated without oxide removal. In the following we will analyze the impact of a contact anneal process on the contact resistivity of contacts plated with the *Reference* and *Easy Plating* sequence. Furthermore, we will show if the silicide formation occurs at the Si/Ni interface and if it decreases the contact resistance. Throughout the paper, the process sequence without the HF-pre-treatment is called "*Easy Plating*", while "*Reference*" refers to the process sequence including a wet chemical treatment before plating.

2 EXPERIMENTAL

All the results shown in this paper are generated using industrially fabricated Cz Al-BSF solar cell precursors. The precursor front-side features a random pyramid texture passivated by a silicon nitride (SiN_x) layer optimized for silver screen printing. The experiment was carried out as shown in Figure 2.

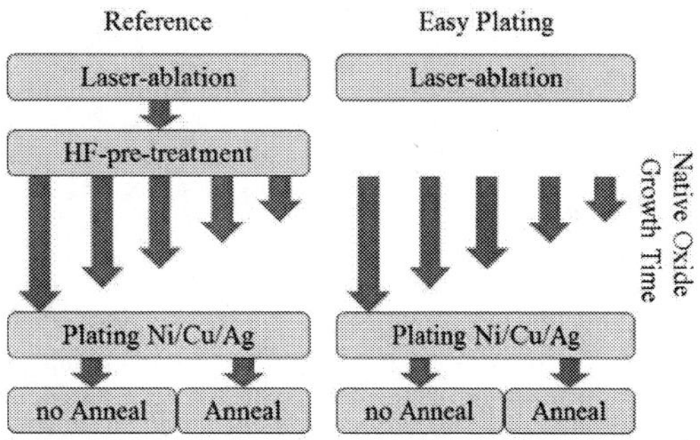

Figure 2: Process scheme of the experiment. Each process sequence features a variation of native oxide growth time before plating in which native oxides are grown at the contact interface

The laser patterning/ablation of the passivation layer was performed using a picosecond-pulsed UV laser system (Coherent Lumera Super Rapid, laser pulse duration < 15 ps, 355 nm). The laser pulse energy was set to result in contact openings with a diameter of 20 µm. After laser-ablation half of the cells follow the *Reference* sequence being transferred in the inline wet chemical pre-treatment (1%-HF, 30 s). This step is followed by a controlled oxide growth on the laser-ablated contact openings varying the delay time (from now on native oxide growth time) between HF-pre-treatment and metal deposition in the range of 20 s – 1440 min. The native oxide growth time variation of cells for the *Easy Plating* sequence starts directly after laser-ablation. In both cases the samples are stored at room temperature under ambient atmosphere. Using light induced plating (LIP), the metal contact consisting of Ni (~1 µm), Cu (~10 µm) and Ag (~ 0.5 µm) was deposited in inline plating tools of the company Rena. To analyze the effect of the thermal treatment certain cells were annealed in an inline furnace under forming gas atmosphere (250°C, 5 min). Each process group featured TLM test structures that were manufactured identically other than the pre-treatment. The characterization of the contact resistance was conducted by using the transmission line measurement. Microstructure analyzes were performed in a Zeiss Auriga SEM.

3 RESULTS AND DISCUSSION

3.1 Oxide influence

To analyze the influence of oxide layers between silicon and nickel the native oxide growth time before metal deposition is varied. Figure 3 shows independently of the process sequence (*Reference / Easy Plating*) increasing contact resistivity for increasing native oxide growth time. Assuming native oxide growth rates on n^+-Si [7] the according native oxide thickness after 1440 min is estimated to be about 8 nm. The results of Fig. 3 illustrate the native oxide-induced contact resistance increase. Moreover, an increased contact resistivity between both process sequences is visible as a result of oxide layers grown while laser-ablation, as the only difference is the HF-pre-treatment.

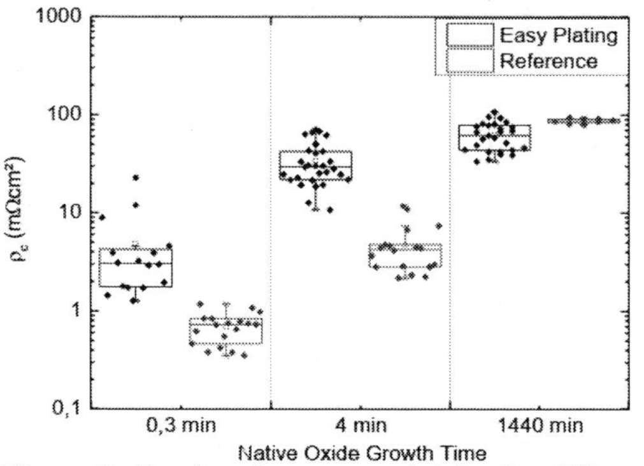

Figure 3: Results of contact resistivity for different native oxide growth times between laser-ablation respectively HF-pre-treatment and metal deposition for the *Easy Plating* (black) and *Reference* (red) process sequence.

As the backside metallization and the plated front-side metal grid are identical on all wafers the contact resistivity is the only influencing factor on the solar cell series resistance. Therefore, Figure 4 displays the interface configuration between silicon and nickel depending on the process and the native oxide growth time. As the surface of the precursor is textured, the input of the laser-pulses is not homogenous respecting to the height of the pyramid. Therefore, the growth rate of the laser-induced oxide might differ over the surface. For a minimal native oxide growth time of 0.3 min no oxide layers are located on the interface for the *Reference* process. With an increase of the native oxide growth time between HF-pre-treatment and plating native oxides occur on the interface. For the *Easy Plating* process sequence, since no HF-pre-treatment is executed, laser-induced oxide layers cover the interface. While the native oxide growth time between laser-ablation and plating increases additional native oxide layers grow on the interface. Higher thickness of oxide layers in the *Easy Plating* process compared to the *Reference* process reveals the higher contact resistivity displayed in Figure 3. After 1440 min the contact resistivity of the *Easy Plating* and the *Reference* process reaches approximately the same value. Thus, the oxide layers for both processes seem to feature equivalent electrical properties regarding the contact resistivity.

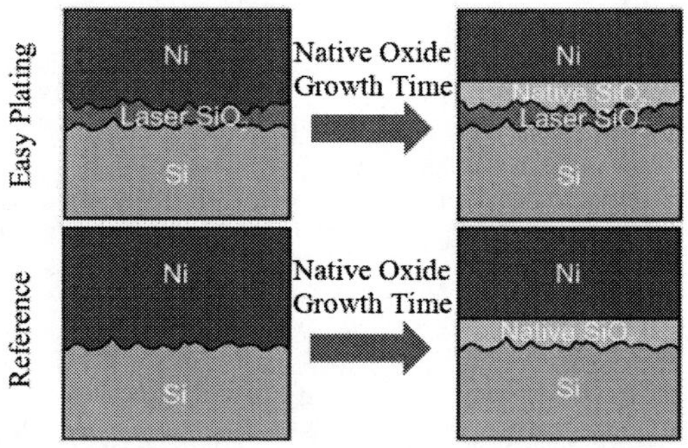

Figure 4: Schematic representation of the interface layers for the *Easy Plating* (up) and *Reference* (down) process and native oxide growth time between laser ablation respectively HF-pre-treatment (left) and plating (right).

To ensure minimized losses of the contact resistivity and a high efficiency a native oxide growth time between laser ablation, HF-pre-treatment and the metal deposition as short as possible is recommended. Furthermore, the reduction of laser-induced oxides has the potential to minimize the increased contact resistivity of the *Easy Plating* against *Reference* process and can make HF-pre-treatment of the *Reference* process obsolete.

3.2 Influence of thermal treatment

Silicides are known from literature to decrease the contact resistivity at a Si-Ni interface. Interface oxides layers can prevent the Ni diffusion into Si and hinder the formation of silicides [9].

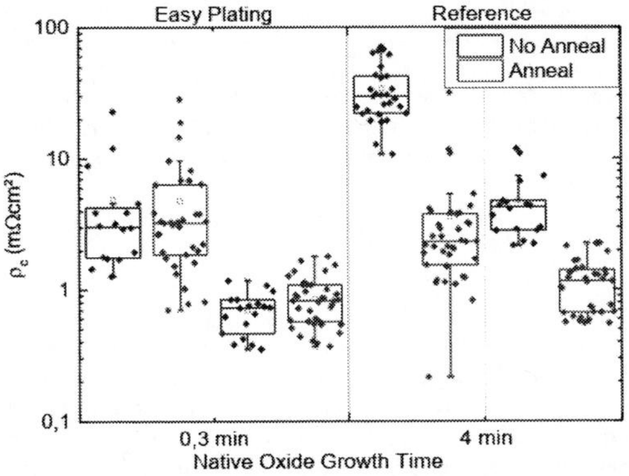

Figure 5: Results of contact resistivity at two different native oxide growth times between laser-ablation and HF-pre-treatment and laser-ablation and metal deposition for *Easy Plating* (left side) and *Reference* (right side), respectively. Process groups with and without anneal are colorized green and blue, respectively.

Figure 5 shows the results of TLM measurements for the *Easy Plating* (left side) and the *Reference* (right side) for a native oxide growth time of 0.3 min and 4 min. In addition, each process results are divided into two groups differentiating before (blue) and after (green) anneal. For the *Easy Plating* process at minimum native oxide growth time of 0.3 min, no appreciable difference of the contact resistivity before and after anneal can be noticed. Also for the *Reference* no noticeable difference can be observed for the contact resistivity before and after anneal for a native oxide growth time of 0.3 min. After a native oxide growth time of 4 min the anneal was capable of reducing the contact resistivity for the *Easy Plating* and the *Reference* group. The *Easy Plating* process sequence is improved at the same level as the contact resistivity after a native oxide growth time of 0.3 min. Equivalent results are obtained for the *Reference* process sequence. Therefore, the anneal process can compensate the negative impact induced by the native oxide layers on the contact interface. Laser-induced oxide layers on the contrary cannot be influenced by the anneal process as shown by the comparison of *Easy Plating* and *Reference* group at a native oxide growth time of 0.3 min. Thus, the structural properties of laser-induced oxides are expected to differ from native oxides.

By SEM investigations the influence of the anneal on contact resistivity can be compared to growth of nickel silicides. According to the SEM images in Fig. 6 the *Reference* process sequence with anneal and a native oxide growth time of 0.3 min reveals the existence of a structure of local areas with increased brightness on the surface assumed to be a local distribution silicides compared to the same process without anneal.

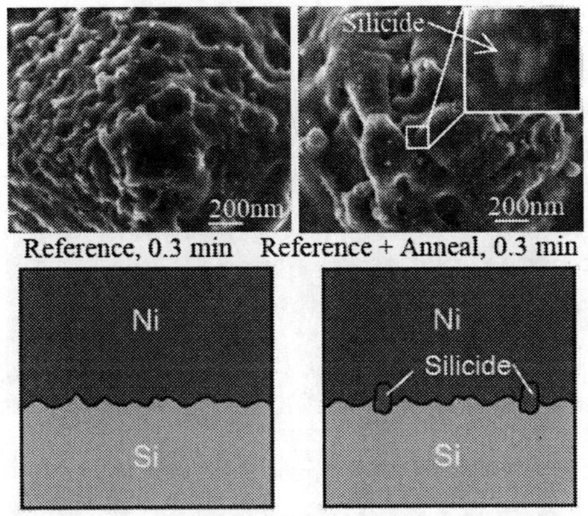

Figure 6: Top View SEM images and of laser ablated surface after metal removal for *Reference* process with a native oxide growth time of 0.3 min without (up left) and with anneal (up right) and the corresponding schematic layer structure before anneal (down left) and after anneal with local silicides (down right)

The diffuse pattern on the surface was characterized by EDX-measurements in the SEM. Figure 7 shows the results of an analysis of two points. Point 1 is located on diffuse structure on the surface while Point 2 is placed on a bare part of the surface. For both points the EDX measurement shows mainly a response at 1.74 eV which stands for silicon. Just a small peak can be noticed for Point 2 at 0.85 eV related to Nickel. At Point 1 the silicon signal reveals to be a bit lower while a Peak rises at the Nickel position. Although a silicide formation occurred for the *Reference* process sequence, no improvement of contact resistivity is observed (Fig. 7).

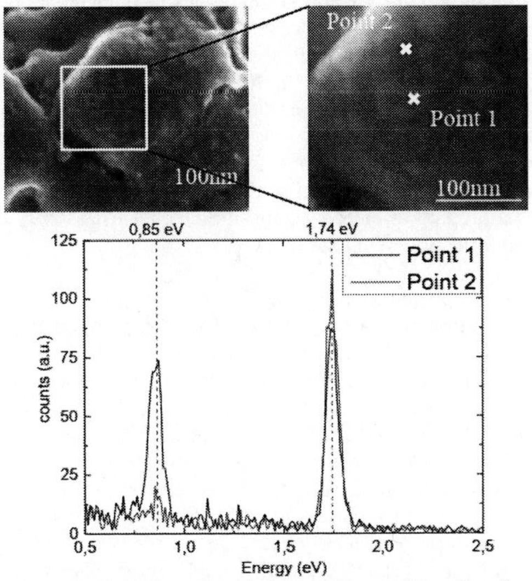

Figure 7: Top View SEM images of laser ablated surface after metal removal for *Reference* process with a native oxide growth time of 0.3 min (up). EDX measurement of two points in an area with and without silicide (down)

For a waiting time of 4 min the SEM images in Figure 8 displays the absence of silicides after the anneal for the *Easy Plating* and the *Reference* process. Therefore, laser-induced and native oxide layers prevent the formation of silicides. However the anneal process sequence manages to reduce the contact resistivity without the occurrence of silicides (Fig. 5). Presumably, as shown in the schematically drawn cross-sections of Fig. 8, local conductive channels are formed due to the anneal that allow the reduction of the contact resistance.

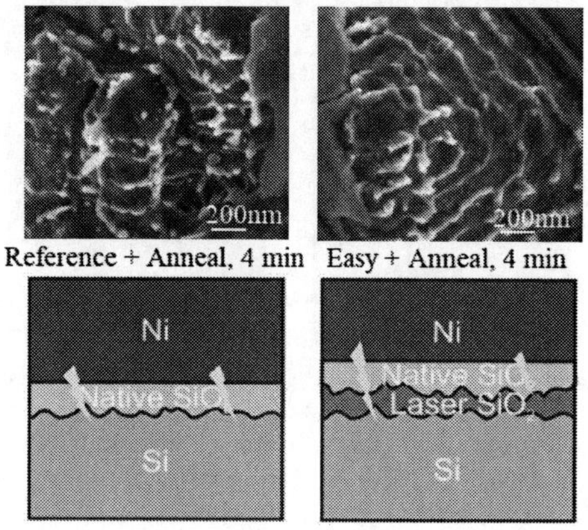

Figure 8: Top View SEM images of laser ablated surface after metal removal for *Reference* (up left) and *Easy Plating* (up right) process with a native oxide growth time of 4 min and the correspondent schematic layer structure (down) with local conductive channels

Thus, for the only process group that showed formation of nickel-silicides (HF-pre-treatment / 0.3 min native oxide growth time) there was no visible improvement of the contact resistivity by the anneal. But the process groups that showed an improvement by the anneal did not show formation of nickel silicides.

To ensure low contact resistivity and high efficiency, oxide layers have a crucial influence. The data suggest that the influence of native oxides on the contact resistivity can be compensated by an anneal for native oxide growth times of at least 4 min. Laser-induced oxides seem to lead to a firm increase of the contact resistivity. This seems in contrast to the theoretical model that predicts an improvement of the contact resistivity by the formation of silicides.

3.3 Optimization of Easy Plating

During the course of this study we applied subtle changes to the processing parameters for both processes (*Easy Plating* and *Reference*) and found that in principle even more favourable values for the contact resistivity could be achievable. As illustrated in Fig. 9 resistivity values under 1 mΩcm² could be achieved with the *Reference* process for the 0.3 min native oxide growth time case. It is remarkable that a very low resistivity value (under 1 mΩcm²) could also be recorded after *Easy Plating* with an anneal which might be indicative of a very favourable amount of laser induced oxide in this specific case. Therefore, the structure of the laser-induced oxides might affect the effect of an anneal. It is however noteworthy for this case that the contact resistivity before anneal was unusually high (above 10 mΩcm²). This observation certainly warrants future studies.

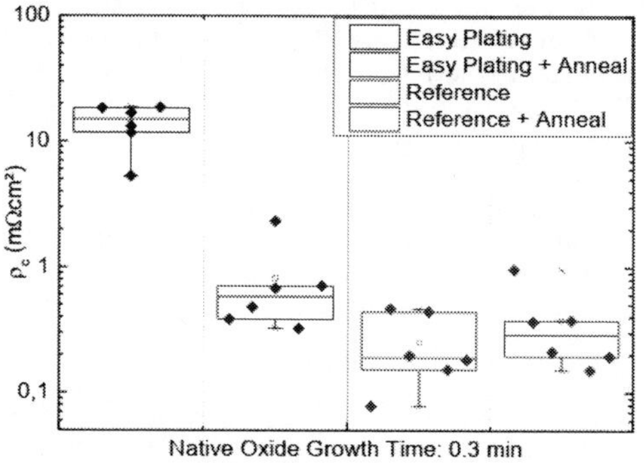

Figure 9: Contact resistivity for Easy Plating and Reference process before (black, red) and after (blue, green) anneal with a minimum native oxide growth time of 0.3 min

5 CONCLUSION

We presented the crucial impact of interfacial oxide layers on the contact resistance that affects the series resistance. As recommended the native oxide growth time between laser-ablation respectively HF-pre-treatment and metal deposition should be kept as short as possible. In the case that the minimum native oxide growth time cannot be applied, a thermal anneal can still compensate the harmful impact of native oxide layers. However, laser-induced oxides layers seem to be the limiting factor for the *Easy Plating* process as they seem to be uninfluenced by that process step. Therefore, the *Easy Plating* process sequence exhibits a slightly higher contact resistance of 3 mΩcm² as the *Reference* process sequence. However further studies seem reasonable as contact resistivities in the range of the *Reference* data were achieved for the *Easy Plating* process. In contrast to the *Reference* process involving parasitic plating, the *Easy Plating* process offers advantages such as a perfect aesthetical appearance, a reduced shaded cell surface and allows to avoid the critical wet chemical pre-treatment step.

6 REFEFERENCES

[1] J. T. Horzel *et al.*, "Industrial Si Solar Cells With Cu-Based Plated Contacts," *IEEE J. Photovoltaics*, vol. 5, no. 6, pp. 1595–1600, http://ieeexplore.ieee.org/ielx7/5503869/7300475/07279053.pdf?tp=&arnumber=7279053&isnumber=7300475, 2015.

[2] S. Kluska *et al.*, "Electrical and Mechanical Properties of Plated Ni/Cu Contacts for Si Solar Cells," *5th International Conference on Silicon Photovoltaics, SiliconPV 2015*, vol. 77, pp. 733–743, 2015.

[3] G. Cimiotti, J. Bartsch, A. Kraft, A. Mondon, and M. Glatthaar, "Design Rules for Solar Cells with Plated Metallization," *Energy Procedia*, vol. 67, pp. 84–92, http://www.sciencedirect.com/science/article/pii/S1876610215005974, 2015.

[4] S. Braun, A. Zuschlag, B. Raabe, and G. Hahn, "The Origin of Background Plating," *Energy Procedia*, vol. 8, pp. 565–570, http://ac.els-cdn.com/S1876610211016924/1-s2.0-S1876610211016924-main.pdf?_tid=0eb15574-9142-11e7-b172-00000aacb361&acdnat=1504510035_0f487e6b5dd656e4443d4f987bddcf43, 2011.
[5] T. E. Orlowski and D. A. Mantell, "Ultraviolet laser-induced oxidation of silicon: The effect of oxygen photodissociation upon oxide growth kinetics," *Journal of Applied Physics*, vol. 64, no. 9, pp. 4410–4414, http://aip.scitation.org/doi/pdf/10.1063/1.341263, 1988.
[6] T.-a. Miura, M. Niwano, D. Shoji, and N. Miyamoto, "Kinetics of oxidation on hydrogen-terminated Si(100) and (111) surfaces stored in air," *J. Appl. Phys.*, vol. 79, no. 8, p. 4373, 1996.
[7] M. Morita, T. Ohmi, E. Hasegawa, M. Kawakami, and M. Ohwada, "Growth of native oxide on a silicon surface," *Journal of Applied Physics*, vol. 68, no. 3, pp. 1272–1281, 1990.
[8] D. J. C. a. E. H. Rhoderick, "Silicide formation in Ni-Si Schottky barrier diodes," *Journal of Physics D: Applied Physics*, vol. 9, no. 6, p. 965, 1976.
[9] A. Büchler *et al.*, "Interface oxides in femtosecond laser structured plated Ni-Cu-Ag contacts for silicon solar cells," *Solar Energy Materials and Solar Cells*, vol. 166, pp. 197–203, 2017.

Industrial solutions for light induced degradation in p-type mc-Si PERC solar cell

Jianwen Dong, Jiaxian Lv, Wei Wang, Quanhua Ye, Yadi Yang, Wenhao Cai, Huiming Zhang, Zhendong Shen, Guidong Chen, Wencao Gu, Xianzhao chen, Jian Sheng, Jian Yang, Chun Zhang, Xiaoxue Zhou, Jiazhen zheng

R&D Center, GCL System Integration Technology Co. , Ltd. Solar BU, 288#, ChenFeng Road, ZhangJiaGang E&T Deve. Zone, SuZhou, JiangSu, China, 215600. dongjianwen@gclsi.com

ABSTRACT: Light-induced degradation (LID) is regarded as a great challenge for both P-type Czochralski (Cz) and multi-crystalline silicon PERC solar cells in mass production. For P-type Cz silicon PERC cell, boron-oxygen related defects are acknowledged to be the main reason for LID, which can be permanently deactivated through a regeneration process. In recent years, a new light and elevated temperature induced degradation effect (LeTID) has been detected in P-type mc-Si PERC cell, which cannot be explained by the activation of boron-oxygen (B-O) complexes or the division of iron-boron (FeB) pairs, there is no physical understanding about the degradation effect.
In this work, industrial solutions for LeTID were evaluated. Under mass production condition, LeTID of partly P-type mc-Si PERC solar cells can be mitigated by treatment-1, the LeTID is between 0.5% and 6% rel.. Treatment-2 looks good for most p-type mc-Si PERC solar cells, the LeTID can be controlled below 2% rel. post treating. Treatment-2 also result in mc-Si PERC cell efficiency increasing ~0.1% abs. Mass production results reveal that treatment-2 is a promising industrial solution for mitigating LeTID in p-type mc-Si PERC solar cells.

Keywords: multi-crystalline silicon, PERC solar cells, light induced degradation

1 INTRODUCTION

Passivated Emitter and Rear Cell (PERC) is a promising high efficiency solar cell technology which have been widely transformed to mass production. International Technology Roadmap for Photovoltaic (ITRPV) predict PERC concepts will become mainstream, and market share will increase to >45% by 2025. Light-induced degradation (LID) is regarded as a great challenge for PERC solar cells in mass production. For P-type Cz silicon PERC cell, boron-oxygen related defects are acknowledged to be the main reason for LID [1,2]. Boron-oxygen LID is traditionally mitigated by minimizing the concentration of boron or oxygen in the silicon bulk [3]. Replacing the boron doping with gallium proves also effective [4]. High-speed regeneration which is achieved at high hydrogen concentration and illumination intensity at T<230 °C [5] also can solve LID.

For P-type mc-Si PERC cell, a new light and elevated temperature induced degradation effect (LeTID) has been detected, which cannot be explained by the activation of boron-oxygen (B-O) complexes or the division of iron-boron (FeB) pairs [6-8]. Currently, there is no physical understanding about the degradation effect and no effective industrial solution to eliminate the LeTID of P-type mc-Si solar cells completely.

In this work, industrial solutions for LeTID were evaluated, it is shown that LeTID of P-type mc-Si PERC cells can be mitigated effectively after treatment.

2 EXPRIMENTS

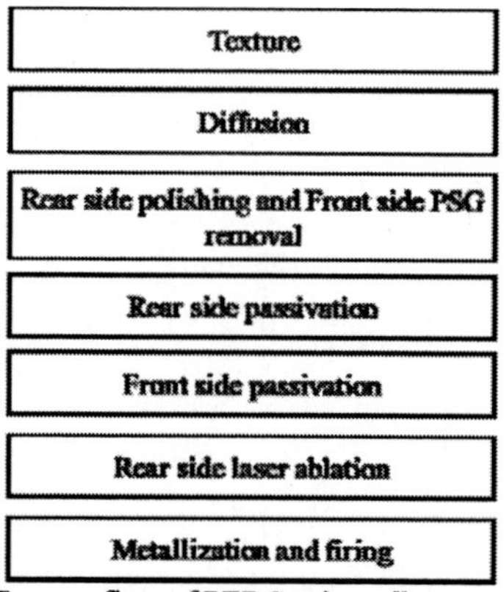

Figure 1: Process flow of PERC solar cells

PERC solar cells were fabricated on 156×156 mm^2 boron doped P-type mc-Si wafers, with bulk resistivity 1Ω.cm ~ 3Ω.cm, and thickness of approximately 180μm. The process flow is illustrated on Fig.1. After IV performance measurement by HALM tester at standard condition (AM 1.5, 25℃, 1000W/m^2), cells were treated under different conditions except reference cells. All cells were subsequently degraded under light exposure (5kwh/m^2) at temperature 45℃. The degradation is monitored by measuring IV performance.

3 RESULTS AND DISCUSSIONS

First, treatment-1 was evaluated for mitigating LeTID of p-type mc-Si PERC solar cells. Under industrial mass production condition, the influences of temperature onLeTID of P-type mc-Si PERC cells and RIE+PERC cells were shown in Fig.2. LeTID of the samples increase at first. For higher temperature, a contrary trend sets in and LeTID decrease. Compared with the samples without treatment, the samples which were treated at the optimized temperature show LeTID decrease 0.5% ~ 1%.

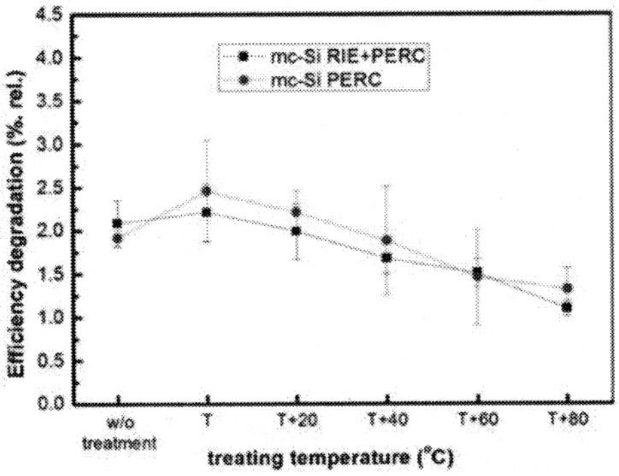

Figure 2: Relative efficiency degradation of P-type mc-Si PERC cells and RIE+PERC cells under light soaking 5kwh/m^2 treated by treatment-1 at different temperature.

Further, treatment-2 was evaluated. As shown in Fig.3, the influences of treating temperature (3a) and time (3b) on the LeTID of P-type mc-Si PERC cells were studied. Under mass production condition, increasing the treating temperature and time is benefit for mitigating the LeTID of the samples. Compared with the samples without treatment, the LeTID decrease ~1% post treating under optimal condition.

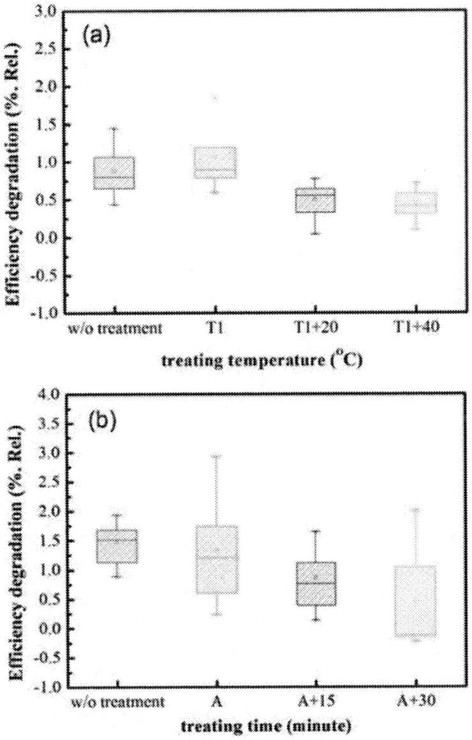

Figure 3: Relative efficiency degradation of P-type mc-Si PERC cells under light soaking 5kwh/m^2 treated by treatment-2 at various temperature (3a) and time (3b).

Several P-type mc-Si PERC cells which were grouped according to cell initial efficiency were treated by treatment-1 and treatment-2 under optimal condition, as shown in Fig.4, LeTID of partly P-type mc-Si PERC solar cells can be mitigated by treatment-1, the LeTID is between 0.5% and 6%, and the average LeTID is about 2%. Treatment-2 looks good for most mc-Si PERC cells, the LeTID can be controlled below 2%, and the average LeTID is about 1% post treating.

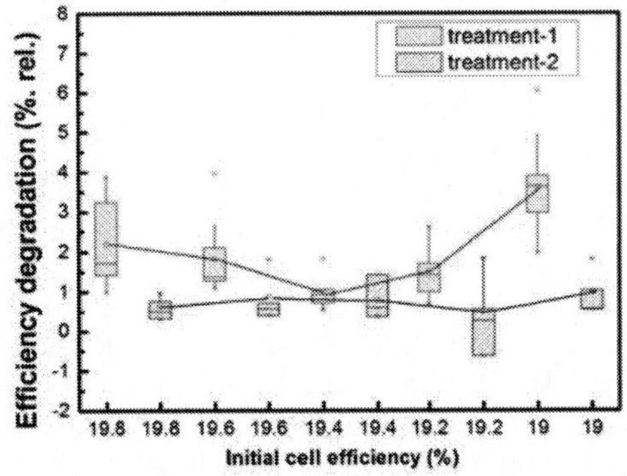

Figure 4: The efficiency degradation of different initial efficiency p-type mc-Si PERC solar cells under light soaking 5kwh/m^2 which regenerated by treatment-1 and treatment-2 under optimal condition.

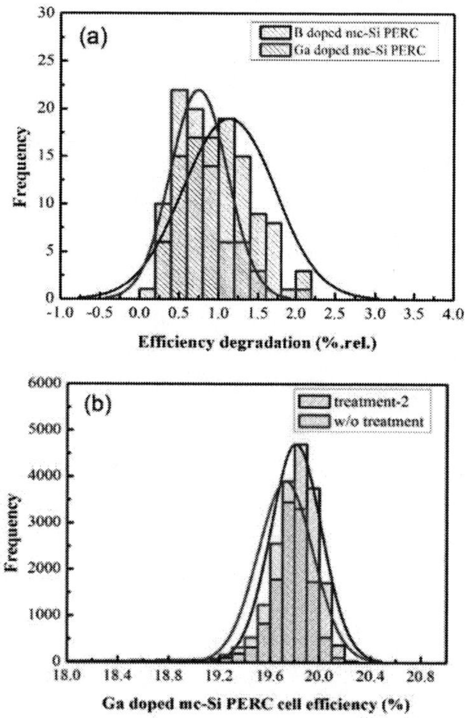

Figure 5: The efficiency degradation of B-doped and Ga-doped mc-Si PERC cells under light soaking 5kwh/m^2 (5a) and efficiency gain of Ga-doped mc-Si PERC cells (5b) treated by treatment-2 under optimal condition.

In this work, treatment-2 have been released to pilot line, Both B-doped and Ga-doped mc-Si PERC cells were treated in mass production. As shown in Fig.5, the LeTID of B-doped and Ga-doped mc-Si PERC cells can be controlled below 2%, and the average LeTID is about 1.1% and 0.8% respectively after treatment (Fig.5a). treatment-2 also result in mc-Si PERC cell efficiency increase of ~0.1% abs..

4 CONCLUSIONS

Under mass production condition, LeTID of partly P-type mc-Si PERC solar cells can be mitigated by treatment-1, the LeTID is between 0.5% and 6%, and average LeTID is about 2%. Treatment-2 looks good for most mc-Si PERC cells, the LeTID can be controlled below 2%, and the average LeTID is about 1% post treating. Treatment-2 also result in mc-Si PERC cell efficiency increase of ~0.1% abs.. Mass production results reveal that treatment-2 is a promising industrial solution for LeTID in p-type mc-Si PERC solar cells.

5 ACKNOWLEDGEMENTS

The authors would like to thank Dr. Zhang and all members of R&D department from GCL System Integration Technology Co., Ltd. MBU.

6 REFERENCES

[1] V.V.Voronkov, R. Falster, Latent complexes of interstitial boron and oxygen dimers as a reason for degradation of silicon-based solar cells, Journal of applied physics 107, 053509 (2010).

[2] Bianca Lim, Karsten Bothe, and Jan Schmidt, Impact of oxygen on the permanent deactivation of boron-oxygen-related recombination centers in crystalline silicon, Journal of applied physics 107, 123707 (2010).

[3] S.W.Glunz et al., Degradation of carrier lifetime in Cz silicon solar cells, Sol. Energy Mater. Sol. Cells 65(2001) 219–229.

[4] S.W.Glunz et al., Comparison of boron- and gallium-doped p-type Czochralski silicon for photovoltaic application, Prog. Photovolt. 7(1999) 463–469.

[5] S.Wilking et al., High speed Regeneration of BO-defects: Improving long-term solar cell performance within seconds, in: Proceedings of the 29th European Photovoltaic Solar Energy Conference and Exhibition, Amsterdam, Netherlands, 2014, pp. 366–372.

[6] K. Ramspeck et al., Light induced degradation of rear passivated mc-Si solar cells, Proceedings of the 27th European Photovoltaic Solar Energy Conference, Frankfurt, Germany, 2012, pp. 861-865.

[7] F. Fertig et al., Light-induced degradation of PECVD aluminium oxide passivated silicon solar cells, Journal of Physical Status Solidi RRL, 1-6 (2014).

[8] F. Kersten et al., Degradation of multicrystalline silicon solar cells and modules after illumination at elevated temperatures, Journal of Solar Energy Materials and Solar Cells 142, 83-86 (2015).

ACCELERATED LIGHT-INDUCED DEGRADATION TEST FOR INDUSTRIAL APPLICATION

Chia-Mei Lin[1,2*], Marcus Glaeser[1,3], Enno Malguth[4], Steffen Uredat[4], Nobert Bernhard[2], Dominik Lausch[1,2]

[1] Fraunhofer Center for Silicon Photovoltaics, Otto-Eißfeldt-Straße 12, 06120, Halle (Saale), Germany
[2] Anhalt University of Applied Science, Bernburger Straße 55, 06366, Köthen, Germany
[3] Martin Luther University, Karl-Freiherr-von-Fritsch-Straße 3, 06114 Halle (Saale), Germany
[4] LayTec, Seesener Str. 10 - 13, 10709 Berlin, Germany
*chia-mei.lin@csp.fraunhofer.de

ABSTRACT: Light-induced degradation (LID) is a well-known problem of industrial silicon solar cells and can lead to severe electrical performance loss through the formation of recombination-active defects during the excess carrier injection by illumination or forward biasing. Thereby an effective LID inspection for industry is essential. Unfortunately, a standard LID test procedure based on the solar cell level is currently missing. Moreover, the maintenance of the traditional degradation source – light is challenging. In order to solve these concerns, a series of experiments are carried out for the development of a reliable and rapid LID test procedure upon a customer friendly LID test machine. Consequently, the commercially available machine, which utilizes current bias instead of light as the degradation source, was developed. Furthermore, four practical experiments which are of great interests to the solar cell suppliers are demonstrated. The high flexibility of operation of this machine was validated

Keywords: LID, LID Scope, boron, oxygen, crystalline silicon solar cells

1 INTRODUCTION

Impurities, such as boron and oxygen, are common in the silicon-based solar cells as boron is a standard p-type dopant and oxygen is inevitable during the Czochralski (Cz)-crystallization processing. Those impurities are known to form recombination-active defects in boron-doped Cz solar cells during illumination or excess carrier injection in the dark, which is known as boron-oxygen related light-induced degradation (BO-LID) [1, 2]. The negative impact of LID on the efficiency of solar cells is unavoidable and can cause up to 10 % efficiency loss (rel.) according to its inherent characteristics. Consequently, the degradation will lead to severe financial losses among owners of solar plants. Therefore, it is of great importance for the solar industry to inspect and control the LID level regularly and precisely.

In 2006 Herguth et al. proposed a three-state model (i.e. annealed (state A), degraded (state B) and regenerated state (state C)) regarding to the BO-LID [3]. In this model, it is reported that the degraded cell performance can be recovered by using prolonged carrier injection at an elevated temperature. The reaction rate of the degradation (i.e. A→B) and regeneration (i.e. B→C) process has been shown to increase with injection carrier density and temperature. The influence of temperature on the velocity can be easily explained through the Arrhenius equation (see equation (1)) [4], as three states present the formation and dissociation of chemical complexes and thus can be described with a kinetic model.

$$\upsilon = \upsilon_{ch} \exp(-\frac{E_A}{K_B T}) \quad (1)$$

In the equation (1), υ is the defect conversion rate between different states, υ_{ch} is the pre-exponential factor, depending crucially on the (exess) carrier density when the activation energy is assumed to be both temperature- and time-independent, E_A is the activation energy, K_B is the Boltzmann constant and T is the temperature. Based on this equation one can see that higher temperature leads to higher kinetic energy and allowing more collisions between molecules, in consequence, the degradation and regeneration reaction rates are accelerated. An increase in excess carrier concentration can cause the LID effect and influence the regeneration rate due to a shift in the Fermi level and a change of the charge states of hydrogen [5] respectively.

Apart from the BO-LID in Cz solar cells, a significant performance degradation due to injection carrier and temperature has been also recently observed on multicrystalline PERC cells, namely LeTID (Light and elevated Temperature Induced Degradation) [6].

According to the aforementioned explanations, it implies that well-controlled temperature and the injected charge carrier density are essential for the investigation or monitoring LID effect. However, it is challenging practically due to the following reasons:

- Maintenance of the traditional degradation source- light
  - Lamp aging
  - Spectrum of light
  - Homogeneity of light exposure over the cell surface
  - Constant carrier injection level during the LID process
- Stability of temperature - the temperature has to be controlled in regimes of 1 K as it has a dramatic influence on the V_{oc}, which reflects the LID level.

Further, the common LID test method based on the module level is costly and time consuming. In order to eliminate the concerns described above, the main subject of this work is to establish a reliable and an accelerated LID test method, which enables us to observe the kinetic behavior of LID effect and to obtain the maximum degradation level simply and precisely based on a customer friendly LID measurement instrument.

2 DEVELOPMENT OF A MACHINE FOR ACCELERATED LID TEST

To achieve our goal, several questions have been made and the corresponding answers are provided by a comprehensive experimental evaluation through various test conditions and facilities.

2.1 How to inject carriers?

The charge carrier density is one of the key dominators for LID effect and can be produced through illumination or current bias. The LID effect is observed as a result of different chemical processes, namely degradation and regeneration, within the solar cell. Those processes are induced and triggered by a shift in the Fermi level due to an increase in excess carrier concentration. In this work, we proposed to utilize a forward bias as the degradation source, so-called current-induced degradation (CID) in accordance to reference [5], which verified the equivalence of CID compared to LID. The carriers can be produced exactly and reproducibly through a forward current via an inexpensive voltage source. Also, the electrical degradation system is supposed to be more robust and less maintenance required when compared to the light source.

2.2 Which electrical parameter should be monitored?

V_{oc} is the common and typical item for LID investigation. This is because the changes of lifetime result in the change of J_o (saturation current) and thus lead to the changes in V_{oc} as shown in equation (2), where n is the ideality factor, q/K_BT is the thermal voltage and J_{sc} is the photo-generated current density.

$$V_{OC} = \frac{nK_BT}{q}\ln(\frac{J_{SC}}{J_0}+1) \qquad (2)$$

However, the presence of defect centers (i.e. B-O complexes) in the bulk can also lead to photo-generated current density losses as the light-generated carriers are recombined and eliminated before reaching the p-n junction.

To determine an appropriate parameter for LID monitoring, V_{oc} and J_{sc} of a solar cell degraded at 140 °C and 1 sun were measured as shown in Fig. 1. The cell is annealed in darkness at 200 °C for 30 minutes in order to deactivate the B-O complexes. During the light-soaking, the solar cell is placed on a hot plate with a fixed temperature of 140 °C. After various periods of light-soaking time, the cell is removed from the light-soaking station and tested via a solar simulator under the standard test condition (STC, 25 °C, 1000 W/m²). It is observed that both parameters have the same tendency and followed the three-state model. Initially, the values of the parameters gradually decrease, after light-soaking for appropriately 15 minutes, the regeneration process occurs and a contrary trend sets in and the values increase. Although both parameters are apparently affected by LID, a special attention must be given to the strongly high fluctuation in J_{sc}.

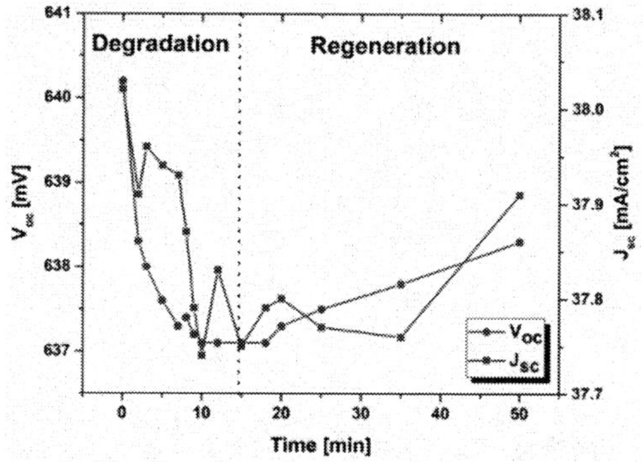

Figure 1: The time-dependent degradation-regeneration behavior of V_{oc} and J_{sc}. The cell is degraded by 140 °C and 1 sun.

To investigate the high fluctuation in J_{sc}, a repeatability test of solar simulator is performed concerning contact issue. Fig. 2 displays the variation comparison of V_{oc} and J_{sc} via a repeatability measurement. In the test sequence, a cell is removed from the test stage after each measurement and is placed and retested on the test stage. This cell is tested under STC for 10 times in total. The variation of each parameter is calculated as shown in equation (3),

$$variation\ [\%] = \frac{Voc\ (N{=}X) - Voc\ (N{=}1)}{Voc\ (N{=}1)} * 100 \qquad (3)$$

where N is the number of measuring times.

The result reveals that J_{sc} has a comparatively larger variation range and variation average compared to V_{oc}. The higher sensitivity to metal contact during the I-V measurement results in a high fluctuated J_{sc} is highly suspected. Based on the analysis above, it can be concluded that V_{oc} is an appropriate parameter for LID monitoring.

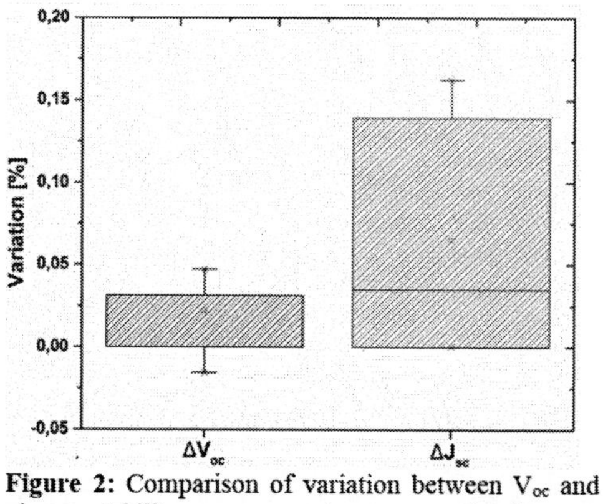

Figure 2: Comparison of variation between V_{oc} and J_{sc} via repeatability measurement over a sun simulator.

2.3 How to enhance the sensitivity of LID-induced ΔV_{oc}?

V_{oc} is temperature-dependent and ΔV_{oc} [%] is reversely proportional to the degradation temperature [4]. In order to avoid data distortion in the high temperature regime, such as unstable degradation temperature during LID process, it is crucial to employ a test condition which raises the sensitivity of ΔV_{oc} to LID process. For a better comparison of the degradation levels between different cells, instead of using the original V_{oc} values, ΔV_{oc} [%] is commonly employed and can be calculate by using equation (4),

$$\Delta Voc\,(t = x)\,[\%] = \frac{Voc\,(t{=}x) - Voc\,(t{=}0)}{Voc\,(t{=}0)} * 100 \qquad (4)$$

where t is the degradation time.

Fig. 3 shows the in-situ monitored ΔV_{oc} with various irradiances of a solar cell during the LID process with 100 °C and 1 sun. It is observed that the degradation level increases as the irradiance decreases. This is because in the lower injection level the cell has a comparatively larger effective lifetime drop than in the high injection level with regard to the interaction between a fixed amount of B-O complexes and a varied amount of free carriers produced by illumination. As a result, the lowest irradiance intensity of 0.1 suns was determined as the standard test condition for LID inspection.

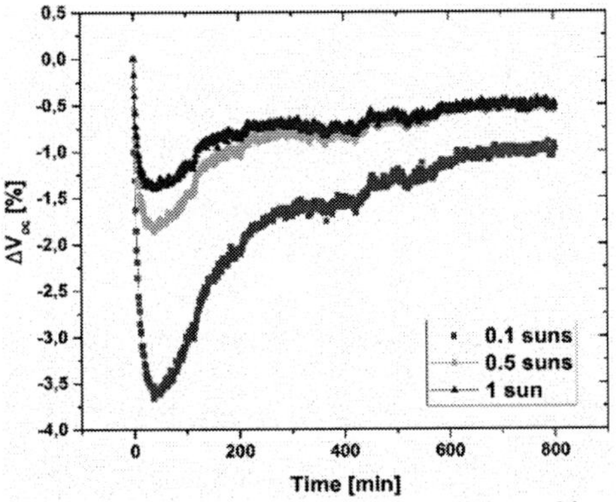

Figure 3: LID-induced ΔV_{oc} over light-soaking time measured under various irradiances during the LID process.

3 INDUSTRIAL APPLICATIONS OF LID SCOPE

3.1 Introduction of LID Scope

Based on the results shown in the section 2, the commercially available machine "LID Scope" (see Fig. 4), which fulfilled our requirements, was developed and manufactured by Laytec AG and Fraunhofer CSP. It uses electrical current as degradation source and a mounted hotplate to achieve various degradation conditions with a high accuracy and repeatability. In addition, it is easy operated and has the function of in-situ monitoring time-dependent V_{oc}, which enables us to analyze data immediately and represents a substantial cost and time saving when compared with ex-situ measurement executed by an operator.

Figure 4: Photo of LID Scope [7]

3.2 Influence of temperature on LID tests

Fig. 5 displays in-situ measured V_{oc} data of four boron-doped Cz Al-BSF halved solar cells as a function of degradation time at various temperatures and at 1 sun (equivalent to 4.7 A current bias). It can be seen that different temperatures result in a different V_{oc} level and that the V_{oc} level is inversely proportional to the degradation temperature. This is because V_{oc} is strongly temperature-dependent; moreover, these V_{oc} curves are in-situ measured. Besides, a significant decrease in the duration of the LID process as well as the time required for reaching the maximum degradation level with rising temperature is observed, which can be comprehend by means of Arrhenius equation as mentioned in the introduction. A lower degradation level and a shorter LID process duration can be achieved once the degradation and regeneration reaction rates are accelerated as well as the ratio of the time constants of degradation and regeneration process is changed [3]. Consequently, a reduction of processing time of 98.5 % can be achieved when we compare the time consumption at the maximum degradation level of the temperatures of 80 °C and 160 °C. However, one should note that with higher temperature also the degradation degree is lower. Consequently one should calculate back the V_{oc} value at standard test conditions by using the empirical data. Based on the promising results observed, an industrial implementation of LID Scope, which offers a rapid and efficient examination of LID level, should be the right approach.

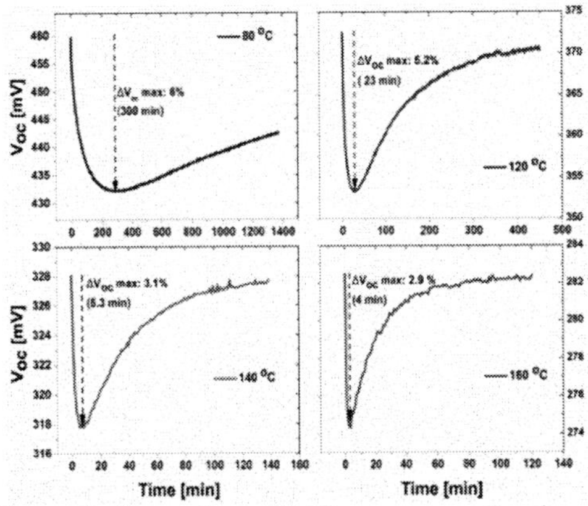

Figure 5: Comparison of LID curves as a function of time under various degradation temperatures at 1 sun (approximately 4.7 A).

3.3 LID test on different material and technology

Fig. 6 depicts in-situ measured V_{oc} data of the p-type solar cells with different material (i.e. mono and multicrystalline) and technology (i.e. Al-BSF and PERC). Special noted that these samples are from different solar cell suppliers. It can be observed that these solar cells have similar degradation-regeneration behavior but very different time span. A maximum degradation level of monocrystalline cells is observed after 5 minutes whereas the multicrystalline cell spent more than 100 minutes. Additionally, a degradation process of the monocrystalline cells is completed within 2 hours whereas it takes approximately more than 30 hours upon the multicrystalline cell. This can be explained by a result of different physical mechanisms. Unlike monocrystalline cells are known to suffer from LID mainly by B-O complexes, the mechanism of multicrystalline cells is much more complicated and cannot be simply explained by the best two known mechanisms: the activation of B-O complexes and the dissociation of Fe-B pairs. In multicrystalline solar cells, another two significant LID effects, sponge LID [8] and LeTID [6] are recently identified. It is even announced that LeTID can be more severe than B-O related degradation in p-type Cz solar cells and is more pronounced on PERC cells than on BSF cells. Interestingly, LeTID can be recovered through illumination and heat treatment as BO-LID, but the processing occurs on much longer time scale. The root cause of this degradation is still under investigation.

In addition, the LID in PERC cells is observed to be more significant than in standard BSF as often reported. This is primarily due to a stronger dependence of efficiency on the bulk lifetime in high efficiency solar cell structures. The purpose of PERC cells is to reduce the recombination on the surface of solar cells. But after LID, the diffusion length of the free carriers is so low that the many carriers cannot even reach the surface, thereby, the impact level of LID on the PERC is comparatively higher than on the BSF cells. To conclude, LID Scope is applicable for different types of solar cells and technologies once these are LID-prone.

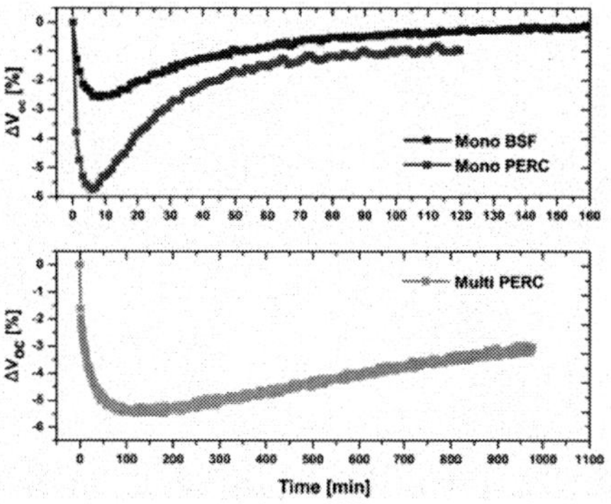

Figure 6: V_{oc} curves of three different types of solar cells during degradation process at 140 °C and 1 sun.

3.4 LID test based on the mini module level

Illuminating a mini module is a typical way to simulate the LID effect under field condition in a laboratory. Fig. 7 shows in-situ measured V_{oc} curves of a solar cell and a mini module at 45 °C and 1 sun. It is noted that neighbor cells are used for the comparison. As can be seen, the tendency of those curves is nearly identical. The difference in the initial V_{oc} values is caused by the difference in the temperature of cell itself. The solar cell assembled in the module received less heat when compare to the single cell under the same thermal setting and thus has higher initial V_{oc} value. To realize the measurement of mini module, the rear side ribbons need to be attached into hot plat through a copper conductive tape. Two front side ribbons are used for the injection of bias current and V_{oc} measurement respectively. Based on this result, the applicability of module level measurement through LID Scope is confirmed.

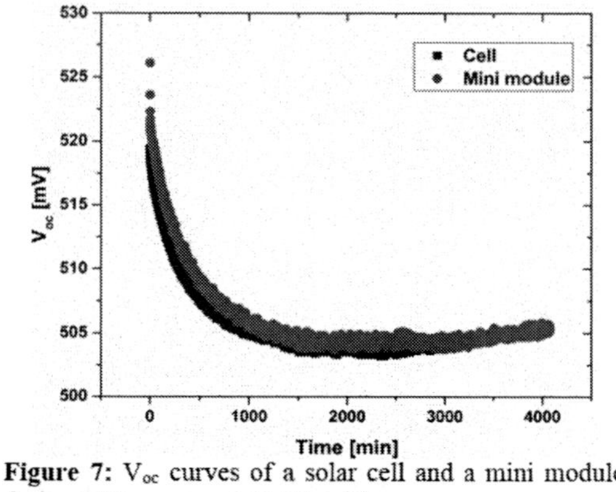

Figure 7: V_{oc} curves of a solar cell and a mini module during LID process at 45 °C and 1 sun.

3.5 Estimation of LID-induced ΔEta

Section 2.2 reveals that V_{oc} could be adopted for monitoring the LID process in a rapid test with respect to a comparatively higher stability during electrical measurement. However, the LID-induced V_{oc} loss is difficult to employ directly into the calculation of the LID-induced loss of a solar system, as well as for companies to compare their competitiveness with others. Accordingly, we attempt to discover a procedure on the base of ΔV_{oc} measurement but it also enables the prediction of the LID-induced ΔEta without building up a complicated measurement system. Fig. 8 displays the correlation between LID-induced ΔV_{oc} and LID-induced ΔEta. Eta and V_{oc} are measured under 1 sun and 0.1 suns through solar cell tester and LID Scope respectively. It is apparent that the correlation is strong, as the correlation coefficient is 0.92, although a slightly higher variation of ΔEta corresponding to the same ΔV_{oc} has been observed. In accordance with the results, we can conclude that our approach is applicable likely within a running production for estimating LID-induced ΔEta even at different temperature regimes.

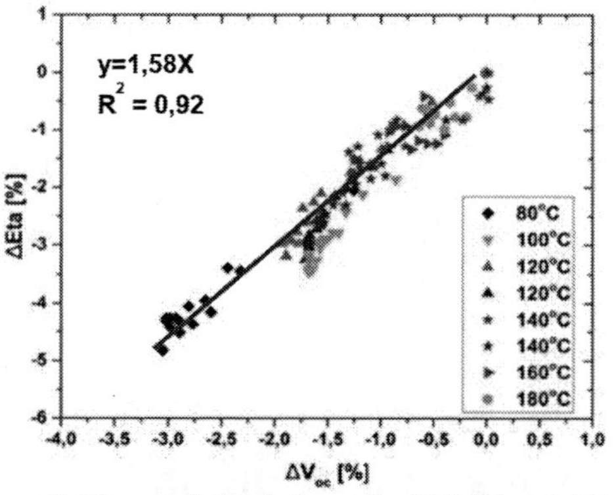

Figure 8: The correlation between the LID-induced ΔV_{oc} and the LID-induced ΔEta of boron-doped Cz Al-BSF solar cells degraded under various degradation temperatures and 1 sun. Eta and V_{oc} are measured under 1 sun and 0.1 suns respectively.

4 CONCLUSIONS

In this work, a reproducible and reliable LID process on solar cell level based on electrical charge carrier injection for mass production is established. A series of experiments have been carried out in order to find the most reasonable and feasible parameters for a customer friendly LID test procedure. Based on these results the LID Scope was developed:

- Utilization of current bias as the degradation source for a stable, user-friendly and reproducible charge carrier injection.
- Integration of a hotplate with automatic control for precise temperature (1 K) from 25 °C to 160 °C.
- In-situ V_{oc} measurement under 0.1 suns.

The proposed test procedure offers various benefits:

- The time needed for reaching the maximum degradation level can be saved by up to 98 % through 160 °C.
- All types of solar cell technology and material can be tested.
- Mini module level-LID test.
- LID-induced ΔEta can be estimated.
- Lower cost for quality management: no module construction is necessary
- Quality control with short response time.
- Easy to use (one click operation)
- More satisfied customers due to trustable quality control.

To conclude, the commercially available machine "LID Scope" is a good alternative for the reliable and accelerated inspection of LID effect on the solar cell level and can be applied for standard test in the solar industry in regard to its simplicity in maintenance, accuracy in measurement and flexibility in operation. Further, a SEMI standard will be offered within the next month to smooth cooperation and communication between partners.

5 REFERENCES

[1] S. W. Glunz et al., Degradation of carrier lifetime in Cz silicon solar cells, Solar Energy Material & Solar Cells, Bd. 65, pp. 219-229, 2001.
[2] K. Bothe und J. Schmidt, Electronically activated boron-oxygen-related recombination centers in crystalline silicon, Journal of applied physics, Bd. 99, p. 013701, 2006.
[3] A. Herguth et al., A new approach to prevent the negative impact of the metastable defect in boron-doped Cz silicon solar cells, in IEEE, USA, 2006.
[4] A. Herguth and G. Hahn, Kinetics of the boron-oxygen related defect in theory and experiment, Journal of Applied Physics, Bd. 108, p. 114509, 2010.
[5] M. Gläser and D. Lausch, Towards a Quantitative Model for BO Regeneration by Means of Charge State Control of Hydrogen, Energy Procedia, Bd. 77, pp. 592-598, 2015.
[6] F. Kersten et al, A new light induced volume degradation effect of mc-Si solar cells and modules, in 31th EUPVSEC, Hamburg, 2015.
[7] http://www.laytec.de/lidscope/
[8] K. Petter et al. 2013, Light induced degradation of solar cells made from high performance multi wafers, in 29th EUPVSEC, Amsterdam, The Netherlands, pp. 551–554.

Comparison of Industrial Solutions to Light Induced Degradation of High Efficiency Cz PERC Cells

Hui Li, Jian Xu, Ke Chen, Hua Fan, Shiling Ma,Chao Yu, Qiang Xu, Xinxiao Ruan
R&D Department, Dongfang Huansheng Photovoltaic (Jiangsu) Co. , Ltd
No. 20, Wenzhuang Road, Economic Development Zone, Yixing, JiangSu, China, 214203. hui.li@dzssolar.com

ABSTRACT: Several industrial solutions to mitigate light induced degradation (LID) is reported for P-type Cz PERC solar cells. The comparison of different ways for regeneration were investigated using P-type Cz PERC solar cells which were fabricated with a bulk resistivity ranging from 0.25 Ω.cm to 3Ω.cm. Illumination with halogen lamp, high-powered LED and laser at elevated temperature were investigated in this paper. Under industrial mass production condition, it is shown that illumination annealing with elevated temperature will be the effective method to mitigate the light induced degradation. Moreover, current injection annealing under industrial mass production condition was also studied and the anti-LID results were compared with that of illuminated annealing. For 2~3Ω.cm P-type PERC cells, the relative efficiency degradation ratio is no more than 1% after treating by industrial current injection annealing tool and the result is no more than 1.5% comparable with that was treated by industrial laser illumination annealing and high-powered LED. For 0.25~0.75Ω.cm P-type PERC cells, current injection annealing is superior to laser illumination annealing , and the relative efficiency degradation ratio is less than 1.5%, the results is the same as 2-3Ω.cm. As the results shown above, it is consider that current injection and laser illumination annealing are the potential industrial solutions to mitigate LID of P-type PERC cells.

Keywords: PERC solar cells, light induced degradation, illumination annealing, hydrogen passivation

1 INTRODUCTION

Light-induced degradation (LID) occurs in both Cz and multicrystalline p-type silicon is one of the biggest challenges currently confronted by the PV industry, especially when the high efficiency PERC solar cells becoming the most promising technology nowadays. The LID in p-type Cz Al-BSF solar cells may be 3–4% [1], compared to the higher degradation up to 4–6% for PERC cells[2], and this value may reach to 10% [3].There are several solutions to mitigate the LID, either by selecting source material to avoid LID, or by mitigating B–O-related LID during cell fabrication[4].

Carrier injection either by light or by current can be used to reduce LID during cell manufacturing for mono PERC cells, as well as mc-Si PERC cells. Illuminated annealing appears to be the most popular approch to mitigate LID, and current injection can be used as an alternative method to provide the carrier injection. Both approaches may have strengths and weaknesses in their suitability for different applications. Franziska Wolny has investigated the LID of PERC cells after illumination annealing with a base resistivity ranging from 0.8Ω.cm to 3.4Ω.cm, it show that the duration of a regeneration procedure must be adjusted according to the base resistivity of the cells[5].

In this work, four different kinds of industrial tools were compared systematically under mass production condition for the first time, and the effective industrial solution was proposed for mitigating LID in mass production.

2 EXPRIMENTS

PERC solar cells were fabricated on 156×156 mm^2 p-type CZ Si wafers of 180um thickness, with a bulk resistivity ranging from 0.25Ω.cm to 0.75Ω.cm. and from 2Ω.cm to 3Ω.cm. The PERC cells were measured IV performance before regeneration, then treated by industrial tools under different conditions, including halogen lamp illumination annealing, high-powered LED illumination annealing, laser illumination annealing and current injection annealing. All cells are subsequently degraded under light exposure 20kwh, 40kwh, and 60kwh at room temperature. The temporal development of the degradation is monitored by measuring the cell IV performance.

3 RESULTS AND DISCUSSIONS

3.1 Regeneration

For illuminated annealing, including halogen lamp illumination annealing, high-powered LED illumination annealing, and laser illumination annealing, light intensity, temperature and time is the key parameters for monitoring regeneration effect. But for current injection, current intensity, temperature and time is the key factor. TableI shows the two different conditions for each regeneration treatment method by using wafers with bulk resistivity of 2-3Ω.cm. And the optimum condition is used to regenerate wafers with bulk resistivity of 0.25-0.75Ω.cm.

Table I: Different regeneration methods for the LID investigations of PERC cells with two group bulk resistivity

Bulk resistivity	Halogen lamp	LED	Laser	Current injection
2-3Ω.cm	C1	C1	C1	C1
	C2	C2	C2	C2
0.25-.75Ω.cm	C2	C2	C1	C1

3.2 Regeneration and LID of PERC cells with bulk resistivity of 2-3Ω.cm

Fig.1 shows the relative decrease in cell efficiency of regenerated Cz PERC cells with bulk resistivity range from 2Ω.cm to 3Ω.cm using halogen lamp illumination, high-powered LED illumination, laser illumination and current injection as regeneration ways. All cells are accepted light exposure duration from 20kwh to 60kwh. After annealing treatment, all the cells has 0.02-0.05%

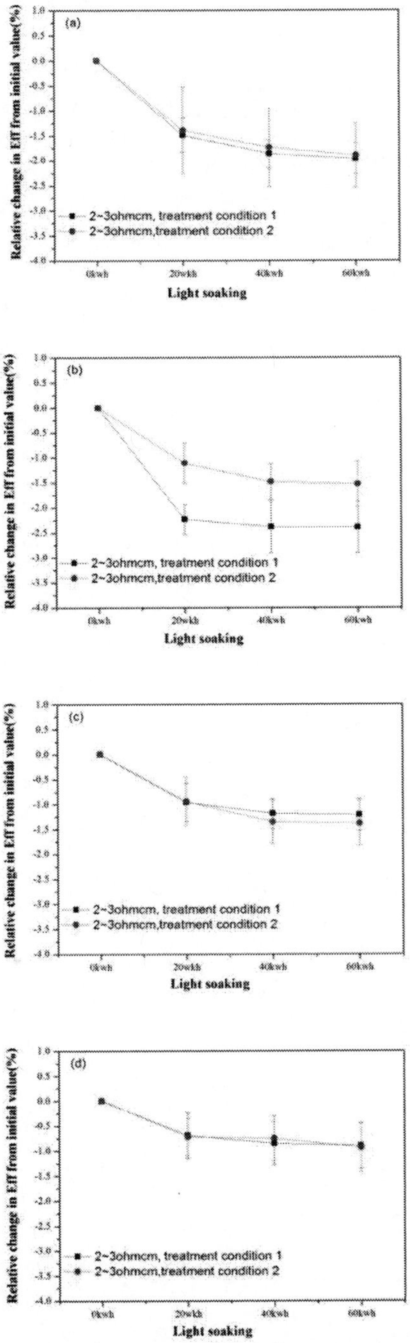

Figure 1: Relative change in efficiency from initial value after regeneration of PERC cells with bulk resistivity of 2-3Ω.cm that are regenerated with two different treatment conditions (a) halogen lamp illumination annealing. (b) LED illumination annealing. (c) laser illumination annealing. (d) current injection annealing.

gain in efficiency, except for halogen lamp illumination regenerated cells which has nearly 0.05% drop in efficiency, and the value can be higher for certain cells. By comparison under increasing light exposure, efficiency has great decrease at 20kWh, and when expused to 40kWh, the degradation tend to be saturated.

As shown in Fig.1, condition 2 is the preferable process for halogen lamp and LED illumination annealing, compared to laser illumination and current injection annealing. From Fig.1(a)-(d), efficiency degrades no more than 1% treating by current injection, and that is less than1.5% by laser and LED illumination annealing after 60kWh light exposure. But the value reaches to 2% for halogen lamp illumination.

Above calculations make the regenerated efficiency as the initial degradation data without consideration the efficiency increase caused by regeneration. Fig.2 shows the relative change in efficiency from initial value after firing. After 60kWh ilumination, the efficiency of cells degrades the same as the Fig1, but for halogen lamp illumination annealing, the relative efficiency degrades more than 2% mainly for the loss after treatment. This loss may caused by the lower light intensity which is less than 10suns for treating.

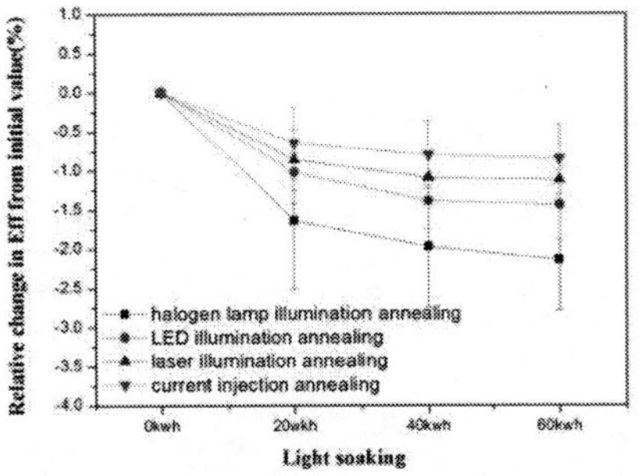

Figure 2: Relative change in efficiency from initial value after firing of PERC cells with bulk resistivity of 2-3Ω.cm that are regenerated with the optimum condition

3.3 Regeneration and LID of PERC cells with bulk resistivity of 0.25-0.75Ω.cm

In order to investigate the effect of the wafer on the regeneration, wafers with bulk resistivity of 0.25-0.75Ω.cm has been treated using different regeneration methods. After treatment, the efficiency of cells has a certain increase of more than 0.05% for current injection

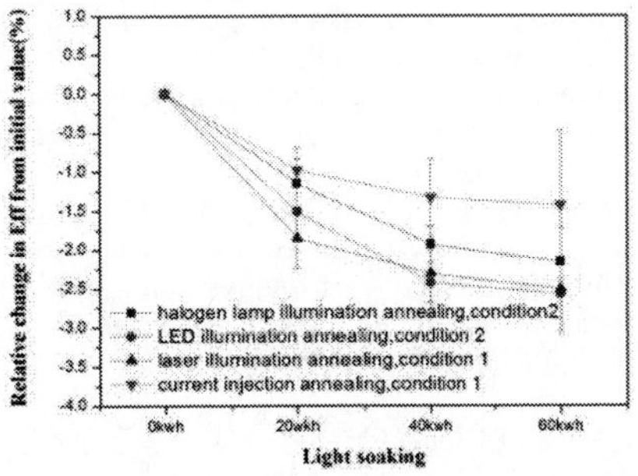

Figure 3: Relative change in efficiency from initial value after regeneration of PERC cells with bulk resistivity of 0.25-0.75Ω.cm that are regenerated with the optimum condition

and LED illunmination annealing, the gain reached 0.14% for laser illumination annealing, but for halogen lamp illumination annealing, the efficiency has dropped 0.04%.

As shown in Fig.3, the relative change in efficiency is no more than 1.5% treating by current injection and 2.5% by other three treatment. However, degradation by halogen lamp illumination annealing seems much lower.

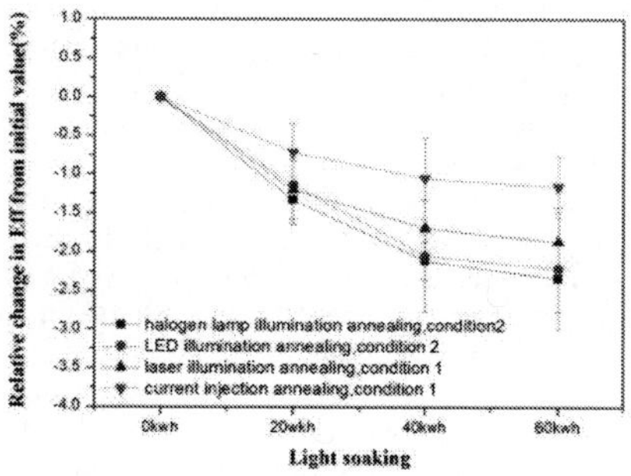

Figure 4: Relative change in efficiency from initial value after firing of PERC cells with bulk resistivity of 0.25-0.75Ω.cm that are regenerated with the optimum condition

In the case that the LID is calculated by making the data after firing as the starting point of degradation as shown in Fig.4.The efficiency drop improves to be less than 2% after laser illumination annealing mainly because 0.14% gain in efficiency after treatment. Meanwhile, the value has a little increase for the cells treating by LED illumination annealing. But, the cells with lower bulk resistivity does not seem to benefit from the halogen lamp regeneration, which shows an efficiency drop of 0.04%, this lead to nearly 0.2% degradation in relative efficiency.

4 CONCLUSIONS

Four different industrial methods for mitigating LID of P-type PERC cells are investigated, including halogen lamp illumination annealing, high-powered LED illumination annealing, laser illumination annealing and current injection annealing. It has shows different performance with the change of bulk resistivity.

For 2~3Ω.cm P-type PERC cells, the relative efficiency degradation ratio is no more than 1% after treating by industrial current injection annealing tool and the result is no more than 1.5% comparable with that was treated by industrial laser illumination annealing and high-powered LED. For lower bulk resistivity wafers, such as 0.25~0.75Ω.cm P-type PERC cells, current injection annealing is superior to laser illumination annealing, and the relative efficiency degradation ratio is less than 1.5%. As the results shown above, it is consider that current injection and laser illumination annealing are the potential industrial solutions to mitigate LID of P-type PERC cells in mass production.

5 ACKNOWLEDGEMENTS

The authors would like to thank Dr. Xinxiao Ruan from Sunpower for the fruitful disscussion and all members of R&D department from Dongfang Huansheng Photovoltaic Co. .

6 REFERENCES

[1] Eunhwan Cho, J.-H. Lai, Young-woo Ok, etal., IEEE 42nd PVSC, USA, June 2015, pp.1-4.

[2] K. Lee, M.-S. Kim, J.-K. Lim etal., 31st EU PVSEC, Germany, 2015, p.1835-1837

[3] J. Knobloch, S.W. Glunz, D. Biro, etal., 25th IEEE Photovoltaic Specialists Conference, USA, June 1996, pp.405-408

[4] Brett Hallam, Catherine Chan, David Payne, etal., Photovoltaics International, Vol.33, Oct 2016, 37-46

[5] Franziska Wolny, Torsten Weber, Matthias Müller, et al., Energy Procedia, 2013, 523-530.

SIMULATION OF THE THERMAL LASER SEPARATION PROCESS (TLS) IN RELATION TO THE CRACK PROPAGATION AT THE WAFER EDGE.

J. Röth[1], C. Belgardt[2], M. Grimm[2], N. Bernhard[1]
[1] Anhalt University of Applied Sciences, Bernburger Str. 55, 06366 Köthen, Germany
[2] 3D-Micromac AG, Technologie-Campus 8, 09126 Chemnitz, Germany

ABSTRACT: Thermal Laser Separation (TLS) is a kerfless wafer dicing technology. This method uses thermal stress to propagate a crack through the substrate. An often noticed vulnerability of this procedure is a crack deviation at the end of the wafer [1]. This study shows a transient opto-thermal FE simulation model. The model allows a prediction of the cracking tip. Thus it is possible to identify the reason for the crack deviations. This approach allows a comparison of different parameters at the wafer edge. Different velocity profiles, beam shapes, laser and cooling powers were manipulated and these calculated results were compared with the experimental measurements of the crack deviation at the wafer edge.
Keywords: Thermal Laser Separation (TLS), Crack Propagation, Dicing, Half Cells

1 INTRODUCTION

TLS is a dicing method which can be used for 1/n-cell production. The greatest advantage of the thermal separation is the superior wafer strength in comparison to other PV dicing methods [2]. This technology has been under development for several years. A vulnerability of the use of TLS is a possible crack deviation at the end of the wafer [1]. This disadvantage can be named as a reason for the low industrial use of TLS. The following study uses an opto-thermal finite volume model to simulate the heat distribution. Thus it is possible to locate the cracking tip and elaborate its behavior next to the wafer edge. The goal of this work is the understanding, describing and reduction of this disadvantage.
The developed model uses the following approach. First, the optical setting of the 3D-Micromac microDice™ research and development tool is calculated, which result in a distinct Gaussian beam shape. The intensity of this beam shape is integrated for each finite volume element of the wafer then. It is to allow the calculation of the temperature dependent absorption including a several time back reflection. After that, the model for the cooling nozzle was adding and the heat conductivity for each time step is simulated. This virtual arrangement is moving over the wafer surface, and the heat distribution can be observed. The total differential of the heat distribution identifies the heavily stressed areas whereby the crack arises. The whole model was validated qualitative by thermographic imaging afterwards. The design is suitable for several process parameter manipulations. These are different velocities profiles, beam shapes, laser and cooling power levels.

2 THEORY AND SIMULATION

2.1 Simulation model and its validation

The following description of the simulation via finite element method was developed with Matlab. To ensure an as accurate as possible translation of the given laser beam (CW 200 W, 1070 nm); a particular beam camera (ophir spiricon) was used. With this setup, the parameter dependent $1/e^2$ dimensions were recorded, and a two-dimensional Gaussian distribution was fitted afterwards. This procedure allows the integration of the intensity for every finite element which results in a high accuracy.
Subsequently, the temperature dependent absorption for each cube and time step was calculated using the Fresnel equations. In addition to this, a several time back reflection (till the change in value < 1%) was added.
With that information, the calculation of the change in the heat for each cube depending on the previous temperature and the time step can be executed. This calculation results in a heat distribution. To this distribution, the model for the process cooling is adding to it. The model uses forced convection on the wafer surface. The heat transfer was assumed according to the literature [5]. Last, the temperature dependent heat conduction between each side of the cube for each time step is added.
The described simulation model results in a transient temperature distribution of the TLS process. To validate the simulated data, two validation steps were carried out.
In the first step, thermographic images from the real process were compared with the simulated temperature profile. The following pictures show such a comparison. Unfortunately, the emissivity of silicon is highly dependent on the temperature. That's why the thermographic image is recorded with an emissivity of one, and the comparison of the different absolute temperatures is not possible. Fortunately this qualitative adjustment also works very fine.

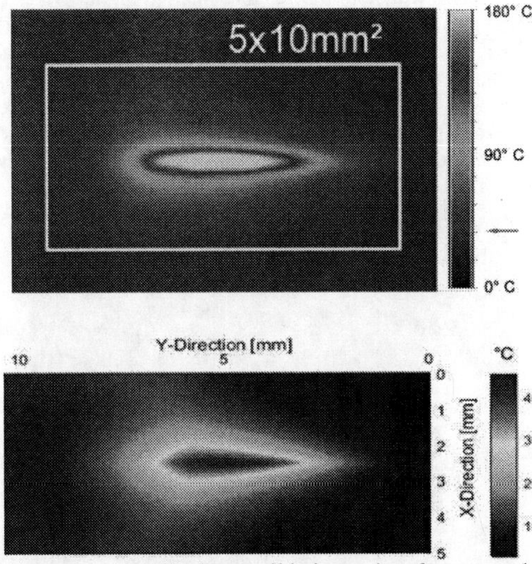

Figure 1: qualitative validation via thermographic imaging (top) and simulation (bottom). Process direction: from the left to the right.

In the second validation step, the position of the cracking tip of a non-fully cleaved sample is measured and compared with the most stressed area of the simulation. It can be assumed that the crack will arise nearby this stress peak [4]. As an indicator for the most stressed area, the maximum negative temperature gradient is chosen. The comparison of these two positions is illustrated in figure 2. It can be pointed out that the cracking tip from the experiment is slightly behind the most stressed area. This difference can be caused by different reasons. First, the cracking tip is hard to detect and second the real cooling is not as homogeneous as in the model for example. However the more important reason is that a crack is opened by tension stress. This stress needs to be in front of the cracking tip to propagate the crack further [3]. In the described simulation model only the position of the tension stress is calculated and an offset for the exact cracking tip is needed. Nevertheless, the difference is not that huge, and with the knowledge of the existence of an offset, the model can be described as useful.

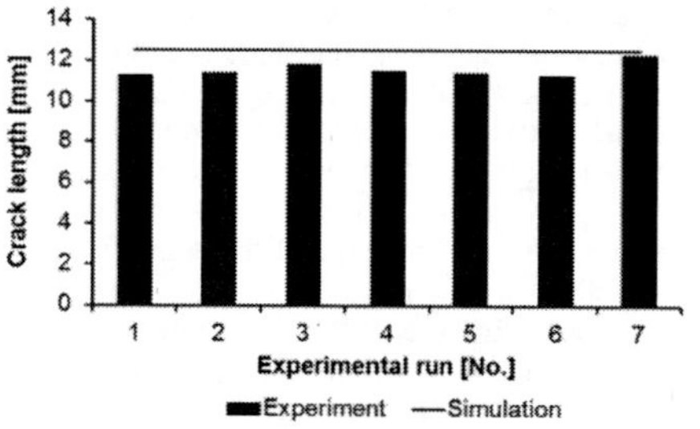

Figure 2: Comparison of the simulated and measured crack lengths. The experimental lengths are slightly behind the simulation.

2.2 Observed phenomenon

This work is focusing on the behavior of the crack around the second wafer edge, at the end of the TLS process. An often recognized phenomenon is a deviation at the end, which forms an unwanted triangle. Such a defect is illustrated in figure 3. A microscope image of such a fracture edge can see it in fig. 4. The process direction is from right to left. The very right mirroring surface is typical for the TLS process. It is remarkable that the pattern changes instantly. Such a change conditioned a suddenly changing stress field [3]. This behavior is unlikely during a continuous process.

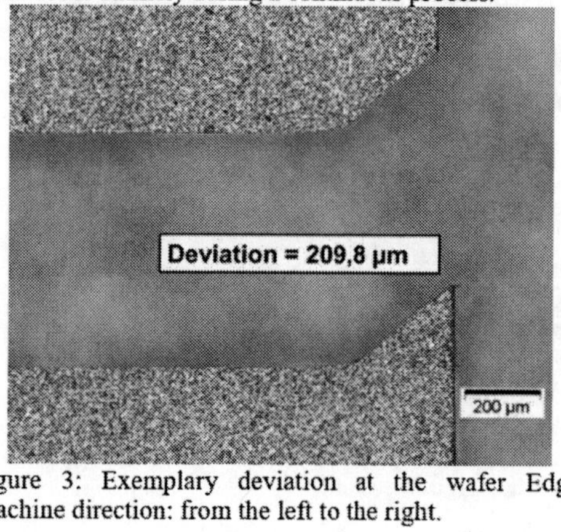

Figure 3: Exemplary deviation at the wafer Edge. Machine direction: from the left to the right.

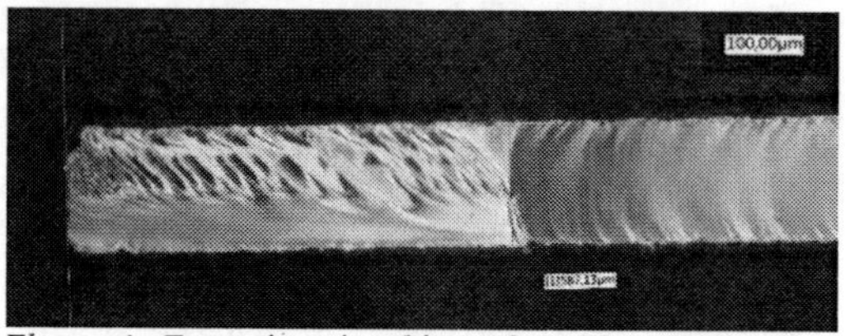

Figure 4: Exemplary breaking edge. The TLS process stops and its cracking path changes abruptly. Process direction: from the right to the left. Picture does not fit to figure 3.

To improve the understanding of the process behavior, the whole process was simulated on a 5 mm by 15 mm sample. During this simulation the position of the highest negative gradient is determined. The use of this value is possible, because of the direct proportionality of the thermal stress level to the temperature gradient [4]. In figure 5 the result of this simulation is illustrated. At the beginning the gradient arises and stabilizes after approximately one mm. At the end of the process the gradient rises again due to a heat accumulation at the wafer edge. Before this negative gradient reaches the wafer edge it collapses and cannot propagate the crack any further. That's why the crack growth stops and the wafer is not fully separated. But the wafer is very fragile now and can be cleaved very easily. This happens during the handling process afterwards and is responsible for the suddenly changing stress field. The separation follows the preferred crystal orientation along the <111> plane and causes the characteristic triangle.

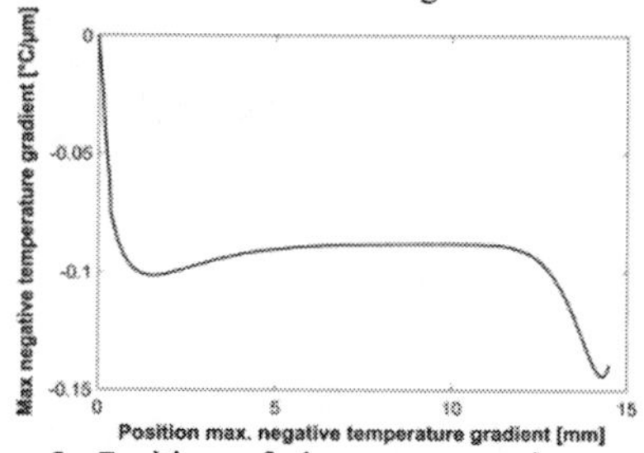

Figure 5: Position of the max. negative temperature gradient on the simulated sample. The temperature gradient collapses right before the wafer edge at 15 mm. That's why the crack propagation stops and the separation is finished due to mechanical stress afterwards. This causes a deviation along the preferred crystal orientation.

3 EXPERIMENTAL

The described simulation procedure is suitable to test several parameter settings. The reduction of the distance between the collapsed gradient and the wafer edge is the aim of this simulation. It can be assumed that a reduction of this distance will propagate the crack further and thus the unwanted deviations will be reduced.

For the simulations, the following parameters were varied: laser power, focus, process velocity, distance and height of the cooling nozzle, heat transfer (cooling) and back side beam shading.

Later on, the parameters of the most promising simulations were transferred to the machine followed by a six times execution on monocrystalline cz (100) wafers. After that, the deviations are measured and compared with the simulated results.

4. RESULTS

Most of the parameters had only slight influence on the distance between the mentioned collapse of the gradient and the wafer edge (< 100 μm). But the beam shading and the heat transfer due to cooling had a remarkable influence on it. In Figure 6 these results are shown. On the first investigation, no beam shading and a low cooling transfer of 2500 W/m²K was chosen. With these simulation parameters, the gradient stopped already 0.7 mm in front of the wafer edge. Adding 25% beam shading reduces the distance dramatically, which is plotted in the middle of figure 6. The laser power is adjusted after the beam shading to its original value. This approach guarantees that the influence results from the beam shape and not from the laser power itself. A further reduction down to 50 μm is possible while using a better cooling transfer of 5000 W/m²K.

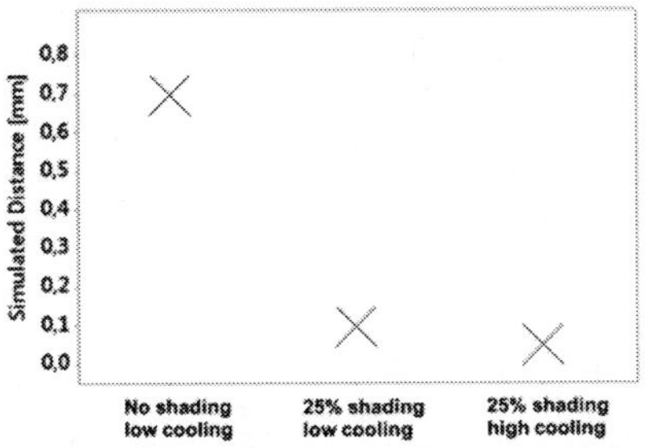

Figure 6: Distance between the simulated gradient and the wafer edge.

The measurements resulting from the real experiment are shown in figure 7. The use of beam shading and the higher cooling power are also beneficial. Thus the same trend is observable.

However, in comparison with the simulation it can be pointed out that the crack stops earlier in the experiment. This difference is already described in figure 2. The tension must be in applied in front of a cracking tip to propagate a crack further. (A tension behind the cracking tip will only cause a compression without any propagation.) A difference in the different tension levels causes different offsets. This statement can be underlined by the following thought. If the 800 μm measurement difference out of figure 2 were added to the simulation results, these fit already quite well but not as good as in figure two.

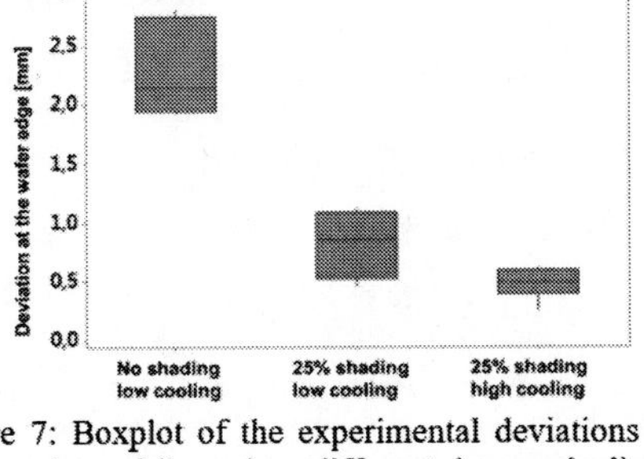

Figure 7: Boxplot of the experimental deviations at the wafer edge while using different beam shading and efficient cooling.

5 DISCUSSION AND CONCLUSION

In this work, a simulation model for the calculation of the most stressed area during the TLS process is presented. With this model, it is possible to identify the reason for the observed crack deviations at the end of the wafer. Also, this model can predict the dimension of such deviations. Nevertheless, these predictions should be handled with care.

Looking at the presented results; all the simulated data underestimates the real experiment.

On the other hand, the comparison of trends between different parameter settings works very well. In this work, remarkable progress has been achieved. The deviations were reduced by a factor of 5. The detection of the various influencing variables was made considerably easier by the model. Thus, e.g., the use of beam shading can be checked quickly.

Remarkably, both improvements were caused by a sharpened temperature gradient. For the crack propagation, a couple of forces (expansion and compression) are necessary. The distance between these two forces can be reduced (sharper gradient) but cannot be zero. Further investigations should be focused on an increased temperature gradient e. g. through the use of a more powerful cooling or a different laser profile. These studies may lead to a repeatable deviation of below 100 μm, which will be suitable for the PV industry.

6 REFERENCES

[1] J. Röth, N. Bernhard, C. Belgardt, M. Grimm, F. Kaule, Thermal Laser Separation (TLS) Dicing Process Study – a New Technology for Cutting Silicon Solar Cells for High-Efficiency Half-Cell Modules, 31st European Photovoltaic Solar Energy Conference and Exhibition, 2015M. Smith, A. Miller, Proceedings 17th European Photovoltaic Solar Energy Conference, Vol. I (2002) 903.

[2] S. Eiternick, F. Kaule, H-U. Zühlke, T. Kießling, M. Grimm, S. Schoenfelder, M. Turek, High Quality Half-Cell Processing Using Thermal Laser Separation, 5th International Conference on Silicon Photovoltaics, 2015

[3] D. Hull, Fractography, Observing, Measuring and Interpreting Fracture Surface Topography, Cambridge University Press, 1999

[4] F. Okamoto & C. Shimizu, "Evaluation of Glass Surface Defect using Thermal Down Shock Test", Glass Symposium 1989 Proceedings, (1989), pp13-14.

[5] F. Puschmann, Experimentelle Untersuchung der Spraykühlung zur Qualitätsverbesserung durch definierte Einstellung des Wärmeübergangs, Otto-von-Guericke-Universität Magdeburg, 2003

METHOD TO COUNTER WARPAGE DUE TO STRINGING FOR BACK CONTACT SOLAR CELLS

Andreas Halm, Engelbert Lemp, Rüdiger Farneda, Jens Theobald, Rudolf Harney
International Solar Energy Research Center - ISC Konstanz, Rudolf-Diesel Str.15 , 78467 Konstanz, Germany

ABSTRACT: In this study, ISC's Zebra cells [1, 2] which are n-type interdigitated back contact solar cells are mechanically bent during ribbon attachment to counter cell warpage. Bending is done mechanically on a specially designed work table in opposite direction of the expectable bowing direction of the cell. In this way, the warpage caused by mechanical stress due to the difference in thermal expansion coefficients between ribbon and cell after ribbon attachment is compensated resulting in a flat cell. Ribbons are attached by soldering. Three ribbon types with different cross section are evaluated in two configurations. Once flat, soldered continuously along the busbar, once stamped with a stress relief structure and soldered locally in-between the stress relieve points. After soldering, single cell modules are produced with the ribbon contacted cells and subjected to extended temperature cycle tests (TC). Results after TC 650 show that even for the extended testing time all single cell modules tested with different ribbon configurations stay well within the limit of 5 % relative power drop after TC 200. Relative FF drop for all single cell modules tested is below 1.3 % relative proving the mechanical stability of the cell-ribbon connection.
Keywords: IBC cell, warpage, bow, interconnection, soldering

1 INTRODUCITON

Up to today, only LG Electronics Inc. manages to industrially fabricate IBC solar cells and modules on 6 inch wafers [3]. One issue in respect to industrial production is module integration. Classical stringing is not feasible due to excessive cell warpage; efforts to adapt stringers to single side soldering or conductive adhesive (CA) gluing are made by several companies. At ISC Konstanz, different routes for module assembly are evaluated targeting to effectively encapsulate and contact ISC's Zebra cells. The Zebra cell is a bifacial back contact back junction solar cell processed on 6 inch n-type Cz wafers featuring eight continuous busbars on the rear side [2] for contacting during module processing. Ribbon interconnection of Zebra cells is especially interesting since it allows bifacial module production. In this work we present a new approach for circumventing cell bowing caused by the themomechanical stress between the interconnection ribbon and the solar cell. For that purpose, Zebra cells are mechanically pre-bent in opposite direction of the anticipated cell warpage prior to ribbon attachment to counter the bowing effect. Standard SnPbAg coated solder ribbons with different dimensions are either directly applied by soldering or stamped with a stress relieve structure in advance. Extensive temperature cycle testing (up to TC 650) is performed on one cell modules containing cells contacted by applying this new method.

2 EXPERIMENTAL

2.1 ribbon preparation

Commercially available SnPbAg coated Cu ribbons of different cross section area are utilized for soldering tests. Table I shows an overview on the ribbon configurations considered in our study:

Table I: solder ribbon configuration

Ribbon	A	B	C
dimension (mm x mm)	1.5 x 0.2	2 x 0.24	4 x 0.15
Cross section (mm²)	0.3	0.48	0.6
Stress relieve (y / n)	y / n	y / n	y
Yield strength (N/mm²)	50	50	?
Pre stretch factor (%)	2	2	0
Solder coating	SnPbAg	SnPbAg	SnPbAg

Ribbons A and B are stretched by 2% before further processing while ribbon C is left non stretched. As additional measure, a mechanically stamped stress relieve structure is investigated in order to reduce mechanical stress between cell and ribbon.

2.2 pre-bending and soldering

Solar cells are mounted on a work station (Figure 1) – called "BOA I" - which allows bending of the Zebra cell with variable radius. The cell is placed on the flat work table, fixed b metal slips and then gradually bent with a adjustable screw. For each ribbon type soldered onto the cells in a first trial the bending radius is adjusted to attain flat cells after ribbon soldering and unmounting the cell.

Figure 1: photo of BOA I, our mechanical pre bending and soldering station with Zebra cell mounted and warped in negative direction to anticipated cell bowing

Soldering is done by hand using commercially available solder flux and a set tip temperature of 320 °C on the soldering iron.

After ribbon soldering, the cell is released and a cross connector is soldered to the individual ribbons. Next, single cell modules are assembled with the ribbon contacted cells, 20 x 20 cm glass sheets, POE encapsulation and transparent backsheet.

3 RELIABILITY TEST

To prove mechanical stability of the interconnection and rule out negative impact of the bending procedure to the structural integrity of the solar cells, temperature cycle tests are performed. The single cell modules in different configurations are placed into a climatic chamber which periodically changes temperature between -40°C and 85°C in 6 hours with a dwelling time of 15 min at the turning points. In contrast to IEC norm

61215, no electrical load is applied during testing at temperatures above 25°C. Total test duration was 650 cycles, single cell modules are characterized initially, after 50, 200, 400 and 650 cycles by means of IV and EL measurements. Figure 2 shows the relative

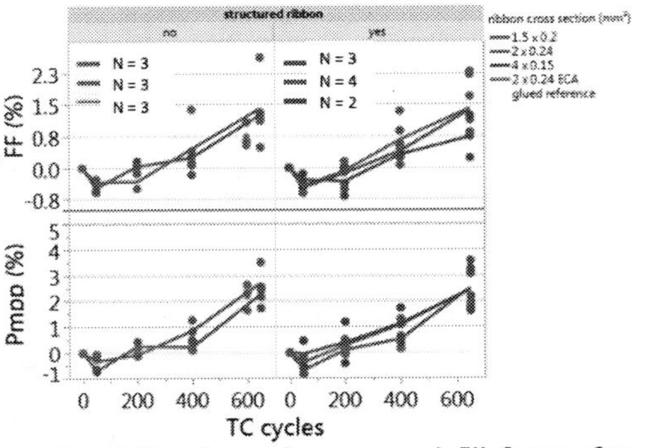

Figure 2: relative change in power and fill factor after temperature cycle testing of single cell laminates containing cells soldered on BOA I – lines show median values and points individual module performance; ribbon A samples in blue, ribbon B in red, ribbon C samples in purple and ECA glued references assembled with ribbon B without pre bending

change in IV parameters after each step. The graph shows relative losses in P_{MPP} and FF of individual minmodules (points) and median values for all different configurations (lines) after TC testing including an ECA glued reference group. Even after 650 cycles the results reveal a stabe connection with median FF losses lower than 1.5 % rel. for all connections evaluated. Median values range between 1.1 and 1.3 % rel. for ribbon A and B independently of structuring and are comparable to the ECA glued, non-bent reference assembled using ribbon B after TC 600; modules with non-stretched 4 x 0.15 mm² ribbon show lower median losses of 0.7 % relative suggesting that the thinner and non-streched ribbon causes lower losses due to higher flexibility. Median

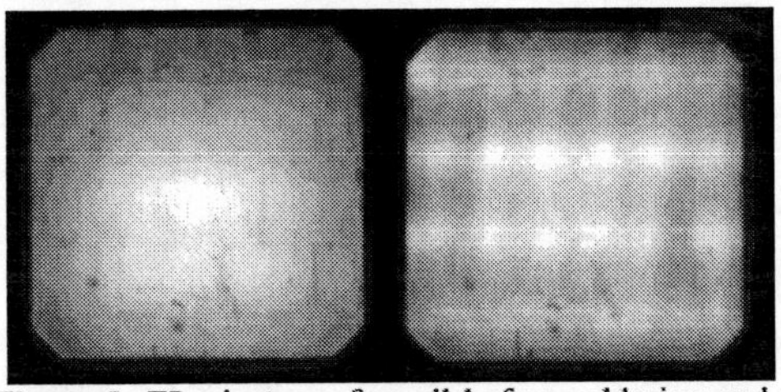

Figure 3: EL picutres of a cell before soldering and the corresponding single cell module after TC 650 reveal no local damage expect one finger interruption (bottom left)

power losses between 2.4 and 2.6 % rel. after 650 cycles lie well below the 5 % limit demanded by IEC test protocol for TC200 including electrical load. An EL picture of an example cell and corresponding single cell module after TC 650 reveal no local defects like cracks or shunts along the ribbon; only one finger interruption on the module (bottom left).

3.1 improved bending procedure

To realize an easier to handle and higher yield bending and releasing process, an improved work station – BOA II – was designed with fixed curvature for one ribbon configuration. The cell is held down by vacuum during soldering; the curved vacuum table turns into a ramp to enable smooth string assembly. The radius at the

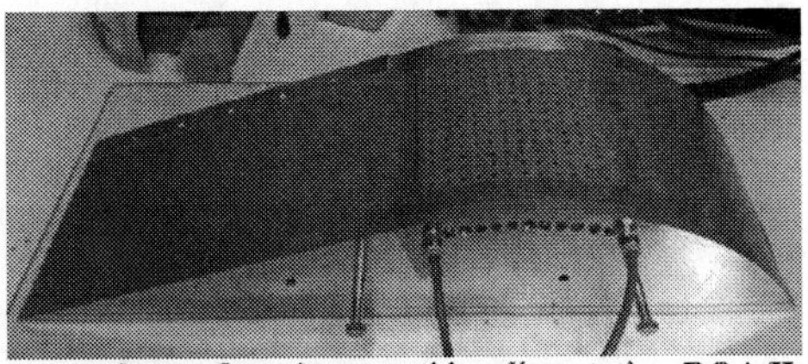

Figure 4: photo of our improved bending station BOA II; cells are hold down by vacuum at the pre bending position with fixed curvature (258 mm radius) followed by a ramp to ease string assembly

solder position is 258 mm which was deduced by experiments performed with BOA I. First tests show the practicability of the approach. Figure 5 shows a 3 cell string assembled on BOA II without any visible warpage.

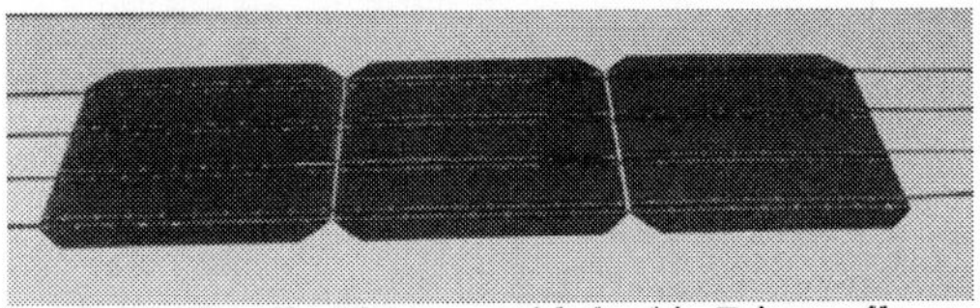

Figure 5: 3 cell string assembled with Zebra cells on BOA II

4 SUMMARY AND OUTLOOK

In this study we investigate a method to counter cell warpage for IBC solar cells with continuous busbars. The assembly procedure does not avoid thermomechanical stress but alters the mechanical state of the ribbon contacted solar cell by bending it opposite to the anticipated warpage direction prior to solder attachment of the interconnection ribbon. Experiments using different ribbon configurations prove the mechanical stability of the connection also after temperature cycle testing up to TC 650 resulting in similar or lower power and FF degradation as for an ECA glued reference after TC 600. Addtionally, no local damages along the ribbon positions are detected in EL leading to conclude that the bending procedure does not negatively influence the structural integrity of the solar cell.

In a next step, an improved work station called BOA II was designed with fixed curvature and a vacuum table for cell clamping. BOA II is successfully applied to assemble a three cell string without any visible bow. Further tests and on string and module level are planned for the near future to prove the viability of the concept.

5 Acknowledgements

This work was supported by the German Federal Ministry of Economic Affairs and Energy (BMWi) within the project PfZ (contract no. 0325840).

6 References

[1] V.D. Mihailetchi et al. "High efficiency ibc solar cells fabricated on large area n-type silicon using industrial available techniques", technical digest, 21 International Photovoltaic Science and Engineering Conference 28.11.2011-2.12.2011, Fukuoka, Japan
[2] G. Galbiati et al., "Large Area Back-Contact Back-Junction Solar Cells with Efficiency Exceeding 21%", 38th IEEE Photovoltaic Specialist Conference, Austin, USA. 1-4 (2012). DOI: 10.1109/PVSC-Vol2.2012.6656700.
[3] www.lgenergy.com.au/products/solar-panels/lg-neon-r-r

ACHIEVING FASTER LAMINATION PROCESS FOR CRYSTALLINE PHOTOVOLTAIC MODULES BY USING LATEST LAMINATION TECHNOLOGIES

Sraisth
Robert Buerkle GmbH
Stuttgarterstrasse 123, 72250, Freudenstadt, Germany

ABSTRACT: The lamination process is the most critical process of the photovoltaic (PV) module manufacturing. It decides the end-product quality of the PV module [1]. The present lamination process time in the PV market for glass-backsheet (GB) and glass-glass (GG) modules with ethylene-vinyl acetate (EVA) encapsulant are around 7.5 to 15 minutes, depending on the single step or two-step process. The module components play an important part in module quality, and one of the main components is the encapsulant. In this paper, EVA encapsulant is used for testing and improving the process, as still more than 90% of PV market is captured by EVA [2]. The paper is focused on methods and fundamentals to reduce the lamination process time by 20% while maintaining or achieving the same or better quality of the module respectively. Several tests like cross-linking, gel content, temperature measurement, electroluminescence test, and edge thickness measurement were performed. Finally, the reduction in manufacturing cost per module due to process time improvement will be calculated and presented for both GB and GG modules.
Keywords: Lamination, Photovoltaic, Gel Content, EVA, Manufacturing Cost, Encapsulant

1 INTRODUCTION

Crystalline Silicon wafer based photovoltaic (PV) modules have market coverage of 93% [3], and presently, 90% of them are glass-backsheet (GB) modules [4]. However, ITRPV has predicted that by 2024 roughly 60% of all crystalline modules will be glass-glass (GG) modules [4], irrespective of the cell type. Crystalline PV module manufacturing has several steps from glass cleaning to module fabrication. Amongst all process steps, stringing of cells and lamination steps are the most vital steps. Presently, the throughput of the stringer is nearly 5% more than the laminator's throughput, and with the continuous development it will increase to 20% by 2019 [5], and this made the lamination process as a bottleneck in the module manufacturing line.

The module is a lay-up of several components, which thermally treated in the process chamber under vacuum, pressure, and temperature parameters [1]. Lamination step bonds all the material together to form a single fabricated module. Mostly in PV market the current lamination process time of GB and GG modules is more than 10 minutes except for few manufacturers like Buerkle [1] and adding an extra minute of non-value added (NVA) operations into the process further reduces the per hour cycles. Therefore, to match the capacity, manufacturers have built multi-stack laminators, or they provide several single stage laminators. However, this will result in huge equipment investment. From the development point of view, the GB module lamination is already an established process, and it does not require intense research except for some tests and fine-tuning to achieve faster process time.

Whereas the GG module lamination is a complex process, and that is why still only 10% of PV modules are GG module in the market despite their proven superior properties over GB modules [4]. Thus, the paper will also discuss the several factors to achieve a high-quality GG module. The different tests like gel-content, temperature uniformity, edge pinching and visual inspection and methods presented and discussed to affirm the module quality for faster process.

1.1 Lamination technologies

There are several latest innovations happened in the lamination process. Amongst all, Buerkle has developed Short Lamination (SL) method three years before and is successfully running in the market [6]. Similarly, a double-sided heating plate for glass-glass lamination is also the latest concept to achieve shorter process time with the defect-free process and has superior quality over the single step membrane lamination process of glass-glass modules [6].

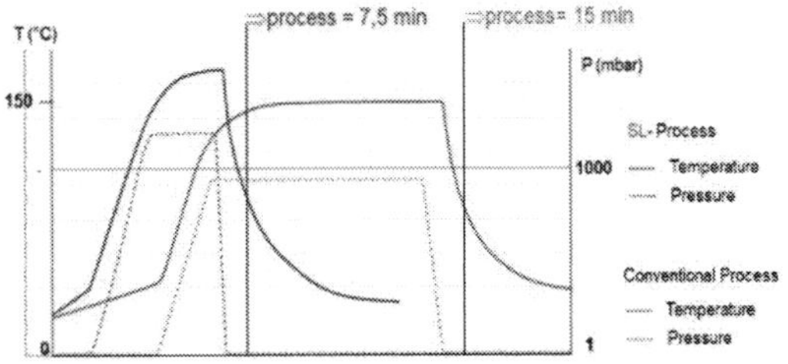

Figure 1: SL process developed by the Buerkle

Fig.1 shows the diagram of comparison of conventional and SL glass-backsheet lamination process. It is visible that almost half of the process time saved in SL lamination.

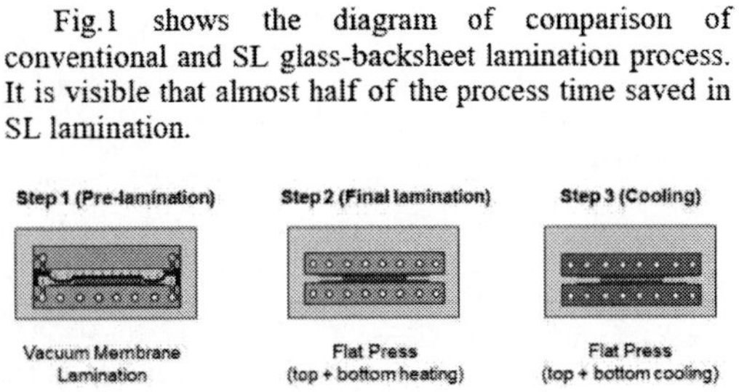

Figure 2: Two-step GG lamination process developed by the Buerkle

Fig.2 shows the two-step lamination process having double-sided heating as the second step. Therefore, a 15-20 minutes lamination process of GG modules can divide in a two-step, and the new process will save half of the time. Cooling press step is an optional step; however, a controlled cooling has a positive impact on the module quality.

2 GLASS-BACKSHEET TRIALS AND RESULTS

Several recipes were tested, and various modules were laminated to get the defect-free result. Full size 60 cells module were laminated and inspected since testing on small samples of size 300mm x 300mm will not lead to the correct interpretation. The recipe developed for the small sample will not work for bigger size module; therefore, full-size modules were tested. Since, the target was to reduce the process time, given that the cross-linking, and temperature uniformity for each new recipe is measured. It helps in understanding the process reality as a minimum value of gel content is required to pass the reliability tests. The gel content was calculated for five different positions in one module.

2.1 DSC test

Fig.3 shows the in-house performed DSC test results. It was important to do the DSC at the facility itself to maintain the correct direction of the trials, which helps to set the right parameters of the process. The graph interprets with respect to fixed process temperature. The process times varied from 5.5 minutes to 6.5 minutes. As seen from the chart for a fixed temperature of 160°C, the cross-linking increases with the increase in process time. Similarly, for other temperatures, it follows the same trend as expected. When the temperature increases, the cross-linking value also gradually rise from the previous lower temperature value [7].

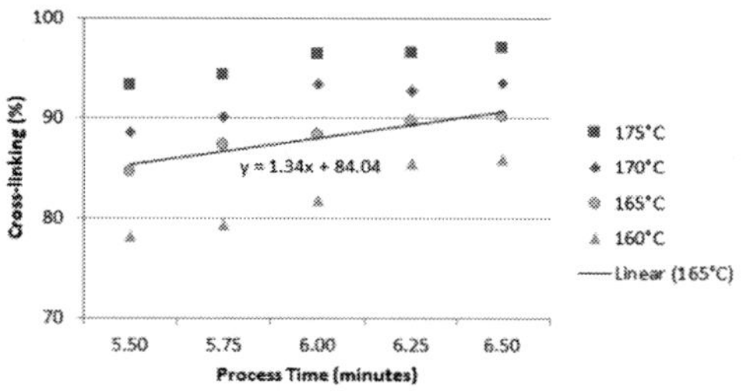

Figure 3: DSC cross-linking results for different processes

From the graph, the gel content value for 170°C at 5.75 minutes process, i.e., 90%, is approximately same as the value for 165°C at 6.5 minutes process. It explains that process time can reduce by increasing the process temperature.

Fig.3 shows the trendline plotted for the fixed temperature of 165°C. Usually, the encapsulant follows logarithmic behavior from zero curing to maximum curing concerning time [8]. In the opening minutes, process time has a significant influence on the gel content value, as gel content value increase rapidly with a slight change in the process time, irrespective of the temperature setting. However, after 70% curing, the graph follows the linear upward trend and later at high temperature, the values stabilized and stay horizontally to the x-axis. Therefore, the linear trendline used for analysis and the equation for the trendline of 165°C curve is:

$$y = 1.34x + 84.04 \quad (1)$$

This equation is in line intercept form, i.e. $y = mx + c$. In the line intercept equation, m is the slope of the line. After, correlating with the trendline equation, the slope of the trendline is, $m=1.34$.

Therefore, with every 15 seconds rise in the process time, the cross-linking increased by 1.34%.

2.2 Soxhlet chemical test results

The soxhlet tests for gel-content determination were performed by Huangzhou First, China at their site. The chemical test was necessary to check the relation between DSC values, where former method is more preferred to establish the final value for a certain process. It also helps in further tests to get an estimation how much the value would be for another test if one of the test values is known.

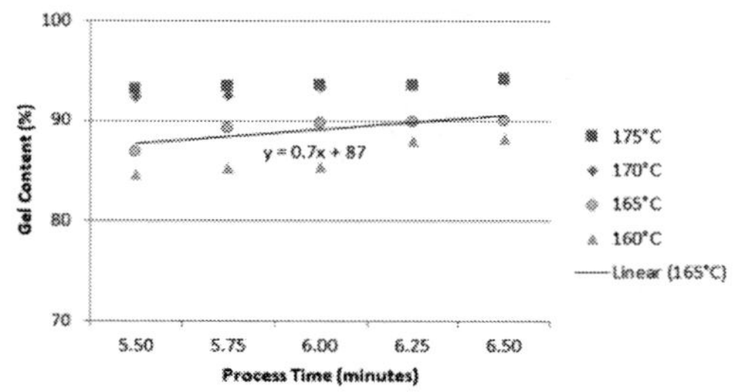

Figure 4: Soxhlet gel content results for different processes

Fig.4 shows the soxhlet gel content results at the fixed process temperatures. The gel content values range from 84% to 95%, whereas, DSC values measured has a bigger range from 78% to 98%. Thus, illustrates that at lower process times DSC shows lower values as compared to chemical test values and shows higher values for high process time. Here, the gel content increases with increasing process time and has lower slope value, $m = 0.7$.

Therefore, with every 15 second rise in process time, the gel content is increased by only 0.7%. It implies that time does not have a significant effect on the gel content value.

The gel content shows stability with increasing process time, and the values at higher temperature show almost similar values signify stable gel content values at higher temperature irrespective of the process time variation. The decreasing slope values with increasing temperature affirmed the statement, see Table I.

Table I: Slope for different temperature plots

Temperature (°C)	Slope (m)
160	1.0
165	0.7
170	0.46
175	0.22

2.3 Visual results

Apart from gel content, temperature uniformity, and electroluminescence test, visual quality is essential. The recipe developed should be defect free during the lamination. There should not be any defects like bubbles in the module, cell breakage or cell shifting. Many modules were laminated. All modules pass through visual inspection and all the flaws, if there were any, were recorded. Defects were categorized into seven different types, that is, Bubbles between the cells, bubbles at the contact area or interconnection area, bubbles on the corner, bubbles on the edges, EVA shrinkage Backsheet optic, and cell breakage (not with EL just visual). During trials, several defects were witnessed, for example, shown in fig.5. However, in the end, the process time optimization achieved without any defects in the module.

Figure 5: Bubbles between the cells in the GB module due to incorrect recipe

2.4 Electroluminescence test

The new process has a pressure more than 1000mbar. Therefore, the module was checked for any micro-cracks during lamination. The EL test was performed with the help of the customer at their site. Fig.6 shows an example of the EL imaging before and after the process for one of the 60 cells module, and there were no cracks witnessed in the laminated module. After that, several more modules are laminated, and the customer is running the line with the same pressure, and no cracks were witnessed due to lamination. It helps to prove that high pressure under certain limit has no role in cell breakage.

a)

b)

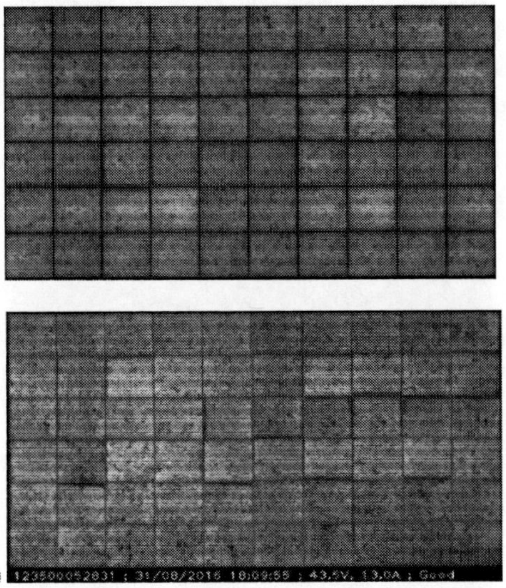

Figure 6: EL image a) before and b) after lamination of the GB module

2.5 Findings

The newly developed recipe is also tested for the temperature homogeneity inside the module. The homogeneity of the heating plate is ±1.5°C. However, to validate that the module follows the same homogeneity, the temperature measurement was performed. The measurement performed using thermocouples at six different points inside the module (4 on corners and 2 in the central area). Fig.7 shows the plot of the temperature profile for the 5.5 minutes process.

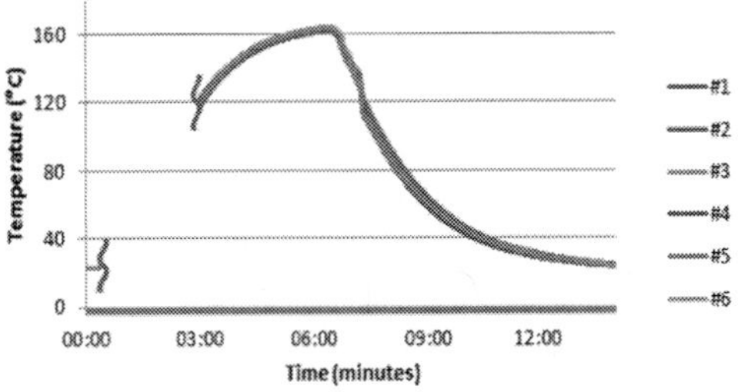

Figure 7: Temperature measurement for 5.5 minutes process, GB

The maximum temperature deviation during curing process measured is 3.1°C and was recorded same during the cooling process. Whereas, minimum deviation measured during curing and cooling processes were 2.1°C and 1.7°C respectively. The temperature uniformity achieved is ±1.5°C.

The process time for EVA material tested should be minimum 5.5 minutes depending upon the overpressure and temperature used in a cycle. Process time shorter than 5.5 minutes, will result in insufficient gel content values. Also, there is a chance of bubbles in the modules because of short evacuation cycle or other factors [9].

Temperature up to 175°C is feasible to use. Also, increase in temperature has more influence in raising the gel content values than the increase in process time.

Overpressure with high temperature used to accelerate the process. Pressure more than 1000mbar can be employed depending on the cycle time and gel content value targeted. Overpressure has no adverse effect on the cells or strings.

3 GLASS-GLASS TRIALS AND RESULTS

Glass-glass lamination is not a simple process to handle, and if the process is in a two-step, then it becomes much more complex. However, it will give shorter cycle time. Because of a two-step process, twice the parameters need to be controlled, and if the recipe is not correct, then several defects occurred during the trials.

The problems associated with GG lamination due to an improper method and incorrect recipe are majorly the defects which are visually detected in the module. The various defects witnessed during glass-glass trials are cell shifting, voids on the edges and the corners, cell breakage and encapsulant overflow, fig.8.

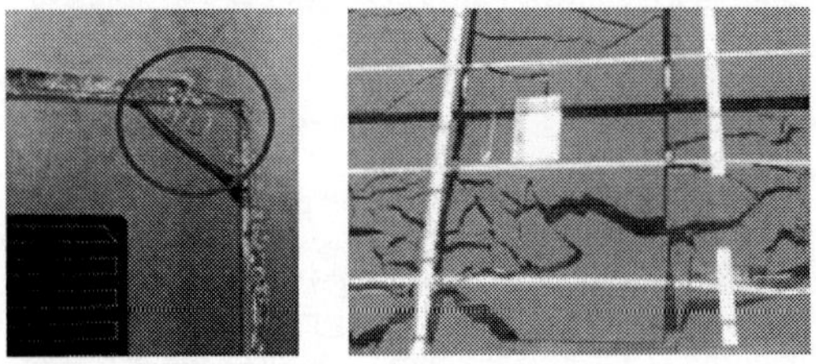

Figure 8: EVA outflow, bubbles/voids and cell breakage defects in GG modules

Another significant problem associated with glass-glass lamination is the edge pinching problem. It is nothing but thinner edge thickness compared to the inner side of the module. Lamination of glass-glass modules has edge pinching problem with the membrane press. It is visible in the fig.9, the sample laminated with the membrane press has deep impression on the edges (a), as compared to the sample laminated with the flat press (b).

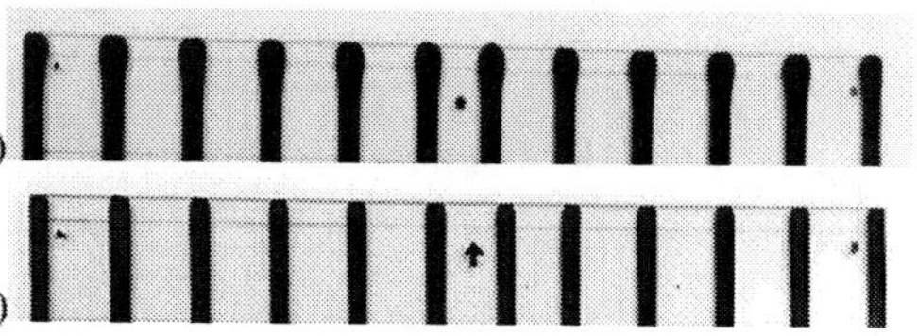

Figure 9: Comparison between a) membrane press and b) flat press lamination

3.1 Gel Content results

The encapsulant should be of the same size of the glass. Otherwise, the shorter size or small pieces just for different positions on the module lead to glass breakage. Gel content measurement is limited in GG trials because a lot of material goes in waste during the tests. Another reason is that the increasing effect of increasing process time or increasing temperature on gel content values has already proved and established for EVA encapsulants in chapter 2. Therefore, only defect-free processes selected for the cross-linking tests.

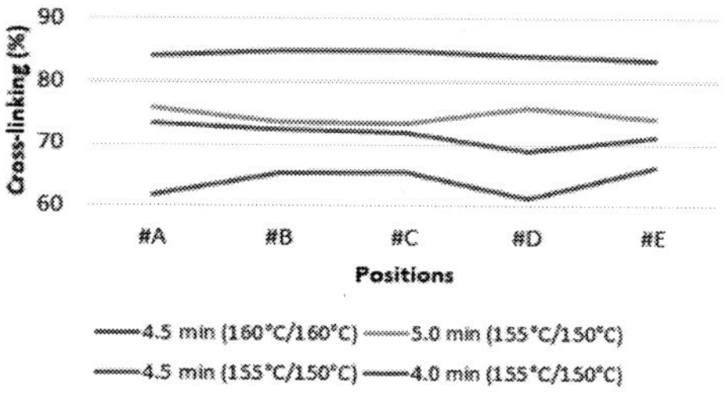

Figure 10: DSC results for GG trials for different processes

Fig.10 explains the importance of the correct process parameters for the lamination. The graph shows the gel content values for various processes measured at five different positions in a module. The deviation in values around the module for 4 minutes process at 155°C (HP1)/150°C (HP2) and for 4.5 minutes process at the same temperature of 160°C in both presses is 5% and 1% respectively. Both the processes are different having different process temperatures and process times, so optimized process parameters lead to uniform cross-linking across the module.

Although, the processes mentioned in fig.11 has good visual quality and gel content values are also in the range of recommended industry values based on practical experiences [10]. However, as per discussion with the Hangzhou First, the minimum gel content value recommended by them is 85% for good reliable quality. Further reliability test, like damp heat test, need to be performed to check if modules having gel content value lower than 85% with the same material is having excellent reliability or not.

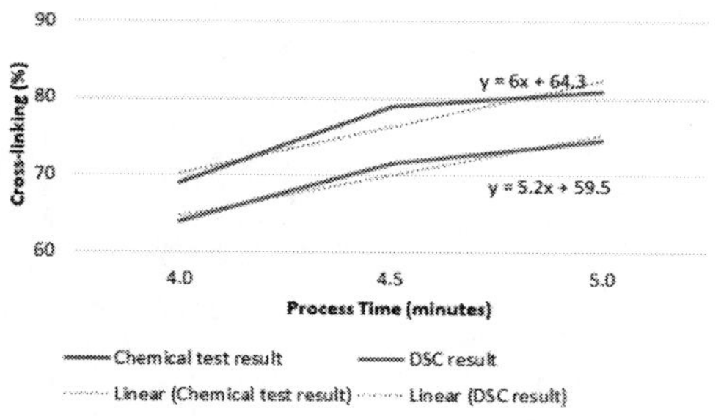

Figure 11: Chemical and DSC test results for GG trials at fixed process temperature

The outcome from the graph here is that it shows the proven fundamental of increase in gel content with the increasing process time. The slope from the chemical test linear trendline is: $m = 6.0$

It indicates the 6% increase in gel content with every half-minute rise in process time. Therefore, with the same process temperatures, the process time of 5.5 minutes can give the reliable module.

As clearly seen, that the chemical test values are 5% to 8 % higher than the DSC test values, which is witnessed in the GB trials with EVA.

Later, the process temperature of both the presses changed to 160°C and measurements carried out for the gel content. Fig.12 shows the increase in gel content for the fixed process time of 4.5 minutes. The chemical results show 87% gel content for the new tested process compared to 79% with previous process temperatures. The value is slightly higher than the recommend gel content value stated by Hangzhou first. It gives flexibility in achieving defect-free reliable module just by the slight increase in temperature and process time up to 6.5 minutes depending upon the capacity requirement.

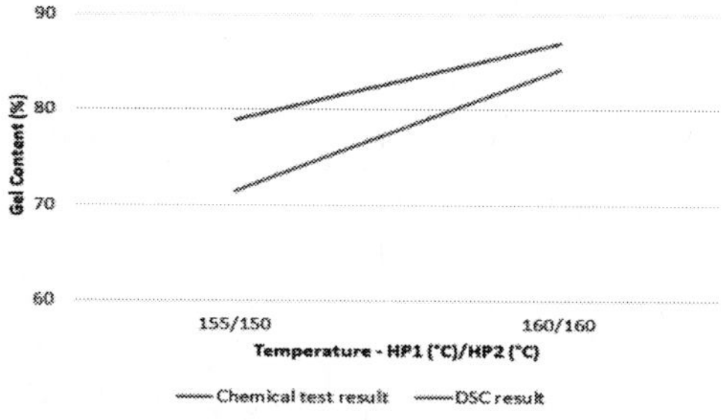

Figure 12: Chemical and DSC test results for GG at process time of 4.5 minutes

3.2 Edge Pinching Results

To check the edge pinching of the module for the 4.5 minutes recipe the module was measured at 16 different points around the module, shown in fig.13, and the values are plotted on a net diagram, shown in fig.14. The values were measured on the edges and 3cm inside from the edges to have a view on edge-pinching. In this module, one side was free from the solar cell, and this helped to evaluate any significance of cells or no cells near edges. Points marked from 1 to 4 in the anti-clockwise direction are no cells side.

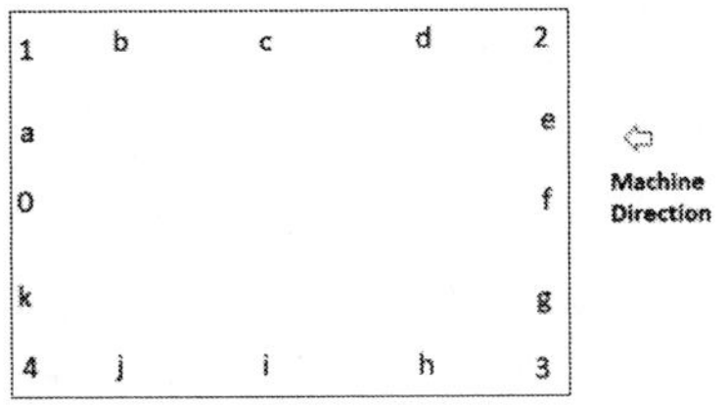

Figure 13: Edge thickness measuring points on the module

Fig.14 shows the Net diagram plotted for the GG module laminated with EVA encapsulant at 160°C (same temperature in both hot-presses) for 4.5 minutes. The values on the edges are smaller as expected. However the maximum deviation between the outside and 3cm inside values is 0.257mm, and the minimum is 0.075mm, and it is at the position marked as 0 which is cell-free side. This small deviation is just because of not having any cells on one edge, giving uniform thickness in that area. Since other three sides covered with cells, which has a thickness of 0.180~0.200mm, therefore, creates little edge thickness difference.

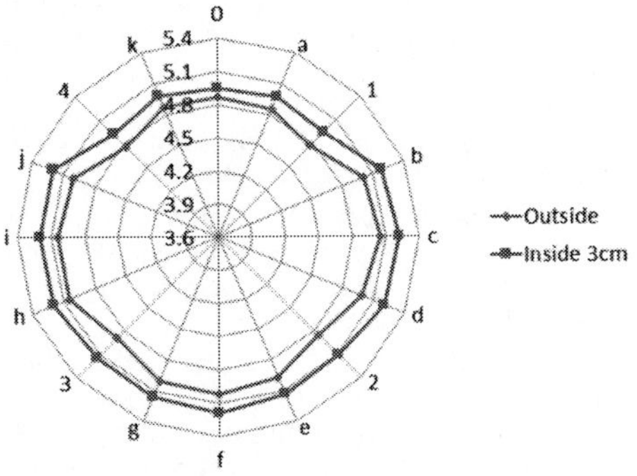

Figure 14: Edge thickness measurement for 4.5 minutes process, GG with EVA

The edge thickness variation is 0.05 – 0.3mm. According to Cattaneo et al. (2014) research, it is an excellent value regarding edge pinch and might consider as no edge pinch or very low edge pinch. The edge thickness variation achieved by other manufacturers and machine suppliers is 0.4 – 0.6mm or even higher up to 1mm. Such a shallow edge pinch value will reduce stresses induced in the module with time and perform for much longer lifetime compared to higher edge pinch value modules [11].

3.3 Findings

Like the GB trial, the temperature measurement is performed for glass-glass trials. Fig.15 shows the temperature uniformity measurement for 4.5 minutes process at 155°C (HP1)/150°C (HP2). The temperature uniformity achieved at the end of curing process followed throughout the cooling process. Here, the second press helps to achieve the required temperature uniformity with the aid of double-sided heating plate and transfer heat more rapidly and steadily from both sides. The minimum deviation achieved between temperatures at the end of the curing process is 1°C, which gives temperature uniformity of ±1°C across the module (considering the device precision). Similarly, the minimum deviation measured at the end of the cooling process is 3°C, which gives a temperature uniformity of ±2°C.

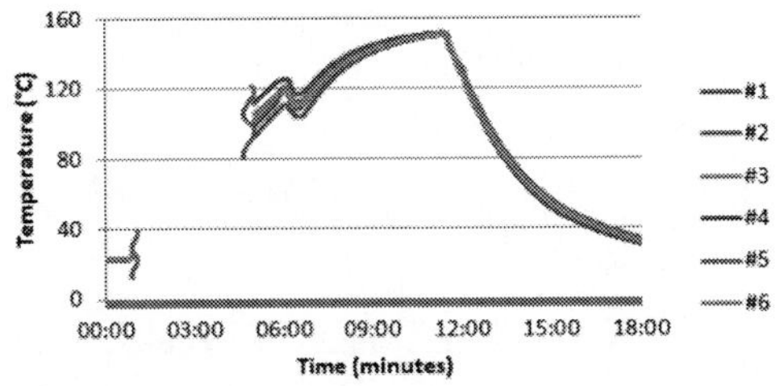

Figure 15: Temperature measurement for 4.5 minutes process, GG

The process time for EVA material tested with GG configuration should be minimum 4.5 minutes depending upon the pressure and temperature used in a cycle. Process time shorter than 4.5 minutes, will result in insufficient gel content values and maybe non-melted portions.

The temperature should be limited to maximum 165°C. Pressure values should be lower than 500mbar (no overpressure). Otherwise, high temperature and pressure can create voids due to EVA outflow from the edges and the corners.

The total process time in the second flat press should be same as first membrane press, it should fulfill the required gel content, or it can also be more than the first press process time. However, it is important to maintain the same process time as it results in good line balancing and having no waiting time for the subsequent process.

4 COST CALCULATION

Manufacturing cost per module calculated for both before and after process optimization. Few assumptions have considered for calculations depending upon the place of manufacturing plant. The assumed manufacturing site is in India. The number of working hours chosen for calculation is 7200. Assuming plant runs for three shifts 8 hours each and for 300 days after removing Sundays and public holidays [12]. 72 cells module considered for the cost comparison because of the maximum size adaptability of the laminator and due to its increasing market demand [13]. The number of cells does not affect the process times. Assuming 72 cells module type has 320W peak power output, irrespective of the module and cell technology, depending upon the different module suppliers and trends in the market [13]. The 200MW capacity of manufacturing plant assumed for calculation. For calculating the manufacturing cost, 60 seconds of NVA operation time added to the process time. Total added times gives the cycle time of the process.

Machine cost is different for SL and VFF laminator and the costs considered for equipment and consumables related to laminator taken from Buerkle sales data. All the equipment and consumables and their costs registered with the defined part number depending upon the type of machine [14]. In this report, the actual prices of the equipment and consumables not mentioned. Since, the selling price of machine equipment and consumables, before and after process optimization, would remain same. Thus, only the impact of process time reduction on the overall costs discussed in following chapters. The process will also change the automation requirement for the module manufacturing.

Table II and III shows the cost comparison before and after the newly developed process for GB and GG modules respectively. Detailed calculation is not presented in this paper.

Table II: Comparison of calculation of manufacturing cost per module before and after the newly developed process for GB 72 cells module

Buerkle SL (Glass-Backsheet) 72 cells		
	Before (%)	After (%)
Equipment cost	100	76.2
Electricity consumption cost	100	76.5
Total manufacturing cost	**100**	**89.8**

Table III: Comparison of calculation of manufacturing cost per module before and after the newly developed process for GG 72 cells module

Buerkle VFF (Glass-Glass) 72 cells		
	Before (%)	After (%)
Equipment cost	100	72.9
Electricity consumption cost	100	73.3
Total manufacturing cost	**100**	**90.1**

5 CONCLUSION

Several recipes were tested and analyzed for both GB and GG type of modules. Shorter process recipes proved and verified by the different laboratory tests. Various laboratory tests performed mainly, DSC measurement, soxhlet chemical test, temperature uniformity test, electroluminescence test and edge pinching test in case of GG module. Tests carried out either at Buerkle research laboratory or supplier's or customer's site. Concluding all the trials the following results achieved:

- GB module with EVA encapsulant: The process time is reduced by 26.7% from 7.5 minutes to 5.5 minutes,
- GG module with EVA encapsulant: The process time is reduced by 30.1% from 7.5 minutes to 4.5 minutes,
- 72 cells GB module: 10.2% reduction in the manufacturing cost per module
- 72 cells GG module: 9.9% reduction in the manufacturing cost per module

As per the ITRPV 2017 report and experiencing the demand from the customers the POE material as an encapsulant is growing in the market. The next research would be on the lamination of GB and GG modules using POE encapsulant. Since POE has superior properties compared to EVA, therefore it brings all new field in the research.

6 REFERENCES

[1] A. Amrani, A.Mahrane, F.Y.Moussa, and Y.Boukennous, "Solar Module Fabrication," *International Journal of Photoenergy, Article ID 27610,* 2007.

[2] ITRPV, "International Technology Roadmap for Photovoltaic (ITRPV): Results 2016. (8th Edition)," 2017.

[3] Fraunhofer ISE, "Photovoltaics Report, updated: 17 November 2016," n. pu., Freiburg, 2016.

[4] ITRPV, "International Technology Roadmap for Photovoltaic (ITRPV): Results 2013. (5th Edition)," 2014.

[5] ITRPV, "International Technology Roadmap for Photovoltaic (ITRPV): Results 2014. (6th Edition)," 2015.

[6] Robert Bürkle GmbH, "Photovoltaikindustrie," 12 September 2016. [Online]. Available: http://www.buerkle-gmbh.de/photovoltaikindustrie.html. [Accessed 14 December 2016].

[7] B.-M. Chen, C.-Y. Peng, J.-L. Cho and G. A. Porter, "Optimization of solar module encapsulant lamination by optical constant determination of Ethylene-Vinyl Acetate," *International Journal of Photoenergy,* vol. 2015, no. Article ID 276404, 7 pages, 2015. doi:10.1155/2015/276404, 2015.

[8] S. Kajari-Schröder, U. Eitner, C. Oprisoni, T. Alshuth, M. Köntges and R. Brendel, "Modeling the Curing Dynamics of Ethylene-Vinyl Acetate," Valencia, Spain, 2010.

[9] L.-E. Perret-Aebi, H.-Y. Li, R. Théron, G. Roeder, Y. Luo, T. Turlings, R. F. M. Lange and C. Ballif, "Insights on EVA lamination process: Where do the bubbles come from?," Valencia, Spain, 2010.

[10] M. D. Kempe, D. C. Miller, J. H. Wohlgemuth, S. R. Kurtz, J. M. Moseley, Q. A. Shah, G. Tamizhmani, K. Sakurai, M. Inoue, T. Doi, A. Masuda, S. L. Samuels and C. E. Vanderpan, "Field testing of thermoplastic encapsulants in high-temperature installations," *Energy Science and Engineering,* vol. 3, no. 6, pp. 565-580, 2015.

[11] G. Cattaneo, A. Faes, H.-Y. Li, F. Galliano, M. Gragert, Y. Yao, R. Grischke, T. Söderström, M. Despeisse, C. Ballif and L.-E. Perret-Aebi, "Lamination process and encapsulation materials for glass–glass PV module design," *PV-tech,* pp. 1-8, March 2015.

[12] International Labour Organization, "India. The Factories Act, 1948 - ILO," 1948. [Online]. Available:https://www.ilo.org/dyn/natlex/docs/WEBTEXT/32063/64873/E87IND01.htm. [Accessed 15 January 2017].

[13] ITRPV, "International Technology Roadmap for Photovoltaic (ITRPV): Results 2015. (7th Edition)," 2016.

[14] Buerkle Sales, "K Numbers: K117809 (For Membrane), K117831 (For Transport Belt), K119382 (For Release Sheet)," n. pu., Freudenstadt, 2017.

ENHANCEMENTS TO CADMIUM TELLURIDE CELL EFFICIENCY

James Sites, Amit Munshi, Jason Kephart, Drew Swanson, Andrew Moore, Tao Song, and W.S. Sampath
Colorado State University, Fort Collins, CO 80523 USA
James.Sites@Colostate.Edu 1-970-491-5850

ABSTRACT: A number of improvements to CdTe solar cells have led to significant efficiency increases in the last few years. The First Solar company has been at the forefront in pushing the efficiency higher at both the cell and module level, but there have also been advances at Colorado State University and other laboratories. The presentation primarily discusses CdTe work at CSU, but does so in the framework of the broader approach to higher-efficiency CdTe and suggestions for further enhancements. The specific focus of our laboratory has been (1) the replacement of the CdS buffer layer with MgZnO, (2) the use of a Te layer in the back contact to reduce the amount of Cu required, (3) higher CdTe deposition temperature leading to larger crystallite CdTe, and (4) incorporation of a CdSeTe layer at the front of the absorber. Taken together, several issues with the front part of the cell, the CdTe absorber, and the back contact have been addressed, and efficiencies have now exceeded 19%.
Keywords: Solar Cells, Cadmium Telluride

1 INTRODUCTION

The cadmium telluride polycrystalline thin-film PV technology, thanks primarily to First Solar, Inc., is highly competitive. That company has produced over 18 GW of installed CdTe modules and has orders for an additional 4 GW. It has achieved efficiencies of 22.1% at the cell level and 18.6% from a module production level. Much of the commercial competitiveness is the speed at which a module can be produced, approximately one module per minute from each production line and less than 3½ hours for a sheet of glass to become a completed module. Currently the company is transitioning from 0.72- to 2.4-m^2 modules. Meanwhile, at least three other companies are building CdTe factories in China to scale their capacity to varying amounts above 100 MW per year.

The road to high-efficiency polycrystalline CdTe cells has had several important milestones. The superstrate structure in common use today was introduced by Bonnet and Rabenhorst in 1972 [1]. In the early 1980's, the use of the CdS/CdTe heterojunction allowed efficiencies of 10% [2]. A few years later, chlorine passivation of the polycrystalline grain boundaries was introduced [3], and subsequent work at the University of South Florida [4] and the U.S. National Renewable Energy Laboratory [5] pushed cell efficiencies to 16.5%.

The highest CdTe efficiencies stayed at this level for over a decade, and there was concern that it might be a permanent plateau. However, in the past five years, work at First Solar and GE has produced efficiencies that now exceed 22%. The factors that have contributed to the recent increases include replacements for the CdS buffer, a ZnTe:Cu back contact, low absorption glass and anti-reflective coatings, and probably most important, the use of a lower band-gap CdSeTe alloy adjacent to the front contact and buffer layers. Others, including our group, have contributed to the different parts of the CdTe cell structure. However, opportunities remain for still higher cell efficiencies, especially if the current CdTe voltage deficit can be reduced as it has with crystalline CdTe .

2 CdTe CELL STRUCTURE

The CdTe cell structure most commonly fabricated at Colorado State University (CSU) is shown in Fig. 1. Much of it is similar to what is done elsewhere, but two somewhat unique features, the MgZnO (MZO) buffer layer between the CdTe absorber and the transparent front contact and the Te layer before the metal contact, will be discussed below.

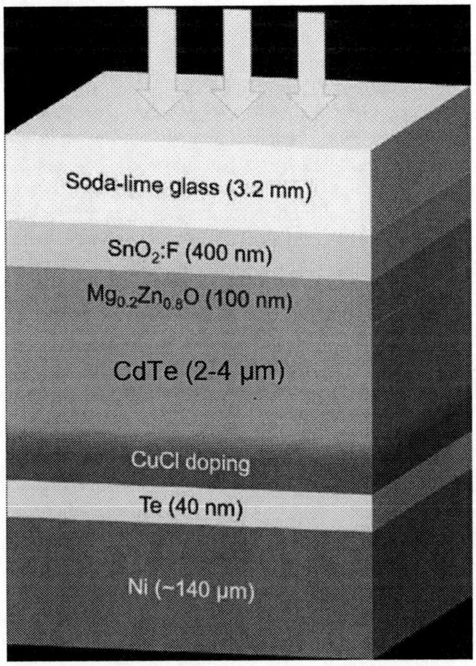

Figure 1. Standard CdTe cell structure at CSU

There are a number of ways that CdTe absorbers have been deposited for high-quality cells [6], but the choice at CSU has been close-space sublimation (CSS). The CSU in-line, single vacuum system is described in Ref. [7]. It has nine stations, so that in addition to depositing CdTe, there are stations for preheating the glass substrate and cooldown of the completed structure and others for depositing $CdCl_2$ for passivation, CuCl to assist with the back contact, and formerly CdS. There is also a two-chamber co-sublimation station [8] for deposition of alloys such as CdMgTe and CdSeTe. The standard size for the glass superstrates is 8 cm by 9 cm, and a typical deposition run takes 30 minutes.

The other layers in Fig. 1 consist of commercially-coated glass (Tec10 from Pilkington), sputtered MZO, evaporated Te, and a sprayed nickel and/or carbon contact layers. Up to 25 cells are delineated on each superstrate with an approximate 0.64-cm^2 area.

3 HIGH-EFFICIENCY CELLS AT CSU

3.1 The MZO layer

The MZO buffer in Fig. 1 serves a similar purpose to the traditional CdS buffer, and it appears to be superior to CdS in three ways. First, it is a higher band-gap material (3.4 eV), which means very little absorption loss for low-wavelength photons and hence a positive impact on the quantum efficiency without requiring extreme thinning. Second, MZO is quite robust when exposed Cl during the passivation process; cross-section micrographs clearly show a very uniform layer both before and after the Cl passivation. Third, and probably most important, variation in the ratio of Mg to Zn allows the conduction-band offset between CdTe and MZO to be adjusted. The best results with approximately 20% Mg appear to correspond to a small spike in the conduction band [9].

3.2 The Te layer

Several groups have observed improvements in efficiency for CdTe cells with a Te layer [10,11] or a Te-rich layer [12,13] at the interface between CdTe and the metallic contact. That strategy has also worked well at CSU. Experimentally, it has allowed us to reduce the amount of copper needed for a good contact by about a factor of ten and produced increases in cell efficiency of 2½% absolute [14]. The contact barrier remained minimal even at quite low temperatures, and the cell voltage increased linearly from 845 mV at room temperature to 1065 mV at 175 K, extrapolating to a value very close to the CdTe band gap at 0 K.

An explanation for the value of the Te layer is given in Fig. 2. The valence-band maximum (VBM) of CdTe is the sum of its electron affinity (4.3 eV) and its band gap (1.5 eV). At a typical carrier polycrystalline CdTe carrier density, the VBM is significantly larger than the workfunction of a typical metal, such as Ni, used for contact. In the absence of the Te layer, the alignment of the CdTe and metal Fermi levels will produce significant band bending in the CdTe and a significant barrier to hole collection.

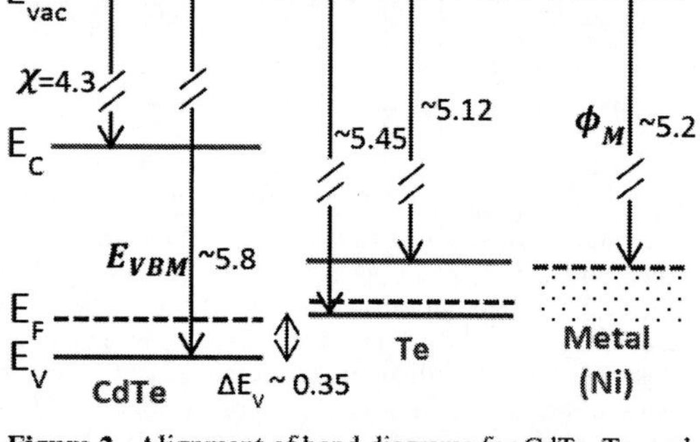

Figure 2. Alignment of band diagrams for CdTe, Te, and a typical metal contact. All values in eV.

With the insertion of the Te layer, there will be a smaller barrier at the CdTe/Te interface and a second one between the Te and the metal. The result, shown in Fig. 3 is a reduction in the CdTe band bending. Effectively the CdTe-to-metal barrier is divided, and the thermal energy of the holes allow them to pass each of the smaller barriers more easily. The values for Te in Fig. 2 are not well-determined independently, and those shown in the figure are those that provided the best fits to experimental current-voltage curves. The thickness of the Te layer is important; it should be relatively thin, but not so thin that it is fully depleted. With the 10^{18} cm^{-3} hole density for the Te deposited at CSU, a thickness of 30-50 nm appears to be optimal.

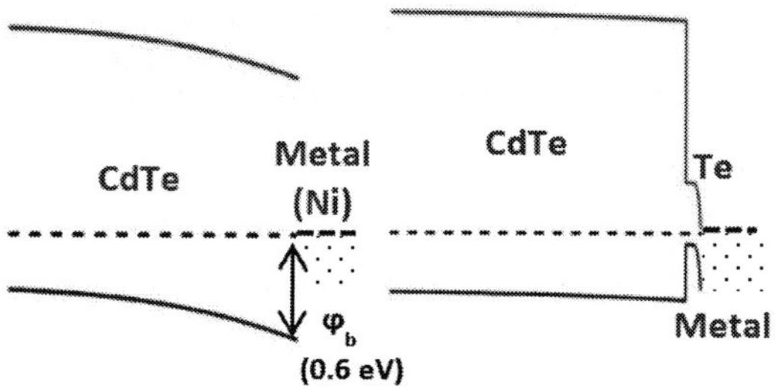

Figure 3. Comparison of rear-interface band bending with and without an intermediate Te layer.

A major advantage of a smaller back-contact barrier is that there is greater flexibility in the use of copper, which has been commonly used to reduce the band bending through heavy doping of the CdTe near the interface. While Cu is also beneficial for greater bulk doping, there is a counter trend of reduced bulk lifetime. The optimal condition seems to be a small amount of Cu, equivalent to about a ½-nm layer, which is about an order of magnitude less that what was formerly used without the Te layer.

3.3 Electron reflection

Not shown in Fig. 1 is the possibility of an electron-reflection layer. Such a layer with a CdMgTe (CMT) alloy to expand the CdTe band gap in the conduction-band direction has been explored in simulation and has shown small increases in current and voltage in preliminary measurements at CSU. Although CMT for this purpose has several attractive features, it also expands the gap in the valence-band direction. Quite possibly though, the combination of the Te layer with CMT electron reflection will allow a flattening of the CdTe bands that is larger than its valence-band expansion.

3.4 Highest-efficiency CdTe cell

In 2016, several high-efficiency strategies were combined at CSU to produce CdTe-absorber cells with efficiencies above 18% [16]. These strategies included the MZO and Te layers, as well as low absorption glass and an anti-reflection coating on its outer surface. The optical enhancements allowed a quantum efficiency above 90% over a broad spectral range, and a current density of 26.8 mA/cm^2. At the same time, the cell voltage reached 863 mV, and the fill-factor was just under 80%.

3.5 Cells with an Se-alloy layer

Although the CdTe-absorber results have been promising, the CSU program has also started to incorporate the Se-containing layer pioneered at First Solar. Such a layer with a reduced band gap at the front of the absorber is shown in Fig. 4. At both First Solar and CSU, the layer is the order of a micron thick and has a graded transition to the CdTe below it.

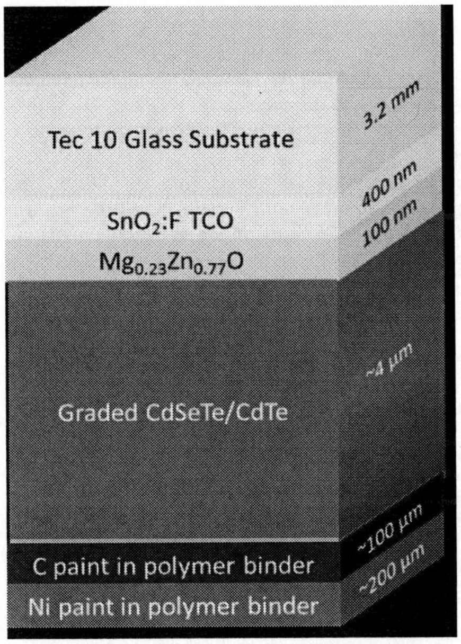

Figure 4. Highest-efficiency CSU cell with the CdSeTe layer in orange and the thinTe layer in blue.

Current-voltage results from the best cell using the Fig. 4 configuration are shown in Fig. 5. The lower-band-gap of the CdSeTe (approximately 20% Se compared to Cd) is clearly seen in both quantum efficiency and photoluminescence and leads to the increased current, even though there was no anti-reflection coating in this case. The highest voltage was slightly below that of cells with all-CdTe absorbers, but with the lower band gap, the voltage deficit was reduced by about 70 mV. The fill-factor here was again approaching 80%.

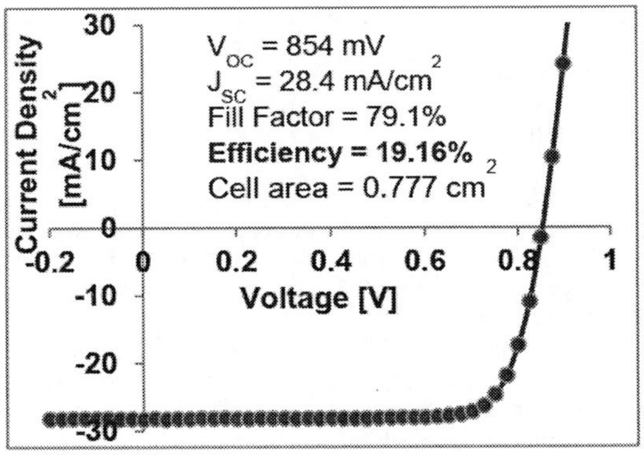

Figure 5. J-V curve of highest efficiency CSU CdTe cell. From Ref. [17].

4 WHAT NEXT FOR CdTe CELLS

The CdTe research community has an informal goal of achieving 25% cell efficiency. To achieve this efficiency, cells will need to incorporate at least some of the following: (1) absorber carrier concentration ~10^{16} cm^{-3} while maintaining a bulk lifetime near 10 ns, (2) back-surface recombination velocity near 10^3 cm/s, (3) an optimal absorber thickness and selenium profile, and (4) incorporation of electron reflection. At the same time to remain cost competitive, the industry needs to maintain module efficiency near 85% of the cell efficiency, high volume and large-size modules, and its record of increasing module lifetime.

5 ACKNOWLEDGMENTS

The authors thank Jennifer Drayton, Alexandra Huss, and Anna Wojtowicz (CSU), and Darius Kuciauskas (NREL) for their contributions and discussions. The work was supported by the U.S. Department of Energy's SunShot Program and by the U.S. National Science Foundation.

6 REFERENCES

[1] D. Bonnet and H. Rabenhorst, Proc. 9th IEEE Photovoltaics Specialists Conf., 129-132 (1972).
[2] Y. Tyan and E. Perez-Albuerne, Proc. 16th IEEE Photovoltaics Conf. 794-800 (1982).
[3] P. Meyers, C. Liu, and T. Frey, U.S Patent 4,710,589 (1982).
[4] J. Britt and C. Ferekides, Appl. Phys. Lett. **62**, 2581-2552 (1993).
[5] X. Wu, J. Keane, R. Dhere, C. DeHart, D. Albin, A. Duda, T. Gessert, S. Asher, D. Levi, and P. Sheldon, Proc. 17th European PVSEC, 995 (2001).
[6] B. McCandless and J. Sites, Ch. 14 in *Handbook of Photovoltaic Science and Engineering*, 2nd Ed., Edited by A. Luque and S. Hegedus (John Wiley and Sons, 2011).
[7] D. Swanson, J. Sites, and W. Sampath, SOLMAT **159**, 389-394 (2017).
[8] P. Kobyakov, A. Moore, J. Raguse, D. Swanson, and W. Sampath, J. Vac. Sci. Technol. **A32**, 021511 (2014).
[9] J. Kephart, J. McCamy, Z. Ma, A. Ganjoo, F. Alamgir, and W. Sampath, SOLMAT **157**, 266-275 (2016).
[10] H. Uda, S. Ikegami, and H. Sonomura, SOLMAT **35**, 293-298 (1994).
[11] W. Xia, H. Lin, H. Wu, C. Tang, I. Irfan, C. Wang, and Y. Gao, SOLMAT **128** 411-420 (2014).
[12] D. Bätzer, A. Romeo, H. Zogg. R. Wendt, and A. Tiwari, Thin Solid Films **387**, 151-154 (2001).
[13] S. Hegedus and B. McCandless, SOLMAT **88**, 75-95 (2005).
[14] A. Moore, T. Song, and J. Sites, MRS Advances (2017).
[15] T. Song, A. Moore, and J. Sites, submitted to IEEE J. Photovoltaics.
[16] A. Munshi, J. Kephart, A. Abbas, T. Shimpi, K. Barth, J. Walls, and W. Sampath, submitted to SOLMAT.
[17] A. Munshi, A. Abbas, J. Raguse, J.-N. Beaudry, J.Sites, K. Barth, J. Walls, and W. Sampath, submitted to IEEE J. Photovoltaics.

EFFECTS OF ULTRA-THIN COPPER LAYER ON THE PERFORMANCE AND STABILITY OF CDTE/CDS SOLAR CELLS

Elisa Artegiani, Daniele Menossi, Claudio Ferrari^ and Alessandro Romeo
Laboratory for Photovoltaics and Solid State Physics, Department of Computer Science, University of Verona, Strada Le Grazie 15, 37134 Verona, Italy
^Istituto dei Materiali per l'Elettronica ed il Magnetismo (IMEM), Parco Area delle Scienze 37/A - 43124 Parma, Italy

ABSTRACT:
CdTe solar cells have shown high efficiency well above 20%. Despite different copper-free back contacts have been successfully introduced this is still required for very high efficiency devices. On the other hand an excessive amount of copper results in degradation of the device and a suitable concentration is very important for a good trade off between conversion efficiency and stability.
A study on how the amount of copper at the back contact affects the performances and the stability of CdTe thin film solar cells is presented. CdTe cells with 0.1 nm Cu layer are compared to devices with 2 nm Cu and the electrical properties are measured. The formation of Cu-Te compounds has been analysed and degradation of cells made with thick and thin Cu layer has been recorded. With different copper thicknesses, different approaches are needed for high efficiencies; moreover devices degrade with different mechanisms.

Keywords: CdTe, Back Contact, Stability

1 INTRODUCTION

CdTe thin film solar cells are the most successful thin film photovoltaic devices in terms of production yield, as attested by the remarkable market performance. Nevertheless the device has also shown to be competitive in terms of efficiency with the latest results from First Solar, where a record efficiency of 22,1 % has been obtained [1].

Because of the high electron affinity of CdTe, one of the most complex points for the development of these devices has been the preparation of the back contact. The insertion of copper has demonstrated to deliver the highest efficiency, moreover recent works have shown that copper is a crucial element for the CdTe doping and consequently for high conversion efficiency, but on the other side copper tends to diffuse into the CdTe bulk degrading the cell.

In this work we analyze superstrate configuration cells made with different copper thickness in order to understand how copper affects the back contact and the CdTe bulk, and in order to address if there is a minimum amount of copper required to perform high efficiencies.

In substrate configuration, Kranz et al. [2] and Perrenoud et al. [3], have shown that a sub-monolayer amount of copper is enough to dope CdTe and reach efficiencies exceeding 13% [2].

2 EXPERIMENTAL

2.1 Preparation of the devices

CdTe thin film solar cells were made in superstrate configuration by low-temperature fabrication process based on vacuum evaporation (VE) with a pressure of 10^{-4} Pa. On an ITO/ZnO stack as front contact, a CdS layer (with a thickness of 100 to 300 nm) is deposited at 150 °C and then annealed at 450 °C in vacuum. Subsequently 6 to 8 micrometers of CdTe are grown in the same vacuum chamber and subsequently activated by $CdCl_2$ applied by wet deposition and annealed in air at 380 °C.

Prior to back contact deposition, the CdTe surface is etched by a bromine/methanol (Br-MeOH) solution in order to remove $CdCl_2$ residues but most important to make with the deposited copper a CuTe compound, which results to be highly stable and performing [4].

Back contact is made by vacuum evaporation of copper (from 0,1 nm to 2 nm), and a 50 nm thick gold layer. Typically a post deposition annealing in air is applied to achieve high efficiency: with solar cells made with 2 nm thick copper layer, the final annealing is made at 200°C in order to deliver efficiencies exceeding 15%.

For 0,1 nm thick copper layer, an additional annealing at 300 °C is applied.

2.2 Characterization

Finished devices have been studied by means of current-voltage (JV). JV measurements are obtained by using a Keithley SourceMeter 2420 at room temperature and 1000 W/m^2. DLCP, CV and AS measurements are carried out by a HP4284A LCR, the temperature is controlled by a Janis cryostat with Lakeshore 325 temperature controller in a vacuum of 10^{-4} Pa and in a temperature range between 100 K and 320 K.

Grazing angle X-ray diffraction (GXRD) was performed at IMEM with a Philips vertical diffractometer with CuKα radiation, Goebel monochromator and with a incident angle of 0.5°.

Accelerated stress tests (AST) were performed in a specific metal chamber, where a rack of halogen lamps and a temperature system control allows to keep the cells under one sun and 80 °C.

3 RESULTS

3.1 JV performance

In order to study the effects of copper on the behaviour of our solar cells, we have changed the amount of copper, depositing the least possible amount, which is about (0,1 ± 0,05) nm.

With the standard fabrication process and the usual annealing of the finished devices at 200°C in air (already mentioned) these cells reach very low efficiencies. From the current-voltage characteristic (see Fig. 1) we can observe a good behaviour in the first quadrant of the graph with no presence of roll-over, a fairly good open circuit potential, but a very low current. This low current may be due to a potential barrier at the back contact or to a low doping of the CdTe bulk.

In order to enhance the current we have applied a second annealing at 300°C in air; which has increased efficiency up to 14 %, as shown in Fig. 1.

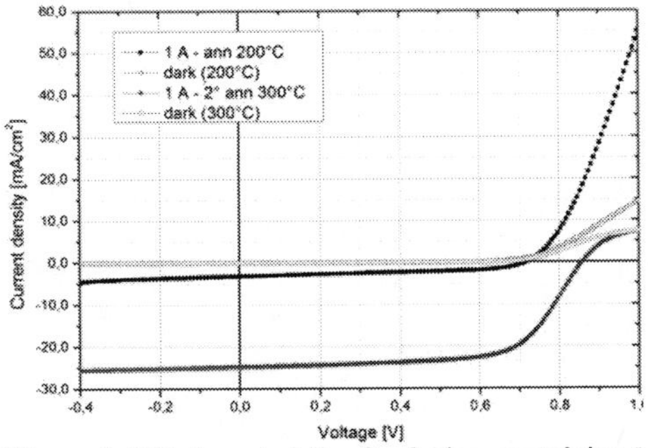

Figure 1: J-V characteristics for devices containing 1 Å of copper: after the first annealing at 200°C and after the second annealing at 300°C.

Table I: Efficiency parameters of devices of Fig. 1.

Annealing temperature	Eff (%)	FF (%)	Voc (mV)	Jsc (mA/cm^2)
200 °C	1,1	48,5	719	3,2
200+300 °C	14,1	66,0	859	25

3.2 Second Annealing

The second annealing improves all the device parameters, but it leads to the presence of a marked roll-over. Probably, only after the second annealing the small amount of copper is diffused into the CdTe, doping CdTe.

In order to have some samples for comparison, together with these cells with 1 Å of copper we have processed also cells with 2 nm thick copper. These have reached efficiencies of about 15 %, just slightly higher than cells containing 1 Å of copper. Comparing the current-voltage characteristics of these cells (see Fig. 2), we notice that the main difference is in the first quadrant of the graph: with 2 nm thick copper layer the roll-over is much less pronounced because most of the amount of copper is still at the back contact improving the performance of the contact.

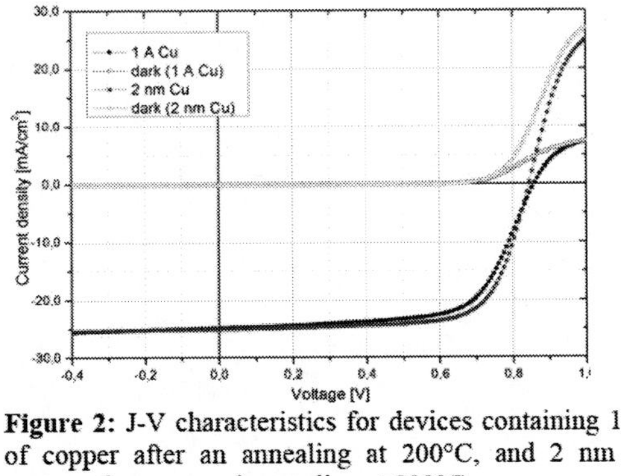

Figure 2: J-V characteristics for devices containing 1 Å of copper after an annealing at 200°C, and 2 nm of copper after a second annealing at 300°C.

Table II: Efficiency parameters of devices of Fig. 2.

	Eff (%)	FF (%)	Voc (mV)	Jsc (mA/cm^2)
2 nm Cu	15,1	71,3	845	25,1
1 Å Cu	14,1	66,0	859	24,9

On the other hand, annealing for a second time at 300°C in air samples with 2 nm thick copper layer, the behaviour is opposite: the efficiency dramatically decreases, as shown in Fig. 3. In particular open circuit voltage and short circuit current decrease, furthermore there is a strong increase of the shunt resistance, which leads to a reduction of the fill factor: probably copper has diffused inside the grain boundaries. After this second annealing there is no presence of roll-over, however seeing the J-V characteristics made in dark, we can observe a lower current flow at high voltages, then, despite the great amount of copper, there seems to be a not perfectly ohmic back contact.

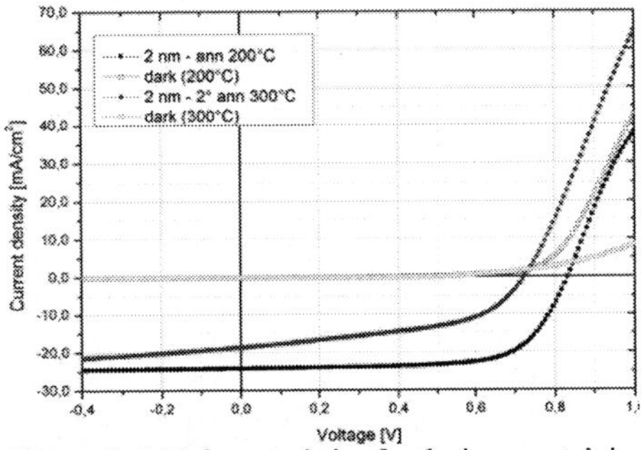

Figure 3: J-V characteristics for devices containing 2 nm of copper: after the first annealing at 200°C and after the second annealing at 300°C.

Table III: Efficiency parameters of devices of Fig. 3

Annealing temperature	Eff (%)	FF (%)	Voc (mV)	Jsc (mA/cm^2)
200 °C	1,1	48,5	719	3,2
200+300 °C	14,1	66,0	859	24,9

3.3 CdTe-back contact interface

GAXRD analysis on copper coated CdTe surfaces has been performed in order to address the formation of Cu-Te compounds. Samples with 0,1 nm Cu (before and after the second annealing at 300°C) have been prepared and their GAXRD patterns have been compared with those of the "2 nm thick Cu case". Surprisingly, the same Cu_xTe_{1-x} compounds have been detected for all the samples, despite the different amount of copper. This result also leads us to conclude that the efficiency improvement after the second annealing for cells with less copper amount is not due to the formation of different Cu-Te compounds at the back contact. .

4 AGING OF THE SOLAR CELLS

A very interesting comparison between cells containing different amount of copper is the degradation during aging tests. The performance degradation of several finished solar cells has been analyzed by measuring their efficiencies at different time steps after having stored under AST conditions as explained in the section 2. A set of cells has been aged in open circuit conditions and a set in short circuit conditions by applying a conductive wire between the front and the back contact.

First of all we have noticed that in devices containing 1 Å thick copper layer, the absorber etching by the bromine/methanol solution does not affect the stability of the cells. Generally cells with or without Br-MeOH etching behave and degrade in the same way, on the contrary cells containing 2 nm of copper need etching to improve the stability of devices [4].

Furthermore, while devices containing 2 nm of copper are more stable in short circuit conditions [4], devices with little amount of copper are more stable in open circuit condition, attesting a totally different degradation mechanism (see Fig. 4).

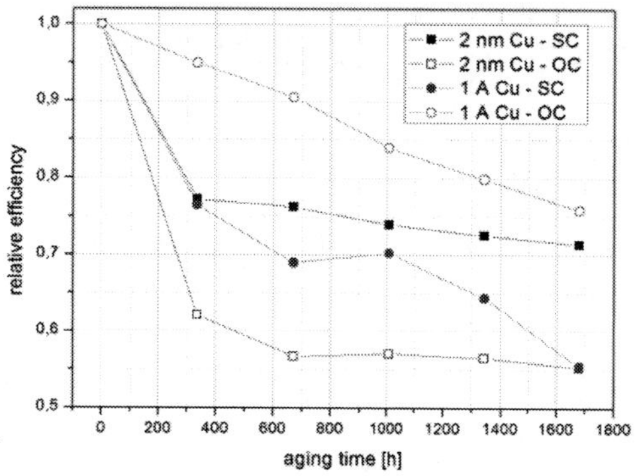

Figure 4: Relative efficiency on aging time (hours). Black dots: samples with 2 nm of copper; red dots: samples with 1Å of copper; aging in open circuit (open dots) and in short circuit conditions (closed dots).

5 CONCLUSIONS

Superstrate configuration CdTe solar cells with 0.1 nm Cu have been prepared for the first time. The average efficiency for the 0.1 nm Cu and 2 nm Cu -contacted cells is in the same range, demonstrating that a very small copper amount is sufficient to dope CdTe. On the other hand a pronounced roll-over is observed for the thin copper case, despite the formation of the same Cu_xTe_{1-x} compounds irrespective of the amount of Cu deposited.

Accelerated stress tests have been applied on both 0,1 and 2 nm Cu contacted cells in short circuit and open circuit condition, showing a very different degradation trend. In the thin copper case, cells degrade faster in short circuit than in open circuit, showing a reduced diffusion of copper.

6. REFERENCES

[1] S. Krum, S. Haymore, First Solar Achieves Yet Another Cell Conversion Efficiency World Record, First Sol. (2016).

[2] L. Kranz, C. Gretener, J. Perrenoud, R. Schmitt, F. Pianezzi, F. La Mattina, et al., Doping of polycrystalline CdTe for high-efficiency solar cells on flexible metal foil, Nat Commun. 4 (2013). http://dx.doi.org/10.1038/ncomms3306.

[3] J. Perrenoud, L. Kranz, C. Gretener, F. Pianezzi, S. Nishiwaki, S. Buecheler, et al., A comprehensive picture of Cu doping in CdTe solar cells, J. Appl. Phys. 114 (2013).

[4] I. Rimmaudo, A. Salavei, E. Artegiani, D. Menossi, M. Giarola, G. Mariotto, et al., Improved stability of CdTe solar cells by absorber surface etching, Sol. Energy Mater. Sol. Cells. 162 (2017). doi:10.1016/j.solmat.2016.12.044.

Performance Characterisation and Extended Reliability Testing of CIGS PV Modules

Peter Lechner, Jonathan Schnepf, Dirk Stellbogen,
Zentrum für Sonnenenergie- und Wasserstoff-Forschung Baden-Württemberg (ZSW),
Meitnerstrasse 1, 70563 Stuttgart, Germany

ABSTRACT: Experimental data from extended reliability testing and from characterisation of the metastable behaviour of CIGS PV modules provide confidence in the bankability of the CIGS PV technology. CIGS modules reveal good stability when subjected to extended DH tests up to 4000 h, by still maintaining 70 to more than 80% of their initial power. The recovery of metastable effects by applying a light-soaking procedure is required after reliability tests; the time dependent behaviour can be fitted with an exponential function. PID chamber tests provide experimental evidence that the degradation of PID-prone CIGS modules is driven by transferred electrical charge. Following this rational, the required test duration of the IEC TS 62804-2 PID test to cover the PID stress load of 25 years field operation can be determined. A case study using different mounting EPDM rubbers for clamping frameless CIGS modules demonstrates that the from-lab-to-field acceleration factor is not unique but must be defined case wise.

Keywords: CIGS Module, metastability, DH, PID

1 INTRODUCTION

Arguments for the competitive bankability of CIGS PV modules need to be supported by results from performance characterisation and extended reliability tests. This holds particularly regarding the following topics.

Metastable changes: the well-known phenomenon of metastability, i.e. module power loss after dark storage at room temperature or after IEC tests like DH and module power gain after light exposure or application of electrical forward bias [1, 2], needs to be demonstrated to be a reversible and reproducible feature of CIGS modules which can be handled without problems. It is also important to know the time constants for metastable power losses and gains, respectively, for properly comparing power monitoring field data to laboratory simulator re-examinations.

Extended damp heat (DH) testing: the IEC 61215 DH test, which is carried out for 1000 h at 85°C and 85% relative humidity (RH), is among the more stringent PV module tests. In the c-Si module community it has already been established to perform extended DH tests [3, 4] to test to failure in order to understand the degradation mode. Extended DH tests can also be employed for comparative reliability assessments to demonstrate that one product or module design outperforms another. Thus, positive outcomes of extended DH tests of CIGS modules could strengthen the confidence in the long term reliability of CIGS modules.

Susceptibility to potential-induced degradation (PID): some types of c-Si and thin film PV modules suffer from PID. A draft for the technical specification IEC 62804-02 TS has been elaborated with the aim to define a test procedure for evaluating PID effects on thin-film PV modules. The chamber conditions have been fixed already at 85°C and 85% RH, while the modules are subjected to maximum system voltage versus ground. However the test duration and thus the severity of the PID stress load is still under discussion and requires more data from PID- tested thin film modules. For CIGS modules we follow the rational proposed from various groups [6, 7, 8] postulating transferred charge as the metric for the stress load and the module degradation during the outdoor field operation and the PID chamber test respectively.

In this paper we present extended chamber experiments with both PID test conditions, close to outdoor field conditions (25°C, 95% RH), and at typical accelerated aging test conditions (60°C, 85% RH). The results suggest further experimental evidence for the equivalence of the PID stress load and the transferred charge ("coulombs").

Further experimental data show that every module type, mounting configuration and climate individually requires the measurement of transferred coulombs in the chamber and in the outdoor field respectively in order to determine the chamber test duration providing equivalent PID stress load as generated by 25 years outdoor operation.

2 EXPERIMENTAL APPROACH

For our experiments we used commercial CIGS glass/glass modules from various manufacturers, both, framed and frameless. For some lab and outdoor PID tests the frameless modules were mounted on a rack with clips using commercially available high resistivity (HR) EPDM rubbers.

The light-soaking tests were performed with a light-exposure chamber at 1000 W/m² and a module temperature at 50°C +/-10°C. For extended DH and PID tests we used an environmental chamber, for I-V curve measurements a pulsed sun simulator class A+AA. After each DH or PID step the modules were stabilized by repeated light-exposure until the module power remained stable within 2%. For the dark storage experiments the modules were kept at room temperature near 25°C and only exposed to 10 ms short light-pulses during each flasher measurements. Electroluminescence (EL) images were taken at forward bias voltage with a current close to the module's Isc.

At the ZSW outdoor test field in Widderstall, Southern Germany, PID test beds are available for monitoring module leakage currents and relevant meteorological data. The bias voltage of -1,000 V (in some cases -500 V) between cell and ground was triggered by the solar irradiance with a switching on/off threshold of 5 W/m². Chamber and outdoor field PID leakage current measurements were performed with a calibrated high precision picoamperemeter.

3 RESULTS AND DISCUSSION

3.1 Extended Damp Heat testing

The behaviour of the normalised power of the DH tested CIGS modules and c-Si benchmark modules, were the data are taken from [5], are revealed in Figure 1. For both module types we observe no or a minor power loss up to 2000 h test duration, while for the CIGS modules 70% to more than 80% of the initial performance is still remaining after 4000 h DH. For comparison, the c-Si modules based on a glass/EVA/TPT construction and used here as a benchmark, develop between 2000 h and 3000 h DH an acetic acid (stemming from the EVA) induced failure mode leading to significant power degradation to 20% to 60% of their initial values after 4000 h DH. According to M. Koehl [5] the required DH test duration in order to cover 25 years outdoor field operation is in the range between about 1000 h and 3000 h DH, depending on the climate where the c-Si modules are deployed and the activation energy for the degradation mode.

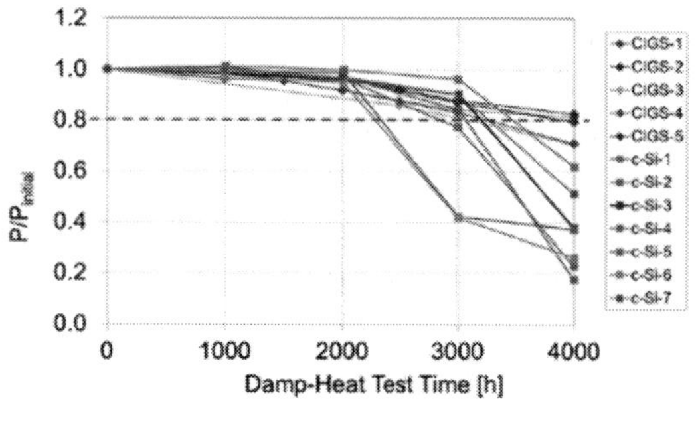

Figure 1: Behaviour of the normalised power of c-Si and CIGS modules after extended DH-test at 85°C/85% RH, data for c-Si are taken from [5]

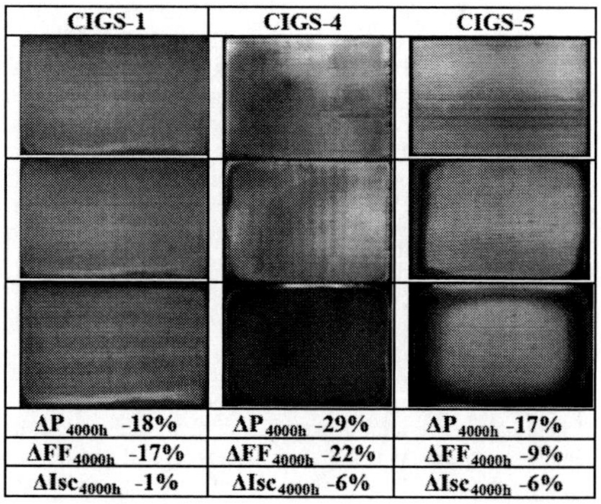

Figure 2: Electroluminescence images of different CIGS modules at the initial state (top) and after 2000 h (middle) DH and 4000 h DH (bottom) and the losses of P, FF and Isc after 4000 h DH

The activation energy for the individual degradation failures still has to be determined for CIGS modules. Nonetheless the fact that CIGS modules outperform standard c-Si modules (which are used at any hot and humid place of the world with a 25 year performance warranty) in a comparative extended DH test is a good precondition for expecting long-term stable operation in hot and humid environments.

It is important to note that the power loss after DH does not always correlate with the EL images. Figure 2 reveals sample CIGS-1 with a power loss of 18%, but an almost unchanged EL image compared with the initial state. In this case the module´s power reduction is based fully on the loss of FF, while the loss of Isc is negligible. In contrast to this reveals sample CIGS-5, which suffers from 17% power degradation coming mainly from 9% FF reduction and 6% Isc loss, an EL image what we intuitively would expect after extended DH testing: inactive and degraded areas near the module edge, probably due to water or humidity ingression. The third example, CIGS-4 with the worst power loss of 29% and an Isc loss of 6% shows an almost completely EL-inactive module area.

3.2 Meta-stable effects

During the extended DH test of CIGS modules a light-soaking procedure is required after each DH step in order to recover from the power loss induced by the dark annealing effect. Figure 3 shows the behaviour of the power recovery after each DH step. For extended DH test durations (2000 h onwards) a full recovery to the 100% initial power is not achieved anymore, the residuum is defined to be the DH induced degradation (as it is plotted in Figure 1).

For the following discussion we define a normalised recovery (R) of the power over time for each individual light-soaking step as:

$$R(t) = (P(t)-P_{min})/(P_{max}-P_{min}), \qquad (1)$$

where P_{min} is the power measured promptly after each DH step before light-soaking and where P_{max} is the power after achieving stabilisation by light-soaking.

The normalised recovery R is shown in Figure 4 in a linear plot and in Figure 5 in a logarithmic plot versus light-soaking time. The recovery significantly slows down and typically takes more than 100 hours for the tested CIGS products. The behaviour of R(t) can be described and fitted with an exponential function:

$$R(t) = 1-\exp-(t/\tau)^{\beta} \qquad (2)$$

where both, the exponent β (best fit with $0<\beta<1$) and the time constant τ need to be adapted for every module product individually.

The experimental recovery data can be checked with the fitted exponential curve whether full recovery was achieved. With this control procedure at hand one can minimise the risk to stop the light-soaking procedure too early (this could happen if one follows the IEC procedure where stabilisation is achieved when the power change is <2% within two LS steps).

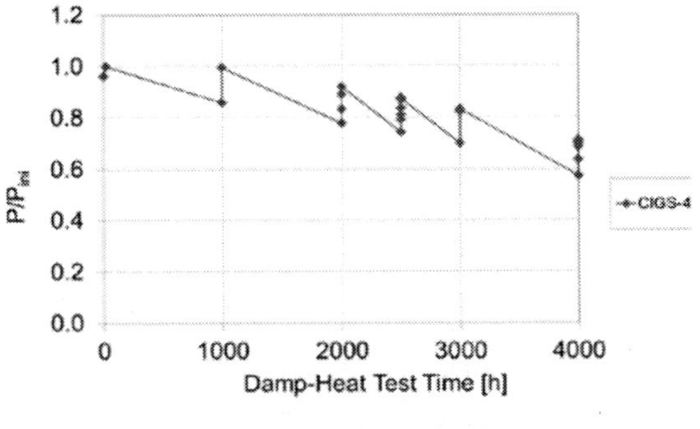

Figure 3: Behavior of the normalized power during the extended DH test of a CIGS module, also including the light-soaking procedure after each DH step

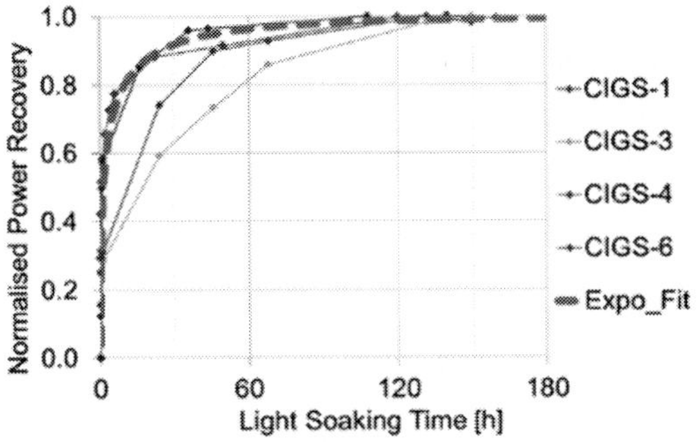

Figure 4: Behaviour of the power recovery R of CIGS modules, experimental data for various manufactures and the exponential fit for CIGS-6

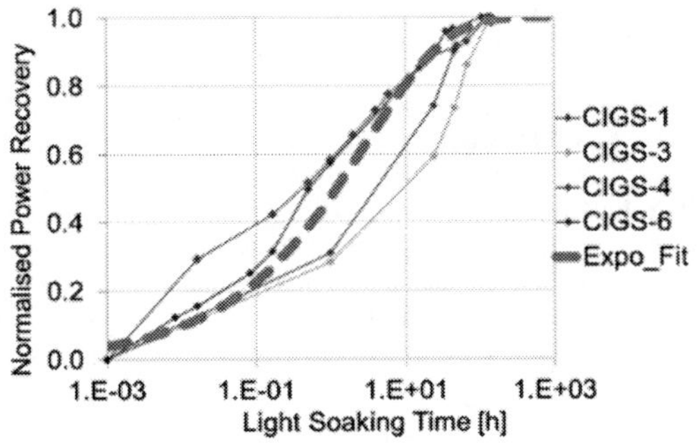

Figure 5: Data from Figure 4 featured in a logarithmic plot

Also the metastability induced power loss of CIGS modules during dark storage is an important issue especially for the practical work with CIGS modules. Figure 6 reveals the power loss over time for different CIGS products. For CIGS-7 we repeated the dark storage procedure and confirmed a very reproducible behaviour. The storage duration required for power loss varies for the tested modules significantly. A power loss of more than 0.5%, which is critical and has a significant impact on flasher measurement uncertainties, takes 20 h for CIGS-6, but only 6 minutes for CIGS-7.

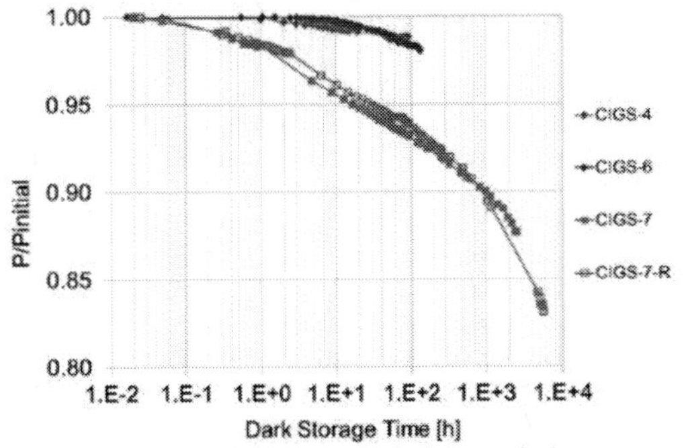

Figure 6: Normalised power during a dark storage experiment for 3 different CIGS products; with CIGS-7 the procedure was repeated after recovery (CIGS-7R)

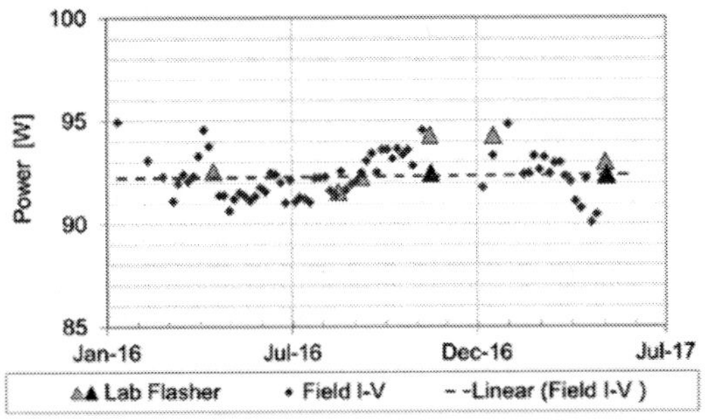

Figure 7: Temperature corrected power of a CIGS module from permanent outdoor I-V monitoring (red dots) and flasher measurement within 2 h (grey triangles) and 1 day (black triangles) after transfer to the lab

It is helpful to know the time constant for the dark storage power loss if fielded CIGS modules are taken to the laboratory for precision flasher measurements, for example in case of power warranty issues. At our module test field in Widderstall I-V curves can be monitored permanently. Figure 7 displays the power of a monitored CIGS module, corrected to 25°C and filtered to irradiance values around 1000 W/m², over the period of almost 1.5 years. This CIGS module, which already has been exposed for several years in the field, does not show a year-to-year degradation, but seasonal fluctuations within one year. The module was re-measured in the lab several times (see also Figure 7); the dwell time for dismantling, transport and measurement was kept short in order to limit the power loss significantly below 0.5%. With this procedure the result of the flasher measurement is found to be in good agreement with the outdoor measurement; also the seasonal power fluctuation can be confirmed. In two cases the lab flasher measurement was repeated the next day, while the module was kept in the dark. By this longer-than-critical dark storage duration a power loss of about 1% to 2% becomes visible.

3.3 Transferred charge as metric for PID

With the following PID experiments we provide further experimental evidence that potential induced degradation of PID-prone CIGS modules is driven by transferred electrical charge ("coulombs"). Two significantly different environmental chamber test conditions were chosen: a "soft", close-to-outdoor condition with 25°C and 95% RH, which resembles cold and humid outdoor situations and an accelerated chamber test condition with 60°C and 85% RH (which is according to one of the IEC TS 62804-1 test conditions for c-Si). Figure 8 reveals the normalised power plotted versus the PID test duration for CIGS A, B and D. It should be mentioned, that after each PID step a light-soaking procedure was executed in order to achieve the light-stabilised state. Obviously product A and B are PID sensitive, while product D is stable against PID stress. As can be seen for CIGS A and B, the power degradation to a certain value takes much longer for the close-to-outdoor condition than for the accelerated test condition. If the data from Figure 8 are plotted versus transferred charge, we come to a different picture (see Figure 9): the transferred coulombs are now a decent metric for the power degradation. At around 5 C both modules A and B have lost about 20% of their initial power, independent whether the test conditions were "soft" or "accelerated".

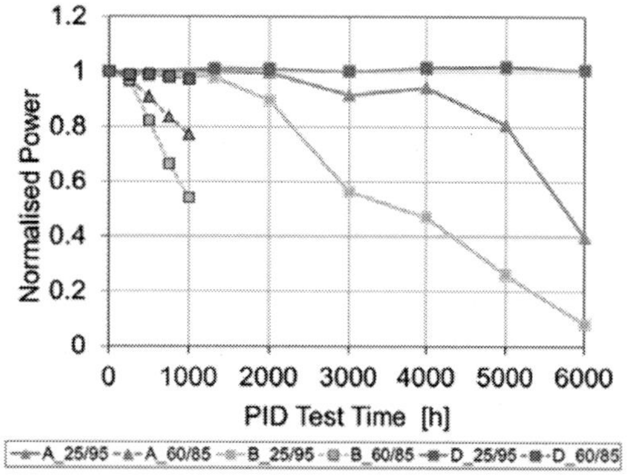

Figure 8: Plot of normalised power versus time of 3 CIGS module types subjected to PID tests at -1000 V and 25°C/95% RH and 60°V/85% RH respectively

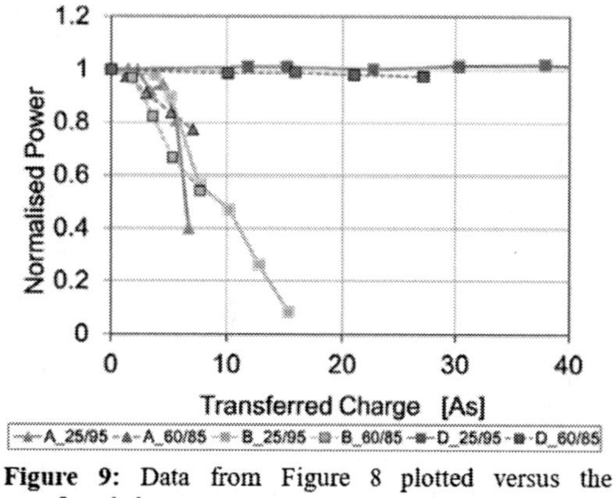

Figure 9: Data from Figure 8 plotted versus the transferred charge

3.4 From-lab-to-field acceleration factor for PID

In this section we show leakage currents and transferred coulombs for frameless CIGS modules, which were mounted on a rack with clips using commercially available high resistivity (HR) EPDM rubbers and standard EPDM rubbers respectively. The HR EPDM rubber should reduce the leakage current flowing from the module's glass surface via the rubbers to ground.

At our outdoor test site at Widderstall we collected detailed leakage current data for both configurations. Figure 10 reveals the time series over one day of the leakage currents for both mounting configurations, with standard and with HR EPDM rubber respectively. In the morning when the modules are still wet from dew, the leakage current shows a peak and is almost similar for both configurations. During the day, when the modules become hot and dry, the standard rubber mounted module shows a second somewhat lower leakage current peak, while the HR EPDM acts as resistor and behaves as intended.

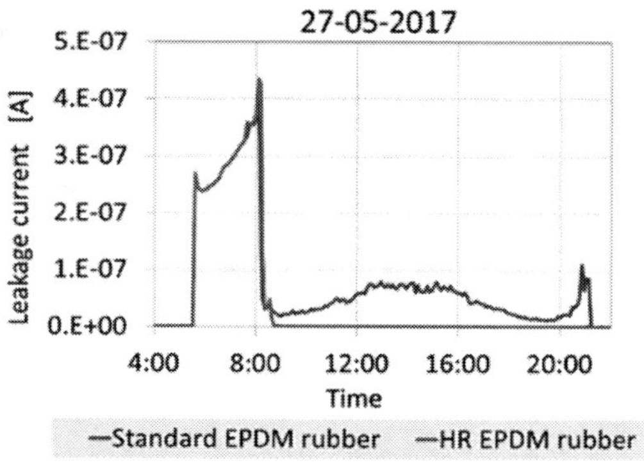

Figure 10: Leakage currents for modules mounted with standard and with HR EPDM rubber over the course of May 27, 2017

In Figure 10 and Figure 11 we show the outdoor leakage current values collected within one year together with the leakage current at 85°C/85% RH (the proposed IEC 62804-2 test condition for thin film) in Arrhenius plots for both, standard and HR EPDM rubber respectively. For the module mounted with standard rubber we observe that the current at IEC condition is significantly higher than the current at any outdoor condition. This means that the IEC PID condition can be expected to be a significantly accelerated PID stress test. For the module mounted with HR EPDM rubber the situation is somewhat different: the leakage current at IEC condition frequently is in the same range as the leakage current at outdoor condition. This behaviour is expected to result in a much lower acceleration factor.

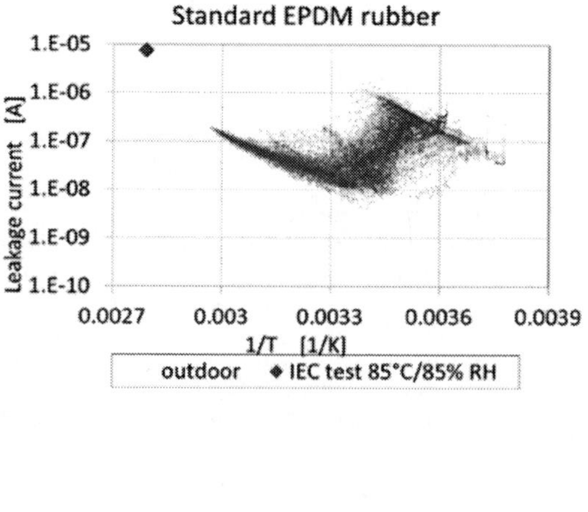

Figure 11: Arrhenius plot of leakage currents from one year outdoor measurement and from IEC chamber test for the standard EPDM rubber configuration

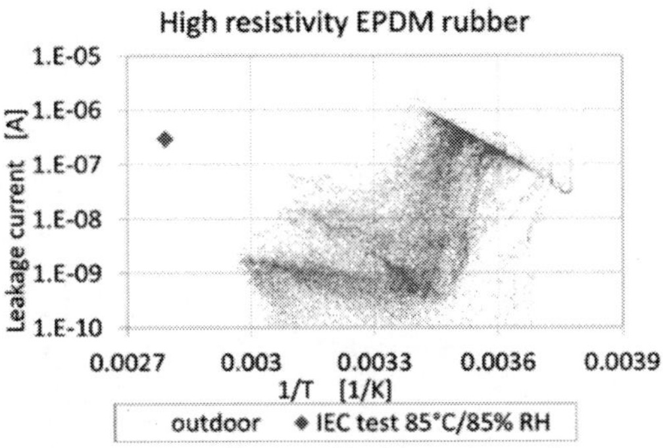

Figure 12: Arrhenius plot of leakage currents from one year outdoor measurement and from IEC chamber test for the HR EPDM rubber configuration

Table I: Transferred charge in different situations

Item	Condition	Standard	HR
1 year outdoor field	Germany; -400V	0.60 C/year	0.40 C/year
1000h IEC test	85°C/85%RH - 1000V	26.9 C	1.06 C
IEC test duration to cover 25years field		550 h	9400 h
Acceleration factor Lab => field		395	23

For the field application of CIGS modules with a maximum system voltage of 1000 V we asume that in the worst case the module at the negative end of the PV module string is subjected to a potential versus ground of -400 V (asume floating system, no grounding applied). Based on outdoor leakage current data shown in Figure 11 and Figure 12 we can calculate, that a charge of 0.60 C/year is transferred in case of the standard EPDM configuration and 0.40 C/year for the HR EPDM configuration respectively (see Table I). During the 1000 h IEC PID test at -1000 V the transferred charge was 26.8 C and 1.1 C for standard and HR rubber respectively.
Based on the rational of equivalent transferred charge, we can further calculate the IEC PID test duration required to cover the PID stress load of 25 years field operation (in Widderstall, Germany) is about 550 h and 9400 h for standard and HR EPDM respectively. The corresponding acceleration factors (from lab to field) are 395 and only 23. This example makes evident that the IEC PID test duration must be individually established for each module type and mounting configuration.

4 SUMMARY

CIGS modules reveal good stability when subjected to extended DH tests up to 4000 h by remaining still 70 to more than 80% of the initial power. EL images show that not the power loss but rather the Isc loss after DH can be correlated to the change of EL features.

Recovery of meta-stable effects by a light-soaking procedure is required after each DH or PID test step. The behaviour of the meta-stable recovery can be described and fitted with a time depending exponential function.

The time constant during for power loss during dark storage was found to be reproducible, however varies significantly for different CIGS module types. It was demonstrated that re-measuring outdoor fielded modules in the lab requires taking into account this critical time.

With PID chamber test experiments, where two significantly different environmental conditions were chosen, one close to outdoor situations and one accelerating condition, experimental evidence was provided that the degradation of PID-prone CIGS modules is driven by transferred coulombs.

Performing both, a one year outdoor field PID experiment and the IEC chamber PID test, we measured the transferred coulombs for frameless CIGS modules, which were mounted on a rack with clips using commercially available high resistivity (HR) EPDM rubbers and standard EPDM rubbers respectively.

In the outdoor experiment a charge of 0.60 C/year is transferred in case of the standard EPDM configuration and 0.40 C/year for the HR EPDM configuration respectively During the IEC PID test the transferred charge was 26.8 C and 1.1 C for standard and HR rubber respectively.

Based on the rational of equivalent transferred charge, we calculate the IEC PID test duration required to cover the PID stress load of 25 years field operation in Germany is about 550 h and 9400 h for standard and HR EPDM respectively. The corresponding acceleration factors (from lab to field) are 395 and only 23.
This example makes evident that the IEC PID test duration must be individually established for each module type and mounting configuration.

5 ACKNOWLEDGEMENT

We would like to gratefully acknowledge the support of a part of this work by the Federal Ministry for Economics and Technology (BMWi), contract number FZ 0325715.

6 REFERENCES

[1] C. Deline, A. Stokes, T. J. Silverman, S. Rummel, D. Jordan, S. Kurtz, Proc. SPIE Optics and Photonics, San Diego, 2012
[2] K. Sakurai, K. Ogawa, H. Shibata, A. Masuda, H. Tomita, D. Schmitz, S. Tokuda, 32nd European Photovoltaic Solar Energy Conference and Exhibition, Munich, 2016, 1690
[3] K. Sakamoto, S. Shimizu, M. Kadowaki, H. Tanaka, K. Suzuki, K. Gobou, C. Yamamoto, T. Doi, A. Matsuda, 29th European Photovoltaic Solar Energy Conference and Exhibition, 2013, 2264
[4] J. Zhu et al., Progress in Photovoltaics: Research and Application 2016; 24,1346
[5] M. Koehl et al., Prog. Photovolt. Res. Appl. 2017, 25: 175-183
[6] P. Lechner, D. Sanchez, D. Geyer, H.-D. Mohring, 27th European Photovoltaic Solar Energy Conference and Exhibition, 2012, 3152
[7] P. Hacke, K. Terwilliger, S. H. Glick, G. Perrin, J. H. Wohlgemuth, S. R. Kurtz, K. Showalter, J. D.

Sherwin, E. Schneller, S. Barkaszi, and R. M. Smith, J. Photonics Energy, Vol. 5, pp. 053083, 2015
[8] T. Weber et. al., Proceedings of the 28th European Photovoltaic Solar Energy Conference and Exhibition, 2013

SEPARATING THE INFLUENCE OF MATERIAL COMPOSITION AND LOCAL DEFECTS ON THE V_{OC} OF CIGS SOLAR MODULES

Johannes Hepp[1,3], Andreas Vetter[1,2], Bernhard Hofbeck[1], Christian Camus[1], Jens Hauch[1], Christoph J. Brabec[1,2]
[1] Bavarian Center for Applied Energy Research (ZAE Bayern)
Immerwahrstr. 2, 91058 Erlangen, Germany
[2] Materials for Electronics and Energy Technology (iMEET)
Martensstraße 7, 91058 Erlangen, Germany: Friedrich-Alexander-University Erlangen-Nuremberg (FAU)
[3] Erlangen Graduate School in Advanced Optical Technologies (SAOT)
Paul-Gordan-Str. 6, 91052 Erlangen, Germany; FAU
E-mail: Johannes.Hepp@ZAE-Bayern.de

ABSTRACT: In spite of the steadily improved quality of co-evaporated $Cu(In,Ga)Se_2$ absorber layers, the existence of local defects affecting the module performance in large area production can still be an issue. Generally, both, lateral composition gradients and point defects can be found in those layers. We show the necessity to separate these two influences and propose two measurement techniques in order to asses the effect of the material and the defects individually. In oder to quantify the theoretical defect-free open circuit voltage (V_{oc}), given by the material composition, we used luminescence spectroscopy. The additional effect on the V_{oc} originating from local defects was determined using dark lock-in thermography. The combination of both these methods enables a local prediction of a module's V_{oc}. In the following, we conduct an indepth analysis of the determination of the material composition and its impact on the resulting V_{oc}.
Keywords: CIGS, Defects, Electroluminescence, Monitoring, Spectroscopy

1 INTRODUCTION

Like any photovoltaic (PV) technology, $Cu(In,Ga)Se_2$ (CIGS) devices strive for even higher efficiencies [1]. Currently, full-size modules are not yet able to reach the record performance values small cells exhibit. In order to transfer these efficiencies to the module level, process development needs to be guided by a quality control tool, which examines performance losses as well as their causes. These losses can be traced back to at least two independent issues. On the one hand the material composition determines the electrical properties of the solar modules. On the other hand, defects like shunts impair the power output of the devices. It is essential to realize that these two factors contribute individually to the module performance. Neither the assessment of merely the material composition, nor the analysis of the defects alone is sufficient for a performance prediction.

In the following, we show the relation between open circuit voltage (V_{oc}) and the luminescence peak wavelength emitted by CIGS photovoltaic samples. Therefore, we compare single cells, for which it is possible to isolate certain properties and defect conditions, with mini modules. In order to provide a scientific relevance of the data we analyzed more than 40 samples for each size.

2 TECHNICAL DETAILS

In order to understand the particular impact of the material composition and the defects individually, we started by analyzing 30 mm x 4 mm PV cells. The advantage is that composition deviations are rather neglectable on such a small area (at least when aiming for assessing performance of a whole module). Furthermore, this approach yields samples of varying defect densities, allowing for separating its impact on cell performance from compositional effects. Subsequently, we moved on to the investigation of mini modules with 16 cell stripes on an 85 mm x 85 mm area. All cell and mini module samples were cut out of a full-size (1200 mm^2 x 600 mm^2) co-evaporated CIGS PV module. All modules were processed in the same way. The structure from bottom to top is glass substrate, a Mo back electrode, a CGIS absorber, a CdS buffer layer, an i-ZnO highly resistive layer, a ZnO:Al front electrode, covered by an encapsulation foile and glass [2].

In order to assure that we only observe the influence of the material composition, we excluded single cells with point defects, using dark lock-in thermography (DLIT) [3][4][5]. The chosen samples exhibited varying ratios of Ga and In. These compositions were determined using X-ray fluorescence spectroscopy (XRF) and are usually expressed as the elemental ratio [Ga]/([Ga]+[In]), called GGI. The used XRF device (custom-built model, Amtec, Germany) requires approximately 30 s for each data point and can only be applied at the non-encapsulated sample.

The electroluminescence (EL) setup used for the investigation of the material influence is a fiber coupled spectrometer (AvaSpecNIR256-1.7, Avantes, Netherlands), sensitive in the NIR range (InGaAs detector: 880 nm – 1620 nm). An additional collimator collects radiation within a measurement spot of ca. 1 mm diameter. The samples are excited using an electrical power source. 1 V was applied per cell (hence, 16 V for mini modules). Additional photoluminescence (PL) measurements were done, using a 532 nm laser as excitation source with a illumination intensity of about 7000 W/m^2. One data point is acquired in less than a second. The evaluation of the spectra is done using an analysis tool programed in Matlab. It fits a polynomial of 4th order around the maxima of the spectra, in order to get a continuous curve and determines its peak wavelength. This helps to increase the accuracy (of determining the peak) even beyond the resolution of the spectrometer which is limited to 3 nm.

3 RESULTS

The relationship between the GGI and the resulting V_{oc} is known for quite some time. In particular, it is a linear correlation within the relevant V_{oc} range (0.46 mV – 0.72 mV) shown in Figure 1 [6].

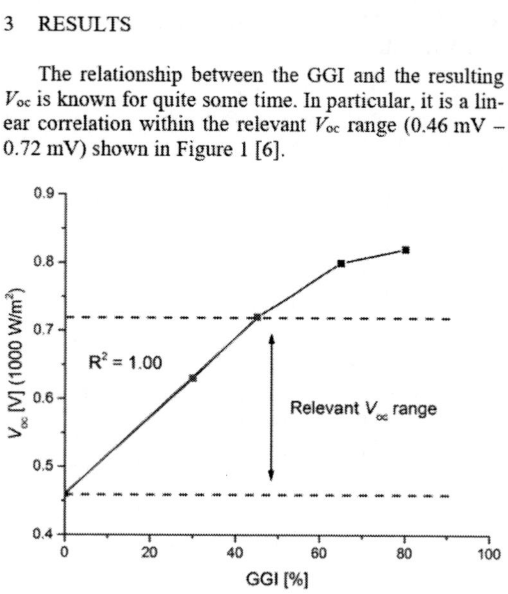

Figure 1: V_{oc} (1000 W/m^2) over GGI (data from [6])

Correspondingly, we also found the relation between the GGI and the position of the luminescence peak to be linear. The results shown in Figure 2 are all data points from one industrial-sized CIGS module. By this means, it is possible to detect the deviations in a commercial product.

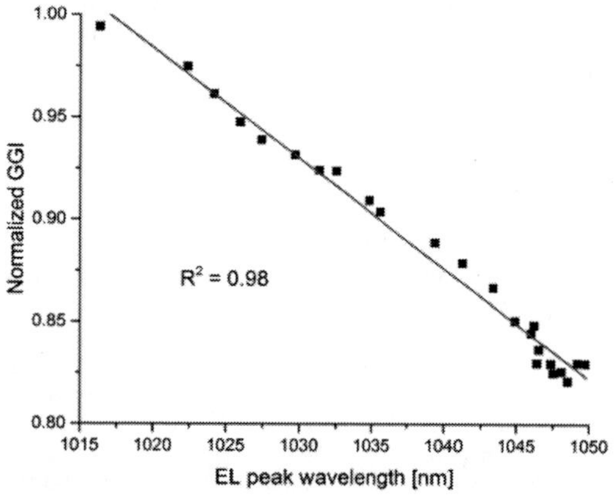

Figure 2: GGI (normalized on maximal value) over EL peak wavelength (quantitative GGI has not been shown due to manufacturer confidentiality)

Whilst luminescence intensity proved not to be a suitable parameter for the sample-to-sample assessment of the material (affected for example by thermal fluctuations and minor local displacements towards the structuring lines or defects), the luminescence peak position is more reliable. Fortunately, neither the encapsulation foil, nor the top glass from the encapsulation process interferes significantly with the EL signal. Moreover, the EL peak wavelength is not affected by close by local defects.

While optimizing the measurement technique, we also compared the influence of the excitation source type on the responding luminescence wavelength. The rather negligible shift between EL and PL for CIGS samples is shown in Figure 3. Thus for the purposes of this study these two types of excitation (photo and electro) are interchangeable.

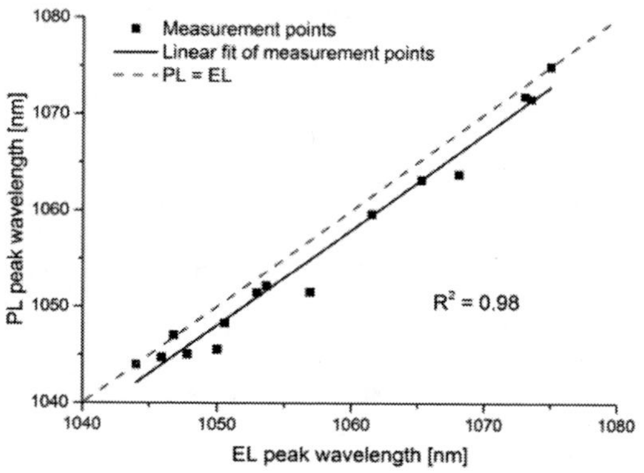

Figure 3: PL peak wavelength over EL peak wavelength

In order to visualize the GGI distribution over an industrial-sized module, we scaned all the mini modules using EL spectroscopy and reassembled them, like a mosaic. An overview of the EL peak wavelength over the whole module (1200 mm x 600 mm) is shown in Figure 4. A steady composition gradient over the module length is clearly noticeable.

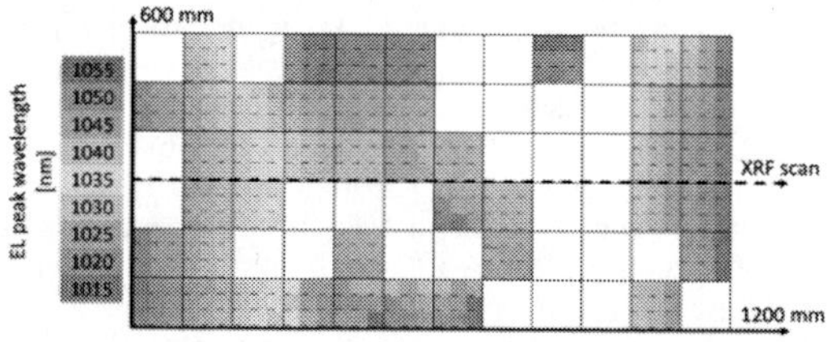

Figure 4: Overview of EL peak wavelength for each mini module within the 1200 mm x 600 mm module (the missing mini modules have been used for other experiments)

Figure 5 displays the relation of luminescence peak wavelength and V_{oc} (1000 W/m^2) for defect-free single cells (hand-picked using DLIT) compared to the values for a regular mini module (including defects). The difference in the quality of the correlation can be seen and quantified by the coefficient of determination (R^2). The mismatch for the small modules is caused by local shunts. This is the reason why, the R^2(mini module) = 0.54 is way lower than R^2(single cell) = 0.86. It is also apparent that, for the same EL peak wavelength (theoretical V_{oc}) the mini modules with local defects exhibit a lower effective V_{oc} than the defect-free single cells. These losses can be accounted for using DLIT and by detecting the intensity of their heat dissipation.

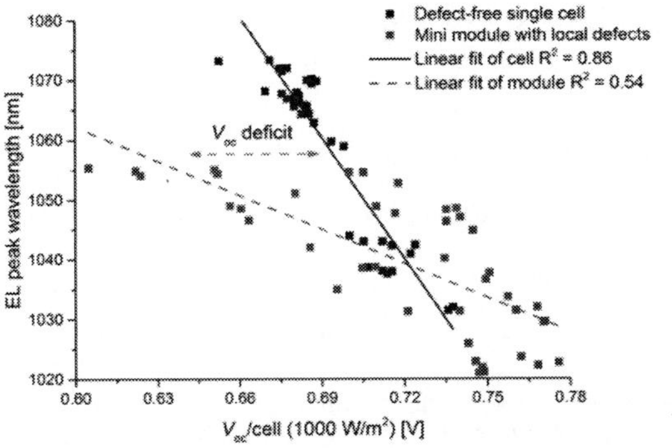

Figure 5: Luminescence peak wavelength over V_{oc} (1000 W/m^2) for defect-free single cells and for mini modules with a varying number of local defects

Moreover, the relation for the mini modules will decline with decreasing illumination intensity (for V_{oc} measurement). That behavior can be explained by the rising effect of the defects on the V_{oc} at lower illumination intensities. Figure 6 shows the comparison between the EL peak wavelength relation to V_{oc} measured at an illumination intensity of 300 W/m^2 and 1000 W/m^2.

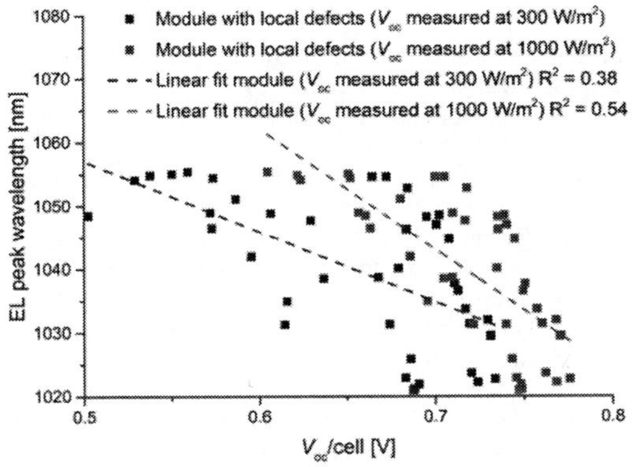

Figure 6: EL peak wavelength over V_{oc} measured at different illumination intensities (300 W/m^2 and 1000 W/m^2)

The extensive results of the defect analysis and combined evaluation of material composition and defect influence are soon to be published.

4 CONCLUSION

To the best of the author's knowledge, this is the first time that the correlation between the EL peak wavelength and the V_{oc} is investigated under the absence as well as under the presence of local defects. It was shown that, the separation of the influence of the material composition and the local defects is necessary in order to determine the V_{oc} of a photovoltaics module correctly. Thanks to the linear correlation between EL peak wavelength and the GGI, the theoretical V_{oc} without point defects can be easily acquired. Neither the absorption of the encapsulation glass, nor close by local defects interfere with this measurement. Further benefits over the GGI acquisition using and XRF device lay in the rapidness and resolution of the measurement at even lower cost. As second measurement technique we propose DLIT, in order to image and quantify the effect of the local defects. The combination of luminescence spectroscopy and lock-in thermography, as well as their correlation to V_{ocs} measured at various illumination intensities, will be detailed analyzed in future work. This method is also expected to work in non-contact mode, using light instead of electricity as excitation source. The large amount of measured sample assures the reliability of the results.

SUMMARY

We investigated the dependency of the V_{oc} and the luminescence peak wavelength. The linear correlation we found is the keystone of our material analyses, and by this means the method to determine the theoretical V_{oc} (V_{oc} without the influence of local defects) of the CIGS photovoltaics modules. This acquisition should work on any CIGS module, regardless of its size and defectiveness. Furthermore, the correlation between GGI, luminescence peak wavelength and V_{oc} was shown in detail.

ACKNOWLEDGMENT

We acknowledge the German Ministry of Economy and Energy for funding (OptiCIGS, 0325724C). Furthermore, Johannes Hepp thanks the Erlangen Graduate School in Advanced Optical Technologies (SAOT) for additional funding.

REFERENCES

[1] P. Jackson, R. Wuerz, D. Hariskos, E. Lotter, W. Witte, and M. Powalla, Phys. Status Solidi - Rapid Res. Lett., Vol. 10 (2016) 583.

[2] T. Feurer, P. Reinhard, and E. Avancini, Prog. Photovolt Res. Appl., Vol 25 (2016) 645.

[3] O. Breitenstein, W. Warta, and M. Langenkamp, Lock-in thermography - Basics and use for evaluating electronic devices and materials, 2nd ed. Springer Verlag, (2010).

[4] A. Vetter, F. S. Babbe, B. Hofbeck, P. Kubis, M. Richter, S. J. Heise, J. Ohland, I. Riedel, and C. J. Brabec, Prog. Photovoltaics Res. Appl., Vol. 24 (2016) 1001.

[5] T. Tran, B. Pieters, M. Schneemann, J. P. Theisen, R. Schäffler, A. Gerber, and U. Rau, 27th Eur. Photovolt. Sol. Energy Conf. Exhib., Vol 1 (2012) 2372.

[6] J. T. Heath, J. D. Cohen, W. N. Shafarman, D. X. Liao, and A. A. Rockett, Appl. Phys. Lett., Vol. 80 (2002) 4540.

TOWARDS AN IMPROVED UNDERSTANDING OF CIGS THIN FILM SOLAR CELLS

*Tetiana Lavrenko**, Ricardo Vidal Lorbada, Dennis Mücke, Thomas Walter,
University of Applied Sciences Ulm, Albert-Einstein-Allee 55, 89081 Ulm, Germany
Balázs Plesz,
Budapest University of Technology and Economics, Magyar tudósok körútja 2, H-1117, Budapest, Hungary
Raymund Schäffler,
Manz CIGS Technology GmbH, Alfred-Leikam-Str. 25, 74523 Schwäbisch Hall, Germany
* Author for correspondence: lavrenko@hs-ulm.de

ABSTRACT: CIGS/CdS thin films and completed devices demonstrate pronounced metastable behavior. On device level, metastabilities can be detected by photoluminescence (PL) imaging. After hot light soaking (HLS) PL intensity of devices increases correlating to the improved Voc. To the contrary, dark annealing degrades PL intensity which corresponds to the reduced Voc values. A question which is still open and being addressed in this contribution is what determines PL intensity on CIGS/CdS thin films. In order to build an appropriate model, PL images are correlated to vibrating Kelvin probe measurements. It has been found that the presence of the CdS layer and induced band-bending at the absorber/buffer interface determine the PL signal from the discussed thin films. The origin of band-bending can be still debated. Furthermore, the impact of band-bending on P1 shunting and the lateral conductivity of the CIGS layer has been investigated. Dark current-voltage measurements on test thin films with a broad P1 scribe (300 μm) indicated that P1 shunting can be modified by the pn-junction formation and certain sample treatments. Moreover, the contribution of P1 shunting to the overall shunt conductance was estimated. It has been shown that shunt conductance of studied devices is not dominated by the P1 shunt and shows perfect stability during long term endurance test.
Keywords: CIGS solar cell, P1 shunting, PL imaging, Kelvin probe measurement, band bending, CIGS lateral conductivity

1 INTRODUCTION

Long term stability of $Cu(In,Ga)Se_2$ (CIGS) solar cells is of crucial importance for commercial success of this technology. It is well-known that CIGS devices demonstrate metastable behavior upon illumination [1] and/or heat treatment [2] which can have positive or detrimental consequences on the overall performance.

One of the measurement techniques to investigate metastable effects of CIGS solar cells is photoluminescence (PL) imaging. [3] A great advantage of this method, in addition to fast data acquisition and high sensitivity, is that PL imaging does not require electrical completion of devices and can be performed already on absorber level. Therefore, a correct interpretation of PL images is very important in order to facilitate deeper understanding of metastabilities in CIGS solar cells.

The interpretation of PL intensity has been under discussion for some time. It has been argued that PL intensity (both steady-state and transient) can be affected by charge carrier lifetimes. Degradation or enhancement of PL intensity would correlate to the decrease or increase in the minority carrier lifetimes, respectively. Another possibility is electric fields inherent to the pn-junction or as a result of band-bending in the absorber layer due to a charge accumulation at the absorber/buffer interface or in the buffer layer. In the literature, a direct correlation of PL intensity and device V_{OC}'s [4], as well as V_{OC}'s and capacitance measurements [2] has been demonstrated. This correlation implies that changes in the charge distribution rather than in the minority carrier lifetimes are responsible for PL metastabilities. A question which is still open and will be addressed in this contribution is what determines PL intensity measured on CIGS/CdS thin films.

As has been mentioned above, one of the possibilities to explain metastable behaviour of PL intensity is induced band-bending. A vibrating Kelvin probe, or vibrating capacitor method [5], [6] allows contactless determination of material work function or surface band-bending, and therefore will be used in correlation to PL images in order to find out what influences PL signal on CIGS/CdS thin films, and build an appropriate model to illustrate underlying mechanism behind recombination kinetics and interpret the PL measurements. It will be demonstrated that PL intensity is determined to a great extent by the presence of the CdS buffer layer and by band-bending induced at the CIGS/CdS interface.

Moreover, the proposed model will be applied to investigate P1 shunting behavior which poses a great interest in the context of weak illumination performance of CIGS modules. The impact of different treatments on P1 conductance will be correlated to the changes in the lateral conductivity of the CIGS layer. Thus, correlating experimental data to the proposed model, the impact of metastabilities on P1 shunting in CIGS/CdS thin films will be discussed. Finally, the contribution of P1 shunting to the overall shunt resistance and its stability during long term endurance test will be presented additionally.

2 EXPERIMENTAL

2.1 Samples under investigation

In this work, CIGS films have been evaporated on the molybdenum back contact sputtered on soda lime glass. As the experiment requires, CIGS absorbers may be covered with the CdS buffer layer (CIGS/CdS - thin film configuration) or completed up to the ZnO bilayer deposition (CIGS/CdS/ZnO - finished device level).

Dark annealing treatment has been performed in the convection oven in the laboratory atmosphere at elevated temperatures. For light soaking treatment, samples are set on the heating plate and exposed to the sulfur plasma lamp illumination. The illumination intensity is adjusted to 100 mW/cm^2. The duration time and temperature will be identified for each experiment individually in the corresponding sections.

The etching treatment will be done in the 5%-solution of the HCl acid for 5 minutes in air.

2.2 Measurement techniques

Current-voltage - (IV) characteristics have been recorded using Agilent 4155 semiconductor analyzer under the Xe-lamp sun simulator. The PL signal was detected with the near-infrared InGaAs camera. For optical excitation, LED arrays with emission wavelengths of 465 nm (blue) and 625 nm (red) are used. For Kelvin probe measurements, a TREK 320C high sensitivity electrostatic voltmeter was used. As the reference probe a graphite electrode with a tip diameter of 100 µm was implemented. A grounded graphite ring was placed around the vibrating electrode to provide additional shielding.

3 RESULTS AND DISCUSSION

3.1 Interpretation of PL images

Fig. 1 represents the PL image of the CIGS/CdS thin films after hot light soaking for 2h at 120°C. The part with an enhanced PL intensity was exposed to light whereas the other part was covered with a foil, and therefore has undergone only dark annealing, resulting in a weaker PL signal.

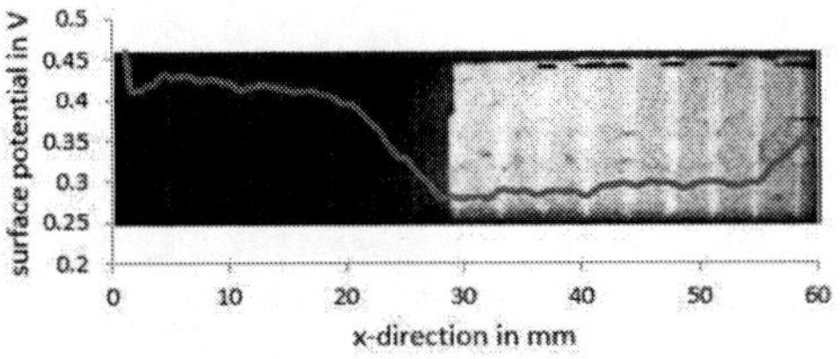

Figure 1: PL image after hot light soaking for 2h @ 120°C. The right part of the sample has undergone light treatment. The left one was covered by a foil, and therefore is only dark annealed. The overlaid green line corresponds to the surface potential scan deduced from the Kelvin probe measurement.

The Kelvin probe measurement – which is reproduced with the green curve overlaid on the PL image in Fig. 1 – points out at the higher values of the surface potentials at the dark-annealed side compared to the values of the light-soaked one. In the discussed arrangement of the Kelvin probe setup, a high value of the surface potential corresponds to a strong band-bending whereas a low value means that a band-bending is reduced.

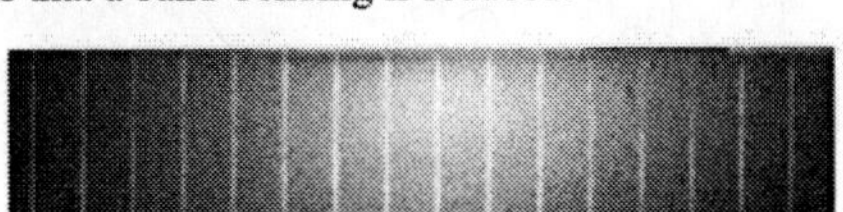

Figure 2: PL image after etching the CdS layer away. The difference in PL intensities between two parts disappears.

The most striking thing in this experiment is that the observed difference in PL intensities disappears after the CdS layer has been removed by etching the sample for 5 min in 5-% solution of hydrochloric acid (HCl). This observation suggests that the CdS layer and detected band-bending have a strong impact on measured PL intensity.

In order to correlate band-bending in the absorber layer and PL intensity a following model is proposed. Fig. 3 shows an energy band diagram simulated in SCAPS 1D. [7] The diagram illustrates two situations: quasi-flat bands (solid lines) and strong band-bending (dashed lines) at the absorber/buffer interface. Under quasi-flat band conditions, upon optical excitation photogenerated electrons can recombine unobstructedly, and thus enhancing PL signal similar to HLS experiment. To the contrary, a strong band-bending would lead to the fast separation of the photogenerated charge carriers and forcing electrons to recombine nonradiatively at the interface. As a result, PL intensity would degrade as has been observed after dark annealing test.

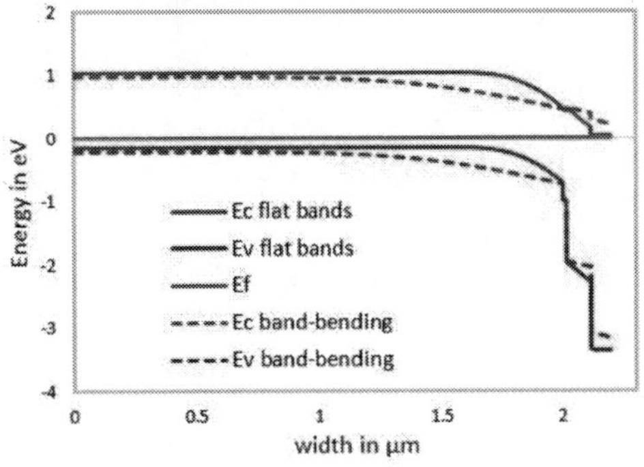

Figure 3: Energy band diagram of CIGS solar cell with quasi-flat bands (solid lines) and enhanced band-bending at the CIGS/CdS interface (dashed lines).

An underlying mechanism for the discussed band-bending can be still debated. A change in the doping concentration in the CIGS layer can modify the width of the space charge region (SCR). Low doping would result in the extended SCR, and therefore in a stronger band-bending at the surface. The opposite is valid for increased doping density. This model is consistent with capacitance measurements on CIGS solar cells discussed in [2]. Another plausible explanation could be a positive charge accumulation at the CIGS/CdS interface or in the CdS layer.

3.2 P1 shunting

If the conductivity of the CIGS material in the P1 scribe increases to a great extent, the parasitic pathways for the light-generated current in CIGS modules can be created leading to significant power losses if not handled appropriately. The effect of reduced shunt resistance is most severe at low light intensities. In [8] it has been shown that the shunt conductance is the dominant loss mechanism at low irradiation conditions.

In [9], the authors investigate P1 scribe behavior correlating PL imaging with Kelvin probe measurements. They observed that PL intensity in the vicinity of the P1 scribe is lower compared to the cell bulk. Moreover, Kelvin probe measurement detected stronger band-bending in the P1 trench. Following the proposed model, low PL intensity must correspond to the extended SCR in the CIGS layer. In the context of P1 shunting, it is worthwhile to consider the impact of the extended SCR on the lateral conductivity of the absorber layer, or, in other words, P1 shunt conductance. For this investigation, a special test structure has been developed. CIGS/CdS thin films have been prepared with a broad P1 scribe: 300 µm instead of reference 90 µm, meaning that the deposited films were continuous and only interrupted by the P1 scribe in the Moly layer where a direct contact of the CIGS semiconductor and the glass substrate was ensured. The schematic of tested samples is drawn in Fig. 4. The P1 shunt conductance has been investigated by measuring current response to applied voltage in the dark while contacting Moly on both sides of P1. The current-voltage measurements in the initial state, after dark annealing for 4h @ 130°C and after etching the CdS layer away are reproduced in Fig. 5.

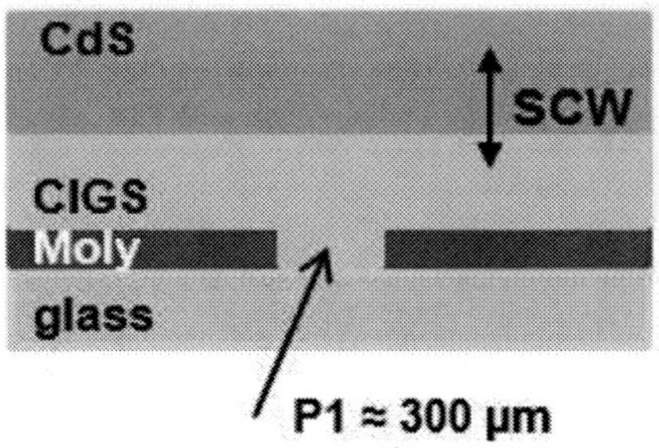

Figure 4: Schematic of a tested sample for investigation of the P1 conductance.

As one can see the current response varies after each treatment. Correlating the experimental data and proposed model, the following inferences can be drawn.

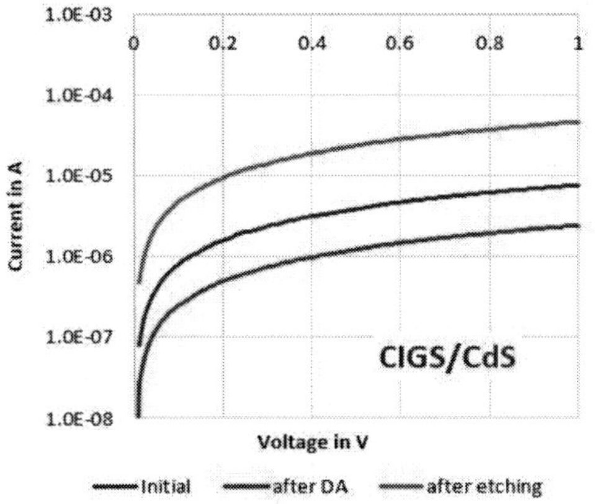

Figure 5: Current-voltage measurements of the CIGS conductance over P1 scribe in the initial state (blue), after dark annealing (red) and etching (green) treatments.

After dark annealing the current flow is reduced compared to the initial state as can be seen in Fig. 5. As mentioned earlier, thermal treatment expands SCR further in the CIGS bulk. Stronger band-bending lowers the CIGS lateral conductivity, and therefore a current flow as can be deduced from the measurement data. After etching the CdS layer away, SCR is eliminated which results in the increased CIGS lateral conductivity and significantly increased current response. This is an important finding which suggests that different treatments and the formation of the pn-junction have a pronounced impact of the CIGS lateral conductivity, and therefore on the P1 shunt conductance.

3.3 Importance of P1 shunting for devices

The influence of shunt resistance becomes superior for the module performance under low light conditions. In this context, the share of P1 shunting in the overall shunt conductance acquires more importance.

In order to estimate the contribution of shunting in P1, dark current-voltage characteristics of single cells (with grid) and interconnected cells (from module) of similar geometry have been measured in the dark. The measurement of two representative cells is reproduced in Fig. 6. The shunting behavior of the two looks similar. There is no visible difference in the current flow in the low voltage regime. The only difference observed for these two devices is increased series resistance for the interconnected cell, which is due to a contacting method from the Mo-to-Mo sides.

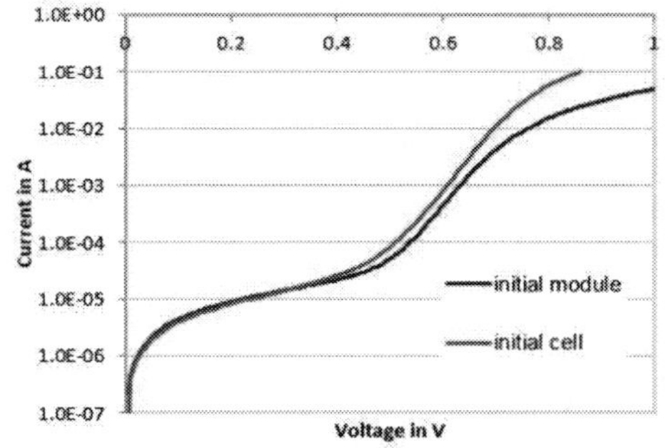

Figure 6: Dark current-voltage measurements of a single and interconnected cell in the initial state.

In addition, shunting around the P1 scribe can also influence the module stability and cause severe losses in the performance. Therefore, its stability is of great importance for the prediction of the lifetimes of CIGS modules. As one can see from Fig. 7 accelerated ageing at 145°C for 1000 h in the laboratory atmosphere proved excellent stability of shunt resistance on the module level.

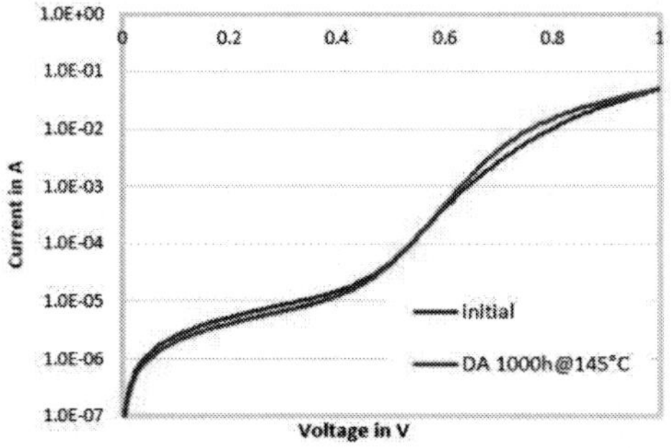

Figure 7: Dark current-voltage measurements of an interconnected cell in the initial state and after 1000h @ 145°C.

All findings considered, one can conclude that within the set of tested samples shunt conductance is not dominated by the P1 shunt. Moreover, the P1 scribe has a minor impact on the module stability during accelerated ageing tests.

4 CONCLUSIONS

CIGS thin films and completed devices demonstrate

pronounced metastable behavior. Such metastabilities can be characterized by PL imaging. On finished device level, PL intensity correlates to the device V_{OC} and can serve as a detection tool for bad cells in a module. In the interim, PL images on CIGS/CdS thin films have to be interpreted carefully as the measured PL intensity is determined by band-bending at the CIGS/CdS interface as has been deduced from vibrating Kelvin probe measurements. Furthermore, the impact of enhanced band-bending on the lateral conductivity of the CIGS layer has been investigated. It has been shown that the CIGS conductance around the P1 scribe depends on sample treatments and is affected by the presence of the CdS layer. On a module level, shunt conductance was not dominated by P1 shunting as has been demonstrated based on comparison between single and interconnected devices. Moreover, shunt conductance of the investigated devices showed a perfect stability during accelerated ageing test.

5 ACKNOWLEDGEMENTS

This study has been funded by the German Federal Ministry of Economic Affairs and Energy within the OptiCIGS project.

6 REFERENCES

[1] T. Lavrenko, F. Schönberger, Y. Wang, M. Teukam, T. Walter, T. Hahn und P. Pistor, „Advanced Luminescence Imaging of CIGS Solar Cells," in *27th European Photovoltaic Solar Energy Conference and Exhibition*, Frankfurt, 2012.

[2] T. Ott, P. Mack, F. Trudel, Y. Schulz, T. Walter, D. Hariskos, O. Klowski und R. Schäffler, „Accelerated Ageing and Luminescence of CIGS Solar Cells," in *26th European Photovoltaic Solar Energy Conference and Exhibition*, Hamburg, 2011.

[3] T. Lavrenko, T. Walter und B. Plesz, „A closer look into metastable effects of Cu(In,Ga)Se2 solar cells," *physica status solidi (c)*, p. 1600197, 23 February 2017.

[4] A. Urbaniak und M. Igalson, „Creation and relaxation of light- and bias-induced metastabilities in Cu(In,Ga)Se2 solar cells," *Journal of Applied Physics*, p. 063720, 2009.

[5] T. Trupke, B. Mitchell, J. W. Weber, W. McMillan, R. A. Bardos und R. Kroeze, „Photoluminescence Imaging for Photovoltaic Applications," *Energy Procedia*, pp. 135-146, 2ß12.

[6] J. Mizsei, „Surface potential transients of ultathin SiO2/Si structures," *Solid-State Electronics*, pp. 951-957, 1991.

[7] T. Lavrenko, K. Marzinzig, T. Walter, B. Plesz und S. Ress, „On the application of the vibrating Kelvin probe method for quality control of Cu(In,Ga)(Se,S)2 thin-film solar modules," *Progress in Photovoltaics*, pp. 1554-1565, 3 February 2016.

[8] B. Litzenburger, S. Pingel, S. Janke, S. Held und R. Alam, „Low light performance of solar cells and modules," in *29th European Photovoltaic Solar Energy Conference*, Amsterdam, 2014.

[9] M. Burgelman, P. Nollet und S. Degrave, „Modelling polycristalline semiconductor solar cells," *Thin Solid Films*, pp. 527-532, 2000.

REVERSE-BREAKDOWN STABILITY OF Cu(In,Ga)Se$_2$ THIN-FILM SOLAR CELLS

Michael Richter*, Mirko Vrenegor, Jürgen Parisi

Laboratory for Chalcogenide Photovoltaics, Department of Energy and Semiconductor Research, Institute of Physics, University of Oldenburg, D-26111 Oldenburg, Germany, *email: michael.richter@uni-oldenburg.de

ABSTRACT: Understanding stability issues of Cu(In,Ga)Se$_2$ (CIGS) modules upon shading is of high relevance regarding applications such as building integrated photovoltaics. With this study, we try to answer the question if the reason for breakdown damage and irreversible defect formation is either a high reverse voltage or a high breakdown current that is imposed to a (shaded) cell in a module. Therefore, reverse current or reverse voltage was applied to a large number of cells and the maximum stress before the cells are damaged was identified to value the resilience. We studied the influence of illumination, light soaking and defects that were already observed in advance. It was found that the voltage, more precisely, the electric field is responsible for the destructive process. We infer this by highly varying maximum current resilience and compared to this, relatively constant maximum voltage values. In case of illuminating the solar cell, the current resilience is drastically increasing. This in turn imply that a total shading of the cell is detrimental for the lifetime of the module. A light soaking treatment is beneficial for the resilience, while minor shunt defects do not harm the stability.

Keywords: CIGS, shunts, shading, stability

1 INTRODUCTION

Thin-film photovoltaic modules with many cells connected in series can imping high reverse voltages to singles cells that are shaded or underperforming – so they are generating less or no current. For solar cells based on Cu(In,Ga)Se$_2$ (CIGS) the stability under reverse stress has not been investigated with respect to the driving force of the irreversible damaging process in much detail [1,2]. It is not clear yet if either the high voltage or the current (burdened from the healthy cells) triggers the breakdown process. To investigate this, we did destruction experiments by applying negative voltage or negative current to single cells until they were damaged. Furthermore, the influence of illumination, pre-conditioning by light soaking, and already existing defects on the resilience were studied by statistical analyzation of the maximum voltage and current that the cells could withstand without permanent damage (e.g., shunting).

2 EXPERIMENTAL DETAILS

Thin-film solar cells with Cu(In,Ga)Se$_2$ (CIGS) absorbers were extracted from a module without encapsulation. The sample's cathode and anode were connected via the adjacent cells by contacting molybdenum with indium patches on the one hand and the aluminum-doped zinc oxide with silver paste on the other hand, respectively. By this, 107 cells with an area of 0.8 cm² have been prepared. All samples were characterized with dark lock-in thermography (DLIT) at +1 V and -3 V to detect any defects and classify the samples to five defect categories: 0 = no defects ... 4 = strong local defect signature within the cell area. Half of the samples were light soaked at 1000 W/m² for 9 hours at 90°C. The increase of charge carrier concentration due to light soaking (LS) [3] was verified with capacitance-voltage measurements.

The destruction experiments were carried out with applying a constant negative voltage or a constant negative current, while the cell's voltage and current was measured with high temporal resolution (~ 30 ms steps). The absolute value for the applied voltage or current was increased incrementally by 10 mV and 0.5 mA after one second, respectively. These experiments were done in dark and under illumination at standard test conditions. By this experimental design eight configurations (V / I, as grown / LS, dark / light) with 10 to 14 cells each could be grouped and compared.

To value the resilience, we identified the measured maximum voltage and current just before an abrupt change in the transient characteristic was observed. Figure 1 shows exemplarily the measured transient IV characteristics of samples that were excited with a voltage under illumination (top) and with current in dark (bottom). Red circles mark the maximum stress–point of maximum resilience–of the cells.

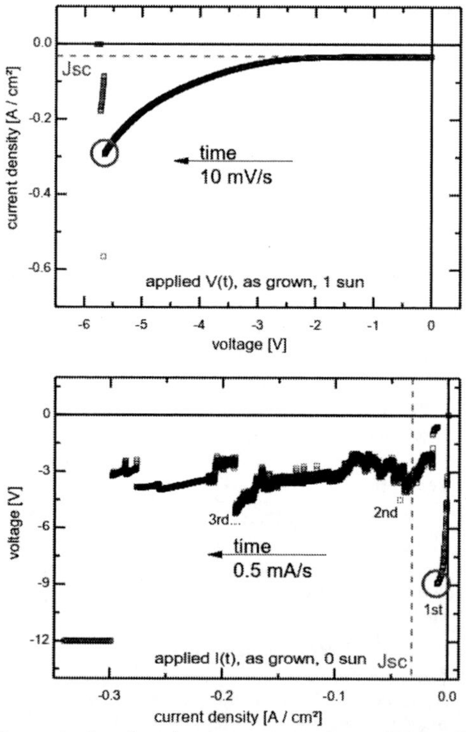

Figure 1: Quasi-stationary current-voltage characteristic of cells with applied voltage (top) and current (bottom).

The applied values were increased after short dwell time. Red circles identify the state just before permanent damage occurs to the cells. The red dashed line mark the short circuit current density of the cells. At the bottom cell, multiple damaging occurs (1st breakdown, 2nd ...).

3 RESULTS AND DISCUSSION

Figure 2 a) and b) show the DLIT image at +1 V and -3 V of an arbitrary sample before the destruction experiment, respectively. This sample exhibits already clear shunt signals but with fairly good performance as the IV characteristic in Figure 2, right shows. After conducting the destruction experiment a "worm" shunt has been formed (Figure 2 c) that is occurring exactly on a spot where a shunt was detected before. However, from all samples with pre-existing DLIT shunt signatures only half of it shows damages on those locations after applying breakdown experiments. We found a higher accordance at higher defect categories. The worm formation leads to a collapsed IV characteristic (Figure 2, right, blue line).

It should be noted that the damage does not happen directly after switching to the next value of voltage or current, but with some delay of around 0.5 seconds. A heating process is likely to play a significant role in this. With respect to the temporal resolution a maximum breakdown current during the destruction process of approximately 1 A was detected, which is in range of a typical module current. The worm is then forming at a local spot and moving to the P1 structuring line, following the electric field in the cell stripe.

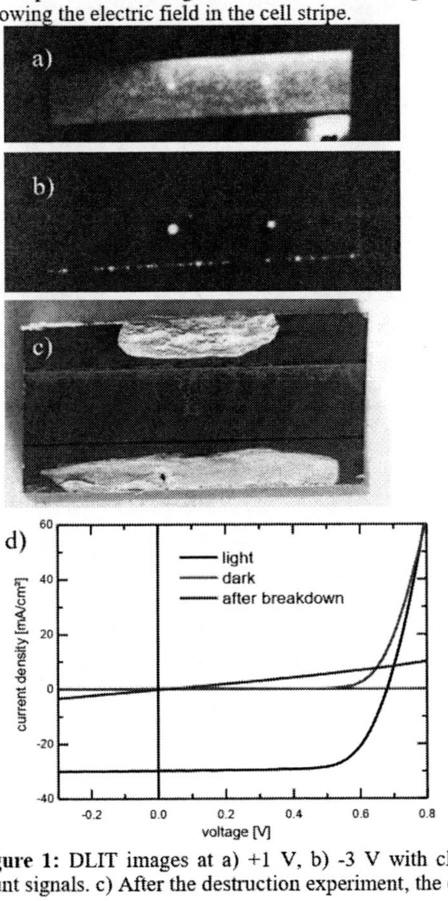

Figure 1: DLIT images at a) +1 V, b) -3 V with clear shunt signals. c) After the destruction experiment, the cell shows worm like damages (P1 line up). d) IV characteristic of the cell after the breakdown measured in dark.

Often, the reverse current is dramatically reduced after the damage occurred (compare Fig. 1, top). Here, we assume an isolation of the shunted region due to evaporation of the transparent conductive oxide layer.

The identified resilience values are shown as box plots in Figure 3. The clustering includes the different defect categories (0 - 4) but not the excitation method V(t) or I(t) as they show similar behavior. The latter reveal that the cell is reacting identically regardless of the excitation with voltage or current. Defect category 3 and 4 contain only few samples.

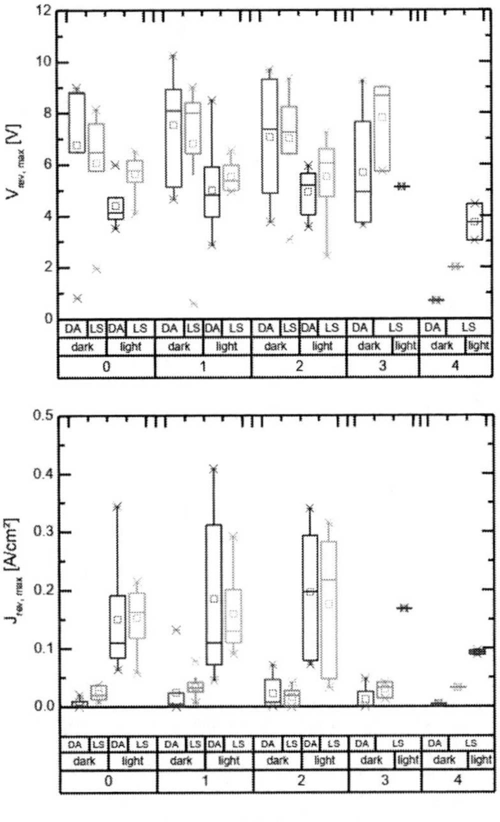

Figure 1: Maximum resilience values of reverse voltage (left) and reverse current density (right) before permanent breakdown damage occurs. The boxes include the min./max. values (x), quartiles (lines) and the mean (square). The samples are clustered by defect category (0 – none ... 4 intense DLIT signature), illumination, and conditioning (LS – light soaking, DA – stored in dark).

A key finding of this experiment is the strong illumination dependence of the resilience. The maximum reverse voltage is slightly lower in dark but the maximum current is much higher with illumination. In some case in dark, the resilience current did not even reach the short circuit current. As the short circuit current is usually not leading to permanent damage and the fact that the resilience voltage is (compared to the current) uniform, we ascribe the voltage or more precise the electrical field to the driving force of the damage process. Furthermore, the applied reverse power at the damage point is much different in dark or light, so a common condition is a high voltage that is either directly applied or realized in the cell by applying the current.

The light soaking lead to higher current resilience in dark (reversible breakdown enhanced), while with illumination no significant change by light soaking is seen. The maximum voltage increases after LS but does not change in dark. It seems as if the LS treatment reduces the strong illumination dependence of the resilience. If the transport at reverse voltage is realized by either (trap-assisted) tunneling, impact ionization, or Poole-Frenkel emission the electric field in the space charge region is the most important factor for the height of the current. We could identify the blue light in the spectrum to be the major cause for the enhanced current transport at reverse voltage. This breakdown trigger effect by blue light is in agreement with many other studies [4,5]. The generation of holes in the buffer layer by blue light could reduce the high density of negative static charge at the absorber surface region (p+ layer) [6]. By this, the type inversion in the absorber is increased leading to a stronger electric field. This process would not only explain the light dependence but also the light soaking effect, which is assumed to increase the negative charge in the absorber's space charge region [7].

The influence of defects is minor. However, the mean values of current resilience slightly increase with defect category. This means that weak defects enhance the reversible breakdown mechanism. Three-dimensional electro-thermal simulations could verify this [8]. Nevertheless, if the shunt signal in DLIT is very strong the resilience is decreasing significantly. Here, we assume a strong localization of the breakdown current leading to high damage probability.

To judge the stability of the cell in a module it becomes apparent that the cells tent to irreversible breakdown if they were completely shaded, not illuminated before (no light soaking) and have no clear defects. In this case, the Impp from the illuminated cells in the module is sufficient to cause damages.

4 CONLCLUSION

Regarding the stability of the cell in a module, we found that cells tend to exhibit permanent defects (break down damage) when they were not illuminated (shaded), not illuminated in advance (no light soaking) and contain strong defect (whereby the defects not necessarily harm the cell performance very much). In case of an operation without illumination and without LS a damage due to the module's Impp is possible. However, a 100% shading of a single cell is not likely in a thin-film photovoltaic module as the aspect ratio is high and the cover glass allows for stray light. The minimum illumination intensity that is required to prevent damage by the module current should be quantified by further studies.

ACKNOWLEDGMENT

We acknowledge the German Ministry of Economy and Energy for funding (OptiCIGS, 0325724C) and R. Schaeffler and J. P. Theisen (MANZ AG, Germany) for providing the solar cell samples. The LCP group gratefully acknowledge financial support by the EWE AG Oldenburg, Germany.

REFERENCES

[1] J. E. Lee, S. Bae, W. Oh, H. Park, S. M. Kim, D. Lee, J. Nam, C. Bin Mo, D. Kim, J. Yang, Y. Kang, H. Lee, and D. Kim, Progress in Photovoltaics: Research and Applications 24, 2016, pp. 1035–1043.

[2] T. J. Silverman, L. Mansfield, I. Repins, and S. Kurtz, IEEE Journal of Photovoltaics 6, 2016, pp. 1333–1338.

[3] S. J. Heise, V. Gerliz, M. S. Hammer, J. Ohland, and J. Keller, Solar Energy Materials and Solar Cells 163, 2017, pp. 270–276.

[4] S. Puttnins, S. Jander, A. Wehrmannc, G. Benndorf, M. Stölzel, A. Müller, H. von Wenckstern, F. Daumea, A. Rahma, and M. Grundmann, Solar Energy Materials and Solar Cells 120, 2014, pp. 506–511.

[5] P. Szaniawski, J. Lindahl, T. Törndahl, U. Zimmermann, and M. Edoff, Thin Solid Films 535, 2013, pp. 326–330.

[6] M. Richter, C. Schubbert, P. Eraerds, J. Parisi, I. Riedel, T. Dalibor, and J. Palm, Solar Energy Materials and Solar Cells 132, 2015, pp. 162–171.

[7] S. Chen, T. Jarmar, S. Södergren, U. Malm, E. Wallin, O. Lundberg, S. Jander, R. Hunger, and L. Stolt, Thin Solid Films 582, 2015, pp. 35–38.

[8] M. Richter, J. Neerken, and J. Parisi, Proceedings 32nd European Photovoltaic Solar Energy Conference, Munich, 2016, p. 1116–1119.

IMAGING OF TCO LATERAL RESISTANCE EFFECTS IN THIN-FILM PV MODULES BY LOCK-IN THERMOGRAPHY AND ELECTROLUMINESCENCE TECHNIQUES

Archana Sinha[A], Martin Bliss[B], Xiaofeng Wu[B], Subinoy Roy[A], Ralph Gottschalg[B], Rajesh Gupta[A*]
[A]Department of Energy Science and Engineering, Indian Institute of Technology Bombay, Mumbai – 400076, India
[B]Centre for Renewable Energy Systems Technology (CREST), Wolfon School Mechanical, Electronic and Manufacturing Engineering, Loughborough University, Loughborough LE113TU, UK
*Email: rajeshgupta@iitb.ac.in, Phone: (+91-22) 2576 7837

ABSTRACT: The lateral sheet resistance of transparent conductive oxide (TCO) electrode in thin-film photovoltaic (PV) modules is a major component of series resistance losses that causes significant reduction in the fill-factor and output power. This paper presents the investigation of TCO lateral resistance effects in the encapsulated thin-film modules by lock-in thermography (LIT) technique, which is predominantly used for shunt investigation in the solar cells. The LIT technique has been employed under both dark and illuminated conditions to compare their spatial sensitivity for imaging TCO resistance effects in a module. The LIT images have also been compared with electroluminescence (EL) images to find a correlation between localized heating and voltage drop across distributed TCO layer resistances, and to determine their advantages and limitations. Experimental results show that both, DLIT and ILIT, exhibit a gradient in thermal signal along the cell width due to variation in power dissipation across the lateral resistance of TCO electrode. However, ILIT appears to be more sensitive for imaging TCO resistance losses due to less junction masking effect. The spatial sensitivity also depends on the width of cell in a module. For narrower cells, DLIT and EL techniques are observed to be more sensitive near the higher potential edge of a cell as compared to ILIT. The study concludes that the LIT technique is also a potential candidate for providing the spatially-resolved characterization of TCO resistive losses in thin-film modules.
Keywords: transparent conductive oxide; lateral sheet resistance; thin-film; photovoltaic module; lock-in thermography; electroluminescence imaging

1 INTRODUCTION

Transparent conductive oxides (TCO) are widely used as a transparent front electrode in most of the thin-film technology photovoltaic (PV) modules due to high optical transmittance (~90%) and low electrical resistivity (~10^{-4} Ωcm). The sheet resistance of TCO material is a key parameter in determining the electrical performance of PV module since the current flows laterally in this layer, as illustrated in Fig. 1. The outdoor degradation of thin-film modules under the moisture and thermal stresses cause a gradual increase in TCO resistivity [1], that increases the overall series resistance of the module and thus a decrease in fill-factor and maximum power [2], [3]. In order to achieve the goal of higher energy yield and improved service lifetime, the TCO lateral resistance losses need to be investigated.

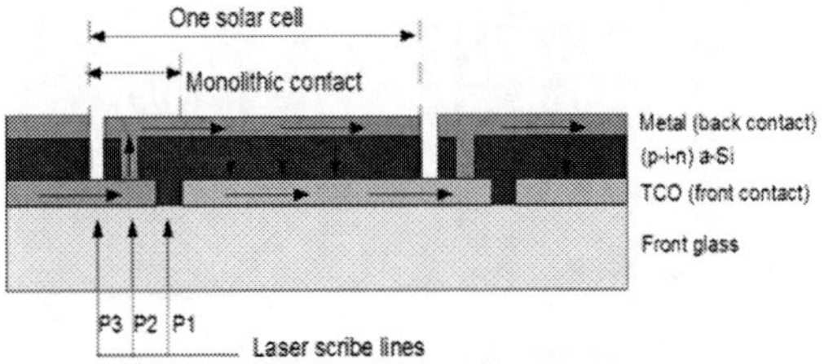

Figure 1: Structure and direction of current flow through the TCO layer of a-Si thin-film solar module

Initial measurements of TCO resistivity were performed by Four probe or Hall effect measurement [4], [5]. However, these methods are only applicable on standalone TCO thin-films. In case of a complete PV device, the camera based electroluminescence (EL) imaging technique has been used as a non-destructive characterization tool for the qualitative and quantitative analysis of TCO lateral resistance effect in coupled with the distributed electrical modelling and simulation approach [6]–[8]. However, EL measurements need a dark environment and higher integration time in thin-film devices for increasing the EL counts (particularly a-Si) [9]. Lock-in thermography (LIT) technique is another popular imaging technique, which characterizes any resistive losses that cause extra power dissipation in the cell. The built-in lock-in feature of this technique allows the detection of weak heat sources with improved spatial resolution by overcoming fast thermal diffusion of heat in the active layer (silicon) and packaging materials. LIT is predominately used for the detection, localization and classification of localized shunts in the c-Si solar cells [10], [11]. Some of the recent studies reported the application of LIT for the detection of processing induced defects or shunts in different thin-film technology modules [12]–[14]. However, it is less utilized to investigate TCO resistance losses. Further, the LIT at the module level is seldom done.

2 METHODOLOGY

In the present study, LIT technique has been employed under both dark and illuminated conditions for the investigation of TCO lateral resistance effects in thin-film PV modules. Dark lock-in thermography (DLIT) is the most common mode of LIT that detects the flow of externally applied current through the cell and enables to identify all the resistive paths contributing to the power losses. In illuminated lock-in thermography (ILIT), the power losses due to series resistance preferentially appear at short-circuit condition, while the dark saturation currents and recombination losses are best observed under open-circuit condition. In view to this, the ILIT investigation has been performed under short-circuit condition. In addition, the EL imaging technique has been utilized as a reference tool for comparison of results obtained from LIT technique, along with their advantages and limitations.

3 EXPERIMENTAL

The experiments were carried out on some single junction amorphous-Si mini-modules, having glass in the front side and a polymer material at the rear side. These modules contain different number of monolithically-connected cells. The electrical parameters of the samples were extracted from their illuminated I-V characteristic curves, which is given in Table 1. It shows that M-3 and M-4 samples have higher FF and efficiency than M-1.

Table 1: Electrical parameters of a-Si mini-modules

Sample	No. of cells	P_{max} (W)	J_{sc} (mA/cm^2)	Cell V_{oc} (V)	FF (%)	η (%)
M-1	1	0.14	10.4	0.90	33.4	3.4
M-3	3	0.12	13.5	0.88	49.9	6.8
M-9	9	0.14	10.7	0.87	53.9	6.0

The LIT investigation was performed by a Stirling-cooled IR camera, consisting of 320 × 256 pixels InSb detector with spectral sensitivity in the range of 3–5 μm. In DLIT, the module was excited by injecting a periodic forward electrical current equivalent to Isc, while in ILIT, the front surface of module under short-circuit condition was periodically irradiated with 850 W/m^2 ± 5%, using 625 nm wavelength (red light) LED source. The temperature modulation was measured at the module rear surface by IR camera. A controller was used to synchronize the power supply excitation with the frames of IR camera to implement lock-in algorithm over the captured images. The lock-in frequency was fixed at 0.5 Hz to assist deeper probing as lock-in frequency is inversely proportional to thermal diffusion length. The thermal images of module were captured at 150 Hz for 200 s. A larger integration time improves the signal-to-noise ratio of the captured images due to reduction in thermal noise level, which is facilitated by averaging over more number of cycles.

The EL imaging was carried out by 1024 x 1024 pixels Si-CCD cooled camera in a dark enclosure. A constant forward current (Isc) was applied into the module and the luminescence radiation was captured through the front glass by EL camera. The integration time was kept 5 mins to increase the EL emission counts from the sample. The image quality was further improved by employing the image subtraction on the EL images of module with and without excitation that reduces the bad pixels of camera and background noise.

4 RESULTS AND DISCUSSION

The DLIT and Jsc-ILIT techniques have been exploited for investigation of TCO sheet resistance effect by correlating it with the extent of power dissipation in the cell. Fig. 2(a) presents the DLIT image and line profile of M-3 mini-module, which shows a characteristic gradient of thermal signal along the cell width (x-direction) with almost no change in signal along the cell length (y-direction). This lateral gradient appears due to the variation in power dissipation along the cell width. The magnitude of lateral external current flow is higher at the injection edge of cell, which decreases as approaching the collection edge of cell due to lateral sheet resistance of TCO layer and back contact, as illustrated in Fig. 3(a). As a result, the power dissipation is higher near positive terminal compared to negative terminal of cell and creates a thermal signal gradient along the cell width. The thermal gradient shows good correspondence to the gradient pattern visible in EL image of same mini-module, as shown in Fig. 2(c) [15]. In this case, the injected current from positive terminal introduces more minority carrier density that causes high radiative recombination at this end and hence high EL intensity. As the position moves towards the other end, the current distribution in the junction changes because of voltage drop across TCO sheet resistance that decreases EL signal.

Fig. 2(b) presents the Jsc-ILIT image and line profile of same mini-module. It also exhibits similar pattern across the cell width, besides little higher thermal signal at the upper edge which is attributed due to small non-uniformity in illumination along the cell length. In this case, almost uniform current generation takes place in the entire module that flows towards the higher potential side of cell and causes increase in the power dissipation at that end, as represented in Fig. 3(b). An interesting observation is found that the decline of thermal signal in Jsc-ILIT is slower than in DLIT. It is because DLIT injects power into the device that is also lost through the junction as heat, while in ILIT, the energy produced can exit the device that is dissipated at short-circuit point and not in the junction, which infers that the lateral gradient is mainly due to TCO losses in ILIT than in DLIT. This suggests that ILIT technique is more appropriate for investigating TCO lateral resistance effects in a PV module as compared to DLIT.

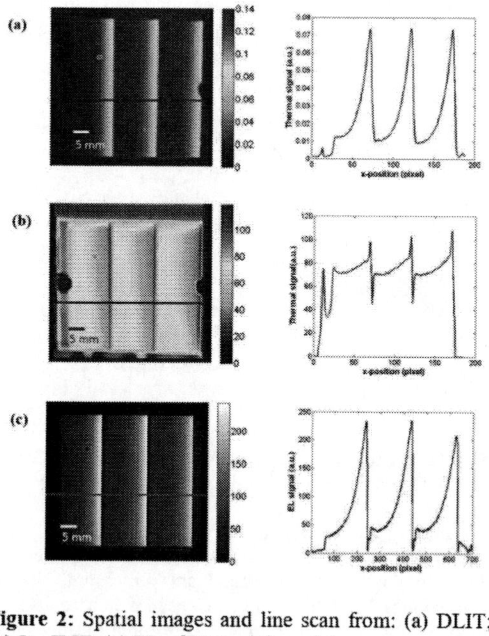

Figure 2: Spatial images and line scan from: (a) DLIT; (b) Jsc-ILIT; (c) EL of M-3 mini-module

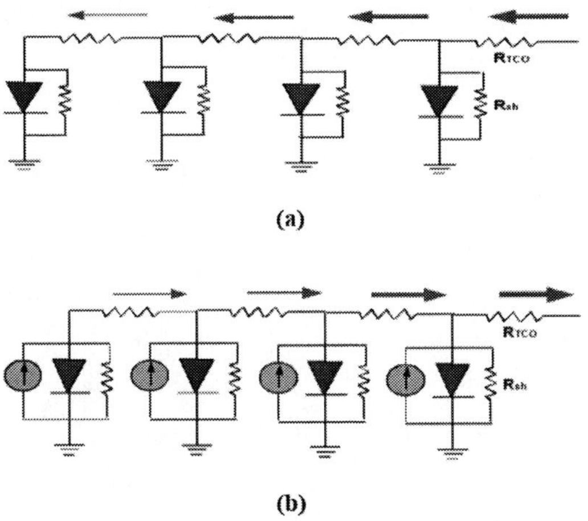

Figure 3: Schematic of distributed circuit of a solar cell indicating the current flow in TCO layer of thin-film module under (a) dark and (b) illumination

The lateral dimension of cell controls the current transport in the module and hence the resistance losses. In order to investigate the effects of cell width on the TCO lateral resistance, both LIT and EL experiments were performed on two other mini-modules of varying cell width. Fig. 4(a)-(c) and Fig. 5(a)-(c) show DLIT, ILIT and EL images of M-1 and M-9 mini-modules respectively. The gradient due to lateral sheet resistance of TCO layer in a wider cell (M-1 sample) is following trend similar to an exponential decay due to logarithmic behavior of junction current to the voltage. In narrower cells (M-9 sample), the gradient appears to be more linear. It is because the resistive path is shorter in a narrower cell, which reduces the voltage shift and makes the current distribution more even. The intensity of EL signal depends on the voltage at that point and the current through the junction. In DLIT, similar effect occurs but the power dissipation is seen from the non-radiative recombination which ends up as heat and I^2R losses through TCO layer; whereas in ILIT, the heating contribution comes from non-radiative recombination at the junction, I^2R losses through TCO layer and the absorbed light energy that can't be converted to current generation. This observation is again in agreement with EL images. By comparing all the images and line profiles, ILIT appears to be more sensitive to the variation of TCO sheet resistance compared to EL and DLIT because of less junction masking. Further, in the case of wider cells, DLIT and EL loses the characterization capability of TCO losses as the output signals get almost saturated.

Also, M-9 mini-module contains some point defects, which are identified as shunts due to increased local heating in the DLIT image. These strong shunts decrease the lateral gradient pattern in both DLIT and EL images since they sink current from surrounding reduce the current flow in lateral direction, resulting in reduced power dissipation and EL emission respectively. The shunting effect is not visible in Jsc-ILIT image because the generated photocurrent flows through the external circuit and bypasses the local leakage paths.

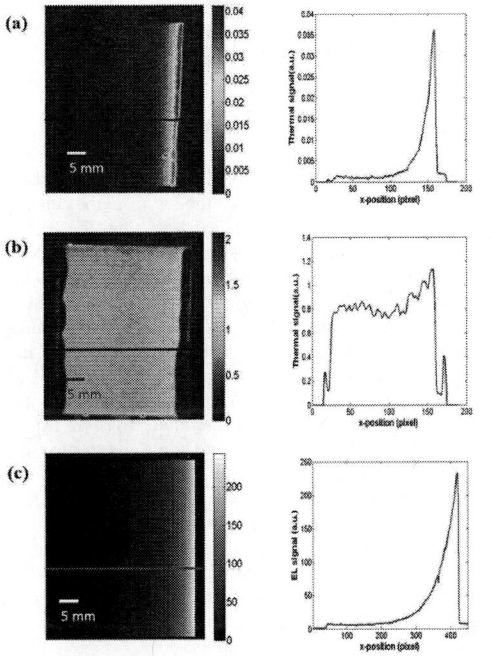

Figure 4: Spatial images along with line scan of M-1 module from (a) DLIT, (b) Jsc-ILIT, and (c) EL.

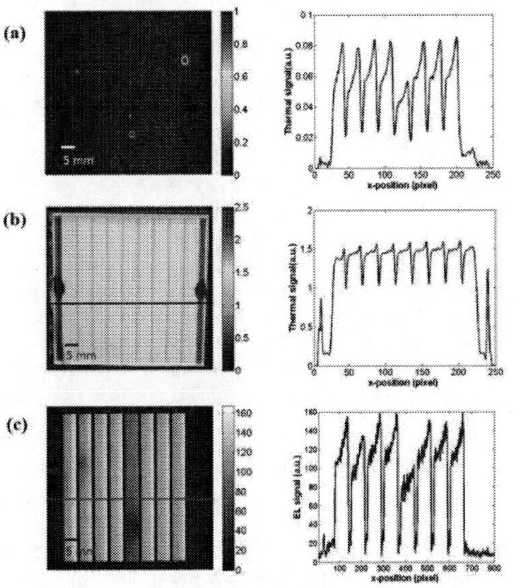

Figure 5: Spatial images along with line scan of M-9 mini-module from (a) DLIT, (b) Jsc-ILIT, and (c) EL.

The results exemplified that LIT method can also be used for investigating the sheet resistance problems of TCO electrode that is detrimental to the health and lifetime of thin-film modules. Although the magnitude and spatial resolution of thermal signal get affected by the encapsulant materials and surrounding conditions, LIT can serve as a diagnostic tool for qualitative investigation of distributed resistive losses in thin-film modules. The quantitative analysis of TCO losses by this technique will be done in future research and is outside the scope of present work.

5 CONCLUSIONS

This paper presents the application of LIT technique under dark and illuminated modes for the electrical investigation of sheet resistance losses in TCO contact layer of thin-film modules. A lateral gradient of thermal signal along the cell width was observed, which can be used to determine the variation in TCO sheet resistance. The gradient pattern observed in DLIT and ILIT images was found to be in close resemblance with that in EL image; however, each technique exhibited different spatial sensitivity for investigation of TCO lateral resistance effect in the module. It is inferred from the study that ILIT was more sensitive near the higher potential edge of cell to the variation of TCO sheet resistance as compared to EL and DLIT methods, particularly for wider cells. Also, this lateral gradient pattern was influenced by presence of strong shunts in a cell that causes significant voltage drop. The study shows that LIT has potential for extending its capability for spatially-resolved characterization of TCO resistance variation in the thin-film modules.

6 REFERENCES

[1] H. M. Lemire, K. A. Peterson, S. Sprawls, K. Singer, I. T. Martin, and R. H. French, *Proc. of SPIE,* 2013, p. 882502.

[2] K. Brecl, M. Topič, and F. Smole, *Prog. Photovoltaics Res. Appl.,* vol. 13, pp. 297–310, 2005.

[3] S. S. Hegedus, S. Albright, F. Jeffrey, T. J. McMahon, and S. Wiedeman, *Prog. Photovoltaics Res. Appl.,* vol. 5, pp. 365–370, 1997.

[4] S. S. Lee, Y. N. Kim, and H. G. Shin, *J. Korean Phys. Soc.,* vol. 53, no. 1, pp. 188–191, 2008.

[5] M. Lapinski, J. Domaradzki, E. L. Prociow, K. Sieradzka, and B. Gornicka, *International Students and Young Scientists Workshop "Photonics and Microsystems,"* 2009, pp. 52–55.

[6] M. Seeland, R. Rösch, B. Muhsin, G. Gobsch, and H. Hoppe, *Energy Procedia,* vol. 31, pp. 167–172, 2011.

[7] X. Wu, M. Bliss, A. Sinha, T. Betts, R. Gupta, and R. Gottschalg, *39th IEEE Photovoltaic Specialists Conference*, 2013, pp. 49–52.

[8] H. Predatsch, U. Heinzmann, and H. Stiebig, *Energy Procedia,* vol. 60, pp. 71–75, 2014.

[9] K. Wang, D. Han, M. Silver, and H. M. Branz, *Sol. Energy*, vol. 30, pp. 219–225, 1991.

[10] S. Huth, O. Breitenstein, A. Huber, D. Dantz, U. Lambert, and F. Atmann, *Solid State Phenom.*, vol. 82–84, pp. 741–746, 2001.

[11] O. Breitenstein, J. P. Rakotoniaina, and M. H. Al Rifai, Prog. Photovoltaics Res. Appl., vol. 11, pp. 515–526, 2003.

[12] H. Straube, J. M. Wagner, J. Schneider, and O. Breitenstein, *J. Appl. Phys.,* vol. 110, p. 84513, 2011.

[13] J. Bachmann, C. Buerhop-Lutz, R. Steim, P. Schilinsky, J. A. Hauch, E. Zeira, J. B. Christoh, *Sol. Energy Mater. Sol. Cells*, vol. 101, pp. 176–179, 2012.

[14] S. Besold, U. Hoyer, J. Bachmann, T. Swonke, P. Schilinsky, and R. Steim, *Sol. Energy Mater. Sol. Cells*, vol. 124, pp. 133–137, 2014.

[15] A. Sinha, *Ph.D. thesis*. Indian Institute of Technology Bombay, 2016.

CHALLENGES FOR HIGH-EFFICIENCY BUFFER-FREE CU(IN,GA)SE$_2$ SOLAR CELLS

Shogo Ishizuka[1], Takashi Koida[1], Noboru Taguchi[2], Shingo Tanaka[2], Paul Fons[3], Hajime Shibata[1]
[1]Research Center for Photovoltaics, National Institute of Advanced Industrial Science and Technology (AIST)
1-1-1 Umezono, Tsukuba, Ibaraki 305-8568, Japan
[2]Research Institute of Electrochemical Energy, National Institute of Advanced Industrial Science and Technology (AIST)
1-8-31 Midorigaoka, Ikeda, Osaka 563-8577, Japan
[3]Nanoelectronics Research Institute, National Institute of Advanced Industrial Science and Technology (AIST)
1-1-1 Higashi, Tsukuba, Ibaraki 305-8565, Japan

ABSTRACT: We report improvements in the energy conversion efficiency in buffer-free Cu(In,Ga)Se$_2$ (CIGS) solar cells. To date, typical CIGS solar cells employ one or two buffer layers. The first buffer layer consists of materials such as CdS, ZnS(O,OH), or In_2S_3, and the second buffer layer consists of materials such as highly resistive intrinsic ZnO or ZnMgO. These buffer layers have been believed to be indispensable components necessary for a proper *p-n* junction to demonstrate high cell efficiencies. In fact, CIGS cell efficiencies demonstrated using buffer-free configurations have been relatively low (about 10%) thus far. In this study, however, we have successfully improved the cell efficiency in buffer-free CIGS solar cells using conventional Mo back contacts and ZnO transparent conductive oxide (TCO) front electrode layers for device component. Independently certified efficiencies exceeding 15% have been demonstrated using the simplified device structure of Mo/CIGS/TCO fabricated on a conventional soda-lime glass substrate. The techniques developed in this study to enhance buffer-free CIGS solar cell efficiency are expected to contribute to a clarification of detailed CIGS device mechanisms and the underlying physics in poorly understood areas. Also, new findings obtained from the simplification of CIGS device structure are expected to contribute to further improvements in CIGS solar cell cost and efficiency regardless of device configurations.
Keywords: Cu(In,Ga)Se$_2$, CIGS, Doping, Silicon, Solar Cell

1 INTRODUCTION

Cu(In,Ga)Se$_2$ (CIGS) solar cells have been fabricated using *n*-type materials such as CdS, ZnS(O,OH), In_2S_3 (first buffer layer) and highly resistive intrinsic ZnO or ZnMgO (second buffer layer) to date. The efficacy of the use of these buffer layers has been proven in enhancing solar cell efficiencies. Today, however, buffer-layer deposition processes generally employ a wet solution process in contrast to the fact that other parts of the device process such as Mo back contact layer deposition and CIGS photoabsorber layer growth basically employ dry processes to demonstrate high cell efficiencies over 20% [1-3]. This fact and the presence of buffer-layers in CIGS device structure often complicate industrial CIGS solar module fabrication processes and the understanding of fundamental device physics. So, challenges for highly efficient solar cells using a buffer-free device structure are also an important factor to contribute to cost reductions in the CIGS solar module and the understanding of material and device physics which have remained unclear because of the complicated device structure of conventional buffer-based CIGS solar cells. In this study, fabrication of completely buffer-free but yet highly efficient CIGS solar cells is attempted. Among lots of approaches which have attempted to eliminate buffer layers to date, CIGS surface modifications with elemental doping using group II or IV elements are expected to be effective [4-8]. Some attempts and successful results using group II elements such as Cd or Zn have already been reported thus far [4-6], whereas there have been no successful reports about the fabrication of a homo *p-n* junction or buffer-free device applications using group IV elements to date. Thus, here we attempted group IV doping using elemental Si to explore a new field. Si-doping effects in CIGS thin films and applications for buffer-free CIGS solar cell devices are reported.

2 EXPERIMENTAL

2.1 Fabrication of Si-doped CIGS thin films

2.0-2.5 μm thick CIGS thin films were grown on Mo-coated soda-lime glass substrates by the three-stage co-evaporation process using elemental Cu, In, Ga, and Se Knudsen cell sources. The substrate temperature used was 350℃ during the first stage where elemental In, Ga, and Se were supplied and was 550℃ during the second and third stages where Cu and Se, and again In, Ga, and Se were supplied, respectively. Si-doping was carried out during the third stage using an elemental Si Knudsen cell source with a source temperature of 1500-1600℃. Detailed experimental conditions can be found elsewhere [9].

2.2 Fabrication of buffer-free CIGS solar cells

Buffer-free CIGS devices were fabricated with a 2.0 μm thick B-doped ZnO (BZO) transparent conductive oxide layer, which was deposited directly on the CIGS film surface by metal organic chemical vapor deposition (MOCVD) at a substrate temperature of 160℃. The electrical properties of the BZO layer used in this study were typically; resistivity ~ 4×10^{-3} Ω cm, mobility ~ 30 cm^2/Vs, carrier concentration ~ 5×10^{19} cm^{-3}. More details about the information of the BZO layer can be found in the literature [9,10]. Grid contacts made of Al or a bilayer of Ni/Al metals were formed on the BZO layer. The designated cell device area used was 0.52 cm^2.

3 RESULTS AND DISCUSSION

Figure 1 shows schematic variations in the elemental composition ratio [Cu]/([Ga]+[In]) (CGI) in growing CIGS films and a substrate temperature used during each stage. In this study, Si-doping in CIGS films was performed during the third stage. As can be seen in Fig. 1, the CGI value in CIGS films decreases across the value of unity during the third stage. Here, we attempted Si-doping using two different timings. One is the term T_{3s}-T_{3e} (from the start to the end of the third stage), where the CGI value is larger than unity and the presence of liquid Cu-Se phases on the growing CIGS surface is expected during the term T_{3s}-T_{cgi1}. The other is the term T_{cgi1}-T_{3e} (only in the latter half of the third stage), where the CGI value is less than unity and thus no presence of liquid Cu-Se phases is expected. Figure 2 compares surface and cross sectional scanning electron microscopy (SEM) images obtained from non Si-doped and Si-doped (performed during the term of T_{cgi1}-T_{3e}) CIGS films. Figure 3 shows secondary ion mass spectrometry (SIMS) depth profiles for elemental Si in these CIGS films.

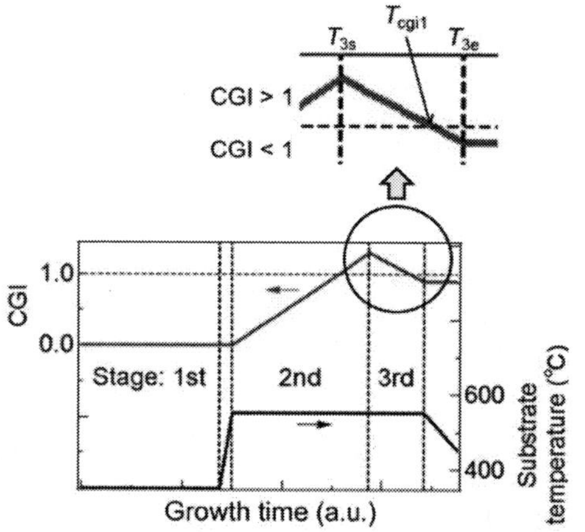

Figure 1: Sequential CIGS film growth profiles regarding CGI and substrate temperature during each stage.

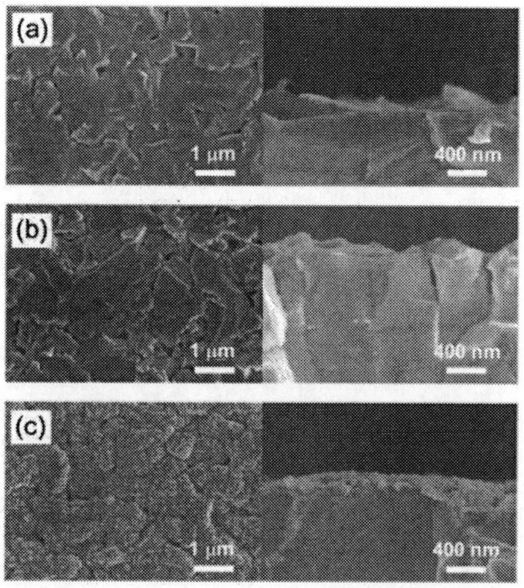

Figure 2: Surface and cross sectional (near surface region) SEM images of (a) non Si-doped (for reference), (b) Si-doped (during T_{cgi1}-T_{3e} at 550℃), and (c) Si-doped (during T_{cgi1}-T_{3e} at 400℃) CIGS films.

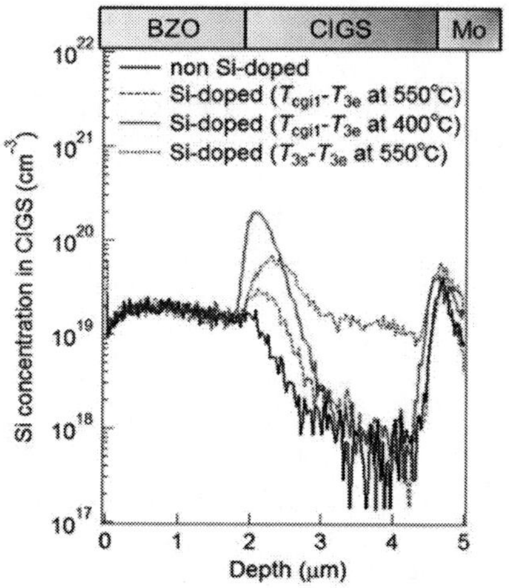

Figure 3: SIMS elemental distribution depth profiles of Si in the CIGS layers with various Si-doping conditions. The Si signal observed in the BZO layer is an artifact attributable to interference from ions such as 16O + 17O + 64Zn + 64Zn (= 28Si + 133Cs = 161). Cs^+ is the primary ion used for the SIMS measurements.

We have found that the surface morphology observed from CIGS films grown with Si-doping performed during the term T_{3s}-T_{3e} is almost the same as a non Si-doped or a Si-doped (during T_{cgi1}-T_{3e} at 550℃) CIGS film shown in Figs. 2(a) and 2(b) [9], and thus not shown here. It should be noted that Si-doping in the CIGS film shown in Fig. 2(c) was carried out using the same timing with the CIGS film shown in Fig. 2(b), but carried out at a relatively low substrate temperature of 400℃ during the term of T_{cgi1}-T_{3e} (the substrate temperature used during the term of T_{3s}-T_{cgi1} was kept at 550℃ and then decreased to 400℃ and Si-doping was performed). This Si-doped CIGS film exhibited the formation of a distinctive thin layer on the CIGS surface as shown in Fig. 2(c), probably containing large amount of elemental Si as can be expected from the SIMS results shown in Fig. 3. Elemental Si depth profiles shown in Fig. 3 suggest that Si-doping performed under a no liquid Cu-Se phase presence condition can concentrate elemental Si at the CIGS surface region. At present, however, Si-doping performed from the start to the end of the third stage (T_{3s}-T_{3e}) has demonstrated the highest buffer-free cell efficiencies among these Si-doping conditions as shown in Figure 4. This result is attributable to the presence of elemental Si diffused into the CIGS layer along CIGS grain boundaries due to the flux effect of liquid Cu-Se phases present during the term of T_{3s}-T_{cigs1}. Elemental Si is expected to have effects of the modification of energy band diagrams at CIGS grain boundaries as well as enhancements in alkali K effects resulting in reduced carrier recombination at grain boundaries and concomitant improvements in carrier transport in the CIGS layer [9].

We have demonstrated over 15% buffer-free cell efficiencies with Si-doping in CIGS layers combined with the modification of CIGS surface [Ga]/([Ga]+[In]) (GGI) composition to date [9]. Further improvements with a precise control of carrier concentration in Si-doped CIGS films as well as defect control near the surface region of CIGS films are expected. Future work will also include a comparative study using buffer-based and buffer-free CIGS devices on buffer-related interface issues which have remained unclear thus far.

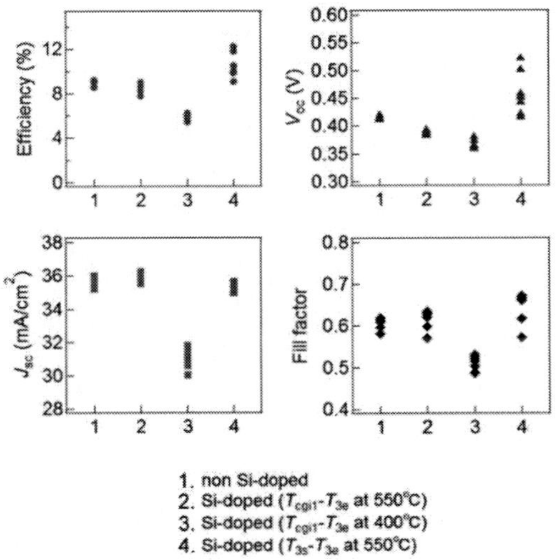

Figure 4: Solar cell parameters obtained from buffer-free CIGS devices fabricated using CIGS photoabsorber layers grown with various Si-doping conditions. No anti-reflection coating was used.

4 CONCLUSION

In summary, the use of properly Si-doped CIGS films as the photoabsorber layer in buffer-free solar cell devices was found to be effective yielding enhanced cell efficiencies in comparison with the use of conventional non Si-doped CIGS films. Although some issues have still remained in buffer-free Si-doped CIGS devices for practical applications, this type of simplified structure devices is expected to be useful for the understanding of fundamental mechanisms and underlying physics in CIGS photovoltaics. Also, effects of elemental Si-doping on the alkali metal distribution profiles in CIGS films observed in our previous work [9] are expected to contribute to developments in alkali diffusion control techniques as well as the clarification of some important parts of the so-called alkali effects on CIGS films and devices.

REFERENCES

[1] T. Kato, Jpn. J. Appl. Phys. 56 (2017) 04CA02.
[2] P. Jackson, R. Wuerz, D. Hariskos, E. Lotter, W. Witte, M. Powalla, Phys. Stat. Sol. RRL 10 (2016) 583.
[3] J. Nishinaga, T. Koida, S. Ishizuka, Y. Kamikawa, H. Takahashi, M. Iioka, H. Higuchi, Y. Ueno, H. Shibata, S. Niki, Appl. Phys. Express 10 (2017) 092301.
[4] K. Ramanathan, F. S. Hasoon, S. Smith, D. L. Young, M. A. Contreras, P. K. Johnson, A. O. Pudov, J. R. Sites, J. Phys. Chem. Solids 64 (2003) 1495.
[5] J. Haarstrich, M. Teichmann, H. Metzner, M. Gnauck, C. Ronning, W. Wesch, T. Rissom, C. A. Kaufmann, H.-W. Schock, W. Scheumann, W. Mannstadt, Sol. Energy Mater. Sol. Cells 116 (2013) 43.
[6] T. Sugiura, S. Chaisitsak, A Yamada, M. Konagai, Y. Kudriavtsev, A. Godines, A. Villegas, R. Asomoza, Jpn. J. Appl. Phys. 39 (2000) 4816.
[7] M. Rusu, S. Wiesner, R. Würz, S. Lehmann, S. Doka-Yamigno, A. Meeder, D. Fuertes Marrón, M. Bär, V. Koteski, H.-E. Mahnke, E. Arushanov, J. Beckmann, K. Höhn, W. Fritsch, W. Bohne, P. Schubert-Bischoff, M. Heuken, A. Jäger-Waldau, A. Rumberg, Th. Schedel-Niedrig, Sol. Energy Mater. Sol. Cells 95, (2011) 1555.
[8] S. Thiru, M. Fujita, A. Kawaharazuka, Y. Horikoshi, Appl. Phys. A 113 (2013) 257.
[9] S. Ishizuka, T. Koida, N. Taguchi, S. Tanaka, P. Fons, H. Shibata, ACS Appl. Mater. Interfaces 9 (2017) 31119.
[10] T. Koida, J. Nishinaga, H. Higuchi, A. Kurokawa, M. Iioka, Y. Kamikawa-Shimizu, A. Yamada, H. Shibata, S. Niki, Thin Solid Films 614 (2016) 79.

HIGH EFFICIENCY CDTE SOLAR CELLS BY LOW TEMPERATURE DEPOSITION WITH MGZNO HRT LAYER

Daniele Menossi, Elisa Artegiani, Francesco Bittau[1], Marco Barbato[2], Matteo Meneghini[2], Gaudenzio Meneghesso[2], Jake W. Bowers[1], John M. Walls[1], Fabio Piccinelli[3] and Alessandro Romeo.
Laboratory for Photovoltaics and Solid State Physics, Department of Computer Science, University of Verona, Strada Le Grazie 15, 37134 Verona, Italy
[1]Centre for Renewable Energy Systems Technology (CREST), Wolfson School of Mechanical, Electrical and Manufacturing Engineering, Loughborough University, Loughborough, Leicestershire, LE11 3TU, UK
[2]Department of Information Engineering, Via Gradenigo 6/B, 35131 University of Padova, Italy
[3]Department of Biotechnology, University of Verona, Verona, 37123, Italy

ABSTRACT: CdTe solar cells have shown high efficiency and the technology is scalable. As a result thin film CdTe modules are competitive with crystalline silicon modules.
Thin film CdTe devices with efficiency above 22% have been reported using high substrate temperatures during the deposition process. It is known that high substrate temperatures result in large grain size with a reduced number of grain boundaries and this is believed to contribute to the high efficiency. However, use of high temperature requires robust substrates and excludes the use of most flexible substrate materials. It also involves higher energy consumption and more complicated machinery.
In this work we present a process for high efficiency solar cells with an improved front contact, by introducing magnesium-doped zinc oxide high resistance transparent layer. By optimizing the fabrication process we have achieved a conversion efficiency exceeding 16%, which is one of the highest reported for substrate temperatures below 500°C.

Keywords: CdTe, TCO Transparent Conductive Oxide, Buffer Layer

1 INTRODUCTION

CdTe solar cells are manufactured on a large scale and with high efficiency by vapour transport deposition (VTD) with a record cell efficiency exceeding 22% [1].

VTD together with Close Space Sublimation has delivered the highest efficiencies using high substrate temperature avoiding re-evaporation of the material.

Although no explanation has been provided for the high current density produced in the champion cell it is speculated that the junction has been optimized by the reduction of absorption in the CdS as well as optimization of the back contact with a back reflector. Back contacts as ZnTe:Cu have been introduced since the beginning of the 2000s [2].

16,7% efficiency by VTD has been obtained by inserting a Cd_2SnO_4 transparent conducting oxide as transparent conductive oxide (TCO) and a Zn_2SnO_4 high resistance transparent (HRT) layer. High efficiency devices exceeding 18% were presented by Sites et al. [3] by introducing a MgZnO buffer layer and other improvements including a broadband antireflection coating and a Te compound as a back reflector.

In this work we study the introduction of a Mg doped ZnO HRT layer to replace our standard insulating ZnO layer. ITO/MZO layers were fabricated at Loughborough with different substrate temperatures and their physical properties have been studied. Finished solar cells have been fabricated on the prepared TCOs and physical and electrical properties have been analysed.

Devices show different performance according to the different substrate temperature deposition. A larger amount of Mg results with higher substrate temperature and consequently a higher band gap is registered.

This optimized process has delivered solar cells with efficiencies up to 16.2%. This results to be the highest value measured in our lab, and would give, if verified in a certified laboratory, the highest efficiency for low substrate temperature deposited CdTe devices.

2 EXPERIMENTAL

2.1 TCO preparation

At Loughborough University, Indium-doped Tin Oxide and Mg doped ZnO thin films were deposited on soda lime glass by Radio-Frequency (RF) magnetron sputtering. The films were deposited using an Orion 8 HV magnetron sputtering system (AJA international, USA) equipped with an AJA 600 series RF power supply.

The 3" diameter ITO target (90% In_2O_3 and 10% SnO_2) and 3" diameter MZO target (11% MgO and 89% ZnO) were rotated at 10 rpm during deposition to assure uniformity. Both targets were sputtered at a constant power density of 3.5 Wcm^{-2} and at an Argon pressure of 133.3 Pa. 1% O_2 in Ar atmosphere was added for the MZO deposition. Substrate temperature was set to 450 °C for ITO deposition, while for MZO deposition substrate temperatures from 20 °C up to 400 °C were applied.

2.2 Solar cell fabrication

At the University of Verona the ITO/MZO substrates were coated with CdS and CdTe by thermal evaporation in a vacuum chamber at a pressure of 10^{-4} Pa with a Edwards XDS10 roughing pump and a Edwards ST-451 turbo-molecular pump.

CdS is evaporated from a tungsten crucible at a deposition rate of 5 Å/sec. During deposition the substrate was kept at 150 °C with halogen lamps. Before and after CdS deposition, the stack is annealed in vacuum at 450 °C for 30 minutes, this allows the polycrystals of ITO/MZO and the deposited CdS to crystallize and increase robustness for the depositions and treatments that follow. CdTe was deposited from a special graphite Knudsen cell at an evaporation rate of 10 Å/sec. The rate was controlled by piezoelectric quartz crystal with an

Intellemetrics IL-150 controller.

The CdTe activation treatment was carried out by deposition of $CdCl_2$ in methanol solution. This is prepared by dissolving the $CdCl_2$ powder (previously dried in a furnace at 0.1 Pa) in methanol to form a saturated solution. The liquid is then deposited in the form of drops on the CdTe surface. A typical amount of 250 µl is calculated. Finally the stack is annealed in air at 380-400 °C for 30 minutes after a 25 minutes ramp from room temperature.

The $CdCl_2$ treatment improves the electrical properties of the absorber (also by the passivation of the grain boundaries), enhances CdS/CdTe intermixing reducing the lattice mismatch and enlarges the grain size by one order of magnitude thereby reducing the density of grain boundaries [4]. In our laboratories we also have developed alternative activation treatment by application of chlorine containing compounds such as difluorochloromethane [5] or $MgCl_2$ [6]. However the $CdCl_2$ wet deposition delivers the highest performance.

The back contact consists of Cu/Au deposited on top of the CdTe layer, which was previously treated with a solution of bromine (50 µl) and methanol (40 ml) to clean the surface and remove $CdCl_2$ residuals to form a Te-rich layer. 2 nm thick Cu and 50 nm thick Au films were subsequently deposited by thermal evaporation at room temperature in a specific chamber with a vacuum of 10^{-3} Pa. Final annealing of the structure for 20 min at 190 °C in air, is necessary to obtain high efficiency.

2.3 Characterization methods

The morphological properties of the CdTe layers were studied by atomic force microscopy (AFM) with a NT-MDT Solver Pro in semi-contact mode.

X-ray diffraction analysis (XRD) of the CdTe layers has been performed by means of a Thermo ARL X'TRA powder diffractometer, operating in Bragg-Brentano geometry equipped with a Cu-anode X-ray source (Kα, k =1.5418 Å) and using a Peltier Si(Li) cooled solid state detector. Current density-voltage (JV) characteristics of the completed cells were measured with a Keithley Source Meter 2420, using a halogen lamp calibrated with a silicon solar cell under an irradiation of 100 mW/cm^2.

Light transmission measurements of TCO layers were performed by a Unicam UV2 UV/Vis spectrometer. Finally external quantum efficiency measurements (EQE) have been performed with a commercial LOANA (PV-Tools GmbH) solar cell analysis system. The EQE were calibrated with a silicon reference sample with known response using an incident spotlight of 1mmx 2 mm area.

3 RESULTS

Different solar cells have been prepared with MZO deposited at different temperatures.

Bittau et al. have already observed that in these layers higher substrate deposition temperatures result in higher amount of Mg into the HRT layer; this provides a larger band gap and, consequently, a higher transparency [7].

Cells with MZO deposited at temperatures above 300°C have resulted in a superior conversion efficiency compared to the standard process with ITO/ZnO solar cells.

Different annealing temperatures of the $CdCl_2$ activation treatment were applied to optimize the fabrication process. Most of the devices had superior efficiency, in the range of 14 % to 16 %. However, the best results were obtained by annealing the CdTe at 390 °C (as described above) and by a reduction of the CdS thickness.

In Figure 1, the current density-voltage (JV) characteristic of a 16.2% efficiency solar cell is shown with a V_{oc}= 852 mV, J_{sc}= 25,8 mA/cm^2 and a F.F.= 74 %. The parameters indicate a higher fill factor and current density, compared to our best ITO/ZnO devices (efficiencies are in the range of 15%) [8].

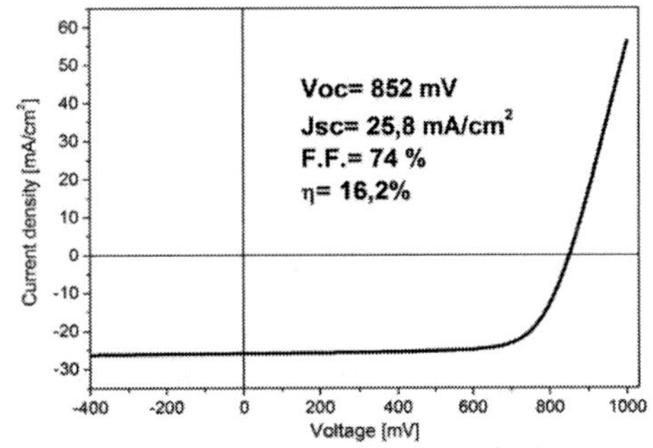

Figure 1: Current density-voltage characteristic of the highest efficiency CdTe solar cell with MZO/ITO front contact.

CdTe grown on CdS/MZO/ITO show a similar morphology irrespective of the MZO temperature and slight different recrystallization (see figure 2). CdTe grain size is almost equal for the MZO (deposited at 200 °C and 400 °C) and for the ITO/ZnO cases. On the other hand the orientation of the crystals shown on the XRD spectra slightly change their preferred orientation.

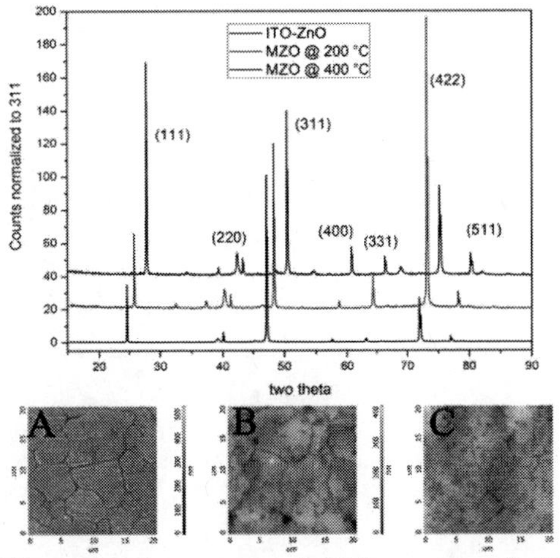

Figure 2: Comparison of XRD spectra of CdTe deposited on CdS and on ITO/ZnO or ITO/MZO (top) and morphology of CdTe deposited on CdS/ZnO/ITO (A), on CdS/MZO/ITO with MZO deposited at 200°C (B) and at 400 °C (C).

As already mentioned the application of MZO/ITO may result in higher transparency of the TCO layer (see figure 3). However, the improved performance is not entirely connected to an increased amount of photons absorbed in the CdTe. In fact quantum efficiency measurement (shown in figure 4) shows a different response according to the different TCOs, and the insertion of the MZO layer improves the response across the entire wavelength range and in particular in the long wavelength region. This is not connected to a reduced absorption in the TCO and CdS layers.

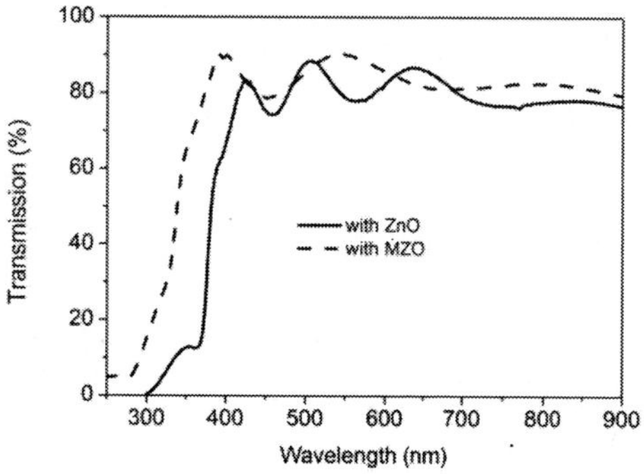

Figure 3: Comparison of transmission spectra of the new ITO/MZO stack with our standard ITO/ZnO stack.

It is possible to explain this phenomenon if we consider that the intermediate layer is not only preventing pin-holes in case of ultra thin CdS layers, or limiting diffusion of impurities from the front contact [9] but also, as mentioned by Klein et al. [10], reducing the band offset between ITO and CdS.
Moreover Rao et al. have shown that MZO has an increased match with CdS compared to ZnO [11] and Kephart et al have demonstrated that it is even possible to tune the MZO band gap and electron affinity in a way to have an optimum match with CdTe, forming MZO/CdTe heterojunction [12].

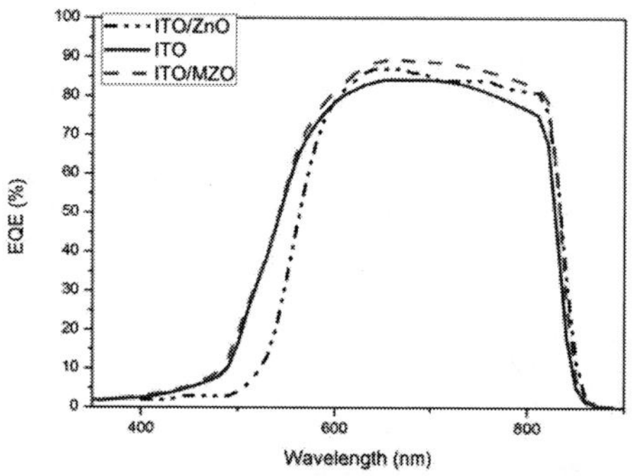

Figure 4: External quantum efficiency of CdTe solar cells made with only ITO and with ITO/MZO front contact (in this case CdS is thicker than record efficiency cells).

4 CONCLUSIONS

Thin film CdTe solar cells have been prepared using a low temperature substrate deposition process by thermal evaporation in vacuum. The process has been optimized for a front contact with insertion of high resistance transparent HRT layer of Mg-doped ZnO.

MZO substrate deposition temperature has been optimized in order to have the ideal band gap and produce a better match with the CdS buffer layer. Various $CdCl_2$ activation treatment temperatures have been used to optimize the fabrication process. The majority of devices have delivered high efficiency (from 14 to 16 %). The highest efficiency of 16,2% is a record for vacuum evaporated CdTe solar cells. The improvement of performance is not justified by the increased transparency of MZO but can be explained by an improved band alignment with the CdS buffer layer, compared to ITO or ITO/ZnO stacks.

5 ACKNOWLEDGMENTS
This work has been supported by Fondazione Cassa di Risparmio di Trento e Rovereto with the TIN-Jet project.

6 REFERENCES.

[1] First Solar, "First Solar Achieves yet Another Cell Conversion Efficiency World Record.," 2016. [Online].http://investor.firstsolar.com/releasedetail.cfm?ReleaseID=956479.

[2] T. A. Gessert, M. J. Romero, R. G. Dhere, and S. E. Asher, "Analysis of the ZnTe:Cu Contact on CdS/CdTe Solar Cells," *MRS Proc.*, vol. 763, 2003.

[3] J. Sites, A. Munshi, J. Kephart, D. Swanson, and W. S. Sampath, "Progress and Challenges with CdTe Cell Efficiency," *43rd IEEE Photovolt. Spec. Conf.*, pp. 3632–3635, 2016.

[4] B. L. Xu, I. Rimmaudo, A. Salavei, F. Piccinelli, S. Di Mare, D. Menossi, A. Bosio, N. Romeo, and A. Romeo, "CdCl2 activation treatment: A comprehensive study by monitoring the annealing temperature," *Thin Solid Films*, vol. 582, pp. 110–114, 2015.

[5] A. Salavei, I. Rimmaudo, F. Piccinelli, P. Zabierowski, and A. Romeo, "Study of difluorochloromethane activation treatment on low substrate temperature deposited CdTe solar cells," *Sol. Energy Mater. Sol. Cells*, vol. 112, pp. 190–195, 2013.

[6] D. Menossi, E. Artegiani, A. Salavei, S. Di Mare, and A. Romeo, "Study of MgCl2 activation treatment on the defects of CdTe solar cells by capacitance-voltage, drive level capacitance profiling and admittance spectroscopy techniques," *Thin Solid Films*.

[7] F. Bittau, E. Artegiani, A. Abbas, D. Menossi, A. Romeo, J. W. Bowers, and J. M. Walls, "Magnesium-doped Zinc Oxide as a High Resistance Transparent Layer for thin film CdS / CdTe solar cells," in *44th IEEE Photovoltaic Specialists Conference*, accepted for publication.

[8] A. Salavei, I. Rimmaudo, F. Piccinelli, D. Menossi and A. Romeo, "Analysis of CdTe activation treatment with a novel approach," in *28th European Photovoltaic Solar Energy Conference and Exhibition*, 2014, pp. 2140–2142.

[9] I. Rimmaudo, A. Salavei, E. Artegiani, D. Menossi, M. Giarola, G. Mariotto, A. Gasparotto, and A. Romeo, "Improved stability of CdTe solar cells by absorber surface etching," *Sol. Energy Mater. Sol. Cells*, vol. 162, 2017.

[10] A. Klein, "Energy band alignment in chalcogenide thin film solar cells from photoelectron spectroscopy," *J. Phys. Condens. Matter*, vol. 27, no. 13, p. 134201, 2015.

[11] G. V. Rao, F. Säuberlich, and A. Klein, "Influence of Mg content on the band alignment

at CdS/(Zn,Mg)O interfaces," *Appl. Phys. Lett.*, vol. 87, no. 3, pp. 1–4, 2005.

[12] J. M. Kephart, J. W. McCamy, Z. Ma, A. Ganjoo, F. M. Alamgir, and W. S. Sampath, "Band alignment of front contact layers for high-efficiency CdTe solar cells," *Sol. Energy Mater. Sol. Cells*, vol. 157, pp. 266–275, 2016.

CHARACTERIZATION AND SIMULATION OF CU2ZNSNS4 ABSORBER LAYERS FABRICATED BY SEQUENTIAL DC MAGNETRON SPUTTERING AND RAPID THERMAL PROCESSING

Maria Zhukova[1]*, Ratan Kotipalli[1], Louise Samain[2], Lionel Fourdrinier[2], Denis Flandre[1]
[1] ICTEAM, UC Louvain, Place du Levant 3, 1348, Louvain-la-Neuve, Belgium
[2] CRM Group, Allée de l'Innovation 1, B57- Quartier Polytech 3, 4000 Liège, Belgium
Phone: +32 (0) 10 47 26 09, e-mail: maria.zhukova@uclouvain.be

ABSTRACT : The thin-film Cu_2ZnSnS_4 (CZTS) solar cell technology is seen as a sustainable and low cost alternative to the CIGS technology. However mastering the CZTS formation is not evident. In particular, the deposition parameters of each layer in the metallic precursor impact on properties of these layers and could influence the final Cu_2ZnSnS_4 (CZTS) absorber quality. To verify and quantify these influences, two kinds of samples with low-density and high-density precursors have been fabricated by sequential DC magnetron sputtering and finalized into CZTS by Rapid Thermal Processing (RTP) with presence of elemental sulfur. In addition to optical and material characterizations, Hall measurements were carried out for estimation of the carrier concentration, mobility and conductivity of fabricated layers. Finally, SCAPS simulations have been run to predict the cell efficiency and compare with experiments. This study demonstrates that appropriate optoelectrical properties : $E_G \sim 1.5\text{-}1.6$ eV, $\alpha > 2 - 3\times10^4$ cm^{-1}, $\mu_h \sim 25\text{-}30$ cm^{2}V^{-1}s^{-1}, $p \sim 5\text{-}6\times10^{15}$ cm^{-3}, rather than absorber composition and morphology, are the success criteria for obtaining high-efficiency (~ 10 %) CZTS-based devices. Therefore, a monitoring of the optoelectrical parameters during the fabrication steps may be particularly relevant to assess the absorber quality in industrial process.
Keywords: Thin film solar cell, CZTS, Sputtering, Rapid Thermal Processing, Optoelectrical properties, Simulation

1 INTRODUCTION

To date, the highest PV efficiency achieved by pure sulfide kesterite Cu_2ZnSnS_4 (CZTS) solar cells fabricated by co-sputtering and annealing of the precursors is of 9.2 % [1], while it still lags behind at 5.2 % only for sequential sputtering of metallic layers in precursor stack [2]. However, this two-step method is considered as the most environment-friendly and industrially viable for large-scale production with future application towards roll-to-roll process on flexible substrates [3]. To try to understand the material and process limitations, many works concerning the characterization and optimization of the CZTS absorber have been focused on structural and optical material properties [4] as well as complete cells [5, 6] but did not report all electronic parameters, i.e., p-type carrier concentration (p) and mobility (μ_h), or discussed the values actually needed for high-efficiency cells.

The aim of the present work is to establish, for the first time, to our best knowledge, a correlation between the process parameters using sequential DC magnetron sputtering and the material and electrical properties of the obtained CZTS layers. The relevance of the results for realizing high-efficiency cells is assessed by means of SCAPS simulations.

2 EXPERIMENTAL DETAILS

2.1 Film deposition, sulfurization and cell finalization

CZTS layers have been fabricated by sequential DC magnetron sputtering (in a Kurt J. Lesker vacuum system) of multilayer precursors from pure (99.99%), 76.2 mm diameter copper (Cu), tin (Sn) and zinc (Zn) targets. The sputtering deposition was carried out at room temperature using Ar as the working gas. The base pressure of the sputtering chamber was about 10^{-6} Torr and 2 mm soda lime glass substrates were used and placed at 110 mm distance to target. This fabrication step allows to adjust composition in metallic precursor easily by deposition time variation. The deposited thicknesses and masses have been controlled by systematic measurements using a Dektak 150 stylus profiler and a 5-decimals microbalance (METTLER AE 200) respectively. For calibration samples, the deposited mass has been double controlled by ICP measurements. Then, recrystallization into polycrystalline thin film and sulfur (S) incorporation are achieved by simultaneous rapid thermal process (RTP) and sulfurization. The sputtering technique also allows for obtaining deposited films with a composition close to that of the source matter. But the deposition parameters impact a lot on the density and adhesion. Two different deposition conditions (Table I) were chosen to obtain low-density (L) and high-density (H) Sn and Zn layers while Cu layer density was kept almost constant close to bulk density. The resulting volume densities, ρ(Cu), ρ(Zn) and ρ(Sn), have been calculated from the measurements of deposited masses Δm (with $\pm$ 0.1 mg precision) and thicknesses (with $\pm$50 nm precision) of thin layers. The densities ρ for Zn and Sn layers have been estimated as one third of the bulk densities for L precursors and one half of the bulk densities for H ones. These results have been achieved by varying mainly the current and working pressure. For each deposition condition, three different metallic stacks were tested with different composition ratios (Table II). The usual way to fix the ratios of the elements in the precursor is to fix their effective thickness, but depending on the deposition parameters, the same deposited mass will result in denser (thinner) or less dense (thicker) layer. So, in the present work, the ratios of precursors are calculated from the extracted densities per unit area $\rho^S = \Delta m/S$ (g/m^2), where S is the 40 $\times$ 40 mm^2 sample area.

All precursors were sulfurized under the same conditions in an AS-ONE 100 RTP system. The samples were placed inside of a graphite box together with ~100 mg of S, ~10 mg of Sn powder and ~30 mg of SnS. This box was placed in the reactor and heated up to 550°C at a rate of 3°C/s and kept at this temperature during 5 minutes. This rapid heating is insured by halogen lamps in Ar atmosphere at a pressure of 0.9 bar. Finally the samples were left to cool down inside the reactor to room temperature, at cooling rate estimated as ~24°C/min. The RTP choice for sulfurization step was done as it showed good results in terms of limiting secondary phase generation and SnS losses [7] as well it is favourable at industrial conditions.

Some preliminary cells were completed in conventional structure SLG/Mo/CZTS/CdS/iZnO/AZO based on the obtained CZTS films. The Mo back contact of about 0.8 µm thick was prepared by DC magnetron sputtering. CdS layer of 40 nm thick was prepared by chemical bath deposition at 60°C during 9 minutes. I-ZnO and AZO were prepared by RF and DC magnetron sputtering with thickness about 100 nm and 350 nm, respectively. The resulted sheet resistance of i- ZnO/AZO measured on calibration samples was about 100 Ohm/sq. Al grids of ~1 µm thick were deposited by evaporation.

Table I: Deposition conditions for Low-density and High-density precursors: $\rho^{bulk}(Cu)$, $\rho^{bulk}(Zn)$ and $\rho^{bulk}(Sn)$ denote the mass densities for copper, zinc and tin solid elements : 8920 kg.m^{-3}; 7140 kg.m^{-3} and 7310 kg.m^{-3} respectively [8]

	Low-density precursors:		$\rho(Cu) \sim \rho^{bulk}(Cu)$ $\rho(Sn) \sim \rho^{bulk}(Sn)/3$ $\rho(Zn) \sim \rho^{bulk}(Zn)/3$	High-density precursors:		$\rho(Cu) \sim \rho^{bulk}(Cu)$ $\rho(Sn) \sim \rho^{bulk}(Sn)/2$ $\rho(Zn) \sim \rho^{bulk}(Zn)/2$
Parameters	Power, W	Current, A	Pressure, mTorr	Power, W	Current, A	Pressure, mTorr
copper layer	230	0.82	3.21	150	0.42	2.44
zinc layer	131	0.62	4.61	150	0.34	2.44
tin layer	50	0.20	2.99	50	0.13	2.44

Table II: Relative mass ratios of the six different stacks compared to target [9] ratios in brackets

Samples names	L1	L2	L3	H1	H2	H3
Stack order	Cu/Zn/Sn	Cu/Zn/Sn	Cu/Zn/Sn	Cu/Sn/Zn	Cu/Sn/Cu/Zn	Cu/Sn/Cu/Zn
Density per unit area, g/m²	0.85/0.98/1.43	1.35/1.29/0.92	1.63/1.22/1.81	1.60/1.32/1.99	-/1.80/1.60/1.20	-/1.5/1.63/1.22
Cu/(Sn+Zn) (0.75)	0.62	0.95	1.96	0.68	0.73	0.76
Zn/Sn (1.22)	1.36	1.36	1.30	1.24	1.22	1.22
Cu/Sn (1.68)	1.47	2.24	2.54	1.51	1.61	1.68
Cu/Zn (1.37)	1.08	1.65	1.96	1.22	1.31	1.37

2.2 Film and cell characterization, simulation

For finished CZTS layers, the normal-incident transmittance spectra in the 250-2500 nm wavelength range were recorded using Perkin Elmer Lambda 950 UV-vis spectrometer, equipped with an integration sphere. These data were used to evaluate absorption coefficients and optical bandgaps for the different samples. Hall measurements were done with Ecopia HMS-3000 system. This equipment uses the Van der Pauw configuration and 0.55 T magnetic field for measuring the resistivity, carrier concentration and mobility in fabricated CZTS layers.

X-Ray Diffraction (XRD) measurements, using CuKα radiation with an acceleration voltage of 30 kV, 30 mA and Bragg-Brentano geometry, were performed for structural examination of the CZTS thin films. The XRD data were collected from 20° to 80° of 2Theta using a high-resolution Bruker D8 Advance diffractometer. Additional Raman spectroscopy measurements were performed using a LabRAM HR 800 instrument from Horiba Scientific. A 514 nm wavelength laser and a 2400 grooves/mm grating were used. The laser was focused onto the sample using a 100x objective with N.A.= 0.95 (spot size ~1 µm).

The surface morphology of the thin films was characterized by Field Emission Scanning Electron Microscope (SEM, ULTRA55). The composition ratio of the thin films was analyzed by Energy Dispersive X-Ray (EDX) spectroscopy integrated with SEM.

For cell measurements, a Newport Oriel IV Solar simulator equipped with a Xenon lamp was used. IV curves were recorded under AM 1.5 illumination and irradiance of 1000 W/m² was calibrated with a reference silicon cell. The sample temperature was not regulated.

Using the CZTS measured optical and electrical parameters the SCAPS [10] simulations have been performed to assess the influence of these parameters and predict perspectives for solar cell performance.

3 RESULTS AND DISCUSSION

3.1. Material CZTS layers characterization

XRD spectra normalized to their maximum peak for all 6 absorber layers (Fig. 1) show the presence of the CZTS phase. For samples L2 and H2, a buckling can be mentioned, which is most probably related to the presence of amorphous phase in the layers at range of 40-50 deg. The bucking at the beginning (~20 deg.) of the scan corresponds more to the amorphous state of the substrate.

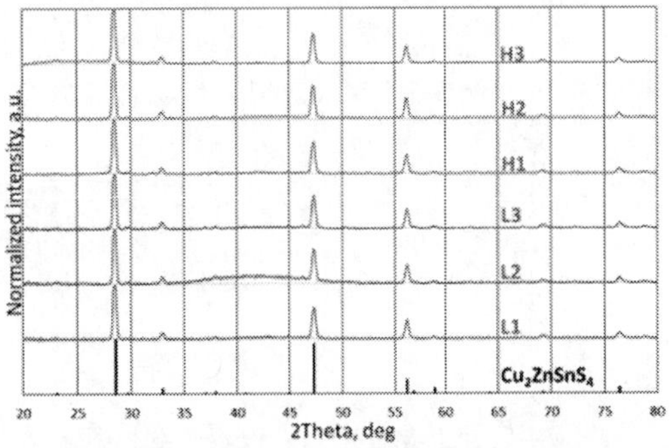

Figure 1: XRD spectra for our 6 CZTS samples and expected peak positions from PDF 26-0575 database

Raman spectra (Fig. 2) confirm the presence of the kesterite CZTS phase with peaks at 288 cm^{-1}, 336 cm^{-1} and 369 cm^{-1} frequencies [11]. But some indications on peaks from the ZnS, SnS or $Cu_{x-1}SnS_x$ (CTS) phases can also be outlined [11]. Samples L2 and L3 feature additional peaks at 468 cm^{-1} corresponding to $Cu_{2-x}S$ phase [11]. This indicates the coexistence of different phases or/and composition and crystal orientation variation. No clear trend with the deposition conditions was identified till now.

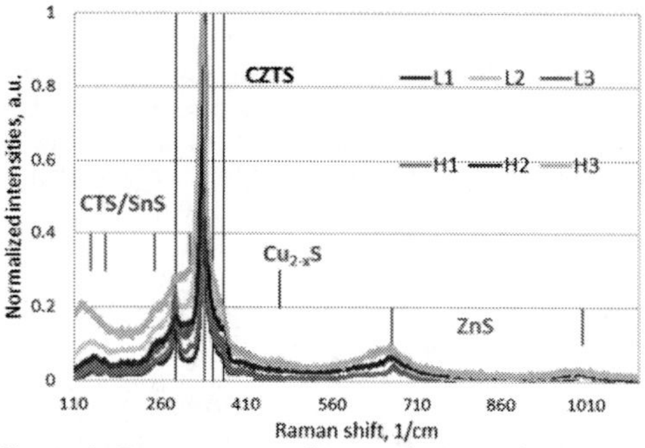

Figure 2: Raman spectra for our six CZTS layers versus expected peak positions for different alloys

After sulfurization, all layers appear grey and present some roughness. Fig. 3 shows top and cross section views of the six layers under discussion. Samples of L-series have inhomogeneous surface morphology. Clearly two different regions can be distinguished: compact islands with round-shaped grains of different sizes (0.2-2 µm) or regions composed of fine grains with round shape (size of tens of nanometres). Sample L1 has additional infusion of rice-shape grain region. Based on literature, we attributed the rice-shaped morphology to the SnS phase and fine-grains region to ZnS phases [11]. The EDX composition analysis confirms these hypotheses by too high concentration in Sn (16 at.%) and S (28 at.%) for SnS and Zn (18 at.%) and S (28 at%) for ZnS regions. This should be due to initial ratios in precursors of these samples. The compact region is mostly present in all samples and was taken into account for composition analysis at 1000 times magnification over 50 measured points (Fig. 4).

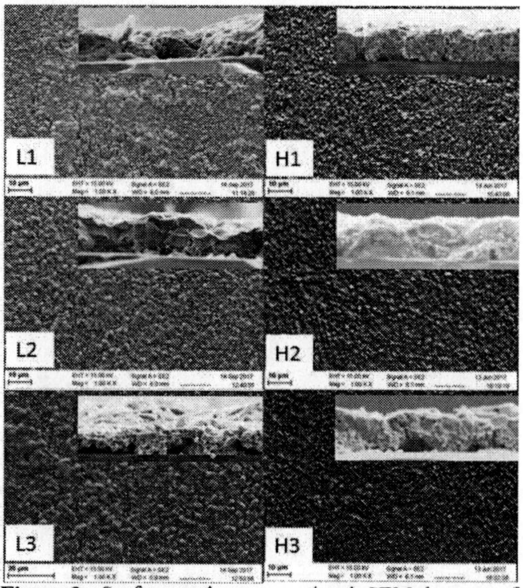

Figure 3: Surface and cross-sectional SEM images of CZTS layers

All films of H series showed homogeneous morphology of compact fine grains. But still two different regions in composition can be outlined. The overall composition is slightly tin-rich and zinc-poor, which is correlated with compositions in precursors.

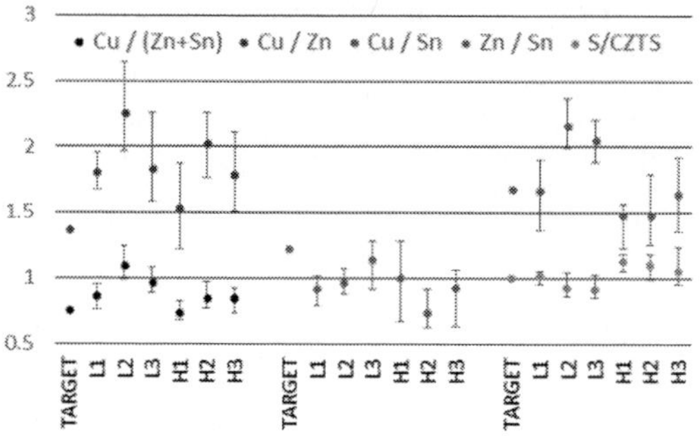

Figure 4: Composition ratios with minimum and maximum values as error bars for all samples compared to target ratios from [9]

In cross-section views, some voids over thickness can be observed for all samples. This could be due to poor adhesion of absorber with substrate. The average thickness is ~1.5 µm for samples of L series and ~2 µm for H samples.

3.2 Optoelectrical properties of CZTS layers

All samples showed relatively good (lower than 60 %) transmittance for all wavelengths, and samples L2 and L3 even lower than 40% (Fig. 5). This can be correlated with the presence of the opaque sub-metallic CuS phase, according to the phase analyses.

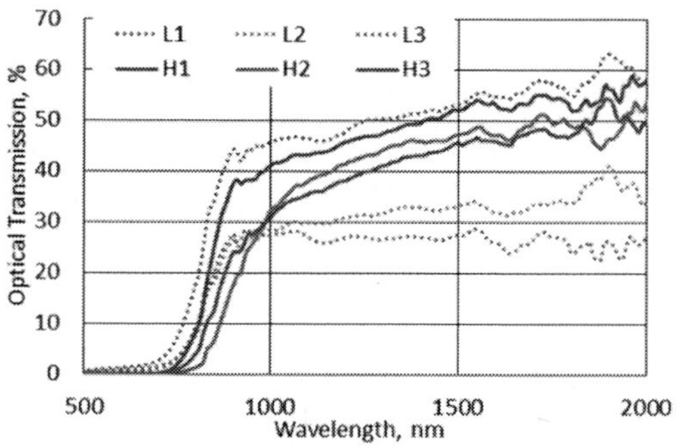

Figure 5: UV-vis transmission results as collected by spectrometer

Bandgaps E_G and averages of absorption coefficient in 250-2000 nm wavelength range are calculated from Fig. 6 and summarized in Table III. They appear in good correlation with literature, i.e. about 1.45 eV and 10^4 respectively [12].

Table III: Band gap, absorption coefficient and Hall measurement results for our 6 CZTS absorber layers

	E_G eV	$\alpha\ 10^4$ cm^{-1}	p cm^{-3}	μ $cm^2 V^{-1}s^{-1}$	ρ Ω.cm
L1	1.53	1.73	1.1×10^{13}	14.04	8.5×10^4
L2	1.43	2.02	2.0×10^{20}	0.19	9.8×10^{-1}
L3	1.49	2.17	5.0×10^{19}	1.04	5.4×10^{-1}
H1	1.58	2.17	8.8×10^{13}	26.41	1.4×10^4
H2	1.39	2.26	7.8×10^{15}	5.52	7.6×10^2
H3	1.49	2.16	1.8×10^{16}	0.33	2.1×10^3

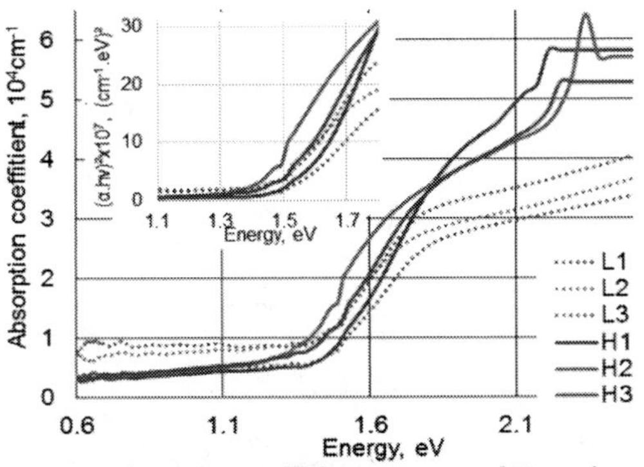

Figure 6: Absorption coefficients spectra and Tauc plots

The bandgap seems to be impacted by the Cu/Sn ratio and increases with increase of Sn concentration, while the absorption coefficient is almost constant (Fig. 7). This in accordance with published discussion about Cu/Sn ratio impact on bandgap [13].

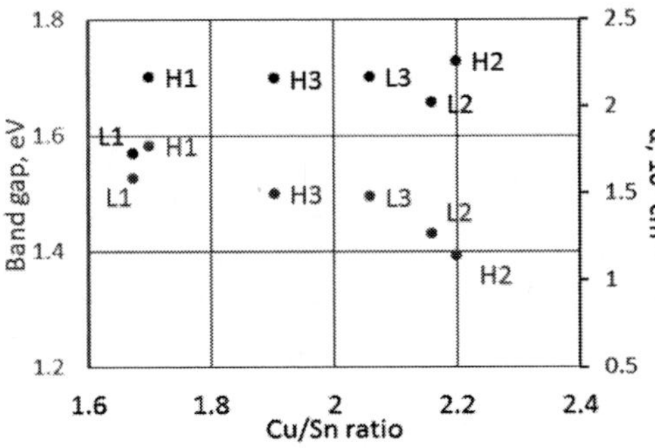

Figure 7: Optical properties versus Cu/Sn ratio.

Reported results of electrical properties widely vary among research groups and fabrication methods: carrier concentration (p) ranges from 10^{16} to 10^{19} cm^{-3}; hole mobility (μ_h) from 0.1 to 30 cm^{2}V^{-1}s^{-1} and corresponding resistivity from 0.1 to 980 Ω.cm [14, 15, 16, 17].

Our Hall measurement results (Fig. 8, Table III) also present large range of electrical properties from low (10^{12} cm^{-3}) to high (10^{20} cm^{-3}) p-type carrier concentration, with a tendency to be inversely proportional to carrier concentration.

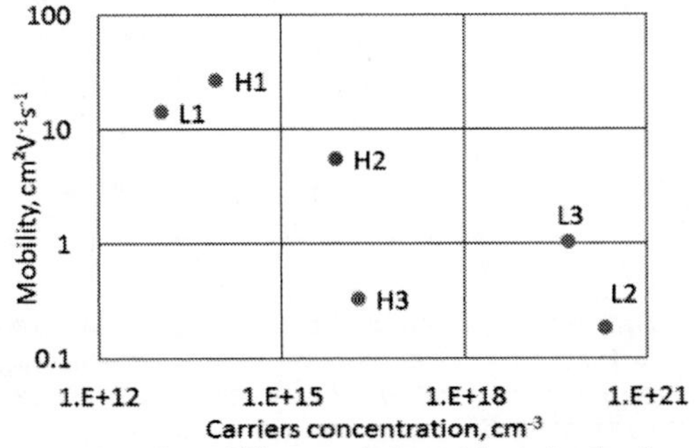

Figure 8: Hole mobility correlation with carrier density.

A linear increase of the carrier concentration with copper concentration in the samples has been observed, similarly to what was reported previously in [18]. Furthermore the mobility increases with tin concentration and the S/metal ratio, except for sample L1. This observation suggests that, for similar ranges of carrier concentration, such as H2 and H3 samples, the tin concentration influences on the mobility. Another possible explanation could be hidden in the unknown effect of sulfur atoms on the conduction mechanisms (Fig. 9).

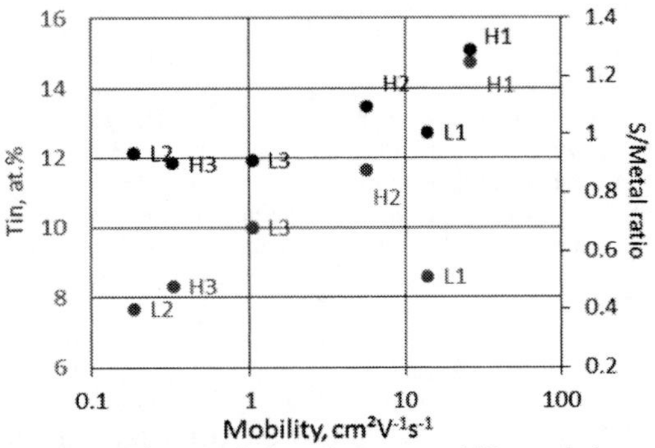

Figure 9: The correlations between mobility and tin or sulfur concentration.

3.3 Simulation optimization

According to SCAPS simulations (Fig. 10), samples H1 and H2 show the most suited p and μ_h values for higher PV efficiency. Note that simulations are limited to 10^{17} cm^{-3} doping levels, below which a rectifying behaviour is obtained, while higher doping levels tend to yield resistive behaviour as observed experimentally before.

The H2-based cell has been used for calibrating predictive SCAPS simulations. Under AM1.5 illumination, the measured photocurrent (I_{PH}) was about 12.7 mA/cm^2 after light soaking of 1.5 hour at negative bias, but extracted efficiency was extremely low (η = 0.78%). Such low cell performance can be related to the S-shape behaviour of J-V characteristics, indicating the presence of Schottky barrier at the rear contact - CZTS/Mo interface (estimated to be about 0.55 eV) and other resistances related problems, namely R_S=313 Ω.cm^2 and R_{SH}=157 Ω.cm^2.

Therefore, to estimate and predict the future roadmap of our H2-based cell process, we have excluded the impact of R_S and R_{SH} by changing their values to experimentally achievable ones [19]. This was followed by the optimization of the key intrinsic optoelectrical parameters such as μ_h, μ_e, p, E_G, α and barrier height. The baseline parameters used for the SCAPS simulations are provided in Table IV.

Impact of R_S, and R_{SH}: Firstly, we estimate the gain in cell efficiency when improving the R_S and R_{SH} from 313 Ω.cm^2 and 157 Ω.cm^2 (measured) to 4 Ω.cm^2 and 2000 Ω.cm^2 (optimization based on experimental values reported in the literature [19]) respectively. Using these values with the measured optoelectronic properties (Fig. 11) as inputs to the SCAPS simulator, a cell efficiency achieved of 3.63 % (i.e. a gain of 2.85 % abs compared to experimental values).

Impact of barrier height at CZTS/Mo interface : In the previous simulations, a barrier height of 0.55 eV have been considered. This level of barrier height at the rear contact is clearly detrimental for the solar cell performances (due to reduced V_{OC} and FF). Therefore, a barrier height of

0.27 eV at the CZTS/Mo interface is next considered. This yields an efficiency of 4.45 % (i.e. a 0.82 % abs gain). These gains in cell efficiencies might look marginal and not so important compared to other optoelectronic parameters. However, the impact of rear contact barrier height gets critically important while moving towards a fully optimized high-efficiency solar cell (where poor FF will significantly impact the cell efficiencies due to barrier height).

Impact of intrinsic μ_h and p : In Fig. 10, we can clearly appreciate the impact of μ_h and acceptor doping concentration (p) on cell efficiency. Therefore, in our SCAPS efficiency-optimization roadmap predictions, we next improve the μ_h from 5.52 $cm^2V^{-1}s^{-1}$ to 25 $cm^2V^{-1}s^{-1}$ and the acceptor concentration from p = 7.8×10^{15} cm^{-3} 5.5×10^{17} cm^{-3} (in all our simulations we have maintained the ratio of μ_h / μ_e to 4 for practicality). This leads to a simulated cell efficiency improvement from 4.45 % to 6.23 %.

Impact of CZTS bandgap (E_G) and absorption (α) : The extracted optical bandgap of H2 absorber is 1.39 eV. It is well known that larger bandgap materials contribute to better V_{OC} and cell efficiencies. Therefore, the next optimization parameter considered for higher cell efficiencies is improved E_G of 1.6 eV in the SCAPS simulations. This eventually led to an increment in the cell efficiency at 6.79 % (i.e. a 0.56 % abs gain). Next, in order to improve the photo-generated current (I_{PH}), it is important to have a higher absorption coefficient of the CZTS films. Therefore, we have increased the absorption coefficient from 2×10^4 cm^{-3} to 3.5×10^4 cm^{-3} in our simulations. This resulted in an improved cell efficiency of 8.44 % (1.65 % abs gain).

Lastly, further optimization of deep defect concentrations (N_{def}) by an order or magnitude (i.e. from 2×10^{18} cm^{-3} to 2×10^{17} cm^{-3}) yields a further improvement in cell efficiencies to 9.27%.

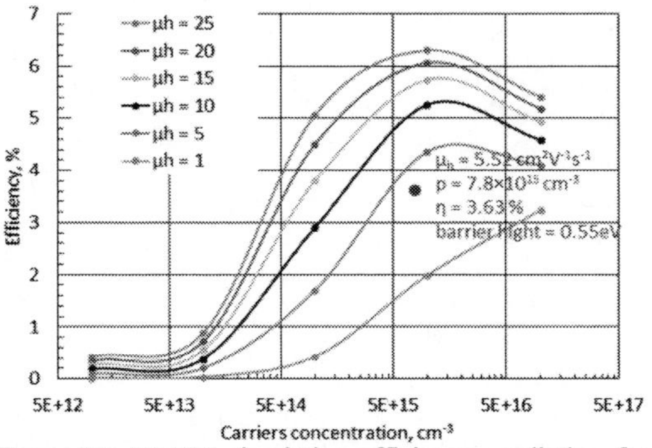

Figure 10: SCAPS simulation efficiency prediction for different hole mobilities (μ_h), carrier concentrations using input parameters from fig. 11 but after optimizing R_S, R_{SH} and barrier height. The red dot corresponds to our experimental cell measurement after R_S and R_{SH} corrections

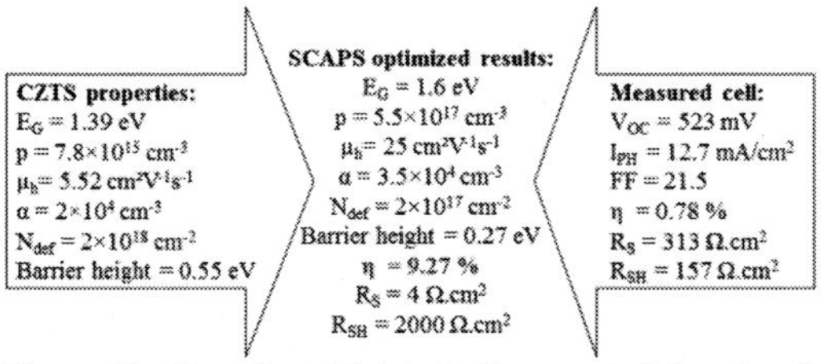

Figure 11: Experimental inputs from material and cell characterizations (left and right) and SCAPS optimization results (middle)

Table IV: Baseline parameters used for modelling CZTS solar cells

Parameter	CZTS	CdS	i-ZnO	ZnO:Al
E_g (eV)	1.39	2.4	3.3	3.3
χ (eV)	4.5	4.45	4.55	4.55
$\varepsilon/\varepsilon_0$	13.6	10	9	9
N_c (cm^{-3})	2.2×10^{18}	1.3×10^{18}	3.1×10^{18}	3×10^{18}
N_v (cm^{-3})	1.5×10^{19}	9.1×10^{18}	1.8×10^{19}	1.8×10^{19}
V_e (cm/s)	3.9×10^7	3.1×10^7	2.4×10^7	2.4×10^7
V_h (cm/s)	1.4×10^7	1.6×10^7	1.3×10^7	1.3×10^7
μ_e (cm^2/Vs)	4*μh	72	100	100
μ_h (cm^2/Vs)	5.52	20	31	3
Doping (cm^{-3})	7.8×10^{15} (a)	5×10^{17} (d)	1×10^{17} (d)	5×10^{19} (d)
Bulk defect properties				
N (cm^{-2})	10^{14} (D)	5×10^{16} (A)	10^{16} (A)	10^{16} (A)
σ_e (cm^2)	10^{-15}	10^{-15}	10^{-15}	10^{-15}
σ_h (cm^2)	10^{-11}	5×10^{-13}	5×10^{-13}	5×10^{-13}

(a) and (d) denote shallow acceptor and donor defect while (A), (D) and (N) denote deep acceptor, donor and neutral defects.

Nomenclature: ΔE_c: Conduction and band offset; E_g: Band gap; $\varepsilon/\varepsilon_0$: Relative permittivity; χ: Affinity; μ_e, μ_h: Velocity of electrons and holes; N_c, N_v: Effective density of states in conduction and valence bands; σ_e, σ_h: Capture cross-section of electrons and holes

4 CONCLUSION

Two groups of metallic precursors have been studied by sequential sputtering, i.e. low- and high-density layers varying composition ratios and/or stack ordering, and finalized into CZTS absorber layers by RTP with presence of S. For all samples material and electrical characterizations have been performed. No significant difference has been observed by XRD and Raman in the crystal structures of the two groups. They all exhibit rough surface with different textures and mainly CZTS phase, coexisting with CTS, $Cu_{2-x}S$ and ZnS phases. From optical characterization, the bandgap energies and absorption coefficients vary respectively from 1.39 to 1.58eV and from 1.73×10^4 to 2.26×10^4 cm^{-1}, suitable for solar cell application and the correlation of the bandgap energy with

Cu/Sn ratio is observed. Carrier concentration and mobility Hall measurements show clear correlations with Cu and Sn concentrations in CZTS layers respectively.

With the measured CZTS optoelectronic properties of H2-based cells as starting inputs to the SCAPS simulator, we have clearly distinguished the impact of each key optoelectronic parameter. Improving these optoelectronic properties, beyond the absorber composition and morphology, is key to achieve above 10%-efficiency CZTS cells. Monitoring of these optoelectrical parameters during the fabrication steps could then be particularly relevant to assess the absorber quality in industrial process.

5 ACKNOWLEDGEMENTS

Authors are grateful to Edgardo Saucedo (IREC, Barcelona) and his team for sharing their experience and valuable advice for this work. We are thankful to Walloon Region and FNRS for financial support. Thanks to Sergio Bernardi for valuable discussions and Ferran Urena Begara for help in Raman interpretation.

6 REFERENCES

[1] K. Sun, et al, Over 9% efficient kesterite Cu2ZnSnS4 solar cell fabricated by using Zn1-XCdXS buffer layer, Adv. Energy Mater. 6 (2016) 1600046-1600052

[2] A. Fairbrother, et al, Development of selective chemical etch to improve the conversion efficiency of Zn-rich Cu2ZnSnS4 solar cells. J. Am. Chem. Soc. 134 (2012) 8018-8021

[3] A.A. Rockett, et al, Current status and opportunities in chalcopyrite solar cells, Curr. Opin. Solid State Mater. Sci. 14 (2010) 143–148

[4] R. Liu, et al, Preparation of high-quality Cu2ZnSnS4 thin films for solar cells via the improvement of sulfur partial pressure using a static annealing sulfurization approach, Sol. Energy Mater. Sol. Cells 157 (2016) 221-228

[5] C. Yan, et al, Boost Voc of pure sulfide kesterite solar cell via a double CZTS layer stacks, Sol. Energy Mater. Sol. Cells 160 (2017) 7-11

[6] Y. Ren, et al, Influence of the Cu2ZnSnS4 absorber thickness on thin film solar cells, Phys. Status Solidi A 212, (2015) 2889-2896

[7] S.M. Pawar, et al, Synthesis of Cu2ZnSnS4 (CZTS) absorber by rapid thermal processing (RTP) sulfurization of stacked metallic precursor films for solar cell application. Mater. Lett. 118 (2014) 76-79

[8] https://www.webelements.com/periodicity/density/

[9] S. Lopez-Martino, et al, Earth-abundant absorber based solar cells onto low weight stainless steel substrate. Sol. Energy Mater. Sol. Celles 130 (2014) 327-353

[10] M. Patel, et al, Enhancement of output performance of Cu2ZnSnS4 thin film solar cells – A numerical simulation approach and comparison to experiments, Physica B 407 (2012) 4391-4397

[11] P.-A. Cormier, et al, One-step synthesis of Cu2ZnSnS4 thin films by reactive magnetron co-sputtering. Acta Materialia 96 (2015) 80-88

[12] K. Ito, et al, Electrical properties of stannite type quaternary semiconductor thin films. Jap. J. of Applied Phys. 27-11 (1988) 2094-2097

[13] C. Malerbe, et al. CTZS stoichiometry effects on the band gap energy. J. of Alloys and Compounds 582 (2014) 528-534

[14] Z. Shi, et al, Kesterite-based next generation high performance thin film solar cell: current progress and future prospects, J. Mater. Sci: Mater Electron 28 (2017) 2290-2306

[15] O. Vigil-Galán, et al, Electrical properties of sprayed Cu2ZnSnS4 thin films and its relation with secondary phase formation and solar cell performance. Sol. Energy Mater. Sol. Cells 132 (2015) 557-562

[16] J.C. Gonzalez, et al, Influence of the sulfurization time on the morphological, chemical, structural and electrical properties of Cu2ZnSnS4 polycrystalline thin films, Sol. Energy Mater. Sol. Cells 123 (2014) 58-64

[17] F. Liu, et al, Insitu growth of Cu2ZnSnS4 thin films by reactive magnetron co-sputtering, Sol. Energy Mater. Sol. Cells 94 (2010) 2231-2434

[18] E.M. Mkawi, et al. Dependence of copper concentration on the properties of the Cu2ZnSnS4 thin films prepared by electrochemical method. Int. J. Electrochim. Sci. 8 (2013) 359-368

[19] W. Wang, et al, Device characteristics of CZTSSe thin-film solar cells with 12.6% efficiency. Adv. Eng. Mater. 4-7 (2014)

PHOSPHOR PARTICLES FOR LUMINESCENT DOWN-SHIFTING IN PHOTOVOLTAICS: DETERMINATION OF COMPLEX REFRACTIVE INDICES

Benjamin Lipovšek[1], Anastasiia Solodovnyk[2,3], Daniel Riedel[3], Andres Osvet[3], Jessica Gast[2], Edda Stern[2], Karen Forberich[3], Miroslaw Batentschuk[3], Christoph J. Barbec[2,3], Janez Krč[1], and Marko Topič[1]
[1] University of Ljubljana, Faculty of Electrical Engineering, Tržaška 25, 1000 Ljubljana, Slovenia
[2] Bavarian Center for Applied Energy Research (ZAE Bayern), Haberstr. 2a, 91058 Erlangen, Germany
[3] Friedrich-Alexander University Erlangen-Nürnberg (FAU), Martensstr. 7, 91058 Erlangen, Germany

ABSTRACT: We demonstrate a novel two-step method for accurate determination of the complex refractive indices of phosphor materials in particulate form. The real part of the refractive index is first measured at a number of selected wavelengths using the Becke Line method; the results are interpolated over a broader wavelength range using Sellmeier's equations. The imaginary part of the refractive index is extracted from measured optical characteristics of polymer layers embedded with phosphor particles; a specialized numerical simulation tool is employed for this purpose. Finally, using the determined complex refractive indices, we numerically investigate the potential of luminescent down-shifting applied to thin-film solution-processed solar cells. We study the influence of the main properties of luminescent down-shifting layers filled with phosphor particles, and indicate the gains in the solar cell performance that can be obtained using this approach.
Keywords: Modelling, Optical Properties, Organic Solar Cell, Photoluminescence, Photovoltaic

1 INTRODUCTION

To improve the conversion efficiency of single-junction solar cells in the short-wavelength region, the concept of luminescent down-shifting (LDS) can be applied [1–4]. This approach is based on the application of photoluminescent (PL) materials that shift the UV and near-UV photons to photons of longer wavelengths where the spectral response of the solar cell is much higher. Among different types of PL materials, phosphor particles show many advantages such as large Stokes shift, high PL quantum yield, and long-term stability [5, 6]. Dispersed in a transparent polymer binder, they can be easily deposited on top of the solar cell in form of an LDS layer [7]. However, due to the relatively large size of phosphor particles (from a few microns to a few tens of microns), scattering at these particles inside the LDS layer can lead to a reduced overall transmission of the incident light into the underlying solar cell [7,8]. Therefore, detailed analysis and careful optimization of the LDS layer composition is required (i.e. the layer thickness, the volume concentration of the phosphor particles, and the size distribution of the phosphor particles), which can be done efficiently using an appropriate optical modelling and simulation tool [9].

The accuracy of numerical simulations, however, depends heavily on the accuracy of the input parameters. In terms of optical simulations of phosphor-based LDS layers, as well as complete solar cells enhanced with such LDS layers, we experienced that the complex refractive indices of the phosphor particles present the most critical input parameter. They are mostly unknown, and, due to the particulate form of the phosphor material, they are also notoriously difficult to determine using the conventional measurement techniques (e.g. ellipsometry).

In this work, we first demonstrate the application of a novel method to precisely determine the wavelength-dependent complex refractive indices ($n(\lambda) - i\cdot k(\lambda)$) of different kinds of particulate phosphor materials [10]. The method consists of two steps that enable seperate determination of the real part, n, and the imaginary part, k (extinction coefficient), of the complex refractive index. Then, using the determined complex refractive indices, we also present the results of complete optical simulations of example solution-processed organic solar cells enhanced with luminescent down-shifting phosphor-filled layers. We investigate the potential of the LDS layers for improved solar spectrum harvesting in these devices, and discuss the optimal parameters of the LDS layers that would be most appropriate for this application.

2 n & k DETERMINATION METHOD

A number of different phosphor materials were included in our study, all of which existed in particulate (powder) form with particle sizes ranging from 10 µm to 20 µm (the precise particle size distributions have been determined experimentally [7]): (i) $Lu_3Al_5O_{12}:Ce^{3+}$, (ii) $Ba_2SiO_4:Eu^{2+}$, (iii) $Ba_3MgSi_2O_8:Eu^{2+},Mn^{2+}$, and (iv) $SrAl_2O_4:Eu^{2+}$. All of these phosphor materials absorb at short wavelenghts (< 500 nm) and emit at longer wavelenghts of the visible spectrum. Example photoluminescence excitation / emission spectra for $Lu_3Al_5O_{12}:Ce^{3+}$ are presented in Fig. 1, together with the SEM image of the phosphor particles.

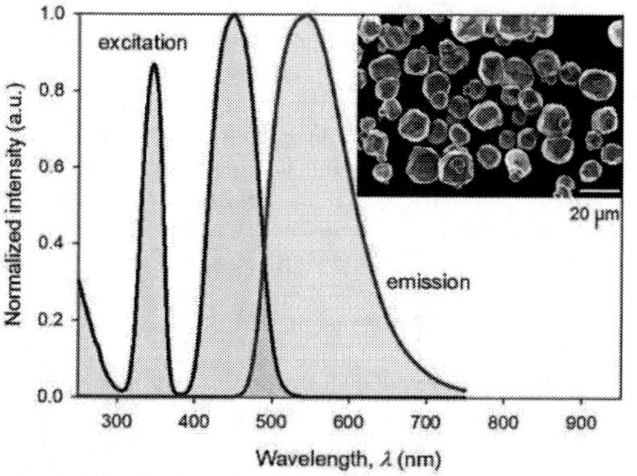

Figure 1: Photoluminescence excitation and emission spectra of $Lu_3Al_5O_{12}:Ce^{3+}$ phopsphor particles. The SEM image of the particles is shown in the inset.

As the first step of the complex refractive index determination, we measured the real part, n, using the established Becke Line method [11]. For this purpose, the phosphor particles were immersed in immersion liquids with varying refractive indices $n_{IL}(\lambda)$. The solutions were then illuminated by a monochromatic light and investigated under an optical microscope, as also demonstrated in Fig. 2. The focus, f, was first set on the particles, and then the distance to the particles, z, was increased by moving the sample table away from the microscope ($z > f$). If the bright halo (Becke Line) moved into the particles (i.e. the particles appeared brighter), the refractive index of the particles was larger than that of the liquid. If the halo moved into the liquid (i.e. the particles appeared darker), the refractive index of the particles was lower than that of the liquid. With changing the liquids (i.e. varying the refractive index of the liquid), we were able to determine the real part of the refractive index of the phosphor particles with great precision (below 0.01) for the given monochromatic illumination. This process was performed separately at six different wavelengths, and the data was finally fitted by a three-term Sellmeier's equation to obtain the dispersion of $n(\lambda)$ throughout the investigated wavelength range of 300 – 800 nm.

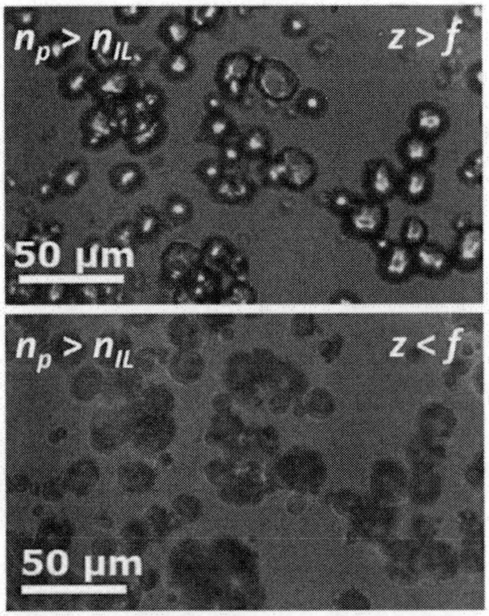

Figure 2: Optical microscopy of $Lu_3Al_5O_{12}:Ce^{3+}$ phosphor particles embedded in an immersion liquid with n_{IL} = 1.7 and illuminated at λ = 595 nm. Top image: distance to the particles, z, is larger than the focal distance, f. Bottom image: distance to the particles, z, is smaller than the focal distance, f. In this example, the refractive index of the particles is larger than that of the immersion liquid.

As the second step, we determined the extinction coefficient, k. For this purpose, the phosphor particles were first embedded in a transparent polymer binder and deposited on top of glass substrate in form of LDS layers with thicknesses between 100 µm and 400 µm. The optical reflectance, R_{tot}, transmittance, T_{tot}, and absorptance, A, of the layers were measured using a spectrally selective angular resolved scattering (SARS) measurement system, which consists of a white illumination source and a goniometer with a line grating as the detector. Using this system, it is possible to measure the "true" optical characteristics (R_{tot}, T_{tot}, A) even in the excitation wavelength region of the phosphor particles, as we can exclude contribution due to the down-shifted light. The measured T_{tot} results for three example LDS layers based on $Lu_3Al_5O_{12}:Ce^{3+}$ phosphor particles are presented in Fig. 3 (full lines).

Next, the measured results were fed into the calibrated three-dimensional optical simulator CROWM enhanced with a recently developed optical model for simulation of LDS layers [9]. Also imported as the input parameters were the previously determined n data as well as various other predetermined properties of the phosphor particles (shape, size distribution etc. [7]). Using a specialized algorithm, we then iteratively determined, separately wavelength-by-wavelength, the $k(\lambda)$ parameters of the phosphor particles that best agree with the measured optical characteristics for all layers with different thicknesses and/or particle volume concentrations. The simulated T_{tot} results of the three example LDS layers based on $Lu_3Al_5O_{12}:Ce^{3+}$ phosphor particles, obtained by using the best fit of the $k(\lambda)$ parameter, are plotted as symbols in Fig. 3.

Both steps of the demonstrated n & k determination method were carefully verified on a number of materials with well-known optical constants (artificial ruby, rock salt, rutile titanium dioxide). We need to point out, however, that the method is especially suitable for characterization of phosphor materials, as it takes luminescence, scattering, and re-absorption of the phosphor particles fully into account.

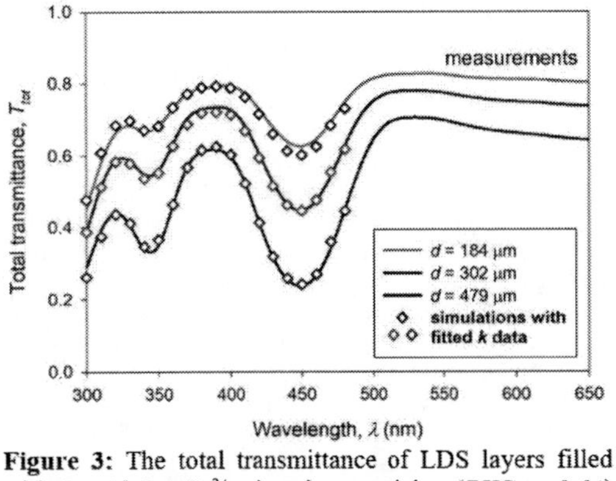

Figure 3: The total transmittance of LDS layers filled with $Lu_3Al_5O_{12}:Ce^{3+}$ phosphor particles (PVC = 6 %) with varied thicknesses, measured by the SARS system (full lines). Also plotted are simulation results (symbols), using previously determined n data and the best overal fit of the extinction coefficient, k (seperately for each wavelength).

3 RESULTS AND DISCUSSION

The demonstrated n & k determination method was applied to all four investigated particulate phosphor materials. The obtained dispersions of n and k are plotted in Figs. 4a and 4b, respectively. It should be noted that the curves of the extinction coefficient, k, correspond well with the photoluminescence excitation profiles of the respective phosphor materials.

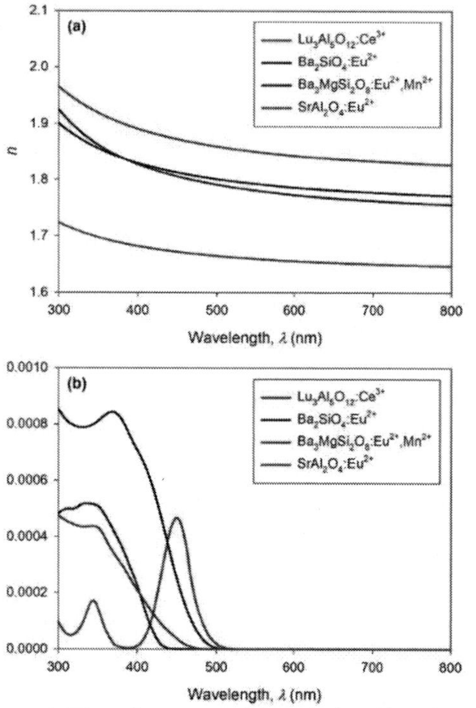

Figure 4: Dispersion curves of the real (a) and imaginary (b) parts of the complex refractive index, as determined for four different phosphor powder materials by means of the presented method.

To evaluate the concept of luminescent down-shifting in thin-film solution-processed solar cells, the determined *n* & *k* parameters of phosphor particles were employed to perform complete optical simulations of solar cells enhanced with LDS layers. For this purpose, we employed the three-dimensional optical model CROWM [12,13]. The model is based on the combination of geometric and wave optics, and it was recently also upgraded with the ability to accurately simulate layers embedded with phosphor particles [9].

The example simulated device in our study was an organic solar cell in superstrate configuration based on the DPP:PC60BM absorber material [13]. The structure was as follows: air / LDS layer / glass (0.5 mm) / ITO (360 nm) / PEDOT:PSS (70 nm) / DPP:$PC_{60}BM$ (120 nm) / Ca (15 nm) / Ag (85 nm). All the input parameters required by the simulations were either determined experimentally or extracted from literature.

First, we simulated the absorptance, *A*, inside the absorber layer for the case of a bare solar cell in the wavelength range of 300 – 800 nm. The results are presented in Fig. 5 (black line). By applying the incident AM1.5 solar spectrum to the simulated *A* results, we also calculated the ideal short-circuit current density, J_{SC}, of 16.14 mA/cm^2. Next, we applied a LDS layer based on the $Lu_3Al_5O_{12}:Ce^{3+}$ phosphor particles on top of the front glass in the simulations, assuming the total layer thickness, *d*, of 100 µm and the phosphor particle volume concentration, *PVC*, of 10 %. The simulated *A* results for this case are also plotted in Fig. 5 (red line). The *A* decrease in the short-wavelength region, due to the absorption of the phosphor particles, and the significant *A* increase in the long-wavelength region can clearly be observed. It should be noted, however, that the observed *A* increase is not only due to the contribution of the down-shifted photons, but also due to the pronounced scattering of light at the phosphor particles inside the LDS layer, which in this case contributes to more efficient light trapping inside the absorber layer of the cell. The calculated ideal J_{SC} of the cell with the LDS layer is 17.09 mA/cm^2, which presents a 5.9 % boost compared to the bare organic solar cell and, thus, indicates promising potential of the LDS concept.

Finally, by taking the characteristics of a typical external quantum efficiency, *EQE*, measurement system into account, we also calculated the »tentative *EQE*« as would be measured by such a system. The results are also plotted in Fig. 5 (dashed blue line). Compared to the *A* curves shown before, the difference in this case is that typical *EQE* measurements are not spectrally selective. Therefore, the contribution of the down-shifted photons to the measured signal is observed at the current (excitation) wavelength of the illumination source, and not at the emission wavelength of the phosphor particles. This *A* to »*EQE*« recalcualtion is important for a clearer future comparison with experimentally determined *EQE* results of fabricated devices.

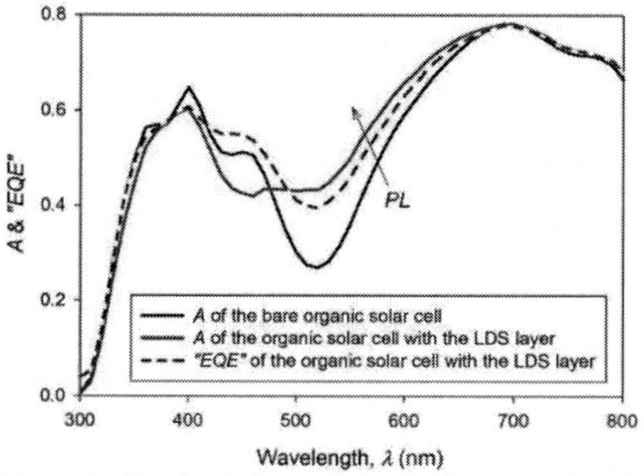

Figure 5: Simulated absorptance of a bare organic solar cell based on the DPP:$PC_{60}BM$ absorber material (black curve), and of the solar cell enhanced with the LDS layer (red curve) based on $Lu_3Al_5O_{12}:Ce^{3+}$ phosphor particles (*d* = 100 µm, *PVC* = 10 %). Also plotted is the recalculated external quantum efficiency that would be obtained through measurements by using a typical non-spectrally-resolved system (dashed blue line).

In the last stage of our work, we performed a series of simulations using the same structure of the organic solar cell, but we systematically altered the properties of the LDS layer. More specifically, we varied the thickness of the layer in the range of 0 – 200 µm, and the phosphor particle volume concentration in the range of 0 – 20 %. For each *d*/*PVC* combination, we simulated the ideal J_{SC} of the organic solar cell. The corresponding values of the relative J_{SC} boost, compared to the solar cell without the LDS layer, are presented in Fig. 6. The results show that both the thickness and the phosphor particle volume concentration of the LDS layer dramatically affect the overal performance of the device. Therefore, careful optimization of such layers, emplying advanced numerical models and using realistic input parameters, is of utmost importance. Finally, the results also support the feasibility of the LDS concept, indicating potential J_{SC} gains of over 6 %.

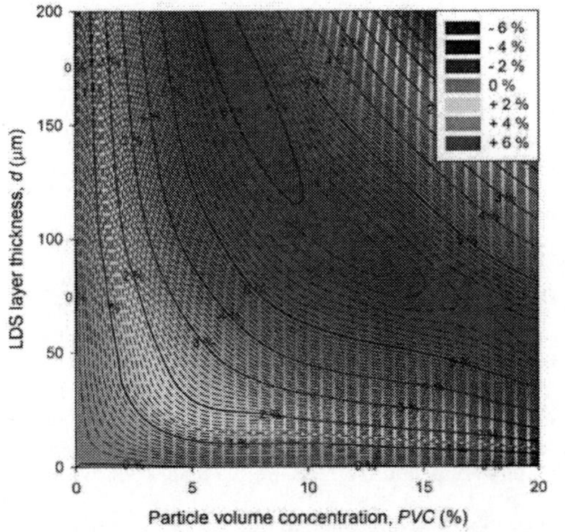

Figure 6: Simulated relative increase of the short-circuit current denisty (in percent) of a thin-film organic solar cell based on the DPP:$PC_{60}BM$ absorber material, as a result of LDS layer application based on $Lu_3Al_5O_{12}:Ce^{3+}$ phosphor particles.

4 CONCLUSIONS

We demonstrated the application of a novel two-step method for complex refractive index determination of particulate phosphor materials. The method consists of n measurement using the Becke Line method, and numerical k extraction from experimentally measured optical characteristics of fabricated layers embedded with phosphor particles. The complete n & k dispersion curves for four commonly used phosphor materials were determined and presented.

Using numerical simulations based on the optical model CROWM, we investigated the potential of luminescent down-shifting layers in organic solar cells based on the DPP:$PC_{60}BM$ absorber material. The results show that proper optimization of the LDS layer properties is crucial for achieving the optimal performance of the device. In the case of LDS layers based on the $Lu_3Al_5O_{12}:Ce^{3+}$ phosphor particles, a short-circuit current density gains of over 6 % can be obtained, which confirms the feasibility of the luminescent down-shifting approach applied to thin-film solution-processed solar cells.

REFERENCES

[1] B. S. Richards, Solar Energy Materials and Solar Cells 90 (2006) 2329.

[2] E. Klampaftis, D. Ross, K. R. McIntosh, B. S. Richards, Solar Energy Materials and Solar Cells 93 (2009) 1182.

[3] E. Klampaftis, M. Congiu, N. Robertson, B. S. Richards, IEEE Journal of Photovoltaics 1 (2011) 29.

[4] T. Uekert, A. Solodovnyk, S. Ponomarenko, A. Osvet, I. Levchuk, J. Gast, M. Batentschuk, K. Forberich, E. Stern, H.-J. Egelhaaf, C. J. Brabec, Solar Energy Materials and Solar Cells 155 (2016) 1.

[5] S. Tanimizu, Fundamentals of Phosphors, CRC Press (2006).

[6] Q. Y. Zhang, X. Y. Huang, Progress in Materials Science 55 (2010) 353.

[7] A. Solodovnyk, B. Lipovšek, D. Riedel, K. Forberich, E. Stern, M. Batentschuk, J. Krč, M. Topič, C. J. Brabec, Journal of Optics 19 (2017) 095901.

[8] A. Solodovnyk, B. Lipovšek, K. Forberich, E. Stern, J. Krč, M. Topič, C. J. Brabec, Optical Materials Express 5 (2015) 1296.

[9] B. Lipovšek, A. Solodovnyk, K. Forberich, E. Stern, J. Krč, C. J. Brabec, M. Topič, Optics Express 23 (2015) A882.

[10] A. Solodovnyk, D. Riedel, B. Lipovšek, A. Osvet, J. Gast, E. Stern, K. Forberich, M. Batentschuk, J. Krč, M. Topič, C. J. Brabec, Optical Materials Express 7 (2017) 2943.

[11] R. A. Carlton, Pharmaceutical Microscopy, Springer (2011).

[12] B. Lipovšek, J. Krč, M. Topič, Informacije MIDEM - Journal of Microelectronics, Electronic Components and Materials 41 (2011) 264.

[13] B. Lipovšek, A. Čampa, F. Guo, C. J. Brabec, K. Forberich, J. Krč, M. Topič, Optics Express 25 (2017) A176.

POWER MATRIX MEASUREMENTS AND ENERGY RATING ANALYSIS OF ORGANIC PV MINI-MODULES

Giorgio Bardizza*, Elena Salis, Ana Maria Gracia Amillo, Thomas Huld, Ewan D. Dunlop
European Commission, DG Joint Research Centre, European Solar Test Installation
Via E. Fermi 2749, I-21027 Ispra (VA) - Italy
* E-mail: giorgio.bardizza@ec.europa.eu - Tel: +39 0332785529

ABSTRACT: In this paper we present accurate measurements of I-V characteristics of two organic PV (OPV) mini-modules, measured indoors following the standard IEC 61853-1. The measurements were performed at a large-area steady-state solar simulator over a reduced range of temperatures and irradiances compared to what the standard IEC 61853-1 requires. The analysis of the electrical parameters dependence on irradiance and temperature shows interesting results. An energy rating evaluation for OPV devices based on the measured power matrix data set has been performed, too. The estimate of the module performance ratio at five different climatic locations shows how the OPV technology has comparable values to some of the most common PV technologies, and also that for some of the considered locations it may even perform relatively better. Spectral and temperature effects are the main contributing factors.
Keywords: Photovoltaic, power matrix, OPV, organic photovoltaic, IEC 61853-1, energy rating

1 INTRODUCTION

Newer PV technologies like organic (OPV) and dye-sensitized (DSC) devices are rapidly evolving and considerable research effort has addressed alternative approaches to increase their efficiency, reaching in the last years promising values [1]. Even if these technologies still need research and development to improve their performances (efficiency and stability) up to a level comparable to well-established and more mature technologies, there are already examples of large-area installations of OPV modules working outdoor at real operating conditions: the Heliatek façade in Nurnberg (Germany) and the DTU solar park in Roskilde (Denmark). The systems are working since more than two years and the energy produced is monitored [2]. This suggests the need of studies about the behavior of these emerging PV devices not only at standard test conditions (STC), but also close to real operating conditions, and the comparison with better established PV technologies (such as c-Si, a-Si, CIGS or CdTe). Experimental results on the performance of OPV devices under different irradiance and temperature conditions are less established than those of Si-based devices. Only few examples are available in the literature [3] [4] and [5] and results are incomplete. In this paper we present accurate measurements of I-V characteristics of two different OPV mini-modules, acquired indoors at a solar simulator over a selection of irradiances and temperatures according to the international standard IEC 61853-1. The measurements were performed at the European Solar Test Installation (ESTI) of the European Commission's Joint Research Centre.

2 DEVICES

The devices measured in this study are two OPV mini-modules (ESTI codes TM801 and QJ87) with 8 cells connected in series (see Figure 1 for a picture of them). Each cell has an active area of about 8 cm^2. The mini-module QJ87 (efficiency about 1%) was produced by the group of Prof. Krebs of the Technical University of Denmark with a roll-to-roll coating technique following the freeOPV design [6].

Figure 1 Picture of the devices under test. ESTI codes TM801 (left) and QJ87 (right).

The device TM801 (nominal efficiency: 4%) was bought more recently from the company InfinityPV. The spectral responsivity of the two devices was measured at ESTI according to the standard IEC 60904-8 [7] and the data are shown in Figure 2.

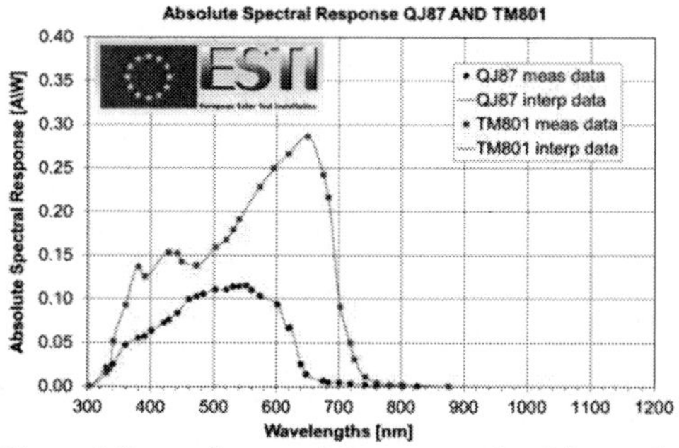

Figure 2 Spectral responsivity measured and interpolated data of QJ87 and TM801.

3 MEASUREMENT SETUP

3.1 Apollo Solar Simulator

The I-V curves at varying irradiance and temperature, used to generate the power matrix of the OPV mini-modules under study, were acquired at the Apollo solar simulator of ESTI [8]. This simulator is a large-area steady-state solar simulator in class AAA according to the standard IEC 60904-9 [9] and it is provided with eleven Xenon lamps. Each lamp illuminates the test plane of 2x2 m^2 and can be controlled independently.

3.2 Power rating measurements
For the measurements presented here, the device under test (DUT) was placed on a vacuum copper plate to maintain it stable and level. The vacuum plate was mounted on a water-cooled plate, so that temperature control and stabilization could be achieved more easily. The DUT temperature was measured by a Pt100 sensor attached to the water-cooled plate and masked towards radiation.
The measurements were performed by setting the temperature of the device through control of the water temperature. The temperature range covered in this study was [15, 45] °C in steps of 5 °C for the device QJ87 and [15, 50] °C in steps of 5 °C for the device TM801. The irradiance was varied between 1000 W/m^2 and 100 W/m^2 with step of approximately 200 W/m^2 down to 200 W/m^2 and set by shuttering a suitable number of lamps. The irradiance level was established and measured by the reading of a calibrated c-Si filtered reference cell (ESTI code PX204A), closely matched to the SR of the devices, mounted close to the DUT holder and kept at (25 ± 1) °C by a Peltier element.
Hence, once the DUT temperature was deemed stable, for each temperature and irradiance combination (T_{meas}, G_{meas}), three I-V curves were acquired by sweeping a voltage across the DUT from short-circuit current (I_{sc}) to open-circuit voltage (V_{oc}). From each of these curves the main electrical parameters were extracted and their dependences on temperature and irradiance analyzed.

4 RESULTS

4.1 Power matrices
A selection of the measured I-V curves of the power matrices of the devices QJ87 and TM801 are shown in Figures 3 and 4 respectively. Three I-V curves were measured at each combination of temperature and irradiance (T_{meas}, G_{meas}) and are reported in the plots. All the irradiances are shown in Figure 3 and 4 for the temperatures 15 °C, 25 °C, 35 °C and 45 °C as example. The electrical parameters extracted from the I-V curves of devices QJ87 and TM801 are the short-circuit current I_{sc}, the open-circuit voltage V_{oc}, the power at maximum power point P_{max} and the fill factor FF.
The average P_{max} values calculated from those extracted from the measurements are shown in Tables 1 and 2 for QJ87 and TM801 respectively. The data sets of all the P_{max} values as function of T and G compose the power matrix in accordance to the definition given in the standard IEC 61853-1 [10], although not all the required combinations were covered in this study in order to avoid severe aging of the device.

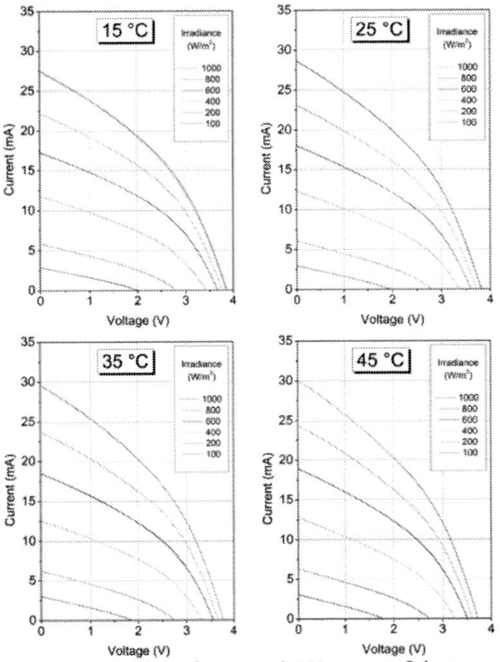

Figure 3 Selection of measured I-V curves of the power matrix of device QJ87. The measurements relative to temperatures of 15 °C, 25 °C, 35 °C and 45 °C and all measured irradiances are shown as example.

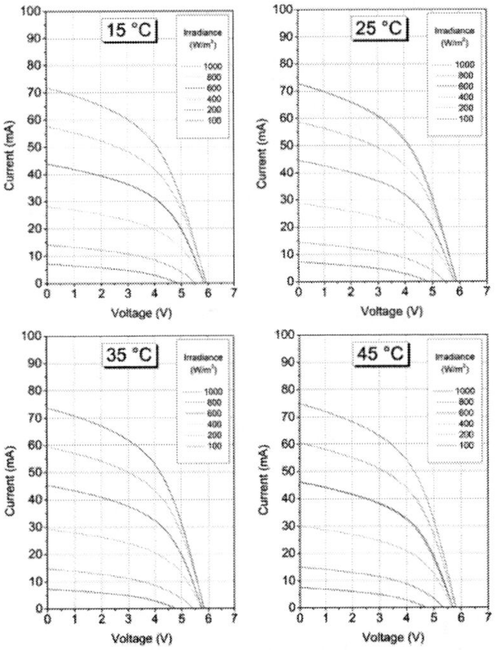

Figure 4 Selection of measured I-V curves of the power matrix of device TM801. The measurements relative to temperatures of 15 °C, 25 °C, 35 °C and 45 °C and all measured irradiances are shown as example. Note the different vertical scale compared to Figure 3.

Table 1 Power matrix average P_{max} values calculated from the I-V curves as a function of T and G for device QJ87.

QJ87 - Maximum power - P_{max} (mW)							
Irradiance (W/m²)	Temperature (°C)						
	15	20	25	30	35	40	45
100	1.640	1.672	1.649	1.647	1.594	1.584	1.541
200	5.307	5.493	5.519	5.510	5.508	5.563	5.540
400	15.027	15.202	15.288	15.370	15.484	15.401	15.095
600	24.798	24.979	25.167	25.274	25.261	25.384	25.203
800	33.250	33.698	33.949	34.217	34.167	33.907	33.839
1000	41.164	42.247	42.281	42.429	42.675	42.663	42.595

Table 2 Power matrix average P_{max} values calculated from the I-V curves as a function of T and G for device TM801.

TM801 - Maximum power - P_{max} (mW)								
Irradiance (W/m²)	Temperature (°C)							
	15	20	25	30	35	40	45	50
100	14.13	14.32	14.41	14.62	14.65	14.84	14.88	15.00
200	34.52	34.84	35.10	35.32	35.51	35.72	36.18	36.44
400	78.74	79.87	80.82	81.27	82.27	82.66	83.45	84.18
600	124.83	125.59	127.33	128.42	129.27	130.05	130.03	130.79
800	167.70	168.91	171.58	171.71	174.05	174.46	175.61	176.21
1000	205.64	206.65	207.76	211.13	212.30	216.05	216.93	218.55

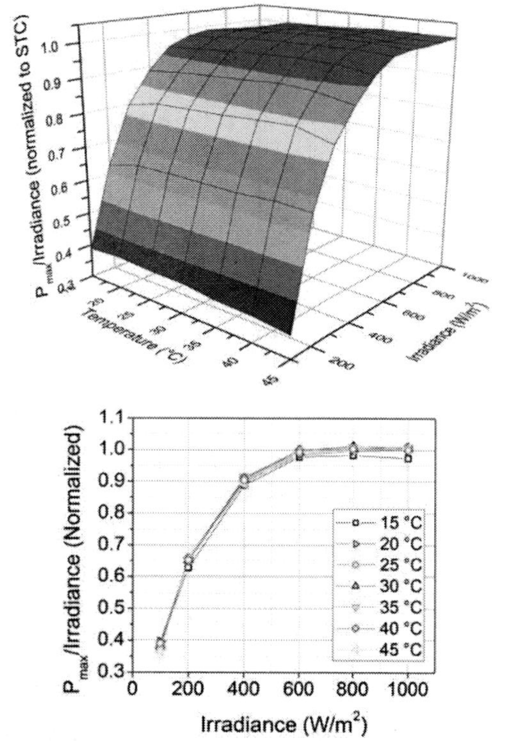

Figure 5 Surface plot (top) showing the dependence of the ratio maximum power divided by irradiance on the temperature and irradiance for QJ87. The average data were normalized to 1 at STC (25 °C and 1000 W/m²). The dependence on the irradiance is shown also separately in the normal plot on the bottom.

As expected, from Figure 3 and 4 one can already see that I_{sc} increases linearly with irradiance while V_{oc} shows a nearly logarithmic dependence.

More interesting to study is the ratio P_{max}/Irradiance as a function of the irradiance level (see bottom plot of Figures 5 and 6). This value is proportional to the efficiency of the devices. It is shown in Figures 5 and 6 for the devices QJ87 and TM801, respectively. The surface plots on the top part of the figures show the simultaneous variation of P_{max}/Irradiance with both temperature and irradiance. The data have been normalized to 1 at 1000 W/m² and 25 °C. In these plots it is clearly noticeable how much the irradiance and the temperature affect the efficiency of the devices under study. As expected the effect of the temperature is relatively smaller compared to the effect of the irradiance, which is however not negligible. This is particularly evident in the measurements of TM801 for which the P_{max}/Irradiance ratio, and thus the efficiency, increases considerably (about 5% in the measured temperature range) at increasing temperature (see bottom plot of Figure 6). This is found at all the irradiance levels included in this study.

The P_{max} temperature dependence is analysed in more details in the next section where also the other electrical parameters are shown.

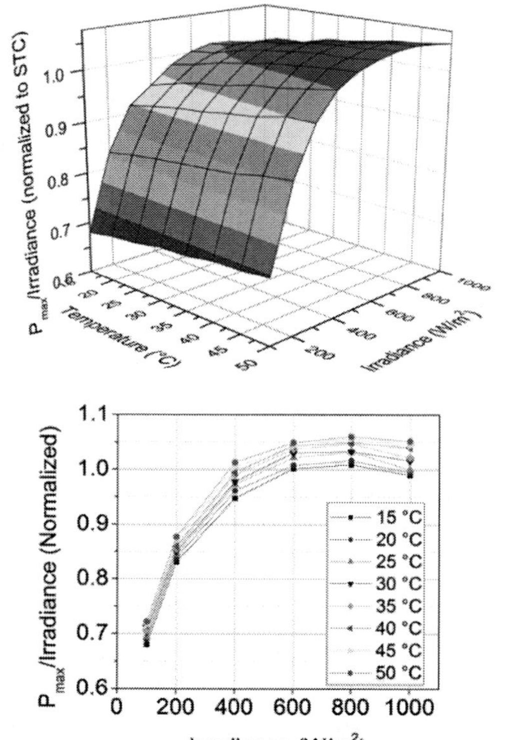

Figure 6 Surface plot (top) showing the dependence of the ratio maximum power divided by irradiance on the temperature and irradiance for TM801. The average data were normalized to 1 at STC (25 °C and 1000 W/m²). The dependence on the irradiance is shown also separately in the normal plot on the bottom.

4.2 Temperature dependences

The temperature dependence of the electrical parameters extracted from the measured I-V curves for QJ87 and TM801 is shown in Figure 7 and 8, respectively. The data corresponding to the measurements performed at 1000 W/m^2, 600 W/m^2, 200 W/m^2 are shown here as an example. The values are normalized to the one at 15 °C for comparison. The error bars represent the maximum relative variation of the parameters in the measured triplet (defined as maximum - minimum divided by the mean value of each triplet).

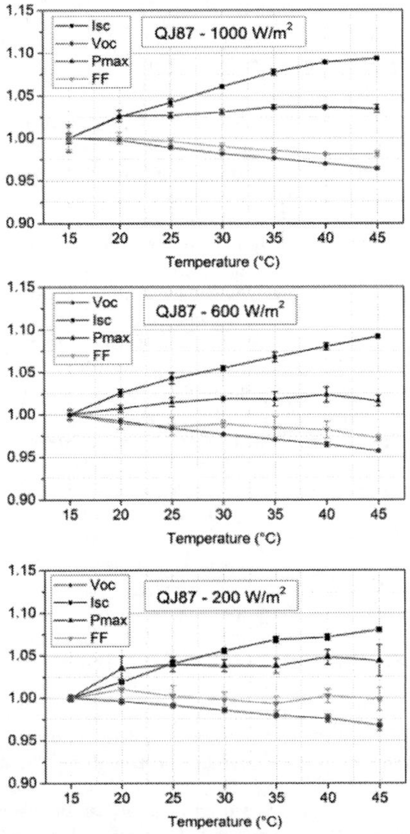

Figure 7 Temperature dependence of the electrical parameters extracted from the I-V curves of QJ87 in the temperature range 15 °C - 45 °C. The mean values and the corresponding maximum relative deviation within each triplet are shown. The values are normalized to the one at 15 °C. The data corresponding to the measurements performed at 1000 W/m^2, 600 W/m^2 and 200 W/m^2 are here shown as example.

It is noted that in general V_{oc} decreases linearly with the temperature and I_{sc} increases linearly, although the ratio between their rates varies with the device. The increase in I_{sc} is about twice compared to V_{oc} for QJ87 and is typically the same for TM801. The variation of FF with temperature is less clear. In the case of QJ87 it decreases at increasing temperature, while the opposite behavior is observed for TM801. In both cases the variation of FF in the measured temperature range is fairly small, below 3%. As a result P_{max} slightly increases up to 35 °C and then starts to decrease for QJ87, while constantly increases in the measured temperature range for TM801. In this second case P_{max} increases more than 5% in the range [15-50] °C.

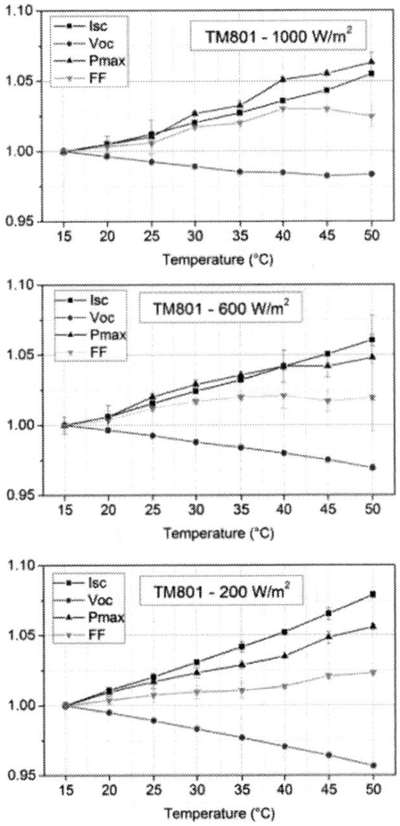

Figure 8 Temperature dependence of the electrical parameters extracted from the I-V curves of TM801 in the temperature range 15 °C - 50 °C. The mean values and the corresponding maximum relative deviation within each triplet are shown. The values are normalized to the one at 15 °C. The data corresponding to the measurements performed at 1000 W/m^2, 600 W/m^2 and 200 W/m^2 are here shown as example.

A comparison of the electrical parameters dependence on temperature for the devices QJ87 and TM801 is shown in Figure 9. The values have been normalized to the value at 15 °C in both cases for comparison. Even if the rate of V_{oc} and I_{sc} variation of the two devices is different, the increase of I_{sc} and decrease of V_{oc} with temperature is confirmed in both cases. Also P_{max} shows an increase with temperature in both cases. Similar behaviors were already reported in previous publications [3] and [4]. For TM801 which is fresher and better performing, P_{max} increases continuously with temperature up to 50 °C, while for QJ87 seems to reach a saturation value. This is an encouraging result, which indicates at least that OPV devices are not negatively affected by high temperatures (50 °C in this case).

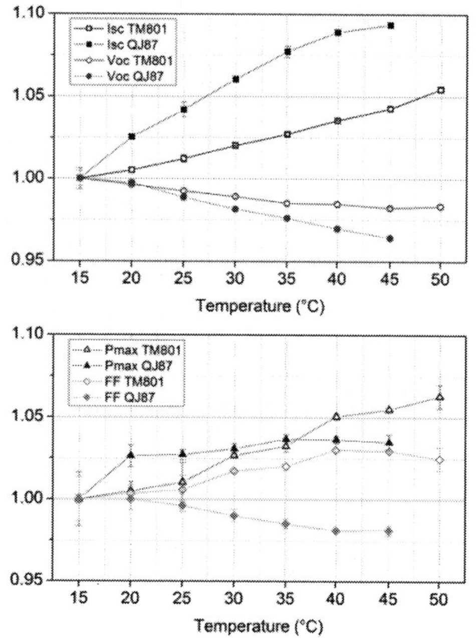

Figure 9 Comparison of the electrical parameters temperature dependency for the devices QJ87 and TM801 for the case at 1000 W/m^2.

4.3 DUTs stability across measurements

I-V curves at STC (1000 W/m^2, 25°C) were performed before and after the power matrix measurement campaign in order to check the stability of the devices during the experiment. The comparison of the initial and final measurements for devices QJ87 and TM801 is shown in Figure 10. Three I-V curves have been performed each time.

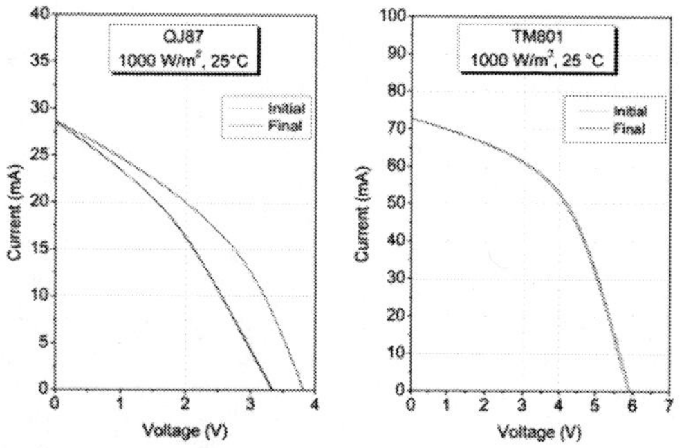

Figure 10 Comparison of the I-V curves (3 for each condition) measured before and after the full power matrix measurement campaign for the devices QJ87 (left) and TM801 (right).

As shown in the plots, QJ87 reported a strong degradation during the measurement campaign affecting the V_{oc} and FF, and consequently P_{max}. This could then partially explain the behavior of P_{max} at high temperature for QJ87. The situation of TM801 is very different. The device well survived to the power matrix measurement campaign and did not show any significant degradation during the measurements. The I-V curves of Figure 10 (initial and final) are well comparable and no remarkable differences can be noted between them.

4.4 Energy rating analysis and results

The measured power matrix was intended to be used as input to the model developed by Huld et al [11], which estimates the performance of PV devices under real working conditions. Then, a comparison with well-established technologies, such as crystalline silicon (c-Si), cadmium telluride (CdTe) and amorphous silicon (a-Si), is carried out. This comparison is based on the yearly Module Performance Ratio (MPR) calculated according to Equation 1.

$$MPR = \frac{\sum PV\ output}{\sum G_T} \cdot \frac{G_{STC}}{P_{STC}} \qquad (1)$$

where the PV output is the power output estimated according to [11]. G_T and G_{STC} are, respectively, the in-plane irradiance received by the modules and the in-plane irradiance under STC. P_{STC} is the power output at STC. The model's coefficients used for the OPV device are obtained fitting the model to the indoor power matrix measurements of QJ87 previously described in this paper. The measurements of TM801 have not been used yet for this calculation for time constrain reasons, but new calculations fitting these data will be executed in the near future. The PV output model requires, also, the in-plane irradiance received by the modules in real conditions and their temperature, T_{mod}. T_{mod} depends on the received irradiance, the ambient temperature and the wind speed, and is calculated using Faiman's model [12]. Regarding the in-plane irradiance, it is calculated from satellite data so that it takes into account the angle-of-incidence effect (some of the irradiance reaching the module's surface is reflected due to a high incidence angle) and the spectral effects. The spectral distribution of the in-plane irradiance varies along the days and with the atmospheric conditions. Besides, different PV technologies use differently the range of wavelengths. The model used in our calculations takes into account this effect in order to estimate the effective irradiance really used by each PV device. For a wider and more complete discussion about the comparison, the different effects and their modelling refer to [13].

The full analysis is based on one year of environmental data. The solar radiation data from satellite are provided hourly with a spatial resolution of about 5 km over large geographical areas. The ambient temperature and the wind speed data are retrieved from the European Centre for Medium-Range Weather Forecast product. Module PV performance for four different technologies has been estimated at five different locations in order to analyze different climatic conditions. The modules are assumed to be installed in a freestanding setting facing the equator with a fixed tilted angle. In every location, the optimal inclination angle is applied so as to maximize the yearly received radiation. The considered locations with the optimal inclination angle are shown in Table 3. The comparison of the calculated MPR values for the considered PV technologies is shown in Figure 11. Looking at these values, the a-Si technology shows the highest MPR values for all the locations (see Figure 11) with the exception of location at 24° N.

Table 3 Coordinates of the five locations considered and optimal inclination angle applied at each of them.

Location	Latitude (°)	Longitude (°)	Optimal angle (°)
Gabon	-1.00	11.00	0
Saudi Arabia	24.91	46.41	25
Spain	38.00	-3.25	34
Slovakia	48.00	18.00	36
Scotland	56.00	-4.00	38

In this location (i.e. in this climate), the OPV outperforms the other technologies with MPR values slightly higher than the one obtained with the a-Si. The OPV reported higher MPR values than c-Si and CdTe in three locations out of five, corresponding to places at low latitude and hot climates. In contrast it has the lowest MPR at the highest latitude, although, in general, the lowest MPR are obtained for the c-Si technology.

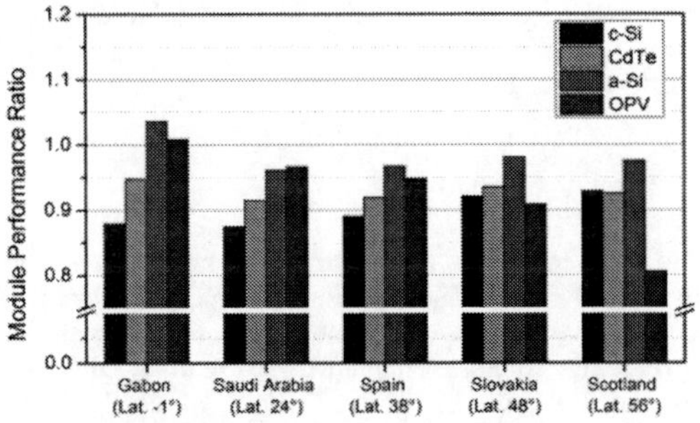

Figure 11 Yearly MPR values calculated for the four technologies at the five different locations.

In general, the performance of the different technologies decreases under low irradiance levels. This effect is even more remarkable for the OPV technology, as it can be observed by its performance in Scotland, where low irradiance levels occur more frequently than in the other considered locations. Indeed, excluding the occurrences with irradiance below 100 W/m^2 the MPR of the OPV at this location increases from 0.80 to 0.93 , which is more similar to the other technologies value: 0.95 for c-Si, 0.98 for CdTe and 1.03 for a-Si.

5 CONCLUSIONS

The power matrix measurements presented in this study have confirmed the unique behaviour of OPV devices with temperature if compared with standard c-Si and thin-film PV technologies already presented by other authors in literature. It is noted that P_{max} is almost constant in the measured temperature range (15 °C - 45 °C) for all the irradiance levels in the case of device QJ87, possibly due to strong aging of the sample, while it increases of about 5% in the case of TM801. The decrease of V_{oc} and FF with temperature is almost compensated by the increase in I_{sc}. The current result of the energy rating evaluation, performed by using only the QJ87 measured power matrix data as input to the calculation, showed that OPV devices better perform in hot tropical climates close to the equator, partially due to an almost lack of variation of P_{max} with temperature. The comparison with more established PV technologies brought to the conclusion that OPV may, in some conditions, be a good alternative candidate for PV deployment, although further research for its stability has to be put in place. Further assessment of the energy rating of the OPV technology has to be considered, though, by inclusion of the newest power matrix measured on the fresher and improved sample. It should be indicated as well that, while the OPV technology is represented by a small device, for the other technologies real commercial modules were used. Therefore caution should be used in deriving too general conclusions on the basis of these results only.

REFERENCES

[1] M.A. Green et al. Solar Cell Efficiency Tables (Version 50) Progress in Photovoltaics: Research and Applications 2017, 25, 668-676.

[2] Carle´ et al., Overcoming the Scaling Lag for Polymer Solar Cells, Joule (2017), http://dx.doi.org/10.1016/j.joule.2017.08.002

[3] Y. Hishikawa, H. Tobita, A. Sasaki, K. Yamagoe, T. Onuma and Y. Tsuno, Current-Voltage Characteristics of Novel PV Devices under Various Irradiance and Temperature Conditions, Photovoltaic Specialists Conference (PVSC), 2013 IEEE 39th, Tampa Bay, Florida

[4] E.A. Katz, et al. Temperature dependence for the photovoltaic device parameters of polymer–fullerene solar cells under operating conditions, J. Appl. Phys. 90 (2001) 5343–5350.

[5] D. Chirvase, et al., Temperature dependent characteristics of poly(3 hexylthiophene)-fullerene based heterojunction organic solar cells, J. Appl. Phys. (2003) 93, 6

[6] F.C. Krebs, et al., Freely available OPV - The fast way to progress, Energy Technology 1 (2013), p. 378-381.

[7] IEC 60904-8 (Ed. 3) Photovoltaic devices – Part 8: Measurement of spectral responsivity of a photovoltaic (PV) device, 2014.

[8] Salis E, Sharlandzhiev I, Field M. Feasibility study for PV measurements at varying irradiances on a large-area steady-state solar simulator, Proceedings of the 33rd European Photovoltaic Solar Energy Conference, EU PVSEC, Sept 25-29, 2017, Amsterdam, The Netherlands

[9] IEC 60904-9 (Ed.2) Photovoltaic devices - Part 9: Solar simulator performance requirements, 2009.

[10] IEC 61853-1 (Ed. 1) Photovoltaic (PV) module performance testing and energy rating - Part 1: Irradiance and temperature performance measurements and power rating, 2011.

[11] Huld T, Friesen G, Skoczek A, Dunlop ED. A power-rating model for crystalline silicon PV modules. Sol Energ Mat Sol Cells 2011;95(12):3359-3369.

[12] Faiman D. Assessing the outdoor operating temperature of photovoltaic modules. Prog Photovoltaics Res Appl 2008;16(4):307-315.

[13] Gracia-Amillo A., Bardizza G., Salis E, Huld T, Dunlop E.D. Energy-Based Metric for analysis of organic PV devices in comparison with conventional industrial technologies. Submitted to *Renewable & Sustainable Energy Reviews*.

"DEVELOPMENT OF A REPRODUCIBLE LASER STRUCTURING PROCESS OF STACKED THIN FILMS ON ULTRA-BARRIER FILMS FOR ORGANIC SOLAR DEVICES"

Niels Friedrich-Schilling [a], Henri Fledderus [b], Frank Peuckert [c], Thomas Kuntze [d], Benedikt Gburek [a], Ahmed Salem [b], Hylke Akkerman [b], Merve Anderson [a]
[a] Heliatek GmbH, Treidlerstraße 3, 01139 Dresden, Germany
[b] Holst Centre, High Tech Campus 31, 5656 AE Eindhoven, Netherlands
[c] 3D-Micromac AG, Technologiecampus 8, 09126 Chemnitz , Germany
[d] Fraunhofer IWS, Winterbergstraße 28, 01277 Dresden, Germany

ABSTRACT: Aim of the work presented in this paper is the development of reproducible and scalable laser processes for selective ablation of stacked thin films on top of an integrated ultra-barrier foil. Industrial manufacturing of functional modules with defined opto-electrical properties requires various coating and structuring processes. Therefore, transparent front contact layer (P1), the absorber layer stack (P2) and the back contact (P3) have to be structured without damage to the ultra-barrier underneath. Focus of the paper is the P1 process as most critical laser process. The laser parameter investigation was performed on sheet-to-sheet samples prepared with a process suitable for upscaling to roll to roll production on 1200 mm web width. A test process was developed in order to find the optimal laser ablation settings for each scribe which include an electrical isolation test and a barrier damage test (optical Ca-test). With this test routine it was possible to identify the process windows for two TCO types without barrier damage and promising solar device performances.

Keywords: Organic Solar Cell, Encapsulation, Laser Processing, Ultra-Barrier foils

1 INTRODUCTION

Organic photovoltaics (OPV) is a promising technology for large area, light weight and integrated devices. A disadvantage of organic materials is their sensitivity against water and oxygen and accordingly the fast degradation under ambient conditions. To guarantee a long life-time of the product, organic solar devices have to be encapsulated in ultra-barriers with very low moisture transmission rates. Best performance can be achieved by encapsulation into glass, which is a kind of "gold standard" for evaluating the achievable life-time. But glass encapsulation also means high weight and complete loss of the flexibility of the organic solar foil.

State of the art encapsulation technology for flexible solar modules is the lamination of a completely processed solar film between two flexible barrier foils under nitrogen atmosphere, see Figure 1. With this technique the ambient moisture is kept outside but residual water in the substrate and the adhesive is also kept inside. In case of Heliatek solar films, PET is used as standard substrate material which shows hygroscopic behaviour. By drying and heating substrate and encapsulation foil the water content is reduced but depending on the process and storage conditions residual water is still present. On the other hand, caused by the currently small but permanently growing market, the price for commercial ultra-barrier foils is still high [1].

Figure 1: Schematic of a solar cell encapsulated between two ultra-barrier foils (left) and the addressed system simplification, called integrated front sheet, with the PV module on top of a barrier film.

2 MOTIVATION

The challenges of the technology for flexible OPV encapsulation are life-time and performance-cost-ratio. In the EU project ALABO both topics are addressed by developing suitable techniques to realize OPV with integrated ultra-barrier layer instead of costly bottom encapsulation. The development of scalable patterning processes of an integrated sheet is an intermediate step on the way to realize devices directly encapsulated with life time comparable to crystalline solar cells, see Figure 1.

In the project the roll-to-roll (R2R) barrier system from Holst Centre is used as substrate [2]. The system shows an extrinsic water vapour transmission rate (WVTR) of 10^{-8}-10^{-6} g*m^{-2}d^{-1}. This value represents the effect when water solely entering through pinholes in the barrier layer. The intrinsic WVTR is below 10^{-6}. These values are sufficient to target long lifetime for OPV devices.

On top of this ultra-barrier, a TCO layer followed by the organic solar layers and a metallic back contact are deposited. The serial interconnection of single solar cells to a complete module is realized by laser ablation, see Figure 2. A critical requirement of this process is to keep the barrier unharmed.

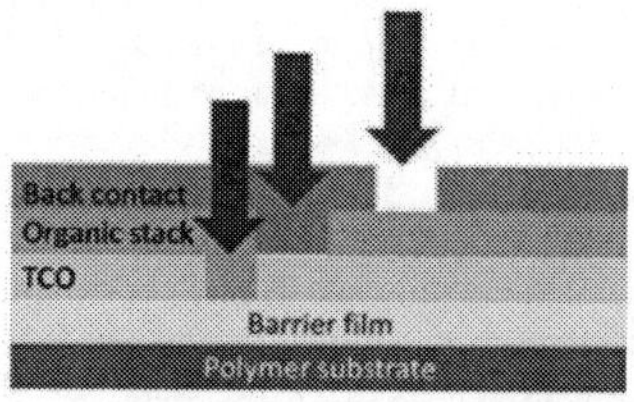

Figure 2: Functional layers of an integrated front sheet and the corresponding laser process steps. P1 to separate the TC layer, P2 for organic removal and P3 for separating the back contact. All of them have to be dimensioned such that the barrier layer is unharmed

Focus of this paper is the structuring of the TCO layer on top of the SiN layer without influencing the barrier performance. To ensure process reproducibility, the investigations were performed by three partners with different laser systems and optical setups. The goal was the development of a scalable ablation process from 6 inch sheet-to-sheet (S2S) to R2R with a web width up to 1200 mm.

3 MATERIAL AND METHOD

3.1 Barrier system and TCO layer

The basic substrate for the integrated front sheet is a R2R produced thin film barrier from Holst Centre [2]. For planarization and particle embedding an organic planarization is coated on top of the PET. This setup allows reproducible WVTR values in the range of 10^{-6} $gm^{-2}d^{-1}$. For the investigations, 6-inch sheets were cut by CO2 laser out of the barrier roll. Subsequent processing, like TCO coating is done in a S2S process. As TCO layers indium tin oxide (ITO) and a dielectric-metal-dielectric (DMD) stack are investigated. The ITO layer is deposited by sputtering with a thickness of 135 nm [2]. For the DMD electrode silver is used as metal layer surrounded by two metal oxide as dielectrics. The overall thickness is in the range of 40 nm to 60 nm. In Figure 3 the overall transmittance and reflection of the TCOs on barrier are shown.

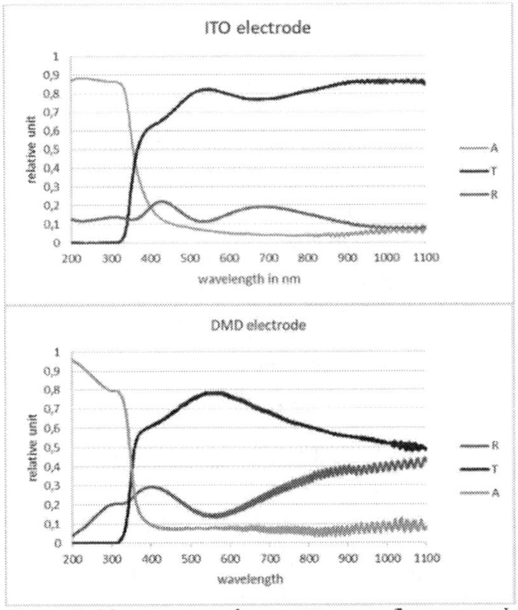

Figure 3: Measurement of transmittance, reflectance and calculated absorbance of ITO (top) and DMD (bottom) electrode on top of the Holst barrier measured with a Shimadzu spectrometer from the electrode side.

3.2 Investigated laser systems and optical setups

Three different laser systems were used to investigate the TCO ablation and the resulting isolation behaviour. For thin film ablation mostly pulsed laser systems are used due to a short interaction time and correspondingly a small thermally influenced area. After some initial test the focus was set on laser systems with a pulse duration in the picosecond (ps) -range. This laser system represents an optimum between achievable ablation quality and a good cost/throughput ratio. Relevant working wavelength for ps-lasers are the infra-red (IR) wavelength around 1030 – 1090 nm and the second and third harmonic of the according wavelength. When looking at the absorption of DMD, IR and visible wavelength are comparable, due to a nearly constant absorption of the silver layer in this stack. For ITO the IR wavelength shows a bit more absorption than for the visible wave length. The barrier shows no absorption in this wavelength range. Both electrodes show high absorption for the third harmonic in the UV range. But in this spectral range, there is also high absorption in the SiN layer and the organic contents, which strongly reduces the process window. Another aspect is also the more critical system stability of a UV laser system with two optical conversion steps for a 24/7 production in R2R.

Because of the discussed risk for using UV wavelength, the focus was set on the basic wavelength in the IR and the second harmonic in the visible wavelength range. The maximum power values and optical configuration of the used laser systems are shown in table 1.

Table 1: Overview of laser parameters and optical setup

	Laser	λ [nm]	Average power [W]	Spot size [µm]	Focal length [mm]
Holst	Talisker Ultra	1064	16W @ 200 kHz	~40 ~ 140	163 566
		532	8W @ 200 kHz	~ 15	163
IWS	Fuego	1064	28 W @ 200 kHz	~ 45	163
		532	18 W @ 200 kHz	~ 20	163
3D-MM	Super Rapid	1064	10W @ 200 kHz	~ 75	255
	Pico-blade	532	30 W @ 200 kHz	~ 10	100

It can be seen that the laser systems and optical configurations of the setups give comparable values. Differences are the optical quality in M^2 and in case of 3D-MM the opportunity to use a so called focus shifter (varioScan from ScanLAB) which allows tracing the focal position in 3-D manner.

3.3 Characterization methods

A large amount of information about the removal quality of the TCO layers and the influence on the barrier performance can be derived through simple characterization methods like optical microscopy and resistance measurement with a multimeter. For a detailed analysis, especially to evaluate influences on the barrier performance, a Ca-test is performed by Holst Centre. The process flow of such a test is sketched in Figure 4, where a barrier with a TC layer is laser scribed with different settings. (4a and 4b). The laser structured foil is laminated on glass/Ca/OCP after which the OCP is cured. If the laser causes damage to the barrier, water can penetrate through the damaged area (4c).

At locations with barrier damage the Ca is converted underneath the hole to $CaO/Ca(OH)_2$ (transparent), creating the white lines in the Ca pad

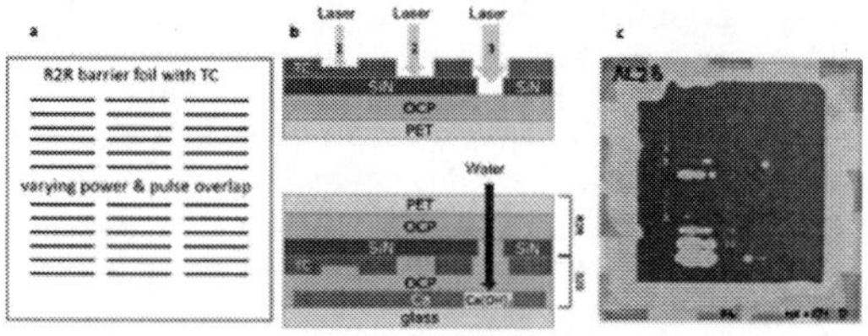

Figure 4: Schematically depicted is the capability of the optical Ca-test to detect local damage to the barrier [2].

The electrical characterization of P1-laser scribes and its influences on the solar cell behaviour can only be determined by building up a complete solar device and measuring the I-V performance. For this a standard tandem stack of Heliatek is used as reference, where a large data base out of R&D and manufacturing can be used for performance comparison.

4 RESULTS

4.1 Single layer ablation behaviour

Starting point of the investigations was the determinations of the ablation thresholds for each individual layer. For this the ablation diameter is measured and the square of it is plotted against the fluence. For the studies the average fluence is used as data base. For calculation of it the pulse energy is divided through the spot size, which was measured with a beam profiler. In Figure 5 the investigations for the Fuego laser at 532 nm wavelength are shown. Table 2: determined ablation threshold values per functional layer at 532 and 1064 nm laser wavelength. The resulting values are given. For DMD as multilayer electrode, it was tried to measure the opening diameter of the silver as most relevant conducting layer in this electrode type.

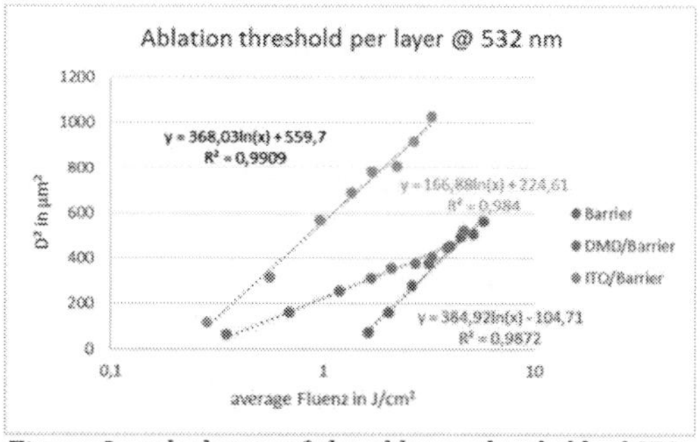

Figure 5: calculation of the ablation threshold of ITO, DMD and the pure barrier with green wavelength (with Fuego setup, 532 nm)

Table 2: determined ablation threshold values per functional layer at 532 and 1064 nm laser wavelength.

layer	wavelength	Threshold
Barrier	532 nm	1,31 J/cm²
	1064 nm	2,57 J/cm²
ITO	532 nm	0,219 J/cm²
	1064 nm	0,295 J/cm²
DMD	532 nm	0,260 J/cm²
	1064 nm	0,160 J/cm²

The ablation threshold values from table 2 indicate a large process window for removal of the electrode without damaging the barrier.

4.2 Laser process

4.2.1 Process window for ITO electrode

The next step is a screening of pulse energy versus pulse overlap to see the ablation characteristics for a fixed parameter combination. Parameter settings where the electrode seems to be removed well enough and the barrier remains intact are then used for isolation measurements. In our case, a square was marked with crossing lines for two purposes: 1) to avoid possible bridges and 2) to test the barrier behaviour at stitching points. Selected laser parameters are as a last step used for a Ca-test sample to determine the barrier performance. Depending on the size of a possible damage, a laser track is quickly visible or takes a while to appear. After 2-3 days at accelerated 60 °C / 90 % relative humidity conditions, all damage can be observed.

For ITO the screening with IR-laser showed a possible process window between 250 mJ/cm² and 1 J/cm² with an overlap between 30 and 90% [2]. One result of the parameter investigation on ITO is shown in Figure 6. Here the C-test results are presented for the 1064 nm ps-laser.

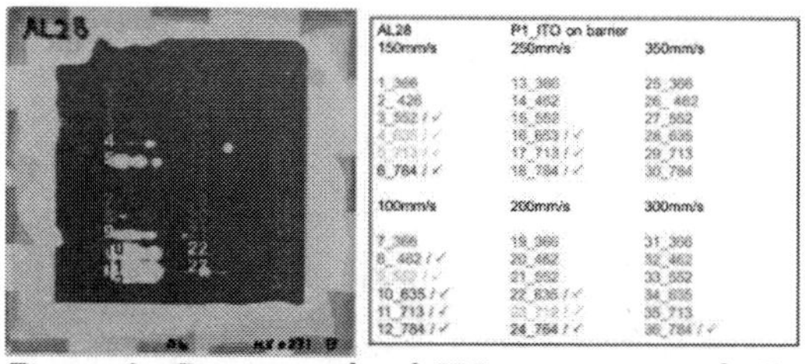

Figure 6: Ca-test result of ITO processing with IR wavelength. The values represent the used fluence. Checkmarks represent isolating parameters.

At fluence around 1 J/cm² a bulge formation, see Figure 7, appears in the track. With higher fluence and/or overlap the bulge formation is enhanced, and at local places, the bulges also open up.

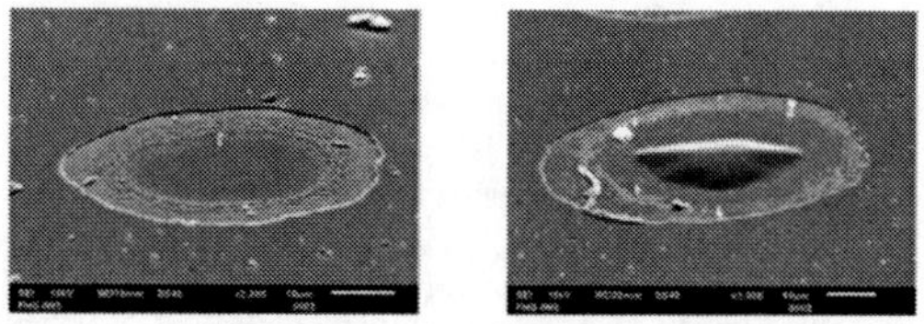

Figure 7: SEM image of single pulses without barrier influence (left) and with formed bubble (right)

A detailed characterization with SEM and Ca-test shows clearly that these bulges are formed underneath the SiN barrier layer. Through this behaviour, the ablation process window gets more reduced than the initial single pulse ablation indicated.

When looking at the ITO electrode and its brittle mechanical behaviour, a clear ablation is expected when working above the ablation threshold and with a sufficient overlap with resulting isolation values larger than the measurement range of the used multimeter. During the laser process investigation also another effect was observed which influences the achievable P1 quality: When scribing with a fixed parameter setting, local ablation variations can be observed in a single track. These ablation variations result in locations with locally reduced scribe width or incomplete ablation and in extreme cases regions with unwanted local barrier damage underneath the laser track.

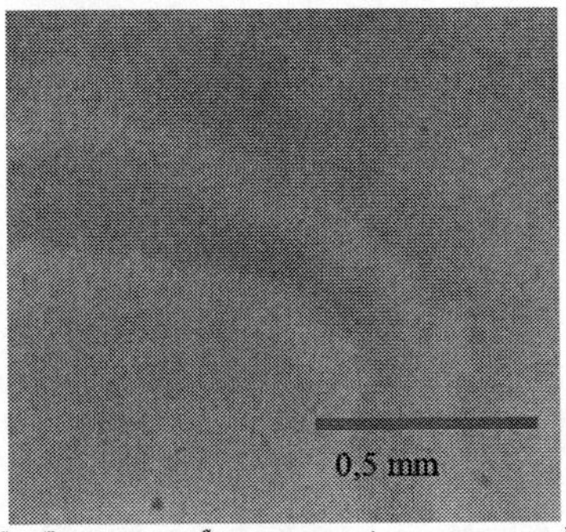

Figure 8: Overview of macroscopic structure which can be observed when structuring the electrode with lines next to each other

When marking multiple scribes close to each other, these local variations seem to form a large scale picture and form a macroscopic picture of a continuous running paint, see Figure 8. Fluctuations in the pulse energy or variations in the height cannot explain such strong local variations. A full detailed study on the ablation inconsistencies during TC ablation on top of barrier foils is in progress and beyond the scope of this paper.

4.2.2 Process window for DMD electrode

The same investigation method was followed for the DMD electrode. A suitable fluence range was identified between 0.3 and 1.1 J/cm² and an overlap range between 15% and 78%. Also for this electrode type the bulge formation in the laser track can be observed around an average fluence of 1 J/cm². In Figure 9 the initial resistance values are shown.

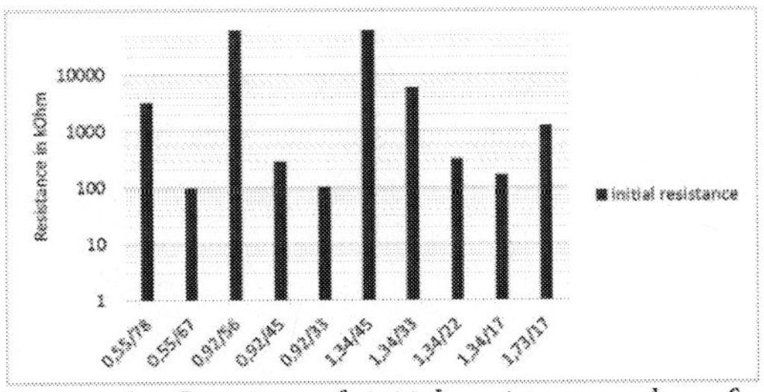

Figure 9: Overview of initial resistance values for different fluence/overlap combinations for processing with green laser wavelength and 163 mm focal length.

The presented results are in a fluence range where the barrier is mostly intact. Most critical are the parameter combination with energy densities of 1.34 J/cm² and higher. It can be seen that only two parameter combinations give an initial resistance above 60 MOhm. All the other settings result in a resistance range of MOhm down to hundreds of kOhm. This behaviour is observed for both tested wavelengths in the same manner.

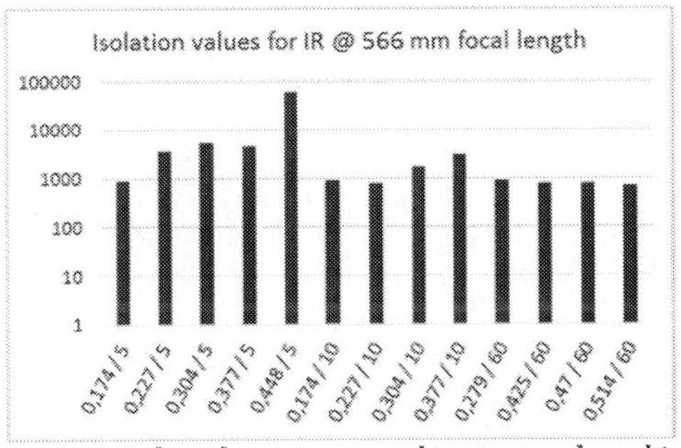

Figure 10: Selected parameter combinations and resulting resistance values for IR processing with a 566 mm focal length lens.

It is expected that the residual conductivity is caused by remaining silver bridges at local points.

Additional to the established S2S processing, where the laser beam is perpendicular to the substrate also the ablation process under an angle was investigated. In this test a focus shifter (Varioscan) from Scanlab is used to keep the substrate in the focal plane. As a result, the test showed an increase in the achievable resistance values compared to the perpendicular structuring, see table 3.

Table 3: Comparison of achieved resistance values for perpendicular and angled laser processing at a fixed overlap for different laser power setting.

Relative power setting	Initial resistance in kOhm	
	Perpendicular	Under angle
55%	410	3200
57,5%	480	3800
60%	740	4500
62,5%	1050	5800
65%	1730	7600

4.3 IV performance

Next step in the process chain is the realization of complete solar cell devices to characterize the electrical performance of the different P1 scribes. In Figure 11 the I-V curve of an ITO/barrier sample processed with 0.78 J/cm² and 78% overlap at 1064 nm is shown.

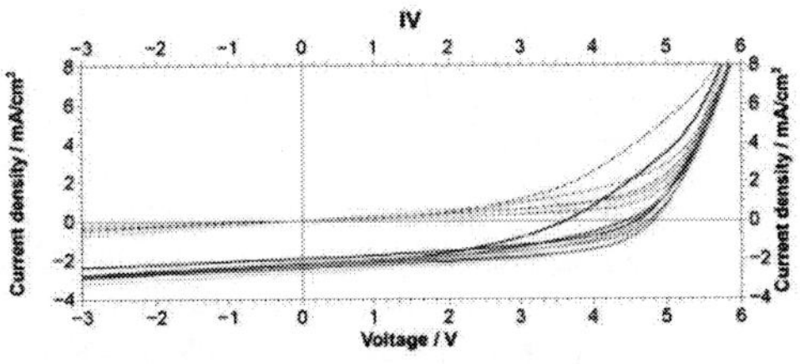

Figure 11: I-V curve of 12 solar modules on an ITO/barrier substrate consisting of 3 interconnected cells. Every cell provides a voltage of 1.7 V. Nearly all devices show the expected full voltage of 5 V. However, all devices show a high leakage current.

On the test substrate only one of twelve samples do not reach the full open circuit voltage of 5 V, which means that the serial interconnection works well. But all samples show high leakage current which is caused by the observed inhomogeneous ablation behaviour.

In Figure 12 the resulting I-V-curves for isolation scribes on DMD are shown. Three laser settings with 0.253 J/cm² and an overlap of 85%, 71% and 57% are evaluated. The resulting initial resistance values are twice 900kOhm and 1200 kOhm respectively.

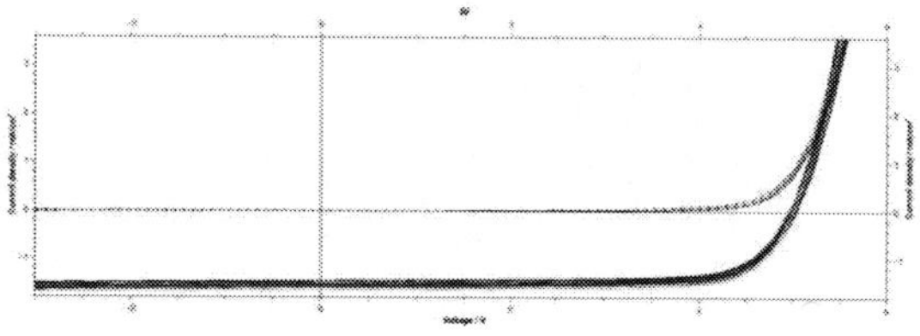

Figure 12: I-V curve of a DMD electrode with three different laser parameter combinations. All samples show full open circuit voltage and a small leakage current.

All tested laser parameters lead to a good module performance. The expected open circuit voltage is reached by every module. Especially highlighted can be the high fill factor of around 75%. The leakage current is also very low and comparable to values achieved on DMD/PET. After the electrical characterization the sheets are tested again in Ca-test to ensure the integrity of the barrier. For the ITO sample local damages were seen related to the P1 scribes. For the DMD substrate no damage related to the P1 process was observed.

A re-measurement of the resistance at functional scribes showed much higher isolation values up to GOhm range. Because of this, a detailed analysis of the DMD electrode was performed.

4.4 Oxidation process

As discussed, the I-V curves indicate an increasing resistance over time. This effect is caused by oxidation of the remaining silver layer after laser processing.

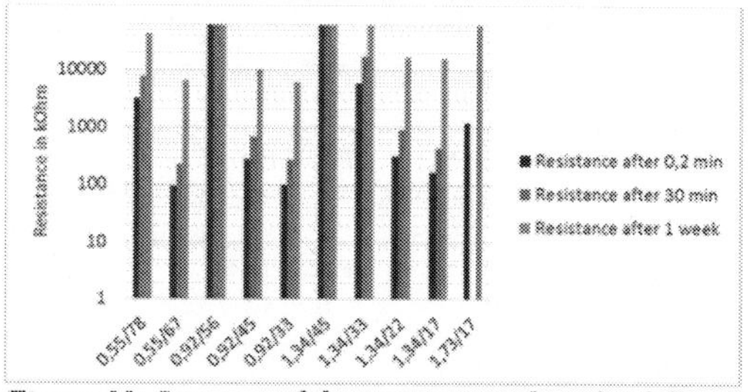

Figure 13: Increase of the resistance values for different laser settings. The resistance is measured directly after processing, after 30 minutes and after 1 week.

A more detailed analysis was done with the green laser wavelength. P1 scribes were patterned with 0.6 J/cm² and different pulse overlaps. Again the initial resistance value is measured directly after laser processing. Due to machine relevant aspects the minimal measurement time for the first data point is 20 seconds. For the next data points a time delay of 30 seconds and more was used.

All samples show an increasing resistance which is expected to be caused by oxidation processed [3] or the loss of percolation paths by reorganization.

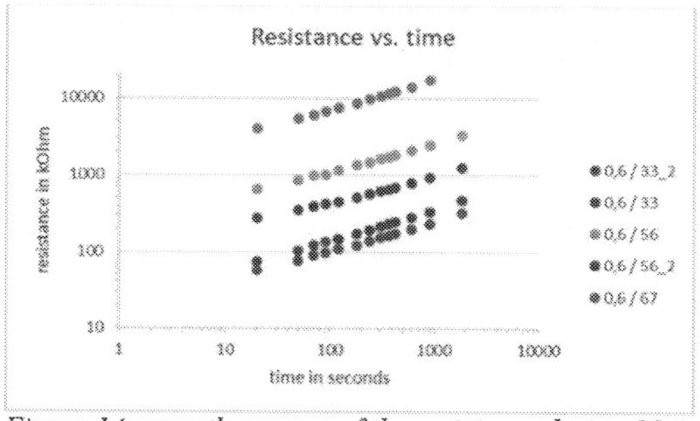

Figure 14: growth process of the resistance during 20 to 1800 seconds after laser processing.

5 SUMMARY AND CONCLUSION

We have investigated a suitable process window for structuring ITO and a DMD as TCO on top of a SiN layer as ultra-barrier. The process windows are confirmed by a special developed test method at Holst Centre to ensure the barrier performance with a good local resolution. Processed solar device on ITO show a high leakage current due to an inhomogeneous ablation behaviour. In contrast, the electrical behaviour of solar devices on a structured DMD electrode is excellent for laser parameters that leave the barrier layer undamaged. This can be explained by the oxidation behaviour of the silver after laser processing enabling effective scribing at moderate fluence combined with a sufficiently smooth edge topography of the laser scribes.

Acknowledgment:
The authors gratefully acknowledge financial support from EU program FP7 ALABO, Grant agreement No: 644026.

References:

[1] Machui, F.; et al. (2014): Cost analysis of roll-to-roll fabricated ITO free single and tandem organic solar modules based on data from manufacture. In: *Energy Environ. Sci.* 7 (9), S. 2792. DOI: 10.1039/C4EE01222D.

[2] Fledderus, H.; et al. (2017): Patterning of organic photovoltaic on R2R processed thin film barriers using IR laser sources. In: Udo Klotzbach, Kunihiko Washio und Rainer Kling (Hg.). SPIE LASE. San Francisco, California, United States, Saturday 28 January 2017: SPIE (SPIE Proceedings), 100920I.

[3] Reinshagen, Holger (2006): Elektrochemie und Grenzflächenkinetik in Niedertemperaturgasplasmen. Universitätsbibliothek, Gießen. Online under: http://geb.uni-giessen.de/geb/volltexte/2006/3877.

EVALUATION EMERGING PV PERFORMANCE RATING UNDER INDOOR LIGHTING SIMULATOR

Yean-San Long[1*], En-Yun Wang[1], Teng-Chun Wu[1], Min-An Tsai[1]
[1]Center for Measurement Standards, Industrial Technology Research Institute, Taiwan
*Corresponding author (mickeylong88@itri.org.tw, Phone: +(886) 3 5911026, Fax: +(886) 3 5821646)

ABSTRACT: The measured I-V hysteresis is complicated by the vast array of different materials, lighting sources, and device architectures. In our study, we used RTOS method for I-V then the results showed better accuracy by eliminating in real time the acceptance effect. We also used this method to compare emerging PV cells with the latter cells exhibited promising power conversion efficiency and no hysteresis behavior under indoor lighting simulator. In this study, we defined a performance ratio to compare between indoor lighting and solar and the results will be show different performance trend for the different type PV in indoor/outdoor, which depends on device. That's will be shown what type PV has the application of energy harvesting for indoor.

Keywords: Organic Photovoltaic, Real-Time Removing Capacitance Effect, One-Sweep Method.

1 INTRODUCTION

In general indoor lighting, a lot of different light sources exist. The International Commission on Illumination (CIE) is responsible for publishing binding standards for different types of light sources, called illuminants. A standard illuminant represents a mathematical table of relative energy versus wavelength, used for colorimetric calculations. It is a (theoretical) source of visible light with a set spectrum, determined by convention and therefore provides a worldwide basis for comparing images under different lighting. Which illuminants are useful standards for indoor illumination? We consider some typical, widely used, light sources for indoor environments (residential and commercial) and relate them to an appropriate illuminant (standard). First, we will propose a standard for the common incandescent light bulb. An incandescent lamp consists of a balloon of (usually) glass in which a filament is placed that is heated to high temperatures by the Joule's effect. To a first approximation the filament can be considered as a black body radiator. In incandescent lamps, the filament usually consists of tungsten heated to a temperature of 2000 to 3000 K. The CIE has agreed upon a standard for the incandescent light bulb, called "illuminant A". This illuminant is intended to represent typical, domestic, tungsten-filament lighting. Its relative spectral power distribution is that of a Planck radiator at a temperature of approximately 2856 K. In other words, a typical light bulb can be characterized by a black body at temperature T = 2856 K. Technically, illuminant A is only defined over the spectral region from 300 nm to 830 nm. However, solar cells often absorb longer wavelengths. Therefore, we will not use illuminant A itself to represent the spectrum of the incandescent light bulb, but we will use Planck's law for a black body at a temperature of 2856 K. In the 300–830 nm range, the difference between Planck's law at this temperature and illuminant A never exceeds 1%. However, In indoor lighting, the International Commission on Illumination is an organization devoted to international cooperation and exchange of information between its member countries on all matters relating to the science and art of lighting. In accordance with an agreement between ISO and CIE, standards are published as double logo standards by ISO. Standards produced by the CIE are a concise documentation of data defining aspects of light and lighting, for which international harmony requires a unique definition. Therefore, the standard follow that the CIE has defined some lighting spectrums for different indoor colorful applications and situations. However, the requirements for color testing do not apply to related indoor energy harvesting under different lighting uniformity and stability, for performance measurement of emerging PV and device. Furthermore, countless types of indoor lighting patterns are required to do performance characteristics measurement for different application contexts. The most common performance indicator is the PV efficiency under standard test conditions (STC). It is the maximum electrical power divided by the total irradiance. For indoor lighting illuminance values are measured used a lux meter or spectradiometer.

(a) lx is a typical unit for measuring indoor lighting.

(b) For outdoors, irradiance is measured in W/m^2 and there is an approximate conversion of 0.0079 W/m^2/ Lux for the sun.

For indoor lighting, fundamental difference between illuminance and irradiance is the weighting of the spectral response. Irradiance includes the power from all wavelengths weighted equally, whereas illuminance weights the power from each wavelength in proportion to the sensitivity of the human (Fig. 1). Although these conditions are seldom achieved in practice (except in the lab), this characterization provides a reasonable guideline for comparing different solar cell types under outdoor conditions. However, the STC are not relevant for indoor applications. Accurate characterization of emerging PV requires level lighting consideration on each very slow temporal response in the I-V curves of the emerging PV are clearly dependent on the voltage sweep direction, even when the sweep time is the order of seconds [1]. Furthermore, the temporal response is dependent on different level lighting consideration. In our study, RTOS (real-time one-sweep) method can real-time remove the capacity effect under solar simulator for I-V successfully and proposes the forward/backward schematic reaction mechanism for DSC [2]. In this study, RTOS method [3, 4] practices for measuring perovskite solar cell performance showed having better accuracy results by eliminating acceptance effect instantly. Additionally, RTOS method will be useful to measuring cell performance more accurately and rapidly when evaluating emerging PV performance in indoor lighting condition. Therefore, we also defined a performance ratio to compare between indoor lighting and solar simulator.

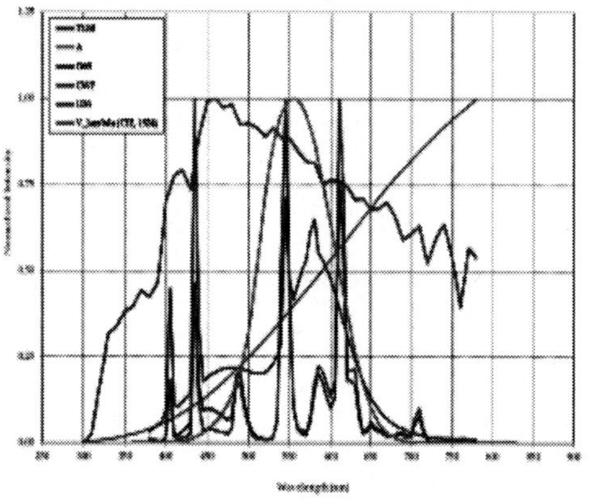

Figure 1: Illumination spectra of CIE Standard Lighting [11, 12]

2 EXPERIMENTS

The samples Sample size ranges from 0.1 cm^2 to 1 cm × 5 cm. These I-V curves were done on a Keithley Source Meter (Model 2460), AM1.5G solar simulator, and Indoor Multi-Light Source System (see Fig. 4). Its meet follow as:

1) Spectral match, available methods are the use of spectroradiometer. Standard lighting spectrum is defined by the deviation from CIE[12]. For three wavelength intervals of interest, the percentage of total irradiance is specified in Table I. Calculate the spectral match for each wavelength interval, which is the ratio of calculated percentage for the CIE spectrum and the lighting simulator spectrum (Table 1). The data comparison with the solar spectrum shall indicate the spectral match classification as per the following as Table 2.

2) Non-uniformity of irradiance in the test plane, the lux meter is recommended to be used as uniformity detector for determining the non-uniformity of irradiance in the test area of the lighting simulator. The uniformity detector shall have a spectral response appropriate for the simulator. The linearity and time response of the uniformity detector shall conform to the characteristics of the simulator being measured. And the designated test area divided by 25. The area covered by the detector measurements should be 100% of the designated test area. The measurement positions should be distributed uniformly over the designated test area. Therefore, the maximum and minimum irradiance are those measured with the detector(s) over the designated test area, likely in Fig. 2. The calculation of non-uniformity is defined as Eq. (1).

3) Temporal instability of irradiance, the lux meter is recommended to be used as temporal instability detector for determining the temporal instability of irradiance in the test area of the lighting simulator. For lighting simulator the temporal instability is related to the irradiance change of measured data sets during the time of data acquisition (Fig. 3 and Table 3).

Table 1: The ratio of calculated percentage for the CIE spectrum

Wavelength	Bandwidth	Percentage of total irradiance (300 nm – 900 nm)				
Band		A	D65	U30	CWF	TL84
1	300 - 450	3.78	20.31	11.40	17.99	15.18
2	450 - 650	40.15	45.35	81.72	76.73	79.13
3	650 - 900	56.06	34.34	6.88	5.29	5.69

Table 2: Spectral definition of lighting simulator classifications

Classification	Percentage
A	1 %
B	3 %
C	5 %

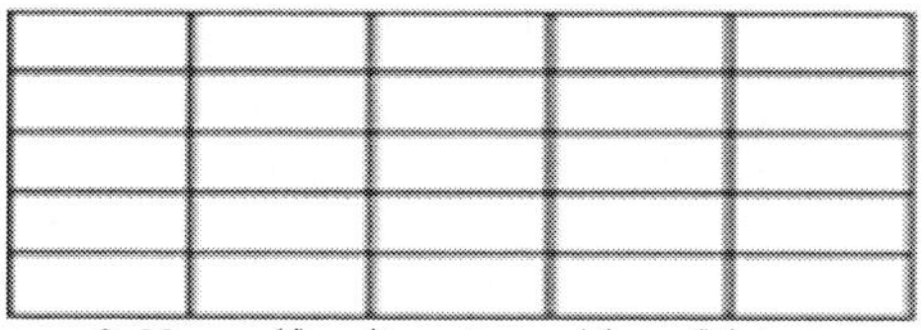

Figure 2: Non-uniformity put a position of detector.

$$Non-uniformity(\%) \equiv \frac{(max-min)}{(max+min)} \times 100\% \quad (1)$$

The Indoor Multi-Light Source included D65 (6500 K, Average North USA sky daylight) / TL84 (4100 K, European shop fluorescent) / CWF (4150 K, cool white fluorescent, shop lighting) / U30 (3000 K, shop lighting) / A (2856 K, typical home lighting), meeting CIE Standard illuminants. The adjustable light intensity (ranging from 0 to 2500 lux) is supplied with an ultraviolet light source for aging testing, with non-uniformity of less than 2% (at 20 cm × 20 cm) and temporal instability of less than 2%. During the I-V measurement in the dark and under Indoor Multi-Light Source System, we followed the testing flowchart in Fig. 6; the scan direction was forward and backward, the sample temperature should be stable at 25 °C with a fluctuation of less than 1 °C, and the irradiance and luminance were determined using a NIST-traceable spectroradiometer (StellarNet) and an NML (National Measurement Laboratory, Taiwan)-traceable lux meter (HIOKI, model FT3424).

Table 3: Temporal instability of indoor lighting simulator.

Item	Average (lux)	Short-term (lux)	Long-term (%)
U30	2709.33	6.35	0.23
	1492.13	1.64	0.11
TL84	2772.08	10.43	0.38
	1507.82	5.61	0.37
D65	1480.16	5.75	0.39
	663.42	2.35	0.35
CWF	2536.80	0.00	0.00
	1199.10	1.91	0.16
A	1902.85	9.01	0.47
	282.92	3.19	1.13

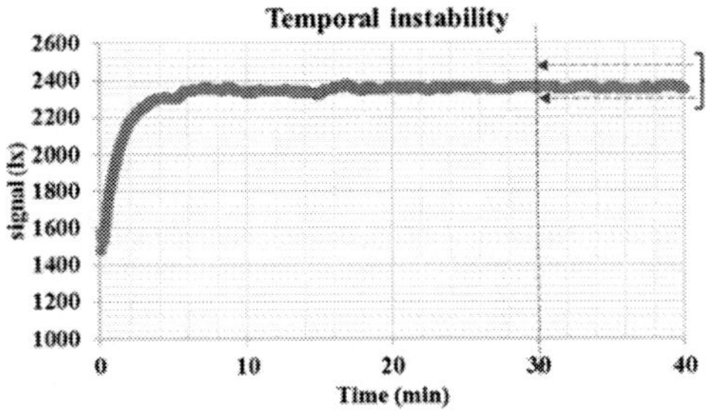

Figure 3: Evaluation of temporal instability.

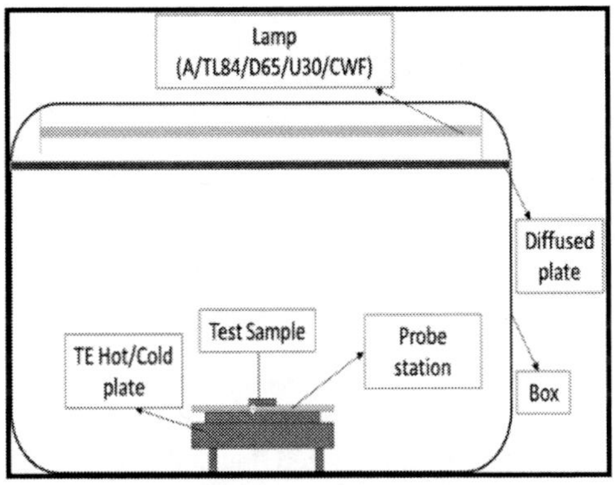

Figure 4: Indoor multi-light source system custom-made by CMS/ITRI for solar cell measurement.

3 ANALYSIS AND RESULTS

In this study, we follow the testing flow [3,4] in the I-V measurement. We used real-time removal of capacity effect; the RTOS method shows the best repeatability (see Fig. 5). A proposed list of steps [2,5] for the measurement of a sample is given in Fig. 8. For most traditional semiconductor PV devices, the current will respond to small changes in bias voltage in less than a millisecond. Sweeping the voltage from open-circuit voltage (Voc) to short-circuit current (Isc) in a few seconds at a constant illumination for such a device will yield accurate currents. Many emerging PV cells respond far more slowly, and consequently, the sweep time must be increased. Fig. 7 shows the I-V for a slowly responding emerging PV cell with different sweep rates.

Therefore, we used the RTOS Method [2] following the testing flowchart (see Fig. 8) in the I-V measurement. applied bias voltage is changed from Isc to Voc (forward scan) or in the reverse direction (backward scan, Voc to Isc). When a capacitance effect occurs, this measurement process produces the phenomenon. In the forward scan, an overshot current appears immediately after an abrupt increase of applied voltage and then the current gradually decreases to an equilibrium state. Sweeping to determine maximum power (Pmax) may not be valid if the device is metastable because of the long times that it sits near Isc or Voc before getting to Pmax shown in Fig. 7(b) to confirm the "near-true" Pmax of emerging PV, especially for samples that depend strongly on sweep direction or rate. This also shows that the asymptotic I-V method may be the only reliable way to obtain efficiency.

In the light intensity under artificial lighting conditions found in offices and factories is less than 5 W/m^2 as compared to 100–1000 W/m^2 under outdoor conditions, depending on the type of light source and its distance. Moreover, the spectrum can be totally different from the outdoor solar reference spectrum (e.g. AM1.5G). The spectrum depends not only on the type of light source, but also on the presence of reflected and diffused light. Therefore, we shown a definition of performance (η) between indoor lighting and solar simulator below as

$$\eta \equiv (P_{max}/Area)/P_{in}. \qquad (2)$$

Where (a) P_{max} means maximum power of device, (b) Area means area of device, (c) P_{in} means Illuminant power of lighting source. Therefore, we also defined a normalized power density (performance ratio, η_n) to compare between indoor lighting and solar simulator below as

$$\eta_n \equiv \eta_{Indoor} / \eta_{AM1.5G} \qquad (3)$$

Where (a) η_{Indoor} means performance of device at each indoor lighting (A/TL84/D65/U30/CWF), (b) $\eta_{AM1.5G}$ means performance of device at solar simulator.

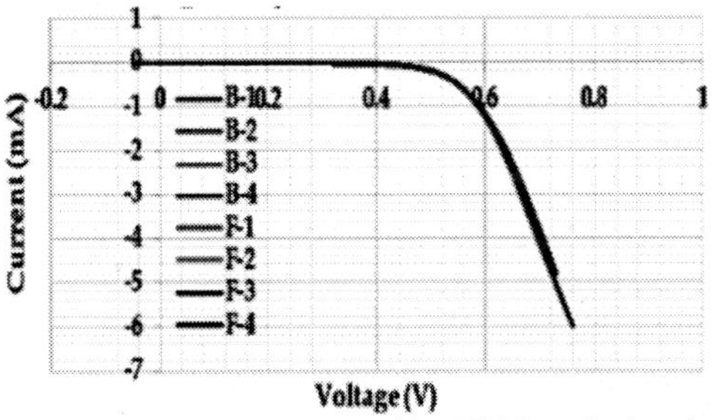

Figure 5: I-V curves measured by RTOS method at dark condition.

4 CONCLUSIONS

The characterization after testing showed that one must measure performance under all typical artificial light sources. Therefore, we propose that each solar cell should be measured varied indoor and outdoor lighting condition for its characterization by RTOS method. In Fig. 8a, we will be compare the Pmax of the different solar cells in light levels in indoor environments to show linearity under near range lighting level, which depends on device. In Fig. 8b shown nearly a constant from 400 lx to 1000 lx, it will be get an average of normalized power density and follow the defined of equation (3). Therefore, the results will be show the different solar cells in light levels in indoor environments to show linearity under near range lighting level, which depends on device. It is also shown the different solar cells in indoor lighting and solar environments to show different performance, ex. DSSC have more 0.26 times for TL84 than D65. We are also known STC under outdoor application, therefore, we also define a STC for indoor application, that is " LS/ 25 °C/ 1000 lx " , LS mean: lighting source, ex. A/D65/TL84/U30/CWF.

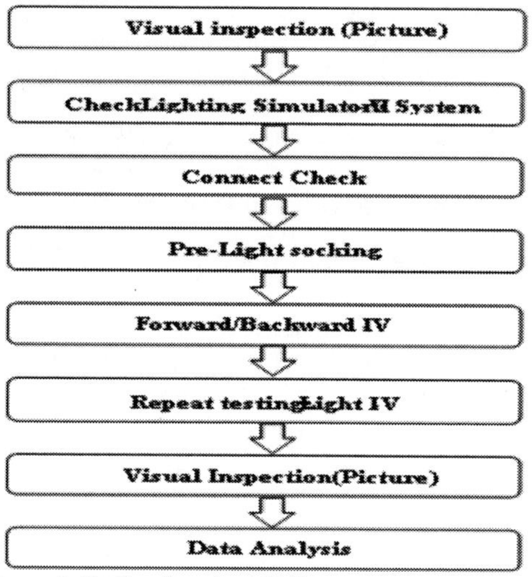

Figure 6: Testing flowchart [2,5].

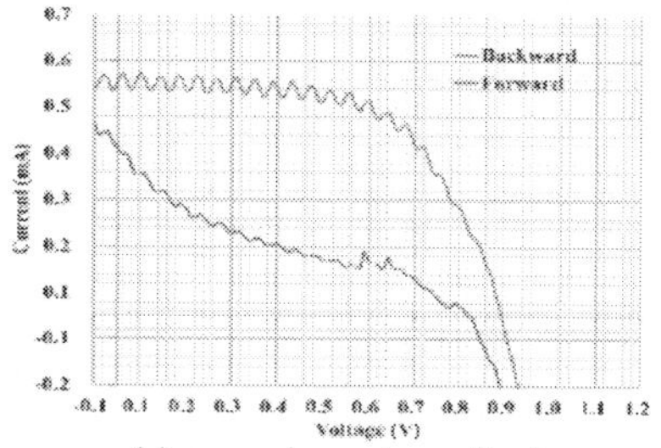

(a) sweeping rate method

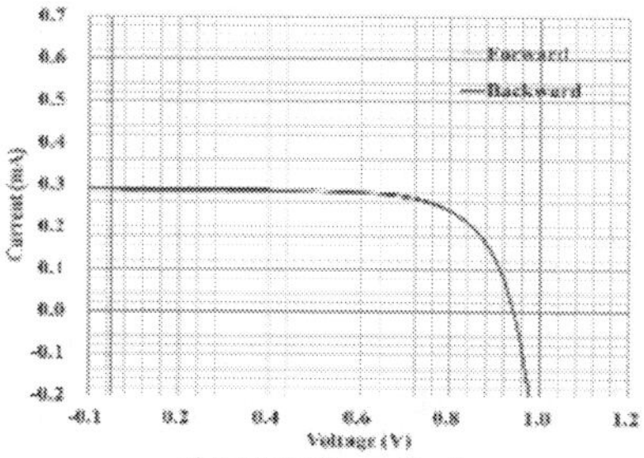

(b) RTOS method

Figure 7: I-V measurement of PSC (at STC) using the (a) sweeping rate method and (b) RTOS method.

In this work, we defined performance and normalized power density, is called performance ratio, shown a comparison for different emerging PV in indoor lighting and solar environments. Therefore, the results also find DSSC have more 0.26 times for TL84 than D65, but OPV/a-Si have less 0.5 times for TL84 than D65. That's shown DSSC have more indoor application than others in this case study.

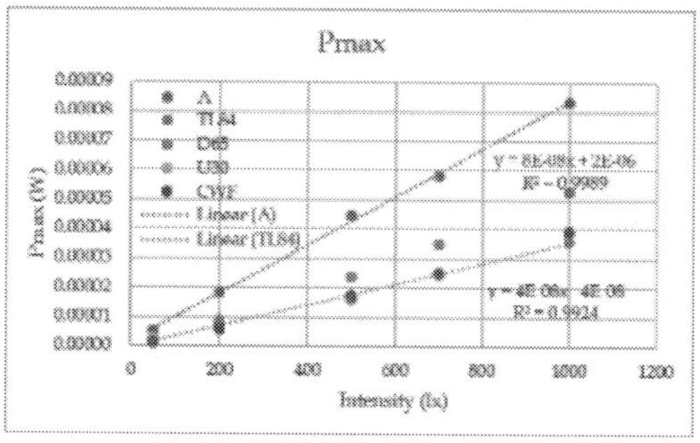

(a) P_{max}

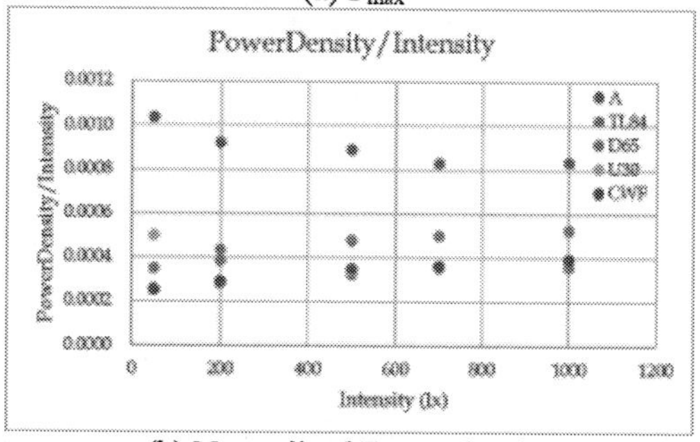

(b) Normalized Power density

Figure 8: Light linearity of DSSC measured by indoor multi-light source system under varied lighting condition (I-V using RTOS method).

When we will be follow through CIE of the defined some lighting spectrums for different indoor colorful applications and situations. However, the requirements for color testing do not apply to related PV indoor performance measurement under different lighting uniformity and stability, for performance measurement of emerging PV and device, and its characteristics will be having a classification. Furthermore, countless types of indoor lighting patterns are required to do performance characteristics measurement for different application contexts. We should be define a different STC than STC in AM1.5G, which we will be define a STC in indoor lighting as below as "I-V measurement for Indoor STC condition: 1000 lx @ 25 °C and Illuminant A". Furthermore, our goal is to determine which light source is relevant for a new international standard test method. Upon the indoor characterization of solar cells, it is the spectrum range of the light source that decides the lighting performance, not the light intensity, which is present and affects the cell equally through different light sources.

REFERENCES

[1] T. C. Wu, S. T. Hsu, and Y. S. Long, "New Set-up Procedures and Integrated Measurement System for Organic Photovoltaic (OPV) Module," *International Photovoltaic Science and Engineering Conference (PVSEC23)*, 2013.

[2] Y. S. Long, S. T. Hsu, and T. C. Wu, "Induction of Internal Capacitance Effect in Performance Measurement of OPV (Organic Photovoltaic) Device by RTOSM (Real-Time One-Sweep Method)," *Journal of Energy and Power Engineering*, Vol 8, pp1059-1066, 2014.

[3] Y. S. Long, S. T. Hsu, and T. C. Wu, "Capacitance Effect from High to Low Level Lighting by Real-Time One-Sweep Method for OPV Device," *IEEE PVSC42*, 2015.

[4] Y. S. Long, S. T. Hsu, and T. C. Wu, "A Study of Capacitance Effect in DSC from High to Low Level Lighting by Real-Time One-Sweep Method," *OPTIC2014*.

[5] SEMI PV57-1214, "Test Method for Current-Voltage (I-V) Performance Measurement of Organic Photovoltaic (OPV) and Dye-Sensitized Solar Cell (DSSC)."

[6] US PTO. 8224598, Method for Forming Optimal Characteristic Curves of Solar Cell and System Thereof.

[7] ASTM E490, "Standard Solar Constant and Zero Air Mass Solar Spectral Irradiance Tables."

[8] Handbook of Photovoltaic Science and Engineering, Chapter 16

[9] IEC 60904-1, "Photovoltaic Devices – Part 1: Measurement of Photovoltaic Current-Voltage Characteristics," 2006.

[10] IEC 60904-3, "Photovoltaic Devices – Part 3: Measurement Principles for Terrestrial Photovoltaic (PV) Solar Devices with Reference Spectral Irradiance Data," 2008.

[11] ISO 10526, "CIE Standard Illuminants for Colorimetry," 2007.

[12] CIE S 014-2, "CIE Standard Illuminants for Colorimetry," 2006.

[13] K. Emery, S. Winter, S. Pinegar, and D. Nalley, "Linearity Testing of Photovoltaic Cells," *4th World PV Conf.*, 2006.

[14] CIE S 014-2, "CIE Colorimetry - Part 2: Standard Illuminants for Colorimetry "

[15] ISO 11664-2:2007, "Colorimetry -- Part 2: CIE standard illuminants"

[16] ASTM D 1729, "Standard Practice for Visual Appraisal of Colors and Color Differences of Diffusely-Illuminated Opaque Materials";

[17] SEMI Doc. 5979, "Specification of indoor lighting simulator requirements for emerging PV "

[18] Handbook of Photovoltaic Science and Engineering, Chapter 16

NUMERICAL OPTICAL OPTIMIZATION OF PEROVSKITE-SILICON TANDEM SOLAR CELLS

K. Jäger,* M. Werth, L. Mazzarella, S. Calnan, F. Ruske, L. Korte, B. Stannowski, B. Rech, and S. Albrecht[§]
Helmholtz-Zentrum Berlin für Materialien und Energie GmbH
Kekuléstraße 5,12489 Berlin, Germany
*klaus.jaeger@helmholtz-berlin.de, [§]steve.albrecht@helmholtz-berlin.de

ABSTRACT: Tandem solar cells with a perovskite top cell and a silicon (c-Si) bottom cell are the most promising route to overcome the theoretical efficiency limit of silicon single-junction solar cells (29.4%), when the low-bandgap bottom cell should be based on silicon. To fully exploit the potential of two terminal perovskite-Si tandem cells, the device must be highly optimized from an optical point of view. In this contribution we present optical simulations for two designs of perovskite-Si tandem solar cells: *regular* structures with the hole-selective contacts of both subcells on the front side, and *inverse* structures with the hole-selective contacts on the rear sides of the subcells. As bottom cell we took a state-of-the-art silicon heterojunction (SHJ) cell with amorphous silicon layers on both sides. We generated absorption profiles with a coherent/incoherent algorithm based on the net-radiation method and used the so obtained absorption spectra of the perovskite and c-Si layers to extract the achievable photocurrent densities for the top and bottom cells.
Keywords: simulation, tandem, optical properties.

1 INTRODUCTION

Perovskite-silicon tandem solar cells aim to surpass the efficiency limit of silicon single-junction solar cells, which is 29.4% [1]. One of the first experimental proofs of this concept showed an efficiency of 18%, hence significantly below that limit [2]. Surpassing the limit significantly requires excellent optical performance of the solar cell, where the following points must be taken into account:

1. Because the overall current density is limited by the subcell generating the lower current density, it is key to distribute the available light between the two subcells such that the generated current densities are equal. This is called *current matching*.
2. The total reflection of the device must be minimized via the following route: first, the interfaces with the largest contributions to the total reflection must be identified. Secondly, suitable anti-reflective measures have to be taken at these interfaces. The most important contributor will be the interface between air and the first layer of the perovskite cell. Since currently it is very demanding to have rough layers in the perovskite cell, reflective losses at this interface have to be tackled via adapting the refractive indices of the respective layers.
3. Parasitic absorption in the contacting layers must be minimized.
4. Because of the indirect bandgap of Si, the bottom absorber absorbs only very weakly for long wavelengths longer than 900 nm. Hence, light trapping becomes very important. As we currently prefer a planer top-cell architecture, light trapping must be enabled at the read side, e.g. by etching the Si wafer such that it exhibits a pyramidal texture. This texture scatters the incident light, which is not absorbed during the first pass, and enhances the path length in Si, such that more light can be absorbed.

In this contribution we present optical simulations of perovskite-silicon tandem cells, which were carried out with the coherent/incoherent net-radation algorithm implemented in the GenPro4 package [3]. We performed layer thickness optimizations for planar cells, where we allowed the thicknesses of the perovskite top cell to vary within experimentally relevant boundaries while we kept the layer thicknesses of the silicon bottom cell constant.

2 METHOD

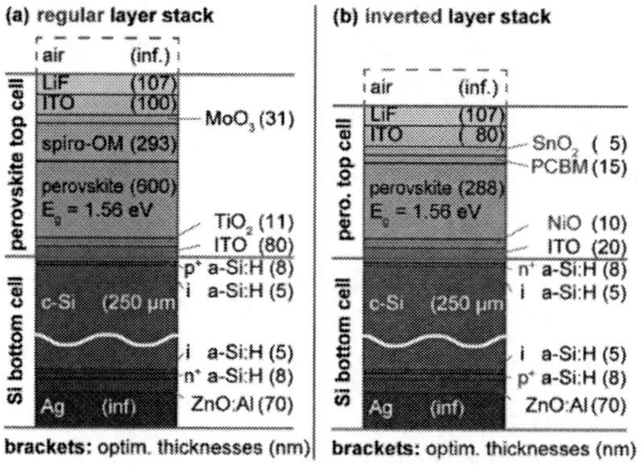

Figure 1: The layer stacks used for the optical simulations: (a) the regular architecture with the hole-selective and *p*-layers on the front of the perovskite and silicon subcells, respectively, and (b) the inverted architecture with the electron-selective and *n*-layers on the front of the perovskite and silicon subcells, respectively. The numbers in brackets refer to the layer thicknesses, in nm, which were derived in the respective optimizations and correspond to Figs. 2 and 3, respectively.

The layer thickness optimizations were conducted with the simulated annealing algorithm [4] as implemented in the MATLAB Global Optimization Toolbox and they allow to tackle points 1-3 listed above: the film thicknesses of the perovskite top cell are adjusted such that reflection losses and parasitic absorption losses are minimized simultaneously while current matching is assured, as previously published [5]. To estimate the PCE of the simulated devices, we started with open circuit voltages (V_{oc}) of state-of-the-art single-junction perovskite and silicon heterojunction (SHJ) cells, being 1.13 V and 0.73 V, respectively. We reduced these voltages in order to take into account that V_{oc} is logarithmically dependent on J_{ph}, $V_{oc} \propto \ln(J_{ph}/J_0)$. The fill factor of the tandem device was estimated to be 81% [5].

We studied the two architectures illustrated in Fig. 1: a *regular* design with the hole-selective contacts of both subcells on the front side, as illustrated in Fig. 1(a) [2], and an *inverse* structure with the hole-selective contacts on the rear side of both subcells, shown in Fig. 1(b) [6]. In contrast to earlier work [5,7], the silicon bottom cell now contains a 70 nm thick ZnO:Al layer, which improves the reflectivity at long wavelength.

The complex refractive index spectra (n, k) used for the simulations were taken from the following sources: LiF [8], spiro-OMeTAD [9], perovskite [10], MoO_3 [11,12], and Ag [13]. For all the other materials, especially all the supporting layers in the inverted design, layers were prepared and characterized in house. For NiO_x [14] and the sputtered ITO layers, ellipsometry and the program RIG-VM was used [15]. The data for the RF-PECVD a-Si:H layers were extracted using SCOUT [16]; more details for the i-layer optimization can be found in Ref. [17].

3 RESULTS

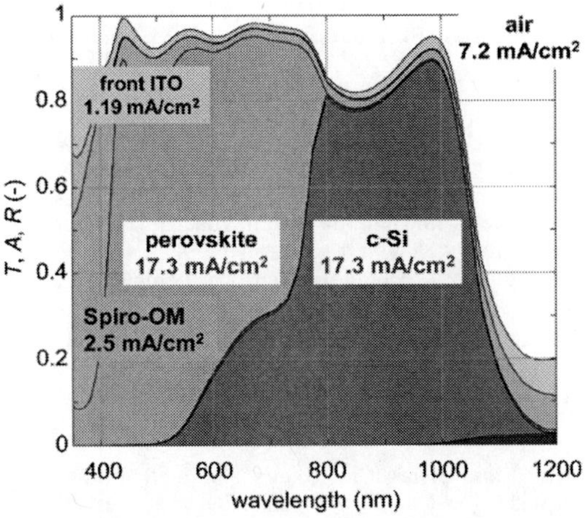

Figure 2: Absorption profile of an optimized tandem device with the *regular* design with the thicknesses shown in Fig. 1(a). The colours correspond to those in Fig. 1(a).

Figure 2 shows the absorption profile for a regular device with the optimized thicknesses shown in Fig. 1(a). This design delivered an optimal matched current density of 17.3 mA/cm² with a 300 nm thick perovskite absorber

As seen in Fig. 3, switching to the inverse architecture drastically reduces the parasitic absorption yielding an optimal matched current density of 18.5 mA/cm² with a 288 nm thick perovskite absorber. This is because in the regular design more than 2.5 mA/cm² are lost in the 150 nm thick Spiro-OMeTAD layer, which is used as hole-selective contact material. In the inverse design, only a 10 nm thin PCBM layer is used at the electron-selective contact, where only about 0.2 mA/cm² are lost.

To enhance light trapping in the long wavelength range, where silicon only is weakly absorbing, we also studied the effect of a pyramidal texture at the rear side of the c-Si wafer, as obtained with standard wet chemical KOH etching of silicon. We assume all interfaces in-between the wafer and the silver back layer to be conformal to the wafer texture. In the simulations, the layer stack is treated with the ray tracing module implemented in GenPro4 (see appendix), for the pyramidal structure the data set provided with the software is used.

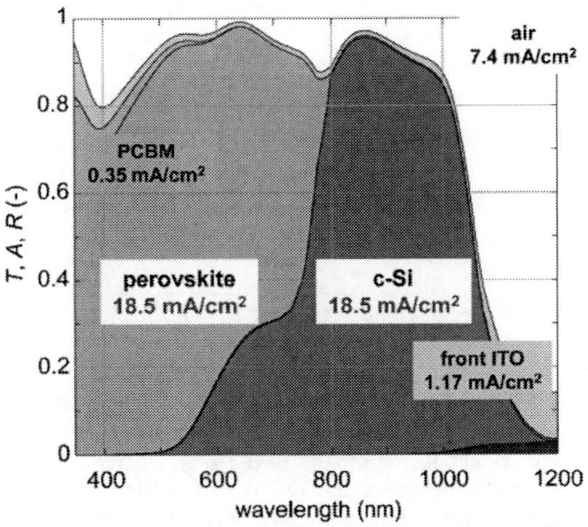

Figure 3: Absorption profile of an optimized tandem device with the *inverted* design with the thicknesses shown in Fig. 1(b). The colours correspond to those in Fig. 1(b).

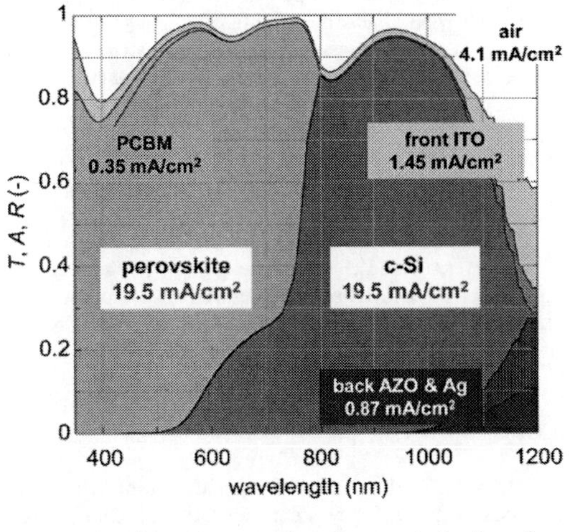

Figure 4: Absorption profile of an optimized tandem device with inverse design and a KOH pyramidal texture on the back side of the silicon wafer. The thicknesses are the same as shown in Fig. 1(b) except the perovskite thickness which is enhanced from 288 nm to 342 nm in this device. The colours correspond to those in Fig. 1(b).

Because simulations with ray tracing take much longer than simulations of complete planar layer stacks, we here omit a full layer thickness optimization with the simulated annealing algorithm but perform a quick optimization using MATLAB's Optimization toolbox, where we only allow the LiF and perovskite layers to vary in thickness.

Figure 4 shows an optimum for the perovskite-silicon tandem solar cell with inverse design and a pyramidal texture at the back. In comparison to the planar design, about 2 mA/cm² can be absorbed in the c-Si wafer additionally above 900 nm. To reach current matching, the perovskite layer thickness had to be increased from 288 nm to 342 nm. The matched current density is 19.8 mA/cm².

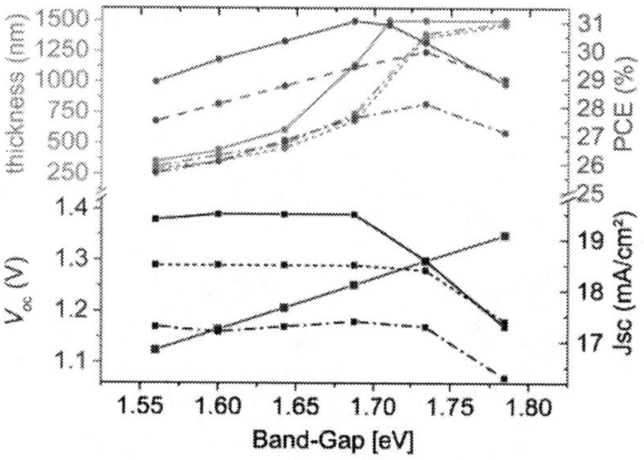

Figure 5: Effect of the perovskite bandgap on the maximum achievable short circuit current density (J_{sc}), perovskite thickness, top cell open circuit voltage (V_{oc}) and potential power conversion efficiency (PCE) for the planar regular design (dashed-dotted lines), the planar inverse design (dashed) and the inverse design with back KOH texturing (full). The perovskite thickness was limited to 1 500 nm.

The PCE of a tandem cell can be further increased by using perovskite absorbers with bandgaps larger than 1.56 eV, because then the top cell voltage can be increased while keeping the current approximately constant. When the upper bound of the perovskite thickness is set to 1 500 nm, the optimal bandgap for the planar devices is around 1.73 eV, as illustrated in Fig. 5 [5,7]. With the planar inverse design, the PCE potential exceeds the c-Si single junction limit of 29.4%.

In the device with back KOH texturing, more current density is generated at long wavelength > 900 nm. Hence, thicker perovskite layers are required to achieve current matching. As a consequence, the optimal bandgap is around 1.68 eV with a potential efficiency of around 31%.

CONCLUSIONS

Perovskite silicon tandem cells are a promising route towards high-efficiency solar cells. In this contribution, we performed coherent-incoherent optical simulations of such solar cells with different designs.

A layer thickness optimization allows to increase the optical performance and to achieve current matching. Further, adapting the perovskite bandgap, the potential efficiency can be increased above the c-Si single junction limit for a fully planar device. With pyramidal back texturing at the back, efficiencies exceeding 30% are feasible.

APPENDIX

In the last part of this manuscript, we simulated the layer stack with pyramidal textures using the ray tracing module implemented in GenPro4. To check the validity of this approach we used GenPro4 to simulate a double textured SHJ cell that was also fabricated at our centre [18]. As shown in Fig. 6, the measured and simulated $1-R$ curves match well. Further, the absorption curve in silicon matches well with the measured external quantum efficiency (EQE).

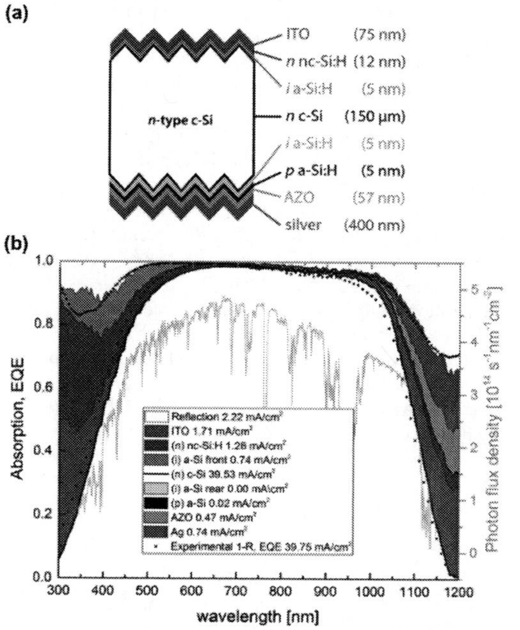

Figure 6: (a) Illustrating a silicon-heterojunction solar cell consisting of a double KOH-textured *n*-type wafer with the emitter on the front side. (b) Simulations of this device performed with GenPro4 with ray tracing. The simulated data agrees well with the measured EQE and $1-R$ curves In addition, the photon flux density according to the solar AM1.5 spectrum is shown.

ACKNOWLEDGEMENTS

We thank Ganna Chistiakova and Steffen Meyer for providung us with the SnO_2 and NiO_x layers, respectively.

We acknowledge the German Federal Ministry of Education and Research (BMBF) for funding via Nachwuchswettbewerb NanoMatFutur (No. 03X5520) and via the project "Materialforschung für die Energiewende" (No. 03SF0540), as well as the German Federal Ministry for Economic Affairs and Energy (BMWi) via the "PersiST" project (No. 0324037C).

REFERENCES

[1] A. Richter, M. Hermle, and S. Glunz, IEEE J. Photovolt. 3 (2013) 1184.

[2] S. Albrecht, M. Saliba, J.-P. Correa Baena, F. Lang, L. Kegelmann, M. Mews, L. Steier, A. Abate, J. Rappich, L. Korte, R. Schlatmann, M. K. Nazeeruddin, A. Hagfeldt, M. Grätzel, B. Rech, Energy Environ. Sci. 9 (2016) 81.

[3] R. Santbergen, T. Meguro, T. Suezaki, G. Koizumi, K. Yamamoto, M. Zeman, IEEE J. Photovolt. 7 (2017) 919.

[4] S. Kirkpatrick, C. D. Gelatt, and M. P. Vecchi, Science 220 (1983) 671.

[5] S. Albrecht, M. Saliba, J.-P. Correa-Baena, K. Jäger, L. Korte, A. Hagfeldt, M. Grätzel, and B. Rech, J. Opt 18 (2016) 064012.

[6] K. A. Bush, A. F. Palmstrom, Z. J. Yu, M. Boccard, R. Cheacharoen, J. P. Mailoa, D. P. McMeekin, R. L. Z. Hoye, C. D. Bailie, T. Leijtens, I. M. Peters, M. C. Minichetti, N. Rolston, R. Prasanna, S. Sofia, D. Harwood, W. Ma, F. Moghadam, H. J. Snaith, T. Buonassisi, Z. C. Holman, S. F. Bent and M. D. McGehee, Nature Energy 2 (2017) 17009.

[7] K. Jäger, L. Korte, B. Rech, and Steve Albrecht, Opt. Express 25 (2017) A473.

[8] H. H. Li, J. Phys. Chem. Ref. Data 5 (1976) 329. Retrieved from https://refractiveindex.info on 2016-03-21.

[9] M. Filipic, P. Löper, B. Niesen, S. D. Wolf, J. Krc, C. Ballif, and M. Topic, Opt. Express 23(2015) A263.

[10] P. Löper, M. Stuckelberger, B. Niesen, J. Werner, M. Filipic, S.-J. Moon, J.-H. Yum, M. Topic, S. De Wolf, and C. Ballif, J. Phys. Chem. Lett. 6 (2015) 66.

[11] S. Roland, S. Neubert, S. Albrecht, B. Stannowski, M. Seger, A. Facchetti, R. Schlatmann, B. Rech, and D. Neher, Adv. Mater. 27 (2015) 1262.

[12] S. Albrecht, B. Grootoonk, S. Neubert, S. Roland, J. Wördenweber, M. Meier, R. Schlatmann, A. Gordijn, and D. Neher, Sol. Energ. Mat. Sol. C. 127 (2014) 157.

[13] E.D Palik, Handbook of optical constants of solids (Academic Press, Orlando, 1985).

[14] J. You, L. Meng, T.-B. Song, T.-F. Guo, Y. Yang, W.-H. Chang, Z. Hong, H. Chen, H. Zhou, Q. Chen, Y. Liu, N. De Marco, Y.Yang, Nat. Nanotechnol. 11 (2016) 75.

[15] A. Pflug, V. Sittinger, F. Ruske, B. Szyszka, and G. Dittmar, Thin Solid Films 455 – 456 (2004) 201.

[16] W. Theiss, Hard and Software for Optical Spectroscopy, Dr.-Bernhard-Klein-Str.110, D-52078 Aachen, Germany (2002) http://www.wtheiss.com.

[17] L. Mazzarella, S. Kirner, O. Gabriel, S. S. Schmidt, L. Korte, B. Stannowski, B. Rech, and R. Schlatmann, Phys. Status Solidi A. (2016) DOI: 10.1002/pssa.201532958.

[18] M. Hendrichs, A. Morales-Vilches, L. Mazzarella, S. Kirner, M. Zelt, H. Rhein, S. Calnan, L. Korte, B. Stannowski, and R. Schlatmann; Proceedings of the 2016 European Photovoltaic Solar Energy Conference (2016) 753.

STUDY OF MICRO-STRUCTURAL PROPERTIES OF ZnO and WO_3 THIN FILMS GROWN BY SPIN COATING

G.Gordillo[1], J.Estrada[2], C.A. Otalora[3] and L.C. Luis[2]

[1] Departamento de Física, Universidad Nacional de Colombia, Bogotá, Colombia
[2]Facultad de Ingeniería, Universidad Nacional de Colombia, Bogotá, Colombia
[3]Departamento de Química, Universidad Nacional de Colombia, Bogotá, Colombia

Micro structural and optical properties of ZnO and WO_3 films prepared by spin coating were studied by XRD (X-ray diffraction) analysis and measurements of transmittance and Urbach Energy. The influence of preparation conditions on the micro structural properties of ZnO and WO_3 films was investigated by X-ray peak broadening. The Williamson Hall (W-H) analysis and size strain plot method were used to study the individual contributions of crystallite sizes and lattice strain on the peak broadening. The physical parameters such as strain, stress and energy density values were calculated for all the reflection peaks of XRD corresponding to both type of samples from the modified form of the W-H plot assuming a uniform deformation model (UDM), uniform stress deformation model (USDM) and uniform deformation energy density model (UDEDM). Further, information regarding the influence of preparation conditions on the formation of structural defects was achieved through Urbach energy measurements.

Keywords: ZnO, WO_3, thin films, PV materials.

1. INTRODUCTION

Thin films of ZnO and WO_3 have been used for many applications including electrochromic devices [1], gas sensing [2], UV lasers [3], light emitting diodes [4], UV detectors [5] and solar cells [6-9]. Besides these applications, recently the compounds ZnO and WO_3 have been used as hole-blocking layer in perovskite based hybrid solar cells [10-14]. The hole-blocking layers, which contribute to transport electron and block hole to electrode, are commonly applied to solar devices to avoid recombination at the interface between the electrode and light absorber. Optimization of the hole-blocking layer has drawn increasing attention as it can significantly improve the performance of perovskite solar cell. These layers are usually prepared by spin-coating based methods or spray-pyrolysis, using a tungsten or zinc precursor solution [15,16].

In the present work, we have investigated the effect of deposition conditions on the formation of structural defects as well as on the structural and optical properties of ZnO and WO_3 films deposited by spin-coating method using tungstic acid (H_2WO_4) in aqueous mixture of 10% H_2O_2 for WO_3 films and Zinc acetate ($Zn(CH_3COO)_2$) in dimethylsulfoxide (DMSO) for ZnO films.

2. EXPERIMENTAL AND THEORETICAL DETAILS

ZnO and WO_3 films were deposited by spin-coating technique and their microstructural properties where improved through correlation of the preparation parameters varied in a wide range, with results obtained from characterization carried out through XRD and Urbach energy measurements; these compounds were grown using the routes described below.

i) The WO_3 films were deposited using a solution of tungstic acid (H_2WO_4) dissolved in an aqueous mixture of 10% H_2O_2 with a concentration of 0.075 M. A typical film of WO_3 is formed depositing successively 10 layers using 100µL of precursor solution, which is spin-coated on a fluorine-doped tin oxide (FTO) conductive glass at 3000 r.p.m. After the deposition, each layer is dried at 150 °C for 10 minutes, followed by a thermal annealing at temperatures ranging from 300 to 400 °C for 30 minutes.

ii) The ZnO films were deposited from solutions prepared using zinc acetate ($Zn(CH_3COO)_2$) in dimethylsulfoxide, with a concentration 0.28M . A typical film of ZnO is formed depositing successively 10 layers using 100µL of precursor solution, which is spin-coated on a fluorine-doped tin oxide (FTO) conductive glass at 3000 r.p.m. After the deposition, each layer is dried at 150 °C for 10 minutes, followed by a thermal annealing at temperatures ranging from 300 to 400 ° C for 30 minutes.

Transmittance and reflectance measurements were performed using a Varian–Cary 5000 spectrophotometer and the film thickness was determined using a Veeco Dektak 150 surface profiler. The XRD (x-ray diffraction) measurements were performed in a Shimadzu-6000 diffractometer.

3. RESULTS

3.1. Influence of post annealing temperature on the optical properties of ZnO and WO_3 films

The influence of annealing temperature TA and film thickness on the optical properties of both ZnO and WO_3 films were evaluated through measurements of spectral transmittance. Fig. 1 show transmittance spectra of ZnO and WO_3 films prepared by spin coating varying TA and film thickness.

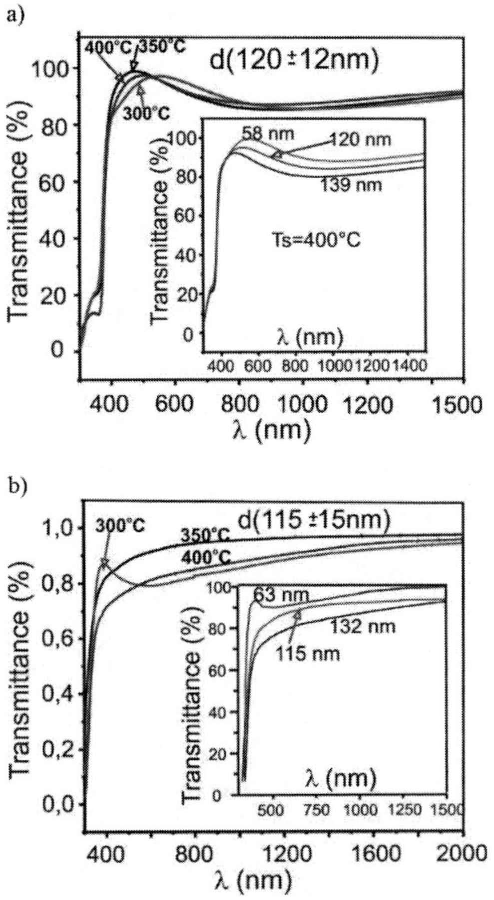

Fig. 1: Influence of post deposition annealing temperature on the transmittance of a) ZnO films and b) WO_3 films deposited by spin coating. In inset are displayed Transmittance spectra of ZnO and WO_3 films showing the effect of film thickness.

From the results shown in figures 1, the following facts can be highlighted.

i) All the transmittance spectra of both ZnO and WO_3 films present the same cutoff wave length, indicating that the energy gap Eg of both types of compound is not affected by the annealing temperature, as well as nor by the thickness. The Eg values obtained for the ZnO and WO_3 films from the intercept with the axis hv of the curve of $(\alpha hv)^2$ vs hv, were 3,3 eV and 3,8 eV respectively.

ii) The ZnO and WO_3 films deposited by spin coating present in general high transmittance (around 90%) and curves of transmittance characterized by having high slope indicating that in both type of samples the absorption of photons occur mainly via fundamental transitions and that both the ZnO and WO_3 films grow with a low density of intrinsic and structural defects.

iii) The slope of the transmittance curves of WO_3 films decreases by increasing the film thickness; this behavior seem to be caused by absorption of photons in states of tails of bands associated with structural defects. This result allow to conclude that increase in the film thickness of the WO_3 films induces an increase in the density of structural defects, generating in this way absorption via states of tail of bands.

Absorption via transitions between states of tails of bands was evaluated through the calculation of the absorption coefficient near the band edge (α_U) that has an exponential dependence on photon energy and Urbach energy [1] given by

$$\alpha_U = \alpha_0 exp\left[\frac{hv - E_i}{Eu}\right] \qquad (1)$$

Where E_U is the Urbach energy, E_1 and α_0 are constant. Thus, a plot of ln α vs. hν should be linear and Urbach energy can be obtained from the slope. In Table 1 are given results that show the effect of post annealing temperature and film thickness on the Urbach energy of both ZnO and WO_3 films. It is observed that annealing temperature and film thickness affects very little the Urbach energy of the ZnO films; however the value of E_U of WO_3 films increases significantly by increasing of both thickness and annealing temperature. These results allow to conclude that crystallographic properties of the ZnO films are slightly improved by increasing of annealing temperature, whereas the crystallinity of the WO_3 films is quite deteriorated by increasing of both, thickness and annealing temperature.

TABLE 1: Influence of post annealing temperature and film thickness on the Urbach Energy of ZnO and WO_3 films

	ZnO					
	Thickness = 120 nm T_A(°C)			T_A=400°C Thickness (nm)		
	300	350	400	62	120	152
E_U (meV)	77,6	66,2	64,3	60,1	64,3	64,9
	WO_3					
	T_A(°C) Film thickness = 115 nm			Thickness (nm) T_A=350°C		
	300	350	400	58	115	149
E_U (meV)	92,5	98,2	119,7	82,4	98,2	117,6

Taking into account that Urbach energy is related to density of localized states in the band gap induced by structural disorder (stress and dislocation) [17], it is possible to conclude that both the samples of ZnO as those of WO_3 deposited by spin coating present an improved crystallinity.

3.2. Influence of post annealing temperature on the structural properties of ZnO and WO_3 films

The structural properties of the ZnO and WO_3 films were also evaluated through XRD measurements. In Fig. 2 are displayed XRD spectra of ZnO and WO_3 films deposited by spin coating on glass substrates coated with a layer of FTO, varying the post deposition annealing temperature maintaining constant the rest of parameters. It is observed that the ZnO films present Bragg peaks of (002), (100), (101), (102), (110), (103) and (112) which belong to the ZnO phase with hexagonal structure (PDF data file # 01-0790208), whereas the WO_3 films present reflections along the (002), (110), (022), (220) and (114) planes which belong to the WO_3 phase with monoclinic structure (PDF data file #01-087-2404).The peaks corresponding to the FTO used as substrate are indicated with an asterisk.

a)

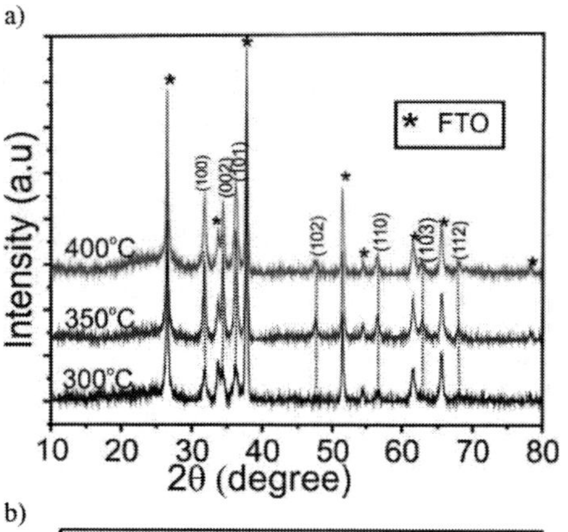

b)

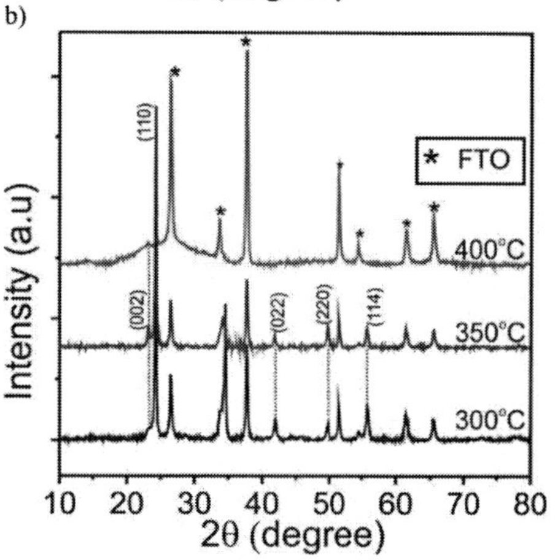

Fig. 2: Influence of post deposition annealing temperature on the XRD spectra of a) ZnO films and b) WO_3 films deposited by spin coating.

It can be seen in Fig. 2 that in the range of studied annealing temperatures (T_A) (300-400°C), the ZnO films grow in polycrystalline form while the WO_3 films show character polycrystalline only at $T_A \leq$ 350°C; at $T_A \geq$ 350°C the WO_3 films do not grow in polycrystalline form, since only exhibit the reflections corresponding to the layer of FTO.

It was found that the full width at half maximum (FWHM) value of the studied ZnO and WO_3 films is affected by the annealing temperature, indicating that this parameter affect their crystallinity. XRD can be utilized to evaluate peak broadening with crystallite size D and lattice strain ε due to dislocation [18]. The particle size of the ZnO and WO_3 can be determined by the X-ray line broadening method using the Williamson-Hall equations [19] given by:

$$\beta_{hkl} * cos\theta_{hkl} = \frac{K\lambda}{D} + 4\varepsilon * sin\theta_{hkl} \quad (2)$$

Where D is the particle size, λ is the wavelength of the radiation for CuKα radiation, k is a constant equal to 0.94, β_{hkl} is the peak width at half-maximum intensity and Θ_{hkl} is the peak position. Therefore, Eq. (2) represents the UDM (Uniform Deformation Modell), where the strain was assumed to be uniform in all crystallographic directions. The term β*cosθ was plotted with respect to 4sinθ for the preferred orientation peaks of ZnO and WO_3 films. Accordingly, the slope and y-intersect of the fitted line represent strain and particle size, respectively.

In the Uniform Stress Deformation Model (USDM), a generalized Hooke's law refers to the strain, keeping only the linear proportionality between the stress and strain as given by σ=Yε, where σ is the stress of the crystal and Y is the modulus of elasticity or Young's modulus. This equation is an approximation that is valid for a significantly small strain. Applying the Hooke's law approximation to Eq. (3) yields:

$$\beta_{hkl} * cos\theta_{hkl} = \frac{K\lambda}{D} + \frac{4\sigma * sin\theta_{hkl}}{Y_{hkl}} \quad (3)$$

The results obtained from the UDM, USDM and models and UDEDM modells, are summarized in Table 2. In this case we used values of Y reported in the literature for ZnO hexagonal (Wurzite) and for WO_3 films grown with monoclinic structure; it was used values of Y= 130 GPa for ZnO [20] and 450 GPa for WO_3 [21].

TABLE 2: Influence of post deposition annealing temperature on the grain size, stress, strain and energy density of ZnO and WO_3 films deposited by spin coating.

		Scherrer	W-H (UDM)		W-H (USDM)	W-H(UDEDM)
	T(°C)	D (nm)	D (nm)	ε x10^{-3}	σ (Mpa)	u (KJ/m^3)
ZnO	300	16.49	11.84	-2.3892	-317.77	379.61
	350	26.19	43.52	1.6491	219.33	180.86
	400	20.47	37.46	2.8056	373.14	523.45
WO3	300	32.43	15.73	-2.9740	-1338.29	1990.0
	350	34.62	26.51	-0.8258	-371.63	153.5

4. CONCLUSIONS

The influence of film thickness and post deposition annealing temperature on the optical and structural properties of ZnO and WO_3 thin films prepared spin coatingwere evaluated through transmittance and X-ray diffraction measurements using Williamson–Hall-isotropic strain model and Urbach energy estimated from absorption coefficient calculated near the band edge. It was found that in general both type of samples grow with a low density of intrinsic and structural defects; however, the study revealed that the increase of film thickness in WO_3 films induces an increase in the density of structural defects, generating in this way absorption via states of tail of bands.

The micro structural behavior investigated by X-ray peak broadening revealed that the particle size of both ZnO and WO_3 films increase as the annealing temperature increases and that in the range of temperature studied (300-350°C) the strain and stress values of the WO_3 films are negative indicating that under these conditions the WO_3 layers present a compressive strain. On the contrary, the ZnO films annealed at 350 and 400°C exhibit expansive strain (**ε positive)** and compressive strain when are annealed at T_A=300°C

It was also found that the energy gap Eg of both types of compound is not affected by the annealing temperature, as well as nor by the thickness. The Eg values obtained for the ZnO and WO_3 films were 3,3 eV and 3,8 eV respectively.

ACKNOWLEDGEMENTS: This work was supported by Colciencias (Contract #184/2016) and Universidad Nacional de Colombia, Sede Bogotá, Facultad de Ciencias, Departamento de Física, Grupo GMS&ES, K30 #45-03,Bogotá DC, Colombia.

REFERENCES

[1] Y. Zhang, S.-H. Lee, A. Mascarenhas, and S. K. Deb, An UV photochromic memory effect in proton-based WO3 electrochromic devices, APPLIED PHYSICS LETTERS 93, 203508 2008

[2] S.H. Lee, H. M. Cheong, P. Liu, D. Smith, C. E. Tracy, J. R. Pitts, and S. K. Deb, hydrogen sensors based on gasochromic oxide thin films, J. Appl. Phys. 88, 3076 2000.

[3] M.H. Huang, S. Mao, H. Feick, H.Q. Yan, Y.Y. Wu, H. Kind, E. Weber, R. Russo, P.D. Yang, Room-Temperature Ultraviolet Nanowire Nanolasers, Science 292 (2001) 1897–1899.

[4] T. Soki, Y. Hatanaka, D.C. Look, ZnO diode fabricated by excimer-laser doping, Applied Physics Letters 76 (2000) 3257.

[5] S. N. Das, K.-Ju Moon, J. P. Kar, Ji-Hyuk Choi, J. Xiong, T. Lee, and Jae-Min Myounga, ZnO single nanowire-based UV detectors, Applied Physics Letters 97, 0221-03 (2010) .

[6] M. Krunks, A. Katerski, T. Dedova, I. Oja Acik, A. Mere, Nanostructured solar cell based on spray pyrolysis deposited ZnO nanorod array, Solar Energy Materials and Solar Cells 92 (2008) 1016–1019.

[7] J. Zhang, Ch. Shi , J. Chen, Y. Wang, M. Li, Preparation of ultra-thin and high-quality WO_3 compact layers and comparision of WO_3 and TiO_2 compact layer thickness in planar perovskite solar cells, Journal of Solid State Chemistry (2016) 238, 223–228

[8] Solar Frontier Press Release. Solar frontier achieves world record thin-film solar cell efficiency: 22.3%, 8. December 2015.http://www.solar-frontier.eu/uploads/media/SFE-Thin-Film-Solar-Cell-Efficiency-EN.pdf

[9] T.K.Todorov, J .Tang, S. Bag, O. Gunawan, T. Gokmen, Y. Zhu, D.B. Mitzi, "Beyond 11% efficiency: Characteristics of state-of-art $Cu_2ZnSn(Se,S)_4$ solar cells", Advanced Energy Materials 2013, *3*, 34–38

[10] Nam-Gyu Park, Organometal Perovskite Light Absorbers Toward a 20% Efficiency Low-Cost Solid-State Mesoscopic Solar Cel, J. Phys. Chem. Lett., (2013) **4**, 2423–2429

[11] J. Duan, J. Wu, J. Zhang , Y. Xu , H. Wang, D. Gao,andP. D. Lund,TiO_2/ZnO/TiO_2 sandwich multi-layer films as a hole blocking layer for efficient perovskite solar cells, Int. J. Energy Res. (2016;) 40, 806–813

[12] E.J. Juarez-Perez, M. Wu βler, F. Fabregat-Santiago, K. Lakus-Wollny, E. Mankel, T. Mayer, W. Jaegermann, I. Mora-Sero.Role of the selective contacts in the performance of lead halide perovskites solar cells, J. Phys. Chem. Lett. 5 (2014) 680–685.

[13] K. Wang, Y. Shi, Q. Dong, Y. Li, S. Wang, X. Yu, M. Wu, T. Ma, Low temperature and solution processed amorphous WOx as electron selective layer for perovskite solar cells, J. Phys. Chem. Lett. 6 (2015) 755–759.

[14] D.Y. Son, J.H. Im, H.S. Kim, N.G. Park, 11% efficient perovskite solar cell based on ZnO nanorod: an effective charge collection system, J. Phys. Chem. Lett. 118 (2014) 16567–16573.

[15] F. Zahedi, R.S.ariani, S.M. Rozati, Effect of substrate temperature on the properties of ZnO thin films prepared by spray pyrolysis, Materials Science in Semiconductor Processing 16 (2013) 245–249

[16] Wu YZ, Yang XD, Chen H, Zhang K, Qin CJ, Liu J, Islam A, Bi E, Ye F, Yin M, Zhang P, Han LY, Peng WQ. Highly compact TiO2 layer for efficienthole-blocking in perovskite solar cells. Applied Physics Express 2014; 7:052301–05230

[17] M.V. Kurik, Review of Urbach's tail. Phys. Status Solidi A, vol. 8, No. 9, 1971.

[18] R. Yogamalar, R. Srinivasan, A. Vinu, K. Ariga, A.C. Bose, X-ray peak broadening analysis in ZnO nanoparticles, Solid State Commun. 149 (2009) 1919.

[19] G.K. Williamson, W.H. Hall, X-ray line broadening from filed aluminium and wolfram, Acta Metall. 1 (1953) 2231

[20] A. Khorsand, W.H. Abd. Majida, M.E. Abrishami, R. Yousefi,X-ray analysis of ZnO nanoparticles by Williamson Hall and size estrain plot methods, Solid State Sciences 13 (2011) 251-256

[21] N.M.G. Parreira, N.J.M, Carvalho and A. Cavaleiro, Synthesis, structural and mechanical characterization of sputtered tungsten oxide coatings Thin Solid Films 510 (2006) 191 – 196.

EARTH-ABUNDANT THIN FILM SOLAR CELLS BASED ON Cu_2MnSnS_4

A. Le Donne[1], S. Binetti[1], M. Acciarri[1] and S. Marchionna[2]
[1]University of Milano-Bicocca, Department of Materials Science and MIBSOLAR center, via Cozzi 55, Milano, Italy
[2] RSE Spa, Ricerca sul Sistema Energetico, Via Rubattino 54, Milano, Italy

ABSTRACT: An attractive solution for the preparation of In and Ga free chalcogenides which may allow terawatt range photovoltaic (PV) applications relies on I_2-II-IV-VI_4 species, such as copper zinc tin sulfide/selenide. A further alternative belonging to this class of materials is Cu_2MnSnS_4 (CMTS), which consists of abundant and non-toxic elements and shows high absorption coefficient and direct band gap suitable for PV applications. In this work, CMTS thin films were grown on Mo-coated soda lime glasses by a two-step process (evaporation in vacuum of the metal precursors followed by annealing in sulfur vapors). SLG/Mo/Sn/Cu/Mn stack structures were sulfurized at 585°C with an initial step at 115°C to enhance the metal intermixing, paying particular attention to the control of Mn/Sn ratio and to the effect of different annealing ramping rate. Some proof of concept PV device based on Cu-poor/Mn-rich CMTS samples were prepared and tested (efficiency 0.6%, open-circuit voltage 303 mV, short-circuit current density 5.1 mA/cm^2, fill factor 38.5%). The beneficial effects of low temperature post-deposition annealing were also investigated, which provided literature record performance (efficiency 0.83%, open-circuit voltage 354 mV, short-circuit current density 5.8 mA/cm^2, fill factor 40%).

Keywords: PV Materials, Manufacturing and Processing, Thin Film Solar Cell, Annealing, Electrical Properties, Photoluminescence

1 INTRODUCTION

As it is well known, in the last decade much attention has been focused on I_2-II-IV-VI_4 thin films as an attractive solution for the preparation of In and Ga free chalcogenides which may allow terawatt range photovoltaic (PV) applications. A further alternative to copper zinc tin sulfide/selenide belonging to this class of materials is copper manganese tin sulfide (CMTS), a p-type semiconductor fully based on Earth-abundant and low-cost elements which shows an important advantage with respect to copper zinc tin sulfide/selenide. As a matter of fact, not only the abundance in the Earth's crust of Mn is two order of magnitude higher than that of Zn (1100 ppm vs 79 ppm), but the amount of Zn produced in 2015 was 4'600'000 tons lower than that of Mn (13'400'000 Zn tons vs 18'000'000 Mn tons) [1]. Therefore, since Mn is definitely cheaper than Zn, optimized CMTS could potentially provide W_p cost definitely lower than copper zinc tin sulfide/selenide, which is crucial for thin film PV applications.

CMTS, which crystallizes into a stannite structure (space group: I-42m), shows high absorption coefficient ($\cong 10^4$ cm^{-1}) and direct band gap [1,2] suitable for PV applications. So far, CMTS was mainly studied as bulk magnetic semiconductor, while, only in the last two years, few papers reported on CMTS thin films for PV applications [1-4]. In particular, our work deals with CMTS thin films grown by a two-step vacuum process, whose first optimization is reported in [1]. The metal precursors deposited on Mo-coated soda lime glass (SLG) by thermal evaporation are annealed in sulfur vapors for 1h at 585°C with an initial 1h step at 115°C to enhance the metal intermixing (ramping rate around 15°C/min). Of the many possible stoichiometries, Cu-poor/Mn-rich CMTS films with Mn/Sn ratio around 1 were chosen in order to prevent the development of both highly conductive (e.g. $Cu_{2-x}S$) and insulating (e.g. MnS) secondary phases.

In this work, some modifications to the previous optimization [1] were adopted to get enhanced device performance. In the first place, a better control of the Mn/Sn ratio has been obtained through a better control of both the Mn and Sn layer thicknesses. Moreover, sulfurization ramping rates around 7°C/min were tested to obtain larger and less defective grains. Finally, the possible beneficial effects of low temperature post-deposition annealing either in air or inert atmosphere between 200 and 275°C on CMTS solar cells were investigated, both in terms of electrical performance and modification of the material properties.

2 EXPERIMENTAL

CMTS thin films with approximate final thickness around 1.5 μm (measured by a Veeco-Dektak 150 profilometer) were grown by a two step process. Metal precursors were deposited by thermal evaporation on 4x4 cm^2 Mo coated SLG in the sequence SLG/Mo/Sn(235 nm)/Cu(176 nm)/Mn(135 nm). The Mo back contact deposited by DC magnetron sputtering was 1 μm thick. All metal precursor layers have been deposited by a 4-sources electron beam (EB) evaporation system (Kurt J. Lesker PVD 75) in a high vacuum chamber with $2x10^{-7}$ mbar base pressure. Both Sn and Cu beads were evaporated with deposition rates of 2.5 Å/sec and 1.5 Å/sec, respectively, while Mn powders were evaporated at 3 Å/sec.

CMTS thin films were then formed by thermal treatment of the metal precursors in sulfur vapors. Such a sulfurization process was performed in a tube furnace, where the stack of metal precursors and a quartz crucible with 0.3 g of sulfur were loaded in the same quartz boat. The metal precursors were then annealed at 585°C in a mixed argon (50 cm^3/min flux) and elemental sulfur atmosphere for 60 min (ramping rates of 7 or 15 °C/min), with an additional initial step at 115°C for 60 min, to enhance the metal intermixing.

Both morphology and chemical composition of the CMTS layers were investigated by a MIRA3 Tescan Scanning Electron Microscope (SEM) equipped for Energy-Dispersive Spectroscopy (EDS). The operating voltage used for all the SEM images was 20 kV, while EDS analyses were performed by a Quantax system (Bruker).

Photoluminescence (PL) spectra were recorded both in the 450-800 nm and in the 850-1700 nm spectral range. PL measurements in the visible range were performed with a spectral resolution of 5 nm by a Fluorolog 3.21 spectrometer provided by Jobin Yvon. The excitation source consists of a 450 W Xe lamp coupled with a 1200 grooves/mm double grating monochromator. The signal detection equipment consists of a 1200 grooves/mm single grating monochromator and of a phototube detector (Hamamatsu R928P). A cooling system consisting of rotary pump, diffusion pump, l-N_2 cold trap and Cryomech PT405 cryorefrigerator was used to perform PL measurements at low temperature up to 7 K. As far as analyses in the IR range are concerned, PL measurements were performed with a spectral resolution of 6.6 nm using a standard lock-in technique in conjunction with a single grating monochromator and a short wavelength enhanced InGaAs detector with maximum responsivity at 1540 nm. A quantum well laser (λ_{exc}=805 nm) was mainly used as excitation source, along with a 532 nm laser sometimes employed for comparison purposes. A cooling system consisting of rotary pump, turbomolecular pump and He closed circuit cryostat was used to perform PL measurements at low temperature up to 15 K.

CMTS solar cells with an active area of 0.15 cm^2 were then obtained as reported in [1].

Post-deposition thermal treatments in air at 200, 225, 250 and 275°C were finally tested to investigate the effect of low temperature annealing on CMTS solar cell efficiency.

The CMTS devices were inspected both by current density-voltage (J-V) measurements and External Quantum Efficiency before and after each annealing. J-V measurements under 1 Sun illumination (100 mW/cm^2) in Air Mass 1.5G conditions were recorded by a Thermo Oriel Solar simulator. External Quantum Efficiency (EQE) measurements were obtained by a SpeQuest quantum efficiency system.

3 RESULTS AND DISCUSSION

In order to get a better control on the Mn/Sn ratio, first of all the metal precursors were grown in the sequence SLG/Mo/Sn/Cu/Mn, instead of the SLG/Mo/Mn/Sn/Cu/Mn sequence tested as PV prototypes in [1]. As a matter of fact, the strict control of the Mn deposition rate being a critical issue, a two-step Mn evaporation increases the error on the final Mn thickness, which in turn affects the CMTS stoichiometry. Furthermore, Mn powders were used for Mn evaporation, as an alternative to the flakes employed in [1], which also enhanced the control of the final Mn layer thickness. EDS (see Table 1) and J-V results (see Fig. 1) showed that these modifications to the previous optimization provided not only enhanced solar cell performance with respect to [1] (efficiency 0.6%, open-circuit voltage 303 mV, short-circuit current density 5.1 mA/cm^2, fill factor 38.5%), but also a good homogeneity of the Mn/Sn ratio.

	1kx	15kx	30kx	50kx
[Mn]/ [Sn]	1.0	1.0	1.1	1.0
[Cu]/ [Sn]+[Mn]	0.8	0.9	0.8	0.9

Table 1. Compositional ratios from EDS analyses performed at different magnifications on CMTS samples obtained by sulfurization of SLG/Mo/Sn/Cu/Mn stacks at 115+585°C (Mn powders for Mn evaporation and 15°C/min ramping rate).

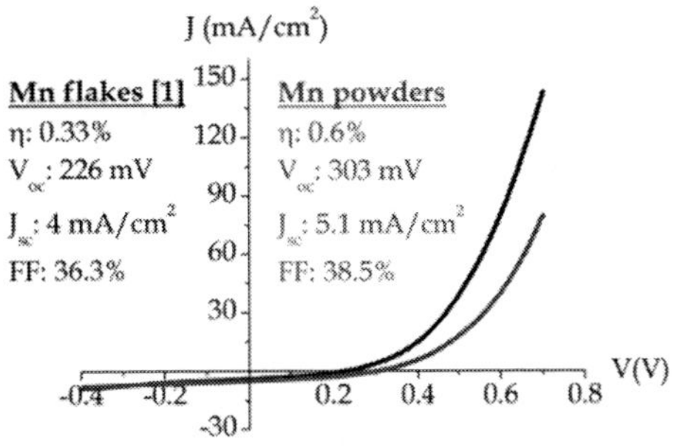

Figure 1. J-V curves under 1 sun illumination performed on CMTS samples obtained by sulfurization of SLG/Mo/Sn/Cu/Mn stacks at 115+585°C (Mn flakes or Mn powders for Mn evaporation and 15°C/min ramping rate).

SEM, EDS and J-V results (see Fig. 2 and Table 2) on CMTS samples grown with sulfurization ramping rates around 7°C/min showed that low sulfurization ramping rate led both to a worse grain packaging and to a significant Sn loss, which in turn provide a performance deterioration. Sulfurization times around 5-10 min will be tested in the case of low ramping rates to reduce the Sn loss.

	15 °C/min	7 °C/min
[Mn]/[Sn]	0.91	1.16
[Cu]/[Mn]+[Sn]	0.93	0.88
[S]/metals	1.01	1.15
Efficiency	0.6%	0.3%

Table 2. Compositional ratios from EDS analyses performed on CMTS samples obtained by sulfurization of SLG/Mo/Sn/Cu/Mn stacks at 115+585°C (Mn powders for Mn evaporation).

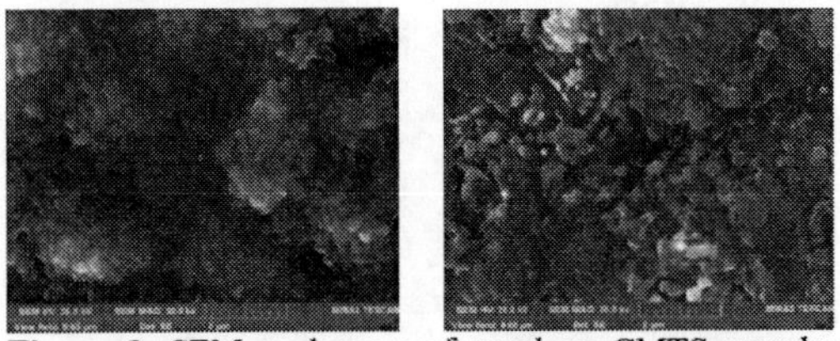

Figure 2. SEM analyses performed on CMTS samples obtained by sulfurization of SLG/Mo/Sn/Cu/Mn stacks at 115+585°C (Mn powders for Mn evaporation): left 15°C/min ramping rate; right 7°C/min ramping rate.

J-V (see Fig. 3), EQE (see Fig.4) and PL results (see Fig.5) showed that post-deposition thermal treatments in air between 200 and 275°C generally reduce recombination losses and increase CdS layer quality, while red-shifting its absorption edge. In particular, a 40 min 225°C annealing in air lead to a significant reduction of recombination losses without a strong increase of CdS absorption, allowing for the best overall performance enhancement (efficiency 0.83%, open-circuit voltage 354 mV, short-circuit current density 5.8 mA/cm^2, fill factor 40%).

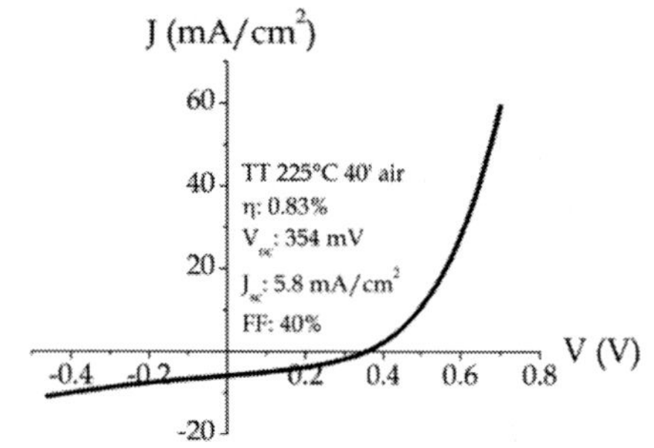

Figure 3. J-V curves under 1 sun illumination performed on SLG/Mo/CMTS)/CdS/i-ZnO/AZO solar cells after a 40' annealing in air at 225°C.

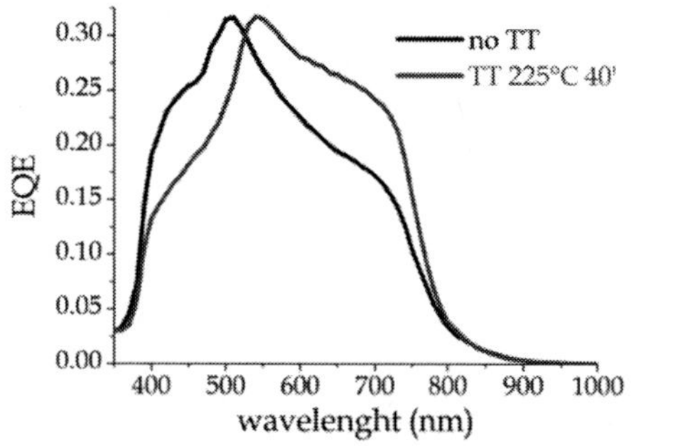

Figure 4. EQE of SLG/Mo/CMTS)/CdS/i-ZnO/AZO solar cells before and after a 40' annealing in air at 225°C.

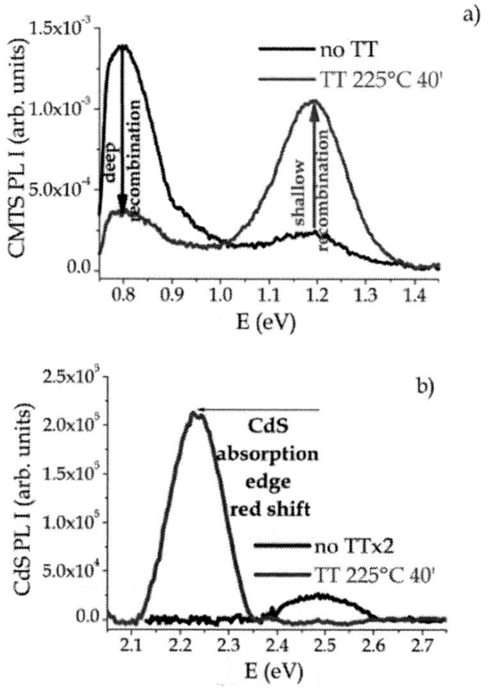

Figure 5. PL spectra of SLG/Mo/CMTS)/CdS/i-ZnO/AZO solar cells before and after a 40' annealing in air at 225°C: a) PL in the IR region due to optically excited CMTS; b) PL in the UV-vis region due to optically excited CdS.

4 CONCLUSIONS

Earth abundant CMTS thin films were obtained by sulfurization at 115°C+585°C with different ramping rates of SLG/Mo/Sn/Cu/Mn stacks grown by thermal evaporation. Cu-poor/Mn-rich CMTS samples with a good homogeneity of the metal compositional ratios were obtained using a 15 °C/min ramping rate and Mn powders for the Mn evaporation. Proof of concept PV devices based on them showed enhanced performance with respect to a previous study (efficiency 0.6% vs 0.33%, open-circuit voltage 302 vs 226 mV, short-circuit current density 5.1 vs 4 mA/cm^2, fill factor 38.5% vs 36.3%). The effects of post-deposition thermal treatments between 200 and 275°C on the same CMTS solar cells were also investigated, both in terms of electrical performance and modification of the material properties. The best annealing at 225°C for 40 min in air allowed for a significant performance enhancement (efficiency 0.83%, open-circuit voltage 354 mV, short-circuit current density 5.8 mA/cm^2, fill factor 40%), leading to the present efficiency record for this promising thin film absorber [5].

6 ACKNOWLEDGEMENTS

Funding was provided by the Research Fund for the Italian Electrical System (Contract Agreement between ERSE and the Ministry of Economic Development-General Directorate for Energy and Mining Resources stipulated on 07/29/2009 in compliance with the Decree of 03/19/2009) and partially by University of Milano-Bicocca (project n. 2016-ATESP-0582).

7 REFERENCES

[1] S. Marchionna, A. Le Donne, M. Merlini, S. Binetti, M. Acciarri, F. Cernuschi, J. Alloys Compd. 693, 95 (2017) and references within

[2] L. Chen, H. Deng, J. Tao, H. Cao, L. Sun, P. Yang, J. Chu, Acta Mat. 109, 1 (2016) and references within

[3] B. Ananthoju, J. Mohapatra, M.K. Jangid, D. Bahadur, N.V. Medhekar, M. Aslam, Scientific Reports 6:35369 (2016) DOI: 10.1038/srep35369

[4] R.R. Prabhakar, S. Zhenghua, Z. Xin, T. Baikie, L.S. Woei, S. Shukla, S.K. Batabyal, O. Gunawan, L.H.Wong, Solar Energy Materials & Solar Cells 157, 867 (2016)

[5] A. Le Donne, S. Marchionna, M. Acciarri, F. Cernuschi, S. Binetti , Solar En. 149 (2017) 125

VACANCY MIGRATION AND ASSOCIATED CHARGE-TRANSFER AT THE ZNS/CZTS INTERFACE

F H Bahrani, J P Goss, P R Briddon and M J Rayson
School of Engineering, Newcastle University, Newcastle upon Tyne, UK, NE1 7RU

ABSTRACT: Secondary phases, such as ZnS in Cu-poor/Zn-rich CZTS, affect the performance of photovoltaics by reducing the absorber volume and decreasing the short circuit current. The band-offset between CZTS and ZnS render the ZnS regions as carrier traps, and the interface between these materials is of significance to CZTS cell performance. The abruptness of the interface between CZTS and ZnS is controlled by the intrinsic diffusion mechanisms, with vacant metal sites playing a key role. To understand the mechanisms underpinning the inter-diffusion between CZTS and ZnS across a [100] interface, we have performed density-functional calculations for two types of inter-diffusion process at the heterojunction interface. For each process the barrier has been estimated, and the processes aligned to a common energy scale based upon the formation energies of the defects at the interface. For some processes the valence-band offset results drives charge transfer across the interface; for example, a Zn vacancy in ZnS extracts electrons from the CZTS valence band. We find mechanism involving V_{Cu} in CZTS are also energetically favorable, which is of significance due to the importance of these defects in p-type doping of the absorber layer.
Keywords: CZTS, Defects, Interfaces, Modelling, PV Materials

1 INTRODUCTION

Cu_2ZnSnS_4 (CZTS) is considered a promising photovoltaic (PV) material, since it is non-toxic, has a direct band gap of about 1.5 eV, is low-cost, and is made up from Earth-abundant elements [1, 2]. As-grown CZTS shows p-type conductivity, most probably due to the shallow acceptor levels of the intrinsic defects, Cu_{Zn} and V_{Cu} [3, 4]. This semiconductor is favorable for photovoltaic devices since it has a high optical absorption coefficient of more than 10^4cm^{-1} in the visible wavelength region [5], although a recently reported efficiency of 8.5% remains inefficient compared to the other materials such as CIGS with reported efficiencies of over 20% [6]. Optimizing the CZTS is includes a focus upon crystallinity, secondary phases, segregation, and inter-diffusion at device interfaces.

CZTS is thermodynamically stable in a relatively narrow region of atomic chemical potentials [4, 7], and secondary phases such as ZnS, $CuSnS_3$ and Cu_2S might become significant contaminants under some growth conditions. However, efficient devices are obtained for Cu-poor/Zn-rich compositions [8, 9], where intrinsic defects (Cu_{Zn} and V_{Cu}) are responsible for the p-type conductivity [8, 9], and then ZnS, a semiconductor with a band gap of 3.6eV [10], is the main competing secondary phase. Secondary phases [11] affect the performance of photovoltaic devices by reducing the volume of absorber layer and decreasing the short circuit current. In addition, lattice misfit between CZTS and the secondary phase has an impact on the PV performance, but the lattice mismatch with ZnS is about 0.5%, so the minority carrier recombination at the interface is low [12].

Typical p-type CZTS solar cells are consecutively deposited layers [13], such as a Mo/CZTS/CdS/ZnO, with ZnS forming as a secondary phase precipitate within CZTS absorber layer. The ZnS is generally most commonly found as a secondary phase close to the interfaces with the contacts and buffer layers [8, 14], but it is also used as an alternative buffer layer to replace CdS [15, 17] due to the toxicity of cadmium. Therefore, the hetero-structure interface between CZTS and the ZnS is of great interest for CZTS cell performance.

The band alignment between CZTS and either ZnS as a precipitation phase or a buffer layer is critical for the conversion efficiency [11, 18]. The valence-band offset is affected by the composition and abruptness of the interface, which is in turn affected by diffusion processes between neighboring materials [19]. For example, the diffusion of Cd and In from the buffer layer is thought to readily diffuse into CZTS, whereas of Zn from a ZnS or ZnO based buffer does not [20]. Conversely, Cu does not seem to diffuse out of CZTS into MoS_2 [21].

Since V_{Cu} and Zn_{Cu} are responsible for the p-type conductivity in CZTS, such defects formed as a result of the intermixing between the buffer layer (or the secondary phase) and the absorber layer are critical for the efficiency of the device. In this paper, the CZTS/ZnS interface is modeled to identify the most stable defects adjacent to the interface, and the energetics by which metal vacancies move between phases.

2 COMPUTATIONAL DETAILS

We performed generalized-gradient approximation [22] density-functional simulations as implemented in AIMPRO [23, 24], with norm-conserving, separable pseudo potentials [25] ($3d^{10}s^1$, $3d^{10}4s^2$, $5s^25p^2$ and $3s^23p^4$ valence sets for Cu, Zn, Sn and S, respectively). The bases consist of independent sets of *s*-, *p*- and *d*-type Gaussian functions of four widths, being 40 functions per atom. Hamiltonian matrix elements are determined using a plane-wave expansion of the charge density and Kohn–Sham potential [26] with a cut-off of 250Ha, leading to well converged total energies with respect to this parameter. The Brillouin-zone is sampled using the Monkhorst-Pack scheme [27]. Structures are deemed optimized when the forces are $<10^{-3}$ a.u. and the structural optimization step reduces the energy by $<10^{-5}$ a.u. Diffusion barriers are obtained via 7-image climbing-nudged-elastic-band calculations [28, 29].

The formation energy of the defect at the interface is

$$E^f(V_y^x) = E^{\text{tot}}(V_y^x) - E^{\text{tot}}(\text{interface}) + \mu_y \qquad (1)$$

where $E^{\text{tot}}(V_y^x)$ and $E^{\text{tot}}(\text{interface})$ are the total energies of the interface with and without a vacancy on site y (Cu, Zn or Sn), x is CZTS or ZnS, and μ_y is the chemical potential of species y. This approach means that the strain energy in the ZnS is not included in the formation energy of the point defect.

3 RESULTS

3.1 Bulk crystal and interface models

It is assumed that the ZnS is a minor component of the composite, and therefore the lattice constants of CZTS are taken to be equilibrium and the ZnS is strained to match. Kesterite CZTS is modeled using a tetragonal, 16-atom cell with the tetragonal axis along [001], and strained ZnS is modeled using a conventional zinc-blende unit cell. All bulk crystals have their Brillouin-zones sampled to converge the energy to better than 10meV/atom. The calculated lattice constants are listed in Table I.

Table I: Calculated lattice parameters of the bulk and strained phases, and the interface supercell. The experimental lattice constants are $a_{CZTS} = 5.426$Å, $c_{CZTS} = 10.81$Å and $a_{ZnS} = 5.45$Å, respectively [30, 31].

	Lattice constant (Å)		
Material	[100]	[010]	[001]
CZTS	5.497	5.497	11.000
ZnS	5.472	5.472	5.472
Strained ZnS	5.435	5.497	5.500
CZTS/ZnS interface	32.832	5.497	11.000

Strained ZnS has the lattice constants fixed to be a_{CZTS} in the [010] direction and $c_{CZTS}/2$ in the [001] direction, both representing a 0.5% strain (Table I). The lattice constant in the [100] direction was obtained by minimizing the energy with the [001] and [010] lattice constants fixed, and represents a −0.7% strain.

In the epitaxial ZnS(100)/CZTS(100) interface is modeled using an orthorhombic, 192-atom supercell made up from three lattice constants of each of CZTS and strained ZnS in the [100] direction, as illustrated in Fig. 1. The Brillouin-zone is sampling using a 1×4×4 mesh. The [100] lattice constant (Table I) is obtained minimizing the energy with the [010] and [001] lattice constants fixed. The supercell [100] lattice constant differs form $3a_{CZTS} + 3a_{ZnS}^{strained}$ only very slightly, the difference arising principally from the detailed differences where the two materials meet.

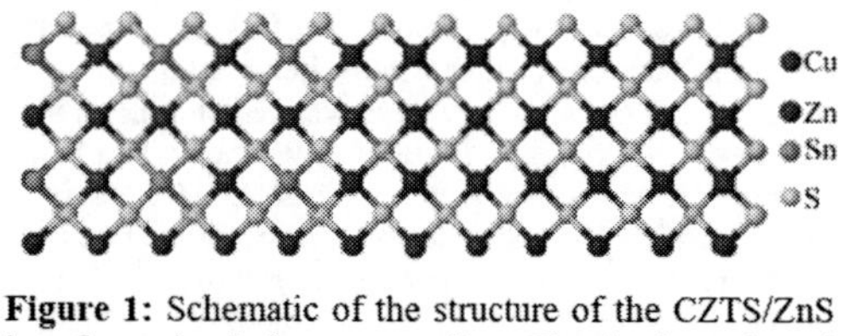

Figure 1: Schematic of the structure of the CZTS/ZnS interface simulation supercell. The horizontal and vertical directions are [100] and [010], respectively, with [001] out of the paper.

Our estimate of the lattice mismatch is in good agreement with the previous work [12] that also obtained that a lattice mismatch of 0.5%.

3.2 Band offset

The valence band offset (VBO) plays a key role in the conversion efficiency of photovoltaic devices. The VBO for the (100) interface has been calculated according to the method outlined in Ref. [32]. Our results, in agreement with the literature [11], indicate that the band offset is of type one with a VBO of 1.05eV (c.f. 0.92eV in Ref.[11]).

However, the band offset alignment of ZnS/CZTS hetero-structure shows that there is a large spike for the valance band. Such a large spike indicating [8, 18] that the ZnS secondary phase in p-type CZTS could be regarded as a resistive barrier to the minority carrier electrons. Therefore, the ZnS could hinder the flow of electrons. Thus, reducing the photo-current (electrons) and blocking the majority carriers' (holes) flow will increase the series resistance. Precipitated ZnS phase [11] consider as an active insulator phase within CZTS but it not acts as a recombination center or carrier trap.

3.3 Inter-diffusion mechanisms

Inter-diffusion of the metal species across the interface impacts p-type conductivity, and is therefore of importance to the device efficiency. A key mode of diffusion occurs when vacant metal sites move through the lattices and across the interface, each diffusion hop being the result of a metal ion moving between nearest sites on the cation sub-lattice. Where there is a vacancy inside CZTS located adjacent to the interface, a Zn ion from the ZnS may migrate from the ZnS into the CZTS, possibly forming misplaced metal ions in the CZTS, i.e. Zn_{Cu} or Zn_{Sn}. Similarly, where the vacancy is initially in the ZnS, metal ions may diffuse out of the CZTS, with the potential for the formation of Cu_{Zn} or Sn_{Zn} substitutions in ZnS.

The two classes of processes are as follows.

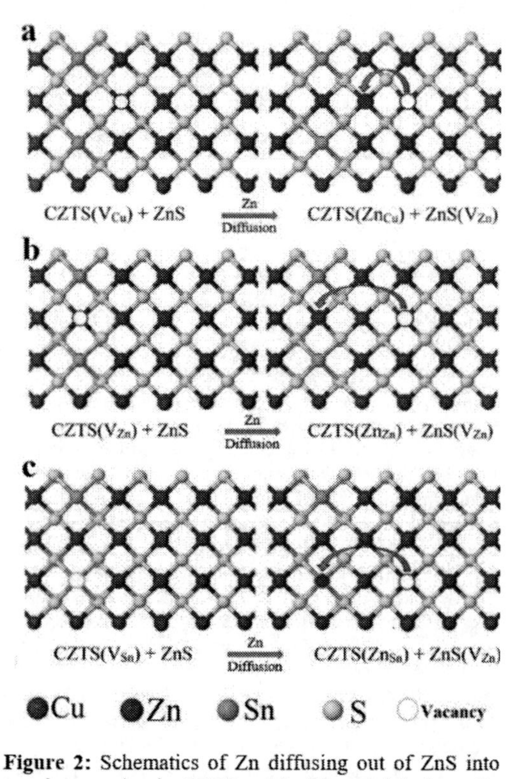

Figure 2: Schematics of Zn diffusing out of ZnS into metal vacancies in CZTS. (a), (b) and (c) show Zn replacing Cu, Zn and Sn, respectively. Colors are as in Fig. 1.

3.3.1 Diffusion mechanism 1: CZTS cation vacancy

In this mechanism, Zn diffuses from ZnS into a metal vacancy in CZTS, leaving V_{Zn} in ZnS, as shown schematically in Fig. 2. The final product of the reaction is Zn_{Cu}, Zn_{Zn} or Zn_{Sn}, in the CZTS, and V_{Zn} inside ZnS. Each case has its own implications regarding the creation and movement of acceptors and donors, and charge transfer across the interface.

First, V_{Cu} in CZTS converting to V_{Zn} in ZnS and Zn_{Cu} in CZTS, Fig. 2(a), is initially a single acceptor, ending as a donor—double-acceptor complex. There is electrostatic cost to separate the charges across the interface, which might be offset due to a combination of the difference in the donor and acceptor levels and the VBO.

Secondly, moving V_{Zn} from CZTS to ZnS (Fig. 2(b)), superficially might be expected to result in a double-acceptor in both cases. However, the VBO means places the double-acceptor level of V_{Zn} in ZnS below the valence band maximum in the neighboring CZTS, and that charge transfer between the point defect in ZnS and the valence band in CZTS might take place.

Thirdly, diffusion of V_{Sn}, ostensibly a quadruple acceptor, results in Zn_{Sn} in CZTS and V_{Zn} in ZnS, Fig. 2(c). Both Zn_{Sn} in CZTS and V_{Zn} in ZnS are double acceptors, so one would not expect any compensation between them, but as with the previous case, one may anticipate charge transfer from the CZTS valence band into the V_{Zn} acceptor levels.

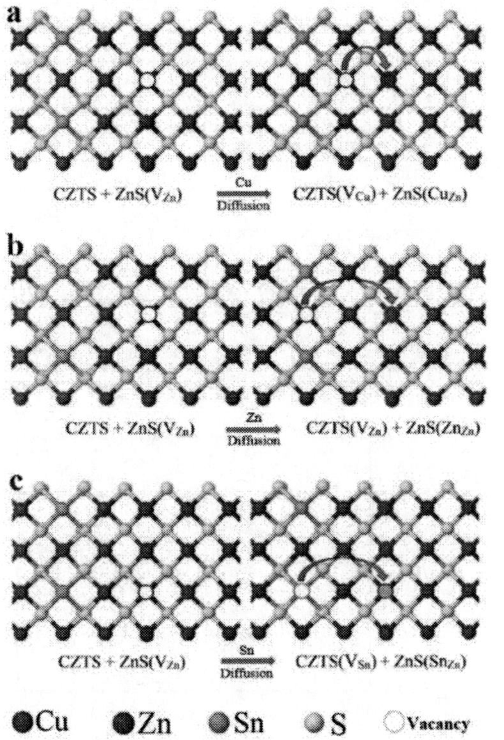

Figure 3: Schematics of V_{Zn} diffusing out of ZnS forming metal vacancies in CZTS. (a), (b) and (c) show Zn being replaced by Cu, Zn and Sn, respectively. Colors are as in Fig. 1.

3.3.2 Diffusion mechanism 2: ZnS cation vacancy

In the second mechanism, there V_{Zn} initially lies inside ZnS, and Cu, Zn, or Sn ions then diffuse out of CZTS across the junction, leaving a cation vacancy behind, as shown in Fig. 3. As with the mechanisms indicated in section 3.3.1, there are nominally three basic diffusion pathways depending upon which of the three metal elements in CZTS are involved, but removal of a Zn atom from CZTS is in practice the reverse of process b in Fig. 2(b). In all cases, the initial state is a double acceptor with the possibility of charge transfer from the CZTS valence band.

First, Cu can diffuse out of the CZTS to substitute for Zn in ZnS (Fig. 3(a)). The products are both expected to be single acceptors, but Cu_{Zn} in ZnS may capture a valence-band electron from CZTS.

Noting that the process involving Zn shown in Fig. 3(b) is the reverse of Fig. 2(b), the other distinct process involves the movement of Sn into ZnS. The products are the double donor Sn_{Zn} in ZnS and the quadruple acceptor V_{Sn} in CZTS (Fig. 3(c)). Here, in contrast to all other cases, it is expected that there will be a net flow of electrons from ZnS to CZTS.

3.4 Formation energies

To identify the most stable defect form and inter-diffusion processes at the hetero-epitaxial interface, both diffusion barriers and the formation energies are required. The formation energies depend upon the atomic chemical potentials; the atomic chemical potentials are linked through the requirement that CZTS is the equilibrium phase, and for this study we have constrained the analysis to Cu-poor/Zn-rich conditions, with $\mu_{Cu} = -0.2\text{eV}$ [7]. Even given this constraint, the values of the remaining species represent an area in the phase diagram (Fig. 4).

Table II: The calculated reference formation energy (eV) of the cation vacancies in the CZTS or ZnS adjacent to the interface for the points P, Q, M, and N shown in Fig. 4.

E^f	P	Q	M	N
$V_{Cu}^{CZTS} \leftrightarrow Zn_{Cu}^{CZTS} + V_{Zn}^{ZnS}$	1.08	1.08	1.08	1.08
$V_{Zn}^{CZTS} \leftrightarrow V_{Zn}^{ZnS}$	1.09	0.95	0.91	1.03
$V_{Sn}^{CZTS} \leftrightarrow Zn_{Sn}^{CZTS} + V_{Zn}^{ZnS}$	2.71	2.15	2.20	2.39
$V_{Zn}^{ZnS} \leftrightarrow Cu_{Zn}^{ZnS} + V_{Cu}^{CZTS}$	1.64	1.51	1.47	1.59
$V_{Zn}^{ZnS} \leftrightarrow Sn_{Zn}^{ZnS} + V_{Sn}^{CZTS}$	1.64	1.51	1.47	1.59

The secondary phases considered in this study are the same as those in Refs. [7, 9].

The formation energies of the cation vacancies in bulk CZTS and bulk, strained ZnS have been estimated using equation 1, and the values are listed in Table II. These energies are based upon supercell calculations using 64-atom, quasi-cubic cells with lattice vectors [200] a_{CZTS}, [020] a_{CZTS} and [001] c_{CZTS}.

Based on the data in Table II, the formation energy to extract the cation from CZTS is lower than that to extract the Zn from ZnS-strained. The cation vacancy inside CZTS is more stable and it can form easily than the V_{Zn} in ZnS-strained. Therefore, in the CZTS/ZnS heterojunction cell, Zn extraction of CZTS is easier than extracting it from ZnS-strained. It is concluded that the single acceptor V_{Cu} has lowest formation energy compared to other defects, it is the dominant defect in the cell.

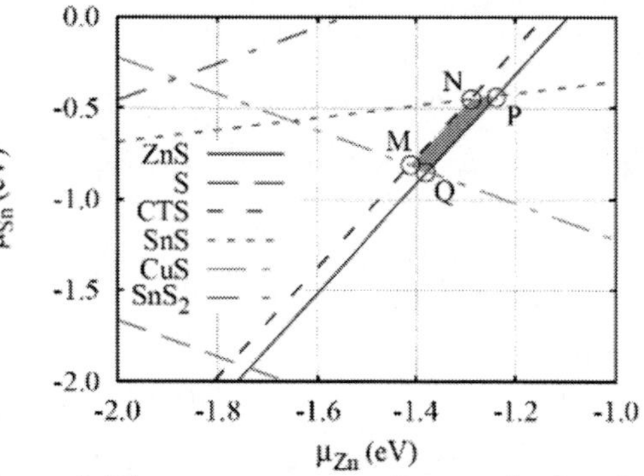

Figure 4: Phase space μ_{Zn}—μ_{Sn} diagram showing the CZTS stable zone for μ_{Cu}=−0.2eV. PQ shows the CZTS/ZnS boundary.

In CZTS [33], Cu-S bonds can be considered as weakly covalent, with Zn-S bonds being more ionic, and a little stronger, and Sn-S ionic bonds stronger still. The bonding and the Cu-poor conditions account for the low formation energy of V_{Cu} and the relatively high formation energy of V_{Sn}, so that it is expected that the Cu vacancy is the dominant cation vacancy in CZTS [33].

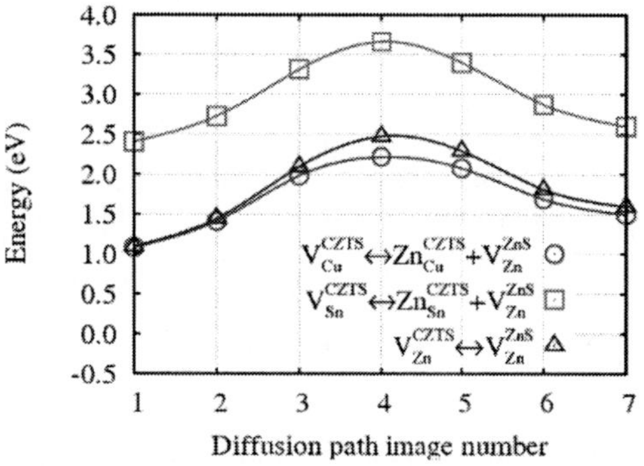

Figure 5: Diffusion barriers calculated for the diffusion of Zn out of ZnS into cation vacancies in CZTS. The energy scale includes the reference formation energy of the cation vacancies in the CZTS adjacent to the interface for the points P in Fig. 4.

3.5 Diffusion barriers

The diffusion of cation vacancies across the interface plays an important role in determining the efficiency of the cell through their impact upon conductivity via both doping and deep carrier traps.

3.5.1 CZTS cation vacancy mechanisms

In the first mechanism, as described in section 3.3.1, initially a cation vacancy lies inside the CZTS and a Zn ion diffuses from the ZnS. The calculated barriers are illustrated in Fig. 5, where the initial energy is based upon the reference formation energy of the cation vacancies in the CZTS adjacent to the interface for the point P in Fig. 4. A summary of the barriers is provided in Table III for the four boundary points labeled in Fig. 4 for the CZTS stable region.

For example, at the point labeled P, the activation energy value of the saddle points with the formation energy of reference defects when there is a V_{Cu}, V_{Zn} and V_{Sn} in the initial state of the CZTS is 2.22, 2.47 and 3.65eV, respectively. However, in all cases the energy is lower for the cation vacancy lying in CZTS than in ZnS, as can be seen in the lower value for all reverse reactions. This is consistent with the inference that in all cases there is a propensity to transfer charge across the interface when the cation vacancy lies in ZnS, which has an associated electrostatic energy cost. In addition, there is a chemical term in the energy difference, associated with the differences in bond-strengths between the metal ions and sulfur.

For the Q, M and N points, the difference between the activation energy relative to P are also listed in Table III. It is concluded that the diffusion barrier of the four points qualitatively have the same behavior for all paths. For the current case of ZnS as a potential secondary phase, it is the boundary line between points P and Q that are arguably the most important. Under these conditions, it is clear in equilibrium conditions the most significant migration path involves the copper vacancy,

Table III: Calculated activation energies including the reference formation energy (eV) of the cation vacancies in the CZTS or ZnS adjacent to the interface for the points P (Fig. 4, Table II), E_P^a. E^{for} and E^{rev} are the forward and reverse migration barriers. For points Q, M and N (Fig. 4), the relative position of the activation energies are lower by ΔE_Q^a, ΔE_M^a and ΔE_N^a, respectively. All energies are in eV.

Reaction	E_P^a	E^{for}	E^{rev}
$V_{Cu}^{CZTS} \leftrightarrow Zn_{Cu}^{CZTS} + V_{Zn}^{ZnS}$	2.22	1.13	0.72
$V_{Zn}^{CZTS} \leftrightarrow V_{Zn}^{ZnS}$	2.47	1.38	0.88
$V_{Sn}^{CZTS} \leftrightarrow Zn_{Sn}^{CZTS} + V_{Zn}^{ZnS}$	3.65	1.25	1.06
$V_{Zn}^{ZnS} \leftrightarrow Cu_{Zn}^{ZnS} + V_{Cu}^{CZTS}$	2.55	0.90	0.64
$V_{Zn}^{ZnS} \leftrightarrow Sn_{Zn}^{ZnS} + V_{Sn}^{CZTS}$	2.93	1.29	1.02
	ΔE_Q^a	ΔE_M^a	ΔE_N^a
$V_{Cu}^{CZTS} \leftrightarrow Zn_{Cu}^{CZTS} + V_{Zn}^{ZnS}$	0.00	0.00	0.00
$V_{Zn}^{CZTS} \leftrightarrow V_{Zn}^{ZnS}$	0.14	0.18	0.06
$V_{Sn}^{CZTS} \leftrightarrow Zn_{Sn}^{CZTS} + V_{Zn}^{ZnS}$	0.41	0.37	0.02
$V_{Zn}^{ZnS} \leftrightarrow Cu_{Zn}^{ZnS} + V_{Cu}^{CZTS}$	0.14	0.19	0.06
$V_{Zn}^{ZnS} \leftrightarrow Sn_{Zn}^{ZnS} + V_{Sn}^{CZTS}$	0.14	0.18	0.06

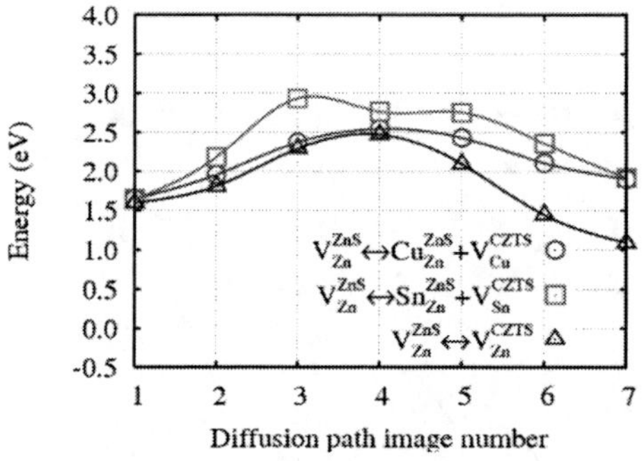

Figure 6: Diffusion barriers calculated for the diffusion of V_{Zn} out of ZnS onto cation sites in CZTS. The energy scale includes the reference formation energy of the cation vacancies in the ZnS adjacent to the interface for the points P in Fig. 4.

although the zinc vacancy is close by in energy.

3.5.2 ZnS cation vacancy mechanisms

In the second mechanism type, described in section 3.3.2, V_{Zn} diffuses out of ZnS. As with the CZTS cation vacancy mechanisms, the energetics are listed in Table III. The barriers calculated for the reference point P (Fig. 4) are shown in Fig. 6. Mirroring the previous cases, the reverse reactions barriers are lower than forward for two of the three reactions that result in impurities (Cu or Sn) being incorporated into the ZnS lattice, with the energy cost including an electrostatic term from charge transferred across the interface, as well as a bond-strength term. However, moving V_{Zn} to form V_{Zn} in CZTS is exothermic, meaning that pre-existing V_{Zn} in ZnS are highly likely to diffuse into the CZTS.

4 CONCLUSIONS

Our calculations show that at a (100)-CZTS/ZnS heterojunction, the energy required to extract a cation from CZTS is lower than that to extract the Zn from the (strained) ZnS. For CZTS under the equilibrium growth conditions considered for this study, extraction of Cu from CZTS is easier than Zn, with Sn extraction being quite unfavorable. This is in line with the p-type doping in as-grown CZTS under these conditions.

The role of charge transfer across the interface is of great significance. In all cases, processes proceed more easily in the direction that reduces or eliminates the transfer of electrons between point defects on either side of the interface, or between a point defect on the ZnS side and the CZTS valence band. However, additionally cation vacancies trapped within the ZnS secondary phase will generate holes in the near-by CZTS, enhancing the p-type doping of the absorber layer.

Finally, we note that under the given conditions the relatively low formation energy of V_{Cu}, and the energetic unfavorability of the migration into the ZnS, means that the secondary phase or capping layer will act as an effective diffusion barrier to these p-type doping centers. This in turn means that the presence of ZnS secondary phases within the absorber layer will affect the diffusivity of copper vacancies, and therefore the operation of PV devices.

6 REFERENCES

[1] Y.-W. Cheng, F.-L. Tang, H.-T. Xue, H.-X. Liu, B. Gao, and Y.-D. Feng, Journal of Physics D: Applied Physics, vol. 49, p. 285107, 2016.

[2] S. A. Phaltane, S. A. Vanalakar, T. S. Bhat, P. S. Patil, S. D. Sartale, and L. D. Kadam, Journal of Materials Science: Materials in Electronics, vol. 28, pp. 8186-8191, 2017.

[3] A. Walsh, S. Chen, S. H. Wei, and X. G. Gong, Advanced Energy Materials, vol. 2, pp. 400-409, 2012.

[4] S. Chen, J.-H. Yang, X. G. Gong, A. Walsh, and S.-H. Wei, Physical Review B, vol. 81, p. 245204, 2010.

[5] S. A. Khalate, R. S. Kate, J. H. Kim, S. M. Pawar, and R. J. Deokate, Superlattices and Microstructures, vol. 103, pp. 335-342, 2017.

[6] M. A. Green, K. Emery, Progress in photovoltaics: research and applications, vol. 23, pp. 1-9, 2015.

[7] S. Chen, A. Walsh, X. G. Gong, and S. H. Wei, Adv Mater, vol. 25, pp. 1522-39, Mar 20 2013.

[8] W. Li, J. Chen, C. Yan, and X. Hao, "The effect of ZnS segregation on Zn-rich CZTS thin film solar cells, Journal of Alloys and Compounds, vol. 632, pp. 178-184, 2015.

[9] P. Sarker, M. M. Al-Jassim, and M. N. Huda, Journal of Applied Physics, vol. 117, p. 035702, Jan 21 2015.

[10] D. A. R. Barkhouse, R. Haight, N. Sakai, H. Hiroi, H. Sugimoto, and D. B. Mitzi, Applied Physics Letters, vol. 100, p. 193904, 2012.

[11] B. G. Mendis, M. C. J. Goodman, J. D. Major, A. A. Taylor, K. Durose, and D. P. Halliday, Journal of applied physics, vol. 112, p. 124508, 2012.

[12] A. Nagoya, R. Asahi, and G. Kresse, Journal of Physics: Condensed Matter, vol. 23, p. 404203, 2011.

[13] O. Vigil-Galán, M. Courel, M. Espindola-Rodriguez, V. Izquierdo-Roca, E. Saucedo, and A. Fairbrother, Journal of Renewable and Sustainable Energy, vol. 5, p. 053137, 2013.

[14] D. Nam, S. Cho, J.-H. Sim, K.-J. Yang, D.-H. Son, D.-H. Kim, et al., Energy Materials and Solar Cells, vol. 149, pp. 226-231, 2016.

[15] M. I. Hossain, P. Chelvanathan, M. M. Alam, M. Akhtaruzzaman, K. Sopian, and N. Amin, in Clean Energy and Technology (CEAT), 2013 IEEE Conference on, 2013, pp. 450-454.

[16] C. Yan, F. Liu, N. Song, B. K. Ng, J. A. Stride, A. Tadich, et al., Applied Physics Letters, vol. 104, p. 173901, 2014.

[17] J. Kim, C. Park, S. M. Pawar, A. I. Inamdar, Y. Jo, J. Han, et al., Thin Solid Films, vol. 566, pp. 88-92, 2014.

[18] W. Bao and M. Ichimura, International Journal of Photoenergy, vol. 2015, 2015.

[19] P. Erhart and K. Albe, Physical Review B, vol. 73, p. 115207, 2006.

[20] T. Kato, H. Hiroi, N. Sakai, S. Muraoka, and H. Sugimoto, in 27th European photovoltaic solar energy conference and exhibition, 2012, p. 2236.

[21] J. Timo Wätjen, J. J. Scragg, M. Edoff, S. Rubino, and C. Platzer-Björkman, Applied Physics Letters, vol. 102, p. 051902, 2013.

[22] J. P. Perdew, K. Burke, and M. Ernzerhof, "Generalized gradient approximation made simple," Physical review letters, vol. 77, p. 3865, 1996.

[23] P. R. Briddon and R. Jones, physica status solidi (b), vol. 217, pp. 131-171, 2000.

[24] M. J. Rayson and P. R. Briddon, Computer Physics Communications, vol. 178, pp. 128-134, 2008.

[25] C. Hartwigsen, S. Goedecker, and J. Hutter, Physical Review B, vol. 58, pp. 3641-3662, Aug 15 1998.

[26] M. J. Rayson and P. R. Briddon, Physical Review B, vol. 80, p. 205104, 2009.

[27] H. J. Monkhorst and J. D. Pack, Physical Review B, vol. 13, pp. 5188-5192, 1976.

[28] G. Henkelman, B. P. Uberuaga, and H. Jónsson, The Journal of chemical physics, vol. 113, pp. 9901-9904, 2000.
[29] G. Henkelman and H. Jónsson, The Journal of chemical physics, vol. 113, pp. 9978-9985, 2000.
[30] H. Katagiri, N. Sasaguchi, S. Hando, S. Hoshino, J. Ohashi, and T. Yokota, Solar Energy Materials and Solar Cells, vol. 49, pp. 407-414, 1997.
[31] K. Wright and J. D. Gale, Physical Review B, vol. 70, p. 035211, Jul 2004.
[32] C. G. Van de Walle and R. M. Martin, Physical Review B, vol. 34, p. 5621, 1986.
[33] T. Maeda, S. Nakamura, and T. Wada, Japanese Journal of Applied Physics, vol. 51, p. 10NC11, Oct 2012.

$(Ag_xCu_{1-x})_2ZnSnS_4$ THIN-FILMS PREPARED BY SPRAY PYROLYSIS

L. Dermenji[1], G. Gurieva[2], M. Guc[1], S. Levcenko[2], N. Curmei[1], L. Bruc[1,*], D.A. Sherban[1], A.V. Simashkevich[1], S. Schorr[2,3], E. Arushanov[1]
[1]Institute of Applied Physics, str. Academiei 5, MD-2028 Chisinau, Republic of Moldova
[2]Helmholtz-Zentrum Berlin für Materialien und Energie, Hahn-Meitner-Platz 1, D-14109 Berlin, Germany
[3]Institute of Geological Sciences, Free University Berlin, Malteserstr. 74-100, 12249 Berlin, Germany
*E-mail: leonid.bruc@phys.asm.md

ABSTRACT: One of the most detrimental problems in the further development of thin film solar cells based on kesterite type compound semiconductors is the limitation in open circuit voltage (V_{OC}). The latter, according to many theoretical and experimental studies, is mostly related to the high concentration of intrinsic defects, mainly with Zn_{Cu} antisites. Recently, a way for overcoming this problem by partial substitution of Cu cations by Ag was proposed. This may lead to a strong decrease of intrinsic defects due to higher formation energy of Zn_{Ag} defects and as result to increase of V_{OC}. Taking this into account, we performed an investigation of $(Ag_xCu_{1-x})_2ZnSnS_4$ thin films deposited by the spray pyrolysis method. The as-deposited thin films with 10, 15 and 20 % of Ag were annealed at 450 °C for 60 min in presence of elemental sulfur. Structural investigations by XRD, as well as Raman spectroscopy studies confirmed the formation of solid solutions. Photoluminescence investigations showed one broad band, which exhibits the blue shift with the increase of Ag concentration. This could be explained by changes in the activation energy of the defect levels involved in radiative transition rather than with the band gap change.
Keywords: Kesterite, Thin Film, Spray Pyrolysis, XRD, Raman Spectroscopy

1 INTRODUCTION

In the past few years $Cu_2ZnSn(S,Se)_4$ (CZTSSe) have been considered as a promising material for the development of cheap and effective thin-film solar cells. In favor of this was the fact that the composition consists not scarce and not toxic elements. The maximum energy conversion factor obtained on solar cells with CZTSSe as an absorber showed 12.6 % in the laboratory [1]. However, further improvements have stopped due to the low open circuit voltage (V_{OC}). The latter, according to many theoretical and experimental studies, is mostly related to the high concentration of intrinsic defects, leading to a low minority carrier lifetime, electrostatic potential fluctuations, tail states etc. [2-4]. The main defect is related with Zn_{Cu} antisites, which leads to creation of the defect acceptor band inside the band gap [5]. Predominance of this defect is provided by the low formation energy, which could be explained by the small difference in the sizes of atoms and chemical properties of Cu and Zn.

Ag was proposed as a promising candidate for substitution of Cu, since it belongs to the same chemical group as copper, but its radius is 16 % larger, which leads to an increase in the formation energy of Zn_{Ag} disorder defects in the cell [6-10]. However, pure $Ag_2ZnSn(S,Se)_4$ (AZTSSe) is a semiconductor with n-type conductivity, which strongly limits its utilization as absorber in thin film solar cell [11]. Therefore, it is necessary to consider the transition series of solid solutions from CZTSSe to AZTSSe in order to overcome the difficulties arising in pure CZTSSe and AZTSSe.

The present study is the first attempt to obtain thin films of $(Ag_xCu_{1-x})_2ZnSnS_4$ (ACZTS) solid solutions by the non-vacuum spray pyrolysis method. In this way the positive influence of Ag on the intrinsic defects of kesterite type compounds was combined with one of the easiest and cheapest deposition methods. Taking this into account, the main purpose of this work was to obtain ACZTS thin films by spray pyrolysis, and to investigate their structural and optical properties for future photovoltaic application. The samples were prepared by spray pyrolysis of an aqueous solution onto soda-lime glass (SLG). The main characteristics of the samples were obtained from the X-ray diffraction (XRD), Raman spectroscopy and photoluminescence (PL) analysis.

2 EXPERIMENTAL DETAILS

The deposition was performed using self-made spray pyrolysis system placed in the glovebox filled with CO_2. The XRD data were collected using a PANalytical X'pertPro MPD diffractometer equipped with CuKα radiation (λ = 1.54056 Å). Detector scans were obtained in Bragg-Brentano geometry, with the step size of 0.02°.

The Raman spectra were obtained with an HR800 Horiba Jobin Yvon spectrometer, matched with a metallographic microscope and a CCD detector. Measurements in backscattering configuration were done using He-Ne laser as excitation source (laser line 632.8 nm). The Raman shift of all measured spectra was calibrated using intercalated spectra obtained from the Si monocrystal reference sample and imposing its main band position at 520 cm^{-1}.

The photoluminescence spectra were measured at room temperature by a 0.5m grating spectrometer coupled with InGaAs detector and excited with a 660 nm laser line with excitation energy about 0.4 kW/cm^2.

3 RESULTS AND DISCUSSIONS

3.1 Solution preparation and deposition

The solution for the deposition of ACZTS thin films was prepared from two precursor solutions. The first initial solution (IS1) consisted of Cu- and Ag- containing complexes, while the second (IS2) one of Zn- and Sn-containing complexes. This separation to IS1 and IS2 is caused by their high stability and homogeneity, these precursor solutions could be stored for a long time (up to 14 days) without being influenced by possible temperature fluctuations. With temperature decrease precipitated crystals could be formed, however they are easily dissolved with a temperature increase.

The first precursor solution IS1 was prepared as described earlier in [12], the only adjustment to the previous routine being the replacement of a part of CuCl by AgCl to ensure the formation of solid solutions. As the result 25 ml of solution IS1 containing water (H_2O), thiourea ($CS(NH_2)_2$ – Th), CuCl and AgCl was obtained. It is worse to be mentioned that chlorides with Cu+ cations were used for the IS1 to ensure the formation of the $[Cu^+(Th)_n]^+$ complex with n > 3, which improves the homogeneity of the solution. While the use of chlorides with Cu^{2+} cations, together with $[Cu^+(Th)n]^+$ complex formation leads to the formation of an additional $1/2(Th)_2^{2+}$(Formamidine disulfide) complex, which dissolves in time with precipitation and has a strong influence on the sulfur concentration in the final solution. Details of the IS2 preparation were described elsewhere [12].

The final solution was prepared just before the deposition process in proportion of: IS1:IS2 = 3:1, which provides a stoichiometric composition of the precursors. The final solution has the concentration of 0.33 M.

The deposition of the final solution was performed in air on the SLG substrate heated up to 420 – 430 °C with a spraying rate ~ 5 ml/min. As result a series of three thin films with 10, 15 and 20 % of Ag was obtained. These concentrations were aimed, based on the previous analysis of the band gap of the ACZTS solid solutions, which was found to decrease slightly for thin films with 20 % of Ag and increase up to 2 eV for the pure Ag_2ZnSnS_4 compound [13]. Finally, to increase the crystalline quality, the as-deposited samples were annealed in vacuum at 450 °C for 60 min in presence of elemental sulfur to prevent it loss.

3.2 Thin films characterization

In order to analyse the quality of obtained thin films X-ray diffraction (XRD), Raman spectroscopy and photoluminescence measurements were applied. In the obtained diffraction patterns (see Figure 1),the most intense Bragg peaks can be attributed to kesterite type structure (space group $I\bar{4}$) [14], wherein the peaks positions are shifted to lower 2θ angles which is due to the incorporation of Ag in the crystal structure forming an ACZTS solid solution. It is important to emphasize that the shift is higher for samples with higher initial Ag concentration, which supports the possibility of control of the Ag/Cu ratio by changing the IS1 composition. However, it should be mentioned that small amounts of secondary phases ($Cu_{2-x}S$, SnS and S) were also found in the samples, while a relatively low signal to noise ratio and wide peaks denotes the relatively low crystalline quality of the thin films. This could be related to the fact that the conditions of sample annealing were selected similar to that previously used for standard Cu_2ZnSnS_4 (CZTS) thin films, while for the ACZTS solid solutions other conditions could be optimal.

The most intense peak in the Raman spectra (Figure 2) at 338 cm^{-1} has a similar position with the main Raman peak of a CZTS kesterite type compound [15]. Most probably this similarity is related with nature of this peak – vibrations of the anion atoms [15], which are almost independent from the Cu/Ag substitution. However, it was found that the peaks at 300 and 368 cm^{-1} increase with an increasing Ag concentrationAccording to the analysis of Raman spectra of CZTS compounds, in these regions there are the E and B symmetry modes [15], which could be related with vibrations of all atoms, including Cu cations. Partial substitution of copper cations to silver ones could be a reason for the observed changes of intensity of the peaks at 300 and 368 cm^{-1}. It should be noted that the additional peaks at 221 and 472 cm^{-1} could be attributed to SnS and $Cu_{2-x}S$ phases, respectively, also found from XRD analysis.

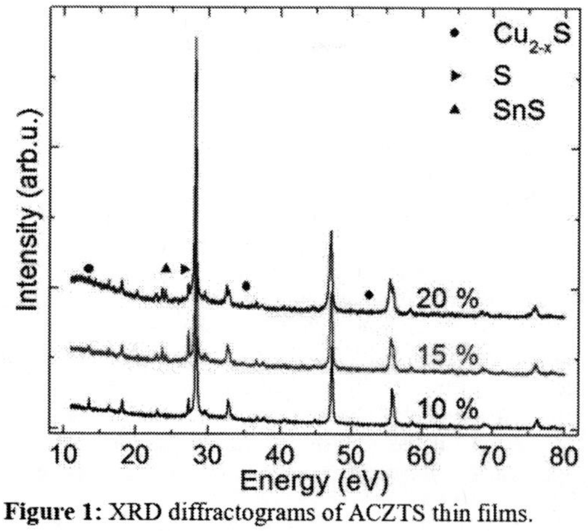

Figure 1: XRD diffractograms of ACZTS thin films.

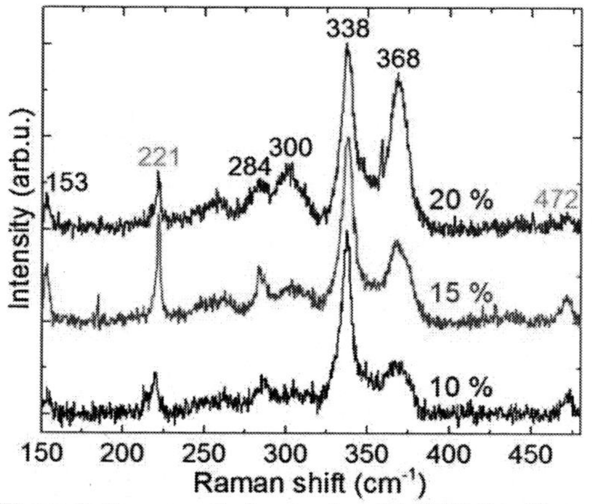

Figure 2: Raman scattering spectra of ACZTS thin films. With orange are marked peaks attributed to the secondary phases.

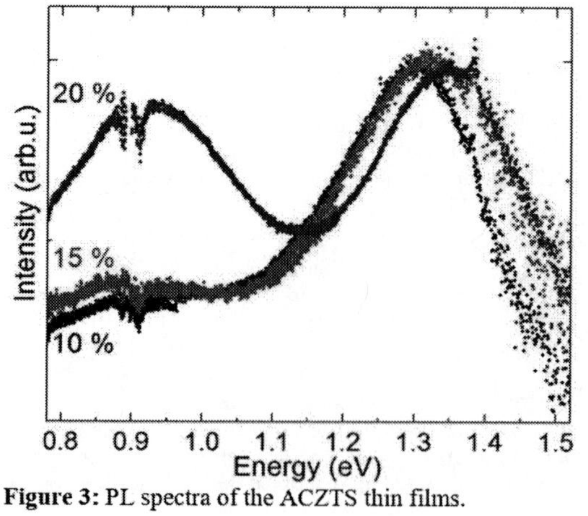

Figure 3: PL spectra of the ACZTS thin films.

In the PL spectra (Figure 3), measured at room temperature a broad band with maximum shifting from 1.30 to 1.34 eV with an Ag concentration increase was observed. Due to the slight band gap decrease in the samples with up to 20 % of Ag [13], most probably this shift is related to a change of the activation energy of defect levels involved in the radiative recombination. At lower energies an additional PL band at about 0.9 eV was also detected, which is probably related to the secondary phases found from the XRD and Raman spectroscopy studies.

4 CONCLUSIONS

From the presented characterization of the ACZTS solid solutions deposited by spray pyrolysis it could be concluded, that this method could be successfully used for the preparation of thin films with a control of Ag concentration. However, additional process optimizations should be performed in order to achieve higher crystalline quality of the thin films as well as to avoid the formation of secondary phases.

ACKNOWLEDGEMENTS
The support of this research by the project STSU 5985 and by Institutional Project CSSDT 15.817.02.04A is acknowledged.

REFERENCES

[1] W. Wang, M.T. Winkler, O. Gunawan, T. Gokmen, T.K. Todorov, Y. Zhu, D.B. Mitzi, Adv. Energy Mater. 4 (2013) 1301465.
[2] O. Gunawan, T.K. Todorov, D.B. Mitzi, Appl. Phys. Lett. 97 (2010) 233506.
[3] I.L. Repins, H. Moutinho, S.G. Choi, A. Kanevce, D. Kuciauskas, P. Dippo, C.L. Beall, J. Carapella, C. DeHart, B. Huang, S.H. Wei, J. Appl. Phys. 114 (2013) 084507.
[4] T. Gokmen, O. Gunawan, T. K. Todorov, D. B. Mitzi, Appl. Phys. Lett. 103 (2013) 103506.
[5] K. G. Lisunov, M. Guk, A. Nateprov, S. Levcenko, V. Tezlevan, E. Arushanov, Sol. Energ. Mater. Sol. Cells 112 (2013) 127.
[6] S. Chen, J.H. Yang, X. Gong, A. Walsh, S.H. Wei, Phys. Rev. B 81 (2010) 245204.
[7] Z.-K. Yuan, S. Chen, H. Xiang, X.-G. Gong, A. Walsh, J.-S. Park, I. Repins, S.-H. Wei, Adv. Funct. Mater. 25 (2015) 6733.
[8] E. Chagarov, K. Sardashti, A.C. Kummel, Y.S. Lee, R. Haight, T.S. Gershon, J. Chem. Phys. 144 (2016) 104704.
[9] T. Gershon, Y.S. Lee, P. Antunez, R. Mankad, S. Singh, D. Bishop, O. Gunawan, M. Hopstaken, R. Haightet, Adv. Energy Mater. 25 (2016) 1502468.
[10] C.J. Hages, M.J. Koeper, R. Agrawal, Sol. Energ. Mater. Sol. Cells 145 (2016) 342.
[11] T. Gershon, K. Sardashti, O. Gunawan, R. Mankad, S. Singh, Y.S. Lee, J.A. Ott, A. Kummel, R. Haight, Adv. Energy Mater. 6 (2016) 1601182.
[12] L. Dermenji, N. Curmei, M. Guc, G. Gurieva, M. Rusu, V. Fedorov, L. Bruc, D. Sherban, S. Schorr, A. Simashkevich, E. Arushanov, Surf. Eng. Appl. Electrochem. 52 (2016) 509.
[13] W. Gong, T. Tabata, K. Takei, M. Morihama, T. Maeda, T. Wada, Phys. Status Solidi C 12 (2015) 700.
[14] G. Gurieva, M. Guc, L.I. Bruk, V. Izquierdo-Roca, A. Pérez Rodríguez, S. Schorr, E. Arushanov, Phys. Status Solidi C 10 (2013) 1082.
[15] M. Guc, S. Levcenko, I.V. Bodnar, V. Izquierdo-Roca, X. Fontane, L.V. Volkova, E. Arushanov, A. Pérez-Rodríguez, Sci. Rep. 6 (2016) 19414.

MICRO CONCENTRATOR CONCEPT FOR COST REDUCTION AND EFFICIENCY ENHANCEMENT OF THIN-FILM CHALCOPYRITE PHOTOVOLTAICS: RESULTS FROM EU JOINT RESEARCH PROGRAM CHEETAH

M. Schmid[1], E. Lotter[2], X. Lin[1], L. Wang[1], R. Klenk[1], K. Eylers[3], F. Ringleb[3], T. Boeck[3], B. Heidmann[1], G. Nenna[4], F. Loffredo[4], F. Villani[4], T. Köhler[1], D. Sancho-Martinez[1], T. Raadik[5], J. Krustok[5], M. Grossberg[5], M.Ch. Lux-Steiner[1]
[1]Helmholtz-Zentrum Berlin für Materialien und Energie, Hahn-Meitner-Platz 1, 14109 Berlin, Germany
[2]Zentrum für Sonnenenergie- und Wasserstoff-Forschung Baden-Württemberg, Meitnerstr. 1, 70563 Stuttgart, Germany
[3]Leibniz-Institute for Crystal Growth, Max-Born-Straße 2, 12489 Berlin, Germany
[4] ENEA Portici Research Center, P.le Enrico Fermi, 80055 Portici (Naples), Italy
[5]Tallinn University of Technology, Ehitajate tee 5, 19086 Tallinn, Estonia

ABSTRACT: Results for research on chalcopyrite micro concentrator solar cells obtained within the framework of CHEETAH joint program are shown. A top-down proof of concept study reveals close to 30% relative efficiency increase under light concentration using inkjet-printed CIGSSe. A novel bottom-up approach for local chalcopyrite absorbers grown from indium islands demonstrates working CISe micro cells. Millimeter-sized lenses are fabricated from PMMA by a casting process to be applied as concentrator optics. For a combined exploitation of direct and diffuse light components an angular splitting concentrator based on chalcopyrite and kesterite absorber material is proposed. The scientific innovation brought will enrich further development of CIGSe solar cells and contribute to their relevance in photovoltaic energy production.
Keywords: Concentrator Cells, Design, Fabrication, Chalcopyrite, $Cu(InGa)Se_2$, Kesterite, CZTS, Lenses

1 BACKGROUND

The European SET-Plan envisages an accelerated development of low-carbon technologies, including photovoltaics (PV). The related aim of reduced material consumption and enhanced efficiency followed in current PV research is reflected in the goals of the joint research program "Cost-reduction through material optimisation and Higher EnErgy output of solAr pHotovoltaic modules - joining Europe's Research and Development efforts in support of its PV industry" (CHEETAH). Within work package 9, task 2 focuses on thin-film chalcopyrite ($Cu(In,Ga)Se_2$ – CIGSe) solar cells and addresses their challenge of containing indium and gallium - elements with high supply risk, which hinders CIGSe to enter production on the terawatt scale. The innovative approach to face this challenge and fabricate CIGSe solar cells with higher efficiency and less material is the application of a micro concentrator concept. By restricting the solar cell dimensions to a few ten to hundred microns in diameter and reducing the coverage with absorber material by a factor of about one hundred, tremendous material saving is possible. Simultaneous concentration of incident sun light onto the reduced absorber areas allows to fully exploit incident radiation and benefit from enhanced performance due to increased illumination energy density. The dimensionality on the microscale enables improved heat dissipation compared to large scale counterparts and opens the path towards fabrication of solar concentrator modules with a compactness comparable to regular thin film products.

In this contribution we show results of CHEETAH joint research on chalcopyrite micro concentrator solar cells: In a top-down proof of concept study, close to 30% relative efficiency increase is achieved by light concentration onto inkjet-printed CIGSSe. Using a novel bottom-up approach for local chalcopyrite growth from indium islands we demonstrate working CISe micro cells. Corresponding millimeter-sized lenses are fabricated from PMMA by a casting process. Finally, we propose an angular splitting concentrator exploiting both direct and diffuse light components by combining chalcopyrite and kesterite absorber material. The related scientific innovation and progress will strengthen further development of CIGSe solar cells and contribute to make them a vital part of photovoltaic energy production.

2 RESULTS AND DISCUSSION

2.1 Top-down micro cells based on inkjet-printed CIGSSe

Material deposition by inkjet-printing offers an ultimate approach for full spatial control over absorber growth. The formation of planar CIGSSe absorbers by inkjet-printing leading to a solar cell efficiency of 11.3% has been reported previously [1]. In brief, a precursor ink containing Cu, In and Ga is formulated, printed by drop-on-demand technology onto a Mo back contact and

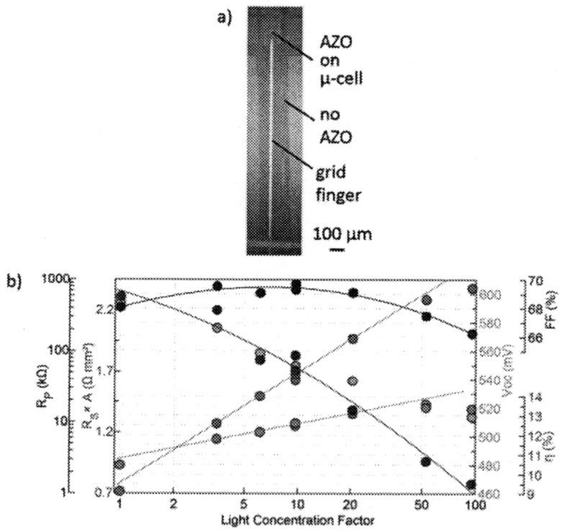

Figure 1: a) Micro solar cell defined on planar inkjet-printed CIGSSe absorber by local etching of ZnO:Al (AZO) front contact, b) corresponding electrical solar cell parameters as a function of light concentration.

subject to a three-stage chalcogenization process in selenium-containing atmosphere and with addition of H_2S. Whilst on the long run a bottom-up approach by local absorber printing is feasible, for the proof of concept study the micro cells were fabricated by the following top-down approach based on planar CIGSSe solar cells: the front part of a Ni-Al grid finger was mechanically separated and a micro cell defined around the tip by chemical etching of the surrounding ZnO:Al front contact, which was also removed along the grid finger. The resulting structure is depicted in Fig. 1a). Fig. 1b) shows the performance of an according micro cell under homogeneously enhanced light illumination. The measured short-circuit current density is assumed to increase linearly with concentration and is used to define the light concentration factor. Up to about 60 times concentration the open-circuit voltage V_{OC} follows the expected logarithmic increase and improves from 460 to 600 mV. Combined with small variations in fill factor *FF* only, the efficiency reaches its maximum at 60 suns as well: $\eta = 13.4\%$ efficiency is obtained which corresponds to a relative increase of 28% compared to $\eta = 10.5\%$ at one sun. Remarkably, the development of series and shunt resistance, R_p and R_s, respectively, reveals a different behavior for different absorber materials – compare the results for micro cells fabricated from co-evaportated CIGSe shown in [2] – and will require further analysis to obtain highly efficient micro concentrator solar cells.

2.2 Bottom-up growth of CISe micro cells from In droplets

True material saving can only be achieved if a bottom-up approach is followed for the micro solar cell fabrication. Addressing in particular a reduced consumption of the rare and expensive element indium, we locally deposit In droplets, which are further processed to CISe absorbers. The growth and regular arrangement of In droplets on laser-patterned substrates has in detail been described elsewhere [3]. Droplets tuned to a diameter of about 40 µm and a thickness of few microns are then overcoated with a 500 nm thick Cu layer. A subsequent selenization process in ramps of 200°C and 550°C leads to the formation of CISe and $CuSe_x$, the latter being removable by selective etching. Fig. 2a) shows an SEM top view of a resulting CISe micro absorber island. In Fig. 2b) the corresponding structure of the micro solar device is represented. Individual absorber islands grown on molybdenum back contact were electrically isolated by SU-8 photoresist, which was deposited by spin-coating and removed on top of the CISe islands by plasma etching. In this way the CdS, i-ZnO ad ZnO:Al front contact layers applied to the entire surface are forming the pn-junction with the local CISe absorber without touching the Mo back contact, present also in between the absorber islands. Resulting dark and illuminated current-voltage characteristics clearly revealing a diode behavior are shown in Fig. 2c). Even though there is still room for improvement regarding the solar cell's performance parameters, the achievement of a working locally grown chalcopyrite solar cell is an important step towards material saving and efficiency enhancement.

2.3 Inkjet-printed PMMA micro lenses

Asides the formation of local CIGSe absorbers the fabrication of adequate lenses is essential to obtain micro concentrator devices. Inkjet-printing mentioned above for

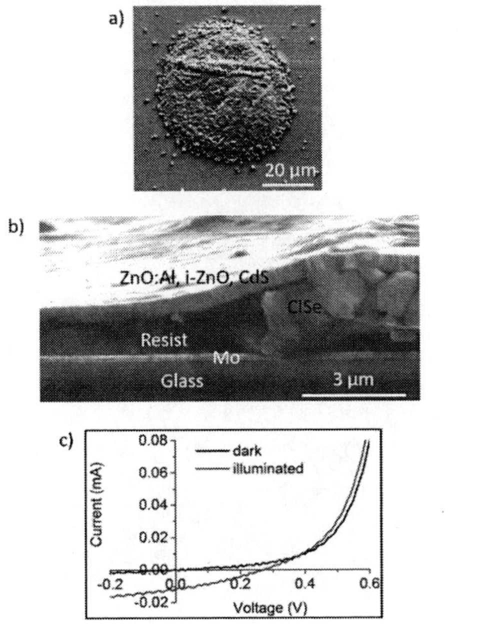

Figure 2: CISe micro solar cell fabricated by a bottom-up process: a) SEM top view of CISe absorber, b) cross sectional SEM image of isolated micro solar cell, c) dark and illuminated current-voltage characteristic.

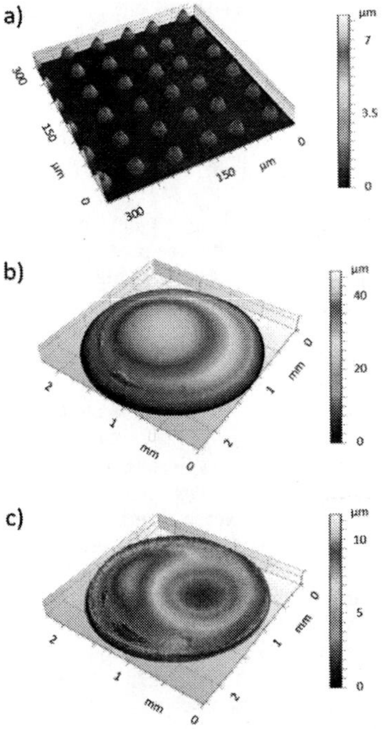

Figure 3: PMMA lenses: a) micrometer-scale array, b) millimeter-scale single optical element revealing deviation from spherical shape as visible in c) corresponding error profile error.

the absorber growth can also be used for the deposition of concentrator optics, enabling on the long run an all-printed device. State-of-the-art inkjet-printed lenses made from poly(methyl methacrylate) (PMMA) usually reach diameters on the order of several tens of microns. An example of a micro lens array deposited by drop-on-demand inkjet printing technology of a high concentration PMMA-based ink onto a 325 x 325 µm^2 substrate is depicted in Fig. 3a). The substrate temperature was kept at 17°C during the printing process and for the subsequent 30 minutes; then the samples were annealed at different temperatures. The results suggest that 250°C is the optimized annealing temperature allowing to reduce the surface roughness and to improve the spherical shape. These findings are significant to produce high-quality optical microstructures. Slightly bigger optical structures than those displayed in Fig. 3a) and having dimensions up to hundreds of microns were manufactured through non-conventional printing technologies (pyro-electrohydrodynamic inkjet printing) by dispensing high-viscosity optical-grade polymer based inks [4]. Having to couple the micro lenses with micro solar cells of approximately 100 µm diameter and reaching concentration factors of 100 -500 times, lens dimensions in the range from 1 - 2.3 mm are needed. Such large dimensions have never been reported for inkjet-printed PMMA lenses in literature due to process constraints. Therefore, micro lenses were manufactured by a casting process via a chemical pipette allowing to deposit the required ink volume in a single droplet. An example of the so-manufactured optical elements having a diameter of 2.2 mm and a maximum height of 46 µm is shown in Fig. 3b). A remaining deviation from the perfect spherical shape, as illustrated by the error profile displayed in Fig. 3c), needs to be addressed in further studies for final application to micro solar cells making integrated concentrator devices feasible.

2.4 CIGSe-CZTS direct-diffuse micro concentrator

As an alternative absorber for micro concentrator solar cells, kesterite Cu_2ZnSnS_4 (CZTS) material was taken into consideration. Testing of CZTS monograin layer solar cells [5] under increased illumination intensities, however, yielded a decrease rather than an increase in efficiency. Fig. 4a) reveals the major reason, namely a significant drop in fill factor (*FF*) with enhanced illumination intensity above one sun, which directly translates to a decrease in efficiency (not light soaked, room-temperature measurements). Therefore, CZTS solar cells are not suitable for operation under concentrated sun light. In contrast, a reduced illumination intensity improves their performance as shown in Fig. 4b). When decreasing the light intensity from 1 to 0.15 suns the *FF* increases by 11% absolute, which finally leads to an increase in efficiency from 4.0% to 4.6% in the example shown here. Kesterite material thus appears predestined to be operated under reduced rather than enhanced illumination intensity. This observation lead to the development of a combined CIGSe-CZTS direct-diffuse micro concentrator concept. The concept takes into account that concentrator optics are only able to focus direct irradiation, whereas the significant part of diffuse light contained in the solar radiation incident in our latitudes cannot be concentrated. Exploitation of both radial parts of the solar spectrum is aspired by placing CIGSe micro cells in the concentration regions and filling the spaces in between the micro cells by CZTS cells such that best performance enhancement of both types of solar cells under increased and reduced illumination intensity, respectively, is achieved. The improved efficiency of CIGSe solar cells with increased illumination intensity was already proven in sec. 2.1; for the striped CIGSe micro cells of the angular splitting concentrator planar co-evaporated absorbers were locally passivation and shaded. Fig. 4c) shows the schematic structure of the combined device and Fig. 4d) the prototype, which could confirm the expected performance behavior.

3 SUMMARY

As the results presented here reveal, significant progress has been achieved with respect to top-down and bottom-up formation of micro absorbers from planar inkjet-printed CIGSSe and sequentially processed CISe based on indium islands, respectively. Furthermore, the development of printed polymer lenses on the millimeter-scale was reported and a novel concept for enhanced exploitation of direct and diffuse sun light proposed. The related joint work within the European project CHEETAH therefore constitutes an important resource of combined forces to advance the development of highly efficient and low cost photovoltaics.

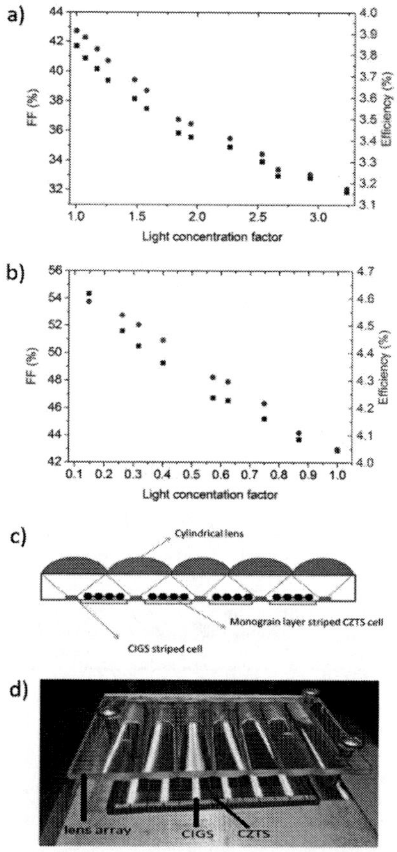

Figure 4: FF and efficiency revealing CZTS solar cell performance under a) enhanced and b) reduced illumination conditions; c) schematic and d) prototype of combined CIGSe-CZTS direct-diffuse micro concentrator concept.

ACKNOWLEDGMENT

This research is part of the EU CHEETAH project and has received funding from the European Union's Seventh Programme for research, technological development and demonstration under grant agreement No 609788.

The authors would like to thank M. Kirsch (HZB) for ZnO sputtering and grid deposition onto CIGSe cells as well as O. Ernst (IKZ) for preparation of resist-based isolating layers. We thank R. Miscioscia (ENEA) for optical profilometer measurements and F. Roca and C. Minarini (ENEA) for advice and support during the course of the activities. T. Varema and M. Kauk-Kuusik (TUT) are acknowledged for support in CZTS solar cell preparation.

REFERENCES

[1] X. Lin, et al., Energy Environ. Sci. 9 (2016) 2037.
[2] E. Lotter, et al., Proc. EU-PVSEC Munich (2016) 1156.
[3] F. Ringleb, et al., APL 108 (2016) 111904.
[4] S. Coppola, et al., APL 106 (2015) 261603.
[5] E. Mellikov et al., in: Copper Zinc Tin Sulfide-Based Thin Film Solar Cells, Wiley-VCH (2015) 289.

GROWTH AND PROPERTIES OF $Cu_2ZnSnSe_4$ FILMS ON FLEXIBLE METALLIC SUBSTRATES

V.F. Gremenok[1], R. Juskenas[2], T.V. Piatlitskaya[3], K. Bente[4], A.V. Stanchik[1],
S.A. Bashkirov[1], R. Giraitis[2], A. Selskis[2], A.N. Pyatlitski[3], V.A. Solodukha[3], C. Berthold[4], K. Nickel[4]
[1]State Scientific and Production Association «Scientific-Practical Materials Research Centre of the National Academy of Sciences of Belarus», 220072, Minsk, P. Brovka str., 19, Belarus, phone: + 375 17 284 12 49,
fax: + 375 17 284 15 58, E-mail:gremenok@physics.by
[2]State research Institute Center for Physical Sciences and Technology, Savanorių Ave. 231, LT-02300 Vilnius, Lithuania
[3]JSC "INTEGRAL" – "INTEGRAL" Holding Managing Company, 220108, Minsk, Kazintsa I.P. str., 121A, Belarus
[4]Applied Mineralogy, University Tübingen, Wilhelmstrasse 56, D-72074 Tübingen, Germany

ABSTRACT: Cu-Sn-Cu-Zn (CZT) precursor films on tantalum (Ta) and molybdenum (Mo) foil substrates were electrochemically deposited in two-electrode cell in galvanostatic mode, further annealed and selenized at elevated temperatures. Modern X-ray diffraction, Raman spectroscopy, energy dispersive X-ray analysis, scanning electron microscopy and photo electrochemical characterization methods have been used for the determination of the physical characteristics of films. XRD measurements showed that phase formation in $Cu_2ZnSnSe_4$ films depends both on the composition ratio in the precursors and on the processing regimes. It has been determined that the as-deposited film contains up to four metallic phases. The pre-annealing of metallic films resulted in the formation of micro-meter size grains of the η-$Cu_{6.26}Sn_5$ and γ-Cu_5Zn_8 phases. XRD studies of Cu-Sn-Cu-Zn precursor layers selenized at temperatures 400 - 550 ^{o}C have shown the formation of $Cu_2ZnSnSe_4$ thin films. However, the selenized films contained ZnSe and CuSe phases. The amount of ZnSe decreased with selenization temperature. The $Cu_2ZnSnSe_4$ films on metallic substrates are densely packed with compact faceted grain structure and without any voids on the surface. The photo electrochemical results showed that all thin films are photoactive and have *p*-type conductivity.
Keywords: Characterisation, CZTS, Electrodeposition, Flexible Substrate, Thin Film

1 INTRODUCTION

The $Cu_2ZnSnSe_4$ (CZTSe) semiconductor compound attracts a great interest due to its potential applications for sustainable thin film solar cell devices [1,2]. CZTSe is a p-type quaternary semiconductor compound most stable in the kesterite structure. This material has the direct energy gap (Eg ~ 1.0 eV) with high inter-band absorption coefficient ($\geq 10^4$ cm^{-1}). The structural and electronic properties of CZTSe are similar to those of $Cu(InGa)Se_2$ (CIGS), successfully used in one of the leading thin film solar cell technologies [3,4]. $Cu_2ZnSnSe_4$ has the additional advantages of having elements which have low costs, low toxicity and are abundant in the earth's crust. This provides a good foundation for the CZTSe-based solar cells, which currently show conversion efficiencies in excess of 11% and makes particularly attractive candidate for large-scale commercial application [5]. CZTSe thin film solar cells deposited on flexible metal foil substrates have advantages of softness, reduced total thickness, light weight, large gravimetric specific power and low cost. These unique advantages offer them more possibilities in such as application fields building integration, space exploration, automotive applications, portable electronics, and etc. [6]. Additionally, metallic flexible substrate utilization eliminates the need of the metallic conductive back contact layer deposition. As a result, the cost of manufacturing and constructing of solar cell modules would be substantially lowered by using flexible metallic substrates. In addition, from the industrial point of view, flexible solar cells enable implementation of the roll-to-roll deposition process which would significantly reduce production costs and potentially raise throughput [7]. A number of methods have been used to obtain CZTSe thin films for photovoltaic applications, including vacuum and non-vacuum methods [1-5,8,9]. Among the different methods for $Cu_2ZnSnSe_4$ production the electrochemical route appears of a great interest because it is easy to control, non-vacuum and low-cost technology, operating at room temperature and using non-toxic solvents and reagents, useful to reduce fabrication cost and to produce large scale CZTSe-based solar cells.

The purpose of this work is characterization of CZTSe films prepared by selenization of electrochemically deposited metallic precursors on Ta and Mo flexible substrates. XRD and Raman spectroscopy data showed that after the annealing at temperature of 500 ^{o}C $Cu_2ZnSnSe_4$ thin films with kesterite or partially disordered kesterite structure was formed. The photo electrochemical results showed that films are photoactive and have *p*-type conductivity.

2 EXPRIMENTAL DETAILS

2.1 Films preparation

The method for obtaining CZTSe films included subsequent electrochemical deposition of Cu-Sn-Cu-Zn precursors on tantalum and molybdenum substrates and selenization of the deposited precursors to form CZTSe film. Flexible Mo and Ta foil substrates were mechanically polished and rinsed subsequently with 96% ethanol and deionized water. The Cu layers were deposited from aqueous solution containing 0.02 M $CuSO_4$ and 0.25 M sodium citrate with pH = 6.3. The Sn layers were deposited from aqueous solution of 0.02 M $SnSO_4$ and 0.25 M sodium citrate with pH = 6.3. The Zn layers were deposited from aqueous solution containing 0.75 M $ZnSO_4$, 0.5 M Na_2SO_4 and organic additives (pH = 3.7). The pH of solutions was adjusted with H_2SO_4. Annealing of the electrochemically deposited CZT layers was conducted in a tube furnace in Ar (purity 99.995 %) atmosphere. A device Omron E5CK-TAA1-500 digitally controlled the temperature regime using an Ntype(Ni–Cr–Si) thermocouple. An accuracy of temperature maintenance was not worse than ± 1.0 ^{o}C. The pre-annealing of CZT precursor layer was performed in a tube furnace in an Ar (N 6.0) 95% þ H_2 5 % gas mixture at a temperature of 350 ^{O}C for 0.5 h. The films pre-annealed were selenized in a home-made quartz container (volume 12.5 cm^3) with 5 mg of powdery Se under 1 bar of Ar gas pressure. Firstly, the tube was evacuated when filled with Ar gas, again evacuated, and afterwards filled with an Ar gas to 1 bar pressure. Selenization was performed at temperatures 450-550 ^{O}C for 30 min. After the selenization samples were naturally cooled down inside the furnace [9,10].

2.2 Sample characterization

XRD characterization of precursors and CZTSe film was performed using a diffractometer D8 Advance (Bruker AXS). CuKa radiation (wavelength 0.154183 nm) was separated by a multilayer Ni/graphite monochromator, which also formed a parallel beam of X-rays. The XR patterns were measured in grazing incidence (GID) step scan mode with a step size of and counting time of 5–10 s. An incidence angle of the primary beam was 0.5^{O}. An appropriate adjustment of the diffractometer was controlled using the certified standard SRM606 from NIST. Phase identification of thin films was performed using the powder diffraction data-base PDF-2(2003 release).

Electron microscopy studies were carried out in SEM EVO- 50EP (Carl Zeiss SMT) arranged with an X-ray energy dispersive spectrometer (EDX), from Oxford Instruments.

Atomic force microscopy (AFM) was used to study the topographic characteristics of thin films. The surface topography was investigated using a NT 206 (Microtestmachines Co., Belarus) high resolution atomic force microscope.

Raman spectra were obtained with a spectral resolution better than 3 cm^{-1} using a confocal Raman spectrometer Nanofinder HE (LOTIS TII). For excitation of Raman radiation a continuous wave solid-state laser with a wavelength of 532 nm was used. Room-temperature Raman measurements were carried out using laser power of 0.2, 0.6, 1.6 and 2.0 mW. The diameter of laser spot on the sample surface was about 0.6-0.7 µm.

The photoelectrochemical (PEC) properties were studied by forming a three-electrode quartz cell with CZTSe films as a photo-electrode, platinum spiral wire as a counter electrode and Ag/AgCl saturated with KCl as a reference electrode. Potentiostat-galvanostat Gill AC (ACM Instruments, UK) was employed to determine photocurrent at constant potential. The 0.1 M Na_2SO_4 (pH 6.5) aqueous solution was used as the electrolyte for photoelectrochemical measurements. The PEC cell was illuminated with a 75 W halogen lamp. Chronoamperometric dependences were recorded in a darkness-lighting (light off/light on) mode.

3 RESULTS AND DISCUSSION

The aim of these studies was to develop a reproducible deposition process of $Cu_2ZnSnSe_4$ thin films on different metallic foils for solar cell technology.

The CZT precursor films of four layers with a uniform glance surface were obtained by electrochemical deposition method. Fig. 1(*a,b*) shows typical XRD patterns of the metallic precursors before and after pre-annealing process on Mo and Ta foil substrates. It has been determined that the as-deposited precursors besides the basic metals (Cu, Sn, Zn) contain a number of intermetallic metallic phases. According to the results, during the layer-by-layer electrochemical deposition process the interaction of metals with the formation of intermetallic phases (η-$Cu_{6.26}Sn_5$, γ-Cu_5Zn_8, $CuZn_5$ etc.) is taken place. The formation of these phases is typical for electrochemical deposition of CZT precursors [9-11]. The Cu_5Zn_8 phase is present in both as-deposited and annealed precursors. One can see that the intensity of peaks for Cu_5Zn_8 phase on the XRD patterns of annealed precursors increases comparing with as-deposited samples. As such, this mechanism can be confirmed indirectly by increase in the grain size of Cu_5Zn_8 phase in the annealed films.

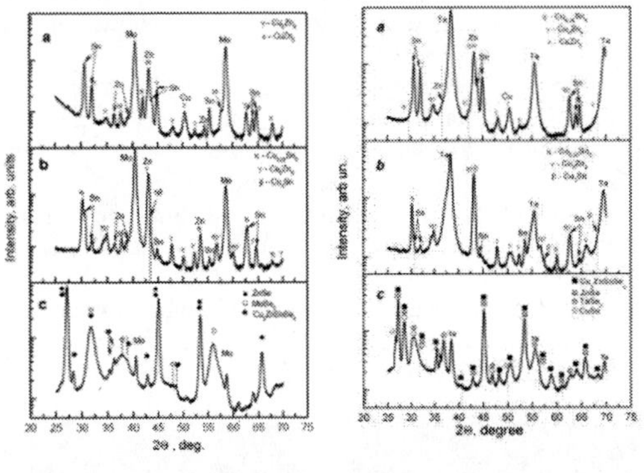

Figure 1: Typical XRD patterns of CZT precursors (***a*** - as-deposited, ***b*** – annealed) and CZTSe thin films (***c***) on Mo and Ta foil substrates

Table I contains the average atomic concentrations of Cu, Zn, Sn and Se determined by EDX method for the all investigated films on metallic foils. The chemical composition was determined by averaging the values from 5 different points on the surface of the same film.

Table I: Chemical composition of CZTSe thin films deposited on different metallic substrates (selenization temperature was maintained at 550 °C for 30 min)

Type substrate	Cu, at. %	Zn, at. %	Sn, at. %	Se, at. %	Cu/(Zn +Sn)
Mo-foil	25.07	15.39	10.14	49.40	0.98
Ta-foil	22.09	18.53	10.83	48.54	0.75

The main tasks of the XRD analysis of selenized CZT were to get evidence for the $Cu_2ZnSnSe_4$ compound formation and to evaluate the phase purity of the absorber layer. Although Cu and Sn selenides can be readily determined by a conventional XRD technique, this is not the case for ZnSe (ICDD # 01-088-2345) and cubic Cu_2SnSe_3 (ICDD # 01-089-287). XRD measurements showed that phase formation in CZTSe films depends both on the composition ratio in the precursors and on the processing regimes.

XRD studies of precursor layers selenized at temperatures 450 - 550 °C have shown the formation of crystalline CZTSe thin films (Fig. 1*c*). When the CZT precursor was annealed at a temperature of 450 °C the peaks of $Cu_2ZnSnSe_4$ phase (ICDD # 04-010-6295) emerged along with those of ZnSe and orthorhombic CuSe. Due to low intensities of the peaks attributable to CuSe it was difficult to establish which one of CuSe phases orthorhombic (ICDD # 01-086-1239) or hexagonal (ICDD # 01-086-1240) was present. The depicted fragments of XRD patterns show that the ZnSe quantity constantly decreases with an increase in the annealing temperature. No ternary selenides were detected. That was corroborated by Raman spectroscopy.

AFM was used to study the topographic characteristics of thin films including grain size, surface roughness, and maximum height of the profile. It is shown that the films obtained on Mo and Ta foil substrates have similar roughness and in both cases a grain structure is formed. The $Cu_2ZnSnSe_4$ thin films show relatively high surface roughness and maximum roughness profile height compared to CZT precursors. The increase in the surface roughness of the films was caused by the growth of grains during annealing and selenization processes.

The surface morphology of the CIGS thin films grown on metallic foils was found to be strongly dependent on the deposition conditions and selenization temperature. It was also found from the SEM micrographs that the uniformity and adhesions of the films on different metallic foils improved with increasing the selenization temperature. The CZTSe thin films are densely packed with compact faceted grain structure and without voids on the surface. The surface SEM micrograph of CZTSe film on Ta foil is shown in Fig. 2*a*.

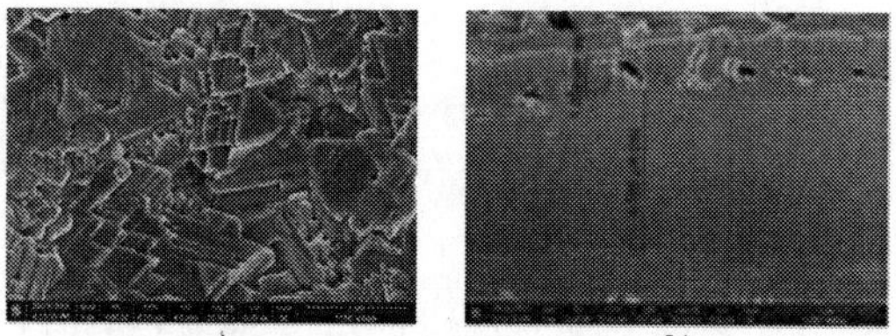

a) *b)*

Figure 2: SEM micrographs of CZTSe film on Ta foil: (*a*) – plane view; (*b*) – cross-section image

The SEM images show large, well-faceted and dense grain structure for the higher selenization temperature ~ 550 °C and the grain sizes of about ~ 0.8 - 1.2 μm, i.e. comparable to the film thickness. The presence of the $TaSe_2$ (or $MoSe_2$) sublayer (additional phase) in interface region can be clearly seen in Fig. 2*b*. Our investigations show that $Cu_2ZnSnSe_4$ thin films with remarkable structural quality can be prepared by selenization method with different morphology. However, we observed that CZTSe films are not always as dense. In some cases the layer can become porous.

Fig. 3 shows the effect of laser power on the Raman spectrum of CZTSe films formed on two different substrates. The spectra of CZTSe films observed from Mo and Ta substrates (Figure 3*a*, *b*) are very similar. Raman bands near 80/81, 172, 196, 235/233 and 386 cm^{-1} are characteristic to CZTSe compound [9,12,13]. The low intensity bands near 247/249 cm^{-1} belong to longitudinal (LO) optical phonon of cubic ZnSe [14,15]. No spectroscopic evidence for presence of $TaSe_2$ and $MoSe_2$ phases was obtained [9,10,15-17]. The wavenumber of the most intense characteristic CZTSe band of A symmetry decreases from 196 to 191 cm^{-1} with increasing the laser power from 0.2 to 2.0 mW (Fig. 3*a*). In addition, the full width at half maximum (FWHM) of this band increases from 6.9 to 9.6 cm^{-1}. It should be noted that frequency of lower intensity 172 cm^{-1} band also decreases at higher excitation powers.

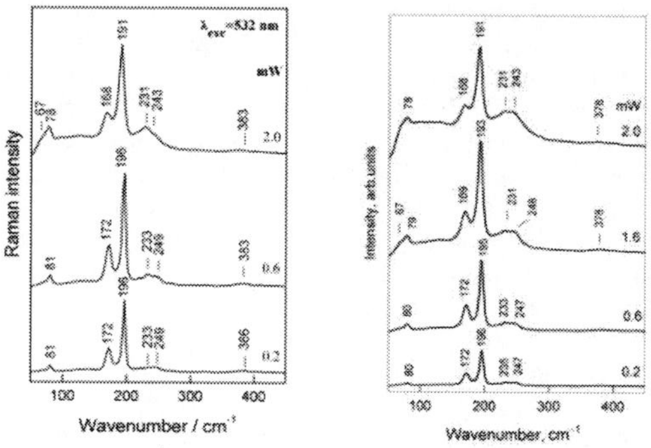

a) *b)*

Figure 3: Raman spectra of CZTSe thin films grown on Mo (*a*) and Ta (*b*) foil substrates observed at different excitation laser (532 nm) powers in the wavenumber regions of 50–450 cm^{-1}

The decrease of the phonon frequency for the A modes of kesterite structure with increase in temperature (laser power) can be explained by the perturbation model according to which the frequency shift is due to the combined effects of thermal expansion and the anharmonic coupling to other phonons [18]. Similar spectral changes were observed for films prepared on the Ta substrate (Fig. 3*b*); the wavenumber of the most intense characteristic CZTSe band decreases from 196 to 191 cm^{-1} with increasing the laser power; the FWHM of this band increases from 5.6 to 8.3 cm^{-1}, indicating that the crystalline material is of high quality without a number of defects [13].

The results of photoelectrochemical measurements of CZTSe thin films under chopped light illumination are presented in Fig. 4.

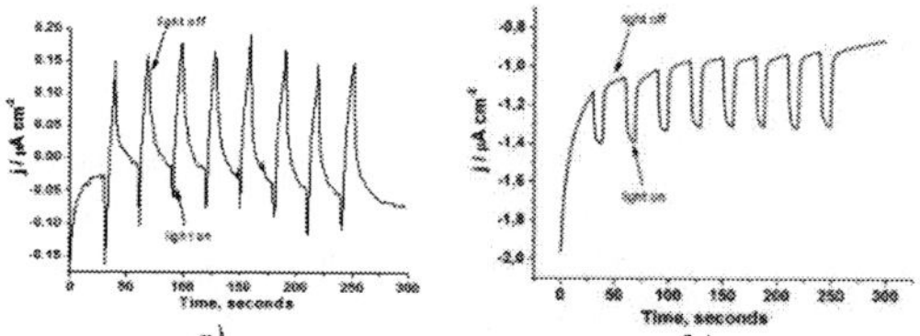

a) *b)*

Figure 4: The dependence of photocurrent on time of thin films prepared by: (*a*) electrodeposition on Mo foil; (*b*) electrodeposition on Ta foil

Chronoamperometric dependences was obtained in the «light off / light on» mode at constant potential, which is determined as stationary potential of the thin films semiconductor in the electrolyte. The photoelectrochemical measurement confirms the photoactivity of films prepared on Mo and Ta substrates. The comparison of results obtained with the films on different substrates shows that all electrodeposited films have *p*-type conductivity. There are positive dark current is observed on Mo substrate without illumination and negative dark current on Ta substrate.

Dependences of the change of photocurrent density per unit surface on time were also different for films deposited on different substrates. On a flexible Mo substrate (Fig. 9*a*), the photocurrent does not reach large values (0.1 μA) and drops sharply after switching off the illumination. After the start of illumination the photocurrent grew unevenly and reached hardly 0.1 μA.

On Ta substrate were obtained higher photocurrents 2 μA and the growth and fall of photocurrent in the «light off/light on» mode is same dependence on time (Fig. 9*b*). The kinetics of current change in the «light off/light on» mode is strictly constant. The current is growing instantly when films were illuminated.

It should be noted, that results we obtained differ from those presented for an annealed mixture of powdery Cu, Zn, Sn and Se [19] as well as for the magnetron sputtered stacked layers of Cu/Zn/Sn [20]. The comparison of our results suggests that fabrication processes strongly depend on the approach of precursor fabrication and conditions of the pre-annealing and selenization. The conditions of this work were the same as commonly used for fabrication of Kesterite absorber layer i.e. a container with CZT precursor and Se powder under Ar atmosphere.

4 CONCLUSIONS

Cu-Sn-Cu-Zn metallic precursors were sequentially deposited on flexible Mo and Ta foil substrate, preliminary annealed and selenized leading to formation of $Cu_2ZnSnSe_4$ thin films Several intermetallic phases (η-$Cu_{6.26}Sn_5$, γ-Cu_5Zn_8, $CuZn_5$ etc.) where observed both in as-deposited and annealed precursors. The usage of GI XRD and a logarithmic intensity scale while analysing the XRD patterns promoted detection of low intensity peaks. The CZTSe films contain undesired ZnSe and CuSe phases. Morphologically, the resulting CZTSe thin films are densely packed with compact faceted grain structure and without voids on the surface. The results obtained by the XRD methods corroborated by Raman spectroscopy. CZTSe films prepared on Mo and Ta substrates the CZTSe and ZnSe phases are detected spectroscopically. There are no observable secondary phases such as Cu_2Se, SnSe and Cu_2SnSe_3. The photoelectrochemical measurement confirms the photoactivity of films prepared on Mo and Ta substrates. The comparison of results obtained with the films on different substrates shows that all electrodeposited films have *p*-type conductivity.

5 ACKNOWLEDGMENTS

This work was supported by the Belarusian Republican Foundation for Fundamental Research (grant F16T/A-006) and the State Programme for Research «Physical material science, new materials and technologies» (MTTEX 1.0.6).

6 REFERENCES

[1] X. Liu, Y. Feng, H. Cui, F. Liu, X. Hao, G. Conibeer, D.B. Mitzi, M. Green, Prog. Photovolt. 24 (2016) 879.
[2] S.A. Bashkirov, R. Kondrotas, V.F. Gremenok, R. Juskenas, International Scientific Journal for Alternative Energy and Ecology, 15-18 (2016) 31.
[3] S. Siebentritt, S. Schorr, Prog. Photovolt. 20 (2012) 512.
[4] M.J. Romero, H. Du, G. Teeter, Y. Yan, M.M. Al-Jassim, Phys. Rev. B 84 (2011) 165324-1.
[5] Y.S. Lee, T. Gershon, O. Gunawan, T.K. Todorov, T. Gokmen, Y. Virgus, S. Guha, Adv. Energy Mater. 5 (2015) 1401372-1.
[6] M. Pagliaro, G. Palmisano, R. Ciriminna, Flexible Solar Cells (Wiley-VCH Verlag GmbH and Co.KgaA, Italy; 2008).
[7] S.Yazici, M.A. Olgar, F.G. Akca, A. Cantas, M. Kurt, G. Aygun, E. Tarhan, E.Yanmaz, L. Ozyuzer, Thin Solid Films 589 (2015) 563.
[8] D. Colombara, A. Crossay, L. Vauche, S. Jaime, M. Arasimowicz, P.-P. Grand, P.J. Dale, Phys. Status Solidi A (2014) 1.
[9] R. Juškėnas, G. Niaura, Z. Mockus, S. Kanapeckaitė, R. Giraitis, R. Kondrotas, A. Naujokaitis, G. Stalnionis, V. Pakštas, V. Karpavičienė, Journal of Alloys and Compounds 655 (2016) 281.
[10] R. Juškėnas, S. Kanapeckaitė, V. Karpavičienė, Z. Mockus, V. Pakštas, A. Selskienė, R. Giraitis, G. Niaura, Solar Energy Materials and Solar Cells 101 (2012) 277.
[11] V. Simć, Z. Marincovič, Journal of the Less-Common Metals 72 (1980) 133.
[12] A. Khare, B. Himmetoglu, M. Johnson, D.J. Norris, M. Cococcioni, E.S. Aydil, J. Appl. Phys. 111 (2012) 083707.
[13] M. Dimitrievska, A. Fairbrother, A. Perez-Rodriguez, E. Saucedo, V. Izquierdo-Roca, Acta Materialia 70 (2014) 272.
[14] D. Colombara, E.V.C. Robert, A. Crossay, A. Taylor, M. Guennou, M. Arasimowicz, J.C.B. Malaquias, R. Djemour, Solar Energy Mater. Solar Cells 123, (2014) 220.
[15] A. Redinger, K. Hönes, X. Fontane, V. Izquierdo-Roca, E. Saucedo, N. Valle, A. Pérez-Rodríguez, S. Siebentritt, Applied Physics Letters 98 (2011) 101907.
[16] V. Izquierdo-Roca, E. Saucedo, C. M. Ruiz, X. Fontané, L. Calvo-Barrio, J.Á. Garcia, P.-P. Grand, J.S. Jaime-Ferrer, A. Pérez-Rodríguez, J.R. Morante, V. Bermudez, Phys. Status Solidi A 206 (2009) 1001.
[17] Z. Yan, C. Jiang, T. R. Pope, C. F. Tsang, J. L. Stickney, P. Goli, J. Renteria, T. T. Salguero, A.A. Balandin, Journal of Applied Physics 114 (2013) 204301.
[18] P.K. Sarswat, M.L. Free, A. Tiwari, Phys. Status Solidi B 248 (2011) 2170.
[19] R.A. Wibowo, W.H. Jung, M.H. Al-Faruqi, I. Amal, K.H. Kim, Mater. Chem. Phys. 124 (2010) 1006.
[20] R.A. Wibowo, S.A. Moeckel, H. Yoo, C. Hetzner, A. Hoelzing, P. Wellmann, R. Hock, Mater. Chem. Phys. 142 (2013) 311.

GROWTH OF P-TYPE CdZnTe THIN FILMS AS PROSPECTIVE ABSORBER LAYER FOR PHOTOVOLTAIC APPLICATION

F.M. Tahzib Enam[1], K.S. Rahman[1], M. Akhtaruzzaman[2], K. Sopian[2], M.A. Islam[3] and N. Amin[1,2*]
[1]Department of Electrical, Electronic and Systems Engineering, Faculty of Engineering and Built Environment, The National University of Malaysia, 43600 Bangi, Selangor, Malaysia
[2]Solar Energy Research Institute, The National University of Malaysia, 43600 Bangi, Selangor, Malaysia
[3]Graduate School of Material Science, Nara Institute of Science and Technology, Ikoma, Nara 630-0101, Japan

ABSTRACT: Cadmium Zinc Telluride (CdZnTe) thin films have been deposited by thermal evaporation technique whether the absorber source is formed by mixing powder of Cadmium (Cd), Zinc (Zn) and Tellurium (Te) particles. Inert gas Argon (Ar) helps to make the CdZnTe (CZT) ternary alloy semiconductor in different annealing conditions. A comprehensive study of the films was executed by X-Ray Diffraction (XRD), Hall Effect Measurement, and UV-VIS-NIR, respectively. The non-stoichiometric composition of CZT ($Cd_{0.4}Zn_{0.6}Te$) was originated which has a direct optimal band gap 1.7 eV with Ar pressure of 1.62 Torr and temperature of 400°C in thermal annealing chamber at a rated current of 28A.
Keywords: Cadmium Zinc Telluride (CdZnTe), thermal evaporation, XRD, band gap.

1 INTRODUCTION

In the last decade, $Cd_{1-x}Zn_xTe$ (CZT) has drawn considerable attention because of its broad applications in optoelectronic devices [1]. This type of ternary alloy material enhances the possibility of tailoring their properties as per requirements in the field of device fabrication. It has various potential properties with zinc blende structure such as the tunability of band gap, lattice parameter with alloy composition, high optical absorption coefficient above the band gap energy and high value of transmittance below the band gap [2]. CZT has a direct optimal band gap of 1.45 eV to 2.26 eV and higher absorption coefficient than CdTe ($\alpha > 5x10^5 cm^{-1}$) [3]. Cadmium Zinc Telluride crystallizes in zinc blende structure showing crystallographic polarity along the <111> direction as shown in Figure 1.

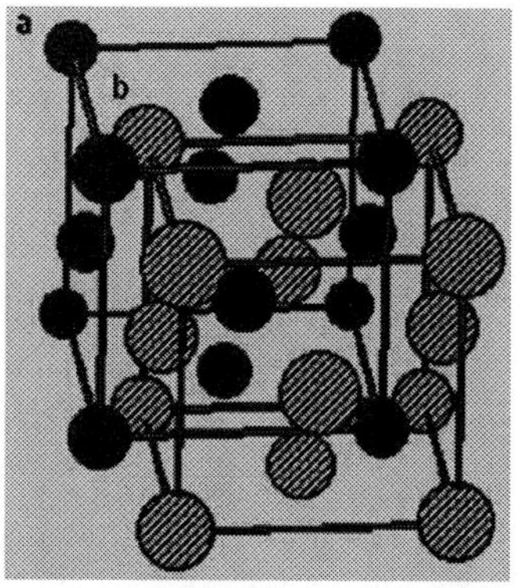

Figure 1: Zinc-Blende Structure

There are two types of {111} alternate planes, one formed by randomly distributed Cd or Zn atoms in (a) and the other formed by Te atoms in (b). The CZT alloy can be described as CdTe crystal with Zn atoms randomly substituting Cd atoms.

Beside solar cells, CZT is also very useful for infrared gamma ray detectors and electro-optical modulators [4]. Generally, it is recognized that large-scale applications must rely on cheap polycrystalline materials. The production cost should be very low in order to make the polycrystalline based devices. The thin film of $Cd_{1-x}Zn_xTe$ can be prepared by vacuum evaporation, magnetron RF sputtering, molecular beam epitaxy, chemical vapor deposition, thermal evaporation, close-spaced sublimation, screen printing etc. [5]. In this investigation, CZT absorber layer is grown by thermal evaporation method where the source material is heated until it reaches the vapor pressure. For large scale polycrystalline application, this process is very simple and cost-effective.

2 EXPERIMENTAL

CZT ternary alloy material can act as a better absorber layer because of its tailoring wide range of band gap. The source is just a mixing powder of three elemental particles of Cd, Zn and Te with their individual molar masses. Firstly, borosilicate glass substrate has been cleaned by ultra-sonic bath process in standard procedure of methanol-acetone-methanol-deionized water (~10 minutes each). Deposition by thermal evaporation happened by heating a metal source until it is hot enough for some atoms to evaporate and then condense on a glass substrate. The source seated on a tungsten filament boat and high current passed through the boat. This heated up the boat and the metal source also. When atoms evaporate, they traveled ballistically in all directions because of the high vacuum like 10^{-4} Torr. In a low vacuum, the layer couldn't stick well enough or adhesion will be poorer because of the collision among the evaporated atoms.

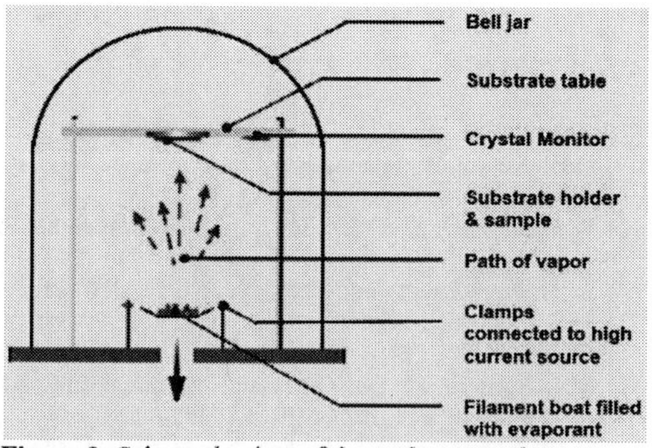

Figure 2: Schematic view of thermal evaporation

In this study, the thickness has been varied by changing the current rating at 26A and 28A. As grown samples are metallic as visible from the Hall measurement. In order to make semiconductor, these two films were annealed in both rapid and conventional thermal annealing in the annealing chamber at 400°C for 15 mins with inert gas Ar at a pressure of 1.62 Torr. This heat treatment removes the metalloid particles Cd, Te and improves the crystallinity also.

3 RESULTS & DISCUSSIONS

Structural and crystallographic properties with the crystal orientation alongside the CZT film's surface were inspected by BRUKER aXS-D8 Advance Cu-Kα diffractometer at room temperature. XRD patterns were documented in the 2θ range from 20° to 80° with a step size of 0.02° using Cu Kα radiation wavelength, λ = 1.5408 Å. The aim of X-ray diffraction analysis was inspecting the composition of the CZT thin film grown by thermal evaporator. Thickness and annealing conditions have been varied to find out the effect on structural and crystallographic properties.

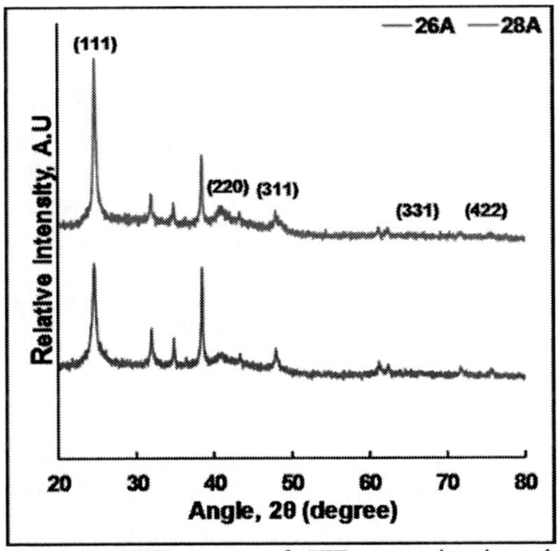

Figure 3: XRD spectra of CZT grown by thermal evaporation

From Figure. 3, it has been seen that the preferential orientation peak (111) is stronger in 28A rather than in 26A and other low relative peak intensities (220), (331), (331) and (422) reduce greatly. There are some other peaks which represent the existence of metalloid particles Cd and Te in the films.

The XRD pattern illustrates the semiconductor characteristics in Figure. 4, where the sharp (111) cubic plane has been found at 2θ=23.8° for 28A rapid annealed sample and two low intensive peaks (220) and (311) are also found at 2θ=39.4° and 2θ=46.6° plane. The diffracted peaks for 26A rapid annealing is not sharp and crystalline. Therefore, it's better to consider the high intensity crystalline peak for 28A where the composition of CZT is $Cd_{0.6}Zn_{0.4}Te$.

In conventional annealing, the crystallinity of 26A sample is poorer than 26A rapid annealed sample. Whereas 28A conventional annealed got very sharp and strong (111) peak than any other previous samples.

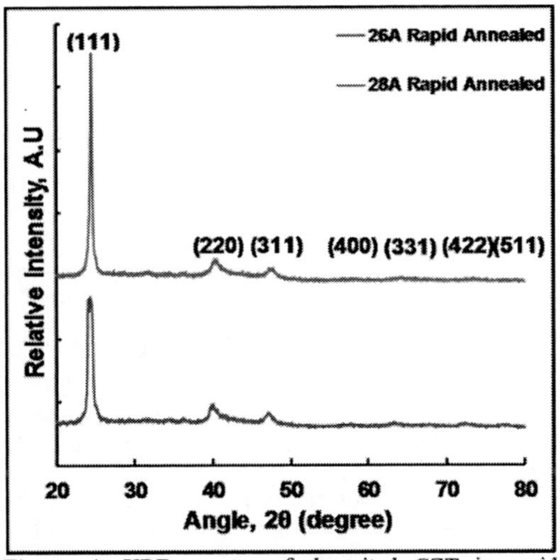

Figure 4: XRD spectra of deposited CZT in rapid annealing

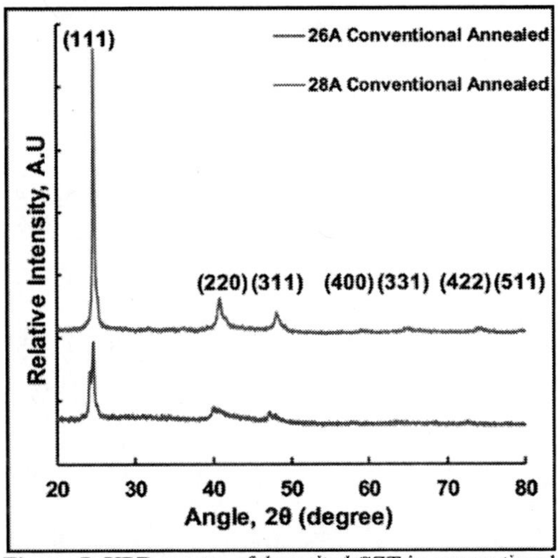

Figure 5: XRD spectra of deposited CZT in conventional annealing

Two more low intensive diffracted peaks of 26A in Figure 5 are quite shaper than 28A rapid annealed sample. The desirable composition of CZT ($Cd_{0.4}Zn_{0.6}Te$) has been achieved from this result, whether Zn has a highest depositing or sticking coefficient than Cd and Te at 28A rated current.

From the calculated values of Table I, average crystallite sizes are in the range of 40.622 nm – 47.791 nm which are significantly large and is suitable for photovoltaic application. Bigger crystallite sizes physically means less grain boundary which might have detrimental carrier recombination centers. Besides, lattice constant of 6.4 Å has been found for all diffracted peaks. The highest micro-strain of 4.121×10^{-3} is obtained for (111) cubic phase for CZT in 26A conventional annealed sample which indicates the highest lattice misfit and dislocation in the film structure. The maximum value of dislocation density $0.606\times10^{11}(cm^{-2})$ is also found for the same sample.

Table I: Calculated values of structural parameters

Sample	hkl	d_{hkl} (nm)	a (Å)	β (radian)	D_{hkl} (nm)	ε [$\times10^{-3}$]	δ [$\times10^{11}$] (cm^{-2})
26A Rapid Annealed	(111)	0.3723	6.449	0.0031	45.134	3.712	0.491
28A Rapid Annealed	(111)	0.3719	6.441	0.0031	45.136	3.707	0.491
26A Conventional Annealed	(111)	0.3720	6.444	0.0035	40.622	4.121	0.606
28A Conventional Annealed	(111)	0.3719	6.442	0.0029	47.791	3.502	0.438

The electrical properties of the CZT thin films prepared in various growth conditions is investigated by Hall-Effect measurements with an integrated resistivity/Hall measurement system (ECOPIA 3000). The magnitude of the magnetic field and the current source was 0.55 T at the maximum and 40 nA, respectively. The measurements were carried out at room temperature. The measured resistivity, carrier concentration, and mobility values are presented in Table II. From the measured values, 26A conventional annealed 28A conventional and rapid annealed films demonstrate p-type polarity.

Table II: Electrical parameters of CZT thin film

Sample	Carrier Concentration (/cm^3)	Mobility (cm^2/Vs)	Resistivity (Ω-cm)	Remark
26A as-grown	1.44×10^{19}	4.19×10^{1}	1.03×10^{-2}	Metallic
28A as-grown	8.82×10^{16}	2.04×10^{2}	3.47×10^{-1}	Metallic
26A rapid annealed	2.98×10^{14}	2.47×10^{-1}	8.45×10^{4}	n-type
28A rapid annealed	1.77×10^{12}	4.65×10^{1}	7.57×10^{4}	p-type
26A conventional annealed	4.90×10^{11}	1.78×10^{2}	7.14×10^{4}	p-type
28A conventional annealed	3.13×10^{13}	3.53×10^{0}	5.65×10^{4}	p-type

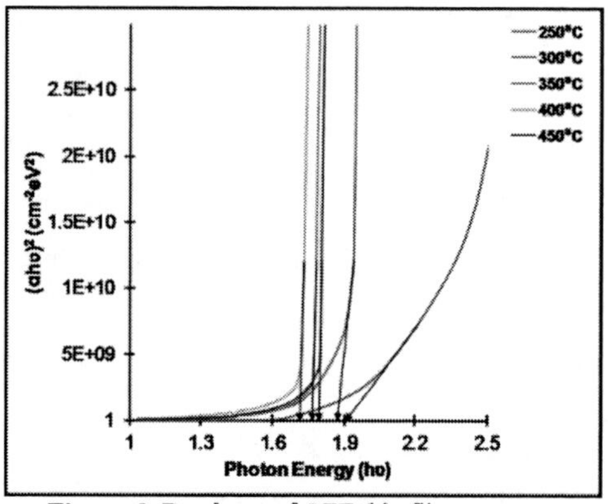

Figure 6: Band gap of CZT thin film

From the UV-Vis analysis, the absorption of CZT has been obtained more than 80%. The band gap of the CZT films are measured by plotting $(\alpha h\upsilon)^n$ versus hυ, where n=2 for the direct allowed transition. The values of the energy band gap are determined by the intercept of the straight line with the hυ axis. In this study, the energy band gap of CZT at 400°C is obtained approximately 1.7 eV.

4 CONCLUSION

CZT ternary alloy composition (Cd0.4Zn0.6Te) has been found by getting preferential orientation peak (111) from XRD analysis. Hot probe and Hall Effect measurement confirms that CZT turns from metallic to p-type semiconductor after post deposition annealing treatment. From UV-Vis analysis, absorption of 80% and band gap of 1.7 eV have been achieved. Therefore, it can be concluded that CZT can act as prospective p-type absorber layer for photovoltaic application by controlling the growth parameters and also the post deposition annealing parameters.

5 Acknowledgment

This work was supported thankfully by Qatar National Research Fund (QNRF) by the grant code of NPRP 6-931-2-382.

6 References

[1] G.N. Panin, C. Diaz-Guerra & J. Piqueras, Semicond, Sci. Technol. 13 (1998) 576.
[2] T.L. Chu, S.S. Chu, F. Friszt, C. Herrington, J. Appl. Phys. 59 (1986) 1259.
[3] G.S. Rekha, Indian Journal of Pure & Applied Physics 48 (2010) 133-135.
[4] B. Anu, P. Rajaram, Mater. Lett. 59 (2005) 3666-3671.
[5] V. Kumar, M.K. Sharma and T.P. Sharma, Eur. Phys. J. Appl. Phys. 50 (2010) 20502.

LIGHT INDUCED MODIFICATION OF COPPER INDIUM GALLIUM DISELENIDE THIN FILMS AND SOLAR CELLS

Torsten Hölscher, Stefan Förster, Thomas Schneider, Matthias Maiberg, Wolf Widdra, and Roland Scheer
Martin-Luther-University Halle-Wittenberg / Institute of physics
Von-Danckelmann-Platz 3, 06120 Halle, Germany
Telephone: +49-345-55-25492, E-Mail: torsten.hoelscher@physik.uni-halle.de

ABSTRACT: We study the light induced degradation of bare $Cu(In,Ga)Se_2$ (CIGSe) surfaces by means of time-resolved photoluminescence (TRPL) measurements. Illumination with 1 sun equivalent white light in air causes a severe reduction of the measured minority carrier lifetime. We assign this to strong recombination at the CIGSe surface. No degradation is observed, when the CIGSe layers were exposed to air without illumination or illumination without air. X-ray photoelectron spectroscopy reveals a correlation between illumination time and concentrations of sodium and oxygen on the CIGSe surface. In comparison to non-illuminated absorber layers, solar cells with light exposed absorber layers show an up to 10% reduction of open circuit voltage, fill factor, and efficiency. However, upon light soaking at 85°C of the complete solar cell, a recovery of J-V parameters for devices with illuminated absorbers is achieved. Both effects, the degradation of bare absorbers and the parameter recovery of complete solar cells, are related to a light induced segregation of Na at the surface/interface.
Keywords: $Cu(In,Ga)Se_2$, Degradation, Light-Soaking, Interfaces, Manufacturing and Processing

1 INTRODUCTION

The material system $Cu(In,Ga)Se_2$ (CIGSe) is well suited for high efficiency thin film solar cells [1].

Nevertheless, the knowledge about the dominant recombination mechanism of CIGSe devices, e.g. limitation of the open circuit voltage by bulk or interface recombination, is still under debate [2]. In addition, a CIGSe layer with its large number of host and doping elements can exhibit some softness in terms of material properties where electrochemical potential gradients can drive diffusion and segregation.

In this work we investigate the impact of illumination in ambient air to the minority charge carrier lifetime of bare CIGSe layers by means of time-resolved photoluminescence (TRPL) [8][9]. TRPL, in principle, can discriminate between several recombination paths [3]. However, due to disturbing effects such as trap states [4], charge carrier separation [5][6], spatially inhomogeneous doping densities, and spatially inhomogeneous minority charge carrier lifetimes [7] an identification of the dominating recombination mechanism from TRPL measurements is hardly possible. Thus, at first instance the TRPL decay time is interpreted as an effective lifetime, $\tau_{n,eff}$, for minority charge carriers [10][11].

Combining the TRPL experiments with X-ray photoelectron spectroscopy (XPS), we find a modified CIGSe surface chemistry for light exposed samples.

We show that solar cells containing such illuminated absorber layers reveal inferior J-V parameters and find a clear signature of interface recombination from temperature-dependent and time-dependent device analysis. Fortunately, the device parameters can be recovered by light soaking (LS) at 85°C.

Our results may be of interest for manufactures since the combination of air and light exposure of bare CIGSe surfaces may well occur in production environment.

2 EXPERIMENTAL

2.1 Sample preparation and light soaking

The 2.8 µm thick absorbers of the investigated CIGSe samples were grown by a co-evaporation 3-stage process, described in Ref. [12]. As substrates, Mo coated soda lime glasses (SLG) were deployed. For incorporation of sodium into the CIGSe layer the glass acts as Na source at the used growth temperature of 625°C. By energy dispersive X-Ray spectroscopy (EDX) the integral elemental ratios $\frac{[Cu]}{[Ga]+[In]}$ and $\frac{[Ga]}{[Ga]+[In]}$ of the grown CIGSe layers were determined to 0.85-0.90 and 0.26-0.29, respectively.

After CIGSe deposition, the 2 × 2 cm^2 samples were investigated by TRPL and stored for 16 ± 2 hours in a light sealed N_2 purged dry box. Following this, one sample was kept in darkness under ambient conditions (air with ~50% relative humidity, 20°C) while another sample was illuminated with 1 sun equivalent white light for 60 min in ambient air at 20°C. The development of the effective minority charge carrier lifetime of the CIGSe layers was recorded by measuring TRPL directly before and after the LS procedure. Then the samples were either transferred to the UHV system of the XPS setup or received the CdS buffer layer. To ensure that the bare absorbers only differ in the yielded light dose we dimmed the light of our laboratory. Further the samples were transported in light-sealed carriers and stayed for an identical period in ambient air. More detailed information is given in Ref. [13].

By sputtering a 50 nm intrinsic and a 350 nm Al doped ZnO layer the samples were completed to solar cells. For electrical characterization a Ni/Al/Ni metal grid was evaporated on top of the layer stack. Three 0.5×1 cm^2 cells were defined by mechanical scribing on each sample. In order to investigate the impact of light soaking at 85°C of complete solar cells the samples were divided into 3 segments with one 0.5×1 cm^2 cell per segment.

2.2 Methods

TRPL measurements were carried out under ambient conditions with a home-built system described in Ref. [14].

To study the chemistry of CIGSe surfaces, a commercial XPS system (*Omicron, Taunusstein*) equipped with a 125 mm hemispherical analyzer was used. For excitation, non-monochromatized Al-K_α radiation was applied. The measurements were conducted at a pass energy of 30 eV.

Depth profiles of the Na distribution were measured using a *SPECTRUMA ANALYTIC* glow discharge optical emission spectroscopy (GDOES) system.

Current-voltage analyses at 25°C and LS at 20°C/85°C were performed with a tungsten halogen lamp. The calibrated photon flux density of the lamp was equivalent to 1 sun. A Mo covered SLG dummy placed next to the samples was used to adjust the temperature of the illuminated side of the devices.

For time dependent open circuit voltage $V_{OC}(t)$ and temperature dependent $V_{OC}(T)$ measurements a *SCIENCETECH Inc.* sun simulator with AM 1.5 filter was used. The red light for the $V_{OC}(t)$ measurements was realized by placing a long pass filter with a cut-off wavelength of $\lambda \approx 665$ nm into the simulator beam path.

3 RESULTS

3.1 CIGSe absorber layers

Fig. 1 gives the transient photoluminescence (PL) intensity of bare absorbers as a function of time after the excitation pulse. The blue decay curve named as "No light" represents the PL decay of CIGSe absorber layers in the initial state, i.e. before starting the LS experiment. Illumination with white light (1 sun equivalent) for 60 min in ambient air (air-light-soak) induces a substantial reduction of $\tau_{n,eff}$ from ~280 ns (blue curve) down to ~4 ns (red curve) as shown in Fig. 1.

In contrast, the exposure of the reference sample for 60 min to the identical ambient conditions but without illumination exhibits no impact on the recombination dynamics. Therefore the TRPL transient of the non-illuminated CIGSe absorber coincides with the blue curve in Fig. 1.

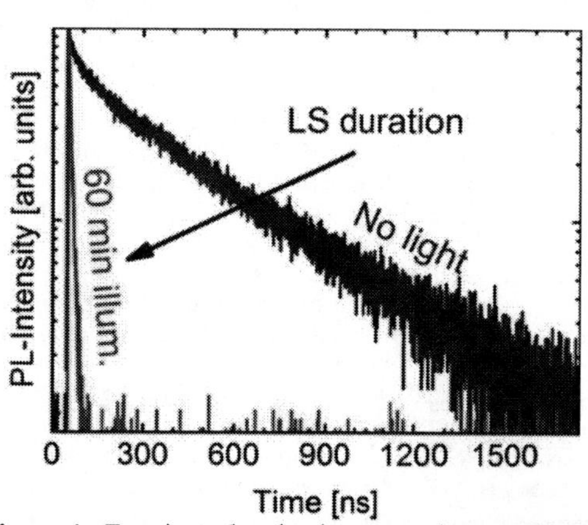

Figure 1: Transient photoluminescence decays of bare CIGSe absorbers before LS and after 60 min air exposition without illumination (blue) as well as after 60 min 1 sun equivalent white light exposure in ambient air (red).

The modification of the CIGSe surface after 60 min illumination in ambient air as revealed by XPS is depicted in Fig. 2 (red line). In comparison to the 60 min air exposed sample without illumination (blue line), the air-light-soaked sample has increased photoemission intensities of Na 1s and O 1s core levels together with peak shifts towards lower binding energies. The Se 3d signal in Fig. 2 shows an additional Se-O [15] peak for the air-light-soaked sample. The formation of Me-O bonds (Me = Cu, In, Ga) was not detected. The systematic of the TRPL and XPS results in Fig. 1 and 2 was perfectly reproduced in a second complete experimental series.

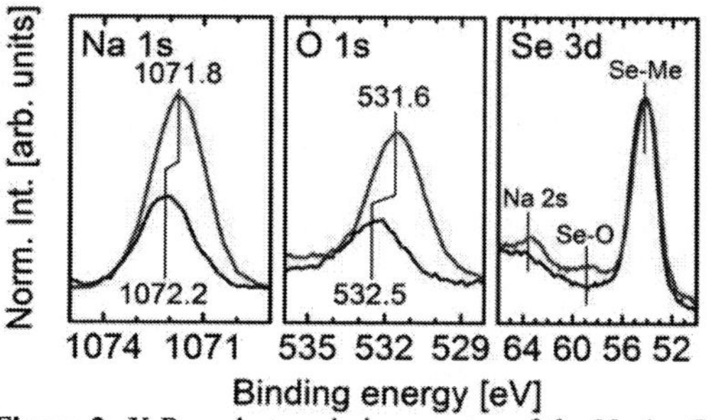

Figure 2: X-Ray photoemission spectra of the Na 1s, O 1s, and Se 3d core levels for a 60 min in ambient air illuminated CIGSe layer (red) and a CIGSe layer exposed to identical ambient conditions without illumination (blue).

3.2 Solar cells

CIGSe layers with and without air-light-soak were fabricated to complete solar cells. The J-V parameters extracted from current voltage measurements are given in Fig. 3. Devices containing air-light-soaked CIGSe absorbers reveal a clearly reduction of V_{OC} and fill factor, FF, resulting in a downsized power conversion efficiency, η_{Cell}. No trend for the short circuit current density, J_{SC}, is observed in Fig. 3.

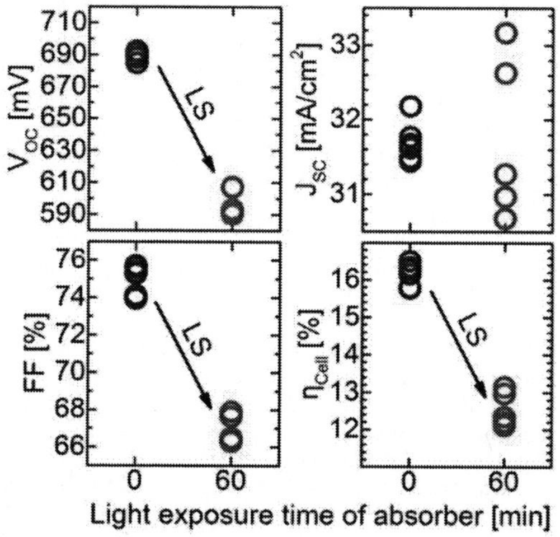

Figure 3: J-V parameter of solar cells containing air-light-soaked CIGSe absorbers (red circles) and non-soaked absorbers (blue circles). The time of air-light-soak is given for discrimination.

However, it is clear that the effect of air-light-soak is persistent during the solar cell fabrication step.

In the next step, the complete solar cells were light soaked at 85°C in air. Fig. 4 shows that V_{OC}, FF, and η_{Cell} recover after 4 hours LS almost up to the level of the reference sample (non air-light-soaked absorber). Only J_{Sc} decreases by ~1 mA/cm² regardless of the absorber treatment. Again, by repeating the LS at 85°C on another set of samples (not shown) all J-V parameter, revealed the identical recovery like in Fig. 4 and converged to a constant level independent from the air-light-soak history of the absorber.

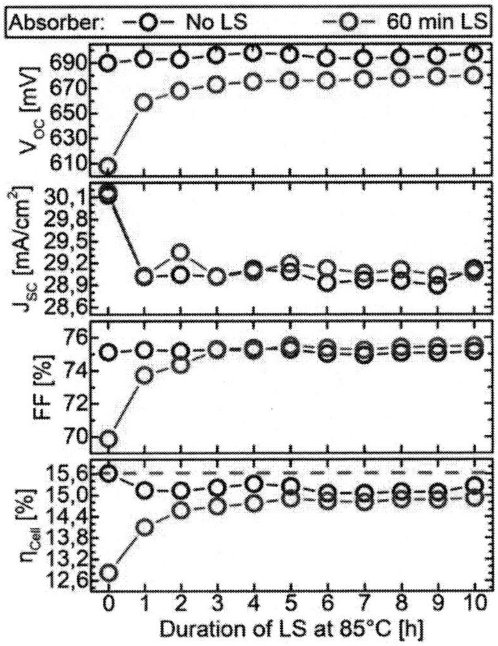

Figure 4: Progression of J-V parameters after light exposure at 85°C of complete solar cells, differing in the LS state of the CIGSe absorber layers. Blue circles: solar cell containing a non-illuminated 60 min air exposed absorber. Red circles: solar cell consisting of an absorber experienced 60 min illumination in ambient air.

4 DISCUSSION

With single wavelength TRPL under linear excitation (like in our case) no distinction between surface and bulk recombination is possible. Thus, we cannot conclude on surface or bulk recombination being the reason for the observed $\tau_{n,eff}$ and V_{oc} reduction by absorber air-light-soak.

To verify surface recombination as the reason for air-light-soak degradation we investigated complete devices by $V_{oc}(T)$ and $V_{oc}(t)$ experiments. The open circuit voltage as a function of temperature for devices based on air-light-soaked and non-air-light-soaked CIGSe absorbers is depicted in Fig. 5. With a linear fit of $V_{OC}(T)$ and extrapolation to T = 0 K the activation energy E_a of the saturation-current density equals $E_a = q \cdot V_{OC}(0K)$, with q the elemental charge. An activation energy of ~1.24 eV and an absorber band gap of ~1.15 eV indicates bulk recombination as the dominant mechanism under open circuit conditions for cells with non-illuminated CIGSe layers (no air-light-soak) [13].

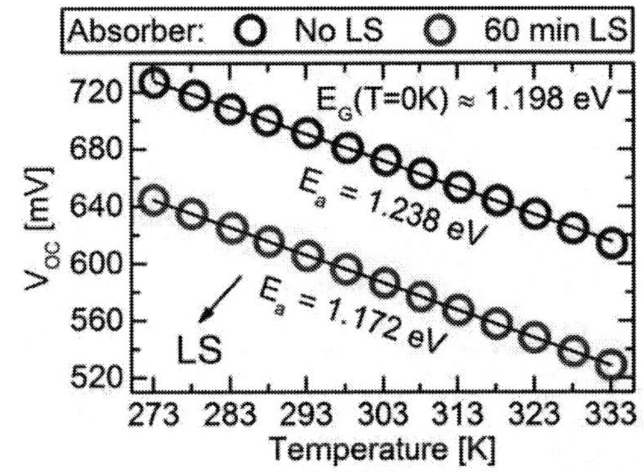

Figure 5: Temperature dependent open circuit voltage of solar cells with CIGSe absorbers illuminated for 60 min in ambient air (red circles) and the non-illuminated reference exposed to identical ambient conditions for the identical duration (blue circles). The activation energy Ea was evaluated by a linear fit of the data (black lines).

In the case of cells with air-light-soaked absorbers we find $E_a \approx 1.17$ eV. According to Ref. [16], E_a is interpreted as an interface recombination limited saturation-current density at V_{oc}. With the assumption of identical band offsets at absorber/buffer and buffer/window interfaces as well as a completely depleted buffer layer, the reduction of E_a could be explained by an increased interface recombination velocity [16].

To confirm our interpretations of $V_{OC}(T)$ measurements, we recorded open circuit voltage transients under red light illumination. The red light effect of CIGSe solar cells is known since long and causes an increased p-type conductivity of the CIGSe absorber [17][21].

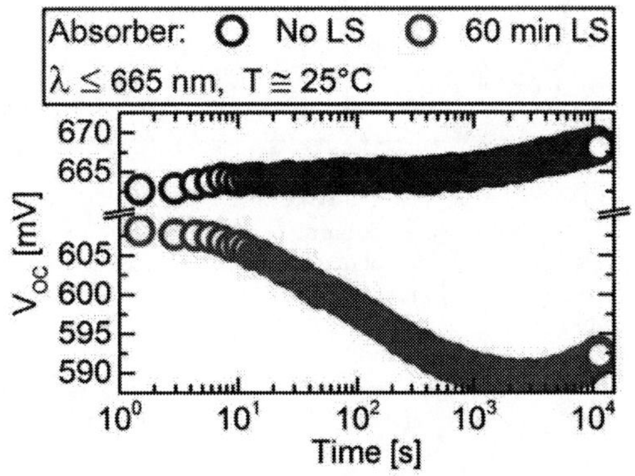

Figure 6: Time dependent open circuit voltage of solar cells with air-light-soaked CIGSe absorbers (red circles) and 60 min air exposed absorbers without illumination (blue circles).

For an inverted absorber layer at the interface, with Eq. 1 after Ref. [17] we can distinguish between an interface recombination or bulk recombination limited solar cell. In the case of interface recombination the time dependent open circuit voltage is given by

$$V_{OC}(t) \propto \frac{1}{N_{A,a}(t)} ln\left(N_{A,a}(t)\right). \tag{1}$$

Here $N_{A,a}(t)$ denotes the doping density as a function of time of the CIGSe absorber. With $dN_{A,a}/dt > 0$ under red light illumination only the red curve for the LS absorber in Fig. 6 fulfills the relation in Eq. 1. Hence air-light-soak of the absorber induces interface recombination. Based on the former experiments, the TRPL degradation of the CIGSe layer in Fig. 1 is explained by an increased surface defect density after illumination.

The XPS measurements in Fig. 2 suggest that the degradation of CIGSe layers is related to a segregation of sodium at the CIGSe surface. To verify the supporting role of sodium, the same experiment as in Fig. 1 was performed on a sample without sodium where the absorber was grown on barrier glass. No degradation was observed after air-light-soak of the CIGSe layer. On the other hand 60 min illumination in high vacuum of a Na containing sample without air exposure revealed also no TRPL degradation. This means the light induced degradation process takes only place when both species Na as well as oxygen (from O_2, H_2O, OH) are present. Therefore, we interpret the degradation of CIGSe layers under illumination in ambient air as sodium induced oxygenation at the very CIGSe surface. Heske *et al.* [18] reported the passivating properties of metallic sodium as well as the increased defect density at the CIGSe surface caused by a reaction of sodium and oxygen (in form of H_2O).

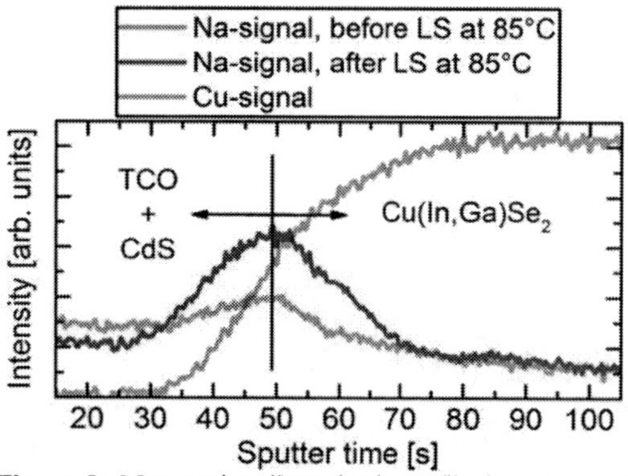

Figure 8: Measured sodium depth profile by means of GDOES for CIGSe solar cells before (orange) and after (green) 10 h illumination with 1 sun equivalent white light at 85°C.

The passivating properties of metallic Na may well explain the improved J-V parameter after LS at 85°C. In Fig. 7, the distribution of Na after LS of the complete cell reveals an increased amount of sodium at the CIGSe/CdS interface. Again we want to emphasize the role of light during the solar cell LS at 85°C. By applying a V_{OC} corresponding forward bias voltage of 0.7 V at 85°C in darkness for 14 hours no improvement of solar cells containing light exposed absorbers was present. Directly afterwards by applying a LS at 85°C for 1 h, a substantial improvement of cell parameters was observed. Therefore we refer the segregation of Na after LS to a light enhanced diffusion mechanism. Our findings are in a good agreement with the results of Chen *et al.* [19]. They reported an identical Na segregation at the CIGSe/CdS interface by applying LS at 25°C to complete solar cells.

5 CONCLUSION

To summarize, air-light-soak of uncoated CIGSe layers induces an increase in surface recombination which degrades the minority carrier lifetime of the bare absorber as well as the solar cell parameters of complete devices.

XPS measurements indicate an enrichment of sodium as well as oxygen at the air-light-soaked CIGSe surface, which is not removed by the CdS chemical bath. Only the combination of sodium, light, and air trigger the surface recombination.

Light soaking at 85°C of complete solar cells based on air-light-soaked absorbers exhibits a recovery of the J-V parameters concomitant with a segregation of sodium at the CIGSe/CdS interface.

Our results point towards a twofold role of sodium, inducing recombination as well as passivation of CIGSe surfaces and interfaces.

6 ACKNOWLEGMENT

The funding by the Deutsche Forschungs-gemeinschaft (DFG) under project SCHE 1745/4-1 is highly appreciated.

7 REFERENCES

[1] P. Jackson, R. Wuerz, D. Hariskos, E. Lotter, M. Powalla, Phys. Status Solidi RRL 10, 8 (2016) 583-586.
[2] H. Mönig, Ch.-H. Fischer, A. Grimm, B. Johnson, C. A. Kaufmann, R. Caballero, I. Lauermann, M. Ch. Lux-Steiner, J. Appl. Phys. 107, 113440 (2010).
[3] R. K. Ahrenkiel, N. Call, S. W. Johnston, W. K. Metzger, Sol. Energy Mater. Sol. Cells 94, (2010) 2197-2204.
[4] M. Maiberg, T. Hölscher, S. Zahedi-Azad, W. Fränzel, R. Scheer, Appl. Phys. Let. 107, 1221104 (2015).
[5] D. Kuciauskas, J. V. Li, A. Kanevce, H. Guthrey, M. Contreras, P. Dippo, K. Ramanathan, J. Appl. Phys. 117, 185102 (2015).
[6] M.. Maiberg, C. Spindler, E. C. Jarzembowski, R. Scheer, Thin Solid Films 582, (2015) 379-382.
[7] M. Maiberg, F. Bertram, M. Müller, R. Scheer, J. Appl. Phys. 112, 085703 (2017).
[8] W. K. Metzger, I. L. Repins, M. Romero, P. Dippo, M. A. Contreras, R. Noufi, D. Levi, Thin Solid Films 517, 7 (2009) 2360-2364.
[9] W. K. Metzger, I. L. Repins, M. A. Contreras, Appl. Phys. Let. 93, 022110 (2008).
[10]I. L. Repins, B. Egaas, L. M. Mansfield, M. A. Contreras, C. P. Muzzillo, C. Beall, S. Glynn, J. Carapella, D. Kusciauscas, Rev. Sci. Instrum. 86, 013907 (2015).
[11]M. Maiberg, R. Scheer, J. Appl. Phys. 116, 123711 (2014).
[12]E. C.Jarzembowski, M. Maiberg, F. Obereigner, K. Kaufmann, S. Krause, R. Scheer, Thin Solid Films 576 (2015) 75-80.

[13]T. Hölscher, S. Förster, T. Schneider, M. Maiberg, W. Widdra, R. Scheer, Appl. Phys. Let. 111, 011604 (2017).
[14]M. Maiberg, C. Spindler, E. C.Jarzembowski, R. Scheer, Thin Solid Films 582, (2015) 379-382.
[15]C. Heske, R. Fink, E. Umbach, W. Riedl, F. Karg, Appl. Phys. Let. 68, 24 (1996) 3431-3433.
[16]H. Wilhelm, H. W. Schock, R. Scheer, J. Appl. Phys. 109, 084514 (2011).
[17]R. Scheer, H. W. Schock, Chalcogenide Photovoltaics: Physics, Technologies and Thin Film Devices, Wiley-VCH, Weinheim (2011).
[18]C. Heske, G. Richter, Z. Chen, R. Fink, E. Umbach, W. Riedl, F. Karg, J. Appl. Phys. 82, 5 (1997) 2411-2420.
[19]S. Chen, T. Jarmar, S. Södergen, U. Malm, E. Wallin, O. Lundberg, S. Jander. R. Hunger, L. Stolt, Thin Solid Films 582, (2015) 35-38.
[20]S.-H. Wei, S. B. Zhang, A. Zunger, J. Appl. Phys. 85, (1999) 7214-7218.
[21]F. Obereigner, N. Barreau, W. Witte, and R. Scheer, J. Appl. Phys., 117, 055704 (2015).

CU(IN,GA)SE₂ THIN FILMS AND MODULES FABRICATED ON POLYIMIDE FOILS BY THE IN-LINE EVAPORATION PROCESS USING THERMAL CRACKED SELENIUM

H Wang[1], F F Liu[2], L Y Yao[1], H Zhang[1], P Gao[1], R B Liu[1], Z B Xiao[1], Q Sun[1]
1.Tianjin Institute of Power Source, Tianjin 300381, PR China
2.The Institute for Photoelectronic Thin Film Devices and Technology Nankai University, Tianjin 300071, PR China
Correspondence email: wanghe632@163.com

ABSTRACT: CIGS films were fabricated by the in-line evaporation process with the thermal cracked Se. The films growth with the thermal cracking Se temperature (T_c) rising to 600°C, had unique properties of large grain size, highly dense smooth surfaces and appropriate carrier concentrations. The improvement on the properties of CIGS films was attributed to the modification of the growth kinetics due to the presence of active Se-radical species. Finally, The monolithic integrated submodule with segments of 14 and the active area of 115.9cm^2 were fabricated on PI foils and achieved the efficiency of 11.3%.
Keywords: CIGS, thermal cracked Se, flexible CIGS submodule, PI foils, growth kinetics

1 INTRODUCTION

Currently, the efficiency of the laboratory-scale cells fabricated by co-evaporation process have been over the polysilicon cells [1]. Moreover, some manufactories in USA and Europe have come true the industrialization of CIGS thin film modules on PI web by in-line co-evaporation process [2]. The quality of crystallization of CIGS films with large area, which is the most critical point to develop the high efficiency solar cells and modules, depended on the deposition temperature and process strongly. However, the dis-matching of the thermal expansion coefficient among polyimide (PI) substrate, Mo contact layer and CIGS film was a serious problem when the substrate temperature (T_{sub}) was of 450oC and above. Lower T_{sub} constrained the atoms deposited onto substrate gaining enough thermal energy, resulting in decreasing the activity of elements and the reaction velocity of the combination of CI GS phase.

To solve this issue, thermal cracked Se technique was applied to grow large area CIGS films by the in-line evaporation process. According to the previous results, utilizing the cracked Se by RF-plasma or high temperature heating led to the significant reduction in the amount of Se material used [3-5]. However, the improvement on the structural properties of CIGS films deposited on the rigid substrate and the performance of solar cells were not prominent for the films deposited at Stub>500°C and/or with the normal Ga content ($0 \leq x < 0.5$) [6-8]. It was possible that cracked Se technique had more contribution for high photovoltaic quality CIGS films deposited at T_{sub}<500°C.

In this paper, CIGS films were grown on PI substrate using the thermal cracked Se and conventional evaporative Se in the in-line co-evaporation process, respectively. The amount of Se material consumed during the two processes was compared. The impact of the thermal cracked Se on the structural and electrical properties of CIGS absorber layer were investigated. Furthermore, we optimized the Tc and fabricated CIGS modules on PI foils.

2 EXPERIMENTS

PI foils are coated with a molybdenum layer of about 1μm by dc sputtering. The process and parameters are optimized to balance the stress to ensure sufficient adhesion among the PI foils, Mo layer and CIGS film, and to gain high enough conductivity. After the Mo laser-patterning process (P1), CIGS layer is deposited by in-line co-evaporation at constant substrate temperatures of ~480°C with and without the thermal cracked Se. The base pressure of the growth chamber was about 3×10^{-4} Pa. The cracked Se source was mostly made up of three parts, including the Se evaporator, a piping system and high-temperature cracked unit. During the process, the temperature of the Se evaporator and the piping system was kept constant at 300°C in order to maintain a stable and sufficient Se atmosphere throughout the whole layer growth and prevent the condensation of Se, while the temperature of the cracked Se unit were varied from 450°C to 750°C. The metal elements of Ga, In, Cu were evaporated onto the substrate sequentially, and formed over-stoichiometric CIGS with segregated phase of CuxSe, then followed by depositing Ga and In to obtain the under-stoichiometry CIGS films. The typical growth rate of CIGS film was 75nm/min and the final thickness was of 1.5μm-1.7μm.

Subsequently, the CIGS films were covered by an approximately 70nm thick CdS layer before the P2 patterning, a high-resistance ZnO layer and an Al-doped high-conductivity ZnO layer as transparent conductive oxide (TCO) were deposited by sputtering with thickness of 50 nm and 600 nm, respectively [9]. After the TCO layer deposition, P3 patterning was performed and completed module structure.

The chemical composition and structural properties of the CIGS films were measured by the XRF-800 x-ray fluorescent spectrometer (XRF) and a Rigaku TTR-III X-ray diffractometer with Cu Kα radiation (XRD). The resolution in theta of xray diffractometer is 0.001°. S-4800 Scanning electron microscopy (SEM) with a cold cathode field emission electron source, was used to characterize the morphology and grain size of CIGS films. The carrier concentration and resistivity of CIGS films were determined by the HL5550 Hall effect measurement apparatus of Accent Optical Technologies, Inc., at room temperature. The current-voltage characteristics of the solar cells were measured under the AM 1.5 spectrum with 1000 W/m^2.illuminations at 25°C.

3 RESULTS AND ANALYSIS

The experiment authenticated that the utilization ratio of raw Se material during the in-line evaporation process was about 10.35% for the thermal cracked Se source and

3.57% for the conventional Se source, respectively. The data in Table 1 indicated that a significant reduction in amount of Se material consumed by using the thermal cracked Se source.

Table I: The comparison of Se material consumption during the process with and without thermal cracked Se technique

	conventional evaporator Se	thermal cracked Se
Consumption	1000g	350g
Process time	20 hour	58 hour
Deposition area of CIGS films	10.8 m^2	31.32m^2
Utilization ratio	3.57 %	10.35%

CIGS films grown under nominally identical conditions, except for difference in the temperature of thermal cracked Se (T_c), have the uniform thickness of 1.5-1.7μm, Cu/(Ga + In) in the range of 0.80～0.90, and Ga/(Ga + In) in the range of 0.31～0.36.

Figure1 shows (112) Bragg peaks of CIGS films grown with Tc varied from 0oC - 750°C. Tc with value of 0oC indicated that a conventional evaporator Se source was used for the growth of CIGS film. Under this condition, a multi-peak structure with some broad shoulders towards higher diffraction angles revealed the presence of a graded CuIn1-xGaxSe2 phase, i.e. the In-rich and Ga-rich phase coexisted in CIGS films. Moreover, additional reflection of secondary-phase In_2Se_3 (110) (JCPDS01-071-0250) had been detected, mainly ascribed to the lower substrate temperature [9-10]. However, narrower FWHM values of (112) peaks and the disappearance of secondary-phase were observed for thermal cracked Se source grown CIGS. As $T_c \geq 600$°C, the multi-peak structure of (112) peak vanished gradually, and almost formed the single peak structure with small shoulder. The result indicated that the graded $Cu(In_{1-x}Ga_x)Se_2$ phase translated into single CIGS phase with the narrower FWHM of (112) peaks by enhancing the cracked temperature of T_c. It implied that using the thermal cracked Se source and higher Tc improved the crystal quality of CIGS films and was likely to affect the diffusion of In and Ga during the film growth.

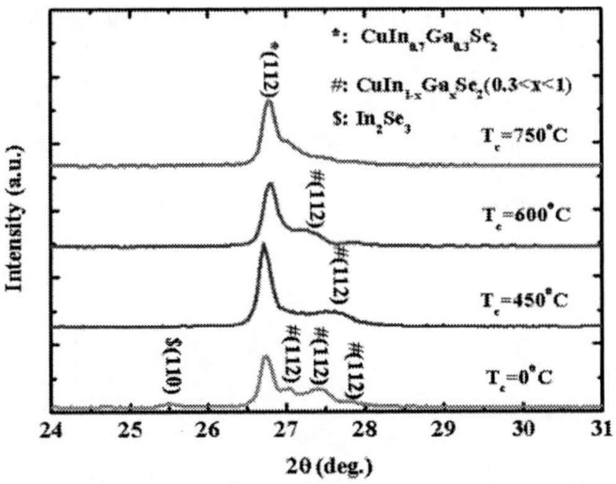

Figure 1: XRD spectra of the (112) peak of CIGS films grown with thermal cracked Se at various T_c. Tc=0°C denoted using the conventional evaporator Se.

Figure2 (a)-(d) show SEM images of surface and cross section of CIGS films as a function of T_c. CIGS films grown with a conventional evaporator Se exhibited discontinuous grains and many crevices on the surface. However, a highly dense and smooth surface was observed from the films grown with a thermal cracked Se source. Furthermore, the cross-sectional image showed that the grain size enhanced with the increase of Tc, and the columnar grains formed at T_c=600 °C with a typical grain size 1-1.5μm along the cross-sectional direction. The morphology of CIGS films had not obvious difference with T_c over of 600°C. The improvement of crystal quality of CIGS films consistent with the results shown in Figure1 was mainly attributed to the modification of the growth kinetics due to the higher reactivity of the active Se species produced and the resulting enhanced migration on the growing film surface, although the substrate temperature was low. With the increase of T_c, more Se radical species promoted the reactions among selenium, the metal elements and/or binary selenide, and formed single CIGS phase. It implied that the active Se radical species might make up for the deficiency of CIGS films deposited at low substrate temperature.

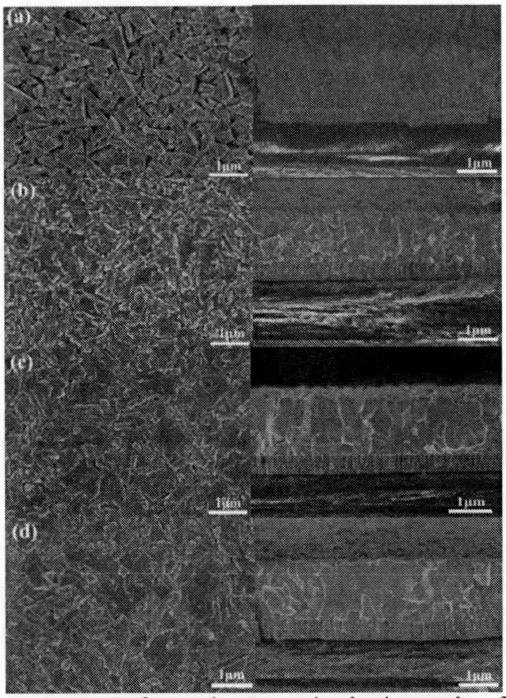

Figure 2: Surface and cross-sectional micrographs of CIGS films deposited on PI/Mo substrates with and without thermal cracked Se: (a) The conventional evaporator Se, (b) with thermal cracked Se and T_c=450°C, (c) T_c=600°C, (d) T_c=750°C.

The alteration of growth kinetics also influenced the electrical properties of CIGS films, as shown in Figure3. The carrier concentration (N_A) increased rapidly as $T_c >$ 450°C, and trended towards the saturation with Tc over of 600°C. The resistivity curve had almost the reverse tendency of N_A with the variation of T_c. The improvement of NA and resistivity of CIGS films probably derived from the passivation of donor defect, i.e. Se vacancies, and the elimination of secondary-phase by using thermal cracked Se source with higher T_c.

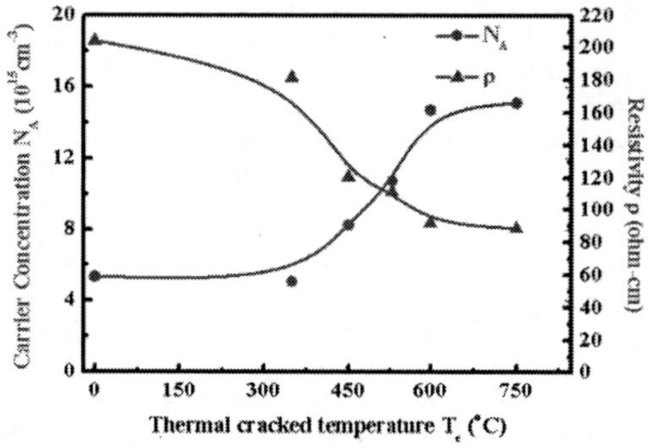

Figure 3: The carrier concentration N_A and resistivity ρ of CIGS films deposited with different T_c.

The melioration of the crystalline quality and electrical properties of CIGS films were expected to be beneficial, especially for improving the performance of low-temperature grown CIGS solar cells and modules. The monolithical integrated submodules with segments of 14 and the active area of 115.9cm^2 were fabricated on PI foils. The I-V curve was shown in figure 4. A submodule efficiency of 11.3% (10cells, open circuit voltage (V_{oc}) 8.02V, short circuit current 262mA, fill factor (FF) 0.625) had been demonstrated with a 480°C-grown CIGS absorber layer by using the thermal cracked Se source with T_c=600°C.The improvement on the structural and electrical properties of CIGS films was attributed to the modification of the growth kinetics due to the higher reactivity of the active Se species and enhanced the reactions between Selenium and elemental metals.

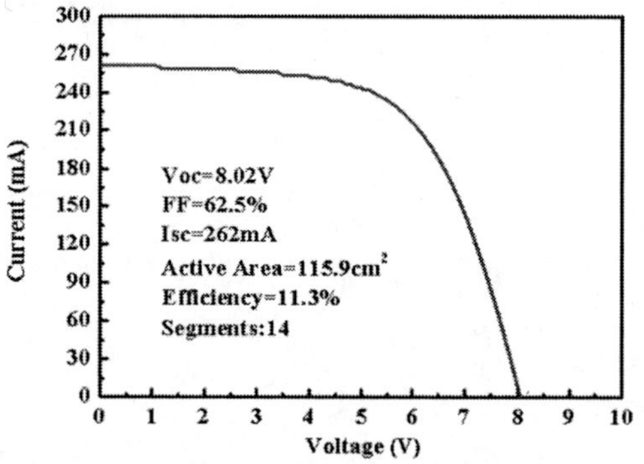

Figure 4: The I-V curve of CIGS module fabricated on PI substrate with thermal cracked Se at T_c=600°C.

4 CONCLUSION

CIGS films were fabricated by the in-line evaporation process at the T_{sub} of 480°C with and without the thermal cracked Se. The films growth using the cracked Se with thermal cracking temperature (T_c) over of 450°C, exhibited single-phase CIGS without secondary phase In_2Se_3. As the Tc rising to 600°C, CIGS films had unique properties of large grain size, highly dense and smooth surfaces. The carrier concentrations of CIGS films also increased with raising the Tc. The improvement on the structral and electrical properties of CIGS films was attributed to the modification of the growth kinetics due to the higher reactivity of the active Se species and enhanced the reactions between Selenium and elemental metals.

Using the thermal cracked Se with T_c=600°C, the monolithic integrated submodule with segments of 14 and the active area of 115.9cm^2 were fabricated on PI foils and achieved the efficiency of 11.3%.

ACKNOWLEDGEMENT

This work was supported by the National Natural Science Foundation of China (61404118), National Natural Science Foundation of China (61504067).

REFERENCES

[1] A. Chirilă, S. Buecheler, F. Pianezzi, P. Bloesch, C. Gretener, A. R. Uhl, C. Fella, L. Kranz, J. Perrenoud, S. Seyrling, R. Verma, S. Nishiwaki, Y. E. Romanyuk, G. Bilger and A. N. Tiwari, NATURE MATERIALS vol.10 (2011) 857.

[2] F. Kessler, R. Würz, S. Spiering, D. Bremaud, Proceeding of 22nd International Photovoltaic Science and Engineering Conference, Hangzhou China, (2012).

[3] Miguel A. Contreras, Ingrid Repins, Wyatt K. Metzger, Manuel Romero and Daniel Abou-Ras. Phys Status Solidi A, vol. 206 (2009) 1042–1048.

[4] Shogo Ishizuka, Akimasa Yamada, Hajime Shibata, Paul Fons, Shigeru Niki. Thin Solid Films, vol. 519 (2011) 7216-7220.

[5] H. Zachmann, S. Puttnins, F. Daume, A. Rahm, K. Otte. Thin Solid Films, vol. 519 (2011) 7317-7320.

[6] Shogo Ishizuka,a Hajime Shibata, Akimasa Yamada, Paul Fons, Keiichiro Sakurai,Koji Matsubara, and Shigeru Niki. Applied Physics Letters, vol. 91 (2007) 0419021-0419023.

[7] Hendrik Zachmann, Andreas Rahm, Karsten Otte, Tobias Eisenbarth, Thomas Unold. Proceeding of 24th European Photovoltaic Solar Energy Conference, Hamburg, Germany (2009) 2425-2428,

[8] Shogo Ishizuka, Akimasa Yamada, Hajime Shibata, Paul Fons, Keiichiro Sakurai, Koji Matsubara, Shigeru Niki. Solar Energy Materials & Solar Cells, vol. 93 (2009) 792–796.

[9] A. Chirilă, P. Bloesch, A. Uhl, S. Seyrling, F. Pianezzi, S. Buecheler, C. Fella, S. Nishiwaki, Y.E. Romanyuk, and A.N. Tiwari. Proceeding of 25th European Photovoltaic Solar Energy Conference, Valencia, Spain, (2010) 3403-3405.

[10] H. Wang, Y. Zhang, X.L. Kou, Y.A. Cai, W. Liu, T. Yu, J.B. Pang, C.J. Li and Y. Sun. SEMICONDUCTOR SCIENCE AND TECHNOLOGY, vol.25 (2010) 055007-055 013.

HIGH-SPEED CURING OF AR COATINGS ON PV MODULES BY LASER IRRADIATION: AN END-OF-LINE APPROACH FOR IMPROVED POWER OUTPUT

Romain Cauchois[1], Dominik Hawelka[2], Norbert Pirch[2], Carsten Johnigk[2], Li Yi[3], Jochen Stollenwerk[2], Hugo Schoot[3]
[1] DSM Materials Science Center | Urmonderbaan 22 | P.O. Box 18 | 6160 MD Geleen | The Netherlands
[2] Fraunhofer Institute for Laser Technology | Steinbachstraße 15 | 52074 Aachen | Germany
[3] DSM Innovation Center | Urmonderbaan 22 | P.O. Box 18 | 6160 MD Geleen | The Netherlands

ABSTRACT: The aim of this work was to up-scale a process for curing AR coatings on cover glass of PV modules based on laser irradiation. To achieve high throughput on a production line, an approach combining the integration of high power laser generators, the technological development of a high-speed laser beam scanner, and the process engineering to tailor the final performance of the AR coating has been conducted. The scientific innovation resides in the development of a high-speed curing technology of silica-based antireflective coatings on the cover glass of PV modules, by applying a high energy density radiation to the AR coating while scanning at high speed (~2x speed of sound). This process was successfully applied to cure the coating on a full-size (0.6 m x 1.2 m) thin-film module with a processing rate of about 18 m²/h, leading to a theoretical throughput of 25-30 modules per hour. This approach is therefore particularly relevant to coat and cure assembled modules in an end-of-line process.

Keywords: Antireflection coating, Laser processing, Module manufacturing, Modelling

1 INTRODUCTION

Anti-reflective (AR) coatings are commonly used in solar applications at the silicon cell level and at the cover glass level. In the latter case, a common approach is to deposit an interference coating based on silica sol-gel chemistry using a large-area liquid-phase deposition technique, such as roll-, aerosol-, or spray-coating.

Conversion of the silica sol deposited on a cover glass requires thermal energy to provide higher mechanical strength and long-term durability performance. While this operation is generally conducted on the production line of the glass manufacturer, it could also be performed at the end of the PV module assembly line. The benefit of applying the ARC at the end of the module line reduces the risk of damaging the coating during harsh module production process steps, especially in thin-film module production, and during module assembly.

In that case, a critical requirement is to prevent degradation of the active layers and the encapsulant. A strict control of the thermal budget can be achieved using a high-intensity CO_2 laser irradiation (λ=10.6 µm) with a high scanning velocity.

2 MODELLING

Finite Element Modeling was conducted to predict the effect of the laser output power, the beam diameter, and the scanning velocity on the extent and temperature profile of the effective Heat Affected Zone. The temperature field is based on a solution of the nonlinear heat equation:

$$\frac{\partial(\rho \cdot C_p \cdot T)}{\partial t} = \nabla[\kappa \cdot \nabla T] + A \cdot I(x(t),\, y(t),\, z)$$

where r is the density of the material, Cp is the heat capacity, k is the thermal conductivity, A is the absorbed fraction of the laser power, I is the intensity distribution of the laser beam, z is the working distance.

As the material properties are temperature-dependent and a layer system is modeled, there is no analytical solution of the heat equation. Thus, a numerical method is applied for the solution.

Figure 1 shows the calculated temperature distributions induced in the workpiece for the first pass of the laser beam. Due to the high scanning speed only a very thin layer (<100 µm) is heated to some 100 °C, whereas the substrate remains cold.

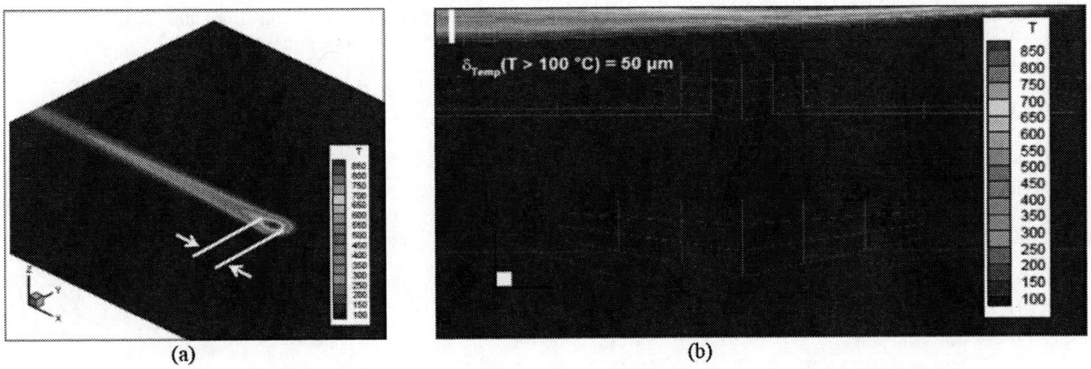

Figure 1: Results of FEM calculation of thermal field along a line scan (a) and corresponding depth profile along z-axis (b).

In **Figure 2** is shown the induced temperature-time cycles on the surface of the coating when scanning multiple overlapping lines at a fixed, but arbitrary position (green line). This way, the temperature is held above the curing temperature within the thin coating layer. The red line shows the maximum temperature induced in the whole substrate.

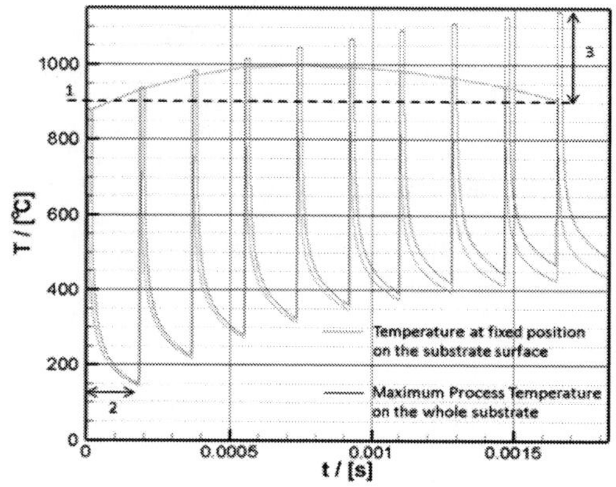

Figure 2: Calculated profile of temperature on the work pieceat a fixed position and on the whole substrate.

The thermal field in the coating generates an additional stress which may lead to cracks formation and propagation if it becomes higher than the coating strength [2,3]. In the presence of a crack in the film, propagation will happen when:

$$\sigma_{fracture} = \sqrt{\frac{2\gamma E l_p}{3\pi a_0 l_c}}$$

where γ is the surface energy per area, l_p is the radius at the crack tip, $2l_c$ is the length of the crack (assumed to be elliptical), E is the Young's modulus, a_0 is the atomic intraplane distance [2].

3 SYSTEM DESIGN FOR HIGH-POWER LASER SOURCE

3.1 High scanning speed using a polygon scanner

In order to achieve high scanning velocity, a standard galvanometer scanner is not sufficient (range of 3 m/s). Instead, a polygon scanner technology [1] using a multifaceted mirror rotating on an air suspended bearing at a speed up to 8000 rpm, has been implemented for this application. This corresponds to a scanning velocity up to 650 m/s.

Unlike the back and forth motion of the galvanometer scanners, polygon scanners rotate at a fixed speed leading to a high number of scanned lines per second (**Figure 3**). The homogeneous conversion of the coating and the high throughput of treatment are then achieved by optimizing the number of overlapping treatments via the scanning speed (y axis) and the axis speed (x axis).

3.2 Optical system for an increased processing width using a 6 kW laser

An optical system was designed and implemented to allow a large scanning area with a width up to 600 mm

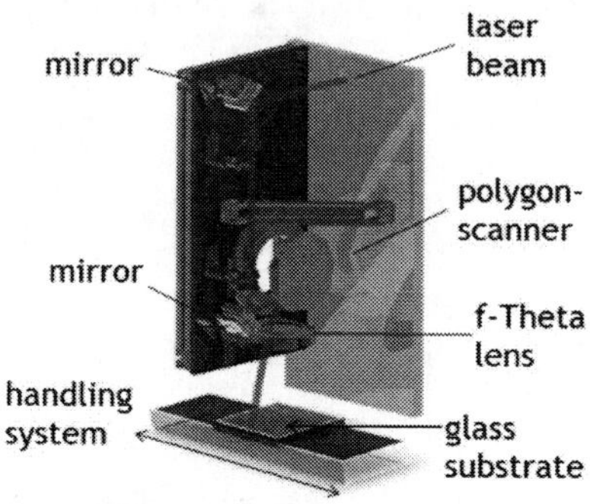

Figure 3: Schematic of the polygon scanner with the faceted mirror providing a linear scanning at high velocity

with a uniform intensity profile and minimal geometric aberrations at the edges. The scanning width mainly depends on the working distance and the focal distance of the focusing optics.

Control of the beam diameter and of the intensity profile homogeneity between the center and the edge of the scanning field is achieved by a zoom telescope using high-quality mirrors. Characterization of the laser parameters after the adaptation indicates negligible loss (ca. 5%) within the beam path.

The complete setup (see **Figure 5**) enables the application of a spatially-localized temperature above 1000 °C with a penetration depth limited to a few microns below the top surface (**Figure 1**). The temperature of the inner layers of the thin-film modules thus remains below a temperature of 30 °C at all time during the process.

4 PROCESS DEVELOPMENT

4.1 Optimization of scanning strategy

In order to realize a sufficient number of overlapping treatments at an industrially relevant processing rate, it can be advantageous to design the laser beam profile (circular or elongated e.g.) to have a shorter dimension parallel to the beam scanning direction (x-axis) and a longer dimension parallel to the substrate movement direction (y-axis). The effect of such strategies when the sample speed is increase is depicted in **Figure 4**.

The mechanical stability of the laser-treated coating depends directly on the number of overlapping treatments. Thus, with a given specific laser spot geometry a specific combination of the polygon speed and the axis speed leads to the required mechanical stability.

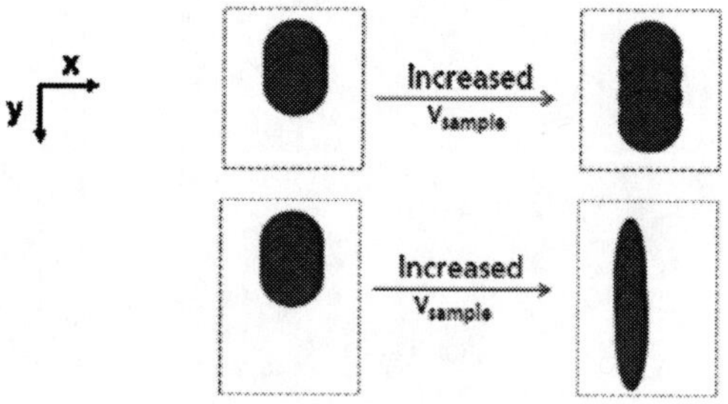

Figure 4: Overlapping of treatment in a scenario of circular (top) or elongated spot geometry (bottom).

Figure 5: Overview of the complete system with high-power CO2 laser and the conveying system for the PV module.

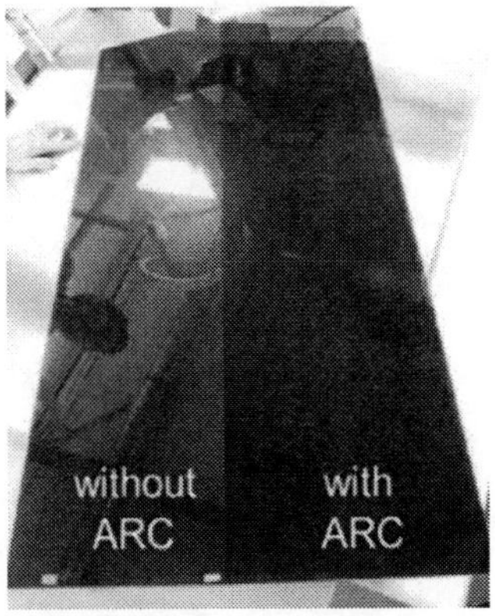

Figure 6: Half-coated module with AR coated and laser cured.

An approach based on a Design of Experiment (DOE) was used to determine the more optimized process for each scanning strategy and lead to process maps like the one presented in **Figure 7**. The main findings of this experiment are:

→ The film cohesion is given by a sufficient transfer of thermal energy, determined by laser treatment overlap and residence time.

→ Defects are initiated as a result from too high and inhomogeneous thermal stresses.

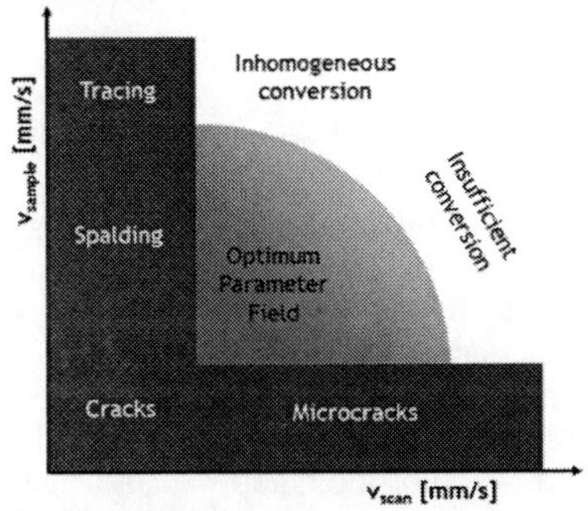

Figure 7: General process map for a given laser power output and scanning strategy.

4.2 Implementation on a full-scale solar module

This high-speed end-of-line process was successfully applied to cure the anti-reflective coating on a full-size module with a processing rate of about 18 m^2/h, leading to a theoretical throughput of 25-30 modules per hour. In the **Figure 6** is depicted a full-size module where only half of the surface has been coated with an AR coating

5 CONCLUSION

The aim of this work was to up-scale a process for curing AR coatings on cover glass of PV modules based on laser irradiation. To achieve high throughput on a production line, an approach combining the integration of high-power laser generators, the technological development of a high-speed laser beam scanner, and the process engineering to tailor the final performance of the AR coating has been conducted.

A 6 kW CO2 laser source was successfully combined with an optimized optical path and a high-speed polygon scanner rotating on air suspended bearing at a speed up to 8000 rpm to cure coatings on glass.

The scientific innovation resides in the development of a high-speed curing technology of silica-based antireflective coatings on the cover glass of PV modules by applying a high-energy density radiation to the coating while scanning at high speed (~2x speed of sound). With an industrial-friendly processing rate of about 18 m^2/s, leading to a theoretical throughput of 25-30 modules per hour. This approach is therefore particularly relevant to coat assembled modules in front-end and end-of-line processes.

REFERENCES

[1] De Loor, "Polygon scanner system for ultra short pulsed laser micro-machining applications", Physics Procedia, vol. 41, 544-551, 2013.

[2] Beresna *et al.*, "Three-dimensional modeling of the heat-affected zone in laser machining applications", Laser Chemistry, vol. 2008, 976205, 2008.

[3] Feit and Rubenchik, ''Mechanisms of CO2 laser mitigation of laser damage growth in fused silica,'' Proc. SPIE, vol. 4932, 91–102, 2003.

STRUCTURAL, MORPHOLOGICAL AND RAMAN SCATTERING STUDIES OF CARBON DOPED ZNO NANOPARTICLES FABRICATED BY PSP TECHNIQUE

R Taziwa*, Luyolo Ntozakhe, and E Meyer

University of Fort Hare, Institute of Technology
Fort Hare Drive, 3rd Floor Chemistry Building, ALICE 5700, South Africa
Tel: +27 (0) 40 602 2086, Fax: +27 (0) 40 653 0665, rtaziwa@ufh.ac.za

Raymond T. Taziwa, University of Fort Hare, Institute of Technology, rtaziwa@ufh.ac.za
Edson Meyer, University of Fort Hare, Institute of Technology emeyer@ufh.ac.za
Luyolo, Ntozakhe, University of Fort Hare Institute of Technology, LNtozakhe@ufh.ac.za

Abstract

ZnO exhibits broad range of properties that depend on doping including a range of conductivity from metallic to insulating, high transparency, piezoelectricity, wide band gap semi-conductivity. Therefore to realize the applications of ZnO for applications in devices and solar cells. It is very important to investigate change in nanomaterials properties due to carbon doping. This paper reveals the effect of carbon doping on the structural, morphological and Raman scattering properties of C-ZnO NPs. In this work Zinc oxide (ZnO) nanoparticles (NPs) were prepared by pneumatic spray pyrolysis technique (PSP) using zin acetate solution as a precursor and tetra-butyl-ammonium as a dopant. The morphological, structural and optical properties of PSP synthesized ZnO NPs were evaluated using SEM, and RS. SEM analysis has revealed that as synthesized NPs have spherical shape. Additionally SEM analysis has revealed that the in-cooperation of C into the ZnO lattice gradually changes the morphology of the NPs changes from mesoporous morphology with have multiple porous network structure to densely packed and agglomerated structure. RS analysis has shown prominent peak at 438.63 cm^{-1}, which is the characteristic peak of $\boldsymbol{E_2^{(2)}}$ $(\boldsymbol{high})$ mode of the Wurtzite ZnO. Additionally Raman phonon shifts due to carbon doping are discussed here in detail.

Keywords: ***Zinc Oxide, Pneumatic Spray Pyrolysis, Carbon doping, Raman Spectroscopy***

1. Introduction

Zinc oxide (ZnO) presents fascinating material system for investigation due to its wide band gap ($E_g = 3.37\ eV\ in\ 300K$) material with a high exciton binding energy ($60\ meV$) [1]. Moreover ZnO is a versatile $II-VI$ semiconductor material with a number of applications including field effect transistors, gas sensors, piezoelectric devices and solar cells [2-5]. Hitherto, its versatile applications has motivated a great deal of research on synthesis and investigation of properties of ZnO nanomaterials [2-5]. In particular modified and un-modified ZnO NPs are expected to become function and integration of nanostructure assembly for dye sensitized solar cells electrodes and nanoscale devices [6, 7]. As a crucial advantage, synthesis of ZnO is more cost effective than other various alternative semiconductors such as $III-V$ semiconductors [8]. Its low price compared with other wide band semiconductor nanomaterials makes it a good candidate for industrial applications. ZnO exhibits a number of properties that depend on doping including conductivity from metallic to insulting, high transparency, piezoelectricity, wide band gap semi-conductivity, room temperature ferromagnetism, huge magneto-optic, and chemical sensing effects [3,5,7]. In the face of its practical prominence, current knowledge of structural, morphological and phonon vibrational properties of doped ZnO nanostructures is rather inadequate. Understanding the specifics of phonon spectrum of ZnO NPs can prove to be pivotal in the development of DSSC employing doped ZnO NPs and nanoscale devices [7]. However most straightforward aspects of doped ZnO nanomaterials are not well studied and the lack of information inhibits further progress. Doping with non-metal elements such as C, S and N has been widely used in literature to improve the structural, morphological and vibrational properties of ZnO. More specifically doping non-metal elements reduces the band gap of wide band semiconductors [9-13]. It has been reported in literature that doping with either N or C enhances a number of properties but not limited to ferromagnetism, magnestotransport properties and *p* type conduction properties

[16]. There are a number of methods commonly used for preparation ZnO NPs ranging from wet chemistry methods [9, 12] to solid phase systems [10]. However solid phase systems for preparation of ZnO NPs outperforms wet chemistry methods because of their simplicity and high yield. Paralleled with solid based methods solution based methods present several challenges in reproducibility, purification and extraction of the nanomaterials prior use. Hence solution methods are not desirable for industrial scaling. Amongst the solid based methods such as laser ablation [14], sputtering coating [14], frequency electron tube sputtering (RFMS) [15], aerosol assisted chemical vapor deposition [16], spin coating [17] and Spray pyrolysis (SP) [18]. Spray pyrolysis methods present a modest and less expensive of fabricating a number of nanomaterials. Specifically SP systems that employ a pneumatic nebulizer/pump technique is a very easy, safe and non-vacuum system and low cost way of producing ZnO NPs. Moreover SP methods provide a relatively simple way to introduce non-metal dopants at ratio required through the proportions of the precursor solutions. Compelled by the aforesaid encouragements, in this paper we aim to understand the effect of carbon doping on the structural, morphological and Raman scattering properties of C-ZnO NPs synthesized by a pneumatic spray pyrolysis technique. The as prepared samples were exposed to a series of structural, morphological and optical evaluation. The results obtained are presented in work.

2. Approach

In typical synthesis procedure 0.1 M of zinc precursor solution was prepared by dissolving 5.4923 g $Zn(O_2CH_3)_2$ in a 250 ml volumetric flask, containing minimum amount of ethanol (99.99 % sigma Aldrich). A few drops of acetic acid (CH_3COOH) were added to the volumetric flask containing zinc ethoxide ($Zn(O_2CH_3)_2$). The $Zn(O_2CH_3)_2$ solution was sonicated on an ultrasonicator at 40 °C for 30 minutes. 3.10 ml of tetrabutylammonium bromide (TBA) $(C_{16}H_{36}BrN)$ was added to the resulting colourless solution of $Zn(O_2CH_3)_2$. The 250 ml volumetric flask containing $Zn(O_2CH_3)_2$ and $(C_{16}H_{36}BrN)$ was topped up to the mark with absolute ethanol. Other precursor solutions containing different level of the carbon dopant $(C_{16}H_{36}BrN)$ were prepared in much the same way. ZnO nanoparticle spray pyrolysis deposition was done on $F{:}SnO_2$ glass substrate at furnace temperatures of 400 °C. The structural, optical and electronic properties of the synthesized nanoparticle were investigated with SEM, XRD, HRTEM and RS.

3. Results and Discussions

Fig 1. (a)- (d) shows the surface morphology of ZnO NPs. SEM analysis has revealed the presence of spherical ZnO NPs. Fig 1. (a) shows the SEM micrographs of the un-doped ZnO NPs shows the presence of spherical NPs with un-even grain size distribution. The un-doped ZnO NPs consist of a mesoporous morphology with a multiple porous network structure. Introduction of carbon Fig 1. (b) shows that the 0.01M C-ZnO NPs is also still mesoporous with un-even grain size distribution. However the introduction of carbon resulted in agglomeration of NPs this is clearly evidence by formation of large spherical NPs surrounded by small NPs. Also a noteworthy feature with the increase carbon dopant concentration Fig 1. (c) and Fig 1. (d) Resulted in loss of the mesoporous morphology with multiple porous network structure and led to formation of dense well packed structure. This suggest that the introduction of carbon catalyses the rate of NPs agglomeration. From the aforementioned discussions, it is therefore safe to conclude that the morphologies of as-prepared ZnO powders depends markedly on the concentration of the carbon dopant.

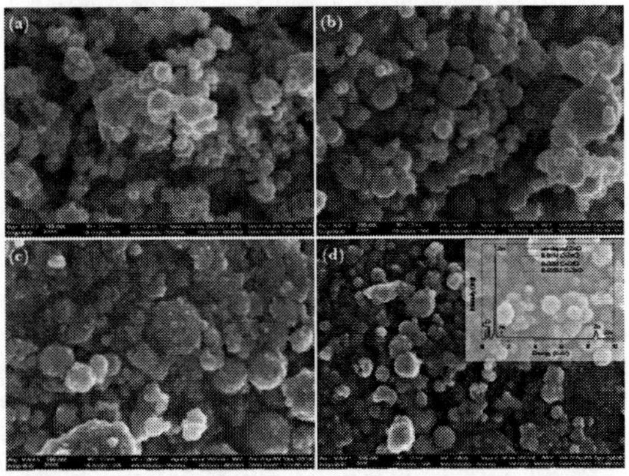

Fig 1. Shows the SEM micrographs of ZnO NPs fabricated by PSP technique (a) un-doped ZnO (b) 0.01M C-ZnO (c) 0.02M C-ZnO and (d) 0.025M C-ZnO. Additionally insert in (d) shows the EDX spectra of the four samples.

Additionally, insert in Fig 1 (d) shows the EDX spectra of all the ZnO samples. The presence of Zn, O and C in the EDX spectra confirms the successful pyrolysis of precursor to form ZnO NPs.

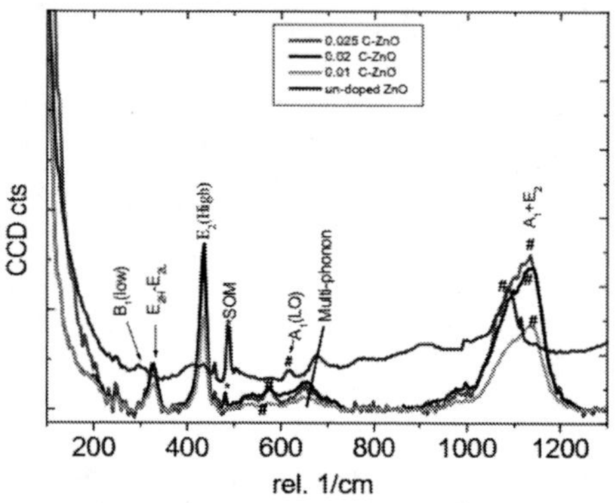

Fig 2: Raman spectra of un-doped and C-ZnO samples fabricated by a PSP technique.

Fig 2. Raman spectra of un-doped and C-ZnO is shown here in Fig 2. The Raman spectra of the un-doped ZnO samples shows the B_1^{low} broad peak at 297.30 cm^{-1}.This mode normally exist in ZnO samples with lattice defects caused by the presence of impurities of either of dopant nature or due to growth conditions [22]. The B_1^{low} mode has been observed in un-doped ZnO films grown by RF sputtering which was attributed to disorder-activated Raman scattering. Hence we can speculate that this mode in our samples is due to the presence of defects due to growth conditions. Surprisingly we did not observed the B_1^{low} mode in our C-ZnO samples. The Raman spectra of C-ZnO samples shows the existence of $E_{2H} - E_{2L}$ multi-phonon mode at the same time this multi-phonon mode does not exist in the Raman spectra of un-doped ZnO samples. The $E_{2H} - E_{2L}$ multi-phonon mode commonly observed at 330 cm^{-1} or 334 cm^{-1} for bulk ZnO samples was red shifted to lower frequencies and was observed for the C-ZnO samples at 325.80 cm^{-1}, 326.3 cm^{-1}, and 325.9 cm^{-1} for 0.01M C-ZnO, 0.02M C-ZnO and 0.025 M C-ZnO NPs respectively. Raman vibration mode of the dominate E_2.(high) of Wurtzite ZnO at 438.63 cm^{-1} commonly observed for bulk ZnO samples at 437 cm^{-1} which was blue shifted to higher frequencies for the un-doped ZnO sample. which was blue shifted to higher frequencies for the un-doped ZnO sample as illustrated here in Fig 3. (a). Fig 3. Shows carbon dopant concentration dependence of Raman peak shift, full width at half Maximum (FWHM), and peak intensity of the E_2 (high) mode of Wurtzite ZnO samples synthesized by PSP technique. It is clearly evident from Fig 3 that the vibrational frequency, Full width at Half Maximum (FHWH) and Intensity are strongly dependent on the concertation carbon dopant. It also evident from Fig 3 (b) that vibrational frequency shifts sharply to lower frequencies in presence of carbon dopant. Our Raman spectra of C-ZnO NPs showed Raman peaks at 434.36 cm^{-1}, 434.37 cm^{-1} and 433.65 cm^{-1} for 0.01 M, 0.02M and 0.025M C-ZnO respectively.

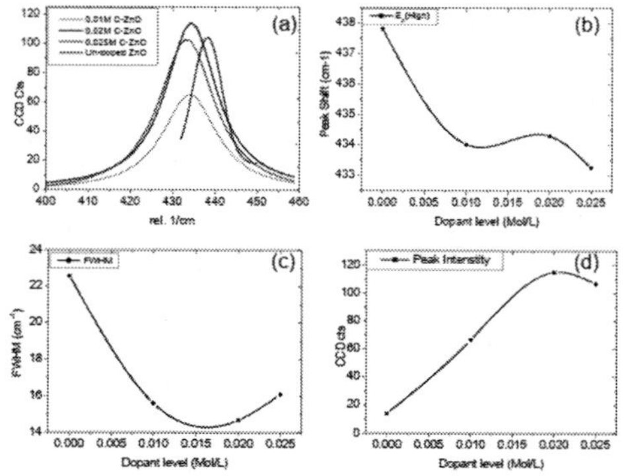

Fig 3: Carbon dopant concentration dependence of Raman peak shift, full width at half Maximum (FWHM), and peak intensity of the E_2 (high) mode of Wurtzite ZnO samples synthesized by PSP technique.

Furthermore it has also been reported that E_2.(high) in ZnO shows a significant change of the vibrational frequency more especially in the presence of foreign species (dopants/impurities). The presence of dopants in the ZnO lattice can cause a significant shift in the Raman vibrational frequency of the E_2.(high) mode. Since the vibrational frequencies of the E_2 (high) mode is related with vibration of oxygen atoms, substitutional doping of the oxygen atoms in ZnO lattice might change the vibrational frequencies of the C-ZnO samples. Hence we can speculate at this stage that the shift to lower frequency numbers can be correlated to the concentration of carbon. We can also conclude that the decrease in phonon frequency is due to both the presence of carbon dopants/impurities and inter molecular stress due to the presence of carbon atoms in the ZnO matrix. On the basis of these observations we conclude C doping in ZnO NPs is also evident through Raman peaks shift to lower frequencies. The dominate E_2.(high) mode. Fig 3. (c) shows the carbon dopant concertation dependence of the Raman line width/ (FWHM). It is clearly evident from Fig 3. (c) that the FWHM decrease sharply with increase in concentration of carbon. It has also been reported that an Increase in FWHM causes weakening of the Wurtzite ZnO structure. Hence the decrease in the FWHM with increase in carbon concentration simply implies that doping with carbon does not weaken or distort the ZnO structure. This is agreement to other experimental observations by Kumar *et al.* [25] were they observed increase in the FWHM with increase in the doping concentration of cobalt, this was attributed ZnO lattice distortions and hence it causes weakening of Wurtzite ZnO structure due to incorporation of metal dopant cobalt. This could be due to preferential substitution of Oxygen (O) atom with carbon (C) atom in the ZnO lattice. Since the C atom 0.26 nm with atomic radii has an almost similar size to the oxygen atom of atomic radii of 0.14 nm. Substitutional doping of O with C is not expected to cause significant lattice distortions as compared to metal dopants like cobalt or nickel with atomic radius of 0.0745 nm and 0.69 nm respectively which preferential substitute the Zn atom in the ZnO lattice. It has been previously reported that for Ni doped ZnO nanostructures that FWHM value for the E_2 (high) mode of less than 11 cm^{-1} signifies that the doped samples retained a good Wurtzite ZnO structure [26]. However for our C doped ZnO samples we observed the FWHM that was estimated to be 22.578 cm^{-1}, 15.590 cm^{-1}, 14.684 cm^{-1} and 16.098 cm^{-1} for the un-doped ZnO, 0.01 C-ZnO, 0.02 C-ZnO and 0.025 C-ZnO samples respectively. Our samples had line shapes greater than 11 cm^{-1} and still retained the perfect ZnO Wurtzite structure. Additionally we observed as shown here in Fig 3 (d) that they is an increase in peak intensity with increase in C dopant concentration.

4. Conclusions

ZnO nanoparticles have been fabricated by pneumatic spray pyrolysis. SEM analysis have revealed the presence of spherical NPs with un-even grain particle size distribution. Additionally SEM analysis has revealed the presence of a mesoporous morphology. The morphological variation due to the presence of carbon in ZnO matrix has resulted in variance in the structural and optical properties of the ZnO nanomaterials fabricated by a PSP technique. The Raman spectra has revealed the presence of the B_1^{low} broad peak at 297.30 cm^{-1}, which is due lattice defects caused by to growth conditions in the Raman spectra of un-doped ZnO samples. The B_1^{low} mode disappeared in spectra of C-ZnO samples. Additionally Raman analysis has revealed the presence of $E_{2H} - E_{2L}$ multi-phonon mode at the same time this multi-phonon mode does not exist in the Raman spectra of un-doped ZnO samples. The $E_{2H} - E_{2L}$ multi-phonon mode was blue shifted to lower frequencies and was observed for the C-ZnO samples at 325.80 cm^{-1}, 326.3 cm^{-1}, and 325.9 cm^{-1} for 0.01M C-ZnO, 0.02M C-ZnO and 0.025 M C-ZnO NPs respectively. Moreover we observed characteristic Raman vibration mode dominate E_2.(high) of Wurtzite ZnO which was red shifted to at 438.63 cm^{-1} for the un-doped ZnO samples and blue shifted to 434.36 cm^{-1}, 434.37 cm^{-1} and 433.65 cm^{-1} for 0.01 M, 0.02M and 0.025M C-ZnO respectively. The red shift of E_2.(high) mode to lower frequency numbers can be correlated to the concentration of carbon. On the basis of these observations we conclude C doping in ZnO NPs is also evident through Raman peaks shift to lower frequencies. The dominate E_2.(high) mode. Additionally the effect of C dopant of peak shift, FWHM and peak intensity has analysed in detail the A1(LO) and intrinsic mode at 678,62 cm^{-1} that is associated with lattice defects. We have observed a decrease in intrinsic defects with increase in carbon dopant level which can be correlated to carbon dopant concentration. Similarly to the E_2(high) mode we have observed Raman peak shifts to lower frequencies with increase in carbon dopant concentration for the A1(LO) mode at 618.68 cm^{-1}. Therefore we can conclude that the decrease in phonon frequency of these modes accompanied by decrease in FWHM is both confirms the presence of carbon dopant (impurities) in our ZnO samples.

5. Acknowledgments

We are grateful for financial support from our sponsors South African National research foundation (NRF), Govani Beki Research and Development centre (GMRDC) of the University of Fort Hare and Sasol Inzalo Foundation. The authors we also like to acknowledge the DST/CSIR Nanotechnology Innovation Centre, National Centre for Nanostructured Materials, CSIR and Electron Microscope Unit, University of Cape Town for characterization of the ZnO NPs.

6. References

1. Rui Zhang, Peng-Gang Yin, Ning Wang, Lin Guo, Photoluminescence and Raman scattering of ZnO nanorods. Solid State Sciences, 11 (2009) 865.
2. B. Miccoli, V. Cauda, A. Bonanno, A. Sanginario, K. Bejtka, F. Bella, M. Fontana, D. Demarchi, One-Dimensional ZnO/Gold Junction for Simultaneous and Versatile Multisensing Measurements. Scientific Reports, 6 (2016) 29763.
3. B. Sang, Y. Nagoya, K. Kushiya, O. Yamase, Solar Energy Mater. Solar Cells. 75 (2003)179.
4. H. Kim, H Yun, H.A. Yoon, S.W. Lee, Integrating ZnO Microwires with Nanoscale Electrodes Using a Suspended PMMA Ribbon for Studying Reliable Electrical and Electromechanical Properties, Adv. Energy Mater. 4 (2014) 1301973.

5. L. Yan, A. Uddin, H. Wang, ZnO Tetrapods: Synthesis and Applications in Solar Cells, Nanomater Nanotechnol, 5 (2015) 19.
6. R. Katoh, A. Furube, T. Yoshihara, K. Hara, G. Fujihashi, S. Takano. S. Murata, H. Arakawa, M. Tachiya, Efficiencies of electron injection from excited N3 dye into nanocrystalline semiconductor (ZrO_2, TiO_2, ZnO, Nb_2O_5, SnO_2, In_2O_3) films, J Phys Chem B. 108 (2004) 4818.
7. O.O. Ogunsolu, J.C. Wang, K. Hanson, Inhibiting Interfacial Recombination Events in Dye-Sensitized Solar Cells using Self-Assembled Bilayers, ACS Appl. Mater. Interfaces. 7 (2015) 27730.

8. J. Yu, X Yu, Hydrothermal Synthesis and Photocatalytic Activity of Zinc Oxide Hollow Spheres, Environ. Sci. Technol. 42 (2008) 4902.
9. J.J. Macías-Sánchez, L. Hinojosa-Reyes, A. Caballero-Quintero, W. de la Cruz, E. Ruiz-Ruiz, A. Hernández-Ramíreza, J.L. Guzmán-Mar, Synthesis of nitrogen-doped ZnO by sol–gel method: characterization and its application on visible photocatalytic degradation of 2,4-D and picloram herbicides, Photochem. Photobiol. Sci. 14 (2015) 536.
10. X.M. Zhang, M.Y. Lu, Y. Zhang, L.J. Chen, Z.L. Wang, Fabrication of a high-brightness blue-light-emitting diode using a ZnO-nanowire array grown on p-GaN thin film, Adv. Mater. 21 (2009) 2767.
11. Guozhen Shen, Jung Hee Cho, Jin Kyoung Yoo, Gyu-Chul Yi, Cheol Jin Lee, Synthesis and Optical Properties of S-Doped ZnO Nanostructures: Nanonails and Nanowires, J. Phys. Chem. B. 109 (2005) 5491.
12. J. Mu, Z. Guo, H Che, H, Electrospinning of C-doped ZnO nanofibers with high visible-light photocatalytic activity, J Sol-Gel Sci Technol, 78 (2016) 3925.
13. A.B. Lavand, Y.S. Malghe, Visible light photocatalytic degradation of 4-chlorophenol using C/ZnO/CdS nanocomposite, Joumal of Saudi Chemical Society, 19 (2015) 471.
14. Keisuke Kawabata, Yasushi Nanai, Seiji Kimura, Tsuyoshi Okuno, Fabrication of ZnO nanoparticles by laser ablation of sintered ZnO in aqueous solution, Appl. Phys. A, 107 (2012) 6745.
15. Xiaohong Yuan, Wenzheng Xua, Fenglin Huanga, Dongsheng Cheng, Qufu Wei, Polyester fabric coated with Ag/ZnO composite film by magnetron sputtering, Applied Surface Science, 390 (2016) 863.
16. V.K. Kaushika, C. Mukherjeeb, T. Gangulic, P.K. Sena, Material characterizations of Al:ZnO thin films grown by aerosol assisted chemical vapour deposition, Journal of Alloys and Compounds, 689, (2016) 1028.
17. P. Raghu, N. Srinatha, C.S. Naveen, H.M. Mahesh, Basavaraj Angadi, Investigation on the effect of Al concentration on the structural, optical and electrical properties of spin coated Al:ZnO thin films, Journal of Alloys and Compounds, 694 (2017) 68.
18. M.A. Rahman, M.R. Phillips, C. Ton-That, Efficient multi-coloured Li-doped ZnO thin films fabricated by spray pyrolysis, Journal of Alloys and Compounds, 691 (2017) 339.
19. A. Bargabas, A. Alshammari, M. FA abpud, H. Kosslick, Room-temperature synthesis of zinc oxide nanoparticles in different media and their application in cyanide photo degradation, Nanoscale Research Letters, 8 (2013) 516.
20. P.K. Giri, S. Bhattacharyya, D.K. Singh, R. Kesavamoorthy, B.K. Panigrahi, K.G.M. Nair Correlation between microstructure and optical properties of ZnO nanostructures synthesized by ball milling, Journal of Applied physics, 102 (2007) 0935151.
21. A.V. Fonoberov, A.A. Balandin, ZnO quantum dos: Physical properties and Optoelectrionic applications, Journal of Nanoelectronics and Optoelectronics, 1 (2006) 19.
22. F.J. Manjon, B Mari, J Serrano, A.H. Romero, Silent Raman modes in zinc oxide and related nitrides
23. M. Tzolov, N Tzenov, D. Dimova-Malinovska, M. Kalitzova, C. Pizzuto, G Vitali, G Zollo, I. Ivanov, Thin Solid Films, 379, (2000) 396.
24. M. Tzolov, N. Tzenov, D. Dimova-Malinovska, M. Kalitzova, C. Pizzuto, G. Vitali, G. Zollo, I. Ivanov, Thin Solid Films, 396 (2001) 274.
25. Y. Kumar, A. Sahai, Sion F. Olive Mendez, N. Goswami, V. Argarwal, Morphological transformation in cobalt doped zinc oxide nanostructures: effect of doping concentration, Ceramics International, 42 (2016) 5184.

REAR-SIDE CONTACTED, LASER-STRUCTURED CIGSE CELLS: A PROOF OF CONCEPT

G. Farías Basulto[a,b], C. Schultz[b], M. D. Heinemann[a], C. A. Kaufmann[a], B. Rau[a], R. Schlatmann[a,b], B. Stegemann[b]

[a] PVcomB/ Helmholtz-Zentrum Berlin für Materialien und Energie GmbH, Schwarzschildstr.3, D-12489 Berlin, Germany
[b] HTW Berlin - University of Applied Sciences, Wilhelminenhofstr. 75a, D-12459 Berlin, Germany

* Corresponding author: E-mail guillermo.farias@helmholtz-berlin.de, Phone: +4930 806215486

This work provides successful proof of a novel rear-side, laser-structured contact concept, as alternative approach for interconnecting chalcopyrite (i.e., $Cu(In_x,Ga_{1-x})Se_2$ or CIGSe) solar cells. This concept is based on employing the rear side molybdenum layer in state of the art CIGSe solar cells not only as back contact, but also as front contact. This is accomplished by creating interdigitating contacts by a P1 laser scribe, isolating the positive and negative contacts. The optimal design was determined by minimizing the power losses depending on the trade-off between dead-area losses and resistive losses. The electrical characteristics of such CIGSe cells were simulated, analyzed and compared with experimental data. The proposed scheme is a promising alternative to conventional interconnection, relying on lower resistive losses within the molybdenum with respect to the transparent conductive oxide (TCO) layers. Thus, this concept allows the usage of thinner TCO layers, which decreases optical losses within the TCO and increases the short-current density j_{sc}.
Keywords: CIGSe, back contact design, monolithic interconnection, laser, patterning

1 INTRODUCTION

Thin film solar cell technology enables a simplified production and a reduced consumption of resources compared to the conventional silicon wafer based technology [1]. One of the essential steps to reduce production costs is the integrated monolithic serial interconnection. It involves the patterning by selective material removal, resulting in precise fine lines in each of the solar cell layers. Typically, three patterning steps, known as P1, P2, and P3, are performed during device fabrication, alternating with layer deposition, in order to achieve monolithic serial interconnection [2]. The first (P1) and the third (P3) scribes determine the width of the cells whereas the second step (P2) creates the back-to-front connection via opening or transforming the absorber layer [3]. The area between the outer edges of P1 and P3 is electrically inactive and cannot be used for energy conversion; therefore it's called "dead area". As illustrated in Fig. 1, the novel concept is based on interdigitating finger-like contacts in the molybdenum back contact layer, separated by the P1 scribe. Thus, two isolated contacts, i.e. the back and "front" contact or the positive and negative contacts, respectively, are formed.

After molybdenum patterning (P1) and the deposition of the absorber, buffer and intrinsic layers, a P2 scribe is executed to provide an electrical path between the transparent conductive oxide (TCO) layer and the molybdenum "front" contact. A final P3 patterning step is required to isolate the solar cell from neighboring cells.

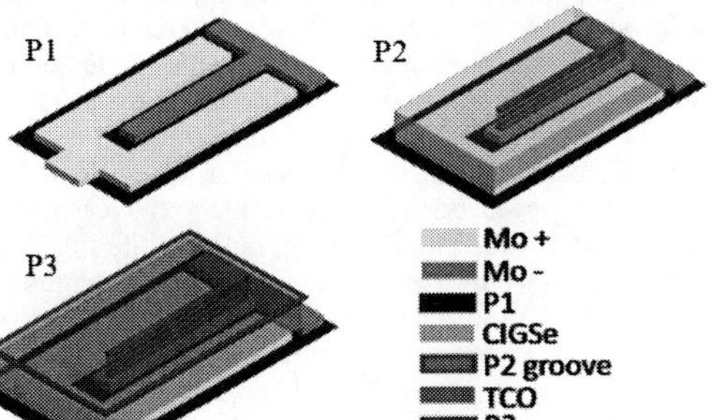

Figure 1: 3-D representation of P1, P2 and P3 scribes for the novel concept.

This patterning scheme may provide advantages with respect to standard designs, such as contacting a cell or even a string from the sides and an increased flexibility of the eligible j-V output parameters by changing the geometry of the contacts. Furthermore, given that molybdenum is used to carry current through the fingers, thinner window layers can be used, enabling the reduction of optical losses in the TCO. Also in comparison to the metal grid used in cell design, shading from top contact grids is eliminated but a higher resistance in the molybdenum finger contacts is present due to the higher resistivity of the material. Finally, a slightly smaller dead-area than in standard modules is possible depending on the particular design [HAAS]. A representation of these three designs can be found in Fig. 2.

In order to optimize device efficiency, a flexible interconnection design scheme needs to be developed, which minimizes the power loss. The power losses included in the model simulations are the area loss and ohmic losses in the finger contacts and the TCO, as well as at their interfaces. By addressing different approaches for the design optimization, the potential and limitations of our novel design will be evaluated. Furthermore, multiple samples were fabricated following this novel concept in order to attain experimental data for comparison.

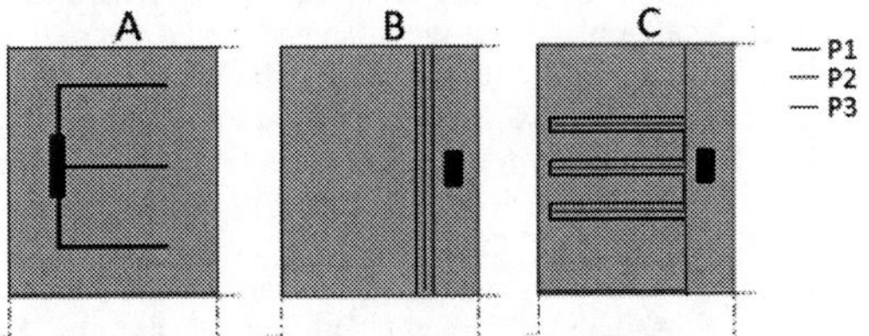

Figure 2: Comparison of three contacting schemes: (A) Top finger grid, cell design. (B) Standard monolithic interconnection. C) Novel concept.

2 EXPERIMENTAL

2.1 Laser patterning setup

For patterning the layers of the solar cell, a high speed motion system was used. This system consists of high-precision linear motor stages for sample translation, which can be moved with velocities of up to 1.2 m/s. The patterning tool is equipped with a picosecond laser source emitting 1064 nm, second harmonic (532 nm) and third harmonic (355 nm) generation with pulse durations of about 10 ps. The maximum pulse energy of 112 µJ can be achieved for 1064 nm radiation, and the repetition rate can be varied from 50 kHz to 8.2 MHz. Furthermore, a nanosecond laser is integrated operating at 532 nm with a maximum pulse energy of 40 µJ and a pulse duration of about 35 ns. The laser is a Q-switched, diode-pumped solid state laser (DPSS) with a pulse-to-pulse energy stability variance less than 1.5 %. Both laser beams possess Gaussian spatial intensity distributions.

Even though all scribing processes have been shown to be possible by applying laser scribing [3, 4], in this study only the molybdenum layer was patterned by means of lasers (i.e. P1 scribe). As it is performed in state of the art CIGS module manufacturing, P2 and P3 patterning were carried out mechanically by a tungsten carbide tip.

2.2 Device fabrication

The samples were prepared according to the R&D process established at the Competence Center for Thin-Film and Nanotechnology for Photovoltaics Berlin (PVcomB). The samples are composed of the following layer sequence: 3.1 mm soda lime glass (SLG) substrate, 150 nm SiO_xN_y barrier layer, 850 nm Mo back contact, ~2.6 µm co-evaporated $Cu(In,Ga)Se_2$ absorber layer, 40 nm CdS buffer layer and transparent conductive oxide (TCO) (i.e., 0. 24 µm ZnO:Al plus ~130 nm i-ZnO).

The co-evaporation process of the CIGSe layer is in detail described by Heinemann et al. [5], similar to the well-known multi-stage process developed at NREL [6]. The sequential co-evaporation of In-Se and Ga-Se at a substrate temperature of 300 °C was followed by the co-evaporation of Cu-Se and In-Ga-Se at a substrate temperature of 530 °C. The resulting CIGSe thin film has a [Cu]/([Ga]+[In]) ratio of 0.88, a thickness of 2.6 µm, a graded band gap depth profile with a minimum band gap energy of 1.1 eV and a surface roughness of around 100 nm [7]. The 20×20 cm^2 substrate was separated into four 10×10 cm^2 sub-panels and covered with CdS, which acts as the n-type emitter, by chemical bath deposition and i-ZnO/ZnO:Al (AZO) as the front contact layer. Finally, onto each cell a metallic grid was evaporated.

In order to proof the concept, the three different schemes presented in Fig. 2 were prepared. Firstly, single squared cells with top grid were used to monitor the spatial fluctuations occurring due to compositional inhomogeneities of the absorber layer (Fig. 2a). Secondly, small cells patterned in state of art module design (consisting of straight fine lines patterned by P1/P2/P3 scribes) show the influence of the TCO–molybdenum interface used to interconnect adjacent cells (Fig. 2b). Thirdly, cells based on the novel concept, interconnected with P1, P2 and P3 patterns, produced in the same chronological order as state of art module design (Fig. 2c). The width of the molybdenum fingers was kept constant at 0.2 mm, since lower widths could possibly induce damages beyond the boundaries of the P1 scribe (i.e. in the active area) due to P2 scribing [8].

Thus, this novel back contact based interconnection scheme can be compared and discussed in multiple ways. All of these devices were patterned and analyzed in order to compare experimental j-V results with those from the simulation.

2.3 Characterization

The j-V measurements were performed using a home built class AAA sun simulator with a Keithley Source Measurement Unit 2425-C equipped with four-point contacts. The j-V parameters of the cells as a function of the investigated cell design were analyzed in order to correlate them with the varying number, length and widths of the contact fingers.

The open-circuit resistance R_s (derived from the slope of the j-V under illumination curve at the point of the open-circuit voltage), the shunt-resistance R_{SH} (derived from the slope of the j-V curve under illumination at the point of the short-circuit current), the fill factor and the efficiency were analyzed. For visual inspections an optical microscope (Axio, Zeiss) and a confocal microscope (Keyence) were utilized. Thermal inspections were carry out by dark lock-in thermography (DLIT) using a TDL640 from Thermosensorik GmbH equipped by an indium antimonide (InSb) focal plane array which is cooled down to ~77 Kelvin.

3 RESULTS

3.1 Equivalent circuit for the novel interconnection scheme

The highest efficiency is obtained when the total sum of power losses is minimized, i.e. a trade-off between area losses and electrical losses is achieved. By assuming a similar electrical behavior as a metallic top grid contact, the power losses through the device´s components, such as TCO, molybdenum fingers and contact interface, can be approximated with well-known formulae.

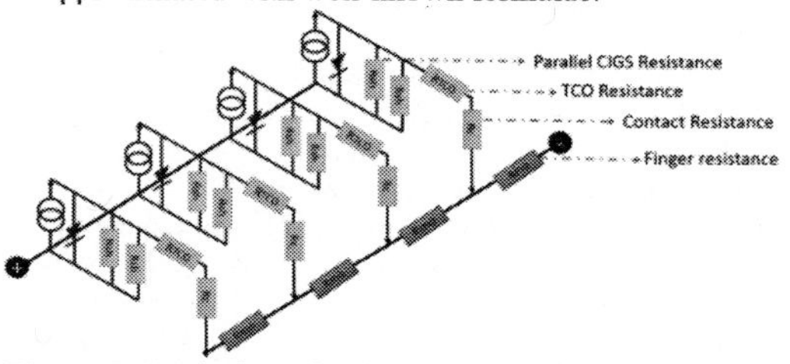

Figure 3: Equivalent circuit of the novel concept.

In Fig. 3, a circuit representing the resistive elements contributing to the overall power loss in the novel contact scheme and, thus, considered in the simulation, is exemplified. In order to facilitate comprehension, the sketch includes only one side of a finger. Firstly, the photo-generated current is collected in the window layer, where the power losses generated due to the resistivity of the TCO are added up at the beginning of the finger. The sum of these series resistances in the TCO is represented in Fig. 3 in green. The power loss in this layer was calculated by a variation of the equation proposed by S. Wenham [9]:

$$P_{Loss\,TCO} = \frac{J_{MPP}^2 * L_f * \rho_\square * S^3}{12} * N \tag{1}$$

where S is the spacing of the fingers in cm, J_{MPP} is the current density at the maximum power point in mA/cm^2,

$\rho_\square$ is the resistivity of the TCO in $\Omega/\square$, L_f is the finger length in cm and N is the number of fingers.

When the current reaches the finger and flows from the tip of the finger to its base, it encounters the resistance of the TCO – molybdenum interface (i.e. contact resistance) and the resistance along each finger which also generates power losses. The interface resistance and finger resistance are represented in Fig. 3 in blue and gray, respectively. These power losses can be approximated according to the method illustrated by A. R. Burgers [10], utilizing the integrated resistances throughout the length of a single finger [11] and adding the contact resistance (R_c) to the equation:

$$P_{Loss\,finger} = \left[\frac{1}{3} * J_{MPP}^2 * S^2 * L_f^3 * \frac{\rho_f}{W_f * D_f} + \left(J_{MPP} * L_f * S\right)^2 * R_c\right] * N \quad (2)$$

where ρ_f is the resistivity of the finger, W_f is the width, D_f is the thickness of the molybdenum and R_c the TCO – molybdenum interface resistance that is assumed to be constant. The resistance of the P2 contact was approximated using transmission line measurement (TLM) structures for typically used mechanical and laser scribing approaches [12]. The finger width cannot be varied without compromising either dead area or back contact losses and therefore was set to a fixed value of 0.2 mm. Even though the area loss is not part of the equivalent circuit, it is important to mention the hypothetical power losses due to the unutilized area. The area loss relate to the total surface required for the finger contact that otherwise would serve to generate photocurrent and therefore is analog to the shading loss introduced top grid fingers or to the dead area in standard interconnection.

3.2 Device simulation

Matlab® was used to calculate the power loss due to the resistive elements in the model and due to the dead area in order to find the minimum power loss for various layouts. Please note that power losses due to the busbar were not taken into account given that they are assumed to be, either part of the neighboring cell or below the aperture area of the panel and the busbar has the same dimensions in all variations. Also, the optical losses due to TCO absorption and the parallel resistances due to material defects were considered indirectly through the values of standard devices used in the simulation (i.e. j_{sc}, V_{mpp}, j_{mpp}) which already include these losses, thus, enabling a valid simulation.

Fig. 4 exhibits a graph of the results obtained through the simulation of the novel design by systematic variation of the total number of contact fingers. The various curves represent the multiple power losses in the device for different layouts, where the power loss due to resistive losses in the contact fingers (gray) together with the dead area losses (red) appears to be the most decisive power losses according to the simulated design; followed by, power losses due to the P2 contact (blue) and finally, window layer losses (green). The latter becomes decisive for designs with a reduced number of fingers. With 3 or less fingers (i.e. with a large spacing) or whenever TCO's sheet resistance is high, the TCO power loss increases significantly.

Thus, according to the simulation, the optimal layout should include 4 fingers, suggesting a simulated minimum total power loss of about 16 %.

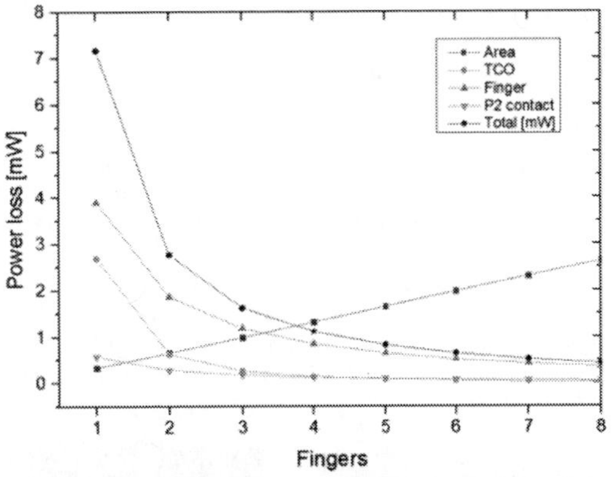

Figure 4: Calculated losses through simulation of the novel concept for CIGSe solar cells with different numbers of contact fingers.

3.3 Experimental results and comparison

Multiple samples were produced following the presented model, varying the number of fingers from 2 to 8 fingers for comparison. The device characterization was performed under standard test conditions (STC). The analysis of the experimental data gathered through solar simulation provides statistical results of the latest comprehensive layout implemented, in order to compare the different contacting schemes and also the diverse layouts for the novel concept.

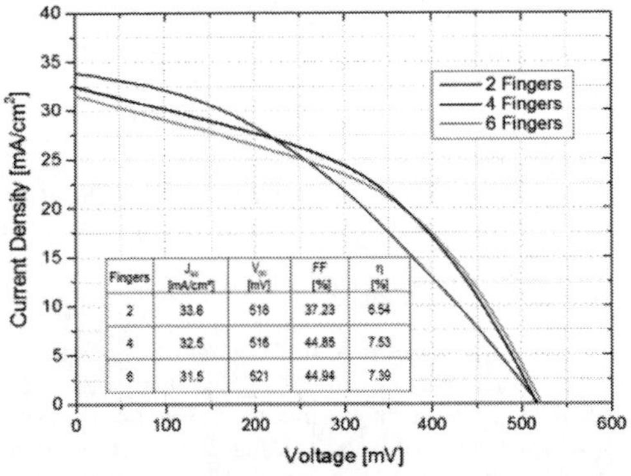

Fingers	J_{sc} [mA/cm²]	V_{oc} [mV]	FF [%]	η [%]
2	33.6	516	37.23	6.54
4	32.5	516	44.85	7.53
6	31.5	521	44.94	7.39

Figure 5: j-V curves of rear contacted CIGSe solar cells.

In Fig. 5, an example of the results from electrical characterization of photovoltaic devices structured with the novel approach is presented. The j-V curves for different layouts (i.e. 2, 4 and 6 fingers) indicate working photovoltaic devices, with the highest efficiency at 4 fingers, which is well in accordance with the results of the simulation (cf. Fig. 4). However, the presence of parasitic resistances affects significantly the performance of the device. On one hand, with a lower number of fingers, the series resistance (R_s) increases due to the increase of current in each finger. On the other hand, by increasing the number of fingers, the overall parallel resistance decreases exponentially according to the measured data.

The power loss due to resistances in the device dissipates in form of heat. According to the simulation, the resistance of the contact fingers is the most significant loss. In order to obtain spatially resolved, quantitative information on these resistive losses, the temperature increase in a cell while in forward bias was visualized through dark lock-in thermography (DLIT) to unveil further possibilities for power loss minimization in the novel design. Both thermographs from rear-side contacted CIGSe cells with 3 and 5 fingers, respectively, shown in Figs. 6a and 6b, present a clear increment of signal intensity from the tip to the base of the fingers. Furthermore, by comparing the two lines in Fig. 6c, it is found that the increase of DLIT signal becomes much more significant with the reduction of number of fingers due to the increase of flowing current. Given the geometry of the device, these behaviors is in accordance with equation (2).

(a) (b)

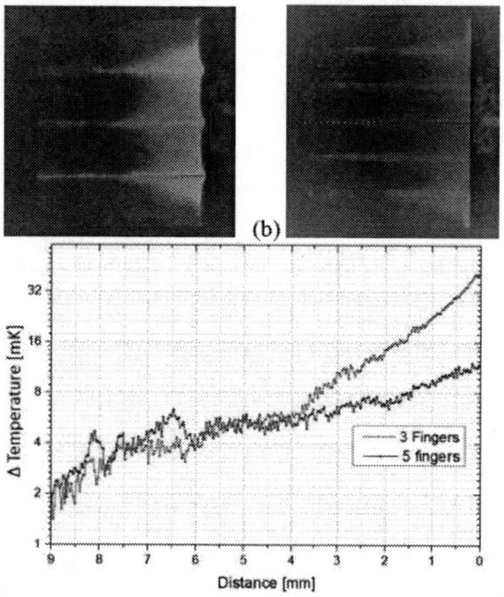

(c)

Figure 6: DLIT images exposing the power losses in the device dissipated as heat. (a) 5 finger sample, (b) 3 finger sample, and (c) characteristic temperature profiles along a contact finger, as indicated in the images above.

The analysis of the DLIT signal of the three-finger cell suggests that around 2/3 of heat dissipated after the breaking point of the green curve in Fig. 6c (i.e. at about 4 mm). Finally, the graph in Fig. 7 presents a summary of the experimental and simulated data obtained from CIGSe cells with different representative layouts (i.e. from 2 to 8 fingers).

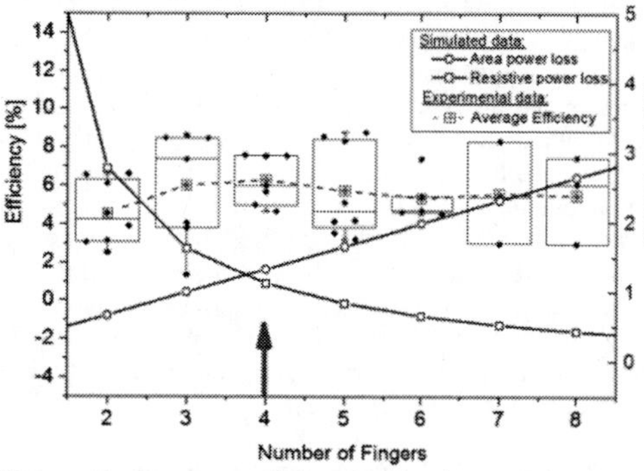

Figure 7: Summary of the simulated power losses and experimentally determined power conversion efficiencies as a function of the number of fingers for the novel rear-side connection scheme.

Thus, the data obtained through solar simulation is in good agreement with the experimental data, suggesting an optimum layout containing 4 contact fingers. Due to the spread of experimental data and possible uncertainties in the material parameters assumed for the simulation one could expect an uncertainty of one finger depending on the properties of the absorber.

4 INTERCONNECTION SCHEME OUTLOOK

The performance of novel concept devices can be further improved to achieve efficiencies comparable to standard module devices by applying corrective measures to decrease the power losses found during this work. These include reducing the series resistances through the back contact and the TCO.

In order to decrease the resistance in the back contact fingers, a combination of multiple actions will be utilized such as: triangular shaped fingers, reduction of finger length and decrease of resistivity and/or increase the depth of the back contact material. By implementing triangular shaped fingers, these power losses can be further minimized not only since the cross-sectional area is reduced but also because this reduces the interface resistance of the contact. Increasing the thickness of the back contact should provide further gains in efficiency by decreasing the sheet resistance of the molybdenum layer and by increasing the cross-sectional area of the finger, thus, reducing the resistance along the contacts.

An improvement of the dead area (DA) is possible by employing this new concept in comparison with the standard process. The scribing patterns of the standard design have a typical width of 300 µm, which causes a 5.66 % area loss for $5\times10mm^2$ cells. For the novel concept, the area loss due to the P1 scribe is more significant given that the P2 scribe is placed within two P1 scribes and the P1 and P3 processes can be overlapped, similar as introduced in [13]. The calculated DA of the novel design is 2.3 mm^2 for a similar cell geometry, which causes a 4.40 % area loss, thus, obtaining a relative decrease of ~22 % of dead area. For longer fingers this dead area is more significant, causing higher dead area losses as the standard interconnection. Therefore, this gain is not yet presented in the cells manufactured in this work given that the length per finger was fixed to 9 mm to match the cells layout.

Due to the lower TCO related power losses, TCO layer thickness might be reduced to minimize optical losses without affecting the performance of the device. Thicker TCOs entail a decrease in transmittance of the material (i.e. Δ transmittance of about 2.5 % at wavelengths between 400 and 800 nm) but a decrease of the average resistivity of this material [14]. Thus by utilizing a thinner cell-TCO of 240 nm, higher efficiencies of almost 3 % relative to a 3.6 times thicker TCO (865 nm) were found. This can be explained by taking into consideration the higher recombination losses of the thicker TCO in comparison to the thinner one. Therefore, this concept allows the usage of thinner TCO layers, which decreases optical losses and increases the short-current density j_{sc}.

5 CONCLUSIONS

A novel concept scheme, relying on interdigitated

rear-side contacts, was effectively applied for CIGSe solar cells with competitive efficiencies, resulting in the fabrication of functional devices with varying number of contact fingers. Furthermore, the electrical behavior was accurately simulated by employing variations of well-known equations. The layout with the best electrical performance represented by the total number of fingers depends largely in both the cell design and the electrical performance of the material at the operating point under standard test conditions (STC). An optimum of 4 (±1) fingers was found in both simulation and experimental data.

The novel concept offers great flexibility in terms of the desired electrical parameters for panel manufacturing, given that the geometry of the contacts can be easily changed and adapted to multiple cell geometries. This provides the possibility of adjusting the design to a multiple set of combinations. The minimization of the dead area is also possible by manipulating this geometry.

The power loss due to the resistive elements that are present in the novel concept are assumed to be the most critical for the elaboration of devices following the layout. The experimental results offer an in-depth insight into the possibility of further manipulating cell design and processes in order to reduce power losses due to resistive elements. By simulating some of this possibilities, a relative decrease of around 60 % of power losses can be achieved. This reduction is enough to make back contacted CIGSe cells competitive to standard module designs.

ACKNOWLEDGEMENTS

The authors gratefully acknowledge the support of the PVcomB team.

REFERENCES

[1] T. Markvart, L. Castaner, "Thin Silicon Solar Cells" in *Solar Cells: Materials, manufacture and operation*, Elsevier, 2004, pp. 122-123.

[2] A.D. Compaan, I. Matulionis and S. Nakade, "Laser scribing of polycrystalline thin films", Elsevier, Optics and Lasers in Engineering Vol. 34, p. 15, 2000.

[3] M. Schüle, C. Schultz, V. Juzumas, K. Stelmaszczyk, M. Weizman, C. Wolf, N. Papathanasiou, B. Rau, R. Schlatmann, V. Quaschning, B. Stegemann, F. Fink, "Laser Patterning of Cigs Based Solar Cells Using Nano- and Picosecond Pulses - Possibilities and Challenges", 28th European Photovoltaic Solar Energy Conference and Exhibition, 3BV.5.39, p. 2302 – 2306, 2013.

[4] C. Schultz, M. Schuele, K. Stelmaszczyk, M. Weizman, O. Gref, F. Friedrich, C. Wolf, N. Papathanasiou, C. Kaufmann, B. Rau, R. Schlatmann, V. Quaschning, F. Fink and B. Stegemann, "Laser-induced local phase transformation of CIGSe for monolithic serial interconnection: Analysis of the material properties", Solar Energy Materials and Solar Cells, vol. 157, pp. 636-643, 2016.

[5] M. Heinemann, R. Mainz, F. Österle, H. Rodriguez-Alvarez, D. Greiner, C. Kaufmann and T. Unold, "Evolution of opto-electronic properties during film formation of complex semiconductors", Scientific Reports, vol. 7, p. 45463, 2017.

[6] A.M. Gabor, J.R. Tuttle, D.S. Albin, M.A. Contreras, R. Noufi, A.M. Hermann, "High-efficiency $CuIn_xGa_{1-x}Se_2$ solar cells made from $(In_x,Ga_{1-x})_2Se_3$ precursor films" in *Appl. Phys. Lett.*, 65, p. 198, 1994.

[7] C.A. Kaufmann, T. Unold, D. Abou-Ras, J. Bundesmann, A. Neisser, R. Klenk, R. Scheer, K. Sakurai, H.-W. Schock, "Investigation of coevaporated Cu(In,Ga)Se2 thin films in highly efficient solar cell devices", Thin Solid Films, vol 515, Issue 15, pp 6217-6221, 2007.

[8] C. Schultz, G. Farías Basulto, S. Ring, C Wolf, R Schlatmann, B. Stegemann, "Revealing and Identifying Laser-Induced Damages in CIGSe Solar Cells by Photoluminescence Spectroscopy" IEEE Journal of Photovoltaics 7 (5), 1442-1449

[9] S. Wenham, M. Green and M. Watt, "Cell Properties and Design" in *Applied photovoltaics*. London: Earthscan, pp. 60-61, 2011.

[10] A. Burgers, "How to design optimal metallization patterns for solar cells", Progress in Photovoltaics: Research and Applications, vol. 7, no. 6, pp. 457-461, 1999.

[11] Pveducation.org. (2016). *Finger Resistance*. [Online].Available:http://pveducation.org/pvcdrom/design/finger-resistance. [Accessed: 12- Apr-2016].

[12] D. Schroder, "Contact Resistance and Schottky Barriers," in *Semiconductor material and device characterization*, 3rd ed. New Jersey: John Wiley & Sons, 2006, pp. 135-138.

[13] S. Haas, A. Bauer, A. Lambertz, U. Rau, "Innovative series connection concept for thin-film solar modules", Proceedings of the 26th European Photovoltaic Solar Energy Conference (Hamburg), 3BO.4.5, p. 2375, 2011.

[14] R. Mereu, S. Marchionna, A. Donne, L. Ciontea, S. Binetti and M. Acciarri, "Optical and electrical studies of transparent conductive AZO and ITO sputtered thin films for CIGS photovoltaics", physica status solidi (c), vol. 11, no. 9-10, pp. 1464-1467, 2014.

PROPERTIES OF CO-SPUTTERED CdSxTe1-x THIN FILMS FOR COMPOSITIONAL OPTIMIZATION IN HIGH PERFORMANCE CdS/CdTe SOLAR CELLS

M. A. Islam[1,3], K. S. Rahman[2], F.M. Tahzib Enam[2], K. Sobayel[2], I. Kamaruzzaman[2], M. Akhtaruzzaman[1], and N. Amin[1,2*]
[1]Solar Energy Research Institute, National University of Malaysia, 43600 Bangi, Selangor, Malaysia
[2]Department of Electrical, Electronic and System Engineering, Faculty of Engineering and Built Environment, The National University of Malaysia, 43600 Bangi, Selangor, Malaysia
[3]Graduate School of Material Science, Nara Institute of Science and Technology, Ikoma, Nara 630-0101, Japan
*Corresponding email: nowshad@eng.ukm.my, aminbgm@gmail.com

ABSTRACT: The formation of CdS_xTe_{1-x} layer in between CdS and CdTe layer is facing a different criticism; whether it has an adverse or advantageous effect on CdS/CdTe solar cell. To understand the effect, CdS_xTe_{1-x} thin film thickness of around 100-120 nm has been prepared by co-sputtering of CdS and CdTe at a 400 °C and the chemical composition, morphology, crystallinity, optical, and electrical properties of the CdS_xTe_{1-x} thin films have been extensively studied and were found strongly dependent on the S concentration. These optical and electrical data found in the experimental studies are then used to model and simulate the CdS/CdTe solar cells. The well-known AMPS-1D simulation software was employed to observe the solar cell performance in presence of CdS_xTe_{1-x} thin film. The S rich $CdS_{0.8}Te_{0.2}$ layer increase band to band matching assist to increase the efficiency, however, S deficient $CdS_{0.3}Te_{0.7}$ layer adversely affect the cell efficiency due to the lowest optical bandgap.
Keywords: CdS_xTe_{1-x}; co-sputtering; optical and electrical properties; AMPS-1D; CdS/CdTe solar cell

1 INTRODUCTION

Despite the successes of CdTe solar cells, however, there are still outstanding issues such as quality of CdTe and CdS thin films, doping of CdTe, CdS/CdTe interface formation, control of CdS_xTe_{1-x} interfacial layer, $CdCl_2$ heat treatment, and back contact, etc. for achieving better performance and cheaper solar cell device. On the other hand, the electrical junction CdS/CdTe solar cells are located in this interfused layer [1]. Among the issues, a full understanding of the formation and behavior of CdS_xTe_{1-x} interfacial layer is believed to improve the cell performance and stability in a further extension.

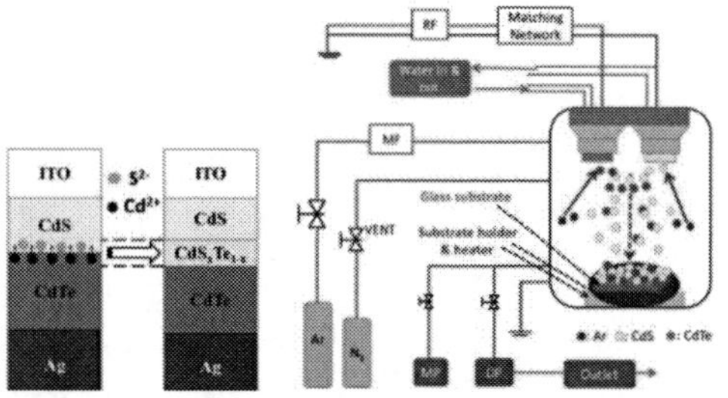

Fig. 1 (left) Schematic of the process of formation of CdS_xTe_{1-x} layer, (right) Schematic of the co-sputtering technique

In this paper, we tried to realize the properties of CdS_xTe_{1-x} layer, its effect on the cell performance. Also, optimum diffusion value of S has been investigated through modeling and simulation. It should be noticed that only bandgap and carrier concentration found in the experiment has been used for simulation, other properties such as electron affinity, density of states, acceptor and donor like defects, and capture cross-section has been taken similar to CdS (for CdS like) and similar to CdTe (for CdS like) films from the literature.

Fig. 1 shows the schematic of S diffusion and of the co-sputtering process. It is well known that CdS_xTe_{1-x} layer is formed unintentionally during the CdS/CdTe cell activation in the interface of CdS and CdTe layer or CdTe layer deposition on CdS layer at high temperature [1, 2]. It can be formed by either diffusion of S into CdTe or the diffusion of Te into CdS [3]. The effect of this layer in the performance of the solar cell is not well understood [4]. It was reported that the formation of this layer is advantageous for CdTe solar cell as it decreases the lattice mismatch between CdS and CdTe layer [5], and it reduces the CdS layer thickness, thus reduces the dark recombination current. Contrary, it was also reported that the formation of CdS_xTe_{1-x} layer decreases the solar cell performance as it has a lower band gap [6]. And the diffusion of Te into the CdS region reduced the band gap of CdS and the diffusion of S into the CdTe region reduces the band gap of CdTe [7]. The structural and optical properties of CdS_xTe_{1-x} grown by different deposition methods are already explained [8]. However, no reports were found investigating the properties of CdS_xTe_{1-x} thin films deposited by co-sputtering technique and the effect of this layer in terms of cell performance is not yet revealed. Thus, it becomes imperative to know the behavior of this CdS_xTe_{1-x} inter-diffused layer in order to make efficient CdTe thin film solar cells.

2 EXPERIMENTAL DETAILS

The CdS_xTe_{1-x} thin films of 100-120 nm is deposited on top of the FTO coated glass substrate at a 400 °C substrate temperature by co-sputtering of CdS and CdTe target supplied by Sigma Aldrich. The CdS and CdTe targets were placed in two different guns. The composition of S and Te in CdS_xTe_{1-x} thin films were varied by changing the RF power ratio of CdS to CdTe ranging from 2.0 watt/cm^2 to 0.0 watt/cm^2. The working pressure of argon ambient was 1.87 Pa during the deposition. Prior to the co-sputtering, the CdS and CdTe are individually deposited by changing the RF power to measure their specific deposition rate. The rate of deposition is varied from 1.14-2.32 Å/sec depending on the RF power of both guns obtained from the thickness profilometer embedded in the sputtering system. After the deposition, the chamber was first cooled down to the room temperature and then the samples were taken out from the chamber.

The compositional analysis was done by Energy-dispersive X-ray spectroscopy (EDX) characterized by Oxford Instruments INCA x-sight SEM detector (Model: 7317). The grain growth and surface morphology is observed by SEM images carried out at an operating voltage 3.0 kV by VPSEM 'LEO 1450 Vp'. The electrical properties such as carrier concentration, mobility, resistivity and hall coefficients are measured by Hall Effect measurement tool 'ECOPIA 3000'. The bandgaps of the films are calculated from the absorption spectra collected by 'Perkin Elmer Instruments Lambda 35' UV-Vis spectrophotometer.

The effects of the bandgap, thickness and electrical properties of CdS_xTe_{1-x} on cell performance were observed by employing 'Analysis of Microelectronic and Photonic Structures (AMPS-1D)' simulation software. The experimentally found optical and electrical properties of CdS_xTe_{1-x} thin films are used for the simulation. The cell output variables such as efficiency, Jsc, Voc, FF, and EQE were investigated to understand the effects of CdS_xTe_{1-x} layer on CdS/CdTe solar cells.

3 RESULTS AND DISCUSSION

The atomic compositions of sulfur (S) in CdS_xTe_{1-x} thin films denoted by 'x' have been calculated from EDX using the following equation, and are listed in Table 1.

$$x = S(\text{atomic }\%)/[S(\text{atomic }\%) + Te(\text{atomic }\%)] \quad (1)$$

The fraction found in the calculation has been simplified to the integer for easy understanding. The SEM images of the CdS_xTe_{1-x} thin films as shown in Fig. 2 are illustrating the evolution of the film morphology with the different S composition. The small nano-sized grains of ellipsoid in shapes are uniformly distributed over the surface and sizes are increased while the S composition is decreased from x=1 to x=0. This is likely due to the nucleation and diffusion rate affected by the S and Te ratios during the deposition process.

Table 1 Atomic ratios of S (fraction has been simplified to integer) in CdS_xTe_{1-x} thin films found from EDX analysis

RF power (CdS:CdTe), watt/cm^2	Composition (x)	RF power (CdS:CdTe), watt/cm^2	Composition (x)
2.0:0.0	1.0	1.5:2.0	0.4
2.0:0.5	0.8	1.0:2.0	0.3
2.0:1.0	0.7	0.5:2.0	0.2
2.0:1.5	0.6	0.0:2.0	0.0
2.0:2.0	0.5		

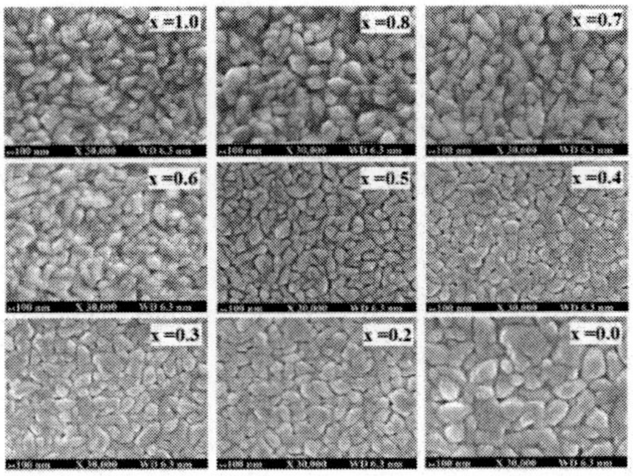

Fig. 2 SEM images of CdS_xTe_{1-x} thin films with different S compositions

The band gap of the CdS_xTe_{1-x} thin films has been calculated from the absorption spectra using the following equation [9],

$$\alpha = [A\,(h\gamma - Eg)^{1/2}]/h\gamma \quad (2)$$

Where α is the absorption coefficients, A is constant, hγ is photon energy and E_g is the bandgap. The observed variation of slopes in the plot of $(\alpha h\gamma)^2$ vs hγ indicating the non-homogeneity of the films associated with the compositional variation. This is likely related to the variation of grain sizes as found in SEM images. With the increase of S composition in CdS_xTe_{1-x} thin films, the bandgap is reduced and the lowest bandgap of 1.3 eV is found for x= 0.3. However, the band gap increases with the further increase of S composition and the highest bandgap of 2.4 eV is found for CdS (x = 1). The bandgap changing with S concentration is in agreement with that of the lattice parameters, supporting the idea of saturation of CdS with tellurium and CdTe with sulfur.

From the variation of band gap with respect to the S composition, a band bowing graph has also been drawn as shown in Fig. 3(c). The band gap variation with the constituent's composition is found to exhibit bowing in a quadratic form according to the relation [10].

$$E_x = E_y + (E_z - E_y - b)x - bx^2 \quad (3)$$

where E_y and E_z are the energy gap of pure compounds CdTe and CdS, E_x is the bandgap of the CdS_xTe_{1-x} thin film with composition x and b is the optical bowing coefficient. The calculated bowing parameter is obtained 1.7 for x = 0.4 which is well matched with the literature [11].

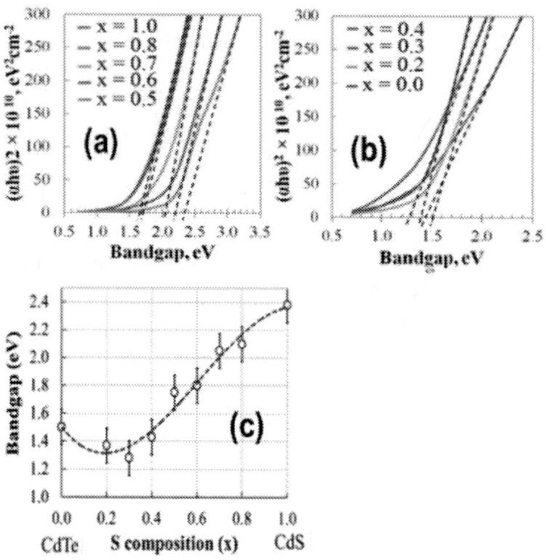

Fig. 3 Graph of (αhγ)2 versus photon energy hγ for CdS_xTe_{1-x} thin films with different S compositions, (a) for x = 1.0 - 0.5 and (b) for x = 0.4 - 0.0, (separating into two groups to see the graph clearly) and (c) variation of the bandgap with S composition in the films (fitting line shows the bowing properties of the film)

Table 2 Values of the electrical parameters of CdS_xTe_{1-x} thin films

Composition (x)	Bandgap, Eg (eV)	Bulk Concentration	Mobility (cm^2/V-s)	Conduction type
1.00	2.4	6.1×10^{15}	2.17	n
0.83	2.2	1.9×10^{16}	4.29	n
0.68	2.0	5.07×10^{16}	1.76	n
0.64	1.8	8.53×10^{16}	1.09	n
0.46	1.8	2.9×10^{16}	2.76	n
0.41	1.4	3.2×10^{15}	25.2	n
0.27	1.3	6.4×10^{15}	13.9	p
0.20	1.4	2.5×10^{16}	3.17	p
0.00	1.5	2.01×10^{14}	50.1	p

The electrical properties of the CdS_xTe_{1-x} layers are shown in Table 2. It has been seen that for the composition x = 0.0 to 0.3, the films are found in p-type conduction. The highest carrier density in the p-type CdS_xTe_{1-x} films is observed for x =0.2 indicating that the small amount of S diffusion to CdTe increases the carrier concentration. Moreover, for the composition x = 0.4 to 1.0, the films found in n-type conduction. In n-type CdS_xTe_{1-x} films, it has been found that the presence of Te into the CdS thin films increased the carrier density of the films at a certain limit. The films resistivity does not change significantly with the variation of S composition. However, the film mobility is found scattered in values.

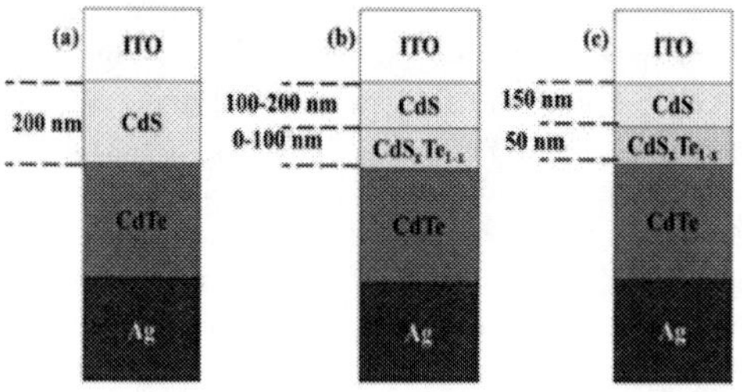

Fig. 4 Structure of $CdS/CdS_xTe_{1-x}/CdTe$ solar cell used in simulation, (a) base line case, (b) with different CdS_xTe_{1-x} thickness (c) with different CdS_xTe_{1-x} band gap, (d) change of Jsc and Voc with the thickness of CdS_xTe_{1-x} layer (x = 0.3, Eg = 1.27 eV)

The CdS/CdTe solar cell structure has been simulated including of the CdS_xTe_{1-x} inter-diffused layer in this study as shown in Fig. 4. There is a baseline case cell (Fig. 4a). The cell with the different thickness of the CdS_xTe_{1-x} thin films with a fixed band gap (1.3 eV for x = 0.3) is shown in Fig. 4b. The cell with the different bandgap of CdS_xTe_{1-x} thin film and corresponding carrier concentration with a fixed thickness 50 nm is shown in Fig.4c. It should be noticed that this inter-diffused layer affects the device performance due to their alteration in the electrical and optical properties [12]. In order to get the real picture of the cell, the effect of CdS_xTe_{1-x} layer has been extensively studied with emphasize given to the change of optical band gap, thickness, and carrier concentrations of the layer. The base parameters used for the other layers are adopted from the different published work [13-15].

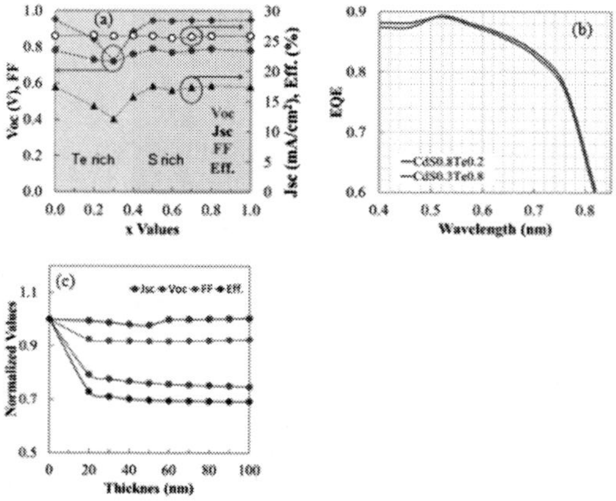

Fig.5 (a) Variation of Jsc, Voc, FF and Eff. with respect to the S concentration on CdS_xTe_{1-x} layer, (b) EQE for two different S concentration and (c) Variation of Jsc, Voc, FF and Eff. with respect to the thickness of $CdS_{0.3}Te_{0.7}$ layer (Eg = 1.3 eV) (normalized with respect to the zero thickness of $CdS_{0.3}Te_{0.7}$ layer)

From the Fig. 5, it has been seen that the efficiency, as well as the Jsc and Voc are strongly dependent on the combined change of band gap and carrier concentration of the CdS_xTe_{1-x} thin films. The highest efficiency of 17.6% is observed for the S rich (x = 0.8), and lowest efficiency of 12.1% is observed for S deficient (x = 0.3) CdS_xTe_{1-x} thin films, respectively. On the other way, the efficiency without CdS_xTe_{1-x} thin film is 17.4% for x=1(CdS) and 17.3% for x =0 (CdTe), both are slightly lower than the obtained efficiency for S rich $CdS_{0.8}Te_{0.2}$ thin films, indicating that S rich and/or CdS like CdS_xTe_{1-x} thin films help to improve efficiency. This may occur due to the improvement of the band to band matching by the CdS_xTe_{1-x} layer, for which the interfacial strain has been reduced, and/or for the higher carrier concentration of CdS_xTe_{1-x} thin films. This matter of fact has also been reported by others [16, 17].

4 CONCLUSION

The co-sputtered CdS_xTe_{1-x} thin films were studied through EDX, SEM, XRD, UV-Vis and Hall Effect measurements. The compositional ratio of S and Te was calculated from the EDX data analysis. The ellipsoid and inhomogeneous grains are increased in size with the decrease of S concentration and/or increase of Te concentration as observed in SEM images. The films are exhibited the direct band gap ranging from 1.3 eV to 2.4 eV which strongly dependent on the composition of the films. The Hall Effect studies revealed that the films are either n (CdS like) or p (CdTe like) type strongly depending on the S and Te compositional ratios. The small amount of S diffusion to CdTe or a huge amount of Te diffusion to CdS can make $CdS_{0.3}Te_{0.7}$, which show high carrier concentration but the lowest efficiency is found in the simulation may be due to the low bandgap. The highest efficiency has been found for S composition ratio of 0.8 of CdS_xTe_{1-x} thin films could be considered as an optimum S diffusion value for CdS/CdTe thin film solar cells.

Acknowledgement

The authors would also like to acknowledge and appreciate part of the contribution of the Ministry of Science and Technology of Malaysia through its research grant with code 0301-02-SF1295 and the grant received from The National University of Malaysia with code DIP-2015-021 and GUP-2016-042.

References

[1] R.G. Dhere, Y. Zhang, M.J. Romero, S.E. Asher, M. Young, B. To, R. Noufi, and T.A. Gessert, Proceedings 33rd IEEE Photovoltaic Specialists Conference 2008 (2008) 1.

[2] M. A. Islam, K. S. Rahman, K. Sobayel, T. Enam, A. M. Ali, M. Zaman, M. Akhtaruzzaman, N. Amin, Solar Energy Materials and Solar Cells, 172 (2017) 384.

[3] Brian E. McCandless and James R. Sites, Cadmium Telluride Solar Cells: Hand book of Photovoltaic Science and Engineering, John Willey & Sons Ltd, 2003, ISBN:0-47149196-9.

[4] D. A. Wood, D. W. Lane, K.D. Rogers, and J.A. Coath, journal of Electronic Materials, 28 (1999) 4433.

[5] I. Matulionis, PhD Thesis, University of Toledo, U.S.A. 2002.

[6] B.E. McCandless et al., Proc. 22th IEEE PVSC, (1991) 967.

[7] G. T pan, Tina F.R. Shen, MN.H lai, R.C. Juang, Thomas C.K. Yang, Solar Energy Materials & Solar cells 95 (2011) 2524.

[8] D.A Wood , K.D. Rogers, D.W. Lane, J.A. Coath, J. Phys. C: Condens. Matter 12 (2000) 4433.

[9] M. A. Islam, M. S. Hossain, M. M. Aliyu, M. R. Karim, T. Razykov, K. Sopian and N. Amin, Thin Solid Films 546 (2013) 367.

[10] N. Muthukumarasamy , S. Jayakumar , M.D. Kannan , R. Balasundaraprabhu, Solar Energy 83 (2009) 522.

[11] K. Wei, F.H. Pollak, J.L. Freeouf, D. Shvydka, A.D. Compaan, J. Appl. Phys. 85 (1999) 7418.

[12] K.R. Murali, P. Thirumoorthy, C. Kannan, V. Sengodan, Solar Energy 83 (2009) 14.

[13] Nowshad Amin, Kamaruzzaman Sopian, Makoto Konagai, solar energy materials & solar cells 91 (2007) 1202.

[14] M. Gloeckler, A.L. Fahrenbruch, and J.R.Sites, 3rd world conference on photovoltaic energy conversion, Japan (2003).

[15] M A Islam, Shrarafat Hossain, Nowshad Amin, M. Mannir Aliyu, Yusuf Sulaiman and Kamaruzzaman Sopian, Advanced Materials Research, 622-623 (2013) 1194.

[16] N. Nakayama H. Matsumoto, and S. Ikegami, Jpn. J. Appl. Phys. 15 (1976) 2281.

[17] D. M. Oman, K. M. Dugan, J. L. Killian, V. Ceekala, C. S. Ferekides and D. L. More, Appl. Phys. Lett. 67 (1995) 1896.

A GROWTH MODEL TO PREDICT THE COMPOSITION OF CADMIUM TELLURIDE FILMS

Xuehai Tan[1], Shou Peng[2,3], Akash Saraf[1], Guogen Liu[1], Shenjiang Xia[2], Jingong Pan[4], Velappan Krishnakumar[5], Bastian Siepchen[5], Alan E. Delahoy[1], and Ken K. Chin*[1]

1. CNBM New Energy Materials Research Center, New Jersey Institute of Technology, Newark, NJ, 07102, USA
2. State Key Laboratory of Advanced Technology for Float Glass, Bengbu Design & Research Institute for Glass Industry, Bengbu, 233000, China
3. China Triumph International Engineering Co. Ltd., Shanghai, 200063, China
4. CNBM (chengdu) Optoelectronic Materials Co. Ltd., Chengdu, 610207, China
5. CTF Solar GmbH, Zur Wetterwarte 50, 01109 Dresden, Germany
** ken.k.chin@njit.edu*

ABSTRACT: In this study, Bube's growth model for cadmium telluride (CdTe) polycrystalline thin film was re-examined with a view to avoiding their assumptions that neglect the vapor pressures of Cd and Te_2 near the film. We proposed a new thermodynamic growth model based on the fact that there is an experimentally verified characteristic ratio (α) of equilibrium partial pressures $P_{Cd}/2P_{Te2}$ that depends on the temperature T and CdTe stoichiometry. By writing $P_{Cd}(0) = 2\alpha(0)P_{Te2}(0)$ and $P_{Cd}(h) = 2\alpha(h)P_{Te2}(h)$, where $\alpha(0)$ is determined by the source stoichiometry, we can solve the equations for $\alpha(h)$ and thereby determine the stoichiometry of CdTe thin film grown under physical vapor deposition (PVD) condition. Simulation was performed to predict the stoichiometry of CdTe thin film as a function of source stoichiometry for various temperature conditions. The results show that for a typical CdTe PVD process with $T_{source} > T_{thin\,film}$: (1) stable deposition, without non-stoichiometric composition shift, can be achieved at the congruent-growth stoichiometry; (2) any stoichiometric deviation from the congruent sublimation point becomes more substantial (in the same direction) in the thin film than in the source; (3) larger ΔT between source and thin film results in more composition shift.
Keywords: CdTe, photovoltaic, growth model.

1 INTRODUCTION

Cadmium telluride (CdTe) thin film photovoltaic technology has demonstrated its commercial success and great potential in the solar energy industry. CdTe is a group II-VI semiconductor with bipolar intrinsic doping capability [1]. The electrical properties of non-doped CdTe depend sensitively on the compound stoichiometry. Non-stoichiometric CdTe film results in either n- or *p*-type conductivity [2]. While excess Cd yields n-type conductivity, excess Te produces *p*-type films.

CdTe thin film solar cells are fabricated as a hetero-junction structure, in which the CdTe layer is usually *p*-type acting as a photo absorbing layer to generate carriers. In the last 30 years, most CdTe photovoltaics (PV) technologies have been using Te-rich CdTe. Some experts in the field believed that at a given temperature, the higher vapor pressure of Cd than that of Te_2 causes loss of Cd during deposition, and the deposited CdTe can be Te-rich even if the source material is stoichiometric [3,4]. The most commonly used extrinsic doping in Te-rich CdTe is group-I dopants, such as Cu. However, the PV technologies based on the use of Te-rich CdTe combined with Cu as dopants and Cl as passivants have failed to produce hole density greater than 10^{15} cm^{-3}. This contributed to the stagnation of open-circuit voltage (V_{OC}) for 30 years.

The recent breakthrough in monocrystalline CdTe solar cells with V_{OC} breaking 1 V barrier demonstrated a new direction of the technology by shifting to a Cd-rich stoichiometry [5]. Through external doping with group V elements, e.g. P or As, Cd-rich CdTe becomes *p*-type with hole densities exceeding 10^{17} cm^{-3}. Potentially, PV technologies based on the use of Cd-rich CdTe may play an extremely important role in the future product market of CdTe solar panels.

For both Te-rich and Cd-rich approaches, substantial gains in V_{OC} cannot be possible with intrinsic doping, but rely on external doping. To effectively introduce external dopants, favorable condition of CdTe stoichiometry is prerequisite. For Te-rich CdTe, the presence of intrinsic defects V_{Cd} enables *p*-type Cu doping by placing Cu onto the Cd sites to form substitutional Cu_{Cd} [6,7,8]. For Cd-rich CdTe, the presence of intrinsic defects V_{Te} enables *p*-type P doping by placing P onto the Te sites to form substitutional P_{Te} [6].

Clearly, stoichiometric effects in CdTe is one important factor not only determining the intrinsic electrical properties, but also correlating to the processing technologies that enable the fabrication of high performance thin film PV devices. The CdTe polycrystalline thin films are usually accomplished by physical vapor deposition (PVD). Until recently, the theory and modeling of the PVD process, either close-space sublimation (CSS) or vapor-transport deposition (VTD) of the CdTe polycrystalline thin film however still has some fundamental flaws. The stoichiometric growth condition was incorrectly expressed in several studies, where the stoichiometric condition of CdTe thin film growth was modeled as $P_{Cd} = 2P_{Te2}$ both near the source and near the thin film [9,10]. As the model only considers the reversible dissociation and recombination of $CdTe(s) \leftrightarrow Cd(g) + 1/2\,Te_2(g)$, such a fixed pressure ratio of 2 lacks theoretical as well as experimental groundings.

The correct stoichiometric condition of CdTe thin film growth was previously expressed as $J_{Cd} = 2J_{Te2}$ by Bube et al. [11], and by Chin et al. [12], where J_{Cd} and J_{Te2} are the diffusive fluxes of Cd and Te_2, respectively, from the source to the superstrate (substrate) on which CdTe thin film is being deposited. In this study, we re-visit the growth models reported in these two works. After reviewing both models and discussing their drawbacks, we propose an improved thermodynamic model by introducing an equation of $P_{Cd} = 2\alpha P_{Te2}$, where α is a characteristic ratio between P_{Cd} and $2P_{Te2}$, determined by CdTe stoichiometry and the temperature. Our growth model achieves an appropriate solution to determine the α value near the thin film. This allows us to predict and thus enables us to control the stoichiometry of CdTe thin film as a function of CdTe source stoichiometry. The results of simulation are shown for various source-film temperature combinations.

2 THEORETICAL MODEL OF CDTE THIN FILM GROWTH

A model of CSS CdTe thin film growth was proposed by Bube et al. in 1984 [11]. In their modeling, the stoichiometric condition was correctly expressed as:

$$J_{Cd} = 2J_{Te2}$$

where J_{Cd} and J_{Te2} are the diffusive fluxes of Cd and Te_2, respectively, from the source at $z = 0$ to the superstrate at $z = h$, and h is the spacing between the source and the superstrate on which CdTe thin film is being deposited. In Bube's approach, there were 4 unknowns:

$P_{Cd}(0)$, $P_{Te2}(0)$, $P_{Cd}(h)$ *and* $P_{Te2}(h)$

but only 3 equations, i.e. the equation of the stoichiometric condition, and the two equations of chemical reaction equilibrium near the source (0) and near the thin film (h):

$$K(0) = P_{Cd}(0)\sqrt{P_{Te2}(0)} = exp\left[\frac{-\Delta G_{CdTe}(T_0)}{RT_0}\right]$$

$$K(h) = P_{Cd}(h)\sqrt{P_{Te2}(h)} = exp\left[\frac{-\Delta G_{CdTe}(T_h)}{RT_h}\right]$$

The steady-state diffusive fluxes of Cd and Te_2, under diffusion-limited transport condition, are described by Fick's first law

$$J_{Cd} = \frac{D_{Cd}}{RhT_{avg}}[P_{Cd}(0) - P_{Cd}(h)]$$

$$J_{Te2} = \frac{D_{Te2}}{RhT_{avg}}[P_{Te2}(0) - P_{Te2}(h)]$$

where T_{avg} is the average temperature of source and thin film in unit of K, and D_{Cd} and D_{Te2} are the diffusion coefficients of Cd and Te_2, respectively, in unit of $m^2 s^{-1}$. As carrier gas, such as He or Ar, has usually been used during CSS CdTe deposition, the system can be treated as a binary gas diffusion system of Cd/carrier gas or Te_2/carrier gas. The diffusion coefficients of Cd and Te_2 in the binary system $D_{i,j}$ can be expressed in the Stefan-Maxwell form [13]

$$D_{i,j} = \frac{3kT_{avg}}{8P_{total}\sigma_{ij}^2}\sqrt{\frac{kT_{avg}}{2\pi}\left(\frac{1}{m_i} + \frac{1}{m_j}\right)}$$

where k is the Boltzmann constant in unit of $J K^{-1}$, P_{total} is the total pressure in unit of Pascal, $\sigma_{ij} = (\sigma_i + \sigma_j)/2$ is the average collision diameter in unit of meter, and m is the molecular (atomic) weight in unit of kilogram.

To resolve the difficulty of 4 unknowns with 3 equations, Bube et al. assumed that over the temperature range typically used during CSS CdTe growth, the vapor pressures of Cd and Te_2 near the thin film can be neglected, i.e. $P_{Cd}(0), P_{Te2}(0) \gg P_{Cd}(h), P_{Te2}(h)$, if the temperature difference between the source and the film (ΔT) is 70 °C or larger. This in turn introduced an additional equation of

$$2 = \frac{D_{Cd,j}P_{Cd}(0)}{D_{Te2,j}P_{Te2}(0)}$$

which allowed them to solve the set of equations and thus yield the CdTe growth rate as a function of growth conditions.

There are two major drawbacks in Bube's work. First, the assumption of $P_{Cd}(0), P_{Te2}(0) \gg P_{Cd}(h), P_{Te2}(h)$ does not hold true, as shown by experimental measurements of Cd and Te_2 partial pressures of $Cd_{1+x}Te_{1-x}$ over a relatively wide temperature range [14]. Second, the work only investigated the effects of deposition condition on growth rate, but not on the stoichiometry of CdTe thin film. Modeling for deposition rate of CdTe have been studied quite extensively, considering both diffusion limited and sublimation limited cases [10]. Considering the advanced technologies nowadays to accurately measure the film thickness both in-situ and ex-situ, it is rather more imperative to model and to predict the CdTe thin film stoichiometry. It is worth noting that a non-stoichiometric composition shift of 0.001 at.% in CdTe may have strong influences on the actual carrier density. Such level of changes in stoichiometry may be beyond the detection limit of any experimental methods.

Recently, our research group has demonstrated a novel and convenient method to resolve the difficulty of 4 unknowns with 3 equations encountered in Bube's work, by treating CdTe with small deviation from the perfect stoichiometry as a solid solution of excess pure Te or Cd solute in congruently grown CdTe solvent [12]. In that growth model, we introduced 1 extra unknown - the excess amount of solute in CdTe thin film $\delta(h)$, but 2 additional equations which are the two polynomial equations of partial pressures based on Henry's law near the source (0) and near the film (h):

$$P_i = H_0(i,T) + \delta(h)H_1(i,T) + \delta(h)^2H_2(i,T)$$
$$P_i = H_0(i,T) + \delta(0)H_1(i,T) + \delta(0)^2H_2(i,T)$$

where i indicates Cd or Te_2, and H's are the coefficients of polynomial that are functions of temperature. Our model was based on the data of experimentally measured equilibrium partial pressures of Cd and Te_2 in binary Cd-Te system as a function of CdTe source stoichiometry, and the coefficients of polynomial for P_{Cd} or P_{Te2} can be obtained by best fitting of experimental data using the polynomial functions. With the excess amount of solute in CdTe source $\delta(0)$ being given - determined by source stoichiometry, we were able to solve the equations and obtain the corresponding value of $\delta(h)$ for CdTe thin film grown under PVD condition.

Perhaps, the most significant advance in our previous modeling work was that the stoichiometry of CdTe thin film under PVD growth was shown to be determined by the stoichiometry of CdTe source as well as the growth condition. This dependence was preliminarily demonstrated for one source temperature and three different thin film temperatures. Following our previous work, we found it worthwhile to present a dedicated work, which models and predicts the stoichiometry of CdTe thin film as a function of the source stoichiometry, for a wider range of temperatures that will be commonly used for a typical CdTe PVD deposition. The potential results should be highly useful to the field of CdTe PV technologies. The scientists and engineers could use the work as a guidance to fine tune the growth parameters and source stoichiometry, aiming to reach a desired CdTe thin film stoichiometry.

In this study, Bube's growth model for CdTe is re-examined with a view of avoiding his assumptions to solve the equations. Our previous assumption of using Henry's law - despite its convenience - will be also avoided, because it has not been experimentally proved that the CdTe deviated slightly from the perfect stoichiometric condition can be depicted as the solid solution obeying the Henry's law. To balance the number of equations and the number of unknowns, we herein introduce an equation correlating the equilibrium partial vapor pressures of Cd and Te_2. This equation is based on the fact that there is a characteristic ratio of $P_{Cd}/2P_{Te2}$ for CdTe, which depends on the temperature T and on CdTe stoichiometry. Considering the source (0) and the thin film (h), respectively, our model provides two additional equations:

$$P_{Cd}(0) = 2\alpha(0)P_{Te2}(0)$$

$$P_{Cd}(h) = 2\alpha(h)P_{Te2}(h)$$

but one extra unknown $\alpha(h)$. With $\alpha(0)$ being given by the source stoichiometry, we are able to solve the 5 equations with 5 unknowns, rendering the dependence of $\alpha(h)$ as a function of $\alpha(0)$. It is interesting to note that the dependence of $\alpha(h)$ on $\alpha(0)$ is equivalent to the dependence of $\delta(h)$ on $\delta(0)$.

In our previous model, we used polynomial functions based on Henry's law to fit the experimental data of partial pressures published by Greenberg [14]. However, the fitting and the extrapolation of polynomial coefficients for vapor pressures at lower temperatures lacks a physical grounding as well as the necessary accuracies. In this study, we instead used a numeric method to fit the experimental data of equilibrium partial pressures of Cd and Te_2 as a function of CdTe stoichiometry. The experimental data at temperatures of 1073K, 1123K, 1173K, and 1223K were digitized from the published figures. The digitized graphs were then tabulated in the range of 49.997 to 50.007 at.% Te with an interval of 0.0001 at.%, using the "smooth spline" function provided by Matlab. Such method does not force the data to follow any analytical formula, but rather simply follow any arbitrary data trend. As shown in Fig. 1, the tabulated data satisfactorily match the Greenberg graph. To extrapolate the vapor pressures at lower temperatures for each tabulated stoichiometry, we used the well-known Antonie equation:

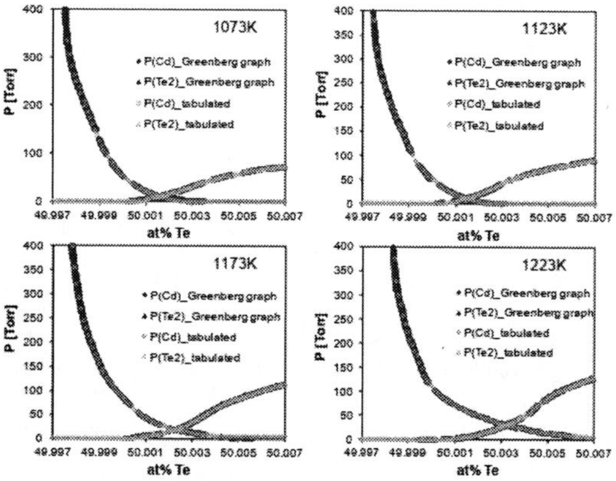

Figure 1: Fitting the experimental data of equilibrium partial pressures of Cd and Te_2 as a function of CdTe stoichiometry using a numeric method. The tabulated data of vapor pressures match the experimental results satisfactorily.

$$log_{10}P = A - \frac{B}{T}$$

The Antoine equation is a class of semi-empirical correlations describing the relation between vapor pressure and temperature for pure components [15]. Here, we treated each stoichiometry of CdTe as a single component.

3 RESULTS AND DISCUSSION

Fig. 2 summarizes the calculated values of $\alpha(h)$ as a function of $\alpha(0)$ for various CdTe source temperatures in the range of 900 to 1180K. For each source temperature, $\alpha(h)$ values were calculated at five different film temperatures to demonstrate the effect of temperature condition. The source-film temperature combinations were selected to be relevant to a typical CSS or VTD deposition of CdTe thin films. For each CdTe source temperature, the five different film temperatures correspond to ΔT $(= T_{source} - T_{thin\,film})$ values of 20, 60, 100, 140 and 180K, respectively.

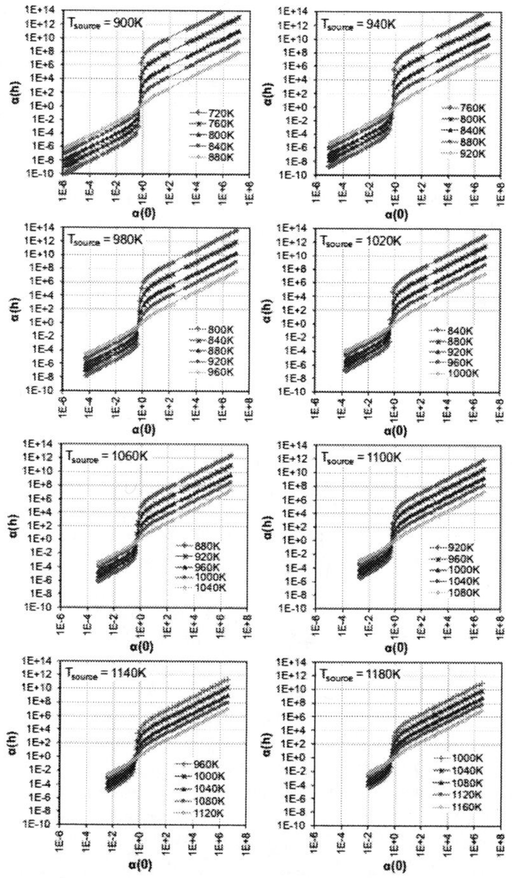

Figure 2: The dependence of $\alpha(h)$ on $\alpha(0)$ for various source temperatures in the range of 900K to 1180K. For each source temperature, $\alpha(h)$ values was calculated at five different film temperatures.

Given the characteristic correlation between the value of α and the stoichiometry of CdTe at a given temperature, the plots of $\alpha(h)$ versus $\alpha(0)$ can be replotted as CdTe film stoichiometry versus CdTe source stoichiometry, with the results being shown in Fig. 3. The characteristic correlations at temperatures of 1073, 1123, 1173 and 1223K were extracted from the experimental data of partial pressures published by Greenberg. The characteristic correlations at other temperatures were obtained by extrapolation of partial pressures using Antonie equation. As shown in Fig. 3, the stoichiometry of CdTe source or thin film was modeled in the range of 49.995 to 50.004 at.% Te for consideration of CdTe thin film PV applications. Further deviation from the perfect CdTe stoichiometry would result in precipitation of elemental Cd or elemental Te phase. The supersaturated solid solution would segregate into a two-phase microstructure with elemental inclusion in the matrix of CdTe compound.

Close inspection of Fig. 2 and Fig. 3 reveals that there is an inflection point in the curve of $\alpha(h)$ vs. $\alpha(0)$ or in the curve of CdTe film stoichiometry vs. source stoichiometry. Near the inflection point, the α ratio of $P_{Cd}/2P_{Te2}$ or the stoichiometry of CdTe shows minimal changes comparing the thin film to the source. This inflection point is positioned on the Te-rich side of stoichiometry in both vapor and solid phases. This observation suggests that the inflection point is near the congruent-growth stoichiometry, rather than near the point of perfect CdTe solid stoichiometry or the point of vapor phase stoichiometric. From the calculated results, stable deposition, without non-stoichiometric composition shift, in principle can be achieved at the congruent-growth stoichiometry. However, as the congruent-sublimation point also shifts with temperature, the CdTe source temperature needs to be controlled in accordance to the source stoichiometry, so that a stable deposition without composition shift can be really achieved.

As shown in Fig. 2 and Fig. 3, on either side of the congruent-growth stoichiometry, any stoichiometric deviation from the congruent sublimation point becomes more substantial (in the same direction) in CdTe thin film than in CdTe source. This modeling result is important for a typical CdTe PVD process with source temperature higher than film temperature. The commonly used CdTe source material with composition close to the perfect CdTe stoichiometry is likely to position on the Te-deficient side of congruent-growth stoichiometry. Consequently, for a typical CdTe PVD process, there is non-negligible shift of CdTe stoichiometry that is however being usually neglected in the field of CdTe thin film PV technologies.

Another point observed from Fig. 2 and Fig. 3 is that larger ΔT between source and thin film during a process of PVD results in more shift in stoichiometry from CdTe source to CdTe thin film. For each source temperature, while the value of ΔT being close to zero, the plot of $\alpha(h)$ vs. $\alpha(0)$ or film stoichiometry vs. source stoichiometry shows a correlation of linear dependency. This linear relation makes sense as the partial vapor pressures of Cd or Te_2 remain similar on either side of source or thin film, leading to minimal changes in CdTe stoichiometry. As the value of ΔT increases, apart from the change in partial vapor pressures, there is also a stronger shift in the characteristic correlation between the value of α and the stoichiometry of CdTe. As a result, any stoichiometric deviation from the inflection point becomes more substantial in CdTe thin film as the temperature difference ΔT increases.

It should be noted that our calculated results have introduced a new considering point among the effects of growth temperature for CdTe thin film PV applications. In general, the thin film growth temperature at the superstrate (substrate) level was believed to affect the quality of CdTe polycrystalline thin film, including the grain size, the defect density, and the film surface morphology [16]. It has also been reported that CdTe thin films deposited at lower substrate temperature show a high degree of porosity between the columnar grains structure [17]. Considering these points, it was argued that a higher growth temperature is more beneficial for growing polycrystalline film with less defects. On the other hand, a lower growth temperature is sometime beneficial from practical viewpoints. For example, a low-temperature solar module fabrication process is needed to accommodate the low strain point of soda-lime-glass superstrate. As demonstrated by the calculated results of current study, the temperature difference ΔT in principle will also affect the stoichiometry of CdTe thin film grown under PVD condition. Since a non-stoichiometric composition shift in CdTe may have strong influences on the actual carrier density, it is imperative to consider the temperature effect on stoichiometry while controlling the temperature condition of a deposition process. As the temperature effect on stoichiometry was usually neglected, this makes our work new and useful to the field of CdTe PV technologies.

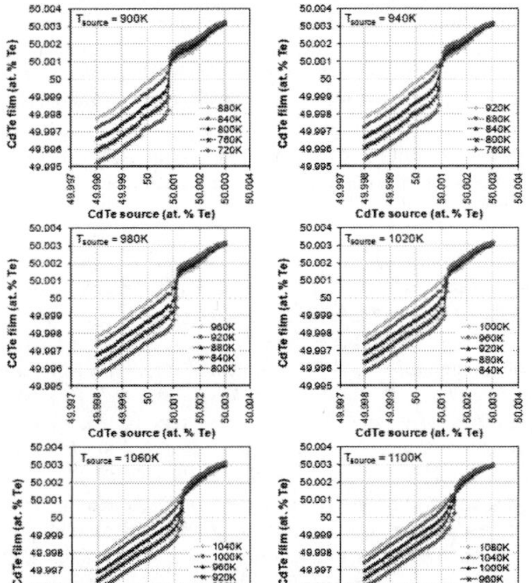

Figure 3: The dependence of CdTe thin film stoichiometry on CdTe source stoichiometry. The results are shown for various source-film temperature combinations.

Derived from the results shown in Fig. 3, we have made a master plot of CdTe composition shift as a function of increasing ΔT, which enables us to predict the stoichiometry of CdTe thin film deposited at a certain source-film temperature combination. As the dependence of composition shift vs. ΔT appears to be linear, the slope and intercept of each line at a given source temperature can be extrapolated using the fitted lines shown in Fig. 4(b) and Fig. 4(c), respectively. Assuming the stoichiometry of CdTe source is known, we can therefore control the stoichiometry of CdTe thin film by fine tuning the temperature condition of a PVD process, using the information provided in Fig. 4.

4 CONCLUSION

The technologies employed for CdTe thin film photovoltaics are strongly dependent on the exact stoichiometry of CdTe thin films. For Te-rich CdTe, Cu-doping is usually used, but there seems to be some fundamental limitations associated. e.g. the doping concentration is limited to less than 10^{15} cm^{-3} due to self-compensation, low carrier lifetime due to Te_{Cd} antisite deep levels, and the stability issue induced by Cu migration. For Cd-rich CdTe, the future seems to be brighter. Experimental study has shown higher V_{OC} for solar cells made from Cd-rich CdTe single crystalline absorber layer. The technologies for Cd-rich material use P- or As-doping, different from that of Te-rich CdTe. Therefore, it is important to foresee the stoichiometry of CdTe films grown under PVD condition, thereby being able to determine the techniques that should be used during the processing steps. As the deviation from the perfect stoichiometry is most likely too small to be accurately measured using experimental methods, a growth model that predicts and controls the CdTe thin film stoichiometry is highly useful.

In this study, we reported an improved thermodynamic model for CdTe thin films grown under PVD conditions. The model considers a characteristic ratio (α) of equilibrium partial pressures $P_{Cd}/2P_{Te2}$ that depends on the temperature T and CdTe stoichiometry. By introducing two equations $P_{Cd}(0) = 2\alpha(0)P_{Te2}(0)$ and $P_{Cd}(h) = 2\alpha(h)P_{Te2}(h)$, where $\alpha(0)$ is determined by the source stoichiometry, we can solve the equations for $\alpha(h)$ and thereby determine the stoichiometry of CdTe thin film. Simulation was performed to predict the stoichiometry of CdTe thin film as a function of source stoichiometry for various source-film temperature combinations. The results show that for a typical CdTe PVD process with $T_{source} > T_{thin\,film}$: (1) Stable deposition, without non-stoichiometric composition shift, can be achieved at the congruent-growth stoichiometry; (2) Any stoichiometric deviation from the congruent sublimation point becomes more substantial (in the same direction) in the thin film than in the source; and (3) Larger ΔT between source and thin film results in more composition shift.

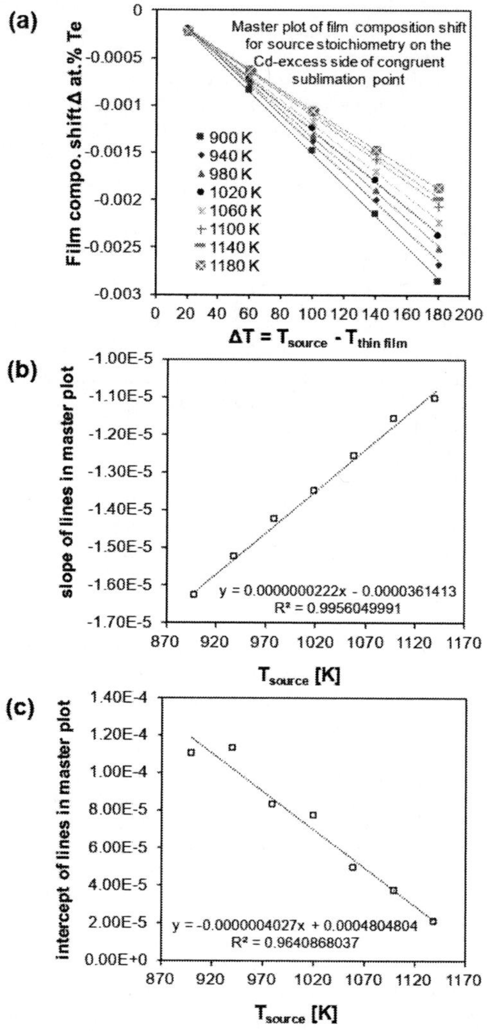

Figure 4: (a) Master plot of CdTe composition shift (% film – % source), on the Cd-excess side of congruent point, as a function of increasing ΔT, for various source temperatures from 900 to 1180K. As the dependence appears to be linear, the slope and intercept of each line (at a given source temperature) shown in master plot can be calculated using the equations shown in (b) and (c), respectively. This allows us to predict the stoichiometry of CdTe thin film deposited at a certain source-film temperature combination.

5 ACKNOWLEDGEMENTS

The authors acknowledge the China Triumph International Engineering Co. Ltd. (CTIEC), Shanghai, China, which offers generous financial support for this work. The authors would like to thank Dr. Zimeng Cheng for the helpful discussions.

6 REFERENCES

[1] F. KrUger and D. de Nobel, *J. Electron.* **1955**, 1: 190-202.

[2] H. Y. Ueng, and S. Y. Yang, *J. Renewable Sustainable Energy* 2012, 4: 011605.

[3] K. Biswas and M. Du, *New Journal of Physics* **2012**, 14: 063020.

[4] M. Khan, V. Evani, S. Collins, V. Palekis, P. Bane, S. Bakhshi, V. Kendre, S. Vatavu, D. Morel and C. Ferekides, *40th IEEE PVSC Proceedings* **2014**, 2343-2347.

[5] J. M. Burst, J. N. Duenow, D. S. Albin, E. Colegrove, M. O. Reese, J. A. Aguiar, C.-S. Jiang, M. K. Patel, M. M. Al-Jassim, D. Kuciauskas, S. Swain, T. Ablekim, K. G. Lynn, and W. K. Metzger, *Nat. Energy* **2016** 1: 16015.

[6] Ji-Hui Yang, Wan-Jian Yin, Ji-Sang Park, Jie Ma, and Su-Huai Wei, *Semicond. Sci. Technol.* **2016**, 31: 083002.

[7] Jie Ma, Su-Huai Wei, T. A. Gessert, and Ken K. Chin, *Physical Review B* **2011**, 83: 245207.

[8] Ken K. Chin, T.A. Gessert, and Su-Huai Wei, *35th IEEE PVSC proceeding* **2010**, 001915-001918.

[9] S. N. Alamri, *Phys. Status Solidi A: Appl. Res.* **2003**, 200: 352.

[10] J. L. Cruz-Campa, D. Zubia, *Sol. Energy Mater. Sol. Cells* **2009**, 93: 15–18.

[11] T. C. Anthony, A. L. Fahrenbruch and R. H. Bube, *J. Vac. Sci. Technol. A,* **1984**, 2, 1296-1302.

[12] K. K. Chin, Z. Cheng, A. E. Delahoy, *J. Cryst. Growth* **2015**, 418: 32-37.

[13] R. Eslamloueyan, M.H. Khademi, *Chemometrics and Intelligent Laboratory Systems* **2010**, 104: 195–204.

[14] J. H. Greenberg, *J. Cryst. Growth* **1996**, 161: 1.

[15] C. Antoine, *Comptes Rendus des Séances de l'Académie des Sciences* **1888**, 107: 681–684, 778–780, 836–837.

[16] J. D. Major, *Semicond. Sci. Technol.* **2016**, 31: 093001.

[17] K. Häfner, H. Morgner, O. Zywitzki, T. Modes, C. Metzner, B. Siepchen, B. Späth, M. Grimm, *26th EUPVSEC proceeding* **2011**, 3052-3054.

ANALYTICAL MODELING OF CZTSE: GE SOLAR CELLS WITH INCOMPLETE GAMMA FUNCTION OF QUANTUM EFFICIENCY

Sanghyun Lee[a], Kent J. Price[b], Edgardo Saucedo[c], Sergio Giraldo[c]
[a]Electronics and Computer Engineering and Technology, Indiana State University, Terre Haute, IN 47809, USA, +1-812-237-3451, Sanghyun.lee@indstate.edu
[b]Physics, Morehead State University, Morehead, KY 40351, USA
+1-606-783-2757, k.price@moreheadstate.edu
[c]Catalonia Institue for Energy Research, Jardins de les Dones de Negre, 1 2 Pl, Sant Adria de Bess 08930, +34-933-562-615, esaucedo@irec.cat

ABSTRACT: We quantitatively develop analytical and numerical models of quantum efficiency for CZTSe: nanolayer Ge (<10 nm) thin film solar cells with incomplete gamma function under the practical linear electric field model, which agrees with empirical data. From capacitance-voltage measurement, the net doping concentration and the depletion width are extracted to use as critical input parameters. As we develop an analytical model, we discuss the effectiveness of the present constant electric field model and compare it to our linear field model. Furthermore, the empirical results are demonstrated for the comparative study. From the measurement, we observe significantly different trends between under forward and reverse bias conditions that are explained with the developed model. An unusual severe carrier collection collapse in >550 nm at forward bias of 0.2 V is observed as compared to the expected spectral response at -0.2 V. Conversely, there is relatively stronger carrier collection between 500 nm and 550 nm at 0.2 V. From the numerical modeling and simulation at 0.2 V, it is found that the alternative explanation is made for the stronger spectral response at the wavelength regions (500 – 550 nm) near the heterojunction interface as compared to dramatic degradation of carrier collection near >550 nm at 0.2 V. This peculiar response is modeled with the linear electric field model with gamma function by inserting thin (< 9 nm) highly defective near-heterojunction interface, which could be eliminated with weak 0.01 Sun (1 mW/cm^2) bias light at 400 nm.
Keywords: thin film solar cells, modeling, quantum efficiency

1 INTRODUCTION

In the past five years, polycrystalline thin film CdTe and Cu(In,Ga)Se_2 (CIGS) solar cells are commercially successful with dramatic conversion efficiency improvements and the cost reduction in module mass production as well as the balance of system (BoS) which is the cost except the module production cost. A champion device from CdTe (Zincblende) and CIGS (Chalcopyrite) solar cells show < 22 % efficiency (< 21.5 %, First Solar and <21.7 %, ZSW respectively). However, some of these materials for thin film solar cells technology are not benign to the environment which causes the toxicity and possible scarcity issues. To fully facilitate the advantage of low manufacturing cost in the polycrystalline thin film solar cells technology, a research on Kesterite solar cells such as Cu_2ZnSnS_4 (CZTS), $Cu_2ZnSnSe_4$ (CZTSe) $Cu_2ZnSn(S,Se)_4$ (CZTSSe) using earth abundant materials gains momentum to pursue low-cost, environment-friendly, and high efficient Kesterite photovoltaic devices with the possible replacement of CdS which contains toxic Cd [1-3]. Although many efforts have been made aiming to improve efficiency, the critical loss mechanism to boost efficiency is not well understood. To address the possible loss mechanism, analytical and numerical models of quantum has been studied by researchers [4] as the spectral responses of quantum efficiency (QE) provides the information about the spatial distribution of the charge recombination path.

In the present work, we fabricated the CZTSe: Ge solar cell devices and developed a practical analytical and numerical model to explain the bias-dependent QE trend, which agrees with empirical QE data. After developing a mathematical expression for QE of CZTSe: Ge nanolayer solar cells, we will compare the result to the computational simulation result.

2 EXPERIMENTAL

The CZTSe: nanolayer Ge thin film was deposited (< 2 um) by direct current (DC) magnetron sputtering onto soda-lime-glass/Mo substrate. The Selenization was performed in a two-step process. The first one at 400 °C (heating ramp of 20 °C/min) during 30 min and 1.5 mbar of Ar pressure, followed by the second step at 550 °C (heating ramp of 20 °C/min) during 15 min and 1 bar of Ar pressure, with a natural cooling down to room temperature. After several chemical etchings to minimize the secondary phases on the surface of the absorber and to passivate, CdS buffer layer (50 nm) was grown by chemical bath deposition. Following 50nm CdS chemical bath deposition, the CZTSe: Ge nanolayer (< 10 nm) device was completed by DC-pulsed sputtering deposition of intrinsic-ZnO (50 nm) and In_2O_3-SnO_2 (ITO, 350 nm) as a transparent conductive window layer. After ITO was deposited, the solar cells were mechanically defined as 5 x 5 mm^2 and complete with Ni/Al front metal fingers and MgF_2 anti-reflective coating.

The electrical characterization was done with Agilent B1500 after calibration with the standard photodiode. The spectral responses were characterized PV Measurements QEX10 QE measurement system with Xenon arc lamp and the spectral distribution of the photon flux at the outlet slit was calibrated with NIST calibrated photodiode G425.

3 RESULT AND DISCUSSION

3.1 Electrical characterization

From the Capacitance-Voltage (C-V) measurement, the net carrier concentration is extracted using the method in [5]. To exclude unwanted effects from mid or deep level defect states, C-V measurements were carried out at low enough frequency, 40 kHz. In Fig. 1, we calculated the average net doping concentration from measurements of

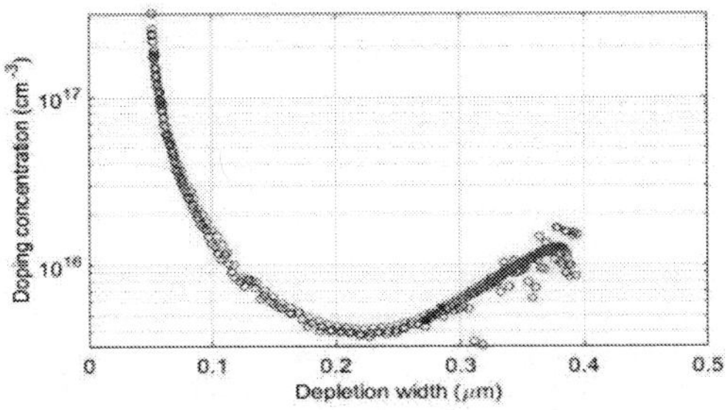

Fig. 1: Net charge carrier profiles of three representative CZTSSe devices calculated from C-V measurements. The blue crosses are measurements at V = 0, indicating the depletion width at V = 0.

three representative devices, 5.0 x 10^{15} cm^{-3} which is a bit lower that reported values elsewhere [6]. The depletion width calculated is 277.5 - 297 nm. In this work, we will use these values for modeling and simulation.

3.2 Analytical modeling

3.2.1 Basic Equations

The analytical model is developed with the carrier collection efficiency model for the CZTSSe solar cells and put in the mathematical expression. The external quantum efficiency, EQE(λ,V) is the ratio of the number of charge carriers collected by the photovoltaic device to the number of photons of a given wavelength or energy onto the device. Since the ITO, ZnO, and CdS layers negligibly contribute to the photocurrent, these only cause optical losses. Hence, the EQE(λ,V) equation is related to the reflection and transmission of each material as below.

$$\mathrm{EQE}(\lambda,V) = \left(1 - R_{ARC}(\lambda)\right) \cdot T_{ITO}(\lambda) \cdot T_{ZnO}(\lambda) \cdot T_{CdS}(\lambda) \cdot IQE_{CZTSSe}(\lambda,V) \cdot h(V) \quad (1)$$

, where $R_{ARC}(\lambda)$ is the reflection of anti-reflection coating. $T_{ITO}(\lambda)$ is transmission of ITO, $T_{ZnO}(\lambda)$ is transmission of ZnO, $T_{CdS}(\lambda)$ is transmission of CdS, and IQE(λ,V) is the internal quantum efficiency (IQE) of the CZTSSe absorber layer, h(V) is the interface recombination function at the CdS/ CZTSSe heterojunction interface [7].

3.2.2 Charge collection from depletion region

By integrating with optical generation function with the collection function of the space charge region, the drift component of IQE is calculated as below [8].

$$IQE_{drift} = \int_0^W \alpha_{CZTSSe} \exp(-\alpha_{CZTSSe} x) CF_{SCR}(x) dx \quad (2)$$

, where α_{CZTSSe} is the absroption coefficient, $CF_{SCR}(x)$ is the charge collection function in the space charge region.

As pointed out Hardrich [8], there is significant interface recombination at the space charge region due to interface defects, but the revised charge collection function was developed with constant electric field in the space charge region. However, the electric field in the space charge regions is not constant for most of the fabricated devices. To reflect a practical situation with an appropriate device model, we develop the constant doping model by assuming constant doping in the space charge region, which is, in particular, applicable to most of the CZTSe: Ge nanolayer devices with narrow space charge region (< 300nm). The electric field in the space charge region linearly decreases along the depth of CZTSSe absorber. Although a constant doping model is nearly close to the practical situation, it also requires assumptions due to the doping profile fluctuation with grain boundaries, fixed charges, impurities, and the dimentionality. We mathematically simply the case with one–dimensional linear electric field at the junction. Hence we are able to develop an alternative expression for the linearly decreasing electric field in the space charge region. By defining the drift time for electron and kappa ($\kappa = \mu_e \tau_e E_{max}$), the collection function under linear electric field is as below.

$$CF_{SCR_{LEF}} = exp\left(-\frac{t_{drift}}{\tau_e}\right) = \exp\left\{-\frac{\int_x^W \frac{dx}{\left(\frac{E_{max}\mu_e}{W}\right)\cdot x}}{\tau_e}\right\}$$
$$= (1 - \frac{x}{W})^{\frac{W}{\kappa}} \quad (3)$$

, where t_{drift} is the drift time, τ_e is the lifetime of the electron, W is the deletion width, μ_e is the mobility of the electron, E_{max} is maxmium electric field at the junction. After integrating with the incomplete gamma function,

$$IQE_{drift} = (\alpha W)^{-\frac{W}{\kappa}} \cdot \Gamma(1 + \frac{W}{\kappa}, \alpha W) \quad (4)$$

3.2.3. Charge collection from quasi-neutral region

For a diffusion component of the photoelectric quantum response, the exact solutions for electron and hole are derived in elsewhere [5] for the case of p-layer in n^+-p junction and we will use this equation to evaluate the diffusion component of IQE in the quasi-neutral region.

$$IQE_{diff} = \frac{\alpha L_n}{\alpha^2 L_n^2 - 1} \exp(-\alpha W) \cdot \left\{\alpha L_n - \frac{\frac{S_b}{D_n}\left[\cosh\left(\frac{d-W}{L_n}\right) - \exp(-\alpha(d-W))\right] + \sinh\left(\frac{d-W}{L_n}\right) + \alpha L_n \exp(\alpha(d-W))}{\frac{S_b}{D_n}\sinh\left(\frac{d-W}{L_n}\right) + \cosh\left(\frac{d-W}{L_n}\right)}\right\} \quad (5)$$

3.3.3 External quantum efficiency

The total IQE is summed up with Eq. 4 and 5 and the external quantum efficiency is calculated with the total IQE in Eq. 1. To avoid reported optical characterization artifacts in [9], we adopted the reported dielectric functions, refractive index, and extinction coefficient to calculate the critical absorption parameters for consistency [10-13] and the electron lifetime is 1-3 ns for this analytical work.

Table I: Key parameters used in this work

CZTSSe layer parameters	Value/Unit
CB effective density of state	2.2 x 10^{18} cm^{-3}
VB effective density of state	1.8 x 10^{19} cm^{-3}
Electron mobility	1.25–50 cm^2/V/s
Hole mobility	1–25 cm^2/V/s
Relative dielectric constant	9

3.3 Characterization and Model comparison

In Fig. 2, the measured band gap from the derivative of the QE curve is 1.11 eV and the total QE-calculated short circuit current is 33.59 mA/cm^2. The QE spectrum for the best fit is analytically simulated with measured model parameters, N_{doping} = 5.0 x 10^{15} cm^{-3} and W = 277 nm to the experimental data. The fitted data at applied voltage

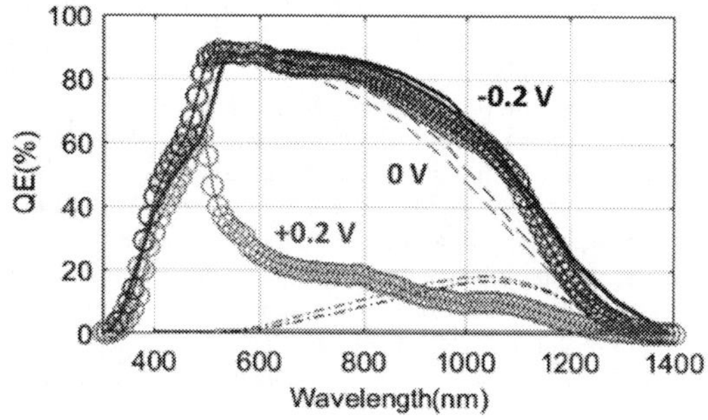

Figure 2: Comparison of experimental EQE curves (dots) with fitted curves (solid lines) at applied voltage, $V_{app} = 0$ V (red) and -0.2 V (blue). Color-coded dashed line lines are drift components and dash-dot lines are diffusion components. EQE data at +0.2 V is added for the reference without fitting.

bias, 0 V and 0.2 V consist of the drift component (dashed line) of carrier collection and the diffusion component of carrier collection (dash-dotted line) as discussed above. Except slight deviation near 500 nm and 900 nm, a good match is found throughout the measured wavelength range region due to the modification of depletion width by applied bias at -0.2 V and no bias. The drift component at -0.2 V increases while the diffusion component decreases as compared to the opposite trend at 0 V.

However, the quantum efficiency measurement at 0.2 V is quite interesting which we could not fit with good error tolerance. With constant uniform doping of the absorber, this setting keeps the shape and the ratio of the quantum efficiency curves at different biases with reference to the near-full collection voltage bias (in this study, V_{app} = -0.2 V) since the shape of the curves in quantum efficiency depends on wavelength and is almost independent of bias voltage [6]. However, we observed the stronger drop at both <500 nm and >550 nm disproportionately whereas it is less near 500 nm. To investigate this further, the voltage bias is held at 0.2 V while we applied very weak (0.01 sun, 1 mW/cm^2) bias lights at three different wavelengths such as 400 nm, 660 nm, and 900 nm in Fig. 3.

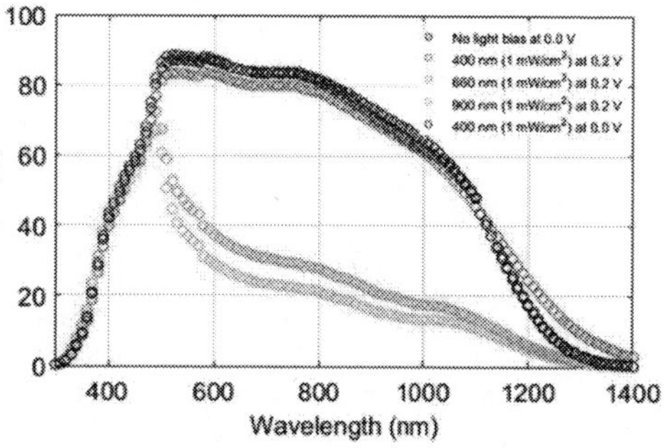

Figure 3: Weak light biased (0.01 sun, 1 mW/cm^2) EQE curves at 0.2 V. A curve in magenta dots is under light bias, 400 nm at 0.2 V, a curve in cyan dots is under light bias, 660 nm at 0.2 V, and a curve in green dots is under light bias, 900 nm at 0.2 V. A curve in black dots is under light bias, 400 nm at 0.0 V. The curve in blue dots without light bias and 0.0 V is added for the reference from the previous figure.

In Fig 3, we measured QE under three different bias light at 400 nm, 660 nm, and 900 nm with the same intensity of light, 0.01 sun (1 mW/cm^2) at 0.2 V along with the reference quantum efficiency without no light and voltage biases. There is no significant difference QE curves under 660 nm (green) and 900 nm (cyan). However, a curve under 400 nm light bias is recovered completely close to the reference QE level at 0 V without light bias. To investigate further, we compared two measured QE curves by normalizing light biased QE divided by QE without no light bias in Fig. 4.

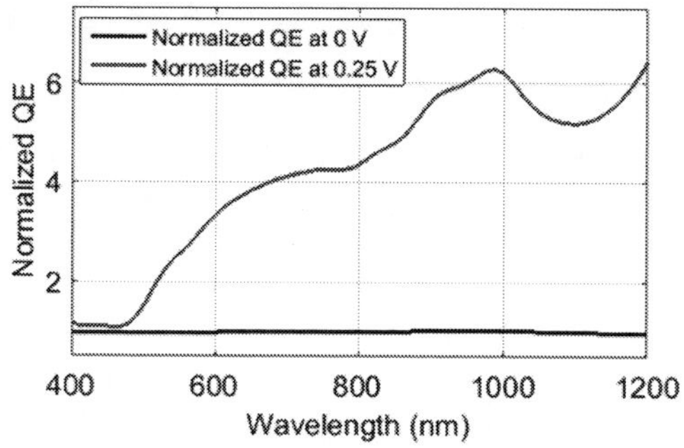

Figure 4: Light bias-dependent normalized QE (QE at 400 nm, 1 mW/cm^2 light bias divided by QE without no light bias) after noise filtering at lower QE response <400 nm and >1200 nm. A curve in blue is normalized QE at 0 V and a curve in magenta is normalized QE at 0.25 V. The normalized QE at 0.25 V increases dramatically at >500 nm after weak 400 nm light is illuminated whereas the normalized QE at 0 V shows almost no increase and kept close to the unity even after 400nm light bias.

In Fig. 4, we normalize QE under 400 nm bias light for 0.25 V (magenta) and 0 V (blue) for the comparison. The normalization is performed by dividing the QE curve under 1 mW/cm^2 light bias at 400 nm by the QE curve without light bias to check the light bias effects on the devices. Apparently, the normalized QE at 0.25 V increases dramatically near >550 nm after weak 400 nm light is illuminated whereas the normalized QE at 0 V shows almost no increase and keeps close to the unity even after 400nm light bias. The traditional explanation for this typical spectral response is the compensation of the acceptor type defects within the CdS layer by holes generated from the valence band [14]. To confirm this scenario, we used a different approach with the highly defective interface layer by simulating with SCAPS simulator.

In SCAPS numerical simulation, as a first step, we simulated the ideal device without interface defects and confirm the ideal response. Then we move onto the simulation with CZTSe: Ge nanolayer devices with the inserted thin (< 9 nm) defective interface layer by performing a serious of acceptor and donor type defects concentration with various energy state in the band gap, confirming the similar severe quantum efficiency drop near > 550 nm with the acceptor type defect concentration, 7.0 x 10^{17} cm^{-3} near mid energy gap energy (+/- 0.3 eV from the mid-gap). By applying voltage bias at 0.25 V under 1 mW/cm^2 light bias at 400 nm, we successfully simulated and normalized the quantum efficiency in Fig. 5. The similar trend of the normlaized QE curve is shown in Fig. 5 which demonstrates strong increase near >550 nm after weak 400 nm light is illuminated at 0.25 V. Therefore, a highly defective heterojunction interface layer is attributed to this typical quantum efficiency response without acceptor type defect within the CdS layer as simulated.

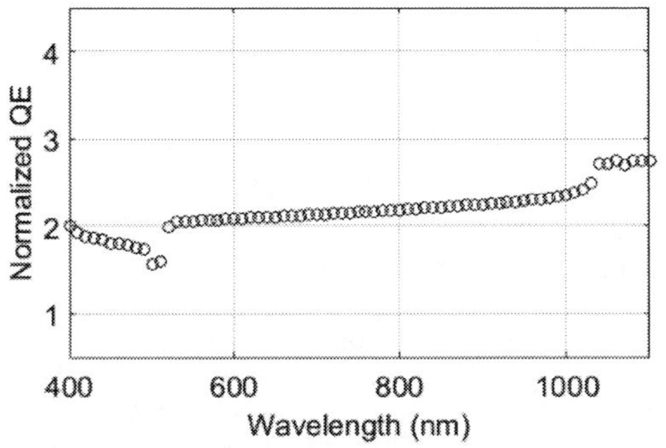

Figure 5: Simulated light bias-dependent normalized QE at 0.25 V (QE at 400 nm 1 mW/cm^2 light bias divided by QE at no light bias) after noise filtering at lower QE response <400 nm and >1100 nm. By inserting a thin interface defect layer with deep acceptor type defect, 7.0 x 10^{17} cm^{-3}. the normalized QE increases dramatically at >550 nm after weak 400 nm light is illuminated which is a similar trend to empirical data.

4 CONCLUSION

The quantum efficiency of CZTSSe devices is analytically modeled with incomplete gamma function to reflect the realistic linear electric field at the heterojunction interface, which agrees with empirical quantum efficiency measurements. The capacitance-voltage measurements illustrate that the net active doping concentration is calculated about 5.0 x 10^{15} cm^{-3} and the depletion width is 277.5 nm. The measured band gap from the quantum efficiency curve is 1.11 eV and the short circuit current by calculating quantum efficiency is 33.59 mA/cm^2. Under the negative voltage bias and no bias, the spectral response, the model well predict the spectral responses. However, under the positive voltage bias at 0.2 V, it is found that unusual severe carrier collection collapse in >550 nm whereas mild carrier collection drop the rest of the wavelength regions. From the numerical modeling and simulation with SCAPS at 0.2 V, we can successfully simulate the relative stronger spectral response at the wavelength regions (500 – 550 nm) near the heterojunction interface at 0.2 V by inserting highly defective near-heterojunction interface defects. By applying weak, 0.01 Sun (1 mW/cm^2) bias light at 400 nm, the stronger drop in quantum efficiency is eliminated and the quantum efficiency is recovered to the similar spectral response at 0 V without light bias.

5. REFERENCE

[1] W.Wang, M. T. Winkler, O. Gunawan, T. Gokmen, T. K. Todorov, Y. Zhu, D. B. Mitzi, Adv. Energy Mater. 4 (2014) 1301465.
[2] D. B. Mitzi, T. K. Todorov, K. Wang, S. Gunha, Sol. Energy Mater. Sol. Cells 95 (2011) 1421.
[3] X. Liu, Y. Feng, H. Cui, F. Liu, X. Hao, G. Conibeer, D. Mitzi, M, Green, Prog. Photovolt: Res. Appl. 24 (2016) 879.
[4] A. Pu, F. Ma, C. Yan, J. Huang, K. Sun, M. Green, X. Hao, Sol. Energ. Mater. Sol. Cells 160 (2017) 372.
[5] S. Sze, Physics of Semiconductor Devices, 3rd ed., Wiley, chap. 13 (2007).
[6] B. Shin, O. Gunawan, Y. Zhu, N.A. Bojarczuk, S.J. Chey, S. Guha, Prog. Photovolt: Res. Appl. 21 (2013) 72–76.
[7] A. L. Fahrenbruch, R. H. Bube, Fundamentals of solar cells: photovoltaic solar energy conversion. Academic Press, (1983).
[8] M. Hadrich, H. Metzner, U. reislohner, C. Kraft, Sol. Energ. Mater. Sol. Cells 95 (2011) 887.
[9] C. Frisk, T. Ericson, S.-Y. Li, P. Szaniawski, J. Olsson, C. Platzer-Björkman, Sol. Energy Mater. Sol. Cells 144 (2016) 364.
[10] A. Nakane, H. Tampo, M. Tamakoshi, S. Fujimoto, K. M. Kim, S. Kim, H. Shibata, S. Niki, H. Fukiwara, J. Appl. Phys. 120 (2016) 064505.
[11] T. Hara, T. Maekawa, S. Minoura, Y. Sago, S. Niki, H. Fujiwara, Phys. Rev. Applied. 2 (2014) 034012.
[12] H. Fujiwara, M. Kondo, Phys. Status Solidi A 71 (2005) 075109.
[13] H. ElAnzeery, O. Daif, M. Buiere, S. Oueslati, K.Messaoudand, D. Agten, G. Brammertz, R. Guindi, B. Kniknie, M. Meuris, J. Poortmans, Phys. Status Solidi A 212 (9) (2015) 1984.
[14] M. Neuschitzer, y. Sanchez, S. Marino, H. Xi, A. Fairbrother, M. Pacidi, S. Haass, V. Roca, A. Rodrogiez, S. Saucedo, Prog. Photovolt: Res. Appl. 23 (2015) 1660

PHYSICAL MODEL OF DEFECT FORMATION IN NON-STOICHIOMETRIC CADMIUM TELLURIDE

Xuehai Tan[1], Shou Peng[2,3], Akash Saraf[1], Shenjiang Xia[2], Jingong Pan[4], Velappan Krishnakumar[5], Bastian Siepchen[5], Alan E. Delahoy[1], and Ken K. Chin*[1]

1. CNBM New Energy Materials Research Center, New Jersey Institute of Technology, Newark, NJ, 07102, USA
2. State Key Laboratory of Advanced Technology for Float Glass, Bengbu Design & Research Institute for Glass Industry, Bengbu, 233000, China
3. China Triumph International Engineering Co. Ltd., Shanghai, 200063, China
4. CNBM (chengdu) Optoelectronic Materials Co. Ltd., Chengdu, 610207, China
5. CTF Solar GmbH, Zur Wetterwarte 50, 01109 Dresden, Germany
** ken.k.chin@njit.edu*

ABSTRACT: In this work, we propose a physical model of defect formation in non-stoichiometric cadmium telluride (CdTe) by detailing the change of free energy for CdTe thin film under growth. The model explains the dependence of various intrinsic point defect densities in CdTe lattice on temperature, on Fermi level, and on vapor pressures of Cd and Te_2 during physical vapor deposition (PVD). The model also establishes a physical foundation for the experimentally observed relationship between the stoichiometry of CdTe films and the vapor pressures of Cd and Te_2 via the formation of various defects. The simulation results based on the model provide theoretical insights and practical guidance for experimentalists to fine tune the carrier concentration in CdTe thin films.

Keywords: CdTe, stoichiometry, carrier concentration, intrinsic defects, defect formation

1 INTRODUCTION

CdTe is one of the most widely used materials for low cost, high efficiency thin film photovoltaic (PV) applications. The intrinsic point defects of CdTe, such as vacancy, interstitial and antisite, play critical roles in the dopant and deep level concentrations. Thus, it is important to optimize the growth conditions of physical vapor deposition (PVD), either close space sublimation (CSS) or vapor transport deposition (VTD), so that the dopant concentration is high, and deep level density is low. The experimentally observed relationship between the stoichiometry of CdTe and the vapor pressures of Cd and Te_2 at various temperatures were published 20 years ago [1], which is illustrated in Fig. 1. There was no theoretical work to explain the relationship of the stoichiometry and the vapor phase pressures until recently, where Chin et al. assume that the non-stoichiometric CdTe thin film behaves as a solid solution [2]. By invoking the Henry's Law, they were able to give an interpretation of the experimental results. Yet, such an interpretation is phenomenological. There is no physical evidence that the non-stoichiometric CdTe polycrystalline thin film is indeed a solid solution.

Alternatively, there may exist a more fundamental correlation between the CdTe stoichiometry and the partial vapor pressures of Cd and Te_2 *via* the formation of various defects with different densities. Therefore, we use the data of formation energies of point defects in CdTe which were obtained by first-principles calculation at zero temperature [3], to develop a physical model of defect formation in non-stoichiometric CdTe. Unlike the theoretical work based on first-principles calculation, our model expresses the chemical potentials as a function of pressure and temperature for practical emphasis. Our model establishes a physical foundation to explain the dependence of various intrinsic defect densities on temperature, on Fermi level, and on vapor pressures of Cd and Te_2 during physical vapor deposition.

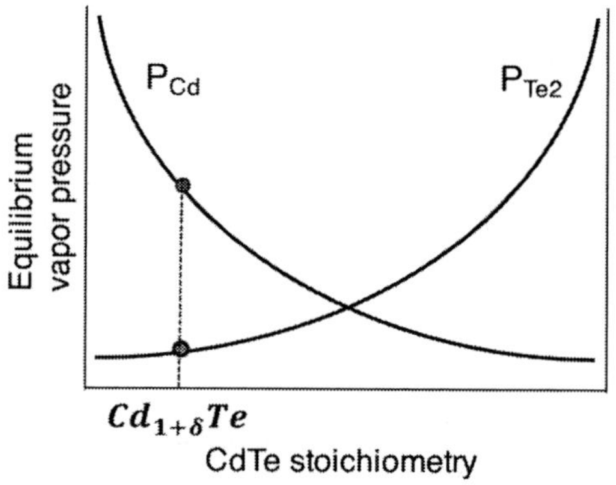

Figure 1: Illustration of the experimentally observed relationship between the stoichiometry of CdTe and the partial vapor pressures of Cd and Te_2.

2 RELATIONSHIP OF CDTE STOICHIOMETRY AND THE POINT DEFECT DENSITIES

As experimentally observed, there are two types of non-stoichiometric CdTe thin films: Cd-rich or Te-deficient, and Cd-deficient or Te-rich. Assuming the deviation from perfect stoichiometric CdTe crystal is small, the two types of slightly non-stoichiometric CdTe can be expressed as

$$Cd_{1+\delta}Te \sim CdTe_{1-\delta} \text{ and } Cd_{1-\delta}Te \sim Cd \qquad (1)$$

respectively.

For Cd-rich CdTe, the dominant point defects are tellurium vacancy V_{Te}, cadmium interstitial Cd_i, and cadmium on tellurium antisite Cd_{Te}, with densities of $\delta_{V_{Te}}$, δ_{Cd_i}, and $\delta_{Cd_{Te}}$, respectively. The formation of these three point defects can be formulated as:

$$Cd + \frac{1}{2}Te_2 \leftrightarrow CdTe$$

$$Cd + CdTe \leftrightarrow Cd_i$$

$$2Cd + \frac{1}{2}Te_2 \leftrightarrow Cd_i \quad (2)$$

$$Cd + \frac{1}{2}Te_2 \leftrightarrow CdTe$$

$$CdTe \leftrightarrow V_{Te} + Te$$

$$Cd \leftrightarrow V_{Te} \quad (3)$$

$$Cd + \frac{1}{2}Te_2 \leftrightarrow CdTe$$

$$Cd + CdTe \leftrightarrow Cd_{Te} + Te$$

$$2Cd \leftrightarrow Cd_{Te} \quad (4)$$

where CdTe is a primitive unit cell of CdTe in the perfect CdTe crystal. The non-stoichiometry index δ of Cd-rich CdTe due to its point defects V_{Te}, Cd_i and Cd_{Te} should satisfy the relationship:

$$\delta_{V_{Te}} + \delta_{Cd_i} + 2\delta_{Cd_{Te}} = \delta \quad (5)$$

Similarly, for Te-rich CdTe, the dominant point defects are cadmium vacancy V_{Cd}, tellurium interstitial Te_i, and tellurium on cadmium antisite Te_{Cd}, with densities of $\delta_{V_{Cd}}$, δ_{Te_i} and $\delta_{Te_{Cd}}$, respectively. The formation of these three point defects can be formulated as:

$$Cd + \frac{1}{2}Te_2 \leftrightarrow CdTe$$

$$\frac{1}{2}Te_2 + CdTe \leftrightarrow Te_i$$

$$Cd + Te_2 \leftrightarrow Te_i \quad (6)$$

$$Cd + \frac{1}{2}Te_2 \leftrightarrow CdTe$$

$$CdTe \leftrightarrow V_{Cd} + Cd$$

$$\frac{1}{2}Te_2 \leftrightarrow V_{Cd} \quad (7)$$

$$Cd + \frac{1}{2}Te_2 \leftrightarrow CdTe$$

$$\frac{1}{2}Te_2 + CdTe \leftrightarrow Te_{Cd} + Cd$$

$$Te_2 \leftrightarrow Te_{Cd} \quad (8)$$

The non-stoichiometry index δ of Te-rich CdTe due to its point defects V_{Cd}, Te_i and Te_{Cd} should satisfy a similar relationship:

$$\delta_{V_{Cd}} + \delta_{Te_i} + 2\delta_{Te_{Cd}} = \delta \quad (9)$$

3 DEPENDENCE OF INTRINSIC DEFECTS DENSITIES ON TEMPERATURE, ON FERMI LEVEL, AND ON VAPOR PRESSURES OF CD AND TE2 UNDER PVD CONDITION

From equations (2) to (4), the free energy of point defect formation in Cd-rich CdTe can be expressed as:

$$2\mu_{Cd} + \frac{1}{2}\mu_{Te_2} \leftrightarrow G_{Cd_i} \quad (10)$$

$$\mu_{Cd} \leftrightarrow G_{V_{Te}} \quad (11)$$

$$2\mu_{Cd} \leftrightarrow G_{Cd_{Te}} \quad (12)$$

We assume that at the growth temperature T of CdTe thin film, thermodynamic equilibrium is reached in the solid-vapor system. By considering the energy change due to vapor heat capacity, the energy change due to pressure, the energy change due to intermixing, the change of formation energy as a function of Fermi level E_f, and the change of configurational entropy due to defect formation, the three equilibrium conditions described in terms of the change of free energy can be detailed as:

$$\int_0^T 2C_p^{Cd}dT + \int_0^T \frac{1}{2}C_p^{Te_2}dT + 2kTln\left(\frac{P_{Cd}}{P_o}\right) + \frac{1}{2}kTln\left(\frac{P_{Te_2}}{P_o}\right) + 2kTln\left(\frac{2}{2.5}\right) + \frac{1}{2}kTln\left(\frac{0.5}{2.5}\right) = \Delta H_{Cd_i} + qE_f + \int_0^T C_p^{Cd_i}dT + T[kln(\delta_{Cd_i})] \quad (13)$$

$$\int_0^T C_p^{Cd}dT + kTln\left(\frac{P_{Cd}}{P_o}\right) = \Delta H_{V_{Te}} + qE_f + \int_0^T C_p^{V_{Te}}dT + T[kln(\delta_{V_{Te}})] \quad (14)$$

$$\int_0^T 2C_p^{Cd}dT + 2kTln\left(\frac{P_{Cd}}{P_o}\right) = \Delta H_{Cd_{Te}} + qE_f + \int_0^T C_p^{Cd_{Te}}dT + T[kln(\delta_{Cd_{Te}})] \quad (15)$$

where the formation energies of the three dominant defects in Cd-rich CdTe are ΔH_{Cd_i}, $\Delta H_{V_{Te}}$ and $\Delta H_{Cd_{Te}}$, respectively, with $\Delta H_{CdTe} = 0$. The formation energies of point defects have been calculated at 0 K using first-principles methods [3].

Considering Cd and Te_2 vapors as monoatomic and diatomic ideal gas, respectively, $C_p^{Cd} = \left(\frac{3}{2} + 1\right)k$, $C_p^{Te_2} = \left(\frac{7}{2} + 1\right)k$, where k is the Boltzmann constant. Since heat capacity of point defect is quite small (~ 0.1 eV) relative to the formation energies of various defects that are on the order of a few eV, the term of point defect heat capacity can be neglected in our calculations. Note that the degeneracy of a point defect is the inverse of the defect density.

Assuming CdTe solid-vapor system is at equilibrium, we have the equation of chemical reaction equilibrium:

$$P_{Cd}\sqrt{P_{Te2}} = exp\left[\frac{-\Delta G_{CdTe}(T)}{RT}\right] \quad (16)$$

Consequently, equations (13) to (15) can be simplified into:

$$6-\frac{\Delta H_{Cd_i}+qE_f+\Delta G_{CdTe}(T)}{kT}+lnP_{Cd}=ln\delta_{Cd_i} \quad (17)$$

$$\frac{5}{2}-\frac{\Delta H_{V_{Te}}+qE_f}{kT}+lnP_{Cd}=ln\delta_{V_{Te}} \quad (18)$$

$$5-\frac{\Delta H_{Cd_{Te}}+qE_f}{kT}+2lnP_{Cd}=ln\delta_{Cd_{Te}} \quad (19)$$

Similarly, for the three dominant point defects in Te-rich CdTe, their equilibrium conditions of defect formation described in terms of the change of free energy can be detailed as:

$$\int_0^T (C_p^{Cd}+C_p^{Te_2})dT+kTln\left(\frac{P_{Cd}P_{Te_2}}{P_o^2}\right)-kTln4$$
$$=\Delta H_{Te_i}-qE_f+\int_0^T C_p^{Te_i}dT+T[kln(\delta_{Te_i})] \quad (20)$$

$$\int_0^T \frac{1}{2}C_p^{Te_2}dT+\frac{1}{2}kTln\left(\frac{P_{Te_2}}{P_o}\right)$$
$$=\Delta H_{V_{Cd}}-qE_f+\int_0^T C_p^{V_{Cd}}dT+T[kln(\delta_{V_{Cd}})] \quad (21)$$

$$\int_0^T C_p^{Te_2}dT+kTln\left(\frac{P_{Te_2}}{P_o}\right)$$
$$=\Delta H_{Te_{Cd}}-qE_f+\int_0^T C_p^{Te_{Cd}}dT+T[kln(\delta_{Te_{Cd}})] \quad (22)$$

where the formation energies are ΔH_{Te_i}, $\Delta H_{V_{Cd}}$ and $\Delta H_{Te_{Cd}}$, respectively. With the equation of chemical reaction equilibrium, equations (20) to (22) can be simplified into:

$$7-2ln2-\frac{\Delta H_{Te_i}-qE_f+\Delta G_{CdTe}(T)}{kT}+\frac{1}{2}lnP_{Te_2}$$
$$=ln\delta_{Te_i} \quad (23)$$

$$\frac{9}{4}-\frac{\Delta H_{V_{Cd}}-qE_f}{kT}+\frac{1}{2}lnP_{Te_2}=ln\delta_{V_{Cd}} \quad (24)$$

$$\frac{9}{2}-\frac{\Delta H_{Te_{Cd}}-qE_f}{kT}+lnP_{Te_2}=ln\delta_{Te_{Cd}} \quad (25)$$

4 MODELING RESULTS AND DISCUSSION

From the equations (17) to (19), the point defect densities in Cd-rich CdTe depend on the growth temperature and on the vapor pressure of Cd. As the formation energies of defects change with the Fermi level E_f, the point defect densities also depend on doping that could change the E_f of CdTe. To demonstrate the dependence of defect densities on growth temperature T, we took equations (17) to (19) to calculate the defect densities as a function of increasing temperature T, while keeping E_f and the ratio of P_{Cd}/P_{Te_2} fixed ($E_f = 0.24$ eV below the conduction band minimum (CBM) and $P_{Cd} \approx 2P_{Te_2}$). DFT study has previously shown that in Cd-rich CdTe Fermi level E_f will be pinned around 0.24 eV below CBM, due to self-compensation, no matter how high the growth temperature is [4]. To demonstrate the dependence on vapor pressure, we took equations (17) to (19) to calculate the defect densities as a function of increasing P_{Cd}/P_{Te_2}, while keeping E_f and the growth temperature T fixed ($E_f = 0.24$ eV below the CBM and $T = 1073K$). The results of modeling are shown in Figure 2. It should be mentioned that the free energy of point defect does not depend on the electric charge of the point defect.

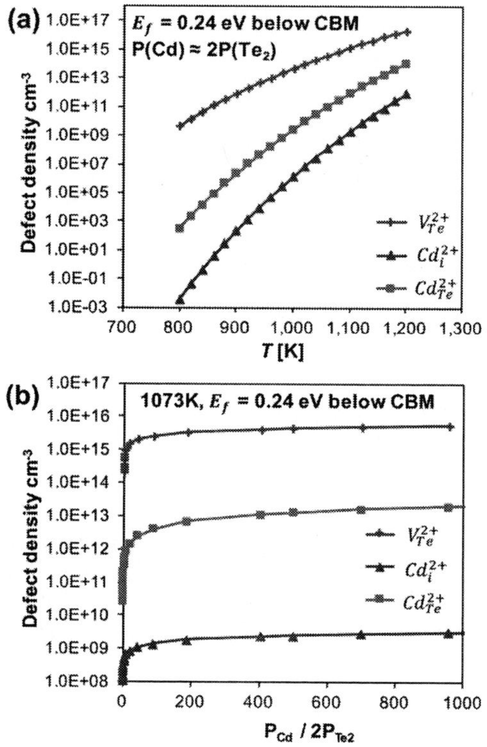

Figure 2: Modeling results of intrinsic point defect density in Cd-rich CdTe as a function of (a) growth temperature, and (b) vapor pressure ratio of P_{Cd}/P_{Te_2}.

From the equations (23) to (25), the point defect densities in Te-rich CdTe instead depend on the vapor pressure of Te_2, different from that in Cd-rich CdTe. To demonstrate the dependence of defect densities on growth temperature T, we took equations (23) to (25) to calculate the defect densities as a function of increasing temperature T, while keeping E_f and the ratio of P_{Te_2}/P_{Cd} fixed ($E_f = 0.67$ eV above the valance band maximum (VBM) and $P_{Cd} \approx 2P_{Te_2}$). DFT study has previously shown that in Te-rich CdTe Fermi level E_f will be pinned around 0.67 eV above VBM, which is the lowest E_f achievable for p-type CdTe under thermodynamic equilibrium growth conditions [4]. To demonstrate the dependence on vapor pressure, we took equations (23) to (25) to calculate the defect densities as a function of increasing P_{Te_2}/P_{Cd}, while keeping E_f and the growth temperature T fixed ($E_f = 0.67$ eV above the VBM and $T = 1073K$). The results of modeling are shown in Figure 3.

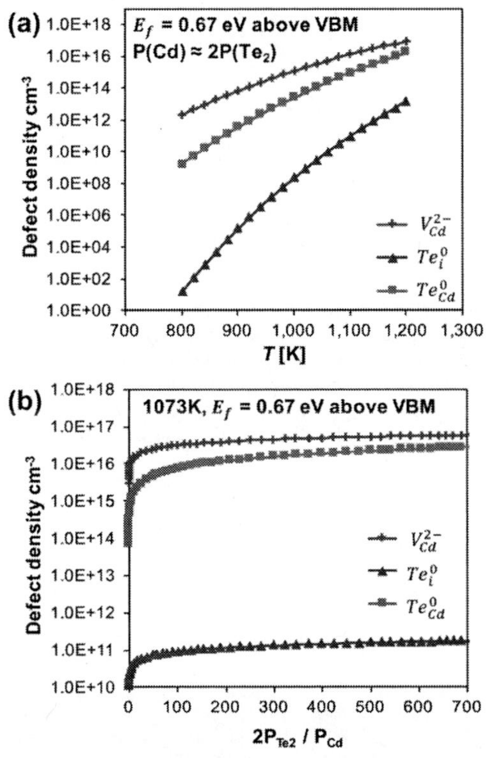

Figure 3: Modeling results of intrinsic point defect densities in Te-rich CdTe as a function of (a) growth temperature, and (b) vapor pressure ratio of P_{Te_2}/P_{Cd}.

As shown in Fig. 2(a) and Fig. 3(a), in either Cd-rich or Te-rich CdTe, all point defect densities increase with higher growth temperature. While the vacancies, i. V_{Te}. and V_{Cd}, are beneficial in Cd-rich and Te-rich CdTe lattice respectively for higher carrier concentration, the interstitial and antisite are usually harmful (deep level defects) compensating the doping efficiencies of vacancy-related defects. As the densities of vacancies are usually one or more orders of magnitude higher than the densities of the other two defects at a given temperature, higher growth temperature seems to be more beneficial for CdTe thin film grown under PVD condition to reach high carrier concentration. After growth, the CdTe thin film is quenched to room temperature. The defect densities formed during growth are thus kept.

As shown in Fig. 2(b) and Fig. 3(b), higher vapor pressure of Cd during PVD increases the point defect densities in Cd-rich CdTe thin film, and higher vapor pressure of Te_2 during PVD increases the point defect densities in Te-rich CdTe thin film. However, based on the relative densities of beneficial defects (vacancies) vs. harmful defects (interstitial and antisite), the vapor pressure dependence under Cd-rich condition is more useful than that under Te-rich condition to fine tune the carrier concentration in CdTe thin films.

It is worth mentioning that as the CdTe thin film growth temperature is around 500 ~ 700° C, the actual bandgap of CdTe during growth may be significantly smaller comparing to the value at room temperature. Such decrease in bandgap at high temperature has not been considered in our modeling. Like the studies using DFT calculation, our modeling still considered standard bandgap value of CdTe, i.e. 1.5 eV. Future experimental or theoretical studies may be needed to investigate the proper bandgap value of CdTe at such high temperatures.

5 CONCLUSION

The non-stoichiometric growth of CdTe thin film is due to the defect formation that determined by, not only the formation energy of the defects, but also the growth temperature and the vapor pressures of Cd and Te_2. In this work, we report a new physical model of defect formation that explains the dependence of intrinsic point defect densities on temperature, on Fermi level, and on vapor pressures of Cd and Te_2 during physical vapor deposition. The proposed model enables the experimentalists to fine tune the growth condition, so that the carrier concentration is high, and deep level density is low.

5 ACKNOWLEDGEMENTS

The authors acknowledge the China Triumph International Engineering Co. Ltd. (CTIEC), Shanghai, China, which offers generous financial support for this work.

6 REFERENCES

[1] J. H. Greenberg, J. Cryst. Growth **1996**, 161: 1.
[2] K. K. Chin, Z. Cheng, A. E. Delahoy, J. Cryst. Growth **2015**, 418: 32-37.
[3] S-H Wei, S. B. Zhang, Phys. Rev. B 2002, 66: 155211.
[4] Ji-Hui Yang, Wan-Jian Yin, Ji-Sang Park, Jie Ma, and Su-Huai Wei, *Semicond. Sci. Technol.* **2016**, 31: 083002.

INVESTIGATION OF KF-TREATMENT INDUCED SURFACE MODIFICATIONS OF $Cu(In,Ga)Se_2$ ABSORBERS AND THEIR CORRELATION WITH DEVICE PERFORMANCE

Isheta Majumdar[*,1,2], Vladimir Parvan[1], Dieter Greiner[1], Christian A. Kaufmann[1], Martha Ch. Lux-Steiner[2], Rutger Schlatmann[1] and Iver Lauermann[**,1]
[1]PVcomB / Helmholtz-Zentrum Berlin für Materialien und Energie, Schwarzschildstraße 3, 12489, Berlin, Germany
[2]Freie Universität Berlin, Fachbereich Physik, Arnimallee 14, 14195, Berlin, Germany
[*]Email: isheta.majumdar@helmholtz-berlin.de, Phone: +49 30 8062 15698
[**]Email: iver.lauermann@helmholtz-berlin.de, Phone: +49 30 8062 15677

ABSTRACT: The influence of a KF-treatment of $Cu(In,Ga)Se_2$ absorbers alone, without the influence of an excessive Se atmosphere and Na from the glass substrate has been studied. K-induced surface modifications differed for varying [K]/([In]+[Se]) (KIS) concentration ratios with a low KIS (0.74) absorber having a K-alloyed CIGSe surface growth and a high KIS (2.16) absorber having a mixed K-alloyed CIGSe and K-In-Se containing small spherical islands growth leading to devices with enhanced efficiencies as compared to the untreated absorber, while a medium KIS (1.73) absorber also having a mixed surface species led to a device with a decreased efficiency than the other two KF-treated absorbers.

Keywords: Alkali treatment, XPS, $Cu(In,Ga)Se_2$, Thin Film Solar Cell

1 INTRODUCTION

Alkali treated $Cu(In,Ga)Se_2$ (CIGSe) absorbers have been well known to result in highly efficient solar cell devices for many years now; the current record-efficiency cell was reported by [1]. One of the reasons for this efficiency enhancement has been attributed to surface modifications in the alkali treated absorbers as compared to the untreated absorbers as reported by [2]-[10] for NaF and/or KF-post deposition treatment (PDT). However, the mechanisms involved are not very clear yet and are subject to further examinations and analysis.

It has been reported in several publications ([5], [9], [10]) that there is a formation of a K-In-Se containing compound as a result of KF-PDT; necessarily with involvement of Na from either the soda-lime glass (SLG) substrate and/or by a NaF-PDT preceding the KF-PDT and a Se atmosphere; whereas no such compound formation has been observed with NaF-PDT alone. Yuan et. al [11] have theoretically predicted the presence of a "Cu-free, K-containing surface layer" using their calculations to state that K dopants have higher formation energies as compared to Na dopants thus giving them less solubility in the CIGSe bulk which in turn leads to formation of a K-enriched surface. A directly observed effect of such a surface modification on the device performance is an increase in the open-circuit voltage (V_{oc}) which has been attributed partly to the band-gap widening at the surface of a KF-PDT absorber ([3], [6]).

In this paper, we present our results on the treatment of CIGSe absorbers with KF, not along the conventional PDT method but in the absence of any Na source and Se atmosphere, to observe surface modifications and their effect on devices prepared with such treatment.

2 EXPERIMENTAL

CIGSe absorbers were prepared as per details given in [12] with a nominal bulk [Cu]/([In]+[Ga]) ratio of 0.90 on Mo-coated SLG substrates with a Na diffusion barrier (SiO_xN_y). These were subjected to varying amounts of KF in an ultra-high vacuum (UHV) chamber (pressure ~ 10^{-10}mbar) with 8, 10 and 12 minutes of KF deposition on CIGSe samples which will be referred to as CIGSe_0.74, CIGSe_1.73 and CIGSe_2.16 further in the text; where 0.74±0.01, 1.73±0.01 and 2.16±0.01 are the respective [K]/([In]+[Se]) (KIS) concentration ratios derived from the quantitative X-ray Photoelectron Spectroscopy (XPS) analysis of the absorbers measured after each treatment. The untreated CIGSe absorber is termed CIGSe_0.

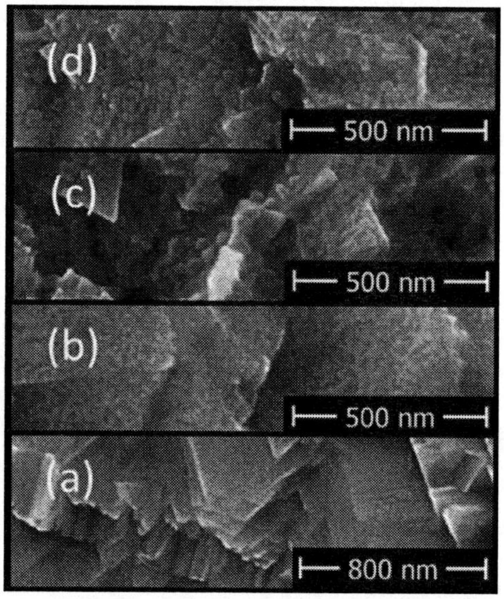

Figure 1: SEM micrographs of KF-treated CIGSe absorber surfaces in (a) CIGSe_0 (b) CIGSe_0.74 (c) CIGSe_1.73 (d) CIGSe_2.16

All samples were KCN-etched for 3 minutes before treatment began. The CIGSe substrates were annealed (in vacuum) at 350°C for about 10 minutes after finishing KF treatment. Scanning Electron Microscopy (SEM) imaging and XPS (using a MgKα source) measurements have been done on all four samples. Solar cell devices were completed from the absorbers with a 60 nm CdS chemical bath deposition, followed by 40 nm i-ZnO / 110 nm ZnO:Al deposition and Ni /Al /Ni front contact grids. Current-voltage (I-V) (under simulated AM 1.5 illumination) and external quantum efficiency (EQE) measurements have been performed.

3 RESULTS AND DISCUSSION

3.1 Surface modifications from KF-treatment

Figure 1(a) shows a top-view SEM image of the untreated absorber; (b)-(d) shows top-view SEM images of the KF-treated absorbers: CIGSe_0.74, CIGSe_1.73 and CIGSe_2.16, respectively. As can be seen clearly from these images, the KF-treated CIGSe absorbers have formed certain surface nanopatterns which are different depending on the varying K content which is not seen on the untreated absorber. On the absorber CIGSe_0.74, a scaly pattern is observed. With increasing KIS values, i.e. for samples CIGSe_1.73 and CIGSe_2.16, little spherical-shaped island growth can be observed, which are denser and seem to coalesce for the CIGSe_2.16 absorber. Such surface nanopatterning with small spherical island growth as a result of KF-PDT has also been observed by Reinhard et al. [4]. However, with sample CIGS_0.74 it is not clear what type of surface pattern is formed as it does not look like spherical islands but rather like an interconnected 2D mesh.

Figure 2 shows the Se 3d XPS spectra measured for the (a) untreated and (b)-(d) KF-treated samples that have been fitted in order to resolve the overlapping spin-orbit components: $3d_{5/2}$ and $3d_{3/2}$, while maintaining a binding energy (BE) difference of 0.86 eV between the two components, a peak area ratio of 3:2 between $3d_{5/2}$ and $3d_{3/2}$ components and the same full width at half maxima (FWHM) values for both components. The red closed circles are the cumulative fits over the raw data (shown in white open circles). Table I tabulates the Se $3d_{5/2}$ peak BE positions obtained after a Shirley background subtraction and fitting with Gaussian peaks.

Table I: Se $3d_{5/2}$ XPS peak fit BE values for various CIGSe samples in eV. The assignments are explained in the text.

Samples	CIGSe	K-In-Se	Se^0
CIGSe_0	53.95	-	-
CIGSe_0.74	53.93	-	-
CIGSe_1.73	54.15	52.10	56.05
CIGSe_2.16	54.14	52.41	55.66

From Figure 2 and Table I it can be observed that the Se 3d peaks fitted for the untreated absorber and the absorber CIGSe_0.74 can be attributed to Se from the CIGSe component (blue closed circles) with a BE of Se $3d_{5/2}$ ~ 53.9 eV. On the other hand, for absorbers CIGSe_1.73 and CIGSe_2.16, there seems to be a shift of the CIGSe contributed Se 3d peak by ~ 0.2 eV towards higher BE. Besides this, two extra contributions can be seen: (i) on the higher BE side, a Se $3d_{5/2}$ component at ~ 56 eV (cyan closed circles) which can be supposed to be coming from elemental Se formation which has also been reported by Lepetit et al. [9] in their KF-PDT absorbers; (ii) on the lower BE side, a Se $3d_{5/2}$ component at ~ 52 eV (green closed circles), which is now attributed to a K-In-Se containing phase present at the topmost surface of KF PDT CIGSe absorbers reported by Handick et al. [10] and also observed by Lepetit et al. [9] in their XPS Se 3d spectra in the case of highly efficient devices formed from such absorbers. What we see here for our absorber surfaces may only be the onset of such a phase formation as with increasing KIS values a shift in BE towards higher values can be seen (52.10 eV to 52.41 eV for K-In-Se component in Table I). Whereas Handick et. al [10] showed a separate K-In-Se surface species formation whose Se 3d contribution is seen mixed and overlapped with the Se 3d contribution from the underlying CIGSe absorber.

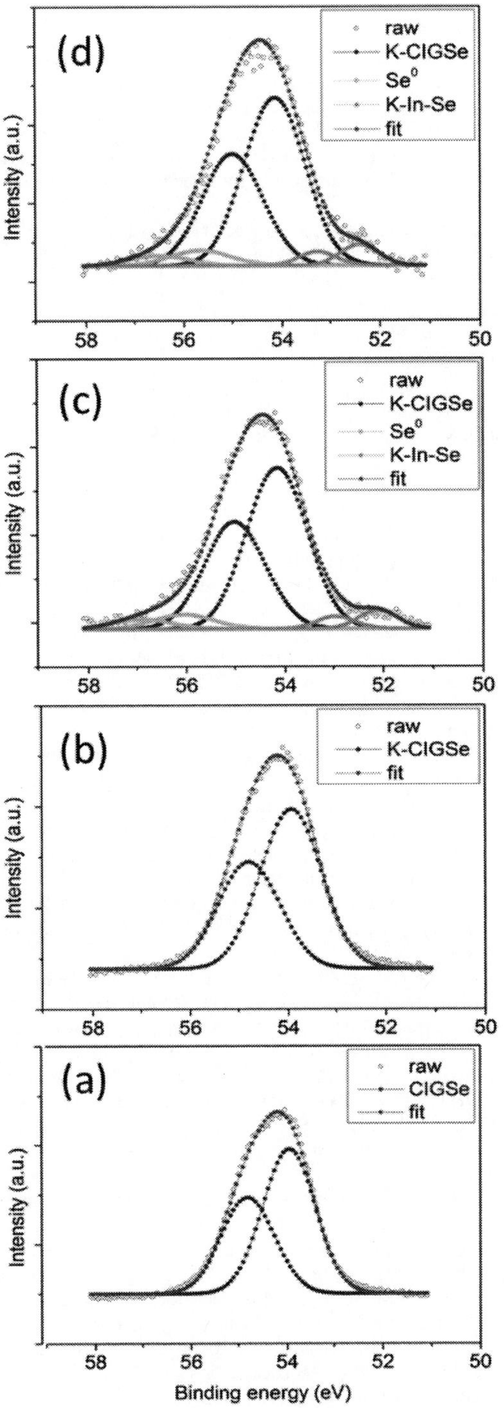

Figure 2: Se 3d XPS peak components fitted for (a) CIGSe_0 (b) CIGSe_0.74 (c) CIGSe_1.73 (d) CIGSe_2.16 absorber surfaces

In our XPS studies, no Cu or Ga signals were detected, except for the bare absorber and very small Cu (1.5 at.%) signal for absorber CIGSe_0.74. Either these elements have been depleted from the surface ([1],[5],[6],[13]) which is a common phenomenon for KF treatments or they have been attenuated by the KF depositions.

Figure 3 shows the In $3d_{5/2}$ XPS spectra measured for the (a) untreated and (b)-(d) KF-treated samples. The red closed circles are the cumulative fits over the raw data (shown in white open circles). Table II tabulates the In $3d_{5/2}$ peak BE positions obtained after a Shirley background subtraction and fitting with Gaussian peaks.

Table II: In $3d_{5/2}$ XPS peak fit BE values for various CIGSe samples in eV. The assignments are explained in the text.

Samples	CIGSe	K-In-Se	In-O
CIGSe_0	444.43	-	-
CIGSe_0.74	444.35	-	445.40
CIGSe_1.73	444.52	443.38	-
CIGSe_2.16	444.46	443.49	-

From Figure 3 and Table II it can be observed that the In $3d_{5/2}$ peaks fitted for the untreated absorber and the absorber CIGSe_0.74 can be attributed to In from the CIGSe component (blue closed circles) with a BE of In $3d_{5/2}$ ~ 444.4 eV for the untreated absorber, while it is slightly shifted for the KF-treated absorber to accommodate an oxidized In component (cyan closed circles) visible on the higher BE side at 445.40 eV that has also been reported in [3], [14], [15]. On the other hand, for absorbers CIGSe_1.73 and CIGSe_2.16, although no oxidation peak is observed, but an extra In component besides the one from CIGSe, is clearly seen in both absorbers. Like in the case of Se 3d peaks, the In $3d_{5/2}$ peak contributions can be seen as: (i) the main CIGSe component, and (ii) the lower BE component (green closed circles) arising from a K-In-Se containing phase that seems to shift towards higher BE with increasing KIS (Table II). Two types of In contribution have also been reported to be found from In MNN Auger peaks for KF-PDT absorbers [9].

Since such modifications are very surface sensitive, X-Ray Diffraction (XRD) and Raman spectroscopy measurements were insensitive to any observable changes in the absorbers. However, by utilizing the Scherrer equation, the crystallite sizes calculated from the FWHM values of the main chalcopyrite (112) XRD peak yielded slightly higher values for the KF-treated absorbers (~ 300 nm) as compared to the untreated absorber (~ 230 nm).

In view of the extra Se and In contributions seen in absorbers with KIS 1.73 and 2.16 and also the nanopatterned surfaces in these absorbers seen from SEM images, it can be concluded that K incorporates into the CIGSe bulk as well as forms a separate phase containing new In-Se bonds that may or may not exist as a separate surface species for these absorbers. On the other hand, the absorber with the least KIS value (0.74) does not seem to exhibit any separate phase formation but is influenced only by K incorporating into the CIGSe chalcopyrite phase. Such a possibility has also been reported by Muzzillo et. al [16], where they intentionally prepared Cu-K-In-Se (CKIS) alloys with low [K]/([K]+[Cu]) (KKC) concentrations and high KKC concentration "$KInSe_2$" surfaces on $CuInSe_2$ (CISe) absorbers and studied the electrical properties of the devices made from such absorbers.

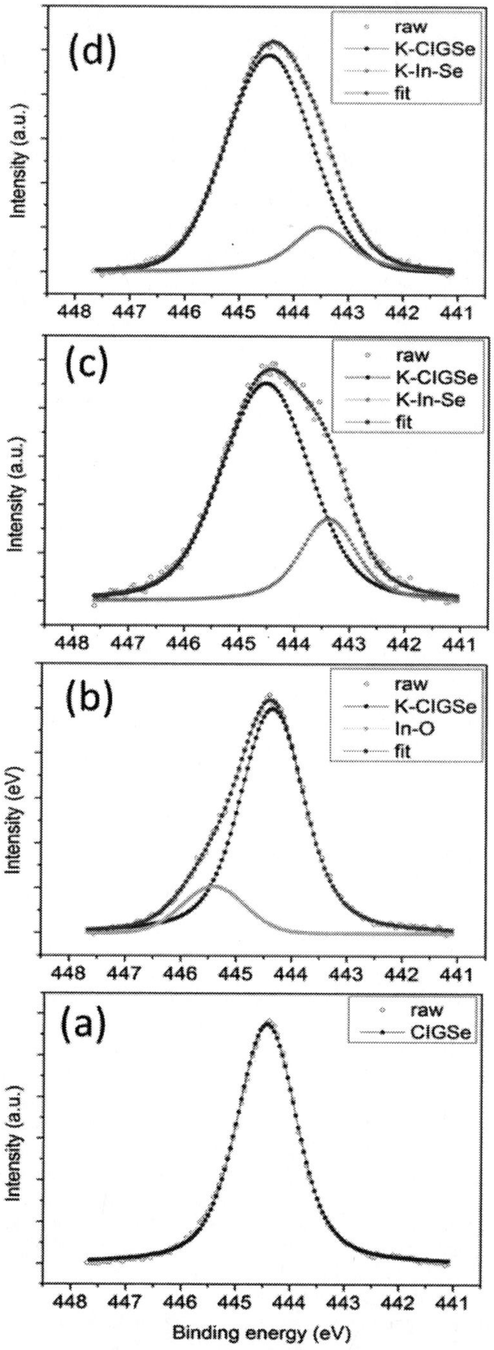

Figure 3: In $3d_{5/2}$ XPS peak components fitted for (a) CIGSe_0 (b) CIGSe_0.74 (c) CIGSe_1.73 (d) CIGSe_2.16 absorber surfaces

3.2 KIS dependent surface modifications and device performance

Figure 4 shows all the cell parameters of the devices made from the untreated and KF-treated absorbers. No anti-reflection coating has been used in the device preparations. The fill factors (FF) of the devices are low, which is an indication of high series resistance, but there has been some improvement with KF-treatment in the case of absorber CIGSe_0.74. The current densities (J_{sc}) have also increased for the devices with KF-treated absorbers and are more pronounced for absorbers CIGSe_0.74 and CIGSe_2.16. The most effective and pronounced improvement has been in the open-circuit voltage (V_{oc}) values with an average 100 mV increase (combining the three KF-treated absorbers) as compared to the untreated absorber, the highest being for the CIGSe_0.74 sample. The efficiency values have also followed a trend similar to the V_{oc} changes and the overall device performance was found to be best for CIGSe_0.74 sample, followed by sample CIGSe_2.16 and then the least one for CIGSe_1.73 sample but still better than the untreated sample.

Table III summarizes the cell parameters obtained for the best cell measured for each of the KF-treatments. It is observed that the best cell of sample CIGSe_0.74 device yields the highest efficiency, V_{oc} and FF values while J_{sc} is highest for the best cell of sample CIGSe_2.16 device and yields the second best efficiency value. Even better device performance could be expected (better FFs) if the absorbers had no Na barrier as the beneficial effects of Na from the SLG substrate is already widely known.

Table III: Best cell parameters of devices made from the KF-treated absorbers

Absorber Sample	ETA (%)	V_{oc} (mV)	J_{sc} (mA/cm^2)	FF (%)
CIGSe_0.74	14.68	626	37.34	62.78
CIGSe_1.73	11.79	600	34.67	56.67
CIGSe_2.16	14.12	601	37.64	62.41

The report by Muzzillo et al. [16] predicts that KF-treatment might benefit device performance by different mechanisms at the surface and the bulk for CISe absorbers. Their devices made out of CKIS absorbers prepared with a KKC value of 0.07 yielded the best efficiency (13.7%), and FF as compared to absorbers with higher KKC values. They attribute such a CKIS absorber with low KKC composition to be a manifestation of K getting incorporated as part of the bulk CKIS alloy that is apparently able to decrease recombination at the absorber/buffer interface. From our set of KF-treated absorbers, CIGSe_0.74 can also be believed to have K incorporated in the CIGSe bulk that also had a low K content. Also from the SEM image of the absorber surface in Figure 2(b), the growth seems to be from within the bulk absorber beneath and not as a separate growth formation as seen in the case of the other two treated absorbers (Figure 1(c), (d)). So the enhanced device performance from CIGSe_0.74 can be attributed to K acting on the CIGSe bulk and we call it a "K-CIGSe" species which is also evident from the Se 3d and In $3d_{5/2}$ XPS peaks seen in (Figure 2(b) and 3(b), respectively). Muzzillo et. al also prepared high KKC composition "$KInSe_2$ surfaces" on CISe absorbers to study the effect of confining K at the buffer/absorber interface. Devices made from such an absorber with KKC value of 0.41 yielded an efficiency of 14.2% as well as enhanced V_{oc} and FF. They claim such an absorber to have a mixed-phase CISe + $KInSe_2$ with high KKC

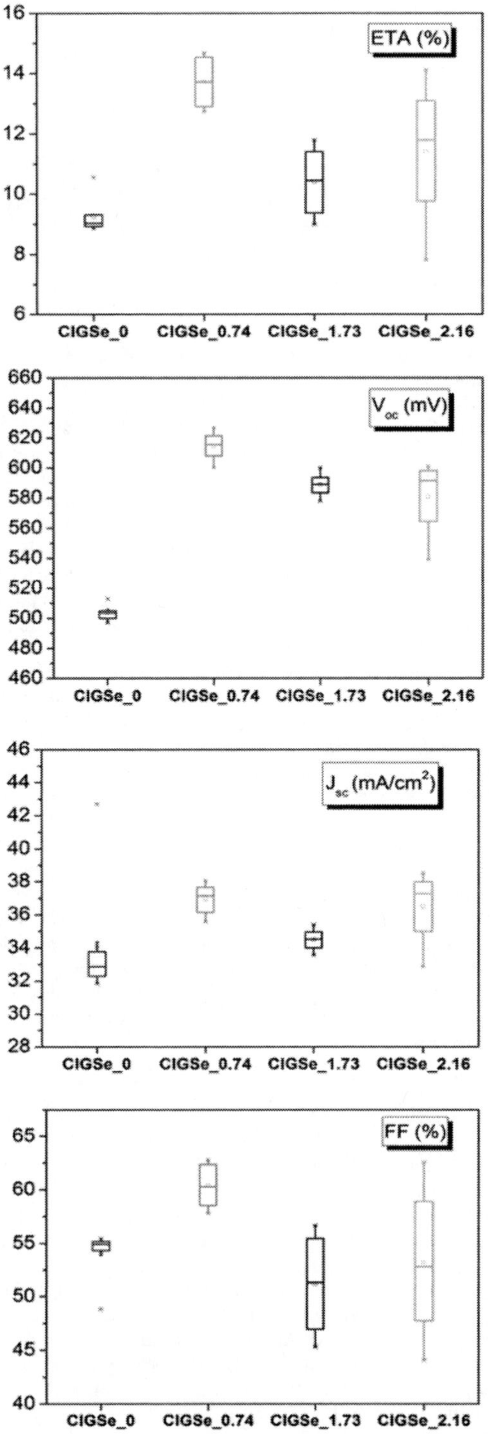

Figure 4: Cell parameters of devices made from untreated and KF-treated CIGSe absorbers

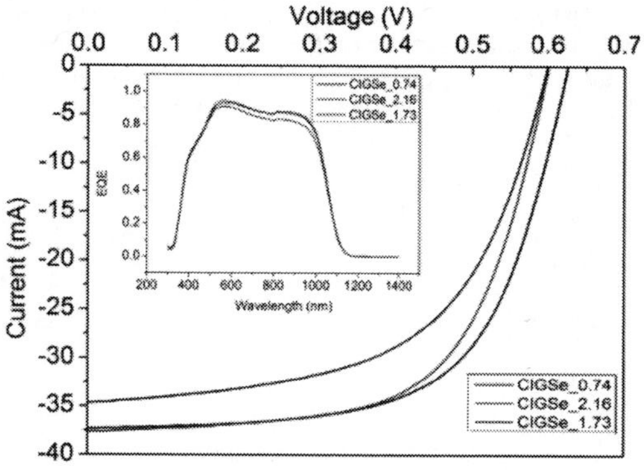

Figure 5: I-V curve for the best cells representing the various KF-treatments.

composition that can decrease recombination at the absorber/buffer interface. From our set of KF-treated absorbers, CIGSe_1.73 and CIGSe_2.16 also seem to have formed a K-In-Se containing phase on top of the CIGSe surface which we have attributed to a "K-In-Se" surface species in our XPS spectra where we have separated them from the "K-CIGSe" components. It is also worth noticing that the CIGSe composition at the surface itself is now modified as compared to the pristine untreated CIGSe surface. This is also very evident from the SEM images of Figure 1(c) and (d) where we clearly see the separate island growths that may contain the K-In-Se phase.

Of the two absorbers with the mixed phases, the one with the higher KIS value, i.e. sample CIGSe_2.16 shows better device efficiency, J_{sc} and FF than CIGSe_1.73, although both show similar V_{oc}. But such a mixed phase surface can be both beneficial and detrimental for device performance as can be seen from the current-voltage (I-V) curves of the best cells from the three type of KF-treated absorbers (Figure 5) and external quantum efficiency (EQE) measurements of the same (inset of Figure 5). However, all KF-treatments led to better device performance as compared to the device made from the untreated absorber. The blue curve representing the lower KIS mixed phase absorber yields the lowest short circuit current as compared to the other two, possibly as a result of increased recombination at the absorber/buffer interface.

4 CONCLUSIONS

These are the first results of KF-treatment on CIGSe absorbers done in our CISSY [17] chamber which is a unique UHV set-up that allows in-situ treatments and in-system XPS measurements on samples. The influence of KF treatments used in the KF-PDT processes for high efficiency CIGSe-based solar cells have been studied except that we did pure KF treatments with no involvement of an excessive Se atmosphere as well as no Na interference from the SLG substrate in order to investigate just the effect of K incorporation in CIGSe. About 8 minutes of KF-treatment alone has increased the V_{oc} by ~ 100 mV and device efficiency from 9% (untreated absorber) to ~ 14%. With three types of absorbers prepared with varying KIS concentration ratios and an untreated absorber for reference, the absorber surfaces reveal different stages of nanopatterned growth with increasing K content or KIS values. K in CIGSe alloying alone as well as the presence of a K-alloyed CIGSe and a K-In-Se containing mixed phase has led to enhanced device performance. Controlled tuning of KIS ratios alone can enhance power conversion efficiencies of CIGSe-based devices. From surface analysis measurements of treated absorbers and electrical measurements of the final devices prepared from them, we tried to find a correlation between the KF-induced surface modifications of the absorbers and their device performance. Although a large number of samples and reproducibility are required to justify the occurrence of any real phenomenon that may relate these two but this is left for testing in future experiments.

ACKNOWLEDGEMENTS

The authors would like to thank Carola Klimm for SEM images, Iris Dorbandt, Bianka Bunn, Katja Mayer-Stillrich, Jakob Lauche, Tim Münchenberg, Michael Kirsch and Jérome Deumer for solar cell device preparation. I.M thanks her research grant Erasmus Mundus EUPHRATES (exchange programme between IIT Bombay, India and FU Berlin, Germany).

REFERENCES

[1] P. Jackson, R. Wuerz, D. Hariskos, E. Lotter, W. Witte and M. Powalla, Phys. Status Solidi RRL, 10, No. 8 (2016) 583.
[2] A. Chirila, P. Reinhard, F. Pianezzi, P. Bloesch, A. Uhl, D. Jaeger, R. Erni, S. Nishiwaki, S. Buecheler, and A. N. Tiwari, Nat. Mater. 12 (2013) 1107.
[3] P. Pistor, D. Greiner, C. A. Kaufmann, S. Brunken, M. Gorgoi, A. Steigert, W. Calvet, I. Lauermann, R. Klenk, T. Unold, and M.-C. Lux-Steiner, Appl. Phys. Lett. 105 (2014) 063901.
[4] P. Reinhard, B. Bissig, F. Pianezzi, H. Hagendorfer, G. Sozzi, R. Menozzi, C. Gretener, S. Nishiwaki, S. Buecheler, and A. N. Tiwari, Nano. Lett., 15 (2015) 3334.
[5] P. Reinhard, B. Bissig, F. Pianezzi, E. Avancini, H. Hagendorfer, D. Keller, P. Fuchs, M. Dobeli, C. Vigo, P. Crivelli, S. Nishiwaki, S. Buecheler, and A. N. Tiwari, Chem. Mater., 27 (2015) 5755.
[6] E. Handick, P. Reinhard, J. H. Alsmeier, L. Köhler, F. Pianezzi, S. Krause, M. Gorgoi, E.Ikenaga, N. Koch, R. G. Wilks, S. Buecheler, A. N. Tiwari, and M. Bär, ACS Appl. Mater. Interfaces, 7 (2015) 27414.
[7] J. A. Aguiar, A. Stokes, C. S. Jiang, T. Aoki, P. G. Kotula, M. K.Patel, B. Gorman, and M. Al-Jassim, Adv. Mater. Interfaces (2016), 1600013.
[8] B. Ümsür, W. Calvet, A. Steigert, I. Lauermann, M. Gorgoi, K.Prietzel, D. Greiner, C. A. Kaufmann, T.Unold, and M. Ch. Lux-Steiner, Phys. Chem. Chem. Phys., 18 (2016) 14129.
[9] T. Lepetit, S. Harel, L. Arzel, G. Ouvrard, and N. Barreau, IEEE J. of Photovoltaics, 6, No.5 (2016) 1316.
[10] E. Handick, P. Reinhard, R. G. Wilks, F. Pianezzi, T. Kunze, D. Kreikemeyer-Lorenzo, L. Weinhardt, M. Blum, W. Yang, M.Gorgoi, E. Ikenaga, D.

Gerlach, S. Ueda, Y. Yamashita, T. Chikyow, C. Heske, S. Buecheler, A. N. Tiwari, and M. Bär, ACS Appl. Mater. Interfaces, 9 (4), (2017) 3581.
[11] Z. K. Yuan, S. Chen, Y. Xie, J. S. Park, H. Xiang, X. G. Gong, and S. H. Wei, Adv. Energy Mater. (2016) 1601191.
[12] C. A. Kaufmann, A. Neisser, R. Klenk, and R. Scheer, Thin Solid Films, 480-481 (2005) 515.
[13] F. Pianezzi, P. Reinhard, A. Chirila, B. Bissig, S. Nishiwaki, S. Buecheler, A. N. Tiwari, Phys. Chem. Chem. Phys. 16 (2014) 8843.
[14] D. Shin, J. Kim, T. Gershon, R. Mankad, M. Hopstaken, S. Guha, B. T. Ahn, B. Shin, Sol. Energy Mater. Sol. Cells, 157 (2016) 695.
[15] G. S. Jeong, E. S. Cha, S. H. Moon, B. T. Ahn, Current Photovoltaic Research, 3 (2015) 65.
[16] C. P. Muzillo, J. V. Li, L. Mansfield, K. Ramanathan, and T. J. Anderson, arXiv:1704.01560v1 [cond-mat.mes-hall].
[17] I. Lauermann, A. Steigert, Journal of large-scale research facilities, 2 (2016) A67.

PROPERTIES OF $Cu_2ZnSn(S_xSe_{1-x})_4$ THIN FILMS OBTAINED BY AN ELECTRODEPOSITION – ANNEALING PROCESS

E.P. Zaretskaya[1], V.F. Gremenok[1], A.V. Stanchik[1], A.N. Pyatlitski[2], V.A. Solodukha[2],
K.A. Urazov[3], M.B. Dergacheva[4], Süleyman Özçelik[5]
[1]State Scientific and Production Association «Scientific-Practical Materials Research Centre of the National Academy of Sciences of Belarus», 220072, Minsk, P. Brovka str., 19, Belarus, phone: + 375 17 284 12 49, fax: + 375 17 284 15 58, e-mail: ezaret@ifttp.bas-net.by
[2]JSC "INTEGRAL" – "INTEGRAL" Holding Managing Company, 220108, Minsk, Kazintsa I.P. str., 121A, Belarus, e-mail: office@bms.by
[3] Center of Chemical Engineering, Kazakh-British Technical University, Tole bi str. 59, Almaty, 050000, e-mail: Kazakhstan u_kazhm@mail.ru
[4]Electrochemistry Technology Department, Sokolsky Institute of Fuel, Catalysis and Electrochemistry, Kunaev, str. 142, Almaty, 050010, e-mail: Kazakhstan m_dergacheva@mail.ru
[5] Gazi University, Photonics Application & Research Center, Central Campus D-Blok 06500 Ankara-Turkey. phone:+903122028465, e-mail:sozcelik@gazi.edu.tr

ABSTRACT: In this study, $Cu_2ZnSn(SxSe_{1-x})_4$ (CZTSSe) absorber films were grown by a two-step process: sequential electrochemical deposition of Cu, Zn, Sn, and S (or Se) components on Mo coated glass followed by annealing in Se(S) atmosphere under N_2 flow. Full characterizations have been carried out using XRD, EDX and Raman scattering measurements. CZTSSe thin film films with band gap energies 1.36 eV and 1.48 eV were grown. The preliminary results demonstrate the applicability of the proposed approach for the producing CZTSSe absorber for thin film solar cells.
Keywords: Kesterite, CZTS, Electrodeposition, Raman Spectroscopy, Thin Film

1 INTRODUCTION

In spite of the great progress that has been achieved in the conversion efficiency $Cu(In,Ga)Se_2$ (CIGS) thin-film solar cells (over 22 %) [1,2], mass production is limited by expensive and rare Ga and In materials. In last decade the researchers have focused on the kesterites like Cu_2ZnSnS_4 and $Cu_2ZnSn(S_xSe_{1-x})_4$ chalcogenide (CZTSSe) materials as a promising candidate for cost effective thin film solar cells [3]. CZTSSe a p-type semiconductor with a high absorption coefficient above 10^4 cm^{-1} and a direct band-gap (E_g = 1.05–1.50 eV depending on the S/Se composition ratio) consists of inexpensive and abundant elements. It could therefore allow to reduce cost production of thin-film solar cells The reported power conversion efficiency of 12.6% had been achieved for the anion-substituted $Cu_2ZnSn(S_xSe_{1-x})_4$ solid solution [4]. Deposition of CZTS thin film absorber layer has been developed by using various vacuum and non-vacuum techniques. One step electrochemical deposition of Cu–Zn–Sn precursors and subsequent sulfurization (or selenization) is considered as a promising approach for production the CZTSSe based thin film solar cells.

This work is devoted to the structural properties analysis of CZTSSe thin films formed by sequential electrochemical deposition of Cu, Zn, Sn, and S (or Se) components followed by annealing in Se(S)-containing atmosphere under N_2 flow. The effect of precursor composition and anion S $\rightarrow$ Se substitution on the CZTSSe thin films growth was studied.

2 EXPERIMENTAL

2.1 Deposition of Coatings.

Electrochemical deposition of Sn-Cu-Zn-S and Sn-Cu-Zn-Se precursor films was performed in a three-electrode electrochemical cell using potentiostat/galvanostat Gill AC (ACM Instruments, UK). Mo/glass substrate was used as a working electrode, platinum – as auxiliary electrode and silver chloride electrode – as a reference electrode.

Sn-Cu-Zn-S thin films were deposited onto the Mo/glass substrates from the electrolyte based on 0.2M tri-sodium citrate, which contains the following salts: 0.01M $CuSO_4{\cdot}5H_2O$, 0.01M $ZnSO_4{\cdot}7H_2O$, 0.005M $SnSO_4$ and 0.05M $Na_2S_2O_3$, all of analytically pure grade. To adjust the pH to 4.6 of the electrolyte was used 0.1M tartaric acid as described in [5]. The electrolysis was performed at the constant potential of – 1.0 V for 30 min at room temperature.

The set of Sn-Cu-Zn-Se thin films were electrodeposited onto the Mo/glass substrates from the electrolyte with pH 1.5, containing 0.1M tartaric acid, 0.005M $CuSO_4{\cdot}5H_2O$, 0.01M $ZnSO_4{\cdot}7H_2O$, 0.01M $SnCl_4$ and 0.005M $NaHSeO_3$, all of analytically pure grade. The electrolysis was performed at the constant potential of – 0.6 V for 30 min at room temperature.

The as-deposited precursors were annealed preliminary at 450°C for 30 min in Ar atmosphere for better crystallization. Then the precursors were reacted to $Se(S)/N_2$ in a tubular furnace under N_2 pressure of 0.8-1.0 mbar inside a partly closed graphite container. Sulfurization was conducted at 500°C and selenization at 550°C for the period 2 hours with subsequent natural quench down to room temperature.

2.2 Investigation Methods.

The phase composition and crystalline structure of the deposited layers were studied by X-ray phase analysis (XPA) with a D8 Advance (Bruker AXS) diffractometer operating with Cu*Kα* radiation (λ = 0.1542 nm) filtered by a multilayered Ni/graphite monochromator. The voltage supplied to the tube was 40 kV, and the current was 40 mA. X-ray diffraction (XRD) patterns were obtained at grazing incidence (GID) in the scanning mode of measurements with a step of 0.04° (on the 2θ scale) and a counting time of 5 s. The angle of incidence of the primary beam was ω = 0.5°. The phases

Table I: CZTSSe thin films chemical composition determined by EDX after annealing in S/N_2 or Se/N_2 atmosphere.

Series, №	Cu, at.%	Zn, at.%	Sn, at.%	S, at.%	Se, at.%	S/Se	Cu/(Zn+Sn)
1a	24.52	3.6	27.25	3.96	16.63	0.23	0.79
2a	32.3	4.7	35.9	5.2	21.9	0.24	0.79
3b	11.91	26.27	5.99	41.18	1.38	29.84	0.36
4b	13.7	25.98	7.30	47.5	1.6	29.68	0.41

were identified by comparing the experimentally determined interplanar spacing *d* with the data of the Joint Council for Power Diffraction Standards (JCPDS), 2013.

The chemical composition of the material of the films was determined using an INCA 350 (Oxford Instruments, UK) energy-dispersive X-ray spectral microanalyzer with a resolution of 1 μm^3 and a sensitivity of 0.1 at.%.

The surface topography of the precursor films and $Cu_2ZnSn(S_xSe_{1-x})_4$ layers was studied using a SOLVER Pro 47 (NT-MDT) high resolution atomic-force microscope (AFM).

The spectral dependences of reflectance of the CZTSSe films were carried out using a spectrophotometer Proscan MC 122. Measurements were carried out for incident angles between 20°and 45° in the spectral range from 300 nm to 2000 nm with a resolution of 3 nm. The Raman spectra were recorded with a Nanofinder High End spectrometer. For the excitation source, we used a solid laser emitting at a wave length of 473 nm. All of the measurements were conducted at a temperature of T = 293 K in the backscattering geometry of recording in the spectral range from 150 to 3000 cm^{-1}.

3 RESULTS AND DISCUSSION

3.1. As deposited precursors.

The morphological changes of the precursors before and after pre-heating in Ar atmosphere were analyzed by AFM. It revealed significant differences in compactness and flatness of their surface after the preheating. The as deposited precursors form a rough surface without voids and microcracks. However, the films adhered well to the SLG substrate and were relatively dense. AFM measurements showed that these voids and uneven surface morphology were reduced when the precursor were preheated at 450°C for 30 min. Root mean square (RMS) of the precursor films decreased from 110 nm (for the as-deposited precursor film) to 60 nm for pre-heated.

It indicates that this stage plays essential role into transformation of the precursor into well mixed alloy and in controlling the morphology of thin film. On the other hand, XRD characterization of pre-heated precursors showed mainly amorphous material with traces of CZTS (or CZTSe) phases and other phases without exact indicating (not shown). Precursors used for selenization experiments had the composition ratio Cu/(Zn+Sn) ≈ 0.6 and content of S = 44.5 at. % while precursors used for sulfurization experiments had the composition ratio Cu/(Zn+Sn) ≈ 0.87 and content of Se = 25.94 at.%.

3.2. Annealed films.

It was found that the phases and the structure of the synthesized CZTSSe thin films to be strongly dependent on their composition as well as processing regimes.

The element composition of the precursors annealed in S/N_2 (or Se/N_2) atmosphere is shown in Table I. During the selenization process sulfur in the precursors is substituted for selenium leading to S/Se ≈ 0.23 whereas during the sulfurization process selenium is substituted for sulfur leading to S/(Se) ≈ 29.84. The values of ratio S/Se of the samples series (***a***) approach to the ternary compound $C_2ZnSnSe_2$, while ratio S/Se of the samples series (***b***) approach to the ternary compound C_2ZnSnS_2. It necessary to mention that the sample series (***a***) are Cu-rich and the sample series (***b***) are Zn-rich. Both series have great deviation from stoichiometry in cation as well an anion sublattice.

The XRD pattern of the Sn-Cu-Zn-S precursor annealed in Se/N_2 atmosphere (series ***a***) measured in GID mode are shown in Fig.1. The logarithmic intensity was used to show better the low intensity peaks.

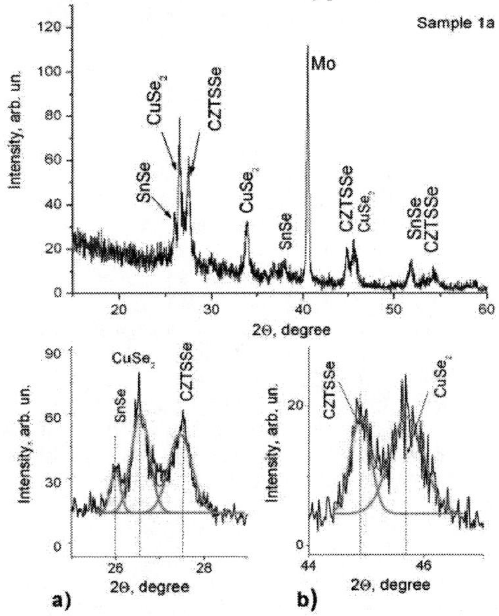

Figure 1: XRD patterns of CZTSSe films series (***a***) with ratio S/Se ≈ 0.23

The existence of the polycrystalline-kesterite CZTSSe phase close to CZTSe (JCDD data file # 96-153-1984) is fixed in the pattern by its main reflections at 27.49°, 45.25 ° and 53.53 ° and other peaks of very low intensity. In addition to the reflection of Mo at 40.27° minor peaks were registered, which coincide or lie very close to those of CZTSe and can be attributed to selenides. Detailed analysis of the samples series (***a***) (where Cu- and Sn are in excess) indicated on the possible presence of secondary SnSe and $CuSe_2$ phases (JCDD data file # 00-01-0984 and # 00-071-004, consequently) (Fig.1, a,b). Due to replacement S to Se and forming CZTSSe solid solutions the (112) peak of CZTS phase is shifted from 26.98° to 27.10°.

Fig. 2 shows XRD pattern of the Sn-Cu-Zn-Se precursor after sulfurization (series ***b***).

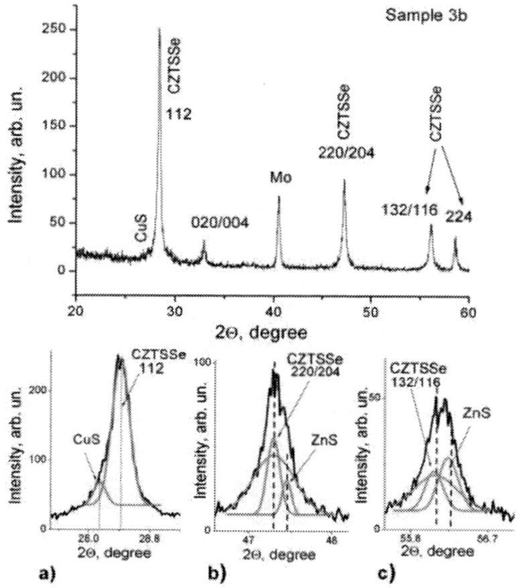

Figure 2: XRD patterns of CZTSSe films series (***b***) with ratio S/Se ≈ 29.84

It was found that all dominant diffraction peaks of this film can be attributed to the peaks of tetragonal structure for CZTS: (112), (020/004), (132/116) at 28.4°, 33.05°, 47.39 ° and 56.15 ° (JCDD data file # 96-900-4751). As it can be seen from XRD fragments low intensity peaks can be attributed to minor sulfide phases CuS (JCDD data file # 00-085-39280) and ZnS (JCDD data file # 00-002-0564). The CZTS phase presents twin peaks 220/204 and 132/116 at 47.29 ° and 56.03°, consequently which confirm the tetragonal structure existence.

All peaks of CZTS phase shifted to larger angles indicating a decrease in the lattice parameters due to the replacement of Se with S. Taking into account these data it may be concluded that the latter film was nearly single-phase and crystallized in partially disordered kesterite structure.

However, it is impossible to determine uniquely crystal structure of CZTSSe films based on XRD data only. To confirm this assumption Raman spectra of grown films were analyzed. All phases present in a solid mixture with CZTSe should be resolved by RS.

Typical RS spectrum of CZTSSe thin films series (***a***) is presented in Fig. 3. In the spectral range of 180-280 cm^{-1} up to five Raman bands were detected at frequencies of about 130, 179, 198, 205 and 234 cm^{-1}. Three components 179, 198 and 234 cm^{-1} can be attributed to $Cu_2ZnSnSe_2$ phase. The major mode at 198 cm^{-1} can be attributed to either the A_1 mode of KS (kesterite) at 193.01 cm^{-1} or the B (TO LO) of KS at 193.22 cm^{-1} or A_1 mode of ST (stannite) 194.6 cm^{-1}. The peak at 179 cm^{-1} can be attributed to the mode A_2 of ST (184.50 cm^{-1}) composed of the vibration of Se anions [6-8]. The peak at 234 cm^{-1} correspond to the vibrations of symmetry E (KS) (233.02) generated by the anion and cation vibrations of $Cu_2ZnSnSe_2$.

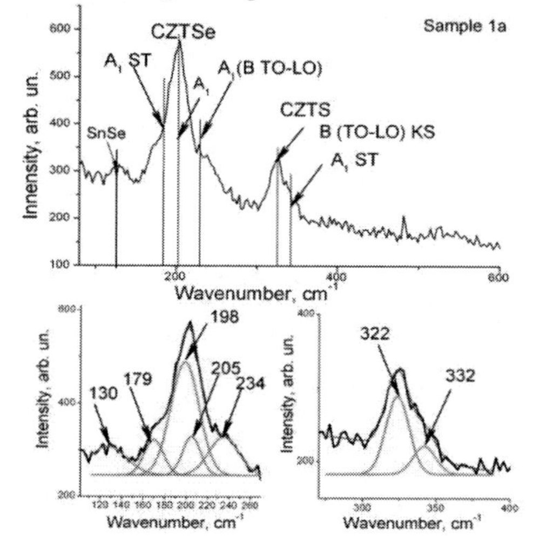

Figure 3: Raman spectrum of the CZTSSe thin film with S/Se ≈ 0.23 (series ***a***)

The position of the dominant A_1 mode is shifted by 3 cm^{-1} to long-wave region from the value to theoretically predicted ordered kesterite $Cu_2ZnSnSe_2$ due to S→ Se anion substitution and disordering of crystal structure [7]. Considerable broadening of the 198 cm^{-1} mode indicates on the possible presence of binary SnSe and CuSe $phases_2$. The low intensity peak visible at 130 cm^{-1} can be attributed to SnSe phase expected from the composition of the films series (***a***) [9]. However, there is no clear resolved peak in the region 260-270 cm^{-1} where vibration mode of CuSe phase should be detected [10]. It may be suggested that this phase statistically distributes in the film material and there is no it's local conglomerates in the mixture.

The appearance of intensive modes in the long-wave region 300-400 cm^{-1}, generated mainly by the S anion vibrations, confirms the forming of CZTSSe solid solutions with two-mode character of vibrations and justifies by other researches [11]. The phonon states at 322 cm^{-1} and 332 cm^{-1} are mainly a result of vibration of symmetry B (TO-LO) KS (309.56 and 313.19 cm^{-1}) and A_1 ST (334.08cm^{-1}) of Cu_2ZnSnS_4 compound [6].

Fig. 4 presents RS spectrum of the CZTSSe thin film with S/Se ≈ 29.84 (series ***b***).

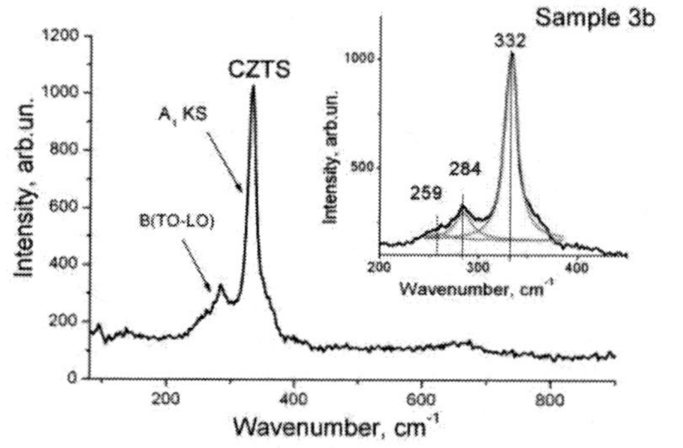

Figure 4: Raman spectrum of the CZTSSe thin film with ratio S/Se ≈ 29.84

A critical effect of annealing in S/N_2 atmosphere is clearly observed. In this case the spectrum is similar to the ordered CZTS kesterite. Three phonon modes appear in the spectral range at 200-400 cm^{-1} with frequencies at 259, 284 and 332 cm^{-1}. Phonon states in the range at 250-300 cm^{-1} are the result of vibration of the Zn cations and S anions with some contribution from the Cu cations.

The main band at 332 cm^{-1} observed in Fig.4 (insert) is supposed to be A-symmetry of KS CZTS or A_1-symmetry of ST [6]. Therefore, it is assumed that the films contain both KS and SN phases. A small shift of the A–peak with respect to more typical value of 338 cm^{-1} can be induced by disordering of crystal structure.

On the other hand, in the RS spectrum shown above (series ***b***) there are no visible bands which might be associated with other phases. The analysis of the Raman data suggests that sulfurization of Sn-Cu-Zn-Se precursors leads to the formation of CZTS phase with quasi-ordered kesterite structure. The obtained result is explained by the high reactivity of elemental sulfur at temperature of 500°C resulting in near complete Se→S substitution and formation of CZTS phase with very low Se content.

To approve this assumption band–gap value of grown films was evaluated from the measurements of reflectance spectra using the Kubelka-Munk (K–M) theory [13-15]. This theory provides the theoretical descriptions of diffuse reflectance spectroscopy. The K–M method is based on the following equation:

$$\frac{k}{s} = \frac{(1-R_\infty)^2}{2R_\infty} = f(R_1)$$

where R_∞ is the reflectance; k – absorption coefficient and s – scattering coefficient.

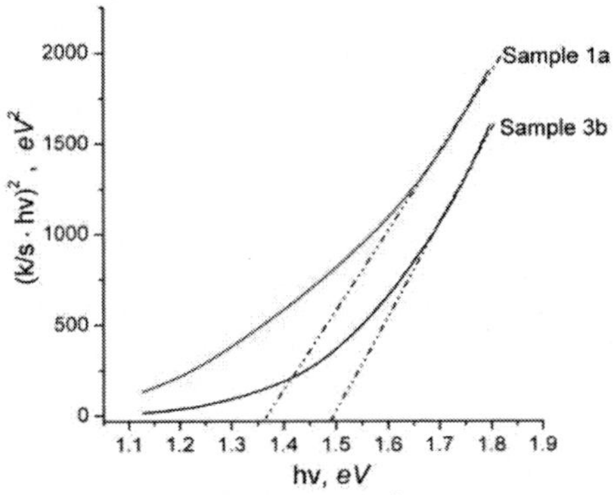

Figure 5. Dependency $[(k/s \cdot h\nu)]^2$ *vs* hν of the sample 1***a*** and 3***b***

From the plot $[(k/s \cdot h\nu)]^2$ *vs* *hν* (Fig. 5), the band gap value of the CZTSSe films material was estimated as 1.36 eV and 1.48 eV for sample ***1a*** (S/Se ≈ 0.23) and ***3b*** (S/(Se) ≈ 29.84), consequently. It is seen that band gap energies are near the one of CZTS and match well to the film compositions.

4 CONCLUSION

The dependence of the structural properties of CZTSSe thin films gown by selenization/(sulfurization) of electrodeposited Cu-Zn-Sn-S(Se) precursors on the S→Se and Se→S substitution has been studied. It was detected that in the process of annealing there is a big loss of Sn and Zn when compared with the initial precursor composition. RS measurements revealed disordered kesterite structure containing both KS and SN phases of grown CZTSSe films. Depending on overall S and Se contents CZTSSe thin films exhibited a shift in band gap from 1.36 to 1.48 eV.

This work demonstrates that it is possible to grow the CZTSSe absorber layer by suggested approach which can be improved by using a more accurate composition control of the precursors and temperature of annealing.

5 AKNOWLEGEMENTS

This work was supported by the Belarusian Republican Foundation for Fundamental Research (grant F16T/A-006) and the Ministry of Education and Science of the Republic of Kazakhstan (grant 0200/GF4).

6 REFERENCES

[1] P. Jackson, D. Hariskos, R. Wuerz, O. Kiowski, A. Bauer, T. M. Friedlmeier, and M. Powalla, Phys. Status Solidi: Rapid Res. Lett., 9 (2015) 28.
[2] M.A. Green, K. Emery, Y. Hishikawa, Y. Warta and E.D. Dunlop, Prog. Photovolt. Res. Appl., 24 (2016) 905.
[3] S. Delbos, EPJ Photovolt., 3 (2012) 35004.
[4] W. Wang, M.T. Winklern, O. Gunawan, T. Gokmen, T.K. Todorov, Y. Zhu and D.B. Mitzi, Adv. Energy Mater., 4 (2014) 1301465.
[5] M.B. Dergacheva, K.A. Urazov and A.E. Nurtazina, Russ. J. Electrochem., 53 (2017) 1.
[6] Khare A., Himmetoglu B., Johnson M., Norris D.J., Cococcioni M., and Audil E.S., J Appl Phys., 111 (2012) 083707.
[7] R. Caballero, E. Garcia-Llamas, J.M. Merino, M. Leorn, I. Babichuk, V. Dzhagan,V. Strelchuk, and M. Valakh, Acta materials, 65 (2014) 412.
[8] Beigom N., Amori M., Postnikov A., Phys. Rew., B 82 (2010) 205204.
[9] Chandrasekhar H.R., Humphreys R.G., Zwick U., Cardona M., Phys. Rev., B 15 (1977) 2177.
[10] Xue C.P., Raptis Y.S., Richter W., Esser N., Siebentritt S., and Lux-Streiner M.Ch., J. Appl. Phys., 96 (4) (2004)1963.
[11] W. Wang, M.T. Winkler, O. Gunawan, T. Gokmen, T.K. Todorov, Y. Zhu, and D.B. Mitzi, Adv. Energy Mater., 4 (2014) 1301465.
[12] M. Dimitrievska, A. Fairbrother, X. Fontané, T. Jawhari, V. Izquierdo-Roca, E. Saucedo, A. Pérez-Rodríguez, Appl. Phys. Lett., 104 (2014) 021901.
[13] Kubelka P., Munk F., Z Techn Phys., 12 (1931) 593.
[14] Kubelka P., J Opt Soc Am., 38 (1948) 448.
[15] Kubelka P., J Opt Soc Am., 44 (1954) 330.

IDENTIFICATION OF TRAP STATES IN HYBRID ORGANIC/INORGANIC PEROVSKITES

G. Gordillo[1], C. A. Otálora[2], E.Romero[2], A. A. Ramirez[3]

[1] Departamento de Física, Universidad Nacional de Colombia, Bogotá, Colombia
[2] Departamento de Química, Universidad Nacional de Colombia, Bogotá, Colombia
[3] Departamento de Ingeniería Química, Universidad Nacional, Bogotá, Colombia

This paper presents results of a study that allowed identifying states of traps in thin films of organolead halide perovskites ($CH_3NH_3PbI_3$) prepared by spin-coating, through thermally Stimulated Current (TSC) measurements. Special emphasis was done in studying the influence of superficial absorption of oxygen from the environment on the depth of traps and on the density of states associated with trapping centers.Deconvolution of the TSC curves obtained from measurements made at a pressure of 200 mBarrevealed the presence of three traps centers located at 0,124, 0,519 and 1,42 eV above the valence band, while in samples exposed to an environment of 2 mBar were identified only two traps centers located at 0.179 and 0,495 eV.

Keywords: Thin films, Trapping and recombination processes, PV Materials .

1. INTRODUCTION

Hybrid organic/inorganic perovskitematerials as light absorbers are currently among the most promising candidates for thin-film photovoltaic (PV) applications. Great advances in terms of efficiency have been achieved with this type of devices, however there are still many doubts about their electrical transport properties and the relationship of these with the efficiency of the devices. Recently, methylammonium lead halide ($MAPbX_3$, X = halogen; CH_3NH_3: MA) and its mixed-halide crystals, have been used as light harvesters for solar cells [1-3]. These devices are cost-effective due to their simple fabrication process andhigh PCE. As a result, these cells have attracted a great dealof attention among researchers in the field.Perovskite solar cells have shown a rapid efficiency improvement (from 2.2% in 2006 to 20.1% in 2014[4]) and in December 2015 – Dyesol Limited (ASX: DYE) announced that a research team headed by Professors Michael Grätzel and Anders Hagfeldt at the EcolePolytechniqueFédérale de Lausanne (EPFL) established a new world record efficiency for its Perovskite Solar Cells, with a certified conversion efficiency (at the laboratories of NewportCorporation) of 21.02%. [5].

Bulk recombination induced by defect states arising from chemical impurities and/or structural defects, have been identified as one of the hurdles for reaching high solar cell performances [6]. Defects may introduce trap and recombination centers, which have a direct influence on the transport phenomena of the charge carriers.Therefore, it is important to study the parameterscharacterizing such states. There arejust a few experimental reports about defect characterization in Hybrid organic/inorganic perovskite materials, howeverthermally stimulated current (TSC) measurements are relatively easy to perform and provide detailed information on trap and/or recombination states [7]. TSC is known to be a very sensitive technique for studying point-defect and traps in semiconductors [8,9] whichpermits a rapid and straightforward survey of the gap states. In the case of organolead halide perovskites films, there are no experimental TSC data available in the literature. In this work, for the first time we present a study of trap in $MAPbI_3$thin films.

In particular the influence of superficial absorption of oxygen from the environment on the depth of traps and on the density of states associated with trapping centers of thin films of $MAPbI_3$deposited in one step by spin-coating using precursor solutions dissolved in DMSO were determined by means of thermally stimulated current measurements.

2. EXPERIMENTAL AND THEORETICAL DETAILS

2.1 Experimental

We studied $MAPbI_3$films deposited on glass slides by spin-coating at 3000 rpm of a mixed halide precursor solution composed of CH_3NH_3I and PbI_2 0,4M in DMSO. The solution was prepared using commercial PbI_2(Sigma Aldrich) and MAI synthesized in our laboratory as precursors. The MAI precursor was synthesized from a mixture at 0° C for 2 hours of 5 mL of hydroiodic acid to 67%, 15 mL of solution 2,0 M of methylamine in methanol and 20 mL of water. The obtained solid was purified by washing it with diethylether up to obtaining the pure and dry compound.

For TSC measurements, electrical contacts of silver were deposited on the sample surface by evaporation, with 1 cm of distance between them. The films were mounted on a cold plate of a cryogenic refrigerationsystem, with a vacuum of 10^{-2}mB. The temperature was monitored by a K type thermocouple and the TSC measurements were performed in the temperature rangefrom 200 to 300K, increasing the temperature at a constant rate using an electronic control system designed and implemented for this purpose. This system includesfacilities to generate ramps of temperature varying at a constant rate in the range of 2-30 °C/min.

The TSC measurements were performed using the following procedure. Initially the sample is cooled until 170 K; after that, the sample is illuminated during 15 minutes with an intensity of 500 W/m^2using a 250 W halogen lampand then the sample isput in dark. After that, temperature is raised at a constant heating rate of 3°C/s keeping a DC voltage of 110 V. The temperature and current are recorded continuously using a virtual Instrument (VI) developed with the software LabVIEW, which allows to visualize in real time the response of current and temperature; in addition it has facilities to generate ramps of temperature at a constant rate and to change the speed of data acquisition.The hardware used includes a USB-600-NI data acquisition card, which allows to acquire the dark-current signal, the voltage signal generated by a thermocouple type K (proportional to temperature) and generate two digital signals used to control the power of the source that supplies current to a heating resistor and the flow of N_2-L that cool the sample.The temperature of heating and cooling is controlled by means of an algorithm of control PID-PWM implemented in the VI.

2.2 Theoretical considerations

The thermally stimulated current (TSC) technique is a method frequently used to identify the presence of trap states in solids; it consists of filling centers after cooling down the material to low temperature and then heating the sample at a constant rate, showing the thermally stimulated current as the centers are emptied. The energy level within the band gap of a particular center is related to the temperature at which it is emptied, while the number of centers contributing to the observed current peak depends on the amount of stored charges which are released.

The thermally stimulated current, due to a single center with an activation energy $E_t$below the conduction band or above the valance band, under slow re-trapping conditions is described by the equation [10]:

$$I = I_0\left[-\frac{E_a}{kT} - \frac{v}{b}\int_{T_0}^{T} e^{-\frac{E_a}{kT}}dT\right] \quad (1)$$

Where T_0 is the temperature at which heating begins after filling the center, k is the Boltzmann constant, and b is the heating rate. The pre-exponential factor I_0 is a weak function of temperature and I exhibits a maximum as a function of temperature. As the temperature increases, the trapped charge carriers are released and this gives rise to current peaks. Consequently, a plot of current versus temperature is called TSC spectrum. If more than one type of center is present, curves obtained from TSC studies may be expected to show several maxima depending upon the activation energy of the centers. In Eq. (1), v is the attempt-to-escape frequency, which is determined by the following expression: [11]

$$v = N_V v_{th} S \quad (2)$$

where N_V is the effective density of states in the valence band, v_{th} is the thermal velocity of a charge carrier and S is capture cross section of the trap or recombination center

The Activation energy can be determined using the initial rise method which is valid for all types of recombination kinetics and it is also independent of the sample heating rate. It is based on the assumption that when centers begin to empty as the temperature is increased, the TSC signal is proportional to $E^{(-\frac{E_a}{kT})}$ [12]. In other words, this approach assumes that current is dominated by the first exponential factor in Eq. (1). Thus, a semi-logarithmic plot of the thermal stimulated current versus 1/kT gives a straight line with a slope value of ($-E_a$).

The center concentration can be obtained taking into account the lower limit of the center density Nt, which can be found by integrating the TSC spectrum over time for each peak according to the following expression [13]

$$\int_{T_0}^{T} I_{TSC}dT = b\int_{t_0}^{t_{final}} Idt = bQ = bQN_tV \quad (3)$$

Where Q is the quantity of charge released during a TSC experiment, which can be calculated from the area under the TSC peak, q is the electron charge and V is the volume of the sample.

3. RESULTS

Several methods have been reported in the literature to evaluate center activation energy from the experimental TSC curves [14-16]. To analyze our results, we have chosen initial rise method described above. Therefore, from the Arrhenius plots, corresponding to the initial rise of the TSC, it is found the activation energies for the observed centers. This method was analyzed by Haake [17] demonstrating that up to a certain critical temperature value corresponding to about 40% of the TSC peak height, this assumption remains valid.

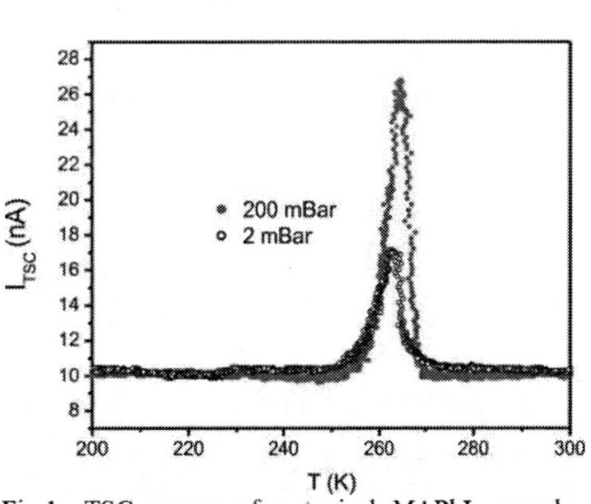

Fig.1: TSC curves of a typical $MAPbI_3$ sample, obtained from measurement performed under an ambient corresponding to pressures of 2 and 200 mBar respectively

Figure 1 shows typical experimental TSC curves of $MAPbI_3$ thin films, obtained using the initial rise method from measurements carried out to samples exposed to an ambient corresponding to a pressure inside the measuring chamber of 20 and 200 mBar respectively.

It is observed from Figure1, that the intensity of the TSC curve obtained from measurements performed at a pressure of 200 mbar is significantly higher than that of the sample exposed at a pressure of 20 mBar, indicating that the density of trapped charge carriers increases when the pressure inside the measuring chamber is increased. This behavior may be attributed to capture of electrons in trap states coming from oxygen molecules chemisorbed on $MAPbI_3$ ($O_{2(g)} + e^- \rightarrow O_2^-{}_{(ad)}$); subsequent releasing of trapped electrons from the sample exposed to a higher density of oxygen molecules gives rise to higher current peaks. Measurements carried out to samples under pressures less than 1 mBar do not present TSC peaks confirming that the behavior of the CTS curves observed in Fig. 1 could be attributed to effects of surface absorption of oxygen.

The peaks of Fig. 1 looks very broad,which can be due to a combination of several peaks, corresponding to recombination center with different activation energy E_A. The deconvolution using Gaussian line shapes, of the TSC curve obtained for a typical $MAPbI_3$ film from measurements carried out at 200 mBar is shown in Fig. 2. The peak positions were not fixed a priori, but were allowed to vary to get the best fit to the experimental spectra; the sum of all the peaks matches the experimental TSC curve everywhere.

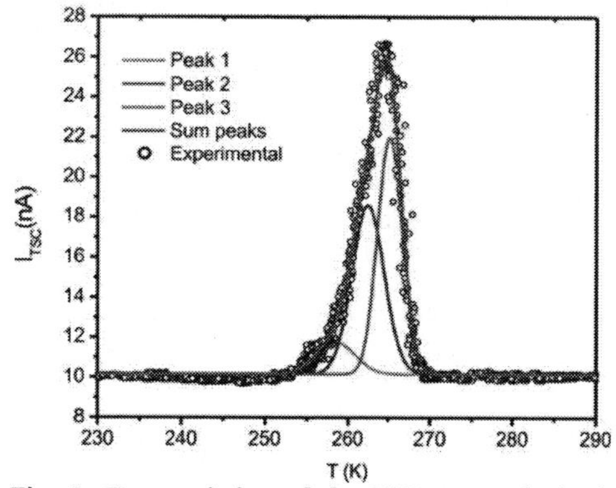

Fig. 2: Deconvolution of the TSC curve obtained from measurements carried out to a MAPbI3 film under a pressure of 200 mBar.The red line is the sum of three peaks used to deconvolute the experimental TSC curve.

In the deconvolution approach, the sum of three peaks matches the experimental TSC curve, indicating that the $MAPbI_3$ film exposed to an ambient of 200 mBar present trap or recombination centers with three different activation energies (see Fig. 3). It was also found that in the first deconvolution approach of the TSC curve of $MAPbI_3$ films exposed to an ambient of 2mBar, the sum of two peaks matches the experimental data, indicating that in this case the $MAPbI_3$ sample present trap or recombination centers with two different activation energies.

Using the initial rise method, the activation energies of $MAPbI_3$ samples were found, assuming that the TSC signal is proportional to $E^{(-\frac{E_a}{kT})}$thus, a semi-logarithmic plot of the current versus 1/kT gives a straight line with a slope value of (-E_a). Fig. 3 shows Activation energy values determined for $MAPbI_3$ filmsexposed to an ambient of 2 and 200 mBar respectively.

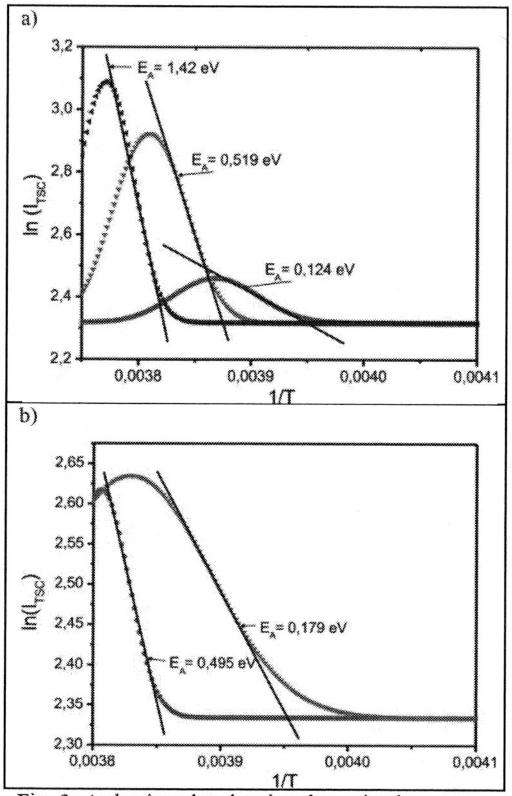

Fig. 3: Arrhenius plot showing the activation energy values of $MAPbI_3$ films exposed to an ambient of a) 200 mBar and b) 2 mBar.

Table 1: values of Activation energy E_A and density of states N_t associated to traps, estimated for $MAPbI_3$ thin films exposed to an ambient of 20 and 200 mBar.

	200mBar		2mBar	
	E_A (eV)	N_t(m^{-3})	E_A (eV)	N_t (m^{-3})
trap 1	0,124	2,8*10^{22}	0,179	1,0*10^{23}
trap 2	0,519	1,0*10^{23}	0,495	3,4*10^{22}
trap 3	1,42	1,0*10^{23}	--	--

In table 1 are listed values of activation energy and density of states $N_t$associated to the traps identified in $MAPbI_3$ films exposed to an ambient of 2 and 200 mBar respectively.

Given the complexity of $MAPbI_3$ films, it is quite difficult to determine the nature of the centers found; however we thing that the observed states of some energy levels can be associated to oxygen adsorbed at the surface of the sample and/or located within the entire crystal. The shallow levels located at 0,124 and 0.179 eV can be considered as traps, whereas the deep levels at 0,519 and 1,42 eV (in samples exposed to an ambient of 200 mBar) and 0,495 eV(in samples exposed to an ambient of 2mBar) can be considered as trap assisted recombination centers.

Similar values of E_A werefoundfor $MAPbI_3$ filmthrough transient photoconductivity measurements carried out at different temperatures and theoretical simulation of the relaxation curves; details of this study is given in Ref. [18]. Using values of the time constants τ obtained from the theoretical simulation of the curves of photocurrent decay measured at temperatures ranging from 25 to 65° C, the activation energy E_A of states corresponding to three differents traps were estimated from the slope of the Arrhenius plots [$\tau = \tau_0 exp(-E_A/kT)$]. It was found that, in general, the photocurrent response is affected by both trap assisted fast recombination processes and traps whose activation process is delayed, being the contribution in the intensity of the photocurrent of the first process greater that the second one. It was found evidence that the $MAPbI_3$ film exhibit a deep trap states at around 459 meV attributed to trap assisted recombination; furthermore, the $MAPbI_3$ films present shallow trap states at 129 and 24 meV that correspond to states of traps whose activation process is delayed.

The presence of deep trap states at around 508 meV were also found in $MAPbI_3$ films from thermally stimulated current (TSC) measurements [19]. Deep defect state (at 0.16 eV above the valence band) have been identified in $MAPbI_3$ films using admittance spectroscopy [20]. According to density functional calculations [21] this defect energy could potentially be ascribed to iodine interstitials.

4. CONCLUSIONS

The thermally stimulated currents technique have been used toidentifythe presence of trap centers in thin films of organic-inorganic lead halide perovskites ($MAPbI_3$) prepared by spin-coating, as well as to calculate both the activation energy and density of stated associated to the traps . It was found that thesuperficial absorption of oxygen from the environment significantly affect the traps and recombination processes;this study revealed that the samples exposed to an ambient of 200 mBar, present traps or recombination centers characterized by one shallow level located at 0,124 eV above the valence band which could be classified as trap and two deep levels at 0,519 and 1,42 eV which could be classified as trap assisted recombination centers.In samples exposed to an ambient of 20 mBar were identified one shallow level at 0.179 eV classified as trap and one level at 0,495 eV considered as trap assisted recombination center.

ACKNOWLEDGEMENTS: This work was supported by Colciencias (Contract #184/2016) and Universidad Nacional de Colombia, Sede Bogotá, Facultad de Ciencias, Departamento de Física, Grupo GMS&ES, K30 #45-03,Bogotá DC, Colombia.

REFERENCES

[1] M. M. Lee, J. Teuscher, T. Miyasaka, T. N. Murakami, H. J. Snaith, Science 2012, 338, 643–647.
[2] H.S. Kim, C.R. Lee, J.-H. Im, K.-B. Lee, T. Moehl, A. Marchioro, S.J. Moon, R. Humphry-Baker, J.H. Yum, J. E. Moser, M. Gratzel, , N.G. Park, "Sci. Rep. 2012, 2, 591:1–7.
[3] J.H. Noh, S.H. Im, J.H. Heo, T.H. Mandal, S.I. Seok, Nano Letters 2013: 13; 1764–1769.
[4] National Center for Photovoltaics at the National Renewable Energy Laboratory, Research cell efficiency records, www.nrel.gov/ncpv/.
[5]http://www.dyesol.com/media/wysiwyg/Documents/2015-asx-announcements/2015-12-08-DYE0397_ _EPFL_achieves_21_efficiency.pdf
[6] S. Siebentritt, Thin Solid Films 535, 1 (2013) 1-4.
[7] M. Courel, O. Vigil-Galán, D. Jiménez-Olarte, M. Espíndola-Rodríguez, and E. Saucedo,Journal of Applied Physics 116, 134503 (2014).
[8] E. Turan, M. Zor, A. S. Aybek, and M. Kul,,Physica B 3 95, 57(2007).
[9] N. S. Yuksek, N. M. Gasanly, and H. Ozkan, Semicond. Sci. Technol. 18, 834 (2003)
[10] N. S. Mohan and R. Chen, J. Phys. D: Appl. Phys. 3, 243 (1970).
[11] N. M. Gasanly and T. Yildirim, ActaPhysicaPolonica A 119, 437 (2011)
[12] R. Chen and S. W. S. Mckeever, Theory of Thermoluminescence andRelated Phenomena (Word Scientific, Singapore, 1997), p. 100
[13] J. Schafferhans, A. Baumann, C. Deibel, and V. Dyakonov, Appl. Phys. Lett. 93, 093303 (2008)
[14] J. M. Wrobel, A. Gubanski, E. Płaczek-Popko, J. Rezmer, and P. Becla, J. Appl. Phys. 103, 063720 (2008).
[15] G. Micocci and A. Tepore, J. Appl. Phys. 80, 894 (1996)
[16] T. Yıldırım and N. M. Gasanly, Solid State Sci. 11, 1562–1566 (2009).
[17] C. H. Haake, J. Opt. Soc. Am. 47, 649 (1957).
[18] G. Gordillo, C. A. Otálora, A.A. Ramirez, Physical Chemistry Chemical Physics, 2016, DOI: 10.1039/C6CP06261J.
[19] A. Baumann, S. Väth, Ph. Rieder, M. C. Heiber, K. Tvingstedt, and V. Dyakonov, J. Phys. Chem. Lett., **18**;6(12), 2350-2354 (2015)
[20] H. Sh. Duan , H. Zhou, Q. Chen, P. Sun, L. Song T. B. Song, B. Bob and Y. Yang, Physical Chemistry Chemical Physics,**17** 112-116 (2015)
[21] M. H. Du, J. Mater. Chem. A, **2**, 9091 (2014)

COMPARISON OF SIMULATION MODELS FOR PEROVSKITE SOLAR CELLS

Santiago Silvestre[1], Elena Mas-Marzá[2], Joaquim Puigdollers[1], Francisco Fabregat-Santiago[2], Valero G. Alfonso[2].
[1] MNT Group, Electronic Engineering Department, Universitat Politécnica de Catalunya (UPC) BarcelonaTech
C/Jordi Girona 1-3,Campus Nord UPC, 08034 Barcelona, Spain.
Tel. +34 93 4017491, email : Santiago.silvestre@upc.edu
[2] Institute of Advanced Materials (INAM), Universitat Jaume I
12006 Castelló, Spain

ABSTRACT: In this work, a comparison of two models for the simulation of the electrical behavior of perovskite solar cells is presented. The two models considered are the well-known one diode model and an analytical model recently presented by Sun et al. These models are detailed in section 3. A perovskite solar cell was fabricated and characterized in order to compare real measurement results with simulation results obtained by using both models. The fabrication and characterization processes are also summarized in the experimental section.
Keywords: Perovskite solar cells, Modeling, Simulation.

1 INTRODUCTION

Perovskite solar cells have shown great potential to achieve high-efficiency solar energy conversion at low cost. In addition, tandem cell structures with a top perovskite solar cell with a tunable higher bandgap than the bottom silicon solar cell with silicon offer good expectations [1-3]. Moreover, an efficiency of 25.2% efficiency was recently reported on a tandem cell structure of perovskites and silicon heterojunction cells [3-4]. Accurate models simulating the electrical characteristics of perovskite solar cells and tandems can help to achieve optimal designs and improve their characteristics and performance.

2 EXPERIMENTAL

2.1 Fabrication Process of the Perovskite solar cell

Main steps of the fabrication process of the perovskite solar cell are the following:

Substrate preparation. Fluorine doped tin oxide (FTO) coated glass substrates were etched with zinc powder and HCl (2M) to obtain 0.224 cm^2 of active electrode area. The substrates were cleaned with soap (Extran® MA 02 neutral) and sonicated for 15 minutes in a solution of water and soap (Extran® MA 02 neutral). Then, the sheets were rinsed with MilliQ water and sonicated for 15 min in ethanol (95%). Finally, the substrates were rinsed with acetone and dried with compressed air.

TiO_2 compact layer deposition. Before deposition a UV/ozone treatment was performed for 15 minutes. Then, a TiO_2 blocking layer was deposited onto the substrates by spray pyrolysis at 450°C, using 5mL of a titanium diisopropoxide bis (75% in isopropanol, Sigma-Aldrich) solution diluted in ethanol, using oxygen as carrier gas. After the spraying process the films were kept at 450 °C for 30 min and left to cool to room temperature. The thickness determined by Scanning Electron Microscopy was ~40 nm.

Mesoporous TiO_2 layer depostion. A mesoporous TiO_2 layer was deposited by spin-coating at 2000 rpm during 10s. After drying at 100°C during 10 min, the TiO_2 layers were heated at 500 °C for 30 min and cooled to room temperature. The thickness determined by Scanning Electron Microscopy was ~200 nm. The perovsite layer was deposited by spin-coating 50 µL of a solution 1.35 M of CH_3NH_3I and PbI_2 in a Mixture of N,N-dimethylformamide and dimethylsulfoxide 9:1 at 4000 rpm during 50s. During the spin-coating procedure, diethyl ether was added to the substrate. Then, the film was heated at 100°C for 3 min.

Spiro-OMeTAD deposition. The spiro-OMeTAD was deposited on top of the perovskite substrates by spin coating. Finally a deposition of 60 nm of gold was performed by thermal evaporation under ultrahigh vacuum conditions, using a commercial MBraun vacuum chamber.

The schematic diagram of the fabricated devide is given in Fig. 1.

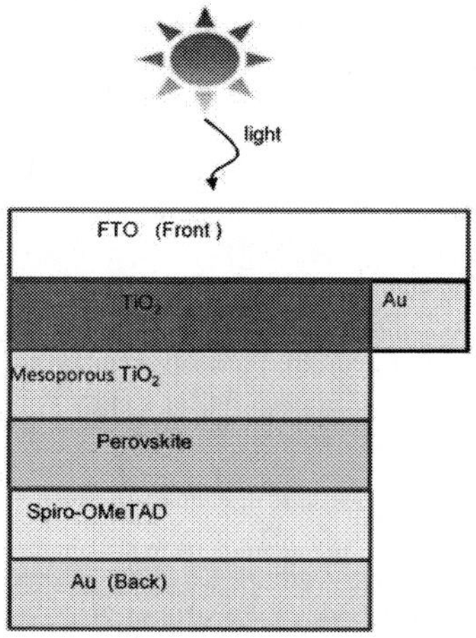

Figure 1: Schematic diagram of the Perovskite solar cell

2.2 Characterization

The External Quantum Efficiency (EQE) was performed using a Xenon lamp power source coupled with a monochromator controlled by a computer; the photocurrent was measured using an optical power meter 70310 by Oriel Instruments.

Current density-voltage (J-V) curves were performed under 1 sun illumination (100 $mW \cdot cm^{-2}$) using a xenon arc lamp simulator (Sun 2000, ABET Technologies) with an AM 1.5 G spectral filter and a Keithley 2400, previously calibrated with an NREL-calibrated Si solar cell. All the measurements were performed with an opaque mask of 0.112 cm^2 and without encapsulation.

The conversion efficiency of the perovskite solar cell is η= 14.11 %. Figures 2 and 3 show the EQE and the I-V characteristics of the perovskite solar cell respectively.

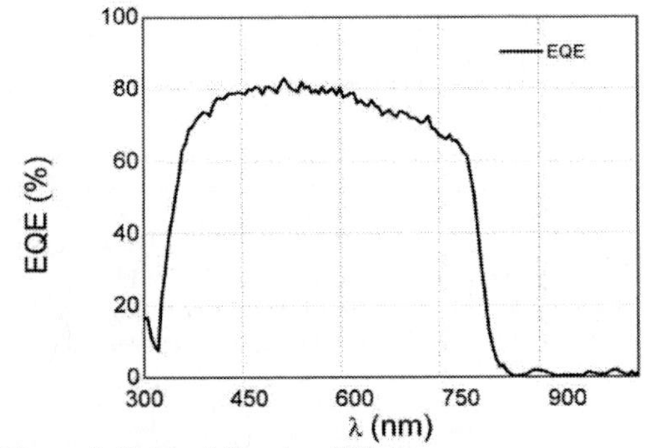

Figure 2: External Quantum Efficiency.

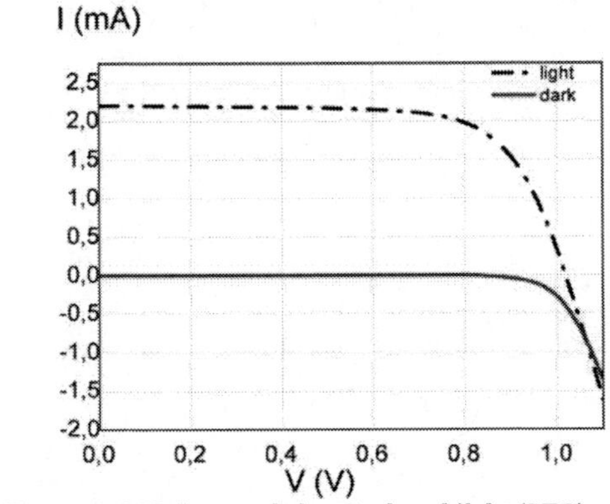

Figure 3: I-V characteristics: Dark and light (STC).

3 SIMULATION MODELS

The first model selected for the simulation for the perovskite solar cell is the model recently presented by Sun et al. [5]. The equations included in the model to calculate the current-voltage characteristics of the perovskite solar cell are the following:

$$J_{dark} = (\alpha_f J_{of} + \alpha_b J_{ob})\left[e^{(\frac{V}{Vt})} - 1\right] \quad (1)$$

$$J_{photo} = q\, G_{max}[A - B\; e^{-m}] \quad (2)$$

$$J_{light} = J_{dark} + J_{photo} \quad (3)$$

where *Gmax* is the maximum absorption, for the perovskite solar cell included in this study $qGmax$ = 19.5 mA/cm^2, *Vt* is the thermal voltage, *Jof* and *Job* are the dark diode current densities recombining at the front and back transport layers respectively, *Jphoto* is the photocurrent density, *Jdark* is the dark current and *Jlight* the solar cell output current density.

Table I shows the equations corresponding to the model parameters included in Eqs (1-2), where Vbi is the build-in potential, D=0.05 s^{-1}cm^2 is the diffusion coefficient [6], s_f and s_b are the effective surface recombination velocities at the front and back interfaces respectively, and to is the thickness of the absorber layer.

The model parameters are functions of the physical parameters of the solar cell.

Table I: Model parameters

Parameter	Equations		
α_f	$\frac{V'}{e^{V'}-1} + \frac{1}{\beta_f}$	$V' = \frac{(V-V_{bi})}{Vt}$	$\beta_f = \frac{D}{to\, s_f}$
α_b	$\frac{V'}{e^{V'}-1} + \frac{1}{\beta_b}$		$\beta_b = \frac{D}{to\, s_b}$
A	$\alpha_f\left(\frac{1-e^{V'-m}}{V'-m} - \beta_f\right)$		$m = \frac{t_o}{\lambda_{ave}}$
B	$\alpha_b\left(\frac{1-e^{V'+m}}{V'+m} - \beta_b\right)$		$\lambda_{ave} = 100\, nm$

The second model selected in this study is the well-known: Five parameters or one diode model, where the relation between the output current and voltage is defined as follows:

$$I = I_{ph} - I_O\left[exp\left(\frac{V+R_sI}{nVt}\right) - 1\right] - \frac{V+R_sI}{R_{sh}} \quad (4)$$

where: *Iph* is the photocurrent, *Io* is the diode reverse saturation current, *n* is the diode ideality factor, R_s and R_{sh} are the series and shunt resistances respectively, *Vt* is the thermal voltage. *I* and *V* are the output current and voltage.

4 RESULTS

Figures 4 and 5 show the comparison between the Current density-Voltage characteristics measured and obtained by simulations in light (STC) and dark conditions respectively.

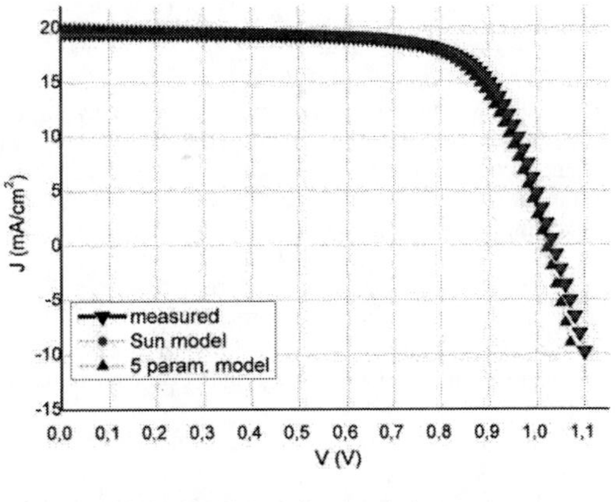

Figure 4: Light (STC) J-V characteristics.

The root-mean-square errors (RMSE) obtained between the measures and simulated output currents are given in Table 1. Finally, Table 2 shows the values of main parameters included in the two models analyzed. These parameters were obtained by fitting the models equations with the measured data by means of parameter extraction techniques [7].

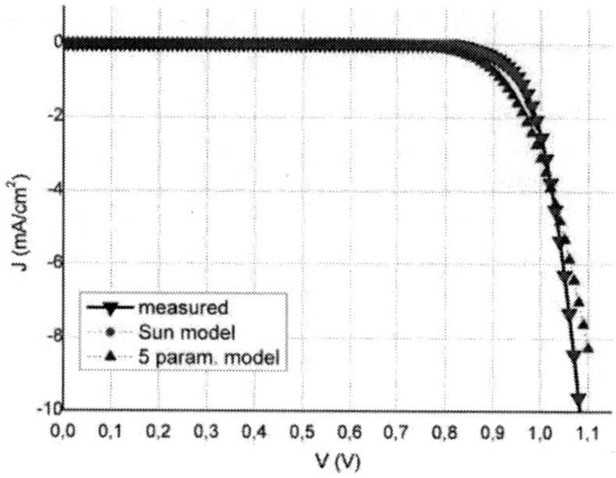

Figure 5: J-V characteristics in dark conditions.

Table II: RMSE obtained for the currents

	RMSE (mA)	
	Sun model	Five Parameter
I light	1.46 10^{-2}	1.32 10^{-1}
I dark	6.95 10^{-2}	3.08 10^{-2}

Table III: Model Parameters

Sun model		Five Parameter	
V_{bi}	0.953 V	n	2.19
to	248 nm	R_s	29.2 Ω
s_f	5.457 10^3 cm/s	R_{sh}	6.723 10^3 Ω
s_b	22.12 cm/s	J_o	1.79 10^{-7} mA/cm^2
J_{fo}	1.85 10^{-16} mA/cm^2	J_{ph}	20 mA/cm^2
J_{bo}	5 10^{-17} mA/cm^2		

5 CONCLUSIONS

The Sun's model allows the best accuracy in the simulations of the perovskite solar cell J-V characteristic in light conditions. Moreover, this model offers relevant information about physical parameters of the cell that can be useful in the design of the devices.

ACKNOWLEDGMENT

This work has been funded by the Spanish Ministry of Economy and Competitiveness under projects number: ENE2014-56237-C4-2-R.

6 REFERENCES

[1] David P. McMeekin, Golnaz Sadoughi, Waqaas Rehman, Giles E. Eperon, Michael Saliba, Maximilian T. Hörantner, Amir Haghighirad, Nobuya Sakai, Lars Korte, Bernd Rech, Michael B. Johnston, Laura M. Herz and Henry J. Snaith. (2016). A mixed-cation lead mixed-halide perovskite absorber for tandem solar cells. Science 351, no. 6269 151-155.

[2] Colin D. Bailie, M. Greyson Christoforo, Jonathan P. Mailoa, Andrea R. Bowring, Eva L. Unger, William H. Nguyen, Julian Burschka, Norman Pellet, Jungwoo Z. Lee, Michael Gratzel, Rommel Noufi, Tonio Buonassisi, Alberto Salleo and Michael D. McGehee. (2015). Semi-transparent perovskite solar cells for tandems with silicon and CIGS. Energy Environ. Sci. 8, 956-963.

[3] The Duong, Niraj Lal, Dale Grant, Daniel Jacobs, Peiting Zheng, Shakir Rahman, Heping Shen, Matthew Stocks, Andrew Blakers, Klaus Weber, Thomas P. White, and Kylie R. Catchpole. (2016). Semitransparent Perovskite Solar Cell With Sputtered Front and Rear Electrodes for a Four-Terminal Tandem. IEEE Journal of Photovoltaics 6 (3) 679-687.

[4] Jérémie Werner, Loris Barraud, Arnaud Walter, Matthias Bräuninger, Florent Sahli, Davide Sacchetto, Nicolas Tétreault, Bertrand Paviet-Salomon, Soo-Jin Moon, Christophe Allebé, Matthieu Despeisse, Sylvain Nicolay, Stefaan De Wolf, Bjoern Niesen and Christophe Ballif. (2016). Efficient near-infrared-transparent perovskite solar cells enabling direct comparison of 4-terminal and monolithic perovskite/silicon tandem cells. ACS Energy Letters 1 (2) 474-480.

[5] Sun, X., Asadpour, R., Nie, W., Mohite, A. D., & Alam, M. A. (2015). A Physics-Based Analytical Model for Perovskite Solar Cells. IEEE Journal of Photovoltaics, 5 (5) 1389-1394.

[6] Stranks, S. D., Eperon, G. E., Grancini, G., Menelaou, C., Alcocer, M. J., Leijtens, T., Laura M. Herz, Annamaria Petrozza, and Snaith, H. J. (2013). Electron-hole diffusion lengths exceeding 1 micrometer in an organometal trihalide perovskite absorber. Science, 342 (6156) 341-344.

[7] Kichou, S., Silvestre, S., Guglielminotti, L., Mora-López, LL., and Muñoz-Cerón, E. (2016). Comparison of two PV array models for the simulation of PV systems using five different algorithms for the parameters identification. Renewable Energy 99 270-279.

LONG TERM THERMAL STABILITY TESTS FOR AIR PROCESSED INKJET INFILTRATED CARBON BASED PRINTED PEROVSKITE SOLAR CELLS

Authors: Syed Ghufran Hashmi *, Alpi Rimppi, David Martineau and Peter Lund
*Corresponding author's email: ghufran.hashmi@aalto.fi, sghufranh28@gmail.com

New Energy Technologies Group (NEW), Department of Applied Physics, Aalto University, P.O. BOX 15100, FI-00076 Aalto (Espoo), Finland

ABSTRACT: We report here long term thermal stability tests of the air-processed inkjet infiltrated carbon-based printed perovskite solar cells (CPSCs) that has been conducted in dark conditions with temperatures as high as up to 80 °C. The long term device performance of these CPSCs was investigated by measuring the J-V curves periodically, incident photon to collected electron efficiency (IPCE) measurements and XRD analysis. The investigations revealed that the thermally exposed devices did not show performance degradation for a period of 3000 hours and retained cent per cent overall initial efficiency. This remarkable durability of fabricated devices presented in this report indicates the potential of the up-scaling of this specific perovskite solar cells geometry by implementing established non vacuum based scalable methods such as screen printing or inkjet printing at significantly lower cost.

Keywords: Perovskite solar cells, stability, carbon counter electrode, inkjet printing, precursor ink, degradation.

1. INTRODUCTION

Solution processed organic–inorganic hybrid perovskite solar cells (PSCs) have drawn significant attention worldwide due to their promising features i.e. rapid increase in solar to electrical conversion efficiency now crossing over 22% [1], abundantly available materials such as titanium dioxide (TiO_2) or Lead (Pb) as well as their easy fabrication schemes [2]. However the key challenge associate presently with these traditional PSCs is their long term performance stability [3-4]. In several reports, the poor stability which consequently caused a performance degradation in the photovoltaic performance have been indicated due to the degradation in perovskite light absorbing layer which remains highly sensitive to air and humidity intrusion [3-6]. As a result, the hydrogen iodide (HI) and methylamine (CH_3NH_2) gradually removes from the perovskite crystals which then transforms back into PbI_2 [3-6]. Hence numerous new materials and new device designs have been adopted to overcome the issue of long term device performance stability. [5, 7-8]. In case of numerous device designs, the holes transporting material (HTM) free and carbon counter electrode based printed perovskite solar cells (CPSCs) looks very promising in terms of scalability, low cost and long term device performance durability [8-10]. We recently have shown a successful demonstration of automated dispensing of the perovskite precursor ink at open air with Dimatix materials inkjet printer for these CPSCs after that the fabricated devices exhibited highly stable photovoltaic performance when subjected to a continuous full sun light soaking test at 35 °C [9] or in very strong continuous UV light soaking for a period of more than 1000 hours in both cases [9-10] proving that the up-scaling of this device geometry is possible without implementing clean room laboratory environment conditions.

For this report, we further examined the CPSCs performance stability at much higher temperatures ranged from 60-80 °C in dark and at open circuit conditions for a period of more than 3000 h. The photovoltaic performance was monitored by measuring periodically the J-V curves and incident photon to collected electron efficiency (IPCE) of the exposed batched of CPSCs. The subjected CPSCs exhibited remarkable stability and did not show any performance degradation in terms of overall conversion efficiency thus ensure the potential of the up-scaling of this specific perovskite solar cells geometry by implementing established non vacuum based scalable methods such as screen printing or inkjet printing at significantly lower cost.

2. EXPERIMENTAL DETAILS

2.1 Measurements setup

The J-V curves of the CPSCs for this experiment were periodically measured in the Xenon lamp based solar simulator (Peccell Technologies, PEC-L01, Japan) full sun light illumination after calibrating it with a reference solar cell (PV measurements Inc) and by using a black tape mask with an aperture area 0.16 cm^2. The long term stability for the fabricated CPSCs was accomplished by keeping the batch of the solar cells under dark conditions for 1896 hours first at 60 °C in an oven (Memmert) and then 1128 hours more in the same oven at 80 °C at open circuit conditions. Moreover, the incident photon to collected electron efficiency (IPCE) spectra were measured with the spectral response measurement unit (QEX7, PV Measurements Inc). The thickness of the screen deposited layers was also measured with a profile meter (Veeco instruments) before the inkjet infiltration of perovskite precuros ink. XRD measurements were performed

2.2 Fabrication of the inkjet infiltrated CPSCs

The CPSCs for this study were fabricated with the following methods:

10x10 cm^2 plates of Fluorine doped tin oxide (FTO) coated glass (Sheet resistance, $R_{SH} = 7$ Ω/Sq, Product code, TCO22-7-Solaronix) were first laser patterned (to define cathode and anode areas) with an automated fiber laser. After that the substrates were subjected to sequential cleaning steps in 1% aqueous solution of Hellmanex, acetone, and isopropanol respectively (15 min each) in an ultrasonic bath and were dried with compressed air.

Then 30-50 nm thin compact TiO_2 layer was grown over the cleaned and patterned FTO glass substrates (by using a glass mask to protect the contact areas) with spray pyrolysis method i.e. by spraying titanium diisopropoxide bis (acetylacetonate) (75% in isopropanol, Sigma-Aldrich) diluted in absolute ethanol (1:10) over a pre-heated (500 °C) hot-plate and the substrate were cooled down to room temperature (RT).

After that the contacts (for anode and cathode) were fabricated by screen printing a low temperature silver paste (Sun Chemical CRSN2442) which were further dried at 150 °C for 15 minutes and were cooled down to RT. Then a semi-transparent mesoporous TiO_2 layer (400-600 nm) was stacked over compact TiO_2 layer by employing a printable titania paste (Ti-Nanoxide T/SP diluted in terpineol, Solaronix) through screen-printing which was first dried at 150°C for 5 minutes and was then sintered at 500 °C for 30 minutes and was cooled down to RT. After that, mesoporous ZrO_2 (Zr-Nanoxide ZT/SP-Solaronix), as an insulating layer (1-2 µm), was deposited by screen-printing over the mesoporous TiO_2 layer and was again dried at 150 °C for 5 minutes and was then sintered at 500°C for 30 min followed by cooling down the FTO Glass substrates to RT.

Then the conductive and highly porous carbon electrode deposited over the ZrO_2 by screen-printing the carbon paste (Elcocarb B/SP, Solaronix), followed by again drying at 150 °C for 5 min, and then sintering at 400 °C for 30 min. The substrates were then again cooled down to RT and were kept in a sealed plastic box before the inkjet infiltration of perovskite precursor ink.

2.3 Recipe of perovskite precursor ink

0.19 g of methyl ammonium iodide (MAI, Dyesol) was first mixed with 0.0176 g of 5-ammonium valeric acid iodide (5-AVAI, Dyesol) and 0.53 g of PbI_2 (TCI Chemicals) in glass vial. After that 1 ml of gamma-butyrolactone (Sigma Aldrich) was introduced in the glass vial and the ingredients were dissolved by stirring the sealed glass vial for 30 min on a preheated hot-plate (70 °C). After 15-20 minutes, a clear yellow solution was obtained which was cooled down to RT. The perovskite precursor ink was then filtered through a PTFE filter and was directly transferred to the tank of the printer cartridge through the disposable syringe.

2.4 Inkjet infiltration of perovskite precursor ink

The inkjet infiltration of perovskite precursor ink was then performed over carbon electrode of the CPSCs by using Dimatix demand on drop materials inkjet printer (DMP-2831, Dimatix-Fujifilm Inc, USA) at 30 °C. A customized waveform with 18V amplitude and 1-8 kHz frequency was used along with 15 µm drop spacing to dispense the liquid which remained very stable during the infiltration. The platen was kept at RT whereas the relative humidity inside the printer hood was ~ 32%. The CPSCs were placed for 3-5 minutes to ensure the liquid penetration within the pores after the automated dispensing of the precursor. Then the CPSCs were carefully transferred in a pre-heated oven (50 °C) where they were kept in a plastic petry dish with lid on and off for 30 and 60 minutes respectively to ensure the complete growth of the crystals of perovskite light absorbing layers. In last, the cells were finally removed from oven and were then stored in vacuum before their characterization.

2.5 CPSCs epoxification

The active and non-active areas (except for silver contacts) of each CPSCs for this study were protected with a two components based commonly available epoxy (Strong Epoxy Professional, Casco-Finland) that was first spread over aforementioned areas manually and was dried overnight in the vacuum conditions before the measurements and exposing them in long term stability test.

3. RESULTS AND DISCUSSIONS

Table 1 represents a batch of CPSCs (4 devices) which were subjected to long term thermal exposure (60 °C) for a period of 1896 hours whereas the trends of photovoltaic performance of this batch are summarized in Figure 1, which reveals exceptionally high stability during the periodic measurements of the exposed CPSCs. A rise (> 25%) in short circuit current densities (J_{SC}) was observed during the periodic measurements whereas the open circuit voltage (V_{OC}) was slightly decrease (□ 6%) which is an expected result as we have observed similar decrement when the viscous epoxy was directly applied over the carbon back contact electrode which might could have affect very slightly the traces of perovskite crystals in the carbon pores [10]. Nevertheless, the V_{OC} after the slight decrease was stabilized and was maintained over the period of time which suggests the slow drying of epoxy over the period of time. Moreover, the overall conversion efficiency was remarkably improved (□ 17%) due to highly impressive decrease (> 48%) in the overall cell resistance (R_{CELL}).

The increase in J_{SC} and efficiency gave an idea of a possible crystal growth through which the inter-particle connectivity could be improved. Although the average fill factor of the exposed batch of CPSC was also remained stable however it was also reduced in the beginning and supports the idea of affecting V_{OC} initially. The improved J_{SC} were also verified initially and at the end of test with incident photon to collected electron efficiency (IPCE) measurements, which also exhibited significant improvements confirming the idea of possible crystals growth in perovskite light absorbing layer.

Table 1: Average initial and aged (1896 h) photovoltaic parameters of a batch of CPSCs which was exposed to 60 °C thermal soaking.

Hours (h)	J_{SC} mA/cm^2	V_{OC} V	FF	Eff %	R_{CELL} Ω
0	13.3±1	0.85±0.02	0.66±0.03	7.5±0.7	69.5±3.6
1896	16.7±0.3	0.80±0.03	0.66±0.02	8.8±0.6	36.1± 3.7

Hence motivated by the these impressive results, we further investigated the long term stability of this batch with same devices under more stressful thermal conditions i.e. 80 °C for a period of 1128 hours.

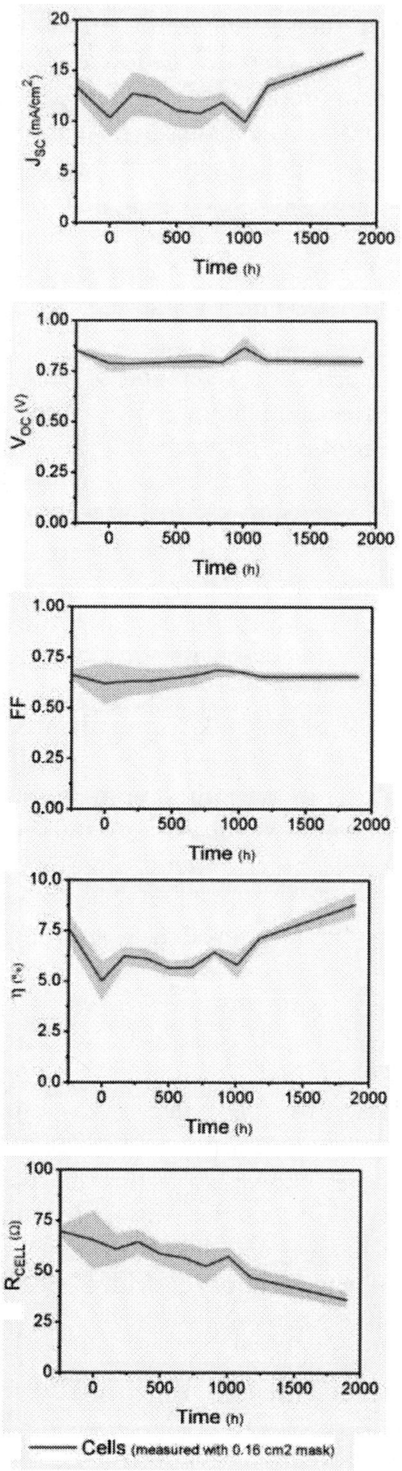

Figure 1: Trends of photovoltaic parameters of a batch of CPSCs exposed to 60 °C.

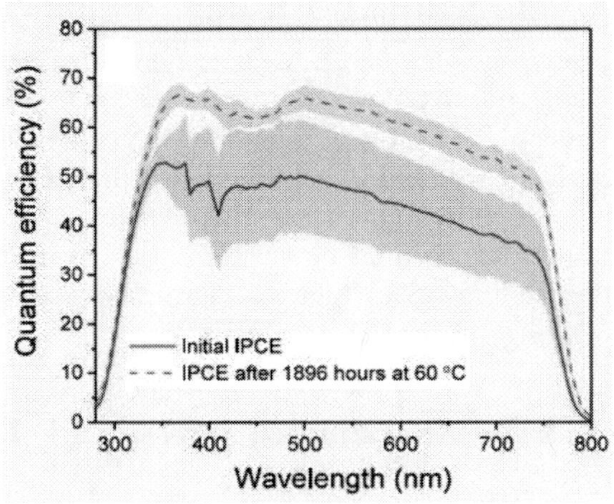

Figure 2: Initial and aged (1896 h) Incident photon to collected electron (IPCE) measurement results of a batch of CPSCs exposed to 60 °C thermal soaking.

The long-term photovoltaic performance trends of the same batch of CPSCs during 80 °C thermal exposure are depicted in Figure 3 stability where the exposed devices again showed remarkable stability and no sign of degradation were seen when the devices were measured periodically for a period of 1128 hours (Table 2). The same IPCE measurements were also performed to cross check the stability of the short circuit densities obtained from J-V curves measurements (Figure 4). Moreover the XRD measurements were also made to verify a possible crystals growth in perovskite light absorbing layer.

Table 2: Average initial and aged (1128 h) photovoltaic parameters of a batch of CPSCs which was exposed to 80 °C thermal soaking.

Hours (h)	J_{SC} mA/cm^2	V_{OC} V	FF	Eff %	R_{CELL} Ω
0	16.7±0.3	0.80±0.03	0.66±0.02	8.8±0.6	69.5±3.6
1896	15.9±0.9	0.80±0.06	0.64±0.04	8.2±1.1	36.1± 3.7

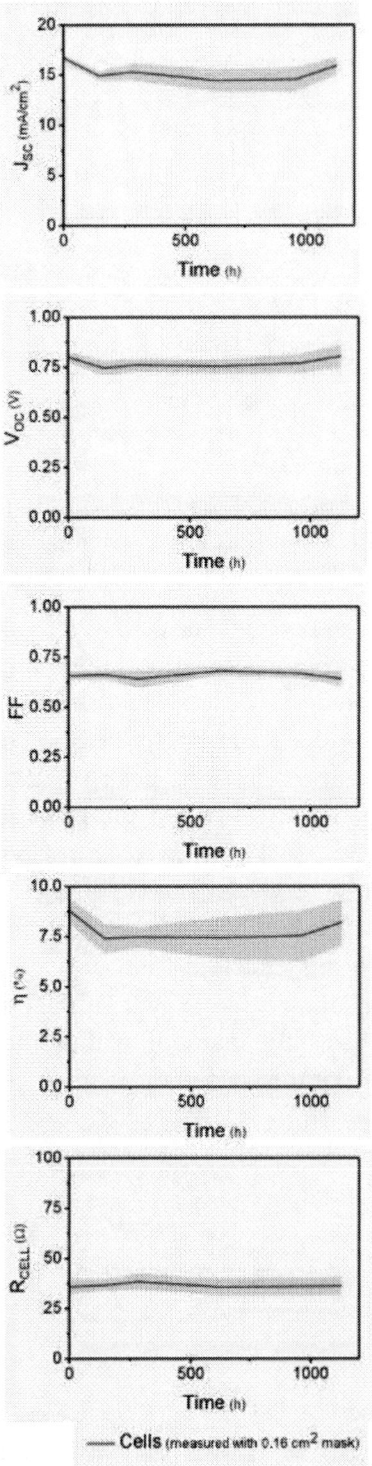

Figure 3: Trends of photovoltaic parameters of a batch of CPSCs exposed to 80 °C up to 1128 hours.

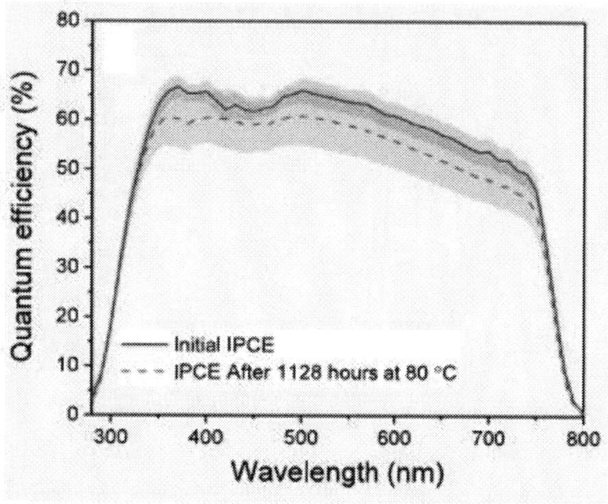

Figure 4: Initial and aged (1128 h) Incident photon to collected electron (IPCE) measurement results of a batch of CPSCs exposed to 60 °C thermal soaking.

Figure 5 represents the XRD data of a CPSC which was compared after 3000 hours with a fresh non aged CPSC. The thermal exposure did not induce any PbI_2 which is confirmed with no appearance of peak at 12.4° which corresponds to PbI_2. On the other hand the peak at 14.1° which corresponds to perovskite crystals stayed stable and was increased in height which supports the idea of increase in integral intensity and thus confirms the crystal growth of the perovskite light absorbing material within the porous structure of the stack [11].

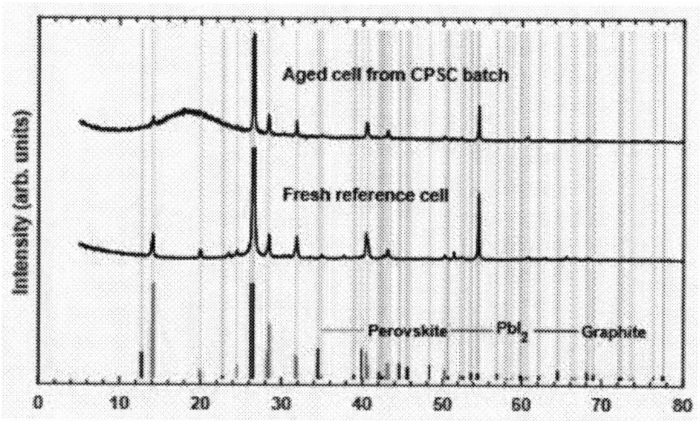

4. CONCLUSIONS

The results presented in this report are extremely remarkable and reveal that these carbon based perovskite solar cell keeps an immense potential for the commercialization. The exposed CPSCs in this study were sealed with a commonly available epoxy. The preliminary sealing exhibited very promising stability and suggest that these device keeps immense potential to even endure the most demanding 85% Relative Humidity and 85 °C long term stability test.

5. ACKNOWLEDGEMENTS

This work was mainly financed through grant number 287641. Ghufran Hashmi grateful to Academy of Finland for the post-doctoral research fellowship and also pays special thanks to Solaronix for hosting his research mobility for the project "Cutting-Edge Third Generation Advanced Photovoltaic Devices". The authors also recognize the provision of facilities and technical support by Aalto University at the OtaNano – Nanomicroscopy Center (Aalto-NMC).

6. REFERENCES

[1] S. S. Shin, Colloidally prepared La doped $BaSnO_3$ electrodes for efficient, photostable perovskite solar cells, Science, 2017.

[2] J. Burschka., Sequential deposition as a route to high-performance perovskite-sensitized solar cells, Nature 2013, 499, 316-9.

[3] Y. Han, S. Meyer, Y. Dkhissi, K. Weber, J. M. Pringle, U. Bach, L. Spiccia and Y. B. Cheng, Degradation observations of encapsulated planar $CH_3NH_3PbI_3$ perovskite solar cells at high temperatures and humidity, J. Mater. Chem. A, 2015, 3, 8139–8147.

[4] D. Wang, M. Wright, N. K. Elumalai and A. Uddin, Sol. Energy Mater. Sol. Cells, 2016, 147, 255–275.

[5] T. Leijtens, G. E. Eperon, S. Pathak, A. Abate, M. M. Lee and H. J. Snaith, Overcoming ultraviolet light instability of sensitized TiO_2

with meso-superstructured organometal tri-halide perovskite solar cellsNat. Commun., 2013, 4 (2885), 1–8.

[6] B. Philippe, B. W. Park, R. Lindblad, J. Oscarsson, S. Ahmadi, E. M. J. Johansson and H. Rensmo, Chem. Mater., 2015, 27(5), 1720–1731.

[7] S. Guarnera, A. Abate, W. Zhang, J. M. Foster, G. Richardson, A. Petrozza, H. J. Snaith, J. Phys. Chem. Lett. 2015, 6, 432.

[8] A. Mei, X. Li, L. Liu, Z. Ku, T. Liu, Y. Rong, M. Xu, M. Hu, J. Chen, Y. Yang, M. Grätzel, H. Han, Science 2014, 345, 295.

[9] S. G. Hashmi, D. Martineau, X. Li, M. Ozkan, A. Tiihonen, M. I. Dar, T. Sarikka, S. M. Zakeeruddin, J. Paltakari, P. D. Lund and M. Graetzel, Advanced Materials Technologies, 2016, 1600183, 1–6.

[10] S. G. Hashmi, A. Tiihonen, D. Martineau, M. Ozkan, P. Vivo, K. Kaunisto, V. Ulla, S. Mohammed Zakeeruddin, M. Grätzel., J. Mater. Chem. A, 2017, 5, 4797-4802.

[11] Syed Ghufran Hashmi, D. Martineau, M. Ibrahim Dar, Teemu T. T. Myllymaki, Teemu Sarikka, V. Ulla, S. Mohammed Zakeeruddin, M. Gratzel., J. Mater. Chem. A, 2017, 5, 12060–12067.

Stability Issues of Perovskite Photovoltaic Cells

Dávid Strachala, Josef Hylský, Jiří Vaněk, Michal Kadlec, Jan Mucha
Brno University of Technology, Faculty of Electrical Engineering and Communication Technologies,
Department of Electrotechnology
Technicka 10, 616 00 Brno, Czech Republic

ABSTRACT: The aim of the work was to find out an optimal way for the preparation of the stable perovskite solar cells. During the manufacture process was created a four solar cell samples with emphasis on stable active perovskite structure. After the perovskite solar cells were prepared, their degradation was observed and analyzed. The influence of humidity and UV radiation began to affect solar cells. The conversion efficiency of the best sample has dropped from $\eta_{MAX} = 7.22$ % to $\eta_{MAX} = 5.48$ % within the 20 hours.
Keywords: Perovskite photovoltaic cell, degradation of perovskite, moisture effect to perovskite structure,

1 PEROVSKITE SOLAR CELL

Well-known scientist Michael Grätzel called perovskite as "unusually cheap" material that could operate successfully in solar cells and combine the seemingly incompatible - low acquisition cost and high efficiency of energy conversion of sunlight into electricity [1]. Advantages of perovskite solar cells lies in their potential use in a broad application such as windowpanes, facade systems and much lower production costs (also availability of materials) in comparison to a traditional silicon solar cells with a relatively easier methods of its preparation. Perovskite solar cells can be considered as promising power source worthy attention.

Perovskite in general is a name for a group of substances with the same crystal structure as $CaTiO_3$. In photovoltaic cells, modified compounds of $CaTiO_3$ are used for absorbing of sunlight. A semiconductor is created as an organic-inorganic hybrid composition as $CH_3NH_3PbX_3$ (X = Cl, I, Br). Solar cells based on perovskite have been continuously developed in the last five years. The first time of publication of this material in the solar cells was in 2009 - energy conversion efficiency of photovoltaic solar cells with liquid electrolyte was 3.13% and 3.81% ($CH_3NH_3PbBr_3$ and $CH_3NH_3PbI_3$) [1]. In the following years, due to optimizations of various parameters of the photovoltaic cells there has been a rapid increase in energy conversion efficiency in a relatively short time [2].

Presented work deals with a hybrid peroskvit solar cell structure of $CH_3NH_3PbI_{3-2}Cl_2/PC_{61}BM$. Fig. 1 shows the general device structure of the perovskite solar cells.

2 DEGRADATION OF PEROVSKITE PHOTOVOLTAIC CELLS

Degradation of perovskite photovoltaic cells is not fully understood. Their main causes are known as the effects of humidity and impact of irradiance, but the physical and interlayer processes influencing the stability still are not clarified.

The degradation of perovskite solar cell can be generally divided into an extrinsic degradation – external influences of oxygen, incident radiation, humidity, temperature and intrinsic degradation – effect of heat and electric field inside the structure. Figure 1. shows a diagram representing the general recombination process, disruption of perovskite bonds, influence of humidity and photo-oxidation process between the interface layer and the HTM contact in the perovskite structure of mesoporous TiO2. In the TiO2 layer is diffusion of oxygen gradually accelerating due to UV radiation. Free oxygen radicals then may react with the HTM layer or perovskite active layer. Oxygen and H2O can enter to the structure through the electrode. Perovskite materials (eg. $CH_3NH_3PbI_3$) are very susceptible to the presence of oxygen and water and can degraded within the minutes, at most hours during the atmospheric influences [3].

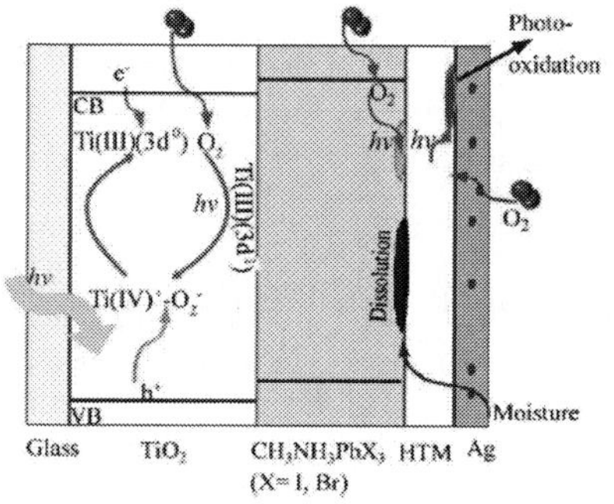

Figure 1: Degradation mechanisms at the perovskite solar cell [4]

One of the main reasons for the material instability lies in disruption of perovskite layer bonds by the atmospheric influences (especially because of humidity and UV radiation). The critical is influence of humidity. The first perovskite structures ($CH_3NH_3PbI_3$ / $PC_{61}BM$) were prone to the humidity. Procedure of perovskite bonds disruption is shown in the following figure 2:

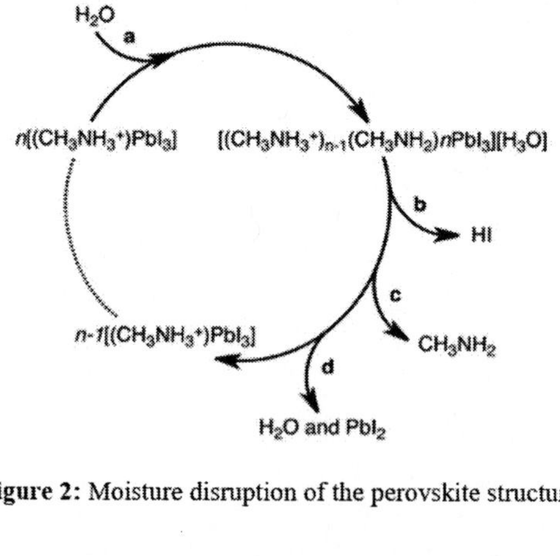

Figure 2: Moisture disruption of the perovskite structure

Degradation of the perovskite photovoltaic cells is already further evident in the bleaching of of originally brown active layer. The cause is the presence of ammonia after bonds disturbance due to H_2O. Reaction of water molecules also causes the formation of hydriodic acid HI, which has strong corrosive influences and distorts the aluminum electrode [5].

3 MANUFACTURING OF REFERENCE PEROVSKITE PHOTOVOLTAIC CELLS

In order to analyze the effects causing degradation of perovskite photovoltaic cells, it was necessary to create the appropriate reference samples. Four samples of perovskite solar cells were manufactured with the structure ITO/PEDOT:PSS/$CH_3NH_3PbI_{3-2}Cl_2$/$PC_{61}BM$/Al electrode and four samples with the structure ITO/PEDOT:PSS/ $CH_3NH_3PbI_3$ /$PC_{61}BM$/Al. Sample 1 and its preparation can be seen in Figure 3.

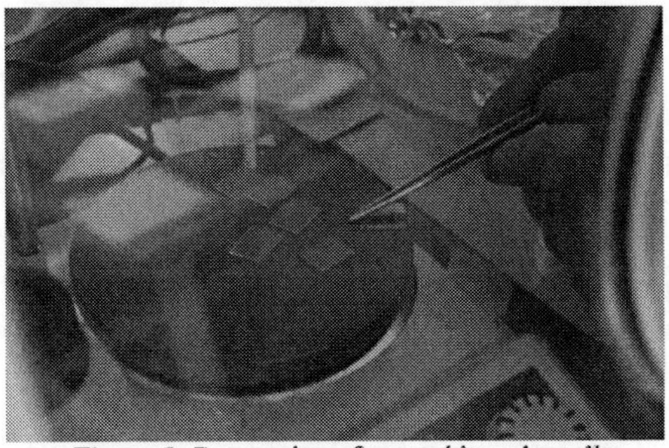

Figure 3: Preparation of perovskite solar cells

The encapsulation through the cover glass and epoxy resin was done to increase the resistiviy of perovskite samples against external influences. Durability of solar cells wasn´t considerably extended and after 20 hours the samples lost significant part of the performance.

The conversion efficiency of perovskite solar cells with the CH3NH3PbI3 active layer was during the initial measurement equal to about 2 %. The highest conversion efficiency have reached samples no.1. (Hmax = 2.46 %); no.2 (ηMAX = 2.85%) and no.4 (ηMAX = 2.56%). Measurements were done on the solar simulator with 1000 W / m2 light source. Current-voltage characteristic were measured in a six hours intervals when the samples were left in the atmospheric environment. The difference in a current voltage characteristic after couple of hours was apparent – decrease of perfomance by 50 % of the initial value. Despite the encapsulation of samples was disrupted the perovskite structure probably due to moisture effect. Moisture contamination into the active layer of cell was also possible during the manufacturing process. Almost at the entire series of the $CH_3NH_3PbI_3$ / $PC_{61}BM$ perovskite solar cell have occurred complete loss of photoelectric effect after 24 hours of operation. Only samples no.3. and no.4 have operated at about 20% of previous performance.

For the perovskite photovoltaic cells with an active layer $CH_3NH_3PbI_{3-2}Cl_2$ were functional only samples no. 1., no. 2., and no. 4. On the other hand the conversion efficiency has reached higher valuse – an average of 5.1 %. Conversion efficiency of the best sample 1 with the active perovskite layer $CH_3NH_3PbI_{3-2}Cl_2$ has reached approximately η_{MAX} = 7. 22 %. With a measurement during illumination the efficiency has increased to a certain value, then again has dropped. The reason is probably in the temperature of the sample during measurements (samples were rapidly heated). Samples were also measured consecutively in short intervals. The gradual degradation of the material, well-known phenomenon of perovskite solar cells, has occurred later. The conversion efficiency of the cells has decreased significantly after 20 hours of its function from η_{MAX} = 7.22 % to η_{MAX} = 5.48 % see Figure 4.

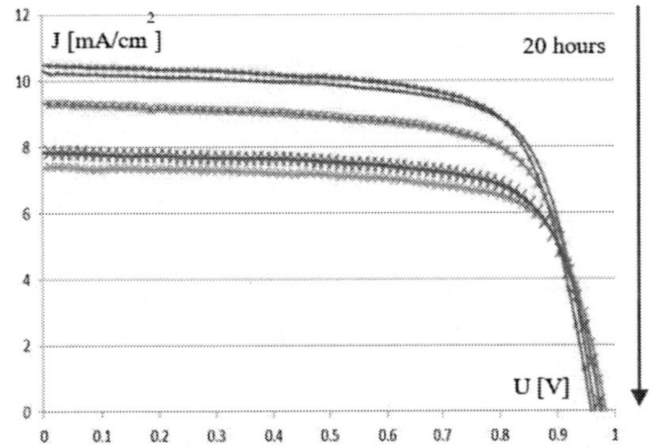

Figure 4: Decrease of perovskite PV cell conversion efficiency after 20 hours

The higher stability of the active layer $CH_3NH_3PbI_{3-2}Cl_2$ is apparently because of the chlorine addition to the structure. Chlorine has a smaller atomic radius and thus apparently creates compact and stable perovskite solar cell by substituting the iodine atoms. Along with the higher efficiency of conversion of solar energy and the stability is better $CH_3NH_3PbI_{3-2}Cl_2$ active layer than $CH_3NH_3PbI_3$.

4 INFLUENCE OF MOISTURE TO AN ACTIVE PEROVSKITE LAYER

Degradation and changes in the parameters of the perovskite sample 1 was most influenced by the moisture, which despite encapsulation may partially penetrate into the layers. An interesting phenomenon has ocurred related to the gradual increasing and lowering of conversion efficiency. The grains have unified leading to the enlargement of its volume. This may reduce the number of grain boundaries which forms a barrier to the current, which could explain the increasing and decreasing of the efficiency. AFM microscope analysis was done because of observation of ongoing processes – Figure 5. This phenomenon will be given additional measurements on perovskite photovoltaic cells.

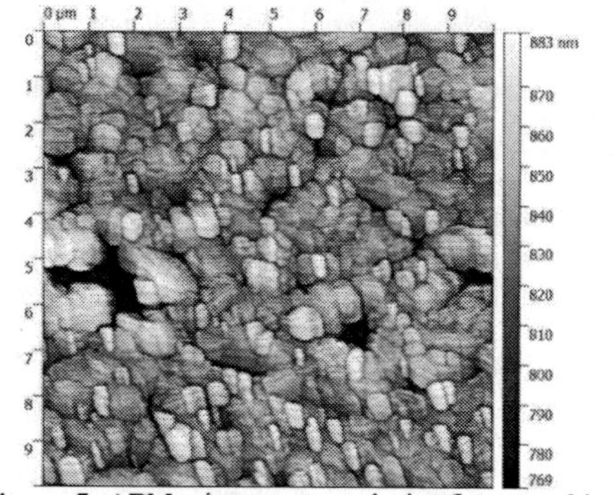

Figure 5: AFM microscopy analysis of a perovskite photovoltaic cell active layer

Exposure of perovskite solar cells to UV radiation has started degradation processes as well. UV rays with moisture disrupted binding of the active perovskite layers, which results into a bleaching. UV radiation also affected halogen atoms and disrupted the stability. Pb2+-X- created an inorganic matrix, which absorbed and emited light. Local irradiation of Pb-X bonds can generate free halide radicals which can decomposed pervoskite to dihalogens I_2, Br_2, Cl_2 (1).

To confirm the assumption of moisture and UV radiation influence was one sample of active perovskite layer exposed to an inert atmosphere and the second one exposed to external influences of moisture and irradiation. The sample exposed to an external influences has significantly bleached, see Figure 6.

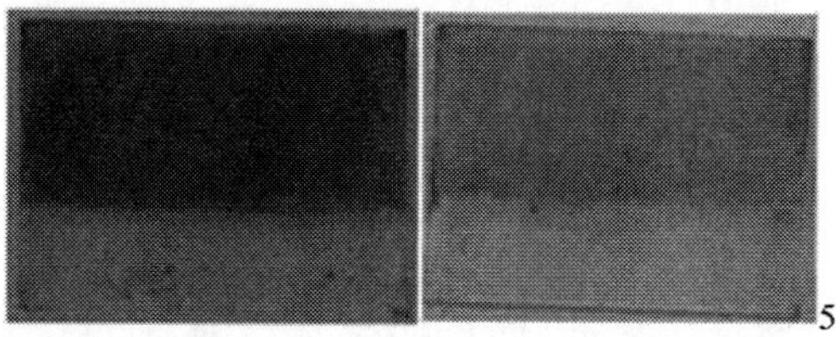
Figure 6. Sample with an active per. layer $CH_3NH_3PbI_{3\text{-}2}Cl_2$ under inert atmosphere, b) sample with an active per. layer $CH_3NH_3PbI_{3\text{-}2}Cl2$ exposed to moisture and UV rays.

5 SUMMARY/CONCLUSION

To observe the degradation of the perovskite solar cells were created four samples of $CH_3NH_3PbI_3$ cells with the highest achieved conversion efficiency η_{MAX} = 2. 85 % and four samples of $CH_3NH_3PbI_{3\text{-}2}Cl_2$ cells with highest achieved conversion efficiency of sample no.1 with η_{MAX} = 7. 22 %. Perovskite solar cell (sample 1) began to degrade after 20 hours, what is its huge disadvantage. The conversion efficiency of measured structures declined significantly within a hours to η_{MAX} = 5.48 %.

An interesting phenomenon has occurred when were samples exposed to the moisture. Conversion efficiency has increased and after some time has decreased. The grains have unified which led to the enlargement of its volume. This may reduce the number of grain boundaries which forms a barrier to the current. Also was observed impact of UV rays and moisture which led to bleaching of an active perovskite layer.

The higher stability of the CH3NH3PbI3-2Cl2 perovskite solar cells were achieved due to chlorine addition into the structure. Chlorine has created compact and stable perovskite solar cell structure by substituting the iodine atoms. Resistance to degradation was at $CH_3NH_3PbI_{3\text{-}2}Cl_2$ structure 3 times greater than in the $CH_3NH_3PbI_3$ structure.

Degradation is currently one of the largest problems associated with the development of perovskite solar cells. Subsequent research will focus mainly on preventing of the degradation by applying the appropriate interlayers between perovskite and its metal contacts. The Cr_2O_3 interlayers will be studied in the context of resistance of the perovskite material against environmental influences (6). The Cr_2O_3 layer will be applied through magnetron sputtering equipment in the Department of Electrical and Electronic Technology at Brno University of Technology.

5.1 References

[1] Kim, H.-S.; Lee, C.-R.; Im, J.-H.; Lee, K.-B.; Moehl, T.; Marchioro, A.; Moon, S.-J.; Humphry-Baker, R.; Yum, J.-H.; Moser, J. E.; Grätzel, M.; Park, N.-G. Nat. Sci. Rep. 2012, 2,591

[2] GRÄTZEL, Michael. *The light and shade of perovskite solar cells*. DOI: 10.1038/nmat4065. ISBN 10.1038/nmat4065. http://www.nature.com/doifinder/10.1038/nmat4065

[3] STRACHALA, D.; HYLSKÝ, J.; KADLEC, M. Organic- Inorganic Hybrid Solar Cells. In *Proceedings of the 22nd Conference STUDENT EEICT 2016*. Brno: Vysoké učení technické v Brně, Fakulta elektrotechniky a komunikačních, 2016. s. 630-634. ISBN: 978-80-214-5350- 0

[4] SONG, Dandan, Jun JI, Yaoyao LI, et al. 2016. Degradation of organometallic perovskite solar cells induced by trap states: degradation and stability. Applied Physics Letters. 108(9), 093901-.DOI: 10.1063/1.4943019. ISBN10.1021/acsami.5b04490. ISSN 0003-6951. http://scitation.aip.org/content/aip/journal/apl/108/9/10.1063/1.4943019

[5] KALTENBRUNNER, Martin, Getachew ADAM, Eric Daniel GŁOWACKI, et al. *Flexible high power-per-weight perovskite solar cells with chromium oxide–metal contacts for improved stability in air*. DOI: 10.1038/NMAT4388. ISBN 10.1038/nmat4388. Dostupné také z: http://www.nature.com/doifinder/10.1038/nmat4388

5.2 Acknowledgement

This research work has been carried out in the Centre for Research and Utilization of Renewable Energy (CVVOZE). Authors gratefully acknowledge financial support from the Ministry of Education, Youth and Sports of the Czech Republic under NPU I programme (project No. LO1210).

SOLUTION-PROCESSED INVERTED ORGANIC SOLAR CELLS: TOWARDS FULLY SPRAY-COATED DEVICES

López-Vicente, Rodolfo [a]*, *Toledo, Carlos* [a], *Padilla, Javier* [b], *Urbina, Antonio* [a] *and Abad, Jose* [b]

[a] *Departamento de Electrónica, Tecnología de Computadoras y Proyectos, Universidad Politécnica de Cartagena, 30202 Cartagena, Spain*

[b] *Departamento de Física Aplicada, Universidad Politécnica de Cartagena, 30202 Cartagena, Spain*

* *Corresponding author: rodolfo.lopezvicente@gmail.com*

Abstract

This work evaluates the possibility of fabricating a full spray inverted organic solar cell. Our experimental approach consists on a sequential substitution of the spin coated or vacuum evaporated layers by spray coated ones. On each layer the different deposition parameters i.e. speed, pressure or distance of spray nozzle from the substrate is optimized, in terms of obtaining the best device performance and stability. The machine who controls the spray coating system is a common 3-axis CNC machine. Different kind of programs has been written for each different experiment. Preliminary results show that the performance parameters of the devices are comparable with those fabricated by spin coating under the same ambient conditions.

Keywords: manufacturing and processing; organic solar cell; ITO; spray coating

1. INTRODUCTION

Humankind is facing a big challenge: the energy problem where resources depletion, sustainable development and environmental impact are key factors [1]. Therefore, renewable energies and especially photovoltaic will play a fundamental role in solving this problem [2]. Among them organic solar cells (OSC) have a great potential due to a low-cost energy production and applications ranging from energy production, to power small devices or in the field of architecture with building-integrated photovoltaic cells (BIPV).

There are currently a wide variety of OSC manufacturing technologies, most of which are well-known printing techniques such as spin coating, inkjet coating, screen printing, doctor blading, or spray coating [3]. The latter is a very promising technique due to no-limitation in substrate dimension and shape, as well as low consumption of materials. The objective of this low-cost approach to OSC manufacturing is to provide energy at a price per kWh which is lower than other competing technologies.

2. OBJECTIVES AND EXPERIMENTAL METHODOLOGY

This work evaluates the possibility of utilizing the spray-coating technique in all the layers of inverted organic solar cells. In the scientific literature there are a few works which successfully demonstrate the application of spray deposition in the different layers of an organic solar cell [4–10]. Inverted architecture is normally used in OSCs due to the better stability as well as to avoid vacuum evaporation of aluminium. The most common structure in this kind of devices is a silver back contact electrode to collect holes, a photoactive layer (P3HT: PCBM) and an electron transport layer (ZnO) on ITO to collect electrons.

Based on previously optimized OSCs fabricated in ambient conditions [11], our experimental approach consists on a sequential substitution of the spin coated or vacuum evaporated layers for spray coated ones. Firstly, only the photoactive layer is spray coated; secondly, the electron transport layer (ZnO) is also spray deposited. Thirdly spray coated Ag back electrode is added, and finally the possibility to replace the ITO transparent electrode by solution processable materials as Au nanoparticles, conducting polymers, graphene or active carbon, among others, will be explored. On each layer the different deposition parameters i.e. speed, pressure or distance of spray nozzle from the substrate is optimized, in terms of obtaining the best device performance and stability. Also temperature of substrate and ink could be controlled A spray coating station has been specifically designed and built for the automated fabrication of inverted OSCs at laboratory scale; specific software for the automation of all the process has been developed and the system has been used to manufacture functional OSCs.

3. RESULTS

3.1. Airbrush scans

Preliminary results for the characterization of inverted OSCs (Ag/P3HT: PCBM/ZnO/ITO) fabricated in our laboratory have been obtained. The sample preparation procedure is not fully spray coated since ZnO layer was spin coated onto ITO/glass substrate while the Ag back electrode was vacuum evaporated. The photoactive layer (P3HT: PCBM) was deposited by spray, varying the number of airbrush scans and the N_2 pressure.

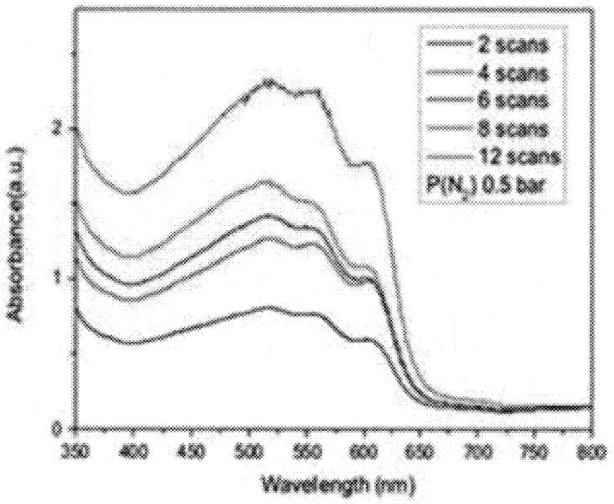

Figure 1. UV-visible absorbance of the P3HT: PCBM photoactive layer on glass as function of the number of airbrush scans.

Figure 1 shows the UV-visible absorbance of P3HT: PCBM photoactive layers as a function of the number of airbrush scans. Absorbance values were correlated with scans number and used to assess the reproducibility of the airbrush technique. These results are very promising considering their fabrication in ambient conditions; however further optimization of the N_2 pressure, airbrush scans and distance between the airbrush and the substrate is needed.

3.2. Photoactive layer by spray coating

In this case 13 substrates with four cells each one were fabricated and only the photoactive layer was deposited by spray coating technique varying the number of scans. A period (1T) is two scans, one way and return. Figure 2 shows I-V curves for the best results.

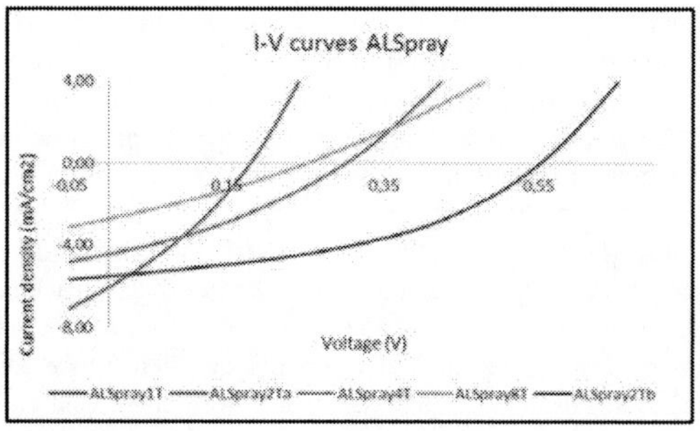

Figure 2. I-V curves of best cells with photoactive layer with different periods fabricated by spray coating technique.

I-V curves shows a power conversion efficiency obtained for the best cell was 1,27% with a fill factor of 43%. The results are shown in the table 1. These results are obtained in ambient conditions. The parameters are obtained by fitting a one-diode model for the fitting of the I-V characteristic of the solar cells under 1 Sun illumination (solar simulator AM1.5G, room temperature).

Name	Scans	Voc (V)	Isc (mA/cm²)	PCE (%)	FF (%)
ALSpray 1T	2	0,2	6,04	0,29	24
ALSpray 2Ta	4	0,27	4,43	0,42	35
ALSpray 4T	8	0,33	3,31	0,34	31
ALSpray 8T	16	0,27	2,72	0,2	27
ALSpray 2Tb	4	0,53	5,51	1,27	43

Table 1. Number of scans, Voc, Isc, PCE and FF for active layer by spray coating.

The distance of the nozzle was the same for all substrates, 11 cm. All devices were fabricated under the same conditions: N_2 pressure of 0.5 bar and 11 cm distance of the nozzle. ZnO layer and Ag electrode were deposited by spin coating and vacuum evaporation technique respectively.

3.3. ZnO and photoactive layer by spray coating

The same conditions were used in this case for both layers, ZnO and photoactive layer. 10 substrates were fabricated this time. Fig 3 shows I-V curves of the best devices.

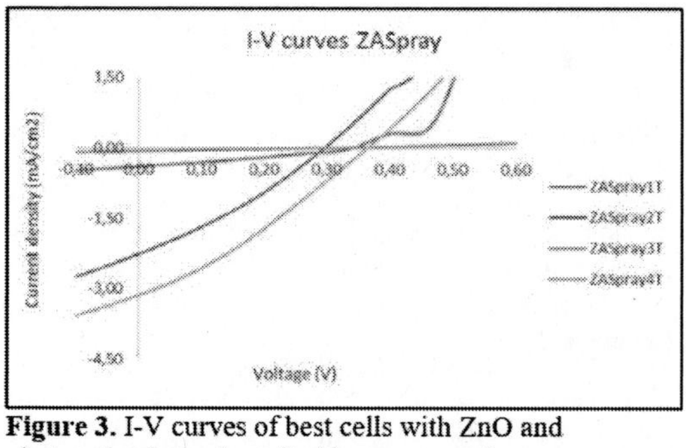

Figure 3. I-V curves of best cells with ZnO and photoactive layer deposited by spray coating.

Figure 3 shows the I-V curves of four best cells with ZnO and photoactive layer by spray coating. Despite the low current obtained, it shows a good filling factor and a moderate efficiency (table 4). The best PCE was obtained by 4T (8 scans) in this experiment.

Name	Scans	Voc (V)	Isc (mA/cm²)	PCE (%)	FF (%)
ZASpray 1T	2	0,27	2,27	0,2	32
ZASpray 2T	4	0,33	0,39	0,04	29
ZASpray 3T	6	0,27	0,06	0	27
ZASpray 4T	8	0,33	3,15	0,35	33

Table 2. Scans, Voc, Isc, PCE and fill factor for best cells with ZnO and active layer by spray coating.

3.4. Stability of photoactive layer

Two devices were manufactured with photoactive layer by spray coating technique in order to compare them against a conventional device manufactured by spin coating. The devices were subjected for 3 days to solar irradiance at standard conditions (1000W/m^2, AM1.5G class AAA) in a solar simulator (ABET Technologies Sun2000).

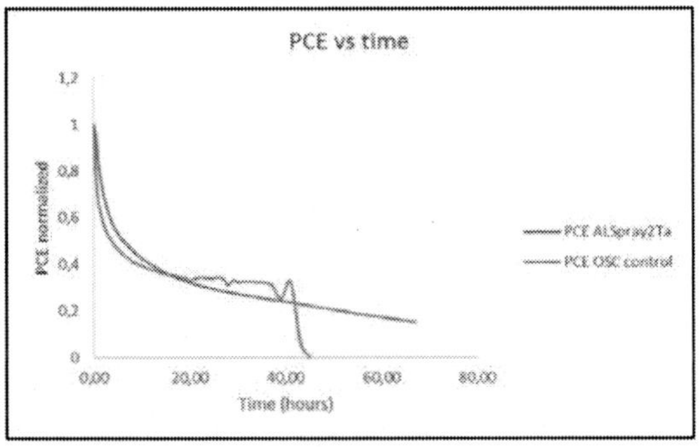

Figure 4. Time evolution of normalized power conversion efficiency of both cells.

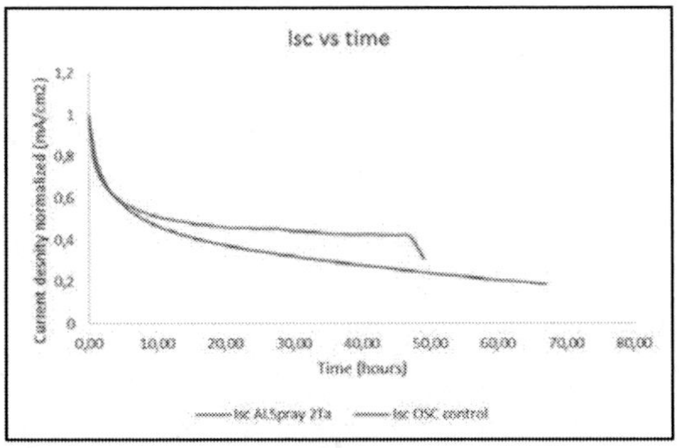

Figure 5. Time evolution of normalized short circuit current density for spray coating and spin coating cells.

Power conversion efficiency (PCE) of best cells is represented as a function of time in figure 4. The cell with photoactive layer deposited by spray coating presents a longer life than control cell fabricated by spin coating still reaching at 67 hours a 0,1% of PCE. On the other hand, in figure 5 the short circuit current is shown as a function of time. This parameter is a better indicator of stability than PCE due the efficiency depends of parameters like V_{oc} and point of maximum power.

4. CONCLUSIONS

Organic solar cells were fabricated using spray-coating technique for the casting of the active layer, which was successfully deposited and optimized by spray coating for 2T periods (4 scans) with promising results reaching at maximum of 1,27% of PCE and 43% of FF.

We deposited ZnO with different scans and best results was obtained with 4T periods (8 scans). The obtained fill factor for all the cells were moderate and efficiencies were modest. Nevertheless, the technique has demonstrated its capability of manufacturing in air. The manufacture by this technique has to be optimized because low currents and low PCE were achieved.

A lower degradation of active layer by spray coating is observed in long term although the stability of control cell is better in short term.

The performance parameters of spray coating devices are comparable with those fabricated by spin coating under the same ambient conditions. The following steps will be to replace the vacuum evaporation and spin coating techniques in fully spray coated devices, since operative "proof of concept" devices have been previously built by an all-spray coated solution processable method.

5. REFERENCES

[1] World Energy Outlook. (International Energy Agency - IEA, 2015).

[2] Global PV Market Report 2016-2020. (The PV Market Alliance - 2016).

[3] F. C. Krebs. Sol. Energy Mater. Sol. Cells 93 (2009) 394–412.

[4] J-W. Kanga, Y-J. Kang, S. Jung, M. Song, D-G. Kim, C. S. Kim, S. H. Kim. Sol. Energy Mater. Sol. Cells 103 (2012) 76-79.

[5] J.-W. Kang, Y.-J. Kang, S. Jung, D.S. You, M. Song, C.S. Kim, D.-G. Kim,J.-K. Kim, S.H. Kim. Org. Electron. 13 (2012) 2940.

[6] J.G. Tait, B.P. Rand, P. Heremans. Org. Electron. 14 (2013) 1002.

[7] T. Wang, N.W. Sacarratt, H. Yi, A.D.F. Dunbar, A.J. Pearson, D.C.Watters, T.S. Glen, A.C. Brook, J. Kingsley,A.R. Buckely, M.W.A. Skoda,A.M. Donald, R.A.L. Jones, A. Iraqi, D.G. Lidzey. Adv. Energy Mater. 3 (2013) 505.

[8] J. Griffin, A.J. Pearson, N.W. Scarratt, T. Wang, D.G. Lidzey, A.R.Buckly. Org. Electron. 15 (2014) 692.

[9] W. Nie, R.C. Coffin, J. Liu, Y. Li, E.D. Peterson, C.M. MacNeill, R.E.Noftle, D.L. Carroll. Appl. Phys. Lett. 100 (2012) 083301.

[10] C. Girotto, D. Moia, B.P. Rand, P. Heremans. Adv. Funct. Mater. 21 (2011) 64.

[11] L. Córcoles, J. Abad, J. Padilla, A. Urbina. Sol. Energy Mater. Sol. Cells 141 (2015) 423-428.

Acknowledgements: F. Séneca (Murcia, 19882/GERM/15) & MINECO (Spain, ENE2016-79282-C5-5-R).

CONDUCTIVE INKS WITH EPOXY RESIN BASED VEHICLES FOR PEROVSKITE SCREEN PRINTING METALLIZATION AS A VIABLE AND LOW-COST ALTERNATIVE TO THERMAL EVAPORATION

C. Montes[1,2], L. Ocaña[1], C. Quinto[1], S. González-Pérez[2], B. González-Díaz[2], M. Friend[1] & M. Cendagorta[1].
[1]Instituto Tecnológico y de Energías Renovables, S. A. (ITER)
Pol. Industrial de Granadilla, s/n.
E 38600 Granadilla de Abona, Spain.
Ph. +34 922 747 700
Fax +34 922 747 701
E-mail cmontes@iter.es
[2]Departamento de Ingeniería Industrial.
Universidad de La Laguna.
Avd. Astrofísico Francisco Sánchez, s/n.
38206 S/C de Tenerife. Spain.
Ph. +34 922 316 502, EXT 6252
E-mail bgdiaz@ull.edu.es

ABSTRACT: In this work, new advances in the research conducted by the solar cells laboratory (SiCelLab) from the Instituto Tecnológico y de Energías Renovables (Canary Islands, Spain) in order to develop inks for screen printing metallization, based on the use of epoxy resin and anhydrous solvent vehicles are presented. This approach has been designed to be applied on thin film perovskite substrates. The methodology applied makes use of an overhead stirrer for mixing and a three roll mill for homogenizing the ink components. From the developed inks, samples have been printed with average volume resistivities measured as low as 3.31 Ω·cm. Preliminary results suggest that a further reduction on these values is possible by way of including a co-binder species that allows reducing the amount of epoxy resin within the mix.
Keywords: Contact, Characterization, Metallization, Screen Printing, Perovskites.

1 INTRODUCTION

Screen printing is a mature, cost-effective and easily scalable metallization technique, widely used nowadays for photovoltaic (PV) cell manufacturing purposes. Thus, from the point of view of its industrial application, due to its relative simplicity and low energy consumption, this technique could become a better alternative to the thermal evaporation processes that are actually being in use for the metallization of perovskite solar cells fabricated on laboratory scale. With this purpose in mind, the Instituto Tecnológico y de Energías Renovables (ITER), has been working on screen printing solutions for perovskite solar cell metallization since 2015, by initially using different commercial pastes and inks. Although these tests were carried out with perovskite-friendly fabrication methods, the aqueous nature of the vehicles used in their formulation, produced a strong degradation in the finished devices [1].

In order to overcome this problem, the laboratory started developing several conductive ink formulas composed by epoxy resins and anhydrous solvents, as vehicles, and different kinds of graphite powders, as the conductive material. Optimal proportions of these three components have been largely studied in order to obtain inks suitable for printing purposes. Different solvents were also analyzed with the aim to increase the quantity of graphite to the mixture and, subsequently, improve the conductivity of the printed contacts [2].

The present paper summarizes the more recent results obtained in this ongoing research, which has been focused on preventing the degradation of perovskite substrates by using non-polar solvents, as well as in extending the kind of conductive powders used, in order to include metallic ones.

2 CONDUCTIVE INK PREPARATION

All the experiments were carried out in an ISO 7 clean room environment. The inks were prepared in three mixing stages: In the first stage, different kinds of conductive powders [3-6] were mixed with the epoxy resin [7] by using an overhead stirrer [8] equipped with a kneading stainless steel beater, until obtaining a conveniently distributed mixture. In the second stage, a three roll mill [9] was used by giving as many passes as it was deemed necessary in order to homogenize the mixture. Finally, in the third stage, the overhead stirrer was used again, albeit this time with a steel beater for lower density blends, in order to add to the mix the hardener [7] and the toluene anhydrous [10].

Table I: Conductive powders used for the performed ink trials.

Conductive powder	Particle maximum size (μm)	Purity (%)
Graphite 1 [11]	<20	n/a
Graphite 2 [12]	<50	99.9
Graphite 3 [13]	<60	99
Graphite 3 [13]/ Ferrocene [14]	<60 / n/a	99 / 98
Graphite 4 [15]	<75	99.95
Aluminium 1 [16]	<75	99

Aluminium 2 [17]	n/a	93
Nickel [18]	<50	99.7
Copper [19]	<45	99.7

Figure 1: Ink being mixed in the overhead stirrer (left) and homogenized at the three roll mill (right).

3 CONTACT PRINTING AND CURING

It is important to indicate that, because of the nature of the vehicle chosen for producing these inks, as soon as the hardener was added to the mixtures, the polymerization reaction was started. Therefore, the inks so prepared had to be not only kept under continuous stirring but also had to be printed at the same time, in order to determine their optimum application moment. Thus, the elapsed time between each impression chosen in an ink trial was set to be of approximately 10 minutes, being the first print performed 10 minutes after adding the hardener, and the last one about 60 minutes later.

The printing itself were performed using screens [20] capable to handle inks with up to 67 micron particle sizes. The screens were mounted on a screen printer [21], provided with a 70 durometer squeegee blade and a dial indicator for fine adjustment of the screen and the squeegee rod to the tool plate, to finely control the pressure exerted and the thickness of the inks and pastes deposited with an accuracy of micrometers.

The contacts were printed over glass substrates of 175mm x 175mm x 2mm, in order to carry out their electrical characterization. In addition, the contacts were also printed on glass substrates of 25mm x 25mm x 2mm with perovskite layer pre-deposited on them, in order to evaluate their effects on these layers.

Figure 2: Contacts being deposited by screen printing.

Finally, the curing process was carried out during 24 hours and at room temperature in the air for the contacts printed on the bigger substrates, and in a vacuum desiccator [22] for the ones printed over the perovskite layered ones, in order to preserve their integrity [23].

4 CHARACTERIZATION

The printed contacts were characterized following the Standard F 1896 Test Method for Determining the Electrical Resistivity of a Printed Conductive Material [24], according to which, the sheet resistance (R_s) can be derived from printed contacts that comply with a length (L) to width (W) geometric ratio of at least 50:1.

$$\mathbf{R_s = R \cdot W/L} \qquad (\Omega / \square)$$

Also, when the thickness (t) of the contact is known, it is possible to obtain its volume resistivity (ρ) by the following formula:

$$\boldsymbol{\rho = R_s \cdot t} \qquad (\Omega \cdot \mathrm{cm})$$

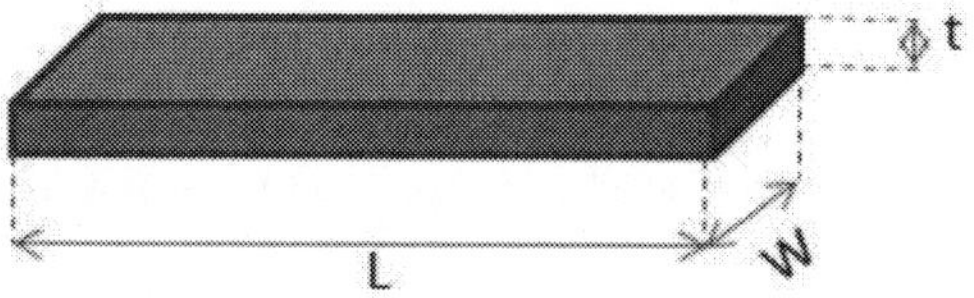

Figure 3: Geometry of a printed contact.

Therefore, screens were designed and produced so as to provide printing patterns that comply with the aforementioned Standard, while being conveniently distributed over the printing area in order to also allow uniformity studies.

The actual measurements were carried out using a semiconductor characterization system [25], configured to operate as an ohmmeter.

Finally, visual inspections of the all the printed samples were carried out by using a digital microscope [26], which allowed us to evaluate the morphology, adhesion level and, for the ones printed over perovskite layers, the degree of interaction between contact and perovskite.

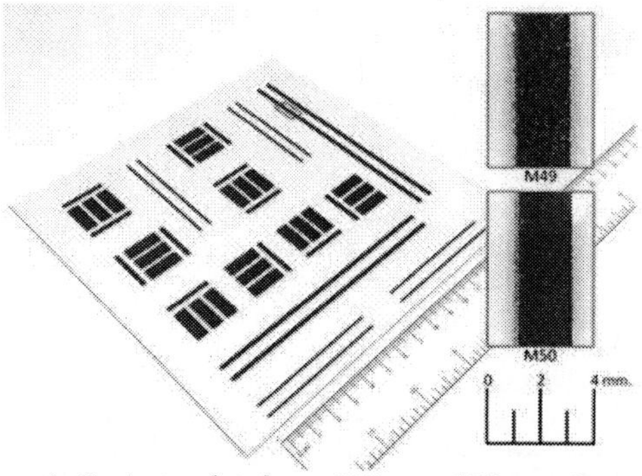

Figure 4: Contacts printed on a 175mm x 175mm x 2mm glass (left), detail of microscopic images of the contact area for different inks (right).

Figure 5: Contacts printed on a 25mm x 25mm x 2mm perovskite covered glass (left), detail of microscopic images with different zoom degrees (right).

5 RESULTS AND DISCUSSION

The methodology followed for producing the ink recipes was as follows: For each of the conductive powders used, a preliminary recipe of ink was sought by way of trial and error until it allowed producing printed contacts with a good appearance and level of adhesion (to the substrate). The ink that complied with both guidelines usually resulted in contacts that had such a high resistance as to be virtually indeterminate. Then, while keeping the proportions of the vehicle used constant, larger amounts of conductive powders were added iteratively, looking for a reduction on the resistances measured in the printed contacts. This process was continued up to reach a state of saturation of powders in the ink, from which it was virtually impossible to obtain a proper impression.

Taking into account that, as explained before, for each ink 6 prints were made, separated by 10 minute intervals, and considering that, in order to impress a proper pattern over the substrates, for many of these prints it was required a fine adjustment on the printing parameters, and that, finally, for each impression, the used screens allowed to produce a set of 12 contacts in total, it was decided to pursue the aforementioned state of saturation by means of looking for the minimum average volume resistivity obtained of all the contacts deposited by each print for any given ink.

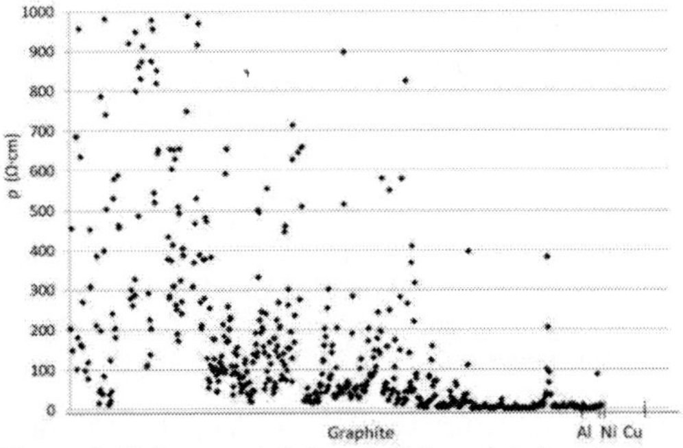

Figure 6: Volume resistivity of all the printed contacts, arranged chronologically and grouped by type of conductive powder.

However, as it can be seen in Figure 6, the volume resistivity values obtained for the printed contacts had a high degree of dispersion, so much so that it became difficult to elucidate whether a variation in a given ink formulation resulted or not in an improvement for said ink. Therefore, the following method was devised as a way of filtering these values and to reduce their dispersion:

Let x1, x2, x3, ..., x12 be the set of volume resistivity values measured for the contacts 1, 2, 3, ..., 12 in a given print. By arranging all the values from the smallest to greatest, it is possible to obtain their median value, $\mathbf{x_m}$.

Let $\mathbf{x_{fi}}$ be defined as:

$$\mathbf{x_{fi} = x_i / x_m}$$

Where i=1, 2, 3, ..., 12 (the number of contacts per print).

A filter for the set of data can be established so as to, from all the possible values, only accept the ones with an x_{fi} below any arbitrary number. If the chosen value for x_{fi} is 10, this is equivalent to filter out all the measured volume resistivity values which resulted to be bigger than an order of magnitude from the set's median value (x_m).

Finally, in order not to lose how objectively representative were the filtered average volume resistivity values ($\bar{x}$), yet another parameter was required: For any given print, let $\mathbf{C_{xf}}$ be defined as the amount of values which were not filtered (that is, which have an $\mathbf{x_{fi}}$ less than or equal to 10). Then, the representativeness or R, of the calculated average volume resistivity ($\bar{x}$) can be defined as:

$$\mathbf{R = C_{xf} / 12}$$

This parameter can be expressed as a percentage (%).

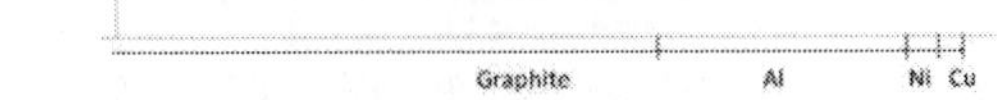

Figure 7: Filtered average volume resistivity ($\bar{x}$) of each print, the size of the balls indicates the representativeness (R) of the plotted value.

Table II shows the best filtered average volume resistivities values obtained from the contacts printed for each conductive powder, together with their representativeness and their corresponding ink recipe expressed in their mass fraction.

Table II: Summary with the best results obtained with their corresponding ink recipes. There is no value of $\bar{x}$ for copper because the obtained one was so high as to be considered off-scale.

Powder	Print time (min.)	$\bar{x}$ (Ω·cm)	R (%)	Recipe (wt%)		
				Conductor	Resin	Solvent
Graphite 2 [12]	50	3.31	100	36.2	54.1	9.7
Aluminium 1[16]	50	3.40	58.3	70.4	25.6	4.0
Nickel [18]	50	6.26	8.3	68.4	26.9	4.7
Copper [19]	30	-	-	78.7	17.8	3.5

Looking at the results summarized in Table II, it is clear that more effort is required in order to improve the quality of the contacts printed with the ink recipes made with aluminium, nickel and copper (particularly for the latter case). However, the results obtained seem to indicate that, contrary to expectations, by simply replacing graphite powders with metallic ones in the ink recipes, the resulting printed contacts do not have lower volume resistivity, despite of having used powders from materials with considerably higher bulk conductivities and densities.

We believe that this is due to an excess of epoxy resin in the formulation of the inks. In order to test such hypothesis, the printed substrates produced with inks that have the heavier conductive powders in their formulations, that is, the ones of nickel and copper, were annealed at 300 ° C for 30 minutes, and the resistivity values of the contacts were obtained again. The results shown in Table III seem to confirm that, possibly due to the evaporation of part of the existing epoxy resin in said contacts, not only did their volume resistivity decreased, but also the representativeness of the obtained values increased.

Table III: Results obtained after annealing the contacts at 300 °C for 30 minutes.

Powder	Print time (min.)	$\bar{x}$ (Ω·cm)	R (%)	Recipe (wt%)		
				Conductor	Resin	Solvent
Nickel [18]	50	0.02	100	68.4	26.9	4.7
Copper [19]	30	$1.02x10^{-3}$	91.7	78.70	17.8	3.5

Unfortunately, annealing substrates with perovskite thin films at 300 °C will result in the loss of its properties as a light-harvesting material [27]. Therefore, it is necessary to look for alternative strategies, in order to reduce the amount of epoxy resin required in the ink formulations by adding some kind of material that acts as a co-binder in the vehicle. Candidates for this purpose could be mineral paraffins such as Nujol and Uvasol [28] or solidifying substances such as phenanthrene or paraffin wax, which become liquid when heated up, so they can be properly homogenized with the conductive powders, and act as solid binders when cooled down [29].

6 ACKNOWLEDGEMENTS

The authors would like to thank Luisa Quintanilla, Sales Manager of JLQ ISMAF, S.L., for having had the courtesy of having ceded much of the graphite used throughout this project.

7 REFERENCES

[1] C. Quinto, A. Linares, E. Llarena, C. Montes, O. González, D. Molina, A. Pío, L. Ocaña, M. Friend and M. Cendagorta, "Screen printing for perovkite solar cells metallization", 31st European Photovoltaic Solar Energy Conference and Exhibition, pp. 1144-1148, 14-18 September 2015, Hamburg, Germany.

[2] C. Montes, A. Linares, E. Llarena, O. González, D. Molina, A. Pío, L. Ocaña, C. Quinto, M. Friend and M. Cendagorta, "Conductive inks with epoxy resin based vehicles for perovskite screen printing metallization", 32nd European Photovoltaic Solar Energy Conference and Exhibition, pp. 1303-1309, 21-24 June 2016, Munich, Germany.

[3] Graphite powder (282863), <20 µm, synthetic from SIGMA-ALDRICH.

[4] Graphite powder SGA 20 99,9 from M. Graphite Kropfmühl GmbH Langheinrichstraße 1, 94051 Hauzenberg, Germany.

[5] Graphite powder SC 20 OS from Graphite Kropfmühl GmbH Langheinrichstraße 1, 94051 Hauzenberg Germany.

[6] Ferrocene 98% (F408-100G). Synonym: Bis(cyclopentadienyl)iron, Di(cyclopentadienyl)iron. CAS Number: 102-54-5. From Sigma-Aldrich

[7] Epoxy resin SR 8500 & hardener SZ 8525 from Sicomin Epoxy System, 31 avenue de la Lardière, 13220 Chateauneuf les Martigues, France.

[8] Overhead stirrer Hei-TORQUE 400 made by Heidolph.

[9] ESSG65 Laboratory Three roll mill, made by Shanghai Espread Industry Co. Ltd.

[10] Toluene puriss (32249). p.a., ACS reagent, reag. ISO, reag. Ph. Eur., ≥99.7% (GC), from SIGMA-ALDRICH.

[11] Graphite powder 282863 Sigma Aldrich. <20 µm, synthetic. 3050 Spruce Street, Saint Louis, MO 63103, USA Website: www.sigmaaldrich.com Email USA: techserv@sial.com Outside USA: eurtechserv@sial.com

[12] Graphite powder SGA 20 OS. Graphite Kropfmühl GmbH Langheinrichstraße 1, 94051 Hauzenberg. www.gk-graphite.com. Acquired via JLQ ISMAF, polígono torrelarragoiti Pabillion 6 F 48170 - Zamudio - Vizcaya. www.ismaf.com

[13] Graphite powder SC20OS. Graphite Kropfmühl GmbH Langheinrichstraße 1, 94051 Hauzenberg. www.gk-graphite.com. Acquired via JLQ ISMAF, polígono torrelarragoiti Pabillion 6 F 48170 - Zamudio - Vizcaya. www.ismaf.com

[14] Ferrocene F408 98%. Sigma Aldrich. 3050 Spruce Street, Saint Louis, MO 63103, USA Website: www.sigmaaldrich.com Email USA: techserv@sial.com Outside USA: eurtechserv@sial.com

[15] Graphite powder FP 99,5. Graphite Kropfmühl GmbH Langheinrichstraße 1, 94051 Hauzenberg. www.gk-graphite.com. Acquired via JLQ ISMAF, polígono torrelarragoiti Pabillion 6 F 48170 - Zamudio - Vizcaya. www.ismaf.com

[16] Aluminium (20093), 99%, powder, 200 mesh, ACROS Organics™

[17] Aluminium (11010) powder fine, >93%. Sigma Aldrich. 3050 Spruce Street, Saint Louis, MO 63103, USA Website: www.sigmaaldrich.com Email USA:

techserv@sial.com Outside USA: eurtechserv@sial.com

[18] Nickel (266981) powder. <50μm, 99.7% trace metals basis. Sigma Aldrich. 3050 Spruce Street, Saint Louis, MO 63103, USA Website: www.sigmaaldrich.com Email USA: techserv@sial.com Outside USA: eurtechserv@sial.com

[19] Copper (357456) powder (dendritic). <45μm, 99.7% trace metals basis. Sigma Aldrich. 3050 Spruce Street, Saint Louis, MO 63103, USA Website: www.sigmaaldrich.com Email USA: techserv@sial.com Outside USA: eurtechserv@sial.com

[20] Stainless steel fabric, 200 mesh, 0.040 D-Ø x 22.5 high tension screens made by SEFAR MAISSA, S.A.U. C/ Alcorcon, 13, 28850 Torrejon Ardoz (Madrid).

[21] MSP-485, Precision Screen Printer. Hary Manufacturing, Inc. Precision Screen Printers and automation equipment (HMI), 24 Cokesbury Road, Lebanon, NJ 08833 USA

[22] Vacuum desiccator model DE02-251-001 by Labbox.

[23] J.Yang, B.D. Siempelkamp, D. Liu, T.L. Kelly, Investigation of CH3NH3PbI3 degradation rates and mechanisms in controlled humidity environments using in situ techniques".. ACS Nano. 24 February 2015, Vol.9(2), pp.1955-1963.

[24] F 1896 – 98 (Preaproved 2004). Test Method for Determining the Electrical Resistivity of a Printed Conductive Material. Test method © by the American Society for Testing Materials (ASTM) International, 100 Barr Harbor Drive, PO Box C700, West Conshohocken, PA 19428-2959, United States.

[25] Semiconductor Characterization System model SCS-4200 by Keithley.

[26] Digital Microscope U500X, focus range 15mm - 40mm, range 50X-500X by Unimake.

[27] Xiao, M., Huang, F., Huang, W., Dkhissi, Y., Zhu, Y., Etheridge, J., Gray-Weale, A., Bach, U., Cheng, Y.-B. and Spiccia, L. A Fast Deposition-Crystallization Procedure for Highly Efficient Lead Iodide Perovskite Thin-Film Solar Cells, Angew. Chem. Int. Ed. 2014 53: 9898–9903. doi:10.1002/anie.201405334

[28] M. Perween; D.B. Parmar, G.R. Bhadu & D.N. Srivastava, "Polymergraphite composite: a versatile use and throw plastic chip electrode" The Analyst, 2014, Vol.139(22), pp.5919-5926.

[29] I. Švancara, K. Vytřas, K. Kalcher, A. Walcarius & J. Wang, "Carbon Paste Electrodes in Facts, Numbers, and Notes: A Review on the Occasion of the 50- Years Jubilee of Carbon Paste in Electrochemistry and Electroanalysis" Electroanalysis, January 2009, Vol.21(1), pp.7-28

STRUCTURAL, OPTICAL AND ELECTRICAL PROPERTIES OF ZNO/PEROVSKITE/CUO/ AL SOLAR CELLS

H.Ait Dads[12*], L. Nkhaili[1], A. Kissani[1], H. El AAkib[12], S. Laalioui[1], M. Ait Ali[2], A. Outzourhit[1]
[1]Nanomaterials Laboratory for Energy and Environmental, Faculty of Sciences Semlalia, Cadi Ayyad University, Marrakech, Morocco.
[2]Laboratory of Coordination Chemistry and Catalysis, Faculty of Sciences Semlalia, Cadi Ayyad University, Marrakech, Morroco

***Corresponding author: h.aitdads@gmail.com**

ABSTRACT: Organolead-halide-perovskite-based solar cells have recently received significant attention due to their excellent photovoltaic performances and low cost. The general formula of the perovskite absorber is ABX3, where A stands for (R-NH3) and B is a metal (Pb or Sn) and X is a halide (Cl, I, or Br).
In this work we report on the fabrication of FTO/ ZnO/ $CH_3NH_3Pb(I_{1-x}Br_x)_3$ /CuO/Al planar heterojunction solar cells. The perovskite acts as an absorber, CuO and ZnO act as a hole conductor and electron transport layers respectively. The perovskite was produced by a two step process in ambient atmosphere. Zinc oxide thin films and copper oxide were fabricated by RF sputtering on indium- tin- oxide covered glass substrates. The structural and optical properties of the films were analyzed by X-ray diffraction, and ultraviolet-visible spectroscopy. The electrical junction properties were investigated by current–voltage (I–V) characteristics. The junction showed a photovoltaic behavior,

Keywords: metal oxide, heterojunction, RF sputtering, current-voltage (I-V), capacitance-conductance measurements.

1 INTRODUCTION

Over the past few years, thin film solar cells based on organic-inorganic lead trihalide perovskite absorbers have experienced an unexpected breakthrough and rapid progress in cell performance. To date, these perovskite thin film solar cells have exclusively employed organic hole conducting polymers which are are normally quite expensive due to complicated synthetic procedure or high- purity requirement[1] [2]. In a quest to explore new inorganic hole conducting materials .

For these perovskite based thin film photovoltaics, we have identified copper oxide as a possible alternative. Here in we demonstrate a device structure consicting of Zno as electron collector, lead halide perovskite as the light harvester and CuO as the Htm. The combination of $CH_3NH_3PB(I_{1-x}Br_x)_3$ with CuO show high PCE of 3,32%..

2 EXPERIMENTAL STUDY

2.1 Synthesis of perovskite $CH_3NH_3PB(I_{1-x}Br_x)_3$

For the perovskite layer, PbI2 and PbBr2 were dissolved in DMF, dropped onto the ZnO film and spin coated followed by annealing at 90 C for 30 min. In the second step, the cell was dipped into the dipping solution of MAI in isopropanol. Following the dipping step, the samples were annealed at 100°C for 30 min, [3]

2.2 Deposition of ZnO Layer

The ZnO thin films were prepared by reactive RF-sputtering. These films were deposited using a high purity Zn metal target with a diameter of 10cm in an argon oxygen gas mixture. Prior to the deposition, ITO substrates were ultrasonically cleaned in acetone and isopropyl alcohol followed by deionized water to remove any impurities and then rinsed with deionized water. The substrates were then mounted at a distance of 10cm above the Zn target (5N: purity, diameter: 10cm). The vacuum chamber was evacuated to a final pressure of 1.5□10-6 mbar. The target was presputtered for 10min, and the substrates were not intentionally heated during the deposition. The reactive sputtering was carried out in an Ar=O2 gas mixture with 30% of O2. The films were deposited using an RF power of 200W for 30min. These parameters were obtained from a previous optimization of the deposition conditions [4][5][6][7].

2.3. Preparation of CuO Films

Copper oxide thin films were prepared by reactive RF sputtering substrates at room temperature from a Cu target (5N purity, 10cm diameter). There active sputtering was carried out in Ar =O2 gas mixture with 30% of O2. The films were deposited at an RF power of 150W for 90min [8] [9][10][11].

The top Al circular electrode was fabricated by thermal evaporation at 2.10-5 mbar base pressure. The thickness of this Al layer is about 100 nm.

2.4. Characterization techniques

The structural properties were investigated by X-ray diffraction with an X-pert MPD diffractometer (using the CuKα radiation. The surface morphology of the as-prepared samples was studied by atomic force microscopy. Optical transmittance measurements were performed using a Shimadzu UV-PC spectrophotometer in the 200 - 3200 nm range.

The current-voltage (I-V) characteristics of FTO/ ZnO/ CH3NH3Pb(I1-xBrx)3 /CuO/Al were measured at room temperature using a Keithley LCZ3000 meter and Keithley 410 programmable ammeter and voltmeter

3 RESULTS And DISCUSSION

The XRD patterns of the Zno thin films present only one peak at 2O=34.1° (figure 1), corresponding to the (002) diffraction peak of the hexagonal wurtzite phase of ZnO. This suggests that the films have a preferred growth orientation in the (002) direction.

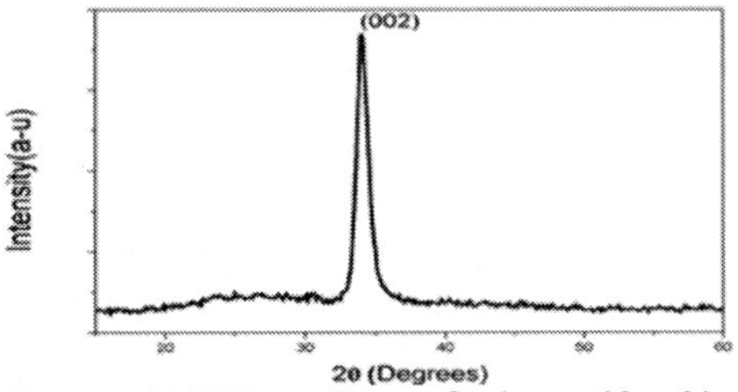

Figure 1: XRD- pattern of zinc oxide thin film deposited on indium tin oxide

Figure 2 shows the optical transmittance spectra of ZnO thin films in the wavelength range from 300 to 2000 nm. The films are highly transparent in the visible range of the electromagnetic spectrum with an average transmittance values up to 84 %.

The optical band gap of the films was determined from the variations of the absorption coefficient with photon energy and by assuming direct transitions. Details concerning the determination of thèse parameters from the optical transmission Spectrum were reported by [8]. The optical band gap Eg were found to be 3.28 eVThe value of the optical gap of our films is in agreement with reported values. [8].

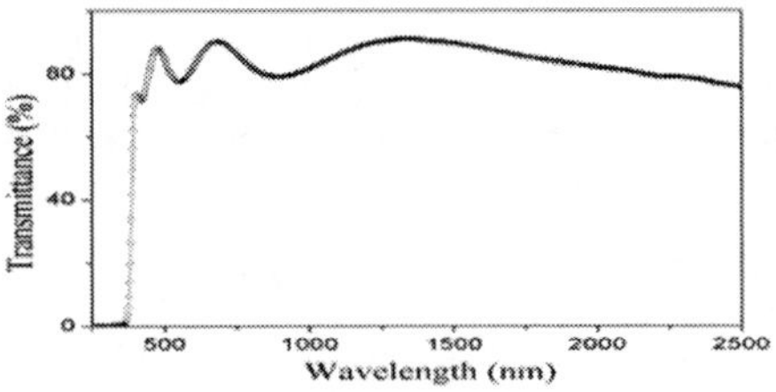

Figure 2: Optical transmission of zinc oxide deposited on indium tin oxide

GIXRD measurement was firstly conducted to confirm the film crystal structure and the diffraction pattern is shown in Figure 3. Besides a tiny signature peak at 12.60° corresponds to a low-level impurity of PbI2, [4] the intense peaks can be respectively assigned to 110, 112, 202, 220, 312, 240, 224 and 314 planes of a perovskite structure. Compared to CH3NH3PbI3, the diffraction peaks of $CH_3NH_3PB(I_{1-x}Br_x)_3$ are shifted uniformly to high angles, which can be caused by the replacement of I by Br. [4][12][13]

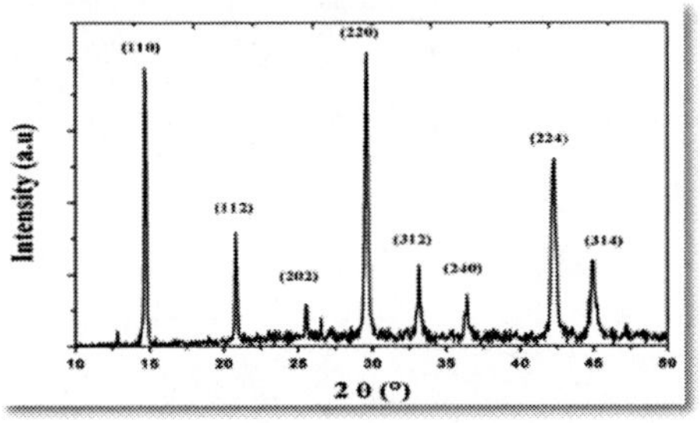

Figure 3: XRD- pattern of perovskite thin film deposited on glass substrate

Figure 4 shows the optical transmittance of perovskite film spin coated on glass substrate. The optical transmission of the as-prepared $CH_3NH_3Pb(I_{1-x}Br_x)_3$ thin films show a sharp drop in the transmission below 600nm. The absorption α coefficient was evaluated from these data. The inset of figure 4 shows the plot of the conventional Tauc equation (α hv)n = A(hv-Eg) where A is an empirical constant, h α is the photon energy and n=2 for the direct band gap semiconductor. By extrapolation of the part of the (α hv)n vs.hv plots. We found that the Eg value for this compound is 1.76 eV. [12][13]

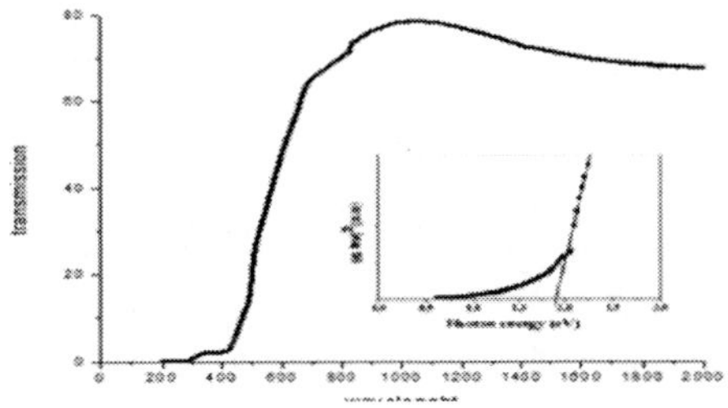

Figure 4: Optical transmission of perovskite deposited on glass substrate

The XRD pattern of CuO thin film was shown in figure 5 and it revealed that the as-deposited film present the (111), (111), (220) and (113) diffraction peaks that are characteristic of the monoclinic phase CuO [24][25][26] .

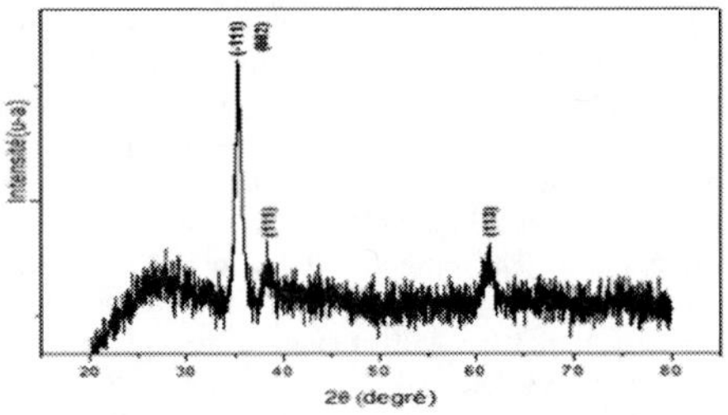

Figure 5: XRD- pattern of CuO thin film deposited on glass substrate

The optical transmission of a typical as-deposited CuO film is shown in Figure 6. In this case, the average transmission in the visible range is around 60%. The presence of interference fringes is an indication of the good quality and the smoothness of the surface of the films. The transmission was used to evaluate the absorption coefficient in the spectral range below the absorption edge. The latter was used to determine the optical gap Eg,by assuming a direct band gap semiconductor. The optical band gap of the films was found to be 1.8eV.[9][10][11]

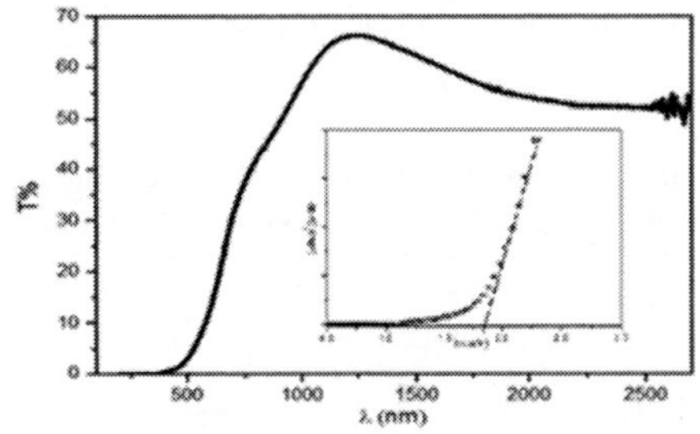

Figure 6: Optical transmission of CuO deposited on glass substrate

Figure 7 presents the J–V curve of the hybrid solar cell with an FTO/Zno/ $CH_3NH_3PB(I_{1-x}Br_x)_3$ / CuO/Al structure. IUnder illumination the solar cell show a photovoltaic behavior and has provided the highest power conversion efficiency (η) of 3.22%, a fill factor (FF) of 0.5, an open-circuit voltage (VOC) of 0.5 V and a lower Is= 3.31 nA. The low ISC suggest that the high resistivity of ZnO and CuO layers is the main factor that limits the cell performance.

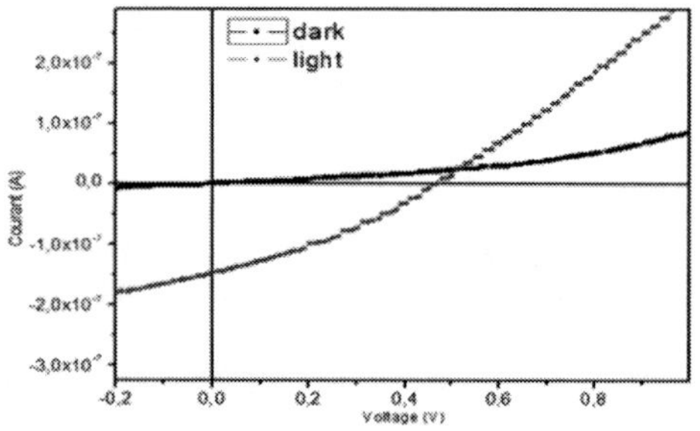

Figure 7: Current -Voltage characteristics of ZnO/ $CH_3NH_3PbI(_{(1-x)}Br_x)_3$ / CuO/Al cells

4 CONCLUSION

$CH_3NH_3PB(I_{1-x}Br_x)_3$ thin films were successfully fabricated by sequential deposition which consist on spin coating a solution obtained by mixing lead iodide and lead bromide PbBr2 on ZnO film, the cell was then dipped into the dipping solution of MAI in isopropanol . X-ray diffraction and optical transmission measurements confirmed the formation of a single-phase mixed perovskite that is suitable for an absorber layer in Tandem cells. The optical band gap of perovskite thin films is 1.76 eV. The as-deposited films of ZnO and CuO were polycrystalline with quasi-uniform grain size. The ZnO films exhibited a (002) preferred growth orientation on glass substrates while CuO films did not exhibit this feature. The average optical transmission of the ZnO films is on the order 80% while its only 60% for CuO.The optical bandgaps of the obtained CuO and ZnO films are found to be 1.8 and 3.28 eV respectively. The high transmittance of the as-prepared ZnO films in the visible range makes them very suitable as window layer in solar cells. The optical bandgap of the as-prepared CuO films, on the other hand, is more suitable for solar cells than that of CuO2.

The ZnO/ $CH_3NH_3PB(I_{1-x}Br_x)_3$ / CuO/Al heterojunction showed a solar cell behavior, The photocurrent was observed under illumination at 480 mW•cm−2 and the solar cell showed a conversion efficiency of 3,22%.The obtained cell parameters low Icc indicating the need of limiting the resistivity of ZnO and CuO films.

References

[1] Ball, J. M., Lee, M. M., Hey, A. & Snaith, H. J. Low-temperature processed mesosuperstructured to thin-film solar cells. Ener Env.Sci.6, 1739–1743 (2013)

[2] Liu, M., Johnston, M. B. & Snaith, H. J. Efficient planar heterojunction perovskite solar cells by vapor deposition. Nat 501, 395–398 (2013)

[3] Burschka, J. et al. Sequential deposition as a route to high-performance perovskite-sensitized solar cells. Nat 499, 316–319 (2013).

[4] H.Ait DADS, S.Bouzit, L. Nkhaili, A. Kissani, A. Outzourhit Structural, optical and electrical properties of planar mixed Perovskite halides/ Al-doped Zinc oxide solar cells. Sol Ener Mater Sol Cells (2015)

[5] Yinzken, W.; Benli, C. Effects of sapphire substrates surface treatment on the ZnO thin films grown by magnetron sputtering. Phys B2008, 403, 1979.

[6] Kuroyanagi, A. Properties of aluminum-doped ZnO thin films grown by electron beam evaporation. App Phys, 28, 219. (1989)

[7] L.Nkhaili, A. El kissani, M. Ait Ali, Y. Ijdiyaou, A.Elmansouri, A. Elkhalfi, A. Outzourhit Effect of RF power on the structural and optical properties of RF-sputtered ZnO thin films
Eur. Phys. J. Appl. Phys. (2014) 66: 30302

[8] Izhizuka, S.; Kato, S.; Maruyama, T.; Akimoto, K. Nitrogen doping into Cu2O thin films deposited by reactive radiofrequency magnetron sputtering. Appl Phys, 40, 2765–2768, (2001)

[9] .B. Balamurugan and B. R. Mehta, "Nanocrystalline Thin Films, Optical Properties. Structural Properties. X-Ray Diffraction," *Thin Sol Fil*, Vol. 396, No. 1-2, (2001), pp. 90-96.

[10] F. Marabelli, G. B. Parraviciny and F. S. Orioli, "Optical Gap of CuO," *Phys Revi B*, Vol. 52, No. 3, (1995), pp. 1433-1436.

[11] J. Ghijsen, L. H. Tjeng, J. V. Elp, H. Eskes, J. Westerink, G. A. Sawatzky and M. T. Czyzyk, "Electronic Structure of Cu2O and CuO," Phys Rev B, Vol. 38, No. 16, (1988), pp. 11322-11330.

[12] W. A. Laban and L. Etgar, Retarding the crystallization of PbI2 for highly reproducible planarstructured perovskite solar cells via sequential deposition. Ener. Environ. Sci., (2013), 6, 3249.

LASER PATTERNING OF PEROVSKITE SOLAR CELLS: PROCESS DEVELOPMENT AND MINI-MODULE FABRICATION

Christof Schultz[a], Felix Schneider[a], Andreas Bartelt[a], Carola Ferber[b], Lukas Kegelmann[b], Steffen Meyer[b], Bernd Rech[b], Rutger Schlatmann[a,c], Steve Albrecht[d], Bert Stegemann[a]

[a] University of Applied Sciences – HTW Berlin, Wilhelminenhofstr. 75a, D-12459 Berlin, Germany
[b] Helmholtz-Zentrum Berlin für Materialien und Energie, Institut für Silizium-Photovoltaik, Kekulestr.5, D-12489 Berlin
[c] PVcomB/ Helmholtz-Zentrum Berlin für Materialien und Energie GmbH, Schwarzschildstr.3, D-12489 Berlin
[d] Helmholtz-Zentrum Berlin für Materialien und Energie, Young Investigator Group for Perovskite Tandem Solar Cells, Kekulèstr. 5, D-12489 Berlin

ABSTRACT: Perovskite solar cells have been rapidly improved over the last years. At small lab-scale sizes efficiencies comparable to crystalline silicon solar cells were achieved. Successful up-scaling to large module sizes requires monolithic interconnection processes. Here we report on laser scribing processes of the P1, P2 and P3 patterning steps to interconnect perovskite solar cells. Particular emphasis is put on the P2 laser patterning step as it is known to induce detrimental material modifications of the absorber layer mostly at the bottom and in the vicinity of the scribed lines. Suitable process windows were established and optimized by systematic variation of the pulse length (ns, ps), wavelength (532 nm, 1064 nm) and the processing side (glass-/layer-side). By systematic analysis of the laser-matter interaction, the selective layer removal was achieved for all three patterning steps. The proper morphology and the electrical functionality, resulted in conditions for successful monolithic serial integration of a first three-cell perovskite mini-module with a power conversion efficiency above 10 % and an interconnection width of 350 μm.

Keywords: Ablation; Perovskite; Laser Processing; Serial Integration

1 INTRODUCTION

Inorganic-organic metal halide perovskites are very promising candidates for absorbers in low cost solar cells [1]. Further interest is driven by the outstanding power conversion efficiencies of more than 22 %, achieved with lab-scale, solution-processed solar cells [2]. Up-scaling from lab-scale to industrial relevant production sizes requires serial interconnection of solar cells to increase the output voltage.

Laser processing has become a key technology in thin-film photovoltaics to achieve monolithic serial interconnection, which involves alternating layer deposition and layer patterning (so called patterning steps P1, P2, P3) [3]. Thereby, the solar cells are patterned by fine lines by selective material removal. The P1 and the P3 scribes determine the width of the cells by isolating the back and front contact, while the P2 scribe enables the interconnection with the adjacent cell via opening the absorber layer. The area between the P1 and P3 scribe cannot be used for energy conversion, thus, it is called *dead area*. In order to achieve a high aperture ratio, the dead area should be as small as possible. In Fig. 1 the patterning of a perovskite solar cell layer stack is schematically illustrated.

For common thin-film technologies, such as thin-film Si and CIGSe, the interaction of the laser irradiation with the specific constituent layers has been extensively studied, and laser patterning is well-established even in industrial production [4, 5]. However, the fabrication of large-area perovskite solar modules requires additional engineering efforts to customize these well-established process parameters and to properly interconnect adjacent cells. A few groups have reported already on successful laser-based serial interconnection and the fabrication of different mini-modules [6-9]. According to these results, particularly the P2 scribe is challenging and might be the origin of a distinct performance drop, when advancing from the cell to module. The rather low open circuit voltages (V_{oc}) and low fill factor (FF) imply, that laser induced effects might be responsible for power losses.

Therefore, the aim of our work is to establish suitable process parameters for the three patterning steps by systematic variation of the laser pulse energy and subsequent comprehensive characterization of the morphological, electrical and chemical properties of the scribed area. While successful P1 and P3 patterning is characterized by a sufficiently large isolation resistance across the corresponding scribe, successful P2 patterning, is obtained by a clean, debris-free bottom of the patterned scribe to enable low contact resistances. Main challenges for P2 and P3 patterning are the thermal sensitivity of the inorganic-organic compound perovskite structure [10, 11]. Thus, in accordance with established patterning processes for thin-film silicon, the laser patterning for the perovskite-based cells is carried out from the glass-side to make use of the advantage of mechanically stress assisted disintegration [12] and to avoid plasma shielding [13]. Successful process development for laser patterning and electrical functionality of the serial interconnection will be proven by a first perovskite mini-module.

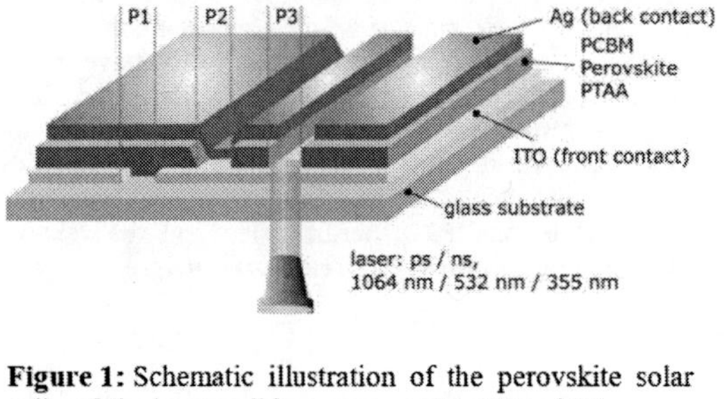

Figure 1: Schematic illustration of the perovskite solar cell and the laser scribing processes P1, P2 and P3.

2 EXPERIMENTAL

2.1 Sample preparation

The perovskite solar cell samples were prepared according to the process established for perovskite single junction and tandem solar cells at the Helmholtz-Zentrum Berlin [14, 15]. The deposition process is carried out by spin coating with top efficiencies of 18 % [14, 15]. For module fabrication we used perovskite solar cells with "inverted" device design (i.e., p-i-n geometry). The formamidinium-cesium based perovskite absorber with 500 nm thickness is embedded in selective contact layers (50 nm PCBM and 5 nm PTAA). These contact layers are deposited onto 1.1 mm glass substrates coated with 150 nm transparent conducting indium-tin-oxide (ITO) layer, acting as front contact. As counter-electrode a 100 nm Ag layer is deposited onto the stack. The whole preparation was carried out under nitrogen atmosphere at atmospheric pressure to avoid absorber degradation [16].

2.2 Laser patterning

A Rofin Baasel patterning tool was utilized for solar cell laser patterning. The system features a high speed, high-accuracy motion system which allows positioning accuracies well below 5 µm and velocities of up to 1.2 m/s. The system is equipped with a nanosecond and a picosecond laser source. The diode-pumped solid state (DPSS) nanosecond laser emits pulses of ~30 ns at a wavelength of 532 nm. Maximum pulse energy of 45 µJ is achieved at a repetition rate of 20 kHz. The repetition rate can be varied up to 400 kHz. The DSSP picosecond laser emits a wavelength of 1064 nm and maximum pulse energy of 114 µJ, the repetitions rate can be chosen separately. Both lasers possess Gaussian intensity distribution. The samples are transferred and processed under continuously flowing nitrogen atmosphere.

2.3 Characterization

Characterization of the P1 scribe done by measuring the isolation resistance between two neighboring cells across the scribe, using a standard source meter with a measuring limit of 40 MΩ, which is considered to be sufficiently large for electrical isolation. The morphology and local conductivity of the P2 scribe were measured by a conductive atomic force microscope (c-AFM, NT-MDT NTEGRA) in current sensing mode using a nitrogen doped tip. The surface was scanned at areas up to 50 µm × 50 µm under a bias voltage of 2 V, and the topography as well as the current signal were recorded simultaneously.

For visual inspections of the laser patterns and modifications of the sample surface a scanning electron microscope (SEM, Hitachi S4100) was used which also enables the local detection of the elemental composition of the sample by means of energy-dispersive x-ray (EDX) spectroscopy. The acceleration voltage was 5 kV.

The results of P3 patterning were evaluated by comparison of the current density-voltage characteristics before and after patterning. A rectangular area was separated by the P3 patterning step and the relative change of the electrical parameters was analyzed.

3 RESULTS AND DISCUSSION

3.1 P1 front contact patterning

Relying on previous studies on laser ablation of transparent oxide layers, P1 laser patterning of the ITO front contact layer with 10 ps pulses was successfully achieved by direct multi-spot ablation at 1064 nm at rather high scribing speeds of 800 m/s (pulse-to-pulse overlap > 90 %). The success of patterning was evaluated by measuring the resistance across the corresponding laser lines, i.e. between two adjacent, separated cells. These lines were patterned with systematically varied fluences, covering a range from 1.6 J/cm² to 6.1 J/cm², corresponding to the onset of material removal and the damage of the substrate, respectively.

In Fig. 2 the isolation resistance as function of the applied laser fluence is shown. The dashed line represents the upper measuring limit of the source meter, i.e. 40 MΩ. It is assumed that at a scribe length of 1 cm this value is sufficiently large for proper cell separation. In order to avoid slight rims and bulging at the scribe edges, which might cause shunts of the very thin absorber layers, patterning was carried out from the layer side. It is found that at a laser fluence below 3.5 J/cm² no significant separation was obtained, whereas above this threshold, sufficient isolation was achieved. Accordingly, for manufacturing monolithically interconnected cells, a laser fluence of ~4 J/cm² was chosen. The so patterned lines show no residues or cracks within the underlying substrate.

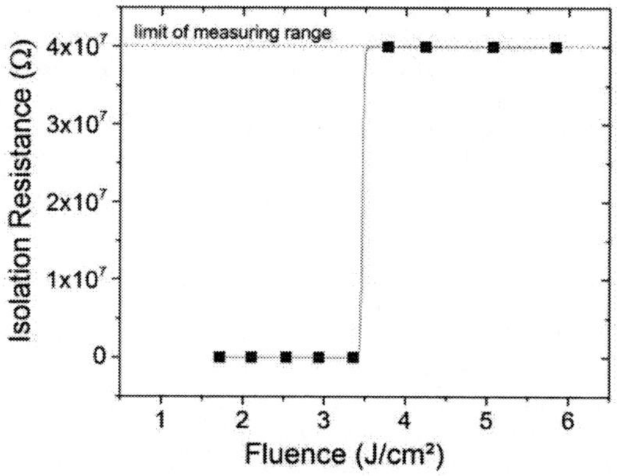

Figure 2: Isolation resistance as function of the applied laser fluence. The dashed line represents the limit of the measuring range of the source meter.

3.2 P2 absorber layer patterning

In contrast to the patterning of the front contact layer, the nanosecond laser emitting a wavelength of 532 nm was used for the absorber opening layer. Laser patterning was carried out from the substrate side, to make use of the high difference between the absorptions coefficients of the front contact layer and the perovskite layer at 532 nm which facilitates mechanical stress-assisted ablation. The applied fluence covered a range from incomplete ablation of the absorber layer (0.5 J/cm²) till the onset of the ablation of the underlying front contact layer (4.2 J/cm²).

For the P2 scribe it is important to create a sufficiently large contact area for the interconnection of adjacent cells to enable low contact resistances. Therefore, a relatively high pulse-to-pulse overlap is necessary. Furthermore, the bottom of the trench must be free of residuals, which might impede low contact resistances down to the front contact layer. Similar to well-established process windows known from common thin-film technologies, such as CIGSe or thin film Si, suitable process parameters were evaluated. Thereby, the laser fluence was systematically varied and the resulting morphological and electrical material modifications were investigated by c-AFM, which is a powerful technique for probing local conductivity variations in heterogeneous samples with high spatial resolution. Information on the local electronic properties is obtained by applying a bias voltage between the tip and the sample and sensing the resulting current. To identify optimal scribing conditions, the morphology of the sample surface as well as the spatially resolved, vertical conductivity was analyzed. A high current corresponds to a high conductivity.

The values of the average conductivity within the grooves for all applied laser fluences are summarized in the graph in Fig. 3. The results show that the conductivity increases within increasing fluence (up to a fluence of ~1.5 J/cm²) due to the ablation of the perovskite and the exposure of the ITO. Beyond 1.5 J/cm² the perovskite layer is completely ablated and the alteration of the ITO sets in, resulting in a decrease of the local conductivity. The inset in Fig. 3 shows an overlay image of the morphology and corresponding current of a laser scribe obtained at a fluence of 1.5 J/cm². This scribe has a homogeneous topography with a constant width of about 35 µm. The bottom of the trench is already smooth as it is preferable for low-ohmic contact resistances. The color bar indicates regions with higher conductivity (red/green), whereas regions with lower conductivity measured are displayed in blue. Accordingly, the conductivity is increased within the scribe due to the exposure of ITO.

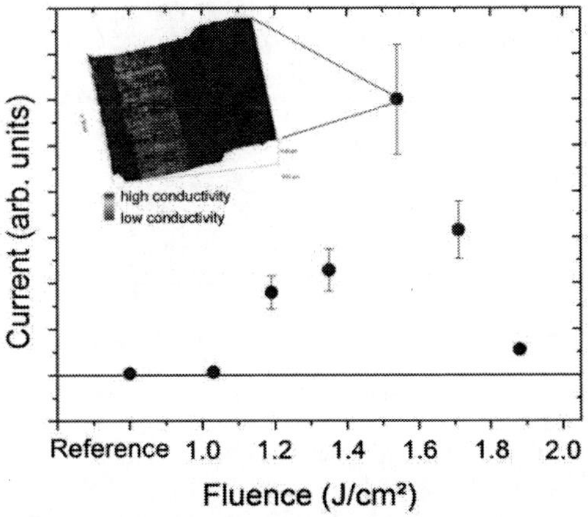

Figure 3: Current between sample surface and c-AFM tip for different laser fluences. The data points correspond to the average value within the laser-treated area. The inset shows an overlay image of the measured morphology and the corresponding local current at a fluence of 1.5 J/cm².

However, a detailed look at the bottom of the trench (inset in Fig. 3) reveals regions with varying conductivity. In order to determine the elemental composition within the scribe EDX measurements were performed. This quantitative analysis relies on measuring the signal intensities for each element in the sample. It allows to determine the present element concentrations.

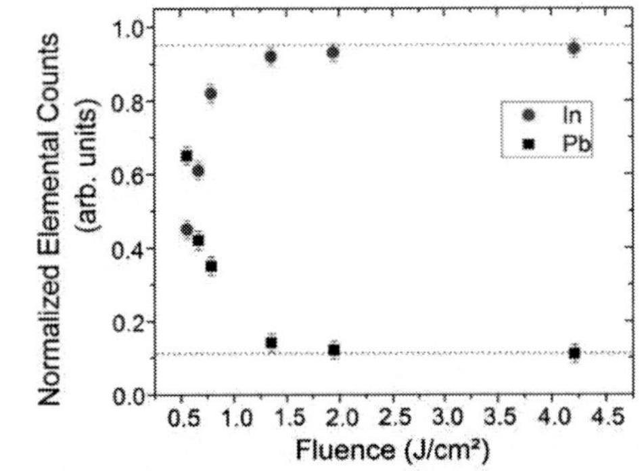

Figure 4: Elemental composition within the P2 scribes as function of the applied laser fluence. Shown are the elements In and Pb.

As elemental references for the material composition within the scribes the elements lead (Pb) and indium (In) were selected. Therefore, Pb represents the absorber material and In the front contact. The results in Fig. 4 show that with increasing laser fluence the perovskite layer is ablated. While the Pb concentration decreases with higher fluence, the In concentration increases in the same way. However, even at high fluence a Pb signal is detected, indicating Pb-containing residuals at the bottom of the scribe. We assume that these lead-rich periodic structures originate from the formation of ripples [13] containing PbI_2 [17, 18], which might impede higher conductivities and, thus, lower contact resistances. Nevertheless, the corresponding process parameters were used for manufacturing the mini-module.

3.3 P3 back contact patterning

The P3 back contact patterning was also carried out with the 532 nm nanosecond laser source irradiated on the sample through the glass substrate to make use of the soft induced-ablation [12]. In contrast to the P2 patterning, where a comparatively high pulse-to-pulse overlap is favorable, the P3 process was carried out with a low pulse-to-pulse overlap to avoid laser-induced damages [19]. Similar to the evaluation of the P2 process parameters, the laser fluence was systematically varied to cover all processes from incomplete separation until laser-induced damaging of the underlying ITO front contact. With the P3 step, a squared area within the initial cell area is separated by laser ablation. The success of the patterning was evaluated by comparison of the j-V parameters before and after laser processing. Beside the relative change of the efficiency and the shunt resistance the comparison focuses on the change in fill factor as integral value for the overall cell performance. In Fig. 5 the comparison of the relative change in fill factor as function of the applied laser fluence is shown. The dashed line represents the threshold between successful and non-successful patterning, values below mean that the fill factor after patterning was comparatively lower. Within the boxes the individual data points (full rhombs) and the median (squares) are shown whereas the box itself displays the standard deviation of each dataset.

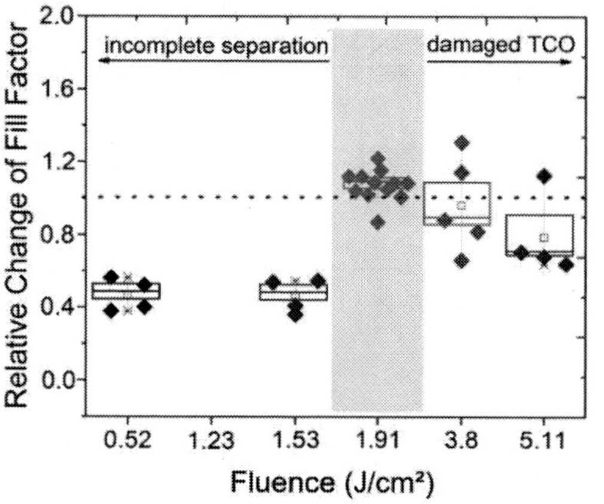

Figure 5: Box-plot of the relative change of the fill factor of P3 laser-patterned solar cells as a function of the applied laser fluence.

As shown in Fig. 5 successful P3 patterning was achieved applying a laser fluence slightly above 1.9 J/cm^2 indicated by a rather constant fill factor before / after patterning. The slightly increased relative change of the fill factor can be ascribed to the smaller cell area after patterning, since it is assumed that the compositional homogeneity increases with smaller cell area. Below 1.9 J/cm^2 incomplete separation was found while above this value the underlying TCO front contact layer and the material around the scribe got damaged.

3.4 Mini-module fabrication

The applicability of all evaluated patterning process parameters for patterning and the electrical functionality were tested by manufacturing a first mini-module. This panel was built on a square inch glass substrate and consists of three interconnected cells (3 × 0.148 cm^2) with interconnection widths of about ~350 µm. The photocurrent density–voltage (j-V) response curve of this mini-module is shown in Fig. 6, the corresponding electrical parameters are provided in Table I.

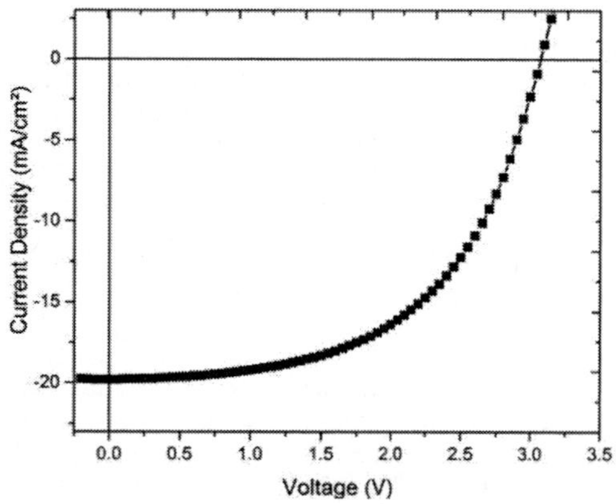

Figure 6: Photocurrent density–voltage (j-V) response of a fully laser-patterned monolithically interconnected mini-module, consisting of three cells (3 × 0.148 cm^2).

Table I: Electrical parameters for the three-cell mini-module from j-V curves in Fig. 6

Parameter		Unit	Value
Short circuit current density	j_{sc}	mA/cm^2	19.30
Open circuit voltage	V_{oc}	V	3.08
Fill factor	FF	%	54.6
Power conversion efficiency	η	%	10.8

These data provide evidence for successful laser processing and electrical functionality of the laser-patterned interconnections. The power conversion efficiency of the fabricated mini-module is above 10 %. The fill factor reflects losses mainly due to the still rather high series resistance, which can be assigned to the Pb-containing residuals found within the P2 scribe (cf. Fig. 4). Furthermore, it is assumed that also the short circuit current density (j_{sc}) is limited by those residuals within, apparently dominating the j-V behavior. However, the rather high V_{oc} value of 3.08 V indicates successful laser patterning.

4 CONCLUSIONS

A fully laser-patterned, series interconnected 3-cell mini-module based on solar cells with efficiencies above 10 % was fabricated. Obtaining an interconnection width of ~350 µm, the laser based patterning for industrial manufacturing has proven to be successful in terms of its electrical functionality. Thereby, adequate isolation resistances for the P1 were achieved. The P2 scribing results are acceptable, however further optimization of the P2 process is part of current efforts to eliminate the residuals within the scribe. The P3 patterning was also successful and allowed patterning without damaging the underlying front contact. Further improvements regarding interconnection losses, optimization of the fill factor and minimization of the dead area are currently in progress.

ACKNOWLEDGEMENTS

The authors gratefully acknowledge the support of the PVcomB team with for the sample preparation. Special thanks to Orman Gref (Technical University Berlin) for the c-AFM measurements. Furthermore, we thank Frank Fink for fruitful discussions.

REFERENCES

[1] E.L. Unger, L. Kegelmann, K. Suchan, D. Sorell, L. Korte, S. Albrecht, Journal of Materials Chemistry A 5 (2017) p. 11401-11409.
[2] N.-G. Park, Nano Convergence 3 (2016) p. 1-13.
[3] A.J. McEvoy, L. Castaner, T. Markvart, Solar cells: materials, manufacture and operation, Academic Press, 2012.
[4] C. Schultz, M. Schuele, K. Stelmaszczyk, M. Weizman, O. Gref, F. Friedrich, C. Wolf, N. Papathanasiou, C. Kaufmann, B. Rau, Solar Energy Materials and Solar Cells 157 (2016) p. 636-643.
[5] B. Stegemann, M. Richter, C. Schultz, H.-U. Pahl, H. Endert, J. Bonse, B. Rau, V. Quaschning, F. Fink, (2011).
[6] S.-J. Moon, J.-H. Yum, L. Löfgren, A. Walter, L. Sansonnens, M. Benkhaira, S. Nicolay, J. Bailat, C.

Ballif, IEEE Journal of Photovoltaics 5 (2015) p. 1087-1092.
[7] G. Mincuzzi, A.L. Palma, A. Di Carlo, T.M. Brown, ChemElectroChem 3 (2016) p. 9-30.
[8] L. Rakocevic, R. Gehlhaar, T. Merckx, W. Qiu, U.W. Paetzold, H. Fledderus, J. Poortmans, IEEE Journal of Photovoltaics 7 (2017) p. 404-408.
[9] B. Turan, A. Huuskonen, I. Kühn, T. Kirchartz, S. Haas, Solar RRL 1 (2017).
[10] A.E. Williams, P.J. Holliman, M.J. Carnie, M.L. Davies, D.A. Worsley, T.M. Watson, Journal of Materials Chemistry A 2 (2014) p. 19338-19346.
[11] E.J. Juarez-Perez, Z. Hawash, S.R. Raga, L.K. Ono, Y.B. Qi, Energy & Environmental Science 9 (2016) p. 3406-3410.
[12] J. Bovatsek, A. Tamhankar, R. Patel, N. Bulgakova, J. Bonse, Thin Solid Films 518 (2010) p. 2897-2904.
[13] D.W. Bäuerle, Laser processing and chemistry, Springer Science & Business Media, 2013.
[14] L. Kegelmann, C.M. Wolff, C.A. Omondi, F. Lang, E.L. Unger, L. Korte, T. Dittrich, D. Neher, B. Rech, S. Albrecht, ACS Applied Materials & Interfaces (2017).
[15] M. Jošt, S. Albrecht, L. Kegelmann, C.M. Wolff, F. Lang, B. Lipovšek, J. Krč, L. Korte, D. Neher, B. Rech, M. Topič, ACS Photonics (2017).
[16] Z. Song, A. Abate, S.C. Watthage, G.K. Liyanage, A.B. Phillips, U. Steiner, M. Graetzel, M.J. Heben, Advanced Energy Materials 6 (2016).
[17] R. Ahuja, H. Arwin, A. Ferreira da Silva, C. Persson, J. Osorio-Guillén, J. Souza de Almeida, C. Moyses Araujo, E. Veje, N. Veissid, C. An, Journal of Applied Physics 92 (2002) p. 7219-7224.
[18] J.S. Manser, J.A. Christians, P.V. Kamat, Chemical Reviews 116 (2016) p. 12956-13008.
[19] C. Schultz, G.A.F. Basulto, S. Ring, C. Wolf, R. Schlatmann, B. Stegemann, IEEE Journal of Photovoltaics 7 (2017) p. 1442-1449.

ON COST-EFFECTIVENESS OF PEROVSKITE/C-SI TANDEM PV SYSTEMS

L.J. Geerligs
ECN Solar Energy
Westerduinweg 3, 1755 LE Petten, The Netherlands

ABSTRACT: The requirements to achieve cost reduction with perovskite/c-Si tandem PV systems relative to single-junction (SJ) perovskite or c-Si PV systems are modeled. Specific cases assume a perovskite band gap of 1.7eV in tandem use and 1.55 eV in SJ use, both having the same module efficiency expressed as a fraction of the Shockley-Queisser limit for the band gap. Compared to SJ perovskite PV systems, current and foreseen c-Si cell manufacturing costs are low enough to achieve cost reduction when adding c-Si cells to a perovskite module, if, in particular, additional loss (other than related to SQ limit as a function of band gap) of power conversion efficiency (PCE) of the perovskite module from SJ to tandem can be mostly avoided (it should be less than about 10-20%). Compared to SJ c-Si PV systems, requirements for cost-effective modification to tandem structure are, not suprisingly, a high PCE and low cost for the added perovskite stack. For example, for reasonable parameter assumptions, a semi-transparent perovskite PCE of 16.4% and a stack fabrication cost of about 35 €/m^2 would result in 10% cost reduction, the same PCE with a stack fabrication cost of 17€/m^2 in 16% cost reduction, for tandem PV systems relative to the c-Si technology. In both tandem-SJ comparisons, if there is a significant increase of module manufacturing cost for, e.g., more expensive junction box, power optimizer, additional encapsulant materials, etc., or balance-of-system costs these should be included in the c-Si cell or perovskite stack costs (translated to the m^2 cost), respectively.
Keywords: hybrid tandem, perovskite, silicon, economic analysis

1 INTRODUCTION

The rapid development of perovskite solar cells to achieving high efficiency, as well as increasingly more stable performance, has resulted in increased attention for hybrid perovskite/crystalline silicon tandem devices. 4-terminal hybrid tandems can potentially provide a unique combination of the attractive aesthetics of thin film (TF) PV, with an energy production that can be higher than the best performing crystalline silicon (c-Si) PV modules. Apart from stability, the cost and performance of such a tandem in comparison to the single junction (SJ) technologies are deciding factors.

This paper analyzes the potential for cost reduction of hybrid TF/c-Si tandem modules compared to the respective single junction modules. Sensitivities to module cost parameters and efficiency are investigated. While the comparison is aimed at perovskite, the framework is valid for other thin film technologies, so the general abbreviation TF will be used for the top submodule technology. The comparison is aimed at c-Si bottom cells, e.g. using the cost for a (156mm)^2 industrial cell as a metric, but with a simple conversion from that cell cost to a m^2 cost, the framework would be applicable to TF/TF tandems as well.

The Shockley-Queisser (SQ) efficiency limit increases from SJ to tandem by about a third. This would roughly be the cap for the cost reduction if the additional cell could be added to the SJ system for zero cost. For a tandem module that builds on a c-Si bottom cell, the potential efficiency increase is even larger than the SQ limit suggests, because c-Si has an Auger-limited efficiency limit significantly below the SQ limit. Therefore, a tandem with an ideal radiatively limited top cell can provide an even larger benefit (e.g., about 16 percentage points in [1]). However, the question is how far cost reduction can approach this potential value in practice. In earlier work, Peters et al.[2], Basore [3], and Bobela et al.[4] reported model analysis for tandem-enabled cost reductions, showing a likelihood of very small cost reductions.

One of the important things these papers pointed out, is that comparison of cost-effectiveness has to be with respect to both constituent single junction technologies, and that best results for the tandem may be expected when those single junction technologies are comparable in cost and performance. Since c-Si cell production presently costs about 1 €/cell which translates to about 35 €/m2, and the typical perovskite stack costs forecasts in literature are also around 30 €/m2 (of course with large uncertainty), and with perovskite module PCE in excess of 15% such comparability of cost and performance for both constituents of the future tandem seems to be a good possibility.

This paper attempts to analyze the cost reduction potential with minimized number and detail of assumptions where possible. It aims to use aggregated real-life numbers where possible. While the model is aimed at 4-terminal tandems (separately contacting TF and c-Si cells), consisting of a TF stack on a glass superstrate, laminated with strings of c-Si bottom cells), monolithically integrated 2-terminal tandems would not be very differently modeled.

At the end of this paper, possible changes in the relation between system cost and levelized cost of energy (LCOE) for SJ versus tandem technology are briefly considered.

2 COST COMPARISON MODEL

2.1 Reference system

The reference system in the model is the single-junction thin film (SJ TF) system. It is parameterized by a system installation cost *C*, and nominal module power output *P*. *C* is thus typically the system Capex, though it could be taken to include additionally an Opex-related term, for the purpose of LCOE calculations. The model assumes losses in inverter and cabling are percentage-wise the same for all technologies.

The relative cost reduction (in €/Wp) for a change from SJ TF to tandem PV system is

$$RCR \equiv -[(C+\Delta C)/(P+\Delta P)-C/P]/(C/P)$$

2.2 Tandem compared to SJ TF system

For the change from SJ TF to tandem, c-Si cells are added to the TF module, resulting in changes of cost and power:

$$\Delta C = C_{cell} + \Delta C_{str}$$
$$\Delta P = P_{cSi,T} + \Delta P_{TF}$$

C_{cell} is the cost of the c-Si cell (wafer and processing) and ΔC_{str} is the sum of all other changes in cost of any other aspect of the PV system, such as additional encapsulant, inverter electronics, cabling, etc (similar to [2]). $P_{cSi,T}$ is the additional power output from the added c-Si bottom cells. ΔP_{TF} is the change of power output from the TF top cell, e.g. due to change of contact materials from non-transparent to semi-transparent, or due to a change of bandgap of the absorber. All parameters are referred to the same (module) area, so area-dependent balance of system (BoS) costs do not change.

It is assumed that ΔC_{str} is composed of the following two parts:

Part 1. balance of system costs that are proportional to the added nominal power output, with a proportionality factor C_{Wp}. These extra costs are thus $\Delta P \cdot C_{Wp}$. This includes inverter and grid connection, as well as a small fraction of the costs of cabling, planning, infrastructure [5].

Part 2. the increased module manufacturing costs, specifically tabbing/stringing and extra encapsulant, as well as possibly increased costs for e.g. the junction box or a module-level power optimizer.

For Part 1 a value of C_{Wp}=0.1 €/Wp is used, which is reasonable for the present and some time into the future [5]. The sensitivity of the results to this value is very small. Part 2 can be combined with the cost per c-Si cell. For example, a change in module-level power optimizer that would add 20€ to the cost of a module would translate into 0.33 €/cell for a 60-cell module.

2.3 Tandem compared to SJ c-Si system

Because the TF system is the benchmark, the comparison of tandem to SJ c-Si system requires additional input parameters and assumptions.

In this paper's model the cost of the c-Si SJ system is related to the cost of the SJ TF system as

$$C_{cSi,SJ} = (1-f)\cdot C + \Delta C + \\ + [P_{cSi,SJ} - (1-f)\cdot P - \Delta P]\cdot C_{Wp}$$

In essence this means that the c-Si SJ generalized system cost is the sum of a fraction (1-f) of the generalized TF SJ system cost, plus the cost of the c-Si cells, with adjustments for the Wp-proportional costs according to the output power difference between TF and c-Si systems. In principle f thus roughly represents the cost of TF stack deposition and monolithic interconnection. Peters et al. [2] have already emphasized the importance of the parameter f for cost-effectiveness of tandems. $P_{cSi,SJ}$ is the power from the encapsulated c-Si cell in SJ use. The relative cost difference between the SJ c-Si and the tandem system, *RCR'*, follows from the difference between $C_{cSi,SJ}/P_{cSi,SJ}$ and $(C+\Delta C)/(P+\Delta P)$.

According to the ITRPV [6], the equivalent of f for c-Si systems is about 0.3 for utility scale PV systems (the c-Si cell cost is ~60% of the module cost, and module cost is slightly below 50% of the utility scale system cost). The most likely application of tandem modules will be in systems where balance of system costs are relatively high, which implies f is reduced below this value of ~0.3. In addition, we assume the perovskite stack production cost will be less than c-Si cell production cost, further reducing f.

3 RESULTS

3.1 Sensitivities

Table 1 shows the center values for the sensitivity analysis. Unless otherwise noted, these center values are also used in the sensitivity plots later in this paper.

Table I: Center values for sensitivity analysis.

Parameter	center value
f	0.15
c-Si PCE SJ	22%
c-Si PCE in tandem	11%
TF PCE SJ	17%
TF PCE in tandem	15.75%
TF PCE loss SJ→tandem	5%
SJ TF system cost	1 €/Wp
c-Si cell cost*	1.2 €/cell
C_{Wp}	0.1 €/Wp

*incl. tabbing/stringing and other extra module conversion costs when c-Si cells are added to the TF module

The c-Si SJ PCE is taken according to the ITRPV for PERx cell types in 2023. The filtered c-Si PCE in tandem is a feasible target for such cells under a perovskite stack with 1.7 eV bandgap [7]. Both c-Si PCEs are for encapsulated situation. The TF SJ aperture area module PCE of 17% is also a realistic target for perovskite of 1.55eV bandgap. The TF tandem PCE of 15.75% corresponds to the same fraction of the Shockley-Queisser efficiency limit (55% of SQ limit), for 1.7eV instead of 1.55 eV bandgap (1.7 eV is a more optimal band gap for application in tandem with c-Si). The TF PCE loss SJ→tandem of 5% is *on top of* this reduction from SJ to TF according to the SQ limit, and this value is probably one of the more challenging ones in this table. The c-Si cell cost can be thought of as composed of 1 €/cell for cell production costs (approximately the present industrial status), plus 0.2 €/cell for extra tabbing/string costs, encapsulant, etc. The value for C_{Wp} of 0.1 €/Wp is on the low end of today's range.

Fig. 1 shows the sensitivities for the cost reduction of tandems with respect to either single junction technology, for variation around these parameters. The sensitivity is calculated as the ratio of the relative variation of the *RCR* divided by the relative variation of the parameter. Note that the value of some parameters will vary by large amounts or have a large uncertainty (e.g. the parameter f, or the TF PCE loss SJ→tandem) while others are much better known or well predictable (e.g. the c-Si PCEs).

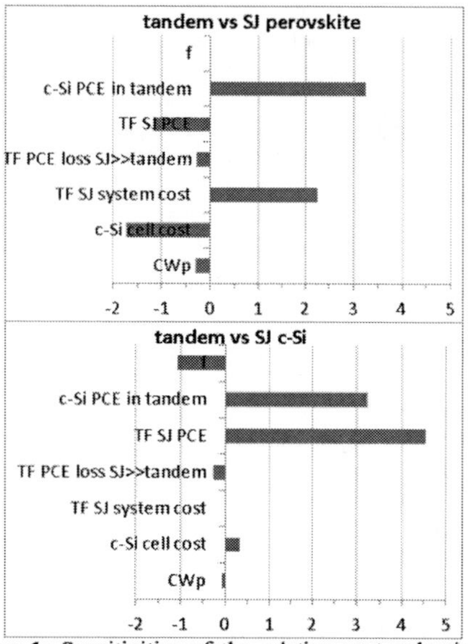

Figure 1: Sensitivities of the relative cost reduction of tandem compared to single junction, for key model parameters. Sensitivity is calculated as defined in the text.

Based on these sensitivities, it makes sense to plot the cost reduction for tandems relative to perovskite as a function of TF SJ system cost, and c-Si cell cost, for both of which the future values are quite uncertain (perhaps by a factor of 2), and the loss in TF PCE from SJ to tandem. The other two important parameters, c-Si PCE in tandem and TF SJ PCE are both relatively less uncertain (the first one more likely to be uncertain by ~10%, the second by perhaps ~30%, both leading to uncertainty in relative cost reduction of ~30%).

Likewise, based on these sensitivities, it makes sense to plot the cost reduction for tandems relative to SJ c-Si as a function of the parameter *f* (determined by the perovskite stack deposition&interconnection cost relative to TF system cost) and the TF PCE.

3.2 Tandems compared to SJ TF

Figure 2 shows the *RCR* for tandem PV systems compared to the bench mark SJ TF PV system, as a function of the cost of the added c-Si cells, the benchmark TF system cost, and the efficiency loss from SJ to tandem use.

Since the current production cost for c-Si cells is around 1 €/Wp, it can be seen that c-Si cells can be added cost-effectively. It should be noted though that all extra module costs, such as junction box, cables, connectors, encapsulant, and also system costs not included in the Wp-proportional costs, should be included in these c-Si cell costs. As an example, 10 €/module for e.g. junction box, connectors, cables, would correspond to 0.17 €/cell, which can be visualized as a left-shift of the curves (a lower cell cost is required to reach the same cost reduction).

Nevertheless, given the progress in reducing cell production costs, a *RCR* of about 10% or more could be achieved, in particular if the extra losses in the TF PCE from SJ to tandem can be largely avoided. The impact of area dependent costs is (implicitly) clearly visible: for expensive system locations, such as residential rooftop, the benchmark system costs will be relatively high, and the *RCR* for a tandem is therefore also high.

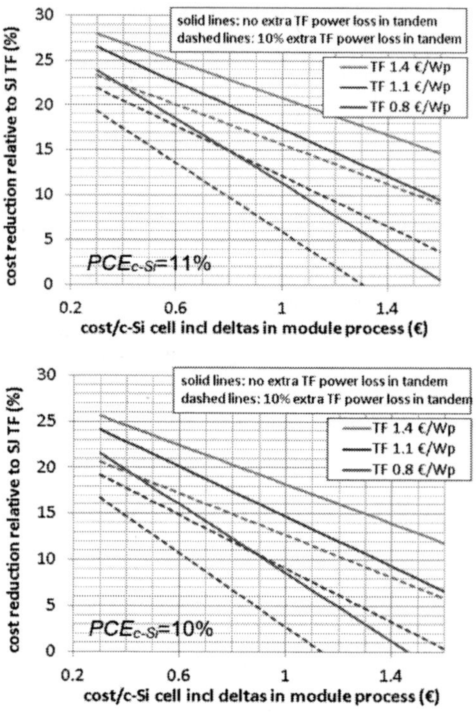

Figure 2: Relative cost reduction of tandem PV systems compared to SJ TF, as a function of the cost of the added c-Si cells. Other parameters as in Table 1, i.e., PCE of 17% for SJ, 15.75% in tandem (solid lines), and 14.2% in tandem (dashed lines). Top: c-Si filtered cell efficiency is 11%, bottom 10%.

3.3 Low cost low performance or high cost high performance c-Si cells?

The fact that the sensitivity to the filtered c-Si cell efficiency is not very large, is illustrated by the comparison of the top and bottom plots in Fig. 1, which are for 11% and 10% filtered c-Si efficiency, respectively. Conversely, this shows that spending c-Si processing costs for higher cell efficiency is less effective for tandem PV systems than SJ c-Si PV systems.

For example, Fig. 3 shows how c-Si cell cost increase for a high efficiency technology affects the system cost for c-Si SJ and TF/c-Si tandem PV systems, for the model parameters as in Table 1. Because the filtered c-Si cell efficiency is only half the SJ c-Si cell efficiency, the allowable cost increase for the high efficiency technology is also halved. In the example of Fig. 3, a SJ PCE increase of 2%$_{abs}$ allows 45% cell cost increase in case of SJ c-Si systems, versus 20% cell cost increase when used in tandems.

The notable exception to this rule of thumb is when changing from Al-BSF c-Si technology to PERx technology, which significantly improves the near-infrared response of the c-Si cells, and therefore has a relatively large benefit in tandem use. This situation should be modeled more precisely. However, a change from PERx to possibly significantly more expensive c-Si technology, e.g. SHJ or IBC, will have only roughly half the cost-effectiveness in tandem application

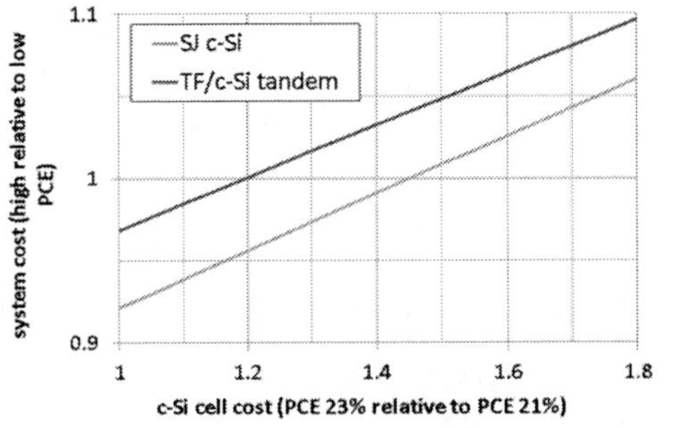

Figure 3: PV system cost for a 23% c-Si cell technology relative to a 21% c-Si cell technology, as a function of the c-Si cell processing cost, for both SJ systems and tandem systems. The break-even (relative system cost = 1) allows a ~2x higher c-Si cell cost increase for the 23% cells in SJ systems than in tandem systems.

3.4 Tandems compared to SJ c-Si

Fig. 4 shows the *RCR'* for tandem PV systems compared to the SJ c-Si PV system based on the same c-Si cells (22% efficiency, cf. Table 1), for two key parameters for the cost reduction: perovskite module efficiency, and fraction *f* of perovskite stack cost in total system cost. As described in section 2.3, since for c-Si cells used in utility scale systems the cell cost is a fraction of ~0.3 of the system cost, it can be expected that in the market segments relevant for tandems (with high BoS costs), for perovskite stacks *f* will be significantly below 0.3.

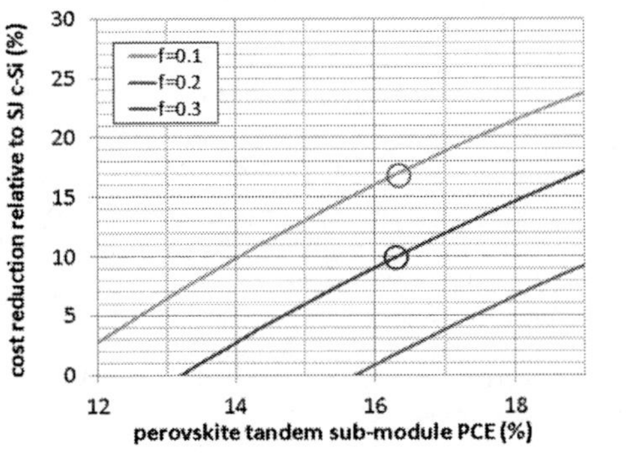

Figure 4: Relative cost reduction of tandems compared to SJ c-Si, as a function of the efficiency of the added semi-transparent TF submodule and the cost fraction of the perovskite stack in the SJ TF system cost. Other parameters as in Table 1.

When assuming a ball park system cost of 1 €/Wp, the parameter *f* can be translated into a perovskite stack cost (including monolithic interconnection, and all other module cost increases from c-Si to tandem). The green circle in Fig. 4 indicates a tandem cost reduction *RCR'*=17% when the stack cost is 18 €/m^2, and the red circle *RCR'*=10% when the stack cost is 35 €/m^2. Such perovskite stack costs seem to be feasible, based on recent literature attempting to estimate future perovskite production costs, e.g. [8], though other references estimate somewhat higher costs [9].

For lower system cost, the allowed perovskite stack cost will be reduced. Also, the increased costs for tandem module and system production (encapsulant, junction box, cabling, etc.) should be effectively included in this perovskite stack cost. Thus, it seems that to obtain cost reduction relative to c-Si SJ systems may be challenging for semi-transparent perovskite module efficiencies (at 1.7eV band gap) with PCE less than about 16%. On the other hand, it should be noted such module PCE is still far below the SQ limit and may have significant room for improvement.

3.5 Effect of module cost components specific for tandem

The previous sections have outlined how module and system costs specific for tandem PV have to be included in the sensitivity plots of Fig. 2-4: For Fig 2-3 (adding c-Si cells to a perovskite TF system) such costs have to be included at a rate of 0.17 €/cell per extra 10 €/module. For Fig. 4 (adding a perovskite stack to a c-Si system), since a standard c-Si module area is ~1.7 m^2, such costs have to be included in the costs for the perovskite stack at about 6 €/m^2 per extra cost of 10 €/module.

The largest part of such costs would probably be related to 4-terminal wiring structure, requiring extra cables, connectors, power optimizer circuits, etc. Therefore, there is significant potential benefit (in cost as well as market acceptance) to be expected from a 2-terminal module with internal voltage matching of the perovskite and c-Si submodules [1,10], if this can be realized at acceptable cost for suitably arranging the c-Si and perovskite submodule architecture.

4 FROM SYSTEM COST DIFFERENCES TO LCOE DIFFERENCES

4.1 LCOE

The levelized cost of electricity is given by

$$LCOE = \frac{Capex + M \cdot Opex}{U \cdot D}$$

where where *Capex* is the system installation cost and *Opex* the annual operational expenses (assumed to be constant), both in €/Wp. *M* is about 1/3 to 2/3 of the economic lifetime of the system in years, depending on the capital discount rate. For utility scale PV, a reasonable assumption is that $M \cdot Opex \approx 0.2$ €/Wp, which can be summed with the *Capex* to arrive at a generalized system cost. For residential rooftop, *Opex* may not be a parameter of interest (i.e., be effectively zero). *U* is the utilization of the system (specific energy yield in kWh/Wp), and *D* is a factor accounting for efficiency degradation as well as financial discount [5]:

$$D = \sum_{t=1}^{n} (1 - Degradation)^t / (1 + WACC_{real})^t$$

where *n* is the economic lifetime of the system in years, *Degradation* the annual degradation of the nominal power of the system, and $WACC_{real}$ the real weighted annual average cost of capital.

LCOE should be a good guideline for the value (and therefore the price) of a high efficiency module, though it has been put forward that other factors, in particular constraints on available area, also play a role in this [11].

4.2 Utilization

Utilization may be a somewhat significant parameter in the cost/benefit evaluation of tandems. One aspect is the temperature coefficient TC of the PCE. The higher band gap top cell can be expected to have a better TC than c-Si [12], as is found for e.g. GaAs cells which have 2x lower TC than standard c-Si cells. For perovskite solar cells, TC is not well established. In addition to effect of differing TCs, also the module temperature in operation can be expected to be slightly less for tandems than for SJ systems because of the reduced thermalization losses. For example, for the parameters as in Table 1, the thermalization power input is reduced by 6%, and therefore the operating temperature increase over ambient temperature would likely be reduced by roughly the same percentage. This can amount to a few degrees in conditions such as for rooftop modules with little ventilation. If power output from perovskite solar cells would have a similar TC as that of GaAs solar cells (about -0.2%/K), this can correspond to an improvement in utilization of a few percents (e.g. 3% better utilization for tandem PV compared to c-Si PV, if a representative operating temperature difference relative to ambient is 25 K).

4.3 Degradation

The degradation rate of tandems incorporating perovskite, might be worse than for c-Si. This will affect the factor D in the LCOE. In Fig. 5 the relative LCOE is plotted as a function of degradation rate, for an economic lifetime of 30 years (the outlook for product warranty on c-Si modules in the ITRPV is 30 years). The curves in Fig. 5 will be slightly less steep for lower economic lifetime: the slope is roughly proportional to system lifetime. Whether or not a residual value after the economic lifetime is included, does not change the figure. The ITRPV outlook for degradation of c-Si systems is about 0.6%/year. Significant changes in degradation rate from c-Si PV system to tandem PV system, of order 1%/year, can negate part or all of an initial cost advantage for tandems.

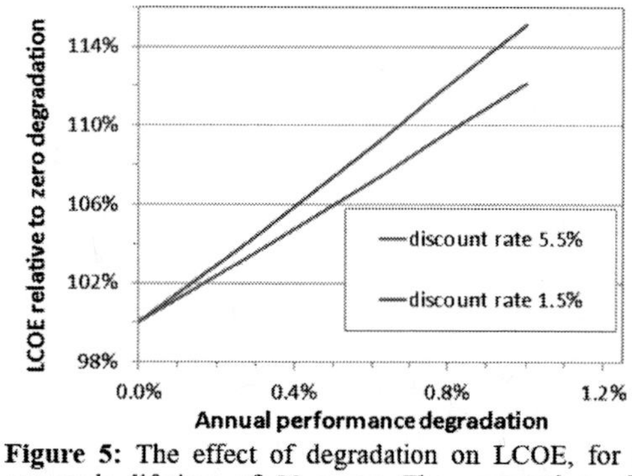

Figure 5: The effect of degradation on LCOE, for an economic lifetime of 30 years. The assumption of a residual value for the system after the economic lifetime does not change this plot. (discount rate = $WACC_{real}$)

5 CONCLUSIONS

This paper analysed under which conditions, using state of the art industrial c-Si production, a cost reduction of tandem systems relative to both SJ systems, of ~10-20% can be achieved.

If perovskite cell stacks can be deposited and interconnected for ~20-30 €/m², and achieve a (semi-transparent, 1.7eV band gap) efficiency of around 16-18%, such significant cost reduction can potentially be achieved when the benchmark PV system cost is around 1 €/Wp. For lower system costs the allowed perovskite stack costs should decrease proportionally. This does not allow much room for added module- or system-level tandem costs (junction box, connectors, MLPE cost increase, etc.) as these have to be included in these stack costs. Higher perovskite PCE improves cost-effectiveness and relaxes these constraints.

For cost-effectiveness of tandems relative to single-junction perovskite modules, PCE losses from non-transparent to semi-transparent perovskite cell will have to be effectively limited to about 10-20%. c-Si cell production costs do not seem to be a foreseeable bottleneck for the cost reduction of tandems with respect to future SJ perovskite modules.

6 REFERENCES

[1] M.H. Futscher and B. Ehrler, ACS Energy Lett. **2**, 2089 (2017).

[2] I.M. Peters et al., RSC Adv **6**, 6691 (2016).

[3] P.A. Basore, in proceedings of the 43th IEEE PVSC, 2016.

[4] D.C. Bobela et al., Progr. Photovolt: Res. Appl. (2016) DOI: 10.1002/pip.2808.

[5] e.g., "PV LCOE in Europe 2014-30", Final report, 23-6-2015. http://www.eupvplatform.org.

[6] International Technology Roadmap for PV, 2017, http://www.itrpv.net/Reports/Downloads/

[7] S.L. Luxembourg et al., these proceedings (33rd EUPVSEC, Amsterdam, 2017).

[8] M. Cai et al., Adv. Sci. **4**, 1600269 (2017).

[9] N.L. Chang et al., Prog. Photovolt: Res. Appl. **25**, 390 (2017).

[10] R. Strandberg, Appl. Phys. Lett. **106**, 033902 (2015)

[11] R.M. Swanson, "The Future of Photovoltaics", Symposium "Photons, Electrons, Bands", Jan 27, 2012, Berkeley, CA, USA. https://www.youtube.com/ watch?v=p12sJYFU6F0

[12] e.g., P. Singh and N.M. Ravindra, Solar Energy Materials & Solar Cells **101**, 36 (2012) and references therein.

PEROVSKITE/CRYSTALLINE SILICON TANDEMS: IMPACT OF PEROVSKITE BAND GAP AND CRYSTALLINE SILICON CELL ARCHITECTURE

S.L. Luxembourg[1], D. Zhang[2], M. Najafi[2], V. Zardetto[3], S. Veenstra[2], L.J. Geerligs[1]
[1]ECN Solar Energy, P.O. Box 1, NL-1755 ZG Petten, The Netherlands
[2]ECN-Solliance, High Tech Campus 21, NL 5656AE Eindhoven, The Netherlands
[3]TNO-Solliance, High Tech Campus 21, NL 5656AE Eindhoven, The Netherlands

ABSTRACT: The combination of a semi-transparent, wide band gap perovskite solar cell (ST PSC) with a crystalline silicon (c-Si) bottom cell in a tandem configuration is considered a promising approach to enhance solar cell efficiencies over the current single junction c-Si cell efficiencies. In this study, we use a simple model based on experimental EQEs and IV-parameters, to predict the 4-terminal tandem efficiency of advanced industrial c-Si cells of different architectures, combined with an experimental 15% ST PSC with a 1.55eV band gap. In addition, we use this model to study the expected effect of the variation of the band gap of the perovskite absorber on the 4-terminal efficiency which results in a positive outlook towards > 25% tandem efficiencies. Furthermore, we demonstrate an experimental 4-terminal tandem efficiency of 24.1% with a MWT-HIT bottom cell.

Keywords: c-Si, Modelling, Perovskite, Tandem

1 INTRODUCTION

With decreasing cell and module prices, balance-of-system (BoS) costs are making up an increasing share of the total costs of photovoltaic systems. Currently, this share is over 50%. Therefore, reduction of BoS-costs is getting increasingly important to lower the cost of solar electricity. Enhancement of solar cell efficiencies is an important driver of BoS-cost reduction. The realization of tandem devices with crystalline silicon bottom cells and wide band gap perovskite top cells is considered a very promising approach towards high efficiency devices. The ongoing increase in the efficiency of perovskite solar cells and the progress being made in terms of their stability draw a lot of attention to this field of research.

Here, we combine an efficient semi-transparent perovskite solar cell (ST PSC) developed within Solliance with several (advanced) industrial crystalline silicon (c-Si) concepts, which have been developed at ECN or in collaboration with external partners. One of these concepts is a silicon heterojunction solar cell, which is generally considered as promising bottom cell technology due to its high overall performance, high V_{oc} and good near-infrared (NIR) response. The specific silicon-heterojunction cell considered here includes metal-wrap-through contacting (MWT-HIT). In a collaborative development with Choshu Industry Co. an efficiency of 23% was realized for this cell concept.[1] In addition, we have included cells with polysilicon (poly-Si) passivating contacts. First, the Passivated Emitter and Rear Polysilicon cell (PERPoly) design with a realized efficiency of 21.5%. PERPoly cells combine a diffused front side emitter with an n++ doped polysilicon (Poly-Si) rear contact, with fire-through (FT) metallization grids.[2] In such a cell structure performance is limited by front surface recombination. In a further development we also consider the implementation of front side poly-Si passivation (two sided poly-Si or 2s-Poly) which reduces front surface recombination. Although, this introduces significant UV-Vis parasitic absorption, this is largely mitigated when implemented as bottom cell in a perovskite/c-Si tandem.[3]

An experimental evaluation of 4-terminal tandem efficiency was performed for a high efficiency MWT-HIT cell. For the other c-Si cell concepts the tandem efficiency was estimated using a simple modelling approach making use of experimental external quantum efficiencies (EQE). The effect of widening of the perovskite band gap was evaluated on the various cell concepts using the same model.

2 METHODOLOGY

2.1 Semi-transparent perovskite solar cell

A semi-transparent p-i-n perovskite solar cell (ST PSC) including a triple-cation perovskite absorber material with a 1.55 eV band gap was used as top cell in this study. The thickness of the perovskite material was around 570 nm, and the cell size 0.09 cm^2. This organic-inorganic trihalide perovskite absorber material consists of a mixture of methylammonium (CH_3NH_3, MA), formamidinium ($HC(NH_2)_2$, FA) and Cs A-site cations which offers enhanced thermal stability and band gap tuning possibilities over the MA tri-iodide absorber [4]. It includes a double (highly transparent) PTAA / NiO_x hole transport layer (HTL) and a double PCBM / ZnO electron transport layer (ETL). More details will be included in a forthcoming publication.

2.2 c-Si bottom cells

Advanced industrial 6 inch n-type c-Si solar cells were prepared in-house or in collaboration with external partners. Details on the 21.9% - MWT-HIT, 21-5% - PERPOLY and 20.0% - 2s-Poly (provided by Tempress, for more information see [5]). SJ IV-measurements and spectral response measurements to determine the external quantum efficiencies (EQE) were performed on these cells. As described below, we have used a simple modelling approach to estimate the 4-terminal tandem efficiencies for the combination of these cells with the ST PSC described above. We derived low-injection level IV-parameters using neutral density filters.

In addition, a 23% MWT-HIT cell which was developed in collaboration with Choshu Industry Co. [1] was used in this study. This cell will be labelled MWT-HIT_I below to distinguish it from the 21.9%-MWT-HIT cell mentioned above, which will be labeled MWT-HIT_{II}. The MWT-HIT_I cell was laminated including front glass and EVA and rear EVA and white reflective back sheet for the purpose of experimentally determining the 4-terminal tandem efficiency in combination with the ST PSC (see below for details of methodology). As unfortunately, no EQE data of the bare cell was available it was not included in our modelling effort.

2.3 Measurement of tandem efficiency

Due to the large difference in scale of the sub-cells we used an indirect method, previously reported by Werner *et al.*, [6] to determine the 4-terminal tandem efficiency. In short: ST PSC EQE and IV-measurements were performed on a black background. In order to determine the J_{sc} of the c-Si bottom cell the EQE of a laminated cell was measured through a perovskite optical filter. The filter was positioned on top of the c-Si cell using index matching liquid to prevent optical losses. Its V_{oc} and FF corresponding to its operation as bottom cell were obtained by adjusting the solar simulator's light intensity to meet the low injection level operation conditions of the bottom cell. The 4-terminal tandem efficiency is then obtained as the sum of the PSC and filtered c-Si bottom cell efficiencies.

2.4 Modelling of tandem efficiency

4-Terminal tandem efficiencies for several advanced industrial c-Si bottom cell architectures combined with the ST PSC described above were determined from a simple modelling approach based on experimental EQEs. A similar approach was outlined previously in Yu *et al.* [7]. The J_{sc} of c-Si bottom cell was derived from its experimentally determined single-junction (SJ) EQE according to:

$$J_{sc,SJ}(\lambda) = q \cdot EQE_{SJ}(\lambda) \cdot \Phi_{incident}(\lambda)$$

$$f_{bottom}(\lambda) = \frac{\Phi_{bottom}(\lambda)}{\Phi_{incident}(\lambda)}$$

$$J_{sc,bottom}(\lambda) = f_{bottom}(\lambda) \cdot J_{sc,SJ}(\lambda)$$

Here, $J_{sc,SJ}(\lambda)$ corresponds to the short-circuit current density per unit of wavelength of the SJ device under AM1.5 illumination. $\Phi_{incident}(\lambda)$ indicates the AM1.5 photon flux, while $\Phi_{bottom}(\lambda)$ corresponds to the photon flux that reaches the bottom cell. Then $f_{bottom}(\lambda)$ refers to the fraction of $\Phi_{incident}(\lambda)$ reaching the bottom cell and $J_{sc,bottom}(\lambda)$ to the bottom cell short-circuit current density per unit of wavelength. $J_{sc,bottom}$ is obtained from the integration of $J_{sc,bottom}(\lambda)$.

Bottom cell V_{oc} and FF are obtained from low-injection level IV measurements as explained above and combined with the modelled $J_{sc,bottom}$ to arrive at the bottom cell efficiencies. The overall 4-T tandem efficiency was obtained by a lossless addition of the top and bottom cells.

To study the effect of the variation of the perovskite top cell's band gap on the overall tandem and bottom cell efficiency a simple approach was taken. The $f_{bottom}(\lambda)$ as defined above was shifted to mimic the presence of a wider band gap top cell, while the shape of the curve and asymptotic value were preserved. By doing so we implicitly assume that an increase of the perovskite band gap will result in a blue shift of its absorption coefficient and that the parasitic absorption (in TCOs, ETL and HTL) is at a constant level over the considered energy (shift) range. Furthermore, we have assumed a constant transmittance below the band gap by taking an spectrally weighted average of the transmittance over the below band gap energy range. Examples of $f_{bottom}(\lambda)$ curves resulting from this procedure are included in Fig. 1.

The J_{sc} of the wider band gap PSCs was obtained from a correction of the 1.55eV band gap J_{sc} through calculation of the current density equivalent of the enhanced transmission. V_{oc} and FF values were derived by assuming constant fractions of the Shockley-Queisser [8, 9] limited values. The magnitudes of these fractions were derived from the experimental ST PSC device with 1.55eV band gap.

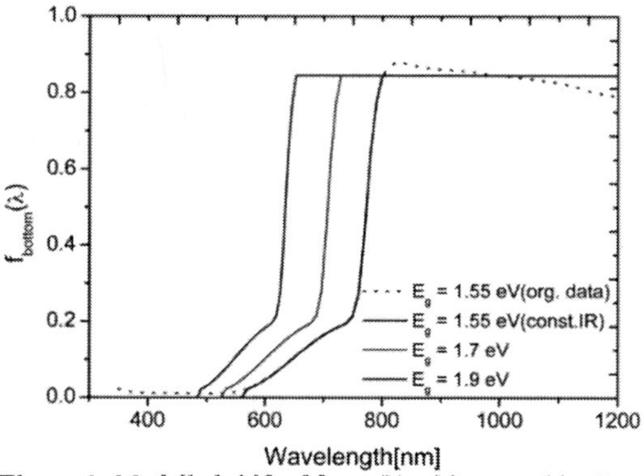

Figure 1: Modelled shift of $f_{bottom}(\lambda)$ with perovskite E_g

3 RESULTS

3.1 Experimental results

The IV-characteristics of the ST PSC are included in Table I. The measurements show low hysteresis, and a stabilized efficiency of 15.0% in a five-minute maximum power point tracking experiment. For preconditioning the cell underwent a 10-minute UV exposure from the glass side. The current ST PSC J_{sc}, V_{oc} and FF values correspond to 0.73, 0.84 and 0.79 of their respective Shockley-Queisser [8, 9] limited values for a band gap of 1.55eV. Below, we use these V_{oc} and FF fractions to estimate PSC and tandem efficiencies for various band gaps. Optical and electrical optimization of the current ST PSC design is expected to result in further efficiency enhancement.

Table I: JV-characteristics of the p-i-n triple-cation ST PSC used as top cell in this study

	J_{sc} [mA/cm^2]	V_{oc} [V]	FF	η [%]
Backward	20	1.056	0.725	15.3
Forward	19.9	1.057	0.716	15
Max. PPT				15

JV-characteristics of the c-Si solar cells which were employed as bottom cell in this study are included in Table II. The variation in V_{oc} among the different devices results from differences in realized recombination currents. As can be expected from the high quality passivation in these devices, the cells with front-and-rear passivating contacts: the MWT-HIT$_{II}$ and the 2s-Poly cells, demonstrate the highest V_{oc} values. The current 2s-Poly cell includes front and rear poly-Si passivating contacts with a thickness of approximately 100 nm. Such thick poly-Si front contact is not optimized for SJ operation due to the associated large UV-Vis optical losses [3, 10], which are the cause of the relatively low J_{sc} of the 2s-Poly cell. In Fig. 2 EQE curves of the PERPoly, MWT-HIT$_{II}$ and 2s-Poly cells are included. From this figure the low wavelength parasitic absorption in the 2s-Poly cell is apparent, The MWT-HIT$_{II}$ cell shows reduced EQE values in the 800 – 1000 nm range which is expected to result from absorption in the not-optimized TCO layers. The MWT-HIT$_{II}$ and 2s-Poly cells demonstrate the highest EQE in the NIR (above 1000 nm).

Table II: Single-junction JV-characteristics of the c-Si cells used as bottom cell in this study. The cells in the first three rows were used in our modelling approach, while the MWT-HIT$_I$ cell was used in the experimental analysis. The results of the latter were obtained on a single-cell laminate including a 3 mm edge of exposed white back sheet. In the last row the realized JV-characteristics of the MWT-HIT$_I$ as bottom cell are included.

	J_{sc} [mA/cm^2]	V_{oc} [V]	FF	η [%]
2s-Poly	38.2	0.685	0.77	20.0
PERPoly	39.7	0.676	0.80	21.5
MWT-HIT	39.3	0.723	0.77	21.9
MWT-HIT$_I$	38.9	0.732	0.77	21.8
MWT-HIT$_{I,bot}$	16.1	0.711	0.80	9.1

The JV-characteristics of the c-Si cells under low-injection level conditions were determined in JV-measurements using a set of neutral density filters to vary the light intensity. These characteristics are used below to calculate the modelled tandem efficiencies below. For the PERPoly and 2s-Poly cells this procedure was performed on laminated cells.

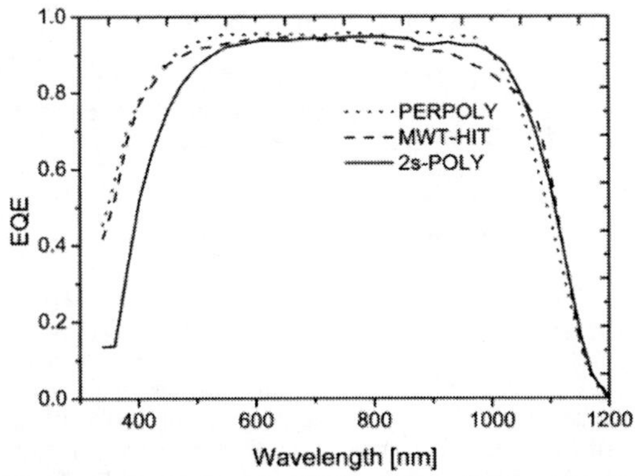

Figure 2: Single-junction EQE curves of the c-Si cells which were evaluated as bottom cell using our modelling approach

An experimental 4-terminal tandem efficiency was determined for the ST PSC joined with the laminated MWT-HIT$_I$ cell obtained as described in the Methodology section. This resulted in a 24.1% tandem efficiency, corresponding to a 2.3%$_{abs}$-efficiency gain over the SJ MWT-HIT$_I$ efficiency measured on the laminated cell. The MWT-HIT$_I$ SJ and bottom cell (MWT-HIT$_{I,bot}$) characteristics are included in Table II. The recorded EQE curves are displayed in Fig. 3. From the comparison of the SJ MWT-HIT$_I$ laminate EQE (J_{sc} = 38.9 mA/cm^2) to the tandem EQE (J_{sc} = 35.6 mA/cm^2) it is apparent that there is significant current loss due to parasitic absorption in the charge collecting layers of the top cell. In addition, there is a clear above band gap tail in the top cell transmission, which may be reduced by light management measures such as front surface texturing. f_{bottom} (λ) was determined from these experimental data as the ratio of the bottom cell EQE measured through the perovskite filter (dashed red curve in Fig. 3) to the SJ EQE (solid black curve). The derived f_{bottom} (λ) was assumed to be valid for all c-Si bottom cells considered in this study. In view of the similarities in their cell design this seems a valid assumption.

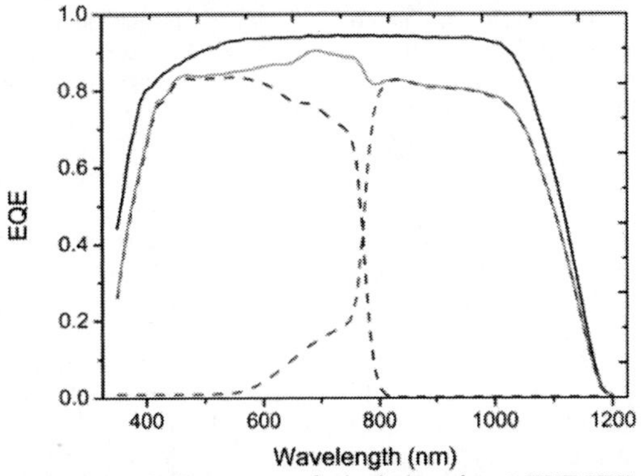

Figure 3: EQE curves of single-junction MWT-HIT$_I$ (solid black) and as bottom cell (dashed red), the ST PSC (dashed blue) and the overall 4-terminal tandem (solid grey)

3.2 Modelled tandem efficiencies

Using the methodology outlined above and the cell parameter inputs as determined in the previous section, we estimated the tandem efficiencies for the c-Si bottom cells included in Table II and the ST PSC described in section 3.1. In addition, we varied the band gap of the perovskite absorber up to 2.0 eV using the methodology described in section 2.4. The resulting tandem efficiencies are included in Fig. 4 as the three lower curves.

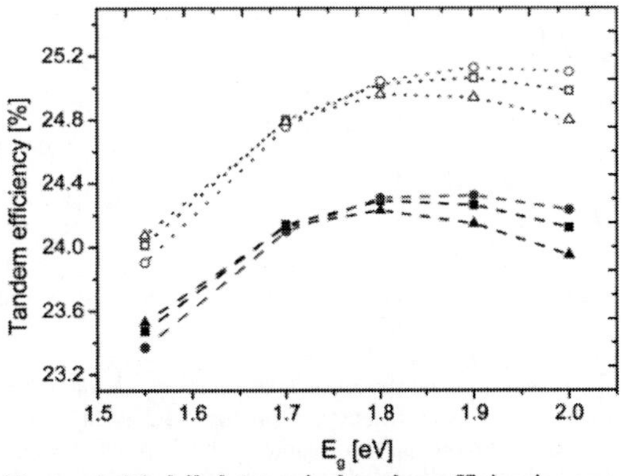

Figure 4: Modelled 4-terminal tandem efficiencies as a function of the ST PSC band gap for a variety of c-Si bottom cells (PERPoly (black), MWT-HIT$_{II}$(red), 2s-Poly (blue)) for the experimentally determined top cell transparency (dashed) as well as for an increased NIR transparency (dotted).

Although, there is a spread of close to 2% in the SJ efficiencies of the c-Si bottom cells, the differences in tandem efficiencies are only very minor. This is in particular the case for perovskite band gap values in the range of 1.55 – 1.8 eV. At 1.55 eV band gap the predicted tandem efficiencies are all between 23.4 – 23.5%. The 2s-Poly cell, which has the lowest SJ efficiency of the considered cells, is even predicted to have the highest tandem efficiency at this band gap, with a tandem over SJ gain of 3.5%. As was mentioned above the SJ efficiency of the 2s-Poly cell is limited by the UV-Vis parasitic absorption in the front poly-Si layer. In the tandem configuration the perovskite top cell filters and uses this low wavelength range effectively. The 2s-Poly cell shows the best overall NIR response which has the largest impact in combination with the lowest band gap perovskite solar cell. In addition, its FF shows a steeper upward trend than the PERPoly cell with decreasing illumination level (corresponding to decreasing perovskite band gap). This is attributed to a higher realized series resistance of this cell. The 2s-Poly cell was selected for the highest V_{oc} and was somewhat underfired, resulting in a higher contact resistance. The FF of the $MWT\text{-}HIT_{II}$ cell studied with the model showed a downward trend of the FF versus illumination level, which is related to a lower shunt resistance.

Under the reduced illumination levels we find for all cells a linear change of V_{oc} with the log of the J_{sc} (see Fig. 5). Since the 2s-Poly cell loses the least in J_{sc} when operated as bottom cell, its reduction in V_{oc} will also be relatively low. The factors mentioned above provide a qualitative insight in the origin of the very similar 4-terminal tandem efficiencies for the different c-Si cells. As for higher perovskite band gap values (and thus higher bottom cell injection levels) the bottom cell JV-characteristics are getting closer to their SJ values. Consequently, the modelled tandem efficiencies are more in line with what would be expected on the basis of the SJ efficiencies (see also Fig. 4).

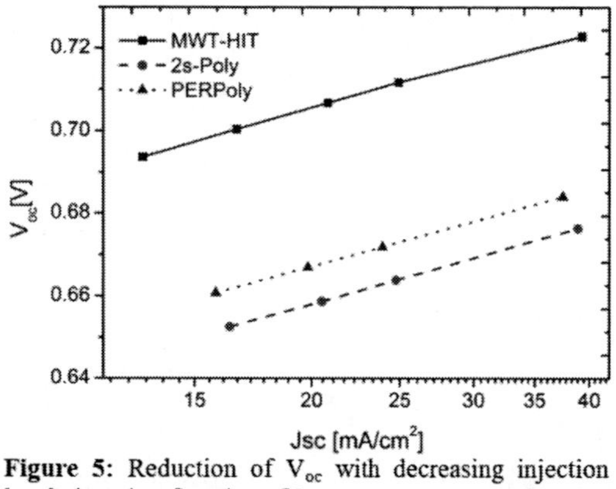

Figure 5: Reduction of V_{oc} with decreasing injection level plotted as function of J_{sc}

Our modelling exercise predicts an approximate 1% gain in tandem efficiency when the PSC band gap is increased to the more optimum values of 1.8-1.9 eV. Here, the model predicts tandem efficiencies of 24.3% for MWT-HIT and PERPoly cell architectures and 24.2% for the 2s-Poly cell, corresponding to tandem over SJ gains of between 2.4% and 4.2%.

The efficiency values in the upper curves have been obtained under the assumption of an increased NIR transparency of 0.9. Such increased transparency is thought to be realized through reduction of parasitic absorption in particular in the PSC TCOs. A possible route to achieve this could be to replace the current ITO layers by more transparent IO:H layers in future devices.

The modelled increase in transparency should enable tandem efficiencies larger than 25% for the considered c-Si solar cells, which are considered to be advanced but industrial cell architectures. Our on-going work on the improvement of the semi-transparent perovskite to cell will further improve its efficiency, which will facilitate 4-terminal tandem efficiencies beyond the values quoted in this study.

4 CONCLUSIONS

We have demonstrated a 24.1% 4-terminal tandem measurement for the combination of 15% semi-transparent perovskite solar cell and a c-Si MWT-HIT bottom cell. With a simple model based on experimental EQEs we have compared the performance of several advanced industrial c-Si solar cell architectures. Although, single-junction efficiencies of these cells differ by approximately $2\%_{abs}$ the modelled 4-terminal tandem efficiencies for a perovskite band gap of 1.55eV become nearly equal. As discussed in the text this is a consequence of the specifics of the individual cells considered. We show that for an optimal perovskite band gap of 1.8 – 1.9 eV and improvement of the NIR transparency of the top cell from 0.85 to 0.9 tandem efficiencies > 25% are expected. This enhanced transparency corresponds to a reduction parasitic absorption. In addition, the current perovskite cell can be improved in terms of light management and will also benefit from reduction of (TCO) parasitic absorption. From electrical modeling the prescence of a significant series resistance in the ST PSC device was identified. In future work we aim to reduce this series resistance by improving the electrode geometry thereby increasing the FF and thus the PCE of the ST-PSC device These on-going improvements will enable further increase of the perovskite/c-Si 4-terminal tandem efficiencies.

ACKNOWLEDGEMENTS

We would like to thank Tempress Systems, Vaassen, The Netherlands, and in particular dr. R. Naber, for kindly providing the front-and-rear polysilicon solar cell (2s-Poly) used in this study. In addition, we would like to thank the people at ECN working on the PERPoly cell and MWT-HIT concepts, in particular G. Coletti, M.K. Stodolny and I. Romijn for providing the PERPoly and MWT-HIT cells and Choshu Industry Co. for providing the $MWT\text{-}HIT_{CIC}$ cell.

REFERENCES

[1] G. Coletti, F. Ishimura, Y. Wu, E.E. Bende, G.J.M. Janssen, K. Hashimoto, Y. Watabe, B.B. van Aken, 32nd EU PVSEC (2016).
[2] M.K. Stodolny, M. Lenes, Y. Wu, G.J.M. Janssen, I. Romijn, J.R.M. Luchies, L.J. Geerligs, Solar Energy Materials and Solar Cells 158 (2016) 24-28.
[3] S.L. Luxembourg, D. Zhang, Y. Wu, M. Najafi, V. Zardetto, W. Verhees, A.R. Burgers, S. Veenstra, L.J.

Geerligs, Energy Procedia 124 (2017) 621-627.
[4] D.P. McMeekin, G. Sadoughi, W. Rehman, G.E. Eperon, M. Saliba, M.T. Hörantner, A. Haghighirad, N. Sakai, L. Korte, B. Rech, M.B. Johnston, L.M. Herz, H.J. Snaith, Science 351 (2016) 151-155.
[5] R.C.G. Naber, M. Lenes, J.R.M. Luchies, 33rd EU PVSEC (2017) 2DO.3.6.
[6] J. Werner, G. Dubuis, A. Walter, P. Löper, S.-J. Moon, S. Nicolay, M. Morales-Masis, S. De Wolf, B. Niesen, C. Ballif, C. Solar Energy Materials and Solar Cells 141 (2015) 407–413.
[7] Z. Yu, M. Leilaeioun, Z. Holman, Nature Energy 1 (2016) 1-4.
[8] W. Shockley, H.J. Queisser, Journal of Applied Physics 32 (1961) 510–519.
[9] S. Rühle, Solar Energy 130 (2016) 139-147.
[10] S. Reiter, N. Koper, R. Reineke-Koch, Y. Larionova, M. Turcu, J. Krügener, D. Tetzlaff, T. Wietler, U. Höhne, J.-D. Kähler, R. Brendel, R. Peibst, Energy Procedia 92 (2016) 199-204.

A ONE-STEP DEPOSITION METHOD ASSISTED WITH NON POLAR WASHING SOLVENT TREATMENT TO PRODUCE UNIFORM THIN LAYERS OF PEROVSKITE VALIDATED THROUGH ELLIPSOMETRY

C. Montes[1,2], L. Ocaña[1], C. Quinto[1], S. González-Pérez[2], B. González-Díaz[2],
C. Hernández-Rodríguez[3], M. Friend[1] & M. Cendagorta[1].
[1]Instituto Tecnológico y de Energías Renovables, S. A. (ITER)
Pol. Industrial de Granadilla, s/n
E 38600 Granadilla de Abona, Spain.
Ph. +34 922 747 700
Fax +34 922 747 701
E-mail cmontes@iter.es
[2]Departamento de Ingeniería Industrial.
[3]Departamento de Física.
Universidad de La Laguna.
Avd. Astrofísico Francisco Sánchez, s/n
38206 S/C de Tenerife. Spain.
Ph. +34 922 316 502, EXT 6252
E-mail bgdiaz@ull.edu.es

ABSTRACT: The present work summarizes the methodology developed by the laboratory of solar cells (SiCelLab) from the Instituto Tecnológico y de Energías Renovables (Canary Islands, Spain), which has been validated by the use of spectral ellipsometry characterization, in order to evaluate the optical properties and the thickness of thin films of methylammonium lead iodide ($CH_3NH_3PBI_3$), deposited from solutions of their chemical precursors, via spin coating over glass substrates by the one-step method, assisted with non polar washing solvent treatment. The application of this methodology has lead to produce samples with a high degree of uniformity and repetitiveness.
Keywords: Characterization, deposition, optical properties, thin film, perovskite.

1 INTRODUCTION

The use of thin film organometal halide perovskites as the light-harvesting layer on solar cell devices, has led to the development of technologies to achieve unprecedented increases on their power conversion efficiencies (PCE) in a very short period of time [1,2].

In the literature have been reported different methods for producing perovskite layers [3-5]. One of the most extended procedures is based on the use of wet chemical formulations of its precursors by dissolving them in polar organic solvents, which then can be deposited by spin coating techniques [6,7].

The laboratory of solar cells (SiCelLab) from the Instituto Tecnológico y de Energías Renovables (ITER), with the support of the University of La Laguna (ULL), has been researching extensively on the one-step deposition method, assisted with non polar washing treatment, in order to produce thin films of perovskite [8,9] and, since the optoelectronic properties of these films strongly depend on their corresponding processing parameters [10,11], a spectral ellipsometer has been used to elucidate and to establish the correspondence between the fabrication parameters and the optoelectronic properties of the resulting devices.

2 EXPERIMENTAL

2.1 Fabrication methodology

The actual sample fabrication required two days and was divided into four subprocesses. All of them were carried out inside an ISO 7 cleanroom, operating at a set temperature of 21 °C and, although our current facilities do not have the means for controlling the ambient relative humidity (RH), this parameter were monitored to guarantee that the fabrication process was carried out at values below 50% RH.

During the first day, the substrates were cleaned and the perovskite precursor solution was prepared. The deposition of the layer to be studied and the curing of the sample were undertaken during the second day.

The substrates, with a size of 25mm x 25 mm x 2 mm, were cleaned in a 4 stages sonication process with temperature during 10 minutes each. In this sense, the substrates were immersed sequentially in a 2% soap solution, in ultra pure water, in acetone and finally in 2-propanol [6]. The rinsing of the substrates between stages was carried out with ultrapure water at a temperature close to the boiling point. This allowed erasing all the active agents that may remain after each cleaning step [12]. Finally, the substrates were dried with nitrogen and stored individually.

The synthesis of the perovskite precursor was performed by stirring PbI_2 and CH_3NH_3I at a molar ratio of 1:1 (0.723 g: 0.25 g) in dimethylformamide (DMF) solvent (1260 μL) at 75 °C [6,7]. After 50 minutes stirring, the velocity was increased (~ 1000 rpm). Finally, the solution was further stirred overnight, this time at room temperature. This allowed for the molecules of the precursors to be adequately suspended and to avoid undesired sedimentation effects.

The perovskite precursor solutions were spin-coated in a one-step deposition method. During this process the relative humidity inside the chamber of the spin-coater was controlled so that the depositions were carried out only when values below 10% RH were reached. The perovskite precursor solutions (100 μL) were statically dispensed on top of a glass substrates, and then the samples were subjected at a spin rate of 5000 rpm for 30 seconds. At the 6th second along the spinning, a non-polar solvent, in this case chlorobenzene, was dispensed in a 3:1 volume ratio with respect to the precursor [6]. This washing treatment helped evacuate the DMF from the precursor solution, accelerating and improving the crystallization of the perovskite [7].

After the deposition process, the samples were cured for 10 minutes inside a stove at a temperature of 100 °C [6]. Finally, the samples were stored in a vacuum desiccator to avoid the degradation of the deposited layers due to the ambient humidity [13].

2.2 Characterization methodology

To get started, all the samples were examined by using a digital microscope [14] just after their deposition, evaluating their macroscopic appearance before proceeding with any further characterization.

Also, in order to elucidate whether the initial deposited layers effectively were of methylammonium lead iodide, X-ray crystallography techniques [15] were applied.

However, the main characterization method applied throughout the project relied on the use of a spectroscopic ellipsometer in order to derive the optical properties of the deposited layers and compare them with those published in the literature [16,17].

With such an equipment, it was possible to obtain the ϕ (PSI) and δ (DELTA) as experimental values for the glass substrate and the deposited thin film considered as a whole and then to develop a fit model from which to derive the optical properties of the studied material [9]. Thus, once set the experimental data to the model, the optical properties of the samples, the refractive index (n) and the extinction coefficient (κ) could be obtained. The absorption coefficient (α) was estimated from the extinction coefficient by the following equation:

$$\alpha = (4 \cdot \pi \cdot \kappa) / \lambda \qquad [18]$$

The values of the deposited thickness derived from the fit model were validated by comparing them with the ones measured by an atomic force microscope (AFM) [9, 19].

The ellipsometer also allowed obtaining experimentally the values of transmittance and optical reflectance of the studied samples, so it was possible to cross-reference them with the values published in the literature [16,20,21] and, in addition, to derive the values of the bandgap of the samples from their optical transmittance [22].

Finally, in order to gain data with a spatial resolution over the surface of the deposited samples, the ellipsometry measurements were performed in a five-point pattern, which will be explained in the following section.

3 RESULTS AND DISCUSSION.

As it can be observed in figure 1, the appearance of a typical sample deposited with the fabrication methodology described in this paper was of a smooth brownish layer that uniformly covers the entire surface of the substrate. Neither discoloration nor defects were appreciable.

Figure 1: Image taken with a digital microscope, depicting a typical perovskite layer deposited on a 25mm x 25 mm x 2 mm substrate with the current fabrication methodology.

The studies conducted by exposing the deposited layers to X-ray diffraction (XRD) showed [8] significant peaks at 14.08°, 28.41°, 31.85°, and 43.19°, which had been assigned to (110), (220), (310), and (330) of pure CH3NH3PbI3 crystals, respectively [23, 24].

As it can be seen in figure 2, the refractive index and extinction coefficient corresponding to the best sample produced (ID number 17027), when plotted together with the ones published by Löper et al [15], Leguy et al [17] and a previous results published by our laboratory [9], adjusted better with the ones from Löper et al [16] (although the obtained values are rather closer to the ones from Leguy et al[17] for the peaks at the refractive index values at 400 and 500 nm and at the extinction coefficient peak values at 350 and 450 nm).

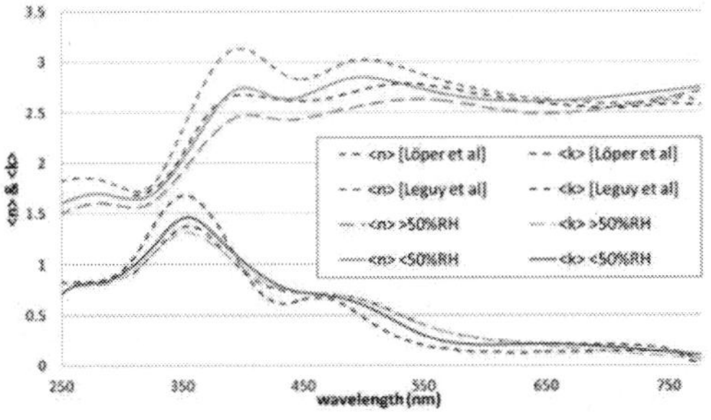

Figure 2: Comparison of the refractive index and the extinction coefficients.

Figure 3 shows a similar adjustment, this time considering the absorption coefficient. Notice that, taking into account our previously published results, the ones communicated in the present paper were a noticeable improvement on the measured absorption in the range of the visible spectrum.

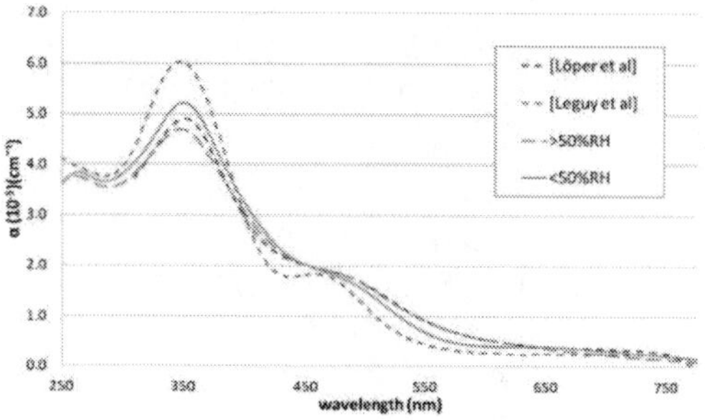

Figure 3: Comparison of the absorption coefficients.

The values obtained for the reflectance and the transmittance were cross-referenced with the literature [16, 25, 26]. Thus, figure 4 shows that the obtained reflectance values are located between 0.2 and 0.4, while the transmittance ones begin from 500 nm on. Also, for the transmittance, it can be observed that there is an inflection point around the 760 nm, which matches with the band gap value.

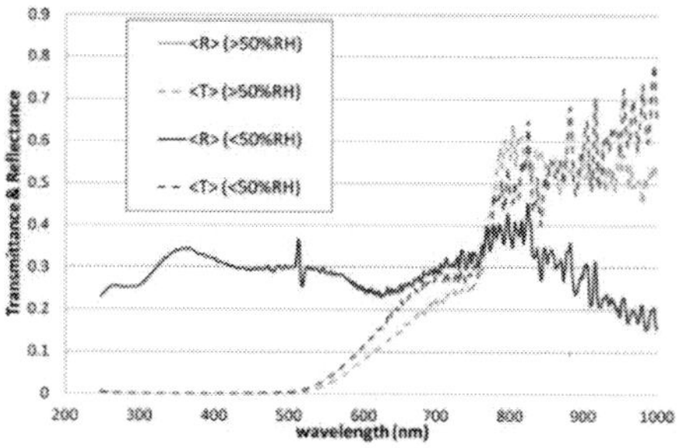

Figure 4: Comparison of the reflectance and the transmittance.

Finally, by using the fit model, the thickness of the deposited layers was obtained and cross-referenced with AFM measurements [9]. Also, by taking measurements distributed over the samples in five locations, the uniformity of such layers was estimated. As it can be seen in figure 5 and Table I, the average thickness of the best sample produced with the current fabrication methodology was around 370 nm, which was a value similar to the one published by Manda Xiao et al [6], improving our previously published results [9].

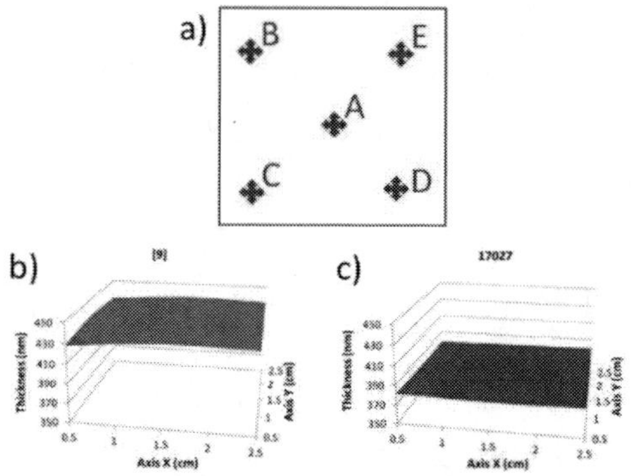

Figure 5: Measurement points (a), perovskite thickness distribution of samples from previously published results (b) and fabricated by current methodology (c).

Table I: Values obtained for the bandgap, thickness, average thickness and standard deviation of the best sample deposited by the current fabrication methodology (17027) and in previously published results (16035).

Sample ID	Band Gap (eV)	Thickness (nm)	Average Thickness (nm)	Standard deviation (nm)
16035_A	1.60	438.78		
16035_B	1.60	428.92		
16035_C	1.60	427.58	432.15	4.37
16035_D	1.60	433.24		
16035_E	1.60	432.25		
17027_A	1.60	377.85		
17027_B	1.60	377.91		
17027_C	1.60	381.26	379.35	1.46
17027_D	1.60	379.95		
17027_E	1.60	379.80		

Thus, in order to conclude, the measurements performed in the samples obtained by the applied fabrication methodology, which has been refined by making use of a characterization methodology that is mainly supported by spectroscopic ellipsometry, seem to lead to a consistent improvement on the quality of the deposited layers.

4 REFERENCES

[1] Akihiro Kojima, Kenjiro Teshima, Yasuo Shirai, and Tsutomu Miyasaka. Organometal Halide Perovskites as Visible-Light Sensitizers for Photovoltaic Cells. Journal of the American Chemical Society 2009 131 (17), 6050-6051 DOI: 10.1021/ja809598r

[2] Research Cell Efficiency Records Chart (webpage). National Center for Photovoltaics, NREL. Acess: February 2017.

https://www.nrel.gov/pv/assets/images/efficiency_chart.jpg
[3] Mehran Habibi, Fatemeh Zabihi, Mohammad Reza Ahmadian-Yazdi, Morteza Eslamian, Progress in emerging solution-processed thin film solar cells – Part II: Perovskite solar cells, Renewable and Sustainable Energy Reviews, Volume 62, September 2016, Pages 1012-1031, ISSN 1364-0321, http://dx.doi.org/10.1016/j.rser.2016.05.042.
[4] Jiandong Fan, Baohua Jia, and Min Gu, "Perovskite-based low-cost and high-efficiency hybrid halide solar cells," Photon. Res. 2, 111-120 (2014)
[5] Zhaoning Song ; Suneth C. Watthage ; Adam B. Phillips and Michael J. Heben "Pathways toward high-performance perovskite solar cells: review of recent advances in organo-metal halide perovskites for photovoltaic applications", J. Photon. Energy. 6(2), 022001 (Apr 15, 2016). ; http://dx.doi.org/10.1117/1.JPE.6.022001
[6] Xiao, M., Huang, F., Huang, W., Dkhissi, Y., Zhu, Y., Etheridge, J., Gray-Weale, A., Bach, U., Cheng, Y.-B. and Spiccia, L. A Fast Deposition-Crystallization Procedure for Highly Efficient Lead Iodide Perovskite Thin-Film Solar Cells, Angew. Chem. Int. Ed. 2014 53: 9898–9903. doi:10.1002/anie.201405334
[7] Kuen-Feng Lin, Sheng Hsiung Chang, Kai-Hung Wang, Hsin-Ming Cheng, Kuo Yuan Chiu, Kun-Mu Lee, Sheng-Hui Chen, Chun-Guey Wu, Unraveling the high performance of tri-iodide perovskite absorber based photovoltaics with a non-polar solvent washing treatment, Solar Energy Materials and Solar Cells, Volume 141, October 2015, Pages 309-314, ISSN 0927-0248, http://dx.doi.org/10.1016/j.solmat.2015.06.002
[8] L.Ocaña, A. Linares, E. Llarena, C. Montes, O. González, D. Molina, A. Pío, C. Quinto, M. Friend and M. Cendagorta, "Adaptation of a Crystalline silicon solar cell laboratory to produce perovskite solar devices", 31st European Photovoltaic Solar Energy Conference and Exhibition, pp. 1138-1143, 14-18 September 2015, Hamburg, Germany.
[9] L. Ocaña, C. Quinto, C. Montes, A. Linares, E. Llarena, O. González, D. Molina, A. Pío, C. Hernández-Rodríguez, S. González-Pérez, R. Guerrero-Lemus, M. Friend and M. Cendagorta, "Optimizing the deposition of thin layers of organic-inorganic hybrid perovskite methylamonium lead iodide (CH3NH3PbI3) on large surfaces through their optical properties", 32nd European Photovoltaic Solar Energy Conference and Exhibition, pp. 1316-1323, 21-24 June 2016, Munich, Germany.
[10] Q. Wang, Q. Dong, Z. Xiao, Y. Yuan and J. Huang, Large fill-factor bilayer iodine perovskite solar cells fabricated by a low-temperature solution-process. Energy Environ. Sci., 2014, 7, 2359–2365. DOI: 10.1039/C4EE00233D
[11] H. Zhou, Q. Chen, G. Li, S. Luo, T.-B. Song, H.-S. Duan, Z. Hong, J. You, Y. Liu and Y. Yang, Photovoltaics. Interface engineering of highly efficient perovskite solar cells. Science, 2014, 345, 542–546. doi: 10.1126/science.1254050
[12] I101 Perovskite Processing Guides (webpage). Ossila Ltd, Kroto Innovation Centre. Access: February 2017. https://www.ossila.com/products/perovskite-ink-air
[X1] J.Yang, B.D. Siempelkamp, D. Liu, T.L. Kelly, Investigation of CH3NH3PbI3 degradation rates and mechanisms in controlled humidity environments using in situ techniques".. ACS Nano. 24 February 2015, Vol.9(2), pp.1955-1963.
[14] Digital Microscope U500X, focus range 15mm - 40mm, range 50X-500X by Unimake.
[15] X-ray diffractometer available at the X-ray Crystallography laboratory, within the Research Support General Service from the University of La Laguna (Canary Islands, Spain).
[16] Löper, P.; Stuckelberger M.; Niesen, B.; Werner J.; Filipič W.; Moon, S.; Yum, J.; Topič M.; De Wolf, S.; Ballif, C. Complex Refractive Index Spectra of CH3NH3PbI3 Perovskite Thin Films Determined by Spectroscopic Ellipsometry and Spectrophotometry, The Journal of Physical Chemistry Letters 2015 6 (1), 66-71, doi: 10.1021/jz502471h
[17] Leguy A.; Hu Y.; Campoy-Quiles M.; Alonso M., Weber O.; Azarhoosh P.; Schilfgaarde M.; Weller M.; Bein T.; Nelson J.; Docampo P.; Barnes P., Reversible Hydration of CH3NH3PbI3 in Films, Single Crystals, and Solar Cells, Chemistry of Materials 2015 27 (9), 3397-3407 doi: 10.1021/acs.chemmater.5b00660
[18] H. Fujiwara, Spectroscopic Ellipsometry: Principles and Applications. 20079,19
[19] Atomic Force Microscope (AFM), model Nanoscope III Multimode by Digital Instrument (Santa Barbara, California) available at the Atomic Force Microscope Service, within the Research Support General Service, integrated in the Univiserity of La Laguna.
[20] The Duong, Niraj Lal, Dale Grant, Daniel Jacobs, Peiting Zheng, Shakir Rahman, Heping Shen, Matthew Stocks, Andrew Blakers, Klaus Weber, Thomas P. White, and Kylie R. Catchpole, "Semitransparent Perovskite Solar Cell With Sputtered Front and Rear Electrodes for a Four-Terminal Tandem". IEEE JOURNAL OF PHOTOVOLTAICS
[21] Yajie Jiang , Martin A. Green, Rui Sheng, Anita Ho-Baillie, "Room temperature optical properties of organic–inorganic lead halide perovskites". Solar Energy Materials & Solar Cells, Volume 137, June 2015, Pages 253–257.
[22] Jayant Dharma (1), Aniruddha Pisal (2) PerkinElmer (3). Simple Method of Measuring the Band Gap Energy Value of TiO2 in the Powder Form using a UV/Vis/NIR Spectrometer. (1)PerkinElmer Technical Center, Global Application Laboratory (2) PerkinElmer and Inc. Shelton, CT USA (3).
[23] M. Liu, M. B. Johnston, and H. J. Snaith, "Efficient planar heterojunction perovskite solar cells by vapour deposition," Nature 501, 395–398 (2013).
[24] Qi Chen, Huanping Zhou, Ziruo Hong, Song Luo, Hsin-Sheng Duan, Hsin-Hua Wang, Yongsheng Liu, Gang Li, and Yang Yang, "Planar Heterojunction Perovskite Solar Cells via Vapor-Assisted Solution Process", Journal of the American Chemical Society 2014,136(2), 622-625
[25] The Duong, Niraj Lal, Dale Grant, Daniel Jacobs, Peiting Zheng, Shakir Rahman, Heping Shen, Matthew Stocks, Andrew Blakers, Klaus Weber, Thomas P. White, and Kylie R. Catchpole, "Semitransparent Perovskite Solar Cell With Sputtered Front and Rear Electrodes for a Four-

Terminal Tandem". IEEE JOURNAL OF PHOTOVOLTAICS

[26] Yajie Jiang , Martin A. Green, Rui Sheng, Anita Ho-Baillie, "Room temperature optical properties of organic–inorganic lead halide perovskites". Solar Energy Materials & Solar Cells, Volume 137, June 2015, Pages 253–257

THE POTENTIAL OF PEROVSKITE SOLAR CELLS IN MOROCCO

Saida LAALIOUI *[1,2], Kawtar BELRHITI ALAOUI [1,2], Aziza CHACHDI [1,3], Houda AITDADS [2], Kassem EL ASSALI[2], Bouchra RAHMOUNI [3], Badr IKKEN [1], Abdelkader OUTZOURHIT [2]

[1]Research Institute for Solar Energy and New Energies (IRESEN), 16, street Amir Sidi Mohamed souissi - Rabat, Morocco
Tel: +212 (0) 537 682 236 | Fax : +212 (0) 537 688 852
[2] Laboratory of Nanomaterials for Energy and Environnement (LNEE), Faculty of Sciences Semlalia, Cadi Ayyad University, PO Box: 2390, Marrakech 40000, Morocco
[3]Faculty of Juridical, Economic and Social Sciences, Km 3 road of Casablanca, B.P784, Settat, Morocco

*laalioui@iresen.org , belrhiti@iresen.org , azizachachdi1@gmail.com , h.aitdads@gmail.com, elassali@uca.ma, bouchra@benhida.ma, ikken@iresen.org, aoutzour@uca.ma

ABSTRACT: The performance of solar cells based on hybrid perovskites has undergone an unprecedented evolution and has stimulated the interest of the scientific and industrial community in the photovoltaics sector.
Nevertheless, no study for the potential of this technology in term of industrial integration and research capability has been performed for Morocco. In order to set up a research roadmap in Morocco for perovskite-based PV, a detailed state of the art of this technology was carried out. Furthermore, a market analysis was performed to in order to gage the interest of the Moroccan industries in this technology. The results show that the tandems cells such using stable perovskite top cell and a silicon heterojunction bottom cell has the highest potential in terms of efficiency, stability and industrial integration. 53% of the companies surveyed expressed their interest in this technology.

Keywords: Tandem, perovskite, Solar Cell Efficiencies, Heterojunction, PV Market.

1 INTRODUCTION

Morocco depends heavily on imports to meet its energy needs, since it imported 98% of its primary energy needs in 2009. Given the dynamics that the national economy is undergoing, the demand for primary energy is raising by an average of almost 5% during the past decade. To alleviate this energy dependency, Morocco launched an energy strategy that is axed on renewable energies. The Moroccan energy strategy aims to raise the part of renewables to 42% of installed capacity in 2020, and to 52% in 2030 [1].

To achieve this objective, Morocco will develop between 2016 and 2030, an additional capacity of electricity generation from renewable sources of more than 10,000 MW distributed as follows: 4,500 MW of solar, 4,200 MW of wind, and 1,300 MW of hydro [1].

The "Moroccan solar plan" is one of the corner stones of this strategy. In addition to set targets of solar capacity, the plan includes the development of technical expertise, the establishment of R&D infrastructure and the promotion of an integrated solar industry. The identification of key PV technologies where the R&D community as well as the local industries can actively contribute is therefore necessary.

Within a few years, the performance of solar cells based on hybrid perovskites has undergone an unprecedented evolution rising from 3.8% in 2009 to 22.10% in 2016 [2], [3]. This has stimulated the interest of the scientific and industrial community in this sector.

Moreover, the abundance and low cost of these materials and their ease of implementation offer a great potential for industrialization to this sector.

Furthermore, perovskite cells have demonstrated their suitability to be used other technologies in the tandem configuration, which constitute a strong impetus for improvement solar cell performance [4].

To build the necessary research roadmap in Morocco, a detailed state of the art of the technology should be established in order identify best cell configurations and key parameters that should be investigated in local R&D projects as well as technological issues that can be adressed by the Moroccan industries for local production.

Morocco has a huge potential for industrial integration of technologies based on thin film solar cells. However, no market analysis has been yet performed to evaluate the real potential of integration of this technology.

2 METHODOLOGY

2.1 State of the art of perovskite-based solar cells

In this work, we aim to establish the state of the art by studying the materials and structures of solar cells based on perovskite, and comparing them to identify the best choices and to develop a device with an excellent suitability for the Moroccan climate.

It is well known that tandem structures increases the performances of the cell device and even exceed the theoretical Shockley-Queisser limit of the simple junction [5]. Particularly, tandem cells combining a perovskite and a silicon sub cells are investigated.

2.2 The potential of perovskite solar cells in Morocco

Local industries were surveyed in order to find out their interest in this new solar cell technology. Four main families of data collection methods can be used: interviews, secondary data analysis, direct observation and questionnaires [6].

Interviews. The interview takes the shape of a conversation, directed objectively by the interviewer, on the object of the study. The interview doesn't make it possible to work on a large scale in terms of time and the costs. This is why it is not adapted for a quantitative treatment of the main object of our study.

The secondary data are data which were collected or

analyzed by other reliable people or official establishments. They are generally data collected at the time of studies or research; available data via official statistics, or produced by survey institutes, presented in the form of a synthetic description, with or without analysis and interpretation.

The direct observation gives access directly to the facts. The data collected through this tool are structured by grids which make it possible to judge and qualify what is observed. On the other hand, this tool is very relevant to study behaviors, attitudes or interactions, which is not paramount for our study.

The **questionnaire:** is used when the respondent is alone to answer the questions asked. It is a fast method to diffuse where the participant answers at his rhythm without the obligation of displacement, which minimizes the time and the cost of the envisaged investigation. It also limits the effects of interviewer's personalities and adapts to respondent schedules.

The present study relied mainly on the questionnaire method.

3 RESULTS

The results of the different studies will be presented in this section, staring by the perovskite study, followed by the tandem cells investigation and finally the results of the survey.

3.1 State of the art of perovskite technology

3.1.1 Choice of the absorber

Research has shown that tin-based perovskites are very unstable, due to the instability of the Sn^{2+} ions which easily transform into Sn^{4+} [5]. In addition, perovskites containing tin like methylammonium tin halide $MASnX_3$ have lower optical gaps and efficiencies than those of the methylammonium lead halide $MAPbX_3$ [3], [4]. This makes lead-based perovskites best suited for the solar cells [7].

Furthermore, the perovskites containing cesium (Cs) and iodide (I) have a good performances and good stability. However, the perovskites containing cesium and bromide (Br) are not good to use in solar cells because they have high band gap up to 2,25 eV [8].

The best single junction solar cells based on perovskites were made of mixed organic cations (methylammonium (MA) and formamidinium (FA)) and mixed halides (Br and I). For example, the perovskite $(MA)_x(FA)_{1-x}PbI_3$ has an efficiency up to 14.9% [4], and the perovskite $MA_{0.17}FA_{0.83}Pb(I_{0.83}Br_{0.17})_3$ offer an efficiency up to 20.2% [8]. However, $Cs_x(MA)_{1-x}PbI_3$ has a lower efficiency which is of the order of 7.68% [4], but the mixture of the cations of Cs and FA has reached an efficiency of 16.5% [8].

One of the interesting approaches is to combine cesium Cs with the two organic cations MA and FA to have a triple cation configuration (Cs / MA / FA). This approach has allowed an efficiency up to 21.1% with the combination$Cs_x(MA_{0.17}FA_{0.83})_{(100-x)}$ Pb $(I_{0.83}Br_{0.17})_3$ [8]. Indeed, this configuration offers a reproducible perovskite films and guaranties a good thermal stability.

When MA is replaced by a larger cation such as ethylamonium EA, the structure of the perovskite changes completely, moving from a three-dimensional structure to a two-dimensional structure that is more resistant to solar radiation and humidity [4].

The use of a single halide in the fabrication of perovskites in the (Br_3 or I_3 or Cl_3) form gave good efficiencies, and the Bromine-based perovskite is known for its high gap, improved solar cell performance and stability [4]. However, the combination of two halides offered better performance [9, 10]. Indeed, perovskites containing iodide and chloride (Cl) have an efficiency of 12.7%, a high diffusion lengths and improved stability compared to those containing only iodide [9]. On the other hand, perovskites combining iodine (I) and bromine (Br) halides such as methylamunium lead bromide-iodide $MAPbI_xBr_{3-x}$ have a maximum efficiency a round 14,9%[4].

3.1.2 Comparison between solar cells based on perovskite

Several combinations of organic and inorganic compounds were investigated by the researchers in this field. The efficiencies of the various perovskite cells are summarized in Table 1 [3], [4], [8], [11].

Table 1: Efficiencis of perovskite solar cells in relation to the absorbers employed

Perovskite	Efficiency
$MASnX_3$	6%
$MAPbX_3$	19.7 %
$MAPbI_3$	19.7 %
$FAPbI_3$	16,60%
$MAPbIxBr_{3-x}$	14,90%
$MAPbIxCl_{3-x}$	12,30%
$CsPbBr_3$	5,95%
$Cs_x(MA)_{1-x}PbI_3$	7,68%
$Cs_x(FA)_{1-x}PbI_3$	16,50%
$(MA)_x(FA)_{1-x}PbI_3$	14,90%
$MA_{0.17}FA_{0.83}Pb(I_{0.83}Br_{0.17})_3$	20,20%
$Cs_x(MA_{0.17}FA_{0.83})_{(100-x)}Pb(I_{0.83}Br_{0.17})_3$	21,10%

3.1.3 Fabrication techniques

Several deposition techniques were used to fabricate op the hybrid perovskite layers. The spin coating is the most widely used technique since it is simple, fast and low cost [12], [3],[13]. The thin film deposition can be carried out either in one or two steps followed by a heat treatment at 100 °C. Thermal evaporation was also used for the fabrication of thin layers of hybrid perovskites [12] in one or two steps. This technique guaranties smooth and uniform films with better quality than spin-coating [3], since it avoids the formation of holes in the film due to the evaporation of the solvents [14], [15].

Another method that has been used is vapor-assisted solution deposition (VASP) where a thin film of the first precursor is deposited on a substrate; the latter is then exposed to the vapor of the second precursor [3]. High efficiencies of up to 20.1% were obtained with this technique [13], [12].

3.1.6 Stability:

The stability of the solar cell based on perovskites depend on several factors: climatic conditions (temperature, humidity, irradiation), the nature of the perovskite absorber, the hole transport layer, electron transport layer and the back contact.

As discussed above, the partial substitution of iodine by Br in MAPb $(I_{1-x}, Br_x)_3$ improves the stability of perovskite [16]. On the other hand, Mckeee et al. [17] showed that the combination of cesium ions with formamidinium (FA) cations in mixed bromide-iodide halide perovskite cells; $FA_{0.83}Cs_{0.17}Pb$ $(I_{0.6}Br_{0.4})_3$, enhances the photo-stability and thermal stability of cells made with this perovskite.

Recently, a new two-dimensional perovskite family has been implemented to overcome stability problems. The two-dimensional perovskites (2D) are derived from the 3D structure $APbX_3$ by the incorporation of voluminous organic cations which play the role of intercalary between the different MX_6 octahedra layers [18]

The pathways to enhance the stability of perovskite cells include the adequate choice of the materials, the structure of the cells and the development of a good encapsulation that protects de device from the precursors of the degradation. So far, the best single junction solar cell containing mixed cations and halide, reached an efficiency of 21, 1% and showed a good stability of up to 1000 hours at 85°C and 85% relative humidity.[19]

3.2 State of the art of tandem cells based on perovskites

3.2.1 Configuration of tandem cells

Generally, the tandem cells are either monolithic or four-terminal. Other configurations are less used because they are more bulky [20]. In the four-terminal configuration, the sub-cells are manufactured independently and are mechanically stacked. Whereas in the monolithic configuration the upper sub-cell is deposited directly on the lower sub-cell thus forming a single block [19], [21]. [22] [2] [23], [24] .

The development of monolithic tandem cells is more demanding than that of four-terminal cells. Indeed, the two subcells must be optimized in such a way as to produce the same current [19]. On the other hand, the upper sub-cell must be able to be deposited at low temperatures in order to avoid damage to the lower subcell [19]. The rear contacts must be replaced by a highly transparent conductive layer [19]. The coupling between the two subcells can be carried out by a tunnel junction of the P^+ N^+ type or by an oxide layer of very small thickness, allowing the passage of the carriers by tunneling.

3.2.2 Tandem cells record efficiencies

According to the principle of the detailed balance, the optimal bandgap is 1.7 eV for the upper cell and 1.1 eV for the lower cell. Several configurations of tandem cells have been developed combining perovskite, CIGS and silicon. The efficiencies of the best combinations are present in the table below [19] [23] [25] [26] [27]:

Table 2: Efficiencies of the best tandem cells

Tandem	Configuration	Efficiency (%)
Perovskite/Perovskite	Monolithic	17
Perovskite/Perovskite	Four-terminal	20,3
Perovskite/ CIGS	Four-terminal	19,5
perovskite/ SHJ	Monolithic	18,1
perovskite/ SHJ	Monolithic	21,2
perovskite/ SHJ	Monolithic	23,6

The perovskite / silicon heterojunction tandem cell has a significant interest because silicon technology is more mature and has achieved efficiencies close to its theoretical limit through the heterojunction structure (SHJ). In these cells, the surfaces of an n-type silicon wafer is passivated with thin layers of intrinsic of hydrogenated amorphous silicon (a-Si: H). A p-type layer of a-Si: H forms is then used to form the heterojunction with n-type crystalline silicon [24].

The efficiency of (SHJ) keeps improving from one year to the next, as it evolved from a value of 23.7% in 2011 to 26.3% in March 2017 [28], [29], [30]. This value is close to the maximum theoretical efficiency a simple junction, which is of the order of 29.4% for the silicon technology [31].

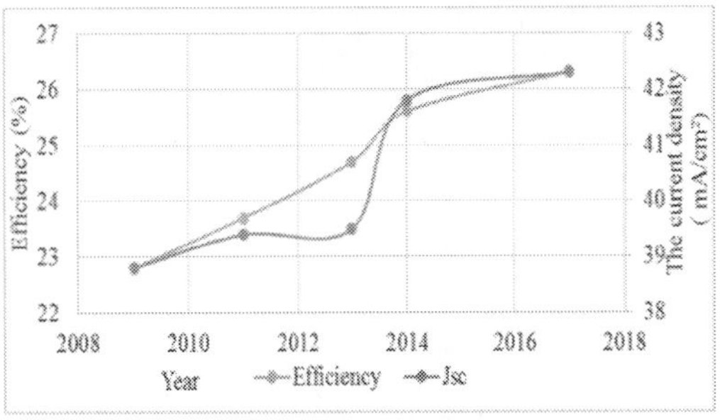

Figure 1: Evolution of performances of SHJ

It is expected that the coupling of the silicon heterojunction (SHJ) cells with the perovskite cells will improve the performance of the resulting devices at a reasonable cost. The efficiencies reported in the literature for this family of tandem cells are summarized in the table 3 [19], [26], [27]. These results suggested that there is still plenty of room for improvement in this technology, as both the pervoskite top cells and the coupling between the two sub cells require further optimization.

Table 3: Performances of the best tandem cells combining Perovskite/ and SJH

Cell structure	surface (cm²)	Scan	Voc (V)	Jsc (mA/cm²)	FF (%)	η (%)
$(MA)_x(FA)_{1-x}PbI_xBr_{3-x}$ / SnO_2/ SHJ	0,25	Inverted	1.785	14	79.5	19.9
		Direct	1.759	14	77.3	19.1
$CH_3NH_3PbI_3$/ PCBM/PEIE/SHJ	1,22	Inverted	1,701	16,1	70,1	19,2
		Direct	1,703	16,1	70,9	19,5
	0,17	Inverted	1,69	15,9	77,6	20,9
		Direct	1,692	15,8	79,9	21,4
$Cs_{0,17}FA_{0,83}Pb(Br_{0,17}I_{0,83})_3$/ NiO/SHJ	1	unknown	1,65	18,1	79	23,6

4 MARKET ANALYSIS

The comapgnies that were surveyed are located in the big cities of Morocco (figure 2). 89% are located in the administrative (Rabat) and economic (Casablanca) capitals of Morocco, whereas 11% are situated Marrakech. The dynamics stimulated by the Moroccan Energy Strategy and in particular the Moroccan Solar Plan, has stimulated the interest of companies working in the photovoltaic sector to invest in new technologies.

As shown in Figure 3, 60% of the actors with activities not related to the photovoltaic sector envisage possible re-orientation towards the photovoltaic sector.

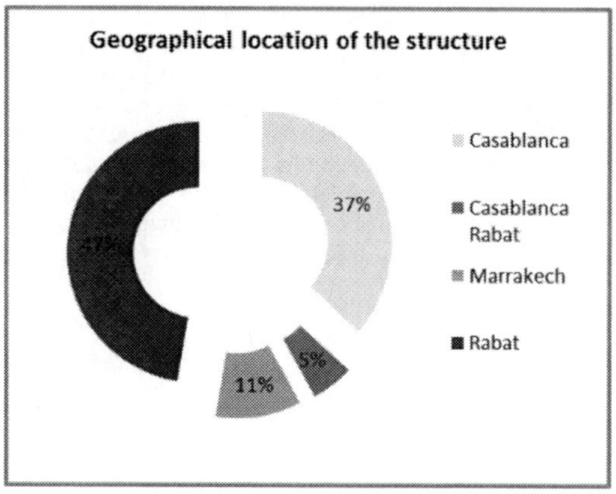

Figure 2: Geographical location of the companies interviewed

This observation is a clear indication of how the photovoltaic industry can be rapidly developed in the coming years in Morocco.

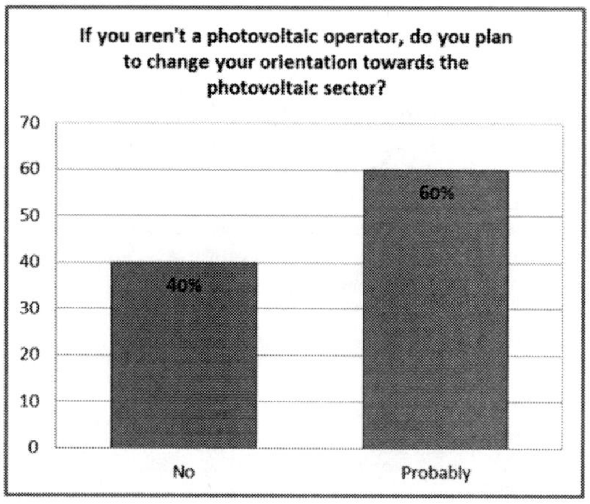

Figure 3: Possibility of the companies to change theirs activity toward the photovoltaic sector

The rapid advances in the perovskite technology in terms of efficiency, stability and lower production costs can be strong impetus and may play a prominent role as alternative or a complement to the photovoltaic silicon solar cells[32]

The cost remains a key factor. As shown in figure 3, 79% of the surveyed companies will opt for this alternative only if it's less expensive than the traditional silicon technology.

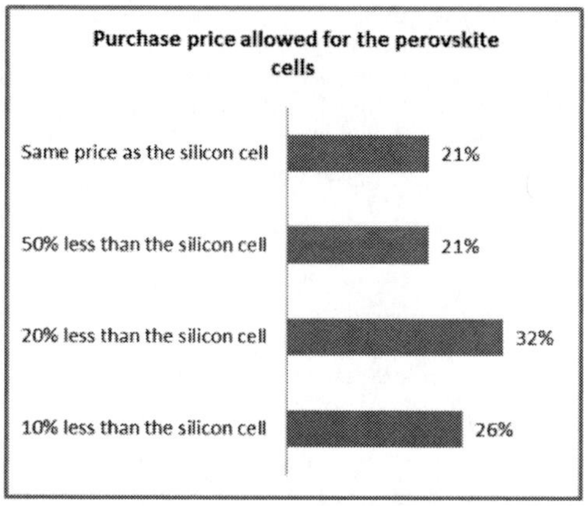

Figure 4: Purchase price allowed for the perovskite cells

Perovskite / Silicon Tandem Technology is now major candidate for the production of cheap photovoltaic panels. 9 out of 10 of the surveyed companies are for the adoption of the Tandem technology.

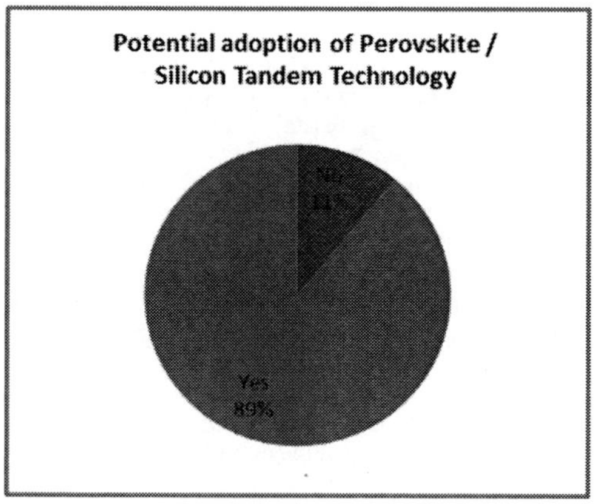

Figure 5: Potential adoption of perovskite/ silicon tandem technology

Regarding the tolerated purchase price for the more efficient Tandem technology, 63% of the companies can readily accept prices that are higher those of the actual silicon technology, while 37% are likely to accept the same price.

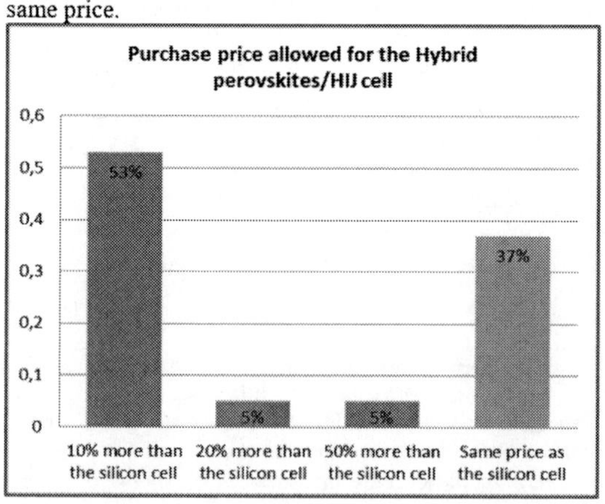

Figure 6: Purchase price allowed for perovskite/ silicon tandem technology

Most of the actors operating in the photovoltaic sector are currently aware that the efficiency of the solar cells of new technologies compete today with those of the traditional silicon cells. According to this survey, 79% of the companies are ready to invest in production units if they are accompanied to produce at an attractive price / quality ratio.

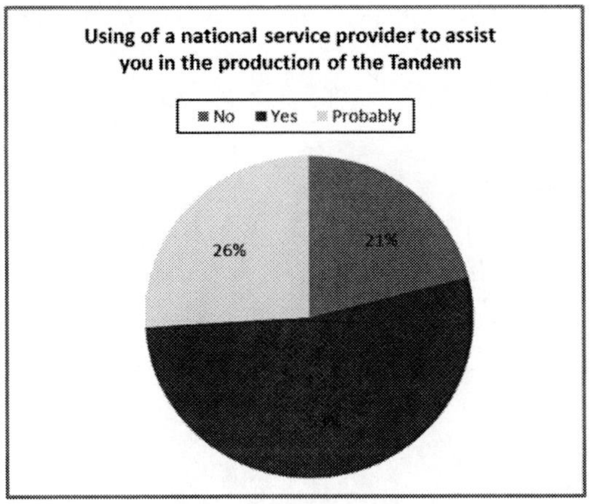

Figure 7: National industries interested by providing a part of the technology from a national reaserch center

5 CONCLUSION

The hybrid perovskite technology is undergoing a continuous development and may be regarded as an attractive alternative to the other technologies and a way to improve their performances.

Indeed, the hybrid perovskite solar cell achieved an efficiency exceeding 22 % while the best tandem cells containing perovskites reached 23, 6%. However, several pathways for improving this technology still need to explore in order to achieve the 46% the theoretical limit of the tandem structures [33].

The overwhelming majority of companies interviewed for the market analysis expressed their interest to opt for this technology either as simple junction or in tandem configurations.

Furthermore, the majority are motivated by the integration of more expensive technologies in there production line instead of conventional silicon solar cells.

The hybrid Perovskite solar cells have a great potential to become a key player in the Moroccan photovoltaic market provided that the pending stability issues are resolved.

5 REFERENCES

[1] [ministry official numbers http://www.mem.gov.ma/].
[2] NREL, "Best Research-Cell Efficiencies," 2017. [Online].Available:https://www.nrel.gov/pv/assets/images/efficiency-chart.png. [Accessed: 14-Apr-2017].
[3] Nam-Gyu Park, "Methodologies for high efficiency perovskite solar cells," nano Convergence, 2016.
[4] Q. Chen, N. De Marco, Y. Michael, T. Song, C. Chen, and H. Zhao, "Under the spotlight: The organic — inorganic hybrid halide perovskite for optoelectronic applications," ", Nano Today, 2015.
[5] D. Wangn, M. Wright, N. K. Elumalai, A. Uddin, "Stability of perovskite solar cells", SOLAR ENERGY MATERIALS & SOLAR CELLS 2015.
[6](http://rb.eclille.fr/l/Analyse_de_donnees/Methodologie_Conception_et_administration_de_questionnaires.
[7] Mohammad Khaja Nazeeruddinand Henry Snaith , Guest Editors, "Methylammonium lead triiodide perovskite solar cells: A new paradigm in photovoltaics". MRS BULLETIN, Volume 40, 2015.
[8] Michael Saliba,a,d* Taisuke Matsui,b Ji-Youn Seo,a Konrad Domanski,a Juan-Pablo Correa-Baena,c Mohammad Khaja Nazeeruddin,d Shaik M. Zakeeruddin,a Wolfgang Tress,a Antonio Abate,a Anders Hagfeldt,c and Michael Gratzel. "Cesium-containing Triple Cation Perovskite Solar Cells: Improved Stability, Reproducibility and High Efficiency", ENERGY AND ENVIRONMENTAL SCIENCE, 2016.
[9] Noh, J. H, Im, S. H., Heo, J. H., Mandal, T. N. & Seok, S. I, "Chemical management for colorful, efficient, and stable inorganic-organic hybrid nanostructured solar cells", Nano Letter. 13, 1764–1769, 2013.
[10] Stranks, S. D. et al. "Electron–hole diffusion lengths exceeding 1 micrometer in an organometal trihalide perovskite absorber", Science 342, 341–344, 2013.
[11] Jae Hui Rhee, Chih-Chun Chung and Eric Wei-Guang Diau, "A perspective of mesoscopic solar cells based on metal chalcogenide quantum dots and organometal-halide perovskites", NPG Asia Materials, 2013.
[12] M. K. Nazeeruddin, H. Snaith, and G. Editors, "Methylammonium lead triiodide perovskite solar cells : A new paradigm in photovoltaics", MRS BULLETIN, pp. 641–645, 2017.
[13] J. Im, I. Jang, N. Pellet, and N. Park, "Growth of CH3NH3PbI3 cuboids with controlled size for high-efficiency perovskite solar cells,", NATURE NANOTECHNOLOGY no. August, pp. 1–6, 2014.
[14] Juliane Borchert, Heidi Boht, Wolfgang Fränzel, René Csuk, Roland Scheer and Paul Pistor, "Structural investigation of co-evaporated methylammonium lead halide perovskite films during growth and thermal decomposition using different PbX2 (X = I, Cl) precursors", Materials Chemistry A, Issue 39, 2015.
[15] C. Wehrenfennig, G. E. Eperon, M. B. Johnston, H. J. Snaith, and L. M. Herz, "High Charge Carrier Mobilities and Lifetimes in Organolead Trihalide Perovskites," pp. 1584–1589, 2014.
[16] H.AitDads . S.Bouzita. L.Nkhailia. A.Elkissania. A.Outzourhit, "Structural, optical and electrical properties of planar mixed perovskite halides/Al-doped Zinc oxide solar cells", Solar Energy Materials and Solar Cells Volume 148, Pages 30-33, 2016.
[17] David P. McMeekin, Golnaz Sadoughi, Waqaas Rehman, Giles E. Eperon, Michael Saliba,1 Maximilian T. Hörantner, Amir Haghighirad, Nobuya Sakai, Lars Korte, Bernd Rech, Michael B. Johnston, Laura M. Herz, Henry J. Snaith, "a mixed-cation lead mixed-halide perovskite absorber for tandem solar cells" Science, ; Vol. 351, pp. 151-155, 2016.
[18] Pablo P. Boix, Shweta Agarwala,Teck Ming Koh, Nripan Mathews, and Subodh G. Mhaisalkar, " Perovskite Solar Cells: Beyond Methylammonium Lead Iodide", Phys. Chem. Lett. , 6, 898–907, 2015.

[19] S. De Wolf, B. Niesen, and C. Ballif, "Efficient Monolithic Perovskite/Silicon Tandem Solar Cell with Cell Area >1 cm 2," Physical Chemistry Letters, pp. 3–8, 2016.
[20] Li, H. Hu, B.Chen, T. Salim, J. Zhang, N. Y. Yuan, J. Ding and Y. Lam, "Reflective Perovskite Solar Cells for Efficient Tandem Application", Mater. Chem. C, 2016
[21] C. D. Bailie et al. Colin D. Bailie, M. Greyson Christoforo, Jonathan P. Mailoa, Andrea R. Bowring, Eva L. Unger,William H. Nguyen, Julian Burschka, Norman Pellet, Jungwoo Z. Lee, Michael Grätzel, Rommel Noufi, Tonio Buonassisi, Alberto Salleo, Michael D. McGehee, "Semi-transparent perovskite solar cells for tandems with silicon and CIGS," Energy Environ. Sci., vol. 8, no. 3, pp. 956–963, 2015.
[22] Y. Hu et al., "Hybrid Perovskite/Perovskite Heterojunction Solar Cells," ACS Nano, vol. 10, no. 6, pp. 5999–6007, 2016.
[23] L. Kranz et al., "High-efficiency polycrystalline thin film tandem solar cells," J. Phys. Chem. Lett., vol. 6, no. 14, pp. 2676–2681, 2015.
[24] N. Marinova, S. Valero, and J. L. Delgado, "Organic and perovskite solar cells: Working principles, materials and interfaces", Colloid Interface Sci., vol. 488, pp. 373–389, 2017.
[25] G. E. Eperon et al., "Perovskite-perovskite tandem photovoltaics with optimized band gaps," Science (80)., vol. 354, no. 6314, pp. 861–865, 2016.
[26] S. Albrecht et al., "Monolithic perovskite/silicon-heterojunction tandem solar cells processed at low temperature," Energy Environ. Sci., vol. 9, no. 1, pp. 81–88, 2016.
[27] K. A. Bush et al., "tandem solar cells with improved stability," no. February, pp. 1–7, 2017.
[28] K. Masuko et al., "Achievement of more than 25% conversion efficiency with crystalline silicon heterojunction solar cell," IEEE J. Photovoltaics, vol. 4, no. 6, pp. 1433–1435, 2014.
[29] M. Taguchi et al., "24.7% Record efficiency HIT solar cell on thin silicon wafer," IEEE J. Photovoltaics, vol. 4, no. 1, pp. 96–99, 2014.
[30] K. Yoshikawa et al., "Silicon heterojunction solar cell with interdigitated back contacts for a photoconversion efficiency over 26%," Nat. Energy, vol. 2, no. March, p. 17032, 2017.
[31] B. P. Rand, J. Genoe, P. Heremans, and J. Poortmans, "Solar Cells Utilizing Small Molecular Weight Organic Semiconductors," Prog. Photovolt Res. Appl., vol. 15, no. February 2013, pp. 659–676, 2015.
[32] Haiming Zhu et al. , "Screening in crystalline liquids protects energetic carriers in hybrid perovskites", Science 353 (6306), 1409-1413), 2016.
[33] William Shockley and Hans J. Queisser, "Detailed Balance Limit of Efficiency of p-n Junction Solar Cells", Journal of Applied Physics, Volume 32, pp. 510-519, 1961.

A FAST AND EASY PEROVSKITE SOLAR CELL SIMULATION TOOL FEATURING ION MIGRATION

A. Fell[1], D. Walter[2], S. W. Glunz[1,3]

[1]Fraunhofer Institute for Solar Energy Systems, Freiburg, Germany
[2]The Australian National University, Canberra, Australia
[3]Department of Sustainable Systems Engineering (INATECH), Albert Ludwigs University Freiburg, Freiburg, Germany

ABSTRACT: Extensions for perovskite solar cells (PSCs) within the solar cell simulation tool Quokka3 are presented, accounting for ion migration. In the "1D detailed solver" the fully coupled equations describing semiconductor carrier transport, ion transport (anions and cations) as well as the general Shockley-Read-Hall formalism (trapping) are implemented, both as steady-state and transient mode. Numerical challenges arising from the narrowness of the Debye layer and the large difference in times scales between semiconductor and ion transport are successfully addressed. Example simulations are presented, showing the influence of pre-bias voltage on fast JV-sweeps, as well as an interaction between ion redistribution and surface recombination that replicates transient open-circuit voltage responses in a $MAPbI_3$ PSC.
Keywords: simulation, modelling, perovskite, ions, quokka

1 INTRODUCTION

Perovskite solar cells (PSCs) are one of the most promising candidates to replace, or via a tandem device enhance, the dominant Silicon photovoltaics (Si-PV) technology. Within several years only, record PSC R&D efficiencies are getting in the range of Si record efficiencies, clearly evidencing the performance potential. However, being a relatively young technology, the understanding of material properties and device physics is largely lacking behind the mature Si-PV technology. Reducing this gap is mandatory to succeed in developing long-term reliable PSCs for industrial application.

A key to gain understanding of PSCs is the ability to model its device physics. Besides significant uncertainties in material properties, one problem the community is facing is that PSCs exhibit a rather special combination of physical effects. In particular, it has been conclusively shown that both semiconductor current transport as well as ion migration play an important role to understand experimental observations and device performance [1–3]. This makes it not straight-forward to use established simulation tools from the semiconductor field and from other photovoltaic technologies, as they are not designed for such an application.

This paper presents the extension of Quokka3 to feature PSC-relevant models. Quokka is a popular "fast and easy" semiconductor simulation tool in the Si-PV community, and up to version 2 its focus was on multidimensional carrier transport in the quasi-neutral bulk [4]. However, Quokka3 features also a one-dimensional solver for the full semiconductor equations [5]. This "1D detailed solver" is where the PSC extensions, mainly ion migration, are included.

2 MODEL

2.1 Charge carrier transport equations

The model for describing carrier transport in the perovskite material is based on the well-established drift-diffusion model for semiconductor carrier transport (electrons and holes), extended by two additional charge carriers: anions and cations [7–9]. It is assumed that ions are neither generated nor "recombine", but a constant integral amount of ions solely redistributes within the perovskite material. Recombination of electrons and holes is considered via intrinsic mechanisms, i.e. Auger and radiative, as well as via the Shockley-Read-Hall (SRH) defect mechanism [6]. SRH can be modelled in its general or simplified form, the general form accounting for the effect of trapping on the charge density and resulting in unequal recombination rates for electrons and holes R_{el} and R_{hol}. The general system of 6 nonlinear differential equations to be solved then comprises the continuity equations for the 4 charge carrier densities

$$q\frac{dn}{dt} = q(G - R_{el}) - \frac{dJ_{el}}{dz}, \quad (1)$$

$$q\frac{dp}{dt} = q(G - R_{hol}) + \frac{dJ_{hol}}{dz}, \quad (2)$$

$$q\frac{dc_{an}}{dt} = -\frac{dJ_{an}}{dz}, \quad (3)$$

$$q\frac{dc_{cat}}{dt} = \frac{dJ_{cat}}{dz}, \quad (4)$$

the continuity equation for the trapped electron density

$$\frac{dn_t}{dt} = R_{el} - R_{hol}, \quad (5)$$

and the Poisson equation for the electric potential

$$\frac{d}{dz}\left(-\varepsilon\frac{d\varphi_e}{dz}\right) = q(N_D - N_A + p - n + c_{cat} - c_{an} - n_t). \quad (6)$$

The current densities due to drift and diffusion can be expressed via the quasi-Fermi potentials (i.e. the electrochemical potential) gradient, and the carrier concentration and mobility as

$$J_X = qc_X\mu_X\frac{d\phi_X}{dz}. \quad (7)$$

To solve those equations, one needs to relate the potentials to the carrier densities via carrier statistics. Fermi-Dirac statistics are applied for electrons and holes

$$n = N_c F_{1/2}\left(\frac{q\phi_{el}+E_c}{kT}\right), \quad (8)$$

$$p = N_v F_{1/2}\left(-\frac{q\phi_{hol}+E_v}{kT}\right), \quad (9)$$

and Boltzmann statistics for ions

$$c_{an} = N_{0,an}\exp\left(\frac{q(\phi_{an}+\varphi_e)}{kT}\right), \quad (10)$$

$$c_{cat} = N_{0,cat}\exp\left(-\frac{q(\phi_{an}+\varphi_e)}{kT}\right). \quad (11)$$

Optical modeling is outside the focus of this paper, and therefore the generation profile G is considered a user-input here.

2.2 Boundary conditions

At the time of writing this paper, Quokka3 supports two different kind of contact types forming the boundary conditions for semiconductor carrier transport: ideal contacts and metal-semiconductor (MS), i.e. Schottky, -type contacts. Both differentiate between an electron and hole contact, which determines the majority carrier type.

The ideal contact model pins the majority quasi-Fermi potential to the potential applied to the contact, and also assumes flat-band conditions, i.e. a zero electric field at the boundary.

The MS contact models majority carrier transport over the boundary via thermionic emission,

$$J_{maj} = A_0 T^2 \exp\left(-\frac{\Phi_B}{kT}\right)\left[\exp\left(\frac{\varphi_{cont}-\phi_{maj}}{kT/q}\right) - 1\right] \pm J_{rec}, \quad (12)$$

and the electric potential is determined via

$$E_{c/v} = q\varphi_{cont} \pm \Phi_B. \quad (13)$$

The barrier-height Φ_B can either be directly defined by the user, or derived via the electron affinity of the perovskite material defining the band-edges E_c and E_v, and the (effective) work-function of the contact material

$$\Phi_B = \pm(E_{c/v} - WF). \quad (14)$$

Note that while fundamentally the contact potential is defined at the boundary, Quokka3 numerically allows the definition of a constant current as well. This is necessary to enable transient open-circuit simulations by setting the boundary current to zero.

Independently of the contact physics, surface recombination is modelled via the simplified SRH mechanism

$$J_{rec} = \frac{pn - n_{i,eff}^2}{S_{n0}(p_1+p)+S_{p0}(n_1+n)}. \quad (15)$$

Via this decoupling, in contrast to some other modeling tools where recombination and current transport over the boundary are essentially one and the same thing, complex contacts can be more easily approximated by their effective properties: effective barrier height and effective surface recombination properties.

As mentioned above, conservation of the integral ion concentration within the perovskite layer is assumed, meaning that the boundary conditions for ion transport is defined by zero current flow

$$J_{cat} = J_{an} = 0. \quad (16)$$

2.3 Implementation in Quokka3

The model detailed in the previous section is implemented within the "1D detailed solver" of Quokka3, using a finite-differences scheme. It supports solving the model both transient and steady-state, i.e. setting all time derivatives to zero. The steady-state solver is also used to find correct start values for the initial operating conditions of a transient simulation.

Despite this was observed to be numerically challenging in previous work [9], the numerical implementation allows a very fine mesh size towards the boundary, enabling the full numerical resolution of the Debye layer.

PSC modeling also benefits from the luminescence modeling implemented in Quokka3 [10]: from the electrical solution at any operating condition spectrally resolved luminescence is straightforward and numerically rapid to calculate from first principles, fully accounting for inhomogeneous carrier profiles and re-absorption effects. With this for example photoluminescence decay experiments accounting for ion migration can well be simulated.

Furthermore, Quokka3 implements two special features useful for modeling PSCs with ion migration: i) a pre-bias functionality of the steady-state solver, and ii) the option to set semiconductor transport to steady-state within the transient solver.

The pre-bias functionality allows the user to define a steady-state operating condition of the device which is solved including ion migration, before running the actual steady-state simulation with a "frozen" ion profile [11]. The primary application are fast JV-measurements after letting the PSC equilibrate at different conditions, as within the fast JV-sweep the slow-moving ions can well be assumed to have a constant distribution. This simplifies this type of simulations considerably, compared to trying to replicate the same within transient simulations.

Allowing to set semiconductor transport to steady-state within the transient solver, i.e. only solving ion-migration transiently, largely speeds up the simulation of the long time-scales required for the slow-moving ions. This is due to some unavoidable numerical inaccuracy resulting in considerable "noise" of the time derivatives of the, relatively, very fast electron and hole transport equations, which puts a limit on the maximum time step for successful convergence. This is resolved by only simulating the steady-state solution for each time step. Due to the large difference in time scales between semiconductor transport and ion migration, this quasi-steady-state approach is a highly valid assumption.

3 APPLICATION EXAMPLES

Two application examples showcasing the PSC functionality of Quokka3 are presented here. The common assumed device properties are listed in Table I. They represent typical values as reported in literature. It is noted however that uncertainties are large, and properties will vary for perovskite materials created by different processes. The exact values within those uncertainties are chosen such that the simulation results show the desired effects for illustrative examples.

Table I: (electrical) PSC device properties, mostly taken from [12].

property	symbol	value
band-gap	E_g	$1.5eV$
electron affinity	E_{EA}	$3.93eV$
relative permittivity	ε_r	30
electron mobility	μ_{el}	$2cm^2V^{-1}s^{-1}$
hole mobility	μ_{hol}	$2cm^2V^{-1}s^{-1}$
conduction band DOS	N_c	$2.2 \cdot 10^{18}cm^{-3}$
valence band DOS	N_v	$1.8 \cdot 10^{19}cm^{-3}$
net doping density	$N_D - N_A$	$0cm^{-3}$
defect density	N_t	$10^{15}cm^{-3}$
electron capture cross section	σ_n	$2 \cdot 10^{-14}cm^2$
hole capture cross section	σ_p	$2 \cdot 10^{-14}cm^2$
defect energy	$E_t - E_i$	0 (midgap)
front and rear electron SRV	S_{n0}	$10^5\,cm/s$
front and rear hole SRV	S_{p0}	$10^5\,cm/s$
hole contact work-function	WF	$5.3eV$
electron contact work-function	WF	$4.0eV$
average ion concentration	$N_{0,an/cat}$	$5 \cdot 10^{18}cm^{-3}$
anion mobility	μ_{an}	0 (immobile)
cation mobility	μ_{cat}	$2 \cdot 10^{-12}cm^2V^{-1}s^{-1}$
uniform generation rate	G	$3 \cdot 10^{21}cm^{-3}$
device thickness	W	$400nm$

3.1 Pre-bias impact on JV-curves

In this example, the pre-bias functionality is showcased. A steady-state solution for the PSC at different bias voltages is simulated, and subsequently a steady-state light JV-curve is simulated with immobile ions. This represents a fast JV-sweep after the different pre-bias conditions, for which redistribution of the slow-moving ions is negligible, an approach often used to characterize PSCs experimentally [9,11].

Figure 1 shows the resulting light JV-curves for pre-bias voltages ranging from 0V to 1.2V. The expected trends from reported experiments can clearly be seen, which are caused by the different ion distributions shown in Figure 2.

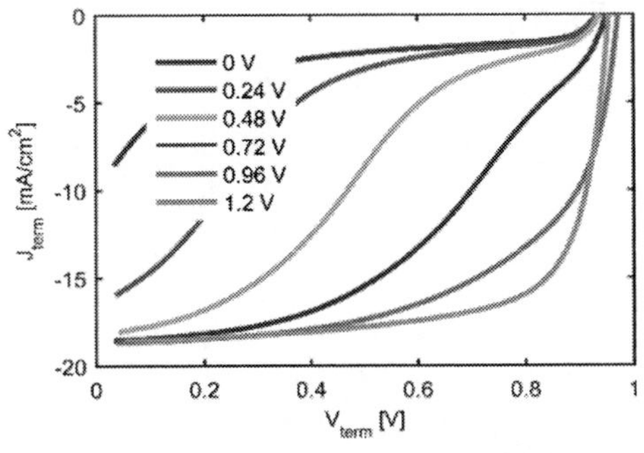

Figure 1: Simulated JV-curves for different immobile cation concentration profiles resulting from different pre-bias voltage steady-state simulations with mobile cations, representing rapid JV-sweeps after different pre-bias conditions.

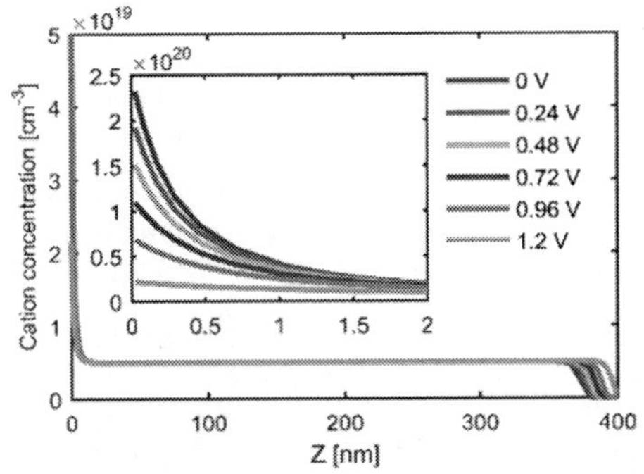

Figure 2: Steady-state cation concentration profiles for the different pre-bias voltages, showing the full numerical resolution of the narrow Debye layer.

3.2 Transient open-circuit voltage during ion relaxation

In the example shown in Figure 3, the influence of transient ion redistribution on the relaxation of open circuit voltage is demonstrated, in comparison to experimental data. A known example of hysteresis in $MAPbI_3$ PSCs is a transient, non-monotonic "overshoot" of the open-circuit voltage when a PSC is abruptly illuminated under open-circuit conditions [1,11]. The cell is initially at steady-state in the dark and then illuminated at t = 0s. In response to the generation of electron-hole pairs, the electric field in the device changes, causing the dark steady-state ion concentration to redistribute towards the illuminated open-circuit steady-state condition. As it does so, the measured voltage is affected by changing recombination currents in the bulk and at the cell surface. With the addition of only a single mobile cation population on top of the standard drift-diffusion model, one of the more counter-intuitive experimental measurements can be qualitatively replicated. Furthermore, the log of the simulated PL signal is plotted, which is approximately proportional to the internal Fermi-Level split or implied V_{OC}. For the case of high surface recombination, it shows a different trend to the (external) V_{OC}, caused by a majority carrier potential drop close to the surfaces which varies over time.

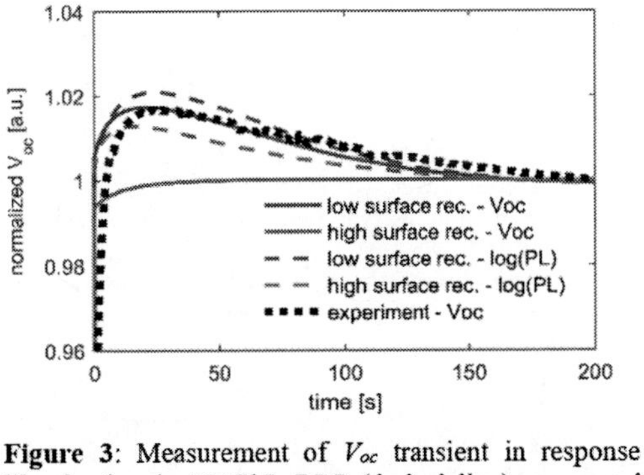

Figure 3: Measurement of V_{oc} transient in response to illumination in $MAPbI_3$ PSC (dashed line), compared to simulated responses with mobile cations. Both the (external) V_{OC} and the log of the PL signal is simulated, showing a different trend for high surface recombination.

4 CONCLUSIONS

Several extensions to the 1D detailed solver within Quokka3 allow the consideration of important effects particular to PSCs, mainly ion migration. A numerical solver for the fully coupled semiconductor and ion drift-diffusion equations is successfully implemented, both for the steady-state and transient case. The solver can account for the effect of trapping within the perovskite material, by implementing the general SRH formalism. The luminescence modelling implemented in Quokka3 further allows for an accurate prediction of spectrally resolved luminescence signal, accounting for non-uniform carrier profile and re-absorption effects.

Two special features of Quokka3 useful for PSC simulations are highlighted by application examples: i) a pre-bias functionality, allowing (fast) steady-state simulations with a fixed ion distribution calculated at a steady-state pre-bias point, useful in particular for representing fast JV-sweeps after different pre-bias conditions; ii) an option to solve semiconductor transport as (quasi-)steady-state within the transient solver, removing numerical limitations due to the large differences in times scales between semiconductor and ion transport.

Currently only the perovskite material layer is solved. Electron and hole contact layers are not resolved but modelled via an effective work-function and surface recombination boundary condition. However, given the large uncertainties and number of unknowns within the perovskite material, resolving the physics within the electron and hole contact materials can be considered a second-order effect.

Nonetheless, work towards implementing multiple semiconductor material layers and interface physics is currently under way, including also (effective) tunneling interfaces. By treating the top-cell as quasi-1D within the skin-concept of Quokka3, this will enable 2D / 3D simulation of tandem cells resolving an unprecedented amount of physical effects.

ACKNOWLEDGEMENTS

Andreas Fell acknowledges funding by the European Commission through the Marie-Curie fellowship "Quokka maturation".

REFERENCES

[1] W. Tress, N. Marinova, T. Moehl, S.M. Zakeeruddin, M.K. Nazeeruddin, M. Grätzel, Understanding the rate-dependent J–V hysteresis, slow time component, and aging in CH 3 NH 3 PbI 3 perovskite solar cells: the role of a compensated electric field, Energy & Environmental Science 8 (3) (2015) 995–1004.

[2] C. Eames, J.M. Frost, P.R.F. Barnes, B.C. O'regan, A. Walsh, M.S. Islam, Ionic transport in hybrid lead iodide perovskite solar cells, Nature communications 6 (2015) 7497.

[3] H.J. Snaith, A. Abate, J.M. Ball, G.E. Eperon, T. Leijtens, N.K. Noel, S.D. Stranks, J.T.-W. Wang, K. Wojciechowski, W. Zhang, Anomalous hysteresis in perovskite solar cells, The journal of physical chemistry letters 5 (9) (2014) 1511–1515.

[4] A. Fell, A Free and Fast Three-Dimensional/Two-Dimensional Solar Cell Simulator Featuring Conductive Boundary and Quasi-Neutrality Approximations, IEEE Transactions on Electron Devices 60 (2) (2013) 733–738.

[5] A. Fell, J. Schön, M.C. Schubert, S.W. Glunz, The concept of skins for silicon solar cell modeling, Solar Energy Materials and Solar Cells (2017).

[6] W. Shockley, W.T. Read Jr, Statistics of the recombinations of holes and electrons, Physical Review 87 (5) (1952) 835.

[7] P. Calado, A.M. Telford, D. Bryant, X. Li, J. Nelson, B.C. O'regan, P.R.F. Barnes, Evidence for ion migration in hybrid perovskite solar cells with minimal hysteresis, Nature communications 7 (2016) 13831.

[8] S. van Reenen, M. Kemerink, H.J. Snaith, Modeling anomalous hysteresis in perovskite solar cells, The journal of physical chemistry letters 6 (19) (2015) 3808–3814.

[9] G. Richardson, S.E.J. O'Kane, R.G. Niemann, T.A. Peltola, J.M. Foster, P.J. Cameron, A.B. Walker, Can slow-moving ions explain hysteresis in the current–voltage curves of perovskite solar cells?, Energy & Environmental Science 9 (4) (2016) 1476–1485.

[10] A. Fell, K.R. McIntosh, M. Abbott, D. Walter, Quokka version 2: selective surface doping, luminescence modeling and data fitting, in: 23rd Photovoltaic Science and Engineering Conference (PVSEC), Taipei, 2013.

[11] D.A. Jacobs, Y. Wu, H. Shen, C. Barugkin, F.J. Beck, T.P. White, K. Weber, K.R. Catchpole, Hysteresis phenomena in perovskite solar cells: The many and varied effects of ionic accumulation, Physical Chemistry Chemical Physics 19 (4) (2017) 3094–3103.

[12] Y. Wu, H. Shen, D. Walter, D. Jacobs, T. Duong, J. Peng, L. Jiang, Y.-B. Cheng, K. Weber, On the Origin of Hysteresis in Perovskite Solar Cells, Advanced Functional Materials 26 (37) (2016) 6807–6813.

RAMAN SPECTROSCOPY AND IMAGING OF TITANIUM DIOXIDE NANOTUBES

R. Taziwa, E Meyer and S. Zinya

Fort Hare Institute of Technology (FHIT), University of Fort Hare, Private Bag X1314, Alice, 5700, Republic of South Africa.
Tel: +27 (0) 40 602 2086, Fax: +27 (0) 40 653 0665
Raymond T. Taziwa, University of Fort Hare, Institute of Technology, rtaziwa@ufh.ac.za
Edson Meyer, University of Fort Hare, Institute of Technology emeyer@ufh.ac.za
Simcelile Zinya, University of Fort Hare, Institute of Technology zinya.sim@gmail.com

Abstract

One-dimensional (1-D) titanium dioxide nanotubes (TNTs) have attracted much attentions as a promising alternative electrode materials for DSSC. As compared to the randomly packed and disordered TiO_2 nanoparticles (TNPs) network with numerous particle-particle interfaces. Vectorial transport of photon generated electrons along the TNTs has been reported to lead to higher charge mobility which is crucial for improvement of DSSC performances. In this work highly cohesive titanium films were deposited on functional substrates (FS) using RF sputtering technique at a sputtering output power of 1KW, operating pressure of 1.5 Pa and at a deposition temperature of 200 °C for 122 minutes to obtain a thickness of 10 µm under an inert argon atmosphere. Subsequently, the RF sputtered titanium films were anodized with ammonium fluoride/glycerol electrolyte solution at room temperature at 60 V for 72 hours. The resulting TNTs on functional substrates (TNTs-FS) were subjected to thermal treatment at 350 °C, 450 °C, 550 °C and 650 °C for 3 hours under oxygen atmosphere. The as prepared and thermally treated TNTs-FS were characterized by SEM, and Confocal Raman Spectroscopy (CRS). CRS Large Area Scan (LAS) and Depth profiling (DP) were employed evaluate the crystallinity phase distribution of TNTs-FS thermally treated at different temperatures. CRS LAS in the XY direction of TNTs-FS has revealed the presence of differently crystallized Anatase phases of TiO_2 with Raman vibrational modes of 159.38 cm^{-1} (E_g), 208.37 cm^{-1} (E_g), 399.67 cm^{-1} (B_{1g}), 514.25 cm^{-1} (A_{1g}) and 641.58 cm^{-1} (E_g) for the samples annealed at 350°C. The effect of annealing temperature on TiO_2 phase evolution was meticulously evaluated using CRS for TNTs-FS for the samples annealed at 350 °C, 450 °C, 550 °C and 650 °C.

1. Introduction

High demand on energy conversion in dye sensitized solar cells (DSSC), necessitates development of well-organized TiO_2 nanotube structures because of their large surface area-to-volume ratio, superior lifetime and providing of optimal passageways for redox electrolyte penetration. Ordered TiO_2 nanostructures, more specifically TiO_2 nanotubes (TNTs) have received a deal of attention in recent years because of their application as effective carrier transporters in DSSCs. Substantial efforts have been focused on fabrication of well aligned TNTs. One dimensional and highly ordered nanotubes architecture offers excellent electrical channel for vectorial charge transfer so that photo induced electron hole pairs can be effectively separated, which minimizes recombination and maximizes photo absorption. In comparison with the randomly packed and disordered TiO_2 nanoparticles (TNPs) networks with numerous nano-particle–nano-particle interfaces. TNTs arrays with highly ordered vertically oriented nanotubular structures provides a larger internal surface area. At the same time introduce a free electron traveling path to reduce carrier recombination possibilities. To further enhance vectorial charge transfer activity TNT arrays, various strategies have been employed such as the improvement of surface morphology, crystal structure, surface area, introduction of heterogeneous structures by decoration and doping with non-metal or metal elements. Amongst these, the enhancement of TNTs crystallinity has been considered as an important and widely investigated approach to achieve better vectorial electron transport in DSSCs. Characterization of TNT arrays continues to grow in importance and to impact the development of efficient electrode materials for DSSCs. The development of efficient defect free Anatase TNT membranes is key to development of DSSC. In addition to the difficulty of fabricating highly ordered TNTs membrane, it must be added the complexity of characterization of these TNTs on top of rigid substrate presents several challenges in obtaining a suitable structural characterization techniques. These difficulties in characterizing the TNT membranes are an existing challenge to understanding crystalline phase distribution on the surface and along the membranes. This paper offers a smart and non-destructive Confocal Raman Spectroscopy technique to evaluate the crystalline phase distribution of TNTs through large area scans (XY) and depth profile (XZ). Raman spectroscopy provides qualitative and quantitative information about the chemical nature and crystalline properties of TNTs. Raman spectroscopy added with Raman imaging provides an excellent tool for investigating crystalline phase distribution. In this work Confocal Raman Spectroscopy coupled with Raman imaging has proven to be a fruitful technique to complement to a number of characterization normally used to probe crystalline phase distribution of TNTs over a large area of the sample.

2. Approach

2.1. RF sputtering of Titanium metal film on glass substrates

Ti films (Titanium target of 99.99%) were sputter coated on FTO glass substrates using RF sputtering system technique. Prior to RF sputtering, the fluorine doped tin oxide (FTO) glass substrates were ultrasonically cleaned using acetone, ethanol and DI water for 15 minutes, respectively. Then the vacuum chamber was pumped down until the base pressure was 5×10^{-4} Pa. The sputtering chamber pressure was kept at 1.5 Pa during the deposition in argon atmosphere. An RF power of 1 KW and substrate temperature of 200 °C was maintained during the depositions to ensure better adhesion between the substrate and titanium film. RF sputtering was done for 122 minutes at deposition rate of 82 nm per minute to obtain a Titanium thin film of about 10 μm (thickness).

2.2. Electro-anodization

The self-organised TNTs were synthesized by anodization of Ti films in the electrolyte consisting of Glycerol + 3.5 vol% deionised water + 0.5 wt% ammonium fluoride (NH_4F). Anodization was performed in a two-electrode configuration connected to a DC power supply (PS 8000 T) with sputtered titanium film as the working electrode and platinum foil as the counter electrode under constant 60V for 72hrs. The as anodized TNTs were annealed at 350 °C, and 450 °C for 3 hours. Samples were collected and prepared for different structural, morphological and topographical evaluation.

3. Results and Discussions

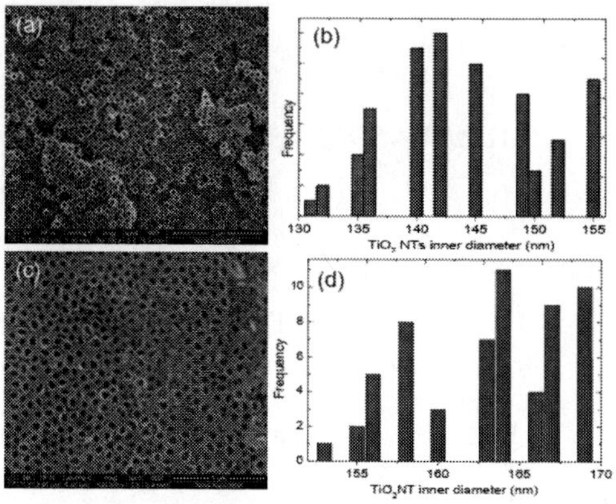

3.1. SEM analysis

Fig 1: Surface SEM images of TNTs on FTO glass substrate annealed at (a) 350 °C (c) 450 °C. additionally, showing histograms of pore size distribution for the TNTs annealed at (b) 350 °C (d) 450 °C

Fig 1 (a) has revealed a that the surface morphology consist of a regular & well orientated tubular layers with TNT pore diameter range of 130 nm-155 nm and modal pore tube diameter of 142.14 nm as shown by the histogram in Fig 1(b).. The observed pore diameter and modal pore diameter for TNTs fabricated on functional substrates is somewhat larger than that reported by Lupiwana et-al [10]. Lupiwana et-al reported internal pore diameter range of 20-70nm and modal pore diameter of 34.56 nm. It is important to note the increased pore diameter of TNTs on FS as compared to TNTs on Ti (foil) is beneficial for DSSC solar cell application as they are likely to retain more of the dye and allow efficient percolation of the redox electrolyte for efficient electron-hole pair generation. Additionally, the cross section SEM image (not shown) has revealed that as fabricated TNTs on FS have ring like structures, which reflect rough walls of the TNTs. The presence of rough tube walls can be ascribed to current oscillations, which also beneficial for DSSC application as it enhances the available surface area for efficient dye loading [9, 10]. Increase in annealing temperature from 350 °C to 450 °C has revealed improved morphology of the TNTs as shown in Fig 1 (c) TNTs annealed at 450 °C revealed an improved surface morphology and with even large pore diameters of the range 150 nm–170 nm and modal pore diameter of 163.35 nm. Yet again, the TNT fabricated on FS have outshined the TNTs fabricated on Ti (foil), which have been reported to have pore size range of 45nm-100 nm and a modal pore diameter of 56.52 nm [10].

Accordingly, the increased pore diameter also results in increased surface area for maximum dye loading, thus increase in photo induced electrons [13]. Simultaneously allowing efficient penetration for redox electrolyte for regeneration of the dye during photo sensitization.

3.2. Confocal Raman Large Area Scans

Confocal Raman spectroscopy was employed to interrogate chemical and structural phase distribution of titanium dioxide nanotubes (TNTs) on functional substrate. TiO_2 has three distinct phases that occur in nature, two metastable phases (brookite and anatase) and stable rutile phase. To evaluate the morphological distribution of the Anatase and Rutile phase of TiO_2 we performed a large area scan (LAS) on the TNTs grown on glass substrate to obtain a conclusive phase distribution on the surface of TNTs annealed at 350 °C covering a large area of the sample. Fig 2(a), (b) and (c) shows the corresponding draw images from basis analysis, a colour coded combined image showing the crystalline phase distribution of titanium dioxide on the surface of the nanotubes as illustrated here in Figure 2(d). The corresponding colour code Raman Spectra is shown here in Fig. 2(e). Fig 2(e) has revealed red area and blue spots with Raman vibrational modes at 160.15 cm^{-1} (Eg), 207.58 cm^{-1} (Eg), 397.54 cm^{-1} (B1g), 512.87 cm^{-1} (A1g) and 637.81 cm^{-1} (Eg) all belonging to the Anatase crystalline phase of TiO_2. Additionally Fig 2(e) has revealed that the surface of TNTs annealed at 350 °C also consist of cyan spots with Raman vibration modes at 160.15 cm^{-1} (Eg), 395.76 cm^{-1} (B1g) and 638.79 cm^{-1} (Eg) all belonging to a differently crystallize Anatase of titanium dioxide. Our Raman measurements over a large area of the sample outclass single Raman single spectra that have been previously reported. As our Confocal Raman large area scan on an area of 500µm x 500µm have given a convincing crystalline phase distribution over a large area of the sample. This can be useful information in the development of electrodes materials for efficient vectorial charge transport in solar cells. Moreover, our Confocal Raman Measurement have also outperformed XRD analysis which do not reveal the exact area of location TiO_2 phase on the sample which is key in producing efficient DSSC employing TNTs. We have also demonstrated in this work the Confocal Raman Spectroscopy can be used as an efficient diagnostic tool to evaluate TNTs on rigid functional substrate without the need for sophisticated sample preparation, more particularly for large scale production of DSSCs employing TNTs substrates.

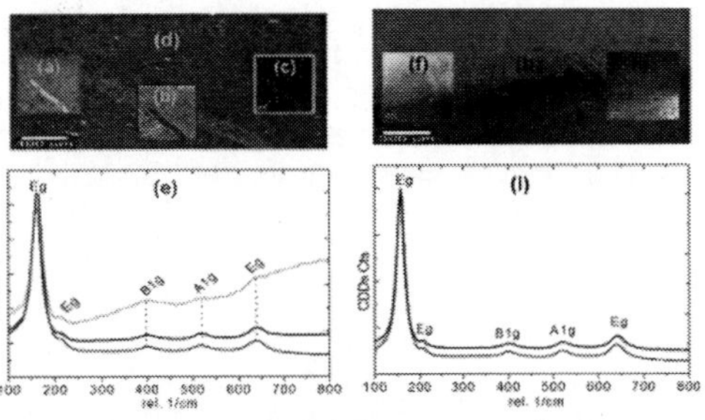

Fig 2: Large area scan of TNTs annealed at 350 °C & 450 °C (a, b, & c) and (f & g) are the resulting draw images showing phase distribution of TiO_2 following basis analysis, respectively. Brighter colours indicate a higher fitting factor and thus higher signal intensity of the corresponding spectrum at that position (d) & (h) are combined images showing different spectra displaying existence of one phase at the surface of the TNTs (e) & (i).

To further evaluate the effect of annealing temperature on the crystalline phase distribution Confocal Raman Spectroscopy was also performed on TNTs on Functional substrates annealed at 450 °C. Figure 2(i) presents the Raman spectra obtained from the large area of the sample. It is clearly evident from the combined colour coded image in Fig. 2(h) that the surface TNTs on Functional Substrates consist of two differently crystallized phases of TiO_2. The corresponding colour coded Raman Spectra in Fig. 2(i) has revealed Raman modes at 156.51 cm^{-1} (Eg), 206.36 cm^{-1} (Eg), 401.97 cm^{-1} (B1g), 520.30 cm^{-1} (A1g) and 638.79 cm^{-1} (Eg) belonging to a well crystallized Anatase phase of TiO_2. Increase in annealing temperature has resulted in improving of the crystallinity of the Anatase phase of TiO_2. These observations are in contention with the literature reporting irregular and collapsed TNTs grown on metal substrate under the same annealing atmosphere [27]. This implies that the nature of the substrate to which TNTs are grown plays pivotal role on the phase composition of the TNTs. These observations are supported by SEM morphology which have revealed a greatly enhanced grain size, large pore diameter and also showing no traces of deformation for the TNTs on functional substrates annealed at 450 °C as shown here in Fig. 1(c). Anatase phase significantly enhance the efficiency of photo-electrodes for catalysis, this is assigned to improved grain size and connections between grain boundaries, which decrease interfacial charge recombination [11]. Anatase is the most favourable in dye sensitized solar cells due to its high bandgap and improved vectorial charge transport of photon generated electrons.

In order to evaluate the crystalline phase distribution of TNTs along the tube walls we performed Confocal Raman depth profile (XZ) on the TNTs on functional substrates. Confocal Raman image depth profiling has revealed that the sample is composed of only Anatase phase of TiO_2 for the sample annealed at 350 °C as shown here in Fig 3 and the Anatase is the most favourable in dye sensitized solar cells due to its high bandgap and high photocatalytic activity. At 550 °C depth profiling has unveiled well-crystallized Anatase phase with no traces of neither rutile or brookite phase. Fig. 3.has revealed Raman vibration modes 156.48 cm^{-1} (Eg), 205.62 cm^{-1} (Eg), 401.40 cm^{-1} (B1g), 517.07 cm^{-1} (A1g) and 638.64 cm^{-1} (Eg).

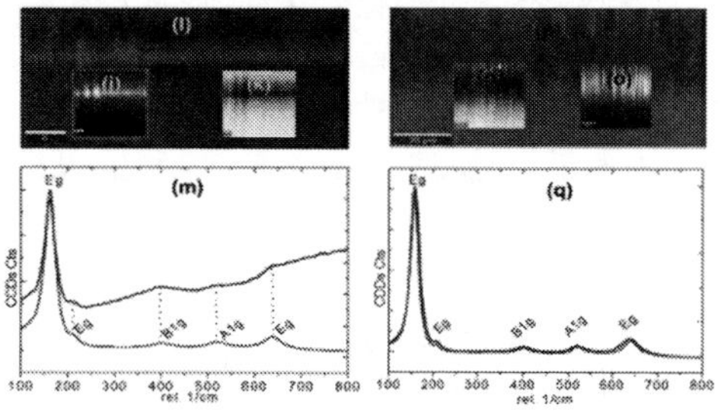

Fig. 3: Depth profiling scan of TNTs annealed at 350 °C & 450 °C (i) & (j) and (n) & (o) are draw images showing phase distribution of TiO_2 following basis analysis, respectively. Brighter colours indicate a higher fitting factor and thus higher signal intensity of the corresponding spectrum at that position (k) & (p) combined image, (m) & (q) show the corresponding colour coded spectra.

At this temperature, all the Raman peaks detected are well resolved and more distinct and these observations are in mutual agreement with SEM micrographs in Fig 1(a) showing the greatly enhanced grain size and large inner diameter with no traces of morphological deformation with increase in annealing temperature.

4. Conclusions

SEM analysis has revealed the presence of porous surface morphology of the TNTs on functional substrates. Additionally, SEM analysis has revealed that the pore size diameter of the TNTs increase with increase in annealing temperature. SEM surface morphology analysis has revealed that the fabricated TNTs on functional substrates showed no signs of deformation as they remained intact and maintained a good surface morphology with increase in annealing temperature. Previous reports have indicated the deformation and collapsing of both surface and cross sectional tube TNTs morphology. The deformation and collapsing of TNTs at was attributed to "substrate effect". Hence the fabricated TNTs on Functional Substrates are more stable and are ideal electrode materials for DSSC. Additionally, in this work we have also utilised confocal Raman Spectroscopy as an ideal non-destructive technique to scrutinize the crystalline phase distribution of TNTs on rigid Functional substrates without the need for complicated sample preparation. Hence Confocal Raman Coupled with Raman imaging can be used as a diagnostic tool to evaluate the crystalline phase distribution of TNTs on rigid functional substrates.

5. Acknowledgments

We are grateful for financial support from our sponsors South African National research foundation (NRF), Govani Beki Research and Development centre (GMRDC) of the University of Fort Hare. The authors would also like to acknowledge the DST/CSIR Nanotechnology Innovation Centre, National Centre for Nanostructured Materials, CSIR and Electron Microscope Unit, University of Cape Town for morphological characterization of Titanium dioxide nanotubes on Functional Substrates. Lastly, we are grateful to the University of Pretoria for assisting in Sputter deposition of titanium film.

6. References

[1] Yang D. J, Park H, Cho S.J, Kim H.G, Choi W.Y, Phys J. Chem. Solids 69, 1272 (2008)

[2] .Mor G.K, Varghese O.K, Paulose O.K, Ong K.G, Grimes C.A, Thin Solid Films 496, 42 (2006)

[3] Gong D, Grimes C.A, Varghese O.K, Hu W, Singh R.S, Chen Z, Dickey E.C, J. Mater. Res. 16, 3331 (2001)

[4] Zhu K, Neale N.R, Miedaner A, Frank A.J, Nano Lett. 7, 69 (2007)

[5] Y. Wang, H. Yang, Y. Liu, H. Wang, H. Shen, J. Yan, H. Xu, Prog. Photovoltaics Res. Appl. 18, 285 (2010)

[6] Y.J. Park, J.M. Ha, G. Ali, H.J. Kim, Y. Addad, S.O. Cho. Nanoscale Res Lett 2015; 10(1): 377.

[7] J.K. Tsai, W.D. Hsu, T.C. Wu, T.H. Meen, W.J. Chong, Nanoscale Res. Lett. 8, 1 (2013)

[8] T. Stergiopoulos, A. Valota, V. Likodimos, T. Speliotis, D. Niarchos, P. Skeldon, G.E. Thompson, P. Falaras, Nanotechnology 20, 365601 (2009)

[9] V. Galstyan, E. Comini, G. Faglia, G. Sberveglieri. Sensors (Basel, Switzerland) 2013; 13(11): 14813-14838.

[10] M. Lupiwana, R. Taziwa, E. Meyer, D. Katwire. Journal of Material Science and Technology Research, 2017, 3, 17-31

[11] M. Jarosz, K. Syrek, J. Kapusta-Kołodziej, J. Mech, K. Małek, K. Hnida, G.D. Sulka. Journal of Physical Chemistry C 119 (2015) 24182.

INFLUENCE OF Pd-DOPED TiO_x ON INVERTED ORGANIC SOLAR CELLS PERFORMANCE

J. G. Sánchez[1], V. S. Balderrama[2], A. Viterisi[1], M. Estrada[2], J. Ferré-Borrull[1], L. F. Marsal[1*] and J. Pallarès[1*]
[1]Departament d'Enginyeria Electrònica Elèctrica i Automàtica, Universitat Rovira i Virgili
Av. Països Catalans 26, 43007 Tarragona, Spain,
Tel. +34 977 55 96 25, Fax. +34 977 55 96 05
[2]Centro de Investigación y de Estudios Avanzados del Instituto Politécnico Nacional (CINVESTAV-I.P.N)
Av. Instituto Politécnico Nacional 2508, 07360, Mexico city, Mexico.
*Corresponding authors: lluis.marsal@urv.cat and josep.pallares@urv.cat

ABSTRACT: We studied the influence of Pd-doped TiO_x layer on the performance of inverted solar cells based on thieno[3,4-b]-thiophene/benzodithiophene (PTB7) and [6,6]-phenyl-C71-butyric acid methyl ester ($PC_{70}BM$). Inverted solar cells were fabricatad using 5 different Pd concentration. As a control analysis, similar devices were fabricated remplacing Pd-doped TiO_x by pristine TiO_x. The solar cells fabricated with lowest Pd-doping concentration show a significantly improvement of the devices performance. Devices with Pd-doped TiO_x exhibit reduction of the shunt losses compared to cells with pristine TiO_x. Finally, this analysis shows that TiO_x with low Pd concentration could improve the performance of solar cells. Nevertheless, high Pd concentrations are detrimental to devices performance.
Keywords: Organic Solar Cell, TiO_2, Doping.

1 INTRODUCTION

Organic solar cells (OSCs) have been considered as promising renewable energy source because of their light weight, high transparency, low-cost and possibility of fabrication on large areas. At present efficiencies up to 11.5% have been achieved [1, 2]. The bulk heterojunction (BHJ) concept is widely used in OSCs based on polymer/fullerene, which provide an automatic selection of electrons from fullerene and holes from the polymer [3]. In recent years, PTBx polymers composed of thieno[3,4-b]thiophene and benzodithiophene alternating units have demonstrated to be promising donor materials due to their wide spectral response and low band gaps [4, 5].

In conventional OSCs, poly(3,4-ethylenedioxythiophene):poly(styrenesulfonate) (PEDOT:PSS) is widely used as interlayer between the ITO and the active layer to improve the electron collection. Nevertheless, the corrosive and hygroscopic nature of PEDOT:PSS can degrade the ITO. Aside the harmful consequences caused by the PEDOT:PSS, low-work-function (LWF) metals such as calcium (Ca) or Barium (Ba) used as top electrodes are susceptible to degradation by oxygen and moisture, which result in detrimental for the OSCs lifetime [6].

In order to overcome these disadvantages, OSCs with inverted structure (iOSCs) have been developed. In iOSCs electrons are collected by ITO (cathode), whereas holes collection is carried out by the metal electrode (anode). PEDOT:PSS is mainly replaced by transition metal oxides such as molybdenum oxide (MoO3) or vanadium oxide (V_2O_5), since TMOs have high work-functions (>5 eV) that allow to obtain low resistance ohmic contacts [7, 8]. Moreover, titanium suboxide (TiO_x) and zinc oxide (ZnO) can substitute the LWF metals. TiO_x has been widely used as ETL in iOSCs, because is transparent in visible spectrum, and can behave as optical spacer, providing stability to OSCs [9]. The TiO_x layer is placed between the cathode and the active layer to can act as electron transporting layer and hole-blocking as well. In order to improve the iOSCs performance, the optimization of the TiO_x layer is required to reduce losses at the TiO_x/organic blend interface.

Doping the TiO_x layer by introducing a small amount of a well-chosen dopant is a promising approach, due to the possibility of modifying the grain size, orientation, the electrical conductivity and Fermi level position [10].

Here we present the synthesis of Pd-doped TiO_x thin film via sol-gel methods using several Pd concentrations. In order to analyse the influence of Pd-TiO_x on the iOSCs performance, we fabricated devices based on BHJ layer of the polymer thieno[3,4-b]-thiophene/ benzo-dithiophene (PTB7) and the fullerene [6,6]-phenyl-C71-butyric acid methyl ester ($PC_{70}BM$), using Pd-doped TiO_x as ETL. Similar devices were fabricated replacing the Pd-doped TiO_x by pristine TiO_x layer for control analysis.

2 MATERIALS AND METHODS

2.1 Pristine and Pd-doped TiO_x synthesis

Pristine and Pd-doped TiO_x layer were synthetized via sol–gel methods using the spin coating technique. Pristine TiO_x solution was prepared by mixing Titanium(IV) isopropoxide $Ti[OCH(CH_3)_2]_4$, 0.68M) in 2-methoxyethanol ($CH_3OCH_2CH_2OH$). Ethanolamine ($H_2NCH_2CH_2OH$, 1.6M) was added as stabilizer. Pd-doped TiO_x solutions were prepared by adding Palladium(II) acetylacetonate ($Pd(C_5H_7O_2)_2$), to the pristine TiO_x solution. All the mixtures were stirred and heated at 120 °C under inert atmosphere. Subsequently, the solutions were diluted in anhydrous methanol with a volume ratio 1:6, respectively, and stirred at room temperature during 1 hour.

2.2 Blend solution preparation

Blend solution was prepared dissolving PTB7 and $PC_{70}BM$ with a weight ratio of 1:1.5 in chlorobenzene (CB) and 1, 8-diiodooctane (DIO) (97:3 v/v), in order to obtain a final solution concentration of 25 mg mL-1. The solution was left stirring and heating under inert atmosphere at 40 °C overnight.

2.3 Device fabrication

The iOSCs were fabricated on ITO-coated glass substrates previously cleaned with detergent and ultra-sonicated in acetone, methanol and isopropanol. Finally, substrates were dried in an oven at 120°C for 20 min and exposed to UV during 15 min. The diluted pristine and Pd-doped TiO_x precursors were spin-cast on ITO substrates at 6000 rpm. In order to convert the precursor in TiO_x by hydrolysis, samples were left in ambient conditions during 1 hour and heated at 400 °C. The heating rate was of 20 °C min−1. Subsequently, the blend solution of PTB7:$PC_{70}BM$ was deposited on top of pristine and Pd-doped TiO_x layers by spin coating at 800 rpm for 30 s, to obtain an active layer with thickness of about 100 nm. Finally, 5 nm of V_2O_5 and 100 nm of Ag were thermally evaporated onto the photoactive layer in a vacuum chamber.

3 RESULTS AND CONCLUSIONS

3.1 Experimental results

The BHJ-iOSCs using Pd-doped TiO_x as ETL were fabricated with the following structure: ITO/Pd-doped TiO_x/PTB7:$PC_{70}BM$/V_2O_5/Ag. In addition, as a control analysis similar, devices were fabricated replacing the Pd-doped TiO_x layer by pristine TiO_x. Figure 1 shows the structure of all devices fabricated.

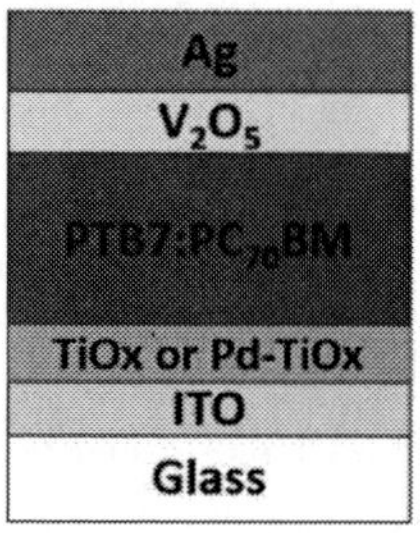

Figure 1: Structure of iOSCs fabricated.

In order to better analyze the influence of Pd-doped TiO_x layer on the iOSCs performance, several devices using 5 different Pd concentration were fabricated. The name of the samples and their Pd concentration used to dope the TiO_x are showed in table 1. The lowest Pd concentration used was 3.5 millmolar (mM) and the highest 56 mM.

Table I: Name and Pd concentration of the samples.

Sample	Pd-doped TiO_x concentration
Pd0	0 (pristine)
Pd3.5	3.5mM
Pd7	7mM
Pd14	14mM
Pd28	28mM
Pd56	56mM

Figure 2 shows the current density-voltage (J-V) characteristics of iOSCs using pristine TiO_x and Pd-doped TiO_x layer. All the the curves were taken under simulated AM 1.5G illumination (100 mW cm-2). The samples Pd0 and Pd3.5 exhibited the highest open circuit voltage (V_{OC}). Moreover, the V_{OC} is slightly reduced as the Pd concentration increased in samples Pd7, Pd14 and Pd28. Nevertheless, Pd56 showed a high V_{OC} decrease. In addition, Pd0 showed the highest short circuit current (J_{SC}). On the other hand, in samples with Pd-doped TiO_x layer, J_{SC} decreased as the Pd concentration increased. However, Pd28 exhibits a J_{SC} similar to than that of Pd3.5.

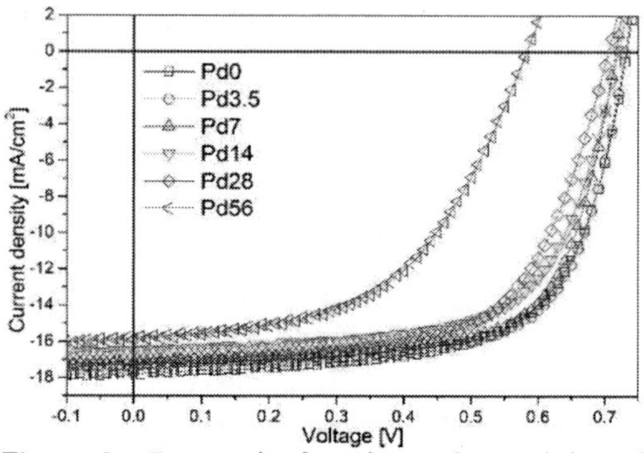

Figure 2: Current density-voltage characteristics of iOSC's based on PTB7-$PC_{70}BM$ with pristine TiO_x and Pd-doped TiO_x with several Pd concentrations under AM 1.5G irradiation.

Table 2 summarizes the performance characteristics of the iOSCs fabricated comprising of V_{OC}, J_{SC}, fill factor (FF), power conversion efficiency (PCE) and both series and shunt resistances (R_S and R_{Sh}). From table 2, V_{OC} is not affected by using low Pd concentration from 3.5 up to 28 mM, while using the highest Pd concentration V_{OC} decreases 20% with respect to the control sample. Samples Pd3.5, Pd7, and Pd28 exhibited a similar J_{SC} decrease (3%), while J_{SC} decreased 7% and 12% in samples Pd14 and Pd56, respectively. FF is slightly reduced in samples Pd7, Pd14 and Pd28, while in sample Pd3.5, FF increased 4%. On the hand, in Pd56 the FF reduced in 20% with respect to control sample. R_{Sh} is increased by 49, 62, 31 and 2 % in samples Pd3.5, Pd7, Pd14 and Pd28, respectively. In sample Pd56 R_{Sh} is reduced by 30%. R_S in not altered in sample Pd3.5, but starts to increase as the Pd concentration increases. Finally, PCE is slightly improved in sample Pd3.5. In samples Pd7, Pd14, Pd28 and Pd56 the PCE decreased as the Pd concentration increased.

Table 2: Performance parameters of iOSC's based on PTB7:$PC_{70}BM$ with pristine TiO_x and Pd-doped TiO_x as ETL

	V_{OC} [mV]	J_{SC} [mA/ cm2]	FF [%]	PCE [%]	R_S [Ω-cm²]	R_{Sh}
Pd0	729	17.7	65.8	8.49	2.2	586
Pd3.5	729	17.2	68.3	8.59	2.2	875
Pd7	719	17.1	65.6	8.21	2.9	950
Pd14	719	16.5	65.3	7.75	3.5	770
Pd28	709	17.3	64.2	7.63	5.4	601
Pd56	579	15.7	52.4	4.81	7.7	408

High Pd concentration in TiO_x (Pd56) results detrimental to iOSCs since it reduces the PCE increased the R_S and reduced V_{OC} and J_{SC}. On the other hand, low Pd concentration (Pd3.5) can improve the performance of iOSCs. In sample Pd3.5 the V_{OC}, J_{SC} and R_S are not affected, while R_{Sh} is increased resulting in the improvement of FF and PCE. Other authors have reported the doping of TiO_x layer. The incorporation of dopant material into the TiO_x layer can reduce the shunt losses and the interfacial charge recombination, resulting in the improvement of the iOSCs performance [10, 11]. Despite the promising results obtained using low Pd concentration (samples Pd3.5), optimization of doping methodology and

Pd concentration are required for improving the iOSCs performance.

3.2 Conclusions

These results have demonstrated that Pd-doped TiO_x affects the performance parameters of iOSCs. In samples with lowest Pd concentration (Pd3.5), the device performance is slightly improved. For Pd concentrations higher than 7mM, R_S increases, which results in PCE and FF reduction. Pd concentrations higher or equal than 56mM are detrimental to iOSCs due to degrade all the performance parameters. Controlled Pd-doping results a promising approach to improve the electrical properties of TiO_x layers, in order to enhance the PCE of iOSCs.

ACKNOWLEDGEMENTS

This work was supported in part by the Spanish Ministry of Economy and competitiveness (MINECO) under grant numbers TEC2015-71324-R and TEC2015-71915-REDT, the ICREA under the ICREA Academia Award, the Catalan authority under project AGAUR 2014 SGR 1344 and the CONACYT Project 237213 in Mexico.

REFERENCES

[1] V. S. Balderrama, F. Avila-Herrera, J. G. Sanchez, J. Pallares, O. Vigil-Galan, L. F. Marsal, and M. Estrada. IEEE J. Photovoltaics, 6, 2,(2016).

[2] M. A. Green, K. Emery, Y. Hishikawa, W. Warta, E. D. Dunlop, D. H. Levi, and A. W. Y. Ho-Baillie. Prog. Photovoltaics Res. Appl., 25, 1 (2017).

[3] G. Yu, J. Gao, J. C. Hummelen, F. Wudl, and A. J. Heeger. Science, 270, 5243 (1995).

[4] P. Han, V. S. Balderrama, A. Mihi, P. Formentin, J. Ferre-Borrull, J. Pallarés, and L. F. Marsal. IEEE J. Photovoltaics, 5, 3 (2015).

[5] Y. Liang, Y. Wu, D. Feng, S. Tsai, H. Son, G. Li, and L. Yu. J. Am. Chem. Soc., 131, 1 (2009).

[6] V. S. Balderrama, M. Estrada, P. L. Han, P. Granero, J. Pallarés, J. Ferré-Borrull, and L. F. Marsal. Sol. Energy Mater. Sol. Cells, 125 (2014).

[7] R. C. I. Mackenzie, V. S. Balderrama, S. Schmeisser, R. Stoof, S. Greedy, J. Pallarès, L. F. Marsal, A. Chanaewa, and E. von Hauff. Adv. Energy Mater., (2015).

[8] M. T. Greiner, M. G. Helander, W.-M. Tang, Z.-B. Wang, J. Qiu, and Z.-H. Lu,. Nat. Mater., 11, 1(2012).

[9] Y. C. Yeh, S. S. Li, C. C. Wu, T. W. Shao, P. C. Kuo, and C. W. Chen. Sol. Energy Mater. Sol. Cells, 125 (2014).

[10] M. Thambidurai, J. Y. Kim, H. Song, Y. Ko, N. Muthukumarasamy, D. Velauthapillai, V. W. Bergmann, S. a. L. Weber, and C. Lee. J. Mater. Chem. A, 2, 29 (2014).

[11] S. Yuan, Y. Zhang, W. Liu, and W. Zhang. Electrochim. Acta, 116 (2014).

CHARACTERISATION OF A MULTIDIMENSIONAL NONLINEAR SOLAR CELL

Thomas Fey, Ingo Kröger, Stefan Winter
Physikalisch-Technische Bundesanstalt (PTB)
Bundesallee 100, 38116 Braunschweig, Germany
Tel. +49 531 592 4185, Fax: +49 531 592 694185, e-mail: Thomas.Fey@ptb.de

ABSTRACT: In general, new and innovative solar cell technologies have spectral responsivities that differ from the spectral responsivities of standard reference solar cells (e.g. c-Si, filtered c-Si). Hence integral calibration methods (e.g. with a solar simulator) require a spectral mismatch correction (IEC 60904-7). Therefore, the relative spectral responsivity of the solar cell is required. The determination of the spectral responsivity for technologies with spectral nonlinearities (e.g. with respect to the irradiance level, like dye-sensitized solar cells (DSC)) is comparatively difficult. However, even though the calibration of the short-circuit current under Standard Test Conditions I_{STC} is performed at an irradiance level E of 1000 W m^{-2}, commonly the spectral responsivity measurement is performed at one irradiance between 300 W m^{-2} and 400 W m^{-2} in accordance to a simplification mentioned in the IEC 60904-8. This simplification can often be used (e.g. for the most Si solar cells), but is not always correct. If this is the case, then the determined spectral responsivity leads to a wrong spectral mismatch correction factor. In this contribution, an example for this behavior is shown with the consequence that for this type of solar cells in general the simplification may not be applied. The solution is the measurement at different irradiance levels with the DSR Method. However, the DSR method uses the combination of a continuous bias radiation and a modulated monochromatic beam. Hence, a dependence of the spectral responsivity of the modulation frequency has to be considered. The most common solar cells (e.g. c-Si) do not show a frequency dependence with respect to the modulation frequency of the monochromatic beam in the range from 0.1 Hz to 99 Hz. Nevertheless, some solar cells such as DSC exhibit a frequency dependence in this range. These multidimensional dependencies of the spectral responsivity as well as its impact and solutions on the determination of the spectral responsivity and the will be presented.
Keywords: Calibration, Characterisation, Dye-Sensitised, Experimental Methods, Spectral Response

1 INTRODUCTION

In general, solar cell calibrations have to be performed at standard test conditions (STC, AM1.5G reference solar spectrum [1], irradiance E = 1000 W m^{-2}, device temperature T = 25 °C). This is often realized using integral calibration methods (e.g. solar simulator). In this case a spectral mismatch correction (IEC 60904-7) have to be performed. A spectral mismatch correction requires the relative spectral responsivity of the solar cell under test and the reference. The absolute spectral responsivity can be determined using the differential spectral responsivity (DSR) method [2,3] (IEC 60904-8). The DSR method is realized using the combination of a continuous bias radiation and a modulated monochromatic beam. The continuous bias radiation sets the solar cell into the working points. The modulated monochromatic beam is used to measure the responsivity of the solar cell for the applied wavelength of the monochromatic beam. If this measurement is performed in the wavelength range of the solar cells and at suited irradiances, the absolute spectral responsivity can be calculated. These sophisticated measurements are time consuming and complicated to realize with sufficient uncertainties. Therefore, the spectral responsivity is often determined using a simplification mentioned in the IEC 60904-8. The simplification allows to perform the differential spectral responsivity measurements at an irradiance level causing a short circuit current of 30 % - 40% relative to the short-circuit current under standard test conditions I_{STC} and assume this differential responsivity as the spectral responsivity at STC. But, the simplification can be wrong if the solar cell exhibit multidimensional nonlinearities of solar cells.

We characterize a multidimensional nonlinear dye-sensitized solar cells. The dye-sensitized solar cell is equipped with a mesoscopic TiO_2 semiconductor electrode. A ruthenium-based dye (Z-907) was applied onto this electrode Here we show the results of complete DSR characterizations at three modulation frequencies of the modulated monochromatic beam and several irradiance levels

2 INSTRUMENTS AND METHODS

2.1 DSR method

The primary Differential Spectral Responsivity (DSR, IEC 60904-4) method [4, 7] uses the combination of a modulated (quasi-) monochromatic beam and a white light bias source (see Fig. 1). The bias radiation E_b sets the solar cell into working points. The corresponding DC output currents I_{SC} are measured with a DC-multimeter. The modulated (quasi-) monochromatic radiations ΔE_{mono} causes additional modulated output currents ΔI_{mono} and are measured with a lock-in amplifier. The DC and modulated output currents are used to perform precise determinations of the spectral responsivity $s(\lambda)$ and the linearity of the solar cell.

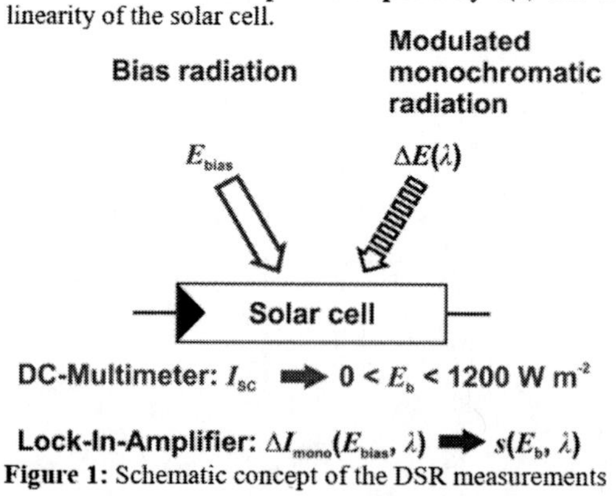

Figure 1: Schematic concept of the DSR measurements

2.2 White light characterization

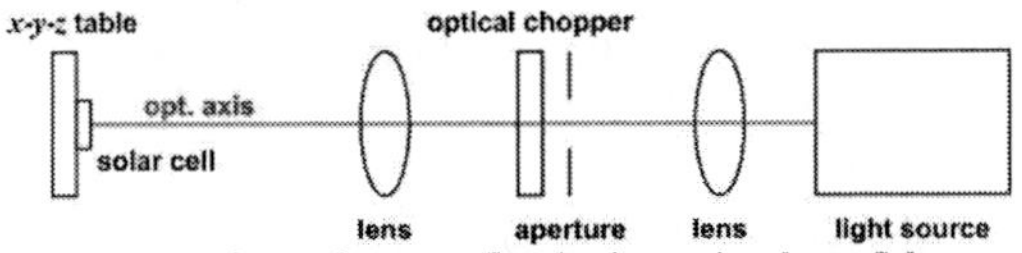

Figure 2: Schematic setup for the investigation of the white light frequency dependence of the dye-sensitized solar cell

The white light frequency dependence was investigated with a setup based on a solar simulator (see Fig. 2). The divergent light of the solar simulator is focused onto a aperture using a lens. The beam cross section is reduced by the aperture. This is necessary to modulate the light of the solar simulator by means of an optical chopper. Behind this spot, the light of the solar simulator is divergent again. However, to be able to carry out investigations at higher irradiances, the light is focused onto the solar cell by means of a second lens. This allows irradiances of up to 1100 W m^{-2} to be reached in the second focus. The irradiance can be adjusted by changing the distance between the solar cell and the remaining setup and determined by means of a reference device. The short-circuit current of the DSC was measured by a lock-in amplifier. To correct the frequency response of the measurement setup (measurement amplifier, lock-in amplifier, wiring), its frequency dependence was determined with a photodiode at very low irradiance.

3 RESULTS

3.1 Spectral characterization

At the beginning, the frequency behavior of the spectral responsivity was investigated. For this purpose, three DSR characterizations were carried out with the modulation frequencies f of the monochromatic irradiance: 2 Hz, 17 Hz and 78.9 Hz. The resulting differential spectral responsivities at different irradiances E_b and frequencies are shown in Fig.3.

The DSC show low differential spectral responsivities $\tilde{s}$ at low irradiances E_b. With increasing the irradiance, the differential spectral responsivity increases in the complete wavelength range. This increase occurs faster for low modulation frequencies (f = 2 Hz, f = 17 Hz) of the monochromatic beam. At high modulation frequencies (f = 78,9 Hz), no saturation can be identified at the applied irradiances. At lower modulation frequencies, a saturation and an additional decrease can be observed at higher wavelengths.

This behavior can be also observed at the AM1.5-weighted differential responsivities as a function of the irradiances E_b. The AM1.5-wheighted differential responsivity increase constantly and no saturation can be observed at the modulation frequency of 78.9 Hz. At f =

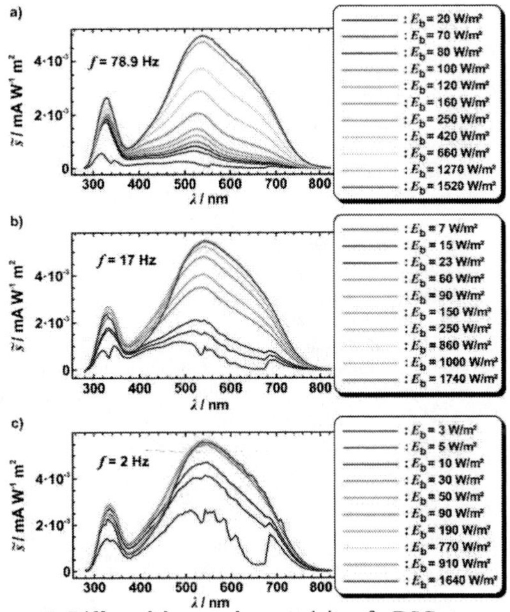

Figure 3: Differential spectral responsivity of a DSC at different bias irradiances and modulation frequencies f of the monochromatic beam.

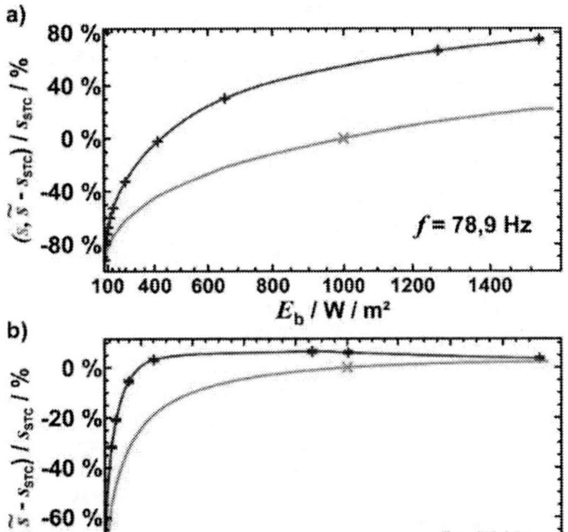

Figure 4: AM1.5-weighted differential responsivities as a function of the corresponding irradiances E_b and modulation frequencies f of the monochromatic beam

17 Hz ($f = 2$ Hz) a saturation can be observed at $E_b = 250$ W m^{-2} ($E_b = 90$ W m^{-2}).

For this strong multidimension nonlinear solar cell the simplification would lead to deviations of 5 % to 20 % at the modulation frequency $f = 78{,}9$ Hz to the value obtained applying the complete DSR method. The deviation decreases in this particular case at lower modulation frequencies. This decrease is related to the integral AM1.5-wheighted linearity of the solar cell. The spectral deviations occur still at the spectral responsivity in dependence of the wavelength.

However, the normalized spectral responsivities are sufficient to calculate the spectral mismatch. To obtain the spectral mismatch, the spectral responsivities were calculated under standard test conditions for the three modulation frequencies and normalized to their respective maximum. The normalized spectral responsivities thus obtained are shown in Fig. 5.

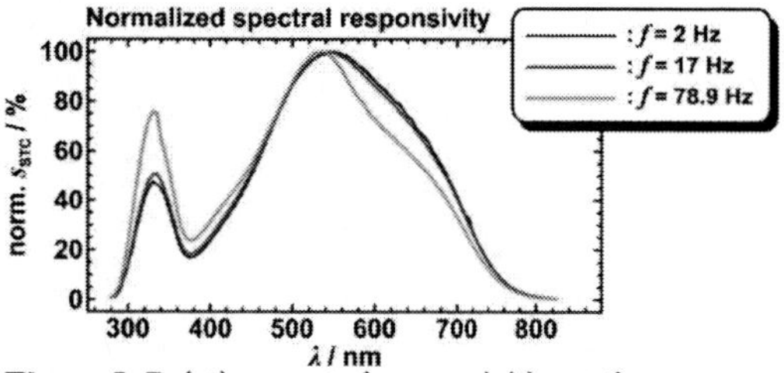

Figure 5: Relative spectral responsivities at the irradiance $E = 1000$ W m^{-2} for the measured modulation frequencies f of the modulated monochromatic beam.

The normalized spectral responsivity of the measurement at a modulation frequency $f = 78.9$ Hz shows clear deviations from the results at $f = 17$ Hz and $f = 2$ Hz. The value at $\lambda = 340$ nm is higher by approx. 30 %, and at $\lambda = 600$ nm it is lower by approx. 15 % than at lower frequencies. The normalized functions of the spectral responsivities at $f = 2$ Hz and $f = 17$ Hz are nearly equivalent. Only at $\lambda = 340$ nm, there is a discrepancy of up to 5 %.

3.4 White light characterization

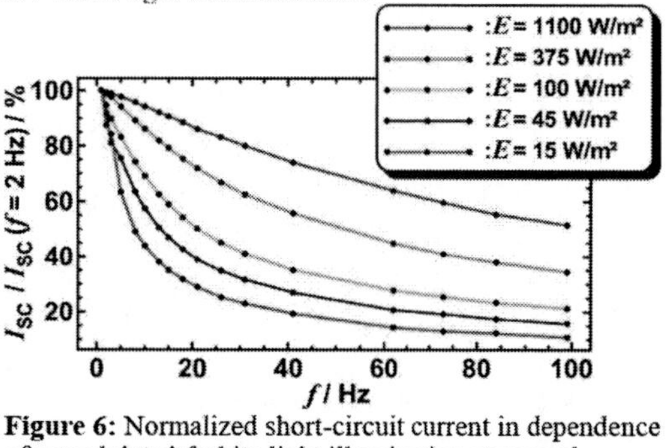

Figure 6: Normalized short-circuit current in dependence of a modulated f white light illumination at several irradiances E

The normalized short-circuit current in dependence of a modulated f white light illumination $I_{sc}(f, E)/I_{sc}(f = 2$ Hz, $E)$ at several irradiances E. The frequency dependence was measured at 1100 W m^{-2}, 375 W m^{-2}, 100 W m^{-2}, 45 W m^{-2} and 15 W m^{-2}. The frequency was varied in the range of 2 Hz to 99 Hz. The results are normalized to the value at $f = 2$ Hz for each irradiance level and shown in Fig. 6. The frequency dependence is also affected by the applied irradiance. The I_{sc} (f, E = 1100 W m^{-2}) drops by approx. 50 % and the I_{sc} (f, E = 15 W m^{-2}) drops by approx. 95 %.

4 SUMMARY

The characterization of a dye-sentized solar cell (DSC) at the frequencies 2 Hz, 17 Hz and 78,9 Hz show that its spectral responsivity exhibits wavelength – dependent multidimensional nonlinearities as a function of the irradiance and the modulation frequency of the monochromatic radiation. Hereby, it turns out that los modulation frequencies are indispensable for a correct determination of the relative spectral responsivity and of the linearity applying the DSR method.

Furthermore, characterization by means of a solar simulator at several irradiance levels is necessary to measure the correct frequency dependence of the solar cell. Only with a comparison and a combination of the results is it possible to make statements on the short-circuit current and the spectral responsivity for DC conditions (f = 0 Hz).

5 LITERATURES

[1] IEC 60904-3 Edition 2, International Electrotechnical Commission, Measurement principles for terrestrial photovoltaic (PV) solar devices with reference spectral irradiance data, Geneva, 2008.

[2] Metzdorf, J. (1987, May). Calibration of solar cells. I. The differential spectral responsivity method. Applied Optics 26, 1701-1708

[3] S. Winter, T. Wittchen, J. Metzdorf: Primary Reference Cell Calibration at the PTB Based on an Improved DSR Facility; in "Proc. 16th European Photovoltaic Solar Energy Conf.", ed. by H. Scherr, B. Mc/Velis, E. Palz, H. A. Ossenbrink, E. Dunlop, P. Helm; Glasgow, (2000) James & James (Science Publ., London), ISBN 1 902916 19 0

6 ACKNOWLEDGMENTS

This work is supported by the European Fund for regional development ("Europa fördert Niedersachsen") and the Federal State of Lower Saxony. Further, my thank goes to Dr. Takagi (Kanagawa Academy of Science and Technology (KAST), Japan) for the stable DSC-sample.

STRUCTURAL AND RAMAN SPECTROSCOPIC CHARACTERIZATION OF C-TIO_2 NANOTUBES SYNTHESIZED BY TEMPLATE ASSISTED SOL-GEL TECHNIQUE

Raymond Taziwa, Edson Meyer and Nwabisa Takata

[1]University of Fort Hare, Institute of Technology
Fort Hare Drive, 3rd Floor Chemistry Building, ALICE 5700, South Africa
Tel: +27 (0) 40 602 2086, Fax: +27 (0) 40 653 2311, rtaziwa@ufh.ac.za
Raymond T. Taziwa, University of Fort Hare, Institute of Technology, rtaziwa@ufh.ac.za
Edson L. Meyer, University of Fort Hare, Institute of Technology emeyer@ufh.ac.za
Nwabisa, Takata, University of Fort Hare, Institute of Technology, ntakata@ufh.ac.za

Abstract

It is of great importance to find ways of improving the charge transport within titanium dioxide nanotubes (TNTs) which will in turn decrease electron-hole pair recombination and subsequently increase the cell efficiency of the dye sensitized solar cells (DSSC). High electron mobility in TNTs relies on the morphology, surface defects and consistence in the crystalline composition. These factors are closely related to the preparation conditions. In this regard this work focuses on synthesis and structural characterization of carbon doped titanium dioxide nanotube. Un-doped and carbon doped titanium dioxide nanotubes (C-TNTs) were synthesized using a template-assisted sol-gel technique employing titanium tetra butoxide precursor and oxalic acid as the dopant. SEM, XRD, and Confocal Raman spectroscopy (CRS) was used to evaluate the morphological and structural properties of the as synthesized TNTs. SEM analysis has revealed that the morphology of TNTs prepared change with introduction of carbon dopant. SEM-EDX spectra has revealed the presence of Ti peaks at 0.45 and 4.9 keV corresponding to $K\alpha 1$ and $K\beta 1$ emission line respectively. Oxygen exhibits a signal at 0.5 keV corresponding to $K\alpha 1$ emission line. XRD analysis has revealed the presence of a mixed Anatase-Brookite phase with diffraction peaks at 2θ angles of 25.49^0, 38.11^0, 40.60^o 48.14 ^{0}, 54.58^0, 63.00^0, 70.11^0 and 75.66^0. CRS large area scan in the XY direction has revealed the presence of Raman active modes at 153.19 cm^{-1}, 205.62 cm^{-1},328.33 cm^{-1}, 404.55 cm^{-1}, 523.26 cm^{-1}, 523.26 cm^{-1} and 648.69 cm^{-1} belonging to a mixed Anatase-Brookite phase. CRS depth profiling in the XZ direction has also validated the presence of a mixed Anatase-Brookite phase with Raman Active modes at153.19 cm^{-1}, 208.87 cm^{-1}, 404.55 cm^{-1}, 523.26 cm^{-1} and 648.55 cm^{-1}.

Key words: Titanium dioxide nanotubes, sol-gel, Carbon doping, Confocal Raman spectroscopy

1. Introduction

In today's society, it is becoming ever important to find alternative sources of energy that are low cost with scientific and commercial relevance. Solar cells have become one of the most widely researched methods of obtaining energy in "***greener***" ways than burning fossil fuels. More particularly, the dye sensitized solar cells (DSSC) employing titanium dioxide nanotubes offer a viable alternative and cost effective solar cells to replace the traditional silicon solar cell technology [1]. Substantial efforts have been focused on fabrication of well aligned TNTs consistent crystalline composition. TNTs architecture offers excellent electrical channel for vectorial charge transfer of electrons [2]. TNTs of different morphologies and geometrical shapes have been prepared by various techniques such as electro-anodization, sonochemical deposition, hydrothermal methods and template assisted sol gel techniques [3, 4, 5]. While many of these fabrication techniques are sophisticated and expensive. Sol gel template assisted methods of fabricating TNTs has attracted considerable interest due to its ability to produce nanomaterials at low temperatures and at low cost. Moreover template assisted sol gel techniques presents numerous advantages such stoichiometric control of the composition at molecular scale which enables production of nanomaterials of a consistent atomic composition [6]. On the other hand before utilization of the fabricated TNTs by any technique for application as electrode material for solar cells it is obligatory to evaluate the crystalline nature of the structure, stability and composition of the crystalline phases. The crystalline nature and composition of the TNTs varies with fabrication techniques and deposition parameters [7]. The stability of the Titanium dioxide phases and hence, potential application in solar cells is entirely dependent on the isomorphic crystalline phase composition present after pre-treatment and deposition on the surface of functional substrate. It is well known that the crystalline structure of TiO_2 also plays a factor in designing the most efficient solar cells. Moreover, it has been well documented that the Anatase polymorph of TiO_2 is preferred in solar cell applications due to the difference in the flat-band potential of Anatase as compared to the Rutile. This is due to the fact that the Anatase polymorph has a conduction band that is 0.2 V more negative than that of rutile counterpart. Hence a superior maximum photo-voltage can be achieved on Anatase than on Rutile provided an identical redox electrolyte is used [7]. Even though, phase transformations are necessary for solar cell applications concerning TiO_2 they have serious implications on the stability of the TNTs more particularly in the presence foreign non-metal and/or metal species [8, 9, 10]. It is well known that addition of foreign species in through non-metal or metal doping in TiO_2 nanostructures can cause structural instability and/or stability which either impedes or catalyses phase transformation [. Several studies have been reported on the effect of annealing temperature on phase transformation in titanium dioxide nanotubes. To the best of our knowledge, no systematic attempt has been made to evaluate the effect of non-metal dopants in phase transformation of titanium dioxide fabricated by a template assisted sol gel technique. This paper presents the effect of carbon doping on the morphological, structural and crystalline phase composition of titanium dioxide nanotubes prepared by a template assisted sol gel technique. In addition to the complexity of fabricating defect free TNT with a uniformly distributed Anatase crystalline phase. It must be pointed out that the difficulty in finding the appropriate characterization technique that is able to map out the crystalline phase distribution on the surface of the TNTs and along the TNTs is a key challenge. Therefore, these difficulties in characterizing TNTs structure are a current challenge that limits the development of solar cells carbon doped titanium dioxide nanotubes (C-TNTs). The development of efficient electrode materials for solar cell applications requires detailed understanding of the morphological,, chemical and as well as structural properties at nanometer scale. Raman spectroscopy provides both qualitative and quantitative information about the chemical nature and crystalline properties of TNTs. Confocal Raman spectroscopy techniques have been demonstrated to reveal unprecedented detail in the vibrational structure of carbon nanotubes. In this work Confocal Raman Spectroscopy coupled with Raman imaging has been used for the first time to successful mapping out crystalline phase composition of C-TNTs and has proven to be a fruitful technique to complement to a number of characterization normally used to probe crystalline phase distribution of TNTs over sample large area. In this work Raman spectroscopy and Raman imaging, has been used for the first time to reveal TiO_2 crystalline phase distribution through Confocal Raman imaging of large area (sample size of 500 x 500 μm). Such Raman measurements will prove to be invaluable for evaluating the crystalline structural uniformity of TNTs for development of efficient electrode materials for solar cell applications will be invaluable for

2. Experimental

2.1. Synthesis *of Sol-gel precursor solutions:*

The precursor solution for un-doped and carbon doped TiO_2 was prepared by mixing (Ti (OBu) $_4$), ACAC, deionized water and EtOH, in molar ratio of 1:1:3:20 respectively. In a typical synthesis two beakers labelled "A" and "B" were used to prepare precursor solutions. 24 cm^3 of ethanol, 2.4 cm^3 of water and 4.5 cm^3 of acetylacetone were transferred to beaker A. To ensure complete mixing Beaker A contents were sonicated for 30 minutes. Beaker B solution was made by mixing 15 cm^3 (Ti (OBu) $_4$) in 24 cm^3 EtOH. After thorough mixing, the beaker "A" solution was added in small quantities using a 10 cm^3 syringe to the solution in beaker B. The yellow mixture was stirred for approximately 2 hours at room temperature. In preparation of the carbon doped precursor solutions just before stirring for 2 hours at room temperature. Different masses of oxalic acid dihydrate were added in the solutions to make the required concentrations for 9 mM to 75 mM.

2.2. Fabrication of TiO_2 nanotubes:

AAM were immersed in the resulting precursor solutions. The membranes were allowed to stand for 72 hours for complete pore filling of the nanopores .The samples then were the dried in air for 10 minutes followed by calcination at 475 °C for 1 hour. The AAM were cooled at room temperature and transferred into a centrifuge tube containing 1M NaOH(aq) solution. The solution was placed in a centrifuge for 1 hour at 10000 rpm for dissolution of the AAM. The sludge was then left to dry at room temperature for 48 hours. The samples were then collected and reserved for analysis.

2.3 Characterization

XRD scans were recorded at several resolutions using a Pan Analytical empyrean XRD system with a Cu anode at Kα radiation wavelength of 0.15418 nm at an operating voltage of 40 KV and 40 mA. Elemental analysis, structural and morphological properties of the synthesized TNTs were acquired using a Zeiss Cross Beam 540 Focus ion beam Scanning Electron Microscope (FIB-SEM). The FIB-SEM is equipped with an Oxford Aztec EDS system with a 50 mm2 silicon drift detector (SDD).The samples were mounted on a double sided carbon sticky tabs and sputter coated with carbon to reduce charging within the SEM. The coating with carbon was performed in an Edwards Carbon Coater. The TNTs were characterized for TiO_2 phase distribution using a using confocal Raman microscopy (Witec ALPHA 300RA).

3. Results and Discussions

3.2. SEM analysis

Fig.1 (a-f) shows the SEM micrographs of un-doped TNTs and carbon doped TNTs (C-TNTs). Fig.1 (a) has revealed that the un-doped TNTs are randomly distributed and clustered with non-uniform length and the TNTs are closely-packed to form of bundles. Fig.1 (b-e) has revealed that the increase in carbon dopant concentration has resulted in formation of loosely packed TNTs. The broken nature of the TNTs is attributed to use of precursor molar ratio of (1:2:3:20 TTB: ACAC: H_2O: EtOH), as opposed to molar ratio of 1:2:3:20 [2] Additionally, the centrifugation method of recovery of TNTs, has induced lateral forces that lead to an irregular lift-off of the nanotubes during detachment from the template surface leading to deformed nanotubes [11] Further increase in carbon dopant concentration has also contributed to deformation of nanotube morphology. Additionally Fig. 1(b-e) has revealed that introduction of the carbon dopant and further increase carbon dopant concentration has resulted in the formation of more broken TNTs. In conclusion the introduction of carbon dopant into the titanium dioxide matrix has cause significant change in the TNTs morphology. The EDX spectra has revealed the presence of strong Ti and O signals. The presence of Ti and O in the spectra confirms the successful hydrolysis of TTB in. The EDX spectra has revealed the presence of Ti peaks at 0.45 keV and 4.9 keV which corresponds to $K\alpha_1$ and $K\beta_1$ emission line respectively. Taziwa *et at* [12] reported the presence of titanium in multiple oxidation states. The 0.52 keV peak of titanium in the EDX spectra was due to the oxidation state of Ti (IV) and the peak at.045 keV arises from an oxidation state of Ti (III). In DSSC, the Ti (IV) oxidation state of semiconductor TiO_2 forms a tight chemical binding interface with dye molecules for improved electron transport between the TiO_2 nanomaterials [13]. We can also observe from Fig. 1 (f) that oxygen exhibits a signal at 0.5 keV corresponding to $K\alpha_1$ emission line. The EDX spectra has revealed the presence of a carbon signal for all fabricated samples. The presence of carbon in the samples can be attributed to sputter coating during sample preparation and the use of a carbon tape during analysis. Additionally, the EDX spectra has revealed the presence of Al peak at 0.99 keV , the observed Al peak may have arose from anodic alumina oxide membranes used to prepare TNTs..

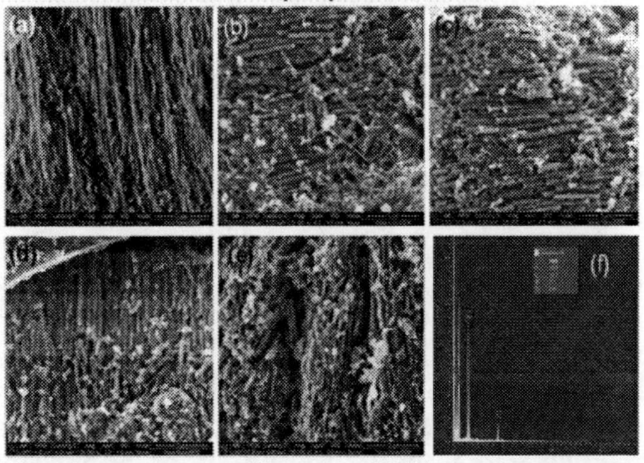

Fig. 1 Cross-sectional SEM micrographs presentation of (a) un-doped TNT, (b) 9 mM C-TNTs, (c) 27 mM C-TNTs, (d) 45 mM C-TNTs, (e) 75 mM C-TNTs and (f) SEM-EDX

3.2. XRD analysis

The crystallographic structure of un-doped and C-TNTs nanotubes fabricated by a template-assisted sol-gel technique using titanium tetra-butoxide as precursor and oxalic acid as dopant are shown here in Fig 2

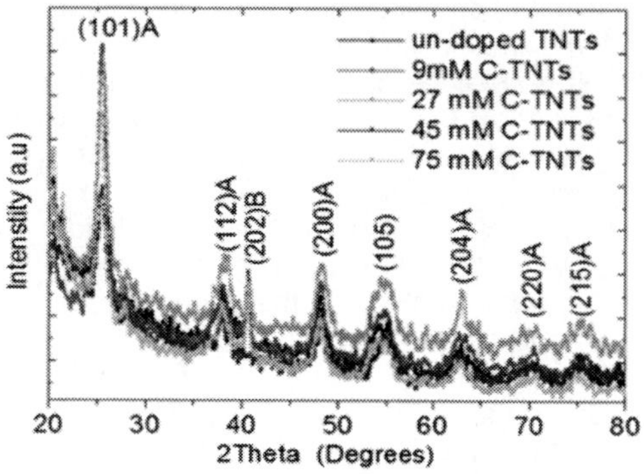

Fig. 2 Powder X-ray diffraction analysis of un-doped and C-TNTs fabricated by a template-assisted sol-gel technique using titanium tetra-butoxide as precursor and oxalic acid as dopant.

XRD analysis has revealed peaks at 2θ angles of 25.49°, 38.11°, 48.14 °, 54.58°, 63.00°, 70.33 and 75.66° all Anatase (JCPDS No. 21- 1272). It is also noticeable that the intensity of the dominate Anatase peak at 2θ angle of 25.62° belonging to the (101) diffraction plane had the highest intensity. This implies that the growth orientation of the un-doped and C-TNTs was in the (101) plane. Moreover, the XRD analysis has revealed the presence of a diffraction peak at 40.60° arising from the [202] crystal plane of Brookite phase of TiO_2 (JCPDS No. 29-1360 TiO_2 [14, 15, 16]. It has been reported that the Brookite content in TiO_2 nano-particles prepared by a sol-gel technique decrease with decrease in titanium precursor to water ratio (Ti-precursor/H_2O) from 1:4 to 1:7. In this work a lower Ti-precursor/H_20 ratio of 1:3 has been utilised. Our lower Ti-precursor/H_2O might have catalysed the formation of the Brookite phase in our samples [15]. Our XRD analysis has revealed that there are no other peaks for amorphous aluminium membrane templates and other impurities suggesting the high purity nature of our samples. The XRD spectra in Fig. 2 have revealed that they are no diffraction peaks that could be attributed to carbon doping. The dopant ions may have been moved in either interstitial or substitutional site of the TiO_2 lattice or either the concentrations were too small to be detected by XRD.

3.3. Confocal Raman spectroscopy: Large area scan (XY)

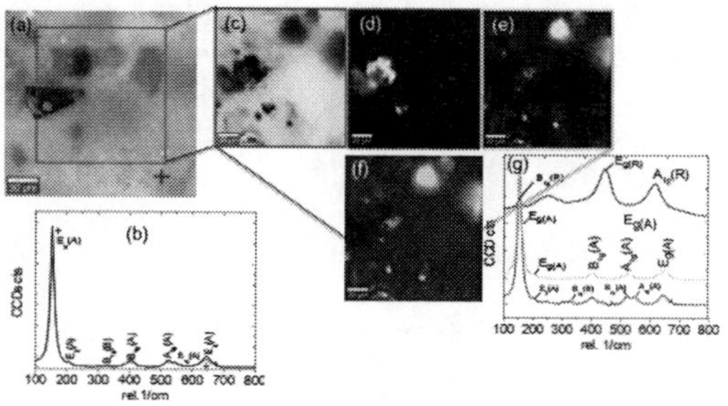

Fig 3 Shows the confocal Raman large area scan (LAS) of un-doped TNTs (a) video image (b) Raman single spectra recorded at the crosses indicated in the image in (a) with corresponding colour coded spectra, (c), (d), & (e) are draw images showing phase distribution of TiO_2 following basis analysis. Brighter colours indicate a higher fitting factor and thus higher signal intensity of the corresponding spectrum at that position, (f) combined image of three different spectra indicating existence of three different phases at the surface of the TNTs pink maroon area less crystallized Anatase phase, blue spots Rutile phase, yellow spots well crystallized Anatase phase (g) shows the corresponding colour coded spectra of (f).

Raman single spectra (RSS) of the un-doped TNTs (both blue and red spectra obtained on the crosses in Fig. 3(a) with corresponding colour coded spectra in Fig 3(b)) have revealed Raman vibrational modes at 153.19 cm^{-1} (Eg-A), 205.62 cm^{-1}(Eg-A), 328.33 cm^{-1}(B1g-B) 404.55 cm^{-1} (B1g-A) 523.26 cm^{-1} (A1g-A), 523.26 cm^{-1}(B1g-A) and 648.69 cm^{-1}(Eg-A) belonging to a mixed Anatase-Brookite phases. This observation is in mutual agreement with XRD analysis which revealed the presence of Brookite diffraction peak at 2θ angle of 40.6°.The presence of Brookite phase of titanium in our samples might be attributed to low titanium precursor/water ratio of 1:3 [15]. Additionally it has been reported that the thermodynamically stable of either Anatase and/or Brookite exist when the crystallite sizes are about 11 nm for Anatase and about 11-35 nm for the Brookite phase. Whilst the Rutile phase is predominately thermodynamically stable in crystallite sizes above 35 nm. Our confocal Raman LAS has revealed that the majority of the sample as depicted by the maroon –yellow area consist of a mixed Anatase-Brookite phase which is mutual agreement with XRD analysis which revealed a crystallite size of 11.39 nm. To obtain a conclusive crystalline phase distribution, a LAS was performed on the video image in Fig 3(a). Confocal Raman imaging LAS scan was performed on the highlighted rectangular area 330 μm x 330 μm shown here in Fig 3 (a). Fig 3(f) shows the combined colour coded image that has revealed existence of about 3 distinct crystalline phase of TiO_2. Fig 3(g) corresponding colour coded spectra of the 3 distinct phases of TiO_2. The majority of the un-doped TNTs sample depicted by the maroon area in Fig. 3 (f), with corresponding maroon Raman spectra, has revealed Raman vibration modes at 153.78 cm^{-1} (Eg-A), 209.08 cm^{-1} (Eg-A), 331.04 cm^{-1} (B1g-B), 403.74 cm^{-1} (B1g-A), 527.16 cm^{-1} (A1g-A), 541.83 cm^{-1} (B1g-A) and 644.59 cm^{-1}(Eg) which consist of Anatase phase of TiO_2 The lowest Eg_1-Anatase mode of the maroon area which is dominant in the un-doped TNT sample has been red shifted by Δ9. 78 cm^{-1} whereas the highest Eg_3-Anatase mode has been blue shifted by Δ3. 59 cm^{-1} compared to the vibrational frequencies the bulk TiO_2. The significant shift is attributed to phonon confinement and decrease in particle size. Additionally Fig 4(f) has revealed that the surface of the un-doped TNTs has blue spots with corresponding blue Raman spectra shown here in Fig 4(g). The blue spots have Raman vibrational frequencies at 150.64.cm^{-1} (B1g-R), 252.14 cm^{-1}(mutli-phonon process), 447.42 cm^{-1} (Eg-R) and 616.09 cm^{-1} (A1g-R) all belonging to a Rutile phase of TiO_2. Whilst the yellow spots, Fig 4(f) have Raman vibrational modes at 153.78 cm^{-1} (Eg-A), 204.72 cm^{-1}(Eg-A), 332.27 cm^{-1}(B1g-B), 521.59 cm^{-1}(B1g-A), 545.51 cm^{-1}(A1g-A), and 644.57 cm^{-1}(Eg-A) all belonging to a mixed Anatase and Brookite phase. Confocal Raman Large area scan of un-doped TNTs prepared by a template assisted sol-gel technique have revealed for the first time that the surface of the nanotubes consist of Anatase, Brookite and Rutile phase of TiO_2. The un-controlled nature of formation and co-existence of the three phase in TNTs membranes might be the main reason why DSSC based on TNTs anodes still have lower efficiency as compared to the TiO_2 nanoparticle counterpart. These results are in mutual agreement with XRD measurements of the un-doped TNTs sample that revealed peaks at 25.62° (101), 38.33°(112), 40.6° (202), 48.40°(200), 54.9°(105), 62.5° (118), 70.7°(220), and 75.1°(215) belonging to mixed Anatase-Brookite phase as shown here in Fig 2. However no additional peaks due Rutile phase of TiO_2 have been detected through XRD analysis. Our confocal Raman LAS measurements have revealed the presence of Rutile phase in our samples. Our method of characterization of TNTs through confocal Raman LAS has outperformed XRD measurements which failed to detect the presence of the Rutile phase in the un-doped TNTs samples.

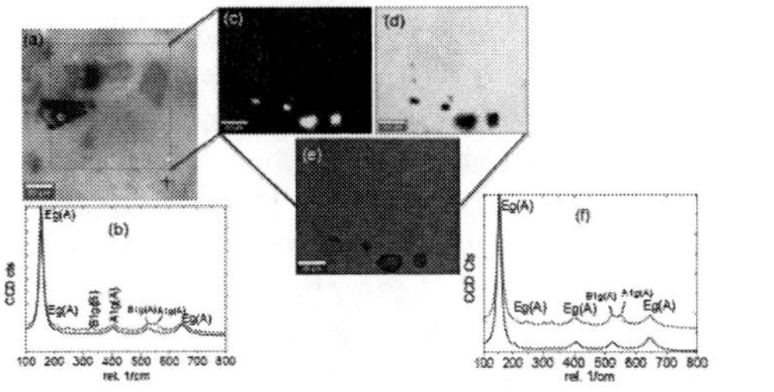

Fig 4 Shows Raman large area scan of 9mM C-TNTs sample (a) video image (b) Raman single spectra recorded at the crosses indicated in the image in (a) with corresponding colours, (c), & (d) draw image results of basis analysis. (e) Combined image, (f) Corresponding colour coded spectra.

Fig 4. Shows the Raman LAS of the 9 mM C-TNTs prepared by template-assisted sol-gel technique. The blue and red RSS recorded, are linked to positions indicated in Fig 4(a) by the blue and red crosses. RSS in Fig 4(b) has revealed Raman vibrational modes at 154.25 cm^{-1} (Eg-A), 208.11 cm^{-1} (Eg-A), 327.53 (B1g-B) 403.43 cm^{-1} (B1g-A) 527.91 cm^{-1} (A1g-A) and 645.49 cm^{-1}(Eg-A) consisting of a mixed Anatase and Brookite phase. Confocal Raman LAS measurements have revealed that the majority of the sample as depicted by the red area in Figure 5(e) consist of a mixed Anatase-Brookite phase with Raman vibrational modes 154.68 cm^{-1} (Eg-A), 207.33 cm^{-1} (Eg-A), 331.92 cm^{-1} (B1g-B), 403.09 cm^{-1} (B1g-A) 529. 66 cm^{-1}(B1g-A), 546.32 cm^{-1} (A1g-A) and 644.54 cm^{-1}(Eg-A). Additionally the blue area in Fig 4(e) have revealed a Raman spectra with vibrational modes at 153.67 cm^{-1} (Eg-A), 205.59 cm^{-1} (Eg-A), 404.55 cm^{-1} (B1g) 547.96 cm^{-1} (A1g) and 645.66 cm^{-1}(Eg) with an Anatase phase only. It is noticeable that introduction of carbon dopant seems to foster formation of a mixed Anatase-Brookite phase [17, 18]. This is probably that the due to the fact that the nucleation and growth mechanism of rutile are inhibited by the introduction of the carbon dopant in TiO_2 crystalline lattice. It is also noticeable that the introduction of the carbon dopant the Raman spectra tapered significantly and the corresponding peak intensities increased. This implies that the crystallinity of the Anatase-Brookite phase was greatly enhanced by doping with carbon atoms. Additionally, our confocal Raman LAS measurements have revealed that the 9 mM C-TNTs do not have Rutile phase TiO_2. The disappearance of the Rutile phase might be due to introduction in TiO_2 matrix which in turn inhibited formation of Rutile phase.

3.4. Confocal Raman spectroscopy: Depth profiling (XZ)

The depth profiling was done using a similar procedure as LAS. Depth profiles were collected by employing the XZ-scan mode starting at the bottom end samples. Confocal Raman depth profiling was performed in the sample area shown by a red line in Fig 5. (a). Fig 5 (b) shows RSS recorded on the blue and red cross on the stitching image in Fig 5(a). RSS the blue and red cross have revealed that the sample consists of a mixed Anatase-Brookite phase with Raman modes at 153.39 cm^{-1} (E_{g1}-Anatase), 204.91 cm^{-1} (E_{g2}-Anatase), 329 cm^{-1} (B_{1g}-Brookite), 404.14 cm^{-1} (B_{1g}-Anatase), 526.76 cm^{-1} (A_{1g}-Anatase), 549.96 cm^{-1} (B_{1g}-Anatase) and 648.07 cm^{-1} (E_{g3}-Anatase).

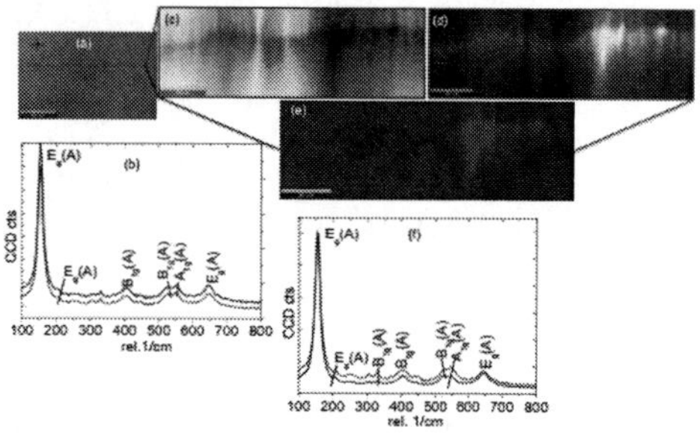

Fig 5 Raman depth profiling of TNTs **(a)** stitching image **(b)** and **(c)** draw images showing phase distribution of TiO_2 following basis analysis. Brighter colours indicate a higher fitting factor and thus higher signal intensity of the corresponding spectrum at that position. **(d)** Combined bitmap image **(f)** shows the corresponding colour coded spectra for the combined bitmap image.

We can safely say that confocal Raman spectroscopy was successfully used to evaluate the crystalline phase composition of TNTs on the surface (XZ) and along the tubes (XZ) via depth profiling and our method of measurements have outperformed RSS spectra and XRD analysis which do not conclusive crystalline phase distribution in TNTs. This information could prove to be vital for the development of efficient electrode materials consisting of non-metal doped TNTs for development of solar cells.

4. Conclusions

SEM micrographs has revealed the formation of closely packed nanotubes. SEM analysis has also confirmed that the morphology of the fabricated TNTs change with increase in carbon dopant concentration. SEM-EDX spectra has revealed the presence of strong Ti and O peaks which endorses the successfully formation of Titanium (IV) oxide. XRD analysis has confirmed the presence of a mixed Anatase-Brookite phase with peaks at 2θ angles of 25.49^0, 38.11^0, 40.60^0 48.14 ^{0}, 54.58^0, 63.00^0, 70.33 and 75.66^0. Confocal Raman large area scan has revealed the presence of Brookite and Anatase for the un-doped TNTs samples. However no additional peaks due Rutile phase of TiO_2 have been detected through XRD analysis. Our confocal Raman LAS measurements have revealed the presence of Rutile phase in our samples. Our method of characterization of TNTs through confocal Raman LAS has outperformed XRD measurements which failed to detect the presence of the Rutile phase in the un-doped TNTs samples. Confocal raman depth profiling has revealed the presence of a mixed Anatase and Brookite phase, no peaks due to Rutile have been detected.

5. Acknowledgements

We are grateful for financial support from our sponsors South African National research foundation (NRF), Govani Beki Research and Development centre (GMRDC) of the University of Fort Hare. The authors would also like to acknowledge the DST/CSIR Nanotechnology Innovation Centre, National Centre for Nanostructured Materials, CSIR and Electron Microscope Unit, University of Cape Town for morphological characterization of Titanium dioxide nanotubes on Functional Substrates.

6. References

[1] S. Bose, V. Somi and K.R. Genwa, "*Recent advances and future prospects for dye sensitized solar cell: A Review*," International Journal of Scientific and Research Publications, vol 5, no 4, pp. 1-9, 2015

[2] M. Zhang, Y. Bando and K. Wada, "Sol-gel template preparation of TiO_2 nanotubes and nanorods," *Journal of Materials Science*, vol. 20, no. 2, pp. 167-170, 2001.

[3] M. Lupiwana, R. Taziwa, E. Meyer and D. Katwire, "*Structural, Morphological, Topographical Characterization of Titanium Dioxide Nanotubes Metal Substrates for Solar Cell Application*," Journal of Material Science and Technology Research, vol 3, no 3, pp. 17-31, 2017.

[4] V. Schipu, A. matei, I. Cernica and C. Podaru, "*Preparation of titanium dioxide films by sol-gel route for gas sensors*,"The International Optical Society for Optical Engineering, Bellingham, 2008.

[5] M.S. Goyat, S. Rana, Sudipta Halder, and P.K. Ghosh, "*Facile fabrication of epoxy-TiO_2 nanocomposites: A critical analysis of TiO_2 impact on mechanical properties and toughening mechanisms*," Ultrasonics

Sonochemistry, vol 40, no A, pp. 861-873, 2017.
[6] L. S. Isley and L. R. Penn, "*Relative Brookite and Anatase content in sol-gel synthesized titanium dioxide nanoparticles*," The Journal of Physical Chemistry B, vol. 110, no. 31, pp. 15134-15139, 2006.
[7] O.K. Varghese, "*Crystallization and high-temperature structural stability of titanium oxide nanotube arrays*," Journal of Materials Research Society, vol 18, no 1, pp 157-165, 2003.
[8] J. Y. Liao, B. C. Lei, D. B. Kuang and Y. C. Su, "*Tri-functional hierarchical TiO_2 spheres consisting of Anatase nanorods and nanoparticles for high efficiency dye-sensitized solar cells*," *Energy and Environmental Science*, vol. 4, pp. 4079-4085, 2011.
[9] N. G. Park, J. Van de lagemaat and A. J. Frank, "*Comparison of dye-sensitized Rutile and Anatase-based TiO_2 solar cells*," Physical Chemistry, vol. B104, pp. 8989-8994, 2000.
[10] C. D. Chen, "*Microwave-assisted topochemical conversion of layered titanate nanosheets to {010}-faceted Anatase nanocrystals for high performance photocatalysts and dye-sensitized solar cells*," Scientific Reports 5, vol. 14, pp. 5801-5811, 2014.
[11] J. Martin, J. Maiz, J. Sacristan and C. Mijangos, "*Tailored polymer-based nanorods and nanotubes by template synthesis*," Polymer, vol. 53, no. 6, pp. 1149-1166, 2012.
[12] . R. Taziwa and E. Meyer, "*Carbon doped nano-crystalline TiO_2 photo-active thin film for solid state photochemical cells*," Advances in nanoparticles, vol. 3, pp. 54-63, 2014.
[13] X. Lu, X. Mou, J. Wu, D. Zhang, L. Zhang, F. Huang, F. Xu and S. Huang, "*Improved-performance dye sensitized solar cells using Nb-doped TiO_2 electrodes; efficient electron injection and transfer*," Advanced Functional Materials, vol. 20, no. 3, pp. 509-515, 2010.
[14] S. P. Burrows, "*Infrared spectroscopic measurement of titanium dioxide nanoparticle shallow trap state energies*," Virginian Polytechnic institute and State University, Blacksburg, 2010.
[15] A. Nosaka and Y. Nosaka, "Characteristics of water adsorbed on TiO_2 photocatalytic surfaces as studied by 1H NMR spectroscopy," *Bulletin of the chemical Society of Japan*, vol. 78, no. 9, pp. 1595-1607, 2005.
[16] V. Schipu, A. matei, I. Cernica and C. Podaru, "Preparation of titanium dioxide films by sol-gel route for gas sensors," in The International Optical Society for Optical Engineering, Bellingham, 2008.
[17] U. Balachandran and N. G. Eror, "*Raman spectra of titanium dioxide*," Journal of Solid state Chemistry, vol. 42, no. 2, pp. 276-282, 1982.
[18] H. A. Hanaor and C. S. Charles, "*Review of the Anatase to rutile phase transformation*," Journal of Materials Science, vol. 46, pp. 855-874, 2011.

ROLE OF THE FABRICATION TECHNIQUE IN THE STABILITY OF CH3NH3PBI3 PEROVSKITE THIN FILMS

Mehran Habibi, Morteza Eslamian
University of Michigan-Shanghai Jiao Tong University Joint Institute
800 Dongchuan Road, Minhang District, Shanghai, 200240, China
Email: mhabibi82@sjtu.edu.cn, morteza.eslamian@sjtu.edu.cn

ABSTRACT: Despite the high power conversion efficiency of the state-of-the-art halide perovskite solar cells, their poor stability against humidity, oxygen, temperature, and light has hindered their commercialization. In this work, it is shown that the nature of the deposition method, and its attributes, is an important parameter that affects the stability of the perovskite films. To this end, the stability of the CH3NH3PbI3 perovskite films deposited by spray and blade coating, as two scalable and commercially viable techniques, and spin coating, as a lab scale technique are compared. To achieve a uniform and defect-free film made by spray and blade coating, the substrate temperature was raised during the deposition process, whereas the anti-solvent treatment was performed in the case of the spin-coated films. It is shown that the perovskite stability is influenced by the morphology of the film which is in turn controlled by the dynamics of the coating process. The stability analysis was performed by obtaining the SEM and XRD graphs of the films exposed to a humid environment (35-60%). Comparison of the ratio of the XRD peak intensity of the lead iodide (PbI2) to perovskite after 10 days revealed that the blade-coated samples are the most stable films. The spray- and spin- coated counterparts, on the other hand, are rough and prone to decomposition. It is also observed that the fabrication of a fully-covered and smooth perovskite film through the anti-solvent engineering can prolong the stability of the spin-coated films. In conclusion, a smooth film with fewer number of gap-free grain boundaries is found to be more stable. In this context, the blade-coated samples and then the spin-coated samples treated by anti-solvent (toluene) showed the best stability.

1 INTRODUCTION

The global energy demand in near future is predicted to be near 1 GW per day [1]. Photovoltaic solar cells as a clean, renewable, and affordable source of energy are expected to have a major contribution in the energy sector to meet the future energy demand. Perovskite solar cell, as an emerging member of the family of the solution-processed solar cells, has proved its potential to be an efficient and cheap source of energy, although it suffers from poor stability and inadequate lifetime [2]. Instability is originated from sensitivity of the perovskite structure to moisture, oxygen, temperature, UV light and solution processing [3]. The methylammonium lead iodide $CH_3NH_3PbI_3$ ($MAPbI_3$) is the common perovskite light harvester, which contains a hygroscopic amine component. Due to the presence of this part, the $MAPbI_3$ may decompose to the perovskite precursors, i.e., PbI_2 and CH_3NH_3I, upon exposure to water and oxygen. After that, a series of chemical reactions may occur resulting in the formation of H_2O, I_2 and H_2, according to the following reactions [4]:

$$CH_3NH_3PbI_3 \overset{H_2O}{\leftrightarrow} CH_3NH_3\,(aq) + PbI_2\,(s)$$
$$CH_3NH_3\,(aq) \leftrightarrow CH_3NH_2\,(aq) + HI\,(aq)$$
$$4HI\,(aq) + O_2 \leftrightarrow 2I_2(s) + 2H_2O$$
$$2HI\,(aq) \overset{h\nu}{\leftrightarrow} I_2(s) + H_2\uparrow$$

To avoid or delay the decomposition of the perovskite structure, several approaches have been reported, to date. Some works have tried to intrinsically stabilize the perovskite active layer by changing its components. For instance, substituting the I ions with the Br ions to tune the stoichiometry of the mixed halide $CH_3NH_3Pb(I_{1-x}Br_x)_3$ perovskites can significantly improve the stability [5]. Also, replacing the methylammonium (MA) cation with formamidinium (FA) cation has shown an improvement in the thermal stability of the perovskite structure [6].

Some attempts have been made to encapsulate the entire device or the perovskite layer. Various encapsulating techniques and materials have been reported in some comprehensive reports [7-9]. In some papers, it is argued that the interface engineering has an encapsulating effect [10-13]. In fact, some interlayers engineered between the electron transporting layer and the metal electrode, e.g., PEIE polymer [12], fullerene amine interlayer (PCBDAN) [11], and MoOx [13], or between the perovskite and the hole transporting layer, e.g., zinc phthalocyanine (ZnPc) [10], not only enhance the electrical performance of the device, but they also protect the perovskite layer against decomposition.

Beside the abovementioned considerations, the fabrication method, through its attributes, is another important parameter that affects the stability. It has been reported that manipulating the process parameters or using alternative casting techniques may alter the thermal and air stability of the perovskite solar cells. Kim and coworkers [14] modified the sequential deposition technique in the fabrication of the $MAPbI_3$ perovskite layer, by changing the steps and duration of the dipping time. Through a two-step CH_3NH_3I (MAI) dipping process they showed that the thermal stability of the devices can enhance. In another work, hybrid chemical vapor deposition was applied which resulted in stable devices. In this case, the fabrication was performed in high temperature, which led to better stability. This was ascribed to the complete evaporation of the excess MAI and water in hybrid chemical vapor deposition [15]. Therefore, in this work to further investigate the aforementioned approach, blade coating, spray coating and spin coating with and without anti-solvent engineering are employed to fabricate the $MAPbI_3$ perovskite films and examine their stability..

2 MATERIALS AND METHODS

Lead iodide (PbI_2, 98.5%), dimethylsulfoxide (DMSO, 99.5 %), γ-butyrolactone (GBL) and toluene (98.8%) were purchased from Sigma-Aldrich, USA.

Methylammonium iodide (MAI, 99.5%) was supplied by Xi'an Polymer Light Technology Co, China. ITO-coated glass substrates (1cm^2) were cleaned by ultrasonication in solution of deionized water and soap, acetone and isopropyl alcohol, sequentially. Then the substrates were treated with UV/Ozone irradiation for 15 minutes. Perovskite films were fabricated by spray, spin and blade coating of $MAPbI_3$ solution. The $MAPbI_3$ solution was prepared by dissolving MAI and PbI_2 powders in GBL and DMSO solvents, with the volume ratio of 7:3, to obtain a solution with the mass concentration of 20 wt%. The solution was preheated to 60 °C and stirred overnight. In the cases of spray and blade coating, the substrate temperatures were kept at 150 °C, during the coating process. Spin coating was performed at 2500 rpm, for 20 s. In the case of spin coating with anti-solvent (toluene) treatment, two step spinning strategy at 1000 and 4000 rpm was followed, with durations of 10 and 25 s, respectively. 100 ml of toluene was poured steadily on the sample, during the second step of spinning. In all four cases, the wet films were annealed on hotplate at 100 °C, for 30 minutes.

The blade coating was performed with a simple blade coater, where the gap between the blade and the substrate was kept near 1 µm. Spray coating was performed for 20 s, by a simple airbrush pen with a flow rate of 10 µl/s and atomizing air pressure of 2 bar [16]. The spray coating resulted in an average film thickness of 405 nm, which is close to the thickness of the blade-coated film (395 nm). The toluene-treated spin-coated film has a thickness of 105 nm, while the spin-coated film without the toluene treatment has an average thickness of 260 nm. The film thickness and roughness were obtained using a stylus profilometer (KLA-Tencor P7, Milpitas, CA, USA).

3 RESULTS AND DISCUSSIONS

To compare the stability of the fabricated films, scanning electron microscope (SEM, Hitachi microscope, Model S-3400N, Tokyo, Japan) and x-ray diffraction (XRD, model D5005, Bruker, Billerica, MA, USA) graphs were taken right after the heat treatment of the fresh films, as well as after 10 days of exposure to the ambient conditions, where the humidity was varying between 35 and 60 %, during those 10 days.

Figure 1 shows the SEM and optical (inset) images of four perovskite films, fabricated by spin coating (a), toluene-treated spin coating (b), spray coating (c) and blade coating (d).

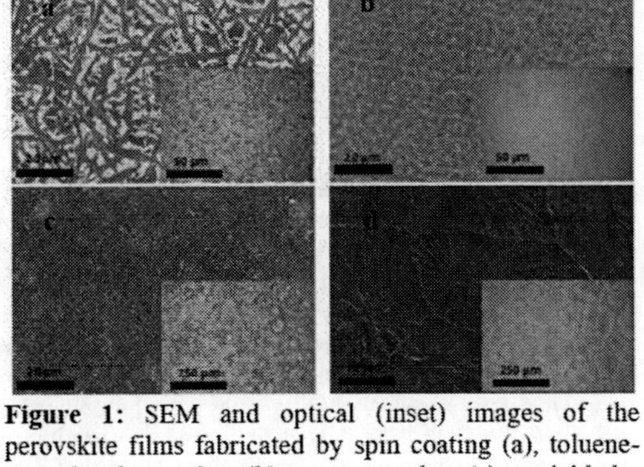

Figure 1: SEM and optical (inset) images of the perovskite films fabricated by spin coating (a), toluene-treated spin coating (b), spray coating (c), and blade-coating (d).

Obviously, different casting methods have caused different morphologies. Huge grain sizes of the blade-coated sample (several tens of µm) in Figure 1-d are much larger than those of the other three counterparts. Spin- and spray- coated films in Figure 1-a and Figure 1-c are significantly rougher (125 nm for both spin- and spray- coated films) than the other two cases (80 nm for the blade-coated film and 15 nm for the spin-coated film with toluene treatment). Figure 2 shows the SEM images of the fresh samples and the same samples after 10 days of exposure to the ambient conditions. Only a minor change is observed in the morphology of the blade-coated and toluene-treated spin-coated films, as shown in Figure 2-a,b and Figure 2-c,d, respectively, while the morphology of the other samples has altered completely, and in some cases new crystals have emerged.

In the next step, new samples were made for the XRD analysis. The XRD graphs of the fresh samples and the same samples after 10 days of exposure to the ambient conditions are shown in Figure 3. Except for the toluene-treated spin-coated sample that has some minor PbI_2 impurity from the beginning (a small peak at 12.7°), in other cases the perovskite main peaks at 14.2°, 28.6° and 32.1° in the as-prepared samples prove the formation of pure perovskite in the tetragonal phase [17].

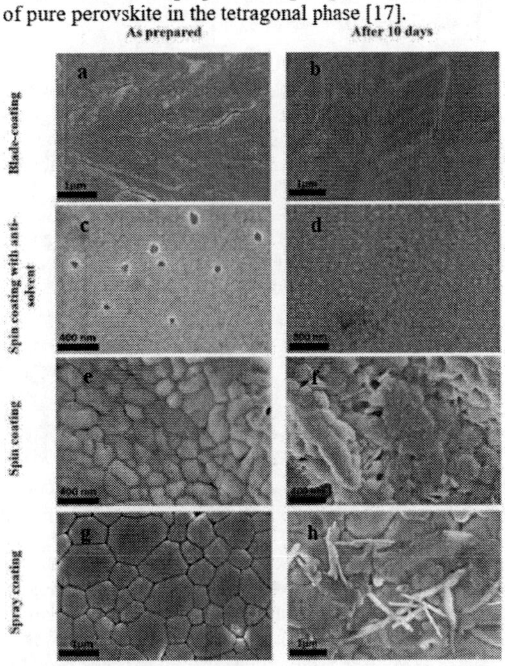

Figure 2: SEM images of the $MAPbI_3$ films; as prepared samples (a,c,e,g) and after 10 days (b,d,f,h) of exposure to the ambient conditions.

After 10 days of exposure to the humid air, a sharp peak appeared at 12.7° only in spin-coated and spray-coated samples (pointed with arrows), which shows the partial decomposition of the perovskite to PbI_2. This peak is absent in the blade-coated sample and is not sharp in the toluene-treated spin-coated sample. It shows the high resistance of the blade-coated film against decomposition, while in toluene-treated spin-coated film, this resistance seems to be smaller. The formation of PbI_2 after 10 days is in accordance with the SEM images in Figure 2, which show the emergence of totally new crystals after exposure to the ambient conditions.

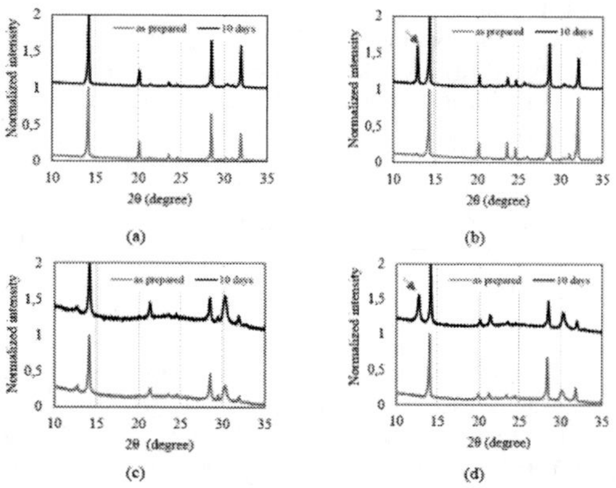

Figure 3: XRD graphs of the $MAPbI_3$ perovskite before and after 10 days of exposure to humid air. The films were fabricated by blade coting (a), spray coating (b), spin coating with toluene treatment (c), and spin coating (d). The intensities were normalized with respect to the highest peak of perovskite at 14.2°. The arrows indicate the appearance of PbI2 at 12.7°.

Defining a peak ratio, as the ratio of the peak intensity of $MAPbI_3$ at 14.2° to the peak intensity of PbI_2 at 12.7° could help us to compare the trend of decomposition. Figure 4 shows the peak ratio of various fresh samples and after 10 days of exposure to the ambient conditions. There is essentially no change in the peak ratio of the blade-coated sample, whereas in the spin- and spray- coated samples, the change is significant. In the case of the toluene-treated spin-coated sample, the change is moderate.

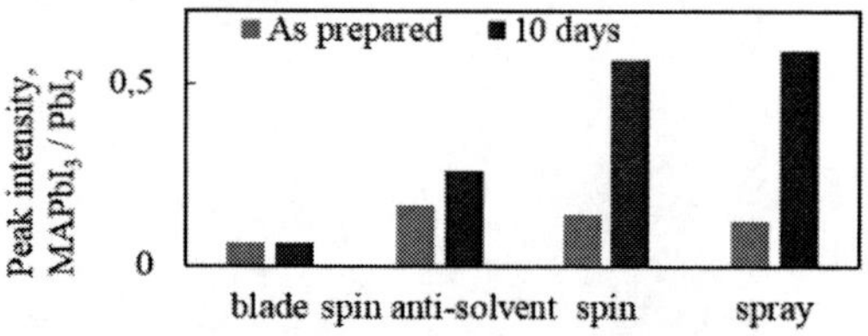

Figure 4: Peak intensity ratio of $MAPbI_3/PbI_2$ extracted from the XRD graphs.

Literature reports have shown that the morphology and crystallinity of the perovskite film strongly influences the electrical performance of the perovskite devices [18]. In fact, morphology strongly depends on the dynamics of the coating process, substrate surface and liquid interaction, and the drying process. Here, we observed that different coating techniques indeed created different morphologies with different stabilities in air. In blade-coating, the movement of the blade presses the liquid film against the substrate surface, which may lead to the formation of self-assembled large domains with few grain boundaries. The formation of these grains has occurred in a short time (less than 3 s) on hot plate. The surface tension forces between the blade and liquid may contribute to the alignment of the large domains [19]. It has been argued that the enhanced stability in blade-coated samples may be originated from the large domains formed in these films [20]. In such blade-coated films, the density of the grain boundaries is low and the lateral grain sizes are large and comparable to the film thickness [19]. Therefore, assuming that the grain boundaries act as the pathways to allow water molecules to infiltrate into the perovskite films, the films with small grain sizes and more grain bouderies (spin-coated films) are more vulnerable against exposure to water and the films with wide grains and therefore lower denisty of grain boundaries have a lower chance to be infiltrated by the water molecules. However, here the spray- and spin-coated films reveal an inconsistency in the above-mentioned argument on the dependence of the stability on grain size only. In spray coating a high substrate temperature was used to shorten the solvent evaporation time and smoothen the surface. However, the roughness of the spray-coated samples is high (125 µm). As it is depicted in Figure 2-g, the grain sizes are smaller than those of the blade-coated sample, but still larger than those of the spin-coated samples. Figure 2-h and also the XRD graphs in Figure 3-b show that the highest decomposition happened in the spray-coated film. In spite of having larger grains (several µm) than the toluene-treated spin-coated sample (tens to hundreds of nm) (Figure 2-c), the decomposition ratio of the spray-coated samples is higher after 10 days (Figure 4). Noteworthy, the gap between the grains in the grain boundaries is high in the spray-coated film (Figure 2-g), which may be responsible for excessive infiltration of water molecules into the film, despite having fewer number of grain boundaries, compared to the spin-coated samples. In the case of spin-coated sample with toluene treatment, the decomposition is considerably smaller than the pristine spin-coated sample. This may be because the toluene-trated film is much more uniform (roughness of 15 nm) than the spin-coated film (roughness of 125 nm). The gap between the grains in the pristine spin-coated film compared to that of the spin-coated film with toluene treatment is also larger (Figure 2-c,e). Although, the grain sizes in these two cases are smaller than those of the spray- and blade- coated samples, we observe a high decomposition only in the pristine spin-coated film after 10 days (Figure 2-f). In other words, the grain size is not the (only) parameter that controls decomposition. It is the number density of the available and wide grain boundaries, which act as the pathways for water molecules, that controls the decomposition.

For an in-depth consideration of the effect of the grain size, we further estimated the crystallite sizes obtained by XRD graphs. This is because individual perovskite grains observed under SEM, may contain several crystalline domains. Here to estimate the average crystallite size (τ) of perovskite films from the XRD pattern, the Scherrer equation is used [21]:

$$\tau = \frac{k\gamma}{\beta \cos\theta} \tag{1}$$

where γ is the radiation wavelength, k is a constant, β is the full width at half maximum (FWHM), and θ is the peak position. According to this equation, an increase in the FWHM, results in a decrease in the crystallite size. To compare the crystallite sizes of the films fabricated by the four mentioned casting techniques, β and θ are calculated from the magnified peaks of the perovskite at 14.2° (Figure 5a). Assuming $k\gamma$ to be equal to 1 in all the cases, the normalized crystallite sizes may be compared as Figure 5b shows. The descending trend of variation of the crystallite sizes from the blade-coated sample to the toluene-treated spin-coated sample is in accordance with the domain or grain size trends observed in SEM images of Figure 2. Therefore, the application of blade coating and then the spray coating not only lead to the creation of large domains, but also cause the formation of large crystallites. Moreover, the spin-coated films treated with anti-solvent not only have the smallest grains, but also have the smallest crystallite size.

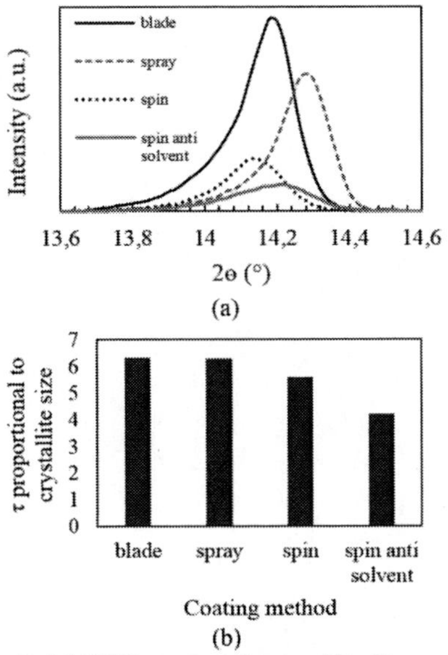

Figure 5: (a) XRD graphs of perovskite films with the magnified peak around 14.2°. (b) τ which is linearly proportional to the average crystallite size of perovskite films estimated by the Scherrer equation.

Table 1 summarizes the morphological differences of the perovskite films fabricated by four different techniques. The qualitative comparaisons of Table 1 are based on the stability of the film against humidity, measured by the peak intensisty ratio of each film, after 10 days of exposure to the ambient conditions. It seems that the film roughness, coverage, and the gap between the grains in grain boundaries play the main roles on decomposition rather than the grain size. In fact, rougher films with lower coverage and thicker cracks and grain boundaries have higher surface contact with air and higher permeability of moisture.

Table 1. Morphological characteristics of perovskite films fabricated by four different casting methods and their stability against humidity

	Blading	Spin with anti-solvent	Spin	Spray
Thickness (nm)	395	105	260	405
Roughness (nm)	80	15	125	125
Coverage (%)	100	100	75	95 (with partial voids)
Grain size range	tens of μm	tens to hundreds of nm	hundreds of nm	several μm
Stability after 10 days	highest	moderate	low	lowest

4 CONCLUSION

Formation of large crystalline domains of perovskite in blade-coating technique had been reported to be the main reason behind an increase in the perovskite stability. To investigate this further, here we fabricated perovskite films by spray coating on hot plate, blade coating on hot plate, spin coating and also spin coating with the anti-solvent treatment. Each method causes the creation of different grain sizes, crystallite sizes and morphologies. Comparison of the roughness values, coverage, grain size and also the gap between the grains in the grain boundaries, led us to conclude that the roughness and the gap between the grains might have the most important roles on decomposition of the films, because the aforementioned effects facilitate the infiltration of water molecules into the perovskite film. Therefore, a more compact and smooth perovskite film is more stable.

5 ACKNOWLEDGMENT

Financial support from the Shanghai Municipal Education Commission via the Oriental Scholar fund and the funding from National Natural Science Foundation of China (NSFC) is acknowledged.

6 REFERENCES

1. Wang, D.; Wright, M.; Elumalai, N.K.; Uddin, A., Stability of perovskite solar cells, Solar Energy Materials and Solar Cells 2016,147, 255-275, doi: 10.1016/j.solmat.2015.12.025.
2. Habibi, M.; Zabihi, F.; Ahmadian-Yazdi, M.R.; Eslamian, M., Progress in emerging solution-processed thin film solar cells – Part II: Perovskite solar cells, Renewable and Sustainable Energy Reviews 2016,62, 1012-1031, doi: 10.1016/j.rser.2016.05.042.
3. Shahbazi, M.; Wang, H., Progress in research on the stability of organometal perovskite solar cells, Solar Energy 2016,123, 74-87, doi: 10.1016/j.solener.2015.11.008.
4. Niu, G.; Guo, X.; Wang, L., Review of recent progress in chemical stability of perovskite solar cells, J. Mater. Chem. A 2015,3, 8970-8980, doi: 10.1039/c4ta04994b.
5. Noh, J.H.; Im, S.H.; Heo, J.H.; Mandal, T.N.; Seok, S.I., Chemical management for colorful, efficient, and stable inorganic-organic hybrid nanostructured solar cells, Nano Lett 2013,13, 1764-1769, doi: 10.1021/nl400349b.
6. Lee, J.W.; Seol, D.J.; Cho, A.N.; Park, N.G., High-efficiency perovskite solar cells based on the black polymorph of HC(NH2)2 PbI3, Adv Mater 2014,26, 4991-4998, doi: 10.1002/adma.201401137.
7. Matteocci, F.; Cinà, L.; Lamanna, E.; Cacovich, S.; Divitini, G.; Midgley, P.A.; Ducati, C.; Di Carlo, A., Encapsulation for long-term stability enhancement of perovskite solar cells, Nano Energy 2016,30, 162-172, doi: 10.1016/j.nanoen.2016.09.041.
8. Wilderspin, T.J.; De Rossi, F.; Watson, T.M., A simple method to evaluate the effectiveness of encapsulation materials for perovskite solar cells, Solar Energy 2016,139, 426-432, doi: 10.1016/j.solener.2016.09.038.
9. Weerasinghe, H.C.; Dkhissi, Y.; Scully, A.D.; Caruso, R.A.; Cheng, Y.-B., Encapsulation for improving the lifetime of flexible perovskite solar cells, Nano Energy 2015,18, 118-125, doi: 10.1016/j.nanoen.2015.10.006.
10. Chen, C.; Li, H.; Jin, J.; Cheng, Y.; Liu, D.; Song, H.; Dai, Q., Highly enhanced long time stability of

perovskite solar cells by involving a hydrophobic hole modification layer, Nano Energy 2017,32, 165-173, doi: 10.1016/j.nanoen.2016.12.009.
11. Xie, J.; Yu, X.; Sun, X.; Huang, J.; Zhang, Y.; Lei, M.; Huang, K.; Xu, D.; Tang, Z.; Cui, C.; Yang, D., Improved performance and air stability of planar perovskite solar cells via interfacial engineering using a fullerene amine interlayer, Nano Energy 2016,28, 330-337, doi: 10.1016/j.nanoen.2016.08.048.
12. Yang, H.; Zhang, J.; Zhang, C.; Chang, J.; Lin, Z.; Chen, D.; Sun, X.; Xi, H.; Han, G.; Hao, Y., Effect of polyelectrolyte interlayer on efficiency and stability of p-i-n perovskite solar cells, Solar Energy 2016,139, 190-198, doi: 10.1016/j.solener.2016.09.048.
13. Sanehira, E.M.; Tremolet de Villers, B.J.; Schulz, P.; Reese, M.O.; Ferrere, S.; Zhu, K.; Lin, L.Y.; Berry, J.J.; Luther, J.M., Influence of Electrode Interfaces on the Stability of Perovskite Solar Cells: Reduced Degradation Using MoOx/Al for Hole Collection, ACS Energy Letters 2016,1, 38-45, doi: 10.1021/acsenergylett.6b00013.
14. Kim, S.; Chung, T.; Bae, S.; Lee, S.-W.; Lee, K.D.; Kim, H.; Lee, S.; Kang, Y.; Lee, H.-S.; Kim, D., Improved performance and thermal stability of perovskite solar cells prepared via a modified sequential deposition process, Organic Electronics 2017,41, 266-273, doi: 10.1016/j.orgel.2016.11.014.
15. Jung, M.C.; Raga, S.R.; Ono, L.K.; Qi, Y., Substantial improvement of perovskite solar cells stability by pinhole-free hole transport layer with doping engineering, Sci Rep 2015,5, 9863, doi: 10.1038/srep09863.
16. Habibi, M.; Rahimzadeh, A.; Bennouna, I.; Eslamian, M., Defect-Free Large-Area (25 cm2) Light Absorbing Perovskite Thin Films Made by Spray Coating, Coatings 2017,7, 42, doi: 10.3390/coatings7030042.
17. Oku, T., Crystal Structures of CH3NH3PbI3 and Related Perovskite Compounds Used for Solar Cells, 2015,doi: 10.5772/59284.
18. Back, H.; Kim, J.; Kim, G.; Kyun Kim, T.; Kang, H.; Kong, J.; Ho Lee, S.; Lee, K., Interfacial modification of hole transport layers for efficient large-area perovskite solar cells achieved via blade-coating, Solar Energy Materials and Solar Cells 2016,144, 309-315, doi: 10.1016/j.solmat.2015.09.018.
19. Deng, Y.; Peng, E.; Shao, Y.; Xiao, Z.; Dong, Q.; Huang, J., Scalable fabrication of efficient organolead trihalide perovskite solar cells with doctor-bladed active layers, Energy Environ. Sci. 2015,8, 1544-1550, doi: 10.1039/c4ee03907f.
20. Kim, J.H.; Williams, S.T.; Cho, N.; Chueh, C.-C.; Jen, A.K.Y., Enhanced Environmental Stability of Planar Heterojunction Perovskite Solar Cells Based on Blade-Coating, Advanced Energy Materials 2015,5, 1401229, doi: 10.1002/aenm.201401229.
21. Ahmadian-Yazdi, M.R.; Habibi, M.; Eslamian, M., Positive effect of the excitation of the wet film by ultrasonic vibration on the characteristics of the perovskite light absorber and the device performance scientific reorts 2017,under review.

PRELIMINARY GUIDELINES FOR ACCURATE I-V MEASUREMENTS ON PEROVSKITE SOLAR CELLS

Ricky B Dunbar, Christopher J Fell, Kenrick F Anderson, Benjamin C Duck, and Gregory J Wilson
CSIRO Energy, 10 Murray Dwyer Cct, Mayfield West, NSW 2304, Australia

ABSTRACT: We propose guidelines for accurate measurement of perovskite solar cells using laboratory solar simulators. The guidelines are designed to assist in determining the cell efficiency in the presence of metastability, degradation and hysteresis.

Keywords: perovskite, measurement, efficiency, accurate, hysteresis

1 INTRODUCTION

Perovskite solar cells (PSC) are a highly attractive approach to solar photovoltaics that has attracted worldwide attention [1,2]. A challenge for the PSC R&D community however, is that laboratory measurements frequently don't reflect the performance that would be observed in a real solar installation [3]. This is because the electrical response of PSC devices is usually sensitive to pre-conditioning (previous exposure to potential difference or irradiance) and sometimes even to degradation during the measurement. When current-voltage (I-V) sweeps are used to measure device efficiency, the curves can exhibit hysteresis even at low sweep rates, indicating the origin of the problem is more complex than simply capacitance. These issues can lead to significant errors in the efficiency extracted from I-V curves.

Here we summarise the progress of a project designed to identify laboratory I-V testing methodologies for PSC, such that these lead to a measured efficiency consistent with the true outdoor performance. This is commonly referred-to as the *steady state* or *stabilised* performance of the device. The project includes consultation with an international peer review committee, and runs 2016-2018. At this stage some preliminary guidelines have been developed, hence we are seeking the widest possible consultation with the PV research community, to ensure the guidelines are practical and consistent with world's best practice.

The draft guidelines are based on our research [4–7] and experiences operating an IEC 17025 accredited cell measurement laboratory for several years, supporting nine institutions that make up the Australian PSC research community. The guidelines also consider other published work in the field, e.g. [8-11], plus our involvement in a working group developing a Draft Technical Report for IEC Technical Committee 82.

The guidelines contain instructions for laboratories at three different levels of technical sophistication – BASIC, suitable for many PV research labs, particularly where the focus is primarily materials development; INTERMEDIATE, a hybrid level for laboratories interested in a higher level of accuracy, and ADVANCED, suitable for dedicated (usually accredited) PV measurement laboratories.

2 PREVIOUS WORK

Naturally, these guidelines are built on the previous work of other authors over several years. An excellent summary of the broader considerations for accurate measurement of all research-scale PV devices is given in Christians et al. [8] and the key points are reproduced here:

- Mask the cell appropriately to avoid stray light, and measure the mask area accurately;
- Calibrate the illumination source carefully;
- Determine the steady-state I_{sc} and V_{oc}, and monitor the stability of these;
- Measure the steady-state current at voltages near P_{mp};
- Record forward and reverse I-V scans at various scan rates and check consistency with stabilised value;
- Cross-check I_{sc} against the spectral quantum efficiency by integrating the latter over the AM1.5 reference spectrum;
- Perform statistical analysis across a batch of samples.

As always, reported device measurements should be accompanied by a detailed description of the above activities.

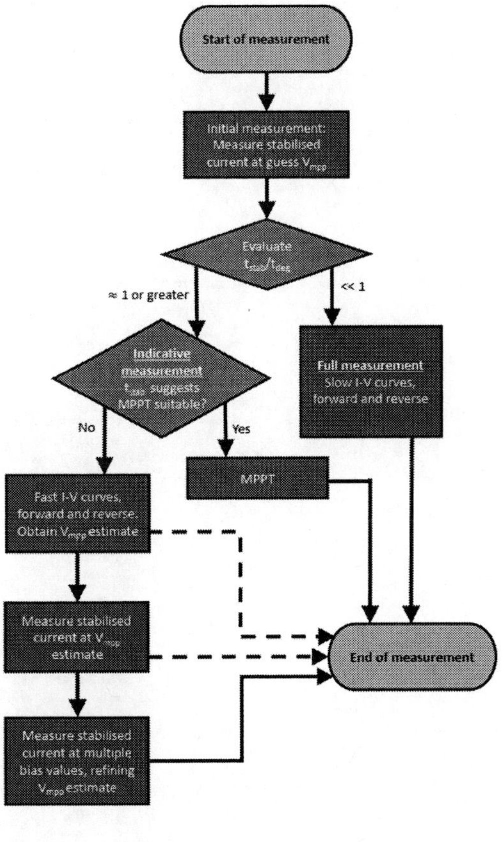

Figure 1: Proposed flowchart for measuring the efficiency of perovskite solar cells

3 DRAFT GUIDELINES

3.1 Laboratory classifications

These Guidelines identify three laboratory classifications based on configurations of equipment and procedures found in research laboratories with different levels of specialisation in PV measurement. The classifications are summarised in Table I.

Table I: Three classes of PV measurement capability

	Basic (BAS)	Intermediate (INT)	Advanced (ADV)
Light source	Non-spectrally matched (e.g. tungsten lamp)	Spectrally matched light source/solar simulator	Spectrally matched light source/solar simulator that undergoes regular characterisation
Irradiance calibration	-	Use of reference devices, typically not spectrally matched. Spectral mismatch typically not accounted for	Range of spectrally matched reference cells, irradiance monitoring during measurement, correction to reference spectrum
Temperature	-	Some consideration of device temperature	Device temperature controlled and monitored, correction to reference temperature according to standard methods
Electronics	Simple	Specialised PV measurement equipment	Specialised PV measurement equipment capable of delivering high accuracy measurements traceable to SI units
Calibration	-	Some items calibrated, typically infrequently	Strict calibration schedule
Methods	Not well developed	Well developed, but usually unverified	Careful adherence to International standard methods. Internally validated and externally reviewed during regular technical assessments
Accuracy	Low, unknown	Moderate, unknown	High, and clearly defined. Expression of result to Standard Test Conditions

3.2 Process flow

A goal of this work is to extend previous efforts to achieve prescriptive guidelines that allow the same level of standardisation in PSC measurements that the community is accustomed to with conventional PV devices. This will likely involve, among other things, a quantitative definitions *steady-state* measurements [5].

The proposed approach is designed to yield a robust efficiency measurement that is appropriate over a range of perovskite cell types. Two parameters are identified as centrally important for identifying the best path – the stabilisation time, t_{stab}, and the degradation time, t_{deg}, as described in Table II.

Table II: Key parameters for the measurement approach

Key parameters	Parameter indicating timescale of process	Description
Stabilisation time	t_{stab}	Following a cell perturbation from the dark state at open-circuit to STC at near V_{mpp}, t_{stab} is the time required to for the measured current to rise or fall to within 1% of the final steady-state value (in the absence of degradation.)
Degradation time	t_{deg}	The time taken for a measurement of current at STC conditions to irreversibly degrade by 1%.

If t_{stab} and t_{deg} are of similar order, it is unlikely that a reliable estimate of the device efficiency can be obtained from a full I-V curve. In this case, I-V curves with higher voltage sweep rates – as slow as possible while still minimising degradation during the scan – can be used to make an estimate of V_{mp}. The remainder of the measurement then involves holding the cell at V_{mp} and recording the current once it stabilises.

If $t_{stab} << t_{deg}$ it is likely that a robust measurement of the steady-state efficiency is possible using the full I-V curve. Both forward and reverse I-V scans should be used, in order to confirm that the measurement is independent of the sweep direction. As per the advice in Section 2, it is also recommended that a final check of the maximum power, P_{mp}, be made using a stabilised measurement of the current at V_{mp}, and also a cross-check of the I_{sc} value against the quantum efficiency (QE) spectrum, integrated over the AM1.5 solar spectrum. A flowchart of the measurement logic is given in Figure 1.

3.3 Pre-conditioning for I-V scans

These guidelines assume that stabilisation criteria for current and voltage are defined in advance. One method for doing this is described in Dunbar *et al.* [5].

i. Forward I-V scan

BAS: Hold the cell at short-circuit for a previously determined time period shown to allow the current to approximately stabilise. INT and ADV: Hold the cell at short-circuit until the measured current stabilises.

ii. Reverse I-V scan:

BAS: Hold the cell at near open-circuit for a previously determined time period shown to allow the current to approximately stabilise. INT and ADV: Hold the cell at near open-circuit until the measured current stabilises.

Alternatively, hold the cell at exactly open-circuit until the measured open-circuit voltage stabilises.

Commence the I-V scan immediately after pre-conditioning is complete.

3.4 Temperature management

Due to the long exposure times typically required for stable or meta-stable measurements, the effect of the light source on the sample temperature needs to be understood. If the time required for the cell to reach a stable temperature under irradiation can be determined, temperature stabilisation can often be achieved in conjunction with pre-conditioning.

BAS: If temperature stabilisation cannot be directly verified, ensure that the cell illumination prior to measurement is for a consistent duration across the devices in the batch. INT: Monitor the cell temperature change with time and determine when stabilisation occurs and the value of the stabilised cell temperature. ADV: Use a temperature control system to achieve a stabilised cell temperature close to the reference temperature, as verified by monitoring and logging the temperature. Again note that temperature stabilisation criteria should be defined in advance.

3.5 Current-voltage measurements

a. Efficiency determination via I-V scans:

It is important to investigate the influence of the voltage scan rate and step size on the resulting I-V curves. A suitable method for this is described by Dunbar *et al.* [7]. The aim is to identify parameters that ensure the dwell time at each bias value is sufficiently long to permit satisfactory stabilisation – this can be verified indirectly by observing that I-V curves become invariant with dwell time for sufficiently long dwell times – without causing significant device degradation.

b. Efficiency determination via stabilised measurements:

The approach for a stabilised measurement is similar to the above. BAS and INT: Repeat measurements of current at fixed hold voltages until stabilised values are achieved. The hold voltage should be selected to be as close as possible to the maximum power point voltage, which can be estimated from I-V scans or multiple stabilised current measurements at different voltages.

ADV: Maximum power point tracking routines can enable an accurate determination of the maximum power point. The final result can depend on the MPPT algorithm and implementation. The performance of the routine should therefore be investigated for the test cell using different routines where possible.

4. CONCLUSIONS

Progress continues to be made in understanding best practices for measuring and reporting the performance of perovskite solar cells. We have presented a workflow that we believe represents a practical and effective approach to the measurement of perovskite cell efficiency and we welcome feedback from the PV research community.

ACKNOWLEDGEMENTS

The authors acknowledge funding support for this project from the Australian Renewable Energy Agency.

REFERENCES

[1] "Perovskite fever," *Nat. Mater.*, vol. 13, no. 9, pp. 837–837, 2014.

[2] N.-G. Park, M. Grätzel, T. Miyasaka, K. Zhu, and K. Emery, "Towards stable and commercially available perovskite solar cells," *Nat. Energy*, vol. 1, no. 11, p. 16152, 2016.

[3] Ricky B. Dunbar, Benjamin C. Duck, Tom Moriarty, Kenrick F. Anderson, Noel W. Duffy, Christopher J. Fell, Jincheol Kim, Anita Ho-Baillie, Doojin Vak, The Duong, Yi Liang Wu, Klaus Weber, Alex Pascoe, Yi-Bing Cheng, Qianqian Lin, Paul L. Burn, Ripon Bhattacharjee, Hongxia Wang and Gregory J. Wilson, "How reliable are efficiency measurements of perovskite solar cells? The first inter-comparison, between two accredited and eight non-accredited laboratories", *J. Mat. Chem. A* (in press).

[4] T. A. Berhe *et al.*, "Organometal halide perovskite solar cells: degradation and stability," *Energy Environ. Sci.*, vol. 9, no. 2, pp. 323–356, 2016.

[5] Ricky B. Dunbar, Walied Moustafa, Alexander R. Pascoe, Timothy W. Jones, Kenrick F. Anderson, Yi-Bing Cheng, Christopher J. Fell and Gregory J. Wilson, "Device pre-conditioning and steady-state temperature dependence of CH3NH3PbI3 perovskite solar cells," *Prog. Photovolt Res. Appl.*, vol. 15, 2016.

[6] Ricky. B. Dunbar, Augustin Barbe and Christopher J. Fell, "An Optical Imaging Method for High-Accuracy Solar Cell Area Measurement," IEEE *Journal of Photovoltaics*, vol 5, no 5, 2015

[7] Ricky B. Dunbar, Timothy W. Jones, Kenrick F. Anderson, Gregory J. Wilson and Christopher J. Fell, "An Investigation Of Perovskite Solar Cell Degradation and I-V Hysteresis for Improved Accuracy Of Efficiency Measurements at Standard Test Conditions," in *31st European Photovoltaic Solar Energy Conf. and Exhibition*, 2015, vol. 1, pp. 1077–1079.

[8] Jeffrey A. Christians, Joseph S. Manser and Prashant V. Kamat, "Best Practices in Perovskite Solar Cell Efficiency Measurements. Avoiding the Error of Making Bad Cells Look Good", *J. Phys. Chem. Lett.*, vol. 6, no. 5, pp. 852–857, 2015.

[9] Henry J. Snaith, "How should you measure your excitonic solar cells?", *Energy Environ. Sci.*, 2012, 5, 6513

[10] Taame Abraha Berhe, Wei-Nien Su, Ching-Hsiang Chen,a Chun-Jern Pan, Ju-Hsiang Cheng, Hung-Ming Chen, Meng-Che Tsai, Liang-Yih Chen, Amare Aregahegn Dubaleb and Bing-Joe Hwang. "Organometal halide perovskite solar cells: degradation and stability", *Energy Environ. Sci.*, 2016, 9, 323

[11] Yoshihiro Hishikawa, Haruya Shimura, Takashi Ueda, Ayumi Sasaki, and Yuki Ishii, "Precise performance characterization of perovskite solar cells", *Current Applied Physics* 16 (2016) 898-904

SNS/CDS THIN FILM SOLAR CELLS BY IONIZED JET DEPOSITION

Daniele Menossi, Simone Di Mare, Elisa Artegiani, Gianpiero Tedeschi[+], Fabio Piccinelli*, and Alessandro Romeo
Department of Computer Science, University of Verona, Verona, 37123, Italy
[+]Noivion S.r.l., Rovereto, 38068, Italy
*Department of Biotechnology, University of Verona, Verona, 37123, Italy

ABSTRACT: Tin sulfide is an excellent candidate for the mass production of solar cells as it is composed of abundant and not toxic elements. From our previous work on thermal evaporation of SnS solar cells, we now present SnS polycrystalline films by ionized jet deposition (IJD). This technology allows transferring atoms from a target of SnS to the substrate due to the generation of plasma from the target, giving enough energy to have a good crystallization of the material also at low temperature. SnS deposited by this technique show an incremented compactness, more robust thin film, which results to be less porous and more crystallized. Moreover the x-ray diffraction technique shows an improved stoichiometry with no secondary phases. Preliminary solar cells were made and indicate that IJD can be a good deposition technique for compounds with reduced congruent evaporation.

Keywords: SnS, thin films, ablation

1 INTRODUCTION

Tin sulfide is an excellent candidate for the mass production of solar cells as it is composed of abundant and not toxic elements. On the contrary materials currently used for the mass production of thin film solar panels (mainly CdTe and CIGS) suffer from limitations related to scarcity issues and/or toxicity of some elements that compose them thus representing a constraint to the spread on a large scale of the cheapest thin film technology. The research in SnS is done actively by major research centers such as MIT and Harvard University confirming the great interest in this material. However, much still needs to be done: efficiencies not greater than 5% [1] have been obtained mainly due to the difficulty of maintaining a proper stoichiometry, avoiding secondary phases and the lack of an ideal n-layer to make a good junction with SnS.

Ionized Jet deposition (IJD), which is an improved version of pulsed jet deposition, allows to control the stoichiometry and to maintain low substrate temperature.

The IJD method is based on the ionization of a gas stream flowing through a metallic nozzle, which serves simultaneously as the auxiliary electrode for the plasma discharge ignition. A pulse at high voltage (up to 25 kV) and a short duration (less than 1 μs) is applied to the cathode. Through a system of trigger and auxiliary electrodes, it causes strong ionization of the gas jet ablating the target surface (see Fig. 1). So a pulsed high power electron beam penetrates approximately for 1 μm into the target, resulting in a rapid evaporation of the target material and its transformation in plasma phase. The non-equilibrium extraction of the target material (ablation) facilitates stoichiometric composition of the plasma. Under optimum conditions, the target stoichiometry is thus preserved in the deposited films. The main feature of the pulsed systems is the ability to generate a high power density of about 10^8 W/cm^2 at the target surface. As a result, the thermodynamic properties of the target material, such as the melting point and specific heat, become irrelevant for the evaporation process. This work follows the previous investigation on the CdTe deposition by IJD [2].

2 EXPERIMENTAL

The The IJD-SnS layers have been deposited in a dedicated vacuum chamber. This is equipped with a IJD gun from Noivion S.r.l., and the vacuum is provided by a roughing and turbo-molecular pumping system, able to reach 10^{-7} mbar.

The distance between target and electron gun is fixed to 4 cm. During the deposition, the chamber pressure is brought between 10^{-4} – 10^{-3} mbar and IJD acceleration voltage is tuned in the range of 13-16 kV. Different substrate temperatures have been tested: from room temperature up to 300 – 400°C.

The SnS solar cells were prepared following the standard procedure mentioned above: on a common soda lime glass, a 300 nm ITO and a 100 nm ZnO layers are deposited and are followed by a deposition of 50 nm of CdS by vacuum thermal evaporation. Finally the back contact is made by depositing a 50 nm gold film, by VTE and with substrate at room temperature. SnS layers have been studied in terms of physical properties by atomic force microscopy (AFM), using a NT-MDT -SMENA A system, equipped with gold coated silicon tips from the same company, in semi-contact mode and by X-ray diffraction (XRD) analysis, processed with a Thermo ARL X'TRA powder diffractometer (in Bragg-Brentano geometry), equipped with a Cu-anode X-ray source (Kα, λ=1.5418 Å) and a Peltier Si (Li) cooled solid state detector). Preliminary finished devices were analyzed by current density–voltage (J-V) analyses, performed with a Keithley Source Meter 2420 at room temperature.

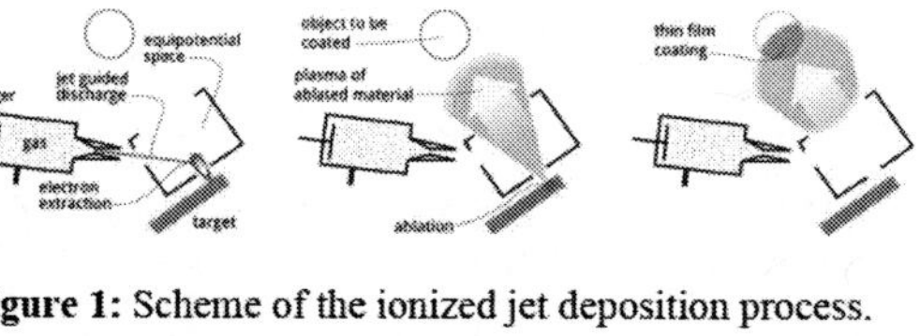

Figure 1: Scheme of the ionized jet deposition process.

3 RESULTS AND DISCUSSION

SnS is deposited by ionized jet deposition on the CdS/ZnO/ITO stacks. Samples have been deposited by applying electron acceleration voltages in the range of 4-5 kV, changing the power from 30 up to 40 W and with a pulse rate in the order of 80 Hz. However even with these parameters, the deposition rate is still low and to reach SnS film thicknesses of about 2 micrometers, we have increased the acceleration voltages up to 12 kV, with power rates of 200-250 W and with similar ablation frequencies of 80 Hz.

Moreover different pressure values during ablation have been tested. When the pressure is lower than 10^{-4} mbar, the plasma is not sufficiently confined and the deposition is minimal; it is necessary to work at pressures around $5*10^{-4}$-10^{-3} mbar to obtain deposition rates of 10 nm/sec.

With these minimal conditions different samples have been processed at different substrate temperatures. Moreover post deposition annealing in vacuum up to 550 °C was also applied. If SnS is deposited at room temperature the layer shows delamination. A possible reason could be the lattice mismatch between CdS and SnS. When substrate temperatures are increased above 200°C the absorber layer results to be more adhesive and compact.

2.1 Morphology

AFM pictures, shown in Fig. 2, reveal a slight decrease of grain size as the substrate temperature is increased from 200 °C to 300 °C, but also a larger roughness (as attested by the large difference in height of the grains observed for the 300 °C in the second picture of Fig.2), which is still increased at 500 °C.

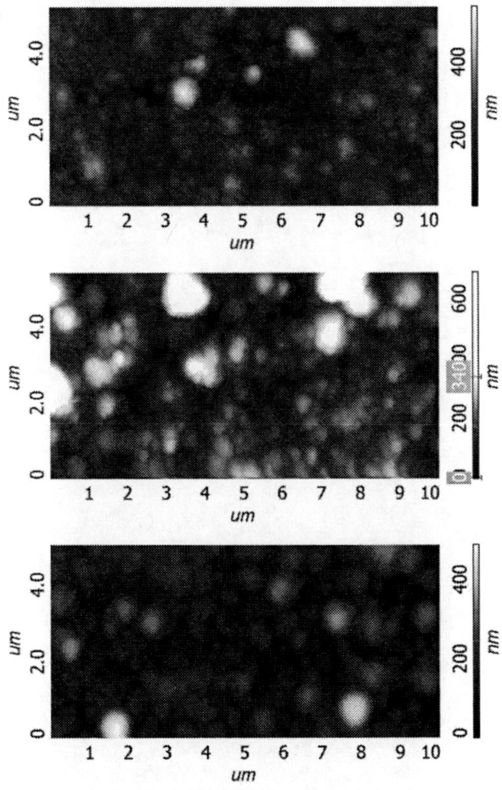

Figure 2: SnS layers deposited by IJD at 200°C (top), 300 °C (center) and 500°C (bottom).

However IJD-SnS layers have a smoother surface compared to VTE-SnS layers. SnS grains grown at a substrate temperature above 400°C are below 1 micrometer, generating a reflective and very smooth surface.

A similar comparison between the samples has been done by XRD analysis. In Fig. 3 are shown XRD patterns of SnS deposited by IJD at two different substrate temperatures. The main reflections, also previously detected for VTE-SnS [3], such as (010) and (111) are observed and very few other peaks are revealed. This might confirm the smooth and uniform low crystallized layer. In fact generally thin films show more peaks as the grains undergo large crystallization, as also attested by the post deposition treated SnS layers reported by the same authors [4]. Moreover the peaks of SnS deposited at 200°C and 300 °C do not show any difference in sharpness demonstrating a very similar crystallization rate.

However no secondary phases are detected from the XRD measurements even at low substrate temperature.

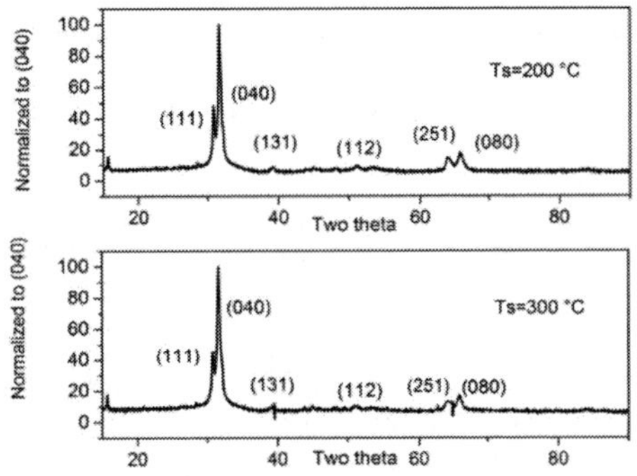

Figure 3: XRD spectra of SnS films deposited by IJD, at different substrate temperatures, namely (black) 200°C, (top) and 300°C (bottom).

2.2 Performance

Prototype solar cells, fabricated with IJD-SnS absorber layer have been tested by current density-voltage measurements. Preliminary photovoltaic parameters of about J_{SC} = 2.1 mA/cm^2, V_{OC} = 0.12 V, FF = 31% and efficiencies of 0.08% have been obtained.

2.3 Pinholes

The low performance is strongly connected with the presence of pinholes generated during the deposition process. As shown in figure 4, a large number of small dots are visible with optical microscope. By increasing the substrate temperature up to 500 °C pinholes are reduced but still present. This phenomenon might be due to deposition of particles during the target ablation and a possible path to overcome this could be to increase the distance between target and substrate since reducing the ablation power did not improve the layer quality.

The low performance of these devices is also connected to the too high-defective interface between CdS and SnS layers, taking into account the large energy band offset between them [5]. So far the best results have been obtained by Steinmann et al., using a Zn(O,S)/ZnO buffer layer structure [6].

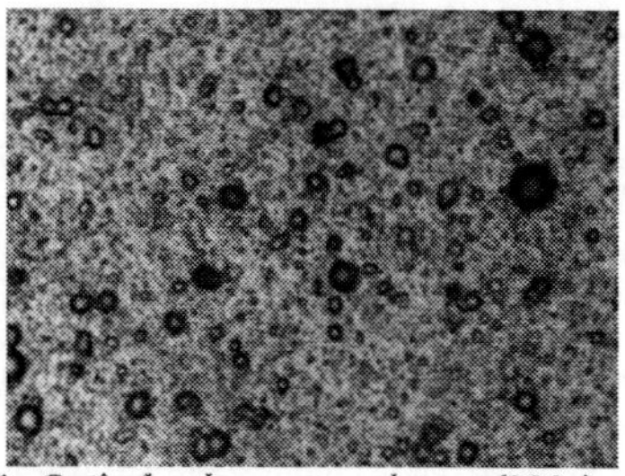

Figure 4: Optical microscope picture (200x) of a SnS layer deposited at 500 °C.

4 CONCLUSIONS

SnS thin film layers have been fabricated for the first time by an improved technique of electron pulsed deposition called ionized jet deposition.
Different vacuum pressures as well as different ablation power values have been tested and the deposition process has been optimized. Comparison of IJD-SnS layers deposited at different temperatures shows the good quality of layers at 300°C temperatures. The morphology of IJD-SnS at high temperature is very different from the VTE one, since its crystal size at these temperatures is smaller. However XRD patterns show that the IJD- layers grown at any temperature between 200°C and 400°C seem not to have secondary phases and confirm the low influence of substrate temperature. These results are promising for optimizing a low-substrate temperature deposition process.

5 ACKNOWLEDGMENTS

This work has been supported by Fondazione Cassa di Risparmio di Trento e Rovereto by funding the TIN-Jet project.

6 REFERENCES

[1] P. Sinsermsuksakul, L. Sun, S.W. Lee, H.H. Park, S.B. Kim, C. Yang, et al., Overcoming efficiency limitations of SnS-based solar cells, Adv. Energy Mater. 4 (2014) 1400496–2. doi:10.1002/aenm.201400496.
[2] A. Salavei, G. Tedeschi, D. Menossi, S. Di Mare, F. Piccinelli, A. Romeo, CdTe thin film solar cells by pulsed electron deposition, in: Photovolt. Spec. Conf. (PVSC), 2016 IEEE 43rd, 2016: pp. 137–140. doi:10.1109/PVSC.2016.7749644.
[3] S. Di Mare, D. Menossi, A. Salavei, E. Artegiani, F. Piccinelli, A. Kumar, et al., SnS Thin Film Solar Cells: Perspectives and Limitations, Coatings. 7 (2017) 34.
[4] S. Di Mare, A. Salavei, F. Piccinelli, A. Romeo, Analysis of SnS Growth and Post Deposition Treatment by Congruent Physical Vapor Deposition, in: 31st Eur. Photovolt. Sol. Energy Conf. Exhib., Hamburg, 2015. doi:10.4229/EUPVSEC20152015-3DV.3.22.
[5] P. Sinsermsuksakul, K. Hartman, S. Bok Kim, J. Heo, L. Sun, H. Hejin Park, et al., Enhancing the efficiency of SnS solar cells via band-offset engineering with a zinc oxysulfide buffer layer, Appl. Phys. Lett. 102 (2013) 53901. doi:10.1063/1.4789855.
[6] V. Steinmann, R. Jaramillo, K. Hartman, R. Chakraborty, R.E. Brandt, J.R. Poindexter, et al., 3.88% Efficient tin sulfide solar cells using congruent thermal evaporation, Adv. Mater. 26 (2014) 7488–92. doi:10.1002/adma.201402219.

EXTERNAL QUANTUM EFFICIENCY MODELING OF GaAs SOLAR CELLS GROWN ON Si: A METHOD TO ASSESS THE THREADING DISLOCATION DENSITY

Arthur Onno*[1], Mingchu Tang[1], Lars Oberbeck[2], Jiang Wu[1], Huiyun Liu[1]

[1]Dept. of Electronic and Electrical Engineering, University College London, London WC1E 7JE, UK

[2]TOTAL Gas, Renewables & Power, 92078 Paris La Défense Cedex, France

*Corresponding author: arthur.onno.13@ucl.ac.uk

ABSTRACT: A method is reported in order to determine an upper bound for the Threading Dislocation (TD) density in experimental GaAs solar cells grown lattice-mismatched on Si. The method is based on the modeling of the devices' External Quantum Efficiency (EQE), using the classic drift-diffusion model, or Hovel model. The model is fitted to experimental EQE measurements, using the diffusion length of minority carriers as the sole fitting parameter. Assuming low surface recombination velocities at both interfaces, a lower bound for the diffusion length of minority carriers is determined. Considering non-radiative recombinations on TDs as the dominant recombination pathway, this lower bound for the diffusion length of minority carriers can be converted to an upper bound for the TD density, using the NTT model. This method is then used to assess the TD density in GaAs solar cells grown on Si by Molecular Beam Epitaxy, using Strained Layer Superlattice (SLS) Dislocation Filter Layers (DFLs) coupled with Thermal Cycle Annealing (TCA) steps in order to reduce the TD density in the active region of the devices. Upper bounds for the TD densities in the low 10^7cm^{-2} are thus extracted from the devices' experimental EQE measurements.

Keywords: Dislocation, Modeling, III-V Semiconductors, c-Si, Epitaxy

1 INTRODUCTION

Direct monolithic integration of III-V materials on silicon in a multijunction architecture, where high-efficiency high-bandgap III-V top cells are epitaxially grown on a crystalline silicon (c-Si) wafer acting as a bottom cell, is an elegant pathway toward the development of silicon-based devices exceeding the ≈29 % theoretical efficiency limit of single junction c-Si solar cells [1]. The main challenge of this approach lies in the difference of lattice parameters between the c-Si substrate and III-V materials of interest, as there is no nitrogen-free III-V material lattice-matched to Si with the required bandgap for a dual junction or triple junction device. As a result of this lattice-mismatch, Threading Dislocations (TDs) nucleate at the III-V/Si interface and propagate upward toward the active region of the device. There, TDs act as non-radiative recombination centers, reducing the minority carriers' diffusion length and lifetime. Hence, the performances of the upper III-V cells are impaired. Demonstration of top cells with a low Threading Dislocation Density (TDD) is thus of paramount importance for the success of the technology.

Rapid and accurate characterization of the TDD in experimental III-V solar cell devices grown on Si is therefore needed. Common characterization methods include Electron Beam Induced Current (EBIC), Cathodoluminescence (CL), Electron Channeling Contrast Imaging (ECCI), Planar-View Transmission Electron Microscopy (PV-TEM), Cross-Sectional Transmission Electron Microscopy (CS-TEM) and Defect Selective Etching (DSE). A review of the pros and cons of each technique has been presented by Yaung *et. al.* [2-3]. Although the best techniques can provide an excellent accuracy in the range of TDDs under investigation (10^4-10^8 cm^{-2}) [2-3], they require advanced microscopy equipment – such as a scanning electron microscopy system (EBIC, CL and ECCI) or a transmission electron microscopy system (PV-TEM and CS-TEM) – and/or advanced sample preparation, leading to the destruction of the portion of sample characterized (PV-TEM, CS-TEM and DSE).

In this contribution, we present an optoelectronic technique in order to calculate an upper bound for the TDD in fully fabricated GaAs solar cells epitaxially grown on Si. Knowing the wavelength-dependent refractive indexes $n(\lambda)$ and absorption coefficients $\alpha(\lambda)$ of the different materials used in the structure, the architecture of the device and its External Quantum Efficiency (EQE) are the only inputs needed to assess the TDD. With further information on the Surface Recombination Velocities (SRVs) of the emitter/window and of the base/back surface field interfaces, the TDD calculation can be refined and an absolute value can be extracted.

We apply this method to experimental p-on-n GaAs solar cells grown on Si using Strained Layer Superlattice (SLS) Dislocation Filter Layers (DFLs), with and without Thermal Cycle Annealing (TCA) steps performed during growth. Lower bounds of 930 nm and 720 nm are extracted for the diffusion lengths of holes in devices grown with and without TCA steps, corresponding to TDDs of 1.5×10^7 cm^{-2} and 2.5×10^7 cm^{-2}, respectively. For the reference sample grown lattice-matched on GaAs, a lower bound of 2760 nm has been extracted for the hole diffusion length.

2 THEORETICAL MODEL

Simulation of the GaAs solar cells' EQE has been carried out using a combination of the minority carrier drift-diffusion model from Hovel and Woodall [4] and the diffusion length model in the presence of TDs developed by Yamaguchi and Amano [5]. A similar model has been developed by McClure *et. al.*, with a good accuracy, to simulate III-V devices grown on polycrystalline substrates [6].

The equations for the contributions of the depletion zone and of the quasi-neutral regions to the photocurrent, as a function of the wavelength, can be found in the extension of the Hovel model presented in Ref. [7]. Reflection at the back of the device – at the interface between the base and the back surface field – has been neglected. The simpler version model from Ref. [7], without back reflection, has consequently been used. In order to calculate a lower bound for the minority carriers' diffusion length, low Surface Recombination Velocities (SRVs) at the emitter/window interface and at the base/back surface field interface have been assumed, with values of 1×10^3 cm.s^{-1} for both interfaces. Reflection on the front surface of the device and absorption in the window and contacting layer have been calculated using OPAL 2 software [8].

Using the diffusion length as the sole fitting parameter, this modeled EQE is matched to experimental measurements. A lower limit for the diffusion length of minority carriers $L_{tot,min}$ – holes in the case of the n-on-p experimental devices presented hereafter – can thus be extracted. According to the NTT model [5], the diffusion length of minority carriers L_{tot} is then given by:

$$\frac{1}{L_{tot}^2} = \frac{1}{L_0^2} + \frac{1}{L_{TD}^2} \quad (1)$$

where L_0 is the diffusion length in the absence of dislocations and L_{TD} is the diffusion length associated with non-radiative Shockley-Read-Hall recombinations on TDs, given by:

$$\frac{1}{L_{TD}^2} = \frac{\pi^3 TDD}{4} \quad (2)$$

Assuming that the diffusion length in the absence of TDs L_0 is much larger than the diffusion length associated with SRH recombinations on TDs L_{TD} ($L_0 \gg L_{TD}$), the contribution to the diffusion length from recombination on TDs dominate and $L_{TD} \approx L_{tot}$. In any case, $L_{TD} \gtrsim L_{tot}$ is assured. An upper limit to the TDD can hence be calculated:

$$TDD \leq TDD_{max} = \frac{4}{\pi^3 L_{tot,min}^2} \quad (3)$$

Knowing the total diffusion length in the absence of TDs L_0, for example from comparison with an identical reference cell grown lattice-matched of GaAs, the calculation can be refined and a better assessment of this upper limit of the TDD can be calculated.

3 EXPERIMENTAL METHODS

3.1 Samples growth

Three samples were grown by solid-source Molecular Beam Epitaxy (MBE): two samples were grown lattice-mismatched on Si, with and without TCA steps; a third sample was grown lattice-matched on GaAs, as a reference. Details of the structure grown on Si are presented in FIGURE 1. The buffer consist of an AlGaAs nucleation layer followed by an AlAs/GaAs superlattice in order to flatten out the growth front and to achieve a smooth growth surface before growth of the four Strained Layer Superlattice (SLS) Dislocation Filter Layers (DFLs). Each DFL consists of a stack of ten alternating InGaAs compression layer and GaAs tension layers; inserted between two thicker GaAs spacer layers. Each DFL is less than 450nm-thick, leading to a total buffer thickness – including the nucleation layer and the AlAs/GaAs superlattice – of less than 3μm. This buffer is replaced with a 200nm-thick n^+-GaAs buffer for the sample grown lattice-matched on GaAs.

A standard cell structure has been used for both absorber materials, including a 30nm-thick n^+-$Al_{0.35}Ga_{0.65}As$ back surface field, a 2000nm-thick n-GaAs base, a 200nm-thick p^+-GaAs emitter, a 30nm-thick p^+-$Al_{0.75}Ga_{0.25}As$ window layer and a 50nm-thick p^+-GaAs contacting and capping layer.

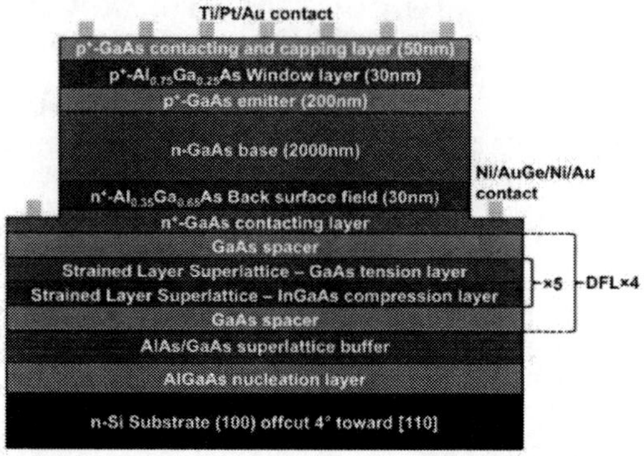

Figure 1: Details of the structure grown on Si. The red/orange layers represent the SLS DFL buffer; the blue layers represent the active layers of the cell. For the sample grown lattice-matched on GaAs, the SLS DFL buffer in orange/red is replaced with a 200nm-thick GaAs buffer.

3.2 Devices fabrication and characterization

Device fabrication was carried out using standard photolithography techniques. The samples were first etched in a H_2SO_4:H_2O_2:H_2O (1:10:80) solution, in order to define and isolate the mesa devices and to access the bottom n^+-type contacting layer. For the contact to the n-type region, a Ni/AuGe/Ni/Au contact structure was thermally evaporated and subsequently annealed at 390°C for 60s. The Ti/Pt/Au contact to the p-type region was then sputtered.

It is to be noted that, in order to avoid oxidation of the underlying Al-rich layers, the top GaAs capping layer has not been etched between the contact pads, leading to a non-negligible absorption in this lower bandgap layer. Moreover, no Anti-Reflection Coating (ARC) has been applied on top of the cells, leading to further reflection losses. Using OPAL 2 software [8], the losses through reflection are evaluated at 12.5mA.cm^{-2} and the losses through absorption in the contacting and window layers are evaluated at 5.0mA.cm^{-2}.

External Quantum Efficiencies (EQEs) were acquired using a ReRa SpeQuest quantum efficiency system, without bias. Absolute EQEs have been extracted by adjusting the measured EQEs to match the cells short-circuit current, measured at room-temperature under AM1.5G illumination from an LOT solar simulator.

4 RESULTS AND DISCUSSION

4.1 TDD extraction from EQE modeling

The experimental (circles) and simulated (solid lines) EQE data are presented in FIGURE 2. From the fitting of the simulated EQEs to the experimental data, lower bounds on the diffusion length of holes $L_{tot,min}$ of 720nm, 930nm and 2760nm have been extracted for the samples grown on Si without TCA steps (blue), grown on Si with TCA steps (red) and grown on GaAs as a reference (black), respectively.

As previously demonstrated with $Al_{0.2}Ga_{0.8}As$ solar cells [9], the use of TCA steps reduces the TDD by improving the mobility of dislocations, thus increasing the probability of dislocations encountering each other and self-annihilating or merging [10]. As a result, the diffusion length of minority carriers is increased, leading to an improvement of the carrier collection efficiency, especially at the back of the cell where lower energy photons are absorbed. This improvement is visible on FIGURE 2, with an enhancement of the EQE at longer wavelengths using TCA steps.

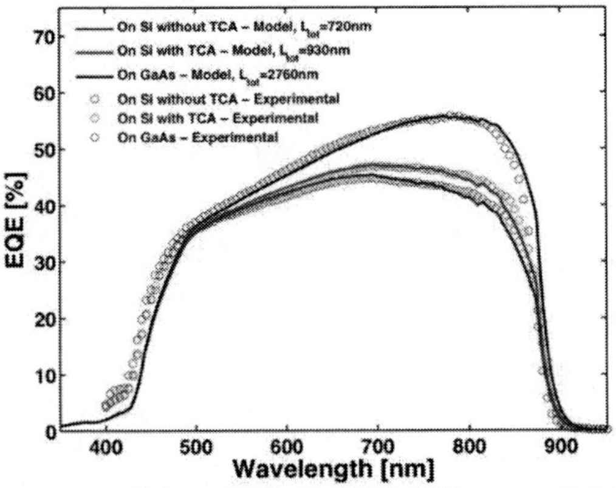

Figure 2: Fitting of the simulated EQE curves (solid lines) on the experimental data (circles). The three samples grown on Si without TCA steps (blue), on Si with TCA steps (red) and on GaAs as a reference (black) are displayed.

Assuming that the diffusion length is strongly dominated by the contribution from SRH recombinations on TDs ($L_{TD} \approx L_{tot}$), upper bounds for the TDD of 1.5×10^7 cm^{-2} and 2.5×10^7 cm^{-2} have been calculated for the samples grown with TCA steps (red) and without TCA steps (blue), respectively. Considering a similar contribution from non-TD-related recombinations for all samples (i.e. L_0=2760nm for the three samples), upper bounds for the TDD of 1.3×10^7 cm^{-2} and 2.3×10^7 cm^{-2} are obtained. The underlying assumptions behind these refined results are that the cell grown lattice-matched on GaAs presents a negligible TDD, that the TD-related and non-TD-related recombination pathways are independent and that, independently of the presence of TDs, the intrinsic material qualities of the three samples are similar.

These results compare with TDDs in the high 10^6 cm^{-2}, extracted from CS-TEM images, previously reported in $Al_{0.2}Ga_{0.8}As$ and GaAs devices grown lattice-mismatched on Si using SLS DFL buffers [9,11]. As expected, the upper bound for the TDD – calculated from EQE modeling – is higher than the CS-TEM measurements. The results are thus consistent. On top of the likely under-evaluation of the diffusion length – related to the assumptions of low SRVs and of independence of the SRH recombination pathways – inhomogeneities in the distribution of the TDD across the grown wafer can also explain the difference between the two calculations.

4.2 Limits of the method

Although a reasonable evaluation of the TDD can be achieved through the reported method, a certain number of hypotheses are necessary. A good confidence in the structure of the cell, especially regarding the thicknesses of the upper layers, is needed to accurately model the spectrum transmitted to the absorber region of the devices. The assumption of negligible reflection on the Back Surface Field appears to be justified, as the difference of refractive index between GaAs and $Al_{0.35}Ga_{0.65}As$ is limited over the range of wavelengths considered [12]. For structures where such a hypothesis would not be justified, implementation of the full model from Ref. [7], including reflection at the rear interface of the cell, would be needed.

The assumption of low SRVs at both interfaces leads to an underestimation of the diffusion lengths. Measurement of the SRV for experimental devices, in the presence and in the absence of TDs, could strongly improve the accuracy of the model. Similarly, assumptions on the contribution L_0 to the diffusion length from non-TD-related recombination pathways, for example from comparison with a reference device grown lattice-matched on GaAs, can increase the precision of the calculation. However, an additional strong hypothesis is then required: that the TD-related and non-TD-related recombination pathways are independent.

Finally, when recombination on TDs is not the dominant recombination pathway – such as in the $Al_{0.2}Ga_{0.8}As$ devices presented in Ref. [11] – our method presents limited interest, as the EQEs of devices grown lattice-matched and lattice-mismatched can be highly similar even though they exhibit a strong difference in TDDs.

5 CONCLUSION

We present an optoelectronic method to calculate an upper bound for the Threading Dislocation Density (TDD) in GaAs solar cells grown lattice-mismatched on Si. The method is based on the fitting of the External Quantum Efficiencies (EQEs) of experimental devices with the drift-diffusion model from Hovel *et. al.* [4]. Assuming low Surface Recombination Velocities (SRVs), a lower bound for the diffusion length of minority carriers is extracted. Using the NTT model [5], an upper bound for the TDD is then calculated. Applied to experimental n-on-p GaAs devices grown on Si using Dislocation Filter Layers (DFLs), a TDD below 1.5×10^7 cm^{-2} has been calculated for the sample grown using Thermal Cycle Annealing (TCA) steps. For the sample grown without such TCA steps, a TDD below 2.5×10^7 cm^{-2} has been extracted.

6 ACKNOLEWDGEMENTS

This work was supported by Total S.A.

7 REFERENCES

[1] A. Richter, M. Hermle, and S. W. Glunz, "Reassessment of the Limiting Efficiency for Crystalline Silicon Solar Cells," *IEEE J. Photovolt.* 2013; **3**(4): 1184-1191. DOI: 10.1109/JPHOTOV.2013.2270351.

[2] K. N. Yaung, S. Tomasulo, J. R. Lang, J. Faucher, and M. L. Lee, "Defect selective etching of $GaAs_yP_{1-y}$ photovoltaic materials," *J. Crys. Growth* 2014; **404**: 140-145. DOI: 10.1016/j.jcrysgro.2014.07.005.

[3] K. N. Yaung, S. Kirnstoetter, J. Faucher, A. Gerger, A. Lochtefeld, A. Barnett, and M. L. Lee, "Threading dislocation density characterization in III–V photovoltaic materials by electron channeling

contrast imaging," *J. Crys. Growth* 2016; **453**: 65-70. DOI: 10.1016/j.jcrysgro.2016.08.015.

[4] H. J. Hovel, and J. M. Woodall, "The effect of depletion region recombination currents on the efficiencies of Si and GaAs solar cells," *Proc. 10th IEEE PVSC* 1973; 25-30.

[5] M. Yamaguchi, and C. Amano, "Efficiency calculations of thinfilm GaAs solar cells on Si substrates," *J. Appl. Phys.* 1985; **58**(9): 3601-3606. DOI: 10.1063/1.335737.

[6] E. L. McClure, Z. S. Bittner, M. A. Slocum, D. V. Forbes, and S. M. Hubbard, "Modeling the Effects of Using Polycrystalline Substrates for Low Cost III -V Photovoltaics," *Proc. 42nd IEEE PVSC* 2015. DOI: 10.1109/PVSC.2015.7355955.

[7] M. P. Lumb, C. G. Bailey, J. G. J. Adams, G. Hillier, F. Tuminello, V. C. Elarde, and R. J. Walters, "Extending the 1-D Hovel Model for Coherent and Incoherent Back Reflections in Homojunction Solar Cells," *IEEE J. Quant. Electron.* 2013; **49**(5): 462-470. DOI: 10.1109/JQE.2013.2252148.

[8] K. R. McIntosh, and S. C. Baker-Finch, "OPAL 2: Rapid optical simulation of silicon solar cells," *Proc. 38th IEEE PVSC* 2012; 265-271. DOI: 10.1109/PVSC.2012.6317616.

[9] A. Onno, J. Wu, Q. Jiang, S. Chen, M. Tang, Y. Maidaniuk, M. Benamara, Y. I. Mazur, G. J. Salamo, N.-P. Harder, L. Oberbeck, and H. Liu, "$Al_{0.2}Ga_{0.8}As$ Solar Cells Monolithically Grown on Si and GaAs by MBE for III-V/Si Tandem Dual-junction Applications," *Energy Procedia* 2016; **92**: 661-668. DOI: 10.1016/j.egypro.2016.07.037.

[10] C. H. Simpson, and W. A. Jesser, "On the Use of Low Energy Misfit Dislocation Structures to Filter Threading Dislocations in Epitaxial Heterostructures," *Phys. Status Solidi (a)* 1995; **149**: 9-20. DOI: 10.1002/pssa.2211490102.

[11] A. Onno, M. Tang, M. Wang, Y. Maidaniuk, M. Benamara, Y. I. Mazur, G. J. Salamo, L. Oberbeck, J. Wu, and H. Liu, "MBE growth of 1.7eV $Al_{0.2}Ga_{0.8}As$ and 1.42eV GaAs solar cells on Si using dislocations filters: an alternative pathway toward III-V/Si solar cells architectures," *Proc. 44th IEEE PVSC* 2017; **manuscript in press**.

[12] D. E. Aspnes, and A. A. Studna, "Dielectric functions and optical parameters of Si, Ge, GaP, GaAs, GaSb, InP, InAs, and InSb from 1.5 to 6.0 eV," *Phys. Rev. B* 1983; **27**(2): 985-1009. DOI: 10.1103/PhysRevB.27.985.

DEVELOPMENT OF III-V ON SI MULTIJUNCTION PHOTOVOLTAICS BY WAFER BONDING

Laura Vauche[1,2], Elias Veinberg Vidal[1,2], Christophe Jany[1,2], Christophe Morales[1,2], Jean Decobert[3], Cécilia Dupré[1,2], Pierre Mur[1,2]
[1]Univ. Grenoble Alpes, Grenoble, 38000, France
[2]CEA, LETI, MINATEC Campus, Grenoble, 38054, France
[3]GIE IIIVLab, F-91767 Palaiseau, France

ABSTRACT: We report the fabrication of GaInP/AlGaAs/Si triple junction (3J) and AlGaAs/Si tandem (2J) solar cells using surface-activated wafer bonding (SAB). We observe maximum power conversion efficiency of 25.2% at 1-sun for a 3J device and 21.8% for a 2J device. The EQE analysis of the cells was carried out, which suggested that the 3J device performance was limited by the J_{SC} generated at the Si subcell. We demonstrate that higher efficiencies are feasible through design and adjustment of anti-reflective coating (ARC), GaAs bonding layer thickness and Si subcell.
Keywords: III-V Semiconductors, Silicon, Multijunction Solar Cell, High-Efficiency, Tandem

1 INTRODUCTION

III-V multijunction solar cells (MJSC) are the state-of-the-art approach for achieving high efficiencies in photovoltaic energy conversion for terrestrial concentrator and space applications [1]. The devices combine multiple p-n junctions of different band gap energies to absorb the spectrum of the sun more efficiently. However, III-V MJSC are expensive and significant cost reduction would be necessary for III-V cells to be cost competitive for low-concentration or one-sun applications, including reuse of the expensive substrate—either GaAs or Ge— and substantial reduction of epitaxial deposition cost [2]. III-V on Si solar cells are promising candidates for next-generation photovoltaics as they combine the low-cost of crystalline silicon (c-Si) and the efficiency potential of III-V/c-Si MJSC. For instance, GaAs/Si and InGaP/Si tandems are potentially cost competitive in scenarios where area-related balance-of-system costs are high, like space-constrained residential applications [3,4]. However, the development of III-V solar cells on Si has historically proven to be challenging. The 4.1% difference in lattice constant between Si and GaAs and the non-polar to polar interface lead to threading dislocations and anti-phase domains, respectively. Wafer bonding has attracted growing interest as it enables the electro-mechanical combination of lattice and thermal expansion mismatched materials with electrically conductive, transparent bonds. High efficiency solar cells using III-V junctions grown lattice-matched to GaAs bonded to InP [1] or Si [5] have been demonstrated in the last years. In the work reported here, we describe the fabrication of III-V on Si multijunction solar cells by wafer bonding. The performance of n-on-p GaInP/AlGaAs/Si triple junction, n-on-p and p-on-n AlGaAs/Si tandem solar cells is discussed.

2 APPROACH

III-V on Si solar cells fabrication main process steps are represented in Fig 1. In a first step (Fig 1.a), III-V epitaxial layers were grown inverted using metalorganic vapor phase epitaxy (MOVPE) on 4" GaAs wafers with release and etch stop layers allowing for the removal of the GaAs substrate after bonding. In the case of 3J GaInP/AlGaAs//Si solar cells, n-on-p top (E_g=1.46 eV) and middle (E_g=1.90 eV) sub-cells were grown at Fraunhofer ISE as reported in [1,5]. For the fabrication of AlGaAs//Si tandem cells, n-on-p (E_g=1.71 eV) and p-on-n (E_g=1.76 eV) solar cells were grown by IQE [6] and III-V Lab [7]. The n-on-p and p-on-n Si homo-junction solar cells were produced by diffusion processes on 4" <100> Si CZ substrates, leading to the formation of emitter at the front surface and back surface field (BSF).

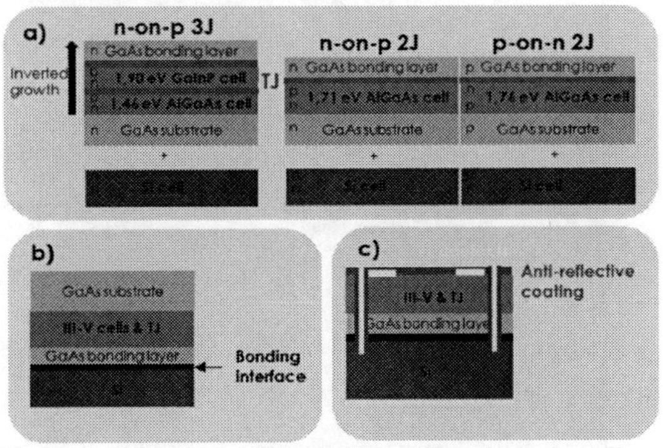

Figure 1: Schematized III-V on Si multijunction solar cell fabrication process. a) MOVPE growth of III-V top and middle subcells, including tunnel junctions (TJ) and fabrication of Si bottom subcells, b) Wafer bonding, c) final solar cell structure.

In a second step (Fig 1.b), Si front surface and GaAs bonding layer surfaces were cleaned and planarized by chemical mechanical polishing (CMP) before being bonded together. We previously reported direct GaAs/Si wafer bonding [7]. In this paper, Si and GaAs were bonded by surface activation bonding (SAB), which is an attractive approach for III-V on Si integration [5,8,9]. Using this technique, the oxide layers formed on GaAs and Si substrate surfaces are removed by Ar beam irradiation and then the surfaces are brought into contact under ultra-high vacuum.
Finally (Fig 1.c), GaAs parent substrate is removed by wet chemical etching. Front metal contacts designed for low concentration (10-20 suns, ~5% shading factor without bus bars [7]) and full sheet back metal contacts are deposited by evaporation. Mesa etching (trenches down to 2 μm inside Si) is performed to partially isolate individual cells of 1 and 2 cm², and a 65 nm SiN_x anti-reflection coating is deposited by plasma-enhanced chemical vapor deposition (PECVD).

3 RESULTS AND DISCUSSION

3.1. Bonding interface properties

The interface properties of wafer-bonded devices by SAB are presented in Fig. 2. The post-bonding quality is ensured through scanning acoustic microscopy (SAM), employing ultrasound. This characterization technique has sub-mm resolution and hence allows to clearly show the areas where the bonding is complete (black) and the ones with voids (white). Both n-GaAs/n-Si and p-GaAs/p-Si bondings using SAB were successful on most of the surface, confirming the interest of this technique for bonding between various III-V designs and Si. Transmission electron microscopy (TEM) and energy dispersive X-ray spectroscopy (EDX) analysis revealed the presence of Ar at the few nm thick bonding interface, as a result of Ar beam irradiation used for the surface activation.

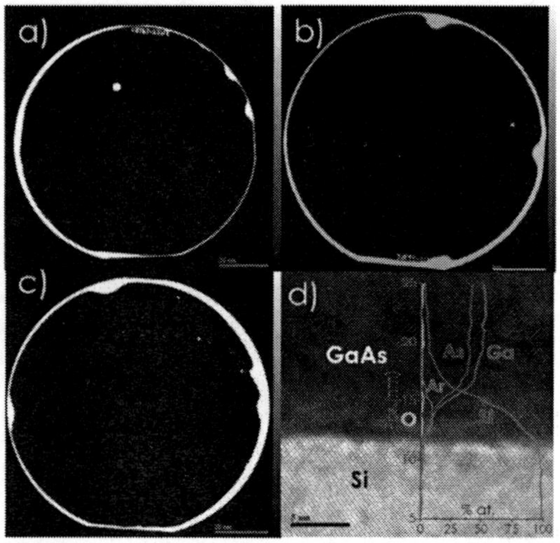

Figure 2: SAM images of SAB wafers (a) n-on-p 3J (b) n-on-p 2J and (c) p-on-n 2J devices showing few voids in white where bonding has not been successful owing to particle contamination on the surface of the wafer. All wafer-pairs are 100 mm diameter. d) TEM image of the bonded n-GaAs/n-Si interface with EDX analysis.

3.2. Triple junction solar cells

The external quantum efficiency (EQE) of n-on-p 3J cells is represented in Fig. 3.

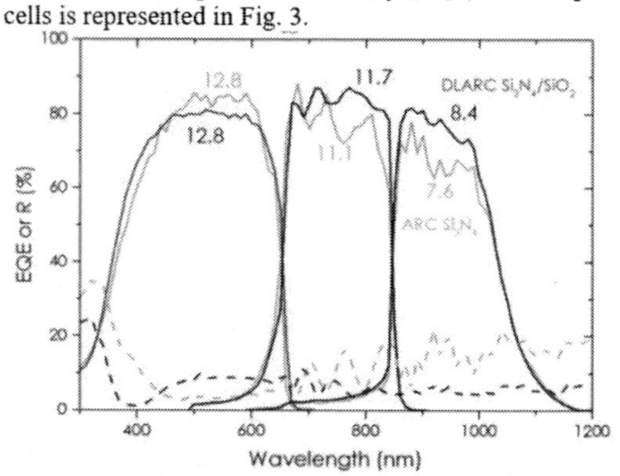

Figure 3 : EQE of n-on-p 3J with single layer ARC (green) and double layer ARC (dark blue). The calculated subcell short-circuit current densities (J_{SC} in mA/cm²) under AM1.5G at 1000 W/m² are shown for both devices. Respective reflection curves are shown in dashed lines.

In 2 terminal MJSC, subcells are connected in series and hence the lowest subcell current drives the total current of the MJSC. The photocurrent mismatch between the top and bottom subcells of the 3J device is quite important (5.2 mA/cm²), and therefore, the bottom Si subcell severely limits the performance of the device. The application of a Double Layer Anti-Reflective Coating (DLARC) consisting of 85 nm of SiO_2 and 65 nm of Si_3N_4 helped to reduce the current mismatch to 4.4 mA/cm². Further reduction is expected from ongoing ARC optimization. In addition, the performance of the 525 µm thick Si cell was limited by the use of low lifetime CZ substrates. Ongoing work on Si cell optimization using PC1D simulations (use of FZ substrates, choice of optimum emitter profile, and integration of rear passivation) will help to obtain maximum efficiency potential from this design. Best performing 3J in literature exhibited only a small current mismatch [5], as reported in Table I. A flash solar simulator with component reference cells (isotypes) has been used to measure I-V curves [10], as reported in [11]. We report a 3J solar cell with 25.2% power conversion efficiency. Similar opto-electronic parameters were extracted from I-V curves measured at FhG-ISE. The V_{OC} of our 3J cells was comparable to the 26% efficient 3J reported in literature [5].

Table I: Performance of 3J III-V on Si under 1 sun under AM1.5g illumination, comparison between this study and literature. *Opto-electronic parameters of our GaInP/AlGaAs/Si DLARC cell are given for AM1.5d illumination *(measured by Fraunhofer ISE CalLab)*.

Cell	Reference	J_{SC} (mA/cm²) top/middle/bot EQE	Current mismatch (mA/cm²)	J_{SC} (mA/cm²) I-V	V_{OC} (mV) I-V	FF (%) I-V	η (%)
GaInP/AlGaAs/Si	this study	12.8/11.1/**7.6**	5.2				
GaInP/AlGaAs/Si DLARC	this study	12.8/11.7/**8.4**	4.4	9.6*(9.8)**	2.95*(2.95)**	88.8*(87.2)**	25.2*(25.2)**
GaInP/AlGaAs/Si	[12]	13.4/12.6/**11.7**	0.9	11.7	3.05	87.5	31.3
GaInP/GaAs/Si	[5]	12.3/**11.6**/12.4	0.8	11.9	3.05	83	30.2
GaInP/AlGaAs/Si	[5]	11.8/12.0/**10.0**	2.0	10.0	2.90	89.8	26.2
GaInP/AlGaAs/Si	[8]	12.1/12.9/**6.5**	6.4	10.0	2.85		~23

3.3. Dual junction solar cells

EQE analysis of 2J devices, as shown in Fig. 4, revealed that the AlGaAs top and Si bottom subcells are almost current matched (current mismatch +0.8 mA/cm² or -1.7 mA/cm²). However, EQE of AlGaAs was lower than expected. We prepared p-on-n (as reported in [6,7]) and n-on-p AlGaAs single-junction solar cells (not shown here) with similar designs, but directly grown on GaAs substrates. These direct-growth devices presented higher performances, with EQE reaching around 80% (versus close to 70% in this paper). Different hypotheses might explain this reduced performance such as inverted growth conditions leading to changes in dopant levels, dopant diffusion [13] or processing issues. The tandem AlGaAs/Si cells V_{OC} were 1.69 V for n-on-p cell and 1.78 V for p-on-n cell, respectively.

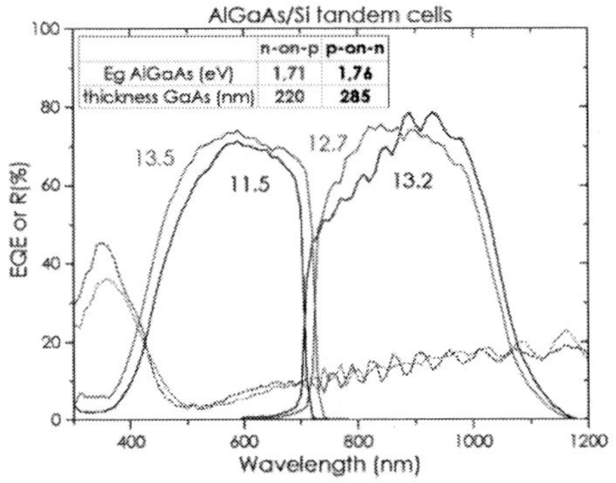

Figure 4: EQE of n-on-p (pink) and p-on-n (black) 2J with single layer ARC. The current densities under AM1.5G at 1000 W/m² are shown for all junctions in both devices. Respective reflection curves are shown in dashed lines. Bandgaps of AlGaAs top cells and thickness of GaAs bonding layer are reported in the insert table.

The use of GaAs as a bonding layer material limits the performance of tandem cells because its 1.4 eV bandgap highly absorbs light in the 700 to 850 nm range and thereby limits the light that should be transmitted to the Si subcell. The severity of this loss greatly depends on GaAs thickness.

First optimizations have been performed on a n-on-p AlGaAs/Si, with a 135 nm GaAs layer (versus 220 and 285 nm for n-on-p and p-on-n cells presented in this study), and using a DLARC, leading a bottom subcell J_{SC} of 16.8 mA/cm². Last optimizations in DLARC, GaAs bonding layer (83 nm) and top cell design have led to 16.4 and 16.2 mA/cm² J_{SC} for top and bottom subcells, respectively. Power conversion efficiency of this 2J cell has been evaluated with a solar flash simulator at 21.8% under 1-sun AM1.5D.

4 CONCLUSIONS

In conclusion, triple junction and tandem cells fabricated by SAB have been demonstrated showing low defective n-GaAs/n-Si and p-GaAs/p-Si bonding interfaces. We observed maximum power conversion efficiency of 25.2% at 1-sun for a 3J device and 21.8% for a 2J device. EQE analysis show that further optimization is needed in order to improve the current matching and devices performance.

ACKNOWLEDGEMENTS

The authors would like to thank Thibaut Desrues for providing the Si homojunction cells, Christophe Lecouvey for GaAs and Si surfaces preparation, Jérémy Da Fonseca and Vincent Rebeyrol for cell processing, SERMA technologies for the TEM and EDX, Clément Weick, Philippe Voarino and Sarah Bernardis for their help with EQE measurements and Franck Fournel, Mathieu Baudrit for valuable discussions.

REFERENCES

[1] T.N.D. Tibbits, P. Beutel, M. Grave, C. Karcher, E. Oliva, G. Siefer, A. Wekkeli, M. Schachtner, F. Dimroth, A.W. Bett, R. Krause, M. Piccin, N. Blanc, M. Muñoz-Rico, C. Arena, E. Guiot, C. Charles-Alfred, C. Drazek, F. Janin, L. Farrugia, B. Hoarau, J. Wasselin, A. Tauzin, T. Signamarcheix, T. Hannappel, K. Schwarzburg, A. Dobrich, NEW EFFICIENCY FRONTIERS WITH WAFER-BONDED MULTI-JUNCTION SOLAR CELLS, in: 29th Eur. Photovolt. Sol. Energy Conf. Exhib., 2014.

[2] J.S. Ward, T. Remo, K. Horowitz, M. Woodhouse, B. Sopori, K. VanSant, P. Basore, Techno-economic analysis of three different substrate removal and reuse strategies for III-V solar cells:, Prog. Photovolt. Res. Appl. 24 (2016) 1284–1292. doi:10.1002/pip.2776.

[3] D.C. Bobela, L. Gedvilas, M. Woodhouse, K.A.W. Horowitz, P.A. Basore, Economic competitiveness of III-V on silicon tandem one-sun photovoltaic solar modules in favorable future scenarios: Economic competitiveness of III-V on on silicon tandem modules, Prog. Photovolt. Res. Appl. 25 (2017) 41–48. doi:10.1002/pip.2808.

[4] S. Essig, C. Allebé, T. Remo, J.F. Geisz, M.A. Steiner, K. Horowitz, L. Barraud, J.S. Ward, M. Schnabel, A. Descoeudres, D.L. Young, M. Woodhouse, M. Despeisse, C. Ballif, A. Tamboli, Raising the one-sun conversion efficiency of III–V/Si solar cells to 32.8% for two junctions and 35.9% for three junctions, Nat. Energy. 6 (2017) 17144.

[5] R. Cariou, J. Benick, P. Beutel, N. Razek, C. Flotgen, M. Hermle, D. Lackner, S.W. Glunz, A.W. Bett, M. Wimplinger, F. Dimroth, Monolithic Two-Terminal III–V//Si Triple-Junction Solar Cells With 30.2% Efficiency Under 1-Sun AM1.5g, IEEE J. Photovolt. 7 (2017) 367 – 373. doi:10.1109/JPHOTOV.2016.2629840.

[6] E. Veinberg-Vidal, L. Vauche, C. Weick, J.D. Fonseca, C. Jany, C. Morales, C. Lecouvey, T. Desrues, P. Voarino, F. Fournel, A. Kaminski-Cachopo, A. Datas, P. Garcia-Linares, M. Baudrit, P. Mur, C. Dupré, Wafer-Bonded AlGaAs//Si Dual-Junction Solar Cells, in: IEEE Photovolt. Spec. Conf., 2017.

[7] E. Veinberg-Vidal, C. Dupré, P. Garcia-Linares, C. Jany, R. Thibon, T. Card, T. Salvetat, P. Scheiblin, C. Brughera, F. Fournel, Y. Desieres, Y. Veschetti, V. Sanzone, P. Mur, J. Decobert, A. Datas, Manufacturing and Characterization of III-V on

Silicon Multijunction Solar Cells, Energy Procedia. 92 (2016) 242–247. doi:10.1016/j.egypro.2016.07.066.
[8] N. Shigekawa, J. Liang, R. Onitsuka, T. Agui, H. Juso, T. Takamoto, Current–voltage and spectral-response characteristics of surface-activated-bonding-based InGaP/GaAs/Si hybrid triple-junction cells, Jpn. J. Appl. Phys. 54 (2015) 08KE03. doi:10.7567/JJAP.54.08KE03.
[9] L. Vauche, E. Veinberg-Vidal, C. Weick, C. Morales, V. Larrey, C. Lecouvey, M. Martin, J.D. Fonseca, C. Jany, T. Desrues, C. Brughera, P. Voarino, T. Salvetat, F. Fournel, M. Baudrit, C. Dupré, Wafer bonding approaches for III-V on Si multi-junction solar cells, in: Photovolt. Spec. Conf. PVSC 2017 IEEE 44th, 2017.
[10] E. Veinberg-Vidal, To be submitted, (n.d.).
[11] C. Domínguez, I. Antón, G. Sala, S. Askins, Current-matching estimation for multijunction cells within a CPV module by means of component cells, Prog. Photovolt. Res. Appl. 21 (2013) 1478–1488. doi:10.1002/pip.2227.
[12] R. Cariou, J. Benick, P. Beutel, N. Tucher, M. Graf, D. Lackner, M. Hermle, S.W. Glunz, A.W. Bett, F. Dimroth, Wafer Bonded III-V on Silicon Multi-Junction Cell Cell with Efficiency beyond 31%, in: IEEE Photovolt. Spec. Conf., 2017.
[13] M.A. Steiner, J.F. Geisz, R.C. Reedy, S. Kurtz, Direct Comparison of Inverted and Non-Inverted Growths of GaInP Solar Cells, in: Photovolt. Spec. Conf. 2008 PVSC 08 33rd IEEE, 2008.

MEASUREMENT OF SUBCELL CAPACITANCE IN TRIPLE JUNCTION SOLAR CELLS WITH PULSED ILLUMINATION

Martin Rutzinger[1,2], Manuel Salzberger[1,2], Helmut Nesswetter[1], Andreas Gerhard[1], Paolo Lugli[2] and Claus G. Zimmermann[1]
[1]Airbus DS GmbH
Robert-Koch-Straße 1 82024 Taufkirchen, Germany
[2]Technische Universität München
Arcisstraße 21, 80333 Munich, Germany

ABSTRACT: The evolution of the subcell capacitance of triple junction solar cells upon 3 MeV electron irradiation is studied. Photocurrent is induced subcell specific using a sequence of spatially homogeneous, monochromatic light pulses. The used method is based on the externally measured transient voltage, which allows to calculate the depletion layer capacitance. Capacitance-voltage curves for each subcell are obtained. From these curves, the base layer effective doping density and thus the carrier removal rate are extracted for the top and middle subcell. Additionally, the built-in voltage and its degradation is measured.

Keywords: III-V Semiconductors, Capacitance, Characterization, Degradation, Multijunction Solar Cell, Radiation Damage, Space Cells

1 INTRODUCTION

Multijunction solar cells (MJSC) are the industry standard solar cell for space applications as they have superior conversion efficiency and mass/power ratios. Their radiation degradation behavior has been studied in prior studies [1,2]. The solar cell key performance indicators, like the open circuit voltage V_{oc} , the short circuit current I_{sc}, and the fill factor FF degrade after electron or proton irradiation. The radiation causes displacements like interstitials, vacancies or more complex defect types in the semiconductor crystal [3]. One reason for the performance degradation is the reduction of minority carrier lifetime as these defects act as non-radiative recombination centers. A second consequence of particle irradiation, which is affecting the cell performance at high irradiation doses, is the permanent localization of majority carriers by radiation induced defect levels. This reduces the effective doping density and is therefore referred as carrier removal (CR) [4]. CR in single junction solar cells due to electron and proton irradiation has been studied previously [5–8]. These studies found, that for one sided n$^+$p junctions the relative effect of CR is more critical for the base layer (p-side) carrier density. The effect was found to have less influence on the more heavily n-doped emitter. For very high fluences, CR can even change the polarity of the base material from p-type to n-type [5]. A polarity change prohibits electrical functioning of the irradiated device as solar cell, since no p-n junction is present anymore. Accurate knowledge about CR characteristics is therefore beneficial for cell optimization, but also for space mission design.

In this work, a novel method to measure the subcell capacitance of MJSC [9] is applied on triple junction solar cells which have been irradiated with 3 MeV electrons. The method is based on the measured external voltage ramp, which is observed, when one subcell is illuminated externally using low intensity monochromatic illumination. With the help of the Mott-Schottky plot, the measured depletion layer capacitance is used to extract the junction's built-in voltage V_{bi} and effective carrier density N_a of the base layer.

2 EXPERIMENTAL

The investigated samples are space grade triple junction (TJ) $In_{0.50}Ga_{0.50}P/Ga_{0.99}In_{0.01}As/Ge$ solar cells with an area of 4 cm². The solar cells have been irradiated with 3 MeV electrons and various fluences ranging from 1E14 cm^{-2} up to 1E16 cm^{-2}. After irradiation, the cells were annealed for 48 h under air mass zero (AM-0) illumination. Additionally, unirradiated begin-of-life (BOL) cells were available.

The experimental setup consists of three light emitting diode (LED) arrays with wavelengths $\lambda_{LED,J1}$ = 450 nm, $\lambda_{LED,J2}$ = 780 nm, and $\lambda_{LED,J3}$ = 1450 nm which are absorbed in the top, middle and bottom junction, respectively. The LEDs are used to provide the measurement pulse with low photocurrent density. Additionally, two lasers with $\lambda_{Laser,J2}$ = 803 nm and $\lambda_{Laser,J3}$ = 1470 nm are used to illuminate the middle and bottom subcells with high photocurrent density. The photocurrent densities $j_{ph,i}$ induced by the light sources in a BOL TJ cell at the used power levels were measured under current limiting conditions [10]. They are given in Table I.

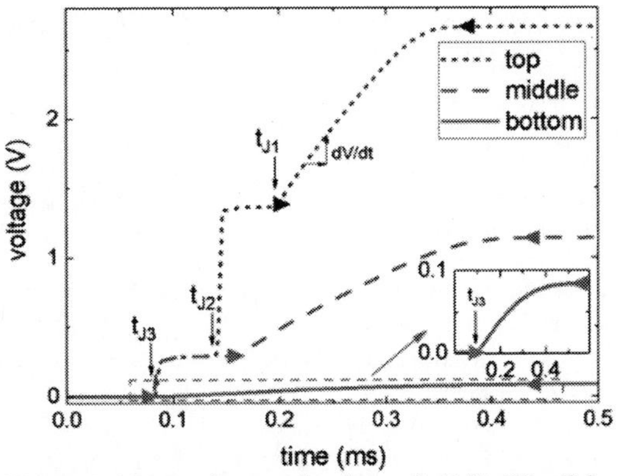

Figure 1: Measured external voltage of a BOL TJ cell for the three pulse sequences top, middle and bottom. The light sources were turned on at the indicated times. The start and end of the region of interest is indicated by triangles.

Table I: LED and laser wavelengths λ and BOL photocurrent density j_{ph} in the respective absorbing subcells.

Junction	LED		Laser	
	λ (nm)	j_{ph} (mA/cm²)	λ (nm)	j_{ph} (mA/cm²)
Top	450	0.58	n/a	n/a
Middle	780	0.29	803	100
Bottom	1450	0.05	1470	150

The setup is enclosed in a dark box to keep out any stray illumination. The solar cell is mounted on a temperature controlled (T=300 K) brass vacuum chuck and is contacted electrically. A Dewetron DEWE-5000 [12] with a 1 MΩ input resistance is used to record the external cell voltage vs. time.

Both, the LEDs and the lasers illuminate the solar cell with spatially homogeneous illumination [11]. This minimizes any in-plane current flow and allows a precise determination of $j_{ph,i}$, which each light source induces in the absorbing subcell i.

The experimental procedure to measure the subcell capacitance is based on a measurement of the subcell voltage as it has been presented in Ref. [14]. There, a MJSC is illuminated using a sequence of monochromatic light pulses. The initially discharged junction capacitances of the un-illuminated subcells take up the very small measurement current, which is in the µA regime. Directly after switching on the light sources, the externally measured cell V_{oc} then equals the sum of the illuminated subcell V_{oc}'s. For small induced photocurrent density in the order of mA/cm², the externally measured voltage is not immediately built up, as first the capacitance of the illuminated subcell needs to be charged. The voltage therefore increases slowly, before it finally saturates at the subcell open circuit voltage as shown in Fig. 1. From the slope of the $V-t$ curves, the C-V curves can be calculated.

The capacitance C is defined as

$$C(V) = \frac{d(Q/A)}{dV} \tag{1}$$

where Q/A is the charge per area and V the voltage. Eq. (1) can be transformed to

$$C(V) = j_c / \frac{dV}{dt} \tag{2}$$

where j_c is the current density, which is charging the capacitance and dV/dt the slope of the externally measured transient cell voltage. Each subcell i of the MJSC solar cell can be modelled as a diode with a recombination current density $j_{d,i}$ and a parallel photocurrent source $j_{ph,i}$. Using Kirchhoff's law, j_c can then be calculated as [9]

$$j_{c,i} = j_{ph,i} - j_{d,i}. \tag{3}$$

$j_{d,i}$ is negligible for low voltages, as the recombination current is exponentially dependent on the voltage across the diode and thus the photocurrent equals the charging current.

Luminescent coupling (LC) is present in the investigated devices [13] and has to be considered. LC describes radiative recombination in one subcell and reabsorption in the subcell below. To eliminate the LC contribution, the reabsorbing subcell is illuminated by a high intensity laser such that the additional LC photocurrent contribution becomes negligible. This is the same concept, which was also used for the extraction of subcell current voltage curves in Ref [14].

For the bottom subcell measurement, the 1450 nm LED is switched on at t_{J3}. The recorded TJ cell voltage is shown in Fig. 1, green solid curve and the inset. No additional laser illumination is necessary here, as the already very low amount of radiative recombination emitted from bottom (Ge) subcell cannot be absorbed by any other subcell. For the measurement of the middle subcell, first the 1470 nm laser is switched on at t_{J3} to flood the bottom subcell. Several ten µs later, after a stable voltage plateau is formed, the 780 nm LED is switched on at t_{J2}. The measured voltage is plotted as red, dashed line in Fig. 1. For the measurement of the top subcell, first, both the 1470 nm and the 803 nm lasers are switched on. Again, after a stable plateau is formed, the 450 nm LED is switched on at t_{J1}, (Fig. 1, black dotted). The voltage plateaus in the middle and top subcell measurement originate from V_{oc}^{bot} and $V_{oc}^{bot} + V_{oc}^{mid}$, respectively [10,15].

The relevant subcell voltage in the top and middle subcell case corresponds to the voltage increase measured from the voltage plateau value before the LED is switched on.

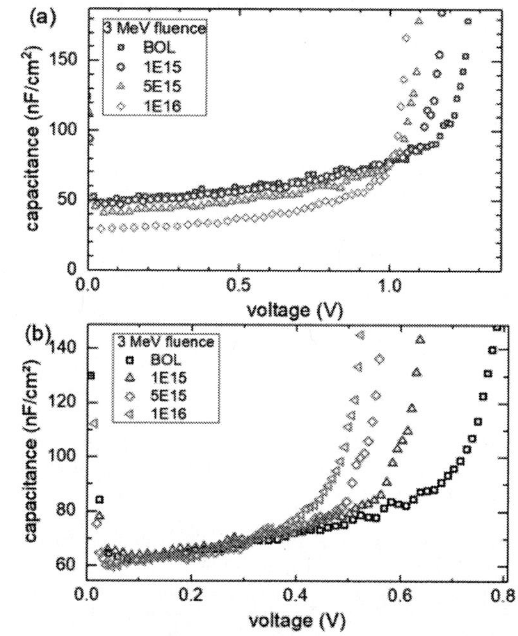

Figure 2: Measured subcell C-V curves of the (a) top and (b) middle subcell for various 3 MeV electron fluences. The capacitance is only measured correctly in the low voltage region. At higher voltages approaching $V_{oc,i}$ the diode recombination is no more neglibible and thus the capacitance is overestimated.

3 RESULTS

The three curves shown in Fig. 1 were measured from a BOL TJ cell. The same measurement was also performed for the irradiated TJ's. The subcell $C-V$ curves could be extracted for the top and middle subcell. The curves of the bottom subcell could not be evaluated, as there the diode current $j_{d,bot}$ is non-negligible, even at low voltages.

Fig. 2 shows the measured subcell $C-V$ curves of top and middle subcell for selected 3 MeV fluences. The results are only valid in the lower voltage region. At higher voltages, approaching the subcell V_{oc}, the assumption, $j_{d,i}=0$ is no longer valid and the capacitance values are overestimated. It can be observed, that the capacitance of the top subcell decreases more drastically than the middle subcell with increasing radiation fluences.

Further analysis of the *C-V* curves is possible, when considering that the pn junction capacitance at low voltage values arises mainly from the depletion layer capacitance C_{dep}. The diffusion layer capacitance C_{diff} only contributes at higher forward voltages, which are not accessible using the approach in this work. C_{diff} is therefore not further considered. For one sided, abrupt n⁺p junctions, C_{dep} can be expressed in terms of the dielectric permittivity ϵ_s, the elementary charge q, the base doping density N_a and the built-in voltage V_{bi} as [16]

$$C_{dep,i}=\sqrt{\frac{\epsilon_s\,q\,N_a}{2(V_{bi}-V-2kT/q)}}. \qquad (4)$$

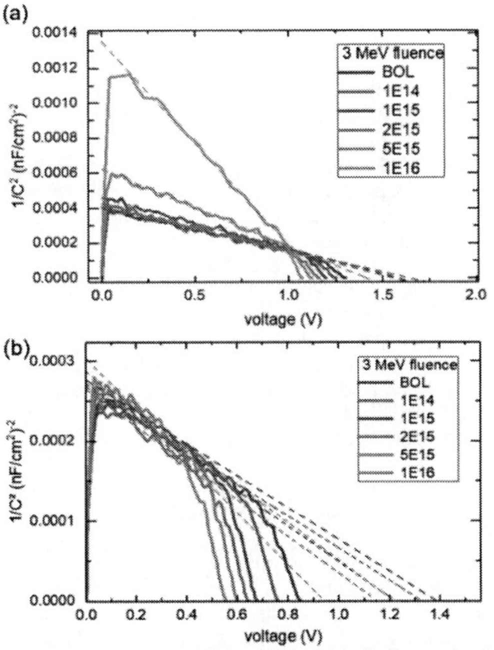

Figure 3: (a) Top and (b) middle subcell Mott-Schottky plots (*1/C² - V*) of irradiated TJ cells. The dashed lines are fitted to the linear plateau region. The slope and voltage axis intersect of the fit curves give access to the effective base layer doping density and built-in voltage.

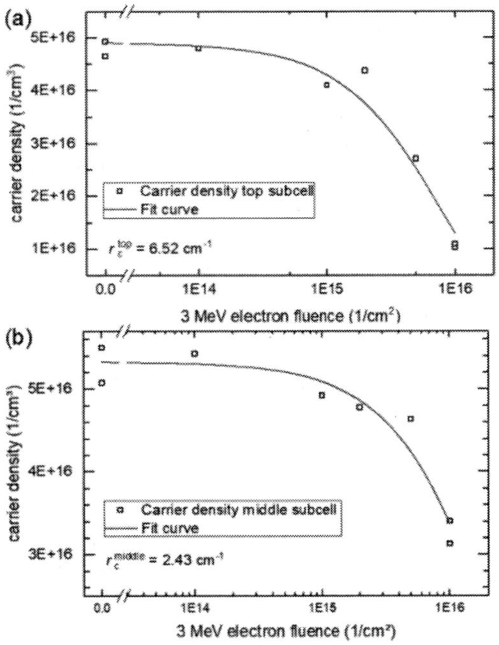

Figure 4: Effective base layer doping density of (a) top and (b) middle subcell vs. 3 MeV electron fluence.

This equation has a linear form in the $1/C^2-V$ form (Mott-Schottky plot). The extrapolated voltage axis intercept is at $V_{bi}-2kT$ and thus allows the extraction of V_{bi}. The slope in the $1/C^2$ form

$$\frac{d1/C_{dep,i}^2}{dV}=\frac{2}{\varepsilon_s\,q\,N_a} \qquad (5)$$

gives access to N_a.

Fig. 3 shows these Mott-Schottky plots of the subcells for all 3 MeV fluences. The dashed lines are numerically fitted to the linear regions. These fit values have been evaluated using Eq. (5) to study the decrease of the effective doping density after irradiation, which is depicted in Fig. 4 (a) for the top and (b) for the middle subcells.

The carrier removal rate r_c is extracted from the data in Fig. 4 by fitting the equation

$$N_a(\Phi)=N_{a0}\exp\frac{-r_c\,\Phi}{N_{a0}} \qquad (6)$$

where Φ is the 3 MeV fluence and N_{a0} the BOL base layer doping density [17]. The obtained values are $r_c^{middle}=2.43$ cm⁻¹ and $r_c^{top}=6.52$ cm⁻¹.

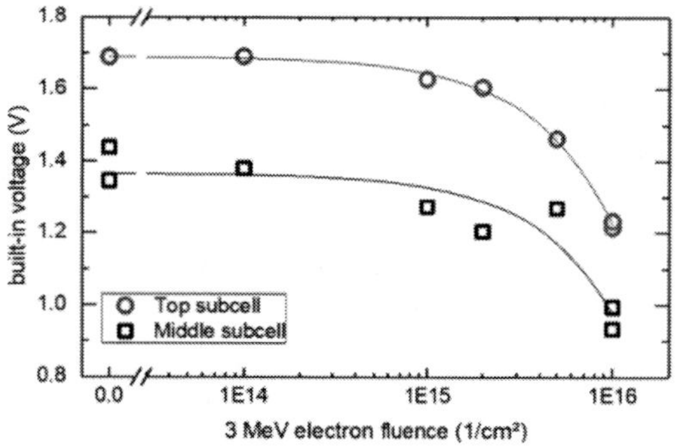

Figure 5: Degradation of the built-in voltage of the top and middle subcell upon 3 MeV electron irradiation. The solid curves represent the linear fit to the data.

Finally, also the degradation of the built-in voltage V_{bi} can be extracted from the voltage axis intersect of the fit in Fig. 3. The results of this evaluation are shown in Fig. 5. We empirically found, that the measured degradation can be described using a linear model

$$V_{bi}(\Phi) = V_{bi,0} - r_v \, \Phi \qquad (6)$$

with the voltage degradation rate r_v. For 3 MeV electron irradition and the given solar cell structure, the obtained values are r_v^{mid} = 3.89E17 V cm²/e- and r_v^{top} = 4.64E17 V cm²/e-.

4 SUMMARY

The experimental approach is based on the slope of the transient voltage upon monochromatic illumination. The slope is interpreted as charging curve of the subcell capacitance. From the capacitance vs. voltage characteristics, the effective base layer doping level and the built-in voltage could be extracted. The carrier removal rate and the degradation of the built-in voltage in triple junction solar cells due to 3 MeV electron irradiation has been investigated. The degradation was found to compensate up to 40% and 80% of the emitter doping density in the top and middle subcell, respectively after electron irradiation with a fluence of 1E16/cm².

BIBLIOGRAPHY

[1] Hoheisel, R., Dimroth, F., Bett, A.W., Messenger, S.R., Jenkins, P.P., and Walters, R.J. (2013) Electroluminescence analysis of irradiated GaInP/GaInAs/Ge space solar cells. Sol. Energy Mater. Sol. Cells, 108, 235–240.

[2] Chen, S., Zhu, L., Yoshita, M., Mochizuki, T., Kim, C., Akiyama, H., Imaizumi, M., and Kanemitsu, Y. (2015) Thorough subcells diagnosis in a multi-junction solar cell via absolute electroluminescence-efficiency measurements. Sci. Rep., 5, 7836.

[3] Khan, A., Yamaguchi, M., Bourgoin, J.C., and Takamoto, T. (2005) Thermal annealing study of 1 MeV electron-irradiation-induced defects in n + p InGaP diodes and solar cells Thermal annealing study of 1 MeV electron-irradiation-induced defects in n ¿ p InGaP diodes and solar cells. 2391 (2002).

[4] Walters, R.J., Warner, J.H., Summers, G.P., Messenger, S.R., and Lorentzen, J.R. (2005) RADIATION RESPONSE MECHANISMS IN MULTIJUNCTION III-V SPACE SOLAR CELLS. IEEE Conf. Photovolt. Spec., 542–547.

[5] Messenger, S.R., Jackson, E.M., Burke, E.A., Walters, R.J., Xapsos, M.A., and Summers, G.P. (1999) Structural changes in InP/Si solar cells following irradiation with protons to very high fluences. J. Appl. Phys., 86 (3), 1230–1235.

[6] Deangelis, N., Bourgoin, J.C., Takamoto, T., Khan, A., and Yamaguchi, M. (2001) Solar cell degradation by electron irradiation. Comparison between Si, GaAs and GaInP cells. Sol. Energy Mater. Sol. Cells, 66 (1–4), 495–500.

[7] Sato, S., Miyamoto, H., Imaizumi, M., Shimazaki, K., Morioka, C., Kawano, K., and Ohshima, T. (2009) Degradation modeling of InGaP/GaAs/Ge triple-junction solar cells irradiated with various-energy protons. Sol. Energy Mater. Sol. Cells, 93 (6–7), 768–773.

[8] Walters, R.J., Messenger, S.R., Cotal, H.L., Xapsos, M.A., Wojtczuk, S.J., Serreze, H.B., and Summers, G.P. (1997) Radiation response of heteroepitaxial n(+)p InP/Si solar cells. J. Appl. Phys., 82 (5), 2164–2175.

[9] Rutzinger, M., Salzberger, M., Nesswetter, H., Gerhard, A., Lugli, P., and Zimmermann, C.G. (2017) Measurement of subcell depletion layer capacitances in multijunction solar cells. Appl. Phys. Lett., submitted.

[10]Nesswetter, H., Jost, N.R., Lugli, P., Bett, A.W., and Zimmermann, C.G. (2014) Determination of subcell IV parameters by a pulsed suns-Voc method including optical coupling. 3–6.

[11]Homburg, O., Mitra, T., and Aschke, L. (2010) Overview: process-optimized beam transformers and their impact on high-power laser applications. Laser Reson. Beam Control XII, Proc. SPIE, 7579 (chapter 2), 75790O.

[12]DEWE-5000 Datasheet. DEWETRON GmbH, Austria, 1–4.

[13]Geisz, J.F., Steiner, M.A., Garcia, I., France, R.M., McMahon, W.E., Osterwald, C.R., and Friedman, D.J. (2015) Generalized Optoelectronic Model of Series-Connected Multijunction Solar Cells. IEEE J. Photovoltaics, 5 (6), 1827–1839.

[14]Rutzinger, M., Nesswetter, H., Lugli, P., Bett, A.W., and Zimmermann, C.G. (2016) Determination of subcell open circuit voltages and Iph-Voc curves in multijunction solar cells by sequentially pulsed, monochromatic illumination. Appl. Phys. Lett., 108 (25), 253902.

[15]Rutzinger, M., Salzberger, M., Nesswetter, H., Gerhard, A., and Zimmermann, C.G. (2016) Determination of subcell capacity in multijunction solar cells by pulsed laser illumination. 251106 (2), 251106.

[16]Sze, S.M. (2007) Physics of semiconductor devices, Wiley-Interscience.

[17]Khan, A., Yamaguchi, M., Bourgoin, J.C., and Takamoto, T. (2002) Thermal annealing study of 1 MeV electron-irradiation-induced defects in n+p InGaP diodes and solar cells. J. Appl. Phys., 91 (4), 2391–2397.

ANALYSIS OF CURRENT GENERATION IN InGaP/GaAs/Ge TRIPLE JUNCTION SOLAR CELLS WITH OPTICALLY NON-UNIFORM LUMINESCENCE COUPLING EFFECT

Bernice Mae F. Yu Jeco[1,2], Katsuhisa Yoshida[1], Ryo Tamaki[1], Nazmul Ahsan[1,2] Yoshitaka Okada[1,2]
1. Research Center for Advanced Science and Technology (RCAST), The University of Tokyo
2. Department of Advanced Interdisciplinary Studies, The University of Tokyo
4-6-1 Komaba, Meguro-ku, Tokyo 153-8904, Japan

ABSTRACT: The luminescence coupling (LC) effect in multijunction solar cells can reduce the current mismatch between subcells and improve the cell performance. The laser beam induced current (LBIC) mapping has revealed, however, that the LC current distribution in a limiting subcell of InGaP/GaAs/Ge triple junction solar cells is spatially mismatched as well. In this paper, the area dependence of the spatial LC current mismatch is evaluated by LBIC mapping and electro-optical modeling. This dependence is evaluated by taking the difference between the quantum efficiencies from the LBIC map fitted into the electro-optical model and by assuming uniform LC current distribution. Quantum efficiency differences obtained were 0.38% and 2.55% for cell areas of 0.31 cm^2 and 0.99 cm^2 having maximum LC factors of 0.69 and 0.61, respectively. These results suggest that larger LC photocurrent generation can be achieved by making it uniform – that it is made spatially matched. When spatial LC current matching is achieved, the conversion efficiency of multijunction solar cells may be further increased.
Keywords: current mapping, luminescent coupling, multijunction solar cell, optical modeling, photon recycling

1 INTRODUCTION

Various ways to increase the power conversion efficiency of multijunction solar cells (MJSCs) have been explored [1]-[2]. While radiative recombination is one of the unavoidable losses in MJSCs, it is possible to reuse the emitted light that this process brings about. This reuse can be categorized as either photon recycling or luminescence coupling (LC) [3]. Photon recycling in MJSCs involves re-absorption of photons within the same junction [4]. On the other hand, the LC effect involves re-absorption of photons emitted from a higher bandgap subcell to a lower bandgap subcell.

The LC effect has been estimated to increase the MJSC conversion efficiency by 1.0% under high sunlight concentration [5]. Moreover, this could lead to reduction of current mismatch between these subcells [6]-[7]. Current mismatch reduction is very important for monolithic MJSCs. It is because whichever subcell produces the least current dictates the output current of this device. Aside from current mismatch among the subcells, the spatial distribution of generated LC current in InGaP/GaAs/Ge triple junction solar cells (3JSCs) is observed to be non-uniform [8], indicating spatial current mismatch. This spatial mismatch possibly originates from radiative emission photons emitted towards the sides of the cell or from their internal reflection [9].

Although in previous works this has been observed for one size of 3JSC [8]-[9], it is unknown yet whether or not this is also observable for other sizes of 3JSCs. In this study, the area dependence of GaAs-to-Ge subcell LC current generation is evaluated by laser beam induced current (LBIC) mapping and analyzed by quasi-2-dimensional (Q2D) electro-optical model.

In the Q2D model, the LC factor and the fraction of non-uniformity were fitted to reproduce the LBIC maps acquired from 3JSC samples having different sizes. Here, the LC factor, α_{LC}, represents the number of radiative emission photons from one subcell successfully converted to LC current in an adjacent lower bandgap subcell. The fraction of non-uniformity, β_{LC}, represents the portion of the local LC current subjected to spatial mismatch. From these parameters, the local net LC current, $J_{LC,net}(x,y)$, is calculated. Knowing the $J_{LC,net}(x,y)$, the mean quantum efficiencies (QEs) of LC current generation are calculated from the fitted LBIC maps considering uniform ($\beta_{LC} = 0$) and non-uniform ($\beta_{LC} \neq 0$, fitted) LC current generation at the Ge bottom cell. Then, the absolute QE difference is derived by subtracting these mean QEs. The mean QE difference is important to know since this may provide an insight whether or not the LC effect would be beneficial for MJSCs of any cell area. Moreover, estimating this can help determine whether or not it is worth addressing the spatial LC current mismatch. If the mean QE is found to be non-negligible, then addressing this can serve as an additional consideration for future improvements of MJSC structures, given that the recent records are yet to reach 50% [2]. Calculations have revealed that the QE difference for a cell area of 0.31 cm^2 having a maximum LC factor of 0.69 is 0.38%. Meanwhile, for a cell area of 0.99 cm^2 having a maximum LC factor of 0.61, the QE difference is 2.55%. These results suggest that for a larger cell area, current non-uniformity becomes more significant. Therefore, finding ways to make the LC current generation more uniform, most especially for larger cell areas, can improve the QE. In turn, the MJSC conversion efficiency may also improve, which is targeted to exceed 50% for achieving grid parity.

2 METHODOLOGY

2.1 Experimental characterization

Commercial monolithic InGaP/GaAs/Ge triple junction solar cells (3JSCs) whose areas are 0.31 cm^2 and 0.99 cm^2 were characterized. In this report, these samples are alternatively referred to as Sample S and Sample L, respectively. For acquiring the spatial LC current generation due to the radiative emission from GaAs middle cell towards the Ge bottom cell, continuous 430-nm and 660-nm LED biases were employed. These LEDs bias the InGaP top cell and the GaAs middle cell. In the setup, the 430-nm LED bias intensity was fixed at 47.5 mW/cm^2. On the other hand, the 660-nm LED bias was varied from 0 mW/cm^2 up to 290 mW/cm^2. This is to vary the GaAs middle cell bias, which in turn varies the GaAs-to-Ge subcell LC factor [5]. Then a pulsating 785-

nm scanning laser was activated to excite the GaAs middle cell. Since the Ge bottom cell was not excited by any light bias, this subcell is limiting the current generation of the 3JSC. Thus, the LBIC maps obtained for cell areas of 0.31 cm^2 and 0.99 cm^2 are considered to show only the photocurrent generated by LC effect. In this setup, the spatial resolution is 23.4 × 20.9 pixels/mm^2.

2.2 Simulation model for non-uniform LC effect analysis

To analyze the measurement results, a Q2D electro-optical model has been developed. Here, the electrical circuit model representing the material properties of the 3JSC samples and the illumination profile of the LC effect were modeled in 2 and 3 dimensions, respectively [9].

The most straightforward definition of LC current generation [7], J_{LC}, is given by

$$J_{LC} = \alpha_{LC,i-1\to i} J_{rad,i-1} \tag{1}$$

where $\alpha_{LC,i-1\to i}$ is the LC factor. It is acquired by light current-voltage (*I-V*) characteristics fitting [5]. This describes the efficiency of converting the photons emitted due to radiative recombination from a higher bandgap or the (i-1)-th subcell into current in an adjacent lower bandgap or the i-th subcell. $J_{rad,i-1}$ is the radiative recombination current at the (i-1)-th subcell, defined by ideal diode equation for solar cells. This definition has been extended into higher dimensions by the equations discussed hereafter, considering the non-uniform profile of LC effect.

The illumination profile of the LC effect in MJSCs was assumed to take a superellipse distribution [8], mathematically defined as,

$$z_{LC}(x,y) = \frac{s^2x^2y^2}{k^4} - \left(\frac{x^2}{k^2} + \frac{y^2}{k^2}\right) + 1 \tag{2}$$

in Cartesian coordinates where s is the sharpness of superellipse corners [10], x and y are the in-plane current map positions and k is defined as half of the side length of the cell sample. For a square or rectangular-shaped solar cell, s is set to 1. This definition follows the criterion $k \geq \pm\, x,\ y$. Then the net local LC current, $J_{LC,net}(x,y)$ becomes

$$J_{LC,net}(x,y) = [\beta_{LC} z_{LC}(x,y) + (1-\beta_{LC})] J_{LC}(x,y) \tag{3}$$

where β_{LC} is defined as the fraction of non-uniformity of the original local LC current, $J_{LC}(x,y)$. Here, β_{LC} is obtained by fitting the LBIC maps into the Q2D model. Simulated variation of β_{LC} is shown in Fig. 1. Here, the LC factor was set to 0.69, which is the maximum α_{LC} fitted from the light *I-V* characteristics of Sample S. As the β_{LC} is increased, the current distribution becomes less uniform. Physically, this non-uniformity may be attributed to photons emitted from the upper bandgap subcell which pass through the escape cone sidewards or subjected to total internal reflection [9].

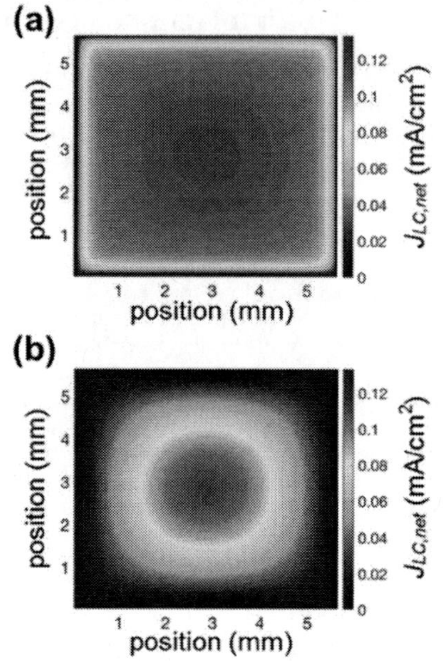

Figure 1: Color map representation of luminescence coupling (LC) current, $J_{LC,net}$, whose fractions of non-uniformity, β_{LC}, are set to (a) 10% and (b) 100%, respectively. The α_{LC} for both LC current distributions is 0.69, which is the maximum acquired for 0.31 cm^2 sample. The spatial resolution of the maps is 23.4 × 20.9 pixels/mm^2 and the applied InGaP/GaAs/Ge triple junction solar cell terminal voltage is 1.5 V.

2.3 Current reduction analysis

For this study, the mean QE of generated LC current, *mLCQE*, has been defined as the averaged ratio of LC current to the incident current, J_{in}, derived from the incident AC power density of the excitation laser. Mathematically, *mLCQE* is given by

$$mLCQE = \frac{\iint J_{LC,net}(x,y)\,dxdy}{J_{in} A_{cell}} \tag{4}$$

where A_{cell} is the cell area and $dxdy$ is the differential area of the InGaP/GaAs/Ge 3JSC sample characterized. Here, larger *mLCQE* indicates larger number of radiative emission photons induced by the input AC power successfully converted to electrons produced by the LC effect. Then the absolute QE difference of generated LC current, Δ*mLCQE*, is calculated using

$$\Delta mLCQE = mLCQE_{uniform} - mLCQE_{fit} \tag{5}$$

where $mLCQE_{uniform}$ and $mLCQE_{fit}$ are obtained from fitted 3JSC *I-V* characteristics considering uniform (β_{LC} = 0) and non-uniform ($\beta_{LC} \neq 0$, fitted) LC current generation in the limiting Ge bottom cell, respectively. Here, larger Δ*mLCQE* indicates larger number of uncollected photons emitted by an upper bandgap subcell towards its adjacent lower bandgap subcell, thereby causing spatial LC current mismatch.

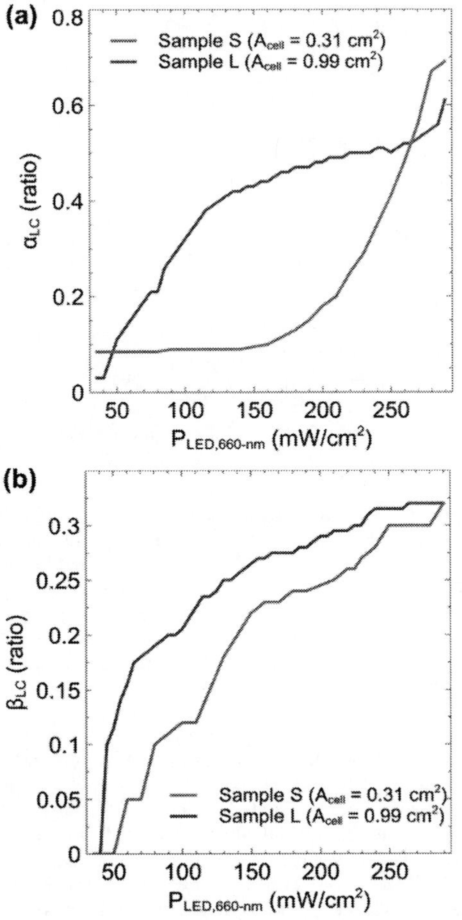

Figure 2: (a) GaAs-to-Ge subcell luminescence coupling (LC) factor, α_{LC}, and (b) fraction of non-uniformity, β_{LC}, as a function of 660-nm LED intensity, $P_{LED,660\text{-}nm}$.

3 RESULTS AND DISCUSSION

3.1 Analysis of luminescence coupling factor and fraction of non-uniformity

To determine how the LC factor varies upon increasing the incident light bias absorbed by an adjacent upper bandgap subcell, the LC factor, α_{LC}, of 0.31 cm^2 and 0.99 cm^2 samples are plotted against increasing 660-nm LED intensity. This plot is shown in Fig. 2(a). As mentioned in Section 2.1, varying the 660-nm LED intensity varies the LC factor, that is, higher α_{LC} indicates larger LC current generation due to radiative emission from the adjacent GaAs middle cell. For both samples, the α_{LC} increases with incident light bias on the GaAs middle cell. This agrees well with the previous findings reported [5],[8].

Also in Fig. 2(a), the difference in the LC factor trend for the samples being studied is apparent. This may be attributed to their differences at which LED intensity will the leakage current from the adjacent GaAs middle cell subjected to low 660-nm LED intensities dominate over LC current generation [11]. This is indicated by different 660-nm LED intensities at which the minimum LC factors are obtained. For Sample S (0.31 cm^2), the

minimum α_{LC}

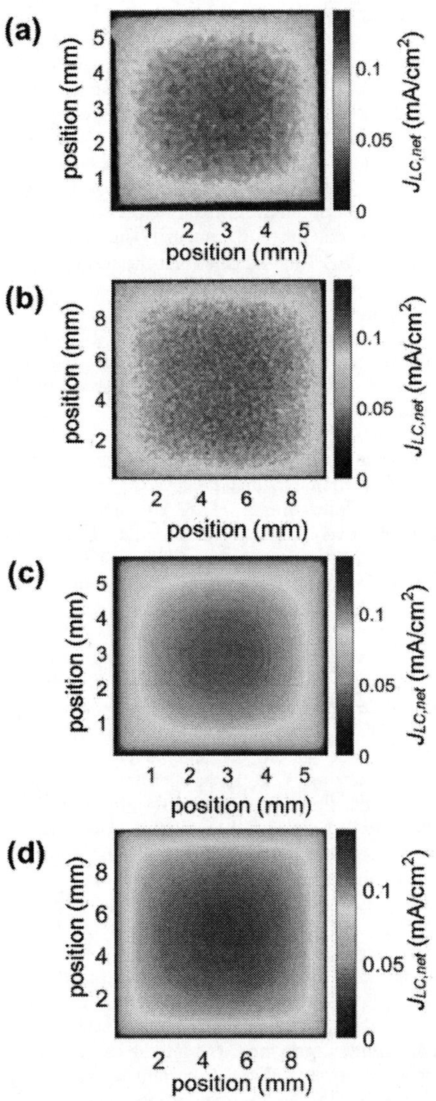

Figure 3: LBIC maps of photocurrent generated by luminescence coupling (LC) effect, $J_{LC,net}$, at the Ge bottom cell of InGaP/GaAs/Ge triple junction solar cell whose areas are (a) 0.31 cm^2 and (b) 0.99 cm^2, respectively. LC current maps of (c) 0.31 cm^2 and (d) 0.99 cm^2 samples reproduced by fitting α_{LC} and β_{LC} into the quasi-2-dimensional model. The LC factors for both samples are at maximum (α_{LC} = 0.69 for 0.31 cm^2 and α_{LC} = 0.61 for 0.99 cm^2 sample) and β_{LC} = 0.32 for both samples. The spatial resolution is 23.4 × 20.9 pixels/mm^2 and the applied terminal voltage is 1.5 V for all maps shown.

obtained is 0.085 at 660-nm LED intensities from 35 to 85 mW/cm^2. On the other hand, for Sample L (0.99 cm^2), the minimum α_{LC} obtained is 0.03 at 660-nm LED intensities from 35 to 40 mW/cm^2. It should be noted that α_{LC} = 0 cannot be obtained due to LBIC measurement artifacts [9]. The leakage current dominance of Sample S that manifested at longer 660-nm LED intensity range may be attributed to larger edge-active region ratio than that of Sample L.

To know the variation of the spatial mismatch of generated LC current upon increasing the incident light bias absorbed by an adjacent upper bandgap subcell, the fraction of non-uniformity, β_{LC}, of 0.31 cm^2 and 0.99 cm^2 samples are plotted against increasing 660-nm LED intensity. This plot is shown in Fig. 2(b). As mentioned in Section 2.2, lower β_{LC} indicates better spatial LC current matching. For both samples, it can be seen that although increasing the 660-nm LED intensity increases α_{LC}, it also increases β_{LC}. This indicates that the stronger the LC effect is, the less uniform the spatial current generation will be. Meanwhile, the value of β_{LC} obtained for both samples have similar trend upon increasing the 660-nm LED bias intensity. From this trend, it can be inferred that the degree of spatial LC current mismatch is independent of the cell area.

3.2 Comparison of the spatial luminescence coupling current distribution for InGaP/GaAs/Ge triple junction solar cell of different areas

To qualitatively compare the amount of uncollected photons emitted by the GaAs middle cell towards the limiting Ge bottom cell, the LBIC maps were acquired using the settings defined in Section 2.1. Figure 3(a) and 3(b) show the acquired GaAs-to-Ge subcell LC current maps from cell areas of 0.31 cm^2 and 0.99 cm^2, respectively. These were acquired at 660-nm LED intensity of 290 mW/cm^2 and an applied MJSC terminal voltage of 1.5 V, which is within the operating region of the 3JSC for Ge bottom cell limited case. Comparing these LC current maps, it can be seen that the larger the cell area is, the longer the region of current uniformity degradation will be. For cell areas of 0.31 cm^2 and 0.99 cm^2, the steep degradation of LC current uniformity occurs within approximately 1 mm and 2 mm, respectively. As reported previously, this degradation towards the cell edges may possibly originate from photons emitted from the GaAs middle cell escaping through the sides of the cell or subjected to total internal reflection [9]. These results qualitatively show that for larger cell areas, there are more uncollected photons which are emitted from the GaAs middle cell towards the cell edges. As a result, less LC current towards the cell edges is produced at the limiting Ge bottom cell.

To analyze the spatial LC current mismatch observed by LBIC mapping, the LC factor, α_{LC}, and the fraction of non-uniformity, β_{LC}, are fitted into the Q2D electro-optical model. The fitted LC current maps for 0.31 cm^2 and 0.99 cm^2 samples at maximum LED intensities indicated in Section 2.1 are shown in Figs. 3(c) and (d), respectively. Here, the estimated LC factors for Samples S and L are 0.69 and 0.61, respectively, and fitted β_{LC} value is 0.32 for both samples at this condition. Comparing the fitted current maps with Figs. 3(a) and (b), the Q2D electro-optical model is able to replicate the LBIC maps acquired from Samples S and L. This suggests that this model is suitable for analyzing the non-uniform LC effect.

3.3 Analysis of absolute quantum efficiency reduction due to non-uniform luminescence coupling effect

To quantify the number of photons emitted by GaAs middle cell successfully converted into LC current in the limiting Ge bottom cell, the mean QE of GaAs-to-Ge subcell LC current was calculated using equation (4).

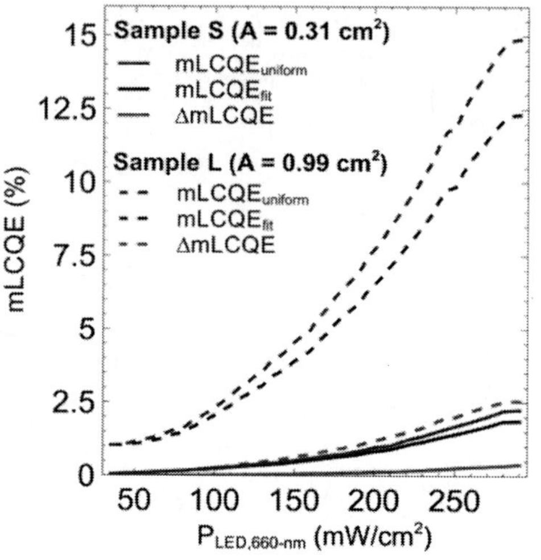

Figure 4: Mean quantum efficiencies derived from non-uniform ($mLCQE_{fit}$, black solid line and black dotted line, fitted β_{LC}) and uniform ($mLCQE_{uniform}$, blue solid line and blue dotted line, $\beta_{LC}= 0$) LC current distribution and their absolute differences, $\Delta mLCQE$, (red solid line and red dotted line) as a function of 660-nm LED intensity, $P_{LED,660\text{-}nm}$ (for cell areas of 0.31 cm^2 and 0.99 cm^2, respectively).

Figure 4 shows the mean QE derived from non-uniform (black solid line and black dotted line) and uniform (blue solid line and blue dotted line) LC current distribution plotted against increasing 660-nm LED intensity (for Samples S and L, respectively). The β_{LC} values used upon deriving the mean QEs of non-uniform and uniform LC current distributions are determined by fitting the LBIC measurements and is set to 0, respectively. For maximum LC factors acquired for Sample S ($\alpha_{LC} = 0.69$) and Sample L ($\alpha_{LC} = 0.61$), the mean QE estimated for non-uniform LC current distribution were 1.87% and 12.32%, respectively. As for the uniform LC current collection, the mean QE estimated are 2.25% and 14.88%, respectively. From these values, it is inferred that the radiative emission photon to electron conversion in the larger cell by LC effect is more efficient than in the smaller cell. This is possibly because the samples belong to different wafers, consequently having Sample L to generate an integrated net LC current of 4.02 A/cm^2, which is about 3.2 times larger than that of Sample S (Integrated LC current = 1.26 A/cm^2).

To compare the uncollected photons emitted from GaAs middle cell towards the limiting Ge bottom cell, the absolute QE differences, $\Delta mLCQE$, considering uniform ($\beta_{LC}= 0$) and non-uniform ($\beta_{LC}\neq 0$, fitted) LC current for Samples S and L were obtained using equation (5). The $\Delta mLCQE$ for Samples S and L are plotted in red solid line and red dotted line in Fig. 4, respectively. Although higher 660-nm LED intensity yields higher mean QE, $\Delta mLCQE$ also increases. At incident 660-nm intensity of 290 mW/cm^2, the values of $\Delta mLCQE$ are 0.38% and 2.55% for 0.31 cm^2 and 0.99 cm^2 samples, respectively. These calculations denote that the stronger the LC effect is, which manifests at high bias intensities of the upper bandgap subcell, the larger the number of uncollected photons emitted by GaAs middle cell and carrier generation towards the center of the sample by internal reflection will be. Nevertheless, these calculations suggest that it will help MJSCs to achieve better current generation by addressing the spatial LC current mismatch, most especially for MJSCs having larger active areas. This could be possibly addressed by implementing light trapping structure on the sides of the cell to prevent photon escape. This, in turn, may improve the MJSC conversion efficiency.

4 CONCLUSION

In summary, the area dependence of GaAs-to-Ge subcell LC current generation has been evaluated by LBIC mapping and analyzed using Q2D electro-optical model. From this study, the following have been observed. First, the Q2D model reproduced well the spatial distribution of produced LC current, suggesting that this model is appropriate for analyzing non-uniform LC current production. In addition, it was also found that although stronger incident light bias yields higher LC factor, the spatial LC current matching degrades. But then, regardless of the strength of the LC effect, the degree of spatial LC current mismatch is found to be independent of the cell area.

The leakage current dominance and the mean QE of LC current generation have been related to cell area. Larger cell was found to have less significant leakage current from the adjacent higher bandgap subcell (GaAs middle cell) even at low incident light bias intensity. Moreover, larger QE was obtained for larger cell area, indicating better conversion of radiative emission photons to electrons produced by the LC effect. However, it was observed that more photons towards the limiting cell get uncollected, which may have escaped through the sides of the cell. Therefore, if spatial LC current mismatch is solved, it can increase QE up to 2.55% for a cell area of 0.99 cm^2. This in turn may increase the MJSC conversion efficiency towards 50% – the current target for the photovoltaics community.

ACKNOWLEDGMENTS

This work is performed under the Research and Development of ultra-high efficiency and low-cost III-V compound semiconductor solar cell modules supported by National Research and Development Agency, New Energy and Industrial Technology Development Organization (NEDO), and Ministry of Economy, Trade and Industry (METI), Japan.

REFERENCES

[1] R.R. King, D.C. Law, K.M. Edmondson, C.M. Fetzer, G.S. Kinsey, H. Yoon, D.D. Krut, J.H. Ermer, R.A. Sherif, and N.H. Karam, Advances in OptoElectronics, (2007) 75384.

[2] M.A. Green, Y. Hishikawa, W. Warta, E.D. Dunlop, D.H. Levi, J. Hohl-Ebinger, and A.W. Ho-Baillie, Progress in Photovoltaics: Research and Applications, Vol. 25 (2017) 668.

[3] M.A. Steiner, J.F. Geisz, T.E. Moriarty, R.M. France, W.E. McMahon, J.M. Olson, S.R. Kurtz, and D.J. Friedman, IEEE Journal of Photovoltaics, Vol. 3 (2013) 879.

[4] S.M. Durbin and J.L. Gray, IEEE Transactions on Electron Devices, Vol. 41 (1994) 239.

[5] T. Sogabe, A. Ogura, C.Y. Hung, V. Evstropov, M. Mintairov, M. Shvarts, and Y. Okada, Applied Physics Letters, Vol. 103 (2013) 263907.

[6] D.J. Friedman, J.F. Geisz, and M.A. Steiner, IEEE Journal of Photovoltaics, Vol. 3 (2013) 1429.

[7] D.J. Friedman, J.F. Geisz, and M.A. Steiner, IEEE Journal of Photovoltaics, Vol. 4 (2014) 986.

[8] B.M.F. Yu Jeco, T. Sogabe, A. Ogura, N. Miyashita, R. Tamaki, and Y. Okada, Proceedings of 43rd IEEE Photovoltaic Specialists Conference (2016) 1229.

[9] B.M.F. Yu Jeco, T. Sogabe, R. Tamaki, N. Ahsan, and Y. Okada, Journal of Photonics for Energy, Vol. 7 (2017) 035501.

[10] M.F. Guasti, A.M. Cobarrubias, F.R. Carrillo, and A.C. Rodríguez, Optik-International Journal for Light and Electron Optics, Vol. 116 (2005) 265.

[11] J.J. Li, S.H. Lim, C.R. Allen, D. Ding, and Y.H. Zhang, IEEE Journal of Photovoltaics, Vol. 1 (2011), 225.

COMPARISON OF NOVEL OPTIMIZATION TECHNIQUES WITH APPLICATION IN MAXIMIZING TANDEM SOLAR CELLS PERFORMANCES

Sherif Michael, Michael Tsutagawa and Silvio Pueschel
Naval Postgraduate School, ECE Dept, Monterey, CA 93943, USA
michael@nps.edu

Abstract. A new method for developing a realistic and accurate physical model of any type of solid state device has been previously introduced by the author. Application to model advanced multi-junction solar cells; Thermo-photovoltaics; sensors; as well as other novel solid state devices were previously presented [1&2]. The primary goal of multijunction solar cell design is to maximize the output power for a given solar spectrum [3-4]. The construction of multijunction or tandem cells places the individual junction layers in series, thereby limiting the overall output current to that of the junction layer producing the lowest current. The solution to optimizing a multijunction design involves both the design of individual junction layers which produce an optimum output power and the design of a series-stacked configuration of these junction layers which yields the highest possible overall output current and voltage. This paper demonstrates and compares the use of different optimization techniques to achieve that goal. The first approach is to use Genetic Algorithm in a two-part process to refine a given multijunction solar cell design for near-optimal output power for a desired light spectrum. Consequently, a Novel Space-Filling Experimental Design optimization technique is also utilized and comparison of the results is presented. These approaches can similarly be utilized to optimize the parameters of any Solid state device to yield any desired performance.

1. THE MODELING SOFTWARE

Two questions arise concerning solar cell modeling and simulation: 1. Can computer modeling and simulation accurately predict solar cell performance? 2. If so, how can solar cell design then be optimized using modeling and simulation? To address the first question, Silvaco ATLAS was used to model an actual Triple junction InGaP/GaAs/Ge solar cell. To address the second question, a genetic algorithm was used to find a more optimal solution for an objective value of maximum efficiency of the triple junction solar cell model.
The optimization routines described in this paper use a solar cell model developed at the Naval Postgraduate School using the ATLAS device simulator by Silvaco International [5]. This model, previously presented [1&2], accurately predicts the electrical characteristics of a solar cell based on virtual fabrication of its physical structure. As an example, an InGaP/GaAs/Ge triple-junction solar cell is modeled, and the device light-induced photogeneration, along with the corresponding spectral response of each layer are illustrated in Figs 1 & 2 as generated from the modeled cell in ATLAS.

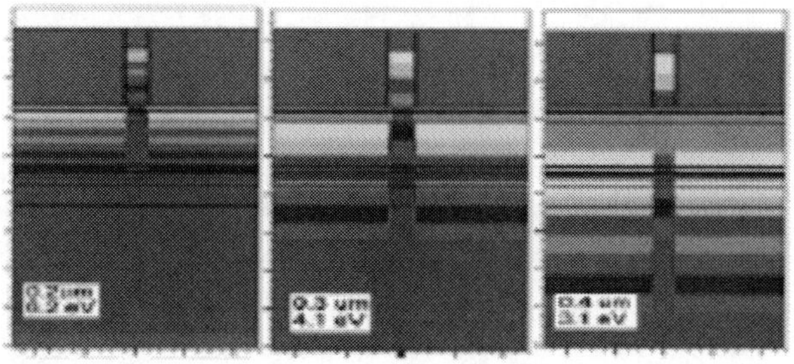

Fig. 1 ATLAS photogeneration rate in a MJ cell.

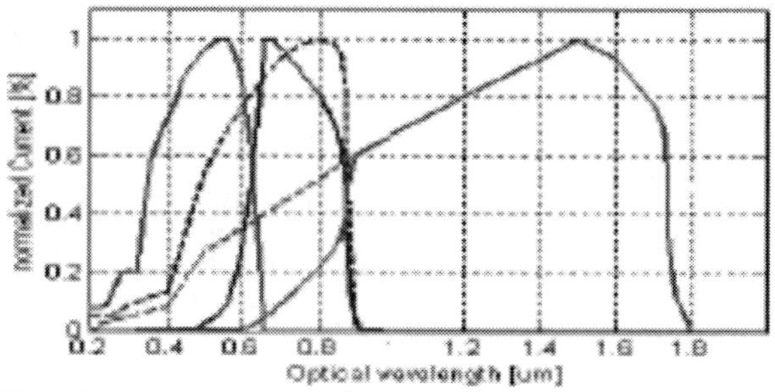

Fig. 2. Spectral response of the individual cells (dashed) and of the stacked cell (continuous).

2. THE GENETIC SEARCH ALGORITHM

The first step in the optimization process is to maximize the output power of each junction layer individually for several layer thicknesses. These optimum configurations will be used to ensure maximum junction layer output power as junction layer thicknesses are changed during the current-matching process. The junction layer optimization process accepts known materials for the window, emitter, base & BSF and determines the ideal thicknesses & dopings for each region. The base thickness was chosen to be a dependent variable to achieve a constant overall junction layer thickness. Thus, seven independent variables remain for each overall junction layer thickness. To search all possible solutions rigorously would require an enormous amount of computational time. Instead, a genetic algorithm [6] was used to search the solution space for the junction layer configuration producing the highest output power. To enact a genetic algorithm for junction layer optimization,

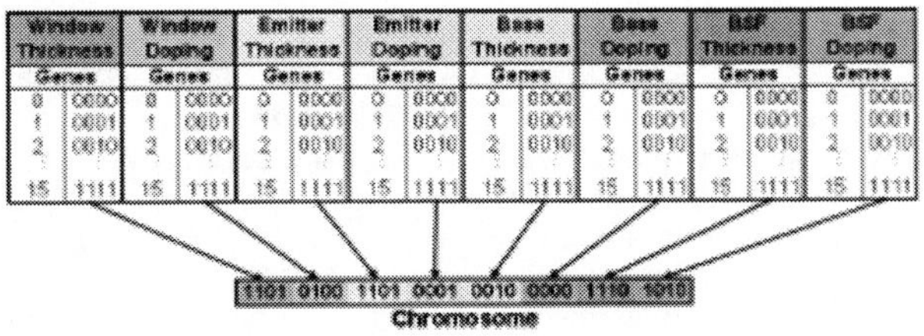

Fig. 3 Chromosome construction from a gene sequence.

each of the eight variable junction layer parameters was encoded into a four-bit binary string.

The encoded binary strings, referred to as genes, were then assembled into 32-bit binary chromosomes (Fig. 3). Each chromosome fully encoded the eight variable properties of a junction layer. A set of 32 randomly selected binary strings made up the initial generation of chromosomes. The encoded properties in each of these chromosomes were used to construct and simulate a junction layer in ATLAS under AM0 illumination. After the simulation of an entire generation of chromosomes, child chromosomes (to make up the next generation) were formed from a mix of the genes from the best performing parent chromosomes. Each parent chromosome was ranked according to the Pmax of its associated junction layer using an elitist roulette wheel selection mechanism. Child chromosomes were formed from parents using a two-point crossover operation (Fig. 4). Bit-inverting "mutations" occurred in each child chromosome according to some small probability to add genetic variety (Fig 5). A relatively small population size was used with recommended crossover and mutation probabilities from [6].

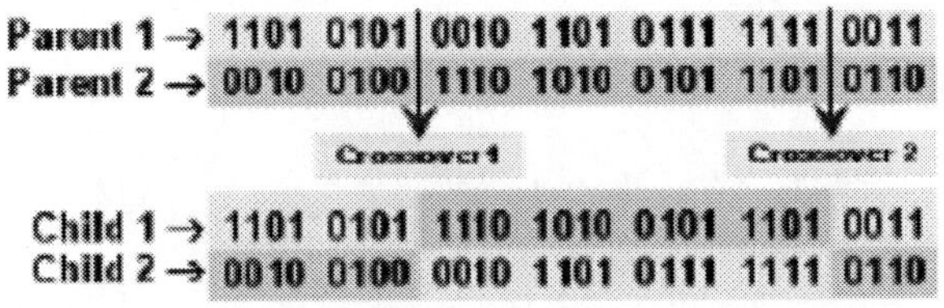

Fig. 4. Two-crossover genetic mixing.

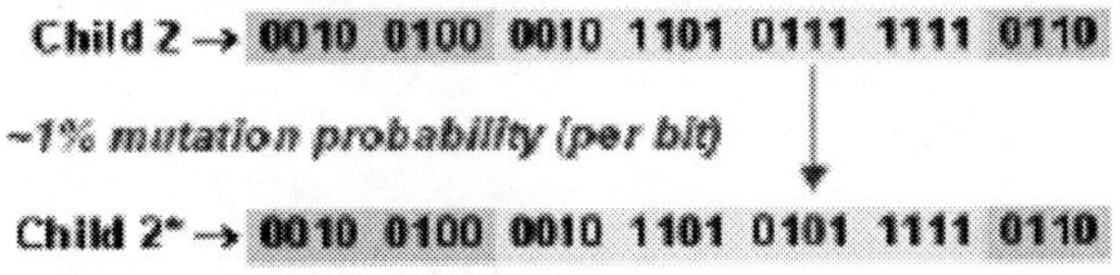

Fig. 5. Bit-inverting mutation.

The genetic search algorithm was allowed to progress for a maximum of 20 generations. This scheme allowed a solution space of over 268 million junction layer designs to be searched to arrive near an optimal junction layer configuration for a specific thickness (Fig. 6). This was repeated for three thickness for each of the three top junction layers (InGaP, GaAs, and InGaNAs).

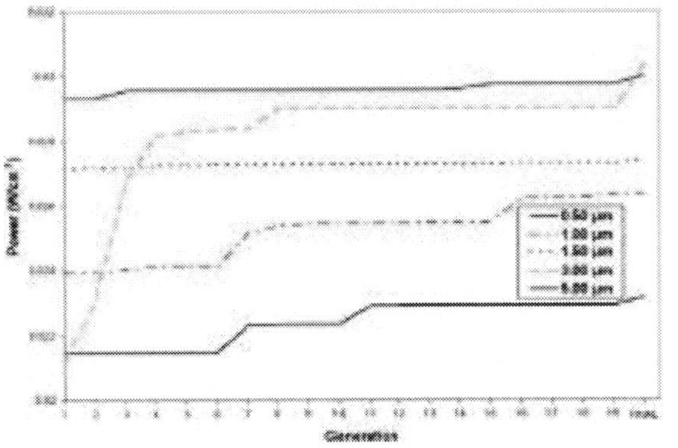

Fig. 6 Power improvement of GaAs junction layer.

3. ITERATIVE CURRENT-MATCHING ROUTINE

To achieve a maximum overall output current from the full multijunction cell, the current produced by each junction layer needed to be matched to the fullest extent possible. To accomplish this, we used an iterative current-matching routine beginning with an ATLAS simulation of the full multijunction cell with each junction layer thickness larger than its estimated final thickness. In this initial state, each junction layer absorbed a large amount of light and little light energy was able to penetrate to each lower junction layer successively.

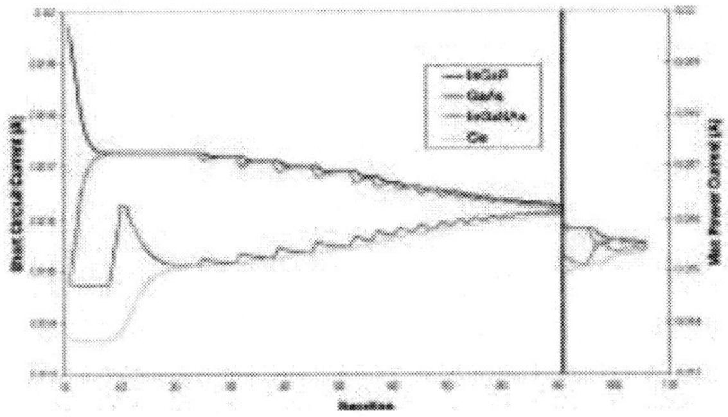

Fig. 7 Progress of iterative current matching.

To match short-circuit currents, junction layers were grouped into pairs from the top junction to the substrate junction (Fig. 7). Short-circuit currents were matched within the pairs by either increasing or decreasing the thickness of the upper junction layer. Then the junction layers were re-grouped with the other adjacent junction layer and the process repeated iteratively until the short-circuit currents of all junction layers were within 99.6% of each other. At this point, the routine changed focus to attempt to match the maximum-power currents of the individual junction layers. The routine ceased when all junction layer maximum-power currents were matched. The optimum cell configuration could then be determined by a review of the output file for the cell with the highest overall maximum power.

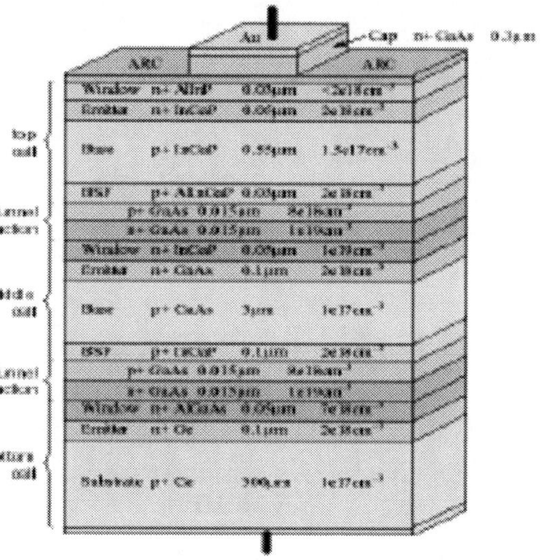

Fig. 8. The new Optimized triple-junction cell structure.

4. THE GA OPTIMIZATION RESULTS

Due to the obvious reasons of propriety, no published reports on the detailed construction of advanced three-junction solar cells could be found. This lack of an experimental baseline complicates any attempt to show the improvement over a tested design obtained by using the optimization procedure proposed in this paper. The advantage of this technique is instead demonstrated using unpublished three-junction solar cell designs. Figures 8&9 illustrate the optimized cell structure and its output, While Table 1 illustrates the achieved improvements.

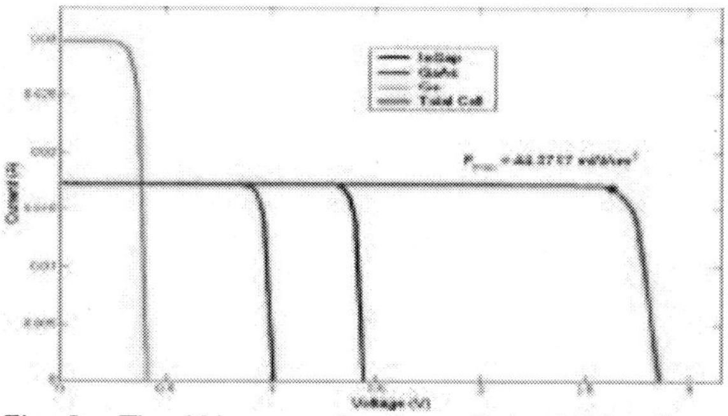

Fig. 9. The I-V curves for the optimized 3-junction cell.

	Layer Material (Layer Thickness)			
	InGaP (0.60μm)	GaAs (1.00μm)	InGaNAs (1.64μm)	Ge (304μm)
P_{max}[3], mW/cm²	22.79	22.61	17.41	13.33
P_{max} optimized, mW/cm²	23.41	28.49	18.75	20.83
% improvement	2.72	26.01	7.70	56.23

Table 1: Resulting improvements after GA optimization.

5. THE NOVEL SPACE-FILLING EXPERIMENTAL DESIGN OPTIMIZATION TECHNIQUE

Next part of the research is to introduce the nearly orthogonal Latin hypercube (NOLH) design of experiments (DoE) as a means for exploring and optimizing solar cell designs in Silvaco Atlas. The general applicability of this approach has been proven, Fig. 10, and compared to the genetic algorithm optimization technique, the following achievements have been made. The pre-generated simulation designs can now be processed in parallel, which drastically reduces the time required to conduct multiple simulations. Moreover, the data generated using the NOLH enabled a better understanding of the simulation input/output relationship and helped to focus the solar cell development by highlighting the design parameters that matter most. Using stepwise regression to build a metamodel helped in finding an optimal design and revealing the interactions among the input parameters. The initial simulation has already yielded promising results as presented in Table 2. below, and has clearly shown the preeminence of the NOLH over the genetic algorithm by identifying a design with greater than 21% more power output than in previous designs. The NOLH DoE should become the new standard for optimizing solar cells in Silvaco Atlas.

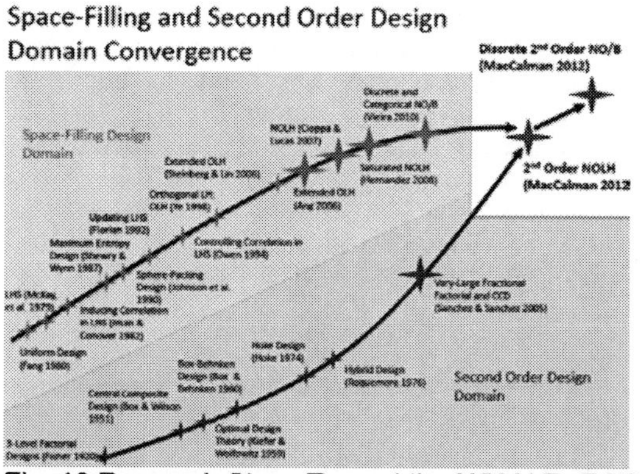

Fig. 10 Research Steps Toward the NOLH DoE [9]

Design parameter	Basis Design	Lin basis	Lin stacked	Second iteration	Third iteration
wnthick	0.33	0.011	0.012	0.007	0.00134
windopconc	2.00e+18	1.0e+19	8.72e+18	5.68e+18	6.13E+18
emthick	0.05	0.024	0.067	0.025	0.0328
emdopconc	2.00e+18	2.42e+18	6.79e+18	3.2e+17	1.50E+17
basethick	0.56	0.343	0.338	0.284	0.357
basedopconc	1.50e+17	3.3e+17	2.4e+17	8.72e+18	4.30E+18
bsfthick	0.03	0.015	0.039	0.034	0.026
bsfdopconc	2.00e+18	6.4e+18	1.65e+18	2.9e+17	1.10E+17
bufthick	0.03	0.04688	0.01391	0.044	0.02746
bufdopconc	1.00e+18	6.75e+18	6.94e+18	8.28e+18	9.44E+18
tunemthick	0.015	0.019	0.013	0.011	0.0131
tunemdopconc	8.00e+18	6.75e+19	3.87e+19	3.8e+17	7.85E+18
tunbasethick	0.015	0.013	0.016	0.02	0.03
tunbasedopconc	1.00e+19	8.35e+18	7.41e+19	5.29e+18	8.00E+18
botwnthick	0.05	0.036	0.021	0.016	0.0106
botwndopconc	1.00e+19	1.77e+18	9.46e+19	1.21e+18	2.44E+18
botemthick	0.1	0.079	0.06	0.126	0.1222
botemdopconc	2.00e+18	1.8e+18	1.88e+18	5.7e+17	1.98E+17
botbasethick	3	4.8	4.8	4.2	5.27
botbasedopconc	1.00e+17	4.2e+17	2.1e+17	4.22e+16	1.88E+16
botbsfthick	0.1	0.106	0.097	0.127	0.1159
botbsfdopconc	2.00e+18	3.7e+17	9.39e+18	1.4e+17	4.72E+17
botbufthick	0.3	0.422	0.325	0.368	0.3422
Botbufdopconc	7.00e+18	3.61e+18	1.15e+19	1.09e+19	4.01E+19
Max power output	35.76	38.58	40.38	42.87	42.86

Table 2: Resulting improvements of the design utilizing the NOLH optimization algorithm.

6. CONCLUSIONS

The combination of the genetic and iterative algorithms presented in this paper and the ATLAS device simulator show great promise for streamlining the design of optimized multijunction solar cells. The complex interactions between the various thicknesses and dopings of the materials available to the solar cell designer make determination of an absolute optimum configuration difficult. The genetic search has shown the ability to search an extremely large solution space to arrive at a near optimal solution without user interaction. The routine is scalable to any large population size and number of generations limited only by computational time. The results of another novel optimization technique, namely the NOLH approach is also presented and comparison of the advantage of each approach is demonstrated and a complete comparison will be introduced in a future presentation.

7. REFERENCES

1. M. Baudrit and C. Algora, "Modeling of GaInP/GaAs Dual Junction Solar Cells Including Tunnel Junction," Proceedings of the 33rd IEEE Photovoltaic Specialists Conference, San Diego, California, May 11–16, 2008.

2. S. Michael and P. Michalopoulous, "A New Technique for the Development of State-of-the-Art Photovoltaic Devices using Silvaco Software," Proceedings of the 6th WSEAS International Multi-conference on Circuits, Systems, Communications and Computers(CSCC 2002), Crete, Greece, July 7-14, 2002. pp. 4121-4125.

3. S. Michael, A. Bates and M. Green, "SILVACO ATLAS as a Solar Cell Modeling Tool," Proceedings of the 31st IEEE Photovoltaic Specialists Conference, Jan 3-7, 2005, Orlando, FL, pp.719-721.

4. S. Michael and M. Tsutagawa, "Improving the Efficiency of Advanced Multi-junction Solar Cells; A new Approach for Design and Optimization," Proceedings of the 2008 Renewable Energy Conference, RE2008, Pusan, Korea, October 12-17, 2008.

5. T. Agui, T. Takamoto, E. Ikeda, and H. Kurita, "High-efficient dual-junction InGaP/GaAs solar cells with improved tunnel interconnect," Indium Phosphide and Related Materials, 1998 International Conference on, pp 203-206, 1998.

6. I. Vurgaftman, J.R. Meyer, L.R. Ram-Mohan, "Band Parameters for III-V Compound Semiconductors and Their Alloys," Journal of Applied Physics, Applied Physics Review, Vol. 89, No. 11, pp. 5840–5841, 2008.

7. K.F. Man, K.S. Tang and S. Kwong, "Genetic Algorithms: Concepts and Applications," IEEE Transactions on Industrial Electronics, Vol. 43, No. 5, pp. 519–534, 1996.

8. A. De Vos, "Detailed Balance Limit of the Efficiency of Tandem Solar Cells," Journal of Applied Physics, Vol. 13, No. 5, May 14, 1980, pp.839–846.

9. S. Püschel, "Optimization of an Advanced Multi-Junction Solar-Cell Design for Space Environments (AM0) Using Nearly Orthogonal Latin Hypercubes," MSEE Thesis, June, 2017.

TEMPERATURE-DEPENDENT PROPERTIES OF AN INVERTED METAMORPHIC FOUR-JUNCTION (IMM-4J) SOLAR CELL

H Zhang, R B Liu, L G Wang, Q Sun, Z B Xiao, Q M Zhang, P Gao, C Xue, M X Jiang, L Shi, Y Tang, L Y Yao, and H Wang
Tianjin Institute of Power Source, Tianjin 300381, PR China

ABSTRACT: The inverted metamorphic four-junction (IMM-4J) solar cell has demonstrated efficiency as high as 33.9% under AM0 at 25 °C. However, the actual operating temperature in space is likely to be as much as 50-80 °C hotter. In order to evaluate the cell performance under the real space operating conditions, we have conducted a measurement of current-voltage characteristics for the IMM-4J cell and the corresponding single-junction subcells as a function of temperature, and calculated the temperature coefficients of the open-circuit voltage, short-circuit current density and fill factor. The Voc temperature coefficient of the IMM-4J cell is the sum of the Voc coefficients of the individual subcells. Jsc for GaInP and GaAs subcells increase linearly with temperature, but Jsc-T dependences for both $Ga_{0.76}In_{0.24}As$ and $Ga_{0.47}In_{0.53}As$ subcells represent the decreasing trend. Besides, Jsc(T) of the individual subcells indicates a transition from GaInP-subcell-limited to $Ga_{0.76}In_{0.24}As$-subcell-limited behavior at 40 °C. It is the transition of Jsc(T) that makes the FF of the IMM-4J cell show the decreasing trend as the temperature increases. Therefore, by adjusting the current of subcells to achieve current-matching and avoiding the GaInAs-subcell-limited behavior for high temperature operation, the IMM-4J cell can potentially produce more energy and lower temperature sensitivity under real operating conditions.
Keywords: inverted metamorphic four-junction (IMM-4J) solar cell, temperature coefficient, current-voltage characteristics

1 INTRODUCTION

The realization of the optimum bandgap combination of the multijunction solar cell for a given solar spectrum can be attained using materials with different lattice constants. Compared with the tandem cells combined through mechanical stacking or wafer bonding, the inverted metamorphic multi-junction (IMM) solar cells pursues expanding the range of bandgaps achievable in multijunction solar cells by using different lattice constants in the same monolithic stack. The focus of IMM performance advancement has shifted from the three-junction IMM (IMM-3J) to IMM-4J device architecture over recent years. To date IMM-4J solar cells have demonstrated the air mass zero (AM0) conversion efficiencies of ~34.2%[1].

IMM-4J solar cell technology is under development, and is considered to be a promising candidate for space applications. However, a key requirement of solar cells intended for space applications is the ability to withstand exposure to the harsh environmental conditions. Besides high-energy particles radiation, there is a considerable range of temperatures that may be encountered for the space use of solar cells. As the spacecraft travels in the orbit, solar cells will experience very large and rapid temperature changes. Typically, in the orbit about earth, the operating temperature of the solar cells is from about 20°C to as high as 85°C [2]. In general, higher operating temperature causes the open-circuit voltage (Voc) to decrease at approximately 2 mV/°C, yet the short-circuit current (Isc) increases slightly due to the rate of diffusion increasing [2]. Also, as temperature is increased, the fill factor (FF) of the I-V curves decreases. Therefore, the overall effect is a reduction in power of a solar cell with an increase in temperature. This change coupled with rapid variations in temperature associated with eclipses can result in power surges that may be problematic.

In this paper, an IMM-4J solar cell, whose structure is GaInP(1.9 eV)/GaAs(1.4 eV)/$Ga_{0.76}In_{0.24}As$(1.0 eV)/$Ga_{0.47}In_{0.53}As$(0.7 eV), and the single junction structures representing each of the IMM-4J subcells have been fabricated using Metalorganic Chemical Vapor Deposition (MOCVD) growth method. In order to study the temperature dependence of IMM-4J solar cells, we conducted a measurement of current-voltage characteristics to show the performance of the IMM-4J cell and the corresponding single-junction subcells as a function of temperature. The focus of this study is a clear understanding of the temperature-induced degradation of the four subcells in the IMM-4J structures.

2 EXPERIMENTS

The typical structure of IMM-4J solar cell shown in Figure.1 was grown in a low-pressure MOCVD reactor on 4-inch GaAs substrate, and processed using standard lithographic technologies applied to an inverted device. During processing, the structure was bonded to a silicon wafer handle and the original GaAs substrate was removed. A bilayer AlxOy/TixOy anti-reflection coating was deposited after fabrication.

To understand the temperature-dependent properties of IMM-4J solar cell, single junction structures representing each of the IMM-4J subcells were fabricated, and the performance of these structures was used to evaluate the viability of the IMM-4J cells. The subcell structures were designed to include the surrounding IMM-4J materials, but the surrounding materials were isotype doped so that they were inactive. Inclusion of the surrounding layers was aimed at producing equivalent optical absorption above the active region, as well as simulating the mechanical strain and thermal loading around the active region. Cell fabrication and characterization of all the subcells were identical with those of the IMM-4J cell.

Table I: AM0 temperature coefficients of Voc, Jsc and FF for the IMM subcells. The bottom row lists the coefficients of the full IMM-4J as derived from the X25 multisource measurements.

Subcell	d(Voc)/dT (mV/°C)	d(Jsc)/dT (μA/cm^2/°C)	d(FF)/dT (°C)
GaInP	-2.16±0.03	12.0±0.8	- (2.64±0.2) ×10^{-4}
GaAs	-2.03±0.02	44.8±2.3	- (6.73±0.1) ×10^{-4}
$Ga_{0.76}In_{0.24}As$	-2.01±0.03	-28.3±1.3	- (1.09±0.05) ×10^{-3}
$Ga_{0.47}In_{0.53}As$	-2.06±0.04	-28.6±1.8	- (3.47±0.20) ×10^{-3}
IMM-4J cell	-8.41±0.06	--	--

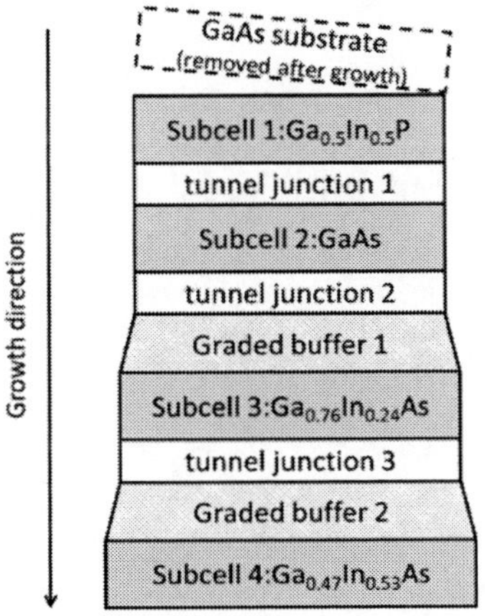

Figure 1: Semiconductor layer structure of the inverted metamorphic four-junction solar cell.

Light current-voltage(J-V) characteristics were measured using simulated AM0 (1367 W/m2) spectrum, and data were collected at nominal ambient temperatures of 25, 40, 55, 70, 80 °C as determined by the thermistor embedded in the copper sample stage.

3 RESULTS AND ANALYSIS

The open-circuit voltages (Voc) of the four single junction subcells varied as a function of temperature are shown in figure 2, and the temperature coefficients of Voc are listed in table 1. Voc for all the four subcells decrease linearly with temperature at a similar rate. The values are consistent with those measured under AM1.5 [3], and the voltage coefficient of the IMM-4J cell whose value of 8.41 mV/°C was derived from the X25 measurement agrees with the sum (8.26 mV/°C) of individual coefficients of the individual subcells.

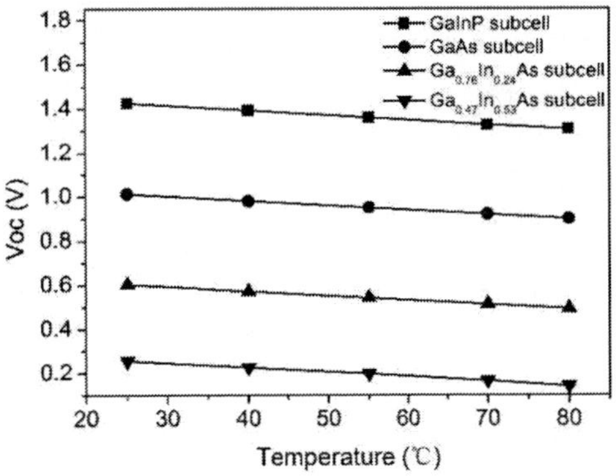

Figure 2: Temperature-dependent Voc of the IMM subcells.

The variations of short-circuit current densities (Jsc) with temperature are shown in figure 3. Jsc for GaInP and GaAs subcells increase linearly with temperature at a rate of 12.0 μA/cm2/°C and 44.8 μA/cm2/°C respectively. However, on the contrary, Jsc-T dependences for both $Ga_{0.76}In_{0.24}As$ and $Ga_{0.47}In_{0.53}As$ subcells represent the decreasing trend, and the decreased rate was 28.3 μA/cm2/°C and 28.6 μA/cm2/°C. The Jsc of GaInP top cell reached 16.95 mA/cm2 at 40 °C, which is still lower than those of other subcells. This suggests that Jsc of the IMM-4J cell is limited by the GaInP top cell until 40 °C. However, with the temperature increasing, the value of Jsc of Ga0.76In0.24As subcell drops rapidly and gradually becomes the minimum. It means when temperature reaches above 40°C, the Jsc of the IMM-4J solar cell becomes Ga0.76In0.24As-subcell-limited and gradually decreases with temperature increasing.

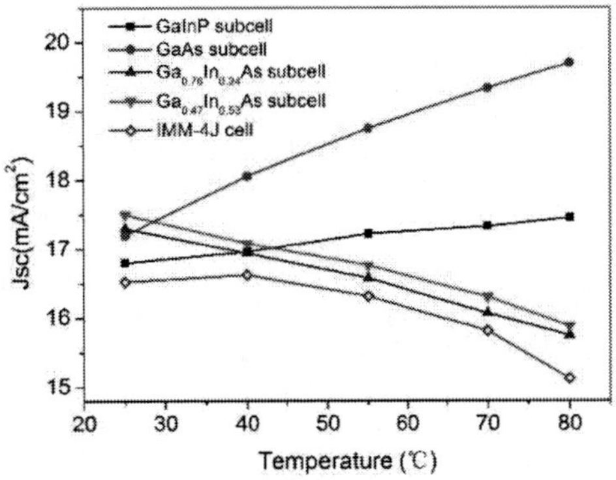

Figure 3: Temperature-dependent Jsc of the IMM subcells and the full IMM-4J cell(open diamond).

The temperature dependence of FF for every subcell and the IMM-4J solar cell is presented in figure 3, and the temperature coefficients of FF are listed in table 1 as well. All the values of FF for the single subcells decrease with temperature. FF for GaInP, GaAs and $Ga_{0.76}In_{0.24}As$ subcells shows a relatively low rate of change with temperature, but FF for $Ga_{0.47}In_{0.53}As$ subcell decreases more quickly with temperature. Furthermore, when the temperature is below 40 °C, because Jsc of the IMM-4J cell is limited by the GaInP top cell, FF of the IMM-4J cell is consistent with that of the GaInP subcell. However, as the temperature increases, the Jsc of the IMM-4J solar cell becomes $Ga_{0.76}In_{0.24}As$-subcell-limited, so FF of the IMM-4J cell decreases rapidly and shows the similar variation trend with that of $Ga_{0.76}In_{0.24}As$ subcell.

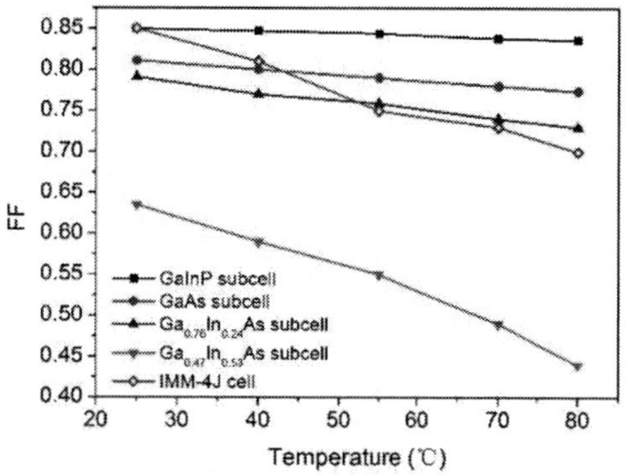

Figure 4 : Temperature-dependent FF of the IMM subcells and the full IMM-4J cell(open diamond).

4 CONCLUSION

$GaInP/GaAs/Ga_{0.76}In_{0.24}As/Ga_{0.47}In_{0.53}As$ IMM-4J solar cell, for which a conversion efficiency of 33.9% (AM0, 25°C) was achieved in our study, has been fabricated using MOCVD growth method. We conducted a series of J-V measurement to show the performance of the IMM-4J cell and the corresponding single-junction subcells as a function of temperature. The Voc temperature coefficient of the IMM-4J cell is the sum of the Voc coefficients of the individual subcells. Jsc for GaInP and GaAs subcells increase linearly with temperature, but Jsc-T dependences for both $Ga_{0.76}In_{0.24}As$ and $Ga_{0.47}In_{0.53}As$ subcells represent the decreasing trend. In addition, Jsc(T) of the individual subcells indicates a transition from GaInP-subcell-limited to $Ga_{0.76}In_{0.24}As$-subcell-limited behavior at 40 °C. It is the transition of Jsc(T) that makes the FF of the IMM-4J cell show the decreasing trend as the temperature increases. Therefore, for the IMM-4J samples in our study, by adjusting the current of subcells to achieve current-matching and avoiding the GaInAs-subcell-limited behavior for high temperature operation, the IMM-4J cell can potentially produce more energy and lower temperature sensitivity under real operating conditions.

REFERENCES

[1] A. B. Cornfeld, P. Patel, J. Spann, D. Aiken, and J. McCarty, "Evolution of a 2.05 eV AlGaInP Top Sub-cell for 5 and 6J-IMM Applications", 38th IEEE Photovoltaic Specialist Conference, Austin, USA (2012), 2788-2791.
[2] A. Luque, and S. Hegedus, Handbook of Photovoltaic Science and Technology. Second Edition, Chichester , West Sussex: John Wiley & Sons Ltd (2011).
[3] M.A. Steiner, J. F. Geisz, D. J. Friedman, W. J. Olavarria, A. Duda, and T. E. Moriarty, "Temperature-dependent measurements of an inverted metamorphic multijunction (IMM) solar cell", 37th IEEE Photovoltaic Specialist Conference, Washington, USA (2011), 2527-2532.

III-V MULTI-JUNCTION SOLAR CELLS UTILISING GROUP IV SIGESN ALLOYS AS A 1.0 EV COMPONENT SUB-CELL

Phoebe Pearce[1], Tom Wilson[1], Andrew Johnson[2], Nicholas Ekins-Daukes[1]
[1] Blackett Laboratory, Department of Physics, Imperial College London, London, SW7 2AZ
[2] IQE plc, Pascal Close, St. Mellons, Cardiff, CF3 0LW

ABSTRACT: Four-junction solar cells can theoretically achieve over 50% efficiency under 1 sun. Optimally efficient four-junction (4J) solar cells require a sub-cell with a bandgap close to 1 eV. Current high-efficiency 4J devices use lattice-mismatched sub-cells to achieve an optimal combination of bandgaps, complicating the fabrication process; thus, novel materials which can achieve a 1 eV bandgap while remaining lattice-matched to Ge are required. Group IV alloy SiGeSn is a promising material system, with functional devices having already been demonstrated. However, further investigation of the fundamental properties of relevant SiGeSn compositions is key to fabricating an efficient 4J device. We present optical measurements (spectroscopic ellipsometry and steady-state photoluminescence) and device characterization of single and triple-junction devices containing SiGeSn sub-cells. Results indicate the desired blueshift of the lowest indirect and direct energy transitions in SiGeSn relative to Ge with the addition of Si and Sn in a composition ratio around 4 to 1. Device measurements indicate good EQE performance in both devices, while the open-circuit voltage is lower than expected based on absorption around 1 eV, indicating the lower indirect edge in SiGeSn limits the voltage.
Keywords: Multijunction Solar Cell, Spectroscopy, Optical Properties

1 MOTIVATION

Multi-junction solar cells are currently the highest-efficiency devices available, with the current record efficiency held by a four-junction (4J) cell [1]. In theory, a 4J device can exceed 50% efficiency without light concentration [2].

An optimally efficient 4J solar cell requires a sub-cell with a bandgap around 1.0 eV. [3]. Currently, the most efficient 4J cells incorporate highly lattice-mismatched 1 eV sub-cells, necessitating the use of inverted metamorphic (IMM) growth [4] or a wafer-bonding process [5], which complicate fabrication and drive up production costs. These processes could be avoided by using a 1.0 eV bandgap material lattice-matched to the other junctions. There is currently a lack of materials which meet these criteria, as shown in Fig. 1. Dilute nitrides [6] have been used, and other options such as dilute bismides have been explored [7]; this work focuses on the Group IV alloy SiGeSn as a possible sub-cell material.

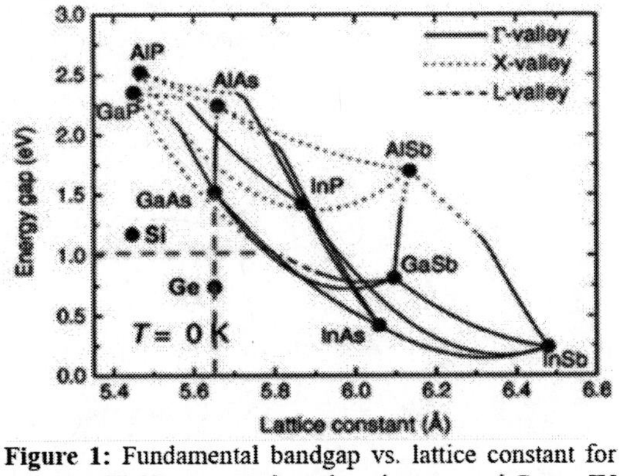

Figure 1: Fundamental bandgap vs. lattice constant for common III-V compound semiconductors, and Group IV Si and Ge. The red lines indicate a bandgap of 1 eV lattice-matched to Ge. Adapted from [8].

Figure 2: Device architecture for a 4J device incorporating a SiGeSn sub-cell.

2 INTRODUCTION

SiGeSn is a Group IV alloy which can be tuned to have a 1 eV direct energy transition while the lattice constant remains equal to that of Ge [9]. Si, Ge and α-Sn (grey tin) are group IV elements with a diamond cubic crystal structure; Si and Ge are semiconductors while α-Sn is a semi-metal (the conduction and valence band have a small overlap).

The SiGeSn alloys of interest are mostly Ge (>85%). Assuming the lattice constants at such compositions can be approximated by linear interpolation between the constituent elements, the ratio of Si:Sn required for lattice-matching to Ge can be predicted using Vegard's law [10]:

$$d_{SiGeSn}(x,y) = (1-x-y)d_{Ge} + xd_{Si} + yd_{Sn}$$

Solving this for $d_{SiGeSn} = d_{Ge}$ gives $x/y = 3.7$, meaning the ratio Si:Sn should be 3.7:1.

While the SiGeSn compositions of interest have a direct transition around 1 eV, the fundamental bandgap remains indirect around 0.8 eV. This bandgap is still 'Ge-like', with the conduction band minimum at the L point, but is blue-shifted relative to Ge. The E_0 transition, between the Γ point maxima and minima in the valence and conduction band, provides the 1 eV direct transition. Adding Si to Ge pushes up the lowest-energy transitions, while adding Sn is necessary to maintain the Ge lattice-match. Table I lists fundamental optical transition energies and lattice constants of Ge, Si and α-Sn.

Fig. 2 shows a possible fully lattice-matched device architecture consisting of InGaP/InGaAs/SiGeSn/Ge. This requires the development of high-quality SiGeSn with the desired direct energy transition, lattice-matched to Ge.

Table I: Optical transition energies and lattice constants of Ge, Si and α-Sn.

Label	*Energy (eV)*		
	Ge [11]	**Si** [11]	**α-Sn** [12]
E_{gap}	0.66	1.1	0
E_0	0.80	4.1	-0.41 [9]
$E_0 + \Delta_0$	1.09	4.1	
E_1	2.05	3.4	1.28
$E_1 + \Delta_1$	2.25	3.4	1.74
	Lattice constant (Å)		
d	5.65	5.43	6.46

In order to measure various optical properties of SiGeSn at different compositions, steady-state photoluminescence (PL) and spectroscopic ellipsometry (SE) were performed on samples grown through metal-organic vapour-phase epitaxy (MOVPE) on Ge substrates. Direct transitions contribute much more strongly to the ellipsometry measurement, while PL is expected to come from the fundamental, in this case indirect, bandgap. Band structure modelling was used to predict key transition energies in SiGeSn alloys and to inform the epitaxial growth. Single and triple-junction devices were produced to demonstrate the feasibility of using SiGeSn as a sub-cell, with light I-V and external quantum efficiency (EQE) measurements presented here.

3 EXPERIMENTAL METHODS

3.1 Sample growth

SiGeSn samples were grown through metal-organic vapour-phase epitaxy (MOVPE) by IQE plc. The samples were all grown on Ge substrates, with an epitaxial layer thickness around 0.5 µm. An ASM Epsilon LPCVD system was used, with tin chloride ($SnCl_4$), germane (GeH_4), and disilane (S_2H_6) precursors for Sn, Ge and Si respectively.

The SiGeSn incorporated into the devices was synthesized using a custom-made ultra-high vacuum chemical vapor deposition (UHV-CVD) tool with Ge_2H_6, Si_3H_8 and SnD_4 precursors and hydrogen carrier gas [13].

3.2 Photoluminescence

PL was measured using excitation from a Spectra-Physics Millennia V Nd:YVO_4 laser with a wavelength of 532 nm (2.33 eV). This relatively high incident energy was chosen to make sure as many carriers as possible were absorbed within a few hundred nanometres at the front of the sample, to ensure carriers were excited in the epitaxial layer rather than the substrate. Assuming that the absorption in the SiGeSn compositions considered here is mostly germanium-like at the high incident photon energy of 2.33 eV, >99% of the incident laser power will be absorbed in the first 500nm of material.

3.3 Spectroscopic ellipsometry

The results in the following section were obtained using a J. A. Woollam Co. V-VASE ellipsometer capable of measuring over a spectral range of 250-1700 nm. The angle of incidence can be varied between 15° and 90°; samples were measured at incidence angles of 75°, 77° and 79°. These angles were chosen because the most sensitive measurements can be obtained around the pseudo-Brewster angle of the sample [14]. The pseudo-Brewster angle is given by $\theta_B = \tan^{-1}(n_1/n_0)$; taking $n_0 = 1$ for air, θ_B is 77° for Ge at 500 nm. The samples measured are not all pure Ge, and θ_B depends on wavelength since n_1 is dispersive, and thus the samples are measured at several values around this angle. Measuring at several incidence angles also allows a simple check of non-ideal behaviour such as depolarization, since calculating the pseudo-dielectric constants should give the same results regardless of incidence angle in the absence of such effects.

3.4 Device measurements

The single and triple-junction devices were characterised using external quantum efficiency (EQE) and current-voltage measurements (I-V). The EQE was measured with a quartz-halogen light source dispersed through a 0.3-m spectrometer. The photocurrent was recorded using a standard lock-in technique. Light I-V characteristics were measured under single-sun illumination conditions using a dual-source TS-Space Systems Class-A solar simulator, which uses both metal halide and quartz halogen lamps to provide a close match to the terrestrial AM1.5G spectrum in the wavelength range 320–2200 nm.

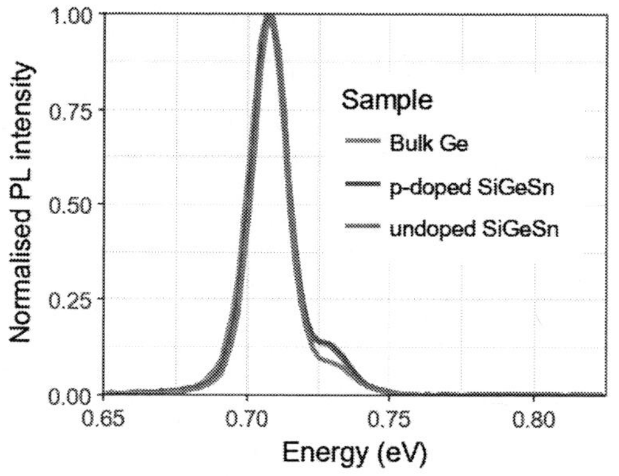

Figure 3: Low-temperature (12K) PL of SiGeSn grown directly on Ge and a piece of Ge substrate.

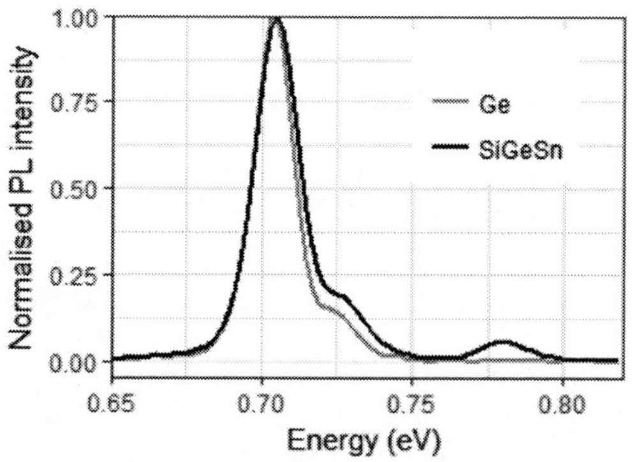

Figure 4: Low-temperature (12K) PL of $Si_{0.06}Ge_{0.925}Sn_{0.015}$ grown on a 10 nm layer of $Si_{0.2}Ge_{0.8}$ on a Ge substrate.

4 RESULTS

4.1 Photoluminescence

Fig. 3 shows low-temperature PL measurements taken from doped and undoped SiGeSn grown directly on Ge substrates, compared to PL from a piece of Ge substrate material. The PL observed from the SiGeSn samples shows an identical peak to the Ge, with no blueshift. A blueshift would be expected from the addition of Si and Sn at the concentrations used. Fig. 4 shows low-temperature PL taken under the same conditions of SiGeSn grown through a similar epitaxial process, but with a 10nm SiGe barrier between the SiGeSn and the Ge substrate. This sample shows an additional peak, not present in the Ge PL, around 0.78 eV.

As the band diagrams in the next section indicate, there is no energetic barrier for carriers excited in the SiGeSn to diffuse into the Ge. Although the incident photon energy was chosen to maximize absorption in the front of the sample, most carriers would still be able to reach the Ge if diffusion lengths are >500 nm. If carriers recombine in the Ge substrate, the resulting PL signal could be detected since SiGeSn is expected to be transparent for photon energies around 0.7 eV. Device modelling results (see section 4.4) have indicated diffusion lengths of at least 5 µm in SiGeSn. The presence of an additional PL peak in a SiGeSn sample in which an energetic barrier has been introduced in the form of an intermediate SiGe layer, and ellipsometry of the undoped SiGeSn sample, indicate that although good-quality SiGeSn has been grown, PL cannot be observed due to carrier diffusion into the substrate.

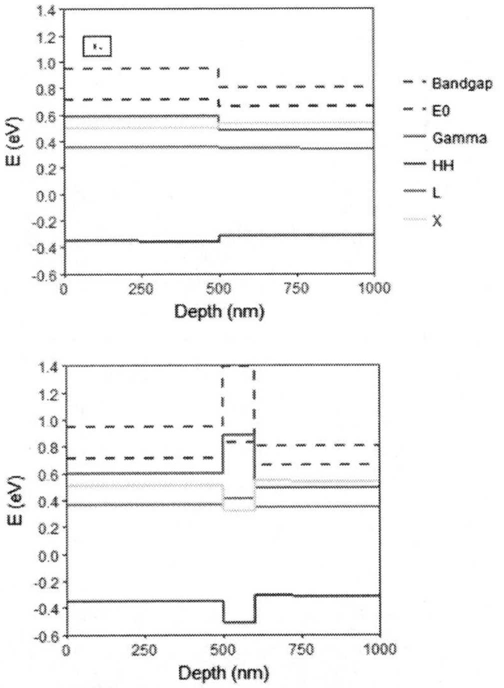

Figure 5: Band diagrams of $Si_{0.06}Ge_{0.925}Sn_{0.015}$ on Ge (top) and $Si_{0.06}Ge_{0.925}Sn_{0.015}$ on $Si_{0.2}Ge_{0.8}$ on Ge (bottom) as predicted by NextNano. The solid lines show the energies at the heavy-hole (HH) conduction band maxima band, and the valence band minima at the L, X and Γ points. The dashed lines shows specific transition energies of interest.

4.2 Band diagrams

NextNano [15] is a semiconductor device modelling tool which can solve both the Poisson drift-diffusion and Schrödinger equations for 1D, 2D or 3D structures. It was used here to understand the band offsets between the SiGeSn and Ge substrates or barrier layers by modelling simple 1D heterostructures. Fig. 5 compares band diagrams of SiGeSn on Ge, and SiGeSn/SiGe on Ge. In the first case, it is energetically favourable for carriers to enter the Ge. In the second case, NextNano predicts that there is still no barrier in the conduction band for electrons due to the X point minimum falling below the L point minimum with the addition of Si to Ge. However, there is a barrier in the valence band for holes travelling to the Ge, and inhibiting either electrons or holes from entering the substrate will reduce recombination there. However, the barrier is not sufficiently high to block all carriers, as the remaining Ge peak in Fig. 4 shows.

Table II shows the fundamental indirect bandgap and direct transition energy of interest for three SiGeSn compositions with a Si:Sn ratio of 3.7:1.

Table II: Bandgap and E_0 energies predicted using NextNano for several SiGeSn compositions lattice-matched to Ge.

Composition	*T = 300K*		*T = 12K*	
	E_{gap}	E_0	E_{gap}	E_0
3% Si, 0.8% Sn	0.69	0.87	0.76	0.96
6% Si, 1.6% Sn	0.71	0.94	0.79	1.03
9% Si, 2.4% Sn	0.73	1.01	0.81	1.10

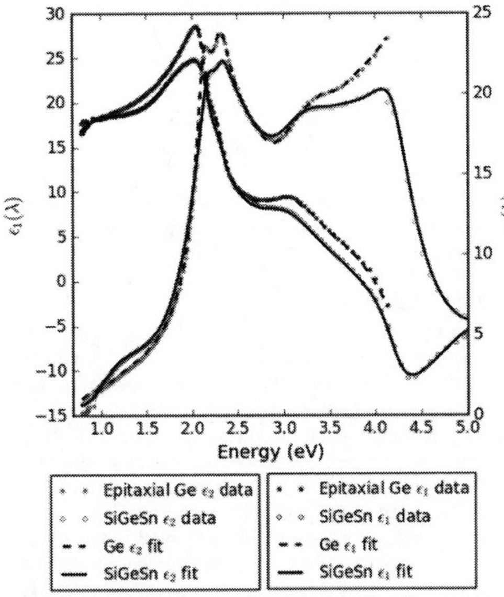

Figure 6: The real and imaginary parts of the pseudo-dielectric constant of an epitaxial Ge sample (green markers) and epitaxial SiGeSn (blue markers), both grown by IQE and with similar layer thicknesses. The dashed black line shows a fit to the Ge data, obtained using the WVASE software, modelling the sample as bulk Ge with a 2.1 nm GeO_2 overlayer. The solid black line shows a fit to the data obtained using a Psemi model [14], [16] for the SiGeSn + 2.1 nm GeO_2. The resulting optical constants fitted for SiGeSn are shown in Fig. 7.

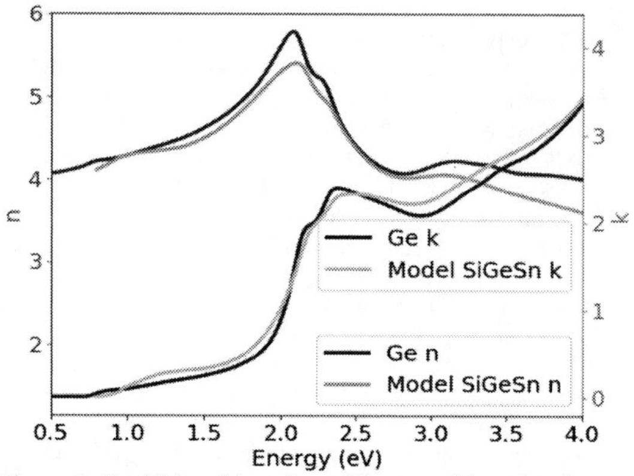

Figure 7: Real (n) and imaginary (k) parts of the refractive index of Ge [17] and SiGeSn, the latter resulting from the fit shown in Fig. 6.

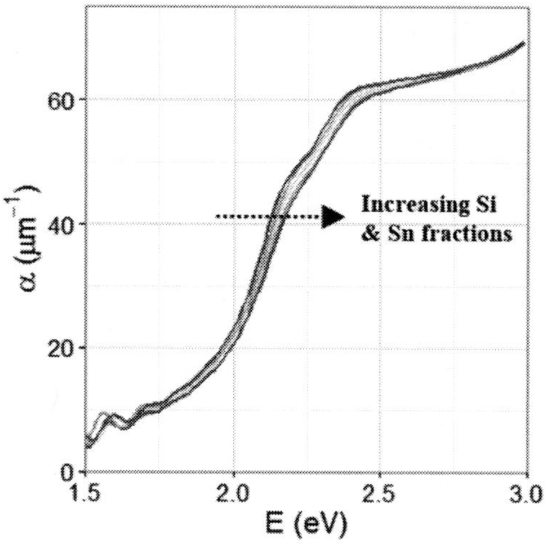

Figure 8: Pseudo-absorption coefficient of SiGeSn barrier samples at various SiGeSn compositions with the same Si:Sn ratio.

4.3 Ellipsometry

Initially, SE was used to confirm the hypothesis that it was not possible to observe the presence of a SiGeSn layer using PL measurements due to carrier diffusion into the Ge substrate. The pseudo-dielectric constants calculated directly from SE data of epitaxial Ge and SiGeSn are shown in Fig. 6. The data used to calculate the results presented here was taken at an incidence angle of 77°, but other incidence angles give identical results. The presence of a feature around 1 eV can be observed in the SiGeSn; this is expected to be the direct transition corresponding to the E_0 energy. This feature cannot be observed clearly in the Ge, since it lies at lower energy (0.8 eV) close to the detection limit of the ellipsometer, and because the signal becomes depolarized below the lowest direct edge of Ge [14].

Since the optical constants of Ge are well-known, the Ge SE data can be fitted by only considering the thickness of an oxide overlayer. The epitaxial Ge data fits very well to a model of bulk Ge with an overlayer of 2.1 nm GeO_2 as described by a Sellmeier model [18]; fitting of the overlayer thickness was done using the WVASE software [14]. Assuming the epitaxial SiGeSn has the same overlayer composition and thickness as epitaxial Ge (as the material is still mostly Ge, and has a similar thickness to the epitaxial Ge sample), a fit to the SiGeSn optical constants can be obtained. This was done using one of the built-in model types in the WVASE software (Psemi, a parametric model for semiconductor optical constants [16]), which allows a flexible definition and manipulation of peak shapes in the dielectric function. The resulting optical constants are Kramers-Kronig consistent, although no physical meaning can be attached to the fitting parameters. The resulting fit to the data is shown in Fig. 6 and the n and k values generated are shown in Fig. 7. The k values show a blueshift in the lowest absorption edge of SiGeSn relative to Ge, as well as a visible shift to higher energies of the E_1 and $E_1 + \Delta_1$ peaks around 2-2.5 eV.

Blueshifting of the E_1 and $E_1 + \Delta_1$ energies was also observed in SE of SiGeSn with a higher-gap barrier material. The absorption coefficient of these samples, calculated from the SE data, is shown in Fig. 8. Lower-energy features are not clearly visible due to strong thin-film interference fringes resulting from reflection of light at the interface between the SiGeSn and barrier material.

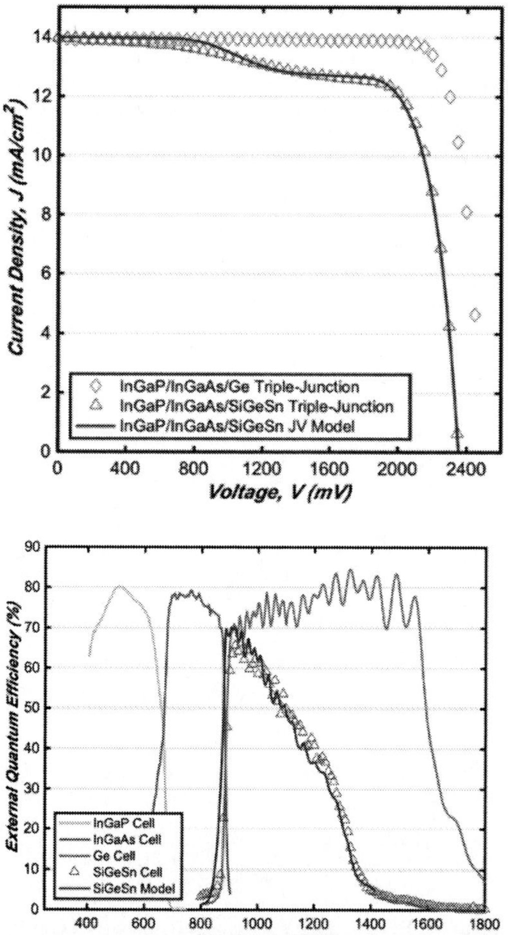

Figure 9: Top: Light-IV curves for a 3J device with a SiGeSn bottom junction and a reference device using a Ge bottom junction. Bottom: EQE of the 3J SiGeSn device. From [19].

4.4 Devices

The EQE of the single and triple-junction cells reached maxima of 80% and 70% respectively. The open-circuit voltage V_{oc} in the triple-junction cell was lower than expected (see Fig. 9, top). While the absorption and thus EQE in the SiGeSn in both devices is expected to be dominated by transitions across the direct gap around 1 eV, carrier thermalisation to the lowest available conduction band states will mean V_{oc} is limited by the indirect gap. This effect, possibly in combination with high non-radiative recombination, will lower the V_{oc}. Fitting the EQE data to an analytic solution of the drift-diffusion equations places a lower bound of 5 µm on the minority carrier diffusions length [13].

5 CONCLUSIONS

PL and ellipsometry measurements in addition to bandstructure modelling confirm the expected blueshift of the bandgap and higher-energy transitions with the addition of Si and Sn in a suitable ratio around 3.7:1. PL measurements were hindered by carrier diffusion into the substrate, which can be mitigated at least partially through the inclusion of a higher-bandgap barrier between the SiGeSn and Ge substrate. Single and triple-junction devices demonstrated the feasibility of using SiGeSn as a sub-cell, although further device optimization and the development of a tunnel junction between SiGeSn and Ge is necessary.

ACKNOWLEDGEMENTS

P. Pearce would like to acknowledge EPSRC CASE sponsorship from IQE plc.

REFERENCES

[1] NREL, 2017. [Online]. Available: http://www.nrel.gov/pv/assets/images/efficiency_chart.jpg. [Accessed: 21-May-2017].
[2] A. De Vos, *J. Phys. D. Appl. Phys.*, vol. 13, no. 5, pp. 839–846, 1980.
[3] R. R. King *et al.*, *24th Eur. Photovolt. Sol. Energy Conf. Exhib.*, pp. 1–7, 2009.
[4] M. Stan *et al.*, *J. Cryst. Growth*, vol. 312, no. 8, pp. 1370–1374, 2010.
[5] T. N. D. Tibbits *et al.*, *29th Eur. PV Sol. Energy Conf.*, no. September, p. 4, 2014.
[6] V. Sabnis, H. Yuen, and M. Wiemer, in *8th International Conference on Concentrating Photovoltaic Systems*, 2012, vol. 14, no. 2012, pp. 14–19.
[7] T. Thomas *et al.*, *Semicond. Sci. Technol.*, vol. 30, no. 9, p. 94010, 2015.
[8] H. Cotal *et al.*, *Energy Environ. Sci.*, vol. 2, no. 2, p. 174, 2009.
[9] P. Moontragoon, R. A. Soref, and Z. Ikonic, *J. Appl. Phys.*, vol. 112, no. 7, 2012.
[10] L. Vegard, *Zeitschrift für Phys.*, vol. 5, no. 1, pp. 17–26, Jan. 1921.
[11] S. Adachi, *Phys. Rev. B*, vol. 38, no. 18, pp. 12966–12976, 1988.
[12] S. Adachi, *J. Appl. Phys.*, vol. 66, no. 2, pp. 813–819, 1989.
[13] R. Roucka *et al.*, *IEEE J. Photovoltaics*, vol. 6, no. 4, pp. 1025–1030, 2016.
[14] J.A. Woollam Co., 2012.
[15] S. Birner *et al.*, *IEEE Trans. Electron Devices*, vol. 54, no. 9, pp. 2137–2142, Sep. 2007.
[16] B. Johs, C. . Herzinger, J. . Dinan, A. Cornfeld, and J. . Benson, *Thin Solid Films*, vol. 313, pp. 137–142, 1998.
[17] D. E. Aspnes and A. A. Studna, *Phys. Rev. B*, vol. 27, no. 2, pp. 985–1009, 1983.
[18] J. W. Fleming, vol. 23, no. 24, 1984.
[19] R. Roucka *et al.*, in *IEEE 43rd Photovoltaic Specialists Conference (PVSC)*, 2016.

INTERNAL-STRESS-ASSISTED EPITAXIAL LIFT-OFF PROCESS FOR THIN FILM GALLIUM ARSENIDE SOLAR CELLS ON METAL FOIL

Youngjo Kim, Sang Hyun Jung, Kangho Kim, Chang Zoo Kim, Hyun-Beom Shin, Kyung Ho Park, Won-Kyu Park, and Ho Kwan Kang*
Korea Advanced Nano Fab. Center, Suwon 16229, Korea
*Correspond author: hokwan.kang@kanc.re.kr, Tel. +82 31 546 6312, Fax +82 31 546 6309

ABSTRACT: High efficiency GaAs thin film solar cells have been fabricated by Al(Ga)As based epitaxial lift-off processes. However, the long-term exposure to the HF etchant could induce surface damages, which are detrimental for device fabrication and substrate reuse processes. Therefore, various ELO techniques have been studied to increase the lateral etching rate of the sacrificial layer. In this study, the influence of the internal stress on the ELO process time has been investigated with the (In)GaAs solar cells, which have different p-n junction structures. The thin film (In)GaAs solar cells are grown by MOCVD on GaAs substrates and transferred to 30 μm thick Au foil by the internal-stress-assisted ELO process. The internal stress is induced by changing the epi-layers from GaAs to $In_{0.015}Ga_{0.985}As$, which has slightly larger lattice constant. As a result, the etching time in the ELO process was reduced from 36 to 4 hours by employing the GaAs/InGaAs heterojunction structure, which has a compressive film stress of -59.0 MPa. Photovoltaic device characteristics of the fabricated thin film solar cells are investigated under AM1.5G illuminations.
Keywords: III-V Semiconductors, Gallium Arsenide Based Cells, Thin Film Solar Cell, Epitaxial Lift-off.

1 INTRODUCTION

GaAs based thin film solar cells promise reliable device performances with higher efficiencies compared to other types of flexible solar cells [1, 2]. Although the device fabrication cost is correspondingly high, it can be lowered by multiple reuses of the GaAs substrate which is the biggest cost driver [3–5].

In general, the thin film GaAs solar cell is fabricated by an epitaxial lift-off (ELO) process, which is based on a selective wet etching of a thin $Al_xGa_{1-x}As$ (x>0.6) sacrificial layer using a solution of hydrofluoric acid (HF) [5–7]. However, the long-term exposure to the HF etchant could induce surface damages of the solar cell epi-layer and the separated GaAs substrate. The high surface roughness and the post-etching residues are known to be the major drawback that should be solved for the substrate reuse process [8–10].

Therefore, various ELO techniques have been studied to increase the lateral etching rate and minimize the detrimental effects on the device performance and the surface quality of the parent substrate [7–12]. Strain-induced ELO process is one of the simple method that can increase the etching rate. Several research groups proposed the stress-assisted ELO processes using various seed metal [7], surface tension [8], and weight [12].

However, there are few studies on the influence of the internal stress in the solar cell epi-layers on the ELO process time. Because the solar cell structure is realized by epitaxy, the multi-layer stress can be controlled by adjusting the growth materials [13, 14]. The strain-induced ELO process using the optimized internal stress of the solar cell epi-layers can be a productive method which is suitable for large-scale industrial utilization.

In this study, we demonstrate the enhancement of the lateral etching rate in the internal-stress-assisted ELO process. (In)GaAs solar cells are inversely grown with an AlAs sacrificial layer (20 nm) by metalorganic chemical vapor deposition (MOCVD). The solar cell epi-structures are separated from the GaAs substrates by employing the internal-stress-assisted ELO process. The fabricated device characteristics are investigated with a class A solar simulator under AM1.5G illuminations.

2 EXPERIMENTAL

2.1 MOCVD growth of (In)GaAs solar cells

Inverted (In)GaAs single-junction solar cell structures are grown by MOCVD (AIXTRON, AIX200/4 RF) on Zn doped p-type GaAs (100) substrates with a misorientation of 6° toward [111]. Fig. 1 shows the schematic diagrams of the solar cells with three different p-n junction structures: GaAs homojunction, InGaAs homojunction, and GaAs/InGaAs heterojunction.

Small amount of In (1.5%) is added in the InGaAs layer, which has a lager lattice constant than GaAs layer, to induce the internal stress. The In compositions of the InGaP window and InGaP back surface field (BSF) layers are also increased to 48.3 and 49.5 % respectively to keep the lattice matching condition between GaAs and $In_{0.015}Ga_{0.985}As$. The window and BSF layers are employed as surface passivation layers to prevent from high surface recombination velocity and Fermi level pinning [15]. Highly doped (>5×10^{18} cm^{-3}) n-type and p-type (In)GaAs layers are grown as capping layers for ohmic contact formation.

Arsine (AsH_3) and phosphine (PH_3) are used as group V precursors, whereas trimethylgallium (TMGa) and trimethylindium (TMIn) are used as group III precursors. Silane (SiH_4) and diethylzinc (DEZn) are used as n-type and p-type dopant sources, respectively. Palladium-diffused ultra-high-purity hydrogen (H_2) is used as a carrier gas with a total flow of 15,000 sccm. The horizontal reactor pressure is fixed at 100 mbar and the reactor temperature is kept at 650 °C during the epitaxial growth.

2.2 Thin film device fabrication

The thin film (In)GaAs solar cells are fabricated by metal foil formation, stress-assisted ELO, and device fabrication processes as shown in Fig. 2. After the epitaxial growth, triple-layer metal films of AuBe/Pt/Au are deposited on the top surface of the solar cell epi-layers as p-type contact and seed metal. In this work, electroplated Au foil (30 μm) is used as an alternative substrate taking advantage of electrical conductivity, flexibility, and HF resistance of the material.

The inverted (In)GaAs solar cell structures are separated from the GaAs substrate by etching the AlAs sacrificial layer in HF (48%) solution. During the ELO process, changes in the lateral etching time are investigated, which depend on the internal stress of the solar cell epi-structures. The separated solar cell film is attached temporarily on a Si substrate using double-sided adhesive tape for following fabrication processes.

AuGe/Ni/Au metal films are patterned on the surface of the n-type GaAs ohmic layer as n-type contact metal. The n-type GaAs ohmic layer is selectively etched in a $NH_4OH:H_2O_2$:DI (2:1:10) solution. Each solar cells are isolated by wet chemical etching. The aperture area of the fabricated solar cell is 5.5×5.5 mm^2. The grid line was designed to be a width of 5 μm and a space of 250 μm. Finally, MgF_2/ZnS double-layers are deposited on the top surface of the solar cell for anti-reflection coating (ARC).

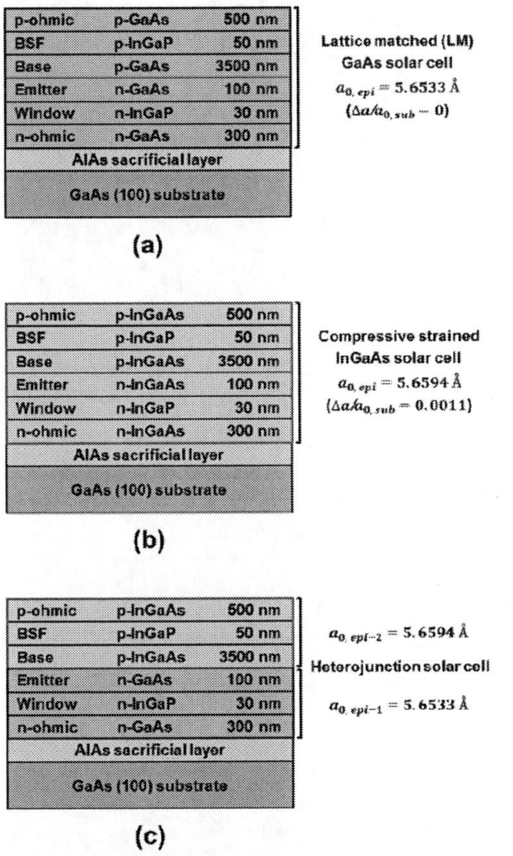

Figure 1: Schematic diagrams of the inverted (In)GaAs solar cells with three different p-n junction structures: (a) GaAs homojunction, (b) InGaAs homojunction, and (c) GaAs/InGaAs heterojunction.

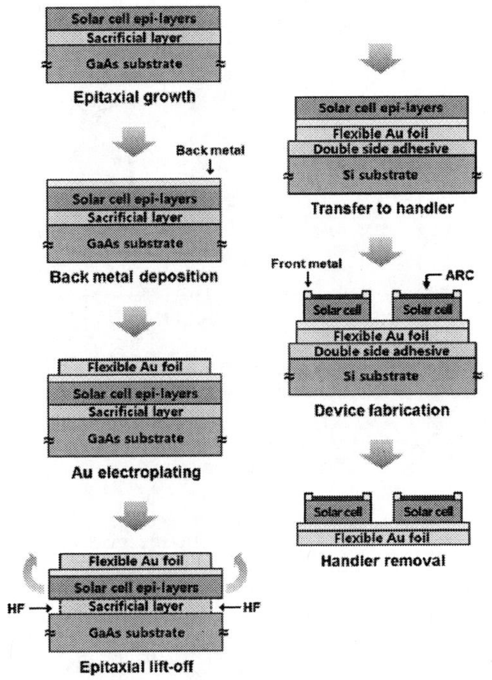

Figure 2: Schematic diagrams of the thin film (In)GaAs solar cell fabrication processes.

3 RESULTS AND DISCUSSION

3.1 Film stress analysis

The film stress depending on the (In)GaAs solar cell epi-structure has been investigated with a stress measurement system (Frontier Semiconductor: FSM 500TC). Fig. 3 shows the measured film stress of the (In)GaAs solar cells with three different p-n junction structures: GaAs homojunction, InGaAs homojunction, and GaAs/InGaAs heterojunction after epitaxial growth, seed metal deposition, and Au electroplating.

The InGaAs homojunction structure shows a compressive film stress of -70.3 MPa, whereas the GaAs homojunction structure shows a tensile film stress of 5.3 MPa. This change is attributed to the increase of lattice constant of the solar cell epi-layers, which induce tensile strain near the top surface of the GaAs substrate [12, 14]. In the same reason, the GaAs/InGaAs heterojunction structure also shows a compressive film stress of -59.0 MPa. This value is a little weaker than the film stress of the InGaAs homojunction structure due to the thinner InGaAs layers.

After the seed metal deposition, the similar trend was observed in the three different solar cell structures. This means that the seed metal structure with a thickness of 250 nm does not contribute to the change of the film stress. After the Au electroplating, in contrast, the film stress was dramatically reduced except the GaAs homojunction structure. 30 μm thick Au foil could contribute to a larger change of film stress because the film stress is inversely proportional to the film thickness.

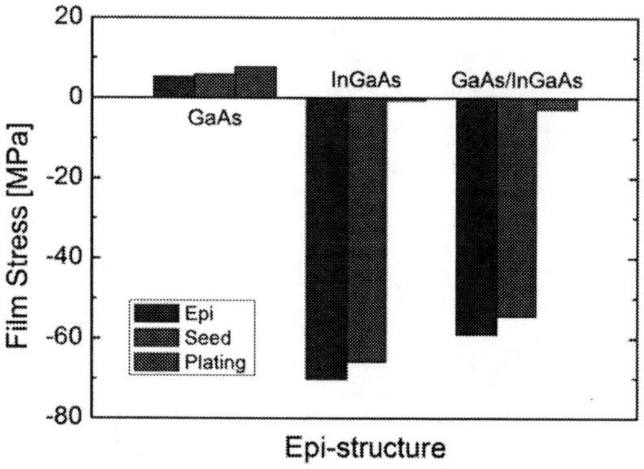

Figure 3: Film stress of the (In)GaAs solar cells with three different p-n junction structures: (a) GaAs homojunction, (b) InGaAs homojunction, and (c) GaAs/InGaAs heterojunction after epitaxial growth, seed metal deposition, and Au electroplating.

3.2 Reduced etching time in the ELO process

The effect of the internal stress on the ELO process time has been investigated within the (In)GaAs solar cells, which have three different p-n junction structures: GaAs homojunction, InGaAs homojunction, and GaAs/InGaAs heterojunction. Fig. 4 shows the photographs of the (In)GaAs solar cell wafers during the ELO process.

The thin film solar cell structures were completely released after 36, 9, and 4 hours for GaAs homojunction, InGaAs homojunction, and GaAs/InGaAs heterojunction wafers, respectively. In the beginning, InGaAs homojunction and GaAs/InGaAs heterojunction wafers showed a similar lateral etching rate under the electroplating contact area (edge ring). After 2 hours, however, the GaAs/InGaAs heterojunction wafer showed much faster lateral etching rate with spontaneous bending. The bend angle of the GaAs/InGaAs heterojunction wafer is larger than that of the InGaAs homojunction wafer as shown in Fig. 5.

It is well known that the strain-induced bending can increase the lateral etching rate in the ELO process [7, 8, 12]. The spontaneous bending is attributed to the internal stress induced by employing InGaAs layers. Therefore, the released thin film solar cell structures have different bending angles as shown in Fig. 4. When the solar cell epi-structure is released, the film stress can be induced in the opposite direction because the substrate replaced from the GaAs substrate to Au metal foil [12, 14].

3.3 Photovoltaic device characteristics

The photovoltaic current density-voltage (J-V) characteristics of the (In)GaAs solar cells have been measured using a class A solar simulator (Wacom: WXS-220S-L2) under AM1.5G illumination. Fig. 6 shows the J-V curves of the epitaxially grown solar cells with three different p-n junction structures: GaAs homojunction, InGaAs homojunction, and GaAs/InGaAs heterojunction. The specific device parameters of the fabricated thin film (In)GaAs solar cells under AM1.5G illuminations are summarized in Table I.

The thin film GaAs homojunction solar cell shows an efficiency of 16.20 % with an open circuit voltage (V_{oc}) of 0.94 V, a short circuit current density (J_{sc}) of 21.89 mA/cm^2, and a fill factor (FF) of 0.7898. On the other hand, the InGaAs homojunction solar cell shows a relatively lower efficiency of 14.77 %. This could be attributed to the FF degraded by employing lower band gap InGaAs material, which is more sensitive to leakage current. An increase of J_{sc} and a decrease of V_{oc} are understandable because the band gap of $In_{0.015}Ga_{0.985}As$ is lower than that of GaAs [16].

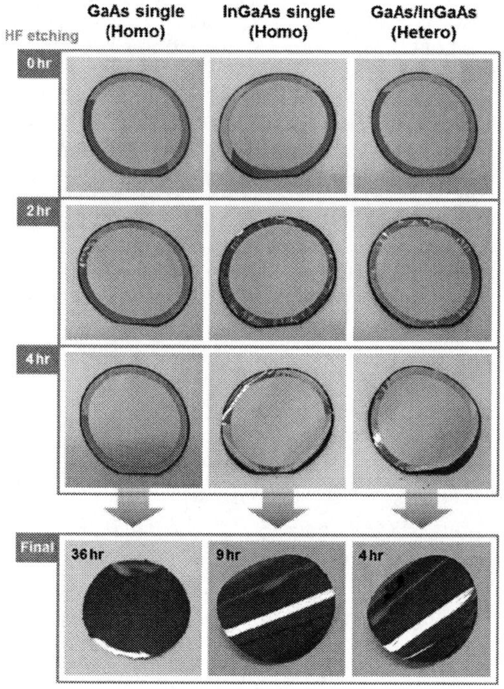

Figure 4: Photographs of the (In)GaAs solar cell wafers with three different p-n junction structures: GaAs homojunction, InGaAs homojunction, and GaAs/InGaAs heterojunction during the ELO process.

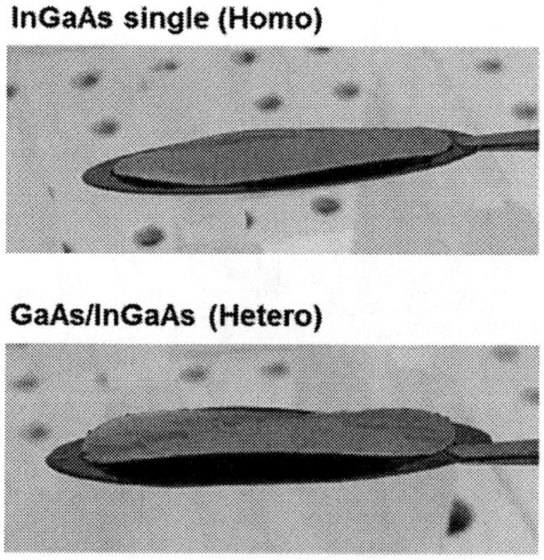

Figure 5: Photographs of the InGaAs homojunction and the GaAs/InGaAs heterojunction wafers after dipping in HF for 3 hours.

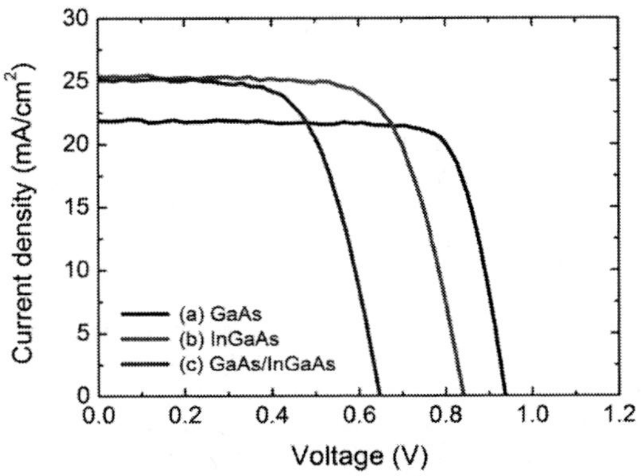

Figure 6: Photovoltaic J-V curves of the (In)GaAs solar cells with three different p-n junction structures: (a) GaAs homojunction, (b) InGaAs homojunction, and (c) GaAs/InGaAs heterojunction.

Table I: Summary of the specific device parameters of the fabricated thin film (In)GaAs solar cells under AM1.5G illuminations.

Junction structure	Efficiency (%)	V_{oc} (V)	J_{sc} (mA/cm^2)	Fill factor
GaAs	16.20	0.94	21.89	0.7898
InGaAs	14.77	0.84	25.41	0.6917
GaAs/InGaAs	10.48	0.65	25.16	0.6443

The GaAs/InGaAs heterojunction solar cell shows the lowest efficiency of 10.48 %. Although J_{sc} and FF were kept at similar values, the V_{oc} decreased by 22.6 % compared to the InGaAs homojunction solar cell. The lower V_{oc} might be attributed to Shockley-Read-Hall recombination, which is induced by defect or trap states [16], at the interface of the GaAs/InGaAs heterojunction.

The device performances of the fabricated (In)GaAs solar cells are not the best, in this work, because we have focused on the effect of the internal stress on the ELO process time. Therefore, further improvements can be achieved by optimizing the growth conditions.

4 CONCLUSION

The effect of the internal stress on the ELO process time has been investigated within the (In)GaAs solar cells, which have different p-n junction structures. The thin film (In)GaAs solar cells have been grown by MOCVD and fabricated on 30 μm thick Au foil by the internal-stress-assisted ELO process. As a result, the etching time in the ELO process was reduced from 36 to 4 hours by employing the GaAs/InGaAs heterojunction structure. Although the device performance is not the best, further improvement is expected to be achieved by optimizing the growth conditions. The simple and fast ELO process can be applied for various III-V based thin film applications.

5 ACKNOWLEDGEMENT

This work was supported by the Korea Institute of Energy Technology Evaluation and Planning (KETEP) and the Ministry of Trade, Industry & Energy (MOTIE) of the Republic of Korea (No. 20163030013980). This research was also supported by the Civil & Military Technology Cooperation Program funded by the MOTIE & DAPA of Korea (16-CM-EN-11).

6 REFERENCES

[1] M. A. Green, K. Emery, Y. Hishikawa, W. Warta, E. D. Dunlop, Progress in Photovoltaics: Research and Application, 24 (2016) 905.
[2] L. S. Mattos, S. R. Scully, M. Syfu, E. Olson, L. Yang, C. Ling, B. M. Kayes, G. He, Proceedings 37th IEEE Photovoltaic Specialists Conference, (2011) 3187.
[3] K. A. Horowitz, M. Woodhouse, H. Lee, G. P. Smestad, Proceedings 11th International Conference on Concentrator Photovoltaic Systems: CPV-11, (2015) 100001.
[4] J. Adams, V. Elarde, A. Hains, C. Stender, F. Tuminello, C. Youtsey, A. Wibowo, M. Osowski, Proceedings 38th IEEE Photovoltaic Specialists Conference, Vol. 2 (2012) 1.
[5] G. J. Bauhuis, P. Mulder, E. J. Haverkamp, J. J. Schermer, E. Bongers, G. Oomen, W. Köstler, G. Strobl, Progress in Photovoltaics: Research and Application, 18 (2010) 155.
[6] K. Lee, J. D. Zimmerman, X. Xiao, K. Sun, S. R. Forrest, Journal of Applied Physics, 111 (2012) 033527.
[7] K. Lee, J. D. Zimmerman, T. W. Hughes, S. R. Forrest, Advanced Functional Materials, 24 (2014) 4284.
[8] C. W. Cheng, K. T. Shiu, N. Li, S. J. Han, L. Shi, D. K. Sadana, Nature Communications, 4 (2013) 1577.
[9] R. H. Horng, M. C. Tseng, F. L. Wu, C. H. Li, C. H. Wu, M. D. Yang, IEEE Transactions on Electron Devices, 59 (3) (2012) 666.
[10] F. L. Wu, S. L. Ou, R. H. Horng, Y. C. Kao, Solar Energy Materials & Solar Cells, 122 (2014) 233.
[11] J. Maeda, Y. Sasaki, N. Dietz, K. Shibahara, S. Yokoyama, S. Miyazaki, M. Hirose, Japanese Journal of Applied Physics, 36 (1997) 1554.
[12] M. M. A. J. Voncken, J. J. Schermer, G. J. Bauhuis, A. T. J. van Niftrik, P. K. Larsen, Journal of Physics: Condensed Matter, 16 (2004) 3585.
[13] A. T. J. van Niftrik, J. J. Schermer, G. J. Bauhuis, J. van Deelen, P. Mulder, P. K. Larsen, Crystal Growth and Design, 7 (12) (2007) 2472.
[14] C. H. Hsueh, S. Lee, T. J. Chuang, Journal of Applied Mechanics, 70 (2003) 151.
[15] J. Plá, M. Barrera, F. Rubinelli, Semiconductor Science and Technology, 22 (2007) 1122.
[16] H. Cotal, C. Fetzer, J. Boisvert, G. Kinsey, R. King, P. Hebert, H. Yoon, and N. Karama, Energy and Environmental Science, 2 (2009) 174.

THE EFFECTS OF SHORT-RANGE ALLOY DISORDER ON THE POTENTIAL VOLTAGE PERFORMANCE IN GAASBI BASED SOLAR CELLS

T. Wilson[1], A. Mellor[1], N. P. Hylton[1] and N. J. Ekins-Daukes[1,2]
[1] The Blackett Laboratory, Imperial College London, London, UK
[2] School of Photovoltaic and Renewable Energy Engineering, University of New South Wales, Sydney, Australia

The large band-gap bowing induced by the incorporation of Bi allows GaAsBi to achieve direct gap energies around 1.0 eV for relatively dilute quantities of Bi, which makes the material attractive as a potential middle sub-cell for future multi-junction solar cells. However, the presence of band-tail states due to inherent short-range alloy disorder may have an adverse effect on voltage performance and optimum sub-cell energies. In this work a range of GaAsBi samples are examined with varying Bi fractions between 0 and 9.7%. A broadening of the fundamental critical point is observed for Bi fractions up to 5.5%, indicative of short-range alloy disorder induced by increased Bi incorporation. Open-circuit voltage performance is observed to degrade in GaAsBi samples with Bi in excess of 4%, which corresponds to band-gap energies in the vicinity of 1.0 eV. Detailed balance calculations incorporating a broadened absorption edge indicate a reduction in maximum efficiency and a blue-shift of the optimum band-gap energy in both single and quad-junction devices. The subsequent blue-shift of optimum bottom cell energy with increased lower-middle cell broadening highlights the potential influence of disordered alloys on optimum multi-junction solar cell design.
Keywords: alloy disorder, multi-junctions, GaAsBi, high efficiency solar cells

1 INTRODUCTION

Multi-junction solar cells offer the most viable route to realizing power conversion efficiencies in excess of the 'Shockley-Queisser' limit, approximately 31% for a single band-gap with energy of 1.3 eV [1] and are the technology of choice for spaced based and concentrated photovoltaic (CPV) applications. By combining different semiconductor materials, where each band-gap energy can be finely tuned, the photogenerated current in each individual sub-cell can be carefully matched to more effectively harness the incident solar radiation, as shown in Fig. 1. Power conversion efficiencies up to 63.8% and 68.8% can be realised for ideal three and four junction solar cells under maximum concentration [2], with today's commercially available triple junction devices producing efficiencies in excess of 40% under concentrated illumination [3]. Making use of the well studied system of GaInP/InGaAs/Ge alloys employed in triple-junction architectures, the addition of a 1.0 eV sub-cell above the 0.66 eV Ge cell in a four-junction device has the potential to produce power conversion efficiencies exceeding 50% [4], a crucial technology milestone to making CPV a cost effective alternative to flat panel Si in the global photovoltaic market [5].

Multi-junction solar cells incorporating a 1.0 eV sub-cell using dilute nitride [6]-[8]and SiGeSn [9], [10] alloys have been demonstrated experimentally, however the use of dilute bismide materials have so far received relatively little attention [11]. It is well established that the valence band anti-crossing interaction (VBAC) splits the degenerate valence bands in dilute bismide alloys into E_+ and E_- sub-bands resulting in a large reduction in band-gap energy for relatively small quantities of Bi [12]. As a result, GaAsBi has been shown to reach a 1.0 eV direct band-gap with as little as 6.1% Bi, and a 0.7% compressive strain mismatch on GaAs [13] making it a promising candidate material for the proposed quad-junction architecture displayed in Fig. 1. However, the large atomic size mismatch between the Bi atom and the GaAs host matrix has been shown to introduce Bi-pair and multi-atom clusters at vacant As sites as the Bi composition is

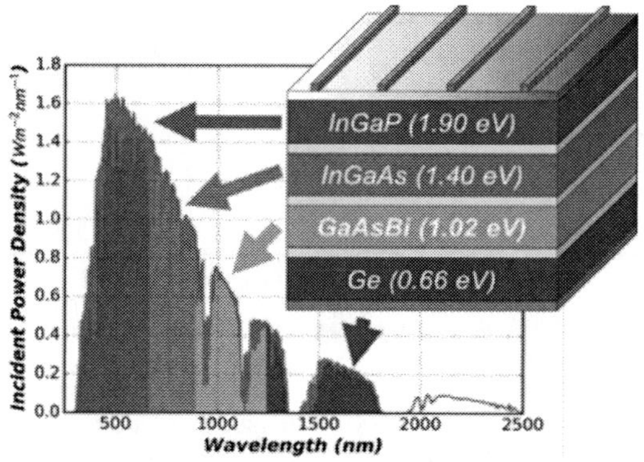

Figure 1: AM1.5D solar spectrum with indicated absorption thresholds as defined by the proposed quad-junction solar cell architecture inset.

increased [14]. This results in, so called, short-range alloy disorder and manifests itself as microscopic fluctuations in energy close to Bi impurity energy levels in the vicinity of the valence band maximum. The effects of these localised energy states are clearly observed as a characteristic red-blue-red shift (s-shape) in the temperature dependence of the peak photoluminescence energy [15] which corresponds with a peak in the spectral full-width at half-maximum [16], [17] and is attributed to the hopping of localised carriers between deep and shallow states [18].

The distribution of localised states has also been shown to introduce an exponential Urbach tail [19] that extends into the band-gap [20], [21], resulting in a broadening of the GaAsBi absorption edge up to 5.5% Bi [22]. The broadening of the absorption edge is indicative of inherent short range alloy disorder as a result of increased Bi incorporation [23]. Such broadening will increase sub band-edge absorption, which may lead to an increase in photogenerated current, however the distribution of extended states will pin the quasi fermi level for holes under illumination, reducing the achievable open-circuit voltage. Rau and Werner have previously demonstrated the adverse affects of alloy disorder in a-Si based solar cells, indicating that the loss of cell voltage can attribute to a reduction of 1.7% absolute cell efficiency [24]. In this work, the effects of short-range alloy disorder on the voltage performance in potential GaAsBi based solar cells are studied. The impact of a broadened absorption edge on maximum efficiency and optimum band-gap energy is assessed in both single and quad-junction devices.

2 EXPERIMENTAL DETAILS

Experimentally determined absorption coefficient data is taken from a previously published work by Masnadi-Shirazi *et al* [22] and is displayed in Fig. 2 as open symbols. The data is taken from a range of bulk $GaAs_{1-x}Bi_x$ films grown via molecular-beam epitaxy on semi-insulating (001) GaAs substrates, with Bi content in the range $0 \leq x \leq 9.7\%$. Further detail on the structure of these samples and growth procedure can be found elsewhere [22], [25].

3 THEORY

3.1 Absorption Coefficient Parametrisation

The absorption coefficient is parametrised using the well-known Critical-Point Parabolic-Band (CPPB) formulism reported by Sadao Adachi [26]-[28], where the complex part of the dielectric function, $\varepsilon_2(\omega)$, is calculated from a number of optical transitions that take place at points of high symmetry in the Brillouin Zone. The real component, $\varepsilon_1(\omega)$, is related to the imaginary component, $\varepsilon_2(\omega)$, via the Kramers-Krönig relations. The software implementation of the CPPB model is reported in more detail elsewhere [29]. As the experimental absorption coefficient data discussed in section 2 is only available in the vicinity of the lowest energy direct band-edge, only the fundamental critical point (CP), describing the E_0 and $E_0 + \Delta_0$ transitions, is considered. The CP centre energy, E_0, the broadening, Γ_{E0}, and amplitude, A, parameters are varied with parameters describing the higher energy transitions assumed to take the values appropriate for fitting GaAs, as listed in [27]. The optical absorption coefficient is calculated from the complex dielectric function and depends on both the real and imaginary components;

$$\alpha(\omega) = \frac{4\pi}{k}\left[\frac{\sqrt{\epsilon_1(\omega)^2 + \epsilon_2(\omega)^2} - \epsilon_1(\omega)}{2}\right]^{1/2} \quad (1)$$

3.2 Estimating Open-Circuit Voltage

The absorptivity of a material can be related to its emitted photon flux by Würfel's generalisation of Planck's radiation law [30] and is given by;

$$\dot{N}(E, T, \Delta\mu) = \frac{2\pi}{c^2 h^3} \int \frac{a(E)E^2}{e^{\frac{E-\Delta\mu}{K_b T}} - 1} . dE \quad (2)$$

Where $a(E)$ represents the spectrally dependent absorptivity and $\Delta\mu$ the chemical potential. The π term corresponds to the etendué for hemispherical emission from the surface of a solar cell into air. Adopting the general detailed-balance result, and treating both solar and ambient radiation as perfect black bodies, the total radiative current density is given by [31];

$$J_{Rad}(V) = q \cdot \left[f_s X \dot{N}(E, T_s, 0) + (1 - X f_s)\dot{N}(E, T_a, 0) - \dot{N}(E, T_a, qV)\right] \quad (3)$$

Where f_s represents the angular range of the sun under direct illumination and takes the value of $f_s = 2 \times 10^{-5}$, X is the concentration factor and T_s and T_a denote the temperature of the sun and ambient respectively. The temperature of the sun is taken to be 5778 K and the ambient temperature is 294 K. The simulated characteristic current-voltage curve is generated by numerically solving Shockley's diode equation, where the V_{OC} is extracted when the total current density equals zero, and the maximum power conversion efficiency is given by;

$$\eta_{Max} = \frac{J_{mpp} V_{mpp}}{P_s} \quad (4)$$

The reciprocity relation connecting the external quantum efficiency (EQE) to the emissivity of a material is used to approximate the absorptivity term in equation (2) [32]. The EQE is calculated using analytic solutions to the semiconductor transport equations for the case of an abrupt n-p junction in the depletion approximation [33], through the software implementation discussed in [29].

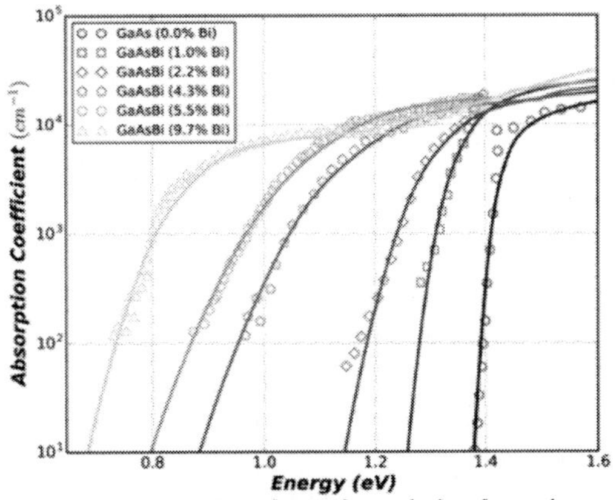

Figure 2: Experimental optical absorption coefficient data taken from [22] (open symbols) and parametrised absorption using the CPPB model (solid lines).

Table I: Fitting parameters used to parametrize the GaAsBi absorption coefficient data shown in Fig. 2.

Sample (Bi Fraction)	E_0 *(eV)*	*G (meV)*	*A* ($eV^{1.5}$)
GaAs (0.0%)	1.425	12	4.5
$GaAs_{1-x}Bi_x$ (1.0%)	1.350	23	6.0
$GaAs_{1-x}Bi_x$ (2.2%)	1.300	45	4.2
$GaAs_{1-x}Bi_x$ (4.3%)	1.145	70	3.0
$GaAs_{1-x}Bi_x$ (5.5%)	1.075	75	3.0
$GaAs_{1-x}Bi_x$ (9.7%)	0.840	45	1.2

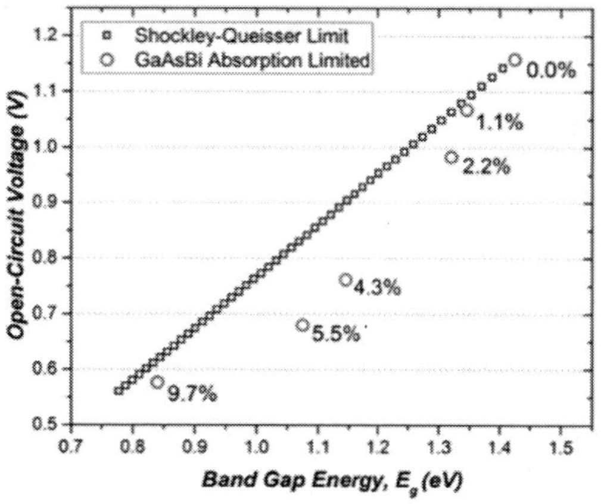

Figure 3: Shockley-Queisser limit and GaAsBi absorption limited calculations of open-circuit voltage.

4 RESULTS AND DISCUSSION

4.1 Parameterisation of GaAsBi Absorption

Experimental GaAsBi absorption coefficient data is plotted as open symbols in Fig. 2. A good agreement is seen between the data and the CPPB model with the parametrised absorption coefficients illustrated by the solid lines. Fitting parameters used in the CPPB model are summarised in Table I. A red-shift of the centre energy of the critical point E_0 is observed, decreasing from 1.300 eV to 0.840 eV for the corresponding increase in Bi content of 1.1% to 9.7%. This is consistent with the VBAC theory [12] and previous results from the literature presented in [13]. The broadening parameter corresponding to the E_0 critical point, Γ_{E0}, increases from 12 meV to 75 meV for the range of Bi fractions spanning 0.0% to 5.5%. This broadening of the absorption edge is indicative of an increase in short-range alloy disorder that arises from the increased incorporation of Bi, where the perceived atomistic randomness at low Bi concentrations decreases as the atoms show an increased tendency to form Bi-Bi pairs and multi-atom clusters [23]. The 9.7% sample clearly indicates a sharpening in the absorption edge, with Γ_{E0} decreasing from 75 meV at 5.5% Bi to 45 meV at 9.7%. This can explained by considering the valence band-edge position in relation to the distribution of extended states. As the Bi incorporation is increased the average compressive strain in the GaAsBi layer also increases, pushing the split valence bands further apart and forcing the heavy hole band deeper into the band-gap. As the heavy hole band shifts upwards it begins to screen out the effects of those localised energy states close to the valence band edge due to Bi-Bi pair and cluster formation [23]. The density of these extended states is understood to decrease exponentially inside the band-gap and so by screening the more abundant distribution close to the band-edge the effects of disorder on the broadened absorption edge becomes less pronounced.

4.2 Calculated Open-Circuit Voltage

The V_{OC} is calculated under both 'Shockley-Queisser' (100% quantum efficiency with an abrupt onset of absorption) and GaAsBi absorption limited conditions, using equations (2) and (3), and is plotted in Fig. 3 as black open squares and red open circles respectively. The parametrised optical absorption presented in Fig. 2 is used to calculate realistic EQE, using appropriate material parameters that produce a good fit to GaAs reference data (not shown here). There is little difference between the V_{OC} calculated in the Shockley-Queisser limiting case and the absorption limited data for the sample with 1.1% Bi, however the samples with 2.2%, 4.3% and 5.5% Bi show a more significant reduction of 71, 154 and 202 mV respectively. The drop in V_{OC} is in good agreement with the increase in Γ_{E0} up to 5.5% Bi reported in Table I and reflects the broadening absorption edge depicted in Fig. 2. This degradation of V_{OC} can be directly attributed to the effects of short-range alloy disorder. It is important to note that Γ_{E0} reaches a maximum for the 5.5% Bi sample, which is not dissimilar to the 6.1% Bi fraction required for a 1.0 eV band-gap energy, and therefore would expect the voltage performance to be similarly degraded in a 1.0 eV sub-cell. The 9.7% Bi sample clearly shows a recovery in V_{OC} that is consistent with the decrease in Γ_{E0} and sharper absorption edge observed in Fig. 2.

The deviation of the absorption limited GaAsBi voltage performance away from that predicted for an ideal 'Shockley-Queisser' device indicates the importance of taking into account the broadened absorption profile when calculating the potential performance limits for disordered materials. This consideration has commonly been overlooked when calculating the performance of 'ideal' materials and in most cases the step function describing absorption above the band-gap is valid. This is illustrated well for the case of GaAs, where the world record power conversion efficiency of 28.8% [34] approaches the Shockley-Queisser limit for a single junction device. However the validity of this approach breaks down when wishing to calculate the performance limits of disordered materials and has the potential to alter the optimum band gap energy for achieving maximum efficiency.

4.3 Evaluating the Effect of Alloy Disorder on Maximum Efficiency in Ideal Single and Multi-Junction Devices

To assess the effect of alloy disorder on ideal single-junction device efficiency and optimum band-gap energy, equation (3) is used to calculate the radiative contributions to current density and the characteristic current-voltage profile is calculated by solving Shockley's diode equation. To simulate the broadened absorption edge due to alloy disorder the conventional step-function absorptivity used in detailed balance calculations is replaced with one that incorporates an Urbach tail to describe sub band-gap absorption;

$$a(E) = \begin{cases} \frac{\exp[E-E_g]}{\sigma}, & E < E_g \\ 1.0, & E \geq E_g \end{cases} \quad (5)$$

Where Eg is the band-gap energy and σ the broadening parameter, analogous to Γ_{E0} used in the parametrisation of the absorption coefficient discussed in section 4.1. This is in good agreement with the absorption edge profile of GaAsBi observed in [20]-[22]. The maximum efficiency of an ideal single-junction device is calculated using equation (4) and is shown in Fig. 4 for a range of increasing broadening energies. For the abrupt absorptivity case (σ = 0 meV) the calculated maximum

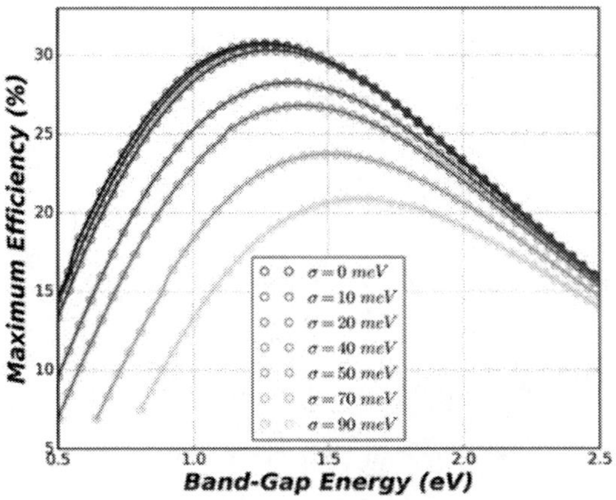

Figure 4: Detailed balance calculations of maximum efficiency for a single junction device considering different degrees of broadening.

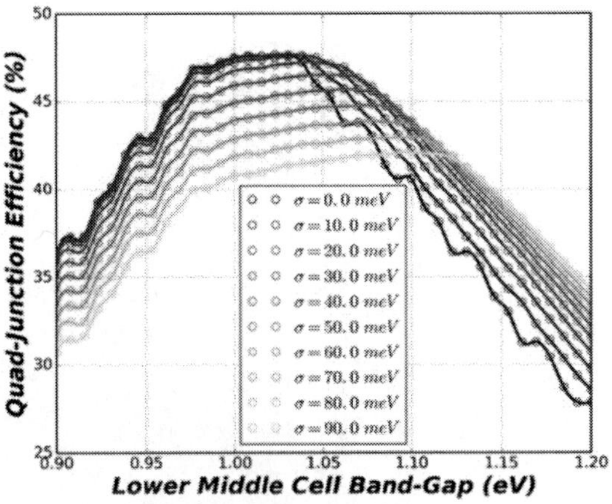

Figure 5: Detailed balance calculations of maximum efficiency in a potential quad-junction structure considering different degrees of broadening.

efficiency of approximately 31% is comparable to that observed by Shockley and Queisser for an optimum band-gap energy of 1.27 eV [1]. As the broadening parameter is increased from 0 to 90 meV, a reduction in maximum efficiency of 10 percentage points is observed with a corresponding blue-shift in the optimum band gap energy from 1.27 eV to 1.61 eV. The effect of σ on maximum efficiency and optimum band-gap energy becomes more significant for values greater than 20 meV, which suggests why GaAs, that is best fitted with a broadening energy of 12 meV as shown in Table I, can be well approximated using an abrupt absorption edge. This is in good agreement with the degradation in V_{OC} shown in Fig. 3 and indicates that the reduction in voltage plays a far greater role in effecting device efficiency than the increase in photogenerated current offered by sub band-gap absorption.

The single-junction calculations clearly highlight the effect of a broadened absorption edge on both the maximum efficiency and corresponding optimum band-gap energy. In a multi-junction device, where the optimum band-gap combinations are carefully selected to achieve the best possible current-matched scenario, a broadened absorption edge will have a pronounced effect on the most desirable sub-cell band-gap energy. Calculated maximum efficiency for a quad-junction device is shown in Fig. 5. The quad-junction band-gap combinations are those used in the potential structure shown in Fig. 1, and the GaAsBi lower-middle sub-cell band-gap energy is varied under different broadening conditions. The maximum quad-junction efficiency decreases from 47.6% to 42.1% as the broadening parameter σ is increased from 0 to 90 meV. A blue-shift in the optimum band-gap energy of the lower middle sub-cell is also observed, moving from 1.01 eV to 1.1 eV. The shape of the curves in Fig. 5 change with increasing σ due to the effect of the broadened band-edge on current matching in the device. Below 0.97 eV the efficiency drops off sharply as the low band-gap of the lower-middle sub-cell makes the bottom cell current limiting. The plateau-like region observed above 0.97 eV indicates the energy range for optimum performance, where the bottom two sub-cells closely match the current generated by the top two sub-cells. The effect of increased photogenerated current absorption as a result of the

Table II: Summary of maximum efficiencies and optimum sub-cell energies for the quad-junction calculations depicted in Fig. 6.

Broadening σ (meV)	η_{max} *(%)*	E_{LM} *(eV)*	E_B *(eV)*
0	47.56	1.012	0.650
30	47.38	1.050	0.700
50	45.98	1.062	0.717
80	43.61	1.087	0.767

broadening band-edge can be clearly observed in the widening of this plateau region. The peak of this region indicates the blue-shifted optimum band-gap energy of the lower-middle sub-cell and above this energy the efficiency rapidly falls away as this sub-cell becomes current limiting. Although the drop in absolute percentage points is limited to 5.5 percentage points for the quad-junction scenario, almost half that observed in the single-junction, the effect of absorption edge broadening on the optimum band-gap energy of the lower-middle sub-cell is still clearly observed. This offers further evidence to show the importance of considering absorption limited performance calculations when assessing the suitability of novel materials that exhibit short-range alloy disorder.

Given that the optimum band-gap energy of the broadened sub-cell has been shown to change it is reasonable to consider this may effect additional sub-cell energies. In the final scenario, the upper two sub-cell energies remain fixed whilst the lower two are varied. As in previous cases, the absorption edge of the lower-middle sub-cell is broadened to simulate the effects of alloy disorder, with energies ranging from σ = 0 meV to σ = 80meV. The results are depicted in Fig. 6 a.) – d.), where the calculated maximum efficiencies (highlighted by the red star) and corresponding optimum sub-cell energies of lower-middle (E_{LM}) and bottom (E_B) sub-cells (indicated by the red dashed-lines) are summarised in Table II. By allowing two sub-cell band-gap energies to vary, the drop in maximum efficiency is reduced further to 3.95 percentage points. Once again the blue-shift of the lower-middle sub-cell optimum energy is observed, however here the optimum band-gap energy of the bottom sub-cell also blue-shifts. E_B shifts from 0.650 eV to 0.767 eV as σ

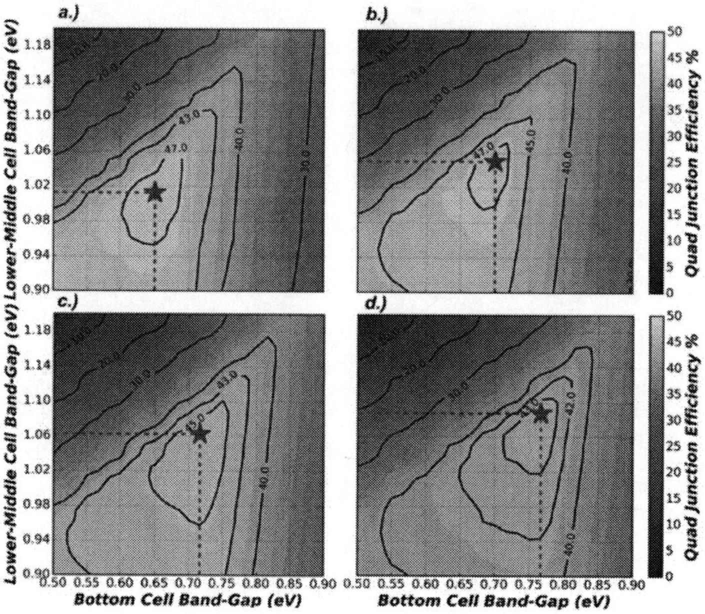

Figure 6: Detailed balance calculations of maximum efficiency for a quad-junction device, where the lower-middle (E_{LM}) and botton (E_B) sub-cell band-gap energies are systematically varied, for broadening parameter values a.) $\sigma = 0$ meV, b.) $\sigma = 30$ meV, c.) $\sigma = 50$ meV and d.) $\sigma = 80$ meV applied to the lower-middle cell.

is increased from 30 to 80 meV indicating the influence of the broadening absorption edge on other sub-cells in the quad-junction. This is a significant result and highlights the importance of considering absorption edge broadening in novel materials when designing future four, or more, junction solar cells.

5. CONCLUSIONS

In this work the effects of absorption edge broadening due to inherent short-range alloy disorder in GaAsBi are evaluated. Experimental absorption coefficient data obtained from the literature is parametrised using a CPPB approach. A clear red-shift of the centre energy E_0 of the fundamental critical point is observed with increasing Bi incorporation, in good agreement with previous results from the literature. The corresponding broadening parameter, Γ_{E0}, increases with Bi fraction up to 5.5% and is attributed to the increased effect of short-range alloy disorder caused by the formation of Bi-pairs and multi-atom clusters. The limiting V_{OC} is calculated using the reciprocity relation that links emissivity to EQE and is observed to decrease with increasing Bi fraction up to 5.5%. This shows that the validity of assuming abrupt 'Shockley-Queisser' like absorption breaks down when wishing to evaluate the performance limits of materials that exhibit short-range alloy disorder, and highlights the importance of considering a broadened absorption edge when calculating optimum band-gap energies.

Detailed balance calculations of a single-junction device with an increasingly broadened absorption edge confirm a reduction in maximum device efficiency from approximately 31% to 21%, which corresponds to a blue-shift in the optimum band-gap energy form 1.27 eV to 1.610 eV. Extending this analysis to the case of a potential quad-junction structure, a smaller reduction in maximum efficiency from 47.6% to 42.1% is observed whilst the optimum energy of the lower middle sub-cell also blue-shifts from 1.01 to 1.10 eV. The increased current density due to sub-band gap absorption results in a widened region of energies where the two upper and two lower sub-cells remain closely current-matched, indicated by a small variation in efficiency. Finally, considering the scenario where the lower-middle and bottom sub-cells are allowed to vary, the maximum efficiency again decreases, from 47.7% to 43.6%, as the broadening of the lower-middle sub-cell is increased from 0 meV to 80 meV. Both lower-middle and bottom sub-cell energies are observed to blue-shift as the broadening is increased, highlighting the role that the broadening of the absorption edge in one sub-cell plays on the optimum band-gap combinations for the whole quad-junction. This shows that considering the role of short-range alloy disorder is of the utmost importance when designing future four or more junction solar cells utilizing novel III-V alloy systems.

REFERENCES

[1] W. Shockley and H. J. Queisser, "Detailed Balance Limit of Efficiency of p-nJunction Solar Cells," *Journal of Applied Physics*, vol. 32, no. 3, pp. 510–519, Mar. 1961.

[2] A. Martí and G. L. Araújo, "Limiting efficiencies for photovoltaic energy conversion in multigap systems," *Solar Energy Materials and Solar Cells*, vol. 43, no. 2, pp. 203–222, Sep. 1996.

[3] "SOLAR JUNCTION TRIPLE JUNCTION SOLAR CELL (TTJ AND SMCC)."

[4] T. Thomas, T. Wilson, M. Führer, D. Alonso-Álvarez, N. J. Ekins-Daukes, D. Lackner, P.

Kailuweit, S. P. Philipps, A. W. Bett, K. Toprasertpong, M. Sugiyama, and Y. Okada, "Potential for reaching 50% power conversion efficiency using quantum heterostructures," presented at the th World Conference on Photovoltaic Energy Conversion, Kyoto, 2014, pp. 1–2.

[5] N. J. Ekins-Daukes, "Cost Requirements for High Volume CPV Manufacturing," presented at the 13TH INTERNATIONAL CONFERENCE ON CONCENTRATOR PHOTOVOLTAIC SYSTEMS (CPV-13), Ottawa, Canada.

[6] D. B. Jackrel, S. R. Bank, H. B. Yuen, M. A. Wistey, J. S. Harris Jr., A. J. Ptak, S. W. Johnston, D. J. Friedman, and S. R. Kurtz, "Dilute nitride GaInNAs and GaInNAsSb solar cells by molecular beam epitaxy," *Journal of Applied Physics*, vol. 101, no. 11, pp. 114916–8, Jun. 2007.

[7] V. Sabnis, H. Yuen, and M. Wiemer, "High-efficiency multijunction solar cells employing dilute nitrides," presented at the 8TH INTERNATIONAL CONFERENCE ON CONCENTRATING PHOTOVOLTAIC SYSTEMS: CPV-8, 2012, pp. 14–19.

[8] A. Aho, A. Tukiainen, V. P. rvi, and M. Guina, "Performance assessment of multijunction solar cells incorporating GaInNAsSb," *Nanoscale Research Letters*, vol. 9, no. 1, pp. 1–7, Feb. 2014.

[9] R. Roucka, A. Clark, T. Wilson, T. Thomas, M. Fuhrer, N. Ekins-Daukes, A. Johnson, R. Hoffman, and D. Begarney, "Demonstrating Dilute-Tin Alloy SiGeSn for Use in Multijunction Photovoltaics: Single- and Multijunction Solar Cells With a 1.0-eV SiGeSn Junction," *IEEE J. Photovoltaics*, vol. 6, no. 4, pp. 1025–1030, May 2016.

[10] T. Wilson, T. Thomas, M. Fuhrer, N. J. Ekins-Daukes, R. Roucka, A. Clark, A. Johnson, R. Hoffman Jnr., and D. Begarney, "Single and multi-junction solar cells utilizing a 1.0 eV SiGeSn junction," presented at the 13TH INTERNATIONAL CONFERENCE ON CONCENTRATOR PHOTOVOLTAIC SYSTEMS (CPV-13), 2016, vol. 1766, pp. 060006–7.

[11] R. D. Richards, A. Mellor, F. Harun, J. S. Cheong, N. P. Hylton, T. Wilson, T. Thomas, J. S. Roberts, N. J. Ekins-Daukes, and J. P. R. David, "Photovoltaic characterisation of GaAsBi/GaAs multiple quantum well devices," *Solar Energy Materials and Solar Cells*, vol. 172, pp. 238–243, Dec. 2017.

[12] K. Alberi, O. D. Dubon, W. Walukiewicz, K. M. Yu, K. Bertulis, and A. Krotkus, "Valence band anticrossing in GaBixAs1−x," *Appl. Phys. Lett.*, vol. 91, no. 5, pp. 051909–4, Jul. 2007.

[13] Z. Batool, K. Hild, T. J. C. Hosea, X. Lu, T. Tiedje, and S. J. Sweeney, "The electronic band structure of GaBiAs/GaAs layers: Influence of strain and band anti-crossing," *Journal of Applied Physics*, vol. 111, no. 11, pp. 113108–8, Jun. 2012.

[14] D. L. Sales, E. Guerrero, J. F. Rodrigo, P. L. Galindo, A. Yáñez, M. Shafi, A. Khatab, R. H. Mari, M. Henini, S. Novikov, M. F. Chisholm, and S. I. Molina, "Distribution of bismuth atoms in epitaxial GaAsBi," *Appl. Phys. Lett.*, vol. 98, no. 10, pp. 101902–4, Mar. 2011.

[15] R. Kudrawiec, M. Syperek, P. Poloczek, J. Misiewicz, R. H. Mari, M. Shafi, M. Henini, Y. G. Gobato, S. V. Novikov, J. Ibáñez, M. Schmidbauer, and S. I. Molina, "Carrier localization in GaBiAs probed by photomodulated transmittance and photoluminescence," *Journal of Applied Physics*, vol. 106, no. 2, pp. 023518–6, Jul. 2009.

[16] S. Imhof, A. Thränhardt, A. Chernikov, M. Koch, N. S. Köster, K. Kolata, S. Chatterjee, S. W. Koch, X. Lu, S. R. Johnson, D. A. Beaton, T. Tiedje, and O. Rubel, "Clustering effects in Ga(AsBi)," *Appl. Phys. Lett.*, vol. 96, no. 13, pp. 131115–4, Mar. 2010.

[17] Y. I. Mazur, V. G. Dorogan, M. Schmidbauer, G. G. Tarasov, S. R. Johnson, X. Lu, M. E. Ware, S. Q. Yu, T. Tiedje, and G. J. Salamo, "Strong excitation intensity dependence of the photoluminescence line shape in GaAs 1−xBi xsingle quantum well samples," *Journal of Applied Physics*, vol. 113, no. 14, pp. 144308–6, Apr. 2013.

[18] O. Rubel, M. Galluppi, S. D. Baranovskii, K. Volz, L. Geelhaar, H. Riechert, P. Thomas, and W. Stolz, "Quantitative description of disorder parameters in (GaIn)(NAs) quantum wells from the temperature-dependent photoluminescence spectroscopy," *Journal of Applied Physics*, vol. 98, no. 6, pp. 063518–8, Sep. 2005.

[19] F. Urbach, "The Long-Wavelength Edge of Photographic Sensitivity and of the Electronic Absorption of Solids," *Phys. Rev.*, vol. 92, no. 5, pp. 1324–1324, Dec. 1953.

[20] C. J. Hunter, F. Bastiman, A. R. Mohmad, R. Richards, J. S. Ng, S. J. Sweeney, and J. P. R. David, "Absorption Characteristics of $GaAs_{1-x}Bi_x$/GaAs Diodes in the Near-Infrared," *IEEE Photon. Technol. Lett.*, vol. 24, no. 23, pp. 2191–2194, Nov. 2012.

[21] C. Gogineni, N. A. Riordan, S. R. Johnson, X. Lu, and T. Tiedje, "Disorder and the Urbach edge in dilute bismide GaAsBi," *Appl. Phys. Lett.*, vol. 103, no. 4, pp. 041110–5, Jul. 2013.

[22] M. Masnadi-Shirazi, R. B. Lewis, V. Bahrami-Yekta, T. Tiedje, M. Chicoine, and P. Servati, "Bandgap and optical absorption edge of GaAs 1−xBi xalloys with 0 < x < 17.8%," *Journal of Applied Physics*, vol. 116, no. 22, pp. 223506–9, Dec. 2014.

[23] M. Usman, C. A. Broderick, Z. Batool, K. Hild, and T. Hosea, "Impact of alloy disorder on the band structure of compressively strained GaBi x As 1− x," *Phys. Rev. B*, 2013.

[24] U. Rau and J. H. Werner, "Radiative efficiency limits of solar cells with lateral band-gap fluctuations," *Appl. Phys. Lett.*, vol. 84, no. 19, pp. 3735–3737, May 2004.

[25] R. B. Lewis, M. Masnadi-Shirazi, and T. Tiedje, "Growth of high Bi concentration GaAs 1−xBi xby molecular beam epitaxy," *Appl. Phys. Lett.*, vol. 101, no. 8, pp. 082112–5, Aug. 2012.

[26] S. Adachi, "Model dielectric constants of GaP, GaAs, GaSb, InP, InAs, and InSb," *Phys. Rev. B*, vol. 35, no. 14, pp. 7454–7463, May 1987.

[27] S. Adachi, "Optical dispersion relations for GaP, GaAs, GaSb, InP, InAs, InSb, Al xGa 1− xAs, and In 1− xGa xAs yP 1− y," *Journal of Applied Physics*, vol. 66, no. 12, pp. 6030–6040, Dec. 1989.

[28] S. Adachi, *Physical Properties of III-V Semiconductor Compounds*. Weinheim, FRG:

Wiley-VCH Verlag GmbH & Co. KGaA, 2005.
[29] D. Alonso-Álvarez, T. Wilson, P. Pearce, and N. Ekins-Daukes, "Solcore: A multi-scale python-based library for the modelling of solar cells and semiconductor materials ," *Submitted to Computer Physics Communications*, pp. 1–21.
[30] P. Wurfel, "The chemical potential of radiation," *J. Phys. C: Solid State Phys.*, vol. 15, no. 18, pp. 3967–3985, Jun. 1982.
[31] G. L. Araújo and A. Martí, "Absolute limiting efficiencies for photovoltaic energy conversion," *Solar Energy Materials and Solar Cells*, vol. 33, no. 2, pp. 213–240, Jan. 1994.
[32] U. Rau, "Reciprocity relation between photovoltaic quantum efficiency and electroluminescent emission of solar cells," *Phys. Rev. B*, vol. 76, no. 8, pp. 435–8, Aug. 2007.
[33] J. Nelson, *The Physics of Solar Cells*, 1st ed. PUBLISHED BY IMPERIAL COLLEGE PRESS AND DISTRIBUTED BY WORLD SCIENTIFIC PUBLISHING CO., 2011.
[34] M. A. Green, Y. Hishikawa, W. Warta, E. D. Dunlop, D. H. Levi, J. Hohl-Ebinger, and A. W. H. Ho-Baillie, "Solar cell efficiency tables (version 50)," *Prog. Photovolt: Res. Appl.*, vol. 25, no. 7, pp. 668–676, Jun. 2017.

A FULL TRANSPARENT ELECTRODE APPLICATION IN III-V COMPOUND SOLAR CELL

Pan Dai[1], Shiro Uchida[2], Ming Tan[1], Wenxian Yang, Meng Xiao[1], Yuanyuan Wu[1], Lifeng Bian[1], Shulong Lu[1]*, and Hui Yang[1]

[1]Key Lab of Nanodevices and Applications, Suzhou Institute of Nano-Tech and Nano-Bionics, Chinese Academy of Sciences (CAS). Suzhou Industrial Park, Ruoshui Road 398, Suzhou, 215123, China

[2]Department of Mechanical Science and Engineering Faculty of Engineering, Chiba Institute of Technology, 2-17-1, Tsudanuma, Narashino, Chiba 275-0016, Japan

*Correspondence E-mail: sllu2008@sinano.ac.cn

ABSTRACT: The application of transparent conducting indium-tin-oxide (ITO) film as full front electrode and hybrid electrode replacing the conventional bus-bar metal electrode in III-V compound solar cell was proposed. High quality ITO film was deposited by electron beam evaporation with low resistivity and high average trasmittance. A high quality non-rectifying contact between ITO and 10 nm N^+-GaAs contact layer was formed, which is benefiting from a high carrier concentration of terrilium-doped N^+-GaAs layer, up to 2×10^{19} cm^{-3}. The absorption, transmittance and reflectance of ITO with different thickness were investigated. A new solar cell structure was designed to simplify the device fabrication procedure. A high efficiency GaAs solar cell was fabricated by using 30 nm ITO hybrid electrode. A good device performance of the GaInP solar cell with the full ITO electrode was observed under concentration. This result indicates a great potential of transparent conducting films in the future fabrication of larger area flexible III-V solar cell.

Keywords: III-V semiconductors, ITO, solar cell, electrode

1 INTRODUCTION

III-V compound multi-junction solar cells have achieved ultrahigh conversion efficiency due to their wide spectral absorption of solar spectrum [1-3]. Numerous studies have been extensively conducted with different epitaxial growth technique, differently designed device structures and various fabrication methods to improve their performances [4-6].One of the issues which limited the improvement of the convention efficiency for the conventional solar cells is the trade-off between the shadow loss and the current extraction capability [7]. To reduce the shadow loss and to increase the current extraction capability, transparent conducting film was proposed to be used as the new electrode to replace bus-bar metals in the III-V compound solar cells [8]. The indium-tin-oxide (ITO) film has been widely utilized as transparent conducting electrode in various optoelectronic devices because of its excellent electrical and optical properties including high electrical conductivity and transparency to visible and infra-light [9-11].

However, it's difficult to form good ohmic contact between ITO film and III-V semiconductor. And a bad contact would increase the series resistance of the device and deteriorate its efficiency. Therefore, a hybrid electrode structure consisting of both metal contact pads and ITO films is universally employed to enhance the device performance, using the metal contact pads to form ohmic contact [12,13]. But such hybrid structure increases the complexity of the device fabrication procedure. A easy way to fabricate the device with only ITO as the electrod is therefore highly anticipated [14,15].

In this paper, we proposed the application of transparent conducting ITO film as the new full front electrode to replace the conventional bus-bar metals in III-V compound solar cell. A new solar cell structure was grown to simplify the fabrication process. The ITO film with low resistivity and high average trasmittance was deposited by electron beam evaporation. The contact resistance between ITO and N-GaAs was investigated. The absorption, transmittance and reflectance of ITO was measured. A high efficiency of 25.6% was obtained for GaAs solar cell with ITO hybrid electrode. The successful operation of the full ITO electrode under concentration not only simplifies the device fabrication process, but also gives great promise to the future application of transparent conducting films in larger area flexible III-V solar cell fabrication.

2 EXPERIMENTS

2.1 Material growth of the solar cells

The epitaxial material of the GaAs and GaInP solar cell were grown on p-type GaAs substrate by all solid-state molecular beam expitaxy (MBE). The structure of GaAs solar cell was grown on the p-type GaAs substrate, which is composed of a p+-GaInP back surface field (BSF), a p-GaAs base layer, an n+-GaAs emitter layer, an n+-AlInP window layer and n+-GaInP contact layer (10 nm, 1×10^{19} cm^{-3}). The structure of GaInP solar cell grown on a p-type GaAs substrate is composed of a p-GaInP base layer (650 nm with the carrier concentration of 3.5×10^{16} cm^{-3}), an n+-GaInP emitter layer (80 nm with 4×10^{18} cm^{-3}), an AlGaInP (30 nm with 4×10^{18} cm^{-3}) as the back surface field layer and an AlInP (20 nm of 4×10^{17} cm^{-3}) as the window layer to decrease the surface recombination. During the GaInP growth, a typical temperature of about 530 ℃ was used. The V/III ratio was approximately 80 and the growth rate was 1 um/h. After the growth of the AlInP window layer, a 10 nm heavily-doped GaAs with the solid-state Te as the n-type dopant was grown to form a good ohmic contact with the ITO film.

2.2 Fabrication of ITO solar cell

The ITO target composition in this work was 95% In_2O_3 and 5% SnO_2 in weight. The evaporation was processed with a constant substrate temperature of 300 ℃. Before the ITO deposition, the surface of the device was cleaned with acetone, isopropanol and deionized water and blown dry using nitrogen gas. The 100-nm-thick ITO film were deposited onto the 10 nm n+-GaAs contact layer of GaInP solar cell via optical thin film coater (OTFC-900). For GaAs solar cell, 170 nm and 30 nm ITO film were deposited onto the contact layer, respectively. During evaporation, the deposition rate was maintained at 0.5 nm/s by controlling the current and voltage. To obtain the best ohmic contact between the ITO film and the n+-GaAs layer, the sample was annealed at different temperatures for 3 min in N2 ambient. The circular transmission line model (c-TLM) patterns made of photoresist were formed using standard photolithography to measure the contacting resistance. The optical properties of ITO film were measured by PekinElmer LAMBDA 750 UV/Vis/NIR spectrophotometer. Devices were processed following the standard III-V solar cell device art. The size of the device was 2.5×2.5 mm. No other anti-reflector coating (ARC) layer was deposited on the test GaInP devices. The typical current-voltage (I-V) characteristics were carried out under air mass 1.5-global (AM1.5G) illumination.

3 RESULTS AND DISCUSSION

3.1 Properties of ITO

Low resistivity of the material is necessary for using as an electrode. It is indicated that annealing can affect the resistivity of ITO material considerably. As the rapid thermal annealing (RTA) temperature increases, the resistivity of ITO film decreases. The resistivity of ITO could be decreased by about 30% after a heat treatment at around 420 ℃ for 150 s under nitrogen ambient. Futher increasing the RTA temperature could do less help in the material resistivity because of the trad-off between material carrier concentration and grain size of the polycrystalline. A resistivity of ITO film of lower than $2\times10^{-4}\,\Omega\cdot cm$ was obtained, which was suitable to applied to a solar cell.

The contact resistance can be significantly affacted by different annealing conditions and different doping densities of materials. During annealing, the indium, tin and oxygen diffusion all play important roles in the contact resistance of ITO/GaAs interface. Figure 1 shows the contact resistance of ITO/n-GaAs under different rapid thermal annealing in N2 ambient. The specific contact resistance decreased with increasing annealed temperature at first, and then increased. With the sample was annealed at 480 ℃ for 3 min, a good ohmic contact between the ITO film and n-GaAs layer with a low resistivity of $7\times10^{-5}\,\Omega\cdot cm^2$ was obtained, which is in the case of the carrier concentration of Si-doped GaAs is 7.6×10^{18} cm^{-3}. In the same annealing condition, the lowest resistance between ITO film and terrilium (Te)-doped GaAs was $2\times10^{-5}\,\Omega\cdot cm^2$.

To achieve a good Ohmic contact between III-V semiconductor and ITO film, high doping concentration of both materials are critical and necessary, so that the carriers can tunnel more easily between GaAs and ITO. Therefore, continuous effort to decrease the contacting resistance relies highly on the increased carrier concentration of the contact layer. The inset in Figure 1 presents the contacting resistance of n-GaAs with ITO for different carrier concentrations of n-GaAs. With increasing concentration, the contact resistance decreased. However, for Si-doped GaAs, the dopant concentration higher than 5×10^{18} cm^{-3} is difficult to achieve, because it suffers from a significant self-compensation problem due to the amphoteric dopant nature of Si. In contrast to Si, terrilium (Te) occupies only the group-V lattice sites and allows a higher carrier concentration. In our work, solid-state Te was used as the dopant source, because it is environmentally friendly. A high carrier concentration of 2×10^{19} cm^{-3} was obtained for GaAs:Te [16]. In this case, the contacnt resistance was as low as $2\times10^{-5}\,\Omega cm^2$, which is comparable to that of mental contact.

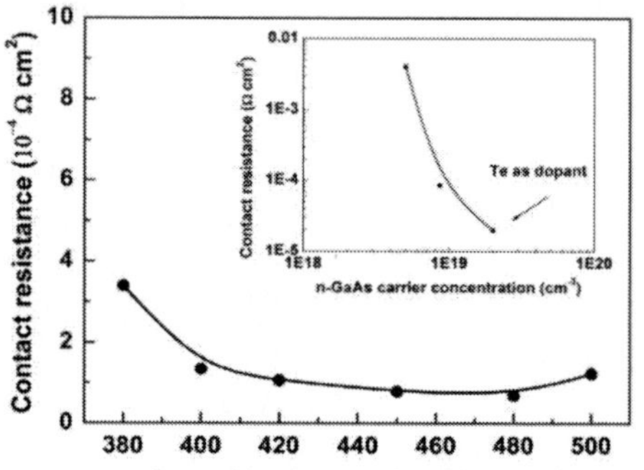

Figure 1: The contact resistance between ITO and GaAs versus the annealing temperature. The inset shows the contacting resistance of n-GaAs with ITO as a function of carrier concentration of n-GaAs.

To avoid the energy loss, high trasmittance and low absorbtion of the ITO film itself is highly demanded. Figure 2 shows the weighted average absorption loss of pure ITO film as a function of the film thickness. The absoption increases with the thickness of ITO increases. A relatively thicker ITO film would absorpt more solar energy. While, the weighted average absorption loss is still lower than 5% when the pure ITO film is less than 200 nm. A 100-nm-thick ITO film was deposited on a glass substrate. The absorption and transmittance are shown in the inset of Fig. 2. It could be found that the average absorption loss is very low of less than 5% and the transmittance is higher than 90% in the visible wave-length range. It indicated that the conductive ITO film is transparent enough to be applied in optoelectronic devices.

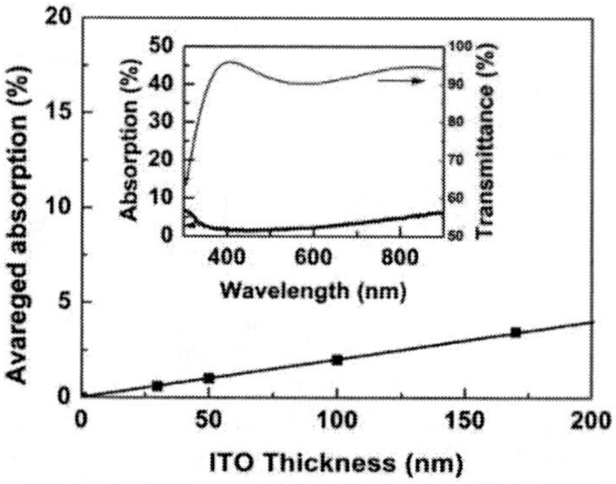

Figure 2: The weighted average absorption loss of pure ITO film as a function of the film thickness. The inset shows the absorption and transmittance of 100-nm-thick ITO film.

3.2 Applicaton in GaAs solar cell

Our first design was to introduce a ITO combined with grid structure which was called hybrid electrode. To enhance the current extraction capability and to simiplify the device fabrication process, a new solar cell structure and fabication mathod were used. Compared to the conventional solar cell with an 300-nm-thickness contact layer, the contact layer of the new structure was only 10 nm. And the thinner contact layer was not etched during the solar cell fabrication process. For GaAs solar cell, GaInP was used as the contact layer to avoide absorption and the doping density is as high as 1×10^{19} cm^{-3} to reduce the contact resistance. The ITO film was directly deposited on the entire surface of the contact layer. The advantage of this design is that ITO film can function as both current spreading layer and antireflection layer. The other antireflection layers were deposited on the ITO film between the intervals among Ti/Pt/Au metal contact pads. The struture of the GaAs solar cell with hybrid electrode was shown in figure 3.

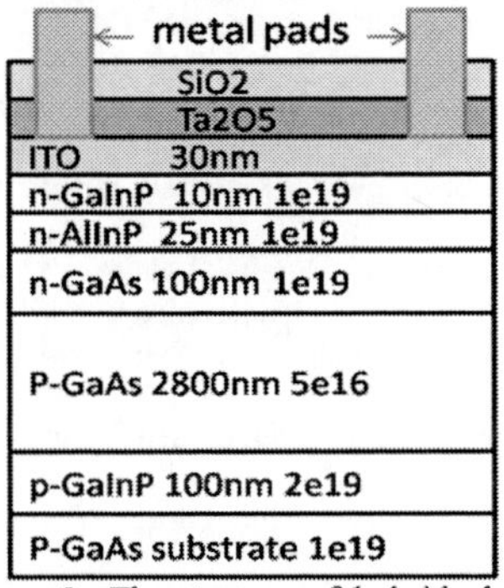

Figure 3: The struture of hybrid electrode GaAs solar cell.

The illuminated currentdensity-voltage (*J-V*) characteristics of the GaAs solar cells with different ITO thicknesses under AM 1.5G were shown in figure 4. For GaAs cell with the 30 nm ITO, a high efficiency of 25.6% is obtained with an open-circuit voltage (V_{oc}) of 1.031 V, a short-circuit current (J_{sc}) of 28.72 mA/cm^2 and a fill factor (*FF*) of 86.50%. The efficiency is comparative to and the V_{oc} is even higher than the best of our conventional GaAs cell because of the high quality ITO film [18]. For 170 nm-ITO trasparent eletrode GaAs cell, the efficiency is much lower of only 18%.

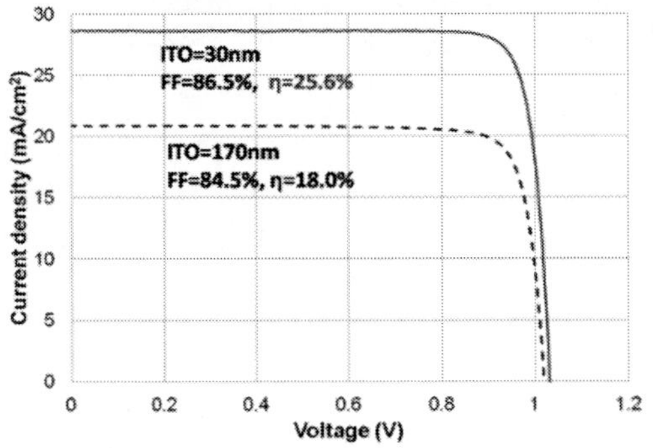

Figure 4: The *J-V* characteristics of hybrid electrode GaAs solar cell with 30 nm ITO (red line) and 170 nm ITO (black line).

The difference mainly stems from the reflection and absorption loss of ITO with different thicknesses. The ITO film combined with Ta2O5 and SiO were deposited as the antireflection coating. For 170-nm-thickness ITO film, the average reflection loss was as high as 15.09%, while it reduced to 2.95% when the ITO was 30 nm. It is induced that thinner ITO is effective for a reduction of reflection. But it is difficult to reduce an reflection loss at the case with thicker ITO, which has the advantage to reduce a lateral resistance. In addition, the weighted average absorption loss of 30 nm ITO film is extremely low of 0.5% compared with that of 170 nm ITO. To apply transparent conducting ITO film as the new full front electrode to replace the conventional bus-bar metals and to improve the solar cell performance, a proper thickness of ITO should be employed.

3.3 ITO as full front electrode in GaInP solar cell

Since GaInP cell is usually on the top of the multi-junction solar cells, the application of ITO to GaInP solar cell was investigated. From our previous study, it is induced that the contact resistance between ITO and GaInP is much higher than that between ITO and GaAs. On the basis of the success contact between GaAs and ITO films, a 10 nm highly Te-doped N+-GaAs was inserted between ITO and AlInP window layer and a GaInP solar cell with the full ITO as the contact layer was fabricated. A high quality non-rectifying contact between ITO and 10 nm N+-GaAs contact layer was formed and the resistance was as low as $2\times10^{-5}\,\Omega$cm^2.

Figure 5 shows the *J-V* characteristics of GaInP solar cell with the full ITO film electrode under AM1.5G illumination condition. The short-circuit current density J_{sc}, V_{oc} and *FF* are 10.51 mA/cm^2, 1.24 V and 80%, respectively, corresponding to a total power conversion efficiency of 10.42%. The lower V_{oc} lies in the imperfect material quality of GaInP as well as its doping [19]. The total photocurrent was lower comparable to our previous reported value in references [17, 18]. Althought the current density was partly benefiting from the nearly zero shadowing loss of ITO film, the high reflectivity of ITO drives down this positive effect. Because the ITO film and highly Te-doped GaAs contact layer was direct contacted without ARC, the upper surface of the device was covered by pure ITO film. Eventhough the ITO film could serve as a part of ARC while being used as the electrode, the reflectivity at about 400 nm is still quite high at about 30% and the average reflectivity is about 14.5%.

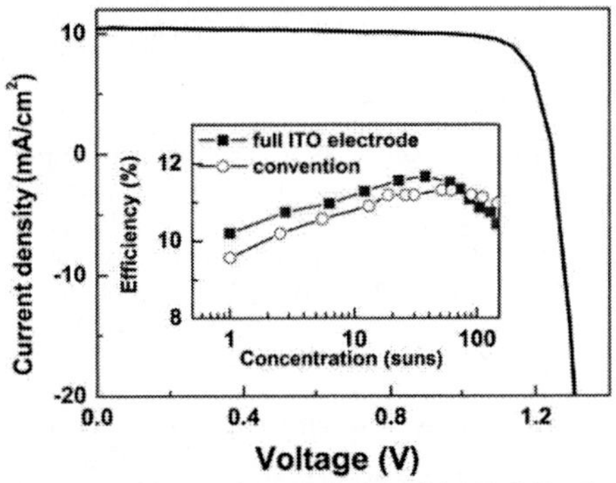

Figure 5: The *J-V* characteristics of the GaInP solar cells with the full ITO film electrode at 1 sun, AM1.5G. The inset shows the efficiency as a function of concentration for GaInP solar cells with the full ITO film electrode (blue square) and with bus-bar metal electrode (black circle).

The inset shows the efficiency results of the GaInP solar cell with the full ITO electrode and with the bus-bar metal electrode under concentration. The device with the full ITO electrode shows a better performance with increasing concentration. This is benefiting from a good lateral current expansion in ITO electrode and none shadow loss. A possible mechanism is that the carriers can easily tunnel across the barrier between the N+-GaAs contact layer and ITO film, because the contact layer is highly doped and thin enough. In the conventional concentration solar cell using metal electrode, in order to decrease the lateral series resistance, the electrode shall be thickened. While by using ITO electrode, carrier were directly collected by ITO film instead of transportation through metal contact pads. At the same time, the shadow loss problem was overcomed. In addition, the easier fabrication of ITO will be helpful to the future fabrication of large area flexible III-V solar cells. A further study to increase the performance of the ITO-based device will focus on optimizing ITO thickness and combining it with the ARC and reducing the reflection loss.

4 CONCLUSION

In conclusion, we successfully applied the ITO film as the hybrid electrode and full front electrode of in III-V compound solar cells to replace the conventional bus-bar metal electrode. An extremely low contact resistance of 2×10^{-5} $\Omega\cdot cm^2$ between ITO and 10 nm N+-GaAs were obtained benefiting from the optimization of thermal annealing and Te-doped N+-GaAs layer. The pure ITO has a low absorption and high trasmittance. A new solar cell structure was designed to simplify the fabrication procedure. A high efficiency of GaAs solar cell was achieved by using 30 nm ITO as hybrid electrode, which had an extremely low average reflection loss. A good device performance of GaInP solar cell with the full ITO electrode under concentration were observed. By using ITO electrode, the lateral resistance can be reduced without increasing its thickness and the shadow loss problem in the conventional metal electrode was avoided. The successful operation of the full ITO electrode not only simplifies the device process, but also gives a great promise to the future application of transparent conducting films in larger area flexible III-V solar cell fabrication.

5 ACKNOWLEDGEMENT

This work is supported in part by the National Natural Science Foundation of China (Grant Nos. 61704186, 61774165, 61534008 and 61376081), the Key Frontier Scientific Research Program of the Chinese Academy of Sciences (Grant No. QYZDB-SSW-JSC014), the Science and Technology Service Network Initiative of the Chinese Academy of Sciences, the Key R&D Program of Jiangsu Province (Grant No. BE2016085) and the External Cooperation Program of BIC, Chinese Academy of Sciences, (Grant No. 121E32KYSB20160071). We are thankful for the technical support from Nano Fabrication Facility, Platform for Characterization & Test of SINANO, CAS.

6 REFERENCES

[1] R.R. King, D.C. Law, K.M. Edmondson, C.M. Fetzer, G.S. Kinsey, H. Yoon, R.A. Sherif and N.H. Karam, Applied Physics Letters 90 (2007) 183516.

[2] R.R. King, D. Bhusari, A. Boca, D. Larrabee, X.Q. Liu, W. Hong, C.M. Fetzer, D.C. Law and N.H. Karam, Progress in Photovoltics: Research and Applications 19 (2011) 797.

[3] W. Guter, J. Schoene, S. P. Philipps, M. Sterner, G. Siefer, A. Wekkeli, E. Welser, E. Oliva, A.W. Bett and F. Dimroth, Applied Physics Letters 94 (2009) 223504.

[4] J.F. Geisz, D.J. Friedman, J.S. Ward, A. Duda, W.J. Olavarria, T.E. Moriarty, J.T. Kiehl, M.J. Romero, A. G. Norman and K.M. Jones, Applied Physics Letters 91 (2007) 023502.

[5] N. Li, K. Lee, C.K. Renshaw, X. Xiao and S.R Forrest, Applied Physics Letters 98 (2011) 053504.

[6] L.W. Lai, J.T. Chen, L.R. Lou, C.H Wu, C.T. Lee, Journal of Electrochemical Society 155 (2008) B1270

[7] K. Nishioka, T. Takamoto, T. Agui, M. Kaneiwa, Y.Uraoka, T. Fuyuki, Solar Energy Materials & Solar Cells 90 (2006) 1308.

[8] C.G. Granqvist, Solar Energy Materials & Solar Cells 91 (2007) 1529.

[9] G. Sberveglieri, Comini E, Faglia G, Ferroni M, G. Jimenez and A.Vomiero, In Conference Record of the Thirty Third IEEE Photovoltaic Specialists Conference, 11 (2008) 1-3.

[10]J.L. Hou, C.H. Wu and S.J. Chang , IEEE Electron Device Letters 36 (2015) 207.

[11]Y. Igasaki and H. Kanma, Applied Surface Science 169 (2001) 508.

[12]C.Y. Tseng and C.T. Lee, Solar Energy 89 (2013) 17.

[13]C.Y. Tseng, C.K. Lee and C.T. Lee, Progress in Photovoltics: Research and Applications 19 (2011) 436.

[14]C.W. Cheng, K.T. Shiu, N. Li, S.J. Han, L. Shi and D. K. Sadana, Nature Communication 4 (2013) 1577.

[15] B.M. Kayes, L. Zhang, R. Twist, I.K. Ding and G.S. Higashi, IEEE Journal of Photovoltaics 4 (2014) 729.

[16]P. Dai, M. Tan, Y.Y. Wu, L. Ji, L.F. Bian, S.L. Lu and H. Yang, Journal of Crystal Growth 413 (2015) 71.

[17]S.L. Lu, L. Ji, W. He, P. Dai, H. Yang, M. Arimochi, H. Yoshida, S. Uchida and M. Ikeda, Nanoscale. Research Letters 6 (2011) 576.

[18]P. Dai, S.L. Lu, M. Arimochi, S. Uchida, T. Watanabe, X.D. Luo and H. Yang Solid State Communication 200 (2014) 9.

DESIGN OF BROADBAND AND OMNIDIRECTIONAL ANTIREFLECTION COATINGS FOR III-V CONCENTRATING MULTIJUNCTION SOLAR CELLS

L.C. Andreani[1,*], M. Liscidini[1], M. Passoni[1], M. Patrini[1], G. Timò[2], F. Trespidi[2]
[1]Department of Physics, University of Pavia, Via Bassi 6, I-27100 Pavia, Italy
[2]Ricerca sul Sistema Energetico (RSE), Strada Torre della Razza, loc. Le Mose, I-29122 Piacenza, Italy
*E-mail: lucio.andreani@unipv.it

ABSTRACT: We present an in-depth theoretical analysis of nanostructured antireflection coatings (ARCs) for broadband and omnidirectional operation in III-V multijunction solar cells. We analyse the performance of nanostructured ARCs within three different models, namely an effective medium with a gradual variation of the average refractive index, a periodic nanostructure, and a fully random nanostructure. We show that the effective medium structure generally outperforms the periodic and random ones, while the three models agree well with each other when the length scale of the nanostructures is much smaller than the wavelength of visible light. We discuss the expected improvement over the performance of the present "standard" ARC structure based on SiO_2/Ta_2O_5 bilayers.
Keywords: III-V multijuncion solar cells, concentration photovoltaics, antireflection coatings, theory and simulation

1 Introduction

III-V multijunction (MJ) solar cells hold the record photovoltaic conversion efficiency values, namely 44.7% for triple-junction cells based on inverted metamorphic InGaP/InGaAs/Ge epitaxial structures, and 46% for four-junction cells realized with epitaxial growth followed by wafer bonding [1]. To improve beyond these values, it is of paramount importance to minimize all sources of optical and electrical losses. In this framework, minimizing reflection losses over the entire solar spectrum and in a wide range of incidence angles is a key task. The relevant range of incidence angles is dictated by the optics of the concentrating system. Various kinds of nanostructured coatings have been grown recently and the study of their optical performance is of great current interest [2-7].

In the present work we theoretically analyse nanostructured antireflection coatings (ARCs) to be grown on top of InGaP/InGaAs/Ge triple-junction solar cells, possibly with an AlInP window layer. Our goal is threefold, namely (i) to identify ARC structures that improve upon the "standard" structure based on SiO_2/Ta_2O_5 bilayers, (ii) to analyse the effect of nanostructuring on the optical response, and (iii) to give guidelines for the realization of improved-performance ARCs by means of epitaxial deposition, evaporation or sputtering, and possibly other nanofabrication techniques like optical or soft lithography.

2 STRUCTURES AND MODELS

The structures investigated in this work are schematically shown in Fig. 1. The standard structure consists of SiO_2, Ta_2O_5 and AlInP layers with optimized thickness values, which are deposited over the InGaP top cell of the multijunction stack. The double composite structure employs two distinct composites, which implement a gradual variation of the refractive index from unity to the InGaP value. Each graded composite is obtained from two materials A and B, with a refractive index that changes gradually from n_A to n_B. To this purpose, the fraction f_B of material B must increase from 0 to 1 across the depth of the graded composite. In a medium that is composed of two materials with complex dielectric functions ε_A, ε_B, the effective dielectric function ε is calculated within effective-medium theory in the Bruggeman approximation:

$$(1-f_B)\frac{\varepsilon_A-\varepsilon}{\varepsilon_A+2\varepsilon}+f_B\frac{\varepsilon_B-\varepsilon}{\varepsilon B+2\varepsilon}=0.$$

This approximation allows to calculate the complex refractive index $N=n+i\kappa=\varepsilon^{1/2}$ of the medium.

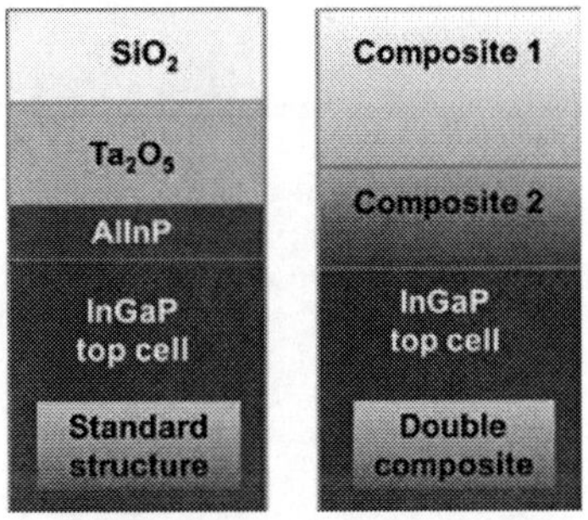

Figure 1: Schematic representation of the ARC structures analyzed in this work. The standard structure employs a SiO_2/Ta_2O_5 bilayer and an AlInP window layer on top of the MJ solar cell. The composite structure employs two graded-index composites with a gradual variation of the refractive index.

It is not trivial to determine the best compositional profile that allows minimizing the reflectance of a graded composite with specified thickness. In this work, each graded composite is represented by a multilayer made of N layers with thicknesses L_1, L_2, …., L_N, each of which is a composite medium in which the filling fraction of material 2 has the value $f_1, f_2, \dots, f_N$. The filling fraction f_j of material 2 is chosen to increase linearly from 0 to 1, while the thickness L_j is taken to be inversely proportional to the corresponding effective refractive index n_j, like in $\lambda/4$ distributed Bragg reflectors. With the condition $L_1+L_2+\dots+L_N=L$, this implies the following choice of the sublayer thicknesses:

$$L_j=\frac{L_j/n_j}{\sum_j(1/n_j)}.$$

This choice of the refractive index profile and of the sublayer thicknesses is assumed in all designs of graded composites structures that are analyzed in the following.

The graded composites are described by means of three different models:

(1) An effective medium model, in which all sublayers are considered to be homogeneous with an effective refractive index that changes gradually from the value of air to that of AlInP. Each sublayer is treated by Bruggeman effective medium theory, Eq. (1), and the thicknesses are determined by Eq. (2).

(2) A periodic model, in which optical nanostructures with a given in-plane period Λ are studied. This model is shown schematically in Fig. 2.

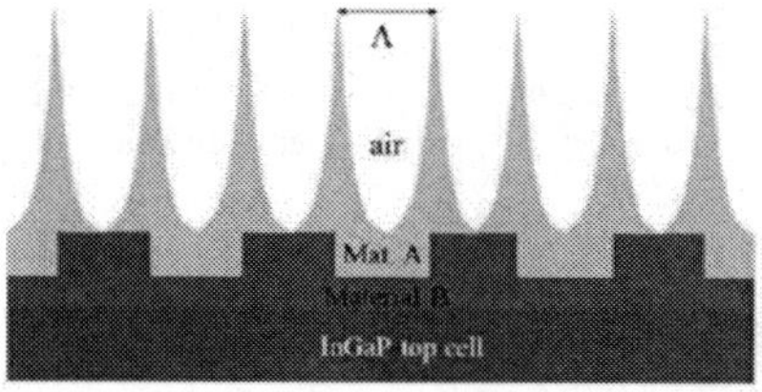

Figure 2: Periodic model of the graded double composite structure. The first composite is made of air and material A, while the second composite is made of materials A and B. The peculiar shape of the protrusions of materials A in the first composite is due to the $\lambda/4$ condition for the sublayer thicknesses, Eq. (2).

(3) A random model, in which the first composite in the ARC is represented by a transition region with Gaussian disorder. The disorder model, taken over from previous studies of light trapping in solar cells [8,9], is characterized by a RMS deviation of the height σ and an in-plane correlation length l_c. This model is shown schematically in Fig. 3.

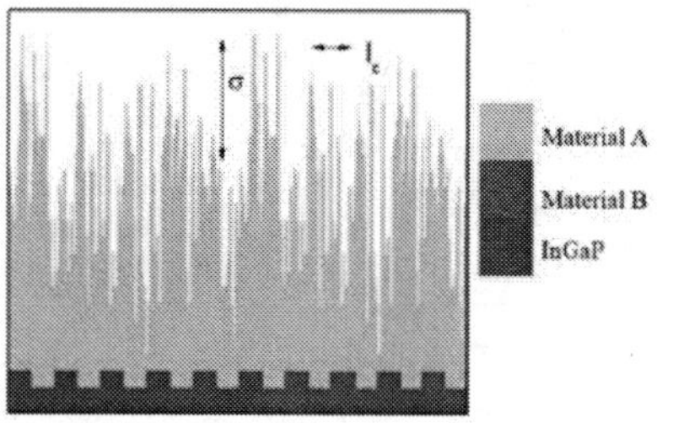

Figure 3: Random model of the graded double composite structure. The first composite is made of air and of material A, while the second composite is made of materials A and B. The first composite is characterized by a Gaussian roughness described by a RMS deviation of the height σ and an in-plane correlation length l_c.

Model (1), which cannot be realized physically, is a limiting case of models (2) and (3) when the relevant length scale (i.e., the periodicity L or the in-plane correlation length l_c) is much smaller than the wavelength of light in the solar spectrum. Model 2 represents a structure that can be realized by epitaxial growth or deposition, possibly followed by lithographic and etching steps. Model 3 is a simple model of a random nanostructure, which can be fabricated e.g. by proper deposition, yielding spiral or elicoidal structures [10].

Model (1) is a simple idealized realization of a graded composite structure and is studied with the transfer-matrix method. Models (2) and (3) are more complex, yet they depend on a small number of parameters, and they are solved by means of Rigorous Coupled-Wave Analysis (RCWA) [11]. In the following we analyse and compare the results of the different models.

3 OPTICAL REFLECTANCE SPECTRA

In Fig. 4 we show optical reflectance spectra of the standard structure, compared with that of the double composite treated in the effective medium approximation. Obviously the reflectance increases with the angle of incidence due to Fresnel laws. Compared to the standard structure, the double composite has lower reflectivity over the visible/IR spectrum and much better angular properties.

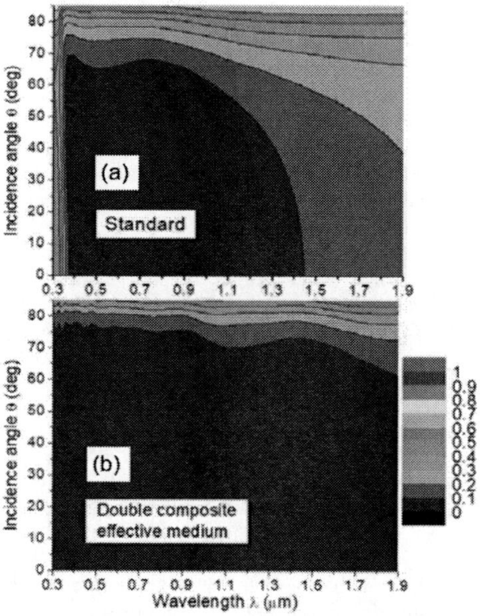

Figure 4: Optical reflectance spectra of the standard structure (a) and of the double composite treated in the effective medium approximation (b) at different values of the angle of incidence. The spectra are averaged over TE/TM polarizations.

In Figure 5 we show the reflectance spectra on an expanded scale (reflectance from 0 to 15%) for the double composite treated as an effective medium (model 1), as a periodic medium (model 2) and as a random medium (model 3). The overall behavior of the reflectance, i.e. the increase with the incidence angle, is similar when calculated with the three models. The sharp oscillations of the reflectance in the periodic and, especially, in the random model, are due to diffraction and diffusion processes in models 2 and 3, respectively. The overall electrical behavior of the MJ solar cell depends on the photogeneration rate that is integrated over the solar spectrum, thus the detailed features of optical spectra are not especially relevant. In the next Sections we discuss the spectrally integrated quantities.

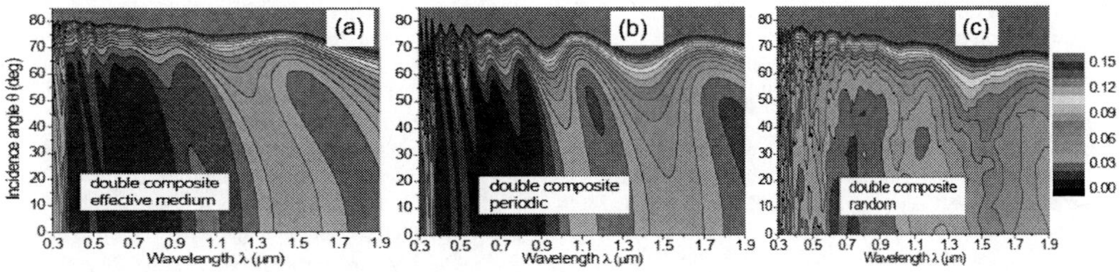

Figure 5: Optical reflectance spectra, averaged over TE/TM polarizations, for the double composite structure treated as an effective medium (a), as a periodic medium (b) or as a random medium (c) at different values of the angle of incidence.

4 SPECTRAL AND ANGULAR INTEGRATION

Optical reflectance (R), transmittance (T), absorbance (A) spectra depend on the wavelength λ and on the incidence angle θ, $I(\lambda,\theta)$. Considering the standard AM1.5G solar spectrum $S(\lambda)$, we calculate the integral of $I(\lambda,\theta)$ weighted with the photon flux in the spectral interval from 300 to 1900 nm:

$$I(\theta) = \frac{\int_{\lambda_1}^{\lambda_2} I(\lambda,\theta)\lambda S(\lambda)\mathrm{d}\lambda}{\int_{\lambda_1}^{\lambda_2} \lambda S(\lambda)\mathrm{d}\lambda}.$$

In a concentrator photovoltaic system, the function $I(\theta)$ should be weighted with the distribution $A(\theta)$ of optical rays that are incident on the multijunction solar cell:

$$\langle I \rangle = \frac{\int I(\theta)A(\theta)\mathrm{d}\theta}{\int A(\theta)\mathrm{d}\theta}.$$

The distribution function here adopted is shown in Fig. 6: it is constant up to θ=40°, then it decreases linearly to zero for θ=70°.

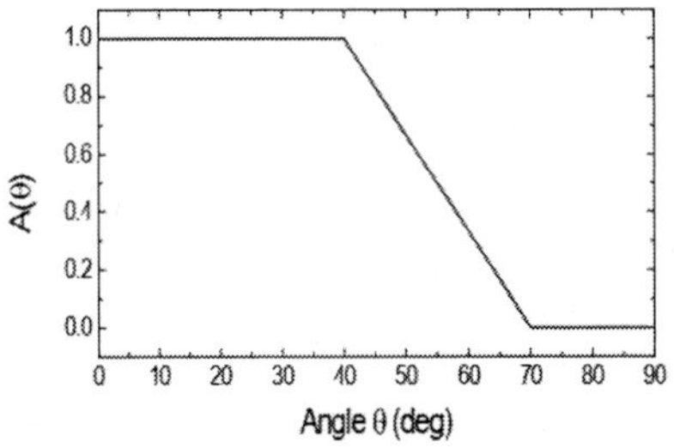

Figure 6: Assumed angular distribution function of optical rays that are incident on the MJ solar cell.

5 RESULTS

In this Section we show the results for the spectrally integrated quantities $T(\theta)$, $R(\theta)$, $A(\theta)$, and for their angular average, referring to the standard and double composite structures defined in Fig. 1.

In Fig. 7 we compare the optical response of the standard structure and that of the double composite treated in the model of an effective medium. The angular dependence is much improved in the double composite, leading to an overall transmittance that increases from 92.9% in the standard structure to 95.7% in the double composite. This results from a strong reduction of the average reflectivity, which more than compensates for a slight increase of absorption in material 2. Also, the polarization dependence is strongly reduced.

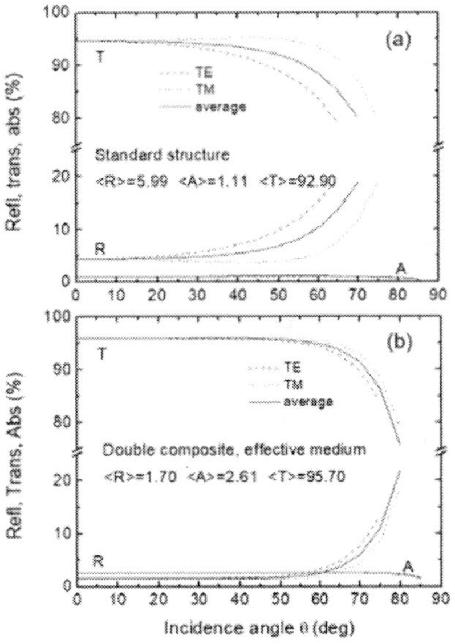

Figure 7: Spectrally integrated reflectance, transmittance, absorbance of the standard structure (a) and of the double composite treated as an effective medium (b).

In Fig. 8 we show the response of the double composite treated in model 2 (periodic medium), for different values of the period. The transmittance is highest when the period is much smaller than the wavelength of visible light. When the period becomes ~λ, optical losses are increased, due to the onset of diffraction processes that increase the spectrally integrated reflectance.

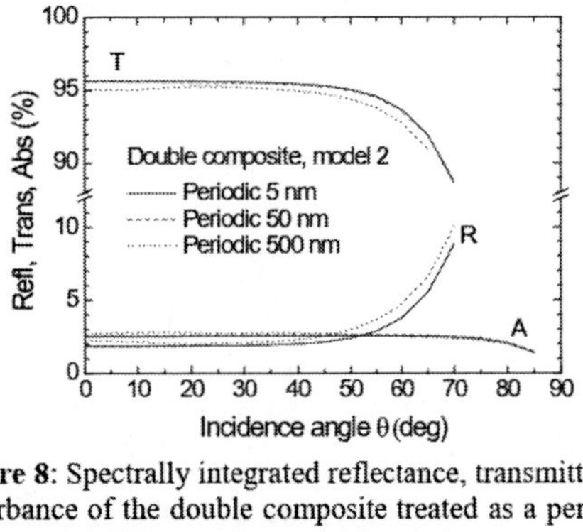

Figure 8: Spectrally integrated reflectance, transmittance, absorbance of the double composite treated as a periodic medium, for different values of the period Λ.

Similar results are obtained in model 3 as a function of the in-plane correlation length l_c, namely, the transmittance is highest when the correlation length is much smaller than the wavelength of visible light. When l_c becomes of the order of λ, diffusion processes produced by the roughness increase the reflectance and reduce the overall transmittance into the solar cell. It should be noticed that the "reflectance" (as well as "transmittance", "absorbance") spectra are not just in the specular or propagation directions, rather they are summed over all scattering angles, thus it is fully understandable that diffraction and diffusion processes may increase the fraction of light that is reflected from the solar cell in all directions.

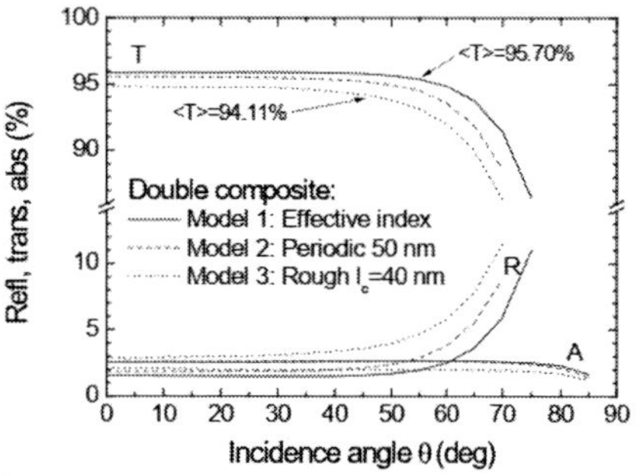

Figure 9: Spectrally integrated reflectance, transmittance, absorbance of the double composite treated as a periodic medium, for different values of the period Λ.

Finally, Fig. 9 compares the results for the double composite treated in the three models. The transmittance is found to decrease when going from the effective index description with period Λ=40 nm (model 1), to the periodic description (model 2), to the rough structure with correlation length l_c=50 nm (model 3). Notice that higher integrated transmittance at normal incidence translates into better angular properties.

We conclude from these results that the effective medium structure has the best performance in terms of overall transmittance, while any nanostructure (periodic or random) degrades the performance unless its spatial scale is much smaller than the wavelength of visible light. We attribute this behavior to the occurrence of diffraction and diffusion processes in the periodic and rough descriptions, respectively. It follows that the crucial feature for optimal ARC performance is the gradual variation of the refractive index profile, which may still be obtained by proper nanostructuring of the constituent layers, but only if the spatial scale of such nanostructures are $<<\lambda$. The optimization criteria for an ARC are very different from those of light-trapping structures for thin-film solar cells, where diffraction and diffusion processes can be usefully employed to increase the optical path of light in the photoactive material.

6 CONCLUSIONS

We have shown in this work that the performance of the standard (bilayer) ARC for MJ solar cells can be substantially improved by using ARCs with a gradual variation of the refractive index. A convenient structure employs two composites that are characterized by a gradual variation of the refractive index. The optimal structure results from a delicate balance between two kinds of optical losses, namely reflection from the solar cell and parasitic absorption in the ARC itself.

The performance of the periodic or nanostructured ARC is always worse compared to the ARC when treated in the effective index approximation. The best performance is obtained when the period Λ or the correlation length l_c is much smaller than the wavelength of visible light. *In the models investigated here, the key ingredient for best ARC performance is a gradual variation of the refractive index.* This can be achieved also by proper nanostructuring of the constituent materials, but only if the spatial scale of such nanostructuring is much smaller than the wavelength of visible light.

Improved transmittance at normal incidence translates into better angular properties at finite θ. The effect of the angular integration is negligible for a properly optimized ARC with gradual variation of the refractive index.

7 ACKNOWLEDGMENTS

This work has also been financed by the Research Fund for the Italian Electrical System under the Contract Agreement between RSE (formerly known as ERSE) , by the Ministry of Economic Development - General Directorate for Nuclear Energy, Renewable Energy and Energy Efficiency stipulated on July 29, 2009 in compliance with the Decree of March 19, 2009.

REFERENCES

[1] M.A. Green et al., Progr. Photov: Res. Appl. 25 (2017) 668.
[2] J. Tommila et al., Solar Energy Materials & Solar Cells 94 (201) 1845.
[3] S. L. Diedenhofen et al., Solar Energy Mat. & Solar Cells 101 (2012) 308.
[4] X. Yan et al., Adv. Funct. Mater. 23 (2012) 583.
[5] F. Galeotti et al., ACS Appl. Mater. Interfaces 6 (2014) 5827.
[6] J. Buencuerpo et al., Opt. Express 23 (2015) A322.
[7] J. Buencuerpo et al., Sci. Rep. 6 (2016) 38682.
[8] P. Kowalczewski et al., Opt. Lett. 37 (2012) 4868.
[9] P. Kowalczewski et al., Opt. Expr. 21 (2013) A808.
[10] A. Barranco et al., Progr. Mat. Sci. 76 (2016) 59.
[11] M. Liscidini et al., Phys. Rev. B 77 (2008) 035324.

ENHANCED PHOTON UTILIZATION IN ULTRATHIN 1.0EV GAINAS SUB-CELL BY SIO_2/AU REFLECTOR

Liyong Yao *, Lirui Liu, Qiming Zhang, He Wang, Heng Zhang, Peng Gao, Qiang Sun
Tianjin Insitute of Power Sources, Tianjin, P.R. China, 300381
e-mail:yaoliyong@yeah.net

ABSTRACT: Multi-junction III-V solar was widely used in space environment. It was meaningful to increase the radiation resistance of III-V solar, which can increase total energy transition and output during several decade years. One of the effective method to increase the radiation resistance of III-V solar was decreasing the thickness, decreased defects and carrier recombination centers can be achieved because the decreased volume and traps quantity. Nevertheless, current decreased because inadequate light absorption. Here, the experiment verified the SiO_2/Au reflector (ODR) was an effective method to improve photon utilization..
Keywords: III-V solar cell, ultrathin, photon cyclic utilization

1 INTRODUCTION

Multi-junction III-V solar was widely used in space environment, because its high photoelectric conversion efficiency exceeding 30% under light intensity AM0[1-3]. However, decay of solar cell performance was inevitable under space environment which was full of high-energy-electron irradiation[4]. It was meaningful to increase the radiation resistance of III-V solar, which can increase total energy transition and output during several decade years. One of the effective method to increase the radiation resistance of III-V solar was decreasing the thickness, decreased defects and carrier recombination centers can be achieved because the decreased volume and traps quantity[5]. Nevertheless, current decreased because inadequate light absorption. Here, the experiment verified the SiO_2/Au reflector (ODR) was an effective method to improve photon utilization.

The 1.0eV GaInAs sub-cell had the poorest radiation resistance. Increase the 1.0eV GaInAs sub-cell can increase the 1.0/1.4/1.9 triple III-V solar cell radiation resistance. The decreased thickness can decrease the solar cell volume, responsively, the defect quantity largely reduced. As a result, the enhanced carrier collection was achieved. So the ultrathin solar cell has high radiation resistance.

Decreasing the 1.0eV GaInAs sub-cell thickness can increase the radiation resistance, however, part of photon penetrate the whole film. Put a mirror at the bottom of the 1.0eV GaInAs sub-cell can reflect the photon and realize photon cyclic utilization[6-8]. Ultra-thin 1.0eV GaInAs sub-cell with a mirror at the bottom was an effective method to increase the radiation resistance without current decrease.

2 EXPERIMENTALS

SiO_2 was selected as the middle because its lower refractive index compare with that of GaInAs, which lead to light trapping effect. Au was selected as the back mirror, because its high reflectivity and high adhesive with SiO_2 layer. The structure of back reflect mirror was shown in Fig.1. The Au and SiO_2 film layers were prepared by electron beam evaporation method,. the 1.0eV GaInAs layer was prepared by MOCVD method on GaAs substrate. The EQE of the prepared solar cells were measured to evaluate the enhancement of photon utilization. The solar cells performance of 1.0eV GaInAs sub-cell were measured under AM0, with the 1.4eV GaInAs cap layer which filter the short wavelength light (<900nm). The SiO_2 layer was selectivity corrosion to ensure ohmic contact of Au with GaInAs.

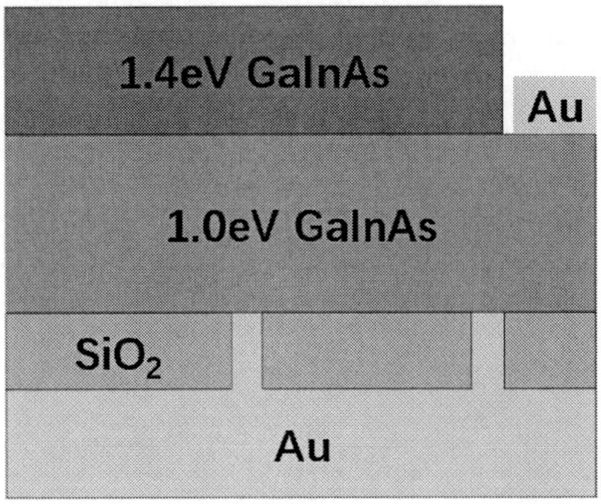

Fig.1 The structure of 1.0eV GaInAs solar with SiO2/Au mirror

3 RESULT AND DISCUSSION

Fig.2 shows the EQE of 1.0eV GaInAs sub-cell with different thickness under AM0. No anti-reflection coating was prepared on the top of solar cell, so the EQE response was low with an peak value of about 75%. The 3μm-thick GaInAs sub-cell exhibit the highest EQE response and the current density was 15.092mA/cm^2 under AM0.

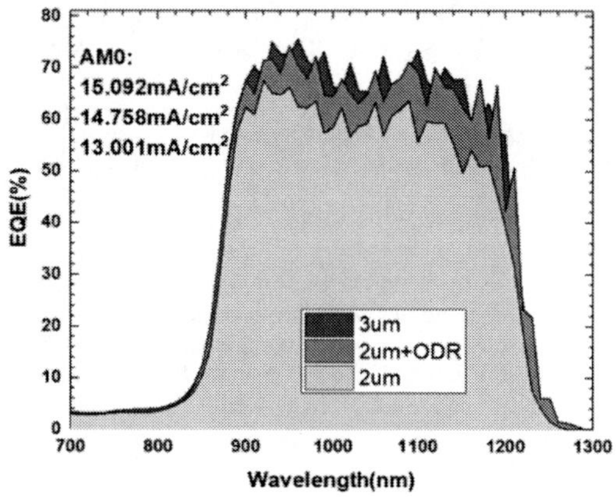

Fig.2 EQE of 1.0eV GaInAs sub-cell with different thickness under AM0

Obviously, the 2μm-thick GaInAs sub-cell exhibit the lowest EQE response and the current density was 13.001mA/cm² under AM0. The 2μm-thick GaInAs sub-cel with an ODR exhibit an high EQE response slightly lower than the 2μm-thick GaInAs sub-cell, and the current density was 14.758mA/cm² under AM0. SiO_2/Au (ODR) can effectively increase the photon utilization.

Fig.3 shows the performance of 1.0eV GaInAs solar with different thickness under AM0. Fig.3(a) shows the current density of 1.0eV GaInAs solar cell with different thickness, 2μm-thick GaInAs sub-cell with an ODR has similar current density with that of 3μm-thick GaInAs sub-cell. Fig.3(b) shows the Voc of 1.0eV GaInAs solar cell with different thickness, 2μm-thick GaInAs sub-cell reveal an higher voltage value than that of 3μm-thick GaInAs sub-cell, owing to the better carrier transport.

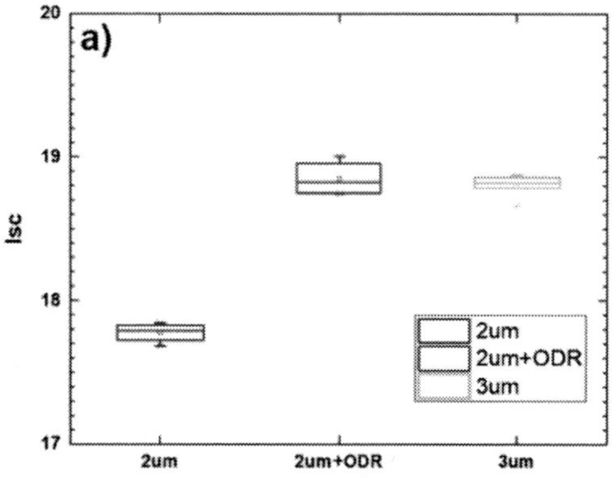

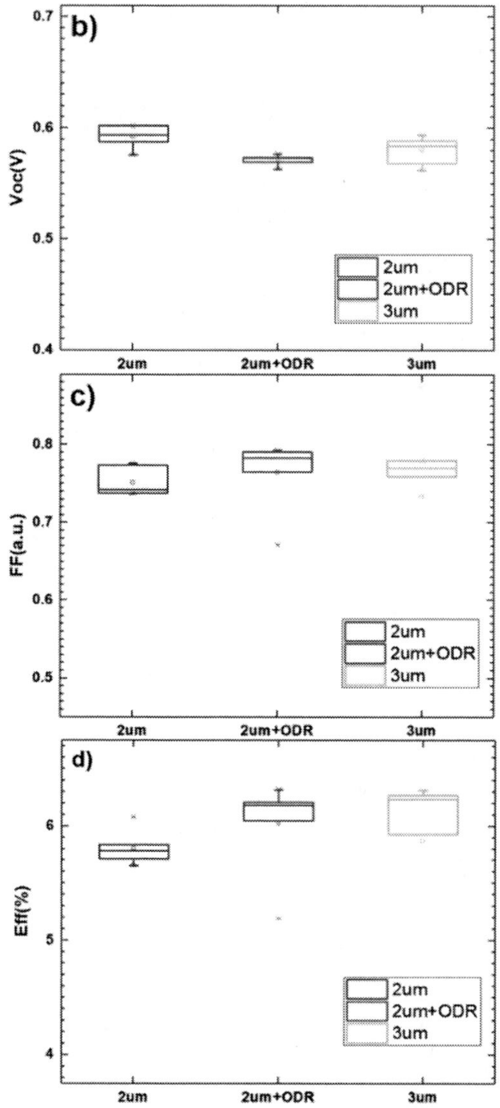

Fig.3 performance of 1.0eV GaInAs solar with different thickness under AM0

Fig.3(c) shows the fill factor of 1.0eV GaInAs solar cell with different thickness, 2μm-thick GaInAs sub-cell with an ODR has the highest fill factor value, reveal the best carrier management. Summing up the above factors, 2μm-thick GaInAs sub-cell with an ODR has similar photoelectric conversion efficiency compared with that of 3μm-thick GaInAs sub-cell, as shown in Fig.3(d).

4 CONCLUSION

SiO_2/Au (ODR) at the bottom of the 1.0eV GaInAs sub-cell can reflect the photon and realize photon cyclic utilization. Ultra-thin 1.0eV GaInAs sub-cell with a mirror at the bottom was an effective method to increase the carrier management without current decrease. Further optimization, like nano structure SiO_2 layer, was under research.

5 REFERENCES

[1] T. Takamoto, T. Agui, A. Yoshida, K. Nakaido, H. Juso, K. Sasaki, K. Nakamura, H. Yamaguchi, T. Kodama, H. Washio, M Imaizumi, M.Takahashi, "World's highest efficiency triplejunction solar cells fabricated by inverted layers transfer process, " in Proc. IEEE PVSC35 2010, pp 000412-000417.
[2] R. R. King, D. Bhusari, D. Larrabee, X. –Q. Liu, E. Rehder, K. Edmondson, H. Cotal, R. K. Jones, J. H. Ermer, C. M. Fetzer, D. C. Law, N. H. Karam, "Solar cell generations over 40% efficiency," Prog. in Photovoltaics: Res. Appl., p. 1255, 2012.
[3] M. A. Green, K. Emery, Y. Hishikawa, W. Warta and E. D. Dunlop, Prog. Photovoltaics, 2015, 23, 1–9
[4] M. Yamaguchi, Radiation resistance of compound semiconductor solar cells, Journal of Applied Physics, 78 (1995) 1476-1480
[5] M. Yamaguchi, "III-V compound multi-junction solar cells:present and future," Sol. Energy Mater. Sol. Cells, vol. 75, pp. 261-269. 2003
[6] M. A. Steiner, J. F. Geisz, I. García, D. J. Friedman, A. Duda, and S. R. Kurtz, "Optical enhancement of the open-circuit voltage in high quality GaAs solar cells," Journal of Applied Physics, vol. 113, no. 12, p. 123109, Mar. 2013.
[7] T. Inoue, K. Watanabe, K. Toprasertpong, H. Fujii, M. Sugiyama, Y. Nakano,, "Enhanced Light Trapping in Multiple Quantum Wells by Thin-Film Structure and Backside Grooved with Dielectric Interface," IEEE J. Photovolt., vol. 5, pp. 697- 703, 2015.
[8] N. Vandamme, H.-L. Chen, A. Gaucher, B. Behaghel, A. Lemaitre, A. Cattoni, C. Dupuis, N. Bardou, J.-F. Guillemoles and S. Collin, "Ultrathin GaAs Solar Cells With a Silver Back Mirror", IEEE J. of Photovoltaics, in print, 2015.

COUNTERACTING PHOTOVOLTAIC EFFECT IN MULTI-JUNCTION SOLAR CELLS

M.A. Mintairov*[1,a], V.V. Evstropov[1,b], S.A. Mintairov[1,c], M.Z. Shvarts[1,d], S.A.Kozhukhovskaia[1,e] & N.A. Kalyuzhnyy[1,f]
1 Ioffe Institute
26 Politekhnicheskaya, St Petersburg 194021, Russian Federation, tel/fax: +7(812) 2972173
A - mamint@mail.ioffe.ru, B – vvevstropov@gmail.com, C - mintairov@scell.ioffe.ru,
D - shvarts@scell.ioffe.rssi.ru, E - kozhukhovskaya@mail.ioffe.ru, F - nickk@mail.ioffe.ru

ABSTRACT: The existence within monolithic double- and triple-junction solar cells of a photoelectric source, which counteracts the basic photovoltaic p-n junctions, is experimentally proved. The paper presents a detailed analysis of the factors affecting the effect. It is established that the counteracting source is tunnel p^+-n^+ junction between GaInP and GaAs subcells. The photoelectric characteristics of samples with different tunnel diodes were investigated and it is shown that when the tunnel p^+-n^+ junction separates the charge carriers, the V_{oc}-J_g dependence has a dropping part, including a sharp jump.
Keywords: Multijunction Solar Cell, Concentrator Cells, Characterization, Photovoltaic Counter Photovoltaic Effect

1 INTRODUCTION

Multi-junction solar cells (MJ SC) are actively used in space and concentrator photovoltaics. The main advantage of these solar cells is the high efficiency of the conversion of solar energy into electrical one, which is achieved due to the use of several p-n junctions in the SC structure. The addition of each p-n junction widens the spectral sensitivity of the device. This determines the tendency to increase the number of p-n junctions in MJ SC. At the same time, the technological process of growing MJ SC structures becomes more complicated with increasing p-n junctions. As a result, effective MJ SCs have a complex structure that contains, in addition to photovoltaic p-n junctions, counter-connected tunnel p-n junctions, which provide electrical connection of MJ SC sub-cells. The presence of such counter-connected p-n junctions can lead to the appearance of a counter photovoltaic effect, i.e. appearance of counter electromotive force. Previously, we reported about possibility of such an effect in GaInP/GaAs/Ge MJ SC [1]. The main reason for this assumption is ether the presence on the V_{oc}-J_g characteristics («open circuit voltage» - «photogenerated current») the maximum or tendency to V_{oc} saturation. This behavior of the Voc-Jg characteristic was also observed in a number of experiments listed in [2]. In [1], we noted that the saturation of the V_{oc}-J_g dependence can be caused by the heating of the subcells p-n junctions by incident radiation. A procedure for obtaining isothermal V_{oc}-J_g dependence was proposed. At the same time, experiments have shown that the counter photo electromotive force is manifested on isothermal dependences at about 50 A/cm^2 photogenerated current densities.

As shown in this paper, this was observed due to the presence of the counteracting photovoltaic effect in the upper tunnel diode (between GaInP and GaAs subcells). In this paper, it has been experimentally shown that the counter photo-effect manifests itself on the V_{oc}-J_g characteristics at a value of J_g close to the magnitude of the peak current of the photoactive tunnel diode. The reasons for a possible increase in the counter photovoltage and photocurrent are also experimentally investigated in this paper.

2 SOURCE OF COUNTERACTING EFFECT AND OBJECTS OF STUDY

The counteracting effect may be connected with carrier separation in counter-connected tunnel p-n junctions. There are two way how carriers may appear in the p-n junctions — the first of them is light absorption and the second one is thermal injection of carriers from subcells p-n junctions. It is worth noting that a GaInP/GaAs/Ge SC structure is designed in such a way that part of the spectrum intended for absorption in the top subcell is transmitted to the middle one. This is required to achieve top and middle subcells photogenerated currents balance. But it also leads to absorption of a transmitted light in the upper tunnel junction. Absorption increases in the case of GaInP/GaAs/Ge SCs intended for use in concentrating installations. In this case, a high peak current of the tunnel diodes is required. This can only be achieved by using the GaAs as a material of n-region. Thus, the n-region of the upper tunnel diode also absorbs a part of the spectrum with a larger than 1.424 eV photon energy. The n-region is rather thin (15 nm), however, it is about 3-5% of middle sub-cell spectrum. The bottom tunnel diode is located behind the GaAs sub-cells and can in principle be considered as not illuminated p-n junction. For this reason, the study of the counteracting photovoltaic effect was carried out precisely for the upper tunnel diode.

To study the effect, double-junction GaInP/GaAs MJ SC structures were used. Such structures contain only one (the upper one of GaInP/GaAs/Ge SC) tunnel diode. In addition to the effect of the light absorption in the tunnel diode, the other two factors influencing the magnitude of the counteracting photocurrent and photovoltage were also studied. Such factors are: the large differential resistance of the tunnel diode and the thermal injection of the charge carriers into the tunnel diode. To study the influence of the above factors, three samples were used, the band diagram of the tunnel part of which is shown in Figure 1. By the term «tunnel part» here and below we mean a set of layers consisting of n- and p-regions layers of tunnel p-n junctions and a layers adhering to them.

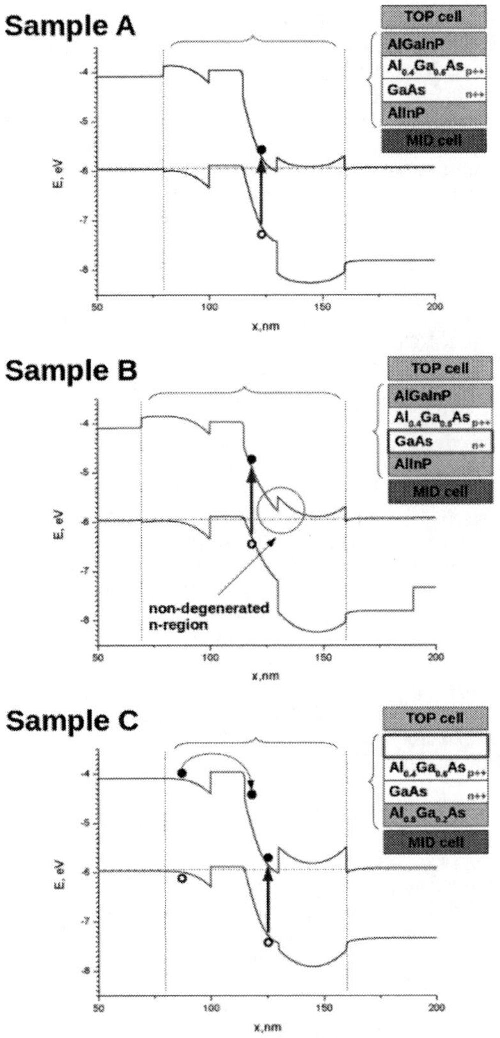

Figure 1: Band diagrams and tunnel part structures of the samples of study. Band diagrams were achieved with help of model described in [3].

All samples were grown by metal-organic vapor phase epitaxy (MOVPE). Sample «A» structure is a standard one for MJ SC devices. Sample «B» has a "defective" tunnel junction with non-degenerated n-region. Sample «C» has no barrier for reducing electron thermal injection. The design of the GaAs and GaInP sub-elements of the structures was slightly different.

3 COUNTERACTING PHOTOVOLTAIC EFFECT IN MJ SC TUNNEL PART

For the three samples studied, a set of light IV-curves recorded at different illumination was obtained (Figure 3, circles). On the basis of these measurements, the isothermal V_{oc}-J_g dependences were constructed (Figure 2). To obtain isothermal dependences, the additional measurements according to the procedure described in [1] were carried out. All measurements were performed on a pulsed solar radiation simulator. The simulator is equipped with devices for monitoring the level of illumination and temperature of the SC under examination, as well as a four-probe system for measuring the IV characteristics. The peak current of the tunnel diodes was registered with the voltage decreasing scanning direction. Maximum concentration of solar radiation reaches about 10,000 suns.

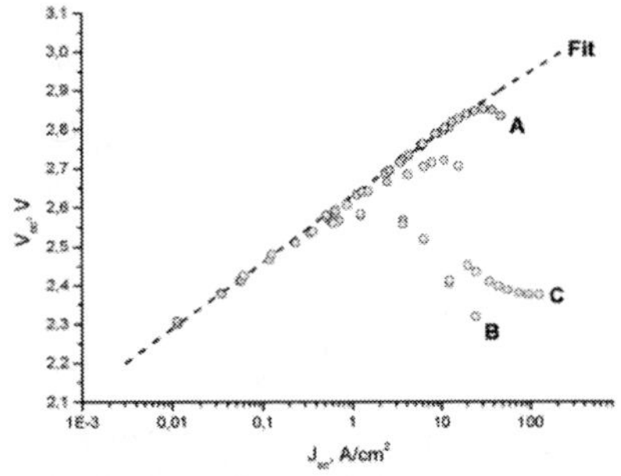

Figure 2: Experimental V_{oc}-J_g isothermal dependencies (circles) of the samples of study. All experimental dependences deviate from a normal dependence (dashed curve). Normal curve obtained by fitting linear parts of the experimental V_{oc}-J_g dependences with formula 1.

The isothermal V_{oc}-J_g dependencies are shown in Figure 2. The linear parts of dependencies were approximated by a two-exponential model using the formula:

$$J_{SC} = J_{02}\left(\frac{V_{OC}}{2kT} - 1\right) + J_{03}\left(\frac{V_{OC}}{3kT} - 1\right) \quad (1)$$

The parameters J_{02}=8.0·10^{-25} A/cm^2, J_{03}=5.5·10^{-16} A/cm^2 were defined. The approximation result is also shown in Figure 2 (dashed line). Parameters J_{02} and J_{03} were used to construct IV-curves, on which there is no influence of the counter photovoltaic effect – «non-counter-effect» IV-curves. The experimental IV-curves (circles), «non-counter-effect» IV-curves (lines) and the result of their voltaic subtraction (triangles) are shown in Figure 3. The results of such voltaic subtraction can be interpreted as the resistive IV-curve of the tunnel part («tunnel part» IV-curve).

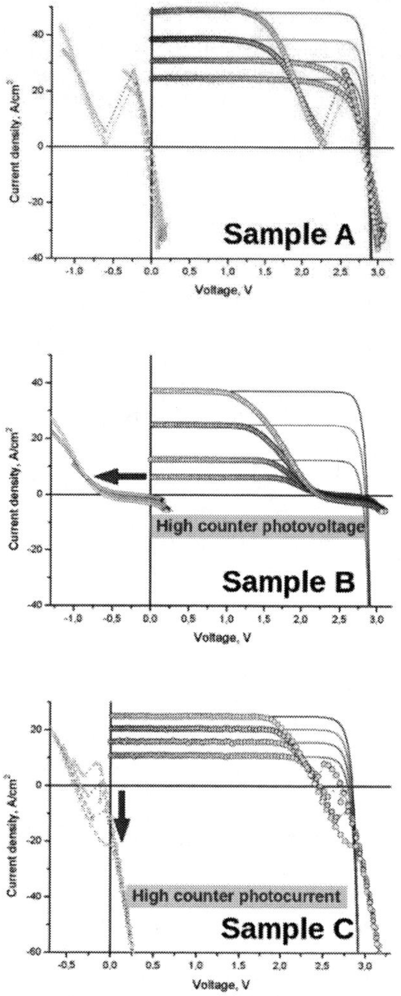

Figure 3: Light IV-curves of the samples of study at different illumination: circles - experimental IV-curves; solid lines – IV-curves constructed from counteracting effect free V_{oc}-J_g dependence (Figure 2, dashed line); triangles - IV curves of resistive tunnel part.

4 RESULTS AND DISCUSSION

It can be seen that tunnel part IV-curve for sample «A» has a highest peak current (of the order of 25 A/cm^2), while also showing a slight (relative to the shift of the full IV-curve) shift of the tunnel part IV-curve downward. This indicates the relatively low value of counteracting photocurrent. The V_{oc}-J_g characteristic for the sample deviates from the counteracting effect free one at $J_g \sim 25$ A/cm^2, which is of the order of magnitude of the peak current. A similar correlation was observed for the other samples.

In the tunnel part of sample «B» there is no degeneration in the n-region (Figure 1). For this reason, the sample «B» tunnel part IV-curve has no peak. and a large differential resistance. Therefore, counteracting photovoltaic effect is much stronger despite the fact that photocurrent magnitude is principle the same as for sample «A». So it produce a high photovoltage even at low light illumination. For this reason, the maximum of V_{oc}-J_g is at current (~ 2 A / cm^2).

In the structure of sample C there is no barrier suppressing thermal injection of carriers from the GaInP sub-cell (Figure 1). This significantly increased the counter photocurrent and tunnel part IV-curve rapidly "shifts down" with illumination increasing. When "peak point" cross 0-axis (at photocurrent about 15 A/cm^2), its V_{oc}-J_g has a sharp V_{oc} jump.

5 CONCLUSION

The counteracting photovoltaic effect in GaInP/GaAs/Ge MJ SC was investigated. It was experimentally shown that such an effect is observed due to the separation of charge carriers by the upper tunnel diode. In the case standard MJ SC structure (sample «A») the effect manifests itself on the V_{oc}-J_g dependence at a photogenerated current close to the peak current of the tunnel diode. For modern GaInP/GaAs/Ge MJ SC the magnitude of such current is quite high (more than 50 A/cm^2). Therefore, at the operating regimes ($J_g \sim$ 5-10 A/cm^2) the influence of the effect on the V_{oc} value is insignificant. Nevertheless, the influence of the effect to maximum power point should occur at lower J_g values. The preliminary estimation we received gave the 0.1 % abs. of efficiency loss for GaInP/GaAs/Ge MJ SC with an efficiency of 37%.

Two extreme cases of manifestations of the counteracting photovoltaic effect were also experimentally investigated. One of them leads to a significant counter photocurrent and the other to a significant photovoltage. Both cases dramatically affect the characteristics of the MJ SC. Thus, suppression of thermal injection of carriers into the tunnel diode and reduction of its differential resistance are an important aspect in the development of monolithic MJ SC for.

6 ANKNOWLEDGMENTS

The work has been supported by the Russian Foundation on Basic Research (grant 17-08-01056-a).

REFERENCES

[1] M A Mintairov, V V Evstropov, M Z Shvarts, S A Kozhukhovskaia, S A Mintairov and N A Kalyuzhnyy, 2016, AIP Conf. Proc. Freiburg (Germany) v 1766 p. 050005

[2] M Stevens, C Downs, D Emerson, J Adler, S Maclachlan and T E Vandervelde, 2015 Proc. of 31 PVSEC, Hamburg (Germany) pp. 1474 – 1477

[3] R. Varache, C.Leendertz, M.E.Gueunier-Farret, J.Haschke, D.Munoz, L.Korte, Solar Energy Materials and Solar Cells, 141, pp. 14-23, 2015

OPTICAL CHARACTERIZATION OF A RED DYE LUMINESCENT SOLAR CONCENTRATOR

Paolo Bernardoni, Michele Tonezzer, Donato Vincenzi, Silvio Fugattini, Micol Boschetti, Vincenzo Guidi
University of Ferrara - Department of Physics and Earth Sciences - Sensor and Semiconductor Laboratory
Via Giuseppe Saragat 1 - Building C, 44122 Ferrara (FE), Italy
Corr.Auth. E-Mail: paolo.bernardoni@unife.it Tel: (+39) 0532 974271

ABSTRACT:
The Luminescent Solar Concentrators (LSC) are innovative devices whose main application is the realization of transparent photovoltaic systems that are particularly interesting in the field of building integrated photovoltaics.
Despite the great potentialities of LSC photovoltaic systems, up to now only a few real world applications exist and most of the scientific work about LSCs concerns only luminescent dyes and small size devices (generally 5×5 cm^2).
The purpose of this work is to investigate the optical behavior of a large (50x50 cm^2) LSC slab that can closely represent the devices used in building integrated applications like roofs or windows.
The output spectrum of the emitted light is measured while the LSC is illuminated in different positions with a round spot. With this setup it is possible to collect information about the contribution of radiative and non-radiative absorption to the performance of the system.
Keywords: Building Integration Photovoltaic, Characterization, Static Concentrators, Photovoltaic Windows, Dye-Sensitized

1 LUMINESCENT SOLAR CONCENTRATORS

Luminescent Solar Concentrators (LSC) are innovative devices that concentrate sunlight on their sides while the main surfaces remain partially transparent (albeit colored depending on the dye): this is a key feature that allows for the development of transparent solar panels particularly suited for building integration.
LSCs concentrate light through two distinct physical phenomena:

- Fluorescence
- Total internal reflection

The first one occurs in materials called fluorophores (organic, rare earth and quantum dot dyes) and consists in the absorption of a photon followed (within nanoseconds) by the emission of another photon at a longer wavelength.
Total internal reflection, instead, takes place when a photon travels from a medium with high refractive index (i.e. glass, PMMA) to a lower one (air): if the angle of incidence is above the so-called critical angle the photon is reflected within the substrate with no transmission occurring.

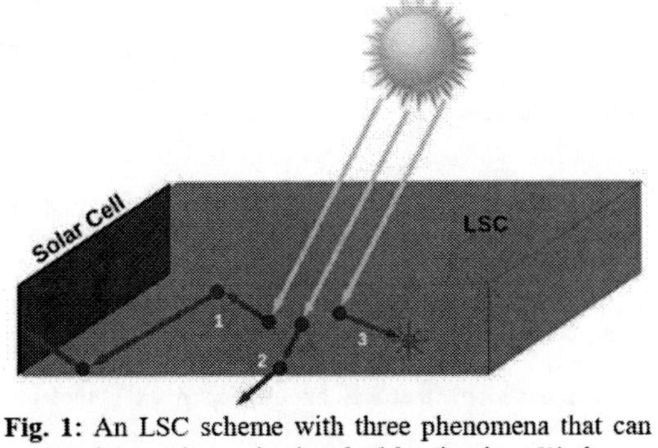

Fig. 1: An LSC scheme with three phenomena that can occur when a photon is absorbed by the dye: (1) the re-emitted radiation is above the critical angle so it is trapped inside the slab and guided to the cells.
(2) The re-emitted radiation is below the critical angle and it is lost.
(3) The emitted radiation is re-absorbed by another dye molecule (self absorption)

In order to achieve light concentration the device is manufactured as a slab or a sheet of transparent material doped with the fluorescent dyes.
When the faces are exposed to sunlight the light emitted by the dyes is partially (~75% for a PMMA substrate with n=1.489) trapped inside the device and guided to its sides by multiple reflections occurring at the LSC/air interface.
This process is affected by multiple efficiency losses some of which are related to the dye (stokes shift, quantum efficiency, self-absorption) and others to the transparent medium (refractive index and flatness of the surface).
The purpose of this work is an optical characterization of an LSC manufactured as a slab of PMMA doped with a 400ppm concentration of Lumogen Red 305 dye (see Fig. 1) as this is the most common configuration in literature [1][2][3].

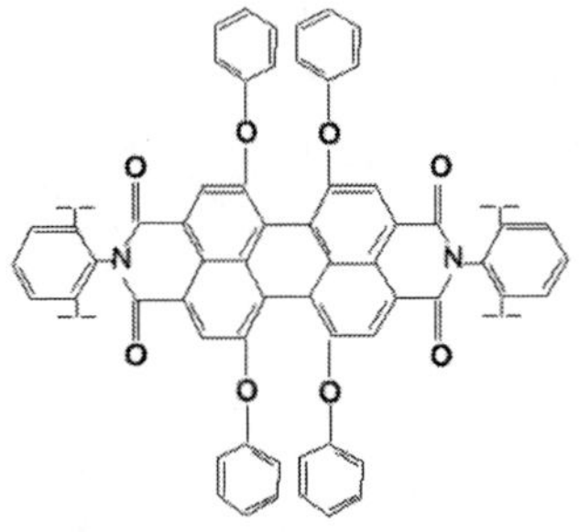

Fig. 2: The molecule of Lumogen Red 305: the organic fluorophore used in this LSC

2 APPLICATIONS

The LSC technology allows the development of transparent solar panels particularly fit for building integration while the low efficiency hinders their use for mass energy production.
These devices can be integrated in smart windows [5], sheds and ceilings [6] that let the light to pass and at the same time generate electricity that can be used by a high efficient illumination system or small appliances.

Fig. 3: The authors installed a canopy with integrated LSCs in Liv-lib' house at Solar Decathlon 2014, the active area was 25m²

3 WHY SIZE MATTERS

In order to use the LSCs as a replacement for the glass slabs adopted in architecture, the devices must be much bigger than the small squares with sides of 5cm to 15cm used commonly [1][2][3][4] for research purposes.

Larger LSCs imply a longer optical path for the photons inside the concentrator and this path is mainly limited by three factors [7]:

- Self-Absorption by the dye (radiative absorption)
- Non radiative absorption by the optical medium and the dye
- Quality of the surface

Since the dye presents a significant overlap between the absorption and emission spectra, the light emitted inside the LSC can undergo multiple absorption-emission processes.

The effect of these processes is called self-absorption and occurs until: the energy of the emitted photon leaves the absorption region, the photon is permanently absorbed in the process, exits the LSC through the escape cone or exits the LSC through the side surfaces.

Among these processes, only the last one is desirable because the photon is available for conversion by the solar cells while the other events lead to an efficiency loss of the device.

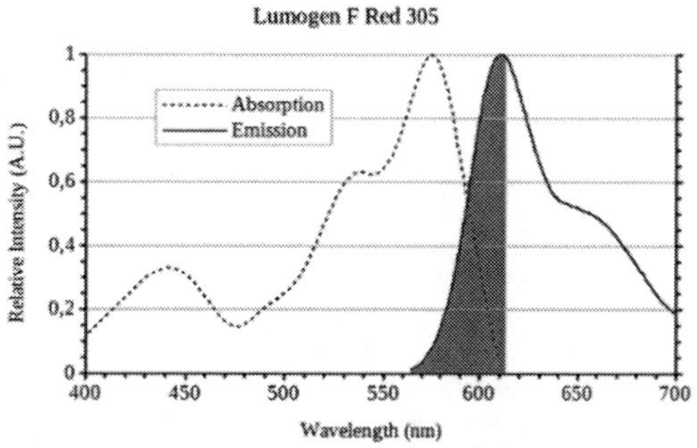

Fig. 4: The absorption and emission spectra of Lumogen Red 305 in CH_2Cl_2 show an overlap (here highlighted in grey), this overlap causes the self-absorption of some emitted photons by the dye itself (Source [8])

4 SETUP

The LSC subject of this study is a square slab with a side length of 50cm and a thickness of 5mm, the slab is made of Poly(methyl methacrylate) (PMMA) as the optical medium doped with 300ppm by weight of Lumogen Red 305.

As a first step, in order to investigate the influence of the optical medium on the emission spectrum of the dye, the output spectrum of the LSC (see Fig. 6) have been measured and compared to the available data about a Lumogen Red 305 solution in dichloromethane (see Fig. 4 [8].

This measure is carried on illuminating the edge of the LSC and analyzing with a spectrometer [9] the light emitted at a right angle from the excitation source.

Then, in order to assess how the optical losses influence the efficiency of the concentrator, the spectrum and intensity of the light emitted by the LSC have been measured when the slab was illuminated only by a round spot moved at different distances from the side.

This was achieved using a black screen behind the slab and another black screen, mounted on rails, in front of the slab.

The top screen had a round circular aperture in the central position with a diameter of 10mm, the screen was more than twice as long as the slab in order to allow for a complete shading of the slab for any position of the aperture.

On one side of the slab, an optic fiber with a diameter of 400um sampled the light emitted by the concentrator; this light was guided by the optic fiber to an Ocean Optics USB4000-XR1-ES USB spectrometer.

The output spectrum is obtained directly from the spectrometer while the intensity is obtained integrating the spectrum.

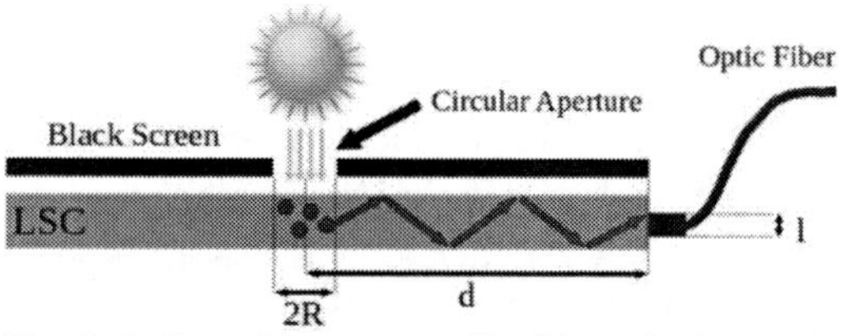

Fig. 5: Outline of the setup used in this work, the LSC is shaded by a black screen with a circular aperture of radius *R*, the light emitted is collected by an optic fiber on one side. The measures were taken at different distances *d* between the circular aperture and the optical fiber.

With this experimental setup both the intensity and the spectrum of the LSC emission were measured at different emission-collection distances by simply moving the position of the screen, and consequently the aperture, then correcting the recorded intensity values by the geometrical coefficient of attenuation.

The expected geometrical attenuation by a completely transparent medium is caused by the smaller angle covered by the fiber when the emission spot is moved at increasing distances and is proportional to the arctangent of the ratio between the half width of the receiver and the distance of the spot.

5 RESULTS

The result about the emission of Lumogen Red 305 in PMMA is displayed in Fig. 6, and is equal, within the resolution of the spectrometer [9], to the spectrum of the dye in dichloromethane solution shown in Fig. 4 and obtained from the datasheet of the manufacturer [8].

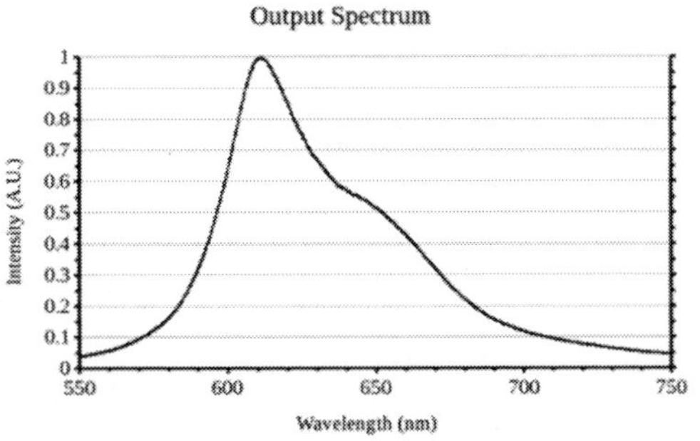

Fig. 6: The emission spectrum of Lumogen Red 305 in PMMA compared to CH_2Cl_2 show a shift in the emission peak lower than the resolution of the spectrometer (~2nm [9]). This spectrum is obtained lighting an edge of the LSC and collecting the light emitted at a right angle from the source in order to remove the effects of self-absorption and reflection from the surface

This means that the interaction between the molecules of the dye and that of the optical medium (PMMA) do not result in a substantial modification of the energy levels of the dye itself [10].
Moving to the results about the output spectrum of the LSC device the curve is, as expected, significantly different from that of the dye alone primarily due to self-absorption.
Lumogen Red 305 has an emission peak at 613nm put this peak lies inside the absorption spectrum (see Fig. 4). This leads to a significant reduction in the light output of the concentrator below 610-620nm due to the so-called self-absorption caused by the dye itself.

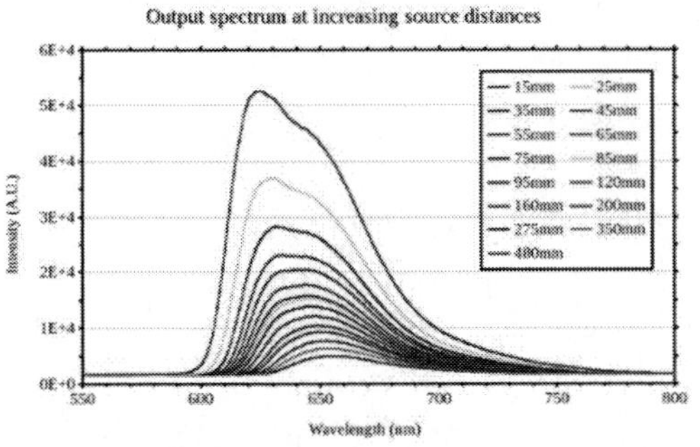

Fig. 7: The output spectrum of an LSC made of PMMA dyed with 300ppm of Lumogen Red 305, the spectrum is measured lighting the LSC at different distances from the side where the light is collected in order to obtain information about the radiative and non-radiative losses occurring in the device

Only the measures at shorter distances (below 10mm) retain the presence of the primary emission peak at 613nm, but this phenomenon ends within few centimeters of optical path from the emission so that at distances above 120mm all the photons lie outside the absorption spectrum and are subject only to the non-radiative absorption by the optical medium.
In the region that falls between 45mm and 55mm the emission peak suddenly shifts from 630 to 642nm compatible with the effect of complete absorption of the 613nm peak by self-absorption while above 120mm the peak slowly shifts from 645 to 650nm with a trend compatible with the absorption by PMMA.
No significant change in the output spectrum is visible beyond 120mm and the light intensity only depends on the geometrical coefficient.
Moving to the plot showing the intensity (number of photons see Fig. 8) versus the distance of the excitation source is evident the presence of three regions where the three absorption regimes (radiative, non-radiative and mixed) are most effective.
In this figure the experimental points are plotted as diamonds while the expected values are obtained fitting the first two experimental points (15 and 25mm) with the geometrical curve

$$I = I_0 \frac{arctan\left(\frac{l}{2d}\right)}{arctan\left(\frac{l}{2d_0}\right)}$$

With l width of the sampling system (in this case the optic fiber), d distance between the center of the spot and the sampling system, I intensity of the signal while the suffix 0 refers to a reference distance but not necessarily to the first value.
This curve represents the expected attenuation in the signal intensity for a system with no optical losses.

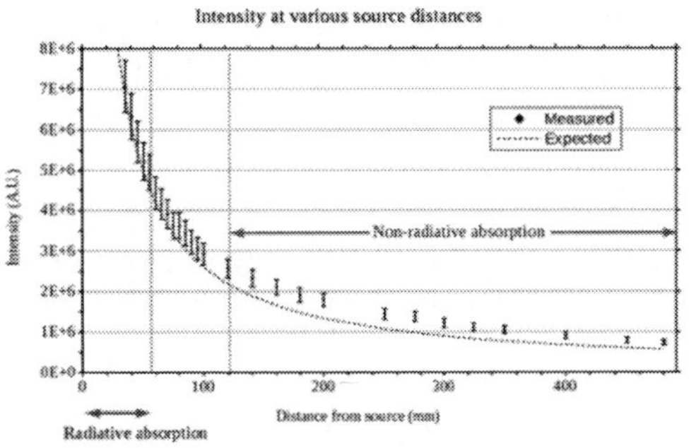

Fig. 8: The expected values are calculated from the measures at the shorter distances fitted with the attenuation expected measuring the intensity of a source with a uniform angular distribution moving away from the receiver.
The two data sets differ when the self-absorption is nearly complete.

6 CONCLUSION

Our results suggest that above 110mm the majority of the photons emitted inside the absorption spectrum is always re-absorbed leaving the non-radiative losses in the transparent medium (PMMA in our sample) as the main contribution to the overall optical losses.
A choice of an optical medium with a low absorbance appears to be more critical than expected in LSCs whose

optical path is longer than 110-120mm (this means a square of about 16-24cm if the cells are placed on all sides).
The most significant outcome of this result is that in large size LSCs-the efficiency losses caused-by self-absorption are only marginally related to their size due to the complete process of self-absorption at the longer optical paths.
The smaller concentrators, instead, present an efficiency that is heavily related to their size due to the short optical length of self-absorption.

7 BIBLIOGRAPHY

[1] Slooff, L. H., et al. *A luminescent solar concentrator with 7.1% power conversion efficiency*. Physica Status Solidi (RRL) - Rapid Research Letters, 2008, 2.6: 257-259
[2] Dienel, T., et al. *Spectral-based analysis of thin film luminescent solar concentrators*. Solar Energy, 2010, 84.8: 1366-1369.
[3] Tonezzer, M. et al. *Luminescent solar concentrators employing new Eu(TTA)3phen-containing parylene films*. Progress in Photovoltaic, 2015, 23: 1037-1044.
[4] Tonezzer, M. et al. *Luminescent Solar Concentrators – State of the Art and Future Perspective*. Solar Cell Nanotechnology, 2014: 293-315.
[5] Aste, N., et al. *Design and performance monitoring of a LSC smart window. Clean Electrical Power (ICCEP)*, 2017 6th International Conference on. IEEE, 2017.
[6] Zarcone, R., et al. *Building integrated photovoltaic system for a solar infrastructure: Liv-lib'project*. Energy Procedia 91 (2016): 887-896.
[7] Debije, M.G., Verbunt P. *Thirty years of luminescent solar concentrator research: solar energy for the built environment*. Advanced Energy Materials 2.1 (2012): 12-35.
[8] BASF Corporation, Lumogen F Red 305 Datasheet
[9] Ocean Optics USB4000-XR1-ES Spectrometer datasheet
[10] Wang X., Wang Z. M., *High-Efficiency Solar Cells - Physics, Materials, and Devices*, Springer 2013

Overview of different characterization techniques used in studying the radiation effect of multijunction solar cells

Uma B.R.[1], Sheeja Krishnan [2], M. Ravindra [1], M. Sankaran [1], N. Raghu[1], Roberta Campesato [3]

[1] ISRO Satellite Centre, Bengaluru, India

[2] Shreedevi Institute of Technology, Manglore, India

[3]CESI, 20134 Milan, via Rubattino 54, Italy

EMAIL ID[1]: uma@isac.gov.in, TEL. NO[1]: +91 8025083531

ABSTRACT — **Photovoltaic Solar array system is a primary source of power generation not only for LEO, MEO, and GEO missions but also for various interplanetary, Lander and rover missions. The most critical characteristic of space solar cells is its radiation hardness. The space radiation environment degrades the long-term performance of the solar cells. Understanding of the performance degradation of solar cells due to radiation is very significant. Different characterization techniques like external quantum efficiency (EQE) vis-à-vis spectral response (SR), Deep level transient spectroscopy (DLTS), Electroluminescence (EL), Photoluminescence (PL) and Ultra fast laser spectroscopy etc are generally used to understand the device parameters in detail. In this paper, an overview of the different characterization techniques used in studying the radiation effect of space grade multijunction solar cells is presented.**

KEY WORDS: **Multijunction solar cells, Electroluminescence, Photoluminescence, DLTS, Ultra fast laser spectroscopy.**

I. Introduction

Photovoltaic Solar array system is a primary source of power generation for all near earth missions as well as for various interplanetary and rover missions because of its robustness, reliability and environment friendliness. Tremendous advances in solar cell technology in the last four to five decades, starting from less than 10% efficient single-crystal silicon solar cell to the state-of-art 30% efficient multijunction solar cells have made it possible to cater high power requirements of spacecraft loads. In a time period of less than four decades space solar arrays have grown in size from less than 1 watt to systems over a few tens of kilowatts.

GaAs-based solar cells replaced Silicon as the technology of choice for critical space missions requiring high operating power because of higher specific power (W/m^2) and high power density (W/kg). Space work-horse GaAs based triple junction solar cell structure (InGaP/InGaAs/Ge), a variant of tandem cells as undisputable leader for space applications. The most critical characteristic of space grade solar cells is its radiation hardness. The space radiation environment degrades the long-term performance of the solar cell. It's very essential to understand the solar cell device behavior in radiation environment.

The solar cell device parameters as well as electrical parameters will change when the solar cells are exposed to radiation environment. To understand the behavior of the solar cells, different characterization techniques are used viz. solar cell dark and illuminated IV measurements, external quantum efficiency (EQE), ElectroLuminescence EL, PhotoLuminescence PL, Deep Level Transient Spectroscopy DLTS and ultra fast laser spectroscopy. Details of the experimental set up, sample preparation, and test results are presented in this paper.

II. CHARACTERIZATION TECHNIQUES

External quantum efficiency (EQE):

External QE measurements are essential for the design and performance evaluation of multijunction solar cells. The EQE measurement gives direct & fast result of radiation effect on Multi Junction Solar Cells by showing the reduction in the EQE spectrum. From the EQE graphs, one can find out the surface recombination velocity & junction depth to diffusion length ratio, for an irradiated solar cell. Basically, the important solar cell performance parameter which can be directly obtained from EQE curve is the short circuit current density (Jsc) of each sub cell that is largely influenced by the radiation. Therefore using EQE measurement it is possible to analyze the radiation resistance of triple junction solar cell and of the sub cells identifying the limiting components as a function of the particles energy and fluency. For example in analyzing the EQE of CTJ30 solar cells [1], it was noted that generally the limiting component cell is the middle junction in InGaAs for all the radiation conditions except for low energy protons where the top cell is highly damaged whereas the middle cell EQE remains substantially unchanged.

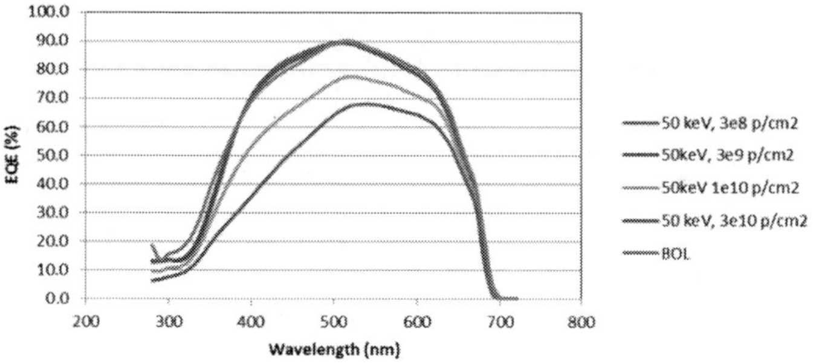

Figure 1: EQE Top cell degradation after 50 keV proton

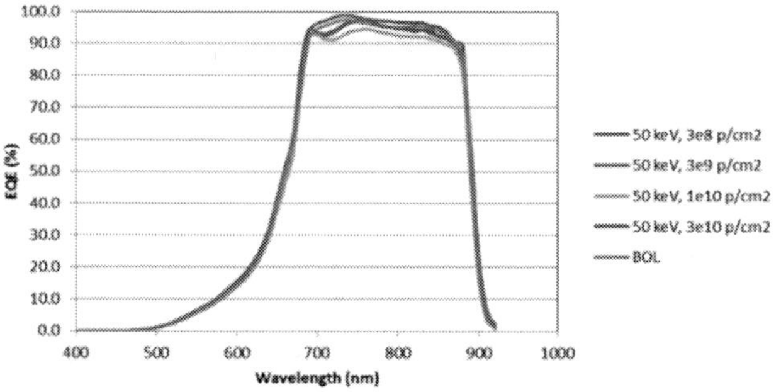

Figure 2: EQE Mid cell degradation after 50 keV proton

Electroluminescence(EL)/photoluminescence(PL):

EL & PL techniques are very powerful tools in determining the defects in the semiconductor [2]. The PL and EL imaging of lattice-matched InGaP/InGaAs/Ge triple junction solar cells is very critical. The electrical and optical coupling of the individual sub-cells presents new challenges in luminescence imaging. Using EL imaging with optical filters, subtle defects on large area solar cells & its sub-cells can be detected and can

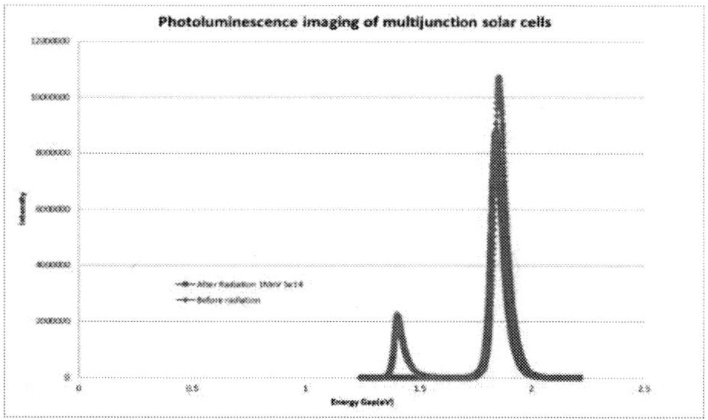

be correlated to the shunts, growth defects, series resistance in the sub-cells.

Figure 3: PL spectra before and after irradiation with electrons 1 MeV energy

In PL, each subcell can be excited homogenously by selecting appropriate light sources in such a way that the excitation light source is only absorbed in the particular subcell, and suitable optics provide constant illumination over the complete cell area. PL is a very useful method to analyze single sub-cell independently and radiation-induced degradation can be identified for the individual sub-cells, by analyzing the PL peaks. Room temperature & Low temperature PL also provide very deep insights of MJSCs vis phonon side band emission & Donor to acceptor pair emission.

Deep Level Transient Spectroscopy (DLTS):

Using this technique an electrically active defects can be determined in the semiconductor [3]. DLTS establishes the defect parameters and measures the concentration of the defects in the material. Room Temperature & Low temperature DLTS also provide deep insights of carrier transport within various energy levels including defect levels of MJSCs, which in turn can be correlated to carrier life time & diffusion length. The limit in DLTS is that it can not be used directly on the triple junction solar cell but only on component cell manufactured in very small size (less than 1 mm area). This is due to the fact that DLTS relies on the measurement of the capacitance at different frequencies. In a triple junction solar cell, the capacitance is very low (sub cells are series connected, therefore the capacitance is given by the equation: $1/C= 1/C_{bot}+1/C_{mid}+1/C_{top}$)
The smaller capacitance (generally the germanium) will dominate the measure thus avoiding the analysis of defects inside middle and top junctions [3].

Electroluminescence (EL) Spectroscopy:

Triple Junction Cell level EL test was performed using EL spectroscopy. EL image of MJ solar cells was taken as (i) without filter, (ii) with 650 nm band pass filter (to study Top Cell-TC) (iii) with 850 nm band pass filter (to probe Middle Cell-MC). However, because of the limitations of Si-detector used, the Bottom Cell (BC) cannot be probed. With this study the hidden defects such as solar cell cracks, micro cracks, local shunts (Rsh), series resistance (Rs), threading defects, growth defects, finger cut and process related defects can be studied. The following are some of the defects captured using EL. By using this technique we can study the effect of radiation on the performance of the solar cells due to these defects. Also we can correlate the reduction in the performance due to radiations to these defects.

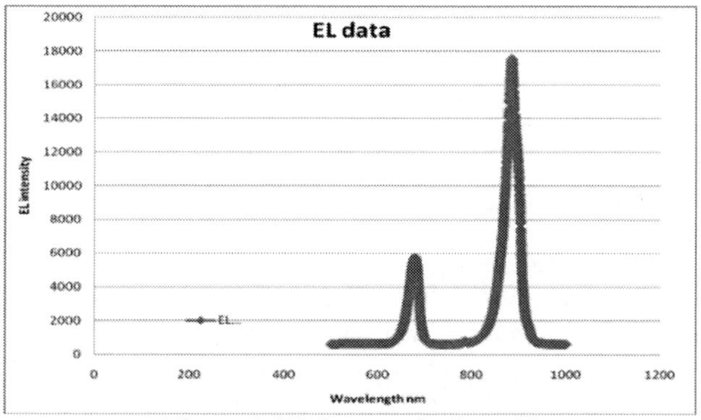

Figure 4. EL image of solar cell

Raw EL image	EL image with 650 nm filter (Top layer image)	EL image with 850 nm filter (Middle layer image)	Inferences
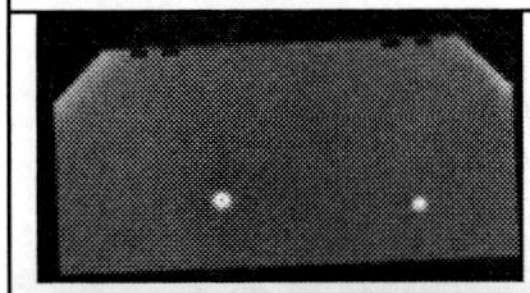	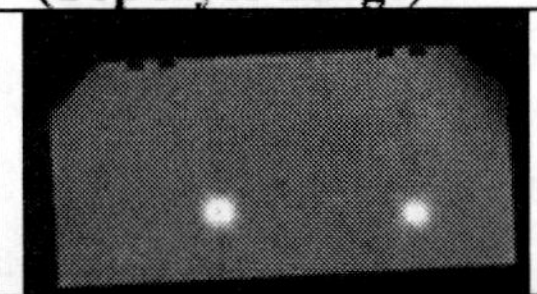	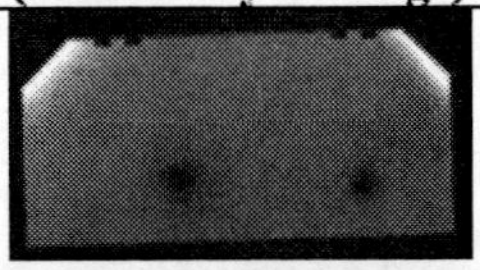	(i) 1 Big bright spot with dark spot inside due to micro crack in TC & MC. Other bright spot due to local shunt in TC. (ii) Local shunt at left & right cropped corners of MC. (iii) Vertical faint black line due to Rs in TC.
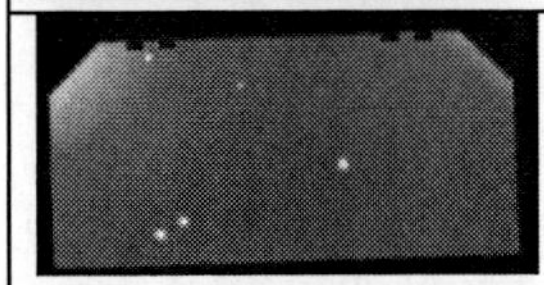	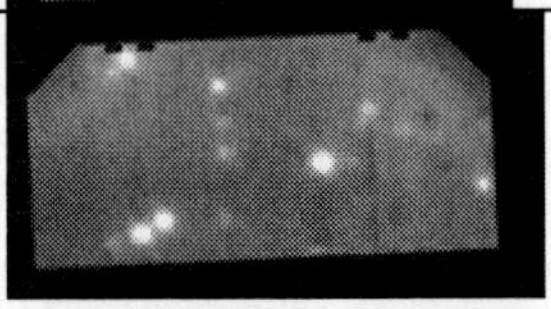	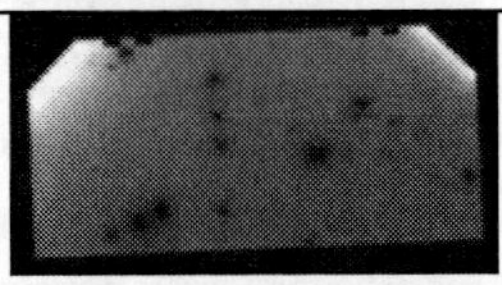	(i) 4 bright spots due to local shunt in TC. (ii) White spots with dark dots inside due to micro crack in TC & MC. (ii) White line linked with 2 dots (complimentary in MC image) due to micro crack in TC. (iv) Local shunt at left & right cropped corners of MC. (v) Vertical faint black line due to Rs in TC.
	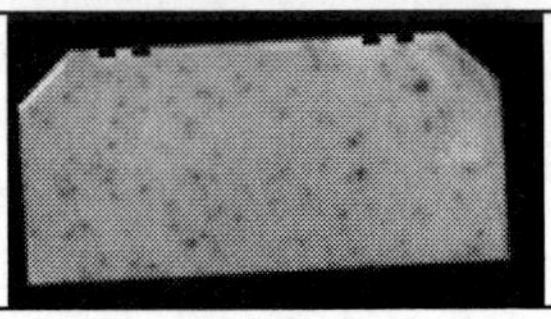	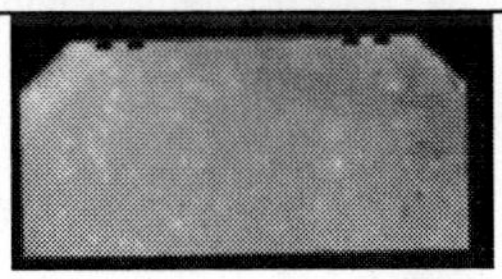	(i) Growth defects in more in MC and less in TC. (ii) Twin dark spots clustered together due to micro crack in MC.

Figure 5. EL Spectroscopy images of solar cell

Ultra fast laser spectroscopy :

This technique is used to investigate the dynamics of carrier life time in femto second range or higher and surface structure of the semiconductors before and after irradiation. This is most suitable for InGaP/InGaAs/Ge multijunction solar cells when the suitable lasers of tunable wavelength excites the MJSC sub-cells and study the carrier decay.

IV characteristics **:**

IV characteristics measurement plays very crucial role in understanding of solar cell behavior due to radiation. From IV measurement we can extract Rsh and Rs of solar cells.

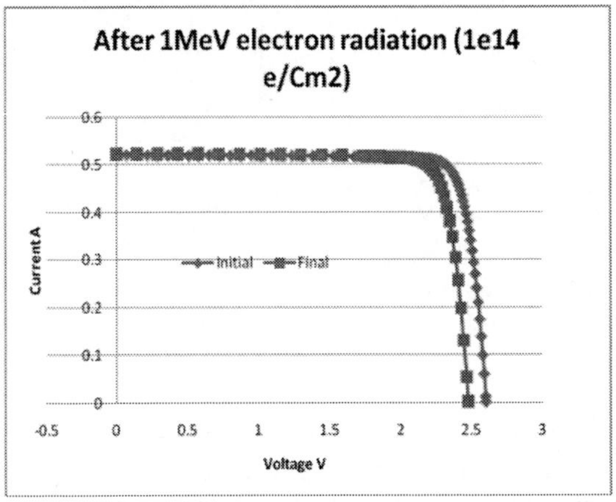

Figure 6. IV characteristics before and after irradiation

I. CONCLUSIONS

To understand the behavior of the solar cells, different characterization techniques are used viz. external quantum efficiency (EQE), EL, PL, DLTS and ultra fast laser spectroscopy. Using these techniques the effect of radiation on the solar cells in particular space grade solar cells can be studied. This will give the information regarding the sub cells behavior which is more vulnerable to radiation.

REFERENCES

[1] R. Campesato, G.Gori, M. Casale, G. Gabetta, M. Sankaran, E.P. Suresh , B.R.Uma, "Radiation Effects on Advanced Multi Junction Solar Cells For Space Mission" 30th EUPVSEC 2016

[2] Helmut Nesswetter, Paolo Lugli, Fellow, IEEE, Andreas W. Bett, and Claus G. Zimmermann, Electroluminescence and Photoluminescence Characterization of Multijunction Solar Cells, IEEE Journal of Photovoltaics , 1-5 (2012)

[3] D. V. Lang, Deep-level transient spectroscopy: A new method to characterize traps in semiconductors, J. Appl. Phys. 45, 3023-3032 (1974).

[4] C. Ghezzi 1; M. Longo 1; R. Magnanini 1; A. Parisini 1; L. Tarricone 1; A. Carbognani 1; C. Bocchi 2; and E. Gombia 2 (2003) Electrical Investigation of Carbon Intrinsically- Doped GaAs Layers Grown by Metalorganic Vapour Phase Epitaxy from TMGa and TBAsin Physica status solidi. C, Conferences and critical reviews

TRANSMITTANCE AND REFLECTANCE MAPS IN 3D-CPCS

Antonio Parretta[1]*, Erica Cavallari[1] and Mario Tucci[2]
[1]University of Ferrara, Physics and Earth Science Department, Via Saragat 1, 44122 Ferrara (FE), Italy
*Phone: +39 3401509953; E-mails: parretta@fe.infn.it; erica.cavallari@student.unife.it
[2]ENEA Research Centre "Casaccia", Via Anguillarese 301, 00123 S. Maria di Galeria (Rome), Italy
E-mail: mario.tucci@enea.it

ABSTRACT: In this paper, we investigate the transmission and reflection properties of photovoltaic 3D-CPCs (Compound Parabolic Concentrators), by applying original simulation methods to obtain the average number of reflections that the transmitted or reflected rays make on the inner wall of the concentrator when it is irradiated by a parallel and uniform beam. The results of the simulations are presented as maps of the entrance opening of the CPC, in which different regions are drawn, each characterized by a different number of internal reflections that the crossing rays undergo before being transmitted or reflected. Maps drawn at different incidence angles of the parallel beam will be presented. The knowledge of these maps allows to optimize the optical design of the concentrator, since, in the real devices, the inner wall reflections are an important source of flux loss, that considerably affect the overall optical efficiency. Preliminary results show that maps of transmittance and reflectance can be obtained by applying a Lambertian source on the exit or entrance opening of the concentrator, respectively. Long-term optical simulations are in progress to improve the angular resolution of these maps.
Keywords: Concentrators, Optical properties, Modelling, Ray Tracing

1 INTRODUCTION

Nonimaging solar concentrators of the type 3D-CPC (Three-Dimensional Compound Parabolic Concentrator) are of considerable interest in the field of CPV in that they allow to reach very high concentration levels, comparable to the theoretical ones, and their optical transmission efficiency is quite constant within a defined angle of incidence of the input beam. A further advantage of these solar concentrators is that they operate with reflective surfaces, that do not induce the spectral dispersion of light and then any chromatic aberration [1-8]. The 3D-CPC is a reflective concentrator with cylindrical symmetry and whose longitudinal cross section profile is obtained by the combination of two parabolas (see Fig. 1a); it is characterized by a step-like transmission efficiency $\eta_\tau(R_w,\theta_{in})$ (see Fig. 1b), allowing the efficient collection of light from 0° to a maximum angle of incidence, the acceptance angle θ_{acc}^{coll}, where the suffix "*coll*" means "collimated" beam irradiation, to be distinguished from "*lamb*" for a Lambertian irradiation. The 3D-CPC used in this work, with θ_{acc}^{coll} = 5°, is the same used in previous communications [9-10].

When a real CPC (R_w < 1) is irradiated in direct mode (that is through its entry aperture) with a parallel beam inclined by an angle of θ_{in}, the input flux Φ_{in} is distributed among three components: a transmitted flux Φ_τ, a back-reflected flux Φ_ρ and an absorbed flux on the inner wall, Φ_α, in such a way that [11]:

$$\Phi_{in} = \Phi_\tau(R_w,\theta_{in}) + \Phi_\rho(R_w,\theta_{in}) + \Phi_\alpha(R_w,\theta_{in}) \quad (1)$$

The transmitted and the reflected beams are suffering an attenuation that is stronger the higher the number of reflections made by the rays on the inner wall of the CPC. Then, a single-ray, for example one of the transmitted ones, with power ϕ_i at input, after $\mathcal{N}_{i,\tau}$ reflections on the inner wall will be attenuated and its output flux will be:

$$\phi_{i,\tau} = \phi_i \cdot R_w^{\mathcal{N}_{i,\tau}} \quad (2)$$

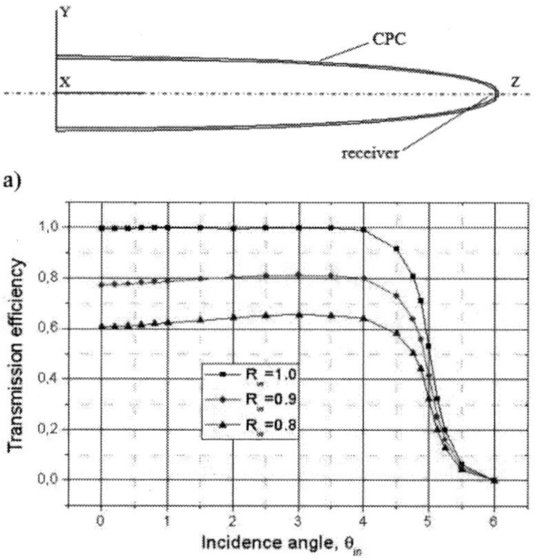

b)

Figure 1: a) Longitudinal cross section of the 3D-CPC simulated in this work. b) Angle-resolved transmission efficiency $\eta_\tau(R_w,\theta_{in})$ of the 3D-CPC with θ_{acc}^{coll} = 5°, irradiated by a parallel beam incident at angle θ_{in} and simulated for three values of the wall reflectivity R_w: 1.0, 0.9 and 0.8.

From Eq. (2) it is easy to get the number $\mathcal{N}_{i,\tau}$ of internal reflections by running two simulations at two values of R_w and measuring the corresponding transmitted fluxes:

$$\mathcal{N}_{i,r}(R'_w, R''_w) \approx \frac{\log\left[\dfrac{\phi_{i,r}(R'_w)}{\phi_{i,r}(R''_w)}\right]}{\log\left[\dfrac{R'_w}{R''_w}\right]} \qquad (3)$$

But the attenuation experienced by a ray, whether it is transmitted or reflected, depends on the impact point P on the entrance opening, because not every ray transmitted or reflected undergoes the same attenuation. Hence the need to distinguish, on the plane of entrance opening, regions crossed by transmitted rays from regions crossed by reflected rays, and to characterize each region by the number of internal reflections that the rays crossing that region make before being transmitted or reflected. It's purpose of this work to obtain maps of the entrance opening carrying, for each point of the map, the information of the number of reflections n made by the transmitted, or m made by the reflected, rays crossing that point of the map. These maps, obtained by simulation codes, are useful to optimize the optical design of the concentrator, since, in real devices, the inner wall reflections are an important source of flux loss, that considerably affect the overall optical efficiency. Maps for different values of the incidence angle of the parallel beam at input will be presented.

2 SIMULATION SCHEMES

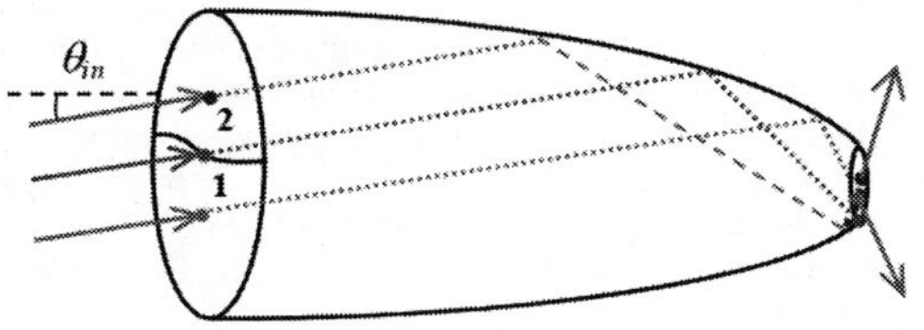

a)

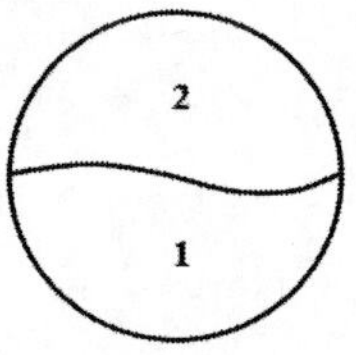

b)

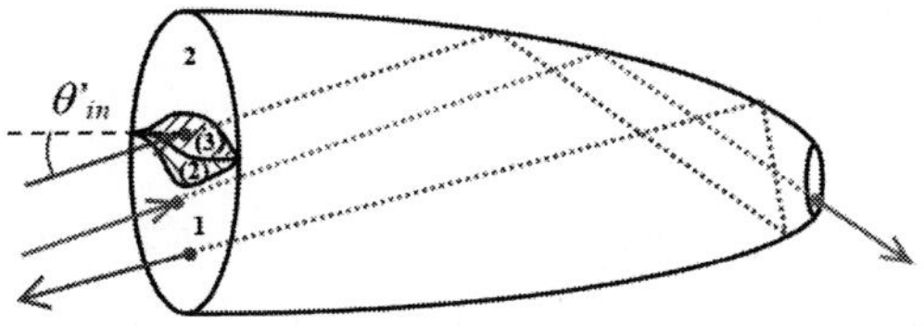

c)

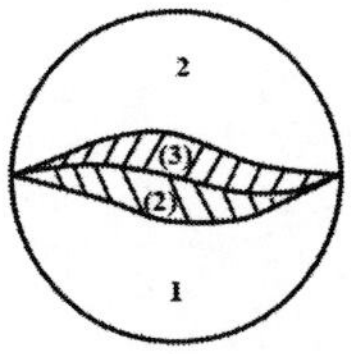

d)

Figure 2: Ray-tracing showing the different regions crossed by rays at the entrance opening of the CPC. a) Here we have two adjacent regions with $n = 1$ and $n = 2$. The green ray, crossing the white region $n = 1$, is transmitted after 1 internal reflection, whereas the blue ray, crossing the white region $n = 2$, is transmitted after 2 internal reflections. The red ray, incident on the edge of the two regions, impacts on the rim of the exit aperture and is scattered. b) It is shown the T map at θ_{in} incidence angle. c) Here the rays are incident at a greater angle on the entrance opening. This determines the splitting of the border line between $n = 1$ and $n = 2$ regions in two hatched regions with $m = 2$ and $m = 3$. The green ray, crossing the white region $n = 1$, continues to be transmitted after 1 internal reflection, whereas the red ray, crossing the hatched region $m = 3$, is back reflected after 3 internal reflections. d) It is shown the T/R map at θ'_{in} incidence angle.

A schematic example of raytracing illustrating the meaning of the different regions in a transmittance/reflectance map (T/R map) is reported in Fig. 2.

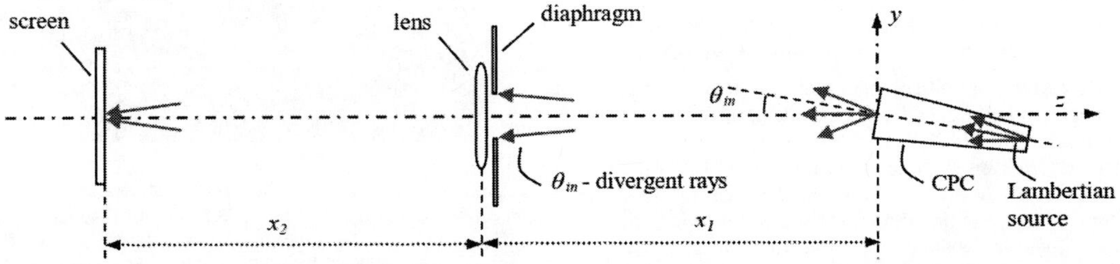

Figure 3: Scheme of the simulation method adopted for obtaining, at θ_{in} angle, the image of the CPC entrance opening when the exit opening is irradiated in a reverse way by a Lambertian beam with 90° angular divergence. The rays transmitted in reverse mode from the entrance opening of the CPC are collected, at θ_{in} angle, by the diaphragm-lens system, which focuses them on the far screen. The blue color of inversely transmitted rays indicates that the red source rays undergo an attenuation inside the CPC.

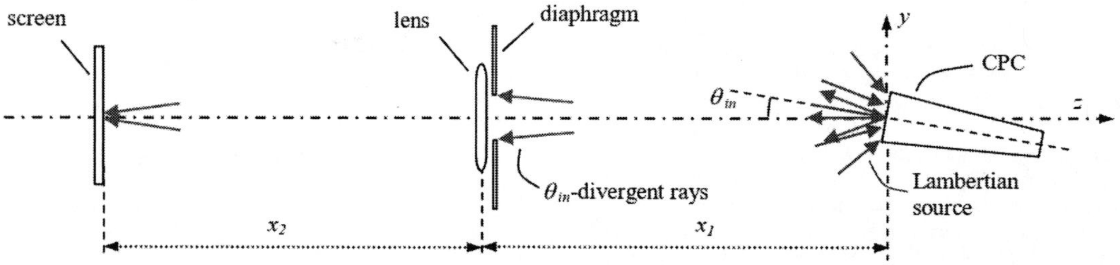

Figure 4: Scheme of the simulation method adopted for obtaining, at θ_{in} angle, the image of the CPC entrance opening when it is irradiated in a direct way by a Lambertian beam with 90° angular divergence. The rays back-reflected by the CPC are collected, at θ_{in} angle, by the diaphragm-lens system, which focuses them on the far screen.

As suggested by Winston [5], we can delineate the different regions of a T map by tracing rays in a reverse way from the exit aperture. The maps of n, in fact, are distorted images of the CPC exit opening viewed from the entrance opening; to get them, therefore, it is sufficient to irradiate the exit opening in reverse mode by a Lambertian source (which emits rays in all directions with the same radiance), and to collect the rays inversely transmitted at a fixed angle θ_{in} (the same angle of incidence of the collimated beam for which we look for T 's map) by a far screen. Fig. 3 shows the scheme adopted to obtain the T map. The CPC is rotated CCW around x axis by an angle θ_{in}. In a T map, Eq. (3) becomes:

$$n(R'_w, R''_w, \theta_{in}, P) \approx \frac{\log\left[\dfrac{I(R'_w, \theta_{in}, P)}{I(R''_w, \theta_{in}, P)}\right]}{\log\left[\dfrac{R'_w}{R''_w}\right]}$$

(4)

where $I(R'_w, \theta_{in}, P)$ and $I(R''_w, \theta_{in}, P)$ are the maps of irradiance measured on the screen when the CPC wall reflectance is R'_w and R''_w, respectively. In Eq. (4), the intensity of the transmitted beam incident on point P at the input opening of the CPC is equivalent, thanks to the reversibility principle [12], to the intensity of the beam transmitted backwards along the same optical path when the CPC is irradiated in reverse mode from the exit aperture. The map of m, number of internal reflections of reflected rays are obtained by irradiating the entrance opening of the CPC by a Lambertian beam and measuring the intensity $I(R_w, \theta_{in}, P)$ on the screen. Combining two maps of $I(R_w, \theta_{in}, P)$ at two wall reflectivities, and applying Eq. (4), the R map of the number m of internal reflections made by the reflected rays can be drawn. The maps of the transmittance and reflectance regions should be built separately. Only after, they can be combined.

3 MAPS OF TRANSMITTED RAYS

Fig. 5 shows the map of number of reflections $n(R'_w, R''_w, \theta_{in}, P) = n(1, 0.8, 0°, P)$ made by the transmitted rays at 0° incidence. The areas with different number of reflections are distinguished by different colors. From Fig. 5 we can see that in most of the central regions of the entrance opening the number of reflections is one and increases going towards the periphery. The small central region with $n = 0$, corresponding to the rays that pass through the CPC without reflections, remains unresolved.

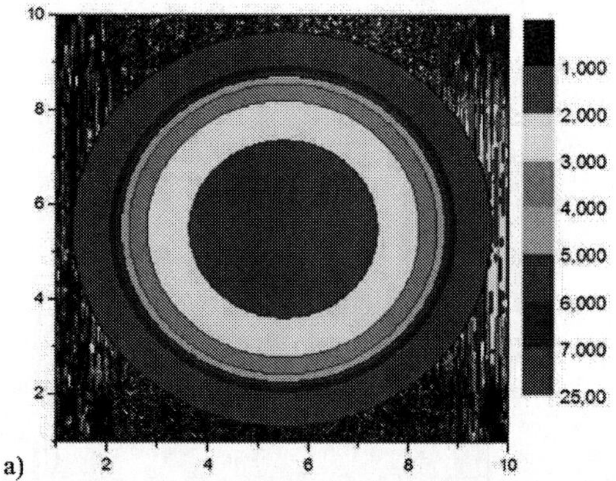

a)

Figure 5: T Map of the local distribution, on the entrance opening of the CPC, of the number of reflections that the rays incident at 0° make on the internal wall of the CPC before being transmitted to the exit opening. The T map is equivalent to the T/R map, as all rays are transmitted.

This is a consequence of the fact that the inverse rays at the output of the CPC that are collected by the system diaphragm+lens+screen (see Fig. 3) have a slight divergence (≈0.7° instead of 0°). The annular regions with $n > 1$ shrink when n increases from 2 to 6, while the annular regions with n between 7 and 25 are grouped in the single green olive region. To obtain the map of n at $\theta_{in} = 5°$, the CPC was rotated by 5° CCW around the x axis, while keeping the center of the entrance opening in the origin of the axes, and two new simulations were made at $R'_w = 1.0$ and $R''_w = 0.8$, following the scheme of Fig. 3. The simulations were carried out with 4M rays and $\approx 12h$ running time.

Figure 6a shows the map of the irradiance obtained on the screen with wall reflectivity $R'_w = 1.0$. The white areas correspond to those of the CPC entrance opening whose incident rays are back-reflected and cannot reach the exit opening. As consequence, those rays do not appear among the inverse rays produced by the Lambertian source placed on the exit opening of the CPC. In the map of Fig. 6a, the most intense regions are those where more rays converge, being that the rays do not undergo optical attenuations.

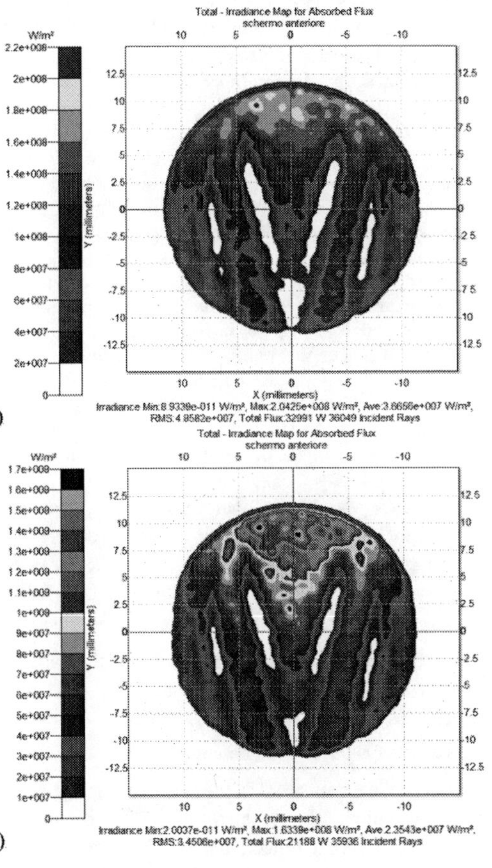

a)

b)

Figure 6: Maps of the irradiance measured on the screen of Fig. 3 when the CPC is oriented at $\theta_{in} = 5°$ respect to z axis. a) Wall reflectivity $R'_w = 1.0$; b) wall reflectivity $R''_w = 0.8$.

This transmission map should show, in principle, a uniform intensity throughout the area affected by the passage of the transmitted rays, because it should reproduce the map of irradiance distribution which could be measured intercepting a direct and uniform, parallel beam, incident on the CPC, by a plane orthogonal to its direction of propagation. The intensity unevenness, in the area affected by the passage of the transmitted rays, is just an effect of the low angular resolution of the simulations. Figure 6b shows the map of the irradiance obtained on the screen with wall reflectivity $R'_w = 0.8$. Differently from the map of Fig. 6a, the intensity of the different zones in the map of Fig. 6b depends on the attenuation that the incident rays undergo as an effect of the optical loss on the CPC inner wall. From the two maps of Fig. 6 we can see that the reduction of the wall reflectivity from 1.0 to 0.8 produces a general reduction of the intensity on the screen, as expected. From the irradiance maps of Fig. 6 we can see that most of rays join the upper part of the screen. As the lens in Fig. 3 reverses the image of the entrance opening, this means that a remarkable decrease of transmission efficiency is observed, in direct mode of irradiation, on that portion of the collimated beam which enters through the upper side of the input aperture. This is justified by the fact that this portion of the beam undergoes a greater number of internal reflections and therefore a greater optical loss [10, 13]. Fig. 6 shows also that the rays contribute to form vertical bands of intensity in the lower part of the map and that the white areas, the one concerning the back-reflected rays, remain well-defined and unaffected by the reduction of R_w.

To obtain the map of number of reflections $n(R'_w, R''_w, \theta_{in}, P) = n(1{,}0.8{,}5°, P)$ made by the transmitted rays at 5° incidence, we have applied Eq. 4, where $I(R'_w, \theta_{in}, P)$ is the irradiance map of Fig. 6a and $I(R''_w, \theta_{in}, P)$ is the irradiance map of Fig. 6b. The map of $n(1{,}0.8{,}5°, P)$ is shown in Fig. 7. The different areas of the map are distinguished by different colors and labeled with the corresponding number of internal reflections.

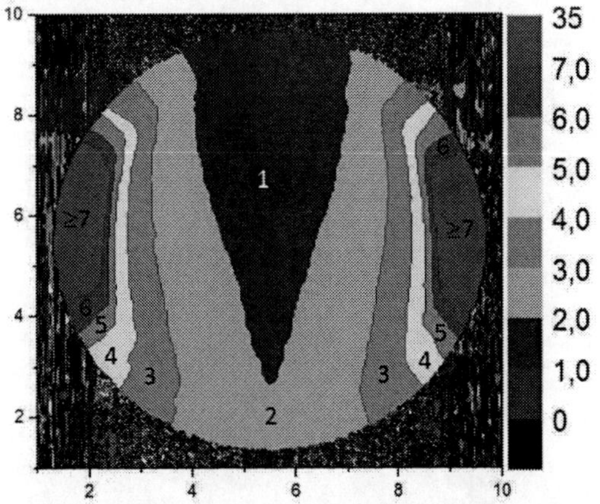

Figure 7: T map of the CPC entrance opening describing the distribution of number of reflections made on the internal wall by the rays incident at 5° before being transmitted to the exit opening.

From Fig. 7 we can see that the 5° rotation of the CPC has had the effect of cutting in two the annular regions with $n > 2$ shown in Fig. 5, separating them, while the region with $n = 2$ remains connected in the lower part of the map and the central region of the map with $n = 1$, is stretched and moved upwards. As in the case of the map of Fig. 5, also here the annular regions with $n \geq 7$ are grouped in the zones of magenta color. The map of Fig. 7 has a specific topological aspect resembling those reported by Winston [5], with the triangular shape of the $n = 1$ region and the V shape of the $n = 2$ region. However, the map of Fig. 7 is not the final map that we look for; this one will be obtained only after the simulation of the maps of reflected rays.

4 MAPS OF REFLECTED RAYS

We now apply the simulation method illustrated in Fig. 4 to obtain the map of reflected rays, the R map. The simplest case, the one with $\theta_{in} = 0°$, is not significant because at this incidence angle there are no reflected rays at all (see Fig. 1b). Fig. 8a shows the map of the irradiance obtained on the screen when the wall reflectivity is $R'_w = 1.0$, whereas Fig. 8b shows the map of the irradiance obtained on the screen when the wall reflectivity is reduced to $R''_w = 0.8$. Both simulations were carried out with 2.5M incident rays.

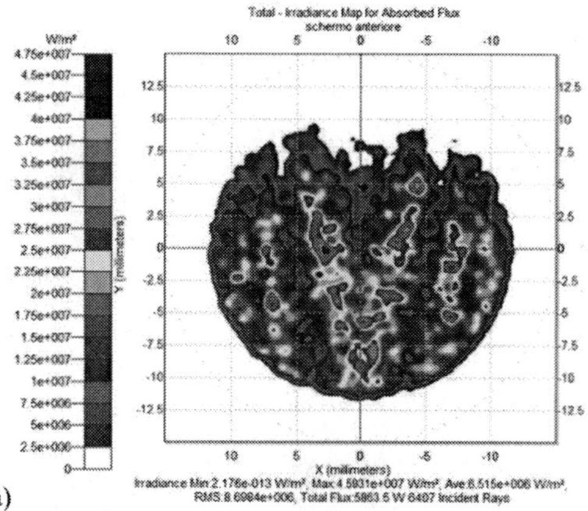

a)

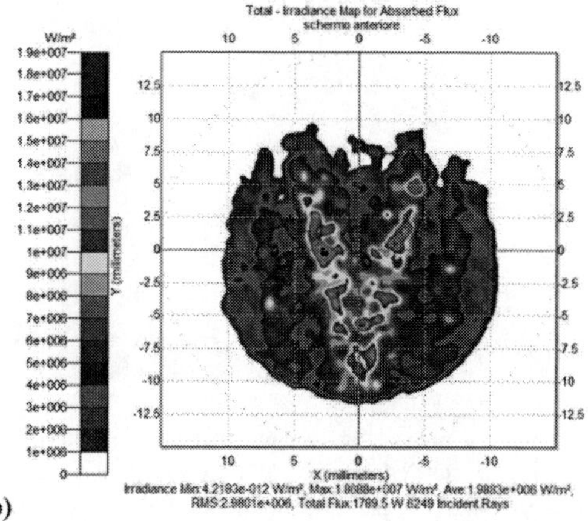

b)

Figure 8: Irradiance maps measured on the screen of Fig. 4 when the CPC is oriented at $\theta_{in} = 5°$ respect to the z axis and the entrance opening is irradiated in direct mode by a Lambertian beam with 10° divergence. (a) Wall reflectivity $R'_w = 1.0$; (b) wall reflectivity $R''_w = 0.8$.

The angular divergence of the Lambertian rays in Fig. 4 was kept low (10°), not 90 ° as in the case of Fig. 3, because now the irradiated aperture is the same from which the reflected rays are exiting; according to Liouville theorem [5], establishing the invariance of the "generalized étendue", the volume occupied by the system in the phase space, we have in fact:

$$A_{in} \cdot \sin^2 \theta_{in} = A_{in} \cdot \sin^2 \theta_{out} \Rightarrow \theta_{in} = \theta_{out} \quad (5)$$

In the present case, as we need to collect only rays back-reflected at 5° (θ_{out}=5°), the application of a θ_{in} = 90° divergent lambertian source at entrance aperture is not necessary, following Eq. (5); we can use therefore a divergence just a little higher than 5°, for example 10°.

The comparison between Fig. 8a and Fig. 8b shows some interesting aspects. First, we note that the V-shaped central region is only slightly attenuated reducing R_w, because this area refers to back-reflected rays characterized by a low number of internal reflections, $m \approx 2-3$ [10], therefore implying a limited attenuation, of $0.8^2 = 0.64$ or $0.8^3 \approx 0.5$. The outermost regions, on the other hand, being crossed by rays that make 5-7 internal reflections, show a stronger attenuation, around 0.3-0.2. But the most interesting comparison is the one between the map of Fig. 6a and that of Fig. 8a. The first concerns the transmitted rays, while the second concerns the reflected rays. The two maps appear complementary. At the top of the map of Fig. 6a we see a very intense region that, in the map of Fig. 8a, is completely devoid of rays. The V-shaped central region in Fig. 6a, which is devoid of any ray, in Fig. 8a is the most intense region of the map. And so forth, observing the lateral regions of the two maps, we can see that the most intense regions of one map correspond to the less intense regions in the other map, and vice versa.

We are now able to obtain the R map, that of the number of reflections made by the back-reflected rays, by applying Eq. (4) in which the quantities $I(R'_w, \theta_{in}, P) = I(1.0, 5°, P)$ and $I(R''_w, \theta_{in}, P) = I(0.8, 5°, P)$ correspond to the irradiances in the maps of Figs. 8a and 8b, respectively. Fig. 9 shows the map of the average number of internal reflections $m(R'_w, R''_w, \theta_{in}, P) = m(1.0, 0.8, 5°, P)$ made by the reflected rays at $\theta_{in} = 5°$ incidence of the collimated beam. The different areas of the map are distinguished by different colors and labeled with the corresponding number of internal reflections. We first note a marked difference with the map of transmitted rays in Fig. 7, that is the average number of reflections made by reflected rays is higher than that made by the transmitted rays. A discussion about this and further results is reported in next section.

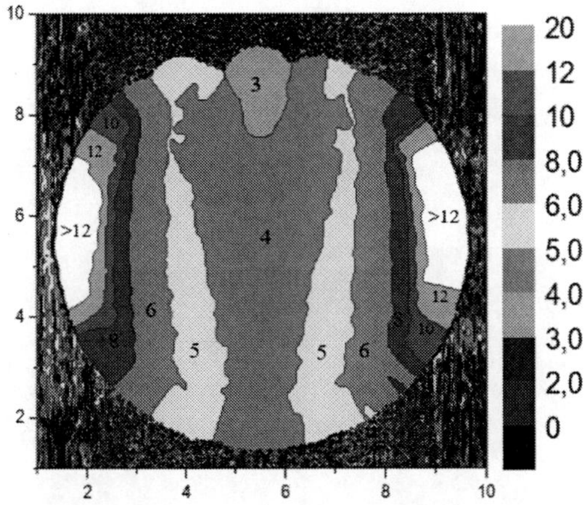

Figure 9: *R* map of the entrance aperture describing the distribution of number of reflections made on the internal wall by rays incident at 5° on the CPC before being back-reflected through the entrance opening.

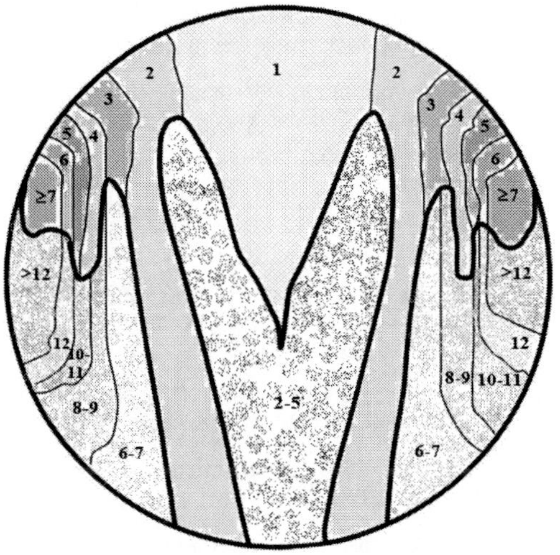

Figure 10: A tentative *T/R* map of the entrance aperture describing the combined distribution of number of reflections n made on the internal wall by the rays incident at 5° on the CPC before being transmitted, and of number of reflections m made on the internal wall by incident rays at 5° on the CPC before being back-reflected.

5 DISCUSSION AND CONCLUSIONS

The *T* and *R* maps of Fig. 7 and 9 should be complementary, that is, they should show only the regions that are crossed by the transmitted rays or by the reflected rays, respectively. On the contrary, the maps are full, and this makes it difficult to interpret the results. The difficulty arises from the fact that the two intensity maps of Fig. 6, from which we obtained the map of n of Fig. 7, do not have an intensity exactly equal to zero in the regions crossed by the reflected rays, due to the low angular resolution of the simulations. Similarly, the two intensity maps of Fig. 8, from which we obtained the map of m of Fig. 9, do not have an intensity exactly equal to zero in the regions crossed by the transmitted rays, also in this case because of the low angular resolution of the simulations. As consequence of this, in the regions in which the intensity should be zero, but it is not, the simulation program extracts an arbitrary value for the number of internal reflections n or m through Eq. (4), even if the intensity is low in both regions. To extract the correct information of n or m from the two maps, therefore, it is necessary to compare them to the intensity maps and neglect the regions in which the intensity is too low. In the end, we got the *T/R* map of Fig. 10, which is only a tentative map.

In this map, the thicker line separates the regions referred to the transmitted rays from the regions referred to the reflected rays. The regions referred to the transmitted rays are full coloured, while those referred to the reflected rays are spray coloured. From Fig. 10 we note that the regions referred to the transmitted rays are well defined, that is adjacent regions differing only by one internal reflection are well separated; instead, the regions referred to the reflected rays contain a range, sometimes large, of number of reflections.

All this, again, is a consequence of the too low angular resolution of the simulations to better distinguish the different regions. As consequence, the regions referred to reflected rays, that are in general thinner than those of transmitted rays, cannot be well resolved. The map of Fig. 10 is then to be considered as a rather approximate map of the real one. We might get a more reliable map improving the angular resolution of optical simulations, but this implies increased processing times, well higher than 12 hours, which are not always tolerated by a code running on a personal computer. We believe, therefore, that the results achieved so far are the best ones that can be achieved as a first attempt and that they should thus be considered as a first, rough result of the proposed method: the map of the average number of internal reflections made by the transmitted and reflected rays of a collimated beam irradiating a 3D-CPC.

REFERENCES

[1] R. Winston, 1974, Principles of Solar Concentrators of a Novel Design, Solar Energy 16, 89-95.

[2] W.T. Welford and R. Winston, 1978, The Optics of Nonimaging Concentrators, Light and Solar Energy, Academic Press.

[3] W.T. Welford and R. Winston, 1989, High Collection Nonimaging Optics, San Diego, Academic Press.

[4] A. Luque, 1989, Solar Cells and Optics for Photovoltaic Concentration, Adam Hilger, Bristol and Philadelphia.

[5] R. Winston, J.C. Miñano, P. Benítez, 2005, Nonimaging Optics, Elsevier Academic Press.

[6] J. Chaves, 2008, Introduction to Nonimaging Optics, CRC Press.

[7] J.J. O'Gallagher, 2008, Nonimaging Optics in Solar Energy, Morgan & Claypool Publishers, Lexington, KY.

[8] A. Antonini, M.A. Butturi, P. Di Benedetto, D. Uderzo, P. Zurru, E. Milan, M. Stefancich, M. Armani, A. Parretta, N. Baggio, *"Rondine® PV Concentrators: Field Results and Developments"*, Progress in Photovoltaics **17**, 451 (2009).

[9] A. Parretta and M. Tucci, *"Models of light collection of 3D-CPC concentrators under Lambertian irradiation"*, 32nd European Photovoltaic Solar Energy Conference, Munich, Germany, June 20-24, 2016.

[10] A. Parretta, R. Chierchia and M. Tucci, *"Models of Light Collection of 3D-CPC Concentrators under Collimated Beams"*, 29th European Photovoltaic Solar Energy Conference and Exhibition, Amsterdam, The Netherlands, Sept 22–26, 2014.

[11] A. Parretta and M. Tucci, *"Optical Models of Light Collection in Solar Concentrators"*, 28th European Photovoltaic Solar Energy Conference and Exhibition, Paris, France, Sept 30 – Oct 4, 2013.

[12] R.C. Jones, *"On Reversibility and Irreversibility in Optics"*, Journal of Optical Society of America, **43**, 138 (1953).

[13] A. Parretta, A. Antonini, M. Stefancich, V. Franceschini, G. Martinelli, M. Armani, *"Characterization of CPC solar concentrators by a laser method"*, Proc. of the SPIE Optics and Photonics Conference, San Diego, California (USA), 26-30 August 2007. Optical Modeling and Measurements for Solar Energy Systems, ed. by Daryl R. Myers, Proc. of SPIE Vol. 6652, 665207, (2007).

EVALUATION OF OUTDOOR PERFORMANCE AND TECHNO-FINANCIAL ANALYSIS OF A STATIONARY HIGH CONCENTRATING PVT SYSTEM

Corry de Keizer*[1], Marnick van de Zande[2], Peter Penning[2], Wiep Folkerts[1]
[1] Solar Energy Application Centre (SEAC), High Tech Campus 21, 5656 AE Eindhoven, www.seac.cc
[2] SunCycle Technology BV, Eindhoven, the Netherlands, www.suncycle.nl
*Corresponding author: dekeizer@seac.cc

ABSTRACT: In this paper, we evaluate the outdoor electrical and thermal performance and the techno-financial feasibility of a HCPV/T system. The SunCycle module is a stationary high concentrating PVT (HCPV/T) concept that was developed and patented by the company SunCycle. The SunCycle module uses internal tracking: a prism and parabolic mirror are rotating independently and thereby concentrating the direct light on a small active cell area. A high efficient III-V cell is used to convert the light to electricity, active cooling is used to produce heat. Since it is a stationary system, it can easily be implemented in an urban environment. In 2015 and 2016, the thermal and electrical output of several prototypes of the SunCycle module have been measured outdoors in the Netherlands over a period of multiple months. A maximum electrical efficiency of more than 16% has been measured in the field. Furthermore, a techno-financial model was implemented to calculate the economic feasibility of the SunCycle system for the Netherlands and Jordan. The calculated financial feasibility for Jordan was good, due to a high share of direct irradiance as well as a high price and net metering for electricity. A further increase in efficiency and functionality of the SunCycle module is needed and expected to be realized over the coming years.
Keywords: concentrators, hybrid, performance, thermal performance

1 INTRODUCTION

The company SunCycle developed an innovative high concentrating PVT (HCPV/T) concept that can be applied in the built environment. Since a large share of the energy demand in the built environment in continental Europe consists of heat, the combination of both solar heat and solar electricity production in one roof is interesting.

The SunCycle module uses an internal tracking mechanism. The prism and parabolic mirror are rotating independently and concentrate the direct light on a small active cell area (see Figure 1). The high-efficiency III-V cell is mounted on a cooling block through which a fluid flows and heat is produced. The module produces both electricity and heat. Since the module is stationary it can easily be implemented in the built environment. The modules have a diameter of approximately 50 cm and it is foreseen that a SunCycle system base unit will consist of 20 units. A concentration ratio of 500 x should be reached.

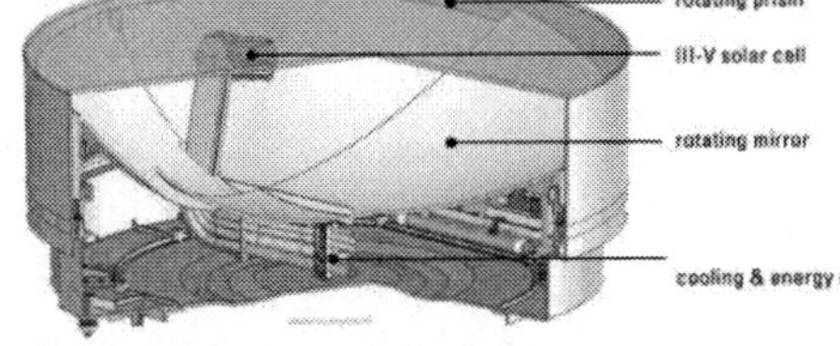

Figure 1: SunCycle module [1]

The work for this paper was carried out in 2015 and 2016. The aim of this paper is to evaluate the electrical and thermal performance of the SunCycle module in an outdoor field test. Furthermore, the goal is to evaluate the techno-financial feasibility for a SunCycle system in the Netherlands and in Jordan.

2 OUTDOOR PERFORMANCE

2.1 Field test set-up

The SolarBEAT facility, see Figure 2, is an outdoor Research & Development infrastructure for innovation on BIPV(T) applications. The facility is a cooperation between SEAC and the Technical University Eindhoven and is located on a large flat south oriented roof with an almost clean horizon. The main use is to measure the performance of BIPV, PVT and solar thermal systems in dependence of several variables (a.o. weather related).

Figure 2: SolarBEAT in 2015

From April 2015 to September 2016 different versions of the SunCycle module have been evaluated (see Figure 3). For every sequence, a single SunCycle module was connected to a pressurized thermal circuit and to a battery set-up. The thermal loop is needed for the required continuous cooling and to determine the thermal response for different inlet fluid temperatures. The fluid inlet temperature could be set between 12 and 80°C. The batteries act as an electrical load.

The following measurement equipment was used:

- Meteorological measurements: DNI measured by a pyrgeometer and ambient temperature. The direct irradiance in the plane of array is calculated by:

$$G_b = DNI \cdot \cos(AOI) \qquad [W/m2]$$

Figure 3: SunCycle module generation 1 (left) and generation 2 (right) on SolarBEAT

- PV performance measurements: Measurements are done at MPP (as determined by the electronics within the module). DC voltage and DC current of both the cell and the module (after DC-DC conversion)
- Thermal performance measurements: input and output temperature of the collector (Pt100, 1/3B) and flow rate (Krohne MID sensor). The heat flow is calculated by:

$$Q_{th} = \rho \cdot c_p \cdot \varphi_v \cdot (T_{out} - T_{in})$$

The thermal performance evaluation was carried out according to the Steady State analysis as described in the norm ISO 9806:2013 [2]. The collector equation for covered collectors is given in the equation:

$$\eta_{th} = \frac{Q_{th}}{A\, G_b} = \eta_0 + a_1 \left(\frac{T_m - T_a}{G_b}\right) + a_2\, G_b \left(\frac{T_m - T_a}{G_b}\right)^2,$$

where

η_0 – zero loss efficiency

a_1 – heat loss coefficient [$W/m^2\,K$]

a_2 – wind speed dependence of the heat loss coefficient [Ws/m^3K]

And the mean collector liquid temperature is defined by $T_m=(T_{out}+T_{in})/2$ and T_{in} and T_{out} the ingoing and outgoing liquid temperature for the module.

2.2 Field test results

The maximum measured electrical efficiency of the first generation module was 6.7%. This is much lower than the stationary measured efficiency (with a solar simulator) of 15%. The differences can be explained by optical losses in the prism and mirror and a low quality concentration, because of non-optimal tracking.

For the second generation module, many improvements have been implemented with regards to cell design, secondary optics, placement of components, etc. This leads to a much higher reached cell efficiency of maximum 16.3% for the second generation. This is still lower than the stationary efficiency of 19.5 % measured with a solar simulator. Further improvements will be implemented.

The thermal efficiency curve for the SunCycle module generation 1 and 2 that was generated with the steady state model is shown in Figure 4. The zero loss efficiency was 27% for the 1st generation SunCycle module. This improved to 31% for the second generation module. Furthermore, the first thermal loss coefficient reduced from 7.8 to 3.8 W/m^2K. The secondary loss coefficient turned out to be not significant. It is expected that the heat losses can be further reduced by better insulation.

There are a large number of measurement points with a low reduced temperature (T_{red}). The large dispersion of values arises due to uncertainties and sensitivity with regards to direct sunlight and electrical performance.

A lot of learning experiences during the field test were made, these will result in a higher quality of modules in the future.

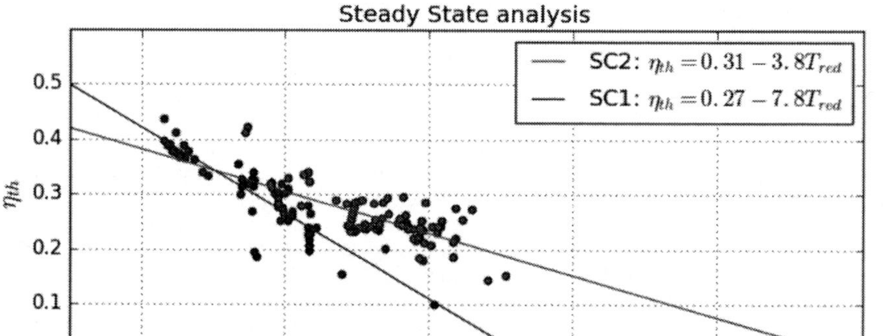

Figure 4: The thermal efficiency curve for the SunCycle module

3 TECHNO-FINANCIAL FEASIBILITY

3.1 Model description

The techno-financial performance of the SunCycle system is important for its commercial success. One of the project goals was to analyze the techno-financial aspects of SunCycle systems. Therefore, a techno-financial model was developed by SEAC in cooperation with Ecofys. The model calculates key financial parameters, like LCOE (levelised cost of electricity), RoI (Return on Investment) and NPV (net present value) from either a consumer or an investor point of view. The model uses technological and financial inputs to calculate key performance indicators. The model has previously been used for analyzing business models for BIPV systems. Since the SunCycle module is a HCPV/T module that produces both power and heat, the model was extended.

A system of 20 modules (approximately 4 m^2), including installation and other system components (inverter, storage, pump, etc.) was used as a reference. The yield was estimated based on the current technology and the expected efficiency increase in 2018 and 2020. The scenarios for efficiency increase are very positive; the yield almost doubles in 2018 and increases with a further 35% in 2020. The increase in efficiency is caused by a low current efficiency of DC-DC and DC-AC conversion (both 80%) and by expected increase in module efficiency.

Like the Netherlands, Jordan currently has a net metering practice. The electricity price depends on electricity usage. We took three scenario's into account: €0.11/kWh, 0.203€/kWh en 0.34€/kWh.

3.2 Results

The case study investigated the financial attractiveness of the ownership of a 20 module HCPV/T system of SunCycle. Using a base case, including future values for yield and costs, and varying critical parameters, several scenarios are made, resulting in the net present value (NPV) and levelized cost of energy (LCoE). The techno-financial analysis is very positive for the situation in Jordan and not positive for the Netherlands. A summary of the results is shown in Figure 5.

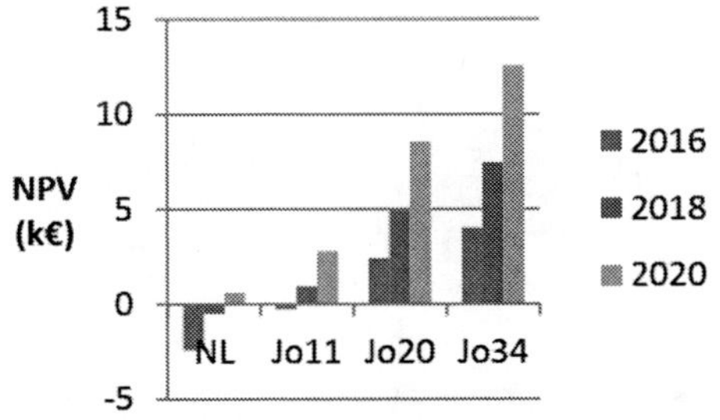

Figure 5: The Net Present Value (NPV) for Netherlands and Jordan, with three different electricity pricing scenarios, for 2016, 2018 and 2020. Jo11 refers to an electricity price of 0.11€/kWh, Jo20 to 0.2 €/kWh and Jo34 to 0.34 €/kWh.

One conclusion is that in the Netherlands the base cases for 2016 and 2018 are not financially beneficial for the user within a reasonable number of years. The initial investment cannot be earned back from the revenue of the yield if a SunCycle system of 20 units is bought and installed in the Netherlands. However, the base case scenarios of 2020 results in a positive NPV, because the yield is expected to increase rapidly and the costs to decrease. One of the reasons for these results is that there is not much direct irradiation in the Netherlands. In Jordan, with almost three times as much direct irradiation, the situation is a lot brighter. In Jordan the SunCycle system has a very positive NPV for all scenarios from 2018 onwards and for most cases in 2016 as well. The payback time of the system is very low in 2020 with 3 to 7 years. The NPV in Jordan is mainly sensitive to the used discount rate and the gas and electricity prices.

3.3 Discussion

The original plan was to use the outcome of the field test for the yield estimation of the techno-financial model. Due to delays, this was not possible. Therefore, estimates have been used. The electrical efficiency of 20% is quite optimistic; it has been measured at certain moments, however, not for longer time frames. Also the improvements in efficiency are ambitious. The yields and costs as forecasted by SunCycle could not be validated, due to a very early market positions.

4 CONCLUSIONS

A summary of the major conclusions follows:

Different versions of the SunCycle module have been measured in 2015 and 2016 outdoors. A maximum operating module efficiency of 16.3 % before DC-DC conversion has been measured. Much knowledge for further increasing the module performance has been generated.

An early techno-financial feasibility study has been carried out for a system with 20 SunCycle modules. A positive outlook exists for Jordan, a country with a high direct irradiance.

ACKNOWLEDGEMENTS

This work was supported by the 'Netherlands Enterprise Agency' (Rijksdienst voor Ondernemend Nederland, RVO) and the Dutch Topteam Energy via the project: 'SunCycle 2nd generation concentrator cell-assembly' with grant number TKIZ01019.

REFERENCES

[1] SunCycle website, www.suncycle.nl, accessed August 2017

[2] ISO 9806:2013. Solar energy – Solar thermal collectors – Test Methods, 2013

NOVEL EPITAXIAL GaAs LIFT-OFF APPROACH VIA VAN DER WAALS INTERFACE IN In_2Se_3 BUFFER LAYER

Nobuaki Kojima, Li Wang, Yoshio Ohshita, Masafumi Yamaguchi
Toyota Technological Institute
2-12-1 Hisakata, Tempaku-ku, Nagoya 468-8511, Japan

ABSTRACT: We propose new epitaxial III-V compond layer transfer technique by peeling from Si substrate via van der Waals interface of $(In_xGa_{1-x})_2Se_3$. To demonstrate this idea, the double layers of In_2Se_3 and GaAs were grown on Si(111) 4° off substrate by molecular beam epitaxy (MBE). Epitaxial growth of α-In_2Se_3 and GaAs on Si(111) can be confirmed by the XRD pole figure measurement. The epitaxial GaAs layer was successfully transferred onto flexible PMMA sheet by just pulling in both directions. It was confirmed that GaAs layer is peeled from Si substrate via van der Waals interface in α-In_2Se_3 layer. We propose this layer transfer technique is applicable to the fabrication of thin film flexible multi-junction solar cells. However, 60° rotation twin formation and island growth of GaAs occurred at the current growth condition. These problems should be solved by optimizing growth conditions.
Keywords: Buffer Layer, Epitaxy, III-V Semiconductors, Thin Film Solar Cell, Tandem

1 INTRODUCTION

The defect zincblende structured III-VI compounds $(In_xGa_{1-x})_2Se_3$ are attractive materials as novel buffer layer for GaAs on Si system. $(In_xGa_{1-x})_2Se_3$ has the lattice constant between that of Si and GaAs, depending on In content, and can have layered structure with loose van der Waals gaps as described below. The layered structured materials are suitable for lattice mismatch and thermal expansion mismatch buffer layers.

The defect zincblende structure is basically a zincblende structure, in which one third of III group atoms are vacant. In this structure, the native vacancies are known to be spontaneously ordered. In Ga_2Se_3, vacancies order along [1-10] direction of the zincblende lattice.[1] On the other hand, in In_2Se_3, vacancies form planes every three indium planes along the (111) plane of the zincblende lattice.[2] Thus, defect zincblende In_2Se_3 exhibits layered structure with van der Waals gaps.

We propose novel buffer layers consisting of $(In_xGa_{1-x})_2Se_3$ compounds for the GaAs on Si(111) system.[3] In this crystallographic orientation, the layered defect zincblende $(In_xGa_{1-x})_2Se_3$ can be grown on Si epitaxially with the van der Waals gaps aligned parallel with the Si(111) substrate surface. The van der Waals interface of the layered structure should absorb any strain caused by lattice mismatch and thermal expansion coefficient difference between Si and GaAs.

We also propose new layer transfer technique for the fabrication of thin film flexible multi-junction solar cells by using $(In_xGa_{1-x})_2Se_3$ buffer layer.[4] The III-V overlayer on $(In_xGa_{1-x})_2Se_3$ buffer layer can be transferred onto a support substrate by peeling from Si substrate via van der Waals interface of $(In_xGa_{1-x})_2Se_3$. Therefore, thin film flexible III-V solar cells can be fabricated more easily than usual epitaxial lift-off (ELO) technique using selective etching by aqueous acid.

However, III-VI compounds have several stable compositions and polytype structures other than the defect zincblende structure. In the case of In-Se system, In_4Se_3, InSe, In_4Se_7 and In_2Se_3 are known as the stable compositions, and α-, β-, γ-, δ-, κ-type In_2Se_3 are known as the polytype structures. The layered defect zincblende In_2Se_3 is included in the α-type structure. Thus, it is necessary to grow α-type $(In_xGa_{1-x})_2Se_3$ preferentially for use of the buffer layer.

In this paper, we report the epitaxial double layer growth of GaAs and α-In_2Se_3 on Si(111) substrate and the demonstration of transfer of GaAs layer onto PMMA sheet by peeling from Si substrate via van der Waals interface.

2 EXPERIMENTAL PROCEDURE

Double layers of In_2Se_3 and GaAs were grown on Si(111) 4° off substrate by molecular beam epitaxy (MBE). The growth temperature and growth time were fixed at 450°C and 1hr for both In_2Se_3 and GaAs layers. VI/III and V/III beam flux ratios for In_2Se_3 and GaAs were 140 and 20, respectively. The typical growth rate for In_2Se_3 and GaAs was 150 to 200 nm/hr.

The crystal structure and crystallographic orientation distribution were characterized by the 2θ-ω scan and pole figure of X-ray diffraction (XRD). Surface morphology of the film was analyzed by scanning electron microscope (SEM).

After the above analysis, the GaAs layer transfer was attempted. The substrate back side was bonded onto glass substrates, and the film top side was bonded with flexible PMMA sheet by adhesive. Then, we pulled glass substrate and PMMA sheet toward the opposite side. The process is illustrated in Fig. 1.

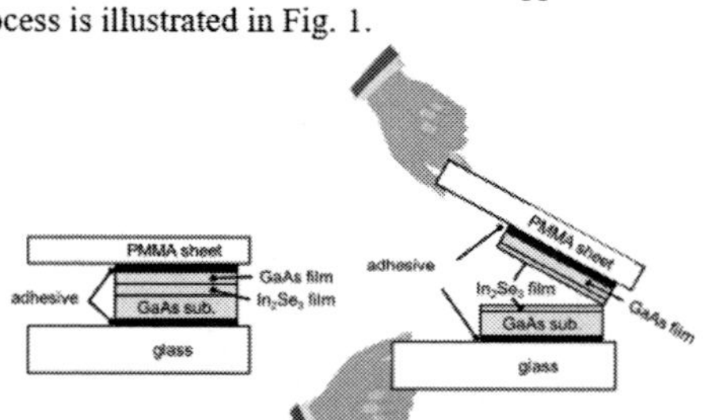

Figure 1: The demonstrated layer transfer process.

3 RESULTS AND DISCUSSION

The crystallographic orientation distribution for each layer was characterized by the XRD pole figure measurement. Figure 2 shows the pole figures for Si(331), α-In_2Se_3(1 1 21) and GaAs(331) plane, respectively. Epitaxial growth of α-In_2Se_3 and GaAs on Si(111) can be

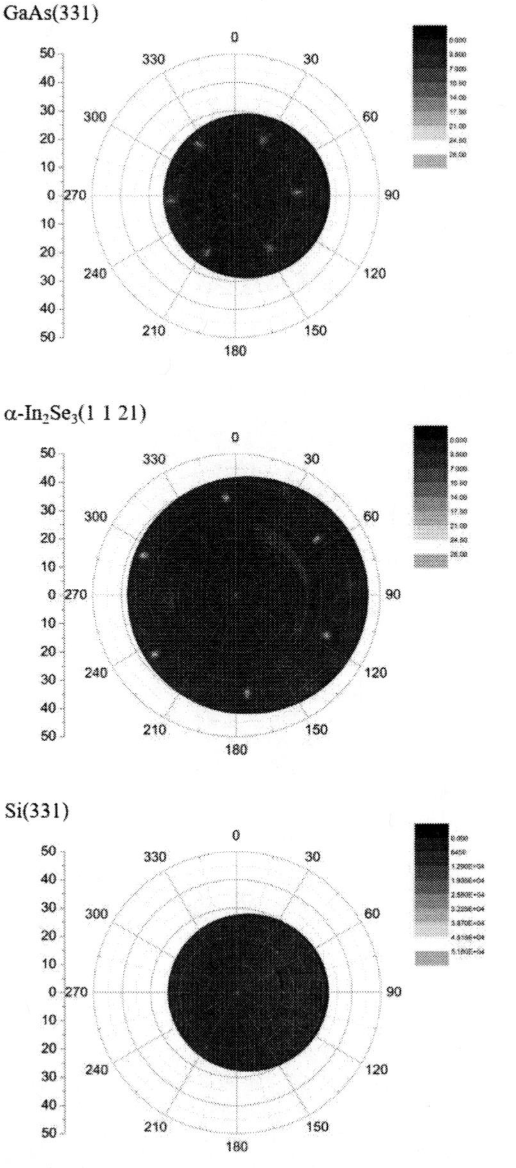

Figure 2: XRD pole figures for Si(331), α-In_2Se_3(1 1 21) and GaAs(331) plane.

confirmed, since the diffraction peaks of α-In_2Se_3 and GaAs are located the particular direction. However, α-In_2Se_3(1 1 21) and GaAs(331) show sixfold symmetry, while Si(331) peaks show threefold symmetry. This result shows 60° rotation twin formation in α-In_2Se_3 and GaAs layers. Peak broadening in α-In_2Se_3 and GaAs is also one of the problems.

Figure 3 shows SEM micrograph of the film surface. GaAs forms islands on α-In_2Se_3 layer. The average island size is around 0.5 μm.

We attempted the GaAs layer transfer onto flexible PMMA sheet by the above mentioned method.

The photograph of the transfered GaAs layer onto PMMA sheet is shown in Fig. 4. We successfully demonstrated the layer transfer by peeling from Si substrate.

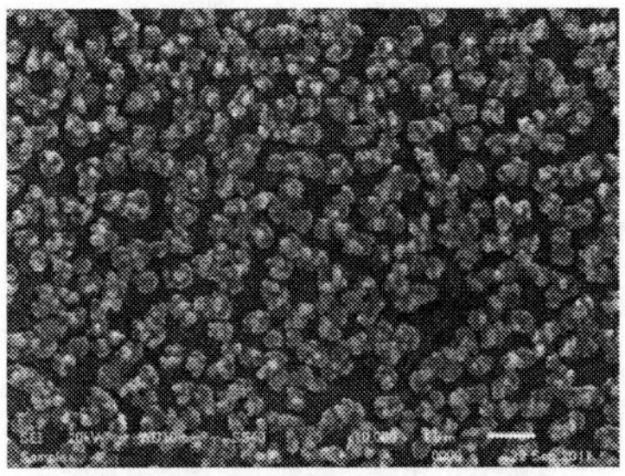

Figure 3: SEM micrograph of the film surface.

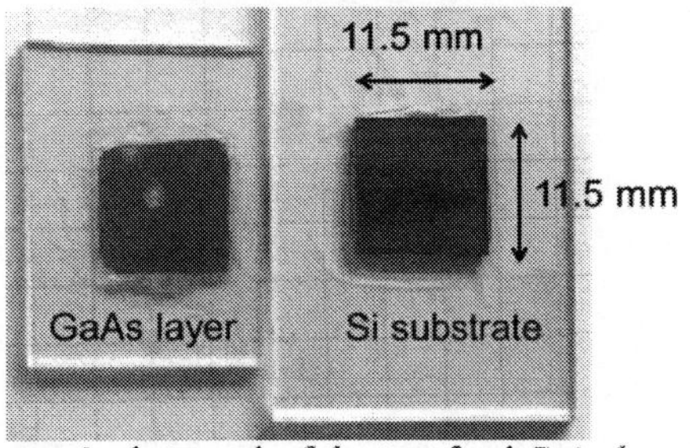

Figure 4: photograph of the transfered GaAs layer onto PMMA sheet.

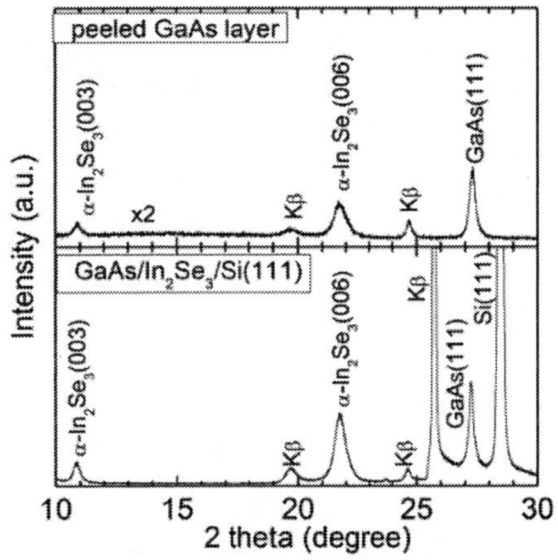

Figure 5: XRD 2θ-ω scanning patterns of $GaAs/In_2Se_3/Si(111)$ and peeled layer

Figure 5 shows XRD 2θ-ω scanning patterns of $GaAs/In_2Se_3/Si(111)$ and peeled layer. The diffraction peaks of α-In_2Se_3 and GaAs are seen in the peeled layer. We confirmed GaAs layer is peeled via van der Waals interface in α-In_2Se_3 layer. We propose this layer transfer technique is applicable to the fabrication of thin film flexible multi-junction solar cells.

4 CONCLUSION

We propose new epitaxial III-V compond layer transfer technique by peeling from Si substrate via van der Waals interface of $(In_xGa_{1-x})_2Se_3$. To demonstrate this idea, the double layers of In_2Se_3 and GaAs were grown on Si(111) 4° off substrate by molecular beam epitaxy

(MBE). Epitaxial growth of α-In_2Se_3 and GaAs on Si(111) can be confirmed by the XRD pole figure measurement. The epitaxial GaAs layer was successfully transferred onto flexible PMMA sheet by just pulling in both directions. It was confirmed that GaAs layer is peeled from Si substrate via van der Waals interface in α-In_2Se_3 layer. We propose this layer transfer technique is applicable to the fabrication of thin film flexible multi-junction solar cells. However, 60° rotation twin formation and island growth of GaAs occurred at the current growth condition. These problems should be solved by optimizing growth conditions.

ACKNOWLEDGEMENT

This work was supported in part by the New Energy and Industrial Technology Development Organization (NEDO) under the Ministry of Economy, Trade and Industry (METI), Japan and by JSPS KAKENHI Grant Numbers JP15K05998 and by the Strategic Research Infrastructure Project, MEXT, Japan.

REFERENCES

[1] K. Ueno et al., Journal of Crystal Growth 237-239 (2002) 1610.
[2] T. Okamoto et al., Journal of Crystal Growth 175/176 (1997) 1045.
[3] N. Kojima et al., AIP Conference Proceedings, 1556 (2013) 38-40.
[4] N. Kojima et al., Proc. 29th EU-PVSEC (2014).

DEGRADE-CPV: A NEW INITIATIVE ON THE DEGRADATION ANALYSIS OF CPV SYSTEMS IN SPAIN AND CYPRUS

Marios Theristis[1,*], Eduardo F. Fernández[2], Juan Pablo Ferrer-Rodríguez[2], Jesús Montes-Romero[2], George Makrides[1], Florencia Almonacid[2], Pedro Pérez-Higueras[2] and George E. Georghiou[1]
[1] PV Technology Laboratory, FOSS Research Centre for Sustainable Energy, University of Cyprus, Nicosia, 1678, Cyprus
[2] Centre for Advanced Studies in Energy and Environment (CEAEMA), IDEA Solar Energy Research Group, Electronics and Automation Engineering Department, University of Jaén, Las Lagunillas Campus, Jaén, 23071, Spain
* Corresponding author: theristis.marios@ucy.ac.cy

ABSTRACT: This paper presents the collaboration initiative by the Universities of Cyprus and Jaén, Spain named as "DEGRADE-CPV". This effort aims in quantifying the degradation rates and modes of different systems based on the concentrating photovoltaic (CPV) technology. Knowledge of degradation is crucial for the reduction of investment risk and also to increase the bankability. The methodology and experimental campaigns conducted at both locations are described and the results of the indoor characterisation using the Helios 3198 solar simulator are presented in this work. The preliminary outdoor characterisation highlights the importance of incorporating corrections on the quantification of the degradation rate.
Keywords: concentrating photovoltaics, degradation, performance

1 INTRODUCTION

One of the most promising research paths for producing clean and sustainable electricity is by using the concentrating photovoltaic technology (CPV) [1]. This technology has demonstrated its ability to produce more cost-effective electricity compared to conventional photovoltaics (PV) at locations with high solar resource [2]. Such systems employ an optical component to concentrate the direct sunlight onto a smaller area made of high efficiency multijunction solar cells (record efficiency of 46% to date [3]). However, although the technology is promising, its maturity and understanding of the behaviour when operating outdoors is clearly lower than conventional PV. Bearing these in mind, important efforts are still needed in order to achieve a better understanding, increase the investors' confidence and therefore promote the market expansion of CPVs as a real alternative energy source to conventional PVs.

A parameter that is well quantified for PV technologies and is directly associated to reliability is the degradation rate. Degradation rate is defined as the power decrease over time and it is a crucial parameter that has not yet been adequately addressed for concentrating photovoltaics. Many studies have been published on the PV degradation [4-7], however, the long-term degradation data of CPV systems are still scarse, mainly because of the young age of this particular technology [8]. The accurate quantification of the CPV degradation is expected to provide reliable lifetime energy yield predictions and thus, to increase the investors' confidence that, in turn, will lead to the market expansion of the technology.

Following this line, a collaboration initiative between the Universities of Cyprus (UCY) and Jaén (UJA) has started in an attempt to answer three open and timely questions concerning different CPV technologies:

1) Why they degrade?
2) How they degrade?
3) By how much they degrade?

In order to answer these questions, outdoor experimental campaigns are conducted in Nicosia, Cyprus and Jaén, Spain where both locations exhibit a high annual direct normal irradiation (see map in Fig. 1) but relatively different atmospheric conditions. Basic geographical, irradiance and ambient characteristics are given in Table 1 for both locations, where DNI_A is the annual direct normal irradiation, T_{air} is the air temperature and RH is the relative humidity.

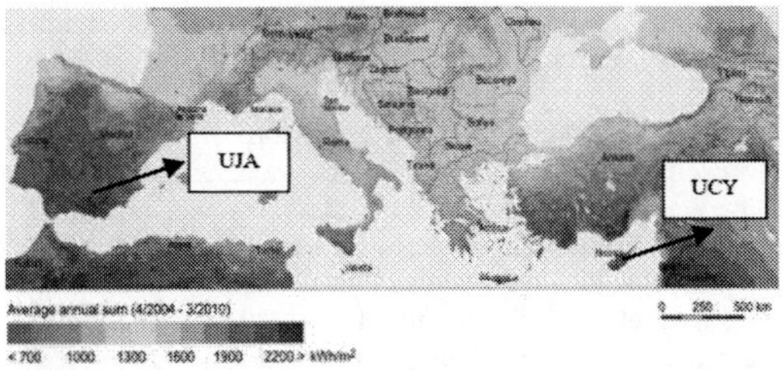

Figure 1: Map of direct normal irradiation (http://solargis.info). The UJA and UCY locations are also shown.

Table I: Annual irradiation and mean meteorological data at each location (data obtained from NASA surface meteorology and solar energy database). The geographical data are also given.

Parameter	UCY	UJA
DNI_A (kWh/m^2/year)	2394	2146
T_{air} (°C)	20.2	16.1
RH (%)	55.4	52.5
Clearness index	0.6	0.56
Latitude (°)	35.15	37.79
Altitude (m)	135	452

2 MATERIALS AND METHODS

Two "as identical as possible" experiments are being conducted at the Universities of Jaén, Spain and Cyprus. Several CPV modules are being tested both indoors and outdoors. The CPV modules consist of a silicon-on-glass Fresnel lens as the primary concentrating optic and a 10 x

10 mm triple-junction solar cell bonded to a direct bonded copper substrate. A secondary optic is attached on the cell's surface for homogenisation purposes.

Initially, all modules were flash tested at Concentrator Standard Test Conditions (CSTC) using a Helios 3198 solar simulator (Fig. 2) in CEAEMA, Jaén. The outdoor experiments have begun on the 1st of April 2017 and consist of grid-connected CPV systems mounted on high accuracy two-axis solar trackers (Fig. 3).

The modules are connected to the same type of micro-inverter (Sunny Boy 240). As such, side-effects due to different maximum power point tracking (MPPT) algorithms are eliminated. The tracking accuracy is monitored by the BPI-TA1 sensor. A pyranometer and a pyrheliometer are used to measure the global and direct normal irradiances (GNI, DNI) respectively. The SolarSIM-D2 device is used for the spectrally resolved DNI, aerosol optical depth (AOD) and precipitable water (PW) measurements. Maximum power point characteristics are recorded on the DC (current and voltage) and AC (power) sides of the system using electrical transducers that are connected to a reliable data logging system. Thermal measurements are taken from the Fresnel lens' surface (close to the edge and outside the active aperture area) and the aluminium heat sinks (as close to the "root" as possible). Meteorological measurements are acquired and stored through a data acquisition platform. Such measurements consist of the ambient temperature, wind speed and direction, relative humidity, pressure and rainfall.

Additional modules are kept at open-circuit conditions and current-voltage measurements are taken in order to have an insight into the transient behaviour of the electrical characteristics of the particular technology. To ensure top quality measurements, all sensors are kept within the calibration period and cleaning is undertaken on a weekly basis. Regular visual inspection is carried out to assess whether any reliability or degradation issues arise.

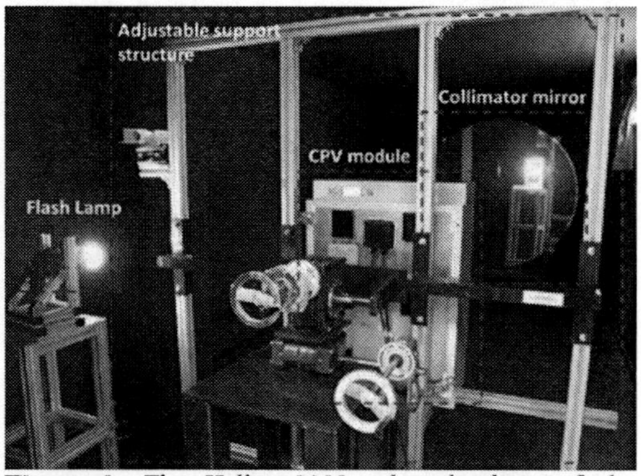

Figure 2: The Helios 3198 solar simulator of the CEAEMA of University of Jaén.

The degradation rate calculated through the outdoor measurements is compared to the one estimated by the CSTC measurements performed annually using the Helios 3198 simulator. In addition to this, outdoor CSTC and Concentrator Standard Operating Conditions (CSOC) are calculated annually in order to observe any deviations from the initial values and the outdoor CSTC (through translation of outdoor I-V measurements [9]) is compared to the corresponding rating obtained indoors. Statistical methods such as Linear Regression (LR), Classical Seasonal Decomposition (CSD), AutoRegressive Integrated Moving Average (ARIMA) or LOcally wEighted Scatterplot Smoothing (LOESS) are also applied on ratings such as the performance ratio (PR) or energy yield. Previous work [5] showed that the LR technique produces results with the highest uncertainty while the most robust methods are the ARIMA and LOESS.

Figure 3: Outdoor experimental setups at the Universities of Jaén (top figure) and Cyprus (bottom figure).

3 PRELIMINARY RESULTS

3.1 Indoor testing

Twelve CPV modules were tested indoors in July 2016. The multi-flash CPV solar simulator "Helios 3198" that incorporates a Xenon lamp was used for the characterisation at CSTC conditions according to the IEC 62670 [10, 11]. Such conditions include irradiance of 1000 W/m^2 and a spectral distribution similar to air mass AM1.5D (according to the ASTM G173-03 [12]).

In order to verify these conditions, a set of component cells was used to quantify the Spectral Matching Ratios (SMR) of top/middle, middle/bottom, top/bottom subcells. The range of all SMR values was kept within 1 ± 0.01 and 0.02 depending on the current ratio of the associated subcells (with an uncertainty of 2.5% on top and middle component cells and 5% on the bottom component cell). Moreover, the room temperature was kept at 25°C ± 0.5°C.

As an example, the current-voltage curve at different irradiance levels and the acceptance angle characteristic of Module SC_05 are shown in Fig. 4 and 5 respectively. The average CSTC efficiency, power output, short-circuit current (I_{sc}) and acceptance angle of the modules were found to be 26.9%, 92.4 W and 0.8° respectively.

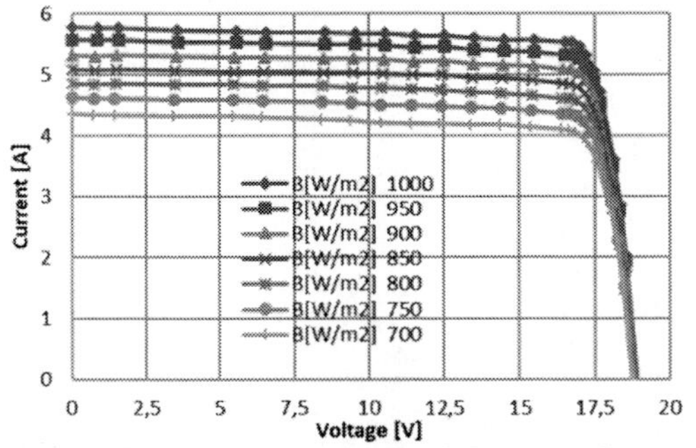

Figure 4: Current-voltage characteristic curve of SC_05 module at irradiance levels from 700 – 1000 W/m^2.

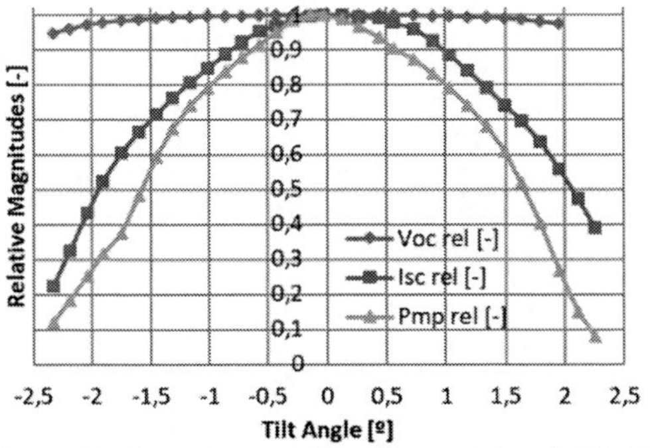

Figure 5: Acceptance angle characteristic of SC_05 module at 1000 W/m^2.

3.2 Outdoor performance

The PR of both systems in April 2017 is shown in Fig. 6 where overcast days with limited number of datapoints were filtered out. The high variability of PR can be explained due to the high aerosol content caused by dust particles that were present in the atmosphere during the experimental campaign. In order to get a better insight, the efficiency was plotted against the AM for days with a different aerosol optical depth (see Fig. 7). As can be seen, the efficiency reduces rapidly with increasing aerosol optical depth. In particular, the efficiency collapses for AM values above the reference on the 22nd of April 2017 in Cyprus due to the relatively high AOD. This behaviour is consistent with studies published elsewhere [13, 14] and highlights the importance of incorporating a spectral correction [15, 16] amongst others (i.e. temperature, etc.) in the degradation analysis.

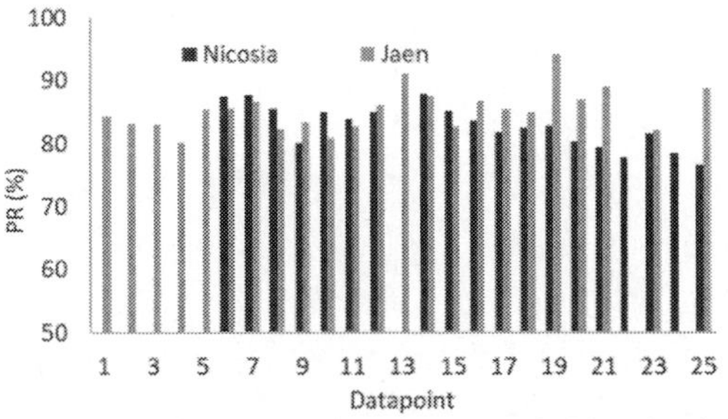

Figure 6: Performance ratio values recorded in Nicosia, Cyprus and Jaén, Spain. Overcast days with limited datapoints were filtered out.

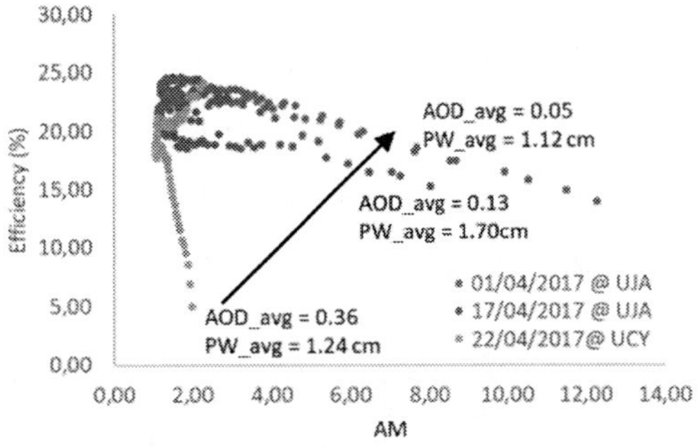

Figure 7: Efficiency against AM for days with different AOD values in Spain and Cyprus.

4 CONCLUSION AND FUTURE WORK

Degradation is an important parameter that has not yet been adequately addressed for the CPV technology. Such parameter can provide a useful insight into the durability of the technology and the lifetime energy yield which is the basis for the reduction of investment risk in a relatively new technology. Following this, DEGRADE-CPV was initiated aiming to investigate the degradation rates and modes of CPV systems. It is open to any interested stakeholders who would like to participate and actively contribute to it. Future work will include the installation of more grid-connected CPV systems. In addition, the statistical methods will be tailored to the needs of the technology and the impact of filtering criteria and corrections will be assessed.

ACKNOWLEDGEMENTS

This work is supported by the EUREKA-Eurostars program (EUREKA/EUSTAR/0116/01) and the Research Promotion Foundation of the Republic of Cyprus. The authors wish to thank Marios Tomazou, Giorgos Panagi and Michalis Papastavrou for their technical support.

REFERENCES

[1] P. Pérez-Higueras and E. F. Fernández, *High Concentrator Photovoltaics: Fundamentals, Engineering and Power Plants*: Springer International Publishing, 2015.

[2] D. L. Talavera, P. Pérez-Higueras, J. A. Ruíz-Arias, and E. F. Fernández, "Levelised cost of electricity in high concentrated photovoltaic grid connected systems: Spatial analysis of Spain," *Applied Energy*, vol. 151, pp. 49-59, 2015.

[3] NREL. (2017, September 18, 2017). *National Center for Photovoltaics. Best research-cell efficiencies chart.* <http://www.nrel.gov/ncpv/images/efficiency_chart.jpg>.

[4] G. Makrides, B. Zinsser, G. E. Georghiou, M. Schubert, and J. H. Werner, "Degradation of different photovoltaic technologies under field conditions," in *Photovoltaic Specialists Conference (PVSC), 2010 35th IEEE*, 2010, pp. 002332-002337.

[5] A. Phinikarides, N. Kindyni, G. Makrides, and G. E. Georghiou, "Review of photovoltaic degradation rate methodologies," *Renewable and Sustainable Energy Reviews,* vol. 40, pp. 143-152, 2014.

[6] A. Kyprianou, A. Phinikarides, G. Makrides, and G. E. Georghiou, "Definition and Computation of the Degradation Rates of Photovoltaic Systems of Different Technologies With Robust Principal Component Analysis," *Photovoltaics, IEEE Journal of,* vol. 5, pp. 1698-1705, 2015.

[7] D. C. Jordan and S. R. Kurtz, "Photovoltaic Degradation Rates—an Analytical Review," *Progress in Photovoltaics: Research and Applications,* vol. 21, pp. 12-29, 2013.

[8] M. Muller, D. Jordan, and S. Kurtz, "Degradation analysis of a CPV module after six years on-Sun," *AIP Conference Proceedings,* vol. 1679, p. 020004, 2016.

[9] M. Muller, S. Kurtz, M. Steiner, and G. Siefer, "Translating outdoor CPV I–V measurements to a CSTC power rating and the associated uncertainty," *Progress in Photovoltaics: Research and Applications,* vol. 23, pp. 1557-1571, 2015.

[10] "Standard IEC 62670-3, "Concentrator Photovoltaic (CPV) - Performance Testing Part 3: Performance Measurements and Power Rating," 2017, Geneva, Switzerland.," ed.

[11] "Standard IEC 62670-1, "Concentrator Photovoltaic (CPV) - Performance Testing Part 1: Standard Conditions," 2013, Geneva, Switzerland.," ed.

[12] "ASTM G173-03(2012), Standard Tables for Reference Solar Spectral Irradiances: Direct Normal and Hemispherical on 37° Tilted Surface, ASTM International, West Conshohocken, PA, 2012, www.astm.org," ed.

[13] M. Theristis, E. F. Fernández, C. Stark, and T. S. O'Donovan, "A theoretical analysis of the impact of atmospheric parameters on the spectral, electrical and thermal performance of a concentrating III–V triple-junction solar cell," *Energy Conversion and Management,* vol. 117, pp. 218-227, 2016.

[14] M. Muller, B. Marion, S. Kurtz, J. Rodriguez, A. W. Bett, R. D. McConnell, G. Sala, and F. Dimroth, "An investigation into spectral parameters as they impact CPV module performance," in *AIP Conference Proceedings,* 2010, p. 307.

[15] M. Theristis, E. F. Fernández, F. Almonacid, and P. Pérez-Higueras, "Spectral Corrections Based on Air Mass, Aerosol Optical Depth, and Precipitable Water for CPV Performance Modeling," *IEEE Journal of Photovoltaics,* vol. 6, pp. 1598-1604, 2016.

[16] M. Theristis, E. Fernández, F. Almonacid, and G. Georghiou, "Spectral Correction of CPV Modules Equipped with GaInP/GaInAs/Ge Solar Cells and Fresnel Lenses," *Applied Sciences,* vol. 7, p. 842, 2017.

QUALIFICATION TESTING VERSUS QUANTITATIVE RELIABILITY TESTING OF PV – GAINING CONFIDENCE IN A RAPIDLY CHANGING TECHNOLOGY

Sarah Kurtz[1], Kent Whitfield[2], Nancy Phillips[3], Tony Sample[4], Christos Monokroussos[5], Edward Hsi[6], Ingrid Repins[1], Peter Hacke[1], Dirk Jordan[1], John Wohlgemuth[7], Peter Seidel[8], Ulrike Jahn[9], Michael Kempe[1], Tadanori Tanahashi[10], Yingnan Chen[11], Bengt Jaeckel[12], Masaaki Yamamichi[13]
[1]National Renewable Energy Laboratory, Golden, CO USA; [2]Underwriters Laboratories, Fremont, CA USA; [3]DuPont, USA; [4]European Commission, JRC, Ispra, Italy; [5]TÜV Rheinland, Shanghai, China; [6]Swiss RE, Switzerland; [7]PowerMark Corporation, USA; [8]First Solar, Mainz, Germany; [9]TÜV Rheinland, Cologne, Germany; [10]National Institute of Advanced Industrial Science and Technology (AIST), Tsukuba, Japan; [11]China General Certification Center, Beijing, China; [12]Underwriters Laboratories, Zeppelinheim, Germany; [13]RTS Corporation, Tokyo, Japan

ABSTRACT: Continued growth of PV system deployment would be enhanced by quantitative, low-uncertainty predictions of the degradation and failure rates of PV modules and systems. The intended product lifetime (decades) far exceeds the product development cycle (months), limiting our ability to reduce the uncertainty of the predictions for this rapidly changing technology. Yet, business decisions (setting insurance rates, analyzing return on investment, etc.) require quantitative risk assessment. Moving toward more quantitative assessments requires consideration of many factors, including the intended application, consequence of a possible failure, variability in the manufacturing, installation, and operation, as well as uncertainty in the measured acceleration factors, which provide the basis for predictions based on accelerated tests. As the industry matures, it is useful to periodically assess the overall strategy for standards development and prioritization of research to provide a technical basis both for the standards and the analysis related to the application of those. To this end, this paper suggests a tiered approach to creating risk assessments. Recent and planned potential improvements in international standards are also summarized.

Keywords: PV reliability, accelerated testing, risk assessment, lifetime prediction

1 INTRODUCTION

As the total investment in the solar industry has increased, persistent questions arise: "What is the degradation rate of the module and system and how do they compare to the component manufacturer's stated warranty and system performance guarantee?" "How much more should I be willing to pay for a module/system that has demonstrated improved reliability?" or, conversely, "What information demonstrates higher confidence to justify a higher price?" While there is ample evidence that PV installations have performed mostly as expected [1-3], there have also been some product recalls and observations of new (unexpected) failures.[1]

Fig. 1 shows our simplistic starting point: certifying a module design by testing 8 engineering samples to the IEC 61215 qualification test [4] is inadequate (and not intended) for making 25-year risk assessments[2]. While IEC 61215 has been demonstrated to be valuable for rapidly uncovering well known failure mechanisms, it is insufficient for assessing long-term risk, evaluating newer or less common materials and designs, establishing field performance degradation, setting insurance rates that must consider power generation over at least 25 years, or selecting the best product for a specific project. Today's rapidly changing PV technology requires many companies to launch new versions of their product every few months while requiring warranties that are decades long.

Today, in addition to qualification testing (IEC 61215 and IEC 61730 [5]) most PV companies require a robust quality management system that controls many aspects of the manufacturing process (incoming materials, processes, etc.) as well as testing beyond IEC 61215. As the PV industry matures, the methods used for quality control (QC) are evolving to utilize new knowledge and to be more consistent, enabling lower QC costs, as with IEC TS 62941 [6]. We expect this evolution to move from pass-fail qualification testing to more sophisticated analyses that provide more quantitative assessment of risk specific to a particular location or type of location, and, thus, enable more quantitative assessment of the value of high-quality components, both in terms of degradation rates and failure rates. One proposed approach to completing a quantitative assessment assigns a Cost Priority Number (CPN) that reflects the cost of repair or loss of revenue associated with a problem [7]. Assignment of a CPN or other rating methodologies [8] relies on being able to link knowledge about the components and system with the anticipated outcomes. The industry has not yet agreed upon the best approaches for gathering and using the information needed for quantifying overall risk.

This paper begins by describing approaches for risk assessment that range from an initial qualification test to the ideal for quantitative assessments. It then proposes a strategy for building on the foundational qualification testing to use standardized data collection from extended stress testing and thoughtful analysis of that data within the context of known field results to provide a useful risk assessment. An overview of a subset of recent and planned international standards development is also presented, highlighting the joint work of the International PV Quality Assurance Task Force (PVQAT) and IEC Technical Committee 82. While this paper is largely focused on PV modules, the concepts must be applied to all components as well as to the entire system.

[1] Here, "failure" broadly refers to any problem that may cause risk for a business partner.

[2] Most investors today plan to obtain return on investment in < 25 years; we refer to 25 years to reflect warranty time lines as well as the needs of second owners.

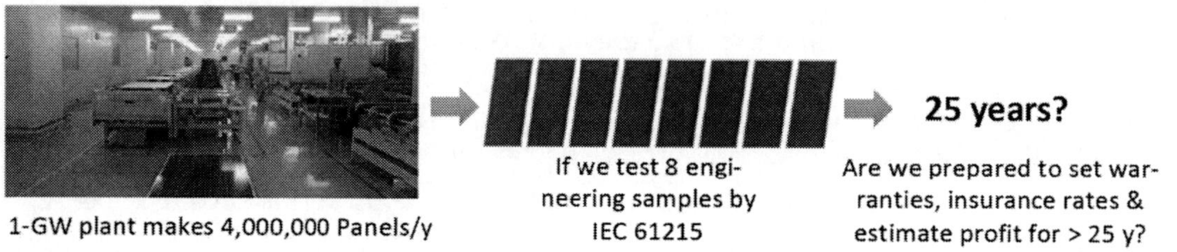

Figure 1: Our challenge with this rapidly changing technology is to use a few months of testing to provide the basis for business decisions that may have impacts for decades. This paper discusses strategies for improving risk assessments.

2 QUALIFICATION vs QUANTITATIVE TESTING

As shown schematically in Figure 2, risk assessments of PV can evolve to be more quantitative. The initial idea (see left side of Fig. 2) is evaluated and analyzed; prototypes are stress tested to qualify a design, and samples from a production line are subjected to extended-stress testing. Finally, a mature product with a specified bill of materials manufactured within a defined process window for a specific application may be tested and modeled quantitatively. From left to right in the simplistic description in Fig. 2, the quantitative nature of the assessment increases. Differentiating the roles and utility of each type of test encounters many complexities, as discussed in this section to set the stage for the strategy discussed in section 3.

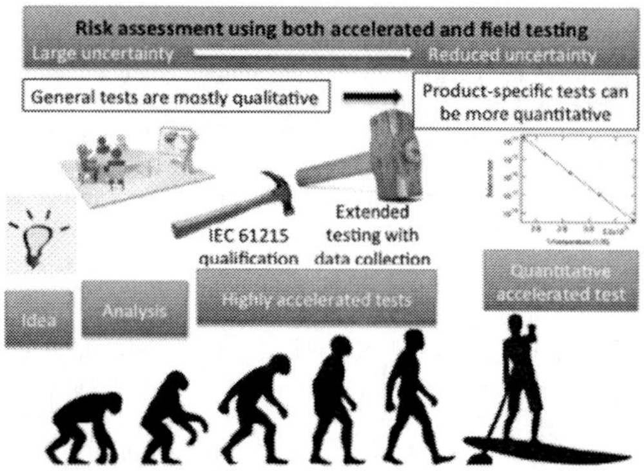

Figure 2: Evolution of maturity of risk assessment as a product evolves from concept to a mature product that can be tested in a more quantitative way because the product design stays fixed long enough to complete the needed testing.

Qualification tests typically provide a pass-fail outcome reflecting whether the test article exceeds a minimum acceptable key indicator. In contrast, a quantitative test is associated with a numerical output providing the user with access to the degree of impact a test has on a key indicator. When multiple data points are measured as a function of varying levels of a stress factor, these may be used to derive a model from which a quantitative prediction may be made. While there is common interest in developing quantitative models (right side of Fig. 2), many such models today fail to provide the desired accuracy and do not include all possible failure mechanisms/stresses; experts are often hesitant to suggest that such models can build confidence over the product's lifetime. While we understand this hesitancy, we assert that we should not let that hesitancy deter us from acquiring and using all relevant data to make increasingly quantitative predictions. It is the purpose of this paper to explore the best strategies to accomplish this goal.

2.1 Can qualification testing be quantitative?

Today, most PV module designs are certified to both IEC 61215 [4] and IEC 61730 [5]; these prescribe a set of stress tests to qualify a prototype with respect to functionality and safety, respectively, before beginning manufacturing. While these two standards are pass-fail by nature, they can provide information beyond a simple pass-fail by:

- Differentiating PV modules with regard to some attributes that are defined on the data sheet, *e.g.* performance, system voltage and mechanical load. As tests mature, additional attributes may be treated in this way.
- Reporting the observations. For example, if one test article fails IEC 61215, the test can be repeated and passed. The failure suggests risk that some samples are defective; such risk can be quantified through statistical sampling from the production population. Many manufacturers decline to share the detailed IEC 61215 test report with their customers even though the data in this report would be valuable toward analyzing the suitability of a module design. Transparency in sharing test results would be beneficial to the community.

2.2 Uncertainty as the metric for quantitative assessment

An accurate model-based prediction of degradation or failure rates of a PV product is very difficult. For example, when a degradation mode involves a multistep process, understanding each step well enough to model it is a challenge. Furthermore, to test a model may require years of work. Often the most difficult part of creating the prediction is assessing the uncertainty of the prediction, because the uncertainty depends on variations in the manufacturing process as well as uncertainty in the kinetics of the failure mechanism and uncertainty in the applied stresses relative to those in the actual use conditions. At the extreme, a prediction of a service life of 25 years ± 25 years may be the result of the study, providing a scientifically rigorous result that has essentially no value.

Studies of acceleration factors to enable prediction of PV degradation or failure rates often lack statistical confidence, or simply are not comprehensive. Typically, the focus is on some details of the prediction (the identification of failure mechanisms, measurement of the kinetic rates, and application of a model to the intended use environment) derived from a small sample set with little analysis of the uncertainty. An unintended impurity in an encapsulant may change its rate of discoloration; variations in the thickness of a solder bond may change its failure rate; mounting of a module in such a way that it overheats could accelerate its degradation, etc. As a result, many experimentally determined degradation rates are bound to have high uncertainties. The accuracy of the output of any model is subject to the degree of accuracy of its input parameters. Compared with the studies of quantifying failure mechanisms, few studies have focused on quantifying and decreasing the uncertainty related to inputs such as the variability of the manufacturing process. Moreover, it is challenging to model the degradation behavior when parallel degradation processes take place. To overcome the difficulties most of the proposed models initiate simplifications and assume that certain processes will dominate under certain environmental conditions and ignore the perplexity of multi-dimensional reaction paths, which further increase the uncertainty of prediction. Similarly, warranties may not specify the use environment, even though the physics predicts that the life of a module varies with use environment.

A general test such as IEC 61215 is not intended for quantitative predictions of lifetime. Efforts to quantify the implications of passing a general test are inherently limited. Although it is clear that different products may show different acceleration factors, preventing confident predictions from a general type approval test, there is an urgent and practical need to better understand and analyze test results in order to provide the basis for setting insurance rates, estimating return on investment, and making other business decisions.

Thus, studies of the mechanisms and associated acceleration factors to make quantitative predictions should be encouraged, but these should always assess the expected uncertainty in the predictions. *In some cases, the uncertainty may be reduced more by the application of tighter control of the production process (when it is known which parameters need to be controlled), or the definition of the use environment than by improved measurements of the kinetics of the failure rates.*

2.3 Is extended-stress testing useful?

Over the years, many protocols have been developed for extended stress testing [9-19]. The number of these tests and the frequency of their use by the community imply that they provide value. Most such tests are designed to apply stresses in a manner that reproduces relevant field stresses, either in combination or sequentially and records a series of data from the extended sequences. But, there is usually a lot of guesswork and some debate about the "best" number of cycles (see section 2.4) and which sequences of testing should be chosen. There is concern that the tests sometimes cause irrelevant failures, adding unnecessary cost, raising unnecessary concerns, or leading to unwise decisions.

We suggest that extended-stress testing can make risk assessment more effective by:

• **Comparisons**: Provides a way to compare two versions of a product, which can be especially useful if a change in the bill of materials or manufacturing process may change field failure rates. Care must be taken to compare products only for test results that are relevant to the use environment. This may be the most common application of extended testing since many companies release new versions of their products every 3-12 months. *The disconnect between the intended lifetime of the product (> 25 years) and the production lifetime (3-12 months) is at the core of our challenge.*

• **More extensive data**: Most extended testing protocols report observations after each step, providing more data than a simple pass-fail at the end of the test. Standardization of the extended-stress test methodology enables comparison of the test data with a larger database of test results, facilitating the comparison of test results from many samples and the correlation of these with field data if accessible records are kept.

• **Identify potential vulnerabilities**: Similar to test-to-failure protocols, overstressing samples can identify vulnerabilities that should be assessed for relevance to the intended use environment. This includes component materials tests that screen out non-durable materials from the design, or identify when mitigation may be required.

• **Use of production modules**: Extended test protocols may be applied to production-line samples, possibly from multiple lines or factory sites, differentiating the extended test from qualification testing of prototypes and potentially leading to an improved understanding of product production controls.

• **Educate the market**: Many experts have reported that buyers and financial institutions often request extended testing, but have minimalistic requirements in terms of test duration, severity and sample size, while expecting that results will provide confidence over the product's expected lifetime. It is therefore important that the scientific community lay the fundamental guidelines for a test protocol with realistic test requirements that can be accomplished in a reasonable timeframe (e.g. 6 months). Such guidelines would provide a basis that will help educate the market over what is realistically achievable and align expectations with reality.

• **Standardization as a means to build a market consensus**: Those who use extended testing routinely report that such testing is very useful toward assessing risk even when the results are not directly associated with quantitative models. But, a plethora of extended test protocols exists. It is often confusing for the buyers, or those who drive financial decisions to understand which one is "best" for their needs. Standardization of the extended-stress test methodology would help build a consensus in the market in terms of both requirements and expectations, and would reduce the time and cost of testing. What the PV community needs is a single, commonly accepted and standardized protocol that provides the information to increase confidence over IEC 61215/61730 qualification and type approval.

2.4 More samples vs. longer stress

For a given budget, one can choose to test more samples for shorter stress or fewer/smaller samples for longer stress. Testing more samples is especially useful when an issue is observed infrequently. On the other hand, extended-stress testing identifies how a product wears out and potential weaknesses in a product, enabling more complete risk assessment, just as test-to-failure is used for risk assessment. While extended-stress testing has become quite popular (see section 2.3) and is offered by most test labs and required by many customers, there is substantial debate about the optimal balance between the number of samples and the length of the stress test. The use of smaller samples (*e.g.* 1- or 4-cell mini modules, or test coupons containing packaging materials with no electrical components or cells) reduces cost (but also introduces new uncertainties), and may be especially useful for screening new materials and for mechanism-specific testing. While it is not well established how to integrate this sort of information, an effective method to screen out materials that are not durable in the application will unquestionably improve the risk assessment. The optimal balance depends on the goal of the testing, as discussed below.

2.5 Risk assessment versus service-life prediction

Although studies often state an ultimate goal of predicting *service life* for PV modules, *failure and degradation rates as a function of time* are much more useful to those who are estimating maintenance costs, setting insurance rates, or quantifying risk of an investment. Failure and degradation rates can be reduced into a single number (service life prediction), but only the use of the time-dependent functions allows calculation of costs and revenues over the lifetime of the system. It's important to describe both the failure rates and degradation rates because failure of a module may require replacement and associated maintenance cost while degradation reduces the electricity generated, both of which can affect the return on investment. In practice, every failure/degradation mechanism must be quantified and the combined effects estimated. Here, we use "risk assessment" as a short way to refer to the evaluation of degradation and failure rates as well as any assessment of risk.

For every risk function, quantifying the uncertainty (which generally increases with time into the project) is essential [7]. As described in section 2.2, quantifying the uncertainty is sometimes more difficult than making the prediction and the challenge is even greater when the uncertainty increases as a function of expected lifetime duration.

Inherent to both risk assessment and service prediction is a means to describe the stresses in the specific application, e.g. the relevant application and climate specifics. The reliability of a given module for 2 different locations, or application types (e.g. field or roof mounted) can be very different, and the relevant parameters (e.g. module temperature, extent of thermal cycling, exposure to salt spray or snow load) need to be understood so appropriate data can be included in the analysis.

2.6 Why should empirical data be included in risk assessments?

Physics-based models (based on understanding the mechanism and quantifying the kinetics) may be required to theoretically explain and understand the development of a degradation process or reaction, but due to the complexity of the processes involved in the field, it is often more practical to rely on experimental observations and correlations of laboratory and field testing. Insurance companies define their rates based on both "exposure rating" (based on physics-based studies) and "experience rating" (based on actuarial-type statistics, which are semi-quantitative and relevant even when they don't reflect the physics of failure).

Using uncertainty as a metric, physics-based models may be inferior to experimental or field observations when the uncertainties in the physics-based predictions are large. In an example of damage caused by hurricanes, historical data may provide a more accurate prediction than a physical model. If the uncertainty of the empirical data is smaller than the uncertainty of the physics-based model, then the empirical data is likely to be superior and should be respected for the value it brings.

On the other hand, experimental data or field observations may be irrelevant if the prediction is made for a product that has different failure mechanisms than described by the so-far observed data. Thus, the best approach is likely to use correlations of accelerated and field test results combined with analyses of how the products differ and how those differences may affect the final results. Comparisons of field data can also provide very useful insight into the effects of design or component selection, if these are analyzed carefully.

3 STRATEGY FOR BETTER ASSESSING RISK

This section proposes a strategy for improving risk assessments by using a tiered approach using standards, knowledge, and processes that consider multiple aspects of product design and implementation.

Goals for future improvements include:

1) Address failures that are being observed in the field,
2) Address new failure mechanisms for new products,
3) Reduce cost by removing unnecessary requirements and standardizing testing protocols, and
4) Reduce the uncertainty in reliability assessments.

3.1 Overall strategy and types of tests

Various types of testing and analysis were described in Fig. 2. We suggest that a robust approach to low-uncertainty risk assessment will benefit from using multiple elements in a systematic way (see Fig. 3).

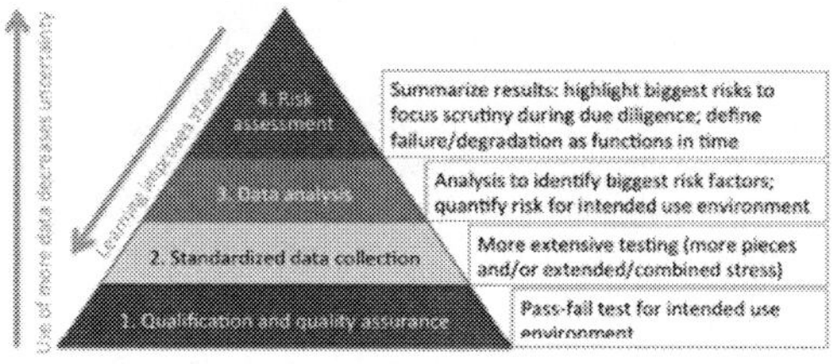

Figure 3: Strategy for improving risk assessments of PV. The colors indicate the time line.

Qualification testing for the intended application provides the foundation (Tier 1 of Fig. 3) of our approach to quantifying risk. The second tier in Fig. 3 describes *standardized data collection for larger sample sets and for extended-stress test sequences that have been found to be useful in identifying product that will be successful in the field*. These *data are analyzed using all available knowledge, including the correlation between laboratory test results and field results for the relevant use environment,* (third tier in Fig. 3) to determine the risk functions that define the final risk assessment (top tier in Fig. 3). The colors in Fig. 3 suggest a time line: Green for Tier 1 suggests that qualification testing is already in place with need for only incremental changes; Yellow suggests completion of a first version of a standard within a year or two; Orange implies that more research is needed before we can standardize our approach. Red implies that a critical confidence has to be reached in observations before making conclusions about risk. The following sections discuss each of these elements.

Table I: Summary of key goals of testing programs and recommended approaches.

Goal	Test element	Approach
1. Identify failure mechanisms for a totally new product	Accelerated test to uncover new and known failure modes; test to failure	Simulate use conditions (including field deployment) to identify expected field failures; test to failure to identify product weaknesses
2. Compare two similar versions of a product	Accelerated test that quickly identifies (known) failures; test to failure	Use extended-stress tests known to cause same failures as in the field, collecting multiple data points; standardize so as to leverage learning from community
3. Quality control	Accelerated test that quickly identifies weakness	Use "smart" (comprehensive) sampling along with tests from IEC 61215, IEC 61730; use standard extended-stress test for fewer samples
4. Reliability assessment	All of the above	Compare data with previously collected data and with literature publications to provide the best analysis possible

3.2 Tier 1: Continued improvements of qualification tests

As products evolve, there is a constant need to update the standards for qualification testing. Based on current reports from PV systems, priority issues include:

1) A range of failures of power electronics associated with both design (IEC 62093) and quality assurance (IEC TS 63157),
2) Potential-induced degradation (PID) of modules,
3) Cracked cells in modules and a range of problems that may be caused during installation,
4) Hot spots, and
5) Quality issues resulting in inconsistent nameplate rating and/or durability.

In response to these and other opportunities, amendments of IEC 61215 and IEC 61730 are being discussed (see section 4). In general, new test methods are first published as IEC Test Specifications (TS), typically without pass-fail requirements. After being used for several years and when well accepted by the community, the test method along with a defined minimum threshold for establishing passage or failure of the test, may be merged into the IEC 61215 and IEC 61730 tests. As an example, IEC 62804-1 was published in 2015 to define how to test c-Si modules for PID. After using the two proposed tests, the IEC committee may now be ready to amend IEC 61215 to include a pass-fail PID test based on IEC 62804-1 experience.

A critical element of starting production of a new design is defining quality assurance procedures. IEC 62941 was developed to guide this process for modules. Methods for consistent implementation of IEC 62941 are being defined by IECRE [20].

Qualification testing provides a "pass" if the conditions of the tests are met, but the conditions may be intentionally varied. For example, the "pass" result may characterize the ability to withstand variable levels of system voltage, snow or wind load, and hail. Now, a Test Specification is being developed for higher temperature operation. If new Test Specifications become well accepted by the community as useful tests, in the future, IEC 61215 may be modified to test for multiple levels for operation at higher temperatures, or extended UV exposure and for robustness to stresses during installation.

3.3 Tier 2: Strategy for additional data collection

While the need for more standardized data is clear, there is not yet agreement on the test methodology and data collection. We suggest that confusion arises from the multiple purposes of the additional data. Table I summarizes key goals and strategies for the data collection.

Outdoor testing in a range of use environments is part of evaluating any product, but getting useful results takes longer than the product-development cycle. Even so, systematic collection of data regarding both the performance and physical changes is essential to analyzing the relevance of the accelerated test data.

If a company is developing an entirely new product and has little information about how the product will fail, it is useful to carefully analyze possible problems, as well as to use accelerated stress testing to simulate conditions of the anticipated use environment [17, 18]; it is also useful to stress the product until it fails so as to identify its weaknesses. (See first item in Table I)

It is more common (second item in Table I) to compare two similar designs (e.g. with different bill of materials or similar products from two manufacturers.) For common PV module designs, the probable failure and degradation mechanisms have been well studied and effective tests exist. Using a standard set of tests and data collection methodology enables consistent comparisons and leverages collective knowledge from the community. While these standardized tests should duplicate failures that are relevant in the field, there can also be value in testing to failure in cases where a change in a design may introduce a new weakness in the product. *Thus, attempting to align the accelerated and field stresses (even if it were possible) may be less useful than cost-effectively collecting an extensive extended-stress data set and evaluating that data set based on field experience coupled with an understanding of the physical processes.*

A third goal of collecting data is for quality control. As part of a quality management system (QMS), companies develop reliability or quality monitoring programs that detect changes in the product. Companies may attempt to confirm product quality through a short (incomplete IEC 61215) set of tests before releasing inventory and then may continue stressing a subset of samples to increase confidence in the consistency of the manufacturing process. Some customers require third party testing of random samples from the lots that are shipped to them. To provide effective quality control, it is essential to sample all production lines and to provide feedback very quickly; we suggest leveraging a quality assurance guide such as IEC 62941 and include:

1) Careful and consistent control of incoming materials/components,
2) "Smart" (rather than purely random) sampling to test appropriately weighted sample populations that represent the production population as a whole with respect to the failure modes of interest,
3) Rapid stress testing such as application of parts of IEC 61215/61730 for modules or IEC 62093 for inverters and other balance-of-system components.

In many cases, application of the standard qualification tests adequately identifies problematic variations in manufacturing. However, extended-stress testing of a subset of samples may also be useful. Although one could apply extended tests to the full sets of samples, practically, there is a tradeoff between the number of samples and the length of the stress tests; we suggest that, especially when quality control is the goal, shorter tests (to give quick feedback to the production process) and broader sampling (to test all production hardware) may be better than using extended testing for all samples.

Developing a useful, standardized set of extended stress tests is critical. For modules, multiple versions of extended-stress testing have already been introduced to the community [9-19]. In many cases, these apply IEC 61215 tests multiple times, recording the results after each cycle [11, 12, 15]. In some cases, different types of stress are applied sequentially [14, 15]. Applying stresses in combination more closely simulates the use environment, so including combined-stress tests is preferable if costs are acceptable. Additionally, it can be useful to measure indicators of changes such as adhesion, leakage current and a polymer's ability to deform without fracturing. *Because acceleration factors can vary by orders of magnitude for different failure mechanisms and use conditions, it will not be possible to design the "perfect" extended-stress test that would identify all relevant failures while avoiding irrelevant failures.* On the other hand, the industry's extensive experience can be used to identify and standardize a useful test sequence that builds on the huge success of IEC 61215 and IEC 61730 to provide more extensive data at minimal cost.

Reliability assessments should use all of the available data in scientifically appropriate ways, as discussed below.

3.3 Tier 3: Strategy for data analysis

The industry will benefit from an increasing engagement by all parties in understanding and analysing data. *This analysis will be easier and more effective if the set of tests and associated data collection methodology are standardized and if research efforts are directed toward comparing the standardized test results to field results in parallel with development of physics-based models, including more detailed understanding of the physical changes.* As a body of knowledge is developed, that knowledge should be used to systematically improve the qualification tests, standardized extended-test methodology and associated data analysis.

The focus of the data analysis should always be on providing useful inputs into the specific risk assessment (section 3.4), which may vary from project to project. For example, a module design that is susceptible to failure under high snow load may be quite acceptable for application in a tropical environment or an inverter that is operated in an air-conditioned enclosure may not function well in a hot ambient with direct sunshine. Also, the data analysis should include considerations of both testing of the design and of consistency of the manufacturing.

The data analysis should reflect both scientific studies that elucidate the details of the failure mechanisms and experimental or field observation studies that indicate the kinetics of the mechanism at stress levels present in the field. For example, potential-induced degradation depends not only on system voltage, but also on temperature, humidity, light, stress history, and electrical loading. There is opportunity for substantial research to better understand the stresses that cause failures/degradation as well as the material and product attributes that may affect the failures and degradation. This understanding, along with knowledge of the process controls that are used in the manufacturing and installation processes, will aid in making more accurate analyses.

Comparison of field experience with accelerated test results is a critical part of verifying any model, yet the number of years required for completion of such studies often prevents clear conclusions and as indicated above, the production lifetime for a fixed bill of materials and production process is often between 3-12 months leading to low manufacturer incentive for lengthy studies. *If records of accelerated test data and installation procedures can be kept alongside of the as-built plant documentation, the meaning of future field outcomes will be clearer.*

A scientific approach must always be taken when analysing both the accelerated test data and field data. The hardware and application being analysed may differ from those for which the data are available. Relevance of the results of extended-stress testing will depend on the site and application being evaluated.

3.4 Tier 4: Useful risk assessments

The highest tier shown in Fig 3 describes the end goal: *a risk assessment that highlights the greatest risks while providing the comprehensive information needed to make business decisions.*

As discussed in section 2.5, service life prediction is useful, but not sufficient: we must quantify the expected degradation and failure rates as a function of time. Whether non-linear degradation occurs mostly near the beginning or end of life can have a large effect on the levelized cost of energy (LCOE) [1]. Thus, reliability *functions* that reflect both degradation and failure rates are much more useful than a single number that is intended to reflect an average lifetime.

Additionally, better data for business decisions will reduce the associated risk, and, potentially, the cost of capital, lowering the overall project cost. This requires more comprehensive information such as a five-part analysis:

1. **Failure/Degradation Mode**
2. **Failure/Degradation Consequence (severity)** – Assign weight to reflect cost associated with the failure.
3. **Failure/Degradation Timing** – This assesses implication on both time value of money as well as the investment horizon.
4. **Failure/Degradation Cause** – This may facilitate risk improvement and management.
5. **Failure/Degradation Scale (occurrence)** – Identifies the fraction of products that could be affected.

All information used in the risk assessment has an associated uncertainty. Part of the risk assessment is estimating this uncertainty and including the uncertainty and its confidence in the conclusions.

3.5 Risk assessments at the system level

The four-tier approach of Fig. 3 should be applied to all of the components and the summary risk assessment completed at the system level, using any available data from similar systems with as many years of experience as possible. *System design, installation, and operation may affect component reliability functions,* as well as the function of the entire system, so are an essential part of the risk assessment.

4 RECENT PROGRESS IN STANDARDIZATION

4.1 IEC and PVQAT efforts

The International Electrotechnical Commission (IEC) Technical Committee 82 (TC82) writes standards for solar photovoltaic energy systems. The International PV Quality Assurance Task Force (PVQAT), initiated in 2011, established a Type A Liaison to IEC TC82 in 2017, supporting discussions and coordinated research to lay the groundwork for new and improved IEC standards.

In 2015, PVQAT-prioritized efforts to address issues identified from failures in the field were summarized (see Table III of [21]). Progress on these and related efforts is summarized in this section.

4.2 Recent progress in PV standards development

Table II summarizes selected recent improvements in IEC documents relevant to this paper.

A key effort of PVQAT has been to improve guidelines for quality assurance. PVQAT initiated discussions of this topic in 2011 and IEC TS 62941 was published in January 2016 to provide guidelines for quality assurance of PV module manufacturing [22, 23]. The IECRE conformity assessment board is implementing IEC TS 62941 [20]. Similarly, opportunities for improved quality assurance were identified for PV system installation and operations/maintenance as described in IEC TS 63049 and implemented by IECRE.

The recent revision of IEC 61215 combined IEC 61215 and IEC 61646 into a single series of documents to align the common testing procedures for silicon and thin-film modules while retaining differentiation of test methods by type of module, where appropriate. The changes to this set of documents were extensive (see Table II for a few highlights).

Two test methods for identifying the susceptibility of silicon modules to potential-induced degradation (PID) were published in 2015 (IEC TS 62804-1).

The cyclic (dynamic) mechanical load test (IEC TS 62782) was published in 2016 with the intent to use, along with thermal cycling and humidity freeze stress, to identify modules that have cracked cells that are likely to degrade in performance after exposure in the field.

Table II: Summary of some recent improvements in IEC standards relevant to this paper

Document	Publication	Description	Value
IEC TS 62941	2016	Quality assurance guideline	Guideline to improve quality of PV modules
IEC TS 63049	2017	Quality assurance guideline	Guideline to improve quality of PV plant installation and operation/maintenance
IEC 61215 series	2016	Hot-spot test measures shunting of every cell	Testing the most vulnerable cells is more likely to identify a problem
		Marking requirements for nameplate and general documentation are better defined	Provides customers important information in a consistent way
		Power output is checked both before and after stress testing	To pass, the modules must perform within specification before *and* after stress
		NOCT (nominal operating cell temperature) replaced by NMOT (nominal module operating temperature)	NMOT reflects the module temperature when biased at the maximum power point
		Robustness of terminations test evaluates both cables and junction boxes	More likely to catch problems
IEC 61730-1 IEC 61730-2	2016	Implemented insulation coordination, overvoltage category, classes, pollution degree and materials groups	Is better aligned with other IEC documents, facilitating better treatment of these concepts
		Definition of creepage, clearance and distance through insulation	Is better aligned with other IEC documents, streamlines introduction of new materials such as edge seals
		Implementation of component qualification	Prequalifying junction boxes and other components reduces time to qualify new design
IEC TS 62804-1	2015	Tests for susceptibility to potential-induced degradation (PID) in Si modules	Provides two tests that may become a basis of a pass-fail test within IEC 61215
IEC TS 62782	2016	Cyclic (dynamic) mechanical load testing	Detects cracked cells in module
IEC TS 62916	2017	Bypass diode electrostatic discharge (ESD) susceptibility testing	Quantifies control needed to avoid damage from ESD; referenced by IEC 62941
IEC 62979	2017	Bypass diode - Thermal runaway test	Tests for thermal runaway when diode switches from forward to reverse bias
IEC 62788 (Materials Testing)	2016-2017	62788-1: Encapsulants 62788-2: Frontsheets/Backsheets 62788-5: Edge Seal 62788-7: Component weathering exposures	Facilitates comparison of candidate materials for use in PV modules and for quality control of incoming materials
IEC 61724 series	2016-2017	Three parts define parameters and procedures for quantifying performance of PV plants	Enables execution of performance guarantee, consistent comparison of systems, and tracking of system health

Two new standards have been published to improve confidence in bypass diodes. The first discusses the control of electrostatic discharge (ESD). Implementation of this standard is one of the key elements in the quality assurance guideline IEC TS 62941. Additionally, to respond to an observed failure mechanism of thermal runaway involving a bypass diode switching from the forward bias state to the reverse bias state while being hot was published as IEC 62979. This failure mechanism is difficult to characterize because it occurs during a quick transient; there is some discussion about whether IEC 62979 is the best approach given the details of the failure mechanism.

A suite of tests (IEC 62788) for characterizing key properties of PV module materials (e.g. encapsulant and back sheet) is being developed with a number of individual documents already published. This set of documents provides consistent data to facilitate comparison of module materials, designed to be very helpful to component suppliers and module manufacturers as they identify ways to reduce cost. These test procedures may also be used for quality control of incoming materials.

The IEC 61724 series describes characterization of PV system performance, including capacity- and energy-test methods. Ultimately, the electricity generated by a PV system is a key metric.

4.3 Planned progress in PV standards development

When the revisions of IEC 61215 and 61730 were published in 2016, there was a decision to delay some changes, but to move forward as quickly as possible with amendments to both documents. Table III summarizes some of the changes that are being discussed. Notably, there is a need for a pass-fail test in IEC 61215 for PID (derived from IEC 62804-1) and there is a plan to introduce the cyclic (dynamic) mechanical load (IEC 62782) as part of a sequence to identify cracked cells in modules. Also, consistent characterization of bifacial and flexible modules will facilitate introduction of these new products into the marketplace. Finally, there has been discussion about modifying IEC 61215's thermal cycling test to include some cycles during which current flows through the bypass diodes instead of the cells. Some failures of bypass diodes have been observed and current flow to cause local heating of thermal fatigued joints accelerates relevant failures without requiring as much stress time as without the current flow. PVQAT Task Group 4 recommends that 50 thermal cycles with current flow through the diodes be added to the 200 thermal cycles that are already done in IEC 61215 with current flow through the cells for a total of 250 cycles. The functionality of the diodes must be verified at the end of this stress.

A suite of climate-specific tests was originally proposed as IEC 62892. In 2017, TC82 decided to revise the approach to climate-specific testing. The document describing extended thermal cycling for a range of applications will be published as a separate document. The stress conditions for encapsulants that have been discussed as IEC 62892-3 is now being discussed as IEC 62788-1-7.

A new effort (IEC 63126) is defining how to apply IEC 61215 and IEC 61730 tests to modules destined for operation at higher temperature. A key element of IEC 63126 may be to increase the thermal endurance testing of bypass diodes; there is a common view that the current thermal endurance test is inadequate for hotter applications. Once established, the option for higher temperature testing may be rolled into IEC 61215 and IEC 61730 much like these currently enable tests to be completed for a range of system voltages.

To address the common problems that are reported for inverters, IEC 62093 is being rewritten to reflect experience with testing inverters; the draft now includes tables differentiating testing of central, string, and micro inverters. A guide for quality assurance of inverter manufacturing (similar to IEC 62941) is also being developed as IEC TS 63157. Given the dominance of inverter problems, a meaningful inverter qualification test, coupled with the planned quality control guideline, should make a substantial difference to overall PV system reliability.

A test for permanent damage resulting from partial shade is being discussed as IEC TS 63140. The procedure includes methods for determining the size of an adverse shadow and for repeatedly applying that adverse shadow. The method is applicable to monolithically integrated PV modules with one series-connected cell group or with multiple series-connected cell groups that are in turn connected in parallel.

While IEC 62759-1 identifies modules that may be damaged during transportation to the installation site, it is widely noted that modules can also be damaged during installation if bounced, dropped, carried on hard hats, walked on, or installed with too much stress. It is anticipated that it would be useful to write a part 2 of IEC 62759 to identify the care that needs to be taken to avoid damage during installation.

Similarly, there is interest in developing tests for various building-integrated PV (BIPV) products. The approach for testing BIPV products is challenging because of the many potential form factors and because of a lack of data identifying all failure mechanisms. A single project team will be discussing the various types of BIPV products to define an approach to tackling this problem.

As discussed above, there is widespread interest in creating a standard for collecting data during and after application of extended stress. Existing extended-stress test sequences will be reviewed to identify commonalities and differences and users will be surveyed to select the test sequences that have been most useful in identifying weaknesses that correlate with field failures. As noted above, it is not possible to define a test that uncovers all field-relevant failures without also causing some failures that will not be found in the specific use environment.

5 CONCLUSIONS

Practicality calls on us to do the impossible: Confidently predicting decades of performance for products that are designed and launched in a few months is a formidable challenge. Yet, billions of euros of investment each year require assessment of the risk in those investments. Here, we propose a multi-tier strategy for collecting and using information in the most effective ways, striving toward meeting this tremendous challenge.

Table III: Summary of priorities for planned amendments of IEC 61215 and IEC 61730 and development of new documents.

Document	Anticipated Publication	Description	Value
IEC 61215 new edition	2019	Add cyclic (dynamic) mechanical load test (IEC TS 62782)	Detect cracked cells in modules
		Add pass-fail requirement for PID (IEC TS 62804-1)	Define consistent way of characterizing PID resistance
		Methods for testing bifacial and flexible modules	These new products are entering the marketplace without consistent ways of communicating their value
		Correction of hot-spot test	Addresses problem with testing certain modules with series-parallel architecture
		Improve simulator requirements	More accurate power output measurements, with less effect from spectral mismatch
		Thermal cycling with current flow through bypass diodes	Thermal fatigue associated with the bypass diodes is sometimes observed; application of current flow through the diodes will increase probability of detecting a problem
IEC 62892	2018	Extended thermal cycling	Provides greater confidence, especially for locations that experience more thermal fatigue
IEC 61730-1 amendment	2018	Allow for insulation thickness to be determined by distance-through-insulation test of 62788-2 or MST-04	Ensures required minimum insulation thickness after lamination
		Addition of weathering requirement for relied-upon insulation	The current test only tests safety of initial conditions; this change gives confidence in safety after weathering
IEC 61730-2	2018	How to deal with bubbles, in particular for a module with a cemented joint	Ensures that reduction in distance between edge of module and active cells is still safe
IEC 62788 (Materials Testing)	2018-2020	62788-1: Encapsulants 62788-2: Frontsheets/backsheets 62788-5: Edge seal 62788-6: Multiple component tests 62788-7: Component weathering exposures	Facilitates comparison of candidate materials for use in PV modules and for quality control of incoming materials
IEC TS 63126	2019	Defines modifications to module qualification tests for modules that will operate at higher temperatures	Useful for modules that will be deployed in the hottest locations and applications
IEC 62093	2019	Updates qualification test for inverters to reflect years of experience	Identify problems with inverter designs
IEC TS 63157	2019	Guideline to improve quality control of power electronics manufacturing	Improve quality of power electronics
IEC TS 63140	2019	Partial shade endurance test	Quantifies permanent changes after part of the module is shaded
IEC 62759-2	2020	Test ability of modules to withstand stresses applied during installation	Rough handling of modules during installation (or walking on after installation) can cause cracks or other damage
Not assigned	2022	Various forms of BIPV will be considered by a single project team	BIPV applications are growing and more comprehensive standards will support growth
TS Extended-stress test	2020	Apply longer stress, periodically reporting changes	Provides additional information about module durability to be analyzed as discussed here

The strategy starts with accelerated tests standardized as qualification tests that quickly identify design problems for the intended application. In manufacturing and installation, the initial qualification testing must be coupled with robust quality assurance programs such as described by IEC TS 62941 and IEC TS 63049. Not only should incoming materials/components be tested, but smart sampling should be used to select finished product from all manufacturing lines and shifts to confirm uniform and appropriate production; during installation, systematic inspections should confirm that all crews have appropriate training and oversight. Qualification tests should be updated to reflect failures observed in the field and should provide options for testing to multiple levels of stress when that is shown to be useful, e.g. for varying snow loads and system voltages.

Building on the tests used for qualification testing and based on experience obtained during deployment of many GWs of PV, standardized extended-stress tests should be developed, widely used, well documented, and well understood. Such testing provides a means to compare different versions of products to facilitate improvements of the bill of materials and other aspects of the products. Standardization and openness of such data collection will facilitate scientific analysis, by leveraging data from the broader community to identify test results that do or do not correlate with field problems in the application of interest. The extended tests should be designed to cost-effectively apply relevant stresses, testing to failure when the added cost is acceptable, and providing preventative measures to reduce technical risks in investments. Predictive models based on a detailed understanding of relevant failure mechanisms can then be applied to both the extended-test and field data to assess the accuracy of the models, identify indications of product weakness that may be relevant to the intended application, and to estimate the uncertainty associated with the risk assessment.

The final risk assessment should use all available data (published in the literature, as well as collected for the product of interest) to facilitate estimation of the anticipated return on investment (ROI) and the associated risk that the ROI may be less. This estimation process will need to include not only an understanding of the physics of failure, but the variability in the manufacturing, installation and maintenance in order to assess the timing, consequence and scale of possible failures and degradation. Many details must be understood to provide the best risk assessments; by pooling data and coordinating studies to better understand failure and degradation, the community will be able to accelerate progress toward the end goal of low-uncertainty predictions.

Standardized qualification and extended-stress testing can play an important role in accelerating this progress. Recent and planned progress in new standards is summarized in section 4.

6 REFERENCES

[1] D. C. Jordan, S. R. Kurtz, K. VanSant, and J. Newmiller, Progress in Photovoltaics: Research and Applications 24 (2016) 978.

[2] U. Jahn and W. Nasse, Progress in Photovoltaics: Research and applications 12 (2004) 441.

[3] D. Jordan and S. Kurtz, 29th European PV Solar Energy Conference and Exhibition, (2014).

[4] IEC 61215 Terrestrial photovoltaic (PV) modules - Design qualification and type approval, 2016.

[5] IEC 61730-1 Photovoltaic (PV) module safety qualification, 2016.

[6] IEC/TS 62941 Guideline for increased confidence in PV module design qualification and type approval, 2016.

[7] U. Jahn, M. Herz, E. Ndrio, E. Janknecht, D. Moser, G. Belluardo, et al., Solar Bankability WP1 Deliverable D 1.

[8] S. M. Shrestha, J. K. Mallineni, K. R. Yedidi, B. Knisely, S. Tatapudi, J. Kuitche, et al., IEEE Journal of Photovoltaics 5 (2015) 174.

[9] D. Cunningham, B. Jaeckel, and A. Roth, PV Module Reliability Workshop, Golden, CO, (2012).

[10] B. Jaeckel, A. Krtschil, D. Cunningham, N. Forney, C. LaMothe, A. Nguyen, et al., Proceedings 26th European Photovoltaic Solar Energy Conference, Hamburg, (2011) 4AV.2.1 3484.

[11] A. Funcell, PV Module Reliability Workshop, Golden, CO, (2012).

[12] J. Meydbray, PV Module Reliability Workshop, Golden, CO, (2012).

[13] D. Meakin, PV Module Reliability Workshop, Golden, CO, (2012).

[14] P. Hacke, K. Terwilliger, S. Glick, R. Smith, G. Perrin, S. Kurtz, et al., Proceedings 40th IEEE Photovoltaic Specialist Conference, (2014) 930.

[15] G. TamizhMani, PV Module Reliability Workshop, Golden, CO, (2012).

[16] "Introducing 'Best of modules' Quality assurance," PV Magazine, 2011 (2011) 92.

[17] K. P. Scott and A. Zielnik, PV Module Reliability Workshop, Golden, CO, (2012).

[18] A. Zielnik, "PV Durability and Reliability Issues," Photovoltaics World, Nov/Dec (2009) 10. http://www.renewableenergyworld.com/rea/news/article/2009/12/pv-durability-and-reliability-issues

[19] J. Wohlgemuth and S. Kurtz, Proceedings 40th IEEE Photovoltaic Specialist Conference, (2014) 3589.

[20] G. Kelly, A. Häring, T. Spooner, G. Ball, S. Kurtz, M. Heinze, et al., Proceedings 43rd IEEE Photovoltaic Specialists Conference, (2016) 1264.

[21] S. Kurtz, T. Sample, J. Wohlgemuth, W. Zhou, N. Bosco, J. Althaus, et al., Proceedings 42nd IEEE Photovoltaic Specialists Conference, (2015).

[22] G. Ramu, M. Yamamichi, W. Zhou, A. Mikonowicz, S. Lokanath, Y. Eguchi, et al., "Updated Proposal for a Guide for Quality Management Systems for PV Manufacturing. Supplemental Requirements to ISO 9001-2008," Technical Report No. NREL/TP--5J00-63742, National Renewable Energy Laboratory, Golden, CO (United States) (2015).

[23] Y. Eguchi, G. Ramu, S. V. Lokanath, M. Yamamichi, S. Kurtz, J. Wohlgemuth, et al., Proceedings 40th IEEE Photovoltaic Specialists Conference, (2014) 3574.

GOOGLE'S LITTLE BOX CHALLENGE AND THE DEVELOPMENT OF THE TRUE AC-MODULE

H. Oldenkamp
OKE-Services
Keizerstraat 121, 2584 BD The Hague, The Netherlands

ABSTRACT: In July 2014 Google started an open competition challenging electronics engineers to design the smallest possible 2 kW inverter. The award for winning this Little Box Challenge (LBC) competition was a $1 000 000 prize. OKE-Services participated in this contest and was selected as one of the 18 finalists.
The paper discusses the procedure, experiences and results of the Little Box Challenge, as well as the possible use of the designs for PV. Further OKE-Services' Little Box inverter design is presented and the latest developments regarding a 'true' ac-module: a PV-module in which the inverter is fully integrated.
Keywords: AC-Modules, Inverter, Building Integrated PV (BIPV)

1 LITTLE BOX CHALLENGE

The assignment of the Little Box Challenge competition was to build a 2 kW stand-alone inverter which converts 400 – 450 Vdc to 240 Vac. Since Henk Oldenkamp has been working on micro inverters since 1994, he registered as team OKE-Services, the name of his one-man company.

Worldwide more than 2000 teams registered of which more than 100 teams applied. 18 teams, including OKE-Services, were selected as finalists and invited to present their inverter and submit it to NREL for testing.

2 REQUIREMENTS LBC AND OKE'S DESIGN SOLUTIONS

The initial inverter requirements of the Little Box Challenge were released on the 22nd of July 2014 and the deadline for submitting an application was one year later.

In the original requirements, an ideal 450 Vdc floating power supply was specified to supply the inverter. The inverter would be loaded with an isolation transformer with a center tab. The center tab and the metal inverter enclosure would be grounded. The maximum ground current through the center tab was specified as 5 mA. Further the maximum peak to peak ripple in the input current was 20% of the average input current. This test setup is presented in Figure 1.

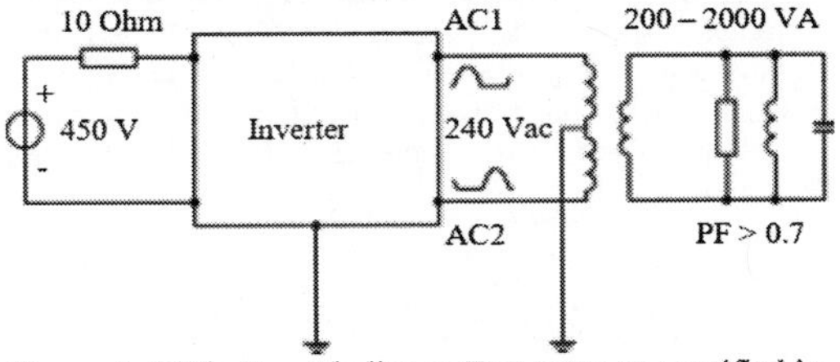

Figure 1: Little Box Challenge Test setup as specified in requirements of July 2014

In the ideal implementation, the voltages on AC1 and AC2 are 120 Vac sinewaves with opposite polarity. In that case the ground current from the center tab will always be zero. To improve the efficiency (and thus reduce the dimensions), I decided to create halve sinewaves and use an unfolding bridge to create the full sinewaves (as shown in the circuit diagram in Figure 1). This saves one large power inductor and reduces the cooling requirements.

This approach will cause some ground current. But, since in my design, the capacity between the electronics and the enclosure of the inverter is kept low (a few nF) the ground current requirement can easily be met. Probably this design could have reached 30 kW/Liter power density.

In March 2015, nearly 9 months after the starting date, a significant change in the requirements became apparent: the power supply was not floating but grounded with 200 nF. Very unprofessional to change it in such a late stage and even more so since this important change was not mentioned in the specifications, but only in the Question and Answers document. Therefore, other teams may not have noticed it. The only reason for changing the requirement was that NREL, who was responsible for testing, did not have such a true floating power supply (….).

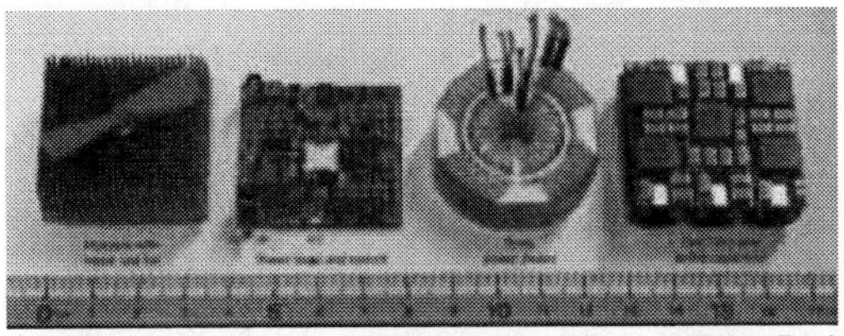

Figure 2: Photo of main sub-assemblies before final assembly of inverter submitted for Little Box Challenge by OKE-Services

Due to this enormous increase of capacity, the ground current of my design will increase to approximately 10 mA which is well above the 5 mA limit. So, this approach can never meet the adapted specifications.

Therefore, I had no choice but to modify my design. The impact is clearly huge: a completely different topology was needed. Originally, my design needed only two GaN-switches with two power inductors (one for the pre-regulator and one to create a rectified sinewave). To meet the 5 mA ground current limit two sinewaves with opposite polarity on AC1 and AC2 needed to be created for which I also needed 5 GaN switches and three power inductors (one GaN switch for the pre-regulator and four GaN switches for a full bridge). On top of that the efficiency will be lower and therefore I needed more cooling capacity. In short: I had to start from scratch…. The new design has much larger dimensions and the power density dropped to approximately 15.6 kW/Liter.

But less than one month before the deadline, the Little Box Challenge Team came up with a new surprise: on the 25th of June 2015 the 5 mA ground current limit was increased to 50 mA. So, I could have used my original design! But the remaining time of only one month was too short to fall back to the first design. All I could do was to use the second design and change the control of the full bridge, optimize the power circuit and reduce the dimensions of the heatsink. This resulted in the 24.4 kW/Liter power density claim (as published on the Little Box Challenge website, Figure 4).

Unfortunately, due to all the extra work caused by the two spec-changes, valuable time for testing was lost. Therefore, it was no surprise that my inverter did not pass the tests.

Figure 3: Photo of inverter submitted for Little Box Challenge by OKE-Services

3 OUTCOME OF LITTLE BOX CHALLENGE

In March 2016 the winner was announced: The Red Electrical Devils with a power density of 8.9 kW/Liter. Number two and three were Schneider Electric Team (6.1 kW/Liter) and Future Energy Electronics Team (3.7 kW/Liter) respectively.

Since the latter two have the lowest energy density, it can be concluded that all other teams (15) failed to pass the tests. These three winning designs are relatively large compared with the other designs. The finalist ranking of the claims are presented in Figure 4.

The outcome of the Little Box Challenge competition can easily be explained by the significant changes in the requirements. The final changes were announced even on the 16th of July 2015, only one week before the deadline.

OKE-Services had to restart the design process after major changes of the requirements announced in March 2015 and again in June 2015. Of course, valuable time for testing was lost. It is likely that other teams with challenging and smaller designs had similar problems.

4 WAS THE LITTLE BOX CHALLENGE USEFUL FOR PV?

None of the concepts designed for this competition are of direct interest for PV applications:

- The Little Box Challenge inverter cannot be used for feeding into the grid
- The maximum dc-current is 5 A and therefore the Little Box Challenge inverter does not match the current of most today's PV-modules.
- The required lifetime to pass the tests is in the order of 100 hours, not 20 years as is common for PV-inverters.
- There was no cost limit for the Little Box Challenge inverter

For PV (micro) inverters just reducing the dimensions makes no sense. Only combining reduction of dimensions, losses and costs will result in an optimal PV-inverter: an integral approach is needed. Further the shape must be optimized for simple installation and for good thermal performance.

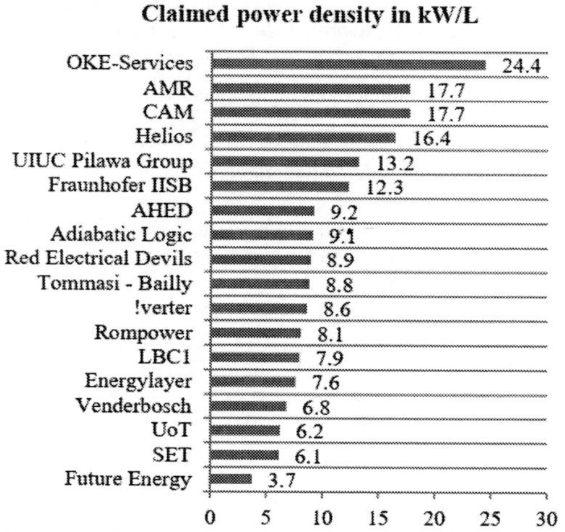

Figure 4: Claimed power density of inverters submitted by the 18 teams for the Little Box Challenge. Source: www.littleboxchallenge.com

5 TRUE AC-MODULE

After completion of the Little Box Challenge competition OKE-Services started developing a 'true' ac-module concept: a PV-module in which the inverter is fully integrated.

For a cost-effective design both the inverter costs as the module costs are reduced. PV-module costs are reduced by eliminating the junction box, the bypass diodes, dc-cables and dc-connectors. The inverter will be roughly 10 times smaller than today's micro inverters with a target efficiency as close as possible to 98%. And, most importantly, for acceptable costs.

For the new inverter design, some of the circuits designed for the Little Box Challenge inverter are being modified and optimized. This results in a significantly smaller micro inverter (compared to today's micro inverters), which can be integrated into the frame of the PV-module.

6 TARGET INVERTER SPECIFICATIONS

The targeted standard features of the inverters are:

- Suitable for 60 x 6" Cells PV module up to 300 W
- Reactive power control down to PF = 0.9
- Peak efficiency in the order of 98% thanks to the GaN switches
- Cost effective
- MTBF >> 1000 years

Some of the innovations that will be integrated are:

- Substring optimizer
- AC-cable current sensing and limitation
- Wifi in each inverter

6.1 Substring optimizer

PV-modules with 60 cells consist of three strings of 20 cells in series. A bypass diode is connected across each substring. This limits the peak reverse voltage across a shaded cell to a safe value. Just like in a string of PV-modules, differences in illumination can occur. Therefore each substring can have a different mpp-current.

Today, all power optimizers and micro inverters optimize the power of the three substrings in series, not the power of each substring. In most circumstances this is far from optimal. Therefore it is decided to include a substring optimizer in the inverter.

The Substring Power Optimizer is realized as a front end with 4 input terminals (the four connections from the PV module) and two output terminals.

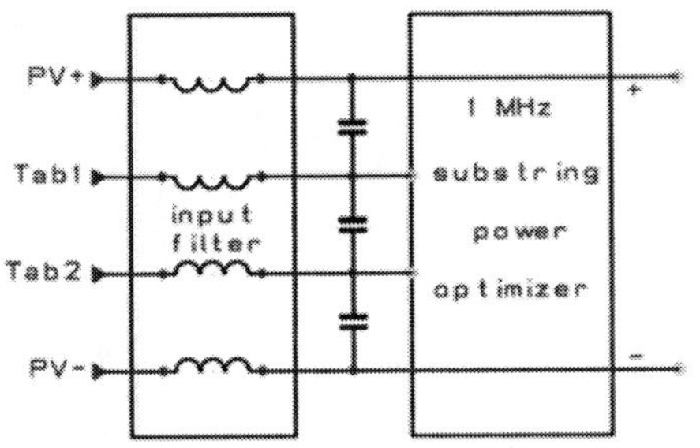

Figure 5: Electronic scheme of Substring Power Optimizer

The advantages of using this Substring Power Optimizer are:

- Highest possible yield
- The output I-V curve is that of a normal unshaded PV-module.
- Even under severe shading conditions each of the three substrings remains forward biased.
- Hot-spot problems are eliminated and thus there is no need for bypass diodes!

6.2 AC-cable current sensing and limitation

For a 300 W inverter the maximum ac-current can be 1.3 Arms at a nominal grid and 1.43 Arms at 10% lower grid. Using an ac-cable with a 16A rating only 11 inverters can be connected on one ac-cable.

However, in practice a PV-module will seldom deliver its maximum power. By monitoring and limiting the current in the ac-cable to 16 Arms the number of inverter can be increased to at least 16 per ac-cable. And, in case of sub-optimal module orientations an even higher number can be connected.

Additionally this will make the system fool proof: it is simply not possible to overload the ac-cable!

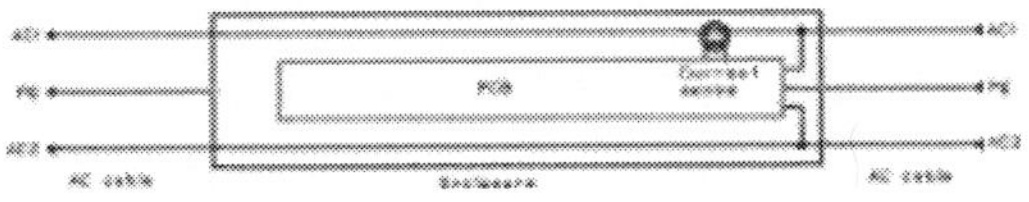

Figure 6: Electronic scheme of AC-cable current limitation

7 COST SAVINGS AND OTHER ADVANTAGES

The following module cost savings can be realized by fully integrating the AC-module inverter into the PV-module frame:

- Junction box
- Bypass diodes
- DC-cables
- DC-connectors
- For single glass modules: thinner back sheet due to low voltage isolation
- For double glass modules: no need to drill a hole in the glass which results in a smaller PV-module and avoids the risks of breaking the glass (due to drilling the hole)

These savings add up to 5 € - 10 € per PV-module, which agrees with 5 - 10% of today's module price. Since the prices of PV-modules will remain under pressure, such a significant cost reduction might be very interesting for module manufacturers.

Furthermore, the ultimate AC-module offers the following advantages for installers and end-users:

- No need to mount and find a place for a central inverter
- Quick cabling
- One type of ac-module for all installations
- Full freedom with system size, orientation and layout, often more modules can be fitted
- No need for dummy modules on shaded spots
- Aesthetically appealing
- Simple system check and module level monitoring
- Fool proof system
- Highest yield
- No junction box: no shading on the rear side. Great for bifacial modules.
- Monitoring at module level: all possible module and inverter failures can be easily detected

8. CONCLUSIONS

To make AC-modules successful, all the advantages that micro inverters inherently offer must be used:

- Get rid of the junction box and the dc-wiring by a new module design
- Develop cost effective flexible ac cables and connectors, both with a small diameter.
- Integrate the micro inverter into the frame

But there is more work to do:

- Adapt module and inverter standards that offer cost reductions for the true ac-module
- Get rid of the relay requirements
- Use the advantage of the low dc-system voltage and the inherent safety

Figure 7: Bifacial double glass module built by CS Wismar with a 4 pin dc connecter integrated in the frame.

To prove the module can be built with standard production equipment a demonstration bifacial double glass module with a four pin connector was built by CS Wismar. This will be shown at the conference. A photo is shown in Figure 7.

9. ACKNOWLEDGEMENTS

OKE-Services wishes to thank the following persons for their contribution to the development of the ultimate AC-module:

- Jan Wittfoth, CS Wismar GmbH
- Erik Eikelboom, Electric Life

This is a pre-peer review version of the paper submitted to the journal Progress in Photovoltaics.
It has been selected by the Executive Committee of the EU PVSEC 2017 for submssion to Progress in Photovoltaics.

THE PRODUCT ENVIRONMENTAL FOOTPRINT (PEF) OF PHOTOVOLTAIC MODULES – LESSONS LEARNED FROM THE ENVIRONMENTAL FOOTPRINT PILOT PHASE ON THE WAY TO A SINGLE MARKET FOR GREEN PRODUCTS IN THE EUROPEAN UNION

Andreas Wade[1], Philippe Stolz[2], Rolf Frischknecht[2], Garvin Heath[3], Parikhit Sinha[4]

[1]First Solar GmbH, Rheinstr 4, 55116 Mainz, Germany, andreas.wade@firstsolar.com,
Deputy-Operating Agent, IEA PVPS Task 12 / Product Group Coordinator PEF Pilot "PV Electricity Generation"
[2]treeze Ltd., Switzerland
[3]National Renewable Energy Laboratory, USA, Operating Agent IEA PVPS Task 12
[4]First Solar Inc., USA

ABSTRACT: Shaping and channeling the transformation of the global energy system requires an understanding of the life cycle environmental impacts of available technology options. With photovoltaic electricity generation expected to be the largest, least cost and most relevant source of energy globally by mid-century, the environmental and resource implications of this technology should be made accessible to a variety of societal, political and scientific stakeholders. The Product Environmental Footprint Guide provides a methodology for assessing the environmental impacts of PV technologies based on life cycle assessment. The establishment of an inclusive multi-stakeholder process with multiple checkpoints, ensuring scientific soundness, societal and political acceptance and methodological quality, is portrayed in this paper and can serve as a model for driving a single market for green products in the EU.

Keywords: Photovoltaics, PV, LCA, environmental footprint, PEF, PEFCR

1 INTRODUCTION

Deployment of photovoltaic (PV) systems in 2016 reached a cumulative capacity of more than 100 GW_p in the European Union and more than 300 GWp worldwide (Schmela, et al., 2017). Continued efforts to decarbonize the global energy system in order to comply with the Paris Agreement will require multiple terawatts of PV capacity additions over the next decades, with solar PV electricity generation expected to be the largest, least-cost and most relevant source of energy globally by mid- century (Breyer, 2017).

Shaping and channeling this transformation of the global energy system through regulatory frameworks requires an understanding of the life cycle environmental impacts of available technology options. This knowledge is a pre-requisite for establishing regulatory frameworks based on a broad societal, scientific and political consensus.

The International Energy Agency Photovoltaic Power System Program (IEA PVPS) aims to enhance the international collaborative efforts to facilitate the role of photovoltaic solar energy as a cornerstone in the transition to sustainable energy systems (PVPS, 2017). Task 12 of this program aims to compile and disseminate reliable environmental, health, safety and sustainability information on the life cycle of photovoltaic technology to the public and policymakers. In line with this goal, the IEA PVPS Task 12 expert group (PVPS, 2017) began participating in the Environmental Footprint Pilot Phase of the European Commission in 2013. Through the formation of a broad coalition of industrial, scientific and societal stakeholders, the Product Environmental Footprint Guide

This is a pre-peer review version of the paper submitted to the journal Progress in Photovoltaics.
It has been selected by the Executive Committee of the EU PVSEC 2017 for submssion to Progress in Photovoltaics.

(European Commission, 2013a) was applied to photovoltaic electricity generation by developing, testing and evaluating Product Environmental Footprint Category Rules (PEFCRs).

2 BACKGROUND AND MOTIVATION

Making informed decisions on the future mix of renewable electricity generation technologies hinges on a thorough understanding of the life cycle environmental impacts and potential trade-offs, based on scientific evidence. This requires decision makers to move away from static one-dimensional approaches based on a singular indicator or individual life cycle stage – and instead evaluate technology options using a holistic, multi-criteria approach based on life cycle assessment (LCA).

The environmental footprint pilot phase, as introduced and recommended by the European Commission (EU Commission, 2013b), provided an opportunity for the joint development and international endorsement of a standardized, scientifically sound, comprehensive and consistent methodology for the life cycle environmental assessment of photovoltaic technologies. The pilot phase also offered the opportunity to confirm the environmental profile of this energy generation technology and communicate the results in a consistent and harmonized way so as to allow conscious policy and consumer decisions to be applied to PV electricity generation.

When compared to already existing and established standards for the assessment of environmental impacts of products in general, e.g. ISO 14044 International Standard on life cycle assessment requirements (ISO, 2016), and guidelines for photovoltaic electricity in particular, e.g. the IEA PVPS Task 12 Methodology Guidelines on Life Cycle Assessment of Photovoltaic Electricity (Fthenakis, et al., 2012)[1] the added value of this approach was the *political mandate given to the European Commission by the EU Member States to develop and test a life cycle assessment based environmental footprint approach that could be used in existing or new EU policies* (Galatola & Pant, 2014). This political mandate, along with the provision of an inclusive, transparent and interactive testing environment during the environmental footprint pilot phase from 2013 to 2016, enabled the establishment of a common basis for measuring and communicating the environmental performance of photovoltaic electricity generation. This methodology can be implemented by all market actors and regulators alike across Europe. This approach will also influence and shape various labels, directives and activities within the European Union and abroad (see Figure 1).

Figure 1 – Policy links of the Product Environmental Footprint, modified from (EU Commission, DG ENV, 2013)

The concept of a pilot phase entails the possibility of discovering inconsistencies and challenges, which leads to recalibrated approaches, means and methods. Though concerns about the inability of the PEF to implement life cycle assessment in European environmental policy due to failures in harmonizing existing methods are potentially valid, it is too early to rule on the application of the newly proposed methodologies and communication channels (Finkbeiner, 2013) as the pilot phase will only be fully evaluated towards the end of 2017.

[1] The most recent version of the LCA methodology guidelines are available in (Frischknecht, R.; et. al., 2016).

This paper aims to provide a detailed overview of the PV electricity pilot process from 2013 to 2017, including details on the scientific basis for the development of the category rules for photovoltaics.

3 APPROACH AND METHODOLOGY

To provide an overview of the PEF pilot process, it is important to understand the setup, pre-requisites and governance structure implemented by the European Commission. Figure 2 below depicts the governance used to manage the 26 participating pilot projects.

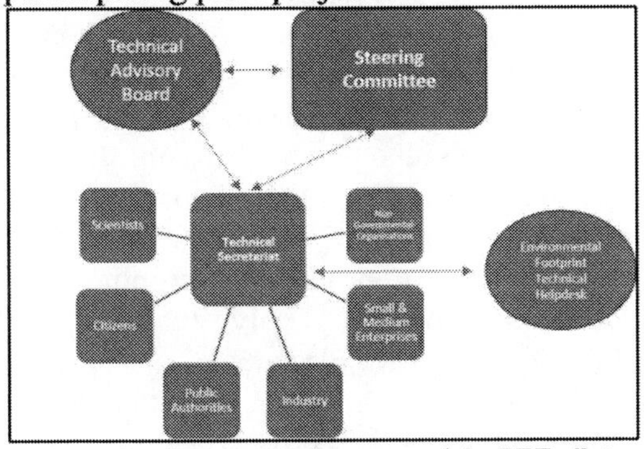

Figure 2 – Governance Structure of the PEF pilot phase

The **Technical Secretariat (TS)** was tasked with drafting the PEFCR for PV, facilitating the harmonization and review process, and executing the screening and supporting studies.

The European Commission set the following pre-requisites for the composition of the TS to ensure market representativeness:

- 75 % of EU market players must be invited,
- all companies contributing to more than 10% of the market must be invited,
- 51% of the market must actively participate in the TS,
- and ensure participation of a wide range of stakeholders, including small & medium enterprises (SMEs), consumers and environmental associations (Non-Governmental Organizations).

The **Steering Committee (SC)** consisted of representatives from all pilot projects, the European Commission, all EU Member States, as well as representatives from one environmental association, one consumer organization and representatives from the **Technical Advisory Board (TAB)**. The SC coordinated the progress of the pilot phase by reviewing, assessing and approving the intermediate results of each pilot and interacting with the TAB on open questions and the need for amendment of work results. The TAB was composed of technical experts appointed by the participating TS.

Figure 3 below depicts the process steps each pilot had to fulfill from 2013 to 2017, leading to the eventual approval and publication of the final Product Environmental Footprint Category Rules by the end of 2017. The remodelling and verification of the PEFCR and screening study are currently being carried out. The individual process steps will be portrayed subsequently.

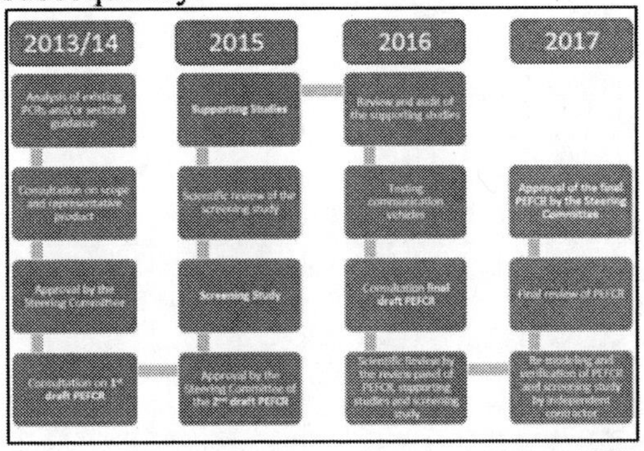

Figure 3 – PEF pilot process steps from 2013 to 2017

3.1 ANALYSIS OF EXISTING PRODUCT CATEGORY RULES AND SECTORAL GUIDELINES

Before starting the development of a new PEFCR, the Technical Secretariat conducted thorough research into whether product category rules (PCRs) had already been developed for the same product category (European Commission, 2016). Table 1 provides an overview of the existing PCRs and sectoral guidelines, which were analysed and screened at the beginning of the process, using the Conformity Assessment Form developed by the Product Category Rule Guidance Development Initiative (Ingwersen & Subramanian, 2013).

Existing PCRs and Sectoral Guidance Documents for the product category "PV Electricity Generation"	Relevance for PEFCR development (H - high / M - medium / L - low)
(1) Methodology Guidelines on Life Cycle Assessment of Photovoltaic Electricity, 2nd edition, IEA PVPS Task 12, International Energy Agency Photovoltaic Power Systems Programme (Fthenakis, et al., 2012) (Frischknecht, R.; et. al., 2016)	H
(2) UN CPC 171 and 173 Electricity, steam and hot/cold water generation and distribution, 2007:08, Version 2.02 (PCR Electricity, steam and hot/cold water generation and distribution, 2017)	M
(3) EPD PCR Basic Module, CPC Division 69, Electricity, Gas and Water Distribution (on own account), Version 1.0, dated 2010-11-30 (PCR Electricity, gas and water distribution (on own account), 2017)	M
(4) EPD PCR Basic Module, CPC Division 17, Electricity, Town Gas, Steam and Hot Water, Version 1.0, dated 2010-08-31 (PCR Electricity, town gas steam and hot water, 2017)	M
(5) NEEDS Deliverable no. 11.2 - RS la, Final report on technical data, costs and life cycle inventories of PV applications, 22.12.2005 (Frankl, Menichetti, & Raugei, 2005)	L
(6) Référentiel d'évaluation des Impacts Environmentaux des systèmes Photovoltaiques par la méthode d'analyse du cycle de vie, ADEME, CYCLeco, 2012 (Payet, et al., 2012)	M
(7) Ökoeffizienzanalyse von Photovoltaikmodulen, bifa-Text Nr. 62, bifa Umweltinstitut, November 2013 (Seitz, Kroban, Pitschke, & Kreibe, 2013)	L
(8) European Standards on Environmental Product Declaration of construction works (EN15804:2013)	H

Table 1 – Existing Product Category Rules and Sectoral Guidance analysed and evaluated in preparation for the development of the Product Environmental Footprint Category Rules (Frischknecht, et al., 2017)

The conformity assessment forms are available to all stakeholders through the environmental footprint wiki pages of the pilot project (EU Commission, 2017b).

3.2 DEFINITION OF SCOPE, REPRESENTATIVE PRODUCT AND FIRST PUBLIC CONSULTATION AND STEERING COMMITTEE APPROVAL

Taking into account the results of the evaluation of the existing category rules and sectoral guidance documents, the functional unit, scope and representative product were defined to further develop the PEFCR following the first public consultation. The **functional unit** is the production of 1 kWh of direct current (DC) electricity from photovoltaic modules (or panels / laminates). The **scope of the PEFCR** includes the manufacturing of the PV module and the balance of system (BOS) components, the operation of the PV system, as well as the dismantling and recycling of material during the end-of-life phase. The use of production equipment and facilities and the supply chain of the materials used are also taken into account while the inverter remains outside the system boundary (Frischknecht, et al., 2017)[2]. The **representative product (RP)** is defined as an average photovoltaic module installed in Europe. The average photovoltaic module is a virtual product composed of the different commercially available PV technologies on the European market including mono- and multi-crystalline silicon, micromorphous silicon, CdTe and CIGS modules, weighted by their respective market shares in the reference year 2012 (see Figure 4). The decision to use a virtual representative product was evaluated as most efficient and meaningful for the further development of the category

[2] The final draft PEFCR is provided as supplementary material.

rules.

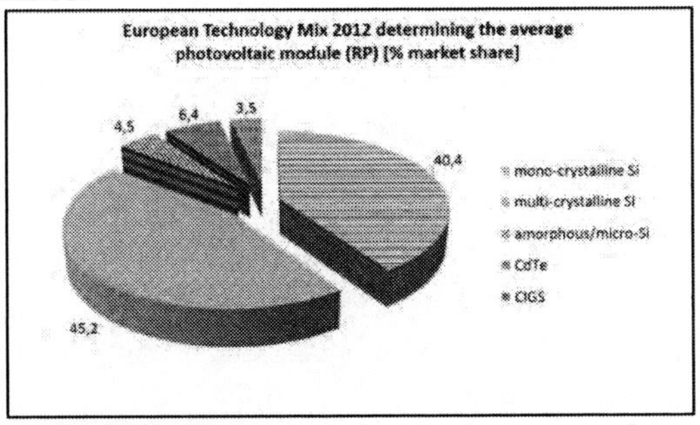

Figure 4 – European photovoltaic technology mix as represented on the EU market in 2012 and used for the definition of the representative product (RP) in the PEF pilot project. Data from (Fraunhofer ISE, 2013)

The statistics on the public consultation on scope and representative product are depicted in Figure 5. Already at this early stage of the process, it became apparent that interest in the project extended well beyond the European Union and across different organization levels, market participants, and representatives of societal and scientific communities.

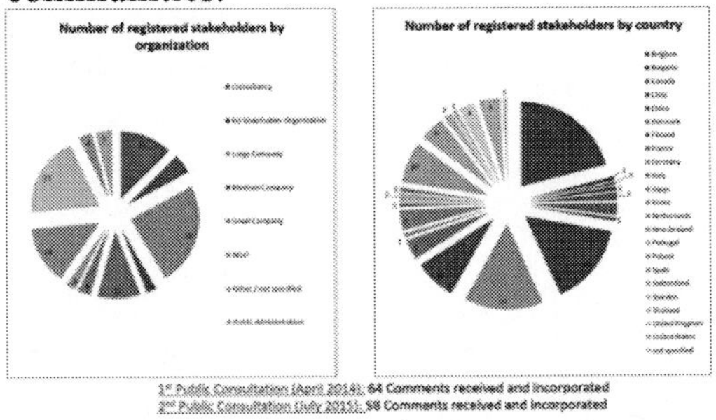

Figure 5 – Statistics on the 1st and 2nd public consultation on the PEFCR "PV Electricity Generation"

Taking into account the feedback received during the public stakeholder consultations, the approved second draft of the PEFCR was used to carry out a screening study, the second deliverable of the pilot process.

3.3 PRODUCT ENVIRONMENTAL FOOTPRINT CATEGORY RULES (PEFCR)

The drafting of the PEFCR followed the template provided in Annex B of the Product Environmental Footprint Pilot Guidance (European Commission, DG Environment, 2016). The PEFCR complement the methodological guidance established through the PEF Guide (European Commission, 2013a) with product specific requirements, which help to ensure methodological consistency and comparability of results. The use of the PEFCR as such is optional for internal applications of the PEF Guide; however, it is recommended for external applications without comparisons or comparative assertions, and is mandatory for external applications with comparisons and comparative assertions.

The PEFCR provides a harmonized set of rules to evaluate the environmental impacts of photovoltaic modules installed in the region of the EU28 and EFTA countries. To achieve this objective, a set of PV specific framework parameters were established and incorporated into the PEFCR. The following default parameters are used to establish the environmental impact of a European PV module:

- Annual yield of the PV system: 975 kWh/kWp (including degradation);
- Annual degradation rate of the PV system: 0.7% per year
- Expected lifetime of the PV system: 30 years.

The annual average yield of optimally oriented modules in Europe was weighted according to the cumulative installed photovoltaic power and corresponds to 1,090 kWh/kWp (excluding degradation effects). This average yield and the expected system lifetime should not be adjusted when carrying out a PEFCR-compliant LCA. A deviation from the default degradation rate is only valid if verification documentation (as defined in the PEFCR) is published along with the assessment. The definition of common framework parameters and default impact categories is of paramount importance since the use of varying and poorly documented framework parameters limits the comparability of the publicly available life cycle assessments of different PV technologies and the applicability of the results for policy or regulatory decisions.

The default environmental footprint impact categories depicted in the PEF Guidance (European Commission, 2013a) were applied throughout the pilot process, in line with various life cycle impact assessment methodologies (JRC/ies, 2010), (JRC/ies, 2011), (JRC/ies, 2010a). In addition to the default impact categories, the TS has included cumulative renewable and non-renewable energy demand and nuclear waste as additional impact categories to be assessed and reported separately. The addition of those impact categories was required to ensure consistency with the previously identified PCRs and sectoral guidance documents depicted in table 1. Table 2 below depicts all impact categories, along with their scientific references, that are used in the PEFCR calculation methodologies.

Impact category	**Indicator**	**Reference**
Indicators required according to the PEF guide		
Climate change	Radiative forcing as Global Warming Potential (GWP100) [kg CO_2 eq.]	(IPCC, 2007)
Ozone depletion	Ozone Depletion Potential (ODP) [kg CFC-11 eq.]	(WMO, 1998)
Human toxicity, cancer effects	Comparative Toxic Unit for humans [CTUh, c]	(Rosenbaum, et al., 2008)
Human toxicity, non-cancer effects	Comparative Toxic Unit for humans [CTUh, n-c]	(Rosenbaum, et al., 2008)
Particulate matter / respiratory effects	Intake fraction for fine particles [kg PM2.5 eq.]	(Greco, Wilson, Spengler, & Levy, 2007); (Humbert, 2009); (Rabl & Spadaro, 2004)
Ionizing radiation, human health	Human exposure efficiency relative to U^{235} [kBq U^{235} eq.]	(Frischknecht, Braunschweig, Hofstetter, & Suter, 2000)
Photochemical ozone formation	Tropospheric ozone concentration increase [kg NMVOC eq.]	(Van Zelm, et al., 2008) as applied in ReCiPe
Acidification	Accumulated Exceedance (AE) [mol H^+ eq.]	(Posch, et al., 2008); (Seppälä, Posch, Johansson, & Hettelingh, 2006)
Eutrophication, terrestrial	Accumulated Exceedance (AE) [mol N eq.]	(Posch, et al., 2008); (Seppälä, Posch, Johansson, & Hettelingh, 2006)
Eutrophication, freshwater	Fraction of nutrients reaching freshwater end compartment (P) [kg P eq.]	(Struijs, Beusen, van Jaarsveld, & Huijbregts, 2008) as implemented in ReCiPe
Eutrophication, marine	Fraction of nutrients reaching marine end compartment (N) [kg N eq.]	(Struijs, Beusen, van Jaarsveld, & Huijbregts, 2008)as implemented in ReCiPe
Ecotoxicity, freshwater	Comparative Toxic Unit for ecosystems [CTUe]	(Rosenbaum, et al., 2008)
Land use	Soil Organic Matter [kg C deficit]	(Milá i Canals, Romanyà, & Cowell, 2007)
Resource depletion, water	Water abstraction related to local scarcity of water [m^3 water eq.]	(Frischknecht, Steiner, & Jungbluth, 2008)
Resource depletion, mineral, fossil, renewable	Scarcity [kg Sb eq.]	(Guinée, et al., 2001)
Additional indicators		
Cumulative energy demand, renewable	Gross energy content of renewable primary energy resources [MJ oil eq.]	(Frischknecht, Wyss, Buesser, Lützkendorf, & Baloutktsi, 2015)
Cumulative energy demand, non-renewable	Gross energy content of non-renewable primary energy resources [MJ oil eq.]	(Frischknecht, Wyss, Buesser, Lützkendorf, & Baloutktsi, 2015)
Nuclear waste	Radiotoxicity index, RTI [m^3 HAA eq.]	(Frischknecht & Büsser Knöpfel, 2013); (Frischknecht & Büsser Knöpfel, 2014)

Table 2: List of life cycle impact assessment indicators considered relevant for the PEFCR

The draft PEFCR provided the basis for carrying out the screening study portrayed below.

The PEF screening study[3] (Stolz, Frischknecht, Wyss, & de Wild-Scholten, 2016) aims to compare the environmental performance of the five selected PV technologies and identify the most relevant life cycle stages and areas of highest impact ('hot spots'). Based on the results of the screening study, the final draft PEFCR (Frischknecht, et al., 2017) also provides guidance on benchmarking and the definition of environmental performance classes for the individual PV technologies in scope. Any footprinting exercise which aims to benchmark an individual technology's environmental impacts per kWh of produced DC electricity must take the geographic location and mounting situation into consideration. It is therefore very difficult to aggregate results into a single metric or label. The importance of these determining factors is illustrated in the PEFCR by comparing a vertically mounted PV system in Sweden (annual yield of 578 kWh/kWp) to an optimally oriented system in Spain (annual yield of 1317 kWh/kWp).

Impact category	Unit	CdTe PV panels			CIS PV panels			micro-Si PV panels			multi-Si PV panels			mono-Si PV panels		
		Minimum	Average	Maximum	Minimum	Average	Maximum	Minimum	Average	Maximum	Minimum	Average	Maximum	Minimum	Average	Maximum
Climate change	kg CO_2 eq.	1.87E-02	2.52E-02	4.75E-02	3.08E-02	4.16E-02	7.84E-02	4.23E-02	5.71E-02	1.08E-01	4.17E-02	5.63E-02	1.06E-01	6.73E-02	9.09E-02	1.71E-01
Ozone depletion	kg CFC-11 eq.	7.47E-10	1.01E-09	1.90E-09	1.34E-09	1.81E-09	3.41E-09	9.09E-10	1.23E-09	2.31E-09	1.33E-09	1.79E-09	3.38E-09	1.80E-09	2.43E-09	4.58E-09
Human toxicity, cancer effects	CTUh, c	6.24E-10	8.42E-10	1.59E-09	9.64E-10	1.30E-09	2.45E-09	1.08E-09	1.46E-09	2.75E-09	9.37E-10	1.27E-09	2.38E-09	1.05E-09	1.42E-09	2.68E-09
Human toxicity, non-cancer effects	CTUh, n-c	3.99E-09	5.39E-09	1.02E-08	6.80E-09	9.17E-09	1.73E-08	7.33E-09	9.90E-09	1.87E-08	7.85E-09	1.06E-08	2.00E-08	1.00E-08	1.36E-08	2.56E-08
Particulate matter	kg PM2.5 eq.	9.92E-06	1.34E-05	2.52E-05	1.36E-05	1.83E-05	3.46E-05	4.40E-05	5.93E-05	1.12E-04	5.07E-05	6.84E-05	1.29E-04	9.82E-05	1.33E-04	2.50E-04
Ionizing radiation, human health	kBq U^{235} eq.	9.44E-04	1.27E-03	2.40E-03	2.62E-03	3.53E-03	6.66E-03	2.02E-03	2.72E-03	5.13E-03	2.16E-03	2.92E-03	5.50E-03	2.83E-03	3.82E-03	7.19E-03
Photochemical ozone formation	kg NMVOC eq.	6.96E-05	9.40E-05	1.77E-04	9.01E-05	1.22E-04	2.29E-04	1.57E-04	2.11E-04	3.99E-04	1.46E-04	1.97E-04	3.72E-04	2.43E-04	3.27E-04	6.17E-04
Acidification	mol H^+ eq.	1.41E-04	1.91E-04	3.60E-04	1.74E-04	2.35E-04	4.43E-04	3.42E-04	4.61E-04	8.69E-04	3.56E-04	4.81E-04	9.07E-04	6.05E-04	8.17E-04	1.54E-03
Terrestrial eutrophication	mol N eq.	2.61E-04	3.53E-04	6.65E-04	3.30E-04	4.46E-04	8.41E-04	5.70E-04	7.69E-04	1.45E-03	5.22E-04	7.05E-04	1.33E-03	8.56E-04	1.15E-03	2.18E-03
Freshwater eutrophication	kg P eq.	3.42E-06	4.61E-06	8.70E-06	4.94E-06	6.67E-06	1.26E-05	4.52E-06	6.10E-06	1.15E-05	5.28E-06	7.13E-06	1.34E-05	5.76E-06	7.78E-06	1.47E-05
Marine eutrophication	kg N eq.	2.31E-05	3.12E-05	5.88E-05	2.86E-05	3.85E-05	7.26E-05	5.07E-05	6.84E-05	1.29E-04	4.78E-05	6.46E-05	1.22E-04	8.49E-05	1.15E-04	2.16E-04
Freshwater ecotoxicity	CTUe	5.99E-02	8.09E-02	1.52E-01	6.87E-02	9.27E-02	1.75E-01	7.08E-02	9.56E-02	1.80E-01	7.47E-02	1.01E-01	1.90E-01	7.81E-02	1.05E-01	1.99E-01
Land use	kg C deficit	2.15E-02	2.90E-02	5.46E-02	4.23E-02	5.71E-02	1.08E-01	3.78E-02	5.10E-02	9.62E-02	4.99E-02	6.73E-02	1.27E-01	6.28E-02	8.47E-02	1.60E-01
Water resource depletion	m^3 water eq.	1.59E-05	2.14E-05	4.04E-05	3.24E-05	4.37E-05	8.24E-05	6.47E-05	8.73E-05	1.64E-04	4.69E-05	6.33E-05	1.19E-04	7.95E-05	1.07E-04	2.02E-04
Mineral, fossil & renewable resource depletion	kg Sb eq.	5.25E-06	7.09E-06	1.34E-05	3.79E-04	5.12E-04	9.65E-04	1.12E-06	1.51E-06	2.85E-06	1.43E-05	1.93E-05	3.64E-05	9.32E-06	1.26E-05	2.37E-05
Single Score	*Pt*	*6.64E-06*	*8.96E-06*	*1.69E-05*	*2.55E-04*	*3.44E-04*	*6.49E-04*	*6.86E-06*	*9.26E-06*	*1.75E-05*	*1.55E-05*	*2.10E-05*	*3.95E-05*	*1.47E-05*	*1.98E-05*	*3.74E-05*

Table 3: Average maximum and minimum environmental impacts of 1 kWh DC electricity produced with mounted mono-Si, multi-Si, a-Si / micro-Si, CdTe and CIS/CIGS PV modules (Frischknecht, et al., 2017)

The values for CIS/CIGS PV modules, portrayed in Table 3, still need to be adjusted as they are based on outdated indium reserves data. In order to enable a more accurate comparison, the methodology will need to be corrected to exclude indium (PVthin; treeze, 2016).

The PEF screening study (Stolz, Frischknecht, Wyss, & de Wild-Scholten, 2016) provides useful interpretation guidance for the LCA practitioner. To support the identification of hot spots in the life cycle, the screening study identifies the most relevant environmental impact categories for the five PV technologies. Based on the weighted results of the screening study, the most relevant impact categories for all PV technologies include:

- Mineral and fossil resource depletion;
- Human toxicity, cancer effects;
- Human toxicity, non-cancer effects;
- Freshwater ecotoxicity;
- Particulate matter potential;
- Acidification potential.

The first three impact categories are largely dominated by the supply chains of semiconductor materials, silver, copper and zinc, which form the hot spots in the production and installation phase. The freshwater ecotoxicity hot spot is identical for all of the analyzed PV technologies and is attributed to the supply chain of electricity for manufacturing which involves the incineration of disposed plastic components of the electrical installation, the disposal of redmud from bauxite digestion, and antimony and chromium VI emissions caused by the extraction of natural gas and beneficiation of iron ore.

[3] The screening study is provided as supplementary material.

The particulate matter and acidification impact categories relate to highly energy intensive production processes such as solar grade silicon and glass manufacturing which rely on electricity produced from hard coal-fired power plants.

3.4 SCREENING STUDY

The PEF screening study is based on life cycle inventory data from ecoinvent v2.2+ (KBOB et.al., 2014) and applies the main parameters of the PEFCR. The five various PV technologies were screened using three standardized mounting types (in roof, on roof, open ground) and two different system scales (3kWp residential-scale and 570 kWp large-scale). Based on these results, the screening study established the baseline environmental profile of the representative product using the prescribed PEF methodology (European Commission, 2013a) and applying the second draft PEFCR. The assessment of the virtual representative product – per 1 kWh of DC electricity (unit of analysis) - serves as the reference for the life cycle impacts of 1 kWp of photovoltaic module power output under standard test conditions..

In addition to the framework parameters depicted in section 3.3, the screening study also evaluated different allocation approaches for assessing the benefits of recycling through a sensitivity analysis. Minimum data quality requirements were included in the screening so that only data with a 'fair or better' quality rating is accepted. All input data for the screening study received a good or higher quality rating. For the environmental impact assessment, the 15 default impact categories as stipulated by the PEF Guide (European Commission, 2013a) have been complemented by three additional impact categories (renewable and non-renewable cumulative energy demand, nuclear waste), which are deemed important for PV electricity. Normalization factors from (Benini, et al., 2014) were used to determine the environmental relevance of the different environmental impact categories.

Figure 6 provides an example of the results of the screening study. The production of the PV modules and the installation and mounting were found to be most important for the majority of the impact categories. The most important impact categories were identified based on the normalized and equally-weighted results of the five PV technologies considered.

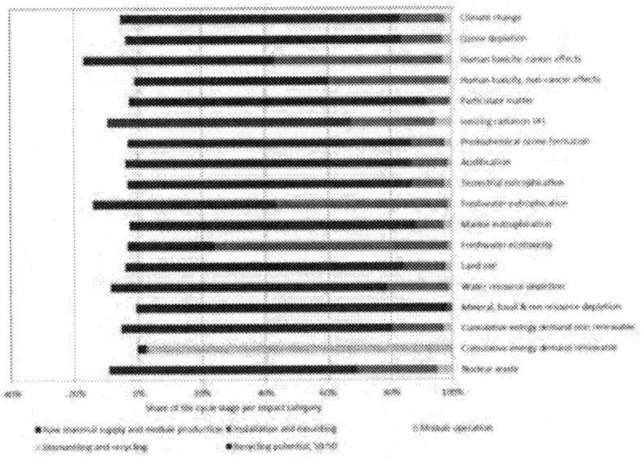

Figure 6 – Environmental impact results (characterized, indexed to 100 %) of 1 kWh of DC electricity produced with a residential scale (3 kWp) PV system with average PV panels mounted on a slanted roof. The potential benefits due to recycling are illustrated relative to the overall environmental impacts from production to end-of-life. (Stolz, Frischknecht, Wyss, & de Wild-Scholten, 2016)

Following the completion of the screening study, industry members of the TS conducted supporting studies on individual photovoltaic module products to complement and benchmark the results of the average representative product. The results and findings of the supporting studies are portrayed in section 3.4. The findings and recommendations from the scientific reviews of the screening study, supporting studies and the draft PEFCR are depicted in section 3.5.

3.4 SUPPORTING STUDIES

The European Commission requires each PEF pilot to carry out at least three supporting studies to ensure the application of the PEFCR to all sub-categories (i.e. PV technologies) of the representative product. The supporting studies were carried out by the industry members of the TS. Each supporting study represented one of the

This is a pre-peer review version of the paper submitted to the journal Progress in Photovoltaics.
It has been selected by the Executive Committee of the EU PVSEC 2017 for submssion to Progress in Photovoltaics.

technologies that were used to compose the average representative product in the screening study, with the exception of the amorphous/micro-crystalline silicon thin-film technology. Due to the decline of amorphous/micro-crystalline silicon thin film PV in the EU market and the lack of commercial manufacturers, this technology was not subject to a supporting study. Table 2 below presents an overview of the products evaluated in the supporting studies.

PV module technology	Product	Organization carrying out the supporting study
Multi-crystalline silicon	REC TwinPeak Series Solar Module	REC Group
Mono-crystalline silicon	SunPower E-Series E20-327 module	TOTAL
CdTe thin film	First Solar Series 4 thin-film module	First Solar
CIGS thin film	Solibro SL2 thin-film module	Solibro

Table 4 – Supporting studies carried out in the context of the PEF Pilot project "PV Electricity Generation"

The objective of all supporting studies was to test the pertinence and practical feasibility of the draft PEFCR, focusing on the following elements:

- Data collection and data quality;
- Identification of the most relevant impact categories (hotspot analysis);
- Validation of normalization and weighting approaches;
- Contribution towards identification of performance classes; and
- Verification requirements.

The supporting studies were submitted to the European Commission for validation. Non-confidential versions of each study are available in the environmental footprint wiki of the pilot project (EU Commission, 2017b).

An important aspect of each supporting study was the validation of the (company- and product-specific) life cycle inventory data in accordance with the data quality requirements depicted in the PEF guidelines (European Commission, 2013a). Table 3 below depicts the minimum data quality required for the supporting study's inventory data.

Data quality requirements for PEF studies	Minimum data quality required	Type of required data quality assessment
Data covering at least 70 % of contributions to each environmental footprint impact category	*Overall "Good" data quality*	*Semi-quantitative*
Data accounting for 20-30 % of contributions to each environmental footprint impact category	*Overall "Fair" data quality*	*Qualitative expert judgement. No quantification required.*
Data used for approximation and filling identified gaps (smaller or equal 10 % of the contribution to each environmental footprint impact category)	*Best available data*	*Qualitative expert judgement. No quantification required.*

Table 5 – Data quality requirements as depicted in (European Commission, 2013a)

Since the accuracy and quality of the input data and adherence to the respective

This is a pre-peer review version of the paper submitted to the journal Progress in Photovoltaics.
It has been selected by the Executive Committee of the EU PVSEC 2017 for submssion to Progress in Photovoltaics.

PEFCR are of paramount importance for a representative and comparable environmental footprint calculation, the European Commission required an additional independent verification step for two out of four supporting studies (mono-crystalline silicon, CdTe thin film). The independent audit concluded that both supporting studies complied with the requirements of the environmental footprint methodology (Ernst & Young et Associés, 2017).

3.5 SCIENTIFIC REVIEWS

Three scientific reviews were carried out during the pilot process, as depicted in Figure 3. The Technical Secretariat appointed a Scientific Review Panel to ensure that the methodology and results presented in the draft PEFCR, screening study, and supporting studies adhere to the PEF requirements (European Commission, 2013a) and support the creation of credible and consistent product environmental footprint profiles. The review panel was tasked with ensuring the fulfillment of the following criteria:

- functional unit, allocation and calculation rules are adequate for the product category under consideration,
- primary and secondary datasets used in the screening study and the supporting studies are relevant, representative, and reliable;
- selected life cycle impact assessment (LCIA) indicators and additional environmental information are appropriate for the product category under consideration;
- the benchmark and performance classes are correctly defined or the lack of performance classes is appropriately justified; and
- both LCA-based data and the additional environmental information prescribed by the PEFCR give a description of the significant environmental aspects associated with the product.

The Scientific Review Panel was composed of Professor Edgar Hertwich (Yale University, USA), Professor Sangwon Suh (Bren School of Environmental Science and Management, UC Santa Barbara, USA) and Ugo Pretato (Studio Fieschi & soci, Italy). In addition to the scientific panel reviews, the European Commission also contracted LCA scientists for a separate independent review.

3.6 TESTING OF COMMUNICATION VEHICLES

The objectives of the environmental footprint pilot were to test the process of developing product specific rules (PEFCRs) and evaluate different verification approaches (scientific review, auditing and certification) and communication vehicles (CVs) for communicating life cycle environmental performance results to business partners, consumers and other company stakeholders (DG ENVIRONMENT, 2017). The Technical Secretariat tested a variety of communication vehicles on selected stakeholder groups and gathered feedback on their effectiveness.–Table 4 below lists the different communication vehicles proposed by the European Commission and highlights (in yellow) the CVs selected and tested by the Technical Secretariat.

Communication Vehicles (CVs)		
Performance label	Loyalty schemes	Campaigns targeting user behaviour
Performance label + QR code	Leaflets, catalogues, etc.	Marketing campaigns/ advertising
Performance improvement label	Instruction manuals	Declaration/ product passport
Barcode	Websites (producer, vendor)	On invoices
Pictogram, POS product advertisement	Websites (3rd party)	PEF external communication report
Consumer receipt – basket comparison	App based on performance	Performance tracking report.

Table 6 - Communciation vehicles suggested for testing in the context of the environmental footprint pilot

This is a pre-peer review version of the paper submitted to the journal Progress in Photovoltaics.
It has been selected by the Executive Committee of the EU PVSEC 2017 for submssion to Progress in Photovoltaics.

phase (Open Evidence et.al., 2016) and (DG ENVIRONMENT, 2016)

The communication of the environmental footprint of electricity generated by a PV system requires an understanding of the system boundaries, the life cycle of the system components, but also – and more importantly – an understanding of the impact of the system installation (geographic location, insulation conditions, tilt, environmental conditions etc.) to evaluate and compare the results in a meaningful way. Therefore, the Technical Secretariat decided to test the CVs using different levels of data aggregation – including the highest aggregation level ('Performance Label') and the most detailed and least aggregated presentation of results ('PEF external communication report'/'Performance tracking report').

To determine whether the targeted stakeholders would be able to understand the environmental performance of one PV system component (the PV panel) based on a fixed set of impact categories (hot spots), the Technical Secretariat invited technical experts and marketing/ communication specialists from selected customer groups to review the different CVs of the industry members.

The results of the different communication tests were recorded in logbooks and have been submitted to the European Commission for further evaluation. A preliminary analysis of all communication tests is available in (LSE & Partners Consortium, 2017).

4 RESULTS AND DISCUSSION

The application of the European Commission's environmental footprint guidance (European Commission, 2013a) to photovoltaic modules provides a commonly accepted methodology for evaluating the life cycle environmental footprint of photovoltaic electricity generation in the European Union. By incorporating societal, scientific, regulatory and industry perspectives while assessing the practical application of the Product Environmental Footprint Category Rules in the screening and supporting studies, the consensus-driven PEFCR development process serves as a leading example of how to establish a foundation for LCA-based environmental regulations and policies for photovoltaic modules in the European Union and abroad.

The PEFCR for PV modules has proven useful in guiding the development of environmental footprint studies by clearly identifying the foreground unit processes which require product specific data versus the background processes which are based on pre-specified data. This delineation limited the scope of work and promoted the comparability of the supporting study's results by restricting the potential for methodological variability. The PEFCR's clearly defined assumptions for the average annual energy yield and lifetime of PV systems (975 kWh/kWp and 30 years respectively) further ensured the comparability of the study results. The PEF screening study of PV modules (Stolz, Frischknecht, Wyss, & de Wild-Scholten, 2016) provided a clear template for documentation and interpretation of the PEF supporting study results and can serve as a basis for future PEF studies. The provision of the background (Ecoinvent 2.2+) database by the authors of the PEFCR also facilitated the reproducibility and comparability of the supporting study.

In addition to providing a framework and basis for future PEF studies, the supporting study and screening studies helped to identify and verify environmental impact categories and hot spots in the life cycle of the PV panel. Developing a better understanding that the impact categories and hot spots are similar in all investigated PV technologies will not only help to inform future decisions on product development, but will also help to shape regulatory measures in the field of ecodesign and eco-efficiency.

Even though the PEFCR provides concrete guidance on how to carry out a PEF-compliant life cycle assessment of PV technologies, the requirements of the Product Environmental Footprint Guide (European Commission, 2013a) and the Pilot Guidance (European Commission, 2016) remain challenging. For instance, the data quality requirements are extensive and demand a substantial effort on behalf of the companies. On the other hand, these requirements offer an opportunity for PV module manufacturers to take their material and energy flows into account in a detailed manner. Another example of the PEF's methodological complexity is the circular footprint formula, which is required to calculate the potentially avoided burdens from recycling. The formula consists of many different variables and terms, rendering its application and implementation in the data models challenging.

The pilot process also helped to identify shortcomings relating to the methodology and data availability and provided suggestions for improvement. For instance, the screening study revealed the dominance of the indium resource in the normalized and weighted results of one specific PV technology (CIS/CIGS). However, these impacts do not reflect this technology's real environmental performance as they are based on outdated indium reserves data and an inconsistent geographical scope for the normalization factors in the ILCD impact assessment method (PVthin; treeze, 2016). We therefore strongly discourage a direct comparison of the environmental impacts of different PV technologies based on the normalized and weighted results until the indium anomaly has been eliminated and a proven and consistent set of assessment factors has been published.

Another challenge that arose during the PV PEF pilot was the modelling of the end-of-life stage. Despite the limited amount of end-of-life modules due to their average 30-year lifetime, PV recycling is already taking place today and life cycle inventories of the current generation recycling processes are readily available (Stolz, Frischknecht, Wambach, Sinha, & Heath, 2016). However, the Product Environmental Footprint Guide (European Commission, 2013a) stipulates that the modelling of the end-of-life stage must be based on current practice. Although the European waste electrical and electronic eqiupment (WEEE) directive requires the recycling of PV modules at their end of life (European Parliament/European Council, 2012), the recycling of waste modules has not yet been documented in EU statistics. As a result, PV modules are assumed to be disposed of in a landfill at end of life in the final PEFCR.

The path towards a single market for green products in the European Union following the conclusion of the PEF pilot phase in 2017 has not yet been communicated. Although the Product Environmental Footprint Category Rules represent a commonly accepted methodology for the life cycle environmental assessment of photovoltaic modules, several methodological and data-related shortcomings still need to be remedied. While the definition of default values for the most important parameters –annual yield, lifetime and degradation rate – ensures the comparability of results in the PEFCR, it is important to thoroughly evaluate the specific location and orientation of each PV system in order to maximize the effective annual yield. The careful selection of the installation site is generally one of the most efficient measures to minimize the environmental impacts of PV electricity. The impact of the location, inclination and orientation of a PV system on the technology's environmental performance can be assessed using the web service ENVI-PV (12, Tech, ADEME, & treeze, 2017) which is based on the life cycle inventories used in the PEF screening study (Pérez-López, et al., 2016). Once the installation site of a PV system has been selected, the PEFCR helps to characterize the environmental impacts of PV technologies based on a consistent and scientifically sound methodology.

5 CONCLUSION

The European Union aims to provide the right incentives for citizens and public authorities to choose the most resource-efficient products through appropriate price signals and clear environmental information (European Commission, 2011). The Single Market for Green Products initiative (EU Commission, 2013b) marks an important step in achieving that goal by establishing a harmonized life cycle assessment-based set of methodologies, which allow for the evaluation and comparison of products based on their life cycle resource use and environmental impacts. The Product Environmental Footprint pilot phase has demonstrated how these product-specific life cycle assessment methodologies can be established on a commonly agreed, harmonized reference as provided in the PEF Guide (European Commission, 2013a). As portrayed in this paper, the establishment of an inclusive multi-stakeholder process with multiple checkpoints to ensure scientific soundness (through scientific reviews), societal and political acceptance (through public stakeholder consultations, steering committee decision gates) and quality (through validation and re-modeling) of Product Environmental Footprint Category Rules is a pre-requisite and can serve as a model on the path towards a single market for green products in the EU.

6 ACKNOWLEDGMENT

The authors wish to acknowledge the support of the International Energy Agency Photovoltaic Power Systems Program Secretariat, which provided invaluable support in managing the Technical Secretariat throughout the multi-year project. Further acknowledgement is given to the members of the Technical Secretariat for providing the financial and human resources to successfully carry out the pilot project, as well as the Scientific Review Panel.

7 REFERENCES

Benini, L., Mancini, L., Sala, S., Manfredi, S., Schau, E., & Pant, R. (2014). *Normalisation method and data for Environmental Footprints.* Ispra, Italy: European Commission Joint Research Centre, Institute for Environment and Sustainability.

Breyer, C. D. (2017). On the Role of Solar Photovoltaics in Global Energy Transition Scenarios: On the Role of Solar Photovoltaics in Global Energy Transition Scenarios. *Progress in Photovoltaics: Research and Applications*. doi:10.1002/pip.2885

DG ENVIRONMENT. (2016). *Background Document for The Testing of Communication Vehicles in the Environmental Footprint Pilot Phase 2013-2016 ver. 1.1.* Brussels, Belgium: European Commission, DG Environment. Retrieved from http://ec.europa.eu/environment/eussd/smgp/pdf/Comm_bgdoc_v1.1.pdf

DG ENVIRONMENT. (2017, June 26). *Environmental Footprint pilot phase.* Retrieved from http://ec.europa.eu/environment/eussd/smgp/ef_pilots.htm

econinvent Centre. (2010). *ecoinvent data v2.2, ecoinvent reports No. 1-25.* Duebendorf, Switzerland: Swiss Centre for Life Cycle Inventories. Retrieved from www.econinvent.org

Ernst & Young et Associés. (2017). *Product and Organisation Environmental Footprint - Verification of embedded impacts and traceability as part of the Environmental Footprint methods implementation.* Brussels: EY et Associés. Retrieved from https://webgate.ec.europa.eu/fpfis/wikis/download/attachments/66622784/EY_PEF_Final%20Report_2017%2003%2024_Vfinal_public%20version.pdf?api=v2

EU Commission. (2013b). *Communication from the Commission to the European Parliament and the Council - Building the Single Market for Green Products - Facilitating better information on the environmental performance of products and organisations COM(2013) 196 final.* Brussels: European Commission.

EU Commission. (2017a, June 15). *DG Environment Product Environmental Footprint Pilot Phase.* Retrieved from http://ec.europa.eu/environment/eussd/smgp/pdf/Fiche_photovoltaic.pdf

EU Commission. (2017b, June 15). *Environmental Footprint Wiki.* Retrieved from https://webgate.ec.europa.eu/fpfis/wikis/spaces/viewspace.action?key=EUENVFP

EU Commission, DG ENV. (2013). *Building the Single Market for Green Products, presentation given at the American Chamber of Commerce to the EU Environment Committee.* Brussels: DG Environment.

European Commission. (2011). *Roadmap to a Resource Efficient Europe, COM(2011) 571 final.* Brussels, Belgium: European Commission.

This is a pre-peer review version of the paper submitted to the journal Progress in Photovoltaics.
It has been selected by the Executive Committee of the EU PVSEC 2017 for submssion to Progress in Photovoltaics.

European Commission. (2013a). *Annex II: Product Environmental Footprint (PEF) Guide to the Commission Recommendation on the use of common methods to measure and communicate the life cycle environmental performance of products and organisations.* Brussels: European Commission.

European Commission, D. E. (2016). *Product Environmental Footprint Pilot Guidance v5.2.* Brussels: European Commission, DG Environment.

European Commission, DG Environment. (2016). *Product Environmental Footprint Pilot Guidance, Guidance for the implementation of the EU Product Environmental Footprint (PEF) during the Environmental Footprint (EF) pilot phase, version 5.2.* Brussels. Retrieved from http://ec.europa.eu/environment/eussd/smgp/pdf/Guidance_products.pdf

European Parliament/European Council. (2012). Directive 2012/19/EU of the European Parliament and of the Council on waste electrical and electronic equipment (WEEE). *Official Journal of the European Union*, 38-71.

Finkbeiner, M. (2013). Product environmental footprint - breakthrough or breakdown for policy implementation of life cycle assessment? *Int J Life Cycle Assess.*

Frankl, P., Menichetti, E., & Raugei, M. (2005). *Final report on technical data, costs and life cycle inventories of PV applications (Deliverable no. 11.2 - RS Ia).* Ambiente Italia (NEEDS project consortium).

Fraunhofer ISE. (2013). *Photovoltaics report.* Freiburg, Germany: FHI-ISE.

Frischknecht, R., & Büsser Knöpfel, S. (2013). *Swiss Eco-Factors 2013 according to the Ecological Scarcity Method. Methodological Fundamentals and their application in Switzerland. Environmental Studies n. 1330.* Bern: Federal Office for the Environment.

Frischknecht, R., & Büsser Knöpfel, S. (2014). Ecological scarcity 2013 - new features and its application in industry and administration - 54th LCA forum. *Int J LCA, 19*(6), 1361-1366.

Frischknecht, R., Braunschweig, A., Hofstetter, P., & Suter, P. (2000). Modelling human health effects of radioactive releases in Life Cycle Impact Assessment. *Environmental Impact Assessment Review, 20*(2), 159-189.

Frischknecht, R., Steiner, R., & Jungbluth, N. (2008). *Methode der ökologischen Knappheit - Ökofaktoren 2006.* Bern: Bundesamt für Umwelt (BAFU).

Frischknecht, R., Stolz, P., Heath, G., Wade, A., Francke, L., Theologitis, I., . . . Tian, J. (2017). *Product Environmental Footprint Category Rules, Photovoltaic Modules used in Photovoltaic Power Systems for Electricity Generation (NACE/CPA class 27.90 "Manufacturing of other electronic equipment").* Brussels: Technical Secretariat PEF Pilot "PV Electricity Generation".

Frischknecht, R., Wyss, F., Buesser, S., Lützkendorf, T., & Baloutktsi, M. (2015). Cumulative energy demand in LCA: the energy harvested approach. *Int J LCA, 20*(7), 957-969.

Frischknecht, R.; et. al. (2016). *Methodology Guidelines on Life Cycle Assessment of Photovoltaic Electricity 3rd Edition.* International Energy Agency Photovoltaic Power Systems Programme, IEA-PVPS T12-08:2016.

Fthenakis, V., Frischknecht, R., Raugei, M., Kim, H., Alsema, E., Held, M., & de Wild-Scholten, M. (2012). *Methodology Guidelines for Life Cycle Assessment of Photovoltaic Electricity, 2nd edition.* International Energy Agency Photovoltaic Power Systems Program (IEA PVPS) Task 12.

Galatola, M., & Pant, R. (2014). Reply to the editorial "Product environmental footprint - breakthrough or breakdown for policy implementation of life cycle assessment?" written by Prof. Finkbeiner (Int J Life Cycle Assess 19(2):266-271). *Int J Life Cycle Assess.*

Greco, S. L., Wilson, A. M., Spengler, J. D., & Levy, J. I. (2007). Spatial patterns of mobile source particulate matter emissions-to-exposure relationships across the United States. *Atmospheric Environment*(41), 1011-1025.

Guinée, J. B., Gorrée, M., Heijungs, R., Huppes, G., Kleijn, R., de Koning, A., . . . Weidema, B. P. (2001). *Life cycle assessment; An operational guide to the ISO standards; Parts 1 and 2.* Den Haag and Leiden: Ministry of Housing, Spatial Planning and Environment (VROM) and Centre of Environmental Science (CML).

Hertwich, E., & et.al. (2016). *Green Energy Choices: The benefits, risks and trade-offs of low carbon technlogies for electricity production. Report of the International Resource Panel.* UNEP.

Humbert, S. (2009). *Geographically Differentiated Life-cycle Impact Assessment of Human Health. PhD Thesis.* Berkeley, California, USA: University of California.

IEA PVPS Task 12, SolarPower Europe (EPIA), International Thin-Film Solar Industry Association (PVthin), Yingli Solar, First Solar, Total, Calyxo, Energy Research Centre of the Netherlands (ECN Solar), Treeze, WWF, BPVA, REC Solar. (2013). *Product Environmental Footprint - On line application form for the Environmental Footprint (EF) pilot phase 2013-2016.* Brussels.

Ingwersen, W., & Subramanian, V. (. (2013). *Guidance for Product Category Rule Development, Version 1.0.* Product Category Rule Guidance Development Initiative. Retrieved from http://www.pcrguidance.org

IPCC. (2007). *The IPCC fourth Assessment Report - Technical Summary.* Cambridge: Cambridge University Press.

ISO. (2016). *ISO 14044:2006 Environmental management -- Life cycle assessment -- Requirements and guidelines.* Geneva: International Organization for Standardization.

JRC/ies. (2010). *ILCD handbook: Framework and requirements for Life Cycle Impact Assessment models and indicators.*

JRC/ies. (2010a). *ILCD handbook: Analysis of existing Environmental Impact Assessment methodologies for use in Life Cycle Assessment.*

JRC/ies. (2011). *ILCD handbook: Recommendation for Life Cycle Impact Assessment in the European context - based on existing environmental impact assessment models and factors.*

KBOB et.al. (2014). *ecoinvent Datenbestand v2.2+; Grundlage für die KBOB-Empfehlung 2009/1:2014: Ökobilanzdaten im Baubereich, Stand April 2014.* Bern, Switzerland: Koordinationskonferenz der Bau- und Liegenschaftsorgane der öffentlichen Bauherren c/o BBL Bundesamt für Bauten und Logistik.

London School of Economics and Political Science (LSE) et.al. (n.d.).

LSE & Partners Consortium. (2017, March 21). *Assessment of different communication vehicles for providing Environmental Footprint information - Communication testing findings.* Retrieved from Environmental Footprint Wiki: https://webgate.ec.europa.eu/fpfis/wikis/download/attachments/66622784/Env_Footprints_20170321_Tab_slides_final.pdf?api=v2

Milá i Canals, L., Romanyà, J., & Cowell, S. J. (2007). Method for assessing impacts on life support functions (LSF) related to the use of 'fertile land' in Life Cycle Assessment (LCA). *Journal of Cleaner Production, 15*(15), 1426-1440.

Open Evidence et.al. (2016). *Assessment of different communication vehicles for providing Environmental Footprint Information.* Barcelona, Spain: Open Evidence/Universitat Oberta de Catalunya. Retrieved from http://www.open-evidence.com/project/assessment-of-different-communication-vehicles-for-providing-environmental-footprint-information/

Payet, J., Blanc, I., Durand, Y., Adra, N., Evon, B., Sié, M., . . . Guermont, C. (2012). *Référential d'évalutaion des Impacts Environmentaux des systèmes Photovoltaiques par la méthode d'analyse du cycle de vie.* Paris: ADMEME, CYCLeco.

PCR Electricity, gas and water distribution (on own account). (2017, June 19). Retrieved from http://environdec.com/en/PCR/Detail/?Pcr=5904

PCR Electricity, steam and hot/cold water generation and distribution. (2017, June 19). Retrieved from http://www.environdec.com/PCR/Detail/?Pcr=5802

PCR Electricity, town gas steam and hot water. (2017, June 19). Retrieved from http://www.environdec.com/en/PCR/Detail/?Pcr=7059

PEF "PV Electricity Generation" - Stakeholder Workspace. (2017, June 19). Retrieved from https://webgate.ec.europa.eu/fpfis/wikis/display/EUENVFP/Stakeholder+workspace:+PEFCR+pilot+Photovoltaic+electricity+generation

Pérez-López, P., Gschwind, B., Blanc, I., Frischknecht, R., Stolz, P., Durand, Y., . . . Blanc, P. (2016). ENVI-PV: an interactive Web Client for multi-criteria life cycle assessment of photovoltaic systems worldwide. *Progress in Photovoltaics: Research and Applications,* 1-15. doi:10.1002/pip.2841

Posch, M., Seppälä, J., Hettelingh, J. P., Johansson, M., Margni, M., & Jolliet, O. (2008). The role of atmospheric dispersion models and ecosystem sensitivity in the determination of characterisation factors for acidifying and eutrophying emissions in LCIA. *Int J LCA*, 477-486.

PVPS, I. (2017, June 14). *International Energy Agency Photovoltaic Power Systems Program.* Retrieved from Who we are: http://www.iea-pvps.org/index.php?id=4

PVthin; treeze. (2016). *Addressing the indium anomaly in the PEFCR for PV electricity generation.* Brussels: The International Thin Film Solar Industry Association (PVthin) / treeze.

Rabl, A., & Spadaro, J. V. (2004). *The RiskPoll software.* retrieved from www.arirabl.com.

Rosenbaum, R. K., Bachmann, T. M., Gold, L. S., Huijbregts, A. J., Jolliet, O., Juraske, R., . . . Hauschild, M. Z. (2008). USEtox - the UNEP-SETAC toxicity model: recommended characterisation factors for human toxicity and freshwater ecotoxicity in life cycle assessment. *International Journal of Life Cycle Assessment, 13*(7), 532-546.

Schmela, M., Döring, T., Gomez, A. P.-B., Arnaud, C., Concas, G., Dunlop, S., . . . Li, M. (2017). *Global Market Outlook For Solar Power 2017-2021.* Brussels, Belgium: SolarPower Europe, Global Solar Council.

Seitz, M., Kroban, M., Pitschke, T., & Kreibe, S. (2013). *Ökoeffizienzanalyse von Photovoltaikmodulen (bifa-Text Nr. 62).* Augsburg, Germany: bifa Umweltinstitut.

Seppälä, J., Posch, M., Johansson, M., & Hettelingh, J. P. (2006). Country-dependent Characterization Factors for Acidification and Terrestrial Eutrophication Based on Accumulated Exceedance as an Impact Category Indicator. *Int J LCA*, 403-416.

Stolz, P., Frischknecht, R., Wambach, K., Sinha, P., & Heath, G. (2016). *Life Cycle Assessment of Photovoltaic Module Recycling.* Uster, CH: treeze Ltd.

Stolz, P., Frischknecht, R., Wyss, F., & de Wild-Scholten, M. (2016). *PEF screening report of electricity from photovoltaic panels in the context of the EU Product Environmental Footprint Category Rules (PEFCR) Pilots v.2.0.* Uster, Switzerland: treeze Ltd. and SmartGreenScans.

Struijs, J., Beusen, A., van Jaarsveld, H., & Huijbregts, M. A. (2008). *Aquatic Eutrophication.* in ReCiPe 2008 A life cycle impact assessment method which comprises harmonised category indicators at the midpoint and the endpoint level. Report I: Characterisation factors.

Van Zelm, R., Huijbregts, M. A., Den Hollander, H. A., Van Jaarsveld, H. A., Sauter, F. J., Stuijs, J., . . . Van de Meent, D. (2008). European characterization factors for human health damage of PM10 and ozone in life cycle impact assessment. *Atmos Environ*, 441-453.

WMO. (1998). *Scientific Assessment of Ozone Depletion.* Geneva: World Meteorological Organisation.

EXTENDED QUALIFICATION TESTING OF 1-CELL CRYSTALLINE SI PV LAMINATES: IMPACTS OF ADVANCED CELL METALLIZATION AND ENCAPSULATION SCHEMES

Jonathan Govaerts[1], Bert Geyer[1,2], Arvid van der Heide[1], Tom Borgers[1], Stefan Hellstrom[3], Bert Broeders[3], Eszter Voroshazi[1], Jozef Szlufcik[1] and Jef Poortmans[1,4]
[1]Imec, Kapeldreef 75, 3001 Leuven, Belgium
[2]Now at ZOEK GmbH
[3]Borealis
[4]Also part-time professor at KULeuven and UHasselt

ABSTRACT: In the presented work we investigate the impact of extended qualification testing on 1-cell laminates with "standard" and "advanced" encapsulation and cell metallization schemes. We evaluate both thermal cycling qualification tests and (extended) damp heat reliability, by determining their impact on the opto-electrical performance of the laminates.
While the thermal cycling testing does not result in a significant degradation of any of the sample types, the result is different for the damp heat reliability testing. Starting from the "standard" buildup using full Al-BSF cells with screenprinted Ag paste metallization and an ethylene-vinyl-acetate (EVA) encapsulant, significant corrosion of the finger metallization is detected after 5000 hours and confirmed through current-voltage (IV), electroluminescence (EL) and series resistance (Rs) measurements. We show that this corrosion mechanism can be addressed in two ways. The first is by avoiding the acetic acid formation that is linked to EVA. This can be achieved through switching to a polyolefin (PO) encapsulant. The second is by making the metallization more corrosion-resistant. We show how this can be achieved through the use of a (densely) plated metallization (as opposed to the more porous screenprinted Ag paste).
Keywords: reliability testing, metallization, encapsulants

1 BACKGROUND AND SAMPLE BUILDUP

To drive down the levelized cost of electricity (LCOE) generated by PV modules [1], one of the important aspects next to increasing (peak) performance is prolonging the operational lifetime of the module, which is closely linked to its degradation behaviour. This latter depends on material selection, production process and of course the installation environment. The first hurdle to tackle for PV technologies in this direction is qualification (e.g. IEC certification) before degradation rate assessment. As shown in Figure 1 the largest failure rates in qualification are typically attributed to thermal cycling and damp heat tests.

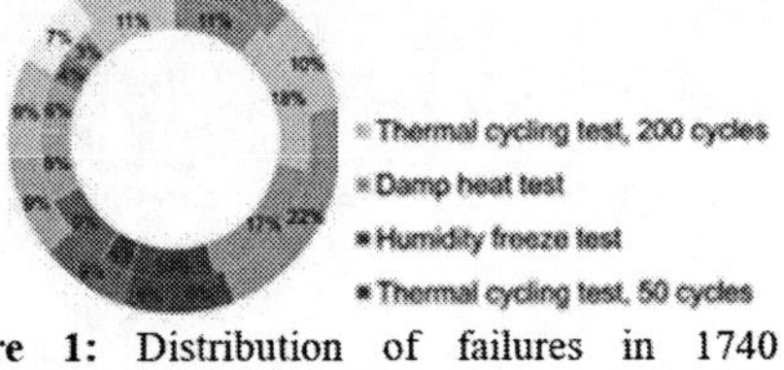

Figure 1: Distribution of failures in 1740 IEC certification tests (after 2000) for crystalline silicon PV modules (inner ring) and thin-film PV modules (outer ring) [2]

The thermal cycling test sequence is designed to detect both adhesion problems between the different layers in the interconnection (contacts) as well as chemical degradation of plastic layers. The damp heat tests on the other hand are mainly intended to look for corrosion issues in the different metallic components and decomposition and delamination issues of encapsulants and backsheets. Currently, improving damp heat reliability is typically tackled by keeping out moisture, e.g. glass-glass and encapsulants with reduced moisture uptake/transmission [3]. As a 2-way mechanism, glass-glass similarly does not allow an easy evacuation of moisture either (though this is potentially only problematic in chamber testing and not relevant in realistic conditions). Consequently, it can result in very different degradation mechanisms [4]. Of course, the presence of a sealant will also have a significant impact on such an effect.

One of the dominating degradation mechanism for the standard module buildup with front glass and plastic backsheet (GB) is the corrosion of the cell metallization triggered by acetic acid formed in the EVA in presence of humidity [2]. To mitigate this process, other options include switching encapsulants to avoid acetic acid formation in presence of humidity and/or adapting the cell metallization to reduce corrosion. This paper reports on experimental results to verify the viability of these different approaches.

In the presented work we investigate the impact of these critical tests on 1-cell laminates with "standard" and "advanced" encapsulation and cell metallization schemes (cf. below). We evaluate both thermal cycling qualification tests and (extended) damp heat reliability, by determining their impact on the opto-electrical performance of the laminates.

Figure 2 shows the schematical buildup of the 1-cell laminates used in the reliability tests and the variations that were investigated together with the used abbreviations. We compared industrial p-type Al BSF (STD) with screenprinted Ag metallization and n-type PERT cells (PERX) with plated metallization [5] with both EVA and PO encapsulants in glass-backsheet (GB) buildup. More details of the cell architecture and metallization layers is given in the respective sections. All cells were soldered with industrial-like equipment and standard SnPbAg-coated Cu tabbing and bussing ribbons, and lamination of the samples was done on an industrial tool. For each sample type 2-5 identical samples were submitted to reliability testing, next to 2 additional identical references were fabricated. These are obviously not exposed to the testing conditions but are simultaneously monitored with the tested samples. All samples before reliability testing have been submitted to 15 kWh/m^2 light soaking to eliminate any effect of light induced degradation. All (4-point) IV-measurements were carried out using a square mask leaving 2 mm spacing around the semisquare cells.

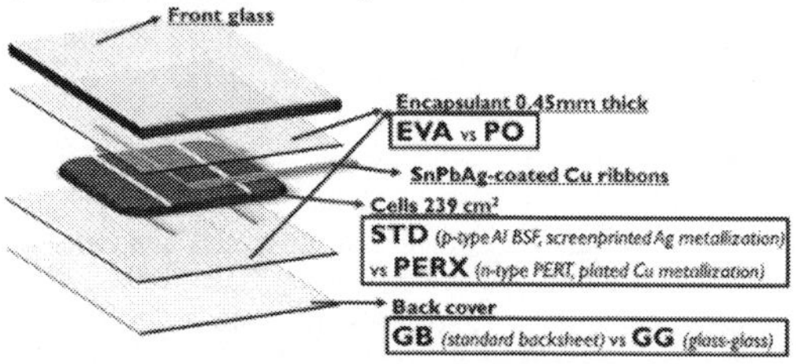

Figure 2: Schematical buildup of the samples under test and variations of the different materials used

2 THERMAL CYCLING

In the first set of tests, we submitted STD-EVA-GB, STD-PO-GB, PERX-EVA-GB and PERX-PO-GB samples to a combined qualification tests based on the standard testing blocks specified in [6]. We applied a sequence of damp heat (DH, 1000 h), light soaking (3xIEC UV, 45 kWh/m2) and thermal cycling (1xIEC TC, 200 cycles). In this test sequence we observed no significant degradation for either type of sample, as illustrated in Figure 3. Using certified and industrial materials, and processes it is not surprising that no delamination, between the layers or chemical change was visually detected and also no power loss was recorded.

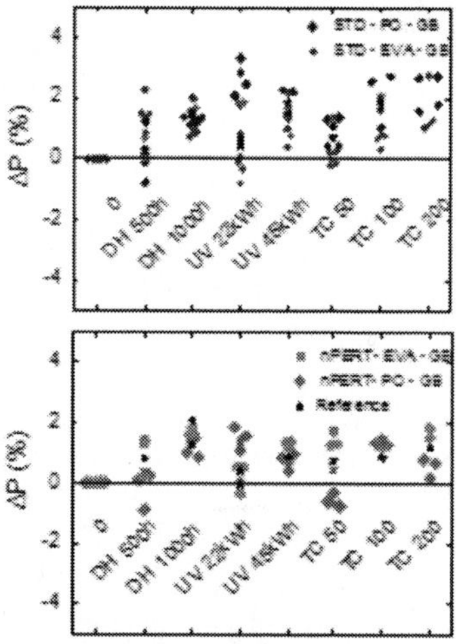

Figure 3: Evolution in performance throughout the combined DH-UV-TC test of STD-EVA-GB and STD-PO-GB samples (top), and PERX-EVA-GB and PERX-PO-GB samples (bottom)

3 DAMP HEAT

Considering the successful pass of the above combined test with short DH, UV and TC, we below focus more on the behaviour of identical 1-cell laminates in damp heat conditions, defined as 85%RH at 85°C [6], where we observe considerable differences depending on the used materials.

3.1 Type 1: STD-EVA-GB

In the first set of samples we consider STD cells, that have a buildup as shown schematically in Figure 4. Particularly important to note is the use of screenprinted Ag paste for the front metallization.

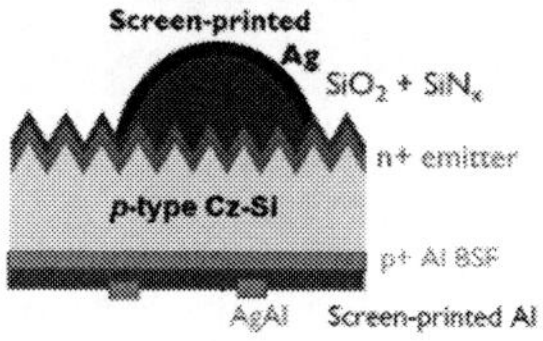

Figure 4: Schematic cross-section of the used STD cells

As indicated in Figure 5, a clear drop in power of ~5%rel can be observed after 5000 hours of exposure linked solely to reduction of the FF. We also note that at 4000 h a significant amount of cracking and delamination occurs in the backsheet, probably accelerating humidity ingress.

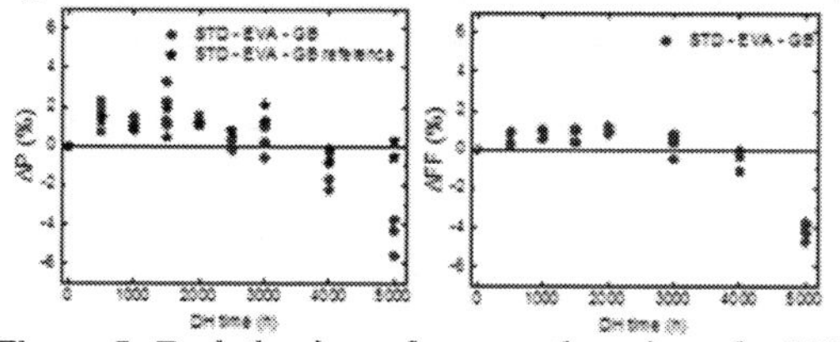

Figure 5: Evolution in performance throughout the DH test of STD-EVA-GB samples

After closer optical inspection and electroluminescence (EL) measurements, a clear degradation footprint is (repeatedly) observed. This pattern is shown in Figure 6 and confirms an increased series resistance (Rs) as determined from the IV-measurements at different illumination intensities [7,8], and points towards corrosion of the finger metallization around the busbars. Also corrosion of the tabbing and bussing ribbons is observed.

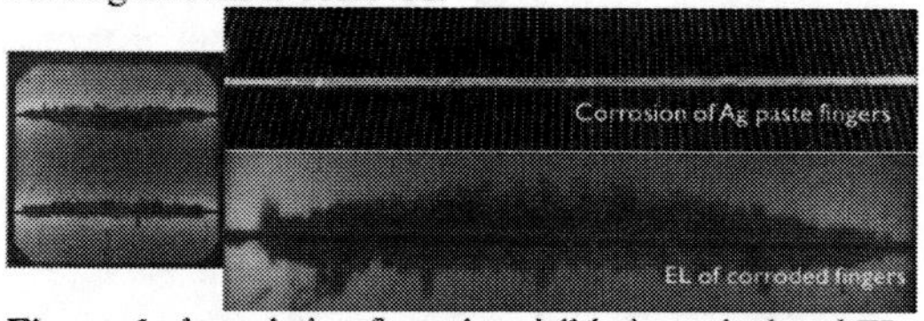

Figure 6: degradation footprint visible in optical and EL imaging: corrosion of the finger metallization around the busbars

This phenomenon is also described in [9], although the root cause of this pattern is not explained. A plausible hypothesis could be that moisture preferentially penetrates along the ribbon-encapsulant interface and Ag paste (glass frit). There it will interact with acetic acid (EVA) which then results in corrosion of the finger metallization [10,11]. But further analysis is still required (and in progress) to confirm this hypothesis.

3.2 Type 2: STD-PO-GB

Alternatively, to avoid/reduce the degradation (mechanism) assumed in the type 1 samples (STD-EVA-GB), using another encapsulant type could be beneficial, both by reducing the moisture transmission, and by avoiding the acetic acid formation present in EVA encapsulants. For these tests, we used a new polyolefine encapsulant from Borealis, QuentysTM BPO882F. No noticeable power loss after 5000h DH testing is recorded, also the FF remains nearly unchanged as shown in Figure 7. In agreement the EL image (Figure 8) of the samples indicates no resistance loss, and probably no corrosion.

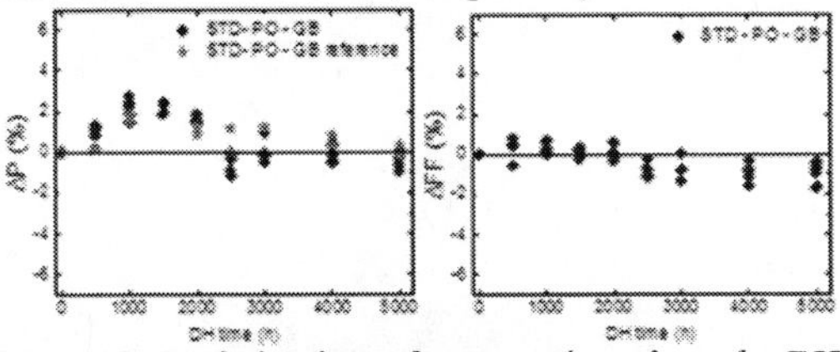

Figure 7: Evolution in performance throughout the DH test of STD-PO-GB samples

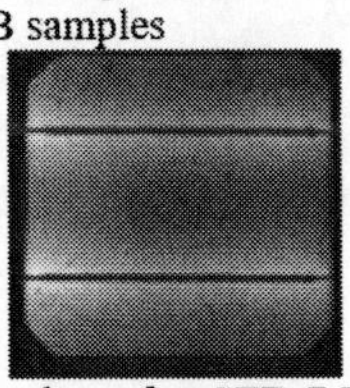

Figure 8: EL mapping of a STD-PO-GB sample after 5000 hours DH exposure (right)

3.3 Type 3: PERX-EVA-GB

Another option to address the degradation (mechanism) assumed in the type 1 samples (STD-EVA-GB), is to use a cell metallization that is more resilient to corrosion. To this end we also submitted PERX samples as in Figure 9, with a plated (NiSi seed/Cu/immersion Ag) finger metallization instead of screenprinted Ag paste.

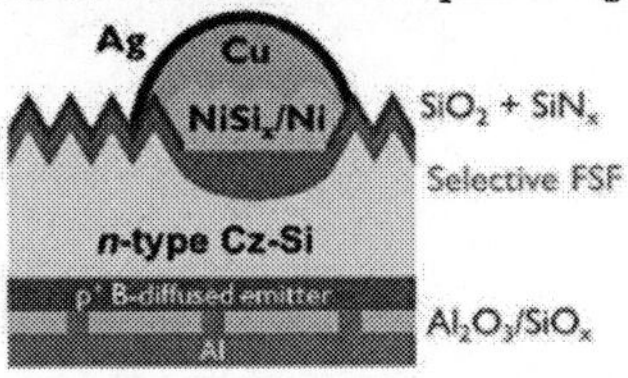

Figure 9: Schematic cross-section of the used PERX cells

Looking at the IV performance results in Figure 10, we observe only 2% power loss during 5000 h, and no change in FF. However, for completeness' sake, it is fair to state that we some degradation in EL inspection (outside the outer busbars), as illustrated in Figure 11. This possibly means that this approach slows down but does not eliminate the degradation mechanism. More likely however, it points towards some slight delamination of the outer fingers.

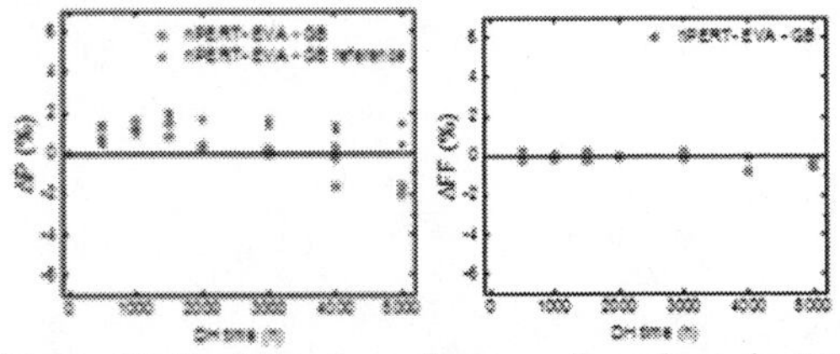

Figure 10: Evolution in performance throughout the DH test of PERX-EVA-GB samples

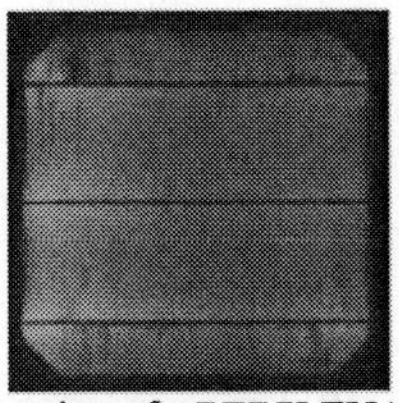

Figure 11: EL mapping of a PERX-EVA-GB sample after 5000 hours DH exposure (right)

3.4 Type 4: PERX-PO-GB

Furthermore we also combined the polyolefin encapsulant and PERX samples with plated metallization and submitted the samples to the same exposure. As expected, we detect no significant degradation in FF yet, as illustrated in Figure 12, though there appears to be some performance drop related to Isc, both for EVA and PO samples. But the degradation mechanism based on corrosion of the finger metallization is clearly not present, as can be determined from the series resistance (Rs) mapping in Figure 13.

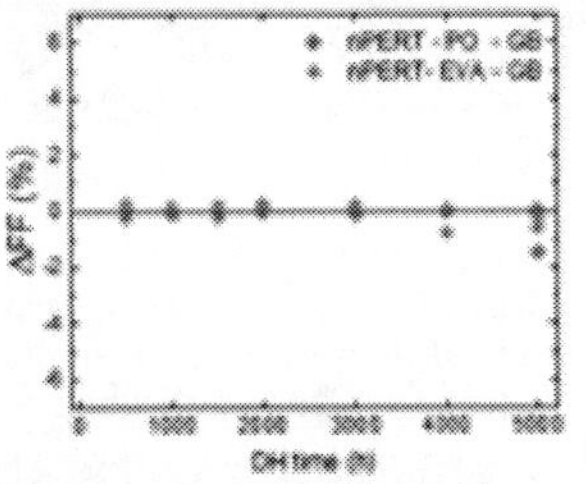

Figure 12: Evolution in performance throughout the DH test of PERX-PO-GB samples

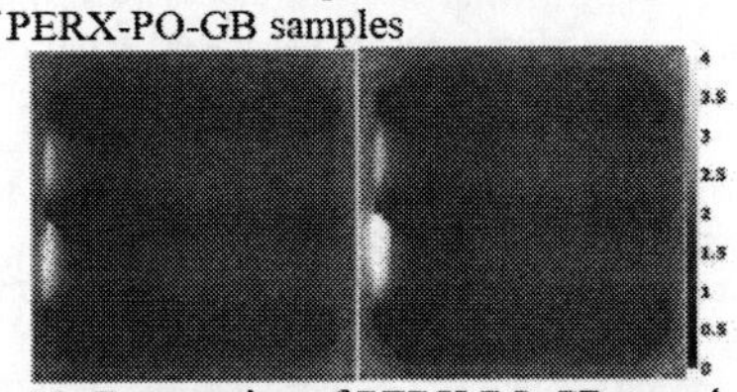

Figure 13: Rs mapping of PERX-PO-GB samples (based on PL-EL measurements): initial (left) and after 5000 hours DH exposure (right)

3.4 Type 5:STD-EVA-GG

Finally we also fabricated samples with a glass-glass buildup instead of glass-backsheet, to check if we can avoid the degradation mechanism by preventing the moisture from coming through the backsheet (that is cracking and delaminating after 4000h). From Figures 14 and 15 it is clear that this does not tackle the degradation mechanism. It seems that replacing the backsheet with glass in itself does not avoid the mechanism as the moisture preferentially penetrates along the tab-encapsulant interface. To address this, this interface would have to be blocked from moisture penetration (e.g. by adding a sealant at the edges).

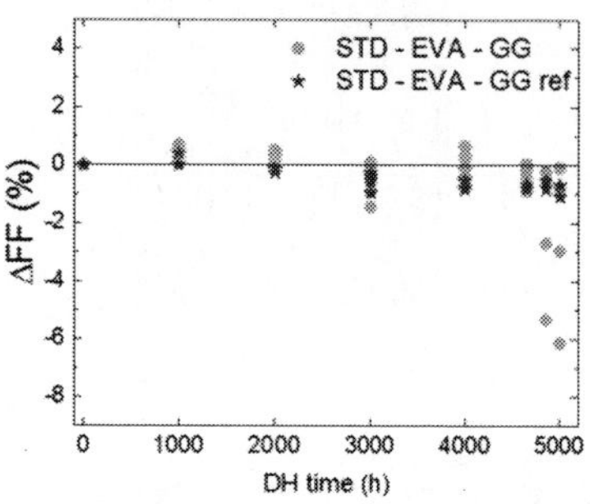

Figure 14: Evolution in performance throughout the DH test of STD-EVA-GG samples

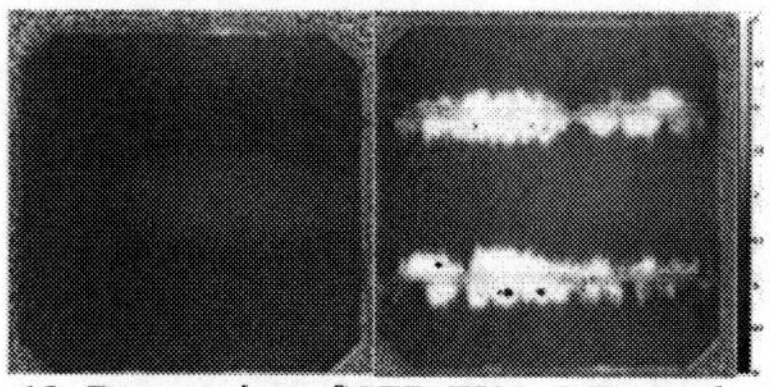

Figure 13: Rs mapping of STD-EVA-GG samples (based on PL-EL measurements): initial (left) and after 5000 hours DH exposure (right)

4 CONCLUSION AND OUTLOOK

In this work we investigate the response of 1-cell laminates with different cell metallizations and encapsulants respond to the qualification tests with the highest failure rates. While we detect no significant degradation for any of the sample types in thermal cycling, extended damp heat testing reveals significant differences depending on used encapsulant and used cell metallization. The corrosion of the cell metallization observed in EVA-encapsulated samples with a screenprinted metallization can be effectively addressed by using a polyolefin encapsulant and/or a plated metallization. Replacing the backsheet with glass in itself however, does not avoid the degradation.

As outlook, more in-depth analysis of the exact corrosion mechanism is in progress.

5 ACKNOWLEDGMENT

The authors want to thank the PERX team at imec (especially Angel Uruena, Emanuele Cornagliotti and Monica Aleman) for supply of the nPERT cells and Reinoud Moors for soldering. We further gratefully acknowledge the partners in imec's SiPV industrial affiliation programme, in particular Borealis for discussions, supply of encapsulant material and financial support. Also, we want to acknowledge the support of the European Union, the European Regional Development Fund ERDF, Flanders Innovation & Entrepreneurship and the Province of Limburg.

6 REFERENCES

[1] J. Poortmans et al., EUPVSEC, 2010
[2] Report IEA-PVPS T13-01:2014
[3] B. Ketola et al., EUPVSEC, 2008
[4] M.D. Kempe et al., Sol.Mat., 2006
[5] E. Cornagliotti, IEEE JPV, 2015
[6] Standard IEC61215, 2005
[7] Bowden et al., EUPVSEC, 2001
[8] Standard IEC60891, 2009
[9] S. Pingel et al., EUPVSEC, 2012
[10] Tanahashi et al., IEEE PVSC, 2016
[11] Kraft et al., IEEE JPV, 2015

ADHESION DEGRADATION OF THE METALLIZATION-ENCAPSULANT INTERFACE

Nick Bosco[1], Peter Hacke[1] and Sarah Kurtz[1]
[1]National Renewable Energy Laboratory, Golden Colorado USA
Jared Tracy[2] and Reinhold H Dauskardt[2]
[2]Stanford University, Stanford California USA

ABSTRACT: Adhesion of the silver metallization/ encapsulant interface was measured through temperature, humidity and voltage bias stress. Results indicate that adhesion at this interface is severely compromised by even short exposures to humidity and quickly plateaus with further, prolonged, exposure. The magnitude of degradation is also identical to that measured through a three-month outdoor, on-sun, exposure. While the addition of a voltage bias also appears capable of further degrading adhesion below this plateau level, in the absence of humidity, adhesion may be similarly degraded.
Keywords: metallization, reliability, screen printing, encapsulation, degradation

1 INTRODUCTION

Delamination in PV modules has been identified as the second most prevalent failure mode seen in the field, according to one survey [1]. However, single factor qualification testing (damp heat, humidity freeze, thermal cycling, potential-induced degradation (PID) stress, etc.) does not readily produce or predict some of the delamination modes seen in the field. For instance, there are numerous examples in the literature of Arco Solar and Siemens Solar modules exhibiting delamination of the encapsulant from the cell near the bus bar and on the grid finger metallization, Fig. 1 [2]. A correlated observation has recently been made during a study of encapsulant to cell adhesion: while the delamination of pristine samples showed no interface preference in the cohesive zone ahead of the delamination front, samples that had been aged at 85°C/ 13.5% relative humidity (RH) clearly demonstrated preferential delamination at the gridlines, Fig 2 [3]. This suggests that the adhesion of the encapsulant to metallization interface degrades faster, or preferentially, to the cell's antireflective coating. While adhesion degradation in the presence of moisture may be the prevalent mechanism, there is also a theory that electrochemical reactions at the metallization's surface may act as an additional mechanism for delamination at this interface [2, 4-6]. Accordingly, there is a current International Electrotechnical Commission (IEC) proposal to investigate serial damp heat (85°C / 85%RH) and voltage bias stress exposure that has been shown to cause similar encapsulant delamination as seen in the field, which originates from the cell's metallization, in some PV modules, Fig 3.

In this work, we isolate the metallization-encapsulant interface and study the mechanisms of adhesion degradation individually to quantify their relative contribution to observed field failures. First, the effect of humidity ingress on adhesion is examined at the silver metallization interface, where bond displacement in the presence of water is the likely degradation mechanism. Next, the processes that occur with the application of voltage bias are examined. This includes the effect of ionic currents that cause Na^+ ions to migrate toward the cell's surface, which has been suggested to also reduce adhesion [7].

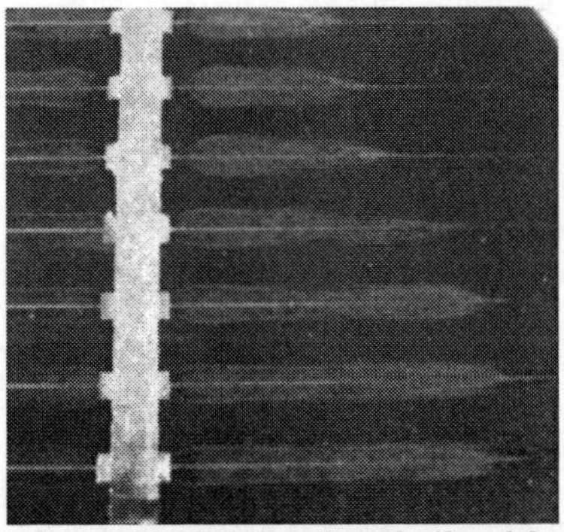

Fig. 1. Image highlighting the selective delamination of the encapsulant/ cell interface of an Arco Solar module.

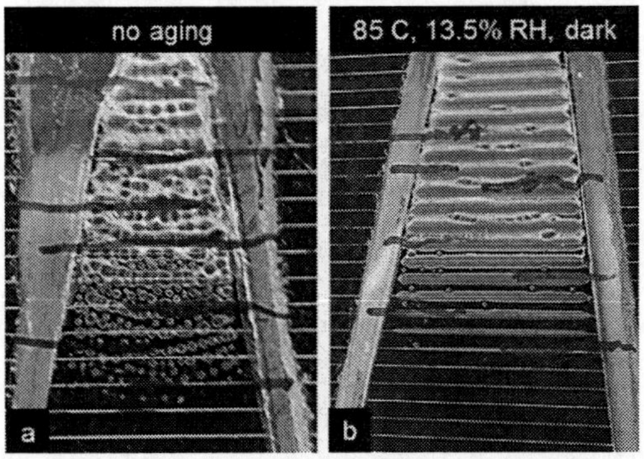

Fig 2. Images of encapsulant/cell interface captured during adhesion tests before (a) and after (b) aging at elevated temperature and humidity. Following aging, cavities in the cohesive zone of the fracture path preferentially form at gridlines. Red lines mark locations of debond extension during the adhesion measurement.

Fig 3. Image of a PV module exhibiting delamination (light blue regions), originating at the cells metallization, following a serial damp heat and PID stress exposure.

2 MATERIALS AND METHODS

Single wafer modules were fabricated for this study. Monocrystalline silicon wafers were first prepared by completely screen printing their front surface with silver paste and then firing in a standard fashion. An interconnect ribbon was then soldered to the Ag front surface of the printed wafer to provide electrical connectivity. These were then laminated to a front sheet of 6 mm thick solar glass with a ~400 µm layer of commercially available ethylene-vinyl acetate (EVA) encapsulant.

The modules were then exposed to environmental aging. Accelerated aging was conducted at 85°C and 0, 13.5, 40 and 85%RH. Both a previously unstressed module, and module which had a pre-conditioning of 1000 hours at 85°C / 85%RH, were also exposed to a -1000 V bias stress between the wafer's front metallization (negative) and an aluminum electrode placed on top of its front sheet of glass (ground). One module received the -1000 V bias stress in a dry heat, 85°C / 0%RH, condition. Finally, one module, without interconnection, was placed outdoors at latitude tilt in Golden, Colorado for three summer months (June–August 2017).

The width-tapered cantilever beam method was used to measure the debond energy (adhesion) of the Ag/EVA interface at intervals through all exposures. This method of sample preparation and testing has been previously presented and is included here for clarity [2]. The metrology employs an elastic width-tapered cantilever beam adhered to the layered structure of interest. When the beam is loaded at its apex, delamination will initiate at the weakest interface and advance upon continued loading. This measurement quantifies the critical value of the energy release rate, G_c, which is the material property of adhesion, and represents the energy required for debond extension, given by:

$$G_c = \frac{P_c}{2\tan(\theta/2)} \frac{\Delta_f}{a_f^2} \qquad (1)$$

where P_c is the critical load plateau at which the debond extends, a_f is a unique value of debond length at the load-line displacement Δ_f, and θ the apex angle of the width-tapered beam.

In this work, width-tapered beams of 20° were fabricated from 1.6 mm thick grade-5 titanium. Multiple Ti beams were adhered to the back of each cell with a two-part epoxy (3M DP420) and any excess wiped clean prior to hardening during its room temperature cure. The beams were therefore present through all environmental exposures. When a module was removed at its designated interval for testing, one beam would be prepared by sectioning the cell around the beam with a sharp carbide scribe. Testing was conducted, at laboratory conditions, within one-hour of removal from its chamber with a small load frame at a constant displacement rate of 10 µm/s while both load and load-line displacement recorded. Because it was also possible to visualize the debond front through the modules front sheet of glass, in-situ measurements of debond length and load-line displacement provided for calculation of G_c as a function of debond length.

3 RESULTS AND ANALYSIS

A summary of the adhesion measurements through temperature and humidity exposure is presented in Fig. 4. Each symbol represents the average debond energy for one width-tapered cantilever beam measurement, and the error bars one standard deviation of that measurement's uncertainty. The initial adhesion of the pristine Ag/EVA interface was measured to be 952±26 J/m^2. The adhesion rapidly drops with temperature and humidity exposure. Degradation is most rapid at the 85°C / 85%RH condition, falling to 134±30 J/m^2 following 300 hours and then decreasing to ~60 J/m^2 following the 1000 hour exposure, which is similar to the module exposed outdoors for only three months (56±3J/m^2). At the 40%RH condition, adhesion degradation is almost as rapid and similarly reaches ~60 J/m^2 following the 1000-hour exposure. At the 13.5%RH condition, the adhesion level following 1000 hours has only dropped to 489±23 J/m^2. The module exposed to the dry heat condition (85°C/ 0%RH) did not exhibit any adhesion degradation following its 1000-hour exposure.

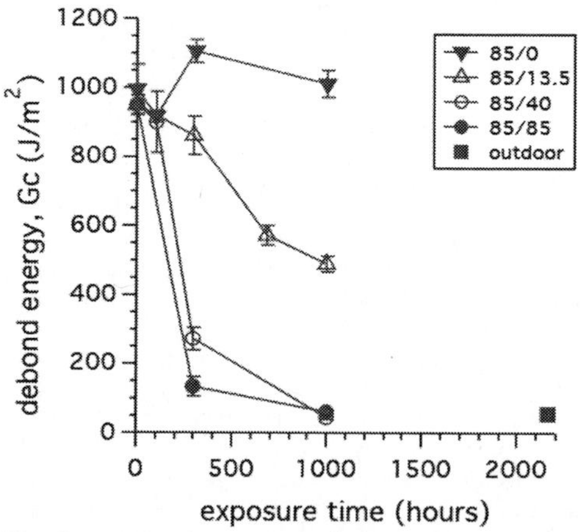

Fig 4. Adhesion measurements of the Ag/ EVA interface through various temperature/ humidity exposures.

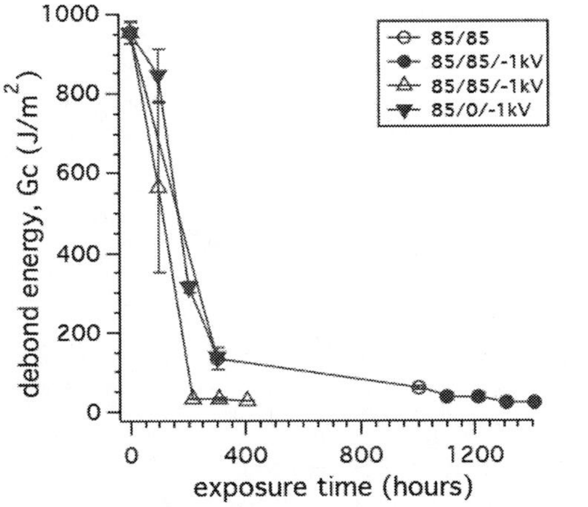

Fig 5. Adhesion measurements of the Ag/ EVA interface through voltage bias exposure.

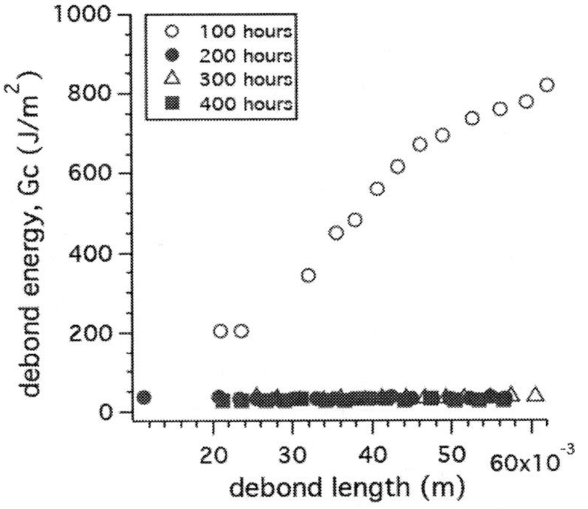

Fig 6. Adhesion measurements of the Ag/ EVA interface as a function of debond length for the non-preconditioned voltage bias exposure (85°C / 85%RH / -1kV).

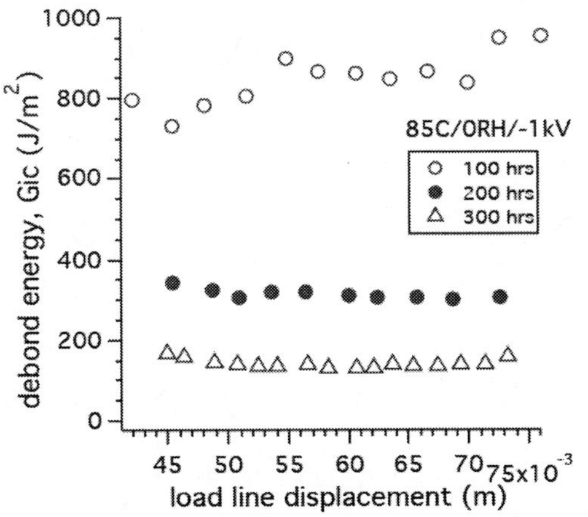

Fig 7. Adhesion measurements of the Ag/ EVA interface as a function of debond length for the dry voltage bias exposure (85°C / 0%RH / -1kV).

A summary of the adhesion measurements through voltage bias exposure is presented in Fig 5. Reproduced in this plot are the results of the measurements through the 85°C/ 85%RH exposure, as this was the pre-condition for one of the voltage bias exposures. The 85°C/ 85%RH/ 1000-hour preconditioned module exhibited a further degradation in its adhesion to 21±2 J/m^2 following an additional 400-hour exposure at 85°C/ 85%RH/ -1 kV. The module that was not preconditioned, however, exhibited a similar degradation in its adhesion (29±2 J/m^2) through only its similar 400-hour voltage bias exposure, suggesting the preconditioning has little or no effect on accelerating adhesion degradation at this interface. The module in the dry heat condition, 85°C/ 0%RH/ -1 kV, degraded similarly to the preconditioned sample following 300 hours of exposure. This condition without bias did not degrade adhesion, clearly demonstrating the independent bias mechanism of adhesion degradation. The large error bar on the 100-hour measurement of the non-preconditioned sample is due to the adhesion of that interface changing significantly from the edge of the cell towards its interior, as illustrated in the evaluation of adhesion with debond length, Fig. 6. The measurements following 100 hours of exposure show that while the adhesion has degraded significantly at the edge of the cell, it approaches its pristine (unexposed) value at the center. This is evidence of two, simultaneous, degradation mechanisms affecting adhesion: moisture ingress and ionic conduction. At the shortest exposure time, moisture absorption from the edge of the cell dominates adhesion degradation, and thus an adhesion profile that qualitatively mimics the expected water content profile. Following the 200-hour exposure, a duration insufficient to transport moisture to the middle of the cell, the ionic conduction mechanism dominates, thus the uniform adhesion with debond length. Adhesion with debond length for the dry bias exposure (85°C / 0%RH / -1kV) is presented in Fig. 7. In this case adhesion is uniform across the area tested, illustrating the expected uniform effect of the bias stress.

4 DISCUSSION

Adhesion degradation of the EVA/ Ag metallization interface has been characterized through temperature, humidity and voltage bias exposures. Results indicate that adhesion at this interface is severely compromised by even short exposures to humidity and quickly plateaus with further, prolonged, exposure. The magnitude of degradation is also identical to that measured through a three-month outdoor, on-sun, exposure, which demonstrates these exposures produce a relevant acceleration. Series and parallel exposures of humidity and voltage stress were also investigated and it was shown the parallel application of stress degraded adhesion to a similar level in a much shorter period of time. Furthermore, in the absence of humidity, the voltage stress alone degraded adhesion at a similar rate, and to a similar extent, as the most humid environment investigated. Previous work found an identical trend in adhesion with a 85°C/ 85%RH exposure and attributed the degradation to dissociation via moisture ingress and the comparatively weak bonding energy of the silane coupling agent to silver in the silver oxide [3].

Conversely, it was also demonstrated in [3] that adhesion between EVA and the cell's antireflective-coated surface was immune to similar temperature and humidity exposures.

It has recently been elucidated that an adhesion energy of greater than 160 J/m^2 is required to ensure encapsulant delamination does not occur in a fielded module over its expected lifetime [8]. While the present study has demonstrated that no expectation should be made on the EVA/ Ag interface retaining an adhesion value above this 160 J/m^2 threshold, immediate and spontaneous encapsulant delamination originating at the cell's metallization is not universally observed. Therefore, even though these areas may serve as nucleation points for encapsulant delamination, significant adhesion degradation at the cell AR coating/ encapsulant interface must also occur before spontaneous delamination is observed.

5 ACKNOWLEDGEMENT

The Authors would like to acknowledge the group of Professor Ajeet Rohatgi of the Photovoltaics Center of Excellence (UCEP) at Georgia Tech University for the fabrication of the silver screen printed wafer.

This work was supported by the U.S. Department of Energy under Contract No. DE-AC36-08GO28308 with Alliance for Sustainable Energy, LLC, the Manager and Operator of the National Renewable Energy Laboratory. Funding provided by the U.S. Department of Energy Office of Energy Efficiency and Renewable Energy Solar Energy Technologies Office.

6 REFERENCES

[1] D. C. Jordan, J. Wohlgemuth, and S. Kurtz, "Technology and Climate Trends in PV Module Degradation," in *27th European Photovoltaic Solar Energy Conference*, 2012, pp. 2411-2415.

[2] N. Bosco, J. Tracy, R. H. Dauskardt, and S. Kurtz, "Development and First Results of the Width-Tapered Beam Method for Adhesion Testing of Photovoltaic Material Systems," presented at the IEEE Photovoltaic Specialist Conference, Portland, OR, 2016.

[3] J. Tracy, N. Bosco, and R. H. Dauskardt, "Evaluation of Encapsulant Adhesion to Surface Metallization of Photovoltaic Cells," *IEEE Journal of Photovoltaics,* 2017 2017.

[4] G. R. Mon, L. Wen, R. G. Ross, and D. Adent, "Effects of temperature and moisture on module leakage currents," in *Proceedings of the 18th IEEE PV Specialists Conference*, Las Vegas, Nevada, USA, 1985, pp. 1179-1184.

[5] J. Wohlgemuth, P. Hacke, N. Bosco, D. C. Miller, M. D. Kempe, and S. R. Kurtz, "Assessing the Causes of Encapsulant Delamination in PV Modules," presented at the IEEE 43rd Photovoltaic Specialists Conference (PVSC), Portland, OR, 2016.

[6] K. Matsuda, T. Watanabe, K. Sakaguchi, M. Yoshikawa, T. Doi, and A. Masuda, "Microscopic Degradation Mechanisms in Silicon Photovoltaic Module Under Long-Term Environmental Exposure," *Japanese Journal of Applied Physics,* vol. 51, 2012.

[7] N. G. Dhere, N. R. Raravikar, A. Mikonowicz, and C. Cording, "Effect of Glass Na Content on Adhesional Strength of PV Modules," presented at the IEEE 29th Photovoltaic Specialists Conference (PVSC), 2002.

[8] N. Bosco, J. Tracy, S. Kurtz, and R. H. Dauskardt, "Defining Threshold Values of Encapsulant and Backsheet Adhesion for PV Module Reliability," *IEEE Journal of Photovoltaics,* vol. in Press, 2017 2017.

DEPTH PROFILING OF OPTICAL, CHEMICAL AND NANOMECHANICAL PROPERTIES OF GLASS/ENCAPSULANT/BACKSHEET PV LAMINATES AGED UNDER DIFFERENT INTENSITIES OF UV LIGHT

Yadong Lyu, Jae Hyun Kim, Andrew Fairbrother, Xiaohong Gu*
Engineering Laboratory, National Institute of Standards and Technology, Gaithersburg, MD 20899, USA

ABSTRACT: Due to the complex multicomponent and multilayer structures of PV modules, it is challenging to quantify the individual components or layers in the modules, hindering the understanding of the root causes of module failures. This work provides a method to effectively investigate the degradation depth profiles of a glass/encapsulant/backsheet laminate system after exposure to different ultra-violet (UV) intensities. Confocal fluorescence imaging, in combination of attenuated total reflection Fourier transformed infrared (ATR-FTIR) imaging, micro-UV-vis spectroscopy and atomic force microscopy-based quantitative nanomechanical mapping (QNM-AFM) were used to investigate the optical, chemical and nanomechanical degradation of each polymeric layer in the laminates. Samples were aged at 85°C and 0 % relative humidity (RH) under 4 UV irradiance levels ranging from 60 W/m^2 to 180 W/m^2 (295 nm – 385 nm) for 3840 h. A non-uniform, depth-depedent, light intensity-depedent degradation behavior was observed in the aged ethylene vinyl acetate (EVA) encapsulant layer based on the the fluorescence imaging. The depth profiles of yellow index, ketone formation and modulus in the encapsulant layer were also found to be similar to that of the fluorescence intensity, following the light intensity decay in the material. The results have suggested that these degradation properties may be cross-correlated and their changes are probably govened by the intensity of the UV light across the sample thickness. In addition, obvious fluorescence was observed in two adhesive layers in the aged poly(terephathalate) (PET) /PET/EVA backsheet, indicating the degradation of the adhesive layers and the potential risk of backsheet inner layer delamination. The study has demonstrated that depth profiling is essential to the understanding of degradation mechanism in the glass/EVA/PPE system, and would provide a scienctific basis for developing standized property testing, weathering and lifetime prediction for PV modules.
Keywords: encapsulant; backsheet; UV light intensity; degradation; yellowing; fluorescence; ATR-FTIR; atomic force microscopy

1 INTRODUCTION

Polymeric photovoltaics (PV) packaging materials, including encapsulant and backsheet, play a critical role to protect modules from weathering in harsh operating environments, and provide mechanical support and electrical insulation [1, 2]. However, these materials are susceptible to degradation under ultra-violet (UV) light, heat, moisture and the cyclic changes of these environmental elements [3, 4], leading to failures such as yellowing, cracking and delamination [5]. These failures can potentially be detrimental to the module performance and raise safety concerns. Therefore, it is important to understand the degradation behaviors of these materials in order to make better material selection and enhance the quality of PV modules.

However, because of the complex multilayer and multicomponent structure of PV modules, the common bulk characterization techniques [6, 7] would not be able to fully reveal the degradation profile of the multilayers, especially when the degradation is not uniform through the thickness. Several spectroscopic imaging techniques, such as Raman imaging and attenuated total reflection Fourier transformed infrared (ATR-FTIR) imaging have been used to examine the depth profile of PV multilayered systems after degradation [8-10]. However, significantly enhanced fluorescence background emitted from aged samples could dominate Raman spectra and mask the spectroscopic vibrational information.

In this work, confocal fluorescence imaging, in combination of FTIR microscopy, micro-UV-vis spectroscopy and quantitative nanomechanical mapping atomic force microscopy (QNM-AFM) were used to investigate the depth-profiles of optical, chemical and mechanical degradation of glass/encapsulant/backsheet laminates after UV exposure under different light intensities. Samples were exposed on the NIST SPHERE *(Simulated Photodegradation via High Energy Radiant Exposure)* at 85 °C and 0 % relative humidity (RH) for 3840 h. The levels of UV irradiance ranged from 60 W/m^2 to 180 W/m^2 (295 nm – 385 nm). Cryo-microtomed cross-sectional samples were prepared and characterized as a function of distance from the exposed surface. The correlation between fluorescence profile and the changes in yellowing index, chemical structures and modulus was investigated. The implication of this work on module performance will be discussed.

2 MATERIALS AND METHODS*

2.1 *Materials*

The glass/EVA/backshee laminates with a size of (180 × 180) mm^2 were prepared using vacuum lamination by a processing procedure with the platen temperature of 150 °C, evacuation time of 4 min, pressing time of 1 min and crosslinking time of 13 min. The glass was a two side-polished fused silica with transmittance of around 95 %. The EVA encapsulant was a ultrafast cure EVA system (Photocap 15420P/UF, STARS) containing a tertbutyl peroxy 2-ethyl-hexyl carbonate (TBEC) curing agent. The backsheet PPE is a polyester based multilayer and mainly composed of pigmented PET outer layer, PET core layer and multiple EVA layers [10]. The details of this backsheet will be discussed later.

** Certain commercial products or equipment are described in this paper to specify adequately the experimental procedure. In no case does such identification imply recommendation or endorsement by the National Institute of Standards and Technology, nor does it imply that it is necessarily the best available for the purpose.*

2.2 *Accelerated Laboratory Exposure*

Accelerated laboratory exposure was performed with the glass-side of the laminates facing UV light source on the NIST SPHERE at 85 °C/0 % RH. The exposure regions on the laminates are at least 30 mm away from the edge of the glass to avoid the oxygen diffusion effect from the periphery. To vary the light intensity of the UV irradiance, the neutral density (ND) filters with nominal transmittances of 40 %, 60 % or 100 % were placed in front of samples during exposure. The measured irradiances on the sample locations through 40 %, 60 % and 100 % ND filters are 60.5 W/m^2, 93.5 W/m^2, and 142.5 W/m^2, respectively, for the wavelength range between 295 nm and 385 nm.

2.3 *Cross-Sectional Sample Preparation*

To conduct cross-sectional analysis, (1 x 5) mm^2 strips under each exposure condition were first cut from the large laminate after exposure using a diamond saw. Then EVA/PPE backsheet part was carefully peeled off from the glass, with minimum polymers remained on the glass. In order to obtrain a smooth surface, cryo-microtoming was performed on the EVA/PPE sample along the thickness direction using diamond blade at - 80 °C. Thin films of 5 µm thick and 2 µm thick of EVA/PPE samples were cut and collected along the cross-section direction for micro-FTIR and micro-UV-Vis microscopy, respectively. The remaining block part of the cross-section samples were used for fluorescence and AFM mapping.

2.4 *Characterization Methods*

Fluorescence measurements Morphology and fluorescence imaging were conducted on the Zeiss LSM510 Meta (Carl Zeiss, Inc., Oberkochen, Germany) laser scanning confocal microscope. For cross-section morphology observation, an objective of the Epiplan-Neofluar 10×/0.3 was used with field view of about (840 × 840) µm^2 and pixel size of (512 × 512) pixels using laser wavelength of 543 nm in reflection mode. The lateral resolution of this objective with 543 nm laser is about 0.9 µm for pinhole size at 1 Airy unit. For fluorescence measurements, an objective of 5x/0.15 was used with diode laser (405 nm) to excite the sample. Scanning area was about (1700 × 1700) µm^2 and the lateral resolution was about 1.4 µm. Both channel mode and lambda mode were used to detect emission signals with long pass (LP) 420 filter. The emission spectra and fluoresence imaging were collected from 417 nm to 748 nm with 10.7 nm increment using the META detector.

UV-Vis-Near IR microspectrophometry (MSP) UV-Vis spectra were recorded using a microspectrophotometer (CRAIC Technologies, Model MSP 121) in the UV-Vis range between 200 nm and 800 nm across the sample in transmission mode. The microscope objective is a Davin reflecting objective 36x, NA 0.5 with aperture size of 7.9 µm × 7.9 µm. Sectioned films with thickness of 2 µm was mounted between quartz microscope slides and cover slips with glycerin as a medium. The micro-UV-Vis spectra were taken, and the yellow index values were calculated based on ASTM E308.

Micro-FTIR measurement Infrared analyses of each layer in EVA/PPE backsheet sample were performed by infrared imaging microscope purged by dry air and equipped with a liquid nitrogen-cooled mercury-cadmium-telluride (MCT) detector. Spectra were collected in transmission mode on the 5 µm thick cross-sectional EVA/PPE slice with an aperture size of (10 x 10) µm^2 and a step size of 10 µm. The spectral ranging from 4000 cm^{-1} to 650 cm^{-1} were recored at 8 cm^{-1} resolution with 64 scans. To ensure the reproducibility, at least three different locations were measured for each data point.

Quantitative nanomechanical mapping atomic force microscopy (QNM-AFM) measurements QNM-AFM was performed on a Bruker Dimension Icon AFM for quantitative information of height, adhesion and modulus of samples at the nanoscale. The modulus was obtained using the Derjaguin-Muller-Toporov (DMT) model following the relative method [11, 12]. Bruker AFM probe of Tap150 was selected based on the estimated Young's moduli of EVA encapsulant. The PDMS-SOFT02012M with nominal modulus of 3.5 MPa was used as reference sample. A (512 × 512) pixels resolution with scan rate of 0.5 Hz was performed at ambient condition.

3 RESULTS AND DISCUSSION

The scanning confocal microscopy of the cross-sectional morphology of fresh EVA/PPE sample is shown in Fig. 1. The thickness of the EVA encapsulant layer is (920 ± 6) µm, and the total thickness of PPE backsheet is (300 ± 4) µm. Combined with our FTIR and Raman results [10], it is found that the PPE backsheet is mainly composed of five layers: EVA outer layer (25 µm), pigmented EVA middle layer (55 µm), EVA inner layer (25 µm), PET core layer (126 µm) and PET outer layer (55 µm). In addition, there are two adhesive layers with thickness around 8 µm and 6 µm between EVA inner and PET core layers, and between PET core and PET outer layers, respectively. (Figure 1).

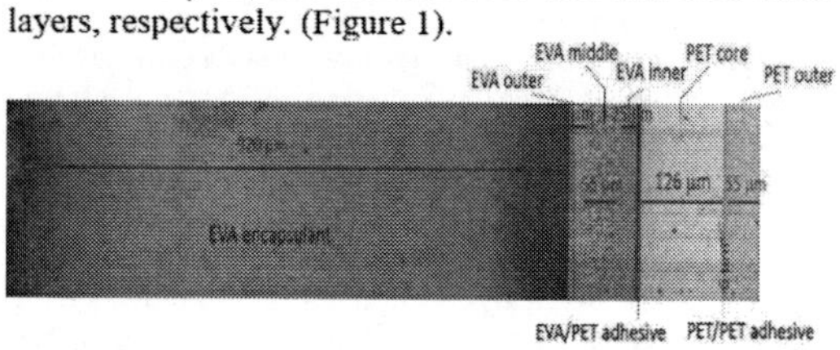

Fig. 1. Cross-sectional morphology of fresh EVA/PPE sample and the corresponding thickness of each layer.

Fig. 2 shows the integral fluorescence emission intensity images ((a)-(d)) and their corresponding line profiles (e) for the cross-sections of EVA encapsulant layers prepared from the glass/EVA/PPE samples before and after exposure to UV/85 °C /0 % RH under different light intensities. The red regions near the exposed side of the specimens have the highest intensity of the fluorescence emission, while the blue interior regions have the lowest intensity. A sequential drop in intensity has been observed as a function of distance (depth) from the exposed side to the interior region. When comparing the results from three UV intensities, one can find that the specimen exposed under 60 % UV has the highest fluorescence intensity. A similar phenomenon was reported by Pern [13] for severely discolored EVA encapsulants. The reason for this could be due to the secondary reactions and depletions of the fluorescent α,β-unsaturated carbonyl species under further UV light. However, the yellowing index shows different behaviors, the higher light intensity, the more yellowing (data not shown). This discrepcy indicates that the origins of yellowing and fluorescence are not always same.

Moreover, the overall degradation depth under three UV intensities is essentially same, as shown in the integral emission line profiles (Fig. 2e). The substantial changes in the fluorescence intensity appear to be within 300 μm from the exposed side of the encapsulant, which can be explained by the exponential light intensity decay in the material according to the Lambert-Beer law [14, 15].

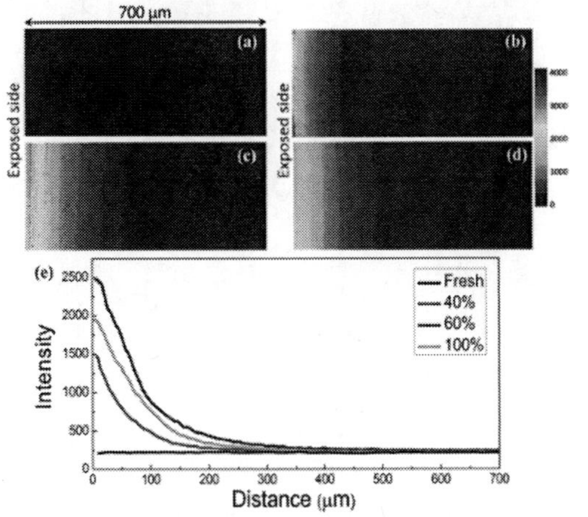

Fig. 2. Integral fluorescence intensity images of the cross-sections of EVA encapsulant layers in the glass/EVA/PPE samples before (a), and after exposure to UV/85 °C/0 % RH under 40 % UV (b), 60 % UV (c) and 100% UV (d). (e) shows their corresponding line profiles of (a)-(d).

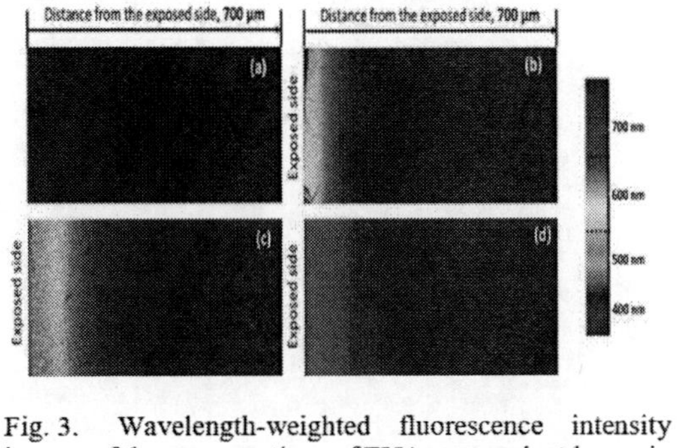

Fig. 3. Wavelength-weighted fluorescence intensity images of the cross-sections of EVA encapsulant layers in glass/EVA/PPE samples before (a), and after exposure to UV/85 °C/0 % RH under 40 % UV (b), 60 % UV (c) and 100% UV (d). (e) shows their corresponding line profiles of (a)-(d). The Red color corresponds to the longest emission wavelength and the purpule corresponds to the shortest emission wavelength.

Fig. 3 shows depth-dependent, wavelength-weighted fluorescence intensity images of the same samples shown in Fig. 2. The different colors represent different emission wavelengths, i.e. the red regions have longer wavelength emission than the blue regions. The whole dark region for the fresh sample indicates that there is no visible fluroscence emission from the fresh EVA encapsulant layer. After exposure, the surface shows more longer wavelength emission compared to the interior region. The short chain α,β-unsaturated carbonyl species probably play an important role for emission at shorter wavelengths, while the relatively long polyconjugated structures lead to emission at longer wavelengths [16]. Further, a progressive shift to a longer wavelength has been observed near the exposed surface region with the increase of the UV intensity. In addition, the higher UV intensity, the longer emission wavelength. The results suggest that the different light intensities not only influence the fluorescece intensity of the exposed surface, but also generate degradation products with different lengths of polyconjugated structures. More details will be discussed elsewhere. [17]

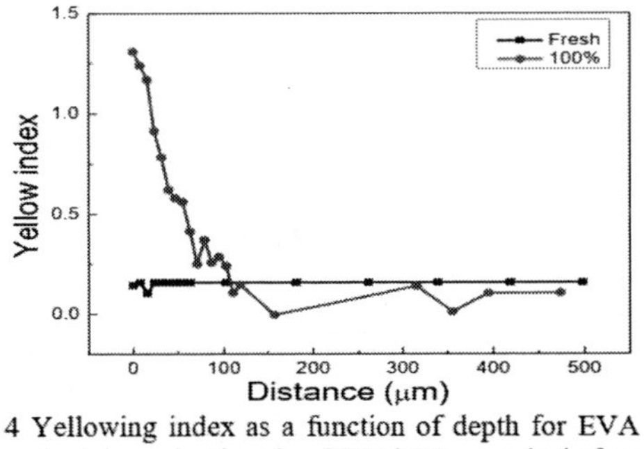

Fig. 4 Yellowing index as a function of depth for EVA encapsulant layer in the glass/EVA/PPE sample before and after exposure to UV/85 °C/0 % RH under 100 % UV. The micro-UV-vis was performed on the 2 μm thick cryo-microtomed cross-sectional EVA/PPE samples before and after UV exposure.

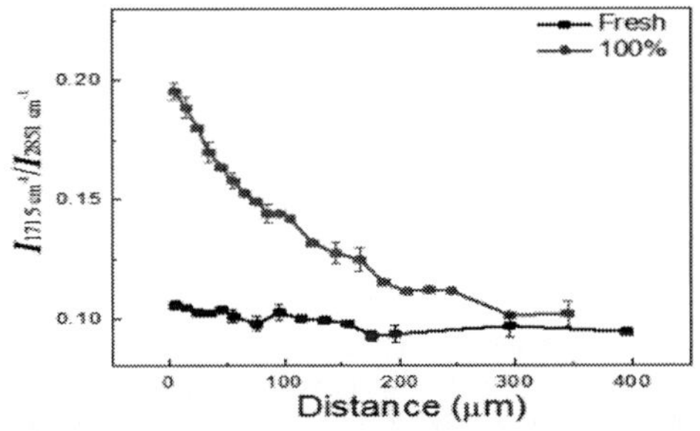

Fig. 5 Ketone formation as a function of depth for EVA encapsulant layer in the glass/EVA/PPE sample before and after exposure to UV/85 °C/0 % RH under 100 % UV. Micro-IR measurement was performed on the 5 μm thick cryo-microtomed cross-sectional The ratio of the absorbance at 1715 cm^{-1} to that at 2851 cm^{-1} is used to represent the relative ketone formation.

In order to understand the relationship between fluorescence profiles and the optical changes of the EVA encapsulant across the thickness after UV exposure, the micro-UV-vis was performed on the 2 μm thick cross-sectional EVA/PPE slices before and after UV exposure. Fig. 4 shows the depth profile of the yellowing index as a function of depth. A sharp decrease in the yellow index has been observed from the UV exposed side to the interior region of the EVA encapsulant, then the value reaches a plateau, which is similar to that of the fresh sample. This trend follows well with the depth profile of the fluorescence intensity. Similarly, a strong depth-dependence of the ketone formation is also observed for the same sample (Fig. 5). The ratio of the absorbance at 1715 cm^{-1} (carbonyl stretching vibration of ketone) to the absorbance at 2851 cm^{-1} (symmetrical stretching vibration of $-CH_2-$) is used as the relative ketone formation index. The ketone products decrease rapidly from the exposed surface, and the degradation depth appears to be consistent with the the evolution of the fluorescence.

The modulus of the EVA encapsulant before and after aging was also measured by QNM-AFM as a function of depth (Fig. 6). For the fresh sample, the DMT modulus of the EVA encapsulant is essentially constant with small variations. However, a substantial drop in the modulus is observed near the exposed encapsulant surface region. This result is consistent with previous work by DMA [18].

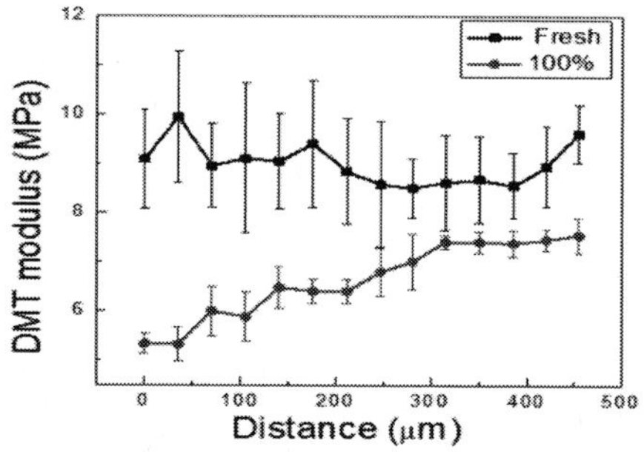

Fig. 6. DMT moduli of fresh and aged (100% UV) EVA encapsulant layers in glass/EVA/PPE samples measured by QNM-AFM. Each modulus data point was averaged from three replicated locations. Error bars represent one standard deviation.

When combining the results from the fluorescence imaging with the ketone formation and the DMT modulus of the aged EVA encapsulant, one can find that a non-uniform degradation behavior has occurred in the encapsulant layer when the UV lights travel through the glass/EVA/PPE laminates. The region near the exposed surface exhibits the highest intensities of fluorescene, the highest amount of ketone formation, but the lowest DMT modulus. The depth profiles of these properties appear to follow the exponential decay of the UV light as a function of depth. Despite the fluorescence is mainly attributed to the interactions between the stabilizing additives such as UV absorbers and peroxides [19-21], while ketone formation and the modulus changes are mainly resulting from the EVA degradation, the results clearly indicate that, in the studied system, the depth-dependent UV intensity across the thickness is a dominant factor for the non-uniform optical, chemical and mechanical changes of the EVA encapsulant. A cross-correlation can also be found among these degradation properties.

Fig. 7 exhibits the fluorescence images and corresponding line profiles of the cross-sections of the PPE backsheets before and after exposure to UV/85 °C/0 % RH under different light intensities. Strong fluorescence emission is observed on the PET outer layer, which could result from the fluorescent whitening agent added in this layer designed to brighten the white color [22]. After degradation, the fluorescence of the PET outer layer decreases with the increase of the UV light intensity. On the contrary, the EVA middle layer shows a non-symmetrical distribution of the fluorescene intensity, which increases with increasing the UV intensity. Moreover, obvious fluorescence emission is observed from two adhesive layers after exposure. The fluorescence intensity of the two adhesive layers appears to be independent of light intensity except for those under 40 % ND filters, indicating that the degradation of adhesive layers is mainly thermal-induced.

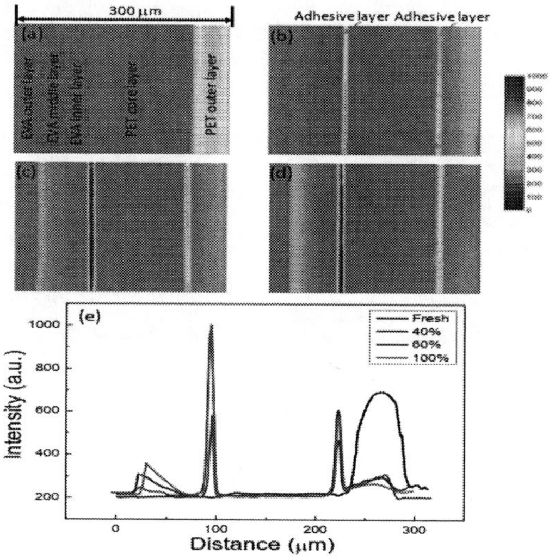

Fig. 7. Integral fluorescence emission images of the PPE backsheet in glass/EVA/PPE before (a), and after exposure to UV/85 °C/0 % RH under 40 % UV (b), 60 % UV (c) and 100% UV (d). (e) shows their corresponding line profiles of (a)-(d).

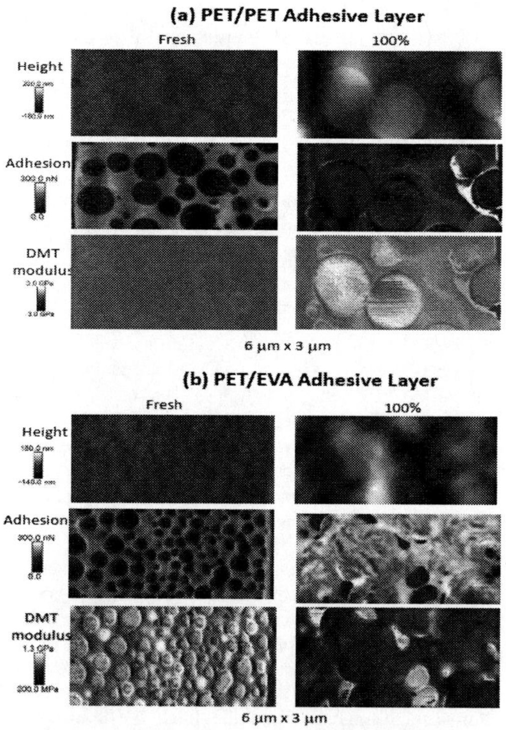

Fig. 8. QNM-AFM height, adhesion and DMT modulus images of (a) PET/PET adhesive layer, and (b) PET/EVA adhesive layer in glass/EVA/PPE samples before and after exposure to UV/85 °C/0 % RH under 100% UV.

The degradation of the adhesive layers has been further confirmed by the AFM-QNM images of the PPE film in the glass/EVA/PPE samples exposed at UV/85° C/0 % RH under 100 % UV (Fig. 8). For the PET/PET adhesive layer, the small nodular structures observed in the fresh sample become larger, probably due to the coalescence of the small latex particles under heat (Fig. 8a). For the PET/EVA adhesive layer, dramatical changes in height, adhesion and DMT modulus are observed (Fig. 8b), indicating that the degradation of this adhesive layer has taken place, leading to microstructural and mechanical changes during exposure. These results suggest a potential risk of inner layer delamination within the backsheet, which could raise potential safety concerns and ultimate reliability of the PV modules.

4 CONCLUSIONS

In this work the degradation depth-profiles of the glass/EVA/PPE laminates exposed to UV/85 °C/0 % RH under different UV intensities have been investigated. The confocal fluorescence imaging, in combination with micro-UV-Vis, micro-FTIR and QNM-AFM were used to characterize the optical, chemical and mechanical properties of the samples as a function of depth. For the encapsulant EVA layer, a non-uniform, depth-depedent, light intensity-dependent degradation profile was found. Similar depth profiles were observed for changes in yellow index, relative ketone formation and the DMT modulus, indicating that these properties can be cross-correlated. Because the degradation depth profiles of these properties followed the exponential decay of the UV light in the material, it is suggested that the UV light across the sample thickness could be a dominating factor governing the depth profile of EVA encapsulant degradation.

In addition, the PPE backsheet in the glass/EVA/PPE laminate before and after UV exposure was also studied. The EVA middle layer revealed the fluorescence emission, and its intensity was increased with the increase of UV intensity. Meanwhile, strong fluorescence emission was observed from two adhesive layers, i.e. PET/PET layer and PET/EVA layer, in the PPE backsheet, and the degradation of these adhesive layers were further confirmed by QNM-AFM imaging. These results suggest a potential risk of inner layer delamination in the backsheet, which could introduce potential safety concerns and be detrimental to the reliability of the PV modules.

5 REFERENCES

[1] G. Jorgensen, K. Terwilliger, J. DelCueto, S. Glick, M. Kempe, J. Pankow, F. Pern, T. McMahon, Solar Energy Materials and Solar Cells 90(16) (2006) 2739-2775.

[2] M. Kempe, Photovoltaics International 9 (2010) 170-176.

[3] F. Liu, L. Jiang, S. Yang, Solar Energy 108 (2014) 88-100

[4] Y. Voronko, G.C. Eder, M. Knausz, G. Oreski, T. Koch, K.A. Berger, Progress in Photovoltaics: Research and Applications (2014)

[5] D.C. Jordan, T.J. Silverman, J.H. Wohlgemuth, S.R. Kurtz, K.T. VanSant, Progress in Photovoltaics: Research and Applications 25(4) (2017) 318-326

[6] K. Agroui, G. Collins, Solar energy materials and solar cells 80(1) (2003) 33-45.

[6] K. Agroui, G. Collins, Renewable Energy 63 (2014) 741-746

[8] Ndiaye A, Charki A, Kobi A, Kébé CM, Ndiaye PA, Sambou V. Solar Energy 2013, 96, 140-151.

[9] Planes E, Yrieix B, Bas C, Flandin L. Solar Energy Materials and Solar Cells 2014, 122, 15-23.

[10] Lin C-C, Krommenhoek PJ, Watson SS, Gu X. Solar Energy Materials and Solar Cells 2016, 144, 289-299.

[11] T. Young, M. Monclus, T. Burnett, W. Broughton, S. Ogin, P. Smith, Measurement Science and Technology 22(12) (2011) 125703

[12] A. Pakzad, J. Simonsen, R.S. Yassar, Composites Science and Technology 72(2) (2012) 314-319

[13] F. Pern, Polymer degradation and stability 41(2) (1993) 125-139

[14] A. Perthué, P. Boutinaud, S. Therias, J.-L. Gardette, Polymer Degradation and Stability 133 (2016) 144-151

[15] P. Gijsman, G. Meijers, G. Vitarelli, Polymer Degradation and Stability 65(3) (1999) 433-441

[16] N.S. Allen, M. Edge, M. Rodriguez, C.M. Liauw, E. Fontan, Polymer Degradation and Stability 68(3) (2000) 363-371

[17] Y. Lyu, J. Kim, A. Fairbrother, X. Gu, Submitted.

[18] J.C. Schlothauer, K. Grabmayer, G.M. Wallner, B. Röder, Progress in Photovoltaics: Research and Applications 24(6) (2016) 855-870

[19] P. Klemchuk, M. Ezrin, G. Lavigne, W. Holley, J. Galica, S. Agro, Polymer degradation and stability 55(3) (1997) 347-365

[20] C. Peike, L. Purschke, K.-A. Weiss, M. Kohl, M. Kempe, Towards the origin of photochemical EVA discoloration, Photovoltaic Specialists Conference (PVSC), 2013 IEEE 39th, IEEE, 2013, pp. 1579-1584.

[21] D.C. Miller, E. Annigoni, A. Ballion, J.G. Bokria, L.S. Bruckman, D.M. Burns, X. Chen, L. Elliott, J. Feng, R.H. French, Degradation in PV encapsulation transmittance: an interlaboratory study towards a climate-specific test, Photovoltaic Specialist Conference (PVSC), 2015 IEEE 42nd, IEEE, 2015, pp. 1-6.

[22] H. Van Aert, I. Srivastava, F. Vangaever, A backsheet for photovoltaic modules, Google Patents, 2014.

COMPARISON OF ACCELERATED UV TEST METHODS WITH FLORIDA EXPOSURE FOR PHOTOVOLTAIC BACKSHEET MATERIALS
EU PVSEC 2017

Emily Parnham, Anthony Whitehead, Sophie Pain, William Brennan
DuPont Teijin Films UK
The Wilton Centre, Redcar, TS10 4RF, UK

ABSTRACT: Photovoltaic (PV) modules currently have lifetime guarantees of 25 years or more. Ideally, all components of the modules would be tested in field conditions for at least 25 years, however this is impractical, so accelerated, laboratory-based testing is required to subject components to an approximation of field conditions. When using accelerated testing it is important to ensure the test does not result in false negatives, hence ruling out good materials, and more importantly, does not give rise to false positives, which could lead to material failures in the field. Current PV qualification standards can only test for "premature" failures and hence are not end-of-life tests. This has resulted in some products passing the qualification tests and subsequently failing in the field within the product lifetime guarantee. This paper compares the mechanical properties of PV backsheet materials weathered outdoors in Florida against those weathered using both ISO 4892-2 (Plastics – Methods of exposure to laboratory light sources – Part 2: Xenon-arc lamps) and the newly proposed IEC 62788-7-2 (Measurement procedures for materials used in photovoltaic modules – Part 2: Environmental exposure - Accelerated weathering tests of polymeric materials).

Keywords: Backsheet, Degradation, Accelerated Weathering

1 INTRODUCTION

Currently, Photovoltaic (PV) modules have lifetime guarantees of 25 years or more, so it is important that the materials used in module construction maintain their functionality for at least 25 years. Throughout the service lifetime of a PV module, it can be exposed to a wide range of environmental stresses that may cause degradation of the module components, which, in turn, can lead to failure of the overall PV module. This paper focuses on just one component of a PV module - the backsheet – and specifically on the effect of accelerated weathering on the outermost polymeric layer of the backsheet. The primary purpose of the outer layer is to protect the inner core PET from UV exposure; hence retaining mechanical integrity of the outermost layer is of fundamental importance to the backsheet lifetime.

Examples of module field failures within the expected service lifetime can be increasingly found in the literature [1]. Some of these failures can be attributed to the polymeric materials [2], illustrating the need to improve the existing standards/qualification tests. There are numerous degradation paths for polymeric materials - hydrolytic, thermal, mechanical and via UV exposure [2-5].

The current testing standard for crystalline silicon modules is IEC 61215. This is a qualification test written to identify flaws in design, materials and processing which could lead to premature field failures [6]. IEC 61215 specifies 15 kW.hr/m^2 exposure in the 320-400 nm region and 3-9% irradiance in the 280-320 nm region, which corresponds to *ca.* 1.5 months Florida direct UV exposure. Assuming an albedo of 10%, this equates to *ca.* 15 months backside exposure, far short of the 25-year lifetime of the module. This illustrates that this test is clearly not designed to simulate outdoor UV exposure for extended periods of time (Figure 1).

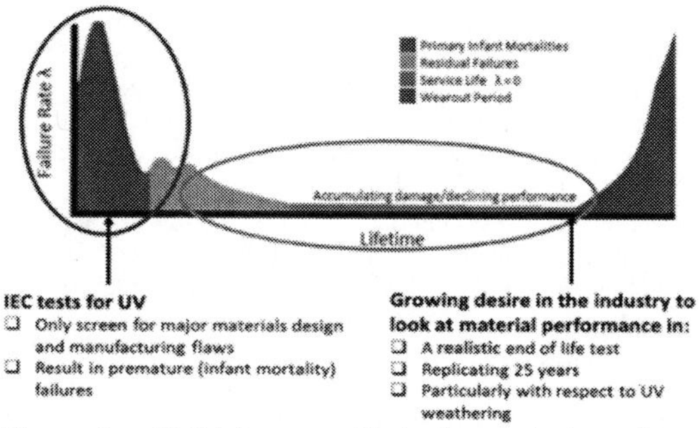

Figure 1: Bathtub curve illustrating typical product failure behaviour

In a research setting, accelerated laboratory-based testing is used to replicate real-life, field conditions. Tests use cycles of high temperatures and irradiances - as well as moisture - in order to degrade materials at a faster rate than would be expected in the field. Materials are then tested to assess the effect of these conditions on material performance/resistance to weathering. These results should be validated against real life outdoor exposure.

Several factors must be considered when determining the harshness of this test. It is commonly agreed that samples should not be weathered at more than 10 times the rate that they would be exposed to in the field [7]. Increasing temperatures and irradiances to levels above those experienced by a sample in the field must also be avoided where possible, as this could induce degradation routes that would not occur in the field [8].

Historically, ISO 4892-2 (Plastics – Methods of exposure to laboratory light sources – Part 2: Xenon-arc lamps) has been widely adopted within the plastics industry. ISO 4892-2 was written with the intention of comparing the relative performance of materials. In order to approximate the weathering effects that occur when materials are exposed in actual end-use environments samples are exposed to repeated cycles of UV irradiance, water spray and humidity at elevated temperatures. UV dosages are matched to 25 years outdoor exposure for any required location (e.g. desert, tropical or temperate regions) and from this, predictions regarding material lifetime can be made.

The PV industry has introduced an alternative accelerated UV test specifically for PV polymeric materials - IEC 62788-7-2 (Measurement procedures for materials used in photovoltaic modules – Part 2: Environmental exposure - Accelerated weathering tests of polymeric materials). IEC 62788-7-2 specifies various parameters, such as a UV light source, irradiance level, temperature and humidity, as outlined in Table 1. The temperatures chosen for acceleration are significantly higher than those in ISO 4892-2 (Table 3, Cycle 1). Arguments for these higher temperature requirements stem from the high maximum temperatures modules can reach in hot climates.

Parameter	ISO 4892-2 (Table 3, Cycle 1)	IEC 62788-7-2 (Condition A3)
Exposure	102 min dry, 18 min water spray	Dry
Light Source	Xe-arc fitted with daylight filters	Xe-arc fitted with daylight filters
Broadband (300-400 nm) / Wm^{-2}	60	81
Narrowband (340 nm) / Wm^{-2}	0.51	0.80
Black Panel Temperature °C	65	90
Chamber Air Temperature °C	38	65
Relative Humidity / %	50	20

Table 1: Test conditions for ISO 4892-2 compared with IEC 62788-7-2

Comparing the IEC 62788-2 test to typical climates will give a first idea of its applicability. IEC 62788-2 specifies an exposure time of 2000 or 4000 hours for both frontsheets and backsheets. This corresponds to 162 and 324 kWh UV irradiation in the 300-400 nm range The latter equates to about 2.7 years direct UV exposure in Florida or about 27 years (assuming an albedo of 10%) [9] backside exposure. The sample in the IEC test, will during this time, see a constant temperature of 65°C or slightly more depending on set-up, sample thickness and sample colour. The actual average air temperatures of hot or even desert climates are in the upper 20°C, lower 30°C due to day/night and summer/winter cycles. The backsheet itself will also be heated by the module cell and reflected irradiation, however cooled by wind or rain. Models calculating backsheet temperature show that peaks of 65°C or more are possible in many southern climates, but these are limited to a few hours per day even in the worst cases [10]. The rate of polymer degradation in IEC 62788-2 is expected to be accelerated relative to real field data.

Figure 2 shows the irradiance spectrums of the two weathering methods. ISO 4892-2 has been matched to AM1.5 sunlight in the region below 400 nm, and has lower irradiance than the solar spectrum across the visible. IEC 62788-7-2 has been matched to AM1.5 sunlight in the visible region, resulting in unrealistically high irradiance levels below 400 nm. This is a second potential cause for over acceleration vs real field exposure.

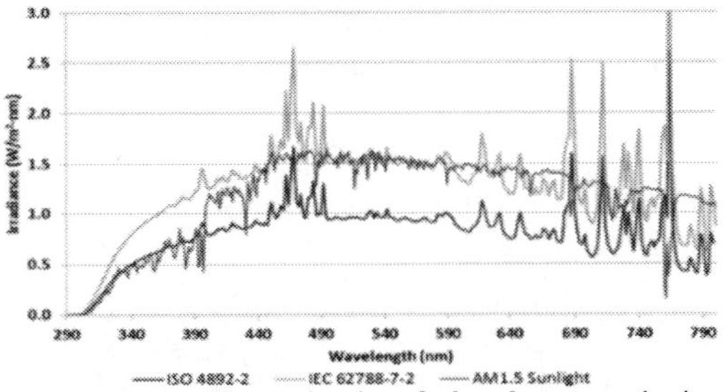

Figure 2: Irradiance levels of the three weathering methods.

Another variable to consider is humidity. IEC 62788-7-2 humidity is set at 20% which is more representative of a desert than a hot and humid climate, so in this case there may be under acceleration of reaction mechanisms such as polymer degradation via hydrolysis.

The choice of temperature, irradiance level and humidity can all result in under or over acceleration. Setting up realistic accelerated weathering tests is a nontrivial task and this highlights the absolute necessity of correlating accelerated weathering data with actual field data.

Assuming an albedo of 10% [9], a PV backsheet specimen mounted with the airside directly facing the sun, 2.5 years direct exposure would give the equivalent of 25 years backsheet UV exposure. Standard practice states that acceleration factors of weathering should be no greater than x10 [7]. This means that direct Florida weathering of backsheets should give an ideal acceleration factor for UV dosage. However as this also accelerates degradation, these accelerated results should be later validated through comparison with materials in application in the field.

This paper will compare the performance of materials for PV backsheets weathered following ISO 4892-2 and the newly proposed IEC 62788-7-2 against direct outdoor Florida weathering.

2 EXPERIMENTAL

Material	Application	Colour	TiO_2
PVF	PV	White	Yes
PET 1	PV	White	No
PET 2	PV	White	Yes
Unstabilised PET	Labels	White	No

Table 2: Description of materials used in this work

Three PV polymeric materials (Table 2), sold for the purpose of use as the outer layer of a backsheet, were studied and compared to an unstabilised white PET. The films were weathered as such and not as part of a full backsheet to fully highlight their material properties.

Materials were weathered in an Atlas Ci-5000 fitted with Type S Boro Inner/Outer optical filters, according to ISO 4892-2 (Table 3, Cycle 1) and IEC 62788-7-2 (Table 1, A3). The filters conform to the ISO standard but not to the requirements of ASTM D7869 as specified in IEC 62788-7-2. This results in a broadband (300-400 nm) of 94 Wm^{-2} rather than 81 Wm^{-2}.

Outdoor weathering of PET 1 and PET 2 was carried out in Miami, Florida (26°N) at a tilt angle of 26° from the horizontal facing south. Specimens were mounted on black-coated aluminium sheets. A thermocouple was fixed to the back of a control sample in order to monitor sample temperature.

Following weathering, mechanical properties of the film were measured. 10 mm wide sample strips were prepared using a JDC precision cutter. Tensile strength (UTS) and elongation to break (ETB %) were measured on an Instron (models 4464, 4301 and 3365), according to the ASTM D882-83 method, with the jaws set at 50 mm separation and the sample pulled at a rate of 25 mm min^{-1}.

SEM was carried out using a FEI Quanta FEG250 Environmental SEM instrument.

3 RESULTS

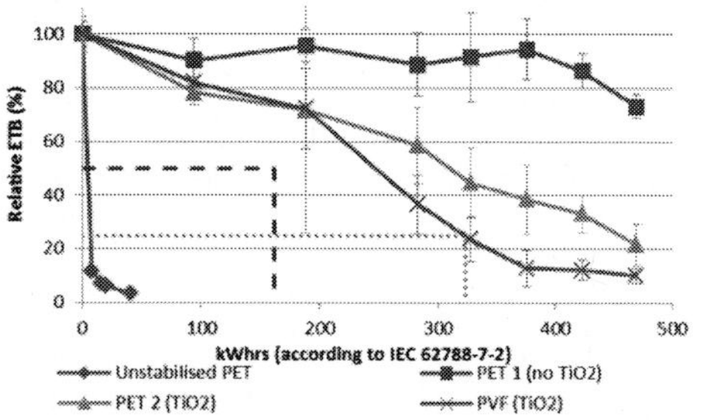

Figure 3: ETB(%) for 3 PV materials and one unstabilised PET material weathered according to IEC 62788-7-2

Figure 3 shows a plot of relative elongation to break (ETB %) against kWh/m^2 UV exposure for each material according to IEC 62788-7-2 (Table 1, A3). The red dotted lines indicate proposed failure thresholds, where a sample should retain 50% ETB by 162 kWh, and 25% ETB by 324 kWh.

As would be expected, the unstabilised PET failed almost immediately with regards to mechanical integrity. PET 1 showed no loss in mechanical properties even after 5000 hours weathering. PVF and PET 2 both showed a reduction in mechanical properties. PET 2 remains well clear of the failure thresholds, while PVF just about passes at 324 kWh. This is surprising as PVF is proven to protect solar panels in the field for more than 25 years [11].

As a second step, we therefore compared the mechanical performance of PET 1 and PET 2 following ISO 4892-2, IEC 62788-7-2 and Florida field weathering [Fig 4].

The mechanical data for PET 1 was consistent irrespective of weathering method followed. PET 2 however shows some differences in terms of mechanical performance over time depending on the weathering method followed. The sample weathered in Florida have the lowest acceleration factor showing no loss in mechanical properties even after 3 years direct UV exposure (which would equate to 30 years backside UV exposure taking an albedo of 10%). The sample weathered according to ISO 4862-2 shows a slight loss in mechanical properties and the sample weathered according to IEC 62788-7-2 shows a more evident loss in mechanical integrity.

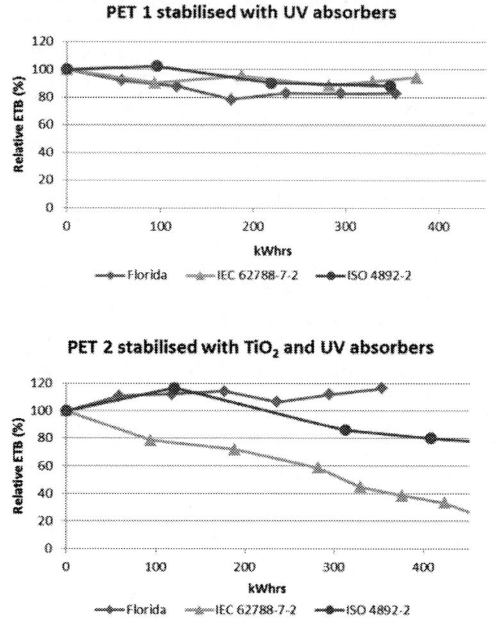

Figure 4: Mechanical data for the PET 1 and PET 2 backsheet materials following weathering

Differences in material composition may explain the difference between Florida and IEC 62788-7-2 data for PET 1 and PET 2.

In order to minimize the loss of mechanical properties, both PET 1 and PET 2 are stabilised. PET 1 contains only UV absorbers, while PET 2 contains both UV absorbers and TiO_2. TiO_2 can protect the binder from direct photochemical degradation, but it is also known to cause indirect degradation of the binder through photocatalysis. The intensity of photocatalytic degradation will depend on TiO2 type and amount and can be tuned by the manufacturer,

As TiO_2 is photocatalytically active at wavelengths below 380 nm, the elevated irradiance levels in this region in the IEC 62788-2 test discussed above may lead to over acceleration of photocatalytic degradation of the polymer binder [12].

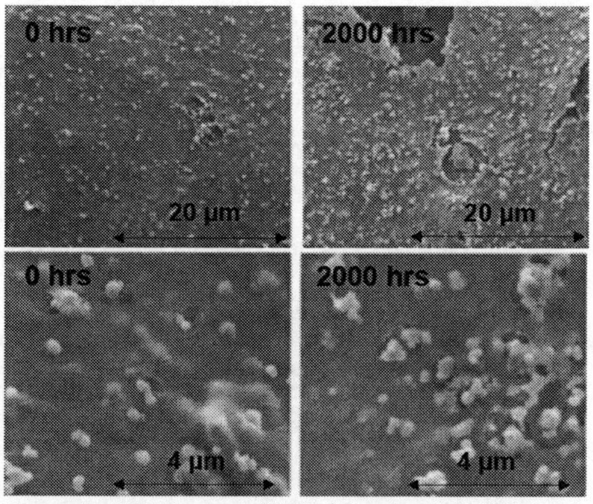

Figure 5: SEM images of PVF following IEC 62788-7-2

Photocatalysis could also explain the rather strong loss in mechanical properties of PVF shown in Figure 3. PVF is reported as being insusceptible to photochemical degradation as it is 'UV transparent'. However, PVF contains *ca.* 20% TiO_2 and so could be subject to photocatalytic degradation, which is possible to see visually as reduction in polymer binder around the TiO_2 particles. The SEM images in Figure 5 confirm this hypothesis, showing reduction of PVF binder directly around the TiO_2 particles.

It has been reported that the rate of photocatalysis has an exponential dependence on temperature, and hence is very sensitive to temperature changes [13-16]. Figure 6 shows the differences between sample temperatures reached following each weathering method. Samples following IEC 62788-7-2 experience significantly higher temperatures than those following other weathering methods, again giving further reasoning for the increased rate of degradation.

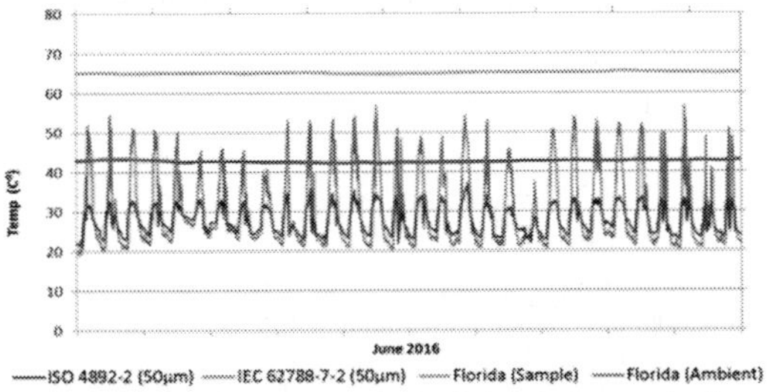

Figure 6: Sample temperatures of the three weathering methods compared to ambient Florida temperatures.

4 CONCLUSION

This paper presents the initial findings to a much larger study. It discusses the newly proposed IEC 62788-2 test conditions versus actual field conditions and analyses deviations in mechanical degradation between the IEC62788-2 test and external Florida weathering for PET and PVF outer layer backsheet films.

Indeed, different stabilized PET samples are shown to degrade clearly less than the accepted failure thresholds. Surprisingly the PVF sample suffers most degradation and does not pass this test well.

The subsequent analysis then implies that the high temperature and increased UV irradiation vs sun levels in the test lead to over acceleration of certain degradation mechanisms, shown most probably here to be predominantly photocatalysis of the polymer binder. PVF is indeed, like most polymers, susceptible to the effects of photocatalysis.

These results raise the understanding of subtle interactions between test conditions and material composition. But in the absence of such awareness, they risk to give false negatives in the evaluation of materials. Further experiments are required in order to determine if this difference in mechanical degradation is a result of elevated temperatures, irradiance or more likely a combination of these two factors.

This paper illustrates again that there needs to be not only consideration of accelerated weathering test procedures and the real maximum irradiance and temperatures these materials will see in application, but also the understanding of the materials under test, their chemistry, composition and possible degradation mechanisms. Once this is achieved, greater understanding and more accurate predictions of material lifetimes should be possible.

6 REFERENCES

[1] D.C. Jordan, J.H. Wohlgemuth, S.R. Kurtz, *Proceedings 27th EUPVSEC,* 2012, 3118-3124
[2] E. Parnham, A. Seaman, A. Whitehead, W. Brennan, E. Ashford, *Proceedings 32nd EUPVSEC,* 2016, 1528-1531
[3] A. Ndiaye, A. Charki, A. Kobi, C.M.F Kébé, P.A. Ndiaye, V. Sambou, *Solar Energy*, 2013, **96**, 140-151
[4] M. Koehl, D. Philipp, N. Lenck, M. Zundel, *Proceedings 24th EUPVSEC,* 2009, 3402-3405
[5] G. Oreski, G.M. Wallner, *Solar Energy*, 2009, **83**, 1040-1047
[6] C.R. Osterwald, T.J. McMahon, *Progress in Photovoltaics: Research & Applications,* 2009, **17,** 11-33
[7] D. Berry, M. Nichols, J. Seebergh, T. Misovski, C. Peters, J. Boisseau, L. Pattison, D. Campbell, J. Quill, J. Zhang, D. Smith and K. Henderson, *Service Life Prediction of Polymeric Materials: Vision for the Future*, California, National Institute of Standards and Technology, 2013.
[8] Atlas Electric Devices Company, *Weathering Testing Guidebook,* Atlas Electric Devices Company, Chicago, Illinois, 2001.
[9] IEC 62788 Measurement procedures for materials used in photovoltaic modules Part 7-2: Environmental exposure – Accelerated weathering tests of polymeric materials
[10] PVWatts Calculator, NREL
[11] http://www.dupont.com/products-and-services/solar-photovoltaic-materials/photovoltaic-backsheet-films.html
[12] T. Egerton and C. King, *J. Oil Colour Chem. Assoc.,* 1979, **62**, 386-391.
[13] S. Costacurta, G.D. Maso, R. Gallo, M. Guglielmi, G. Brusatin, P. Falcaro, *ACS Appl. Mater. Interfaces*, 2010, **2**, 1294-1298
[14] Q. Hu, B. Liu, Z. Zhang, M. Song, X. Zhao, *Journal of Wuhan University of Technology-Mater. Sci. Ed.*, 2010, **25**, 210-213
[15] M. Matsuoka, Y. Ide, M. Ogawa, *Phys. Chem. Chem. Phys.*, 2014, **16**, 3520-3522
[16] R. B. Cundall, R. Rudham and M. S. Salim, *J. Chem. Soc. Faraday Trans.,* 1976, **72**, 1642-1651 (DOI:10.1039/F19767201642).

ANALYSIS OF UV DEGRADATION OF PV BACKSHEETS USING ARRHENIUS FORMALISM TO EXTRACT INTRINSIC MATERIAL CHARACTERISTICS AND MODEL LIFETIME PERFORMANCE

Axel Borne[1], Stephan Padlewski[1], Kaushik Roy Choudhury[2], Thomas Felder[2], T.-John Trout[2], Michael Koehl[3]
(1) Du Pont de Nemours International S.A., Geneva, Switzerland; (2) DuPont, Wilmington, DE, USA; (3) Fraunhofer Institute for Solar Energy Systems, Freiburg, Germany)

ABSTRACT: Photovoltaic need to operate for at least 25 years to make PV today as one of the most competitive source of energy and provide good returns on investment. PV panels are deployed in most regions across the world where they need to endure a broad spectrum of climatic conditions, with stress factors such as the UV irradiation, humidity, and temperature that can degrade the materials used in panels. The incumbent IEC quality standard is not designed to predict the long-term performance of the various materials used, and does not consider the different stresses that exist in the field. The backsheet component is designed to protect the PV panel over its expected lifetime from the stresses of the environment. Preserving its integrity is critical for the sake of electrical safety protection and to maximize the performance and yield of the PV module. This study uses an Arrhenius approach to model the rate of degradation of the backsheet as function of the UV irradiation and temperature. The Arrhenius formalism provides access to the activation energy. The activation energy, considered as thermo-dynamical fingerprint of materials, together with the Arrhenius formalism provide the ability to model degradation and lifetimes under various temperatures. The results show that the presented approach allows realistic life time estimation for materials of PV-modules for specific climatic conditions using accelerated ageing tests on material level.

Keywords: Arrhenius formalism, materials, module, durability, reliability

1 ACCELERATED UV TESTING

Polymeric backsheets are key components to assure functional and safety integrity of PV modules. To provide a safe and productive PV module lifetime of 25 years the PV module must have a durable backsheet. Present testing procedures for IEC standard 61215 approval do not provide adequate durability assurance. In this context, industrial partners, testing bodies and research institutes have explored more extensive testing. In this paper we present accelerated UV exposure results from Fraunhofer, in the frame of a Round-Robin campaign performed under the European SOPHIA project, as well as results from DuPont and NIST, and analyze these results with Arrhenius formalism to calculate and predict backsheet degradation.

Over an expected 25 year service life, a photovoltaic panel is exposed to significant UV irradiation, with a cumulative energy in a desert environment that can exceed 2,300kWh/m^2 on the front through direct exposure and 275kWh/m^2 on the back side of the panel through albedo exposure (10%-20%). [1] With a pre-conditioning of only 15kWh/m^2 on the panel front side and no exposure on the back, the IEC 61215 quality standard largely ignores the significant long term effects of UV exposure [2]. Equipment for full-size UV module exposure is expensive, but valid results can be obtained from component tests, in our case consisting of backsheet coupons irradiated on the backsheet airside.

UV exposures were conducted on PET-based and Tedlar® PVF-based backsheets at three different laboratories, with three different exposure systems.

Table 1: UV exposure lamp and irradiance levels at each testing site.

Laboratory	UV lamp	Irradiance (W/m2) 300nm-400nm
Fraunhofer	metal halide	>150
NIST	microwave discharge	170
DuPont	Xenon arc	85

The Fraunhofer experiments used a PET-based backsheet designated "A", while NIST and DuPont each used the same PET-based backsheet (different from "A") designated "B".

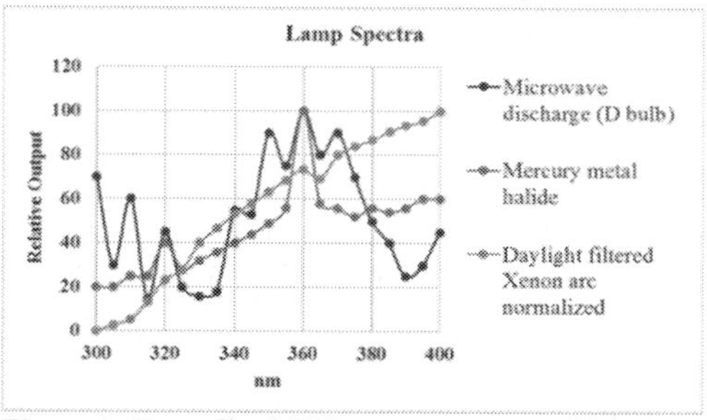

Figure 1. Normalized UV output spectrum of lamps used in this study.

As can be seen in Figure 1, the spectra are similar over the region of interest, and no issues with reciprocity were seen. The microwave discharge lamp used at NIST provides slightly higher UV dose than the metal halide or xenon arc lamps in the short wave 300nm-320nm range. Backsheet samples were irradiated up to 235kWh/m^2 and yellowness in YI units was measured at intervals. Samples were held between 45°C and 85°C during exposure to determine temperature effects on YI values. The data was then used to calculate activation energy (Ea) for yellowing.

2 ARRHENIUS FORMALISM

UV exposure stimulates a wide range of chemical reactions [3] and can damage polymeric materials. In many cases the degradation leads to yellowing and yellowing can be used to track and model the degradation. Starting with a nominal "white" color backsheet material for instance, we can track the polymer degradation by monitoring the coloration change and the related change in Yellowing Index (YI).

An Arrhenius kinetic rate approach [4] can be used to model the thermally activated YI variation under UV exposure. The kinetic rate is given by the generic Equation 1 with the rate constant K as function of the Activation Energy E_a, the frequency factor A and the temperature T.

$$K(T) = A\exp\left[-\frac{E_a}{k_b T}\right] \qquad (1)$$

Equation 2 presents the linear form of the Arrhenius equation (Arrhenius plot) for graphical presentation, and can be used to extract E_a and A.

$$\cdot\ \ln[K\ (T)] = -\frac{E_a}{k_b T} + \ln A \qquad (2)$$

A linear relationship can be obtained between the inverse of temperature (in degrees Kelvin) and the natural log(ln) of the measured rate constant (K) as shown in Figure 4.

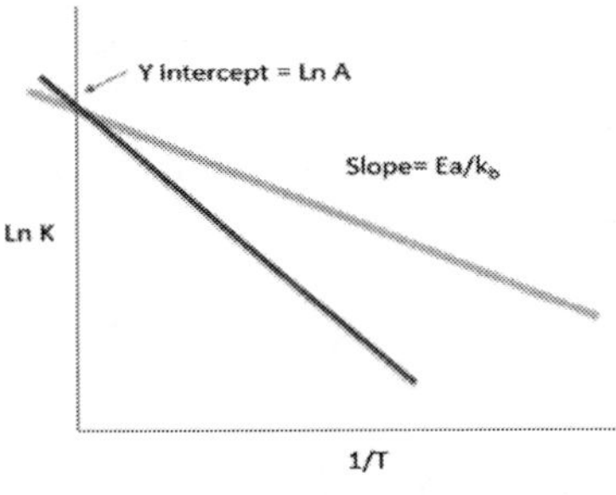

Figure 2: Plot of the Arrhenius equation to extract E_a and A.

The activation energy Ea is calculated from the slope of the graph and the frequency factor A is derived the intercept of the graph with the y-axis. The knowledge of the activation energy E_a and frequency factor A is key to the interpretation of accelerated data for the predictions at lower temperatures. The Arrhenius kinetics model can be utilized across the temperature range where a constant relationship between the effect of temperature on the reaction rate is maintained, i.e. where the graph is linear. The reaction mechanism should not change over the temperature range studied. Linearity is often lost when crossing a phase change, such as freezing, glass-phase transition of the material or when other dominant reactions with similar rates compete with the main degradation process.

The kinetic rate is related to the change in YI according to Equation 3, showing the variation of YI as a function of time and temperature. Here we assumed that each backsheet material is characterized by an initial yellowing index value, hereby noted YI_o, and a saturation yellowing index value after complete degradation of the backsheet noted as YI_{max}.

$$YI(time,T) = YI_0 + (YI_{max} - YI_0)(1 - \exp[-K(T)\cdot time]) \qquad (3)$$

In this manuscript, the activation Energy and the frequency factor is derived from experimental test results derived from a Round-Robin campaign performed under the European project SOPHIA [5], as well as independent experiments performed at NIST and DuPont.

3 YELLOWING DATA

For each UV exposure dataset obtained at a particular exposure temperature, an exponential fit was applied to the dataset as described in the earlier section.

3.1 Sophia dataset

Spectral measurements were performed on PET-based backsheet "A" and Tedlar® PVF-based backsheet, taken under 60°C and 80°C exposure temperatures as listed below in Table 2. YI were extracted out of the spectral data as mentioned above.

Table 2: Measurement points in kWh/m².

	0	40	80	90	120
60°C (333 Kelvin)	x			x	x
80°C (353 Kelvin)	x	x	x		x

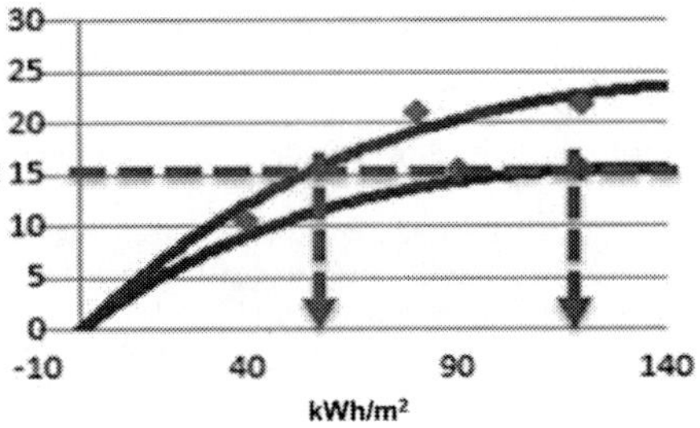

Figure 3: Example of YI vs UV dose for sample "A" for Fraunhofer dataset. The lower graph is measured at T_1 = 60°C and the upper one at T_2 = 80°C.

In Figure 3 the red fit curve of sample "A" represents the yellowness index for temperature 80°C (T_2) as a function of dose (kWh/m²) and the blue fit curve represents the yellowness index as a function of dose (kWh/m²) for a temperature of 60°C (T_1). For a given yellowness index one can determine the required dose. For example, for a YI=ΔYI value of 15.7 for sample "A" (PET sample), we determine a dose of 57 kWh/m² for 80°C (T_2) and a dose of 120kWh/m² for 60°C (T_1).

3.2 NIST dataset

Spectral measurements were taken on PET-based backsheet "B" under 65°C and 85°C. Delta YI vs exposure is plotted in Figure 4.

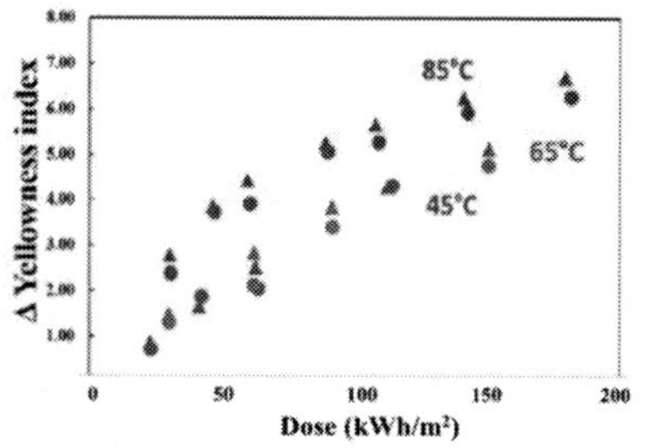

Figure 4: Example of YI vs UV dose for sample "B" for different temperatures for NIST dataset.

3.3 DuPont dataset

Spectral measurements were performed on PET-based backsheet "B" and Tedlar® PVF-based backsheet, taken under 65°C and 85°C. YI vs exposure is plotted in Figure 5.

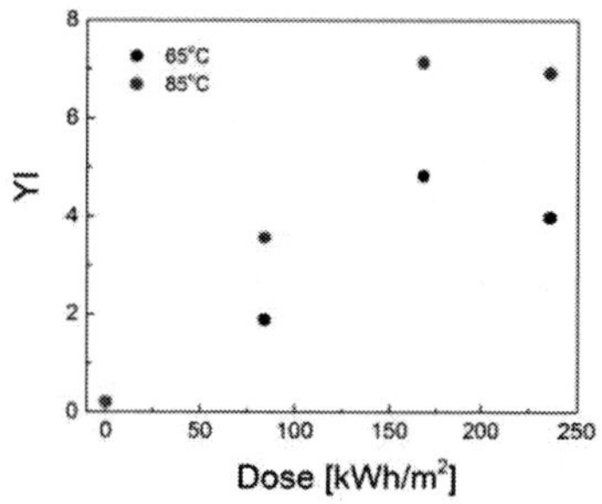

Figure 5: Example of YI vs UV dose for sample "B" for different temperatures for DuPont dataset.

All the PET data showed observable yellowing with increasing exposure and with increasing temperature.

For the PVF-based backsheets, the change in yellowing was minimal with increasing exposure and increasing temperature, as shown in Figure 6.

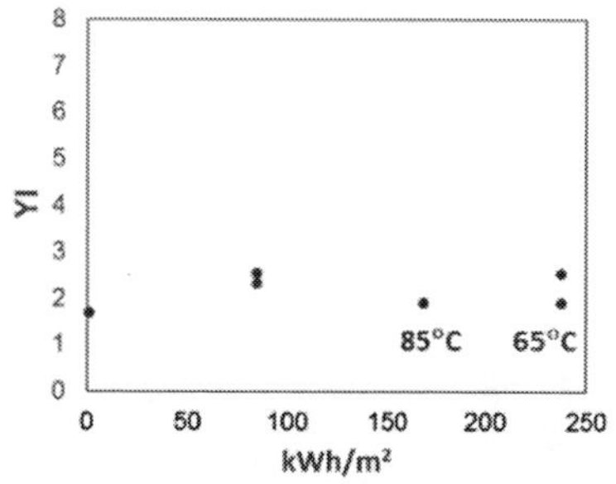

Figure 6: Tedlar® PVF data from DuPont showing minimal change in yellowing. Blue dots are 65°C and red dots are 85°C.

4 ACTIVATION ENERGY EXTRACTION

Table 3 lists the activation energies E_a extracted from the Arrhenius plots of the yellowing data as outlined in section 2. E_a for the PET-based backsheet sample "A" from the Sophia data is estimated at ~36 KJ/mol. This is different from the E_a value extracted for sample "B" from the NIST dataset, at ~14 KJ/mol and that extracted from the DuPont dataset, at ~18 KJ/mol. This difference can be attributed to the fact that the backsheet samples used in the NIST and DuPont studies are identical ones, and are different from that used in the SOFIA study. The difference also shows that even for backsheets fabricated from nominally the same material, PET in this case, the exact material composition of the backsheet outer layer can vary and thereby can lead to different performance of the resulting backsheet. As seen in Figure 6, the PVF-based backsheet shows very little change in yellowness as exposure increases and temperature increases. The Ea value for the PVF-based backsheet is estimated to be >66 KJ/mol.

Table 3: Activation energy Ea in kJ/mol for the three datasets.

Ea (kJ/mol)			
Laboratory	**PET A**	**PET B**	**PVF**
Fraunhofer	36.4		>66
NIST		13.5	
DuPont		17.5	>66

5 RATE OF YI DEGRADATION

The extracted activation energies can be used to calculate YI values using the kinetic rate equation for any given temperature. Using the calculated YI values, the yellowing of PVF- and PET-based backsheets at 20°C is plotted in Figure 7 as a function of UV dose. As is clearly evident from the plot, significant yellowing is observed for the PET-based backsheet sample. In contrast, the PVF-based backsheet has minimal change in yellowness. Interestingly, the plot in Figure 7 compares favorably with DuPont measured field data, as shown in Figure 8.

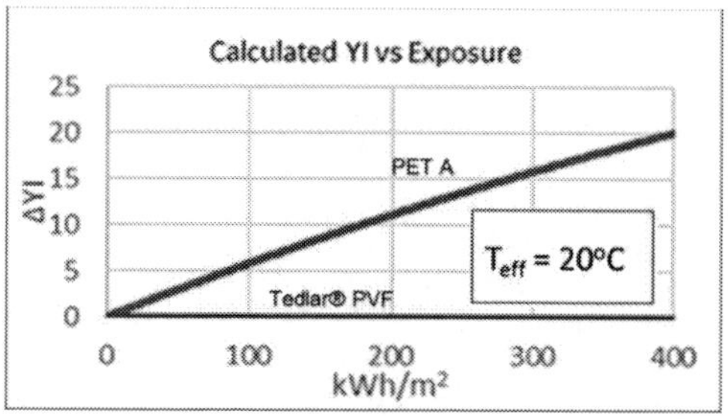

Figure 7: Change in YI calculated with Equation 2 at 20°C using Ea of 36 KJ/mol for PET and 66 KJ/mol for PVF.

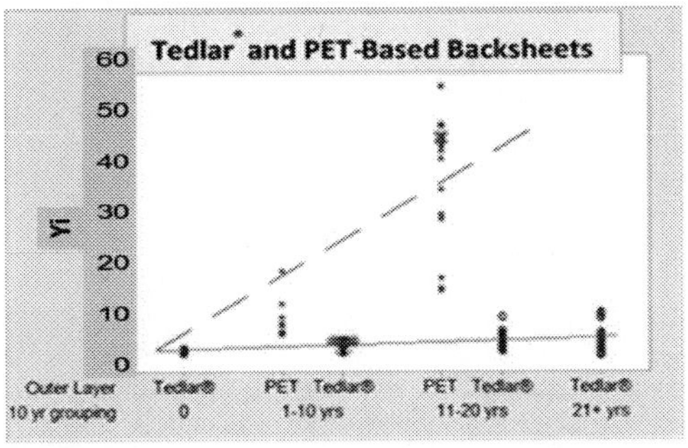

Figure 8: DuPont field data showing change in YI vs years in service for PET (blue) and Tedlar® PVF (red).

6 MECHANICAL LOSS vs YELLOWING

UV exposure of the polymeric backsheet can lead to degradation that can eventually compromise the mechanical integrity of the backsheet in PV modules. Yellowing of the backsheet is believed to be an indicator of such degradation. DuPont has found a correlation between airside yellowing and loss of elongation of backsheets. [6] A wide variety of backsheet coupons were exposed to UVA for a total dose of $195 kWh/m^2$. The yellowing YI and % elongation of the backsheets was measured at the start and conclusion of the exposure, and the loss in % elongation was plotted vs YI. Next, modules which had been in the field for 5-30 years were obtained and the backsheets removed. Yellowness YI and % elongation were measured for the fielded backsheets, and these data points were included with the accelerated exposure data. The plot of all the data is shown in Figure 9. In both the field and accelerated UV exposure data, yellowing higher than a YI value >11 results in near total loss of % elongation for all backsheets. This loss of mechanical properties leads to cracking, delamination, and loss of module protection.

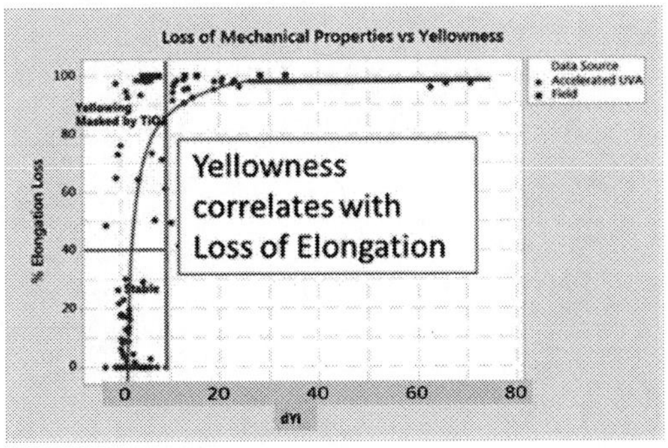

Figure 9: Loss of backsheet maximum elongation as function of the variation of YI.

7 ARRHENIUS PREDICTION OF BACKSHEET USEFUL LIFETIME

From Section 6, a YI value of 11 can be used as the critical point where significant backsheet mechanical loss is observed. Using Equation 3, in conjunction with Figure 7, we can estimate the time needed to reach this limiting YI value, when serious concerns about the integrity of the backsheet can be raised. In Table 4, we calculate this critical time needed by the two types of backsheets to reach the limiting YI value of 11. As expected, the results demonstrate a strong temperature sensitivity of the YI degradation for the PET-based backsheet. It would take far less time (and hence far less UV energy) for the PET-based backsheets to lose mechanical properties, compared to PVF. The model predicts that it would take between 11 and 20 years for PET to reach this YI value at operating temperatures ranging from 20°C to 30°C, whereas it would take >50 years to reach that level of degradation with PVF.

Table 4: Time in years to reach YI of 11.

years	20°C	30°C	40°C	50°C	60°C	70°C
PET	18.2	11.2	7.0	4.7	3.0	2.0
PVF	>50	>50	>50	>50	>50	29.9

These values match well the field observations and accelerated testing results correlating elongation loss with YI values, as demonstrated in Figure 9.

8 CONCLUSIONS

Preserving the integrity of the backsheet in PV panels is critical for the sake of electrical safety and to maximize the performance and yield of the PV module. UV exposure of the polymeric backsheet can lead to chemical changes and polymer degradation that can eventually compromise the mechanical integrity of the backsheet in PV modules. In this study UV exposures on PET-based and Tedlar® PVF-based backsheets were conducted at three different laboratories, with three different exposure systems, and yellowing of the backsheets as an indicator of degradation was monitored. We use an Arrhenius approach to model the rate of yellowing of the backsheet as a function of the UV irradiation and temperature. The results show that PET–based backsheets have significantly lower activation energy for yellowing than Tedlar PVF-based backsheets. This leads to a much higher change in yellowness index for the PET-based backsheets, a finding that correlates favorably with accelerated and field data. We further use the model to estimate the time needed to reach a value of YI when significant loss of mechanical properties is observed in accelerated tests and in the field. PET degradation is predicted to reach a critical loss value between 11 and 20 years, while PVF®-based backsheets maintain mechanical integrity >50 years, indicating the superior UV resistance of Tedlar.

We gratefully acknowledge Xiaohong Gu of NIST for supplying data used in this study.

9 REFERENCES

[1] Solar Power Europe: Global Market Outlook 2017
[2] IEC 61215 "Crystalline silicon terrestrial photovoltaic modules- design qualification and type approval" http://www.iec.ch/ (2016)
[3] [N.J. Turro and J.C. Scaiano, "Principles of Molecular Photochemistry"], University Science Books (2009)
[4] "Arrhenius Equation", https://chem.libretexts.org/Core/Physical_and_Theoretical_Chemistry/Kinetics/Modeling_Reaction_Kinetics/Temperature_Dependence_of_Reaction_Rates/The_Arrhenius_Law/Arrhenius_

[5] M. Kohl, Ultraviolet radiation round-robin testing of various backsheets for photovoltaic modules", IEEE 42nd Photovoltaic Specialists Conference (PVSC), 2015
[6] W. Gambogi, et. al. "Comparison of Accelerated and Outdoor Weatherability of PV Backsheets and Modules" IEEE PVSC (2016)

COMPARISON OF TRACEABLE CALIBRATIONS FOR FILTERED REFERENCE CELLS

Harald Müllejans[1], Ingo Kröger[2], Willem Zaaiman[1], Diego Pavanello[1], Elena Salis[1], Roberto Galleano[1]
[1]European Commission, Joint Research Centre, Directorate C – Energy, Transport & Climate, Energy Efficiency and Renewables Unit, European Solar Test Installation (ESTI), Via Fermi 2749, I-21027 Ispra (VA), Italy
Email: harald.muellejans@ec.europa.eu
[2]Physikalisch-Technische Bundesanstalt (PTB), Bundesallee 100, 38116 Braunschweig, Germany

ABSTRACT: Results from traceable calibrations of PV reference cells are compared. Two primary calibration methods were used, namely the differential spectral responsivity (DSR) method at the Physikalisch-Technische Bundesanstalt (PTB) and the direct sunlight method (DSM) at the European Solar Test Installation (ESTI). Furthermore, secondary calibrations were also made at ESTI against the World Photovoltaic Scale (WPVS). The calibration of c-Si reference cells is well established. Here we investigate the extension to filtered c-Si reference cells. Calibrations were performed on three cells. One unfiltered c-Si reference cell was used for control, confirming the good agreement in the calibration of such devices reported previously. Furthermore, two filtered c-Si PV reference cells were calibrated using the same methods. The agreement between the methods is studied and possible sources for discrepancies are investigated. The spectral mismatch correction is identified as crucial factor. The variation in the spectral mismatch correction with variation in measurement of spectral responsivity is found to be within uncertainty limits, whereas the measurement of spectral irradiance of the simulator can contribute much larger uncertainties due to artefacts in the spectral region around 1050 nm. A correction of the measured data alleviates the effect, so that the equivalence of all methods is demonstrated.
Keywords: Calibration, Reference Cell, Traceability, World Photovoltaic Scale (WPVS)

1 INTRODUCTION

Traceability of PV reference cell calibration [1] is essential to guarantee reliable results. The international standard IEC 60904-4 [1] describes several calibration methods. In all of them the PV reference cell for the measurement of solar irradiance is calibrated against instruments (which are not PV devices) measuring irradiance traceable to the World Radiometric Reference (WRR) or the SI irradiance scale. Furthermore calibration of a PV reference cell against another PV reference cell is also possible and considerably less complex [2]. The latter involves an additional transfer step and therefore has in general a higher uncertainty (UC) associated to it.

The current state-of-the-art best measurement capability for primary calibrations is just below 0.5% (*k=2*) [3]. However, recent work of combining results from calibration with different independent methods on the same PV reference cell has been able to lower the UC of the calibration value assigned to such cells to 0.23% (*k=2*) [4]. Such cells are the embodiment of the World Photovoltaic Scale (WPVS) with the lowest available UC to date [4]. Consequently it is then possible to perform secondary calibrations of other PV reference cells against such a cell representing the WPVS with an UC below 0.5% (*k=2*), i.e. with a measurement UC of a similar magnitude as for a single primary calibration.

While the equivalence of all different calibration methods for c-Si PV reference cells is well established [4], much less attention has been given to the calibration of filtered c-Si reference cells, required for the calibration of PV technologies other than c-Si.

Therefore, here we compare the results of PV reference calibration using two primary calibration methods, directly traceable to WRR and SI irradiance scales, and the secondary calibration traceable to WPVS. The methods are applied to three PV reference cells, one unfiltered c-Si reference cell (for control) and two filtered c-Si reference cells.

2 METHODS

2.1 Direct Sunlight Method (DSM)

The DSM at ESTI has been described in detail recently [3]. In brief the direct solar irradiance is measured, the total with a cavity radiometer (traceable to WRR) and its spectral irradiance distribution with a suitable spectroradiometer, calibrated against the SI irradiance scale. Furthermore, the short-circuit current of the reference cell is measured in parallel with the above mentioned parameters, while the cell is temperature-controlled to 25 °C. The field-of-view of both the spectroradiometer and the reference cell is limited to the same angle as for the cavity radiometers by suitable collimation tubes. Measurements are repeated a number of times on each day and on at least three different days and results are averaged.

2.2 Differential Spectral Responsivity (DSR)

The DSR at PTB has been described previously [5]. The monochromatic irradiance is generated by a xenon/halogen lamp system and a double-monochromator system followed by an imaging optics generating a uniform illumination of the reference cell. The cell is simultaneously illuminated with white light bias generating bias irradiance levels from 0 W/m² - 1100 W/m². The absolute differential spectral responsivity of the reference cell is measured at different bias irradiance level in order to determine the non-linearity of the DSR. The temperature of the cell is controlled to 25 °C. The spectral responsivity as well as the short-circuit current are derived mathematically from the DSR measurements using the tabulated AM1.5G spectral irradiance [6]. The reference to measure the irradiance of the monochromatic light is a calibrated photodiode traceable to SI via the cryogenic radiometer at PTB.

2.3 Secondary calibration against WPVS

The short-circuit current of the PV reference cell is measured under simulated sunlight from a steady-state solar simulator, while the cell temperature is controlled to

25 °C. The irradiance of the simulated sunlight is measured with another PV reference cell (of the ESTI reference cell set [4]) calibrated previously and representing the WPVS. A spectral mismatch correction is applied based on the measured spectral irradiance distribution of the simulated sunlight.

3 RESULTS

Here all calibration results currently available for the three cells are presented. Further measurements have been taken but data analysis is not completed yet. They will be presented when available. The SR of the three cells is shown in Fig. 1.

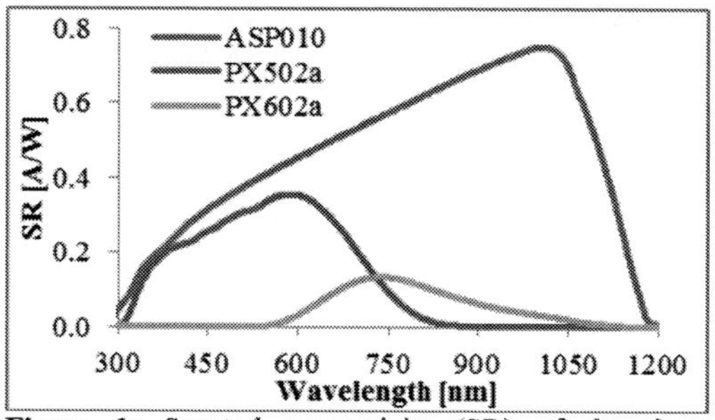

Figure 1: Spectral responsivity (SR) of the three reference cells as measured at PTB.

3.1 ASP010

The reference cell ASP010 (manufactured by ESTI) consists of a 2cm×2cm c-Si high-efficiency non-encapsulated cell which is covered by a circular precision aperture of area 200 mm^2. Results are available from DSM for three years, although only for 2014 do they represent a full calibration of the required three days [1], whereas for 2013 and 2015 they are the data for a single day each (with respective higher UC). The latter were included to show the high repeatability. The DSR calibrated in 2016 agrees with these results as do all calibrations against the WPVS since 2009. The latter show also the stability of the reference cell and the reduction of measurement UC with time due to improvement of methods.

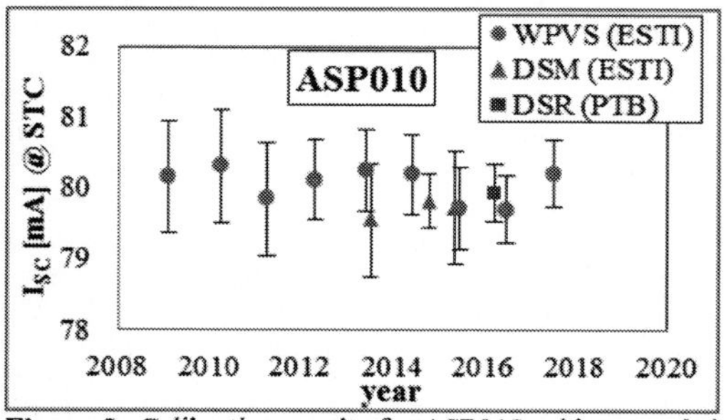

Figure 2: Calibration results for ASP010 with expanded UCs (*k=2*).

3.2 PX502a

The reference cell PX502a (manufacturer PRC Krochmann) is a c-Si cell of 2cm×2cm encapsulated with a filter blocking the transmission of light with wavelengths above 850 nm. Results from DSM are not yet available. The DSR calibrated in 2016 agrees with calibrations against the WPVS from 2009 to 2014. However, in the recent years (2015-2017) the secondary calibrations showed noticeably higher values which are not consistent within declared measurement UCs. This is not attributed to instability of the reference cell but rather increased UC in spectral mismatch correction as presented within the discussion below.

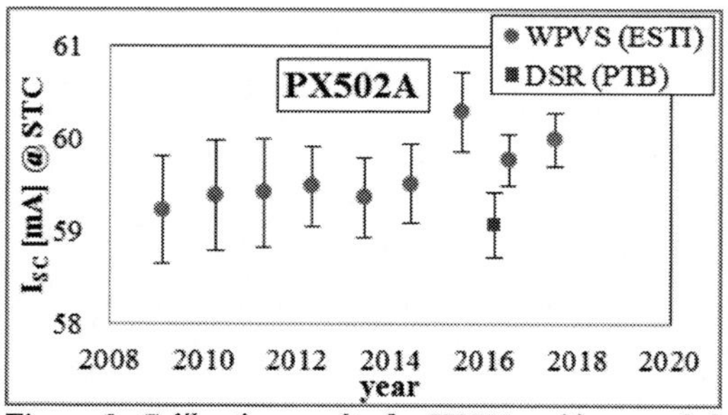

Figure 3: Calibration results for PX502a with expanded UCs (*k=2*).

3.3 PX602a

The reference cell PX602a (AK110, manufacturer Konica Minolta) is a c-Si cell of 2cm×2cm encapsulated with a filter blocking the transmission light with wavelengths below 500 nm and reducing transmission above 800 nm. Additionally the encapsulation is textured to minimize back-reflections. Results are available from DSM for 2013 (alas for a single day, i.e. no full calibration), DSR in 2016 and secondary calibrations since 2012 (when the reference cell was acquired). There is higher variability in the results although most still overlap within declared measurement UC.

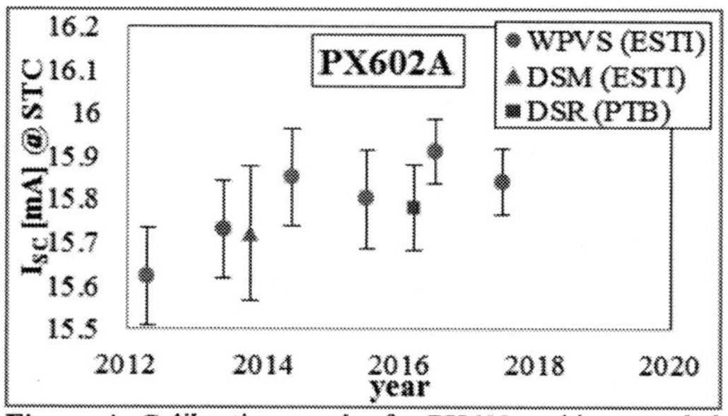

Figure 4: Calibration results for PX602a with expanded UCs (*k=2*).

4 DISCUSSION

4.1 Spectral mismatch (SMM) correction

We investigate the influence of the SMM correction on the results, in particular those of the secondary calibration. For each of the cells the SR was measured by PTB (Fig. 1) but also by ESTI at each annual calibration yielding 9 data sets (from repeated measurements) for ASP010 and PX502a and 6 for PX602a. Furthermore the spectral irradiance of the steady-state solar simulator used for secondary calibration has been measured during every yearly secondary calibration campaign, yielding 9 different spectra. With an in-house batch routine the SSM for all combinations of SR and spectral irradiance was calculated, for fixed SR of the c-Si reference cell (PX301C) and the reference spectrum AM1.5G [6]. In the following we present representative results. Please note that for historic reason the SMM at ESTI is defined as the inverse of the MM as defined in [7].

For any given spectral irradiance, the standard deviation of the SMM with different measurements of SR is typically 0.15% for all cells. As an example the data for the spectral irradiance of 2017 and all SRs of ASP010 is presented in Fig. 5.

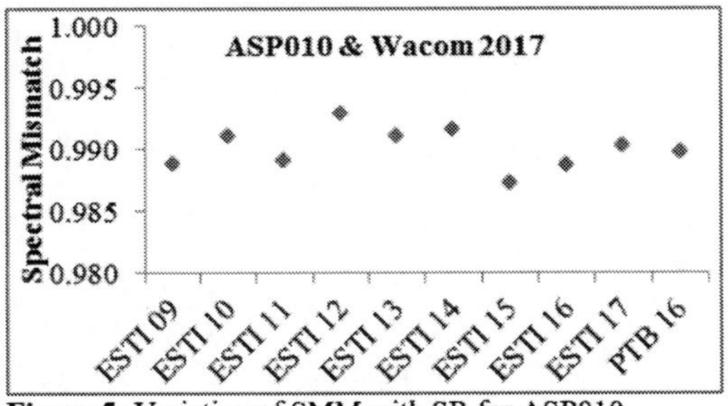

Figure 5: Variation of SMM with SR for ASP010.

For the SMM calculated for a given SR and with varying spectral irradiance the standard deviation for ASP010 is of the same size (0.15%) (not shown) whereas for PX502a it is much higher (Fig. 6). From the level in 2009-2011 the SMM dropped dramatically (-2.5%) in 2012 and 2013 and then went back to original levels from 2014. From Fig. 3 it is clear that the change in SMM did not cause a change of calibration value, as the results from 2009-2014 are basically identical. On the other hand, the change in calibration value for 2015-2017 (Fig. 3) is not related to a change in SMM (Fig. 6). Therefore, the question arises what cause there might be. For completeness we note that for ASP010 the SMM is constant (not shown) as the SMM between to c-Si cells is less sensitive to variations in spectra. For PX602a a variation of SMM was noted but on a smaller scale (ca. +1%) (not shown).

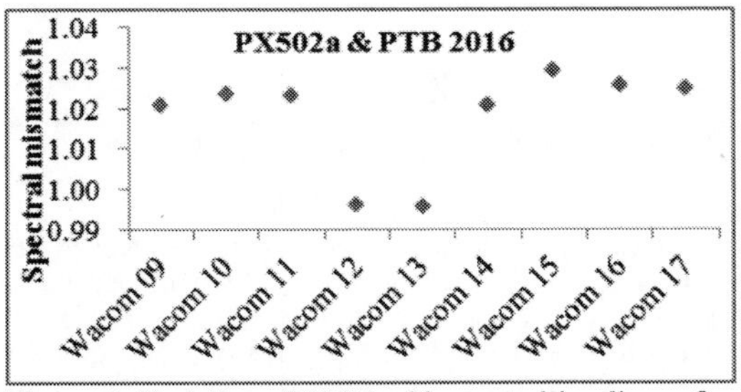

Figure 6: Variation of SMM with spectral irradiance for PX502a.

4.2 Variation of SMM with time

First we investigate why the SMM dropped in 2012 and 2013. Essentially the spectral irradiance of the solar simulator was similar in all years except 2012 and 2013. As example between 2013 and 2015 (Fig. 7) a marked difference occurred, notably between the UV-visible and the infrared wavelength range. This is due to a different adjustment of the solar simulator. Correctly therefore the SMM varied (Fig. 6), and the SMM corrected calibration value (Fig. 3) remained constant (up to 2014). This is an encouraging result, as it indicates that the method in itself is working satisfactorily.

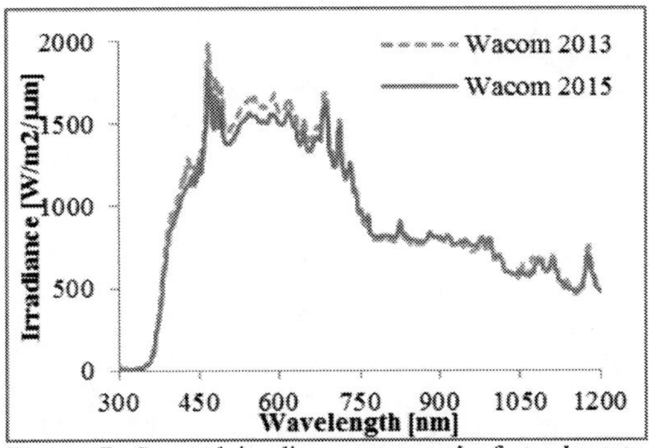

Figure 7: Spectral irradiance measured of steady-state solar simulator in 2013 and 2015 showing a different balance between UV-visible and infrared wavelength range.

4.3 Variation of SMM due to spectral artefacts

Now we investigate the spectra from 2015-2017. Based on observations that the spectra of natural sunlight measured with the same spectroradiometer sometimes showed a discontinuity at the wavelength of 1050nm (the spectrum of natural sunlight is rather featureless in this wavelength range facilitating the detection), the spectra of the steady state solar simulator were examined in detail around 1050 nm. This task is made difficult as the spectrum of the simulator has a peak just above 1050 nm and in general the measured spectra have a certain statistical noise. However, upon close inspection it was found that up to 2014 the spectra were continuous at 1050 nm whereas since then they showed a feature at this wavelength. For demonstrating this, three spectra are plotted on top of each other (Fig. 8). Below 1050 nm artificial high values are observed in 2015 and 2017 (and also 2016, not shown) with respect to previous years (example of 2012 shown). The observed step is about +6% at 1050 nm for 2015 and 2017 with respect to 2012.

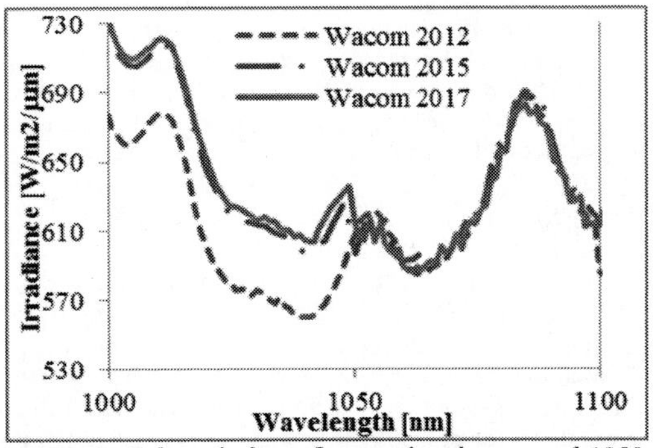

Figure 8: Enlarged view of spectral region around 1050 nm for years 2012, 2015 and 2017. Spectrum in 2012 was continuous at 1050nm whereas in the other years exhibited a marked step.

This was also observed in previous work comparing the OL spectroradiometer used here with a compact array spectroradiometer (CAS) from PTB [8]. A similar step is observed (Fig. 9) and therefore this is classed as a recurrent artefact of the spectroradiometer. In the previous work a correction for this has been applied which eliminates the effect (Fig. 9) and resulted in spectra of both instruments to be fully consistent within declared measurement UCs. The correction was based on a comparison of spectra of natural sunlight measured simultaneously with the OL spectroradiometer and another spectroradiometer at ESTI permanently mounted at the outdoor calibration facility and traceably calibrated.

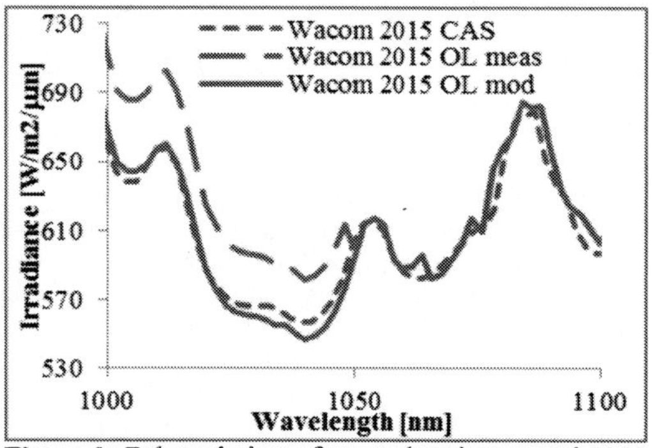

Figure 9: Enlarged view of spectral region around 1050 nm for measurements in 2015 with spectroradiometer CAS (PTB) and OL (ESTI). Measured data for ESTI show a step at 1050nm which is corrected in the modified data.

4.4 Variation of SMM with modified spectra

We now investigate the variation of SMM based on various possible corrections of this artefact. The spectra of 2015 and 2017 were manually modified in the following two ways:

- multiplication in the fixed wavelength range 832 nm – 1048 nm with a constant factor (0.90, 0.92, 0.94, 0.96 and 0.98). This is based on previous work [8] where 0.94 was used.

- multiplication with a factor varying linearly from 1 to 0.94 from a starting wavelength (600 nm, 700 nm, 800 nm, 900 nm and 1000 nm) up to 1048 nm. This is based on the assumption that the size of the artefact is probably a continuous function of wavelength.

All SMMs for these spectra were calculated. The effect of modified spectra on SMM for ASP010 is small (up to +0.34% for the first type and +0.19% for the second type). For PX502a the effect is much larger (up to -2.54% for the first type and -1.39% for the second type). For PX602a the respective values are -0.24% and +0.36%.

For the cells ASP010 and PX602a all variations are essentially within repeatability and reproducibility and declared measurement UCs. For PX502a they are far beyond these limits and the exact correction is important. The most likely candidates for suitable correction are a constant factor of 0.94 between 832 nm and 1048 nm (as used previously [8]) or a linear correction from value 1 at 700 nm decreasing to 0.94 at 1048 nm. The former gives a variation of SMM for PX502a of -1.52% and the latter -1.17%, i.e. of similar magnitude. A further indication that the flat correction of 0.94 is meaningful was obtained by calculating the SMM for the two spectra of CAS (PTB) and OL modified (Fig. 9) for PX502a which yielded an agreement to better than 0.01%.

The average correction for the two most likely correction functions are +0.2% for ASP010, -1.3% for PX502a and -0.1% for PX602a. Applying these corrections to the calibration values presented above (Fig. 2, Fig. 3 and Fig. 4) for the years 2015-2017 makes the results of these years fully consistent with all other results (Fig. 10, Fig. 11 and Fig. 12).

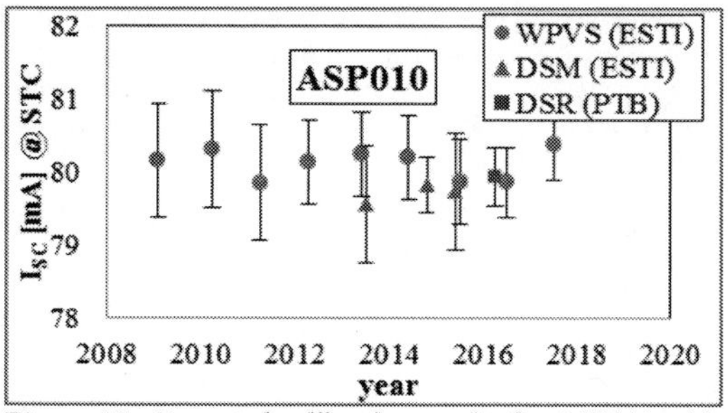

Figure 10: Corrected calibration results for ASP010 with expanded UCs (*k=2*).

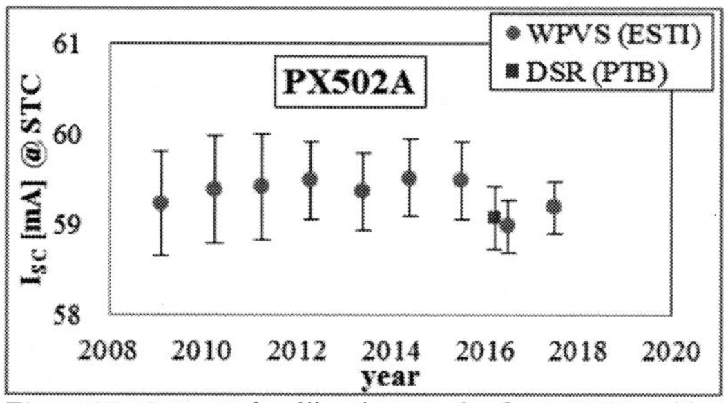

Figure11: Corrected calibration results for PX502a with expanded UCs (*k=2*).

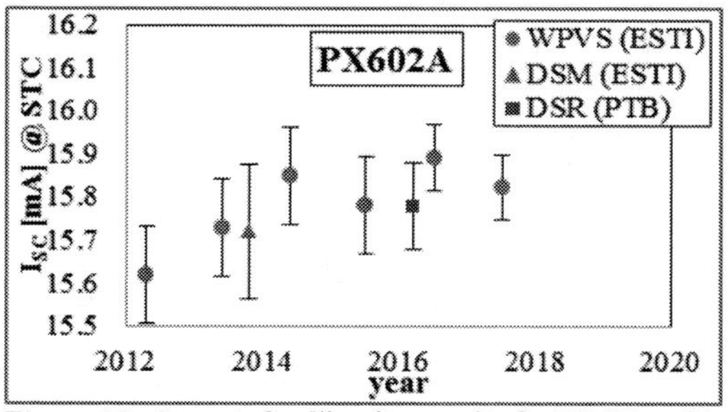

Figure 12: Corrected calibration results for PX602a with expanded UCs (*k=2*).

This paper intended to investigate qualitatively the causes of the observed variations and to give an estimate of the size of the effect. For a complete correction it is planned to characterize the artefact of the OL spectroradiometer quantitatively by comparing it to two further spectroradiometers, another OL with similar characteristics and a CAS recently acquired at ESTI, which is similar to the instrument held by PTB. Based on these measurements a full correction function should be obtained and applied to measured spectra.

5 CONCLUSIONS

The results for the unfiltered c-Si reference cell (ASP010) are fully in line with previous experience, i.e. the calibration of all methods and traceability chains are fully consistent within the quoted UCs. For one of the filtered cells (PX602a) the three methods agree well; however, the secondary calibrations show a somewhat larger scatter over time, which is attributed to the method and not to the instability of the device. For the other filtered cell (PX502a) the result of the DSR calibration agrees well with the secondary calibrations up to 2014. Since then, the secondary calibrations were significantly higher. This is attributed to the method and not to the instability of the reference cell. It was traced to an artefact in the measurement of the spectral irradiance of the solar simulator, whereas the variation of repeated measurements of spectral responsivity was insignificant. Based on plausible corrections of the artefact, it was shown that calibration values after correction are consistent. The determination of the exact correction function will be subject of further work.

These findings underline the importance and the impact of deviations in spectral irradiance measurements on the spectral mismatch correction factors and hence the calibration values of PV devices, in particular for filtered reference cells. A systematic deviation (100% correlated in wavelength over a distinct wavelength range) in spectral irradiance measurements can lead to changes in the SMM of up to 2.5%. Therefore the implementation of correlations of UCs of spectral irradiance measurements and SR measurements in the calculation of SMM correction factors is important.

6 ACKNOWLEDGEMENTS

The research work leading to this article was carried out within the EMRP ENG55 project "Towards an energy-based parameter for photovoltaic classification". The EMRP is jointly funded by the EMRP participating countries within EURAMET and the European Union.

7 REFERENCES

[1] IEC 60904-4 "Photovoltaic devices - Part 4: Reference solar devices - Procedures for establishing calibration traceability" ed.1 (2009).
[2] IEC 60904-2 "Photovoltaic devices - Part 2: Requirements for photovoltaic reference devices" ed.3 (2015).
[3] H. Müllejans, W. Zaaiman, D. Pavanello, E. D. Dunlop "Traceable calibration of photovoltaic reference cells using natural sunlight" *J. Phys. Conf. Ser.* (2017) submitted.
[4] H. Müllejans, W. Zaaiman, E. D. Dunlop "Reduction of uncertainties for photovoltaic reference cells" *Metrologia* **52** (2015) 646–653.
[5] S. Winter, T. Wittchen, J. Metzdorf: "Primary Reference Cell Calibration at the PTB Based on an Improved DSR Facility" *Proc. 16th European Photovoltaic Solar Energy Conf.* (2000) 2198-2201.
[6] IEC 60904-3 "Photovoltaic devices - Part 3: Measurement principles for terrestrial photovoltaic (PV) solar devices with reference spectral irradiance data".ed.3 (2016).
[7] IEC 60904-7 "Photovoltaic devices - Part 7: Computation of the spectral mismatch correction for measurements of photovoltaic devices" ed.3 (2008).
[8] R. Galleano, I. Kröger, F. Plag, S. Winter, H. Müllejans "Traceable spectral irradiance measurements in photovoltaics: Results of the PTB and JRC spectroradiometer comparison using different light sources" *Measurement* (2017) in print http://dx.doi.org/10.1016/j.measurement.2017.09.007
.

EXTENDING THE SPECTRUM CHARACTERISATION OF SOLAR SIMULATORS FROM 300 nm TO 1200 nm: CHALLENGES ON SPECTRAL MEASUREMENTS IN THE UV AND NIR

G. Belluardo[1*], R. Galleano[2], W. Zaaiman[2], M. Pravettoni[3], M. Halwachs[4], R. Fucci[5], A. Drobisch[6], M. Friederichs[7], E. Haverkamp[8], A. Phinikarides[9], G. Friesen[10]
[1]Institute for Renewable Energy, EURAC Research, 39100 Bolzano, Italy;
[2]Energy Efficiency and Renewables Unit, European Commission DG JRC Dir. C, 21027 Ispra, Italy;
[3]Solar Energy Research Institute of Singapore, National University of Singapore; [4]Austrian Institute of Technology GmbH, 1210 Vienna, Austria; [5]ENEA-Agenzia Nazionale per le Nuove Tecnologie l'Energia e lo Sviluppo Economico Sostenibile, 80025 Napoli, Italy; [6]Photovoltaik Institut Berlin, 10997 Berlin, Germany; [7]PV Lab Germany GmbH, 14482 Potsdam, Germany; [8]Institute for Molecules and Materials, Radboud University Nijmegen, 6525 AJ Nijmegen, The Netherlands; [9]Photovoltaic Technology Laboratory, Department of Electrical and Computer Engineering, University of Cyprus, 1678 Nicosia, Cyprus; [10]Institute of Applied Sustainability to the Built Environment, University of Applied Sciences and Arts of Southern Switzerland, CH-6952 Canobbio, Switzerland.
[*]Phone: +39 0471 055626; Fax: +39 0471 055699; E-mail: giorgio.belluardo@eurac.edu

ABSTRACT: Innovative photovoltaic technologies with spectral sensitivity exceeding the 400 to 1100 nm limits (as currently defined by the international standard IEC 60904-9) are nowadays available on the market. This poses new challenges in the correct measurement of the spectral content of solar simulators and natural sunlight in those wavelength bands that lie outside these limits. This study proposes an extension of the IEC 60904-9 bandwidth by adding two bands in the UV (300-400 nm) and NIR (1100-1200 nm) regions. This new proposed extension is analysed in terms of spectral match, using spectral measurements of Global Normal Irradiance (GNI) acquired during the 6th European Spectroradiometer Intercomparison by eight independent laboratories. A laboratory is selected to provide reference spectra, and the spectral match of the other ones is calculated, both on a single-measurement level and on a daily average level. The intra-day and inter-day variations are evaluated as well. Results show that all investigated laboratories are capable to assure a spectral match well below the ±25% limit corresponding to class-A simulators. When the more stringent, informal class-A+ corresponding to the ±12.5% limit is considered, four out of seven laboratories are still compliant with it.
Keywords: IEC 60904-9; spectroradiometer; spectral match

1 INTRODUCTION

The development and market-diffusion of innovative photovoltaic technologies with spectral sensitivity between 300 nm and 1200 nm or above poses new issues in the measurements of spectral content of solar simulators and natural sunlight. Spectroradiometers have become more and more reliable and accurate in the spectral range defined by the standard IEC 60904-9 [1] (400 nm to 1100 nm). However, a forthcoming extension of it would create new challenges for instruments calibration and for spectroradiometers' detector stability and reliability, an increased impact of internal stray-light error and the need for double-detector instruments, just to mention some.

The purpose of this work is to present and discuss some results of the 6th European Spectroradiometer Intercomparison as a real-world, case study on the increased measurement uncertainty when extending the spectra characterisation bandwidth of a solar simulator, as defined by the present IEC60904-9, from 400-1100 nm to 300-1200 nm. The spectral matches of simultaneous measurements of seven different participating spectroradiometers is tested according to the actual IEC 60904-9 bandwidth definition and to a possible future scenario of adding two new bands in the UV (300-400 nm) and NIR (1100-1200 nm) regions. The participating spectroradiometers have different calibration reference chains to SI units and different manufacturers, thus, well representing the variety of the most used instruments in the PV community.

In these terms, this work aims at contributing to the ongoing discussion at the IEC Standardisation Committee, in an attempt to extend the actual band subdivision of IEC 60904-9 to include part of the UV and NIR content for the characterisation of solar simulators' spectra. On the other hand, a modification of the actual standard extending the bandwidth definition seems nowadays necessary to keep pace with the advancements and the new needs of the PV industry (e.g. high-efficiency c-Si, new ARC, colored PV, multi-junction,...).

2 EXPERIMENTAL APPROACH

The sixth European spectroradiometer intercomparison of solar spectrum took place at the "ENEA – Agenzia Nazionale per le Nuove Tecnologie l'Energia e lo Sviluppo Economico Sostenibile" in Trisaia (MT) Italy between the 23rd and the 27th of May 2016. Documentation on the results of the previous editions can be found in [2-5]. Fourteen laboratories participated with a total of twelve spectroradiometers representing both poly and mono chromator technologies. Spectroradiometers have been calibrated prior to the event, either internally by the laboratory itself or externally by accredited calibration laboratories.

The instruments were mounted close to each other on solar trackers, assuring a pointing accuracy of 0.5° peak to peak. A set of cavity radiometers were also installed at the measurements site as reference instruments for broadband irradiance. The solar

Table I: Main characteristics of the spectroradiometers participating the sixth European spectroradiometer intercomparison.

Laboratory	Instrument name	Instrument type	Wavelength band (nm)	Calibration & traceability path
AIT	Ocean Optics	Polychromator, two CCD array detector	300–1600	In house, standard lamp
ENEA	Stellarnet	Polychromator, two CCD array detectors	300–1700	External accredited cal. lab
JRC	EKO wiser system	Polychromator, two CCD array detectors	300–1700	In house, standard lamp
PI Berlin	Tec5	Polychromator, CCD array detector	300–1700	In house, standard lamp
Radboud University	EKO MS711	Polychromator, CCD array detector	300–1100	Manufacturer
PV Lab	Instrument System CAS 140CT 156	Polychromator, CCD array detector	300–1100	Manufacturer
SUPSI	AVANTES (customized)	Polychromator, two CCD, one PDA detectors	300–1700	In house, standard lamp
University of Cyprus	Stellarnet	Polychromator, two CCD array detectors	300–1700	In house calibration

spectrum measurements and validation followed a precise procedure. The measurement time was divided into 1 minute sessions and measurements were considered 'stable' when the total irradiance remained within 1% or better during the measurement time window.

Finally, in order to compare measurements from different laboratories, the acquired spectra were convolved using a Gaussian function in order to increase and harmonize the spectral bandwidth to 4-nm full-width half-maximum, and interpolated to 1 nm.

All participating laboratories were able to perform spectral measurements of global normal incidence irradiance (GNI), and only a few ones performed also spectral measurements of direct normal incidence irradiance (DNI) by applying collimating tubes to the instruments. For the purposes of this study, only GNI measurements from eight participating spectroradiometers were considered. Their main characteristics - including the instrument name and type, typical wavelength range covered by the instruments, and the calibration source – are reported in Table I. According to a confidentiality non-disclosure agreement amongst participants, the results are presented in an anonymous way.

3 RESULTS

Figure 1 shows a comparison of GNI measurements performed by the participating laboratories. In the upper graph the spectra simultaneously-acquired by the reference laboratory (Lab A) and by the others (Lab B to L) on the 24th May at 10:45 (UTC0) are reported. The Lab A's spectroradiometer was chosen as reference because it is calibrated to the SI units with a metrological unbroken chain, and affected by low calibration uncertainty, fully evaluated. In the lower graph, the corresponding deviation from the reference laboratory, normalized to its peak irradiance, is also shown. It is evident, especially in the VIS band (300-700 nm), that the spectral irradiance measured by Lab H is lower than the other laboratories but with no apparent spectral distortion. This could be related to systematic effects in the calibration process. Also the underestimation of spectral irradiance by Lab E at around 1100 nm seems to be related to a loss of accuracy around the higher sensitivity limit of the instrument.

In order to evaluate the spectral match of the measurements with respect to a reference spectrum, the procedure described in IEC 60904-9 is applied with the sole difference that the spectra measured by the reference laboratory (Lab A) are considered as reference in place of the AM 1.5 Global standard spectrum defined in IEC 60904-3 [6]: this is to stress the deviations between instruments rather than the deviation of the measured spectra from the standard reference one. For each interval specified by the IEC 60904-9, the integrated spectral irradiance is divided by the total spectral irradiance between 400 nm and 1100 nm. This percentage is then divided by the corresponding percentage calculated for the reference laboratory, to give the spectral match. Table II shows the spectral match of the measurements acquired on the 24th May at 13:55 (UTC0) by the participating laboratories, with respect to the reference spectrum. The corresponding Air Mass value at this time of the day was 1.38, the highest one during the daily measurements session.

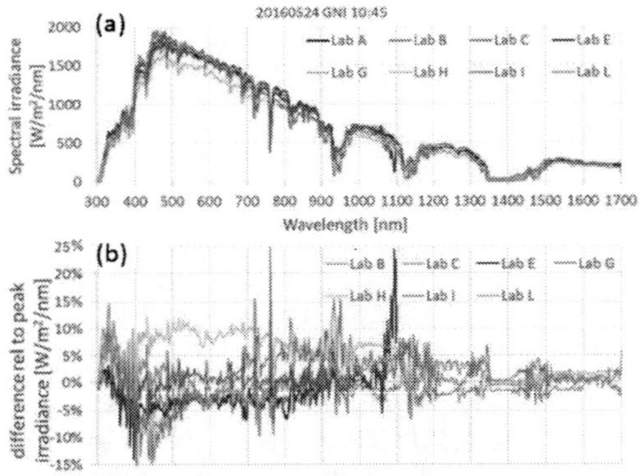

Figure 1: a) GNI spectra measured simultaneously by eight laboratories on the 24th May at 10:45 (UTC0).
b) Deviation of measured performed by Lab B to L from the corresponding measurement by Lab A, normalized to its peak irradiance.

In addition to the bandwidth subdivision of the actual IEC 60904-9, also a new proposed extension having two additional bands in the UV and NIR is presented, and the values of the spectral match are reported in Table II as well (in italic). This extension is a worst case scenario compared to a more conservative bandwidth subdivision where the six bandwidths system is maintained and each band is widen to include the new wavelength limits (see [7]). The spectral match within the new proposed additional bands (300-400 nm and 1100-1200 nm) is worse than in the other bands, as expected. However, the

Table II: Spectral match with respect to the reference lab's measurement on 24th May 2016 at 1:55PM (AM=1.38). Values in italic correspond to the range IEC 60904-9 extension to 300-1200 nm.

Range (nm)	Lab B		Lab C		Lab G		Lab H		Lab I		Lab L		Lab E	
300-400		*8.37*		*-3.76*		*9.07*		*2.85*		*-3.17*		*8.19*		*-1.12*
400-500	-0.14	*-0.48*	-5.16	*-5.00*	-0.70	*-1.74*	-2.22	*-2.62*	-2.13	*-1.80*	-8.82	*-9.76*	-0.43	*-0.54*
500-600	0.55	*0.12*	-0.55	*-0.45*	-3.38	*-4.53*	-0.42	*-0.89*	0.80	*1.02*	-3.31	*-4.30*	0.56	*0.47*
600-700	-0.77	*-1.21*	0.72	*0.83*	-3.76	*-4.92*	-1.19	*-1.66*	0.59	*0.81*	0.87	*-0.08*	-0.29	*-0.38*
700-800	-1.36	*-1.80*	3.30	*3.41*	-2.63	*-3.77*	-0.07	*-0.54*	0.68	*0.90*	7.88	*7.00*	-1.22	*-1.31*
800-900	-0.69	*-1.12*	3.21	*3.31*	2.77	*1.68*	-0.70	*-1.18*	0.63	*0.86*	11.63	*10.79*	-0.83	*-0.92*
900-1100	2.14	*1.71*	1.07	*1.18*	10.22	*9.21*	5.44	*4.99*	-0.13	*0.10*	-1.34	*-2.31*	1.90	*1.81*
1100-1200		*-4.60*		*3.63*		*11.49*		*6.28*		*-0.52*		*8.56*		*4.40*

Table III: daily average values of spectral match with respect to the reference lab's measurement on 24th May 2016. Values in italic correspond to the range IEC 60904-9 extension to 300-1200 nm.

Range (nm)	Lab B		Lab C		Lab G		Lab H		Lab I		Lab L		Lab E	
300-400		*8.74*		*-4.22*		*10.31*		*3.45*		*-2.84*		*3.51*		*-0.92*
400-500	0.21	*-0.09*	-5.25	*-5.07*	0.24	*-0.89*	-3.01	*-3.64*	-1.84	*-1.54*	-11.14	*-12.09*	-0.59	*-0.71*
500-600	0.90	*0.51*	-0.74	*-0.60*	-2.49	*-3.74*	-1.57	*-2.28*	0.88	*1.07*	-4.92	*-5.92*	0.58	*0.48*
600-700	-0.47	*-0.86*	0.49	*0.62*	-3.49	*-4.75*	-1.35	*-2.06*	0.58	*0.77*	0.23	*-0.72*	-0.18	*-0.29*
700-800	-1.08	*-1.48*	3.14	*3.27*	-3.13	*-4.39*	0.04	*-0.66*	0.66	*0.85*	7.49	*6.60*	-1.14	*-1.24*
800-900	-0.86	*-1.26*	3.07	*3.20*	1.11	*-0.10*	-0.91	*-1.61*	0.44	*0.63*	10.47	*9.62*	-0.85	*-0.95*
900-1100	0.75	*0.36*	1.89	*2.02*	9.12	*8.02*	8.36	*7.73*	-0.43	*-0.24*	5.63	*4.73*	1.88	*1.78*
1100-1200		*-5.85*		*3.70*		*11.22*		*10.44*		*-0.34*		*15.43*		*4.20*

values of spectral match remain below 10%, except for the 1100-1200 nm band of Lab G.

For each laboratory listed in Table II, the spectral match differs in the common wavelength bins because the considered total irradiance is integrated in the wavelength range 400-1100 nm for the left-hand side column and in the range 300-1200 nm for the right-hand side column.

The mean values of spectral match for the spectra measured on the 24th May are reported in Table III. Comparing with Table II, it is clear that the intra-day variability depends on both spectral range and laboratory, with maximum absolute differences between single measurement and daily mean close to 7% (see Lab E in the 900-1100 nm band).

In order to better evaluate the behavior of the different laboratories' measurements in terms of spectral match during a specific day, and between different days, the daily average values of spectral match are calculated for the 24th and 26th May, and shown in Figure 2. Black-dotted lines indicate the ±25% limit corresponding to class-A simulators, which in this case refers to a limit on the quality of the spectroradiometers, under the assumption that the reference spectroradiometer (Lab A) acquires the "true" spectrum. For the extended spectral range graphs, also a red-dotted line is displayed corresponding to the ±12.5% "Class A+" limit. Error bars are plotted, corresponding to the ± standard deviation, useful to evaluate the intra-day variation of the spectral match. Results from Lab E and L on the 26th May are missing because these laboratories performed only spectral DNI measurements on this day.

All the considered instruments show spectral match well within the ±25% limits of class-A, even in the new added bands. This good figure of measurement equivalence should assure, also in the extended wavelength range, a correct simulators' characterization, where limitation on signal-to-noise conditions due to short integration time may worsen the measurements accuracy. This represents a sound improvement with respect to the results of the previous spectroradiometer intercomparison [7], where differences up to 30% had been observed in the new bands.

However, the spectral deviation of spectroradiometers increases in the bands 300-400 nm and 1100-1200 nm. For these bands, also the standard deviation of spectral matches is larger for some laboratories (H, L and G), and in some cases is higher than the value of spectral match itself, denoting a possible instrument intra-day drift or instability.

The trend of spectral match is similar in both considered days, which indicates that deviations are reproducible during different days, i.e. they may be affected by systematic errors.

4 CONCLUSIONS

A new bandwidth subdivision extending the IEC 60904-9 spectral irradiance coverage from 400-1100 nm to 300-1200 nm is analysed in this work in terms of spectral match.

As it has often been proposed in the last years, two additional bands in the UV (300-400 nm) and NIR (1100-1200 nm), nowadays necessary to keep pace with the advances and the new needs of the PV industry and market, where several innovative products span spectral sensitivity between 300 nm and 1200 nm - and above.

A real-world case study is presented, where the spectral match is evaluated for seven instruments that took part to the 6th European Spectroradiometer Intercomparison in 2016. The reference spectrum is provided by a reference laboratory, whose spectroradiometer has a fully-evaluated low calibration uncertainty, and is linked to the SI units via a metrological unbroken chain.

By comparing the spectral match values among intercomparison's partners, a figure of merit of measurements equivalence can be evaluated both in the present and in possible new wavelength limits for the

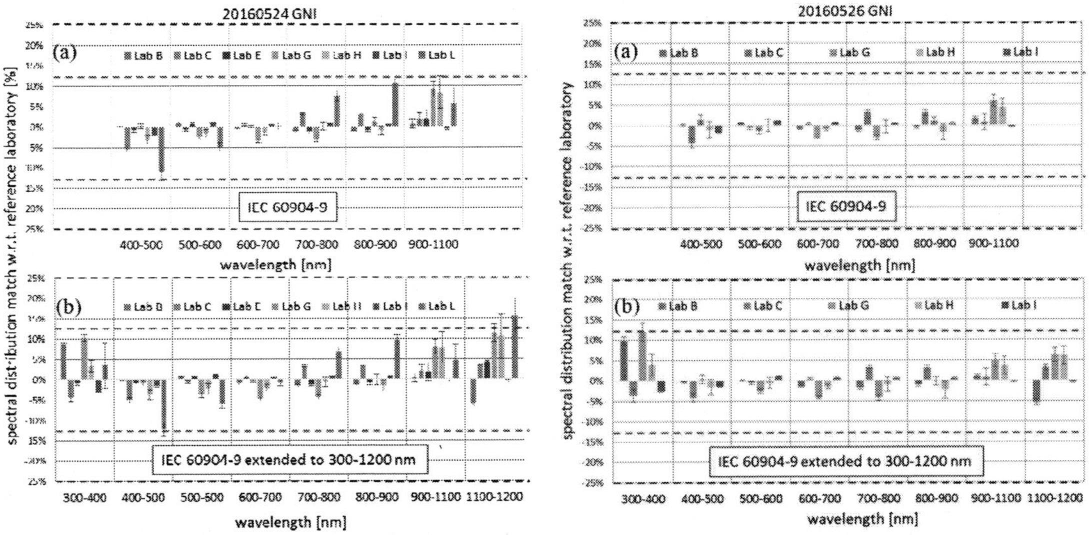

Figure 2: Spectral match for the seven laboratories participating the 6th Spectroradiometer Intercomparison, with respect to the reference laboratory (Lab A), on two different days. a) 6-bands bandwidth definition as defined by the standard IEC 60904-9. b) New proposed 8-bands bandwidth definition, with the addition of two bands in the UV (300-400 nm) and NIR (1100-1200 nm) regions. Black dotted lines: ±25% limit (class-A). Red dotted lines: ±12.5% limit (proposed class-A+).

IEC 60904-9 standard.

Results show that all participating laboratories providing spectral GNI measurements are well-below the ±25% limits of class-A, thus foreseeing a correct simulators' characterization also in the extended wavelength range. This represents a sound improvement with respect to the previous edition of the spectroradiometer intercomparison, where values of spectral match above 30% were found. The use of a class-A+, with a more stringent limit of ±12.5%, is also analyzed and discussed. In this case only one out of seven laboratories shows a daily average spectral match above this level, in the band 1100-1200 nm. Furthermore, when considering intra-day variability, three laboratories have ± standard deviation error bars exceeding the class-A+ limits, denoting a possible instrument intra-day drift or instability.

This work aims at contributing to the ongoing discussion at the IEC Standardisation Committee in an attempt to extend the actual band subdivision of IEC 60904-9 to include part of the UV and NIR content for the characterisation of solar simulators' spectra. New possible bandwidth subdivisions may be explored and assessed, where the six-bands subdivision is kept relevant limits are widened to accommodate the 300-1200 nm broadband range.

5 REFERENCES

[1] Photovoltaic Devices - Part 9: Solar Simulator Performance Requirements, International Organization for Standardization, ISO Central Secretariat, Geneva, Switzerland, ISO/IEC 60904-9:2007.

[2] Galleano, R., Zaaiman, W., Virtuani, A., Pavanello, D., Morabito, P., Minuto, A., Spena, A., Bartocci, S., Fucci, R., Leanza, G., Fasanaro D., Catena, M. Intercomparison campaign of spectroradiometers for a correct estimation of solar spectral irradiance: results and potential impact on photovoltaic devices calibration. Progress in Photovoltaics, vol. 22 no. 11 (2014), p. 1128-1137.

[3] Galleano, R., Zaaiman, W., Strati, C., Bartocci, S., Pravettoni, M., Marzoli, M., Fucci, R., Leanza, G., Timò, G., Minuto, A., Catena, M., Aleo, F., Takagi, S., Akiyama, A., Nuñez, R., Belluardo, G. Second international spectroradiometer intercomparison: results and impact on PV device calibration. Progress in Photovoltaics, vol. 23 no. 7 (2015), p. 929-938.

[4] Pravettoni, M., Galleano, R., Jungst, G., Balenzategui, J., Bartocci, S., Bogeat Sanchez-Piqueras, J.A., Del Bolgia, M., Fabero, F., Fucci, R., Gadermaier, J., Gomez Rodriguez, T., Hoogendijk, K., Lanconelli, C., Marzoli, M., Minuto, A., Rennhofer, M., Serrano Perez, A., Zaaiman, W. Results of the fourth international spectral measurement intercomparison of a steady-state AM0 solar simulator. Proc. International PVSEC-25, Korea, 2015.

[5] Galleano, R., Zaaiman, Alonso-Alvarez, D., Minuto, A., Ferretti, N., Fucci, R., Pravettoni, M., Halwachs, M., Friederichs, M., Plag, F., Friedrich, D., Haverkamp, E. Results of the Fifth International Spectroradiometer Comparison for Improved Solar Spectral Irradiance Measurements and Related Impact on Reference Solar Cell Calibration. IEEE Journal of Photovoltaics, vol. 6 no. 6 (2016), p. 1587-1597.

[6] Photovoltaic Devices - Part 3: Measurement principles for terrestrial photovoltaic (PV) solar devices with reference spectral irradiance data, International Organization for Standardization, ISO Central Secretariat, Geneva, Switzerland, ISO/IEC 60904-3:2016.

[7] Pravettoni, M., Galleano, R., Zaaiman, W. Spectral Measurements in the UV and NIR: Evidence from Previous International Spectral Measurement Intercomparisons. Proc. International PVSEC-26, Singapore, 2016.

SPECTRAL AND ANGULAR CORRECTION - A MULTIDIMENSIONAL APPROACH TO MODEL MEASUREMENTS UNDER OUTDOOR CONDITIONS

F. Plag*, S. Riechelmann, I. Kröger and S. Winter

[1]Physikalisch-Technische Bundesanstalt (PTB), Bundesallee 100, D-38116 Braunschweig, Germany

*corresponding author: Tel. +49 531 592 4148, fax: +49 531 592 694148, e-mail: Fabian.Plag@ptb.de

ABSTRACT: For the determination of a PV modules energy output under different realistic outdoor conditions the influences of spectral and angular effects have to be considered and quantified. The standard IEC 60904-7 describes a procedure to correct for spectral deviations [1]. This work introduces a novel multidimensional model bringing spectral angular effects into the field of PV metrology with focus on quantifying their impact on high accuracy performance measurements and PV module energy rating. Beside the spectral responsivity of a photovoltaic device the model also considers the angular responsivity with respect to the incident spectral irradiance. In this work, the spectral and angular effects are expressed as a combined correction factor. This study aims on the calculation of correction factors for clear sky conditions, since high accuracy measurements or calibration measurements are usually performed under this conditions. We compared the angular losses determined with our model to results of a current state-of-the-art model for several types of solar cells and under different horizontal outdoor irradiation conditions as case studies. We observed differences of more than 3.8 % in the angular loss of the photocurrent generated by the diffuse irradiance. This method is also applicable for modeling indoor measurements using solar simulators containing diffuse irradiance, as well as for the consideration of spectrally dependent albedo and bifacial modules at outdoor conditions. The authors suggest to develop an international standard addressing one of the most important key findings of this study: the angular mismatch correction procedure.

Keywords: energy rating, characterization, calibration, outdoor measurements and monitoring

1 INTRODUCTION

It is presently under discussion how to handle angular losses of PV modules in an applicable manner in order to determine its energy output under various irradiance conditions (energy rating) [1]. A widely-spread approach is the angular loss model after Martin and Ruiz (MR-model), which assumes an isotropic sky radiance, isotropic ground reflectance as well as a spectrally independent and azimuthal symmetric PV module angular dependent responsivity [2, 3]. In order to investigate the impact of these simplifications, we developed a comprehensive multidimensional model. We performed calculations with an anisotropic spectral sky radiance field and PV devices characterized thoroughly for spectral and directional properties. The idea of studying spectral and directional properties combined was already introduced in the late 80ies [4, 5] but according to our knowledge it has never been studied in the context of PV module energy rating. We calculated the global in-plane spectral irradiance for clear sky condition by the means of the radiative transfer software package libRadtran [6]. In addition to spectral irradiances, libRadtran provides the angular distribution of spectral radiance by solving transfer equations including scattering and absorption of the sunlight in the atmosphere and on the ground. The spectrally resolved angular responsivities of different solar cells are measured with our recently developed multispectral laser-based differential spectral responsivity facility [7, 8]. We compare the short circuit current output of two characterized solar cells under global horizontal sunlight relatively for different solar zenith angles and their corresponding calculated spectral sky radiance, in order to obtain the spectral angular mismatches referring to standard test conditions with direct irradiance only. In a second part, we calculated the mismatch factors between an ideal detector following the cosine law with the angle of incidence and a real solar cell to compare them with the MR-model.

However, calculations performed with this new approach require an immense characterization and computation efforts. Hence, the authors have no intention to substitute state-of-the-art models to perform PV module energy rating. Furthermore, we focus on the localization of systematic uncertainties and their magnitude for typical scenarios in order to improve solar PV metrology in the area of energy rating and for high accuracy calibration using global natural sunlight.

2 THEORETICAL BACKGROUND

To investigate the influence of spectral and angular losses for the determination of PV modules energy output, a combined metric is introduced, accounting for the sky's anisotropic spectral radiance and the modules spectral and angular responsivity, which cannot generally be treated as independent detector quantities. Our method enables us to overcome systematic uncertainties introduced by assumptions and idealizations made when using state-of-the-art approaches. The following subsection will introduce our novel approach and will summarize a widely used state-of-the-art model.

2.1 Multidimensional model

In order to describe the output signal of a detector that is receiving input optical power from various directions, it is necessary to understand how the detector with a responsivity s accumulates the input. An ideal detector with a flat surface is accumulating incoming radiance over all infinitesimal solid angles $d\Omega$ within its field of view (FOV), following the cosine law with the angle of incidence (AOI). Regarding the output photocurrent I_{SC} of a PV device with an individual angular responsivity $s(\theta, \varphi)$ in A m² W^{-1}, the following general relation to the radiance L in W m^{-2} sr^{-1} can be found:

$$I_{SC} = \oiint_{\Omega} L(\theta,\varphi) \cdot s(\theta,\varphi)\mathrm{d}\Omega \qquad \text{eq. 2.1}$$

where, $\mathrm{d}\Omega = \sin\theta\, \mathrm{d}\theta \mathrm{d}\varphi.$

To visualize the directionality for a typical scenario during an outdoor secondary calibration, a radiative transfer calculation providing the sky radiance L was performed with the radiative transfer software package libRadtran. In the given example, the position of the sun is at a zenith angle of 48.2° (AM 1.5) and at an azimuth angle of 180°, indicated as yellow star in Figure 1. It shows the radiance distribution of a clear sky providing information of the directional non-uniformity of the sky, in particular for the circumsolar and horizontal regions.

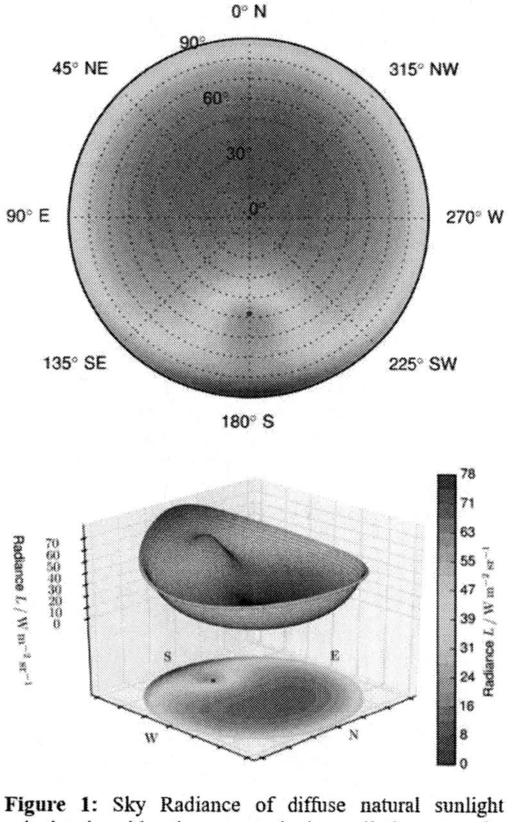

Figure 1: Sky Radiance of diffuse natural sunlight calculated with the atmospheric radiation transfer software package libRadtran for the visible wavelength range. Top: Contour plot of the diffusive components at clear sky condition with a solar zenith angle of 48.2°. Bottom: Surface plot of the same situation. The suns position is shown as a yellow star.

Equation 2.1 remains valid only if the light source radiance L used for the solar cell measurement is spectrally independent, which is usually not the case (particularly for solar simulators with multiple light sources and for natural sunlight at clear sky conditions).

The Earth's atmospheric transmissivity for solar radiation results from a combination of wavelength-dependent scattering, absorption and reflection processes. In order to visualize the resulting spectral effects, the previously calculated spectral radiance L_λ was rendered into RGB-colors for each solar zenith angle across the solar principle plane, representing the meridian from north to south along the position of the sun (see Figure 2).

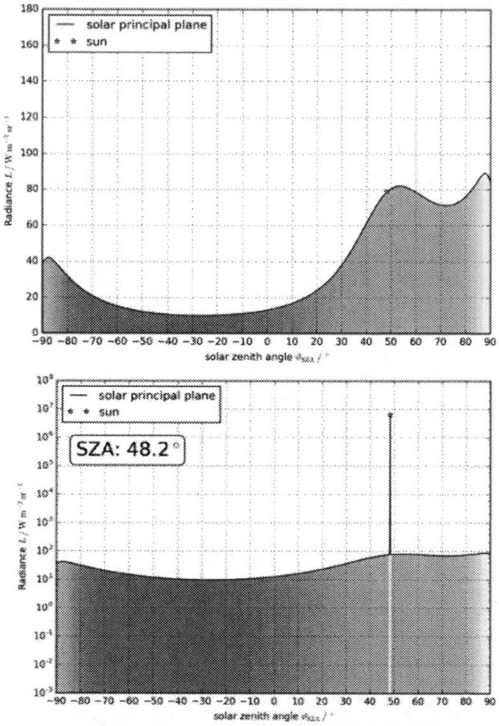

Figure 2: Sky radiance along the solar principal plane (meridian from north to south) taken from simulated sky radiance of Fig.1. The surface color is calculated by a conversion of the spectral radiance in the visual wavelength range to RGB-colors and indicates the sky's bluish coloration and the color of the sun. The suns position is marked by a yellow star at the solar zenith angle. Top: Linear scaled plot of the diffuse radiation components. Bottom: Logarithmic scaled plot of direct and diffuse radiance components.

If the angular responsivity of the investigated PV device provides also spectral dependencies, independent corrections of the spectral mismatch and angular mismatch are not applicable any more, when the device is exposed to an anisotropic spectral radiance source (e.g. a clear sky). For a comprehensive modeling, the spectral radiance $L_\lambda(\lambda, \theta, \varphi)$ as source property and the spectral angular responsivity $s(\lambda, \theta, \varphi)$ as detector property have to be included. The output photocurrent I_{SC} of a PV device is then defined as:

$$I_{SC} = \int_{\lambda} \oiint_{\Omega} L_\lambda(\lambda, \theta, \varphi) \cdot s(\lambda, \theta, \varphi)\mathrm{d}\Omega\, \mathrm{d}\lambda \qquad \text{eq. 2.2}$$

To derive a factor correcting for spectral angular mismatch analog to the spectral mismatch defined in the standard IEC 60904-7 [1], a reference spectral radiance distribution needs to be defined. For a better comparability to irradiance conditions at primary calibrations, the reference spectral radiance field is assumed hereto be a point light source. Therefore, the incident spectral radiance $L_{\lambda,\mathrm{ref}}(\theta, \varphi)$ is only apparent at θ

= 0°, with $L_{\lambda,\mathrm{ref}}(\theta, \varphi) = 0$ for incidence angles $\theta \neq 0°$. The spectral angular mismatch AMM_λ for devices with differing responsivities $s(\theta, \varphi)$ is defined as the product of two output current relations. Each under individual irradiation conditions, primary (reference) and secondary (including diffuse radiance). It can be expressed as:

$$\mathrm{AMM}_\lambda = \frac{I_{\mathrm{Ref,ref}}}{I_{\mathrm{Source,ref}}} \Bigg/ \frac{I_{\mathrm{Ref,DUT}}}{I_{\mathrm{Source,DUT}}} = \frac{\int_\lambda \oiint_\Omega L_{\lambda,\mathrm{ref}}(\lambda, \theta, \varphi) \cdot s_{\mathrm{ref}}(\lambda, \theta, \varphi) d\Omega \, d\lambda}{\int_\lambda \oiint_\Omega L_{\lambda,\mathrm{Source}}(\lambda, \theta, \varphi) \cdot s_{\mathrm{ref}}(\lambda, \theta, \varphi) d\Omega \, d\lambda} \cdot \frac{\int_\lambda \oiint_\Omega L_{\lambda,\mathrm{Source}}(\lambda, \theta, \varphi) \cdot s_{\mathrm{DUT}}(\lambda, \theta, \varphi) d\Omega \, d\lambda}{\int_\lambda \oiint_\Omega L_{\lambda,\mathrm{ref}}(\lambda, \theta, \varphi) \cdot s_{\mathrm{DUT}}(\lambda, \theta, \varphi) d\Omega \, d\lambda} \quad \text{eq. 2.3}$$

The measured short circuit current of a device under test I_{meas} exposed to diffuse irradiance can be corrected to reference conditions for spectral angular mismatch AMM_λ with the correction factor $f_{\mathrm{AMM},\lambda}$ according to :

$$I_{\mathrm{corr}} = \frac{I_{\mathrm{meas,DUT}}}{\mathrm{AMM}_\lambda} = I_{\mathrm{meas,DUT}} \cdot f_{\mathrm{AMM},\lambda} \quad \text{eq. 2.4}$$

$$\text{with } f_{\mathrm{AMM},\lambda} = \frac{1}{\mathrm{AMM}_\lambda}$$

Equation 2.4 already includes the spectral mismatch. The spectral angular mismatch factor can be used for calculations of the mismatch between an ideal detector with cosine behavior and a real device or between two real detectors. The latter is more relevant for high accuracy outdoor calibrations of PV devices using global sunlight.

2.2 Martin and Ruiz (MR-)model

The future energy rating standard IEC 61853-3 [9] proposes the use of an angular loss model introduced by Martin and Ruiz [2,3]. For crystalline and amorphous silicon PV modules Martin and Ruiz proposed the use of the following analytical expression to describe the angular transmission $\tau(\theta)$ with an angular loss coefficient a_r:

$$\tau(\theta) = \frac{1-\exp\left(-\frac{\cos(\theta)}{a_r}\right)}{1-\exp\left(-\frac{1}{a_r}\right)} \quad \text{eq. 2.5}$$

The value of a_r has to be determined by a nonlinear fit and it decreases for better angular responsivities. It can be estimated by using measured values $\tau(\theta)$. The relative angular loss F_D of the short circuit current at diffuse sky irradiance condition can be determined by the expression given in Equation 2.6:

$$F_D(a_r, \beta) = \exp\left\{-\frac{1}{a_r}\left[c_1\left(\sin\beta + \frac{\pi - \beta - \sin\beta}{1+\cos\beta}\right) + c_2\left(\sin\beta + \frac{\pi - \beta - \sin\beta}{1+\cos\beta}\right)^2\right]\right\} \quad \text{eq. 2.6}$$

with the coefficients $c_1 = \frac{4}{3\pi}$ and

$$c_2 = 0.5 \cdot a_r - 0.154$$

β is the angle of the modules surface normal with respect to the zenith, which is set to zero in this work.

3 CHARACTERIZATION OF THE SOLAR CELLS ANGULAR DEPENDENT SPECTRAL RESPONSIVITY

We measured the angular responsivities $s(\lambda, \theta, \varphi)$ of several types of reference solar cells in WPVS design at our Laser-DSR facility by the use of a spectrally tunable modulated monochromatic light source and a two-axis goniometer (θ and φ). To minimize irradiance non-linearity effects, we irradiated the samples with steady state bias white light (mounted at fixed position relative to the solar cells on the goniometer) and measured the modulated monochromatic light with lock-in technique. Polarization effects were compensated by measuring angular dependent spectral responsivity of the sample at two orthogonal orientated polarization states. Therefore, we used polarizers optimized for the wavelength range of interest.

To test the samples on their azimuthal symmetry (φ-axis) we changed their orientation with an increment of 15° between 0° and 90°. A wavelength range from 300 nm – 1150 nm was covered in 50 nm step width. The angular resolution of the zenith-axis (θ-axis) was 5°. The resulting values were interpolated on high resolution grids for angle and wavelength (1 degree and 1 nm, respectively) and subsequently used for computing angular losses with the model proposed in this work. To extend the responsivity data sets onto the complete hemisphere, the three missing quadrants of the solar cells field of view were assumed to behave symmetric compared to the measured first quadrant. To compare the resulting losses for diffuse irradiance between the loss models, a_r coefficients for the MR-model were derived for two different samples (see Figures 3 and 4).

While the determination of the fitting parameter a_r works well for the measured broadband angular transmission of the encapsulated reference cell (Figure 3 top), the fitting algorithm is cannot properly consider super cosine behavior of the device at 1050 nm (Figure 3 bottom). Differences of 0.01 in the angular loss coefficient a_r were found for both devices at different azimuthal orientations.

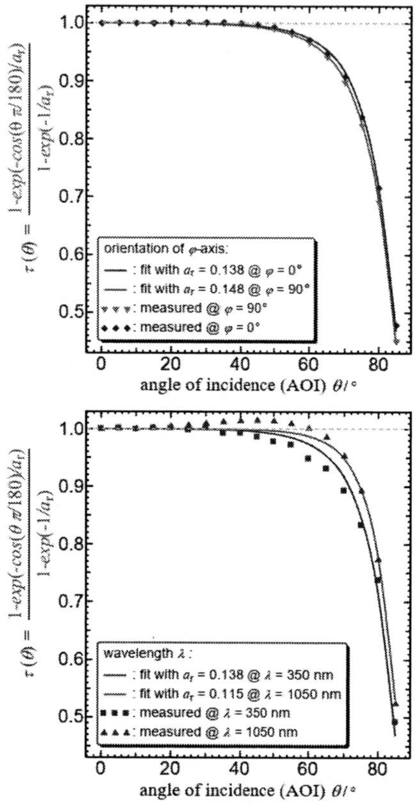

Figure 3: Angular loss coefficients a_r for an encapsulated reference solar cell, determined with a Levenberg–Marquardt algorithm. Top: "broadband" angular transmissions of two different orientations of the device (φ-axis) and for an AM1.5G-weighted spectrum. Blue diamonds indicate measured values at 0°, green triangles indicate measured values at 90°. Dark blue and dark green lines show the analytic expressions with corresponding a_r-values respectively. Bottom: spectrally resolved angular transmission of the encapsulated device for measurements taken at 350 nm (violet squares) and at 1050 nm (dark red triangles). Blue and red lines show the analytic expressions with corresponding a_r-values for selected wavelengths.

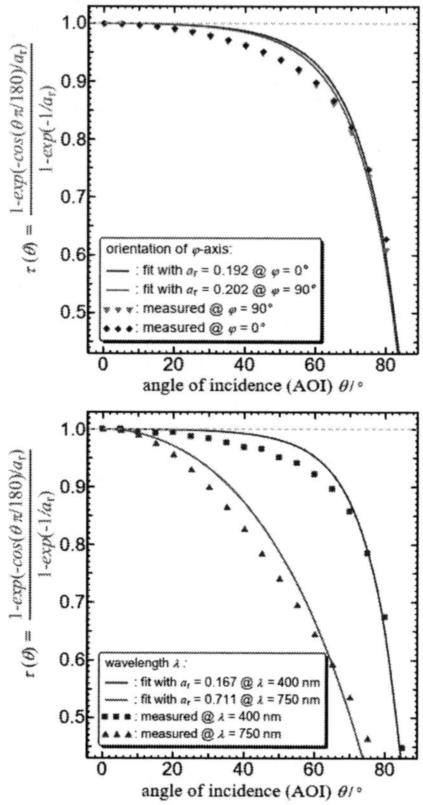

Figure 4: Angular loss coefficients a_r for an IR-filtered reference solar cell, determined with a Levenberg–Marquardt algorithm. Left-hand side: "broadband" angular transmissions of two different orientations of the device (φ-axis) and for an AM1.5G weighted spectrum. Blue diamonds indicate measured values at 0°, green triangles indicate measured values at 90°. Dark blue and dark green lines show the analytic expressions with corresponding a_r-values respectively. Right-hand side: spectrally resolved angular transmission of the device for measurements taken at 400 nm (violet squares) and at 750 nm (dark red triangles). Blue and red lines show the analytic expressions with corresponding a_r-values for selected wavelengths of the IR-filtered reference cell.

The evaluation of the measured IR-filtered reference solar cell results in a poor agreement with Equation 2.5. In the MR-model optical absorption properties of the encapsulation material are neglected, resulting in a larger disagreement of the derived values for a_r (see Figure 4). The consequence is an underestimation of angular losses for isotropic diffuse irradiance, compared to the direct use of the measured and interpolated angular dependent responsivity together with the new approach proposed in this work.

For further calculation, the average a_r values (mean value over 7 azimuthal orientations) for the examined devices were determined to be 0.143 and 0.197 for the encapsulated and the IR-filtered cell respectively.

4 APPLICATION OF THE MULTIDIMENSIONAL MODEL AND COMPARISON TO THE MR-MODEL

This section includes calculations of the spectral and angular mismatch of two different reference solar cells under global horizontal solar irradiance (calculated with libRadtran for varying solar zenith angles θ_{SZA}.

The identification of differences in the angular loss of the direct beam under varying angle of incidence is trivial, when comparing the measured angular transmission directly with the curve determined with a_r. The evaluation of angular losses under diffuse light is more demanding. In the second part of this section we show a comparison between angular loss models for diffuse global horizontal irradiance.

4.1 Computation of spectral angular mismatch between two different reference solar cells for a measurement under global sunlight

The approach presented here, additionally allows a separation between diffuse (sky) radiance and direct irradiance contribution to mismatch. In a first step the spectral angular mismatch at a horizontal device orientation between two reference solar cells was calculated for varying solar zenith angles. The solar cell considered to be the reference detector was encapsulated and the cell considered as device under test was not encapsulated. The mismatch indicates the systematic deviation due to the differences in spectral and angular distribution of the measurement condition compared to the reference condition (direct beam only in the direction of the surface normal with a defined reference spectral irradiance [10]).

While the contribution of the direct irradiance to the angular mismatch between the two investigated samples is nearly zero for solar zenith angles around 0°, it strongly increases for solar zenith angles larger than 30° (see Figure 5). This trend reverses for solar zenith angles higher than 50° due to individually differing detector properties. A maximum effect of 5.8 % spectral angular mismatch for the global horizontal case was found for this example calculation at a solar zenith angle of 50°.
Larger solar zenith angles lead also to an increase of the spectral angular mismatch contributed by diffuse irradiance due to a proportional increase of sky radiance near to the horizon. Nevertheless, we expect a larger total uncertainty of the mismatch factors calculated for large zenith angles. The calculation includes the conventional spectral mismatch. Thus, it can be separated for obtaining information about the pure angular mismatch. More conclusions about the uncertainties in the method presented can be obtained by further studies.

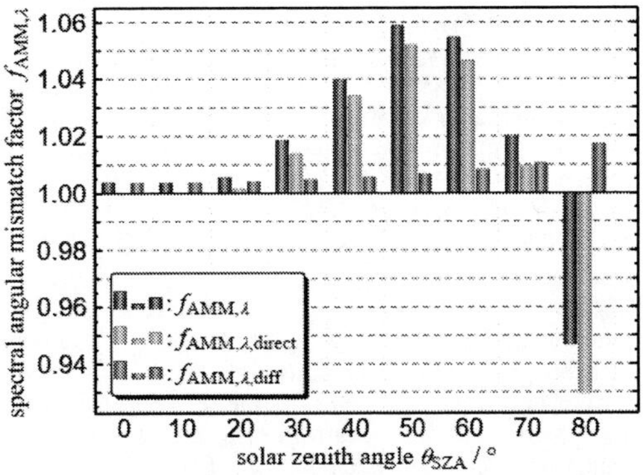

Figure 5: Spectral angular mismatch $f_{AMM,\lambda}$ between an encapsulated and a non-encapsulated reference solar cell under global horizontal irradiance conditions for a variation of solar zenith angles θ_{SZA} in steps of 10° (from zenith to horizon). A separation between the influence of the sky radiance (diff) and direct irradiance on the spectral angular mismatch was performed, indicated here in green and red bars respectively.

4.2 Comparison of angular loss models

We compare the existing state-of-the-art model by Martin and Ruiz (MR-model) with a new approach to study the influence of approximations and on the resulting angular losses.

A substantial difference between the models is that the consideration of the sky's anisotropic radiance. While for the MR-model the simplification to an isotropic sky radiance is obligatory, the new model is able to perform angular loss calculations for both situations: isotropic (ISO) and anisotropic (ANISO) sky radiance. Two different anisotropic sky radiance situations were compared to investigate discrepancies in the calculated angular losses: The first condition considers a solar zenith angle θ_{SZA} of 0°, and the second of 50°. Table I summarizes the differently calculated angular losses for two reference solar cells for diffuse sunlight and horizontal orientation.

Table I: Comparison between angular loss models for diffuse global horizontal irradiance

	Angular loss compared to an ideal detector / %			
model:	MR model ISO	New model ISO	New model ANISO	New model ANISO
θ_{SZA}	-	-	0°	50°
PV device				
Encapsulated cell	-3.93	-3.97	-4.42	-6.23
IR-filtered cell	-6.81	-7.56	-7.79	-10.64

The angular losses under diffuse irradiation conditions at θ_{SZA} = 0° of an encapsulated reference cell and an IR-filtered cell calculated by the MR-model differ from those of the model studied here by 0.04 % and 0.75% respectively. The comparison has been performed for a horizontal device orientation and clear sky condition. The limitation in the analytical function for $\tau(\theta)$ (Equation 2.5) to determine an appropriate angular loss

coefficient a_r for the filtered solar cell was found to be the cause of a disagreement between the models.
A comparison between the new anisotropic model and the MR-model discovers differences of 0.5 % and 1 % for the encapsulated and for the IR-filtered solar cell, respectively at $\theta_{SZA} = 0°$. For larger solar zenith angles ($\theta_{SZA} = 50°$), a larger difference of 2.3 % for the encapsulated device and 3.83 % for the IR-filtered device is observed.
Differences of more than 3 % angular loss were found when using an isotropic and an anisotropic sky radiance distribution as input parameter for our new model. Discrepancies are expected to further increase for varying solar zenith positions due to a resulting increase of the anisotropy of the sky radiance.

5 DISCUSSION AND OUTLOOK

The study on different models for the angular losses of PV modules using different approximations enables the understanding of influences contributing to total uncertainty of mismatch correction factors.

The presented approach allows us to perform case studies with a higher degree of freedom and overcomes limitations due to simplifications and approximations (e.g isotropy of the sky radiance and determination of angular loss coefficient with large discrepancies). Further investigation and a comprehensive sensitivity analysis will enable a discussion on the impact of measurement uncertainty on the angular losses obtained with different approaches.

The model proposed here allow detailed investigations of the sensitivity to the quality of input parameters, such as the need for high accuracy angular dependent responsivity measurements for a large number of azimuthal orientations.

Also, the analysis of measurement uncertainty for the global sunlight procedure described in IEC 61853-2 [11] to determine the angle of incidence effect of PV modules (and a_r values) is subject to further studies. Furthermore, our model is capable to consider for tilt and rotation of modules, spectrally dependent albedo e.g. for studying influences on BIPV or bifacial PV modules. This method additionally enables the advanced modeling of indoor measurements using solar simulators containing diffuse irradiance. The findings of this research will be transferred to standardization bodies to stimulate the implementation of an angular mismatch correction procedure into the IEC 60904-X standard series.

6 SUMMARY

We present a novel multidimensional approach for the computation of spectral angular mismatch. To motivate the need of such a correction, we show that the sky's solar irradiance is not isotropic and that PV devices spectral and directional properties can differ significantly.

We used an example calculation to demonstrate the impact of spectral angular mismatch between two different reference solar cells under global horizontal sunlight. A contribution of the mismatch of more than 5.8 % was found for a global horizontal case. A detailed comparison between the MR-model and our new approach obtained absolute differences between the angular losses of the diffuse sunlight contribution of more than 3.8 %. This differences are caused by an inaccurate fitting procedure used in the state-of-the-art model and the approximation of an isotropic sky radiance. Our investigation enables the understanding of individual influences contribution to the uncertainty of the spectral angular mismatch.

7 ACKNOWLEDGEMENTS

The research work leading to this article was carried out within the EMRP ENG55 project "Towards an energy-based parameter for photovoltaic classification" (PhotoClass). The EMRP is jointly funded by the EMRP participating countries within EURAMET and the European Union.
This work is supported by the Braunschweig international graduate school of metrology (igsm).

8 REFERENCES

[1] IEC 60904-7 Edition 3, Computation of the spectral mismatch correction for measurements of photovoltaic devices, Geneva (2008).
[2] N. Martin and J.M.Ruiz, Calculation of the PV modules angular losses under field conditions by means of an analytical model, Solar Energy Materials & Solar Cells, 70, pp. 25-38, (2001).
[3] N. Martin and J.M.Ruiz, A new model for PV modules angular losses under field conditions, Int. J. of Solar Energy, Vol 22(1), pp. 19-31, (2002).
[4] K. A. Emery and C. R. Osterwald, Efficiency Measurements and Other Performance-Rating Methods, Chapter 4 in Current Topics in Photovoltaics Vol 3, (1988).
[5] Shimokawa R. et al., Possible Errors due to Deviation from the Cosine Response in the Reference Cell Calibration under Global Irradiance, Jap. J. of Appl. Physics, Vol. 25, No. 2, pp. L102 L104, DOI: 10.1143/JJAP.25.L102, (1986).
[6] Emde, C. et al., The libRadtran software package for radiative transfer calculations (version 2.0.1), Geosci. Model Dev., 9, 1647-1672, doi:10.5194/gmd-9-1647-2016, (2016).
[7] S. Winter et al., Design, realization and uncertainty analysis of a laser based primary calibration facility for solar cells at PTB, Measurement, 51, pp. 457 - 463, (2014).
[8] Kröger et al., Spectral angular responsivity calibration facility at PTB, presented at EU PVSEC, (2017).
[9] Future IEC 61853-3 Edition 1 (draft), Photovoltaic (PV) module performance testing and energy rating – Part 3: Energy rating of PV modules, Geneva, (2016).
[10]IEC 60904-3 Measurement principles for terrestrial photovoltaic (PV) solar devices with reference spectral irradiance data. Geneva, (2008).
[11]IEC 61853-2 Edition 1, Photovoltaic (PV) module performance testing and energy rating – Part 2: Spectral responsivity, incidence angle and module operating temperature measurements, Geneva, (2016).

REPRODUCIBLE OUTDOOR I-V CURVE MEASUREMENT BY THE USE OF PV MODULE IRRADIANCE SENSORS AND COMPARISON WITH INDOOR RESULTS

Y. Hishikawa, T. Doi, M. Higa, T. Takenouchi, H. Ohshima and K. Yamagoe
National Institute of Advanced Industrial Science and Technology (AIST), Research Center for Photovoltaics (RCPV)
Central 2, Umezono 1-1-1, Tsukuba, Ibaraki, 305-8568 Japan

ABSTRACT: Precise outdoor I-V curve measurement of PV modules is investigated. Experimental results are analyzed for clarifying the factors which affect the reproducibility, in comparison with the the indoor measurement results. It is confirmed that the procedure of the present study, which uses PV module irradiance sensor (PVMS) and fast I-V measurements within 0.2 s, etc., enables the reproducibility of the P_{max}, I_{sc}, V_{oc} and FF within 1% of standard deviation, based on data including many cloudy days when the solar irradiance is very unstable. Their averaged values also agree with the indoor results typically within ±1.5%. However, when we look into the details of the data, larger deviation sometimes appears, which are important for practical outdoor measurement when the available period is limited. The present results indicate that the spatial nonuniformity of solar irradiance and spectral mismatch between the PVMS and module under test (MUT), affect the reproducibility and precision. Other factors such as the module temperature measurement and differences in the alignment and angular dependences of PVMS and MUT are also relevant. Clarifying those factors are important for further improving the precision and measurement opportunity.

Keywords: Qualification and Testing, Reference Cell, PV Module, Performance, Photovoltaic, outdoor

1 INTRODUCTION

There is growing need for the precise outdoor performance measurement of photovoltaic (PV) modules for low-cost onsite performance measurement, monitoring, and diagnosis etc.. Recently, the authors demonstrated that the measurement procedure including the fast I–V curve measurements in 0.2 s and irradiance measurement using a PV module irradiance sensor PVMS etc., enables significant improvement of the reproducibility to a standard deviation better than 1% even on cloudy days when the irradiance is unstable [1][2]. In this study, the results are analyzed for quantifying the factors which affect the small residual deviation.

2 EXPERIMENT

The outdoor *I-V* curves of commercial crystalline silicon PV modules, which is hereafter denoted as modules under test (MUT), were measured using ADC 4601 *I-V* meters at a sweep time of 0.2 s and intervals of 2.5 -5 s throughout the daytime. The solar irradiance was measured by using a PVMS, which includes an active 5 inch c-Si cell and eight dummy cells around it. The module temperature T_m was measured by resistive temperature detectors (Pt100). The indoor *I-V* curves of the MUT under four irradiance and temperature conditions were precisely measured, and were used to calculate the *I-V* curves at various irradiances and temperatures by the linear interpolation method [3][6], for comparison with the outdoor results. Other details are described in [2].

3 RESULTS AND DISCUSSION

3.1 Reproducibility of the outdoor I-V curve measurement

Typical outdoor experimental I-V curves of the present study are shown in Fig. 1. The statistical standard deviation σ of their P_{max} is a small value of 0.5% (1σ), although the majority of measurements was carried out on cloudy days when the irradiance is unstable. Other parameters such as I_{sc}, V_{oc} and FF also showed similar reproducibility, although not shown in the figure. The averaged value of those parameters agreed with the indoor results within ±1.5%. The result is consistent with the previous studies [1][2]. If we pay attention to a day when the irradiance is very unstable as shown in Fig. 2(a), however, the σ tends to become large, and the portion of outlying data increases (Fig. 2(b). It is important to quantify the origin of the scatter of the data for practical outdoor measurement when the measurement period is limited. It is also useful for further improving the precision of the outdoor I-V curve measurement.

3.2 Spatial nonuniformity of irradiance

Partial shading of sunlight by the clouds and the reflection by nearby objects probably result in nonuniform irradiance [1]. The effect of clouds in a PV systems scale has been recognized [7][8], and combined spatial and temporal nonuniformity of sunlight has been also recently pointed out [9]. However, the nonuniformity in a module scale has not been quantitatively studied yet.

Figure 3 shows the experimental setup of the present study, where three PVMS's were used for investigating the nonuniformity. The distances among the PVMS's were about 1 - 2 m, which are comparable to the size of commercial PV modules. On clear sunny days (Fig. 4(a)), the relative values of irradiance were nearly constant throughout the daytime, and their variation in 10 hours was smaller than about ±1% (Fig. 4(b)). In a shorter time scale of 1 minute, their variation was within about ±0.3% (Fig. 4(c)). It is noted that the results are also affected by the noise of the experimental setup. It is also noted that the U-shaped (green line) and S-shaped (blue line) small variation of the relative irradiance of about ±1% reflect the slight angular misalignment of a few tenths of a degree among the three PVMS's [10].

On a cloudy day as shown in Fig. 5(a), the relative variation was large, and frequently exceeded ±2% (Fig. 5(b)). It even exceeded 10% when the irradiance very rapidly changed (Figs. 5(c) and (d)). It is more correlated with the temporal variation of irradiance rather than the absolute value of irradiance or time of the day. Also, it is larger for PVMC-5b, whose distance from the reference (PVMC-5c) is larger than PVMC-5a. These results suggest that the variation of the relative irradiance of PVMS's in Fig. 5(d) is mainly due to the spatial nonuniformity of solar irradiance, although further study based on longer period of time is necessary for statistical discussion.

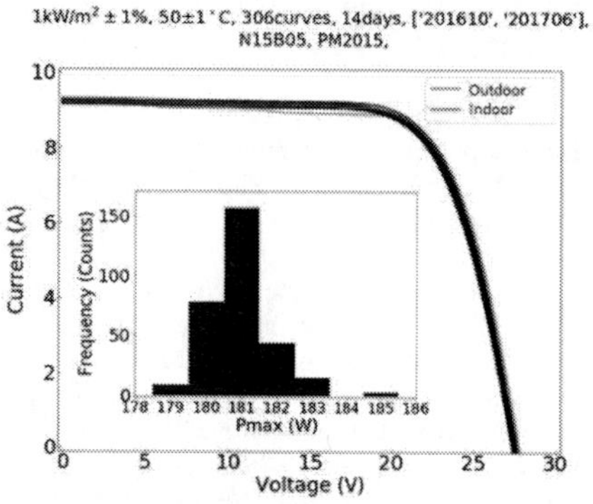

Fig. 1 Typical *I-V* curves of an MUT measured in October 2016 and June 2017 (306 I-V curves in total). The irradiance and module temperature are 1.0 ± 0.01 kW/m² and 50 ± 1 °C, respectively. The curves are corrected for irradiance to 1 kW/m² [3]. The histogram of their P_{max} is shown in the inset.

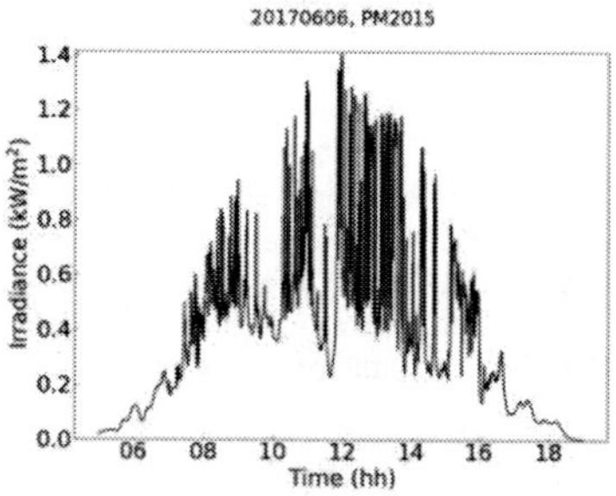

(a)

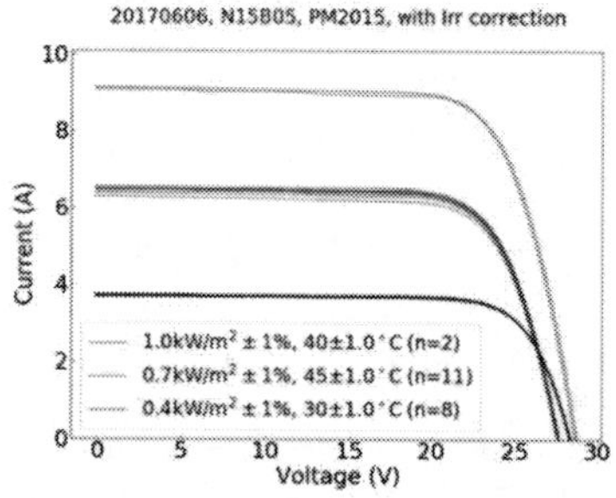

(b)

Fig. 2. (a) Solar irradiance on a day when its temporal variation is significant. (b) I-V curves measured under the irradiance and module temperature of 1.0 ± 0.01 kW/m² and 40 ± 1 °C (red lines), 0.7 ± 0.007 kW/m² and 45 ± 1 °C (green lines), and 0.4 ± 0.004 kW/m² and 30 ± 1 °C (blue). The curves are corrected for irradiance.

Fig. 3. experimental setup of the present study, for investigating the nonuniformity of solar irradiance in a module scale. PVMC-5a, 5b, and 5c denote the PVMS's used for the measurement.

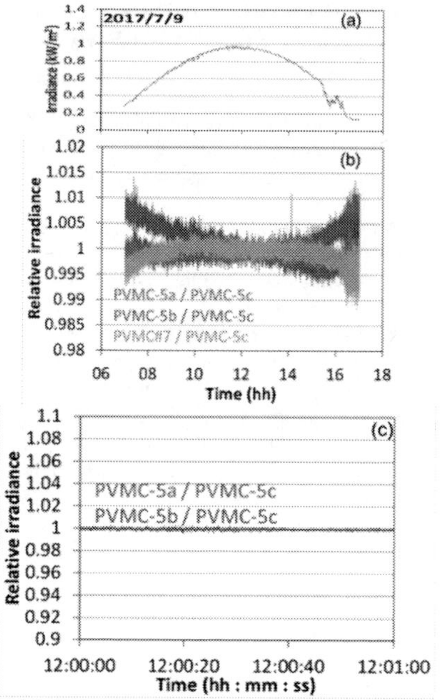

Fig. 4. (a) Solar irradiance on a clear sunny day. (b) Variation of the relative irradiance measured by different PVMS's as shown in Fig. 3, normalized by PVMC-5c in 10 hours and (c) in 1 minute.

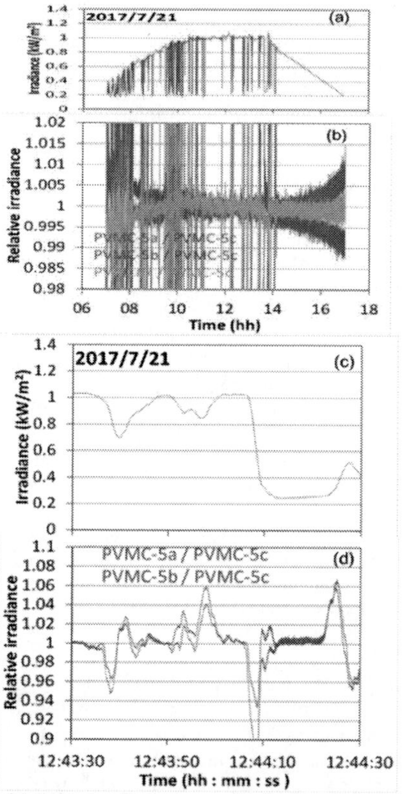

Fig. 5. (a) Solar irradiance on a cloudy day. (b) Variation of the relative irradiance measured by different PVMS's as shown in Fig. 3, normalized by PVMC-5c in 10 hours and (c) in 1 minute.

3.3 Spectral mismatch

The solar spectrum, or the spectral irradiance, is also expected to vary especially on cloudy days, being affected by the scattering and reflection of light by the moving clouds. The spectral mismatch error [11] is much smaller when the irradiance is measured by a PVMS than when measured by a thermal pyranometer [12], thanks to the very similar spectral responsivities of MUT and PVMS. However, it is possibly not negligible even when PVMS is used, if the variation of spectrum is very large. The spectral variation of sunlight on cloudy days has not been well investigated so far.

The spectral mismatch is dependent on the spectral responsivities of PVMS and MUT. The present study assumed the spectra as shown in Fig. 6. On a clear sunny day, the relative solar spectrum was not very different from the standard reference spectrum [13], and, the spectral mismatch (MM) [12] between the PVMS and MUT was within 1 ± 0.002 throughout the daytime in Fig. 7. On the other hand, the relative spectrum was very blue-rich on a cloudy day when the sunlight is shaded by the cloud, as shown in Fig. 8(b), mainly due to relatively large contribution of diffused light. The resultant MM up to about 1.5% is estimated, as indicated in the figures. It is noted that the effect of MM is probably negligible in the results of Figs. 4(c) and 5(b), since the spectral responses of the PVMS's are almost identical.

3.4 Other sources which affect the reproducibility

In addition to the those discussed in the previous sections, there are factors to be considered for fully understanding the scatter of the outdoor I-V curves such as Figs. 1 and 2(b), and further improving the practical precision of the outdoor measurement. The distribution of temperature in an MUT and its precise measurement are the most important factors to affect the output voltage. The in-plane distribution and perpendicular distribution of the temperature are both important, and are much affected by the wind speed and direction [14]. The differences in the angular dependence and irradiance-versus-I_{sc} nonlinearity between the PVMS and MUT are also important for comparing different module technologies. The light reflection from the objects near the MUT also affects the irradiance at MUT for many outdoor measurement setups. The details of those factors are not clearly established yet, and further investigation is desired.

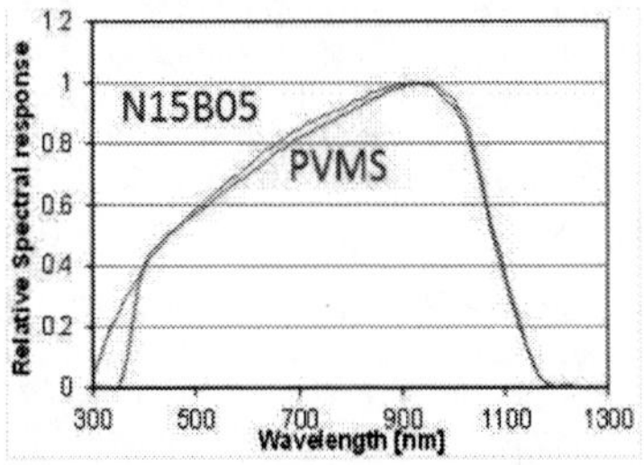

Fig. 6. Examples of the spectral responsivities of PVMS and MUT used in the present study.

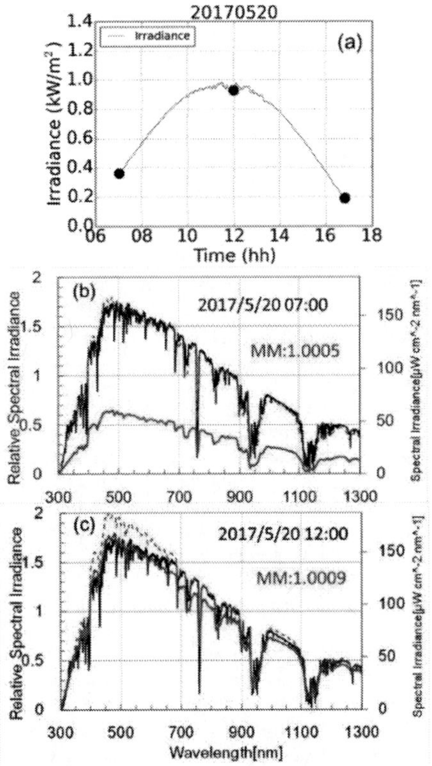

Fig. 7. (a) Examples of the variation of irradiance and (b)(c) spectral irradiance on a clear sunny day.

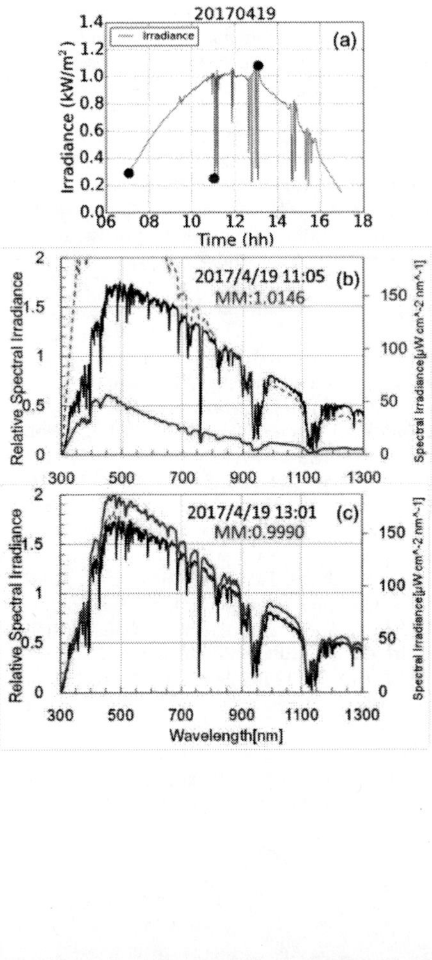

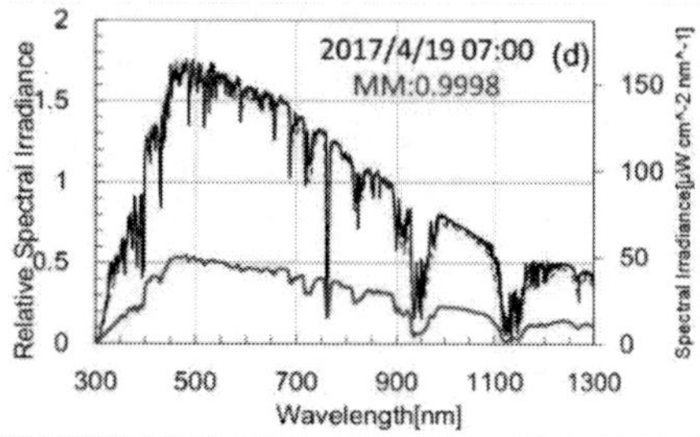

Fig. 8. (a) Examples of the variation of irradiance and (b) -(d) spectral irradiance on a cloudy day when the irradiance is unstable.

3 SUMMARY

The use of PVMS and fast I-V measurement within 0.2 s, etc., effectively improved the measurement precision, which enabled reproducibility (σ) of the measured P_{max} within 1% even on cloudy days when the solar irradiance is very unstable, which agrees with the results of [1] and [2]. However, when we look into the details of the measured data, outlying data with larger deviation sometimes appear.

Since the distance between the PVMS and MUT is comparable to the size of MUT, the present result suggests that the nonuniformity of irradiance within an MUT is also not negligible, in a short period of time when the temporal variation of irradiance due to the cloud is large. It is probably one of the major factors which affect the residual small deviation of the precise outdoor I-V measurement such as Figs. 1 and 2(b). Although the spectral mismatch possibly exceeds 1% in some conditions, it is usually expected to be within ±0.5% for MUT's of crystalline silicon technology, unless the solar irradiance is low due to the shading of sunlight by the cloud. The optical properties of the cloud is not well understood yet. Other important factors, whose details are not discussed in this study, include distribution of temperature in MUT and its precise measurement, and different angular dependence and linearity of PVMS and MUT, as well as the light reflection from the objects near the MUT.

Quantitative clarification of those factors are inevitable for further improving the precision of practical outdoor measurements, and for implementing the precise measurement technology to large scale PV module, string, and array characterization with high throughput [15][16].

ACKNOWLEDGEMENT

This work is supported by NEDO under METI. The authors Thank Prof. T. Kobayashi of Gifu University, Prof. K. Nishioka of Miyazaki University, Prof. T. Minemoto of Ritsumeikan University, Prof. Y. Ueda of Tokyo University of Science, and K. Masuda of JET for fruitful discussions.

References

[1] Y. Hishikawa, T. Doi, M. Higa, K. Yamagoe, and H. Ohshima, "Precise Outdoor PV Module Performance Characterization Under Unstable Irradiance" IEEE J. Photovol. 6-5 (2016) 1221-1227 DOI: 10.1109/JPHOTOV.2016.2571620.

[2] Yoshihiro Hishikawa, Takuya Doi, Michiya Higa, Hironori Ohshima, Takakazu Takenouchi and Kengo Yamagoe, "Effects of synchronous irradiance monitoring and correction of I-V curves on the outdoor performance measurements of PV Modules", Jpn. J. Appl. Phys. 56-8S2 (2017 Jul 27) 08MD07.

[3] IEC 60891:2009 "Photovoltaic devices - Procedures for temperature and irradiance corrections to measured I-V characteristics".

[4] B. Marion, S. Rummel and A, Anderberg, "Current-Voltage Curve Translation by Bilinear Interpolation" Prog. Photovolt: Res. Appl. 12 (2004) 593-607.

[5] Y. Tsuno, Y. Hishikawa and K. Kurokawa, " Modeling of the I–V curves of the PV modules using linear interpolation/ extrapolation", Solar Energy Materials & Solar Cells 93 (2009) 1070–1073.

[6] Y. Tsuno and Y. Hishikawa, "Comparison of Curve Correction Procedures for Current–Voltage Characteristics of Photovoltaic Devices", Jpn. J. Appl. Phys. 51 (2012) 10NF02.

[7] E. C. Kern Jr., E. M. Gulachenski, and G. A. Kern, "Cloud effects on distributed photovoltaic generation", IEEE Trans. Energy Conversion, 4 (1989) 184-190

[8] U. Sangpanich, J. Thongpron, K. Kirtikara and C. Jivacate, "Study of Moving Cloud Effects on the Poper Quality of Grid-Connected Photovoltaic System Compared with Simulation of PV Power Using a DC Generator", Technical Digest of the International PVSEC-14, Bangkok (2004) 959-960

[9] J. Zhang, T. Kobayashi and J. Yoshino, "Characteristics of Solar Irradiance Fluctuations and Corresponding Weather Conditions" Proceedings of JSES/JWEA Joint Conference (2016), Matsuyama, 443-446 (in Japanese)

[10] T. Doi, Y. Hishikawa, M. Higa, T. Takenouchi, H.Ohshima, K. Yamagoe, "Effects of deviation of the normal vector of a plane on the solar irradiance measurement using PV module irradiance sensor" Proceedings of JSES/JWEA Joint Conference (2016), Matsuyama, 37-40 (in Japanese)

[11] IEC 60904-7:2008 "Photovoltaic devices - Computation of the spectral mismatch correction for measurements of photovoltaic devices "

[12] J. Chantana, H. Mano, Y. Horio, Y. Hishikawa, T. Minemoto, " Spectral mismatch correction factor indicated by average photon " energy for precise outdoor performance measurements of different type photovoltaic modules", Renewable Energy 114 (2017) 567-573

[13] IEC 60904-3:2016 "Photovoltaic devices - Measurement principles for terrestrial photovoltaic (PV) solar devices with reference spectral irradiance data"

[14] K. Okumoto, K. Sakamoto, Y. Ota, K. Nishioka" The measurement and examination of temperature distribution for a photovoltaic module in windy conditions " Proceedings of JSES/JWEA Joint Conference (2016), Matsuyama, 247-248 (in Japanese)

[15] Y. Hishikawa, A. Masuda, T. Doi, Y. Ueda "Application of Precise PV Module Performance Measurement Technology to System Measurement and Diagnosis" Proceedings of JSES/JWEA Joint Conference (2016), Matsuyama, 293-296 (in Japanese)

[16] K. Masuda, "Study of outdoor performance-measurement of photovoltaic modules in an array" Proceedings of JSES/JWEA Joint Conference (2016), Matsuyama, 345-346 (in Japanese)

SMART PV MODULE BATCH TESTING: REDUCTION OF PERFORMANCE MEASURMENT UNCERTAINTY BY UPTO 50%

Bengt Jaeckel[1], Blagovest Mihaylov[2], Ralph Gottschalg[2], and Juergen Arp[3],

[1] UL International GmbH, Admiral-Rosendahl-Strasse 9, 63263 Neu-Isenburg (Zeppelinheim), Germany, Bengt.Jaeckel@gmail.com, Tel. +49-162-2621901
[2] Centre for Renewable Energy Systems Technology, Loughborough University, B.V.Mihaylov@lboro.ac.uk, R.Gottschalg@lboro.ac.uk, LE11 3TU, UK
[3] PV LAB Germany GmbH, Gartenstrasse 36, 14482 Potsdam, Germany, j.arp@pv-lab.de

ABSTRACT:
Measurement uncertainty may bias verification measurements for PV systems, e.g. delivered DC capacity for a PV power plant or establishing warranty claims. The problem is present when evaluating modules both on site in a PV plant and in a laboratory. This becomes quite critical if the module's STC performance is borderline in terms of compliance with stated/expected power levels as minimum acceptable power levels are minimum power minus measurement uncertainty.
The measurement uncertainty equates directly to a financial risk in the due diligence of larger plants. This is even more obvious as PV module prices have fallen by 30% in 2016 alone. The cost of capital plays an increasingly important role in the overall systems' cost and is strongly affected by the perceived risk.
To reduce this risk of false DC rating, for large enough projects – typically in the range of several MW, a batch of modules is tested to confirm DC baseline performance.
Typical measurement uncertainties are from 2.5% to 3% with 95% confidence for crystalline PV modules. This contribution presents an approach to nearly halve the uncertainty in the measurement of batches of modules compared to the uncertainty of a single laboratory. This is achieved via measuring a sub-sample of these modules by two additional laboratories with different traceability chains. A GUM-compliant uncertainty analysis is presented for this approach and an example is given for the three measurement laboratories involved in this work. These laboratories have measurement uncertainties (k=2) in the range of 2.5-3% and the resulting uncertainty is less than 1.5%. The presented approach is a cost-effective method to verify large numbers of modules in field tests with very low uncertainty, which is comparable to expensive high precision individual PV module measurements.

Keywords: electroluminescence, micro cracks, probability, power loss

1. INTRODUCTION

With PV module prices falling, by 30% [1] in 2016 alone, the cost of capital plays an increasingly important role in the overall systems' cost. Financing costs are correlated with the perceived risk. While there are numerous risk contributors in the lifecycle of a PV project [2], the most straightforward is associated with the performance estimates based on the DC rating of the modules. To reduce this risk, a batch of modules should be tested independently to confirm baseline performance, at least for large enough projects. The number of samples can be derived e.g. using ISO 2859-1 [3]. Similarly, a batch of modules is tested in case of having to investigate a PV module warranty issue.

Reputable PV characterization laboratories have established measurement procedures that ensure high repeatability and reproducibility of their measurements. This is documented as part of an ISO 17025 accreditation. Repeatability, however, is only a fraction of the overall uncertainties. Typical overall measurement uncertainties for PV module STC [4] measurements are from 2.5% to 3% with 95% confidence for crystalline silicon PV modules. International inter-laboratory comparisons suggest that the uncertainty estimates are not too conservative [5]-[8]. Recent announcements claim for individual PV module measurements uncertainties as low as 1.5% [9]. State-of-art measurement costs per single module are in the range of >1000€ and therefor are prohibitively expensive for large batches of PV modules that need to be tested to get a representative result.

In contrast, field measurements are much cheaper, but have larger uncertainties. It is however possible for field measurements to have measurement uncertainties comparable to typical laboratory ones, if special care it taken for the setup. This includes a thermally controlled environment in the mobile laboratory and onsite control of all measurement equipment, including the homogeneity of the light source and positioning of the irradiance sensors. Portable I-V curve tracers using the sun will typically have much higher uncertainties in the range greater than 5% when using specific reference modules and larger than 7% without such measures.

This paper focuses on the assessment of larger batches of PV modules taken from the field and not on individual "golden" modules. An approach to reduce the financial risk while keeping cost reasonable is described within this contribution.

2. APPROACH OUTLINE

A metrological approach is used to minimize the performance risk equivalent to the reduction of measurement uncertainty. In short this is achieved via multiple laboratory measurements, reducing the measurement uncertainty of the mean value of a batch of PV modules. The proposed methodology results in a lower uncertainty than that of a single laboratory while still being cost effective. This is especially the case when applied to large PV fields.

The proposed procedure is sketched in Figure 1, where an example of a mobile laboratory (lab1) with excellent repeatability is used in conjunction with two further calibration laboratories (lab2 and lab3). During the time of the field measurement, the Job Specific Reference Modules (JSRMs) are measured onsite, carefully packed and sent to the other calibration laboratories to generate the calibration file for the measurements with reduced uncertainties.

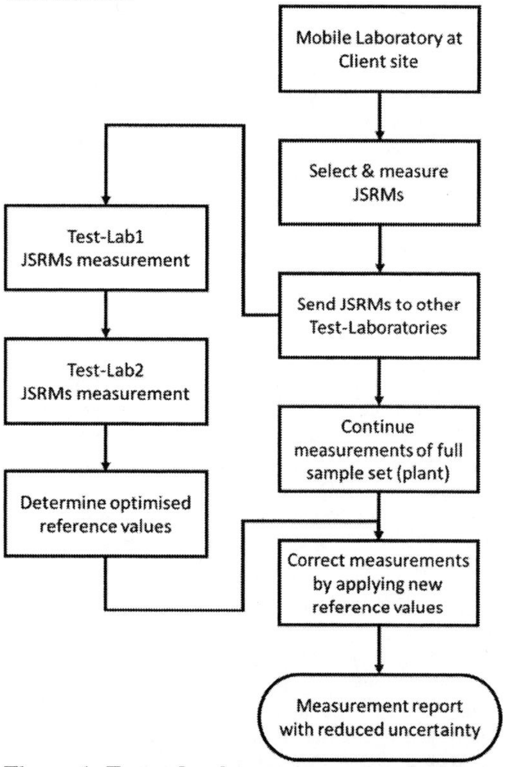

Figure 1: Example of a measurement of a larger sample as e.g. a warranty claim for a large PV system.

With a comprehensive uncertainty analysis of each PV measurement laboratory, it is possible to separate the uncertainty contributions as [10]:

- Persistent (i.e. systematic) between consecutive measurements
- Volatile (i.e. random) for each measurement
- A combination of both with an estimate of what part might be persistent.

From the detailed uncertainty analysis, it can be concluded that the unknown contribution of persistent effects such as reference cell calibration uncertainty, irradiance homogeneity and mismatch factor uncertainty are dominating the overall measurement uncertainty. For a single laboratory, it is not possible without substantial investment, to decrease this uncertainty.

Different laboratories have varying measurement setups (thus different spectrum of the solar simulator); operating procedures; reference cells, data acquisition calibration and traceability chains. It can therefore be assumed that the measurements are independent. As such, the mean of PV modules measured on more than one setup can have smaller uncertainties than the systematic uncertainty of the single laboratory.

A sensitivity study in [15] illustrated the impact of the number of PV modules and of laboratories for typical setups as illustrated in Figure 2. Increasing the number of PV modules leads to a reduction of approximately 10%, whereas the involvement of only 3 laboratories already reduces the uncertainty by up to further 40%.

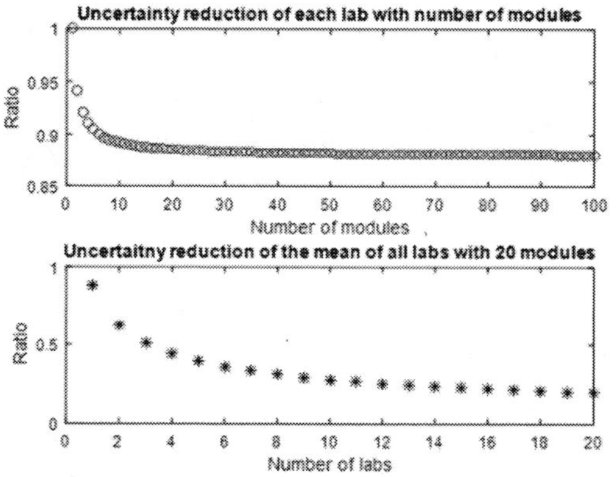

Figure 2: Impact of number of modules (top) and impact of laboratories performing the measurements (bottom) assuming equal uncertainties [15].

3. UNCERTAINTY ESTIMATION

In the following, the uncertainty calculation is briefly outlined. More details can be found in [15]. For brevity, the focus is on P_{MAX} measurements, but it is equally applicable to I_{SC} and V_{OC}.

First the uncertainty of each laboratory is estimated following a consistent framework, such as the one in [14], resulting in estimates of the uncertainties due to random, systematic and mixed effects.

The uncertainty of the mean maximum power of l modules, $\overline{P_l}$, can be calculated as a function of the uncertainties due to random, $u_r(R)$, and systematic, $u_r(S)$, effects [11][12]:

$$u_r^2(\overline{P_l}) = u_r^2(S) + \frac{u_r^2(R)\left(\sum_{j=1}^{l} P_j^2\right)}{\left(\sum_{j=1}^{l} P_j\right)^2} \quad (1)$$

In metrology, it is common practice to measure a standard by multiple laboratories and to assign the weighted average as the reference value. In this paper, the approach is extended to measuring the mean values of the performance characteristics of a batch of modules by multiple partners. The smallest uncertainty is achieved via using a weighted mean, $\overline{P_w}$, of all laboratories. The weights are chosen to be reciprocal to the variance of each laboratory [12].

Since it is impractical to measure all modules by multiple partners, the method is further modified so that only a small number of samples m, $m < l$, of the batch of modules is measured by multiple laboratories.

A correction can then be applied to the mean value of all PV modules measured by the principal laboratory. This correction is based on the weighted mean of the sample PV modules measured by all laboratories, divided by that of the principal laboratory alone. For simplicity, it is assumed that all correlations cancel out resulting in a conservative estimate [15]. The relative uncertainty associated with the corrected mean is:

$$u_r^2(\overline{P_{i,l}cor}) = \frac{u_r^2(R_i)}{m} + \frac{u_r^2(R_i)}{l} + u_r^2(\overline{P_w}) \quad (2)$$

4. COST IMPLICATIONS

The approach and uncertainty estimation was applied to a case study and the results were presented in [15] in detail. For convenience, a summary is reported here. A sample batch of 17 PV modules was measured by three laboratories. The principal laboratory measured the full batch of 120 PV modules. So to summarise we had for the calculations presented in [15] the following numbers: n=3, m =17 and l =120.

The three laboratories that participated were:

- CREST, UK
- UL, DE
- PV LAB, DE

A summary of the results is presented in Table 1 and 2.

Table 1: Summary of laboratory uncertainty

Laboratory	Relative P_{MAX} Uncertainties at k = 2 in %		
	Total, $U_r(P_i)$,	Systematic, $U_r(S_i)$,	Random, $U_r(R_i)$,
CREST	2.75	2.42	1.31
PV LAB	2.59	2.17	1.40
UL	3.02	2.79	1.17

Table 2: Summary of laboratory measurements

Laboratory	$\overline{P_{i,m}}$, W	STD, W	$U_r(\overline{P_{i,m}})$,%
CREST	323.56	2.61	2.44
PV LAB	323.51	2.44	2.20
UL	323.32	2.33	2.80
Weighted Mean	$\overline{P_w}$, =323.48		$U_r(\overline{P_w},)$ =1.41

The excellent agreement between the measurement partners is shown in Figure 3, where data from the JSRMs measurements is reported. The results are normalized to the nameplate rating and the deviation from the principal laboratory is reported on the secondary axis. Due to the excellent agreement, the correction that has to be applied to the principal laboratory is almost negligible, however the uncertainty of the corrected mean is reduced to 1.44 % at k=2. Therefore, the uncertainty of the batch mean value was reduced by almost half.

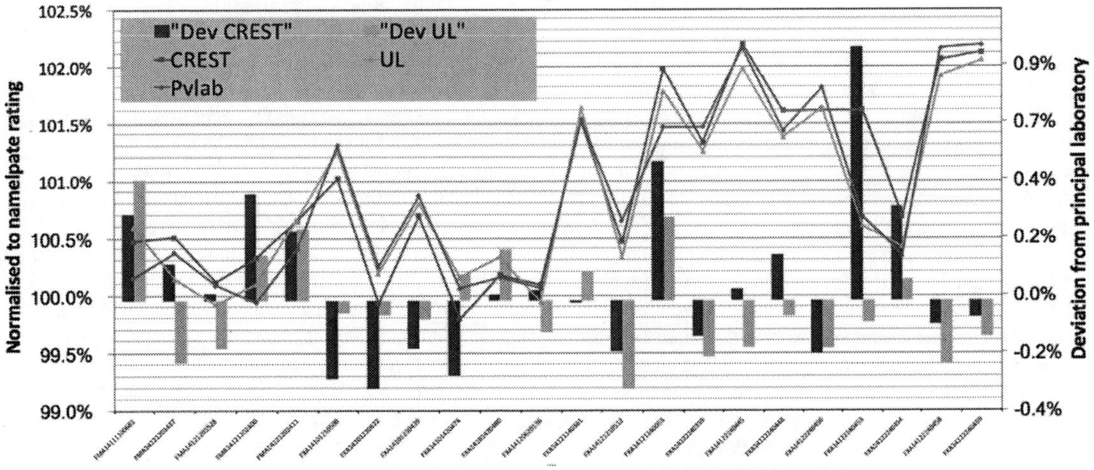

Figure 3: Normalised Pmax measurements of the JSRM modules.

To better understand the financials and technical benefit of the proposed procedure example costs for the case study are presented in Table 3. Within the three-laboratory test approach, with a sub-set of modules tested (JSRFs) in each laboratory, it is possible to reduce the total uncertainty of all 120 modules and the mean of these modules is finally evaluated with a measurement uncertainty of ~1.5%. The costs now have to be compared with a reasonable number of modules that can be shipped to a laboratory with a similar low uncertainty. There the total costs including shipping of e.g. 20 modules would be in the range of 10-20000€. With that we have now similar costs thus similar financial considerations. But the statistical and technical relevance is different. With the three-laboratory approach, it is possible to measure in this case 6-times more modules and with that the confidence in the delivered DC power is much greater compared to just the smaller single-high precision laboratory.

Table 3: Summary of measurement costs

	No. of PV modules	Costs, pm €	Setup cost, €	Total, €
Lab 1	120	40	2,000	6,800
Lab 2	17	120	NA	2,040
Lab 3	17	120	NA	2,040
		Transportation costs		3 x 2000
			Grand Total:	16,880

5. LIMITATIONS

Based on the used assumptions and the chosen example some points needs to be discussed.

As described in [16] when using a weighted mean the measurand, in this case i.e. the mean of the sample PV modules, has to be stable. It is a reasonable assumption that if care is taken during shipping of PV modules, the mean of c-Si PV modules measured in a short timespan would be stable. Technologies exhibiting metastabilities would require an additional validation step of the stabilization procedure (see e.g. MQT 19.1 of IEC 61215) and potentially an additional uncertainty contribution term.

The approach assumes no correlation between measurements at different laboratories. In practice, there could be some level of correlation between partners, if for example employing the same type of simulator and reference cell or having the same traceability chain. These points could lead to some correlations that are currently not considered. This and the magnitude of the correlations could be assessed but are assumed to be small.

The approach is only valid if robust and comprehensive uncertainty analyses are performed by all partners and if these analyses are consistent with one another. It, therefore, requires disclosure and transparency between partners.

6. CONCLUSIONS

The approach presented in this paper offers clear opportunities. Uncertainties of measurements performed in the field of a large sample batch of PV modules can now have significantly lower uncertainty if a sub-sample is measured by other laboratories. This can be useful during commissioning, but also for estimating degradation, which can be smaller than typical measurement uncertainties and thus is hard to prove. The applicability of the procedure depends on the size of the project. It is estimated that the most cost-effective uncertainty reduction can be achieve if up to 20 PV modules are measured by three laboratories. For the case study, 120 modules were measured, 17 of which by three laboratories. This resulted in a mean of Pmax measurement with uncertainty calculated as 1.44% at k=2. This is approximately half of the single module measurement uncertainty of the principal laboratory and comparable to state-of-art single module precision measurements. The cost was estimated to be approximately double that of the principal laboratory measurements alone, and comparable to state-of-art measurements of a much smaller sample of modules, i.e. 20 modules. This approach can therefore be applied when the lowest risks associated with the DC rating of a power plant is required due to the significantly higher statistically relevance of the determined DC power.

7. REFERENCES

[1] B. Gallagher, "US PV System Pricing H2 2016: System Price Breakdowns and Forecasts," 2016.

[2] T. Lowder, M. Mendelsohn, B. Speer, R. Hill, "Continuing Developments in PV Risk Management: Strategies, Solutions, and Implications" no. February. 2013.

[3] ISO 2859-1, 1999, Sampling procedures for inspection by attributes - Part 1: Sampling schemes indexed by acceptance quality limit (AQL) for lot-by-lot inspection

[4] IEC 60904 series: Photovoltaic devices

[5] D. Dirnberger, U. Kräling, H. Müllejans, E. Salis, K. Emery, Y. Hishikawa, and K. Kiefer, "Progress in photovoltaic module calibration: results of a worldwide intercomparison between four reference laboratories," Meas. Sci. Technol., vol. 25, no. 10, p. 105005, 2014.

[6] W. Herrmann, S. Zamini, F. Fabero, T. Betts, N. Van Der Borg, K. Kiefer, G. Friesen, H. Muellejans, and D. Fraile, "PV Module Output Power Characterisation In Test Laboratories And In The PV Industry – Results Of The European Performance Project," in 25th EUPVSEC/WCPEC-5, 2010, pp. 3879–3883.

[7] Y. Hishikawa, H. Liu, H. Hsieh, T. Inoue, K. Kim, C. Limsakul, S. Kim, R. Ninae, and K. Morita, "Round-robin measurement intercomparison of c-Si PV modules among Asian testing laboratories," Prog. Photovoltaics Res. Appl., vol. 21, no. 5, pp. 1181–1188, 2012.

[8] B. Mihaylov et al., "Results of the Sophia Module Intercomparison Part-1: STC, Low Irradiance Conditions and Temperature Coefficients Measurements of c-Si Technologies," in 29th EUPVSEC, 2014, pp. 2443–2448.

[9] Press release Fraunhofer ISE, 2017: "PV Modulkalibrierung", https://www.ise.fraunhofer.de/de/geschaeftsfelder/photovoltaik/photovoltaische-module-und-kraftwerke/modulkalibrierung/pv-modulkalibrierung.html

[10] A. Possolo, "Simple Guide for Evaluating and Expressing the Uncertainty of NIST Measurement Results", NIST Technical Note 1900, 2015

[11] J. L. Gardner, "Uncertainties in Photometric Integrals," NMI Technical Report 9, 2005

[12] E. R. Woolliams, "Determining the uncertainty associated with integrals of spectral quantities," EMPR-ENG05-1.3.1, 2013

[13] H. Müllejans, W. Zaaiman, and E. D. Dunlop, "Reduction of uncertainties for photovoltaic reference cells," Metrologia, vol. 52, pp. 646–653, 2015

[14] B. Mihaylov, T. R. Betts, A. Pozza, H. Muellejans, and R. Gottschalg, "Uncertainty Estimation of Temperature Coefficient Measurements of PV Modules," IEEE J. Photovoltaics, vol. 6, no. 6, pp. 1554–1563, 2016

[15] B. Mihaylov, B. Jaeckel, J. Arp, R. Gottschalg, „Reduced measurement uncertainty in PV module batch testing", IEEE Washington 2017

[16] M. G. Cox, "The evaluation of key comparison data," Metrologia, vol. 39, no. 6, pp. 589–595, Dec. 2002

HIGH-PERFORMANCE AR COATING ON GLASS APPLIED USING HIGH-PRESSURE MOLDING

J. Jong[1], E. Brouwer[1], V. Rosca[2]*, A.R. Burgers[2], A.J. Carr[2], L.A.G. Okel[2]
[1] TOWA Europe B.V., Geograaf 14, 6921 EW Duiven, The Netherlands
[2] ECN Solar Energy, Westerduinweg 3, 1755 LE, Petten, The Netherlands
*Corresponding author. E-mail: rosca@ecn.nl

ABSTRACT: Herein we report on the development and testing of a high-performance structured AR coating applied to solar glass by means of TOWA's high-pressure molding technology. This technology allows the application of optically-active structures on glass with unprecedented freedom of geometry and precision. The structured layer with AR functionality were based on a theoretically defined geometry, which then was translated into a mold design. As deduced from the systematic I-V measurements at different angles of incidence, the developed ARC provided a ca. 2.1% rel. gain in I_{SC} at normal incidence and greatly improved light capturing at high angles of incidence as compared to flat glass, but also compared to a typical commercial structured solar glass. The experimentally-derived incident angle dependence was then used to estimate the impact of ARC on annual yield for a typical residential system. Depending on the location and the system orientation the ARC improves the annual yield by 5-7% rel. when compared to a flat glass-black background configuration and by 2-2.5% rel. when compared to a flat glass-white background configuration. Aside from greatly light trapping, the coating could offer additional advantages, like improved aesthetics, protection against mechanical impact, and electrical safety of a module and could benefit such applications as BIPV, automotive PV, portable PV, space-based PV.
Keywords: Light trapping, antireflection coating, PV module, PV system, Manufacturing and processing

1 INTRODUCTION

Past experience has shown that anti-reflective coatings (ARC) on glass can improve light coupling, particularly under high angles of incidence. This is especially interesting for panels with a fixed mounting, which are the vast majority of modules. An ARC will enhance energy yield when the sun is not at an optimum position relative to the panel, which, for some systems is the majority of the time. In this paper we report the main results of a study aimed at using high-pressure molding technology, developed by TOWA, to form a high-performance ARC on glass and in this way increase the energy yield of the module at minimum cost increase.

TOWA's molding technology is suitable for the encapsulation of LEDs and electronic components in general. The LEDs are encapsulated as a complete wafer whereby the encapsulant can be formed into the shape of a lens at the location of each individual LED. The shape of the lens can be tuned by changing the design of the mold used during the encapsulation process. Lenses can be made to focus the light produced by the LED or make it diffuse. The key advantage of this molding technology is the great precision and freedom of geometry of the molded structures (see Fig. 1). It is thought that the high-pressure molding technology can be scaled up to the size required for coating a sheet of glass with the dimensions of a full-sized module (approximately 1×1.6 m).

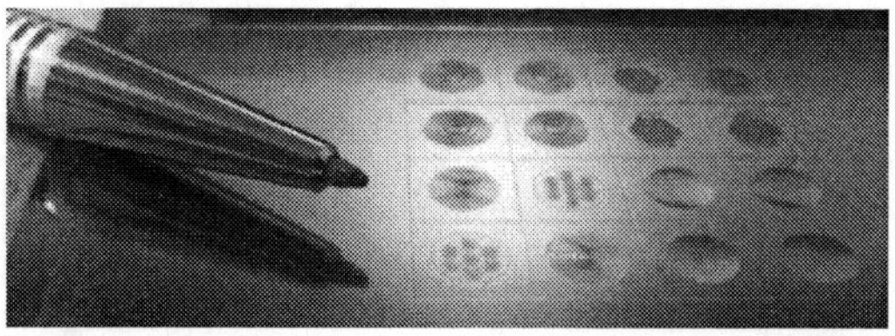

Figure 1: Examples of structures that can be produced using TOWA's high-pressure molding technology.

In this project we made use of the design flexibility offered by TOWA's molding technology to develop a high-performance ARC on glass, which would allow capturing most of the light that is normally lost in state-of-the-art PV panels. In this project we took the complete route from defining optically-active structures by means of ray tracing and then translating the theoretically-defined geometries into a mold design, to producing coatings on glass, the experimental evaluation of the developed coatings, and finally calculation of the impact of the developed structured on the annual yield for a typical residential PV system. Herein we describe main results and conclusions of the project, with focus on manufacturing of the test samples and evaluation of the optical performance of the developed AR coating. The evaluation of the optical performance included systematic I-V measurements at various angles of incidence and annual yield calculation for typical residential systems using PVsyst simulation software.

2 METHODS AND PROCEDURES

In this section we describe the procedure for application of the optically active coatings on glass substrate, manufacturing and testing of single-cell modules, and the annual yield calculations methodology. The geometry of the structures responsible for the AR effect was thoroughly optimized by means of optical modelling, but the results of this theoretical work will be reported elsewhere.

2.1 Application of AR coating

The optically-active coatings on glass were obtained by means of high-pressure molding process using TOWA's CPM 1080 molding system. A dedicated modular mold, carefully selected separation foil, and a commercial silicone compound had been used to produce coated solar glass samples. The molding process closely resembled an industrial high-pressure molding process for encapsulation of micro-electronic devices. An additional ex-situ post-curing step was introduced in order to ensure the optimal degree of cross-linking of the compound. The glass substrate was a 200×200×4mm solar glass sheet and an area of approx. 145×145mm area was covered by optically-active structures. Therefore, the area above the (6-inch) cell was not completely covered with ARC. The resulting coating consists of an array of dome-shaped lenses (see Fig. 2). The lenses are not stand- alone objects: there is a very thin underlying continuous layer of molding compound connecting all lenses.

Figure 2: A close-up photograph of TOWA's structured AR coating on a solar glass substrate.

2.2 Manufacture of single-cell modules

In order to experimentally evaluate the effect of the TOWA coating, the coated glass samples were used to manufacture single-cell PV modules. In principle any type of PV cell and module interconnection method could be used to evaluate the effect of the coatings. However, we decided to use the interdigitated back-contact (IBC) cells. The main advantage of using IBC cells is the absence of metallization on the front of the cell (sunny side). In this way complications related to the presence of the front metallization are avoided. This allowed a more straightforward evaluation of the beneficial effect of the coating.

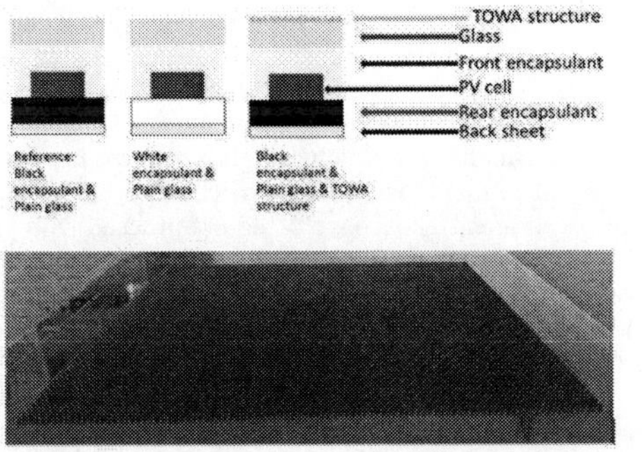

Figure 3: Single-cell laminate lay-outs manufactured and tested (top) and photograph of an IBC module having black rear encapsulant and glass coated with TOWA's AR coating (bottom).

The modules used for evaluation of the performance of AR coating were manufactured using the pilot back-contact module assembly line supplied by Eurotron B.V. (The Netherlands) using commercially available module materials (conductive back-sheet foil, electrically conductive adhesive, and EVA encapsulants) and a standard module manufacturing process practiced at ECN. Cells were ECN's Mercury IBC cells with Sirius interconnection design [1].

Prior to module manufacturing, the stability of the molded layer was tested by means of repeated lamination trials. Although thermal stability of the layers was not expected to be an issue, there might be a risk of some deformation of the layer under pressure during the lamination step. Therefore, especially designed Teflon spacers were developed in order to limit the pressure applied to the coating, but still ensure good mechanical contact and heat exchange with the whole glass area.

Three types of modules were manufactured and tested: (1) modules having black rear encapsulant and flat glass, (2) modules having white rear encapsulant and flat glass, and (3) modules having black rear encapsulant and glass with TOWA's AR coating. See Figure 3. The I-V characteristics of all modules were measured under various angle of incidence, as described in the next section.

2.3 I-V measurements

The optical performance of the developed AR coating was evaluated by means of systematic I-V measurements of the single-cell IBC modules. The I-V measurements under standard test conditions (STC) and at varied angle of incidence were carried out using PASAN IIIb flash tester (class-A according to IEC 60904-9). Firstly, all modules were measured under STC and then cell-to-module losses were calculated from cell I-V data acquired with same flash tester and under same conditions. Secondly, the I-V data was acquired under various angle of incidence. The angle of incidence was controlled by means of a dedicated in-house developed setup. This setup was thoroughly tested and adjusted to provide reliable and repeatable measurements [2]. To ensure significance of the effect of the angle of incidence on the module power output, at least three modules were measured per group.

Importantly, the modules were masked using black tape in order to expose the same aperture area for all modules. Tape was applied at a distance of 1.25 mm from the cell edge, which is typical inter-cell distance for back-contact modules.

2.4 Annual yield calculations

In order to assess the potential impact of the developed AR coating on the system level and compare to other module configurations, annual yield calculations were carried out using the PVsyst software [3]. PVsyst applies an Incidence Angle Modifier (IAM) based on the ASHRAE model to each module, unless a specific IAM has been defined for the module [4]. The IAMs for each module lay-out were calculated from experimentally obtained current values after normalization and cosine correction. Annual yield calculations were performed for three different system orientations and three locations (see Table 1).

Table 1: System variations: location, tilt angle and orientation.

Location	Tilt angle	Orientation
Amsterdam	38° (optimum)	South
	60°	South
	90°	West (façade)
Kyoto	30° (optimum)	South
	60°	South
	90°	West (façade)
Doha (Qatar)	25° (optimum)	South
	60°	South
	90°	West (façade)

For the PV component a 5kW system has been chosen consisting of 18×280W SolarWorld Sunmodule SW280 panels and a Fronius International 5kW Primo 5.0-1 inverter. This module was used as the basis for the flat glass-white back sheet module. We have then adjusted the module parameters to obtain realistic versions of the module with the flat glass and the black encapsulant, and for the TOWA ARC and black encapsulant configurations. These calculations were based on the measured results at STC (at normal incidence) of the single-cell modules. The modules also include an IAM particular to the given module.

3 RESULTS AND DATA ANALYSIS

3.1 Characteristics of AR coating on glass

Figure 2 shows a picture of a solar glass sheet with AR coating. The coating showed exceptionally good adhesion to glass and aesthetically improved appearance. The parameters of the molding process as well as the separator foil had to be optimized in order to obtain defect-free coating. The coating consisted of ca. 12000 individual lenses.

Due to some practical considerations, the roughness of the surface of individual lenses was sub-optimal and that could have negatively impacted the optical effect. Furthermore, due to practical consideration, the cell area could not be completely covered, which diminished the overall effect.

So far no reliability studies have been carried out on the modules built with TOWA's AR coatings. However, we have carried out TC200 and DH1000 tests (IEC61215) on glass samples coated with a thin (0.25 mm) continuous flat layer of the same compound on glass. The coated glass samples were exposed to TC200 and DH1000 tests (IEC61215), with visual appearance, adhesion (cross-cut), and transmission measurements (integration sphere) as characterization methods before and after accelerated stress tests. For DH1000 samples we observed very limited loss of transmission, hardly any change in visual appearance, but some loss of adhesion (delamination for 20-25% of cross-cut area). For TC200 samples we have observed full retention of adhesion, no visual changes and very limited change in light transmission. Screening of various molding compounds to improve the coating weathering characteristics is in order.

Table 2: Comparison of cell-to-module Isc change for modules with the three configurations.

Code	Glass / background	Rel. cell-to-module Isc change (%)*	
		Avrg	SD
B	Flat / black	0	0.06
B-AR	AR coat. / black	+2.13	0.07
W	Flat / white	+3.38	0.10

* Compared to modules with flat glass and black rear encapsulant.

3.2 Optical performance at STC

Table 2 compares cell-to-module (CTM) change for the modules having three configurations at STC conditions (normal incidence). CTM losses were calculated from the experimental data for at least three modules. The cell data measured under identical conditions and same flash tester. The module in which AR coating is combined with black background shows a significant I_{SC} gain, which is still ca. 30% lower than the gain observed for flat glass-white background configuration. Note, however, that the AR coating was optimized to improve light collection under a broad range of angles of incidence and ultimately improve annual yield.

3.3 Optical performance at varied incident angles

The optical performance of various module configurations at varied angles of incidence can be compared using derived IAM values for each module configuration. Such a comparison is shown in Fig. 4.

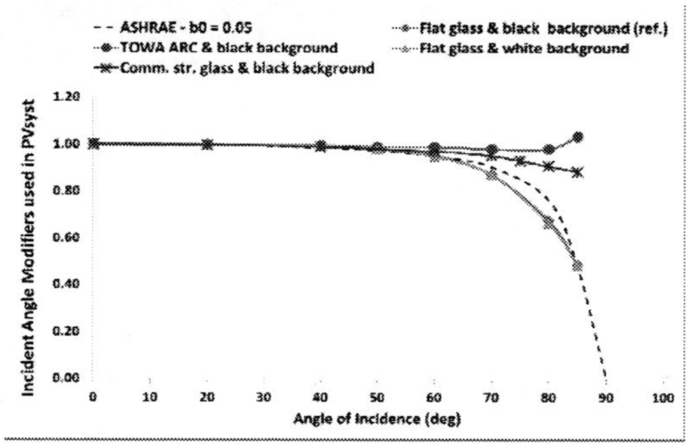

Figure 4: Measured Incident Angle Modifiers (IAMs) for a fixed azimuth, compared to the standard ASHRAE – b_0=0.05 used in annual yield simulations; also included is a previously measured IAM for a commercial structured glass.

TOWA's structured AR coating shows a great improvement when compared to flat glass, but also a significant improvement when compared to a very good commercial structured glass recently tested at ECN. Note that the commercial structured glass was measured using different cell and glass (at an earlier stage and not part of this experiment), hence we could not proceed with proper comparison on the system level. Interestingly, IAM values measured for TOWA's coating at grazing angles exceeding the value of one. The effect is reproducible and is currently being investigated.

The AR effect of the TOWA's coating is determined primarily by the geometric effect. At the same time, the refractive index of the compound (as given by the supplier) is 1.41. which means that an index-matching effect should also have a contribution.

Finally, note that the performance of TOWA's ARC is even more impressive if we take into account that the coating was not fully optimized from both geometry and processing points of view (see subsection 3.1).

3.4 Effect on annual yield

Before we proceed with presentation and discussion of annual yield data, it is important to mention that in this study we focused on evaluation of the optical effect of the developed ARC itself. For that purpose flat glass-black background configuration was chosen as reference. The flat glass – white background configuration was another configuration we have included because (i) this configuration provides a simple and inexpensive way to boost the module current and (ii) we could prepare and test the required samples without difficulties. Although we had planned benchmarking on the system level against commercial structured glass products and other AR coatings on flat glass, it was not possible to obtain required samples in the required format (1-cell module) and determine IAM for the same cell and module bill of material.

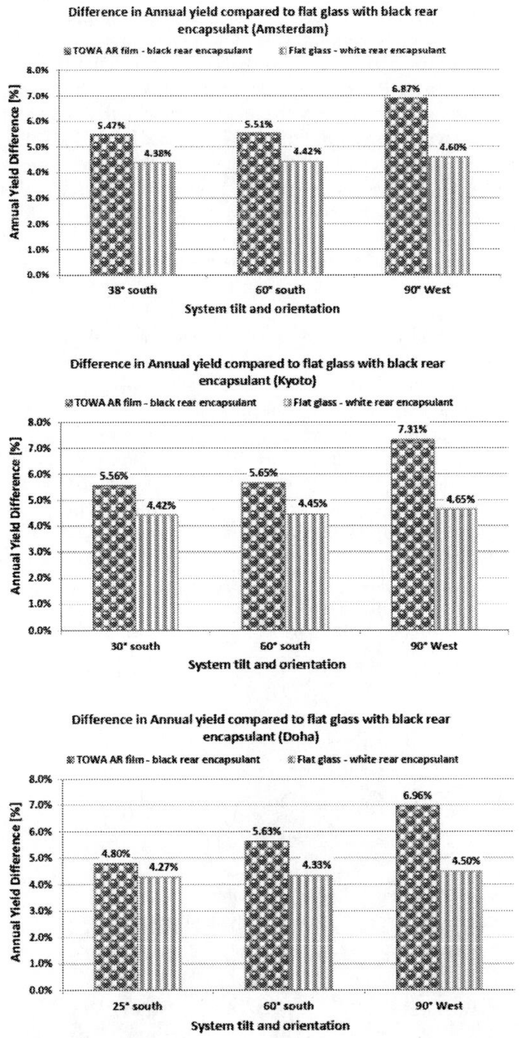

Figure 5. Overview of relative difference in annual yield for three glass-encapsulant combinations, three locations and three orientations.

Turning now to presentation and analysis of the results, the relative differences in annual yield calculated for the three module configurations are presented in Figure 5. The flat glass - white background configuration brings an angle-independent improvement over the flat glass - black background reference, whereas TOWA's AR coating gives an improvement over both black- (5-7% rel.) and white background (2-2.5% rel.) configurations for all locations and orientations, with a prominent angle-dependence effect, as expected. Accordingly, from the monthly yield data (not shown) for the 38° and 60° orientation, TOWA's ARC improves yield significantly in spring-autumn period, whereas for the 90° facing a dramatic increase in yield is observed for winter months over the white background configuration. In other words, the further from optimal tilt angle for given location, the higher the gain and beneficial effect of the coating.

As for potentially interesting practical applications of the proposed technology, TOWA's AR coating should allow replacement of the flat glass - white background configuration with TOWA ARC – black background configuration (required from aesthetical considerations, for example) without loss and even with a gain in annual yield. Furthermore, colored compounds can be used to improve the visual appearance of the module with minimum performance loss. Naturally, the TOWA's ARC could be an interesting alternative to commercial structured glasses or AR coatings on flat glass and can be combined with white background to achieve exceptional cell-to-module current gains.

4 CONCLUSIONS

Herein we reported on development and testing of high-performance AR coating applied to solar glass by means of TOWA's high-pressure molding technology. This technology allows the application of optically-active structures on glass with unprecedented freedom of geometry and precision. As deduced from the systematic I-V measurements at different angles of incidence, the developed ARC is responsible for ca. 2.1% rel. gain in I_{SC} at normal incidence and greatly-improved light capturing at high angles of incidence as compared to flat glass, but also to a typical commercial structured glass. The experimentally-derived incident angle dependence was then used to estimate the impact of ARC on the annual yield. The TOWA's structured ARC improves the annual yield by 5-7% rel. as compared to flat glass-black background configuration and by 2-2.5% rel. as compared to flat glass – white background configuration, depending on the location and system orientation. Importantly, the above results although impressive already were obtained for a not fully optimized coating. Aside from greatly enhanced light trapping on module level, the coating could contribute to improved aesthetics, protection against mechanical impact, and electrical safety of the module and could benefit such application as BIPV, automotive PV, portable PV, space-based PV.

ACKNOWLEDGEMENTS

The work reported in this paper had been carried out in the framework of CALIFORNIA project (TKI Solar Energy initiative), financially supported by the Dutch Ministry of Economic Affairs.

REFERENCES

[1] I. Cesar et al., Energy Procedia 55 (2014) 633.
[3] L.H. Slooff, A.J. Carr P.M. Sommeling, R. van de Voort, 32nd European Photovoltaic Solar Energy Conference (2016) 903.
[3] PVsyst V6.43, PV "Study of Photovoltaic Systems". PVsyst SA, Satigny, Switzerland. www.pvsyst.com.
[4] J.H. Fatehi, K.J. Sauer, 40th IEEE Photovoltaic Specialist Conference (2014) 1335.

PV MODULE SAND ABRASION TESTING

G. Mathiak, N. Pfeiffer, L. Rimmelspacher, W. Herrmann, J. Althaus, F. Reil,
TÜV Rheinland Energy GmbH, Am Grauen Stein, 51105 Cologne, Germany.
C. Holze, toughtrough GmbH, Wiener Str. 5, 28359 Bremen, Germany.
A. Morlier, Institut für Solarenergieforschung GmbH Hameln, Am Ohrberg 1, 31860 Emmerthal, Germany.
Email: Gerhard.Mathiak@de.tuv.com

ABSTRACT: Due to the high solar irradiation and the large number of sunny days the use of large PV systems in desert areas is very profitable. However, the PV modules are exposed to harsh environmental conditions such as high temperatures, high UV radiation or sandstorms. Currently there is no suitable standardized method for testing the abrasion resistance of PV modules and solar glass under sandstorm conditions. Qualification testing of PV modules for arid climate conditions will require to include a sand abrasion test and a dry cleaning resistivity test. In order to be able to assess the individual influencing factors in more detail, this work examines the sand trickling tests defined in DIN 52348 [1] and ASTM D968-05 [2] in detail. Furthermore, we developed a test facility, which can be used to simulate the impacts of desert sand on the surface of PV modules and solar glasses. Via transmittance and reflectance measurements on PV modules and solar glass sheets before and after sand abrasion tests, the samples are characterized and their resistance to desert conditions is assessed.

Keywords: Qualification and testing, Sand abrasion, Trickling sand, Glass coating, Transmittance and reflectance

1 Introduction

An important aspect for the use of photovoltaic modules and large PV systems in desert areas, beside the high intensity and irradiation, is the number of sunny days in the period of one year, which additionally leads to a large energy gain. This is a major advantage for the use of photovoltaic systems, as they are directly dependent on solar energy. Finally, the great advantage for desert regions is that the high space requirement of PV systems does not pose a problem. However, there are some environmental situations in desert areas that can have a negative effect on the efficiency of PV modules. In addition to dust and sand contamination (soiling), resistance to sand drifts and sandstorms, higher UV radiation, cleaning and higher temperatures are the main challenges for the PV module design and product qualification testing. Sandstorms can have a great influence on the glass surface of the modules and thus strongly influence their performance. In the case of photovoltaic modules that are installed relatively close to the ground, the kinetic energy of the sand particles accelerated by wind gusts and sandstorms is sufficient to cause damage to the glass surface of the modules. In most cases, standardised ageing tests are also available for the evaluation of PV modules especially the front glasses at certain locations. To date, no accelerated ageing guideline has been developed for the degradation of PV modules due to sandstorm impact, in order to realistically take into account performance losses caused by sand particle abrasion. In general, the type of stress on solid-state abrasion of materials is described by the properties of the erosive particles (size, shape, hardness and density) and the external conditions (impact velocity, angle of incidence and temperature). The numerous parameters that can occur at the same time, which define the sandstorm erosion of materials in desert areas, show the complexity of this observation. The threat of erosion effects on front and rear side of PV modules leads to a growing expectation for a suitable test procedure. With the help of test procedures, plant operators could adapt and develop an optimized risk analysis for performance and cost calculations to the expected service life of PV modules.

More and more PV systems are installed in deserts with arid (hot-dry) climate. Higher module temperatures, ultraviolet radiation, sand storms and soiling by dust are the main challenges for the special module design and the module test. Performance testing for arid climates is under development in new IEC test standards. A new approach is made in various parts of IEC 62892 [3], where test programs are defined to differentiate the PV modules performance in multiple climates, including arid climates.

Fig.1: Photo of the test setup according to DIN 52348 with funnel and plexiglass tubes. Under the tube it is possible to position full-size PV modules and glasses.

A combination of sand and dust tests with UV exposure is recommended to determine the resistance of PV components against abrasion effects from wind-blown sand. Each location of a PV power plant in a desert should be analyzed for the local sand type (hardness, shape, grain size distribution), its wind conditions (speed, frequency) and its history of sand and dust storms [4].

2 Sand Abrasion Test Standards

For sand abrasion testing typically sand trickling tests and blowing sand tests are applied. In this study two different sand trickling abrasion tests were studied. The sand trickling tests according to DIN 52348 and ASTM D968-05 are usually used in the industry to investigate the abrasion resistance of glass and surface coatings. The DIN test setup (see Fig. 1) consists of a funnel for the sand and a tube with two sieves to guide the free fall of the sand. The sieves homogenize the particle concentration over the exposition area. The final tube piece is 1.025 m high and has a diameter of 120 mm. The maximum speed of the free falling sand particles is 4.4 m/s. Standardized natural silica sand (ASTM 20/30 sand) with 95 % SiO_2 and grain diameter of 0.5 mm to 0.7 mm is mainly used for this test. Sand with grain diameter of 0.7 to 0.9 mm was used for comparison. Silica sand produces a higher rate of abrasion compared to the most natural desert sands. The test is adapted to tempered glass sheets and modules in original size and is without continuous sample rotation (Tab. 1). The ASTM-test setup has a thinner tube and no sieves.

Sand Trickling Tests				
	DIN 52348		**ASTM D 968-05**	
		adapted		**adapted**
Sand mass (typical)	3 kg	3 kg	3 kg	3 kg
Test duration (approx.)	7 min	7 min	7 min	7 min
Exposition area Ø [mm]	120	120	25	50
Tube height/ free fall [cm]	102.5	102.5-146	98	98-202
Particle speed (calcul.) [m/s]	4.46	4.46-5.4	4.22	4.22-6.6
Temperature	25°C	25°C	25°C	25°C
Impact angle	45°	45°	45°	45°
Mass per area [kg/m²]	265	265	6114	1529

Tab. 1: Test parameters for typical sand trickling test with sand mass of 3 kg. The duration of such an experiment is approx. 7 minutes. For the research tests some parameters were adapted due to the size of the samples and varied. The main difference of the adapted versions of ASTM D968-05 is the bigger tube diameter of 50 mm.

The sand blowing test according to MIL STD 810G was originally developed for environmental tests of military materials and components. The test method "510.5 Procedure II: Blowing Sand" [5] uses as abrading sand silica sand with at least 95 % SiO_2 particles with grain diameters in the range 150 µm to 850 µm. The test is performed inside a closed wind tunnel chamber. The air temperature can be chosen between ambient and high temperature up to 70°C. During the test, wind speeds of 18 m/s are capable of blowing large particle sand, whilst wind gusts can reach up to 29 m/s. The test sand concentration for natural conditions is 0.18 g/m³. Sand concentration of 2.2 g/m³ is applied for material likely to be used close to helicopters operating over unpaved surfaces. Typical test duration for PV applications is 1.5 hours per vulnerable area i.e. on the front glass and the backsheet.

The sand trickling test is simpler and faster than the more realistic sand blowing test. In the applied configuration the exposition area of the sand trickling test is smaller than the solar cell size.

After completion of the sand abrasion test PV modules and solar glasses with and without anti-soiling or anti-reflective coatings (ARC) are optically characterized.

3 Results of Sand Trickling for Solar Glass

3.1 Diagnostic Method

The aim of the various diagnostic measurements is to determine changes in the optical transmittance through the solar glass to a laminated solar cell behind the glass. Therefore, three types of hemispherical measurements were performed with a 10 mm diameter direct light beam and an integrating sphere (diameter =150 mm) in a wavelength range from 300 nm to 1600 nm (see Fig. 2):
1) Total transmittance τ_{tot} through the solar glass sheet
2) Total reflectance ρ_{tot} of the solar glass with multiple reflections at both surfaces and
3) Reflectance at the front surface r_1 using a light trap behind the glass.

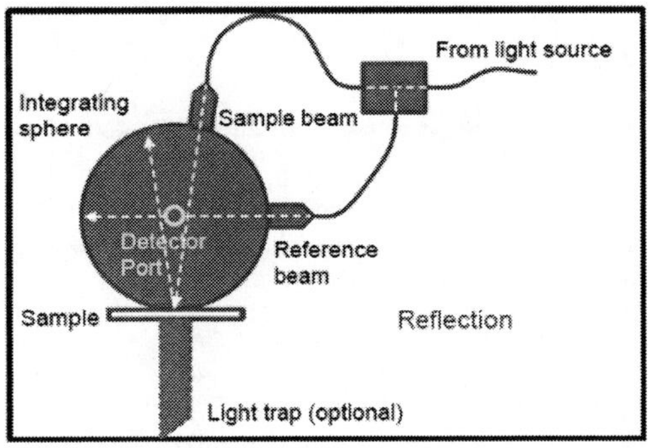

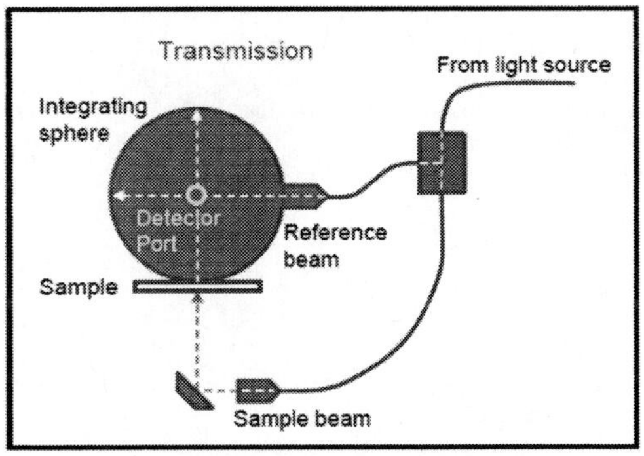

Fig. 2: Diagnostic setup with integrating sphere for reflectance (optional with light trap) and transmittance measurements.

With the results from these measurements, it is possible to calculate using an evaluation method developed at the University of Rapperswil [6]:
4) Reflectance at the rear surface r_2,
5) Internal transmittance t and
6) Interface-corrected transmittance τ_Kor,G2.

This parameter is preferred for the evaluation of the sand abrasion test because it corresponds to the transmittance to an optical coupled solar cell behind the glass without reflectance at the glass rear surface. The transmittance curve can be integrated over the spectral range under consideration of the sun spectrum and the spectral response of the PV module to obtain the effect on the PV module. For higher simplicity, the change of the parameters were compared at the chosen wavelength λ=500 nm.

Fig. 3 shows an example of three transmittance and three reflectance curves. A solar glass sheet with anti-reflective coating was measured after an ASTM-test with 9kg coarse sand and long tube. Slight discrepancies in the glass transmittance and reflectance evaluation were found. Especially, the calculation of the reflectance at the rear surface r_2 shows variations. This failure can be caused by light losses to the glass edges due to diffuse reflection.

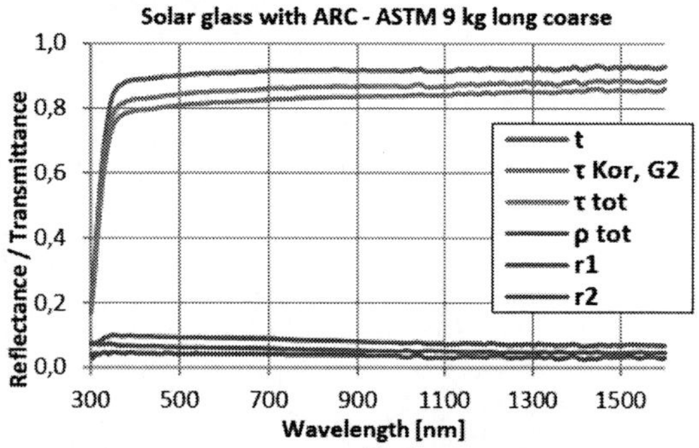

Fig. 3: Reflectances and transmittances of a one-side AR-coated PV glass. The most relevant result for PV applications is the interface-corrected transmittance τ_Kor,G2.

3.2 Parameter study for transmittance

The deterioration of the transmittance τ_Kor,G2 was determined by sand trickling tests according to the DIN and ASTM standards with different grain sizes and tube lengths as listed in Tab. 1. In Fig. 4 as a first example the degradation of the transmittance τ_Kor,G2 was determined after sand trickling test with 3 kg, 6 kg and 9 kg sand mass. The resulting transmittance loss is approx. 0.5 % per test mass of 3 kg.

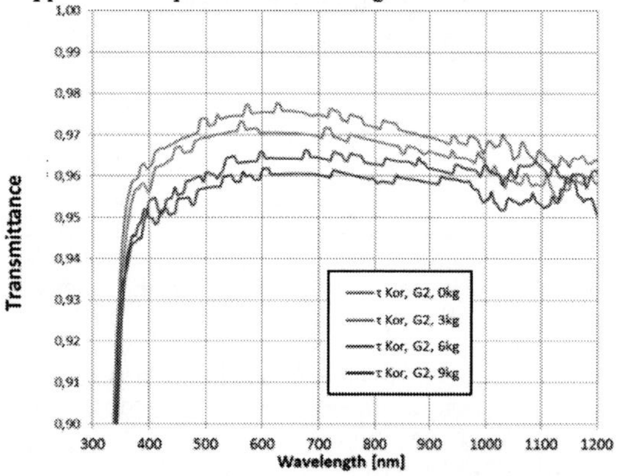

Fig. 4: Degradation of corrected spectral transmittance τ_Kor,G2 of the solar glass with ARC after DIN test with 3 kg (0.5 % transmittance loss), 6 kg (1.0 % loss) and 9 kg (1.5 % loss).

Fig. 5 and 6 show the transmittance change in dependence of the loaded sand mass. The DIN standard tests show almost a proportional relation between sand mass and transmittance change. The ASTM standard tests show high transmittance losses due to the higher mass per area load. In the initial phase the ARC is slightly removed or the glass without ARC is slightly deteriorated, in the final phase the basic glass is deteriorated strongly and the deterioration goes to saturation with a value dependent of the test parameters. In Fig. 5 the transmittance curve of a slightly structured glass without ARC is shown. The average grain size factor for the DIN test between fine and coarse sand is 1.81. For the ASTM test, this is slightly lower at 1.44. Fig. 6 shows a glass with ARC. The average grain size factor for DIN is 2.44. For the ASTM test, it is 1.26.

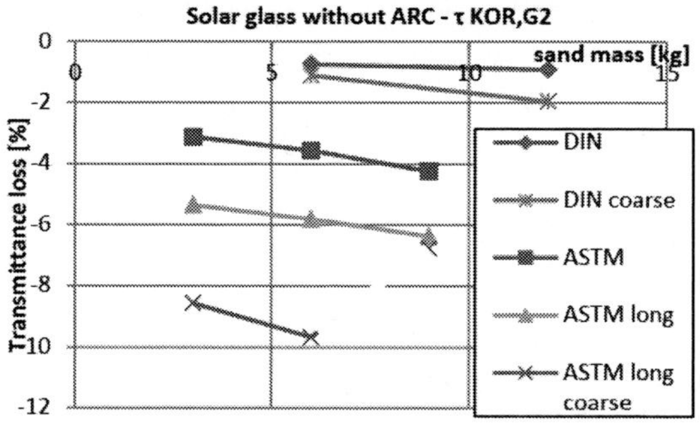

Fig. 5: Transmittance changes (at 500 nm) of interface-corrected transmittance τ_Kor,G2 with variation of the sand mass. Separated into standard tests and associated load changes – glass without ARC

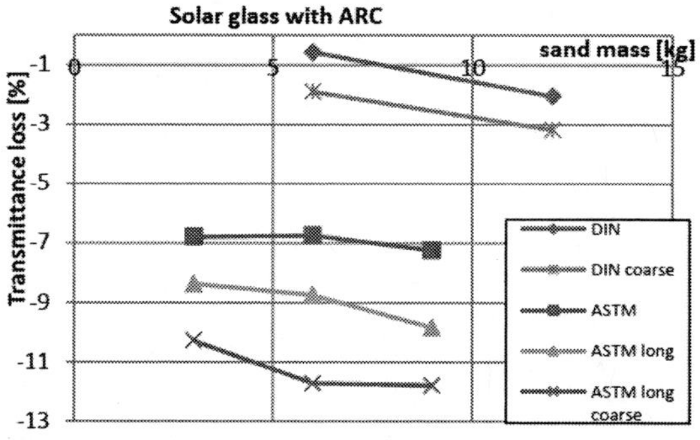

Fig. 6: Transmittance changes (at 500 nm) over the sand mass. Separated into standard tests and associated load changes. The glass with ARC was tested at the new test setup.

3.3 Parameter study for reflectance at the front surface

Several solar glass types of different manufacturers were tested during two test campaigns at toughtrough [7] in Bremen and with the new set up in Cologne.

In Fig. 7 and 8 reflectance curves of comparable glasses at two different setups are shown. The abrasion effect of the setup in Cologne is stronger. Fig. 9 shows the reflectance curves of a glass with another ARC type and Fig. 10 of a glass without ARC of the same manufacturer.

Fig. 11 and 12 show the reflectance of solar glass sheets at a wavelength of 500 nm over the loaded sand mass. As in the case with the transmittance changes, smaller changes were observed for solar glasses without ARC. Both figures show an increase in reflectance for DIN standard stress and ASTM test with extended tube and coarse sand. In Fig. 12 the ASTM test (long tube, coarse sand) increases the reflectance to a maximum value of 6.4%.

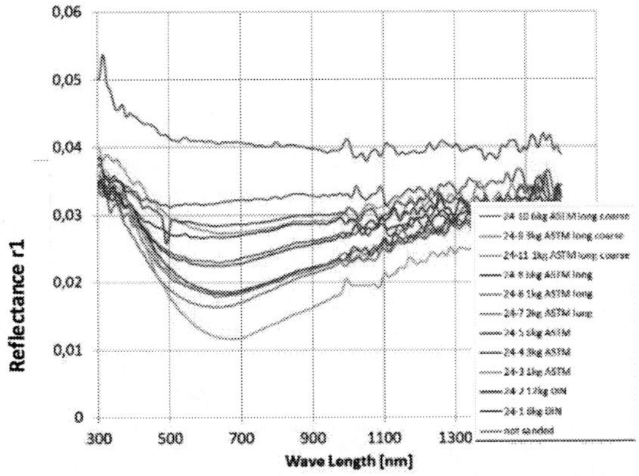

Fig. 7: Reflectance at the front surface in the range from 300 nm to 1600 nm of solar glass with ARC during the campaign in Bremen. A test with 6 kg sand in the ASTM-setup with long tube and coarse sand leads to the complete removement of the ARC with reflectance above 4% over the whole range.

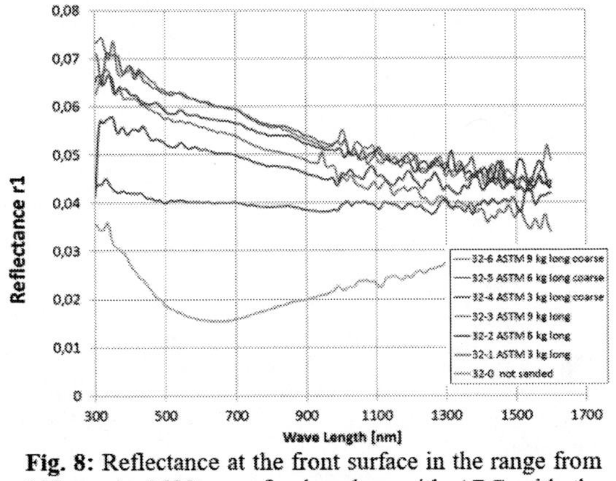

Fig. 8: Reflectance at the front surface in the range from 300 nm to 1600 nm of solar glass with ARC with the setup in Cologne. All ASTM tests with the long tube lead to reflectances above 4% over the whole range independently of using fine or coarse sand.

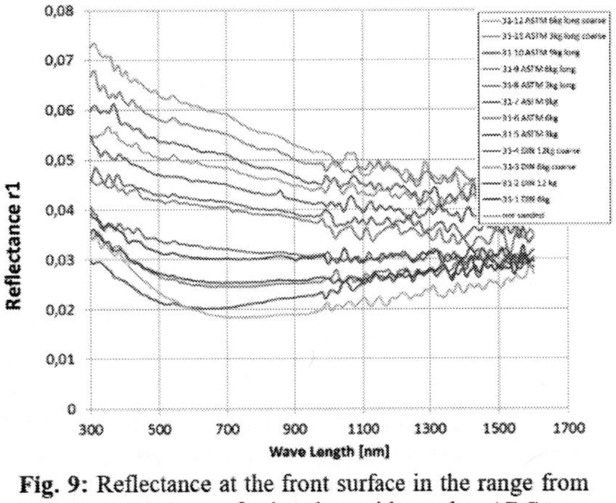

Fig. 9: Reflectance at the front surface in the range from 300 nm to 1600 nm of solar glass with another ARC type with the setup in Cologne. All ASTM tests with the long tube lead also to reflectances above 4% over the whole range independently of using fine or coarse sand.

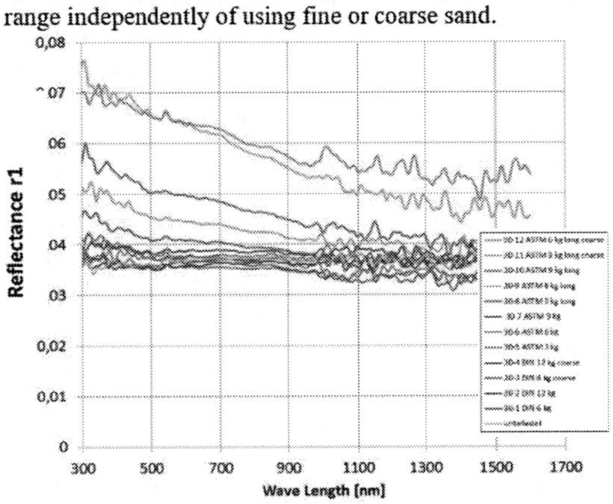

Fig. 10: Reflectance at the front surface in the range from 300 nm to 1600 nm of solar glass without ARC with the setup in Cologne.

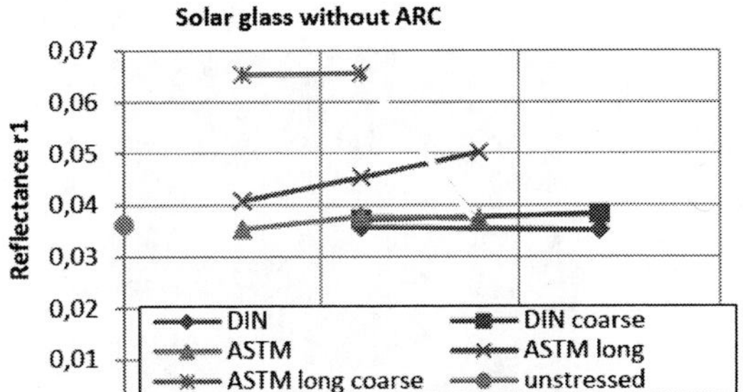

Fig. 11: Reflectance r1 of a solar glass without ARC at a wavelength of 500 nm as a function of sand mass excerpted from Fig. 10. Transmittance in Fig. 5

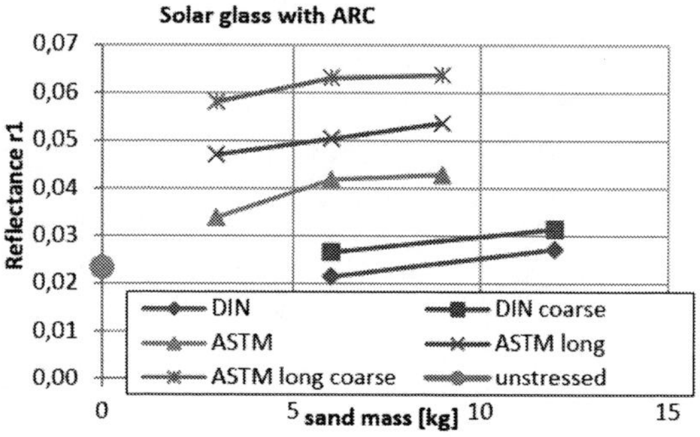

Fig. 12: Reflectance r1from a solar glass with ARC at a wavelength of 500 nm applied over the sand mass. of a solar glass excerpted from Fig. 9. Transmittance in Fig.6.

4 Results of Sand Trickling for PV Modules

4.1 Diagnostic Method

The transmittance of the glass cover of a full-size PV module is characterized by the short-circuit current measured with a pulsed solar simulator. Due to the size of the exposition area of the sand abrasion test, only single cells can be measured, either directly contacted through the backsheet foil or in a non-destructive way with partially shading of the cell under test. The short-circuit current of the partially shaded cell Ish is measured directly contacted to the cell under test after opening of the backsheet or non-destructive as described in Fig. 14.

In order to demonstrate the equivalence of both test methods (direct and shaded), measurements can be done simultaneously with a glass-foil module. The non-destructive partially shading method is the only method suitable for glass-glass modules. For a 6-inch cell Ish is approx. 46% (120 mm circle mask, see Fig. 13) or 8% (50 mm circle mask) of the Isc of the fully illuminated cell.

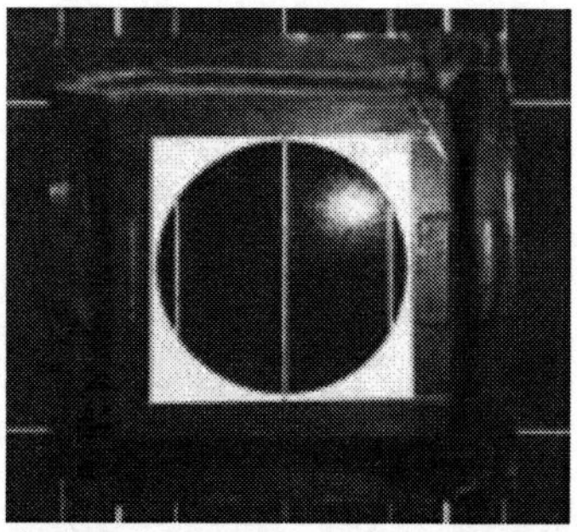

Fig. 13: The image shows a partially shaded cell. The tape around the mask shades the neighbor cells to avoid the entry of light from the side.

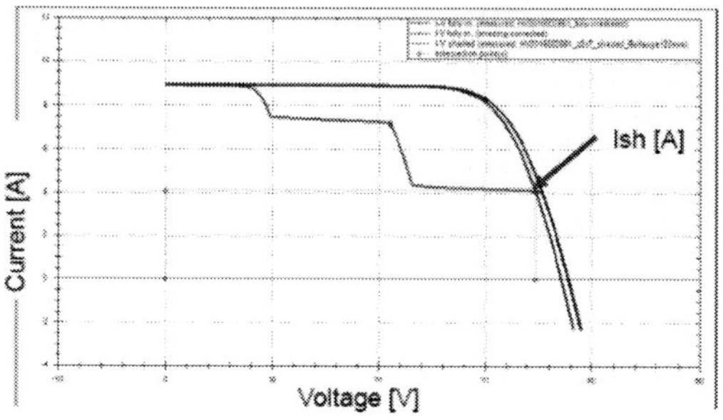

Fig. 14: The short-circuit current of the partially shaded cell Ish is determined by the comparison of the IV-curve of the shaded module and the IV curve of the unshaded module calculated for 59 cells [8].

4.2 Parameter study

The measurement principles were applied to glass-glass and glass-foil modules, which have been subjected to DIN and ASTM test procedures with different sand types and tube lengths to analyze the impact of particle size and drop velocity on the module surfaces.

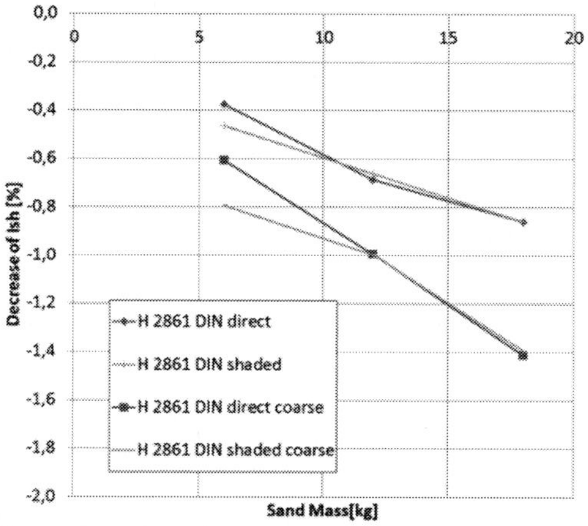

Fig. 15: Short-circuit current changes of DIN configuration tested module. Direct and shaded measurements with standard and coarse sand are compared.

The DIN configuration was used to compare the direct contact method with the shading method. Our measurements have demonstrated that both methods deliver comparable results (Fig. 15). The non-destructive method is more suitable as standard test method due to higher flexibility and less effort of this method. For the following tests the partially shading method was used.

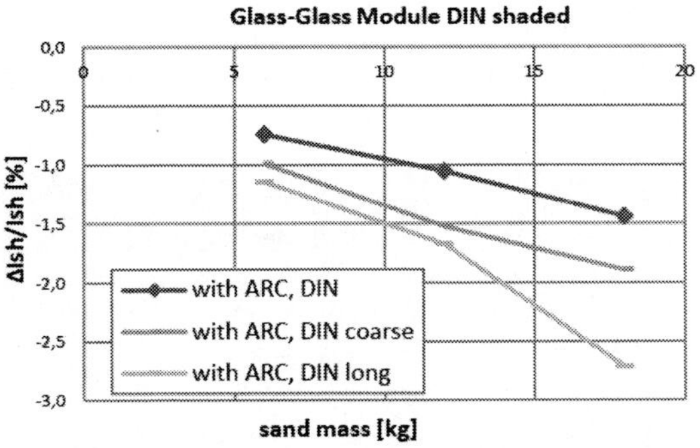

Fig. 16: Change of short-circuit current of glass-glass module with ARC for various DIN tests as function of the sand mass.

Fig. 16 shows short-circuit current changes of various DIN tests for variable sand mass. It can be seen that an increase of the sand grain size leads to an increase of ΔIsh/Ish. The grain size is increased from 0.5-0.7 mm to 0.7-0.9 mm. The grain size factor is 1.67 in average. In a further test the tube has been extended from 1015 mm to 1360 mm, which leads to an increase of the calculated particle speed from 4.7 m/s to 5.4 m/s. In average for the three stress levels a 23 % reduction of ΔIsh/Ish was found.

Fig. 17 shows short-circuit current changes of a PV module with ARC and a module without ARC of various ASTM tests for variable sand mass. In comparison to Fig. 16, higher transmission losses are observed due to the higher mass per area impact. The module without ARC shows a smaller sensitivity to sand abrasion than the module with ARC. Overall, the transmission loss ΔIsh/Ish of the module with ARC, compared to the module without ARC is 118% in average. In a direct comparison of the PV modules with and without ARC, when stressed with coarse sand, the average transmittance loss is 52.4%. The PV module with ARC has a grain size factor of 1.41. In contrast to this, the PV module without ARC has a grain size factor of 2.04.

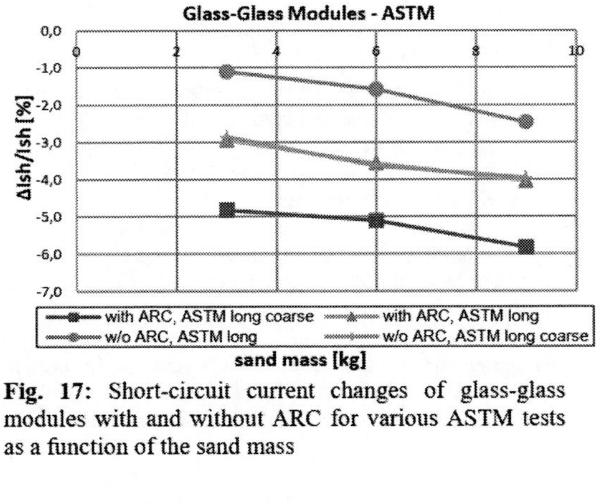

Fig. 17: Short-circuit current changes of glass-glass modules with and without ARC for various ASTM tests as a function of the sand mass

4.3 *Impact on angular response*

The angular dependency of the transmittance according to IEC 61853-2 [9] is determined by Isc measurements in the angle of incidence range -85° to +85°. The incidence angle modifier (IAM) is calculated by normalization to normal incidence and division by the value of the cosine function for the angle of incidence.

Six solar cells were investigated on three PV modules. It has been observed that although ΔIsh/Ish changes, the IAM does not change between before and after sand abrasion tests (Fig. 18).

Previous measurements with other PV modules have led to a significant change of IAM at higher angle of incidence. After the sand blowing test (MIL STD 810G) of glass samples, the short-circuit current of a unshaded optically coupled cell was measured in dependence of the test duration and the angle of incidence (Fig. 19). The improvement of angular response in the angular range above 60° can be explained by roughening of the front glass, which generates diffuse light. The difference between both experiments is that during the sand trickling test the sample angle is 45°, while during the blowing sand test the sample is nearly normal to the sand beam.

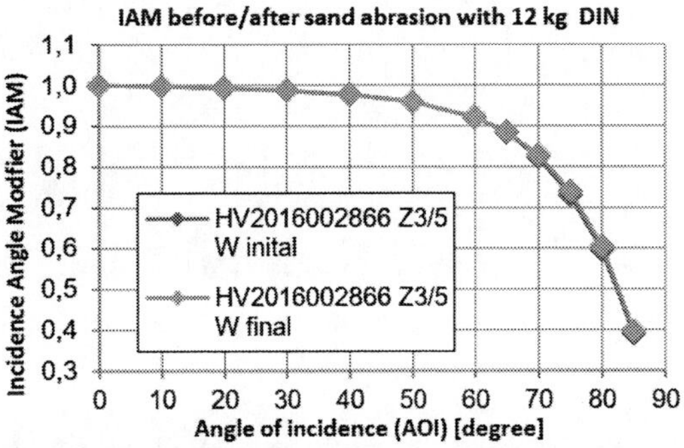

Fig. 18: Incidence angle modifier (IAM) over the angle of Incidence of a glass-foil module with ARC. The effect by sand abrasion with 12 kg sand according to DIN on the IAM is very low.

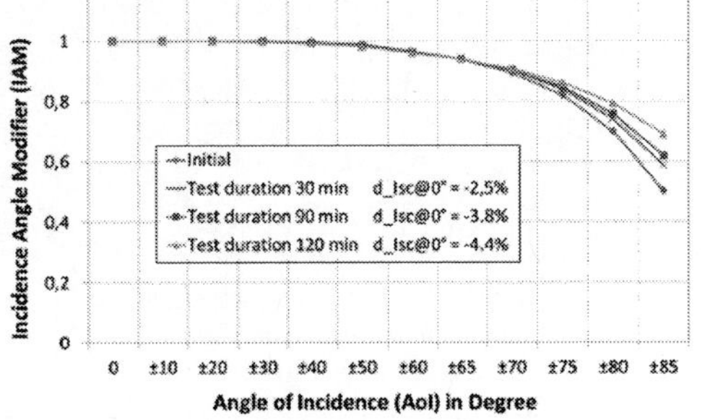

Fig. 19: Incidence angle modifier (IAM) over the angle of incidence. The loss of Isc at normal incidence as a function of test duration is given in the insert.

5 Microscopic and profilometric measurements on solar glasses

For the understanding of the abrasion effects on solar glass it is necessary to perform microscopic and surface roughness measurements.

5.1 Microscopy

Sand abrasion is caused by formation of lateral cracks, which develop almost parallel to the surface in order to split off. The lateral cracks are mainly responsible for the removal of the glass material. Fig. 20 shows a microscopic image of a solar glass with ARC, which has been subjected to ASTM test with 6 kg coarse sand and a long tube. Here the destructive effects of sand impact on the glass surface and glass coating are clearly visible. Fig. 21 shows a microscopic image of a solar glass with ARC, which has been subjected to standard ASTM test with 6 kg standard sand. On both images, a preferred direction of the deterioration is not detectable.

Fig. 20: Microscopic image of a solar glass with ARC after 6 kg ASTM setup (long tube, coarse sand, Fig. 6)).

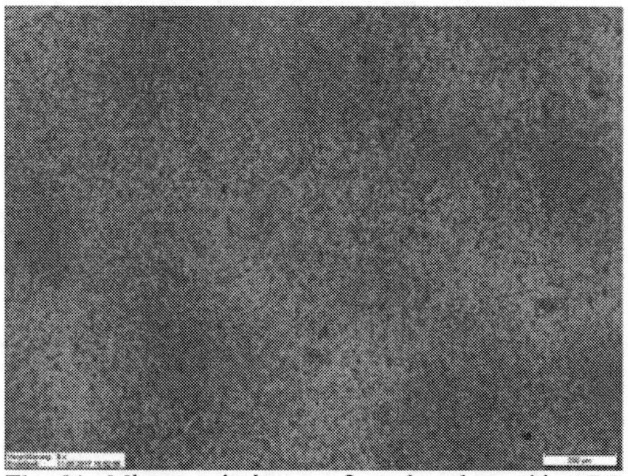

Fig. 21: Microscopic image of a solar glass with ARC after 6 kg ASTM setup (standard tube and sand, Fig. 6).

5.2 Profilometric measurements

Finally, profilometer images were taken with the Veeco Wyko NT9100 optical profilometer. Four images were taken for each analyzed sample. The measurement represents an area of 0.84 mm x 5 mm. For illustration purposes, a blue mask has been added to show the abrasions more clearly. The dark areas mark the deepest impacts. Fig. 22 and 23 corresponds to Fig. 20. Fig. 22 shows the profilometer image of a solar glass with ARC after 6 kg ASTM-test with long tube and coarse sand. In Fig. 23 the 3D view of the measured profilometer image is shown. Fig. 24 and 25 corresponds to Fig. 21.

Fig. 22: Profilometer image of a solar glass with ARC after 6 kg ASTM setup (long tube, coarse sand, Fig. 20))

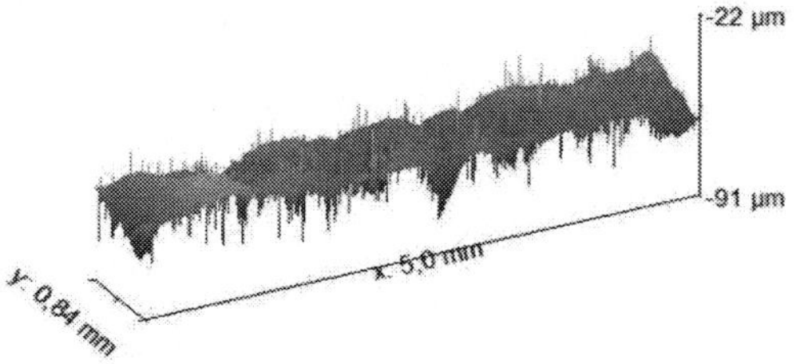

Fig. 23: 3D view of the measured profilometer image of Fig. 22.

Fig. 24: Profilometer image of a solar glass with ARC after 6 kg ASTM setup (standard, Fig. 21))

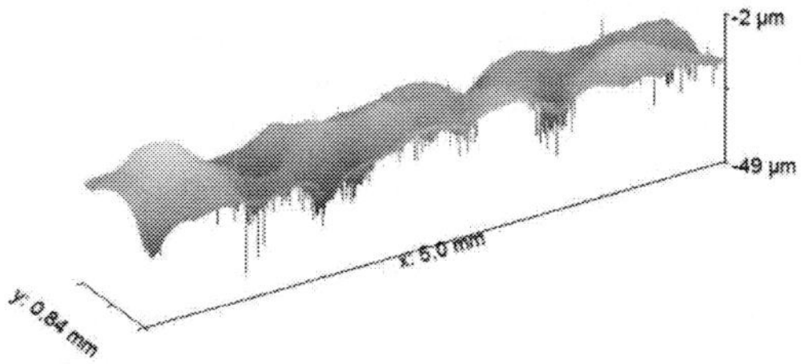

Fig. 25: 3D view of the measured profilometer image of Fig. 24.

6 Conclusions

With the new sand abrasion test setup PV modules and solar glass were tested according to DIN 52348 and ASTM D968-05. In addition, the sand particle speed (due to tube length), the impact angle at the module carriage and the grain sizes can be varied.

For PV modules, the DIN standard is more practical as the ASTM test, due the larger test area and the homogeneous sand distribution.

An extension of the tubes with a change in particle velocity leads with the same sand to significant transmittance losses, both for PV modules and for solar glass.

A significant factor with regard to transmittance losses is the sand grain size distribution. A size factor of 2 was determined for the change from the standard sand (d = 0.5-0.7mm) to the coarser sand (d = 0.7-0.9 mm)

PV modules and solar glass sheets with anti-reflective coating react more sensitive to the abrasion stress. It was found that the transmission loss ΔIsh/Ish is proportional to the sand mass in accordance with the DIN standard. For ASTM tests with higher sand masses per area, ΔIsh/Ish decreases steeply and approaches quickly to saturation when stress level is further increased. PV modules without coating can even gain transmittance by minor roughening of the surface.

Both measurement methods for PV modules (direct and non-destructive by partial shading) deliver comparable results in angular responsivity and normal incidence. The non-destructive method is suitable as standard test method.

Slight discrepancies in the glass transmittance evaluation according to the University of Rapperswil were found. This failure can be caused by diffuse light losses to the glass edges. The reflectance of the front surface r1 is therefore the more stable parameter.

Due to the angular characteristic measurements on PV modules, no changes in the angular properties could be observed. This represents a discrepancy for older investigations.

A new standard for the sand abrasion test of solar glass is currently discussed in the PVQAT working group 12 "Soiling and Dust" as a part of NP IEC 62788-8-2 [10]. In this standard small glass samples are tested. Differences in the hardening of the glass samples and the full-size glasses may lead to differences in the test results. Verification measurements with full-size modules and a laboratory round robin test with different glass samples shall be performed.

From real weather data and reported PV module damages in the field, appropriate test parameters for standard test procedures have to be deduced. More data from fielded PV modules and from laboratory tests need to be collected to define pass/fail-criteria on a scientific basis. This concerns the transmittance of the front glass with durable coating and the safety aspects for backsheet insulation.

7 References

[1] DIN 52348: 1985:02: Testing of glass and plastics, abrasion test, sand trickling method.

[2] ASTM D 968-05:2010: Standard Test Methods for Abrasion Resistance of Organic Coatings by Falling Abrasive.

[3] IEC 62892 (series): Testing of PV modules to differentiate performance in multiple climates and applications

[4] J. P. Thornton: The Effect of Sandstorms on PV Arrays and Components, NREL/TP-411-4787, 1992.

[5] MIL STD 810G: Department of Defense Test Method Standard: Environmental Engineering Considerations and Laboratory Tests, 2008. – Method 510.5, Proc. II: Blowing Sand.

[6] F. Ruesch, L. Omlin, S. Brunold: "Solarglas PV. Zertifizierungsverfahren für Gläser im Einsatz in c-Si Fotovoltaikmodulen", Schlussbericht BFE-Projekt SI/500614, 2012.

[7] G. Mathiak, M. Hansen, M. Schweiger, L. Rimmelspacher, W. Herrmann, J. Althaus, F. Reil: „PV Module Test for Arid Climates including Sand Storm and Dust Testing" Proc. 32nd European Photovoltaic Solar Energy Conference and Exhibition, 2016, 5BO.11.4, 1568-1572.

[8] W. Herrmann, M. Schweiger, L. Rimmelspacher: "Solar simulator Measurement Procedures for Determination of the Angular Characteristic of PV Modules" Proc. 29th EU PVSEC 2014, 5CO.16.2, 2403-2406.

[9] IEC 61853-2:2016: Photovoltaic (PV) module performance testing and energy rating - Part 2: Spectral responsivity, incidence angle and module operating temperature measurements.

[10] IEC 62788-8-2 (NP) Measurement procedures for materials used in photovoltaic modules - Part 8-2: Materials and coatings for the irradiance incident surface of photovoltaic modules or similar solar devices: Abrasion and environmental testing.

8 Acknowledgement

The work presented has been elaborated in the frame of the project 'Fidelitas' (FKZ: 0325735A) supported by the German Federal Ministry for Economic Affairs and Energy.

EVALUATION OF ANTIREFLECTION AND ANTISOILING COATINGS FOR PV MODULES IN THE ATACAMA DESERT

Daniel Díaz Almeida[1], Francisco Araya[1], Pablo Ferrada[1], Asier Sanz[2], Naiara Yurrita[3], Oihana Zubillaga[3]

[1]Centro de Desarrollo Energético Antofagasta, Universidad de Antofagasta, Av. Angamos 601, Antofagasta, Chile
[2]TECNALIA Research & Innovation, Parque Tecnológico de Bizkaia, Derio E-48160, Spain
[3]TECNALIA Research & Innovation, Paseo Mikeletegi 2, San Sebastian E-200009, Spain
Phone: +56 55 2 513530; Email: daniel.diaz@uantof.cl

ABSTRACT: An outdoor evaluation of antireflection and anti-soiling coatings for photovoltaic (PV) modules was performed. The main idea was to answer the question how coatings for PV glass behave under extreme environmental conditions where soiling and high irradiation levels are present. In order to study the performance of these glass and glass coatings, the current-voltage characteristics of laminated solar cells with these materials were measured outdoor at different exposure times. The observables for comparison were the short circuit current (I_{sc}) and the efficiency (η). Four groups (G1 to G4) were defined according to the PV glass used for laminating solar cells. The modules were tested on field, under the conditions of soiling and solar exposure, in the Atacama Desert. To evaluate the different PV glass and glass treatments for the groups, measurements M0 to M5 (corresponding to different measurement times) were carried out every 6 weeks. It was found a decrease of about 6.6% in the I_{sc} within this period and about 0.2% daily loss in absolute efficiency. It came out that the anti-reflection coating (ARC) with the anti-soiling treatment type 2 used in the G4 group led to a better performance under outdoor conditions with regard the other samples.
Keywords: Photovoltaic, Performance, PV Module, Soiling, Environmental Effect, Solar Cell Efficiencies.

1 INTRODUCTION.

Soiling can be understood as the climate-dependent aeolian transport of dust particles from the ground and atmosphere onto a surface of a PV module. When dust is present on the surfaces, the overall performance of a PV plant is degraded. Some of the direct effects of soiling are power and efficiency losses due to the decrease in the amount solar radiation available for solar cells. Climatic conditions can play an important role in PV performance and dust deposition [1]. Costa et al 2016 reviewed the soiling impacts related to energy systems. It is pointed out that some soiling mechanisms are still unclear. On this line, places with increasing PV installations and abundant solar resource such as north of Africa, Middle East, India and deserts in China require detailed quantification of soiling impacts. Another place of interest is the Atacama Desert in Chile where the installed capacity up to August 2017 reached 1.7 GW [2].

The Atacama Desert exhibits highest surface irradiation values on Earth [3], a high mean altitude, a large number of days with clear skies and low absorption ozone and water vapor columns. The abundant solar resource as well as political and economical situation pushes the implementation of PV and solar technologies in this place. Therefore, the unique environmental conditions of this region are challenging for technology developments in PV. Progress into the development of a PV module with module materials adapted to environmental conditions of Atacama Desert (AtaMo) as well as the impact of soiling and abrasion on several AtaMo configurations were shown in [4] and [5].

The effect of dust on PV performance can be determined by measuring the current-voltage (IV) characteristics of the PV modules. As dust accumulates, it is expected that the short circuit current (I_{sc}) and the current at the maximum power point will be the most affected. Setting up an experiment where modules are exposed to the environment for a long period of time, IV curves of clean and dusty modules can be analyzed. A study conducted in Dakar (Senegal) [6] showed that the maximum power output losses were 18% and 78% for a mc-Si module and mono c-Si module, respectively. The loss of I_{max} increased from 23% to 80% for respectively mc-Si and mono c-Si modules. Conversely, the maximum voltage (V_{max}) and the open-circuit voltage (V_{oc}) were not affected by dust. The fill factor (FF) reduced from 2% for the mc-Si module to 17% for the mono c-Si module.

Depending on the environmental conditions, soiling distribution will be affected [7]. The different surface structures and anti-reflection coatings applied to the PV glass may turn non-effective for increasing the overall gain of the PV module and transmittance may decrease [8]. According to the aforementioned, the study of the PV systems performance affected by soiling phenomena is of high interest since it has a direct impact over the PV modules power conversion as well as over the economic factor and technology development. In order to identify best performing materials for PV modules, long-term measurements performed on field, under local conditions, are required. The soiling process, which potentially leads to power losses in PV modules, can be addressed by the implementation of featured surfaces and designs to diminish both dust accumulation and optical losses. The use of special coatings on the glass have a potential reduction in power loss caused by dust settlement [9].

In this work, we performed a study and assessment of antireflection and different anti-soiling coatings through measurements on laminated solar cells installed in the Atacama Desert. Thus, the performance of these module materials, was performed at module level. The main idea was to answer the question how coatings for PV glass behave under extreme environmental conditions where soiling and high irradiation levels are present.

2 MATERIALS AND METHODS.

2.1 Samples.

Four different module groups were defined according

to the PV glass used. Group G1 corresponds to a standard glass, while G2 receives an antireflection coating (ARC). Groups G3 and G4 exhibit both an ARC but differentiate each other by the anti-soiling treatment. For each kind of glass, 3 samples were used in combination with same encapsulant and solar cells produced by Sunways AG. Consequently, each group (G1 to G4) is made up of 3 laminated solar cells.

Monocrystalline silicon solar cells of 12.5x12.5 cm^2, 260-310 μm thickness with Ag/Al busbars and, 17% efficiency were used. These solar cells were encapsulated into mini-modules with the different glasses described above.

2.2 Methodology

In order to study the performance of these glass and glass coatings, the IV characteristics of laminated solar cells with these materials were measured outdoor at different exposure times. The defined observables for comparison were the short circuit current (I_{sc}) and the efficiency (η).

The measurements M0, M1, M2, M3, M4 and M5 (corresponding to different measurement times) have been carried out every 6 weeks. The period of evaluation started on October 2016 and is to be carried out for at least 12 months.

The measurements performed before any dust accumulation define a reference and starting point denoted as a measurement zero (M0). The location for outdoor measurements is the Solar Platform of the Atacama Desert (PSDA) in Chile, belonging to University of Antofagasta, and located at coordinates 24°05'24.1"S 69°55'44.2"W (see Fig. 1). Outdoor measurement conditions such as those at the PSDA can strongly differ from standard conditions (STC) used to rate solar cells and PV modules. Therefore, IV data from outdoor characterization were corrected to STC. The IV characteristics of PV modules performing on field at PSDA were recorded with a Cetis PV-CTL1 IV tracer from *h.a.l.m.* GmbH.

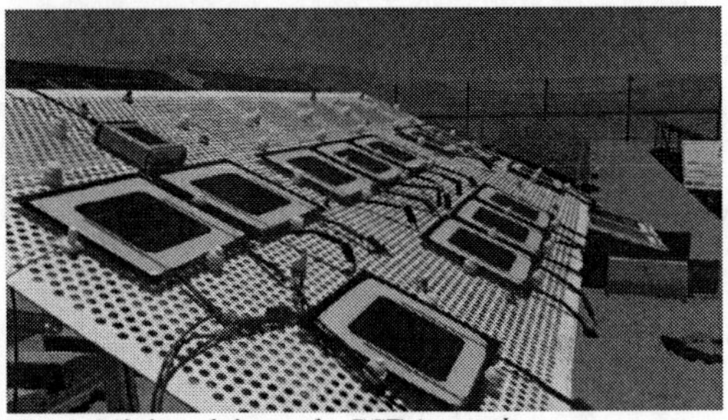

Fig. 1. Mini modules at the PSDA test site.

The mini-modules-samples are placed over a 20° tilted grid at the PSDA distributed in a matrix array isolated with no shadowing and with the terminals of the modules easy to reach. All outdoor measurement have been performed near noon time with irradiances between 900 W/m^2 and 1050 W/m^2, after one and half month between them, without removing any layer of deposited dust.

3 ANALYSIS AND RESULTS

3.1 Measurements at clean conditions

Table 1 summarizes the IV results at M0 carried out indoor at STC with a Sun Simulator and outdoor corrected to STC with the IV curve tracer. The table includes the absolute values for I_{sc}, η and the standard deviation denoted as d I_{sc}, and d η. A lower current and lower efficiency on field compared to indoor conditions were found (up to 0.5 A and 2.8% absolute, respectively). A deeper analysis reveals that the resistive effect due to cables and contacts is higher at outdoor conditions compared to indoor environment. This effect can be shown as the fill factor (FF) was affected most with a relative difference up to 10% (outdoor with respect to indoor).

Table I: Comparison among indoor (in) and outdoor (out) measurements carried out for the four groups.

Samples		Isc (A)	d Isc (A)	η (%)	d η (%)
G1	in	5.4	0.01	14.2	0.1
	out	5.0	0.09	11.5	0.1
G2	in	5.5	0.03	14.3	0.1
	out	5.0	0.03	11.9	0.1
G3	in	5.5	0.01	14.6	0.1
	out	5.2	0.05	12.3	0.3
G4	in	5.6	0.03	14.7	0.1
	out	5.1	0.06	11.9	0.1

3.2 Measurements performed on field

3.2.1 Short Circuit Current measurements.

The measurement M0 to M5 of all modules has been performed near noon time with irradiances between 900 W/m^2 and 1050 W/m^2, after one and half month between them, without removing any layer of soiling deposited. Measurement defined as M0 is the reference or starting point for further comparisons

Figure 2 shows I_{sc} for each group of study comparing the data obtained from the measurement zero (M0 out) to the sixth one, M5, with soiling exposure in the Atacama Desert. At measurement M0, G3 and G4 achieved highest I_{sc} values (5.19 A and 5.14 A, respectively) compared to those of G1 and G2. Under Atacama Desert conditions, G3 and G4 at M2 were the leading groups with a better performance and higher I_{sc} (5.02 A both) as well as for M3 measurement, but with a small decrease of the short circuit current for the G4 group (5.00 A, respectively), compared to the other groups (G1 and G2). It means that the ARC with anti-soiling treatment 2 led to a good performance under outdoor conditions.

For measurement M4, a different behavior of the samples is observed since there was an increment of I_{sc} compared to M3 measurements due to an unusual raining condition that took place the day before the data were recorded, which means the dust was removed from the samples. Under this condition, both G2 and G4 produced higher I_{sc} values with 5.08 A and 5.07 A, respectively, compared to the groups G1 and G3, which implies an increment of ~4% for G2 and ~2% for G4 with regard to the later measurement.

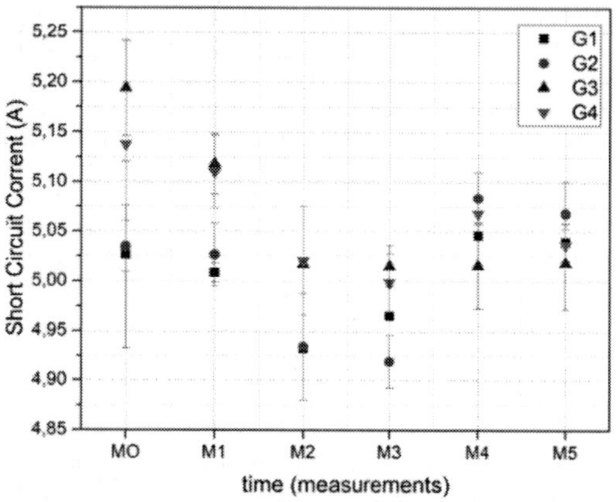

Fig. 2. Short circuit current measurements M0, M1, M2, M3, M4 and M5 performed outdoor for the samples from G1, G2, G3 and G4 corrected to standard testing conditions (STC).

It can be observed that the performance is similar between the samples, following an expected tendency, e.g., for M0 measurement the value of the I_{sc} are closer to those measured indoor for all samples (Table I). Period M5 presented a different condition compared to the later measurements due to low temperatures, high humidity and winter season that took place at the PSDA, which produced crystallized water and dew over the front glasses as well as at the terminals of the samples. As a result, the accumulated dust was removed. Therefore, measurements corresponding to M4 and M5 exhibit a current generation similar to no soiling conditions. Nevertheless, after 5-6 measurements (approximately a year of exposure) the group G2 and G4 led to the higher Isc, 5.07 A and 5.04 A, respectively, and slower degradation than G1 and G3. The evaluation shows that samples from G3 and G4, whose only difference is the anti-soiling treatment, produced the best performance during measurements M0-M3 followed by G2, meanwhile, for M4 and M5, G3 showed a mayor degradation than the others with an ARC and anti-soiling treatment.

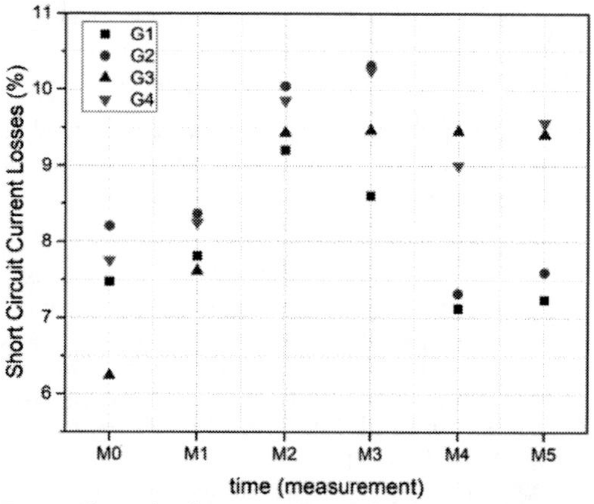

Fig 3. Short circuit current losses for the samples from G1, G2, G3, and G4, at PSDA outdoor soiling exposure.

Figure 3 shows I_{sc} losses for each group relative to the M0 indoor measurement. It can be observed that from M0 to M3 measurements, the samples with a lower loss rate were group G1 with standard glass and G3, with a minimum loss of 7.5% and 6.2% at M0, and 8.6% and 9.5% at M3, respectively. In addition, the group G3 maintains almost a steady I_{sc} loss tendency from M3 to M5 in spite of the change of the soiling conditions mentioned above. On the other hand, for G4 suffered an overall increase of 9.1% (on average) in I_{sc} losses for all the measurement periods despite its ARC and anti-soiling treatment.

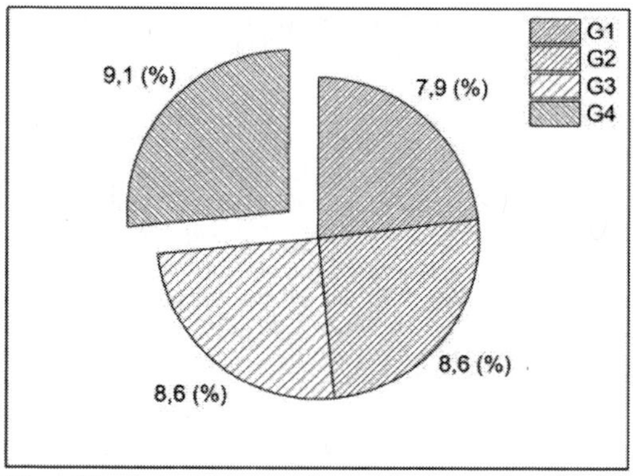

Fig 4. Short circuit current losses for groups G1, G2, G3 and G4.

From Fig 4, it is observable how I_{sc} was affected due to the exposure in Atacama Desert conditions with the its losses increased by an average of 8.6% with regard to M0 (indoor), 1.4%/period, and 0.2%/day. Comparing the I_{sc} losses and I_{sc} G2 and G3 have the same average current losses rate but in Fig 2 and Fig 3 it can be seen that G3 samples with ARC and anti-soiling treatment type 1 show an overall better performance over G2 samples with ARC only. Even though, G4 was the best group between those with ARC and anti-soiling treatment.

3.2.2 Efficiency measurements

Under these extreme environmental conditions at the PSDA, the efficiency of the modules was hardly affected during the measurement time. So far, it could be observed, in an unexpected way, G1 samples to exhibit a better performance in terms of efficiency showing a slower decay during the period of analysis, whereas G3 showed a faster η decay.

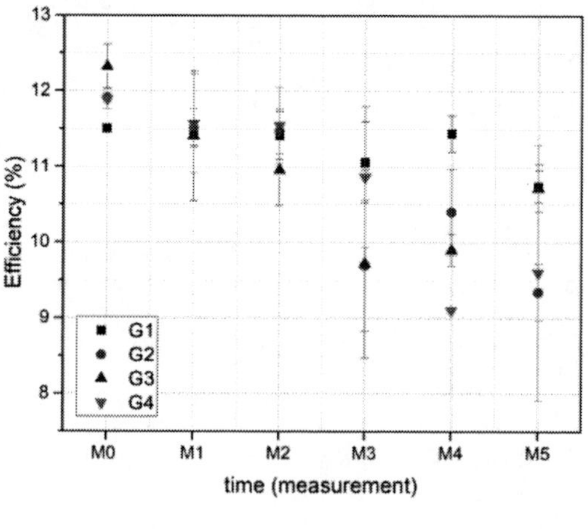

Fig 5. Efficiency measurements M0, M1, M2, M3, M4 and M5 performed outdoor for the samples from G1, G2, G3 and G4 corrected to standard testing conditions (STC).

Figure 5 shows the comparison of efficiencies along the measurement period. At the beginning (M0), samples from G3 reached the best efficiency, exhibiting a significant decay at M3 and recovering at M4 to be above G4, and above both G2 and G4 at M5. The samples with ARC and anti-soiling treatment together were expected to have overall better performance.

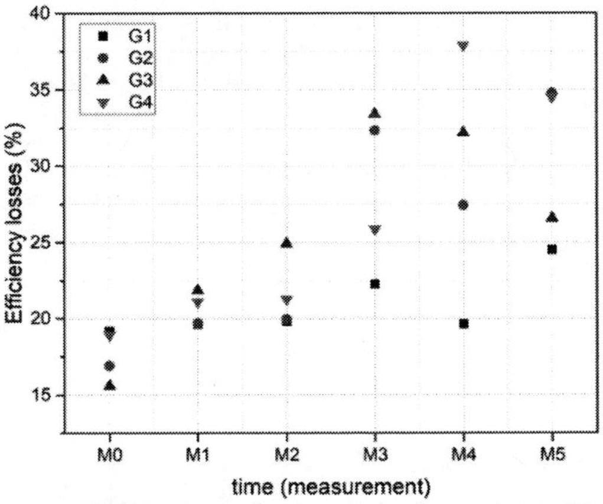

Fig 6. Efficiency losses for the samples from G1, G2, G3, and G4, at PSDA outdoor soiling exposure.

Figure 6 depicts the efficiency losses of the samples after 6 measurements. Regarding to M0 (indoor), samples from group G3 had an increasing tendency compared to the other samples, starting with 15.6% η loss at M0 and 33.4% at M3. Nevertheless, this loss decreased considerably for M4 to M5. Group 1 with the standard glass led to lower efficiency losses over the periods M2-M5 compared to groups G2-G4. Note that at clean conditions, the difference of the measurement performed M0 outdoor respect to M0 indoor, showed in Table I, exhibited an efficiency loss mainly due to the resistive issues when metal contacts and connectors are exposure to outdoor conditions in contrast to STC. It is noticed that for more than six months of exposure to the Atacama Desert, samples from G4 were more degraded and hence had a higher efficiency loss for the last two measurements, which corresponds to a clean condition of the module glass.

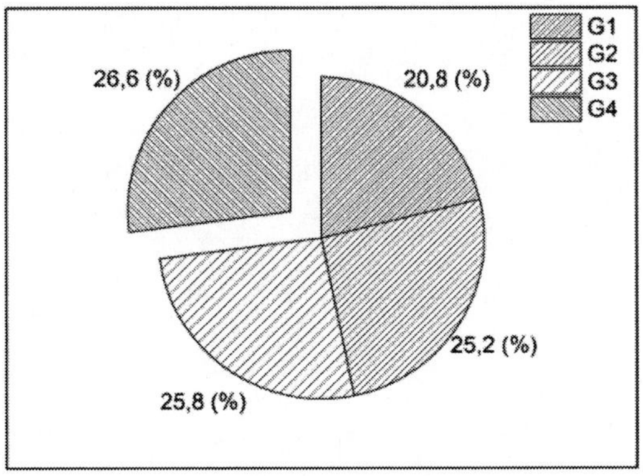

Fig 7. Efficiency losses for groups G1, G2, G3 and G4.

Figure 7 presents a comparison of the efficiency losses between the samples and can be noticed that as well as for the I_{sc} G4 had the highest losses with an average of 26%/period respect to the reference M0 (indoor) in contrast to G1 that had the lowest losses, 20.8%/period, and both G2 and G3, 25.2%/period and 25.8%/period, respectively, which had similar performance and overall average losses of 24.6%/period. On the other hand, when a comparison among groups per measurement was considered they had an efficiency loss about 5.5% in between, and 0.6%/day.

The first estimation of the degradation rate was done at the first three measurements, M0-M2, and it was found for G2 and G4 a similar degradation value for the efficiency about 0.14%/month and 0.11%/month, respectively, being G4 the group with the best behavior in the analyzed periods since it had the lowest degradation rate. Nonetheless, after 11 months of outdoor exposure, precisely the samples of G4 suffered higher losses due to corrosion at the contacts. This issue can be associated to the chemistry of dust [10], humidity, high temperatures during the day plus high solar irradiance. Despite the later fact, this group with ARC and anti-soiling treatment type 2 was the one with higher average short circuit current generation with 1.6%/period current drop as well as G3 with ARC and anti-soiling type 1 among samples as shown in Table II.

Table II. Comparison of I_{sc} values.

Samples	I_{sc} avg. (A)	I_{sc} drop avg. (A/period)
G1	**5.04**	**0.06**
G2	**5.05**	**0.06**
G3	**5.09**	**0.08**
G4	**5.09**	**0.08**

So far, a performance comparison of the samples regarding the measurement at STC condition, M0 (in), has been carried out. In addition, a comparison between samples respect to the outdoor reference M0 (out) has been done to estimate the amount of current lost since the very first measurement at the PSDA.

Table III. Comparison of I_{sc} losses respect to M0 (out) for 11 months of measurements.

Samples	I_{sc} avg. (A)	I_{sc} drop avg. (A/period)
G1	**5.04**	**0.02**
G2	**5.05**	**0.02**
G3	**5.09**	**0.04**
G4	**5.09**	**0.03**

Table III shows a comparison of I_{sc} losses respect to M0 (out) which exhibit that samples of group G1 and G2 have a similar performance in terms of short circuit current losses with an average loss of 0.02 A/period compared to G3 and G4 with 0.04 A/period and 0.03 A/period of average loss, respectively. Therefore, samples achieved an overall average of 0.03 A/period of current lost.

4 OUTLOOK

In order to obtain a complete picture with regard to the performance of PV glass and glass treatments at module level, field measurements need to be carried out for at least one year. There is some aspect of the procedure that will be applied to the current and future measurements. For instance, the measurement of the module temperature will be used to correct data to STC by using IEC 60891 procedure [4]. Finally, the results obtained after the measurements carried out for a year will be compared with a similar test with the same samples done under the conditions of Atlantic weather for 18 months period.

5 CONCLUSIONS

Photovoltaic (PV) mini modules consisting of laminated solar cells with standard PV glass (group G1) and featured glass such as antireflection coating ARC (G2), ARC with anti soiling treatment 1 (G3) and ARC with anti soiling treatment 2 (G4) were evaluated for almost one year (11 months) at the Solar Platform of Atacama Desert (PSDA) in Chile. The current-voltage (IV) characteristics were carried out every 6 weeks at the PSDA test site and corrected to standard testing conditions (STC).

The samples showed a decrease about 8.6% in the short circuit current (I_{sc}) generation respect to STC conditions used at reference for PV technology development, in the period of 11 month of measurements which were divided in 6 weeks intervals. Furthermore, samples showed an overall average of 0.03 A/period I_{sc} lost, and at least 0.2% daily loss. Groups G3 and G4 resulted in a better response to extreme soiling exposure producing high I_{sc} despite the corrosion on metal contacts, connectors and the environmental stresses.

Among the four groups, G1 with standard glass still resulted to be a good option for environments without such extreme conditions. However, in terms of performance, those glasses combing ARC and anti-soiling treatment may be more feasible for better performance, reliability and endurance in desert environments.

ACKNOWLEDGEMENTS

The authors acknowledge TECNALIA Research & Innovation for providing the mini-modules to be characterized in the Atacama Desert.

The authors acknowledge the support provided in part by FONDECYT Project Number 3160190 and CONICYT/FONDAP/15110019 "Solar Energy Research Center" SERC-Chile.

REFERENCES

[1] S. C. S. Costa, ASAC Diniz ASAC, L. L. Kazmerski. Renew Sust Energ Rev 2016;63;33-61.
[2] Reporte Mensual ERNC, CNE Vol. 12, August 2017, "*Reporte Sector Energético* Agosto 2017". https://www.cne.cl/wp-content/uploads/2015/06/RMensual_ERNC_v201708-1.pdf.
[3] R. Escobar, C. Cortés, A. Pino, E. Bueno Pereira, F. Ramos Martins, J. M. Cardemil. Renewable Energy 2014; 71: 324-332.
[4] E. Cabrera, A. Schneider, J. Rabanal, P. Ferrada, R. Cordero, E. Fuentealba, R. Kopecek. 31st EU PVSEC, Hamburg (Germany), 14-18 September 2015.
[5] E. Cabrera, A. Schneider, E. Wefringhaus, J. Rabanal, P. Ferrada, D. Thaller, F. Araya, A. Marzo, M. Trigo, D. Olivares, J. Haas, E. Fuentealba, R. Kopecek. 32nd EU PVSEC, Munich (Germany), 20-24 June 2016.
[6] A. Ndiaye, CMF Kébé, P.A. Ndiaye, A. Charki, A. Kobi, V. Sambou. International Journal of Physical Sciences 2013;8(21);1166-1173
[7] M. Gostein, B. Stueve, et al., IEEE 43rd Photovoltaic Specialists Conference (PVSC), (2016) 3129-3131.
[8] E. Klimm, T. Lorenz, K. A. Weiss, Proceedings 28th European Photovoltaic Solar Energy Conference, (2013) 3099-3102.
[9] A. Reinhart, et al., Solar Energy 96 (2013) 283-291.
[10] D. Olivares et al. Energy Procedia 124 (2017) 547–553.

LONG TERM STATISTICS ON MICRO CRACKS AND THEIR IMPACT ON PERFORMANCE

Bengt Jaeckel[1], Thomas Franke[2] and Jürgen Arp[2]

[1] UL International GmbH, Admiral-Rosendahl-Strasse 9, 63263 Neu-Isenburg (Zeppelinheim), Germany, bengt.jaeckel@gmail.com

[2] PV LAB Germany GmbH, Gartenstrasse 36, 14482 Potsdam, Germany, j.arp@pv-lab.de

ABSTRACT:
Over six years the change in isolated cell areas due to micro cracks and the change in STC power are analyzed and brought in context with type of crack and its potential PV module power loss. Intermediate results about four years of operation were presented in 2015 [6]. This work this is an update to include collected data after 6 years of operation which are split in two sub periods of three years each. Other than in the first presentation we now have three data points and can identify a first trend.

The data is analyzed on cell and on module level. On cell level every cell is characterized by the type of micro crack, the percentage of potentially isolated area and of isolated area. The analysis compares the development of the micro crack status of the different types of cracks with respect to isolated areas and potential power loss. 143 cells in three categories have been specifically selected for detailed monitoring and analysis. Changes in the percentage of the isolated areas were investigated. In the three-year period between 2011 and 2014 10,9% of the areas which have been endangered to become isolated have really been electrically separated from the live circuit. This results in 3,6% per year. In the following three-year period between 2014 and 2017 the amount of newly isolated areas adds up to 8,25%, which equals 2,75% per year.

For the whole time-span of six years the amount of isolated areas adds up to 19,1% or 3,2% per year in average.

The STC power measurement in 2015 shows a power loss which exceeds the expected natural degradation with an average annual power loss of 1,5%, but was within the uncertainty of measurement. The measurement in summer 2017 did not confirm this trend of development of the STC power degradation. The average annual power loss has been reduced to 1,0% per year.

Keywords: electroluminescence, micro cracks, probability, power loss

1. MOTIVATION

In recent years there has been extensive research on the topic of micro cracks. For instance, Marc Köntges [1]-[4] delivered ground breaking work several years ago. Especially his theory of potential power loss caused by micro cracks serves as a basis for the analysis and evaluation of micro cracks and their potential impact on the performance of the solar cell/module. The simulation basically delivers the quantification of the risk. What has been missing in the past is information about the probability that this scenario occurs and in particular when under normal outdoor operating conditions. This probability information is crucial to better assess the impact of micro-cracks on the long term reliable performance of a PV power plant asset.

2. INTRODUCTION

In wafer based, crystalline silicon PV module micro cracks are found regularly. Electroluminescence [5] made it easy to detect and characterize micro cracks with very high spatial resolution.

Based on publication on Electroluminescence (EL) it is not clear to what extends the power output of PV modules is influenced by cracks in the short and long term. Some worst-case scenarios (end-of life) can be easily created [2]-[4] to sort out defective modules but this may overdo it in a lot of cases.

The work from Koentges delivered the possibility to calculate the possible power loss due to a specific crack. Koentges approach is based on a almost worse case simulation and does not include the important time dependence. To get a better understanding of PV power plant performance the time evaluation of cracks is important to understand. The aim of this project was to have a first look into this difficult to assess topic because it just takes a lot of time actually years, better decades. This is the first update and continues the data sharing from 2015 [6]. Within this paper we provide data for 2014 and 2017 with respect to the particular used installation where the module now exposed to the outdoor environment for around 6 years in Potsdam/Germany. The data points 2011, 2014 and 2017 were chosen to have two directly comparable time periods.

The setup and details are described in the next sections.

3. EXPERIMENTAL DETAILS

PV Lab started a long-term experiment on the outdoor development of micro cracks in late 2011. In 2014, 2015 and 2017 the micro cracks and their changes have been characterized

The system contains 130 multi crystalline PV modules with each 60 6" cells, and therefore 7800 cells in total. The installed PV modules have been pre-selected to:

- show in EL a lot of cell cracks
- all different crack types are covered
- no other defects like delamination

All modules have been tested with Electroluminescence (EL) and infrared Thermography (IR) as received in 2011, in 2014 in 2015, and this year, 2017, so after 6 years of outdoor exposure in Potsdam in a roof installation with totally free air ventilation, similar to open rack field mounted module. The system is shown in Figure 1. Within this contribution no distinguishing is done on the two different module tilt angles for simplicity and as there was till now no obvious difference. .

Figure 1: PV system investigated within this project. It contains 130 modules with two different inclinations.

The cracked cells were classified using known understanding [7] by using EL. Special attention was paid in the repeated measurements to those classified cells to investigate cell crack propagation. But all other cells in the modules were checked if new cracks appeared.

EL is the main tests to measure cell crack propagation. A brief look at IR images was done to check for critical hotspot heating. Since no defects e.g. browning of backsheet was observed also after 6 years of outdoor exposure this paper will not further discuss IR herein. Nevertheless, performance testing was done outdoors to determine power degradation using a state of the art portable, fully calibrated IV curve tracer.

4. EXPERIMENTAL EL DETAILS

The first measurement campaign in 2011 was done utilizing standard laboratory equipment. The presented measurements from 2015 and 2017 were done outdoors, resulting in a slightly lower resolution and quality of the images. But this is not critical since the effects are still clearly visible. The used outdoor procedure provides a significant advantage compared to indoor measurements: it is much more realistic! This is due to the fact that the PV modules were measured without unmounting and transporting them. Typically these steps result in extra cracks or other defects that are "artificially" introduced to the PV module and are not necessarily from the environmental stress.

Initially and as presented in [6] we separated the cracks into two main divisions as:

a) single cracked and
b) multiple cracked cells.

For both divisions it is clearly visible that some of the cracks are very stable and the EL image basically does not change in the period of 6 years of outdoor exposure. On the other side, some images clearly show the risk potential of cracked cells and their change. Either the EL contrast darkens in the crack effected areas or turning completely dark. For both cases that at least results in a higher serial resistance resulting in current mismatch and therefor power loss.

For a more clear picture a more simplified analysis of the isolated areas was done usingthe following two decisions:

1. Only completely black areas are classified as isolated areas.
2. Only cells which show a clearly defined potential area for isolation have been taken into account.

"black&white"-approach:

Hier muss noch das mit der Änderung rein… so wie du es auf den folien hat. Was ist ein change und so. Das ist die vorraussetzung, um dieT später zu verstehen

This is a change to the previous published data but results in a more clear result. The cells have been assigned to five different categories as below. Within the contribution only the categories 1-3 were included in the evaluation.

Category 1:	Category 2:	Category 3:
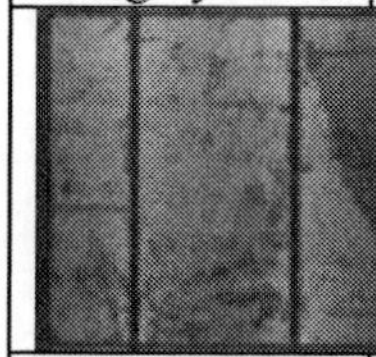		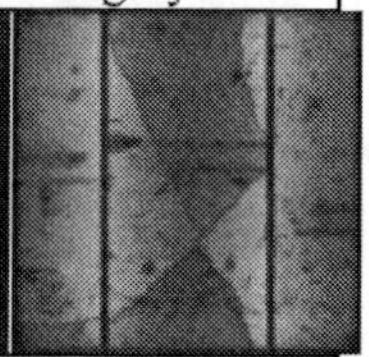
Single Crack at the corner of a cell.	Single cracks along the busbar outside the busbar.	Cracks between the busbars surrounding a clearly defined area.

The cells that fall into the categories 4 and 5 are excluded as there is typically not a clear change visible within the used strict black&white-approach.

Category 4:	**Category 5:**
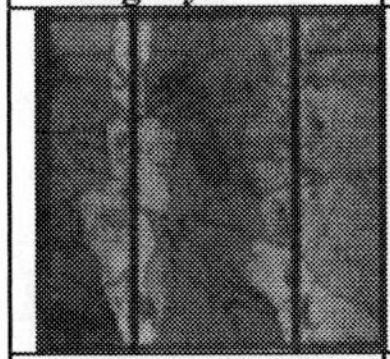	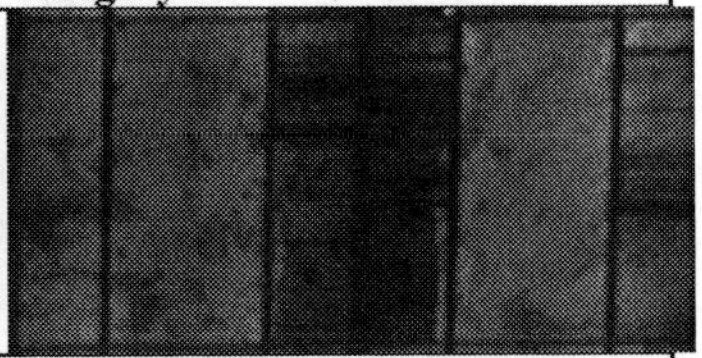
Multiple and dentritic cracks.	**Cracks below the busbar.**

In Table 1 the number of cells for the different categories and the percentage of change are shown for the two time periods under investigation. Significant difference means, that the part of the cell which has been potentially separated by the micro crack is separated and appears black in EL imaging taken at short circuit current (Isc). If simply divided by the six years of outdoor exposure on average over all 3,2% of the potentially isolated area becomes isolated per year.

Table 1: Cell crack statistics

	Category 1	Category 2	Category 3	Total
Affected cells (2011-14)	10/33	32/81	7/26	49/140
Area loss [%] 2011-14	12,0%	11,2%	8,3%	10,9%
Area loss [% p.a.] 2011-14	4,0%	3,7%	2,8%	3,6%
Affected cells (2014-17)	13/33	23/81	3/26	39/140
Area loss [%] 2014-17	16,7%	4,9%	8,1%	8,3%
Area loss [% p.a.] 2014-17	5,6%	1,6%	2,7%	2,8%
Affected cells (2011-17)	19/33	42/81	10/26	71/140
Area loss [%] 2011-17	28,6%	16,1%	16,4%	19,1%
Area loss [% p.a.] 2011-17	4,8%	2,7%7/	2,7%	3,2%

Hier muss noch etwas Text rein... was lernen wir daraus und so

5. STC POWER DETAILS

To further investigate the impact of the cracks and to quantify the DC powerloss IV curves were taken outdoors in 2014 and 2017. The data is compared to the flash-test data from the manufacturer.

The measured average power of the modules is summarized in Table 2. It must be considered that the data from 2014 and 2017 is taken outoddors. The estimated measurement uncertainty is determined to be in the range of 6%. Within that in mind the power of the modules did not change much in the time period from 2014 to 2017. It must be noted that the standard deviation is increasing. This can have different reasons including factors as time of the year (2014: June, 2017: August), time of date (morning vs afternoon), deviation from 1000W/m², module temperature and also soiling as the modules were not particularly cleaned prior to the measurements. Including those impact factors it is not possible to draw a clear conclusion from the measurements so far. That may change within the next years as more cracks may get active and a more significant power loss may develop.

Table 2: Performance data from the 130 modules normalized to STC.

	2011	2014	2017
Averages	224.5	206.8	210.9
Standard dev	8.4	10.9	13.8

6. DISCUSSION

Hier muss noch input rein!!!

7. CONCLUSION

Hier muss noch input rein!!!

8. REFERENCES

[1] M. Köntges, S. Kajari-Schröder, et al., Cell Cracks measured by UV fluorescence in the field, 27th EUPVSEC, Frankfurt, Germany, 2012

[2] M. Köntges, K. Bothe., Elektrolumineszensmessung an Photovoltaik-Modulen. Photovoltaik Aktuell supplement in Elektro Praktiker 7/8 2008, p. 36-40

[3] M. Köntges, I. Kunze, et al., Quantifying the risk of power loss in PV modules due to micro cracks, 25th EUPVSEC, Valencia, Spain, 2010

[4] M. Köntges, S. Kajari-Schröder, et al., Crack statistic of crystalline silicon photovoltaic modules, NREL Workshop, PV Module Reliability Workshop, Golden, Colorado, USA, 2011

[5] T. Fuyuky, H. Kondo, T. Yamazaki, Y. Takahaschi, Y. Uraoka, Photographic surveying of minority carrier diffusion length in polycrystalline silicon solar cells by electroluminescence, Applied Physics Letters 86 (2005) 262108

[6] J. Arp, B. Jaeckel, Long term statistics on micro cracks and their impact on performance, 30th EUPVSEC, Hamburg, Germany, 2015

[7] B. Jaeckel, J. Arp, Blick in die Zukunft, PV Magazine 01/2013

Single cracks:

#	Crack-type	EL image before outdoor exposure	EL image after 4 years of outdoor exposure	EL image after 6 years of outdoor exposure	comment
1	C01 – single crack			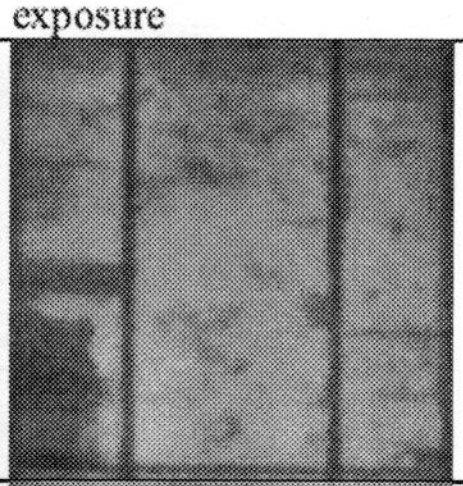	Single crack that is complete. No obvious change in 4 years of outdoor exposure.
2	C01 – single crack, complete		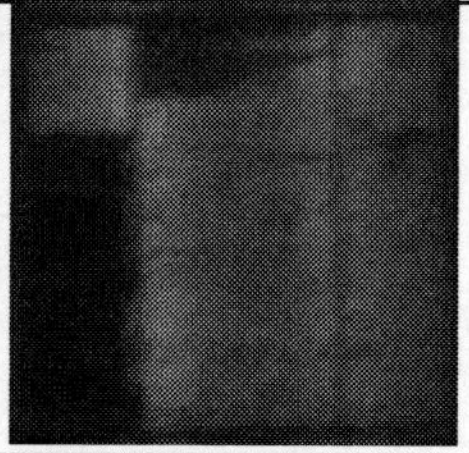		Two single cracks starting at cell edge and ending at busbar (could also be just a single crack that is going through the busbar). After 4 years of outdoor exposure no major change in EL crack appearance
3	C01 – single crack, complete				Single cracks starting at Busbar and ending at busbar. After 4 years of outdoor exposure no major change in EL crack appearance
4	C01 – single crack, complete		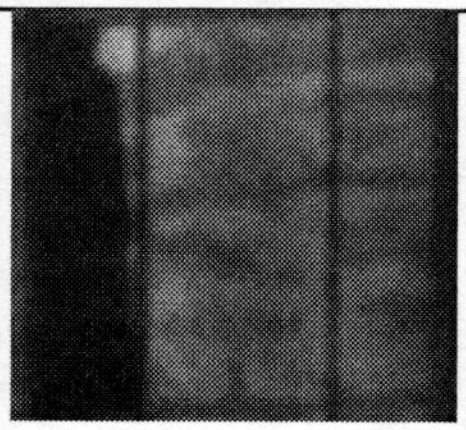	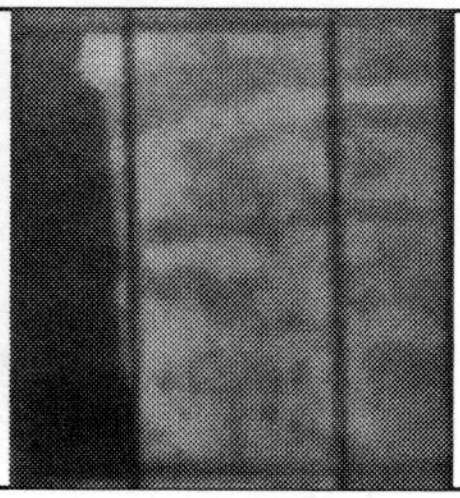	Single cracks starting at cell edge and ending at busbar. After 4 years of outdoor exposure no major change of crack in EL visible. EL contrast is slightly more intense.

5	C01 – single crack, complet e	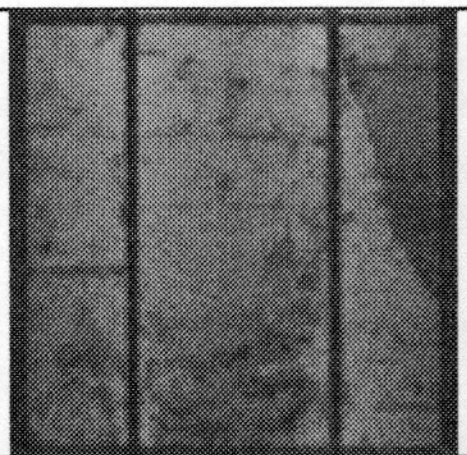	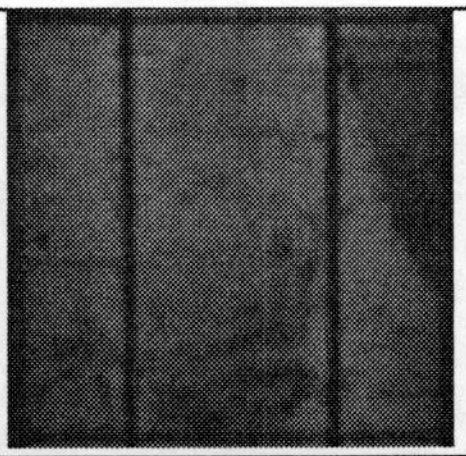	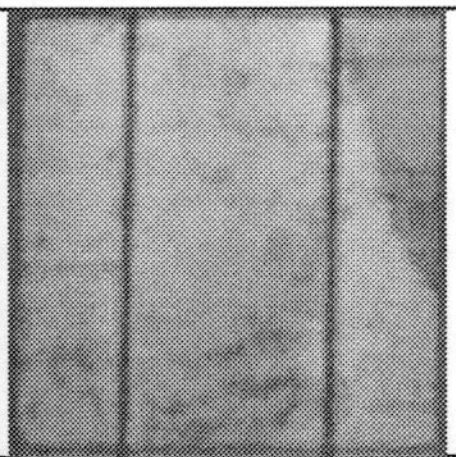	Single cracks starting at cell edge and ending at busbar. After 4 years of outdoor exposure no major change of crack in EL visible. EL contrast is slightly more intense.
6	C01 – single crack, complet e				Single cracks starting at cell edge and ending at cell edge After 4 years of outdoor exposure a strong increase in EL contrast is visible. Additional grid finger interruptions are also visible.

Multiple cracks:

	Crack-type	EL image before outdoor exposure	EL image after 4 years of outdoor exposure	EL image after 6 years of outdoor exposure	comment
7	C07-multipl e cracks	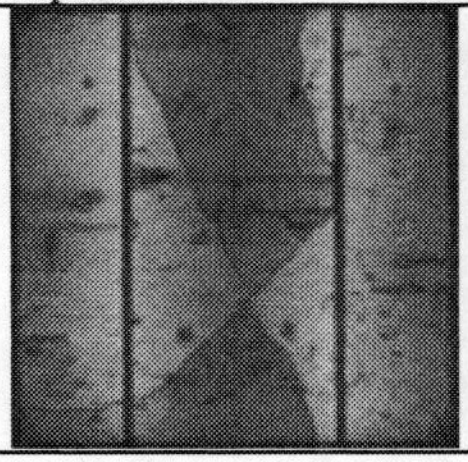		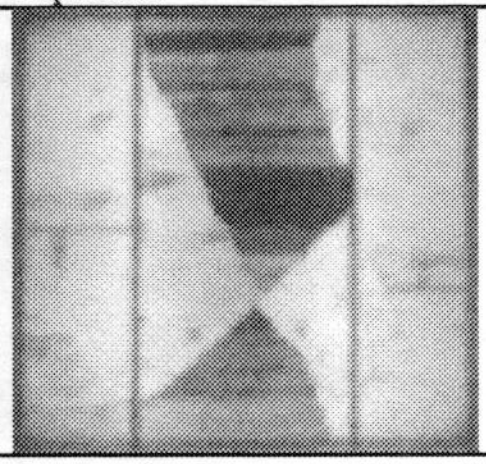	Multiple cracks that show the potential of islanding parts of the cell. After 4 years of outdoor exposure most of the potential islanded area are disconnected from the active cell circuit

8	C07-multiple cracks	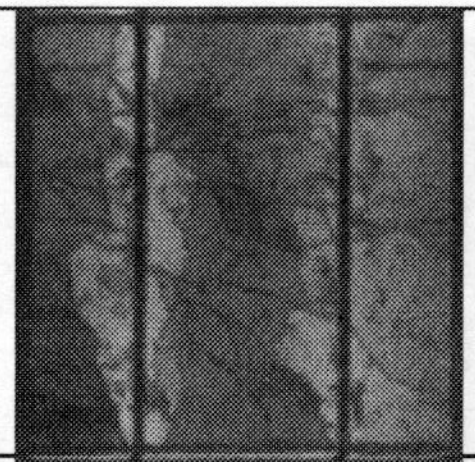	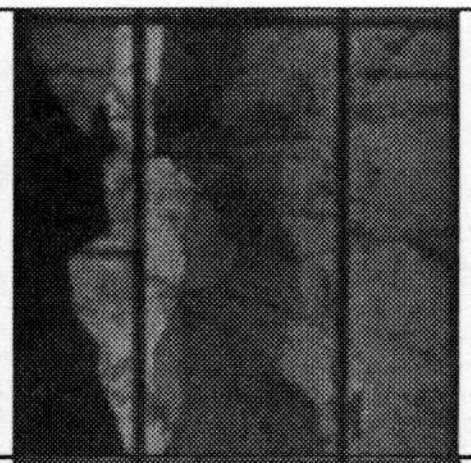	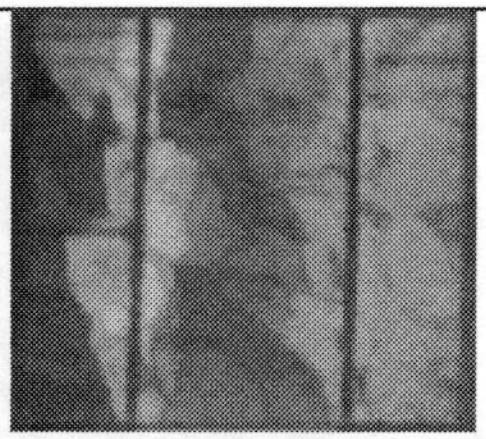	Multiple cracks that show very high potential of islanding large parts of the cell. Interestingly after 4years of outdoor it does not look as severe as one could imagine. Still in the center of the cell large areas are still active.
9	C07-multiple cracks	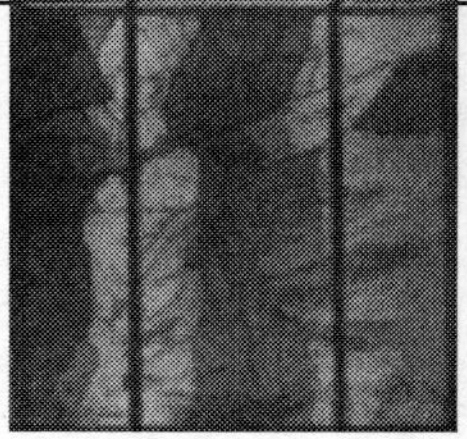			Multiple cracks that show very high potential of islanding large parts of the cell. Interestingly after 4years of outdoor it does not look as severe as one could imagine. Still in the center of the cell large areas are still active.

VOLTAGE DEPENDENCE OF POTENTIAL-INDUCED DEGRADATION AND RECOVERY ON PHOTOVOLTAIC ONE-CELL LAMINATES

J. Carolus[1,2], J. Govaerts[2], E. Voroshazi[2], W. De Ceuninck[1,2], and M. Daenen[1,2]
[1]Hasselt University, Martelarenlaan 42, 3500 Hasselt, Belgium
[2]imec, Kapeldreef 75, 3000 Leuven, Belgium

ABSTRACT: A high voltage degradation mechanism on photovoltaic (PV) modules called potential-induced degradation (PID) has gained a lot of attention since 2010 when Solon published their findings about PID [1]. The research conducted on PID mainly focused on industrial scale PV systems where potential differences between the solar cell and the grounded frame in the order of 1000 V and even more are not uncommon [2–5]. In this work, a stress test for PID according to the foil method as described in IEC 62804-1 and a recovery test in the same conditions on a set of five one-cell laminates is performed. However, by stressing the one-cell laminates at different voltages (-200 V, -600 V and -1000 V) and curing them at +500 V, which is compliant with residential voltages in the PV market, and +1000 V as a reference, more insight on the impact of stress and curing voltage on PID is gained. This is the basis for a voltage dependent model for PID. Furthermore, it is shown that the impact of the stress voltage on PID as a function of time exhibits a complex behaviour. At last, the impact of PID at different stress voltages on the different operational parameters of the one-cell laminates and their physical meaning is described.
Keywords: Degradation, Diffusion, PV Module, PID, Sodium

1 INTRODUCTION

Research on PID has gained attention since 2010 when Solon published their findings on a degradation mechanism which is caused by a high potential difference between the grounded frame and the solar cell of a photovoltaic (PV) module, a result of connecting multiple PV modules in a series connection [1]. Naumann *et al.* described the root cause of this phenomenon. PID of the shunting type is caused by sodium ions (Na^+) migrating from the soda lime glass through the encapsulant into the anti-reflective coating (ARC) under the influence of a significant electrical field across the stack [6]. From the ARC Na^+ diffuses into stacking faults through the PN-junction of the solar cell creating direct shunt paths. This results in a substantial lowering of the shunt resistance, degrading the power output of the PV module. PID of the shunting type occurs on PV modules when Na^+ migration is promoted by the combination of the soda lime glass, the encapsulant and the ARC on the solar cell. The working principle of PID of the shunting type is shown in Figure 1. Solutions for new PV modules (the so-called PID-free modules) as well as cures for affected PV modules have already been published [1, 7]. On a cell level, PID can be prevented by engineering the ARC in such a way it is conductive, and thus cancelling the electrical field across this layer [1, 2, 8]. On a module level, PID can be prevented by using a sodium free front material or an encapsulant which does not promote sodium ion migration [9]. On a system level, the grounding of the system can be executed in such a way the voltage across the PV module stack is driving the sodium ions away from the solar cell. Curing PID affected PV modules is mainly done by reversing the polarity of the voltage between the frame and the solar cell at night [1, 2, 5].

In the last decade, a lot of research with respect to potential- induced degradation (PID) on PV modules has been conducted. The work performed in this area mainly focused on industrial scale PV systems where potential differences between the solar cell and the grounded frame in the order of 1000 V and even more are not uncommon [2]–[5]. However, residential scale PV applications are commonly constructed out of a limited number in series connected PV modules, resulting in a lower stress voltage with respect to PID. Furthermore, curing applications are forced to limit their curing voltage due to inverter warranty.

In this article, the results of a PID stress test according to the foil method as described in IEC62804-1 and a recovery test in the same conditions are performed. However, to simulate the lower stress voltages as observed by residential PV applications, lower stress voltages (-200 V and -600 V) and the commonly used stress voltage (-1000 V) are applied. Moreover, a PID recovery test at a lower curing voltage (+500 V), which is compliant with residential voltages in the PV market, and at +1000 V as a reference are performed.

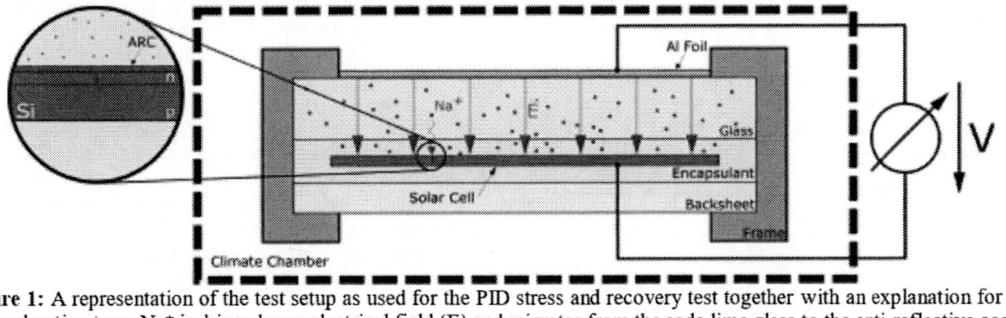

Figure 1: A representation of the test setup as used for the PID stress and recovery test together with an explanation for PID of the shunting type. Na^+ is driven by an electrical field (E) and migrates from the soda lime glass to the anti-reflective coating on the solar cell where it diffuses into stacking faults through the PN-junction.

2 EXPERIMENTAL

Two commercially available p-type aluminium back surface field (Al BSF) solar cells with two different wafers (two mono-crystalline and three multi-crystalline) were laminated into five one-cell laminates. It is important to note that the stack of layers and the cell processing was different between the two sample types. Apart from solar cell type and their processing, the one-cell laminates were identical to each other and composed of a soda lime glass front, an ethylene-vinyl acetate (EVA) encapsulant and a white polyvinyl fluoride (PVF) back sheet. Each of the three multi-crystalline single-cell laminates were PID stressed at a different voltage (-200 V, -600 V and - 1000 V) for more than 1000 hours. The two mono-crystalline single-cell laminates were first PID stressed at -200 V and -600 V for 380 hours and later at -1000 V for 130 hours. The multi-voltage single-module stress test was performed according to the foil-method as described by the PID standard IEC 62804-1: "test methods for detection of potential- induced degradation of crystalline silicon photovoltaic (PV) modules". Next to the PID stress tests, PID recovery tests were conducted as well. For this experiment, the one-cell laminate stressed at -1000 V was cured at +1000 V in order to set a reference. Next to this sample, two additional one-cell laminates were prepared. Note that these laminates are identical to the previous laminates. However, the EVA encapsulant originated from a different batch. The two one-cell laminates were stressed at -1000 V and -500 V and both one-cell laminates were subsequently cured; first thermally at room temperature for about 950 hours and afterwards with a curing voltage of +500 V. An overview of the experiment with the one-cell laminates as included in this test together with their stress and curing voltage is shown in Table 1. The stressing and curing of the one-cell laminates was conducted inside a climate chamber at 60 °C and 60%RH in order to obtain higher stress and recovery rates. A voltage source generated the voltage difference between the short circuited solar cells and the aluminium foil during the stress and recovery test. The test setup as used while testing the one-cell laminates for PID is schematically shown in Figure 1. It is to be noted that the one-cell laminates used in this experiment were not equipped with a frame.

Table 1: An overview of the experiment with the seven one-cell laminates as included in this test with their stress voltage, their PID after stressing, their curing voltage and their PID after curing.

Sample name	Stressing Voltage [V]	PID after Stress [%]	Curing voltage[V]	PID after Curing [%]
Multi01	1000	92,13	1000	13,17
Multi02	200	96,69	-	-
Multi03	600	94,90	-	-
Multi04	1000	37,45	500	0,40
Multi05	500	6,15	500	-1,82
Mono01	600/1000	6,07	-	-
Mono02	200/1000	4,38	-	-

3 RESULTS

3.1 Impact of stress voltage on PID

In what follows, the degradation caused by PID is presented by $\%PID_S$ and is calculated as follows:

$$\%PID_S = 100 * \left(1 - \frac{P_S}{P_I}\right)$$

with P_S the maximal power output of the one-cell laminate after the stress test and P_I the initial maximal power output as measured before the stress test.

The $\%PID_S$ of the different one-cell laminates is plotted as a function of time in Figure 2. From this graph, it is clear that the mono-crystalline one-cell laminates included in this test are not PID sensitive for both the low (-200 V/-600 V) and the high stress voltages (-1000 V). Therefore, the mono-crystalline one-cell laminates are excluded for further investigations in this paper. The multi-crystalline one-cell laminates included in this test show a high PID sensitivity; at -1000 V, a degradation level of more than 60% after 48 hours of PID stress has been reached. After 600 hours of PID stress at -1000 V, a degradation level of 99% was reached. At -600 V, a lower degradation rate can be observed; after 48 hours, a degradation level of 20% was reached. After 1000 hours at -600 V, a degradation level of 95% was reached. For the single-cell laminate stressed at -200 V, an S-shaped degradation curve which has an incubation time of about 100 hours is shown. 1000 Hours of PID stress at -200 V results in a PID level of about 97%. From here it is clear that the different stress voltages result in a different incubation time while the final degradation level is the same for all stress voltages while Hattendorf *et al.* [10] performed similar experiments in which they observed a difference in final degradation level of the output of the PV module while the incubation time showed no significant differences. It is to be noted that these results were also obtained using the foil-method. When only the frame is grounded instead of an electrode covering the front, a difference in final degradation level is expected for the different stress voltages due to the limited conductivity of the soda lime glass. In such a case, PID will mostly affect the solar cells at the perimeter, depending on the relative humidity.

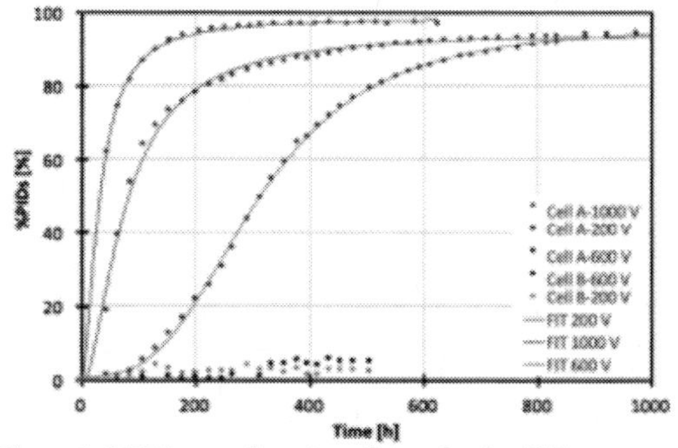

Figure 2: $\%PID_S$ as a function of time for the different one-cell laminates at different stress voltages together with the fits of the logistic function.

When $\%PID_S$ is plotted as a function of the stress voltage, as shown in Figure 3, a changing behaviour as a function of time can be observed. At about 64 h of PID stress, a linear behaviour can be observed, which is in accordance with Annigoni *et al.* [11]. They showed a linear behaviour after 96 h of PID stress. However, when varying PID stress durations are investigated, an upward bending curve can be perceived clearly for stress durations of less than 64 h whereas long degradation durations (over 64 h) show a downward bending curve. From this, it can be stated that the impact of the different stress voltages is somewhat more complex due to the S-shaped curvature.

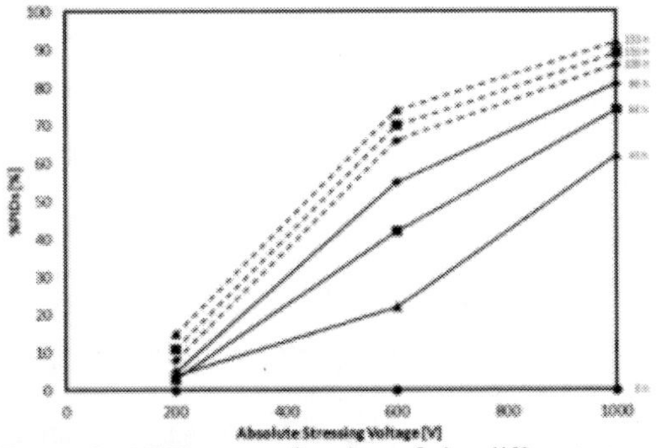

Figure 3: %PIDs as a function of the different stress voltages at different stress durations.

The S-shaped degradation curves of all three one-cell laminates are fitted with a function as described by Berghold *et al.* [4]. They proposed to apply a logistic function which describes the degradation accurately:

$$f(x) = \frac{A_1 - A_2}{1 + \left(\frac{x}{S_1}\right)^{p_1}} + A_2$$

with A_1 the initial PID level, which is normally 0, A_2 the stabilized PID level after degradation, x the time, S_1 the halfway time of the S-shaped curve and p_1 a shape factor. The fits of this logistic function for all one-cell laminates are presented in Figure 2. The different parameters as obtained to fit this logistic function are presented in Table 2.

Table 2: Overview of the different fit parameters for the different stress voltages. Note that the final degradation level (A2) does not show a significant difference.

Stressing Voltage [V]	A1 [%]	A2 [%]	S1 [-]	p1 [-]
200	0	96,93	310,12	2,91
600	0	94,88	72,60	1,73
1000	0	98,81	32,09	1,67

This overview verifies that the final degradation level for the three different stress voltages show no significant differences. Furthermore, it shows a drastically degrading halfway time (S_1) by increasing the stress voltage. When plotting S_1 as a function of the stress voltage, as shown in Figure 4, an asymptotic behaviour can be observed. This is not a surprising result because when the stress voltage tends to go to zero, the halfway time will tend to go to infinity, whereas an increasingly higher stress voltage results in a halfway time approaching zero. Apart from the halfway time, we observe a significant difference in shape factor p_1, as plotted in Figure 4, which verifies our suggestion that the stress voltage does not only impact the time scale during the total degradation period, but the shape of the S-shaped curve as well. When a high voltage stress is applied, a quite aggressive transition from the incubation phase to the stressing phase was observed while the transition from the stressing phase to the stabilisation phase was much smoother. When a lower stress voltage is applied, the transition between the different phases happens smoother which shows in the symmetry of the curves. Nevertheless, in order to link the stress voltage to S1 and p1, experiments with more intermediate voltages are needed.

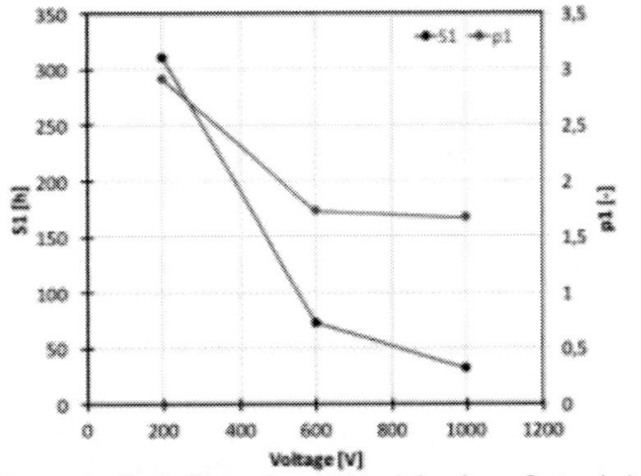

Figure 4: The halfway time (S_1) and the shape factor (p_1) as a function of stress voltage.

As already mentioned, the degradation rate is a function of both time and stress voltage. This is shown in Figure 5, in which the derivatives of the previously discussed fits are shown. We can clearly observe a difference in peak shape and peak intensity. Peak shape in this case is the consequence of a difference in the stressing and the stabilizing phase. Peak intensity gives us more insight in maximal degradation rate.

When the maximal degradation rate is plotted as a function of the stress voltage, a quadratic relationship can be observed. This is shown in Figure 6.

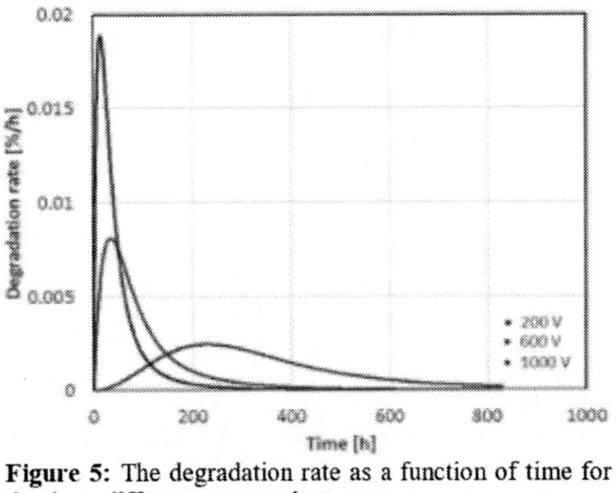

Figure 5: The degradation rate as a function of time for the three different stress voltages.

3.2 Impact of curing voltage on PID

After stressing a single-cell laminate at -1000 V for 150 hours, the electrical field was reversed for 150 hours in order to cure this one-cell laminate. It is notable that the curing process stopped while about 15% of PID was still present. This is compliant with our earlier research in which we performed a PID stress test according to the foil-method as described in the PID standard IEC 62804-1 and recovery test in the same conditions on 49 full-size PV modules [12] (multi-module single stress voltage experiment). In the results, we described a recovered and an unrecovered part contributing to PID. The results are summarized in Figure 7, in which the recovered percentage PID ($\%PID_R$) is plotted as a function of the percentage of PID after stress ($\%PID_S$). PV modules with a $\%PID_S$ of less than 85% approach the ideal full recovery

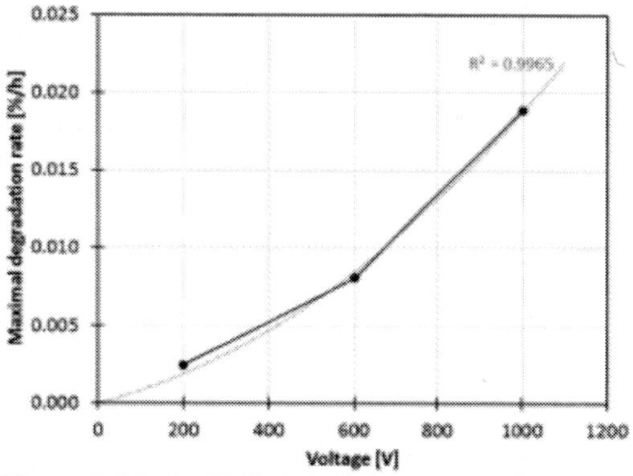

Figure 6: Maximal PID degradation rate as a function of stress voltage shows a quadratic relationship.

(1:1 linear recovery relation), whereas some of the PV modules with a %PIDs of more than 85% seem to recover slower, or in worst case, do not recover at all. However, no intermediate measurements were performed while recovering the PV modules for PID. It cannot be determined whether the asymptotic trend towards a stabilized state was reached or not in the full module test while we are sure that the curing of the single cell modules was stabilised.

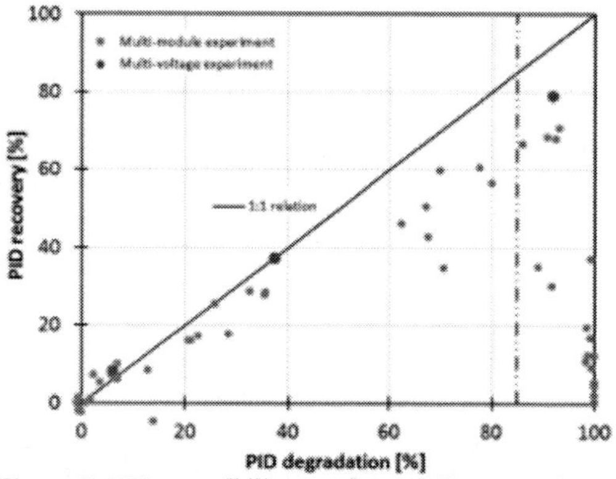

Figure 7: PID reversibility: an almost full recovered part for full- size PV modules with a PID level less than 85% and a non-recovered part for some PV modules with a PID level over 85%. The red dots in this plot show the reversibility as observed in this experiment.

After being stressed for 168 hours according to the foil method as described in IEC62804-1, the sample under test at -500 V shows a degradation level of about 6% whereas the sample under stress at -1000 V shows a degradation level of about 37%. Note that the stress level after 168 hours of stress at -1000 V is still less than the sample as included in the previous experiment when stressed for 48 hours at -1000 V. After thermally recovering the one-cell laminates at room temperature for about 950 hours, the sample stressed at -1000 V shows a recovery of 14%, resulting in a remaining PID level of 23%. The sample stressed at –500 V shows a recovery of 3% after the thermal curing cycle, resulting in a remaining PID level of 3%. When both samples under stress were cured at +500 V for 144 hours, a total recovery could be observed. This shows that, for low PID levels, no significant secondary degradation mechanisms are activated by the higher voltage stress. The results are summarized in Figure 8.

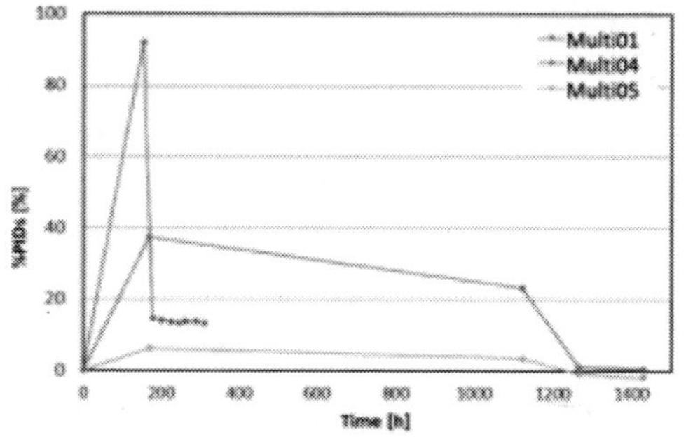

Figure 8: Different stress and curing voltages combined. Total recovery of the PV modules is possible by using a lower curing voltage than stress voltage.

3.3 Impact of PID on FF

When the fill factor (FF) is plotted as a function of %PIDs, as shown in Figure 9 (coloured points), a linear behaviour can be observed for PID levels of less than 70%. Once this value is reached, the FF tends to stabilize at around 25%. This is due to the fact that the one-cell laminate is stressed to a degree where the IV-curve has changed into a straight line. This is caused by a decrease in shunt resistance to such a degree the total solar cell is shunted. The shunt resistance in this case is in the order size of, or even smaller than, the series resistance. This is in accordance with our earlier work on full-size PV-modules, from which the FFs as a function of %PIDS are represented by the black dots in Figure 9 [12]. Note that the fill factor degraded according to the same trend for all the different voltages applied to the one-cell laminates as well as for all different 49 PV modules stressed at -1000 V. This indicates that the FF is a reliable quantification for PID stress, unregarded the PV module type. Furthermore, it shows that no significant secondary degradation mechanisms are activated when higher stress voltages are applied.

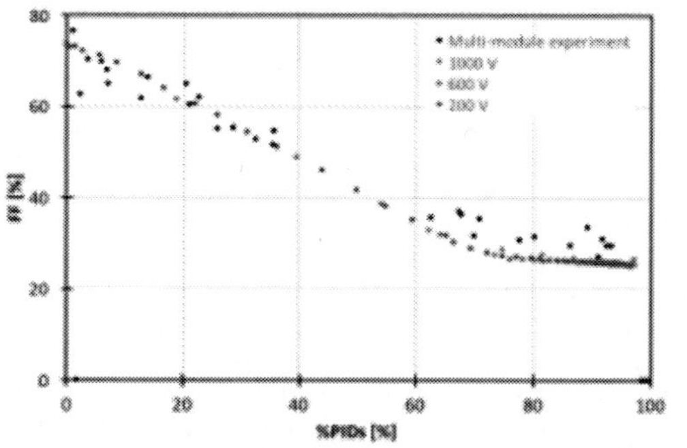

Figure 9: The fill factor (FF) as a function of % PIDs. A linear trend line can be observed for PID levels less than 70%.

3.4 Impact of PID on V_{OC}

When the open circuit voltage (V_{OC}) is plotted as a function of %PIDs, as shown in Figure 10 (coloured dots), a clear trend can be observed. From this plot, it is clear that the V_{OC} degrades at the same rate for the different applied stress voltages. For PID levels of less than 60% and more than 75%, V_{OC} degrades linearly while the region in between shows a non-linear behaviour. This is the moment the IV-curve degrades from an IV-curve into a straight line. This is in accordance with our earlier research conducted on full scale PV modules where no significant changes for the V_{OC} at PID levels of less than at least 40% were concluded, the results of this multi-module experiment are shown as black dots in Figure 10 [12]. The different stress voltages do not show a different behaviour in V_{OC} degradation, which means no significant secondary degradation mechanisms are activated for the higher stress voltages.

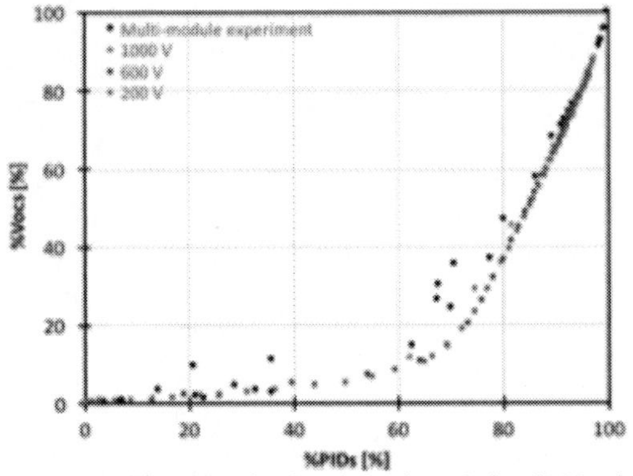

Figure 10: The open circuit voltage degradation (%V_{OCS}) as a function of %PID_S. The V_{OC} does not show a significant degradation for PID levels less than 60%.

3.5 Impact of PID on I_{SC}

When the short circuit current degradation (%I_{SC}) is plotted as a function of %PID_S, as shown in Figure 11 (coloured dots), a clear trend can be observed. From this plot, it is clear that the I_{SC} degrades at the same rate for the different applied stress voltages. For PID levels lower than 70%, the I_{SC} degrades linearly whereas the I_{SC} degrades dramatically for PID levels over 80%. This is in accordance to the moment the shunt resistance has degraded to a value near to the value of the series resistance. The same behaviour with respect to I_{SC} degradation was observed in our research on full-size PV modules, presented by black dots in Figure 11 [12]. Just as the degradation of FF and V_{OC}, the degradation of I_{SC} does not show a significant difference for the different applied stress voltages.

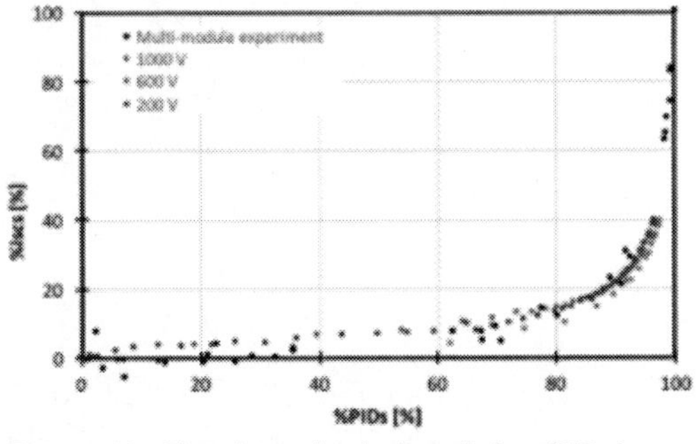

Figure 11: The short circuit degradation (%I_{SC}) as a function of %PID_S. The I_{SC} does not degrade significantly for PID levels less than 70%.

4 CONCLUSIONS

In this paper, we presented the impact of the stress voltage on the behaviour of PID. We performed PID stress tests according to the foil method as described in IEC62804-1. Three different stress voltages (-200 V, -600 V and - 1000 V) were applied and intermediate characterisation measurements were conducted. The results showed an S-shaped degradation curve for the laminates with multi-crystalline solar cells while the laminates with mono-crystalline solar cells did not degrade at all. Nevertheless, in earlier experiments we executed [12], there was no significant relation between wafer type and PID. Interestingly, the stabilized degradation level seems to be the same for all stress voltages and tends to approach 100% degradation. Due to the S-shaped degradation curve, we observed a logistic behaviour of the degradation due to PID as a function stress voltage. Only for one certain stress time, a linear relationship between PID degradation and stress voltage can be observed. The impact of the stress voltage is an important parameter which should be described very clearly in order to make valid predictions about PID as a function of time The S-shaped degradation curves were fitted with a logistic function as proposed by Berghold *et al.* [4]. The fitting parameters of this logistic function for different stress levels show that the voltage is not only affecting the time scale factor of the degradation, but it has an impact on the symmetry of the S-shaped curve as well. When a high voltage stress is applied, a quite aggressive transition from the incubation phase to the stressing phase was observed while the transition from the stressing phase to the stabilisation phase was much smoother. When a lower stress voltage is applied, the transition between the different phases happens smoother which shows in the symmetry of the curves. The same can be observed when the derivative, which is the degradation speed, of the S-shape curve is plotted as a function of time. Next to peak shape, the peak intensities differ for the different voltages. The maximal degradation rate shows a quadratic relationship with the stressing voltage. However, more experiments with more intermediate stress voltages are planned.

Next to the stress tests, we performed recovery tests as well by reversing the polarity of the high voltage. We have shown that a full recovery is possible when a lower curing voltage than the stress voltage is applied. This is promising for the residential scale PV installations since the maximal curing voltages are often limited due to inverter warranty. However, at high degradation levels we observed a non-recovered part of degradation. This is in accordance with our earlier work where we reported on the irreversibility for PV modules with PID levels over 85% [12].

At last, we presented the impact of PID on the different PV parameters such as FF, V_{OC}, I_{SC}. The degradation of the different PV parameters as a function of PID follow the same trend. This is interesting since it indicates that no significant secondary degradation mechanisms are activated by applying a higher stress voltage. From these results, it is clear that the shunt resistance degrades to a value near the value of the series resistance. Once this happens, the FF, V_{OC} and J_{SC} are affected dramatically. Furthermore, we see that these trends are very clear for all different PV modules as well [12]. This enables the possibility to detect PID. Although, PID is not isolated from other degradation mechanisms when measuring only 1 PV parameter.

5 REFERENCES

[1] S. Pingel, O. Frank, M. Winkler, S. Daryan, T. Geipel, H. Hoehne, and J. Berghold, "Potential induced degradation of solar cells and panels."

[2] J. Berghold, O. Frank, H. Hoehne, S. Pingel, B. Richardson, and M. Winkler, "Potential Induced Degradation of solar cells and panels."

[3] S. Koch, C. Seidel, P. Grunow, S. Krauter, and M. Schoppa, "Polarization effects and test for crystalline silicon cells," in *Photovoltaik Institut Berlin*, 2011, pp. 1–6.

[4] J. Berghold, S. Koch, S. Pingel, S. Janke, A. Ukar, P. Grunow, and T. Shioda, "PID: from material properties to outdoor performance and quality control counter measures," 2015, p. 95630A.

[5] R. Swanson, M. Cudzinovic, D. DeCeuster, V. Desai, J. Jürgens, N. Kaminar, W. Mulligan, L. Rodrigues-Barbarosa, D. Rose, D. Smith, A. Terao, and K. Wilson, "The surface polarization effect in high-efficiency silicon solar cells," 2005.

[6] V. Naumann, D. Lausch, A. Hähnel, J. Bauer, O. Breitenstein, A. Graff, M. Werner, S. Swatek, S. Großer, J. Bagdahn, and C. Hagendorf, "Explanation of potential-induced degradation of the shunting type by Na decoration of stacking faults in Si solar cells," *Sol. Energy Mater. Sol. Cells*, vol. 120, pp. 383–389, 2014.

[7] D. Lausch, V. Naumann, A. Graff, A. Hähnel, O. Breitenstein, C. Hagendorf, and J. Bagdahn, "Sodium Outdiffusion from Stacking Faults as Root Cause for the Recovery Process of Potential-induced Degradation (PID)," *Energy Procedia*, vol. 55, pp. 486–493, 2014.

[8] S. Koch, D. Nieschalk, J. Berghold, S. Wendlandt, S. Krauter, and P. Grunow, "Potential Induced Degradation Effects on Crystalline Silicon Cells with Various Antireflective Coatings," pp. 1985–1990, 2012.

[9] P. Hacke, K. Terwilliger, R. Smith, S. Glick, J. Pankow, M. Kempe, S. Kurtz, I. Bennett, and M. Kloos, "System Voltage Potential- Induced Degradation Mechanisms in PV Modules and Methods for Test Preprint," 2011.

[10] J. Hattendorf, R. Löw, W.-M. Gnehr, L. Wulff, M. C. Koekten, D. Koshnicharov, A. Blauaermel, and J. A. Esquivel, "Potential Induced Degradation in Mono-Crystalline Silicon Based Modules: An Acceleration Model," *27th Eur. Photovolt. Sol. Energy Conf. Exhib.*, pp. 3405–3410, Jan. 2012.

[11] A. Eleonora, J. Marko, G. Federico, Y. L. Heng, P.-A. Laure-Emmanuelle, T. Marko, S.-M. Fanny, V. Alessandro, and B. Christophe, "Modeling potential-induced degradation (PID) in crystalline silicon solar cells: from accelerated-aging laboratory testing to outdoor prediction," in *32nd European Photovoltaic Solar Energy Conference and Exhibition*, 2016, pp. 1558–1563.

[12] J. Carolus, W. De Ceuninck, and M. Daenen, "Irreversible damage at high levels of potential-induced degradation on photovoltaic modules: A test campaign," in *IEEE International Reliability Physics Symposium Proceedings*, 2017.

6 ACKNOWLEDGEMENTS

This project receives the support of the European Union, the European Regional Development Fund ERDF, Flanders Innovation & Entrepreneurship and the Province of Limburg. Moreover, the authors would like to thank Jan Mertens and Reinoud Moors for their help during the measurements and the preparation of the one-cell laminates and pidbull nv for both their financial support.

INVESTIGATION OF CORRELATION BETWEEN FIELD PERFORMANCE AND INDOOR ACCELERATION MEASUREMENTS OF POTENTIAL INDUCED DEGRADATION (PID) FOR C-SI PV MODULES

Yifeng Chen[1,*], Kaitlyn VanSant[2], Yong Sheng Khoo[3], Zigang Wang[1], Wei Luo[3,4], Chris Deline[4], Peter Hacke[4], Jing Chai[3], Li Yin[1], Yan Wang[3], Armin G. Aberle[3], Yang Yang[1], Pietro Altermatt[1], Zhiqiang Feng[1], Sarah Kurtz[4] and Pierre J. Verlinden[1]

[1] State Key Laboratory of PV Science and Technology, Trina Solar, ChangZhou, 213031 China

[2] Colorado School of Mines, 1500 Illinois St, Golden, CO, USA.

[3] Solar Energy Research Institute of Singapore (SERIS), Singapore 117574

[4] National Renewable Energy Laboratory (NREL), 15013 Denver West Parkway, Golden, CO, USA

[*] Corresponding author: phone: (+86) 519 8158 7508, e-mail: yifeng.chen01@trinasolar.com

ABSTRACT: The correlation between outdoor field testing and indoor chamber measurement for potential-induced degradation in crystalline silicon PV modules is unclear. The motivation of this work is to build up a direct comparison and investigate the correlation between indoor and outdoor performance. First of all, four PID indoor tests were defined from basic to premium, based on the difficulty of the test. Six groups of modules were designed and fabricated in Trina Solar, with different grades of PID resistance from weak to strong. The outdoor field test systems were built in three different climates: tropical, subtropical, and semi-arid climates in Singapore, Changzhou (China) and Golden (CO, USA) respectively, in a collaborative testing campaign involving Trina, SERIS and NREL. The system in Golden has been on-sun for more than 13 months where <4% power loss which includes PID and other reliability losses was detected and power loss of <2% is shown for all six groups of modules after 21 weeks in the field in the subtropical climate in Changzhou. However, PID of 23% was observed after 18 weeks in Singapore for the most PID-susceptible module, while the other PID module types showed degradation of only less than 5%.

Keywords: Potential Induced Degradation (PID), Crystalline Silicon, Modules, Chamber tests, Field testing

1 INTRODUCTION

Potential induced degradation (PID) is one of the most severe reliability problems for today's crystalline silicon PV industry. The power of a module could degrade over 50% shortly after the installation of a PV system, although this behavior is typically found in hot and humid climate areas. To date, a complete understanding of the PID mechanisms has yet to be established, although some models have been proposed, mostly based on indoor simulation investigations [1-4]. A good comprehensive review was presented by Luo et al. [5].

The relationship between the real field performance and indoor chamber tests is still not very clear. As a result, it is difficult to predict the module lifetime under high-voltage stress in field with the indoor accelerated tests. Some previous investigations were carried out to reveal the interrelationship of PID and environmental parameters. For instance, Hacke et al. showed that the PID of modules is strongly correlated to the voltage, temperature and relative humidity (R.H.) [6-7].

As the correlation between indoor and outdoor PID behavior is not very clear, the PV industry is seriously concerned with the design and quality control of modules that will exhibit PID resistance in various different climates over the course of the product's warranty.

In this work, we build up a direct comparison of indoor chamber measurements and outdoor field tests. First of all, we developed 6 types of PV modules with different grades of PID resistance from weak to strong. Second, we built up three outdoor test systems via an international collaboration involving Trina Solar, NREL and SERIS. The systems operate under three different climates: tropical in Singapore (by SERIS), subtropical in Changzhou (China, by Trina Solar) and semi-arrid, in Golden, CO (USA, by NREL). External voltage is applied between the poles and the frame of the modules. We recorded the environmental parameters such as relative humidity, temperature, irradiance, etc. The modules are removed from the field for indoor IV characterization regularly in Changzhou and Singapore whereas outdoor I-V characterization is regularly conducted in Golden. The indoor accelerated experiments and the outdoor field results are compared. The result of leakage current comparison was reported in our previous publication [8]. The ultimate goal of this project is to determine the minimum indoor PID testing necessary to ensure a 25-year module lifetime in a given climate.

2 INDOOR CHAMBER TESTS

Before the chamber tests, we designed several test conditions for accelerated indoor chamber measurements, by changing the relative humidity, temperature, applied voltage and duration of time in the chamber. We also considered a case that involved metal foil covering the surface of the module. By varying these parameters, we developed 4 groups of test conditions, starting from the basic test conditions defined by IEC 62804-1 [9], to progressively more aggressive conditions.

According to Ref. [6], the leakage current density of modules would be 10 times higher than in ambient relative humidity of 85%, if the module is covered with metal foil, like copper (Cu). One may reasonably suggest that special modules passing the premium PID test could demonstrate much stronger PID resistance in the field than the modules that just pass the basic and medium PID test defined in Table I.

Table I: Potential induced degradation indoor chamber tests levels

Test conditions	Basic	Medium	Hard	Premium
Relative humidity [%]	85	85	85	85
Temperature [°C]	60	85	85	85
Applied voltage [V]	±1000	±1000	±1000	±1000
Test time [h]	96	96	600	600
Metal foil Covered	No	No	No	Yes

The solar cells for this project were fabricated with full area Al-BSF on p-type multi-crystalline silicon substrates with a resistivity from 0.5-3 Ωcm. The refractive index and the thickness of the SiN_x:H deposited by Plasma Enhanced Chemical Vapor Deposition (PECVD) were adjusted to yield different PID resistance at the cell level. Meanwhile, ethylene-vinyl acetate (EVA) sheets with different bulk resistivity were used during module fabrication. The modules are standard-sized 60-cell modules. The most PID-resistant group of modules was fabricated with the best cells into a double glass module. With the combination of the PID-resistant cell designs and encapsulation materials, 6 different types of modules (from A to F) with varying PID-resistance were fabricated ranging from very weak to very strong, as shown in Table II.

Table II: Modules divided in groups (from A to F) based on PID test results

Module	Encap.	Basic	Medium	Hard	Premium
Type A	Backsheet	NP	NP	NP	NP
Type B	Backsheet	PASS	PASS	NP	NP
Type C	Backsheet	PASS	PASS	PASS	NP
Type D	Backsheet	PASS	PASS	PASS	PASS
Type E	Backsheet	PASS	PASS	PASS	PASS
Type F	Double Glass	PASS	PASS	PASS	PASS

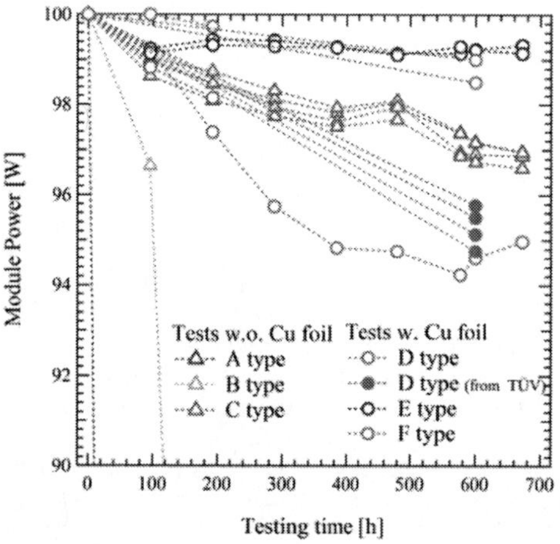

Figure 1: Normalized module power as a function of indoor PID chamber test (under 85%R.H./85°C/-1000V) time for 6 types of modules, where type A module is completely PID vulnerable. The triangle symbols represent the accelerated measurements without Cu foil, while the circle symbols denote the cases with Cu foil.

The difficulty of the indoor chamber tests from basic through to premium were defined in Table I. During the fabrication of the modules, the same backsheet was used from Type A to Type E modules, while Type F modules were encapsulated with double glass. NP is abbreviation for Not Pass and represents a degradation >5% after the PID chamber test. Therefore, A type modules are considered to be very susceptible to PID when they operate in the field. The result of the indoor chamber test was plotted in Fig.1.

As illustrated in Fig.1, the Type A module showed very quick degradation during the measurement. Type B module degraded <4% for the first 96h PID test, but shows an abrupt degradation after extended testing. With careful design, backsheet modules (type D and type E) are also able to pass the most aggressive chamber test, which was independently confirmed by TÜV Rheinland in Shanghai (light blue solid circles). F-type modules are double-glass modules which prevent moisture from penetrating into the module and feature excellent PID-resistant performance. We expect that a module capable of passing such an aggressive indoor test would not suffer from PID degradation over its 25-year lifetime under any real-world conditions.

3 OUTDOOR FIELD EXPERIMENTS

As can be seen from the indoor test results, the severity of PID is greatly affected by the humidity, temperature and whether the surface is covered with metal foil. For field testing, we chose Changzhou (China), Singapore and Golden (2,426m of altitude), CO (USA) to represent the typical PID test conditions found in subtropical, tropical and warm summer continental (Dfb) climates [11], respectively. The average temperature and humidity are quite different in these three climates, as shown in Table II, which will yield a different severity of PID that occurs in the field. The measurement in Singapore will be performed by SERIS, while that in Changzhou and Golden will be done by Trina and NREL respectively.

Table III: Climate data of three field test locations [10,11]

City	Avg. T [°C]	Avg. R.H. [%]	Climate
Singapore	34.2	84.1	Tropical
Changzhou	37.8	80	Subtropical
Golden	31.9	47.8	Dfb

Table IV: The number of modules tested in voltage from 0 V to 1500 V

Modules	0 V.	400 V	600 V	1000 V	1500 V
Type A	2 pcs	3 pcs	3 pcs	4 pcs	0
Type B	2 pcs	3 pcs	3 pcs	4 pcs	0
Type C	2 pcs	3 pcs	3 pcs	4 pcs	0
Type D	2 pcs	3 pcs	3 pcs	4 pcs	0
Type E	2 pcs	0	0	4 pcs	3 pcs
Type F	2 pcs	0	0	4 pcs	3 pcs

The experiment plan was shown in Table IV. Each testing location has 66 modules categorized as groups A to F. The modules are connected to a fixed resistor to make sure that the modules perform close to their maximum power point during operation. The aluminium frames are grounded and the positive poles of the modules are connected to an external voltage source, applying 0V, -400V, -600V, -1000V or -1500V to the modules. The 0V modules are installed in the field without any applied voltage as references to monitor any degradation unrelated to PID. During the field measurements, all the sites collected irradiance, wind speed, R.H., and module temperatures of each module type. During the night time or low irradiance conditions of less than 10 W/m^2, the external voltage source is disconnected and data collected under these conditions were not taken into consideration. All the modules in the three test fields were installed at the local best tilt angles to simulate the real field installation. At the Singapore site, the modules were mounted flat, which leads to accumulation of water that remains until it is sufficiently heated and evaporation can occur, a particularly stressful condition.

At regular time intervals, the modules are either removed from the field for indoor IV characterization or their I-V characteristics are measured in the field. The power of the modules with 0 V bias is also measured as a reference.

The normalized module power as a function of outdoor test time was plot in Fig.2, for Singapore (a), Changzhou (b) and Golden (c), respectively. Note that the power shown in Fig.2 is the true power measured, the impact of other aging is not taken into account, namely we did not remove the effect of long term reliability by subtracting the power loss of the reference modules.

The testing among the three labs was supposed to start at the same time. However, due to issues associated with building out the test fields , the connection of the outdoor system was delayed in Singapore and Changzhou. By the end of August, 2017, the system in Singapore and Changzhou have been in operation for 38 weeks and 21 weeks, respectively. The system in Golden had been on-sun for 13 months, while NREL experienced a 2.5 month system shutdown between when the IV characterization was completed at 40 weeks and when the IV characterization was conducted at 57 weeks. The NREL IV data was collected using a Daystar DS-1000 PV I-V curve tracer, and which show a bit variability in the data. The outdoor measurements at NREL also reflect up to about 2% of loss from soiling. The colors in Fig. 2 represent the modules with different strengths of PID resistance, as shown in Table II.

As illustrated in Fig. 2, the overall module power generally shows a decrease as a function of test time. In each location, the degradation of Type A modules is the largest, the second largest degradation could be the Type B modules, while Type F modules show the least degradation. By in general, it is hard to see the statistical significance of this result, except at SERIS, suggested by the error bars. This is consisient with the indoor test results shown in Fig.1. For modules from Type C to Type F, the degradation is quite small and may be impacted by other long term degradation effects such as light induced degradation, corrsion of front anti-reflection coating of glass, corrosion of the metal contacts, etc. and it is therefore difficult to determine whether the degradation of Type C is higher than Type D and E or vice versa.

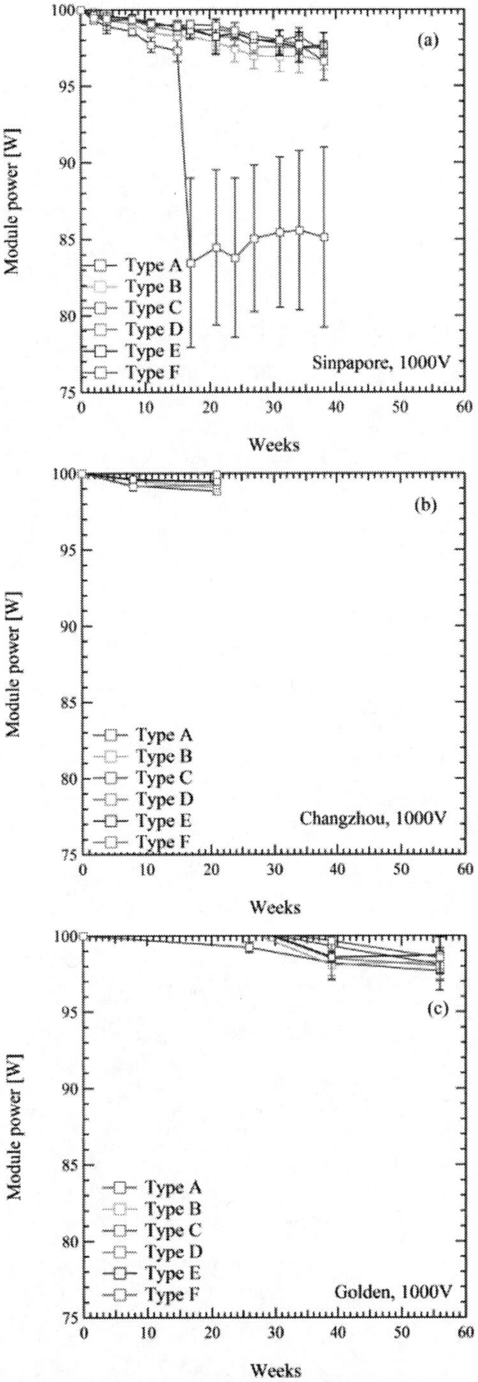

Figure 2: The measured module power as a function of outdoor testing time for the case in Singapore (a), Changzhou (b) and Golden (c), respectively. The different colors represent the different modules types.

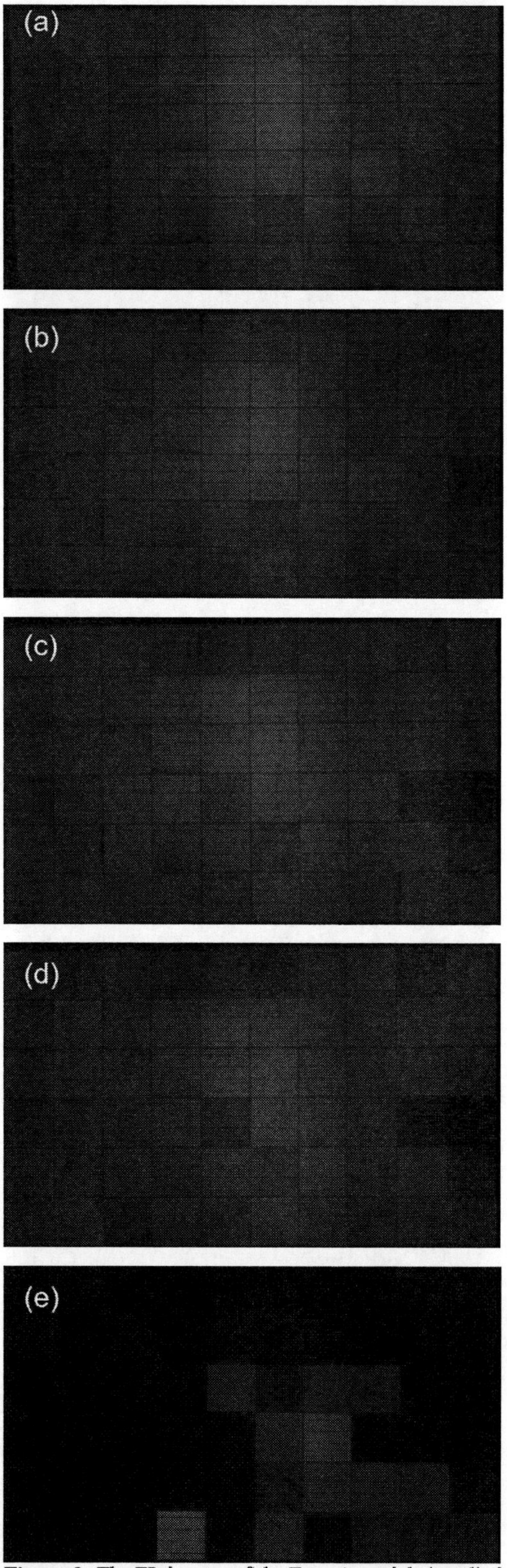

Figure 3: The EL images of the Type A module installed in Singapore that has the biggest degradation, before (a) and after outdoor operation of 8 weeks (b), 11 weeks (c), 15 weeks (d) and 18 weeks (e).

Degradation of <5% was shown for most of the modules in all three locations, except for the Type A module in Singapore. This module type exhibited a sudden drop of maximum 23% of power which was detected after only 18 weeks of operation in the field. This abrupt drop of power was also shown in Fig.1.

For a better understanding of the degradation mechanisms observed during field testing, we also performed electroluminescence (EL) measurements. EL measurements were taken using a current of 1 A in order to clearly identify shunts. Fig. 3 shows the EL evolution of the Type A module installed in Singapore that shows the largest degradation in power. The EL image of the module before the field test is shown in Fig. 3 (a), with a power output of 259W after light soaking preconditioning. After 8 weeks of outdoor testing, the module was removed from the field and sent to the Lab in SERIS for EL and IV measurements. It appears that the EL image stays normal as illustrated in Fig.3 (b), but the power has decreased to 254W. After 11 weeks' operation in the field, dark cells start to appear from the edge of the module as shown in Fig. 3(c), while the power of the module dropped to 251W. After another 3 weeks, the edge cells became even darker as shown in Fig. 3(d), indicating more severe PID which was verified with IV measurements that showed another 2W was lost, bringing the power output down to 249 W. After 18 weeks, a sudden significant decrease of the power to 194W (25% degradation) was detected. The EL image associated with 18 weeks is shown in Fig. 3 (e). As can be seemed, the EL images confirm that most of the cells in the module became greatly impacted by the PID, and not by another degradation mechanism.

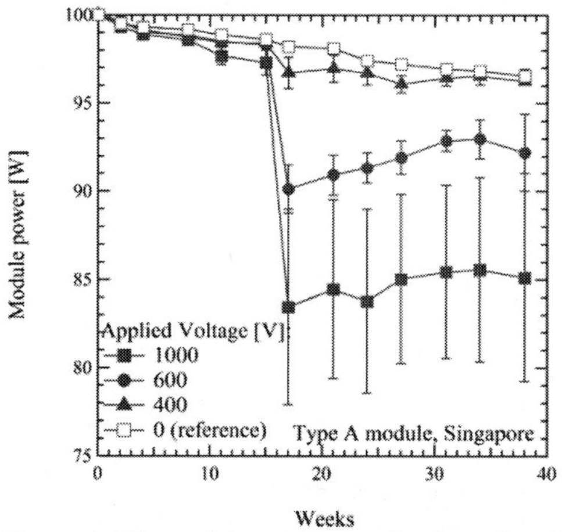

Figure 4: The module power as a function of outdoor testing time for Type A module tested in Singapore. The different symbols represent different applied voltage, while the open square symbols are the reference modules that are installed outdoor with zero bias voltage.

Fig. 4 plots the module power in Singapore as a function of test time, with applied voltage from 0 V to 1000 V. The 0 V samples installed in the field connected with a resistor but without any bias voltage, which shows 3-4% degradation after 38 weeks of outdoor testing. As expected, the degradation increases with an increase in applied voltage. The 1000V group shows a medium degradation of 15%, while the 600V group shows a medium degradation from 8-10% and that of 400V group is about 6-8%. Interestingly, the abrupt drop of power can be seen to varying degrees for all three voltage groups.

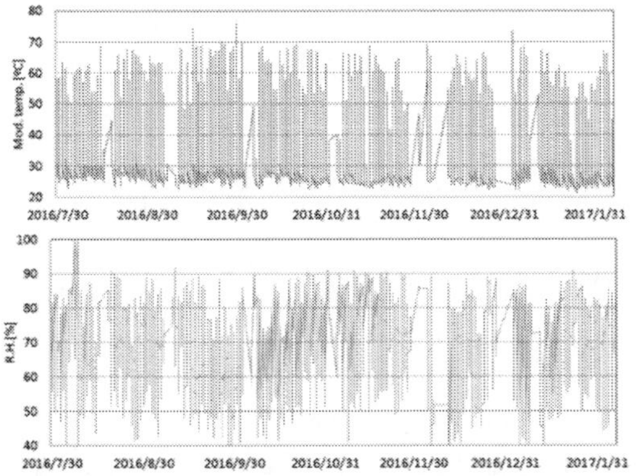

Figure 5: The measured module temperature (up) and relative humidity (down) as a function of date in Singapore. Each bar in the figure represents and value measured in each day, the top of the bar is the maximum value for the day while the bottom is the minimum value. The connection point between bar to bar is the daily average value from 7:00 to 19:00. The system in Singapore started from end of July 2016. Severe degradation was detected in Type A module at the end of Jan 2017.

Was there something special about the weather before the 18 weeks data points were taken? As it appears that most of the degradation took place in that short time, we suspected that the module may see a sudden weather change. To answer this question, the measured module temperature and R.H. in the environment were plotted in Fig.5, covering the time from the system was connected to the end of Jan. 2017, when the Type A modules degraded rapidly. Each bar in the figure represents and value measured in each day, the top of the bar is the maximum value for the day while the bottom is the minimum value. The connection point between bar to bar is the daily average value from 7:00 to 19:00. The highest module temperature per day could go from 60 to 78 °C while the R.H. varied from 50-90%. Though there is no sudden humidity and temperature change in December 2016 and January 2017 comparing to other months, the rainy season in Singapore is typically from December to February. So the reasons that cause the sudden drop of Type A modules regardless of applied voltage could be rain and stagnating water, or even extreme weathers like lighting, or bright sunshine in the rain. Another possible explanation could be that the module will degrade when the charge to a critical point. At such point, the accumulated charge density is bigger enough to create strong electric field which may 1) impact the minority carriers at the surface and 2) drift the particles like Na+ to pass through the junction. However, it is still not very clear to conclude the dominating impact to the sudden drop of module power in Fig. 4, which needs further investigations.

4 CORRELATION BETWEEN INDOOR AND OUTDOOR PERFORMANCE

Comparing the indoor testing and outdoor testing, there are several conclusions that can be deduced: 1) the outdoor performance as a function of test time shows some consistency with the indoor accelerated measurements. Modules with the strongest PID resistance (Type F, double glass) show the best outdoor performance, while modules that are the most vulnerable to PID (Type A) show substantially larger degradation than the other groups; 2) when PID happen in Type A and Type B modules, both indoor and outdoor degradation starts with a sudden drop of module power; 3) Indoor chamber test shows acceleration when comparing the indoor and outdoor results. For example, in the chamber test, the power of Type A module drops very quickly and a final degradation of >90% was detected only after 96 h measurement of 60 °C and 85% R.H. with bias of -1000V, while it takes 18 weeks for the same type of modules to show degradation of <25% in Singapore, and insignificant degradation was observed in this module type in Golden after a year and one month. However, the precise model to bridge the indoor and outdoor performance is still unclear.

5 CONCLUSIONS

Three field test systems were built in tropical, subtropical, and semi-arid climates in Singapore, Changzhou (China) and Golden (CO, USA) respectively to investigate the relationship between field performance and indoor chamber measurement of PID with 6 types of modules with PID-resistance ranging from very weak to very strong. The strongest module type degraded less than 5% after the most aggressive PID test (85% R.H., 85 °C, 600h, -1000V, and with Cu foil in front of the glass), while the weakest one degraded over 90% after the basic PID test (85% R.H., 60 °C, 96h, -1000V). During the outdoor field tests, most of the modules show <5% degradation to date, except Type A modules installed in Singapore. A PID of over 20% was observed after 18 weeks for the most PID-susceptible module in Singapore, confirmed by the dark EL images. It was shown that degradation increases with the applied voltage for these modules, this shows some consistent with the indoor measurement results.

6 ACKNOWLEDGEMENTS

This work is financially supported by the National High-tech R&D Program of China (863 Program, No. 2015AA050302).

This work was partially supported by the U.S. Department of Energy under Contract No. DE-AC36-08GO28308 with the National Renewable Energy Laboratory. Funding provided by U.S. Department of Energy Office of Energy Efficiency and Renewable Energy Solar Energy Technologies Office.

The U.S. Government retains and the publisher, by accepting the article for publication, acknowledges that the U.S. Government retains a nonexclusive, paid-up, irrevocable, worldwide license to publish or reproduce the published form of this work, or allow others to do so, for U.S. Government purposes.

REFERENCES

[1] R. Swanson, M. Cudzinovic, D. DeCeuster, V. Desai, J. Jürgens, N. Kaminar, W. Mulligan, L. Barbarosa, D. Rose and D. Smith, Proceedings of 15th International Photovoltaic Science and Engineering Conference, Shanghai, China (2005) 410-411.
[2] S. Pingel, O. Frank, M. Winkler, S. Daryan, T. Geipel, H. Hoehne and J. Berghold, Proceedings of the 35th IEEE Photovoltaic Specialists Conference, Honolulu, HI, United States (2010) 2817-2822.
[3] P. Hacke, K. Terwilliger, S. Glick, D. Trudell, N. Bosco, S. Johnston and S. Kurtz, Proceedings of the 35th IEEE Photovoltaic Specialists Conference, Honolulu, HI, United States (2010) 000244-000250.
[4] V. Naumann, D. Lausch, A. Hähnel, J. Bauer, O. Breitenstein, A. Graff, M. Werner, S. Swatek, S. Großer and J. Bagdahn, Solar Energy Materials and Solar Cells (2014) 120, 383-389.
[5] W. Luo, Y. S. Khoo, P. Hacke, V. Naumann, D. Lausch, S.P. Harvey, J. P. Singh, J. Chai, Y. Wang, A. G. Aberle and S. Ramakrishna, Potential-induced Degradation in Photovoltaic Modules: A Critical Review, Energy & Environmental Science, vol. 10 (2016) 43-68.
[6] P. Hacke, K. Terwilliger, R. Smith, S. Glick, J. Pankow, M. Kempe, S. K. I. Bennett and M. Kloos, Proceedings of the 37th IEEE Photovoltaic Specialists Conference, Seattle, WA, United States (2011) 000814-000820.
[7] P. Hacke, R. Smith, K. Terwilliger, S. Glick, D. Jordan, S. Johnston, M. Kempe, and S. Kurtz, IEEE Journal of Photovoltaics, 3 (3) (2012) 246-253.
[8] Y. Chen, P. Hacke, Y. Khoo, K. VanSant, Z. Wang, W. Luo, J. Chai, C. Deline, Y. Wang, A.G. Aberle, P.P. Altermatt, Z. Feng, S. Kurtz, and P.J. Verlinden, Proceedings of the 44th IEEE Photovoltaic Specialists Conference, in print, 2017
[9] IEC TS 62804-1, Edition 1.0, 2015.
[10] data from Wikipedia, https://en.wikipedia.org/wiki/Wiki
[11] The Köppen Climate. https://en.wikipedia.org/wiki/K%C3%B6ppen_climate_classification

POTENTIAL-INDUCED DEGRADATION OF PHOTOVOLTAIC MODULES COMPOSED OF INTERDIGITATED BACK CONTACT SOLAR CELLS OBSERVED IN AN ACTUAL PHOTOVOLTAIC SYSTEM

Tetsuyuki Ishii*[1], Ritsuko Sato[2], Sungwoo Choi[2], Yasuo Chiba[2], and Atsushi Masuda[2]
[1]Central Research Institute of Electric Power Industry, 2-6-1 Nagasaka, Yokosuka-shi, Kanagawa 240-0196, Japan
[2]National Institute of Advanced Industrial Science and Technology, 807-1 Shukumachi, Tosu-shi, Saga 841-0052, Japan
*TEL: +81-70-5548-8047 / FAX: +81-46-856-5571, E-mail: tetsu@criepi.denken.or.jp

ABSTRACT: Crystalline silicon (c-Si) photovoltaic (PV) modules are the dominant PV technology with about 90% of the market share at present. Recently, various kinds of high-efficient PV solar cells are commercialized. However, there is little information available on the reliability and durability of these high-efficient PV modules. In this study, we investigate the annual degradation rates of two kinds of PV modules composed of single crystalline silicon (sc-Si) and interdigitated back contact (IBC) solar cells using three indices. The results indicate that the annual degradation rates of the sc-Si and IBC PV modules were 0.0%/Year and 0.7%/Year, respectively. The high annual degradation rates showed by IBC PV modules would be due to the effects of potential induced degradation (PID).
Keywords: PV Module, c-Si, Back Contact, Reliability, Degradation

1 INTRODUCTION

In recent years, the amount of photovoltaic (PV) installations has increased acceleratingly. The cumulative installations of PV systems reached about 302 GW at the end of 2016. PV systems are dispersed power generation plants. Therefore, the reliability and durability of PV systems are very important. Because electric power supply from solar systems are not stable, electric companies must control electric power supply using mainly thermal power plants, considering the electric power supply from the PV systems connected to the electric power grid. If PV systems did not show performance degradation, we could estimate the power supply from PV systems based on the nominal capacity of the installed PV systems. However, if the PV systems were largely degraded, the electric power supply estimation would become very difficult. On this point, the reliability and durability of PV systems is important. In addition, the electric power generation costs depend on the usable years of PV systems. Moreover, the credibility of the PV module manufacturers will increases with increasing the reliability and durability of their PV modules.

Interdigitated back contact (IBC) solar cells are one of the single crystalline silicon (sc-Si) solar cells with high efficiencies. IBC solar cells eliminate shadow losses of the fingers and busbars on the top surface of solar cells, because all metal electrodes in the IBC solar cells are located on the back side. The purpose of this study is to investigate the reliability and durability of PV modules consisting in IBC solar cells. Here, we investigate we investigate the annual degradation rates of two kinds of PV modules composed of conventional p-type aluminum back surface field (Al-BSF) sc-Si and IBC solar cells using three indices.

2 EXPERIMENTAL METHOD

2.1 Outdoor test field for PV modules

Figure 1 shows a mosaic image of the outdoor test field at AIST Kyushu Center in Saga Prefecture in Japan taken by a drone. The outdoor test field can be divided into nine sections, which are named E-1, E-2 ..., etc. A PCS and I-V curve tracer were installed in each section.

Table I: Configuration of the installed c-Si PV modules in the outdoor test field at AIST Kyushu Center

	Kind	Total Pmax [kW]	Array Configuration	Install Mon/Year
E-1A	p-type sc-Si	5.04	7S × 1P × 4A	9/2010
E-1B	p-type sc-Si	4.9	5S × 1P × 4A	12/2012
E-2A	p-type pc-Si	5	6S × 1P × 4A	9/2010
E-2B	p-type pc-Si	5	5S × 1P × 4A	12/2012
W-2A	SHJ	4.8	5S × 1P × 4A	12/2012
W-2B	IBC sc-Si	4.68	6S × 1P × 4A	12/2012
M-3C	SHJ	1.55	5S × 1P × 1A	06/2016
M-3D	n-type sc-Si	1.89	7S × 1P × 1A	06/2016
W-1A	SHJ	1.25	5S × 1P × 1A	06/2016
W-1B	IBC sc-Si	1.05	5S × 1P × 1A	06/2016
W-1C	PERC sc-Si	1.45	5S × 1P × 1A	06/2016
W-1D	PERC sc-Si	1.475	5S × 1P × 1A	06/2016

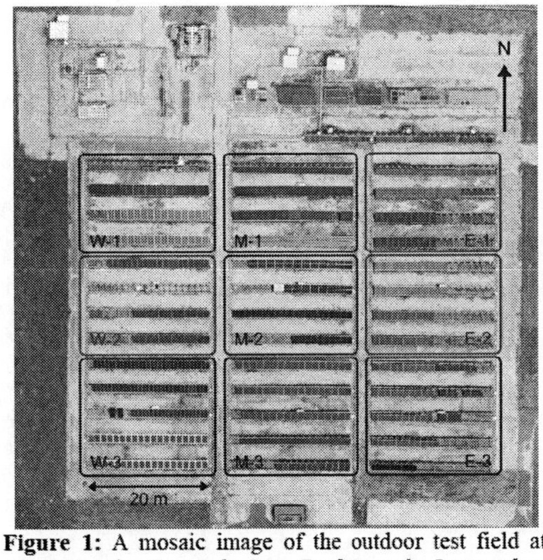

Figure 1: A mosaic image of the outdoor test field at AIST Kyushu Center in Saga Prefecture in Japan taken by a drone

In the outdoor test field, we have investigated 11 kinds (22 types) of PV modules composed of both c-Si and thin-film technologies, which include almost all commercially available kinds of PV modules. These include 6 kinds (12 types) of c-Si PV modules shown in Table I. The details of the above six types of c-Si PV modules in Table I were investigated in the paper [1]. In

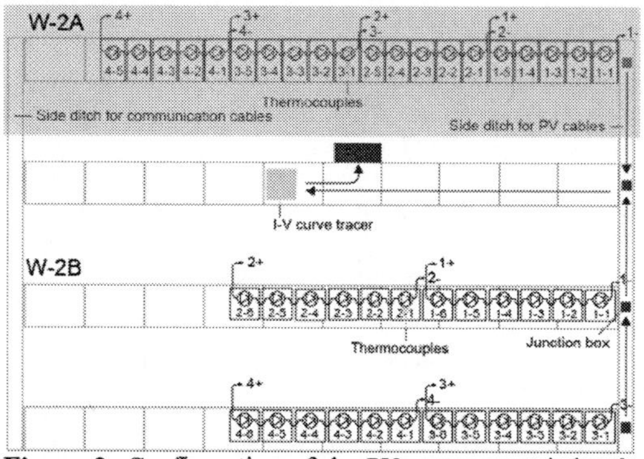

Figure 2: Configuration of the PV system consisting in the IBC PV modules installed in W-2B

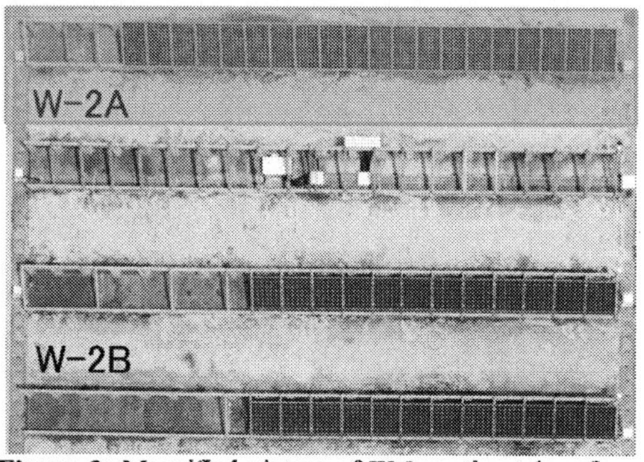

Figure 3: Magnified picture of W-2 section taken from above by a drone

this study, we focus on the two types of PV modules composed of the p-type sc-Si (E-1A) and IBC sc-Si (W-2B) solar cells.

2.2 Configuration of the investigated PV systems

Figure 2 shows the configuration of the PV system consisting in the IBC PV modules installed in W-2B. One string is composed of six PV modules. There are four PV strings. As shown in Table I, the total nominal power outputs of the sc-Si and IBC PV arrays are approximately 5.04 kW and 4.68 kW, respectively. The multiple (eight) string PCS with high frequency transformer is used in the system. In addition, DC circuit is not grounded, which is said to be floating. In the section of W-B, two types of PV modules are connected to one PCS. Both four PV strings of silicon hetero junction (SHJ) installed in W-2A and IBC installed in W-2B are connected to the PCS. In this study, we does not discuss SHJ (W-2A) because of the space constraints. The magnified picture taken from above by a drone is shown in Figure 3. In this picture, we can identify each solar cells. Therefore, the resolution of the picture is quite high (about a few centimeter).

2.3 Outdoor measurement parameters

I-V curve of every PV string, in-plain global solar irradiance, and module temperature are measured at 10 time intervals. Every PV strings are generally connected to electric power grid. Communication and PV cables are set to the left side ditch and PV cables are set to the right side ditch.

2.4 Three indices to evaluate the PV performance

We use three indices in order to evaluate the annual

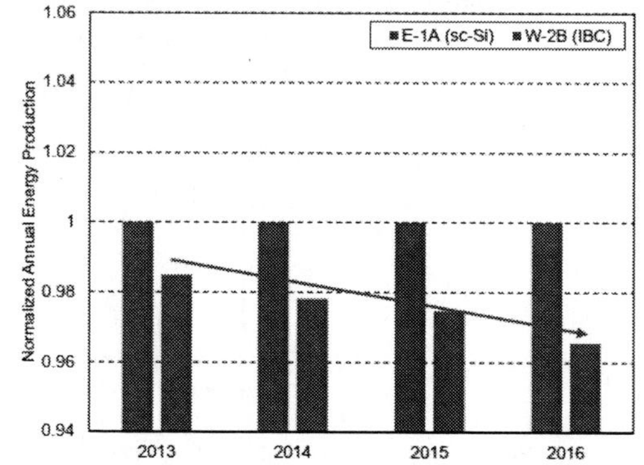

Figure 4: The normalized annual energy production of IBC arrays per 1 kW

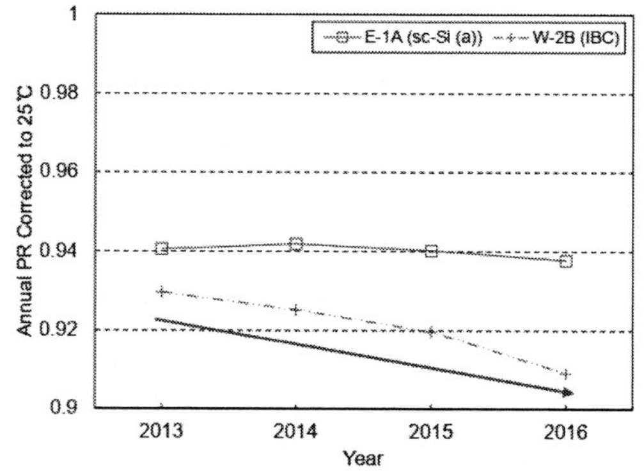

Figure 5: performance ratio corrected to 25℃ ($PR_{T=25}$)

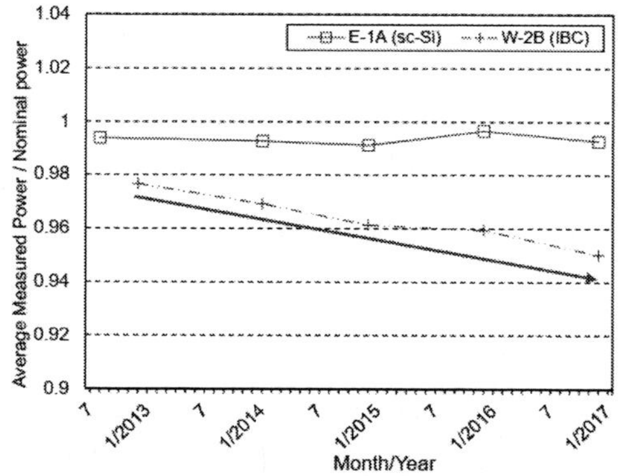

Figure 6: Measured powers normalized by the nameplate powers defined by the module manufacturers

degradation rates of sc-Si and IBC PV modules in this study. Two indices and one index are based on the outdoor and indoor measurements, respectively.

First index is the annual energy production. We can never eliminate missing data in particular in outdoor measurements. Therefore, in order to remove the effect of missing data, we only select the time intervals at which the powers of all the PV strings are more than 0 W.

Second index is performance ratio corrected to 25℃ ($PR_{T=25}$). When the fluctuation of solar irradiance is large, the precision of $PR_{T=25}$ becomes quite low. Therefore, we use only the data which were measured under stable and high solar irradiance conditions.

The third index is based on indoor measurements. Once a year, we removed all the c-Si PV modules form the mounting and washed the surface of the PV modules. Then, all the PV modules were measured by a pulsed solar simulator under the standard test conditions (STC). The measured powers are normalized by the nameplate power defined by the module manufacturers. The third index is the average values over the normalize values of all the PV modules. The light source is Xe lamp. The seep direction and time are from I_{SC} to V_{OC} and 100 ms, respectively. Until now, we have conducted five times indoor measurements of all the PV module.

3 RESULTS AND DISCUSSION

3.1 Annual degradation rates of sc-Si and IBC PV arrays

The normalized annual energy production per 1 kW in Figure 4. Because the total solar irradiation was different from every year, we normalized the annual energy production of IBC arrays by that of sc-Si arrays, which decreased by about 2% from 2013 to 2016. Here, we assumed that the annual energy production sc-Si arrays did not show any degradation and were used as a reference arrays.

$PR_{T=25}$ is shown in Fig. 5. As is the case with the annual energy production, The $PR_{T=25}$ of IBC arrays decreased by about 2% from 2013 to 2016. However, the $PR_{T=25}$ of sc-Si arrays showed no or little degradation.

Figure 6 shows the average normalized indoor power measured by the solar simulator under STC. The numbers of sc-Si and IBC PV modules are 28 and 26, respectively. All the indices show the same trend. Therefore, the results of this study indicate that the annual degradation rates of sc-Si and IBC modules are 0.0%/year and 0.7%/year, respectively.

3.2 Relationship electrical potential and degradation

Why the IBC PV modules show high degradation rates? Here, we discuss the relationship between module locations in a string and PV cell potentials. Because the DC circuit is not grounded, the electrical potential of 1-1, 1-2, and 1-3 modules is negative, although that of 1-4, 1-5, and 1-6 module (Fig. 2) is positive as shown in the upper part of Fig. 7. The lower part of Fig. 7 shows that the performance of IBC PV modules decreased only under positive potential conditions.

We also confirmed the performance degradation by the change in the I-V curves of IBC PV modules. The shapes of I-V curves only under positive potential conditions were varied.

The three modules would be at negative potentials showed no degradation, while the other three modules would be at positive potentials showed clear signs of degradation. The magnitude of the degradation increases with increasing electrical potential. We refer to this phenomenon literally as "potential-induced degradation (PID)" of n-type IBC solar cells. Only the performance of IBC PV modules was influenced by the electrical potential, whereas the performance of sc-Si PV modules was almost homogeneous and does not seem to be influenced by the electrical potential. Furthermore, we cannot identify clear changes in both I-V curves and EL images of sc-Si (E-1A) PV modules.

3.3 Mechanism of PID of n-type IBC solar cells

The PID of the n-type IBC solar cells coincides with

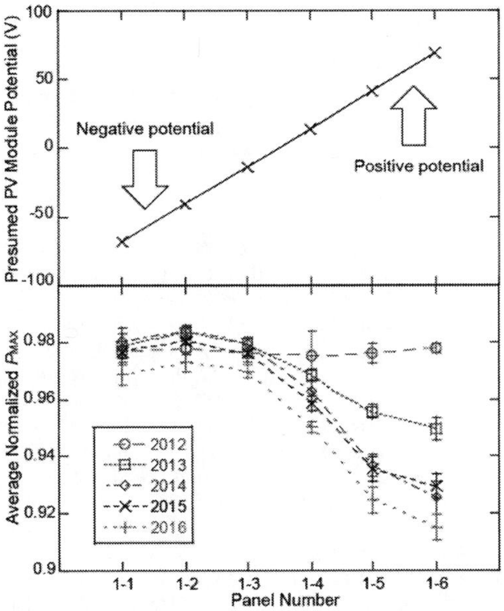

Figure 7: Relationship presumed electrical potential and degradation of IBC PV modules

SunPower's report, in which the degradation was referred to as "surface polarization" [2]. They proposed the following mechanism. First, the surface of IBC solar cells under positive potential conditions is negatively charged by leakage currents because the electrical resistivity of the hydrogenated silicon nitride (SiNx:H) layer would be very high. The SiNx:H layer, which would be charged negatively, thus attracts photoexcited holes, which results in recombination. Consequently, the diffusion length of minority carriers (holes) in the n-layer would be shortened [2].

The mechanism of the PID of p-type solar cells is considered to be different from that of n-type technologies. The PID of p-type Si technologies has been vigorously investigated [3–9] since the discovery of this phenomenon around 2010. The most probable mechanism is cations transport to the p-n junction region [3–7], although one study suggests that sodium ion diffusion is not a sufficient condition for PID occurrence [9].

4 SUMMARY

We investigated the annual degradation rates of sc-Si (E-1A) and IBC (W-2B) PV modules installed in Saga prefecture in Japan using three indices. The results are consistent among all the indices. The annual degradation rates of sc-Si (E-1A) and IBC (W-2B) are 0.0% and 0.7%, respectively. The performance of IBC (W-2B) modules decreased only under positive potential conditions. This would be due to "PID" of n-type IBC solar cells.

5 ACKNOWLEDGEMENT

This study was supported by New Energy and Industrial Technology Development Organization (NEDO) under Ministry of Economy, Trade and Industry (METI).

6 REFERENCES

[1] Ishii, T., Masuda, A., Annual degradation rates of recent crystalline-silicon photovoltaic modules under actual operating conditions, *Progress in Photovoltaics: Research and Applications*, Available online, DOI: 10.1002/pip.2903.

[2] Swanson, R., Cudzinovic M., DeCeuster D., Desai V., Jürgens J., Kaminar N., Mulligan W., Rodrigues-Barbarosa L., Rose D., Smith D., Terao A., Wilson K., The surface polarization effect in high-efficiency silicon solar cells, *Proceedings of the 15th International Photovoltaic Science & Engineering Conference 2005*; Shanghai, China, pp. 410-411.

[3] Naumann V., Lausch D., Hähnel A., Bauer J., Breitenstein O., Graff A., Werner M., Swatek S., Großer S., Bagdahn J., Hagendorf C., Explanation of potential-induced degradation of the shunting type by Na decoration of stacking faults in Si solar cells, *Solar Energy Materials and Solar Cells* 2014; **120**(Part A), pp. 383-389, DOI: 10.1016/j.solmat.2013.06.015.

[4] Lausch D., Naumann V., Breitenstein O., Bauer J., Graff A., Bagdahn J., Hagendorf C., Potential-induced degradation (PID): Introduction of a novel test approach and explanation of increased depletion region recombination, *IEEE Journal of Photovoltaics* 2014; **4**(3), pp. 834-840, DOI: 10.1109/JPHOTOV.2014.2300238.

[5] Hara K., Ichinose H., Murakami T. N., Masuda A., Crystalline Si photovoltaic modules based on TiO2-coated cover glass against potential-induced degradation, *RSC Advances* 2014; **4**, pp. 44291-44295, DOI: 10.1039/c4ra06791f.

[6] Hacke P., Terwilliger K., Glick S., Tamizhmani G., Tatapudi S., Stark C., Koch S., Weber T., Berghold J., Hoffmann S., Koehl M., Dietrich S., Ebert M., Mathiak G., Interlaboratory study to determine repeatability of the damp-heat test method for potential-induced degradation and polarization in crystalline silicon photovoltaic modules, *IEEE Journal of Photovoltaics 2015*; **5**(1), pp. 94-101, DOI: 10.1109/JPHOTOV.2014.2361650.

[7] Kapur J., Stika K. M., Westphal C. S., Norwood J. L., Hamzavytehrany B., Prevention of potential-induced degradation with thin ionomer film, *IEEE Journal of Photovoltaics 2015*; **5**(1), pp. 219-223, DOI: 10.1109/JPHOTOV.2014.2365465.

[8] Jonai S., Hara K., Tsutsui Y., Nakahama H., Masuda A., Relationship between cross-linking conditions of ethylene vinyl acetate and potential induced degradation for crystalline silicon photovoltaic modules, *Japanese Journal of Applied Physics* 2015; **54**(8S1), 08KG01, DOI: 10.7567/JJAP.54.08KG01.

[9] Masuda A., Hara Y., Jonai S., Consideration on Na diffusion and recovery phenomena in potential-induced degradation for crystalline Si photovoltaic modules, *Japanese Journal of Applied Physics* 2016; **55**(2S), 02BF10, DOI: 10.7567/JJAP.55.02BF10.

PERFORMANCE LOSS INDUCED BY LeTID IN THE FIELD

Friederike Kersten[a], Fabian Fertig[a], Kai Petter[a], Bernhard Klöter[a], Evelyn Herzog[a], Matthias B. Strobel[a], Johannes Heitmann[b], Jörg W. Müller[a]
[a]Hanwha Q CELLS GmbH, Sonnenallee 17-21, 06766 Bitterfeld-Wolfen, Germany
[b]TU Bergakademie Freiberg, Institute of Applied Physics, Leipziger Straße 23, 09599 Freiberg, Germany

ABSTRACT: If not adequately suppressed, Light and elevated Temperature Induced Degradation (LeTID) has been shown to cause severe degradation of crystalline silicon solar cells and modules with passivated emitter and rear cell (PERC). Within this work, the system performance of LeTID-sensitive multicrystalline silicon (*mc*-Si) modules is investigated when operated in temperate (Germany) and Mediterranean (Cyprus) climates, and a correlation to predict the observed field performance based on accelerated laboratory testing is presented. Severe degradation of LeTID-sensitive modules of more than 7% in maximum output power is detected after one thousand hours of laboratory testing, which is shown to correspond to ~ 3 years of field installation time in Cyprus. In contrast, LeTID-sensitive modules installed in Germany show a degradation of 2.5% within the same time period. Due to lower irradiance and module temperatures in Germany LeTID is less pronounced than in Cyprus. Hence, a significant dependence of system performance and energy yield loss of LeTID-sensitive modules on installation site is shown, which affects one-to-one the levelized cost of electricity (LCOE). By applying Hanwha Q CELLS' Q.ANTUM technology, it is shown that LeTID in the field is suppressed, independently of installation site.
Keywords: Multicrystalline Silicon, PV Module, Degradation, Performance

1 INTRODUCTION

In 2012, *mc*-Si solar cells with plasma-enhanced chemical vapor deposited (PECVD) aluminum oxide (AlO_x) passivated rear side and *mc*-Si aluminum back-surface field (Al-BSF) solar cells have been shown to degrade more at elevated temperatures [1,2]. Both studies excluded the observed behavior to be due to boron-oxygen (BO) defect formation or iron-boron (FeB) pair dissociation, which has been speculated to be the root cause of the light-induced degradation (LID) observed at lower temperatures in *mc*-Si in previous studies, e.g. [3-8]. Since then, many groups have investigated the degradation of *mc*-Si, confirming cells with dielectrically passivated rear side to be significantly more affected than Al-BSF solar cells [9], showing local inhomogeneities of the defect characteristics [10-12], and more recently, ways to mitigate excessive LeTID [13-20]. Furthermore, it has been shown that LeTID also occurs within passivated emitter and rear cells (PERC) processed on Czochralski-grown silicon substrates [x]. As the observed LID is significantly more pronounced at elevated temperatures, Hanwha Q CELLS has introduced the term LeTID for "Light and elevated Temperature Induced Degradation" [14] as denomination for this new degradation mechanism at the time. If not suppressed, LeTID can lead to a loss in relative conversion efficiency of more than 10% [14]. LeTID is relevant under field conditions [14] and, therefore, needs to be suppressed to enable mass production of solar cells with PERC technology. A method to suppress LeTID in mass production of dielectrically passivated solar cells and modules has been developed by Hanwha Q CELLS [14] as part of its Q.ANTUM technology [22]. In this work, performance loss measurements for LeTID-sensitive modules in system installations in temperate and Mediterranean climates and a correlation between field operation time and accelerated laboratory testing are shown. Furthermore, it is shown that Q.ANTUM technology suppresses LeTID in the field, independent of installation site.

2 LeTID ON MODULE LEVEL IN THE LABORATORY AND FIELD

2.1 Experimental

LeTID-sensitive PERC modules were manufactured on standard industrial *mc*-Si substrates. Furthermore, neighbored *mc*-Si substrates were used to process modules with Q.ANTUM technology [22]. The module degradation experiments in the laboratory were performed at a temperature of 75°C in a climate chamber, with excess carriers injected by current (CID, current-induced degradation). The current was adjusted to simulate the excess charge carrier density during operation at maximum power point (MPP) and 1 sun illumination, which was previously proposed as standard accelerated test conditions for LeTID [14]. With this method, a high number of modules can be tested in parallel at moderate cost in the laboratory. LeTID-sensitive PERC and LeTID-suppressing Q.ANTUM modules were installed on outdoor test fields in Thalheim, Germany (DE) and Nicosia, Cyprus (CYP). The irradiance (G_{mod}) in module plane and module temperature (T_{mod}) were logged and averaged every 5 min. The energy yield of each system was averaged in 15 min intervals. All night values were eliminated by using a data filter which excluded all values with $G_{mod} < 10$ W/m^2 due to increased measurement uncertainties. Once every three months, the modules installed in Germany and Cyprus were removed from the system racks and tested by means of current-voltage measurements under standard test conditions (STC) in the laboratory. After each testing cycle, the modules were re-installed outdoors.

2.2 Field performance of LeTID-sensitive modules in temperate and Mediterranean climates

Fig. 1 shows the annual irradiance distribution measured at the module installation sites in Germany and Cyprus. Each G_{mod} bar shows the sum of irradiated energy for one year in 100 W/m^2 steps. For Thalheim (temperate climate), lower irradiance values with a maximum of

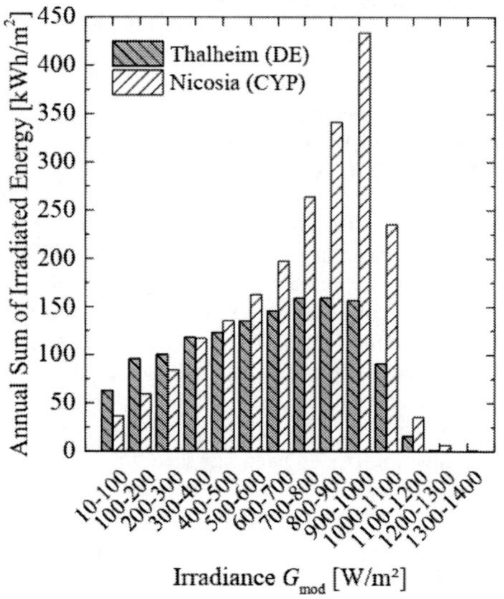

Fig. 1. Annual irradiance distribution of irradiated energy measured in module plane in Thalheim (DE) and Nicosia (CYP).

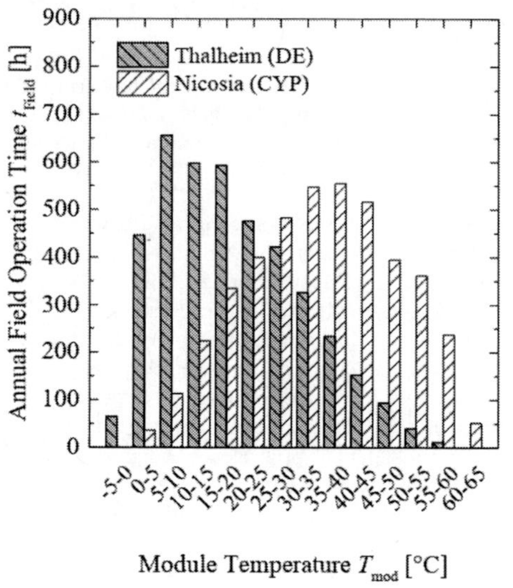

Fig. 2. Annual module temperature distribution measured in Thalheim (DE) and Nicosia (CYP).

150 kWh/m^2 for the intervals 700 W/m^2 < G_{mod} < 1000 W/m^2 compared with Nicosia (Mediterranean climate) are measured. While in Thalheim, the modules operate for ~30% of all operation time in comparably low-light conditions with G_{mod} < 400 W/m^2, in Nicosia the highest irradiated energy value of 434 kWh/m^2 was measured in the range of 900 W/m^2 to 1000 W/m^2. As shown in Fig. 2, these relatively high irradiance values around 1000 W/m^2 in Cyprus resulted in comparably high module temperatures. In contrast to the depicted irradiation profile in Fig. 1, the module temperature measured in Nicosia showed a Gaussian distribution curve. The weighted average T_{mod} of 35°C to 40°C is significantly higher temperatures compared to Thalheim with 5°C to 10°C. In Nicosia, elevated temperatures with T_{mod} > 50°C occured for ~25% of field operation time per year. Hence, these climatic conditions in Cyprus are well-suited for the characterization of LeTID in comparison to the temperate climate in Germany with annually ~70% of field operation time with T_{mod} < 25°C.

2.3 Comparison of field data and accelerated degradation tests in the laboratory

As illustrated in the previous subsection, the irradiation in module plane and module temperatures over several months and years at two different test fields in Cyprus and Germany were measured. Together with a model for the irradiance- and temperature-dependence of LeTID kinetics, a test-site-specific time constant corresponding to the proposed accelerated laboratory test performed at 75°C is deduced; e.g., 290 h laboratory test time corresponds to 1 year of field installation time in Cyprus. Fig. 3 shows as filled symbols the relative module power loss due to LeTID for LeTID-sensitive PERC and LeTID-suppressing Q.ANTUM outdoor-tested modules in Cyprus and Germany, respectively, referring to the lower x-axis (field operation time t_{Field}). Furthermore, the corresponding values for degradation tests in the laboratory are shown as open squares. The time-dependent relative module power loss under laboratory testing at 75°C with CID in MPP mode refers to the upper x-axis (lab degradation time t_{Lab}). It can be seen field performance can be modelled well by laboratory tests, see Fig. 3. LeTID-sensitive PERC modules showed a significant loss in module power of more than 7% due to LeTID, both in the field and under accelerated aging in the laboratory. Equivalently installed modules in Germany showed a degradation of ~2.5% for the same time period. Due to lower irradiance values and module temperature in Germany, LeTID rate is less pronounced as in Cyprus. Fig. 4 shows the relative loss of specific yield on the left and LCOE on the right y-axis compared with a reference system during 3 years of field installation in Cyprus. The loss in system output power due to LeTID shown in Fig. 3 affects one-to-one the energy yield and the LCOE. It can be seen that the LeTID-suppressing Q.ANTUM modules show a higher specific yield and, hence, higher revenue than the reference system. The lower specific yield after approximately 2 years field installation time results from soiling of the Q.ANTUM system in Cyprus due to nearby construction, which was removed at the end of October. As shown in Fig. 3-4, applying the optimized defect-engineering Q.ANTUM process, LeTID could be fully suppressed in the field. Fig. 5 shows that Q.ANTUM modules also do not degrade after extended laboratory testing of up to a time of several thousands of hours. With the developed model and test-site-specific time constant the laboratory degradation time of 4300 h can be calculated to correspond to 15 years of field installation in Mediterranean climate.

3 CONCLUSION

The presented results show performance loss measurements due to LeTID in system installations with LeTID-sensitive PERC modules in different climates. Long-term degradation tests in a climate chamber with LeTID-sensitive PERC modules resulted in performance loss of more than 7%, which was also observed in the field in Mediterranean climate (Cyprus) within the first 3 years after installation. Equivalent modules installed in Germany showed a degradation of 2.5% within the same

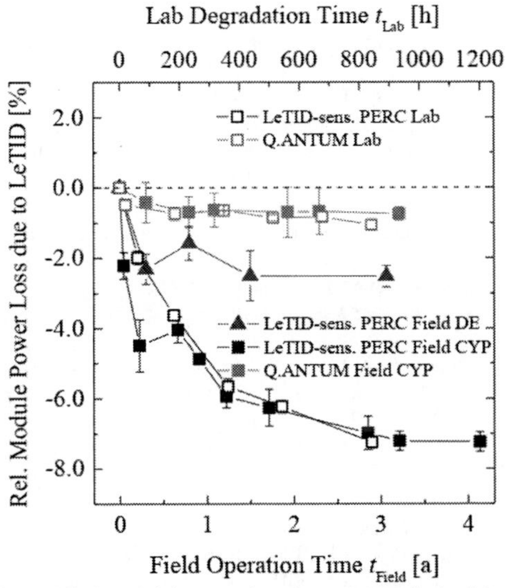

Fig. 3. Relative module power loss due to LeTID during laboratory degradation time (upper x-axis) and field operation in Thalheim (DE) and Nicosia (CYP) (lower x-axis).

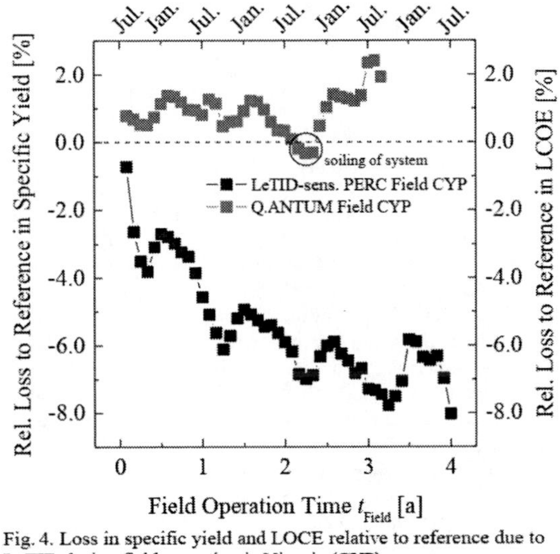

Fig. 4. Loss in specific yield and LOCE relative to reference due to LeTID during field operation in Nicosia (CYP).

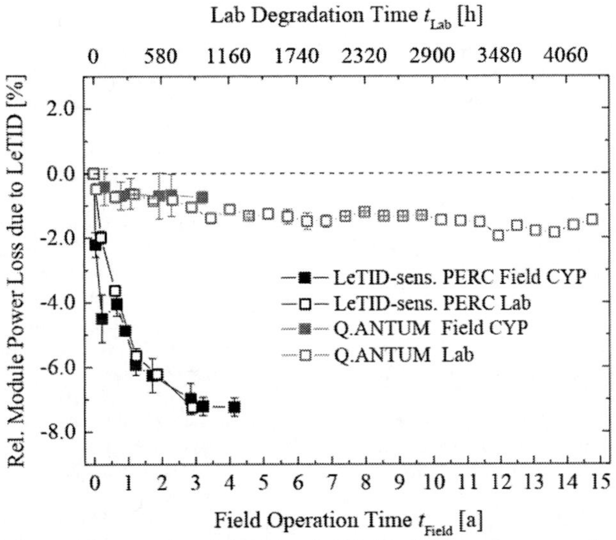

Fig. 5. Time-dependent relative module power degradation due to LeTID. The lower x-axis corresponds to field operation time in Nicosia (CYP) and the upper x-axis to CID laboratory time in MPP mode at 75°C in a climate chamber.

time period. This demonstrates the high LeTID susceptibility of PERC modules without LeTID suppression in the field when being installed in areas with comparably high module temperature such as in Mediterranean climates. Furthermore, it is shown that the observed degradation behavior in the field can be adequately predicted by using a test-site-specific time constant, which is easily adaptable to any climate zone. Hanwha Q CELLS' Q.ANTUM modules are shown to suppress LeTID, which is validated by during outdoor field operation data and long-term laboratory testing in climate chambers, which corresponds to a field installation time of 15 years in Mediterranean climate. Beyond that, the benefit of Q.ANTUM technology in energy yield and LCOE were shown in field installations.

ACKNOWLEDGEMENTS

The authors would like to thank George Makrides and George Georghiou from PV Technology Laboratory at University of Cyprus and the entire team of the Reiner Lemoine Research Center, Pilot Line, Energy Yield and Module Test Center at Hanwha Q CELLS for their contribution to this work.

REFERENCES

[1] K. Ramspeck, S. Zimmermann, H. Nagel, A. Metz, Y. Gassenbauer, B. Birkmann, A. Seidl. Light induced degradation of rear passivated *mc*-Si solar cells. Proc.

27th EU PVSEC 2012, Frankfurt, Germany, pp. 861–865.

[2] F. Fertig, K. Krauß, I. Geisemeyer, J. Broisch, H. Höffler, J. O. Odden, A.-K. Soiland, S. Rein. Fully Solderable Large-Area Screen-Printed Al-BSF *p*-Type *mc*-Si Solar Cells from 100% Solar Grade Feedstock Yielding $\eta > 17\%$: Challenges and Potential on Cell and Module Level. Proc. 27th EU PVSEC 2012, Frankfurt, Germany, pp. 1031–1038.

[3] S. De Wolf, P. Choulat, J. Szlufcik, I. Périchaud, S. Martinuzzi, C. Häßler, W. Krumbe. Light-induced degradation of very low resistivity multi-crystalline silicon solar cells. Proc. 38th IEEE PVSC 2000, Austin, Texas, pp. 53–56.

[4] B. Damiani, K. Nakayashiki, D. S. Kim, V. Yelundur, S. Ostapenko, I. Tarasov, A. Rohatgi. Light induced degradation in promising multi-crystalline silicon materials for solar cell fabrication. Proc. 3rd World Conference on Photovoltaic Energy Conversion 2003, Osaka, Japan, pp. 927–30.

[5] S. Dubois, N. Enjalbert, J. Garandet. Slow down of the light-induced-degradation in compensated solar-grade multicrystalline silicon. Applied Physical Letters 2008;93:103510.

[6] J. Junge, A. Herguth, G. Hahn, D. Kreßner-Kiel, R. Zierer. Investigation of degradation in solar cells from different *mc*-Si materials. Energy Procedia 2011;8:52–57.

[7] K. Peter, P. Preis, P. E. Díaz-Pérez, J. Theobald, E. Enebakk, A.-K. Soiland, A. Savtchouk, M. Wilson, J. Lagowski. Light induced degradation in multicrystalline solar grade silicon solar cells evaluated using accelerated LID. Proc. 26th EU PVSEC 2011, Hamburg, Germany, pp. 1856–1858.

[8] K. Ounadjela, O. Sidelkheir, C.S. Jiang, M. M. Al-Jassim. Light-Induced Degradation in Upgraded Metallurgical-Grade Silicon Solar Cells. Proc. 38th IEEE PVSC 2012, Austin, Texas, pp. 2739–2743.

[9] F. Fertig, K. Krauß, S. Rein. Light-induced degradation of PECVD aluminium oxide passivated silicon solar cells. Physical Status Solidi RRL 2014;9:41–46.

[10] K. Krauß, F. Fertig, D. Menzel, S. Rein. Light-Induced Degradation of Silicon Solar Cells with Aluminium Oxide Passivated Rear Side. Energy Procedia 2015;77:599–609.

[11] M. Selinger, W. Kwapil, F. Schindler, K. Krauß, F. Fertig, B. Michl, W. Warta, M. C. Schubert. Spatially Resolved Analysis of Light Induced Degradation of Multicrystalline PERC Solar Cells. Energy Procedia 2016;92:867–872.

[12] T. Luka, S. Großer, C. Hagendorf, K. Ramspeck, M. Turek. Intra-grain versus grain boundary degradation due to illumination and annealing behavior of multi-crystalline solar cells. Solar Energy Materials and Solar Cells 2016;158:43–49.

[13] S. Frigge, H. Mehlich, T. Grosse. LID free module made from high performance PERC solar cells. Proc. 31st EU PVSEC 2015, Hamburg, Germany.

[14] F. Kersten, P. Engelhart, H.-C. Ploigt, A. Stekolnikov, T. Lindner, F. Stenzel, M. Bartzsch, A. Szpeth *et al.* Degradation of multicrystalline silicon solar cells and modules after illumination at elevated temperature. Solar Energy Materials and Solar Cells 2015;142:83–86.

[15] C. E. Chan, D. N. R. Payne, B. J. Hallam, M. D. Abbott, T. H. Fung, A. M. Wenham, B. S. Tjahjono, S. R. Wenham. Rapid Stabilization of High-Performance Multicrystalline *P*-type Silicon PERC Cells. IEEE Journal of Photovoltaics 2016;6:1473–1479.

[16] A. Zuschlag, D. Skorka, G. Hahn. Degradation and Regeneration Analysis in *mc*-Si. 43rd IEEE PVSC, Portland, Oregon, 2016.

[17] D. N. R. Payne, C. E. Chan, B. J. Hallam, B. Hoex, M. D. Abbott, S. R. Wenham, D. M. Bagnall. Rapid passivation of carrier-induced defects in *p*-type multi-crystalline silicon. Solar Energy Materials and Solar Cells 2016;158:102–106.

[18] K. Krauß, A. A. Brand, F. Fertig, S. Rein, J. Nekarda. Fast regeneration processes to avoid light-induced degradation in multicrystalline silicon solar cells, IEEE Journal Photovoltaics 2016;6:1427–1431.

[19] D. Bredemeier, D. Walter, S. Herlufsen, J. Schmidt. Lifetime degradation and regeneration in multicrystalline silicon under illumination at elevated temperature. AIP Advances 2016;6:035119–9.

[20] K. Nakayashiki, J. Hofstetter, A. E. Morishige, T.-T. A. Li, D. B. Needleman, M. A. Jensen *et al.* Engineering Solutions and Root-Cause Analysis for Light-Induced Degradation in *p*-Type Multicrystalline Silicon PERC Modules. IEEE Journal of Photovoltaics 2016;6:860–868.

[21] F. Fertig, R. Lantzsch, A. Mohr, M. Schaper, M. Bartzsch, D. Wissen, A. Mette, S. Peters *et al.* Mass production of *p*-type Cz silicon solar cells approaching average stable conversion efficiencies of 22 %. Energy Procedia 2017;124:338–345.

[22] P. Engelhart, D. Manger, B. Klöter, S. Hermann, A. Stekolnikov, S. Peters, H.-C. Ploigt, A. Eifler *et al.* Q.ANTUM - Q-CELLS next generation high-power silicon cell and module concept. Proc. 26th EU PVSEC 2011, Hamburg, Germany, pp. 821–826.

Shadows from People and Tools Can Cause Permanent Damage in Monolithic Thin-Film Photovoltaic Modules

Timothy J Silverman and Ingrid Repins
National Renewable Energy Laboratory
Golden, Colorado, United States

ABSTRACT

Monolithic thin-film modules, including CIGS, CdTe and a-Si products, can be damaged by shadows that occur during service as a result of operations and maintenance activities. We show results from commercial modules exposed to shadows from human bodies and from cleaning tools. These shadows all resulted in extensive formation of shunts and permanent loss of power. Permanent loss of P_{mp} measured at standard test conditions ranged from 1% for single exposures to human shadows to > 30% for multiple exposures to cleaning shadows. In every case the first event caused more damage than each subsequent event. In some cases uniform light exposure after shadow events caused damage to worsen.

Keywords: thin film, module, performance, reliability, shunts

1 Introduction

Monolithic thin-film photovoltaic (PV) modules, including CIGS, CdTe and a-Si products, typically contain a large number of series-connected cells. When a shadow falls on some of these cells while others are illuminated, the shadowed cells pass current by operating in reverse bias. This causes localized heating at heterogeneities in the thin-film device, leading to the formation of shunts that have a permanent effect on performance [1–3]. Because of the large number of cells between the module terminals, there is limited opportunity to protect against reverse bias using bypass diodes. It is possible to incorporate dozens of bypass diodes, but this adds cost and decreases unshaded performance. Typical commercial monolithic thin-film products either lack a bypass diode or have a single bypass diode to protect over 100 cells.

The PV module type qualification test standard contains a test, the "hot spot endurance test", where shadows are intentionally applied to an otherwise illuminated module. This test screens for major visual defects, loss of insulation resistance, and complete loss of electrical function. As such it is primarily a safety test and does not screen for loss of efficiency resulting from shadows.

Product documentation sometimes forbids certain types of shadows, but neither the thoroughness of these guidelines nor 100% compliance by everyone present at an installation can be guaranteed. In this work we present case studies quantifying the permanent loss of efficiency that may occur in commercial monolithic PV modules due to shadows. We applied shadows from human bodies and from cleaning tools. Human body shadows are practically impossible to avoid in the field and the shadows we applied were not forbidden by manufacturer guidelines. Shadows from cleaning tools are in some cases forbidden by manufacturer guidelines but can still be observed in the field. Cleaning shadows can be of greater concern because they are often intentionally applied to an entire system.

2 Method

2.1 Specimens

We tested four commercially available monolithic thin-film modules from multiple manufacturers. The modules included CIGS and CdTe products. These specimens had no field or test exposure prior to the start the tests described here. The modules are designated with letters A–D.

2.2 Performance stabilization

We used an indoor light chamber to stabilize the modules' electrical states (sometimes called light soaking or preconditioning) at 1000 W m^{-2} and 50°C until P_{mp} changed less than 1% per 20 h according to in-situ I-V curve measurements [4]. 30 min after the light exposure we measured the room-temperature I-V curve using a pulsed solar simulator. This stabilization and pulsed I-V curve measurement procedure was repeated in some cases after step 1 of stress and in every case at the end of the experiment. Every I-V curve and P_{mp} result shown in this work results from this stabilization and indoor measurement procedure. I-V curves are normalized to initially measured I_{sc} and V_{oc}, designated $I_{sc,0}$ and $V_{oc,0}$.

2.3 Outdoor testing

We deployed modules outdoors on a latitude-tilt rack. Because the current in a string of modules remains approximately constant when a shadow is cast on a small portion of the string, we tested individual modules under constant-current conditions. Modules were connected to a DC load configured to pass a constant current of I_s, with

$$I_s = E_{POA}\frac{I_{mp,STC}}{E_{STC}},$$

where E_{POA} is the plane-of-array irradiance, $I_{mp,STC}$ is the current at maximum power point under standard test conditions (STC) and is taken from the product nameplate, and E_{STC} is the irradiance at STC (1000 Wm^{-2}). For products without a bypass diode, passing I_s can require reverse voltage and we provided up to 50 V for this purpose.

Between steps of outdoor shadow testing, we measured I-V curves in situ. We use the ratio I_{mp}/I_{sc} to show the loss of shunt resistance outdoors. This ratio is less sensitive to changes in temperature and irradiance than other indirect measures of shunt resistance such as *FF*.

2.3.1 Human shadows

On modules A and B we cast shadows exclusively from human bodies.

On module A, step 1 was the application of multiple shadows from student volunteers during a tour. The event lasted for approximately five minutes. The volunteers were not specially trained regarding the purpose of their shadows. Irradiance was variable but included some sunny periods. Steps 2 through 10 were events lasting less than one minute of intentional shadows from arms, head and torso when irradiance was > 800 Wm^{-2}. Step 11 was a three-day period where the module was made available to volunteers to cast human-body-only shadows.

On module B, steps 1 through 9 were events lasting less than one minute of intentional shadows from arms, head and torso when irradiance was > 800 Wm^{-2}. The load was modulated during these steps in aid of in situ photoluminescence (PL) imaging (see Section 2.4 for details).

2.3.2 Cleaning shadows

On modules C and D we cast shadows using a cleaning tool with a 25-mm-diameter, 1800-mm-long handle and a 50-mm-wide blade. The blade's long dimension exceeded the modules' maximum dimension.

On module C, steps 1 through 5 were cleaning events occurring when irradiance was > 800 Wm^{-2}. The blade of the cleaning tool was held parallel to the module's integration scribe lines and passed across every cell in < 15 s. The load was modulated during these steps in aid of in situ PL imaging (see Section 2.4 for details).

On module D, steps 1 through 10 were cleaning events occurring when irradiance was > 800 Wm^{-2}. The blade of the cleaning tool was held parallel to the module's integration scribe lines and passed across every cell in < 15 s. Steps 11 and 12 were each ten additional cleanings of the same type. Step 13 was a two-day period where the module and cleaning tools were made available to volunteers to cast any shadow.

2.4 Luminescence imaging

Before stress and after each shadow in the outdoor test, we either performed electroluminescence (EL) imaging indoors or illuminated PL imaging outdoors. For indoor EL imaging, we forward-biased the module to pass its nameplate I_{sc} and collected EL emission using an InGaAs camera. Outdoor PL imaging was done using the OCPL technique with the same camera and modulating between I_s and V_{oc} at 22 Hz. The appearance of shunts in OCPL images is indistinguishable from that in indoor EL images [5]. Although shadows were applied to the entire surface of each module, in this work EL and PL images are cropped to show a square section of each module.

3 Results and discussion

Every shadow we applied resulted in the formation or strengthening (reduction of resistance) of localized shunts, visible in luminescence images, and efficiency loss due to the permanent loss of shunt resistance. Because this efficiency loss was due to a loss of shunt resistance, the loss of efficiency would be greater when measured at low irradiance.

3.1 Human shadows

Figure 1 shows EL images and I-V curves from before step 1, after step 1 and the end of the human shadow test on module A. A single shadow event caused many new shunts to appear in the EL image (Figure 1b) and a 1% permanent loss in P_{mp}. Subsequent events both created new shunts and strengthened existing ones (Figure 1c), leading to a 9% permanent loss in P_{mp}. Figure 1e shows that I_{mp}/I_{sc} declined after each event, illustrating that repeated exposure to shadows causes additional damage.

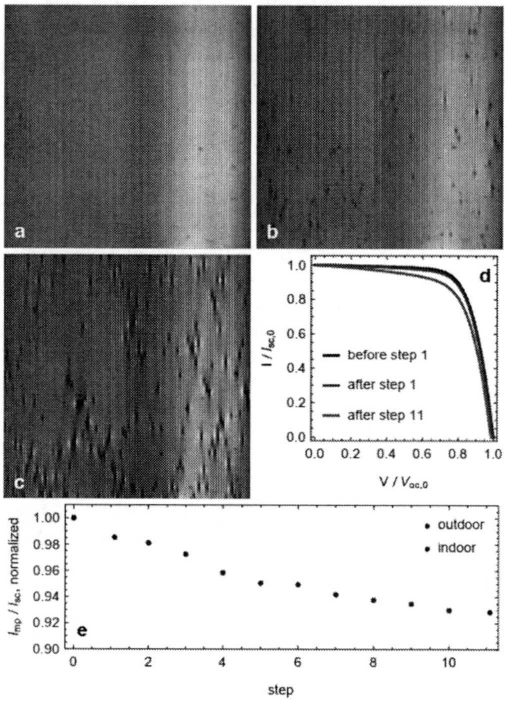

Figure 1: Results from human shadow testing on module A are shown: EL images from before the test (a), after step 1 (b), after step 11 (c); stabilized and normalized indoor STC I-V curves from the same three stages (d); and I_{mp}/I_{sc} from each step of the test.

Figure 2 shows PL images from before step 1, after step 1 and after step 9 and I-V curves from the beginning and end of the human shadow test on module B. A single shadow event created many new shunts (Figure 2b) and subsequent events created a small number of new shunts (Figure 2c), leading to a 10% permanent loss in P_{mp}. Figure 2e shows that I_{mp}/I_{sc} declined after each event, but in contrast to Module A, the first shadow event did much more damage than subsequent events. This illustrates that different products accumulate shunt damage at different rates. Figure 2e also shows that the stabilization process, which is simply further light exposure at P_{mp}, caused the damage done by shadows to worsen.

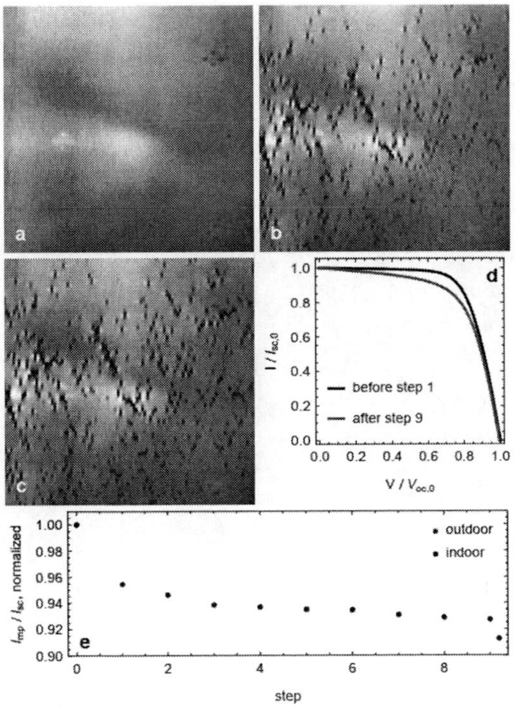

Figure 2: Results from human shadow testing on module B are shown: PL images from before the test (a), after step 1 (b), after step 9 (c); stabilized and normalized indoor STC I-V curves from before step 1 and after step 9 (d); and I_{mp}/I_{sc} from each step of the test.

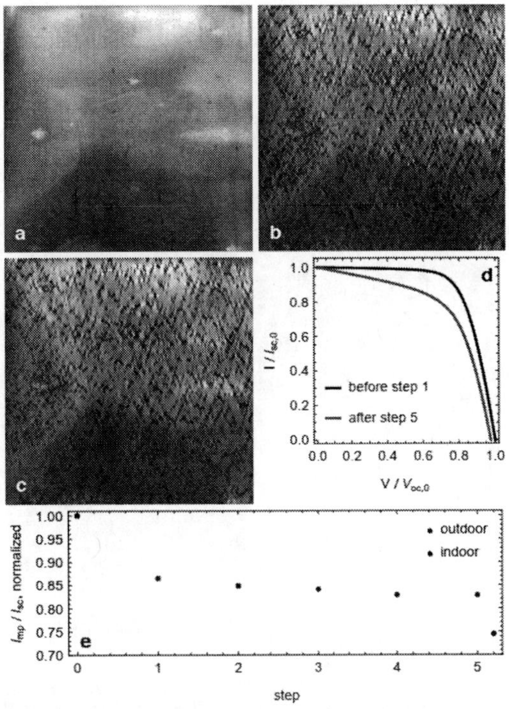

Figure 3: Results from cleaning tool shadow testing on module C are shown: PL images from before the test (a), after step 1 (b), after step 5 (c); stabilized and normalized indoor STC I-V curves from before step 1 and after step 5 (d); and I_{mp}/I_{sc} from each step of the test.

3.2 Cleaning shadows

Figure 3 shows PL images from before step 1, after step 1 and after step 5 and I-V curves from the beginning and end of the cleaning shadow test on module C. Cleaning the module one time introduced thousands of new shunts (Figure 3b and Figure 4) and the decline in I_{sc}/I_{mp} (Figure 3e) shows that this first event caused a major loss of shunt resistance. This damage resulted from a 15-second event, showing that these shunts can form very quickly. Subsequent events did minor additional damage and the final decline in P_{mp} was $> 30\%$. A substantial fraction of this reduction occurred after the last step of testing, during stabilization. Because the shadow steps of this test were conducted with in situ PL imaging (to create Figure 4), the load was not held steady at I_s, but was modulated between I_s and V_{oc}. Whether this modulation can affect the extent of damage is now under investigation.

Figure 5 shows EL images and I-V curves from before step 1, after step 1 and after step 13 of the cleaning shadow test on module D. The first event created many new shunts (Figure 5b), resulting in a 4% permanent loss of P_{mp} and subsequent tests mainly strengthened existing shunts (Figure 5c), leading to a 15% permanent loss of P_{mp}. Figure 5e shows that the first cleaning event caused the most damage, however the subsequent > 30 cleaning events combined to approximately triple the damage of this first event.

4 CONCLUSION

Commercial monolithically integrated thin-film PV modules can be severely and permanently damaged by the same type of shadows that may be incident on them during service due to operations and maintenance activities. The mechanism causing this damage is only active when current is flowing through the module. These shadows include those cast by human bodies, which during our testing never contravened manufacturer recommendations, and those cast by cleaning shadows, which in some cases did contravene manufacturer recommendations. The first shadow event caused the most damage to unexposed modules, with subsequent events causing less additional damage in most cases. In some cases, exposing the module to uniform light for pre-measurement stabilization caused the damage already done by shadows to worsen.

ACKNOWLEDGMENT

This work was supported by the U.S. Department of Energy under Contract No. DE-AC36-08GO28308 with Alliance for Sustainable Energy, LLC, the Manager and Operator of the National Renewable Energy Laboratory. Funding provided by U.S. Department of Energy Office of Energy Efficiency and Renewable Energy Solar Energy Technologies Office.

The U.S. Government retains and the publisher, by ac-

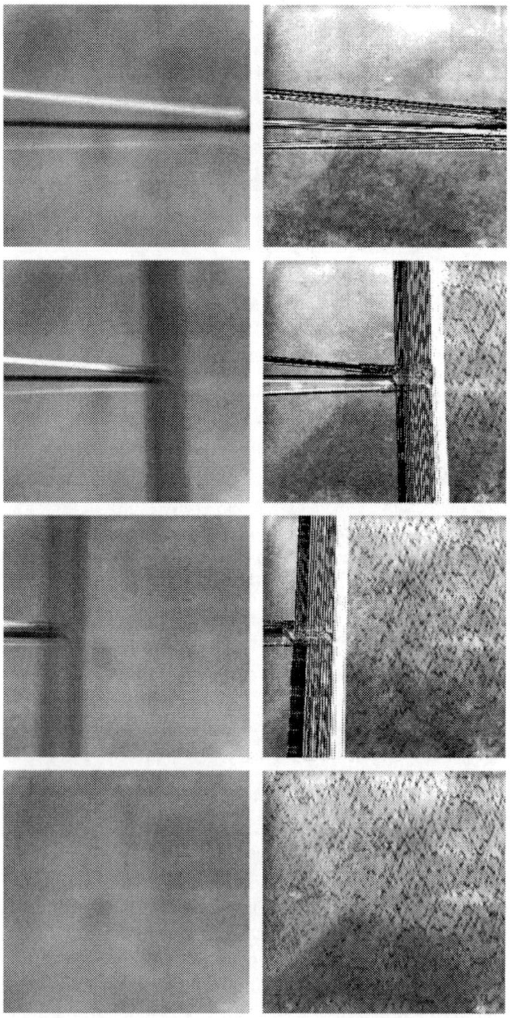

Figure 4: Images from a video (left column) and in situ PL images (right column) taken during the first cleaning step on module C. The motion of the cleaning tool is blurred because each image is an average of several frames.

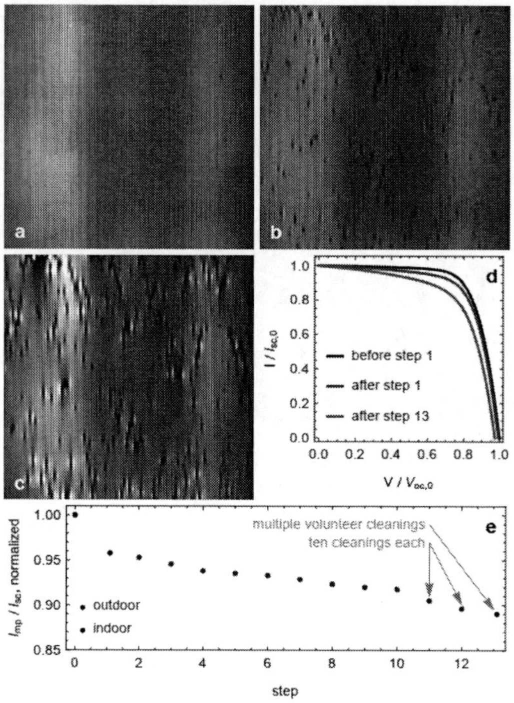

Figure 5: Results from cleaning tool shadow testing on module D are shown: EL images from before the test (a), after step 1 (b), after step 13 (c); stabilized and normalized indoor STC I-V curves from the same three stages (d); and I_{mp}/I_{sc} from each step of the test.

cepting the article for publication, acknowledges that the U.S. Government retains a nonexclusive, paid-up, irrevocable, worldwide license to publish or reproduce the published form of this work, or allow others to do so, for U.S. Government purposes.

REFERENCES

[1] P. O. Westin, U. Zimmermann, L. Stolt, and M. Edoff, "Reverse bias damage in CIGS modules," in *24th European Photovoltaic Solar Energy Conference and Exhibition*, 2009, pp. 2967–2970.

[2] T. J. Silverman, M. G. Deceglie, X. Sun, R. L. Garris, M. A. Alam, C. Deline, and S. Kurtz, "Thermal and electrical effects of partial shade in monolithic thin-film photovoltaic modules," *IEEE Journal of Photovoltaics*, vol. 5, no. 6, pp. 1742–1747, 2015.

[3] J. E. Lee, S. Bae, W. Oh, H. Park, S. M. Kim, D. Lee, J. Nam, C. B. Mo, D. Kim, J. Yang, Y. Kang, H.-s. Lee, and D. Kim, "Investigation of damage caused by partial shading of $CuIn_xGa_{(1-x)}Se_2$ photovoltaic modules with bypass diodes," *Progress in Photovoltaics: Research and Applications*, 2016.

[4] M. G. Deceglie, T. J. Silverman, B. Marion, and S. R. Kurtz, "Robust measurement of thin-film photovoltaic modules exhibiting light-induced transients," in *SPIE*

Optics+ Photonics for Sustainable Energy. International Society for Optics and Photonics, 2015, pp. 956 108–956 108.

[5] T. J. Silverman, M. G. Deceglie, K. VanSant, S. Johnston, and I. Repins, “Illumianted outdoor luminescence imaging of photovoltaic modules,” in *Proceedings of the 44th IEEE Photovoltaic Specialists Conference (PVSC)*, 2017.

SINGLE SIDE VERSUS DOUBLE SIDE ILLUMINATION METHOD IV MEASUREMENTS FOR SEVERAL TYPES OF BIFACIAL PV MODULES

S. Roest, W. Nawara, B.B. Van Aken and E. Garcia Goma
Eternal Sun Group, Den Haag, Zuid-Holland, Wolga 11 2491 BK, The Netherlands
ECN Solar Energy, Noord-Holland, Westerduinweg 3 1755 LE, The Netherlands
PVMD TU Delft, Delft, Zuid-Holland, Mekelweg 4 2624 CD, The Netherlands

ABSTRACT: As the deployment of bifacial PV modules power plants becomes a reality, it is more and more evident that an accurate power rating of such devices is necessary to give right performance predictions for end users. For this reason, the international community is already developing a standard to rate performance of bifacial PV modules, which is in its draft stage. This standard considers two testing methods for such devices, the double side illumination and the single side equivalent illumination method. The present work will experimentally perform these two methods on several types of bifacial PV modules as well as on bifacial PV modules of the same production bin. After analyzing the results, it is concluded that for design qualification in R&D laboratories, a double side measurement is indeed necessary on top of the single side characterization, as the single side method can yield maximum power differences up to +1.8% and down to -1.2% compared to double side measurements. However, in modules of the same production line these differences reduce to +0.4% and -0.5%, and are expectable from observing the rear side behavior separately.
Keywords: Bifacial, standard, characterization, single side, double side.

1 INTRODUCTION

As the cost of PV modules decreases dramatically, more attention is faced towards the reduction of balance of system costs. One approach to achieve such reduction is to obtain more electrical power and annual energy per unit of area. Bifacial PV modules, which can efficiently convert light received from the front side as well as from the rear side, represent an excellent alternative to fulfil such approach. Furthermore, the manufacturing costs of producing such modules has reduced significantly through the years. As an example, the recent deployment of large bifacial PV power plants provides empirical proof of the cost competitiveness of the technology against traditional monofacial PV devices. However, the commercial power rating of bifacial PV is still based on its front performance. In order to greater boost the implementation of bifacial PV and by extension of solar energy in general, it is crucial to establish a performance rating standard that considers the rear side contribution of the devices.

Ideally, the indoor performance rating of bifacial PV modules would be carried out reproducing the exact same operating conditions found outdoor and that affect the performance of the module. This includes of course the ability to controllably illuminate the module from both front and rear side. This represents, however, a technical and logistical challenge for many due to the associated cost of using two controlled light sources instead of one. Therefore, such standard also aims to consider an alternative single side equivalent illumination method which can effectively reproduce the electrical effects of any double side simultaneous illumination situation. Furthermore, such method should consider means to effectively reduce any light falling into the rear side to prevent any error. This can be achieved for instance by covering the non-illuminated side with a black antireflective cover.

Figure 1 shows the schematic of both aforementioned methods, where G_F, G_R and G_E represent front, rear and equivalent irradiance, being the latter employed in the single side method and the first ones applied in the double side method. The full description of these methods as well as the correlations necessary to compare them between themselves are being carried out by the international community and an initial draft of such standard has already been submitted [1].

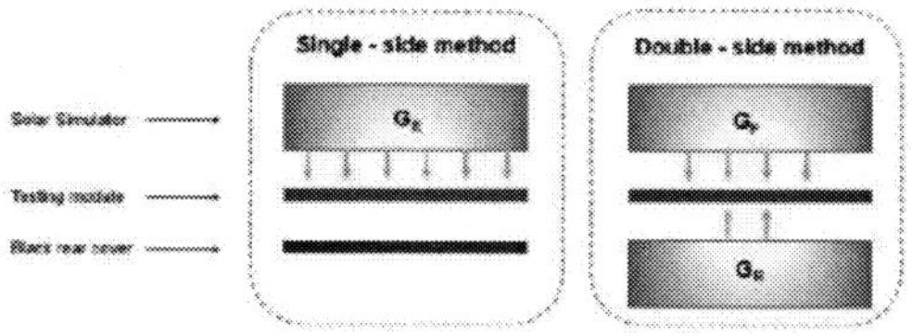

Figure 1: Schematic of single and double side illumination methods

Previous work performed by the present authors [2] describes in detail the development of an indoor setup that combines two solar simulators and which is suited to perform double side illumination performance measurements. The same work presents an initial assessment of the comparison of single and double illumination methods performance output for a single type of bifacial PV module.

As a natural continuation of that work, the research presented in this manuscript aims to experimentally compare the performance rating obtained employing both methods for several types of bifacial PV modules in order to assess in which sorting situations the single side method alone is representative of bifacial performance between modules and in which situations an additional double side method characterization is recommended.

To do so, different sequential steps are followed. First, the general IV curves of every module under front, rear and combined irradiance are analyzed in order to understand the factors that determine the electrical output in these three situations. Secondly, the bifacialities of the modules under different irradiances are determined and discussed. Finally, every module is measured under both single and double side illumination conditions, and the output of both methods correlated as established in the standard draft.

2 EXPERIMENTAL METHOD

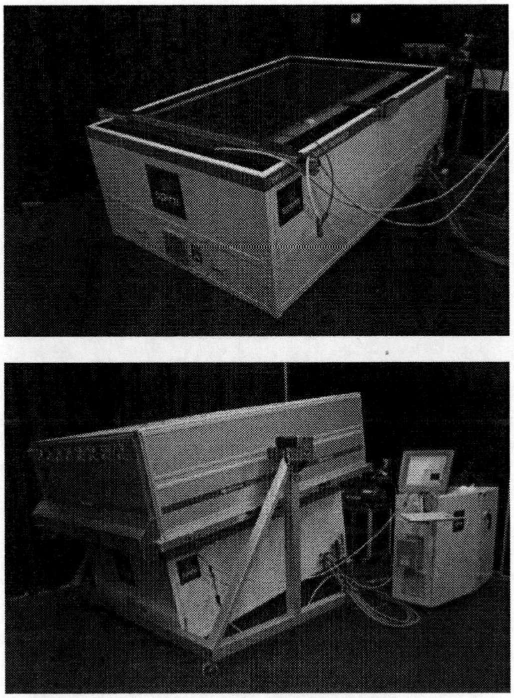

Figure 2: Single side and double side setup

2.1 Description of the testing setup and procedures

The experimental setup employed in this paper is shown in Figure 2 and it consists of a Single Long Pulse (SLP) flash solar simulator for single side illumination IV measurements and the addition of a steady state simulator on top of it in order to provide rear irradiance for the double side illumination measurements. The light uniformity provided by the flash simulator is Class A and the light uniformity of the rear steady state simulator at low irradiances, such as 100 or 200 W/m^2 is Class B, as rated according to the standard IEC 60904-9. Additionally, no spectral correction per different spectral response of front and rear side is applied. More information of the setup can be found in the previous work of the authors [2].

For the initial IV curve characterization, every module is characterized under 1000 W/m^2 on the front and rear side, separately, and afterwards characterized with simultaneous front irradiance at 1000 W/m^2 and rear irradiance at 200 W/m^2. Secondly, the bifacialities of the modules along different irradiances are analyzed.

Bifaciality is described as the ratio between performances of a module when it is illuminated under the same irradiance on the rear and on the front side separately. Every electrical parameter has its own bifaciality, however the ones relevant for the standard draft [1] are short circuit bifaciality $\varphi_{ISC,STC}$ and maximum power bifaciality $\Phi_{PMAX,STC}$, which are described in Equation (1) and (2). The standard draft, however, considers the usage of bifaciality under Standard Test Conditions (STC), which are 1000 W/m^2 and 25 C°, to be employed to correlate both methods. This however might have an effect when comparing both methods at different irradiances. Therefore an assessment of these bifacialities per every module under different irradiances, from 200 to 800 W/m^2 is carried out.

Figure 3: From top left to bottom, details of HJT framed module, PERT not junction box shaded and PERT white junction box shaded module

$$\varphi_{ISC} = \frac{I_{SC,rear}}{I_{SC,front}} \qquad (1)$$

$$\varphi_{PMAX} = \frac{P_{MAX,rear}}{P_{MAX,front}} \qquad (2)$$

Finally, the performance of every module is measured employing single and double side measurements, and the maximum power ratings compared. For the single side measurements, irradiances form 200 to 1000 W/m^2 changing in steps of 100 W/m^2 are employed. For the double side measurements, the front side is kept at G_F 800 W/m^2 and the rear side G_R set at three different low irradiance levels, ranging between 50 to 300 W/m^2. In order to correlate both methods output, the guidelines suggested in the standard draft [1] are followed, where the double side irradiance conditions are translated to single side equivalent irradiance G_E via Equation (3).

$$G_E = G_F + G_R * Min(\varphi_{ISC,STC}, \varphi_{PMAX,STC}) \qquad (3)$$

2.2 Commercial bifacial PV modules

Due to the broader reception that bifacial PV modules receive, more manufacturers are launching their own modules with their own specific design. Material wise, most of the commercial bifacial PV modules are based in PERC, PERT and heterojunction technology (HJT) solar cells. Besides this aspect, modules can have a frame that protrudes from one of the sides or be frameless. Finally, the junction box location is determinant as it can partially shade some of the cells. This junction box can be either black or white. During this research the modules employed are a 60 small area cell HJT framed module without junction box shading, a 40 cell PERT frameless module also without junction box shading and a 60 cell PERT frameless bifacial module with a white junction box partially shading some cells. Figure 3 shows some detail of the aforementioned modules. All the modules have three bypass diodes which divide the modules in three strings each, being all equal except for the 40 cell module, which has one string of 8 cells and two of 16 cells.

Regarding modules with the same design and even from the same bin of a production line, these can still present some alterations, as for instance the current mismatch between cells when irradiated from the rear side. This happens because manufacturing lines traditional sort cells based on their front performance, and thus the rear performance of the cells is not taken into account. This can be observed in Figure 3 where the cells of a commercial module show different shades of blue when observed form the rear side. In order to understand this effect, two modules of the same bin are analyzed.

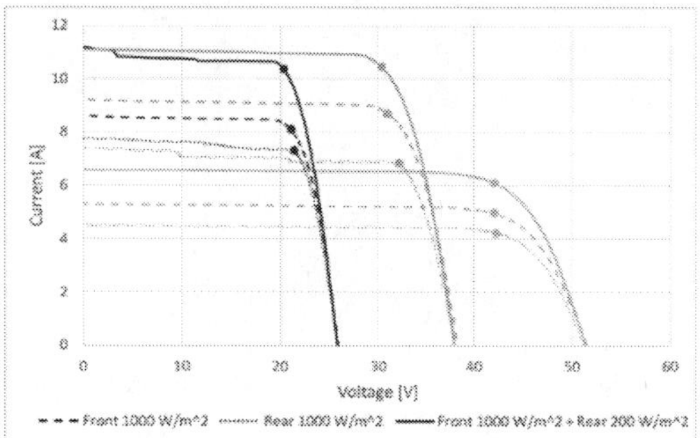

Figure 4: IV curves for different bifacial modules for front, rear and combined illumination HJT framed module (yellow), PERT not junction box shaded (purple) and PERT white junction box shaded module (green). Dots represent maximum power points

3 RESULTS AND DISCUSSION

3.1 IV curve dependence on the irradiated side

Figure 4 shows the IV characteristics for the three different types of modules tested in this research. It can be observed that the three modules present a very smooth IV curve when irradiated from the front side, which is related to a low mismatch current between cells, proving that the three modules have undergone an adequate front side cell sorting procedure.

The HJT module not junction box shaded shows at the same time a very uniform rear performance, consequence of the natural bifacial behavior of heterojunction cells, which cross section is very symmetric. As a consequence, when the module is irradiated by both sides simultaneously, its IV curve still shows a very smooth behavior.

When observing the PERT frameless module without junction box, however, the rear performance shows plateaus, which are caused by non-uniform current generated between the strings. This situation limits the maximum power and therefore is relevant for applications such vertical bifacial modules. When under double side illumination, this effect is also visible, although in terms of maximum power is less intense as the maximum power point is found before the plateau begins.

For the PERT frameless module with white junction box partially shading some of the cells, the rear side performance is limited by the shading of the junction boxes and rear cell mismatch. However the double side illumination curve shows a more uniform shape than the addition of front and rear curves, as it could be expected based on the output from the previous module. This is due to an added front current generated by the cells that are on top of the white junction boxes, due to reflection. This effect occurs during the single front side illumination but it is not visible in the curve since the vast majority of the cells operate at low current.

From observing this curves it is clear that depending on the design of the module the double side behavior will keep a different relation respect their front and rear separate performance.

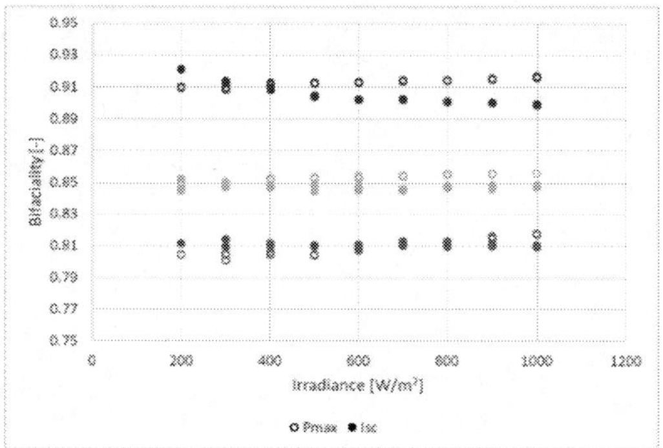

Figure 5: Bifacialities for different PV modules at 25 C° under different irradiances: HJT framed module (yellow), PERT not junction box shaded (purple) and PERT white junction box shaded module (green)

3.2 Bifacialities under different irradiance levels

The same three modules are analyzed regarding bifacialities at different irradiance levels, as shown in Figure 5. Under STC conditions, for all the modules the I_{SC} bifaciality has a lower value than the P_{MAX} bifaciality, therefore being the first one the value chosen for calculating the equivalent irradiance in a single side performance measurement as indicated by Equation (3). It is to be noted however, that if these two values are not equal, it is expected to generate differences in the electrical output of the single side versus the double side methods, since the equivalent irradiance G_E will be such to generate the same I_{SC} as in the double side method, but not the same P_{MAX}. Additionally, this offset between bifacialities at STC is greater for different modules, peaking in this research for the PERT Frameless module without unction box shading, which has a worse rear cell mismatch, as seen earlier.

Regarding the change in bifacialities among irradiances, it remains fairly constant within one per cent respect to the values at STC, except for the PERT frameless module without junction box shading. For this module the I_{SC} bifaciality increases significantly at low irradiance.

Therefore, bifacialities over irradiance behave differently per module depending on its design and also on the sorting of rear cells.

3.3 Maximum power rating per method

Figure 6, 7 and 8 show the maximum power output for the three modules using double side illumination method and single side illumination method. For the double side, G_F is kept as 800 W/m^2 while G_R is kept set different low level irradiances. Taking into account the short circuit bifaciality determined in the previous section, the corresponding G_E to which this double side irradiance condition correspond is calculated. For the single side measurements, G_E values of 800, 900 and 1000 W/ m^2 are employed.

For the HJT framed module without junction box shading, the single side measurement generates 1.2% lower Pmax at an equivalent irradiance of 1000 W/m^2. This is consequence of using I_{SC} bifaciality while applying Equation (3), and not employing P_{MAX} bifaciality is higher. In such case, the single side method G_E will underestimate the true bifacial maximum power.

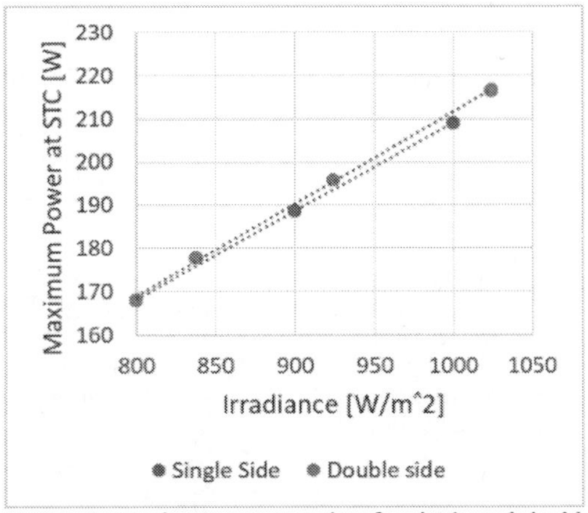

Figure 6: Maximum power rating for single and double side method for HJT framed module at 25 C° under different irradiances

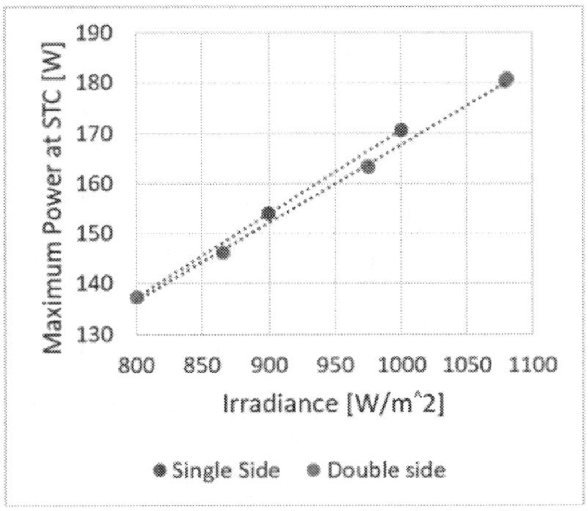

Figure 7: Maximum power rating for single and double side method for PERT not junction box shaded module at 25 C° under different irradiances

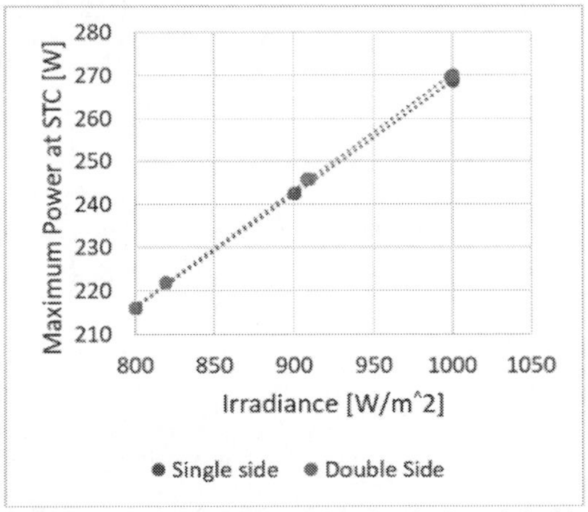

Figure 8: Maximum power rating for single and double side method for PERT white junction box shaded module at 25 C° under different irradiances

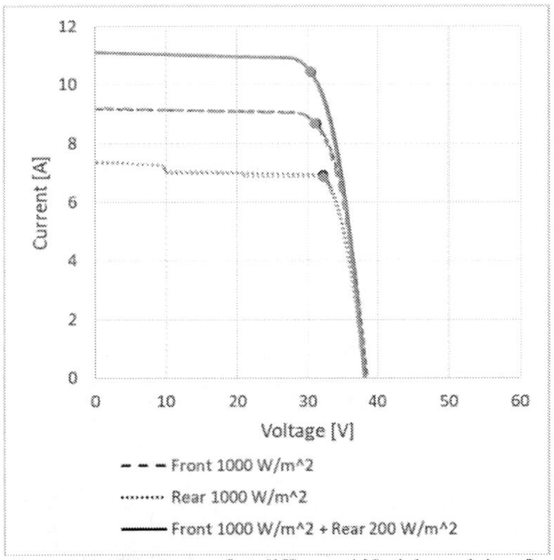

Figure 9: IV curves for different bifacial modules for different serials

For the PERT Frameless module without junction box shading, however, the same difference does not appear, although the bifaciality of I_{SC} is also used. For this module the single side method over estimates the double side power by 1.8%. This might be consequence of the mismatch current of rear cells, which affects the shape of the IV curve when under double side illumination, as shown in Figure 2, therefore altering the relation between methods.

Finally, the PERT frameless module with white junction box shows an underestimation of power of 0.5% when applying the single side method, which is more in line with the expected situation of single side method under estimating the maximum power. This can relate to the fact that for this design the double side IV curve shows a shape more similar to the front side curve than the previous module, as shown in Figure 4.

Overall, in this research, applying single side method measurements to different module designs can either overestimate the real double side illumination power 1.8% or under estimate it up to 1.2%. This represent an absolute range of 3.0 % between methods.

3.4 Analysis of modules of the same bin

The previous points are also analyzed for two modules with the same design, type and production bin, in order to understand if the differences found hold also for such situation.

Figure 9 shows the IV curve of both modules for front, rear and combined irradiance performance. It can be seen that while the front side behaves similarly, consequence of efficient cell sorting, the rear side shape varies, accentuating one plateau more for one of the modules. This difference however is not clearly visible in double illumination performance as it represents a small change since the rear contributes with 200 W/m^2.

When observing the single versus double side performance of the second module, shown in Figure 10, the single side method over estimates by 0.4% the maximum power.

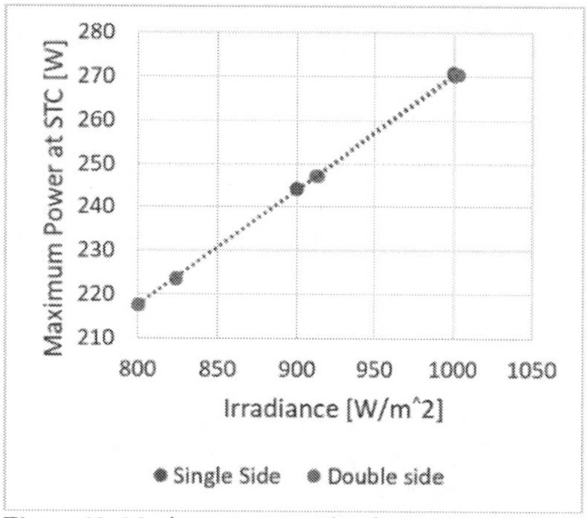

Figure 10: Maximum power rating for single and double side method for another serial of PERT white junction box shaded module at 25 C° under different irradiances

4 CONCLUSIONS

This research has compared the performance of different types of bifacial modules applying both the methods contemplated by the current IEC standard draft in terms of power rating of bifacial PV modules, which are double side illumination and single side illumination. It has been observed that maximum power using single side method can deviate up to +1.2% or -1.8% depending on the module design, which is associated to how the double illumination IV curves behave. This double side illumination behavior is not fully predictable by observing front and rear side performances separately. Therefore for design qualification, in R&D laboratories, it is advisable to perform double side measurements on top of single side measurements in order to understand what kind of offset might be between methods and in which direction.

Concerning manufacturing purposes, for modules of the same bin, however, the offset between methods in maximum power varies less, from +0.4% to -0.5%, and it is associated to a change in rear side IV curve behavior. Therefore for production it is advisable to do sampling every certain amount of modules of front and rear side performance separately, in order to calculate bifacialities, while the rest of modules in the production line can be tested only from the front side with single side equivalent irradiance.

5 ACKNOWLEDGEMENTS

Special thanks to Dr. Rudi Santbergen from PVMD TU Delft and to Tempress for their support.

6 REFERENCES

[1] V. Fakhfouri, "IEC 60904: Photovoltaic Devices Part 1-2: Measurement of current-voltage characteristics of bifacial photovoltaic (PV) devices," Proposal 82/1044/NP, July 28, 2015.

[2] S. Roest, W. Nawara, B. B. Van Aken, E. Garcia Goma, Proceedings 44th IEEE Photovoltaic Solar Conference, 2017.

Comparison of electrical performance of bifacial silicon PV modules

Juan Lopez-Garcia*, Alberto Casado, Diego Pavanello, Tony Sample
European Commission, DG JRC, Directorate C - Energy, Transport and Climate, Energy Efficiency and Renewables Unit, Via Fermi 2749, 21027 Ispra (VA), Italy
*Tel: +39 0332783720, Fax: +39 0332789268, e-mail: juan.lopez-garcia@ec.europa.eu

ABSTRACT: The interest on the n-type crystalline silicon bifacial PV modules has increased in the last years since they can increase the performance in comparison to traditional PV modules. As the PV modules are generally sold according to the nominal power output and efficiency under standard test conditions STC, the lack of standards for bifacial modules leads each manufacturer to claim different amounts for the "added value" of the bifaciality that makes it difficult for customers to directly compare between bifacial manufacturers. In this work, we have compared the electrical performance of framed and frameless c-Si bifacial modules from 5 different manufacturers utilising rear covers of different reflectivity and different measurement methods such as a single-sweep flash simulator, the so-called multi-flash (MF) method, a large-area steady state sun simulator and under natural sunlight for the accurate measurement of the module parameters. It was observed that depending on the measuring method, some modules presented STC front power out of the tolerance range given by the manufacturer. The design of the module also plays an important role on the rear power contribution.

Keywords: Bifacial; c-Si; PV Modules; n-type; Energy performance.

1 INTRODUCTION

In order to reduce the cost of solar power generation, the photovoltaic (PV) industry is focused on improving the solar cells and module efficiency, lifetime, cost and PV module's energy yield. The interest on the c-Si bifacial PV modules has increased in the last years since they can increase the performance in comparison to traditional PV modules [1-3]. This is due to the fact that both sides of the cells, front and rear, absorb solar radiation, collecting not only incident light on the front surface but also sunlight scattered or reflected from the ground or environment onto the rear surface. To assess their performance and quality, PV modules are measured and characterized in a controlled environment under standard test conditions (STC) as defined by the International Electrotechnical Commission (IEC) [4]. Although the IEC is currently working on the development of a draft technical specification [5] for bifacial photovoltaic devices IEC 60904-1-2, there is currently no standard for bifacial modules. As the PV modules are generally sold according to the nominal power output and efficiency under STC conditions, the lack of bifacial standards leads each manufacturer to claim different amounts for the "added value" of the bifaciality which makes it difficult for customers to directly compare between bifacial manufacturers. There are currently around 10-12 manufacturers selling framed and frameless bifacial silicon modules with different type of cells such as HJT, PERC, n-PERT, BSF, etc. The development of a measurement method to appropriately assign the nameplate rating is considered one of the main challenges for the bifacial PV modules in order to increase the acceptance of the customers. Different groups and laboratories have proposed different methods to characterise bifacial modules under simulated sunlight such as double-sided simulators, single side flash simulators or the use of mirrors [1, 6]. Additionally, the indoor testing of high-efficiency and highly-capacitive bifacial PV modules with a standard flash solar simulator may be affected by strong measurement artefacts and lead, depending on the pulse duration and sweep speed, to under or over-estimations of the measured power. Different indoor methods with long pulse solar simulator or alternative methods are necessary in order to accurately determine the module parameters [7-9].

In this work, the electrical performances of framed and frameless c-Si bifacial modules from 5 different manufacturers have been determined using four different simulated and natural light methods. The object was to compare the different methods and the divergence from the nominal rated power.
The impact of using different rear covers of varying reflectivity was also investigated

2 EXPERIMENTAL DETAILS

2.1.Module characteristics

Different commercially available bifacial n-type crystalline Si 4 glass/glass and 1 glass/foil modules (both framed and frameless) from 5 manufacturers were characterized indoor and under natural sunlight in the outdoor test field at the European Solar Test Installation (ESTI) under STC conditions. The modules consisted of 60 or 72 cells in three sub-strings and they were light soaked to eliminate possible light induced degradation (LID) following the standard IEC 61215-2 [4]. Figure 1 and Table I show the front and rear images and general information of the bifacial modules under test including the maximum power bifaciality φ_{Pmax}. As shown in table I, module D and E are frameless while modules A, B, C are framed modules. It should be noted that the module A and B, have a lip of the frame covering a small part of the rear of the cells. Whereas, module C has wider edge spacing avoiding any shade to the rear cell surface. Modules A, B, and E have junction boxes partially covering the rear side cells. In the case of the module E, the three junction boxes are covering cells in each of the three sub-strings.

2.2. I-V measurement setup.

I-V curve measurements were performed with a class AAA PASAN IIIB flash solar simulator (with sweep times or flash duration in the order of 10 ms) at 1000 W/m^2 and 25.0 ± 0.1°C. In addition to the standard forward sweep single flash method, the so-called multi-flash (MF) method was also used as described in [10]. In this method, 35-40 individual points are taken to construct the final I-V curve. In this case, each point of the I-V curve was obtained by applying a constant voltage to the module which was then subjected to a flash from the solar simulator at the desired irradiance level. The resultant module I-V curve was then obtained by performing an interpolation of the measured points [7, 8].

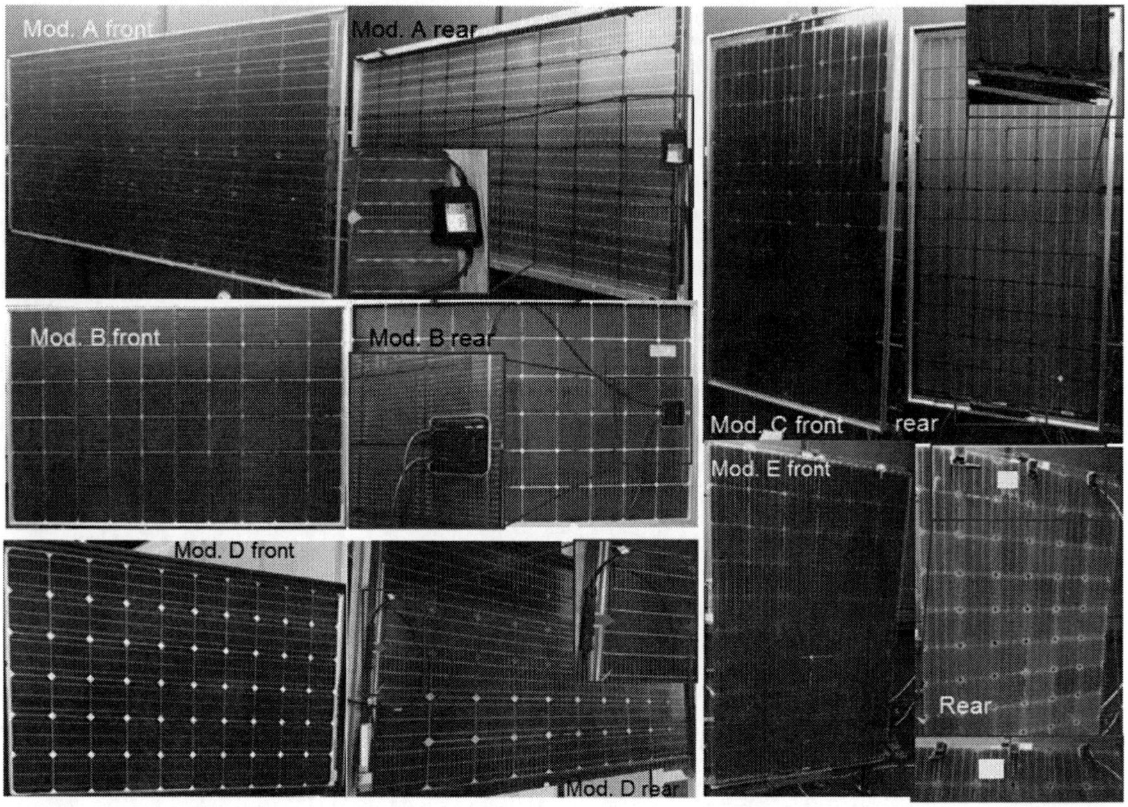

Figure 1 Front side and rear side images of the c-Si bifacial modules characterised. A zoom of the junction box area is included.

Table I List of c-Si bifacial modules tested and corresponding maximum power bifaciality, φ_{Pmax},

Module	Frame	φ_{Pmax} (%)	N° Cells
Module A	Yes	58	60
Module B	Yes	66	60
Module C	Yes	86	72
Module D	No	97	60
Module E	No	76	60

With this method, capacitive effects of the highly capacitive modules are avoided by considering only stabilized current and voltage output values for each measurement. Additionally, a class AAA continuous large-area or steady state solar simulator (Apollo) with single I-V sweep times of around 1 second, at temperature of 25.0 ± 0.1 °C was also used to measure the I-V curves. Finally, I-V curves were taken under natural sunlight in the outdoor ESTI facility at 25.0 ± 0.5 °C. I-V curves of all of the modules were corrected to STC. The front and rear side of the module has been defined by the manufacturer, in practice it means that the rear side is that incorporating the junction box or labels. The electrical characterization of the modules was carried out with a rear cover, with the same dimensions as the module, placed in direct contact to the non-illuminated surface of the module, thus only light that is transmitted through the module can be scattered or reflected back to the module. Three rear covers with reflectance 4%, 54% and 93%, respectively, on average in the visible spectrum 380-750nm, consisting of a black ("non-reflective") painted wooden panel, a stainless steel panel of 0.5mm of thickness and a gloss white reflective panel were used. A Perkin–Elmer LAMBDA 9 spectrophotometer was used to determine the optical reflectance at room temperature of the rear panels. I-V curves have been obtained by illuminating alternatively the front or the rear side while covering the rear with the different covers.

The spectral responsivity of representative modules were measured using a differential spectral responsivity technique with a large-area flash solar simulator equipped with a number of bandpass filters to obtain illumination of modules with quasi-monochromatic light [11].

3 RESULTS AND DISCUSSION

3.1. Different I-V measurement methods.

The electrical parameters of the bifacial silicon modules were measured by means of four different methods. Figure 2 shows the I-V curves at STC conditions of framed and frameless modules from different manufacturers, namely from A to D, measured with a flash simulator using a multi-flash method with a non-reflective black rear cover. Both front and rear illuminations were tested. It should be noted that the design of module, in particular, the junction box location and frame have a strong influence on the I-V curves of the rear side. For example, modules with frames or junction boxes partially covering cells of the rear side lead to kinks in the I-V curves which reduce the rear I_{mpp} and P_{max}. Modules A, B and E show up to three marked kinks in the rear I-V curves due to the partial shading of the cells by the junction box (all three cases) and the frame (modules A and B). It is especially relevant the case of the module B with a big junction box covering almost half of two cells of a sub-string and the lips of the frame covering partially the cells at the edge of the module.

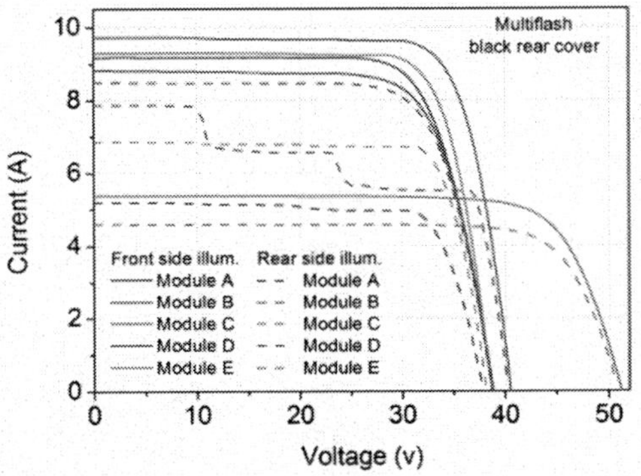

Figure 2 I-V curves of modules commercially available Si bifacial modules measured with a multi-flash method, with a black rear cover.

Conventional flash light simulators are generally used for the measurement of module performance due to the possibility to have very controlled environmental conditions and their ease of use. Moreover, class AAA solar simulators nowadays lead to a spectral mismatch factor close to unity for standard c-Si modules. However, for high-efficiency c-Si modules with increased capacitance, the effects of the highly-capacitive devices can lead to inaccurate measurements of the PV module electrical parameters. In order to test the accuracy of characterisation methods, a direct comparison of the P_{max} change among different measuring methods (flash simulator with a forward sweep (FS), multi-flash method (MF), a steady state simulator (SS) or under natural sunlight (outdoor)) was carried out considering the Multi-flash method as the reference point. The results for front side and rear side illumination are shown in figures 3 and 4, where an opaque black rear cover was used. In general, all the modules showed a lower P_{max} measured with the continuous light simulator, natural sunlight or single flash forward sweep, with respect to the Multi-flash method. Single I-V measurements made, on the nominally front surface of the modules, as a forward sweep (≤10ms) at 1000W/m^2 exhibited lower P_{max} values ranging from -0.5% to -7.4%, in comparison to the multi-flash method (mainly due to a decrease in the FF). Steady state simulator exhibited the lowest differences with respect to the multi-flash method, below -1% with the exception of the module C. A similar result was obtained with the natural sunlight measurements with a decrease in the P_{max} in the range between -0.5% and -1.3% for the front illumination. However, the decrease in the continuous light simulator or natural sunlight is also due to a decrease in the V_{oc}, which in this case is associated, among other things, to the temperature control.

3.2. Comparison with nominal rated power.

The difference between the P_{max} measured indoor and outdoor at the ESTI laboratory and the nominal rated STC front P_{max} given by the datasheet of the manufacturer is plotted in the figure 5, where the respective tolerance ranges given by the manufacturers are shown as error bars. Modules A, B and E claimed positive tolerance ranges of 0/+2%, 0/+3%, 0/+2%, respectively, meanwhile module C and D claimed -5%/+10% and -2/+3%, respectively. The declared ESTI P_{max} uncertainties for pulsed solar simulator, steady state and outdoor system (1.1%, 1.7% and 1.7%, respectively) are well below standard producers uncertainties.

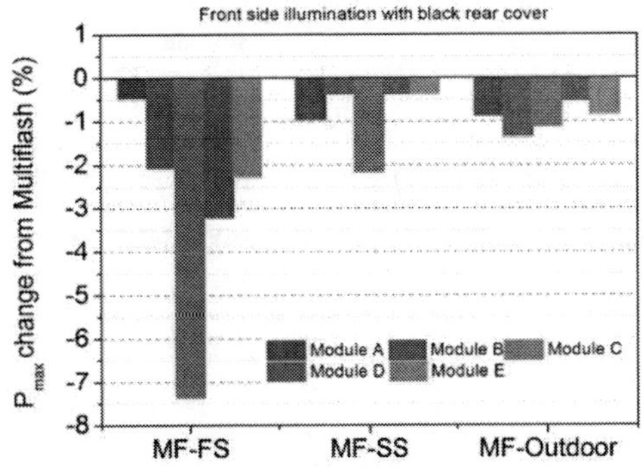

Figure 3 P_{max} change for bifacial modules measured by means of a flash simulator with a forward sweep method, a steady state simulator and under natural sunlight with respect to the multi-flash method for front illumination.

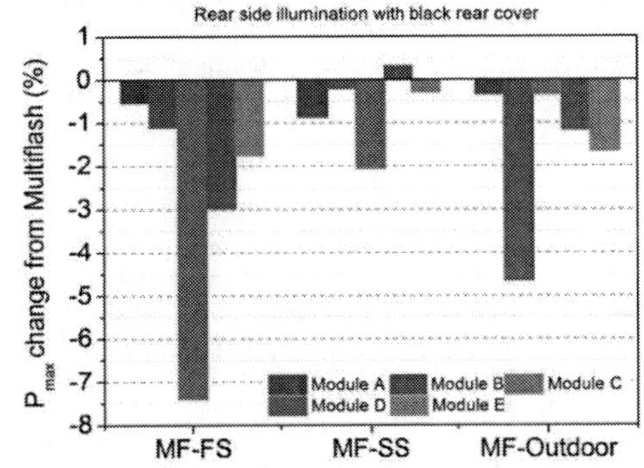

Figure 4 P_{max} change for bifacial modules measured by means of a flash simulator with a forward sweep method, a steady state simulator and under natural sunlight with respect to the multi-flash method for rear illumination.

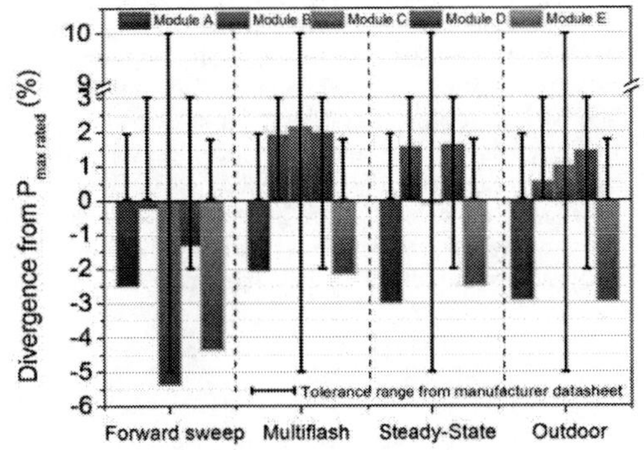

Figure 5 Divergence of the P_{max} between the measurements at the ESTI lab and the nominal rated STC front power. The tolerance range is also indicated with error bars.

As mentioned before, a notable influence of the measuring method was found. Assuming the forward sweep method in a flash simulator as the most common method for the electrical characterisation, the divergence from the P_{max} rated was out of the tolerance range under STC conditions for the front side for modules A, B, C and E, with negative divergences between -0.2% and -5.5%. Modules A and E also exhibited a front P_{max} out of the tolerance range for measurements with Multi-flash, Steady-state and natural sunlight, between -2% and -3%. The correct labelling of the PV modules is important also from the economic point of view for large systems.

Most of the manufacturers claimed in their datasheet an additional power contribution of up to 25-30% considering the power gain from the rear side depending on the mounting conditions and albedo but without any other specification. Only one manufacturer from the devices under test provided in the datasheet, some general specification in the installation guide for maximum system yield and optimum performance ratio.

3.3. Effects of the rear side cover.

Normally, bifacial Si modules are tested indoors with a black rear cover to ensure that no stray light (for example due to reflection from the wall/measurement structure) can enter the module [1]. In this section we present the effects of a white and stainless steel cover at the rear side. Figure 6 shows the optical reflectance of three rear covers consisting on a black ("opaque or non-reflective") painted wooden panel, a stainless steel panel of 0.5mm of thickness and a gloss white reflective panel. The average reflectance value in the visible spectrum 380-750nm is also shown. Assuming the black rear cover as the reference or monofacial configuration, the P_{max} or the I_{sc} change (P_{max} or I_{sc} gain) due to the other more reflective covers (steel and white) can be calculated by means of the equation:

$$P_{max}\text{change (\%)} = \frac{P_{max\ cover} - P_{max\ black\ cover}}{P_{max\ black\ cover}} \times 100 \quad (1)$$

Where $P_{max\ cover}$ is the P_{max} of the steel or the reflective white sheet covers. Figures 7 and 8 show the change in the P_{max} and I_{sc} with respect to the black rear cover or monofacial configuration for a white and a steel rear cover for the front side and rear side illumination, respectively, with a flash simulator using a multi-flash method. Even though the rear irradiance contribution due to reflected or diffuse light is very low (the rear panel is placed in direct contact to the module) a gain in P_{max} from 1% to 2.7% for the front side illumination and a gain between 2% and 3.8% for the rear side illumination were obtained for a reflective white cover placed behind the module as observed in figure 7 and 8. The extent of the reflected light by the cover depends also on the spacing of cells and backsheet properties (glass, foil, textured, etc). Preliminary measurement of the rear irradiance for the reflective white cover at a distance of 5cm from the rear surface (not published yet) indicates values below 28 W/m^2 on average that this is roughly the 2.8% of the front irradiance of 1000 W/m^2. This is in agreement with the P_{max} gain due to the white rear cover.

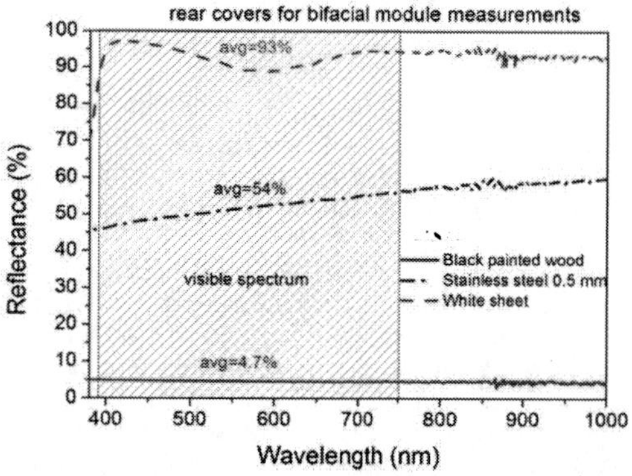

Figure 6 Reflectance as a function of the wavelength of different materials used as rear cover.

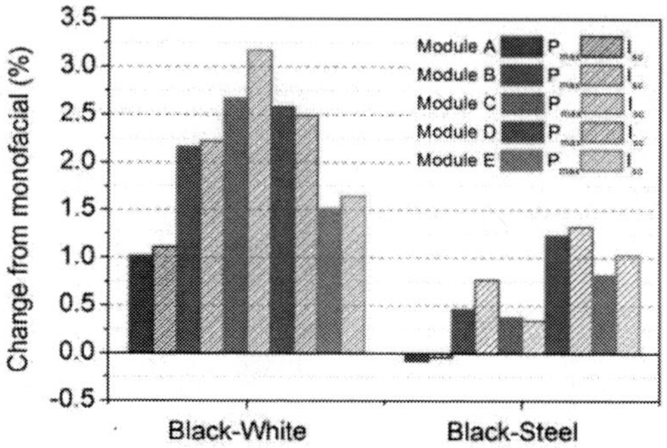

Figure 7 Change in the P_{max} and I_{sc} of bifacial modules for covers with different reflectance with respect to the black rear cover illuminating the front side.

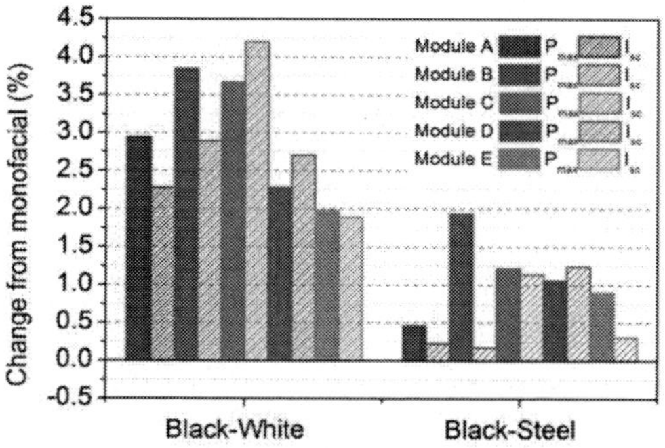

Figure 8 Change in the P_{max} and I_{sc} of bifacial modules for covers with different reflectance with respect to the black rear cover illuminating the rear side.

The P_{max} gain for the stainless steel cover was found between 0.5 and 1.2% for the front side and between 0.5 and 1.9% for the rear side.

The increase in P_{max} is mainly due to a similar increase in the I_{sc} and I_{mpp}. However, as can be observed in figure 8, the increase in the P_{max} for modules A, B and E in the rear side (framed modules with partial "self-shading" by the frame and junction box) is not directly related to the increase in the I_{sc}, due to partial shading of the frame leading to kinks in the rear side I-V curves that reduces the I_{mpp} and P_{max} while maintaining the I_{sc}, as will be discussed in the next section.

The obtained gain in I_{sc} is in good agreement with reported values for similar experiments ranging from +1.3 to +3% with white sheets at 0cm from the rear module surface [3, 12-15].

It could be expected that the gain in P_{max} was proportional to the reflectance of the rear cover. However, some modules are not producing the expected power due to losses of diffuse light or because they are not colleting all the light reflected by the cover. For the front side, the P_{max} gain with the steel cover is approximately 45-52% the gain with the white cover for modules D and E. However, Modules A, B and C exhibit a relatively low gain with the steel with respect to the black cover (below 80% of the gain with the white cover). Whereas the rear side exhibits a more expected behaviour and the modules B, C, D and E shows a gain with the stainless steel cover of 51%, 33%, 46% and 45% below the gain with the reflective white cover.

3.4. Effects of partial "self-shading".

As describe in the IEC standard draft for bifacial modules IEC 60904-1-2, Bifaciality refers to the ratios between the main I-V characteristics of the rear side and the front side of a bifacial device, typically at Standard Test Conditions (STC). It is quantified with reference to bifaciality coefficients, namely the short circuit current bifaciality coefficient φ_{Isc}, the open-circuit voltage bifaciality coefficient φ_{Voc} and the maximum power bifaciality coefficient φ_{Pmax}. This last one can be defined as:

$$\varphi_{Pmax}(\%) = \frac{P_{\max \text{Rear}}}{P_{\max \text{Front}}} \times 100 \qquad (2)$$

where $P_{max\ Rear}$ and $P_{max\ Front}$ are the module output power measured when illuminating only the rear and the front side at STC, respectively, with a black rear cover. Table I shows the maximum power bifaciality φ_{Pmax} of the modules under test and shows a wide range of values ranging from 58% to 97%. A consequence of the partial "self-shading" in the module B (around 4% of the total active area), for example, due to the lips of the frame, the junction box and the label, is that the bifaciality for P_{max} is 66% while the bifaciality for I_{sc} is 80% and the I-V curve of the rear exhibits the kinks (one for each sub-string covered) shown in figure 2.

Manufacturers of standard Silicon monofacial PV modules used their production lines, module design and new bifacial cell technologies to produce bifacial PV modules. However, the design of the framed modules needs to be considered as a priority to take the maximum advantage of the reflected light on the rear side and profit of the bifacial technology. For example, in the case of module B, a question arises: What would be the real contribution of the rear side with an appropriate frame, junction box and nameplate design and location to avoid partial self-shading? Assuming that the bifaciality is maintained constant with the irradiance and with the area covered, the percentage and P_{max} contribution of the rear side without covering the cells can be calculated as follow:

- The front side of the module B was covered with tape in the same area and position that the rear side (as shown in figure 9) and the I-V curve was measured.
- The maximum power bifaciality φ_{Pmax} of the covered side can be calculated as:

$$\varphi_{Pmax\ covered}(\%) = \frac{P_{\max \text{Rear}}}{P_{\max \text{Front covered}}} \times 100 \ (3)$$

Where $P_{max\ rear}$ is the normal P_{max} of the already covered rear side.

As the bifaciality is an intrinsic property of the solar cells, the P_{max} of the rear side without cover or partial self-shading, $P_{max\ rear\ uncover,}$ can be obtained from the equation:

$$P_{\max rear\ uncovered} = \varphi_{Pmax} covered(\%) \times P_{\max front} \quad (4)$$

Figure 10 shows the I-V curves of the front side, rear side and front side partially covered with tape in the same place and area (~4% of the active area) for several irradiances obtained with a flash simulator. The front side covered I-V curve exhibits the same kinks at the same voltages that the rear side I-V curve without variation on the V_{oc}. In the Table II are listed the values of the maximum power bifaciality and P_{max} values for the module B at 1000W/m^2, 400 W/m^2 and 200 W/m^2, where this last one is a characteristic rear irradiance in outdoor conditions. The maximum power bifaciality of the uncovered sides was found to be 86%, similar to the short circuit current bifaciality. An increase in the P_{max} rear without self-shading of the frame, junction box and nameplate around 30% was obtained with additional absolute values for the rear at 1000W/m^2 and 200W/m^2 of 58 W and 13 W, respectively.

Figure 9 Front side of the module B covered in the same place and with the same area than the rear side.

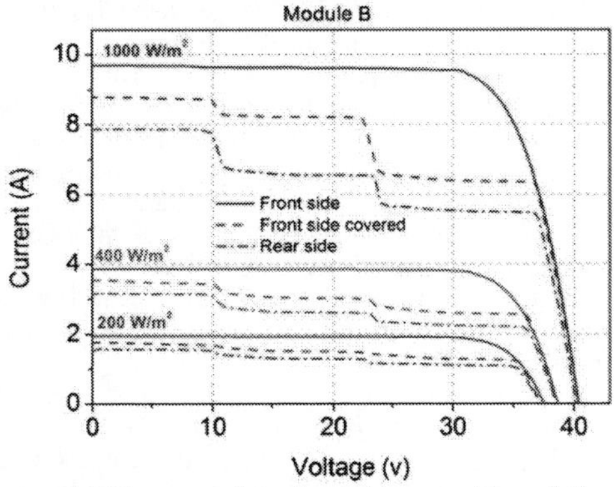

Figure 10 I-V curves of the front side, rear side and front side partially covered of the module B with a flash simulator at several irradiances.

4 CONCLUSIONS

Crystalline silicon Bifacial PV modules from 5 different manufacturers were characterised indoor and outdoor by mean of a flash simulator with a single flash forward sweep and multi-flash method, a steady state simulator and under natural sunlight at the ESTI facilities.

The capacitive effects in highly-efficiency bifacial modules need to be taken into account in order to use the appropriate characterisation system and method to determine the electrical performance. In comparison to the multi-flash method (used as the reference characterisation method) a single flash forward sweep method yielded lower or much lower values of P_{max} from -0.5% to -7.4% (mainly due to a decrease in the FF).

A steady state simulator and natural sunlight exhibited similar behaviour with respect to the Multi-flash method. The steady state simulator exhibited differences of below -1% (with the exception of the module C), with respect to the multi-flash method. The natural sunlight also showed slightly lower values of P_{max} in the range between -0.5% and -1.3%, with respect to the multi-flash method. The observed lower P_{max} values are due, in part, to a decrease in the V_{oc}, which in this case is associated, among other things, to the temperature control of the module.

Table II Irradiances, P_{max} values for the front, rear and front side covered and the calculated P_{max} for the rear side uncovered of the module B at several irradiances.

Irradiance (W/m^2)	$P_{max\ front}$ (W)	$P_{max\ rear}$ (W)	φ_{Pmax} (%)	$P_{max\ front\ covered}$ (W)	Calculated $\varphi_{Pmax\ uncovered}$ (%)	$P_{max\ rear\ uncovered}$ (W)	Increase $P_{max\ rear}$ (%)
1000	298.1	**198.4**	66	230.6	86	**256.5**	29.3
400	118.4	**78.4**	66	91	86	**101.9**	30.1
200	58.5	**37.8**	65	43.8	86	**50.4**	33.5

The design of module, in particular, the junction box location, nameplate and frame had a strong influence on the I-V curves of the rear side with partial "self-shading" in some modules (A, B and E) leading to kinks in the I-V curves which reduce the rear I_{mpp}, and P_{max}.

Depending on the measuring method, some modules presented STC front power out of the tolerance range given by the manufacturer in the nameplate or datasheet with negative divergences between -0.2% and -5.5%. This highlights the need for all modules independently of the bifaciality to quote the STC_{front} and STC_{rear} values for the module together with the measurement setup, ie type of rear cover (if used), type of simulator. The correct labelling of the PV modules is also relevant from the economic point of view.

The effects of placing a white reflective and a stainless steel cover in direct contact to the rear side of the modules were also studied. It was observed in agreement with the draft of the IEC 60904-1-2, that the parasitic reflections from the rear side of the device under test can increase the P_{max} and then the uncertainty in measurements. Therefore, according to the standard, the use of an opaque black cover to limit the reflections to irradiance values below 3W/m^2 is recommended in order to obtain reproducible and reliable performance parameters.

The bifacial design of the module plays an important role in order to make the most of the bifacial technology and special care needs to be taken by the manufacturers. In this study module B showed the impact of partial "self-shading" of the lips of the frame, the nameplate and the junction box (4% total active area) can lead to a "loss" of ~30% in the P_{max} of the rear contribution.

5 REFERENCES

[1]J.P. Singh, A.G. Aberle, T.M. Walsh, Electrical characterization method for bifacial photovoltaic modules, Sol. Energy Mat. Sol. Cells, 127 (2014) 136-142.

[2] J. Appelbaum, Bifacial photovoltaic panels field, Renew. Energy, 85 (2016) 338-343.

[3] J.P. Singh, S. Guo, I.M. Peters, A.G. Aberle, T.M. Walsh, Comparison of Glass/Glass and Glass/Backsheet PV Modules Using Bifacial Silicon Solar Cells, J. Photovoltaics, 5 (2015) 783-791.

[4] International Electrotechnical Commission, IEC 61215-2 ed 1.0 Crystalline silicon terrestrial photovoltaic (PV) modules: design qualification and type approval, 2016.

[5] International Electrotechnical Commission, IEC 60904-1-2 TS: Measurement of current-voltage characteristics of bifacial photovoltaic (PV) devices, (it is being discussed by the TC82).

[6] K. Emery *et* al., Rating photovoltaics, 39th IEEE Photovoltaic Specialists Conference, Tampa, FL, 2013, pp.1-6.

[7] N. Ferretti, Y. Pelet, J. Berghold, V. Fakhfouri, P. Grunow, Performance Testing of High-Efficient PV Modules Using Single 10 ms Flash Pulses, 28th European Photovoltaic Solar Energy Conference and Exhibition, Paris, France, 2013, pp. 3184 - 3187.

[8] A. Virtuani, G. Rigamonti, P. Beljean, G. Friesen, M. Pravettoni, D. Chianese, A fast and accurate method for the performance testing of high-efficiency C-Si photovoltaic modules using A 10 Ms single-pulse solar simulator, 38th IEEE Photovoltaic Specialists Conference, Austin, TX, 2012, pp. 496-500.

[9] H.A. Ossenbrink, W. Zaaiman, J. Bishop, Do multi-flash solar simulators measure the wrong fill factor?, 33rd IEEE Photovoltaic Specialists Conference, 1993, pp. 1194-1196.

[10] J. Lopez-Garcia, B. Haile, D. Pavanello, A. Pozza, T. Sample, Characterisation of n-Type Bifacial Silicon PV Modules, 32nd European Photovoltaic Solar Energy Conference and Exhibition, Munich, Germany, 2016, pp. 1724 - 1729.

[11] R. Van Steenwinkel, Measurements of spectral responsivities of cells and modules, 7th European Photovoltaic Solar Energy Conference and Exhibition, Sevilla, Spain, 1986, pp. 325-329.

[12] B.B. Van Aken, M.J. Jansen, N.J.J. Dekker, Reliability and energy output of bifacial modules, 39th IEEE Photovoltaic Specialists Conference., Tampa, FL, 2013, pp. 1610-1614.

[13] C. Erben, J.B. Heng, X. Zhigang, J. Fatehi, B. Yang, High efficiency bifacial modules in outdoor testing, 42nd IEEE Photovoltaic Specialist Conference, 2015, pp. 1-3.

[14] L. Kreinin, N. Bordin, A. Karsenty, A. Drori, D. Grobgeld, N. Eisenberg, PV module power gain due to bifacial design. Preliminary experimental and simulation data, 35th IEEE Photovoltaic Specialists Conference, Honolulu, USA, 2010, pp. 2171-2175.

[15] H. Mehlich, A. Waltinger, J. Zhao, Y. Yao, B. Bonnet-Eymard, Pilot production of Bifacial Heterojunction cells and modules, 2nd Hercules workshop, Berlin, 2016.

ACKNOWLEDGEMENTS

Authors would like to thank Dr. Cecilia Guillen for the reflectance measurements. The project leading to this application (PV-ENERATE) has received funding from the EMPIR programme co-financed by the Participating States and from the European Union's Horizon 2020 research and innovation programme.

ENERGY RATING OF C-SI AND MC-SI COMMERCIAL PV-MODULES IN ACCORDANCE WITH IEC 61853-1,-2,-3 AND IMPACT ON THE ANNUAL YIELD

C. Monokroussos[1], X. Y. Zhang[1], M. Schweiger[2], D. Etienne[1], S. Liu[1], A. Zhou[1], V. Feng[1], Y. Zhang[1] and C. Zou[1]
1 TÜV Rheinland (Shanghai) Co., Ltd., No.177, Lane 777, West Guangzhong, 200072, Shanghai, P.R. China
2 TÜV Rheinland Energie und Umwelt GmbH, Am Grauen Stein, 51105, Cologne, Germany

ABSTRACT: With the market expansion and increase of global installation capacity, a requirement for energy rating of the photovoltaic (PV) modules has been raised in the PV community. The IEC 61853 standards series aims to provide a complete methodology for performance testing and energy rating of PV modules. This paper analyzed measurement results of 8 monocrystalline (c-Si) and 6 multicrystalline (mc-Si) PV-module types in terms of the light-induced degradation, irradiance dependent behavior, thermal behavior and angular response according to IEC 61215-1,-1-1,-2 and IEC 61853-1,-2 draft. Using synthetic hourly meteorological data of 5 sites from MeteoNorm in PVsyst, energy rating was carried out according to IEC 61853-3 draft and compared with the results calculated in PVsyst. Minor differences on the mean annual specific energy yields (<0.9%) were observed between the two calculation methods for c-Si and mc-Si modules. Generally, the coefficient of variation (CoV) varied within 2.5% (k=2) for both c-Si and mc-Si module types with differences between PVsyst and IEC 61853-3 staying within 0.5%. Overall, it is shown that maximum differences as large as 5% in terms of PV-module's specific yield are to be expected between mainstream silicon-based PV-modules.
Keywords: crystalline silicon PV modules, LID, angular response, energy rating, PVsyst, IEC 61853

1 INTRODUCTION

In 2016 the PV community has witnessed several record breakthroughs with a global expansion of 75 GW, leading to an installed global photovoltaic capacity exceeding 300 GW.[1] In such a context, rating of the PV products is vital to the whole global market and industry. Currently, power rating under Standard Test Conditions (STC: device temperature of 25 °C and incident irradiance of 1000 W/m^2 with the spectral distribution AM1.5G as defined in IEC 60904-3 [2]) provides not only a globally accepted tool for comparison between different PV technologies and various module types, but also is an important reference for the pricing of PV modules. However, with the rapid increment of PV installation capacity, the PV community, especially the owners and investors of PV power plants, realize the limitation of power rating under STC in describing and predicting the performance of a certain PV module type in the field. Many studies on the outdoor performance and theoretical energy yield prediction as well as comparison of different simulation software have taken place, aiming to find an effective and reliable approach for the energy yield prediction and energy rating of PV modules. [3] It is well established that the site-specific factors like insolation, climate and surrounding environment conditions have a direct impact on the energy yield and performance of PV modules in a PV power plant, thus on the levelized costs of electricity (LCOE).

To comprehensively understand the energy yield, essential properties that drive the outdoor performance of PV modules have to be characterized and analyzed, for example, the spectral mismatch [14], the light-induced degradation (LID)[18] , irradiance dependent behavior [20], thermal behavior[21] and incident angle dependency [23]. The IEC Technical Committee 82 (TC 82) have proposed the IEC 61853 series of standards [24-27], which firstly describe the requirements for a holistic characterization of PV-module performance under conditions that are seen in the field and secondly provides the basis for an energy rating of PV modules with standard reference climatic profiles. To be more specific, the irradiance dependent behavior and thermal behavior of the PV modules are evaluated through power rating over a range of irradiances and temperatures in IEC 61853-1, while the test procedures for measuring the performance effect of angle of incidence, the module operating temperature and its dependence to irradiance, ambient temperature and wind speed, as well as the impact of spectral response on energy production are described in IEC 61853-2. [24-25] IEC 61853-3 CD provides a methodology of the energy rating calculation using the climatic profile defined in IEC 61853-4 CD. [26-27]

As an important factor for the bankability of a PV power plant, energy yield is usually predicted through in-depth simulations. A common reference in the industry is PVsyst software [4,9], which is a useful tool for the simulation and analysis of PV systems. With an input of the crucial performance aspects of a PV-module such as STC performance, irradiance dependent behavior, angular response, temperature dependence and LID, energy yield can be calculated in combine with the meteorological data, installation conditions and information of other components in a PV system.

In this paper, we have performed PV-module characterization under IEC 61853 and analysis on 14 commercially-available crystalline-silicon PV module types (8 c-Si and 6 mc-Si) in terms of the LID, irradiance dependent behavior, thermal behavior and angular response. Based on the data available, projections and comparisons on energy rating were carried out through PVsyst and the methodology described in IEC 61853-3 with the synthetic hourly meteorological data of 5 sites from MeteoNorm database.

2 METHODOLOGY

2.1 Module characterization

This work presents the comparison of 14 module types (8 c-Si and 6 mc-Si) sampled from 8 PV-module manufacturers located in Asia-Pacific region between September 2016 and March 2017. For each PV-module type 4 full-sized PV-modules and 1 spare were sampled from production by independent engineers. In addition to these samples encapsulated solar cells of the same bill of materials as the PV-modules were available for the test specimens and were used for determining the spectral response of the test modules. The test specimens were subsequently transported for testing to the ISO/IEC 17025 accredited PV-laboratory of TÜV Rheinland in

Shanghai. Initially all PV-modules were verified to meet the requirements of IEC 61215-1 for visual defects, electrical circuitry and safety. [28] The test protocol was illustrated in **Figure 1**. Prior to their characterization in accordance with IEC 61853-1,-2, PV-modules were electrically stabilized in accordance with IEC 61215-1,-1-1,-2 standards.[28] In addition electroluminescence (EL) imaging was performed for all modules, which showed that all samples were in good condition. The test items that will be reported in this work are: i) light-induced degradation (LID), 4 samples per module type; ii) irradiance-temperature-efficiency (GTE) matrix, 3 samples per module type; iii) spectral response, 1 encapsulated cell sample with the same bill of materials (BOM) of each module type; iv) angular response, 1 sample per module type; and v) temperature coefficient, 1 sample per module type. All 14 PV-module types are reported here anonymously to ensure the confidentiality of all parties involved. The outdoor spectral mismatch effect is not taken into account here because the spectral irradiance data were not available in the meteorological database of PVsyst. Research has shown that the impact of spectral effects on the annual energy yield was less than 2% due to the similar spectral responsivities of c-Si and mc-Si [12,31], therefore this is a reasonable assumption. However, in laboratory measurements all the performance data have been corrected by spectral mismatch factor.

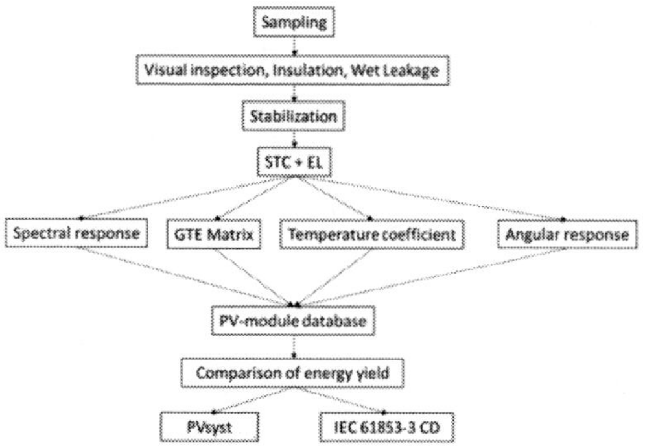

Figure 1: The test and simulation protocol used for calculation of energy yield.

All power measurements are performed with a state-of-the-art commercially available A+A+A+ pulsed solar simulator in a dark chamber especially designed to reduce the impact of indirect light. The level of irradiance is calibrated for each measurement against a primary reference cell. The used primary reference cells are calibrated in duplicates biyearly at Physikalisch-Technische Bundesanstalt (PTB). The temperature reference cell and tested PV module are controlled within ±0.5°C. The measurement reproducibility is confined within ±0.5%, k=2 and spectral mismatch uncertainty within ±0.35%, k=2. The described experimental setup shows a combined expanded measurement uncertainty, which is less than ±2.0%, k=2 under the commonly referred STC. Low irradiances were adjusted using different meshes without altering the spectral distribution. Temperature and irradiance corrections of the measured IV data were carried out according to the procedures defined in IEC 60891.[32]

For the angular response measurement, a rotatable open rack was employed with a ball-point rotating unit mounted on a calibrated goniometer. The equivalent optical distance between the test sample and the light source was approximately 8m. PV modules under test were specially prepared by accessing directly the electrical terminations of the measured solar cell located at the middle of the PV module to confine the non-uniformity of irradiance variation on volume of rotation of the target cell. The non-uniformity of irradiance reached 0.6% in the test plane and 1.9% on the volume of rotation of the test cell. Three successive baffle layers were employed to suppress the diffusive, in order to restrict the irradiance on the test plane within a field of view of 20° and non-reflective walls to eliminate secondary reflections.

Total measurement uncertainties were calculated for a DUT according to the principles detailed in the "Guide to the expression of uncertainty in measurement" (GUM).[33]

2.2 Energy yield simulation

Projections of the performance data measured in the laboratory to annual specific energy yield were performed using PVsyst software and a calculation method derived from IEC 61853-3 CD. Five test sites that represent different climates were chosen: i) Chennai in India with a tropical wet and dry climate, ii) Cologne-Bonn in Germany with a temperate-oceanic climate, iii) DaTong in China with a continental monsoon-influenced steppe climate, iv) Kumamoto in Japan with a humid subtropical climate, and v) Los Angeles in USA with a subtropical, hot-summer climate, classified as Mediterranean. A summary of the selected sites and the climatic conditions is shown in **Table I**.

Table I: Summary of test sites and abstract information used in specific energy yield comparison

Sites	Chennai India	Cologne-Bonn Germany	DaTong China	Kumamoto Japan	Los Angeles USA
Coordinate	13°N 80.18°E	50.7°N 7.15°E	40.1°N 113.33°E	32.82°N 130.72°E	37.62°N, 118.45°W
Köppen-Geiger Climate	Aw	Cfb	BSk	Cfa	Csa (inland) Csb (coast)
PV-module Tilt	15°	38°	38°	30°	32°
Global Irrad. [kWh/m²]	1941.8	979	1480.9	1377.7	1820.7
Diffuse Irrad. [kWh/m²]	942.8	544.6	670.4	790.1	668.6
Mean Temp. [°C]	28.2	11	8	17.3	17.3

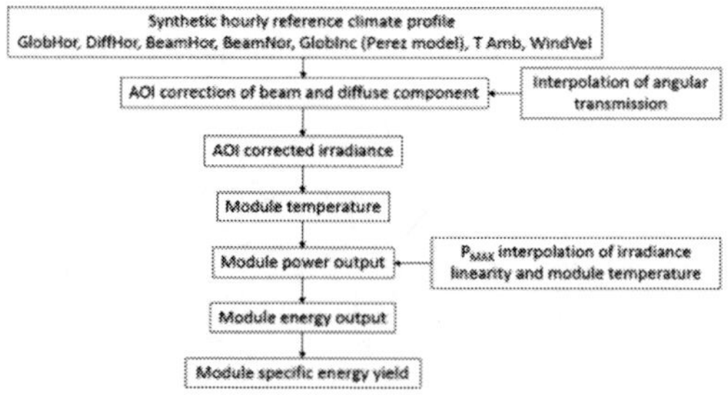

Figure 2: A flow chart of calculation procedure based on EC 61853-3 draft standard series.

The hourly meteorological data of these five sites

used in both simulation methods were synthetized through MeteoNorm embedded in PVsyst software. The methodology employed in this work is based on a relative comparison of specific energy yield between different PV technologies. Therefore, the comparison will not be significantly impacted in spite of the high uncertainties of the Meteonorm insolation data, as the same data will be used for all modules. In addition, spectral effect and wire loss were not considered in both calculations in PVsyst and IEC 61853-3 method for consistency in comparison. Inverter sizing losses were kept below 0.1% at PVsyst, while additional losses such as soiling or annual degradation were not taken into consideration in the PVsyst simulation. Wind effects on the thermal behavior of PV modules in the field was neglected, while a heat transfer coefficient of 20 W/m²·K was used as a default setting. A flow chart of the calculation procedures based on IEC 61853-3 was given in **Figure 2**. Formulas can be found in IEC 61853-2, -3. [25]

3 RESULTS AND DISCUSSION

3.1 Light-induced degradation

LID describes a phenomenon that PV modules can undergo a performance and power degradation due to the exposure under illumination. Several degradation mechanisms have been proposed, for example, boron-oxygen complex activation (B-LID), iron-boron pair dissociation (FeB-LID) and mc-PERC LID, sponge LID. [15-17] Among them, B-LID is the best known LID mechanisms which was first observed in boron-doped Czochralski-grown c-Si devices in 1970s because of the high oxygen concentration inside.[15] While mc-PERC LID has been seen in p-type mc-Si solar cells and can be induced by high temperature. It is more severe in PREC devices and can lead to an efficiency loss up to 10%. [16] Most industrial crystalline silicon solar cells and modules are suffering from some type of LID. Even 1% drop in power will result in considerable energy and capital losses.

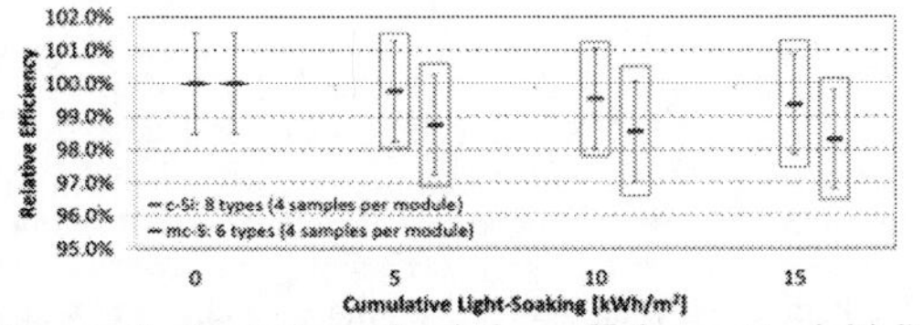

Figure 3: Variation of relative efficiency (to initial) against cumulative irradiation dose for different c-Si and mc-Si module types at STC. Boxes represent the standard deviation of technology variation (k=2) and error bars the measurement uncertainty (k=2). All samples considered in this work passes the stabilization criteria in accordance with IEC 61215-1,-1-1,-2 standards. It is shown that module type variability may exceed the mean values for both technologies.

In accordance with IEC 61215-1,-1-1,-2 standards [28], we electrically stabilized all PV-modules before any further measurement. The results of relative efficiency (to initial) against cumulative irradiation dose for different c-Si and mc-Si module types at STC are illustrated in **Figure 3**. The LID relative efficiency in most cases stayed within 3% of the initial values. Monocrystalline samples, the majority of which were made by Cz-BSF cells, showed a mean LID of 0.64%. Polycrystalline, mostly made of BSF cells, showed on average 1.69% relative efficiency decrease after 15 kWh/m² of light-soaking. These results are in agreement with the results of Ramspeck et al. [17], who studied LID for different technologies including Cz-BSF c-Si and BSF mc-Si solar cells. The coefficient of variation (CoV) exceeded for both technology types the bounds of relative measurement uncertainty CoV=1.91%, k=2 for c-Si and CoV=1.82%, k=2 for mc-Si. Interestingly, the results show a significant variation within technology groups; rather than technology type c-Si or mc-Si. The dispersion is attributed to both wafer quality and manufacturing processes employed.

3.2 Effect of irradiance on module efficiency

IEC 61853-1 provides a matrix of module performance with respect to temperature and irradiance as a useful tool to characterize the low light behavior and thermal behavior of the PV module under test. Several typical irradiances and temperatures are selected in this matrix: 1) 100, 200, 400, 600, 800, 1000 and 1000 W/m²; 2) 15°C, 25°C, 50°C and 75°C. Results in the Irradiance - Efficiency are sensitive to the cell technology. **Figure 4** presents abstractly the linearity of relative efficiency with irradiance for c-Si and mc-Si samples using the performance under STC as the reference.

The results showed no significant variation of relative efficiency in respect to their STC performance in the range of 600-1100 W/m² for both c-Si and mc-Si PV modules. However, the c-Si samples were generally performed better than mc-Si ones in the low irradiance range from 100 W/m² to 800 W/m². Performance higher than that under STC can be observed in the range of 600-800 W/m² for c-Si samples, which is attributed to the lower losses over the series resistance of the PV-module

Significant differences were observed in lower irradiances for both PV technologies, indicating the performance can be affected by technology types for irradiances below 400W/m². The mc-Si samples are more affected by the low irradiance with about 9% drop compared to their STC, while c-Si samples showed 2.4% higher mean relative efficiencies than that of mc-Si. Furthermore, the variability of modules of the same technology group can be significant for irradiances below 200W/m². The CoV exceeded the bounds of relative measurement uncertainty of both technology types, with CoV=2.32%, k=2 for c-Si and CoV=1.86%, k=2 for mc-Si at 100 W/m².

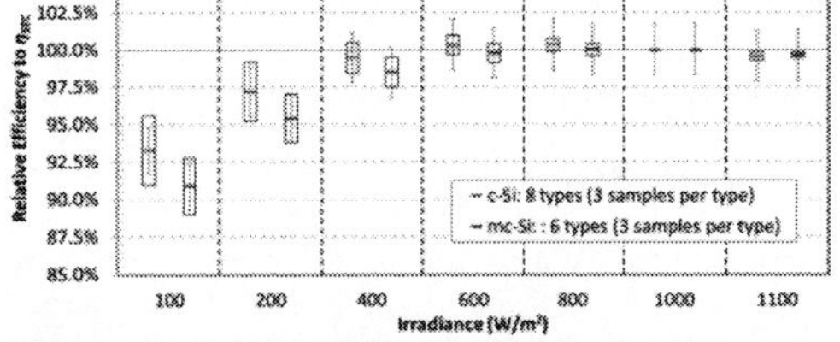

Figure 4: Relative efficiency deviation (to STC) against irradiance for different c-Si and mc-Si module types at 25°C and spectral distribution of AM1.5G. Boxes represent the standard deviation of technology variation (k=2) and error bars the measurement uncertainty (k=2).

3.3 Effect of temperature on module efficiency

Thermal behavior plays an important role in the photovoltaic performance of solar cells and their energy output, which decrease with the increasing operating temperature. This is because the internal carrier recombination is accelerated due to the enhanced carrier concentrations. [19] A photovoltaic (PV) module in the field is rarely working in the same condition as STC. Most of the incident solar energy which fails to be converted into electricity, dissipates as heat, leading to an increased module temperature and worsened performance. Generally, a linear dependency of both the electrical performance and the power output of a photovoltaic (PV) module on its operating temperature are expected for c-Si module types.[19][20]

The temperature coefficients, α of I_{SC}, β of V_{OC} and δ of P_{MAX}, for c-Si and mc-Si technology types are displayed and compared in **Figure 5**. The temperature coefficient of I_{SC} is the most sensitive parameter to both the variability of module types and less likely to the cell technology, with the mean temperature coefficient $\alpha = 0.045\%\cdot K^{-1}$ for c-Si PV-modules and $\alpha = 0.049\%\cdot K^{-1}$ for mc-Si. In the other hand, the mean temperature coefficients of V_{OC} and P_{MAX} were approximately $\beta = -0.30\%\cdot K^{-1}$ and $\delta = -0.40\%\cdot K^{-1}$ for both c-Si and mc-Si samples. The temperature coefficient of P_{MAX} was the most stable parameter which indicate the power output values of c-Si and mc-Si PV modules can be almost equally affected by the module operating temperature. It is noted however, that high-efficiency modules were not included in this examination such as HJT, which are known to exhibit different thermal behavior.

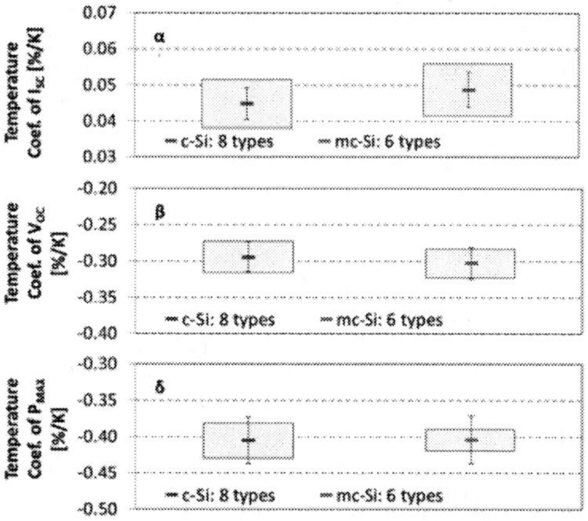

Figure 5: Temperature coefficients of I_{SC}, V_{OC} and P_{MAX} for c-Si and mc-Si technology types. Boxes represent the standard deviation of technology variation (k=2) and error bars the measurement uncertainty (k=2). Differences overlap with the uncertainty in measurement.

3.4 Angular response

A laminated structure with different layers is usually adopted by the PV modules. Optical losses happen either by means of reflection or when the light transmits through inactive materials such as glass or EVA before being absorbed by the solar cells. [22] Therefore, modules' photovoltaic performance can be affected by not only the module's optical characteristics but also the angle of incidence (AoI, angle between incident sunlight and module normal). Typically the STC power measurement is performed in laboratories with solar simulators that have strong specular components [22], thus most of the sunlight will be normal to the front surface of PV modules. When operating outdoors, however, PV-modules are subject to diffuse angular profiles that may introduce significant optical losses. Several methods have been applied to the modules to reduce the reflection loss, for example, using anti-reflection (AR) layers and deeply structured surface for light trapping. [34] All the c-Si and mc-Si samples, which were investigated here used standard float glass with AR coating.

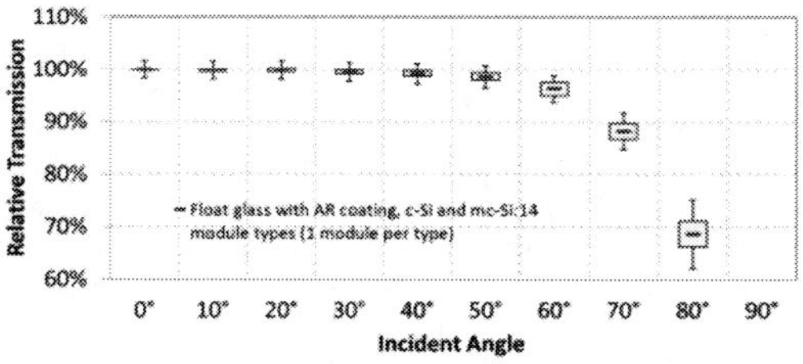

Figure 6: Variability of relative angular transmission for c-Si or mc-Si module types. Boxes represent the standard deviation of technology variation (k=2) and error bars the measurement uncertainty (k=2).

The I_{SC} values measured at different incident angles were translated into the relative transmission and the results are presented in **Figure 6**. The CoVs at the measured incident angles for both mc-Si and c-Si are much smaller than their corresponding measurement uncertainties, which indicate that the angular responses of the samples are predominantly dependent on the glass types rather than the solar cell technology. Most transmission deviations were observed at oblique angles above 60°. The measurement uncertainty and CoV of the angular response also have strong angular dependencies. Although, the measurement uncertainty is significant for oblique angles exceeding 50°, higher variations were observed for different module types. The CoV seen at 70° and 80° reached ±1.5% (k=2) and ±2.4% (k=2), respectively.

3.5 Energy yield

Energy yield (EY) is important for PV projects, especially for PV power plants, since it has a direct influence on the levelized cost of electricity (LCOE) and revenue. However, energy yield is subject to local, seasonal and year-to-year variations, for example the insolation, climate, and installation conditions, which makes the prediction even more complicated. [3] The specific energy yield (kWh/kWp) represents the total net energy yield generated per kWp of installed PV. Specific energy yield is regarded as a useful parameter in comparing the power generation ability of different PV devices under a certain condition, such as the location or system design, without influenced by their nominal power, or total installed power. Therefore, the specific yield was used as a diagnostic tool to examine the annual yield of our 14 PV module types at 5 different sites. The results were analyzed in terms of solar cell technologies and calculation methods.

Figure 7 displays the variations of specific energy yield over 5 sites and coefficient of variation for c-Si (blue) and mc-Si (red) module types calculated through PVsyst and IEC 61853-3 CD, respectively. Maximum differences up to ±2.5% in terms of PV-module's specific yield are observed in the two mainstream silicon-based

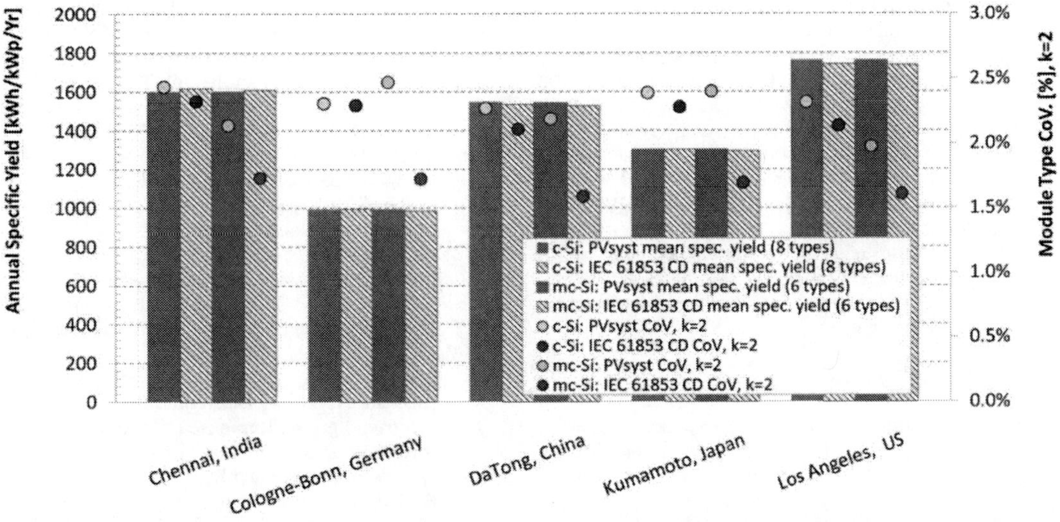

Figure 7: Variation of specific energy yield over 5 sites and coefficient of variation for c-Si (blue) and mc-Si (red) module types calculated through PVsyst and IEC 61853-3 CD, respectively. It is noted that relative differences between module types do not significantly differ per site. However, variations with standard deviations exceeding ±2%, k=2 are to be expected within c-Si and mc-Si technology types.

PV-modules at different sites. Although the mean differences between c-Si and mc-Si were not significant (0.9%) the technology variation within each category was significantly higher (2.5%).

PVsyst simulations showed that the relative differences between module types generally do not differ considerably per site. CoV of c-Si PV modules varied within 2.27%-2.44% (k=2), while the CoV for mc-Si PV samples varied within 1.97% (Los Angeles) to 2.47% (Cologne-Bonn). In general higher CoVs were observed in sites where diffuse light had higher contributions, such as Cologne-Bonn and Kumamoto.

The mean specific energy yield values calculated according to IEC 61853-3 CD [26] are close to that from PVsyst. Agreements between the two models stayed within 0.5%. It is noted however, that relative differences in terms of specific energy yield of mc-Si PV modules turned out to be somewhat smaller than their counterpart using PVsyst by 0.3-0.5% with with CoV varying from 1.59% (k=2) at Datong to 1.73% (k=2) at Chennai. This indicates some minor calculation differences between the two approaches, which is reasonable considering PVsyst simulations are based on equivalent diode modeling, while the IEC 61853-3 employs interpolation methods. The simulations also showed that differences may also depend on the interpolation scheme used, which is a point that requires further evaluation by the standard. Another thing to be noted is that the differences of CoVs between c-Si and mc-Si are also slightly bigger in Cologne-Bonn and Kumamoto, where the diffuse light has a higher contribution in the isolation volume. This probably was caused by the theoretical model of the angular light transmission used in IEC 61853-2 introduces assumptions that simplify the angular profile to a single parameter. According to IEC 61853-2, a_r is used as a single fitting parameter in the angular light transmission formula. [26] Furthermore, PVsyst simulations are based on equivalent diode modeling, which also introduces uncertainties. The differences between the simulated value and measured value may have led to the underestimation of the variation in terms of the specific energy yield of mc-Si PV modules. In any case all differences observed were fairly small and showed excellent agreement (within 0.9%).

4 CONCLUSIONS

Energy rating in accordance with IEC 61853 is critical and can lead to a paradigm shift of industry's practices. It was shown that IEC 61853 testing may reveal maximum differences as large as 5% in terms of PV-module's specific yield between mainstream silicon-based PV-modules.

This paper, a project was initiated with the purpose of providing comprehensive information on the variability of silicon-based PV-modules, which are available in the market. 14 mainstream crystalline silicon-based PV modules types (8 c-Si and 6 mc-Si) sampled from 8 PV-module manufacturers were measured in terms of spectra response, light-induced degradation, irradiance dependent behavior, thermal behavior and angular response.

Polycrystalline sample was found more subject to LID than monocrystalline ones, even though all the LID stayed within 3% of the initial value after 15 kWh/m^2 of light-soaking. The coefficient of variation also indicated the variation of LID may vary more within module types than technology (c-Si or mc-Si) types.

Generally better performance under low irradiance was observed in c-Si samples compared to the mc-Si ones. When the irradiance was below 200W/m^2, a dependency on module types started to emerge in both PV technologies.

The sensitivity of temperature coefficients to module variability is not particularly significant for P_{MAX}, V_{OC} and I_{SC} for both c-Si and mc-Si PV samples. Their mean temperature coefficients of I_{SC}, V_{OC} and P_{MAX} were approximately $\alpha = 0.050\%\cdot K^{-1}$, $\beta = -0.30\%\cdot K^{-1}$ and $\delta = -0.40\%\cdot K^{-1}$, respectively with variations overlapping with the measurement uncertainty.

The angular response measurements of the samples displayed a predominant dependency on the glass types rather than the solar cell technology. Most transmission deviations were observed at oblique angles above 60°.

In the end, projections of specific energy yield in 5 different sites were carried out based on the results of abovementioned measurements for all 14 PV module types. Minor differences of specific energy yields were observed between the two calculation methods of PVsyst and IEC 61853-3 draft.

Lastly maximum relative differences as large as ±2.5% regarding to PV-module's specific yields of different module types were observed in the mainstream silicon-based PV-modules.

ACKNOWLEDGEMENTS

The authors gratefully acknowledge the contribution of Dr. Werner Herrmann (TÜV Rheinland Energie und Umwelt GmbH) and Mr. Kengo Morita (TÜV Rheinland Japan Ltd) for the helpful discussions and data contribution.

5 REFERENCES

[1] Snapshot of Global Photovoltaic Markets 2016. Report IEA PVPS T1-31(2017)

[2] IEC 60904-3: 2016, Photovoltaic devices – Part 3, Measurement principles for terrestrial photovoltaic (PV) solar devices with reference spectral irradiance data

[3] M. Schweiger, W. Herrmann. Energy Rating Label for PV Modules for Improving Energy Yield Prediction in Different Climates. 31st European Photovoltaic Solar Energy Conference and Exhibition, Hamburg (2015)

[4] F. Carigiet, F. Baumgartner, J. Sutterlueti, N. Allet, M. Pezzotti, J. Haller. Verification of Measured PV Energy Yield versus Forecast and Loss Analysis. 28th European Photovoltaic Solar Energy Conference and Exhibition, Paris (2013)

[5] B. Müller, L. Hardt, A. Armbruster, K. Kiefer, C. Reise. Yield Predictions for Photovoltaic Power Plants: Empirical Validation, Recent Advances and Remaining Uncertainties. 29th European Photovoltaic Solar Energy Conference and Exhibition, Amsterdam (2014)

[6] D. Buß, M. Strobel, M. Koentopp. Influence of Technological Changes on the energy yield of PV Modules; An Outdoor Study. Energy Procedia 55 (2014) 38

[7] T. A. Guerra, J. A. Guerra, B. O. Tabernero, G. de la Cruz García. Comparative Energy Performance Analysis of Six Primary Photovoltaic Technologies in Madrid (Spain). Energies, 10 (2017) 772

[8] J. Muñoz, O. Perpiñán. A Simple Model for the Prediction of Yearly Energy Yields for Grid-Connected PV Systems Starting from Monthly Meteorological Data. Renewable Energy, 97 (2016) 680

[9] G. Makrides, A. Phinikarides, E. Herzog, M. B. Strobel, G.E. Georghiou. Outdoor Performance and Modelling of Innovative Crystalline Silicon Photovoltaic Modules under Hot Climate Conditions. 32nd European Photovoltaic Solar Energy Conference and Exhibition. Munich (2016).

[10] T. Huld, E. Dunlop, H. G. Beyer, R. Gottschalg. Data Sets for Energy Rating of Photovoltaic Modules. Solar Energy, 93 (2013) 267.

[11] D. Dirnberger, B. Müller, C. Reise. PV Module Energy Rating: Opportunities and Limitations. Progress in Photovoltaics, 23 (2015) 1754.

[12] D. Dirnberger, G. Blackburn, B. Müller, C. Reise. On the Impact of Solar Spectral Irradiance on the Yield of Different PV Technologies. Solar Energy Materials and Solar Cells. 132 (2015) 431.

[13] B. Zinsser, M. B. Schubert, J. H. Werner. Spectral Dependent Annual Yield of Different Photovoltaic Technologies. 26th European Photovoltaic Solar Energy Conference and Exhibition, Hamburg (2011)

[14] M. Schweiger, M. Ulrich, I. Nixdorf, L. Rimmelspacher, U. Jahn, W. Herrmann. Spectral Analysis of Various Thin-Film Modules Using High Precision Spectral Response Data and Solar Spectral Irradiance Data. 27th European Photovoltaic Solar Energy Conference and Exhibition. Frankfurt (2012)

[15] B. Sopori, P. Basnyat, S. Devayajanam, S. Shet, V. Mehta, J. Binns, J. Appel. Understanding Light-Induced Degradation of c-Si Solar Cells. 2012 IEEE Photovoltaic Specialists Conference, Austin (2012)

[16] T. Luka, C. Hagendorf, M. Turek. Multicrystalline PERC solar cells: Is light-induced degradation challenging the efficiency gain of rear passivation? Photovoltaics International. 32 (2016) 37.

[17] K. Ramspeck, S. Zimmermann, H. Nagel, A. Metz, Y. Gassenbauer, B. Birkmann, A. Seidl. Light Induced Degradation of Rear Passivated mc-Si Solar Cells. 27th European Photovoltaic Solar Energy Conference and Exhibition. Frankfurt (2012)

[18] G. E. Bunea, K. E. Wilson, Y. Meydbray, M. P. Campbell, D. M. De Ceuster. Low Light Performance of Mono-Crystalline Silicon Solar Cells. 2006 IEEE 4th World Conference on Photovoltaic Energy Conference, Waikoloa (2006)

[19] S. Dubey, J. N. Sarvaiya, B. Seshadri. Temperature Dependent Photovoltaic (PV) Efficiency and Its Effect on PV Production in the World-A Review. Energy Procedia, 33 (2013) 311.

[20] M. Schweiger, S. Michalski, U. Jahn, W. Herrmann, U. Rau. Non-Linearity of Temperature Coefficients, Equivalent Cell Temperature and Temperature Behaviour of Different PV-Module Technologies. 28th European Photovoltaic Solar Energy Conference and Exhibition, Paris (2013)

[21] M. Ebert, H. Stascheit, I. Haedrich, U. Eitner. The Impact of Angular Dependent Loss Measurement on PV Module Energy Yield Prediction. 29th European Photovoltaic Solar Energy Conference and Exhibition, Amsterdam (2014)

[22] C. Monokroussos, D. Etienne, J. Theophil, J. Lau, L. Xu, S. Kera, K. Morita, C. Dreier, R. Struwe, C. Zou, T. Hasegawa. Effect of Glass/EVA/Backsheet Encapsulation Package on the Industrial Rating of c-Si PV Modules. 31st European Photovoltaic Solar Energy Conference and Exhibition, Hamburg (2015)

[23] A. Yusufoglu, T.M. Pletzer, B. Min, H. Kurz. Angle-of-Incidence Losses and Their Influence on the Annual Energy. 29th European Photovoltaic Solar Energy Conference and Exhibition, Amsterdam (2014)

[24] IEC 61853-1: 2011, Photovoltaic (PV) module performance testing and energy rating - Part 1: Irradiance and temperature performance

measurements and power rating
[25] IEC 61853-2: 2016, Photovoltaic (PV) module performance testing and energy rating - Part 2: Spectral responsivity, incidence angle and module operating temperature measurements
[26] IEC 61853-3 CD: 2015, Photovoltaic (PV) module performance testing and energy rating - Part 3: Energy rating of PV modules
[27] IEC 61853-4 CD: 2015, Photovoltaic (PV) module performance testing and energy rating - Part 4: Standard reference climatic profiles
[28] IEC 61215-1: 2016, Terrestrial photovoltaic (PV) modules - Design qualification and type approval - Part 1: Test requirements
[29] IEC 61215-1-1: 2016, Terrestrial photovoltaic (PV) modules - Design qualification and type approval - Part 1-1: Special requirements for testing of crystalline silicon photovoltaic (PV) modules
[30] IEC 61215-2: 2016, Terrestrial photovoltaic (PV) modules - Design qualification and type approval - Part 2: Test procedures
[31] M. Schweiger, W. Herrmann, A. Gerber, U. Rau. Understanding the Energy Yield of Photovoltaic Modules in Different Climates by Linear Performance Loss Analysis of the Module Performance Ratio. IET Renewable Power Generation, 11 (2017) 558.
[32] IEC 60891: 2009, Photovoltaic devices - Procedures for temperature and irradiance corrections to measured I-V characteristics
[33] Evaluation of Measurement Data - Guide to the Expression of Uncertainty in Measurement. JCGM 100 (2008)
[34] H. K. Raut, V. A. Ganesh, A. S. Nairb, S. Ramakrishna. Anti-reflective coatings: A critical, in-depth review. Energy & Environmental Science. 4 (2011) 3779.
[35] M. A. Green. Two New Efficient Crystalline Silicon Light-Trapping Textures. Progress in Photovoltaics: Research and Applications. 7 (1999) 317.
[36] P.I. Widenborg, A. G. Aberle. Polycrystalline Silicon Thin-Film Solar Cells on AIT-Textured Glass Superstrates. Advances in OptoElectronics. 2007 (2007) 24584.

A SYSTEMATIC COMPARISON OF 12 EMPIRICAL MODELS USED FOR ENERGY YIELD PREDICTIONS VS PV TECHNOLOGY

Steve Ransome[1] (SRCL, UK) and Juergen Sutterlueti[2] (Gantner Instruments, AUT)
[1] steve@steveransome.com ; +44 (0)7515 565010 ; http://www.steveransome.com
[2] Gantner Instruments Environment Solutions GmbH, 08297 Zwoenitz/ Germany

ABSTRACT: Twelve different empirical models have been compared by fitting indoor matrix IEC 61853-1 data (PV module performance at different irradiances and module temperatures). Models are used to characterize PV modules and predict energy yield using site climate data (e.g. hourly irradiance and ambient temperature for a year). The empirical models were found to have some coefficients that could give unphysical fits (e.g. wrong temperature coefficients or unfeasible performance at either low or high light levels). Therefore, a new optimized "Mechanistic Performance Model" (MPM) was developed with only physically significant coefficients. The MPM has been tested with both outdoor (from Gantner Instruments) and indoor matrix data from a variety of institutes and has proven to be more robust and behave better than the empirical models tested with inputs some of which had random added noise.
Keywords: Energy Rating; Energy Performance; Modelling

1 INTRODUCTION

Twelve different empirical energy yield prediction models have been compared against each other. They characterise PV module performance (e.g. power "P", efficiency "η" or the dc performance ratio "PR_{DC}") versus weather parameters including the plane of array irradiance G (kW/m²), module temperature T (C) and WS windspeed (ms⁻¹).

These models usually sum 3-7 empirical fit coefficients C_i multiplied by a variety of weather functions $func_i$ (G, T and WS) as in equation <1>

$$PR_{DC} = \sum_{i=1}^{N} C_i \times func_i(G, T, WS \dots) \quad <1>$$

The optimised fit coefficients C_1-C_N can be derived from a simple tool such as Excel solver or Python SciPy curve fit [1] to minimise the rms differences between the measured and modelled P_{MAX}.

The twelve models here have been compared using

1) Internal measured matrices provided by 3rd parties such as ASU, Sandia, CFV and NIST (all USA); JRC ESTI (Italy/EU), TUV Rheinland (Germany) and SUPSI (Switzerland) on a variety of PV module technologies including "standard" and advanced "e.g. HIT and ABC" c-Si and Thin Films such as CdTe
2) Outdoor logged data from Gantner Instruments test facility (OTF4) in Tempe, AZ [2].

2 SHAPE OF USUAL PV PERFORMANCE CURVES

Figure 1 illustrates the usual shape of a PV module's relative efficiency $\eta_{MEASURED}/\eta_{STC}$ (i.e. PR_{DC} or MPR) vs. plane of array irradiance G (x axis) and module temperature T (curves). The dots are measured data and coloured lines are smooth fits at different temperatures.

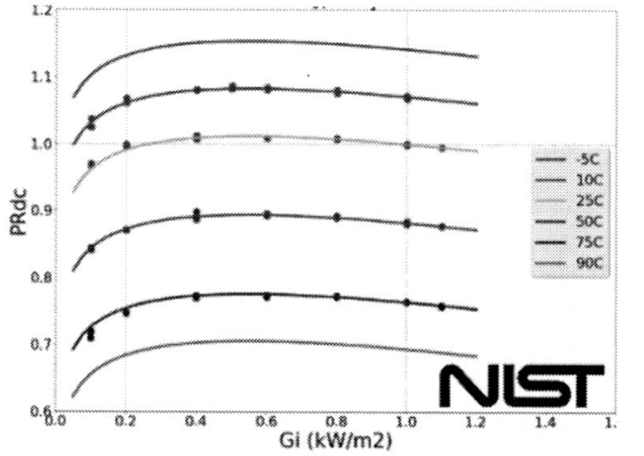

Figure 1: Typical shape of normalised module efficiency PR_{DC} (y axis) vs. T_{MODULE} (coloured lines and dots) and irradiance kW/m² (x axis) courtesy NIST/CFV.

The twelve models studied are listed in Table I. They contain various selections of weather functions of irradiance, module temperature or other parameters to fit including the following

Irradiance:	G_I G_I^2 G_I^{-1} $\ln(G_I)$ $\ln(G_I)^2$
Temperature:	T_{MOD} T_{MOD}^2 T_{AMB} T_{AMB}^2
Combinations:	$T_{MOD}*\ln(G_I)$; $T_{MOD}*T_{AMB}$;
Other:	SolarHeight, WindSpeed ...

The models are code named A to P in this work and are deliberately not in the same order as listed in the introduction.

Some of the models derive their own Gamma values from their fits, others temperature correct by the datasheet value and these are identified in column <D>.

3 MATRIX MEASUREMENTS (IEC 61853-1)

The matrix method as described in IEC 61853-1 [3] defines the measurement of PV performance at 23 different pairs of irradiance (G=0.1, 0.2, 0.4, 0.6, 0.8, 1.0 and 1.1 kW/m²) and module temperature (T=15, 25, 50 and 75C) [4]. Five of the points with the "hottest+lowest irradiance" and "coldest+highest irradiance" that rarely occur can be ignored. These measured points are plotted as dots in figure 1 which is a c-Si module ID=40 measured by CFV. In this graph, all 28 points have been taken plus an extra one at (0.5, 10).

Table I: Comparison of twelve empirical models (some models may have more than one version)

Columns: <A>	**<B>**	**<C>**	**<D>**	**<E>**
Model Name [Reference]	Number of Coefficients C_N	Are coefficients normalised?	Is T_{MOD} modelled?	Further comments
CREST [5]	5	☒No	☑Yes	
HEYDENRICH [6]	3	☒No	☒No	
IEC 60891[7]	5?	☒No	☑Yes	Complicated calculation procedure
LFM 2013[8]	6	☑Yes	☒No	Simplest 2 LFM factors only nI_{DC} and nV_{DC}
Mother PV[9]	5	☒No	☒No	Self-referenced Isc
POLYNOMIAL 4th[10]	4	☒No	☒No	Simple $P = G_I * [\Sigma_N C_N * G_I^N] * T_{CORR}$, N=4
PVCOMPARE[11]	7	☒No	☑Yes	Includes T_{AMB} and SolAlt terms
PVGIS[12]	6	☒No	☑Yes	Also known as "simplified King model"
PVUSA[13]	4	☒No	☑Yes	Only use G>0.7kW/m^2 as poor low light fit
PVUSA+[14]	5	☑Yes	☑Yes	Extra coefficient to improve fit at low light
SRCL2014 [15]	5	☑Yes	☑Yes	Physical coefficients e.g. LLEC, R_S
New mechanistic model "MPM"				
Model "N" [1][4]	5-6	☑Yes	☑Yes	New model in this paper

4 COMPARING MODELS vs. PV TECHNOLOGY

It was previously shown [4] that some of the empirical models used coefficients which could give unphysical effects such as poor performance extrapolation at low or high light levels, non-constant or just incorrect temperature coefficients.

Partial differentiation of their PR_{DC} equations by T_{MOD} gives the P_{MAX} coefficient (gamma) behaviour vs. temperature by

$$gamma = \frac{\partial PR_{DC}}{\partial T_{MOD}} \quad <2>$$

proving some of their coefficients can be seen to result in unphysical behaviour if their C_N is non-zero.

Because of these limitations a new model is needed with only physical coefficients. Table 2 lists some the most important effects on PV performance vs. weather input which all need to be modelled correctly.

Table 2: Most important PV performance dependencies on weather parameters.

Dependency	Comments
1.$I_{MAX} \propto$ $\mathbf{G_I}$	neglecting SR, AOI
2.$V_{MAX} \propto$ $\mathbf{log(G_I)}$	from 1 diode model
3.$T_{MOD} \sim T_{AMB} -$ $\mathbf{fn(WS)}$	due to NOCT/NMOT
4.$P_{MAX} \propto (1+\gamma*(T_{MOD}-25))$	P_{MAX} temp. coefficient
5.$\Delta P_{MAX} \propto \mathbf{I_{MAX}^2} * R_{SERIES}$	Eff falls with G_I^2
6.$R_{SHUNT} \propto$ $\mathbf{1/exp(G_I)}$	varies by technology

A new "Mechanistic Performance Model" (MPM) [4] was developed choosing coefficients that guarantee sensible behaviours (i.e. physically significant curve fits) using the equations in Table 2 so it has 6 coefficients C_1 to C_6 as defined in equation <3>.

MPM formula

$$PR_{DC} = C_1 + C_2 \times dT_{MOD} + C_3 \times Log_{10}(G_I) + C_4 \times G_I + C_5 \times WS + C_6/G_I \quad <3>$$

Where

$dT_{MOD} = (T_{MOD} - 25)$;

G_I = reference sensor plane of array irradiance (kW/m^2)

WS = wind speed (ms^{-1}).

The MPM can be used to fit both IEC 61853 or outdoor data.

5 MODEL FIT ROBUSTNESS

Robust statistical model fits should not depend on outliers.

It has been found [4] that a "perfect" IV curve (e.g. from the 1 diode model) could be fitted reasonably well by empirical models even with "unphysical coefficients" because their coefficients C_N tended to be near 0 or else self-compensated.

However as soon as the data became imperfect (e.g. bad point(s), atypical behaviour or noise) this could cause the unphysical coefficient's C_N to rise in an attempt to fit the data points better resulting in unphysical fit.

Figure 2 shows graphs fitted by 3 models, A(top), D(middle) (both are empirical which contain unphysical coefficients) and the MPM (bottom row).

The data in left hand column comes directly from an SAPM model [16] so it will be noiseless and should be able to be fitted well by any model.

The other two columns have had a random noise component added (rms error of 2%) to the points resulting in much worse scatter that would normally be seen. This allows us to check the robustness of fitting from each model, how reproducibly it can be fitted. (Note the same noisy data is modelled in each of the right two columns and there are two seeds to give separate random data).

Even though the data is very noisy the MPM N (bottom row) gave sensible fits every time – they are all parallel curves with constant temperature coefficients similar to figure 1 (note that the temperatures of these lines are not linearly spaced) and reasonable extrapolated behaviour at both high light levels (almost constant) and low (dropping somewhat).

The empirical models A and D fit well only when data is perfect (left column) but any scatter makes the fits' unphysical coefficients appear as one of

1) wrong extrapolated performance at low light

2) wrong extrapolated performance at high light
4) wrong gamma reducing at low light
5) wrong gamma increasing at low light

Note that 4) and 5) mean that the P_{MAX} temperature coefficient is not constant with light level when it should be.)

The MPM is therefore a more robust model which fits noisy data as well as it can.

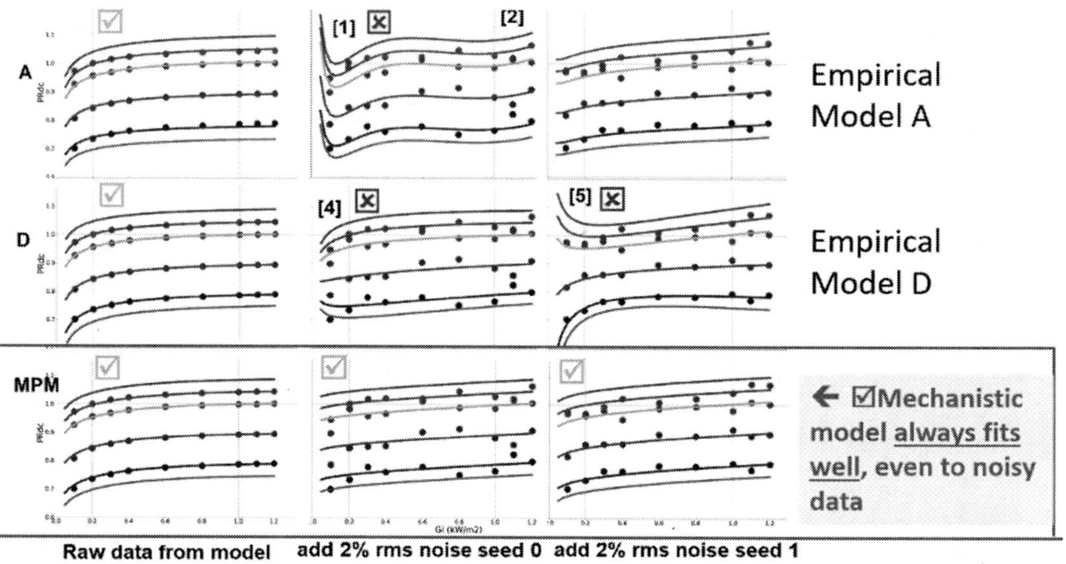

Figure 2: Model fits to "perfect" and "added noise" SAPM model data. Left column is raw "noiseless" data (which all models can fit well), other columns have added 2% rms noise (two different seed values) to test curve fitting for empirical models A(top), D(middle) and MPM (bottom row). Artefacts 1-5 which only affect models A and D and not the MPM are described in the text.

The coefficients of a mechanistic model are also more meaningful than those of an empirical model. For example, the MPM coefficients give

C_1 → P_{MAX} (Overall quality = module rating)
C_2 → Gamma (P_{MAX} temperature coefficient)
C_3 → V_{OC} (Low light logarithmic drop)
C_4 → R_{SERIES} (High light linear dependence)
C_5 →Wind Speed (From NOCT)
C_6 → R_{SHUNT} (varies by technology)

Figure 3 compares the first 5 coefficients from an Empirical model (top) and the MPM (bottom) for 13 modules of varying technologies (right column) measured by SUPSI.

The values of the empirical model's coefficients seem random whereas the MPM's coefficients (lower half) are systematic, normalised and are reasonable. For example, C1+C4 will give the module $P_{MAX}/P_{NOMINAL}$ rating and the C2 gives the calculated gamma temperature coefficient which can be seen to be -0.42 to -0.48%/K for the standard c-Si and also believable lower values for the thin films.

Technology	ID	C_1	C_2	C_3	C_4	C_5	rms
c-Si	60)	-42.3	53.9	-10.7	-32.9	-8.0	0.22%
c-Si	62)	127.2	159.8	-31.7	-97.8	-23.9	0.22%
c-Si	64)	-71.5	90.5	-18.0	-55.3	-13.5	0.09%
c-Si	66)	-93.4	117.6	-23.4	-72.0	-17.6	1.84%
c-Si	67)	100.2	125.8	24.6	75.6	18.4	0.24%
c-Si	68)	-69.5	87.9	-17.5	-53.8	-13.1	0.16%
c-Si	70)	-37.3	131.4	-98.7	21.9	23.5	0.10%
c-Si	71)	6.4	-6.8	1.4	4.1	0.9	0.07%
c-Si	72)	60.7	132.5	75.7	10.1	-10.2	0.59%
c-Si	73)	53.8	-68.9	16.3	38.4	8.7	0.09%
TF a-Si	65)	0.2	1.1	-0.3	-0.5	-0.1	0.94%
TF a-Si	74)	90.8	121.1	31.9	62.8	13.2	0.32%
TF CdTe	63)	-0.6	2.2	-0.6	-1.2	-0.3	0.27%
c-Si	60)	96.2%	-0.45%	8.3%	-2.1%	0.0%	0.07%
c-Si	62)	109.6%	-0.42%	20.5%	-10.0%	0.0%	0.09%
c-Si	64)	106.4%	-0.45%	8.5%	-6.4%	0.0%	0.09%
c-Si	66)	107.7%	-0.48%	11.9%	-7.7%	0.0%	0.08%
c-Si	67)	115.2%	-0.46%	18.2%	-15.4%	0.0%	0.11%
c-Si	68)	107.6%	-0.47%	10.4%	-7.5%	0.0%	0.09%
c-Si	70)	103.7%	-0.46%	3.9%	-4.3%	0.0%	0.08%
c-Si	71)	113.7%	-0.46%	24.4%	-12.4%	0.0%	0.08%
c-Si	72)	99.6%	-0.44%	0.7%	1.2%	0.0%	0.20%
c-Si	73)	109.4%	-0.45%	17.1%	-9.2%	0.0%	0.09%
TF a-Si	65)	112.2%	-0.11%	31.6%	-11.9%	0.0%	0.21%
TF a-Si	74)	122.7%	-0.22%	39.5%	-23.1%	0.0%	0.33%
TF CdTe	63)	121.5%	-0.25%	19.6%	-20.2%	0.0%	0.16%

Figure 3: Comparing coefficients for an empirical model (top half) vs. those from the MPM (bottom half) for 13 modules of different technologies as measured by SUPSI.

6 MPM MEASUREMENTS OF 3rd PARTY MATRICES

Many institutes such as ESTI, TUV Rheinland, SUPSI and CFV can measure indoor matrices IEC 61853-1. ASU can measure outdoor too with meshes. [4]

Figure 4 illustrates the fit by the MPM to typical measurements from 7 institutes/companies and comparing data from the SAPM and 1 diode model used by PVSYST. The Gantner Instruments outdoor data is added for comparison and will be described in section 7.

The RMS error is usually < 0.3% indoors (if low scatter measurements) for all technologies. The scatter and rms error may be worse for institutes measuring a wider irradiance range e.g. "100-1100W/m²" rather than "300-800W/m²". SUPSI and TUV now measure wider irradiance ranges than shown.

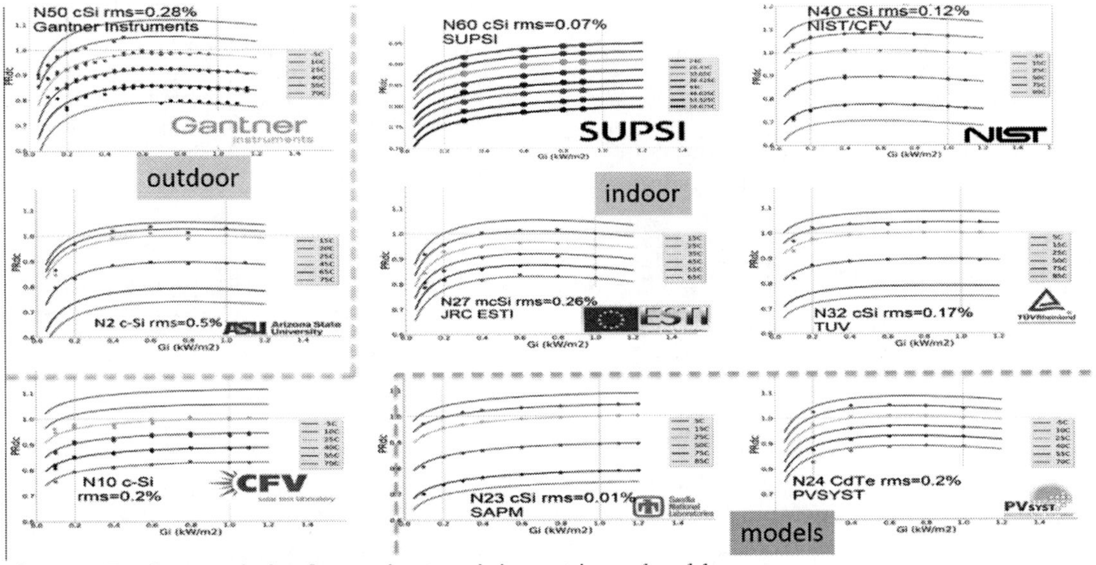

Figure 4: MPM fits to matrix data from outdoor tests, indoor matrices and models.

7 MPM FITTING OUTDOOR MEASUREMENTS

Gantner Instruments [2] sells an Outdoor Test Facility (OTF) capable of 500W (200V, 20A) PV Module IV scans and has a vast amount of outdoor IV and weather measurement data since 2010 for many technologies including: standard and high efficiency c-Si; CdTe; a-Si:uc-Si and CIGS at their US locations Tempe (AZ) [2]. The OTF was used to determine how accurately MPM fits could be made to well measured outdoor data. Figure 5 plots the number of minutely measurements by irradiance and T_{MODULE} bins for over 300,000 IV curves measured every minute for 12 months during 2016 for a standard 200Wp crystalline Si module vs. a reference cell.

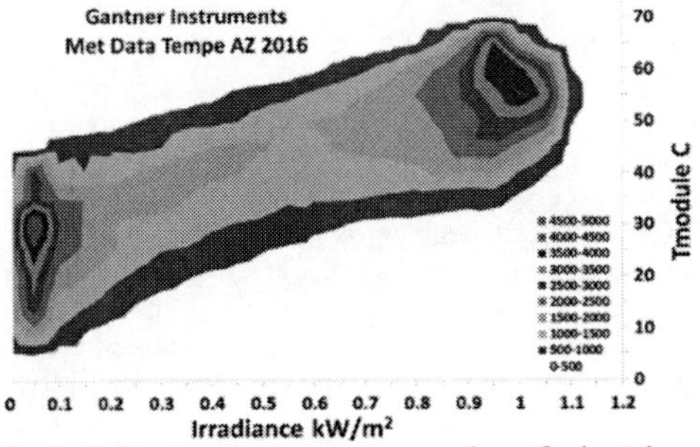

Figure 5: Gantner Instruments – number of minutely measurements per bin for 1 year 2016.

The MPM was used to fit the measurement data and present the results compensated to 15C bins in figure 6. As expected from figure 5 points are mostly in the "dull+cool" (top left) to "bright+warm" (bottom right) ranges. The MPM fits this data very well with its smooth and parallel lines and has an rms error of only 0.28% which is low considering there have been no spectral, dirt, ageing or AOI corrections.

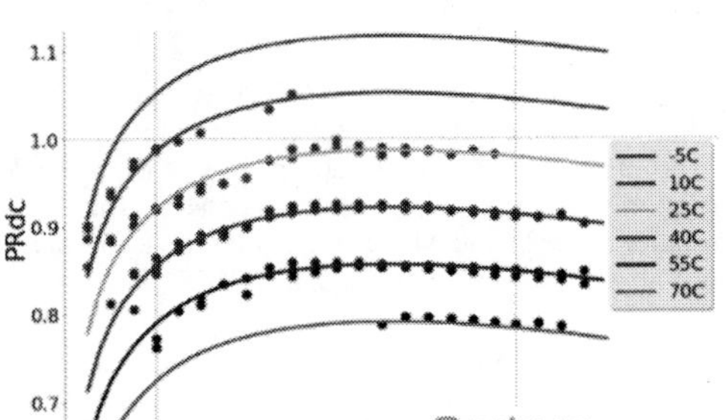

Figure 6: Good MPM Fit to measured cSi OTF4 MID ID#12 data for Gantner Instruments data at Tempe in AZ

Figure 7 also illustrates the MPM fit versus time. To prove the model's capability it was optimised to fit data from one year and then tested against data from another.

The fit to measured residuals are mostly within 2% but within the middle of the day with clear skies (when most energy will be generated) the residual is within 1%. Note that no corrections have needed to be made for effects such as spectrum, soiling, angle of incidence, degradation.

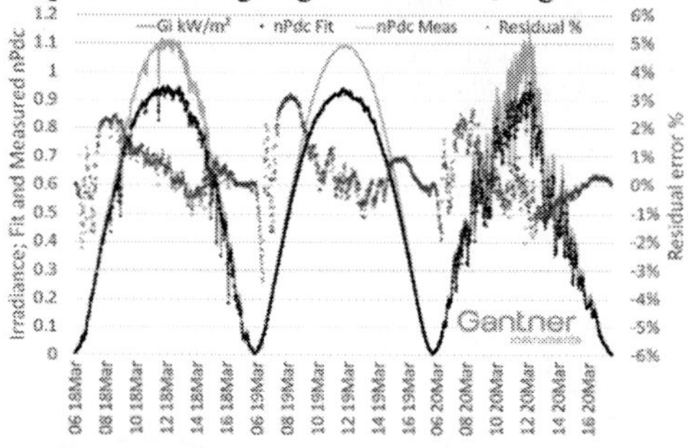

Figure 7: MPM Fit to measured cSi OTF4 MID ID#12 data vs. time for Gantner Instruments data at Tempe in AZ.

Thus the MPM can fit well measured outdoor data too.

8 MPM ANALYSIS VS. PV TECHNOLOGY

The MPM can quantify any P_{MAX} tolerance, Gamma, low light efficiency drop and R_{SERIES} differences between module manufacturers and technologies.

Figure 8 plots graphs of data and fits performed by SUPSI on different PV technologies - c-Si (2 left), CdTe (top right) and a-Si (bottom right). The c-Si modules (left) show normal behaviour fitted well. The CdTe module appears to have a high series resistance meaning that its efficiency at high irradiance (right) will be lower than at lower irradiance (left) due to its high $I^2.R_{SERIES}$ loss. The a-Si module has a low P_{MAX} temperature coefficient but poor low light level response.

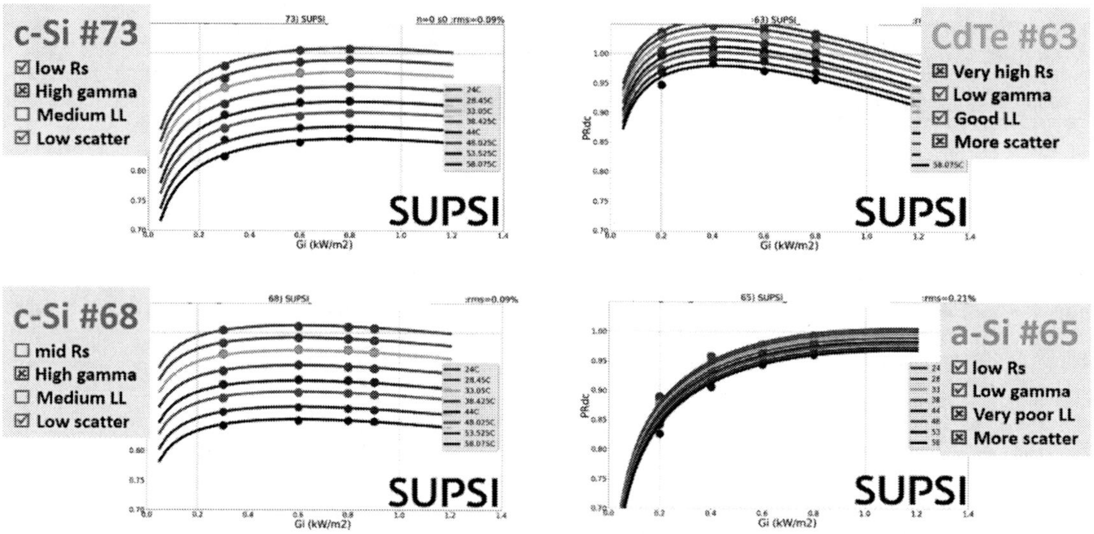

Figure 8: Internal matrix measurements and MPM fits for four different manufacturers measured by SUPSI.

10 ENERGY YIELD

The variability and bad fits of some of the empirical models (e.g. extrapolation problems or wrong temperature coefficients) will cause variability in energy yield predictions at sites differently depending on their insolation distribution vs. Temperature and Irradiance This will be discussed in [17] in Nov 2017.

11 CONCLUSIONS

Twelve empirical models were studied – they all include "unphysical" coefficients and cannot repeatably fit imperfect or noisy data (extrapolation or gamma errors), they are not robust, their coefficients may be random and are not useful in performance understanding.

A new optimised Mechanistic Performance Model (MPM) has been introduced which

- Fits data from 3rd parties ASU, ESTI, SUPSI, TUV Rheinland, CFV (Sandia, NIST)
- Replicates predictions from existing models such as SAPM and 1-diode (PVSYST)
- Fits outdoor data with accurate measurements from Gantner Instruments

The MPM has been added to Gantner Instruments' gantner.webportal SaaS platform [2] for advanced analysis and fault detection.

12 REFERENCES

[1] https://docs.scipy.org/doc/scipy/reference/generated/scipy.optimize.curve_fit.html
[2] Gantner Instruments www.gantner-webportal.com
[3] Mani et al Photovoltaic Module Power Rating per **IEC 61853-1**Standard
[4] Ransome et al, 44th PVSC 2017 Washington DC "How to Choose the best Empirical Model for Optimum Energy Yield Predictions"
[5] Alhusna "Influence of Spectral Variations on Photovoltaic Module Energy Rating" PVSAT 13 Bangor 2017

[6] Heydenrich et al, "Describing the world with 3 parameters" 23rd PVSEC Valencia 2008 4DO 9.4
[7] IEC 60891:2009 "Photovoltaic devices - Procedures for temperature and irradiance corrections to measured I-V characteristics"
[8] Stein et al "Outdoor PV performance evaluation of three different models: single-diode, SAPM and Loss Factor Model" 28th PVSEC 2013 Paris, France
[9] Montgareuil et al "MotherPV method From watt-peak to watt-hours" 1st Sophia Workshop on PV Performance Modelling
[10] e.g. $PR_{DC} = C_0 + C_1*G + C_2*G^2 \ldots + C_n*G^n$ where n=4
[11] Jardine et al "PV-COMPARE: Direct Comparison of Eleven PV Technologies at Two Locations in Northern and Southern Europe" 17th PVSEC Munich 2001
[12] Yordanov thesis 2012
[13] Whitaker et al "Application and validation of a new PV performance characterization method" 26th PVSC 1997
[14] Ransome, "Comparing PV simulation models and methods with outdoor measurements" Hawaii PVSC 35, 2010
[15] Ransome "How Simulation Program kWh/kWp Predictions Depend on PV Model Discrepancies" 29th PVSEC 2014 Amsterdam
[16] Bruce King et al "Simplifying methods to calibrate the Sandia Array Performance Model: elimination of the traditional thermal test SAPM" 6th PVPMC Oct, 2016
[17] "7MoO.5.4 Optimised fitting of indoor (e.g. IEC 61853 matrix) and outdoor PV measurements for diagnostics and energy yield predictions" PVSEC-27 Nov 2017 Shiga Japan – To be presented

Acknowledgements:
ASU, CFV, Gantner Instruments, NIST, PVGIS, TUV Rheinland, SANDIA, SUPSI staff for data and discussions.

PERFORMANCE ANALYSIS OF PRE-CRACKED PV-MODULES AT REALISTIC LOADING CONDITIONS

Claudia Buerhop[1], Thilo Winkler[1], Frank W. Fecher[1], Andreas Bemm[2], Jens Hauch[1], Christian Camus[1], Christoph J. Brabec[1,3]

[1]ZAE Bayern e.V., D-91058 Erlangen, Immerwahrstraße 2,

[2]Allianz Risk Consulting GmbH, 81737 München, Fritz-Schäffer Straße 9

[3]i-MEET, FAU Erlangen-Nürnberg, 91058 Erlangen, Martensstraße 7

ABSTRACT: Damaged modules, especially with existing cell cracks, are observed quite often in PV-installations. Little knowledge exists about the stability of such pre-cracked modules under real operating conditions. Previous investigations and existing standards focus on the degradation of new, defect-free modules. This work highlights a twofold approach for performance study of pre-cracked modules: 1) artificial stressing of 20 representative pre-cracked modules with a novel load test set-up simulating snow and wind loads, 2) simulating IV-curves of cracked cells. The experimental results are discussed with respect to field exposure data. EL-images of stressed PV-modules indicate that new cracks are only initiated once a certain threshold is exceeded. Below this threshold formerly unseen cracks open, as can be seen by small changes in the EL-intensity. Power measurements yield a strong power reduction at the loaded state. Due to the solar cell being under tension the cracks open and the separated areas lose their electrical contact. Considering a break resistance in the IV-curve, simulations lead to good agreement with the observed experimental results.

It is highlighted that the electric contact after unloading for the previously unloaded module recovers visibly in the EL-image as well as in the power output. Simulating real weather (operating) conditions by moderate static loading tests reveals no measurable changes in the EL-images affecting the power output.

Keywords: EL-imaging, pre-cracked modules, cracks, degradation, performance.

1 INTRODUCTION

Quality control of PV-installations is of great interest today. Imaging methods, such as thermography (infrared) IR-imaging and electroluminescence (EL) imaging, are commonly used to visualize defective modules and their location [1-3].

Fig. 1 shows an aIR-image of a PV-plant. aIR-imaging means a drone-mounted IR-camera is used for the inspection [4]. The cells with homogeneously increased temperatures are typical for cell fracture [5]. Using EL, cracks as well as fractures can be detected/visualized in detail [6, 7]. Statistical analysis reveals that many PV-plants suffer from modules with cell fracture [1].

How cracks and fractures influence the lifetime and future performance under real operating conditions is still not known completely, yet. At operating conditions, the modules are exposed to various stresses, e. g. temperature cycles, wind and snow loads.

Existing standards and standardized test procedures do not provide the relevant information [8] for lifetime prediction of pre-cracked modules.

Therefore, the crack evolution in PV-modules for "realistic" weather conditions is studied. A one-year field study in Southern Bavaria [9] shows the mean daily wind speed is approx. 32 km/h, the most frequent wind speed is about 28 km/h and a maximum wind speed of about 140 km/h was observed only in a single storm.. The snow coverage can be rather low. Daily module temperature cycles range up to 60 K while the air temperature differs for the same period by maximum 20 K. Applying existing standards for structures [10, 11] corresponding wind loads, $p(v{=}30\ \text{km/h}) \approx 150$ Pa and , $p(v{=}140\ \text{km/h}) \approx 1300$ Pa are investigated. Respectively, for snow load results $p(\text{height of fresh snow} = 1\ \text{m}) \approx 500$ Pa.

Fig. 1: aIR-image of PV-plant with several modules showing locally increased temperatures due to cell fracture

For a better understanding of the future lifetime of pre-cracked modules we decided on a twofold approach: combining field exposure with lab testing and additional electrical simulations - the latter two being highlighted in this paper by computer simulating realistic loading conditions and monitoring crack growth simultaneously. The load test facility as well as the first results will be presented in this paper.

2 EXPERIMENTAL PROCEDURE

For this investigation pre-cracked, crystalline PV-modules (230 W) from a landslide damage were investigated exhibiting differing numbers of cells with cracks and fracture. Many cracks were identified in parallel and in vicinity to the busbars.

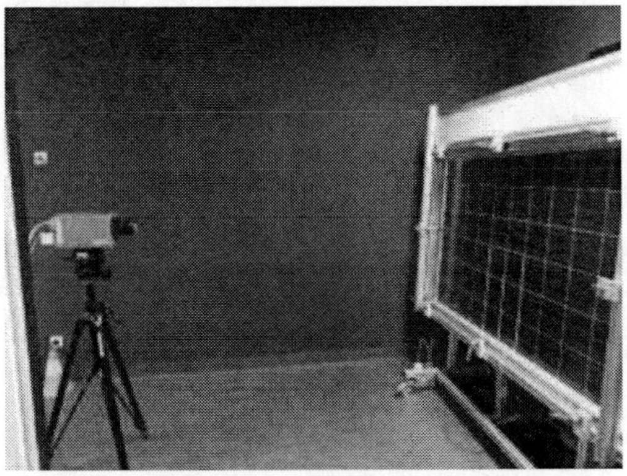

Fig. 2: Test facility for mechanical loading of PV-modules by vacuum and overpressure, EL-imaging, and illuminated IV-curve measurements. (Mobile sun simulator not shown in this picture.)

2.1 Loading experiments

Twenty modules were available for artificial load testing. For the first time, simultaneous EL-imaging during loading of the modules is presented here. The test facility applies the static pressure by vacuum pumps and air blowers. Thus, underpressure as well as overpressure can be used to simulate snow, wind loads and wind suction. We applied loads up to 5000 Pa. The test procedure includes a stepwise increase of the pressure by 200 Pa with a holding time of 30 min, afterwards followed by a relaxation period of 5 min without load. At the loaded and unloaded stage, EL-images were recorded. Standard EL-images are recorded using the EL-measurement system GE 2048 2048 DD NIR (greateyes GmbH, Germany). In addition, IV-curves were measured at an illumination of approximately 780 W/m² with a halogen lamp setup and the peak power P_{pk} calculated for standard test conditions (25°C, 1000 W/m², 1.5AM). In total there were fifteen loading and unloading steps. After the entire test sequence of fifteen loadings, the modules were loaded again with a fairly small load of 100 to 400 Pa simulating normal moderate operating conditions. We call this *ex-post loading*. Strain gauges recorded the displacement. The test routine was completed by initial (before first loading) and final (after last loading) IV-measurements with a solar simulator at standard test conditions. Fig. 1 shows a photography of the set-up with a mounted module. Here, a mobile self-made solar simulator measured the IV-curves at mechanically loaded and unloaded states.

2.2 Electric simulations

A simple equivalent electric circuit model was set up with LTspice for a 60-cell PV-module with cracked and fractured cells. In order to consider varying electric contacts of the separated cell parts by a crack, a break resistance R_B is introduced [6], which is high for completely electrically disconnected parts and it is rather low, if the cell areas are still electrically connected. The broken cell area was varied in the simulation as well as the number of broken cells. The module had three bypass diodes. The cell parameters for the one-diode model used for the cell simulations under nominal operating conditions (of irradiance = 800 W/m²) are: photocurrent I_{ph} = 6.84 A, diode saturation current I_s = 26.63 * 10-9 A, diode ideality factor n = 1.122, series resistance R_s = 7.92 mΩ and shunt resistance R_{shunt} = 612.5 Ω.

3 RESULTS AND DISCUSSION

3.1 Crack propagation

Applying homogeneous mechanical loads to PV-modules can initiate cracks if the solar cell is at tensile stress and the material's fracture toughness is exceeded [12]. Simulating wind suction overpressure is applied. At these conditions the solar cell are at compressive stress, such that cracks are not initiated. However, applying underpressure (in analogy to wind pressure or snow load) the solar cell is under tension. Cracks are able to grow or can be initiated. Fig. 3 shows the initiation and stepwise propagation of a crack in slow motion within 260 ms.

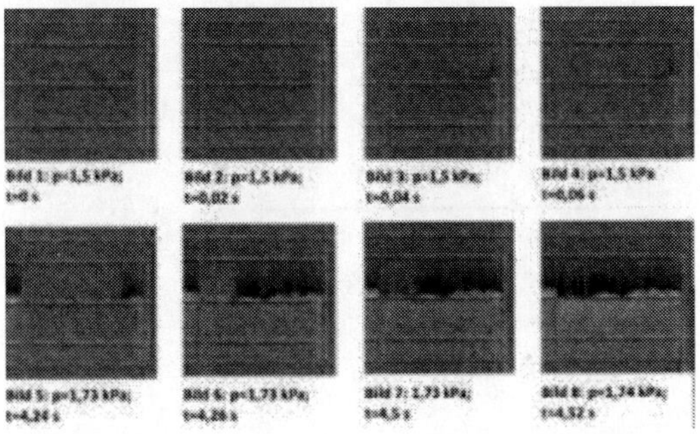

Fig. 3: Visualizing stepwise crack propagation through a solar cell within 5 s at loading with 1500 Pa and 1740 Pa, sequence recorded using an InGaAs-detector with an integration time of 20 ms (IRCAM, Germany), here difference EL images are shown, for which the image of the unloaded state was subtracted from the actual EL-image.

3.2 Simulated electrical performance

Cracks, visualized in fig. 3, can severely affect the electrical performance of the PV-modules. If cracks are not going through the entire thickness of the solar cell, an electrical contact still exists. The variety and quality of the contact is simulated by means of break resistance [13]. A large R_B leads to a stepped IV-curve with a plateau due to the constantly reduced current in the substring with the cracked cell. If the crack does not separate the fractions completely, a lower R_B is suitable. Fig. 5 depicts the change of the IV-curves and the shift of maximum power point mpp with changing R_B.

$n = 0$	$n = 3$	$n = 10$	$n' = 10$	$n = 15$	$n' = 15$	$n = \text{end}$
initial	Loaded with 600 Pa	Loaded with 2000 Pa	Unloaded after 2000 Pa	Loaded with 3000 Pa	Unloaded after 3000 Pa	Final
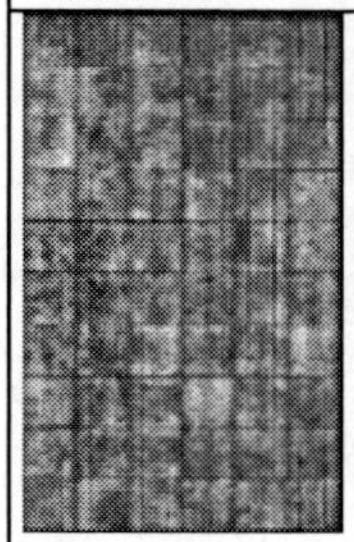	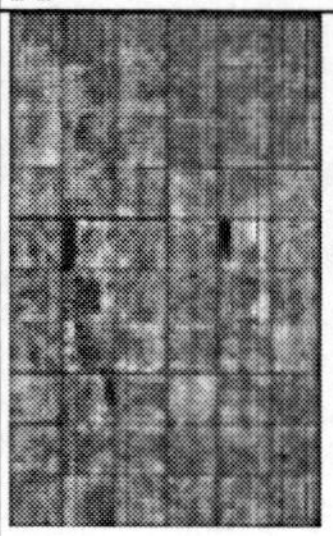	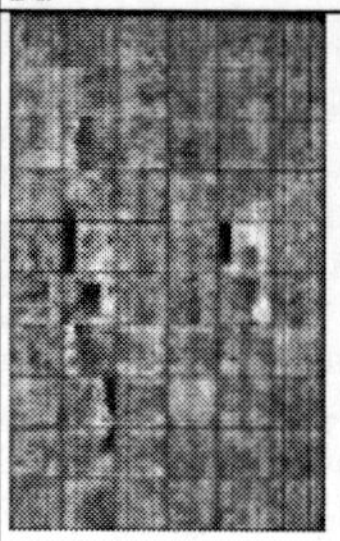	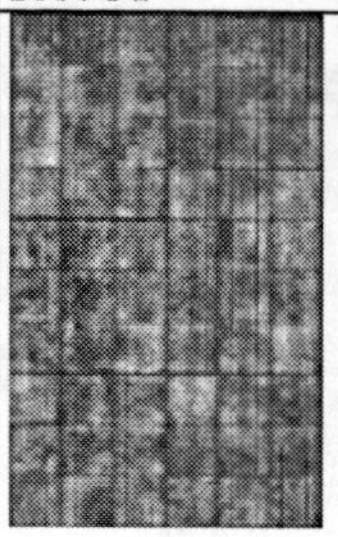	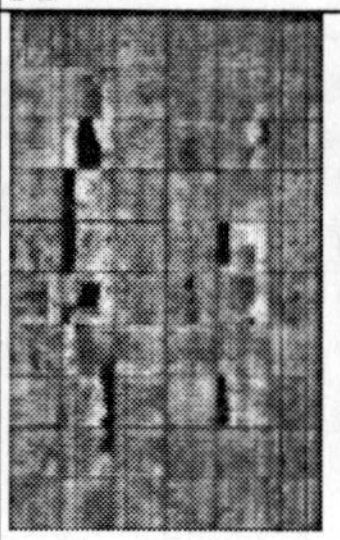	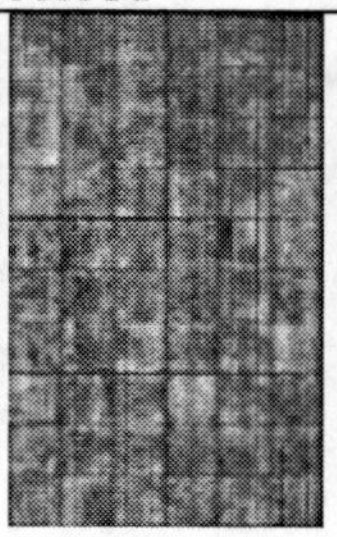	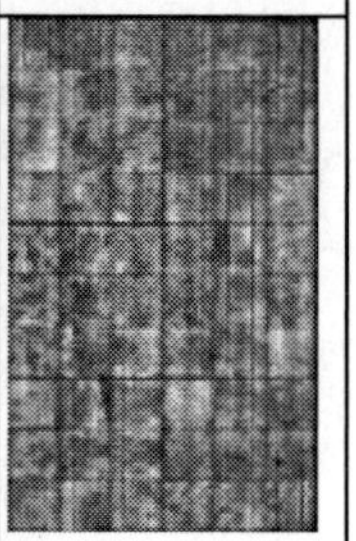
$P = 229.7$ W						$P = 227.9$ W

Fig. 4: EL-images for various loaded and subsequent unloaded stages as well as initial and final power measurements of module ID 922, n is the number of the experiment in ascending load order, n' describes the corresponding unloaded stage

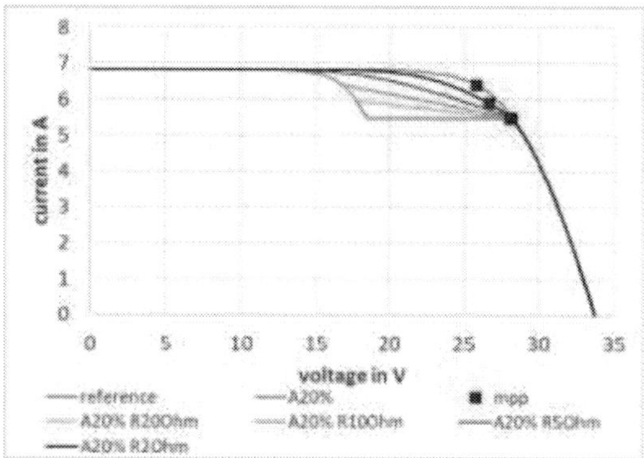

Fig. 5: Simulated IV-curves for **one** cracked cell, where the crack separates an area of 20%, with varying the break resistance between 2 Ω and infinity

Frequently, more than one cell has a crack in a module. In general, the break resistances of the cracked cells are not identical. Fig. 6 shows several combinations of fracture size (A = 10% or 20%) and R_B ranging from 2 Ω to high values (10 k Ω).

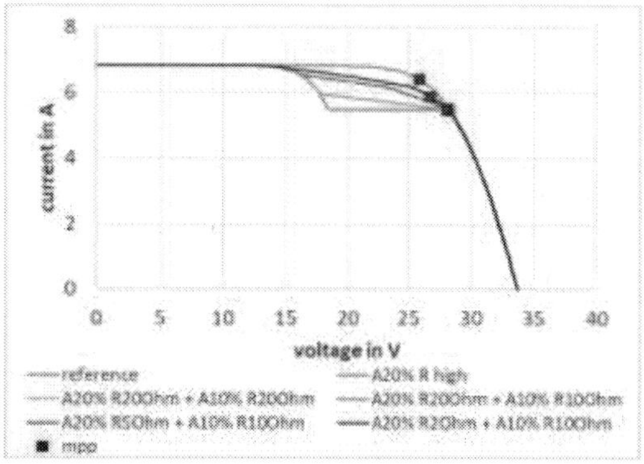

Fig. 6: Simulated IV-curves for **two** cracked cells in the same substring and varying separated area (A = 10 – 20%) and break resistance (R = 2 – 20 Ω)

For rather low R_B-values an IV-curve shaped similarly to the "uncracked" curve results. The mpp-values (power, current) change related to the fractured area and the break resistance. It is also visible that the cell with the largest cracked cell area in combination with the highest break resistance has the strongest and the dominating impact on the IV-curve.

3.3 Electrical performance at mechanical loading

Deflection and stress evaluation

The loading causes considerable bending of the PV-module. Deflection up to 3 cm is measured, see Table 1. Compressive stresses up to 27 MPa are measured in the center of the glass plate. In the solar cell corresponding tensile stresses are expectable. Finite-element-method (FEM)-simulation reveal that the highest stresses are present in the cells of the second long row and the corner cells [12, 14].

Table 1: Characteristic parameters measured for different loadings, deflection and principal stresses σ_1 and σ_2

Load In Pa	*Deflection in mm*	*Stress σ_1 in MPa*	*Stress σ_2 in MPa*
400	8.4	-10.4	1.26
1000	14.2	-17.6	2.9
3000	24.1	-25.3	9.3
5000	29.7	-27.5	16.0

EL-imaging

Demonstrating the influence on the loading procedure on the tested modules, exemplary EL-images of module ID 922 at different loading conditions are presented in Fig. 4. Initially, one cell obviously shows a fractured cell area visible as darkened cell areas, which are (partially) electrically disconnected. At a closer look, at least five additional cells with cracks in parallel to the busbars can be detected. Applying a low load of 600 Pa to the module, three cells with dark areas are obvious. The previously seen fractured cell appears much darker than before which is probably related to the loss of electrical contact at the crack. Increasing the external load induces more stress in the solar cells. Under a load of 2000 Pa and 3000 Pa, the EL-images reveal more crack openings, visible by their dark cell areas. Surprisingly, upon unloading some of the previously dark areas in fractured cells disappear. In Fig. 4 it is obvious, that freshly induced fractures are able to reconstruct the electric contact, whereas "old" fractures, already visible in the initial image, do not close. These openings and closures of cracks have also been described by Gabor [15] and Paggi [16]. Even more interesting is how pre-cracked modules perform under moderate operating conditions. Therefore, the modules are loaded again with moderate loads up to 400 Pa. First analysis revealed that *ex-post loaded* modules perform well although they were loaded previously up to 5000 Pa and accordingly contained many cracked cells [17].

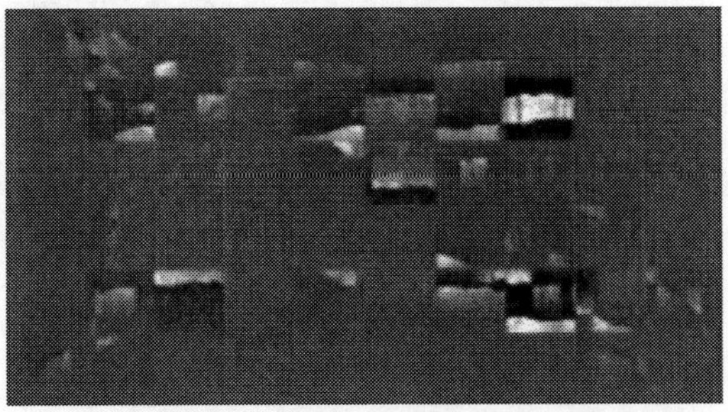

Fig. 7: Difference EL-image between loaded (1600 Pa) and unloaded initial stage for module ID 484

The impact of different loading steps is quantified by counting the number of cells with changes in the EL-image compared to the previous image. Therefore, difference EL-images are created by subtracting two EL-images, EL at the loaded stage and EL at the last unloaded initial stage, see Fig. 7. Changes become visible as white and dark areas. Many cracked cells are visible with cracks orientated parallel and diagonally to the cell edges depending on their location in the module.

IV-curves

IV-curves of loaded modules are shown in Fig. 8. As seen in the simulated graphs (Fig. 5 and Fig. 6) the shape changes constantly. Referring to the former results the break resistance seems to increase during loading as well does the fractured cell area. According to the IV-curves relevant fractured cells are present in one string because of the step-like character.

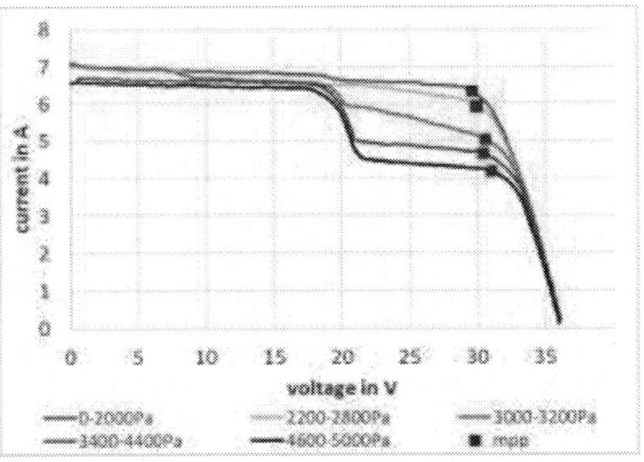

Fig. 8: Measured IV-curves at loaded states representative for certain temporary stable crack conditions are shown, module ID 922.

For increasing loads, the measured power at maximum power point (mpp) decreases continuously by 33%. In this case, the power decrease is related almost linearly to the decrease of I_{mpp}.

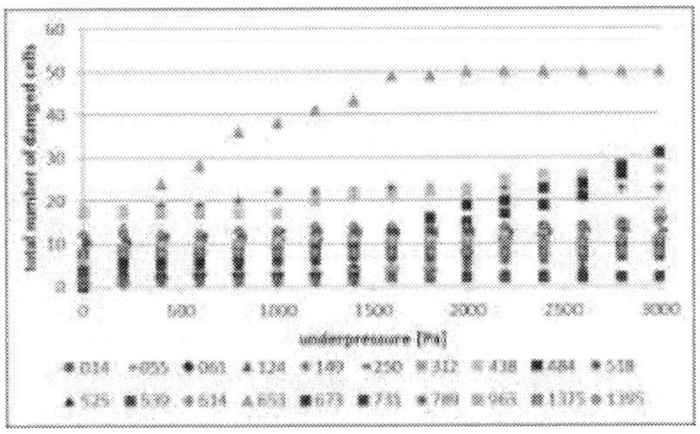

Fig. 9: Increase of newly cracked cells with increasing underpressure for all investigated modules [18]

Fig. 9 reveals how all investigated modules behave under load. With increasing underpressure the number of damaged cells increases as visualized by EL at the loaded stage. The increase is very different for the various modules. Most of them have a constant number of cracked cells up to approximately 1000 Pa. Then, the fracture strength of some up till then undamaged cells is probably exceeded and cracks are initiated. Module ID 653 is outstanding because of the extremely large number of fractured cells. In this case, a large number of hidden cracks seem to be present which become visible under loading.

Performance evolution under loading cycles

It is of great interest how the modules perform electrically under loading. Fig. 10 presents first data of measured peak power P_{pk} with the mobile solar simulator. The unloaded data (at 0 Pa) are measured directly before the subsequent loading. For better comparison the values "unloaded" are plotted at a 0 Pa close to the real previous and subsequent loading parameters. Up to ~1500 Pa the loading has no impact on the determined power output. At pressures above 3000 Pa, the power output under load is significantly lower than at the unloaded stage. Under load, the cracks are open and cause a power loss. Unloaded, the cracks are closed and the electrical contact is recovered yielding high, unreduced power output.

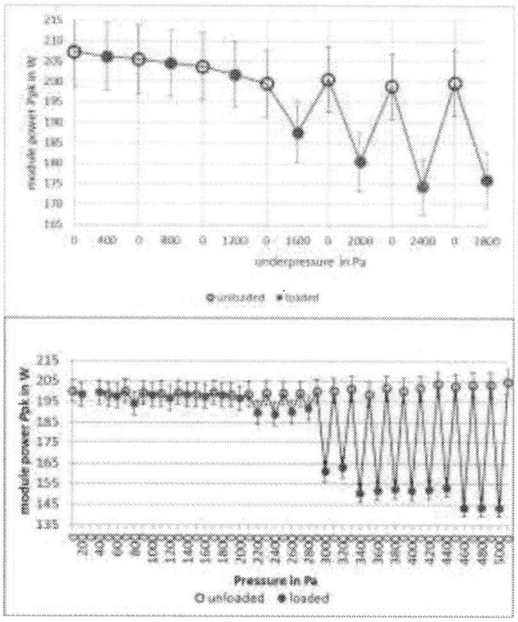

Fig. 10: Measurements of peak power P_{pk} of module ID 484 (top) and ID 922 (bottom) for different mechanical load states (at the test facility Fig. 2). The line indicates the order of the measurements.

Analyzing module power and EL-intensities of the fractured areas (between light grey to deep black) of background-corrected EL-images has to be done carefully. The impact of the non-contact area is recognizable by its grey-to-black appearance in the EL-image. Fig. 4 and Fig. 11 visualize the change of EL-intensity of certain cell areas for the loading qualitatively. Fig. 11 (bottom) shows a quantitative analysis of fractured areas as a function of the loading. For this first approach, the size of the fractured area is not taken into account, therefore the area does not change, but only its EL-intensity. Since the cracked cells are located mainly in two rows, only dominating cells of two strings are evaluated. For comparison, data of an unchanged cell (S2 D5) is plotted as well. While the intact cell has an EL-intensity of 2600 counts, really black, disconnected areas have an intensity of less than 500 counts in average. It is obvious that dark appearing cells and their continuous intensity reduction have minor impact on the power output (S1 B5, S3 E5). However, as soon as a certain threshold (here 340 counts) is underrun, a power drop results (see S3 E3). For distinct cells a spontaneous change is observed, e. g. S1 B3, S3 E3, S3 F8. These drops are related to measurable power reductions. For the measured power drop at 3000 Pa, cell S3 E3 has the dominating impact. Thus, only cell areas with EL-intensities below 500 counts have the potential to decrease the power if it is the largest affected cell area in the substring. E. g. cell S3 F8 does not cause further power reduction at $p > 4200$ Pa because the black area of S3 E3 is much larger. Furthermore, more broken cells with black area do not instantaneously lead to a power reduction. Therefore, the maximum black area of a cell within one substring has to increase.

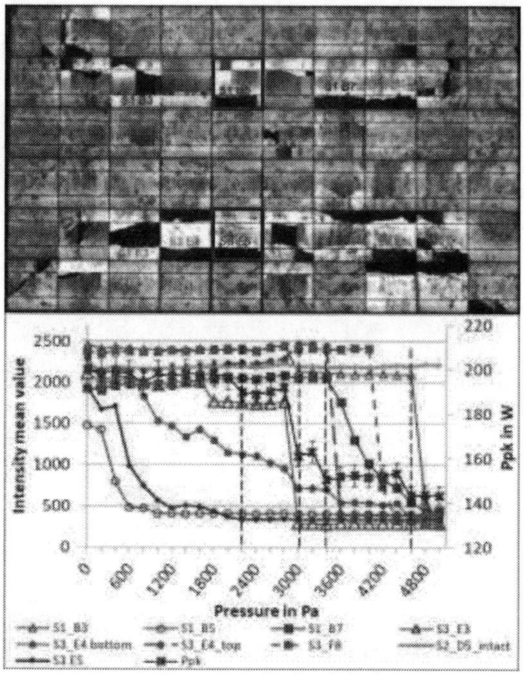

Fig. 11: Evaluation of EL-intensity of cracked areas of certain cells: EL-image of module ID 922 at loading with 5000 Pa (top), mean EL-intensity of the separated, dark area of the indicated cells (bottom). Framed cells showed initially dark areas in the EL-images, see Fig. 3. Note, S3 E4 is evaluated separately for both cell areas separated by the two cracks.

Module power

The influence on the power output in the unloaded state is shown in Fig. 12. Since the initial and final power measurements are carried out with the solar simulator, the module power was measured for the mechanically unloaded module. The data clearly reveal that no significant difference of the module power can be measured for the initial and final stage (except for 653). The values differ within the measurement accuracy marked by the error bars. Although nearly all modules suffer from damage due to the loading procedure, no influence on the power at the unloaded stage is measured. Module ID 653 with the extremely high number of defective cells behaves extraordinarily as it exhibits a significant drop in power.

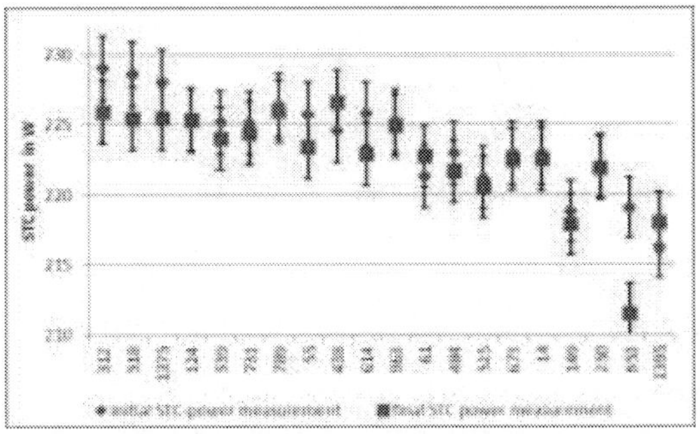

Fig. 12: STC-power of the tested pre-cracked modules before and after the complete loading test series

While Fig. 12 exhibits the module power before and after the loading procedure, Fig. 13 shows the power measured and number of cracked cells during loading. Fig. 13 reveals that the number of visibly cracked cells starts to increase continuously at a load of 1400 Pa whereas the power typically starts to drop at p = 2000 Pa. Since new cracks are generated and crack opening increases, most modules lose 20 to 30% of their power at the loaded state. Rarely modules withstand the loading. Nearly all modules show power reduction at $p > 3000$ Pa.

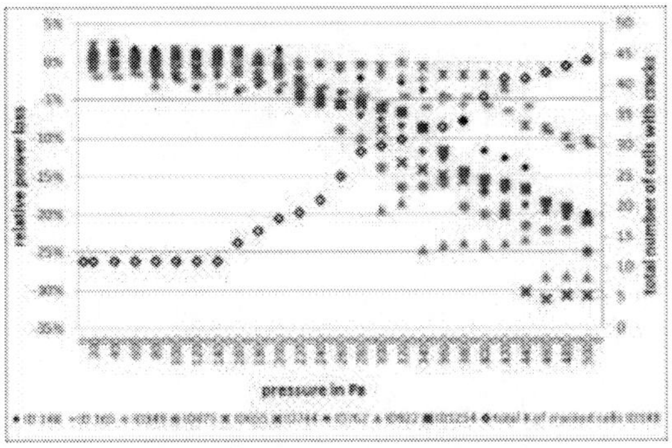

Fig. 13: Measured power loss as well as number cracked cells (representatively shown for module ID 148, compare fig. 9, too) with increasing loading for various modules (at the loaded state). $p = 0$ Pa indicates the unloaded state between the loadings.

These experimental results and findings agree well with field-exposed modules under real operating conditions [9]. Many cells with cracks were detected but the performance was stable at a high level and no further degradation in terms of new cracked cell could be visualized.

4 CONCLUSION

The lab results and one-year field findings yield that at moderate stresses up to 1200 Pa a negative impact on the performance of pre-cracked modules is not likely. However, the loading experiments clearly reveal that the modules at higher mechanically loaded stages have a reduced power output compared to the subsequent unloaded stage. Strikingly, opened cracks mostly close again when the load is withdrawn. Thus, modules in field, that experienced extreme load events, e. g. storm, can operate almost undisturbed after those extreme events.

5 ACKNOWLEDGEMENT

ZAE Bayern gratefully thanks the German Federal Ministry for Economic Affairs and Energy (BMWi) for financial funding of this project.

6 REFERENCES

[1] C. Buerhop, et al., SPIE 9938 Optics and Photonics, Optics + Photonics for Sustainable Energy (2016) 99380L.

[2] J. A. Tsanakas, et al.; Renewable and Sustainable Energy Reviews 62 (2016) 695.

[3] S. Koch, et al., 32nd EU PVSEC (2016) 1736 .

[4] C. Buerhop, et al., Third Southern African Solar Energy Conference SASEC (2015) 224.

[5] C. Buerhop, et al.; SOLMAT 107 (2012) 154.

[6] M. Köntges, et al.; Solar Energy Materials and Solar Cells 95 (2011) 1131.

[7] M. Sander, et al.; Solar Energy Materials and Solar Cells 111 (2013) 82.

[8] Standard IEC 61215 1995, 2005 Crystalline silicium terrestrial photovoltaic (PV) Modules - Design qualification and type approval 10.16 mechanical load test.

[9] C. Buerhop, et al.; Progress in Photovoltaics (2017) submitted.

[10]Standard DIN EN 1991-1-4 2010 Eurocode 1: Actions on structures - Part 1-4: General actions - wind loads, 2010 Allgemeine Einwirkungen - Windlasten.

[11]Standard DIN EN 1991-1-3 2010 Eurocode 1 - Actions on structures - Part 1-3: General actions - Snow loads, 2010 Allgemeine Einwirkungen - Schneelasten.

[12]C. Buerhop-Lutz, et al., 44th IEEE (2017).

[13]J. Käsewieter, et al.; Energy Procedia 55 (2014) 469.

[14]S. Kajari-Schröder, et al., 26. Symposium Photovoltaische Solarenergie (2011) 050.

[15]A. M. Gabor, et al., 43th IEEE, Photovoltaic Specialists Conference (PVSC) (2016).

[16]M. Paggi, et al.; scientific reports 4 (2014) 4506.

[17]C. Buerhop-Lutz, et al., SPIE 10370 Optics and Photonics, Optics + Photonics for Sustainable Energy (2017).

[18]S. Wirsching (2016) Master Thesis, FAU Erlangen-Nürnberg.

IN-LAMINATE FATIGUE OF SOLAR CELL INTERCONNECTORS: EXPERIMENTAL INVESTIGATION OF SENSITIVITIES

Matthias Pander, Rico Meier, Sebastian Schindler, Ulli Zeller, Matthias Ebert
Fraunhofer Center for Silicon Photovoltaics, Otto-Eissfeldt-Str. 12, 06120 Halle (Saale), Germany
phone: +49-345-5589-5215, fax: +49-345-5589-5999, matthias.pander@csp.fraunhofer.de

ABSTRACT: Fatigue of solar cell interconnectors is a serious failure mode for solar modules, as their interruption can lead to severe loss in module power as well as safety issues (e.g. arcing). Currently early failure and design issues are investigated by time consuming thermal cycling tests. Cyclic mechanical load testing accelerates this failure mode.

Within this work a coupon-based in-laminate fatigue test setup is introduced and evaluated for different load/test conditions. This includes load amplitude, frequency and test temperature. The relation between load amplitude and cycles to failure can be described by a power law model. The results show a significant reduction of cycles to failure at elevated temperatures. The test frequency showed a significant influence between 0.5 and 5 Hz, where higher deformation speeds resulted in higher cycles to failure. However, between 0.05 and 0.5 Hz there were no obvious differences in fatigue behavior. The velocity/frequency behavior needs to be considered when wind gust effects are discussed and for the design of accelerated fatigue tests. Another crucial influencing factor is introduced by processing conditions. Most significant for the achievable cycles to failure is the pre-deformation of the cell interconnectors, e.g. crimp. A crimp is found to significantly reduce cycles to failure. The comparison of cycles to failure for two tabber-stringers shows significant differences. Therefore, a reliable module design needs to consider the tabber-stringer process parameters.

Keywords: Ribbons, Solar Cell Interconnector, Cyclic Mechanical Loading, Lifetime Modeling, Fatigue

1 INTRODUCTION

The driving force for interconnector fatigue is a relative displacement of the solar cells. This movement can be either introduced by temperature changes or mechanical loads and differs from each other in amplitude, occurrence and by direction [1]. The cyclic loading may result in material fatigue and electrical failure of certain ribbons, strings (Figure 1) or even the whole modules. Therefore, fatigue of interconnectors is identified as one of the key failure modes in PV modules [2].

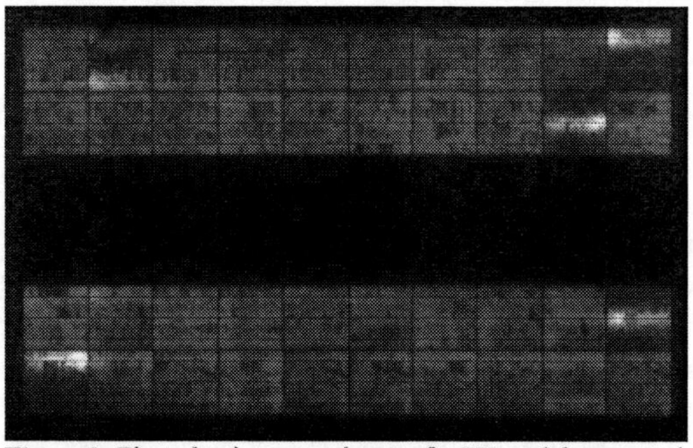

Figure 1: Electroluminescence image of a PV module extracted from a PV installation after five years [5]

Currently time-consuming thermal cycling tests are state of the art for regular type approval. Temperature cycling produces an almost equal strain distribution within the cell gaps across the module, so all ribbons are tested at the same time. But the regular test time is around 40 days, corresponding to 200 thermal cycles according to IEC 61215. A test to failure is even more expensive with respect to climate chamber usage and energy demand. Therefore, a cyclic mechanical load test was introduced to accelerate ribbon fatigue testing [3] and incorporated in a technical standard (IEC TS 62782:2016, [4]). The test has shorter test times but also limitations. Due to an unequal distribution of ribbon strain inside the module only certain positions depending on the mounting conditions are prone to fatigue [5]. Testing of a statistical relevant amount of samples leads to high experimental effort for installation and high material expenses. There are also technical limits regarding possible investigation of high-frequent environmental conditions (as wind) or for further acceleration of the test. The usage of different temperatures is not addressed by the standard but important for the investigation of the temperature dependent fatigue behavior.

Therefore, certain demands for the test setup for small samples were formulated:

1. Well defined, reproducible loading and mounting
2. Wide range of test parameters
    a. temperature range representing environmental conditions in the field (-40 to 85°C)
    b. loads representing snow or wind loads (+/-3 kN)
    c. accelerated testing (frequency > 0.1 Hz, which is used in IEC TS 62782:2016)
3. Cost effective equivalent to full size modules

2 TEST SETUP AND SAMPLE DEFINITION

The chosen test setup is a four-point bending setup due to the constant strain between the load rollers. The samples have a structure equivalent to current modules and are assembled with standard components [6]. In the technical standard the load amplitude of ±1000 Pa for full-sized module testing is suggested [4]. By numerical simulation it was found that this load corresponds to a relative change in gap of about 2.82 %. As a starting point for our test the load amplitude was set to reach the same change in gap width resulting in ribbon failures occurring between 1167 and 6384 cycles [6].

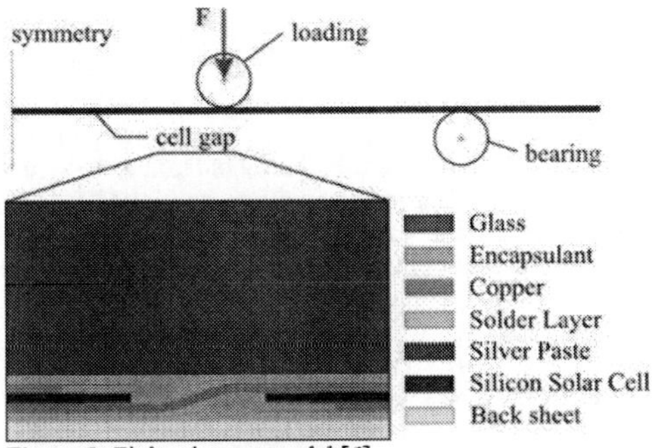

Figure 2: Finite element model [6]

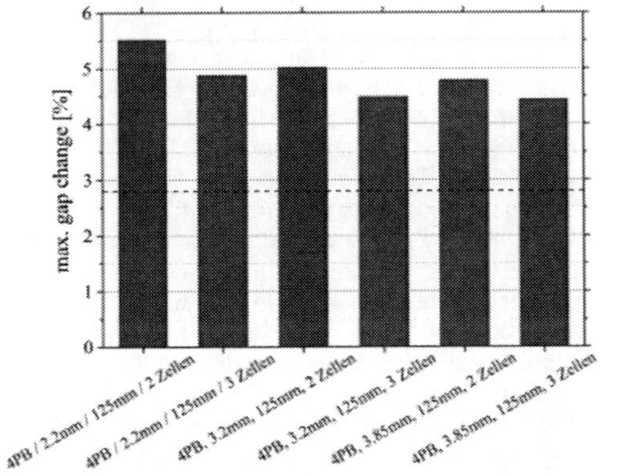

Figure 3: Maximum gap changes simulated for different setups

The experiments were conducted with a servo-hydraulic testing machine MTS Landmark. Two setups, one with (maximum sample size 750 mm x 200 mm) and one without climate chamber (maximum sample size 250 mm x 200 mm) were evaluated to achieve the desired gap change. The maximum bearing distance was fixed to 250 mm. The load roller distance for 4PB was fixed to 125 mm. Half cells were chosen for a suitable laminate. The initial cell gap was always 3 mm. A finite element simulation model was used to analyze two and three cell laminates under four-point (4PB) bending (Figure 2). The simulation was evaluated with respect to the achievable maximum cell gap change, which was limited by the glass strength (120 MPa was used in the simulation as fracture strength for fully tempered glass). In Figure 3 the maximum achievable changes in cell gap for different setups are shown. All setups were found to achieve a gap change larger than 2.8 % as required. 2.2 mm glass allowed the highest gap change but was excluded due to rare availability and strength requirements. Results of 3.2 mm and 3.85 mm glass (nominal 4 mm) had only minor differences. For reasons of availability and better heat strengthening behavior 4 mm glass was chosen. The decision for 3 half-cell laminates had the practical advantage of a higher number of possible breakage observations for better statistics (max. of 5 compared to 3). In summary the following test setup and samples were chosen:

- Setup: four-point bending with 125 mm distance of load rollers and 250 mm distance of support rollers
- 4 mm thick fully tempered glass, length: 280 mm, width: 200 mm and 180 mm (for comparison)
- One cell string with three half cells and 3 mm cell gap
- Standard EVA encapsulant and TPE backsheet

3 EXPERIMENTAL PROCEDURE

The nominal force amplitude to achieve the same gap change compared to [5] was 1.35 kN (width: 200 mm) and 1.22 kN (width: 180 mm). For accelerating the test with respect to comparison and sensitivity studies the test should be finished within 1000 cycles, therefore the force amplitude was increased to 2 kN (width: 200 mm) and 1.8 kN (width: 180 mm).

The complete test setup is shown on the left side of Figure 4. On the right side the four-point bending setup with sample is presented in more detail. The load was alternated between pre-force and chosen load amplitude with specified frequency. The sample was connected to a power supply that operates the mini-module in forward-bias. Impedance sensors were used to detect the induced current that results from an AC current signal from a signal generator (Figure 5a and b). Representative induction signals are shown in Figure 5c. The signal amplitude is similar at the beginning of the test. When an interconnector breaks its induction signal drops to zero and the current flow is redirected to the closest remaining interconnection(s) increasing the corresponding current through that ribbon (compare Figure 5).

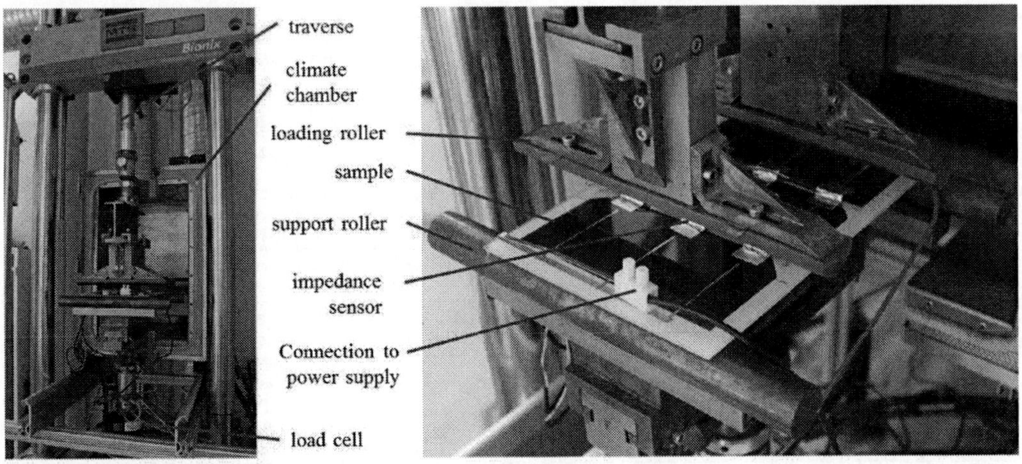

Figure 4: Experimental setup for In-Laminate fatigue testing

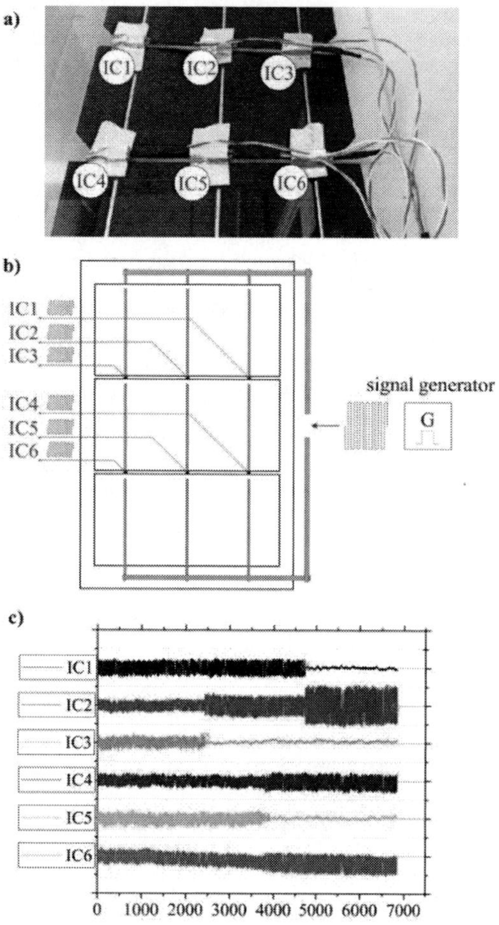

Figure 5: Ribbon breakage detection,
- a) sample with sensors,
- b) scheme for interconnectors and signal generator
- c) induction signal during a representative test

For the experiments three batches were produced in our module production line (overview in Table 1). The interconnectors were crimped (mechanically pre-deformed into s-shape) except for batch 3 and soldered by industrial tabber-stringer with infrared heating. The main differences between these samples were the interconnector batch and the glass width. At least three samples were used for each parameter set giving a minimum of 9 up to 15 breakage events. The standard test sequence was a triangular shaped cyclic force controlled loading between a minimum force (initial load for sample contact) and a maximum force with frequency of 0.5 Hz. The measurement was stopped when all ribbons between two cells were broken.

4 RESULTS AND DISCUSSION

4.1 Temperature dependence

It was found that an increase in temperature to 60 °C significantly reduces the cycles to failure (CTF) by factor 4.5 compared to room temperature (RT, Figure 6). Low temperatures such as -10 °C increase the cycles to failure by factor 8, whereas the deviation in CTF is not affected by temperature. The effect is attributed to the change of stiffness of the encapsulant [8]. Due to lower stiffness at 60°C the bending of the laminate is higher and the cells can move further resulting in a higher change of cell gap during each cycle. This leads to higher load on the ribbons and consequently lower cycles to failure (CTF). At -10°C this behavior is inverted. Additionally, the fatigue behavior of the copper material itself is influenced by temperature. In [7] it was shown that the fatigue strength of the copper ribbon is reduced at elevated temperatures.

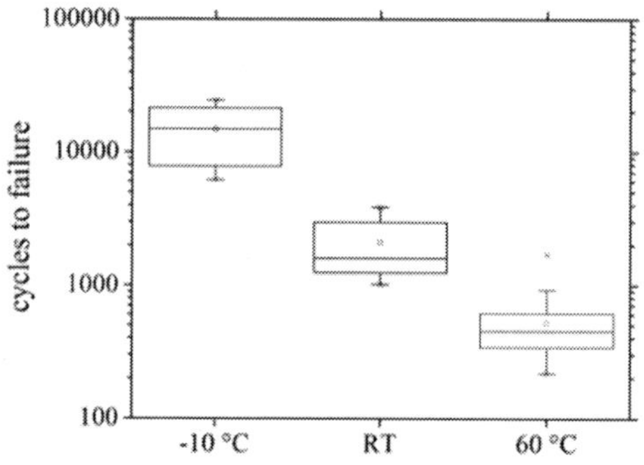

Figure 6: In-laminate fatigue, dependence on temperature, 50/2000 N, 0.5 Hz

Table I: Overview of produced samples and test parameters

Sample batch	number of samples	Objective	Parameters
280 mm x 200 mm x 3.85 mm Glass/EVA/EVA/TPE, interconnector batch 1	29	Temperature dependence	50/2000 N, 0.5 Hz, -10°C, 50/2000 N, 0.5 Hz, RT (23°C), 50/2000 N, 0.5 Hz, 60°C
		Frequency dependence	0.05, 0.5, 5 Hz
280 mm x 180 mm x 3.85 mm Glass/EVA/EVA/TPE, interconnector batch 2	33	Load dependence	45/900N, 45/1150 N, 45/1450N, 45/1800 N, 45/2000 N at RT, 0.5 Hz
		Load dependence at elevated temperature	same force amplitudes, 60°C, 0.5 Hz
280 mm x 180 mm x 3.85 mm Glass/EVA/EVA/TPE, interconnector batch 3	9	Manufacturing equipment and crimp effect	45/1800 N, RT, 0.5 Hz 3 samples with crimp on semi-automatic tabber-stringer, 3 samples without crimp on semi-automatic tabber-stringer, 3 samples with crimp on full-automatic tabber-stringer

4.2 Load frequency dependence

Another parameter where viscoelastic behavior of polymers (encapsulant and backsheet) plays an important role is the speed of loading / unloading (loading frequency). Since stiffness of the polymer normally increases for higher frequencies the cell displacement is reduced. This leads to an increase in CTF at 5 Hz by factor 1.6 (Figure 7). On the other hand, the test time is effectively reduced by factor 5. A reduction of frequency to 0.05 Hz has no significant effect on the mean CTF. The force amplitude variation was below 1 % and therefore had no significant impact on the observed cycles to failure. This investigation was repeated at 60°C. Here, the higher frequency slightly increases the CTF by a factor of 1.7 (Figure 7).

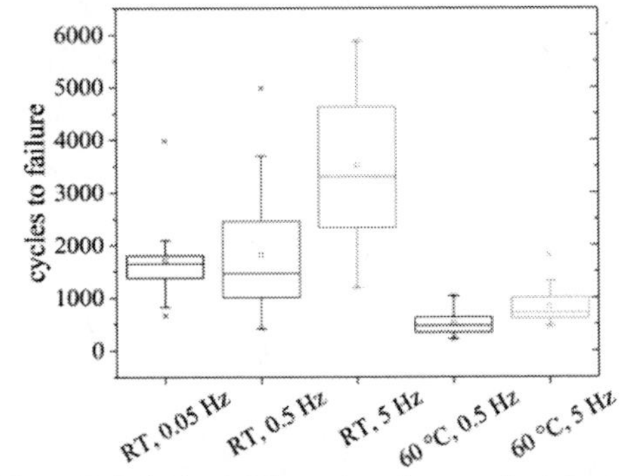

Figure 7: In-laminate fatigue, dependence on load frequency, 50/2000 N, at room temperature (RT) and 60°C

4.3 Load amplitude dependence

The load amplitude dependence follows a power law, both, for room temperature as well as 60 °C (Figure 8). The slope changes significantly between the two temperatures because of the temperature and load frequency depended behavior of the polymer materials. Since the load frequency was constant at all load steps the load rate was reduced for lower load amplitudes.

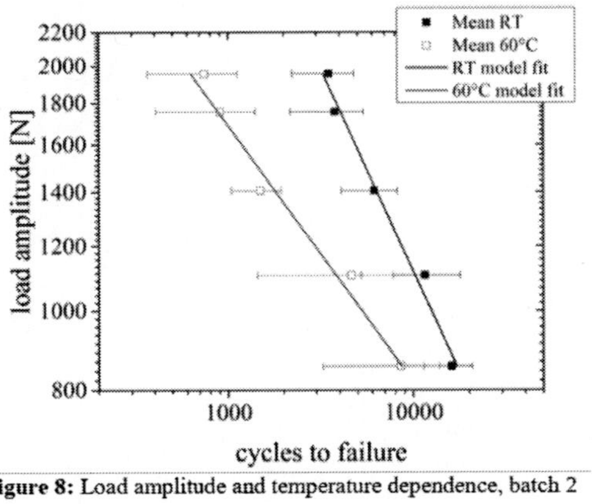

Figure 8: Load amplitude and temperature dependence, batch 2

4.4 Manufacturing equipment and crimp effect

The crimping process is known to have a significant impact on the resulting ribbon lifetime [6]. This was confirmed by the measurements shown in Figure 9. Due to pre-damage there is a significant reduction of mean cycles to failure (factor 10). Samples with no preliminary crimping show a higher scattering of failure incidents for stringer 1. Stringer 2 has better scattering but some early failures seem to occur due to unfavorable variations during lay-up and lamination.

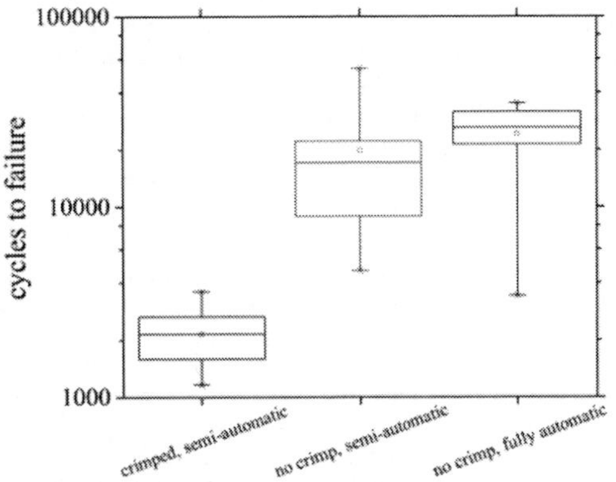

Figure 9: Manufacturing equipment and crimp effect, 45/1800 N, 0.5 Hz

5 SUMMARY AND OUTLOOK

The proposed test setup is a faster and more cost-effective alternative to TCT and full-size module mechanical load testing. Testing times can be reduced from months in thermal cycling tests to a few hours in cyclic mechanical load tests. With the introduced setup it was possible to investigate temperature dependence and frequency dependence of solar cell interconnector fatigue under module-like conditions with minimum effort compared to conventional full-size testing. It is found that temperature is a crucial factor and elevated temperatures strongly accelerate the fatigue. Higher testing frequencies (5 Hz) increase the cycles to failure, which needs to be considered regarding further acceleration. This behavior is attributed to temperature and rate-dependent mechanical properties of the polymer. The relationship between load amplitude and cycles to failure can be described phenomenologically with a power law model. Further results show a significant influence of material batch (yield strength, breakage strain) as well as the treatment during manufacturing (stringer and crimping).

This accelerated in-laminate fatigue based on small samples can be used in research and development and quality assurance. Applications with respect to interconnector fatigue are:

- Interconnector quality testing and benchmarking
- Comparison of different encapsulants and backsheets
- Monitoring and optimization of manufacturing processes (stringing and lamination)

Another application mainly for research is the validation of simulation models / results.

6 ACKNOWLEDGEMENTS

Manuel Schulze and Uwe Müller are gratefully acknowledged for sample preparation as well as Frank Wenger for testing the samples.

The financial support by the German Federal Ministry for Economic Affairs and Energy within project "LAURA" (contract no. 0325716B) is gratefully acknowledged.

7 REFERENCES

[1] R. Meier et al., "Reliability of copper-ribbons in photovoltaic modules under thermo-mechanical loading", 35th IEEE Photovoltaic Specialists Conference (PVSC), 2010

[2] W. Herrmann et al, "PV Module Degradation Caused by Thermo-Mechanical Stress: Real Impacts of Outdoor Weathering versus Accelerated Testing in the Laboratory", Proc. of SPIE Vol. 7773, 2010

[3] N. Bosco et al, Evaluation of Dynamic Mechanical Loading as an Accelerated Test Method for Ribbon Fatigue, 39th IEEE PVSC, 2013

[4] IEC TS 62782:2016, Photovoltaic (PV) modules - Cyclic (dynamic) mechanical load testing, March 2016

[5] Pander et al, Fatigue Analysis of Solar Cell Interconnectors due to Cyclic Mechanical Loading, Proc. 32nd EU PVSEC, 2016

[6] M. Pander et al, "Lifetime Estimation for Solar Cell Interconnectors", 28th EU PVSEC, 2013

[7] R. Meier et al, "Microstructural optimization approach of solar cell interconnectors fatigue behavior for enhanced module lifetime in extreme climates", Silicon PV 2016

[8] M. Pander et al, "Thermo-mechanical assessment of solar cell displacement with respect to the viscoelastic behaviour of the encapsulant", 12th EuroSimE, 2011

AUTHOR INDEX

AUTHOR INDEX

AUTHOR INDEX

AUTHOR INDEX

AUTHOR INDEX

AUTHOR INDEX

AUTHOR INDEX

AUTHOR INDEX

AUTHOR INDEX

AUTHOR INDEX

AUTHOR INDEX

AUTHOR INDEX

AUTHOR INDEX

AUTHOR INDEX

AUTHOR INDEX

AUTHOR INDEX

AUTHOR INDEX

AUTHOR INDEX

AUTHOR INDEX

AUTHOR INDEX

AUTHOR INDEX

AUTHOR INDEX

AUTHOR INDEX

AUTHOR INDEX

33rd European Photovoltaic Solar Energy Conference and Exhibition (EU PVSEC 2017)

Amsterdam, The Netherlands
25-29 September 2017

Volume 3 of 4

Editors:

A. Smets
P. Helm
N. Taylor

ISBN: 978-1-5108-5397-3

Printed from e-media with permission by:

Curran Associates, Inc.
57 Morehouse Lane
Red Hook, NY 12571

Some format issues inherent in the e-media version may also appear in this print version.

Printed by Curran Associates, Inc. (2018)

For permission requests, please contact WIP – Renewable Energies
at the address below.

WIP – Renewable Energies
Sylvensteinstr. 2
81369 Munchen
Germany

Phone: +49 89 72012735
Fax: +49 89 72012791

wip@wip-munich.de

Additional copies of this publication are available from:

Curran Associates, Inc.
57 Morehouse Lane
Red Hook, NY 12571 USA
Phone: 845-758-0400
Fax: 845-758-2633
Email: curran@proceedings.com
Web: www.proceedings.com

33rd European Photovoltaic Solar Energy Conference and Exhibition

Proceedings of the International Conference held in Amsterdam, The Netherlands

25 September – 29 September 2017

Edited by:

A. SMETS
Delft University of Technology
Professor Solar Energy
Delft, The Netherlands

N. TAYLOR
European Commission Joint Research Centre
Ispra, Italy

P. HELM
WIP Renewable Energies
Munich, Germany

Coordination of the Technical Programme:
European Commission Joint Research Centre
Via E. Fermi 1
21020 Ispra (VA)
Italy

Institutional Support:
European Commission
UNESCO – United Nations Educational, Scientific and Cultural Organization, Natural Sciences Sector
WCRE – World Council for Renewable Energy

Supporting Organisations:
ARE – Alliance for Rural Electrification
EREF – European Renewable Energies Federation
ESA – European Space Agency
EUFORES –The European Forum for Renewable Energy Sources
EUREC – The Association of European Renewable Energy Research Centres
European Technology & Innovation Platform PV
IEA PVPS – International Energy Agency, Photovoltaic Power Systems Programme
REN21 – Renewable Energy Policy Network for the 21st Century
TKI Urban Energy - Topsector Energie

Institutional PV Industry Cooperation:
SolarPower Europe
SOLARUNITED

Supporting Associations:
AIE – European Association of Electrical Contractors
AREA – African Renewable Energy Alliance
CABA – Continental Automated Buildings Association
EPIC – European Photonics Industry Consortium
GENSED - Turkish Solar Energy Industry Association
MPIA - Malaysian Photovoltaic Industry Association
SASIA - Saudi Arabia Solar Industry Association

EU PVSEC realised by:
WIP
Sylvensteinstr. 2, 81369 München, Germany
Tel: +49 89 720 12 735, Fax: +49 89 720 12 791
Email: pv.conference@wip-munich.de
www.photovoltaic-conference.com
www.wip-munich.de

Proceedings produced and published by:
WIP
Sylvensteinstr. 2, 81369 München, Germany
Tel: +49 89 720 12 735, Fax: +49 89 720 12 791
Email: pv.conference@wip-munich.de
www.photovoltaic-conference.com
www.wip-munich.de

ISSN 2196-0992

33rd EUROPEAN PHOTOVOLTAIC SOLAR ENERGY CONFERENCE AND EXHIBITION
25 SEPTEMBER – 29 SEPTEMBER 2017
AMSTERDAM, THE NETHERLANDS

EU PVSEC COMMITTEES

P.A. Basore, PV Specialist Services, Solana Beach, USA
S. Bowden, Arizona State University, Tempe, USA
T.M. Bruton, Progress in Photovoltaics, Woking, United Kingdom
A. Danel, CEA, Le Bourget du Lac, France
B. Dauksher, Arizona State University, Tempe, USA
M. Debucquoy, imec, Leuven, Belgium
P. Delli Veneri, ENEA, Portici, Italy
T. Desrues, CEA, Le Bourget du Lac, France
D. Dimova-Malinovska, Bulgarian Academy of Sciences, Sofia, Bulgaria
S. Dubois, CEA, Le Bourget du Lac, France
T. Dullweber, ISFH, Emmerthal, Germany
P. Fath, RCT-Solutions, Konstanz, Germany
S. Gall, HZB, Berlin, Germany
I. Gordon, imec, Leuven, Belgium
J. Haschke, EPFL, Neuchâtel, Switzerland
K. Hesse, Wacker Chemie, Burghausen, Germany
J. Horzel, CSEM, Neuchâtel, Switzerland
J.-R. Huang, Motech Industries, Taoyuan, Taiwan
J. John, imec, Leuven, Belgium
D.-W. Kang, Cheongju University, Republic of Korea
S.D. Kim, San Jose, USA
A. Meeder, Geo-En Energy Technologies, Berlin, Germany
J. Meier, Meier Technologies, Berlingen, Switzerland
A. Metz, Alzenau, Germany
C. Molpeceres, UPM, Madrid, Spain
J.W. Müller, Hanwha Q CELLS, Bitterfeld-Wolfen, Germany
Y. Ohshita, Toyota Technological Institute, Nagoya, Japan
S. Peters, Hanwha Q CELLS, Bitterfeld-Wolfen, Germany
K. Petter, Hanwha Q CELLS, Bitterfeld-Wolfen, Germany
R. Preu, Fraunhofer ISE, Freiburg, Germany
J. Rentsch, Fraunhofer ISE, Freiburg, Germany
E. Sauar, Brighterlite, Oslo, Norway
H. Schlemm, Meyer Burger, Hohenstein-Ernstthal, Germany
J. Schmidt, ISFH, Emmerthal, Germany
M.C. Schubert, Fraunhofer ISE, Freiburg, Germany
B. Terheiden, University of Konstanz, Germany
M. Vetter, IPHT, Jena, Germany
G.P. Willeke, Fraunhofer ISE, Freiburg, Germany
D. Wood, Hemlock Semiconductor, Hemlock, USA
A. Yamada, Tokyo Institute of Technology, Japan
D.L. Young, NREL, Golden, USA
M. Zeman, Delft University of Technology, The Netherlands
J. Zhao, Meyer Burger, Hohenstein-Ernstthal, Germany

Topic 3 – Thin Film Photovoltaics

C.J. Brabec, University of Erlangen-Nuremberg, Germany
A.N. Tiwari, EMPA, Dübendorf, Switzerland
S.C. Veenstra, ECN, Eindhoven, The Netherlands
W. Wischmann, ZSW, Stuttgart, Germany

G. Bardizza, European Commission JRC, Ispra, Italy
R. Birkmire, University of Delaware, Newark, USA
D. Cheyns, imec, Leuven, Belgium
S. Cros, CEA, Le Bourget du Lac, France
T. Dalibor, AVANCIS, Torgau, Germany
B. Dimmler, Manz CIGS Technology, Schwäbisch Hall, Germany
K. Durose, Stephenson Institute for Renewable Energy, Liverpool, United Kingdom
S. Huang, UNSW Australia, Sydney, Australia
S.J.C. Irvine, Swansea University, St. Asaph, United Kingdom
C.-F. Lin, NTU, Taipei, Taiwan
M. Lira-Cantú, CIN2, Barcelona, Spain
M.Ch. Lux-Steiner, HZB, Berlin, Germany
M. Meuris, imec, Leuven, Belgium
J. Parisi, University of Oldenburg, Germany
R. Po, eni spa, Novara, Italy
M. Powalla, ZSW, Stuttgart, Germany
A. Redinger, University of Luxembourg, Luxembourg
A. Romeo, University of Verona, Italy
V. Sittinger, Fraunhofer IST, Braunschweig, Germany
A. Virtuani, EPFL, Neuchâtel, Switzerland

Topic 4 – Concentrator and Space Photovoltaics

G. Flamand, imec, Leuven, Belgium
C. Signorini, European Space Agency, Noordwijk, The Netherlands

I. Antón, UPM, Madrid, Spain
A.W. Bett, Fraunhofer ISE, Freiburg, Germany
R. Campesato, CESI, Milan, Italy
V. Khorenko, Azur Space, Heilbronn, Germany
R. King, Arizona State University, Tempe, USA
F. Rubio Berenguel, Soitec Solar, Freiburg, Germany
G. Siefer, Fraunhofer ISE, Freiburg, Germany
G. Timò, RSE, Piacenza, Italy

Topic 5 – Performance, Reliability and Sustainability of Photovoltaic Modules and Balance of System Components

M. de Wild-Scholten, SmartGreenScans, Groet, The Netherlands
R. Gottschalg, Loughborough University, United Kingdom
N.M. Pearsall, Northumbria University, Newcastle upon Tyne, United Kingdom
M. Perrin, CEA, Le Bourget du Lac, France
T. Sample, European Commission JRC, Ispra, Italy
K. Wambach, Wambach-Consulting, Aindling, Germany

K. Baert, KU Leuven, Heverlee, Belgium
G. Beaucarne, Dow Corning, Seneffe, Belgium
R. Einhaus, Apollon Solar, Lyon, France
G. Friesen, SUPSI, Canobbio, Switzerland
V. Fthenakis, Columbia University, New York, USA
W.J. Gambogi, DuPont, Wilmington, USA
L. Gomes de Freitas, Federal University of Uberlândia, Brazil
G. Graditi, ENEA, Portici, Italy
Y. Hishikawa, AIST, Tsukuba, Japan
U. Jahn, TÜV Rheinland Energy, Cologne, Germany
S. Krauter, University of Paderborn, Germany
A.R. Lagunas, CENER, Sarriguren-Navarra, Spain
C. Monokroussos, TÜV Rheinland, Shanghai, China
M.A. Muñoz-García, UPM, Madrid, Spain
H. Nagel, Fraunhofer ISE, Freiburg, Germany
H. Nussbaumer, ZHAW, Winterthur, Switzerland
S. Oktik, Sisecam, Istanbul, Turkey
G. Oreski, PCCL, Leoben, Austria
W. Palitzsch, Loser Chemie, Zwickau, Germany
J.N. Roy, IIT Kharagpur, India
R. Sinton, Sinton Instruments, Boulder, USA
M. Van Iseghem, EDF R&D, Moret-sur-Loing, France
E. Voroshazi, imec, Leuven, Belgium
A. Wade, First Solar, Mainz, Germany

Topic 6 – PV System Performance and Integration

F.P. Baumgartner, ZHAW, Winterthur, Switzerland
F. Bonemazzi, ENEL, Rome, Italy
F. Frontini, SUPSI, Canobbio, Switzerland

P. Lechner, ZSW, Stuttgart, Germany
O. Mayer, GE Global Research, Garching, Germany
F. Montanari, ENEL Green Power, Rome, Italy
C. Protogeropoulos, EEPS, Athens, Greece
A. Scognamiglio, ENEA, Portici, Italy
W.G.J.H.M. van Sark, Utrecht University, The Netherlands
I. Weiss, WIP - Renewable Energies, Munich, Germany

C. Alonso-Tristán, UBU, Burgos, Spain
J. Binder, ZSW, Stuttgart, Germany
P. Blanc, MINES ParisTech, Paris, France
P. Bonomo, SUPSI, Canobbio, Switzerland
F. Bourry, CEA, Le Bourget du Lac, France
I. Da Silva, Strathmore University, Nairobi, Kenya
B. Gaiddon, Hespul, Lyon, France
M. Grottke, WIP - Renewable Energies, Munich, Germany
I.B. Hagemann, Architekturbüro Hagemann, Aachen, Germany
C. Helmke, Clake Energy, Erfurt, Germany
T. Huld, European Commission JRC, Ispra, Italy
E. Kaijuka-Okwenje, Rural Electrification Agency, Kampala, Uganda
S. Koopman, CSIR, Pretoria, South Africa
P.A. Leicester, Loughborough University, United Kingdom
P. Malbranche, CEA, Le Bourget du Lac, France
G. Mütter, Alternative Energy Solutions, Vienna, Austria
C. Nyman, Soleco, Porvoo, Finland
K. Peter, ISC Konstanz, Germany
T. Reijenga, BEAR-iD, Gouda, The Netherlands
J. Remund, Meteotest, Bern, Switzerland
K.T. Roro, CSIR, Pretoria, South Africa
M. Schroedter-Homscheidt, German Aerospace Center, Wessling, Germany
M. Sengupta, NREL, Golden, USA
D. Stellbogen, ZSW, Stuttgart, Germany
J. Sutterlueti, Gantner Instruments, Schruns, Austria
S. Tselepis, Pikermi, Athens, Greece
X. Vallvé, Trama TecnoAmbiental, Barcelona, Spain
M. Zehner, Rosenheim University of Applied Sciences, Germany

Topic 7 – PV Economics, Markets and Policies

C. Breyer, Lappeenranta University of Technology, Finland
M. Getsiou, European Commission DG RTD, Brussels, Belgium
G. Masson, Becquerel Institute, Brussels, Belgium
T. Nordmann, TNC Consulting, Feldmeilen, Switzerland
S. Nowak, NET Nowak Energy & Technology, St. Ursen, Switzerland
E. Perezagua, Consultores de Energía Fotovoltaica, Madrid, Spain

S. Caneva, WIP - Renewable Energies, Munich, Germany
I. Kaizuka, RTS, Tokyo, Japan
D. Mayer, MINES ParisTech, Paris, France
P. Mazzucchelli, EUREC, Brussels, Belgium
S. Szabó, European Commission JRC, Ispra, Italy

SUBJECT INDEX

Topic Code	Session Type	Day Codes
1 New Materials and Concepts for Photovoltaic Devices	P = Plenary Session	A = Monday, 25 September 2017
2 Silicon Photovoltaics	O = Oral Session	B = Tuesday, 26 September 2017
3 Thin Film Photovoltaics	V = Visual Session	C = Wednesday, 27 September 2017
4 Concentrator and Space Photovoltaics		D = Thursday, 28 September 2017
5 Performance, Reliability and Sustainability of Photovoltaic Modules and Balance of System Components		E = Friday, 29 September 2017
6 PV System Performance and Integration		
7 PV Economics, Markets and Policies		

e.g. 3CP.1 ⇒ 3=Thin Film Photovoltaics, C=Wednesday, P=Plenary Session, 1=Session 1

TABLE OF CONTENT
of EU PVSEC 2017 Proceedings Papers

Silicon Photovoltaics

Visual PRESENTATIONS 2AV.3 Heterojunction Solar Cells

Thin Film Photovoltaics

Plenary SESSION 3CP.1 Thin Film PV: Pushing the Limits with Breakthroughs in Industry and Research

Oral PRESENTATIONS 3AO.7 Optical Losses and TCO`s

Oral PRESENTATIONS 3AO.8 Module Stability and Characterisation

Oral PRESENTATIONS 3BO.9 Manufacturing and Performance Improvements

Concentrator and Space Photovoltaics

Oral PRESENTATIONS 4DO.4 III-V-Based Devices for Terrestrial and Space Applications

Visual PRESENTATIONS 4CV.4 III-V-Based Devices for Terrestrial and Space Applications

Performance, Reliability and Sustainability of Photovoltaic Modules and Balance of System Components

Plenary SESSION 5DP.1 Performance, Reliability and Sustainability of Photovoltaic Modules and Balance of System Components

Oral PRESENTATIONS 5EO.1 Sustainability and Recycling

Visual PRESENTATIONS 5BV.4 PV Module Performance and Reliability (I)

PV System Performance and Integration

Plenary SESSION 6DP.2 PV System Performance and Integration

Oral PRESENTATIONS 6BO.7 Advanced Field Performance Estimation

Oral PRESENTATIONS 6BO.8 Failure Modes and Degradation

Visual PRESENTATIONS 6BV.3 Solar Resource and Forecasting / Building, Infrastructure and Landscape Applications / Grid and Energy System Integration

PV Economics, Markets and Policies

Oral PRESENTATIONS 7DO.8 Global PV Economics and Market Trends

Oral PRESENTATIONS 7DO.9 Innovative National PV Market Economics Business Cases

Oral PRESENTATIONS 7EO.3 PV-Related Policies, Strategies and Societal Issues

Visual PRESENTATIONS 7DV.1 PV Economics and Markets / PV-Related Policies, Strategies and Societal Issues

FOREWORD

The EU PVSEC 2017 in Amsterdam concluded with the strong message that the photovoltaic revolution is just getting started. Research and innovation across the sector, together with fast deployment and numerous application options, are paving the way to putting PV at the centre-stage of sustainable energy systems. This year 2,516 participants from 89 countries took part in this most inspiring platform. It provided a vital forum for information exchange in the field of PV solar energy research, demonstration and applications. In the exhibition 64 companies from all parts of the world welcomed visitors and presented their products and services.

The EU PVSEC 2017 Proceedings give a comprehensive overview of the photovoltaic solar sector, its current status and future prospects in science, research, innovation, development and deployment on more than 4,500 pages. Selection for inclusion in the conference was made by paper review experts and topic organisers (see the listing on pages III-V), to whom we express our sincere gratitude for their comprehensive review work and overall contribution to the success of the conference. In addition to the 619 submitted papers, the proceedings include 70 presentations (slides) shown during the plenary and oral presentations as well as 232 poster files of the visual presentations. In total this amounts to 921 publications.

The 2017 EU PVSEC Proceedings maintain our commitment to providing quick and open access to high quality scientific results. They constitute a powerful tool for targeted and quick information search and retrieval, enabling you to search by keyword, paper title and authors. We are sure that these features will help to simplify the use and exploit the full potential of this extensive source of information.

The Conference Proceedings are published as downloadable DVD file and are also fully accessible online. A DOI code (Digital Object Identifier) has been assigned to each paper. This ensures unequivocal and permanent identification and full citability. The papers can be viewed and downloaded in a full free open access from the EU PVSEC Proceedings website www.eupvsec-proceedings.com.

We are confident that these Proceedings will play an important role in providing a comprehensive overview of the current actors and activities in the global PV sector and that they will disseminate information on the state-of-the-art of technologies and applications. This can generate further research, add momentum to innovation and promote interest in PV worldwide.

We would like to cordially thank all authors and participants of the EU PVSEC 2017 for their contributions and look forward to welcoming you in 2018 at the 35th EU PVSEC 2018 in Brussels, Belgium.

The Editors

EU PVSEC 2017 25 - 29 September 2017
33rd European PV Solar Energy Conference and Exhibition
RAI Convention & Exhibition Centre • Amsterdam • The Netherlands

Selected Highlights

Nigel Taylor, Technical Programme Chair

.... with help from the session chairs and others

European Commission

Silicon

European Commission

2nd Fabrication Run of n-Type HP mc Solar Cells
IV Results

Fraunhofer ISE

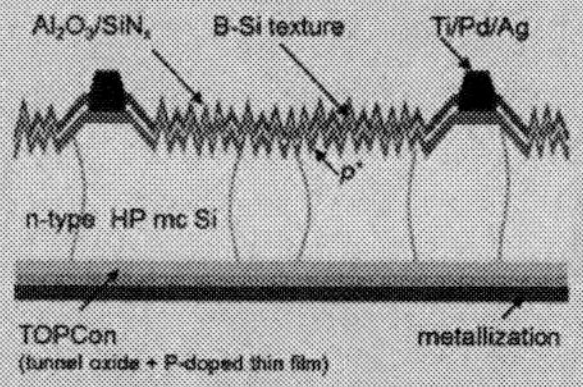

V_{oc} [mV]	J_{sc} [mA/cm²]	*FF* [%]	*pFF* [%]	η [%]
674.2	41.1	80.5	81.6	22.3*

Cell area 2×2 cm², aperture area (*) certified by Fraunhofer ISE CalLab

- 2nd cell batch
- Applying of „lessons learned" in cell fabrication
- Same material as previous solar cells

➔ New record efficiency of 22.3% for a mc silicon solar cell

2DO.3.1 Approaching 22% Efficiency with Multicrystalline n-Type Silicon Solar Cells
J. Benick et al, Fraunhofer ISE, Freiburg, Germany

European Commission

Q.ANTUM technology on p-Cz substrates in mass production

Hanwha Q CELLS

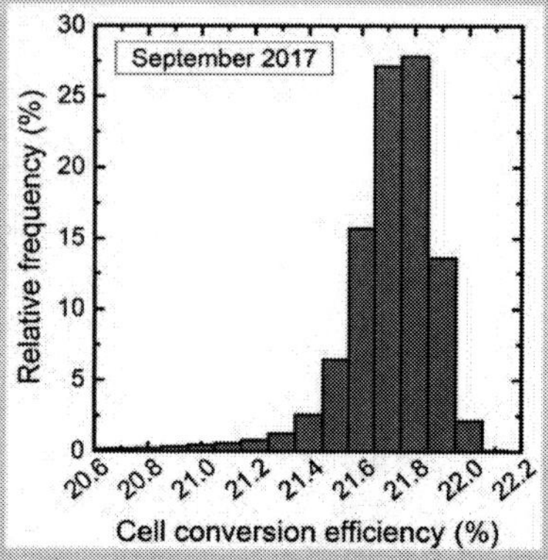

Platform change to Q.ANTUM[1] secures:

- Efficiency increase of +0.6%abs. per year
- Evolutionary development process[2] and upgrade of existing tool set
- Silver consumption: -50%
- Throughput increase
- **Avg. cell efficiency of 21.7% in production (max. >22%)**

2CO.9.6 Mass Production of Q.ANTUM Solar Cells and Modules on p-Type Cz Silicon Substrates
F. Fertig et al, Hanwha Q CELLS, Bitterfeld-Wolfen, Germany

European Commission

24.13% efficiency IBC cell

Trinasolar

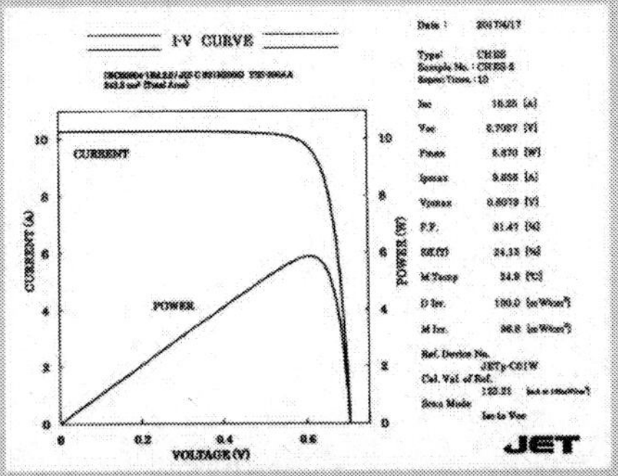

Compared to 23.5% cell (2016):

- 10 mV gain in V_{oc} by using selective emitter structure;
- J_{sc} is also improved by using less absorption in SiN_x;
- Higher FF is due to the reduced R_s, mainly from lower ρ_c on emitter.

Area	V_{oc}	I_{sc}	V_{mpp}	I_{mpp}	FF	Eff
[cm²]	[mV]	[A]	[mV]	[A]	[%]	[%]
244.9	702.7	10.25	607.9	9.66	81.47	24.13

* Independently measured at Japan Electrical Safety & Environment Technology Laboratories (JET)

2CO.11.5 Large-Area (6 Inch) Screen-Printed IBC Solar Cells with Efficiency Approaching 24% without Passivated Contacts

G. Xu et al, Trina Solar Energy, Changzhou, China

European Commission

Advantages of a-Si/c-Si heterojunction (HJ) back contact (BC) Si solar cell

a-Si/c-Si HJ : High V_{oc} and FF by excellent a-Si/c-Si interface

Back contact : High J_{sc} and FF by small optical and resistive loss

Si wafer-based: Reasonable production cost

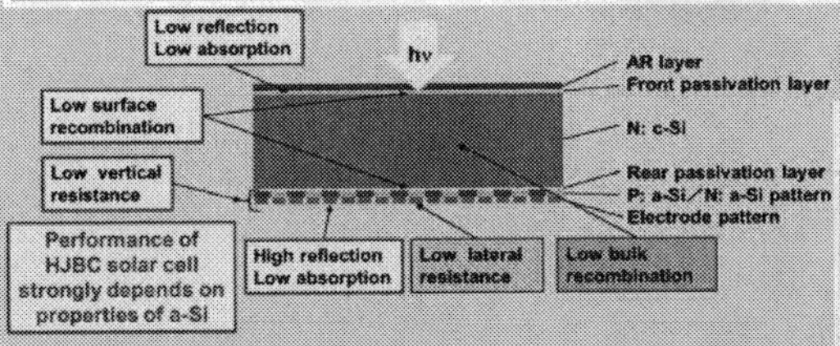

kaneka

High efficiency c-Si solar cell; PV parameters

*da: designated area *ta: total area *ap: aperture area

	V_{OC} (V)	J_{SC} (mA/cm²)	FF (%)	Eff (%)	Area* (cm²)	Cell type
Panasonic	0.740	41.8	82.7	25.6	143.7 (da)	HJBC [1]
SunPower	0.737	41.3	82.7	25.2	153.5 (ta)	IBC [2]
Fraunhofer ISE	0.725	42.5	83.3	25.7	4.017(da)	TOPCon [3]
Kaneka	0.738	40.8	83.5	25.1	151.9 (ap)	HJ [4]
Kaneka	0.744	42.3	83.8	26.3	180.4 (da)	HJBC [5]
Kaneka	0.740	42.5	84.7	26.6	179.7 (da)	HJBC [6]
Kaneka	0.738	42.7	84.9	26.7	79.0 (da)	HJBC
Theoretical limit	0.753	43.6	89.2	29.3		150um-thick wafer [7]
Ratio (26.7%/limit)	0.980	0.979	0.952	0.911		

- High V_{oc}, J_{sc} and FF obtained by excellent PECVD passivation and structure design
- Challenge for improvement of FF is a key to achieve >27% efficiency.

[1] K. Masuko et al., IEEE Journal of Photovoltaics, 4, 1433 (2014), [2] "SunPower 2015 Analyst Day" (2015), [3] A. Richter et al., Silicon PV proceedings (2017), [4] D. Adachi et al., Applied Physics Letters, 107, 233506 (2015), [5] K. Yoshikawa et al., Nat. Energy 2, 17032 (2017) [6] K. Yoshikawa et al., Silicon PV proceedings (2017) [7] A. Richter et al., IEEE J. Photovolt. 3, 1184 (2013).

2BP1.1 Record-Breaking Efficiency Back-Contact Heterojunction Crystalline Si Solar Cell and Module

K. Yamamoto et al, Kaneka, Osaka, Japan

J.L. Hernández, Kaneka, Westerlo-Oevel, Belgium

European Commission

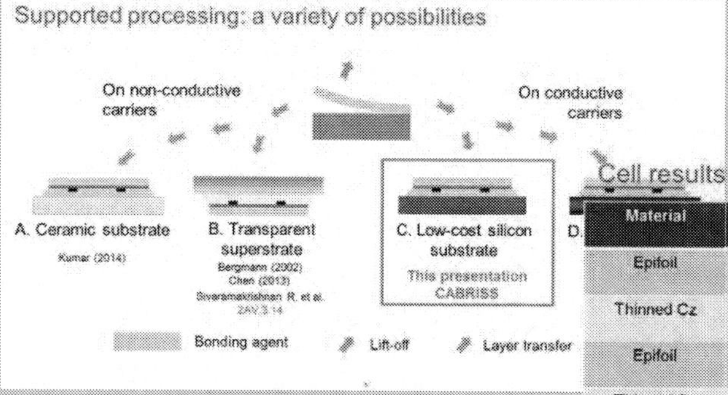

imec

TU WIEN · SINTEF · PyroGenesis · CABRISS · INKRON

Cell results

Material	Configuration		J_{sc} [mA/cm²]	V_{oc} [mV]	FF [%]	Eff. [%]	
Epifoil	Bonded to p⁺ Cz substrate	**Best** Avg.	**33.2** 32.6	**681** 676	**69.2** 69.1	**15.7** 15.3	*This work*
Thinned Cz	Bonded to p⁺ Cz substrate	**Best** Avg.	**33.2** 32.7	**738** 739	**73.0** 72.1	**17.9** 17.3	*This work*
Epifoil	Freestanding	**Best** Avg.	**33.5** 32.9	**693** 656	**73.2** 72.1	**17.0** 15.6	*Bearda et al. [2AV.3.26]*
Thinned Cz	Freestanding	**Best**	**34.0** 33.9	**744** 723	**74.4** 71.6	**18.8** 17.6	*Bearda et al. [2AV.3.26]*
Comparison wit) literature (cells bonded to conductive substrates)							
Epifoil (18 μm)	Bonded to steel	**Best**	**34.5**	**632**	**77.2**	**16.8**	*Amberwave/UNSW*
Slim-cut (25 μm)	Bonded to steel	**Best**	**33.6**	**580**	**76.7**	**14.9**	*Astrowatt*

2CO.12.2 Solar Cells on < 50μm Thick Epitaxial Foils Conductively Bonded to Low-Cost Si Carrier

H. Sivaramakrishnan Radhakrishnan et al, imec, Leuven, Belgium
N. Bednar & N. Adamovic, Vienna University of Technology, Austria
R. Roozeman & J. Heikkinen, INKRON, Espoo, Finland
A. Ulyashin & M. Syvertsen, SINTEF, Oslo, Norway

7

European Commission

Thin Film

8

European Commission

Key techniques implemented into the record sub(mini)modules

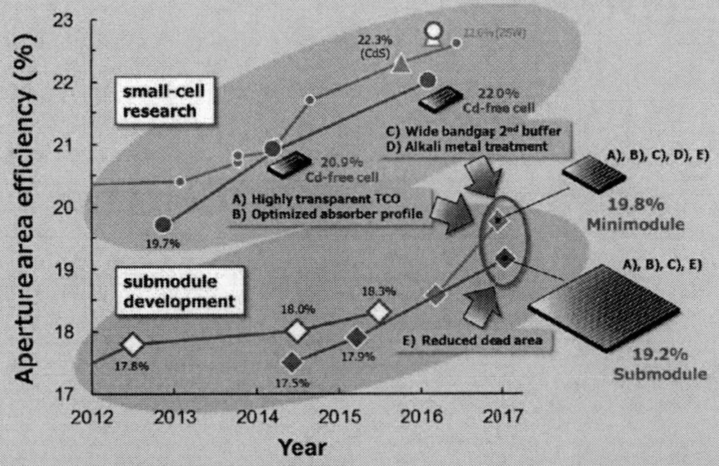

3CP.1.1 Present Status of Solar Frontier Cu(In,Ga)(Se,S)2 Record Efficiencies and Overall Progress

V. Bermudez et al, Solar Frontier, Atsugi, Japan

9

European Commission

HIGH EFFICIENCY MODULES

- Current in-house 30×30cm^2 champion module with 18.2% efficiency
- high Ga/high S absorber, wide bandgap buffer - highly Na doped In_xS_y:Na/i-$Zn_{0.85}Mg_{0.15}O$

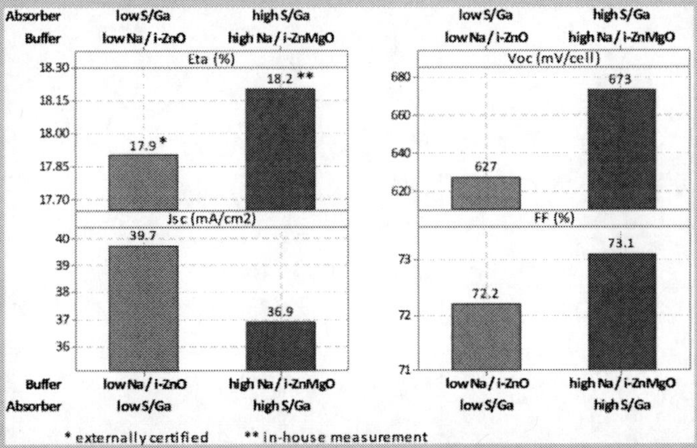

3BO.9.1 Wide Bandgap Sequential Absorber with Tunable Buffer Bandgap for CIGSSe Solar Modules at 18% Efficiency

R. Lechner et al, Avancis, Munich, Germany

10

European Commission

SOLIBRO Hi-Tech
A Hanergy Company

FULL SIZE: CHAMPION MODULE GEN 3

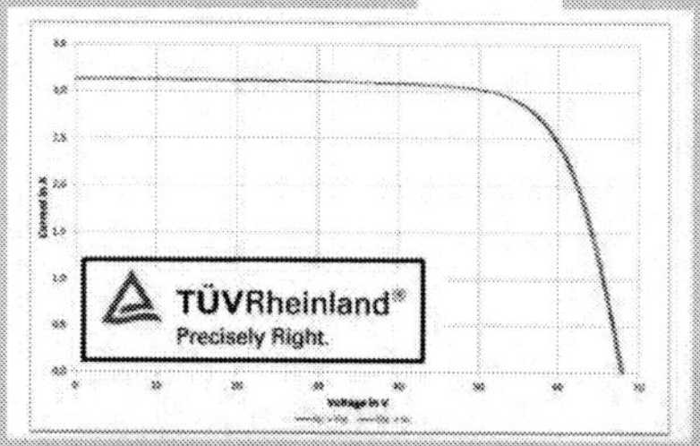

P_{mpp} [W]	159.4
V_{oc} [V]	67.9
I_{sc} [A]	3.12
FF [%]	75.2
η	17.0 %

- 17.9% Aperture Area Efficiency
- Externally verified
- Processed in SL2 production
- Module with future target process GEN3

17.0% @ 0.94m²
totale area efficiency
New world record

CELL AND MODULE DESIGN

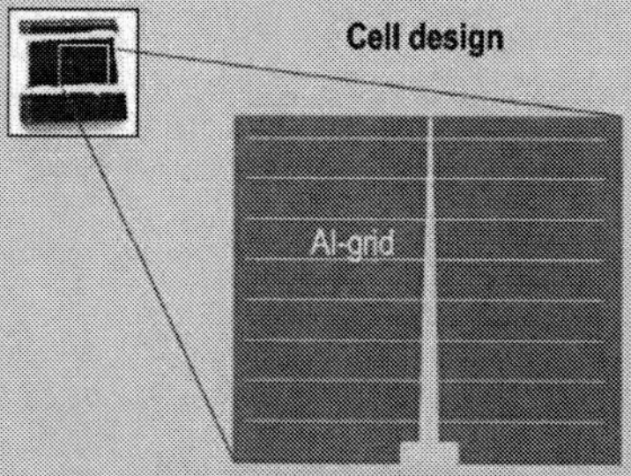

- Sheet resistance: 70 Ohms/sq
- Cell area: 10x10 mm²
- Dead area loss: ~ 3%
- Encapsulation: MgF_2

3CP.1.2 **17% Total Area Efficiency at Commercial Size CIGS Module**

P. Kratzert et al, Solibro Research, Uppsala, Sweden

Double Vision

Light Trapping

IV Results of III-V//Si Solar Cells

- Increase in J_{sc}
  - 11.6 → 12.7 mA/cm²
- Efficiency: 33.3%
- Record efficiency for 2T multi-junction solar cell

V_{oc} [V]	J_{sc} [mA/cm²]	FF [%]	η [%]
Gen 1			
3.046	11.7	87.5	31.3
Gen 2			
3.125	11.6	86.5	31.4
Gen 3			
3.127	**12.7**	83.8	**33.3**

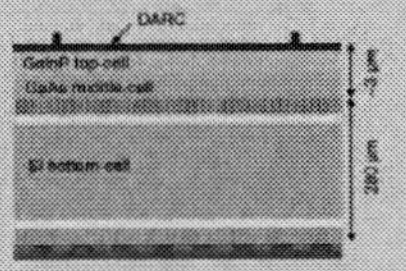

1AP.1.3 Monolithic III-V//Si Multi-Junction Solar Cell Exceeding an Efficiency of 31%

Fraunhofer ISE, Freiburg, Germany

Surplus collection of carriers at 3rd terminal injected from top cell

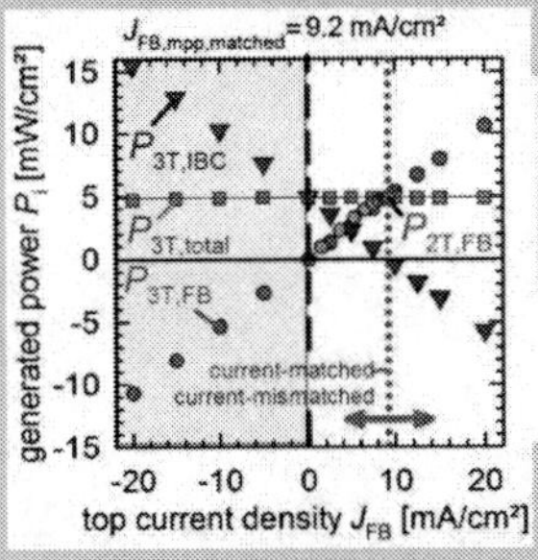

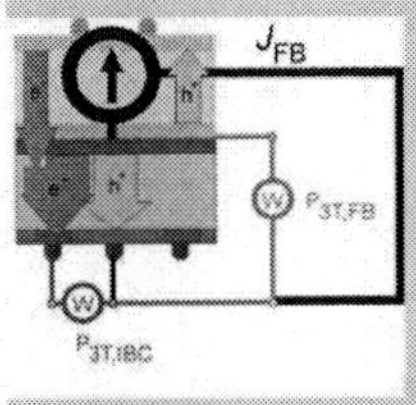

1AP.1.2 : Maximum Power Extraction Enabled by Monolithic Tandems Using Interdigitated Back Contact Bottom Cells with Three Terminals

ISFH and NREL

European Commission

13

22.7% Perovskite-CIGS thin film tandem solar cells

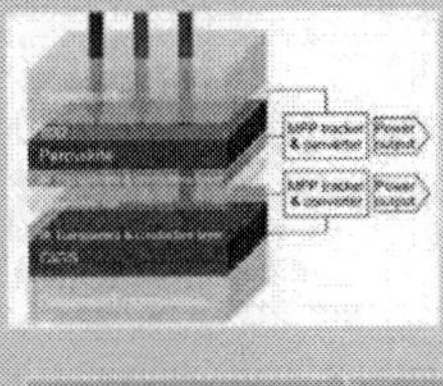

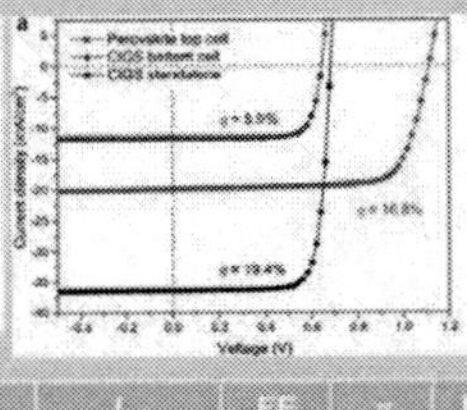

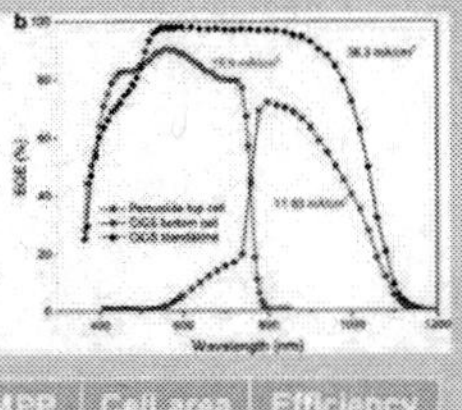

Solar Cells	V_{OC} (V)	J_{SC} (mA/cm²)	FF (%)	η (%)	MPP (%)	Cell area (cm²)	Efficiency gain
CIGS (stand alone)	0.684	36.4	78.2	19.4	19.4	0.213	
CIGS bottom cell	0.645	11.6	78.1	5.9	5.9	0.213	
Perovskite top cell	1.116	19.9	75.7	16.8	16.8	0.273	
Perovskite-CIGS 4-terminal tandem					22.7	0.213	3.3

Further improvement in Perovskite-CIGS thin film tandem solar cells via bandgap engineering, optical coupling etc.

Laboratory for Thin Films and Photovoltaics

13

3DO.7.4 Efficient and Stable NIR-Transparent Perovskite Solar Cells Prepared by Partial Ion Ex-Change Method for All-Thin-Film Tandem Applications

F. Fu, S. Pisoni, T. Feurer, A. Wäckerlin, S. Nishiwaki, A.N. Tiwari & S. Buecheler

EMPA, Dübendorf, Switzerland

European Commission

Performance

European Commission

~~MWp~~ MWh/MWp

Energy rating uncertainty

Total-order sensitivity analysis

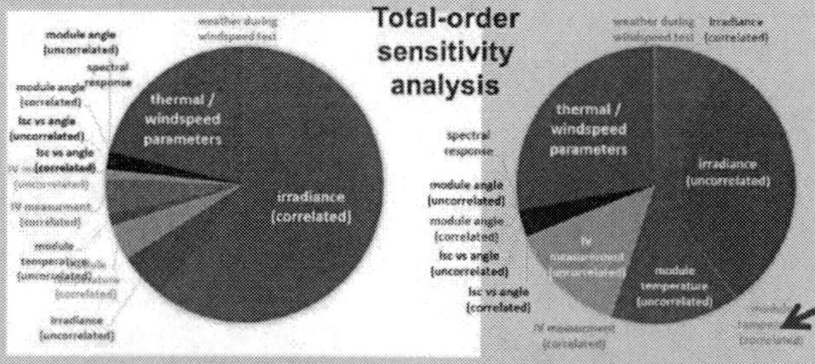

Uncertainty of $E_{mod,year}$
2.16 % (95% conf.)

Uncertainty of *CSER*
1.84 % (95% conf.)

Simulation of 1000 modules, mixed types.
Uncertainty model based on survey of accredited test labs.

IEC 61853 Energy rating standard

Part 1 – Measure performance matrix
Measure Isc, Pmax, Voc, Vmax at vs. irradiance and temperature

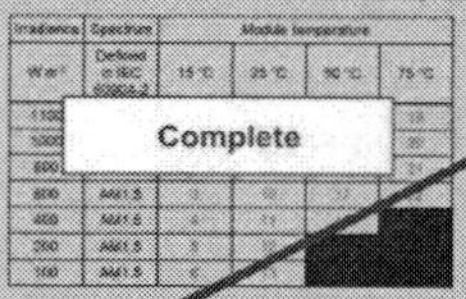

Complete

Part 2 – Other measurements

- Ang...
- Sp...
- Thermal & wind speed parameters

Complete

Part 3 – Standardised model

Final stages

CSER = **Climate specific energy ratio**
Performance relative to STC as %

Part 4 – Reference climates
Several (currently 6) profiles
Hourly values for:

- Ambie...
- Wind ...
- Sun el...
- Sun incidence angle
- Global horizontal irradiance
- Direct horizontal irradiance
- Global in-plane irradiance
- Direct in-plane irradiance
- Spectral irradiance

Negotiating which climates to include

5CO.7.4 Progress in Energy Rating Standards: Accuracy and Optimisation
J.C. Blakesley, NPL, United Kingdom
T. Huld & H. Müllejans, European Commission JRC, Ispra, Italy

PhotoClass PV-Enerate

European Commission

Bifacial impact in the field

Bifacial Single Axis Tracker (NM)

Exceptional service in the national interest

Sandia National Laboratories

- Daily Potential Bifacial Energy Gains were estimated from front and back irradiance data using reference cells.
- Potential gains increase when tracker is off-track, but yield decreases

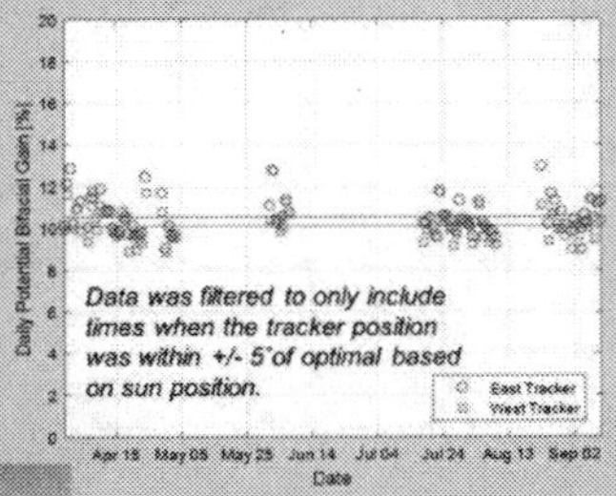

16

6CO.13.3 Outdoor Field Performance from Bifacial Photovoltaic Modules and Systems

J.S. Stein et al, Sandia National Laboratories, Albuquerque, United States
C. Deline, NREL, Golden, United States
F. Toor, University of Iowa, United States

17

European Commission

Light and elevated Temperature Induced Degradation

Piercing the Puzzle Together

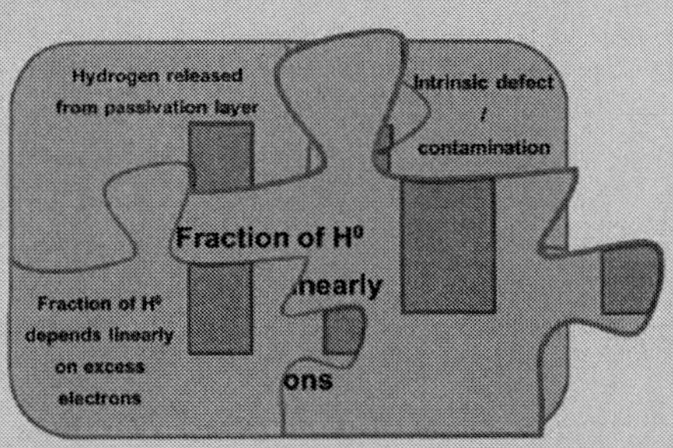

Model
Degradation

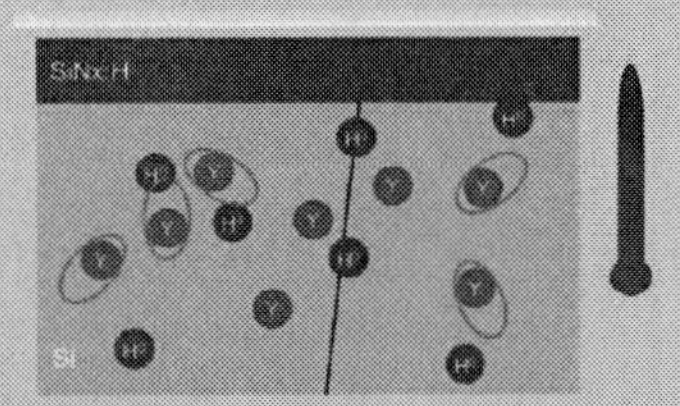

- LeTID observed in all p-type Si materials
- Defect activation only possible after fast firing of SiN_x:H-layer
- LeTID defect involves one fast-diffusing species
- Defect activated by inducing a change in charge state

→ Key to controlling LeTID: Manipulation of hydrogen

- Promising route for investigations: Float-zone silicon

2BP.1.4 Understanding Light-Induced Degradation in Multicrystalline Silicon: Possible Complex Formation Mechanisms

F. Schindler et al, Fraunhofer ISE, Freiburg, Germany

18

European Commission

Big data can help (1)

More than one million PV systems < 30 kWp have been installed in Germany.

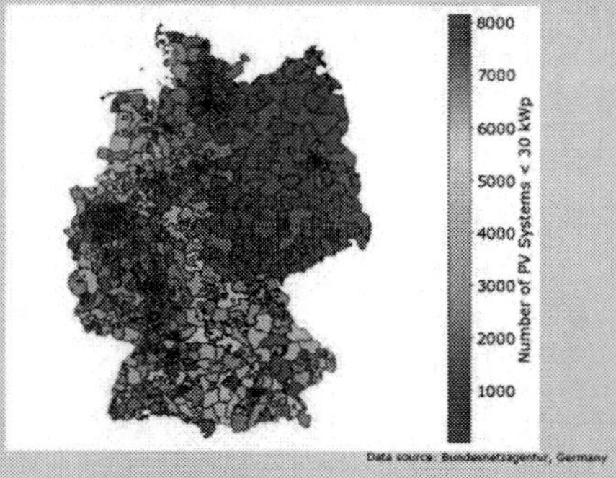

There is an annual loss of €450m to €500m by residential roof-top PV systems in Germany.

Year	Annual electric energy production in TWh	Annual yield loss in TWh	Monetary loss in million EUR
2014	36.1	1.46	€460m
2015	38.7	1.59	€489m
2016	37.6	1.58	€460m

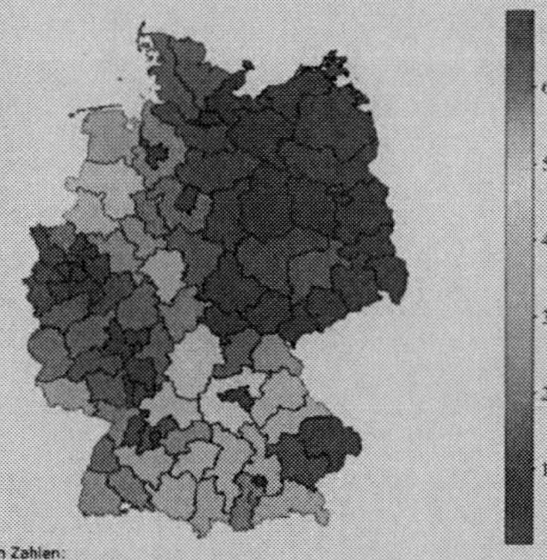

Source for mean Feed-In Tariff: BMWI, EEG in Zahlen: Vergütungen, Differenzkosten und EEG-Umlage 2000 bis 2017.

6BO.7.1 Survey on Yield of PV Systems in Germany 2014 to 2016
H. te Heesen & M. Rumpler
Trier University of Applied Science, Neubrücke, Germany
V. Herbort Ulm University of Applied Sciences, Germany

19 European Commission

Big data can help (2)

Model development: Model Architecture

i-EM

Two parallel approaches

- **Generic Fault Prediction (Supervision-Diagnosis Model):** *predict inverter deviations from nominal behavior*
- **Specific Fault Prediction (Short-Term Fault Prediction Model):** *predict a specific fault class*

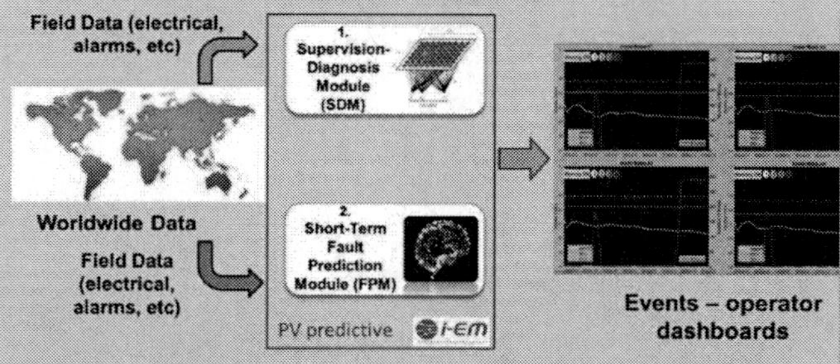

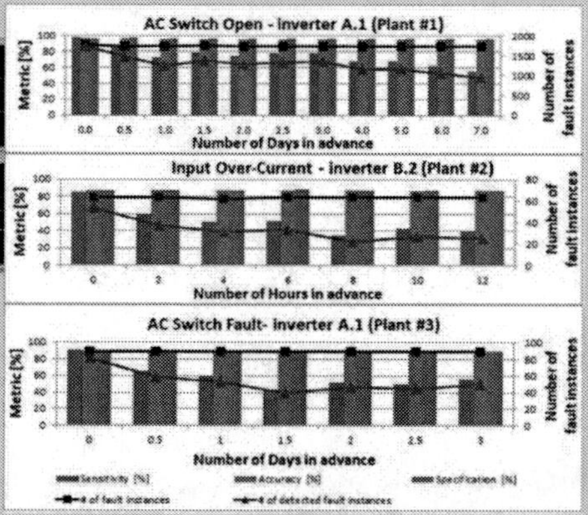

6DP.2.4 Predictive Maintenance in Photovoltaic Plants with a Big Data Approach
Alessandro Betti et al, i-EM (Livorno, ITALY), Enel Green Power (Rome, Italy)

20 European Commission

Non-destructive testing increasingly quantitative

Indication for PID

PID
- Homogeneous T increase
- Located at one string end
- Trend of PID-ratio increase towards the end
- < 50% of modules affected
- Incubation time
- PID-cells close the frame

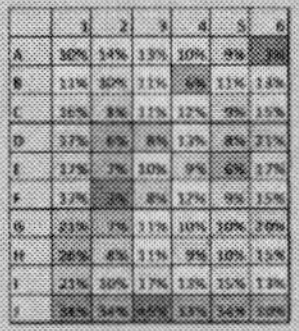

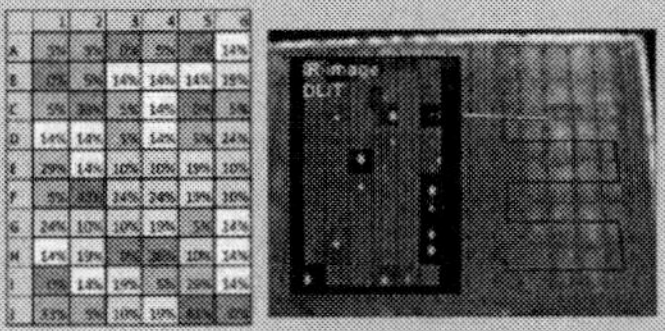

- Homogeneous T increase
- Arbitrarily distributed
- many modules may be affected
- Close to commissioning
- Affected cells everywhere

6BO.8.2 Quantitative Study of Potential Induced Degradation of a Roof-Top PV-Installation with IR-Imaging

C. Buerhop-Lutz et al, ZAE Bayern, Erlangen, Germany
C. Zetzmann Rauschert, Pressig, Germany
C.J. Brabec University of Erlangen-Nuremberg, Germany

European Commission

Which module is an insurance case?

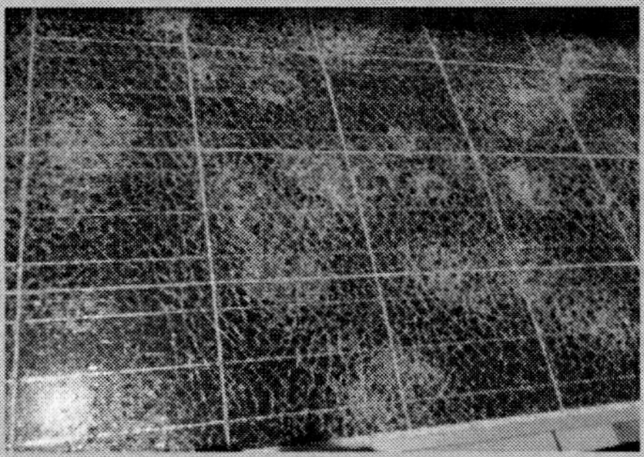

6BO.8.3 Scientific Investigation of a PV Generator After Hail

W. Mühleisen et al, CTR, Villach, Austria
M. Spielberger PVSV, Guttaring, Austria
H. Sonnleitner ENcome, Klagenfurt, Austria
Y. Voronko OFI, Vienna, Austria

European Commission

Product Environmental Footprint Category Rules (PEFCR) for PV

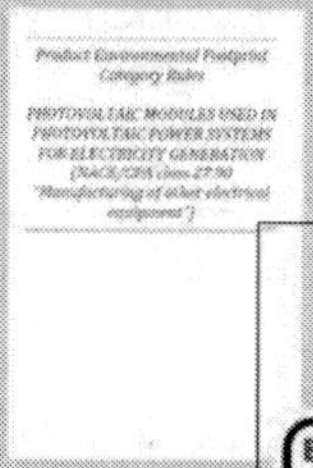

The PEFCR provides a harmonized set of rules to evaluate the environmental impacts of electricity produced from photovoltaic modules installed in the region of the EU28 and EFTA countries.

Circular Economy

Is it feasible to aggregate the complex results of a Product Environmental Footprint evaluation into a single score?

Communication Vehicles (CVs)

Performance label	Loyalty schemes	Campaigns targeting user behaviour
Performance label + QR code	Leaflets, catalogues, etc.	Marketing campaigns/ advertising
Performance improvement label	Instruction manuals	Declaration/ product passport
Barcode	Websites (producer, vendor)	On invoices
Pictogram, POS product advertisement	Websites (3rd party)	PEF external communication report
Consumer receipt – basket comparison	App based on performance	Performance tracking report.

Communication vehicles tested during the PEF Pilot Phase for the product category of PV electricity production

5DP.1.4 The Product Environmental Footprint (PEF) of Photovoltaic Modules – Lessons Learned from the Environmental Footprint Pilot Phase on the Way to a Single Market for Green Products in the European Union

A. Wade, First Solar, Mainz, Germany
P. Stolz & R. Frischknecht Treeze, Uster, Switzerland
G. Heath, NREL, Golden, United States
P. Sinha, First Solar, Tempe, United States

European Commission

Product Innovation

European Commission

BIPV / VIBV / IIPV

Solar module for automotive applications

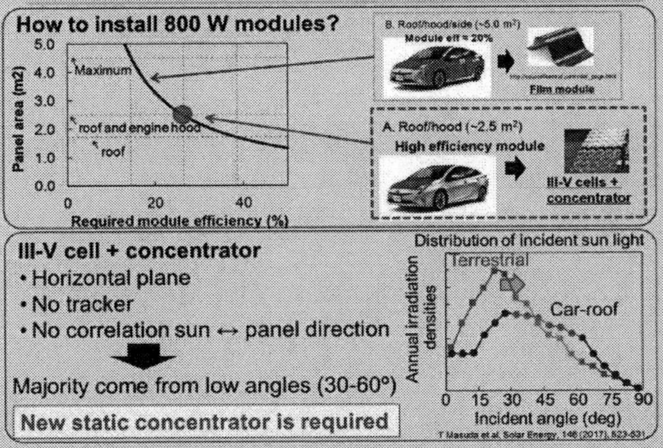

- New static low concentrator was proposed for the automotive application.

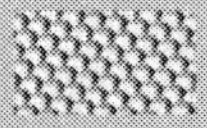

- Prototype module was fabricated by using cutting tool for lens processing.

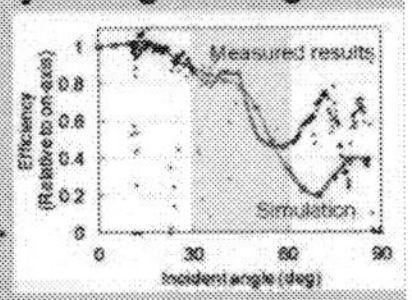

- Characteristics of lens is in good agreement with calculated results.

4DO.4.6 Solar Powered Vehicles with Static Concentrator Photovoltaics

T. Masuda et al, Toyota, Susono, Japan

K. Araki & M. Yamaguchi, TTI, Nagoya, Japan

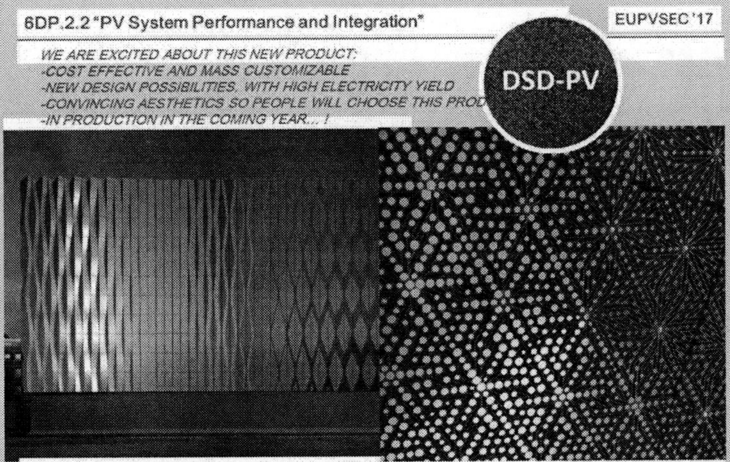

6DP.2.2 Aesthetics and Performance of PV

T. Minderhoud, UNStudio, Amsterdam, Netherlands

Terrawatt

European Commission

Different Phases of VRE Integration – Towards Sector Coupling

Phase	Description	Country Examples
1	VRE capacity is not relevant at the all-system level	Most countries, incl. Mexico, Indonesia, South Africa
2	VRE capacity becomes noticeable to the system operator	Brazil, China, India, Sweden, Texas
3	Flexibility becomes relevant with greater swings in the supply/demand balance	Italy, Germany, Portugal, Spain, UK
4	Stability becomes relevant. VRE capacity covers nearly 100% of demand at certain times	Ireland, South Australia, Denmark
5	Structural surpluses emerge; electrification of other sectors becomes relevant	
6	Bridging seasonal deficit periods and supplying non-electricity applications; seasonal storage & synthetic fuels	

Electricity only accounts for around one fifth of total final energy demar
renewables will require multiplying their uses in buildings, indus

Acceptance
Landscapes
Clean Energy for All

EUPVSEC 2017, Opening, Paolo Frankl, IEA

European Commission

Open access proceedings

29

European Commission

Thanks!

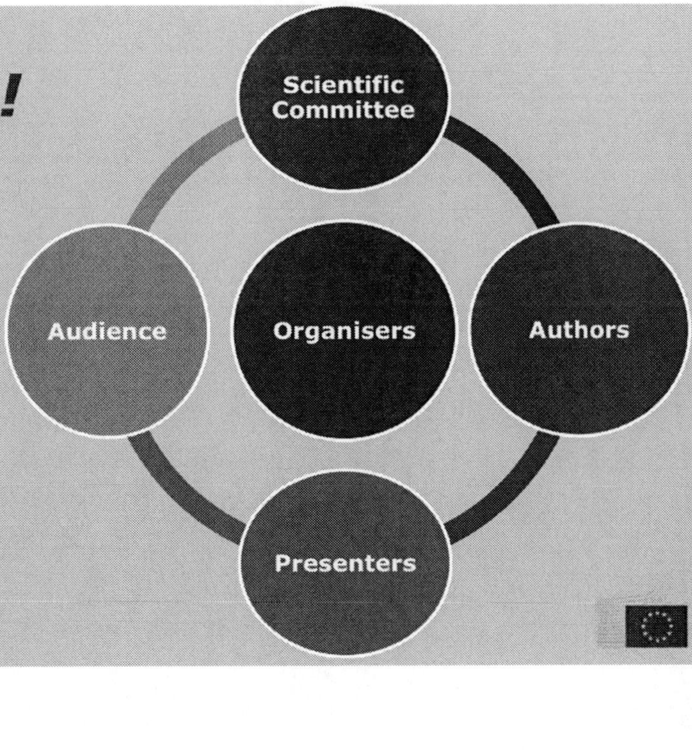

30

European Commission

LIST OF EXHIBITORS
(in alphabetical order)

3D-Micromac	DE
Abet Technologies	US
Advanced Silicon Group	US
ASYS	DE
Bentham Instruments	UK
BERGER Lichttechnik	DE
Berlin-Brandenburg Energy Network	DE
CSEM SA	CH
Delft University of Technology	NL
Dutch Energetics Management	NL
ECN	NL
Engineered Materials Systems	US
ETA Florence Renewable Energies	IT
EUREC	BE
European Commission, DG JRC	BE
FME	NL
greateyes	DE
h.a.l.m. elektronik	DE
Heraeus Photovoltaics	DE
Holland Solar	NL
HydroPV Technologies	NL
IEA PVPS	CH
InnoLas Solutions	DE
Ionx Cleaning Facilities	NL
ISC Konstanz	DE
IZOVAC Technologies	BY
Jonas & Redmann	DE
JRT Metallization Lines	DE
Kipp & Zonen	NL
KOPEL / KYOSHIN ELECTRIC	JP
Luvata Pori Oy	FI
Newport Spectra-Physics	DE
pv magazine group	DE
pv-tools	DE
Quantum Zurich	CH
RENA Technologies	DE
Semilab	HU
SINGULUS TECHNOLOGIES	DE
Sinton Instruments	US
Sisecam Flat Glass	TR
Solar Swiss Connect	CH
SOLARC Innovative Solarprodukte	DE
SOLARUNITED	DE
Solaxess SA	CH
Solibro	DE
Solliance	NL
Stäubli Electrical Connectors	CH
Sunprojects	NL
SUPSI ISAAC	CH
SVCS	CZ
Tempress Systems	NL
TFSC-Instrument	FR
TKI Urban Energy	NL
University of Ljubljana, LPVO	SI
Valentin Software	DE
Van der Valk Solar Systems	NL
Vela Solaris	CH
VITRONIC	DE
VON ARDENNE	DE
Wiley	US
WIP Renewable Energies	DE
ZSW Baden-Württemberg	DE

We thank the EU PVSEC 2017 Sponsors

BERGER Lichttechnik

LONGi Solar

RENA

SINGULUS

DEGRADATION BEHAVIOR WITH ACETIC ACID IN CRYSTALLINE SILICON PHOTOVOLTAIC CELLS

Tadanori Tanahashi, Yukiko Hara, and Atsushi Masuda
National Institute of Advanced Industrial Science and Technology (AIST)
Central 2, 1-1-1, Umezono, Tsukuba, Ibaraki 305-8568, JAPAN

ABSTRACT: It has been understood that the degradation behavior duiring corrosion process in crystalline silicon photovoltaic (PV) modules under extended damp heat (DH) stress test does not accord with that in fielded PV modules, because the electroluminescence (EL) appearance of fielded PV modules does not coincide with that in PV modules during extended DH stress test. However, we found some fielded PV modules with the similar EL appearance with that after extended DH stress test, in which the dark EL area is spread in the perimeter on PV cells. Furthermore, in a durable PV cell to acetic acid atmosphere, the distribution of dark EL area is limited in the center region on PV cell, even when PV module with this PV cell model is exposed to long-term DH stress. By the detail kinetic analysis on the degradation behavior of these PV cells / modules, we suggest that the similar degradation mechanism(s) works in both PV modules with different EL appearance.
Keywords: PV Module, Degradation, Metallization

1 INTRODUCTION

Long-term reliability of photovoltaic (PV) modules deployed in fields under various climate conditions is a crucial issue to assure the advance in levelized cost of electricity (LCOE) in the PV systems. We focus on the corrosion induced by hygrothermal stress, because the degradation with this cause is observed at the late stage of their lifetime, not in "infant-mortality" period [1]. Corrosion induces the current-collection failure in the pathway from surface of crystalline silicon (c-Si) PV cell to front electrodes, and it is facilitated by acetic acid (HAc) which is liberated by hydrolysis of ethylene vinyl acetate (EVA), when moisture is penetrated into PV modules from ambient environment.

So far, it has been understood that the electroluminescence (EL) images observed after extended damp heat (DH) stress test, as an indoor hygrothermal stress test, are dissimilar with those observed in fielded PV modules for a couple of decades, i.e., the dark EL areas are spread in center of respective cells of these veteran PV modules (Fig. 1a), although the degradation identified by EL darkening in PV modules with extended DH stress test is confirmed only on the perimeter of each cell (Fig. 1b [2]) [3], [4]. Therefore, it has been concluded that the degradation mechanism(s) working in PV modules during extended DH stress test would not be the same with that functioning in the fielded PV modules [5].

However, we found the veteran PV modules exposed in the Cfa climate for ca. 20 to 30 years, in which the dark EL areas were spread on the perimeter of each cell (Fig. 2). Indeed, the distribution of EL dark area in each cell was quite similar with that observed in PV modules after extended DH stress test (Figs. 2c and 2d [6]). These results suggest that a common degradation mechanism(s) works in fielded- and DH stressed-PV modules, although the EL appearance (distribution of dark EL area) is different between them.

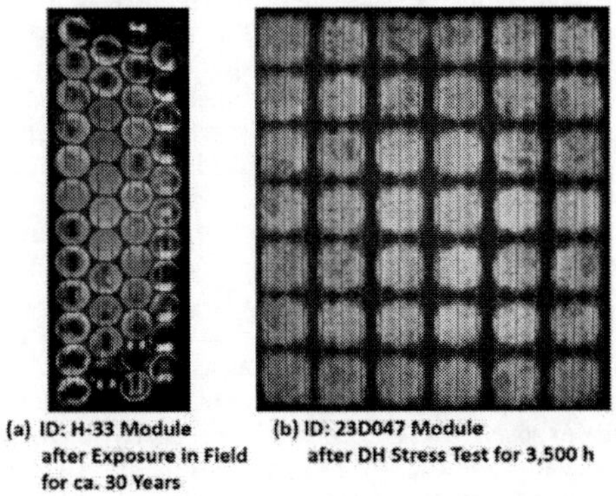

Figure 1: Typical EL images in fielded PV module (a) and PV modules after extended DH steress test (b).

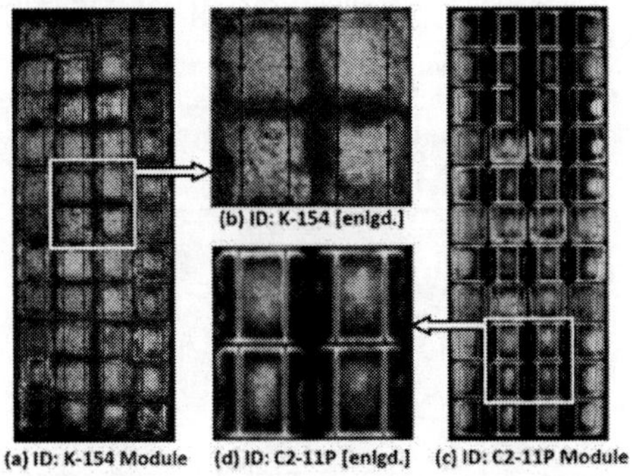

Figure 2: EL images of PV modules with dark EL area in the perimeter on PV cells. Subfigure (b) and (d) are the enlarged images of subfigure (a) and (c), respectively.

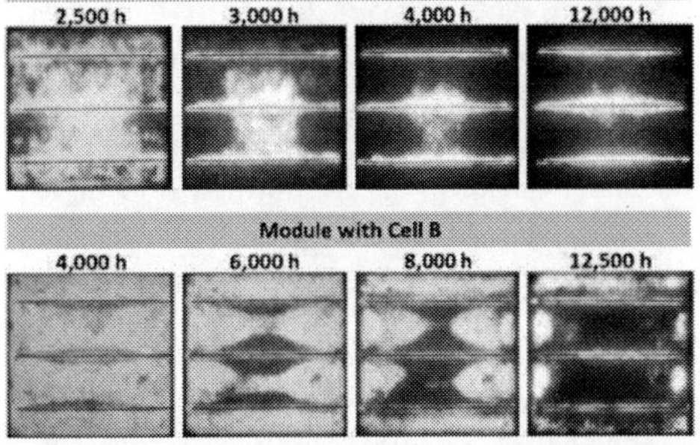

Figure 3: EL images of PV mini-modules with Cell A or Cell B during DH stress test. The duration of test is indicated on top on respective EL images.

We attempt to confirm the manifestation of this putative common mechanism(s) in PV modules during extended DH stress test, because the distributions of dark EL area in respective PV modules with two models of PV cell corresponded to those of fielded PV modules, i.e., the dark EL area was spread in the perimeter on PV cell in "Module with Cell A", although the central region of PV cell in "Module with Cell B" was degraded (Fig. 3).

2 EXPERIMENTAL PROCEDURES

2.1 PV cells and PV modules

Two models ("Cell A" and "Cell B") of c-Si PV cell (156 x 156 mm) were purchased from the distinct cell-manufactures, which were distributed as PV cells to fabricate the commercially available PV modules. Using these PV cells, PV mini-modules (180 x 180 mm, 1 cell/module) were assembled with common materials (semi-tempered front glass, EVA, backsheet [polyvinyl fluoride/polyethylene terephthalate/ polyvinyl fluoride), as previously described [7].

2.2 Determination of degradation behavior in PV cells and PV modules

Bare PV cells were exposed to HAc vapor under the hygrothermal conditions (85°C / 80% relative humidity [rh]), as reported [8]. DH stress test of PV modules with the respective models of PV cell was carried out at 85°C / 85% rh, in accordance with IEC standard [9]. Periodically, PV parameters (maximum power [P_{max}], open-circuit voltage [V_{oc}], short-circuit current [I_{sc}], and fill factor [FF]) of them were determined using a solar simulator under the standard test conditions. Simultaneously, the degradation behavior was confirmed by visual inspection, EL imaging.

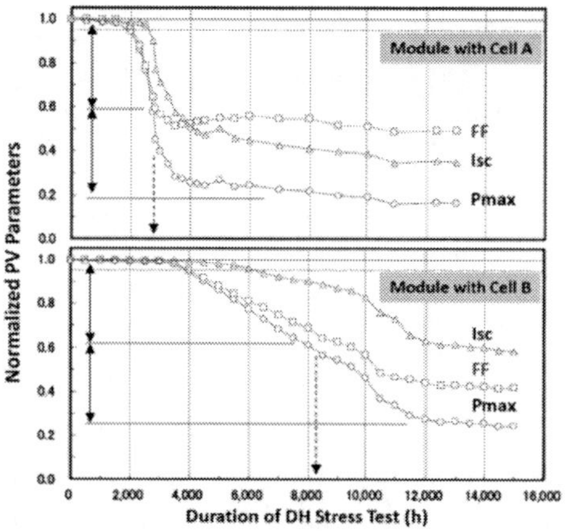

Figure 4: Degradation profiles of PV parameters in PV mini-modules with Cell A or Cell B during DH stress test.

3 RESULTS

3.1 Degradation behavior observed in PV parameters

As described in "INTRODUCTION" section, the distribution of dark area in EL images was dissimilar between the respective PV modules with different PV cell-model, when the extended DH stress was applied to these PV modules (Fig. 3). The whole shape of degradation profiles observed in the electrical characteristics was nearly similar in both PV modules, although the time-frames of these profiles were not comparable between both of them (Fig. 4). Indeed, the duration of lag phase in P_{max}-reduction (this duration was defined as a period required in 5% reduction of each PV parameter) was ca. 2,000 h and ca. 4,000 h in "Module with Cell A" and "Module with Cell B", respectively. After that, the reduction of FF prior to that of I_{sc} was observed in both PV modules, and the decrease in P_{max} depended on these reduction of them (Fig. 4), irrespective of cell model laminated in the PV modules. The extents of P_{max} in "Module with Cell A" and "Module with Cell B" were saturated at ca. 6,000 h and ca. 12,000 h, respectively. And, the periods required in 50% degradation (hereafter, this period is called as "Median Degradation Period [DP_{50}]") was roughly estimated approximately 2,800 h and 8,500 h in "Module with Cell A" and "Module with Cell B", respectively. In the bare PV cells exposed to HAc vapor, similar degradation behavior with early FF-reduction followed by I_{sc}-decrease was confirmed, as reported previously [8], [10]. These results indicate that the degradation caused by the corrosion with HAc may be accomplished by a common mechanism in both model of PV cell.

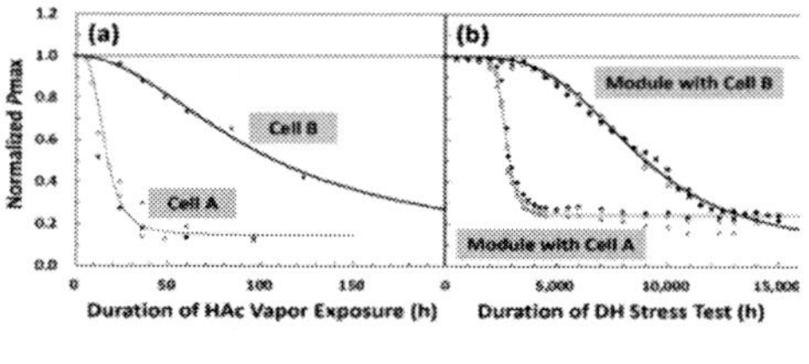

Figure 5: Degradation of P_{max} in bare PV cells (Cell A / Cell B) exposed to HAc vapor (a) and in PV mini-modules with Cell A or Cell B during DH stress test. The lines indicate the fitted results to logistic models, in respective samples and test conditions.

3.2 Kinetic analysis of degradation

The kinetics in degradation behavior described above was modeled by the fitting to the four-parameter logistic equation:

$$\frac{y\,(t)}{y\,(0)} = \frac{U\text{-}L}{1 + e^{S(t-K)}} + L\,,$$

where y (0) is each PV parameter at an initial state, y (t) is each of them at t (duration of HAc-vapor exposure or DH stress test), U is maximum y value that can be estimated, L is minimum y value that can be estimated, K is duration for inflection point (it is same with DP_{50}), and S is the steepness parameter. The extents of P_{max}-degradation in bare PV cells exposed to HAc vapor and in PV modules under DH stress test were well fitted to this model, as shown in Fig. 5. And, the fitting of the extents of FF- and I_{sc}-reductions to this logistic equation has succeeded in both types of stressed tests (data not shown).

Based on the DP_{50}s in respective PV parameters, which were extracted from the degradation profiles observed in PV cells exposed to HAc vapor and in PV modules under DH stress tests, the acceleration factor between both test procedures was assessed, at first. As shown in Fig. 6, DP_{50}s of respective parameters in both PV cells exposed to HAc vapor were nearly completely correlated to those of PV modules under DH stress conditions, and the acceleration factor of degradation under HAc vapor exposure against that under DH stress conditions was estimated ca. 62 (with ca. 2,000 h offset period). These results indicate that, if we can pay due regard to the acceleration factor, the kinetic data obtained in HAc-vapor exposure test are applicable to the prediction of degradation behavior under DH stress test, and vice versa.

The degradation rates of I_{sc} and FF were estimated by

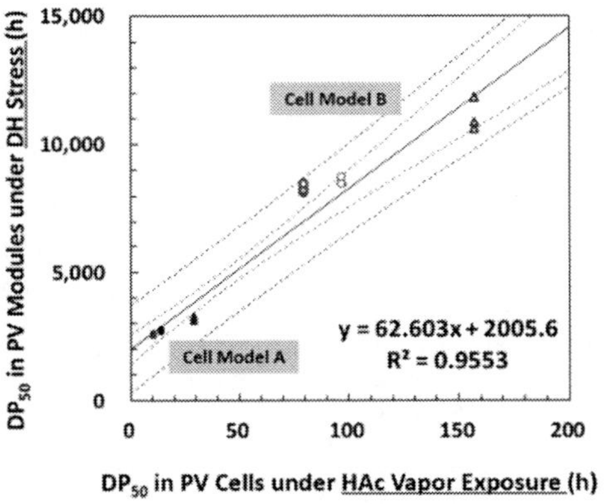

Figure 6: Correlation of DP50s in HAc exposure of PV cell and DH stress test of PV modules assembled using the same model of PV cell. Black-, red-, and bule-symbols indicate the DP50s of P_{max}, FF, and I_{sc}, respectively.

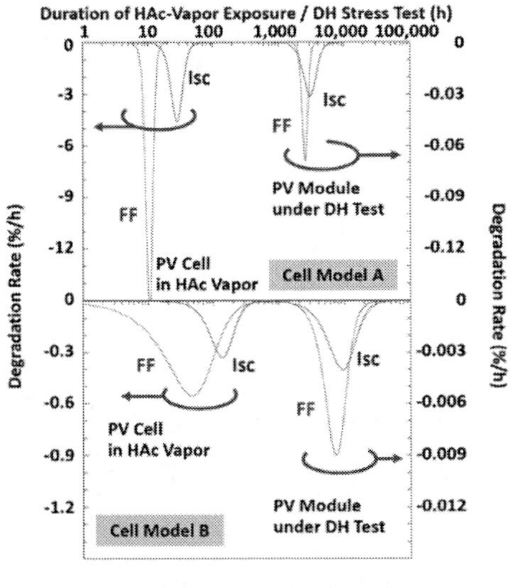

Figure 7: Hourly degradation rates observed in PV cells exposed to HAc vapor (left axis) and in PV mini-modules with these cells under DH stress (right axis).

the numerical calculation of derivative form of the four-parameter logistic equation (Fig. 7). In the comparison between the degradation rates in PV modules with Cell A or Cell B during DH stress test, the maximum degradation rate in PV module with Cell A was ca. 10 times as high as that in PV module with Cell B (see the magnitude of scale indicated in right axis of upper / lower panels). However, in both PV modules under DH stress test, some similar features were observed as follows; 1) after the rapid and potent change of degradation rate in FF, the slow and broad peak of that in I_{sc} was induced, 2) the maximum rate in I_{sc} was ca. 45% of that in FF, 3) since these waveforms of FF and I_{sc} were nearly completely overlapped, the respective contribution of these parameters to power-loss was not clearly divided. And, most features observed in PV modules under DH stress test were also confirmed in the comparison between the degradation rates in bare PV cells (Cell A or Cell B) with exposure to HAc vapor, although the contribution of FF-reduction to power-loss was nearly separated from that of I_{sc}-decrease (especially, in the case of Cell A under HAc vapor exposure). These findings reveal that, in also the degradation rate analysis, the corrosion by HAc in both cell models would work by a common mechanism.

We further evaluated the relationship of steepness in degradation phase between both PV cells (Table I). In HAc-vapor exposure, the ratio of I_{sc}-steepness in Cell B to that in Cell A was 0.52, and that for FF-steepness is 0.15. And, those of I_{sc}- and FF-steepness in DH stress test were 0.49 and 0.26, respectively. These results indicate that the decreasing in I_{sc} of Cell B is more gradual than that of Cell A in both stress conditions, since this characteristic is nearly quantitatively kept in both tests (0.52 vs. 0.49), as well as it is observed in those of FF-steepness (0.15 vs. 0.26). The ratio in FF-steepness between Cell A and Cell B is much smaller than that in I_{sc}-steepness, irrespective of the applied stress conditions. Therefore, it is suggested that the power-losses induced by I_{sc}- and FF-decreasing are subjected to the distinct degradation mechanisms, although these degradation-steps progressing in DH stress test are mimicked in HAc-vapor exposure test.

Table I: Estimated steepness of degradation observed in the PV cells exposed to HAc vapor and in the PV modules under DH stress conditions.

	I_{sc}			FF		
Cell / Module	A	B	B/A	A	B	B/A
HAc Vapor	7.45	3.84	0.52	13.6	2.05	0.15
DH	6.75	3.29	0.49	16.1	4.18	0.26

4 DISCUSSION

We demonstrate by the kinetic analysis of the degradation behaviors that, even when the distinct EL images are induced in PV modules with hygrothermal stress (essentially, it is degradation with acetic acid generated by moisture-ingress within PV modules), the degradation mechanism(s) is highly likely common to both PV modules.

The degradation with hygrothermal stress in PV modules has been evaluated by the DH stress test, and this significant degradation is induced in the metallization on the perimeter of cell (spreading of dark EL area) [11]–[13]. However, some recent reports suggest that this pattern of EL darkening has not been observed in the PV modules exposed in fields [3]–[5]. In this study, we found that the peripheral-degraded PV cells were identified in some fielded PV modules for long-term. In addition, using the PV cells with different degradation behavior with acetic acid, we suggest that the difference in EL patterns is subjected to a common degradation mechanism(s), although the intensities of kinetic parameters are not comparable, i.e., only the rate of degradation is different between these cells / modules. Therefore, it is concluded that the most susceptible region of metallization on PV cell seems to be degraded in each cell / module, however the precise cause(s) should be elucidated in future works.

It is also found that the degradation in PV modules with hygrothermal stress (under DH stress test) is mostly mimicked by that in PV cells exposed to HAc vapor with high acceleration factor. Interestingly, the duration required in the initiation of degradation in PV modules under DH stress test, which corresponded to time-zero in HAc-vapor exposure test of PV cells, was ca. 2,000 h. This period almost coincides with that required in the entire penetration of moisture within a whole module[4], [14]. Then, in the HAc-vapor exposure test of PV cells, the degradation is initiated without delay caused by moisture penetration, and the isotropic degradation on the PV cell is observed. These findings suggest that the degradation-mode observed during the hyper-accelerated degradation with HAc-vapor exposure is completely correlated with that observed in DH stress test of PV modules, reflecting the period of moisture ingress into PV modules. And, the different distribution of dark EL area between these PV modules under DH stress test would be explained by a term as dynamics including anisotropic distribution of moisture within a PV module. Furthermore, we observed that the peaks of FF- and I_{sc}-reduction rates were near fully separated in the HAc-vapor exposure test of PV cells, but not in the DH stress test of PV modules, although the sequence of these reductions was not switched. These results suggest that the degradation mechanism in FF-induced power-loss is likely different from that in I_{sc}-induced power-loss; however, both mechanisms seem to be closely related to each other.

For the prediction of service lifetime of PV modules exposed in fields, recent three reports based on the DH stress tests should be taken a deep interest [5], [12], [13]. In these reports, the authors proposed the respective equations to predict the service lifetime of PV modules deployed under various climate zones, although the degradation mechanism(s) with separated FF- and I_{sc}-reductions has not been discussed in details. Then, the argument including the respective mechanisms of FF- and I_{sc}-reduction during HAc-vapor exposure test of PV cells is necessary to establish the precise prediction of service lifetime of PV modules in fields [10].

5 CONCLUSIONS

We have found that the expression of dark EL area on PV cell within PV module is altered with the durability of PV cells to HAc vapor, i.e., the EL darkening in the center region of PV cell within PV module was induced in the durable PV cells. In addition, the dark EL area on the perimeter of PV cells was also identified in the long-term fielded PV modules. From these results, we suggest that the degradation with corrosion is subjected to a common mechanism(s) which is comprised of FF- and I_{sc}-reductions, irrespective of the appearance in EL images.

6 ACKNOWLEDGEMENTS

We wish to thank the support by New Energy and Industrial Technology Development Organization, Japan.

7 REFERENCES

[1] IEA-PVPS Task13 Team, "Review of failures of photovoltaic modules," International Energy Agency, Paris, 2014.

[2] "Interim Progress Reports in the Consortium Study II on Fabrication and Characterization of Solar Cell Modules with Long Life and High Reliability (in Japanese)," National Institute of Advanced Industrial Science and Technology (AIST), Tsukuba, Japan, 2012.

[3] P. Bentz, "Standardization requirements: An industrial point of view," presented at. *SOPHIA Workshop PV Module Reliability 2012*, 2012.

[4] J. H. Wohlgemuth and M. D. Kempe, "Equating damp heat testing with field failures of PV modules," in *Proc. 39th IEEE Photovoltaic Spec. Conf.*, 2013, pp. 0126–0131.

[5] M. D. Kempe and D. C. Jordan, "Evaluation and modeling of the potential effects of a module manufacturing anomaly," *Prog. Photovoltaics: Res. Appl.*, Jul. 2017, DOI: 10.1002/pip.2908.

[6] "Progress Reports in the Consortium Study II on Fabrication and Characterization of Solar Cell Modules with Long Life and High Reliability (in Japanese)," National Institute of Advanced Industrial Science and Technology (AIST), Tsukuba, Japan, 2014.

[7] S. Suzuki, T. Tanahashi, T. Doi, and A. Masuda, "Acceleration of degradation by highly accelerated stress test and air-included highly accelerated stress test in crystalline silicon photovoltaic modules," *Jpn. J. Appl. Phys.*, vol. 55, no. 2, p. 22302, Feb. 2016.

[8] T. Tanahashi, N. Sakamoto, H. Shibata, and A. Masuda, "Electrical detection of gap formation underneath finger electrodes on c-Si PV cells exposed to acetic acid vapor under hygrothermal conditions," in *Proc. 43rd IEEE Photovoltaic Spec. Conf.*, 2016, pp. 1075–1079.

[9] *Crystalline Silicon Terrestrial Photovoltaic (PV) Modules: Design Qualification and Type Approval*. IEC 61215: 2005, 2005.

[10] T. Tanahashi, N. Sakamoto, H. Shibata, and A. Masuda, "Cause of current-collection failure observed in I_{sc}-reduction phase of PV cells and modules exposed to acetic acid," in *Proc. 44th IEEE Photovoltaic Spec. Conf.*, 2017, in press.

[11] C. Peike, S. Hoffmann, P. Hülsmann, B. Thaidigsmann, K. A. Weiβ, M. Koehl, and P. Bentz, "Origin of damp-heat induced cell degradation," *Sol. Energy Mater. Sol. Cells*, vol. 116, pp. 49–54, Sep. 2013.

[12] M. Koehl, M. Heck, and S. Wiesmeier, "Modelling of conditions for accelerated lifetime testing of humidity impact on PV-modules based on monitoring of climatic data," *Sol. Energy Mater. Sol. Cells*, vol. 99, pp. 282–291, Aug. 2012.

[13] M. Koehl, S. Hoffmann, and S. Wiesmeier, "Evaluation of damp-heat testing of photovoltaic modules," *Prog. Photovoltaics: Res. Appl.*, vol. 25, no. 2, pp. 175–183, Feb. 2017.

[14] P. Hülsmann and K.-A. Weiss, "Simulation of water ingress into PV-modules: IEC-testing versus outdoor exposure," *Sol. Energy*, vol. 115, pp. 347–353, May 2015.

QUANTIFICATION OF SOLAR CELL FAILURE SIGNATURES BASED ON STATISTICAL ANALYSIS OF ELECTROLUMINESCENCE IMAGES

Sergiu Spataru[1], Harsh Parikh[1], Peter Hacke[2], Gisele A. dos Reis Benatto[3], Dezso Sera[1], and Peter B. Poulsen[3]
[1]Aalborg University, Department of Energy Technology
Pontoppidanstraede 111, DK-9220 Aalborg, Denmark, Tel: +45 5264 7520, Email: ssp@et.aau.dk
[2]National Renewable Energy Laboratory (NREL)
Golden, CO, 80401, United States
[3]Department of Photonics Engineering, Technical University of Denmark,
Frederiksborgvej 399, 4000, Roskilde, Denmark

ABSTRACT: We propose a method to identify and quantify the extent of solar cell cracks, shunting, or damaged cell interconnects, present in crystalline silicon photovoltaic (PV) modules by statistical analysis of the electroluminescence (EL) intensity distributions of individual cells within the module. From the EL intensity distributions (ELID) of each cell, we calculated summary statistics such as standard deviation, median, skewness and kurtosis, and analyzed how they correlate with the type of the solar cell degradation.

We found that the dispersion of the ELID increases with the size and severity of the solar cell cracks, correlating with an increase in standard deviation and decrease in kurtosis. For shunted cells, we found that the ELID median is strongly correlated with the extent of cell shunting. Last, cells with damaged interconnect ribbons show current crowding and increased series resistance regions, characterized by increased dispersion and skewness of the ELID.

These cell-level diagnostic parameters can be used to quantify the level of mismatch between the solar cells in the module, which can represent the extent of the module degradation, due to transportation, installation, or field operation. The method can be easily automated for quality control by module manufacturers or installers, or can be used as a diagnostic tool by plant operators and diagnostic service providers.

Keywords: Characterization, Degradation, Electroluminescence, Failure detection, Photovoltaic Module, PID, Cell cracks, Statistical parameters

1. INTRODUCTION

Electroluminescence (EL) imaging has become an indispensable diagnostic tool for PV modules, allowing identification of a wide range of solar cell failure and degradation modes, such as cell cracks, shunting, broken cell fingers and interconnects, and other cell defects [1]. Most often, EL imaging is used as a qualitative/visual diagnostic tool however, it presents significant potential to quantify the magnitude or severity of degradation and defects, through model based or image analysis methods.

Previously proposed EL quantification approaches vary significantly in methodology depending on the failure mode analysed, such as: series resistance mapping [2], voltage mismatch mapping [3], detection and quantification of solar cell micro-cracks [4, 5] and fractures [6]. Several research groups, including ours have published on physics-based approaches for analysing series and shunt resistance losses; however, some outstanding issues, such as knowledge of voltage drop over each cell and determination of shunt resistance losses from EL make it difficult to perform on industrial scales (manufacturing lines, power plants).

A more generic method for quantifying the diagnostic information contained in EL images was proposed in the new draft IEC 60904-13 EL imaging standard draft [7] and in [8], based on statistical analysis of the EL signal. In this work, we performed an experimental investigation into the solar cell ELID signatures, corresponding to different degradation modes, such as cell cracks, shunting, and cells with damaged interconnects affecting the series resistance. The purpose of this work is to find the statistics and metrics that correlate with the magnitude of the degradation, for use in quantifying solar cell or PV module degradation.

We performed the investigations on standard 60-cell crystalline silicon PV modules, which were artificially degraded to induce cell cracks, shunting and interconnect damage in a controlled manner. The PV modules were imaged by EL, before and after each degradation cycle. The ELID of individual solar cells was analysed for identifying patterns correlating with the failure mode type and magnitude. Summary statistics, such as standard deviation (STD), median and kurtosis were calculated for each solar cell ELID, and the statistics of healthy cells were compared to those of damaged solar cells, to quantify the level of mismatch due to the degradation.

We found that each of the three types of degradation modes investigated has a specific signature and quantifiable impact on the ELID of the solar cell with use of appropriate distribution statistics, which are well correlated with the magnitude of the degradation. These parameters can be used to rate solar cell or module damage after manufacturing or transportation (such as cell cracks or grid finger breaks), or degradation due to weathering stresses. The calculation and analysis of these parameters can be automated and used in quality control for module manufacturers or installers, or can be used as a diagnostic tool by plant operators and diagnostic service providers.

2. METHODOLOGY

2.1. Solar cell-level EL intensity distribution analysis

All solar cells in the test modules were analysed individually, by automatically extracting the cell-level EL images from that of the module. From each solar cell sub-EL image, the cell-level ELID $p_{EL-Cell}(k, i)$, was calculated as (1):

$$p_{EL-Cell}\left(k,i\right)=\frac{n_i^k}{n^k},\ 0\le i<L, 1\le k\le N_c \qquad (1)$$

Where k is the solar cell number, N_c is the number of solar cells in the module, n_i^k is number of occurrences of gray level i in cell k, n^k is the total number of pixels in the image of cell k.

From the cell-level ELID, EL distribution parameters are calculated for each solar cell, such as STD, mean, median, skewness, and kurtosis, for the different stages of degradation of the cells. These parameters are presented in the next sub-section.

2.2. Diagnostic parameters derived from the ELID

Mean and median are the basic statistical parameter which can be derived from the the EL intensity distributions. This parameters are important as they are directly correlated with the luminesence of the cell. Thus from the luminescence, an indication of the cell voltage (compared with the rest of the cells) can be estimated. Mean of a distribution is defined as:

$$\mu_{Cell}\ k = \frac{1}{L}\cdot\sum_{i=0}^{L-1} p_{EL-Cell}(k,i) \qquad (2)$$

Standard deviation is a measure of dispersion that indicates the spread of the data. In the case of cell ELID, the increases in the dispersion of the luminescence is correlated with the defects of the module and to the power loss depending on the severity and fault type. Since new/"healthy" solar cells tend to have a more uniform luminescence distribution occurring across the cell surface, resulting in a tighter ELID. Whereas, EL images of degraded/defected solar cells, will vary in luminescence across the solar cell surface, depending on the defect type, resulting in a wider ELI distribution. For example, for ribbon damage type faults, the luminescence increases from in the defective region due to current crowding and the change in the series resistance of the cell. The STD corresponding to and ELID is calculated as:

$$\sigma_{Cell}(k) = \sqrt{\frac{1}{L-1}\sum_{i=0}^{L-1}\left[p_{EL-Cell}(k,i) - \mu_{Cell}\ k\ \right]^2} \qquad (3)$$

Skewness characterizes the degree of asymmetry of a distribution around its mean [9]. It is a statistical parameter, which essentially measures the relative size of two tails of a dataset. Positive values for the skewness indicate data that are skewed right and the right tail is relatively long, more observations are below the mean and the mean of the curve is greater than the median [9]. Negative values for the skewness indicate data that are skewed left, it denotes small number of low observations and a large number of observations with higher value, and median is greater than mean in this case. Skewness of an ELID is calculated as:

$$\gamma_1(k) = \frac{1}{L}\sum_{i=0}^{L-1}\left(\frac{p_{EL-Cell}(k) - \mu_{Cell}(k)}{\sigma_{Cell}(k)}\right)^3 \qquad (4)$$

Kurtosis is a measure of the combined weight of the tails relative to the rest of the distribution [9]. It quantifies whether the shape of the data distribution matches the Gaussian distribution, and how outlier-prone a distribution is, as illustrated in Fig. 1. Certain cell failures, such as cell cracks, series resistance, or ribbon damage are charactezied by localized cell regions with strongly decreased/increased luminesence, relative to other cell regions. These corelate with outliers in the ELID and can be characteized by the kurtosis.
For an ELID, the kurtosis is calculated as:

$$\kappa(k) = \frac{1}{L}\sum_{i=0}^{L-1}\left(\frac{p_{EL-Cell}(k) - \mu_{Cell}(k)}{\sigma_{Cell}(k)}\right)^4 \qquad (5)$$

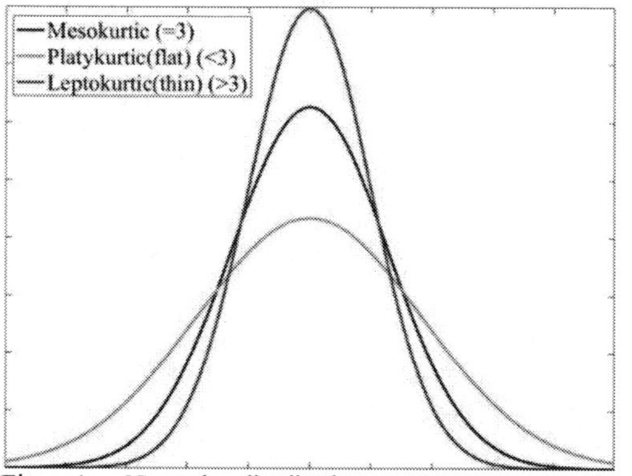

Fig. 1: Normal distribution curves representing Mesokurtic, Platykurtic (more outlier-prone) and Leptokurtic distributions (less outlier-prone).

Inactive cell area is an empirical parameter, previously proposed in [6], for quantifying the cracked area of modules or cells. This parameters quantyfies the darker regions of the cell or module, which would correspond to inactive or defective cell/module areas. Calculating $I_{EL-cell}(k)$ for a cell requires the determination of a module levels intensity theshold value *TH*, as depicted in Fig. 2. Methods for calculating *TH* are discuss in [6, 10].

$$I_{EL-Cell}(k)[\%] = 100\sum_{i=0}^{TH} p_{EL-Cell}(k,i) \qquad (6)$$

The $I_{EL-cell}$ *(k)* can be used to quantify both: (i) the percent of damaged/degraded solar cell regions in the module: (ii) as well as cell defects, low lifetime and dark regions in the cell EL image.

ELID full-width at x% of peak is an empirical parameter, shown in Fig. 3, that characterizes the spread of the ELID, in different regions of the distribution.
Its sensitivity can be adjusted by the choice of the *x% of peak* parameter, which will determine the points of the ELID where the width is calculated. Full-width of a histogram at any given [%] is calculated as:

$$PK(k) = \max\left[p_{EL-Cell}(k,i)\right] \qquad (7)$$

$$FW_x(k) = \max\{i \mid p_{EL-Cell}(k,i) = \frac{x}{100}PK(k)\} - \min\{i \mid p_{EL-Cell}(k,i) = \frac{x}{100}PK(k)\} \qquad (8)$$

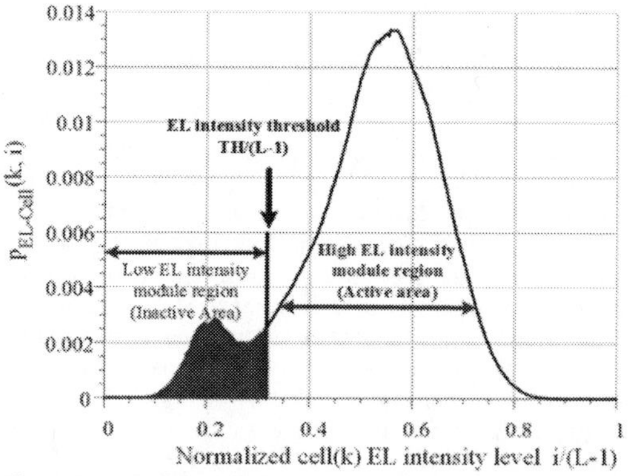

Fig. 2: ELID of the a PV module highlighting the inactive area in the histogram

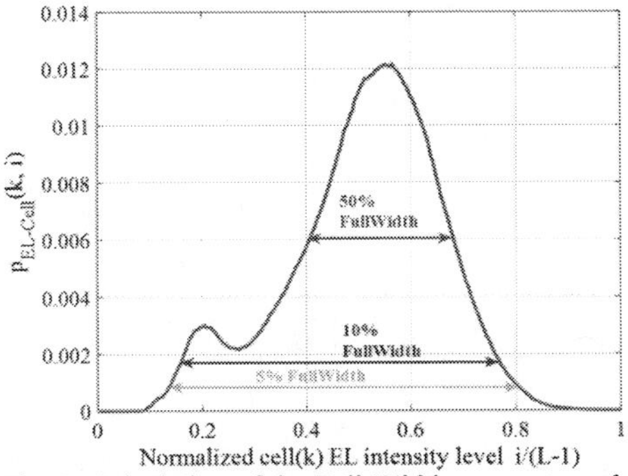

Fig. 3: Calculation of the Full-Width parameter calulated at 5%, 10% and 50% of ELID peak.

3. EXPERIMENT

Several new, mono- and poly-crystalline silicon PV modules were artificially degraded through:

(i) Mechanical loading, thermal cycling, and humidity freeze cycles – to show the progression of solar cell cracking in 5 stages of degradation:

(ii) Voltage stress at high temperature and humidity – for causing potential-induced degradation by cell shunting (PID-s).

(iii) Mechanical damage to the cell interconnects – to replicate solar cell regions with increased series resistance as occurs in the field by thermomechanical fatigue.

The PV modules were imaged before and after each degradation cycle, in a dark imaging studio with a high-resolution near IR CCD camera, under high current bias conditions (close to the STC maximum power-point current - I_{mp}). The experiments were performed under constant exposure and current bias.

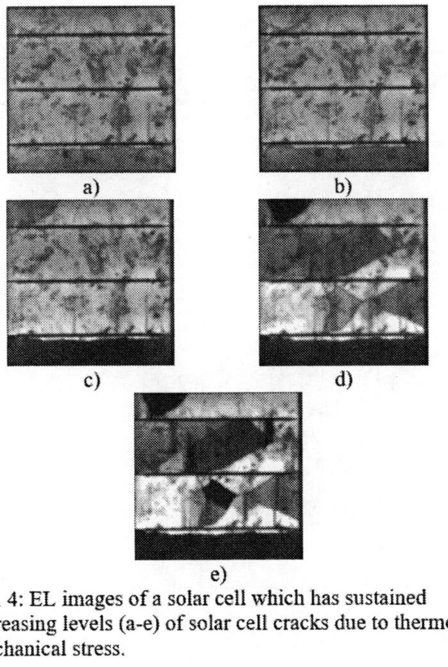

Fig. 4: EL images of a solar cell which has sustained increasing levels (a-e) of solar cell cracks due to thermo-mechanical stress.

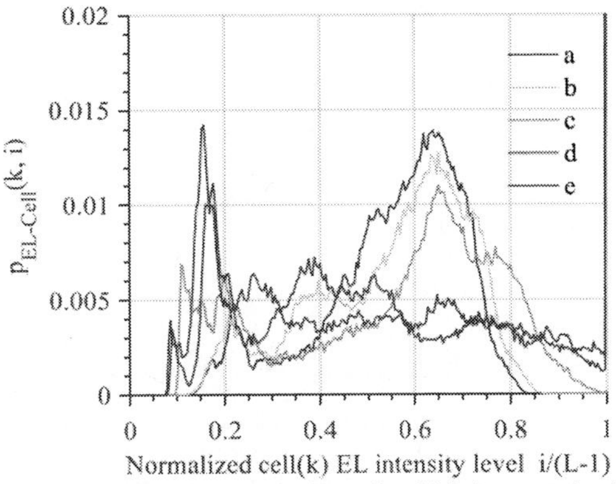

Fig. 5: ELID of the same solar cell, which has sustained increasing levels (a-e) of cell cracks

4. RESULTS

4.1. Analysis of ELIDs of cracked solar cells

To demonstrate the application of ELID statistics to quantify cell degradation, we use first use it to analyze a poly-Si solar cell, shown in Fig. 4.

The cell is part of a standard 60-cell module which has sustained several cycles of thermo-mechanical stress, causing the progression of solar cell cracking, from minor-micro cracks (in Fig. 4a) to severe cell fractures (in Fig. 4e). To note that the module was imaged under the same bias current and exposure time conditions.

Next, we calculate the cell ELID $p_{EL\text{-}Cell}$ for each state of degradation (a to e), according to (1). The ELIDs corresponding to each state of degradation are shown in Fig. 5 starting from a bell shaped bi-modal distribution, and progressively increasing in dispersion in both luminescence intensity directions. Since cell cracks are associated with series resistance [11], they can cause current crowding and voltage drops in the cells, appearing as higher or lower EL intensity regions.

The increased dispersion of the ELID can be quantified by several statistics, such as the STD and kurtosis, shown in Fig. 6 to correlate well with the increased severity of the cell cracks.

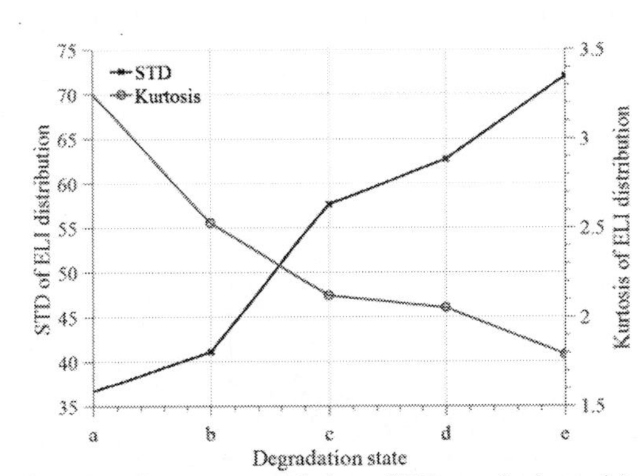

Fig. 6: Summary statistics (STD and kurtosis) corresponding to the ELID of a cell in five states of mechanical degradation (cell cracks). More and larger cell cracks cause an increase in the STD, and a decrease in the cell ELID kurtosis.

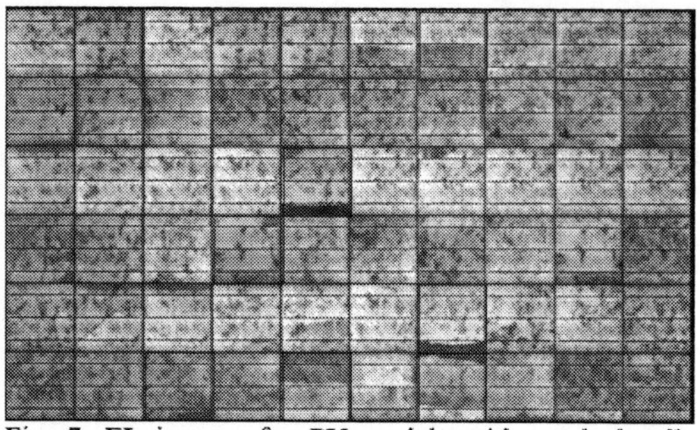

Fig. 7: EL image of a PV module with cracked cells (corresponding to degradation c). The top 7 cells with the lowest ELID kurtosis are highlighted in color and their ELIDs are shown in Fig. 8, compared to the other cells (black).

Since the absolute values of the summary statistics depend also on the imaging conditions, as well as module's physical properties, it can be difficult to interpret them independently. Considering that the solar cells within a PV module are sorted during manufacturing based on similar performance and electrical characteristics, we can quantify the ELIDs of degraded cells, relative to the rest of the cells within, if not all cells are degraded at the same level.

To exemplify, Fig. 7 shows a module where the cells with the lowest kurtosis are colored. Their ELIDs are shown in Fig. 8, in corresponding colors, compared to the ELIDs of the rest (53) of the cells (shown in black). Here, we can observe how their ELID is affected by the cracks, relative to the rest of the cells.

Furthermore, Fig. 9 shows the ELID kurtosis for each cell in this module, where darker colors represent lower Kurtosis values. From these statistics we could further quantify the degradation by calculating a cell kurtosis-mismatch factor, or a distribution of the cell kurtosis in the module, where the module degradation would be evaluated based on the dispersion of this cell-kurtosis distribution, similar to the method presented in [8].

4.2. Analysis of ELIDs of shunted solar cell

We performed a similar analysis for a mono-Si module, which has sustained PID-s, due to high voltage stress applied under high temperature and humidity.

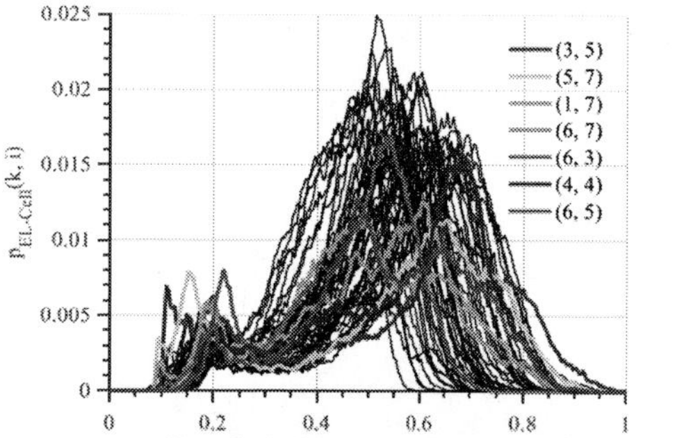

Fig. 8: ELIDs corresponding to the 60-cells of the module in Fig. 7, affected by cell cracks. The colored distributions, showing the row and column in the module, correspond to the highlighted cells in Fig. 7, and have the lowest Kurtosis within the PV module.

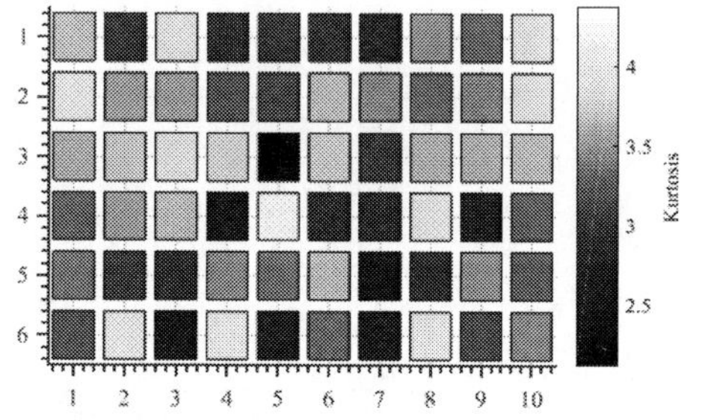

Fig. 9: PV module color map showing the ELID kurtosis of each solar cell. Darker colors represent lower Kurtosis values, correlating with more severe cell cracks.

The shunting can be observed in the lower EL intensity cells, located closer to the module frame. The module was imaged under high current bias (STC I_{mp}), however the effect of shunting is even more pronounced under lower bias [1]. The coloured cells in Fig. 10 have the lowest ELID median within the module. Their corresponding ELIDs are shown in Fig. 11 in the same colours, compared to the rest of the cells (shown in black). Here we can observe that the PID-s is associated with a shifting of the cell ELID towards lower intensities. However, it does not increase the dispersion of the ELID. This effect can be observed in other cells in the module as well, which are affected (to a lower extent) by PID-s.

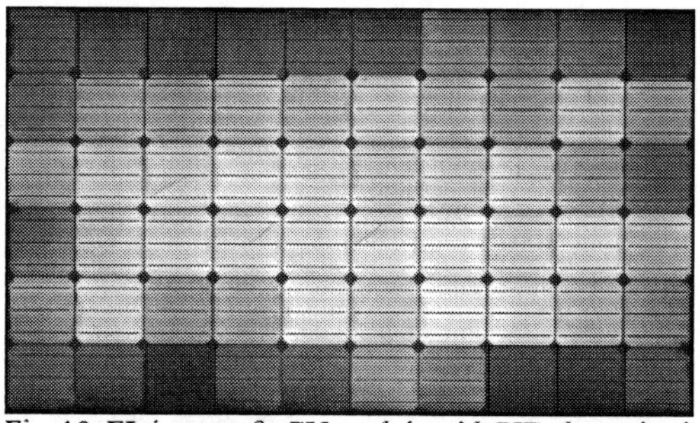

Fig. 10: EL image of a PV module with PID shunted solar cells. The top 7 cells with the lowest ELID median are highlighted in color.

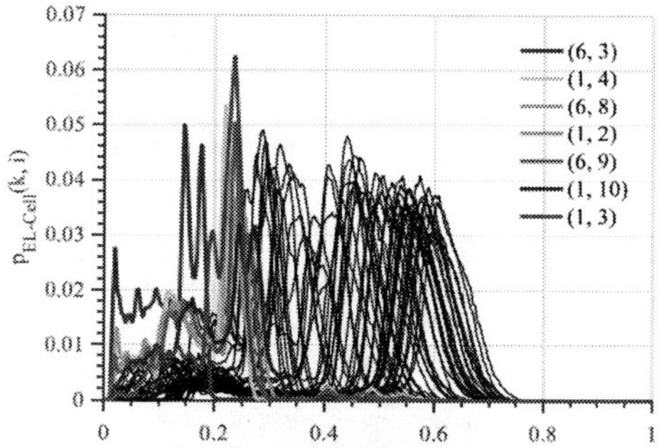

Fig. 11: ELIDs corresponding to the 60-cells of the module in Fig. 10, with shunted cells. The colored distributions correspond to the highlighted cells in Fig. 10, and have the lowest ELID median in the module.

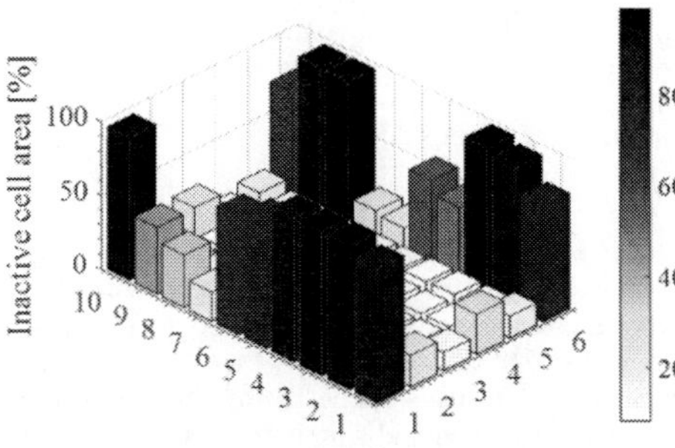

Fig. 12: 3-dimensional graph corresponding of the PV module with PID shunted solar cells. The different colored distributions correspond to the inactive area.

Fig. 12, shows a 3-dimensional graph corresponding to the PV module with PID shunted solar cells and it can be seen from the graph that the inactive area of the cells closer to the module frame increases a lot due to significant decrease in the luminescence. On further analysis, it was observed that the inactive area mismatch factor of the module increases from 10.2% to 33.2% due to damaged cells resulting in increase in power loss. A significant increase in cell-STD mismatch factor (from 1% to 33.5%) was also evaluated, whereas the median of the severely degraded cells is seen to be shifted left with respect to the healthy ones as shown in Fig. 11.

The sensitivity of the spread of the dispersed data due to the defective cells compared to the healthy ones can be characterized with the full width at 10% peak mismatch factor inferring in a decrease in the parameter for the shunted cell as shown in Fig. 13.

Fig. 14 **Error! Reference source not found.**shows the cell ELID median for all 60 cells within the degraded module, where darker colours correspond to lower median values. It can also be seen from Fig. 11 that the median of the damaged cells have shifted to the left of the distribution curve compared to that of the healthy cells. Here we can observe the typical PID-s pattern and quantify the cell mismatch due to shunting.

4.3. Analysis of ELIDs of solar cells with damaged solar cell interconnect ribbons

The last degradation mode we analysed with ELID statistics was broken cell interconnects, which can occur due to bad soldering and thermal stress. Although this failure does not cause significant power loss by itself, unless all ribbons of a cell are broken, it can cause arcing and localized hotspots [1], further damaging the PV module. We replicated this failure mode, by interrupting a ribbon per cell, for several cells, from the module backside.

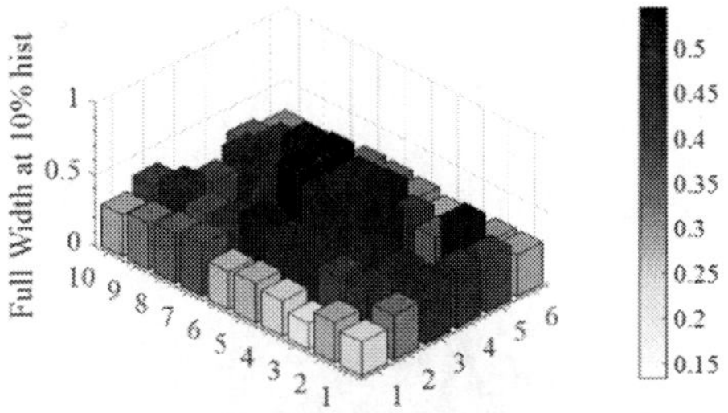

Fig. 13: Fullwidth at x% peak representing the dispersion of the shunted cells.

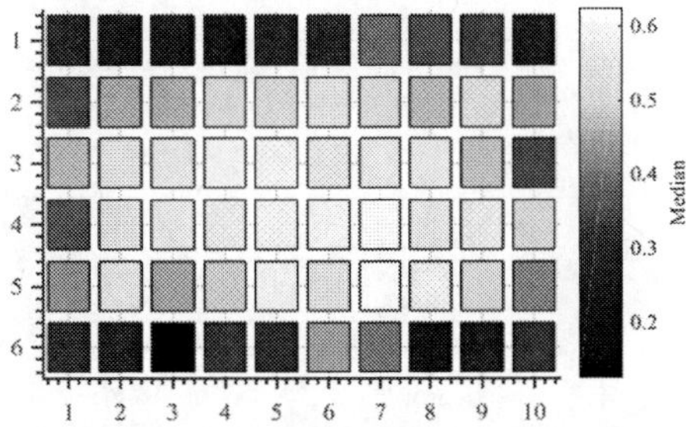

Fig. 14: PV module color map showing the ELID median of each solar cell. Darker colors represent lower median values, correlating with lower ELI due to shunting of the cells.

As can be observed from Fig. 15, this causes current crowding around the undamaged ribbon – associated with increased EL intensity, as well as change in the cell series resistance, characterized by voltage drops and lower EL intensity around the broken ribbons. These two effects compound into an increased dispersion of the cell ELID. Correspondingly, the top seven cells with the highest STD are highlighted in color in Fig. 15 and their corresponding ELIDs are shown colored in Fig. 16.

As shown in Fig. 16, the ELIDs of the damaged cells differ significantly from the undamaged cells (shown in black), having an increased dispersion as well as a skew of the distribution toward lower EL intensity values.

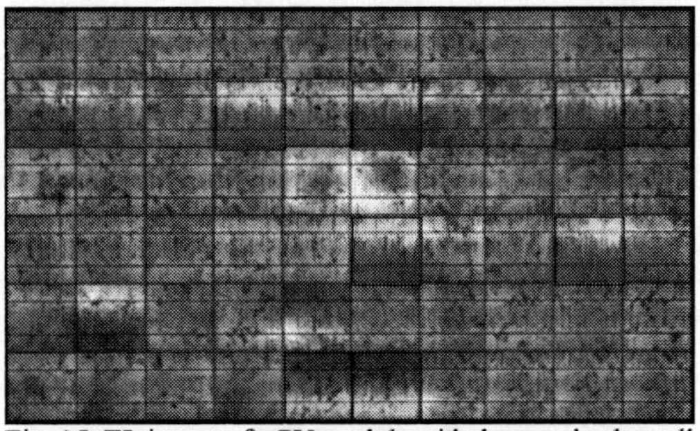

Fig. 15: EL image of a PV module with damaged solar cell interconnects. The top 7 cells with the highest STD are highlighted, and their corresponding ELIDs are colored in Fig. 16.

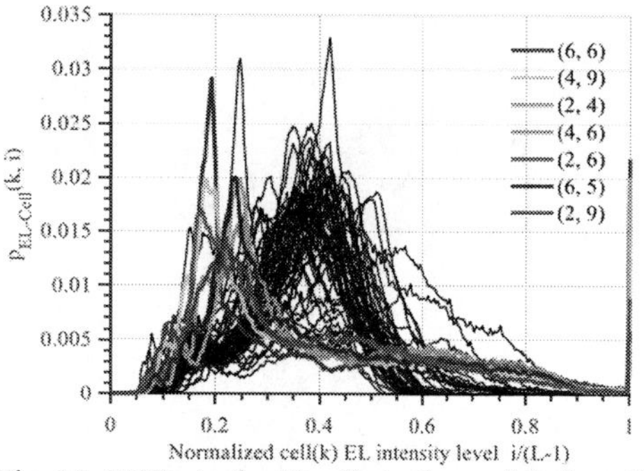

Fig. 16: ELIDs to the 60-cells in the module from Fig. , where the colored distributions correspond to the highlighted cells, and have the highest ELI STD within the PV module.

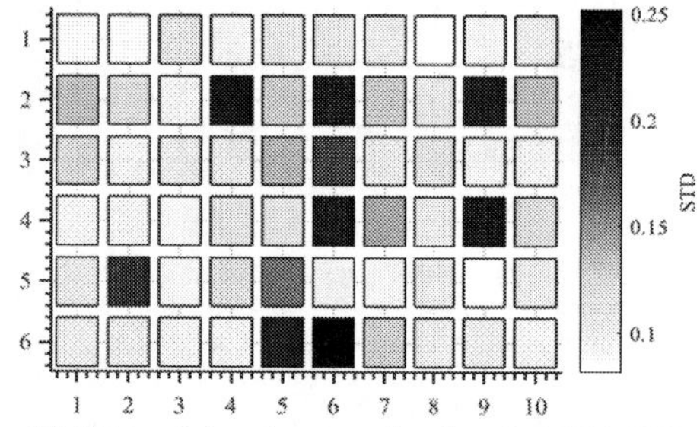

Fig. 17: PV module color map showing the ELID STD of each solar cell. Darker colors represent higher STD values, correlating with higher ELI dispersion due to the series resistance regions and current crowding.

These changes in the ELID are best characterized by statistics such as the STD and skewness, or the ELID distribution with at 10% of peak (as proposed in the EL imaging standard draft [7]), all three parameters increasing for the damaged cells. Fig. 17 shows the cell ELID STD for all 60 cell of the module, where darker colors are associated with higher STD values. Here, we can observe that the cells with damaged ribbons, or series resistance problems have a higher ELID STD. Based on these parameters, we can calculate a cell STD mismatch factor, as the difference between the most frequent cell STD and the cells with the highest STD, for example.

The sensitivity of the spread of the dispersed data can be characterized with the full width at 10% peak mismatch factor corresponding to an increase in the parameter for the damaged cell as shown in Fig. 18.

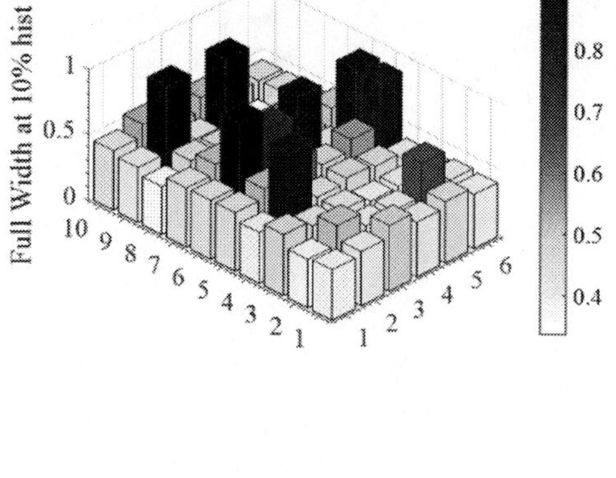

Fig. 18: The colored ELID correspond to the highlighted cells in Fig. 15. Fullwidth at 10% peak representing the dispersion of the defective cells can be seen.

5. SUMMARY AND CONCLUSIONS

In this work we presented the ELID signatures of three main solar cell degradation modes: cell cracking, PID-s shunting and broken interconnect ribbons causing series resistance increase. Moreover, we showed the potential of cell-level ELID statistics to quantify these failures and the cell mismatch they cause. The main advantage of this method is its computational simplicity, making it suitable for automation and large-scale analysis of EL images in module manufacturing or plant inspection. Its main disadvantage is that the statistical parameters do not have a direct and univocal physical meaning (as series resistance for example), and must be interpreted in the context of other ELID statistics. Comparatively, analytical EL analysis methods, such as series resistance or voltage mapping, can give a direct insight into the underlying physical degradation mechanism of the solar cells. However, large-scale application of such methods for module EL analysis is still in development, due to their complexity.

Table 1 Statistical parameters sensitivity to faults

Types of faults	Mismatch-parameter factor					
	Median	γ_1	μ	σ	k	FW_x
Cell breaks			S	S	S	
RD		S		S		S
PID	S			S	S	S

*S – parameter is sensitive

Depending on the type of faults, different statistical parameters are useful in identifying the level of degradation of the PV modules and are helpful in module characterization and diagnostics. The sensitivity to faults of these parameters can be summarized as in Table 1.

These cell-level diagnostic parameters can be used to quantify the severity and distribution of the cell degradation/damage in the module, as well as determine the level of mismatch between the solar cells in the modules, which can be used as an indicator of module quality and different degradation modes can be investigated based on the obtained interpretations.

By analyzing the ELID signatures of three main solar degradation modes with different statistical parameters, it is to be noted that the skewness and kurtosis appear to be very dependent on the sample size. One of the disadvantage of using skewness and kurtosis for shape statistics is that it needs a large dataset for computation for reducing the uncertainty for estimating the location, dispersion and shape parameters. On the other hand, statistical EL methods such as Kurtosis, Full-width and STD have proved to be very useful, indicating the broad range of dispersed values, allowing in a proper estimation of detecting the level of degradation/damage occurring in each cell.

ACKNOWLEDGMENT

This work was partially supported by Aalborg University, Innovation Fund Denmark within the research project "DronEL – Fast and accurate inspection of large photovoltaic plants using aerial drone imaging", project number 6154-00012B, as well as by the U.S. Department of Energy under Contract No. DE-AC36-08-GO28308 with the National Renewable Energy Laboratory.

REFERENCES

[1] M. Köntges, S. Kurtz, C. Packard, U. Jahn, K. A. Berger, K. Kato, T. Friesen, H. Liu, and M. Van Iseghem, *Review of Failures of Photovoltaic Modules*, International Energy Agency, 2014.

[2] D. Hinken, K. Ramspeck, K. Bothe, B. Fischer, and R. Brendel, "Series resistance imaging of solar cells by voltage dependent electroluminescence," *Applied Physics Letters,* vol. 91, no. 18, pp. 182104, 2007.

[3] M. Köntges, M. Siebert, D. Hinken, U. Eitner, K. Bothe, and T. Potthof, "Quantitative analysis of PV-modules by electroluminescence images for quality control," in 24th European Photovoltaic Solar Energy Conference, Hamburg, Germany, 2009, pp. 3226 - 3231.

[4] M. Köntges, I. Kunze, S. Kajari-Schröder, X. Breitenmoser, and B. Bjørneklett, "Quantifying the risk of power loss in PV modules due to micro cracks," in 25th European Photovoltaic Solar Energy Conference and Exhibition / 5th World Conference on Photovoltaic Energy Conversion, Valencia, Spain, 2010, pp. 3745 - 3752.

[5] S. Spataru, P. Hacke, and D. Sera, "Automatic Detection and Evaluation of Solar Cell Micro-Cracks in Electroluminescence Images Using Matched Filters " in 43rd IEEE Photovoltaic Specialists Conference Portland, OR, 2016, pp. 1-6.

[6] S. Spataru, P. Hacke, D. Sera, S. Glick, T. Kerekes, and R. Teodorescu, "Quantifying Solar Cell Cracks in Photovoltaic Modules by Electroluminescence Imaging," in 42nd IEEE Photovoltaic Specialist Conference, New Orleans, 2015, pp. 8.

[7] I. E. Commission. "PNW/TS 82-901 Ed. 1.0 Photovoltaic devices - Part 13: Electroluminescence of photovoltaic modules (proposed future IEC TS 60904-13," 29 October, 2014; http://www.iec.ch/.

[8] S. Spataru, "Characterization and Diagnostics for Photovoltaic Modules and Arrays," Department of Energy Technology, Aalborg University, Aalborg, Denmark, 2015.

[9] R. Shanmugam, and R. Chattamvelli, "Skewness and Kurtosis," *Statistics for Scientists and Engineers*, pp. 89-110: John Wiley & Sons, Inc, 2015.

[10] S. Spataru, P. Hacke, and D. Sera, "Automatic Detection of Inactive Solar Cell Cracks in Electroluminescence Images," in 44th IEEE Photovoltaic Specialists Conference, Washington D.C, 2017.

[11] M. Köntges, I. Kunze, S. Kajari-Schröder, X. Breitenmoser, and B. Bjørneklett, "The risk of power loss in crystalline silicon based photovoltaic modules due to micro-cracks," *Solar Energy Materials and Solar Cells,* vol. 95, no. 4, pp. 1131-1137, 2011.

NON-DESTRUCTIVE EVALUATION OF DELAMINATION IN PHOTOVOLTAIC MODULE BY THERMOGRAPHY

Archana Sinha*, Humaid Mohammad, Rajesh Gupta
Department of Energy Science and Engineering, Indian Institute of Technology Bombay, Mumbai – 400076, India
*Email: archana.sinha@iitb.ac.in, Phone: (+91-22) 2576 7837

ABSTRACT: Delamination is one of the most commonly observed structural defects in the photovoltaic (PV) modules that leads to the hotspots formation and reduction in the PV performance. This paper presents a thermographic method for the non-destructive evaluation of delamination in PV modules. In this method, the module gets thermally excited by a uniform heat flux and the surface temperature response has been evaluated for investigating the presence of sub-surface delamination defect. The influence of different excitation waveforms, such as step, rectangular and half-sine on the thermal contrast evolutions created by delamination of varying thickness have been studied by a 3-dimensional (3-D) electro-thermal modelling and simulation approach, which is also supported by an experimental activity. A characteristic parameter peak contrast time has been considered in a thermal contrast profile for comparing these waveforms as it is less affected by the heat flux intensity and other ambient conditions. Advantages and limitations of each waveform for this investigation have been compared and discussed. Results show that the rectangular heating provides faster detection of delamination, while half-sine waveform possesses better characterization ability. Such investigation paves the way of cost-effective solution for identifying the severity of delamination defect and its impact on the module performance.
Keywords: PV module; delamination; electro-thermal modelling and simulation; thermal contrast; thermography

1 INTRODUCTION

Long-term performance and reliability of PV modules in the field condition are of paramount significance in order to build the confidence in the large-scale PV installation for the residential and commercial applications. The structural integrity of PV modules is impeded by the presence of delamination at any interface, lead to power degradation and reduced service lifetime. Delamination is defined as the breakdown of the bonds between material layers that constitute a module laminate. [1] Improper lamination procedures at production units and harsh outdoor field conditions such as high ambient temperature, thermal excursions between day and night, exposure to UV radiation, moisture and pollutants are the main contributors of interfacial adhesion loss [2]–[4]. The void created due to delamination reduces the rate of heat transfer across the module layers to the surrounding that leads to the formation of local hotspots [5], [6]. As a result, the cell temperature increases that deteriorate the electrical performance and encapsulation material quality. It also facilitates the moisture ingression that generates some catastrophic failures, such as corrosion of metallic contacts and decrease in electrical insulation resistance [7]. Identification of such losses at an early stage is needed to mitigate the loss in performance and associated potential risks involved.

Delamination in any interface of front module layers is visible through the transparent glass, which can be identified in the form of white patches across the metal gridlines and module edges [8]. On the other side, backside delamination cannot be easily visualized due to high opacity of backsheet, unless it grows to a significant size. Fig. 1 shows a representative case of backside delamination in the module, which is present at the solar cell/EVA interface. The traditional approach to investigate the adhesion loss between the solar cell and encapsulant layers is peel strength test; however, it is destructive in nature [9]. Thus, it is utmost important to detect and characterize delamination in PV modules by using some non-destructive testing (NDT) methods. Infrared (IR) thermography is a popular NDT method, particularly after the advancement in thermal cameras and related electronics, which has revolutionized its applicability in various fields. It is widely used for defect investigation in any structure due to superior features, such as fast, non-invasive, non-contact, large area inspection, portable, and easy to implement. The present work is aimed to exploit this technique for the non-destructive evaluation of delamination in the PV modules.

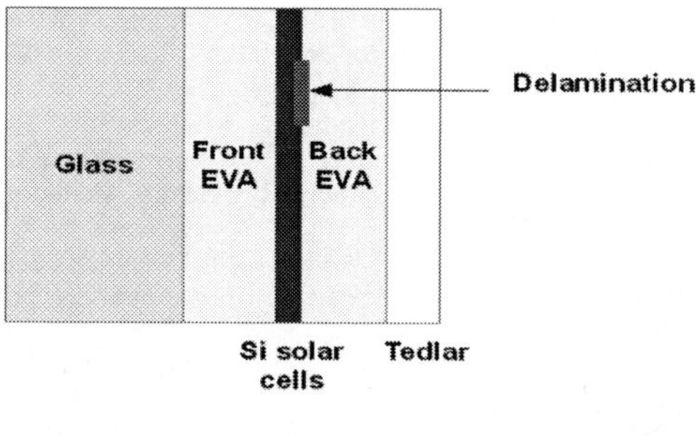

Figure 1: Cross-sectional view of PV module indicating the backside delamination

2 METHODOLOGY

In active IR thermography, a part of the module is thermally excited by irradiating it with uniform heat flux and the subsequent surface temperature response is measured by an IR camera [10]. In the presence of any delamination, the heat propagation is perturbed inside the module that brings the change in surface temperature over the delaminated region, which is quantified in terms of thermal contrast. Absolute contrast is the difference of temperature over delaminated and non-delaminated regions of the module. Running contrast is another way of representing the contrast, which is the ratio of absolute contrast to the surface temperature over non-delaminated region. It has practical significance since it is independent of heat flux intensity. Different excitation waveforms, such as step, rectangular and half-sine have also been employed to realize some improvement in the conventional inspection method.

Delamination investigation by IR thermography method is a heat transfer problem in which the thermal signature is affected by various parameters and environmental factors. To perform this study under a controlled condition, modelling and simulation provides relatively fast and effective approach. In the present study, a 3-D electro-thermal modelling technique has been exploited to simulate the thermographic method for investigation of delamination in PV modules. Further, the results obtained from simulation have been supported by conducting thermography experiments over a delaminated module.

3 ELECTRO-THERMAL MODELLING AND SIMULATION

Electro-thermal modelling and simulation technique is useful in translating the heat transfer problem to equivalent electrical circuits, which is easy to model. It is based on the correspondence between the fundamental laws of heat transfer and electricity, as given below. [11]

$$Q = \frac{\Delta T}{R_{th}} \leftrightarrow I = \frac{\Delta V}{R_{el}}$$

In the above relation, the current (I) is equivalent to the rate of heat flux (Q), voltage (V) plays the role of temperature (T) and electrical resistance (R_{el}) corresponds to the thermal resistance (R_{th}) of a material layer. Fig. 2(a) shows the schematic of 1-D model of a delaminated PV module under thermal investigation. However, in the present study, a 3-D electro-thermal model (shown in Fig. 2(b)) has been used to simulate the heat conduction across the module layers that takes care of lateral heat diffusion phenomenon and facilitates fairly accurate estimation of temperature rise.

In 3-D simulation, the module area (x-y direction) has been divided into a discrete number of sections depending the resolution, while the division of the entire thickness (z-direction) has been decided by considering the thermal diffusivity of each layer. A current source with equivalent shape and magnitude of the signal has been used to model the heat source, whereas the nodal voltage at backsheet has been taken as the surface temperature. Thermal properties of each module layer were assumed to be independent of temperature and the heat flux intensity to be uniform over the module surface, which is a reasonable consideration.

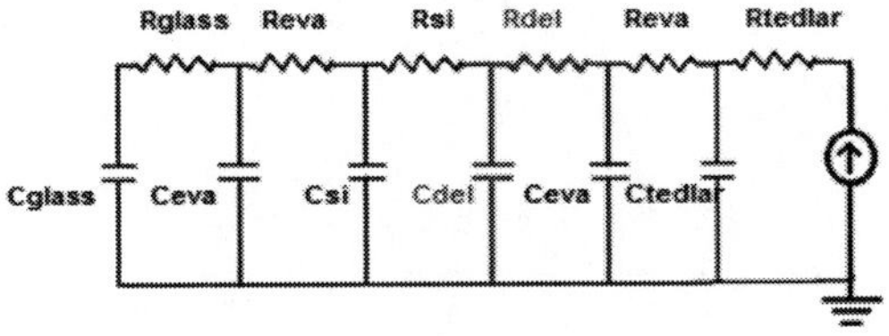

(a)

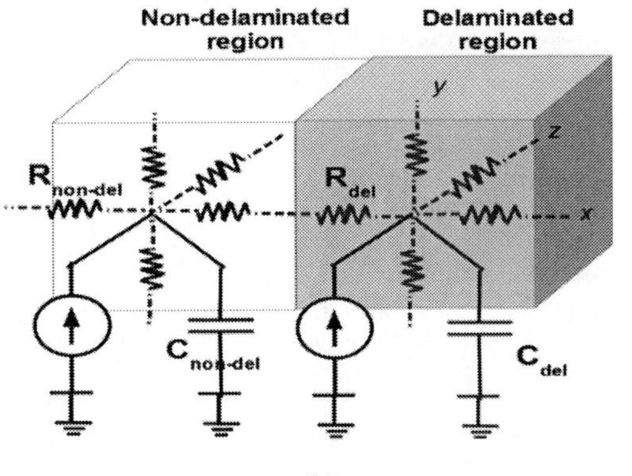

(b)

Figure 2: Schematic of (a) 1-D and (b) 3-D models consisting of equivalent R-C network

4 EXPERIMENTAL

In order to support the simulation results, IR thermographic experiments were carried out on old field-exposed delaminated c-Si PV module. It consists of a string of 36 serially-connected mono-crystalline solar cells encapsulated with glass/ EVA at the front side and EVA/ backsheet at the backside.

The thermography investigation on PV module was performed by a Stirling-cooled IR camera, consisting of 320 × 256 pixels InSb detector with spectral sensitivity in the range of 3–5 μm. The heat source used for this application was a single 1500 W halogen lamp, which imparted the average heat flux of 800 W/m^2 on the module back surface with non-uniformity less than 5%. Different excitation waveforms like step, rectangular and half-sine were employed using function generator connected to the heat source, whose shapes are shown in Fig. 3. The module temperature at backsheet was measured by the IR camera.

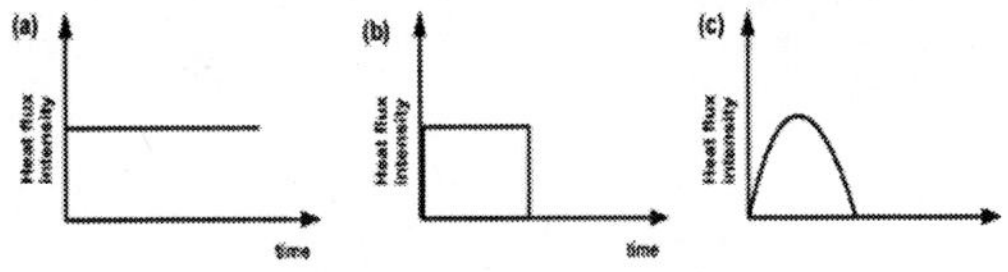

Figure 3: Excitation waveforms used for investigation (a) step, (b) rectangular, and (c) half-sine

5 RESULTS AND DISCUSSION

Fig. 4 shows the thermal image of part of a PV module, where extra hot region is identified as a delamination defect. The delaminated interface creates a thermal mismatch to the inward movement of heat flow that is reflected back to the surface, resulting in temperature rise. An image analysis has been performed to evaluate the area affected by delamination defect, which indicates its severity.

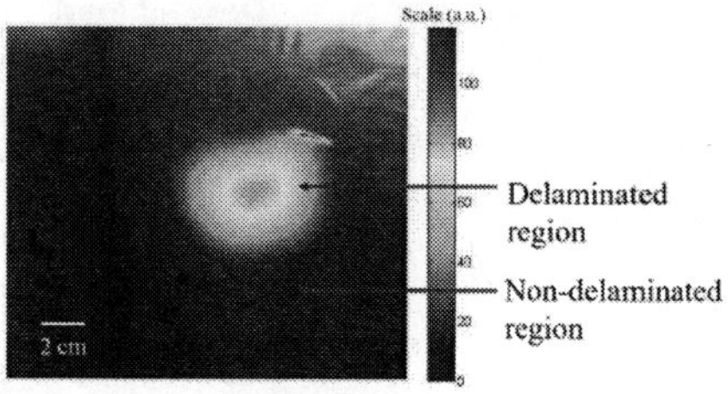

Figure 4: Thermal image of a part of module, showing delamination as a hot region

The simulated plots of thermal contrast profiles due to this delamination defect from different excitation waveforms are given in Fig. 5-7. Fig. 5(a) shows that the temporal evolution of absolute contrast to a step change in heat flux, which increases linearly at the beginning and then gradually reaches its saturation value. The maximum value of contrast is related to the delamination thickness; however, it is largely affected by the heat flux intensity and other ambient conditions. Another characteristic parameter 'time constant' has been considered in this evolution curve, which is defined as the time taken for the absolute contrast due to delamination to reach 63.2% of its saturated value [5]. It is independent of heat flux intensity and only function of thermal properties (thermal resistance and thermal capacitance) of delamination defect, which is advantageous in providing the information about the severity of the defect. It is also evident from the graph that the time constant increases with the increase in delamination thickness and hence thicker delamination takes longer time for its characterization. However, for its estimation, the knowledge of the saturation value of absolute thermal contrast is needed for which one has to wait for longer time. Fig. 5(b) shows that the temporal evolution of running contrast, which first increases and then progressively decreases. The peak in this contrast profile is another important parameter, which can be used to characterize delamination. As the delamination thickness increases, the peak contrast time increases; however, it becomes broader that would lead to high uncertainty in its estimation.

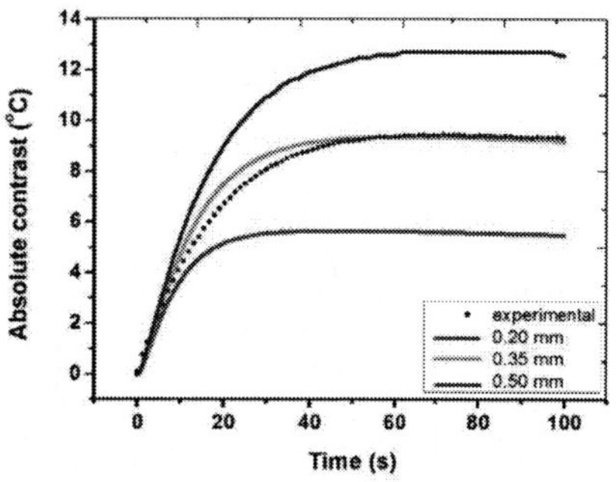

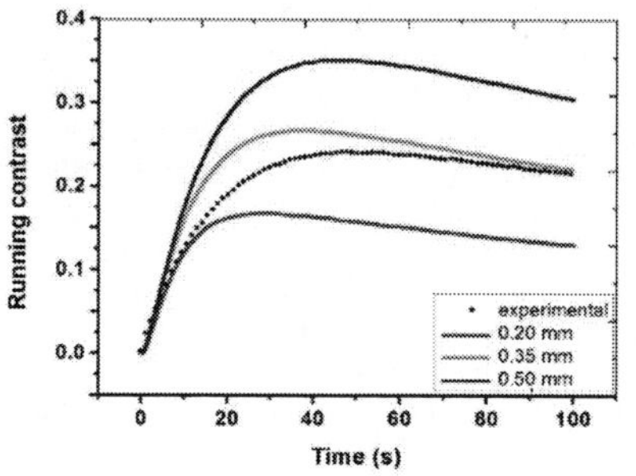

Figure 5: (a) Absolute and (b) running thermal contrast for different delamination thickness under step heating.

Fig. 6(a) and (b) show the simulated evolutions of absolute contrast and running contrast from rectangular heating respectively. In this case, absolute contrast first increases, reaches its maximum near the end of heating and then starts declining non-linearly. The time corresponding to maximum contrast reaches earlier as compared to in the step heating, which signifies that the rectangular heating provides faster investigation of delamination. In case of thinner delamination, the absolute contrast attains its saturated value before the end of heating time. This shows that the lower heating time is sufficient for investigating thinner delamination. The evolution of running contrast shows a peculiar peak just after the end of heating time, which tends to be prominently visible with the increase in delamination thickness. This time corresponds to better visualization of delaminated region in the thermal image due to higher thermal signal w.r.t. background.

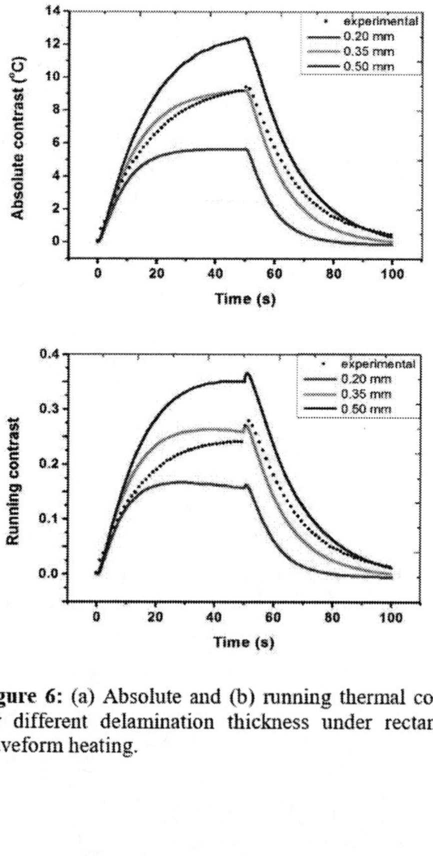

Figure 6: (a) Absolute and (b) running thermal contrast for different delamination thickness under rectangular waveform heating.

Fig. 7(a) shows the simulated plot of absolute contrast from half-sine heating, which shows a peak; however, the peak contrast value is lower than in case of rectangular heating, which is ascribed to the lower input energy transferred to the module. As delamination thickness increases, the peak is shifting towards higher time and becoming sharper. In contrast to rectangular heating, this waveform has an added advantage of attaining the peak before the heating of module terminates but at the cost of lower absolute contrast value. Fig. 7(b) shows the evolution of running contrast, which also exhibits a peak. Interestingly, this peak contrast time varies with delamination thickness and demonstrates higher sensitivity for its characterization as compared to rectangular heating where the peak contrast time occurs at the end of heating.

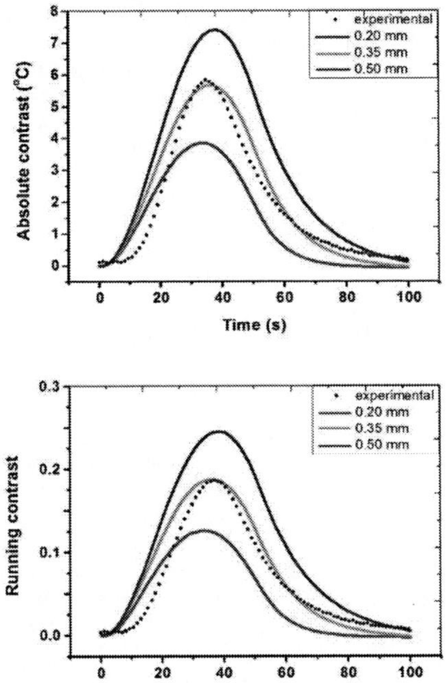

Figure 7: (a) Absolute and (b) running thermal contrast evolution for different delamination thickness under half-sine waveform heating.

The temporal contrast obtained from the experiments for a delamination having thickness of 0.35 mm has also been plotted in Fig. 5-7. It shows that the experimental curves are closely following the simulated curves with a small level of deviation, which could be due to assumptions and values of thermal properties of individual module layers used in the simulation. The present study demonstrated the potential of IR technique for investigation of delamination in PV modules in fast and non-contact manner without destroying them, which would be useful for inline and on-site field testing.

6 CONCLUSIONS

In this paper, a non-destructive approach using IR thermography was presented for the detection and characterization of backside delamination in the PV modules. The application of different excitation waveforms was presented and discussed for fast and better evaluation of delamination defect. Peak running contrast time was considered to be a more appropriate parameter for comparing the appropriateness of each waveform since it is independent of heat flux intensity and other surrounding variables. The results showed that the rectangular waveform is more suitable for faster delamination detection, while half-sine heating waveform provides efficient characterization of delamination in PV modules. This low-cost method can also be applicable to any other module structure where delamination is a common defect that causes loss in the long-term performance and reliability.

7 REFERENCES

[1] M. A. Quintana, D. L. King, T. J. McMahon, and C. R. Osterwald, *29th IEEE Photovoltaic Specialists Conference*, 2002, pp. 1436–1439.
[2] A. Ndiaye, A. Charki, A. Kobi, C. M. F. Kebe, P. A. Ndiaye, and V. Sambou, *Sol. Energy*, vol. 96, pp. 140–151, 2013.
[3] V. Sharma and S. S. Chandel, *Renew. Sustain. Energy Rev.*, vol. 27, pp. 753–767, 2013.
[4] E. E. van Dyk, J. B. Chamel, and A. R. Gxasheka, *Sol. Energy Mater. Sol. Cells*, vol. 88, pp. 403–411, 2005.
[5] S. Kaplanis and E. Kaplani, *Simul. Model. Pract. Theory*, vol. 19, pp. 1201–1211, 2011.
[6] I. Geisemeyer, F. Fertig, W. Warta, S. Rein, and M. C. Schubert, *Sol. Energy Mater. Sol. Cells*, vol. 120, pp. 259–269, 2014.
[7] N. C. Park, J. S. Jeong, B. J. Kang, and D. H. Kim, *Microelectron. Reliab.*, vol. 53, pp. 1818–1822, 2013.
[8] S. Chattopadhyay et al., *IEEE J. Photovoltaics*, vol. 4, no. 6, pp. 1470–1476, 2014.
[9] G. Oreski and G. M. Wallner, *Sol. Energy Mater. Sol. Cells*, vol. 89, pp. 139–151, 2005.
[10] A. Sinha, O. S. Sastry, and R. Gupta, *Nondestruct. Test. Eval.*, vol. 31, pp. 1–16, 2015.
[11] R. Gupta and S. Tuli, NDT&E Int., vol. 38, pp. 11–19, 2005.

UPDATE ON RANKINGS OF CONVERSION EFFICIENCIES AND ENERGY YIELD OF MICRO-INVERTERS, INCLUDING INVERTERS FOR TWO PV MODULES

Stefan Krauter and Jörg Bendfeld
University of Paderborn, Electrical Energy Technology – Sustainable Energy Concepts
UPB, EET – NEK, Pohlweg 55, D-33098 Paderborn, Germany

ABSTRACT: To compare efficiency and yield of most micro-inverters available on the world market in 2014–2017, an in- and outdoor test laboratory at the University of Paderborn have been set up. The inverters have been fed by identical and calibrated crystalline silicon PV modules of 215 W_p. To monitor accurately AC power output and yield, each of the micro-inverters has been equipped with a calibrated precision electricity meter. For micro-inverters requiring control units for grid-feeding that has been acquired also. The comparison includes energy yield over one to two years and efficiency-load characteristics. Purchase costs vary considerably between the models in comparison, sometimes inverter costs are higher than module costs, particularly if an additional grid-connection or interface device is necessary for operation. The weighted conversion efficiency according to EU and CEC standards has been measured and compiled. While some inverters have been optimized for high irradiance levels, they ranked better at the CEC efficiency, others performed very well also for low irradiance levels, thus ranking higher at in the EU-efficiency tables. These results deviated from the actual energy yield measurements, which showed a slightly different ranking. One device showed an extreme deviation of the efficiency rating from the yield results. An accurate but very slow MPPT algorithm that hardly could follow quickly changing irradiance conditions has possibly caused this effect. Apparently, some inverters are been optimized to show excellent EU and CEC efficiency ratings. The new inverters featuring two inputs did not show an exceptional performance at the EU- and CEC-ratings (ranking 5–8 only), but they achieved rank 1 and 2 at the AC energy yield ranking. For the customer the AC yield is a major performance indicator of the micro-inverter and should be included in the datasheet. Further investigations are on the way, including degradation and temperature effects.
Keywords: Inverter, Yield, Small Grid-connected PV Systems, Power Conditioning

1 INTRODUCTION

Micro-inverters are inverters that are connected principally to a single PV module (sometimes to two modules, as indicated in the tables), so each module-inverter combination acts as an independent power plant. The micro-inverter consists of a maximum power point tracker (MPPT), the DC-AC inverter, and an islanding protection unit (see e.g., [1]). To achieve higher power outputs, several module-inverter combinations are interconnected in parallel on the AC output side. This configuration offers various advantages: Easier planning and easier installation, easy up- and downscaling of the power plant, including extensions or repair that could be carried out during power plant operation. Effect of shadowing is very limited, and due to low system voltages, potential induced degradation (PID) does not occur. Logistics is simplified. An excellent overview of the development and the advantages of micro-inverters has been compiled by H. Oldenkamp [2].

However, costs of power plants based on micro inverters are about 20% higher. Some of the inverters cannot be operated by themselves and require a control unit (often combined with a remote shutdown option and a monitoring system), or a protective device for grid interfacing (depending on national regulations), thus adding extra costs. Also, conversion efficiency may not be as high as for central inverters. Due to smart master-slave concepts centralized solutions with multiple but relatively large inverters may offer higher yields under weak light conditions. [3] is giving a performance comparison of a micro inverter, a power-optimizer and a central inverter. Unfortunately, the only micro-inverter used in that comparison has been one with a quite poor performance (according to our comparisons [4] and [5]), and therefore the comparison results conceivably do not favor the category "micro-inverter" in general.

2 INDOOR TESTS

Due to the reproducible test conditions in the indoor lab, the inverters have been examined individually with predefined and controlled input data. Input has been a PV module simulator with data set according to the modules used in the outdoor test. The principal output data recorded has been the delivered AC power of the inverters. Besides input power, output is also a function of input voltage. If input voltage is getting too low, the inverters even stop operating. The following examinations are based on the possible range of input data (including voltage) given the specific PV module also used for the outdoor investigation

2.1 Peak, EU- and CEC conversion efficiency

Peak efficiency is often reached close to the maximum load of the inverter. Peak efficiency in data sheets is not a helpful value since most of the time the inverters operate in the range of 20% to 30% of their rated power. Consequently, an adequately weighted efficiency is a more adequate measure. One type of weighted efficiency is the so-called "European Efficiency" η_{Euro} – it is calculated according to:

$$\eta_{Euro} = 0.03 \cdot \eta_{5\%} + 0.06 \cdot \eta_{10\%} + 0.13 \cdot \eta_{20\%} + 0.1 \cdot \eta_{30\%} + 0.48 \cdot \eta_{50\%} + 0.2 \cdot \eta_{100\%} \quad (1)$$

The other is the "CEC efficiency" by the California Energy Commission (CEC). CEC efficiency is computed as an average value of DC-AC conversion efficiencies at six pre-defined outputs between 10% and 100% of the rated power (with an emphasis on higher irradiance levels) is determined by:

$$\eta_{CEC} = 0.04 \cdot \eta_{10\%} + 0.05 \cdot \eta_{20\%} + 0.12 \cdot \eta_{30\%} + 0.21 \cdot \eta_{50\%} + 0.53 \cdot \eta_{75\%} + 0.05 \cdot \eta_{100\%} \quad (2)$$

For the European Efficiency the weighting factors are lower for the high relative power values.

2.2 Test arrangement and results

Ten different output power values from zero up to the maximum operational power of the PV modules have been considered (they are roughly identical for all seven micro inverters). The values used for the single input inverters (adjusted by controlling the current) are: 10 W, 25 W, 50 W, 75 W, 100 W, 125 W, 150 W, 175 W, 200 W, and 225 W. The averaging time for the power meter used (Zimmer LMG 670) was set to 500 ms. One micro inverter showed an unstable AC power output for values below 100 W, for that device the averaging time had to be extended to two seconds.

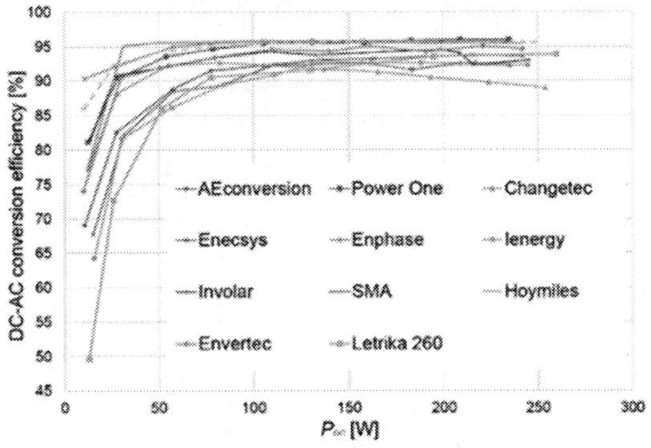

Figure 1: Measured DC-AC conversion efficiencies as a function of DC input power for eleven micro-inverters with single PV module inputs.

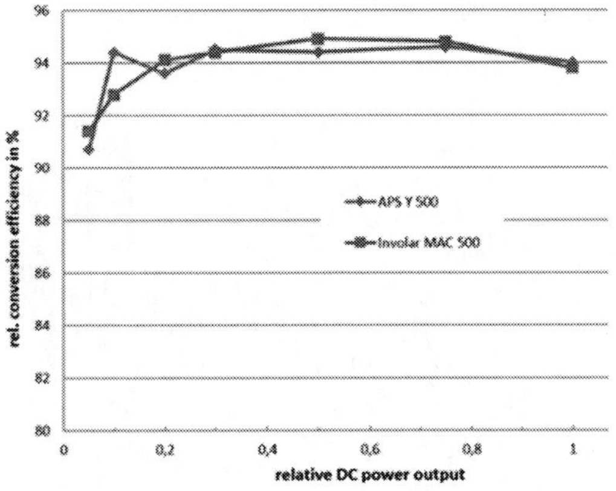

Figure 2: Measured DC-AC conversion efficiencies as a function of relative DC input power for two inverters that have inputs for two PV modules (2 · 250 W).

For each power-step 25 values have been recorded. The measured DC-AC conversion efficiencies of all inverters are shown in Fig. 1 and Fig. 2. Based on those measurements, the European (EU) efficiencies and the CEC efficiencies for the micro inverters have been calculated according to (1) and (2), eleven micro-inverters are designed for single modules, two inverters have inputs for 2 modules: Involar MAC 500 and the APS Y 500.

The ranking considering the European conversion efficiency is shown in Table I: AEconversion/Aptronic and Enecsys have the same EU-efficiency, thus sharing rank number ten.

Table I: Ranking of all tested micro-inverters by "European conversion efficiency", according to (1), same types and models as in Table III.

Rank	Manufacturer	European efficiency	Relative eff. vs #1
1	SMA	95.4 %	100.0 %
2	Enphase	95.2 %	99.8 %
3	Hoymiles	95.0%	99.5 %
4	ABB/Power One	94.6 %	99.2 %
5	Involar MAC 500	94.3 %	98.8 %
6	APS Y 500	94.1 %	98.6 %
7	Envertech	93.2 %	97.7 %
8	Involar MAC 250	92.7 %	97.2 %
9	Changetec	90.9 %	95.3 %
10	AEconversion	90.3 %	94.7 %
10	Enecsys	90.3 %	94.7 %
11	Ienergy	89.9 %	94.3 %
12	Letrika 260	88.7 %	93.0 %

The ranking for the CEC-conversion-efficiency is given in Table II, here the Letrika 260 and the Ienergy achieved the same results, therefore they are sharing rank number ten.

Table II: Ranking of all micro-inverters by CEC conversion efficiency, according to (2), same types and models as in Table III.

Rank	Manufacturer	CEC efficiency	Relative eff. vs #1
1	Enphase	95.6 %	100.0 %
2	ABB/Power One	95.5 %	99.9 %
3	Hoymiles	95.4 %	99.8 %
4	SMA	95.1 %	99.5 %
5	Involar MAC 500	94.6 %	99.0 %
6	APS Y 500	94.5 %	98.9 %
7	Envertech	94.1%	98.4 %
8	Involar MAC 250	93.9 %	98.2 %
9	Enecsys	92.0 %	96.3 %
10	Letrica 260	91.5 %	95.7 %
10	Ienergy	91.5 %	95.7 %
11	AEconversion	91.2 %	95.5 %
12	Changetec	90.9 %	95.1 %

3 OUTDOOR TESTS

3.1 Test arrangement and measurements

For the PV outdoor test lab (see Fig. 3) installed on the roof of the University of Paderborn (51.707° N, 8.771° E), a specific test system comprising of eight photovoltaic modules with eight micro-inverters has been employed. To examine thirteen inverters, five of them have been interchanged.Ten different output power values from zero up to the maximum operational power of the PV modules have been considered (they are roughly identical for all seven micro inverters). The values used for the single input inverters (adjusted by controlling the current) are: 10 W, 25 W, 50 W, 75 W, 100 W, 125 W, 150 W, 175 W, 200 W, and 225 W. The averaging time for the power meter used (Zimmer LMG 670) was set to 500 ms. One micro inverter showed an unstable AC power output for values below 100 W, for that device the averaging time had to be extended to two seconds.

Figure 3: PV outdoor laboratory at the University of Paderborn for comparison of micro-inverters using eight equal, calibrated PV modules.

Each PV module (QC-CO2 by Hanwa Q-cells) consists of 60 solar cells. In stated plant, these equal and calibrated modules are the input for each micro inverter. The goal of this investigation is to analyze the performance of the inverters in the lab and under real operating conditions comparing their energy yield simultaneously at the same climatic conditions and solar irradiance. The climatic conditions have been monitored during the whole test period. The meteorological monitoring equipment consists of two calibrated pyranometers in the plane of the module (CMP 21 and SP2 lite by Kipp & Zonen), a 3-D ultrasonic anemometer (by Thies), and a thermo-hygro sensor (by Thies).

Fig. 4 shows the electrical measurement setup. Each micro inverter has been directly connected to a calibrated electrical energy meter with a S0-interface. All S0-interfaces have been connected to an efficiency recording data server.

Fig. 5 shows an example of the collected data for single day (data recording every 5 min). Due to the high resolution in time, power and energy output, some specific characteristics could be observed: E.g., starting behaviors, MPP tracking (accuracy and speed), dropouts and performance.

Fig. 6 shows an example of the collected data for single day during spring of the year 2015. It is visible that some inverters have difficulties to follow the quick change of irradiance levels during that day (e.g. IEnergy, and Hoymiles – see also Fig. 7).

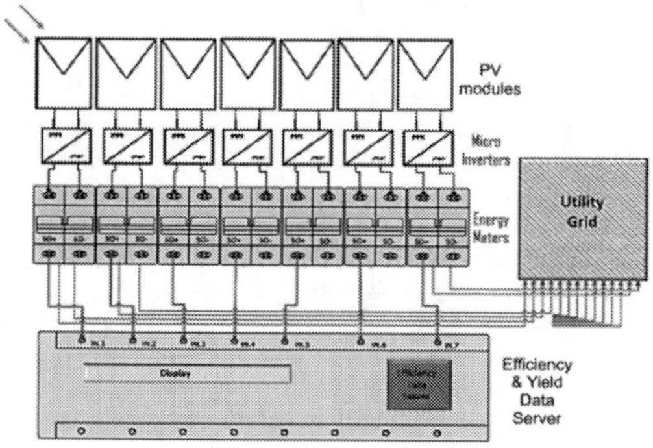

Figure 4: Scheme of the outdoor test setup for the micro-inverters (MI) investigated.

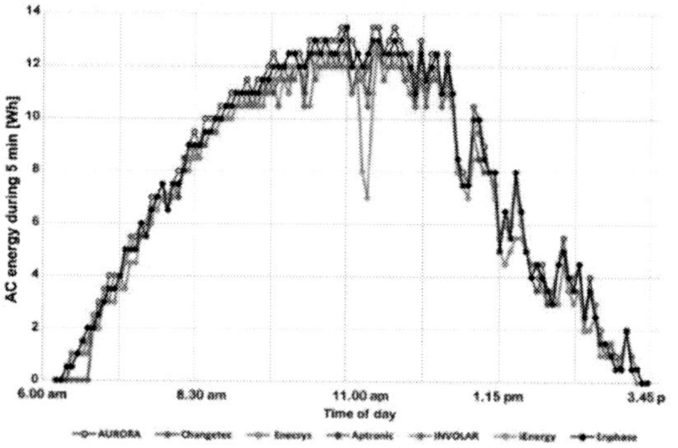

Figure 5: Actual recorded data of AC energy generation (over an integral of 5 min) for seven micro-inverters during the day of 10/31/2013.

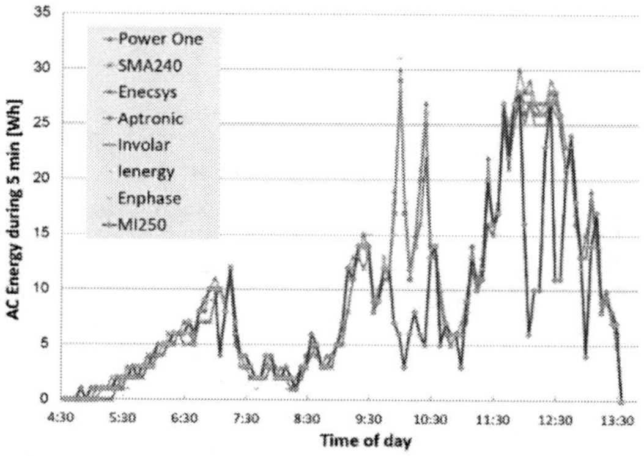

Figure 6: Actual recorded data of AC energy generation of seven micro-inverters (over an integral of 5 min) during the day of 4/6/2015 with changing irradiance. Some inverters have not been able to follow a quick increase of irradiance.

A ranking list according to the total AC energy yield of the micro inverters during the common operation period is shown in Table III (July 2015 until June 2016, (Envertech 6/1/2016 until 6/15/2016 only) relative energy yield vs. former #1 [4]. To make future progress in yield improvements more transparent, that #1 reference will be kept (as 100%). Eleven micro-inverters are designed for single modules; two new inverters have inputs for 2 modules: Involar MAC 500 and the APS Y 500. The

Letrika 260 has been at the outdoor test laboratory for insufficient time to achieve qualified yield results, but it seems to stay in the lower average of the range of inverters examined. It is equipped to perform active anti-islanding directly, so it does not need any additional device as a grid interface - even in countries that have require protection devices according to IEC 62116.

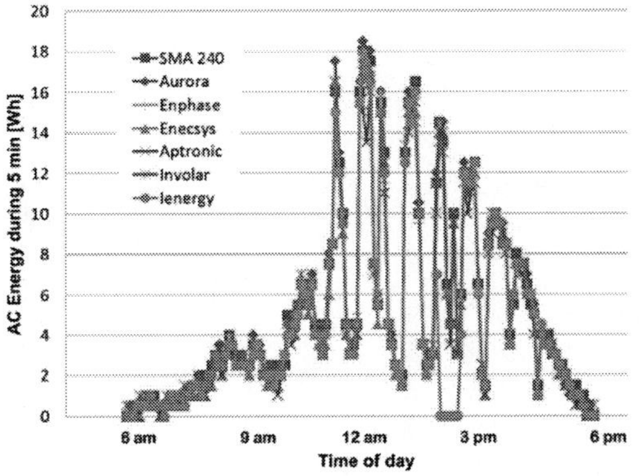

Figure 7: Actual recorded AC-energy generated within a 5 min interval during a day with changing irradiance levels. One inverter adapts quickly (blue line: ABB/Aurora), the other not (red line: Hoymiles MI 250).

Table III: Ranking of all micro-inverters by measured AC energy yield (in Paderborn 2015–2017), referenced to best performer of last examination [4] [5]. Update on 20/09/2017.

Rank	Manufacturer	Type, Model	rel. yield vs. ABB (Ex-#1) [4]
1	Involar	MAC 500	100.7 %
2	APSystems	Y 500	100.5 %
2	ABB/Power One/Aurora	Micro-0.25-i	100.0 %
4	SMA	Sunny Boy 240	95.5 %
5	Enphase	M 215	95.9 %
6	Involar	MAC 250	93.8 %
7	IEnergy	GT 260	93.4 %
8	Envertech	EVT 248	92.1 %
9	Lektrika	SMI 260	91.5%
10	AEconversion /Aptronic	INV 250-45	90.3 %
11	Enecsys	SMI-S-240W	88.7 %
12	Hoymiles	MI 250	78.4 %
13	Changetec	ELV 300-25	75.6 %

4 RESULTS, DISCUSSION AND CONCLUSION

4.1 Results

Some of the results could be expected by taking a close look at the efficiency curves of Fig. 5. For example: If highest performance occurs at relatively low input power (and the corresponding low irradiance levels), the ranking for the European Efficiency (which puts a higher weight on lower irradiance levels) is higher, while the CEC ranking is worse due to the higher weighting of high irradiance levels (this is the case for the Changetec inverter). The best in the CEC ranking has been the Enphase inverter, which was second in European Efficiency ranking. For the SMA inverters the ranking positions have been the other way round: SMA is optimized for European irradiance conditions, Enphase for Californian conditions. While the conversion efficiency rating of the new micro-inverters with two inputs, Involar MAC 500 and the APS Y 500, has not been exceptional (ranks 5-8), they achieved rank 1 and 2 at the energy yield ranking. The layout of your paper should have exactly the same format as these notes.

4.2 Correlation of yield with conversion efficiency

Some of the rankings by yield (as shown in Table III) cannot be explained by the curves of Fig. 1 and 2 and the European or the CEC efficiency. We observed significant differences of the MPPT devices in their capabilities to adjust to an increase of irradiance quickly (reduction of internal load resistance). Best in this discipline has been the Power One / Aurora inverter, which has not been the best in the static conversion efficiency measurements, but in energy yield. The device MI 250 showed an extreme deviation of its efficiency rating from the yield results. An accurate but slow MPPT algorithm that had difficulties to follow changing irradiance conditions has possibly caused this effect.

Fig. 8 shows the relative energy yield as a function of European efficiency (which is often an important criterion for the purchase decision by a potential customer). As can be observed, there is just a quite vague correlation between those two values with high deviations. For the example indicated by the numbers 1,2,3 at Fig. 8: A customer wants to buy an efficient inverter and he usually takes the datasheet values (or even actual measurements) of the European conversion efficiency as an indicator. According to that, he would prefer Inverter (1), offering a slightly EU-efficiency: 95% instead of 94.6% (2), resp. 94.5% (3).

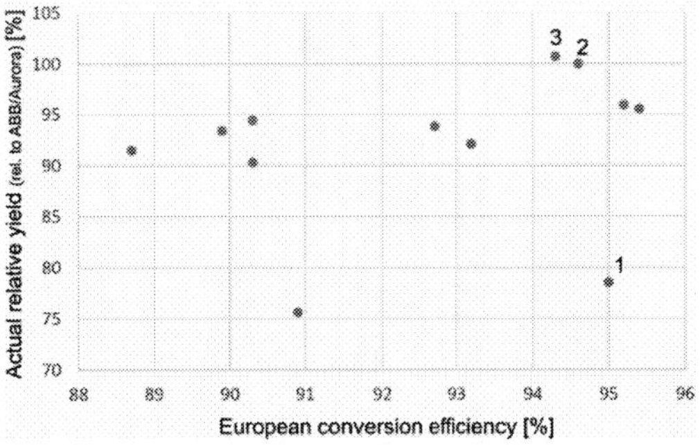

Figure 8: Electrical energy yield as a function of European efficiency: The user cannot expect a higher yield if he purchased an inverter with a higher European efficiency – the yield can be significantly lower, see also [5].

After purchase and during operation he will find out that his inverter will have a relative AC yield of 78.4% only, compared to 100% (2) or even 100.5% (3). This is unsatisfying; therefore, the actual on-site AC yield values (also considering the dynamic behavior of the inverters) should be the major indicator for a potential customer and should be given in the datasheet.

4.3 Conclusion

For the time being, no standard is able to overcome the constraints by the ranking via weighted efficiencies, so the on-site measurement of electrical energy yield is the best method to achieve reliable results that are helpful and relevant for the customer – or at least for customers with similar climatic conditions.

4.4 Outlook

To overcome the considerable deviation between yield and efficiency measurement results, a dynamic test will be deployed to find a better indicator for the relative yield to be expected by the different inverters. A statistics of the frequency of irradiance changes, its amplitudes, and the velocities of changes will be translated into a test procedure.

While the lab tests have been made at laboratory temperature conditions (20°-22°C) and the yield measurements at actual operating temperature for moderate Paderborn conditions (which did not cause a detectable degradation after 2 years [6]), additional efficiency tests at elevated temperatures (e.g. 40°C and 60°C) will be carried out (see also [7]).

5 REFERENCES

[1] W-F. Lai, S-M. Chen, T-J. Liang, K-W. Lee, A. Ioinovici, "Design and Implementation of Grid Connection Photovoltaic Inverter" in *IEEE Energy Conversion Congress and Exposition (ECCE),* 2012, pp. 2426–2432.

[2] H. Oldenkamp, I.J. de Jong, "The return of the AC-module inverter" in *24th European Photovoltaic Solar Energy Conference (EUPVSEC)*, 2009, pp. 3101–3104.

[3] D. Stellbogen; P. Lechner; M. Senger, "Field and laboratory performance characterisation of microinverter and Power optimizer systems" in *32nd European Photovoltaic Solar Energy Conference (EUPVSEC),* 2016, pp. 1654–1659.

[4] S. Krauter; J. Bendfeld, "Cost, performance, and yield comparison of eight different micro-inverters", in *42nd IEEE Photovoltaic Specialist Conference (PVSC)*, 2015.

[5] S. Krauter; J. Bendfeld, "Deviations of results for energy yield from efficiency rankings of micro-inverters", in *32nd European Photovoltaic Solar Energy Conference (EUPVSEC)*, 2016, pp. 1508–1511.

[6] S. Krauter; J. Bendfeld, "Long-term performance of PV micro-inverters", in *33rd European Photovoltaic Solar Energy Conference (EUPVSEC)*, 2017.

[7] M. A. Hossain; Y. Xu; T. J. Peshek; L. Ji; A. R. Abramson; R. H. French: "Microinverter Thermal Performance in the Real-World: Measurements and Modeling", in *PLOS ONE*
DOI:10.1371/journal.pone.0131279 July 6, 2015.

PERFORMANCE ANALYSIS OF THE PROPOSED SINGLE-STAGE TRANSFORMERLESS BOOST GRID-CONNECTED MICROINVERTER (STBM) UNDER PARTIAL SHADING CONDITIONS

Fernando C. Melo, Rodolfo R. de Godoi, Luiz C. Freitas, Ernane A. A. Coelho, João B. Vieira Jr., Valdeir J. Farias, and Luiz C. G. Freitas

Núcleo de Pesquisa em Eletrônica de Potência (NUPEP)
Faculdade de Engenharia Elétrica (FEELT) - Universidade Federal de Uberlândia (UFU)
Uberlândia, Minas Gerais, Brazil 38400-902
e-mails: fernando_cmelo@outlook.com; lcgfreitas@yahoo.com.br

ABSTRACT: Microinverters applied to photovoltaic systems can increase the energy generation by reducing the effects of partial shading compared to string inverters, since the energy harvesting of each PV module is controlled individually. In this context, the STBM single-stage transformerless grid-connected Boost microinverter 500 W prototype was evaluated experimentally under an irradiance step, steady state operation and two different partial shading conditions. Besides that, an experimental comparison of a commercial string inverter (2 kW) and a microinverter (250 W) was presented proving that the latter has a higher kWh/kWp (specific production) performance under the same operating conditions.
Keywords: Grid-Connected Inverter, Microinverter, Photovoltaic, PV Module, String Inverter.

1 INTRODUCTION

String inverters are based in series connections of PV modules through dc cables (strings). These systems can present disadvantages such as: mismatching losses if the PV modules have different electrical characteristics; generally, there is only one MPPT for a large group of PV modules and due to the high dc voltage and long cabling, there is risk of electric arc [1].

For solving these problems, Microinverter for low power systems can be applied since its efficiency is greater when compared to string inverters by reducing the losses by mismatching (gain of 3 to 5%), by partial shading (gain of 5 to 25%) and by the reduction of faults due to short dc cable (gain of 0 to 15%). The MPPT is performed individually by the microinverters for each PV module, which can result in a 3 to 10% of efficiency gain. This can increase the solar energy extraction for roofs that are orientated differently, therefore is possible to install a microinverter for each PV module.

Finally, there is a safety improvement and high area utilization, since it is possible to install PV modules with different inclinations [2]-[5]. Nevertheless, microinverters can increase cost per module or cost per Wp compared to string inverters. Figure 1 (a) presents the performance of string inverters under a partial shading of one PV module and is possible to verify that in this condition, the entire energy produced by the string is reduced. In Fig. 1 (b) is shown that this problem can be solved by installing a microinverter for each PV module [2].

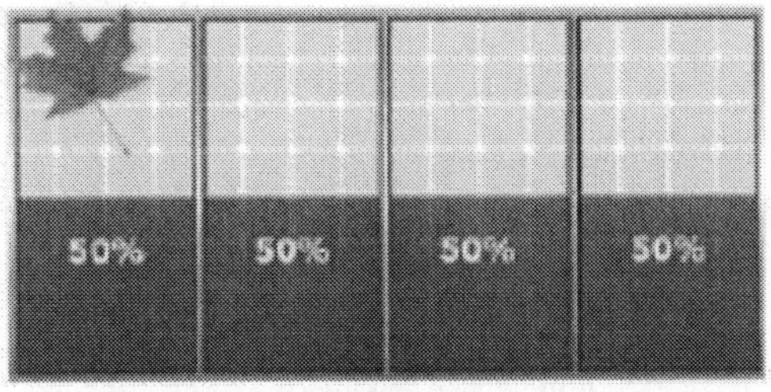

(a)

(b)

Figure 1: (a) Partial shading on a string inverter and its respective reduction of the overall energy production and (b) using microinverters for each PV module the partial shading losses are reduced.

For commercial microinverters (MIC) topologies generally, isolated converters are used, lowering the efficiency, lifetime and increasing the costs. On the other hand, the so called transformerless inverters can operate with high switching frequencies, reducing reactive elements and heat dissipation. Then, one can produce converters with lower weight, volume and costs [5]-[7].

Different topologies of single-stage transformerless PV inverters have been proposed for obtaining a more compact and efficient structure. In the literature are found various single-stage topologies that were reviewed in [8], such as non-isolated Boost inverter with four switches. Another topology uses two Buck-Boost converters that provide each half-cycle of the output voltage from a dc symmetrical source, allowing the MPPT operation. One can observe another single-stage topology with four switches and zero current switching (ZCS) named as Resonant Buck-Boost.

In [9] is presented two single-stage inverters: one for step-down, composed by a typical Voltage Source Inverter (VSI) and the other is for step-up applications (Current Source Inverter - CSI). Moreover, are presented topologies that operate as step-down and step-up, such as the Z-Source Inverter and the Natural Soft-Switching Inverter. A new topology of voltage inverter is presented in [10], operating as a step-up or a step-down converter single-stage and transformerless.

The pursuit for increasing efficiency and lowering costs, which can be achieved by reducing the number of solid-state devices, has motivated the development of the *Single-Stage Transformerless Boost Microinverter*

(STBM) [11]-[14]. The structure is portrayed in Fig. 2(a) and the main operational feature observed is the output voltage gain, since it affords Boost operating mode with totally controlled sinusoidal output voltage and input inductors currents, without using any transformer. Additionally, it has reduced number of switches and diodes. The schematic diagram of the microinverter with the Modified P&O "Gama" MPPT algorithm generating the variable I_{REF}* and the grid current control loop composed by the Proportional-Resonant (P-Res) controller is presented in Fig. 2(b). In this figure, one can verify that the reference (V_{REF}) is adjusted by the P-Res to inject a sinusoidal current into the grid. More details concerning the STBM operating principle are presented in [13]-[15].

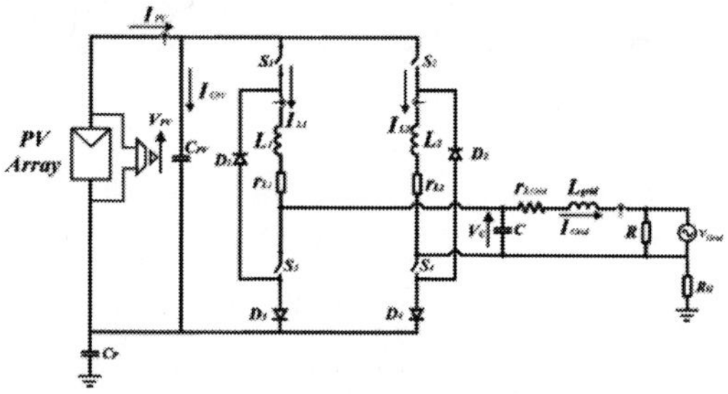

(a)

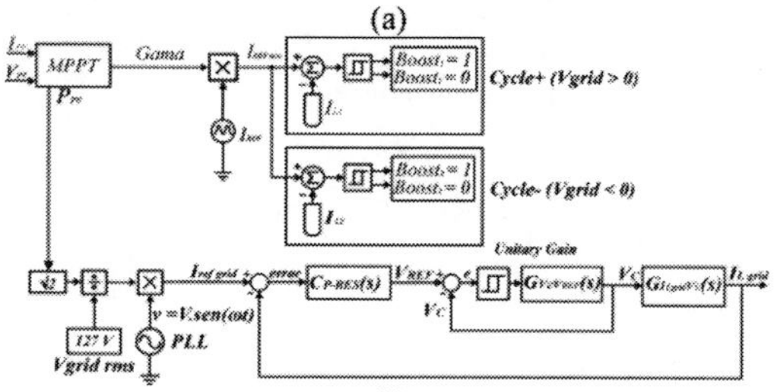

(b)

Figure 2: (a) Single-stage Transformerless Boost Microinverter (STBM) and (b) the control diagram of the Modified P&O MPPT technique and the grid current control.

In this paper, experimental results of the STBM for an irradiance (W/m^2) step maintaining the temperature at 25°C and steady state condition (830 W/m^2 and modules temperature at 57°C) are presented. Also, the STBM was evaluated for two partial shading conditions for testing the Modified P&O "Gama" MPPT reliability as shown in Section 2. Furthermore, to validate the advantages of microinverters over string inverters, an experimental comparison of a commercial string inverter (2 kW) and a microinverter (250 W) was presented in real operating conditions using the Sun-Earth PV modules installed at the laboratory roof. All the acquired data in the period evaluated is also presented in terms of kWh/kWp (specific production) in Section 3.

2 STBM EXPERIMENTAL RESULTS

The single-stage inverter prototype built in laboratory has the specifications and general data presented in Tab. 1. For implementing the switching logic, Modified P&O "Gama" MPPT algorithm, synchronism control loop to the grid and the current injection control was used the DSP microcontroller Texas Instruments® TMS320F28335 [16]-[222].

Figure 3 shows the prototype developed in laboratory, highlighting the power circuit board and the Solar Array Simulator (SAS) E4361A from Keysight® used during the initial stage of tests and on the final tests was used the SAS from Regatron® model TopCon Quadro (Programmable DC Power Supply of maximum power of 10 kW). The final tests were carried out using a PV array composed by two Amerisolar (250Wp) connected in series.

The transient irradiance test was made for a positive variation from 600 W/m^2 to 1000 W/m^2 (using Regatron SAS), presented in Fig. 4, increasing the injected current amplitude and the available PV power, proving the satisfactory operation of the proposed MPPT algorithm.

Table I: STBM Prototype Specifications

Design Specifications
RMS Output Voltage, $V_{O(rms)}$ = 127 Vrms
Input Voltage, V_{MP} = 59.6 V – Input power, P_{PV} = 490 Wp
Maximum switching frequency, f_s = 25 kHz
Microinverter Power Circuit Specifications
Inductors, L_1 and L_2 = 360 μH, 40 A
Grid Inductor, L_{grid} = 9 mH
Input Capacitor, C_{PV} = 3 mF
Output Capacitor, C = 10 μF
Switches, $S_1 - S_2$, IRFP4868
Switches, $S_3 - S_4$, IRFP4768
Diodes, $D_1 - D_4$, STTH200L04TV
Microcontroller, DSP TMS320F28335

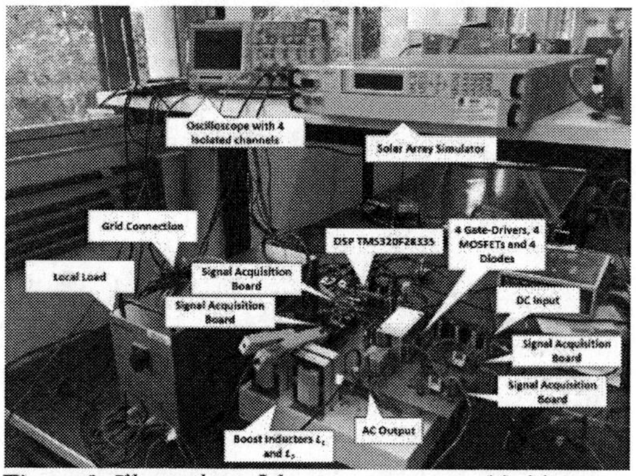

Figure 3: Illustration of the prototype assembled in laboratory and the measuring instruments used.

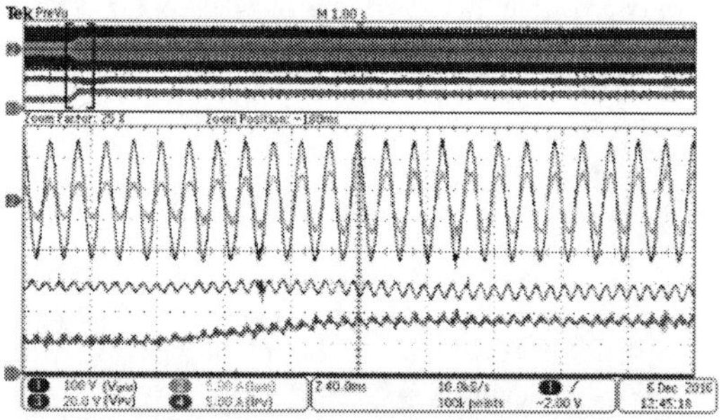

Figure 4: Irradiance step from 600 W/m^2 to 1000 W/m^2, showing the grid voltage and injected current. Below, are shown the PV array voltage and current.

Figure 5 shows the experimental test under real operating conditions of irradiance and PV module temperature for two Amerisolar PV modules installed on the laboratory rooftop (irradiance at 830 W/m², PV modules temperature at 57°C and ambient temperature at 29°C). This figure presents the grid voltage and current waveforms and Boost inductors currents. One can observe that the injected current waveform is in accordance with the standard IEEE-1547 [23].

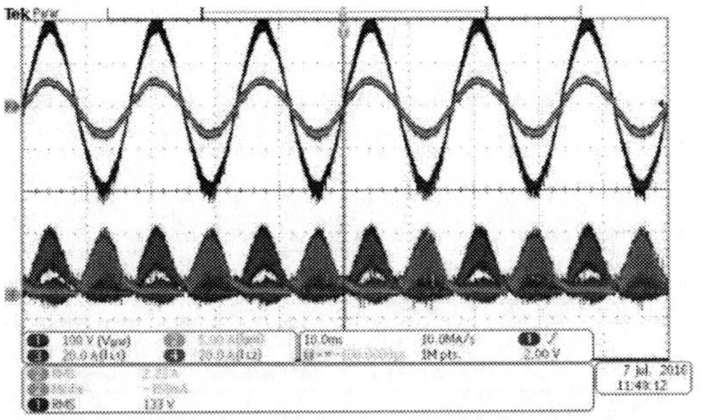

Figure 5: Grid voltage and current and below are the Boost inductors currents, for operation of the PV array at 830 W/m², modules temperature at 57°C and ambient temperature at 29°C.

Partial shadings can be caused by nearby buildings, trees, among others. For validating the proposed Modified P&O "Gama" MPPT algorithm, was performed a partial shading in one of two Amerisolar (250 Wp) PV modules of the array, as shown in Fig. 6 (a) and the resulting I-V and P-V curves is illustrated in Fig. 6 (b). The modules were characterized using the Solar I-V from HT Instruments®, then the collected points were inserted in the SAS (Solar Array Simulator) from Regatron® TopCon Quadro (*Programmable DC Power Supply*) for testing this condition on the STBM.

Moreover, the Fig. 6 (b) shows the inverter extracting the maximum power at 1033 W/m² and modules temperature at 55.1°C (red dot on the I-V curve, P_{max} = 309.05 W and the measured power is equal to 300 W, resulting in a MPPT efficiency of 97.1 %). Another test of partial shading was implemented in both PV modules, as illustrated in Fig. 7 (a), and the resulting I-V and P-V curves are illustrated in Fig. 7 (b), proving also that the inverter is extracting the maximum power at 1018 W/m² and modules temperature at 56.2°C (P_{max} = 194.19 W and the measured power is equal to 182 W, resulting in a MPPT efficiency of 94 %).

(a)

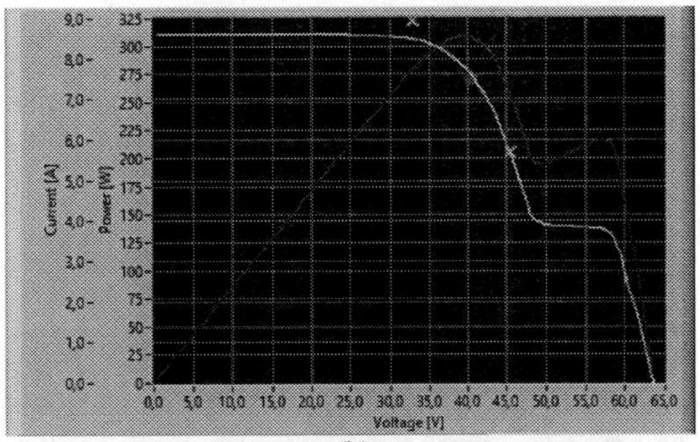

(b)

Figure 6: (a) Partial shading in one Amerisolar PV module of the array and (b) the resulting I-V and P-V curves, showing the inverter extracting the maximum power at 1033 W/m² and modules temperature at 55.1°C.

(a)

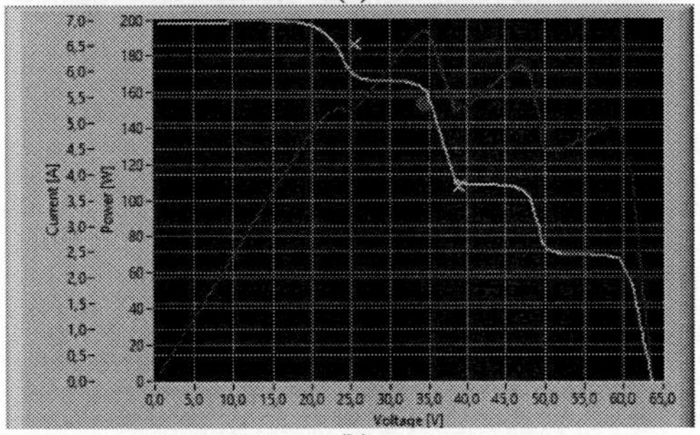

(b)

Figure 7: (a) Partial shading in both PV modules and (b) the resulting I-V and P-V curves, showing also the inverter extracting the maximum power at 1018 W/m² and modules temperature at 56.2°C.

3 PERFORMANCE COMPARISON OF COMMERCIAL STRING INVERTER AND MICROINVERTER

For validating the presented theory concerning the advantages of microinverters over string inverters, experimental tests were carried out with the Sun-Earth PV modules of 235 Wp installed on the laboratory roof with the same horizontal inclination of 20°, azimuthal orientation of 13° and ambient conditions (soiling, temperature and irradiation) as shown in Fig. 8. One can observe that 7 PV modules, composing an installed power of 1,645 kWp, were connected to the string inverter Fronius IG 2000, which specifications are presented in Table II. The generation data has been saved in the *Fronius*

Datalogger Web (acquisition time interval).

The test with the Hoymiles MI-250 microinverter, which specifications are illustrated in Table III, was performed with a 235 Wp PV module and its respective generation data were obtained with the HT Solar 300 N (*Solar System Analyzer*).

Figure 8: Sun-Earth 235 Wp PV modules. In red, PV string connected to the Fronius inverter and in yellow is the PV module connected to the Hoymiles microinverter.

Table II: Fronius IG 2000 Specifications

DC Input	
Recommended PV panel power	1500 - 2500 Wp
Maximum DC input voltage	500 V
Operating MPPT DC Voltage Range	150 - 450 V
Maximum DC input current	13.6 A
AC Output	
Maximum output power at 40°C	2000 W
Rated output AC voltage	220 V
AC mains voltage range	180 - 264 V
Maximum AC current	8.35 A
Maximum return current	0 A
Operating frequency range	59.3 to 60.5 Hz
Total Harmonic Distortion	< 5 %
Power Factor	1
General Data	
Maximum Efficiency	95.2 %
Consumption during operation	7 W
Stand-by consumption	<0.15 W (night)
Ambient temperature range	-25 to 50 °C

Table III: Hoymiles MI-250 Microinverter Specifications

DC Input	
Input Power	200 W- 310 W
MPPT Voltage Range	27 - 48 V
Maximum Input Voltage	60V
Nominal MPPT Efficiency	99.8%
AC Output	
Output Power	250 W
Power Factor	> 0.99
Current Total Harmonic Distortion	< 3%
Maximum Efficiency	96.7%
Dimensions (Length x Height x Depth)	18.3 x 16.4 x 2.8 cm
Weight	1.65 kg

For comparing the performance of different PV systems, one can use the ratio of kWh generated by a given period per kWp. As an example, was observed that on July 7, 2017, the string inverter and the microinverter processed practically the same amount of energy as shown in Table IV. During this day, the generation of solar energy showed high intermittency due to the great cloudiness, as illustrated in the output energy curves of both analyzed systems in Figs. 9 (a) and (b).

It was also verified that on July 13, 2017, the string inverter processed 4.63 kWh/kWp and the microinverter was around 4.7 kWh/kWp, presenting a gain of 1.5 % of the energy extracted. One can observe that the energy measurements for this day occurred with clear skies, as shown in the output energy curves of Figs. 10 (a) and (b).

Table IV: Performance of Hoymiles MI-250 Microinverter and Fronius IG 2000 String Inverter (2 days)

Date	*Hoymiles MI-250 (kWh/kWp)*	*Fronius IG 2000 (kWh/kWp)*	*Hoymiles/ Fronius (%)*
07/07/2017	3.52	3.514	**0.17**
07/13/2017	4.7	4.63	**1.5**

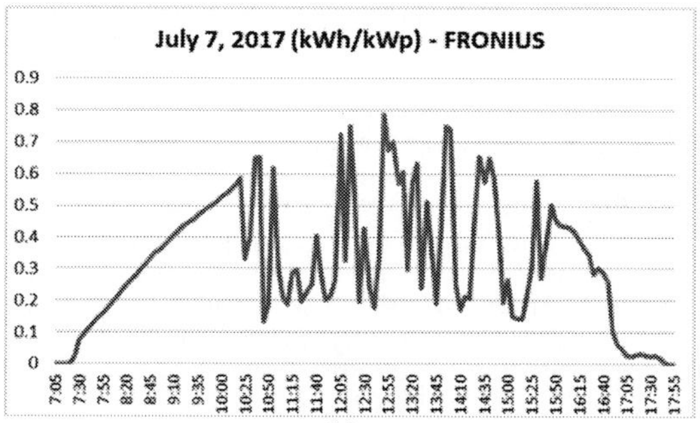

(a)

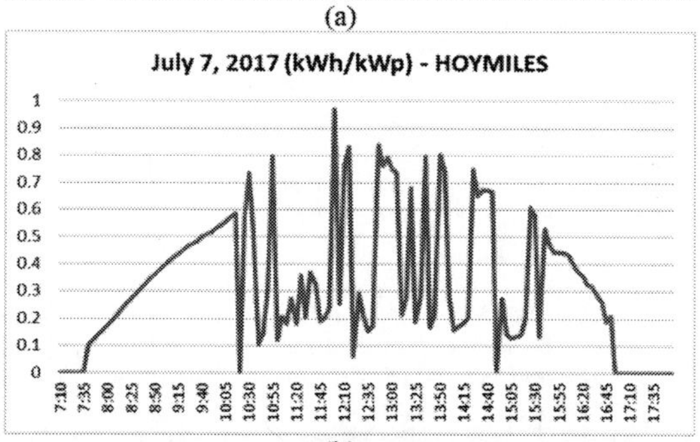

(b)

Figure 9: Output energy curve obtained showing the kW/kWp along the same day (July 7) (a) from the string inverter and (b) from the microinverter.

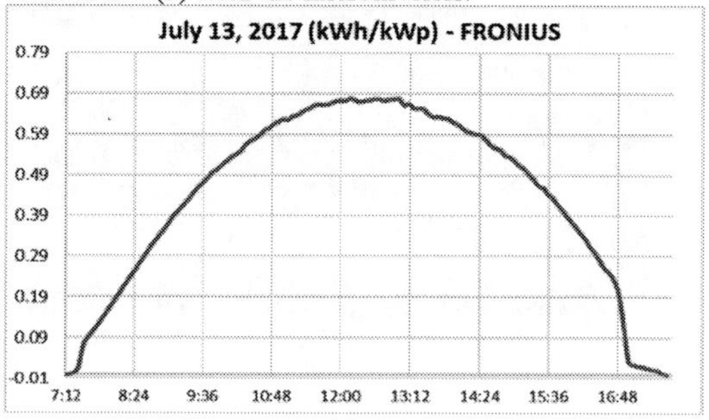

(a)

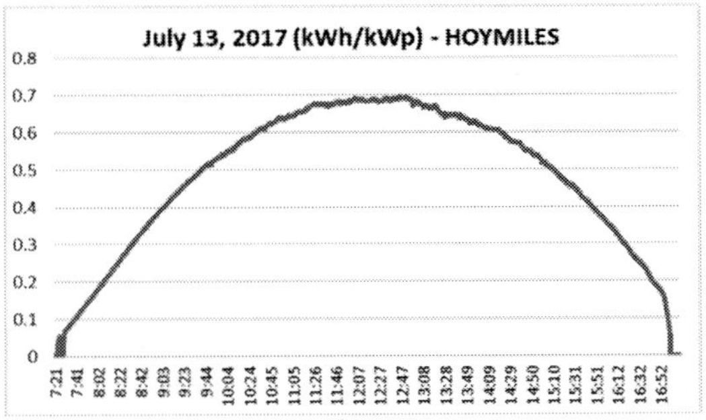

(b)

Figure 10: Output energy curve obtained showing the kW/kWp along the same day (July 13) (a) from the string inverter and (b) from the microinverter.

All the obtained results in the evaluated period consisted between the end of June and early August are presented in Table V, in which one can observe that the generation in terms of kWh/kWp by means of the microinverter is greater than the one obtained by the string inverter, being 4.36 % greater in average. This result can prove the theory previously discussed, mainly due to the fact the MPPT algorithm implemented by the microinverter is dedicated to an individual PV Module.

Table V: Performance Comparison of Hoymiles MI-250 Microinverter and Fronius IG 2000 String Inverter

Date	*Hoymiles MI-250 (kWh/kWp)*	*Fronius IG 2000 (kWh/kWp)*	*Hoymiles/ Fronius (%)*
06/27/2017 to 06/30/2017	12.39	12.26	**0.11**
07/06/2017 to 07/13/2017	32.85	28.82	**13.98**
07/15/2017 to 07/22/2017	33.15	32.99	**0.48**
07/25/2017 to 08/01/2017	40.00	38.88	**2.88**
Average	-	-	**4.36**

4 CONCLUSIONS

The STBM topology was evaluated for a positive irradiance step (W/m^2) using the SAS, steady-state operation under real climate conditions and two cases of partial shading (for one and both PV modules). In all cases and conditions, the STBM MPPT algorithm P&O Modified "Gama" can extract the maximum power presenting efficiency results of 99% proving its reliability.

This paper also presented a contribution of comparing experimentally a commercial string inverter (2 kW) and a microinverter of 250 W. They were submitted to real operating conditions in terms of weather, being evaluated during the end of June and early August 2017. The PV modules under test were the Sun-Earth 235 Wp, installed at the laboratory roof. During the energy assessment of two specific days, one cloudy and the other with clear skies, the microinverter presented a slightly higher performance than the string inverter (0.17 and 1.5%, respectively). The acquired generation data in the period evaluated in terms of kWh/kWp, showed that the microinverter had an average productivity gain of 4.36% higher than the string inverter.

This can evidence its higher performance for low power application, for example, for residential consumers that demand a reduction in the electrical bill without making a high initial capital investment.

If the application requires an installed power higher than 1 kW, in terms of cost benefit, the string inverter can be considered, because its cost $/Wp is lower than the microinverter.

5 ACKNOWLEDGMENTS

The authors wish to thank the financial support for this project provided by CNPq (Process 406845/2013-1, 472457/2013-6, 304252/2013-1, 304307/2013-0) and FAPEMIG (Process APQ-01219-13, TEC-PPM-00031-14).

6 REFERENCES

[1] T. Salmi, M. Bouzguenda, A. Gastli, A. Masmoudi, "A Novel Transformerless Inverter Topology without Zero-Crossing Distortion." in International Journal of Renewable Energy Research, vol. 2, no. 1, pp. 140-146, 2012.

[2] J. Tao, V. Xue, "Grid-Connected Micro Solar Inverter Implement Using a C200 MCU." Application Report SPRABT0 – Texas Instruments, January 2013.

[3] Q. Li and P. Wolfs, "A Review of the Single Phase Photovoltaic Module Integrated Converter Topologies with Three Different DC Link Configurations," in IEEE Trans. on Power Electronics, vol. 23, no. 3, pp. 1320-1333, May 2008.

[4] C. T. Rodriguez, D. V. de la Fuente, G. Garcera, E. Figueres, J. A. G., "Reconfigurable Control Scheme for a PV Microinverter Working in Both Grid-Connected and Island Modes," in IEEE Trans. on Industrial Electronics, vol. 60, no. 4, pp. 1582 – 1595, Apr. 2013.

[5] D. Meneses, F. Blaabjerg, Ó García and J. A. Cobos, "Review and Comparison of Step-Up Transformerless Topologies for Photovoltaic AC-Module Application," in IEEE Trans. on Power Electronics, vol. 28, no. 6, pp. 2649-2663, June 2013.

[6] H. Hu, S. Harb, N. H. Kutkut, Z. J. Shen, I. Batarseh, "A Single-Stage Microinverter Without Using Eletrolytic Capacitors," in IEEE Trans. on Power Electron., vol. 28, no 6, pp. 2677 – 2687, Jun. 2013.

[7] S. Jiang, D. Cao, Y. Li, F.Peng, "Grid-Connected Boost-Half-Bridge Photovoltaic Microinverter System Using Repetitive Current Control and Maximum Power Point Tracking," in IEEE Trans. on Power Electron., vol.27, no.11, pp.4711,4722, Nov. 2012.

[8] D. Meneses, F. Blaabjerg, Ó García and J. A. Cobos, "Review and Comparison of Step-Up Transformerless Topologies for Photovoltaic AC-Module Application," in IEEE Trans. on Power Electronics, vol. 28, no. 6, pp. 2649-2663, June 2013.

[9] W. Wu, J. Ji and F. Blaabjerg, "Aalborg Inverter - A New Type of "Buck in Buck, Boost in Boost" Grid-Tied Inverter," in IEEE Transactions on Power Electronics, vol. 30, no. 9, pp. 4784-4793, Sept. 2015.

[10] Piazza, G.L.; Barbi, I., "New Step-Up/Step-Down DC–AC Converter," in IEEE Trans. on Power Electronics, vol.29, no.9, pp.4512,4520, Sept. 2014.

[11] L. S. Garcia, L. C. Freitas, G. M. Buiatti, E. A. A. Coelho, V. J. Farias, L. C. G. Freitas, "Evaluation of a single-stage current source inverter with high-voltage gain supplied by a polymer electrolyte membrane fuel cell," in IET Power Electronics, v. 5, n. 9, Nov. 2012, pp. 1834-1846.
[12] L. S. Garcia, L. C. Freitas, G. M. Buiatti, E. A. A. Coelho, V. J. Farias, L. C. G. Freitas, "Dual Transformerless Single-Stage Current Source Inverter with Energy Management Control Strategy," in IEEE Trans. on Power Electronics, vol.28, no.10, pp.4644,4656, Oct. 2013.
[13] Fernando C. Melo, Lucas S. Garcia, Luiz C. de Freitas, Gustavo M. Buiatti, Ernane A. A. Coelho, Luiz C. G. Freitas, "Novel Transformeless Single-Stage 4-Switches Buck-Boost Inverter," in Proc. of IEEE Applied Power Electronics Conference and Exposition 2013, pp. 2811-2816;
[14] F. C. Melo, L. S. Garcia, E. A. A. Coelho, V. J. Farias and L. C. G. Freitas, "Transformerless Microinverter for PV Systems," in Proc. of 28th European Photovoltaic Solar Energy Conference and Exhibition, 2013, pp. 2902-2907.
[15] F.C. Melo, L. S. Garcia, L. C. Freitas, E. A. A. Coelho, V. J. Farias and L. C. G. Freitas, "Proposal of a Photovoltaic AC-Module with a Single-Stage Transformerless Grid-Connected Boost Microinverter (STBM)," in IEEE Transactions on Industrial Electronics, vol. PP, no. 99, pp. 1-1.
[16] Kjaer, S.B.; Pedersen, J.K.; Blaabjerg, F., "A review of single-phase grid-connected inverters for photovoltaic modules," in IEEE Trans. on Industry Applications, vol.41, no.5, pp.1292,1306, Sept.-Oct. 2005.
[17] H. Hu, S. Harb, N. Kutkut, I. Batarseh and Z. J. Shen, "Power decoupling techniques for micro-inverters in PV systems-a review," in Proc. of 2010 IEEE Energy Conversion Congress and Exposition, Atlanta, GA, 2010, pp. 3235-3240.
[18] L. P. Sampaio, M. A. G. de Brito, L. G. Junior, G. de A. e Melo and C. A. Canesin, "Single-phase current-source-boost inverter for renewable energy sources," in Proc. of IEEE International Symposium on Industrial Electronics, 2011, pp. 1118-1123.
[19] M. A. G. Brito, L. P. Sampaio, L. G. Junior, C. A. Canesin, "Evaluation of MPPT techniques for photovoltaic applications," in Proc. of IEEE International Symposium on Industrial Electronics 2011, pp.1039-1044.
[20] Sachin Jain, Vivek Agarwal, "New current control based MPPT technique for single stage grid connected PV systems", in Proc. of Energy Conversion and Management, Volume 48, Issue 2, February 2007, pp. 625-644.
[21] B. S. Prasad, S. Jain and V. Agarwal, "Universal Single-Stage Grid-Connected Inverter," in IEEE Trans. on Energy Conversion, vol. 23, no. 1, pp. 128-137, March 2008.
[22] Ciobotaru, M.; Teodorescu, R.; Blaabjerg, F., "Control of single-stage single-phase PV inverter," in Proc. of 2005 European Conference on Power Electronics and Applications, pp.10.
[23] IEEE Standard Conformance Test Procedures for Equipment Interconnecting Distributed Resources with Electric Power Systems," in IEEE Std 1547.1-2005, pp.1-62, July 1, 2005.

This is a pre-peer review version of the paper submitted to the journal Progress in Photovoltaics.
It has been selected by the Executive Committee of the EU PVSEC 2017 for submission to Progress in Photovoltaics.

Verifying defective PV-modules by IR-imaging and controlling with module optimizers

Claudia Buerhop[1], Frank W. Fecher[1], Tobias Pickel[1], Adrian Häring[2], Tim Adamski[2], Christian Camus[1], Jens Hauch[1], Christoph J. Brabec[1,3]

[1] Bavarian Center for Applied Energy Research (ZAE Bayern), Immerwahrstraße 2, D-91058 Erlangen, Germany
[2] SolarEdge Technologies GmbH, Werner-Eckert-Straße 6, 81829 Munich, Germany
[3] Institute Materials for Electronics and Energy Technology (i-MEET), Friedrich-Alexander-University Erlangen-Nürnberg, Martensstraße 7, 91058 Erlangen, Germany

Abstract Quality control and efficient operation of roof-top, residential PV-installations is an economic and technical challenge because these PV-systems often suffer from various roof orientations and inclinations as well as partial and temporal shading. Module optimizers, which track the operating point individually for each module group, are good solutions providing also additional electrical information. The purpose of this work is verifying the identification and localization of defective PV-modules with IR-imaging and with module optimizers as well as quantifying the impact on the performance. Several PV-plants were inspected and carefully analyzed. The findings using IR-imaging can clearly be classified to substring failures due to open circuit and short circuit, bypass-diode faults, cell defect, and some undefined optimizer problems. The electric data reveal that these findings are related to certain power loss and monthly or annual yield reduction. Interesting is the temporal behavior of particular defective sites, which show short-term recovery. In summary, there is a good agreement between the occasionally performed IR-findings and the continuously monitored electrical data.

Keywords: thermography, module optimizer, IR-imaging, yield, monitoring, defects.

INTRODUCTION

Infrared (IR)-imaging is widely used for quality inspection of photovoltaic (PV)-plants [1-3]. Defective modules are reliably detected at operating conditions. The benefit of this method would be increased even more if the impact on the module performance could be quantified. Data originating from the solar simulator are of limited usefulness since operating conditions cannot be mimicked. Operating conditions means that the modules are at elevated temperatures, have a varying temperature distribution, are connected in series and therefore interact and influence each other. Defective modules within a string configuration impact current and voltage of the overall string maximum power point and accordingly the defect temperature [4]. In contrast, module optimizers control one or two modules, thus, a limited interaction of the modules is possible.

As a first approach, a test facility was built up in order to study the correlation between IR-measured defect temperatures and module performance [4]. This motivated a study on the assessment for IR-inspection and efforts on system-design and improvement of real PV-installations [5].

In a cooperative study the experiences are now used to verify both, IR-inspection and module-level monitoring in terms of defective module detection and quantification of their impact on the performance.

Module level control systems, so-called module optimizers, are frequently installed in residential roof-top PV-plants. Here, the modules are exposed to varying solar irradiance, so that string inverters are not the best solution. Shading, differing roof orientations and inclinations are some of the reasons. In order to avoid performance losses of the PV-system due to partial shading under series-connection conditions, module optimizers can limit the losses [6].

This paper, for the first time, sheds light on IR-inspections of real PV-systems with a quantitative defect assessment based on module optimizers. Of interest is the impact of the maximum power point (MPP)-tracking of the module optimizers on the defect temperature difference and the resulting yield loss. Due to reduced mismatch these numbers should be lower than for standard string solutions. Thus, for the same defect lower impact is expected. For the first time IR inspection results are compared to power and yield data from module-level-optimized monitoring data, thus answering the question how defects detected in IR images affect power and yield of the modules. Comparative electrical circuit simulations complete the results for failures like shaded cells and broken cells for single modules with module-optimizers and for module strings.

EXPERIMENTAL PROCEDURE

Ten roof-top residential PV-installations equipped with module optimizers were investigated to compare the detection of defective modules by IR-imaging and module optimizers, respectively. The average peak power of the PV-plants is 25 kWp. They were operating for one to six years. An important peculiarity of these installations is the presence of various roofs with differing angle of inclination and orientation. Table 1 gives an overview of the investigated PV-plants and their key figures including the found failures.

Table 1 Characteristic data of the investigated PV-systems with module optimizers. The designation of the PV-plants is based on anonymized data of the PV-system specification. (Abbreviations: c-Si = crystalline silicon, TF = thin-film (CIGS-technology), SCM = short-circuited module (due to 3 defect bypass-diodes), SCO = short-circuited optimizer, OCM = open-circuited module, HCM = module with hot cell, SCS = module with one short-circuited substring, OCS1 = module with one open-circuited substring, OCS2 = module with two open-circuited substring, GB = module with Glass breakage, - = no failure.)

PV-plant	Module technology	Size (in kWp)	Number of modules	Duration of operation (in years)	Number of modules per optimizer	Number of found failures
B	c-Si	20.75	83	5.3	1	-
G	c-Si	23.92	92	1.75	1	-
H	c-Si	16.5	60	2.3	2	1 SCS
L	TF	27.39	168	1.2	3	-
P	c-Si	99.75	406	3.1	2	2 SCO or 4 SCM (reason unknown)
R	c-Si	20.5	109	2.1	2 (1x single module)	7 OCS1, 5 OCS2
SC	c-Si	29.7	110	3.3	2	-
SO	c-Si	21.86	93	5.6	1	1 OCM, 1 SCM, 1 HCM
ST	c-Si	18.02	68	2.75	2	1 OCS1
W	TF	29.96	120	1.6	3	20 GB

The applied module optimizers control one, two or three modules depending on the site design. They are tracking the MPP independently for each module (two or three modules, respectively). Simultaneously, they are logging power, voltage and yield data. By analyzing the module-level data one can see how power and yield are affected by different influences such as defects (found by IR), ambient conditions, module orientation, shading, module position in the array and others.

Fig. 1: Measurement system for aIR-PV-check of PV-plants

For characterizing the installed PV-modules or PV-plants, drone mounted IR-imaging systems are used [7]. The measurement system, as shown in Fig. 1, consists of an unmanned drone (Multicopter), a lightweight IR-camera PI 450 (optris), a visible camera GoPro and navigating equipment. Solar irradiance E and ambient temperature T_{amb} are recorded, too. The measurements were carried out on suitable days in November 2016, March 2017 and July 2017.

Time-resolved monitoring data for energy, power, voltage and current of the module(s) per module optimizer were provided by SolarEdge via their web portal, as well as the ambient temperature, wind speed and relative humidity delivered from a weather data service [8].

The data evaluation combines singular IR investigations and historical monitoring data of the entire operation period.

RESULTS

This is a pre-peer review version of the paper submitted to the journal Progress in Photovoltaics.
It has been selected by the Executive Committee of the EU PVSEC 2017 for submission to Progress in Photovoltaics.

Simulations

Whereas in normal module strings typically up to 22 or 24 modules are connected in series, systems with module optimizers have only one or two modules connected in series. The resulting electrical differences between both systems become important when module failures occur or operating conditions vary for the different modules. This will be shown by simulating and comparing the *I-V* and *P-V* curves of a single module and a string with 20 modules. The modules consist of 60 cells with 3 bypass diodes, i.e. 20 cells per substring. The curves were simulated for standard test conditions (STC) of irradiance $E = 1000$ W/m² and module temperature $T = 25$°C. As failure type we assumed the shading of only one cell in the module and one cell in the whole module string. It is seen in Fig. 2, depending on the relative insolation of the shaded cell, the I-V curve changes and the current in the substring is limited by the shaded cell. In both cases (module and module string) the curves behave the same, but due to the higher number of substrings the relative impact on the *I-V* curve is lower for the module string. Consequently the impact of a failure can more easily be detected and does not disappear in the noise of monitoring data (as seen in Fig. 8). The *P-V* curves with the shaded cell exhibit two maxima. In module strings with a shaded cell with relative insolation of 80% and lower, the global maximum is found for the lower voltage and the bypass diode of the substring with the shaded cell is conducting. In this working point the string current is the same as it would be without shading and the whole power of the substring with the shaded cell is lost. In contrast, in the single module the bypass diode becomes conductive for much lower relative insolation of 50% and lower. This fact might be the explanation why the shaded cells in Fig. 3 (left) do not heat up and appear cool despite the shading by the dormer. When the right maximum is tracked and the bypass diode is inactive, all cells in the module work on the lower current and become warmer than in a module without failures or shading. However, the temperature rise should be less than 1 K when assuming a reasonable relation between produced electrical power and temperature of a crystalline silicon cell of 1 W/K. This temperature rise is hardly detectable by IR inspection. When the bypass diode is conducting, the shaded cell is pushed onto a negative bias that compensates the sum of positive biases of the unshaded cells plus the activation voltage of the diode (i.e. 19 * 0.5 V + 0.6 V = 10.1 V). The grade of cell shading determines the current in the substring and, therefore, the dissipated power in the shaded cell. For instance of 50% shading 45.45 W (= 4.5 A * 10.1 V) are dissipated as heat in the shaded cell. If more cells are shaded in one substring, the negative bias is shared among the shaded cells. Hence, less energy is dissipated into heat per shaded cell. For instance, for two cells with 50% shading the dissipated power is 21.6 W (= 4.5 A * 9.6 V / 2).

As the electrical behavior of shaded cells and cells with completely electrically isolated broken cell parts are almost the same, the results above are transferable on modules with cell breakage. Here, the relative insolation corresponds to the relative area of unbroken cell part.

This is a pre-peer review version of the paper submitted to the journal Progress in Photovoltaics.
It has been selected by the Executive Committee of the EU PVSEC 2017 for submission to Progress in Photovoltaics.

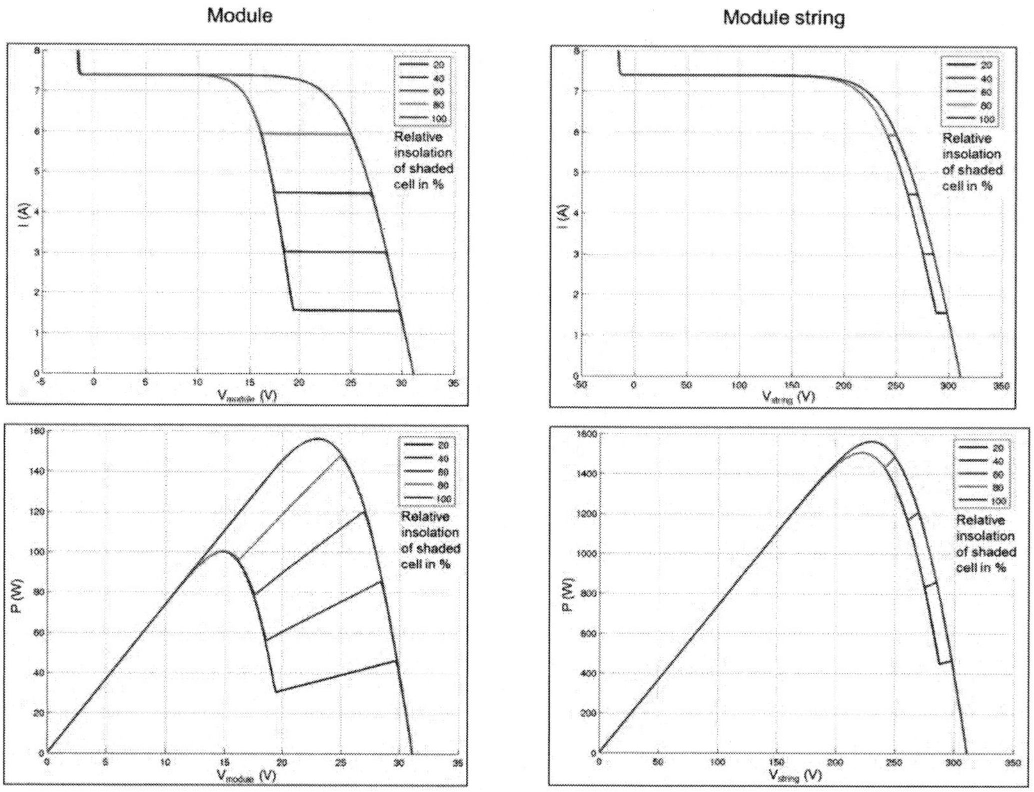

Fig. 2: Simulated *I-V* (top) and *P-V* (bottom) curves of a single module (left) and a module string (with 20 modules in series) (right) with one shaded cell for different relative insolations.

IR-Investigations

Many private houses have roof segments with differing roof inclinations, e.g. due to dormer windows. In IR-images, PV-modules located at such dormers appear cooler than the other roof areas, compare Fig. 3. The IR-camera looks nearly perpendicular on the main roof area. These modules have an apparent temperature of 34°C whereas for the dormer modules a temperature of 30°C is measured, as shown in Fig. 3 right. Due to the larger observation angle the emissivity is lower. Thus, more radiation from the cool sky is reflected and pretends a lower module temperature. Frequently, found temperature differences are between 4 – 10 K. The influence on the energy yield due to the different roof angles and the shading is shown in Fig. 4 for PV-plant B and R for the days 19th – 21st June 2017.

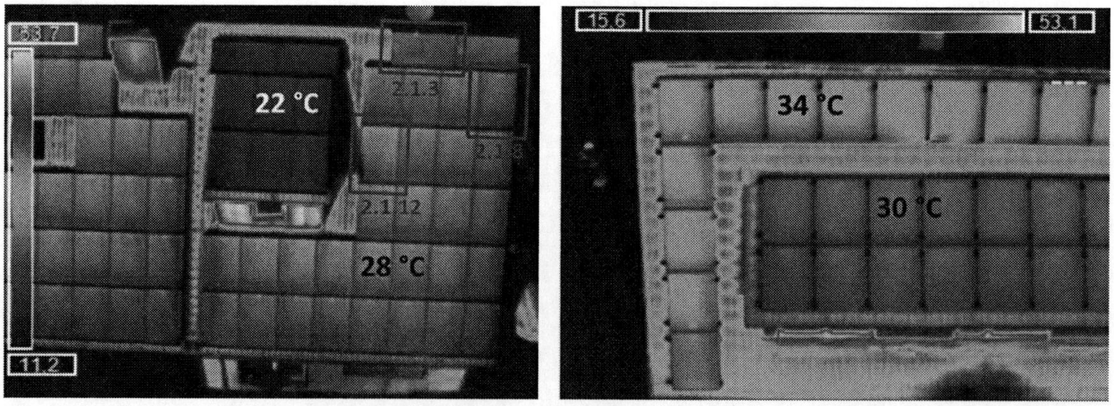

Fig. 3: PV-plant B (left) and PV-plant R (right), Franconia, Germany, 29th November 2016.

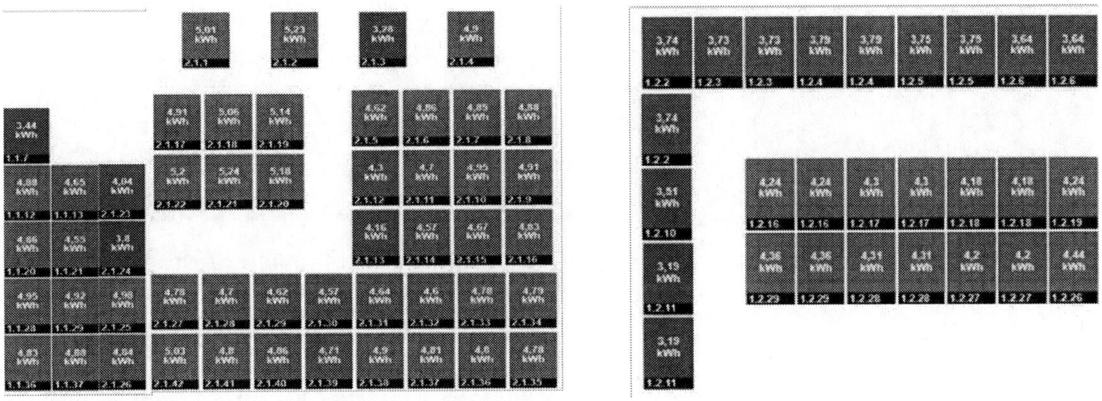

Fig. 4: Parts of PV-plant B (left) and PV-plant R (right): Cumulative energy yield for 19th – 21st June 2017

The different inclination of the modules also affects the yield, see Fig. 5. During summer the modules on the flater roof (dormer) produce more energy, whereas in winter the modules on the steep roof perform slightly better. This is due to the slightly higher irradiance in summer for the flatter roof as angle-resolved irradiance data show [9].

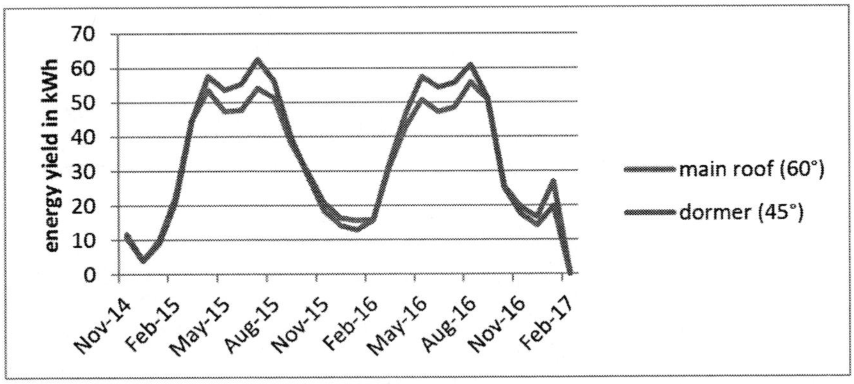

Fig. 5: Monthly mean energy yield of the modules with same direction for the two different angles on the roof of PV-plant R, as seen in Fig. 3 right.

Comparative data analysis

Temperature abnormalities in IR-images are a strong indication for defective modules, see Fig. 6 and Fig. 7. In PV-plant H, a single module has a short-circuited substring (SCS) due to a defective bypass diode, Fig. 6. This defect is characterized by cells with differently increased temperatures within one substring [10]. While defect-free cells have temperatures of roughly 12°C, the warmest cell reaches a temperature of 39.6°C. The monitoring data of the modules of optimizer 1.1.9 yield a power output of 4.62 kWh. The neighboring modules with optimizer 1.1.8 have an output of 5.55 kWh for the same period. Thus, about 1/6 of the energy is missing corresponding to one substring as the two optimized modules have six substrings together. Analyzing the monitoring data yields that, as expected, one third of the voltage of the module is missing. The historical data proves that this defect has been present since the installation of the PV-plant.

This is a pre-peer review version of the paper submitted to the journal Progress in Photovoltaics.
It has been selected by the Executive Committee of the EU PVSEC 2017 for submission to Progress in Photovoltaics.

Fig. 6: Inspection of an array of PV-plant H. Optimizer 1.1.8 without defective (black framed) modules and optimizer 1.1.9 with one defective module with short circuited-substring (SCS) (red framed), $E \approx 400$ W/m², $T_{amb} \approx 5$°C, 29th November 2016, left: IR-image, right: module layout with energy yield from 19th – 21st June 2017.

The yield of the same modules for different periods are listed in Table 2. It is obvious that the daily data are not representative for the yield of the defective module. Analyzing a longer period a much lower yield ratio is obtained [11].

Table 2: Yield and ratio of defective module group (1.1.9) versus defect-free module group (1.1.8) for different time scales of plant H.

energy yield in kWh		defect-free module 1.1.8	defective module 1.1.9	ratio $yield_{def.}/yield_{def\text{-}free}$
daily yield	29th Nov 16	0.96	0.83	87.0%
weekly yield	week no. 47, 2016	3.09	2.49	80.4%
	week no. 46, 2016	3.09	2.45	79.2%
	week no. 45, 2016	3.13	2.49	79.6%
monthly yield	Jun 16	62.51	51.56	82.5%
	Nov 16	16.15	13.16	81.5%
yearly yield	2016	494.92	410.31	82.9%

Substring failure is a module failure frequently found [12]. In PV-plant R about 12% of the modules have substring failures, see in Fig. 7. The good, defect-free modules have temperatures of about 37°C. In general, a substring failure leads to 7 - 8 K temperature increase [13]. Two bypassed substrings show a temperature increase of about10 K due to limited heat conduction.

For the power output or energy yield a reduction according to the number of bypassed substrings is expected. Yield values of the monitoring system confirm that one substring failure in two modules produces 1/6 less than two good modules.

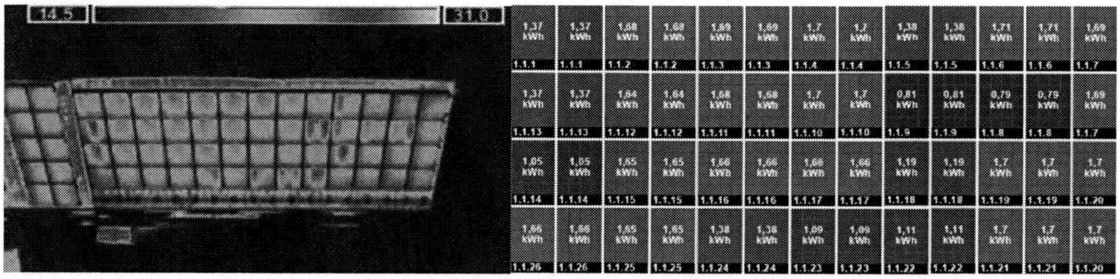

Fig. 7: IR-image of a rooftop PV-plant R with several modules with one or two open-circuited substrings (OCS1 or OCS2) (left), monitoring chart showing the yield on module-level (right), $E \approx 950$ W/m², $T_{amb} \approx 5$°C.

This is a pre-peer review version of the paper submitted to the journal Progress in Photovoltaics.
It has been selected by the Executive Committee of the EU PVSEC 2017 for submission to Progress in Photovoltaics.

The analysis of the historical monitoring data is interesting for investigating failure history. The impact on the energy yield and the voltage are presented in Fig. 8. The module data are classified by the IR-image according to the number of bypassed substrings: none (green), 1 substring (yellow), 2 substrings (orange), and 3 substrings (red). It is obvious that several modules suffered from two and three defective substrings since their installation. Other modules degraded stepwise during lifetime. There are also modules which showed short-term recovery. That means the voltage recovered for a certain period of time. Loose contacts are a possible explanation. Temperature changes may influence the material contact significantly. The yield is determined by the module quality. Whereas in 2015 less modules with defective substrings led to a high yield of 9.3 MWh, in 2016 the increased number of defective modules reduced the energy production (8.7 MWh) significantly, as seen in Fig. 8. Due to the fact that PV-plant R had already many inactive substrings from the time of installation and since the insolation fluctuates over the years, the additional power reduction by new inactive substrings is not clearly visible when looking on the energy yield of the PV-plant over the last years (cf. Fig. 8 dashed line). Since solar irradiance data are not available for the location of this PV-plant, the calculation of the more meaningful performance ratio is not possible.

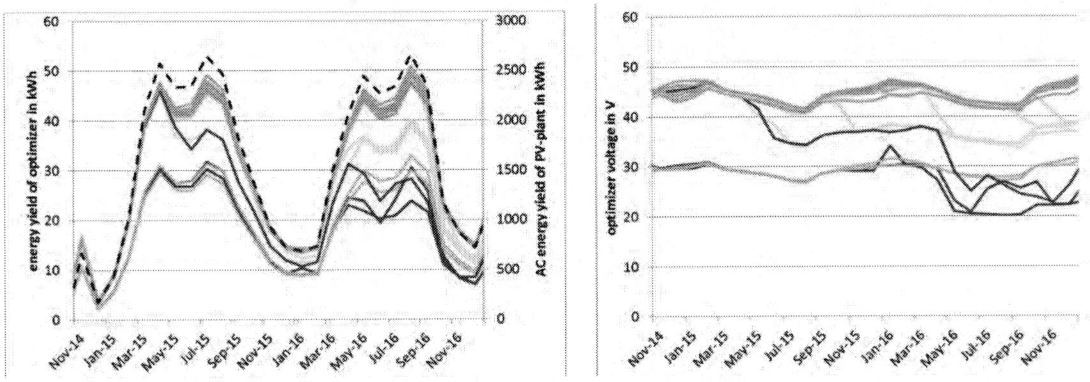

Fig. 8: Monthly energy yield of the modules (left) and monthly averaged measured MPP voltage of the modules (right) shown for PV-plant R in Fig. 7. In all cases, measurements add data of two modules. The green trace is for the non-suspicious modules, the yellow for the modules with one suspicious substring, the orange for two suspicious substrings and the red one for three suspicious substrings. The black dashed line indicated the AC energy yield of the plant in left plot.

The IR-image of PV-plant SO revealed that module 4.1.14 exhibited a cell that was 4-6 K warmer than the other cells. The hot cell module (HCM) was in total about 2 K warmer than the adjacent modules because of the other MPP of the individually tracked modules (cf. Fig. 12). The SolarEdge web portal revealed a power reduction of 8.3 % at the time of the IR-measurement and a long term loss in energy yield of 6.6 % compared to its adjacent module 4.1.5.

In PV-plant W CIGS-modules of size 1.7 m x 0.7 m are installed. In the IR-image we detected stripes of increased temperature (5-7 K) almost perpendicular to the center of the long side of several modules (see Fig. 9). Evaluation of the visual images gives evidence of glass breakages at these sites. The loss in energy yield is depicted in Fig. 9. The energy data yield a reduction of 10 % for each broken module. This forces the current to take a "detour" along the crack resulting in higher current densities and temperature [14].

This is a pre-peer review version of the paper submitted to the journal Progress in Photovoltaics.
It has been selected by the Executive Committee of the EU PVSEC 2017 for submission to Progress in Photovoltaics.

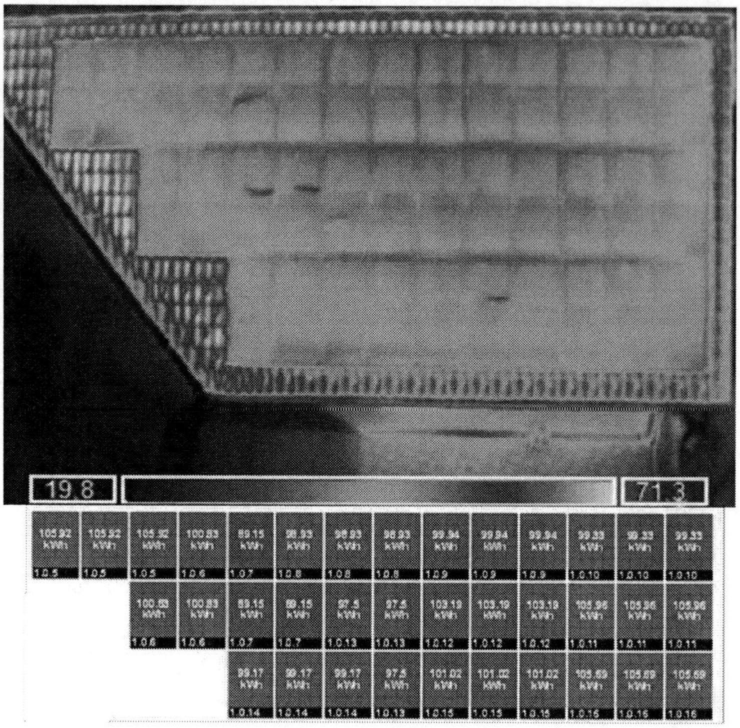

Fig. 9 Part of PV-plant W with CIGS thin-film modules. Stripes of higher temperatures indicate glass breakage in the IR-image (on 16th March 2017 at 13:30 h) (top) and lower energy yield of optimizers connecting 3 modules (1st January – 13th July 2017) (bottom). E = 850 W/m², T_{amb} = 16 °C.

Results and discussion of fluctuating effects

The module monitoring reveals situations which would not be detected by singular imaging measurement. On two consecutive days in the summer (19th and 20th June 2017) the weather was clear sky. This can be seen in the power-time and current-time curves of module 2.1.8 of PV-plant B (cf Fig. 10). They show ideal outdoor behavior. However, some modules are affected by shading. For instance, module 2.1.12 is shaded by a dormer in the afternoon which would not be noticed when doing the imaging earlier in time (cf. monitoring curves in Fig. 10 and IR-image in Fig. 3 left). However, the shaded cells in this module are cold and not hot or the whole module has a temperature compared to the neighboring ones higher as expected from the simulations in Fig. 2 when the module is working at the lower or higher voltage maximum power point due to the shading, respectively. The reason is seen in the monitoring data of this module. Just three minutes before taking the IR-image the module was shut down by the module optimizer as it produced too low power due to the shading.

But also other variable effects are not seen by random imaging measurements. Although the two measurement days in summer were almost identical by the weather data (cf. in Fig. 10), module 2.1.3 reveals totally different behavior in the mornings. This module is shaded by the receiver arm of a TV satellite dish on the roof ridge. In the morning of the 19th June 2017 the bypass diode of one substring was active, as seen in Fig. 10 by reduction of the power and the voltage compared to the other modules. In the afternoon the voltage was higher and the current was lower compared to module 2.1.8. The behavior can be explained by the two maxima in the power-voltage curve as simulated in Fig. 2 and explained before. Another explanation would be an unstable electrical contact affected by ambient conditions. Even more interesting is that on the 20th June 2017 module 2.1.3 remained the whole day in the higher voltage maximum power point although the day was similar to the one before by irradiance and weather – only the ambient temperature was higher by about 2 K.

This is a pre-peer review version of the paper submitted to the journal Progress in Photovoltaics.
It has been selected by the Executive Committee of the EU PVSEC 2017 for submission to Progress in Photovoltaics.

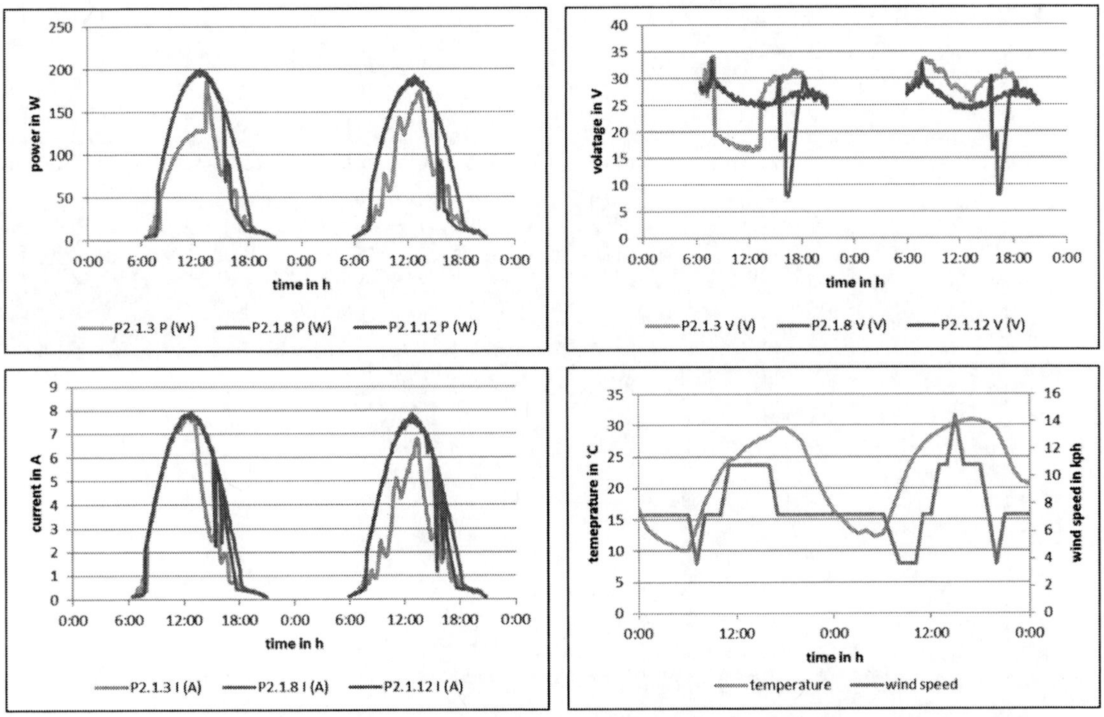

Fig. 10: Power, voltage and current curves over time for 19th and 20th June 2017of modules 2.1.3, 2.1.12 and 2.1.8 of PV-plant B and the temperature and wind speed during this time.

Besides the detection of defects in the PV-plant, the analysis of monitoring data on module level also reveals other information which would not easily be available otherwise. In the schematic energy yield of PV-plant ST the influence of a shading tree in the lower right corner of the house is seen (cf. Fig. 11) where also the highest reduction in energy is measured. By plotting the daily power curves of the single module optimizers over time the vanishing of the shadow over almost the whole roof to lower right corner can be followed as the modules start to generate power.

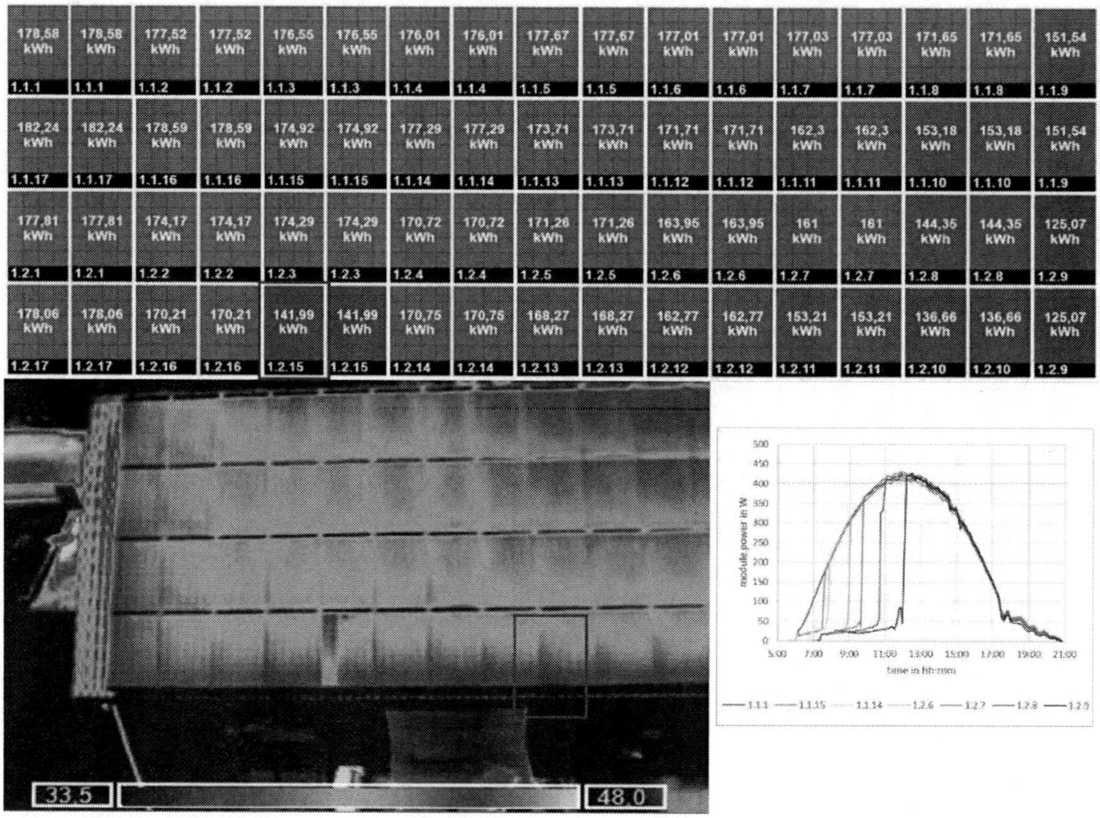

Fig. 11 Energy layout (top) of PV-plant ST, IR-image (bottom, left) of the left part, and power output of selected modules across the PV-plant during one day (bottom, right). An inactive substring (OCS1) in the left module of the optimizer 1.2.15 (red frame) is seen in the IR-image and noticed in the energy yield. The energy yield decreases from the top-left to the bottom right of the roof-top plant due to the moving shadow of a tree during the day.

Effect of convective cooling

In IR-images the modules on the edges or corners of module fields or arrays can have lower temperatures due to better convection and more wind exposure as seen in Fig. 12. Consequently, the lower module temperature should lead to a higher power generation which could be detected with one module optimizers. Those edge modules (cf. module 4.1.18) are typically 4 K cooler which results in a power increase of 2 % (for a temperature coefficient for the module power of 0.5 % / K). However, as the maximum output power of the modules varies by up to 5 % due to variation in production, the increased power effect of cooler modules is not resolvable by evaluation of the power or energy yield module layout map of a certain time.

Therefore, we compared the power ratios of an edge module with a module inside the module array on the roof of PV-plant SO for two time situation: First, the time of the IR-measurement (6th July 2017), when the temperature difference is known, and second, the cumulative generated energy over a few months (1st January -11th July 2017). At the latter ratio the effect of the cooler module should be flattened/equalized as the wind direction turns and the ratio tends to approach one. By evaluating modules 4.1.18 and 4.1.9 by this way we got ratios of 1.02 for the moment and 0.99 for the long term, which reveals the 3 % increase in power production for the 4 K cooler module at the time of IR-measurement. By this a low performing module can outperform a better module when it is cooled and tracked separately. When tracking many modules in a long string this effect cannot be discovered as it is compensated by the module mismatch.

This is a pre-peer review version of the paper submitted to the journal Progress in Photovoltaics.
It has been selected by the Executive Committee of the EU PVSEC 2017 for submission to Progress in Photovoltaics.

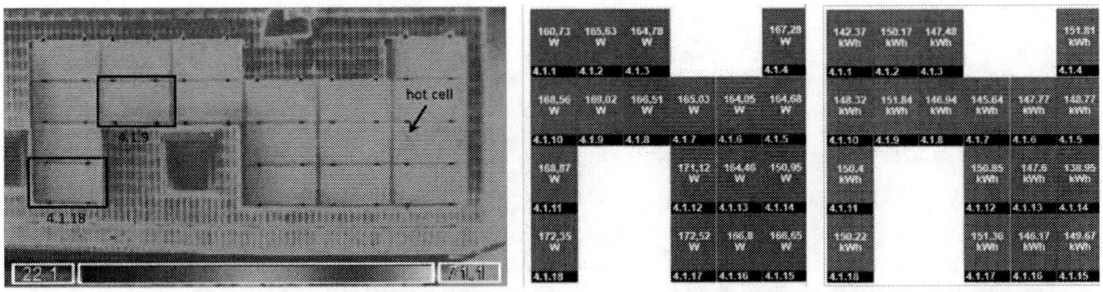

Fig. 12 Part 1 of PV-plant SO: IR-image (left) and module layout with power generation (middle) on 6th July 2017 at 12:30 h. Module layout with energy yield (right) for 1st January – 11th July 2017. In the IR-image module 4.1.14 shows a damaged hot cell (HCM) (as seen by arrow and mentioned in section "defect detection").

Out of power modules

In PV-plant SO we found one short-circuited module (SCM) and one open-circuited module (OCM) (cf. Fig. 13 top). The SCM showed the typical IR pattern with several hotspots in the module, while the OCM had a homogeneous higher temperature of about 3-5 K. In the SolarEdge web portal both modules gain no yield as seen in the plant layout with energy yield in Fig. 13 (bottom) for the modules 2.1.22 and 2.1.18, respectively. Whereas from the OCM a signal with a non-zero voltage was randomly received, the SCM gave no monitoring data at all. The reason for the SCM is still not answered – possibly all three bypass-diodes are or the module optimizer is short-circuited (SCO).

Usually, the OCMs are easily to detect. However, in PV-plant SO several other modules showed the same higher temperature distribution as the OCM 2.1.18, but they generated normal power. In this case the IR-image failed with the interpretation. Yet another module 2.1.14 was not suspicious in the IR, but indicated 14 % less power than its neighbors. The same situation was found for module 1.1.17 with a loss of 12 %. There are two possible explanations: wrong signal transfer or modules belonging to a lower power class.

This is a pre-peer review version of the paper submitted to the journal Progress in Photovoltaics.
It has been selected by the Executive Committee of the EU PVSEC 2017 for submission to Progress in Photovoltaics.

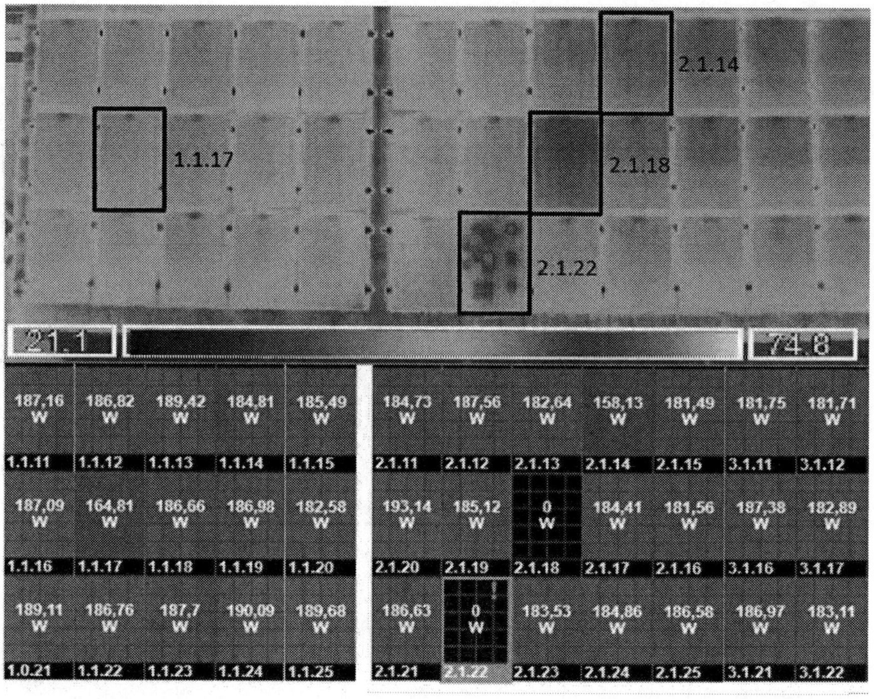

Fig. 13 Part 2 of PV-plant SO: IR-image (top) and module layout with power generation (bottom) on 6th July 2017 at 12:30 h. Modules 1.1.17 and 2.1.14 are unsuspicious in the IR-image but shows less power in the monitoring. Module 2.1.18 is open-circuited (OCM). Module or optimizer 2.1.22 is short-circuited (SCM or SCO) or all three bypass-diodes are defective.

CONCLUSION

The present results confirm that the findings in the IR-images agree to the ones in the monitoring. IR-images show a snap-shot of the present quality under the local operating conditions. Defects can be identified and precisely localized. Due to temporal changes defective sites may be active and visible or not. Historical monitoring data allow for root-cause analysis. The IR-findings agree to the monitored energy yield.

The investigation has shown, that both methods identified the failures at the PV-modules. The advantage of the module-level monitoring, when available in the plant, is to find failures continuously during operation. Further, by the possibility to see the impact to power and yield, operating and maintenance (O&M) staff can decide to react immediately or correct later during a scheduled maintenance. The thermography imaging has the advantage, that more information is given to find the source of the failures, e.g. by pattern of the module image. In addition the thermography can identify fire risks outside the module, e.g. damage of connectors and cables.

The combination of the thermography and the use of a system offering module-level-monitoring and module MPPT allowed to investigate thermal effects and their impact to power and yield output, e.g. 1/6 for substring failures, as well as the effects of cooling by wind or the impact of an overheated cell within a PV module under real conditions.

REFERENCES

1 King DL, Kratochvil JA, Quintana MA, and McMahon TJ. Applications for infrared imaging equipment in photovoltaic cell, module, and system testing. in Photovoltaic Specialists Conference, 2000. Conference Record of the Twenty-Eighth IEEE. 2000. pp. 1487-1490.

2 Jahn U, Buerhop C, Hoyer U, Lerche B, and Wittmann S. Infrarotmessungen an Photovoltaik-Modulen. *ep Photovoltaik aktuell* **2008(1/2)**: 32-38.

3 Tsanakas JA, Ha L, and Buerhop C. Faults and infrared thermographic diagnosis in operating c-Si photovoltaic modules: A review of research and future challenges. *Renewable and Sustainable Energy Reviews* **2016**; **62**: 695-709, DOI: 10.1016/j.rser.2016.04.079.

4 Buerhop C, Scheuerpflug H, and Pickel T. Defect Analysis of installed PV-Modules - IR-Thermography and In-String Power Measurement. in *31st EU PVSEC*. 2015. Hamburg, Germany, pp. 1692-1697, DOI: 10.4229/EUPVSEC20152015-5BO.12.6.

5 Häring A, Henne T, and Dobler S. ASSESSMENT FOR IR INSPECTION CYCLES AND EFFORTS RELATED TO SYSTEM DESIGN. in *32nd EU PVSEC*. 2016. Munich, Germany, pp. 1750-1753, DOI: 10.4229/EUPVSEC20162016-5DO.12.5.

6 Hanson AJ, Deline CA, MacAlpine SM, Stauth JT, and Sullivan CR. Partial-Shading Assessment of Photovoltaic Installations via Module-Level Monitoring. *IEEE Journal of Photovoltaics* **2014**; **4(6)**: 1618-1624, DOI: 10.1109/JPHOTOV.2014.2351623.

7 Buerhop C and Scheuerpflug H. Inspecting PV-plants using aerial, drone-mounted infrared thermography system. in *Third Southern African Solar Energy Conference SASEC*. 2015. Kruger National Park, South Africa, pp. 224-229.

8 WeatherBug. weather forecast. date of access July 12th, 2017.

9 renewables.ninja. Weather forecast. date of access July 9th, 2017.

10 Buerhop C, Schlegel D, Nieß M, Vodermayer C, and Brabec CJ. Reliability of IR-imaging of PV-plants under operating conditions. *SOLMAT* **2012**; **107(0)**: 154-164, DOI: 10.1016/j.solmat.2012.07.011.

11 Buerhop-Lutz C, Winkler T, Wirsching S, Pickel T, - Bemm A-, Camus C, Hauch J, and Brabec CJ. Analyzing the degradation of pre-damaged PV-modules in *SPIE Optics and Photonics*. 2017. San Diego, USA.

12 Buerhop C, Scheuerpflug H, and Camus C. Statistical overview of findings in IR-imaging of PV-plants. in *SPIE* 2016. San Diego, CA, USA, pp. 99380L-99380L-9, DOI: 10.1117/12.2237821.

13 Teubner J, Kruse I, Scheuerpflug H, Buerhop-Lutz C, Hauch J, Camus C, and Brabec CJ. Comparison of drone-based IR-imaging with module resolved monitoring power data. *Energy Procedia* **2017**.

14 Helbig A, Kirchartz T, Schaeffler R, Werner JH, and Rau U. Quantitative electroluminescence analysis of resistive losses in Cu(In, Ga)Se2 thin-film modules. *Solar Energy Materials and Solar Cells* **2010**; **94(6)**: 979-984, DOI: DOI: 10.1016/j.solmat.2010.01.028.

ELECTRICAL AND THERMAL MODELING OF JUNCTION BOXES

Max Mittag, Christoph Kutter, Stephan Hoffmann, Pascal Romer, Andreas J. Beinert, Tobias Zech
Fraunhofer Institute for Solar Energy Systems ISE
Heidenhofstr. 2, 79110 Freiburg, Germany
max.mittag@ise.fraunhofer.de

ABSTRACT: The power of photovoltaic modules is the product of single gain and loss factors. These factors influence the cell-to-module (CTM) ratio and final power of modules. We extend an existing CTM-methodology by presenting a complete model to calculate the losses attributed to junction boxes and cabling. We find the total junction box losses to be small (< 1 W) compared to the power of common photovoltaic modules. Electrical losses in cabling are the dominant loss factor (> 80%) for junction boxes. We simulate the thermal behavior of a junction box using the finite element method and analyze the temperatures of bypass diodes. The model is verified using infrared thermography. We find the diode temperature in the analyzed setup to be below critical temperatures for a thermal runaway. From our FEM-model we find that diode temperatures can be reduced by 13K using potting material with increased thermal conductivity (0.8 W/m*K compared to 0.2 W/m*K). Electrical losses of the junction box increase with elevated temperatures but are still comparably low. We perform an economic analysis and consider costs of power loss as well as material costs for cabling. We find the optimal cable cross section to be 4 mm² at Standard Testing Conditions. At irradiations other than 1000 W/m² the optimal cross section is found to be different (6 mm² at 1200 W/m², 2.5 mm² at 600 W/m²).

Keywords: CTM, Cell-to-Module, Power Prediction, Junction Box, Diode, FEM

1 INTRODUCTION

The integration of solar cells into photovoltaic modules causes a change of power and reference area which defines the efficiency of solar devices. The cell-to-module power ratio (CTM) describes the ratio of the module power after integration of the solar cells relative to the sum of the power of the solar cells before integration [1]-[3]. Optimization and therefore detailed knowledge of the influence factors for CTM is necessary to further increase the power and efficiency of PV modules and to eliminate unnecessary losses.

A unified methodology was presented by Haedrich et al [4] summarizing previous work and offering a comprehensive toolbox to analyze the CTM ratio and its contributing factors. This model does not include losses in junction boxes. Thus parameters in CTM calculation models are unable to precisely evaluate commercially available photovoltaic modules and components.

2 METHOD

We extend the existing methodology by presenting a new loss factor for junction boxes and cabling. Cable dimensions, bypass diode losses, contact resistance in plugs and jacks as well as internal ohmic losses in the junction box itself are considered. Modeling is validated by measurements on several industrial available junction boxes.

We perform a cost analysis and calculate an optimal cable thickness for solar modules reducing the total costs (including power loss and material costs) of solar cables and junction boxes for different irradiations.

We simulate the thermal 3D profile of a junction box connected to a PV module by finite element modeling (FEM). We present a 3-dimendional FEM model, which is based on a detailed CAD model reproducing the junction box exactly. The FEM model is adjusted by thermographic measurements.

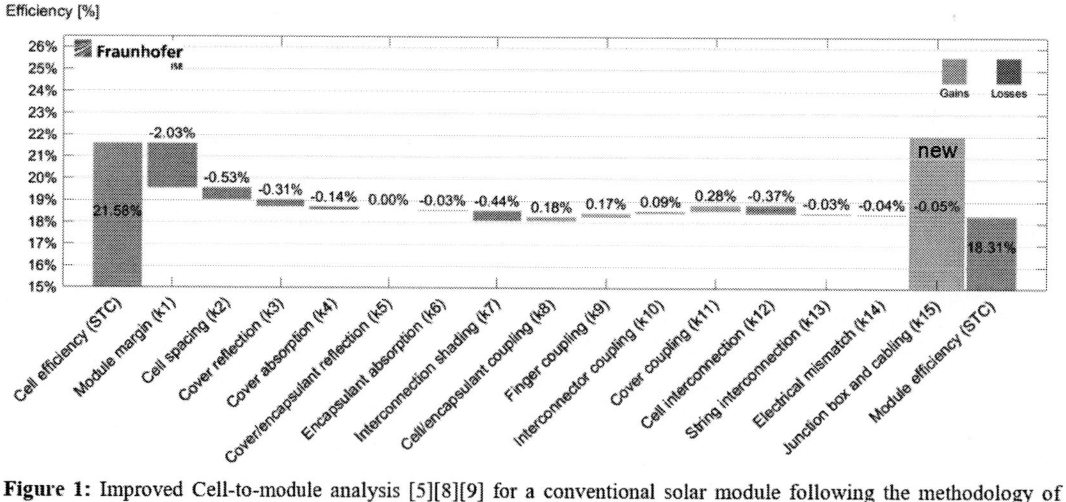

Figure 1: Improved Cell-to-module analysis [5][8][9] for a conventional solar module following the methodology of Haedrich [2]

Temperatures in the junction box and the bypass diode as a function of the module current and ambient temperature are evaluated. We analyze the temperature dependent electrical losses in the bypass diode and estimate their relevance for losses in the junction box. We use the FEM-simulation results to analyze the risk of elevated temperatures within junction boxes (risk of thermal runway) and validate our simulation with measurement results.

3 MODELING OF JUNCTION BOX AND CABLING LOSSES

Electrical losses in junction boxes and cabling are caused by the ohmic resistance of electrical conductive components at the I_{MPP} of the module. The total junction box loss comprises five different power loss mechanisms:

a) ohmic resistance of the cable
b) contact resistance between jack and plug of the junction box cables
c) ohmic resistance of the internal circuits of the junction box
d) contact resistance between the string interconnector ribbon and junction box connectors
e) reverse leakage current in bypass diodes

$$P_{loss} = P_{cable} + P_{plugs} + P_{internal} + P_{string\,connector\,joints} + P_{diode}$$

As shown in Figure 2 commonly used Schottky diodes have a reverse leakage current of around 5 µA at 25 °C. Depending on the number of strings, solar cells per string and the number of bypass diodes, losses of type a) vary significantly. Nonetheless the bypass diode losses are usually negligible for solar modules due to the small currents. At significantly elevated operating temperatures (120 °C) reverse leakage current increases to approx. 10 mA for Schottky diodes (Figure 2) and therefore losses in typical modules may reach 0.1% of the nominal power.

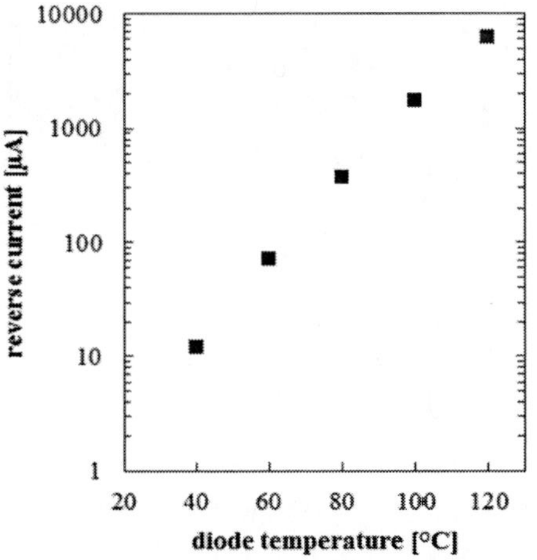

Figure 2: Measurement results of reverse current of a Schottky bypass diode

The ohmic losses in the connecting cables are dependent on the second power of the module current, on cable length, specific resistance and cross section. Commercially available cables are made by using tin coated copper (specific ohmic resistance of approx. 0.0175 Ω*m/mm²) and feature typical cross sections between 2.5 and 6 mm².

Common modules are equipped with two connecting cables that feature one jack or one plug each. Cables are connected to form module strings. Each jack is usually connected with one plug and therefore contact resistance of jack and plug can be measured together. We use a transfer length method (TLM) to measure the contact resistance [8]. Figure 3 shows the result of these TLM-measurements for two different jack-plug combinations. The measured resistance of jacks and plugs ranges around 0.3 mΩ. Results are in good accordance to datasheet specifications (< 0.5 mΩ).

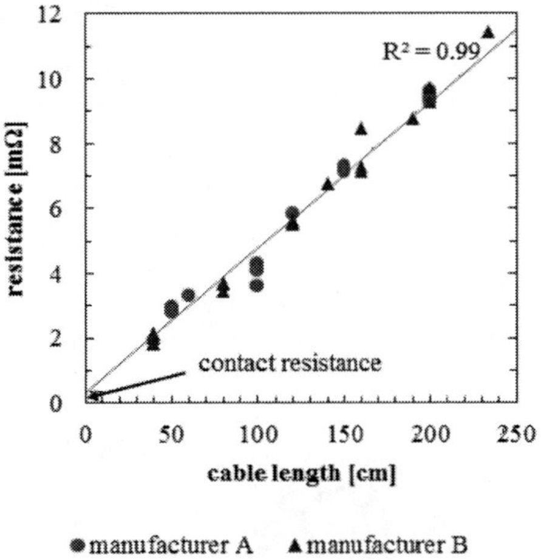

Figure 3: Resistance of two jack&plug combinations with cables of different lengths, linear fit for manufacturer A

Losses from contact resistance between the string interconnector ribbon and junction box connectors as well as the losses from the ohmic resistance of the internal circuits of the junction box are both measured together. Separating the measurement is possible but offers no advantage for practical applications. We measure the internal losses of different junction boxes by again using a transfer length method. String interconnector ribbons of variable length are connected with the junction box using soldering or plug-in connection. Figure 4 shows the results of resistance measurements on four different junction boxes. By doing a linear fit for every measured junction box, we obtain the sum of internal resistance and contact resistance between junction box and string connector ribbon. Resistances range from 0.27 mΩ to 1.5 mΩ.

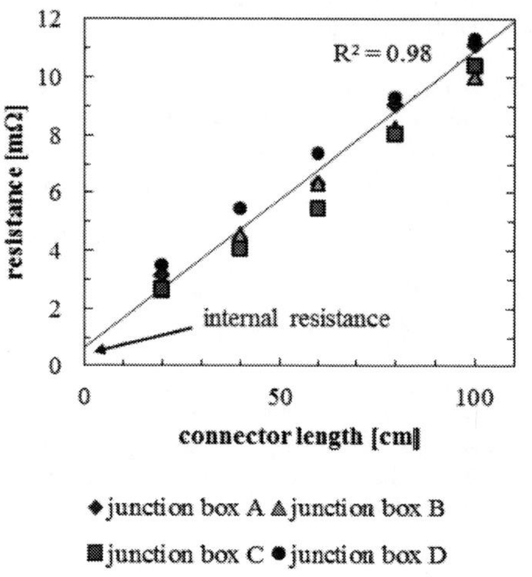

Figure 4: Internal resistance of junction boxes with different string connector lengths, linear fit for box A

We implement all sub-models to calculate the complete junction box losses and combine all single factors. Calculation for a common module (9.1 A I_{MPP}, 303 Wp) with cables of 1 m length (4 mm², copper core) results in losses of approx. 0.84 W which represent 0.28% of the module power. Most important factor for junction box losses are the ohmic losses in the connecting cables (Figure 5).

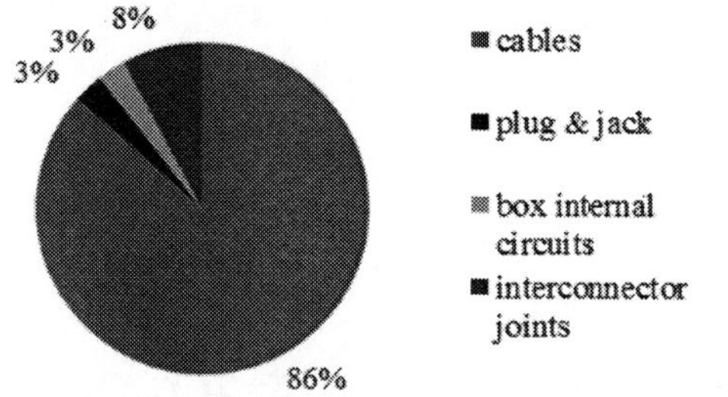

Figure 5: Share of different loss mechanisms for a common module and junction box setup

The junction box loss calculation model has been integrated into SmartCalc.CTM [7][9] by Fraunhofer ISE which is a software to analyze cell to module losses for photovoltaic modules.

4 ECONOMIC ANALYSIS OF JUNCTION BOX LOSSES

Assuming a module price of 0.40 €/Wp [10] the calculated losses (0.84 W at STC) caused by junction boxes equal 0.34 €. For other cable cross sections, the losses change and therefore their costs (Figure 6, red square). We assume the price of solar cable to be 0.50 €/m for 6 mm², 0.385 €/m for 4 mm² cross section and 0.265 €/m for 2 mm² based on market research. Calculating the junction box losses and using these cost information, we find a minimum of the total costs at STC (including cabling material and losses) for a cable cross section of 4.0 mm² (Figure 6, black circles). Other costs such as junction box material, plugs etc. are kept constant and are not included.

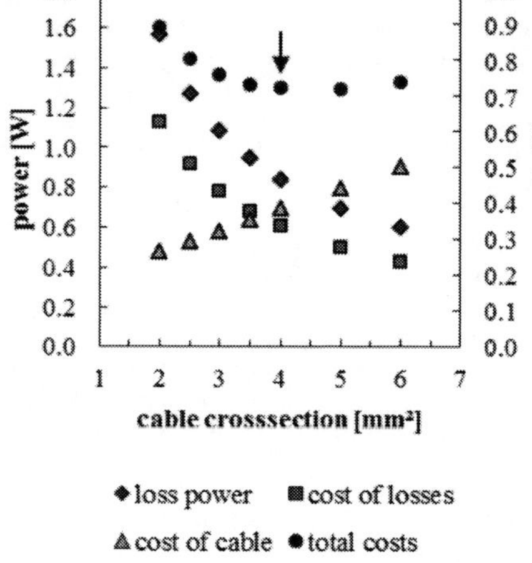

Figure 6: Power losses (@STC) and respective costs, cable and total costs

It is well known that laboratory testing at STC only provides limited information about outdoor performance. We therefore extend our model and calculate the total costs of power loss – including electrical losses as well as material costs for cabling – for different levels of irradiation. We assume a linear change of the current with irradiation. Results are displayed in figure 7. While at STC, a cable cross section of 4 mm² leads to minimal total costs, different cross sections are favorable at other irradiations. We therefore conclude that based on the environmental conditions a customized design of the junction box and cabling may reduce Levelized Costs of Electricity.

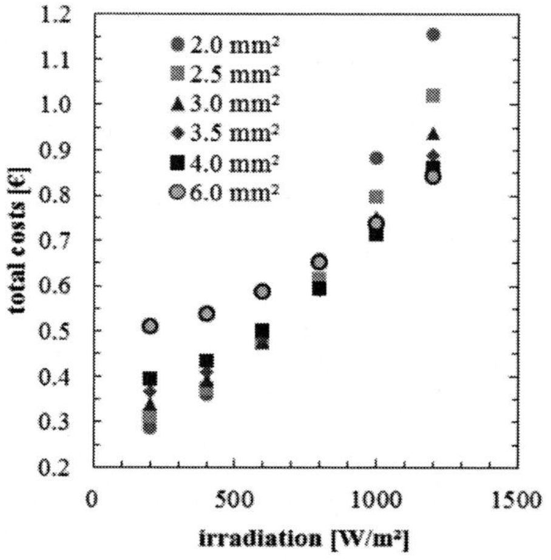

Figure 7: Total costs of junction box and cabling losses including costs for cables of various cross sections at different irradiation

5 INFLUENCE OF ELEVATED JUNCTION BOX TEMPERATURES ON LOSSES AND SAFETY

Since losses in junction boxes not only depend on module currents but also on temperature, we investigate the temperature profile of a junction box as a function of the current and thermal conductivity of potting materials. We also put particular interest in the temperature of the diode as the reverse current increases exponentially with increasing temperature (Figure 2).

We set up an electro-thermal FEM-model using CAD-files of the geometry of the junction box (Figure 8).

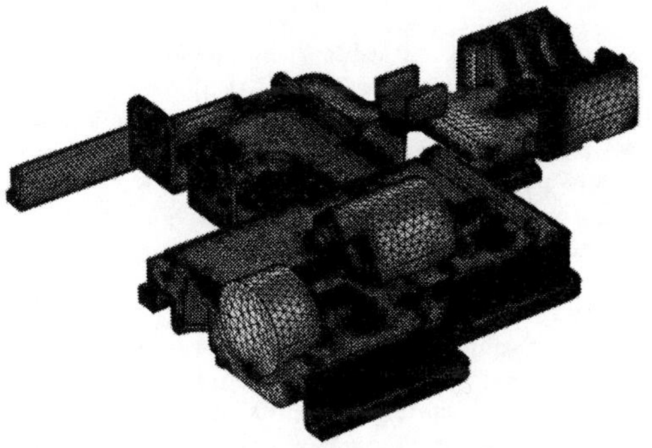

Figure 8: Symmetry element of a junction box with mesh for FEM-modelling

We perform measurements on the diodes of the junction box and find the forward voltage drop over the Schottky diodes to be approx. 0.45 V at 9 A (Figure 9). Using these values we are able to calculate the losses dissipated as heat during bypass operation to be 4 W per diode. A measurement of the temperature dependency of the threshold voltage (forward voltage drop) shows a decrease with higher temperatures (Figure 9).

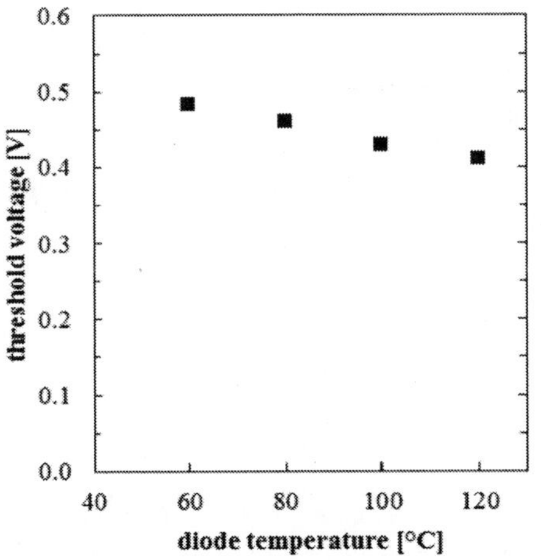

Figure 9: Temperature dependency of the threshold voltage of an exemplary Schottky diode

We feed the loss power of the bypass diode (4 W) as boundary condition into the FEM-model and simulate the heat spread within the junction box (Figure 10).

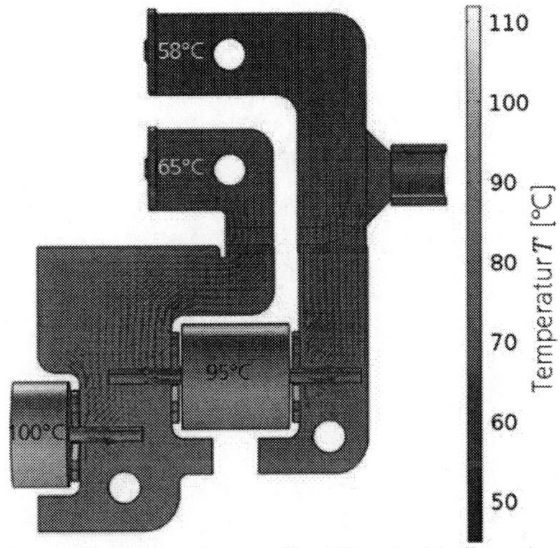

Figure 10: Temperature profile of the electrically active parts of the junction box

We balance the FEM simulation with infrared thermography measurements (Figure 11) and adapt the coefficient of convection to match the infrared-measurement results.

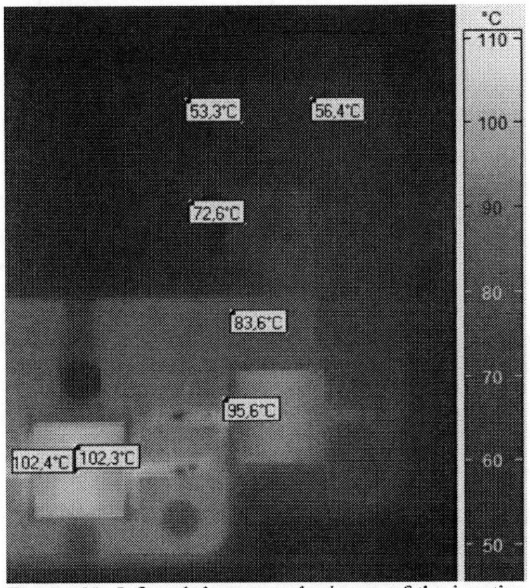

Figure 11: Infrared thermography image of the junction box of a module in failure mode at 9 A.

We find the highest temperature to be at the symmetry plane in the center diode and obtain a temperature of about 100 °C at a forward current of 9 A (failure mode of bypass diode).

When exceeding the stationary point, where the cooling capacity is equal to the dissipated heat of the diode, the thermal dependency of the reverse current is amplified by the so called thermal runaway effect which may lead to damage of junction boxes [11][12]. By simulating the heat spreading within the junction box using a FEM analysis, we obtain the diode temperature in dependency of the module current and are capable of evaluating the risk for thermal runaway as well as investigating electrical losses.

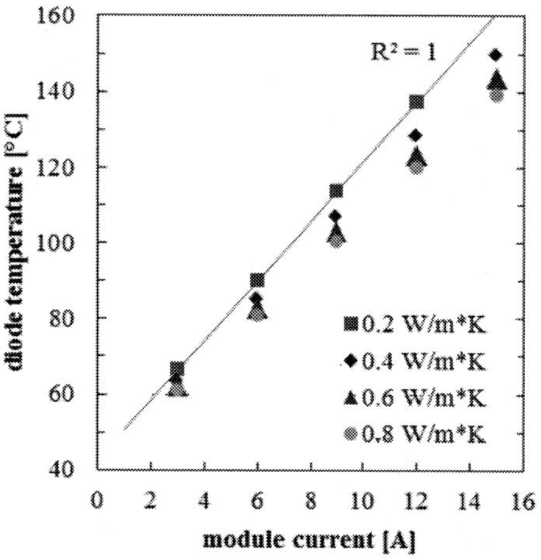

Figure 12: Simulated diode temperature versus module current for different thermal conductivities of the junction box potting material, linear fit for 0.2 W/m*K

The FEM model shows that the diode temperature increases linearly with increasing module current (figure 12). Assuming common solar cells and modules with currents below 10 A, we find the bypass diode temperature within the simulated junction boxes to be lower than the critical thermal runaway temperature (130 °C, depending on diode and manufacturer) [11]. Nonetheless risk of module and diode damage increases for solar cells with higher currents such as bifacial cells and elevated temperatures in general.

We vary the thermal conductivity of the potting material in the FEM simulation and evaluate the influence on diode temperature.
Improving the thermal conductivity from 0.2 to 0.8 W/(m*K) leads to decrease in diode temperature of 13 K (at 9 A). Improving the potting material therefore leads to lower temperatures of bypass diodes (Figure 12).

As shown in figure 2, the elevated temperature within the junction box increases the reverse current of the bypass diode. Even with a significantly increased reverse current (1.8 mA @ 100 °C compared to 5 µA at 25 °C) the bypass diode losses are small compared to other loss factors such as losses within cables.

Using the total junction box loss power (0.84 W) and the module current (9.1 A) we are able to calculate the total junction box resistance and find it to be 10 mΩ. Assuming the temperature coefficient of the resistance to be similar to copper ($\alpha = 3.93 \times 10^{-3}$ K^{-1}) we are capable to calculate the junction box losses at elevated temperatures. Junction box losses of a module at 60 °C increase only slightly assuming the cables to have the same temperature as the junction box itself. We find the total junction box resistance to be approx. 1.5% higher at 60 °C compared to STC.

6 SUMMARY

We present a model to calculate the electrical losses in junction boxes and cabling and implement this model into the software SmartCalc.CTM [7][9] to calculate CTM-losses for crystalline photovoltaic modules. We find the total losses to be low compared to the total module power output (0.25 - 0.30% at STC) and the losses within the cable to be the most important single loss factor regarding junction box losses (> 80%).

A cost calculation and a sensitivity analysis are performed showing an optimal cross section of module connection cables to be at 4.0 mm² to reduce total costs at STC. At irradiations different from 1000 W/m² or for modules with currents significantly different from common module technologies other cable cross sections are favorable.

We present a detailed electro-thermal FEM model of a junction box. The model is adjusted by a thermography measurement and simulates the thermal profile of the junction box. We calculate the diode temperature in dependency of the PV module current. A diode temperature of approx. 100 °C is obtained for a module in failure mode at 9 A.

We find the electrical losses in bypass diodes to be insignificant at normal operation for STC and elevated temperatures. The losses in connection cables and internal wiring within the junction box increase with elevated temperatures but are still low (< 1 W) compared to the power output of common photovoltaic modules.

7 ACKNOWLEDGEMENTS

We would like to thank the German Ministry of Economic Affairs and Energy ("Mosbit" FKZ 0324036A) for their funding.

We would like to thank Matthieu Ebert for his valuable contributions to SmartCalc.CTM and the photovoltaics module department of Fraunhofer ISE.

8 REFERENCES

[1] Jang, J. Y., et al, "*Research on decrease of cell to module loss for crystalline silicon photovoltaic module*", Journal of Renewable and Sustainable Energy 5, 2013

[2] Woehrle, N., et al, "*Solar cell demand for bifacial andsingulated-cell module architectures*", Photovoltaics International, vol. 36, 2017

[3] Mittag, M., et al, "*Analysis of Backsheet and Rear Cover Reflection Gains for Bifacial Solar Cells*", 33rd European PV Solar Energy Conference and Exhibition, 2017

[4] Haedrich, I., "*Unified methodology for determining CTM ratios: Systematic prediction of module power*", Solar energy materials and solar cells 131 (2014), pp.14-23, ISSN: 0927-0248, SiliconPV, 2014

[5] Wirth, H. et al, "*Photovoltaic Modules: Technology and Reliability*", ISBN 978-3-11-034827-9, 2016

[6] Mittag, M., et al, "*Cell-to-Module (CTM) Analysis for Photovoltaic Modules with Shingled Solar Cells*", 44th IEEE Photovoltaic Specialists Conference, 2017

[7] Mittag, M., et al, "*Systematic PV module optimization with the cell-to-module (CTM) analysis software*", Photovoltaics International, vol. 36, 2017

[8] Reeves, G.K. et al, "*Obtaining the Specific Contact Resistance from Transmission Line Measurements*", IEEE Electron Device Letters Vol 3 Ed 5, 1982

[9] www.cell-to-module.com

[10] www.pv-magazine.com/features/investors/module-price-index/

[11] Shiradkar, N. S. et al, "*Predicting thermal runaway in bypass diodes in photovoltaic modules*", IEEE 40th Photovoltaic Specialist Conference (PVSC), DOI: 10.1109/PVSC.2014.6924881, 2014

[12] Fahrenbruch, S.A. et al, "*Solar Bypass Diodes Then and Now*", PV Management Magazine, 2010

BEYOND WASTE – THE FATE OF END-OF-LIFE PHOTOVOLTAIC PANELS FROM LARGE SCALE PV INSTALLATIONS IN THE EU THE SOCIO-ECONOMIC BENEFITS OF HIGH VALUE RECYCLING COMPARED TO RE-USE

Andreas Wade[1], Parikhit Sinha[2], Karen Drozdiak[2], Eddie Brutsch[3]
[1]First Solar GmbH, Rheinstr 4, 55116 Mainz, Germany, andreas.wade@firstsolar.com
[2]First Solar Inc., USA
[3]University of California, Santa Barbara, USA

ABSTRACT: This case study investigates the economics of potential re-use, (informal) recycling and (illegal) disposal scenarios for end-of-life (EoL)1 photovoltaic panels from large-scale photovoltaic systems2 and compares these with the economics and socio-economic impacts of high value recycling.
Stakeholders have raised concerns, that (illegal) shipments of end-of-life photovoltaic panels ('PV waste trafficking'), be it for re-use, (informal) recycling or (illegal) disposal from today's major photovoltaic markets into world regions with no appropriate regulatory framework and infrastructure for the management of end-of-life photovoltaic panels could lead to environmental and health risks.
Given the fast evolution of PV technologies – reflected in efficiency improvements and cost reductions and the correspondingly small proportion of total system cost attributed to the PV module over the last couple of years – as well as in PV recycling technologies, the study demonstrates that the economic incentives for re-use, of end-of-life photovoltaic panels from large scale PV systems are not sufficient to induce a large volume stream towards those options and motivate PV waste trafficking, which has been a topic of raising concern with regard to other WEEE fractions. We therefore conclude that the likelihood of a thriving secondary market for PV panels is very low, given that future new panels will be relatively inexpensive and of higher efficiency, which in turn will drive down BOS costs, making secondary panels unattractive from a system cost perspective.

Keywords: Photovoltaics, PV, end-of-life, re-use, recycling

1 INTRODUCTION

Deployment of photovoltaic (PV) systems in the European Union has reached a cumulative capacity of more than 100 GW_p in 2016 (SolarPower Europe, 2017). Multiple terawatts of PV capacity additions will be needed over the next decades to help decarbonize the global energy system, with solar PV electricity generation expected to be the largest, least cost and most relevant source of energy globally by mid- century (Breyer, On the Role of Solar Photovoltaics in Global Energy Transition Scenarios, 2017).
At the end of 2016, cumulative global PV waste streams are expected to have reached 43,500-250,000 metric tonnes. This represents 0.1%-0.6% of the cumulative mass of all installed panels (4 million metric tonnes). Meanwhile, PV waste streams are bound to only increase further. Given an average panel lifetime of 30 years, large amounts of annual waste are anticipated by the early 2030s. These are equivalent to 4% of installed PV panels in that year, with waste amounts by the 2050s (5.5-6 million tonnes) almost matching the mass contained in new installations (6.7 million tonnes) (IEA PVPS Task 12; IRENA, 2016).

2 BACKGROUND AND MOTIVATION

This case study investigates the economic and technical feasibility of potential re-use, recycling and disposal scenarios for end-of-life (EoL) photovoltaic panels from large-scale photovoltaic systems. For the purposes of this study, EoL is defined as the point in time when the owner of a PV system decides to dismantle the system and dispose of the system components. This does not necessarily correspond with the physical end-of-life of all system components. Instead, a PV system's EoL could be triggered by regulatory or economic framework conditions (e.g. by the end date of a granted Feed-in-Tariff or by an incentive to re-power an existing installation due to the availability of more efficient and lower cost PV panels).
Stakeholders have raised environmental and health concerns regarding potential shipments of end-of-life photovoltaic panels ('PV waste trafficking'), be it for reuse, recycling or disposal, from today's major photovoltaic markets into world regions with no appropriate regulatory framework and infrastructure for the management of end-of-life photovoltaic panels.
Since 2012, the member states of the European Union have adopted a comprehensive regulatory framework that addresses the recycling and disposal routes for PV panels and other components from residential and utility-scale installations alike. The various national transpositions of the WEEE Directive (2012/19/EC) implement detailed requirements for the collection, treatment, recycling and depollution of PV panel waste streams, while also defining conditions under which potential waste could be exported from the European Union. (Wambach, 2012). The export of used photovoltaic modules as products out of the EU into developing economies can be considered as a potential business model for today's operation and maintenance service providers, re-sellers of PV system portfolios, when considering future repowering of old installations in the EU. *Figure 1* depicts a decision tree for the end-of-life management of PV systems.
The analysis of economic and technical implications of dismantling, shipping, re-installing and re-use of photovoltaic panels and the comparison to the economics of new PV system installations will provide a first insight on the feasibility of re-use business cases and the likelihood of waste trafficking scenarios under the umbrella of re-use outside of the EU.

Figure 1 Decision tree for the end-of-life management of a PV system installed in the European Union

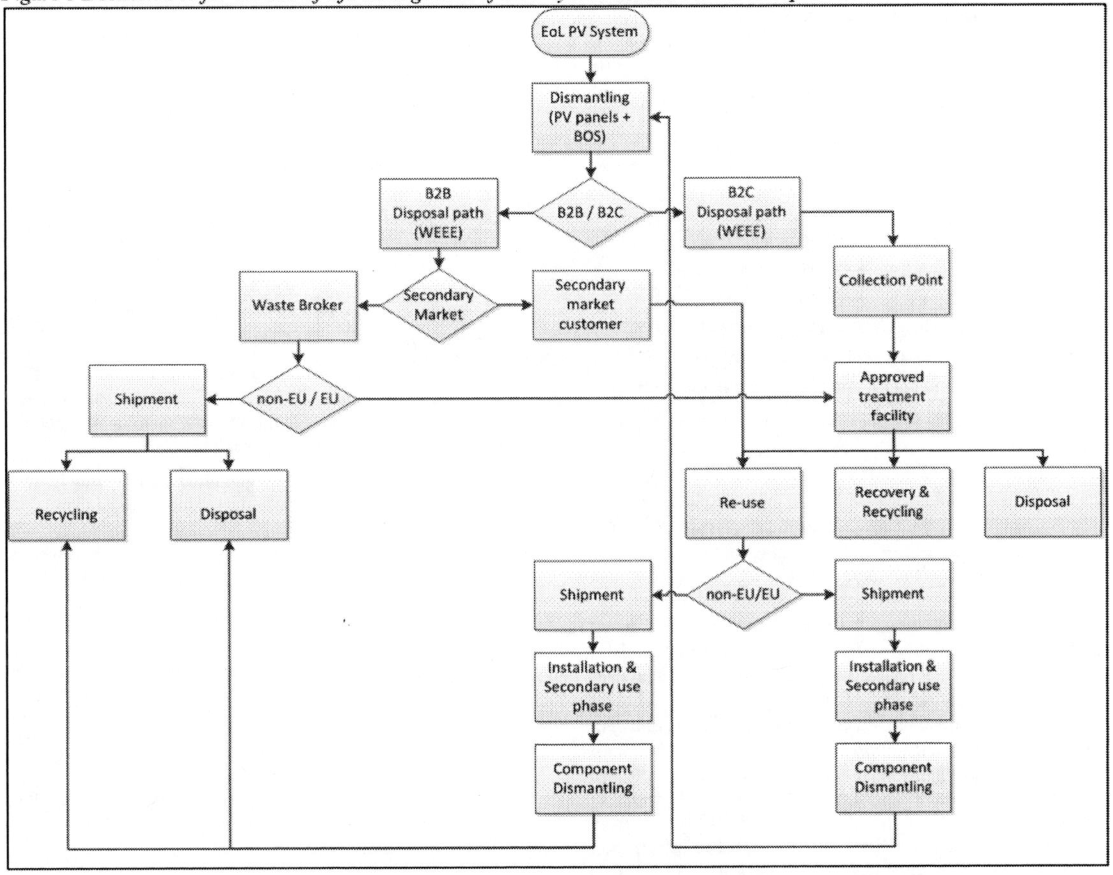

3 APPROACH AND METHODOLOGY

To evaluate the economic and technical feasibility of a re-use scenario for end-of-life photovoltaic panels outside the European Union, the different technical steps leading to an installation of used panels in a secondary PV system outside the EU are portrayed and evaluated. Based on these technical framework conditions, which determine the parameters for the economic analysis, the feasibility of this scenario is evaluated through a probabilistic comparison of potential levelized costs of electricity (LCOE) from a secondary PV system to a new PV system installed at a given location.

3.1 REGULATORY FRAMEWORK

End-of-life management of photovoltaic panels from large-scale installations typically follows the logic of a business-to-business (B2B) waste management scenario – unless it is regulated differently[1]. For this analysis, the B2B case is more relevant for two reasons. Firstly, the amount of end-of-life PV panels expected from utility-scale PV installations will be much higher than from small-scale private household installations. Secondly, the B2B approach provides a greater degree of freedom to the owner of the system on how to handle the system components at end-of-life, which will be the basis for this analysis. It can also be assumed that private household photovoltaic systems will eventually end up in a B2B context as the dismantling would have to be done by a trained professional linked to some form of reverse logistics. The decision tree depicted the different options for dismantled PV systems which will be investigated further in this case study.

Given the deployment rates in recent decades and the expected PV system lifetime of 25 to 30 years, most of the scenarios are still somewhat hypothetical. Most of today's end-of-life PV panels are related to early failures (thus covered by producer warranties) or stem from "pioneer" systems and are hence in very low quantities. A more comprehensive overview on the expected volumes, regulatory framework conditions and an overview of state-of-the-art PV recycling technologies is provided by (IEA PVPS Task 12; IRENA, 2016) and (Komoto & Lee, 2017) and will not be presented in this case study.

3.2 HIGH-VALUE RECYCLING SCENARIO ASSUMPTIONS

Recycling technology for PV panels can be categorized as either bulk recycling (recovery of high-mass fraction materials such as glass, aluminum, and copper) or high value recycling (recovery of both bulk materials and semiconductor and trace metals). Based on current average

[1] E.g. through the transposition of the WEEE Directive in the EU. Some EU Member States define end-of-life photovoltaic panels as so called „dual-use" equipment which could potentially be used in both private households as well as in a B2B context. In this case, the regulator foresees that this equipment has to be dealt with as consumer equipment, and collection and recycling has to be taken care of by the public authorities (e.g. through municipal collection points and treatment by approved recycling facilities).

PV panel compositions, the following recoverable material fractions can be depicted for 2030.

Figure 2 Average recoverable material fractions of PV Panels in 2030 based on **(Weckend, et al., 2016)**

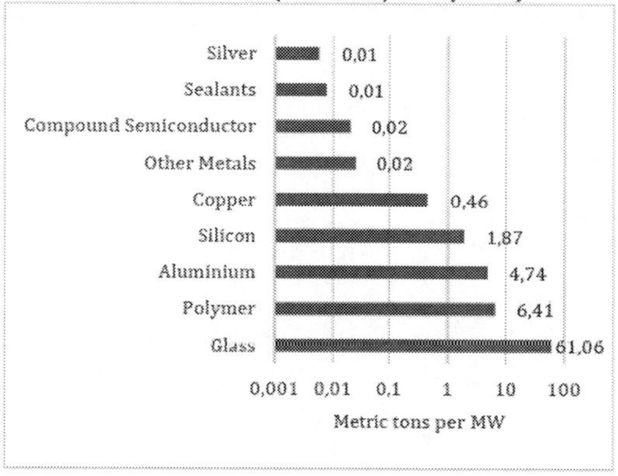

In Europe, the current raw material recovery rate for PV module recycling is 65%-70% by mass and is in line with the EU WEEE Directive. CENELEC, the European Committee for Electrotechnical Standardization, has developed a supplementary standard specific to PV panel collection and treatment (EN50625-2-4 & TS50625-3-5) to assist treatment operators. The standard specifies various administrative, organizational, and technical requirements aimed at preventing pollution and improper disposal, minimizing emissions, promoting increased material recycling and high value recovery operations, and impeding PV waste shipments to facilities that fail to comply with standard environmental, health and safety requirements. The standard includes specific depollution requirements whereby the content of hazardous substances in output glass fractions shall not exceed the following defined limit values:

- 1 mg/kg (dry matter) cadmium (Si-based PV) / 10 mg/kg (dry matter) cadmium (non-Si-based PV);
- 1 mg/kg (dry matter) selenium (Si-based PV) / 10 mg/kg (dry matter) selenium (non-Si-based PV);
- 100 mg/kg (dry matter) lead.

In addition to the PV panels, the balance-of-system (BOS) components also offer a potential secondary raw material value stream after decommissioning. Based on common decommissioning plans for utility scale installations in the US (Desert Stateline, LLC; LSA Associates, Inc., 2014) and a case study on the decommissioning of a 10 MW PV project in Germany (Schimpf, G. et.al.; Gehrlicher Solar AG, 2011), this study will present the present value of the net cost to decommission (NDC_{PV}) a utility scale PV power plant using Eq. 1.

$$NDC_{PV} = \frac{[(DC_T + IC_T + MR_T + LF_T) - SV_T - LV_T]}{(1+r)^T} \tag{1}$$

where,

$DC_T + IC_T =$	Direct cost (labor, equipment) and indirect cost of PV plant de-installation, demolition, recovery, and land reclamation in year T
$MR_T =$	PV module recycling cost in year T.
$LF_T =$	Landfill disposal cost in year T, including landfill tipping fees and hauling, of non-salvageable material.
$SV_T =$	Scrap value of steel, copper, and aluminum recovered during PV solar field and power equipment removal and sold to recyclers at prices prevailing in year T.
$LV_T =$	Value of reclaimed land in year T.
$r =$	Rate of annual discount applied to costs and revenues realized in year T (expected value = 2.91 %).

In order for the net costs to be negative (i.e. profitable), the scrap metal value and/or land value must exceed the decommissioning costs.

The NDC_{PV} model is used to evaluate the proposed Little Bear Project located in Fresno County, California, which has the plant characteristics shown in Table 1. The parameters of Eq. 1 vary based on different decommissioning scenarios shown in Table 2, such as whether cabling is above or below-ground and whether PV modules are recycled or landfilled at end-of-life.

3.3 RE-USE SCENARIO ASSUMPTIONS

To evaluate whether there would be an economic incentive for the shipment of used photovoltaic panels for re-use, *Table 3* depicts the parameters used for the probabilistic evaluation of the LCOE of a new and a secondary PV installation at a given location. An economic incentive would only exist if the levelized cost of electricity of the secondary system would be below the LCOE of a newly installed system at the same location. The calculation of the LCOE (USD/MWh) is based on (ETIP PV - PV LCOE Working Group, 2015) and depicted in equations 2 and 3.

$$LCOE = \frac{CAPEX + \sum_{t=1}^{n}[OPEX(t)/(1+WACC_{nom})^t]}{\sum_{t=1}^{n}[Utilization_0 \times (1-Degradation)^t/(1+WACC_{real})^t]} \quad (2)$$

$$WACC_{real} = (1+WACC_{nom})/(1+Inflation)-1 \quad (3)$$

Table 1 – NDC$_{PV}$ model's default PV plant characteristics **(Cates & Stifter, 2017)**

Characteristic	Units	Model Value
Project Life	Years	25
Distance to Landfill	Miles Roundtrip	72
Land Type	NA	CROPLAND
Land Area	Acres/MW-ac	7.1
dc/ac Conversion	dc/ac	1.165
Module Area	m^2/module	2.47
Mounting	NA	SINGLE-AXIS
Module Weight	kg/module	35
Module Efficiency	kW-dc/m^2	0.174
Module Capacity	kW-dc	0.430
Module Capacity	W-ac	369
PV Modules/MW-ac	Number	2711
Steel (Structure)	kg/m^2/module	15.790
Copper (Structure)	kg/m^2/module	1.1660
Aluminum (Structure)	kg/m^2/module	0.1876
Other Material (Structure)	kg/m^2/module	0.5053
Steel (Equipment)	kg/MW-dc	15,557
Copper (Equipment)	kg/MW-dc	5,001
Aluminum (Equipment)	kg/MW-dc	1,234
Other Material (Equipment)	kg/MW-dc	11,880
Cable Length (Trays)	LF/MW-ac	1,700
Steel (Trays)	lbs/LF	4.25
Aluminum (Trays)	lbs/LF	2.755

Table 2 – NDC$_{PV}$ model's PV plant decomissioning scenarios and base parameters ($/MWac) corresponding to Equation 1 **(Cates & Stifter, 2017)**

	Scenario	1a	2a	1b	2b	3	4
	Cable Installation	Below-Grade	Below-Grade	Below-Grade	Below-Grade	Above-Grade	Above-Grade
	Cable Removal	Remove	Remove	Abandon	Abandon	Remove	Remove
	PV Module EOL Treatment	Recycle	Landfill	Recycle	Landfill	Landfill	Recycle
DC	Direct Cost	113,272	113,272	92,655	92,655	92,655	92,655
IC	Indirect Cost	37,775	37,775	30,917	30,917	30,917	30,917
MR	Module Recycling Cost	32,394	-	32,394	-	-	31,670
LF	Landfill Cost	5,978	16,807	6,918	17,747	16,791	6,204
SVSs	Steel	(104,009)	(104,009)	(104,009)	(104,009)	(106,872)	(106,872)
SVSc	Copper	(96,513)	(96,513)	-	-	(103,937)	(103,937)
SVSa	Aluminum	(4,665)	(4,665)	(4,665)	(4,665)	(8,945)	(8,945)
LV	Land Value	(362,099)	(362,099)	(351,548)	(351,548)	(362,099)	(362,099)

Table 3 – Parameter overview for LCOE evaluation of a re-use PV system compared to a new PV system in 2020 corresponding to Equations 2 and 3

Parameter	New Utility-Scale PV System in 2020	2020 Utility-Scale PV System with Reused Panels from 2010	Notes and References
t = time (yr)			Panel lifetime 2010 & 2030 according to the IEA PVPS Task 12 methodology guidelines for PV LCA (Frischknecht, Heath, Sinha, Raugei, & de Wild-Scholten, 2016)
n = economic lifetime of system (yr)	30	20	
CAPEX = total investment expenditure of the system, made at t=0 ($/kWp)	780	924	New system CAPEX 2020 according to (Gallagher, Ben; gtm research, November 2016); Secondary system CAPEX based on BOS cost + Shipping cost + Dismantling cost; Normal distribution for new system CAPEX (mean=780; stdev=40)
Panel cost ($/kWp)	280	0	According to (Gallagher, Ben; gtm research, November 2016)
Dismantling cost for panel reuse ($/kWp)	0	210	Based on average quotes received from PV maintenance service providers active in Europe, for system dismantling in France, Germany and Spain Normal distribution (mean=210; stdev=50)
Shipping cost for panel reuse ($/kWp)	0	55	Average shipment cost based on quotes received from international forwarding companies; upper bound: 80$ per kWp for a hypothetical shipment of one container from Kitzingen (Germany) to Lusaka (Zambia) / lower bound: 30$ per kWp for a hypothetical shipment of one container from Potenza (Italy) to Delhi (India). Triangular distribution (min=30; likely=55; max=80)
BOS cost ($/kWp)	500	659	Normal distribution for new system BOS (mean=500; stdev=20); BOS cost for reuse system based on new system adjusting for area-dependent fraction (49%) of BOS cost by ratio of new (17.4%) to reused (10.6%) panel efficiency. Panel efficiency degradation for 2010 panel was assumed at 0.5% p.a., starting with an initial panel efficiency of 11.1% (80W, FS Series 2). Assumed panel efficiency for average 2020 PV panel according to (ETIP PV - PV LCOE Working Group, 2015) based on the average panel efficiency of 2014 (15%) and an average annual rate of efficiency increase of 0.4% p.a..
OPEX(t) = operation and maintenance expenditure in year t ($/kWp/yr)	15	15[2]	Triangular distribution (min=10; likely=15; max=30) and adjusted for expected annual rate of OPEX decrease (2%) according to (ETIP PV - PV LCOE Working Group, 2015)
$WACC_{nom}$ = nominal weighted average cost of capital (per yr)	7,10%	7,10%	Eq. 3
WACCreal = real weighted average cost of capital (per yr)	5%	5%	Triangular distribution (min=2%; likely=5%; max=8%)
Inflation = annual inflation rate	2%	2%	Based on EU average (2001-2014)
$Utilisation_0$ = initial annual utilisation in year 0 without degradation (kWh/kWp/y)	1617	1617	Based on average plane of array irradiation in Ivory Coast, Ghana, Zambia, Ethiopia, India and triangular distribuution of performance ratio (min=0.75; likely=0.80; max=0.825); average values based on (Breyer & Schmid, 2010)
Degradatio n = annual degradation of the nominal power of the system (per yr)	0,5%	0,5%	Triangular distribution (min=0.2%; likely=0.5%; max=0.8%)

[2] this value has the potential to fall within a larger range based on actual degredation rates of the older system

The major differences between the new system and the reuse system relate to the shorter lifetime of the reuse system (20 years compared to 30 years), the lower conversion efficiency of the reuse panels due to vintage (10.6% compared to 17.4%) and 10 years of field degradation, and the related higher BOS costs needed in order to achieve a similar nameplate power output. Another major influencing factor when it comes to the CAPEX of the reuse system are the labor costs related to the careful dismantling of the old system so that the panels can be recovered fully intact for reuse in a secondary installation. The assumption that the reuse panels are obtained at no cost may underestimate the resulting LCOE – if a maintenance service provider is considering the dismantling, shipment and re-installation of used panels, it is almost inevitable that a markup would be added on those and that the CAPEX portrayed above would be higher than just the sum of dismantling and shipment.

4 RESULTS AND DISCUSSION

In order to evaluate the feasibility of the different end-of-life management pathways depicted in the decision tree (Figure 1), the net economic benefits of a high-value recycling scenario have been evaluated through a probabilistic cost optimization model using Monte Carlo analysis (Cates & Stifter, 2017) and compared to a recent decommissioning cost optimization model developed by (Fthenakis, Zhang, & Choi, 2017).
To evaluate the economic feasibility of a re-use scenario, the levelized costs of electricity of a secondary PV system reusing 10-year old PV panels have been compared to the LCOE of a new PV system for the year 2020. Sensitivities on the model parameters have been evaluated though probabilistic Monte Carlo analysis.

4.1 DECOMMISSIONING AND HIGH VALUE RECYCLING ECONOMIC RESULTS

The analysis indicates that the commercial scrap value of PV power plant decommissioning exceeds decommissioning costs – which mainly relates to the scrap value of the recovered steel and copper. This result clearly incentivizes recycling over disposal and is in line with the findings from (Fthenakis, Zhang, & Choi, 2017) which estimated a net profit of up to $1.58 per module area. As shown in Figure 3 below, the Monte Carlo analysis conducted by (Cates & Stifter, 2017) indicates 100% confidence of net profit from PV plant decommissioning when land value was included, and approximately 90% confidence in a net profit when land value was excluded. Figure 3 shows the results for the high-value recycling scenario 4 in which cabling is above-ground and removed at end-of-life for copper recovery and PV modules are recycled at end-of-life. Scrap value of copper from cables and steel from mounting structures provide the bulk of the revenues from PV plant decommissioning, along with land value when available.

Figure 3 – Net Discounted Decommissioning Costs ($/MWac) for Scenario 4 including the value of land (top) and excluding the value of land (bottom) **(Cates & Stifter, 2017)**

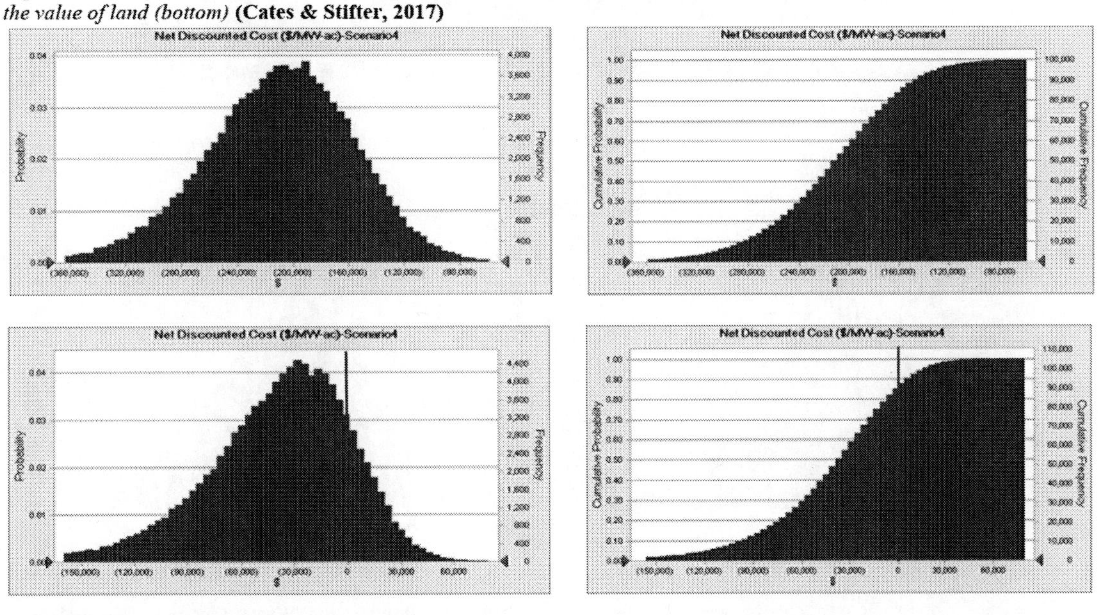

4.2 COMPARATIVE RESULTS OF THE RE-USE SCENARIO

The Monte Carlo analysis on the levelized cost of electricity for a new PV system and a secondary PV system re-using 10 year old panels indicates a 90% confidence level that the LCOE of the system consisting of previously used panels is higher than the LCOE that could be realized by a system using new panels in a given location. Table 4 and Figure 4 below summarizes the results which were generated using Oracle Crystal Ball v. 11.1.2.4.850. The results can be attributed to the shorter system lifetime of the re-use system (20 years versus 30 years for the new system), the higher system CAPEX due to the lower efficiency of the re-use system which requires more area and hence more BOS structures in addition to the added cost of module dismantling and shipping. Last but not least, the energy yield of the re-use system is lower due to reused panel degradation.

Table 4 – Results of the Monte Carlo Analysis of the LCOE of a reuse system compared to a new PV system

	Base case	Mean	Standard Deviation
Control case: LCOE ($/MWh) of new utility-scale system in 2020 (50MWp system deployed in Malaga with 5% real WACC) (ETIP PV - PV LCOE Working Group, 2015)	41	-	-
LCOE ($/MWh) of new utility-scale system in 2020 (10MWp system in this study; similar insulation conditions as in the control case)	40	43	7
LCOE ($/MWh) of utility-scale system in 2020 with reused panels from 2010 (9.5 MWp system in this study; similar insulation conditions as in the control case)	55	58	9
Probability of $LCOE_{reuse} > LCOE_{new}$	>90%		

Figure 4 – Comparative LCOE results of a new PV system compared to a re-use PV system at a given location

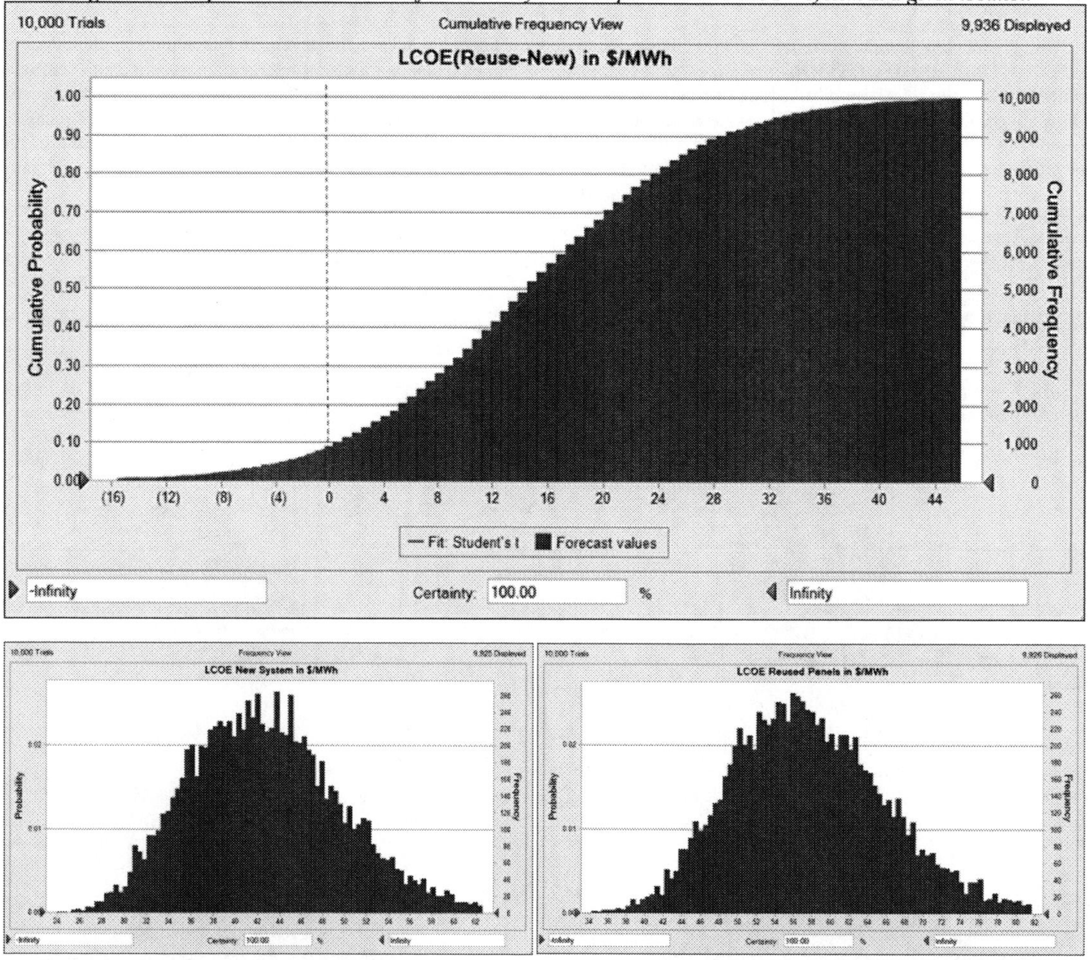

5 SUMMARY

When evaluating the different options for end-of-life management of PV systems in the European Union, the regulatory framework already requires the collection, treatment and recycling of PV panel components and inverters through the Waste Electrical and Electronic Equipment Directive (WEEE) 2012/19/EC, which has been implemented in the various member states of the EU through national legislation. This study has shown that high-value recycling of the complete system beyond the regulated components offers an economic incentive for the asset owner as the net costs of decommissioning would be negative. A potential alternative to decommissioning and high-value recycling of an end-of-life PV system would be the dismantling for re-use. This topic has gained attention in recent policy discussions around the reusability of electrical and electronic equipment in general and photovoltaic panels in particular, as several research studies are underway to demonstrate the technical feasibilty of reuse. Addressing this discussion and stakeholder concerns that end-of-life module streams would be exported out of the European Union into developing economies, this study investigated the economic feasibility of such a scenario using the perspective of an operation and maintenance service provider, tasked to dismantle a PV system in the EU for re-powering. Even under very conservative assumptions it becomes clear, that the high learning rate realized in photovoltaic modules in recent years as well as the ongoing cost reductions make it very unlikely, that a secondary PV system would be able to achieve a levelized cost of electricity below the LCOE of a new system installed at the same location.

Summarizing the findings of both analyses, it can be concluded that a scenario of large amounts of EoL PV panels being shipped out of the European Union into developing economies for the purpose of re-use can be considered very unlikely at this point in time, given the economics of new PV systems in these markets and the underlying net revenues from the decommissioning and high-value recycling of PV plants in the European Union.

6 REFERENCES

Breyer, C. (2017). On the Role of Solar Photovoltaics in Global Energy Transition Scenarios. *Progress in Photovoltaics: Research and Applications*. doi:10.1002/pip2885

Breyer, C., & Schmid, J. (2010). Population Density and Area Weighted Solar Irradiation: Global Overview on Solar Resource for Fixed Tilted, 1-Axis and 2-Axis PV Systems. *25th EU PVSEC*. Valencia, Spain.

Cates, A., & Stifter, R. (2017). *PV Power Plant Net Decommissioning Cost Model Technical Report.* San Francisco, CA: ERM.

Desert Stateline, LLC; LSA Associates, Inc. (2014). *Closure, Decomissioning, And Reclamation Plan - Stateline Solar Farm Project San Bernadino County, California.* Needles, San Bernadino County, California, USA: U.S. Bureau of Land Managment.

ETIP PV - PV LCOE Working Group. (2015). *PV LCOE in Europe 2014-30, Final Report.* Brussels, Belgium: European Technology and Innovation Platform Photovoltaics. Tratto da http://www.etip-pv.eu/publications/other-publications/pv-costs.html

Frischknecht, R., Heath, G., Sinha, P., Raugei, M., & de Wild-Scholten, M. (2016). *Methodology Guidelines on Life Cycle Assessment of Photovoltaic Electricity - 3rd Edition.* Ulster, Switzerland: IEA PVPS Task 12.

Fthenakis, V., Zhang, Z., & Choi, J.-K. (2017). Cost Optimization of Decommissioning and Recycling CdTe PV Power Plants. *Proc. IEEE PVSC.* Washington D.C.: IEEE.

Gallagher, Ben; gtm research. (November 2016). *U.S. PV System Pricing H2 2016 - System Pricing, Breakdowns and Forecasts.* gtm research.

IEA PVPS Task 12; IRENA. (2016). *End-of-Life Management: Photovoltaic Panels.* Munich: IEA PVPS Task 12, IRENA.

Komoto, K., & Lee, J.-S. (2017). *End-of-Life Management of Photovoltaic Panels: Trends in PV Module Recycling Technologies.* IEA PVPS T12-10:2017.

Schimpf, G. et.al.; Gehrlicher Solar AG. (2011). *Untersuchungen zum Recycling einer konkreten PV Anlage.* Berlin, Germany: Forum Solarpraxis.

SolarPower Europe. (2017). *Global Market Outlook For Solar Power (2017-2021).* Brussels: SolarPower Europe, Global Solar Council.

Wambach, K. (2012). PV Module Take Back and Recycling Systems in Europe - New Challenges under WEEE. *Proceedings of the 27th EU PV Specialist Conference.* Frankfurt am Main, Germany.

Weckend, S., Wade, A., Heath, G. A., Wambach, K., Currás, T. A., Zhang, J., . . . Wuester, H. (2016). *End-of-Life Management: Photovoltaic Panels.* Munich, Germany: IRENA, IEA PVPS Task 12.

TECHNOLOGY TRENDS IN PV MODULE RECYCLING FROM VIEWPOINTS OF PATENTS AND R&D PROJECTS

Keiichi Komoto[1], Jin-Seok Lee[2], Andreas Wade[3] and Garvin Heath[4]
[1]Mizuho Information & Research Institute, 2-3 Kanda-Nishikicho, Chiyoda-ku, Tokyo 101-8443, Japan,
tel:+81-3-5281-5286, fax:+81-3-5281-5466, e-mail: keiichi.komoto@mizuho-ir.co.jp
[2]Korea Institute of Energy Research (KIER), Republic of Korea, jslee@kier.re.kr,
[3]First Solar/SolarPower Europe, Germany/Belgium, andreas.wade@firstsolar.com,
[4]National Renewable Energy Laboratory, USA, garvin.heath@nrel.gov

ABSTRACT: PV technology is the one of the most environmentally friendly technologies among all electricity generation technologies. The environmental aspects are evaluated from a life-cycle viewpoint including end-of-life management. In the case of PV modules, recycling has become an important emerging topic and various discussions and activities have been conducted and developed by governments, organizations, and companies. While the recycling of waste PV modules has already begun to be commercialized, various technologies for PV module recycling are under development in order to improve process efficiency, economics, recovery and recycling rates, and environmental performance. To meet the needs for future recycling and recovery operations, further efforts, including the acceleration of technology R&D, are expected. In this report, an overview of technology trends in relation to PV module recycling is given from the perspectives of patents and R&D projects, and expected upcoming issues related to PV module recycling technologies are addressed.
Keywords: PV Module, Recycling, Sustainable

1 INTRODUCTION

The photovoltaic (PV) market is growing rapidly, and further market expansion is expected all over the world. In addition to its positive impacts on energy security and climate change, PV technology is also among the most environmentally friendly technologies of all electricity generation technologies, particularly when evaluated from a life-cycle viewpoint, including end-of-life management. This means that proper end-of-life management is an indispensable issue for 'clean' energy technologies.

All technologies eventually degrade to where they enter their end-of-life stage, eventually replacement. PV technology, especially the PV module, has a useful lifespan of approximately 30 years or more. With PV deployment increasing exponentially, the number of PV modules that reach the end of useful life will also greatly increase, accumulating exponentially, after the time lag of operation, as waste. A report published by International Energy Agency, Photovoltaic Power Systems Programme (IEA PVPS) Task12 and the International Renewable Energy Agency (IRENA) in 2016 [1] projected waste PV modules to amount to 1.7–8.0 million tons cumulatively by 2030 and to 60-78 million tons cumulatively by 2050.

Generally, sustainable waste management offers opportunities known as the 3Rs: reduce, reuse, and recycle. When a product cannot be repaired or reused, recycling is the next best option. In anticipation of the large volume of waste PV modules, and to retain PV's position as a clean energy technology, PV module recycling has become an important emerging topic, and various discussions and activities have been conducted and developed by governments, organizations, and companies. Discussions on the topic of PV module reuse are considered to be less mature.

This study aims to provide an international survey of trends related to the development of PV module recycling technology from viewpoints of patents and R&D projects, and was implemented under the IEA PVPS Task12.

2 PATENT ANALYSIS OF PV MODULE RECYCLING TECHNOLOGY

Many resources are available for an analysis of PV recycling technology. However, patents may be a particularly practical resource for technology trend analyses in that these documents represent commercial interests in specific technology areas.

In this study, patents issued in Europe, Germany, France, the UK, the U.S., China, Japan, Korea, and the PCT (patent cooperation treaty), were searched by the online WIPS (worldwide intellectual property service) system as of December 2016, screened for relevance to recycling of PV modules (and not just selected module components or balance of system), and augmented by input from experts within IEA PVPS Task12. The following results are based on analysis of information extracted from each patent.

2.1 Patent trends for crystalline Si PV module

Forty-five percent of identified patents focus on separating module components by removing encapsulants (mostly EVA). Methods classifies as (solely) mechanical account for 40% of the total. Many patents have been filed for the separation of module components and not to recover every material from PV modules.

Figure 1 shows recovered materials from crystalline Si PV module by processing method.

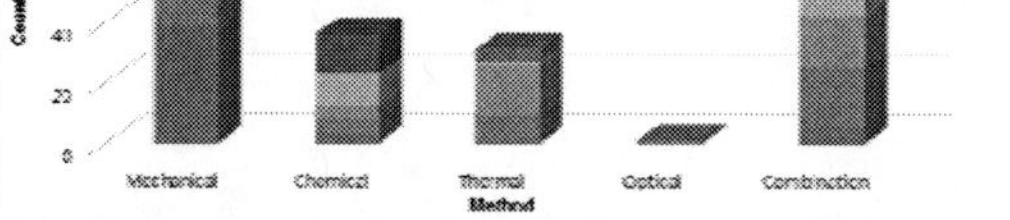

Figure 1: Recovered materials from crystalline Si PV module by processing method based on analysis of PV module recycling patents [2]

A large number of mechanical methods involved apparatuses or processes for separating the frames from the crystalline silicon PV modules. Although various materials can be recovered through solely mechanical means from waste PV modules, aluminum is the most common material recovered using relatively simple mechanical methods. There were also many patents for recovering glass and solar cells by mechanically crushing the modules.

Among chemical methods, the majority of patents involved the processing of solar cells recovered from modules. There were also patents of methods that chemically swell or dissolve EVAs in PV modules for the separation of the modules. Many materials can be recovered using chemical methods, and the largest number of patents was for recovering silver, followed by patents for silicon. Several patents were for recovering the silver and aluminum used as electrodes in the front and back contacts of solar cells. These were followed by patents for methods that chemically swell or dissolve EVAs to recover glass and solar cells.

The majority of patents involving thermal methods were thermal decomposition of EVAs for solar cell recovery because solar cells can be recovered through the thermal processing of EVAs. There were also patents that included the additional recovery of glass.

Patents for combinations of two or more methods which resulted in the widest variety of recovered materials.

The focus of material recovery in analyzed patents ranked in the order of patent counts, are glass, solar cells, silicon, aluminum, copper, followed by silver.

2.2 Patent trends for thin-film compounds PV module

The patent technologies related to compound PV module recycling are more inclined to be commercialized as the patent assignees are corporations in most countries. Of these, 53% of identified patents address encapsulants and 47% for semiconductor materials. The patents for thin-film compounds PV module claim a total-recycling process that covers all recycling steps, from module separation to material recovery. Technologies that combine two or more methods account for 64% of the total. An analysis of patents with regard to recovered materials shows that the recovery of semiconductor materials is the most important concern in compound PV module recycling technology patents.

Figure 2 shows recovered materials from thin-film PV by processing method.

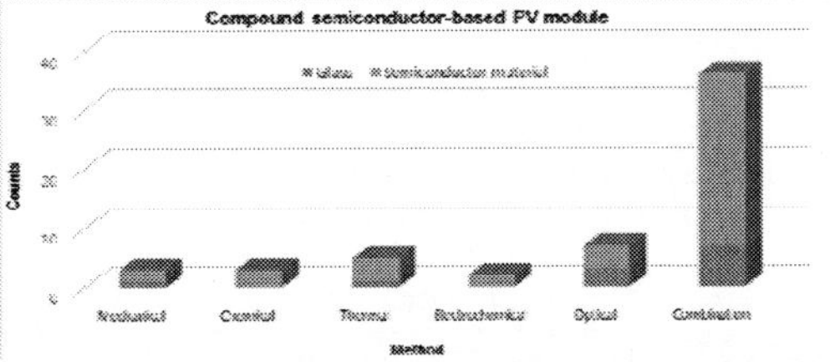

Figure 2: Recovered materials from thin-film compounds PV module by processing method based on analysis of PV module recycling patents [2]

Semiconductor materials were the materials targeted for recovery in about 80% of the thin film recycling patents. Patents for recovering semiconductor materials were found for all processing methods, while the number of patents for glass recovery was relatively small for all processing methods.

It can be seen that from the application for the first patent to the present, the focus has been on the recovery of semiconductor materials and the improvement of existing technologies, rather than on the development of wholly new technologies.

3 TECHNOLOGY R&D FOR PV MODULE RECYCLING

We collected and examined governmental R&D roadmaps, funding documents, policies and other announcements in Japan, Korea, China, Europe, etc.

Recycling technologies for crystalline Si PV modules and thin-film compounds PV modules have different aspects owing to the differences in module structures and the metals contained in them. Currently, both types of PV modules are recycling by proven technology, although recycling facilities for crystalline Si are not designated to treat PV modules. However, to improve process efficiency, recovery and recycling rates, cost effectiveness, and environmental performance, several technology R&D projects have been implemented.

3.1 Recycling R&D for crystalline Si PV modules

Processes for crystalline Si PV module recycling can be roughly divided into those that eliminate the encapsulant from the laminated structure and those that recover the metals from the Si cells.

Figure 3 shows possible processes for crystalline Si PV module recycling.

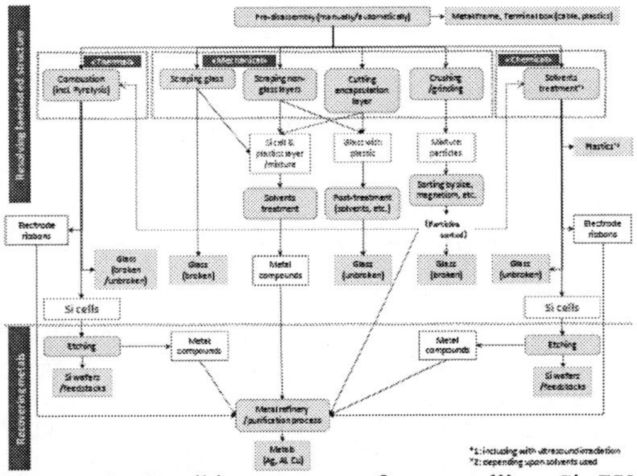

Figure 3: Possible processes for crystalline Si PV module recycling [2]

Eliminating the encapsulant from the laminated structure is one of the most difficult and important target of recycling technology R&D. Thermal, mechanical and chemical approaches can be used.

The thermal approach is a combustion process and the expected materials recovered are glass, Si cells, and electrode ribbons. Under certain conditions, glass and Si cells can be recovered without breakage, which is the benefit of this approach. The recovery rates of Si cells without breakage depend upon the thicknesses of the cells/wafers and the burning conditions.

As a mechanical process, cutting the encapsulation layer, scraping non-glass layers, scraping glass and crushing/grinding technologies are examined. For glass of a higher quality and recovery rate, processes that are without breakage are superior, and may extend to other materials than glass. These mechanical technologies are often combined with a post-treatment step, typically a chemical treatment, to separate Si chips and other metals from the remaining mixture.

Although the chemical processes are technologically feasible and enable the recovery of Si cells, such processes require long treatment times and require liquid waste treatment steps as well. They may suitable as on-site, small-scale treatments, akin to combination of thermal and mechanical processes for the recovery of Si cells and metals.

Recovering metals from Si cells can be achieved by chemical approaches with an acid or alkali hydroxide, for example. Another method is a direct treatment by a metal refinery company.

3.2 Recycling R&D for thin-film compounds PV modules

With regard to the recycling of thin-film PV modules, processes are roughly divided into those that eliminate the encapsulant from the laminated structures and those that recover the metals and substrate glasses.

Figure 4 shows possible processes for thin-film compounds PV module recycling.

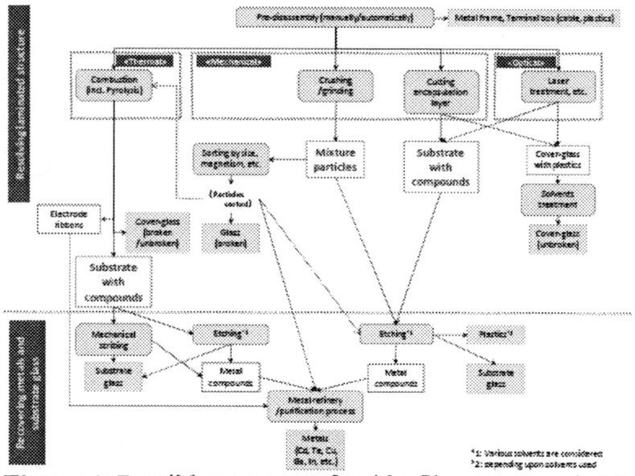

Figure 4: Possible processes for thin-film compounds PV module recycling [2]

To eliminate the encapsulant from the laminated structures, combustion as a thermal approach, cutting the encapsulation layer as a mechanical approach, and a laser treatment as an optical approach have been developed, in addition to the already proven mechanical crushing technology. These processes will enable the separation of double-glass structures and recovery of cover glass components without any damages or contamination. Under certain conditions, substrate glass with semiconductor layer will be recovered while also maintaining shape. Considering the treatment speed and yield during the next step for the recovery of metals from substrates, crushing the substrate will also be an effective approach. However, to increase the quality and value of the recovered glass, non-damaged glass and glass cullet components with a larger particle size are attractive.

With regard to recovering semiconductor metals from substrate, chemical approaches, such as etching are promising. However, treatments that generate gases and waste liquids are critical management issues. If substrate is recovered without breakage, mechanical scraping may be an alternative process.

One important difference between approaches to recycling thin film as compared to c-Si modules is that the objective of eliminating the encapsulant from the laminated structure of thin-film PV modules is to recover both the cover glass and the substrate glass with the semiconductor layer, whereas the objectives for c-Si modules are separating and recovering glass, Si cells, and other metals.

In preparation for mass-treatment methods for waste PV modules in the future, it is expected that the pace of R&D projects will accelerate, allowing researchers to resolve remaining issues and contribute to the development of proper schemes for PV module recycling businesses, as well as for the end-of-life management of PV modules.

4 EXPECTED ISSUES RELATED TO PV MODULE RECYCLING TECHNOLOGY

Growing PV panel waste represents a new environmental challenge, but also opportunities to create and pursue new economic avenues. End-of-life management could become a significant component of the PV value chain and can spawn new industries, supporting considerable economic value creation [3].

Currently, the recycling of both thin film and crystalline Si types of PV modules is commercialized. However, to improve process efficiency, recovery and recycling rates, cost effectiveness, and environmental performance, several activities have been implemented. In the near decades, the need for end-of-life management strategies for waste PV modules will obviously be accelerated.

To meet this need, further efforts (as shown in Table I) including the acceleration of current R&D projects are expected to produce opportunities for increased performance and benefits.

Table I: Opportunities for increased performance and benefits

<u>Technical Innovation</u> ➢ Higher recovery/recycling rate: close to 100% ➢ More economical, more environmentally friendly processes ➢ Higher quality and value of materials recovered ➢ Recycled materials recovered for use as closed-loop feedstock materials for new PV modules ➢ Scaling up: from laboratory to demonstration and commercial-scale operations
<u>Recycling Businesses</u> ➢ Capacities and capabilities for a sustainable business model, including mass-treatment and on-site processing ➢ Securing a certain amount of waste PV modules for suitable operations, collection networks ➢ Collaboration with entities that use recovered/recycled materials

REFERENCES

[1] IRENA/IEA PVPS Task12, End-of-Life Management: Solar Photovoltaic Panels (2016).

[2] IEA PVPS Task12, End-of-Life Management of Photovoltaic Panels: Trends in PV Module Recycling

Technology, Report IEA PVPS T12-10:2017 (to be published).
[3] P. Sinha, Global Policy on PV End of Life Care, 44th IEEE Photovoltaic Specialists Conference (2017).

ECO-SOLAR FACTORY: ENVIRONMENTAL IMPACT OPTIMISATION OF PV PRODUCTION

K. Wambach[1], R. Peche[1], M. Seitz[1], M.P. Bellmann[2], G.S. Park[3], J. Denafas[4], F. Buchholz[5], R. Einhaus[6], G. Noja[7], B. Ehlen[8], R. Roligheten[9], P. Romero[10], A. Bollar[11]

[1]bifa Umweltinstitut GmbH, Am Mittleren Moos 46, 86167 Augsburg, Germany;
phone: +49 821 7000-198, fax -100, e-mail: kwambach@bifa.de
[2]SINTEF Materials and Chemistry, Trondheim, Norway
[3]Norsun AS, Oslo, Norway
[4]UAB Soli Tek R&D, Vilnius, Lithuania
[5]International Solar Energy Research Center - ISC Konstanz, Konstanz, Germany
[6]Apollon Solar, Lyon, France
[7]Garbo Srl, Cerano, Italy
[8]Boukje.com Consulting BV, Bleiswijk, The Netherlands
[9]Steuler Solar Technology AS, Porsgrunn, Norway
[10]Aimen Laser Application Centre, Porrino, Spain
[11]Ingesea Automation SL, Elgoibar, Spain

ABSTRACT: This conference paper deals on one hand with the monitoring of the current degrees of achievement of the targets of the European H2020-FOF-13-2015 Eco-Solar project regarding project targets in material reduction. On the other hand the environmental impacts like carbon footprint, acidification or resource depletion of the current processing status of the individual work packages are considered in relation to the environmental impacts of the manufacturing processes of state of the art PV modules (baseline).

In the halftime assessment, by implementing all previous project developments into the production process chain (ingot → wafer → cell → module) the project target reduction demand of aluminium can be achieved totally for both, mono- and multi-Si based PV modules, the project target reduction demand of argon gas can be achieved totally for mono-Si based PV modules only. Most of the other project targets benefit as well, but without achieving the target values yet.

Compared to the results of state of the art PV modules (baseline) environmental advantages exist for almost all examined impact categories. The only exception is eutrophication terrestrial of the improved production process of multi-Si based PV modules. In the case of climate change, the greenhouse emissions are reduced by more than 11 % for mono-Si based PV modules and more than 15 % for multi-Si based PV modules. The improvements trace back to the ingot production as well as module production mainly and to a lower proportion to the wafer production.

Keywords: LCA, LCI, carbon footprint, material reduction, resource depletion, mono-Si based PV modules, multi-Si based PV modules, eco-efficiency analysis

1 AIM AND APPROACH USED

The European H2020-FOF-13-2015 Eco-Solar project that started in October 2015 envisions increasing resource and energy efficiency over the entire photovoltaic value chain, while simultaneously maximising recycling and remanufacturing efficiencies, by introducing design for recovery, repair and reuse, and collaborating for improvements in waste reduction. Reusing materials and reducing the consumption of raw materials will make solar cell panels both cheaper and greener.

One of the main aims of work package 5 “Environmental impact and repurposing of waste products” is to provide a comprehensive analysis of the progress made with respect to environmental impacts. First results from the initial assessment at start of the project (benchmark processes) and the midterm assessment are presented. The implementation of the new eco-friendly production methods requires an ecological validation and has to be economically viable as well. bifa therefore carried out an eco-efficiency analysis according to figure 1. The reference module is a standard module (glass-EVA-Tedlar laminate, nominal power of 270 Wp, 60 cells with an area of 156 x 156 mm each).

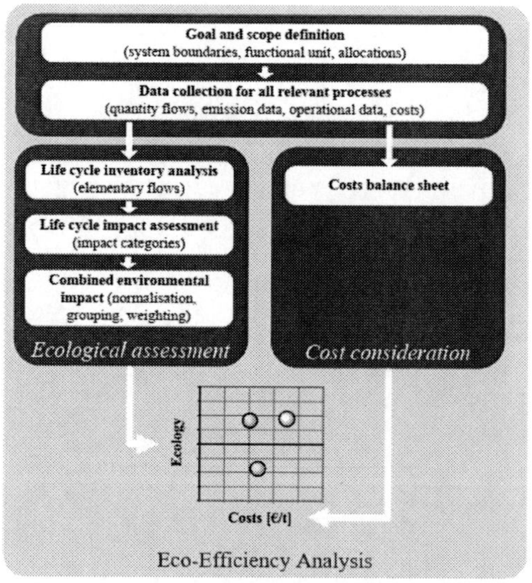

Figure 1: Eco-efficiency analysis carried out by bifa

2 SCIENTIFIC INNOVATION AND RELEVANCE

The input data collection for the initial LCI was based on a state of the art module design: standard EVA laminated module (60 6-inch solar cells, about 270Wp) with front glass and polymer back sheet including aluminium framing. All relevant production steps starting from the ingot casting were modelled using LCI data provided by all partners. Data gaps were filled by using preferably recently published data and expert assumptions. Preliminary baseline data of the PV-module production as well as data sets for standard processes like energy supply, transports etc. from professional data bases are used as basis in the material and energy flow modelling [1 - 5]. The model structure is designed to facilitate a sectoral evaluation of the results of the LCA (Life Cycle Assessment). The LCI (Life Cycle Inventory) baseline of the individual sectors of the PV-module production (ingots, wafer, cell and module) is finished.

This conference paper deals with first monitoring results of the current degrees of achievement of the project targets regarding material reduction. Most important reduction potentials of waste and resource consumption are shown in figure 2. The environmental impacts of the current processing status of the individual work packages are considered in relation to the environmental impacts of the manufacturing processes of state of the art PV modules, based in LCI baseline. For this, a midterm, preliminary LCA has been prepared on the basis of data from the project partners regarding the progress in their work packages.

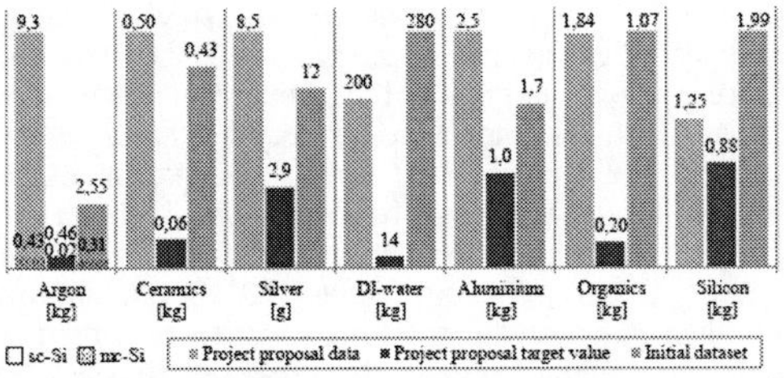

Figure 2: Comparison of proposal targets and initial values at project start

3 STATUS OF SELECTED INDIVIDUAL PROJECT DEVELOPMENTS REGARDING ENVIRONMENTAL IMPACT AND MATERIAL REDUCTION

Basis for the midterm, preliminary LCA are data from the project partners regarding the progress in their work packages. These data were integrated in the material and energy flow model and evaluated. The selection of impact categories for the LCIA (Life Cycle Impact Assessment) has been made by following the ILCD/PEF (International Reference Life Cycle Data System / Product Environmental Footprint) recommendations [6],[7].

3.1 Argon gas recovery

The first step of the value chain included is the production of cz-crystals and multicrystalline ingots. During this process step, pure argon gas, used to remove contaminants, is currently being vented into the air. Though perhaps argon is abundantly available in the earth's atmosphere (approximately 1 % of mass), there are still (ecological and financial) costs involved in the production of pure gas. For instance, the argon consumption in NorSun's wafer facility is responsible for 3-5 % of the wafer cost.

In other industries argon gas recycling has become already state of the art but in PV-production this was not yet the case. A relatively simple and cost efficient technique – developed by GR2L – is based on a chemical looping combustion process to convert combustible species in the exhaust gas stream to carbon dioxide and water, followed by removal of carbon dioxide and water in reclaimable reactor beds. Such a recycling system has been integrated in NorSun's factory collecting and recycling exhaust gas from eight crystal pullers. First results from crystals grown with argon recycling show similar material performance compared to the standard process without recycling.

Environmental impact and material reduction on the example mono-Si based PV module

The approach considers that used argon gas will be treated and afterwards meets the requirement of argon gas for up to 90 % by weight. The data for modelling the argon recovery process are provided by project partner NorSun. The result regarding material reduction shows, that one project target – reduction of argon gas demand to 0.46 kg per PV module – can be achieved totally (see figure 3) by implementing of the argon gas recovery process. Compared with the results of the baseline-scenario environmental, although not large, advantages exist for all examined impact categories. The proportion range goes from 0.1 % for depletion of mineral, fossil and renewable resources to 1.8 % for climate change and depletion of water resources. Saving of over 95 % could be realised if the waste argon is purified and reused and a full implementation of the Ar recovery system is planned.

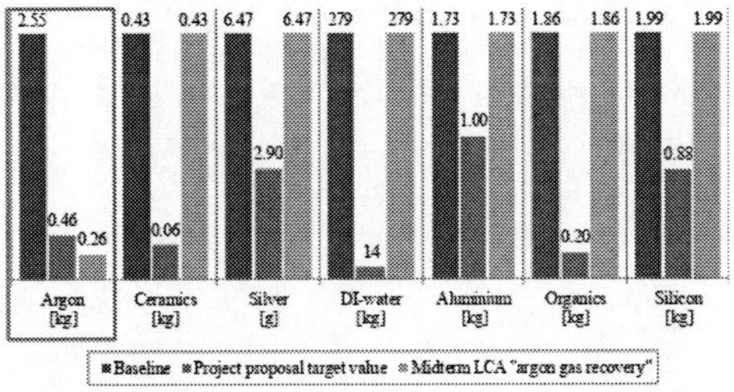

Figure 3: Argon gas recovery - reduction of waste and resource consumption per production of 1 mono-Si based PV-module compared with the baseline process and the project targets.

3.2 New wire sawing process with thinner diamond wire and silicon kerf recovery process from sawing machines coolant

After crystallisation, side-, top- and bottom-parts are cut off before the ingot is cut into blocks and latter wafered. A considerable amount of silicon is lost in form of submicron saw dust (kerf-loss) during squaring and wafer cutting processes. Up to 50 % of the initial silicon amount is lost and can be recovered from the coolant depending on the sawing conditions. In addition to the possibility to use thinner diamond saw wire associated with a decrease in silicon kerf-loss, the ability to recycle the kerf-loss for solar ingot production will have beneficial effects in terms of savings of precious poly-silicon consumption and of waste reduction. Therefore, Garbo has patented and implemented a silicon recycling process which removes contaminations and brings silicon to 99.999 % purity. Purified silicon is then dried, compacted and packed under vacuum in order to obtain a stable silicon powder that, however, might need to undergo a high temperature oxygen degassing process before being used as feed material in the standard production of solar ingots and cells.

The approach considers two aspects:

Firstly a new wire sawing process developed by the project partner Norsun utilises thinner diamond wires (from 80 to 60 µm) and allows more wafers to be sawn out from one brick. The additional yield achieved is about 7.92 %. Further the thinner diamond wire leads to a decrease in silicon kerf and wire waste generation.

Secondly, a newly developed silicon kerf recovery process from sawing machines coolant makes it possible to re-use at least part of the silicon. In the LCA analysis it is assumed that 10 % of secondary silicon from kerf can be added to the necessary silicon demand for wafer production without the need for additional preparation of the secondary silicon during melting. This approach is not recommended for mono-Si wafer production because of high oxygen content and therefore is limited to multicrystalline ingot production. The data for modelling the kerf recovery process are provided by project partner Garbo.

The material reduction result shows, that several project targets benefit of implementing the new sawing process and the recovery process, but with small changes only. Compared with the baseline-scenario the demand of argon gas, ceramics and silicon can be reduced by approx. 18 % and the demand of DI-water by about 1 %. Compared with the results of the baseline-scenario environmental advantages exist for all examined environmental impact categories. The proportion range goes from 0.8 % for depletion of mineral, fossil and renewable resources to 8.2 % for climate change (figure 4).

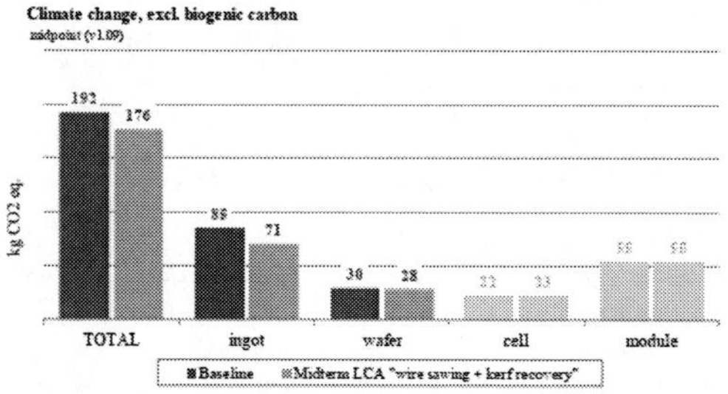

Figure 4: Wire sawing and kerf recovery, climate change – greenhouse gas emissions at the beginning of the project and after implementing a new wire sawing process and a silicon kerf recovery process, calculated for the production of 1 multi-Si based PV-module

Within Eco-Solar, the consortium is looking at three factors on kerf-loss optimisation that will give both environmental and monetary gain. Garbo is working on the reduction of chemicals consumption in Si kerf cleaning step and improving compact density of powders, while Norsun is carrying out sawing tests with increased kerf-loss saturation level in sawing coolant for Garbo to study if (how) the saturation level in the coolant affects the coolant recycling process and the recycled coolant quality.

3.3 New cell process

In crystalline solar cell manufacturing several wet chemical etching and cleaning steps are implemented, which have negative impact on the environmental footprint. By reducing and avoiding certain chemicals or metals and using less pure or recycled chemicals, waste can be reduced and the ecologic footprint of solar cells and panels can be improved.

The approach considers that the standard cell process is replaced by the current status of the cell process at Soli Tek. This cell process is for multi-Si based PV module only. Though better cell efficiencies due to process modifications were obtained no changes in the LCA analysis occurred with respect to material reduction compared to the baseline-scenario, yet. Compared with the results of the baseline-scenario for climate change and depletion of water resources environmental advantages exist amounting to 0.8 % and 2.7 %. The proportion range of the environmental disadvantage goes from 1.7 % for depletion of mineral, fossil and renewable resources to 16 % for eutrophication terrestrial.

Especially silver is an important element, as it is used in most solar cells currently produced, providing the metal contact that "collects" and "drains" the current from the solar cell. In the Eco-Solar project, ISC-Konstanz is developing different solar cell architectures, to minimise the use of silver. In addition, an innovative interconnection scheme enabled by Apollon Solar's module design, avoids soldering of a contact tape onto the rear side silver pads and the front side busbars. Only the small contact fingers on the front side are needed. In addition, ISC-Konstanz and Soli Tek develop an alkaline method for saw damage removal from diamond wire sawing, texturisation and cleaning, that shall replace the current state of art concentrated HF:HNO3, and will investigate advanced processes for emitter formation that will enable higher throughput, resource efficiency and avoid the needs for chemical phosphorous glass layer removal (PSG-removal).

Aimen is working on advanced laser treatment of solar cells, in order to minimize the cut-off area at the wafer edges to a negligible fraction of the whole wafer area, to avoid the wet chemistry normally used to separate the front and rear of the solar cells.

3.4 Usage of an EVA-free glass/glass frameless NICE module (Work package 4 "module design for remanufacturing")

After completion of the cells, they are electrically series connected and assembled into a solar module. As Apollon Solar's NICE-modules are already designed for disassembly Apollon Solar focuses on further reduction of the module BOM (Bill of materials) focussing on a frameless version of the PV module, and integration of electronics in the junction box, to monitor the performance in the field.

The approach considers that the standard EVA laminated modules with front glass and organic backsheet including an aluminium framing shall be replaced by EVA-free glass/glass frameless NICE modules. The data for modelling of a NICE module at the present state of development are provided by Apollon Solar. The material reduction result shows, that one project target – reduction of aluminium demand up to 1 kg per PV module– can be achieved totally. With organics, a second project target benefits as well, but here 74 % of the target value is reached only. Compared with the results of the baseline-scenario environmental advantages exist for all examined impact categories. The proportion range goes from 0.3 % for ozone depletion to more than 55 % for depletion of mineral, fossil and renewable resources. The improvements shown refer to the production step module production only (see figure 5).

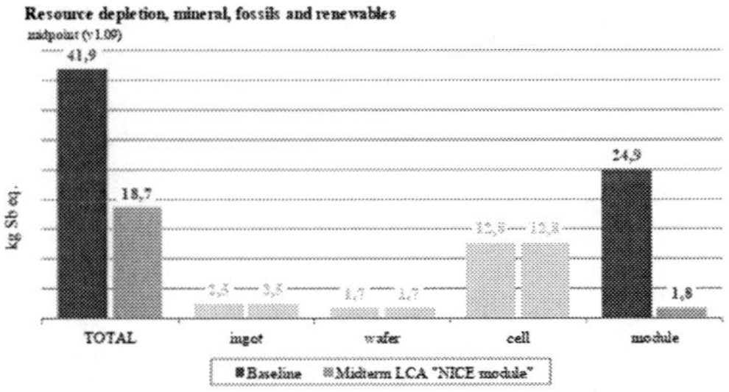

Figure 5: NICE module, resource depletion - consumption of mineral, fossil and renewable resources at the beginning of the project and after replacing the standard EVA laminated module by NICE module, calculated for the production of 1 mono-Si based PV-module

Apollon Solar demonstrates that module assembly of the Eco-Solar cells result in comparable module output as with current (less eco-friendly) state of the art module technology. Furthermore first NICE modules from cells without busbars are made as small laboratory samples for testing and data generation to be used in the LCA. Apollon Solar is working on the disassembly of modules into their major components (glass sheets, copper wires, cells) allowing for their recovery as entire pieces, preserving the value and strong potential for re-use (preferably in PV or other industries).

4 MIDTERM LCA: FIRST EXEMPLARY RESULTS AFTER IMPLEMENTATION OF ALL SELECTED PROJECT DEVELOPMENTS INTO MATERIAL AND ENERGY FLOW MODEL

4.1 Mono-Si based PV module

This approach considers the project developments:

- argon gas recovery,
- new wire sawing process with thinner diamond wire
- and usage of an EVA-free glass/glass frameless NICE module are implemented into the production process chain for a mono-Si based PV module.

The material reduction result shows, that two project targets – reduction of argon gas demand up to 0.46 kg and reduction of aluminium demand up to 1 kg per PV module – can be achieved totally by implementing all project developments. Except demand of silver and DI-water the other project targets benefit as well, but without achieving the target values yet: the demand of ceramics can be reduced by approx. 7.9 %, the demand of silicon by approx. 7.9 % as well and the demand of organics by more than 74 %.

Table 1: Comparison of the environmental burdens and benefits of the impact categories of baseline-scenario and midterm LCA "all project developments" for the production of 1 mono-Si based PV-module (

Impact category	Baseline	Midterm LCA	Environmental impact of midterm LCA compared to baseline
Climate change [kg CO_2 eq.]	316 kg	282 kg	Benefit of 34 kg → 11 % emission reduction
Acidification [Mole of H+ eq.]	2.02 Mole	1.82 Mole	Benefit of 0,21 Mole → 10.2 % emission reduction
Eutrophication terrestrial [Mole of N eq.]	2.92 Mole	2.72 Mole	Benefit of 0,21 Mole → 7 % emission reduction
Ozone depletion [mg CFC-11 eq.]	28.1 mg	26.3 mg	Benefit of 1,8 mg → 6.6 % emission reduction
Photochemical ozone formation [kg NMVOC]	0.91 kg	0.82 kg	Benefit of 0,09 kg → 10.4 % emission reduction
Resource depletion water [m^3 eq.]	7.79 m^3	6.98 m^3	Benefit of 0,81 m^3 → 10.4 % resource use reduction
Resource depletion mineral, fossils and renewables [g Sb eq.]	41.9 g	18.3 g	Benefit of 23.6 g → 56.2 % resource use reduction

Compared with the results of the baseline-scenario environmental advantages exist for all examined impact categories. The proportion range goes from 6.6 % for ozone depletion to more than 56 % for depletion of mineral, fossil and renewable resources (see table 1).

4.2 Multi-Si based PV module

This approach considers that the project developments argon gas recovery, new wire sawing process with thinner diamond wire, kerf recovery process, new cell process and usage of an EVA-free glass/glass frameless NICE module are implemented into the production process chain for a multi-Si based PV module.

The result regarding material reduction shows, that one project targets – reduction of aluminium demand up to 1 kg per PV module – can be achieved totally by implementing all project developments. Except demand of silver and DI-water the other project targets benefit as well, but without achieving the target values yet: the demand of argon gas can be reduced by approx. 92 %, the demand of ceramics by approx. 18 %, the demand of silicon by approx. 7.9 % as well and the demand of organics by more than 74 %.

Table 2: Comparison of the environmental burdens and benefits of the impact categories of baseline-scenario and midterm LCA "all project developments" for the production of 1 multi-Si based PV-module

Impact category	Baseline	Midterm LCA	Environmental impact of midterm LCA compared to baseline
Climate change [kg CO_2 eq.]	192 kg	162 kg	Benefit of 30 kg → 15.4 % emission reduction
Acidification [Mole of H+ eq.]	1.31 Mole	1.20 Mole	Benefit of 0,11 Mole → 8.5 % emission reduction
Eutrophication terrestrial [Mole of N eq.]	2.00 Mole	2.16 Mole	Burden of 0,16 Mole → 8.1 % additional emission
Ozone depletion [mg CFC-11 eq.]	16.80 mg	15.44 mg	Benefit of 1.36 mg → 8.1 % emission reduction
Photochemical ozone formation [kg NMVOC]	0.597 kg	0.590 kg	Benefit of 0,007 kg → 1.2 % emission reduction
Resource depletion water [m^3 eq.]	3.6 m^3	2.9 m^3	Benefit of 0,6 m^3 → 17.5 % resource use reduction
Resource depletion mineral, fossils and renewables [g Sb eq.]	45.0 g	22.2 g	Benefit of 22.8 g → 50.7 % resource use reduction

Compared with the results of the baseline-scenario environmental advantages exist for all examined impact categories except for eutrophication terrestrial. The proportion range goes from 1.2 % for photochemical ozone formation to almost 51 % for depletion of mineral, fossil and renewable resources. The environmental disadvantage for eutrophication terrestrial is approx. 8 %

due to nitrogen oxide emissions from the new cell process (see table 2).

5 CONCLUSION

The results of the midterm assessment clearly show that some projects targets are already achieved (e.g. Ar gas recycling, reduction of Al consumption) and can be fully implemented. The optimisation of solar cell processes, water consumption and saving and substitution of chemicals has just started. The tests with thinner diamond wires showed the higher yield potential of wafers und minimisation of kerf losses. The reclamation process of kerf is mostly developed and shows the potential of use as secondary feedstock in ingot growing though the relative high oxygen content may limit the share in the feedstock mix or may require a special pre-treatment step.

Europe wants to reduce its needs for raw materials and raise the level of recycling of resources in the solar power industry. Our target is that after the successful completion of this project the greenhouse gas emissions from solar panel manufacturing will be reduced by 25 to 30 per cent and the waste generated will be decreased by 10% minimum.

REFERENCES

[1] R. Frischknecht, R. Itten, P. Sinha, M. de Wild-Scholten, J. Zhang, V. Fthenakis, H. C. Kim, M. Raugei, M. Stucki, 2015, Life Cycle Inventories and Life Cycle Assessment of Photovoltaic Systems, International Energy Agency (IEA) PVPS Task 12, Report T12-04:2015.

[2] P. Dold, 2015, Silicon Crystallization Technologies, Semiconductors and Semimetals, Volume 92, Pages 1-61, ISSN 0080-8784.

[3] n.a., 2017, International Technology Roadmap for Photovoltaic (ITRPV) 2016 Results, 8th Edition March 2017

[4] N. Jungbluth, M. Stucki, R. Frischknecht, 2009, Photovoltaics, in R. Dones (Ed.) et al., Sachbilanzen von Energiesystemen: Grundlagen für den ökologischen Vergleich von Energiesystemen und den Einbezug von Energiesystemen in Ökobilanzen für die Schweiz, ecoinvent report No. 6-XII, Swiss Centre for Life Cycle Inventories, Dübendorf 2009

[5] H.J. Möller, 2015, Wafering of Silicon, Semiconductors and Semimetals, Volume 92, Pages 63-109, ISSN 0080-8784.

[6] European Commission-Joint Research Centre-Institute for Environment and Sustainability: International Reference Life Cycle Data System (ILCD) Handbook-Recommendations for Life Cycle Impact Assessment in the European context. First edition November 2011. EUR 24571 EN. Luxemburg. Publications Office of the European Union; 2011

[7] European Commission, Joint Research Centre, Institute for Environment and Sustainability. Characterisation factors of the ILCD Recommended Life Cycle Impact Assessment methods. Database and Supporting Information. First edition. February 2012. EUR 25167. Luxembourg. Publications Office of the European Union; 2012.

ACKNOWLEDGEMENT

This project has received funding from the European Union's Horizon 2020 research and innovation programme under grant agreement No. 679692.

CABRISS
RECYCLING OF SI-KERF FROM PV

Terje Halvorsen[1], Monica Moen[1], Knut Mørk[1],
Delphine Grosset-Bourbange[2], Pascal Rivat[2], Hamza Hajjaji[3], Virginie Brizé[3], Fabrice Coustier[3]

[1] ReSiTec AS, Setesdalsveien 110, NO-4617 Kristiansand, Norway.
[2] FerroPem, 517 avenue de la Boisse, 73000 Chambéry, France.
[3] Univ. Grenoble Alpes, INES, F-73375 Le Bourget du Lac, France.
CEA, LITEN, Department of Solar Technologies, F-73375 Le Bourget du Lac, France.

Corresponding authors:
terje.halvorsen@resitec.no +47-90-029-911

Authors e-mails:
terje.halvorsen@resitec.no; monica.moen@resitec.no; knut.mork@resitec.no; fabrice.coustier@cea.fr;
pascal.rivat@pemsil.com; delphine.grosset-bourbange@pemsil.com; virginie.brize@cea.fr

ABSTRACT: The European Horizon 2020 project CABRISS [1] aim to implement a circular economy for solar cells based on recycled, reused and recovered materials. One important pillar in this project is the recycling of silicon kerf from diamond wire cutting of wafers and blocks. The first objective is to optimize Si-kerf environment in order to facilitate silicon recycling. The second is to demonstrate the full purification of Si-kerf at industrial scale. A recycling process has to be flexible since the silicon kerf will vary in shape, size, contamination level and concentration from source to source. CEA, FerroPem and Resitec have achieved interesting results on this topic within the CABRISS project. Important topics for investigation have been the influence of additives and sawing conditions [2], wet and dry mechanical separation methods and further purification by pyro metallurgical processes. Si-kerf has been recycled into 2-4N silicon powder in pilot scale and industrial scale with an acceptable oxygen content. This material has been successfully melted, purified and used for ingot production. There are also several safety topics that need to be investigated for scale up of the recycling process. Recycling of Si-kerf has been demonstrated at an industrial scale.

Keywords: Silicon, Recycling, Sustainable, PV Materials

1 INTRODUCTION

The European Horizon 2020 project CABRISS [1] aim to implement a circular economy for solar cells based on recycled, reused and recovered materials. This paper is related to the activity of recycling the Si-kerf from diamond wire cutting of wafers and blocks. Traditional slurry with silicon carbide and glycol has not been a topic of this project. The CABRISS project concerns also other parts of the value chain for PV production such as recycling of broken cells or end of life panels, recycling of indium and silver, and other topics [1].

The process of cutting silicon ingots into wafers suitable for solar cells has a kerf loss of approximately 40% by weight silicon. There is progress to reduce thickness of diamond wire, which will reduce the kerf loss. The kerf loss from block cutting is in the range of 5%. Diamond wire is today more commonly used for mono silicon, but in the CABRISS project, both mono, mono like and multi silicon have been considered.

Several R&D project have aimed to recycle Si-kerf from PV. Several processes have also been established and some patented. To our knowledge no process have been successfully industrialized so far. This has mainly been a result of recycling cost and investment cost in combination with low prices for PV silicon in the last years. Resitec has focused on a cost efficient process for recycling the Si-kerf into 2-4N silicon for use in novel solar cell concepts or alternative markets. FerroPem and Silicio FerroSolar have focused on further melting and purification of the recycled kerf to produce ingots. CEA has focused on Si-kerf recycling and mechanisms to passivate the Si-kerf. Recycled silicon kerf has been used in conventional and novel solar cell concepts within the CABRISS project.

Recycling of Si-kerf is energy efficient. Considerable amounts of energy have been consumed to produce PV quality silicon. Resitec estimates its process to give an energy saving of >15MWh pr 1000kg Si-kerf compared to conventional production of similar Si-powder. Through the CABRISS project, we have also seen that recycled Si-kerf have applications outside of the PV industry.

2 EXTRACTION OF SI-KERF

Si-kerf from diamond wire cutting exists in highly diluted coolant (water + additives) from the cutting process. There are variations in particle size of the silicon particles depending on the method for cutting, wire thickness, wire speed etc. A typical particle size distribution of Si-kerf from diamond wire wafering is shown in figure 1. The average particle size is 1.7 μm and a 90% point at 4.3 μm. These are small particles that are challenging to process.

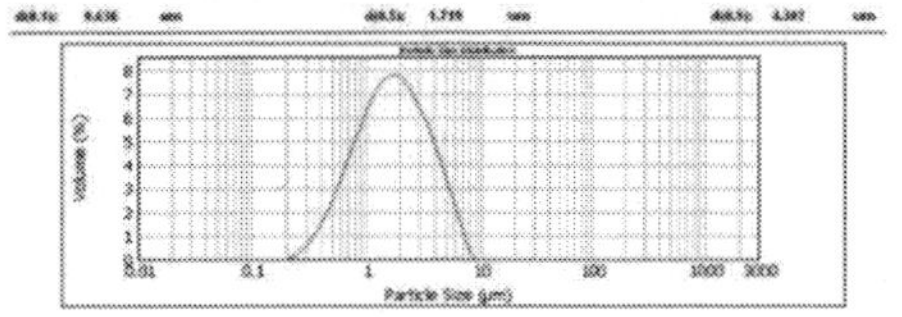

Figure 1: Particle size distribution of Si-kerf from diamond wire wafering measured by Malvern Mastersizer at Resitec.

Due to the fine particles and large volumes of cutting fluid with a low content of silicon particles (typically 0.5-2%), it is important to have an efficient method for dewatering the particles. Suitable equipment for efficient dewatering in two steps have been identified and tested in pilot and industrial scale.

Diamond wire cutting is used normally with water as fluid with the addition of a coolant additive. Several types of additives are available commercially and there are recommendations from the producers of the cutting equipment. Alternative additive systems have also been developed. In CABRISS, CEA has investigated different types of additives and their effect on oxidation of the kerf [2]. It is important to avoid oxidation of the kerf as much as possible during recycling to keep its quality high. Figure 2 shows results on oxidation for different types of coolant used by CEA.

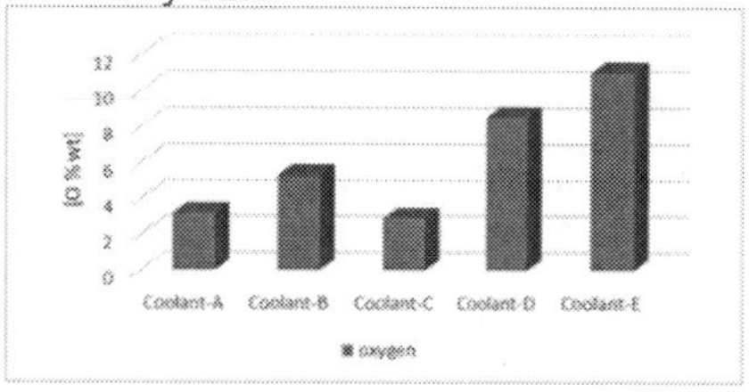

Figure 2: Oxygen in recycled Si-kerf with different coolant additives from investigations done by CEA.

It is essential that the Si-kerf is passivated before recycling. Unless passivated, the kerf will start oxidizing due to the oxygen available in water. This reaction is exothermic, and in a storage tank of Si-kerf /water slurry, there will be a temperature increase depending on the solids concentration in the water. The oxidation reaction develops large amounts of hydrogen. The oxidation which develops hydrogen results in a deterioration of the silicon quality and creates a serious safety problem. Reaction rate will depend on particle size (available surface area), temperature and other parameters. This is a safety risk during scale-up from lab to industrialization. Reliable passivation methods have been developed and tested in pilot and industrial scale. Hydrogen concentrations need to be monitored at all times.

3 RECYCLING OF SI-KERF

Si-kerf has been recycled from diamond wire wafering and block cutting. The particle size may vary and the wafer kerf is normally finer particles compared to the kerf from block cutting. Material from mono- and multi silicon has been recycled.

CEA has studied recycling of Si-kerf from a conventional process and from an optimized process. The recycling was done in connection with a diamond wire wafering saw using a water based coolant and achieved a high purity Si powder with the chemical analysis shown in figure 3. This is material recycled in pilot scale from a standard industrial type diamond wire wafering saw.

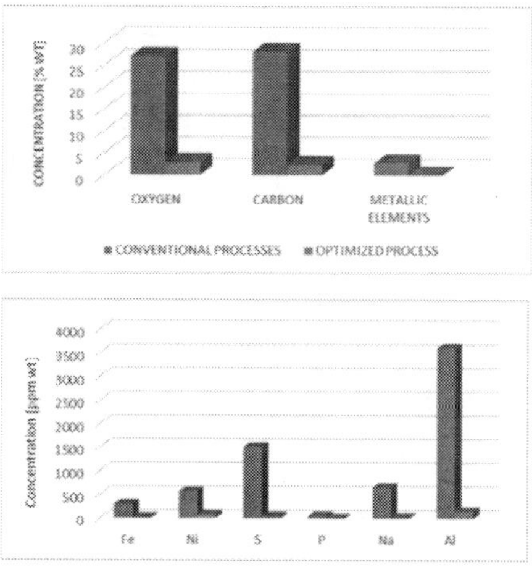

Figure 3: Results from recycling of Si-kerf at CEA from a conventional process and an optimized process.

The results from figure 3 show a significant reduction in carbon, oxygen and metallic elements, particularly aluminum. These are important parameters for using the material to produce ingots or other use in PV.

Resitec have recycled Si-kerf from block cutting and wafering of both mono and multi silicon. The sources for Si-kerf have been Fraunhofer THM and other sources outside the CABRISS consortium. The recycling is done in pilot scale of 10-100kg and also demonstrated in industrial scale > 1000kg batches. Resitec has aimed to have a cost efficient process to produce a cheap substrate which is used further in the CABRISS project or to produce silicon powder with purity suitable as feedstock for further purification steps for PV use. Purity of 99,5-99,9% have been achieved in industrial scale and 4N purity in pilot scale. Figure 4 shows results for 99,5% silicon powder recycled from a block cutting process. This material may need further purification steps prior to PV ingot production or it can be in alternative PV applications.

Si-type	Purity [wt%]	Particle size distribution [µm]			Fe [%]	Al [%]	Ca [%]	Other elements [%]
		D10	D50	D90				
Fine, poly	>99,5	1	5	15	<0,02	<0,2	<0,02	<0,01
Coarse, poly	>99,3	10	25	110	<0,06	<0,7	<0,02	<0,01

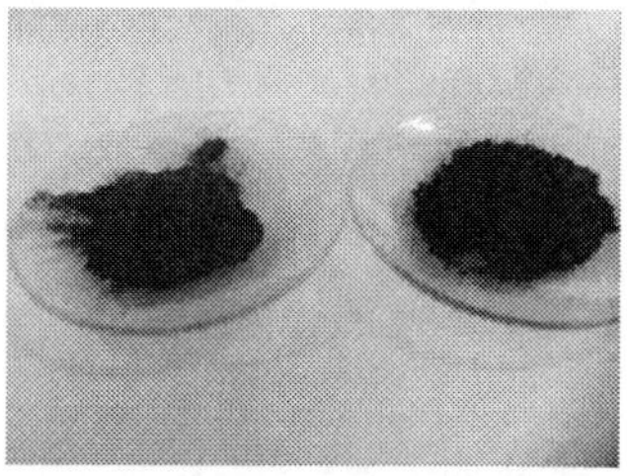

Figure 4: Results of recycled Si-kerf from a block cutting process and a picture of the dry silicon powder.

Several tests were performed at different scale to optimize the process. A recycling process for Si-kerf at industrial scale has to be flexible to fit into the existing production system of an industrial source for Si-kerf. Figure 5 shows the particle size distribution for recycled

Si-kerf from wafering of silicon by diamond wire. This test was done in pilot scale. Scale up of this process is currently on going at Resitec.

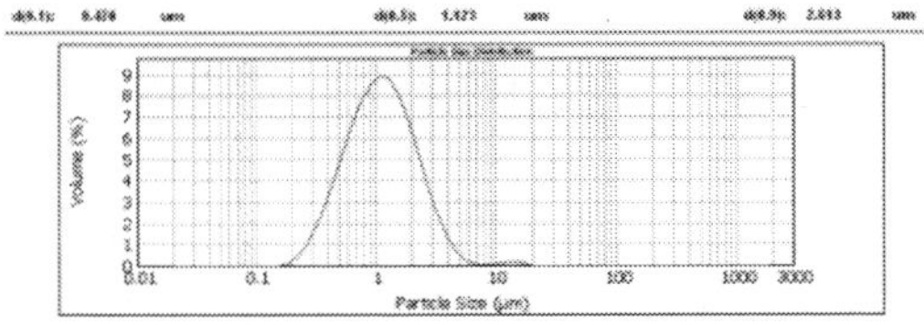

Figure 5: Particle size distribution measured by Malvern Mastersizer at Resitec.

Figure 6 shows an analysis of elements in the recycled Si-kerf from wafering with diamond wire saw. This is the same material as shown in figure 5. The analysis are done with GDMS by EAG Laboratories in France. Materials from the different recycling tests have been further used in the CABRISS project as feedstock for conventional routes for ingots or in novel routes as a silicon powder.

Element	Concentration [ppm wt]	Element	Concentration [ppm wt]
Li	< 0.005	Pd	< 0.005
Be	< 0.005	Ag	< 0.01
B	0.09	Cd	< 0.01
C	-	In	Binder
N	-	Sn	< 0.1
O	-	Sb	< 0.05
F	1.1	Te	< 0.01
Na	6.7	I	< 0.001
Mg	8.0	Cs	< 1
Al	3.4	Ba	0.25
Si	Matrix	La	0.17
P	5.7	Ce	0.24
S	13	Pr	0.08
Cl	2.1	Nd	0.10
K	2.9	Sm	< 0.005
Ca	62	Eu	< 0.005
Sc	0.04	Gd	< 0.005
Ti	0.31	Tb	< 0.005
V	0.01	Dy	< 0.005
Cr	0.35	Ho	< 0.005
Mn	0.19	Er	< 0.005
Fe	7.4	Tm	< 0.005
Co	0.05	Yb	< 0.005
Ni	62	Lu	< 0.005
Cu	2.7	Hf	< 0.05
Zn	2.7	Ta	< 1
Ga	0.10	W	0.25
Ge	1.1	Re	< 0.01
As	0.17	Os	< 0.005
Se	< 0.05	Ir	< 0.005
Br	< 0.01	Pt	< 0.01
Rb	< 0.01	Au	< 0.01
Sr	0.24	Hg	< 0.05
Y	0.01	Tl	< 0.005
Zr	0.06	Pb	0.71
Nb	0.02	Bi	< 0.005
Mo	0.09	Th	0.01
Ru	< 0.001	U	0.007
Rh	< 0.001		

Figure 6: Analytical results of elements in recycled Si-kerf from wafering of silicon at an industrial site. The oxygen content was 4,3% and the carbon content was 0,56%. The elements are measured by GDMS and the O and C level are measured by IGA, both analysis are performed by EAG Laboratories in France.

4 MELTING AND PURIFICATION

The recycled Si-kerf was supplied within the CABRISS project for multiple different PV applications. These are melting, hot pressing, plasma spray and further purification steps.

FerroPem in France did several small scale tests for melting the material and these tests resulted in a scale-up for > 300 kg tests at Silico FerroSolar in Spain. Both companies are a part of Ferro Atlantica. Analysis of melting tests by FerroPem is shown in figure 7.

Sample	Analysis	Al	Ca	Fe	Ti	Cu	Mn	Ni	P	B	Cr	V	W
359	before melting (ppmw)	213,8	77,2	94,1	5,6	9,5	<2	<2	2,5	<2	2,6	<2	<5
359	after melting (ppmw)	207,0	12,6	77,0	3,5	5,8	<2	2,1	2,7	<2	2,3	<2	<5
327	before melting (ppmw)	821,1	197,1	370,5	3,8	63,8	<2	3,1	4,6	2,8	4,9	<2	<5
327	after melting (ppmw)	626,0	13,5	224,0	2,0	42,8	<2	3,1	2,4	2,9	3,8	<2	<5

Sample	Analysis	C (%)	O (%)	S (%)
359	before melting (ppmw)	0,045	1,855	< 0,05
359	after melting (ppmw)	0,015	0,17	<0,05
327	before melting (ppmw)	0,0525	2,265	< 0,05
327	after melting (ppmw)	0,028	0,19	< 0,05

Figure 7: Analysis of Si-kerf after melting tests at FerroPem.

Figure 7 shows a reduction in several of the metallic elements and oxygen and carbon. Figure 8 shows similar results from trials in larger scale at Silicio FerroSolar.

Sample	Kg in	Al	Ca	Fe	Ti	P	B	Cu	Cr	K	V	Ni	Na	Kg out
Resitec initial kerf	1000	173,0	66,3	108,2	3,8	1,1	<1	15,1	4,0	7,1	0,0	2,6	14,0	827
TES 1-2	230	433,1	1,4	139,1	3,6	1,1	<1	14,1	5,9	2,4	<1	2,7	0,7	129
TES 3-4	150	833,8	40,7	296,6	11,6	2,3	<1	10,1	6,7	4,8	0,6	2,8	2,3	138
TES 5-9	365	1355,9	79,0	359,6	19,7	2,3	<1	16,6	10,2	3,0	0,4	4,1	2,7	320
TES 10-13	255	773,7	60,7	852,9	8,6	2,7	<1	12,3	8,4	6,5	0,5	3,4	2,0	240

Figure 8: Results of metallic elements (ppmw) after large scale melting tests. ICP-OES measurements are done by Silicio FerroSolar.

The results show that the material did not reach PV purity, but needs further purification to reach the traditional PV specifications. The further purification step to reach the solar grade quality can be done through the confidential Silicio FerroSolar process. This is currently being tested. Figure 9 shows the purification route suggested.

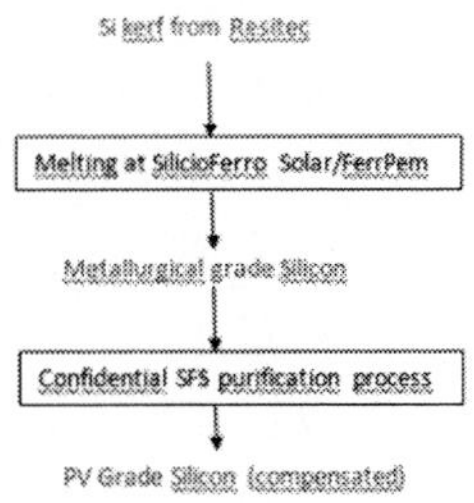

Figure 9: Suggested route for purification of Si-kerf to PV grade.

In figure 10, we can see a picture of the melted material with its origin in recycled Si-kerf. This material is from the tests by Silicio FerroSolar in Spain in larger scale as shown with chemical analysis from figure 8.

Figure 10: Picture of silicon chunks from melted and solidified material with its origin in Si-kerf.

Sintef worked on the further purification of Si-kerf by melting and these results are presented separately [3].

5 CONCLUSIONS

Through the CABRISS project a method for recycling of Si-kerf has been developed. Work by CEA has shown in pilot scale how an optimized process for Si-kerf recovery can give a high purity (4N) with low oxygen content. Resitec has developed a process for recycling of Si-kerf which has been demonstrated in pilot and industrial scale for producing 2-4N Si-powder. FerroPem and Silicio FerroSolar have used the recycled material for melting and solidification with good yield and have shown further purification in melted stage. Further up-scaling tests are in progress.

6 REFERENCES

[1] https://www.spire2030.eu/cabriss

[2] Hajjaji, H., Coustier, F., Brizé. V., Benayad, A., Benmansour, M., Chabli, A.: "Recycling of Kerf-Loss Silicon Powder from Diamond-Wire Cutting Without Chemical Treatment.", EU PVSEC 2017.

[3] Syvertsen, M., Halvorsen, T., Mørk, K., Kaden, T., Ulyashin, A.:" Remelting and purification of Si-kerf for PV wafers, EU PVSEC 2017.

Development of a modular cradle to cradle process-chain for c-Si-PV panel recycling

J. Glatthaar[1], , E. Kamdje[2], J. B. Barnickel[1], M. Dax[3], V. Schaub[4], H. G. Stevens[5], B. Jehle[6], U. Ricklefs[2], E. A. Stadlbauer[1], H.Weigand[1]
[1]Kompetenzzentrum ZEuUS,Technische Hochschule Mittelhessen (THM), Wiesenstr. 14, D-35390 Giessen; +49-641-309-2548; joerg.glatthaar@zeuus.thm.de
[2]Kompetenzzentrum NanoP,Technische Hochschule Mittelhessen (THM), Wiesenstr. 14, D-35390 Giessen
[3]Ruehl Solar, Konrad Becker Str.1, D-35102 Lohra-Kirchvers
[4]Abfallwirtschaft Lahn-Dill (AWLD), Am grauen Stein, D-35614 Asslar
[5]SM InnoTech GmbH&Co.KG, Vennweg 18, D-46395 Bocholt
[6]ZME Elektronik Recycling GmbH, Auf dem langen Furt 17, D-35452 Heuchelheim

ABSTRACT: End-of-life photovoltaic modules are part of the electric and electronic equipment addressed by the WEEE-directive. Thus, defined recycling rates need to be fulfilled for this particular waste stream. As opposed to largely destructive recovery technologies available on the market this project aims at developing a comprehensive c-Si-PV-panel recycling concept capable of conserving technological intelligence inherent to end-of-life PV modules. The approach is characterized by a tailor-made repair/recycling process for the individual modules based on a reliable failure analysis. Therefore, visual inspections of the end-of-life modules are complemented by electroluminescence measurements and the determination of current-voltage characteristics.

The most appropriate recycling procedure is assigned to each module. Ideally refurbishment can be achieved by eliminating panel defects in single repair steps (e.g. exchange of broken glass or wafers or defect connection boxes), which restore full function. Photovoltaic modules with irreparable damages are successively dismantled into components. All intact parts are recovered for the assembly of 2nd life PV modules. Broken and destroyed parts are fed into industrial raw material repair/recycling process chains. This cascaded recycling approach is expected to be more environmentally beneficial compared to existing raw material recycling in terms of energy demand and resource consumption.
Keywords: PV modules, Silicon Solar Cells, Recycling, , Encapsulation

1 INTRODUCTION

Since 2012 photovoltaic panels (PV) have been included in the EU WEEE Directive [1]. Therefore, collection and recycling of discarded end-of-life PV-modules to an extent of 80% is mandatory. On the other hand, the photovoltaics industry has started voluntary efforts on PV recycling, particularly regarding the organization of collection, transport and treatment (recycling) of out-of-spec, damaged and End-of-Life-photovoltaic panels by PV CYLCLE.

A typical photovoltaic panel consists of a laminate stabilized by an aluminum frame joint by silicone-based edge sealant or double-sided adhesive tapes preventing ingress of moisture. In 1st generation modules the laminate consists of the Si-cells embedded in layers of polymer encapsulant, typically an ethylene vinyl acetate copolymer (EVA) which bonds to a front glass and a polymer back sheet.

Current technologies mainly recycle the bulk materials glass, aluminium and copper by destructive methods completely disuniting the module compound structure [2].

Obviously. there is a need for a recycling technology that conserves function and value of module components. Moreover, there is an increasing demand for fully operative, refurbished or reproduced older models of photovoltaic panels at the growing second hand market for photovoltaic components. The first integral approach was shown by Wambach et.al. who introduced a thermal dismantling technique [3]. Later on the applied pyrolysis techniques were improved and combined with dissolution of the laminate in organic solvents or strong acids by other research groups. However, afore methods are energy-consuming, especially if working at elevated temperatures. Therefore, successful business cases are rather scarce. The refurbishment in many cases is limited to exchange of damaged connecting cables and junction boxes or reframing of panels to restore water-tightness.

Recently, the photovoltaics industry has started to consider recyclability in the latest design of photovoltaic panels [4].

Combining these concepts, in this report we present an integral approach fathoming the feasibility to widely conserve the chemical and technological intelligence inherent in spent PV modules by a stepwise repair/ recycling process (Figure 1).

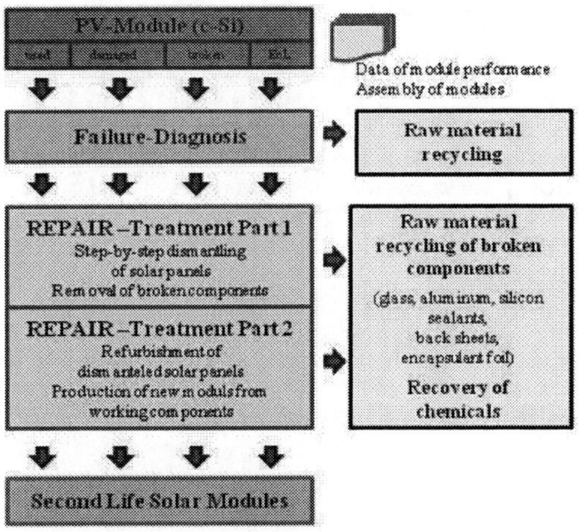

Figure 1 The PV-Rec recycling concept

Based on the experimental failure analysis, each module is assigned the most appropriate recycling procedure. In accordance with the EU-WEEE requirements the ideal case is refurbishment: PV modules with minor damages end up in tailor-made repair procedures to be specified below, which restore full function and result in 2nd life modules in a most value-conserving manner. Modules with severe or irreparable damages are successively dismantled into components used in the assembly of the 2nd life PV modules. Broken and destroyed parts are fed into industrial raw material recycling process chains.

The variety of module design (size, silicon cell types, frame, glass, back sheets, sealants and encapsulants) is the main challenge both for the dismantling and the repair.

Essential steps of this modular cradle to cradle process-chain in c-Si-PV panel recycling are:

i. Failure analysis
ii. Chemo-mechanical removal of the aluminum frame
iii. Chemo-thermal removal of the glass and back sheet
iv. Dissolution of the encapsulant, mainly ethylene vinyl acetate EVA

2 EXPERIMENTAL

2.1 Failure Analysis

A prerequisite to this approach is the availability of a reliable failure detection and analysis. Different types of defects are known in spent photovoltaic panels ranging from problems in insulation, open circuits, contacts, glass and wafer fracture to partial delamination [5,6]. In a first step, these defects are localized and classified mainly by visual inspection while still being installed. This preliminary assessment is complemented by testing both the performance of the entire module as well as single cells in a c-Si panel. Measurements applied are: (a) image analysis concerning the geometry of conductors and single cells, (b) powered by an external current source modules and single cells show heating and electroluminescence. These effects are analyzed by image analysis (c) current-voltage characteristics with and without of illumination, (d) illumination with modulated high power light sources on single cells as well as the whole module

2.2 Chemo-mechanical removal of the aluminum frame

Starting point for the development of a non-destructive mechanical removal of the aluminum frames is a force analysis of the sealant joint. Tensile strength tests, plain strain tests and tensile shear strength tests were conducted on an Inspect table BLUE 5k 222 EDC testing machine. The specimen were prepared from HelioBond® PVA 205 (Koemmerling), a typical 1k silicone sealant (Figure 2) [7, 8]. The joint surfaces were cleaned and pretreated with Koerabond HG 83 (Koemmerling). Mechanical or chemical pre-treatment routes were derived from the results of the force analysis and tested with pressed or bolt Aluminum frames.

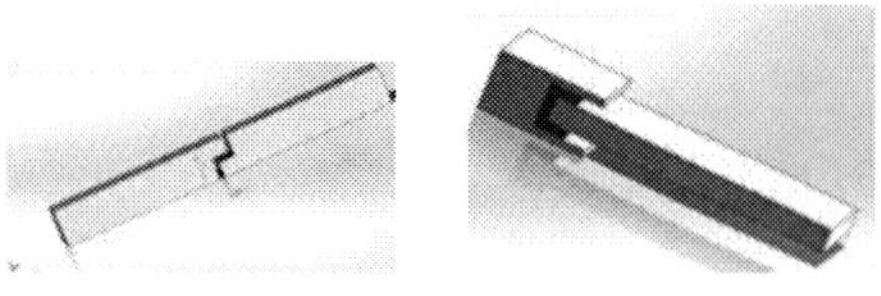

Figure 2 Examples of specimen used in different experiments; Left: Lap shear strength Right: Specimen modeling the u shape joint

Special, catalytically active solvent mixtures reduce the adhesion of the silicon sealant or are able to dissolve it. Suitable mixtures were identified through immersion experiments [9].

2.3 Chemo-thermal removal of the glass and back sheet

The temperature dependence and the influence of selected chemicals on the stiffness of the encapsulant were tested using rectangular specimen (length:160mm, width:10mm, thickness:0,5mm) prepared from samples of ethylene vinyl acetate copolymer foil PHOTOCAP 15585P HLT (Specialized Technology Resources), laminated under industrial standard conditions at 140 °C and having a gel content of 85%) [10,11]. The tensile strength tests were conducted on a standard universal testing machine. A heating unit was applied which consists of a stainless steal tube with an inner diameter of 13mm, equipped with a tailor-made heating collar (Horst GmbH), power regulator and PT 100 thermocouple.

In a first test series separation techniques were applied to small test samples (30mm × 30mm) with PET type back sheet (BP Solar) or with PVF type back sheet (Hyundai) cut from spent mono crystalline Si solar panels. The tests were expanded to whole intact wafers and to whole solar panels with different defects (e.g. broken front glas, intact front glas and broken wafers).

2.4 Dissolution of the encapsulant,mainly ethylene vinyl acetate EVA

Swelling and polymer dissolution properties of selected suitable chemicals were tested in immersion experiments [9].

In a first test series small test samples (30mm × 30mm) with PET type back sheet (BP Solar) or with PVF type back sheet (Hyundai) were cut from spent mono crystalline Si solar panels as well as from almost intact modules. Chemical cocktails were tested in terms of dissolution rates. In a second series of tests, intact wafers (125mm × 125mm) were immersed into these chemical cocktails.

3 RESULTS AND DISCUSSION

3.1 Failure Analysis

Especially mechanical defects like cracks and scratches in glass and sometimes on conductors, broken cells etc. could easily be detected and identified by image analysis. (Figure 3).

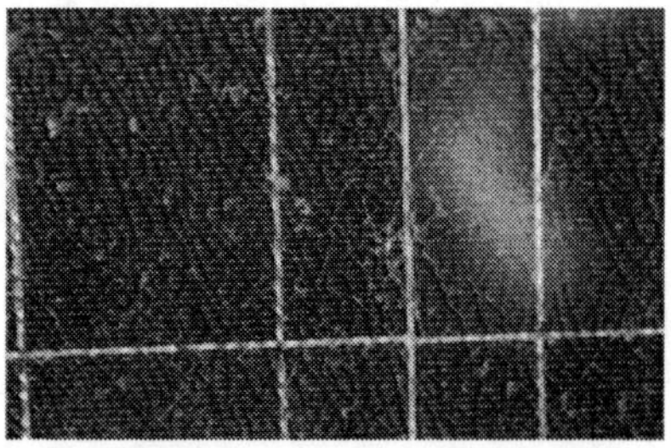

Figure 3 Image analysis detects totally broken glass although the single cells show full efficiency

At externally powered modules images were taken in the NIR (~ 1 µm) spectral range. These images show heated areas. In these areas the cells work well. Dark areas do not contribute (Figure 4)–. Shorts show strong spots. From these images defect cells and shorts can be localized.

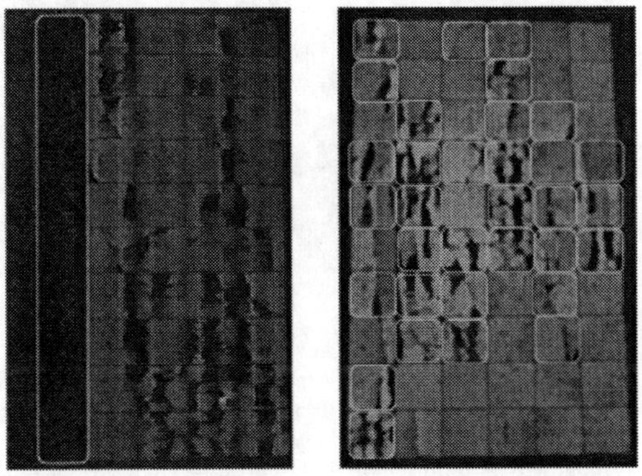

Figure 4 Externally powered modules Left: Defect string (marked) and defect single cells, black areas reduce the energy captured by that cell; Right: Marked defect single cells

To capture these images three CMOS-cameras were tested. The "Solar Cam for NIR" from MBJ gave best results. On the other hand this camera is expensive (~ 4T€). The "UI-3240LE-NIR-SL" of IDS and the "Raspberry PI 2-V2" (~ 30 €) show a very low sensitivity in the NIR. The images of the modules are captured in a dark room at maximum time of exposure (20 ms). Then the mean is taken on ten from these images. In most cases the resulting noisy mean images would be sufficient to detect black areas too but with a very cheap setup.

Under the test conditions used none of these cameras was able to detect electroluminescence at a wavelength of 1.3 μm.

A single cell in a module illuminated by a modulated light source (frequency <1 kHz) gave small signals to the external contacts. In this case all photodiodes in the chain behave as capacitors connected in series. A technique to qualify the cell is under development.

A frame with integrated high power illumination and camera is under development as well. The goal of this setup is to compare different modules of the same type. To complete this measurement in an additional automated setup the current-voltage-characteristic of single cells can be taken under different levels of illumination.

3.2 Chemo-mechanical removal of the aluminum frame

The aim of running the tensile strength tests for specimen modeling the u joint between Al frame and laminate is to simulate the change of forces within the joint, if the sealant is partially removed through chemical dissolution or mechanical treatment. We expect the holding force to be strongly reduced. Therefore the sides of the aluminum plate, that is sticking into the sealant, was covered with Vaseline to prevent adhesive bonding to sealant during the preparation only allowing adhesion on the head of the plate. The force field should correspond to the one of a simple cross tension test. Indeed, this is the case. The maximum force is reduced from 200-250 N down to 60 N. Instead of a cohesion rupture a adhesion break is observed in the upper parts (Figure 5).

At the head a mixed adhesion cohesion break is observed. This phenomenon is object of further studies and simulations.

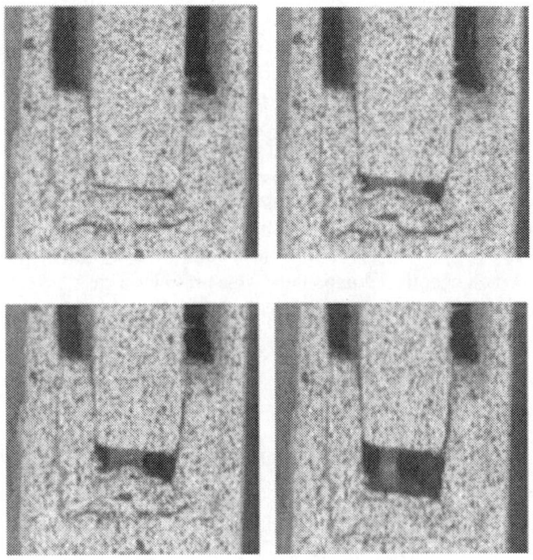

Figure 5 Course of failure of a specially pretreated u shaped joint during tensile testing. Upper left: prior to start; Upper right: Early stage, Lower left and right: advanced stages.

The result of the tensile testing can be directly applied to the removal of the aluminum frames:

Pre-treatment by exposure to silicone-dissolving catalysts weakened adhesion and cohesion forces. Suitable chemical mixtures for application under ambient conditions were developed.

As an alternative to chemical treatment, the sealant layer could be partially destroyed by mechanical pre-treatment.

In case of bolted aluminum frames the metallic parts are easily removed. Pressed aluminum frames have larger holding forces and need both mechanical and chemical treatment as well as special separation techniques.

The exposed glass/wafer/back sheet-sandwich is not affected by the removal of the frame. However immediate stabilization is necessary to prevent fracturing. Dismantled pressed and bolt frame parts are intact and can be re-used in the 2nd life PV panel assembly.

3.3 Chemo-thermal removal of the glass or back sheet materials

At EUROTRON and ECN a recycling process for newly designed modules with new thermoplastic encapsulants (TPO) was developed. At about 200°C, below the point of encapsulant decomposition, the frameless laminate is cut through the encapsulant layer using a wire saw. By this process front glass is separated from intact Si cells [4].

Encouraged by this report we have developed a new approach to separate the laminate components glass, encapsulant, Si cells and back sheet of old standard type PV modules with standard EVA encapsulant. The tensile strength behavior of EVA foil during the crosslinking process is known for ambient temperature and 80°C [12]. In our studies, we covered the temperature range from

23°C to 140°C, the lamination temperature. The results are shown in figure 6. At moderate temperatures the cross-linked EVA loses its elastic properties. For T > 80 °C the curves drop and the tensile strength is strongly reduced from 15 N/mm² down to 0.5 N/mm². At the same time the maximum strain is reduced from over 500% at room temperature to 100% at 120°C. This suggests that a series of new repair options can be performed in a temperature window from 80 to 120°C. Options include:

Exchange of broken Si cells
Nondestructive removal of back sheets
Separation of front glass and of intact Si cells
Exchange of damaged EVA encapsulant layers

Similar effects as observed for the temperature dependence of EVA were also found for several of the tested chemical compounds. This provides a case-by-case alternative to the thermal treatment of the encapsulant.

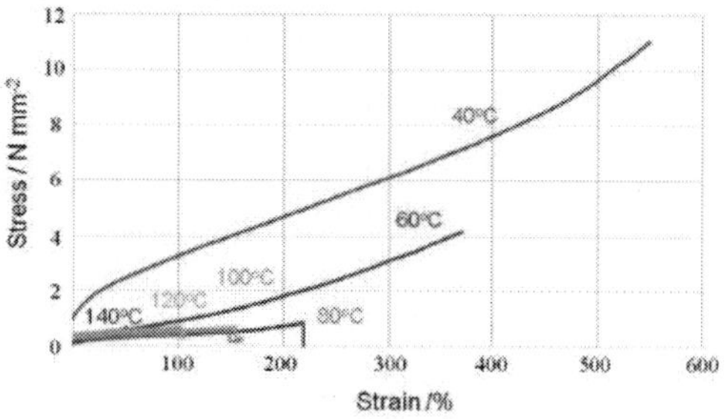

Figure 6 Stress-strain behavior of EVA at varied treatment temperature

When the front glass of the modules is broken neither the above mentioned nor established wire saw cutting can be applied because the glass-encapsulant-Si-cell interface is too uneven. For this case, we have identified a new method to separate broken front glass directly from the laminate (Figures 7, and 8). Therewith, the glass particles could be removed from a single Si cell within less than 60 s without the aid of chemicals (see 3.4). The method is currently under optimization. After chemical removal of encapsulant residues and back sheet material this will provide a high-purity glass fraction suited for recycling.

Figure 7 Laminate with broken front glass before treatment (single Si cell)

Figure 8 Laminate after treatment

3.4 Dissolution of the EVA encapsulant

Reported dissolution capabilities of chemicals for cross-linked EVA foils with a gel-content of about 80 to 85 % are limited to pure solvents like Trichloroethylene, Toluene or mixtures of Toluene and Dichlorobenzene [13]. The dissolution is accelerated at elevated temperatures of up to 100 °C and with simultaneous ultrasound treatment. However, under these treatment conditions the EVA swells faster than it dissolves. The strain generated thereby causes a breakdown of the Si cells. Clamping the laminate in between two glass plates during the solvent treatment prevented the destruction of Si cells [14]. The use of mixtures of ether derivatives of phenol and benzyl alcohol and even more complex mixtures with special chemicals like tetradecene and other long chained ester and polyether derivatives is reported [15]. Many of the applied chemicals are hazardous therefore requiring safety measures when used in the disassembly of PV-modules.

Ongoing work is related to the identification of chemical mixtures capable to cause an adhesion rupture at the glass or Si cell interface rather than dissolution of the encapsulant. We were able to find chemicals with comparable swelling tendencies, comparable dissolution abilities but less hazardous properties compared to toluene. Mixtures of these chemicals with basic solvents were found to remove PET-based back sheets within 1 to 2 hours of treatment at 80°C and PVF based back sheets within 5 to 8 hours at 80°C. Additionally, in case of multi-layered back sheets, these mixtures are able to separate the PET or PVF layers from the EVA layers within the back sheets. These mixtures are able to liberate intact Si cells from the EVA encapsulant residues.

In case of broken front glass 20 to 50% of the broken glass particles were removed without extra effort, but needed very long treatment times >24h. Since we have found a better way of removal of broken glass (see 3.3) we focus onto the chemical removal of encapsulant and back sheets. The tests are now extended step-by-step towards the immersion of complete sets of PV panels.

4 CONCLUSIONS

We have investigated new approaches for a function-conserving recycling.

The image analysis is a powerful tool for the detection and localization of cells, circuits and main defects in the visible spectral range. In the NIR spectral range it can be applied to detect shorts and areas of considerable function losses in the cells.

In most cases image acquisition and analysis can be done at low cost and the setup can be used both for the VIS and NIR.

An easy non-destructive removal of pressed or bolt aluminum frames is accomplished by mechanical or combined mechanical and chemical pretreatment routines, derived from tensile strength studies of the silicone sealant. Suitable mixtures of silicone-dissolving catalysts and solvent were identified and successfully tested. After the frame removal, the remaining laminate becomes quite flexible and needs extra stabilization.

Two new routines, (i) the thermo-mechanical treatment and (ii) the chemical treatment, have been developed for the failure-diagnosis-based stepwise dismantling of the laminate. Both routines need only moderately elevated temperatures.

(i) After warm-up to temperatures between 80°C to 140°C the back sheet is easily removed from the laminate. Broken Si cells, identified in the failure diagnosis, can be removed in a next step accomplishing a final replacement with an intact Si cell of the same type. Intact front glass can be separated e. g. through wire saw cutting. A new technique allows the separation of broken front glass.

(ii) Back sheets can chemically removed in dissent time from the laminate. The chemical mixtures identified from immersion tests, accomplish the liberation of Si cells from the EVA residues as well as the separation of the different layers (EVA/PET/PVF) within multi-layered back sheets. Broken front glass parts can be removed only partially by chemical treatment.

5 ACKNOWLEDGEMENT

This project (HA project no. 497/16-09) is funded in the framework of Hessen ModellProjekte, financed with funds of LOEWE – Landes-Offensive zur Entwicklung Wissenschaftlich-ökonomischer Exzellenz, Förderlinie 3: KMU-Verbundvorhaben (State Offensive for the Development of Scientific and Economic Excellence).

6 REFERENCES

[1] European Commission, "DIRECTIVE 2012/19/EU OF THE EUROPEAN PARLIAMENT AND OF THE COUNCIL on waste electrical and electronic equipment (WEEE) of 4 July 2012," 24 July 2012.

[2] see methods cited in Latunussa et al Solar Energy Materials & Solar Cells, 156 (2016) 101

[3] E. Bombach, I. Röver, A. Müller, S.Schlenker, K. Wambach, R. Kopecek, E. Wefringhaus, Proceedings 21th European Photovoltaic Solar Energy Conference, (2006)

[4] M.A.A. Goris, V. Rosca, L.J. Geerligs, B. de Gier, Proceedings 31th European Photovoltaic Solar Energy Conference, (2015) 1925

[5] IRENA and IEA-PVPS (2016), "End-of-Life Management: Solar Photovoltaic Panels," International Renewable Energy Agency and International Energy Agency Photovoltaic Power Systems.

[6] IEA-PVPS (2014a), Review of Failures of Photovoltaic Modules, International Energy Agency - Photovoltaic Power Systems T13, Bern, www.iea-pvps.org (viewed May 2015)

[7] Bachelor Thesis, Oezun Durgut, Technische Hochschule Mittelhessen (THM), September 2017

[8] DIN-EN-ISO-527-2:2012-06, Plastics – Determination of tensile properties Part 2: Test conditions for moulding an extrusion plastics

[9] DIN-EN-ISO-175:2011-03, Plastics – Methods of test for the determination of the effects of immersion in liquid chemicals

[10] DIN-EN-ISO-527-3:1995-07, Plastics – Determination of tensile properties Part 3: Test conditions for films and sheets (ISO-527-3:1995 + Corr 1:1998 + Corr 2:2001) (includes corrigendum AC:1998 + AC:2002)

[11] Master Thesis, Julian K. Barnikel, Technische Hochschule Mittelhessen (THM), August 2017

[12] M. Narkis, J. Miltz, J. Appl. Polym. Sci. 21 (1977), 703

[13] S. Kang, S. Yoo, J.Lee, B. Boo, H. Ryu, J. Korean Inst. Resources Recycling 20 (2011), 45

[14] T. Doi et.al., Solar Energy Materials & Solar Cells 67 (2001) 397

[15] S. Kimura, Jpn. Kokai Tokkyo Koho (2014), JP 2014105253 A Jun 09, 2014. | Language: Japanese, M. Mizuta, Jpn. Kokai Tokkyo Koho (2014), JP 2014094321 A 20140522,

A REVIEW OF SEMI EMERGING PHOTOVOLTAIC STANDARDS: 2013–2017

Shu-Tsung Hsu*, Yean-San Long, Teng-Chun Wu, Hsin-Hsin Hsieh
Center for Measurement Standards, Industrial Technology Research Institute
Rm.401, Bldg.53, No. 195, Sec. 4, Chung Hsing Rd., Hsin-Chu 31040, Taiwan, R. O. C.
*Corresponding author, andersonhsu@itri.org.tw, Phone: + (886) 3 5917970, Fax: + (886) 3 5837801

ABSTRACT: Emerging photovoltaic, including of organic solar cell and dye-sensitized solar cell, are high-potential PV products used for energy harvesting, especially in the indoor illumination applications. This paper aims to examine the standardization research of pivot SEMI emerging PV standards, which were developed by SEMI OPV and DSSC task force in Taiwan during 2013 to 2017. This work consists of three international standards in detailed specifications, including SEMI PV57, SEMI PV69 and SEMI PV76. PV57 is designed to evaluate solar cell's characteristics without capacity effect for I-V measurement. PV69 is designed to provide a suitable statistic method for SR evaluation by solving the slow response time problem, and proposes a guideline to correct performance data measured on an isothermal platform under STC. Therefore, after harmonizing both PV57 and PV69 in solving the hysteresis problem due to capacity effect and getting the reliable I-V, PV76 improves the existing durability tests, e.g., IEC 61646 for thin-film product, towards a more complete simulation or real operating conditions of emerging PV to inspect the product's defect or weakness in early stage. Consequently, these standardizations have provided the contributions to alleviate the performances of emerging PV, and influence the relevant reliability aspect in a long-term perspective.
Keywords: Emerging Photovoltaic, Organic Solar Cell, Dye-Sensitized Solar Cell

1 INTRODUCTION

Emerging PV includes organic solar cell (OPV) and dye-sensitized solar cell (DSSC). The operation principle of OPV/DSSC is using layers of organic molecules subject to lighting after excitation electronic then pass to the inorganic/organic layer of the wide energy gap nano-layer and voltage. The major differences between OPV/DSSC and p-n junction solar cells are spectrum, absorption range, photoelectric conversion response time and standard test condition (STC), etc.

OPV/DSSC are the high-potential products used for energy harvesting, especially in the context of indoor illumination applications. It is important to enhance the quality and reliability of such products, and to overcome the measurement error and hysteresis problem caused by capacity effect. PV industry needs these specifications to get the reliable measurement. Therefore, it is necessary to standardize I-V (current – voltage) and SR (spectrum response) test methods for OPV/DSSC.

The detailed documents of industry standards for OPV/DSSC have been developing by SEMI OPV and DSSC task force (TF) in Taiwan since 2013. ITRI coordinated this TF to set up a series of experiments [1-3] and discussions, which focused on the performance evaluation method of I-V (current – voltage), SR (spectrum response) and durability for OPV/DSSC. The work of this research has successfully developed three test standards in detailed specifications, including SEMI PV57 [4], SEMI PV69 [5] and SEMI PV76 [6]. Consequently, OPV/DSSC device makers and buyers, or any other party interested, can thus have the common testing standards to refer to when desired.

2 SEMI PV57 [7]

This work presents a new test method, which differs with IEC 60904-1 [8] (Table I) addressed on single and multi-junction samples, for determining I-V performance of OPV/DSSC, including the indoor application in low levels' lighting. Results were applied for SEMI Doc. 5597 and released as SEMI PV57 by voting in 2014.

Table I: I-V test method for photovoltaic

Item	SEMI PV57	IEC 60904-1
test sample	OPV and DSSC	single and multi-junction devices
lighting source	simulated sunlight (steady state)	natural or simulated sunlight
reference spectrum	AM1.5G refer to IEC 60904-3 [9]	
temperature	± 1 °C	
reference device	refer to IEC 60904-2 [15], and list table of the filtered c-Si reference cell	refer to IEC 60904-2
test procedure	adding light soaking and check criteria for sweep direction	NA
test report	with title list and adding report form	with title list

2.1 Test sequence and test items

To be differential with standard test condition (STC, 1000 Wm^{-2}) used for c-Si devices, there are two new indexes were defined in SEMI PV57 for the demand of low lighting. One is very low irradiance condition named VLIC (AM1.5G, 25 °C, 60 Wm^{-2}), and the other is ultra-low irradiance condition named ULIC (AM1.5G, 25 °C, 1 Wm^{-2}). The testing procedure is written as follows: (1) visual inspection (2) measured active area (3) check I-V/IPCE (incident photon-electron conversion efficiency) measurement system (4) check wire connection (5) light soaking (6) forward/backward I-V at STC, VLIC and ULIC (7) I-V at STC, VLIC and ULIC (8) visual inspection (9) data analysis.

Therefore, the advantage of this standardization can reduce the measurement error due to capacitive effect, and evaluate the relationship between short-circuit current (Isc) and spectral correction factor (spectral mismatch factor, MMF). OPV/DSSC needs define the specific basis of reference to make correction for spectral issue and reserve extra time for I-V test as well.

2.2 Experiment and discussion

Performance characterizations of OPV/DSSC have been investigated, in order to clarify how to determine their performance accurately. Such characterization requires considering different levels lighting on very slow temporal response in the current-voltage (I-V) curves, which are clearly dependent on the voltage sweep direction, even when the sweep time is the order of seconds.

Real-time one-sweep (RTOS) method [10] or SEMI PV57 C9.7.3 is a good test way in measuring cell performance more accurately and rapidly when evaluates performance of OPV/DSSC. RTOS method highlights the temporal stability of the output current is measured at a fixed bias voltage as shown in Fig. 1, in order to get the stable P_{max}, especially for its data influenced in different sweep direction without capacity effect (hysteresis phenomenon). This mean backward sweep has saturation in HUMO (highest occupied molecular orbital), and forward sweep has saturation in LUMO (lowest unoccupied molecular orbital), and its response time is shown to increase while decreasing levels lighting in the forward and to keep constant in the backward. This phenomenon also demonstrated the influence from capacity during the measurement of I-V curves by using a steady-state simulator.

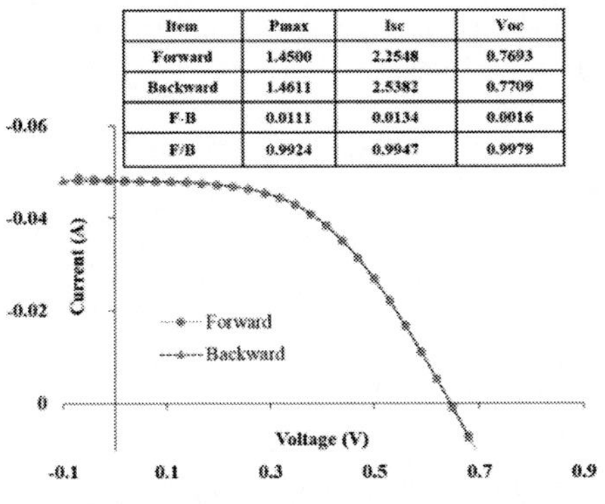

Figure 1: The stable I-V measurement by RTOS method for ionic-type DSSC in forward and backward sweep without hysteresis phenomenon

3 SEMI PV69 [11-13]

This work presents a new test method, which differs with IEC 60904-8 [10] (Table II) addressed on single and multi-junction samples, in determining SR performance of OPV/DSSC. Results were applied for SEMI Doc. 5647 and released as SEMI PV69 by voting in 2015.

Table II: SR test method for photovoltaic

Item	SEMI PV69	IEC 60904-8
test sample	OPV and DSSC	linear and non-linear for single-junction devices
lighting source	simulated sunlight (steady state)	
reference spectrum	AM1.5G refer to IEC 60904-3	
temperature	± 1 °C	
reference device	refer to IEC 60904-2 [15]	
test procedure	(1) adding light soaking (2) including real-time removing capacity effect in AC/DC mode (3) the final performance data shall have deviations 1.5 % or less.	(1) before the spectral response measurement of thin-film devices, the device under test shall be stabilized, as specified in the light soaking test procedure (2) linear and non-linear for single-junction devices (3) no check required.
test report	with title list and adding report form	with title list only

3.1 Test sequence and test items

Five test samples required for SR test are made by the same raw materials and process at least. General test sample assembled need include seal, package and four well-attached wires (or terminal wires). Terminal wires are used to multiply set designed for the safety of the rated current and length is 5 cm.

The performance data of test sample is highly responsive to the external measurement conditions. Reliable evaluation results require the measurement to be performed under proper conditions. The testing procedure is written as follows: (1) visual inspection (2) measured active area (3) light soaking (4) check SR/IPCE measurement system (5) check wire connection (6) SR checks AC/DC mode at STC or NTC (normal test conditions) (7) SR at STC or NTC (8) visual inspection (9) data analysis.

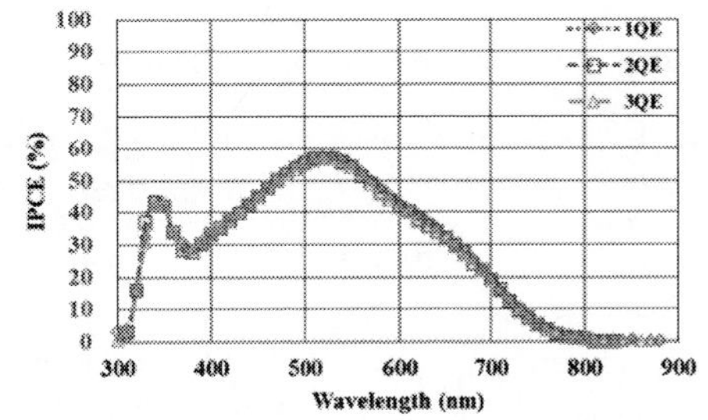

Figure 2: SR curve measured by SEMI PV69 and repeatability less than 1.5 %. (Ionic DSSC 1cm x 5 cm, DC mode)

3.2 Experiment and discussion

Following the test procedures in 3.1, three SR curves (see Fig. 2) were defined by both DC mode and RTOS method. Due to low response time characteristics of OPV/DSSC, RTOS method can remove the capacitive effect instantly, and amend the spectral response of OPV/DSSC.

4 SEMI PV76 [16-18]

By harmonizing both SEMI PV57 (I-V) and SEMI PV69 (SR) in solving the hysteresis problem due to capacity effect first, the result can get the reliable I-V and SR characteristics successfully. In addition, this work focused on the detailed guidelines of durability test including severity I & II for OPV/DSSC. The main findings of this work were applied for SEMI Doc. 5598, and successfully released as SEMI PV76 in Jan. 2017.

4.1 Test sequence and test items

Emerging PV has the portfolio for energy harvesting in low-light application. The current technology challenge for OPV/DSSC products still focused on a serious packaging problem that certainly appears in subtropical environments. In general, environmental degradation factors (humidity, temperature and light) could seriously make OPV/DSSC with poor performance and reliability, and shorten its lifetime as well. SEMI PV76 aims to improve the existing durability tests, such as IEC 61646 [19] for thin-film PV product, towards a more complete simulation or test requirements for emerging PV. This work has proposed three aging tests (TC, LS and DH) and one combined test (TC50 + DH), according to the specific test condition (see Table III) designed for durability test with two severities (I & II), and compared with IEC 61646.

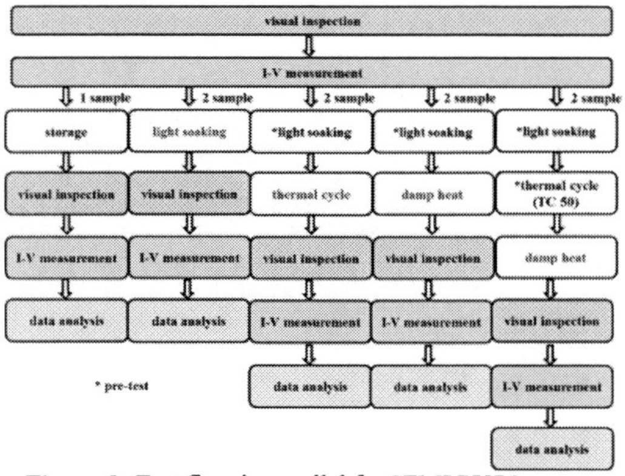

Figure 3: Test flow in parallel for SEMI PV76

Table III: Durability test method for PV cell and module

Item	SEMI PV76	IEC 61646
Test sample	OPV and DSSC (cell and module)	thin-film module
Pre-LS	1 h*	NA
Severity	I & II	NA
LS (simulated by lighting exposure model)	1000 Wm^{-2} (**300 Wm^{-2}) ; 60 °C ± 1 °C ; 1000 h(I)/ 3000 h(II)	1000 Wm^{-2} ; 50 °C ± 10 °C ; at least 43 $kWhm^{-2}$
DH (simulated by inverse power model)	65 °C ± 2 °C ; 65 %RH ± 10 %RH ; 600 h(I)/ 1800 h(II)	85 °C ; 85 %RH ; 1000 h
TC (simulated by Arrhenius model)	-10 °C ~ 60 °C ; TC50 (I)/ TC150 (II)	-40 °C ~ 85 °C ; TC50/ TC200
Combined test	TC50 + DH	NA
Life-time estimated	3 Yr* (I)/ 10 Yr (II)	20-25 Yr
Criteria	$\triangle P_{max}$ < 5 %	$\triangle P_{max}$ < 5 %
I-V requirement	SEMI PV57	IEC 60904-1

Note: * h (hour), Yr (year); **only for low-lighting

4.2 Experiment and discussion

In this study, test samples of DSSC used two cells (3cm x 4cm) composed of gel dye, electrolyte and FCO (Face-Centred Orthorhombic). The major equipment includes a Keithley source meter (Model 2460) for I-V measurement, an AM1.5G solar simulator for light soaking test, and a chamber for damp heat test. Generally, the characteristic resistance of a solar cell is the output resistance of the solar cell at its maximum power point. If the resistance of the load is equal to the characteristic resistance of the solar cell, then the maximum power is transferred to the load, and the solar cell operates at its maximum power point. These parameters (resistance, power, etc.) are very important in solar cell analysis, particularly in examining the impact of parasitic loss mechanisms. Resistive effects in solar cells could reduce the efficiency of the solar cell by dissipating power in the resistances. The most common parasitic resistances are series resistance (R_s) and shunt resistance (R_{sh}). Fig. 4 showed the same decay trend during a light soaking test while monitoring maximum power (P_{max}) and normalized resistance (R_s/R_{sh}). There are different decay start points within this degradation curve. In Fig. 5(a), the T80 happened at 1100 minute (or min) and the other decay start from 1100 min to 2200 min or after 2200 min. Consequently, Fig. 5 showed the same decay trend during a damp heat test while monitoring P_{max} and R_s/R_{sh}. In Fig. 5(a) the T80 happened at 600 min and the other decay start from 600 min to 900 min or after 900 min. In Fig. 6(b) and Fig. 5(b), R_s/R_{sh} also shown the similar degradation trend for OPV/DSSC.

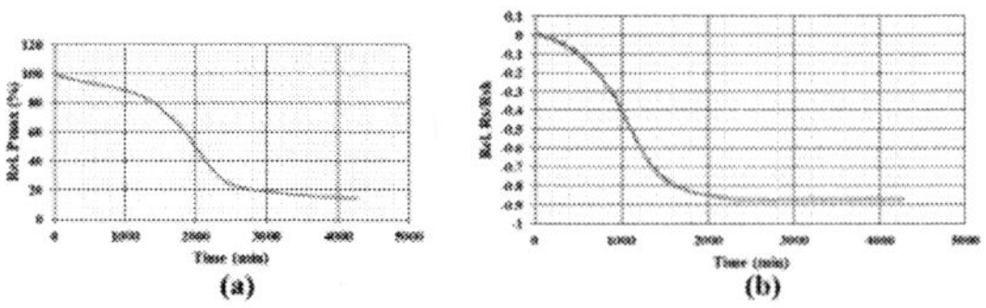

Figure 4: Light soaking test at severity I for relative (a) P_{max} (b) R_s/R_{sh}

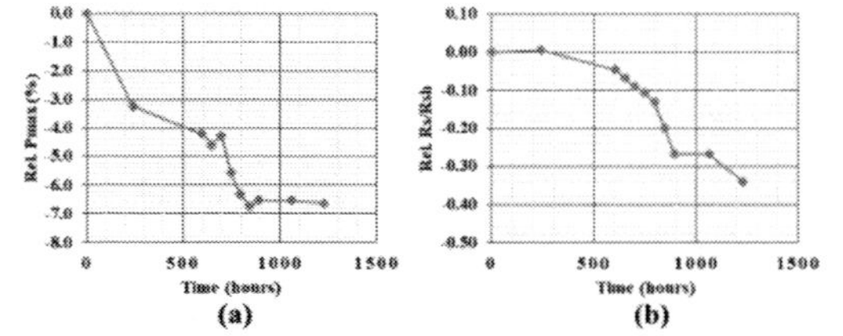

Figure 5: Damp heat test at severity I for relative (a) P_{max} (b) Rs/R_{sh}

5 CONCLUSIONS

This paper aims to examine the standardization research of pivot SEMI emerging PV standards during 2013 to 2017 in Taiwan. This research consists of three test methods in detailed specifications, including current-voltage (I-V), spectrum response (SR) and durability. In addition, this study standardized a new durability test method named SEMI PV76 for OPV/DSSC, which leveraged previous work in I-V (SEMI PV57) and SR (SEMI PV69). SEMI PV76 can improve the existing aging tests, towards a more complete simulation or real operating conditions of emerging PV, in order to determine its performance accurately and inspect the product's defect or weakness in early stage. Consequently, the findings of these standardizations make contributions of emerging PV, e.g., RTOS method and multi-indoor light system [20-22], and influence the relevant reliability aspect in a long-term perspective.

For other PV technologies, the proposed standards would be helpful to refer to in similar indoor application like OPV/DSSC. Upon for the low-light energy harvesting application and differentiating with standard test condition (STC, 1000 Wm^{-2}) used for c-Si devices, there are two new indexes VLIC and ULIC were defined in SEMI PV57. Base on the low-light application indoors, the efficiency of OPV/DSSC module shall be defined as (P_{max}/module area)/input power. The value of input power depends on the choice of irradiation type or artificial lighting source, and keeps 1000 Wm^{-2} in using natural sunlight or simulated solar irradiation. Furthermore, there are two ongoing drafts are developing by OPV and DSSC TF since Jan. in 2017, to define the specification of the indoor lighting simulator (SEMI Doc. 5979 [23-24]) and current-voltage (I-V) measurement in indoor lighting (SEMI Doc. 5980 [25]). Doc. 5979 has been joining global cycle voting in June 2017.

6 ACKNOWLEDGEMENTS

The authors appreciate sincerely for all the discussions and contribution from SEMI OPV and DSSC TF in Taiwan.

7 REFERENCES

[1] S. T. Hsu, Y. S. Long, T. C. Wu, H. H. Hsieh, A Case Study of Developing SEMI OPV/DSSC Standards in Taiwan, 32 nd EU PVSEC, 3DV.2.59 (2016).

[2] S. T. Hsu, Y. S. Long, T. C. Wu, Measurement and Standardization for Emerging Photovoltaic, IEEE International Instrumentation and Measurement Technology Conference (2016).

[3] S. T. Hsu, Y. S. Long, T. C. Wu, A Case Study of Developing SEMI PV Standards in Taiwan (2009~2014), *Applied Mechanics and Materials*, ISSN: 1662-7482, Vol. 870, pp 231-236, 2017. doi:10.4028/www.scientific.net/AMM.870.231.

[4] SEMI PV57, Test Method for Current-Voltage (I-V) Performance Measurement of Organic Photovoltaic (OPV) and Dye-Sensitized Solar Cell (DSSC).

[5] SEMI PV69, Test Method for Spectrum Response (SR) Measurement of Organic Photovoltaic (OPV) and Dye-Sensitized Solar Cell (DSSC).

[6] SEMI PV76, Test Method for Durability of Low Light Intensity Organic Photovoltaic (OPV) and Dye Sensitized Solar Cell (DSSC).

[7] S. T. Hsu, Y. S. Long, T. C. Wu, "Standardization of Current - Voltage Test Method for DSSC Products", *Applied Mechanics and Materials*, ISSN: 1662-7482, Vol. 870, pp 263-268, 2017. doi:10.4028/www.scientific.net/AMM.870.263.

[8] IEC 60904-1:2006, Photovoltaic Devices - Part 1: Measurement of Photovoltaic Current-Voltage Characteristics.

[9] IEC 60904-3:2016, Photovoltaic Devices - Part 3: Measurement Principles for Terrestrial Photovoltaic (PV) Solar Devices with Reference Spectral Irradiance Data.

[10] Y. S. Long, S. T. Hsu, T. C. Wu, Induction of Internal Capacitance Effect in Performance Measurement of Organic Photovoltaic Device (OPV) by Real-Time One-Sweep Method (RTOSM), *Journal of Energy and Power Engineering*, 8 (2014) 1059-1066 .

[11] S. T. Hsu, Y. S. Long, T. C. Wu, M. A. Tsai, The Spectrum Response Measurement for OPV and DSSC. Part 1 - Apparatus and Test Procedures, The 43rd IEEE Photovoltaic Specialists Conference (2016).

[12] S. T. Hsu, Y. S. Long, T. C. Wu, M. A. Tsai, The Spectrum Response Measurement for OPV and DSSC. Part 2 – Calculations and Standardization, The 43rd IEEE Photovoltaic Specialists Conference (2016).

[13] S. T. Hsu, Y. S. Long, T. C. Wu, Jay Lin, Standardization of Spectrum Response Measurement for Organic Photovoltaic and Dye Sensitized Solar Cell, SNEC 10th (2016) International Photovoltaic Power Generation Conference.

[14] IEC 60904-8:2014, Photovoltaic Devices - Part 8: Measurement of Spectral Responsivity of a Photovoltaic (PV) Device.

[15] IEC 60904-2:2015, Photovoltaic Devices – Part 2: Requirements for Photovoltaic Reference Devices.

[16] S. T. Hsu, Y. S. Long, T. C. Wu, J. C. Chou, Durability Test for Organic Photovoltaic (OPV) and Dye Sensitized Solar Cell (DSSC), PVSEC26, Oct, 2016

[17] S. T. Hsu, Y. S. Long, T. C. Wu, J. C. Chou, Standardized Durability Test for Emerging Photovoltaic in Low Light Intensity Environment, 4th IMETI, 2016

[18] Shu-Tsung Hsu, Yean-San Long, Teng-Chun Wu, Hsin-Hsin Hsieh, Requirement of Durability Test for DSSC and OPV, 32 nd EU PVSEC, 2016

[19] IEC 61646:2008, Thin-Film Terrestrial Photovoltaic (PV) Modules - Design Qualification and Type Approval.

[20] Y. S. Long, S. T. Hsu, T. C. Wu, Performance Ratio of PV Compare between Indoor Lighting and Solar Simulator, Optics & Photonics Taiwan International Conference (2016).

[21] Y. S. Long, S. T. Hsu, T. C. Wu, Energy Harvesting Characteristics of Emerging PV for Indoor and Outdoor, The 43rd IEEE Photovoltaic Specialists Conference (2016).

[22] T. C. Wu, Y. S. Long, S. T. Hsu, Performance Rating of PV Compare between Indoor Lighting and Solar Simulator, SNEC 11th (2017) International Photovoltaic Power Generation Conference.

[23] SEMI Doc. 5979, Specification of Indoor Lighting Simulator Requirements for Emerging PV.

[24] Y. S. Long, S. T. Hsu, T. C. Wu, Evaluation Emerging PV Performance Rating under Indoor Lighting Simulator, 33 rd EU PVSEC, 3DV.2.44 (2017).

[25] SEMI Doc. 5980, Test Method of Current-Voltage (I-V) Measurement in Indoor Lighting for DSC and OPV.

NOTE ON COLE-COLE DIAGRAMS OF PHOTOVOLTAIC MODULES EVALUATION

Ladislava Černá[1], Tomáš Finsterle[1], Pavel Hrzina[1,2] and Vítězslav Benda[1]

[1]Department of Electrotechnology, Faculty of Electrical Engineering, Czech Technical University in Prague, Technicka 2, 166 27 Prague, Czech Republic.

[2] University Centre for Energy Efficient Buildings, Czech Technical University in Prague, Trinecka 1024, 273 43 Bustehrad, Czech Republic.

Phone: +420 224 352 090; E-mail: ladislava.cerna@fel.cvut.cz

finsttom@fel.cvut.cz, hrzinap@fel.cvut.cz, benda@fel.cvut.cz

ABSTRACT: There exist many method used for photovoltaic (PV) modules' and systems' diagnostics. One of the less conventional is impedance analysis of PV cells or modules which gives additional information about measured cell, module respectively. Due to high amount of possible sources of uncertainties and low repeatability when performed under samples' irradiation, the evaluation of obtained results is relatively complicated. In this paper, the comparison of conventional methods and impedance analysis and its possible evaluation and limits are presented.
Keywords: Photovoltaic, Electroluminescence, Capacitance, Flash Test, Impedance Spectroscopy, Biasing, Series Resistance

1 INTRODUCTION

There exist many methods for defects identification in photovoltaic devices. Most of them are based on measurement of static parameters (namely I-V curve measurement, e.g. [1]) or by defects' visualisation (thermography, electroluminescence (ELCD) – e.g. [2]). Defects apparent from ELCD sometimes don't influence the static characteristics but represent the performance drop in the future. It is advantageous to detect such possible failures also by electrical characterization that can be easily performed on-site without demanding measurement means. These prerequisites can be fulfilled using impedance spectroscopy which requires LCR meter only and can be performed independently on weather. But the problem with obtained results' evaluation still remains.

2 IMPEDANCE SPECTROSCOPY

Every cell is characterized by its basic static parameters which are usually obtained by measuring volt-ampere characteristics - short circuit current I_{SC} (A), open circuit voltage V_{OC} (V), shunt resistance R_{SH} (Ω) which represents defects inside the cell, series resistance R_S (Ω), which represents ohmic losses, maximum power P_{MAX}, also *MPP* (W) / maximum power point, fill factor *FF* (-), $FF = P_{MAX}/(I_{SC} \cdot V_{OC})$ and efficiency η (%) given by the ratio of P_{MAX} and cell's irradiation P_{IN}. Moreover, there exist also other parameters based on the AC equivalent circuit like transition capacitance (C_T) and diffusion capacitance (C_D). A solar cell (module) can be modelled using the equivalent circuit shown at Fig. 1 [3]:

When combining C_D with C_T and R_D with R_{SH} the simplified dynamic model of PV cell, module respectively, can be obtained as shown on Fig. 2 [3].

PV module consists of cells which are serially connected in order to reach a sufficiently high voltage output, so the module can be modelled using same circuit and parameters which values represent the serial connection of the parameters of individual cells.

The AC parameters can be measured using frequency domain technique – impedance spectroscopy. The Impedance Spectroscopy is characterized by the measurement and analysis of some or all impedance related functions of an electronic device. In impedance spectroscopy, the complex impedance $Z(\omega) = R(\omega) + jX(\omega)$ of a device is measured directly within a large range of frequencies. A purely sinusoidal voltage with varying frequency is applied to the terminals of the device under test, and the phase shift and amplitude of the voltage and current signals are measured. The ratio between the applied voltage and the resultant current is calculated and this gives the impedance $Z(\omega)$ of the device under test. The plotting of $R(\omega)$ and $X(\omega)$ on a complex plane, in function of the varying frequency, gives the impedance spectrum of the device. In the photovoltaic area, this method is used for many years, mainly for AC parameters estimation such a R_P determination (e.g. [4], [5]), but in [6] is presented also the usage for the donor concentration determination. For measurement, the LCR meter is usually used.

Cole-Cole Diagram

The graphical representation of $X(\omega)$ vs. $R(\omega)$ is called the "Cole-Cole diagram" of the solar cell, and it gives the impedance arcs from which the different AC parameters can be extracted. The construction of Cole-Cole diagram allows simple evaluation of parameters and also the quality of the tested cell, module respectively [3]:

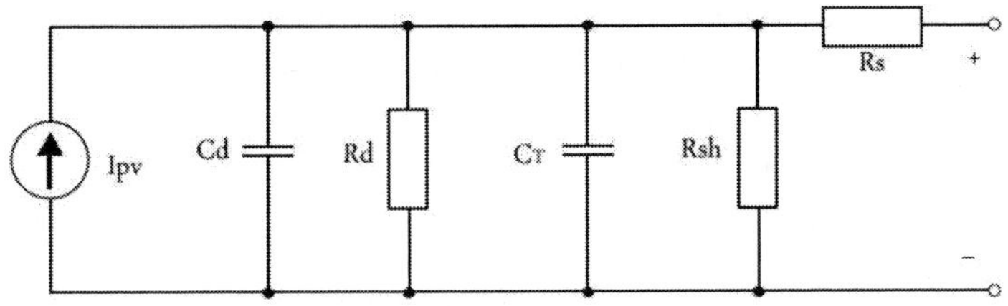

Figure 1: Dynamic equivalent circuit of PV cell (module)

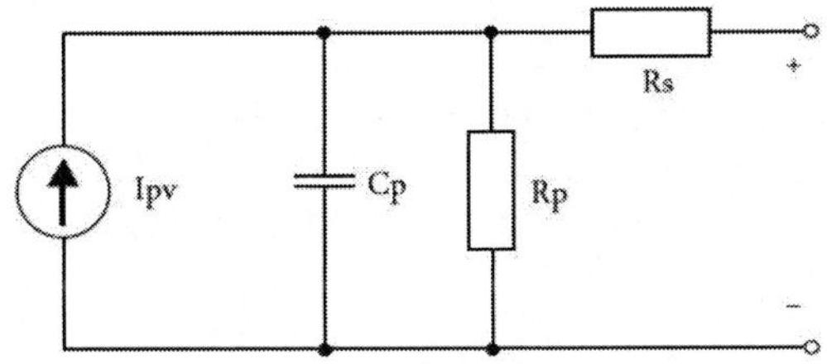

Figure 2: Simplified dynamic equivalent circuit of PV cell (module)

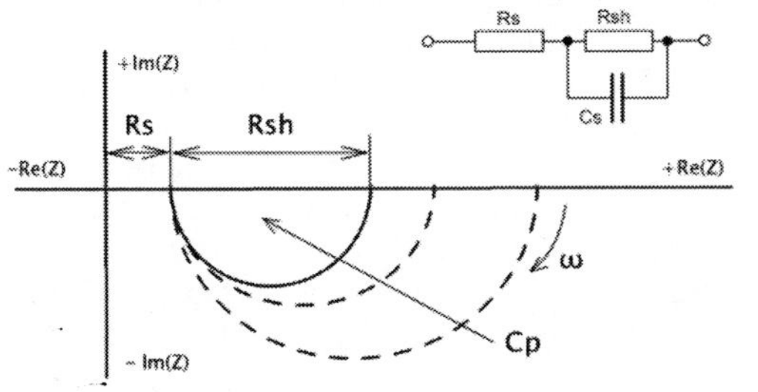

Figure 3: Cole-Cole diagram of PV cell (module)

The problem of measurement the Cole-Cole diagrams lays in many factors that can influence the resulting curves. The Cole-Cole circle's diameters correspond with the PV cell's parameters, but also with other aspects like irradiance level and its spectrum, DC biasing, test signal level and cable length.

The most problematic parameter is the capacitance which is the representation of physical capacitances inside the structure, particularly space charge capacitance of depletion region and capacitance due to minority carrier oscillation in response to the AC signal. Especially the second capacitance represents relatively difficult problem because it shows strong dependence on injection level. Diffusion capacitance is defined as $C_D = dQ/dV$, where Q represents injected positive charge and V the voltage [7]. For lower frequencies can be expressed by [8]:

$$C_D = \frac{e}{kT}\left(\frac{eL_p p_{n0}}{2} + \frac{eL_n n_{p0}}{2}\right)\exp\left(\frac{eV}{kT}\right) \qquad (1)$$

where e is electron charge, k Boltzmann constant, T temperature, L_p and L_n diffusion lengths of holes and electrons and p_{n0} and n_{p0} initial minority carriers concentrations.

From the equation above it is apparent that injection level is tightly connected with minority carrier generation, it means, in the case of PV structures, also with irradiance. Because the irradiation of structure is never sufficiently constant, different results can be obtained when evaluating these parameters, so the results than are unrepeatable. For this reasons, it can be recommended to perform the AC measurement in dark conditions.

The AC power voltage value (measuring signal) has, if usual range is used (tens of millivolts), no significant effect on the measured values, but DC biasing is changing the operational point of the cell and thus influence the diameter and area of measured Cole-Cole diagram. Extracted parameters are than strongly influenced by this value as can be seen on Fig. 4. The reason is in properties of the P-N junction inside the cell. If the junction is closed, also the capacitance of the junction is relatively low. If the voltage is applied than the width of depletion region is lowered and consequently capacitance rises – analogy with variable capacitor where the capacitance is controlled by the distance between electrodes can be found. With rising capacitance, the complex part of impedance decreases which propagates in circle dimensions.

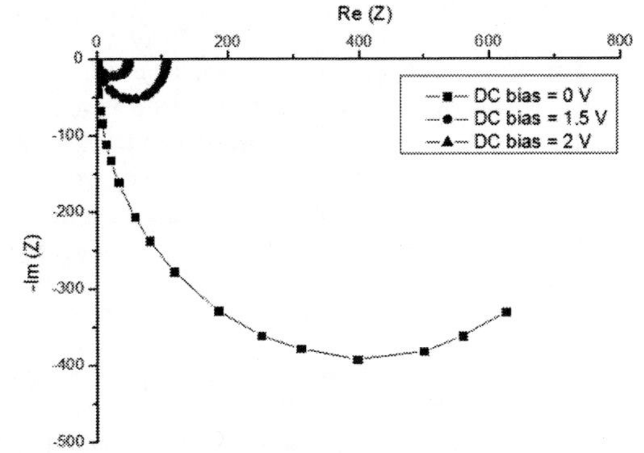

Figure 4: Cole-Cole diagram of 4 multicrystalline PV module

DC biasing also influence the value of shunt resistance. When the junction is opened than also R_{SH} decreases – the capacitor's representation of the junction behaves like resistor when the frequency is equal to zero thus it doesn't behave like open circuit and lowers final R_{SH} value.

The question is than, which value of DC bias is could be set to achieve parameters' values corresponding to values obtained by standard measurement methods like

I-V curve measurement under STC (standard testing conditions).

3 RESISTANCES EXTRACTION METHODS

To find the proper configuration of Impedance Spectroscopy method it is essential to determine the correct values of resistances. There exist the following methods for R_{SH} and R_S determination: determination from the slope of the line around the V_{OC} (R_S) or I_{SC} (R_{SH}) point, standard method according to IEC 60891 [9], approximation method e.g. [10] and derivation method e.g. [11]. Although R_S and R_{SH} are usually considered to be constant and independent on irradiance level, there is a strong exponential dependence on irradiance of both resistances. All of mentioned methods are performed under irradiation, therefore modification should be used.

R_S determination

For verification of measured results under zero biasing it is necessary to measure the dark series resistance. The R_S dependence on irradiation is given by

$$R_S(G) = R_{S\infty} + AG^{-B} \qquad (2)$$

Where A and B are constants specific for the given PV cell, module respectively and $R_{S\infty}$ states for theoretical value of R_S when the irradiance is infinite (R_S decreases with the rising irradiance)

The value of R_S under dark conditions can be given by fitting measured results under different irradiance conditions.

R_{SH} determination

For R_{SH} determination the calculation from dark current measurement can be used. The dark *I-V* curve can be divided into three parts. In the first part, where the voltage is as small as possible and linear (approx. lower than 40 mV per one cell), the shunt resistance influence predominates and equation for *I-V* curve can be reduced to:

$$I \approx \frac{V}{R_{SH}} \qquad (3)$$

4 EXPERIMENTS

There was a comparison of impedance spectroscopy (IS), flash test (FT) and resistance determination used. For measurement, the multicrystalline Si modules consisting of 4 serially connected PV cells were used. For IS measurement, the LCR meter HP 4284A has been used and FT have been performed at PASAN Sun Sim 3c.

From the presumptions mentioned above it is obvious that IS performed under zero voltage DC bias should give the value of R_{SH} corresponding to value measured under dark current measurement method. From the graph at Fig. 4 can be found that the value for zero biasing will be approx. 800 Ω. From the dark current measurement, the value of approx. 750 Ω has been calculated (see Fig.5). The value of dark shunt resistance than can be considered as corresponding to measured at zero voltage bias.

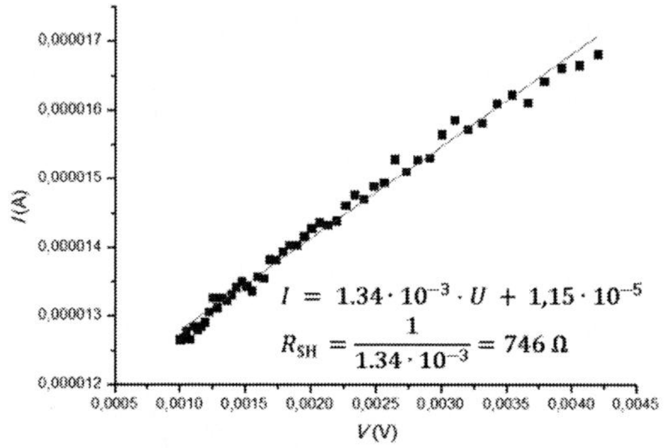

Figure 5: Shunt resistance determination

The value of series resistance is in this case too low, so every measurement obtained by IS will be burden by relatively high uncertainty, so it has not been performed in this case nor further examined. The comparison of R_{SH} values obtained by FT and IS is apparent from the following table:

Table 1: Dependence of R_{SH} at injection level controlled by irradiance G and DC biasing

DC Biasing (V)	G (W/m^2)	R_{SH} – IS (Ω)
0	0	800
1.5	450	108
2.0	1000	48

5 CONCLUSION

The basic description of Impedance spectroscopy method and its properties have been presented. There was an influence of different conditions discussed.

From the comparison of measured values it can be stated that it doesn't matter if the P-N junction is opened by biasing or by irradiation. After biasing of approx. 0.5 V per one cell, the results obtained by IS should be corresponding to those obtained by STC measurement. On the other hand, when lower injection is used, lower biasing can relatively well serve as the simulator of this condition and provide the possibilities of studying the defects behaviour under AC conditions like they would be under constant irradiation. In the future, there will be investigated other aspects like temperature and different defects propagation.

6 ACKNOWLEDGMENT

This work has been supported by project of University Centre for Energy Efficient Buildings, SGS SGS16/074/OHK3/1T/13, European Regional Development Fund-Project "Centre of Advanced Photovoltaics" (No.CZ.02.1.01/0.0/0.0/15_003/0000464) and Laboratory of Photovoltaic Systems Diagnostics.

7 REFERENCE

[1] IBRAHIM, A., Athapol KITIYANAN, Takashi FUYUKI a Ayumi TANI. 2006. *Analysis of Electrical Characteristics of Photovoltaic Single Crystal Silicon Solar Cells at Outdoor Measurements*. Editor Frank Kreith, D Goswami. New York: Nova Science Publishers, ix, 256 p. DOI 10.4236/sgre.2011.22020.

[2] HOYER, Ulrich; BUERHOP, Claudia; JAHN, Ulrike. Electroluminescence and infrared imaging for quality improvements of PV modules. In: *23rd European Photovoltaic Solar Energy Conference and Exhibition, Valencia*. 2008.

[3] CHEVNIDHYA, D., K. KIRTIKARA a C. JIVACATE Dynamic Impedance Characterization of Solar Cells and PV Modules Based on Frequency and Time Domain Analyses. *Trends In Solar Energy Research*. Hough, T.P. New York: Nova Science Publishers, 2006, s. 21-45. ISBN 1-59454-866-8; LCCN: 2005034740.

[4] THONGPRON, J.; KIRTIKARA, K. Voltage and frequency dependent impedances of monocrystalline, polycrystalline and amorphous silicon solar cells. In: *Photovoltaic Energy Conversion, Conference Record of the 2006 IEEE 4th World Conference on*. IEEE, 2006. p. 2116-2119.

[5] JOHNSON, Jay, et al. Creating dynamic equivalent PV circuit models with impedance spectroscopy for arc fault modeling. In: *Photovoltaic Specialists Conference (PVSC), 2011 37th IEEE*. IEEE, 2011. p. 002328-002333.

[6] AL ABDULLAH, K., et al. Low Frequency and Low Temperature Behavior of Si, Solar Cell by AC Impedance Measurements. *Energy Procedia*, 2012, 19: 183-191.

[7] LUQUE, Antonio; HEGEDUS, Steven (ed.). *Handbook of photovoltaic science and engineering*. John Wiley & Sons, 2011.

[8] FRANK, H.; ŠNEJDAR, V. *Principy a vlastnosti polovodičových součástek*. SNTL-Nakladatelství technické literatury. Prague 1976.

[9] IEC 60891:2009 ed.2. *Photovoltaic devices - Procedures for temperature and irradiance corrections to measured I-V characteristics*. IEC:2009.

[10] A general and precise method to evaluate the series resistance of photovoltaic cells from I-V characteristics. Chibbaro, C.; Zimbone, M.; Litrico, G.; Baeri, P.; Lo Trovato, M. L.; Aleo, F. *Journal of Applied Physics*. Aug2011, roč. 110, č. 4, p044505. 9p. DOI: 10.1063/1.3622617.

[11] K. Ishibasi, Y. Kimura, and M. Niwano, J. Appl. Phys.103, 094597 (2008)

[12] CERNA, L., V. BENDA a Z. MACHACEK. A note on irradiance dependence of photovoltaic cell and module parameters. 2012 28th International Conference on Microelectronics Proceedings [online]. IEEE, 2012, , 273-276 [cit. 2017-02-03]. DOI: 10.1109/MIEL.2012.6222852. ISBN 978-1-4673-0238-8. Dostupné z: http://ieeexplore.ieee.org/document/6222852/

Concept of a photoluminescence measurement system

Rita Ebner, Gusztáv Újvári, Bernhard Kubicek, Martin Halwachs
AIT Austrian Institute of Technology, Center for Energy, www.ait.ac.at, Giefinggasse 2, 1210 Vienna, Austria, T +43 50550-6628, F +43 50550-6390, rita.ebner@ait.ac.at

ABSTRACT: Camera-based measurements, such as electroluminescence (EL), photoluminescence (PL), and infrared (IR) technologies, offer excellent possibilities for determining production failures or defects in solar modules which cannot be detected by means of standard power measurements. These types of optical measurement provide high resolution images with a two-dimensional distribution of the characteristic features of PV modules. This paper focuses on the concept of a LED operated PL-measurement system for PV-modules. Furthermore the performed PL-images of different module technologies were compared with EL-images.
Keywords: Photoluminescence, Measurement system

1 Introduction

Photoluminescence (PL) imaging, is compared to EL imaging, a contactless and non-destructive measurement technique [1]. The combination of PL imaging with EL imaging of finished cells gives further information about cell performance losses (e.g. resistance losses). In PL imaging systems commonly high power lasers are used to generate excess charge carriers, and an InGaAs- CCD camera works as detector. But as this equipment has limited resolution an alternative equipment was searched for. Light emitting diode (LED) arrays turned out to be a very cost efficient alternative excitation source [2]. The PL system built up at AIT consists of a sensitive Si-CCD-camera (300 to 1000 nm wavelength window) and different innovative LED arrays with peak wavelengths between 525 nm and 840 nm were used (see Fig.1) [3, 4].

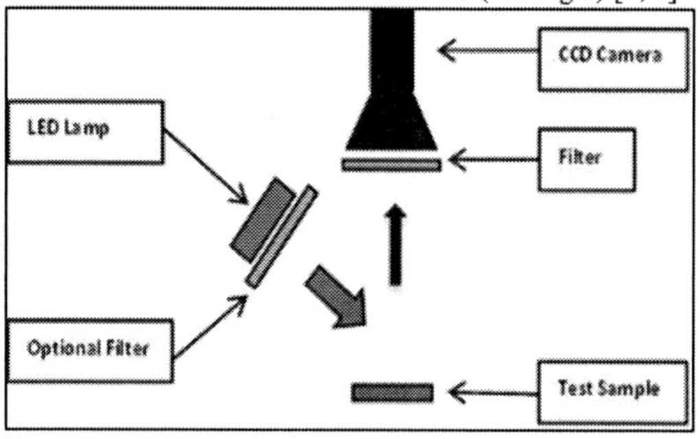

Fig.1 Concept of photoluminescence measurement system.

2 Concept of the PL measurement system

Depending upon the materials and the technology used, solar cells are better or worse at converting the different color bands of sunlight into electricity. The spectral sensitivity describes the wavelength range in which a cell works most efficiently and influences the efficiency under different irradiance conditions.

While crystalline solar cells are particularly sensitive to long wavelength solar radiation, thin-film cells utilize the visible light better. Amorphous silicon cells can absorb short wavelength light optimally. In contrast, CdTe and CIS are better at absorbing medium wavelength light.

Because of the different sensitivity of thin-film cells, for PL-measurements of these technologies, also different LEDs have to be used (see Fig.2). For aSi-modules LED arrays with a peak wavelength of 525 nm, for CdTe-modules arrays with a peak wavelength of 625 nm and for CIS/CIGS arrays with a peak wavelength of 840 nm are used.

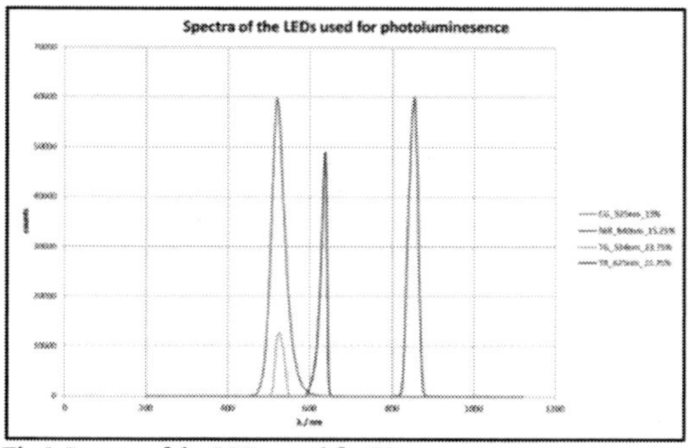

Fig.2 Spectra of the LEDs used for PL.

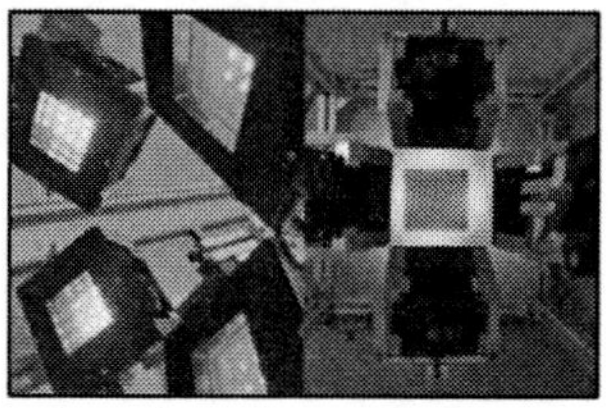
Fig.3 shows the PL-setup with 4 green lamps

The LED spectrum each was chosen to have a negligible overlap with the luminescence spectrum. In our setup the camera is placed on the same side as the LED arrays. This requires a good filtering as a large portion of the excitation light is reflected on the module and strikes the camera. Additionally a cooling of the LEDs is necessary (see Fig.4). The green LEDs also need a water cooling.

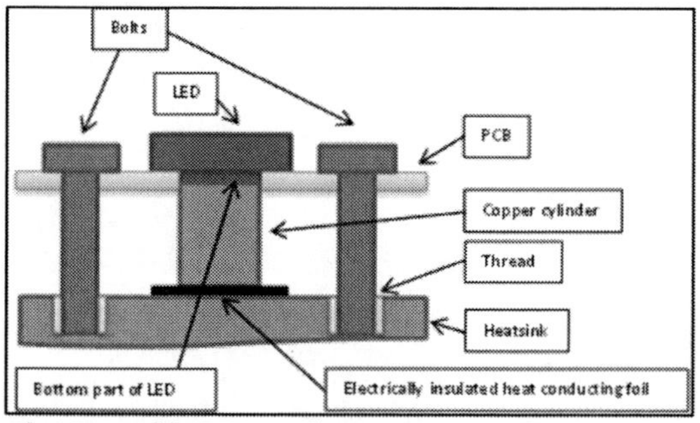

Fig.4 Setup of the LEDs-cooling system.

3 Experiments

PL-measurements of CIS, CdTe and aSi-modules and of a multicrystalline Si solar cell were performed and compared with EL-images.

PL measurements of CIGS and CIS modules performed with NIR 840 nm lamps delivered very good results. The powerful green 525 nm lamps illuminating the whole CdTe modules also delivered very good results. Red 625 nm lamps did not deliver any results applied on CdTe or aSi modules. The powerful green lamps (525 nm) delivered only weak results on aSi modules. To get good results for aSi modules, it is necessary to use multiple excitation wavelengths of 465 nm (blue), 525 nm (green), 625 nm (red) and 740 nm (near infrared).

3.1 Characterization of a CdTe-module

Comparison of electroluminescence (EL) and photoluminescence (PL) images of a broken CdTe-module:
A section of the EL and the same section of the PL-image of a CdTe-module is shown in Fig.5). The module was broken after transport. The PL-image clearly gives more information about the material properties and the position of the defect.

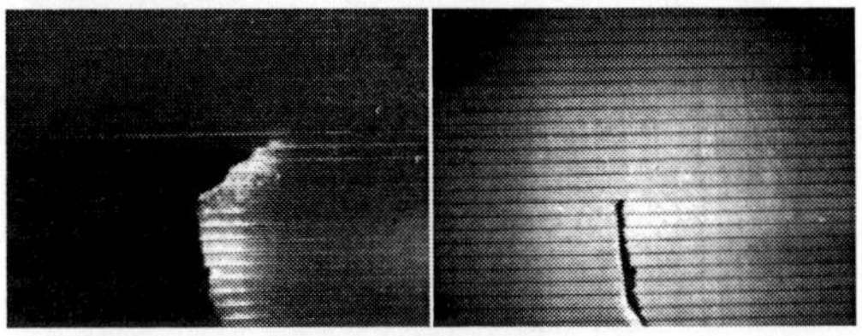

Fig.5 CdTe, left: EL-image, right: PL-image.

Fig.6a and Fig.6b show the EL-images of a CdTe-module measured in "cold" and "warm conditions". Only the EL-image taken in "warm condition" supplies enough information and is comparable to the PL-image.

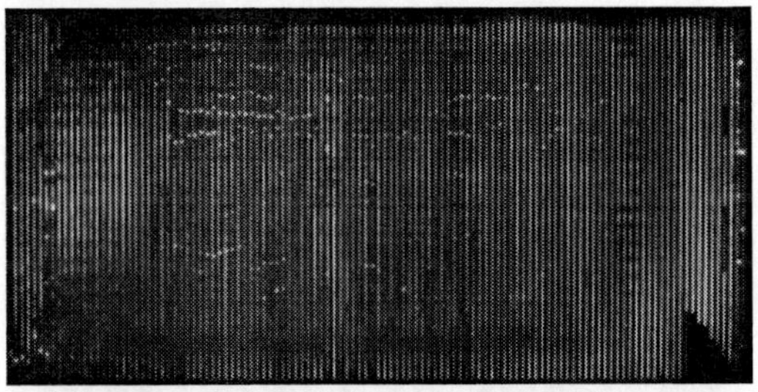

Fig.6a CdTe-module, EL-image "cold condition".

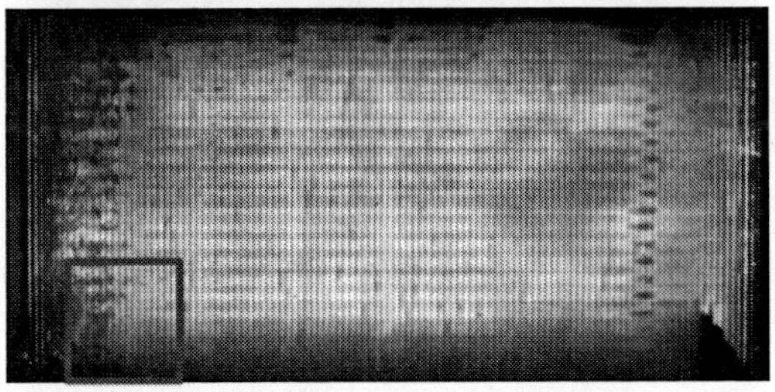

Fig.6b CdTe-module, EL-image "warm condition".

Fig.7 shows a section of the EL-image of the CdTe (Fig.6b, marked with a red rectangle) and the PL-image of the same section.

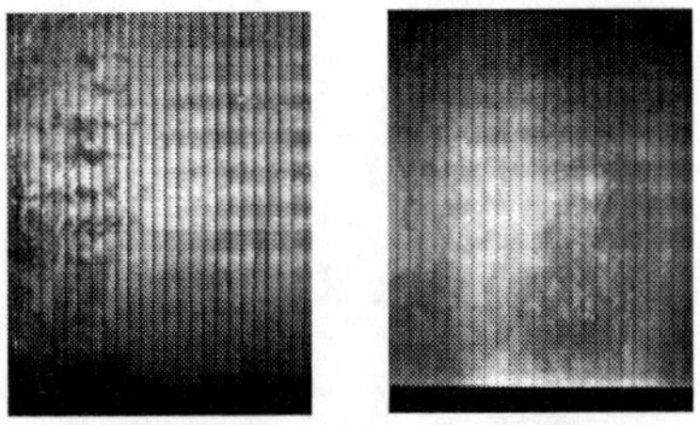

Fig.7 CdTe, left: EL-image ("warm"), right: PL-image.

3.2 Characterization of a CIS module

Photoluminescence and electroluminescence measurements were executed of a defect CIS-module (see EL-image, Fig.8). EL and PL images of one section of a CIS-module were performed and compared. It can be clearly seen that the defects are given in more detail in the PL-images. In the EL-images there are only dark areas with no exact position of the defect. In the PL-image the exact position is determinable.

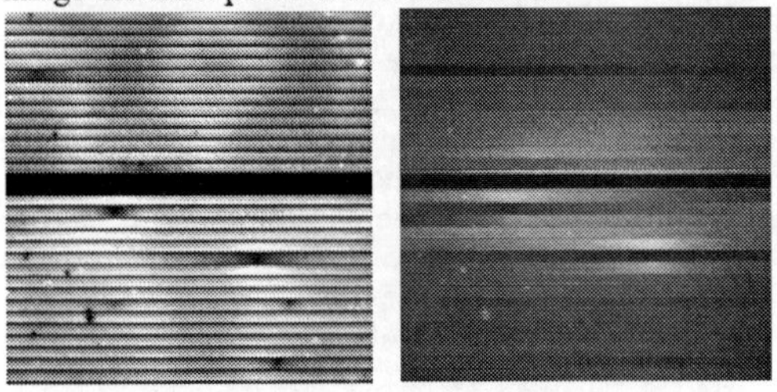

Fig.8 CIS, left: EL-image, right: PL-image.

3.3 Characterization of an aSi-module

Fig.9 shows a section of the EL and the same section of the PL image of an aSi-module without defects.

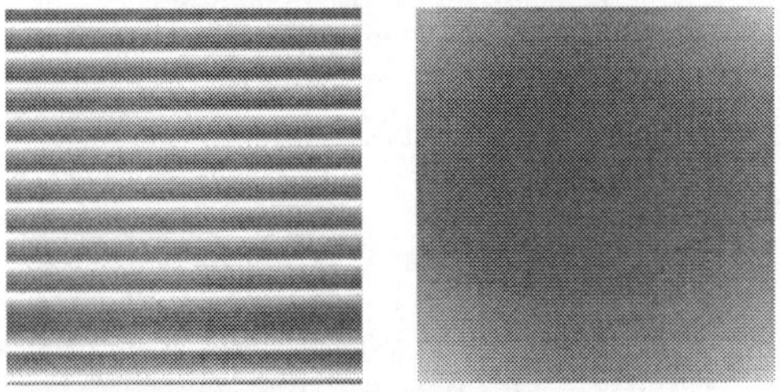

Fig.9 aSi, left: EL-image, right: PL-image.

Fig.10 shows a section of the EL and the same section of the PL-image of an aSi-module with defects. The module was broken after transport. The PL-image clearly gives more information about the material properties and the position and the origin of the defect. It is thus a fast and contactless method to investigate defect thin film modules.

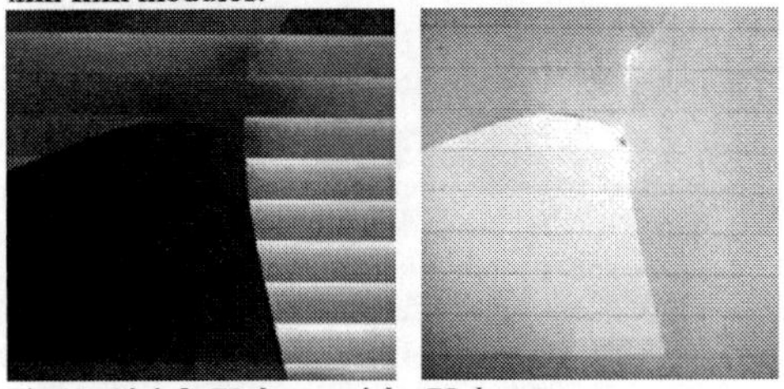

Fig.10 aSi, left: EL-image, right; PL-image.

3.4 Characterization of a mc-solar cell

Fig.11 shows the EL-image and PL-image of a multicrystalline solar cell. In this case the PL-image again clearly gives more information about the material properties and the position of the cracks.

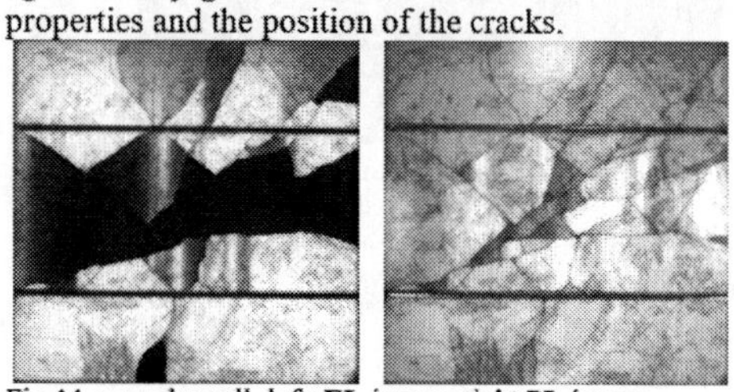

Fig.11 mc-solar cell, left: EL-image, right PL-image.

4 Conclusion

PL measurements can be used to detect defects throughout the entire solar cell manufacturing process. In a previous work PL imaging was used for analyzing solar cells in PV modules and for the investigation of contact properties [5], [6], [7]. PL images help to find out, if material defects or contacting problems are the reason for power losses in solar cells and modules. Compared to EL images, PL imaging has the advantage that contacting failures do not influence the contrast of the PL images. Concerning thin film modules, the investigation with PL helped to identify the origin of many defects.

A further advantage of PL-measurements compared to EL measurements is, that CdTe modules could be measured faster and in "cold condition". EL-measurements of CdTe modules always have to be performed in "warm condition" (specific temperature of the module) in order to get enough information.

Furthermore, it could be demonstrated that PL imaging of solar cells and modules could be carried out succesfully by means of LED excitation instead of hazardous and costly laser excitation.

5 References

[1] T. Trupke. R.A. Bardos, M.D. Abbott, F.W. Chen, J:E. Cotter, A. Lorenz, "Fast photoluminescence imaging of Silicon wafers", IEEE, 4th World Converence on PV Energy Conversion (2006).

[2] A. Lawerenz, K. Lauer, M. Blech, A. Laades, M. Zentgraf, „Photoluminescence lifetime using LED arrays exitation source", 25th PVSEC, Valencia, Spain (2010).

[3] R. Ebner, B. Kubicek, G. Újvári, „Photoluminescenc measurement system for thin film modules", 30th PVSEC, Hamburg, Germany (2015)

[4] R. Ebner, B. Kubicek, G. Újvári, „Photoluminescenc measurement system for thin film modules", PV-Tagung, Villach, Austria (2016)

[5] R. Ebner: "Non-destructive techniques for quality control of PV-modules: infrared thermography, electro- and photoluminescence imaging", IECON 2013, 39th Annual Conference of the IEEE Industrial Electronics Society (2013).

[6] T. Trupke: "Fast photoluminescence imaging of Silicon wafers", IEEE, 4th World Conference on PV Energy Conversion (2006).

[7] R. Ebner, B. Kubicek, G. Újvári, „Zerstörungsfreie-Messverfahren zur Untersuchung von Solarzellen und Modulen: Elektrolumineszenz, Photolumineszenz und Infrarotthermografie", 29. PV-Symposium, Photovoltaische Solarenergie, Bad Staffelstein, Deutschland (2014)

This investigation was performed as part of the FFG Emission Project AMSEL.

COMPARISON OF PRIMARY AND SECONDARY SOLAR CELL CALIBRATIONS

Thomas Fey[1], Ingo Kröger[1], Stefan Winter[1], Tom Betts[2], Willem Zaaiman[3], Diego Pavanello[3], Harald Müllejans[3]
[1]Physikalisch-Technische Bundesanstalt (PTB)
Bundesallee 100, 38116 Braunschweig, Germany
Tel. +49 531 592 4185, Fax: +49 531 592 694185, email: Thomas.Fey@ptb.de
[2]Loughborough University (LU), Centre for Renewable Energy Systems Technology (CREST)
[3] European Commission, Joint Research Centre, Directorate C - Energy, Transport & Climate, Energy Efficiency and Renewables, European Solar Test Installation

ABSTRACT: Results from calibrations to determine the power conversion efficiency of photovoltaic (PV) devices can differ if they are obtained using different calibration methods or different facilities. However, every calibration value has a measurement uncertainty. The consistency of the calibration results around the world is a major task of metrology institutes and calibration laboratories and is ensured by regular comparisons. However, one important parameter of performance measurements of PV devices is the short-circuit current under standard test conditions I_{STC}. Therefore, the I_{STC} of an ensemble of reference solar cells were calibrated at the Joint Research Centre (JRC), at the Physikalisch-Technische Bundesanstalt (PTB) and at Centre for Renewable Energy Systems Technology (CREST) of Loughborough University (LU, United Kingdom). The determined I_{STC} of CREST and PTB are consistent within the uncertainties. Additionally, the spectral responsivities of the reference solar cells were determined at the facilities of JRC and PTB and the results discussed.
Keywords: Calibration, Characterisation, Experimental Methods, Spectral Response

1 INTRODUCTION

Results from measurements to determine the power conversion efficiency of photovoltaic (PV) devices can differ if they are obtained using different calibration methods or different measurement facilities. In general, the comparability of the results is ensured by using the standard test conditions (STC, AM1.5G reference solar spectrum [1], irradiance E = 1000 W m^{-2}, device temperature T = 25 °C). Whenever the measurement conditions differ from the STC, data post-processing [2, 3] has to be performed to correct these differences. Especially using white light calibration methods, the spectral mismatch (SMM) correction has to be applied. For SMM correction the spectral responsivity $s(\lambda)$ of the device under test (DUT) and the reference device (Ref) is needed. Therefore, both, the measurement of the relative spectral responsivity and the measurement of the I_{STC}, are crucial for calibration of PV devices. However, every calibration value has a measurement uncertainty. The consistency of the calibration results taking into account the measurement uncertainties is a major task of metrology institutes and calibration laboratories. Therefore, worldwide comparisons are regularly performed.

The calibration facility of the Physikalisch-Technische Bundesanstalt (PTB) is based on the successful application of the differential spectral responsivity (DSR) method [4] and serves as a qualified World Photovoltaic Scale (WPVS) laboratory [5]. Recently, PTB has developed a multifunctional laser-based DSR-facility for the characterization and primary calibration with lowest uncertainties of the I_{STC} ($U(I_{STC})$ = 0.4 %, k = 2) [6], spectral responsivity and linearity. However, the DSR method uses the combination of a modulated (quasi-) monochromatic beam and a white light bias source. The bias irradiation (E_b) sets the solar cell into the working points. The size of the PV device calibrated using the DSR-method is limited by the signal to noise ratio related to the power of the (quasi-) monochromatic beam. The Centre for Renewable Energy Systems at Technology (CREST, Loughborough University) developed a workaround for determination of the spectral responsivity. CREST uses a polychromatic spectral responsivity (SR) facility to obtain a better signal to noise ratio. The polychromatic spectral responsivity (SR) facility consist of a solar simulator, different optical filters and a spectrometer. Finally, an approximation of the spectral responsivity can be calculated. Additionally, the solar simulator can be used to perform a secondary calibration of the I_{STC} by comparing the DUT with a calibrated reference solar cell.

The Joint Research Centre (JRC), is able to calibrate PV devices using spectral and integral calibration methods. The here shown measurement results are performed with the Direct Sunlight Method (DSM), the "Oriel CSS facility" and the "Wacom facility". The Direct Sunlight Method (DSM) is an integral calibration method for calibration of solar cells (e.g. reference solar cells). Generally, the aim is to determine I_{STC} of the DUT. The DSM realise the illumination with the direct natural sunlight. The Oriel CSS facility is used to measure the spectral responsivity and the Wacom facility to measure the linearity.

Here we show a comparison of I_{STC}, spectral responsivity and linearity of an ensemble of reference solar cells measured using the facilities of JRC, PTB and CREST.

2 INSTRUMENTS AND METHODS

2.1 DSR method

The primary Differential Spectral Responsivity (DSR, IEC 60904-4) method [4, 7] uses the combination of a modulated (quasi-) monochromatic beam and a white light bias source (see Fig. 1). The bias radiation E_b sets the solar cell into working points. The corresponding DC output currents I_{SC} are measured with a DC-multimeter. The modulated (quasi-) monochromatic radiations ΔE_{mono} causes additional modulated output currents ΔI_{mono} and are measured with a lock-in amplifier. The DC and modulated output currents are used to perform precise determinations of the spectral responsivity $s(\lambda)$, the linearity of the solar cell and the I_{STC}.

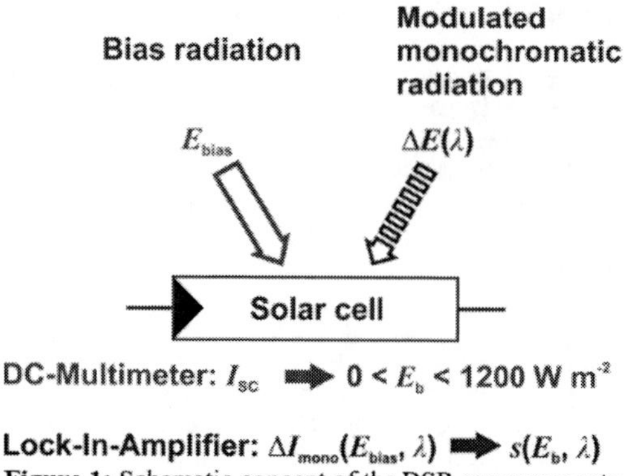

Figure 1: Schematic concept of the DSR measurements

2.2 Direct Sunlight-Method

The Direct Sunlight-Method (DSM, IEC 60904-4) is a precise integral calibration method for calibration of the short circuit current under standard test conditions I_{STC} of solar cells. The DSM realizes the illumination of the reference solar cells with direct natural sunlight. The corresponding output currents are measured using DC-multimeters, the spectra of the sunlight are measured using an EKO Wiser system, the irradiances are measured using absolute cavity radiometers (directly traceable to WRR-PMOD - Davos, Switzerland) and the temperature with an integrated temperature sensors. However, the direct natural sunlight doesn't match the reference spectrum AM1.5G. Therefore, a spectral mismatch correction (IEC 60904-7) have to be applied. The spectral mismatch correction requires the spectral responsivity of the solar cell in addition to the measured direct natural sunlight spectrum. The spectral responsivity $s(\lambda)$ was measured using the Oriel CSS (SR2) system at JRC.

2.3 Polychromatic Spectral Responsivity-Method

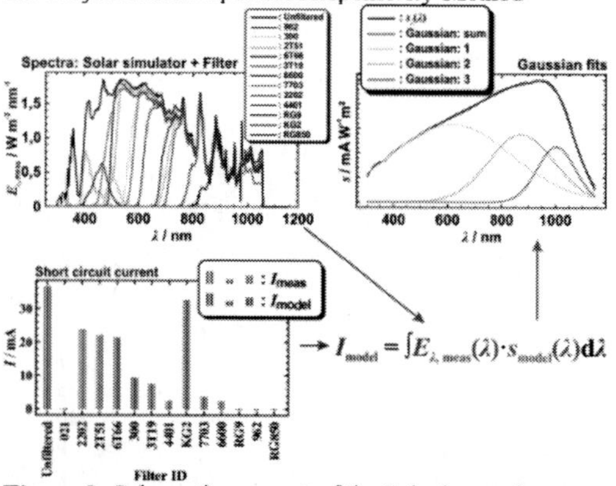

Figure 2: Schematic concept of the Polychromatic Spectral Responsivity-Method

CREST uses a PASAN 3b solar simulator for solar cell calibrations of e.g. the I_{STC}. Furthermore, the spectrum of the solar simulator can be changed by polychromatic filters. The polychromatic spectra $E_{\lambda,meas}$ and the related currents I_{meas} are used to optimize a spectral responsivity model (see Fig. 2). For example, a fifth order Gaussian distribution function can be chosen as a simulated spectral responsivity $s_{model}(\lambda)$ [8, 9]. This simulated spectral responsivity and the measured spectra are used to calculate estimated short circuit currents I_{model}. The calculated and the corresponding measured currents can be compared. Finally, the parameters of the Gaussian distributions have to be optimized with an iterative fitting algorithm to minimize the difference between the calculated and the corresponding measured currents.

3 RESULTS

A set of two unfiltered and one filtered mono c-Si reference solar cells were calibrated, using the Differential Spectral Responsivity (DSR) method at PTB, the Direct Sunlight Method (DSM) at JRC and the polychromatic spectral responsivity method at CREST. The List of calibrated reference solar cells is summarized in Table I.

Table I: List of used reference solar cells

	Technology	Size	Type
Sample #1	c-Si	4 cm^2	unfiltered, encapsulated
Sample #2	c-Si	4 cm^2	unfiltered, encapsulated
Sample #3	c-Si	4 cm^2	NIR filtered, encapsulated

3.1 Linearity

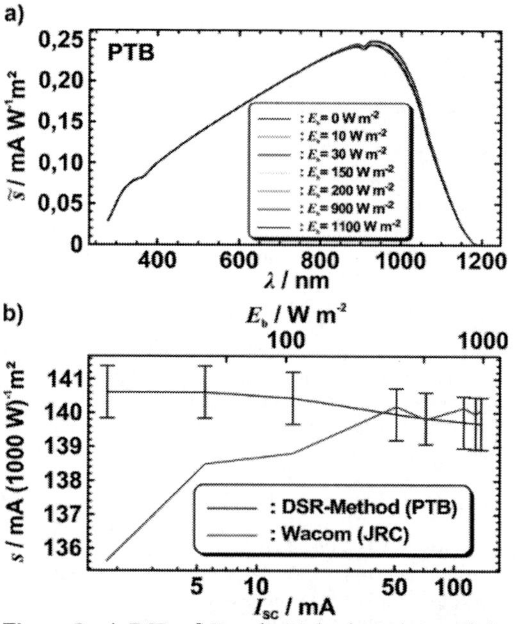

Figure 3: a) DSR of Sample #2 in dependence of the irradiance. b) Linearity of Sample #2 in the irradiance range between 0 W m^{-2}and 1000 W m^{-2}

The linearity of the solar cells was calculated using the DSR method at PTB and measured using the WACOM facility at JRC. The preliminary results of Sample #2 are shown in Fig. 3.

The irradiance varies in the range between 0 W m^{-2} and 1000 W m^{-2}. The linearity within the uncertainties agree above 500 W m^{-2}. The reason for the disagreement below 500 W m^{-2} needs to be investigated.

3.2 Spectral Responsivity

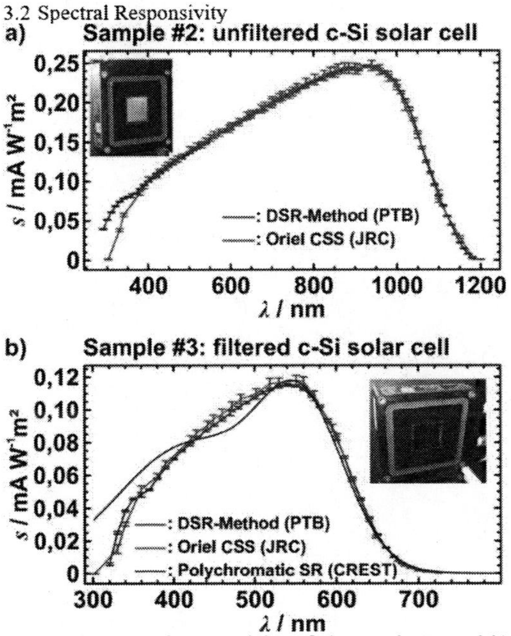

Figure 4: Spectral responsivity of a) Sample #2 and b) Sample #3 measured at JRC, CREST and PTB.

The spectral responsivity was determined using the DSR method at PTB, the Oriel (SR2) facility at JRC and approximated using the polychromatic method at CREST. The preliminary results of Sample #2 and #3 are shown in Fig. 4.

The results of the unfiltered c-Si reference solar cell Sample #2 (see Fig. 4 a)) at JRC and PTB agree in the wavelength range above 400 nm. A disagreement below 400 nm can be observed. Furthermore, an approximation of the spectral responsivity was determined for the filtered Sample #3 (see Fig. 4b)) using the polychromatic spectral responsivity method at CREST. The polychromatic method achieved a good approximation above 550 nm.

3.3 Short-circuit current under standard test conditions

The short-circuit currents under standard test conditions I_{STC} of the ensemble of reference solar cells was calibrated at PTB using the DSR method and at CREST using the PASAN III b solar simulator.

The I_{STC}, CREST is measured using the unfiltered Pasan III b and a calibrated unfiltered c-Si reference solar sell. With respect to the limitation of the spectrometer (See Fig. 2), no spectral mismatch corrections were applied. Furthermore, we calculate the En-number (ISO 17043) as a conformity check using eq. (1). The results are summarized in Table II.

$$En_i = \frac{\left|I_{STC,LU} - I_{STC,PTB}\right|}{\sqrt{\left(U\left(I_{STC,LU}\right)\right)^2 + \left(U\left(I_{STC,PTB}\right)\right)^2}} \qquad (1)$$

Table II: The short-circuit currents under standard test conditions I_{STC} for the ensemble of reference solar cells. The measurements were performed at PTB and at CREST. Additionally, the En-numbers were calculated using equation (1).

	$I_{STC,PTB}$ / mA	$I_{STC,CREST}$ / mA	En-n.
Sample #1	141.11 ± 0.53 %	141.2 ± 2.8 %	0.03
Sample #2	139.73 ± 0.53 %	140.3 ± 2.7 %	0.15
Sample #3	35.70 ± 0.56 %	36.3 ± 2.7 %	0.59

4 SUMMARY

An ensemble of three reference solar cells was characterized (linearity, spectral responsivity $s(\lambda)$ and short-circuit current under standard test conditions I_{STC}) at JRC, CREST and PTB using the Direct Sunlight-Method (JRC), Oriel CSS (SR2) facility (JRC), the Differential Spectral Responsivity-Method (PTB) and a Solar simulator (CREST).

The differential spectral responsivities were measured using the DSR method at PTB and used to calculate the AM1.5 weighted differential spectral responsivity. Sample #2 showed the biggest nonlinearity of 0.6 % within the measured irradiance range. The determined linearity within the uncertainties at JRC and PTB agree above 500 W m^{-2}. The reason for the disagreement below 500 W m^{-2} needs to be investigated.

The measurements of the spectral responsivity at JRC and PTB shows agreements in the wavelength range above 400 nm within the uncertainties and a discrepancy in the wavelength range below 400 nm. The reason for this deviation needs to be investigated. The determined spectral responsivity for the filtered Sample #3 using polychromatic spectral responsivity measurement achieve a good approximation above 550 nm.

The I_{STC} of the ensemble were calibrated at PTB, CREST and JRC. The data analysis at JRC is currently in progress. The determined I_{STC} by PTB and CREST were used to calculate the En-numbers. The En-numbers are smaller than 1 and demonstrate the very good agreement of the calibration values and reasonable stated uncertainties.

5 ACKNOWLEDGEMENTS

The research work leading to this article was carried out within the EMRP ENG55 project "Towards an energy-based parameter for photovoltaic classification - PhotoClass". The EMRP is jointly funded by the EMRP participating countries within EURAMET and the European Union.

6 LITERATURE

[1] IEC 60904-3 Edition 2, International Electrotechnical Commission, Measurement principles for terrestrial photovoltaic (PV) solar devices with reference spectral irradiance data, Geneva, 2008.
[2] IEC 60891 Edition 2, International Electrotechnical Commission, Procedures for temperature and irradiance corrections to measured I-V characteristics, Geneva, 2009.
[3] IEC 60904-7 Edition 3, International Electrotechnical

Commission, Computation of the spectral mismatch correction for measurements of photovoltaic devices, Geneva, 2008.
[4] Metzdorf, J. (1987, May). Calibration of solar cells. I. The differential spectral responsivity method. Applied Optics 26, 1701-1708
[5] Osterwald C.R. et al. (1999). The Results of the PEP'93 Intercomparison of Reference Cell Calibrations and Newer Technology Performance Measurements: Final Report, National Renewable Energy Laboratory Tech. Rep. NREL/TP-520-23477
[6] Winter, S. et al. (2014). Design, realization and uncertainty analysis of a laser-based primary calibration facility for solar cells at PTB. Measurement 51, 457-463
[7] S. Winter, T. Wittchen, J. Metzdorf: Primary Reference Cell Calibration at the PTB Based on an Improved DSR Facility; in "Proc. 16th European Photovoltaic Solar Energy Conf.", ed. by H. Scherr, B. Mc/Velis, E. Palz, H. A. Ossenbrink, E. Dunlop, P. Helm; Glasgow, (2000) James & James (Science Publ., London), ISBN 1 902916 19 0
[8] Sara, I. D., Polychromatic determination of spectral response of PV devices, doctoral dissertations, Loughborough University: Doctoral Thesis, 2013.
[9] Alhusna, H. et al. "Spectral Response Measurements of Photovoltaic Devices using a Pulsed Source Solar Simulator," in PHOTOVOLTAIC SCIENCE APPLICATIONS AND TECHNOLOGY 11, University of Leeds, Leeds, UK, 2015.

FEASIBILITY STUDY FOR PV MEASUREMENTS AT VARYING IRRADIANCES ON A LARGE-AREA STEADY-STATE SOLAR SIMULATOR

Elena Salis*, Ivan Sharlandzhiev, Michael Field
European Commission, DG Joint Research Centre, European Solar Test Installation
v. E. Fermi 2749, I-21027 Ispra (VA), Italy
*Tel: +39 0332 785 467, Fax: +39 0332 789 992, e-mail: elena.salis@ec.europa.eu

ABSTRACT: Reliable prediction of photovoltaic (PV) module performance under real operating conditions requires several steps beyond the current comparison at Standard Test Conditions (STC). The standard IEC 61853-1 establishes the requirements for power matrix measurements, which involve variation of the module's temperature as well as of the irradiance level with respect to STC. The large-area steady-state solar simulator (LACSS) of the European Solar Test Installation (ESTI) can be used for PV device characterisation at conditions other than STC, e.g. power matrix and linearity measurements. For both cases, the irradiance level should be varied without significantly affecting the spectral content and the spatial non-uniformity (NU) on the test plane. At ESTI this can be achieved by an appropriate selection of the LACSS lamps. However, as this procedure implies a possible change in the NU, a better characterisation of the solar simulator and an automation tool for irradiance prediction were deemed necessary to increase confidence in the process. Initial comparison between predictions and measurements demonstrated satisfactory agreement typically within ±1% for irradiance distribution and NU. Also, the result of the solar simulator assessment is positive for advanced PV measurements, although care must be taken in the selection of lamp combinations and in their constant update.
Keywords: Characterisation, Energy Rating, Irradiance Non-uniformity, Performance, Solar Simulator

1 INTRODUCTION

Reliable prediction of photovoltaic (PV) module performance under real operating conditions needs several steps beyond the measurements at Standard Test Conditions (STC) as required by the standard IEC 60904-1 [1]. These additional steps are the subject of the standard series IEC 61853 concerning energy rating, where the standard IEC 61853-1 [2] deals with power matrix measurements. In order to perform such measurements, one should vary the module's temperature as well as the irradiance level on the test plane over the relevant ranges (see for example [3]). While varying the temperature of a PV device may be done without much difficulty at a steady-state solar simulator, the variation of the irradiance level without significantly affecting the spectral and spatial distribution of the light may be tricky. Additionally, a variation of the irradiance level with respect to STC implies the need to reliably measure it with a reference device that, as requested by the standard IEC 60904-4 [4], has to be linear according to the standard IEC 60904-10 [5].

Among the facilities available at the European Solar Test Installation (ESTI) of the Joint Research Centre (JRC), the large-area steady-state solar simulator Apollo has in principle the capability to be used for a wider range of PV device characterisations than just STC measurements; some of them are power matrix and linearity measurements. In both cases, the irradiance level should be changed without significantly affecting spectral content and the spatial non-uniformity (NU) on the test plane.

As the Apollo solar simulator consists of eleven xenon lamps that can be controlled and shuttered individually, a method to vary the irradiance on the test plane is by selection of the necessary lamps to achieve the required irradiance level by superposition of their light beams, each lamp providing approximately 100 W/m² at full lamp power.

We present here the first results of a feasibility study that was run to check firstly whether the NU requirements can be met for all the relevant irradiance levels and, secondly, to find the combination of lamps that would achieve the lowest spatial NU on the test plane.

2 EXPERIMENTAL SETUP

2.1 The Apollo Solar Simulator

The solar simulator characterised in this study is a multi-source large-area steady-state solar simulator and from now on it will be referred to as Apollo solar simulator or simply Apollo.

The Apollo solar simulator consists of eleven xenon lamps that can be individually controlled both in power and shuttering. Each of them produces about 100 W/m² on the test plane, for a total achievable irradiance of about 1050 W/m² at full lamp power. Figure 1 shows it with all the lamps on and open. The test plane for which the Apollo is a class AAA solar simulator according to the standard IEC 60904-9 [6] covers an area of 2 m × 2 m at about 8 m distance from the light sources.

Figure 1. A photo showing the eleven lamps of the Apollo solar simulator with their shutter open.

The standard target irradiance of 1000 W/m^2 on the test plane is obtained by superposition of the eleven lamp beams, each of which is differently aligned towards the plane. Every lamp has two sets of multi-lens optics that distribute its light on the entire test plane. However, due to the different optical alignment between each lamp and the test plane itself, every light source of the Apollo produces a specific irradiance spatial distribution. For PV measurements involving all the eleven lamps, the class A conformity to the relevant IEC standard [6] for the irradiance spatial distribution is assured by periodic maintenance measurements of the Apollo performance. Typical value of its spatial non-uniformity (NU) at 1000 W/m^2 is 1.2% over the entire test plane.

In order to use the Apollo for irradiances lower than 1000 W/m^2, there are two main options. The first is to lower the power to the lamps until the irradiance level is achieved. This may be done as far as the power reaches about 60% of the total, because at power lower than this the xenon discharge inside the lamp's bulb is not stable anymore. This option has the advantage that the spatial uniformity on the test plane is in principle not altered significantly, as no optics is modified in this procedure and only negligible variations in the spatial distribution due to change of spectral content might be expected. However, it is exactly this change in the spectral distribution – due to the reduced electrical power to the lamp – that represents a major negative effect. This has to be avoided as much as possible for both power matrix and linearity measurements, whenever it is not possible to properly deal with it by spectral mismatch calculation.

The second option to achieve lower irradiances on the Apollo test plane consists in shuttering some of the lamps or even to filter the light by the interposition of appropriate neutral mesh filters, according to the target irradiance level to be achieved. This method does not alter significantly the spectral content of the light, as all lamps have very similar spectrum. It may instead alter the spatial distribution of the light on the test plane, depending on the optics of the single lamps and on which lamps are combined to obtain a certain irradiance level. Therefore, an extended characterisation of the solar simulator is necessary in order to establish whether the unavoidable change in spatial distribution may be acceptable or not on the basis of the requirements set in the standards IEC 60904-9 [6] and IEC 60904-10 [5]. Additionally, this characterisation can give important input to the calculation of the overall uncertainty for all those measurements where the irradiance level is changed by the method discussed here.

3 METHODOLOGY

3.1 Measurement methodology

The spatial irradiance non-uniformity of a solar simulator has to be measured according to one of the methods described in the standard IEC 60904-9 [6]. As the temporal stability of the Apollo was not an issue, we chose to use only one calibrated large cell moved across the test plane, which has been divided in 10 × 10 squares of 20 cm × 20 cm area each. The cell is an encapsulated crystalline-silicon shunted reference cell with active area of 15 cm × 15 cm (ESTI code: PX702C). It is masked so that only the active area is exposed to light and it is annually calibrated at ESTI. Additionally, the cell had been previously verified to be linear according to the standard IEC 60904-10 [5] in the irradiance range between 50 W/m^2 and 1000 W/m^2 (unpublished data).

In this way, the spatial non-uniformity matrix (obtained by acquiring the cell's reading at the 100 positions on the test plane) could be straightforwardly transformed into an irradiance matrix describing the test plane for each specific lamp selection, thus giving an irradiance value associated to each of the 20 cm × 20 cm squares and based on the following applicable equation:

$$G_i = 1000 \times I_i / I_{STC} \qquad \text{Eq. 1}$$

where G_i is the calculated irradiance at a specific position *i* for a specific lamp selection, I_i is the short-circuit current of the reference cell measured at that position for that selection, and I_{STC} is the calibration value of the reference cell at STC.

A rigid structure, which covered about two thirds of the height and the full width of the test plane, was mounted on the support that is usually used to hold the PV modules to be measured. As this support can be vertically adjusted, the structure for non-uniformity measurements could be shifted up or down to cover the entire test plane. The reference cell was mounted on a holder that could slide horizontally along the rigid structure to each of the 10 required positions. The reference cell output was measured by a calibrated voltmeter and acquired by a PC through a Labview program.

As said, the purpose of this feasibility study was to characterise the individual lamps and to validate a tool to easily estimate *a priori* the irradiance spatial distribution (and so the spatial non-uniformity) for any selection of the Apollo lamps as well as the best lamp combination for a given irradiance.

Therefore, we measured the irradiance spatial distribution produced by each lamp and by a representative number of lamp combinations. The single-lamp measurements were performed twice in order to have some statistics, as one complete set (11 lamps) of measurements could not be done in a single day. Although, when a lamp combination required only two or three lamps we performed those measurements, including the combination, within the same day. The non-uniformity of the irradiance on the test plane is calculated according to the standard IEC 60904-9 [6] with the following equation:

$$NU = (max_{IRR} - min_{IRR})/(max_{IRR} + min_{IRR}) \qquad \text{Eq. 2}$$

where max_{IRR} and min_{IRR} are, respectively, the maximum and the minimum irradiance values measured on the test plane. NU is given as percentage.

3.2 The predictive software

As we dealt with irradiance matrices, and not just with non-uniformity matrices consisting in deviation values from a reference point, it was possible to consider them as physical and not just as mathematical entities and to apply a simplified superposition approach. In other words, we assumed that for each position of the test plane we could simply sum the irradiance values as measured for single lamps in order to estimate the irradiance value achievable when a selection of those lamps would be used.

A Visual Basic script was developed in order to combine the single-lamp measurements for any selection of lamps and to find the optimum lamp combination for a given irradiance level on the basis of the calculated expected non-uniformity values.

The script used the eleven single-lamp irradiance matrices to generate a new matrix of irradiance spatial distribution on the basis of the lamps selected by the user through a graphic interface. From such new matrix, the NU value associated to it was calculated, too. In addition, the script was programmed to evaluate all the lamp combinations that would achieve a desired irradiance level within a given range and to rank the results by smallest non-uniformity value. This additional option would be very useful at mid and low irradiance, when few lamps are necessary but at the same time the non-uniformity worsens with possible significant differences in the irradiance spatial distribution depending on the lamp selection.

A sequential approach was used to validate the script. As first step, the correct combination of single-lamp measurements into a new irradiance matrix was tested for a selection of lamp matrices and successively compared to an actual measurement performed with the same selection of lamps. This was done for some lamp combinations. Once the correctness of the mathematical combination of the single-lamp matrices was assured and a satisfying consistency of the calculated output with actual measurements verified, the optimisation script was evaluated. In order to do this, we calculated all the combinations of lamps that would produce a certain irradiance level within a reasonable range. As we were dealing with irradiance matrices, all the positions in the calculated matrix should show a value within the selected range in order to be included in the ranking.

4 RESULTS

Before opting for the single-cell method, we had checked that no significant negative influence was introduced by the fact that the active area of the cell was slightly narrower than the considered square area on the test plane. The measurements taken by positioning the cell within the target area had the effect of slightly smoothing the irradiance curve that would describe the continuous space when the cell was moved across the test plane on consecutive, and not discrete, positions.

4.1 Single-lamp measurements

Figure 2 reports an example matrix of irradiance spatial distribution as measured for one of the Apollo lamps. The values are given in W/m².

As expected, the spatial non-uniformity produced by a single lamp is not as good as with all eleven lamps illuminating the test plane (see Figure 3). Indeed, the NU value on the full test plane is about 3.1% (class B in [6]) in the case shown in Figure 2, while it is 1.2% with all the lamps (Figure 3). However, for PV devices that are smaller than the test plane a better NU value is calculated for the same lamp; it is about 2.5% (class B according to [6]) on an area of 1.6 m × 1.6 m concentric to the first and 1.6% (class A according to [6]) on a concentric area of 1.2 m × 1.2 m.

The colour map applied to these two matrices highlights the deviations around the mean value (identified by the white colour) of the whole matrix itself and between the minimum (blue colour) and the maximum value (red colour). As these three values differ for the two matrices shown in Figure 2 and Figure 3 and they do not even correspond to the same position of the test plane, in this case the colour map is only used as a simple way to visualise the irradiance variations within the test plane and do not aim at comparing the two matrices.

The case displayed in Figure 2 is the best NU result (considering all the three concentric areas) among the single-lamp measurements performed in January 2017. The power to the lamp was in this case about 95%. These measurements have also been used to give indications for spatial distribution improvements during the recent periodic maintenance of the Apollo solar simulator, and a comparison with the new spatial distributions will be possible when the new measurement campaign will be completed.

The single-lamp measurements were performed twice in order to build some statistical data for the solar simulator. In this paper, though, we refer mainly to the first set of measurements with only few mentions to some comparison with the second data set. More extended information will be published when sufficient data collection and analysis will be available.

91.4	92.2	93.7	94.2	92.3	91.8	91.6	90.6	89.2	89.6
91.4	91.5	93.0	92.7	92.3	91.7	90.7	91.0	89.7	89.1
91.4	91.0	91.8	91.1	91.8	91.9	90.5	91.4	90.6	89.6
91.0	90.3	90.8	90.2	90.8	91.1	90.1	91.2	90.6	89.4
91.0	90.0	90.9	90.3	90.7	91.8	90.9	91.5	90.8	89.5
91.3	90.1	90.1	90.1	90.6	92.4	90.9	91.3	90.5	89.3
91.8	91.5	90.5	90.7	90.5	91.7	91.3	90.8	90.0	89.1
92.8	92.7	93.0	91.1	90.0	91.4	90.7	91.2	90.7	89.7
92.0	90.3	89.3	88.7	88.5	90.5	89.1	88.7	89.0	89.8
91.5	93.5	93.5	92.4	91.6	92.2	91.0	90.9	91.1	91.8

Figure 2. Example of irradiance matrix measured for one lamp of the Apollo. The values are in W/m². NU value for the whole test plane is 3.1%

4.2 Lamp combinations

Some comparisons were done between the software prediction, based on the single-lamp measurements, and the actual measurement of the irradiance spatial distribution for the selected lamp combination. The number of lamps involved was between 2 and 11. The deviations between the combined irradiance, estimated on the basis of one measurement for each lamp, and the lamp combination as measured are always of the order of -1% when all eleven lamps are considered, thus introducing a small but systematic underestimation of the irradiance distribution. They are well within ±1% for each of the 100 positions of the test plane when only a selection of lamps is considered, with two cases of the 14 considered predictions within (-0.5% , +1.6%) and only one case that reached (+0.8% , +2.5%) range. In particular, the latter and one of the other two negative results were connected to temporary issues with one of the three lamps selected for this particular combination.

1028.7	1028.7	1027.2	1022.5	1026.4	1025.7	1026.6	1024.7	1020.5	1024.7
1028.0	1030.1	1029.1	1020.0	1025.5	1026.0	1026.0	1024.4	1020.0	1016.1
1019.9	1020.6	1019.5	1016.4	1018.2	1023.8	1022.1	1026.2	1020.8	1013.0
1015.6	1017.0	1018.6	1016.4	1017.0	1015.1	1017.9	1021.5	1014.2	1006.2
1017.8	1017.0	1018.2	1014.8	1015.4	1014.0	1016.3	1023.1	1019.0	1010.8
1022.5	1019.6	1018.8	1016.3	1018.3	1015.3	1014.9	1022.0	1019.9	1011.9
1017.4	1015.6	1011.1	1009.5	1018.6	1013.7	1016.6	1020.6	1022.9	1015.1
1018.6	1015.6	1013.8	1011.6	1018.5	1013.1	1018.3	1021.7	1022.7	1019.4
1010.2	1010.2	1007.6	1007.0	1007.8	1009.3	1010.1	1016.3	1014.3	1017.9
1009.1	1017.6	1014.3	1012.1	1016.9	1020.6	1011.4	1013.3	1013.3	1014.7

Figure 3. Measured irradiance distribution on the Apollo test plane with all eleven lamps contributing to it. The values are in W/m^2. NU value is 1.2% on the whole test plane and 0.82% in the central area of 1.2 m × 1.2 m.

When the average irradiance matrix is estimated from the pair of measurements performed for each lamp, and then compared to the actual measurement for the corresponding selection of lamps, all estimates are well within ±1% with the exception of the average related to the problematic combination that is within -0.3% and +1.3%.

For example, Figure 4 shows the irradiance spatial distribution matrix measured for the combination of lamps 5 and 7. The colour shades are centred around the average irradiance value of the overall matrix (which represents the test plane), and bounded by the minimum and the maximum values. So, the darkest red cell shows the maximum value of the measured irradiance and the darkest blue the minimum one. The values for NU are 2.2% if the entire test plane is considered and 1.8% for both the concentric areas. The former corresponds to class B and the latter to class A for spatial non-uniformity according to IEC 60904-9 [6]. The average irradiance of the matrix shown in Figure 4 is 176.7 W/m^2 (at about 95% lamp power).

Figure 5 is the corresponding estimated irradiance matrix, as calculated by the software on the basis of the single-lamp measurements for lamps 5 and 7, which were performed on the same day of the 5+7 combination measurement. The average irradiance is 177.0 W/m^2.

A look at the colour map shows some differences, which are driven by the individual values as well as by the average, minimum and maximum values of the two matrices. Even though, the main features of the irradiance spatial distribution are kept and can be used to estimate the suitability of a lamp selection compared to another one. Figure 6 displays the individual deviations between the measured and the predicted irradiance matrices shown in Figure 4 and Figure 5, which are all within ±1%.

This verification allows us to use the script in order to roughly estimate the average irradiance and the NU value that we can expect for any combination of the Apollo lamps.

177.4	177.9	177.0	177.4	179.3	179.4	177.6	179.2	177.0	176.6
175.7	177.3	176.6	177.0	177.1	177.2	176.7	178.6	176.7	175.2
174.7	176.5	175.8	176.4	178.0	177.0	175.6	178.6	176.6	173.6
174.3	174.2	174.1	175.6	177.7	176.9	174.9	178.2	176.1	172.7
174.2	175.3	176.0	176.6	178.2	176.6	175.6	178.8	176.4	173.8
175.2	176.1	176.6	177.4	179.2	176.6	175.7	179.5	176.9	174.2
176.7	176.6	177.0	176.5	180.2	178.2	176.3	179.3	178.3	175.1
177.8	177.7	175.8	177.6	180.3	178.6	176.5	179.2	178.9	176.1
177.4	176.9	177.2	176.5	175.2	176.1	176.2	179.2	175.4	175.1
176.2	175.9	174.7	174.5	174.3	176.8	176.9	179.1	177.0	175.2

Figure 4. Measured irradiance matrix for lamp combination 5+7. The values are in W/m^2. NU value is 2.16% on the entire test plane, and 1.75% on both the concentric areas.

177.8	178.4	177.5	177.8	179.5	179.1	177.7	178.4	177.9	176.9
175.9	177.2	176.7	177.1	178.3	178.3	176.5	178.8	177.1	173.8
174.8	176.6	176.0	176.4	178.4	177.4	175.5	178.7	176.5	172.4
173.9	175.2	175.5	176.0	178.1	177.5	175.1	178.2	176.2	173.0
175.2	176.0	176.7	177.5	178.8	177.2	176.1	178.8	175.7	174.4
175.8	176.5	177.1	178.2	180.4	177.9	176.6	180.4	177.7	175.1
175.8	175.2	177.3	176.6	180.3	179.1	177.0	179.7	178.7	175.7
177.7	178.8	177.5	176.2	179.0	178.6	176.5	179.2	178.7	176.7
177.0	176.4	176.7	176.0	174.9	176.3	176.6	179.0	175.5	175.2
175.4	175.6	174.6	174.5	174.7	177.7	177.4	179.6	177.6	175.7

Figure 5. Estimated irradiance matrix based on the first single-lamp measurement for lamps 5 and 7, which were performed immediately before the lamp combination. The values are in W/m^2.

0.2%	0.3%	0.3%	0.2%	0.1%	-0.2%	0.0%	-0.4%	0.5%	0.2%
0.1%	0.0%	0.1%	0.1%	0.7%	0.6%	-0.2%	0.1%	0.2%	-0.8%
0.0%	0.1%	0.1%	0.0%	0.3%	0.2%	-0.1%	0.1%	-0.1%	-0.7%
-0.2%	0.6%	0.8%	0.2%	0.2%	0.4%	0.1%	0.0%	0.1%	0.2%
0.6%	0.4%	0.4%	0.5%	0.4%	0.3%	0.3%	0.0%	-0.4%	0.4%
0.3%	0.3%	0.3%	0.4%	0.6%	0.7%	0.5%	0.5%	0.4%	0.5%
-0.5%	-0.8%	0.1%	0.0%	0.1%	0.5%	0.4%	0.3%	0.3%	0.4%
-0.1%	0.6%	1.0%	-0.8%	-0.7%	0.0%	0.0%	0.0%	-0.1%	0.3%
-0.2%	-0.3%	-0.3%	-0.3%	-0.2%	0.1%	0.3%	-0.1%	0.1%	0.0%
-0.4%	-0.2%	-0.1%	0.0%	0.2%	0.5%	0.3%	0.3%	0.3%	0.3%

Figure 6. Percentage deviations of the estimated values of Figure 5 versus those of Figure 4.

4.3 Optimisation algorithm

An optimisation algorithm was included in the Visual Basic script in order to estimate which lamp combination would achieve the lowest NU value on the test plane for an irradiance level within a range specified by the user. As for the combination algorithm, this optimisation process is based on actual single-lamp measurements, which have to be kept updated with the periodic maintenance of the solar simulator.

Table I and Table II show two examples of lamp combinations that could be selected to produce, respectively, an irradiance level of 600 W/m^2 and 900 W/m^2 on the test plane. The list reported in these tables is limited to the first ten combinations only, but the algorithm calculates and evaluates all the possible combinations for the specified irradiance range and gives as output the first thirty ones. In this way, it is possible to evaluate *a priori* the combination or combinations of lamps that would be the best choice for PV measurements at a specified irradiance.

Table I. Ranked list of lamp combinations that could be used to achieve a target irradiance of 600 W/m^2. The ordering criterion is based on NU value.

Target irradiance: 600 W/m^2		
Ranked list	**Lamp selection**	**NU (on 2m×2m)**
1	2, 4, 5, 6, 7, 9, 11	1.359%
2	1, 2, 4, 5, 7, 9, 11	1.362%
3	4, 5, 6, 7, 9, 10, 11	1.454%
4	1, 2, 5, 7, 8, 9, 11	1.458%
5	1, 4, 5, 6, 7, 9, 11	1.469%
6	1, 2, 4, 5, 9, 10, 11	1.488%
7	1, 5, 6, 7, 8, 10, 11	1.501%
8	1, 2, 4, 5, 7, 10, 11	1.501%
9	1, 2, 4, 5, 6, 9, 11	1.504%
10	2, 4, 5, 7, 8, 9, 11	1.516%

Table II. Ranked list of lamp combinations that could be used to achieve a target irradiance of 900 W/m^2. The ordering criterion is based on NU value.

Target irradiance: 900 W/m^2		
Ranked list	**Lamp selection**	**NU (on 2m×2m)**
1	1, 2, 4, 5, 6, 7, 8, 9, 10, 11	1.372%
2	1, 2, 3, 5, 6, 7, 8, 9, 10, 11	1.507%
3	1, 2, 3, 4, 5, 6, 7, 9, 10, 11	1.521%
4	1, 2, 3, 4, 6, 7, 8, 9, 10, 11	1.556%
5	1, 3, 4, 5, 6, 7, 8, 9, 10, 11	1.559%
6	1, 2, 3, 4, 5, 6, 7, 8, 9, 10	1.567%
7	1, 2, 3, 4, 5, 7, 8, 9, 10, 11	1.569%
8	1, 2, 3, 4, 5, 6, 7, 8, 9, 11	1.609%
9	2, 3, 4, 5, 6, 7, 8, 9, 10, 11	1.677%
10	1, 2, 3, 4, 5, 6, 8, 9, 10, 11	1.742%

7 CONCLUSIONS

A feasibility study has been initiated to assess the suitability of the Apollo solar simulator for PV performance measurements at irradiance level other than the standard 1000 W/m^2. Some of the first results of the irradiance spatial distribution as measured for individual lamps as well as for combinations of lamps are shown. The latter are compared to the estimate calculated by a Visual Basic script on the basis of the single-lamp measurements and an acceptable agreement, typically within ±1%, is found between them. This agreement allows to exploit the script to predict the irradiance distribution given by any lamp combination and consequently its corresponding NU value, without the crucial need of measuring every single combination of lamps for every required irradiance level.

The optimisation algorithm is an additional step that gives the possibility of selecting the best lamp combination for a given irradiance range on the basis of the best NU value. Additional measurements of the irradiance spatial distribution produced by other selected lamp combinations will allow us to increase the confidence in such predictions, as for the moment only a few cases have been checked by comparison to real measurements.

The study here presented highlights also that careful choice of the lamp selection has to be made especially at the lowest irradiances and depending on the size of the PV device to be measured.

8 ACKNOWLEDGEMENTS

The authors thank H. Müllejans and E. Dunlop from ESTI for scientific discussion and support.

The research work leading to this article was partly carried out within the EMRP ENG55 project "Towards an energy-based parameter for photovoltaic classification". The EMRP is jointly funded by the EMRP participating countries within EURAMET and the European Union.

9 REFERENCES

[1] IEC 60904-1 Photovoltaic devices - Part 1: Measurement of photovoltaic current-voltage characteristics. (Geneva: International Electrotechnical Commission)

[2] IEC 61853-1. Photovoltaic (PV) module performance testing and energy rating - Part 1: Irradiance and temperature performance measurements and power rating. (Geneva: International Electrotechnical Commission)

[3] Kenny R P, Viganó D, Salis E, Bardizza G, Norton M, Müllejans H and Zaaiman W 2013 Power rating of photovoltaic modules including validation of procedures to implement IEC 61853-1 on solar simulators and under natural sunlight *Prog Photovoltaics Res Appl* **21** 1384-99

[4] IEC 60904-4. Photovoltaic devices - Part 4: Reference solar devices - Procedures for establishing calibration traceability. (Geneva: International Electrotechnical Commission)

[5] IEC 60904-10 Photovoltaic devices - Part 10: Methods of linearity measurement. (Geneva: International Electrotechnical Commission)
[6] IEC 60904-9. Photovoltaic devices – Part 9: Solar simulator performance requirements. (Geneva: International Electrotechnical Commission)

A CAMERA-BASED CHARACTERIZATION METHOD FOR SOLAR SIMULATORS

Riechelmann, Stefan[1] and Plag, Fabian[1]
[1]Physikalisch-Technische Bundesanstalt (PTB), Bundesallee 100, D-38116 Braunschweig, Germany

ABSTRACT: Differences in the angular distribution of the light field emitted by various solar simulators that are used along the PV calibration chain can lead to systematic deviations in measured PV device's performances. To evaluate the influence of this effect, the light field of a simulator must be characterized. Therefore, we developed a CCD camera-based approach, where a camera is placed in the test plane which captures pictures of the light source. The distribution of the pixel values of the captured picture is used to derive a radiance distribution. We applied our technique to a xenon-based solar simulator and compared the results to measurements derived by a scanning method, where the radiance has been measured with a spectroradiometer. The opening angle, where 80 % of the irradiance that reaches the test plane originates from, is derived by both methods for this solar simulator. With the camera-based system an opening angle of 17.2° has been detected, compared to an opening angle of 17.7°, derived with the scanning system. The difference of 0.5° equals to about 3 % difference between both methods.

Keywords: Solar cell, qualification and testing, utilities, characterisation, experimental methods

1 INTRODUCTION

A secondary calibration of PV devices is commonly performed with solar simulators that show a wide range of different designs. All of them have in common, that a light source is irradiating the PV device under test, but shape and size of that light source varies widely, dependent on application purpose and simulator technology. Therefore, the angle of incidence in which light is received by a device under test also varies widely with the used solar simulator. A possible approach to characterize the radiance field emitted by the solar simulator is to carefully sample the radiance emitted by the light source point by point with a scanning detector. Those spectrometer-based light field characterizations provide spectral and spatial information about the light source of a solar simulator and have been recently developed [1] and are currently evaluated at the PTB. However, those measurements are time consuming and the necessary equipment is expensive. Therefore, we propose a CCD camera-based measurement technique for the determination of the radiance field emitted by solar simulators. Capturing the light source with a commercial grade camera provides a high angular resolution of the light field, while the spectral information compared to a spectrometer-based approach is limited to three broadband color channels in the visible wavelength range. As an example, the angular distribution of the light field of a xenon-based solar simulator measured with the proposed method is presented. The proposed method is addressed to solar simulator manufacturers and may help to optimize the design of light sources by a fast and cost-effective method compared to spectrometer-based scanning techniques.

2 APPROACH

A CCD-based camera is placed in the test plane of the solar simulator oriented towards the light engine. During the measurement, the camera is focused on the surface of the light source, e.g. a diffusing plate. We choose a high exposure time and flash the solar simulator manually during that period. This procedure requires a minimal synchronization effort. By using neutral density filters an overexposure of the captured picture is avoided. The use of the RAW picture format is also crucial, since the conversion in other formats like compressed JPG applies a highly non-linear gamma correction, which complicates further evaluation. In a first approach, the mean count values of the red, green and blue color channels of the RAW pictures are used to derive qualitative information about the angular distribution of the light field. An exemplary capture of the light source of a xenon-based solar simulator is shown in Figure 1. The most relevant emitting element of the light source is the diffuser plate of the simulator. In the middle of the diffuser several shadow rings are visible that ensure a better homogeneity of the irradiance in the detection plane.

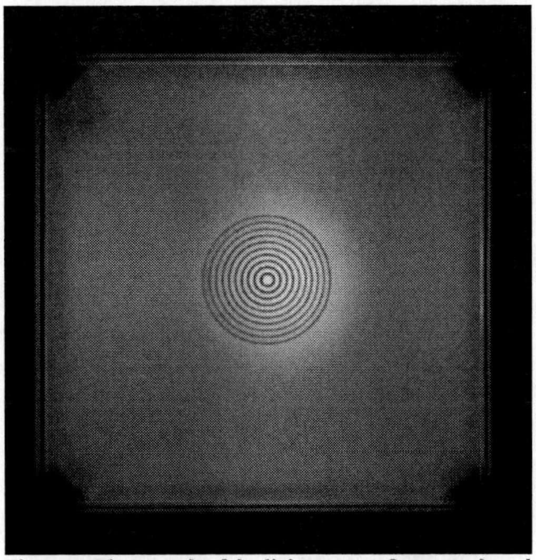

Figure 1: Photograph of the light source of a xenon-based solar simulator. Compared to regular JPEG picture no gamma correction has been applied since this would introduce a strong non-linearity and consequently distorting the measured radiance distribution.

We use the spatial information contained in the pixel values by performing a mapping where corresponding zenith and azimuth angles are assigned to each camera pixel. This is achieved by gridding the camera pixels to Cartesian coordinates, which are scaled according to the known size of the light engine and its distance to the testing area. Afterwards, the coordinates of the Cartesian grid are converted into the polar coordinate system:

$$r_i = \sqrt{x_i^2 + y_i^2 + z^2}$$
$$\theta_i = \arccos\left(\frac{z}{r_i}\right)$$
$$\varphi_i = \arctan2\left(\frac{y_i}{x_i}\right)$$

where r_i, θ_i and φ_i is the radius, zenith and azimuth angle of pixel i in a polar grid and z is the distance between light source and the test plane. The conversion of the regular Cartesian grid results in an irregular polar grid, which consequently is interpolated to a regular polar grid with a step size of 0.05° to provide a high sampling resolution at the edges of the light engine.
Since the camera pixel values are now on a polar grid with a high resolution, we integrate the measured pixel values P over the hemisphere to derive an integral value proportional to the irradiance E received by the test plane:

$$P_{int} = \iint_{\Omega} P(\theta, \varphi) \cdot \cos\theta \, d\Omega$$

where $\Omega = \sin\theta \, d\theta \, d\varphi$ is the solid angle and P_{int} is the integrated pixel value in counts. When we assume that the incident irradiance E in the test plane equals to 1000 W m^{-2}, an according scaling factor for the integral pixel value P_{int} can be determined:

$$E = SF \cdot P_{int} := 1000 \text{ Wm}^{-2}$$

By multiplying $P(\theta, \varphi)$ with the scaling factor SF we obtain the angular distribution of the radiance received in the test plane under the assumption that its integral equals to 1000 W m^{-2}:

$$P(\theta, \varphi) \cdot SF = L(\theta, \varphi)$$
$$\iint_{\Omega} L(\theta, \varphi) \cdot \cos\theta \, d\Omega = 1000 \; W \, m^{-2}$$

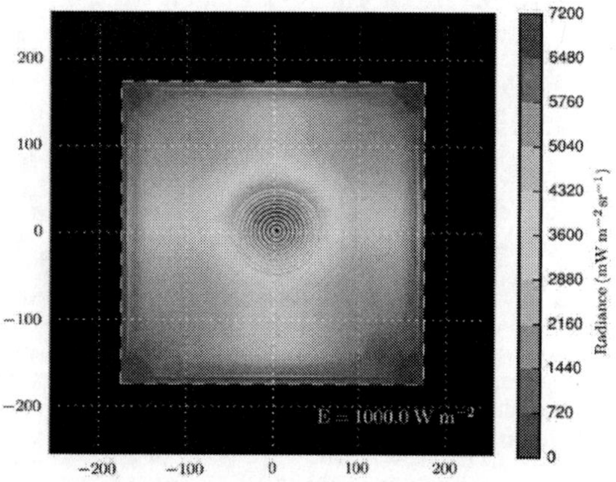

Figure 2: Contour plot of the radiance distribution of the examined xenon-based solar simulator.

However, since commercial CCD cameras have spectral responsivities in the visible part of the spectrum like the human eye, the method only works well for solar simulators with little spectral changes in the angular distribution of their emitted radiance, which might be the case for xenon-based solar simulators. This can be examined by evaluating the count values of the red, green and blue color channel of the camera separately. If no crucial shift in the angular distribution is apparent, this indicates a wavelength-independent behavior at least in the visible range of the emitted radiance. A more thorough examination of the spectral behavior can be achieved by applying bandpass filters to the camera setup. Other technologies like LED-based solar simulators, do also not provide a wavelength-independent angular distribution. This is due to their different light source design – instead of a xenon lamp emitting the whole spectrum simultaneously, the light emitting elements are spatially separated in the form of LEDs with different colors. While this also results in a homogeneous irradiance distribution in the test plane due to imaging techniques, a strong wavelength dependency of the angular distribution of the emitted light field can be assumed. However, this behavior can also be characterized by the proposed method by flashing each LED type subsequently and therefore performing virtually a spectral measurement of the angular distribution.

3 MEASUREMENTS AND RESULTS

The emitted radiance field of a xenon-based solar simulator has been examined with the proposed characterization method. Figure 1 shows the picture that was captured during a single flash of the light source. Apparent features are shadow rings where the xenon lamp has its brightest spot. Furthermore, the holding elements of the diffuser plate appear as dark triangles at the edges and an aluminum profile edging provides also a contribution to the overall radiance received in the test plane. By applying the described algorithm, the angular distribution of radiance $L(\theta, \varphi)$ emitted by the light source has been derived (Figure 2). For this, the assumption has been made, that the integral over $L(\theta, \varphi)$ equates to 1000 W m^{-2}. This number can be adjusted to the actual irradiance detected at the test plane. These results now enable the user to optimize the light engine in respect to the angular distribution of the emitted light field. The camera method can be also applied for the estimation of the irradiance content within a certain angle of incidence. For this, the irradiance originating from a region around the center of the solar simulator is integrated in slices of 1° intervals, resulting in a histogram showing the irradiance per angle of incidence (Figure 3). Since the emitting area around the center is very small due to geometrical reasons, most radiation reaches the test device with an angle of incidence of 5 - 20°. To evaluate our results, we performed the same calculation based on measurements obtained with a point-by-point scanning method based on a spectrometer with a field of view of 1° [1]. Compared to the scanning method, the camera-based method underestimates the irradiance between 5° and 16° and overestimates between 18° and 26°. By a step-wise integration of the irradiance with higher angles of incidence, one can derive the irradiance originating from within a certain angle of incidence (Figure 4). As an example, we evaluated that 80 % of the irradiance originates from within an angle of

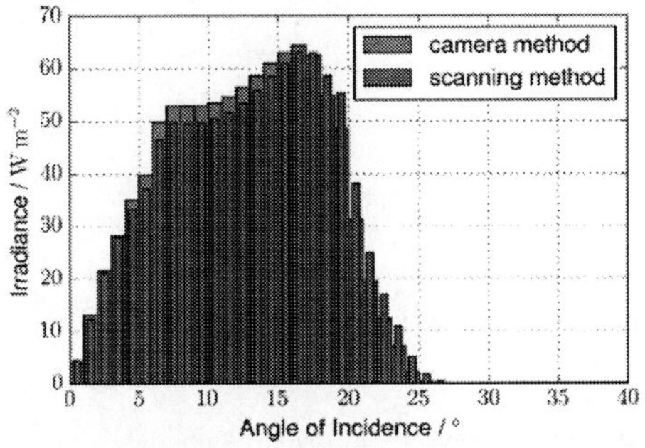

Figure 3: Histogram that shows how much irradiance originates from a certain angle of incidence. The bin size of each corresponding bar is 1°. The results of the camera-based method is shown as blue bars, the results of the scanning reference method is shown as orange bars.

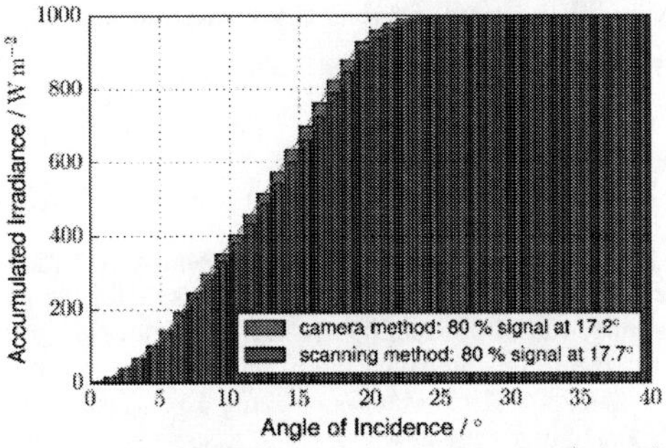

Figure 4: Histogram that shows how much irradiance originates from a certain angle of incidence. The irradiance shown in Figure 3 is accumulated from angle to angle, indicating how much irradiance comes from within a certain angle of incidence.

incidence of 17.2° according to the camera-based method, compared to 17.7° according to the scanning method. The difference of 0.5° may be due to camera non-linearity and limitations concerning the spectral responsivity of the camera. One can also estimate the contribution of specific features to the overall irradiance, e.g. the contribution of the reflecting aluminum profile edging has been estimated to about 1.1 % of the total irradiance received by the testing plane by integrating the radiance field after setting all radiance values inside the area of the diffuser plate to zero.

4 SUMMARY

A camera based method for the determination of the angular distribution of the radiance emitted by a solar simulator is presented. We validated the results by a comparison against a scanning method. Without further evaluation of non-linearity and specific optical properties the results agreed reasonably well within 3 %.

5 ACKNOWLEDGEMENT

The authors acknowledge financial support of the MNPQ-Program provided by the German Federal Ministry of Economic Affairs and Energy. This work is supported by the company h.a.l.m. elektronik GmbH providing a laboratory grade solar simulator for solar cell characterization.

6 REFERENCES

[1] F. Plag et al., Comprehensive Analysis of a Pulsed Solar Simulator to Determine Measurement Uncertainty Components. Amsterdam: (2014). Proceedings of 29th European Photovoltaic Solar Energy Conference and Exhibition. pp. 2435 - 2442.

[2] S. Winter et al., Camera-based Uniformity Measurements of Irradiation Fields. Daejeon: 2008 Proceedings of 10th International Conference New Developments and Applications in Optical Radiometry. pp. 189 - 191.

MULTIFUNCTIONAL LED-BASED FACILITY: INTEGRAL AND SPECTRAL CHARACTERIZATION OF SOLAR CELLS

Antje Schweitzer[1], Florian Witt[1], Stefan Riechelmann[1], Tom Schulze-Bubert[2], and Stefan Winter[1]
[1]Physikalisch-Technische Bundesanstalt (PTB), Bundesallee 100, 38116 Braunschweig, Germany
[2]Now with: Newport Spectra-Physics GmbH, Stahnsdorf
Tel.: +49 531 592 4189, fax: +49 531 592 69 4183, e-mail: Antje.Schweitzer@ptb.de

ABSTRACT:

The PTB developed a spectrally tunable LED-based facility, which combines integral and spectral measurement methods for the performance characterization of solar cells. In this study we present the facility and its measuring capabilities. Within the usage as solar simulator arbitrary AM.X spectra up to the AM0 spectrum can be simulated. Due to the precise controllability of the LEDs it is suitable for the simulation of diurnal variation of solar irradiance and for the calibration of complex solar cell technologies. Thereby requirements are met solar simulators which offer flexibility for the adjustment of irradiance and spectral distribution at the same time. As the LEDs can be controlled separately, spectral measurement methods are available in addition to it. We performed short circuit current measurements, a simplified spectral responsivity measurement method and spectral mismatch correction with our facility.

Keywords: calibration, characterization, spectral responsivity

1 INTRODUCTION

The characterization and calibration of solar cells and modules is performed, for worldwide comparability, under defined standard test conditions including irradiance, spectral distribution and temperature. These standard test conditions (STC) are defined by the IEC 60904-3 standard with an irradiance of 1000 W/m^2, an AM1.5G spectrum and a device temperature of 25°C. Commonly halogen and xenon based solar simulators are used for this purpose. They are optimized for the AM1.5G spectrum.

However, STC are not suitable for the evaluation of performance of solar modules at real operating conditions as these conditions differs strongly from STC. Hence energy based metric characterizations are gaining importance for the solar research sector. Therefore, a flexible adjustment of spectral irradiance is necessary to create AM.X spectra related to a variety of different outdoor irradiation conditions.

Equally the characterization of complex solar cell technologies results in special requirements regarding to a flexible adjustment of spectral irradiance. For the calibration of multi-junction solar cells (MJSC) an AM0 spectrum with an irradiance of 1360 W/m^2 is required [2]. The spectral ranges of the spectrum covering the spectral responsivity of the different layers of a MJSC need to be varied precisely in defined steps so that each cell creates a corresponding short-circuit current. Since the spectral distribution of conventional xenon and halogen based systems can be varied marginally, these systems are not suitable for energy based metric investigations or the characterization of MJSC. Out of these reasons solar simulators are required which offer flexibility for the adjustment of irradiance and spectral distribution at the same time. Thus, the PTB developed a spectral tunable LED based facility that comply with these requirements. Every LED channel is separate selectable, thus high flexibility for the adjustment of spectral irradiance can be achieved. Furthermore, each channel can be operated in steady or modulated mode. Hence, the facility is suitable for spectral characterizations of solar cells. This enables the application of our facility as solar simulator (integral) and for spectral responsivity measurements.

Integral as Solar simulator:

- I_{SC} and IV-curve measurements of WPVS and multi-junction solar cells (MJSC)
- Simulation of the diurnal variation of solar irradiance

Spectral:

- Spectral responsivity (SR) measurements
- mismatch correction

2 EXPERIMENTAL

The facility consists of an integrating sphere and is equipped with 27 different LED types with center wavelengths from 405 nm up to 980 nm and a halogen lamp to cover the infrared area. The LEDs and the lamp are thermal stabilized to ensure steady operating conditions. The barium sulphate coated sphere ensures a spatially uniform mixing of all colors. The homogenous irradiated measurement area is located directly at the outlet of the sphere has a size of 5x5 cm^2.

Each LED-type can be operated in a continuous or modulated mode. The spectrum is adjusted and corrected with a feedback loop continuously with an array spectroradiometer and a monitoring photodiode. By means of the measured deviation between the target spectrum and actual spectrum the optimal power level of each LED type is calculated by an an optimization algorithm. The spectrum is readjusted continuously and instabilities like for example changes of the spectrum due to the spectral shift and power loss during the warm-up time of the LEDs are compensated. The schematic layout of the facility is depicted in Figure 1.

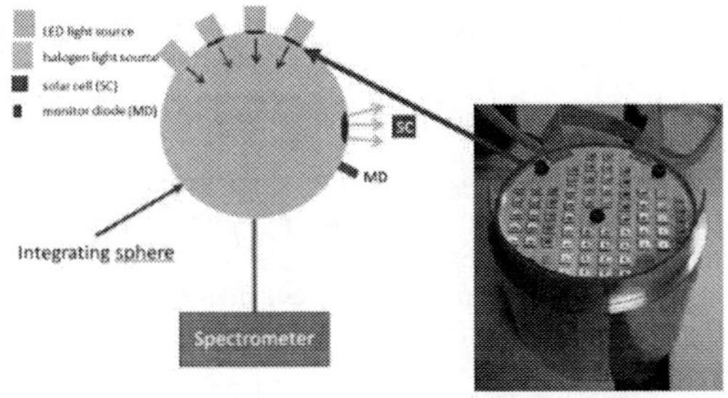

Figure 1 shows a charting of the facility consisting of an integrating sphere, LED and halogen light sources, a monitoring photodiode and an array spectroradiometer. A photograph of the thermal stabilization and the LEDs is shown.

2.1 Solar simulator functionality

The power of every LED-type can be controlled separately and tuned continuously, hence the spectral distribution of irradiance E_λ can be adjusted with high flexibility. The facility is equipped with a sufficient amount of LEDs, so that it is possible to reach total irradiance levels as required for the AM1.5G standard spectrum only with LEDs as light source. This enables our facility for IV and I_{SC} measurements at standard test conditions. Due to the flexible and precise controllability of the LEDs arbitrary AM.X spectra can be simulated (Figure 2). Hence, the facility is suitable for the simulation of the diurnal variation of solar irradiance for realistic performance prediction of solar cell technologies.

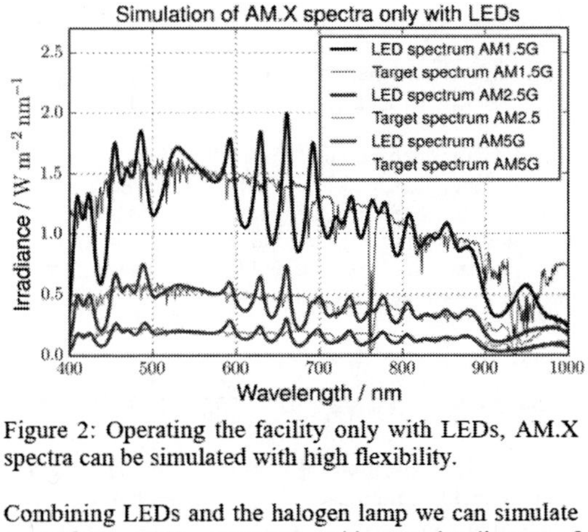

Figure 2: Operating the facility only with LEDs, AM.X spectra can be simulated with high flexibility.

Combining LEDs and the halogen lamp we can simulate a gapless AM0 spectrum with an irradiance of 1360 W/m² which is required for the calibration of MJSC, see Figure 3. The spectral irradiance of the solar simulator needs to be varied precisely in defined steps under consideration of the responsivities of the different layers of a MJSC, so that each cell generates it's short-circuit current for the adjusted spectrum.

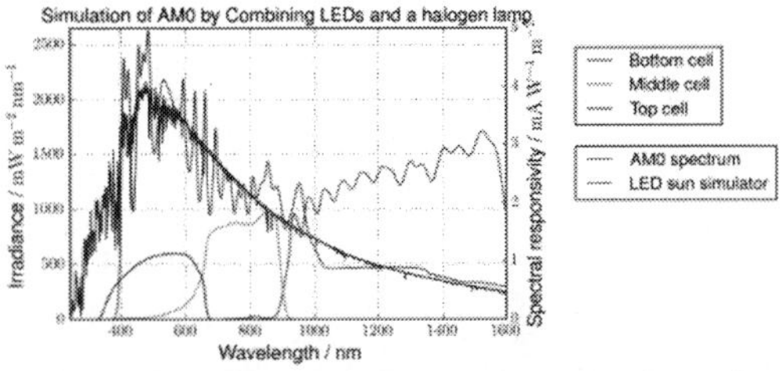

Figure 3: Combining LEDs and a halogen lamp a gapless AM0 spectrum can be simulated with an irradiance of 1360 W/m². This is necessary for the calibration of MJSC. The SR curves of a triple MJSC are depicted in the graph concerning to the bottom, middle and top cell.

2.1.1 I_{SC} measurements of WPVS solar cells at STC

The relative spectral distribution of the AM1.5G spectrum is adjusted with the spectrometer and then scaled absolute with I_{SC} of a reference cell (REF). Afterwards the reference solar cell gets removed and the device under test (DUT) is positioned in the same measuring plane. We used as reference (REF), DUT 1 and DUT 3 three unfiltered WPSV solar cells and as DUT 2 one IR filtered WPVS solar cell. Subsequently we placed the DUTs one after the other in the measuring plane and determined the short circuit current $I_{SC,measured}$(see Table 1). For the unfiltered cells the relative deviation lays between 2.2 % and 3.3 %, for the filtered cell it is due to the higher spectral mismatch nearly 12 %.

Solar cell	$I_{SC,measured}$ [mA]	$I_{SC,STC}$ [mA]	Relative deviation [%]
REF	125.74	125.76	0.02
DUT 1	144.50	149.48	3.33
DUT 2	40.02	35.84	11.66
DUT 3	135.61	138.77	2.28

Table 1: The AM1.5G spectrum is adjusted by the means of the spectrometer and I_{SC} the reference solar cell. The short circuit current of three different DUT solar cells is determined and compared to $I_{SC,STC}$.

2.2 Spectral measurements

2.2.1 Spectral responsivity measurements

The PTB calibrates solar cells performing the differential spectral responsivity (DSR) method [1]. The solar cell is irradiated at the same time with cw bias light and modulated monochromatic light. For the bias irradiation halogen lamps are used and for the monochromatic light a spectral tunable laser system or a halogen/xenon lamp and a monochromator. By the means of a lock-in amplifier the signal of the solar cell with the frequency of the modulated monochromatic light is determined and thereout the spectral responsivity derived. In this study this method is referred to as conventional DSR method.

We applied with our system a simplified SR measurement method. We irradiated a solar cell with the halogen lamp as steady bias radiation and modulated every LED-type one after each other. With a lock-in amplifier we measured the AC-current signal of the solar cell. In a first step, we measured a calibrated reference solar cell, subsequently we measured a solar cell which is considered as device under test (DUT. We multiply the ratio of the monitor corrected AC-signal of the DUT and the reference solar cell with the absolute spectral responsivity of the reference solar cell and determine hence the absolute spectral responsivity of the DUT:

$$SR_{\mathrm{DUT}} = \frac{Signal_{\mathrm{DUT,LED}}}{Signal_{\mathrm{REF,LED}}} \cdot SR_{REF}$$

We chose for this measurement as our device under test the filtered WPVS solar cell DUT 2 and the as the reference solar cell the device indicated as REF.
In Figure 44 (top) the monitor corrected AC signal of the REF (red) and the DUT (blue) and the ratio DUT/REF (green) are depicted. Every dot corresponds to the center wavelength of a LED_i with the channel number i. The SR curve of the DUT determined with the conventional DSR method (pink) is in close match to the SR curve determined with the LED-facility (blue) with the exception of the measuring point at 567 nm, see Figure 4 bottom. The deviation is due to a broad spectral bandwidth of this LED-type and that the SR of the REF is increasing while the SR of the DUT is decreasing.

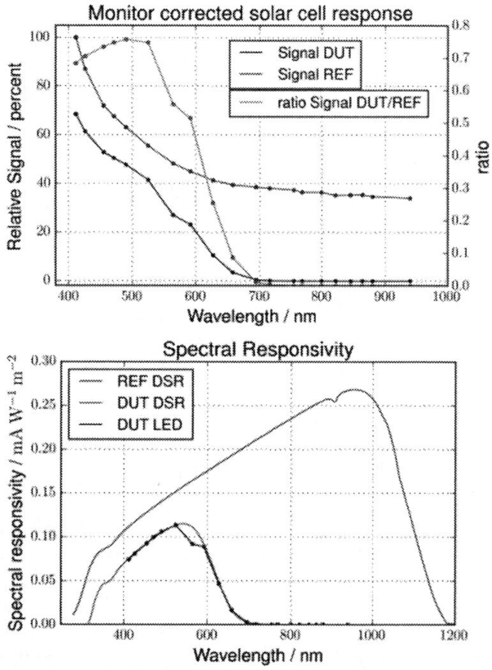

Figure 4 Top: Monitor corrected AC-Signal of the REF and DUT solar cell and the ratio of both signals which is equivalent to the relative SR curve of the DUT. Bottom: The absolute SR curve of the REF (red) and the DUT (pink) determined with the conventional DSR and the absolute SR curve determined with the LED-facility (blue).

2.2.2 Mismatch correction

Any simulated spectrum is non-perfect and deviates from the target spectrum. In order to determine the correct I_{SC} at STC the spectral mismatch MMF consistent of the SR_DUT, the SR_REF, the reference spectrum $E_{AM1.5G}$ and the spectrum of the solar simulator E_{LED} has to be taken into account. This is considered by the means of the spectral mismatch factor:

$$MMF = \frac{\int SR_{Ref}(\lambda) \cdot E_{AM1.5G} d\lambda}{\int SR_{Ref}(\lambda) \cdot E_{LED} d\lambda} \cdot \frac{\int SR_{DUT}(\lambda) \cdot E_{LED} d\lambda}{\int SR_{DUT}(\lambda) \cdot E_{AM1.5G} d\lambda}$$

We compare the spectral mismatch factor of our DUT solar cell determined with the conventional DSR method ($\mathrm{MMF_{DSR}}$) and with the LED-facility ($\mathrm{MMF_{LED}}$). We chose the integral limits at 400 nm and 980 nm. To derive the mismatch factor for the LED-facility we substituted in the formula above DUT LED (blue) and for the conventional DSR method DUT DSR (pink), see Figure 5. As reference spectrum, we chose the standard AM1.5G spectrum (violet) and for the source spectrum the simulated AM1.5G spectrum we used in chapter 2.1.1 for the I_{SC} measurements (green).

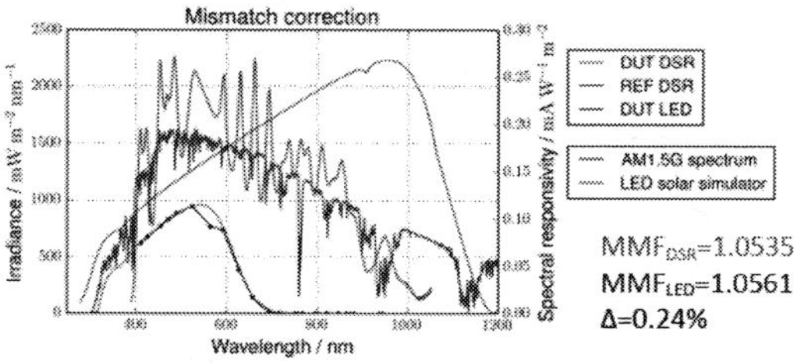

Figure 5 shows in violet the AM1.5G reference spectrum, the spectrum of the LED solar simulator (green), the SR curve of the REF (red) and the DUT (pink) determined with the conventional DSR method and the SR curve of DUT determined with the LED-facility (blue).

The deviation between both mismatch factors is less than 0.3 % from which follows that the LED facility is suitable for spectral mismatch corrections within a reasonable tolerance. As final step, we use now the spectral mismatch factor ($\mathrm{MMF_{LED}}$) to correct the measured short circuit current of the filtered device DUT 2 from chapter 2.1.1:

$$I_{\mathrm{SC},corrected} = \frac{I_{\mathrm{SC},measured}}{MMF_{LED}} = \frac{40.02\ mA}{1.0561} = 37.89\ mA$$

The relative deviation of the measured short circuit current to the calibrated $I_{\mathrm{SC,STC}}$ = 35.84 mA is thereby reduced to 5.7 %.
The remaining deviation is due to the very diffuse radiance distribution of the LED-facility in combination with the diverse angular dependent responsivity of both solar cells. The characterization of the angular dependent radiance distribution of the solar simulator [3],the

knowledge about the angular and spectral dependent responsivity of the measured solar cells [4] and the thereout arising ASMM factor [5] may be an approach to correct this remaining deviation.

3 CONCLUSION

We presented the measuring capabilities and application possibilities of our newly developed LED-facility and showed preliminary results.
We performed with our facility integral and spectral characterizations of solar cells. We show that the facility is suitable for I_{SC} measurements and for the determination of the spectral responsivity to correct for spectral mismatch. Further investigations and characterizations of the facility are planned and the extension of the setup with more LEDs. The implementation of additional LEDs in the UV and IR range would extend the range of possible applications. A narrower LED in the range of 570 nm can further improve the spectral responsivity measurements.

4 ACKNOWLEDGEMENT

The research work leading to this article was partly carried out within the EMRP ENG51 project "SolCell". The EMRP is jointly funded by the EMRP participating countries within EURAMET and the European Union.

5 REFERENCES

[1] S. Winter et al, Proceedings. EU PVSEC Munich, Design, realization and uncertainty analysis of a laser based primary calibration facility for solar cells at PTB.,2014
[2] F. Witt et al, Recent Advances in synthetic Calibrations of Multi Junction Solar Cells and their corresponding Component Cells, ESPC 2016
[3] I. Kroeger et al, Spectral Angular Responsivity Calibration Facility at PTB, EU PVSEC 2017
[4] F. Plag et al, Spectral and Angular Correction- a Multidimensional Approach to Model Measurements under Outdoor Conditions, EU PVSEC 2017
[5] S. Riechelmann et al, A Camera-based Characterization Method for Solar Simulators, EU PVSEC 2017

TOWARDS ACCURATE, HIGH-FREQUENCY I-V CURVE MEASUREMENTS OF PHOTOVOLTAIC MODULES APPLYING ELECTRONIC LOADS

K. Spiliotis[1,2,4], G.H. Yordanov[1,2], G. Van den Broeck[1,2,4], H. Goverde[2,3], K. Baert[1,2], J. Driesen[1,2]

[1]KU Leuven, ESAT-ELECTA, Kasteelpark Arenberg 10, 3001 Leuven-Heverlee, Belgium
[2]EnergyVille, ThorPark 8310, 3600 Genk, Belgium
[3]IMEC, Kapeldref 75, 3001, Leuven-Heverlee, Belgium
[4]EIT InnoEnergy, J.F. Kennedylaan 2, 5612 AB Eindhoven, The Netherlands

ABSTRACT: We present a new measurement methodology for accurate, high-frequency outdoor I-V characterization of PV modules. A hybrid sweep is performed in order to obtain more detailed I-V curves of the PV module, hereby combining the advantage of a voltage sweep (for good resolution in the near-horizontal parts of the curve) with the advantage of a current sweep (giving precise information about the very steep parts). The new approach is compared to the state-of-the-art current and voltage sweep methods. Furthermore, an advanced data-point distribution algorithm is discussed, which ensures sufficient information in the ranges of interest, while keeping the total amount of required measurement points as low as possible to minimize the data storage, sweep time, and computational effort. The improved technique provides accurate, high-resolution characterization about the maximum power point, MPP without necessarily requiring too many measurement points. The new methods enable I-V measurement frequencies up to 1 Hz (which is essential for accurate energy yield estimations) while maintaining low computational effort.

Keywords: PV Module, Electrical characterization, Hybrid I-V sweep, Measurement automation.

1 INTRODUCTION

An important part of the PV modules development at EnergyVille is the outdoor performance testing. To be able to estimate and model the energy yield also during highly varying weather conditions, high frequency (1 Hz) and accurate I-V characterization is required [1]. However, at this sampling rate, the amount of data produced can be enormous, hereby strongly enlarging the storage demand and limiting data processing capabilities due to a big computational effort required to lookup the measured data. Next to that, state-of-the-art current and voltage sweeping methods produce detailed information between the MPP and open circuit voltage as well as from short circuit current up to the MPP, respectively. Accurate and detailed information about the complete IV curve is required for analyses of both the degradation and the energy production.

2 STATE OF THE ART

The I-V curves of PV modules provide valuable information for their performance, such as the values of short-circuit current (I_{SC}), open-circuit voltage (V_{OC}) and the MPP location. Common ways to sweep the I-V curve include the use of variable resistor [2, 3], capacitive load [4], electronic load [5-7], bipolar power amplifier [8], four-quadrant power supply [9], and DC-DC converters [10]. Multichannel electronic loads are characterized by high degree of flexibility and modularity [11]. That makes them well suited for technical and scientific research. Some experimental setups use electronic loads for I-V curve estimation and MPP tracking [12-14]. Electronic loads are based on MOSFETs and use current or voltage ramps in order to sweep the I-V curve of the connected PV module. Non-monotonic voltage control waveforms are used to measure accurate I-V curves of highly capacitive PV modules with short-pulse (~10 ms) sun simulators. Studies concerning energy yield, ageing, control and balancing among others require accurate I-V characteristics to ensure reliable results. Often, the means of improving the I-V curve measurement is by increasing the number of set points, decreasing the stepping time and dynamically estimating their distribution. These approaches, however, come along with certain drawbacks. An increased number of set points results in high volumes of data; dynamic measurement point setting requires continuous and fast computations; while decreased time intervals between I-V point measurements require costly high-performance load controllers or prolonged operation away from the MPP which results in higher operating temperatures of the PV cells.

3 METHODOLOGY AND NOVEL APPROACHES

As part of an outdoor measurement campaign for testing innovative façade BIPV modules, KU Leuven and IMEC carry out experiments involving 9-cell c-Si modules without bypass diodes [13-15]. Within this campaign, the present work proposes two new approaches for I-V curve sweeping developed in such a way as to avoid the aforementioned complications. A voltage-controlled sweep would obtain more information on the flat part of the I-V curve, while a current-controlled sweep emphasizes the rest. The proposed approaches are designed to obtain high detail on the entire I-V without extra computing cost. They have been conceptualized and implemented experimentally.

3.1 Variable-step current sweep

The I-V curve measurement specification requires $N = 50$ data points per curve, each being an average of 20 voltage and current measurements. Because of the good quality cells used in the modules, the shunt resistance is very high and the lack of bypass diodes results in I-V curves that are very flat between short circuit and the kink near the maximum power point (MPP); see Fig. 1. Kinks result when one cell limits the current output of the entire string, which results in a cut-off. Often the MPP is located at the kink, meaning that the associated P-V curve is not differentiable at this point. Therefore, a very good resolution is mandatory for achieving low uncertainties of the power, current and voltage at the MPP.

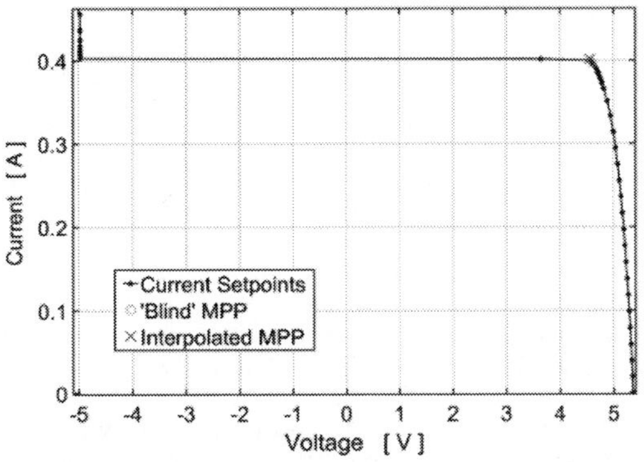

Figure 1: I-V curve of a 9-cell c-Si PV module without bypass diodes measured with the described in section 3.1

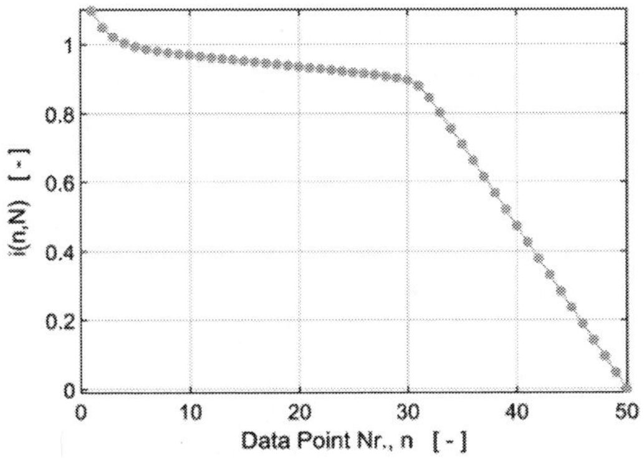

Figure 2: The current setpoints function $i(\mathbf{n},\mathbf{N})$ calculated for $\mathbf{N}$ = 50 data points per I-V curve according to the method described in section 3.1.

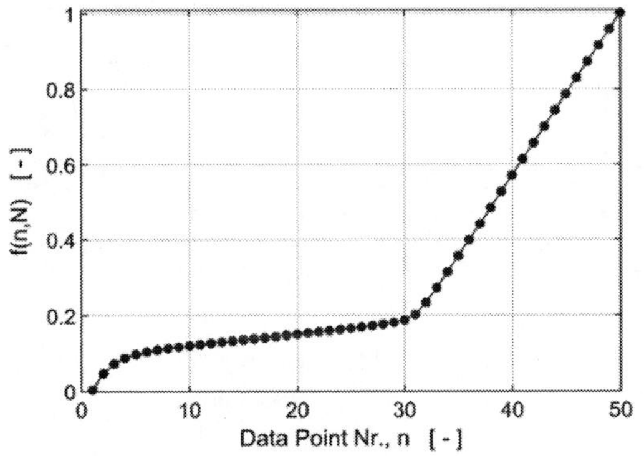

Figure 3: The auxiliary function $f(\mathbf{n},\mathbf{N})$ calculated for $\mathbf{N}$ = 50 data points per I-V curve.

Because of the low number of data points being recorded for each I-V curve, we had to distribute these points optimally so that the best resolution is achieved about the MPP. For c-Si PV cells and modules, the latter is usually achieved between 92% and 95% of the short-circuit current, $\mathbf{I_{SC}}$ [16]. However, when there is a cut-off due to a cell that significantly underperforms the others, the MPP may be at the kink and the $\mathbf{I_{MPP}}$ is almost equal to the $\mathbf{I_{SC}}$. Therefore, we aimed for best resolution (densest data points) between about $0.90\mathbf{I_{SC}}$ and $1.00\mathbf{I_{SC}}$. The short-circuit current is measured before the I-V curve. Then, we define the current setpoints for the electronic load control between $1.10\mathbf{I_{SC}}$ and 0 A in case the in-plane solar irradiance increases in the meantime. This particular multichannel electronic load is operated in constant current mode because all the other modes require unacceptably long measurement times for a complete I-V curve.

We need a scalable solution that works for an arbitrary, but not too low, number of data points per I-V curve, $\mathbf{N} \geq 50$. For a given $\mathbf{N}$, our data-acquisition (DAQ) software calculates a vector of $\mathbf{N}$ elements containing the desired setpoints in terms of parts of the $\mathbf{I_{SC}}$; see Figure 2. The current setpoints are then obtained by multiplying that vector by the measured $\mathbf{I_{SC}}$. To obtain these optimal values along the I-V curve, we take the following approach. We define a dimensionless function $i(\mathbf{n},\mathbf{N}) = 1.10(1 - f(\mathbf{n},\mathbf{N}))$ where $\mathbf{n}$ is the data point number ($\mathbf{n}$ = 1 corresponding to 110% of $\mathbf{I_{SC}}$ and $\mathbf{n} = \mathbf{N}$ corresponding to open circuit) while $f(\mathbf{n},\mathbf{N})$ is an auxiliary function (also dimensionless) which satisfies $f(1,\mathbf{N}) = 0$ and $f(\mathbf{N},\mathbf{N}) = 1.00$. Examples for $i(\mathbf{n},\mathbf{N})$ and $f(\mathbf{n},\mathbf{N})$ are given in Figs. 2 and 3, respectively for $\mathbf{N}$ = 50. We model the auxiliary function $f(\mathbf{n},\mathbf{N})$ as a hyperbola which undergoes translation and rotation with respect to the origin of the coordinate system, and subtract an exponentially decaying term to obtain the required curving at the low $\mathbf{n}$ values. We require that, independent on $\mathbf{N}$, approximately a tenth of all the data points in the I-V curve lie between $1.10\mathbf{I_{SC}}$ and $1.00\mathbf{I_{SC}}$; half of all points lie between $1.00\mathbf{I_{SC}}$ and $0.90\mathbf{I_{SC}}$; and the rest are more or less evenly distributed from $0.90\mathbf{I_{SC}}$ down to open circuit ($\mathbf{I}$ = 0 A). We use these requirements to parameterize the hyperbolic and exponential parts of the auxiliary function $f(\mathbf{n},\mathbf{N})$ as well as the initial translation and rotation of the hyperbola. For brevity, the involved equations are omitted in the present paper. With this approach, we obtain satisfactory resolution of the measured I-V curves and P_{max} accuracy within 0.6%. Fig. 1 shows an example 2-quadrant I-V curve for one of the tested modules. The circle and the cross represent the MPP location as determined from the original and fitted curve, respectively.

3.2 Hybrid sweep

A hybrid sweeping scheme is employed, where two sweeps are performed (in voltage and current control mode) to emphasize different parts of the I-V curve. The voltage sweep gives enough points in flat near-horizontal regions, while the current sweep does so in near-vertical parts. Specifically, the method includes a current sweep with k equidistant current set points between 0 A and 92%-95% of $\mathbf{I_{SC}}$ [16], followed by equidistant voltage set points starting from 0 V and increasing to $\mathbf{V_{OC}}$. We can see in Fig. 4 an illustration of the proposed sweep strategy. We can clearly observe that in case of a current sweep for example, the resolution of our current set points has to be drastically increased close to $\mathbf{I_{SC}}$ or the local level of detail will suffer. Similarly, much finer voltage steps would be required between $\mathbf{V_k}$ and $\mathbf{V_{OC}}$. Sweeping between 0 A and $\mathbf{I_{SC}}$ and 0 V and $\mathbf{V_{OC}}$, with few equidistant current and voltage set points accordingly may provide accurate enough I-V characteristic without increasing the number of data points. $\mathbf{I_{MPP}}$ and $\mathbf{V_{MPP}}$ are approximated through $\mathbf{I_k}$ and $\mathbf{V_k}$, respectively. I-V curves of partially shaded modules with bypass diodes require complete current and voltage sweeps since the MPP location is generally unknown a priori. Partially shaded modules with integral bypass diodes in each cell (e.g. with SunPower Maxeon Gen III IBC cells) require many more data points for greater detail. Fig. 5 depicts our proposed hybrid I-V curve sweeping algorithm.

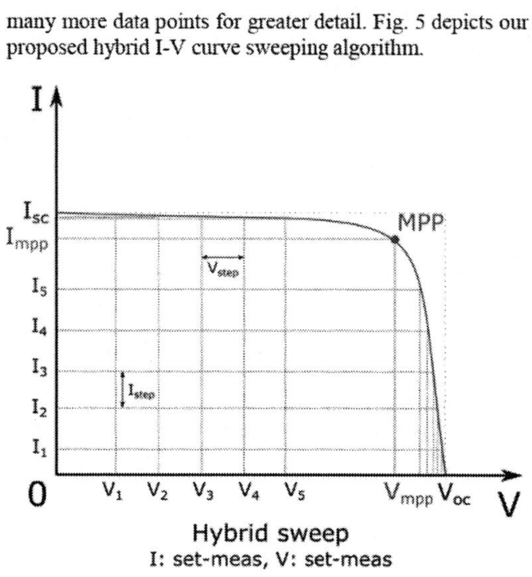

Figure 4: Hybrid sweeping on the I-V curve.

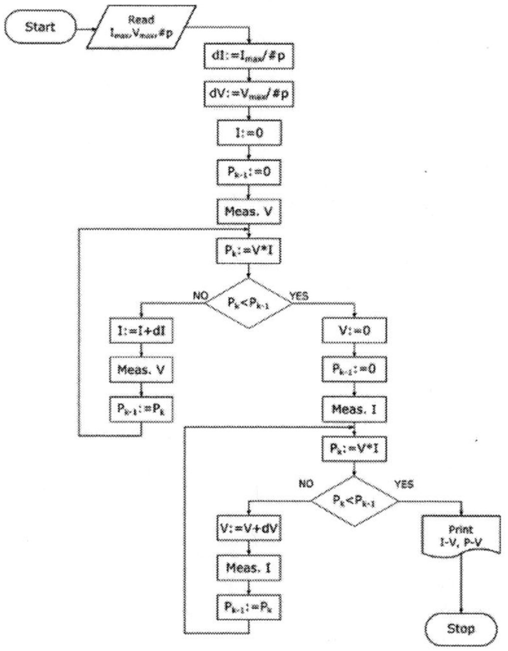

Figure 5: Flowchart depicting the hybrid sweeping algorithm.

4 RESULTS AND DISCUSSION

The voltage and current sweeps alongside the novel methodology are implemented in practice using 9-cell c-Si PV modules without bypass diodes, Höckerl & Hackl PMLI and PLI-606 electronic loads, and National Instruments compactRIO and compactDAQ controllers programmed with LabVIEW-based software.

Figs. 6 and 7 and illustrate the performance of the hybrid sweep compared to the voltage and current sweeps for a certain number of measurement points. Specifically, the sweeps' performance is tested using between 50 and 300 measurement points. We see that with a hybrid sweep even as few as 50 points may suffice to determine the **MPP**, **I_{SC}**, and **V_{OC}**. It is clear that, at a low number of data points, the region close to **V_{OC}** cannot be resolved satisfactory when performing an equidistant voltage sweep. Similarly, the flat part of the I-V curve needs to be interpolated when applying a current sweep. Hybrid sweeping may help to estimate more accurately the series and shunt resistance, **R_S** and **R_P**, as it provides a lot of information about the flat parts of the curve.

Fig. 8 shows P-V curves of seven BIPV modules in the MPP vicinity. The corresponding I-V curves were obtained with the variable-step current sweep algorithm. Although the optimal data point density is not always achieved at the MPP, the finer steps around it allows for accurate curvilinear fitting for MPP interpolation.

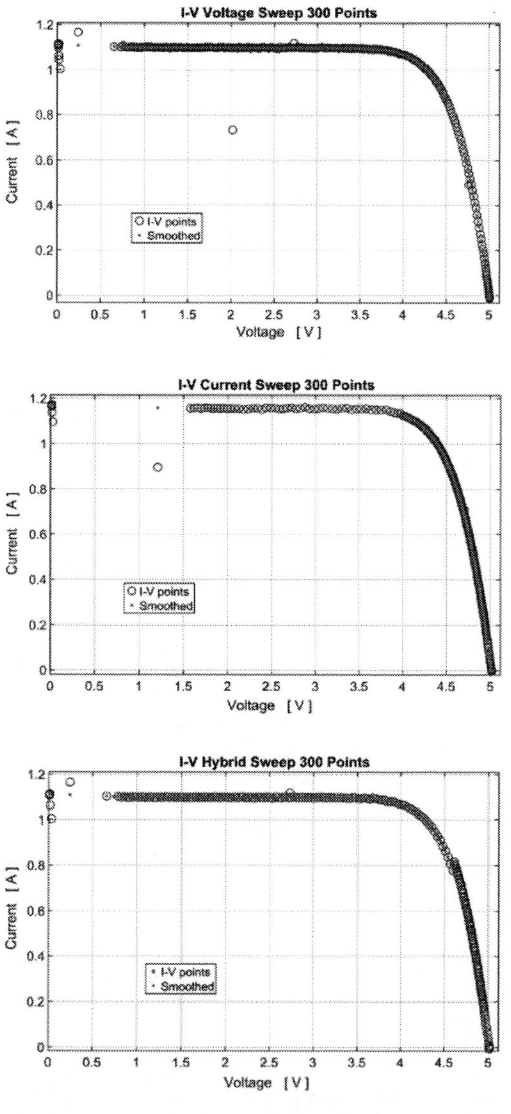

Figure 6: Voltage, current and hybrid sweeping (from top to bottom) for 300 data points.

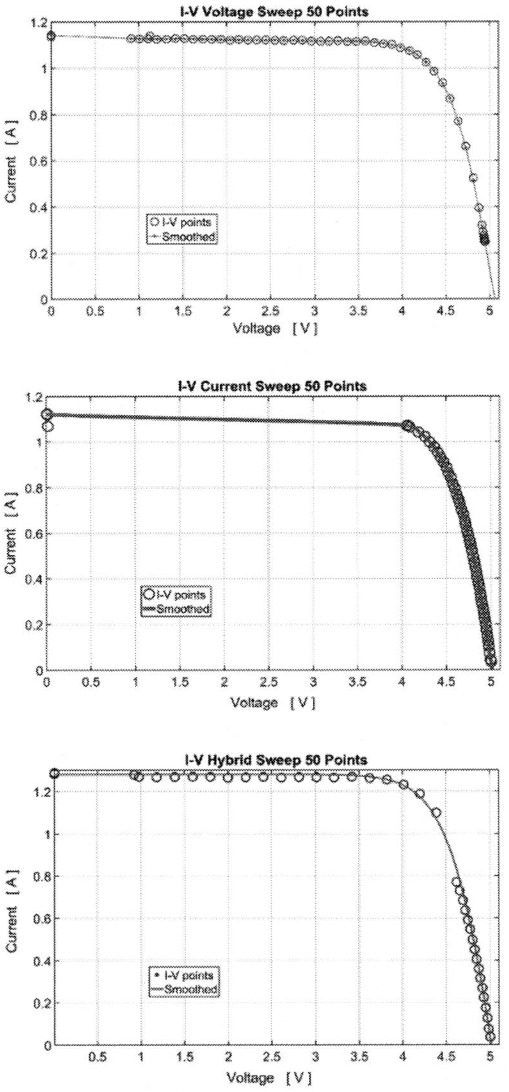

Figure 7: Voltage, current and hybrid sweep (from top to bottom) for 50 data points

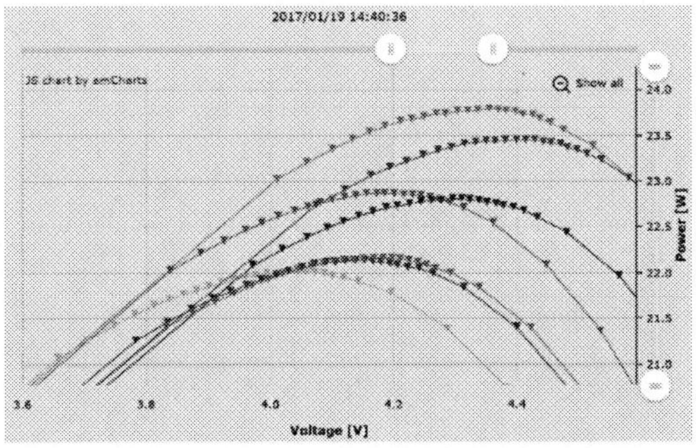

Figure 8: A screenshot from the data-querying web interface showing P-V curves of seven BIPV minimodules near the MPPs.

Finally, Fig. 9 shows results from simulated hybrid sweep of a stepped I-V curve of a 60-cell c-Si PV module with three bypass diodes and different degree of shading in the three sub-strings. The simulations were done in Matlab using Simulink.

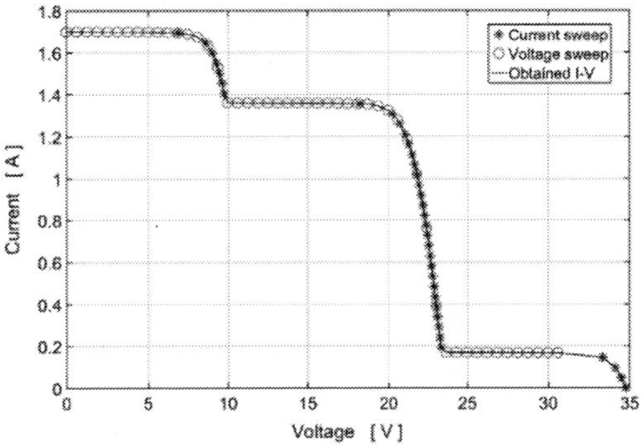

Figure 9: Results from a hybrid I-V sweep simulation.

The simulations show that it is it is possible to obtain satisfactory resolution in all parts of the I-V curve even with a modest number of points. In fact, the small gap seen near the lowest knee was due to not covering the entire voltage range in the voltage sweep all the way to the **V_{OC}**. Fine estimation of the MPP is then possible by fitting locally the I-V curve in the vicinity of the 'blind' MPP (the measured MPP). Same applies to the **V_{OC}** and **I_{SC}**.

From practical experience, the dynamic behavior of the electronic load used to sweep the I-V curve are of big importance. It may have a very fast response in constant current regulation mode but be very slow in constant voltage mode, thus making the hybrid sweeping too long.

Furthermore, some loads as the one used to obtain the results in Figs. 6 and 7 can have a 'jumpy' behavior due to saturation at either the open-circuit or short-circuit end. These I-V curves took 0.1 s to sweep. The observed behavior is clearly due to the load and not to the tested low-capacitance c-Si minimodule. Such load-related issues can be tackled by programing more sophisticated control waveforms, e.g. longer time steps near the saturation point(s). Any remaining outliers can be post-processed by using fitting techniques that discard them.

5 CONCLUSIONS

The results in the present work indicate significant accuracy improvements in the electrical characterization of PV modules. Specifically, a major reduction of the points number per I-V curve can be achieved. Dynamic point setting is not necessary which saves time and computational effort. Equidistant points suffice for the case of hybrid sweeping. More accurate determination of the shunt and series resistances (**R_P** and **R_S**, respectively) is possible. And finally, the load's response time and resolution may not be superior thus preventing high equipment costs. We compared the presented methods in practical terms and implemented the variable current step sweeping in outdoor testing experiments. The proposed methodology achieves improved measurement accuracy and reduced data storage costs.

ACKNOWLEDGEMENTS

This project receives the support of the European Union, the European Regional Development Fund ERDF, Flanders Innovation & Entrepreneurship and the Province of Limburg. Our colleagues Willem Bertels, Wout Parys, and Patricia Elsen from KU Leuven's Civil Engineering department deserve our special thanks for installing the tested PV modules.

REFERENCES

[1] H. Goverde, D. Goossens, J. Govaerts, F. Catthoor, K. Baert, J. Poortmans and J. Driesen, "Spatial and Temporal analysis of wind effects on PV modules: Consequences for electrical power evaluation", Solar Energy, vol. 147, pp. 292-299, 2017.

[2] E. E. Van Dyk, A. R. Gxasheka and E. L. Meyer, "Monitoring current-voltage characteristics and energy output of silicon PV modules," Renewable Energy, vol. 30, pp. 399-411, 2005.

[3] F. Recart, M. Mäckel, A. Cuevas and R. Sinton, "Simple data acquisition of the current-voltage and illumination-voltage curves of solar cells," in IEEE WCPEC, 2006.

[4] M. M. Mahmood, "Transient analysis of a PV power generator charging a capacitor for measurement of the I-V characteristics," Renewable Energy, vol. 31, pp. 2198-2206, 2006.

[5] Y. Kuai and Y. S, "An electronic load for testing photovoltaic panels," Journal of Power Sources, vol. 154, pp. 308-313, 2006.

[6] N. Forero, J. Hernandez and G. Gordillo, "Development of a monitoring system for a PV solar plant," Energy Conversion and Management, vol. 47, pp. 2329-2336, 2006.

[7] J. Salmon and R. Phelps, "Solar cell measurement system for NPS Spacecraft Architecture and Technology Demonstration Satellite NPSAT1," in 17th Annual AIAAIUSU Conference on Small Satellites, 2003.

[8] M. G. Gunvench and C. Gurcan, "Solar simulator and I-V measurement system for large area solar cell testing," in Proceedings of the 2004 American Society for Engineering Education Annual Conference & Exposition, 2004.

[9] J. Fernandez-Reche and I. Canadas, "PSA Solar furnace, A facility for testing PV cells under concentrated solar radiation," Solar Energy Materials & Solar Cells, vol. 90, pp. 2480-2488, 2006.

[10] E. Duran, J. Galan, M. Sidrach-de-Cadrona and J. M. Andujar, "A new Application of the Buck-Boost-Derived Converters to obtain the I-V curve of PV Modules," in 38th IEEE PESC, 2007.

[11] E. Duran, M. B. Ferrera, J. M. Andujar and M. S. Mesa, "I–V and P-V curves measuring system for PV modules based on DC-DC converters and portable graphical environment," in Industrial Electronics (ISIE), 2010 IEEE International Symposium on, 2010.

[12] M. Bliss, T. R. Betts, R. Gottschalg and D. Infield, "Improved outdoor monitoring of photovoltaic modules," Proc. PVSAT-2, 2005.

[13] G. Van den Broeck, W. Parys, H. Goverde, J. Poortmans, J. Driesen, K. Baert and D. Saelens, "Experimental Analysis of the Performance of Facade-Integrated BIPV in Different Configurations," in 32nd European Photovoltaic Solar Energy Conference and Exhibition, Munich, 2016.

[14] G. Van den Broeck, B. Herteleer, H. Goverde, F. Catthoor, K. Baert and J. Driesen, "Substring-level Energy-yield Assessment of Photovoltaic Modules subject to Partial Shading Conditions," in European PV Solar Energy Conference and Exhibition, Hamburg, 2015.

[15] W. Parys, G. Van den Broeck, H. Goverde, K. Baert, D. Saelens, "Whole building energy analysis of a BIPV curtain wall." Advanced Building Skins, edition 11, Bern, 2016.

[16] G.H. Yordanov, "The PVGIS relative efficiency model: Equations for k_1 through k_6." Proceedings 29th EU PVSEC, 2014, Amsterdam, pp. 3297–3301.

EVALUATION OF A COMPREHENSIVE I-V OUTDOOR-CHARACTERIZATION METHOD FOR PHOTOVOLTAIC MODULES

L. Gottschalk, D. Pflaum, S. Schneider, A. Schulze+, F. Becker*, M. Queck*, B. Hüttl

University of Applied Sciences Coburg, Department of Electrical Engineering and Computer Sciences, 96450 Coburg, Friedrich-Streib-Str. 2, Germany, Tel.: +49 (0) 9561 317 202, bernd.huettl@hs-coburg.de
+ University of Applied Sciences Rosenheim, Department of Applied Natural and Social Sciences, Germany, 83024 Rosenheim, Hochschulstr. 1
* Calyxo GmbH, Germany, 06766 Bitterfeld-Wolfen, Sonnenallee 1a

ABSTRACT: Outdoor I-V measurements on photovoltaic (PV) modules are advantageous because light soaking or light-induced degradation can be studied under natural conditions. Such outdoor analyzes can achieve precision of indoor flasher measurements if the self-reference algorithm is applied. This new concept delivers I-V characteristics for much more test conditions than required by IEC 61853 standard. Low-light curves of all relevant electrical I-V parameters can be generated for specific temperatures. We report about experimental results of a 6-day outdoor measurement campaign on PV modules. We evaluate the opportunities determining temperature coefficients, STC and NOCT data and to deduce I-V curves as a basis for fitting analyzes. Finally, the accuracy of the new outdoor method is investigated.

Keywords: Characterization, Module, Monitoring, Performance

1 INTRODUCTION

Precise power ratings of PV modules and generators are a prerequisite for economic operation of PV facilities. Outdoor I-V analyzes of photovoltaic modules offer some advantages in comparison to indoor characterizations:

- Effects like light soaking [1] or light-induced degradation can be quantified by just in time electrical measurements on natural conditions.
- I-V measurements are done slowly (within 0.2 to 2 seconds) and electrical transient errors [2] can be avoided for highly capacitive modules.
- The natural offer of meteorological conditions delivers a wide range of irradiance (E) and temperature (T) conditions [3]. Hence, the I-V measurements can be performed simply for all required IEC 61853 test conditions and for much more.
- Outdoor analyzes do not need time and logistics for module transfer to indoor lab.

Fig. 1 presents a sketch of an usual outdoor test set-up for PV modules.

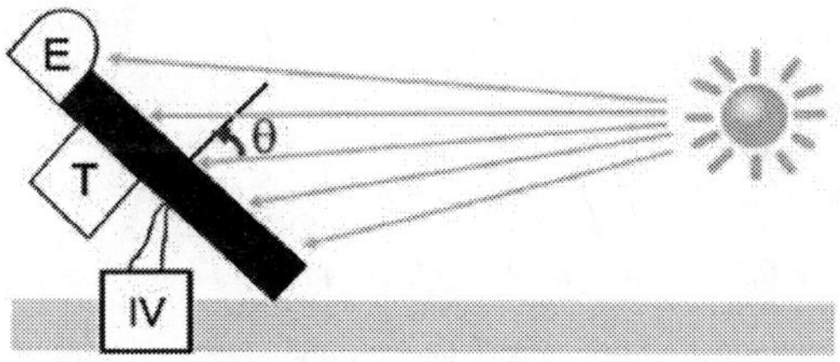

Figure 1: Sketch of an outdoor set-up for electrical I-V characterization of tilted PV module, equipped with sensors for detection of irradiance (E) and temperature (T)

The irradiance in the tilted module plane (scaled by θ as shown in fig. 1) is typically detected by thermopile pyranometers. Temperature sensors on the backside of modules are usually thermocouples or resistor-type thermometers. Passive electric loads are applied commonly for I-V measurements. The set-up in our use is equipped with a planar silicone rubber heater mounted in close thermal contact to module's backside to achieve an extended temperature range up to 100°C for outdoor characterization [4].

For example, fig. 2 shows available test conditions for irradiance and temperature of a 6-day measurement campaign in Coburg (Bavaria, Germany) in July. Note: E_{mod} and T_{mod} refer now to module's tilted plane irradiance and backside temperature, respectively. During the measurement campaign were logged presumably 42.000 different I-V curves of the module. In each minute were measured six curves wherein one of them is stored as a complete table. For other five curves the I-V parameters (I_{sc}, V_{oc}, I_{mpp}, and V_{mpp}) are only logged.

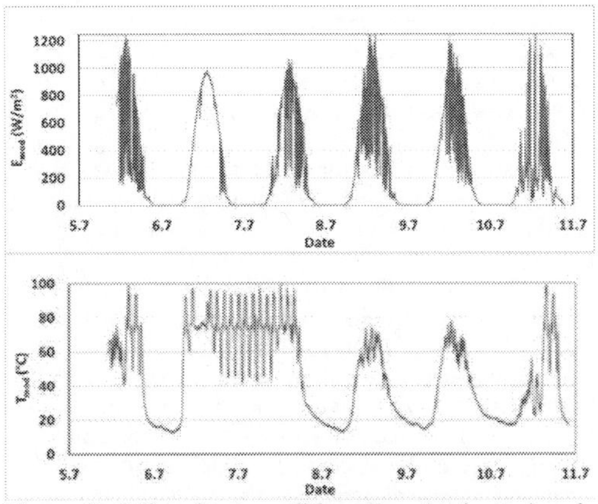

Figure 2: Offer of test conditions E_{mod} (at the top) and T_{mod} (at the bottom) in PV module plane for 6-day measurement campaign in Bavaria (Germany)

The T_{mod} distribution (fig. 2, bottom) shows clearly the impact of used heater: The temperature of module could varied with cycling time of 2 hours between around 50 and 95°C if the heater was turned on (four days and one night within the measurement campaign). In this way, I-V measurements could be accomplished even at high temperatures.

Unfortunately, the accuracy of outdoor measurements of electrical parameters – e.g. power at maximum power point (P_{mpp}) or efficiency (η) – is very poor if they are represented vs. E_{mod} for certain T_{mod}. Such low-light characteristics are shown for our measurement campaign on a PV module in fig. 3: The low light curves of normalized P_{mpp} and η are represented for three different backside temperatures T_{mod} (25, 50 and 75°C). It turns out that it is nearly impossible to determine reliable values for P_{mpp} and η.

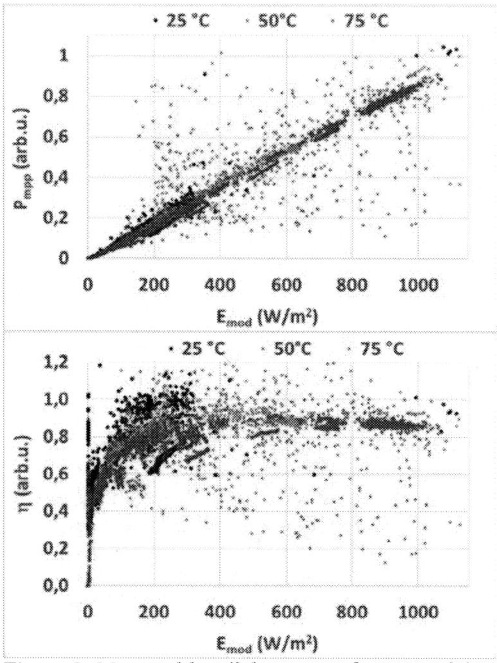

Figure 3: Measured low-light curves of a PV module: P_{mpp} and η vs. measured irradiance E_{mod} for three backside temperatures T_{meas} (25±2°C; 50±2°C and 75±2°C); note: power and efficiency are normalized by STC values of P_{mpp} and η, respectively

What are the reasons for the low reliability of such type of outdoor analyzes? Generally, outdoor measurements suffer on low definiteness of test conditions:

- The detected backside temperatures differ usually from module's cell temperatures due to impact of wind cooling, sun heating and thermal capacities.
- The irradiance detected by pyranometer in tilted module plane is mostly different to those reaching the cells. This is due specific angle dependent reflection of modules according Maxwell's law, which is different to directional response of pyranometers.
- Lastly, the natural-sunlight spectra deviate usually from AM 1.5 G spectrum mainly due to varying absorptions and aerosol extinctions.

In addition, there are also technical measurement reasons resulting in additional uncertainty: The response time of thermopile pyranometers is with 15 to 30 s much longer than I-V measurement time (1-2 s). In case of fast irradiance changes as observed in our measurement campaign (fig. 2 at the top) significant fluctuations for low-light characteristics occur as shown in fig. 3.

For these reasons, state of the art I-V-analyzes require solely clear-sky or partly sunny steady state conditions in order to determine reliably I-V parameters of modules [5, 6]. The I-V data for any irradiance and temperature (as well as STC) are obtained then by means of effective extrapolation parameters, which can be derived by at least two or three measurements within a certain limited range of E and T test conditions. However, such concepts fail if the I-V behavior is nonlinear, for example for rather low irradiances.

Very recently, the so-called self-reference algorithm (SRA) was introduced [3, 4]. The application of SRA on measured electrical outdoor data, e.g. on data of mentioned measurement campaign (fig. 2 and 3), improves the quality of low-light characteristics significantly as shown by fig. 4 which should be compared with fig. 3.

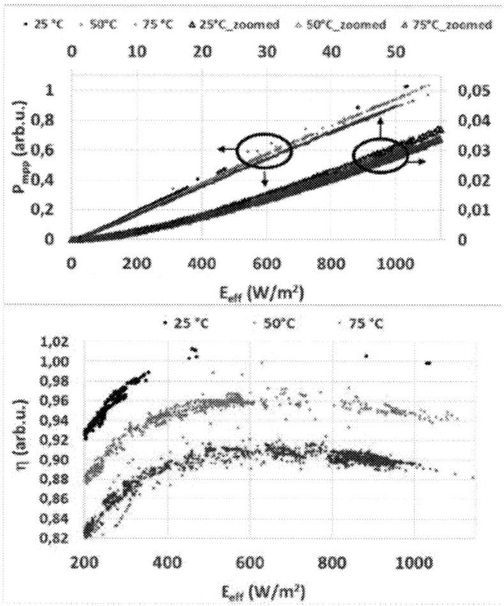

Figure 4: SRA based low-light curves of a PV module: P_{mpp} and η vs. effective irradiance E_{eff} for three effective temperatures T_{eff} (25±2°C; 50±2°C and 75±2°C); note: power and efficiency are normalized by STC values of P_{mpp} and η, respectively

The low-light curves for normalized P_{mpp} and η (fig. 4 top and bottom, respectively; please note the zoomed scale of fig. 4!) are very smooth and the uncertainty is reduced prominently. It becomes feasible to prove a nonlinear behavior of P_{mpp} - E low-light curves for very low irradiances (0 - 50 W/m²). This can be observed easily for the three zoomed curves in fig. 4 (zooming factor: 20).

In this paper, we consider the concept and opportunities of SRA and evaluate the accuracy of the new method.

2 CONCEPT AND OPPORTUNITIES OF SELF-REFERENCE ALGORITHM

2.1 Concept and calibration of SRA

The improvements of SRA are achieved by replacing the measured test conditions E_{mod} and T_{mod} by effective data (E_{eff}) and (T_{eff}) by means of self-referencing to module's measured short circuit current

($I_{sc,meas}$) and open circuit voltage ($V_{oc,meas}$), respectively [3, 4, 7]. Moreover, there is the advantage that response time or synchronization issues of E and T sensors are no longer relevant for data acquisition.

The effective temperatures T_{eff} are obtained within "i" distinct irradiance intervals (interval width: ΔE_i) by determination of distinct temperature coefficients for V_{oc} (TCV_i) and V_{oc}'s at 25°C ($V_{oc,25°C,i}$) according equation (1):

$$T_{eff} = \left(\frac{V_{oc,meas,i}(T_{eff})}{V_{oc,25°C,i}} - 1 \right) \cdot \frac{100\%}{TCV_i} + 25°C \quad (1)$$

The different $V_{oc,meas,i}$ are the real time detected $V_{oc,meas}$ in each interval ΔE_i. The TCV_i's and $V_{oc,25°C,i}$'s are determined by linear regression which is applied on all measured data pairs of $V_{oc,meas,i}$ and T_{mod} for each ΔE_i. The effective irradiation E_{eff} depends on $I_{sc,meas}$ according to equation (2):

$$E_{eff} = \frac{1000\, W/m^2}{I_{sc,STC}} \cdot \frac{I_{SC,meas}(T_{eff})}{1 + \frac{TCI \cdot (T_{eff} - 25°C)}{100\%}} \quad (2)$$

Equation (2) needs knowledge of the temperature coefficient for I_{sc} (TCI) and of I_{sc} at standard conditions ($I_{sc,STC}$). Hence an exact prediction of $I_{sc,STC}$ is needed for calibration of SRA. An preferred outdoor $I_{sc,STC}$ calibration has to consider spectral, thermal, irradiance and directional response issues, reported elsewhere [8]. We use a factorization model [7] which considers these aspects. Finally, $I_{sc,STC}$ can be deduced from measured $I_{sc,meas}$ according equation (3):

$$I_{sc,STC} = I_{sc,meas} \cdot k \cdot t \cdot m \cdot w \quad (3)$$

Factor k is a spectral correction factor depending on spectral response of module and on actual spectrum of solar radiation. The temperature impact, modeled by t, is taken into account by a temperature measurement and the use of temperature coefficient [7]. Factor m corrects linearly deviations of actual irradiation with respect to 1000 W/m². Finally factor w models the $I_{sc,meas}$ as a function of angle of incidence which can be measured by a tracker system.

For our measurement campaign we calibrated the system with I_{sc} measurement at clear sky with T_{mod} = 48 °C, E_{mod} = 970 W/m² and according equation (2). For SRA processing a width of irradiance interval of ΔE_i = 50 W/m² was chosen. The low-light curves were drawn for certain temperatures X including a width of acceptance of ΔT_{eff} = 4°C, hence T_{eff} = X ± 2°C.

2.2 Low-light characteristics of I-V parameters

SRA method allows the representation of low-light characteristics for all relevant parameters of I-V curves (I_{sc}, V_{oc}, fill factor) as for example shown for V_{oc} in fig. 5.

One advantage of SRA with respect to common outdoor analyzes becomes visible: It is feasible to determine electrical parameters in complete range of naturally available test conditions, and these are typically much more than those required by standard IEC 61853.

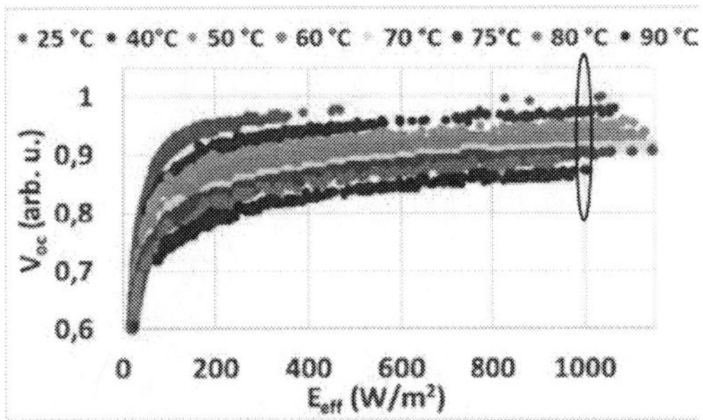

Figure 5: Low-light curves of a PV module: V_{oc} vs. effective irradiance E_{eff} for eight different effective temperatures T_{eff} (X±2°C); note: voltage is normalized by STC value of V_{oc}

2.3 Temperature coefficients and prediction of I-V parameters for specific test conditions (STC, NOCT)

The huge number of data points of electrical parameters vs. E_{eff} and T_{eff} allows simply and with high precision the determination of temperature coefficients for voltage (TCV), current (TCI) or power (TCP) at user-defined irradiances under outdoor conditions (e.g., STC: 1000 W/m² or NOCT: 850 W/m²). For example, TCV was determined by means of V_{oc} data at 1000 W/m² and regression line analysis. This is depicted in fig. 6. The TCV was predicted to be TCV = - 0.2 ± 0.004 %/K.

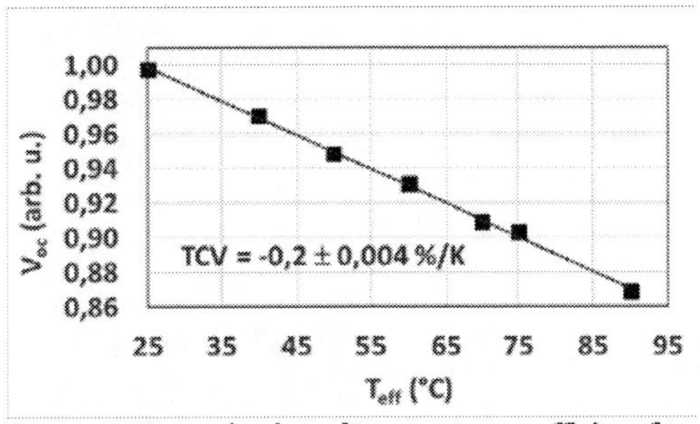

Figure 6: Determination of temperature coefficient for V_{oc} (TCV = - 0.2 ± 0.004 %/K) at irradiance of E_{eff} = 1000 W/m²; note: voltage is normalized by STC value of V_{oc}

The knowledge of temperature coefficients enables reliable predictions of V_{oc}, I_{sc} and P_{mpp} (also for V_{mpp} or I_{mpp}) for STC or NOCT conditions even though the rather low temperatures of 25°C or 20°C could not reached naturally within measurement campaign. Note: in our measurement campaign we could measure electrical data also for T_{eff} = 25°C by happenstances.

2.4 I-V and P-V curves

SRA uses electrical data, which were deduced from detected I-V-characteristics. According to the SRA concept only measured test conditions for irradiance and temperature E_{mod} and T_{mod} were replaced by actually acting E_{eff} and T_{eff}. Hence, I-V and power-voltage (P-V) curves can simply been presented with reallocated test conditions as shown in fig. 7. This figure represents I-V and P-V curves of module tested within measurement campaign for fixed E_{eff} = 400 W/m² and variable temperatures (T_{eff}: 25, 50 and 75°C).

Detailed analyzes of I-V curves by fitting routines allow access on further parameters of PV modules which are suitable for I-V modelling and module characterization. For example the resistances (series and parallel), ideality factor and saturation current are a matter of interest. We will not state fitting results in this contribution.

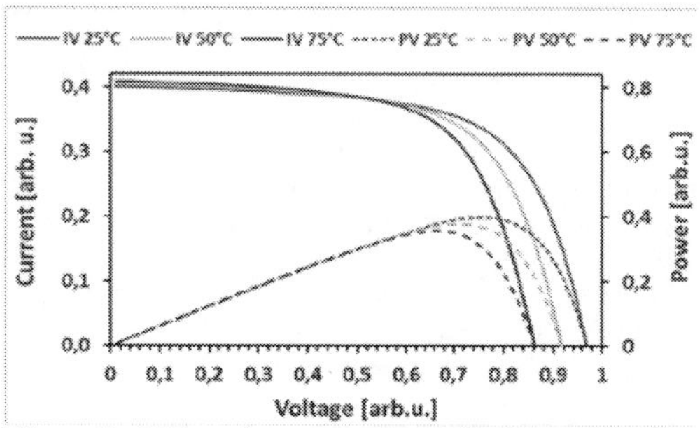

Figure 7: I-V and power-voltage (P-V) curves for effective irradiance of $E_{eff} = 400$ W/m^2 and three effective temperatures $T_{eff} = 25°C, 50°C, 75°C$; note: current, voltage and power are normalized by STC values of I_{sc}, V_{oc} and P_{mpp}, respectively

3 PRECISION OF SRA RESULTS

Finally, the accuracy of SRA results is evaluated. The normalized experimental standard deviations of η and P_{mpp} ($s_η$ and s_P, respectively) are considered by means of low-light curves as presented in fig. 4. A suitable image enlargement of the η curve reveals the spreading of η values (fig. 8, at the top).

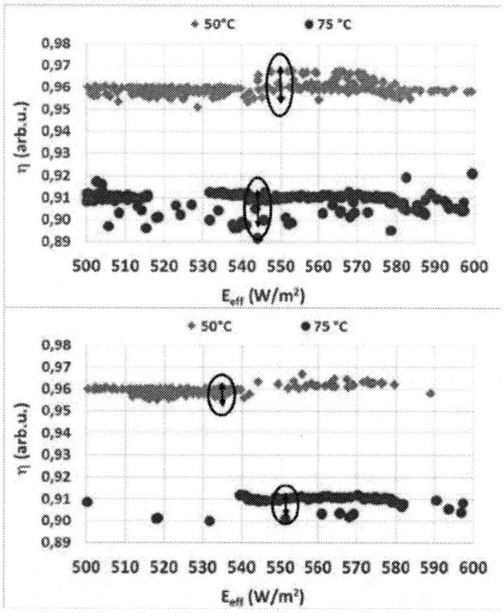

Figure 8: Enlarged low-light curves of η (zoomed in from fig. 4), at the top: SRA algorithm calculated and presented with intervals of $ΔE_i = 50$ W/m^2 and $ΔT_{eff} = 4°C$; at the bottom: reduced intervals with $ΔE_i = 20$ W/m^2 and $ΔT_{eff} = 2°C$

The standard deviation of η can be determined by using the data of fig. 8. We found a standard deviation of about $s_η = 1.1\%$ (fig. 8 top) and one for power (not presented in fig. 8) of about $s_P = 0.9\%$. The spreading of data points is not primary caused by the measurement system (passive electrical load) but by SRA concept itself. This is due to the application of intervals of irradiance ($ΔE_i$) and temperature ($ΔT_{eff}$) for data processing and presentation. We used generally processing intervals of $ΔE_i = 50$ W/m^2 and $ΔT_{eff} = 4°C$. It is a proper way to reduce the width of intervals in order to decrease the experimental standard deviations down to the level, which is caused by measurement equipment. For example the application of $ΔE_i = 20$ W/m^2 and $ΔT_{eff} = 2°C$ gives rise for reduced standard deviations with $s_η = 0.7$ % and $s_P = 0.6$ %, respectively. However, the interval reduction results in a loss of data points as well, shown in fig. 8 at the bottom. This is disadvantageously for keeping the measurement uncertainty as low as possible.

Nevertheless, the total measurement uncertainty of SRA is mainly caused by calibration itself, meaning the determination uncertainty for $I_{sc,\,STC}$. A rather extended inaccuracy calculation reveals for our set-up an uncertainty of η of about ± 6%. The precision could be further increased by means of more reliable short circuit measurements and application of more precise spectrometer and pyranometer.

Finally, we consider experimentally the consistence of P_{mpp} data at standard test conditions ($P_{mpp,\,STC}$) determined by SRA. For this purpose we have determined $P_{mpp,\,STC}$ by means of three different methods:

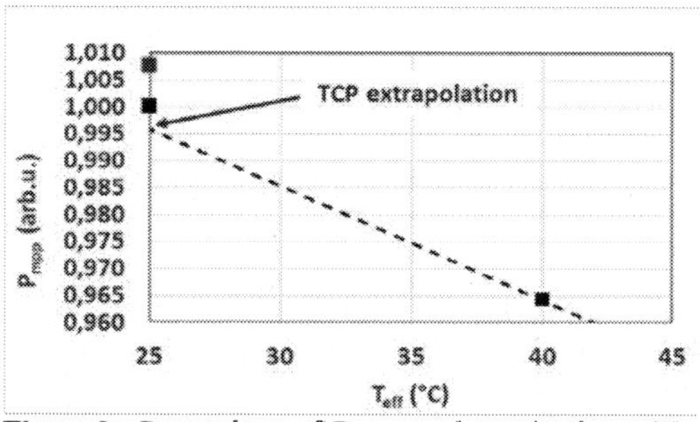

Figure 9: Comparison of $P_{mpp,\,STC}$ determinations: (1) direct measurement of $P_{mpp,\,STC}$ by indoor flasher (red box), (2) direct outdoor measurement by SRA (black box at 25°C) and (3) outdoor measurement by SRA and regression line extrapolation

(1) By indoor measurement using a calibrated flasher system (fig. 9, highlighted by red box),

(2) By outdoor measurement with applied SRA for $E_{eff} = 1069$ W/m^2 and $T_{eff} = 27°C$ including linear correction towards STC values (fig. 9, highlighted by black box at $T_{eff} = 25°C$), and

(3) By outdoor measurement at $E_{eff} = 1000$ W/m^2, with variable temperatures T_{eff} in the range of 40 to 90°C and by linear extrapolation of TCP regression line according section 2.3. This concept is also shown in fig. 9 and the intersection at 25°C is emphasized too.

Note: for all three methods the same $I_{sc,STC}$ calibration was is use.

The $P_{mpp,\,STC}$ values of all methods vary by less than

1.5% and prove the high reliability of SRA method. It turns out that electrical outdoor analyzes nowadays have the same precision as standard indoor measurements.

4 SUMMARY

Electrical outdoor measurements on PV modules can achieve precisions like those performed with calibrated indoor flasher technique if the self-reference algorithm (SRA) is applied. We have concluded a measurement uncertainty of SRA based outdoor analyzes by about ± 6 % for module efficiency. An experimental comparison of in- and outdoor determination of $P_{mpp, STC}$ on the same module reveals a conformity of the values by about 1.5 %.

Low-light characteristics of all relevant I-V parameters can be measured by SRA within the wide range of naturally offered irradiance and temperature. Electrical data of PV modules can be produced for much more test conditions than required by standard IEC 61853.

The temperature coefficients of P_{mpp}, V_{oc} and I_{sc} can be determined very precisely by linear regression analysis because a huge number of I-V data for specific irradiances is available. Hence, all electrical parameters of modules can be predicted accurately for STC and NOCT if extrapolation of regression lines is used.

SRA concept replaces the sensed test conditions for irradiance (by pyranometer) and temperature (by resistor-type sensor) by actually acting effective values. Accordingly, I-V and P-V curves are recorded with re-allocated and valid test conditions. Special fitting analyzes of I-V curves can deliver reliable electrical parameters for I-V modelling (series resistor, parallel resistor, ideality factor, saturation current) although "only" outdoor data were used.

5 REFERENCES

[1] M. Gostein, L. Dunn; "Light Soaking Effects On Photovoltaic Modules: Overview And Literature Review", Conference Record of the IEEE Photovoltaic Specialists Conference (2011), 003126–003131

[2] T. Roth, D. Wichmann, K. Meyer, M. Orlob; "In-depth analysis of transient errors of inline I-V measurements", Energy Procedia vol. 8 (2011), pp. 82–87.

[3] H.-D. Mohring, D. Stellbogen, D. Geyer, P. Lechner; "Comparative Assessment of Different Thin Film Technologies from Outdoor I-V Characteristics", 26th European Photovoltaic Solar Energy Conference, (2011), 4AV.2.46, pp. 3625 - 3628.

[4] D. Kaiser, M. Simmler, M. Wachter, T. Faber, F. Becker, M. Sayala, T. Kaden, A. Schulze, B. Hüttl; "Concept for Fast and Precise PV Module Outdoor Characterization", 31th European Photovoltaic Solar Energy Conference, (2015), 5AV.6.10, pp. 1956 - 1958.

[5] Ch. Bendel, A. Wagner; „Photovoltaic Measurement Relevant to the Energy Yield" 3rd World Conference on Photovoltaic Energy Conversion (2003), 7P-B3-09

[6] Y. Hishikawa, H. Ohshima, M. Higa, K. Yamagoe, T. Doi "Precise Determination of the STC I-V Curves by Wide-Range Linear Extrapolation of Outdoor I-V Curves on Partly Sunny Days"; 32nd European Photovoltaic Solar Energy Conference and Exhibition (2016), 5DO.11.4

[7] M. Wachter, L. Gottschalk, M. Simmler, A. Schulze, F. Becker, M. Sayala, B. Hüttl; "Short Circuit Current Measurements at Clear-Sky Conditions on Photovoltaic Modules: Basic for a Reliable Self-Reference Algorithm"; 32nd European Photovoltaic Solar Energy Conference and Exhibition (2016), 5BV.4.16

[8] Th. Fey, I. Kröger, F. Witt, St. Winter; Comprehensively Characterized Solar Cells: Impact of Angular, Spectral, and Non-linear Effects"; 32nd European Photovoltaic Solar Energy Conference and Exhibition (2016), 5DO.11.3

6 ACKNOWLEDGEMENT

The authors gratefully acknowledge the German Federal Ministry of Education and Research (BMBF) for funding the project "PV-Präzis" under support code FKZ 03FH018PX4.

PROPOSAL AND INVESTIGATION OF NOVEL PORTABLE DEGRADATION DIAGNOSIS SYSTEM FOR PV MODULE IN ACTUAL OPERATION

T. Tanaka[1], T. Nagayama[1], T. Hayashi[1], T. Yanagidaira[1], Y. Inui[2]
1: Ibaraki University
4-12-1 Nakanarusawa, Hitachi, Ibaraki 316-8511, Japan
2: The University of Shiga Prefecture
2500 Hassaka, Hikone, Shiga 522-8533, Japan

ABSTRACT: In this study, the authors proposed a novel portable degradation diagnosis system that can detect a specific degraded PV module among a number of actually operated sound PV modules. At first, in order to confirm basic performance of the proposed system, numerical simulation for the system were carried out. From the simulation results, it was made clear that the performance evaluation of PV module is feasible by using the proposed system. Next, some demonstration tests were executed after a test equipment based on the proposed system was assembled. Through the experiments, it was successfully demonstrated that the proposed system has sufficient ability to detect degradation of PV module and the required time for diagnosis per module is very short. Furthermore, the proposed diagnosis system is portable and can be utilized everywhere. Therefore, the proposed system is a promising new candidate for on-site degradation diagnosis system that can detect a specific degraded PV module among a number of actually operated sound PV modules.
Keywords: PV module, portable, degradation, diagnosis

1 INTRODUCTION

Recently, renewable energy is a key to solve environmental and energetic problems. From this point of view, photovoltaic (PV) system is a promising technology to utilize renewable energy effectively. So far, many researches on PV system have been carried out [1-6]. Owing to these efforts, PV systems are widely commercialized now.

On the other hand, degradation of PV modules has been reported in many papers when PV systems are operated outdoors for long time [7-10]. The degradation provokes the decrease in the environmental and economic benefits of the PV system owners. The degradation of PV module is also a serious problem for the utlity because the utility has to compensate the the reduction of power generation by the PV systems in the utlity grid to keep the balance between electric generation and electric demand properly. Therefore, the degradation of PV modules damages the stability of the utility grid and its general users are also affected by the degradation. This means that the degradation of PV modules has a widespread impact on many stakeholders and so researches on maintenance of PV modules are very important.

Various degradation diagnosis methods of PV modules have been proposed and investigated [11-14]. Some researchers have developed a diagnosis method that evaluates power generation of every PV string in which a number of PV modules are connected in series. However, detection of a specific degraded PV module from a lot of PV modules installed in PV system is difficult.

To reduce maintenance cost of PV systems through minimization of refurbishing parts, some degradation diagnosis systems that can detect a specific degraded PV module among a number of actually operated PV modules is required. Especially, since PV system is installed in huge wide area in the case of general PV power plant and it often occurs that PV modules are put in not easily accessible location in the case of PV system for domestic use, portability of the degradation diagnosis system is also needed. However, no such kind of system with both the detectability of a specific degraded module and the portability has been developed yet.

In this study, considering the above backgrounds, the authors at first proposed a novel portable degradation diagnosis system that can detect a specific degraded PV module among a number of actually operated sound PV modules. Next, to confirm the basic performance of the proposed system, numerical simulations for the system were executed. Then, a prototype of the experimental equipment of the proposed system was assembled and its operating characteristics were experimentally confirmed.

2 PROPOSED DEGRADATION DIAGNOSIS SYSTEM

2.1 Configuration and operation

Schematic diagram of the proposed novel portable degradation diagnosis system for PV module is shown in Fig. 1. This system is composed of a xenon lamp, a Li-ion battery, a boost-up chopper circuit, a trigger circuit, and a series circuit of diode, resistance and capacitor. This system is divided into two parts: a light source part and an evaluation part.

With respect to the process of the degradation diagnosis, a flash lighting from the xenon lamp exposes a target PV module. The electric charges generated by the flash lighting is stored in the capacitor, and the terminal voltage of the capacitor rises in proportion to the

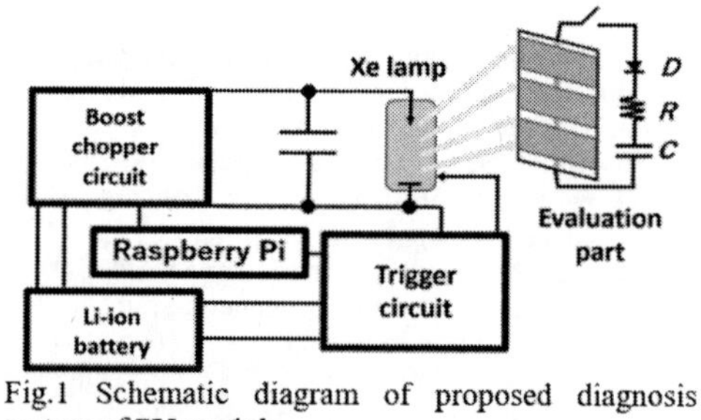

Fig.1 Schematic diagram of proposed diagnosis system of PV module.

generated electric charges. Therefore, degraded PV module can be identified by measuring and comparing the capacitor voltage because the generated electric charges by degraded PV module is less in comparison with sound PV modules.

2.2 Features

The features of the proposed system are as follows. Although solar simulators using xenon lamp are popular, commercialized solar simulators with high accuracy are usually too large to use them at the actual sites of PV power plants. In addition, large capacity power sources are needed to drive such solar simulators. Therefore, the utilization site of the conventional solar simulators is limited inside laboratories. In contrast to the conventional ones, the key advantage and originality of the proposed system are employment of a Li-ion battery and a step-up chopper circuit as the lighting circuit of the xenon lamp. Since they realize large reduction in weight and size, the light source part of the proposed system becomes portable. In fact, maintenance service person can carry it anywhere and use it at actual sites of PV power plants in the same manner as the solar simulator installed in laboratory. In addition, the flash lighting is selected and so the battery capacity can be saved. Thus, the equipment assembled based on this proposal has high portability. The proposed degradation diagnosis system, therefore, has features of both the detectability of a specific degraded PV module among a number of operated sound PV modules and portability. Thus, using the proposed system can minimize repairing parts of PV power plant, leading to suppression of the maintenance cost.

3 SIMULATION

In the proposed system, the *RC* series circuit is employed as the evaluation part, and applicability of this circuit to diagnosis of PV module was not made clear. Therefore, at first, in order to confirm that performance of PV module can be evaluated by the *RC* series circuit employed in the proposed system, a series of diagnosis procedure was simulated. In this section, the numerical method, conditions and results are discussed.

3.1 Method

Current–voltage (*I–V*) curve of a PV module is simulated by the following equation.

Table I: Parameters used in simulations.

PV module	Set value
I_0	6.19×10^{-7} A
q	1.6×10^{-19} C
m	36
k	1.38×10^{-23} J/K
n	1.63
T	298 K
R_{SH}	200 Ω
R_S	0.05, 5.05 Ω
I_{SC}	8A, 7A, 6A
RC series circuit	**Set value**
C	500 μF
R	1 Ω

$$I = I_{SC} - I_0\left[\exp\left\{\frac{q(V+R_SI)}{mnkT}\right\}-1\right]-\frac{V+R_SI}{R_{SH}} \tag{1}$$

where I_0 is the saturation current, q the charge of an electron, m the number of PV cells in a PV module, n the diode coefficient, k the Boltzmann constant, T the temperature, R_S the internal series resistance of the PV module, and R_{SH} the internal parallel resistance of the PV module.

Using both the current and the voltage of PV module obtained from equation (1), the capacitor voltage in the series *RC* circuit existing in the evaluation part can be calcuated by the following eauations.

$$RC\frac{dV_C}{dt}+V_C=V \tag{2}$$

$$\frac{V-V_C}{R}=I \tag{3}$$

where V_C is the capacitor voltage, R and C the resistance and capacitance of the *RC* series circuit.

3.2 Simulation condition

The numerical setting values of the PV module and the *RC* series circuit are listed in Table I.

With respect to the assumption of degradation of PV module, the authors took into account two degradations of PV module. One is the increase in the series resistance due to corrosion of internal wire. To realize the degradation in the simulations, the authors considered two internal resistances of 0.05 Ω as the normal and 5.05 Ω as the degradation. Besides, the other is the reduction of the short circuit current related to yellowing, whiting and delamination on the surface of PV module. In the simulations, three short circuit currents were taken into consideration as shown in Table I.

As was described in Section 2, flash light is irradiated to a target PV module in the proposed degradation diagnosis system. Therefore, some setting of flash light is also needed. In this study, a pulse flash lighting with a duration of 2 ms was assumed. To satisfy this lighting condition, while the short circuit current was set to 0 A under the condition of no flash light, the short circuit current indicated in Table I was set just for 2 ms when the flash light is irradiated to PV module.

3.3 Numerical results

The authors computed time variations of capacitor voltage when three flash lights were irradiated to target PV module and the internal series resistance was set as 0.05 Ω

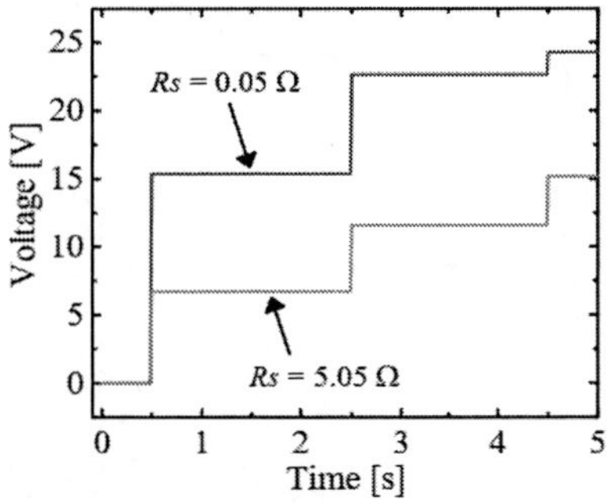

Fig.2 Comparison of numerical results of capacitor voltage with and without degradation of increase in internal series resistance.

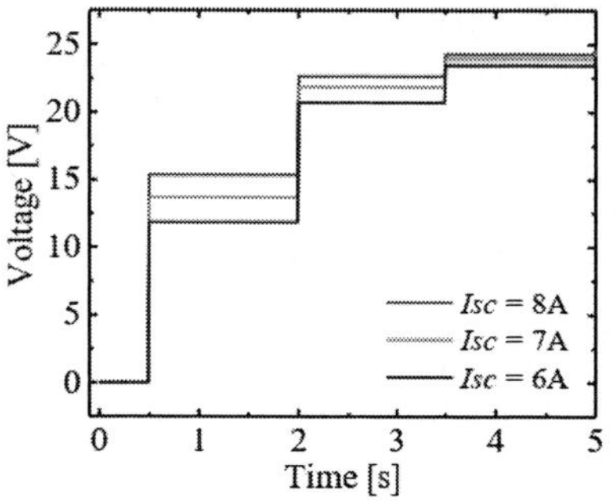

Fig.3 Comparison of numerical results of capacitor voltage with and without degradation of decrease in short circuit current.

and 5.05 Ω. The obtained results are depicted in Fig. 2.

From this figure, it can be observed that the capacitor voltage comes at about 16 V after the first flash light in the case of the internal series resistance of 0.05 Ω. On the contrary, the capacitor voltage remains at about 7 V in the case of the internal series resistance of 5.05 Ω. Thus, from this simulation, it can be said that the performance evaluation related to the degradation of PV module attributed to the corrosion of internal wire is expected.

Next, the authors calculated time variations of the capacitor voltage when three flash lights were irradiated to the target module and the short circuit current were changed as indicated in Table I. In this case, the internal series resistance is fixed as 0.05 Ω. The simulation results are shown in Fig. 3.

From this figure, it is found that the highest capacitor voltage is obtained in the case of the highest short circuit current. Then, the capacitor voltage is reduced as the decrease in the short circuit current. This result implies that the proposed diagnosis system has possibility to detect degradation of PV module related to change of the light transmission rate due to yellowing, whiting and delamination on the surface of PV module.

It can be confirmed from the above that the proposed system has sufficient diagnostic ability of degraded PV module associated with the increase in the series resistance and the decrease in the light transmission rate.

4 DEMONSTRATION TEST

Since good results were obtained through the above simulations, a prototype of the experimental equipment was assembled based on the proposal and some demonstration tests were carried out. In this section, the configuration and operation of the prototype equipment and results of the tests are described.

4.1 Demonstration equipment

The circuit structure of the prototype equipment assembled based on the proposal is shown in Fig. 4. Since Li-ion battery with terminal voltage of 180 V is utilized as power source of the xenon lamp, the light source part of the proposed diagnosis system is portable. With respect to the operation, the capacitor C_{11}, which is connected to xenon lamp in parallel, is charged until 450 V. After that, a trigger signal of about 3 kV is provided to the xenon lamp. Hereby, since electric avalanche phenomenon occurs in the lamp, the charge of the capacitor C_{11} is released in a moment. At this time, flash light is irradiated to target PV module, and power generation attributed by the flash light is stored in the capacitor of the *RC* series circuit. The capacitor voltage becomes an index of the performance evaluation of target PV module.

4.2 Test conditions

With reference to the simulation conditions, the cement resistance of 1 Ω and the electrolytic capacitor of 470 μF were employed in the *RC* circuit of the evaluation part. As for the terget module, the commercialized PV module (SOLAR CENTER, Model SCM 30W) was used

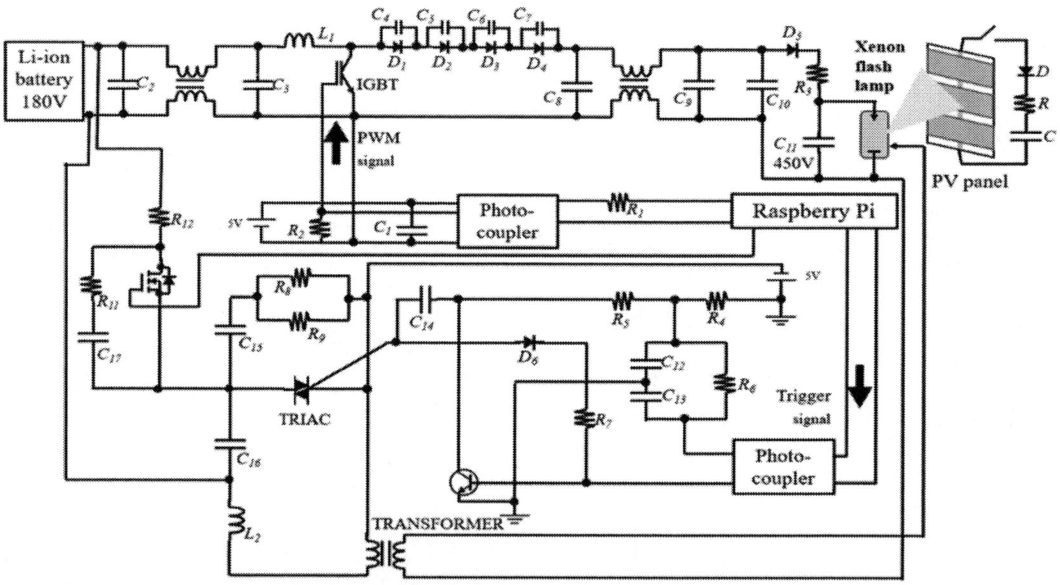

Fig.4 Circuit structure of proposed novel portable degradation diagnosis system for PV module.

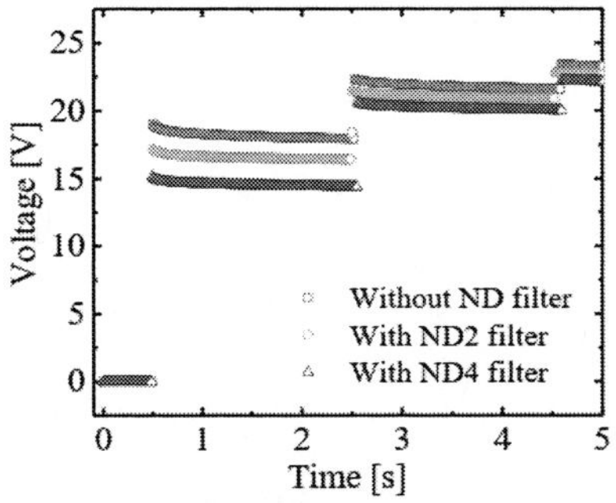

Fig.5 Comparison of experimantal results of capacitor voltage with and without ND filter.

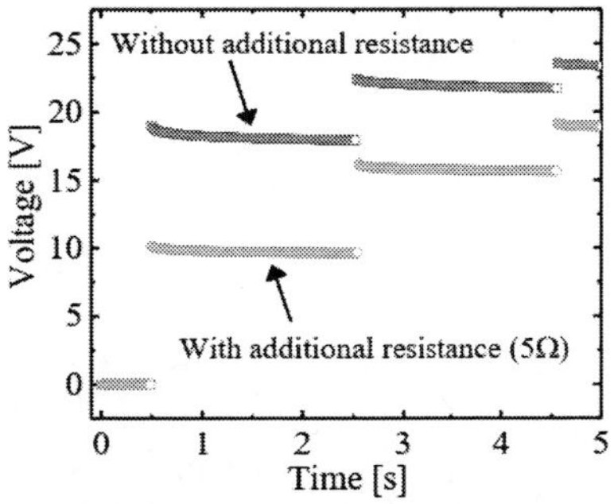

Fig.6 Comparison of numerical results of capacitor voltage with and without additional resistance.

in the demonstration tests.

With respect to the degradation of PV module, in the same manner as the above described simulations, both the increase in the internal series resistance due to the corrision of internal wire and the decrease in the short circuit current due to the yellowing and delamination on the surface were taken into account. For this purpose, the additional resistance of 5 Ω is connected between the terget module and the *RC* series circuit in the case of degradation related to the corrision of internal wire. On the other hand, in oder to simulate the actual reduction of the light transmission rate and the reduction of the short circuit current related to the yellowing and delamination on the surface, two types of neutral-density filters of ND2 and ND4 (Cokin, recutangler ND filter) were put on the xenon lamp.

4.3 Test results for change of short circuit current

Fig. 5 depicts the time variations of the capacitor voltage when three flash lights were provided to the target modules. In the test, two types of ND2 and ND4 filter were put on the xenon lamp. To compare, the capacitor voltage is also plotted in this figure when no ND filter was used.

From this figure, it is made clear that, in the case of no ND filter, the capacitor voltage becomes about 19 V after the first flash light. On the other hand, the capacitor voltage reaches only 17 V in the case of ND2 and 15V in the case of ND4. Therefore, the capacitor voltage is reduced depending on the filtering level. The difference in capacitor voltage can be distinguished only by one-time irradiation of flash light. Therefore, the proposed system can be used to identify the lowering degree of the light transmission rate of PV module. In addition, its diagnosis time per module is very short.

4.4 Test results for change of internal resistance

Fig. 6 shows time variations of the capacitor voltage when the flash lights were irradiated to the target modules combined with and without the additional resistance of 5 Ω.

From these experimental results, it is seen that the capacitor voltage is relatively low when the additional resistance is added to the target module. In particular, the capacitor voltage after first flash light was approximately 19.0 V in case the additional resistance is not connected while that reached only about 10.0 V in case of the additional resistance. The largest difference on the capacitor voltage is indicated after the first flash light. From the above, it can be confirmed that the degradation related to the change of the internal resistance due to the corrosion of internal wire of PV module can be also detected by the proposed system and the required diagnosis time per module is also very short.

According to the former study [15], generated electricity of PV system is dramatically reduced when only one limited module has significant degradation among a lot of sound PV modules. Therefore, the proposed system is considered to be a powerful tool of onsite diagnosis system of PV power generation system because maintenance service person freely carries the xenon light equipment of the proposed system and checks the performance of PV module one by one in the wide site of PV power generation plant by using the proposed diagnosis system.

5 CONCLUSIONS

In this study, the authors proposed a novel portable degradation diagnosis system that can detect a specific degraded PV module among a number of actually operated sound PV modules. The proposed system has the feature of both the detectability of a specific degraded PV module among a number of operated sound PV modules and portability.

In the proposed system, the *RC* series circuit is selected as the evaluation part. To confirm that the performance of PV module can be evaluated by the *RC* series circuit employed in the proposed system, some simulations and experiments were carried out. From the numerical results, it was made clear that the performance evaluation of PV module is feasible by the proposed system. Some demonstration tests were also executed by using the prototype equipment assembled based on the proposal. Through the experiments, it was successfully demonstrated that the proposed system has sufficient ability to detect degradation of PV module and the diagnosis time per module is very short.

From the above, the proposed system is considered to be a powerful tool of onsite diagnosis system of PV power generation system because maintenance service person freely carries the xenon light equipment of the proposed system and check the performance of PV module one by one in the wide site of PV power generation plant. By using the proposed system, a specific degraded PV module can be found among a number of sound PV modules. This means that repairing module can be minimized. As a result, the proposed system contributes to reduce the maitenance cost.

Finally, the authors would express their appreciation to the financial support by International Technological Interaction, TEPCO Memorial Foundation.

REFERENCES

[1] K. Kawamura, K. Naka, N. Yonekura, S. Yamanaka, H. Kawamura, H. Ohno, K. Naito, Solar Energy Materials & Solar Cells 75 (2003) 631
[2] P. Mohanty, G. Bhuvaneswari, R. Balasubramanian, N.K. Dhaliwal, Renewable and Sustainable Energy Reviews 38 (2014) 581
[3] V. Dash and P. Bajpai, Sustainable Energy Technologies and Assessments 9 (2015) 68
[4] R. Xu, K. Ni, Y. Hu, J. Si, H. Wen, D. Yo, Energy Conversion and Management 148 (2017) 100
[5] C.S. Ruschel, F.P. Gasparin, E.R. Costa, A. Krenzinger, Solar Energy 133 (2016) 35
[6] M. Ito, M. Kudo, M. Nagura, K. Kurokawa, Progess in Photovoltailcs: Research and Application 19 (2011) 878
[7] M.A. Munoz, M.C. Alonso-Galcia, N. Vela, F. Chenlo, Solar Energy 85 (2011) 2264
[8] A. Sayyah, M.N. Horenstein, M.K. Mazmber, Solar Energy 107 (2014) 576
[9] P. Sanchez-Friera, M. Piliougine, J. Carretero, M.S. de Cardona, Progress in Photovoltailcs: Research and Application 19 (2012) 658
[10] N. Kahoul, M. Houabes, M. Sadok, Energy Conversion and Management 82 (2014) 320
[11] S. Silvestre, S. Kichou, A. Chouder, G. Nofuentes, E. Karatepe, Energy 86 (2015) 42
[12] A. Chouder, S. Silvestre, Energy Conversion and Management 51 (2010) 1929
[13] E. Garoudja, F. Harrou, Y. Sun, K. Kara, A. Chouder, S. Silvestre, Solar Energy 150 (2017) 458
[14] M. Dhimish, V. Holmes, Solar Energy 137 (2016) 236
[15] T. Hayashi, T. Nagayama, T. Tanaka, Y. Inui, Proceedings of the International Conference on Electrical Engineering (ICEE2017), 201702200000050 (2002) 1698.

OUTDOOR CHARACTERIZATION OF CDTE TECHNOLOGY AND SEASONAL PERFORMANCE ANALYSIS AT DIFFERENT LATITUDES IN EUROPE

Cristina Cornaro[#], Marco Pierro[#*], David Moser*, Gustavo Nofuentes[°], Miguel Alonso-Abella[§], Chris Gueymard[&]
[#]Department of Enterprise Engineering, University of Rome Tor Vergata,
Via del Politecnico, 1 – 00133 Rome, Italy, cornaro@uniroma2.it, pierro@uniroma2.it
*EURAC Research, Viale Druso, 1, 39100 Bolzano, Italy e-mail: david.moser@eurac.edu, marco.pierro@eurac.edu
°Grupo de Investigación y Desarrollo de Energía Solar, Escuela Politécnica Superior, University of Jaén, Campus de Las Lagunillas, s/n. 23071-Jaén, Spain, gnofuen@ujaen.es
§Photovoltaic Solar Energy Unit (Renewable Energy Division, CIEMAT), Avda.Complutense 40, 28040 Madrid, Spain, miguel.alonso@ciemat.es
&Solar Consulting Services, P.O. Box 392, Colebrook, NH 03576, Chris@SolarConsultingServices.com

ABSTRACT: The aim of this work is to improve knowledge about the observed seasonal behavior of CdTe modules under real operating conditions. The study consists of a characterization of the CdTe photovoltaic (PV) technology based on the application of the Sandia Array Performance Model (SAPM) to two PV fields in Italy (Viterbo and Bolzano) and two PV test modules in Spain (Madrid and Jaén). SAPM is a semi-empirical power estimation model that takes into account the four main phenomena that affect the Performance Ratio (PR) of PV plants or modules: temperature, angle of incidence, irradiance and spectral effects. New outdoor experimental procedures were set up at the ESTER station of University of Rome Tor Vergata to validate the temperature coefficients, the angle of incidence coefficients and the spectral mismatch coefficients of CdTe modules to be used in the SAPM model implementation. A detailed analysis of the deviation between the experimental PR data and the modeled PR estimated by SAPM reveals that it is possible to characterize the residuals and attribute them to phenomena not considered by the model.
Keywords: PV, CdTe, outdoor characterization, performance ratio, spectral irradiance.

1 INTRODUCTION

There is still some scientific debate regarding the seasonal behavior of CdTe technology under real operating conditions. Some preliminary results that illustrate different estimates of the seasonal effect's magnitude can be found in the literature. In particular, it has been observed that this seasonal effect is highly variable, depending on the PV modules' location, with various explanations for its causality [1] [2] [3] [4] [5] [6]. For example, in [7] [8] [9] [10] the better performance during summer was attributed to spectral effects associated with a positive correlation between the spectral mismatch factor (MMF) and the water vapor content of the atmosphere. Other authors ascribe the seasonal variability of the CdTe performance to meta-stability effects [11] that could be related to the cell materials [12] [13] [14] [15].

It is emphasized that the seasonal variability in PR is difficult to evaluate because it is usually masked by other perturbing effects.

Considering the rapid development of CdTe technologies, the present study uses an original experimental procedure to fully characterize PV modules under outdoor (real) conditions. A detailed methodology is developed here to isolate the performance seasonality by comparing the results of the theoretical Sandia Array Performance Model (SAPM) [16] to experimental observations. Moreover, the method is used to analyze the differences in seasonal performance patterns observed at various sites, and to elucidate the origin of the seasonal effect and its relationship with latitude and/or climatic features of some atmospheric constituents.

2 MATERIAL AND METHODS

2.1 Methodology

The method consists of two steps. In the first step, three experimental CdTe modules are tested at the ESTER station of University of Rome Tor Vergata, Italy. Their thermal coefficients, along with the coefficients needed to describe the absolute air-mass (AM) spectral dependence, as well as the glass reflectance effect, are evaluated and compared with those obtained at Sandia Laboratory in Albuquerque, New Mexico, using a different procedure. This experiment is designed to validate the CdTe module characterization according to SAPM, verifying whether the coefficients used in the parametric model depend on procedure or location. In the second step, these experimental coefficients are used to evaluate the performance ratio (PR) using SAPM at four different locations, including Tor Vergata. The deviation observed between the monthly experimental and theoretical performances is assumed to be mainly due to three effects not considered in SAPM: (i) initial degradation; (ii) degradation rates of the nominal power; and (iii) performance seasonality. Based on this assumption, the PR modeled with SAPM can be corrected using three additional factors, one for each of these effects.

2.2 Experimental activity and data

The outdoor characterization is carried out using three CdTe modules manufactured by First Solar, model FS 275. The experiment is performed at the ESTER lab (lat. 41.86°N, long. 12.62°E). The laboratory is equipped with a complete meteorological station, a tracker stand, as well as both variable and fixed-tilt stands that can host PV modules of various technologies [17]. More precisely, this station measures the global and diffuse horizontal irradiance and the reflected irradiance with three Kipp&Zonen CM21 pyranometers, as well as direct normal irradiance with a Kipp&Zonen CH1 pyrheliometer. In addition, air temperature and relative humidity are monitored by a Rotronic thermohygrometer, atmospheric pressure by a Skye Instruments Ltd air pressure sensor model SKPS-810, and precipitation by an EM Ltd model ARG-100 gauge. All the data are recorded by a Campbell Scientific CR100 data logger with 1-minute time step. A number of additional variables are measured on the tracking, variable-tilt and fixed-tilt stands: back-of-module (BOM) temperature

(T_{bom}) with a PT100, global tilted irradiance (*GTI*) measured in the plane-of-array (POA) with a CM21 pyranometer, as well as the maximum power point (MPP) electrical values and the I-V curves with a MPPT 3000, kindly provided by the SUPSI laboratory, Switzerland. Finally, the incident spectrum is measured on the tracking stand by two EKO spectroradiometers, models MS-710 (350–1000 nm) and MS-712 (900–1700 nm). The BOM temperature, POA irradiance and MPP data are recorded every minute, whereas the spectral data and I-V curves are acquired every 10 minutes. Finally, a portable station is used for the evaluation of temperature coefficients and angle of incidence. This portable station measures the horizontal and POA irradiances with CM21 pyranometers, the air and BOM temperatures with PT100 sensors, and the module's I-V curves. For this setup, the observations are recorded every 20 seconds with a Campbell Scientific CR1000 data logger. All data streams are filtered to remove the unphysical or inconsistent data points caused by monitoring problems or instrument malfunction. Overall, a stringent quality control procedure [18] is applied to the database, considering both a plausible-value check and a time-consistency check. Finally, a check of coherence between the various instrument measurements is applied.

In addition to this extensive experimental setup, PV production data have been obtained at four different locations in two countries: Viterbo and Bolzano (Italy), and Jaén and Madrid (Spain). The PV plant near Viterbo (lat. 42.45°N, long. 12.08°E) relies on 9000 First Solar modules, model FS180 rated at 80 W_p, with a plant nominal power of 720 kW_p. Power production data sampling was done every 15 minutes from 3/2013 to 2/2014.

In parallel, four years of experimental data, from 1/2011 to 12/2014, are available for the outdoor test facility of Airport Bolzano Dolomiti (ABD) (lat. 46.46°N, long. 11.33°E) located in South Tyrol, Italy. The hourly-average power generation of a 662 kW_p commercial PV plant with CdTe thin-film technology was measured during the whole test period. The modules are oriented 8.5° west of south and tilted at 30°. The DC and AC electrical parameters were measured every 15 minutes by commercial inverters. Moreover, an independent meteorological station provided measurements of global horizontal and diffuse irradiance, POA irradiance, wind speed, air temperature and BOM temperature with a frequency of 1 minute, averaged to a 15-min time step.

One CdTe module from First Solar, model FS270, was also tested under outdoor lab conditions in Spain, at both Jaén (lat. 37.77°N, long. 3.79°W) and Madrid (lat. 40.45°N, long. 3.73W) during 2013. Irradiance at all locations was measured on POA using pyranometers.

2.3 Sandia Array Performance model

The SAPM [16] estimates the module power production through the two following equations:

$$G_{eff} = GTI\ MMF\left[\frac{IAM(AOI)DNI\cos(AOI)+DHI}{GTI}\right] \quad (1)$$

and

$$P_m = I_m V_m \quad (2)$$

with

$$I_m = I_{m0}\left[C_0 G_{eff} + C_1 G_{eff}^2\right]\left[1 + \alpha_{Im}(T_{cell} - T_0)\right]$$

$$V_m = V_{m0}\left[1 + \beta_{Vm}(T_{cell} - T_0)\right] + C_2 N_s \delta(Tc)\ln G_{eff} + C_3 N_s\left[\delta(Tc)\ln G_{eff}\right]^2$$

where *GTI* is the POA irradiance, α_{Im} and β_{Vm} are the maximum current and voltage thermal coefficients, C_i(i = 1 to 3) are irradiance coefficients, N_s is the number of cells in series, and $\delta(Tc)$ is the thermal voltage. *DNI* and *DHI* are direct normal irradiance and diffuse horizontal irradiance, respectively. G_{eff} is the effective POA irradiance that reaches the PV cells, and P_m is the maximum power produced by the module under real operating conditions. MMF is the spectral mismatch factor, which describes the spectral effect on module performance whenever the incident spectrum is different from the reference spectrum:

$$MMF = \left(\frac{\int SR_{PV}(\lambda)\,E_{meas}(\lambda)\,d\lambda}{\int SR_{pyr}(\lambda)\,E_{meas}(\lambda)d\lambda}\right) / \left(\frac{\int SR_{PV}(\lambda)\,E_{ref}(\lambda)d\lambda}{\int SR_{pyr}(\lambda)\,E_{ref}(\lambda)d\lambda}\right) \quad (3)$$

where SR_{PV} and SR_{pyr} are the spectral responses of the module and pyranometer, respectively, E_{meas} is the measured spectral irradiance in the POA, and E_{ref} is the reference spectrum at air mass (AM) 1.5 (AM1.5G, according to IEC 60904-7 Ed.3.0.2008). Finally, The cell temperature is estimated as:

$$T_{cell} = T_{bom} + DT\left[{GTI}/{1000}\right] \quad (4)$$

where DT=3°C is the difference between the cell and the BOM temperatures, measured at 1000 W/m^2, from Sandia data base.

In Eq. (1), IAM is the angle of incidence modifier that describes the increase in irradiance reflection due to the module transparent cover:

$$IAM = \tau(AOI)\cos(AOI) \quad (5)$$

where AOI is the angle of incidence and $\tau(AOI)$ is the module cover's transmittance coefficient.

The SAPM method approximates MMF by a 4th-order polynomial function (F1) of AM:

$$MMF \approx F_1(AM) = \sum_{n=0}^{4} A_n\,(AM)^n \quad (6)$$

The five coefficients A_n were originally obtained empirically by Sandia during very clear conditions. Thus, function F_1 describes the impact on the module performance of the spectral mismatch related to fine weather conditions, and is referred to as "Air-Mass Performance Ratio", PR_{AM}.

In Eq. (5), IAM is approximated by a 5th-order polynomial function (F_2) of AOI:

$$IAM \approx F_2(AOI) = \sum_{n=0}^{5} B_n\,(AOI)^n. \quad (7)$$

Hence, the effect of varying AOI on the module performance is described by the "angle-of-incidence Performance Ratio", PR_{AOI}:

$$PR_{AOI} = \left[\frac{IAM(AOI)DNI\cos(AOI)+DHI}{GTI}\right]. \quad (8)$$

Therefore, application of the SAPM method requires the determination of three main parameters, namely the thermal coefficients and the F_1 and F_2 coefficients.

The original irradiance coefficients, C_i, provided by Sandia were used here without further verification for two reasons: the impact of this effect is usually very small and, moreover, it appears only when irradiance is low and shading effects are dominant.

3 OUTDOOR CHARACTERIZATION

3.1 Temperature coefficients

The modules' thermal coefficients are measured outdoors following the CEI EN 61215 procedure. At the Rome site, one module is placed on the tracking stand and kept at normal incidence. During sunny days with POA irradiance (GTI) ≈ 1 kW/m^2, the module is intentionally kept completely shaded until T_{bom} reaches the ambient temperature value. The module is then uncovered and the I-V curves are recorded every 20 s until T_{bom} stabilizes. Solar irradiance, air and BOM temperatures, as well as wind speed, are simultaneously recorded too. Table 1 reports the electrical parameters thus measured at the ESTER lab in comparison with those provided by Sandia and by the module datasheet.

From Table 1, the tested module apparently underwent significant degradation, producing 59 W_p rather than the anticipated 75 W_p. As expected, this degradation mainly affects the Fill Factor since the MPP values undergo greater changes with respect to I_{sc} and V_{oc}. On the other hand, the measured thermal coefficient agrees with both the datasheet and the Sandia database values. The bigger difference is found in the coefficients related to variations in current. Indeed, since the current depends very weakly on temperature, it is much more difficult to evaluate this dependence. Moreover, the greater noise in current measurements is due to the high sensitivity to irradiance variations.

Table 1: CdTe electrical parameters from the module datasheet, the Sandia database and ESTER outdoor measurements.

	Data Sheet	Sandia DB	ESTER	95% confidence bounds	
I_{sc} [A]	1.23	1.20	1.20	—	—
V_{oc} [V]	89.6	92	87.1	—	—
I_m [A]	1.1	0.970	0.97	—	—
V_m [A]	68.2	69.4	60.8	—	—
P_m [W]	75	67.30	59.1	—	—
α [%/°C]	0.04	0.04	0.06	0.05	0.08
β [%/°C]	-0.25	-0.25	-0.24	-0.24	-0.23
α_1 [%/°C]	np	0.04	0.03	-0.04	0.10
β_1 [%/°C]	np	-0.29	-0.27	-0.33	-0.21
γ [%/°C]	-0.25	-0.25	-0.24	-0.25	-0.23

3.2 Angle of incidence effect

The AOI performance ratio defined by Eq. (7) can be derived experimentally from:

$$\mathrm{PR}_{AOI} = \frac{(I_{sc}(AOI)/GTI)}{I_{sc0}/G_0} \quad (9)$$

where $I_{sc}(AOI)$ and I_{sc0} are the short-circuit currents measured at a specific AOI, and AOI=0, respectively. G_0 is equal to 1000 W/m^2. The short-circuit currents are temperature-corrected to make them correspond to T_{cell}=25°C. Moreover, to avoid spectral influences, all current measurements are done near AM1.5. In particular, I_{sc0} is precisely measured at AOI=0° and AM=1.5.

The test module, SN2246, is placed on a variable-tilt stand (Fig. 1). The tilt is varied progressively so that AOI decreases from 87° to 0°. For each AOI, the I-V curve is recorded every 20 s during 10 min, along with the POA and horizontal irradiances, the air and BOM temperatures, and wind speed. During the same 10-min period, the diffuse and direct normal irradiances are recorded too, but as a 1-min average only. Finally, the 20-s measurements are averaged and the 1-min IAM factor is calculated using Eq. (7).

Figure 1: Variable-tilt stand at the ESTER lab.

Table 2 compares the IAM fit parameters measured at ESTER to those provided by Sandia, showing a very good agreement.

Table 2: IAM coefficients for Eq. (7) empirically obtained at Sandia and ESTER.

Coefficient	Sandia	ESTER
B_0	1	1.008
B_1	-0.002438	-0.00334
B_2	3.10E-04	3.66E-04
B_3	-1.25E-05	-1.42E-05
B_4	2.11E-07	2.33E-07
B_5	-1.36E-09	-1.42E-09

3.3 Spectral effect

The spectral mismatch defined by Eq. (6) can be obtained experimentally from:

$$\mathrm{MMF} \cong \frac{(I_{sc}(AM)/GTI)}{I_{sc0}/G_0}, \quad (10)$$

where $I_{sc}(AM)$ and I_{sc0} are respectively the short-circuit currents measured at a specific AM within the interval [1.45–1.55] and at AOI=0°, while GTI varies within the interval [950–1050] W/m^2. A temperature correction is applied to translate them to the reference T_{cell}=25°C.

For additional tests, two CdTe modules are placed on a tracking stand (Fig. 2). The I-V curve, incident spectrum, POA irradiance, together with the horizontal irradiance, air and BOM temperatures, wind speed and direction, relative humidity, diffuse and direct normal irradiances, are recorded every 10 minutes. The spectral mismatch fit coefficients are calculated for the periods 4/2014–5/2014 and 7/2014–8/2014.

Table 3 reports the mean values of I_{sc0} obtained for two test modules over the test periods. In turn, these values can be used in Eq. (10) to calculate MMF from the corresponding values of the Average Photon Energy (APE) and AM.

Table 3: Mean measured values of I_{sc0} over the test period at the ESTER lab.

Serial number	I_{sc0} [A]	APE [eV]	AM [—]
FS1946	1.21±0.1	1.606±0.007	1.50±0.2
FS2271	1.20±0.1	1.606±0.007	1.50±0.2

Figure 2: Tracking stand with the CdTe modules for MMF evaluation.

The current measurements used to evaluate MMF are filtered to remove outliers. Moreover, clear-sky conditions (diffuse/global ratio <10%) are selected for the tests described in what follows. Figure 3a shows experimental results of the calculated MMF under AM values in the range 1–7.5, as well as their comparison with a simple fit of the same data and with the values obtained using the fitting coefficients provided by Sandia. A significant scatter in the MMF experimental data can be observed. The outliers at AM<4 can be due to a time constant difference between the thermopile radiometer (a few seconds) and the nearly instantaneous PV module current during periods when the irradiance is changing rapidly [19]. For AM>4, a double trend in the experimental data is visible. This seems related to the fact that the data points come from two different measurement campaigns and seasons. In any case, at low air masses, a good agreement between the two fits is found. The main difference that appears for AM>4 may be due to the dependence of the spectral effect of CdTe modules on water vapor (which varies seasonally), as pointed out in [19]. Nevertheless, this difference has a very small impact on the daily or monthly PR.

Figure 3b additionally displays the contemporaneous solar spectrum observations obtained with the two EKO spectroradiometers (MS-710 and MS-712) mounted in the POA (Fig. 2). For AM<4 at least, this experiment validates the CdTe module characterization according to SAPM, thus suggesting that the coefficients used in the parametric model do not depend on procedure or location. Table 4 reports the coefficient of the fits used to approximate MMF under clear-sky conditions. As above, a very good agreement between the fit coefficients measured in Rome and Albuquerque is found.

Table 4: Empirical MMF coefficients for Eq. (6).

Coefficient	Sandia	ESTER
A_0	0.9362	0.9417
A_1	8.10E-02	6.52E-02
A_2	-3.18E-02	-2.02E-02
A_3	4.90E-03	2.19E-03
A_4	-2.86E-04	-9.07E-05

4 PERFORMANCE ANALYSIS

Figure 4 displays the monthly measured PR values and their calculated counterpart, using SAPM at all four locations. It is stressed that the performance seasonality is only weakly visible at all locations, and that a non-negligible bias appears between the observed and modeled PR in Bolzano, Jaén and Madrid. Moreover, the seasonal patterns of the measured monthly-mean performances at Bolzano and Viterbo significantly differ from the calculated performances.

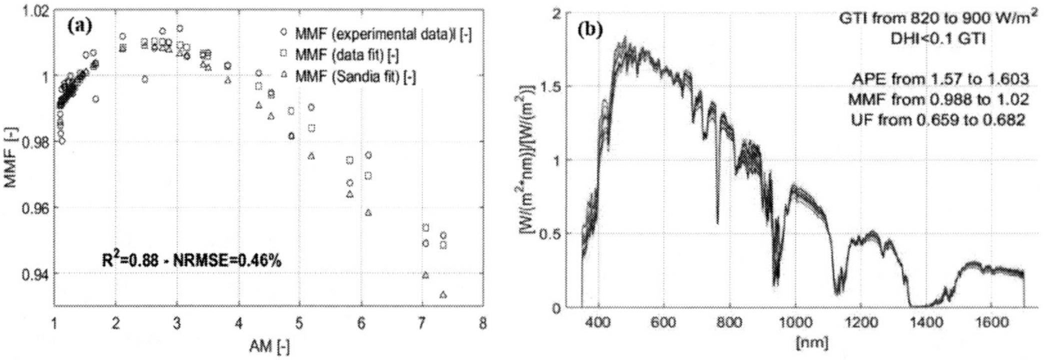

Figure 3: (a) Experimental measurements of spectral mismatch factor (MMF) during clear-sky conditions, and comparison between the data fitted with the local ESTER coefficients (red) and the data fitted with the Sandia coefficients (blue); (b) Spectral observations corresponding to the MMF measurements shown in the left plot.

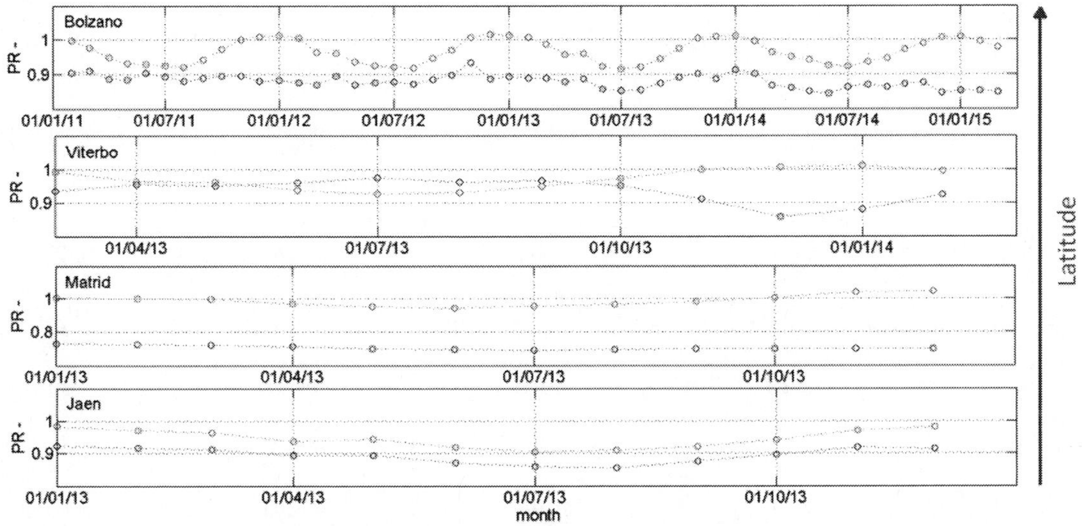

Figure 4: Monthly-mean measured PR (black dots) and calculated PR with SAPM (black dots) at four different locations.

From the above observations, it seems that the main difference between observed and modeled performance could be related to time variations in the nominal power with respect to the nominal nameplate power (P_n). Thus, defining P_m as the power output measured at maximum power point (mpp), and P_{m0} as the ideal power measured under STC conditions for a given day, the observed production could be modeled as:

$$P_m(obs) = \left(\frac{P_m0}{P_n}\right) P_m(SAPM) = RP_n\, P_m(SAPM) \quad (11)$$

The correction factor RPn accounts for the time variation in plant capacity, and can be estimated by obtaining the mean daily ratio between the power measured when GTI is in the range 750–1100 W/m^2 and the estimated power from SAPM:

$$RP_n = \langle \frac{P_m(obs\ @\ GTI \approx G_o)}{P_m(SAPM\ @\ GTI \approx G_o)} \rangle_{day}. \quad (12)$$

This correction represents the mean daily value of the power measured near 1000 W/m^2, after conversion to STC according to SAPM. Therefore, this metric should be independent from irradiance, cell temperature, angle of incidence and AM.

The ratio, R_{Pn}, relating the experimental and theoretical productions (near STC) can be explained in terms of the effects not considered in SAPM, such as the initial power loss or the degradation rate of the nominal power and performance seasonality.

Thus, R_{Pn} can be decomposed into three performance factors:

$$RP_n = RP_{n0}\, PRF_d\, PRF_s. \quad (13)$$

where RP_{n0} is the initial power loss, i.e. the ratio between the nominal power at the measurement's starting time and the nominal power provided by the manufacturer, PRF_d is the power degradation factor, and PRF_s is the seasonal performance factor. The first two factors can be modeled by a linear fit of the monthly-mean value of RP_n :

$$RP_{n0}\, PRF_d(month) \cong PLR_0 - PLR(month-1) = PLR_0\left(1 - \frac{PLR}{PLR_0}(months-1)\right), \quad (14)$$

where PLR is the power loss rate that accounts for the monthly degradation of the module capacity. The second factor, which accounts for performance seasonality, can be modeled by a periodic fit function of the day number:

$$PRF_s(day) = \frac{RP_n\,(day)}{PLR_0 - PLR(month-1)} = Fit(\tfrac{2\pi}{365} day + \varphi). \quad (15)$$

Figure 5 reports the monthly averages of the seasonal performance factor and of the periodic fitting function (second-order Fourier function) at the four locations. Interestingly, the performance seasonality is now visible for Bolzano and Viterbo, with higher performance in summer and lower in winter. Moreover, PR is greater in Viterbo (10%) than in Bolzano (7%), in spite of the lower latitude of the former. Conversely, seasonal variations are only slightly visible in Madrid, and almost negligible in Jaén. The reason for this is still not fully understood. Note, however that differences in seasonality magnitude were also found by Schweiger [1] under differing climates. Although SAPM appears to take spectral effects into account (through MMF), their modeling is actually done only as a function of AM (Fig. 3), and only for clear-sky conditions (since under overcast conditions DNI is very weak and MMF does not depend on AM anymore). This incomplete modeling of spectral effects, previously noticed in, e.g., [20], [21], [22] combined with the performance seasonality of the nominal power due to meta-stability of the cell junction, can be addressed by the empirical PRF_s factor shown in Fig. 5.

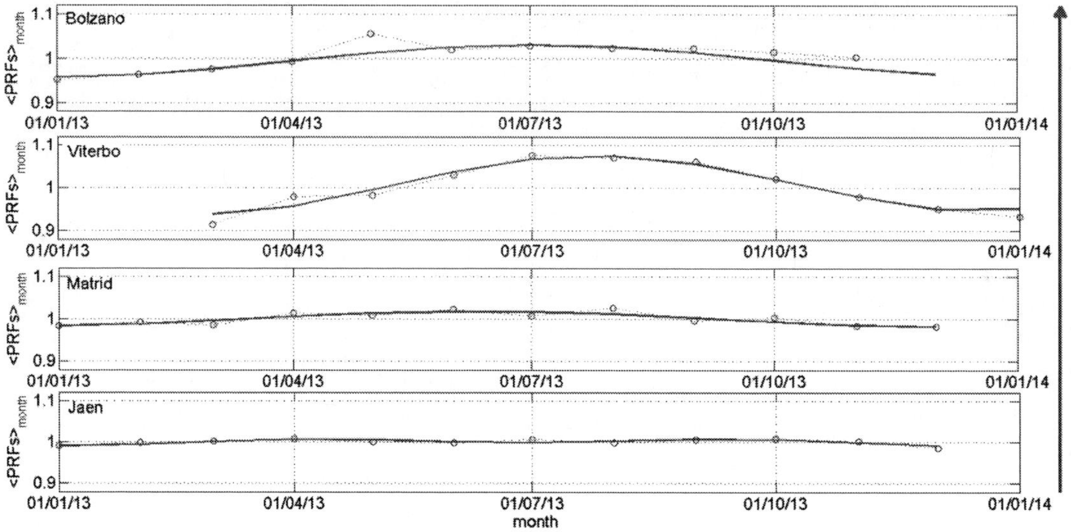

Figure 5: Monthly-mean seasonal Performance Ratio Factor (PRF) at four locations (black dots: observations; red line: periodic fits). The vertical arrow indicates the direction of increasing latitudes.

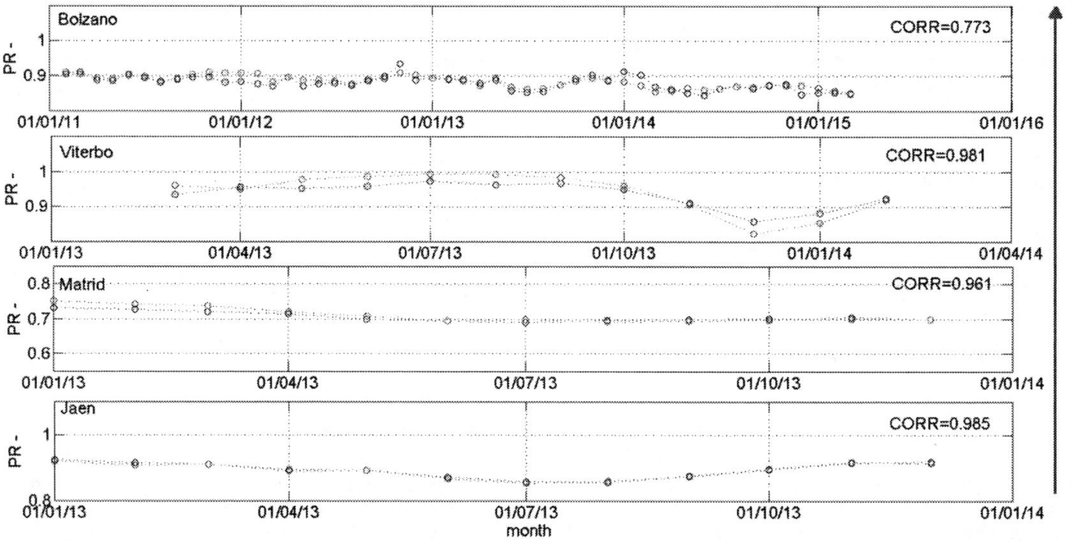

Figure 6: Monthly PR measured (black dots), and calculated with the corrected SAPM model (red dots), at four locations. The vertical arrow indicates the direction of increasing latitudes.

Figure 6 compares the experimental PR to the corresponding calculated values derived from the corrected SAPM (according to Eq. 11), where the seasonal performance factor is calculated by the periodic function described above. An excellent agreement between the monthly-mean experimental and modeled performances can now be observed. It is stressed that the determinations of PR_{SAPM} obtained from SAPM should describe the performance behavior of CdTe modules independently of plant location. Conversely, the performance factors defined by Eq. 13 must be derived from measurements of the power output of each specific plant. Hence, they provide a site-dependent correction, which apparently depends on latitude, climate, or both.

5 DISCUSSION

After additional consideration for power degradation and performance seasonality, the corrected SAPM model provides a reliable estimate of the performance of CdTe modules, as can be observed in Fig. 6. Nevertheless, SAPM simplifies the modeling of spectral effects by considering MMF as a function of air mass only. Indeed, the MMF model can be used only under clear-sky conditions. Moreover, the spectral correction in SAPM does not include the impact of water vapor or aerosol optical depth. Thus the seasonal performance factor could depend on incomplete spectral modeling as well as meta-stability due to transient modification of the CdTe cell.

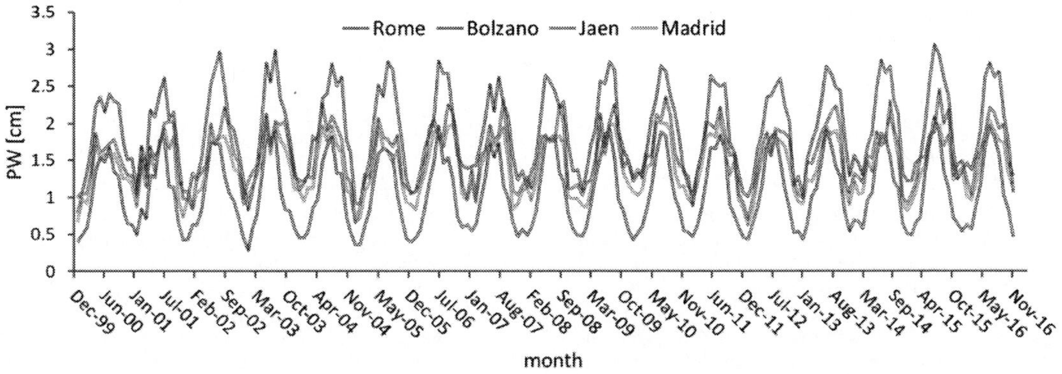

Figure 7: Sixteen years of *PW* time series at the four test locations (data from NASA's MERRA2 reanalysis).

Physically, the impact of precipitable water (*PW*) on the spectrum is outside the spectral response of CdTe, so that the higher *PW*, the bluer is the spectrum, and the larger the mismatch between the usable incident spectrum and the reference AM1.5 spectrum. Since the seasonal maximum *PW* normally occurs during summer and the minimum during winter, the module performance should be higher in summer than in winter (see Fig. 7). This explanation is consistent with the PRF_s' seasonal trend. Indeed, a remarkably high correlation is found between *PW* and PRF_s, as can be observed in Fig. 8 in the case of Bolzano, for instance. This could indicate a possible dependence of PRF_s on the spectral effect [7] [8] [9]. However, the spectral effect should then increase with latitude [23], but this is not supported by the experimental data, since the performance seasonality is found greater in Viterbo than in Bolzano (see Fig. 5).

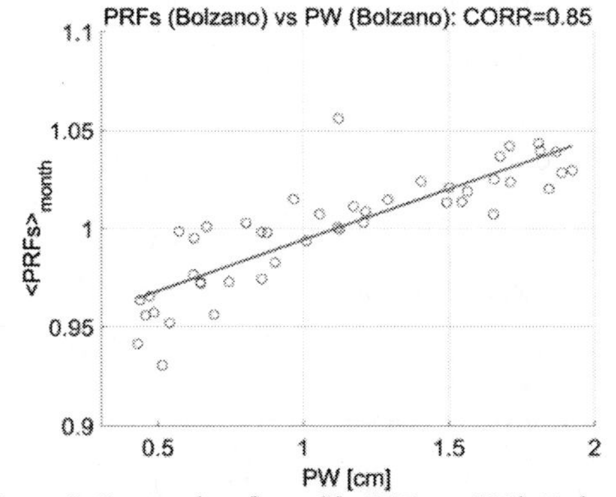

Figure 8: Scatter plot of monthly PRF_s vs. *PW* in Bolzano.

Nevertheless, a significant correlation between PRF_s and air temperature can be observed, as shown in Fig. 9. This could indicate that the observed seasonality could also depend on reversible meta-stability of the cell that could be related to modification of the back contact under light and temperature stress [12] [13] [14] [15].

Keeping in mind that the above correlations do not mean causality, it should be remarked that *PW* is directly related to both air temperature and relative humidity (see, e.g., [24]). Thus the correlation between PRF_s and *PW* could simply reflect the correlation between PRF_s and T_{air}. Moreover, the transient meta-stability phenomenon produces seasonal changes of the actual maximum power at STC, and PRF_s precisely attempts to take this periodic fluctuation of the PV capacity into consideration, per Eq. (13). In general, meta-stability affects both the open circuit and mppt voltage @STC, as well as V_{oc0}, V_{m0}, I_{sc0} and I_{m0} (short circuit and mppt currents @STC). For this reason, the cell fill factor (@STC) results differ from that of the nameplate one.

Figure 10 shows the monthly trends in RI_{m0} and RV_{m0} in Bolzano, as obtained experimentally similarly to RP_n:

$$RI_{m0} = \langle \frac{I_m(obs\ @\ GTI \approx G_0)}{I_m(SAPM\ @\ GTI \approx G_0)} \rangle_{month}. \quad (16)$$

$$RV_{m0} = \langle \frac{V_m(obs\ @\ GTI \approx G_0)}{V_m(SAPM\ @\ GTI \approx G_0)} \rangle_{month}. \quad (17)$$

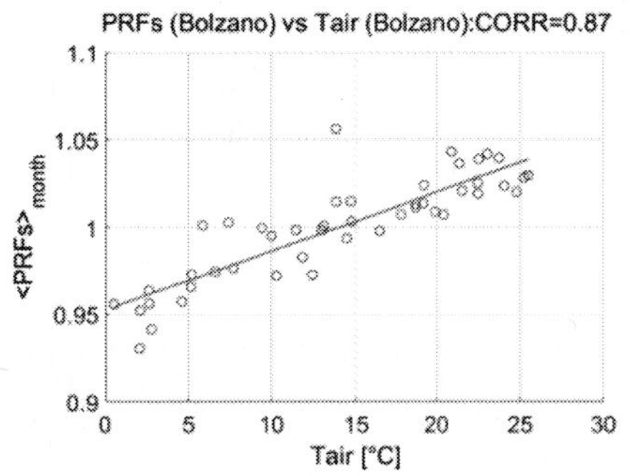

Figure 9: Scatter plot of monthly PRF_s vs. T_{air} in Bolzano.

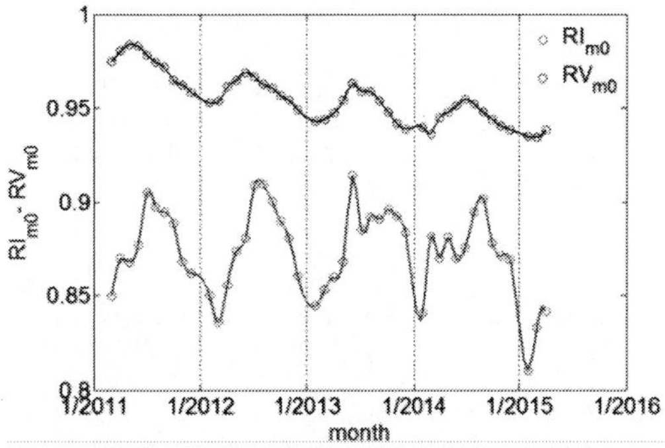

Figure 10: Seasonal variation of I_{m0} and V_{m0} in Bolzano.

It can be observed that the initial power loss mainly depends on current degradation (with an average RI_{m0} of ≈87%). The power degradation mainly results from the voltage degradation, whereas both V_{mo} and I_{mo} contribute to the performance seasonality.

This evidence suggests that meta-stability effects may contribute to the performance seasonality.

6 CONCLUSIONS

An outdoor experimental procedure for CdTe modules characterization was developed, and the main coefficients of the SAPM model were evaluated. The empirical coefficients derived experimentally in Rome are in close agreement with those originally obtained in Albuquerque. Nevertheless, a significant difference in the semi-empirical spectral function is found here for low-sun conditions (AM>4). This could be due to the spectral modelling of the SAPM model not considering the spectral dependence on *PW*.

A method to correct the SAPM model for power degradation and performance seasonality was developed and tested using data from four plants or test modules at locations in Italy and Spain, with latitudes in the range 37–47°N. Based on these specific corrections, good agreement was found between the corrected SAPM model and the experimental data. Significant performance seasonality was observed for the plants at Viterbo (10% variation) and Bolzano (7% variation), but not for the test modules at Jaén and Madrid. The reason why this seasonal effect occurs only at some locations is still an open question. Does this seasonality depend on spectral effects, such as the *PW* influence on the solar spectrum, and/or on meta-stability of the CdTe cell junction that are not considered in the SAPM model?

At least, it is found that *PW* and T_{air} are well correlated with the monthly seasonal performance factor, thus indicating that these variables could both influence the seasonal behaviour. Further analysis is now underway to better understand this intricate situation.

REFERENCES

[1] Schweiger, et al., 2017, "Understanding the Energy Yield of Photovoltaic Modules in Different Climates by Linear Performance Loss Analysis of the Module Performance Ratio," IET Renew. Power Gener., **11**, pp. 558–565.

[2] Schweiger, et al., 2017, "Performance Stability of Photovoltaic Modules in Different Climates," Prog. Photovoltaics Res. Appl. DOI: 10.1002/pip.2904

[3] Cañete, C., et al., 2014, "Energy Performance of Different Photovoltaic Module Technologies under Outdoor Conditions," Energy, **65**, pp. 295–302.

[4] Gottschalg, R., et al., 2003, "Experimental Study of Variations of the Solar Spectrum of Relevance to Thin Film Solar Cells," Sol. Energy Mater. Sol. Cells, **79**(4), pp. 527–537.

[5] Nofuentes, G., and Casa, J., 2013, "Solar Spectral and Module Temperature Influence on the Outdoor Performance of Thin Film PV Modules Deployed on a Sunny Inland Site," Int. J. Photoenergy, **2013**.

[6] Singh, Y. K., et al., 2015, "Effect of Spectrum and Temperature on Seasonal Performance of CdTe Technology in Composite Climate of India," 2015 IEEE 42nd Photovolt. Spec. Conf. PVSC 2015, pp. 5–7.

[7] Nelson, L., et al., 2013, "Changes in Cadmium Telluride Photovoltaic System Performance due to Spectrum," IEEE J. Photovoltaics, **3**(1), pp. 488–493.

[8] Lee, M., and Panchula, A., 2016, "Spectral Correction for Photovoltaic Module Performance Based on Air Mass and Precipitable Water," *43rd IEEE 43rd Photovoltaic Specialists Conference*, pp. 1351–1356.

[9] Lee, M., et al., 2015, "Understanding next Generation Cadmium Telluride Photovoltaic Performance due to Spectrum," *2015 IEEE 42nd Photovoltaic Specialist Conference, PVSC 2015*.

[10] Mikofski, M., et al., 2016, "Evaluation and Correction of the Impact of Spectral Variation of Irradiance on PV Performance," *Conf. Rec. 2016 IEEE Photovolt. Spec. Conf.*, pp. 1357–1362.

[11] Schweiger, M., and Herrmann, W., 2016, "Electrical Stability of PV Modules in Different Climates," *Conf. Rec. of the 2016 IEEE Photovoltaic Specialists Conference*, pp. 3685–3687.

[12] Gostein, M., and Dunn, L., 2011, "Light Soaking Effects on PV Modules: Overview and Literature Review," NREL PV Modul. Reliab. Work., pp. 1–25.

[13] Batzner, D. L., et al., 2000, "Study of the Back Contacts on CdTe/CdS Solar Cells," Thin Solid Films, **361**, pp. 463–467.

[14] Dobson, K. D., et al., 2000, "Stability of CdTe/CdS Thin-Film Solar Cells," Sol. Energy Mater. Sol. Cells, **62**(3), pp. 295–325.

[15] Virtuani A., et al., 2011, "Comparison of Indoor and Outdoor Performance Measurements of Recent Commercially Available Solar Modules," Prog. Photovoltaics, **19**, pp. 11–20.

[16] King, D. L., et al., 2004, "Photovoltaic Array Performance Model," Sandia Rep. No. 2004-3535, **8**, pp. 1–19.

[17] Spena, A., et al., 2008, "Outdoor ESTER Test Facility for Advanced Technologies PV Modules," *Conf. Rec.of the IEEE Photovoltaic Specialists Conference*.

[18] Zahumensky, I., 2004, "Guidelines on Quality Control Procedures for Data from Automatic Weather Stations," Wmo, (8), pp. 1–10.

[19] Emery, K. A., 2012, "Photovoltaic Module Calibration Value versus Optical Air Mass : The Air Mass Function," Prog. Photovoltaics Res. Appl., **22**, pp. 560–573.

[20] Theristis, M., et al., 2016, "Spectral Corrections Based on Air Mass , Aerosol Optical Depth , and Precipitable Water for CPV Performance Modeling," ieee J. photovoltaics, **6**(6), pp. 1598–1604.

[21] Duck, B. C. and Fell, C. J., 2016, "Improving the Spectral Correction Function," *43rd IEEE Photovoltaic Specialist Coference (PVSC)*, pp. 2647–2652.

[22] Lee, M., et al., 2016, "Variation in Spectral Correction of PV Module Performance Based on Two Different Precipitable Water Estimates," *43rd Photovoltaic Specialist Conference (PVSC)*, pp. 2692–2697.

[23] Huld, T., et al., 2016, "Photovoltaic Energy Rating Data Sets for Europe," Sol. Energy, **133**, pp. 349–362.

[24] Gueymard, C. A., 2014, "Impact of on-Site Atmospheric Water Vapor Estimation Methods on the Accuracy of Local Solar Irradiance Predictions," Sol. Energy, **101**, pp. 74–82.

LIGHT-INDUCED DEGRADATION IN MULTICRYSTALLINE MODULES AND SYSTEMS UNDER FIELD CONDITIONS

A.T. Alasfour and F.G. Alzubi
Kuwait Institute for Scientific Research
Safat 13109, Kuwait
Tel.: +965 99484624, E-mail: alasfour.wahab@gamil.com

ABSTRACT: While recently there has been a growing interest in LID exhibited by multicrystalline silicon (mc-Si), little research has been undertaken to investigate this phenomenon under field conditions. In this study, we evaluate the effect of sun soaking on commercially available mc-Si modules and further onto PV systems. After prolonged sunlight exposure, the relative efficiency of the modules degraded by 1%. However, on a system level, LID was found to nearly double as a result of current and voltage non-uniformities between modules. These preliminary findings show that LID in the field has two types of losses, one that is seen on a module level, and another stemming from increased parameter mismatch.
Keywords: light-induced degradation, mc-Si, PV systems, reliability

1 INTRODUCTION

Light-induced degradation (LID) is a serious issue in crystalline silicon as it causes a significant loss in solar cell efficiency upon initial light exposure. The phenomenon was discovered back in the early 1970s [1, 2], and since then, extensive research has been done mono-crystalline silicon cells grown by Czochralski or Float-Zone processes. Not until 1997 was LID measured in multi-crystalline silicon (mc-Si) cells [3]. A decade later, Pingel et al. reported a severe drop in relative efficiency of about 11% [4], generating a wide interest in LID exhibited by the technology. As of today, studies have pointed to 3 competing theories explaining the cause of LID: (1) the formation of boron-oxygen (BO) complexes [5-7], (2) the existence of 3rd transition metals in the silicon material [8-12], and (3) the type of passivation layer/cell fabrication process used [13-17]. In all theories, LID is a result of an increase and/or enhancement of recombination centers located in the bulk or the surface of the wafer.

Although considerable research has been devoted to mc-Si wafers/cells studied in laboratories, rather less attention has been paid to modules tested under field conditions; only a few researchers have investigated LID outdoors[4, 15, 18, 19]. From a designer and operator perspective, the extent of LID on a PV system remains unclear. For instance, many studies have shown different degradation levels amongst wafers cut from the same ingot [11, 20, 21], which will almost certainly lead to parameter mismatch losses when connected; thus, rendering LID to increase on a system level. Understanding the extent of these losses is becoming ever more important as the global share of mc-Si have reached 70% of the annual total production in 2016 [22].

In this study, we subject commercially available mc-Si modules, of the same model, to natural sunlight in order to further understand LID expected to occur under field conditions. In addition, we investigate the mismatch loss between the samples before and after sun soaking to establish any differences between LID on a module and system level.

2 EXPERIMENTAL

The experiment was performed on 7 commercial mc-Si modules of the same model composed of 60 Al-BSF cells that have not been previously exposed to sunlight. In an effort to minimize LID measurement errors, the samples were first stored in the dark for 24 hr in order to restore any dissociated iron-boron pairs that might lead to lower-than-actual LID [11]. Prior to sun soaking, an IV characteristic measurement of each module was carried out in accordance with IEC 60904.

After this initial benchmark, the modules were taken outside for sun soaking in an open-circuit mode and a tilt angle of 30°. To ensure power stability, the samples were kept outside until they received an accumulative solar radiation of 400kWh/m^2; this criterion resulted in an exposure of 78 days. Irradiance data was collected using a calibrated reference cell with an HOQ-filter mounted on a flat surface with the same tilt angle as modules. Since the experiment started in January, the min, max, and average ambient temperatures were 4°C, 32°C, and 20°C respectively throughout the soaking process.

Once the modules received the required insolation, they were brought back into the lab for their final IV measurement. This was done immediately after light-soaking such that any metastable defects that lead to LID are not reversed. The electrical parameters presented throughout this paper were measured at STC using an indoor class AAA pulsed solar simulator.

In an effort to understand the extent of LID in the field, a preliminary investigation of its impact on mismatch losses was carried out. Only 3 cases were examined by comparing mismatch losses before and after sun soaking due to a large variety of possible configurations. In all cases, a 7x7 sized array consisting of the modules tested in this experiment was considered.

3 RESULTS AND DISCUSSION

Table I presents the averaged electrical parameters of all the modules before and after sun soaking. As can be seen, prolonged illumination reduced the relative efficiency by 1%. The stronger degradation seen in Isc, compared with Voc, is consistent with literature. Moreover, the fill factor (FF), i.e. a quality measure of the cell fabrication process, showed little change, indicating no major contribution to LID.

Table I: Averaged electrical parameters of 7 mc-Si modules before and after of sun soaking.

Parameters	Before	After	Relative Δ
Voc [V]	37.57	37.51	-0.16%
Isc [A]	8.54	8.47	-0.80%
Vmp [V]	30.37	30.27	-0.31%
Imp [A]	7.98	7.92	-0.71%
Pmp [W]	242.20	239.75	-1.01%
Fill Factor	75.52%	75.47%	-0.06%
$\eta_{\%Rel}$	14.78%	14.62%	-0.99%

Figure 1 shows a box plot that details the impact of LID on each module. Sun exposure clearly does not degrade modules equally; 2 samples had a few parameter outliers worth discussing. In one case, the module's Voc and Vmp dropped 3 and 6 times more than the rest of the group. The severe degradation is likely a consequence of strong LID in most of the connected cells, leading to a low accumulated voltage.

On the other hand, in a different module, Vmp increased while Imp significantly decreased. It is highly likely that one of the cell's current degraded significantly more than the other 59 cells resulting in a strong current mismatch. This is believed to be a result of poor cell processing, as that sample was observed to have the largest reduction in FF. As for the surprising increase in Vmp, more modules are undergoing sun soaking to determine if this behavior was simply an outlier.

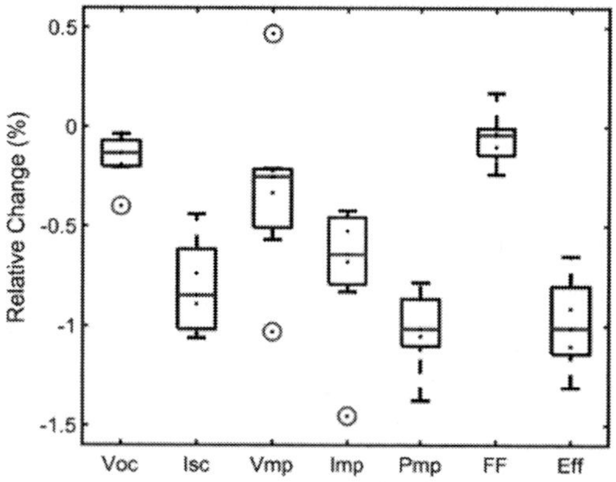

Figure 1: Box plot showing the relative changes of the electrical parameters before and after sun soaking for each of the 7 mc-Si modules.

As previously noted, disparities amongst modules are expected where many researchers have reported different degradation behaviors and severity levels even between wafers cut from the same ingot. However, deviations in Vmp and Imp are likely to increase mismatch losses (MML) on a system level. To test this hypothesis, 3 cases were examined by comparing MML before and after sunlight exposure. A 7x7 array consisting of modules measured in the study was considered for all cases.

In case 1, the current and voltage MML were isolated for a better understanding of how each parameter's mismatch changed as a consequence of LID. Isolating current MML was done by connecting 7 identical series (S) strings consisting of all the modules in parallel (P). Whereas voltage MML was isolated by connecting 7 identical P strings in S. On the other hand, case 2 represents a more practical design layout, where the modules were randomly arranged within the array. Like case 1, both S-to-P and P-to-S configurations were examined. As for case 3, the modules were connected in the worst possible way to benchmark the strongest LID footprint in MML.

Figure 2 presents the findings of all 3 cases before and after sun soaking. MML resulted in an increase as a result of LID in all but one array configuration. Before sun exposure, MML ranged from 0.5% to 1.7%; whereas after the experiment, it ranged from 0.8% to 2.3%.

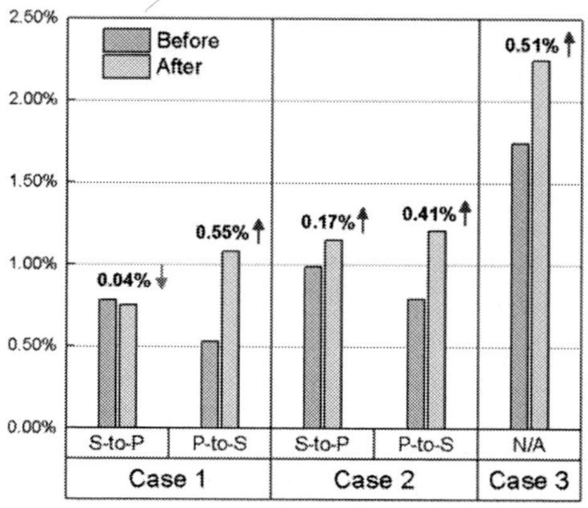

Figure 2: MML before and after LID for a 7x7 array consisting of the modules tested in this experiment examined under 3 different cases.

When isolating current and voltage mismatch, i.e. case 1, a slight improvement in Imp uniformity between the modules was observed, while Vmp deviations nearly doubled after illumination; figure 3 further illustrates this point. Furthermore, case 1 reveals that before LID, it was best to have an array with a P-to-S configuration by taking advantage of low voltage mismatch and suppressing high current mismatch. However, after LID, when current mismatch was lower, an S-to-P configuration became the preferable choice.

Conversely in case 2, optimizing the connections rendered ineffective as MML after sun exposure was about 1.2% for both S-to-P and P-to-S setups. It is important to note that these results reflect one of many random possible arrays; hence a larger difference between the two configurations is possible with a different module arrangement. Additional random MML scenarios need to be studied to verify that an S-to-P configuration is likely to yield lower MML values, and to what extent.

Finally, for the worst possible array layout, i.e. case 3, MML was 2.2%, whereas before LID it was capped at 1.7%. A 0.5% reduction in power is solely a result of increased module mismatch after sun exposure and not a side effect of spatial irradiance/temperature variabilities or shading. Therefore, at a system level, LID is aggregated with losses from both individual modules as well as enhanced mismatch parameters. In other words, if LID is 1% on a module level, it can be as high as 1.5% on an array level.

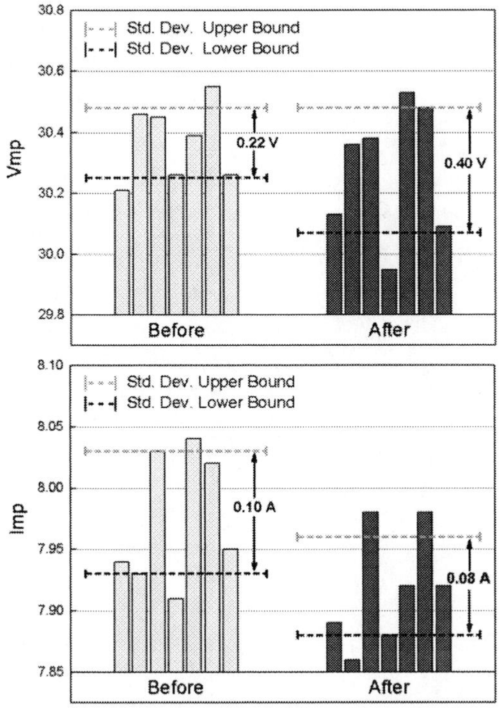

Figure 3: Current and voltage deviations between all the 7 mc-Si modules before and after sun soaking. Highlighted by standard deviations.

The impact of LID can be even greater on larger systems through cascading effects as shown in figure 4. When simulating a random scenario for a 1MWp plant, with only one inverter, power loss was seen to be about 2%. These additional losses will lead to either a reduction in energy yield or an increase in cost and reliability issues, i.e. by either integrating more inverters with maximum power point trackers or include power optimizers in each module.

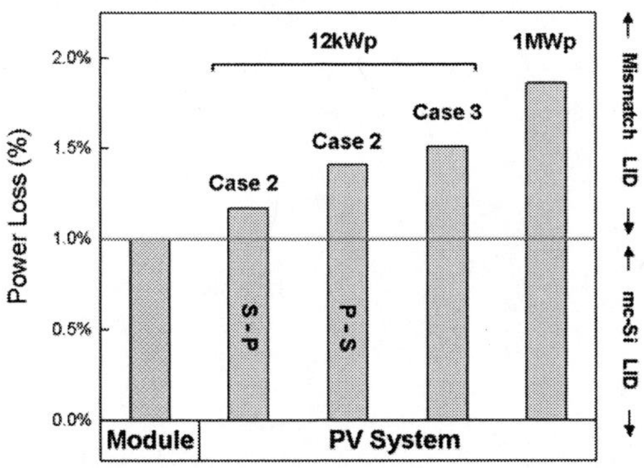

Figure 4: Power loss comparison between mc-Si modules and different PV systems as a result of LID.

With that said, these results are only based on the modules studied in this experiment. Very little is known about the MML changes due to LID; deviations amongst the same mc-Si devices before and after illumination are rarely reported. Rather than averaging the findings of future LID studies, including the details, particularly that of voltage and current, of all the samples investigated is valuable information for understanding the extent LID in a PV system.

4 CONCLUSIONS

Commercially available mc-Si modules of the same model were subjected to prolonged natural sunlight. A loss of 1% in relative efficiency was seen after illumination. Although this was an average value, parameter disparities amongst the samples were observed. While this was expected, seeing a very strong increase in voltage deviations and no change in current deviations between the modules was surprising. It should be noted that the lack of change in current deviations was regarded to be coincidental as each sample's current degraded differently.

Extending the work onto a system level, a preliminary investigation of MML for a 7x7 array was carried out based on the parameter mismatches observed in this study. The effect of LID on a system was found to nearly double in comparison to a module as a consequence of increased mismatch. We thus conclude that LID in the field has two combined losses, one that is seen on a module level, and another that stems from the increased mismatch. For a better understanding of LID in PV systems, further outdoor studies reporting on parameter mismatches are needed.

5 ACKNOWLEDGMENTS

The work was financially supported by the Kuwait Foundation for the Advancement of Sciences (KFAS). Facility and equipment support was provided by the Kuwait Institute for Scientific Research (KISR). The authors wish to thank A. Al-Qattan for his continuous support and encouragement. A. Alquennah is acknowledged for her assistance in characterizing the PV modules.

6 REFERENCES

[1] R. L. Crabb, "Photon induced degradation of electron irradiated silicon solar cells," in *9th IEEE Photovoltaic Specialists Conference* Silver Springs, MD, USA, 1972, pp. 243-249: IEEE.

[2] H. Fischer and W. Pschunder, "Investigation of photon and thermal induced changes in silicon solar cells," in *10th IEEE Photovoltaic Specialist Conference*, Palo Alto, CA, USA, 1973, pp. 404-411.

[3] H. Nagel, J. Schmidt, A. G. Aberle, and R. Hezel, "Exceptionally High Bulk Minority-Carrier Lifetimes in Block-Cast Multicrystalline Silicon," in *14th European Photovoltaic Solar Energy Conference*, Barcelona, Spain, 1997.

[4] S. Pingel *et al.*, "Initial Degradation of Industrial Silicon Solar Cells in Solar Panels," presented at the 25th European Photovoltaic Solar Energy Conference and Exhibition Feria Valencia, Spain 2010.

[5] J. Schmidt and K. Bothe, "Structure and transformation of the metastable boron- and oxygen-related defect center in crystalline silicon," *Physical Review*, vol. 69, pp. 024107-1, 2004.

[6] K. Bothe, R. Sinton, and J. Schmidt, "Fundamental Boron– Oxygen-related Carrier Lifetime Limit in Mono- and Multicrystalline Silicon," *Progress in*

Photovoltaics: Research and Applications, vol. 13, 2005.
[7] V. V. Voronkov and R. Falster, "Light-Induced Boron-Oxygen Recombination Centres in Silicon: Understanding their Formation and Elimination," *Solid State Phenomena,* vol. 205-206, pp. 3-14, 2013.
[8] S. Naito and T. Nakashizu, "Electric degradation and defect formation of silicon due to Cu, Fe, and Ni contamination," (in 652), *Material Research Society* vol. 262, 1992.
[9] J. H. Reiss, R. R. King, and K. W. Mitchell, "Characterization of diffusion length degradation in Czochralski silicon solar cells," *Applied Physics Letters,* vol. 68, no. 23, pp. 3302-3304, 1996.
[10] W. B. Henley, D. A. Ramappa, and L. Jastrezbski, "Detection of copper contamination in silicon by surface photovoltage diffusion length measurements," *Applied Physics Letters,* vol. 74, no. 2, pp. 278-280, 1999.
[11] D. H. Macdonald, L. J. Geerligs, and S. Riepe, "Light-Induced Lifetime Degradation in Multicrystalline Silicon " 2002.
[12] H. Savin, M. Yli-Koski, and A. Haarahiltunen, "Role of copper in light induced minority-carrier lifetime degradation of silicon," *Applied Physics Letters,* vol. 95, no. 15, 2009.
[13] K. Krauss, F. Fertig, D. Menzel, and S. Rein, "Light-induced Degradation of Silicon Solar Cells with Aluminiumoxide Passivated Rear Side," *Energy Procedia,* vol. 77, pp. 599-606, 2015.
[14] F. Kersten, J. Heitmann, and J. W. Müller, "Influence of Al2O3 and SiNx Passivation Layers on LeTID," *Energy Procedia,* vol. 92, pp. 828-832, 2016.
[15] K. Nakayashiki *et al.*, "Engineering Solutions and Root-Cause Analysis for Light-Induced Degradation in *p*-Type Multicrystalline Silicon PERC Modules," *IEEE Journal of Photovoltaics,* vol. 6, no. 4, pp. 860-868, 2016.
[16] D. Bredemeier, D. Walter, S. Herlufsen, and J. Schmidt, "Lifetime degradation and regeneration in multicrystalline silicon under illumination at elevated temperature," *AIP Advances,* vol. 6, no. 3, 2016.
[17] A. Zuschlag, D. Skorka, and G. Hahn, "Degradation and regeneration in mc-Si after different gettering steps," *Progress in Photovoltaics: Research and Applications,* vol. 25, no. 7, pp. 545-552, 2017.
[18] M. A. Munoz, F. Chenlo, and M. C. Alonso-García, "Influence of initial power stabilization over crystalline-Si photovoltaic modules maximum power," *Progress in Photovoltaics: Research and Applications,* vol. 19, no. 4, pp. 417-422, 2011.
[19] F. Kersten *et al.*, "Degradation of multicrystalline silicon solar cells and modules after illumination at elevated temperature," *Solar Energy Materials and Solar Cells,* vol. 142, pp. 83-86, 2015.
[20] S. D. Wolf *et al.*, "Light-induced degradation of very low resistivity multi-crystalline silicon solar cells," in *28th IEEE Photovoltaic Specialists Conference,* Anchorage, AK, USA, 2000: IEEE.
[21] J. Junge, A. Herguth, G. Hahn, D. Kreßner-Kiel, and R. Zierer, "Investigation of degradation in solar cells from different mc-Si materials," *Energy Procedia,* vol. 8, pp. 52-57, 2011.
[22] S. Philipps and W. Warmuth, "Photovoltaics Report," Fraunhofer Institute for Solar Energy Systems and PSE AG2017.

AN INEXPENSIVE AND ACCURATE SOLAR IRRADIANCE SENSOR BASED IN A SMALL CALIBRATED PV MODULE

Nouredine ERRAISSI*, Noura AARICH, Mohamed AKHSASSI, Mustapha RAOUFI and Amin BENNOUNA

NEE Laboratory, Faculty of Sciences Semlalia, Cadi Ayyad University, BP 2390 Marrakech, Morocco
(*Corresponding author mail erraissi.noureddine@gmail.com)

ABSTRACT: In this paper, we have proposed and realized an inexpensive and accurate sensor to measure the solar irradiance, the sensor is based in a small calibrated PV module. The method of measure consists in acquiring the short circuit current I_{SC} which is proportional to the incident solar irradiance by measuring the voltage across a precision shunt resistor connected to the reference module, then the final value is corrected according to ambient temperature to take into account the variation of the short circuit current I_{SC} with cell temperature. The reference module was calibrated following the recommendations stated in the international standards IEC 60904-2 and IEC 60906-4, it was used to measure in-plane and horizontal irradiance for more than one year, acquired data were compared with data from commercial and highly accurate sensors, and the results have shown that the accuracy was better than 3%.

Keywords: Solar irradiance measurements, Pyranometer, Small PV module, Calibration, Low-cost.

1 INTRODUCTION

In order to keep continuous development and farther improvement in the photovoltaic sector, accurate and consistent evaluations of the installed PV systems are significant, for this, several parameters which related to the PV system performance have to be monitored. Solar irradiance is the first and the most influencing parameter in the performance of PV systems [1]. In addition the output energy of PV installation is directly proportional to the irradiance reaching the surface of the PV modules, this irradiance is the dominant contributors to uncertainty in the models predicting the daily energy as reported in [2], and the efforts to reduce the uncertainty in PV system output predictions may yield the greatest improvements by focusing on the in-plane irradiance accuracy. Many studies related to the PV monitoring and PV performance evaluation as a function of solar irradiance can be found in the literature [3–6]. As a consequence the measurement of solar irradiance should follow strict recommendations in term of accuracy and reliability.

The measurements of the solar irradiance are carried out indirectly by the interpretation of the satellite images, a number of methods for estimating solar radiation from satellite data have been developed in the last decades [7–10]. Or can be measured directly by solar irradiance sensors, most sensors used to measure solar irradiance are thermopile pyranometers, photodiode pyranometers, and photovoltaic reference cells.

Pyranometers are widely used; they are the most accurate sensors technologies, they have a constant spectral response in the whole solar spectral range, they can cover the wavelength in the interval of 290 nm and 2800 nm [11]. But their main disadvantages are their high cost, their long response time up to 15s for Secondary Standard Class and 30s for First Class ones [12], their significant errors for instantaneous measurements (clear, cloudy) in addition to their low intensity signal output which must be properly amplified and protected from noise.

Contrariwise, calibrated reference cells are less expensive, they have much faster response time (order of microseconds) and they have the same selective spectral sensitivity as the modules comprising the PV array. Their main disadvantages are limited spectral response and the temperature influence. In general the yearly solar radiation measured by calibrated reference cells is less than that measured by pyranometers [13].

Pyranometers are highly used for solar irradiance measurements, but in some cases they are unsuitable solution because of their high cost, especially for small and medium PV installations. To solve this problem of cost, we have proposed and realized a low-cost solar irradiance sensor with a satisfying accuracy that complies with the standard IEC 61724 (better than 5%) [14]; the sensor is based in a small PV module calibrated with a commercial pyranometer.

2 METHODOLOGY

The measurement of solar irradiance must follow some recommendations described in the IEC 61724 standard, first the solar irradiance should be measured in the same place where the PV array is installed, the accuracy of this measurement must be better than 5% regardless of the sensor chosen, and in the particular case of the reference PV devices, a calibration method is necessary before their uses, the two standards IEC 60904-2 and IEC 60906-4 [15,16] describe the different calibration steps that must follow. In this work we have selected a PV module based in crystalline silicon technology because it has a good stability as declared in [13].

The solar irradiance is obtained by measuring the short circuit current I_{SC} of a PV module which is proportional of the incident solar irradiance reaching its surface. However, it is difficult to measure the short circuit current directly, a precision resistor with a small value is connected to PV module, and the resistor value must be small enough to remain satisfactorily near the short circuit condition, this value should ensures that the product of short circuit condition and shunt resistor must be less than 0.03 the open circuit voltage [17]. The short circuit current is then calculated as the ratio between the shunt resistor used and the voltage measured across it.

Firstly the sensibility factor of the PV module is determined as the ratio between the output voltage measured across the shunt resistor and the solar irradiance measured by accurate pyranometer in Standard Test Conditions (STC). These conditions are rarely achieved, as a consequence, we have made several test and experiments in different days with clear sky at noon where the solar irradiance is supposed to be maximal (around 1000w/m^2) to ensure that we are close to STC.

Then the solar irradiance is calculated using the output voltage across the shunt resistor and the temperature ambient. The final value of the solar irradiance must be corrected to take into account the variation of the short circuit current I_{SC} with cell temperature. In order to reduce the cost of the system we have used only the ambient temperature measurement, the temperature cell is then calculated by the model presented in the Eq 1:

$$T_m = T_a + \gamma * G \quad (1)$$

The short circuit current and the solar irradiance are related by the relation given in Eq 2:

$$I_{SC} = \frac{G}{G_{ref}} * I_{SC,ref}[1 + \alpha_{SC}((T_a + \gamma * G) - T_{ref})] \quad (2)$$

A precision resistor of 0.5Ω is used, it has a temperature coefficient expressed in [ppm/1°C] (parts per million per centigrade degree) which indicate that its resistance varies with its temperature, as the resistor is placed in the back surface of the PV module inside the J-Box, we can approximately consider that its temperature is the same as the PV module temperature, and then its resistance can be expressed (Eq 3):

$$R_{Sh} = R_{Sh,ref}[1 + \alpha_R((T_a + \gamma * G) - T_{ref})] \quad (3)$$

The voltage across the shunt resistor which is equal to the product of the short circuit current and the shunt resistor is given in Eq 4:

$$V_{Sh} = \frac{G}{G_{ref}} * I_{SC,ref} * R_{Sh,ref}[1 + (\alpha_{SC} + \alpha_R)((T_a + \gamma * G) - T_{ref}) + \alpha_{Sh} * \alpha_R((T_a + \gamma * G) - T_{ref})^2] \quad (4)$$

The resolution of the Eq 4 gives the final solar irradiance as expressed in Eq 5:

$$G = \frac{-[1+\beta(T_a - T_{ref})] + \sqrt{[1+\beta(T_a - T_{ref})]^2 - \frac{4\beta\gamma V_{Sh}}{S_{ref}}}}{2\beta\gamma} \quad (5)$$

Where:

- G is the solar irradiance in [W/m^2];
- R_{Sh} is the shunt resistor in [Ω];
- I_{SC} is the short circuit current of the PV module in [A];
- $R_{Sh,ref}$ is the shunt resistor for 25 °C in [Ω];
- $I_{SC,ref}$ is the short circuit current of the PV module in STC in [A];
- T_m is the module temperature in [°C];
- α_{SC} is the temperature coefficient of I_{SC} in [%/°C];
- β is the sum of α_{SC} and α_R;
- T_{ref} is the ambient temperature in STC in [°C];
- S_{ref} is the sensibility factor in [V/W/m^2];
- α_R is the temperature coefficient of shunt resistor in [ppm/1°C];
- V_{Sh} is the output voltage in [V];

3 DESCRIPTION OF THE PROPOSED SENSOR

The system is composed by the two sensors, solar irradiance and ambient temperature; the signals coming from them are converted by an Analog to Digital Converter (ADC) and transmitted to the controller board, the final value of the two parameters and then calculated and stored in a secure SD card memory before transferred to the users via internet, Figure 1 shows the bloc diagram of the proposed system.

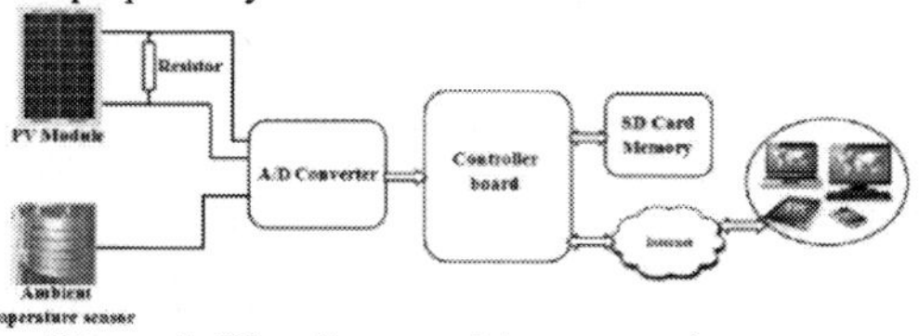

Figure 1: Bloc diagram of the proposed system

3.1 PV module

The PV device used for solar irradiance measurement is a small PV module of 20W based on polycrystalline cell technology, it is characterized by a short circuit current of 1.30A and an open circuit voltage of 22V, the current and the voltage in the maximum power are 1.18A and 17V consecutively.

3.2 Analog to Digital Converter

An Analog to Digital Converter (ADC) with a resolution of 16-bits is used to convert the signal coming from the sensor, it has also a programmable gain amplifier making the full-scale range of the voltage on their input programmable from +5V to +0.265V, and this provide a resolution less than 0.1W/m^2 for solar irradiance.

3.3 Controller board

A micro-computer board running in Linux operating system is used, the collected data are stored in CSV file and a web server implemented in the board is used to allow to the users the access to data via internet. Figure 2 presents a picture of the realized system.

Figure 2: Picture of the realized system

4 RESULTS AND DISCUSSION

A comparison of our measurements against those obtained with a pyranometer and a commercial calibrated reference cell has been carried out, for this all data were measured simultaneously and recorded with an interval of 5 minutes. Figure 3 gives the comparison between the daily solar irradiance measured by our sensor and a commercial reference cell for both day with clear sky (a) and day with cloudy sky (b).

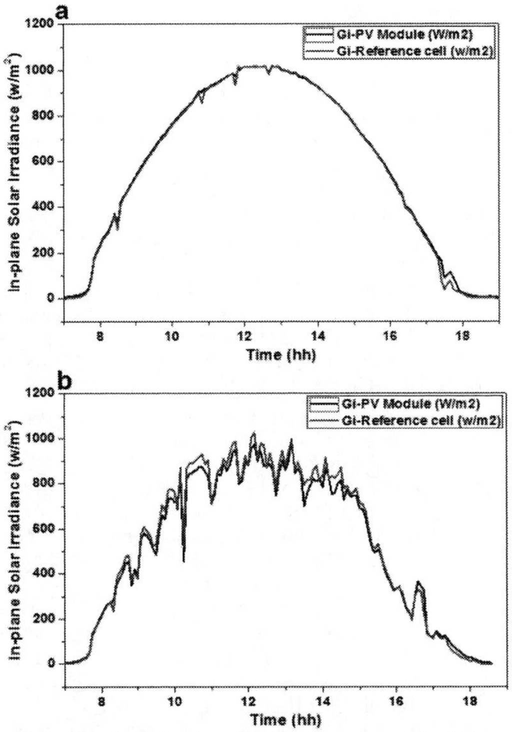

Figure 3: Daily in-plane irradiance measured by PV module and in-plane irradiance measured by commercial reference cell (a) day with clear sky (b) day with cloudy sky.

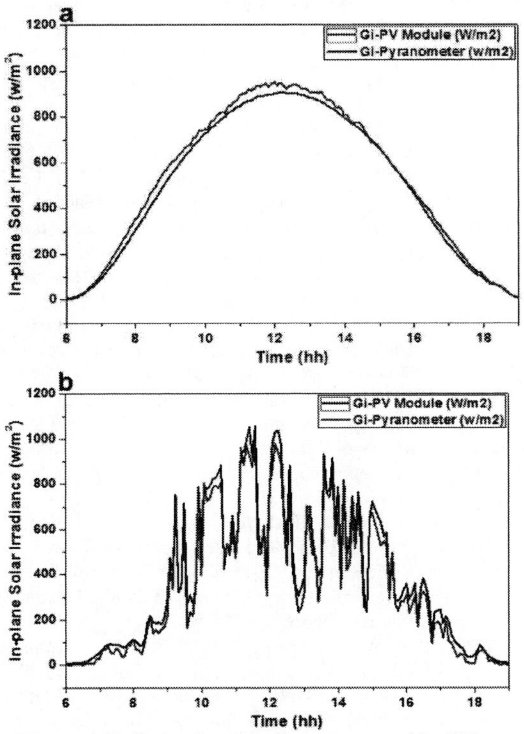

Figure 4: Daily in-plane irradiance measured by PV module and in-plane irradiance measured by pyranometer (a) day with clear sky (b) day with cloudy sky.

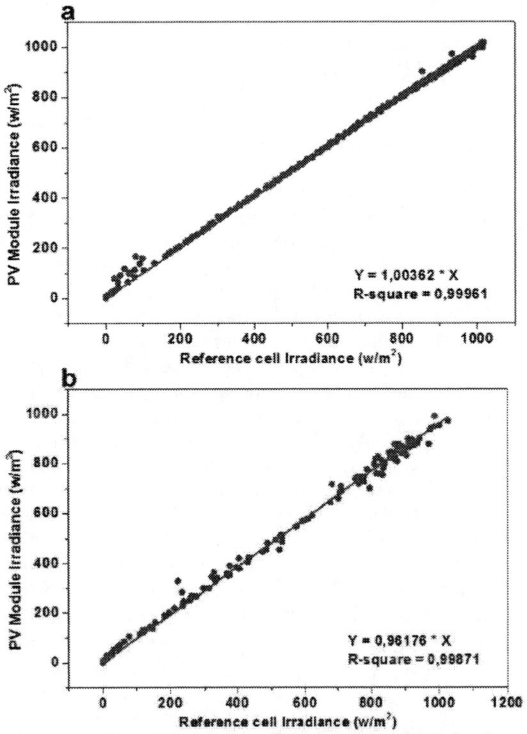

Figure 5: Correlation between the irradiance measured by PV module and the irradiance measured by commercial reference cell (a) days with clear sky (b) days with cloudy sky.

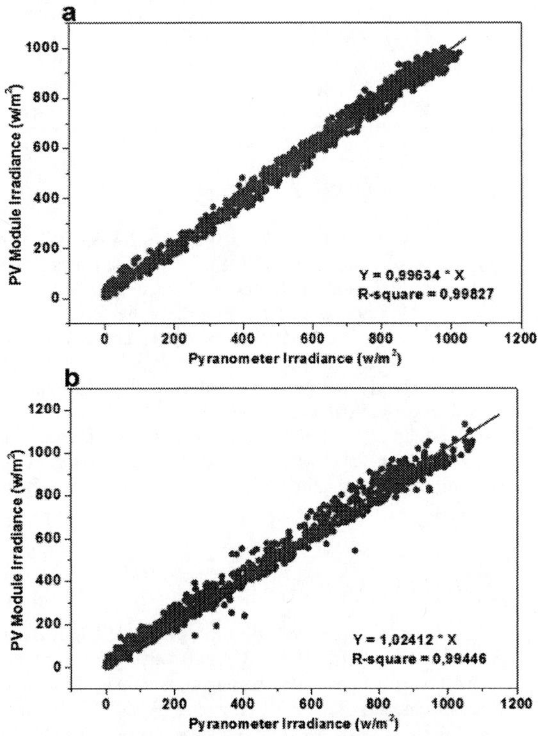

Figure 6: correlation between the irradiance measured by PV module and the irradiance measured by pyranometer (a) days with clear sky (b) days with cloudy sky.

The result shows that the daily profile of the two sensors is similar, with a small difference observed in the cloudy day, and this because the reference cell is placed far from our sensor by 10 meter.

As the same, the measurement of our sensor are compared with these obtained with a Kipp and Zonen pyranometer in Figure 4, the result shows that the two curves are similar both in clear and cloudy day, we can see that in clear day the solar irradiance measured with pyranometer is higher than that measured with the PV module which can be explained by the difference between their reflectivity and spectral response.

We have presented the solar irradiance measured by the PV module as a function of the solar irradiance measured by the commercial reference cell (Figure 5) and pyranometer (Figure 6) both for sunny and cloudy days. We have also calculated the Root Mean Square Error (RMSE) and Mean Bias Error (MBE); the result is given in the Table 1. The good correlation coefficient is obtained in sunny days with the commercial reference cell (0.99961), a small value of RMSE is also obtained in this case (0.87%), thus because the PV module and the reference cell have the same spectral sensitivity. In the case of pyranometer a good correlation coefficient in sunny days (0.99827) and a RMSE better than 3% are obtained. In cloudy days, these coefficients are less good but our proposed sensor still providing accuracy better than 5%.

Table 1: RMSE and MBE obtained from the comparison of solar irradiance measured with our calibrated PV module with those measured with a pyranometer and a reference cell for sunny and cloudy days.

		Pyranometer	Reference cell
Sunny days	RMSE (%)	2.54	0.87
	MBE (%)	-0.16	0.53
Cloudy days	RMSE (%)	4.46	3.36
	MBE (%)	3.31	-3.51

5 CONCLUSION

This work discusses the design of a low cost sensor for solar irradiance measurement; it is based in a small calibrated PV module. It has been used to measure in-plane and horizontal irradiance for more than one year.

The results exhibits that the total accuracy of the sensor was better than 3% for sunny days and better than 5% for cloudy days. In general it is found to be in accordance with that expected by IEC 61724 standard (better than 5%) in all conditions. Also the sensor was found to be reliable as it was tested in different extreme environmental conditions, and it showed satisfying results.

6 ACKNOWLEDGMENTS

This work is part of the "PROPRE.MA" project: sponsored by IRESEN for 3 years starting on February 2014, carried by 20 higher education Moroccan institutions and one private company (RESING). Each of the 20 cities host the same 3x2kWp grid connected PV plants based on m-Si, p-Si and a-Si technologies. "Propre.Ma" will draw grid-connected PV yield maps for Morocco with ground calibration in 20 different cities.

7 REFERENCES

[1] C. Rus-Casas, J.D. Aguilar, P. Rodrigo, F. Almonacid, P.J. Perez-Higueras, Classification of methods for annual energy harvesting calculations of photovoltaic generators, Energy Convers. Manag. 78 (2014) 527–536. doi:10.1016/j.enconman.2013.11.006.

[2] C.W. Hansen, C.M. Martin, Photovoltaic System Modeling: Uncertainty and Sensitivity Analyses, 2015.

[3] A. Driesse, W. Zaaiman, Characterization of global irradiance sensors for use with PV Systems, in: 2015 IEEE 42nd Photovolt. Spec. Conf. PVSC 2015, 2015. doi:10.1109/PVSC.2015.7356004.

[4] D. Thevenard, S. Pelland, Estimating the uncertainty in long-term photovoltaic yield predictions, Sol. Energy. 91 (2013) 432–445. doi:10.1016/j.solener.2011.05.006.

[5] D. Dirnberger, G. Blackburn, B. Müller, C. Reise, On the impact of solar spectral irradiance on the yield of different PV technologies, Sol. Energy Mater. Sol. Cells. 132 (2014) 431–442. doi:10.1016/j.solmat.2014.09.034.

[6] D. Dirnberger, B. Müller, C. Reise, On the uncertainty of energetic impact on the yield of different PV technologies due to varying spectral irradiance, Sol. Energy. 111 (2015) 82–96. doi:10.1016/j.solener.2014.10.033.

[7] C. Rigollier, M. Lefèvre, L. Wald, The method Heliosat-2 for deriving shortwave solar radiation from satellite images, Sol. Energy. 77 (2004) 159–169. doi:10.1016/j.solener.2004.04.017.

[8] C. Schillings, R. Meyer, H. Mannstein, Validation of a method for deriving high resolution direct normal irradiance from satellite data and application for the Arabian Peninsula, Sol. Energy. 76 (2004) 485–497. doi:10.1016/j.solener.2003.07.037.

[9] F.R. Martins, E.B. Pereira, S.L. Abreu, Satellite-derived solar resource maps for Brazil under SWERA project, Sol. Energy. 81 (2007) 517–528. doi:10.1016/j.solener.2006.07.009.

[10] S. Janjai, A method for estimating direct normal solar irradiation from satellite data for a tropical environment, Sol. Energy. 84 (2010) 1685–1695. doi:10.1016/j.solener.2010.05.017.

[11] M. Paulescu, E. Paulescu, P. Gravila, V. Badescu, Solar radiation measurements, Weather Modeling and Forecasting of PV Systems Operation (Chapter 2), in: Green Energy Technol., 2013. doi:10.1007/978-1-4471-4649-0.

[12] M. Sengupta, A. Habte, S. Kurtz, A. Dobos, S. Wilbert, E. Lorenz, T. Stoffel, D. Renné, C. Gueymard, D. Myers, S. Wilcox, P. Blanc, R. Perez, Best Practices Handbook for the Collection and Use of Solar Resource Data for Solar Energy Applications, National Renewable Energy Laboratory (NREL) 2015.

[13] A. Woyte, M. Richter, D. Moser, S. Mau, N. Reich, U. Jahn, Monitoring of Photovoltaic Systems: Good Practices and Systematic Analysis, in: 28th Eur. PV Sol. Energy Conf. Exhib. 30 Sept. – 4 Oct., Paris, France, 2013: pp. 1689–1699. doi:10.1017/CBO9781107415324.004.

[14] IEC 61724, Photovoltaic System Performance Monitoring-Guidelines for Measurement, Data Exchange, and Analysis, 1st ed., International Electrotechnical Commission (IEC), Switzerland, Geneva, 1998.

[15] IEC 60904-2, Photovoltaic Devices: Requirements for Reference Solar Devices, 2nd Ed, International Electrotechnical Commission (IEC), Switzerland, Geneva, 2007.

[16] IEC 60904-6, Photovoltaic Devices: Requirements for Reference Solar Modules, 1st Ed., International Electrotechnical Commission (IEC), Switzerland, Geneva, 1994.

[17] M. Fuentes, M. Vivar, J.M. Burgos, J. Aguilera, J.A. Vacas, Design of an accurate , low-cost autonomous data logger for PV system monitoring using Arduino ™ that complies with IEC standards, Sol. Energy Mater. Sol. Cells. 130 (2014) 529–543. doi:10.1016/j.solmat.2014.08.008.

DO THIN FILM PV MODULES OFFER AN ADVANTAGE UNDER PARTIAL SHADING CONDITIONS?

Chris Tzikas[1], Gabriela Gómez[1,2], Menno van den Donker[1], Klaas Bakker[2,3], Arno H.M. Smets[2], Wiep Folkerts[1]

[1]Solar Energy Application Centre (SEAC), High Tech Campus 21, Eindhoven, The Netherlands
[2]Delft University of Technology, PhotoVoltaic Materials and Devices group, Delft, The Netherlands
[3]Solliance Solar Research, Eindhoven, The Netherlands

ABSTRACT: Building Integrated PhotoVoltaics (BIPV) is one of the fastest growing segments of the photovoltaic industry worldwide. More than 25 years development has brought to the market diversified products for seamless integration into the building envelope by using a variety of PV technologies. One of the biggest loss factors, though, for such applications are the partial shading phenomena caused by shading elements on the roof or in close proximity such as chimneys, dormers, trees, poles etc. These shadings may have a detrimental effect mostly on the performance but sometimes also on the reliability of the PV modules used in BIPV applications.
Keywords: Building Integrated PV (BIPV), Thin Film, Shading, Reliability, CIGS

1 INTRODUCTION

This paper focuses on the effect shading has on the performance and reliability of various PV technologies used in Building Integrated Photovoltaics (BIPV). Building Integrated Photovoltaics are photovoltaic materials that are used to replace conventional building materials in parts of the building. The current world market share of BIPV systems is only 3% [1] and can therefore be considered a niche market.

In order for BIPV to move out of the niche market certain challenges must be addressed. One of the main technical challenges is to assure an adequate performance under non-ideal conditions. For the scope of this paper, non-ideal conditions are defined as high operating temperatures and light obstructions. Such conditions can be very difficult to be avoided in these systems and have a drastic impact on the performance and lifespan of the PV installation.

Due to their nature, BIPV applications are often subject to shading caused by obstruction elements such as: trees and poles in close proximity, chimneys, exhaust pipes, dormers or surrounding buildings (see Figure 1). This partial shading effect has detrimental impact on the system's performance but could also sometimes affect the reliability of the PV modules over time.

Figure 1: Light obstruction on rooftop PV systems

In order to study the effects of shading on the performance and the reliability of various PV technologies, a comprehensive range of experiments was designed that included partial shading experiments, EL imaging of the PV modules before and after these experiments and IV tracing under different shading scenarios.

2 EXPERIMENTAL SETUP

The experiments presented in this paper were conducted at the Photovoltaics Laboratory of TU Delft using the following equipment:

- Steady State Solar Simulator, Eternal Sun.
- 650W DC load, BK Precision.
- I-V tracer, ReRa IVtraQ.
- Electro-luminescence (EL) camera with high pass IR filter, Nikon D7200.
- Optical Microscope.
- Resistance Data Logger with temperature and voltage channels, Pico PT-104.
- Opaque cardboard for shading experiments.

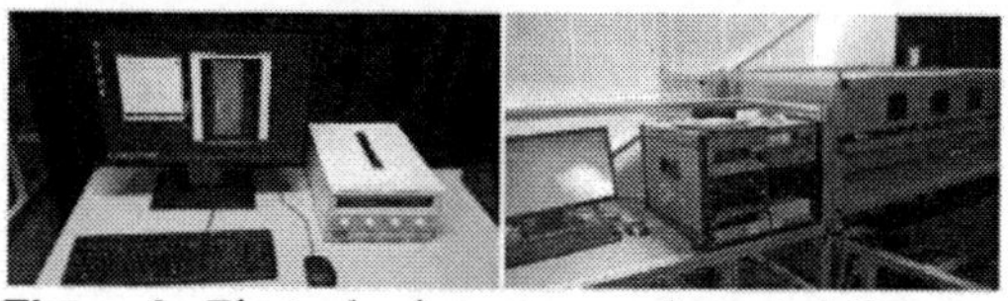

Figure 2: Electro-luminescence equipment, I-V tracer and solar simulator used for the experiments

For this study, 11 commercially available PV modules of 3 different PV technologies (crystalline Silicon, Cadmium Telluride, Copper Indium Gallium Selenide) were characterized:

- 2 c-Si modules (c-Si_A, c-Si_B).
- 2 CdTe modules (CdTe_A, CdTe_B).
- 7 CIGS modules (CIGS_A, CIGS_B, CIGS_C, CIGS_D, CIGS_E, CIGS_F, CIGS_G).

The steps of the experimental procedure was followed the steps of which are listed below:

1. Initial electro-luminescence (EL) imaging to corroborate the initial module condition.
2. Unshaded IV tracing to establish initial nominal power.
3. IV tracing with progressive shading stress, depicted in Figure 3 and Figure 4.
4. Unshaded IV tracing to assess possible damage done to the module by the shading stress.
5. Final EL imaging to determine module condition after shading stress.

The IV curves obtained were analyzed and the Nominal Power (P_{max}) was calculated. The maximum power was corrected for module temperature according to equation (1):

$$P_{corr} = \frac{P_m}{1+a(T_m - T_{STC})} \quad (1)$$

where: P_{corr} is the corrected power output, P_m is the measured power, α is the temperature coefficient for P_{max} according to the manufacturer's datasheet, T_m is the module temperature at the time of the measurement and T_{STC} is the temperature at Standard Test Conditions.

The shading stress was performed using a black opaque cardboard, imposing direct shading, with increments of 5% of the total module area in landscape (see Figure 3) and portrait orientation (see Figure 4). It is important to point out the different cell orientation for the CdTe and CIGS_A modules compared to the rest of the thin-film (TF) modules.

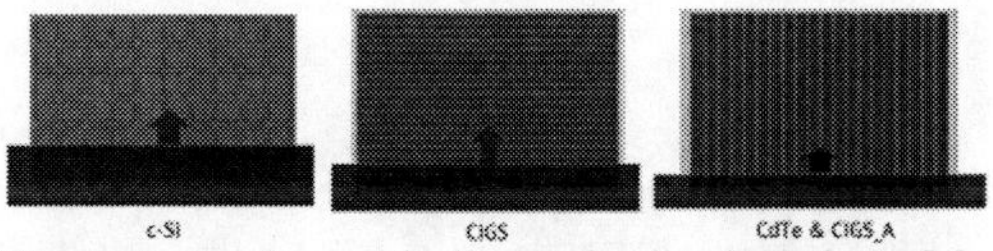

Figure 3: Landscape shading and cell orientation of the PV modules tested

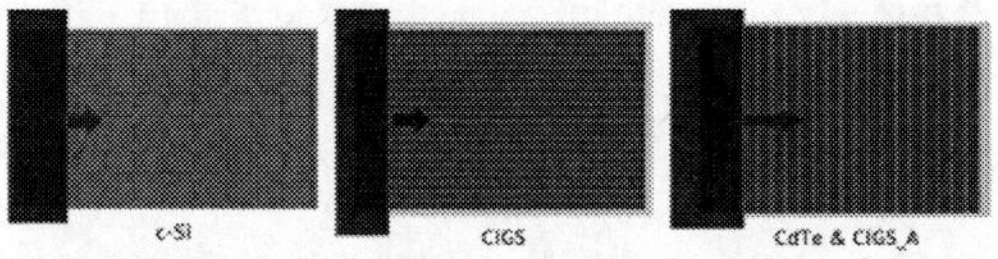

Figure 4: Portrait shading and cell orientation of the PV modules tested. The arrow shows the direction of the progressive shading stress applied.

The performance study aims to investigate if thin-film (TF) modules perform better under partial shading conditions. This is a claim made also by TF manufacturers and attributed to the monolithic integration and long geometry of their solar cells [2].

For the reliability study, the shading pattern was selected in order to simulate the worst case shading scenario for a PV module. It was found from the experiments and literature [3] [4] [5] [6] that covering the whole length of a small amount of TF solar cells causes the most damage as it creates the highest power dissipation for the shaded cells. This shading pattern corresponds to landscape shading orientation for CIGS and portrait shading orientation for CdTe and CIGS_A modules. The size of the cardboard mask was selected so there could be the highest dissipation possible on the maximum number of solar cells. Figure 5 shows a characteristic IV curve for a TF module under shading conditions. In order to obtain highest dissipation, the shaded characteristic curve must have a short circuit current (I_{sc_shaded}) higher than the current at maximum power point (I_{mpp}) of the unshaded module. The 10% module shading percentage was selected for this study.

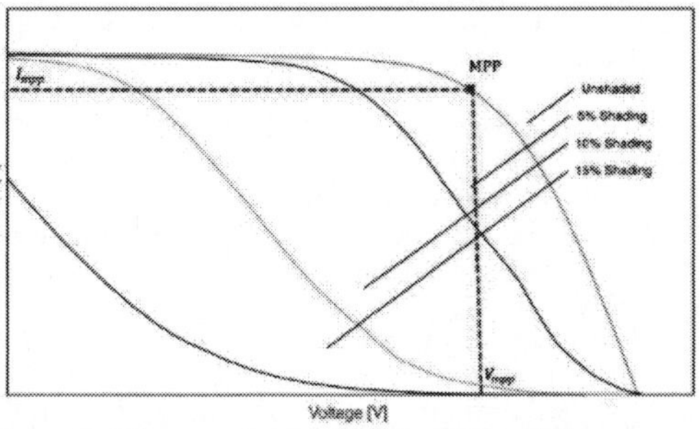

Figure 5: IV curve of a shaded thin-film module

The PV modules were shaded with the selected mask and short circuited while being illuminated by the solar simulator. The experiment was conducted for 10 minutes while current and temperature of the module were being recorded. At the end of the illumination period, an unshaded IV tracing was performed and an EL image was taken.

3 PERFORMANCE RESULTS

In this paragraph the results of the landscape and portrait shading experiments are presented. Both the relative power of the PV modules and the module shading percentage are normalized so that all results are directly comparable regardless of the power rating or the size of the PV module.

The results of the landscape shading experiment are presented in Figure 6. The c-Si modules show a typical behavior with 2 plateaus forming every time a by-pass diode is activated, consequently shutting off one part of the module. The difference between c-Si_A and c-Si_B resides in the number of cells connected to the bypass diodes. The two CdTe and the CIGS_A show the best performance in these conditions with performance reducing linearly as all cells get equally shaded. The CIGS modules lose all power at around 20% module shading as whole cells were shaded and forced into reverse bias.

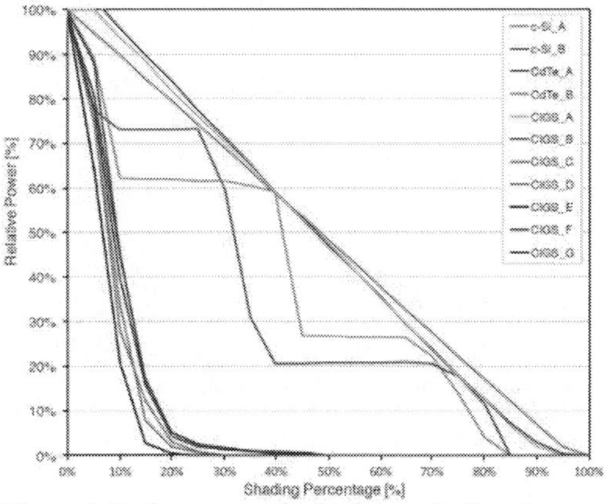

Figure 6: Performance under landscape shading stress

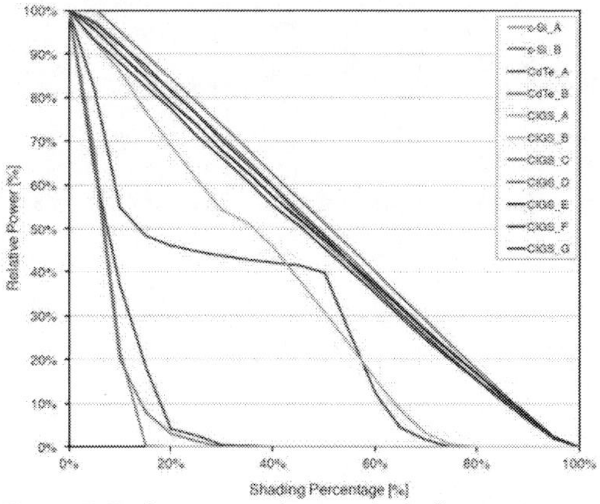

Figure 7: Performance under portrait shading stress

In the portrait shading experiment (see Figure 7), the opposite behaviour is observed for all PV modules apart from the CdTe_A and CIGS_A modules. The CdTe_A presented a plateau from 20% to 50% confirming the presence of two substrings connected in parallel allowing disconnection of the first half of the solar cells when the module is shaded past 20%. The CIGS_A module on the other hand is equipped with multiple bypass diodes allowing it to retain power by shutting down parts of the module up to almost 80% shading.

The overall results show a higher tolerance of the TF modules under partial shading stress compared to c-Si ones. Nevertheless a careful study of the shading patterns for BIPV applications is essential for the proper installation that will maximise the energy yield. In addition to this, modules with integrated bypass diodes (CIGS_A) or more than one substring (CdTe_A), show a superior performance under any shading conditions.

4 RELIABILITY RESULTS

In this paragraph, the reliability investigation is presented. The PV modules were subjected to partial shading with opaque cardboards under short circuit conditions. This allowed to achieve the worst case operating scenario for these modules. The emphasis will be on CIGS modules as both the c-Si and CdTe didn't show any clear signs of damage.

The results on CIGS modules show damage visualised as dark spots in the EL images. These dark spots are shunts located at the areas where the panel was masked for shading (see Figure 8 - Figure 14).

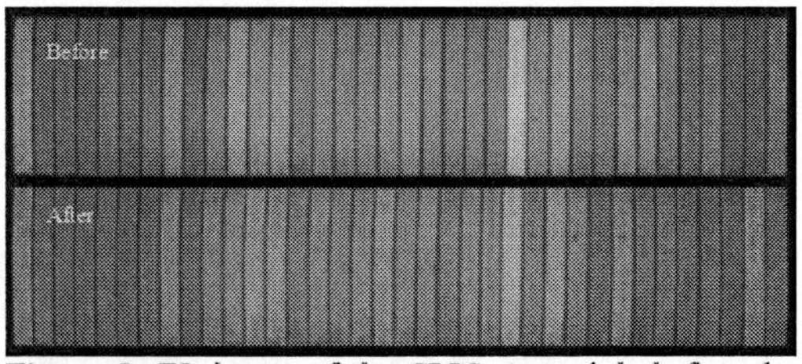

Figure 8: EL image of the CIGS_A module before the shading stress was applied (top image) and after the shading stress was applied (bottom image).

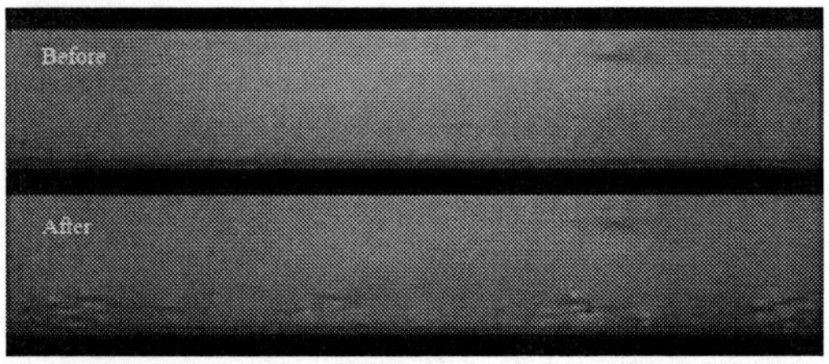

Figure 9: EL image of the masked side of the CIGS_B module before and after the shading stress respectively

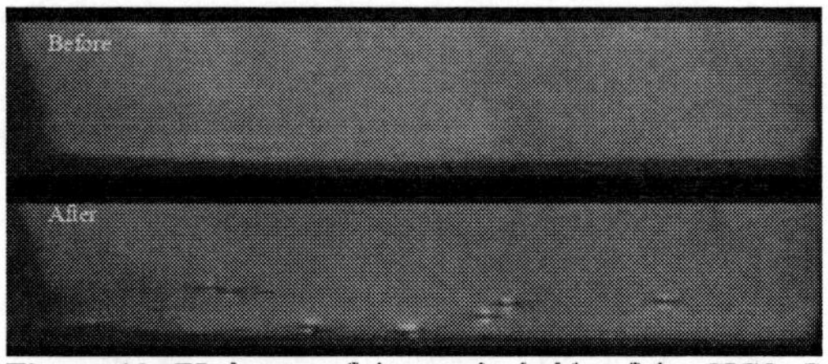

Figure 10: EL image of the masked side of the CIGS_C module before and after the shading stress respectively

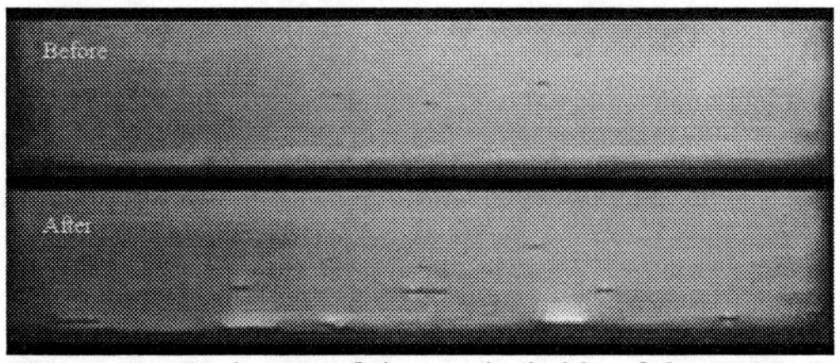

Figure 11: EL image of the masked side of the CIGS_D module before and after the shading stress respectively

Figure 12: EL image of the masked side of the CIGS_E module before and after the shading stress respectively

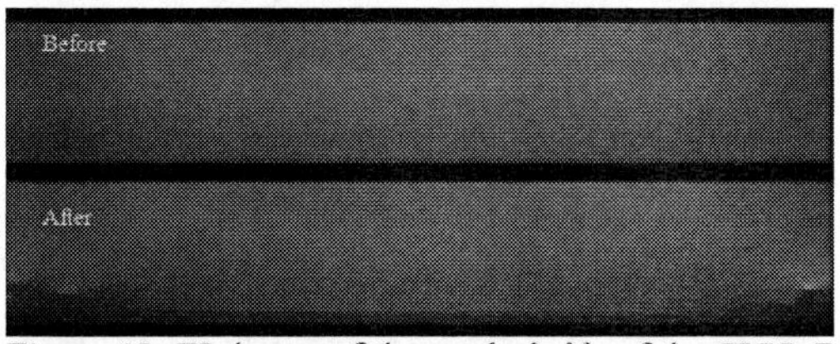

Figure 13: EL image of the masked side of the CIGS_F module before and after the shading stress respectively

Figure 14: EL image of the masked side of the CIGS_G module before and after the shading stress respectively

Shunting due to reverse bias in the masked cells was evident in the majority of the CIGS modules. The most severe effect was observed for the CIGS_B, CIGS_C, CIGS_D and CIGS_F modules. CIGS_A didn't show any signs of shunting while also the effect on CIGS_E was very small.

The outstanding tolerance of CIGS_A module can be attributed to the different solar cells' interconnection design. Contrary to the rest of the CIGS modules of this study, which use a monolithic integration, the solar cells' interconnection in the CIGS_A module rely on a conductive metal grid (also known as shingle overlap). The shingled interconnection allows for a distribution of current on a wider area. This reduces the stress due to reverse bias. Moreover, the integrated bypass diodes of this module help in avoiding a high reverse bias stress on the shaded cells. Along with the CIGS_A, also the CIGS_E module appears to be more tolerant to partial shading. This is due to the partiotioning of the cells which helps reducing the effect of reverse bias on the shaded cells.

In some cases, the shunts created due to partial shading produced a visible defect to the naked eye. Figure 15 shows the visible worm-like defect. These defects are localised close to the scribing lines.

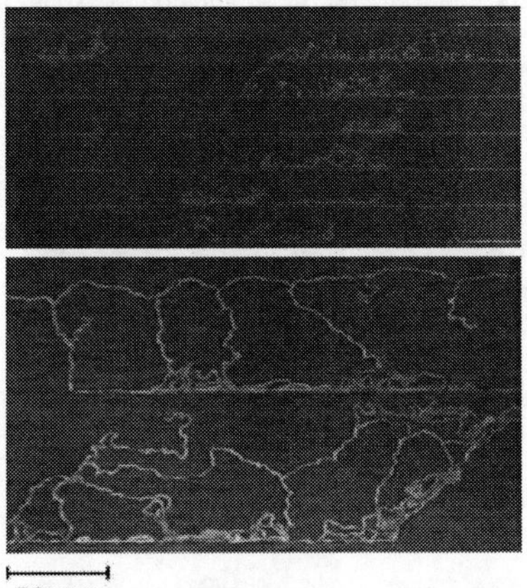

Figure 15: Worm-like defects that were created by applying shading stress on CIGS modules

After the reliability experiments, unshaded IV tracing was performed for each module to determine potential performance losses due to shunting. The losses were limited (2% - 8%) and affected mostly the heaviest shunted modules (CIGS_B, CIGS_F). Furthermore, a partial healing phenomenon was observed after storing these modules in dark for a couple of weeks.

5 CONCLUSIONS

TF modules have in general better performance under partial shading due to their unique cell layout. The TF modules consisting of more than one string (CdTe_A) or having more than one bypass diode (CIGS_A) have superior performance and improved reliability.

Despite the aforementioned advantages of TF modules in terms of performance, in extreme partial shading conditions, cell defects were evident, potentially causing shunt paths in areas where power is dissipated. Also in some modules worm-like defects visible to the naked eye were formed. Module partitioning (CIGS_E) and multiple strings (CdTe_A) can mitigate the problem.

6 REFERENCES

[1] A. Reinders, P. Verlinden, W. van Sark and A. Freundlich, Photovoltaic Solar Energy: From Fundamentals to Applications, Wiley & Sons, 2017.

[2] Solar Frontier, "High shadow tolerance," Solar Frontier, [Online]. Available: https://www.solar-frontier.eu/en/cis-technology/high-shadow-tolerance/. [Accessed 2017 April 13].

[3] T. J. Silverman, M. G. Deceglie, C. Deline and S. Kurtz, "Partial Shade Stress Test for Thin-Film Photovoltaic Modules," NREL, California, USA, August 2015.

[4] J. E. Lee, S. Bae, W. Oh, H. Park, S. M. Kim, D. Lee, J. Nam, C. B. Mo, D. Kim, J. Y. Yang, Y. Kang, H.-s. Lee and D. Kim, "Investigation of damage caused by partial shading of CuInxGa(1-x)Se2 photovoltaic modules with bypass diodes," *Progress in Photovoltaics,* vol. 24, no. 8, pp. 1035-1043, April 2016.

[5] T. J. Silverman, L. Mansfield, I. Repins and S. Kurtz, "Damage in Monolithic Thin-Film Photovoltaic Modules Due to Partial Shade," *IEEE Journal of Photovoltaics,* vol. 6, no. 5, pp. 1333-1338, July 2016.

[6] P.-O. Westin, U. Zimmermann, L. Stolt and M. Edoff, "Reverse bias damage in CIGS modules," in *24th European Photovoltaic Solar Energy Conference,* Hamburg, DE, September 2009.

ESTIMATION OF LOCAL DETERIORATION FACTOR IN CRYSTALLINE SI PV MODULE BY PARTIAL SHADING

Tomohito TANASE[1], Yasuhito TAKAHASHI*[1], Koji FUJIWARA[1]
[1]Dept. of Electrical Engineering, Doshisha University
1−3 Tataramiyakodani, Kyotanabe, Kyoto 610−0321, JAPAN
Phone: +81−774−65−6266, Fax: +81−774−65−6266
E−mail: ytakahashi@mail.doshisha.ac.jp (*Corresponding author: Yasuhito TAKAHASHI)

ABSTRACT: Accurately diagnosing failures of a photovoltaic (PV) module is important for improving its reliability. In this paper, the authors investigate a method to diagnose faults of a crystalline silicon PV module from its current-voltage (*I-V*) curves. Specifically, the authors try to detect the failure PV cell by shading one PV cell in the PV module and measuring the *I-V* curve of the PV module. As a result, local deterioration caused by shunt resistance can be detected. Furthermore, based on the fact that the influence of deterioration due to the shunt resistance does not appear near the open-circuit voltage in the PV cell's *I-V* characteristic, the authors propose the method to extract the *I-V* characteristic of the deteriorated PV cell from the PV module's *I-V* characteristic. As a result, it is confirmed that the extracted and measured *I-V* curve of the deteriorated PV cell agrees well.
Keywords: crystalline silicon PV module, diagnosis method, partial shading

1 INTRODUCTION

This paper aims at developing a method which can estimate local deterioration in a photovoltaic (PV) module with high accuracy. The PV power generation system has been said that it is maintenance-free for a long time. However, in fact, some troubles due to long-term operation have been reported [1]. In the worst case, some failures may cause eventual large-scale accidents such as smoke emission and fire. In order to prevent a decrease in output power and a major accident, a method for diagnosing a failure in a PV module is required.

Electro Luminescence (EL) image diagnosis and thermal image diagnosis are widely used for detecting failures in PV modules. These image diagnosis methods have the advantage of detecting faulty points in a PV module directly, although they cannot evaluate decrease of output power quantitatively. By contrast, the measurement of *I-V* characteristic is also carried out as a fault detection method. The measurement of the *I-V* curve enables us to evaluate the change of output power quantitatively. However, as a disadvantage, it is difficult to detect faulty points and estimate cause of the fault when a PV module includes a local deterioration.

In this paper, the authors investigate a method to estimate local deterioration factors in crystalline silicon PV modules based on the *I-V* curve measurement under partial shading of a PV cell. The effectiveness of the presented method is verified by numerical simulation and experiment of a PV module with artificially-generated local fault.

2 *I-V* CHARACTERISTICS OF PV MODULE INCLUDING DETERIORATED PV CELLS

This section examines *I-V* characteristics when a deteriorated PV cell is included in a PV module by numerical simulation. Fig. 1 shows schematic diagrams of the PV module and PV cells investigated in this paper and Fig. 2 shows the *I-V* characteristics of each PV cell in the PV module. Basic PV module consists of ten basic PV cells, PV module A consists of nine basic PV cells and one PV cell A, and PV module B consists of nine basic PV cells and one PV cell B. Here, PV cell A includes deterioration due to shunt resistance R_{sh} (for

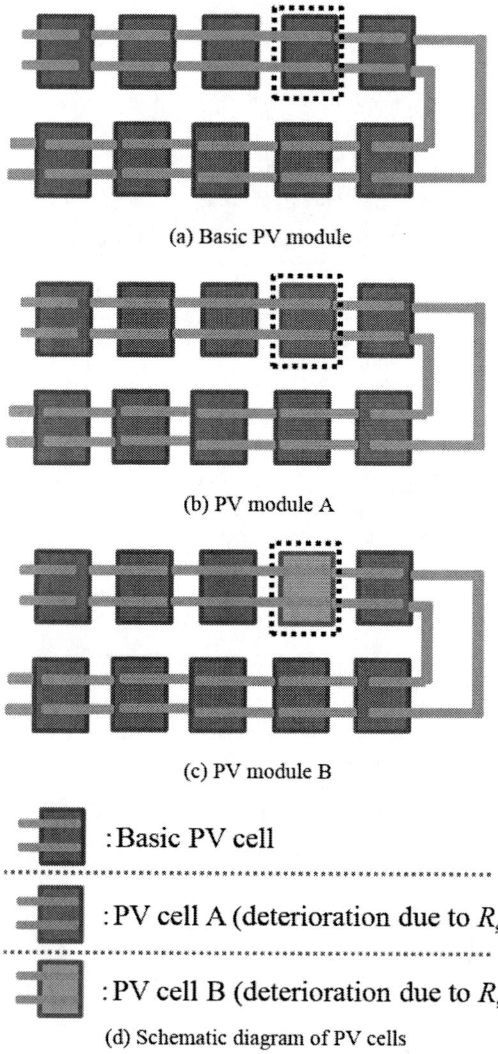

Figure 1: Schematic diagram of PV modules and PV cells used in this study

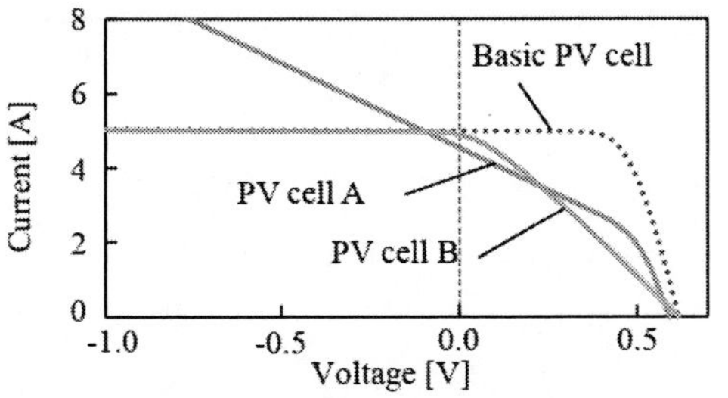

Figure 2: *I-V* characteristics of PV cell shown in Fig. 1

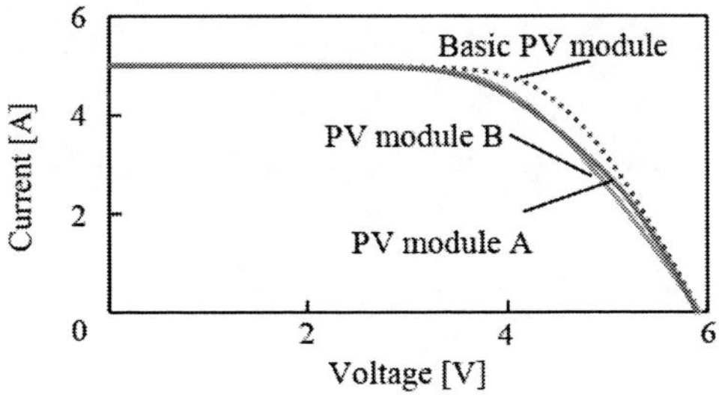

Figure 3: *I-V* characteristics of PV module shown in Fig.1

instance, a failure caused by leakage current), and PV cell B includes deterioration due to series resistance R_s (for instance, failure in solder joints).

In this verification, we investigate the change of the *I-V* characteristic when deterioration due to R_{sh} or R_s occurs in a PV module. Fig. 3 shows the *I-V* characteristic of each PV module. As shown in Fig. 3, the *I-V* characteristics of PV module A is almost the same as that of PV module B, which results in the difficulty to detect the cause of the local deterioration based on the shape of the *I-V* curve. Although it is possible to identify equivalent circuit parameters of each PV cell in the PV module and reproduce the *I-V* curve of the PV module by serially composing *I-V* curves of every PV cell, the accurate estimation of the equivalent circuit parameters in every PV cell is not easy. Therefore, we propose a simple and easy method to judge local deterioration factors in a PV module.

3 ESTIMATION OF LOCAL DETERIORATION FACTOR BASED ON *I-V* CHARACTERISTICS WITH PARTIAL SHADING

In Fig. 2, it can be seen that the *I-V* curve of PV cell A has a large slope in the second quadrant. This feature is unique to a deteriorated PV cell due to shunt resistance. However, in a general PV module, since the bypass diodes are connected, the measurement of the *I-V* curve in the second quadrant cannot be performed. In addition, even if the measurement can be performed, the influence of the local deterioration in a PV cell on the *I-V* curve of the PV module is small as shown in Fig. 3. As a result, the influence of local deterioration cannot be observed clearly. Therefore, in order to extract the information of local deterioration efficiently, we attempt to reflect the distinguish feature of the *I-V* curve in the second quadrant of the deteriorated PV cell to the *I-V* characteristic of the PV module. In order to realize this, we shade a deteriorated PV cell with a metal mesh for lowering the current value by limiting incident irradiance.

The basis of this approach is explained using the example in the previous section. In the *I-V* characteristic of the PV module shown in Fig. 3, the current value is within the range of 0 A to 5 A. However, the difference in the *I-V* curves between PV cells A and B appears especially in the second quadrant (5 A or more) as shown in Fig. 2. Here, we consider the change of shape of the *I-V* curve when shading the deteriorated PV cell in a PV module. Fig. 4 shows the *I-V* characteristics when the incident irradiance is decreased by 40% in each PV cell shown in Fig 1(d). Fig. 5 shows the *I-V* characteristics of each PV module when the incident irradiance is reduced by 40% by shading the PV cell which are surrounded by black broken lines in the PV module shown in Fig. 1. In Fig. 5, the current value drastically decreases when the basic PV cell and PV cell B are shaded, whereas the current value does not decrease greatly in PV cell A. In particular, before and after shading, the short-circuit current does not change significantly.

The reason of the distinguish change of the *I-V* curve of the PV module when shading the deteriorated PV cell can be explained as follows. The *I-V* characteristic of the PV module is obtained by serial combination of the *I-V* characteristics of each PV cell, that is, by summing up voltage at the same current in the *I-V* curve of each PV cell. For simplicity, we consider the serial composition of the voltage at the short-circuit current point. The short-circuit current of the basic PV cell is about 5 A. The corresponding voltage in PV cell A with shading is about -0.5 V as shown in Fig. 4. This is because the slope of the *I-V* curve of PV cell A in the second quadrant is steep. Therefore, the short-circuit current of the PV module A with shading is almost the same as its original value. On the other hand, the voltage corresponding to the short-circuit current of 5 A in PV cell B and basic PV cell with shading seems to be much smaller than -1 V as shown in Fig. 4. For this reason, the short-circuit current of the PV module B and basic module with shading is much smaller than their original value.

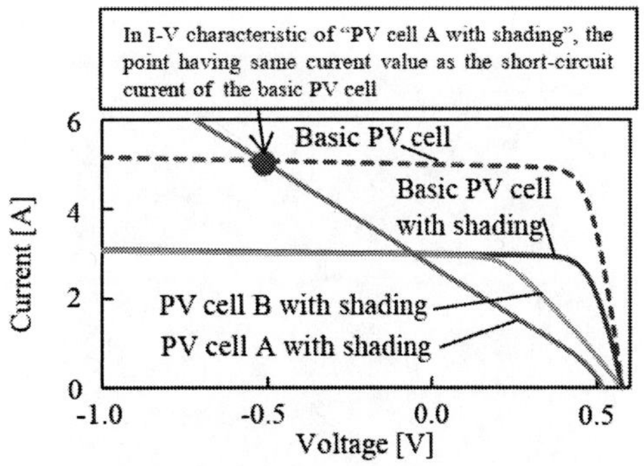

Figure 4: Simulated *I-V* characteristics of each PV cell when reducing incident irradiance by shading the cell.

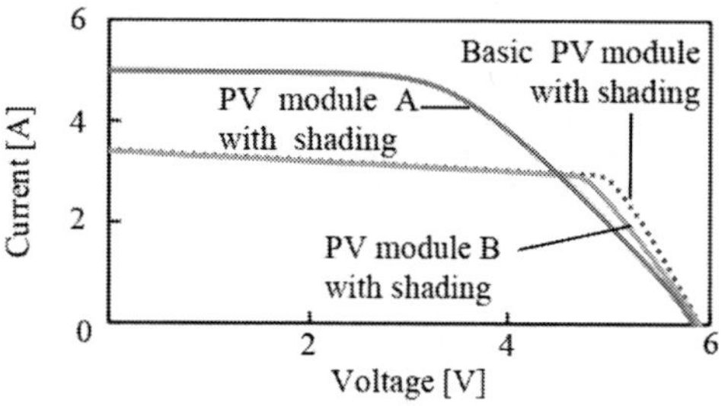

Figure 5: Simulated *I-V* characteristics of each PV module when reducing incident irradiance by shading one PV cell.

As mentioned above, when one PV cell in the PV module is shaded, the distinguish feature of the *I-V* curve in the second quadrant can be reflected to the *I-V* curve of the PV module clearly. This principle enables us to detect the local deterioration factor due to R_{sh} in the PV module. The procedure is as follows:

i) Measure the *I-V* characteristic of the PV module with and without shading one deteriorated PV cell.

ii) If the short-circuit current value does not change from the non-shading state, it can be estimated that the cause of the deterioration is due to shunt resistance R_{sh} in the shaded PV cell. If the short-circuit current value changes drastically before and after shading, deterioration due to shunt resistance R_{sh} does not occur in the PV cell.

We verify the validity of the presented approach with a real PV module shown in Fig. 6. In this experiment, the serially-connected 16 PV cells surrounded by red broken line shown in Fig. 6 (a) are defined as Basic PV module. The shunt resistor of 0.51 Ω is connected to one of the PV cell in parallel from the back side of the module as a local deterioration. This state of the module is regarded as PV module with a deteriorated PV cell. Then, the deteriorated PV cell in the PV module is shaded by a metal mesh whose opening ratio is 64 %. The PV module in this state is defined as PV module when partially shading a deteriorated PV cell. Fig. 7 shows the measurement results. Since the short-circuit current of the PV module when shading the deteriorated cell does not decrease, it can be judged that the deterioration due to the shunt resistance has occurred in this PV cell.

(a) front side

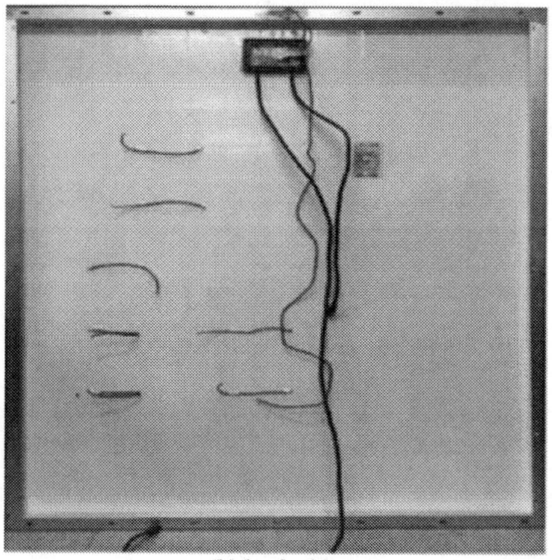

(b) back side

Figure 6: PV module used for experiment (using 16 PV cells indicated by red dotted line)

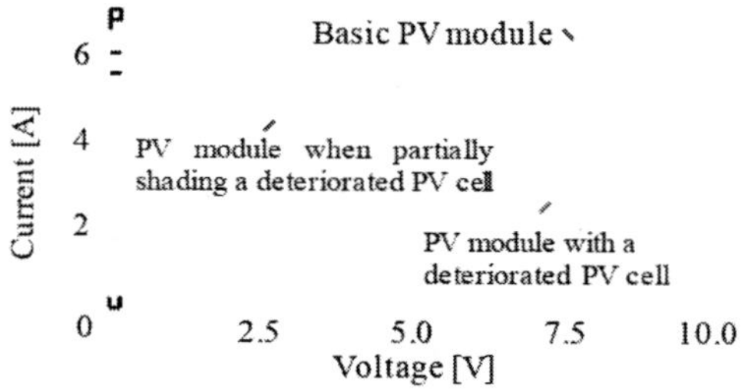

Figure 7: Measured *I-V* characteristics of PV module

4 EXTRACTION OF *I-V* CHARACTERISTIC OF DETERIORATED PV CELL IN PV MODULE

In this section, we discuss a method for extracting the *I-V* characteristic of a single deteriorated PV cell from the *I-V* characteristic of a PV module including the deteriorated PV cell.

Fig .8 shows schematic diagrams of PV modules used in this verification. A PV module composed of *n* basic PV cells is called basic PV module, and the PV module with a deteriorated PV cell due to shunt resistance is called PV module α. If we have the basic PV module, the *I-V* characteristic of a single deteriorated PV cell can be easily obtained. However, in general, we only have PV module α when estimating local deterioration. Here, the *I-V* characteristic of PV module α' is defined by multiplying the voltage of the *I-V* curve of PV module α by (n - 1) / n. From Fig. 7, the influence of the deteriorated PV cell appears near the maximum power point in the *I-V* curve of the PV module. In other words, the *I-V* curves of the basic PV module and PV module α are almost the same when the voltage is higher than the maximum power point as shown in Fig. 7. Therefore, in this region, PV module α' can be regarded as the *I-V* characteristic of the PV module consisting of *n*-1 basic PV cells. Furthermore, the *I-V* characteristics obtained when shading the deteriorated PV cell is defined as PV

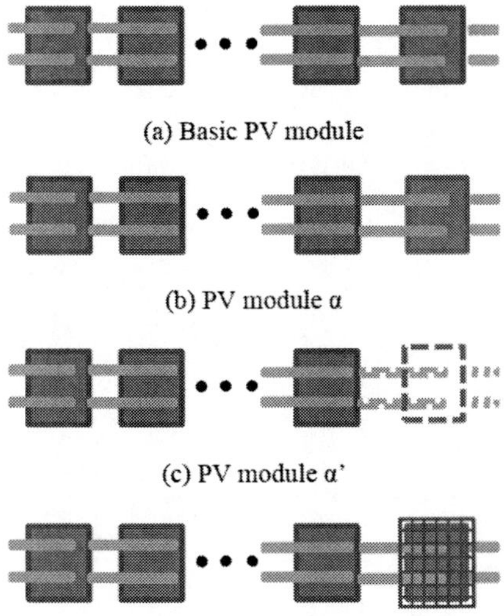

Figure 8: Schematic diagrams of PV module used for extracting *I-V* characteristic of the PV cell deteriorated due to shunt resistance from that of the PV module α

Module β.Then, by subtracting the *I-V* curve of the PV module α' from that of the PV module β, we can extract the *I-V* characteristic of one deteriorated PV cell with shading. Fig. 9 shows a schematic diagram of the above procedure.

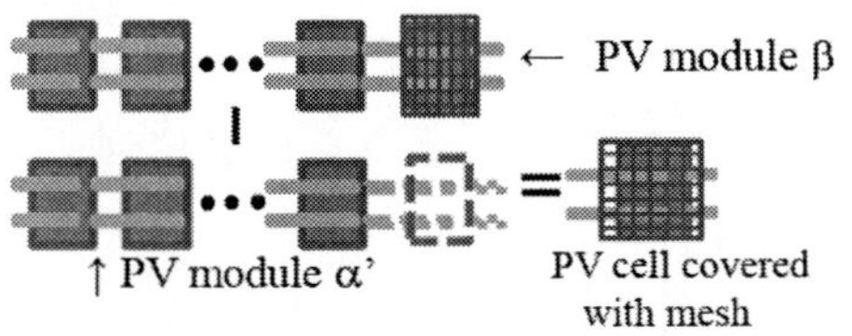

Figure 9: Method for extracting the *I-V* characteristics of deteriorated PV cell due to shunt resistance from that of PV module β and PV module α'

Using the above procedure, the *I-V* characteristic of the deteriorated PV cell investigated in the previous section is extracted. Fig. 10 shows the *I-V* characteristics of PV module α' calculated from PV module α. Then, by subtracting the PV module β from the PV module α', the *I-V* characteristic of the deteriorated PV cell are calculated.

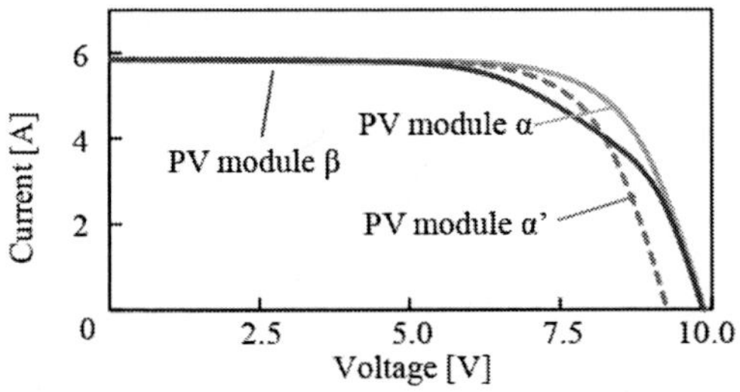

Figure 10: *I-V* characteristics PV module α' calculated from PV module α in the experiment of Section 3

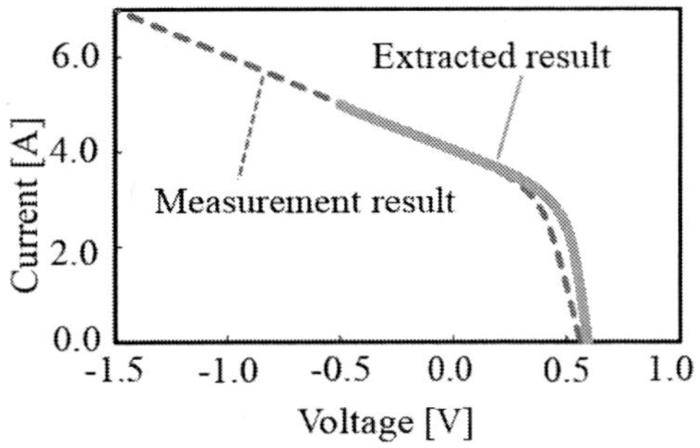

Figure 11: Comparison between extracted and measured result of the *I-V* characteristics of the PV cell deteriorated due to shunt resistance with shading by metal mesh

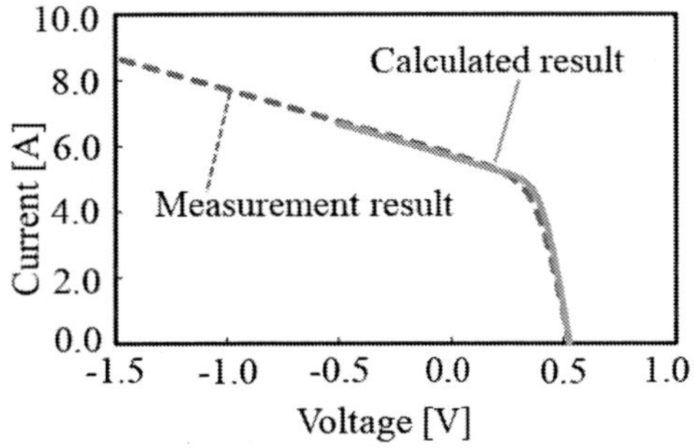

Figure 12: Comparison between finally calculated and measured result of the *I-V* characteristics of the PV cell deteriorated due to shunt resistance with no shading

For comparison, the *I-V* characteristic of one deteriorated PV cell covered with a metal mesh was directly measured from the back side of the PV module. Fig. 11 shows these results. The extracted *I-V* curve agrees well with the measured one, which indicates the validity of the presented method.

Using the *I-V* characteristic of the deteriorated PV cell with shading, the *I-V* characteristic without shading is estimated. First, the equivalent circuit analysis is performed from the obtained characteristics, and four equivalent circuit parameters (photocurrent I_{ph}, series resistance R_s, shunt resistance R_{sh} and diode factor n) are obtained [2]. Then, by multiplying the photocurrent by the reciprocal of shading rate (64 %) of the metal mesh used in this experiment, the value of the photocurrent without shading is obtained. Finally, the *I-V* characteristic of the deteriorated PV cell without shading is calculated using the corrected photocurrent and three other equivalent circuit parameters. Fig. 12 shows the obtained results. For comparison, the measurement results of the *I-V* characteristics of the deteriorated PV cell without shading are also shown. From Fig. 12, it is confirmed that the sufficiently accurate *I-V* characteristics can be estimated.

5. CONCLUSION

This paper proposed the method to evaluate the local deterioration in the PV module. As a result, when deterioration due to shunt resistance occurred, it was possible to detect the occurrence of deterioration from the *I-V* characteristic of the PV module by shading the PV cell. It was also possible to extract the characteristics of deteriorated due to shunt resistance PV cell from the *I-V* characteristic of the deterioration PV module by utilizing the fact that the *I-V* characteristic of basic PV cell and deteriorated PV cell due to shunt resistance are not significantly different under a certain current value. In future, the presented method will be applied to other deteriorated cases and its effectiveness will be examined.

References

[1] K. Kato, A. Miller, The Institute of Electrical Engineers of Japan, Vol.134, No.10, (2014).
[2] Y. Kimura, Y, Takahashi, K. Fujiwara, K. Masuda, S. Nishikawa, H. Kato, Grand Renewable Energy 2014 Proceedings (2014).

A COMPARATIVE STUDY OF PV MODULES PERFORMANCE BETWEEN PREDICTION MODELS AND EXPERIENCE IN THE GREEN ENERGY PARK: CRYSTALLINE TECHNOLOGY

C. Hajjaj[1, 2], M. Benhmida[2], S. Sahnoun[2], A. Alami Merrouni[1], A. Ghennioui[1], A. Benlarabi[1,4], B. Ikken[1], Z. Naimi[1], A. Bennouna[3], M. Maaroufi[4]
[1]Institut de Recherche en Énergie Solaire et en Énergies Nouvelles (IRESEN), Rue Abou Marouane Essaadi, Quartier Administratif Rabat Instituts 6208, Agdal - Rabat, Morocco
[2]Laboratory of Electronics, Instrumentation and Energetic, Faculty of Sciences, Chouaïb Doukkali University, B.P 20, El Jadida, Morocco
[3]Université Cadi Ayyad, Boulevard Abdelkrim Al Khattabi, Marrakech 40000, Maroc
[4]Ecole Mohammedia des Ingénieurs, Avenue Ibn Sina, Rabat 10000, Maroc
hajjaj.c@ucd.ac.ma

ABSTRACT: This paper aims to give a critical overview of the existing prediction models for PV modules performance over a long term. Indeed, each model takes into account a number of meteorological parameters (cell temperature, ambient temperature, solar radiation and solar spectrum), and depends on regressive coefficients to weight each term of the equations, yet, each model won't be giving the exact performances.
We will compare the efficiency calculated by Evans and Durish prediction models with the experimental data collected from the outdoor PV test platform at the GEP and highlight the relevance of each model.
Keywords: prediction, PV performance, statistical indicators.

1 INTRODUCTION

Photovoltaic (PV) array/module/cell efficiency predictions are essential to accurately assess its electrical performance. However, the PV modules technologies are tested according to the IEC standards that stipulate strict conditions: Solar Radiation of 1000 W/m², a temperature of 25°C, and a solar spectrum of 1,5 [1].

Several models had been put in place to approach the real output through equations of estimation taking into account the solar radiation, the temperature of the cell, and the air mass [2]. Nonetheless, all those models remain far from the reality. In this study, we will proceed to a critical overview of the existing models, confronting them to real measured efficiency applied on Poly-crystalline technologies installed in outdoor conditions in green energy park, the research platform of IRESEN.

2 FORECAST MODEL

To model the electricity production of a region or a simple photovoltaic installation over several days, various factors such as weather forecasts, sun irradiation and the efficiency of photovoltaic modules must be taken into consideration. The main parameters used to compare the two prediction models are shown in Figure 1.

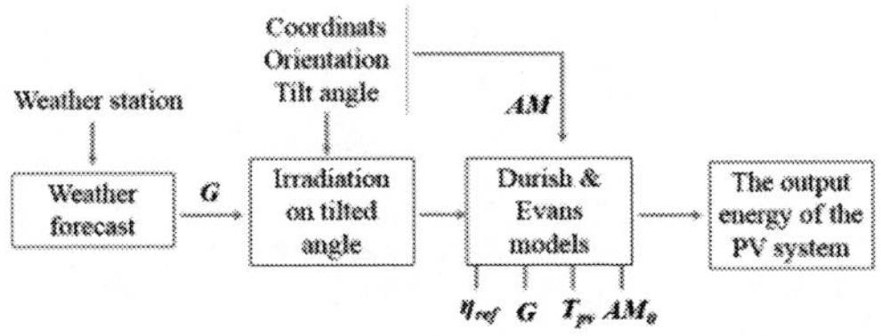

Figure 1: Schematic diagram of the PV production forecast.

3 EFFICIENCY MODELS

In our study, we have tested two energetic models proposed by Durish and Evans to predict the photovoltaic modules production [3,4] from meteorological data measured on the green energy park (GEP) site at the green city of benguerir in Morocco [5].
The efficiency model of the photovoltaic model proposed by Evans [3]:

$$\eta_{PV} = p\left[q\frac{G_\beta}{G_{\beta,o}} + \left(\frac{G_\beta}{G_{\beta,o}}\right)^m\right] \times \left[1 + r\frac{\theta_{cell}}{\theta_{cell,0}} + s\frac{AM}{AM_0} + \left(\frac{AM}{AM_0}\right)^u\right]$$

Where $\eta_{PV,ref}$, T_{PV}, G, β_{ref} and γ are respectively the nominal electrical efficiency at the reference temperature (T_{ref} = 25 °C), solar irradiation intensity (W/m²), efficiency correction coefficient for temperature and efficiency correction coefficient for solar radiation.
The efficiency model of the photovoltaic model proposed by Durish [4]:

$$\eta_{PV} = p\left[q\frac{G_\beta}{G_{\beta,o}} + \left(\frac{G_\beta}{G_{\beta,o}}\right)^m\right] \times \left[1 + r\frac{\theta_{cell}}{\theta_{cell,0}} + s\frac{AM}{AM_0} + \left(\frac{AM}{AM_0}\right)^u\right]$$

Where p, q, m, r, s, and u are constants, specific for each technology and which must be defined experimentally.

G_β : Tilted angle of PV module

$G_{\beta,0}$: Standard insolation in sea level.

T_{cell} : Temperature of the cell.

$T_{cell,0}$: Standard temperature of the module.

AM : Air mass

AM_0 : Air mass in normal conditions

4 Statistical methods and performance indicators

Standardized performance evaluation techniques facilitate forecast benchmarking; Common metrics proposed in the literature include Mean Bias Error MBE, Mean Squared Error MSE, Root Mean Squared Error RMSE, Standard Deviation SD, Mean Bias Error, and finality the most adapted for solar forecasting the normalized Root Mean Squared Error NRMSE. These are defined as follows:

$$e_x = y_{meas,x} - y_{pred,x}$$

Where $y_{meas,x}$ and $y_{pred,x}$ are respectively the ith measured and predicted value.

$$MBE = \frac{1}{N}\sum_{i=1}^{N} e_x$$

$$RMSE = \sqrt{\frac{1}{N}\sum_{i=1}^{N} e_x^2}$$

$$NRMSE = \left(\frac{1}{N} \sum_{i=1}^{N} \frac{|e_x|}{y_{meas,x}} \right) \times 100$$

$$TS = \sqrt{\frac{(N-1)MBE^2}{(RMSE^2 - MBE^2)}}$$

$$SD = \sqrt{\frac{N(RMSE^2 - MBE^2)}{(N-1)}}$$

MBE gives the average forecast error; RMSE gives more weight to the largest errors. As for SD, it captures the part of the RMSE that is not due to systematic error and provides an indication of the RMSE that can be achieved once the bias is zero according to this equation $RMSE^2=MBE^2+SD^2$.

5 EXPERIMENTAL SETUP

5.1 Green Energy Park

GEP's Photovoltaic power plant have two main goals, first clean electricity generation, the second is being a test platform. This power plant was designed by combining a multitude of test-set-ups in order to investigate and characterize various PV modules technologies in harsh weather conditions. Also to get valuable data about their performance in situ. IRESEN is aiming to implement a newly adapted certification process for the outdoor measurement and characterization. Also, it was created to assess PV technologies and to develop novel adapted technologies. (Figure 2).

Figure 2: 3D view of Green Energy Park power plant

5.1 Poly-Si photovoltaic panel installation

An experiment with a Poly-Si PV panel (Figure 3) with nominal output power 252.8 W at STC was conducted from 11 sept 2016 to 11 sept 2017 (12 month time period).

The PV panel is tilted 32° from the horizontal and is facing to the south. A high-quality solar irradiance sensor is mounted close to the PV panel and is placed in the same plane as the PV panel. A single temperature probe is glued on the back of the PV panel.

The PV panel power, the incident solar irradiance and the PV panel temperature are recorded with a one-second time step using a real-time embedded industrial controller.

Figure 3: Pictures of the PV Poly-Si panel in Green Energy Park

4 RESULTS AND DISCUSSION

In the light of the performances indicators (MBE, RMSE, NRMSE and SD) calculated in table 1 for each model we conclude that Evans model gives a better prediction than Durish model.

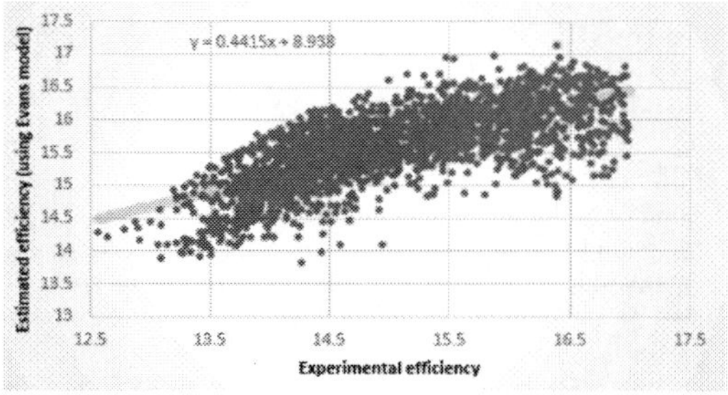

Figure 4: Correlation between the data efficiencies given by Evans model and the experimental results of efficiency

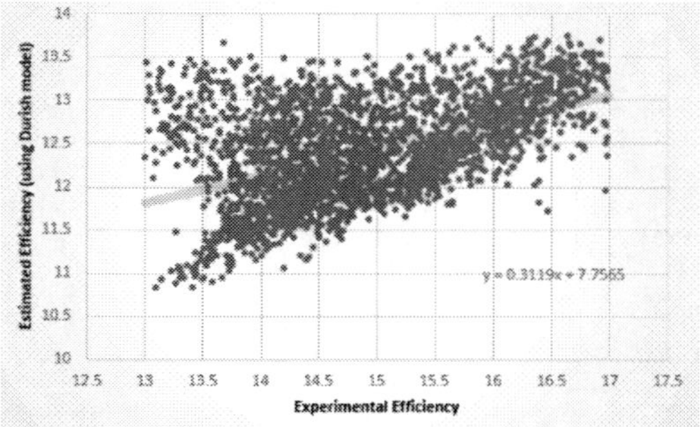

Figure 5: Correlation between the data efficiencies given by Durish model and the experimental results of efficiency

Table I: Statistical evaluation of different performance indicators

	Evans	**Durish**
MBE	0.5659	2.3558
RMSE	0.8493	2.5804
NRMSE	0.0566	0.1744
TS	44.9408	118.5617
SD	0.6335	1.0529
R^2	0.5222	0.1058
Rank	**1**	**2**

4 CONCLUSION

In the light of the performance indicators, we have concluded that Evans model gives a better prediction than a Durish model

Reference

[1] A. Q. Jakhrani, S. R. Samo, S. A. Kamboh, J. Labadin, A. R. H. Rigit, "An Improved Mathematical Model for Computing Power Output of solar Photovoltaic Modules", International Journal of Photoenergy, September 2014.

[2] I. Bërdufi, P. Berberi, D. Mitrushi, V. Muda, D. Topçiu, U. Buzra, "The Performance of a Grid Connected Photovoltaic System Part I: Durisch and Evans Energetic Models".

[3] Evans DL. Simplified method for predicting photovoltaic array output. Solar Energy 1981.

[4] Durisch W, Bitnar B, Mayor JC, Kiess H, Lam KH, Close J. Efficiency model for photovoltaic modules and demonstration of its application to energy yield estimation. Solar Energy Materials and Solar Cells 2007.

[5] GREEN ENERGY PARK Platform. URL: http://www.greenenergypark.ma/

ILLUMINATION HOMOGENEITY OF BIFACIAL SYSTEMS – OUTDOOR MEASUREMENTS WITH SYSTEMATICALLY VARIED INSTALLATION CONDITIONS

Thomas Baumann*, Markus Klenk, Nico Keller, Hartmut Nussbaumer, Franz Baumgartner
ZHAW, Zurich University of Applied Sciences, SoE, Institute of Energy Systems and Fluid Engineering
Technikumstrasse 9, CH-8401 Winterthur, Switzerland, www.iefe.zhaw.ch
*phone: +41 (0) 58 934 71 87; e-mail: thomas.baumann@zhaw.ch; www.zhaw.ch/=bamt

ABSTRACT: The optimum mounting condition for bifacial modules is a trade-off maximizing the performance of the front and backside output obtained for the given installation conditions. At the backside of bifacial PV modules there are more complex conditions concerning the illumination intensity and illumination homogeneity, compared to the frontside. In extended PV plants, with parallel rows of bifacial modules, there are additional factors, such as reduced ground albedo due to shading by adjacent rows, which have to be considered.
At the ZHAW, we implemented an outdoor test system to analyse the performance of bifacial arrays at differing mounting conditions by a new methodical measurement approach. The tilt angle is adjusted periodically between 0° and 90° in twelve positions within a one minute interval. In this paper we focus on outdoor measurement results concerning the illumination homogeneity at the front and backside and its impact on the total output. Not only the module data itself is analysed, but also a set of multiple small light sensors is attached at the rim of the bifacial centre module in order to reveal the illumination homogeneity in more detail. This setup enables a detailed analysis of the sensitivity at different tilt angles as well as more general yield results from long-term measurements.
Keywords: Bifacial, PV Array, Shading, System Performance, Evaluation

1 INTRODUCTION

The potential for an improved module power output of bifacial PV systems, compared to standard monofacial ones, was repeatedly demonstrated by simulations [1]–[3], measurements on single modules [4]–[8] or installations [9] with specific installation conditions. In spite of the considerable potential the installed capacity of bifacial systems is still marginal. A major obstacle for investors is the limited predictability of the bifacial PV system output. Even in the PV community there is considerable uncertainty about the real benefit due to bifaciality, as reflected by the numerous publications dealing with this issue.

For monofacial standard systems the simulation by software tools is a proven means and state-of-the-art since long. Due to the large amount of installed monofacial systems all over the world exists a huge amount of comparable data which allows an appraisal of projected arrays. Thus, for standard monofacial systems, investors can rely on a meaningful appraisal based on experience and simulations.

Because of more complicated physical conditions the predictability of bifacial PV systems is considerably less straightforward and reliable. The bifacial power gain is inherently based on the utilization of the radiation which is impinging on the modules backside. This backside illumination, reflected light from the modules surrounding, will show an inhomogeneous distribution over the backside area, dependent on multiple factors. Optimised mounting conditions for bifacial modules are a trade-off maximizing the performance of the front and backside output obtained for the given ground reflectance, the installation height, inclination and orientation. These more complex conditions concerning the illumination intensity and illumination homogeneity, compared to the frontside, hamper the development of reliable simulation software tools for bifacial modules. In real, extended systems, the arrangement of multiple modules and the specific mounting conditions will have additional effects. Shading or the reflectance caused by adjacent rows has to be taken into account. This is particularly true for some mounting conditions which are of interest for bifacial systems, such as vertical installation. Also the reduced albedo due to shading of the ground has to be considered.

A systematic analysis of a bifacial system in a real, extended array is therefore an appropriate means to improve the described obstacles, as it produces data to test and improve respective physical models and algorithms.

2 APPROACH

At the Zurich University of Applied Science (ZHAW) we currently put an array for the systematic measurements of bifacial systems with differing mounting conditions to the test. The trials are carried out at a 3x3 module array based on large, commercially available, 60-cell modules with a continuous, automated variation of the tilt angle. Other parameters like the distance between adjacent rows, the installation height or the ground reflectivity are manually adjusted. Due to the continuous variation of the tilt angle trends and interdependencies are revealed and several mounting situations are simultaneously analysed.

The basic concept of the test rig is shown in Figure 1. It offers the option to include shading and light reflection effects by adjacent rows of modules in different configurations. This is a major advantage compared to test stands with single, free standing modules. Three rows of modules, with manually adjustable distance between the rows, are mounted on vertically adjustable pillars in order to position the modules in different heights. An important feature of the measurement set-up is the automatic variation of the tilt angle in certain steps. For each step an I/V-curve of the centre module is measured in order to get the power as a function of the tilt angle of the module over a long period. All panels change their tilt angle continuously and coordinated with the central row.
By a change of the orientation of the whole measurement test array one can measure the power output of a typical flat roof or ground mounted, free field PV system, mainly

oriented to the south (northern hemisphere) or a tracking system changing the orientation from east to west. In the latter case, the measurement system can be used as tracking system itself. In addition, the influence of reflective grounds with differing reflectivity can be analysed.

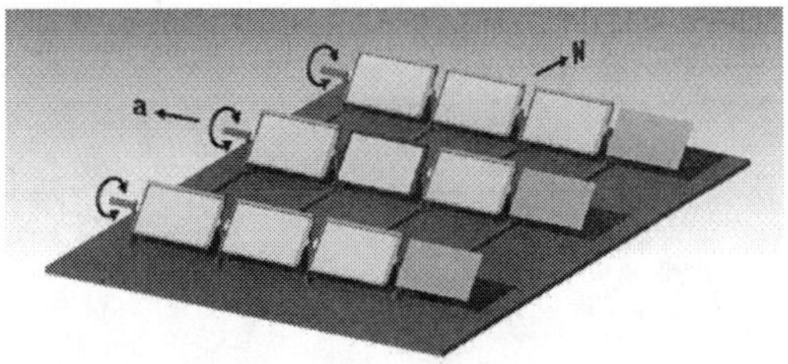

Figure 1: Measurement setup with permanently revolving modules, indicated with round arrows. Height, distance between rows and reflecting ground can be changed manually. The most relevant module in the centre, which is best suited to represent the actual conditions in real installations, is marked red.

Due to the bifaciality, the used PV modules are also sensitive to indirect shading by reason of reduced diffuse reflectance from the ground. Therefore, the panel in the centre of such an array is best suited to reflect the general conditions for a typical module in an extended bifacial power plant. By analysis of the other modules in the array the effect as a result of the limited extension and the validity of simulation tools can be tested.

An advantage of the new methodical bifacial test setup is that even small changes in yield at different tilt angels will be detected, compared to fixed mounted bifacial PV module rows at different tilt angles. This is due to the fact that measurement values of the same module at the identical position are used, which excludes drift effects of individual modules and local variations of the ambient.

3 OBJECTIVES

An important objective of the test rig, called Bifacial Outdoor Rotor Tester (BIFOROT), is the generation of data in long term measurements to test and improve existing prediction algorithms for bifacial systems. The variation of the mounting conditions allows the implementation of varying parameters (installation height, albedo, row distance) which results in an extended range of validity. The long term measurements will also reveal the energy yield and the impact of shading for specific mounting situations. For all mounting situations and lighting conditions the optimized tilt angle will be determined due to the continuous revolving of the modules.

Long term measurements are however no prerequisite to obtain meaningful results. Also at shorter time scales important conclusions about the general properties of bifacial systems can be drawn. This is especially true if certain parameters, such as the albedo from the surrounding ground, can be replaced at virtually unchanged lighting conditions. Since the adjustment of new installation conditions can be lengthy for such a large array, we additionally implemented a miniaturized version. Comparison of yield analysis from the BIFOROT and the miniaturized BIFOROT test array are subject to prior publications [10].

4 MEASUREMENT SETUP

A matrix of 3x3 modules may not be sufficient to reflect the shading effects in extended arrays for specific installation conditions, such as vertical installation. This is particularly true for small distances between rows. Therefore additional shading elements are respectively applied to one side of the rows, a further extension in the large array is not feasible due to the limited available space on the roof of the building.

Figure 2 shows the test rig, equipped with commercially available bifacial modules from the cell and module manufacturer Megacell (MBF-GG60-270). The white sheet under the test array from the manufacturer Sika (Sarnafil TS 77-20 RAL 9016) is commercially used for roof waterproofing. The measured albedo factor (measured at axis height) after a year installation and regular cleaning is 0.51, whereas the albedo factor of the surrounding concrete slabs is 0.17, also measured at axis height.

Figure 2: South orientated installation of the BIFOROT test array on the roof of the ZHAW in Winterthur. Aerial view on the roof and test array from southwest direction. The module tilt angle is changing permanently in a synchronized manner within an interval of one minute. Module 1 to module 3 (M1 –M3) are the devices under test (DUT).

The respective output versus the tilt angle is recorded for a specific set of tilt angles. The duration of a cycle is one minute, in which one cycle means the movement from the 0° horizontal position to the 90° vertical position (for south orientation of the frontside). There are 12 tilt angles / measurement positions in each cycle (0°, 10°, 15°, 18°, 21° 25°, 30°, 35°, 40°, 45°, 60°, 90°).

Multiple small light sensors are attached in vertical and horizontal positions at the centre module (M2) front (Figure 3) and backside (Figure 4). East of the centre module a rotating pyranometer and ISE reference cell is installed. Several environmental sensors are located on the north side of the test array and on top of the roof. These sensor measurements include the wind speed, wind direction, ambient temperature and the global horizontal irradiation.

Table I represents the used specifications for the BIFOROT, which were used for the measurements. All results in the following chapter are based on these specifications.

Figure 3: Small irradiance sensors - crystalline silicon cells - (red marked) enable a detailed analysis of the mapping of illumination intensity and homogeneity on the module frontside (south side). The pyranometer and silicon reference sensors ISE cell (green marked) are mounted on the east side of the module mounting frame, moving synchro with the DUT.

Figure 4: Small irradiance sensors - crystalline silicon cells - (red marked) on the module backside (north side).

Table I: Used specifications of the BIFOROT for the analysis in this paper

Name	Value/Description
Azimuth Angle	0° (north-south orientation)
Axis height	0.75 m (from ground to axis centre)
Axial spacing	2.86 m (from the centre of the axis to the centre of the axis)
Ground albedo	0.51 (measured at axis height)
Module height	lower edge depending on tilt angle (Figure 11), module centre always at axis height (0.75 m)
Module 1 (M1)	frontside covered for $I_{SC,back}$ measurement
Module 2 (M2)	I/V-curve measurement ($I_{SC,bifacial}$)
Module 3 (M3)	backside covered for $I_{SC,front}$ measurement

5 DATA EVALUATION

The long-time measurement started on 5th of October 2016 and was planned to continue for at least one year. Unfortunately, on 28th of March 2017, two construction cranes of a larger construction site were erected in the south of the test array. These cranes caused significant shading on the test array. Therefore, the measurement data could only be evaluated until 27th of March 2017.

In order to exclude shading from the building and reflection effects from the glass pane (Figures 2 and 5), the used data are limited to an azimuth angle range of -86.6° to +85.1° (respectively 93.4° to 265.1°), as well as to a minimum sun elevation of 0°. Figure 6 shows the sun positions during the measuring period with the corresponding angles of incidence. The white gaps correlate with the measurement outages by reason of maintenance, heavy snowfall or icing of the bearings.

Figure 5: Reflection of the sunlight in the glass pane of the roof structure.

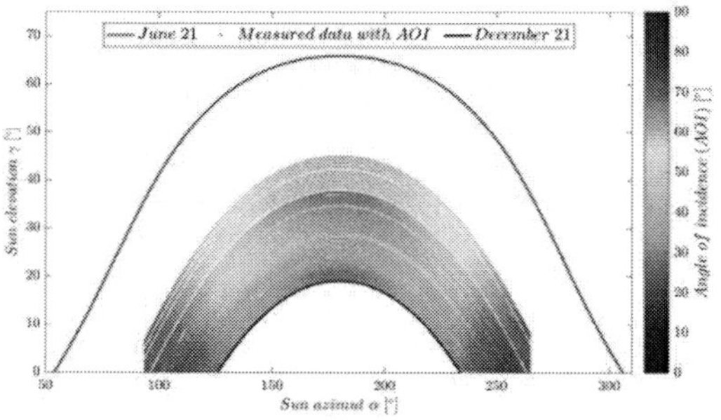

Figure 6: Solar positions at the measuring location during the measurement period with the angle of incidence (colored points), solar positions on December 21 (blue) and June 21 (red). In addition, the graph shows the interruptions (e.g. due to maintenance, heavy snowfall or frost) by the white gaps.

The used centre module (M2) has a bifaciality B of 0.694, which is calculated according to Formula (1). This value is low compared to other bifacial modules because of the shading on the modules backside, caused by the junction box. The MPP values of each side were measured with the ZHAW LED Flasher [11], whereas the not measured side was covered by a cardboard.

$$B = \frac{P_{mpp,back}}{P_{mpp,front}} = \frac{188.5\ W}{271.44\ W} = 0.694 \qquad (1)$$

5.1 Energy yield

The energy yield of the centre module (M2) is shown in Figure 7 (dark blue) for the whole measurement period (winter season). The maximal energy yield for the given setup was achieved at a 40° tilt angle. In Figure 7, the monthly energy yields from October 2016 to March 2017 are shown as well. The yield of January is low due to heavy snowfall and icing of the rotating axis.

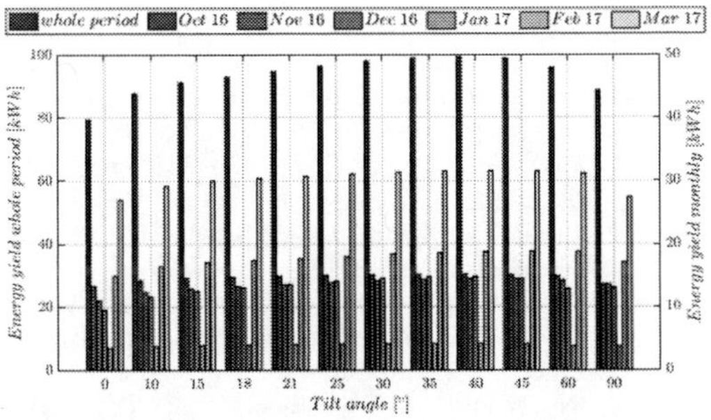

Figure 7: Energy yield (dark blue, left y-Axis) over the whole measurement period (05.10.2016 until 27.03.2017) and energy yield per month (coloured, right y-Axis) per tilt angle of module 2 (M2). The yield of January is low due to heavy snowfall and icing of the rotating axis.

5.2 Irradiance on module front and backside

Figure 8 shows the energy yield of M2 compared to the total irradiation measured with the ISE cell, which is installed in the module frontside plane. Additionally, the total irradiation on the module frontside of M3 is shown. The total irradiation is respectively obtained by integrating the I_{SC} measurement values. The courses of the total frontside irradiations (ISE and I_{SC}) are almost identical, whereas the energy yield has a different course due to the bifacial gain.

Figure 9 includes the sum of the short circuit current of the frontside (M3), backside (M1), their sum and bifacial (M2) per tilt angle. It is evident that the sum of the short circuit current on the backside of the module is nearly constant over all tilt angles. The backside contribution (Bc) is calculated according to Formula (2) and shown on the right y-axes of Figure 9.

$$Bc = \frac{\sum I_{SC,bifacial}}{\sum I_{SC,front}} - 1 \quad (2)$$

The relative contribution of the backside to the total illumination in Figure 9 is in a range between 0.15 at 45° and 0.35 at 0° dependent on the tilt angle.

The irradiation on the module backside, measured at different positions, is shown in Figure 10. The minimal irradiation on the backside of the module limits the bifacial gain. The red lines in Figure 9 (Bc) and Figure 10 (minimal value) have an identical course.

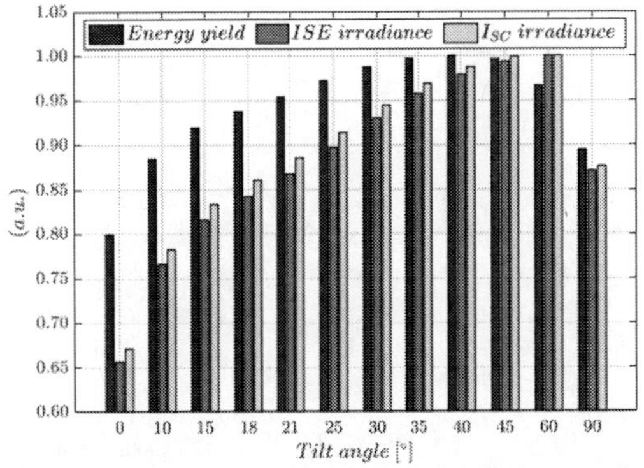

Figure 8: The course of the total irradiation on the module frontside (M2) measured with the ISE cell, as well as the total irradiation on the module frontside of M3 is shown in comparison to the energy yield per tilt angle of module 2. All values are normalized to the maximum per category.

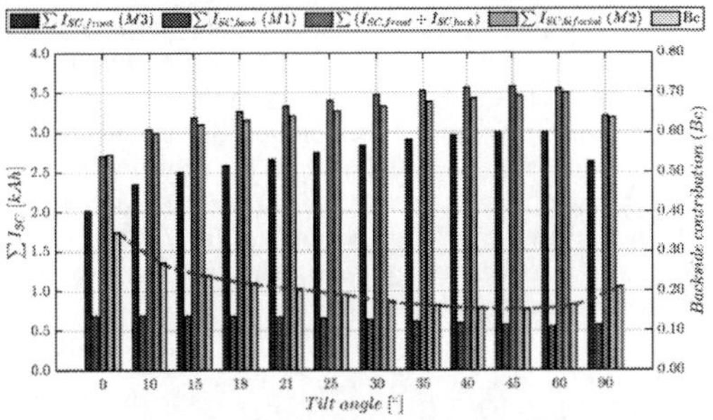

Figure 9: Total irradiation (integrated I_{SC}) over the whole measurement period for frontside (M3), backside (M1), sum of front and backside and bifacial (M2) per tilt angle. The contribution of the backside (yellow, right y-Axis) was calculated according to Formula (2).

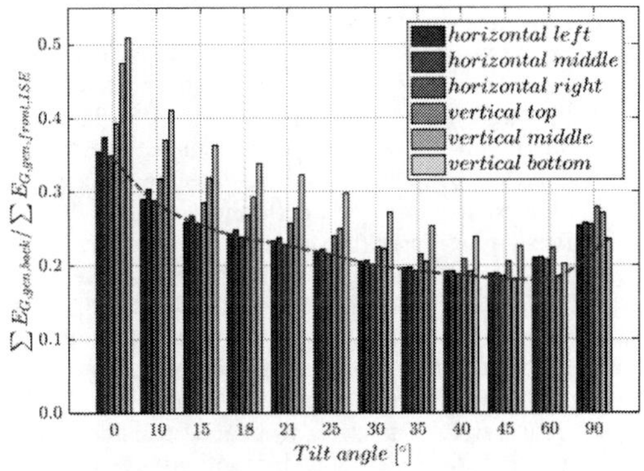

Figure 10: The irradiation over the whole measurement period on the module backside of module 2 (M2) is measured at different positions and related to the irradiation on the module frontside (ISE cell)

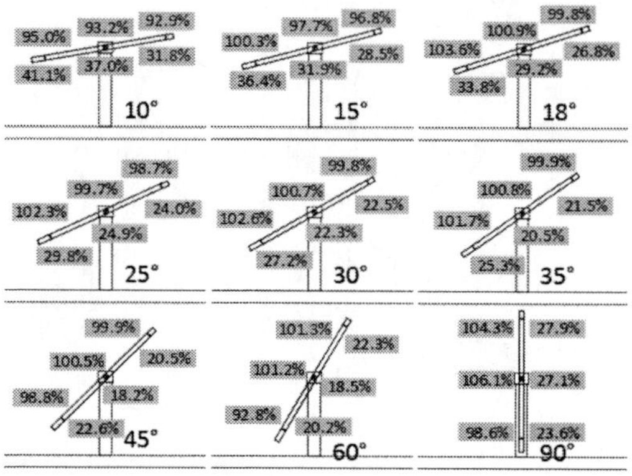

Figure 11: The vertical total irradiation for the front and backside of module 2 (M2) over the whole measurement period. The irradiation is related to the irradiation measured on the module frontside plane by the ISE cell.

The vertical irradiation distribution for three sensor heights (front and backside) and for different tilt angles is shown in Figure 11. The values are normalised to the ISE irradiation on the module frontside. The further the distance to the ground, the more diffuse irradiation reaches the backside of the module and contributes to the energy yield. The lowest value on the backside of the module limits the bifacial gain. A higher mounting height results in a higher energy yield, which has already been shown in other publications [8].

6 CONCLUSION AND OUTLOOK

The new test system was successfully operated during the first long term measurement period. The data provide valuable information about the distribution of irradiation on the module backside at different tilt angles. These data are intended to be used for the development of an energy yield simulation model.

The maximal energy yield for the measurement period (winter season) and the given setup was achieved at a tilt angle range between 35° and 45°. The total backside illumination was found to be almost constant for all tilt angles. This could indicate that the planning of bifacial systems can be performed similar to the monofacial procedures, but this has to be proved. The relative contribution of the backside to the total illumination was measured between 0.15 at 45° and 0.35 at 0° depending on the tilt angle. The minimal irradiation on the backside of the module limits the bifacial gain.

Further studies will include optimisations of the measuring device. The already obtained and future data will be used to develop and proof an energy yield simulation model in cooperation with the ISC Konstanz.

ACKNOWLEDGMENT

All analyses made for this paper by the leading author were performed as part of a project work during his Master's degree studies.

REFERENCES

[1] H. Nussbaumer *et al.*, 'PV Installations Based on Vertically Mounted Bifacial Modules Evaluation of Energy Yield and Shading Effects', 2015.

[2] U. A. Yusufoglu *et al.*, 'Simulation of Energy Production by Bifacial Modules with Revision of Ground Reflection', *Energy Procedia*, vol. 55, pp. 389–395, 2014.

[3] U. A. Yusufoglu, T. M. Pletzer, L. J. Koduvelikulathu, C. Comparotto, R. Kopecek, and H. Kurz, 'Analysis of the Annual Performance of Bifacial Modules and Optimization Methods', *IEEE J. Photovolt.*, vol. 5, no. 1, pp. 320–328, Jan. 2015.

[4] J. P. Singh, T. M. Walsh, and A. G. Aberle, 'Performance Investigation of Bifacial PV Modules in the Tropics', WIP, 2012.

[5] Z. Wang *et al.*, 'A Study on Electrical Performance of N Type Bifacial Module Using Transparent Backsheet', WIP, 2013.

[6] T. Nordmann, T. Vontobel, and R. Lingel, 'Exploiting High Solar Irradiation in Alpine Regions Using Bifacial PV Modules - First Results of Bifacial Module Technology in Alpine Conditions for High Energy Yields and Outlook for further Benefits for High Grid Penetration', presented at the 28thEUPVSEC2013, 2013.

[7] F. P. Baumgartner *et al.*, 'Bifacial modules used in cable based PV carport applications', presented at the World PV Conference, Kyoto, 2014.

[8] C. Comparotto *et al.*, 'Bifacial n-Type Solar Modules: Indoor and Outdoor Evaluation', WIP, 2014.

[9] K. Sugibuchi, N. Ishikawa, and S. Obara, 'Bifacial-PV Power Output Gain in the Field Test Using EarthON High Bifaciality Solar Cells', WIP, 2013.

[10] H. Nussbaumer *et al.*, 'Influence of Low Concentration on the Energy Harvest of PV Systems Using Bifacial Modules', presented at the 32nd European Photovoltaic Solar Energy Conference and Exhibition, München, 2016.

[11] D. Schär and F. P. Baumgartner, 'Portable LED Flasher with Implemented Bypass Diode Tester', WIP, 2014.

POTENTIAL FOR PHOTO-GENERATED CURRENT FOR BIFACIAL PV MODULES IN THE ATACAMA DESERT

Pablo Ferrada[1], Francisco Araya[1], Aitor Marzo[1], Jayaprasad Arumughan[2], Paulo Ayala[3], Michael Rogoll[4], Patricio Valdivia Lefort[3,5], Enrique Cabrera[2]
[1]Universidad de Antofagasta, Centro de Desarrollo Energético Antofagasta, Av. Angamos 601, Antofagasta, Chile
[2]International Solar Energy Research Center (ISC Konstanz), Rudolf Diesel-Str. 15 D-78467 Konstanz, Germany
[3]Fraunhofer Chile, Center for Solar Energy Technologies (CSET), Av. Vickuña Mackena 4860, Santiago, Chile
[4]University of Applied Sciences Upper Austria, Campus Wels, Stelzhammerstraße 69, 4600 Wels, Austria
[5]Universidad Técnica Federico Santa Maria, Departamento de Ingeniería Eléctrica, Santiago, Chile

Phone: +56 55 2 513530; Email: pablo.ferrada@uantof.cl

ABSTRACT: Theoretical calculations of the photo-generated current density and outdoor measurements were performed for bifacial photovoltaic (PV) modules. The main idea was to find the potential of crystalline silicon (c-Si) based bifacial PV technology considering solar spectra in the Atacama Desert, Chile. An average air mass (AM) at noon for this location averaged 1.15 and the PV modules tilt angle was 18°. In order to compute the current density for a variety of glass-encapsulant combinations, the transmittance of glass-encapsulant-glass structures and the spectral response of the front as well as the rear side of the bifacial solar cell were measured. Integrating the product of the spectral irradiance, transmittance and spectral response, it came out that the thermoplastic material as encapsulant led to a gain in current obtained with Atacama AM1.15 compared to that with the reference AM1.5 spectrum. For a vertical installation and before noon, the current density was highest with the glass featured with antireflection coating combined with ethylene vinyl acetate. Conversely after noon, the glass with antireflection coating combined with the thermoplastic material as encapsulant led to the maximum current density.
Keywords: bifacial, silicon solar cell, module, spectral response, solar radiation

1 INTRODUCTION

Photovoltaic (PV) generation based on monofacial modules has reached an important technological maturity. Today, the continuity of its development in terms of efficiency and cost may be at an inflection point. The state of the art opens alternative routes to achieve lower prices and increased efficiency. One of these routes is the development of bifacial PV modules. Further reductions of the levelized cost of energy (LCOE) will be also driven by reduction of the balance of system (BOS) cost. An important issue for lowering BOS cost is the enhancement of the energy yield (kWh/m^2) of PV modules. Under favorable installation conditions (e.g., high albedo of the ground at the installation site) the rear side contribution to the total energy production can reach 30% [1]. Bifacial solar cells are suitable to be processed in standard module manufacturing lines and is becoming more relevant as many new advanced cell concepts are inherently bifacial.

Bifacial PV implemented on locations such as deserts, white-snowy reflecting grounds, or on water, allows for using more solar energy which otherwise would be lost with monofacial modules. Particularly, deserts can reach an albedo of 40% [2], from which bifacial PV can make a significant profit.

One of the regions with high potential for energy generation from the sun is the Atacama Desert in Chile. The Atacama Desert receives the highest radiation levels on earth. Some of the reasons are the high mean altitude, clear skies and low absorption ozone and water vapor columns found in this region. As a result, the global horizontal irradiation (GHI) exceeds 2500 kWh/m^{2}year [3] (2800 kWh/m^{2}year at the plane of array).

The fall in prices of PV modules combined with Chile's attraction as the right place for investors, resulted in a first Chilean growth in PV applications that led the private sector to plan and build several solar farms in the north of the country. Chile's cumulative installed capacity for PV installations stood at 1.7 GW at the end of August 2017, according to the report of the National Energy Commission (CNE), 2017 [4].

The increasing number of PV systems installed in the Atacama Desert has revealed a number of issues related to the operation of PV technology. In order to maximize the performance, solar spectrum effects [5], module materials and solar cell concept for the modules must be taken into consideration. Together with the possibility of bifacial PV modules to take advantage from the front side and rear side illumination, they can incorporate improved module materials and be installed using different geometrical configurations. Advancements in the development of a PV module adapted to climate conditions of Atacama Desert (AtaMo) and the impact of soiling and abrasion on several AtaMo configurations were shown already in the framework of Solar ChilD BMBF-project [6] and [7].

In this work, the potential of bifacial PV modules considering a representative solar spectrum of Atacama Desert, the use of different module materials such as PV glass and encapsulants, and the spectral response of a bifacial solar cell are addressed. Moreover, the field measurements of a commercial bifacial technology installed in the Atacama Desert are reported.

2 EXPERIMENTAL

Float-glass as standard (STD) and float glass with antireflection coating (ARC) were used for this study. These PV glass types were 3.2 mm and 1.5 mm thick and with an area of 5×5 cm^2. Encapsulants were ethylene vinyl acetate (EVA, E), a thermoplastic material (TM) and a modified EVA with a low UV cutoff (U) of thicknesses 450 μm, 450 μm and 400 μm, respectively. The PV glass and encapsulants were combined to obtain glass-encapsulant-glass. The rear side glass was 1.5 mm thick and it was used as support plate for preparation. Lamination was performed at 145 ± 5 °C in 16 min.

The wavelength-dependent transmittance (*T*) and reflectance (*R*) of glass, encapsulants, and glass-encapsulants structures. was measured with a Perkin Elmer Spectrophotometer. The Current-Voltage (IV) characteristics were measured with a Flasher device from h.a.l.m. elektronik. The Quantum Efficiency and reflection of the metallized solar cells were determined with an IQE Scan of PV Tools GmbH.

Outdoor measurements of a bifacial PV module were carried out using an IV curve meter from Papendorf Software Engineering GmbH. Data collected comprises the period from February 2nd to March 10th, 2017. This single monitored bifacial module (Enduro B270 from Si Module GmbH) belongs to a larger installation at the *Solar Platform of the Atacama Desert* (PSDA) located at coordinates 24°05'24.1"S 69°55'44.2"W. The set up is at 20° with azimuth 0° (pointing north).

3 METHODOLOGY

The performance calculation required: the spectral response (SR) of the bifacial solar cells, the transmittance of glass-encapsulant-glass structures (*T*) and the spectral irradiance for the Atacama Desert (Φ_{in}). The integral with regard to the wavelength of the product between Φ_{in}, *T* and SR results in the maximum theoretical photo-generated current density (J_m).

3.1 Calculation of J_m

We used the number of photons (n_{ph}) impinging on an area *A* within a time interval Δt. The value of n_{ph} is obtainable from the ratio of the optical energy $E_\lambda(\lambda)$ of the radiation at a given λ to the energy of a single photon $E_{ph}(\lambda)$. The number of photons depends on the spectral irradiance $\Phi_{in}(\lambda)$ given in W/m^{2}nm. Equation 1 incorporates the transmittance *T* of glass-encapsulant-glass structures and SR to compute J_m [8]:

$$J_m = \int_{\lambda_0}^{\lambda_1} \Phi_{in}(\lambda)\cdot T(\lambda)\cdot SR(\lambda)\cdot d\lambda \qquad (1)$$

Where the spectral response SR is related to the external quantum efficiency (EQE) through Eq. 2.

$$SR(\lambda) = \frac{q\cdot\lambda\cdot EQE(\lambda)}{h\cdot c} \qquad (2)$$

Where *q* is the elementary carrier charge, *h* is the Planck's constant, and *c* is the speed of light.

3.2 Global spectrum in the Atacama Desert

The Global Horizontal, Direct Normal and Diffuse Horizontal Spectral Irradiances (GHIλ, DNIλ and DHIλ), were calculated for a representative region between latitudes 13° and 30° south, which includes Perú, Bolivia, Argentina and Chile. In order to come out with these spectra, the temperature, relative humidity, precipitable water, ozone, carbon dioxide, turbidity, albedo, air mass (AM), and surface tilt angle were considered as inputs to be introduced in SMARTS v. 2.9.2 code [9,10]. These inputs are atmospheric and astronomical parameters from satellite databases, model retrievals and equations, which were averaged in the defined area, and individually changed with respect to ASTM G173-03 ref spectra [11].

A resulting mean global spectral irradiance incident on a tilted surface is shown in Fig. 1. Solar spectra for the front side of the solar cell at difference tilt angles (15°, 20°, 30°, 45° and 90°) were calculated from the obtained mean GHI, DNI and DHI, according to the models defined in [9,12]. The difference was less than 1%, according to a ground measurement comparison. The spectra for the rear side were computed in a similar way as for the front side. The calculation was performed considering an anisotropic atmosphere for the diffuse radiation, except in the ultraviolet range [9]. The reflected component from the ground was considered isotropic.

3.3 Average values of *T* and SR

The transmittance of glass-encapsulant-glass structures and SR of the bifacial solar cell were weighted with the reference spectra ASTM G173-03 AM1.5 spectrum to obtain an average value according to Eq. 3:

$$X_{avg} = \frac{\int \Phi_{in}(\lambda)\cdot X(\lambda)d\lambda}{\int \Phi_{in}(\lambda)d\lambda} \qquad (3)$$

Where *X* is the variable to be replaced by the λ-dependent *T* or SR and integrated within the spectral range of interest.

4 RESULTS

4.1 The global spectrum

A mean value of AM for noon was calculated along the year for the Atacama Desert. The value obtained was 1.15 and a tilted surface at 18° was considered. There are differences between Atacama and reference spectra, especially for the short wavelength, both ultraviolet (UV) and visible (VIS) ranges. According to the spectral range, the GTI$_\lambda$ calculated for Atacama Desert can be 139% higher for UV-A (315<λ<400 nm) and 107% higher for VIS (400<λ<780 nm) compared to reference spectrum.

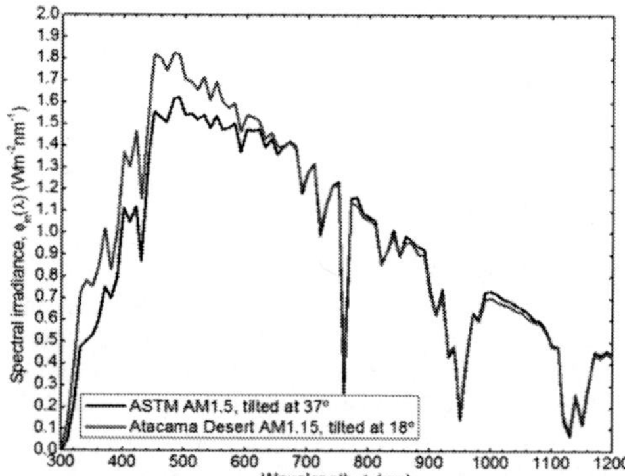

Fig. 1. Global tilted spectral irradiance (GTIλ).

4.2 Transmittance of glass-encapsulant-glass structures

A comparison of the averaged transmittance of the glass-encapsulant-glass structures in the wavelength range of 400 nm to 1100 nm resulted in 87.7% for STD-E-STD, 86.1% for STD-TM-STD, 87.6% for STD-U-STD, 90.6% for ARC-E-STD, and 87.1% for ARC-TM combinations. Standard deviation of the transmittance kept below 1.5% for STD glass and 2% for ARC combinations. The wavelength-dependent transmittance of glass-encapsulants structures is shown in Fig. 2.

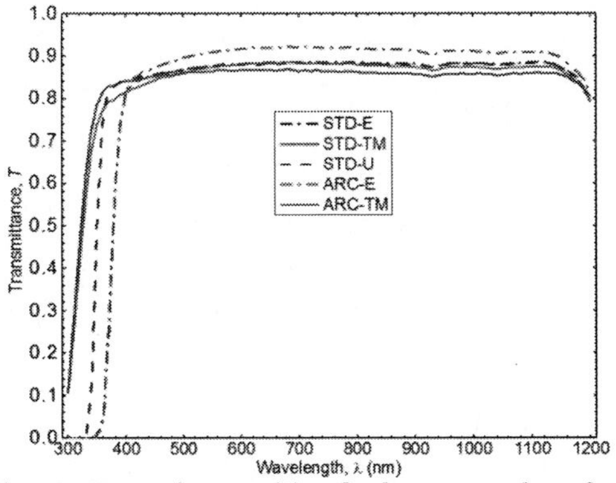

Fig. 2. Transmittance (T) of glass-encapsulant-glass structures. The rear side always exhibits a STD glass.

4.3 Current-voltage and spectral response

The open circuit voltage (V_{oc}), short circuit current density (J_{sc}), fill factor (FF), and efficiency (η) were extracted from current-voltage (IV) characteristics of the bifacial solar cell, measured from its front and rear side, these values are summarized in Table 1.

Table 1: Current-voltage measurements under standard testing conditions (1 kW/m^2, AM1.5 and 25°C).

Side	V_{oc} (mV)	J_{sc} (mA/cm^2)	FF (%)	η (%)
Front	653.1	39.2	78.3	20.0
Rear	649.7	34.6	78.2	17.6

The spectral response, shown in Fig. 3, was calculated based upon the external quantum efficiency, measured from the front and rear side of the solar cell, and its reflection.

In the UV, SR averaged 0.21 A/W for the front side and 0.13 A/W for the rear side of the bifacial solar cell. Likewise, in the VIS, average SR values were 0.44 A/W for the front side and 0.4 A/W for the rear side. Finally, in the IR, SR resulted in 0.59 A/W for the front side and 0.51 A/W for the rear side. The overall SR in the 300<λ<1200 nm range resulted in averages of 0.47 A/W for the front and 0.41 A/W for the rear side of the bifacial solar cell.

The value of the short current density (J_{sc}) obtained from IV measurements at STC can be compared with the integrated photo-generated current density J_m of the bare solar cell, for the 300 to 1200 nm wavelength range without considering the glass-encapsulant structures. This calculation was carried out as an indication to validate quantum efficiency measurements. The difference defined as $(J_m\text{-}J_{sc})/J_{sc}$ resulted in 0.9% for the front side and 0.8% for the rear side of the bifacial solar cell.

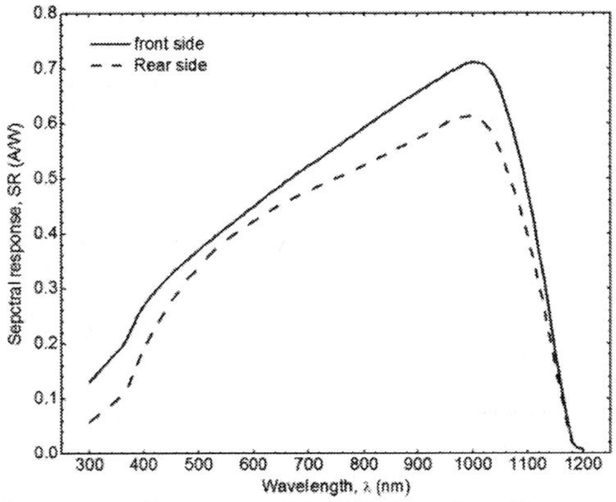

Fig. 3. Spectral response of the bifacial solar cell.

4.4 Potential of bifacial modules as a function of spectra

Figure 4 shows a comparison of J_m values obtained with one encapsulant (TM and U) with respect to J_m obtained with the referent encapsulant E, with the same glass type and same solar spectrum. This calculation is performed assuming normal illumination and same solar spectrum for front and read side of the PV device. The obtained result showed that for the front side, the TM led to a 0.6% loss, whereas, U led to a 0.7% gain with respect to E encapsulant combined with STD glass under AM1.5. The same calculation for glass with ARC indicates that TM produced a 2.8% loss with respect to E encapsulant under AM1.5. Using AM1.15, TM reduced the loss to 0% and the U encapsulant increased the gain to a 1% with respect to E encapsulant combined with STD glass. Similarly, for the glass with ARC the loss is reduced to 2.3% with respect to E encapsulant. A comparison of J_m obtained with AM1.15 with respect to J_m obtained with AM1.5, with same glass and same encapsulant, is shown in Fig. 4, which indicates that the TM encapsulant combined with STD glass produced the maximum gain (3.9%) and 3.8% when TM is combined with ARC glass.

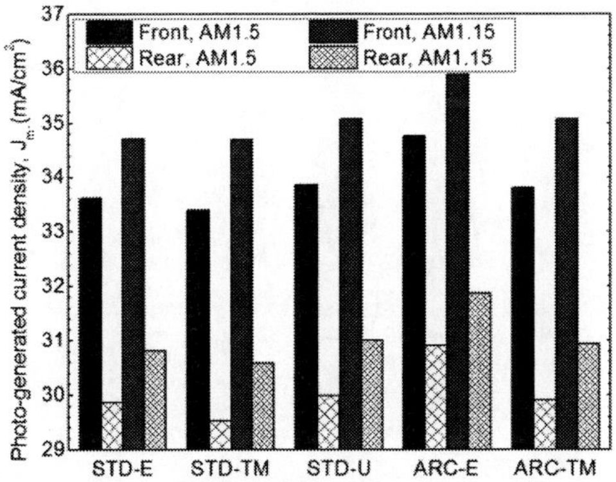

Fig. 4. Photo-generated current density (J_m) as a function of glass type and encapsulant for ASTM G173-03 reference AM1.5 and Atacama Desert AM1.15 spectrum.

4.5 Potential of bifacial modules as a function of angle

Figure 4 depicts the incorporation of different solar spectra when a module is installed at several tilt angles. It means that for a given tilt angle, front and rear side solar spectra are different. It came out that for all glass-encapsulant combinations, a tilt angle of 20° produced the maximum J_m. This angle is close to 18°, derived from Atacama Desert AM1.15. Thus, as the angle deviates from 20°, lower J_m values are obtained for front and rear side of the bifacial solar cell. For STD glass, the combination with U encapsulant led to the higher J_m gain (4.3%) compared to J_m at 45°. Similarly for ARC glass, the combination with TM encapsulant resulted in the higher gain (3.8%) compared to J_m at 45°. The case of 90° module tilt angle with 0° azimuth is not shown in Fig. 5.

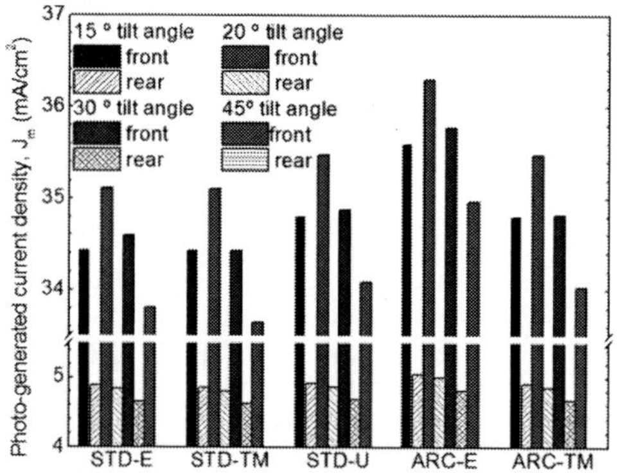

Fig. 5: J_m values as a function of glass type and encapsulant for AM1.15 with varying module installation angles.

4.6 bifacial modules at vertical configuration

Figure 5 depicts the configuration at 90° module tilt and 90° azimuth angle, i.e. east-west orientation. Before noon, the current density J_m is highest with ARC glass combined with the E encapsulant. After noon, J_m is maximized with ARC glass combined with the TM encapsulant.

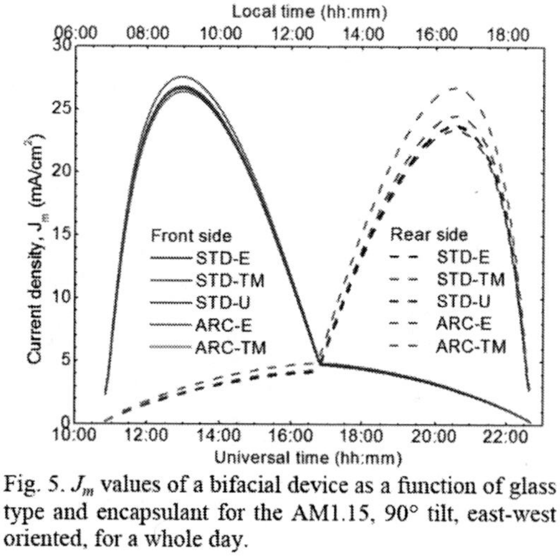

Fig. 5. J_m values of a bifacial device as a function of glass type and encapsulant for the AM1.15, 90° tilt, east-west oriented, for a whole day.

4.7 Field measurements at real conditions

Data corresponding to the period from February 2nd to March 10th 2017 of the monitored bifacial PV module at the test site (PSDA) in the Atacama Desert was evaluated. The ratio of the bifacial short circuit current I_{sc} and the solar irradiance (received in the front side (Irr_{front}) of the PV module was calculated. The value of I_{sc}/Irr_{front} was plotted in Fig. 6 as a function of albedo and back side module temperature. This value represents a measure of the generation, which is independent of Irr_{front}. Note that albedo was defined as the ratio of reflected irradiance (Irr_{ref}) and Irr_{front}, both measured with silicon photocells. These irradiance sensors are mounted at a bifacial module, which is not the module monitored with the IV curve meter.

Figure 6 shows an increasing tendency for I_{sc}/Irr_{front} as albedo grows. Mornings and afternoons are associated with a larger albedo and lower temperature resulting in higher I_{sc}/Irr_{front} compared to those when albedo is low. In this situation, at lowest albedo, highest temperature is reached. As a consequence, higher values of the generated current are mostly caused by the reflected irradiance.

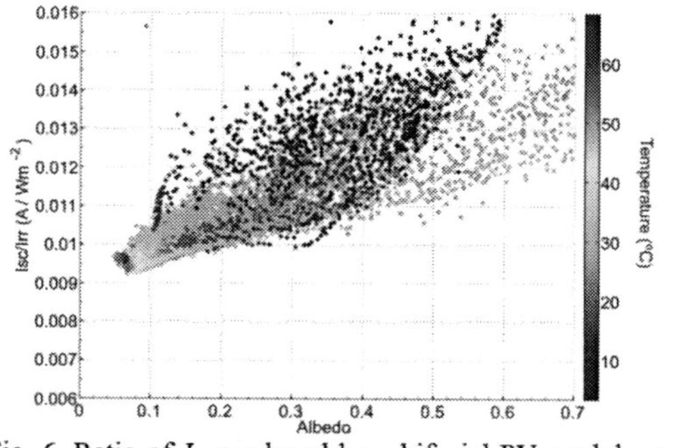

Fig. 6. Ratio of I_{sc} produced by a bifacial PV module on the field at PSDA in the Atacama Desert to the solar irradiance (Irr_{front}) measured by a photo cell at the front side as a function of albedo and module temperature (T_{mod}).

5 CONCLUSIONS

The calculation of a representative solar spectrum for Atacama Desert, the measurement of the optical properties module materials, as well as the spectral response of a bifacial solar cell were used to study the influence of air mass (AM), module tilt angle and orientation of PV modules for a variety of glass-encapsulant combinations. The results were that the thermoplastic material (TM) was able to produce a gain in current obtained with Atacama AM1.15 compared to that with the reference AM1.5. A comparison of J_m obtained with AM1.15 with respect to J_m obtained with AM1.5 with same glass and same encapsulant indicates that the TM encapsulant combined with standard (STD) glass produced the maximum gain (3.9%) and 3.8% when TM is combined with ARC glass.

The study for a whole day for a 90°-tilt angle and east-west orientation led to the conclusion that before noon, the current density is highest with the glass featured with antireflection coating (ARC) combined with ethylene vinyl acetate (EVA). After noon, the current density is maximized with ARC glass combined with TM as encapsulant.

A single bifacial module fixed at 20º, belonging to a larger bifacial PV plant, was monitored for a period of 2 months at the Solar Platform of Atacama Desert (PSDA). The ratio of the bifacial short circuit current (I_{sc}) to the solar irradiance received at the front side of the module (Irr_{front}) increased linearly with albedo. Thus, for the installation used, higher values of the generation were mostly caused by the reflected irradiance.

ACKNOLEDGEMENTS

The authors acknowledge the support provided in part by FONDECYT/CONICYT Project Number 3160190, CONICYT/FONDAP/15110019 "Solar Energy Research Center" SERC-Chile and in part by the Federal Ministry of Education and Research of Germany (BMBF) Solar Collaboration between Chile and Deutschland (Solar ChilD) project number 01DN14005.

REFERENCES

[1] R. Kopecek, I. Shoukry, J. Libal. Photovoltaics International 30 (2015).
[2] Li and Garand. Journal of Geophysical Research 99, N1 D4 (1944), 8335-8350.
[3] R. Escobar, C. Cortés, A. Pino, E. B. Pereira, F. Ramos Martins, J. M. Cardemil. Renewable Energy 71 (2014) 324-332.
[4] Reporte Mensual ERNC, CNE Vol. 12, August 2017, "*Reporte Sector Energético* Agosto 2017". https://www.cne.cl/wp-content/uploads/2015/06/RMensual_ERNC_v201708-1.pdf
[5] M. Simon, E. L. Meyer .Prog. Photovolt: Res. Appl. 19 (2011) 1–10.
[6] E. Cabrera, A. Schneider, J. Rabanal, P. Ferrada, R. Cordero, E. Fuentealba, R. Kopecek. 31st EU PVSEC, Hamburg (Germany), 14-18 September 2015.
[7] E. Cabrera, A. Schneider, E. Wefringhaus, J. Rabanal, P. Ferrada, D. Thaller, F. Araya, A. Marzo, M. Trigo, D. Olivares, J. Haas, E. Fuentealba, R. Kopecek. 32nd EU PVSEC, Munich (Germany), 20-24 June 2016.
[8] P. Ferrada, A. Marzo, E. Cabrera, H. Chu, V. del Campo, J. Rabanal, D. Diaz-Almeida, A. Schneider, R. Kopecek. Sol. Energy 144 (2017) 580-593.
[9] C. A. Gueymard. Sol Energy 71 (2001) 325-346.
[10] C. A. Gueymard. SMARTS, A Simple Model of the Atmospheric Radiative Transfer of Sunshine: Algorithms and Performance Assessment. SMARTS 1995.
[11] A. Marzo, P. Ferrada, F. Beiza, J. Alonso-Montesinos, J. Ballestrín, R. Román. 11th ISES EUROSUN Conference, Palma de Mallorca (Spain) 11-14 October 2016.
[12] E. Lorenzo. Energy Collected and Delivered by PV Modules. In: A. Luque, S. Hegedus, editors. Handbook of Photovoltaic Science and Engineering, Chichester - England: John Wiley & Sons Ltd. 2003, p. 905–970.

SHADING EFFECTS OF THE MOUNTING STRUCTURE OF BIFACIAL PV MODULES AND IMPACT TO THE BYPASS DIODE LIFETIME

Sven Muenter, Viktor Wesselak, Sebastian Voswinckel, Lukas Gerstenberg
University of Applied Sciences Nordhausen, Institut für Regenerative Energietechnik (in.RET)
Weinberghof 4, D-99734 Nordhausen/Germany
sven.muenter@hs-nordhausen.de; Phone: +49 3631 420857

ABSTRACT: This paper investigates the effects of partial shading and high temperature stress on bypass diodes of bifacial Photovoltaic modules. An indoor test to proof the reliability of bypass diodes under a high thermal- and electrical stress level was realized. The indoor test was carried out in a climatic chamber under constant ambience conditions. Furthermore, an outdoor Test set up was designed and developed to proof the reliability of the bypass diodes under atmospheric conditions with various shading situations at the surface of a bifacial photovoltaic module. The results and methods of the different tests will be discussed.
Keywords: bifacial, hot spot, shading, bypass diode

1 INTRODUCTION

Bifacial photovoltaic technologies are the next big step for high performance photovoltaic modules on the market. Bifacial photovoltaic modules/cells are able to generate a higher output energy compared to standard photovoltaic technologies. The concept of bifacial photovoltaic cells exists since the 1960s [1]. With this cell concept also the reflected light on the backside of the module is used to generate electricity. For bifacial photovoltaic technologies the shading effects are a higher problem in comparison to monofacial photovoltaic technologies.
Bypass diodes are reverse connected diodes to the sub strings of a photovoltaic module. They are used to prohibit reverse voltage across the cells in due of an asymmetrical shading of the photovoltaic module sub strings. Shading effects of a photovoltaic module can be generated by a lot of reasons like growing up vegetation or constructions. Also contamination by dirt can cause the same effect. Furthermore, modules in photovoltaic systems are often mounted on metal racks. The mounting structure consists of different rails on which the module is attached with module clamps. The typical mounting rails for monofacial photovoltaic modules are located behind the solar cells. In case of bifacial photovoltaic modules this leads to an inhomogeneous irradiation caused by a diffuse shadow on the backside. Followed by this, the bypass diode can be activated and a part of the module current will flow through the diode. In addition to the power losses continuous load is applied to it. This load can lead to accelerated aging and failure of the bypass diode.

2 BYPASS DIODE FAILURES

Bypass diodes can fail in two different types of failure mode. The first type of failure mode is a short circuit of the bypass diode. In this case the concerned substring of the module gets short circuited by the bypass diode. For a photovoltaic module with three substrings and one bypass diode per substring a power loss of 1/3 per failed bypass diode, is the consequence. The other possible type of failure mode is an open circuit situation. An open circuit failure at one bypass diode doesn't immediately affect the power output of the module. However, if the concerned substring gets a shading situation, a reverse bias across the cells of the substring is the consequence. As reaction of the reverse bias across the substring overheating, cell breakdown and fire hazard are possible consequences.
The possible failure mechanism can be categorized into two types: wear-out- and catastrophic failure mechanism. Catastrophic failures are failure mechanisms as a result of a one-time event. However, wear-out failure mechanism are the results of a gradual degeneration of the bypass diodes. Expected typical photovoltaic catastrophic failure mechanisms are arcing, electrostatic discharge (ESD) and thermal runaway [2]. Arcing can appear as a result of a poor electrical connection of the bypass diode to the rest of the circuit. Therefore, arcing is mainly a failure mechanism caused by an electrical connecting issue and not by a failure of the bypass diode.
ESD events at bypass diodes are a critical source of error by the manufacturing process of photovoltaic modules. Production steps like handling, flash testing and junction box attachment get a high potential for triggering ESD events. A study in 2012 showed that ESD based failure mechanisms are the most frequent cause of failure for bypass diodes in photovoltaic modules [3]. Thermal runaway is called a process by destroying the bypass diode in succession of overheating. In case of a longer term shading the bypass diode obtains a thermal equilibrium. If the shading breaks off rapidly, the diode turns to reserve bias. If the ability for heat transfer from the junction box lower then the power dissipation in the reverse bias, the bypass diode temperature begins to increase. Subjected to the temperature the power dissipation of the diode increases exponentially. This behavior of power dissipation leads to a continues rise of temperature till the bypass diode is completely destroyed by overheating.
Classified as typical wear-out failure mechanism is high temperature forward bias (HTFB) operation and high temperature reverse bias (HTRB). HTFB operation describe the operation mode of a bypass diode if the diode becomes forward biased by a high temperature. The operating temperature of the bypass diode depends on the current passing the diode and the ambient temperature. Especially bypass diodes of photovoltaic modules operating under permanent partial shading are susceptible for failures as a result of HTFB operations. In the unshaded state of photovoltaic modules, bypass diodes work in reverse bias. However, leakage currents in combination with high temperatures is known to cause localized breakdowns which can eventually cause a complete breakdown of the device. This process is known as HTRB [2].

3 EXPERIMENTAL

The used photovoltaic modules in the discussed test are commercially available bifacial photovoltaic modules. The cells in the photovoltaic modules are serial connected in three sub strings. Every substring is reverse connected with one bypass diode (Figure 1).

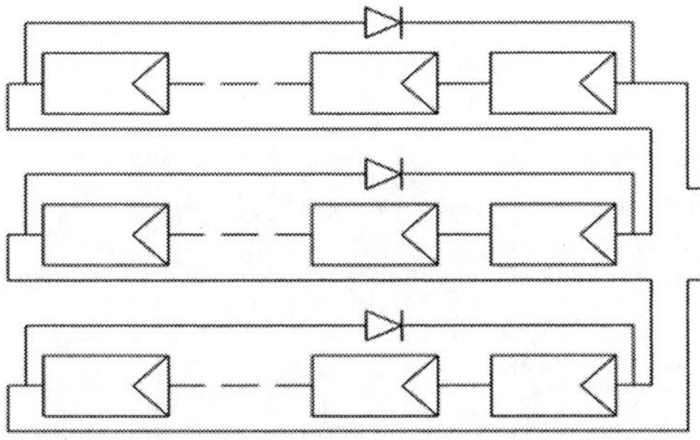

Figure 1: Connection substring and bypass diodes

Previously to the experiments the test modules were prepared for use in the test. The junction box was opened and the heat transfer paste was removed to get direct access to the bypass diodes.

The discussed exercise test of the bypass diodes is classified in two sections. The first test section is assumed from the bypass diode test in IEC 61215. This test section was realized in a climatic chamber under static conditions. The target of the first section was to proof the performance of the bypass diodes under a high thermal- and electrical stress level. The second test section was carried out at an outdoor test field at environmental conditions. During these test section the influence of different shading events at the front of the module was investigated.

3.1 CLIMATIC CHAMBER

Purpose of the test was to determine the junction temperatures of the bypass diodes under high temperature condition and power consumption. The test setup was build up in a climatic chamber at a constant ambient temperature of 75 °C. During the test the bypass diodes get forward bias by a constant current. The test execution is split in two parts, each of them lasts 60 minutes.

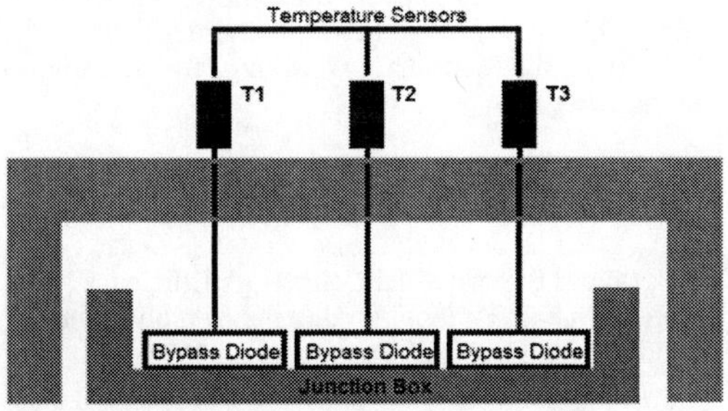

Figure 2: Pt100 temperature sensors over JB

During the first part of the test the current was adjusted for 60 minutes to the open circuit current (I_{OC}) specified by manufacture. For the second part the current was raised up to 1.25 times I_{OC} for the next 60 minutes. Over the whole time of the test execution the surface temperature and the forward voltage of the bypass diodes were logged. To minimize the heat conduction by the temperature measurement, pt-100 temperature sensors with a small bearing area were used (figure 2). The final temperature values after 60 minutes of every part of the test are used to calculate the junction temperature of the bypass diodes. Figure 3 shows the first part of the test and figure 4 the second part. The junction temperature was compared to the maximum operation junction temperature of 175 °C specified by producer. To calculate the junction temperature, equation (1) was used.

$$T_J = T_{CASE} + R_{THJC} \cdot U_D \cdot I_D \tag{1}$$

Table 1: Symbols to equal (1)

Symbol	Parameter	Unit
T_J	Junction temperature	°C
T_{CASE}	Surface temperature	°C
R_{THJC}	Junction to case	°C/W
U_D	Forward voltage	V
I_D	Forward current	A

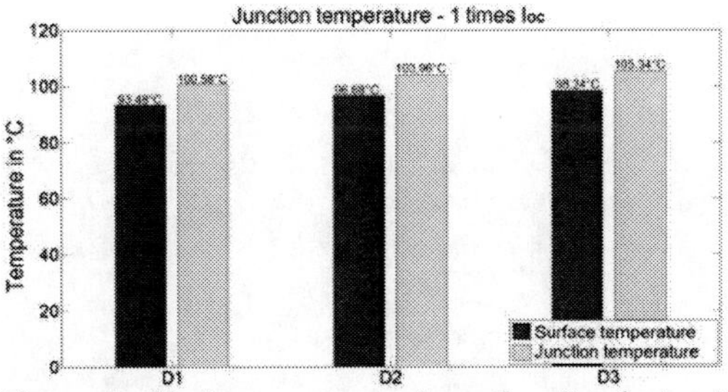

Figure 3: Final temperature values after 60 minutes 1 times I_{OC}

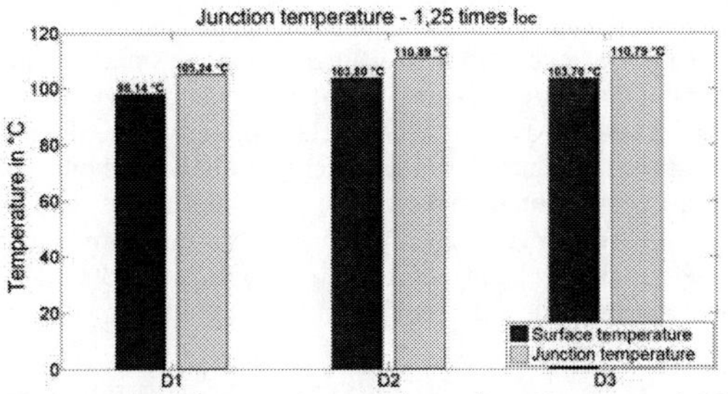

Figure 4: Final temperature values after 60 minutes 1,25 times I_{OC}

Before and after the test, the test module was characterized concerning the open circuit voltage, the insulation resistance, the functionality of the bypass diodes and the maximum power point. These data are the reference values for evaluating possible degeneration effects during the experiment.

3.2 OUTDOOR TEST

The purpose of the outdoor test was to investigate the reliability of the bypass diodes under different shading events at the module surface. The test module was mounted in a south orientation with a pitch of 30 °C.

The mounting situation of the module was realized without any rails behind the photovoltaic module, to avoid shadows at the backside of the module. For the test execution five different shading situation at the front surface of the module were realized (figure 5).

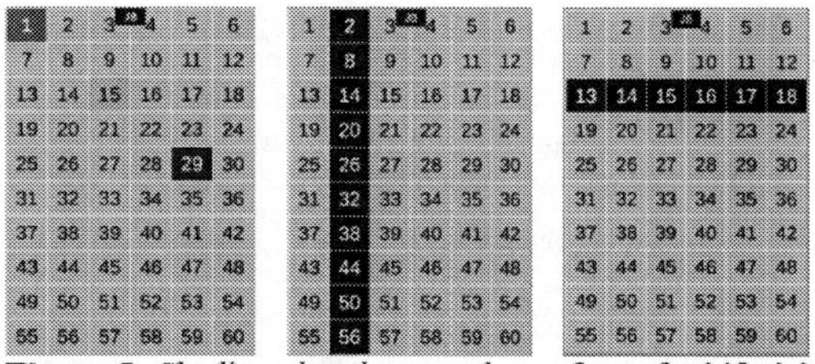

Figure 5: Shading situations on the surface of a bifacial module. Situation 1-3 (left), Situation 4 (middle), Situation 5 (right)

The shading situation were carried out as diffuse and hard shadow. For the simulation of diffuse shadow events a textile mesh with an absorptance of 76 % was used. For realizing the hard shadow situations black molliton with an absorptance close to 100 % was used (figure 6).

Figure 6: Shading situation 4 as hard shadow (left) and shading situation 5 as diffuse shadow (right)

Every test cycle with a different shading situation and type of shadow (diffuse, hard) was maintained for 72 hours. Before starting a test cycle and afterwards, the test module was characterized concerning the functionality of the bypass diodes, signs of hot spots and the maximum power point. The generated data were compared after every test cycle to identify degeneration of the cells, power losses or occurring bypass diode failures.

4 RESULTS AND DISCUSSION

The analysis of the climatic chamber test shows that the final junction temperatures after both parts of the test were below the maximum operation junction temperature of 175 °C specified by the producer. The maximum junction temperature determined after the first part of the test with a 1 times I_{OC} current amounted 105.34 °C after the second part of the test with a current of 1.25 times I_{OC} the maximum temperature amounted 110.89 °C. Furthermore, the values of insulation resistance and maximum power output of the module showed no change after the test. Also the bypass diodes were fully functional after the test. Therefore, the climatic chamber test doesn't show any weak spots relating to the performance of the bypass diodes. The results of the outdoor test relating the reliability of the bypass diodes under the realized shading events are positive. All shading tests were executed on the front side of the module under the assumption that the shading events show the same result on both sides of the module. After all shading situations the bypass diodes are still fully functional. Furthermore, during the test there were no signs of hot spots or other types of damages at the test module detected. Also there were no indications for a power loss detected.

Worth mentioning is that an asymmetrically shading situation of one sub string (shading situations 1-5) results in a thermic stress for the related bypass diode.

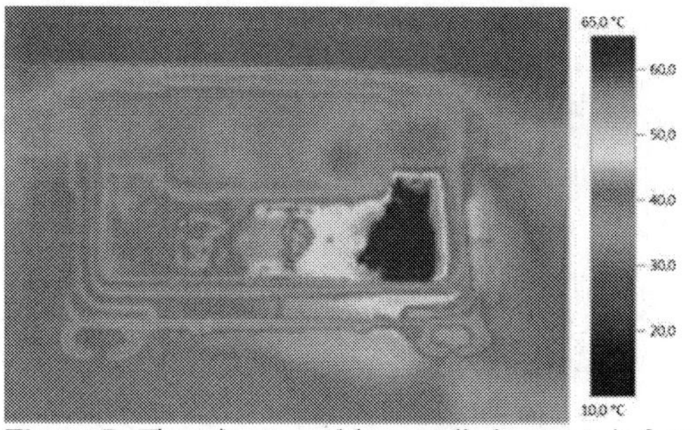

Figure 7: Thermic stressed bypass diode as result from an asymmetric shading

Figure 7 shows an infrared image of the open junction box of the test module during shading situation 1 carried out as hard shadow. If an asymmetrical shading situation like the simulated situation comes to be permanent at a module in the field, diode wear out failure mechanism like HTBF is expected.

5 OUTLOOK AND FUTURE WORK

The outdoor test will be intensified regarding long time partial shading test at the front side of the module. Also, based on the shading effects of different mounting systems, there will be modeled shading situations on the backside of the bifacial photovoltaic modules investigated.

6 ACKNOWLEDGEMENT

This paper has been supported by the European Social Fund (ESF) and the Free State of Thuringia (Thüringer Ministerium für Wirtschaft, Wissenschaft und Digitale Gesellschaft).

7 REFERENCES

[1] C. Duran: Bifacial Solar Cells: High Efficiency Design, Characterization, Modules and Applications. Universität Konstanz, Dissertation, 2012

[2] N. Shiradkar: Predictive modelling for Assessing the Reliability of Bypass Diodes in Photovoltaic Modules. University of Central Florida, 2015

[3] J.V. Loo: Bypass Diodes for PV Modules: 10 years of field experience.
Intersolar Europe Conference, Munich, 2012

IMPACT OF INHOMOGENEOUS IRRADIANCE AT THE REAR OF BIFACIAL PANELS ON MODELLED ENERGY YIELD

G.J.M. Janssen* (janssen@ecn.nl), R.S.R. Gali, K. de Groot, A.J. Carr, B.B. Van Aken, I.G. Romijn
ECN Solar Energy, P.O. Box 1, NL-1755 ZG Petten, The Netherlands
*Phone: +31 88 515 4803

ABSTRACT: The inhomogeneity of the rear irradiance of bifacial PV panels and its effect on the power loss due to current mismatch was investigated by modelling. We have found that the inhomogeneity of the rear irradiance depends on orientation, elevation, and tilt angle of the panel and on irradiance conditions. Typically the inhomogeneity of the rear irradiance is largest at a high fraction of direct light. Conversely, the inhomogeneity of the total irradiance is smallest when there is a large direct light component. The simulations indicated that the relative standard deviation (RSD) of the total irradiance will not exceed 5% even at albedo as high as 0.8, and is usually much smaller. Simulations mimicking near-field shading also showed that the RSD will be below 10% unless the beam component is small, i.e. at low light conditions. LTspice circuit simulations with an inhomogeneous current distribution showed that a current distribution with an RSD below 10% will not lead to cells going into reverse bias. This means the risk for hot-spots is small. A simple parametrization for the power loss as function of the RSD of the current distribution was established, which can be used for energy yield calculations.
Keywords: bifacial, module, simulation, shading, current mismatch

1 INTRODUCTION

By converting light falling on the rear as well as on the front of a solar panel, bifacial PV systems can produce 10-30% more electricity than conventional PV systems [1]. However, the irradiance on the rear of a solar panel is usually more inhomogeneous than on the front, i.e. the side facing the sun-light. This is first of all due to the shade created by the direct light falling on the panel, or its adjacent panels in a shed, which reduces the ground reflected irradiance on the rear [2,3]. In addition there may be near-field shade of objects at the rear such as the frame, the support of the panels, junction box, other equipment or panels in the next row of a system. These objects will block ground-reflected light as well as diffuse light falling on the rear.

Besides reducing the generated current, these shading effects induce a current mismatch between the individual cells of the panel, resulting in additional power loss. In monofacial systems the current mismatch is mostly caused by shading of the direct irradiance and is, therefore, often quite large. To avoid hot-spots, a whole group of cells is then by-passed by a diode, thereby substantially reducing the energy output of the system. Recent experimental results obtained by ECN indicated that the inhomogeneity in rear side irradiance of an equator-facing, tilted panel did not lead to cells going into reverse bias, i.e. did not cause conditions at which hotspots can occur [4]. It was found that the inhomogeneity in rear irradiance, in that study caused by variations in ground-reflected light as well as by objects at the rear, was small with respect to the total irradiance on the panel.

In this paper we first investigate the inhomogeneity of the irradiance for a wider range of conditions. Therefore, we carried out simulations with the recently developed ECN model designed to calculate the bifacial energy yield [5]. This model includes shading aspects affecting the irradiance on the rear, such as reduction of the ground-reflected irradiance by self-shading, limitations of the view factors for diffuse and reflected irradiance, and shading of the direct light on the rear. Next, we present LTspice [6] calculations to show the impact of inhomogeneous irradiation of bifacial photovoltaic (PV) panels on the expected power output. We find a simple parametrization, coupling the simulated irradiance inhomogeneity to the modelled power loss.

2 DESCRIPTION OF THE MODEL

2.1 Irradiance model

We calculated the effect of inhomogeneous irradiance on the rear for a landscape oriented, free-standing panel in Amsterdam. Meteorological data with one-hour resolution for Amsterdam were synthetically generated from monthly average values in the database PVGIS using PVSyst [7,8]. Two configurations were considered: south-facing with a tilt angle of 38° and an east-orientated vertical module. The edge of the module was 0.5 m above the ground, unless specified otherwise.

The irradiance G_{front} on the front side of the free-standing module with tilt angle β can essentially be written as [9]:

$$G_{front} = (GHI - DHI) \cdot \frac{\cos\theta}{\cos\theta_z} + DHI \cdot \frac{1+\cos\beta}{2}$$

$$+\gamma \cdot DHI \cdot \frac{1-\cos\beta}{2} +$$

$$\gamma \cdot (GHI - DHI) \cdot \left[\frac{1-\cos\beta}{2} - VF_{front-shade}\right] \quad (1a)$$

Here GHI and DHI are the global and the diffuse irradiance on a horizontal plane, θ is the angle of incidence of the beam on the tilted plane, θ_z is the zenith angle of the sun, and γ is the ground-reflection coefficient or albedo. $VF_{front-shade}$ is the view-factor of the front to its self-shade; this term reduces the total ground-reflected irradiance. $VF_{front-shade}$ will only deviate from zero when the beam is incident on the rear of the panel, i.e. it will usually be negligible for an equator-facing panel. For an east-facing system $VF_{front-shade}$ is non-zero as soon as here is direct irradiance from the west, i.e.50% of the time. Note that, like most other models, the ECN software actually uses a more complex equation to calculate the contribution of diffuse irradiance on the tilted plane but such details are not relevant for this study.

For the irradiance G_{rear} on the rear side a similar

equation applies:

$$G_{rear} = (GHI - DHI) \cdot \frac{\cos\theta}{\cos\theta_z} + DHI \cdot \frac{1-\cos\beta}{2}$$

$$+\gamma \cdot DHI \cdot \frac{1+\cos\beta}{2} +$$

$$\gamma \cdot (GHI - DHI) \cdot \left[\frac{1+\cos\beta}{2} - VF_{rear-shade}\right] \quad (1b)$$

In this case θ is the angle of incidence of the beam on the rear of the tilted plane. $VF_{rear\text{-}shade}$ is the view-factor of the rear to its self-shade, and this term can be quite significant. As shown by Yusufoglu et al. [2,3] $VF_{rear-shade}$ does not only depend on the position of the sun with respect to the module and the tilt angle, but also on the elevation of the module. The $VF_{rear-shade}$ varies over the module surface. This makes the rear irradiance intrinsically inhomogeneous.

It must be noted that Eq. 1a and 1b only include self-shading caused by the beam. For a free-standing panel this is probably a valid assumption, unless the tilt angle and elevation are very low. In the case of a PV system consisting of many modules, the diffuse irradiance hitting the ground will also be reduced by the presence of adjacent modules or modules in adjacent sheds.

A convenient way to characterize the inhomogeneity of the irradiance of n cells in a panel (or cell-string) is by means of the relative standard deviation RSD:

$$\text{RSD} = \frac{\sqrt{\frac{1}{n}\sum_{i=1}^{n}(G_i - G_{mean})^2}}{G_{mean}} \quad (2)$$

Where G_i is the irradiance on cell i and G_{mean} the average over all n cells. The RSD can be calculated for the total irradiance, but also for front or rear irradiance only.

2.2 LTspice model

To investigate the effect of G_{mean} and RSD on the power loss due to current mismatch, we constructed irradiance distributions with a random generator. We selected six different values for G_{mean} and seven RSD values between 1 and 20%. An irradiance distribution for 60 solar cells was made to match each (G_{mean}, RSD) combination. This irradiance distribution was then fed to LTspice to simulate a 60-cell 240 Wp front efficiency module. The short-circuit current of each cell was taken proportional to its incident radiation. Furthermore, the module was divided in three 20-cell strings and each string was protected with a by-pass diode.

3 RESULTS

3.1 Inhomogeneity due to self-shading

Fig. 1 shows examples of inhomogeneous irradiance on the rear of a panel. The most remarkable feature is that the inhomogeneity decreases rapidly with elevation, due to the quadratic dependency of $VF_{rear-shade}$ on the distance between the panel and its shade on the ground. Smaller distances not only lead to a larger$VF_{rear-shade}$, but also to larger variations of the value over the panel. [3]. We also found that the RSD of portrait-orientated panels is smaller than for landscape-orientated ones that

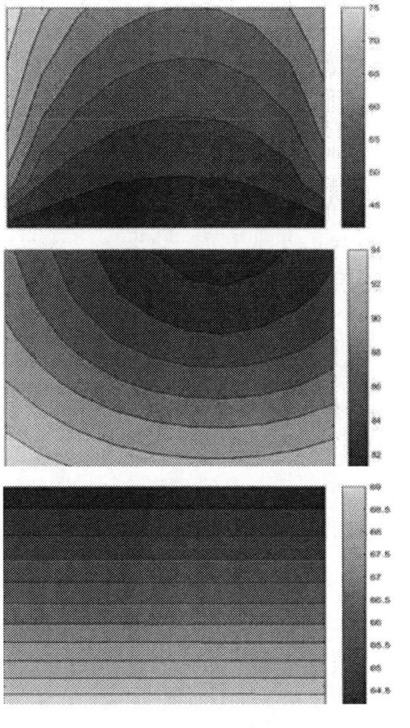

Figure 1: Examples of inhomogeneous irradiance at the rear of a 10x6 cell landscape-oriented, south-facing panel in Amsterdam, with tilt angle 38° and albedo 0.2.The scale is rear irradiance inWm^{-2}. The top and center graph are free-standing modules with elevation 0 and 0.5 m, respectively. The bottom panel is in the center of a 19 panel-wide shed, at 2 m elevation. The RSD of the rear irradiance was 17.9% (top) 4.6% (center), and 2.6% (bottom).

we consider in this study, since portrait -orientated panels have more solar cells at larger distance from a shade on the ground.. Other system lay-out parameters that affect the inhomogeneity are the tilt angle, i.e. at lower tilt angle the relative difference in distances to the shade are larger, and the position of the panel in the system. On panels that are not located near the edges of the sheds the rear irradiance has a regular pattern, which can be used when grouping cells in strings (Fig. 1, bottom). The second factor determining the inhomogeneity is the irradiance. As will be shown in more detail below, the inhomogeneity on the rear will increase when the ground reflected contribution increases and the diffuse contribution decreases.

The RSD of the rear irradiance of the south-facing, free-standing panel with 38° tilt at 0.5 m elevation can be as large as 13%, depending on the albedo. The RSD increases with the total irradiance G_{tot} incident on the panel (see Fig. 2). Conditions of high total irradiance imply a large contribution of the beam component (see Fig. 3) and therefore a more pronounced self-shading effect. However, it must be noted that also the position and size of shade affects these variations, the highest RSD is often obtained at times when the sun is just before or after its most southern position.

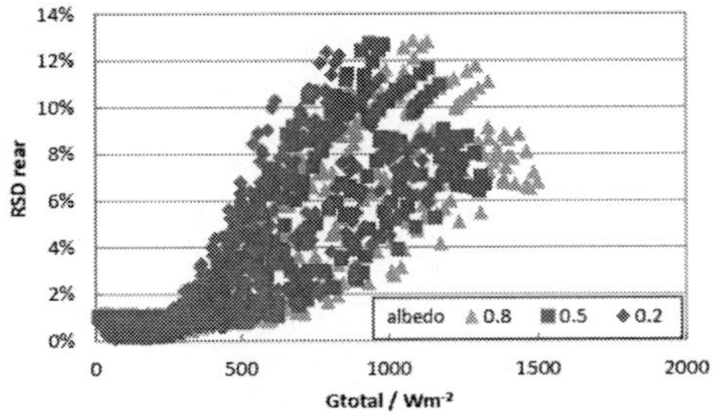

Figure 2: The calculated RSD of the rear irradiance of a free standing, south-facing panel in Amsterdam during the month of April.

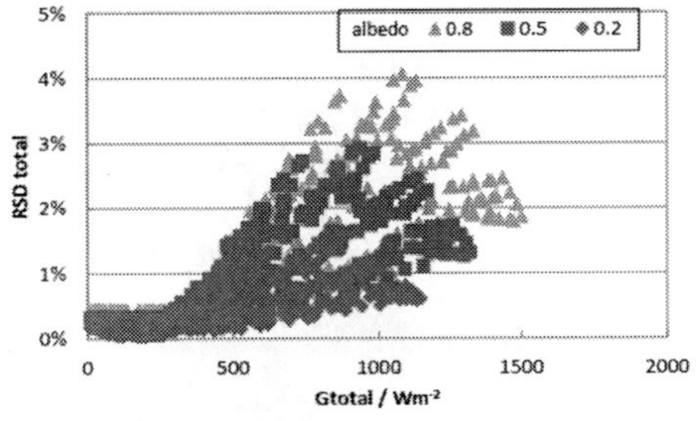

Figure 4: The calculated RSD of the total irradiance of a free standing, south-facing panel in Amsterdam during the month of April.

The dependency of the rear RSD on the albedo is small. This is because the irradiance on the rear is dominated by the albedo, i.e. the numerator and denominator in eq. 2 both essentially scale with the albedo. The calculated RSD values shift to higher values of G_{tot} because G_{tot} increases with albedo.

Although RSDs in the order of 13% of the rear irradiance may seem quite significant, the impact on the module or string output is determined by the variations occurring in the total generated current in each cell, i.e. variations in the total incident irradiance. In the cases under consideration the RSD of the total irradiance is much smaller, ranging from 1.5% at albedo 0.2 to 4% at albedo 0.8, as Fig. 4 shows.

The ratio G_{rear}/G_{front} determines what the contribution of the rear RSD will be on the RSD of the total irradiance. This ratio depends on the irradiance conditions and, as Fig. 5 shows, decreases with the total irradiance. When the *diffuse* component dominates, as is often the case at low G_{tot}, the ratio is given by:

$$\frac{G_{rear}}{G_{front}} = \frac{(1-\cos\beta)+\gamma(1+\cos\beta)}{(1+\cos\beta)+\gamma(1-\cos\beta)} \quad (3)$$

This means that at tilt angle β=0° (and high enough elevation) G_{rear}/G_{front} is equal to the albedo γ; at 90° tilt angle the ratio is equal to one. At 38° tilt angle the ratio is 0.31, 0.58 and 0.83 at albedo 0.2, 0.5 and 0.8, respectively. This is in good agreement with the data at low G_{tot} shown in Fig. 5.

At high G_{tot} the *beam* component is dominant and G_{rear}/G_{front} is determined by (neglecting the self-shade):

$$\frac{G_{rear}}{G_{front}} = \frac{\gamma(1+\cos\beta)/2}{\cos\theta/\cos\theta_z+\gamma(1-\cos\beta)/2} \quad (4)$$

Now G_{rear}/G_{front} depends on the position of the sun with respect to the panel. At Amsterdam conditions in April, the ratio for a panel with 38° tilt angle decreases to values of half the albedo (Fig. 5).

Comparison of Fig. 2 and 5 shows that when the irradiance on the rear is most inhomogeneous, the fraction of rear irradiance is small. In most south-facing configurations the G_{rear}/G_{front} will be lowest at high irradiance conditions, and this leads to RSD values of the total irradiance that are considerably lower than the RSD of the rear irradiance, even at high albedo.

The module tilt angle is usually such that maximum use is made of the beam component, i.e. that the annual sum of $(GHI - DHI)\cos\theta/\cos\theta_z$ is highest. However, bifacial panels can also be advantageously used in an east-west configuration. In this case the tilt angle is 90°, i.e., the modules are placed vertically, to make best use of beam components coming from either the east or the west. Although this would lead to lower $\cos\theta/\cos\theta_z$ values and thus higher G_{rear}/G_{front} compared to the south-facing configuration, the self-shade of the panel is usually further away from the rear (in this case the rear is the side not receiving direct light) and the RSD of the G_{rear} is smaller. Fig. 6 (top) shows the RSD of the west side. The

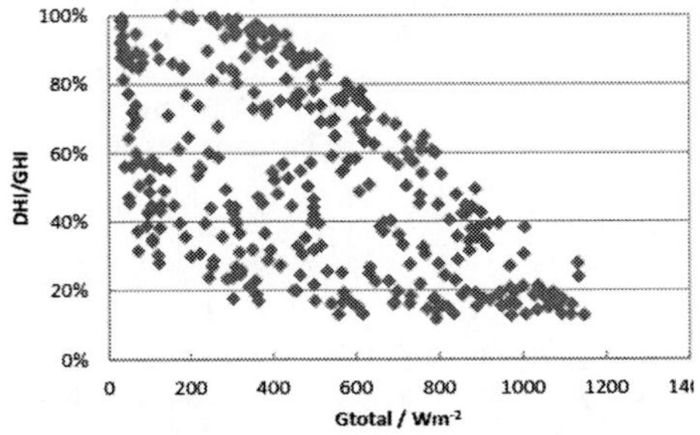

Figure 3: The ratio DHI/GHI in Amsterdam during the month of April plotted versus the total irradiance on a free standing, south-facing panel in Amsterdam.

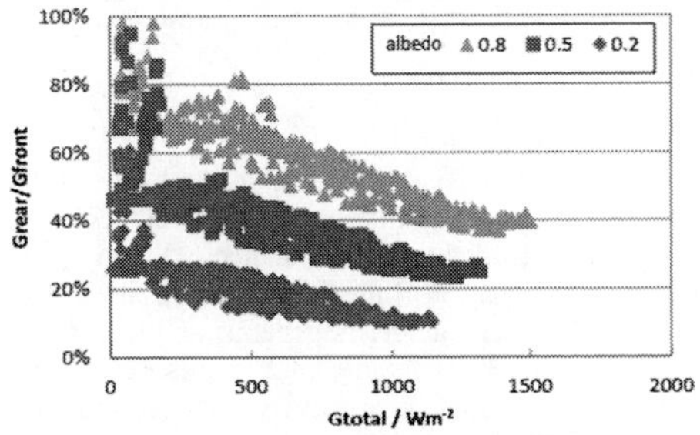

Figure 5: The ratio G_{rear}/G_{front} for a free standing, south-facing panel in Amsterdam during the month of April.

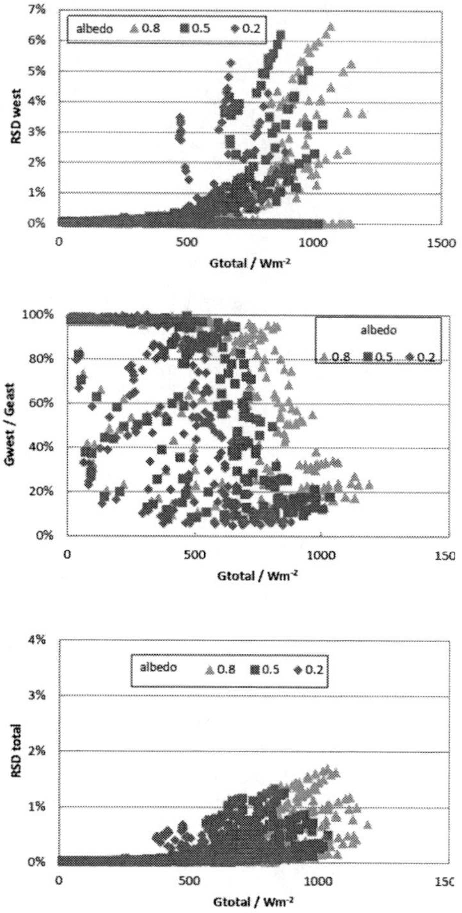

Figure 6: The calculated RSD of the irradiance on the west side (top), the ratio G_{west}/G_{east} (center) and the RSD of the total irradiance (bottom) on a free standing, vertical, east-west facing panel in Amsterdam during the month of April.

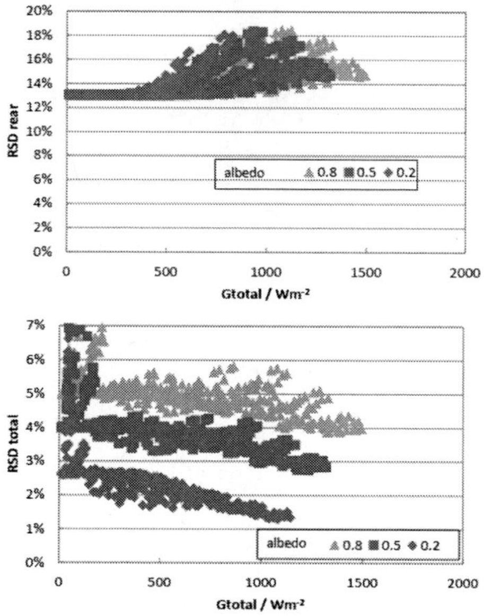

Figure 7: The calculated RSD of the rear irradiance (top) and of the total irradiance (bottom) on a free standing, south-facing panel in Amsterdam during the month of April. In the calculation it was assumed that one of the cells does not receive any rear irradiance.

inhomogeneity is either very small, i.e., when the direct light is from the west, or up to values of about 7%. The G_{west}/G_{east} is now 100% when diffuse light prevails, and reduces to values much lower than the albedo at high G_{tot} values (Fig. 6, center). A similar picture applies for the east side of the panel and in the resulting RSD values for the total irradiance are < 2% even at high albedo (Fig. 6, bottom).

3.2 Near-field shading

Shading by objects such as junction boxes at the rear can easily lead to the situation where several cells receive only a fraction of the rear irradiance [4]. Using Eq. 2 it can be calculated that if the irradiance on the rear of one or two cells is zero and homogeneous and non-zero on all other cells, the RSD for the rear irradiance of a 60-cell module will be 13%, respectively 19%. In the case that these two cells receive half the irradiance the other cells receive, the RSD will be 9%. In practice, there will also be differences in self-shading as discussed in 3.1, but these will yield a smaller increase in RSD. Moreover, it will not matter much which cells are completely shaded.

Note, that in the experimental, outdoor study at ECN, which included near-field shading but where no cell was completely blocked, the rear irradiance of a portrait-oriented module had an RSD of 13.8%

Fig. 7 shows the calculated RSD of the rear and total irradiance, under identical conditions as Fig. 2 and Fig. 4, with the addition that one cell receives no rear irradiance. Because of the fairly constant RSD of the rear irradiance, the RSD of the total irradiance now decreases with the total irradiance, in agreement with the decreasing ratio of G_{rear}/G_{front} (Fig. 5). At Amsterdam conditions, the RSD of the total irradiance of a 60 cell module with one cell completely shaded at the rear will then be 2-4% at moderate albedo and about 5% at high albedo (Fig. 7).

3.3 Impact of current mismatch on electrical output.

The inhomogeneous irradiance on the rear leads to variation in the photocurrents of the solar cells, which are connected in series. LTspice was used to calculate the power output of modules with inhomogeneous irradiance distributions, characterized by combinations of G_{mean} and RSD, as explained in section 2.2. The power loss due to the current mismatch can be quantified by the parameter f:

$$f = 1 - \frac{P_{inhom}}{P_{hom}}, \qquad (5)$$

where P_{hom} is the sum of the power at MPP of 60 cells with a homogeneous irradiance G_{mean}; P_{inhom} is the output of a string of 60 cells, which have an irradiance distribution with average irradiance G_{mean} and an RSD > 0.

Fig. 8a shows the mismatch fraction f calculated with each distribution as a function of the mean irradiation G_{mean}. The mismatch fraction f is not strongly dependent on G_{mean}. However, as Fig 8b shows there is a clear correlation with the RSD of the distribution, especially for RSD < 10%.

For RSD < 10% the relative power loss f depends only on RSD and not on the total irradiance. In the majority of realistic irradiance-view factor-albedo conditions the RSD is significantly smaller than 10%; even with one cell fully blocked for rear irradiance and at albedo 0.8, the RSD of the total irradiance is < 6% (see Fig. 7 bottom).

At RSD > 10% the loss factor f increases more rapidly with RSD and exhibits a large variation, in the order of ±5%, for very similar RSD values. In these cases the RSD alone is not sufficient to accurately predict the loss.

4 DISCUSSION

The results for the free-standing panel in Amsterdam show that the inhomogeneity of the total irradiance due to self-shading effects will be limited to < 5%, even at high albedo (Fig. 4). The main cause is that the homogeneous front side irradiance will dominate, especially at high irradiance conditions where the ratio between front and rear irradiance will approach half the ground-reflection coefficient.

Such small variations in irradiance, or equivalently in current, will not lead to hot-spots, as simulations by LTspice showed. This result is in agreement with observations in the literature that inhomogeneities in cell power up to 15% over a string can still be tolerated without large mismatch losses [10].

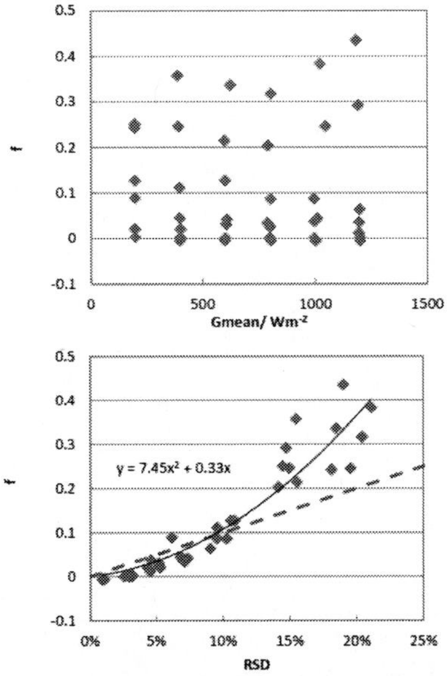

Figure 8: The calculated power mismatch fraction f as a function of the mean irradiance (top) and as a function of the relative standard deviation RSD (bottom). In the bottom figure the dashed, red line indicates f=RSD, the black line is the best fit through all data points with the quadratic equation given.

Near-field shading, as long as it is not shading the direct light, will in most cases not give rise to a large inhomogeneity in the overall irradiance, except when the front and rear irradiance are similar. This can only happen in the absence of a direct beam component and thus at low total irradiance; but in that case, hot-spots will not occur because the energy dissipation is too low.

Consider a possible "worst-case scenario" where the front side is covered by snow and sun light is reflected of fresh snow below the bifacial modules. In this case the RSD of the overall irradiance will be identical to that of the rear. However, in this case the module's rear side must face down. Due to the sun's position in the sky in the winter, i.e. low and in the Southern hemisphere, the total (rear) irradiance due to diffuse and ground-reflected light will not be high. Thus the dissipated energy will usually be limited.

The calculations here were done for Amsterdam at 52° north. At locations closer to the equator lower tilt angles are used and more inhomogeneity in the rear irradiance is expected. But at these locations the beam component will be (much) stronger, which will limit the effect of the inhomogeneous rear irradiance on the total irradiation.

For energy yield predictions the power losses due to current inhomogeneity that is not caused by direct beam shading can be approximated by a linear dependence when the RSD does not exceed 10%. For higher RSD values, a quadratic expression as found for the 60-cell case in section 3.3 can be used. Note, that these relationships were obtained for a 6o-cell circuit that is a single module under MPP tracking. Larger module strings are not within the scope of this paper, but we expect a similar relationship to be valid as long as the variations in the irradiance are in the same range.

5 CONCLUSIONS

- Due to self-shading the rear irradiance of a PV-panel in Amsterdam shows relative standard deviations up to the order of 13% at an elevation of 0.5 m. Larger RSD values for the rear irradiance are obtained for higher beam component, but the maximum values are not sensitive to the albedo value. The RSD decreases with elevation and tilt angle.
- The effect of the inhomogeneous rear irradiance on the variation of the total radiance depends on the ratio of the front and rear irradiance. The presence of a strong beam component on the front, and thus a low G_{rear}/G_{front} ratio, reduces the RSD of the total irradiance to < 2% at albedo 0.2, and < 5% at albedo 0.8. For vertical east-west facing vertical panels even smaller values were found.
- A near-field shade that fully blocks reflected and diffuse light on the rear can cause an inhomogeneity in the rear irradiance exceeding 13%. But the occurrence of hotspots is not very likely. At low light intensity, the induced reverse voltage in the shaded cell is not accompanied by a significant power dissipation, whereas at high light intensity the inhomogeneity in the total irradiance is much reduced by the dominating, homogeneous front irradiance.
- Because the inhomogeneity in the total irradiance rarely exceeds 10%, the relative power loss will be about the same as the RSD of the total irradiance.

ACKNOWLEDGEMENTS

This work was supported by the Dutch Ministry of Economic Affairs within the TKI Urban Energy TKI-toeslag project BING.

REFERENCES

[1] Electric Power Research Institute EPRI, "Bifacial Solar Photovoltaic Modules - Program on Technology Innovation", (2016).
[2] U. A. Yusufoglu, T. M. Pletzer, L. J. Koduvelikulathu, C. Comparotto, R. Kopecek, and H. Kurz, IEEE Journal of Photovoltaics 5 (2015) 320-328.
[3] U. A. Yusufoglu, T. H. Lee, T. M. Pletzer, A. Halm, L. J. Koduvelikulathu, C. Comparotto, R. Kopecek, and H. Kurz, Energy Procedia 55 (2014) 389-395.
[4] K. M. de Groot and B. B. Van Aken, Energy Proc. (2017) Proceedings SiPV 2017, in press.
[5] G. J. M. Janssen, B. B. Van Aken, A. J. Carr, and A. A. Mewe, Energy Procedia 77 (2015) 364-373.
[6] LTspice, http://www.linear.com/designtools/software/#LTspice, (2017).
[7] PVGIS, http://re.jrc.ec.europa.eu/pvgis/apps4/pvest.php#, (2016).
[8] PVsyst V6, www.pvsyst.com, (2016).
[9] J. A. Duffie and W. A. Beckmann, *Solar Engineering of Thermal processes,* 2nd ed.; John Wiley &Sons, 1991.
[10] A. Massi Pavan, A. Mellit, D. De Pieri, and V. Lughi, Prog.Photovolt: Res.Appl. 22 (2014) 332-345.

IV MEASUREMENT OF BIFACIAL MODULES: BIFACIAL VS. MONOFACIAL ILLUMINATION

Alexandra Schmid, Gonca Dülger, Ghassan Baraah, Ulli Kräling
Fraunhofer ISE, Heidenhofstr.2, 79110 Freiburg
alexandra.schmid@ise.fraunhofer.de

ABSTRACT: Bifacial modules are able to use light incident on both sides und thus have the potential for significantly increasing the yield of PV power plants. However, as current testing procedures do not respect gains from rear side illumination, the bifacial technology comes along with new challenges for a reproducible and comparable module characterization. Currently a draft for a new standard is in preparation, including different measurement methods with front side only and bifacial illumination. In this paper, we compare measurement results of these different approaches to characterize bifacial modules. The differences between single sided and bifacial illumination are investigated. The role of the light incident side and the influence of module properties on the resulting measured power are analyzed. Additionally, as most labs do not have the possibility to measure the IV curve under bifacial illumination, a method is presented to calculate the power under bifacial illumination based on monofacial measurements.
Keywords: Bifacial, Calibration, Characterisation, PV Module, Performance

1 INTRODUCTION

For the measurement of bifacial modules different methods are currently under discussion. As the existing standards for the IV measurement of photovoltaic devices do not consider gains from rear irradiation, nor define the measurement conditions for the module rear side, it is often not clear how the nominal power of commercial modules is determined. The labeled values are often extrapolated from frontside STC measurements, assuming a linear power boost, or measured with an undefined reflector behind the module. In any case, the comparability of datasheet values is poor.

A new standard for the IV measurement of bifacial modules is already in development, allowing for different options for the assessment of the bifacial gain: frontside measurement under elevated irradiance, measurement under bifacial illumination and outdoor measurement with different rear intensities.

In this paper, measurement results of different indoor approaches for the determination of power of bifacial modules are compared: measurements under single sided illumination, under elevated front side illumination as proposed in [3], and under simultaneous bifacial illumination. It is analyzed in which cases the results are in good agreement, and which module properties cause deviations between the different approaches, with respect to irradiance intensity regimes.

2 METHODS

2.1 STC under single sided illumination

For a basic characterization of bifacial modules, it is necessary to measure each side separately at STC, while the other side has to be protected from incident stray light. For the measurements in this work, the module's rear side was covered by a black curtain. The spectrally weighted reflection of the material is 4.3% and rather constant over the relevant wavelength range between 300-1200nm. Along the module's long edges a mask prevents light from passing by the module, so the incident light on the rear cover is reduced to transmission through the module and light passing by the short egdes of the module. This way the electrical module parameters of each side are determined with maximum precision. With these results the bifaciality φ of current and power can be calculated:

$$\varphi_{Isc} = \frac{I_{sc,rear}}{I_{sc,front}} \qquad (1)$$

$$\varphi_{Pmpp} = \frac{P_{mpp,rear}}{P_{mpp,front}} \qquad (2)$$

The bifaciality is needed for the calculation of the elevated irradiance Ge, as described in the next section. The single sided STC parameters are also the basis for the calculations in section 3.3, and for yield simulations of bifacial PV systems [7].

For framed modules, special care has to be taken for the module position in the rear side measurement. The module surface is shifted by the frame width behind the normal measurement plane. Thus the intensity has to be adjusted to obtain one sun in the plane of the rear module surface.

2.2 Front side under elevated irradiance

Since the module's behavior in bifacial operation and the bifacial gain cannot be directly assessed from the single sided STC measurements, Fakhfouri et al. [3] propose additional front-side measurements at higher irradiances $G_{e,i}$. $G_{e,i}$ is calculated by the short-circuit current bifaciality coefficient φ:

$$G_{e,i} = 1000W/m^2 + \varphi \cdot G_{r,i} \qquad (3)$$

At least three different irradiances $G_{r,i}$ have to be measured and the results interpolated to G_R = 100 and 200 Wm^{-2}. With this method, the increased resistive loss due to series resistance is considered, as the generated current is equal to the current under the respective bifacial illumination.

2.3 Bifacial illumination

As bifacial modules in real applications operate under simultaneous front and rear side illumination, it is advisable to characterize this module type under bifacial illumination For determination of parameters under bifacial operation and the bifacial gain, a new measurement set up was developed at CalLab PV Modules, which enables bifacial illumination of full size modules up to 1x2m in irradiation quality better than AAA. Two mirrors in 45° angle to the lamp are directing the light of the solar simulator simultaneously on both sides of the module, like described in [6] for cells. The mirrors are made from a silver coated reflector sheet, with a reflectance over 95% in the wavelength range 300-1200nm, so the reflected spectrum remains A+ quality. The reflector sheets are attached to glass panes to get a smooth surface for maintaining irradiance homogeneity.

Figure 1 shows a schematic of the set-up. The lamp power can be adjusted between 100-1000W/m^2, and by inserting attenuation filters the rear intensity can be reduced. With this variable light intensity and variable front-to-rear intensity, typical irradiation conditions for different installation geometries (e.g. south, east-west) can be simulated.

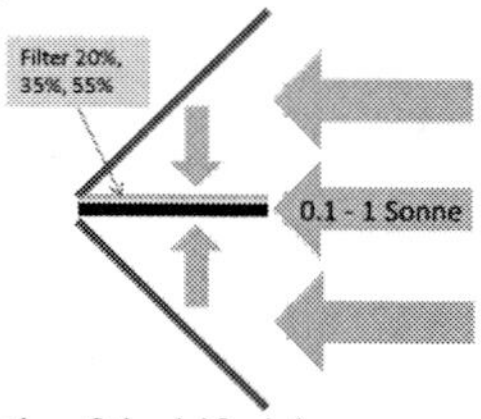

Figure 1 schematic of the bifacial measurement set-up: 2 mirrors in 45° angle to the lamp are directing the light on both sides of the module

The attenuation filters used for this work are made of woven wire mesh. This material shows spectrally neutral transmission and good spatial homogeneity on big areas, as shown in [1]. The currently available transmissions are 20%, 35% and 55%. For reaching lower transmissions of 10-20%, representing the range of realistic installation conditions, new filters of different materials are currently under design.

As this mirror set-up is installed in the same sun simulator where the normal single sided measurements are performed, all the general measurement procedures are not affected. All measurements are carried out as section and hysteresis measurements.

2.4 Devices under test

The described measurement procedures were compared for different commercial bifacial modules, consisting of 60 or 72 cells. Since a module's fill factor (FF) is always influenced by different effects of the serial connection of single cells, like current mismatch, standard size commercial modules as well one-cell modules were investigated for a closer FF analysis.

3 RESULTS

3.1 single sided vs. bifacial measurement

In the first step, single sided measurements over a wide intensity range of 100W-1000W/m^2 are compared to bifacial measurements. In the following figures, the irradiance is always symmetrical (frontside irradiance = rearside irradiance). In figure 2, the sum of Isc for front side irradiance and for rear side irradiance is compared to the Isc measured under bifacial irradiance. The deviation is below ±0.3% for all intensities. From this good agreement, it can be concluded that in this intensity regime the assumption that the short circuit current adds linearly is applicable. Further, the bifacial illumination quality is sufficient for reproducing the expected current based on the single sided precision measurements.

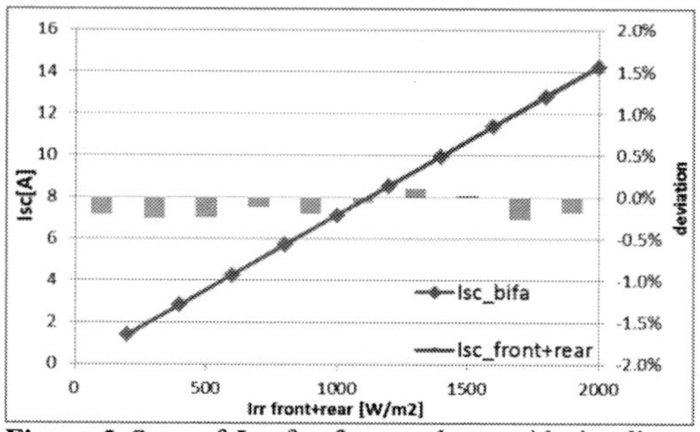

Figure 2 Sum of Isc for front and rear side irradiance compared to Isc for bifacial irradiance

Figure 3 shows the Voc measured for frontside, rearside and bifacial irradiance. The x-axis refers here to Isc instead of irradiance, in order to compensate the bifaciality. At the same current level, Voc is independent of light incident side in the complete irradiance range.

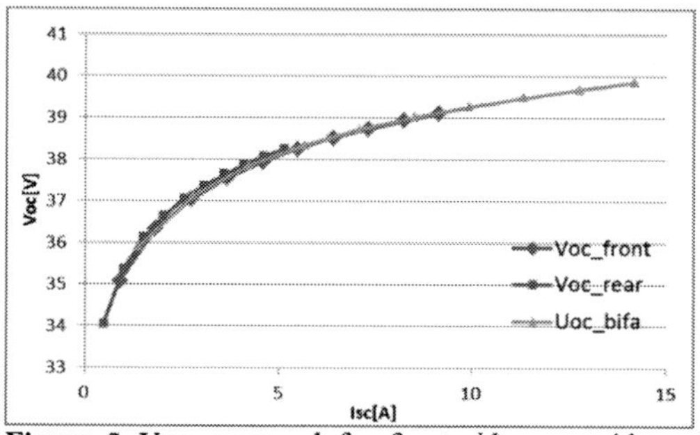

Figure 3 Voc measured for front side, rear side and bifacial irradiance, with respect to Isc

While Isc and Voc for bifacial operation can be calculated based only on the single sided measurements for any given irradiance condition, like shown in [4], the bifacial Pmpp is affected by resistive losses due to series resistance and shunt resistance. Figure 4 shows the deviations of the bifacial Pmpp and the sum of Pmpp for front and rear side irradiance.

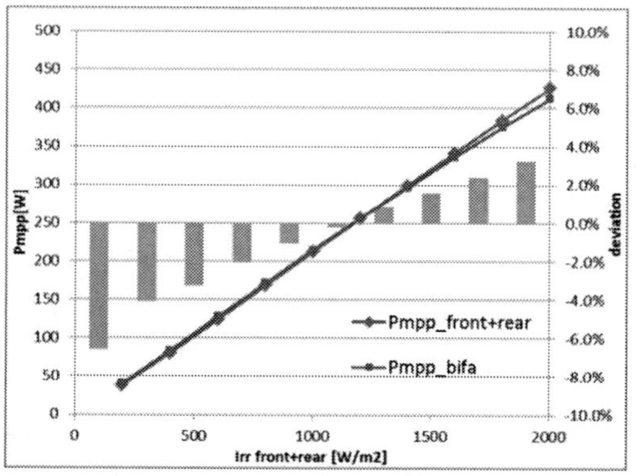

Figure 4 Sum of Pmpp for front and rear side irradiance compared to Pmpp for bifacial irradiance

For low irradiances, the Pmpp of the single sided measurements is more affected by the shunt resistance as under bifacial irradiation, as the module operates at a higher injection level because of the simultaneous illumination. For high irradiances, the measured power for simultaneous illumination is lower than the sum of the single sided measurements, due to the increased series resistance loss. These effects cause a deviation of linear superposition in Pmpp of up to 6% for extreme irradiance combinations.

3.2 Ge method vs. bifacial measurement

For comparing single sided and bifacial measurements of commercial modules, it is important to consider the shape of the IV curves. Most bifacial modules have distorted rear I-V-curves, due to partial shading by the junction box, cabling, frame or label, or due to cell sorting by frontside current only.

As in the Ge method a modules IV curve is measured only under elevated front irradiance, the distortion of the rear IV curve will not be detected, as can be seen in figure 5. While the IV curve of the Ge measurement is as smooth as the frontside STC, the IV curve for the bifacial measurement is affected by the partial shading of the module´s rearside.

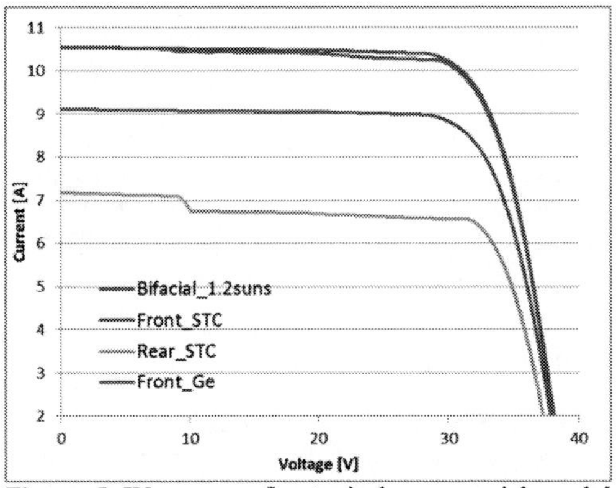

Figure 5 IV curves of a typical commercial module under front side irradiance and under rear side irradiance (STC condition), compared to bifacial and Ge irradiance corresponding to 200W rear irradiance

Depending on the rear intensity and the severity of the distortion, this leads to a deviation between measured power and FF for Ge and bifacial measurement. Figure 6 shows measured FFs for different illumination conditions for a typical module. FF_bifa refers to symmetrical irradiance for front and rear side. Additionally, the FFs for the Ge and bifacial measurement, corresponding to 1000W/m^2 front and 200 W/m^2 rear intensity are shown.

For all bifacial modules measured with the bifacial set-up and with the Ge-method, the deviation of measured Pmpp was calculated. Table 1 shows the mean deviations for commercial modules and for single cell modules.

Table 1: mean deviation of measured Power for bifacial and Ge measurement, for irradiance condition 1000W/m2 front and 200 W/m2 rear:

	Commercial Modules	One cell modules
Mean deviation	0.54%	0.15%
stdev.	0.24%	0.20%

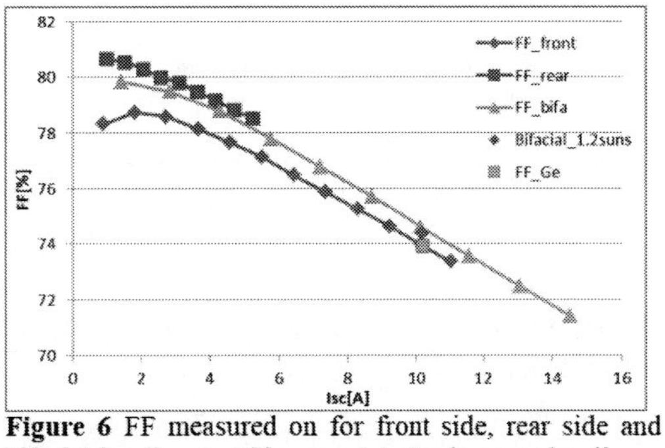

Figure 6 FF measured on for front side, rear side and bifacial irradiance, with respect to Isc in same irradiance condition. Additionally the FF for the Ge and bifacial measurement with 200W rear intensity are shown

For the measured commercial modules a mean deviation of 0.5% in Pmpp between Ge and bifacial measurement was found for the irradiance condition 1000W/m2 front and 200 W/m^2 rear intensity.

In order to have a closer look at the FF and to separate the effects of the serial interconnection of the cells from the single cell behavior, we used one cell modules for further investigation of the FF. Three different bifacial cell technologies were investigated.

For the measurement under single sided illumination on front and rear side, we found a slightly higher FF for the rear side illumination even for the single cell modules. Figure 7 shows the FFs for rear and front irradiance for different intensities.

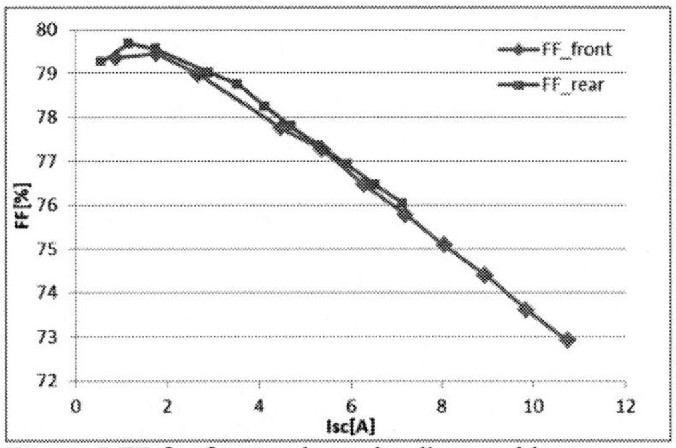

Figure 7 FF for front and rear irradiance with respect to Isc for a one cell module.

To calculate the FF difference at the same current level, three one cell modules were measured under a front intensity $G_{r,STC}$ defined as

$$G_{r,STC} = 1000W/m^2 \cdot \varphi_{Isc} \quad (4)$$

Under this intensity the same current is generated on the front side as on the rear side at STC. In Table 2 the results are listed. The FF for rear illumination is found to be 0.3% abs. higher than for frontside illumination. This result suggests that the FF is dependent on the light incident side. The cause for this effect is currently under further examination.

Table 2: differences in FF for front side and rear side irradiance for 3 one cell modules:

	Isc_[A]	FF_[%]	delta FF
Mod1_front_$G_{r,STC}$	5.881	77.966	
Mod1_rear STC	5.882	78.122	**0.20%**
Mod2_front_$G_{r,STC}$	5.916	76.677	
Mod2_rear STC	5.914	76.936	**0.34%**
Mod3_front_$G_{r,STC}$	7.785	76.991	
Mod3_rear STC	7.791	77.344	**0.46%**
mean:			**0.33%**
stdev:			0.13%

3.3 Calculation of bifacial power

As most labs and module producers do not have the possibility to measure a module`s IV curve under bifacial illumination, it is desirable to get reliable results for the bifacial gain based on monofacial measurements.

Singh et al. introduced in [2] a method to calculate the bifacial efficiency based on monofacial STC measurements of front and rear side. For different commercial modules, the results of this method were found to underestimate the measured bifacial power. Therefore, instead of using the ideal value for n=1, and an empirical relation for the pseudo fill factor pFF as proposed by Singh, these values were determined from a SunsVoc measurement [5] of the front side.

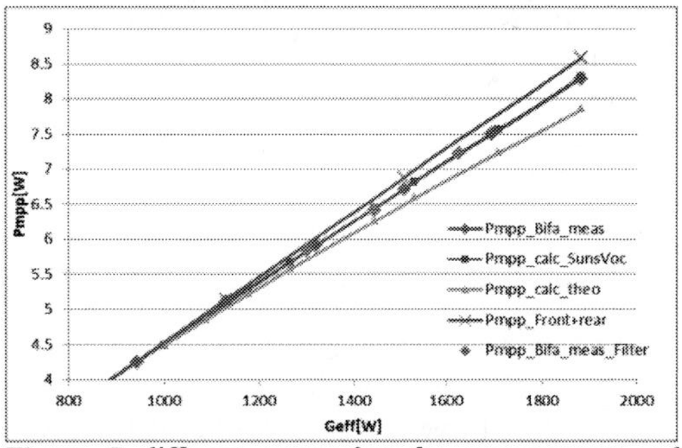

Figure 8 different approaches for measurement and calculation of bifacial Pmpp

With this modification, very good agreement with the results of the bifacial measurements could be reached, as is shown in figure 8 and Table 3. For the extreme case of 1000W/m^2 on front and rear side the mean deviation of measured and calculated power is 0.2%.

Table 3: mean deviation of measured bifacial power and calculated values for 1 sun front and rear intensity:

	meas-calc SunsVoc	meas-calc theo
Mean deviation	0.2%	-3.7%

4 CONCLUSIONS

For the basic characterization of bifacial modules, single sided STC measurements are required to assess the module parameters for each side. A comparison of different methods for the determination of the bifacial gain shows slightly different results. For modules with severe distortion of the rear IV curve, the measurement under real bifacial illumination is recommended to assess the correct IV curve for realistic operating conditions. In the Ge-method, the effects of partial shaded rear sides on the IV curve are not detected.

However, for low rear irradiances up to 200W/m^2, the different approaches are in good agreement of about 0.5%.

Further, it is shown that based on single sided STC measurements and an additional SunsVoc measurement, the bifacial efficiency can be calculated. The results are in very good agreement with the measurement under bifacial illumination.

5 ACKNOWLEDGEMENTS

This work was partly funded by the EMPIR programme co-financed by the Participating States and from the European Union's Horizon 2020 research and innovation programme, and by the German Federal Ministry of Economic Affairs and Energy in the project MOSBIT (FKZ 0324036A).

6 REFERENCES

[1] A. A. Santamaria, Proceedings 29th European Photovoltaic Solar Energy Conference, Vol. I (2014) 3214.

[2] J. P. Singh et al., A new method to characterize bifacial solar cells, Prog. Photovolt: Res. Appl. 2014; 22:903–909

[3] V. Fakhfouri et al., IEC 60904-1-2: Measurement of current-voltage characteristics of bifacial photovoltaic devices, Presentation at 3rd bifi PV workshop, Miyazaki, Japan, (2016).

[4] J. P. Singh et al., Electrical characterization method for bifacial photovoltaic modules, Sol. Energy Mater. Sol. Cells, 127 (2014), pp. 136-142

[5] R.A. Sinton, and A. Cuevas, A Quasi-Steady-State Open-Circuit Voltage Method for Solar Cell Characterization, 16th European Photovoltaic Solar Energy Conference, 2000

[6] A. Edler et al., Flasher setup for bifacial measurements, Presentation at 1st bifi PV workshop, Constance, Germany, 2012.

[7] C. Reise, A.Schmid, Realistic Yield Expectations for Bifacial PV Systems – An Assessment of Announced, Predicted and Observed Benefits, 31st European Photovoltaic Solar Energy Conference, pp.1775-1779

OUTDOOR PERFORMANCE ANALYSIS OF THE SI-HETEROJUNCTION MODULES WITH DIFFERENT CELL AND MODULE DESIGNS

K. Emtsev[1], D. Malevskiy[2], D. Andronikov[1], A. Abramov[1], S. Yakovlev[1], A. Titov[1], E.Terukov[1,2], D. Orekhov[1], B. Bulygin[3], A. Dubrovskiy[3]
[1] R&D Center for Thin Film Technologies in Energetics (TFTE), St.-Petersburg, Russia
[2] Ioffe Institute, St.-Petersburg, Russia
[3] Hevel LLC, Novocheboksarsk, Russia

ABSTRACT: The results of the annual outdoor monitoring of different Si-heterojunction (HJT) modules are presented. Standard 60 cell modules were assembled in glass/backsheet and bifacial glass/glass configurations. Further, HJT cells for modules were produced in monofacial and bifacial designs. This allows us to compare performance of the monofacial and bifacial modules side by side. In particular, we evaluate bifacial gain for different seasons and weather conditions. In addition, the behavior of HJT modules with respect to the standard mc-Si module is assessed.

Keywords: heterojunction, bifacial, performance, monitoring, reliability

1 INTRODUCTION

Silicon heterojunction technology (Si-HJT) enables very high cell and module efficiencies [1] and according to the ITRPV roadmap is expected to grow its market share steadily over the next years [2]. Several new mid-scale production facilities, indeed, have been announced recently. Further, new cell interconnection technologies for HJT cells are becoming available commercially. Needless to say, such cells naturally support bifacial modules that will further increase their output efficiency. Since 2014 there are ongoing activities to develop HJT cells at the TFTE R&D center in Saint-Petersburg, Russia [3]. Technology utilizes former a-Si micromorph equipment. HJT 6" cell efficiency of 22.7% on production equipment has been achieved. Recent results of upscaling the HJT module production at the Hevel LLC, Russia based on these cells are reported at this conference (see paper 2AV.3.12) [4]. In this work, we compare the outdoor performance of different HJT module designs throughout a year. In particular, performance of monofacial *vs.* bifacial modules at various weather conditions is assessed. In addition, we compare the behavior of HJT modules with respect to the standard mc-Si module.

2 EXPERIMENTAL SETUP

Outdoor monitoring stand is located in Novocheboksarsk, Russia (N56°07′, E47°30′) and includes overall 9 modules (see Fig. 1). The collected data for each module are P_{mpp}, I_{mpp}, V_{mpp} DC values at the maximum power point, in plane solar irradiance via Kipp&Zonen pyranometer, module and ambient temperatures. The data are collected every 30 sec. The monitoring begun in August 2016 and continues thereafter.

The standard 60 cell modules for this study were produced from HJT cells of different designs via smart wire contacting technology (SWCT) [5]. The first batch of modules was assembled at the Meyer Burger from the cells produced at the TFTE, Saint-Petersburg. Starting from April 2017, five modules of different classes fully produced at Hevel LLC, Russia were installed. The highest STC (standard test conditions) output power of the HJT module was 297 W. The power rating of the modules was independently confirmed by Fraunhofer ISE. The overall description of the modules is given in Table 1. There is one module based on monofacial cell design. Bifacial cells were assembled in two different module configurations, namely, glass/backsheet and glass/glass. Bifacial module was measured with the black back cover. As reference, commercial mc-Si module was monitored. In parallel, indoor climatic tests were carried out on the HJT modules produced at Hevel. Thermocycling (TC200) and damp heat tests (DH1000) as per IEC 61215 has passed successfully.

Solar irradiation and ambient temperatures at the site are shown in Fig. 2. The coldest month of monitoring January also exhibits the highest temperature variation.

Figure 1: Outdoor monitoring stand with 8 HJT modules and one reference mc-Si module located in Novocheboksarsk, Russia (N56°07′, E47°30′).

Table 1: Monitoring modules with different cell and module designs.

No.	Module	Cell type	Cell Connection	Module type	Pmpp, W
1	mc-Si	mc-Si	3 BB	Glass/Backsheet	235
2	HJT	bifacial	SWCT	Glass/Backsheet	291
3	HJT	bifacial	SWCT	Glass/Backsheet	282
4	HJT	monofacial	SWCT	Glass/Backsheet	273
5	HJT	bifacial	SWCT	Glass/Backsheet	261
6	HJT	bifacial	SWCT	Glass/Backsheet	262
7	HJT	bifacial	SWCT	Glass/Backsheet	297
8	HJT	bifacial	SWCT	Glass/Glass	263
9	HJT	bifacial	SWCT	Glass/Backsheet	272

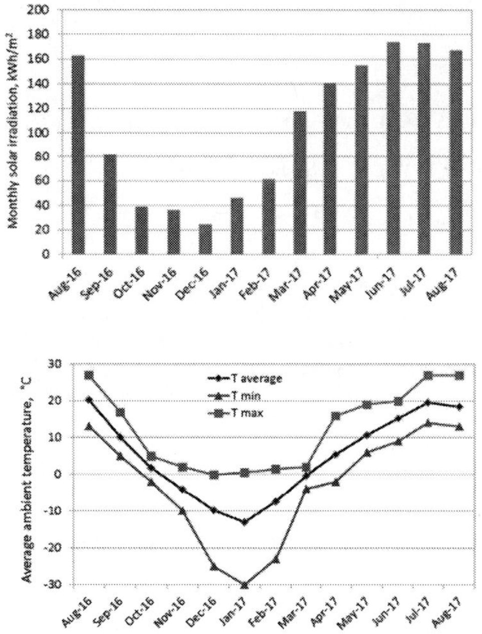

Figure 2: Monthly solar irradiation (top) and ambient temperatures (bottom) for the monitoring period.

Daytime temperatures down to -30°C occurred in winter. At the same time, the winter warmest days were close to 0°C. Generally, however, winter season is of very low irradiance and, in addition, the performance was partially affected by snow. In this respect, the period from March (still with snow coverage) till August appears most useful for data evaluation.

3. RESULTS AND DISCUSSION

Fig. 3 compares energy yield defined as generated DC power output in kWh divided by the STC nominal power in kW for monofacial glass/backsheet HJT and standard mc-Si modules. The HJT data represents average of the several modules as, essentially, no difference in energy yield between modules of different classes occurred. Also, modules based on monofacial cells performed equally to the ones based on bifacial cells in glass/backsheet design. As can be seen in Fig. 3, the HJT modules perform better in summer, while for colder months energy yield of the mc-Si is somewhat higher. For the hottest month the relative gain of HJT over mc-Si module of +3.3% is detected. At the same time, in March with the ambient temperatures slightly above 0°C the yield of both technologies is the same. This difference is due to lower temperature coefficient of the output power for the HJT cells as opposed to mc-Si. Based on the selected outdoor data for clear days with the irradiance close to STC value of 1000 W/m^2 we estimate temperature coefficients to be around -0.31%/°C and -0.45%/°C for HJT and mc-Si, respectively. To illustrate this effect further, in Fig. 4 we show the normalized output power of different types of module technologies for hot summer and cold winter days. The highest gain of around 6.2% of monofacial HJT module over mc-Si module is observed at the solar noon when the module

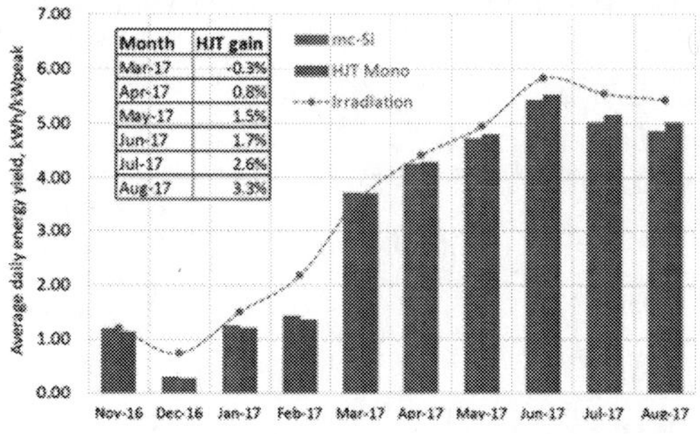

Figure 3: Average daily energy yield for the monofacial HJT and mc-Si modules. Inset shows the energy gain of the HJT over mc-Si for different months. Winter data are partially affected by snow.

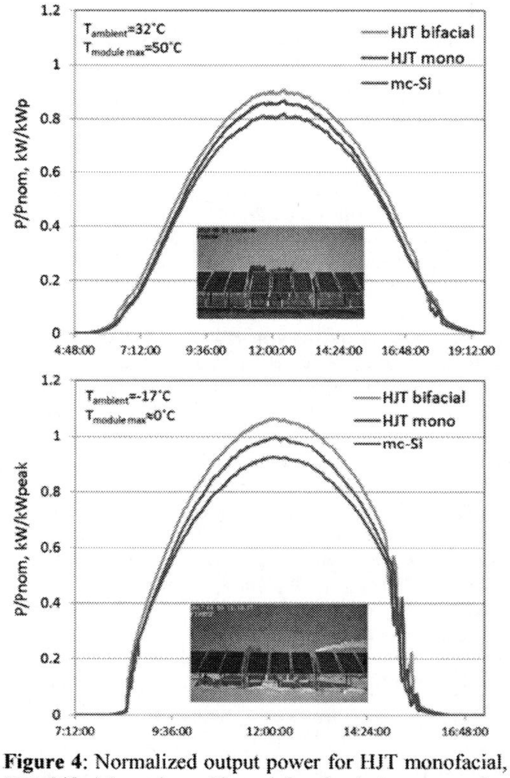

Figure 4: Normalized output power for HJT monofacial, HJT bifacial, and mc-Si modules for hot summer day 21.08.2017 (top) and cold winter day 30.01.2017 (bottom).

temperature is also the highest. The situation is reversed on cold winter day (Fig. 4, bottom). With the module temperature of ≈0°C output of the mc-Si module is significantly higher than that of the mono HJT module.

In Fig. 5 we plot energy yield of monofacial glass/backsheet vs. bifacial glass/glass HJT modules for the full year period. The bifacial module is seen to outperform a monofacial HJT module at any time of the year. In total, annual energy yield of the bifacial module exceeded that of the monofacial HJT by 13%. The bifacial gain is due to collection of the reflected light by module's backside. Therefore, it depends on many factors: 1) the bifaciality of the cell in the module itself,

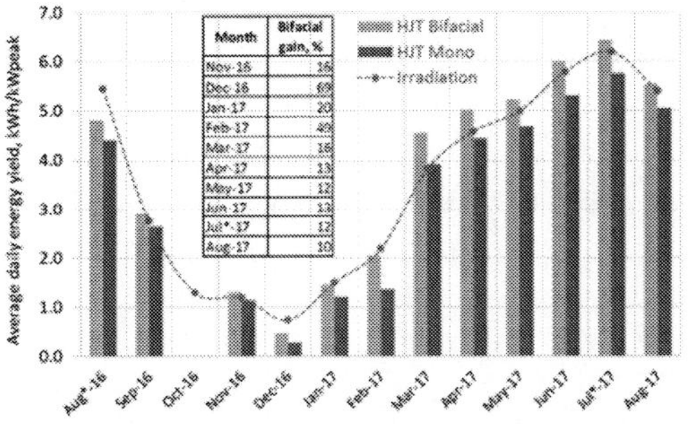

Figure 5: Average daily energy yield of the monofacial and bifacial HJT modules. In total, 13% annual gain of bifacial HJT over monofacial HJT is observed. Winter data are partially affected by snow.

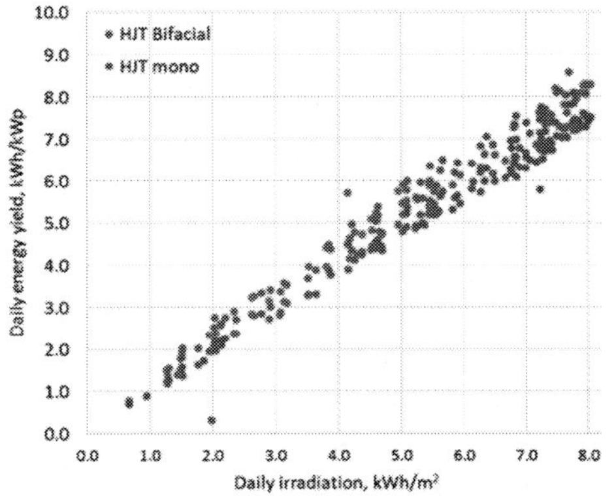

Figure 6: Comparison of the bifacial and monofacial HJT modules daily energy yield as a function of daily irradiation. Data from April to August 2017.

module design (back glass, encapsulant, and junction box type), weather conditions such as direct and diffused irradiance ratio, ground reflectivity (albedo), installation height and frame back shading of the module etc. In Fig. 6 daily energy yield of the two module types as a function of daily irradiation is shown. At any weather conditions there is substantial bifacial gain due to backside light collection. From Fig. 4(top), it is interesting to note that power yield of bifacial module exceeds that of monofacial modules at any time of the day, also at low light and grazing incidence conditions. At the coldest time, bifacial gain overcompensates a relative loss with respect to mc-Si module due to temperature coefficients (see Fig.4, bottom). In Fig. 7 we plot bifacial gain as function of daily irradiation for different season and ground conditions. On sunny days bifacial gain ranges from 7 to 12% with little difference between summer and winter. As irradiation goes down at cloudy or mixed weather conditions, bifacial gain is seen to increase reaching 15-20% and even 25% for strongly overcast days. This fact can be understood in terms of enhanced diffused light scattering conditions on such days. Our data show little dependence between different seasons in terms of reflective ground conditions with perhaps spring values characteristic of bare ground being somewhat lower than those of summer and winter.

The reported bifacial gain for HJT modules in literature is in line with our data. Bifacial HJT modules made by Meyer Burger installed in Lugano, Switzerland

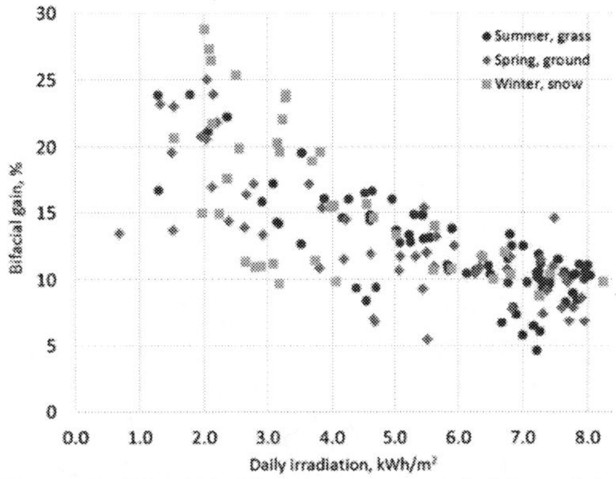

Figure 7: HJT bifacial gain over monofacial module design as a function of daily irradiation for different seasons and ground conditions.

Table 2: The STC indoor measurements after outdoor exposure of the HJT modules produced at Hevel.

ID	Initial data Pmpp, W	Exposed to 180 kWh/m² 1.5 months	Exposed to 800 kWh/m² 5 months	Delta, %
7	297	296	296	-0.3%
2	291	292	292	0.4%
3	282	283	283	0.4%
9	272	276	273	0.7%
5	261	263	260	-0.1%

showed 10-14% gain with respect to mc-Si modules [6]. Similarly, higher bifacial gain was observed for low irradiance cloudy conditions as compared to sunny days (13.6% vs. 16.7%).

The HJT modules fully manufactured at Hevel were remeasured indoor at the STC after 1.5 and 5 months of outdoor exposure. The respective data are presented in Table 2. No degradation (i.e. <1%) of the HJT module power was detected.

4. SUMMARY

In this work the annual outdoor performance of different HJT modules is evaluated in direct comparison with each other and with respect to the reference mc-Si module. Power gain of the HJT monofacial modules over mc-Si is shown to increase with the ambient temperature and reach +3.3% for the hottest periods at this monitoring site. Bifacial glass/glass HJT modules outperform monofacial HJT modules in all weather and seasonal conditions. In total, annual energy gain of 13% is obtained. On sunny days bifacial gain is of the order of 7-12% and increases at low light cloudy conditions up to 15-25%. The HJT modules fully manufactured at Hevel show no degradation after 6 months of outdoor exposure.

5. ACKNOWLEDGEMENTS

We thank Meyer Burger, Thun and Meyer Burger Research, Neuchatel for help in assembling the first batch of modules for the present study.

REFERENCES

[1] M. Taguchi, A. Yano, S. Tohoda, K. Matsuyama, Y. Nakamura, T. Nishiwaki, K. Fujita, E. Maruyama, IEEE Journal of Photovoltaics, 4, 96(2014).
[2] The International Technology Roadmap for Photovoltaic (ITRPV): http://www.itrpv.net/
[3] S. Abolmasov et al., "Feasibility Study on the Use of Gen5 PECVD Reactors for Manufacturing of High-Efficiency Silicon Heterojunction Solar Cells", Proc. of the 31st EUPVSEC, pp. 1046–1049, 2015.
[4] D. Andronikov et al., "A successful conversion of silicon thin-film solar module production to high efficiency heterojunction technology", paper 2AV.3.12, this conference.
[5] A. Faes et al., "Smart wire solar cell interconnection technology", Proc. of the 29st EUPVSEC, pp. 2555–2561, 2014.
[6] T. Soderström et al., "Low cost high energy yield solar module lines and its applications", Photovoltaic Specialist Conference (PVSC), IEEE 42nd, 2015. DOI: 10.1109/PVSC.2015.7356431.

COMPARISON OF BIFACIAL MODULE LABORATORY TESTING METHODS

Bonna Newman[1], Anna J. Carr[1], Koen de Groot[1], Nico Dekker[1], Bas van Aken[1], Ard Vlooswijk[2], Bas van de Loo[2]
[1]ECN Solar Energy, P.O. Box 1, NL-1755 ZG Petten, The Netherlands
[2]Tempress Systems, Inc., Radeweg 31, 8171 MD Vaassen

ABSTRACT: Bifacial modules, capable of additional power generation from the rear side, are a promising technology for increased energy yield compared to standard PV (monofacial) modules and with a low added cost. In order to provide customers and end-users confidence in the bifacial products now available on the market a clear laboratory test is needed that is both accurate and feasible with the current characterization equipment. A draft bifacial standard test has been proposed and suggests options for both full laboratory testing and characterization for modules from a manufacturing line. The authors conducted a survey of various parties in the PV value chain including institutes, module manufacturers, and equipment suppliers, gathering more information about the advantages and disadvantages of different test protocols and to use as a guide for recommendation for best practices after future laboratory evaluation.

Keywords: bifacial modules, characterization, LCOE

1 INTRODUCTION

Many module manufacturers are now offering bifacial modules, capable of additional power generation and annual energy yield due to light falling on the rear side, in addition to the standard monofacial modules. This is due to the increased production of p-PERC and n-PERT cell technologies that are amenable to bifacial applications. The new bifacial modules can be manufactured for similar costs as the monofacial variety but offer increased energy yield and therefore potential for lower levelized cost of electricity (LCOE) [2]. However, an official standard for characterizing and reporting performance for bifacial modules is not yet available although a draft standard has been proposed (IEC 60904) [1]. More uniform test definitions and conditions is of great importance for certification and comparison between laboratories and manufacturing locations. However, the added complexity of bifacial module testing and the attempt to simulate the outdoor environment in the laboratory space is a key challenge to widespread adoption of bifacial modules.

A bifacial module test standard should fulfil a number of useful criteria. It should be accepted by the community as an accessible and standardized way to measure performance of modules in a laboratory environment; it should measure the necessary parameters for quantifiable comparison between modules; it should result in parameters necessary for input for accurate energy yield modelling; and have a predictable relationship to outdoor performance in common applications (neglecting degradation). The purpose of this work is to evaluate various methods of indoor standardized testing of bifacial modules based on (1) community acceptance and opinions, (2) accuracy and ease of testing methods, and (3) utility for energy yield modelling and outdoor performance. In this contribution, we report on the first part, a survey conducted with participants from along the PV value chain to gather information about the community opinions on what is feasible, economic, and reasonable for bifacial module characterization. Further evaluations in the laboratory and outdoor will be reported in future publications.

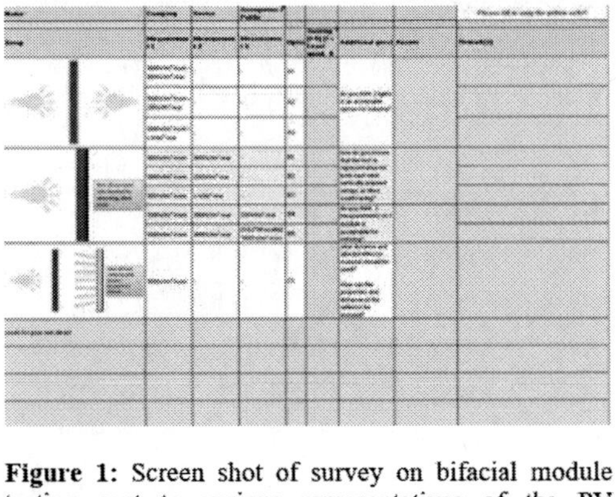

Figure 1: Screen shot of survey on bifacial module testing sent to various representatives of the PV valuechain.

2 BIFACIAL TESTING SURVEY

To gather input we surveyed various parties along the PV module value chain. We divided these parties into three types: module manufacturers, research institutes, and equipment suppliers. Many of the module manufacturers are also playing the role of system developers for many bifacial module systems today. While this end user group is not specifically included in this survey due to the technical nature of the testing methods, their opinions are also of significant value.

The survey is arranged as a worksheet with 3 basic equipment setups and variations of measurements that can be made with each of these equipment setups, resulting in nine specific test configurations. A screen shot of the survey is shown in Figure 1. The participants are asked to rate each proposed method on a scale of 1-5 with 5 as the best. In addition a few specific questions are also asked regarding the specific test setups. There was also space for the respondent to write-in other ideas and methods for bifacial testing. We received 13 responses from the survey with five representatives each from institutes and module manufacturers and three from equipment suppliers. In the following sections, we describe each of three equipment setups and the general responses and comments received.

2.1 Setup A: Two Light Sources

The first equipment setup requires two light sources that are flashed simultaneously (Figure 2). One will be on the front of the module and the other illuminating the rear

Figure 2: Schematic of Setup A, with two light sources illuminating the front and rear simultaneously. One measurement can be sufficient.

of the module. This setup has the benefit that it requires only one measurement but the logistics of timing two flashes, controlling the reflection of the environment, and the added cost of more equipment make it difficult. Industry commented that the current equipment is not suitable for this setup and requires significant and potentially expensive modifications. One respondent pointed out that the rear illumination could potentially be of lower quality in order to ease costs. Another drawback identified was the possible physical footprint of the test setup for uniform illumination.

There were three specific test setups that were rated as described in Table 1 . The general feedback was that method A1 would result in irrelevant values as a module would never be in this situation outside. Method A2 was considered a reasonable test but all respondents believed

Table 1: Setup A proposed methods. All methods are a single measurement with simultaneous front and rear side illumination. XX indicates a tunable or variable value where respondents could enter a suggested value.

	Front Source Power	Rear Source Power
A1	1000 W/m^2	1000 W/m^2
A2	1000 W/m^2	200 W/m^2
A3	1000 W/m^2	XX W/m^2

that tuning a flash tester to such low power might cause problems. Method A3 was favored among the first equipment setup with the idea that the rear illumination source would not be fixed in power but rather could vary from 100 W/m^2 to 300 W/m^2 to evaluate linearity and also provide useful input for annual yield modelling.

2.2 Setup B: Single Light Source and Multiple Measurements

The second equipment setup, shown in Figure 3, consists of only one illumination source making it more compatible with standard monofacial module testing apparatus most common today. In this setup, the non-illuminated side of the module, during any specific measurement is covered by a black absorbing material, such that all measurements are effectively monofacial. Therefore, bifacial data collection requires multiple measurements.

Different series of multiple measurements were proposed as summarized in Table 2. For B1, B4, and B5, the bifaciality factor (BF) can be calculated after the first two measurements M1 and M2, such that,

$$BF = P_{mpp}(M2)/P_{mpp}(M1).$$

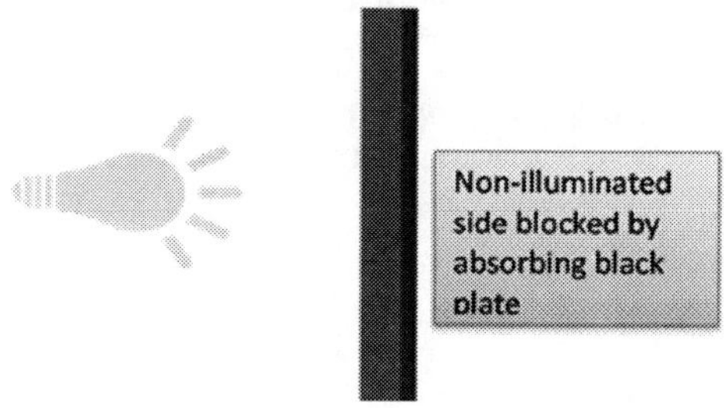

Figure 3: Schematic of Setup B with one light source and a non-reflecting black absorbing material is placed against the module on the non-illuminated side. Multiple measurements are necessary.

The bifaciality factor here is calculated in terms of maximum power, but a similar term can be calculated from measurements M1 and M2 for each of the module parameters of short-circuit current, open-circuit voltage, and fill factor. Each of these calculations could also be provided on a module specification sheet or used as input for modelling.

Overall, Setup B was preferred by many research institutions as it gives the most amount of data. However, the need for multiple measurements for each module made it impractical for most module manufacturers. Test B1 was somewhat more practical with fewer measurements and can be used to calculate performance in various installation configurations, such as fixed tilt (south), vertical east-west, or even tracking. However, the impact of higher current density in the module is not well characterized with B1. Therefore the results for half-cell modules would probably be more representative than the results for full-cell modules due to interconnection series resistance.

B2 and B4 both offer more realistic characterization of the rear side performance with diffuse albedo light, but suffer the issue identified in A2, in that it is difficult to get a uniform distribution and good spectrum at the low intensity of 200 W/m^2. For B3, various intensities were suggested ranging from 100 – 300 W/m^2.

Test setup B5 was identified as one of the best methods for testing (and is similar to the draft standard) but many module manufacturers suggested that three measurements

Table 2: Series of multiple measurements proposed for Setup B. Illumination intensity and illumination side (F – front, R – rear) is indicated. The bifaciality factor (BF), calculated from the ratio of M2 to M1, is used to determing illumination intesity in B5.

	Measurement		
	M1	M2	M3
B1	1 kW/m^2 –F	1 kW/m^2-R	-
B2	1 kW/m^2 –F	200 W/m^2-R	-
B3	1 kW/m^2 –F	XX W/m^2-R	-
B4	1 kW/m^2 –F	1 kW/m^2-R	200 W/m^2-R
B5	1 kW/m^2-F	1 kW/m^2-R	(1+0.2*BF)* 1 kW/m^2-F

on a single module is too many. An equipment supplier mentioned the fact that some module test equipment could do both measurements M1 and M3 at the same time.

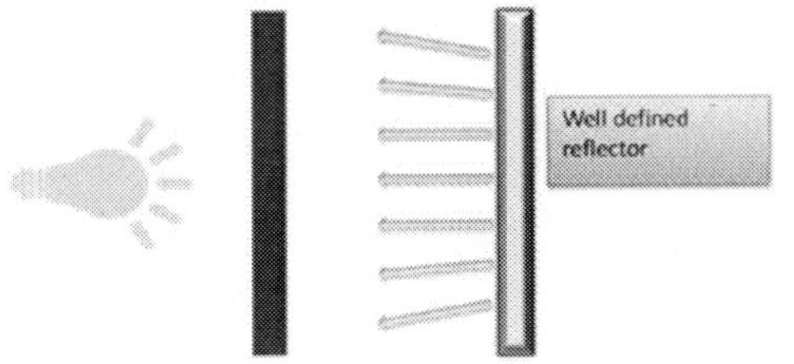

Figure 4: Schematic of Setup C with one light source illuminating the front of the module and a well-defined diffuse reflector behind the module at some distance.

Setup B also does not sufficiently account for self-shading on the rear of the module such as, large or misplaced junction boxes. In this case, the rear measurements and bifaciality factor would not reflect the real performance of the module. A significant shade on the rear of a bifacial module only contributes as a much smaller partial shade for the whole module when illuminated from the front under real outdoor conditions. In these cases, simulations of energy yield would likely be very inaccurate if based upon Setup B test results.

2.3 Setup C: Single Light Source and Defined Reflector

The third equipment setup consists of a single measurement from the front of the module with a diffuse reflector placed a specific distance behind the module with a known reflectivity, as shown in Figure 4. In this Setup, light that does not illuminate the module directly, or passes through the inactive and transparent areas of the module is reflected diffusely by the reflector behind and illuminates the rear side of the module at the same time as the front side and with only one measurement. This method was praised by respondents for potentially simulating the real outdoor environment of direct light, albedo, and self-shading. But it was also recognized that that ensuring uniformity of illumination, specifications of the reflector, and measurement setup would result in a difficult implantation and limit the useful comparison between various measurements on different setups. This would pose a large challenge to Setup C.

2.4 Write-in Suggestions

In addition to the scoring of the proposed tests A-C, the respondents were also asked for any other ideas. Many module manufacturers suggested that for each module bill of materials (including specific cell types and bifaciality), only one B1 or B5 type measurement should be made. For all following modules, a simple standard monofacial flash test could be used to predict module performance in bifacial conditions. For institutions, it was suggested that a B5 test should include a 4th measurement (M4) at lower illumination intensities, i.e. approximately 800 W/m^2 in order to establish the module characteristics with less than one sun. A third suggestion was also made that multiple measurements with equipment arranged in Setup C with varying distances between the module and reflector to simulate different installation and albedo intensity conditions.

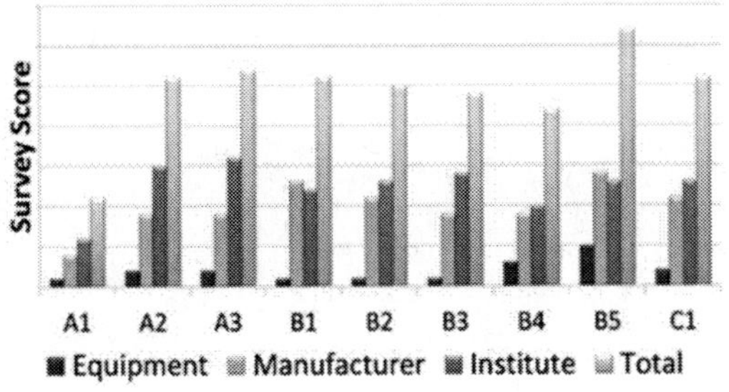

Figure 5: Histogram of survey results by respondent type and in total.

3 RESULTS AND CONCLUSIONS

A histogram of the responses from each sector and total is seen in Figure 5. Test setup B5 is preferred for almost all sectors and overall as it provides the most information. Research institutes also indicated a preference for A2 and A3, despite the measurement difficulty of two light sources as it would clearly mimic the outdoor performance in the most controlled way. Research institutes also scored Setup C highly as a possible way to clearly mimic real-world conditions. Manufacturers preferred B1 for a reduced number of measurements and B5 in order to gather more module performance parameters.

4 RECOMMENDATIONS AND OUTLOOK

It is crucial for the large-scale adoption of bifacial modules that a well demonstrated standardized test procedure is defined such that it is indicative of real world performance. It should also be a relatively simple test procedure that provides quantitative information for energy yield modelling and calculations. Finally, it should also be a relatively simple test procedure that can be reproduced at various institutions for comparison purposes.

Based on the survey results, the overall PV community is still uncertain of what test will be sufficient to address the various needs of each part of the value chain. For the end user and systems developers, the gain or value for bifacial modules is still often unclear [3]. LG has published a bifacial users guide explaining how different module characteristics and installation choices such as distance from the ground, albedo light of rear surfaces, orientation, and inverter or optimizer choice will impact the peformance in a bifacial system [4]. Module manufacturers prefer to use fewer measurements and then calculate peformance while institutes prefer either multiple measurements or methods that illuminates both sides of the module at the same time, most closely related to real-world applications. Overall, it is clear however that tracking the behavior of modules with various illumination intensities, including lower light behavior, is preferred by all respondents to verify the assumed linear response.

The results of the community survey for bifacial module characterization reported in this paper will provide a basis for a laboratory experiment to compare the highest ranked testing procedures as outlined above along with the proposed standard test method IEC 60904, Parts 1-2 Draft A [1]. Further, outcomes of these tests will be used for annual yield and performance modelling of bifacial systems [5]. These simulations and laboratory tests will be compared with outdoor bifacial module data in order to establish an empirical justification and to provide a clear recommendation for practical bifacial module and characterization in both industry and research.

5 REFERENCES

1] IEC 60904, Parts 1-2 Measurement of current-voltage characteristics of bifacial photovoltaic (PV) devices (Draft A), 2017.
[2] IRENA (2016), *The Power to Change: Solar and Wind Cost Reduction Potential to 2025*, ISBN 978-92-95111-97-4
[3] Private Communications with various PV installers and systems developers.
[3] LG Bifacial Module User Guide. Intersolar Munich Germany, 2017.
[4] Janssen, G.J.M.; Aken, B.B. van; Carr, A.J.; Mewe, A.A.; Outdoor performance of bifacial modules by measurements and modelling *Energy Procedia (Elsevier), 2015, Ed.77, p.364-373.*

A NOVEL HEAT DISSIPATING MATERIAL FOR ENHANCING THE PERFORMANCE OF PHOTOVOLTAIC PANELS

Min-An Tsai[1], Chien-Yuan Chen[2], Lian-Chih Chen[2], Yung-Tsung Chen[2], Chen-Hsing Liu[2], Hung-Sen Wu[1] , Teng-Chun Wu[1]
[1]Center for Measurement Standards, Industrial Technology Research Institute, Taiwan
[2]Big Green Environmental Technology. Co. Ltd., New Taipei City, Taiwan.

ABSTRACT: The Performance of PV panel has a strong dependent on the temperature of PV panel (More precisely is the temperature of the cell, whether it is made by Si, CIGS or other material). The performance can be enhanced by reducing the temperature of the PV panel. Here we provide a novel heat dissipating material, which is made of ceramic and metal powder. The substances are mixed with special solvent, and printed on backsheet to increase the dissipation area and the ability of thermal conductivity. In this work, we demonstrated the improved result of heat dissipation by using the novel material in the PV panel. And we also arranged the relative test of reliability of the novel material, such as the NOCT, Thermal Cycle, Damp Heat, UV exploration and the performance of system level, which will be shown in the future.

Keywords: Heat Dissipating,

1 Introduction

Photovoltaics represent a clean, silent, renewable alternative energy source, potentially beneficial to the environment by preventing greenhouse gases from being produced and entering the atmosphere. With increasing energy demand and environmental pollution, better utilization of solar energy is considered to be the best way for energy generation [1, 2]. In fact, only a small fraction of the incoming sunlight striking the cell is converted into electrical energy (a typical efficiency value for crystalline-Si cells is 14–17%) ; most of the absorbed solar radiation energy by solar cells will be converted into thermal energy and may cause the cells junction temperature to rise unless the heat is efficiently dissipated to the environment [3, 4]. The photovoltaic cell efficiency decreases with increasing cell temperature because of the negative temperature coefficient of crystalline-Si (about −0.4 ~ −0.5%/K). As the temperature of the solar cell increases, the band gap of the device decreases, resulting in an increasing short circuit current and a decreasing open circuit voltage, as reported previously. This phenomenon occurs because the thermally excited electrons begin to dominate the electrical properties of the silicon cells [5]. The detailed physics of the temperature impact on the performance of solar cells can be found in other references [6-8].

Additionally, decreasing the temperature of the module can increase the reliability of the PV module because some of the materials in PV modules are polymers, such as EVA (Ethylene-vinyl acetate), PET (Polyethylene terephthalate), PVF (Polyvinyl fluoride), PO (Polyethylene), and PVDF (Polyvinylidene fluoride), and these materials degrade rapidly with higher temperature, as reported in other references [9-12]. To decrease the temperature of the PV module, different types of approaches have been used, such as applying structured glass in the front side of the module and a V shape fin in the back side or filling the thermal conductivity filler in encapsulant to enhance the heat dissipation [11,13,19,20]. Other approaches include combining other systems, such as a thermal solar panel for heatingwater and a cooling PV module, or devices such as a thermoelectric device for generating electricity from the temperature difference between the airside and the PV module [21-27]. Here we provide a novel heat dissipating material, which is made of ceramic and metal powder. The substances are mixed with special solvent, and printed on backsheet to increase the dissipation area and the ability of thermal conductivity. In this work, we demonstrated the improved result of heat dissipation by using the novel material in the PV panel.

2 Experiments and details

To enhance the heat dissipation in the PV panel, the backsheet was printed by a highly thermal conducting material. To compare the impact of the highly thermal conducting material on the performance of the PV panel, one of the typical structures of a backsheet without novel material was used as a reference. The configuration of the PV panel for the backsheet with and without highly thermal conducting is shown in Fig.1.

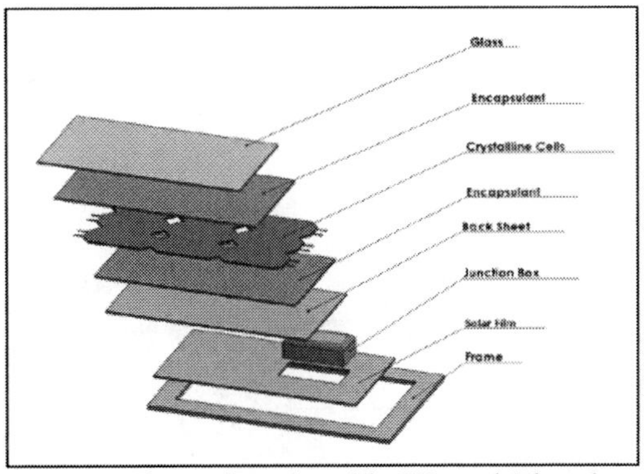

Figure 1: The configuration of the PV panel (The solar film reprensents the novel heat dissipating material)

The PV panels were put on the indoor light soaking system for one hour to compare the temperature and the performance. The figure 2 was shown the position of the temperature-sensors and the result. Furthermore, in order to make sure the additional material didn't influence the reliability of the PV panel, here also prepare the PV panel with novel material for UV 、Damp Heat and Thermal Cycle test according to IEC 61215.

3 Results

As explained in the introduction, illuminating the solar cell by a solar simulator induces a temperature rise in it, as shown in Fig. 2(T1~T3). The energy that is not used to generate electricity in the solar cell induces a temperature rise. The temperature of the solar cell increases as the exposure time increases. The position of T3 is near the junction box, so the temperature is influence by the solar cell and the bypass diode. Therefore the final temperature of the PV panel with and without novel material was similar. For T1 the heat-source is solar cell, the temperature of PV panel with novel material is low than without novel material. For T2 the effect of the reduce temperature is obvious, because the novel material show the good thermal diffusion to diffuse the heat energy to the edge. The different of temperature is about 10 degree.

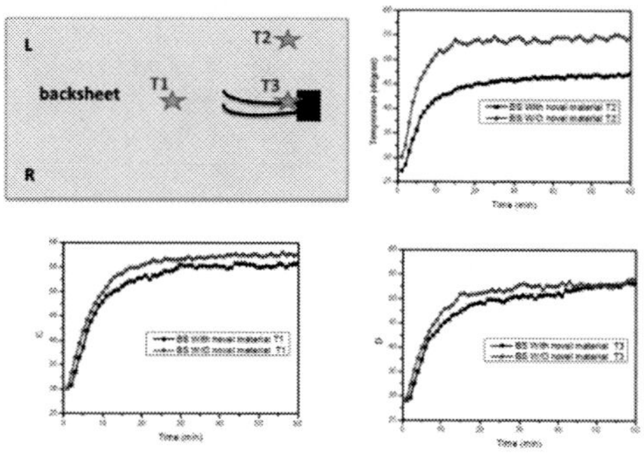

Figure 2: the comparsion of the temperature for PV panel with/without novel heat dissipating material in three different position(T1, T2, and T3).

For the performance of the PV panel, the maximum power decrease with the illumination time. Due to the temperature increased. So when the PV panel with the novel material shows the lower temperature compared with reference PV panel, the performance is show the same behavior, as show as figure 3.

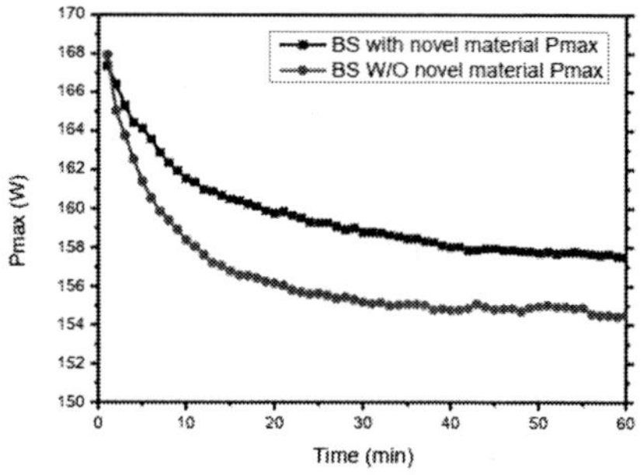

Figure 3: the comparsion of the performance for PV panel with/without novel heat dissipating material.

Here we also demonstrated the reliability of the novel material when it is combined with the PV panel, as showed as figure 4. The test result of the UV explosion(60 kWh/m^2) 、Damp Heat (1000 hours) and Thermal Cycle (200 cycles) were showed in table 1.

Table 1: Test results of the reliability

	UV 60	DH1000	TC200
degradation rate	-0.97%	-1.65%	-0.94%
	-0.45%	-1.74%	-1.59%

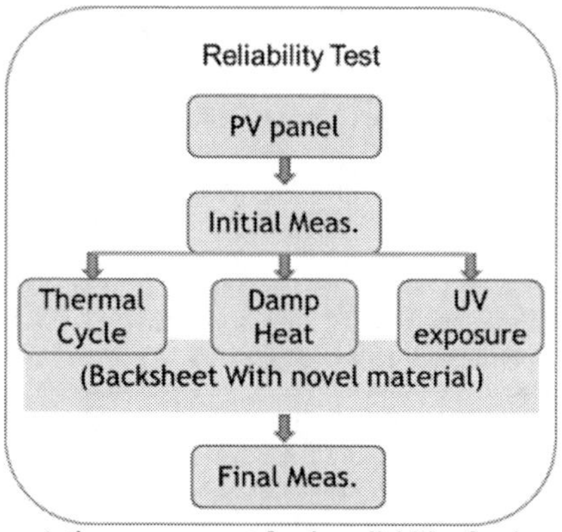

Figure 4. the arrangement for the reliability for the novel material.

4 Conclusions

The preliminary results shows that the maximum power of PV panel can be improved by reducing the temperature of the PV panel. The temperature is reduced by using our novel heat dissipating material which reveals a highly thermal conducting characteristic. The material have a strong weather resistant due to most of its ingredients are inorganic materials. The novel material have a highly potential to increase the performance of a PV panel, even the PV system.

5. References

[1] Sun S-S., Sariciftci N. S.: Organic photovoltaics: Mechanisms materials and devices. CRC Press, Boca Raton (2005).
[2] Bube R. H.: Photovoltaic materials. Imperial College Press, London (1998).
[3] Tripanagnostopoulos Y., Nousia T., Souliotis M., Yianoulis P.: Hybrid photovoltaic/thermal solar system. Solar Energy, 72, 217–234 (2002).
[4] Royne A., Dey C. J., Mills D. R.: Cooling of photovoltaic cells under concentrated illumination: A critical review. Solar Energy Materials and Solar Cells, 86, 451–483 (2005).
[5] Skoplaki E, Palyvos JA. On the temperature dependence of photovoltaic module electrical performance: a review of efficiency/power correlations. Sol Energy 2009;83:614-24.
[6] Armstrong S, Hurley WG. A thermal model for photovoltaic panels under varying atmospheric conditions. Appl Therm Eng 2010;30:1488-95.
[7] Segado PM, Carretero J, Sidrach-de-Cardona M. Models to predict the operating temperature of different photovoltaic modules in outdoor conditions. Prog Photovolt 2015;23:1267-82.
[8] Skoplaki E, Palyvos JA. Operating temperature of photovoltaic modules: a survey of pertinent correlations. Renew Energ 2009;34:23-9.
[9] Teo HG, Lee PS, Hawlader MNA. An active cooling system for photovoltaic modules. Appl Energ 2012;90:309-15.
[10] Bahaidarah H, Subhan A, Gandhidasan P, Rehman S. Performance evaluation of a PV (photovoltaic) module by back surface water cooling for hot climatic conditions. Energy 2013;59:445-53.
[11] Solanki CS, Sangani CS, Gunashekar D, Antony G.

Enhanced heat dissipation of V-trough PV modules for better performance. Sol Energ Mat Sol C 2008;92: 1634-8.

PERFORMANCE OF MULTI-BUSBAR PV MODULES

Yan Xie, Shu Zhang, Yifeng Chen, Hongwei Huang, Jianmei Xu, Zhiqiang Feng and Pierre J. Verlinden
State Key Laboratory of PV Science and Technology, Trina Solar
No.2 Trina Road, Trina PV Park, New District, Changzhou, Jiangsu, China, 213031, yan.xie@trinasolar.com

ABSTRACT: A detailed investigation of the reliability of PV modules using Multi-Busbar (MBB) technology with manual soldering interconnection process is presented. The MBB technology is compared to the conventional four-busbar (4BB) soldering process. The MBB technology increased the output power (P_{mmp}) of the PV modules by 2.9%. The Fill Factor (*FF*) of MBB modules is enhanced by 2.2% compared to that of 4BB modules. The relative I_{sc} gain for MBB module compared to 4BB modules is 0.9%. The corresponding shading factor of the wires is 53%. The results of qualification tests such as thermal cycling (TC200), humidity-freeze (HF10), damp heat (DH1000) and mechanical loading tests show that, compared to conventional stringing processes, MBB technology exhibits a comparable if not better reliability, thanks to better distributed stress. As a result of a lower series resistance, the results of a 4-month field test show that MBB modules have an improved energy production when the irradiance is higher than 450 W/m^2. Hence, the MBB technology is proved to be feasible and promising in mass production of PV modules.
Keywords: multi busbar, module, *I-V* properties, reliability, field performance

1 INTRODUCTION

Typical PV module assembly processes are based on solar cell interconnection followed by lamination using a protective cover, encapsulant and backsheet. Solar cell interconnection is conventionally performed by soldering 2 to 5 copper ribbons to the solar cell metallization grids. To achieve PV modules with higher efficiency and reliability, new kinds of interconnection technologies have been proposed. A low temperature conductive film interconnection technology combined with light capured ribbons (LCR) increased the module power by 1.83% [1]. The Multi-Busbar (MBB) technology appears as another novel cell interconnection technology to increase the module power.

MBB technology has gradually been a hot research in recent years due to the following advantages. First, the ribbons of the MBB modules are substituted by copper wires that requires narrower or no busbars on the cells and, therefore, results in reduced silver paste consumption. Second, the round shape of the wires used for interconnection in MBB modules generates additional reflection gains from the air/glass interface to the cell which brings on improved light absorption and improved generated current. Third, the large number of wires reduces the requirement for lateral transport which results in lower series resistance. Finally, a more homogenous busbar design of the cell leads to higher tolerance to cracks.

There are several approaches to achieve MBB interconnection such as soldering [2, 3], low temperature alloy lamination (Smart wire) [4] and the prefabricated interconnected grid method (Merlin). Among them, the MBB technology of soldering is the closest to the traditional cell interconnection technology. The current module equipment can be upgraded to assemble MBB modules. Schmid company was the first one to develop a special stringer machine for MBB modules. Now, several Chinese or Korean equipment manufacturers are developing this kind of tool. The main challenges of the MBB technology are: 1. maintaining a high power output and a high manufacturing yield due to the increased number of soldering points; 2. how to measure accurately the performance of MBB cells when the number of busbar increases.

In this paper, MBB cells were processed, characterized and sorted by dense probe cell tester, which was produced by Trina Solar. The sorted cells were assembled into MBB modules by manual soldering. MBB modules, composed of 60 cells, are compared to standard industrial 4 busbar (4BB) modules in terms of *I-V* properties, reliability and field performance. The results show great superiority of the MBB modules.

2 EXPERIMENTAL

All the cells were p-type 156×156 mm^2 industrial multicrystalline (mc-Si) cells with a thickness of 190 µm ± 20 µm. 1000 standard solar cells in each group were processed using two metallization patterns, one with 4 busbars for the reference and the other with 12 busbars for the new group. The front side pattern of the MBB cells is shown on Fig. 1. The busbar width of the 4BB cells was 1.0 mm. For the MBB front grid, each busbar contains 2 lines and 3 rectangular pads with 500 µm × 900 µm dimensions which were designed to enhance the quality of contact between the wires and fingers. 4BB cells were characterized by a standard industrial measurement frame while MBB cells were characterized by another measurement frame with a density of probes, which was produced by Trina Solar. To avoid mismatch, cells within a module were sorted according to their *I-V* parameters respectively.

Several PV modules were fabricated for each group. For module reliability testing, experiments were performed on modules containing 60 cells. The interconnection of MBB cells were processed by manual soldering while 4BB cells were processed by an automatic stringer machine. The ribbons applied for the 4BB cells and the wires for the MBB cells were both made of copper with a Sn63Pb37 solder coating made with a hot-dip process. For the 4BB cells, the applied ribbons have a rectangular cross-section with the dimension of 0.25 mm × 1.10 mm. The MBB cells were interconnected with 12 round wires with a diameter of 390 µm. The other materials (glass, encapsulant, backsheet, J-box, cable, connectors, ..) used in the module assembly were identical.

Before and after the reliability testing, the

performance of the modules at STC was tested using a Pasan solar simulator. The reference module used here is a mc-Si PV 4BB module with the same spectral response as modules in the experiment and has been calibrated by Fraunhofer ISE. The reliability testing was done according to IEC 61215.

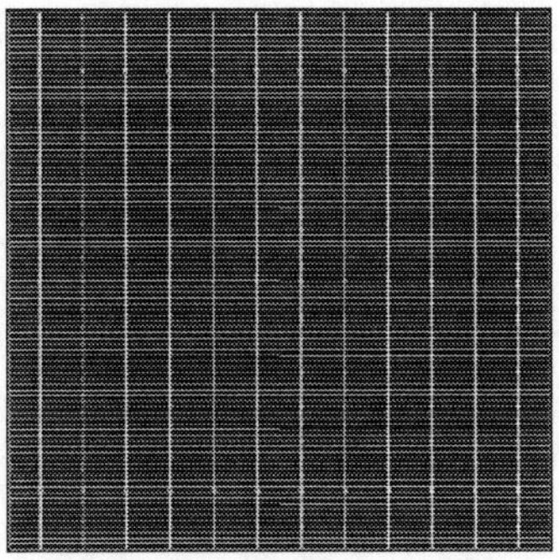

Fig.1: Front side pattern for MBB cells

3 RESULTS AND DISUSSION

3.1 Performance characteristics of cells

As mentioned in the introduction, the technology used to make MBB cells can save silver paste. In this experiment, the silver paste consumption of MBB cells was reduced by 20 mg (16.7%) compared to that of 4BB cells.

The median electrical parameters of both cell groups are shown in Table I. The short-circuit current (I_{sc}) and fill factor (*FF*) data of MBB cells are not comparable because we used another measurement frame with dense probe to measure MBB cells, creating additional shading on the cells. The open-circuit voltage (V_{oc}) data of MBB cells should also be a little lower than for the other group because of wide shading area of the probes.

Table I: Median IV parameters of the two groups of solar cells, containing 1000 cells each.

Group	Eff. (%)	V_{oc} (mV)	I_{sc} (A)	FF (%)	R_s (mohm)
MBB	18.21	631.5	8.795	79.77	1.6
4BB	18.36	633.1	8.894	79.36	2.1

3.2 Initial I-V characteristics of modules

The median electrical parameters of the two groups of PV modules are listed in Table II. As expected, the series resistance (R_s) is decreased with the increasing number of busbar. While the series resistance of the cells is reduced from 2.1 mohm to 1.6 mohm, a 23.8% reduction thanks to a significant reduction of the finger length, the series resistance R_s of MBB modules is 14.8% lower than that of 4BB modules. The decrease in R_s is close to 12.6% as similarly demonstrated by Hoon et al [5]. Due to the reduced R_s, *FF* of MBB modules is enhanced by 2.2% compared to 4BB modules.

The relative I_{sc} gain for MBB module compared with 4BB modules is 0.9%. Busbar shading of 4BB cell is 2.8% with 1.1 mm-wide ribbons. While for MBB cell, busbar shading is 3.0% by using 12 wires with a diameter of 390 μm. Thanks to the geometry of the round wire with MBB technology, this shading can be reduced to a lower level that results in the above I_{sc} gain. The measured corresponding shading factor of the wires is 53%, a bit higher than 35.7% as Braun et al [6] simulated. The shading factor can be further reduced to obtain more I_{sc} gain by optimizing the wire geometry and the wire reflectivity. This will be investigated in future work.

The open-circuit voltage V_{oc} of MBB modules is almost the same as 4BB modules due to the similar recombination in both cells.

Generally, all these above factors result in a 2.9% relative gain in output power P_{mmp} for the MBB technology.

Table II: The median electrical parameters of the two groups of PV modules, containing 10 pieces each.

Group	P_{mmp} (W)	V_{oc} (V)	I_{sc} (A)	*FF* (%)	R_s (ohm)
MBB	271.3	37.86	9.236	77.57	0.402
4BB	263.6	37.94	9.152	75.91	0.472

3.3 Thermal cycling test

Thermal cycling test is designed to determine the ability of the modules to withstand thermal mismatch, fatigue and other stresses caused by repeated changes of temperature. It is critical to verify the reliability of interconnections. During each TC cycle, the temperature cycle ranged from -40 to 85℃. Dethlefsen et al [7] did TC200 tests to MBB modules with 20 pieces cells, and no degradation was observed. The full size MBB modules were tested in this paper. The electrical parameter degradation of MBB modules and 4BB modules after TC200 tests are shown in Table III. The result shows that the MBB module is as reliable as 4BB module.

Table III: The median relative change in electrical parameters for two groups of PV modules after TC200 test (relative to initial parameters)

Groups	ΔP_{mmp} (%)	ΔI_{sc} (%)	ΔR_s (%)	ΔV_{oc} (%)	ΔFF (%)
MBB	-0.84	-0.31	0.47	-0.48	0.00
4BB	-0.80	-0.06	0.43	-0.39	-0.39

3.4 Humidity freeze test

The purpose of this test is to determine the ability of the module to withstand the effects of high temperature and humidity followed by sub-zero temperatures. This is not a thermal shock test. Humidity freeze testing is vital to evaluate the reliability of interconnections. During the testing, the relative humidity is maintained at 85% ±5% R.H. and the temperature is 85 ±2℃. The moisture can easily extend into the module, including the interconnection parts. When the temperature falls down to -40℃, the moisture freezes and expands, thus stress in interconnections appears. The results after HF10 tests are visible in Table V. The result shows that MBB modules suffer from almost the same power degradation ratio as 4BB modules.

Table V: Median relative change in electrical parameters of two groups of PV modules after HF10 test (relative to initial parameters)

Groups	ΔP_{mmp} (%)	ΔI_{sc} (%)	ΔRs (%)	ΔV_{oc} (%)	ΔFF (%)
MBB	-0.98	-0.42	3.68	-0.03	-0.55
4BB	-1.06	-0.51	4.08	-0.05	-0.50

3.5 Damp heat test

The purpose of this test is to detemine the ability of the module to withstand the effects of long-term penetration of humidity. The high amount of humidity for a long time may cause metal corrosion and contact failure. The high temperature accelerates permeation of water vapor into the module. DH1000 tests were conducted to both MBB and 4BB modules. The relative change in electrical parameters of the PV modules after DH1000 tests is illustrated in Table VI. Thus, the MBB technology was considered to be able to provide the highly reliable modules.

Table VI: Median relative change in electrical parameters of two groups of PV modules after DH1000 test (relative to initial parameters)

Groups	ΔP_{mmp} (%)	ΔI_{sc} (%)	ΔR_s (%)	ΔV_{oc} (%)	ΔFF (%)
MBB	0.11	0.19	-0.48	-0.08	0.00
4BB	-0.49	-0.67	-1.71	-0.11	0.26

3.6 Mechanical loading test

The purpose of this test is to determine the ability of the module to withstand wind, snow, static or ice loads. Table VII shows the results of ML test. It was observed that the MBB techonology was a reliable solution. This is easily understandable, as stress is better distributed for MBB technology.

Table VII: Median relative change in electrical parameters of two groups of PV modules after ML test (relative to initial parameters)

Groups	ΔP_{mmp} (%)	ΔI_{sc} (%)	ΔR_s (%)	ΔV_{oc} (%)	ΔFF (%)
MBB	-0.15	0.37	2.06	-1.32	-0.39
4BB	-0.49	-0.08	1.96	-1.58	-0.26

3.7 Field test

The field test of both MBB modules and 4BB modules was examined in Changzhou, China. After 4 months of continuous testing from September to December, it can be found that the average energy generation of MBB modules is 2.09% higher than that of 4BB. Figure 2 plots energy generation gain of MBB group for varying irradiance. According to the simulated curve, $y = 2\times10^{-05}x + 0.0155$ (Here define irradiance as x, and y means energy generation gain), we conclude that energy generation gain of MBB modules goes up with higher irradiance. The energy generation gain reaches to 3.6% when the irradiance is 1000 W/m^2. This is easily understandable since the power output of MBB modules measured at STC conditions (1,000W/m^2) is 2.9% higher than that of 4BB modules, mostly due to a reduction in series resistance. The benefit of MBB is mostly visible at high intensity due to lower series resistance loss. It is normal that at low intensity the energy generation is lower for MBB.

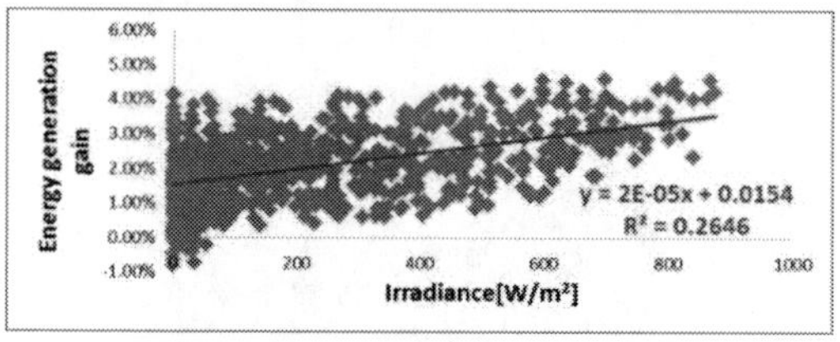

Fig.2: Energy generation gain with varying irradiance for 4months in Changzhou, China

3.8 Levelized cost of electricity (LCOE)

LCOE is a measure of a power source which attempts to compare different methods of electricity generation on a consistent basis. It is an economic assessment of the average total cost to build and operate a power-generating asset over its lifetime divided by the total energy output of the asset over that lifetime. The LCOE can also be regarded as the average minimum cost at which electricity must be sold in order to break-even over the lifetime of the project.

The LCOE is calculated based on the following assumptions. Firstly, we assumpt 2nd resource areas where total year irradiation is 586~670 $\times10^4$ kJ/cm^2•a. Secondly, the discount rate keeps to be 7% every year. Thirdly, there is a 0.05% energy yield gain per year due to high reliability.

As shown in Figure 3, LCOE of MBB module is 1% lower than 4BB based on 1000V system voltage. When the system voltage increases to 1500V, there is still 1% gap between 4BB and MBB modules.

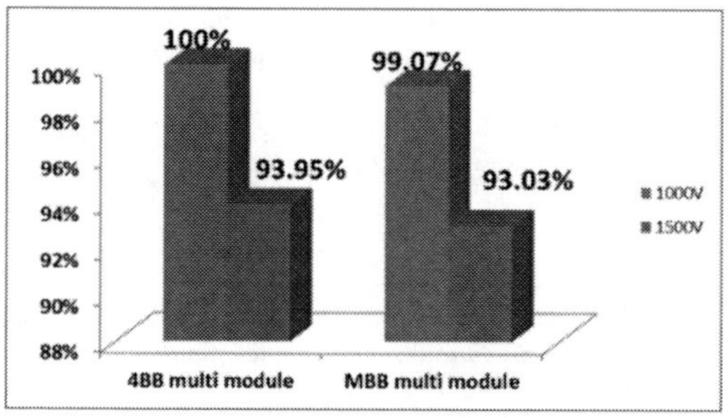

Fig.3: Normalized LCOE of the two groups of PV modules

4 CONCLUSION

It was demonstrated that the MBB technology shows its advantage on module level. The MBB technology by manual soldering increased the P_{mmp} of the PV modules by 2.9%. The result of a sequence of reliability testing shows that the power degradation of MBB modules is comparable to that of conventionally 4BB modules regarding thermal cycling, humidity ingress and thermal mechanical stress, demonstrating the high-reliability of MBB modules. The results of field test shows MBB module is 2.09% higher than that of 4BB. LCOE of MBB module is 1% lower than 4BB. Hence, the MBB technology is proved to be feasible and promising in mass production of PV modules.

ACKNOWLEDGEMENT

This work is supported by the National "863" Project under the Project Number of 2015AA050302 and the Changzhou Sci&Tech Program with the Grant Number of CC20150005.

REFERENCES

[1] S. Zhang, et al. Reliability of low temperature conductive film interconnection process for PV modules, 32th EU PVSEC, 1598-1602 (2016)
[2] S. Schindler, et al. Soldering process and material characterization of miniaturized contact structures of a newly developed multi busbar cell metallization concept, 28th EU PVSEC, 480-483 (2013)
[3] S. Braun, et al. Multi-busbar solar cells and modules: high efficiencies and low silver consumption, Energy Procedia, 334-339(2013)
[4] T. Söderström, et al. Smart wire connection technology, 28th EU PVSEC, 495-499 (2013)
[5] Multi-wire ribbon technology evaluation and contact design optimization for commercial application, 29th EU PVSEC, 218-220 (2014)
[6] S. Braun, et al. Solar cell improvement by using a multi busbar design as front electrode, Energy Procedia, 227-233(2012)
[7] A.F. Dethlefsen, et al. Reliability Testing of Multi-Busbar Modules, 29th EU PVSEC, 3323-3326 (2014)

5BV.4.45 ACCURATELY SIMULATING PV ENERGY PRODUCTION: EXPLORING THE IMPACT OF MODULE BUILD-UP

H. Goverde[1,2], D. Anagnostos[1,2,4], J. Govaerts[1,2], P. Manganiello[1,2], E. Voroshazi[1,2], K. Baert[2,3], J. Szlufcik1[2], F. Catthoor[1,2,3], J. Poortmans[1,2,3,5], J. Driesen[2,3]
[1]Imec, Kapeldreef 75, 3001 Leuven, Belgium
[2] EnergyVille, Dennenstraat 7, B-3600 Genk, Belgium
[3] KULeuven, Oude Markt 13 bus 5005, 3000 Leuven, Belgium [4] National Technical University of Athens (NTUA), Herron Polytechneiou 9, Zographou Campus 157 80 Athens
[5]University Hasselt, Martelarenlaan 42, 3500 Hasselt, Belgium
*corresponding author: hans.goverde@imec.be, Tel: +32-16-28 85 23, Fax: +32-16-28 10 97

ABSTRACT: A holistic approach for energy yield evaluation of photovoltaic (PV) modules has been developed. The optical, thermal and electrical behaviour of PV modules is integrated into the simulation framework, and as such is based on bottom-up and physics-based models. Incorporating a fine-grain temporal and spatial resolution, the framework enables accuracy and versatility even during highly varying and non-uniform weather conditions. Due to its physics-based nature, the approach can be extrapolated beyond its initial calibration. In contrast to black-box models, it can therefore be safely used for the energy yield exploration of novel PV module technologies and other system-specific (e.g. environmental) conditions. In this paper, an overview of current state-of-the-art energy yield modelling approaches will be discussed. The accuracy of our approach is compared to the other state-of-the-art approaches. Finally, several simulation results are shown where optical, thermal or electrical module characteristics are varied, to demonstrate the physics-based, exploratory simulation capabilities.

1 INTRODUCTION

The power output of PV modules depends on many parameters like module technology, system configuration, thermal state of the system, weather conditions etc. All those parameters need to be taken into account for an accurate evaluation of the energy production of PV modules. State-of-the-art energy yield evaluation approaches, reported in the literature, are able to estimate the energy yield with an accuracy between 5 – 20%, depending on the weather conditions and evaluation period [1]. Especially during days, which are characterized by highly varying weather conditions, the estimation error increases significantly. Next to that, most of the approaches described in literature use parametric black or grey-box energy yield models, and therefore cannot be reliably extrapolated beyond calibrated measured points and their intermediate ranges. Hence, they are not able to simulate other module technologies or system configurations than the ones for which the model was calibrated and optimized. To perform a true what-if exploration, e.g. to evaluate the potential LCOE reduction of a certain PV module technology variation, it is necessary to extrapolate.

In this paper, we give an overview of the state-of-the-art modelling approaches. After that, the developed modelling approach will be described and benchmarked versus the state-of-the-art approaches. In the last and major part of the paper, the effect of various encapsulants, replacement of the backsheet by glass and the effect of the temperature coefficient of a solar cell on the energy production of PV modules will be investigated. Those examples will demonstrate a number of the what-if exploration capabilities of this bottom-up, physics-based energy yield simulation approach.

2 COMPARISON TO THE STATE-OF-THE-ART

The current state-of-art modelling approaches can be divided into three main classes: bottom-up, circuit-based and parametric modelling approaches. Table 1 shows the strengths and the weaknesses of each class. The first type consists of mostly finite element modelling (FEM) that is used to simulate the optical, thermal and electrical properties of the PV module [2,3]. These approaches are complete and accurate, however, their major disadvantage is that the simulation requires a large computational effort (much slower than real-time). The second class which is circuit-based describes the thermal and electrical behaviour of PV modules [4,5]. These equivalent circuit-based models have limited exploratory possibilities and coupling between e.g. the electrical and thermal behaviour. The last and most common class is black- or grey-box modelling using an empirically-extracted module response to calculate the energy production [6,7]. This third class of models is also not capable of performing exploratory simulation, neither can they predict the energy production accurately for sub-optimal operational conditions (e.g. partial shading conditions).

Table 1: Overview of strengths and weaknesses of the different modelling classes.

	FEM modelling	*Circuit-based*	*Black or grey-box*
Extrapolation	+	-	--
Temporal variations	+/-	+	+
Non-uniform irradiance	+	+	-
Fast	-	+	++
Versatile	-	+/-	--
Physics – based	++	+/-	-
Accurate	++	+/-	+/-

The Electrical, Optical and Thermal (EOT) modelling approach used in this paper uses meteorological data (ambient temperature, irradiance and wind speed and direction) as input for the environmental conditions. Material properties (optical, thermal and electrical constants, thickness, etc.) and cell and module technology parameters (cell performance, temperature coefficients, EQE, etc.) serve as input to represent the PV module technology under evaluation. A full description and demonstration of this modelling approach can be found in [8,9].

In the approach, every solar cell in the PV module is modelled individually to be able to simulate the effect of partial shading on the energy production [10]. The individual sub-models are both thermally and electrically coupled to create the model of a full module. The optical part of the model calculates the light absorption, heat and carrier generation in each layer of the PV module. The result is used by the thermal and the electrical parts of the model. The thermal part of the EOT model consists of an RC equivalent thermal network that incorporates the heat capacity of each element and the conduction, convection and radiation of heat both within the module and from the module to the environment. Next to that, the thermal circuit also takes into account reduction of heat generation due to the dissipation of electrical energy in the electrical load and the thermal state of the PV module. Thermal resistances and capacities in the thermal circuit are equal to the physical value of each layer in the module. By incorporating the thermal capacity of each layer, thermal state effects are integrated. The thermal network is able to calculate the solar cell temperature as a function of illumination, ambient temperature, sky temperature, wind speed and direction, electrical operation point and thermal state. The well-known (1- or 2-) diode model is used to describe the electrical part of the PV module. Parameters of the diode model are being varied depending on solar cell temperature and illumination.

3 VALIDATION AND ACCURACY

Validation of our fully-coupled and physics-based energy yield simulation approach is performed on a residential PV system in Western European climate (full size module, facing South and 20° inclination angle and managed by Oldenburg University). Next to ours, also the PVsyst and Sandia models, applied according to best practice, were used to simulate the energy production of the PV system [6, 11]. The energy production of the system was calculated by the various models using the same weather input data: in-plane irradiance, ambient temperature, wind speed and wind direction. All the parameters were measured every second and these data were injected into the approaches. The results were based on 9 months of weather data recorded from Oldenburg. The relative RMSE difference for the complete 9 month dataset between simulated and measured energy production for different time intervals (from daily up to 10 s) is displayed by Fig. 1 [12]. Note that PVsyst only supports coarse-grain intervals up to 15 min .

The EOT modelling approach is clearly the best performing method for all the evaluation intervals. Due to the incorporation of correct wind behaviour, extracted from wind tunnel measurements and FEM CFD simulations, and incorporation of the thermal state, the approach is able to evaluate the energy production even during 10 s time intervals [13-15]. The error increase at shorter time intervals is mainly induced by the measurement uncertainties of e.g. the pyranometer. Larger evaluation intervals are less affected by these uncertainties due to averaging. Note that the Sandia model also achieved a relatively good accuracy. However, this method cannot be used for physics-based exploratory simulations due to the parametric modelling approach.

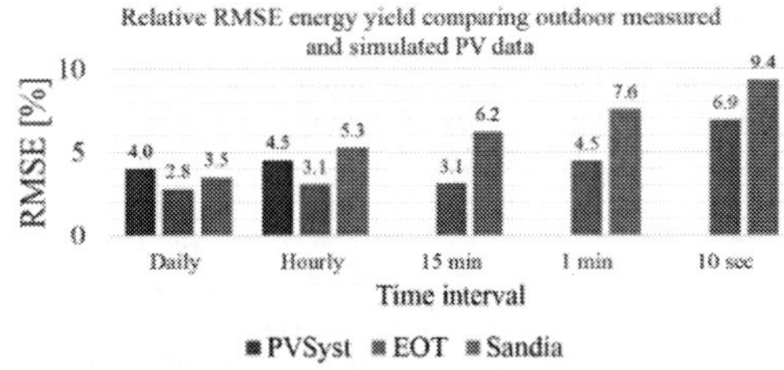

Fig. 1: The relative RMSE between the measured and simulated energy for different time intervals for the Oldenburg dataset.

5 RELATIVE STDV COMPARISON

In this section we want to demonstrate how our approach can be applied to do simulations for PV modules with alternative optical, thermal and electrical behaviour that requires extrapolation of the validated model.

A. Impact of a UV-transparent front-side encapsulant

The effect of a UV transparent front-side encapsulant on the energy production was simulated. As this layer transmits more light, the conversion efficiency increases. Ray-tracer simulations, in combination with one-cell laminated power measurements, were used to build and calibrate a model of a full size 60-cell PV modules (1.0 x 1.6 m^2, 6" full Al-BSF *c*-Si, white backsheet) with either a standard EVA encapsulant or a UV-transparent alternative. By replacing the standard EVA layer by a UV-transparent encapsulant, the power of the full module increases 1.8 % relative (Fig. 2).

The wave-length dependent optical properties of both the UV transparent and the EVA encapsulant were integrated into the modelling approach. The measured weather data, recorded in Oldenburg, was again used as input. Note that by injecting this data into the model, it was assumed that both modules were mounted in the identical configuration as the module used in Section III. The energy production of both modules for the 9 months of weather data is displayed in Fig. 2.

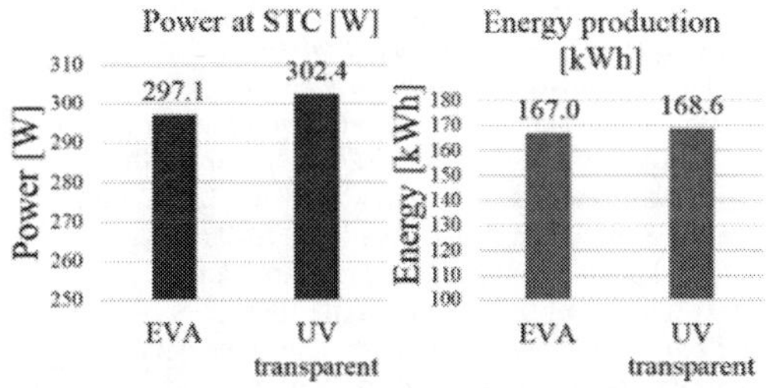

Fig. 2: (left) PV module power (at STC) for a white backsheet module containing UV transparent encapsulant or EVA. (right) Simulated energy production over 9 months of both module at Western EU climate

As expected, the energy production of the PV module with a UV-transparent encapsulant is higher compared to the one with EVA under an identical environmental profile. The additional UV light enhanced the photo-generated current which resulted in an increase of the energy production. However, despite a power gain of 1.8

% relative, the energy production increased only 1.0 % relative .

Due to the bottom-up modelling, our analysis also allows to get detailed insights in the underlying mechanism for these results. The deviation between the power and the energy production increase is caused by the additional heat generated in the solar cell. The energy of photons in the UV range is much bigger than the bandgap energy of silicon. When a charge carrier is created by those photons, it is injected well above the band gap (assuming one photon creates only one charge carrier). The charge carrier will thermalize back toward the bandgap energy, hereby releasing its excessive energy. Consequently, the solar cell heats up due to this energy, hereby reducing its conversion efficiency. Therefore, a power gain of 1.8 % relative will only result in an energy production increase of 1.0 % relative.

B. Impact of a glass-glass vs glass-backsheet build-up

Because of various reasons (e.g. improved reliability and for bifacial modules), more and more module manufacturers are focusing on the production of glass-glass PV modules instead of the traditional glass-backsheet build-up. It is expected that the thermal resistivity and conductivity of the glass are influencing the energy production as both values affect the thermal inertia, and hereby the thermal response of the PV module.

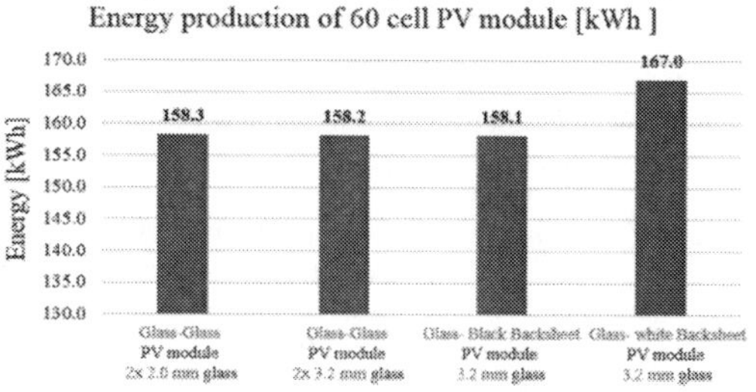

Fig. 3: Simulated energy produced by the various glass-glass and glass-backsheet PV modules with black rear encapsulant

To verify the effect of the thermal properties of a glass-glass module on the energy production, three different models of a full size module (60 cell 1.0 x 1.6 m, 6" full Al-BSF *c*-Si) were "constructed" assuming standard EVA as front encapsulant. The first one contained the thermal properties of a glass-glass module with 2x 2.0 mm thick glass plates and a *black* rear encapsulant, the second model 2x 3.2 mm thick glass plates and a *black* rear encapsulant and the last one a 3.2 mm glass on the front and *black* backsheet on the rear. The same data applied in the section above were used as input and the energy production of each module was calculated (Fig. 3).

One can see that the energy production is not significantly influenced by replacing the backsheet layer by a glass plate. The glass-glass module containing 2x 2.0 mm glass was the best performing module. Detailed analyses showed that the thermal response is mainly dominated by the effect of wind. Therefore, the variation of the thermal properties has only a minor impact on the *annual* energy production. However, during highly-varying weather conditions, the *short-term* energy production is significantly affected by the thermal properties of the glass as shown in [9]. This conclusion can vary in different installations and climate conditions (work in progress).

In addition the effect of a black vs. white backsheet is also displayed in Fig. 3. Due to the increased reflectance and less parasitic heat generation in the back layer, a PV module containing a white backsheet produces 6.6 %rel more energy. Further analysis learns that this increase is mainly caused by the photo-generated current gain due to the additional reflection. As the operating temperature was mainly dominated by the wind effects, no significant deviation in temperature was observed in these conditions.

C. Impact of cell technology on energy yield

It is well known that the conversion efficiency is strongly influenced by the operating temperature of the solar cell. However, this temperature dependency can vary between – 0.41 and -0.29 %rel/°C depending on the solar cell technology [16]. The effect of this temperature coefficient on the energy yield was investigated by developing various models for PV modules containing different cell technologies. These simulations demonstrate the capabilities of the modelling approach to perform also electrical optimization simulations. The measured Oldenburg weather data was used again as input.

As the different cell technologies vary considerably in performance in this section the energy *yield* is considered, meaning that the energy production of one kW_p was investigated. Thus, the energy production was normalized according to their conversion efficiency. Using this method, the effect of the temperature coefficient on the annual energy yield was investigated (Fig. 4). For these modules front glass (3.2 mm), white backsheet and standard EVA encapsulant are assumed.

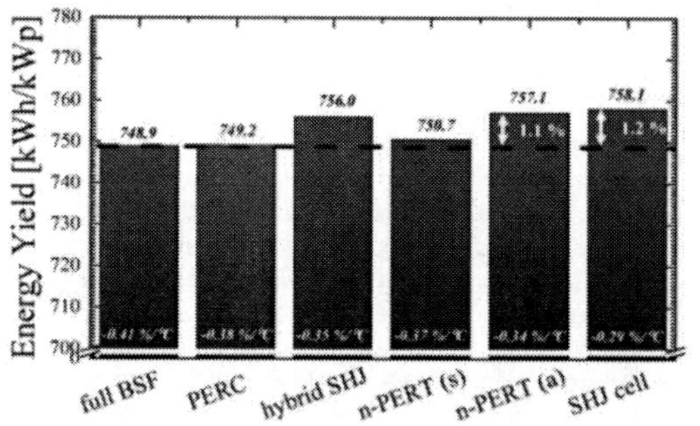

Fig. 4: Simulated energy yield for the various solar cell technologies. The full Al-BSF solar cell was taken as reference.

According to the simulations, the PV module containing SHJ solar cells will have the highest energy yield. This technology will produce 1.2 % relative more energy compared to the full Al-BSF solar cells. This energy yield gain is related to the temperature coefficient, as this was the only parameter varied. The conversion efficiency of the SHJ cell was less reduced at elevated operating temperatures, consequently leading to the highest energy yield when subjected to equivalent weather conditions. Further analysis learns that for the Oldenburg location, the energy yield will increase 70 kWh/kW_p in case the efficiency temperature coefficient would decrease 1.0 % relative.

D. Other applications and future extensions

In all of the described simulations, weather data recorded in Oldenburg were used as input of the modelling approach. Obviously, other weather conditions recorded at different climates can be injected and the energy production can be simulated as a function of those conditions. Currently, the implementation of other climates is being developed and will be available soon.

Thanks to the bottom-up approach, the modelling approach can and will be extended by including physics-based degradation mechanisms. Due to this extension, the approach, in combination with manufacturing, financing and operating costs, can then be used for detailed, exploratory, bottom-up LCOE calculation. It will serve as a tool, which will not only optimize the LCOE by including cell and module technology, but also will optimize the LCOE according to location-specific electrical load profiles, environmental conditions including wind cooling, and degradation behaviour, hereby clearing the path towards an extensive regional optimization of PV modules. This feature is clearly not available in current state-of-the-art modelling approaches.

6 CONCLUSIONS

In this paper, a holistic, accurate and bottom-up energy yield modelling approach was presented. State-of-the-art simulation methodologies were discussed as well as the accuracy of the approach demonstrating that it was the most accurate for all evaluation intervals. The proposed modelling approach was able to estimate the energy production accurately during intervals up to 10 s.

The capability of performing exploratory simulations was demonstrated by the a number of use cases in which optical, thermal and electrical properties were varied. Those specific models were then used to investigate the impact of the variation on the energy production. In this paper, the specific properties were varied separately. However, the approach is also capable of simulating a large number of combinations, in order to find a global optimum in function of e.g. the climate. This feature is not accessible by the other models and experimentally testing all possibilities requires too much time and resources.

7 ACKNOWLEDGEMENTS

This project receives the support of the European Union, the European Regional Development Fund ERDF, Flanders Innovation & Entrepreneurship and the Province of Limburg. The authors like to acknowledge the funding agencies and the industrial partners of the imec Industrial Affiliation Program in SiPV and KFAS under project code: P115-15EE-01 for their support and for providing the materials used in this work. Furthermore, the authors gratefully acknowledge B. Aldalali, D. Goossens and Oldenburg University for their support and advice.

8 REFERENCES

[1] G. Makrides, B. Zinsser, M. Schubert, G. Georghiou, *Energy Yield Prediction Errors and Uncertainties of Different Photovoltaic Models*, Progress in photovoltaics: research and applications, 21, 500-516

[2] B. Kubicek, K.A. Berger, *"Ab initio simulation of the pv energy production dependence on module materials and layering"*, 26th EUPVSEC, 272-274, 2011

[3] H. Tian, F. Mancilla-David, K. Ellis, E. Muljadi, P. Jenkins, *"A cell-to-module-to-array detailed model for photovoltaic panels"*, Sol. Energ., 86, 2696-2706, 2012.

[4] V.L. Brano, A. Orioli, G. Ciulla, A.D. Gangi, *"An imporved five-parameter model for photovoltaic modules"*, Solmat, 94,1359-1370, 2010.

[5] A. Mermoud, T. Lejeune, *Performance assessment of a simulation model for PV modules of any available technology"*, 25th EUPVSEC, 4786-4791, 2010.

[6] C.W. Hansen, D.M. Riley, M. Jaramillo, *"Calibration of the Sandia array performance model using indoor measurements"*, 38rd IEEE PVSC, pp. 3067-3072, 2012.

[7] W. De Soto, S.A. Klein, W.A. Beckman, *Improvement and validation of a model for photovoltaic array performance*, Sol. Energ, 80, 78-88, 2006.

[8] H. Goverde, B. Herteleer, D. Anagnostos, G. Köse, D. Goossens, B. Aldalali, J. Govaerts, K. Baert, F. Catthoor, J. Driesen and J. Poortmans, *"Energy Yield Prediction Model for PV Modules Including Spatial and Temporal Effects"*, 29th EUPVSEC, 2014

[9] H. Goverde, D. Anagnostos, B. Herteleer, J. Govaerts, K. Baert, B. Aldalali, F. Catthoor, J. Driesen and J. Poortmans, *"Model requirements for accurate shot term energy yield predictions during fast-varying weather conditions"*, 31st EUPVSEC, Germany, 2015

[10] P. Manganiello, M. Baka, H. Goverde, T. Borgers, J. Govaerts, A. Van Der Heide, E. Voroshazi, Francky Catthoor, *Bottom-up energy generation simulation to compare PV module topologies under non-uniform and dynamic operating conditions,* submitted to IEEE PVSC 2017

[11] PVsyst photovoltaic software, available at http://www.pvsyst.com/en/

[12] D. Anagnostos, H. Goverde, B. Herteleer, F. Catthoor, D. Soudris, J. Driesen, J. Poortmans, *"Demonstration and validation of an energy yield prediction model suitable for non-steady state non-uniform conditions"*, Proceedings 6th WCPEC, 2014

[13] H. Goverde, G. Van den Broeck, D. Anagnostos, B. Herteleer, J. Govaerts, D. Goossens, E. Voroshazi, K. Baert, F. Catthoor, J. Driesen, J. Poortmans, *Impact of Wind on Intra-Module Energy Yield Variations*, Proceedings 32nd EUPVSEC, pp. 1665-1668, 2016

[14] H. Goverde, D. Goossens, J. Govaerts, V. Dubey, F. Catthoor, K. Baert, J. Poortmans and J. Driesen, *"Spatial and temporal analysis of wind effects on PV module temperature and performance"*, Sustainable Energy Technologies and Assessments, 11, 36-41, 2015

[15] H. Goverde, D. Goossens, J. Govaerts, F. Catthoor, K. Baert, J. Poortmans, J. Driesen, *"Spatial and temporal analysis of wind effects on PV modules; consequences for power evaluation"*, Accepted to Solar Energy, 2016

[16] J.P. Seif, J. Haschke, J. Cattin, S. De Wolf, L. Tous, P. Choulat, M. Aleman, E. Cornagliotti, A. Uruena, R. Russell, F. Duerinckx, J. Szlufcik, A. Abdallah, B. Aissa, M.M. Kivambe, N. Tabet, *"Impact of solar cell architecture on the temperature dependency of electrical performance"*, 32nd EUPVSEC, 2016.

THE EFFECT OF NON-UNIFORM TEMPERATURE DISTRIBUTION IN PV CELLS AND THEIR INTERCONNECTIONS

Petr Wolf[1], Vítězslav Benda[2]
[1]University Centre for Energy Efficient Buildings, Czech Technical University in Prague
Třinecká 1024, 273 43 Buštěhrad, Czech Republic
[2]Faculty of Electrical Engineering, Czech Technical University in Prague
Technická 2, 166 27 Prague 6, Czech Republic

ABSTRACT: The paper is focused on effects caused by non-uniform temperature distribution in photovoltaic (PV) arrays. The influence of non-uniform operating conditions on electrical parameters has been studied on both individual solar cells and on their parallel interconnection. To determine the temperature effects, simulations, as well as measurements on crystalline silicon PV cells, has been carried out. The results can be used to give an estimation on how the temperature would affect the overall electrical yield. This can be useful in designing a PV system where the non-uniform temperature can occur in an effective way.

Keywords: Energy performance, Non-uniform, Photovoltaic, Solar Cell Efficiencies, Temperature gradient

1 INTRODUCTION

It is generally known that the efficiency of photovoltaic cells drops with increasing temperature. As the temperature effects mostly the cell voltage, non-uniform temperature distribution becomes critical when the cells are connected in parallel [1]. Similar effects that occur at interconnected cells can be observed within a single cell if operated with a temperature gradient. A high-temperature gradient within a cell is typical for concentrating PV systems due to non-uniform irradiation and cooling. For such systems, the non-uniformity is mostly expressed in form of Gaussian function or its combinations and the problems have been discussed in [2], [3]. Non-uniformity in temperature distribution affects the PV system in two ways: (1) cells may experience an efficiency loss, (2) temperature variation induces thermal fatigue. This causes further irreversible damage to solar cell due to excessive localized heating and reduces the reliability of the system [4].

2 SCALE OF INSPECTION OF NON-UNIFORM TEMPERATURE

There are two effects of power loss caused by increased temperature: (i) common power loss due to temperature increase and (ii) power loss due to MPP mismatch. This applies to all situations where the non-uniform temperature can occur: within PV generators, PV modules or single solar cells. For many applications of non-concentrated PV systems, the mean temperature of the whole PV generator with a minimal error of the power output can be assumed. When high non-uniformity of temperature is suspected, both power losses (i), and (ii) should be considered for simulation to achieve accurate results.

On PV generator scale, non-uniform temperature can occur due to dissimilar heat transfer. Upper modules are surrounded by preheated air; therefore they are operating at a higher temperature. The same situation is typical for modules situated in the middle of PV array which have worse heat transfer to the ambient air. The situation is more critical in the case of building integrated systems with no free ventilation on the back side of modules. For such cases, it might be more suitable to consider a temperature profile across the module (still considering uniform temperature on cells). The non-uniform temperature on the scale of PV cells becomes critical in case of concentrated PV systems.

3 PARALLEL INTERCONNECTION OF PV ELEMENTS

The aim of this work was to inspect the effect of common power loss due to temperature increase and power loss due to MPP mismatch separately.

The first case, presents a parallel connection of PV elements operating at different temperature (each of the element has a uniform temperature). These elements can, in general, be either PV cells, PV modules or interconnected modules in a series (strings). To assure the high precision of experimental validation we have focused on simulating and measuring two parallel connected cells. This can be done on a small experimental setup where irradiance can be kept uniform and temperature easily adjusted.

The simulation was based on getting the initial data by measuring the I-V curves of two 4'' c-Si cells in the range of 30 – 100 °C with 5 K step. Pasan Cell Tester IIc was used; measuring was performed at 1000 Wm^{-2}. The obtained data was smoothed and interpolated on the I-V curve as well as in the temperature range. As a result, high-resolution I-V curves with 1 K temperature resolution were obtained. Simulation of the parallel operation of two cells on different temperature was done using curve superposition technique, presented in [5] for simulation of a partially shaded PV array. As no other elements like bypass diode were used, the technique is based on simply adding the current of each cell at the same voltage point

$$I(V) = I_1(V) + I_2(V), \quad (1)$$

where I and V states for the parallel connected current and voltage, I_1 and I_2 for the current of single cells 1 and 2, respectively.

Fig.1 shows the simulation results. As expected, the power loss due to temperature increase was linear, indicating 0.229 % / K (half of the typical c-Si temperature power coefficient, as only one cell, was considered to be heated). Power loss due to MPP mismatch was found to be insignificant in case of low-temperature difference. At a temperature difference of

40 K, the power loss due to MPP mismatch was over 5 % and showed the same slope as the power loss due to temperature increase. The overall loss can be easily estimated as the sum of the two types of power losses, Fig. 2.

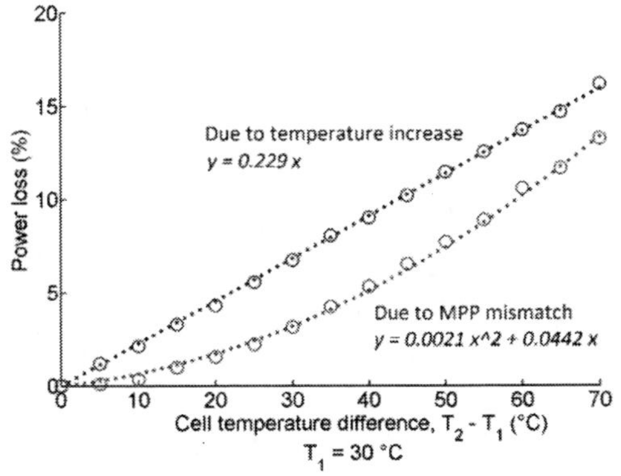

Figure 1: Efficiency drop of two parallel connected cells of different temperatures

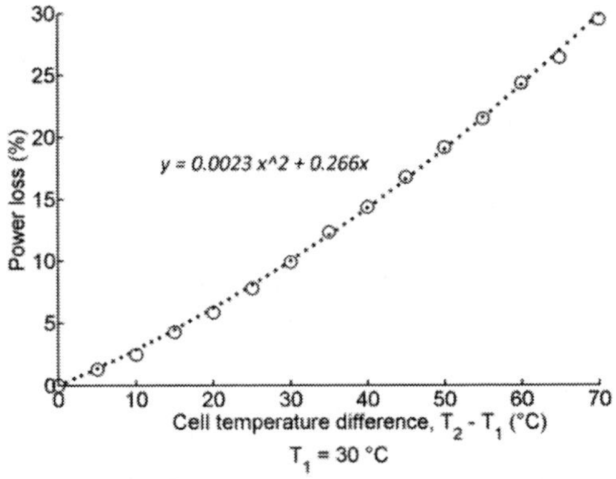

Figure 2: Overall power loss of two parallel connected cells of different temperatures

High-temperature difference rarely occurs on non-concentrating PV systems. On the other hand, the problems caused by MPP mismatch should not be neglected on many types of installations, such as building integrated photovoltaic (BIPV) systems. As found in [1] on a back-ventilated PV façade the temperature difference on modules can reach up to 20 K. This can according to our estimation lead to 1.5 % power loss caused by MPP mismatch, (Fig. 1).

The verification of the estimation of power loss caused by parallel connection of cells operating under different temperatures was validated. Two cells were connected in parallel and one of them was heated. A good agreement with the simulated result was obtained. At temperature difference of 37 K, simulation showed 4.66 % of power loss due to MPP mismatch, where almost 4.3 % was obtained experimentally.

4 NON-UNIFORM TEMPERATURE WITHIN ONE SINGLE PV CELL

The model of a PV cell operating with temperature gradient was based on the assumption that the I-V curve can be estimated as a parallel interconnection of cell elements, each working at a different well-defined uniform temperature. The basis of the methodology can be found in [6]. The experimental setup consisted of a stainless steel, gold-coated plate, heated from one side and cooled from the other side. The PV cell was mechanically attached on the top of this plate. On the contrary to previous measurements published in [6], this setup was improved and the temperature was measured using the infrared camera (Testo 882). This enabled to get a high-definition 2D temperature profile of the PV cell surface, Fig. 3. The emissivity of used 4'' c-Si cell was set by comparing with contact measurement (0.85 on active cell parts and 0.11 on busbars and fingers).

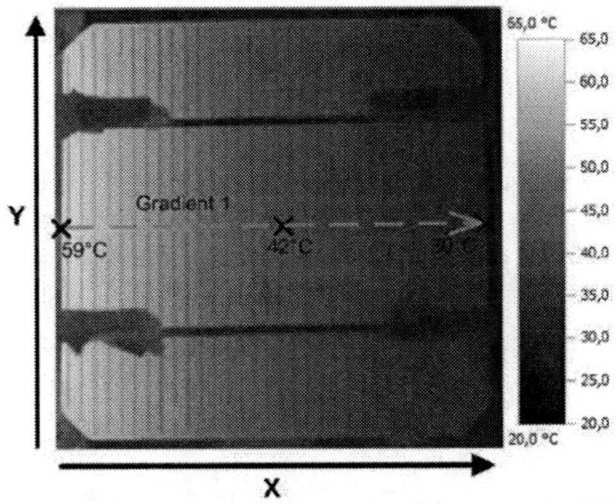

Figure 3: Temperature distribution within one cell with high-temperature gradient

Two temperature gradients were obtained: Gradient 1 (59 – 30 °C), Gradient 2 (81 – 36 °C). As the thermovision proved high uniformity of temperature on the Y-axis (maximal difference on Y-axis was found 1.5 K), for the simulation only the temperature profile across X-axis was considered. The temperature was calculated based on relative position on X-axis using the 2nd-degree polynomial function, as a good match to the surface temperature was found, Fig 4.

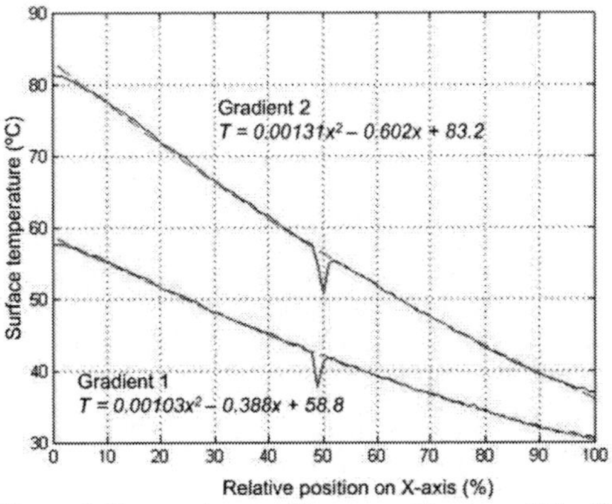

Figure 4: Temperature profiles across X-axis on PV cell

As the next step, the simulation was performed based on the method of parallel interconnection of cell slices. This method was similar to the method described in Section 3. For simulation purpose, the cell was divided into 100 parts (slices), each of them was assumed to have a uniform temperature. To get the I-V curve of the cell, these pieces were simulated as being parallel connected. The power loss due to MPP mismatch was found 2.3 % and 4.4 % for Gradient 1 and Gradient 2, respectively.

As seen in Fig. 5 and also in Fig. 6, using this simulation method we got an I-V curve and MPP that fitted much better to the measured results than by using single mean temperature only. The error of estimated power of PV cell was reduced from 2.3 % to 1.2 % (Gradient 1), from 4.4 % to 1.2 % (Gradient 2).

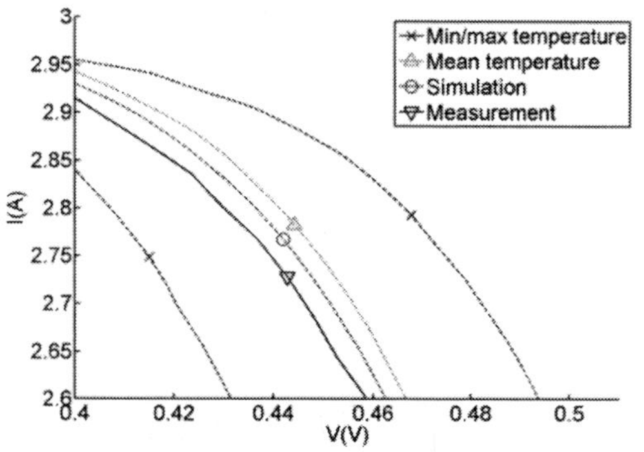

Figure 5: Measured and simulated I-V characteristics (Gradient 1)

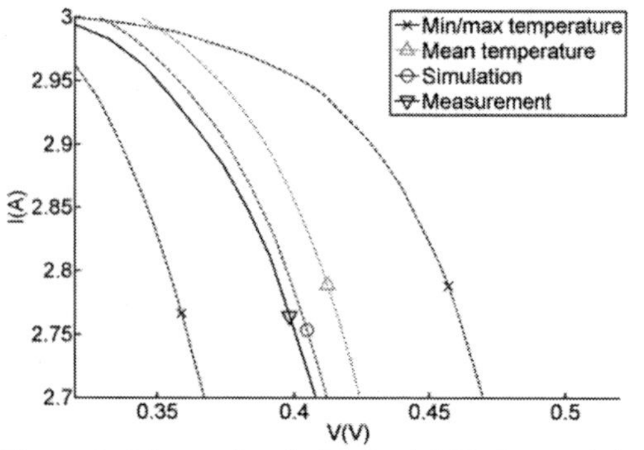

Figure 6: Measured and simulated I-V characteristics (Gradient 2)

5 CONCLUSION

Non-uniform distribution of temperature on the PV generator causes power and yield losses. There are many situations, where these losses cannot be neglected as they have a similar impact on the performance of PV systems like other losses (caused by the uniform increase of temperature, electric losses in cables and interconnections, ageing,...).

To find out the relevance of the MPP mismatch losses caused by non-uniform temperature, two study cases were chosen: non-uniform temperature on two parallel interconnected cells (Section 3) and single cell with a temperature gradient (Section 4). In both cases, simulations based on the addition of current of parallel interconnected parts together with measurement validations were performed.

When two cells with different temperature were connected in parallel (first case) it was found that power losses by MPP mismatch were negligible by low temperatures, but became quite significant when the temperature difference was high (at 15 K temperature difference causes almost 1 % power loss by MPP mismatch).

In the second case, it was found by simulation that in case of the high-temperature gradient using the mean temperature for power estimation leads to wrong results (inaccuracy of 2.3 % and 4.4 % in case of Gradient 1 and Gradient 2, respectively). By using the method of parallel interconnection of parts with the same temperature it was possible to estimate the power of the cell with much higher precision than by using only the mean temperature.

6 ACKNOWLEDGEMENTS

This work has been supported by the Ministry of Education, Youth and Sports within National Sustainability Programme I (NPU I), project No. LO1605 - University Centre for Energy Efficient Buildings – Sustainability Phase and by the Operational Programme Research, Development and Education of the European Structural and Investment Funds, project CZ.02.1.01/0.0/0.0/15_003/0000464 Centre for Advanced Photovoltaics.

7 REFERENCES

[1] Benda, V., Macháček, et al. Effect on Non-uniform temperature distribution in the photovoltaic array on efficiency of photovoltaic systems. In ISES Solar World Congress: Conference proceedings, 2009.

[2] Domenech-Garret, J.L. Cell behavior under different non-uniform temperature and radiation profiles using a two dimensional finite element model. Solar Energy, Vol. 85, 2011, s. 256-264.

[3] Franklin, E., Coventry, J. Effects of Highly Non-uniform Illumination Distribution on Electrical Performance of Solar Cells. In 40th ANZSES Conference: Proceedings, Newcastle, Australia, 2002.

[4] Bahaidarah, H.M.S., Baloch, A.A.B. et al. Uniform cooling of photovoltaic panel: A review. Renewable and Sustainable Energy Reviews, Vol. 57, 2016, 1520-1544.

[5] Patel, H., Agarwal, V. MATLAB-based modelling to study the effects of partial shading in PV array characteristics. In IEEE Trans. Energy Convers. 23(1), pp. 302-310, 2008.

[6] Wolf, P. - Benda, V. - Macháček, Z. The effect of surface temperature distribution in PV cells on the overall power generation. In MedPower: Proceedings. Thessaloniki, Greece. 2008.

COMPARISON OF OPTICAL GAINS AND ELECTRICAL LOSSES IN MODULES WITH DIFFERENT DESIGNS OF PARTIAL CELLS IN DESERT REGIONS

Hamed Hanifi[1,2]*, David Daßler[1], Marko Turek[1], Jens Schneider[1], Joerg Bagdahn[1]
[1]Fraunhofer-Center for Silicon Photovoltaics CSP, Otto-Eissfeldt-Str. 12, 06120 Halle (Saale), Germany
[2]Anhalt University of Applied Sciences, Bernburger-Str. 55, 06366, Koethen, Germany
*Corresponding author: +49 345 5589 5515, hamed.hanifi@csp.fraunhofer.de

ABSTRACT: Modules with partial cells have shown better performance compared to full-cell modules due to reduced electrical losses and increased optical gains. In this paper, we have compared different module concepts of full-cell, half-cell and third-cell with different tab widths of 0.8 mm and 1.5 mm in moderate and desert conditions. First, we present the statistical evaluation of indoor measurements results of the modules. The modules were measured with a mask which limits the spacing of the cells and keeps the total inactive area of the mini-modules constant. The efficiency and fill factor of the modules are evaluated to analyze the electrical and optical losses of the modules. We show that the modules with partial cells show better fill factor and efficiency compared to the similar full-cell modules. The modules were measured under different irradiation levels to evaluate the performance of the modules under different light conditions. Based on the measured data, the energy yields of the modules are simulated in Morocco and Germany as desert and moderate climates. We show that extra energy yield can be harvested by choosing the module concept and optimized tab width for each climate. The optimized half-cell modules benefit of 4.12% and 4.97% extra annual energy yield compared to full-cell modules in moderate and desert climates correspondingly. These values increase respectively by transition to third-cell modules to 5.2% and 5.68% at moderate and desert environments.

Keywords: Partial cells, Half-cell module, Third-cell module, Desert Module, PV Module, Module Manufacturing, Ribbons, Design, Electrical Properties, Experimental Methods, Optical Losses

1 INTRODUCTION

Modules in desert regions benefit from high irradiation levels which lead to higher energy production. However, high irradiation leads to excessive electrical losses in form of heat in the cells and module interconnections [1, 2]. The electrical losses in interconnection have a quadratic relation with the current and thus the irradiation level. This means that reduction of nominal short-circuit current of solar cells by cutting them can decrease the electrical losses and consequently the temperature of the modules [1, 2, 3]. Although the cutting process decreases the efficiency of the solar cells [4, 5] , this loss will be overcompensated after module fabrication due to extra optical gains and reduced electrical losses [1, 3].

Increased market share of modules with partial cells from almost 2% of the share of module production in 2016 to about 5% in 2017 shows the attraction of these modules especially for various kinds of applications such as shading resistant modules and high power modules [6, 7, 8]. Based on the International Technology Roadmap for Photovoltaics, the market share of modules with partial cells is expected to be increased to 40% by 2027 [7].

In this work, we compare the electrical characteristics and efficiency of mini-modules with full-cells, half-cells and third cells to investigate the performance of these modules under high and low irradiation levels. The modules have optimized and standard tab dimensions to compare the optical and electrical performance of the modules. The module area is kept constant for all module concepts while the spacing between cells was modified. In the next step, we have compared the energy yield of such modules in Morocco and Germany to show the advantages of modules with partial cells in desert regions.

2 EXPERIMENTAL SETUP

The modules were fabricated of mono crystalline ALBSF cells with 3 busbars of very narrow width. The solar cells were cut with an optimized laser process to half- and third-cells. The solar cells have been characterized before and after cutting process by using the LOANA cell measurement system. The measurement device combines I-V-measurements with spectral resolved analysis such as quantum efficiency measurements and spatially resolved analysis such as light-beam-induced-current and luminescence imaging.

In the next step, the mini-modules were fabricated using standard materials as shown in Table 1.

Used materials	
Cell type	Full-cells (156×156 cm^2) Half-cells (156×78 cm^2) Third-cells (156×52 cm^2)
Tab width	0.8 mm 1.5 mm
Laminate	Bridgestone EVA SKY S88 Krempel Backsheet Interfloat rolled solar glass with no anti reflection coating (200 mm × 200 mm × 3.2 mm) Bruker-Spaleck interconnecting tab with SnPb coating

Table 1: Overview of used materials and their dimensions for mini-module fabrication

The tab width for each concept of module is chosen based on the optimized tab width for half-cell and full-cell modules [1].

Figure 1: Module layout with (left) full-cell, (middle) half-cells, (right) third-cell with the covering mask

Figure 2: A third-cell module with 0.8 mm (left) and 1.5 mm (right)

The total spacing between the cells inside each laminate was considered to be constant. The modules were fabricated in 200 mm × 200 mm laminates. Since the total inactive area of the module is to be constant, the spacing between the cells connected in series is varied from 6 mm for the full-cell module to 3 mm for the third-cell module. After module fabrication, the modules were characterized electrically with a mask with the inner dimension of 168 mm × 162 mm. By using the mask, it is expected that mini-module represents the performance of a large module. The spacing from the sides of the mini-module is considered to be constant as 3 mm for all mini-modules. The modules were characterized under 100, 250, 500, 750 and 1000 W/m^2 to estimate the performance of each concept under different irradiation for energy yield calculation purposes. The module layout for each concept of modules follows as Table 2.

Module type	Spacing	Tab width	Quantity
Full-cell	6 mm	0.8 mm	5
		1.5 mm	5
Half-cell	4 mm	0.8 mm	5
		1.5 mm	5
Third-cell	3 mm	0.8 mm	5
		1.5 mm	5

Table 2: Module layout and quantity for each concept of cells size

3 RESULTS

3.1 Indoor measurement

The optical losses in module increase linearly by increasing the tab width [1]. The tabs with 1.5 mm and 0.8 mm cover 2.88% and 1.53% of the active area of a full size cell respectively. The extra spacing between partial cells is expected to influence the current generation in the relevant modules [3].

Due to the same thickness and length of the tabs, the electrical resistance of them has an inverse relation with the tab width. The electrical losses through the tabs have a square relationship with the current passing through the tabs.

$$P_{ele}[W] = \rho[\Omega mm]\frac{L_{tab}[mm]}{W_{tab}[mm] \times d_{tab}[mm]} I_{tab}^2[A] \quad (1)$$

Where P_{ele} is the electrical losses through each tab, ρ is the electrical resistivity, L_{tab}, W_{tab} and d_{tab} are the length, width and thickness of the tab respectively and finally, I_{tab} is the current passing through tabs. It should be mentioned that current passing through tab increases linearly over its length. Therefore, total electrical losses of the tabs laid on the cells differ from the losses in the spacing tabs [1, 3]. The theoretical electrical losses for the half-cell module and third-cell module are expected to be four and nine times less compared to the similar full-cell module.

Figure 3 demonstrates the measured fill factor values of 30 modules covered by mask under standard test conditions. The fill factor increases from half-cell to third-cell due to the reduced current through tabs. Furthermore, wider tab width shows better fill factor due to the lower total impedance.

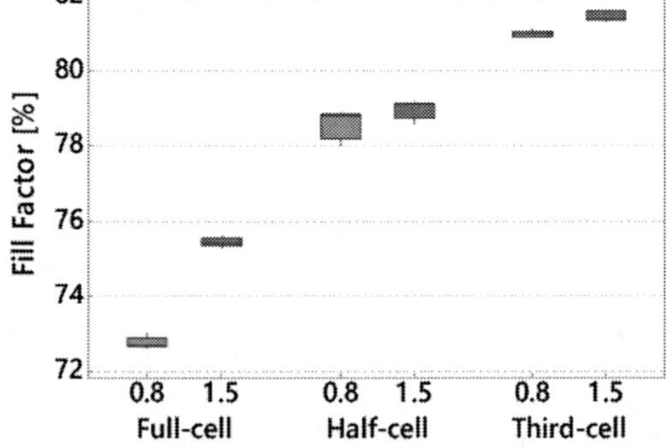

Figure 3: Measured fill factor values of 30 mini-modules with mask under standard test conditions

Figure 4 demonstrates the measured efficiency of the modules under standard test conditions. The efficiency of modules increases systematically by cutting the cells to half and three parts respectively. Full-cell modules with wider tab width show better efficiency compared to the efficiency gained from the module with narrower tab width due to lower Ohmic losses and better balance between optical and electrical losses of the mini-module. However, half-cell and third-cell modules with narrower tab width show better efficiency compared to the modules with wider tab width due to lower current passing through tabs and consequently lower electrical losses. There is an absolute gain in median efficiency of 1% from optimized full-cell module to half-cell module and 1.4% from optimized full-cell module to third-cell module with narrower tab width.

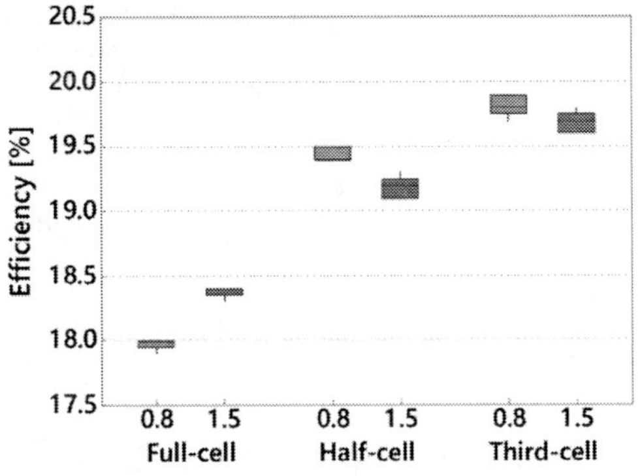

Figure 4: Measured efficiency values of 30 mini-modules with mask under standard test conditions

Figure 5 shows the average efficiency of full-cell, half-cell and third-cell modules with the 0.8 mm and 1.5 mm tab widths under low light, mid light and high irradiation conditions.

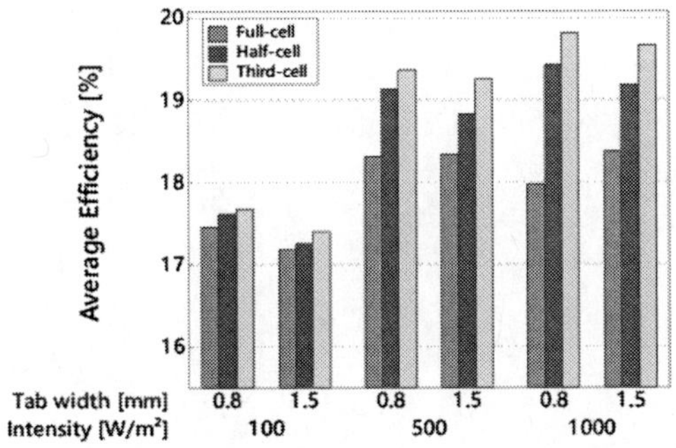

Figure 5: Average efficiency comparison of three different module concepts with 0.8 mm and 1.5 mm tab widths under different intensity conditions

The same trend of better performance of half-cell to full-cell and third-cell to half-cell concept follows by all intensity ranges. Regarding to the tab widths, half-cell and third-cell modules with narrow tab widths show systematically better performance under both high intensity and low light conditions. The reason is that the optical losses are the main loss mechanism in modules with partial cells in both low light condition and high illumination level. On the other hand, full-cell module with 1.5 mm tab width show better performance compared to the same module with 0.8 mm at high irradiation levels while at low light conditions, the module with 0.8 mm have better efficiency compared to the wider tab width. The reason for this change is a turnover between optical and electrical losses as the dominant loss mechanism. At high irradiation cases, the major loss mechanism in full-cell modules is the electrical losses and therefore, the wider tab width show better efficiency. On the other hand, at low light conditions, optical losses are the main loss mechanism and module with narrower tab width shows better performance due to the less coverage of active area and low electrical losses compared to high irradiation conditions.

3.2 Energy yield calculation

The energy yield of all samples was simulated for Germany as moderate climate and Morocco as desert climate by INSEL software. The average hourly irradiance over three years from 2012 to 2014 for both climates is simulated under five different irradiation classes from 100 W/m² to 1000 W/m². The number of hours with related irradiance within these classes is determined.

Figure 6 shows the energy yield simulation results for full-cell, half-cell and third-cell module concepts with 0.8 mm and 1.5 mm tab widths in Morocco as desert environment and Germany as moderate climate. The calculation is based on the measured weather data in Berlin and Marrakech within three years from 2012 and 2014.

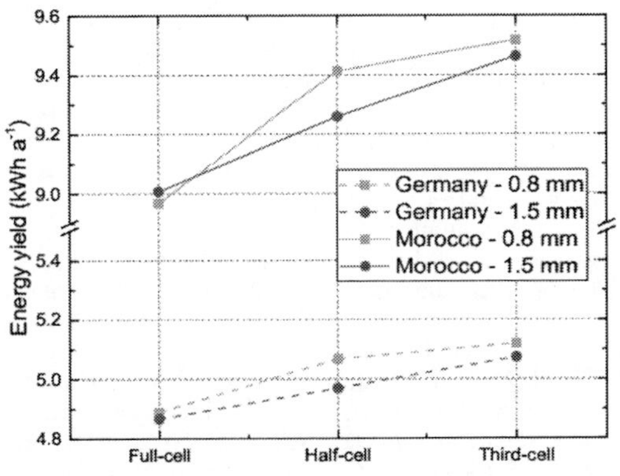

Figure 6: Energy yield calculation of the three module concepts with different tab widths between 2012 and 2014 in Morocco as desert climate and Germany as moderate climate

Comparison of the energy yield results of Morocco and Germany show an average of 45% lower harvested energy in Germany compared to Morocco for all similar modules.

The results show similar trend as Figure 4 for energy yield under high irradiation level. The energy yield for full-cell module with 1.5 mm tab width in Morocco is higher than the similar module with 0.8 mm due to high electrical losses in the ribbons. On the other hand, the energy yield of the full-cell module with 0.8 mm in Germany is higher than similar module with 1.5 mm due to the change of loss mechanism from electrical losses to optical losses in low light conditions. This shows that the tab width for full-cell modules in moderate climates can be reduced to half which leads to considerable reduction in material consumption and costs. By changing the module concept from full-cell to half-cell, the modules with 0.8 mm show higher energy yield in both Germany and Morocco since the optical losses are the dominant loss mechanism for this concept of module in both moderate and desert climates. The similar trend is valid for third-cell module. The rate of increase in energy yield of third-cell module is slightly lower compared to half-cell module. The optimized tab width of third-cell module could be narrower than 0.8 mm but due to the fabrication limits and width of the busbar print on the solar cell, production of such module was challenging.

The optimized module with half-cell concept shows relatively 4.12% and 4.97% extra annual energy yield compared to the optimized full-cell module in moderate and desert environment respectively. The relative extra energy yield of the optimized third-cell module compared to the optimized full-cell module at moderate and desert climates are 5.2% and 5.68% correspondingly.

4 CONCLUSION AND OUTLOOK

In this paper, we show that modules with partial solar cells benefit of increased module efficiency due to the reduced electrical losses and increase optical gains. The electrical losses and consequently the efficiency of modules can be optimized by optimization of module interconnections.

The losses of cell cutting are overcompensated by after module production.

The behavior of modules under different irradiation conditions are different and full-cell modules with narrower tab width show better performance in low light condition compared to high irradiation level due to the change of the dominant loss mechanism from electrical losses to optical losses. This means that in places with moderate climates the tab width of full-cell modules can be reduced to 0.8 mm which leads to less material consumption and higher energy yield.

Modules with partial cells show higher efficiency and power compared to full-cell modules in any irradiation levels. This gain in efficiency 1% and 1.4% by moving from optimized full-cell module to optimized half- and third-cell modules respectively.

The half-cell modules benefit of 4.12% and 4.97% extra annual energy yield compared to full-cell modules in moderate and desert climates correspondingly. These values increase respectively by transition to third-cell modules to 5.2% and 5.68% at moderate and desert environments.

These solar cells used for this paper were 3 busbar solar cells and a comparison between modules with partial cells and cells with more busbars will be investigated in the future.

5 ACKNOWLEDGEMENTS

The authors thank the Federal Ministry for Economic Affairs and Energy (BMWi) and Federal Ministry of Research and Education (BMBF) for financial support under contracts of "Wuestenmodule" with reference number of 03FH014IX4 and "CTM100+" with the reference number of 0324033.

6 REFERENCES

[1] H. Hanifi, D. Dassler, J. Schneider, M. Turek, S. Schindler und J. Bagdahn, „Optimized tab width in half-cell modules," *Energy Procedia,* 2016.

[2] S. Malik, D. Dassler, J. Fröbel, J. Schneider und M. Ebert, „Outdoor Data Evaluation of Half- / Full-Cell Modules with Regard to Measurement Uncertainties and the Application of Statistical Methods," in *EU PVSEC*, Amsterdam, 2014.

[3] S. Guo, J. Schneider, F. Lu, H. Hanifi, M. Turek, M. Dyrba und I. M. Peters, „Investigation on the short circuit current increase for PV modules using halved cells," *Solar Energy Materials and Solar Cells,* pp. 240-247, 2014.

[4] S. Eiternick, K. Kaufmann, J. Schneider und M. Turek, „Loss Analysis for Laser Separated Solar Cells," *Energy Procedia 55C,* pp. 320-324, 2014.

[5] S. Eiternick, F. Kaule, H.-U. Zühlke, T. Kießling, M. Grimm, S. Schoenfelder und M. Turek, „High Quality Half-Cell Processing Using Thermal Laser Separation," *Energy Procedia,* Bd. 77, pp. 340-345, 2015.

[6] H. Hanifi, J. Schneider und J. Bagdahn, „Reduced Shading Effect on Half-cell Modules – Measurement and Simulation," in *31th European Photovoltaic Solar Energy Conference and Exhibition*, Hamburg, 2015.

[7] „International Technology Roadmap for Photovoltaics," IRTPV, September 2017.

[8] H. Hanifi, J. Schneider und J. Bagdahn, „Reduced shading effect in half-cell modules - measurement and simulation," in *31st EUPVSEC*, Hamburg, 2015.

CELL TO MODULE GAINS FOR HIGH EFFICIENCY BACK CONTACT CELLS

Nicolas Guillevin[1], Bonna Newman[1], Evert Bende[1], Lars Okel[1], Mark Jansen[1], Nico Dekker[1], Wilma Eerenstein[2]
[1] ECN Solar Energy, P.O. Box 1, NL-1755 ZG Petten, The Netherlands
[2] Exasun B.V., Laan van Ypenburg 122, 2497 GC Den Haag, The Netherlands

ABSTRACT: Back contact cells architectures, especially Interdigitated Back Contact (IBC), offer the potential for the highest efficiencies, best aesthetics, and application flexibility. Foil-based back contact module interconnect for these architectures has been shown to be a high throughput, industrially viable, and reliable solution for both IBC and Metal Wrap Through (MWT) cells. By collecting current from distributed points on the cell, foil reduces series resistance in the module interconnection, leading to lower cell-to-module (CTM) FF loss. This specific feature makes the conductive foil the interconnection technology of choice for high current back contact solar cells. In addition, the capacity of the conductive foil to tolerate higher Isc in the module with limited series resistance increase offers the opportunity to further improve module power based on light management techniques and increase of the module Isc. In this work, we focused on increasing the CTM Isc and power gains by improving capture of the light inside the module by using a reflective "Intra-Module" Foil (IMF) placed in between back contact cells interconnected with back contact foil. We present results on single-cell and 2x2 back contact laminates manufactured with IBC and MWT cells interconnected based on ECN's conductive foil technology. CTM power gain above 5.1% is demonstrated with Isc gain more than above 8% on 2x2 laminates.

Keywords: IBC, MWT, modules, CTM, materials, light management

1 INTRODUCTION

In standard PV modules, interconnection of cells into modules results in a decrease of both power and total area efficiency. These cell-to-module losses are typically a result of resistance losses (fill factor) from the cell interconnections while the efficiency is decreased due to the existence of inactive areas in between the cells and around the edges of the module. Typical cell-to-module losses, measured on the power output, are on the order of 2%-3% and efficiency losses are often 5% or greater [1].

Back contact cells architectures, Interdigitated Back Contact (IBC) and Metal Wrap Through (MWT), offer the potential for the highest efficiencies, best aesthetics, and application flexibility. However, the higher currents in these modules also increase the risk of higher fill factor losses. Back contact foil interconnection technology is based on a metallized back sheet that is patterned with two different electrode areas to interconnect all back contact cells. Among other benefits, back contact foil interconnection improves on other back contact module interconnection technology as it is not dependent on the cell to carry current. Thereby higher currents can be generated in the cell and efficiently extracted without increased series resistance when multiple cells are interconnected in series and laminated into modules. This increases the module fill factor and decreases CTM losses. Back contact foil module interconnect technology for high efficiency cell architectures has already been shown to be a high throughput, industrially viable, and reliable solution for all back contact cells [2].

Use of back contact foil and the ability to carry higher currents with fewer resistive losses also allows optical optimization of a back contact module. A highly reflective intra-module foil (IMF) can be placed between cells and along the edges in order to reflect light incident on inactive module areas back onto the high efficiency cells. With high reflectance, module efficiency losses can be countered by making inactive module areas more active. Similar materials have been integrated into standard H-pattern stringing and tabbed modules [3] but have not been applied to back contact modules.

In this work, we focused on increasing the CTM by gaining current I_{sc} by improving capture of the light inside the module by using a reflective IMF placed in between the back contact cells connected by back contact foil. We present results on single-cell and 2x2 back contact laminates manufactured with IBC and MWT cells. Through the combination of the IMF and back contact foil technology for module interconnection, we demonstrate greater than 5% CTM power gains and decreased cell to module efficiency losses for 2x2 mini-modules.

2 OPTIMIZED LIGHT MANAGEMENT

Inactive areas make up more than 5% of the module

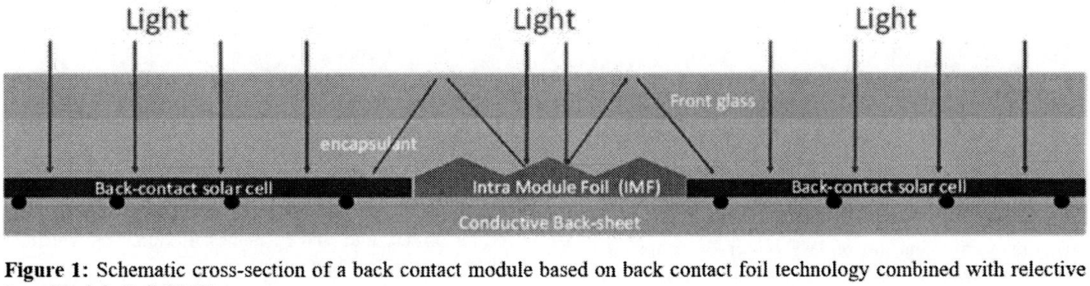

Figure 1: Schematic cross-section of a back contact module based on back contact foil technology combined with relective Intra-Module Foil (IMF).

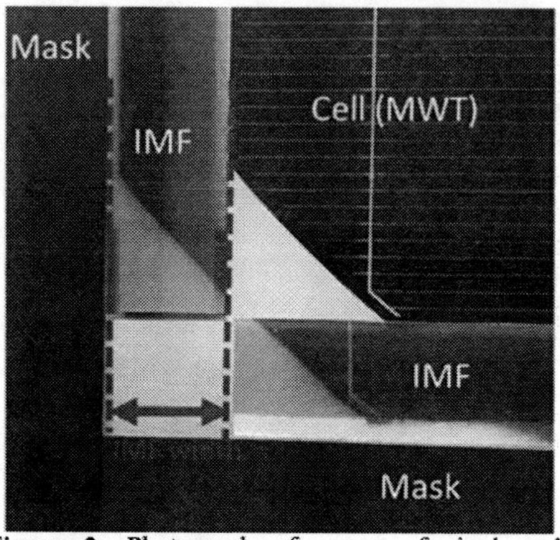

Figure 2: Photograph of corner of single cell laminate showing IMF, mask to determine the effective IMF width, and an MWT cell.

area. By optimizing the light trapping inside the module for light incident on these areas, a large I_{sc} increase can be achieved. The module structure was re-designed to allow redirecting the light falling onto the electrically inactive area of the module (in between the cells and module edges) towards the cells by mean of reflection. For high efficiency IBC and MWT cells, this light can be efficiently converted into currents. Reflection of the light is assured by the reflective IMF, a structured foil, placed in between the cells inside the module. Figure 1 shows a schematic of a cross section of the IMF integrated with back contact cells and foil. In this way, the IMF allows an effective use of the regions between the cells as it efficiently redirects more than 80% of the light toward the cells, increasing current. The IMF pattern is optimized using ray tracing modelling to maximize the internal reflection of light that falls between the cells. This 'results in maximum light trapping and thereby increases the short circuit current of the module. The conductive foil interconnection technology mitigates the potential increase of the series resistance resulting from higher I_{sc}. Thereby FF loss is mitigated.

3 RESULTS

3.1 Single-Cell Laminates

In order to quantify the potential current and power gain as a function of IMF width around the cell, we made a series single cell laminates using MWT and IBC cells with 1 cm IMF strips on all sides. Figure 2 shows a close up of the corner of one of the single cell laminates. As can be seen here, no special attention was given for the corners of the pseudo-square cells. Similar single cell laminates were also made with white encapsulant (EVA) for comparison. Each of the single cell laminates was masked with tape to determine the amount of IMF or white encapsulant exposed around the cells. In order to measure the impact of the IMF and white encapsulant specifically, flat glass with no anti-reflection coating was used in this experiment. The results for current and power are show in Figure 3 and Figure 4 respectively.

With standard cell spacing of 1.25 mm (minimum spacing measured), both MWT and IBC laminates have cell to module losses in current, probably due to the glass. The MWT laminates have slightly less current loss than the IBC laminates due to the natural light scattering off of the front side metallization. The gain in current for both cell types is approximately linear with the width of IMF or white encapsulant. With only 2.5 mm of IMF width between cells, CTM current loss is 0. With 10 mm IMF width, MWT laminates show more than 10.6% current gain and IBC laminates show more than 9% current gain. This is almost 8% more current gain than what is measured for white encapsulant.

Similar results are seen for power in **Error! Reference source not found.**. With 1.25 mm IMF or white encapsulant, the laminates have almost 5% CTM loss. At 5 mm cell spacing, there are no CTM power losses and with 10 mm cell spacing there is almost 5% gain in power from CTM. For white encapsulant, gains in CTM power are not realized for 10 mm cell spacing. We note that the slope of both the power CTM and current CTM as a function of cell spacing are approximately the same. This suggests that the module FF is not decreased due to the increased current.

3.2 2x2 Mini-Modules

To further optimize the IMF performance, IBC 2x2 mini-modules were made with anti-reflection coated glass on the front. The cell spacing was set at 8 mm to maximize the impact IMF and minimize module size. A picture of one of these mini-modules is seen in Figure 5. For the mini-modules, IMF was also integrated into the corners between cells and oriented to maximize current gain. The photograph in Figure 5 is taken at an angle so that the IMF can be seen between the cells. When photographed directly, the IMF is more difficult to see as it reflects the all-black surface of the IBC cells. This all black appearance may be useful in aesthetic applications for high power modules.

Two sets of mini-modules were made. One with a standard glass/back contact foil structure and another

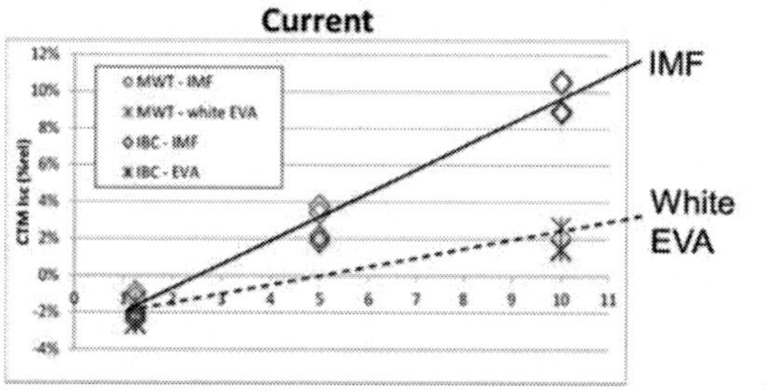

Figure 3: Cell to laminate (CTM) short circuit current gains and losses as a function of IMF width as compared with white encapsulant.

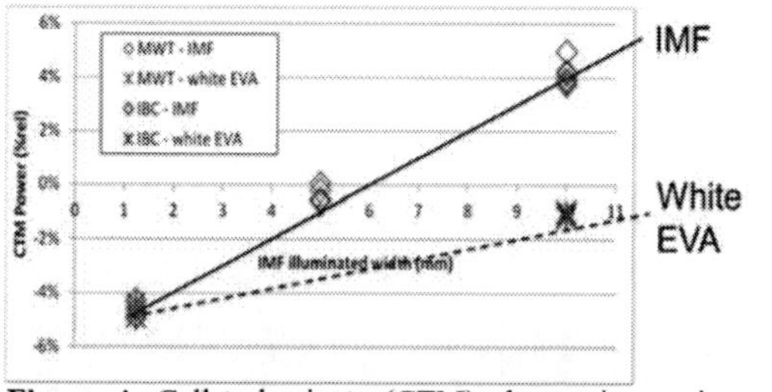

Figure 4: Cell-to-laminate (CTM) change in maximum power of IBC and MWT single cell laminates with IMF and compared to white EVA.

Figure 5: IBC mini-module with 8mm wide IMF integrated along cell edges and corners.

with glass/back contact foil/glass structure. Table 1 shows the CTM results for these two constructions. The double glass module suffered some problems during lamination due to slight shifts in the IMF such that it was about 1 mm away from the edges of the cell in some places. This decreased the current gain in CTM and resulted in 4.8% CTM power gain. However, the glass/foil module had 5.1% cell to module power gain due to a CTM gain in short circuit current of more than 8%. For both material stacks, the FF CTM loss was slightly higher than expected. This is likely due to alignment issues in the lamination process. Further optimization is ongoing and will be applied in full-size modules integrated with IMF.

Table 1: Cell-to-module relative changes for 2x2 IBC modules. All measurements are masked.

Laminate Architecture	CTM			
	P_{mpp}	I_{sc}	V_{oc}	FF
Glass/foil/glass	+4.8%	+7.8%	-0.4%	-2.5%
Glass/foil	+5.1%	+8.4%	-0.1%	-2.9%

4 CONCLUSIONS

We demonstrate increased module power and efficiency by improved module design. In this case, integration of IMF with back contact foil and high efficiency IBC cells result in a 5.1% cell to module power gain for mini-modules. We will expand this concept to full-size modules to demonstrate similar CTM gains in 60 cells.

Module technology should enhance cell performance rather than decrease cell performance. With proper light management and electrical optimization, this is possible for all back contact cell structures. Such innovation can also be of benefit in residential rooftop, area-limited, or BIPV applications where module power should be maximized in order to realize the most cost competitive electricity production combined with aesthetic.

ACKNOWLEDGEMENTS

ECN acknowledges the cooperation with DSM within the project CHAMP, funded by Dutch TKI Urban Energy.

REFERENCES

[1] D. Smith *et al.*, "Silicon Solar Cells with total area efficiency above 25%," 2016 IEEE 42nd Photovoltaic specialist Conference (PVSC), Portland, OR, 2016.
[2] V. Rosca, *et al.*, European PV Solar Energy Conference and Exhibition, Hamburg, Germany, 14-18. September 2015.
[3] S. Zhang *et al.*, "335Watt world record P-type mono-crystalline module with 20.6 % efficiency PERC solar cells," 2015 IEEE 42nd Photovoltaic Specialist Conference (PVSC), New Orleans, LA, 2015.

THE STUDY ON THE IMPACT OF THE WVTR OF THE BACKSHEET TO THE ANTI-PID PERFORMANCE OF THE MODULE

Xia Cai[1]*, Zhichun Ni[1], Feng Qian[1], Shuping Wang[1], Leizhen Hu[1], Qiangming Zhang[1], Guoqing Chen[1], Ruofei Zhao[2], Heyu Chen[2]
[1] Suzhou Talesun Solar Technologies Co., Ltd. Changshu, Jiangsu, 215542, China
[2] DSM Sunshine Solar Technology (Suzhou) Co., Ltd. Suzhou, Jiangsu, 201203, China
* Corresponding Author. E-mail: xia.cai@talesun.com

ABSTRACT: The backsheet is the key material of the module, which need have reliable insulation, water resistance and aging resistance. The WVTR (water vapor transmission rate) is the critical parameters which represent the water resistance of the backsheet. The backsheet WVTR has significant impact on the anti PID (potential induced degradation) performance of the module, which has been studied in this paper. The study shows that the higher backsheet WVTR, the more serious of the PID phenomenon of the module, and the higher power degradation of the module in the field application.
Keywords: backsheet, PID (potential induced degradation), WVTR (water vapor transmission rate)

1 INTRODUCTION

The PID phenomenon has aroused the great attention of the whole PV industry due to the huge influence on the long-term reliability of the module. The main external factors affected the PID phenomenon of the module are the environment temperature, the environment humidity and the voltage potential. The WVTR of the backsheet plays an important role in the process of PID phenomenon of the module.

The backsheet is the key material for protecting the module, which isolate cells and encapsulation material with the atmosphere environment. The backsheet need tolerance all sorts of environment stress, and need have reliable insulation, water resistance and aging resistance. The backsheet is crucial to the outdoor reliability, power degradation and life time of the module.

The water-resistance of the backsheet determines the effect of the environment humidity on the module. The typical value of the WVTRWVTR of the backsheet is usually less than 2g/m•²d (38℃, 90%RH), but the WVTR of the backsheet changes with temperature, and the temperature of the backsheet is higher than 38℃ in the field application. So, it's incomplete to evaluate the initial WVTR of the backsheet. In order to choose the suitable backsheet, the changes of the WVTR of the backsheet in the field application of the module, the changes of the WVTR of the backsheet with different temperature, and the influence of the WVTR of the backsheet on the PID phenomenon of the module need be studied.

2 EXPERIMENTAL METHOD

Two module samples are choosed and PID tested in the conditions of 85℃, 85% RH and -1000V bias voltage. The schematic diagram of the PID test is shown in Figure 1. The materials of the two samples are same except the bachsheet. The variation of the WVTR of the backsheet after the long-term PID test affected the power degradation is studied, and the varying trend of the WVTR of the backsheet after PCT, Damp Heat, UV and with different temperature is studied at the same time.

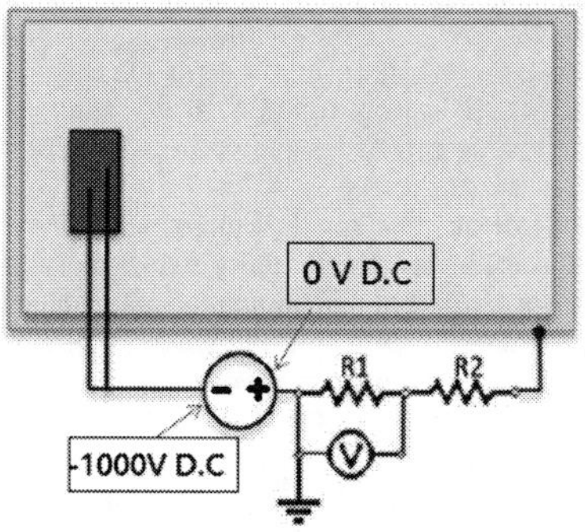

Figure 1: The schematic diagram of the PID test

3 EXPERIMENTAL RESULTS

Table I: The PID test of the different module

Serial number	PID test（85℃, 85%RH, -1000V）								
	Time/h	0	192	384	672	768	960	1056	1248
PH67271008031115	Power vaule/W	310.76	308.89	307.91	308.91	308.70	309.30	305.02	309.84
	Attenuation ratio/%	0.00	0.60	0.92	0.60	0.66	0.47	1.85	0.30
PH67271008041115	Power vaule/W	306.79	306.67	305.72	305.86	300.72	295.02	287.81	279.76
	Attenuation ratio/%	0.00	0.04	0.35	0.30	1.98	3.84	6.19	8.81

Notes: The module PH67271008031115 uses the backsheet S, the module PH67271008041115 uses the backsheet J, and the rest materials of the module are same.

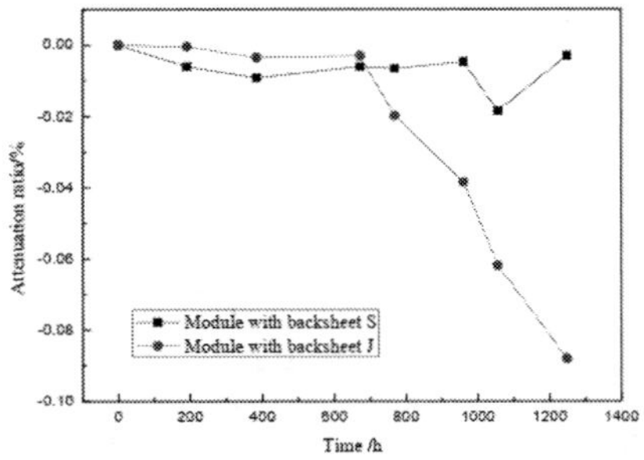

Figure 2: The PID test result of different module

The PID test result of different module is shown in Figure 2. The power degradation of the module with the backsheet S is less than the one with backsheet J during whole test time. During PID test process, the gap of the power degradation of this two modules grown larger after 672h. The power of the module with backsheet keeps constant, and the module with backsheet J decreases sharply . The power degradation ratio of the module with backsheet S is less than 1% when the PID time achieves 1248h. The power value of the module with backsheet J is better than the one with backsheet S in the first half of the test cycle, but the power of the module drop seriously in the second half time. The power degradation ratio of the module with backsheet J is 8.81%. The key difference come from the backsheet S has less WVTR. Which can be seen in Figure 3.

4 EXPERIMENTAL ANALYSIS

The WVTR of the backsheet in different temperature is shown in Figure 3.

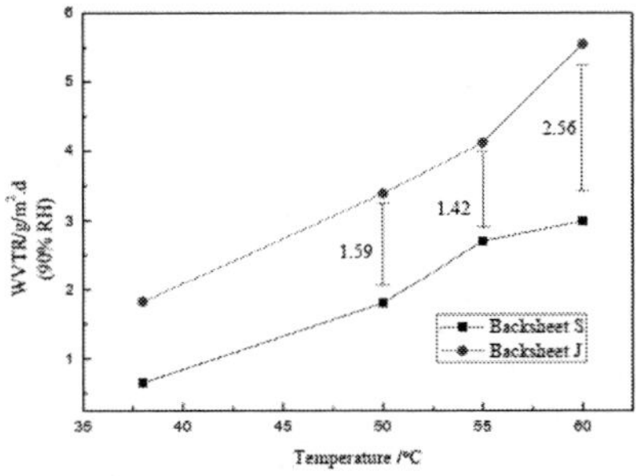

Figure 3: The WVTR of the backsheet in different temperature

The WVTR of the backsheet S and backsheet J increase with the increase of temperature. At the same time, the gap between them is larger. The WVTR value of the backsheet J increases from 1.82g/m²•d to 5.55g/m²•d as the temperature of the backsheet increases from 38 ℃ to 60 ℃, while the WVTR value of the backsheet increases from 0.65g/m²•d to 2.99g/m²•d. The WVTR value of backsheet J is much higher than the WVTR value of backsheet S in the high temperature.

Table Ⅱ: The WVTR of the backsheet after PCT

PCT time	WVTR (g/m²•d, 38℃&90%RH)	
	Backsheet S	Backsheet J
0h	0.64	1.82
48h	0.60	2.40
72h	0.59	2.40
96h	0.61	2.30
144h	0.61	Cracks

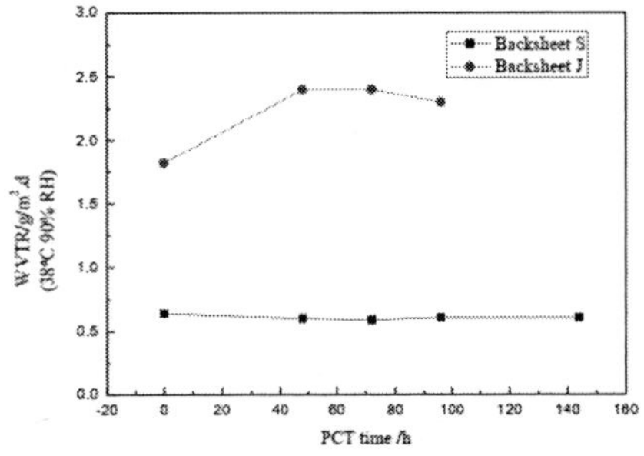

Figure 4: The WVTR of the backsheet after PCT

The WVTR of the backsheet after PCT is shown in Figure 4. The WVTR of the backsheet S keeps stable after PCT, while the WVTR of backsheet J increases by 26.37%, from 1.82g/m²•d to 2.3g/m²•d as the test time from 0h to 96h. When the PCT test time increases to 144h, the backsheet J is crack. The WVTR of the backsheet S is lower than the WVTR of the backsheet J after PCT.

Table Ⅲ: The WVTR of the backsheet after Damp Heat

Damp Heat time	WVTR (g/m²•d, 38℃&90%RH)	
	Backsheet S	Backsheet J
Initial	0.64	1.82
300h	0.60	1.84
672h	0.59	2.03
1000h	0.57	2.20
1100h	0.62	2.24

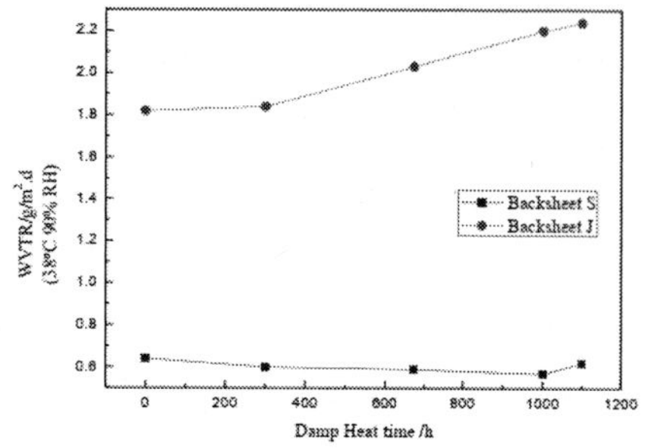

Figure 5: The WVTR of backsheet after damp heat

The WVTR of backsheet after damp heat is shown in Figure 5. The chamber air temperature of the apparatus is 85℃, and the chamber relative humidity of the apparatus is 85% RH. The WVTR of the backsheet S keeps stable after Damp Heat, while the WVTR of the backsheet J increases by 23.07%, from 1.82g/m^2•d to 2.24g/m^2•d as the test time from 0h to 1100h. The ratio of the backsheet J WVTR increase by 23.07%. The WVTR of the backsheet S is less than the WVTR of the backsheet J after damp heat.

Table V, The WVTR of the backsheet after UV

UV time	WVTR（g/m^2•d, 38℃&90%RH）	
	Backsheet S	Backsheet J
Initial	0.64	1.82
300h	0.59	1.80
672h	0.62	1.98
1100h	0.60	2.13

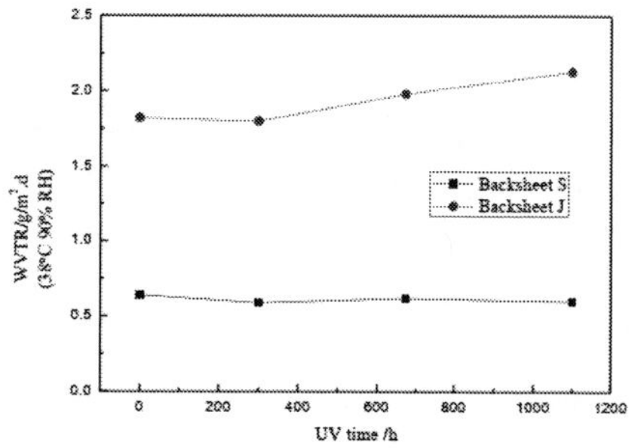

Figure 6: The WVTR of the backsheet after UV

The WVTR of the backsheet after UV is shown in Figure 6. The WVTR of the backsheet S is stable after UV, while the WVTR of the backsheet J increases from 1.82g/m^2•d to 2.13g/m^2•d as the test time from 0 to 1100h. The ratio of the backsheet J WVTR increase by 17.03%. The WVTR of the backsheet J is higher than the WVTR of the backsheet S.

5 CONCLUSIONS

The WVTR of the backsheet is a very important indicator of the evaluation to the backsheet. The WVTR of the backsheet S is more stable than the backsheet J in different temperature and aging conditions, such as PCT, Damp Heat, UV. Which means the water resistance performance of the backsheet S is better than backsheet J.

The WVTR of the backsheet has significant impact on the anti PID performance of the module. The higher WVTR of the backsheet, the more seriousof the PID phenomenon of the module, and the higher power degradation of the module in the field.

6 REFERENCES

[1] ISO 15106-3:2003 Plastics—Film and sheeting—Determination of water vapour transmission rate—Part 3: Electrolytic detection sensor method.
[2] IEC 62804 Ed. 1.0 DRAFT Rev D System voltage durability test for crystalline silicon modules –design qualification and type approval.
[3] IEC 60068-2-78 Environmental testing-part 2-78: Test-Test cab: Damp heat, steady state.
[4] Kim H B, Jung T H, Kang G H, et al. The Effect of PID Generation by Components of the PV Module [J]. Journal of the Korean Solar Energy Society, 2013, 26(10):760-765.
[5] Chang M, Chen H, Chen C, et al. Water cooking for backsheet and PV module endurance evaluation[C]// Photovoltaic Specialist Conference. IEEE, 2014:1997-2000.
[6] Gong H, Wang G, Gao M. Degradation Mechanism Analysis of Damp Heat Aged PV Modules[C]// Eu-Pvsec. 2012:3518-3522.
[7] Kim N, Park N, Han C. Developing accelerated life test for backsheet used in PV module and correlation study between field and accelerated conditions[C]// Photovoltaic Specialists Conference. IEEE, 2012:003162-003165.

ELECTRICAL PERFORMANCE EVALUATION OF C-SI SOLAR CELL SUBJECTED TO POTENTIAL INDUCED DEGRADATION

Zeel Purohit, Manoj Kumar, Brijesh Tripathi
Department of Science, School of Technology, Pandit Deendayal Petroleum University, Gujarat, Gandhinagar – 382007 (India)

ABSTRACT: Potential-induced degradation (PID) is a severe degradation phenomena that leads to massive decline of power output over time in solar photovoltaic (PV) power plants. The potential difference between PV cell matrix and grounded aluminum frame leads to the movement of charge species through the available leakage path resulting into degradation of the panel. In order to understand the cell level defects contributing to the PID, a c-Si solar cell has been subjected to high voltage stress condition (-800 V) for 24 hours. The two-diode model has been further used to investigate the recombination effect under PID. The alternating current (AC) impedance spectroscopy (IS) has been used to investigate the effect on charge accumulation process under PID effect. A significant change in impedance pattern indicates inductive behavior which serves better insight into physical process associated under PID effect.

Keywords: c-Si solar cells, reliability, Potential induced degradation (PID), two-diode model, Impedance spectroscopy (IS)

1 INTRODCTION

The PV installation projects are emerging exponentially under clean energy cumulative target over the last few years where reliability and sustainability have become important. There are various factors which affect the performance of c-Si module on the field over the long operation of period. The reliability of PV modules is mainly affected by the parameters like climatic change (humidity & temperature), high system voltages stress and encapsulation materials fatigue over the long periods of operation [1-2]. Potential induced degradation (PID) is identified as the major reliability issue which prompted mid-life failures and leads towards enormous decline of power output due to high voltage stress between the aluminium frame and the solar cells [1-5]. Although, some of the solutions have been tried through power electronics systems at the cost of extra power consumption but this phenomena needs to be addressed at the cell level for better performance of the PV system. There are various reports on the performance parameters of c-Si solar panels subjected to PID [6-11]. However, an in-depth study in terms of cell level performance has not been presented.

The present study provides an improved understanding of reduction in electrical performance of c-Si solar cells subjected to PID. The experimental *J-V* characteristics are analysed and theoretically fitted using the two-diode model to investigate the recombination mechanism associated to PID. Hereafter, the characteristics of *p-n* junction subjected to high voltage stress was further evaluated with Impedance spectroscopy (IS) to decouple various physical processes.

2 EXPERIMENTAL PROCEDURE

2.1. Description of set-up

The c-Si cells (1 cm^2) consisting SiN_x ARC coating, EVA (ethylene-vinyl acetate polymer) as the encapsulation layer along with tempered soda lime cover glass was used for PID test. The polyvinyl fluoride (PVF) material sheet serves as the back sheet. The PID tests were carried out with the encapsulated c-Si solar cells which subjected at -800 V high voltage stress by connecting the positive electrode to the aluminum foil which is placed on the front cover glass and negative electrode to shorted terminals of the c-Si cell as shown schematically in Fig. 1. The PID tests were carried out with controlled condition of temperature maintained at 80±2 °C with relative humidity maintained at 65 %RH.

2.2. Description of characterization

The J-V electrical response was measured before and after PID under one sun-illumination (1000 W/m^2, AM1.5G spectrum) using solar simulator (SS80AAA, Photoemission Tech., USA) and source measuring unit (U2722A, Agilent). The surface morphology were studied using field emission scanning electron microscope (FESEM) (Carl Zeiss, Ultra-55, 1-20 kV voltage range) equipped with EDAX (Oxford Instruments) for PID affected solar cells. The Impedance measurements of the solar cells were done before and after PID at 0 V bias under dark, using Electrochemical workstation (CHI 660D) equipped with a frequency response analyzer with 5 mV AC perturbation signal in the frequency range of 0.1 Hz to1 MHz.

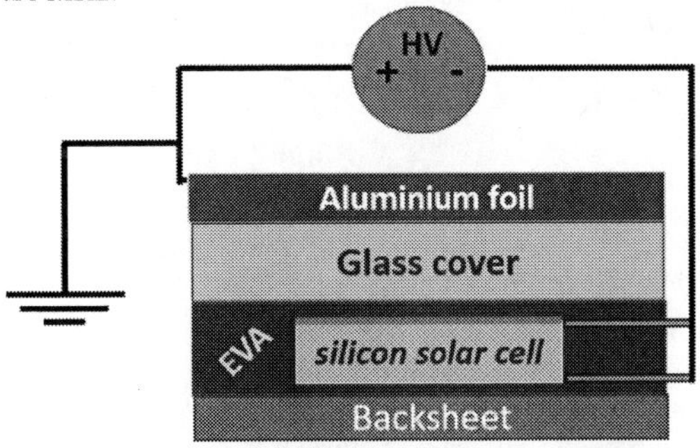

Figure 1: The schematic representation of the potential induced degradation (PID) test setup.

3 THEORETICAL DESCRIPTION

The two-diode based equivalent circuit of the solar cells is shown in Fig. 3. The saturation current densities of the first and second diodes are denoted as J_{01} and J_{02} and ideality factors of the first and second diode are represented as n and m, respectively. The parallel/shunt resistance (R_{Sh}), and series resistance (R_S) are included to quantify the deviation of the current density-voltage (J-V) characteristics from the ideal behaviour. The two-diode model equation for illuminated solar cell can be expressed as [12]:

$$J = J_L - J_{01}\left\{exp\left[\frac{q(V+JR_S)}{nk_BT}\right] - 1\right\} - J_{02}\left\{exp\left[\frac{q(V+JR_S)}{mk_BT}\right] - 1\right\} - \frac{V+JR_s}{R_{Sh}} \quad (1)$$

where, J_L represents light generated current, q represents the charge of an electron, k_B represents the Boltzmann constant, and T represents the absolute temperature. The two-diode is used to fit experimental data using the Matlab/Simulink code to estimate R_S and R_{Sh}, n and m of solar cell subjected to 24 hours PID.

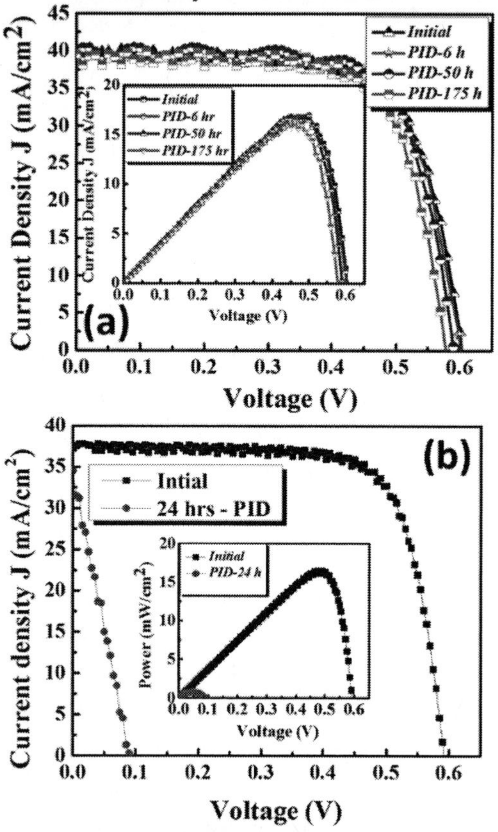

Figure 2: The J-V characteristic of c-Si solar cells subjected to -800 V at room temperature (25 °C) for 175 h exposure time (ET).

4 RESULT AND DISCUSSION

4.1 The effect of temperature on PID

Fig. 2(a) shows the J-V characteristics of the PID affected c-Si solar cell subjected to high voltage stress -800 V. The exposure time (ET) was varied to an extent of 6 h, 50 h and 175 h. The values of electrical performance parameters extended upto maximum ET (175 h) under PID such as V_{OC}, J_{SC}, FF and η are noted to be in the range of 0.57-0.6 V, 39.52-40.26 mA/cm^2, 70%-71% and 16.0 %-16.8% respectively as shown in Fig. 2(a). It is observed that, there is marginal change after 175 h in the performance parameters values as compared to the initial conditions at ambient temperature (25 °C). Inset of Fig 2(a) shows the P-V curve before and after PID performance which indicates no significant degradation in power for various ET at ambient temperature. However, after 24 h PID test at 80° C/60-65%RH demonstrated drastic reduction in J-V performance as shown in Fig.2 (b).The drastic reduction in V_{OC}, J_{SC}, FF and η parameters noted to be from 0.59 to 0.09 V, 37.23 to 34.50 mA/cm^2, 74% to 24% and 16.5% to 0.83% ,respectively. This results signifies the impact of temperature on PID test which plays an important role in the reduction of performance.

4.2 J-V performance evaluation with two-diode model

The evaluation of J-V performance further carried out by employing two-diode model in order to explore the various electrical performance parameters for initial and PID affected cell. As observed in Fig. 3, the theoretical curve fits well with the experimental results before and after PID by using Eq. 1. Table 1 displays the theoretical extracted parameters such as n, m, R_S and R_{Sh} before and after PID. The R_S values show a marginal difference from 0.1 to 0.15 ohm. In contrast, the R_{Sh} values decrease by the factor of 100 after PID. The decrement in the R_{Sh} values result in the increase of the leakage current across the junction under high voltage stress, which is due the increase in the impurity levels in the junction. Moreover, ideality factors n and m are found to increase from 1.5 to 2.5 and 3.5 to 4.5, respectively. The result suggest that the recombination increases through scape charge region (SCR) and quasi neutral region (QNR). In addition, the recombination current J_{01} and J_{02} increases from 5.25×10^{-9} to 1.89×10^{-6} A/cm^2 and under PID which gives an indication of a local intermediate defect states which possibly initiated by increase in impurities level under high voltage stress.

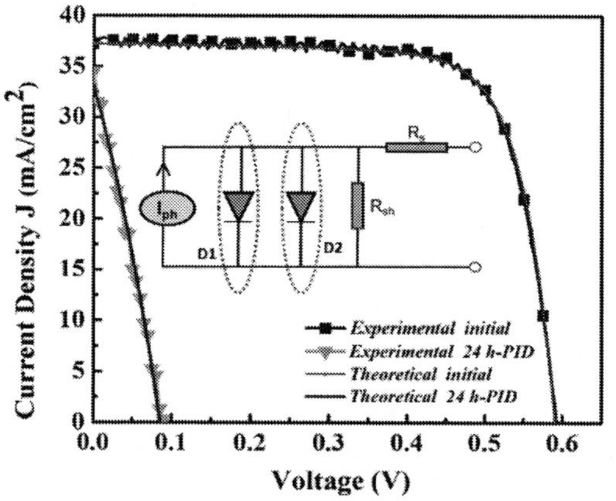

Figure 3: The J-V characteristic of c-Si solar cell at -800 V high voltage bias with -80° C/60-65% RH under illumination. The two-diode model to explain J-V characteristics of the c-Si solar cell affected by PID

4.3 FESEM and EDAX analysis of PID affected cell

The surface morphological characterization have been done for PID affected c-Si solar cells to investigate high voltage stress induced effect on surface of PID affected solar cells by using field emission scanning electron microscopy technique (FESEM). Fig 4(a) and (b), show the surface morphology and the Energy dispersive x-rays spectroscopy (EDAX) analysis of c-Si solar cells before and after PID treatment. It is clearly seen that, the deformation has been found on PID affected cell as seen in marked region. The EDAX elemental investigation reveals the 4 at% Na^+ impurities presence at deformed surface regions. These Na^+ impurities mainly responsible for the micro-structural surface deformation and contributes towards the shunt path [8-10]. The Na^+ diffused impurities creates the local trap states within junction and leads to increases recombination in SCR and QNR regions which can be corroborated by increasing value of ideality factors of n and m, respectively. This increase in recombination values responsible for the all-over reduction in the J-V performance with the decrease in the FF from 74 % to 24 %.

Table 1: The electrical performance parameters using two-diode model for c-Si solar cell subjected to PID at field condition (85°C / 65 %RH)

Parameters	**Initial**	**PID**
J_{SC} (mA/cm^2)	37.23	34.5
V_{OC} (V)	0.59	0.09
J_{MAX} (mA/cm^2)	33.68	18.48
V_{MAX} (V)	0.49	0.05
FF (%)	74	24
R_S (Ω)	0.1	0.15
R_{Sh} (Ω)	3000	10
n	1.5	2.5
m	3.5	4.5
η (%)	16.5	0.83

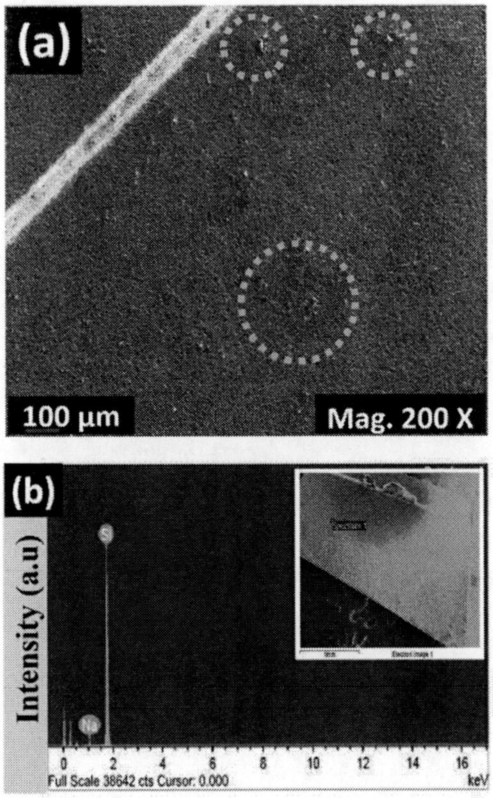

Figure 4: (a) The FESEM micrographs of the PID affected cell (b) EDAX elemental spectra confirming the presence of Na^+ impurities in PID affected cell.

4.4 Impedance Spectroscopy (IS) of PID affected c-Si cell

The Impedance spectroscopy technique has been widely used to investigate various mechanisms associated with physical processes. The impedance patterns (Z'-Z") has been obtained for initial and PID affected cells to explore physical phenomena involved in PID. Fig.5 (a) and (b) demonstrates the Nyquist curves of the initial and PID affected cell in dark at 0 V bias, respectively. The initial Nyquist pattern shows the capacitive behavior with high R_{Sh} (3000 Ω). In contrast, PID affected cell impedance at 0 V bias dominated by deleterious inductive pattern at higher frequency with negligible R_{Sh} value, as shown in Fig. 5(b). The high R_{Sh} value mainly indicates the low leakage current across p-n junction and low recombination for initial cell condition along with capacitive behavior which signifies good quality of SCR region. The inductive behavior under PID effect denotes the effect of Na^+ impurities driven through micro structural deformation under high voltage stress. Moreover, the Na^+ impurities mainly affect the junction quality and charge accumulation process by recombination which results into the reduction of η from 16.5% to 0.83%.

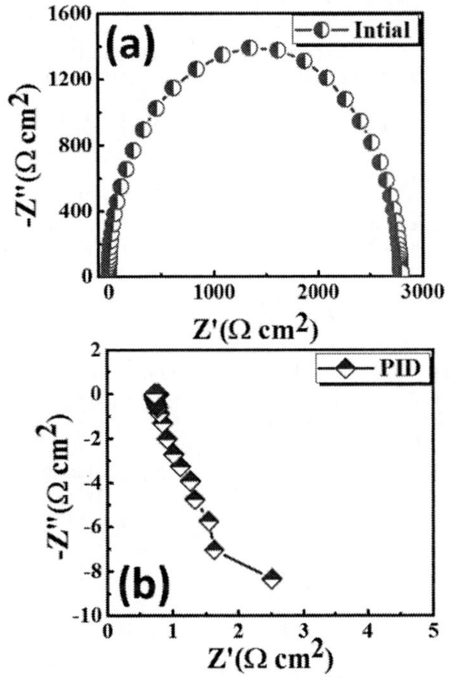

Figure 5: Nyquist plots describes the impedance behavior before and after PID under 0V bias in dark

5 CONCLUSION

Potential-induced degradation (PID) has been investigated by applying high voltage stress of up to -800 V for 24 h under the field conditions (80°C/65% RH). The J-V response of c-Si solar cells has been recorded before and after PID to analyze corresponding effect on electrical performance parameters such as V_{OC}, J_{SC}, FF and η drastically decrease under PID. The values of FF and η reduced from 74% to 24% and 16.5% to 0.83%, respectively. The two-diode model has been used to investigate the recombination mechanism under PID at cell level. The surface morphological analysis of PID affected solar cell confirms Na^+ impurities incorporation, observed at the structural deformed region which developed due to high voltage stress. The Impedance patterns reveal the inductive behavior of the PID affected due to presence of Na^+ impurities driven through high voltage stress.

References

[1] W. Luo, Y. Khoo, P. Hacke, V. Naumann, D. Lausch, S. Harvey, J. Singh, J. Chai, Y. Wang, A. Aberle and S. Ramakrishna, Energy Environ. Sci. 10 (2017) 43-68.
[2] M. Schutze et al., 37th IEEE Photovoltaic Specialists Conference (2011) 821-826.
[3] S. Pingel, O. Frank, M. Winkler, S. Daryan, T. Geipel, H. Hoehne and J. Berghold, 35th IEEE Photovoltaic Specialists Conference (2010) 2817-2822.
[4] S. Hoffmann and M. Koehl, Progress in Photovoltaics: Research and Applications 22 (2012) 173-179.
[5] P. Hacke et al.,37th IEEE Photovoltaic Specialists Conference (2011) 814-820.
[6] M. Koentopp, M. Krober and C. Taubitz, IEEE Journal of Photovoltaics 6 (2016) 252-257.
[7] D. Lausch, V. Naumann, O. Breitenstein, J. Bauer, A. Graff, J. Bagdahn and C. Hagendorf, Journal of Photovoltaics, 4 (2014) 834-840.
[8] V. Naumann, D. Lausch, S. Großer, M. Werner, S. Swatek, C. Hagendorf and J. Bagdahn, Energy Procedia, 33 (2013) 76-83.
[9] V. Naumann, D. Lausch, A. Hähnel, J. Bauer, O. Breitenstein, A. Graff, M. Werner, S. Swatek, S. Großer, J. Bagdahn and C. Hagendorf, Solar Energy Materials and Solar Cells,12 (2014) 383-389.
[10] S. Harvey, J. Aguiar, P. Hacke, H. Guthrey, S. Johnston and M. Al-Jassim, IEEE Journal of Photovoltaics, 6 (2016) 1440-1445.
[11] M. Oprea, S. Spataru, D. Sera, P. Poulsen, S. Thorsteinsson, R. Basu, A. Andersen and K. Frederiksen, IEEE 43rd Photovoltaic Specialists Conference (PVSC), (2016).
[12] M. Oprea, S. Spataru, D. Sera, P. Poulsen, S. Thorsteinsson, R. Basu, A. Andersen and K. Frederiksen, IEEE 43rd Photovoltaic Specialists Conference (PVSC), (2016).

FORECASTING POWER LOSSES DUE TO POTENTIAL-INDUCED DEGRADATION (PID)

J.Slamberger, M.Schwark
AIT Austrian Institute of Technology, Center for Energy
Giefinggasse 2, 1210 Wien, AT
T: +43 (0) 50550-6477 , F: +43 (0) 50550-6311, jan.slamberger@ait.ac.at

ABSTRACT: In this work, the models for forecasting power losses due to potential-induced degradation (PID) are presented. The PID is one of the most considerable types of degradation, in which case power losses may even exceed 30%. Additionally, the current standard (IEC 62804) applicable to PID testing specifies no threshold values for PID-tested modules. Usually, power losses of 5% under STC are accepted by PID testing institutes. However, at low-irradiation in many cases may lead to power losses of 20% and above. Therefore a good knowledge of PID processes is needed to quantify the energy yield losses. PID process is strongly dependent on strength of the electric field, temperature, relative humidity, conductive soiling and time.
Keywords: reliability, degradation, PV module, shunt

1 INTRODUCTION

One of the most considerable types of degradation of PV modules is potential-induced degradation (PID), in which case power losses may even exceed 30% [1]. The PID mechanism at p-type cells is supposed to be a reduction of the shunt resistance and an increase of reverse saturation current and the diode ideality factor due to the recombination in the space-charge region. The main suspects for this mechanism are sodium ions (Na^+), which drift due to negative electrical field from the sodium rich glass through anti reflection (AR) coating and penetrate crystal defects crossing the p-n junction. This process depends on several factors such as the combination of materials, the strength of the electric field (EF), temperature, relative humidity (RH), conductive soiling and time. PID is a reversible process, and thus can, to a large extent, be diminished by increased temperature and/or an oppositely charged EF.

Restricted to disclosing, the power losses under Standard Testing Conditions (STC), existing standards applicable to PID testing (IEC 62804) specify no threshold values for PID-tested modules. Usually, power losses of 5% under STC are accepted by PID testing institutes. However, at low-irradiation may in many cases lead to power losses of 20% and above.

In order to quantify the magnitude of power losses, a model for Degradation and Regeneration (PIR) is required. In literature, some models for the simulation of PID and PIR can be found [2]-[3]. These models were developed as interpolations for polynomic or exponential functions, which do not include all components.

In this work, the model for thermal (TR) and potential-induced regeneration, presented in [3], was tested. In addition, a PID model was created. To verify existing models and in order to create new ones, the power characteristics of a PV module over time and in dependence of temperature, relative humidity and strength of the EF need to be known. Parallel to the PID and PIR tests, the dark curves were measured, which were subsequently used to calculate the characteristics of light curve and maximum power outputs at multiple points in time during the stress test.

2 MEASUREMENTS AND RESULTS

The high voltage stress test with different voltages (1000V, 800V, 600V, 400V and 200V) between the module contacts and the frame has been conducted in the climatic chamber at different relative humidity (15%-92%) at five different temperatures (30°C, 40°C, 50°C, 60°, 70°C and 85°C). Before and after each test the I-V characteristics at STC and low irradiance have been measured with the pulsed light simulator and the electroluminescence images at Isc and 0.1Isc have been captured. Moreover the in-situ dark I-V characteristics measurements with the frequency of 5 minutes have been performed to monitor the progression of the degradation. The degeneration was limited to 30% of the nominal power output, which allowed a full recovery (>95%) of the power output. Figures 1 and 2 present the summary of the measurement results at temperatures of 60°C and 40°C and relative humidity of 85%., whereby $p(t)$ is relative power. The processes and values of voltages between cells and frame for all sections are presented in Table I.

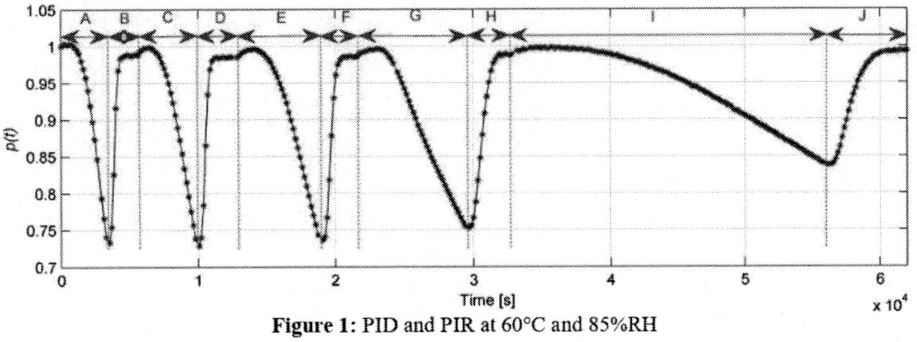

Figure 1: PID and PIR at 60°C and 85%RH

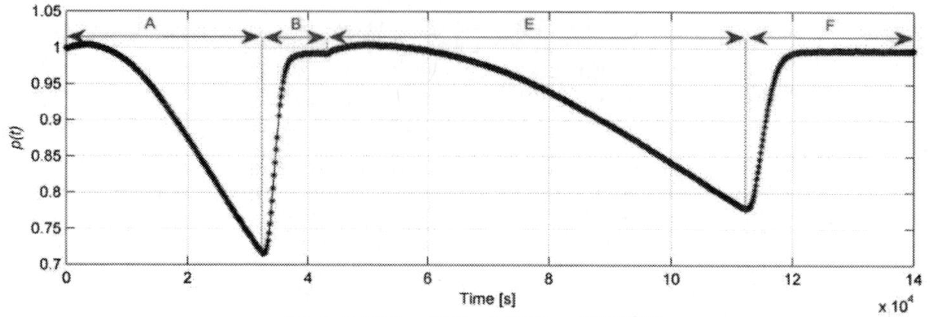

Figure 2: PID and PIR at 40°C and 85%RH

Table I: Definition of sections

Section	Process	Voltage [V]
A	PID	-1000
B	PIR	1000
C	PID	-800
D	PIR	800
E	PID	-600
F	PIR	600
G	PID	-400
H	PIR	400
I	PID	-200
J	PIR	200

The PIR process appeared to be faster than the PID process at the same temperature, relative humidity and voltage (absolute), whereby the PIR process does not recover the module's power to initial values but slightly lower (approximately 98-99%). The module's power recovers also at the beginning of the PID process and reaches values even higher as initially.

The verification of models for TR and PIR, presented with following equations

$$p(t) = \frac{P(t) - P_{min}}{P_{max} - P_{min}}$$

$$p(t) = 1 - e^{-\left(\frac{t}{\tau}\right)^{\beta}}$$

where p(t) is normalized power, P(t) power, P_{min} minimal power, P_{max} maximal power and β, τ parameters for PIR/TR fit, shows suitable fit to the measurements points (Figure 3) with an exception that the parameter β is not constant for different temperatures and voltages, as stated. The parameters are presented in Table II.

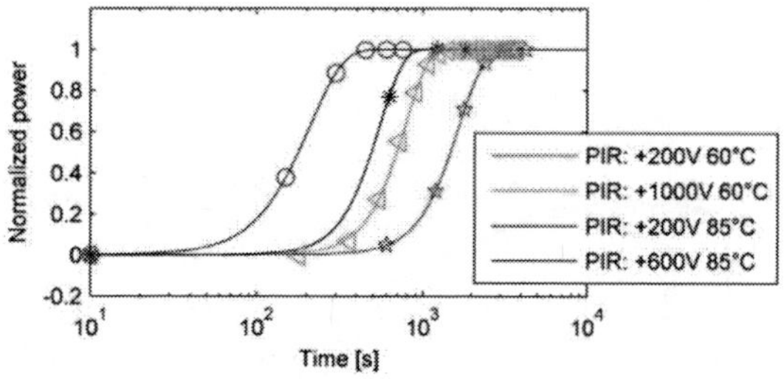

Figure 3: Normalized power during PIR processes (RH=85%)

Furthermore, also the new developed model for PID process, presented with following equations

$$p(t) = \frac{P(t)}{P_{max}}$$

$$p(t) = a \cdot e^{-\left(\frac{t-b}{c}\right)^2}$$

where a, b and c are fit parameters, shows appropriate fit to the measurement points (Figure 4 and Table III).

Table II: Parameters for PIR fits (RH=85%)

Temperature [°C]	Voltage [V]	β	τ
60	200	2,869	1693
60	1000	3,119	780,1
85	200	3,12	549,8
85	600	2,201	210,5

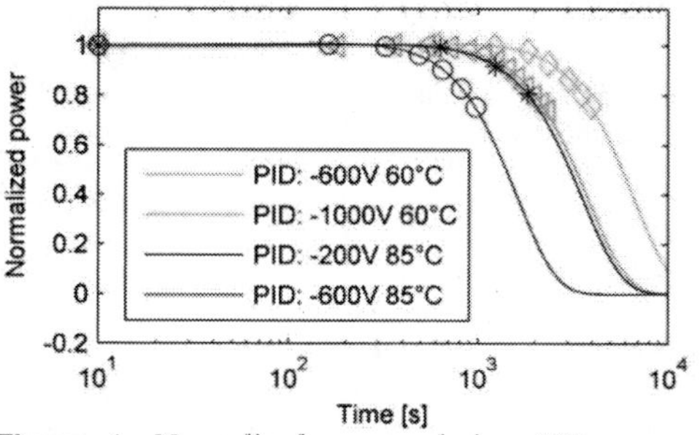

Figure 4: Normalized power during PID processes (RH=85%)

Table III: Parameters for PIR fits (RH=85%)

Temperature [°C]	Voltage [V]	*a*	*b*	*c*
60	600	1,006	714,7	6054
60	1000	1,005	295,3	3589
85	200	1,005	224,9	3409
85	600	1,009	151,3	1505

Both processes are strongly dependent on the strength of the EF, temperature and relative humidity. Furthermore, the velocity of both processes is scalable with transferred charge (Figure 5), what leads to research and measurements of the dependence of leakage current on the strength of the EF, temperature, relative humidity and time. The measurements results show the linear dependency on leakage current on the strength of the EF (Figure 6), exponential dependency on temperature, exponential dependency with limitation (S-Form) on relative humidity (Figure 7) and power dependency with time (Figure 8).

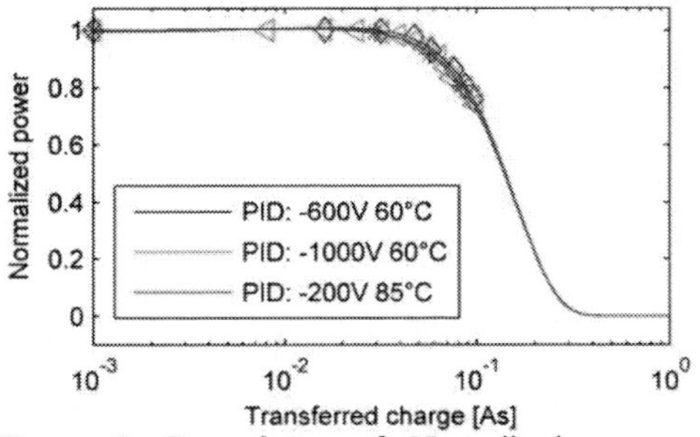

Figure 5: Dependency of Normalized power on transferred charge

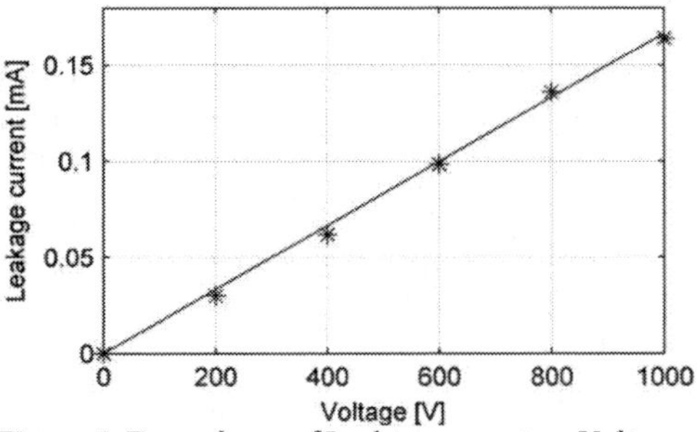

Figure 6: Dependency of Leakage current on Voltage

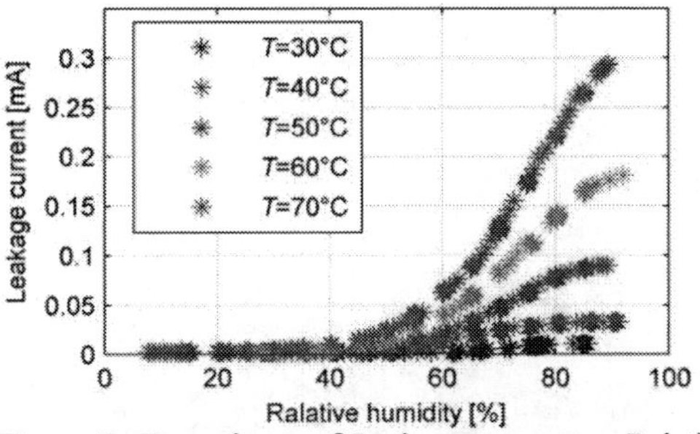

Figure 7: Dependency of Leakage current on Relative humidity and Temperature

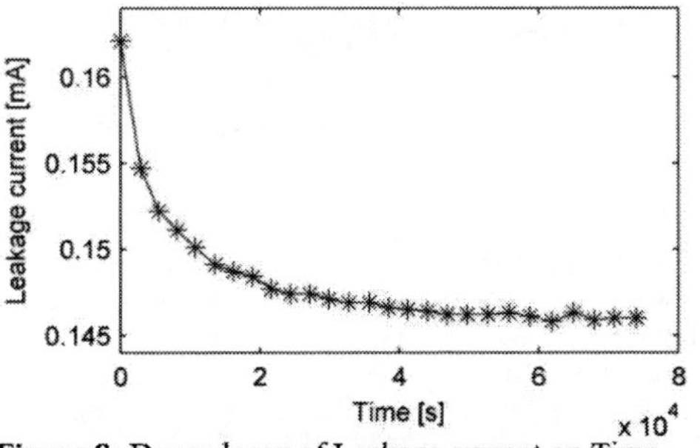

Figure 8: Dependency of Leakage current on Time

Additionally, the reproducibility of PID and TR process at a temperature of 60°C, relative humidity of 85% and voltage of 1000V (0V for TR) were made (Figures 9 and 10). The results show big difference of process velocity in case of PID and no significant difference in case of TR. It has been found that the PID process depends also on the history of the module. 1st run was made after two months of dark storage at 25°C, 2nd, 3rd and 4th runs were made immediately after PIR process and 5th and 6th run after TR process.

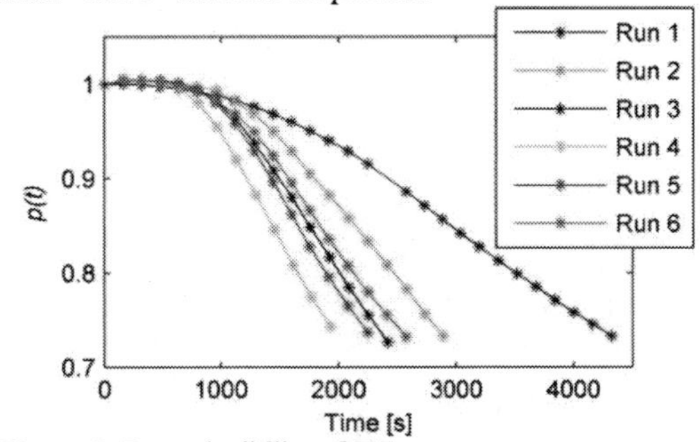

Figure 9: Reproducibility of PID process

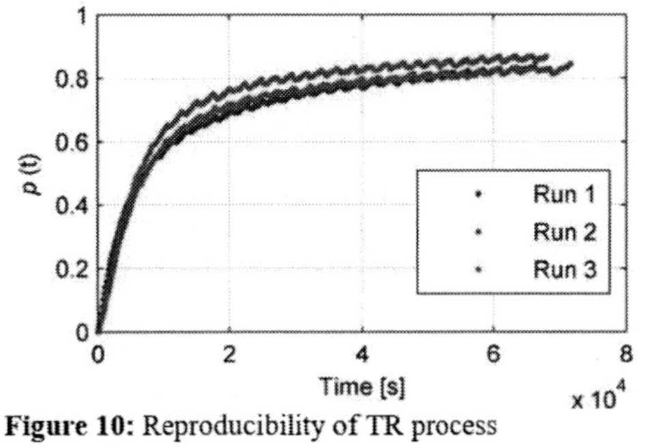

Figure 10: Reproducibility of TR process

3 CONCLUSION

This work, presents a suitable model for forecasting the power losses due to PID. Furthermore, the extension of the models with aforementioned factors is planned and thereafter, the comparison and validation of the models with the monitoring data from the real power plant.

Additionally, a reproducibility issue has been detected and will be researched in detail in the ongoing work. It is supposed that in case of immediate switch from PIR to PID process, some sodium ions stay on the surface of the cells and do not need the same amount of energy as sodium ions located in glass to penetrate to p-n junction of the cell, what accelerate PID process.

4 ACKNOWLEDGEMENT

This work was conducted as part of the Austrian "Energy Research Program" project INFINITY. Funded by the Austrian Climate and Energy Fund and the Austrian Research Promotion Agency (FFG), both are gratefully acknowledged.

5 REFERENCES

[1] W. Vaaßen *"Qualität von PV-Kraftwerken"* PV-Symposium 2016 Kloster Banz, Bad Staffelstein

[2] P. Hacke, et al., *„Accelerated Testing and Modeling of Potential-Induced Degradation as a Function of Temperature and Relative Humidity"*, IEEE Journal of Photovoltaics, vol 5, no.6, pp 1549-1553,2015

[3] P. Lechner, et al., *"Evaluation of Recovery Methods after Potential Induced Degradation of PV Modules"*, 31st European Photovoltaic Solar Energy Conference and Exhibition, Hamburg, 2015

PV SILICON MODULE DEGRADATION UNDER HIGH POSITIVE VOLTAGE BIAS

Kristijan Brecl, Matevž Bokalič, Marko Topič
Faculty of Electrical Engineering, University of Ljubljana
Tržaška 25, SI-1000 Ljubljana, Slovenia
E-mail: Kristijan.Brecl@fe.uni-lj.si

ABSTRACT: The photovoltaic (PV) modules in PV systems are connected in series and therefore exposed to high voltage stress. To study and analyze the behavior of the PV modules during their lifetime it is essential to expose them to accelerated tests. In our work we evaluate performance and electroluminescence of in-house developed PV modules exposed to damp heat (RH = 85%, T = 60 °C) and connected to positive voltage bias of 1000 V. The modules exhibited a severe and almost linear degradation over time. We could identify three degradation mechanisms: cell degradation, silver corrosion and encapsulation foil evaporation. After 1200 h of accelerated test the modules degraded for 11%.
Keywords: crystalline silicon, degradation, high positive voltage, Potential Induced Degradation

1 INTRODUCTION

The knowledge of degradation mechanisms in photovoltaic (PV) modules and systems is crucial to understand and predict the behavior of a PV system during its lifetime. PV modules in PV system are normally connected in series to reduce Joule losses in the cables and are therefore exposed to high voltages. Currently, the system voltages are up to 1000 V, but there are already discussions going on to rise the system voltage up to 1500 V. Usually, the PV systems are using transformless inverters, thus the PV module strings are not grounded. The ground is connected to the PV module frames, which means that a PV module can feel up to the half of the system voltage in either positive or negative direction. High voltage stress can cause solar cell degradation known as potential induced degradation (PID) [1]. The degradation is normally caused by ion migration between glass, aluminum frame and solar cells. The intensity of the degradation depends strongly on the materials used in the PV module and sole solar cell structure. Additionally, the degradation effects are also temperature and humidity dependent, thus higher temperature and humidity values accelerate the degradation.

The PID effect was revealed in the 1970s [2] but it raised the interest of researchers in the last decade after the discovery of the SunPowers' polarization effect [3]. Until now, several studies of the PID effect have been published [4]–[6]. However, the majority of the publications deal with the degradation mechanisms of p-type solar cells under negative voltage. The defects in p-type solar cells occur mostly due to sodium ions [7] that introduce shunts over the NP junction. This type of degradation is recoverable by applying reverse polarity to the PV module.

If a positive high-voltage bias stress is applied to a p-type solar cell, a corrosion of the silver conduction paths can occur [8], [9].

In the paper, we are presenting the behavior of the PV modules with p-type solar cell exposed to a positive high-voltage bias stress.

2 METHODOLOGY

To investigate the influence of the leakage current in PV modules under high positive voltage bias stress, we developed small area crystalline silicon PV modules with dimension of 23 x 33 cm^2 (Figure 1). The module fabrication procedure followed a typical commercial manufacturing process, only on a smaller size and manually assembled. Commercially available pseudo-square high efficient p-type monocrystalline solar cells with a textured front surface covered by a SiN_x antireflective coating and with no special treatment or production steps to make the solar cells PID resistant were used. The cells were cut to form small-area solar cells with a dimension of 78 x 39 mm^2 by a laser cutter (Trotec Speedy 300). The laser cuts the cells up to the half of the total depth from the back side; afterwards the cells are broken apart. Each module consists of 12 small-area solar cells connected in series. Due to the small dimension of the module, we used a thinner aluminium frame to seal them and a smaller junction box usually used in thin-film technologies. For the front module's surface, we used a commercially available textured solar white glass (Saint-Gobain Albarino), an ethylene vinyl acetate (EVA) foil from Hanwha Chemical and a backsheet from Coveme.

Figure 1: Photo of in-house developed test PV module: a) front site, b) back side.

Before the test, we characterized all PV modules electrically to obtain the initial module parameters and investigated them by the electroluminescence imaging technique [10]. For the experiment, modules were exposed to accelerated test conditions DH85/60 in a climate chamber (Kambič KK2310-CHLT) at relative humidity (RH) and temperature of 85% and 60 °C, respectively. The climate conditions were chosen with regard to the defined PID test requirements in the IEC-62804 standard [11]. During the test, the high positive voltage bias was applied between the short circuited module terminals and the grounded frame. The leakage current of each individual module was measured during the test with a datalogger [12].

The modules were periodically taken out of the chamber to perform electrical measurements and electroluminescence imaging.

2.1 Initial characterization tests

In total we measured 4 PV modules in two runs. Before the test the power of the modules was measured under sun simulator to obtain the initial performance parameters. In additional to the tested modules we produced also a reference module which was kept in the dark and unbiased at room temperature during the whole test. The PV module parameters are shown in table I.

Table I: Initial parameters of the modules under test.

	P_{STC} (W)	η (%)	I_{sc} (A)	V_{oc} (V)	FF (%)
Module1	7.27	9.65	1.28	7.29	78.0
Module2	6.91	9.18	1.27	7.33	74.1
Module3	7.05	9.28	1.24	6.73	84.4
Module4	6.26	8.25	1.23	6.63	76.7
Reference module	7.13	9.47	1.28	7.25	76.8

In addition to the electrical measurements, we also inspected the modules with the electroluminescence (EL) imaging technique. Modules in the first round were investigated by EL just before and at the end of the whole test, while in the second round EL characterization was made also during the test period. In this paper we will present the results of the second round. The results of the first test round are presented in [13]. EL images of two modules before the test are presented in Figure 2.

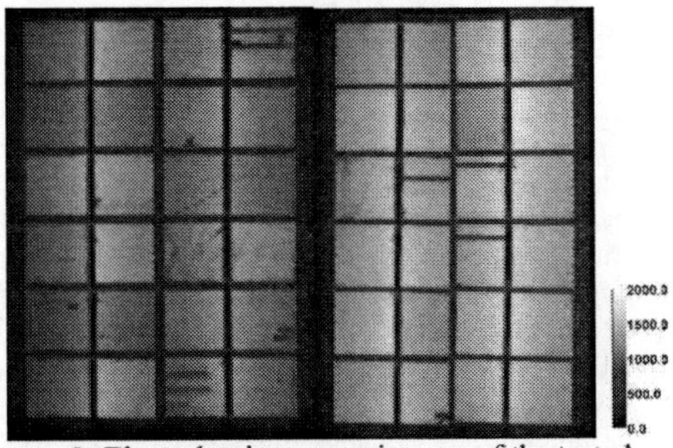

Figure 2: Electroluminescence images of the tested modules with 12 cells in series before the test (room temperature and $I = I_{SC\_STC}$)

3 RESULTS

3.1 Leakage current

During the whole test, we measured the leakage current between the short-circuited PV module connections and grounded aluminum frame. The leakage current was measured in 10 s intervals and is shown in Figure 3

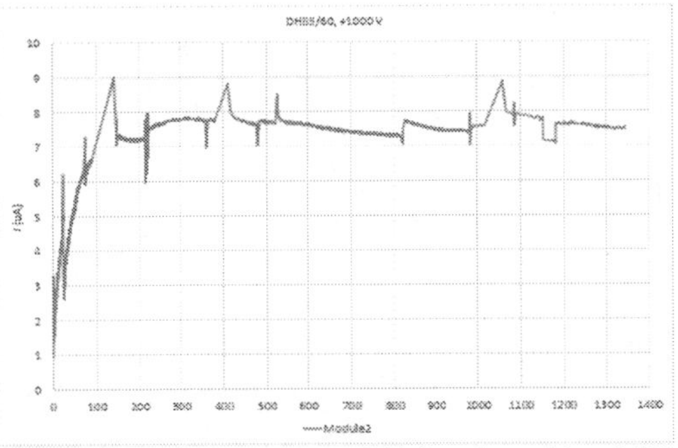

Figure 3: Leakage current of high voltage biased PV module in a climatic chamber under DH85/60 during the test.

As it is evident from the Figure 3 the leakage current rises over the time to a "saturated" value. The "spikes" and short term deviations are due to the opening of the climatic chamber. The absolute value of the leakage current at a positive bias of 1000 V was around 8-10 μA.

3.2 Power output

The power output of the tested modules starts to drop from the very beginning of the test. At the first measurement after 210 hours, the modules exhibit a 1% power loss. The 5% loss pass criteria [11] milestone was reached after around 670 hours of DH85/60 test (Figure 4). Following the IEC 62804 standard, these PV modules could be declared as PID free modules since the requested test period is only 96 h. Anyhow, our result show that the PV modules start to degrade from the beginning and degrade very rapidly.

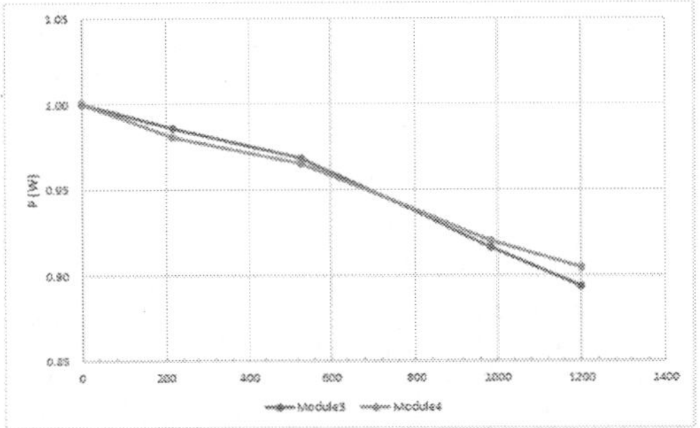

Figure 4: Normalized output power of tested modules.

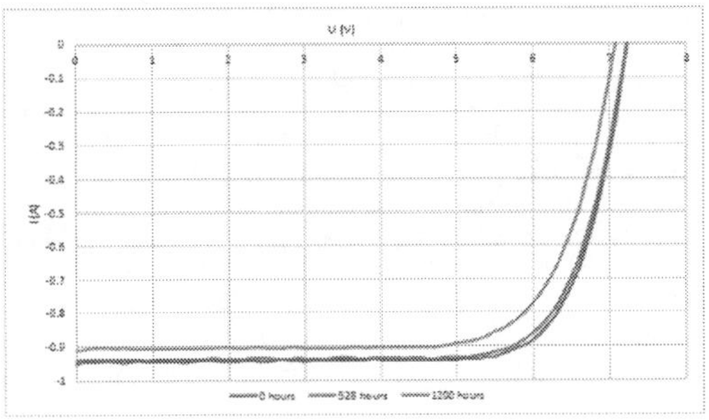

Figure 5: I-V characteristics of Module 3 during DH85/60 under conditions close to standard test conditions.

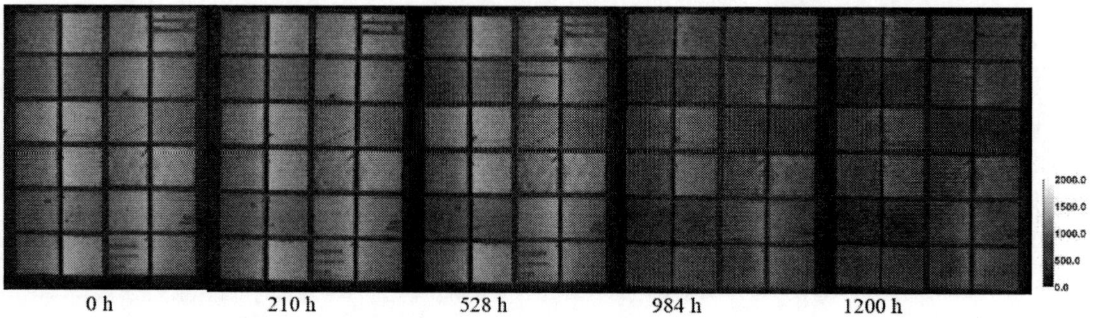

Figure 6: Electroluminescence images of Module3 during the DH85/60 test.

Figure 5 shows the *I-V* curves of the Module 3 after 528 h and 1200 h of DH85/60 test. After 528 h the main influence is seen around the maximum power point (MPP) point thus the fill factor of the PV module is affected. After 1200 h of DH85/60 test all characteristic points exhibit a strong deterioration. The total power degradation after 1200 h of DH85/60 test is 11%.

3.3 Electroluminescence response

The degradation of the PV modules due to the DH85/60 test was investigated also by EL inspection. The EL inspection was made each time the modules were taken out of the chamber for the electrical tests. All modules exhibit a similar EL degradation. In Figure 6 the EL images of the Module3 during the DH85/60 test is presented. After 210 hours the degradation of EL is almost invisible. In the third image from the left taken after 528 h of the test we could already define which cells degrade. The last two images (taken after 984 h and 1200 h) exhibit a very severe degradation in the EL response. Not all cells are affected the same. The weaker cells (weaker response in the first EL image) degrade faster and more evident. When a single solar cell within the PV module is inspected, the highest degradation can be seen in the center of each active part of the cell. At the end the EL intensity was reduced by 21%.

3.4 Visual inspection

Beside the degradation / deterioration of measured parameters the changes were observed also by the naked eye (Figure 7). The silver fingers and also the busbars became colorized due to corrosion (Figure 7 – a and b). Additionally, after 1000 h the EVA foil start to evaporate and build bubbles which are trapped in the pyramidal structure of the glass (Figure 7 c).

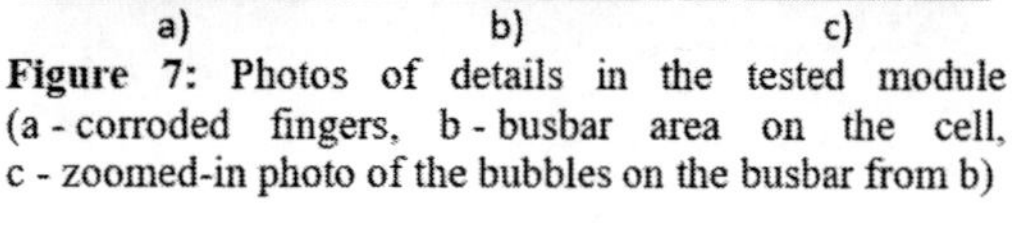

a) b) c)

Figure 7: Photos of details in the tested module (a - corroded fingers, b - busbar area on the cell, c - zoomed-in photo of the bubbles on the busbar from b)

4 DISCUSSION

The degradation of PV modules with *p*-type solar cells under positive bias occur due to two effects. The first one is the degradation of the sole cell, which is proved by EL inspection. The degradation mechanism is currently unknown. The ion migration, which causes the PID under negative bias stress is not present here since due to the voltage potential the positive sodium ions are pushed away from the cell. The second effect is the corrosion of silver lines. With a positive bias, the ionization of the silver with the reaction of $Ag \rightarrow Ag^{+} + e^{-}$ is stimulated and accelerated. High corrosion of silver contact lines results in a higher series resistance of the module. Additionally, after longer period the EVA around silver lines reacts and evaporates.

5 CONCLUSSION

Custom in-house assembled *p*-type monocrystalline silicon modules were tested at high positive voltage stress under damp heat DH85/60 conditions. The modules start to degrade from the beginning and degrade to the standard pass criterion of 5% after around 640 hours of DH85/60. The degradation is exhibited in the sole solar cell degradation, silver line corrosion and also EVA evaporation at longer periods of DH85/60 test. Despite the well-known PID under negative bias stress of *p*-type silicon modules, we could observe a very severe degradation of the modules also under high positive bias stress.

The main effect which starts early is the corrosion of silver lines. After a very long exposure to DH85/60 or more severe conditions the silver line could corrode completely and break the conduction lines.

ACKNOWLEDGMENT

The authors acknowledge the financial support of the Slovenian Research Agency (Research Programme P2-0197, Project J2-7105).

6 REFERENCES

[1] S. Pingel *et al.*, "Potential Induced Degradation of solar cells and panels," in *2010 35th IEEE Photovoltaic Specialists Conference*, 2010, pp. 002817–002822.

[2] Anvanced Energy, "Understanding Potential Induced Degradation." .

[3] R. Swanson *et al.*, "The surface polarization effect in high-efficiency silicon solar cells," in *15th PVSEC*, 2005.

[4] J. A. del Cueto and T. J. McMahon, "Analysis of leakage currents in photovoltaic modules under high-voltage bias in the field," *Prog. Photovolt. Res. Appl.*, vol. 10, no. 1, pp. 15–28, Jan. 2002.

[5] H.-C. Liu, C.-T. Huang, W.-K. Lee, and M.-H. Lin, "High Voltage Stress Impact on P Type Crystalline Silicon PV Module," *Energy Power Eng.*, vol. 05, no. 07, p. 455, Sep. 2013.

[6] N. G. Dhere, N. S. Shiradkar, and E. Schneller, "Evolution of Leakage Current Paths in MC-Si PV Modules From Leading Manufacturers Undergoing High-Voltage Bias Testing," *IEEE J. Photovolt.*, vol. 4, no. 2, pp. 654–658, Mar. 2014.

[7] J. Bauer, V. Naumann, S. Großer, C. Hagendorf, M. Schütze, and O. Breitenstein, "On the mechanism of potential-induced degradation in crystalline silicon solar cells," *Phys. Status Solidi RRL – Rapid Res. Lett.*, vol. 6, no. 8, pp. 331–333, Aug. 2012.

[8] M. K. P. Hacke, "Characterization of Multicrystalline Silicon Modules with System Bias Voltage Applied in Damp Heat." 2011.

[9] Z. Xiong, T. M. Walsh, and A. G. Aberle, "PV module durability testing under high voltage biased damp heat conditions," *Energy Procedia*, vol. 8, pp. 384–389, Jan. 2011.

[10] M. Bokalič, K. Brecl, and M. Topič, "Imaging Inspection System and Methodology for Evaluation of Inhomogeneities in PV Modules - A case of Multicrystalline Silicon," presented at the 33rd European Photovoltaic Solar Energy Conference and Exhibition, Amsterdam, 2017.

[11] IEC, "System Voltage Durability Qualification Test for Crystalline Silicon Modules." International Electrotechnical Commission, Geneva, Switzerland, Standard No. IEC-62804, 2013.

[12] LPVO, "LPVO Datalogger." Faculty of Electrical Engineering, LAboratory of Photovoltaics and Optoelectronics.

[13] K. Brecl, M. Bokalič, and M. Topič, "Examination of Photovoltaic Silicon Module Degradation Under High-Voltage Bias and Damp Heat by Electroluminescence," *J. Sol. Energy Eng.*, vol. 139, no. 3, pp. 031011-031011-6, Mar. 2017.

A REVIEW OF POTENTIAL INDUCED DEGRADATION IN THIN-FILM PLANTS

T. Weber, C. Hinz, M. Leers, P. Grunow, L. Podlowski
PI Photovoltaik-Institut Berlin AG, Wrangelstr. 100, 10997 Berlin, Germany
weber@pi-berlin.com; phone +49 30 8145264-111, fax +49 30 8145264-101

ABSTRACT: Thin Film (TF) plants which were installed within the past 10 years often suffered from quality issues like underperformance and poor module quality. PID could be identified as one of the major defect types, which sometimes results in an underperformance far beyond the warranty terms. Affected projects end up in legal confrontations and quite often require technical experts to evaluate the situation and to determine and validate the failure. Finally, in many cases the modules must be replaced.

In recent years, results of PID affected PV plants have been collected by the authors showing underperformance. Results of three investigated plants are presented in which the performance of the technologies CIGS, CdTe and μc-Si was studied. Investigation and characterization in the field and in the laboratory, including PID tests, were performed. The CdTe modules during field-investigations show a clear degradation towards the negative string end. The PID test proved their sensitivity to the phenomenon and revealed TCO-corrosion. For μc-Si TF modules a PID sensitivity could also be proven but here the power evaluation and failure distinguishing to the "white spot" phenomenon was challenging. The CIGS modules under investigation showed also a very clear PID degradation towards the negative string end which could be proven by PID tests and electroluminescence on complete strings in the laboratory.
Keywords: Thin Film, Reliability, Degradation, 62804-2, PID

1 INTRODUCTION

Nowadays, there is a worldwide production capacity of about 5 GW of thin film (TF) module technology. In total, an estimated cumulative installed capacity of 15 to 24 GW exists (5-8% of 300 GW installed worldwide in 2016). TF plants which were installed within the past 10 years often suffered from quality issues like underperformance and poor module quality. PID could be identified as one of the major defect types, which sometimes results in an underperformance far beyond the warranty terms. Affected projects end up in legal confrontations and quite often require technical experts to evaluate the situation and to determine and validate the failure. Finally, in many cases the modules must be replaced.

This work does not focus on the development of a TF test standard [1,2] or the mechanisms behind PID in TF modules [3-6]. When results of TF PID were presented in the past, they almost always focussed on results gained in the laboratory. In this work, the authors present real data gained from the field, which could be identified by outdoor investigation techniques. This is supported by data gained from the lab on the same modules to investigate the issues in more detail and to validate the field findings.

First, a definition of the terms and methods related to PID for TF is essential. "Voltage potential that exists between the active circuit and the grounded module surfaces can lead to module degradation by multiple mechanisms including ionic transport in the encapsulant, superstrate or substrate; hot carriers in the cell; redistribution of charges that degrade the active layer of the cell or its surfaces; failure of adhesion at interfaces, and corrosion of module components. These degradation mechanisms in thin-film modules caused by voltage stress and promoted by high temperature and humidity have been labelled potential-induced degradation, polarization, electrolytic corrosion, bar-graphing [, TCO-corrosion,] and electrochemical corrosion. They are most active in wet or damp environments and in environments prone to soiling of modules with conductive, acidic, caustic, or ionic species that lead to increased conduction on the module surfaces. In the field, modules have been observed to degrade in positive as well as negative polarity strings depending on the cell construction, module materials, and design." [2] The cited paragraph is out of the technical specification for TF PID testing (IEC TS 62804-2). This specification describes the test methods to determine module PID susceptibility under positive and negative bias directions. Correlating stress levels from tests to PID durability in a natural environment is important but a completely different topic. [4,7]

In recent years, results of PID affected PV plants were gained by the authors showing underperformance. Results of three investigated plants are presented in which the performance of the technologies CIGS, CdTe and μc-Si was studied. Investigation and characterization in the field and in the laboratory as well as PID tests were performed.

2 METHODOLOGY

2.1 General

Once in operation PV plants have to be regularly evaluated, inspected and tested to ensure that the investment is providing the expected revenue and is operating in a safe and reliable manner. The operation and maintenance (O&M) teams and the installed monitoring systems are able to provide the first hints of performance ratio (PR) losses. In-depth analysis is necessary to address the origin of the failures. Therefore, external experts often work in the field to troubleshoot and to advise the plant owners in finding the right solution to identify and solve the issues. The variety of services inter alia range from monitoring data and PR analysis, documentation check, on-site inspections and module selection for laboratory testing. For the PV plants considered in this work, all of those approaches were used to get a complete overview. The focus for PID-analysis on TF within this work is on the on-site inspection and laboratory testing at the module level. Not all results are presented. The picture is not a complete one due to the reason that the results are not representative for a particular module type, manufacturer or technology.

2.2 On-site Investigations

To investigate the modules on-site, engineers conduct IV-curve tracing at module and string level, infrared thermal imaging and electroluminescence tests. Visual inspections are also play an important role to evaluate the status and workmanship of the plants. For the PV plants presented here, the following equipment was used:

- Power measurements were performed with a calibrated HT I-V400 peak power measuring device and IV-curve tracer with a measurement uncertainty of +/-5%. Calibrated irradiation and temperature sensors determine the actual conditions enabling corrections to STC conditions.
- By means of an electroluminescence (EL) camera EL pictures were taken. The modules were powered string-wise with the short circuit current. The pictures reveal shunts and dark areas enabling the evaluation of the module status. Further information can be found elsewhere [8,9].

2.3 Laboratory Testing

In PI Berlin's accredited test laboratory (ISO 17025) tests are performed according to IEC 61646 Ed.2.0 (2008) and 61215-X Ed.1.0 (2016) [10,11], among others. The PID tests were performed acc. to TS 62804-2 TF (simulating the original mounting construction and applying maximum system voltage in a climate chamber of 85°C and rel. humidity of 85%). They were divided in the categories of sensitivity test (selection of non-affected modules), risk assessment (selection of medium-affected modules) and recovery tests (selecting strongly degraded modules and reversal of the test voltage). Before and after each exposure the IV curves have been recorded using a Class AAA flasher Pasan SSIIIb. The maximum power (Pmax) has been extracted from the IV-curve under standard test conditions (STC). The measurement uncertainties are for single junction ±2.9% (CdTe and CIGS) and ±5% for double junction (µc-Si). Furthermore, visual inspection and EL have been performed before and after the stress tests to evaluate the results [8]. EL has been performed using a Nikon D 800 camera with the IR-filter removed from the high-sensitivity 36 Mpixel CCD-chip. A 50 mm f/1.4 high-precision IR-optimised optic from Zeiss was used.

2.3 Selected TF-plants

Three different TF PV plants were chosen composed of the module technologies CIGS, CdTe and microcrystalline silicon tandem (µc-Si). A list of the basic properties can be found in **Table 1**. All plants are located in central Europe and are ground-mounted installations.

Table 1: Overview of the PV plants investigated

	I	II	III
Technology	CIGS	CdTe	µc-Si
Plant age in years	4	6	2
Size in MW	21	2.3	2.0
Nominal Power in W	105 & 110	67.5	128
Grounding Instruction	None, APT (Anti PID Technology)	None, not specified	Neg.-pole grounding

3 RESULTS AND DISCUSSION

3.3 Plant I

For the CIGS plant (I) under investigation, laboratory measurements were performed. 20 strings à 10 modules were selected randomly out of all plant areas to ensure a representative result distribution. **Figure 1** shows the power measurement results for all modules string-wise. The deviations of power were determined by comparing the actual power to the nominal value. One can see that all strings show module degradations for a few modules up to -40% to -90%. If one looks at the individual string and correlates the power loss to the module position in the string, it can be seen that modules at the negative string end degrade most (see **Figure 2**). The corresponding EL pictures show also a darkening of the cells towards the negative string end. These evenly distributed results lead to the suspicion of a potential induced degradation as a reason for the power deviations.

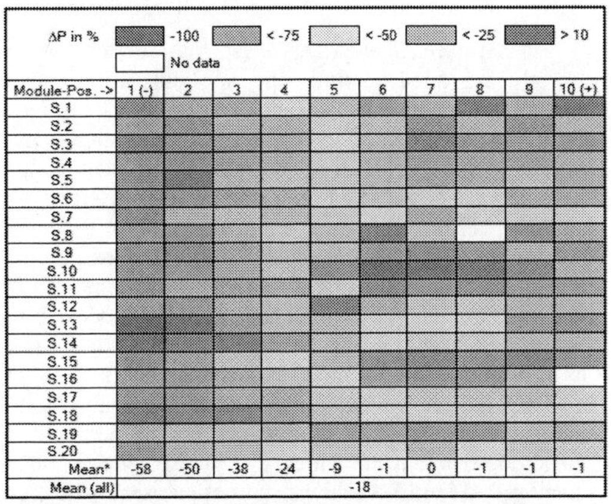

Figure 1: Degradation of power output (compared to label) string-wise for a CIGS plant under investigation

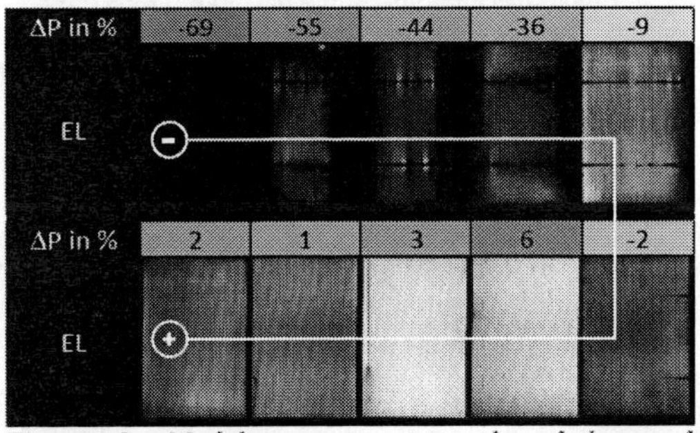

Figure 2: Module power output degradation and corresponding EL pictures for modules of one investigated string. Left top side negative end of the string and left bottom side positive end

To verify this suspicion, PID tests were performed to evaluate the modules in terms of general sensitivity, the further risk during operation and the regeneration potential. The sensitivity tests were performed on five modules under each polarity. Only the least affected modules were selected for the 96h long PID test. **Figure 3** shows the results. Depicted is the module power evolution (incl. the standard deviations) for modules tested under positive bias direction (blue) and under negative bias conditions (orange). The initial power for all modules was determined as 7% above nominal value. The group of modules tested under positive bias direction degraded slightly whereas the modules stressed under negative bias direction degraded about -86%.

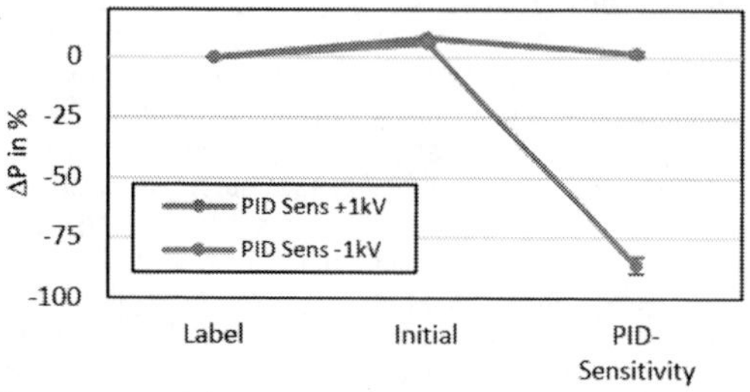

Figure 3: Module power evolution for PID-Sensitivity Tests. Blue +1kV and orange -1kV PID-Sensitivity tested for 96h

Furthermore, a recovery test was performed on strongly degraded modules by applying the opposite polarity to the modules during the PID test. **Figure 4** shows the results as power deviation to the nominal value. Initially the modules showed a degradation of - (58 +/- 17)%. After 96 h PID recovery test the modules recovered to a value of - (17 +/- 9)%. An additional 96 h further improved the power of the modules to a final deviation of - (8 +/- 5)%. If the module type is recoverable, then as a consequence one could redesign the system layout and set the voltage to positive values only by grounding, for example. In the specific case of this plant and module type, it was not recommended to do this because of other types of degradation which could be already determined and which is to be expected for future operation. **Figure 5** shows the appearance of those degradation which were provoked by the PID regeneration test as an example for the module type under investigation on one module. One could find a discoloration and bubbles in the edge sealing leading to a reduced insulation. Furthermore, worm-like delamination has occurred over the semiconductor itself. The backside of the module showed strong glass corrosion on the edges and in front of the rails of the mounting structure. If one take a closer look to the power deviations in the strings, one can find also a hint of a different degradation mechanism on the positive voltage string-end (see **Figure 2**, visible in power loss and EL). The highest power of an individual module in the string is around module position seven (counted from left, the negative string-end). Towards the positive string-end, one can see slight decreases for all strings.

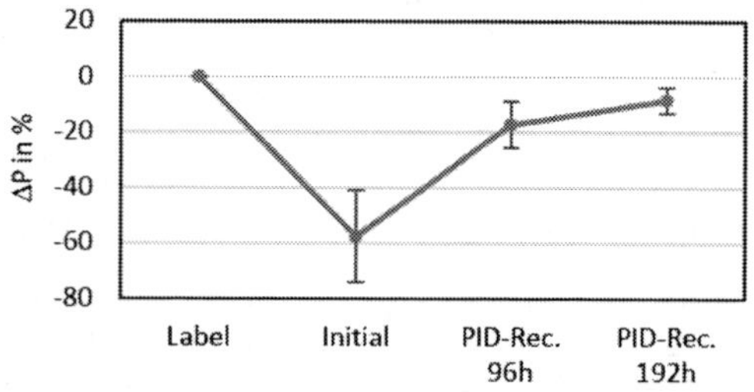

Figure 4: Module power evolution for PID-Recovery Tests

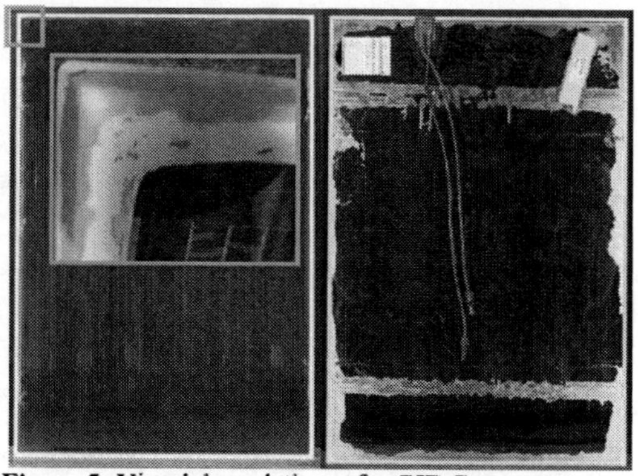

Figure 5: Visual degradations after PID-Recovery test +1kV, 196h on module front (left) and back (right)

The results for PV plant III can be summarized as

- Strong PID determined in the strings investigated towards negative string-end.
- PID-Sensitivity tests validate this result
- Regeneration under positive bias conditions is not possible because of superimposed degradation mechanism leading to new module failures and degradations

Finally, all modules in the plant were replaced by new modules.

3.1 Plant II

In PV plant II, 12 strings with 20 modules each were investigated with IV-curve measurements and electroluminescence. **Figure 6** shows the STC corrected data for each module in its string position. The 67.5W CdTe modules were in the 6th year of operation, which results in a minimum deviation of -12.6% to the nominal power the producer guaranteed (90% guaranteed for 10 years). This value includes acc. EN 50380 (2003) a measurement uncertainty of 2.9% and the 90% guaranteed power. In the figure, the colour code ranges from green, representing above guaranteed values, to yellow and orange to red, which indicates the modules with the lowest power. Some modules were not able to be measured due to a total failure (grey) and four were not measured due to glass breakage.

Only two modules pass the evaluation meaning that 99% of the modules failed. The mean deviation is -26%. Interestingly, one could determine a mean deviation across all strings for the first five modules from the negative string end of 40% and a mean deviation of -21% for the last five modules in the string at the positive polarity end. This indicates most likely a potential induced degradation superimposed by a "normal" degradation, which is also beyond the guaranteed values.

Figure 7 shows the EL-signal of four modules supplied with a current in forward mode. One can see inhomogeneous semiconductors with many failure patterns impaired the module from the edges inwards. For this module type and its current state it is hard to distinguish from production induced inhomogeneities and field-induced failure patterns. **Figure 8** shows a picture of a table in the plant. One can see cracked modules with the naked eye.

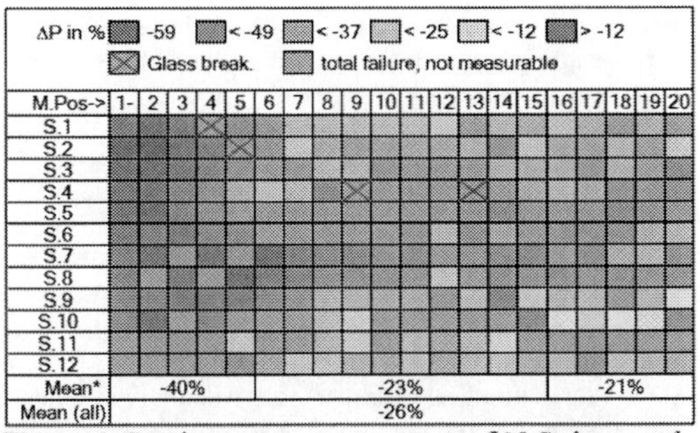

Figure 6: On-site power measurements of 12 Strings, each with 20 modules. Some modules show already in the sixth year of operation a total failure (grey). Depicted are for each module position in the string the power deviations to label acc. the colour code. Mean* power deviations were determined for the marked sections and all modules

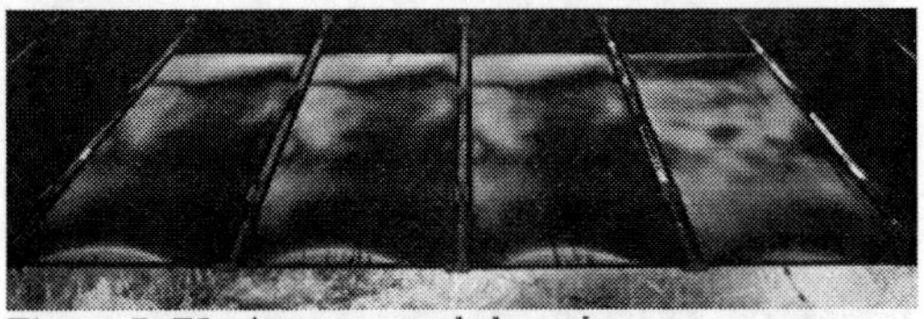

Figure 7: EL pictures recorded on-site

Figure 8: Picture of a table in plant II

During the field measurement campaign 20 modules were selected for further in-depth laboratory analysis. To summarize the results:

- All investigated modules fail the guaranteed power (90% after 10a) already in the sixth year significantly with a mean deviation of -54% determined in the laboratory (incl. all failures)
- Burn marks (18% of all modules affected) on the current collector straps inside the modules indicate a too low specified or too low reverse current overload protection (RCOP)
- 20% of all investigated modules (n=300) in the plant show glass breakage, most likely induced by the insufficient RCOP
- A plant design failure supports that, by interconnecting six strings in parallel without string protecting diodes (knowing that this is common but not feasible for that module type).

Moreover, in climate chamber tests PID, TCO-Corrosion was proven on all modules, which developed around the clamps. **Figure 9** shows the test results as an evolution of the power deviation to label. The modules were taken from different string positions. Two modules from the negative string end (black) and three from the positive string end (red, depicted as -/+End). After the initial laboratory power measurement, four modules were treated in PID tests under negative bias direction and one module as reference in damp heat (DH, dotted line). The modules from the negative string end have a mean initial deviation of -43%, whereas the modules from the positive string end have a mean initial deviation of –17%. Those modules from the negative string end degrade further (A and B). The modules from the opposite string end regenerate despite the negative potential during the exposure. The DH module (E) behaves like the modules from the negative string end. One can conclude that the modules are sensitive to PID due to further power reduction and the TCO-Corrosion. The power improvements can be explained most likely not by recovery but an improvement of the semiconductor due to stabilization by high temperatures of 85°C during the climate chamber treatment (some manufacturers prescribe pre-conditioning before power measurements after dark storage in heat environment, but this was not the case here). Finally, the plant owner was advised to replace all modules in order to assure the economic viability of the plant.

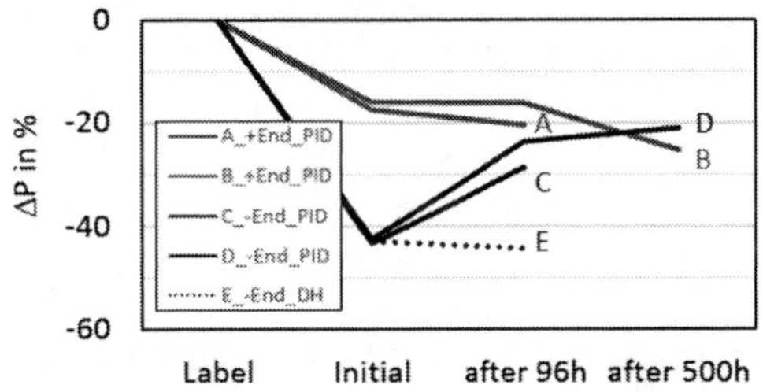

Figure 9: Module power evolution of CdTe modules from producer/plant II. The modules were taken from the negative and the positive string end. After initial laboratory power measurement the modules were treated in PID tests under negative bias direction and one module as reference in damp heat (DH)

3.2 Plant III

For PV plant III, equipped with µc-Si modules, laboratory measurements were performed. The modules showing abnormalities with suspicion for TCO-corrosion (orange) and white areas (blue) indicating a low shadowing tolerance are shown in **Figure 10** (left). To further investigate this issue and to validate the suspicion, eight modules were selected: two from the positive end of a string (A,C) and two from the negative end (B,D) all showing no TCO-corrosion. Two more modules from the negative string end were selected showing TCO-corrosion already after two years of operation (E,F). The modules were tested as shown in **Table 2**.

Table 2: Module properties and performed tests

	Free of TCO-Corrosion		Showing already TCO-C.
	Take from + string end	Take from - string end	
(+) PID	A	B	
(-) PID	C	D	E,F
DH, HS	G,H		

Figure 10: Left: Pictures of µc-Si module (E) of producer III with initial visual degradations showing TCO-corrosion (orange) and "white spots" (blue). Right: TCO-corrosion after 500h PID(-) test

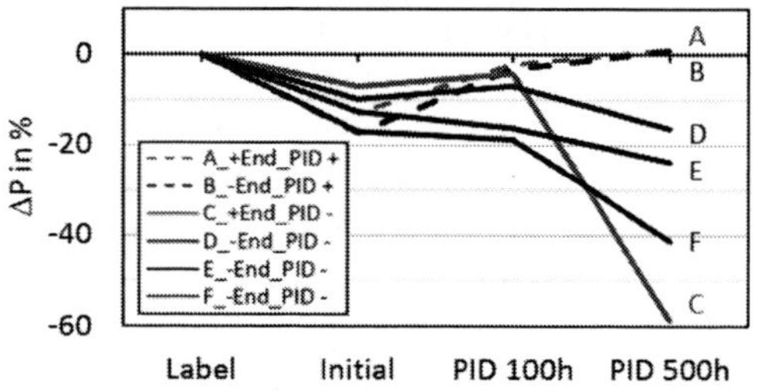

Figure 11: Module power evolution of µc-Si modules from producer II. The modules were taken from different string positions. After initial laboratory power measurement, PID tests were conducted under positive and negative bias direction

The Hot-Spot Tests revealed that the µc-Si TF modules are highly sensitive to shading. A brief partial shading leads to irreversible spots on the semiconductor (hot-spots) visible as grey marks (white spots). In the test no significant power loss could be found however temperatures of max. 80°C were measured. The failures reported from the field could be reproduced in the test performed.

The PID-Test results are presented in **Figure 11**. The modules from a negative string position in the plant show partially visible TCO corrosion and an initial power deviation up to 18% (E,F). All investigated modules show a significant further power degradation under negative bias conditions and reveal power losses up to -55% (C to F). Furthermore, TCO-corrosion could be proven as can be seen in the corner of the right hand side picture of **Figure 10**. **Figure 12** show an example of an EL picture of module E after 500h PID test under negative bias conditions. The module shows dark areas at the edges and corners in the EL picture. However, modules tested under positive bias conditions reveal no power related degradation processes but show visually a change of the semiconductor layers (fog like). In contrast, under positive bias conditions a power increase (up to labelled power) was observed for the tested modules coming from a negative string position in the field due to a temperature regeneration effect (A,B). Accordingly, it can be stated that PID for these modules (such as TCO-Corrosion) can be stopped, by avoiding negative bias by applying a proper grounding or by reversing the voltage into positive bias. Not shown are the results of modules G and H after the DH test. Both modules behaved in the same way as the positive PID tested modules indicating again that positive bias is not degrading the modules and that the modules regenerate under warm conditions (85°C).

It is challenging to differentiate between degradation and power improvement for Si TF modules. Moreover, for evaluating the status of a module in the right way it is necessary to measure the power of the module using the right procedure. Consequently, for µc-Si modules it is suggested to follow the test specification from the producer to evaluate power degradation. In this case, the procedure prescribes a regeneration by heat for 48 hours at 90°C (e.g. damp heat chamber) followed by light induced degradation. This enables an evaluation of the power-state of the modules before all treatments. Furthermore, it is important to distinguish between the "white-spot" phenomena and the TCO-Corrosion. The authors presented this already elsewhere on the same module type. [12]

Figure 12: Electroluminescence pictures of µc-Si modules of producer III (same module as in above figure) after PID-Test under negative bias conditions revealing dark areas around the edges indicating TCO-corrosion

4 SUMMARY AND CONCLUSION

Field measurements (IV-curve tracing, EL and IR) enable one to investigate large quantities of modules in the range of hundreds. These characterization methods suffer from instable measurement conditions (intraday and seasonal) and high measurement uncertainties (5 to 10% for power) [13]. On the other hand, it can reveal degradation at the module level during operation and provide insights into the reasons for the degradation. One big advantage is that the modules are already stabilized in terms of power and no pre-conditioning like in laboratory measurements is necessary.

In most cases, laboratory measurements are limited due to financial or resource constraints – therefore in most cases only low module quantities (low double-digit range) are feasible. Nevertheless, using a representative and string-wise sample allows for a validation of PID related degradation. Despite higher accuracy (in this case 2.9%), it will be necessary to use pre-conditioning procedures due to dark storage (due to transport) or seasonal power variations (Steabler-Wronski effect in Si TF). PID tests enable one to test for general module sensitivity to PID, regeneration potential and the further risk to operate the modules in the plant. **Table 3** summarizes these advantages and disadvantages between field and laboratory measurements.

Table 3: Evaluation of field and laboratory measurement advantages and disadvantages

	Field	Laboratory
Sample number	+ High	- Low
Costs	- Low	+ High
Measurement cond.	- Instable	+ Stable
Accuracy	- Low	+ High

A combination of field and laboratory tests allowed us to draw the right conclusions and give more precise and confident results. This enables the parties involved to take measures against PID, e.g. negative pole grounding, regeneration or module replacement. A complete module replacement was necessary in two of the three investigated projects.

The following recommendations are proposed to improve inspection methods of modules in operating plants:

- Further automatization is needed in failure detection and analysis in the field. Monitoring service providers implement this already but it also necessary for module testing like electroluminescence. This becomes even more critical as plant sizes continue to increase toward the GW range. The results provide evidence that a correlation between power drop and EL signal would lower inspection time, increase the number of inspected modules and lower the costs.
- Intensive work towards understanding the mechanisms behind module failures for TF is necessary. Therefore, knowledge exchange along the whole value chain is essential.

To support this the PEARL project was established: "Performance and Electroluminescence Analysis on Reliability and Lifetime of Thin-Film Photovoltaics". The PEARL TF-PV project aims to reduce the cost of electricity produced by thin-film PV power plants, by improving plant reliability, yield, and prediction of overall plant lifetime. To achieve this, large and small thin-film photovoltaic plants will be inspected by using in particular electroluminescence imaging. Electroluminescence imaging observes the light emitted by solar cells during the application of a voltage and provides spatially resolved information about a solar cell or module quality. During the project, the applicability and understanding of electroluminescence imaging methods to large-scale measurements on thin-film solar cells and modules will be improved with the objective to obtain knowledge about the appearance, behaviour and progression of failure mechanisms in thin-film solar cells. The obtained information will be used to increase the long-term profitability of thin-film photovoltaic projects by increasing operating yield, reducing operating and maintenance costs, improving accuracy of investment models, and increasing bankability.

The project is an international collaboration of industrial partners and research centres from Germany, Austria and the Netherlands, which are brought together via the Solar-era.net framework, namely: Forschungszentrum Jülich, Helmholz-Zentrum Berlin für Materialien und Energie (PVcomB), PI-Berlin, Austrian Institure of Technology, Crystalsol, TNO, ECN, TU-Delft, Solar Tester, KiesZon, eigenenergie.net and Straightforward.

5 ACKNOWLEDGEMENT

Many people at PI Berlin have contributed to the present work. The authors particularly acknowledge the input by Poonam Kothari, Stefan Wendlandt, Nicoletta Ferretti, Alexander Preiss, Carsten Leighton, Sören Rindert, Stefan Janke, Simon Koch and Steven Xuereb.

6 REFERENCES

[1] P. Lechner, J. Schnepf, D. Geyer, R. Schaeffler, R. Wächter, T. Repmann, "DOES THE NEW IEC 62804-2 PID TEST PROCEDURE COVER A SERVICE LIFE OF CIGS PV MODULES?", 32nd EU PVSEC, Hamburg, 2015

[2] Technical specification IEC TS 62804-2 Ed.1, 82/1238/CD:2017-01

[3] S. Yamaguchi, S. Jonai, K. Hara, H. Komaki, Y. Shimizu-Kamikawa, "Potential-induced degradation of Cu(In,Ga)Se2 photovoltaic module", Japanese Journal of Applied Physics, 2015

[4] P. Lechner "PID Failure of c-Si and Thin-Film Modules and Possible Correlation with Leakage Currents", NREL PV Module Reliability Workshop, 2013

[5] M. Atsushi, H. Yukiko, "Potential-induced degradation of thin-film Si photovoltaic modules", Japanese Journal of Applied Physics, 2017

[6] M. A. Contreras, P. Hacke and I. Repins,"Development of Cu (In,Ga)Se2 Test Coupons for Potential Induced Degradation Studies", Photovoltaic Specialists Conference (PVSC), 2016 IEEE 43rd, 2016

[7] T. Weber, J. Berghold, F. Heilmann, M. Roericht, S. Krauter, P. Grunow, "TEST SEQUENCE DEVELOPMENT FOR EVALUATION OF POTENTIAL INDUCED DEGRADATION ON THIN-FILM MODULES", 28th EU PVSEC, Paris, 2013

[8] T. Weber, E. Benfares, S. Krauter, P. Grunow., "Electroluminescence on the TCO-Corrosion of Thin Film Modules", 25th EU PVSEC, Valencia, 2010

[9] S. Koch, T. Weber, C. Sobottka, A. Fladung, P. Clemens, J. Berghold, "OUTDOOR ELECTROLUMINESCENCE IMAGING OF CRYSTALLINE PHOTOVOLTAIC MODULES: COMPARATIVE STUDY BETWEEN MANUAL GROUND-LEVEL INSPECTIONS AND DRONE-BASED AERIAL SURVEYS", 32nd EU PVSEC, München, 2016

[10] IEC 61646 Ed. 2.0 (2008)

[11] IEC 61215-2, Ed.1.0 (2016)

[12] S. Wendlandt, S. Berendes, T. Weber, J. Berghold, S. Krauter and P. Grunow· „SHADOWING INVESTIGATIONS ON THIN FILM MODULES", 32nd EU PVSEC, München, 2016

[13] A. Preiss, U. Siegfriedt, T. Röder, P. Grunow, "Aktuelle Betrachtung der Messunsicherheit der Leistungsmessungen an PV-Modulen unter Feldeinsatzbedingungen, oder: Sind die 5% Unsicherheit (Pmax) wirklich die untere Grenze?", PV Symposium Bad Staffelstein, 2017

EARLY POTENTIAL INDUCED DEGRADATION (PID) DETECTION IN THE FIELD: VOLTAGE MEASUREMENT METHODS

Michalis Florides*, George Makrides and George E. Georghiou
PV Technology Laboratory, FOSS Research Centre for Sustainable Energy, Department of Electrical and Computer Engineering, University of Cyprus.
Panepistimiou 1 Avenue, P.O. Box 20537, Nicosia, 1678, Cyprus.
* florides.michalis@ucy.ac.cy

ABSTRACT: This paper examines the use of two different voltage measurement methods which could be used to detect Potential Induced Degradation (PID) in the field before any significant power loss occurs. The two methods are the open circuit voltage and the forward bias voltage measurement under light and dark conditions respectively. A theoretical analysis shows that the lower is the photocurrent or the forward bias current, the earlier PID can be detected. The experimental study of the aforementioned methods has validated the theoretical results and also revealed that the forward bias voltage method is a better and more accurate way to detect PID due to the controlled applied forward bias current. On the other hand, the measurement uncertainty of the pyranometer at low irradiance levels makes the open circuit voltage method a less promising technique for early PID detection.

Keywords: Photovoltaics (PV), Potential Induced Degradation (PID), Detection in the Field

1 INTRODUCTION

During the last few years, Potential Induced Degradation (PID) has become a major degradation mechanism for silicon wafer-based photovoltaic systems (PV) as a result of the use of increased system voltage up to 1000 V DC. This failure mode is more pronounced nowadays because of the high potential developing between the modules and ground and the utilisation of transformer-less inverters, which do not allow for grounding of the negative terminal of a PV module string. In such cases, high voltage is developed between each module terminals and its grounded frame within the string, with half of the modules under positive bias and the other half under negative bias. The modules under negative bias, and especially the ones at the end of the string, are susceptible to PID [1]. In addition, it is commonly accepted that sodium ions (Na+) drift under the influence of the electric field that results from the negative bias voltage, which subsequently causes shunting of the cells by the Na+ ions [2], [3].

Even though the cause of PID is known, there is still no widely accepted detection mechanism, which can be used in the field to sense PID development in the PV modules under operation. The most common method for detecting PID is electroluminescence (EL) imaging, but this is time consuming and costly as the modules have to be dismantled for testing. Another method reported for the detection of PID is the use of a drone copter with a thermal camera to identify PID affected modules [4].

A lot of work has been done on PID, but the main focus so far has been the cause of PID and methods to prevent it. Instead, our work [5] focuses on early PID detection in the field before any significant power loss occurs. This is important since early detection will flag a warning to take recovery measures or to replace the PV modules before a significant amount of energy (and consequently revenue) is lost. The increasing system voltages used in PV systems render the development of a reliable and affordable method for PID detection a very timely issue [6].

This work examines electrical quantities, which can be used for detecting PID, with an inexpensive and less time consuming manner compared to methodologies applied at present. PID affects the performance of a PV cell/module by decreasing its shunt resistance (R_{sh}) and consequently its fill factor (FF) and at a later stage its open circuit voltage (V_{oc}) [1], [7], [8]. On the other hand, the short circuit current (I_{sc}) is also reduced, but this occurs when PID is very severe [1]. The use of the open circuit voltage (V_{oc}) under light conditions and the forward bias voltage (V_{fb}) under dark conditions are proposed as the electrical quantities to monitor for the identification of PID. The benefit of using the aforementioned electrical quantities is that the detection algorithm can be implemented with affordable equipment and/or can be embedded in the PV inverter software.

Initially, a simulation of the open circuit voltage (V_{oc}) with photocurrent (I_{ph}) was performed to demonstrate the concept of early PID detection at low irradiance. The same concept applies for the forward bias voltage (V_{fb}) at low forward bias current (I_{fb}). Then, an experiment was set up outdoors to verify the simulation results.

2 EXPERIMENTAL SETUP

2.1 Outdoor Infrastructure

For this test, an outdoor experimental setup has been built as shown in Fig. 1 and multi-crystalline silicon cells have been tested. The cells are PID prone with a cell area of 243.4 cm^2, 3 mm thick glass, x-23-105 EVA and TPT Akasol PV1000V back-sheet.

A metallic structure was used to mount the solar cells and the various devices such as the pyranometer and ambient temperature and relative humidity sensors which were used during the experiment. Electrical boxes housing the electrical circuits and the data logger (Fig. 1, inset A) were mounted on the structure as well. A high voltage power supply (HV PSU) was used to apply -1000 V DC bias voltage to the solar cells, which were isolated from the metallic structure with rubber holders.

Figure 1: PID experimental setup: metallic structure with solar cells, sensors (pyranometer, ambient temperature and relative humidity) and data logger.

2.2 Indoor Testing

For the indoor tests a Flasher was used to assess the power of the cell during PID progression and a Source Meter for performing the dark IV measurements. In addition, the cell temperature during the dark IV test was monitored with a k-type thermometer.

The equipment and its measuring accuracy are summarized in Table I.

Table I: Equipment and measurement accuracy

Equipment	Accuracy
Outdoor infrastructure	
HV PSU	10%
Pyranometer (Hukseflux SR11)	First Class
T_{amb} sensor (Vaisala HMDW110)	0.2%
RH sensor (Vaisala HMDW110)	4%
Data logger (TopMessage)	16 bits
Indoor test	
Flasher (Pasan)	3%
Source Meter (Keithley 2430)	< 0.1%
k-type thermometer (Fluke 52II)	< 1.5%

2.3 Electrical Connection

The positive and negative wires of the cell were left at open circuit and the open circuit voltage (V_{oc}) was measured by the data logger. Aluminium tape was applied at half of the perimeter of each cell frame (Fig. 1, inset B) to emulate the module frame. The HV PSU negative terminal (-ve) was connected to the cell frame (aluminium tape) and the positive terminal (+ve) to the positive terminal (+ve) of the cell.

3 METHODOLOGY

The following methodology was used for the simulation of the open circuit voltage (V_{oc}) with photocurrent (I_{ph}) and to experimentally study the progression of PID in the samples under test which was then related to the open circuit voltage (V_{oc}) and forward bias voltage (V_{fb}).

3.1 Open Circuit Voltage vs Photocurrent Simulation

In order to demonstrate that the open circuit voltage (V_{oc}) is an appropriate electrical quantity to use for PID detection, a model was developed based on the one-diode model for c-Si PV cells. Since the shunt resistance (R_{sh}) value decreases with increasing PID [7], [8], the PID progression was modelled by different values of shunt resistance (R_{sh}). The value of the shunt resistance (R_{sh}) was varied exponentially from 1 mΩ to 1 kΩ. For the simulation, the reverse saturation current (I_{rs}) at a cell temperature of 25 °C was set equal to 0.1 nA and the ideality factor (n) equal to 1.

PID appears to be more severe at low irradiance (i.e. low photocurrent) [9], [10], hence the open circuit voltage (V_{oc}) was simulated for various photocurrent (I_{ph}) values to identify the best current range for the assessment of the occurrence of PID.

In addition, the power loss (P_{loss}) was estimated based on the ratio of the output power change with shunt resistance (R_{sh}). The output power was calculated by finding the maximum power operating point of the cell. This was undertaken in order to assess whether it is possible to detect PID before any significant power loss occurs by using the open circuit voltage (V_{oc}) measurement method. The ambient temperature (T_{amb}) was set to 25 °C and the cell temperature (T_{cell}) was calculated for different photocurrent values (i.e. irradiance) by assuming a 3.125 °C/A temperature rise from ambient. Both the reverse saturation current (I_{rs}) and the thermal voltage (V_T) of the cell were varied with temperature according to the p-n junction behaviour to obtain more realistic results.

3.2 Outdoor Test

The HV PSU output voltage was set to -1000 V DC continuously. All the sensors were connected to the data logger, which was sampling at every second and averaging over a period of 1 minute.

3.3 PID Progression

The PID progression was monitored by indoor Flash tests of the cell at standard test conditions (STC). The recorded power was related to the open circuit voltage (V_{oc}) and forward bias voltage (V_{fb}), in order to see if there is any correlation between them as PID was progressing at outdoor conditions. The cell power (P_{cell}) was also calculated as a percentage by dividing the recorded power with the initial power of the cell recorded before the test had started.

3.4 Open Circuit Voltage Measurement

The open circuit voltage (V_{oc}) of the cell was measured continuously by the data logger even when the cell was biased at -1000 V DC. This was possible via an analog isolation signal conditioner which was isolating the cell high voltage side from the data logger low voltage side.

The data logger was also recording the global plane of array irradiance (G_{POA}), the ambient temperature (T_{amb}) and the relative humidity (RH). The cell temperature (T_{cell}) was not measured but was estimated based on the following empirical formula:

$$T_{cell} = T_{amb} + G_{POA} * 243.4 * 10^{-4} \quad (1)$$

The G_{POA} coefficient has units of m²°C/W.

3.5 Forward Bias Voltage Measurement

The cell was forward biased under dark conditions and the forward bias voltage (V_{fb}) was measured at various forward bias currents (I_{fb}) at room temperature. The current was varied from 0.1 mA to 3 A. This is useful for establishing the change of the forward bias voltage with PID progression at outdoor conditions and to identify the best forward bias current range to be used for PID detection. The cell temperature (T_{cell}) was monitored as well in order to eliminate any uncertainties due to temperature variation during the test.

4 RESULTS

Continuous monitoring of the cells had been taking place and the days on the x-axis of the plots correspond to the days since the start of the relevant experiment. The x-axis tick is placed at 12:00 o'clock of each day. Only the results of one cell are presented.

4.1 Open Circuit Voltage vs Photocurrent Simulation

Fig. 2 shows the open circuit voltage (V_{oc}) and cell power loss (P_{loss}) with shunt resistance (R_{sh}) for various values of photocurrent (I_{ph}, i.e. solar irradiation). The PID progress is modelled by decreasing R_{sh} as described in the methodology. In order to assess PID occurrence, a voltage of 0.4 V was set as the threshold for PID detection, i.e. when the cell voltage drops below 0.4 V PID has occurred.

From the results, it can be seen that at low photocurrent (less than 0.1 A, and hence irradiance lower than < 12.5 W/m^2), PID can be detected before the power reduces by more than 1%. On the other hand, if the open circuit voltage (V_{oc}) is measured at higher photocurrent values, then PID can only be detected after a significant power loss occurs. Therefore, for a photocurrent higher than 4 A, and hence high irradiance (higher than 500 W/m^2), PID can only be detected after the power has been reduced by more than 40%.

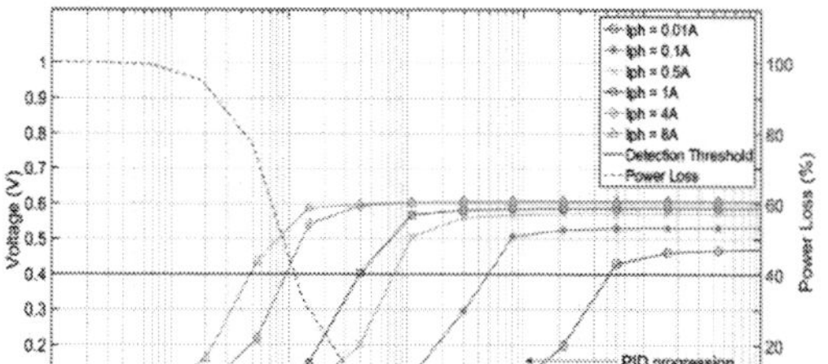

Figure 2: Open circuit voltage (V_{oc}) and percentage cell power loss (P_{loss}) vs shunt resistance (R_{sh}) for various values of photocurrent (I_{ph}, i.e. solar irradiation). The cell output power before any power loss occurs is 2.6 W. The more the PID progresses the lower the shunt resistance is.

Therefore, for an early detection of PID with the proposed method, the open circuit voltage (V_{oc}) has to be measured at low irradiance. In this way, a detection algorithm can be embedded into the inverter which will run early in the morning when the irradiance is low in order to measure the open circuit voltage and assess PID occurrence. The same idea of low bias current applies to the forward bias voltage (V_{fb}) method under dark conditions. Both electrical quantities have been tested experimentally with the results presented in the following sections.

4.2 Outdoor Measurements

The ambient conditions, the measured open circuit voltage (V_{oc}) and the estimated cell temperature (T_{cell}) during the test are shown in Fig. 3 in order to get an insight into the outdoor test environment conditions. These conditions are valid for both the open circuit voltage and forward bias voltage tests since the same cell was used.

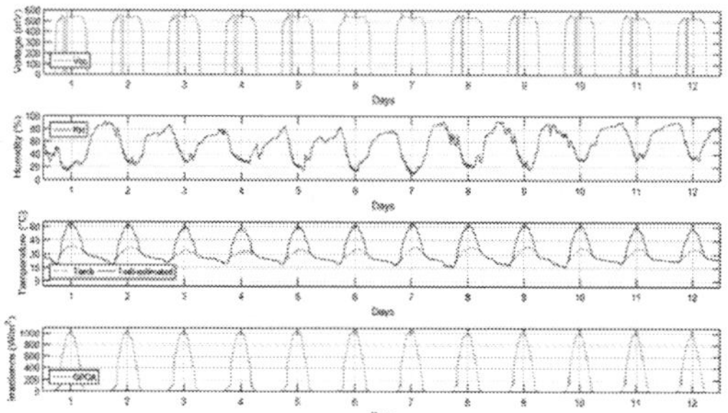

Figure 3: Open circuit voltage (V_{oc}), relative humidity (RH), ambient temperature (T_{amb}), estimated cell temperature (T_{cell}) and irradiance (G_{POA}). Rapid drops of Voc down to 0 V are due to the removal of the cell for indoor testing.

The open circuit voltage (V_{oc}) of Day 1 early in the morning was recorded as a reference for comparison with the next days in which the high voltage bias (-1000 V DC) was applied to the cell. The high voltage was applied on the first day just after the indoor test. After the indoor test on Day 11, the high voltage bias was switched off to study the thermal recovery of the cell under the sun light.

4.3 Indoor Measurements

From the indoor measurements, the cell power (P_{cell}) and its forward bias voltage (V_{fb}) were recorded while PID was progressing (Fig. 4). Although the cell power did not change by more than 5%, the forward bias voltage had significant changes. According to the simulation results, this is possible since the shunt resistance (R_{sh}) has to change to a high degree for significant amount of power to be lost. The forward bias voltage is analysed in more detail in section 4.5. It was plotted along with cell power in Fig. 4 to demonstrate that it changes significantly before any power loss occurs.

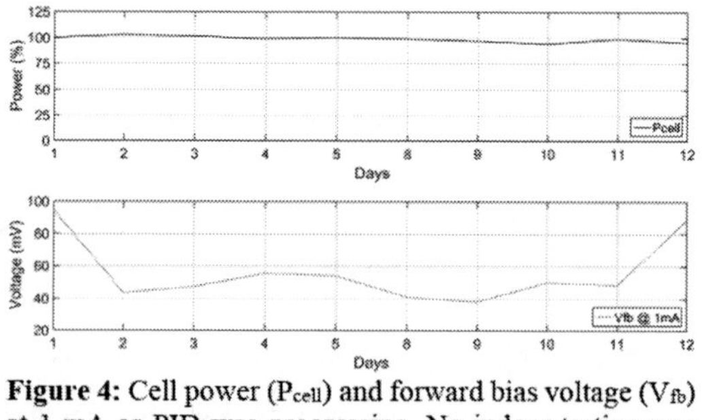

Figure 4: Cell power (P_{cell}) and forward bias voltage (V_{fb}) at 1 mA as PID was progressing. No indoor testing was done on Days 6 and 7.

From Fig. 4 it is also seen that the forward bias voltage (V_{fb}) drops on Day 2 and then partially recovers and drops again. On Day 12 the voltage almost fully recovers because the -1000 V bias voltage was switched off on Day 11. The recovery of the voltage is due to the increase of the cell shunt resistance (R_{sh}) due to the thermal recovery of PID. The thermal recovery on some days was because the weather conditions were such that the recovery rate was higher than the PID progression rate.

4.4 Open Circuit Voltage vs PID Progression

The open circuit voltage (V_{oc}) of the cell was monitored during the test and a comparison (Fig. 5) was made for values early in the morning to check whether it is affected by PID progression.

The reference open circuit voltage (before the -1000 V was applied) was recorded on Day 1. As PID was progressing from Day 1 to Day 9, the voltage was decreasing as expected due to the reduction of the shunt resistance (R_{sh}). For the rest of the days, which are not shown in the graph, the weather conditions were such that the cell thermally recovered partially from PID as shown in Fig. 4.

From the irradiance levels in Fig. 6, the effect is more pronounced at irradiance values lower than 10 W/m^2 (i.e. photocurrent values lower than 8 mA) as predicted by the simulation results in section 4.1.

The decrease in voltage can be mostly attributed to the PID progression (R_{sh} reduction) and not to differences in the irradiance on each day. For example, if Day 2 and Day 8 are compared, according to the irradiance values in Fig. 5, Day 8 open circuit voltage (V_{oc}) should be higher than Day 2. However, Day 2 voltage is higher due to the more severe PID on Day 8. This discrepancy is due to the inability of the pyranometer to accurately measure very low irradiance levels.

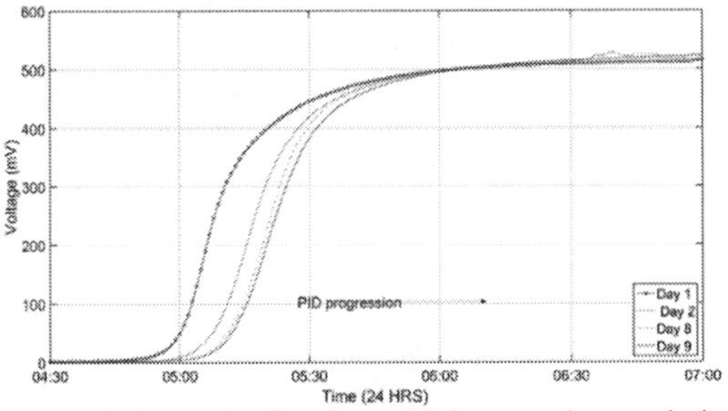

Figure 5: Open circuit voltage (V_{oc}) comparison early in the morning with PID progression. Day 1 is the reference day and Day 9 was the worst PID day.

To alleviate any effects of temperature on the open circuit voltage (V_{oc}) change with PID progression, the estimated cell temperature (T_{cell}) was plotted on the respective days of comparison as shown in Fig. 6. For example, if Day 1 and Day 8 are compared, Day 1 voltage should be lower than Day 8 due to the higher cell temperature (T_{cell}) as expected from the p-n junction diode behaviour. However, Day 1 voltage is higher due to the more severe PID on Day 8. Hence, the changes in the open circuit voltage are mostly attributed to PID progression.

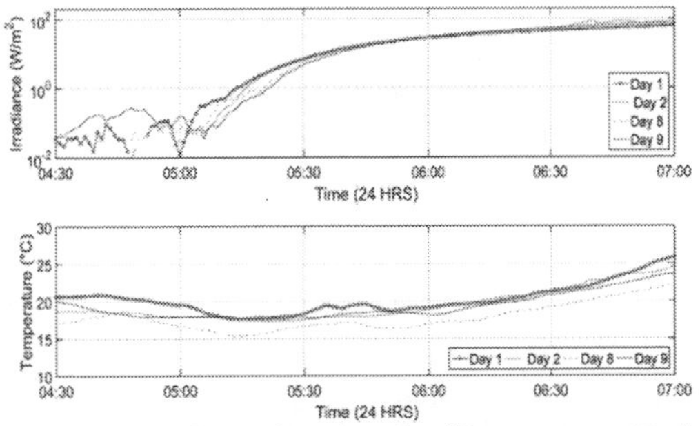

Figure 6: Irradiance (G_{POA}) and cell temperature (T_{cell}) early in the morning during PID testing. PID was increasing from Day 1 to Day 9.

It has been shown that the open circuit voltage (V_{oc}) early in the morning changes with PID progression before significant power loss (P_{loss}) occurs. In this test the results were as expected because of the short duration of the test and hence the very similar irradiance values on each day (clear sky, see Fig. 3). For a long time of testing, if the irradiance is highly different on each day due to weather conditions, an accurate way of measuring it is required otherwise false PID indication could be alarmed. It has been seen that a pyranometer cannot measure accurately very low levels of irradiance and hence a more predictable method for PID detection like the forward bias voltage (V_{fb}) method described in section 4.5 is required.

4.5 Forward Bias Voltage vs PID Progression

The forward bias voltage (V_{fb}) method under dark conditions was another PID detection technique to examine since it is affected by the cell shunt resistance (R_{sh}) in a similar way to the open circuit voltage. From the results in Fig. 7, it can be seen that for bias current values of less than 10 mA, the cell voltage was changing to a high degree with PID progression. This happened even before significant power was lost (lower than 5%, Fig. 4) as predicted from the simulation results in section 4.1. On the other hand, for bias current values greater than 100 mA no change in the cell voltage had been observed. The higher the bias current level, the more severe the PID degradation needed to observe any voltage changes. Hence low bias current levels are recommended for early PID detection.

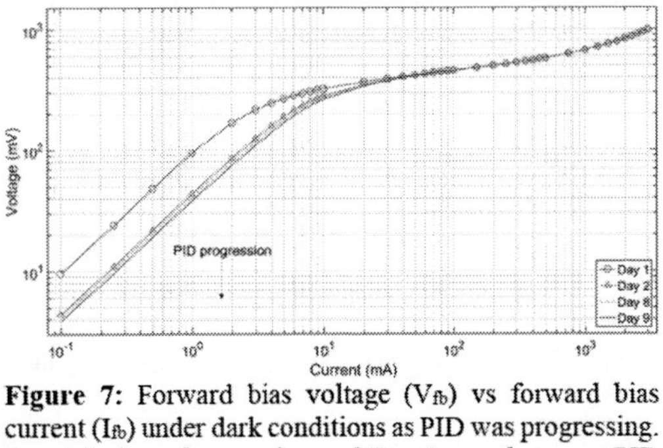

Figure 7: Forward bias voltage (V_{fb}) vs forward bias current (I_{fb}) under dark conditions as PID was progressing. Day 1 is the reference day and Day 9 was the worst PID day.

The reference forward bias voltage (before the -1000 V was applied) was recorded on Day 1. As PID was progressing from Day 1 to Day 9, the voltage was decreasing as expected due to the reduction of the shunt resistance (R_{sh}). The rest of the days, which are not shown in the graph, are days in which the weather conditions were such that the cell thermally recovered partially from PID as shown in Fig. 4.

The decrease in voltage can be attributed to the PID progression (R_{sh} reduction) and not to differences in the cell temperature on each day. For example, if Day 1 and Day 2 are compared (Fig. 8), Day 1 voltage should be less than Day 2 due to the higher cell temperature (T_{cell}) as expected from the p-n junction diode behaviour. However, Day 1 voltage is higher due to the more severe PID on Day 2. Hence, the changes in the forward bias voltage are mostly related to PID progression.

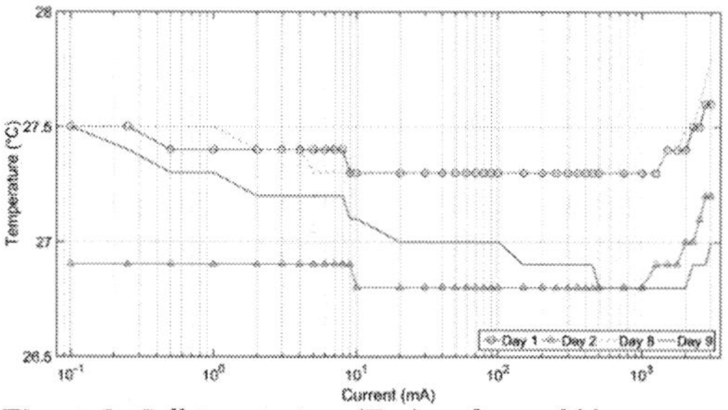

Figure 8: Cell temperature (T_{cell}) vs forward bias current (I_{fb}) during the dark IV measurements.

Both voltage measurement methods demonstrated similar results, however, the forward bias voltage (V_{fb}) method for PID detection seems better than the open circuit voltage (V_{oc}) method since the applied forward bias current can be controlled accurately and does not depend on the irradiation (i.e. photocurrent, I_{ph}) measurement which is highly affected by the pyranometer uncertainty at very low irradiance levels.

5 CONCLUSIONS

This work presented two possible methods for early PID detection in the field, the open circuit voltage (V_{oc}) method and the forward bias voltage (V_{fb}) method under light and dark conditions respectively. Initially, a simulation was performed and it has been shown that for early PID detection before any significant power loss occurs, the bias current, whether it is the photocurrent (I_{ph}) or the applied forward bias current (I_{fb}), has to be very low (< 100 mA). An outdoor experiment with multi-crystalline cells was set up and the two aforementioned detection methods have been tested. The open circuit voltage changed with PID progression early in the morning for photocurrent values lower than 10 mA. The forward bias voltage method demonstrated similar results for bias current values lower than 10 mA as well. During the experiment, the cell power (P_{cell}) did not change by more than 5%, however, the two proposed methods showed significant voltage changes as predicted by the simulation results and can be used for early PID detection.

6 ACKNOWLEDGEMENTS

This work was funded through the IPERMON project which is co-financed by the SOLAR-ERA.NET Transnational Calls of the European Union under the grant agreement number SOLAR-ERA.NET/1214/08. The authors would also like to acknowledge the financial support from the Cyprus Research Promotion Foundation.

Simon Koch from PI Berlin is gratefully acknowledged for his help with the PID prone cells.

7 REFERENCES

[1] S. Pingel, O. Frank, M. Winkler, T. Geipel, H. Hoehne, and J. Berghold, "Potential Induced Degradation of Solar Cells and Panels," in *35th IEEE Photovoltaic Specialists Conference (PVSC)*, 2010, pp. 002817–002822.

[2] V. Naumann, C. Hagendorf, S. Grosser, M. Werner, and J. Bagdahn, "Micro Structural Root Cause Analysis of Potential Induced Degradation in c-Si Solar Cells," *Energy Procedia*, vol. 27, pp. 1–6, 2012.

[3] S. P. Harvey, J. Aguiar, P. Hacke, S. Johnston, and M. Al-Jassim, "Sodium Accumulation at Potential-Induced Degradation Shunted Areas in Polycrystalline Silicon Modules," in *43rd IEEE Photovoltaic Specialists Conference*, 2016.

[4] T. Kaden, K. Lammers, and H. J. Moller, "Power Loss Prognosis from Thermographic Images of PID Affected Silicon Solar Modules," *Sol. Energy Mater. Sol. Cells*, vol. 142, pp. 24–28, 2015.

[5] M. Florides, G. Konstantinou, V. Venizelou, G. Makrides, and G. E. Georghiou, "Potential Induced Degradation (PID) Correlation to Leakage and Reverse Bias Currents," in *44th IEEE Photovoltaic Specialits Conference (PVSC)*, 2017.

[6] W. Luo *et al.*, "Potential-induced degradation in photovoltaic modules: a critical review," *Energy Environ. Sci.*, vol. 10, pp. 43–68, 2017.

[7] M. Bahr and K. Lauer, "Analysis of Activation Energies and Decay-time Constants of Potential-induced Degraded Crystalline Silicon Solar Cells," *Energy Procedia*, vol. 77, pp. 2–7, 2015.

[8] A. Masuda, Y. Hara, and S. Jonai, "Consideration on Na Diffusion and Recovery Phenomena in Potential-Induced Degradation for Crystalline Si Photovoltaic Modules," *Jpn. J. Appl. Phys.*, vol. 55, no. 2S, 2016.

[9] B. Braisaz and K. Radouane, "PID Results at Low Irradiances on c-Si Modules," in *40th IEEE Photovoltaic Specialist Conference (PVSC)*, 2014, pp. 1993–1996.

[10] S. Spataru, D. Sera, P. Hacke, T. Kerekes, and R. Teodorescu, "Fault Identification in Crystalline Silicon PV Modules by Complementary Analysis of the Light and Dark Current–Voltage Characteristics," *Prog. Photovoltaics Res. Appl.*, vol. 24, no. 4, pp. 517–532, 2016.

IMAGING INSPECTION SYSTEM AND METHODOLOGY FOR EVALUATION OF INHOMOGENEITIES IN PV MODULES – A CASE OF MULTICRYSTALLINE SILICON

Matevž Bokalič, Kristijan Brecl, Marko Topič
University of Ljubljana, Faculty of Electrical Engineering
Tržaška 25, SI-1000 Ljubljana, Slovenia
E-mail: Matevz.Bokalic@fe.uni-lj.si

ABSTRACT: Aiming for comprehensive imaging inspection of photovoltaic (PV) modules we have constructed a computer controlled imaging system comprising visual, UV phosphorescence and near-infrared (NIR) as well as short-wave-infrared (SWIR) electroluminescence (EL) inspection. The imaging system is computer controlled via custom developed LabVIEW® application. For image inspection we developed a software package, which displays all images side by side. The software package includes an inhomogeneity database assembled from previous project and literature that guides the image evaluation process. EL images of PV modules reveal a number of inhomogeneities including disconnected areas, cracks, contact finger interruptions, dark dots, bad edges, improper ribbon interconnection. The distribution of the inhomogeneities across the modules varies significantly. Cracks and disconnected areas are distributed most unevenly. Correlation between cracks and non UV-phosphorescent areas is strong. Random compression cracks are often indicated by a small dark circle on a UV phosphorescence image. These cracks often appear in combination with a longer normal crack. Accurate inhomogeneity identification and evolution either during accelerated testing or field operation of multicrystalline silicon PV modules are presented as case studies of temporal changes.
Keywords: Electroluminescence, Imaging, Multicrystalline Silicon, Characterisation, Inhomogeneities, PV Module

1 INTRODUCTION

Inhomogeneities have negative impact on performance of photovoltaic (PV) modules. They also reveal manufacturing deficiencies and indicate possible further degradation. Inhomogeneities are detected using various spatially resolved characterization methods [1,2], which allow their identification and make quality estimations possible.

Among spatially resolved methods for complete PV modules, the electroluminescence (EL) [3,4] became the most widely used as it offers fast and detailed examination of the modules. EL spectrum is primarily defined by the bandgap of the semiconductor and is usually in range between 1000 and 1300 nm for crystalline silicon solar cells. Two camera types cover this wavelength range. A cost effective silicon based sensors provide high resolution in range between 300 and 1100 nm, i.e. visual and near infrared (NIR) part of the spectrum. A more specialized short-wave infrared (SWIR) InGaAs based sensor covers range between 900 and 1700 nm. InGaAs sensors allow shorter acquisition times, but have limited resolution to below one megapixel. Use of both sensors offers additional image analysis, such as minority carrier diffusion length in silicon solar cells [5] or bandgap imaging in CIGS PV modules [6]. Complementary, ultraviolet phosphorescence (UV-ph) relies on phosphorescence of the ethylene vinyl acetate (EVA) foil. If EVA is sealed from oxygen, it becomes phosphorescent after exposure to UV light, and thus enables crack detection in the modules [7,8].

In this paper we will first describe the imaging inspection system at Laboratory of Photovoltaics and Optoelectronics (LPVO) within the University of Ljubljana, and continue with the methodology for evaluation of inhomogeneities in PV modules, which includes a custom program that leads the process based on the inhomogeneity database. We will then use the system to detect and identify various inhomogeneities based on different measurements, and present some temporal changes that we detected in inspection of 17 kW solar power plant.

2 METHODOLOGY

2.1 Imaging Inspection System

Imaging inspection system at LPVO is installed in a dark room. The imaging takes place in an air conditioned chamber separated with black curtains from the dark room. All imaging devices are mounted at fixed positions on a solid aluminium post (Fig. 1 a), whereas the inspected module is positioned on a wall-mounted holder (Fig. 1 b). Behind the module, there is a spring loaded insulated temperature sensor.

The imaging system consists of three cameras in different spectral ranges. An unmodified DSLR camera is used to obtain a high-resolution visual image of the inspected module and, combined with an in-house-developed UV-LED illumination, a UV-ph image. Zoom level and focus are set constant for all images. Aperture, sensitivity and exposure times are predefined based on illumination conditions for visual and UV-ph image.

Electroluminescence image acquisition takes place in the dark by two different cameras. An air-cooled high-resolution 16-megapixel silicon CCD based FLI ML16803 camera acquires a NIR part of luminescence spectrum up to 1100 nm. An air-cooled 640×512 pixels InGaAs plane-array based Photonic Science PSL VGA 15 µm camera acquires a SWIR part of luminescence spectrum (900 – 1700 nm). Acquisition time is set on per-module per-excitation basis to obtain the best ratio between available dynamic range of the signal and acquisition time. Electroluminescence is excited by a power supply, that ensures enough power to excite electroluminescence in full-size crystalline PV modules as well as commercial thin-film PV modules. Temperature of the module is measured by a digital temperature sensor and read by an in-house developed 1-wire to USB interface.

The whole system is controlled via a custom developed modular Lab-View program wrapper that controls and synchronizes 4 subprograms, where each of them controls one instrument (Fig. 2). The program allows easy manual control of all instruments from a single screen, and also offers automated measurements of predefined queue of single measurements.

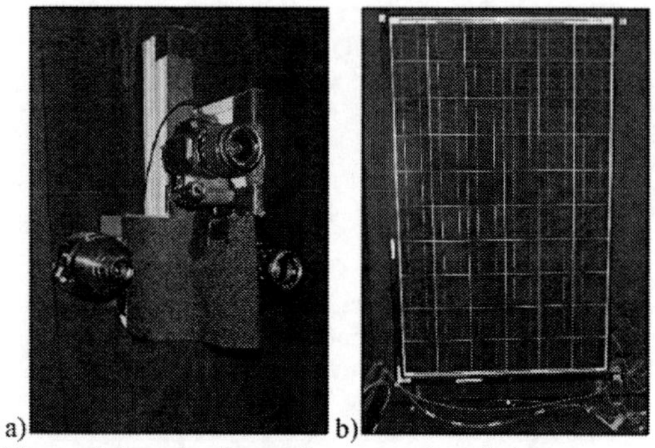

Figure 1: Imaging inspection system in a dark room
a) Three cameras at fixed positions on a post
b) Wall-mounted module holder

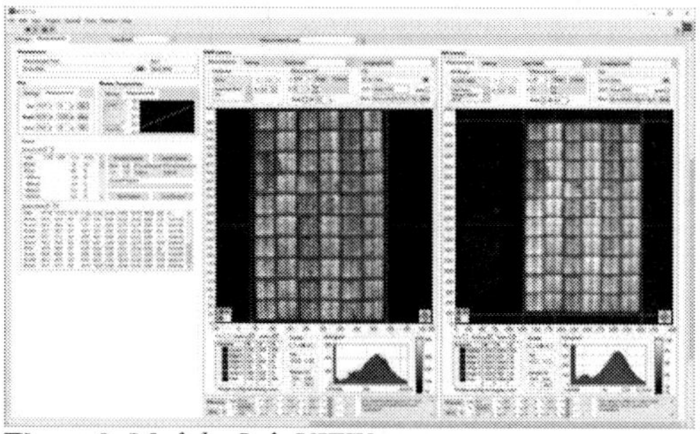

Figure 2: Modular Lab-VIEW program

After acquisition, images are post processed to reflect actual electroluminescence emission. ImageJ [9] based program package Fiji [10] is used for post processing. Standard post processing steps are organized in macros and include background subtraction, outliers removal and vignetting correction.

2.2 Evaluation Program

For image inspection we developed a software package, which displays visual, UV-ph and EL (NIR and SWIR swappable) images side by side (Fig. 3). Synchronized zoom and pan together with triple cursor and quantitative per-cell EL evaluation allows systematic comparison and evaluation of the images. The program contains a configurable inhomogeneity database, which includes a description of each inhomogeneity including instructions for its identification and an example image. Inhomogeneities from the database are listed and described more detailed in the next sub-section.

The module evaluation process is as follows. The operator selects a module from the list on the left. The images and the data of the module is displayed. The inhomogeneity database then leads the operator over all predefined inhomogeneities to be evaluated on the module. To enhance visibility of certain inhomogeneities, different false-colouring options are available, e.g. UV threshold mask, fire, blue. For each inhomogeneity type the following data should be evaluated: the number of cells affected, evaluation of intensity of the inhomogeneity (usually a fraction of area affected by the inhomogeneity, which may be evaluated using calculate area tool) and a comment if required. To ease the evaluation process, the operator should annotate inhomogeneities on the images using drawing tools. The entered data is stored to the database for reference and statistical evaluation.

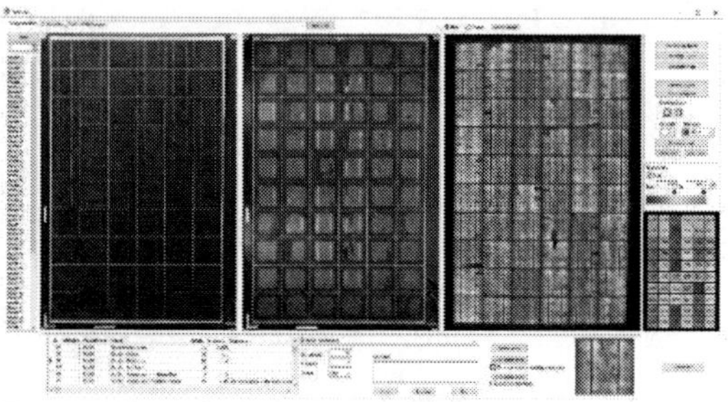

Figure 3: UV-vis-EL program's front panel

2.3 Classification of inhomogeneities

Inhomogeneity database was constructed from our previous work and literature [11–14]. Inhomogeneities are classified into nine main categories with subcategories:

- Inactive areas
- Cracks
  - Active, semi active, inactive, compression
- Contact finger interruptions
  - Compete, strong, weak
- Manufacturing marks
  - Dark dots (perpendicular, parallel, shading)
  - Belt marks
- Silicon brick mark
  - Bad edges (parallel, perpendicular, frame, corner)
- Reduced lifetime
- Ribbon related failures
  - Solder flux visible, corrosion, improper interconnect
- UV exhibited inhomogeneities
  - Average area, cracks, along ribbon, dots
- Other

Each inhomogeneity is described by its characteristic fingerprint, which is most often found in EL image and in case of cracks also in UV-ph image. In rare cases (e.g. visible solder flux and ribbon corrosion) the fingerprint is also found in visual image. The detailed description of the inhomogeneities and their fingerprints can be found in the literature [11–14].

3 RESULTS

A selection of inhomogeneities that were detected during the inspection of multi-crystalline Silicon PV modules from 17 kW PV power plant is presented. The inspected power plant was installed on the roof of the faculty's building (Ljubljana: N46°02.69', E14°29.38', orientation: 25° east, inclination: 30°) in late 2010. It has been in constant operation for almost 7 years. The only interruptions in operation were: a two-months shutdown in 2016 when it was completely disassembled and all the modules were characterized, and several shorter shutdowns for maintenance, experimental work and regular testing of selected modules.

A majority of the categorized inhomogeneities from the preceding section was detected in the modules. The exceptions are belt marks, which originate from certain types of cell manufacturing lines, and visible solder flux, which is a consequence of inaccurate module assembly.

The distribution of the inhomogeneities between the modules varies significantly. Cracks and disconnected areas are distributed most unevenly. Fig. 4 presents three modules with varying presence of (induced)

inhomogeneities. Fig. 4 a displays a module with minimal inhomogeneities. Some minor weak finger interruptions are visible and in the second row some cells are affected by silicon brick marks (darker edges of the cells). About 75% of cell in module in Fig. 4 b are affected by contact finger interruptions, ranging from weak (green - i), through strong (blue - ii), to complete (red - iii). Module displayed in Fig. 4 c is heavily affected by cracks. All distinguishable types of cracks can be found: inactive (green - i), semi active (blue - ii), active (dashed red – iii-a), active leading to inactive areas (solid red – iii-b), and compression (orange - iv).
During inspection of our power plant, UV-phosphorescence imaging revealed very good agreement between non UV phosphorescent areas and crack locations. Two examples of obtained correlations are presented in Fig. 5. Cell borders are labelled with orange, while cracks identified from EL images are labelled with red.

Fig. 5 a shows correlation between measurements on different types of cracks: top right cell: inactive, bottom right cell: semi active, bottom left cell: active crack propagated into disconnected area. In this case all EL-identified cracks have corresponding non-UV-phosphorescent areas which propagate away from the cracks in approximately equal distances in all directions. There are additional non-UV-phosphorescent areas that propagate along the marked green-dashed lines as they would propagate along the cracks. All these lines run along the busbars; however, no cracks nor contact finger interruptions (as a consequence of a crack under the busbar) can be observed in EL image. Such behaviour is not unique to this example as we have observed it on many other cells. This suggests that in some cases oxygen may travel along the ribbons from cell edges towards the central part of the cell and extinguishes the UV-phosphorescence along the way.

In Fig. 5 b, an example of compression crack is presented. Such a crack usually appears as a cross or an asterisk. Because it is very small, it is very difficult to observe in EL image, especially in case of multi-crystalline silicon cells, where grain boundaries may form very similar patterns. However, in UV-ph image, compression crack usually results in a non UV-phosphorescent area around it, which is easy to observe.

Several inhomogeneities change over time. Among others, we observed crack formation, either from existing cracks or compression crack, or anew, contact finger interruption formation and healing, and ribbon to busbar contact changes. In rare cases we also observed inspection induced damage which resulted in extremely severe crack and compression crack formation. Please note, that the left-hand side images in Fig. 6 were taken in 2013 with our old EL setup with lower resolution silicon CCD based camera [15].

Fig. 6 presents two examples of temporal changes of inhomogeneities. In Fig. 6 a, a crack formation is presented. In this example, the new crack originates from an older existing crack. Note that the old crack might have been initiated by the compression crack in the middle of the crack. In Fig. 6 b, ribbon to busbar contact degradation is evident. Accurate examination reveals, that the area along the busbar in the left-hand side figure (in dashed-green rectangle) emits higher EL signal than the area along the busbar in the right-hand side figure (in dashed-red

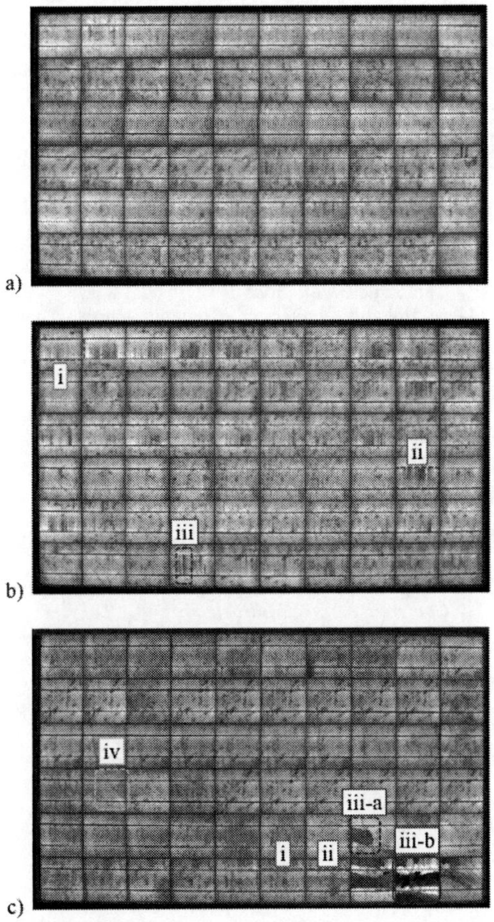

Figure 4: Modules showing varying presence of inhomogeneities:
a) Minimal induced inhomogeneities
b) Multiple finger interruptions
c) Multiple cracks

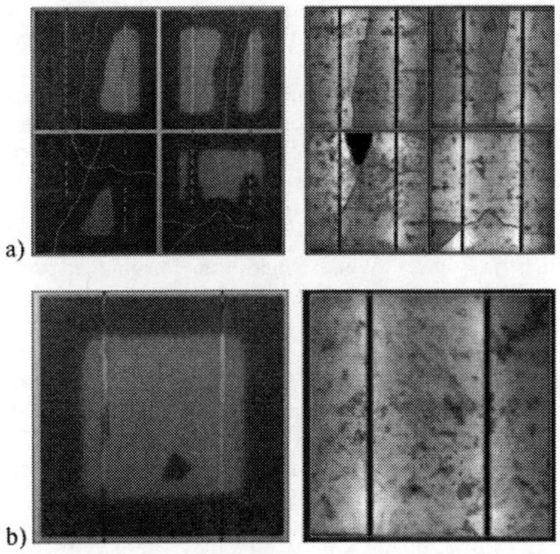

Figure 5: Correlation between UV phosphorescence and electroluminescence images
a) Crack correlation
b) Compression crack correlation

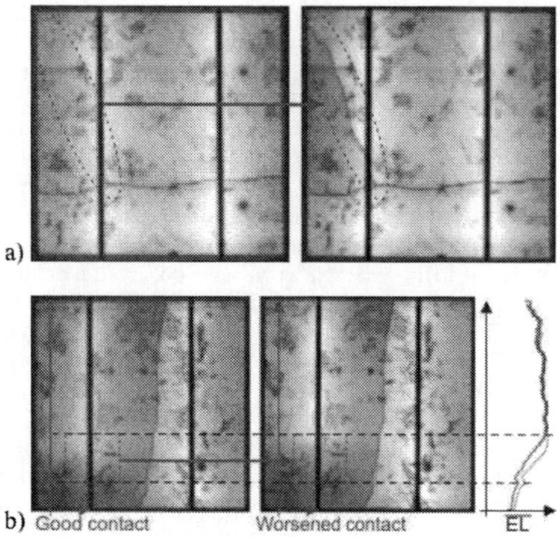

Figure 6: Temporal changes of inhomogeneities
a) Crack formation
b) Ribbon to busbar contact degradation

rectangle). The change is best observed on a normalized averaged EL linescan over the arrowed rectangles' area. The most probable cause is a newly formed disconnection between the ribbon and the busbar in the centre of the dashed rectangles. Note that in both figures the bottom part around the left busbar is darker than the bottom part around the right busbar, indicating an already present bad solder joint between the ribbon and the busbar.

4 CONCLUSION

We have presented the methodology for evaluation of inhomogeneities in PV modules. The presentation includes setup description including hardware and software, evaluation program description, and a list of inhomogeneities. The image analysis is done manually, primarily to gain knowledge about properties and development of inhomogeneities. In results section we presented examples of inhomogeneities, their distribution between modules, UV-ph to EL correlation and temporal dynamics.

Complete comparative image analysis eases image inspection and makes inhomogeneity identification more reliable. EL image provides the most information, while UV phosphorescence is a good indicator of crack presence. Usually, visual image provides very little to no information, however it may provide essential information in certain cases. While single time imaging inspection provides qualitative information about the modules, a comparison with sequence of older images, preferably starting from the production line or day 0 of operation, provides the temporal information, which could lead to estimation of inhomogeneity induced performance degradation.

Organized inhomogeneity data collection is important, as it enables evaluation verification, tracking of inhomogeneity development, statistical evaluation, and link between module inhomogeneity and module performance. These will be our future focus and a mean to establish a model of the power plant.

5 Acknowledgments

The authors acknowledge the financial support from the Slovenian Research Agency (Research Programme P2-0197, Project J2-7105).

6 References

[1] M. Bokalič, M. Topič, *Spatially Resolved Characterization in Thin-Film Photovoltaics*, Springer International Publishing, Cham, **2015**.
[2] D. Abou-Ras, T. Kirchartz, U. Rau, Eds. , *Advanced Characterization Techniques for Thin Film Solar Cells*, Wiley-VCH Verlag GmbH & Co. KGaA, **2011**.
[3] T. Fuyuki, H. Kondo, T. Yamazaki, Y. Takahashi, Y. Uraoka, *Appl Phys Lett* **2005**, *86*, 262108.
[4] U. Rau, *Phys Rev B* **2007**, *76*, 85303.
[5] P. Würfel, T. Trupke, T. Puzzer, E. Schäffer, W. Warta, S. W. Glunz, *J Appl Phys* **2007**, *101*, 123110.
[6] M. Bokalič, B. E. Pieters, A. Gerber, U. Rau, M. Topič, *Prog. Photovolt: Res. Appl.* **2017**, *25*, 184.
[7] J. Schlothauer, S. Jungwirth, M. Köhl, B. Röder, *Sol Energ Mat Sol Cells* **2012**, *102*, 75.
[8] M. Arnaud, M. Siebert, I. Kunze, G. Mathiak, M. Köntges, in *44th IEEE Photovoltaic Specialist Conference (PVSC)*, IEEE, Washington, DC, USA, **2017**, In print, #377.
[9] C. A. Schneider, W. S. Rasband, K. W. Eliceiri, *Nat. Methods* **2012**, *9*, 671.
[10] J. Schindelin, I. Arganda-Carreras, E. Frise, V. Kaynig, M. Longair, T. Pietzsch, S. Preibisch, C. Rueden, S. Saalfeld, B. Schmid, J.-Y. Tinevez, D. J. White, V. Hartenstein, K. Eliceiri, P. Tomancak, A. Cardona, *Nat. Methods* **2012**, *9*, 676.
[11] M. Bokalič, Conversion Efficiency Inhomogeneities of Photovoltaic Devices, University of Ljubljana, **2013**.
[12] M. Köntges, S. Kurtz, C. Packard, U. Jahn, K. A. Berger, K. Kato, T. Friesen, H. Liu, M. Van Iseghem, *Review of Failures of Photovoltaic Modules*, IEA, **2014**.
[13] *IEC NP 60904-13 Electroluminescence of Photovoltaic Modules (82/901/NP)*, **2014**.
[14] K. Bedrich, et. al, EL Round-Robin Project Draft, **2016**.
[15] M. Bokalič, K. Brecl, M. Topič, in *50th International Conference on Microelectronics, Devices and Materials and the Workshop on Ceramic Microsystems, MIDEM 2014 Proceedings*, Ljubljana, Slovenija, **2014**, pp. 291–295.

DIGITAL IMAGE PROCESSING ALGORITHMS FOR QUALITY-ENHANCEMENT OF ELECTROLUMINESCENCE PICTURING IN PV-FIELDS

G. Behrens[1], A. Domnik[1], K. Mertens[2], A. Arnds[2] , M. Diehl[3]

[1]Solar Computing Lab - University of Applied Sciences Bielefeld, Artilleriestraße 9, 32427 Minden, Germany
[2]PV-Testlab – University of Applied Sciences Münster, Department of Electrical Engineering and Computer Science, Stegerwaldstraße 39, 48565 Steinfurth
[3]Photovoltaikbüro, Ternus und Diehl GbR, Schönauerhofstraße 27, 65428 Rüsselsheim

Telephone +49.571.8385-184, email: gbehrens@fh-bielefeld.de

ABSTRACT: The focus of this work is on the development of digital image processing algorithms for electroluminescence pictures and videos of a PV-field taken under different lightening conditions. Algorithms are optimized for contraction of the exposure time of picture taking and for enhancement of the quality of the taken video-sequences.

Keywords: PV-Modules, Failure diagnosis, Electroluminescence measuring, image processing algorithms

1 AIMS

Electroluminescence Measuring enables a deep insight into photovoltaic modules. Module failures can be detected and analyzed before measurable losses of energy yield would have been occured.
This work analyses the low-cost video-taking system of electroluminescence measuring of Mertens et al. [1] by applying image processing algorithms with the aim of optimization for use under field conditions.
The exposure time of EL-picture taking should be contracted and the quality of EL-video-sequences should be enhanced.
Optimized quality of EL-measuring should enable field-work of qualified PV-experts under daylight conditions without necessity of night-time work.
Therefore different algorithms of digital image processing are investigated and implemented into the EL-measuring- system. Furthermore the algorithms are parameterized on different extraneous lighting conditions.
The software-system for EL-image and EL-video capturing should work independent from the powering device "pvServe" [2] and recognize automatically the cycles of the modules powering current.
The in Mertens et al. [3] described "optical lock-in-effect" should be developed and implemented.

2 APPROACH

The software-system includes the programming interface of "pvServe" powering-device (figure 1). Thus the cycles of modules powering current could be parameterized in frequency and intensity and can optimally adopted on respective extraneous lighting conditions by the developed algorithms.

```
gtr
scr
scr,i,5,0
scr,u,50
scr,run
scr,loop
scr,i,5,0
scr,delay,3000
scr,i,2,0
scr,delay,2000
mode,script
sb,r
```

Figure 1: Example of programming script to control the powering-device

For SW-realization of optical lock-in-effect by capturing EL-video-sequences an image algorithm of three steps is developed:
At first the powered module status and the unpowered module status are detected. Secondly all frames for powered subsequence and all frames of unpowered subsequence are arithmetically averaged. And thirdly the averaged result-images are subtracted and enhanced.
For detection of powered status and unpowered status for all frames of the sequence the image features "intensity", "intensity–gradient" and "normalized histogram", which are calculated by using the OpenSource C++-library "OpenCV" [4]. The features were evaluated by decision-function based on different threshold-methods [5] like "LevelTop", "LevelBottom", LevelTopBottom", Average" and "Median" every time for different possible parameter sets for modules powering current control, for camera settings and for extraneous lighting conditions.
The algorithm for the optical lock-in-effect uses a self-adoption on different intensity distributions within the frames by the CLAHE-method "Contrast Limited Adaptive Histogram Equalization". As a result the frame-contrast can be enhanced for different illumination scenarios, which can occur under field conditions (see in figure 2).

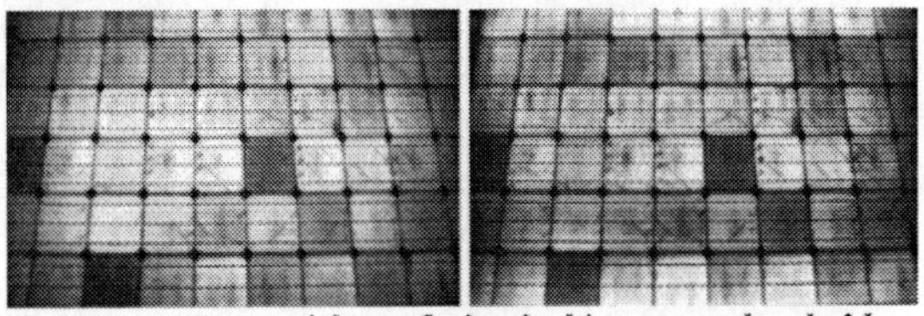

Figure 2: Comparision of simple histogram threshold algorithm (left site) an CLAHE- algorithm (right site)

3 RESULTS AND CONCLUSIONS

Sequences of experiments were performed on a moderately damaged PV-module (micro-cracks) and an intensively damaged PV-module (cell-cracks) under different extraneous lightening conditions. In all

experiments a camera Canon EOS 700D has had the same settings in video-mode with a framerate of 7 frames per second and a light sensitivity of ISO-3200. The cycles of modules powering were set as 3000ms for the powered module state and 2000ms for the unpowered module state. Best technology for power state detection in static experiments (static camera) was a threshold on basic image feature "intensity".
At the following figures 3, 4 and 5 the same four frames are shown respectively: upper left – the last frame of the unpowered subsequence, upper right – the last frame of the powered subsequence, under right – the difference image, calculated from the averregd frames of the two different subsequences and enhanced by CLAHE-algorithm.
At first, a moderately damaged PV-module was inspected inside the laboratory with ceiling lighting only, also seen in Figure 3. In this condition in the enhanced difference image of EL-measurement video all little damages and module failures like micro cracks and fingers were very good to see.

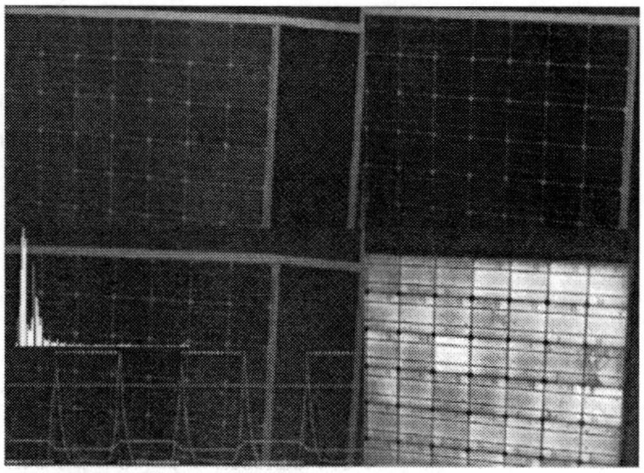

Figure 3: EL-measurement inside laboratory with ceiling lighting only

Figure 4 shows a measurement of the before named module under strong solar light influence passing through nearby windows and open doors of the laboratory at midday. The level of detail in the enhanced difference image of the EL-measuring video was reduced in relation to Figure 3, only bigger damages of the PV-module like cell cracks were visible in this experiment. Figure 5 shows a measurement-result also under solar light conditions plus ceiling lighting for a stronger damaged PV-module. Damages like cell-cracks could be very well visualized in the enhanced difference image of EL-video-sequences.
Results of experiments show that Low-cost-Electroluminescence for PV-modules through an enhanced difference image of EL-video-sequences is a practicable technique also under daylight conditions. The developed system allows full field-implementation of the method in twilight conditions. Under stronger lightening the level of detail decreases in the result images and only strong damages could be recognized.
In the future, cameras with higher-light-sensitivity in the wavelength spectrum of the electroluminescence effect could be used and the development and implementation of new image processing algorithms could much more optimize the low-cost outdoor EL-measurement with higher detail levels in the result images also under daylight conditions at cloudy days.

Figure 4: EL-measurement inside laboratory with strong solar light passing through the windows and open doors around noon.

Figure 5: EL-measurement inside laboratory with ceiling lighting and daylight passing through windows and open doors on a cloudy day.

4 SCIENTIFIC INNOVATION AND RELEVANCE

The scientific innovation of our work is that we implemented a software system realizing the "optical Lock-in effect" on video-sequences for low-cost-outdoor electroluminescence measurements with simple camera-types. With our new SW-system the EL- measurement can be provided in the PV-fields not only at night under dark lighting conditions, but also under conditions with some extraneous lighting like we have it in laboratories with ceiling lights and outdoor in time near twilight.
The SW-System works independent from the powering system and can be used very easily and cost-efficient.

5 REFERENCES

1. K. Mertens, H. Kösters, M. Diehl "LowCost-Outdoor-EL: Cost-Efficient Extensive On-Site Quality Analysis of Solar Modules", Proceedings of 31st European Photovoltaic Solar Energy Conference and Exhibition, Hamburg, 2015

2. M. Diehl "Photovoltaikanlagen untersuchen mit dem pvserve", [Online]. Available:

http://pvserve.de/pvserve-photovoltaikanlagen-thermografie-elektrolumineszenz. [18 April 2016].

3. K. Mertens, A. Arnds, G. Behrens, A. Domnik, "Low-cost-Outdoor-Electroluminescence : Significant Improvements of the method", Proceedings of 32st European Photovoltaic Solar Energy Conference and Exhibition, Munich, 2016

4. „OpenCV 3.1.0," [Online]. Available: http://docs.opencv.org/3.1.0. [03 Juli 2016].

5. P.S. Heckbert „Graphics Gems IV", Oxford: Elesevier LTD, 1994

EFFICIENT DETECTION OF FINGER INTERRUPTIONS FROM PHOTOLUMINESCENCE IMAGES

Iskra Zafirovska[1], Mattias K. Juhl[1], and Thorsten Trupke[1,2]
[1] The University of New South Wales, Sydney, Australia
[2] BT Imaging Pty Ltd, Sydney, Australia

ABSTRACT: Finger interruptions or finger breaks are a common occurrence during manufacturing of screen-printed solar cells and modules, and may result in a decrease in power output due to an increase in effective series resistance. Electroluminescence imaging is generally used for the detection of finger interruptions. This paper demonstrates contactless detection of finger interruptions using line scan photoluminescence imaging. Simulations and experimental imaging results are used to demonstrate the difference between line scan photoluminescence images and conventional luminescence imaging techniques. The advantages of using line scan photoluminescence imaging over electroluminescence imaging for detection of finger interruptions are presented. In line scan photoluminescence images finger interruptions cause an increase in luminescence intensity, whilst regions of high recombination cause a decrease in intensity. In electroluminescence images both types of defects result in decreased luminescence intensity. This contrast inversion enables more effective identification of defect types, in particular when using automated image processing algorithms. Analysis of line scan photoluminescence imaging results demonstrates the impact of finger interruptions as a function of location and illumination intensity. The increase in luminescence intensity from a finger interruption is found to have a linear correlation with the power loss.

Keywords: Characterisation, Defects, Electroluminescence, Photoluminescence, PV Module, Simulation

1 INTRODUCTION

Finger interruptions (also known as finger breaks) are induced during manufacture of screen-printed silicon solar cells and modules. This primarily occurs during cell metallisation and during module interconnection and lamination [1]. An efficient method for detection of finger interruptions is desirable for manufacturers, as their presence may alter the current path through the front metallisation, resulting in an increase in effective series resistance (R_S) and a decrease in power output from the sample [2].

During cell manufacturing detection of finger interruptions is generally performed using high-resolution optical vision systems, however electroluminescence (EL) imaging [3] is becoming more common for this purpose in module manufacturing. Luminescence imaging techniques are increasingly being used due to their ability to provide spatially resolved data at high speeds. Unlike EL imaging open-circuit photoluminescence (PL) imaging [4] is contactless, but it cannot detect finger interruptions since the sample is uniformly illuminated. All local diodes essentially generate the same voltage, and thus there is no driving force for lateral current flow, which is the basis of visualising R_S features in luminescence images [5]. This paper evaluates the efficiency of line scan PL imaging as a method of detecting finger interruptions on silicon solar cells and modules, and as a contactless alternative to EL imaging. Using line scan PL imaging for detection of finger interruptions was demonstrated in prior work [6]. It should be noted that the concepts presented in this paper also apply to other features that result in an increase in effective R_S.

Line scan PL imaging is performed by restricting the camera field of view to a thin line spanning the width of a sample, and localising photoexcitation to within this line. Individual "line" images are continuously acquired in sync with the sample motion, enabling "on-the-fly" imaging. The final image is generated by combining the acquired line images. The localised line illumination creates a voltage difference within the sample as only a fraction of the cell area is illuminated at any one time. This results in lateral current flow between the illuminated and non-illuminated regions [7-9], which enables finger interruptions and other R_S features to be visualised.

2 METHODOLOGY

2.1 Line scan imaging simulations

The efficiency of detecting finger interruptions using line scan PL imaging was quantified using simulations performed with *Griddler 2.5 Pro* [10]. The cell model used in all simulations was a uniform, monocrystalline silicon, three busbar, full area Al-BSF solar cell with parameters representative of an industrial screen-printed cell manufactured in 2014 [11]. Key parameters of this model are shown in Table 1. Each simulation represents an instantaneous view of the line scan imaging process, in which sample movement is modelled by changing the location of a line shaped Gaussian illumination profile along the length of the cell model. A line region of interest representing the camera field of view is extracted from each simulation and combined to generate the final image. This process performed with sample movement perpendicular to the front metal fingers of the cell (vertical sample movement) is demonstrated in Fig. 1.

Input parameter	Value	Units
Finger width	60	μm
Finger height	12	μm
Finger pitch	2.08	mm
Busbar width	1300	μm
Busbar height	15	μm
Metal resistance	3.75	mΩ/□
Contact resistance	2	mΩ-cm^2
Emitter sheet resistance	75	Ω/□
Passivated area recombination	201	fA/cm^2
Metal contact recombination	204	fA/cm^2

Table 1: Key optical and electrical parameter values of the monocrystalline cell model used for all *Griddler* simulations [11].

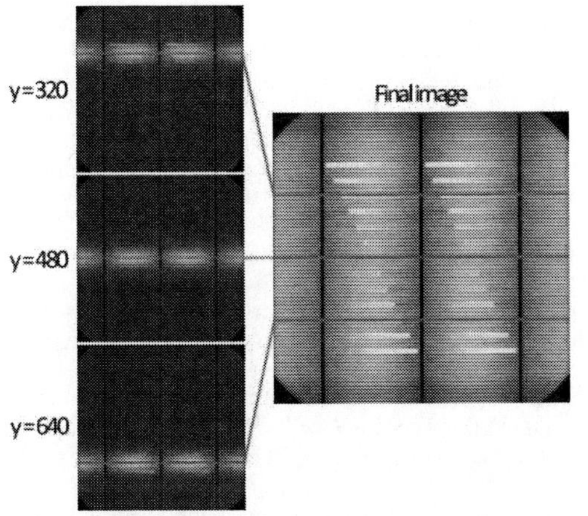

Fig. 1: Process of simulating a line scan PL image, demonstrated using three of the numerous simulations required. The green lines represent the centre of illumination and camera field of view, which is extracted from each simulation and combined to generate the final image.

2.2 Experiments

To validate the simulations, line scan PL images were acquired on the cell level using an iLS-W3 line scanning PL system from BT imaging and on the module level using a prototype imaging tool developed at UNSW [9]. The prototype imaging tool is also capable of acquiring line scan EL images. Specifics of the tool's image acquisition process are published elsewhere [6].

Two monocrystalline silicon, full area Al-BSF cells were used for cell level experiments, shown in Fig. 2 and denoted as samples A and B, respectively. The specific finger interruptions that were considered in our analysis are highlighted in red. These samples were chosen due to their similarity to the cell model used in the *Griddler* simulations.

A 60-cell, multicrystalline silicon PV module manufactured in 2015 was used for module level experiments, a line scan PL image of which is shown in Fig. 3. This module had undergone mechanical load testing and therefore contains multiple cells with cracks and finger interruptions. The finger interruptions result in increased luminescence intensity, and are clearly visible against the cell cracks and recombination active grain boundaries that result in lower luminescence intensity. This is a key advantage of using line scan PL imaging for detection of R_S defects from individual images, as will be discussed in more detail below.

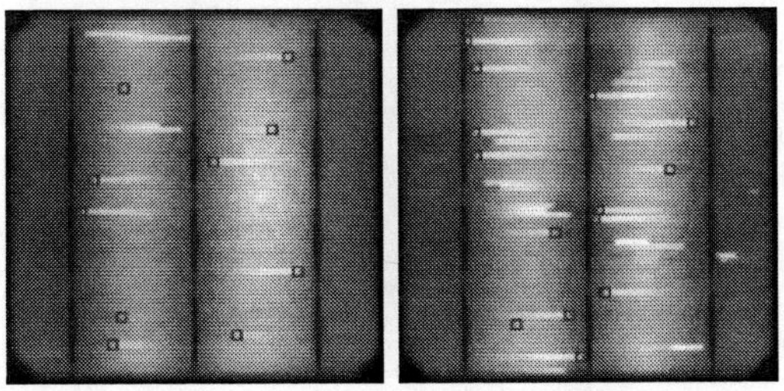

(a) (b)

Fig. 2: Line scan PL images with vertical sample movement of monocrystalline cells (a) sample A and (b) sample B. Peak illumination intensity was approximately 2.3 Suns. Finger interruptions considered in the analysis are identified with red boxes.

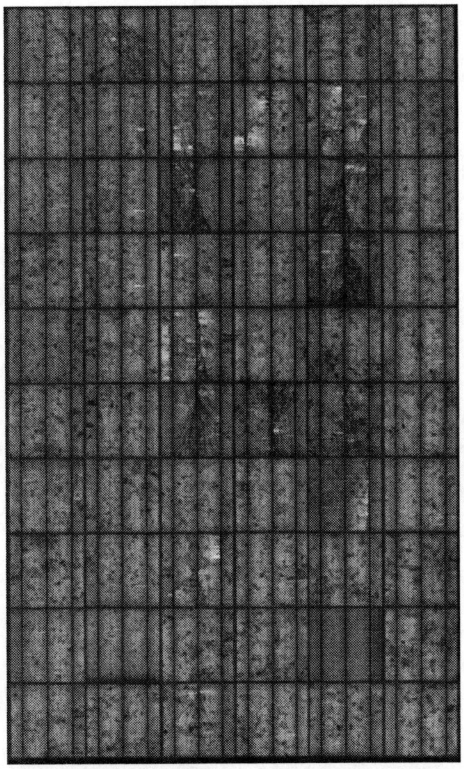

Fig. 3: Line scan PL image of a multicrystalline module that had undergone mechanical load testing. Peak illumination intensity was approximately 2.2 Suns.

2.3 Quantifying changes in luminescence intensity

Efficient detection of finger interruptions requires a large difference in luminescence intensity to be observed between the region impacted by the interruption and the adjacent unaffected regions. Large luminescence contrast between those areas enables more reliable detection of finger interruptions using automatic feature detection algorithms. We define a metric (ΔPL) that quantifies luminescence intensity changes around R_S features in line scan PL images:

$$\Delta PL = \frac{PL_{Rs} - PL_n}{PL_n} \quad (1)$$

where PL_{Rs} is the average PL intensity of the affected region (outlined in yellow in Fig. 4) and PL_n is the average PL intensity of the closest unaffected region (outlined in green in Fig. 4). PL_{Rs} is calculated from the area equal to the finger length multiplied by twice the finger spacing and centred over the interrupted finger being considered. PL_n is calculated from the closest area of the same size that does not overlap PL_{Rs}.

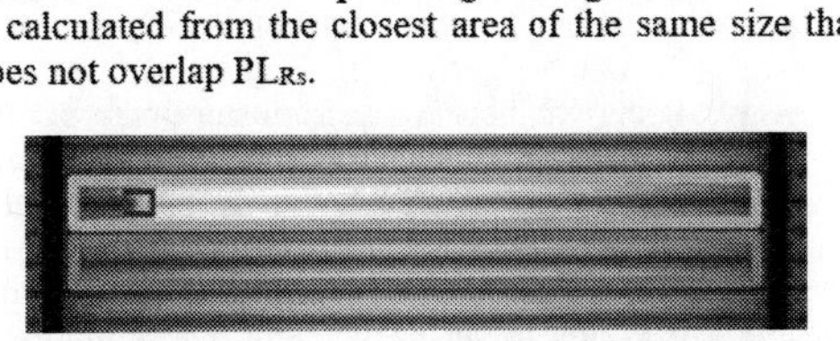

Fig. 4: Close up of a single finger interruption within the cell model showing the regions used to define ΔPL. The red box highlights the location of the finger interruption. The yellow box outlines the region averaged to determine PL_{Rs}. The green box outlines the region averaged to determine PL_n.

3 RESULTS AND DISCUSSION

3.1 Comparing luminescence imaging techniques

Simulated and experimental data are used to investigate how finger interruptions (and any other R_S features) and variations in bulk lifetime (such as regions of high recombination) appear in line scan PL images when compared to conventional luminescence imaging techniques. We will demonstrate that only line scan PL imaging can easily distinguish between R_S features and high recombination sites using a single image. This is shown in the comparison of the simulated luminescence images (open-circuit PL image, line scan PL image and EL image) in Fig. 5. Finger interruptions and recombination features were introduced to the model, namely a region of increased recombination similar in shape to a grain boundary and centred over a region affected by five finger interruptions (highlighted by a red box), a single finger interruption (highlighted by a blue box), a finger with two interruptions (highlighted by a green box), and a rectangular region of increased recombination centred over a finger (highlighted by a yellow box).

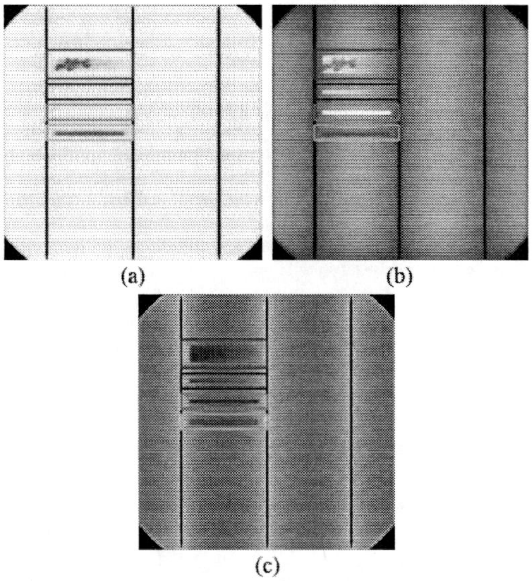

Fig. 5: Comparison of simulated luminescence images containing both finger interruption and recombination defects (see text for details). (a) Open-circuit PL image with 2.25 Suns uniform illumination intensity. (b) Line scan PL image with vertical sample movement and peak illumination intensity of 2.25 Suns. (c) EL image with 650 mV forward bias.

In the open-circuit PL image the impact of finger interruptions is not visible, because the uniform full area illumination does not provide a driving force for lateral current flow. In the line scan PL image areas affected by finger interruptions result in higher luminescence intensity, as current flow away from the interruption is impeded due to increased R_S and hence a larger concentration of carriers remain within the affected region. Note that the luminescence intensity over the entire cell area is higher in the open-circuit PL image compared to the line scan PL image, despite the same peak illumination intensity being used for both images. This is caused by carriers moving away from the imaging field of view and recombining in non-imaged areas during the line scan process. In the EL image finger interruptions result in lower luminescence intensity, because the current flow into cell regions with higher R_S is impeded.

In the EL image it is more difficult to reliably distinguish between high recombination sites and the finger interruptions within the red box, as they both exhibit reduced luminescence intensity [3, 5]. However in the line scan PL image the distinction between the two defect types is clear due to the inverted luminescence contrast [6, 7, 9]. This contrast inversion is valuable as it enables efficient defect detection regardless of variations in bulk lifetime particularly prevalent in multicrystalline material [5].

A clear distinction between defect types, as simulated above, is also observed in the experimental line scan PL image of the multicrystalline module in Fig. 3. To further highlight this effect, one cell was cropped from Fig. 3 and also from the corresponding EL image on the same module (not shown here) that was acquired using 10 A forward current with the prototype imaging tool. The cell images cropped from the line scan PL image and line scan EL image are shown in Fig. 6(a) and Fig. 6(b), respectively. Finger interruptions are significantly easier to identify in the line scan PL image than in the EL image. An example of automated image processing is shown in Fig. 6(c), which shows the results of proprietary algorithms developed by BT Imaging performed on the line scan PL image from Fig. 6(a). The severe finger interruptions within the cell are detected correctly. The dark region on the bottom edge of the cell provides another example for the distinction between R_S and recombination defects. It is unclear in the EL image whether this region is attributed to high R_S or reduced lifetime. In contrast, from the line scan PL image it is clear that this is a region of reduced lifetime, in this case caused by higher impurity concentrations as the wafer was cut from an edge brick.

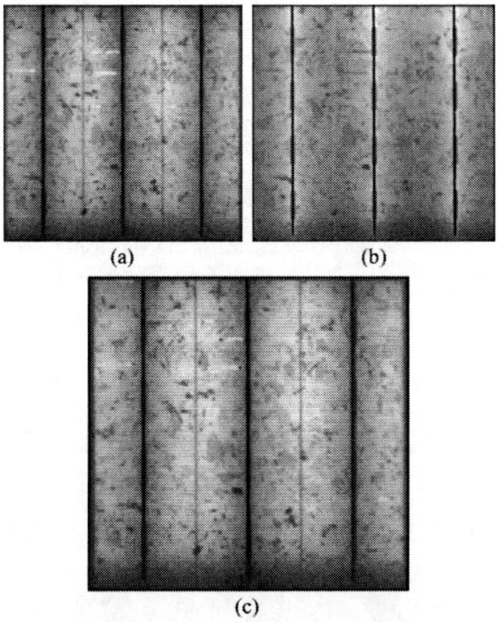

Fig. 6: (a) Line scan PL and (b) EL images of a cell cropped from module line scanning PL and EL images. (c) line scan PL image from (a) shown with a colour overlay from an automated image processing algorithm, where purple diamonds identify the location of finger interruptions and red highlights regions of enhanced luminescence due to an interruption.

3.2 Enhanced detection via image combination

More sensitive detection of finger interruptions and other R_S defects can be achieved through the combination of line scan PL and EL imaging, rather than relying on an individual image. An effective way to observe the combined results is to divide the line scan PL image intensity by the EL image intensity to form a ratio image. Since high recombination sites appear with lower luminescence intensity in both images, their impact will be suppressed or completely cancelled out. On the other hand, the impact of even very minor R_S effects will be amplified due to the contrast inversion. This is demonstrated in Fig. 7, which shows the ratio images of two cells cropped from the module images. Fig. 7(a) is the ratio of images shown in Fig. 6(a) and Fig. 6(b). We observe that the dark region of lower lifetime on the bottom edge of the cell seen in the line scan PL and EL images is no longer visible in the ratio image. The ratio image also reveals a number of very subtle R_S features, which cannot be seen or detected from the individual images.

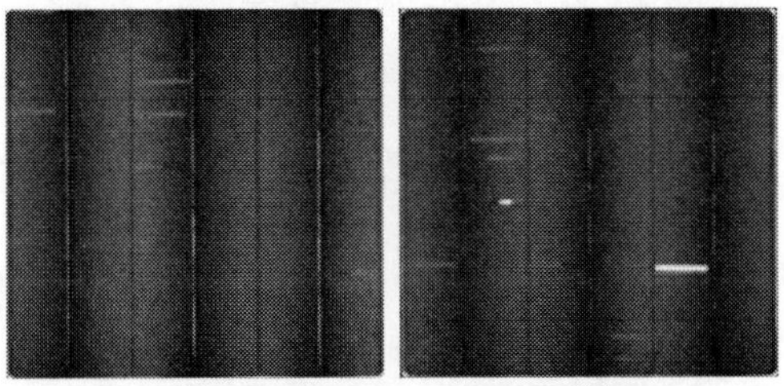

(a) (b)

Fig. 7: Ratio images of cells cropped from experimental module images. (a) is the ratio of images shown in Fig. 6(a) and Fig. 6(b). (b) is the ratio of a different cell from the same module.

3.3 Impact of finger interruption location

The location of a finger interruption relative to the busbars has a strong impact on the luminescence intensity contrast in line scan PL images. The closer the interruption is located to the busbars, the larger the contrast. ΔPL was simulated for 13 finger interruptions at varying distances from the busbar. Experimental ΔPL data were calculated for the finger interruptions highlighted with red boxes in Fig. 2. These results are plotted as a function of distance from the busbar nearest to the left side of the finger interruption in Fig. 8. Experimental values are in good agreement with simulated values. It can be seen that ΔPL approaches zero for interruptions located at the midpoint between the busbars, and increases with proximity to either busbar [2].

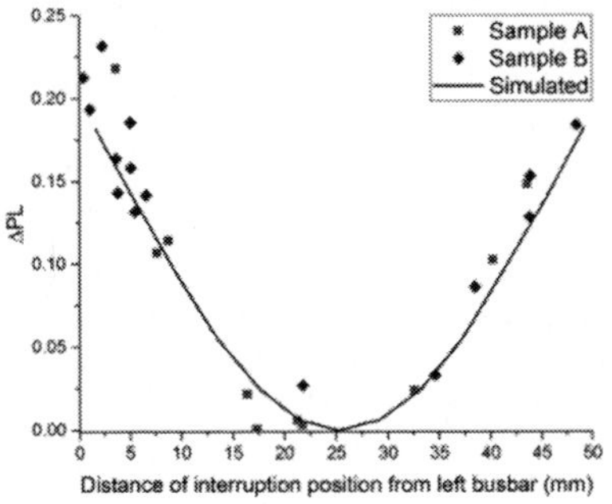

Fig. 8: Impact of the position of finger interruptions on ΔPL, calculated from simulated and experimental line scan PL images. Busbar spacing was 50.7 mm.

The higher luminescence intensity observed around finger interruptions in line scan PL images is caused by carriers that are not extracted from the location in which they are generated, which causes a power loss. The relationship between ΔPL and the power loss is shown in Fig. 9. The increased luminescence observed from regions of high R_S scales linearly with power loss, and can therefore in principle be used to quantify the impact of the high R_S region on sample performance.

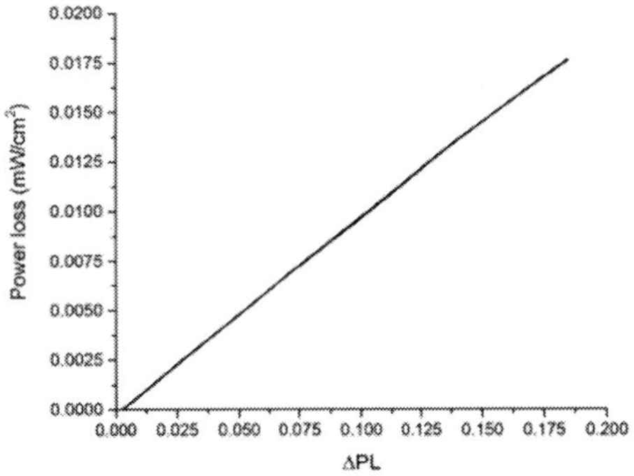

Fig. 9: Power loss at the maximum power point as a function of the simulated ΔPL values shown in Fig. 8. The power loss was simulated for a single finger interruption under uniform 2.25 Suns illumination, with respect to a cell with no finger breaks. $R^2 = 0.9997$.

3.4 Impact of illumination intensity

It has previously been shown that the illumination intensity used during line scan PL imaging influences the luminescence intensity contrast around R_S features [12]. Varying the illumination intensity changes the number of carriers generated within the illuminated region, which has an impact on the effectiveness of lateral carrier flow to surrounding non-illuminated regions. We quantitatively demonstrate this effect through analysis of the relationship between ΔPL and peak illumination intensity. Three finger interruptions were introduced into the cell model at different locations relative to the left busbar. Line scan PL images were simulated at peak illumination intensities varying from 0.05 Suns to 10 Suns. Experimental ΔPL values were also calculated for six of the interruptions highlighted in Fig. 2, at illumination intensities ranging from 0.2 Suns to 13.4 Suns.

The simulated and experimental ΔPL results are shown in Fig. 10. The results show that an optimal illumination intensity range exists for which ΔPL exhibits a maximum. Defect detection will be most efficient on images acquired at this intensity. For the simulated cell this occurs at approximately 2.5 Suns, whilst for the experimental cells it occurs between 1 Suns and 1.2 Suns.

The reduction in ΔPL at high and low illumination intensities is a result of the difference in current required to flow laterally at different illumination intensities. At very low illumination intensities the impact of R_S is negligible, hence carriers can effectively redistribute throughout the entire cell area regardless of lateral variations in R_S. There is thus no difference in carrier density between defective and non-defective regions. As the illumination intensity increases, the current flow between illuminated and non-illuminated regions increases. This results in increasing ΔPL, as the higher R_S in the defective region starts to impede the redistribution of carriers.

For even larger illumination intensities, even the R_S in the non-defective region begins to impede the redistribution of carriers, causing the PL intensity from the non-defective cell region to "catch up" with the defective region, resulting in a peak in ΔPL and then a reduced ΔPL.

For very high illumination intensities essentially no carriers leave either the defective or non-defective region, producing a unity ΔPL. The line scan PL image is then comparable to a conventional open-circuit PL image, i.e. one in which R_S features are not visible.

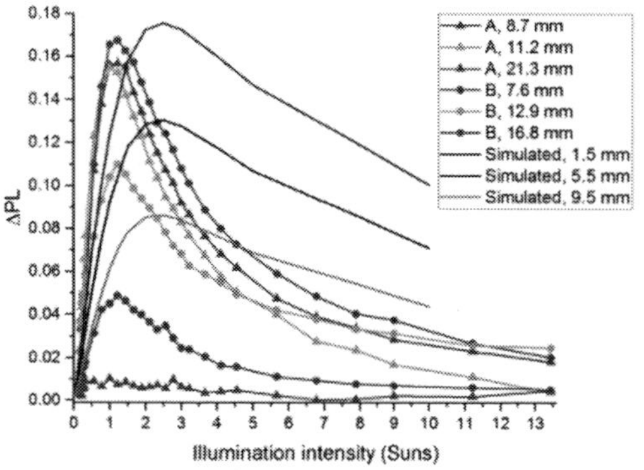

Fig. 10: Simulated and experimental ΔPL as a function of peak illumination intensity. Simulated finger interruptions are shown for three locations and identified by the distance from the left busbar. Experimental finger interruptions are identified by the name of the cell sample and distance from the busbar closest to the left side of the interruption.

4 CONCLUSIONS

The use of line scan PL imaging for detection of finger interruptions and other R_S defects was demonstrated for silicon solar cells and modules. We showed that R_S defects appear with inverted contrast compared to high recombination sites in line scan PL images, whereas both types of defects appear with reduced luminescence intensity in EL images. This contrast inversion enables more reliable defect detection using individual line scan PL images, and improves the accuracy of automatic defect detection algorithms. By combining line scan PL and EL images via calculation of ratio images, high recombination sites can be suppressed to enable very sensitive R_S defect detection. A good correlation between the luminescence contrast in line scan PL images associated with R_S features and the power loss of a cell was demonstrated. It was also shown that the detection of R_S defects can be optimised by varying the illumination intensity used during line scan PL imaging, as this impacts the luminescence contrast. Detection of finger interruptions (and other R_S defects) is more reliable using line scan PL images than EL images, especially in the case of multicrystalline material.

5 ACKNOWLEDGEMENT

This program has been supported by the Australian Government through the Australian Renewable Energy Agency (ARENA) grant RND009. Responsibility for the views, information or advice expressed herein is not accepted by the Australian Government.

6 REFERENCES

[1] P. Chaturvedi, B. Hoex, and T. M. Walsh, "Broken metal fingers in silicon wafer solar cells and PV modules," *Solar Energy Materials & Solar Cells*, vol. 108, pp. 78-81, January 2013.

[2] R. De Rose, A. Malomo, P. Magnone, F. Crupi, G. Cellere, M. Martire, *et al.*, "A methodology to account for the finger interruptions in solar cell performance," *Microelectronics Reliability*, vol. 52, pp. 2500-2503, September-October 2012.

[3] T. Fuyuki, H. Kondo, T. Yamazaki, Y. Takahashi, and Y. Uraoka, "Photographic surveying of minority carrier diffusion length in polycrystalline silicon solar cells by electroluminescence," *Applied Physics Letters*, vol. 86, pp. 262108-1 - 262108-3, June 2005.

[4] T. Trupke, R. A. Bardos, M. C. Schubert, and W. Warta, "Photoluminescence imaging of silicon wafers," *Applied Physics Letters*, vol. 89, pp. 044107-1 - 044107-3, July 2006.

[5] T. Trupke, R. A. Bardos, M. D. Abbott, F. W. Chen, J. E. Cotter, and A. Lorenz, "Fast Photoluminescence Imaging of Silicon Wafers," in *4th World Conference on Photovoltaic Energy Conversion (WCPEC)*, Waikoloa, Hawaii, USA, 2006, pp. 928-931.

[6] I. Zafirovska, M. K. Juhl, J. W. Weber, J. Wong, and T. Trupke, "Detection of Finger Interruptions in Silicon Solar Cells Using Line Scan Photoluminescence Imaging," *IEEE Journal of Photovoltaics*, vol. PP, pp. 1-7, August 2017.

[7] M. Kasemann, L. M. Reindl, and B. Michl, "Contactless Qualitative Series Resistance Imaging on Solar Cells," *IEEE Journal of Photovoltaics*, vol. 2, pp. 181-183, April 2012.

[8] J.-M. Wagner, J. Carstensen, A. Berhane, A. Schütt, and H. Föll, "Serial resistance analysis with the shaded luminescence technique," in *26th European Photovoltaic Solar Energy Conference and Exhibition (26th EU PVSEC)*, Hamburg, Germany, 2011, pp. 1569-1575.

[9] I. Zafirovska, M. K. Juhl, J. W. Weber, O. Kunz, and T. Trupke, "Module Inspection Using Line Scanning Photoluminescence Imaging," in *32nd European Photovoltaic Solar Energy Conference and Exhibition (32nd EU PVSEC)*, Munich, Germany, 2016, pp. 1826-1829.

[10] J. Wong, "Griddler: Intelligent computer aided design of complex solar cell metallization patterns," in *IEEE 39th Photovoltaic Specialists Conference (PVSC)*, Tampa Bay, Florida, USA, 2013, pp. 933-938.

[11] A. Fell, K. R. McIntosh, P. P. Altermatt, G. J. M. Janssen, R. Stangl, A. Ho-Baillie, *et al.*, "Input Parameters for the Simulation of Silicon Solar Cells in 2014," *IEEE Journal of Photovoltaics*, vol. 5, pp. 1250-1263, July 2015.

[12] T. Trupke, K. R. McIntosh, J. W. Weber, W. McMillan, L. Ryves, J. Haunschild, *et al.*, "Inline Photoluminescence Imaging for Industrial Wafer- and Cell Manufacturing," in *22nd Workshop on Crystalline Silicon Solar Cells & Modules: Materials and Processes*, Vail, Colorado, USA, 2012.

APPLICABILITY OF HIGHLY ACCELERATED THERMAL CYCLING TESTING FOR MULTIPLE TYPES OF POLYCRYSTALLINE SILICON PHOTOVOLTAIC MODULES

Masaaki Fujimori*†, Toru Kohno*, Yuki Tsuno**, Kengo Morita**
*Center for Technology Innovation – Electronics, Hitachi, Ltd.,
1-280 Higashi-Koigakubo, Kokubunji, Tokyo 185-8601, Japan,
†masaaki.fujimori.xc@hitachi.com
**TÜV Rheinland Japan Ltd., 4-5-24 Chigasaki-higashi, Tsuzuki-ku Yokohama 224-0033, Japan

ABSTRACT: To estimate long-term reliability of silicon photovoltaic modules within a short period of time, a new accelerated testing based on thermal cycling stress have been developed. Degradation modes induced by the testing or a time to induce a certain amount of degradation was estimated by using only one kind of photovoltaic modules. In this paper, we present results of analysis for several kinds of silicon photovoltaic modules. The analysis was performed for failure modes observed in current-voltage characteristics or electroluminescence images, and microscopic observation of degradation induced by the testing. The results shows the developed testing induces the same failure modes as same as those induced by thermal cycling testing defined for silicon photovoltaic modules in International Electrochemical Commission (IEC) qualification. This indicates that the developed testing is capable of accelerating the thermal cycling testing defined in the IEC qualification and is available for almost all silicon photovoltaic modules of conventional structure.
Keywords: PV Module, Polycrystalline, Silicon, Reliability, Qualification and Testing, Degradation

1 INTRODUCTION

Photovoltaic (PV) modules are used under severe conditions such like sunlight exposure, suffering large temperature changes, or moisture ingress for long time. Long-term reliability of the modules is critical to stable electric power generation with high efficiency under those conditions throughout their life time. To examine the long term reliability of PV modules, accelerated testing is generally used. The most standard accelerated testing for PV modules is a testing set described in the International Electrochemical Commission (IEC) qualification. In spite of its standard position in module qualification, the IEC qualification does not guarantee long-term reliability of qualified modules. The reason that the qualification cannot guarantee long-term reliability is lack of enough stress in each accelerated testing to examine the long-term reliability of PV modules. Thus the accelerated testing described in the IEC qualification is often performed repeatedly when one want to evaluate long-term reliability of a PV module. However, repetition of the accelerated testing is thoroughly time consuming process, therefore many testing procedures have been proposed to shorten the testing time. A typical example of such attempts is to perform different kinds of accelerated testing sequentially or alternate repetition of them. Although some of those attempts succeeded in making testing time shorter, those also make understanding a relationship between stresses and degradation difficult. In addition, we cannot compare results of those testing with accumulated data obtained by using the IEC qualification testing.

We have been developing efficient accelerated testing toward short-term evaluation of long-term reliability of PV modules. In the development, we took following three strategies: (1) Narrowing targets to mono- and poly-crystalline silicon PV modules. Because the most demand of PV modules is supplied by mono- and poly-crystalline silicon PV modules (Si-PV modules), the narrowing target is reasonable. (2) Equivalence of failure modes with those of accelerated testing of IEC qualification. We may quickly degrade testing modules if we impose excessive stress to the modules during accelerated testing, but obtained result would not probably reflect long-term reliability of the testing modules. Thus we should avoid imposing overstress on testing modules. A sure way avoiding overstress is applying stress that induces the same failure modes with those produced by IEC qualification tests. (3) Confinement target testing to thermal cycling (TC) testing. We focused our development on TC testing because the TC stress induces mainly degradation related to interconnects including busbars and fingers that is known to be the most frequent degradation modes in the field [1-3]. From those prerequisites, IEC 61215 TC (IEC-TC) test protocol is a measure of our development of TC testing.

At the first, we made testing by using light irradiation and high frequency of TC in exchange for decreasing temperature difference [4]. Then, we examined effects of light irradiation, and temperature difference in TC on degradation of PV modules. Based on the obtained results, we optimized stress imposed in TC testing and finally proposed highly accelerated TC (HATC) testing [5]. The HATC testing yields an acceleration factor of 2.3 compared to the IEC-TC testing. If we take into account cycle time difference between the HATC (2 h) and the IEC-TC (usually 6 h) testing, the HATC can induce the same degradation modes nearly five times faster than the IEC-TC testing. This makes possible to evaluate long-term reliability of PV modules in a short period of time. During the development, we have used only a kind of PV module purchased from a major vendor, which are utilized in commercial use. In this study, we investigated how much the HATC testing is effective on various types of Si-PV modules.

2 EXPERIMENTAL

2.1 Accelerated testing

The TC pattern of the HATC is depicted in figure 1 together with that of the IEC-TC for comparison. For both the testing, an electric current corresponding to the maximum power point current is applied to testing modules over temperature of 25 °C. Specific features of

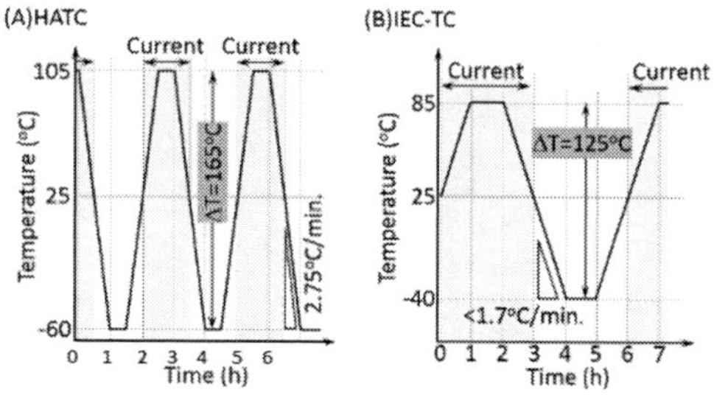

Figure 1: Thermal cycling patterns for (A) HATC and (B) IEC-TC testing. The HATC is characterized by a large temperature difference and a quick cycle. An electric current corresponding to the maximum power point current is applied to a PV module over temperature of 25 °C (yellow colored regions).

the HATC are a large temperature difference and a quick cycle. This results in a large ramp rate. In addition, the maximum and the minimum temperatures of the HATC are both exceeded those of the IEC-TC. Although it is not unveiled that how large ramp rate is allowed in TC testing for PV modules, the rate remains quite a small value compared to that of the TC testing defined as semiconductor devices mechanical and climatic test method in IEC 60749-25. We also decided the maximum and the minimum limits of temperature in TC by taking into account physical properties of constituent materials of PV modules such as glass transition temperatures, softening temperatures, compressive strength and tensile shear adhesion strength. Thus we assumed the HATC does not impose over stress on testing modules. This assumption was actually confirmed by results of the HATC testing.

Seven kinds of commercial Si-PV modules purchased from major six manufacturers were used as samples (Table 1). We applied 400 cycles of the HATC testing to all types of the modules except for the module of the vendor A, type-1. This module was performed only 200 cycles of the HATC due to physical analysis. The IEC-TC testing of 1600 cycles was performed only for one type of the module (Vendor A, type 7) as a reference.

2.2 Analyzing

For both the TC testing, we measured current-voltage (I-V) characteristics of the PV modules at every 200 cycles. The measurements were performed under standard test conditions (STC). Specific values such as the maximum power of a PV module (P_{max}) were obtained from the I-V curves. We will discuss and compare degradation modes getting from those values. Electroluminescence (EL) images were also taken before and after accelerated testing with a constant module current of 8 (A) and used to examine degradation modes of the testing modules. We also carried out physical analysis to obtain microscopic views of connectivity at around interconnects, especially states of soldering. In addition to accelerated testing listed in table 1, we also performed 400 cycles of the IEC-TC testing using the vendor A's PV modules of type 1 for this purpose. After 400 cycles of both the IEC-TC and the HATC testing, the PV modules were disassembled carefully and observed surfaces and sections of PV cells with optical and scanning electron microscopy (SEM).

Table 1: Sample modules used in the study. All modules are polycrystalline silicon PV modules. Two modules are prepared for each types of module except for Vendor-C Type-3' (No.3). Interconnects information on a cell of each module (numbers of busbars and fingers) and performed accelerated testing are also shown.

Vendor	Type	No.	# of busbar	# of finger	Testing
A	1	1,2	3	86	HATC
B	2	1,2	3	71	HATC
C	3	1,2	4	91	HATC
C	3'	1	4	101	HATC
D	4	1,2	3	75	HATC
E	5	1,2	3	71	HATC
F	6	1,2	3	85	HATC
G	7	1,2	3	72	HATC
G	7	3,4	3	72	IEC-TC

3 RESULTS AND DISCUSSIONS

Figure 2 shows characteristic values of I-V curves for the PV modules performed IEC-TC testing. The values can be roughly divided into two groups: one is a group that shows small changes with increasing thermal cycles except initial decrease (short circuit current I_{sc}, the maximum power point current I_{pm}, and open circuit voltage V_{oc}), the other group shows continuous decrease with increasing thermal cycles, resulting in relatively large changes (the maximum power point voltage V_{pm}, the maximum power P_{max}, and change of fill factor ΔFF). We can find a change in shape of the I-V curve from behavior of those parameters because the behavior reflects the change in the shape of the I-V curves. In this case, the behavior of the characteristic values means that a slope at around V_{oc} in I-V curves becomes gentle. This is caused by increase of series resistance of a PV module. That is, degradation induced by the IEC-TC is originated in increase of the series resistance of a testing module.

Figure 3 shows changes in the characteristic values obtained from I-V curves of seven kinds of PV modules after HATC testing of 400 cycles. Similar trends to the IEC-TC were also observed. One of the trends is formed by the group of I_{sc}, I_{pm}, and V_{oc}, and the other is formed by V_{pm}, P_{max}, and ΔFF. The former remains nearly constant regardless of increasing in thermal cycles. On the other hand, decreasing of the parameters in the latter progresses with increasing thermal cycles. Because degree of degradation varies with reliability of each module, variation appears in amount of the changes in the parameters.

EL images of a part of the each module are shown in figure 4. A pair of the images taken before and after accelerated testing is presented for each PV module. The upper image is for before and the lower is for after testing. We show reference EL images of the IEC-TC using vendor A in Figure 4(A). Brightness clearly changes after the IEC-TC testing. Characteristic changes are roughly summarized in three modes: (a) decreasing brightness along busbar, (b) decreasing brightness in local regions around finger electrodes, and (c) ununiformed brightness spreading on relatively wide area in a cell. The (a) is probably caused by damaged soldering beneath a busbar. The (b) is obviously due to disconnection of finger electrodes. The (c) seems a preliminary step of (a) or (b). Anyway, it is caused by ununiformity in current flow due to uneven resistance. Those modes are all directory originated from increase in electric resistance.

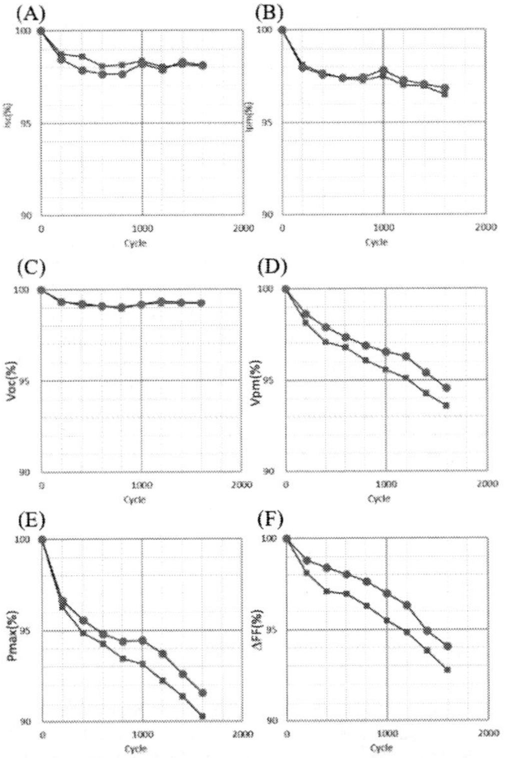

Figure 2: The changes of characteristic values observed in I-V curves for the PV modules performed IEC-TC testing (Vendor A). (A) Short circuit current I_{sc}, (B) Maximum power point current I_{pm}, (C) Open circuit voltage V_{oc}, (D) Maximum power point voltage V_{pm}, (E) Maximum output power P_{max}, (F) Change rate of fill factor ΔFF. I_{sc}, I_{pm}, and V_{oc} show small changes but V_{pm}, P_{max}, and ΔFF shows large changes, relatively.

In the case of HATC testing, we can observed the same modes in EL images. The failure modes (b) and (c) are seen in all EL images and the failure mode (a) is clearly in the EL images of types 1, 2, 6 and merely in the image of type 7. The amount of decreasing in brightness for the type 3, 3', 4, and 5 is relatively low. (This is consistent with relatively small decreasing in P_{max} for those types.) All failure modes produced by the HATC testing are consistent with those produced by the IEC-TC testing.

This was also confirmed in microscopic views. We showed optical microscopic images at sections along busbars in figure 5. The sections were chosen from dark and bright areas in the EL images of the PV modules after the IEC-TC and the HATC testing. Whereas no damage was observed in the bright area (figure 5 (A)), voids and crack formation in solder alloy were clearly seen between busbars and a cell at the dark area in the EL image (figure 5 (B)) after 400 cycles of the IEC-TC testing. Similar cracks are observed between busbars and cells after the HATC testing, but the crack formation at the bright area (figure 5 (C)) is different from the case of the IEC-TC. This crack formation in the bright area regardless of imposing the same number of TC's is probably due to higher stress in the HATC. Electric current path maybe formed outside of the view such as perpendicular to the section. We also examined the failures at finger electrodes observed as darkness in EL images from microscopic view

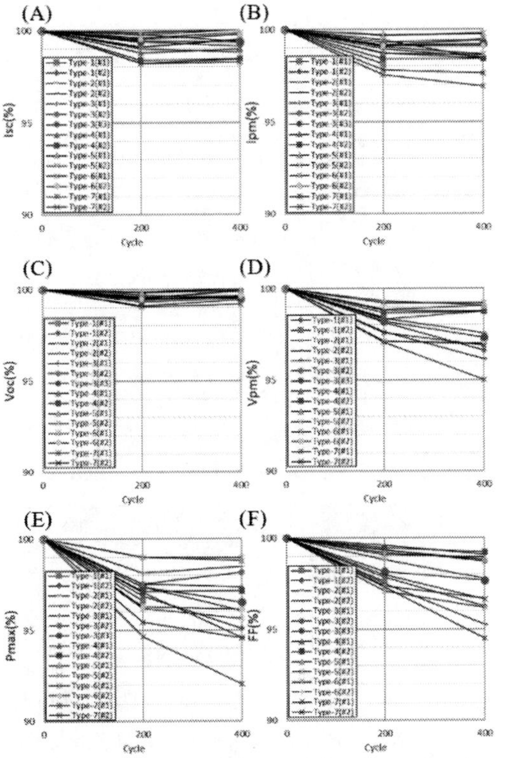

Figure 3: The changes of characteristic values taken from I-V curves for PV modules performed HATC testing (from Vendor A to F). (A) I_{sc}, (B) I_{pm}, (C) V_{oc}, (D) V_{pm}, (E) P_{max}, (F) ΔFF.

with SEM. Breaking of finger electrodes were found in surface images of SEM in both the case of the IEC-TC and the HATC testing.

From those results of electrical properties, EL images, and microscopic analysis, the failure modes and their origins are equivalent between the IEC-TC and the HATC testing. That is, we can conclude that the HATC testing accelerates the same failure modes as those induced by the IEC-TC testing. As previously estimated by using only one kind of a PV module, an acceleration factor of the HATC testing was at 2.3 [5]. In addition, if we take into account of cycle time (6 hours typically for the IEC-TC and 3 hours for the HATC testing) the HATC testing can be carried out to induce the same degrees of degradation nearly five times faster than the IEC-TC testing. We may expect that the HATC testing provides nearly the same acceleration to other Si PV-modules of conventional structure and materials.

We would, then, give a brief consideration of the PV modules by comparing the results of the HATC testing. We can see high uniformity of brightness in the EL images for the case of the four-busbar modules (the vendor C, type-3 and 3'). Those modules shows relatively small degradation in the P_{max}. This indicates that more busbars is resistant to the stress by thermal cycles. A number of busbars, however, is not always the most important factor affecting degradation of a PV module. Actually, there are the three-busbar PV modules that are lower degradation in P_{max} (the vendor E, type-3). This is the same for a number of finger electrodes. By comparing the P_{max} of the type-3 and the type-3' modules of the vendor C, it can be seen that the type-3' module degrades faster than the type-3

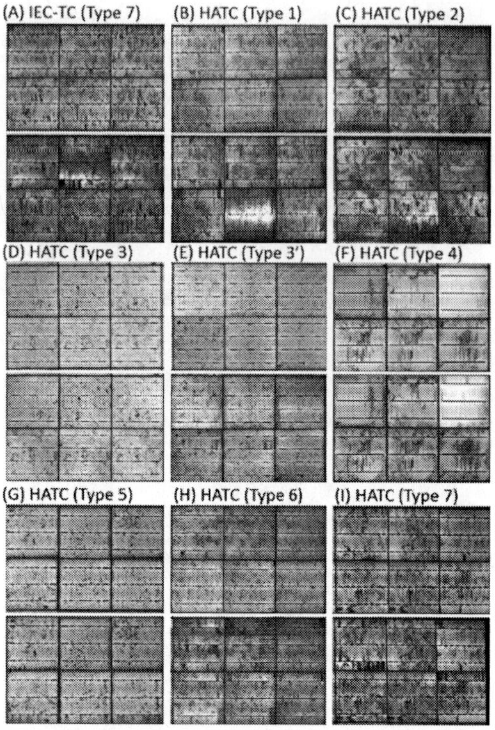

Figure 4: Comparison of EL images typically observed in each PV module. The images were taken before (upper) and after (lower) accelerated testing. An EL image of one module in two samples is shown for each type of modules. (A) Vendor A's modules of type 7 are used for IEC-TC as a reference. From (B) to (I), images for each vendor's module used for HATC.

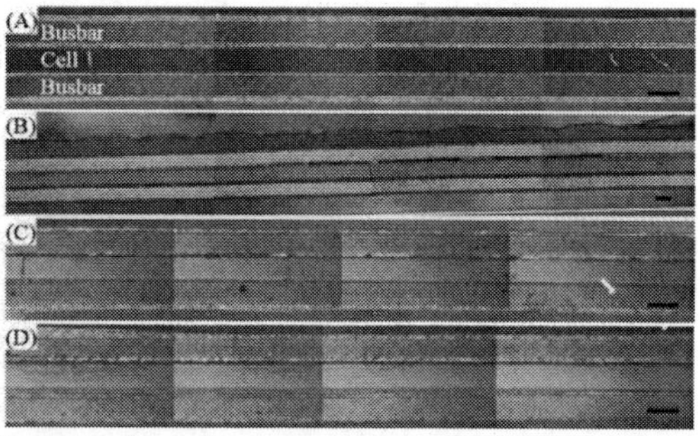

Figure 5: Optical microscopic images of a section along a busbar. (A) At bright area in an EL image after 400 cycles of IEC-TC testing. (B) At dark area in an EL image after 400 cycles of IEC-TC testing. (C) At bright area in an EL image after 400 cycles of HATC testing. (D) At dark area in an EL image after 400 cycles of HATC testing. All scale bars at lower right of each figure are 200 (μm). The vendor A's PV modules of type-1 were used.

module despite more finger electrodes. This shows that the finger number is not always a dominant factor of degradation. Because PV modules with lower degradation in P_{max} show higher uniformity in brightness of EL images regardless of busbar numbers, more important factor to affect degradation of PV modules in TC testing is firm soldering between the busbars and the cell. Note that our results give only the view on the TC stress. The TC testing is irresponsible for other stresses such as thermal stress. Thus, we should take several different perspectives into account when we consider the total reliability of a PV module.

The simple average of P_{max} for the fourteen modules after 400 cycles of the HATC testing is 96.4 (%) and standard deviation of 1.95. Our preliminary result shows 200 cycles of HATC correspond to 7.9 years in field use [6]. If we assume this correspondence of the cycle to field use, degradation of 0.23 % per year would be expected to P_{max} for 20-year use. This is an acceptable small value but based on assumption and results from small amount of the samples. Therefore, we should properly consider the correspondence of the cycle to field use and a statistical treatment if we predict absolute values of output power and so on. Though such problems still remain, the HATC testing can be used to estimate long-term reliability of PV modules within a short period of time.

4 CONCLUSION

Based on the several kinds of analysis for the seven kinds of polycrystalline silicon PV modules, we confirmed equivalence of failure modes obtained by the HATC and by the IEC-TC testing. No inconsistency was observed in the failure modes observed in the I-V characteristics and in the EL images. Both the testing also shows the same origin of the failures in the microscopic view. Because both the testing induce the degradation at connectivity between busbars and cells, and at finger electrodes, we can expect that the HATC testing is available for almost all Si-PV modules of a conventional structure. Thus we conclude the HATC testing can be used as an accelerated version of the IEC-TC testing.

REFERENCES

[1] J. H. Wohlgemuth, *et al.*, "Long Term Reliability of Photovoltaic Modules", 4th World Conference on Photovoltaic Energy Conversion, (2006).
[2] D. DeGraaff, *et al.*, "Degradation Mechanisms in Si module Technologies Observed in the Field; Their Analysis and Statistics." NREL 2011 Photovoltaic Module Reliability. Workshop, Golden, CO, (2011).
[3] M. Kntges, *et al.*, "Review of Failures of Photovoltaic Modules", Report IEA-PVPS T13-01:2014, p.5 (2014).
[4] M. Fujimori, *et al.*, "Rapid Thermal Cycle Test Consisting of Repeated Stress based on IEC Qualification Test", Technical Digest in the proceedings of the 6th World Conference of Photovoltaic Energy Conversion, (2014).
[5] M. Fujimori, *et al.*, "Highly Accelerated Thermal Cycling Test for Short Term Examination of Photovoltaic Module Reliability", in the proceedings of the 31st European Photovoltaic Solar Energy Conference and Exhibition, p.1911 (2015).
[6] K. Morita et al., "New acceleration testing method "highly accelerated thermal cycling test": Acceleration method through degradation mode analysis", in the proceedings of the 31st European Photovoltaic Solar Energy Conference and Exhibition, p.2515 (2015).

NON-UNIFORM MECHANICAL LOADS DUE TO WIND EFFECT ON PHOTOVOLTAIC MODULE

Shu-Tsung Hsu*, Hsin-Hsin Hsieh
Center for Measurement Standards, Industrial Technology Research Institute
Rm.401, Bldg.53, No. 195, Sec. 4, Chung Hsing Rd., Hsin-Chu 31040, Taiwan, R. O. C.
*Corresponding author, andersonhsu@itri.org.tw, Phone: + (886) 3 5917970, Fax: + (886) 3 5837801

ABSTRACT: The issue of typhoon has received considerable critical attention due to the associated strong winds generally damage PV modules severely. Previous IEC standards have examined the effect of static uniform-loads (IEC 61215-2:2016) or dynamic uniform-loads (IEC TS 62782:2016), but overlooked the wind loads with moment effect or non-uniform loads on PV module. In this paper, we develop an analytical model about a mean extended wind load (MEWL) test method, and a non-uniform dynamic mechanical load (NUDML) system, which can simulate the wind action on PV module by uneven mechanical loads. Results of this study also provide a severity test considering the environmental condition such as wind velocity (V), wind direction angle (β) and inclined angle (α) of PV module.
Keywords: Non-uniform, Photovoltaic module, Wind

1 INTRODUCTION

Typhoon is one of the world's most frequently occurring natural disaster that causes severe damage. The losses caused by typhoons are not only related to the strength and structure of a particular typhoon, but also to the population density, home range, and type of economy in the affected area. Taiwan, situated between 22-25° N and 120-121° E, is centrally situated in the main path arc of typhoons generated in the Northwest Pacific Ocean. On average, 3~4 typhoons approach or make landfall in Taiwan yearly. The severity of a typhoon is associated with its wind strength and moisture content. Thus, the installation of a photovoltaic (PV) module is of necessity concerned with the wind uplift produced by typhoons. Breakage of the glass covers of PV modules and frames is mainly due to the strong wind uplift and the associated large deflection. In Fig. 1, MERANTI typhoon had numerous PV modules or trackers damaged at Southern Taiwan in September 2016. The major damages of modules due to wind loads in open-rack mounting configurations generally include broken glass, crack or micro-crack of cell, failures of metal frame, fastener or support system, etc. Therefore, a good simulation of the wind resistance by mechanical load is necessary for preventing such damage.

In this work, a new concept of test method [2] was introduced for uneven surface-loads on single PV module due to wind pressure, in order to simulate the non-uniform loads induced by wind flow through along the surfaces of PV module, especially at some worst cases with torsion moment effect under specific wind direction and inclined angle of PV module.

Figure 1: MERANTI typhoon damaged solar trackers at Southern Taiwan in 2016 [1]

2 TEST STANDARDS OF MECHANICAL LOADS FOR PV MODULE

2.1 IEC 61215-2 [3]

IEC 61215-2 is designed for the type approval of PV modules, and provides the static mechanical load test in C11.16. This test defined three cycles of 2400 Pa uniform load, applied for one hour to front and back surfaces of PV module in turn. Optional snow load of 5400 Pa during the last front cycle. To pass the tests, the modules need to withstand a uniform test stress of 2400 Pa, and even 5400 Pa in cases of higher snow load classes.

2.2 IEC TS 62782 [4]

IEC TS 62782 provides a test method (± 1000 Pa, 1000 cycles, 1 to 3 cycles per minute) for performing a (cyclic) dynamic mechanical loading test in which the module is supported at the design support points and a uniform load normal at the module surface is cycled in alternating negative and positive directions. This test may be utilized to evaluate if components within the module including solar cells, interconnect ribbons and/or electrical bonds within the module are susceptible to breakage or if edge seals are likely to fail due to the mechanical stresses encountered during installation and operation.

2.3 IEC 62938 [5]

IEC 62938 recommends a consistent pulling force of 2400 Pa or 5400 Pa in three one-hour cycles. The load to be applied to the PV module and its distribution by separate weight elements are determined as a function of the characteristic snow load Sk. High snow load frequently causes damage to photovoltaic systems. Systems on slanted roofs are most affected, due to the uneven distribution of the overall snow load. When snow slides down the PV module and accumulates on the lower frame, a particularly critical snow load occurs, which could lead to an overload of the entire mounting system. Until now, standards for the type approval of photovoltaic systems only included load tests carried out horizontally.

Consequently, the above IEC standards still lack of documents defined the test method of non-uniform loads due to the wind effect on PV modules.

3 NBE-AE 88 [6] WITH INCLINED ANGLE

The Spanish standard NBE-AE 88 defines the whole range of tilt angles (0°-90°), with values close to those from European standard (Eurocode) [7, 8], for canopies with tilt angles up to 30 degrees. The tilt angle mentioned in all standards is actually the wind angle (angle between wind direction and structure surface) considering the horizontal direction of the wind. In addition, NBE-AE 88 presents a model of wind action on inclined open surfaces very similar with the solar panel [9, 10], even if it is not especially dedicated. Values of the limit pressure coefficients (C_{p1} and C_{p2}) are given for different wind direction (according to a loading diagram in Fig. 2) and wind angles α (see Table I). Therefore, we assume the tilt angle equal to the inclined angle for PV module. The dimensionless pressure coefficient (C_p) in Eq 1 generally can also be simulated from wind-tunnel experiment or CFD.

$$C_p = (P-P_\infty)/(0.5\rho V^2) \qquad (1)$$

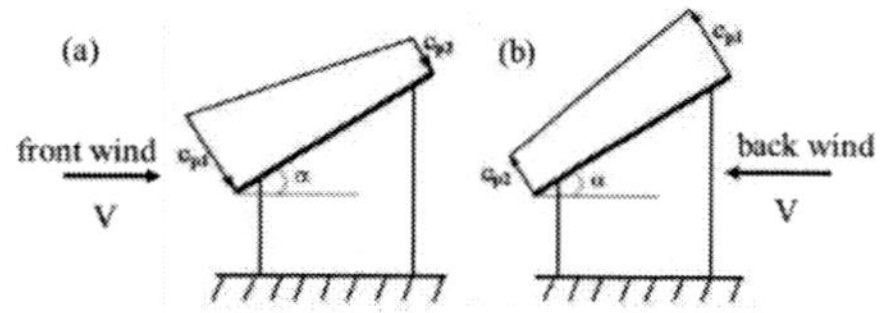

Figure 2: Pressure coefficients C_p and load distribution diagram on PV platforms (a) front wind ($\beta = 180°$) (b) back wind ($\beta = 0°$)

Table I: Pressure coefficient C_p for open platforms [10], extended coefficient C_α and F_I/F_{II} [2]

α (°)	C_{p1}	C_{p2}	C_{p1}/C_{p2}	F_I/F_{II}	C_α
20	1.2	0.4	3	1.0	2.3
30	1.6	0.8	2	0.5	1.7
40	1.6	0.8	2	0.5	1.7
50	1.4	1.0	1.4	0.2	1.3
60	1.2	1.2	1	NA	1

4 MEAN DESIGN WIND LOAD (MDWL) AND MEAN EXTENDED WIND LOAD (MEWL) [2]

In this paper, we developed a simulation method of wind-load on PV module due to wind action with the moment effect [10, 11, 12]. The wind pressure estimated along the long axis (or length) on PV module, no matter the wind direction either from front side (load case 1, see Fig. 2(a)) or back side (load case 2, see Fig. 2(b)). For the general case of trapezoidal distribution of wind pressure, a torsion effect appears. For example, the calculated moment M_I equals to the total force F_I (triangle I) multiplied by arm distance e (see Fig. 3). F_{II} means the total force of rectangle II. F_{III} stands for the total force of rectangle III, and insteads of triangle I due to the same moment M_I. L is the length of PV module. Eqs. (2) - (8) define each load in detail. A_{ref} means the surface area of solar panel

$$F_{III} = 2/3 \times F_I \qquad (2)$$

$$F_B = F_{III} + 1/2 \times F_{II} \qquad (3)$$

$$F_A = 1/2 \times F_{II} \qquad (4)$$

$$C_\alpha = F_B/F_A \qquad (5)$$

$$W_p = 0.5\rho V^2 \qquad (6)$$

$$P_A = C_{p2} \times W_p \times SF = F_A/(0.5A_{ref}) \qquad (7)$$

$$P_B = C_\alpha \times P_A = F_B/(0.5A_{ref}) \qquad (8)$$

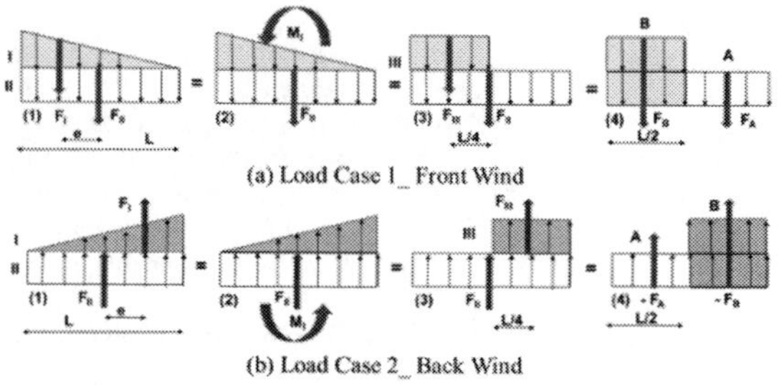

Figure 3: Wind action decomposed diagram (a) load case 1 (front wind) (b) load case 2 (back wind)

Consequently, if we consider using F_A and F_B to simulate wind-loads on both half-side of the module (Figs. 3(a)(4), 3(b)(4)) coherently, the force F_A is the mean design wind load (MDWL) covering safety factor (SF) 3 and C_{p2}, and then F_B is the mean extended wind load (MEWL) covering the moment effect and the extended coefficient C_α. Upon the mechanical loads test of IEC 61215 and IEC TS 62782, the design wind loads consider SF 3 as a minimum requirement to fulfil the general application. The value of SF presents the different safety concerns. C_α means the ratio of F_B/F_A (or P_B/P_A) considering the severity of inclined angle α. The more value of C_α mean the more moment effect due to wind load need module to withstand. This also is helpful in design support platform since distributed pressure along a module's surface can be transferred to some fasteners of customized support directly.

5 SEVERITY TEST WITH UN-EVEN WIND LOADS FOR PHOTOVOLTAIC MODULE [2]

Angle of orientation and wind velocity are studied together to get the optimum design for installation of the PV modules. Table I and Table II shows the variations of design wind-pressure P_A and extended wind-pressure P_B, for different inclined angles α and wind speeds V. The effect of wind-pressure on the surface of PV module can be simulated by both P_A and P_B, which are determined in the choices of fixed wind velocity and inclined angle. In addition, the designed wind-pressure P_A is determined in C_{p2}, inclined angle α, wind pressure W_P and safety factor 3. Base on the moment effect, the extended wind-pressure P_B is determined in an extended coefficient C_α, inclined angle α and radio of C_{p1}/C_{p2}. In Table II, the value of C_α decreases while inclined angle (α) increases (e.g., α = 40°, P_B = 1.7P_A). Otherwise, the value of C_α increases while inclined angle α decreases (e.g., α = 20°, P_B = 2.3P_A). P_A and P_B strongly rely on the environmental condition like velocity (V) and inclined angle (α). Therefore, the chosen severity test should fulfill the environmental condition including module installation. Low inclined angles significantly reduced the effects of wind uplift and drifting [11].

Table II: Wind pressure (P_A(Pa), P_B(Pa)) with severity of inclined angle α, wind velocity V (m/s) and wind pressure W_P (Pa) at BEAUFORT scale (BS) 8-17. [2]

α(°)	BS	8	9	10	11	12
	V	20.7	24.4	28.4	32.6	36.9
	W_P	277	385	521	687	880
20	P_A	332	462	625	824	1056
	P_B	764	1063	1438	1895	2429
30	P_A	665	924	1250	1649	2112
	P_B	1131	1571	2125	2803	3590
40	P_A	665	924	1250	1649	2112
	P_B	1131	1571	2125	2803	3590
50	P_A	831	1155	1563	2061	2640
	P_B	1080	1502	2032	2679	3432
60	P_A	997	1386	1876	2473	3168
	P_B	997	1386	1876	2473	3168

α(°)	BS	13	14	15	16	17
	V	41.4	46.1	50.9	56.0	61.2
	W_P	1108	1373	1674	2026	2420
20	P_A	1330	1648	2009	2431	2904
	P_B	3059	3790	4621	5591	6679
30	P_A	2659	3295	4018	4862	5808
	P_B	4520	5602	6831	8265	9874
40	P_A	2659	3295	4018	4862	5808
	P_B	4520	5602	6831	8265	9874
50	P_A	3324	4119	5022	6078	7260
	P_B	4321	5355	6529	7901	9438
60	P_A	3989	4943	6026	7294	8712
	P_B	3989	4943	6026	7294	8712

Notes: $\rho = 1.2923$ kg/m^3, temperature = 25 °C, SF=3.

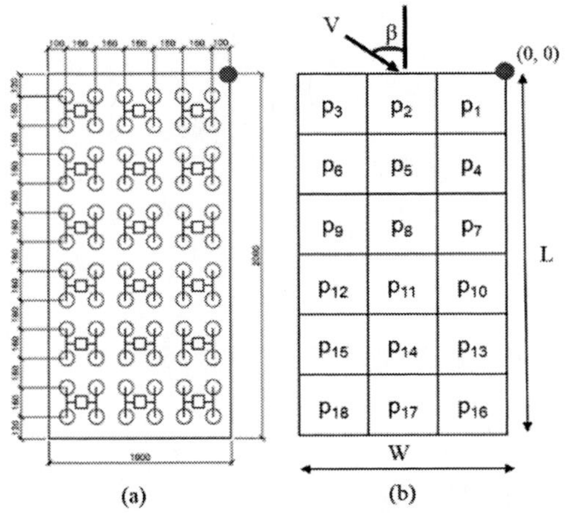

Figure 4: (a) Diagram of NUDML system with 18 (3x6) pistons, and each piston having four suction cups (b) Diagram of eighteen independent mean surface-pressure loads. Note: V is wind velocity; β is wind direction angle. L: length; W: width; •: relative original point (0, 0).

6 NON-UNIFORM DYNAMIC MECHANICAL LOAD (NUDML) TEST

NUDML is a new test capacity to simulate the uneven wind loads on PV module in test lab. The major failures triggered by NUDML generally include broken glass, crack or micro-crack of cell, failures of frame, fastener and support system (option). The other advantage for such a test is to monitor performance degradation of PV module during the testing process.

Fig. 4 shows the design of a non-uniform dynamic mechanical load system. The system requirement of air pressure is adjustable independently by each piston with four vacuum suckers. Output range should be up to ± 10000 Pa or more, in order to meet the wind intensity speed over 60 m/s at least for BEAUFORT scale level 17 (strong typhoon). The space between suckers is 16-20 cm to fulfil the specifics of IEC TS 62782 as well. Test module is supported at the design support points.

This technical specification provides a test method for performing a dynamic mechanical loading test in which the module is supported at the design support points and a non-uniform load normal to the module surface is cycled in alternating negative and positive directions, as following example for different wind velocity range as followings.

A. Choose wind velocity V.
B. Select extended coefficient C_α with severity of inclined angle α (see Table I).
C. Calculate design wind pressure P_A (see Table II).
D. Calculate extended wind pressure P_B (see Table II).
E. Apply the dynamic mechanical load at a cycle rate, cycles being defined as one positive and one negative load application.
F. Each single cycle is composed of the first half-cycle (pressure mode) and the second half-cycle (suction mode), due to both the front and back wind (see Fig. 2, Fig. 5).

(1) On the left-half module (see Fig. 6(a)): Force the extended wind pressure P_B on the first half-cycle for load case 1; force the negative design wind pressure P_A on the second half-cycle for load case 2.

(2) On the right-half module (see Fig. 6(b)): Force the design wind pressure P_A on the first half-cycle for load case 1; force the negative extended wind pressure P_B on the second half-cycle for load case 2.

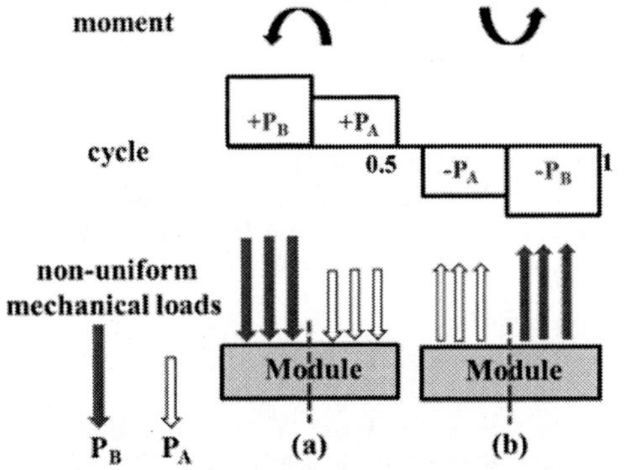

Figure 5: A schematic drawing of the non-uniform dynamic mechanical loads distribution along the long axil of test module for (a) load case 1 (b) load case 2.
Note: '+' means pressure, '-'means negative pressure

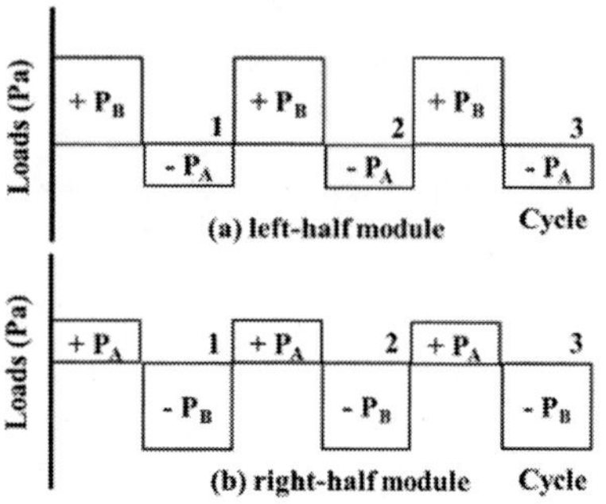

Figure 6: Non-uniform dynamic mechanical loads within three cycles (a) on the left-half module (b) on the right-half module

7 CONCLUSIONS

Taiwan (22-25° N, 120-121° E) situate in the main path of typhoons generated in the Western North Pacific Ocean (19-28° N, 117-125° E). On average, 3~4 typhoons approach or landfall in Taiwan yearly. The issue of typhoon has received considerable critical attention since the associated strong winds generally damage PV modules severely. Previous IEC standards have examined the effect of static uniform-loads (IEC 61215-2:2016) or dynamic uniform-loads (IEC TS 62782:2016) on PV module, and have overlooked the wind loads with moment effect or non-uniform loads on PV module indeed.

In this paper, we have developed an analytical model about mean design wind load (MDWL) and mean extended wind load (MEWL), according to mean design wind pressure (P_A) and mean extended wind pressure (P_B) (see Fig. 3). Results of this study revealed that the result of MDWL/MEWL test strongly rely on the environmental condition of wind velocity (V), wind direction angle (β) and inclined angle (α) of PV module (see Table I and Table II). In addition, the test requirement of a non-uniform dynamic mechanical load (NUDML) test was also defined (see Figs. 4-6). Therefore, this work contributes to ongoing discussions about the test condition of wind actions on PV platform, and fulfills the PV industry or any other party interested, can thus have the common testing method to refer to when desired.

8 ACKNOWLEDGEMENT

This work was under the support by the Bureau of Energy, Ministry of Economic Affairs, Taiwan, R. O. C.

9 REFERENCES

[1] http://news.ftv.com.tw/NewsContent.aspx?ntype=class&sno=2016917U13M1.
[2] S. T. Hsu, T. C. Wu, S. Y. Ting, Simulated Wind Action on Photovoltaic Module by Non-uniform Dynamic Mechanical Load and Mean Extended Wind Load, SNEC 11th (2017) International Photovoltaic Power Generation Conference.
[3] IEC 61215-2:2016, Crystalline Silicon Terrestrial Photovoltaic (PV) Modules - Design Qualification and Type Approval.
[4] IEC TS 62782:2016, Photovoltaic (PV) Modules – Cyclic (Dynamic) Mechanical Load Testing.
[5] IEC 62938 Ed. 1, Non-uniform Snow Load Testing for PV Modules.
[6] NBE-AE/88, Actions on Structures, 1988 (in Spanish).
[7] EN 1991-1-4:2005, Eurocode 1: Actions on Structures - Part 1-4: General Actions -Wind Actions.
[8] ASCE/SEI 7-05, Minimum Design Loads For Buildings and Other Structures.
[9] I. Scaletchi, I. Vişa, R. Velicu and M. Moldovan, Torsion Moment from Wind Action on PV Platforms, 26th European Photovoltaic Solar Energy Conference and Exhibition, pp 3556-3559 (2010).
[10] Puneeth Kumar H P, S B Prakash, CFD Analysis of Wind Pressure over Solar Panels at Different Orientations of Placement, *International Journal of Advanced Technology in Engineering and Science*, Volume No. 02, Issue No. 07 (2014).
[11] K. C. Su, K. M. Chung, S. T. Hsu, Numerical Simulation of Wind Loads on Solar Panels, The Seventh International Symposium on Physics of Fluids (2017).
[12] R. Velicu, G. Moldovean, I. Scaleţchi and B.R. Butuc, Wind Loads on an Azimuthal Photovoltaic Platform. Experimental Study, RE&PQJ, Vol.1, No.8, pp 427-432, April 2010.

EFFECT OF LIGHT IRRADIATION TREATMENT ON HYGROTHERMAL DEGRADATION OF CRYSTALLINE SILICON PHOTOVOLTAIC MODULES

Yoshiyuki Kobayashi[1], Hideyuki Morita[1], Kentaro Mori[1], Atsushi Masuda[2]
[1] Toray Industries, Inc., [2] National Institute of Advanced Industrial Science and Technology (AIST)
[1] Seta Plant, 1-1, Oe 1-chome, Otsu, Shiga 520-2141, JAPAN
Tel: +81-77-544-3531, FAX: +81-77-544-3534
E-mail:Yoshiyuki_Kobayashi@nts.toray.co.jp

ABSTRACT: We conducted combined test of xenon light irradiation test and subsequent damp heat (DH) test on crystalline silicon photovoltaic (PV) modules using backsheets (BSs) with different water vapor transmission rates (WVTRs), and then compared the results of the combined test with that of simple DH test. We found that in the combined test WVTR of BS had little influence on the required time for beginning of maximum power (P_{max}) drop, and degradation began at the similar time for any modules, which is unlike the result of simple DH test. Moreover, we found that deteriorated region in the combined test spreads in different way compared with the case of simple DH test.
Keywords: Degradation, Durability, PV Module, Reliability

1 INTRODUCTION

It is known that light irradiation and hygrothermal stresses are one of the causes of deteriorating long-term reliability for crystalline silicon photovoltaic (PV) module. It was reported that degradation is caused by corrosion of metal materials due to acetic acid generated by hydrolysis of ethylene vinyl acetate (EVA) encapsulant [1], and that in damp heat (DH) test, degradation rate of PV module is varied with barrier property of backsheet (BS) toward water vapor [2]. On the other hand, effect of light irradiation on reliability of PV module is not sufficiently clarified. Although it was reported that maximum power (P_{max}) reduces earlier in the case that UV light irradiation test and subsequent DH test are carried out than the case that DH test is carried out alone [3], effects of barrier property of BS have not been clarified. This work is focused on investigating correlation between barrier property of BS and performance in PV modules.

2 EXPERIMENTAL PROCEDURE

Major materials which constitute the sample (1 cell module) are as follows; multicrystalline silicon solar cell (3 bus bar, 156 x 156 mm, t=0.18 mm), EVA (fast cure type t=0.45 mm), tempered and embossed cover glass (190 x 190 mm, t=3.2 mm). Three kinds of BS, polyethylene terephthalate (PET)/Al/PET, silica-evaporated PET (SiO_x BS), polyvinyl fluoride (PVF)/PET/PVF, whose water vapor transmission rates (WVTRs) are shown in Table 1, were used. One sample was fabricated without BS. The conditions of UV exposure test are as follows; light source is long arc xenon lamp with intensity of UV 90 W/m^2 at 300-400 nm, chamber temperature 63°C, humidity less than 30%RH.

As the first step, samples were exposed to UV for 1500 h or 2000 h. As the next step, samples of the above mentioned are exposed to DH condition of 85°C, 85% RH. In addition, terminal condition of module during the test was short circuit.

Table 1: WVTR of BS employed for the test modules.

Backsheet	WVTR at 85 °C, 85%RH (g/m^2/day)
PET/Al/PET	extremely low
SiO_x BS	1
PVF/PET/PVF	30
without	extremely high

3 RESULTS AND DISCUSSION

Figure 1 shows P_{max} retention after DH test using BS with different WVTRs [2]. Figures 2 and 3 show the result after combined test of xenon light irradiation test (1500 h or 2000 h) and subsequent DH test. Figure 1 suggests that in the case of DH test, required time for beginning of P_{max} drop is varied with WVTR of BS [2]. In the case of combined test, the results described below were obtained. First, P_{max} of every module after the combined test began to drop in a shorter DH test time than the case of DH test alone, as shown in Figs. 2 and 3. Second, required time for beginning of P_{max} drop was almost the same regardless of the difference in WVTR of BS. This behavior may suggest that light irradiation treatment generate substance which causes or accelerates P_{max} drop, and the substance is produced with no relation to WVTR of BS. Third, the rate of P_{max} decrease after initiation of P_{max} drop was almost the same among the tested modules by extending light irradiation test time. From these results, we are considering that the following phenomena might occur in combined test. Under the light irradiation, some substances which accelerate hydrolysis of EVA might be generated. In DH treatment, acetic acid might be generated by the effect of such substance with small amount of water vapor in EVA. It is known that carboxylic acid is generated by photodecomposition of ester (Norrish type II reaction) and hydrogen cation (H^+) is catalyst for hydrolysis reaction of ester. We suppose that the substance which accelerates hydrolysis of EVA might be H^+ generated from EVA.

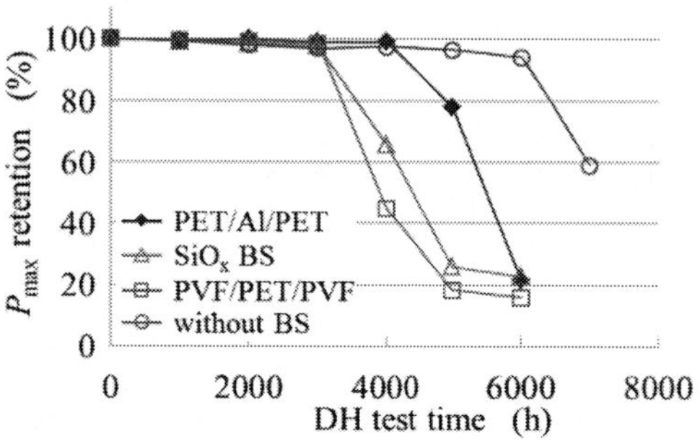

Figure 1: P_{max} retention after DH test [2]. Legends show type of BS.

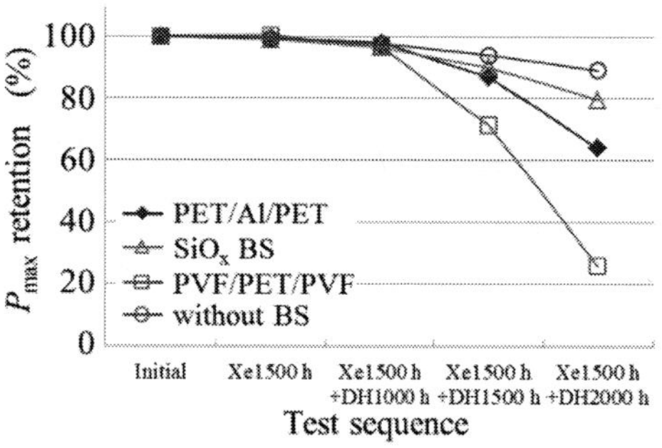

Figure 2: P_{max} retention after combined test of xenon light irradiation test (1500 h) and subsequent DH test. Legends show type of BS.

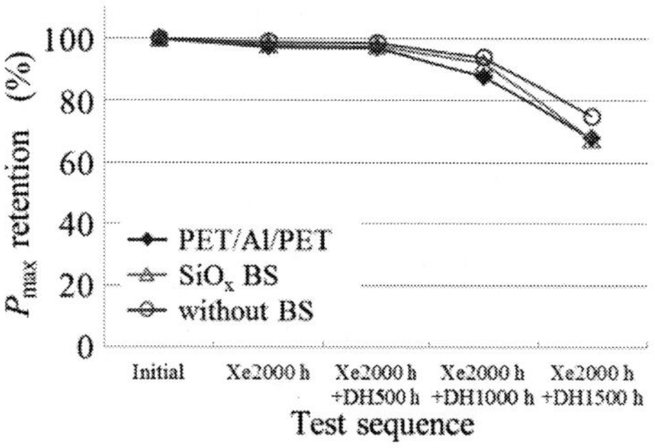

Figure 3: P_{max} retention after combined test of xenon light irradiation test (2000 h) and subsequent DH test. Legends show type of BS. The result of tested module using PVF/PET/PVF was not obtained due to breakage of module.

Figures 4, 5 and 6 show electroluminescence (EL) images of tested modules. Dark area in the cell corresponds to deteriorated region of the cell. In the case of simple DH test, dark area spreads mainly from the edge of the cells for the modules with BS. We consider that deterioration of the cells during DH test occurs by the following mechanism. First, water vapor diffuses from the edge of the cell passing through BS and concentration of moisture in EVA becomes higher at the edge of the cell than that at the center of the cell. We consider that main route of moisture ingress is BS and moisture cannot pass through cells. Second, hydrolysis of EVA occurs depending on concentration of moisture in EVA and acetic acid is generated. Concentration of acetic acid becomes higher at the edge of the cell than that at the center of the cell due to the difference in concentration of moisture. Finally, finger electrode on cell is corroded by acetic acid from the edge of the cell. Rate of corrosion of finger electrode depends on concentration of acetic acid. The difference in concentration of moisture brings higher concentration of acetic acid in EVA at the edge of the cell, resulting in beginning of corrosion of finger electrode at the edge of the cell. On the other hand, when combined test is carried out, dark area expands mainly from the center of the cells. We are considering that the phenomena described below might occur in combined test. At first, light irradiation to EVA might generate substance, which causes or accelerates hydrolysis of EVA, over whole areas of EVA in front side. Then, during DH treatment, acetic acid might be generated at the whole area on the cell by such substance. At the edge of the cell, acetic acid is generated, however also goes to outside. On the other hand, at the center of the cell, acetic acid is generated and stagnates. Finally, concentration of acetic acid at the center of the cell becomes larger than that at the edge of the cell, and finger electrode on the cell is corroded from the center of the cell.

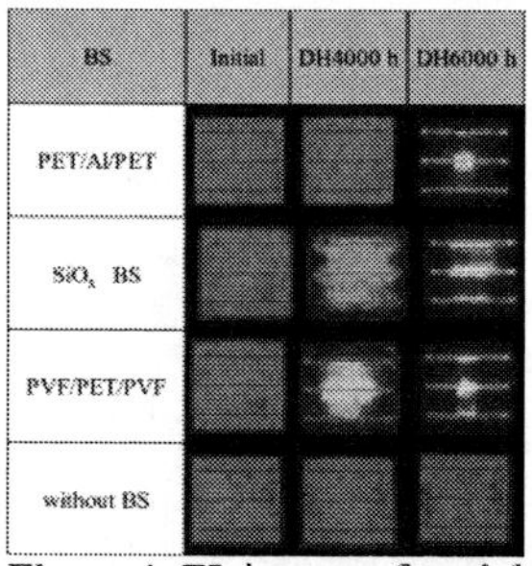

Figure 4: EL images of modules after DH test.

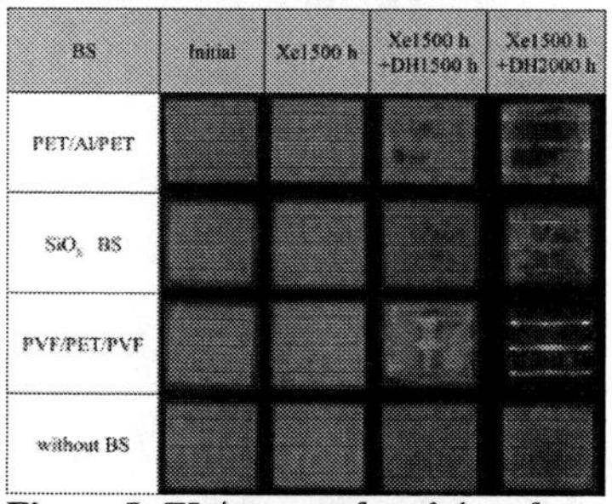

Figure 5: EL images of modules after xenon light irradiation test (1500 h) and subsequent DH test.

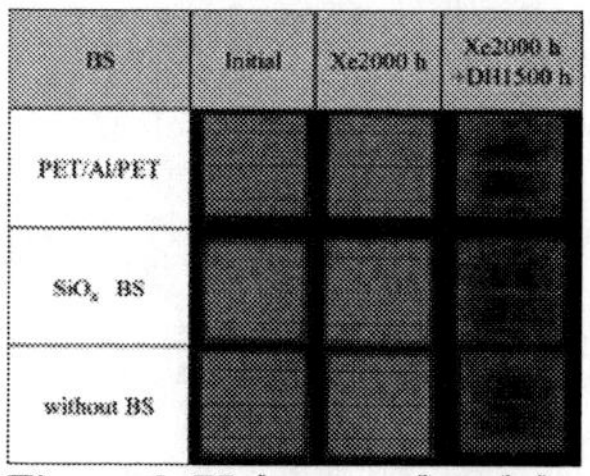

Figure 6: EL images of modules after xenon light irradiation test (2000 h) and subsequent DH test.

4 CONCLUSIONS

WVTR of BS is less effective on reduction of P_{max} in the case of combined test, where UV light irradiation test and subsequent DH test are carried out, than the case of DH test alone. The way that deteriorated region spreads in the case of combined test differs from that in the case of DH test alone.

ACKNOWLEDGEMENT

This work was supported by New Energy and Industrial Technology Development Organization (NEDO) under the Ministry of Economy, Trade and Industry.

REFERENCES

[1] M. D. Kempe, G. J. Jorgensen, K. M. Terwilliger, T. J. McMahon, C. E. Kennedy, T. T. Borek, Sol. Energy Mater. Sol. Cells **91**, 315 (2007).
[2] M. Miyashita and A. Masuda, Proc. 28th European Photovoltaic Solar Energy Conf. and Exhib., 2013, p. 2828.
[3] T. Ngo, Y. Heta, T. Doi, and A. Masuda, Jpn. J. Appl. Phys. **55**, 052301 (2016).

RELIABILITY INVESTIGATION OF THE REAR SIDE METALLIZATION OF PERC CELLS

Tobias Urban[a], Marko Turek[b], Stephan Großer[b], Christian Hagendorf[b] and Johannes Heitmann[a]

[a] Technische Universität Bergakademie Freiberg, Institute of Applied Physics, Leipziger Straße 23, 09599 Freiberg, Germany

E-mail: tobias.urban@physik.tu-freiberg.de; Fax: +49 3731 394004; Tel: +49 3731 392779

[b] Fraunhofer Center for Silicon Photovoltaics CSP, Otto-Eißfeld-Straße 12, 06120 Halle (Saale), Germany

ABSTRACT: The rear side passivation of PERC (passivated emitter and rear) solar cells implies a modified contacting approach in comparison to Al back surface field cells. This substantial modification of the rear side leads to higher cell performance but can also imply new failure and degradation mechanisms. In our work, we present a systematic approach for a quantitative evaluation of individual components of the rear side metallization. We propose an experimental method to separate the major contributions to the series resistance of the rear side of PERC solar cells and show their impact on cell reliability during temperature cycling stress tests.

The local contact between metallization and silicon, especially at voids, shows no degradation due to temperature cycling nor electromigration after high current injection. The degradation of the lateral resistance of the rear side metallization shows a good correlation with the measured series resistance increase.

Keywords: PERC, module, degradation, rear contact, metallization

1 INTRODUCTION

During the last years, the photovoltaic industry changed from conventional BSF technology to the more advanced PERC concept [1]. The charm of this concept is the introduction of only two additional production steps leading to an increased cell efficiency: the rear side passivation followed by locally opening of these layers. However, this fundamentally affects the rear side metallization process as the full area contact (BSF) changes to a local one. The optimization of the cell efficiency has been studied in detail [2, 3] while the reliability and degradation behavior is not yet fully understood.

It is known that PERC rear contacts can show voids, which form locally in the former AlSi-eutectic [4]. Studies have shown that the structural properties and the electric impact of the voids can be quantified [5]. The size of a void is in the range of µm up to cm in length and can dominate the area fraction of the microstructure for the electrical connection of Al-paste to the Si-substrate. On cell level, no eminent increase in the resistance is reported [6]. The electrical contact between the Al-rear metallization and the Si-substrate results from the contact of the Al-paste to the residual AlSi-eutectic which remains at the edges of the whole voided contact. However, the contact area is smaller than for a filled contact and could be more prone to mechanical stress resulting in an increase of the series resistance. Mechanical stress to the rear side metallization of PERC solar cells arise in solar modules. Thus, at module level, the rear side metallization is affected by thermo-mechanical stress due to the interaction of many different materials, for example the difference between thermal expansion coefficients of silicon wafer and metallization.

The presented work focusses on the influence of temperature cycling test (TCT, cycling between -40°C and 85°C) on the lateral conductivity of the rear side metallization and the interface between aluminum and silicon. An approach to separate contributions of rear contacts and the metallization layer to the total series resistance is presented. Furthermore, the reliability of voided contacts with respect to current injection is tested. Based on a systematic study, electrical parameters were measured and discussed in terms of degradation mechanism.

2 METHOD

The series resistance of a solar cell is an important contribution to the losses which reduce the total cell power. In order to test the long-term reliability of PERC solar cells in modules, three reliability tests based on accelerated stress tests have been applied. Two tests are based on a standard module assembly and test the electrical performance of the whole setup. The third test is based on a special test structure and allows the separation of the total series resistance and the lateral rear side resistance.

For the standard samples, only commercially available materials like glass, EVA, backsheet, ribbons and PERC cells (void affected) were used and manufactured as single cell mini-modules. These mini-modules were subject to a thermal cycle test (TCT) including 200 cycles between -40 °C and 85 °C in order to evaluate the impact of the voids on the total series resistance [7]. To this end, dark and illuminated *IV*-curves were measured before and after lamination, and after TCT. A reference sample which has not been stress tested has also been measured in parallel to identify the effects caused by the TCT. Furthermore, a second set of samples has been subject to current injection as a further reliability test. For 21 days, two groups of mini-modules have been exposed to a current injection of 8A. One group has been tested at 25°C and the other group at elevated temperatures of 85°C.

The degradation of the metallization layer was analyzed on a different set of cells with an adapted rear metallization layout and a cell dimension of 104 x 52 mm^2 performed by laser cutting of 6" cells. A special interconnection scheme has been employed which allows to extract two separate information: the resistance of rear side metallization as one part of overall series resistance and the series resistance of the full cell in dark. On the one hand, this enables the measurement of the resistance between both rear side ribbons, see Figure 1 and on the other hand, it is possible to measure the dark *IV*-curve if the module is interconnected in traditional way. Based on a diode model, the series resistance can be extracted from the *IV*-curve [8]. The cross-linking of the rear side ribbons therefore lies outside the module and enables the switching between both configurations due to simple soldering/resolving during the degradation process (area within the module marked by black dashed line in Figure 1). Hereby it is possible to determine the rear side resistance R_{rear} and series resistance r_s at the same degradation state.

$$R_{rear} \sim R_{tab} + R_c + R_{Ag} + R_{AgAl} + R_{Al} \qquad (1)$$

The resistance measurement gives an integral value for many resistance contributions, Eq. 1. The interconnection has their proportion by the ribbon resistance (R_{tab}) and the contact resistance at the solder joint (R_c), both are assumed to be constant during TCT [9]. The metallization part is divided into a contribution of the silver resistance at the solder pad (R_{Ag}), the alloy formation between silver and aluminum (R_{AgAl}) [11] and the resistance of the aluminum rear side metallization layer.

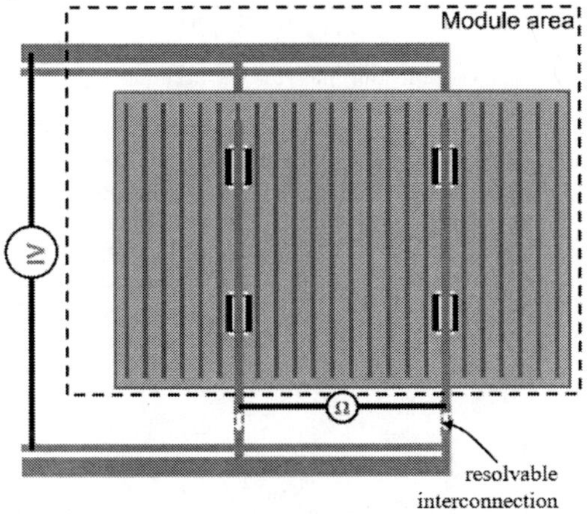

Figure 1: Interconnection scheme for solar cells with two different measurement possibilities, at one side after resolving the outside interconnection the resistance measurement between rear side ribbons (A) and a classical *IV*-measurement (B) and.

In this special one-cell interconnection, the resistance of the copper ribbon (R_{tab}) and the contact resistance (R_c) can be assumed as constant [9, 11] and therefore the change in R_{rear} is only due to the rear side metallization including all other mentioned resistance contributions. Furthermore, it has been tested whether the degradation is constant for the usual operating temperature of a solar module. To this end, the *IV*-curve and rear resistance was measured in a temperature range between -40 °C up to 85 °C.

3 RESULTS AND DISCUSSION

The TCT stress tests on the void-affected one-cell mini-modules resulted in rather small changes in the short circuit current and open circuit voltage only, see Figure 2. However, the fill factor showed a degradation, which is responsible for an efficiency loss, which was not observed for the pseudo fill factor. Hence, this degradation can be attributed to an increase in series resistance of more than 3% which has not been observed in the reference sample. However, based on this experiment, a separation of different contributions leading to this increased series resistance is not possible.

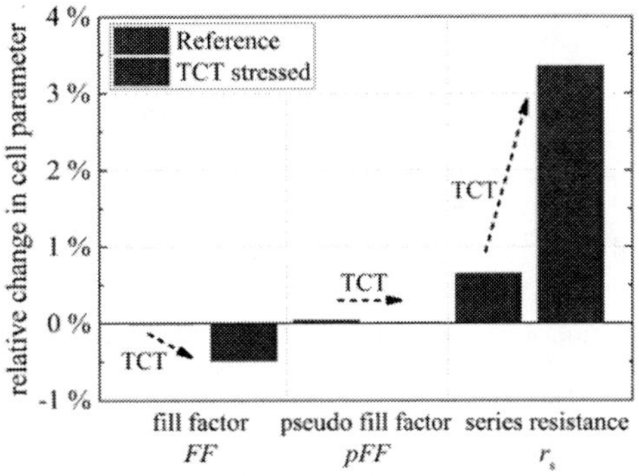

Figure 2: Relative change in fill factor, pseudo fill factor, and series resistance due to TCT-stress test of mini-modules.

A cross-section of the cell metallization in a mini-module after TCT-test has been prepared. The microstructure of a laminated rear contact with void formation is shown in Figure 3. One can see that voids in a PERC cell are still present and not filled with the encapsulant during the lamination process. The electric connection of the wafer to the metallization layer is present at the void's rim, which is encircled (green) in Figure 3. No interface modification (metallization layer to local rear contact) or damage has been identified which can be attributed to thermo-mechanical stress. Therefore, no indication of void related impact on the total series resistance has been found.

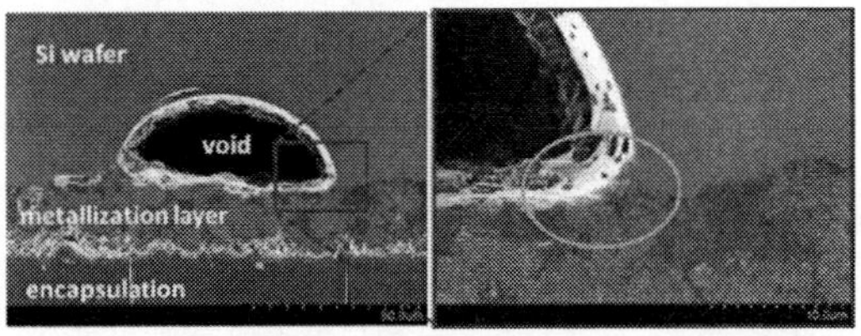

Figure 3: Cross section of a PERC cell in a mini-module after TCT 200. The position of the electric contact is green encircled in the magnification on the right hand side.

The stress test due to the application of a current to the cell leads to a completely different result. In this case, the major contribution to the efficiency loss is due to a degradation in I_{sc} and V_{oc} which can be explained by a major increase in saturation current J_{01}. While this degradation is clearly visible at elevated temperatures but not at room temperature, we attribute this effect rather to the mc-LeTID degradation which is caused by carrier injection at elevated temperatures [12-14] than to PERC-voids or a electromigration related degradation of the metallization and contacts.

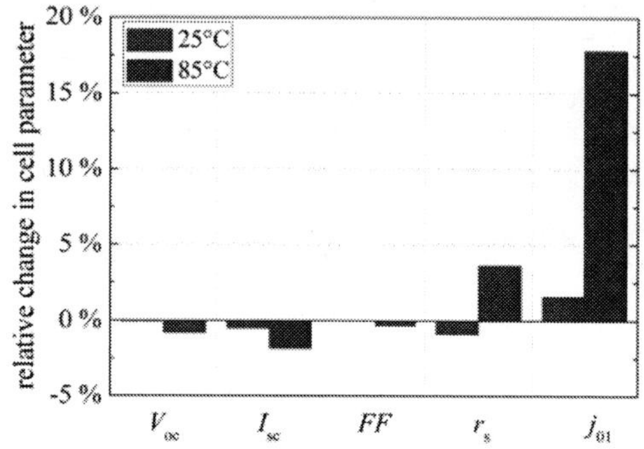

Figure 4: Relative change in open circuit voltage, short circuit current, fill factor, series resistance, and saturation current due to current induced stress test.

The third experiment is based on a special test structure and allows the separation of the total series resistance and the rear side series resistance.

The degradation measurements reveal an increasing resistance of the metallization resistance R_{rear} during TCT, which is nearly linear for the investigated period, as example for one module/cell plotted in Figure 5. However, the deviation of the measured resistance increases with temperature cycle counts (TCC). After 200 cycles the resistance is ~ 70% higher in comparison with the initial state. The soldering contact resistance and ribbon resistivity are assumed as constant [10, 11]. Therefore, the resistance increase has to be caused by the rear side metallization. It can be concluded, that the degradation of the aluminum metallization with their porous structure is the obvious weak point of the system.

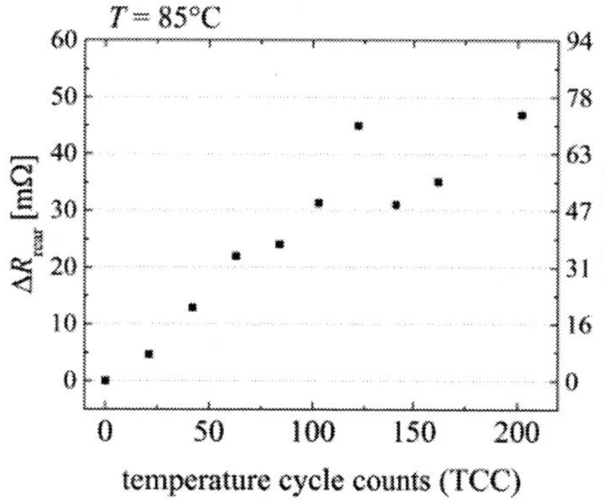

Figure 5: Plot of resistance increase of the rear side metallization in relation to the starting point.

The influence of this resistance increase on the overall series resistance of the solar cell is well visible in Figure 6. The series resistance of the same one cell module was determined by fitting of the dark IV-curve at temperature of 85°C to a classical diode model [8]. A correlation between R_{rear} and r_s can be established and is in agreement with the estimation (red line), however the measurement variations increase with degradation time, see Figure 6.

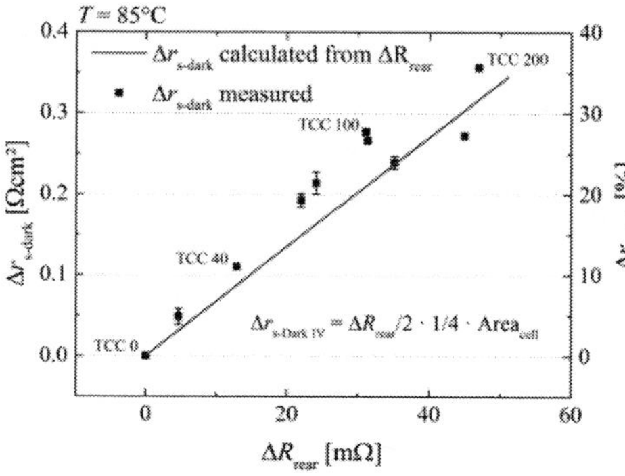

Figure 6: The relation between series resistance (from dark IV-curve) increase in relation to the starting point plotted vs. the corresponding ΔR_{rear} values, TCT count marked at data points. Calculation due to area weight resistance approach.

Furthermore, the series resistance depends on the measurement temperature during the *IV*-measurement, see Figure 7. The absolute series resistance increases due to the temperature dependency of metal resistances. In our experimental setup, this approach of measuring the series resistance at varying temperatures simplifies the data acquisition but has to be accounted for when evaluating the findings.

In Figure 7 the trend is displayed for three temperatures. It can be described by a fit function of type $\Delta r_{s\text{-Dark}} = a - b \cdot c^{TCC}$. Here, the free fit parameters a, b and c have been introduced. For the limit of many temperature cycles TCC, it reaches the asymptotic value a. On the other hand, for short degradation times, it follows that $\Delta r_{s\text{-Dark}} = (a - b) - b \cdot \ln(c) \cdot TCC$ which implies b=a since $\Delta r_{s\text{-Dark}} = 0$ for TCC $\rightarrow$ 0.

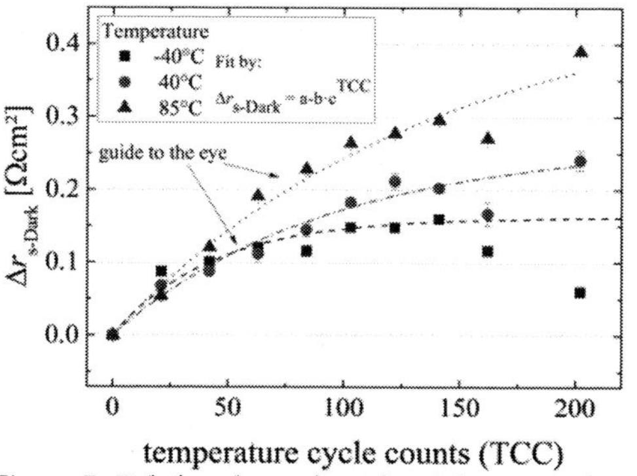

Figure 7: Relative change in series resistance during TCT of one module for 3 different temperatures, with regard to initial value for each temperature.

Based on our experimental data, the free parameters have been obtained by fitting the proposed fit function to the measurement results. The determined fit parameters, see Table 1, show an increase in the saturation level (described by parameter a) with increasing temperature. This implies directly the same behaviour for parameter b as $\Delta r_{s\text{-Dark}} = 0$ for TCC = 0. This effect can be explained by the temperature dependency of the resistance, therefore the degradation phenomenon seems to be the same for all investigated temperatures.

Table 1: Summary of the fit coefficients from curves displayed in Figure 7, whereby a = b for the constrain of TCC=0.

T [°C]	a [Ωcm²]	c [-]
-40	0.16	0.98
40	0.27	0.99
85	0.47	0.99

CONCLUSION

In this paper a measurement approach with 3 different setups have been presented to test the rear side contacts with respect to the reliability of PERC cells in modules. It was found out that thermo-mechanical stress led to an increase in the total series resistance. Simple standard setups using single cell minimodules reveal that PERC rear contacts and voids show no indication for degradation due to fractures at the electric contact interface. Also no indication for electromigration was found. By using a specific test structure the separation of the rear side series resistance from the total series resistance was achieved. A clear correlation of total resistance increase to the rear side resistance of the PERC metallization layer has been found. Therefore the observed degradation could be attributed to the rear side metallization layer even though the root cause is not understood yet and under further investigation. However, this study shows that the long term reliability of PERC solar cells in modules under thermo-mechanical load is an issue and must be addressed with rapid test methods and a deep understanding of the failure mechanism for reliable PERC modules.

ACKNOWLEDGMENT

This work was funded by the German Federal Ministry for Economic Affairs under the acronym "AdmMo" with identification code 0325775(E/F).

REFERENCES

[1] ITRPV: *International Technology Roadmap for Photovoltaic: 2017 Results including maturity reports*. Eighth Edition. September 2017. URL: http://www.itrpv.net/Reports/Downloads/

[2] FISCHER, Bernhard: *Loss Analysis of Crystalline Silicon Solar Cells Using Photoconductance and Quantum Efficiency Measurements* : Cuvillier Verlag Göttingen, (2003)

[3] KRANZ, C. et al.: *Simulation-based Efficiency Gain Analysis of 21.2%-efficient Screen-printed PERC Solar Cells*, In: Energy Procedia 92 (2016), p. 109-115

[4] URREJOLA, E. et al.: *Silicon diffusion in aluminum for rear passivated solar cells*. In: Applied Physics Letters 98 (2011), Nr. 15, p. 153508

[5] GROßER, S. et al.: *Quantification of Void Defects on PERC Solar Cell Rear Contacts*. In: Energy Procedia 92 (2016), p. 37–41

[6] LOTTSPEICH, F. et al.: *Investigation of Rear Contact Resistance of Line Contacted Industrial PERC Solar Cells*. 31st European Photovoltaic Solar Energy Conference and Exhibition (2015), p. 485–488

[7] TUREK, M.: *Current and illumination dependent series resistance of solar cells*. In: JOURNAL OF APPLIED PHYSICS 115 (2014), Nr. 14, S. 144503

[8] RIßLAND, S. ; BREITENSTEIN, O.: Considering the Distributed Series Resistance in a Two-diode Model. In: Energy Procedia 38 (2013), S. 167–175

[9] URBAN, T. et al.: Analysis of Ohmic Losses Due to Solder and Pressure Interconnection and Related Interface Resistances for Solar Cells. In: Energy Procedia 77 (2015), S. 420–427

[10] URBAN, T.et al.: Influence of Silver-Aluminium Alloy at Solar Cell Rear Side on Series Resistance and Open Circuit Voltage. In: Energy Procedia 92 (2016), S. 236–241

[11] MEIER, R. et al.: Microstructural Optimization Approach of Solar Cell Interconnectors Fatigue Behavior for Enhanced Module Lifetime in Extreme Climates. In: Energy Procedia 92 (2016), p. 560–568

[12] KERSTEN, F. et al.: Degradation of multicrystalline silicon solar cells and modules after illumination at elevated temperature, In: Solar Energy Materials and Solar Cells 142 (2015), p. 83-86

[13] LUKA, T. et al.: Multicrystalline PERC solar cells: Is light-induced degradation challenging the efficiency gain of rear passivation? In: PV International 2016, Nr. 32, p. 37

[14] LUKA, T. et al.: Intra-grain versus grain boundary degradation due to illumination and annealing behavior of multi-crystalline solar cells. In: Solar Energy Materials and Solar Cells 158 (2016), p. 43–49

LONG TERM DEVELOPMENT OF PHOTOVOLTAIC MODULE FAILURES DURING ACCELERATED AGING TESTS

C. Hirschl[1], G.C. Eder[2], L. Neumaier[1], W. Mühleisen[1], Y. Voronko[2], R. Ebner[3], B. Kubicek[3], K.A. Berger[3]

[1] CTR Carinthian Tech Research AG, Europastr.12, 9524 Villach/St.Magdalen, Austria (lukas.neumaier@ctr.at, wolfgang.muehleisen@ctr.at);
Corresponding author: E-Mail: christina.hirschl@ctr.at; T: +43 4242 56300 260
[2] OFI Austrian Research Institute for Chemistry and Technology, Arsenal Object 213, Franz-Grill-Str. 5, 1030 Vienna, Austria, (yuliya.voronko@ofi.at, gabriele.eder@ofi.at)
[3] AIT Austrian Institute of Technology, Giefinggasse 2, A-1210 Vienna, Austria (rita.ebner@ait.ac.at, bernhard.kubicek@ait.ac.at, karl.berger@ait.ac.at)

ABSTRACT: A series of test-modules – intact and with deliberately generated failures (micro-cracks, cell-cracks, glass breakage and soldering defects) - were exposed to various stress conditions (temperature, humidity and irradiation) in order to see the stress-induced effects and potential propagation of the failures on the performance. The modules were thoroughly electrically characterized before and after the accelerated ageing and outdoor tests by electroluminescence and IV-curve measurements. In addition, the test-modules were subjected to time resolved characterisation during the ageing tests by UV-fluorescence imaging. The induction period for the fluorescence effects of the polymeric encapsulant to show up and the characteristic fluorescence patterns formed upon damp-heat storage, irradiation and outdoor exposure were analysed in detail. The effect of cell-cracks and glass breakage on the local extinction of the fluorescence is discussed.

Keywords: crystalline PV, Aging, failures, UV-fluorescence, Operation and Maintenance

1. INTRODUCTION

With the enhanced competition in the photovoltaic (PV) market over the last years, the PV industry experienced tremendous pressure to reduce production costs and at the same time improve module efficiency, reliability and long-term performance. In order to be competitive in the market, PV module manufacturers have to warrant operational lifetimes of 25+ years at a maximum total yield loss of 25 %.
One of the major concerns especially for operation and maintenance (O&M) companies is the forecast how module failures develop over time and affect the yield under different climatic conditions in the future [1]. In the Austrian flagship research project INFINITY this topic is addressed. An investigation focusing on the propagation of various failures (glass breakage, micro cracks and soldering defect) under various stress conditions (temperature + humidity and irradiation) is presented. Their effect on the module performance and the encapsulation material is monitored by different measurement techniques (power comparison, electroluminescence (EL) and ultraviolet fluorescence (UV-F).

2. APPROACH

For this investigation, identical 6-cell PV modules (poly-crystalline Si-cells, ethylene vinyl acetate (EVA)-encapsulation and non-gastight polyethylene terephthalate (PET)- based backsheet) were manufactured (i) without failure, (ii) with micro-cracks and (iii) with soldering defects. At a time, two sample modules of each type (i-iii) were stored under accelerated ageing conditions such as damp heat (DH) at 85 °C, 85 % r.H. and irradiation at 1000 W/m (artificial sunlight, Xe-arc lamp, 300-2500 nm) as well as under outdoor conditions for comparison. All samples were characterized before and after a light stabilization with 20 kWh and after the aging procedure by power and EL measurements. In addition, the effect of failure development over time under the different stress conditions was visualized by measuring UV-F of the polymeric encapsulant every 100 h of accelerated aging and every month of outdoor weathering.

3. RESULTS

Two test-modules of each of the three groups (modules without failure = reference, with micro-cracks and soldering defects) were stored under DH, irradiation and outdoor conditions.

3.1 Measurement of the reference modules

As expected, the reference modules did not show strong degradation effects as determined by performance and EL measurements (Fig. 1) irrespective of the stress conditions applied. The power of the modules was 24.5 W under STC conditions and decreased between 1 and 2 % upon 1000 h of irradiation or 2000 h of DH treatment.

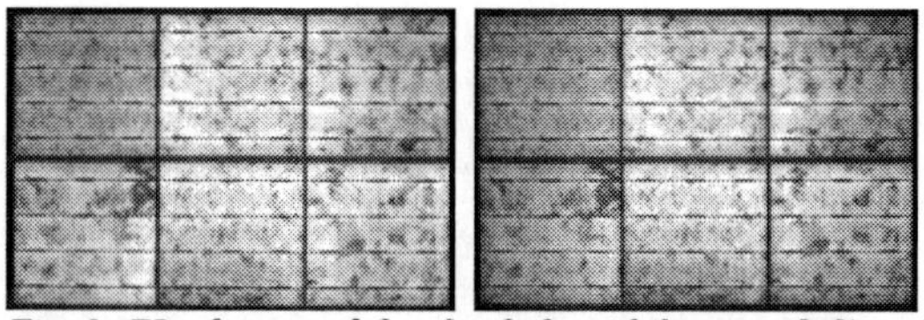

Fig. 1: EL of test-module after light stabilization (left) and after 2000 h DH (right)

The original modules do not show a specific fluorescence behavior after manufacturing (after light stabilization, Fig. 2 middle). Upon subsequent DH storage, however, the polymeric encapsulation shows beginning fluorescing effects (Fig. 2, right). These effects are not homogeneously distributed over the whole module but are stronger above the backsheet and the rims of the cells [2]. The broadness increases with increasing storage time.
In contrast, upon irradiation with artificial sunlight in the EVA of the modules fluorescing groups are formed (Fig. 2, left). The UV-fluorescence is evenly distributed in the polymeric encapsulation and increases in intensity with increasing irradiation dose. Between and at the rims of the cells the fluorescence is extinguished by oxygen (=quenching) permeating into the encapsulation via the polymeric backsheet.

Fig 2: UV-F images of test-modules after 1000 h of irradiation (left), after light stabilization (middle) and after 2000 h DH (right)

In order to see (i) which effect deliberately generated failures (micro-cracks, cell breakage, glass breakage, and soldering defects) have on the developing UV-fluorescence patterns and (ii) if a propagation of such failures occurs with storage time, defect test-modules were treated in the same way as the reference modules described above.

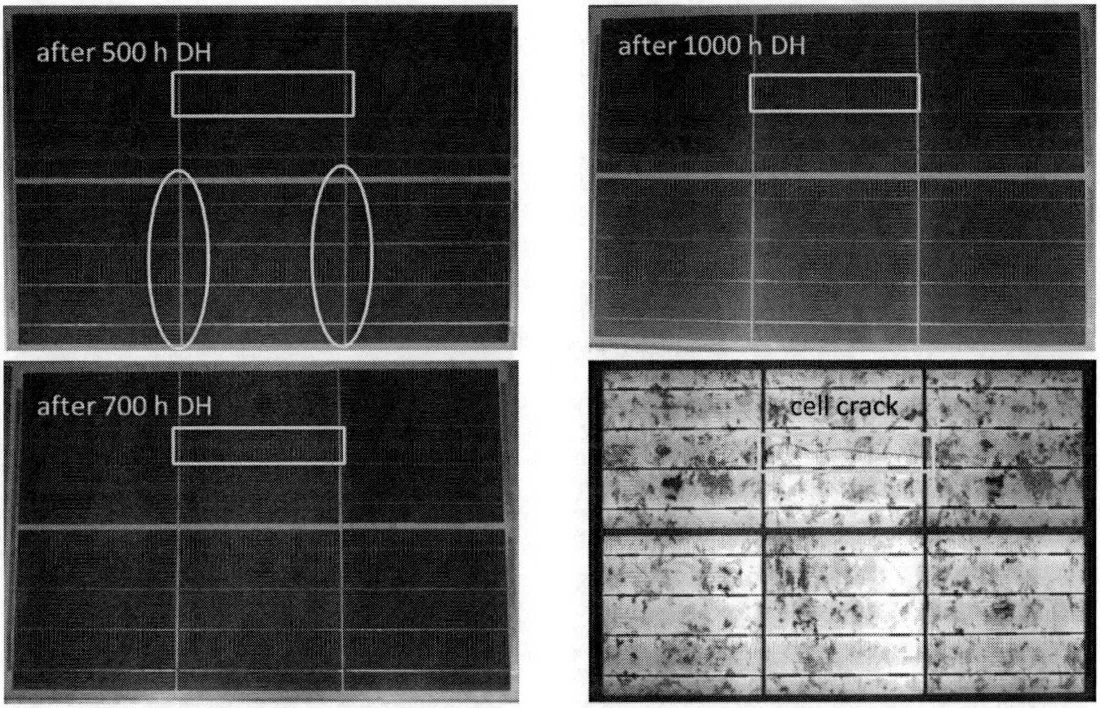

Fig. 3: UV-F images of a test-module with a cell-crack (EL-image right, bottom) after DH storage for 500, 700 and 1000 h

3.2 *Measurement of test-modules with cell-cracks*

A test-module showing a cell crack in the EL image gives UV-fluorescence images upon DH-storage as shown in Fig. 3. In addition to the fluorescence formed with increasing storage time above the backsheet and the rims of the cells, additional fluorescence is observed above the cell-crack making it clearly visible to the observer.

After 1000 h and 2000 h DH, neither the performance nor the EL-image of the test-modules were changed drastically upon artificial ageing. The measured power loss was ~1 % (after 1000 h DH) and 1-2 % (after 2000 h DH) for all modules, irrespective of the failure modes.

Upon irradiation with artificial sunlight, the evolving fluorescence pattern of test modules with micro-cracks (see Fig. 4) differed clearly from that of the test-modules without failures.

The UV-F images taken every 100 h of weathering again showed the continuous increase in intensity of fluorescence above the cells. However, there were numerous dark spots where no fluorescence was generated. By comparison with the EL-image of that module a clear correlation with the positions of the micro-cracks in the cells could be obtained (see Fig. 4). The effect could first been noticed after 300 h storage time. This is in good agreement with findings reported in literature on the induction period for the visible detection of UV-fluorescence [3].

It should be noticed that the length of the micro-cracks (as visualized by the EL-images did not increase upon accelerated ageing. The loss in power of the test-module with numerous micro-cracks was ~ 3.5 % upon weathering, which was twice as high as for the intact module.

Another test-module had a glass breakage coming up at the beginning of the irradiation test most probably due to thermal tensions. It was first noticed after 100 h storage time (see Fig. 5) and increased in length reaching its final magnitude after 400 h. It could be observed that the fluorescence formed homogeneously in the polymeric encapsulation over the cells and disappeared beneath the glass breakage after 300 h of further irradiation. The EL image showed that the glass breakage induced by thermal stress in the first phase of the weathering test did not cause a crack in the underlying cells. The disappearance of the fluorescence beneath the glass breakage is attributed to an oxygen quenching effect. The incoming oxygen can interact (e.g. energy transfer, collision, reaction) with the activated and fluorescing sites of the polymer leading to a decrease of its fluorescence intensity. The glass breakage and quenching had no impact on the performance of the test-module (loss < 1 %).

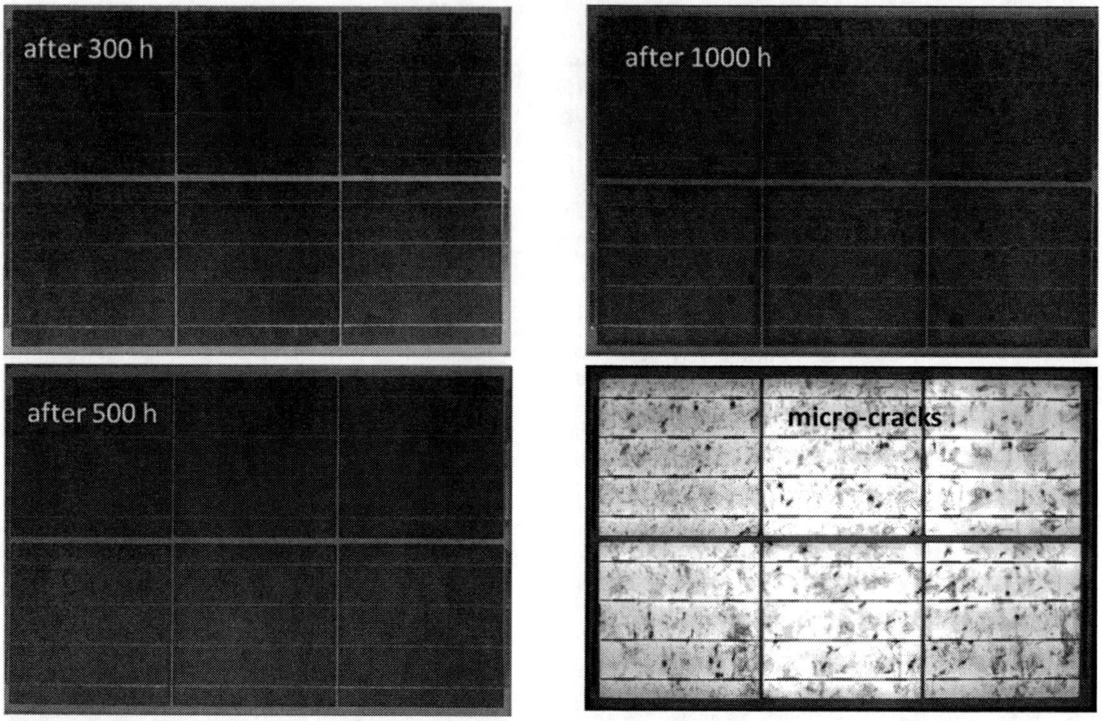

Fig. 4: UV-F images of a test-module with numerous micro-cracks in the cells (EL-image right bottom) after irradiation for 300. 500 and 1000 h

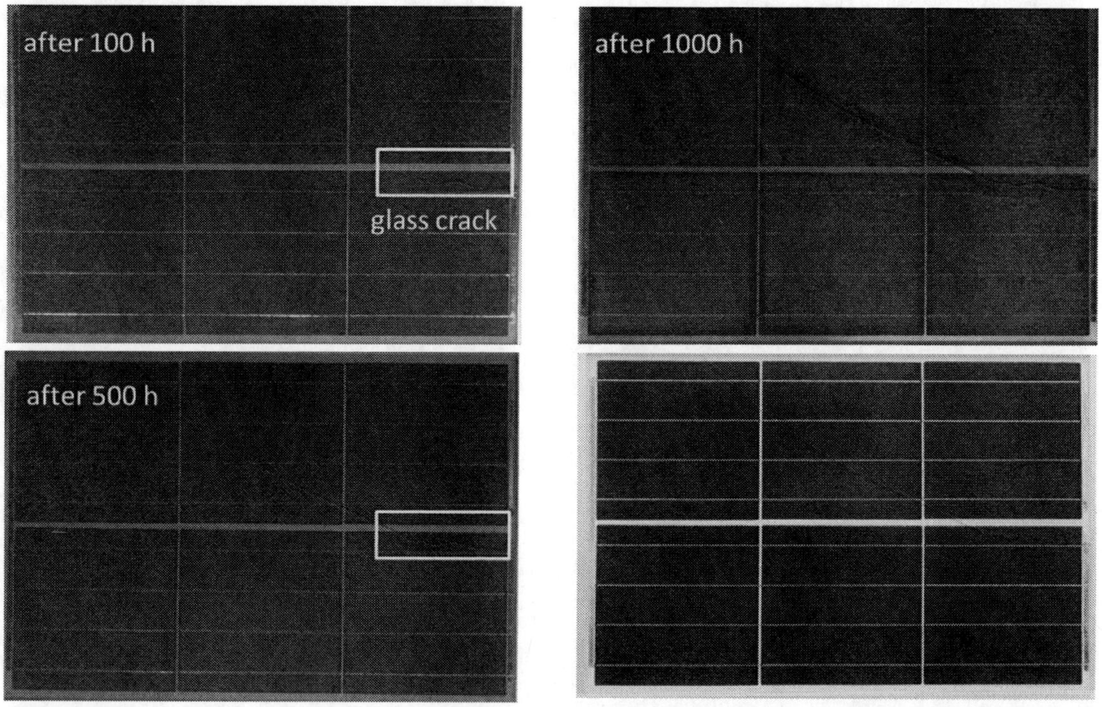

Fig. 5: UV-F images of a test-module glass breakage (photo right bottom) after irradiation for 100, 500 and 1000 h

3.3 Measurement of test-modules with soldering failures

The EL images of test-modules with soldering-defects all show a darker region of one cell caused by insufficient contact of the cell to the conducting ribbon (see Fig. 6). One of the ribbons was cut there to produce the soldering defect. This effect visible in the EL image was detected in the original state, after accelerated ageing under DH or irradiation conditions and outdoor testing.

Fig. 6: EL-image of a test-module with soldering failure

The performance of the modules with soldering defect was slightly lower (-1 to -1.5 %) compared to the intact modules. This effect was due to a lower fill factor, while U_{OC} and I_{SC} were unaffected. Upon ageing, the test modules with a deliberately generated soldering defect behaved like the intact modules, no stress-induced propagation of the failure could be observed in the test program applied.
The UV-F images of the test modules with soldering failures were identical to those of the intact reference modules, irrespective of the stress treatment.

4 SUMMARY

The results presented show a good robustness of the performance of PV-modules against stress even when failures such as micro cracks, cell-cracks or glass breakage are present. However, numerous micro-cracks led to a higher rate of degradation, the power loss upon irradiation (1000 W/m^2 for 1000 h) was twice as high as for the intact reference module. Soldering defects caused a lower fill factor of the test modules leading to a lower power. The relative decrease in power upon accelerated ageing though was not affected by the failure.
It could be demonstrated that the innovative and non-invasive characterization technique UV-fluorescence imaging is a perfect tool to detect and visualize cell cracks, micro-cracks and glass breakage [3,4,5,6]. The fluorescence formed in the polymeric encapsulant upon weathering / irradiation is extinguished in the presence of oxygen. Oxygen permeating into the encapsulant via a glass breakage or a polymeric backsheet can react as quencher for fluorescence as the oxygen molecules can interact (e.g. energy transfer, collision, reaction) with the activated and fluorescing sites of the polymer leading to a decrease of its fluorescence intensity.
The fluorescence effects observed upon DH storage (no irradiation) above permeable backsheets are correlated to the incoming water vapor and its interaction with the polymeric encapsulant. The intensity of this type of fluorescence is much lower and fluorescence spectroscopic measurements [2,5,7] have shown that the absorption maximum of water induced fluorescence lies at lower wavelength than for irradiation induced fluorescence.

5 ACKNOWLEDGEMENTS

This work was conducted as part of the Austrian "Energy Research Program" project INFINITY. Funded by the Austrian Climate and Energy Fond and the Austrian Research Promotion Agency (FFG), both are gratefully acknowledged.

6 REFERENCES

[1] A.W. Czanderna, F.J. Pern, "Encapsulation of PV modules using ethylene vinyl acetate copolymer as a pottant: A critical review", Solar Energy Materials and Solar Cells 43(2):101-181, 1996
DOI:10.1016/0927-0248(95)00150-6
[2] B. Roeder, J, Schlothauer, „Luminescence spectroscopy as powerful tool for non-destructive inspection of PV module encapsulants", Sophia Workshop 2015
[3] M. Köntges, S. Kajari-Schröder I. Kunze, "Crack statistics for waferbased silicon cell modules in the field" Sophia Workshop PV Module Reliability, 2012 Lugano.
[4] M. Köntges, A. Morlier, M. Siebert, „UV-Fluoreszenz Messtechnik für die Modulinspektion im Außeneinsatz", TÜV Rheinland, 12. Modul-Workshop 2015
https://www.tuv.com/media/germany/10_industrialservices/download_landingpage_solar/12_workshop_photovoltaik_modul/33_Koentges_UV-Fluoreszenz_Messtechnik_fuer_Modulinspektion_im_Ausseneinsatz.pdf
[5] J. Schlothauer, K. Grabmayer, G. Wallner, B. Roeder, „Correlation of spatially resolved photoluminescence and viscoelastic mechanical properties of encapsulating EVA in differently aged PV modules", Progress in photovoltaics Volume 24, Issue 6, Pages 855–870, 2016, DOI: 10.1002/pip.2734
[6] W. Mühleisen et al., „Outdoor Detection and Visualization of Hailstorm Damages of Photovoltaic Plants", submitted 09.2017
[7] J. Schlothauer, "Evaluierung von EVA-Degradations-prozessen in Si-Photovoltaikmodulen mittels 2D-Lumineszenz", PhD Thesis, HTU Berlin 2016

Monitoring of moisture ingress in PV module laminates during accelerated aging tests

L. Neumaier[1], W. Mühleisen[1], C. Hirschl[1], G.C. Eder[2], M. Aichinger[3], L. Plessing[4], A. Martin[4], A. Zimmerman[5]

[1] CTR Carinthian Tech Research AG, Europastr.12, 9524 Villach/St.Magdalen, Austria (christina.hirschl@ctr.at)
Corresponding author: E: lukas.neumaier@ctr.at; T: +43 4242 56300 206
[2] OFI Austrian Research Institute for Chemistry and Technology, Franz Grill Str. 5, 1030 Vienna, Austria (gabriele.eder@ofi.at)
[3] ertex solartechnik GmbH, Peter Mitterhofer Str. 4, 3300 Amstetten, Austria (martin.aichinger@ertex-solar.at)
[4] crystalsol GmbH, Simmeringer Hauptstrasse 24, 1110 Wien, Austria (lukas.plessing@crystalsol.com, amaury.martin@crystalsol.com)
[5] Sunplugged GmbH, Affenhausen 80, 6413 Wildermieming, Austria (andreas.zimmermann@sunplugged.at)

ABSTRACT: Within this work miniature digital temperature sensors were utilized to monitor the moisture ingress in photovoltaic modules during accelerated aging. The investigated modules were developed within the research project PV@Façade, which deals with the integration of PV modules in the building envelope to obtain long-term resistant multi-material composites. Hence, eleven samples, varying in manufacturing procedure, material composition and/or material supplier were analyzed while stored under damp-heat conditions (85°C, 85 % r.H.). As expected, the glass-glass setup with edge sealing performed best in regard to moisture resistance. For the non-glass modules, samples using TPO as encapsulation material showed good results. Nevertheless it was found, that already small differences in the distance of the sensor to the module edge of the different samples strongly influence the outcome of this experiment.
Keywords: Moisture ingress, monitoring, accelerated aging

1 INTRODUCTION

During their productive lifetime, photovoltaic (PV) modules are exposed to various environmental stresses, such as temperature changes, humidity fluctuations and UV-radiation according to the climatic environment of the installation site. Especially in building integrated photovoltaic (BIPV) applications heat accumulation and water condensation can cause additional stress for the utilized PV technology. Within the project PV@Façade, such specific challenges, raised by the integration of PV modules in the building envelope, are addressed to obtain a long-term resistant multi-material composite. Thus, various PV-active materials (crystalline Si, CIGS and CZTS) were characterized and ideally integrated into BIPV-elements by (i) lamination with Polyvinylbutyral (PVB) between glass and a façade panel (compact laminate) or by (ii) adhesion of a flexible semi-finished PV-product (various compositions) to the façade panel. Within this work, miniature digital humidity and temperature sensors were utilized to monitor the micro climate within different PV modules, varying in their manufacturing procedure, material composition and/or material supplier, during accelerated aging tests [1,2].

2 APPROACH

For monitoring the moisture ingress into the PV modules an "Arduino Due" based analysis unit comprising 25 miniature digital humidity and temperature sensors was developed. A simple graphical user interface allows the operator to adjust log interval as well as other presets. First, several commercially available humidity sensors were evaluated according to their performance and practicability for the planned measurements (see Figure 1a-c).

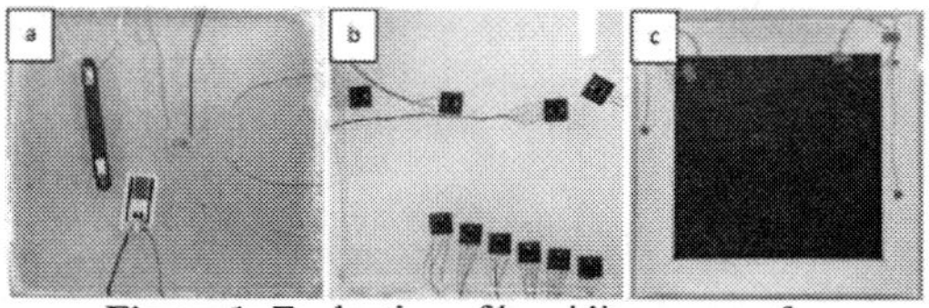

Figure 1: Evaluation of humidity sensors for experimental testing and sample assembly

The most suited sensor (Sensirion SHT25) was subsequently integrated in various PV modules, which differ in their assembly and connection (lamination, adhesion) to the façade panels (see Figure 2).

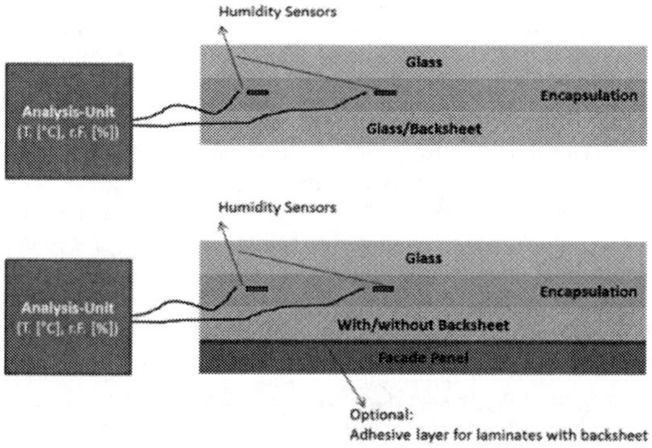

Figure 2: Experimental setup and sample variations; encapsulation layer contains PV technology

In total 11 configurations/laminates (L1-L11) were manufactured and evaluated (see Figure 3).

Component	L1	L2	L3	L4	L5	L6	L7	L8	L9	L10	L11
Sunny Side											
Front-Glass	X	X	X			X	X				
Front-Sheet ETFE				X							
Front-Sheet PET+ETFE					X			X	X	X	X
Back Side											
Back-Glass	X	X									
Backsheet with Al Layer				X	X			X	X	X	X
Backsheet without Al Layer							X				
Attachment method											
Adhesive attachment on facade panel				X	X		X	X	X	X	X
Direct lamation on facade panel			X			X					
Encapsulation											
EVA							X				
PVB	X		X			X					
TPO								X	X	X	
Ionomer				X	X						X
Edge Sealing											
Butyl		X									

Laminate # (L#)

Figure 3: Sample overview

Before investigating the moisture ingress in an accelerated aging test (damp heat), the initial status (after lamination) of all samples was determined and compared. Next, the modules were stored in a climate-chamber (Vötsch VC3 4018) and exposed to elevated humidity and temperature (85°C, 85 % r.H., damp heat testing). To ensure sufficient accuracy, the measuring interval was set to 2 minutes. Additionally, a detailed analysis of the moisture ingress was conducted by correlating the sensor distance to the module edge and the respective relative humidity value.

3 RESULTS

3.1 Initial Humidity

The initial humidity content inside the PV modules was determined directly after the manufacturing procedure (see Figure 4).

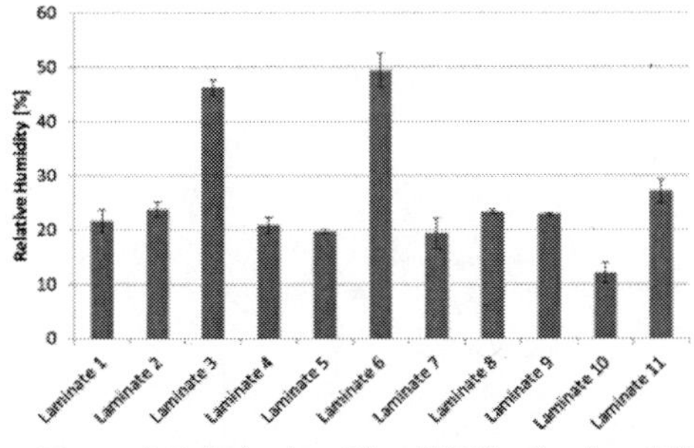

Figure 4: Relative humidity (25°C) of various PV module laminates directly after the manufacturing procedure

Most of the samples show humidity between 20-30 % r.H. at room temperature, whereas two samples (L3 & L6) mismatch this behavior. This can be explained by the setup of these samples. Here the fundamental difference to the other samples is the direct lamination on the façade panel, hence a backsheet is missing and the encapsulant is in direct contact with the composite material of the panel. It can be assumed that residual humidity which is stored in the façade panels directly diffuses into the encapsulation layer during the manufacturing process. Laminate L10 has the lowest initial humidity content. This sample was put on a heating plate for 5 minutes before the lamination procedure to ensure a proper melting of the TPO. Hence, the sample could be dried out, which would explain the lowest humidity content of all laminates.

3.2 Accelerated Aging

For determining the moisture ingress of the laminates during accelerated aging, the samples were stored in a climate chamber for 300 hours. In a first step, the increasing relative humidity content was monitored for laminate 7 (standard PV Module setup, see Figure 5).

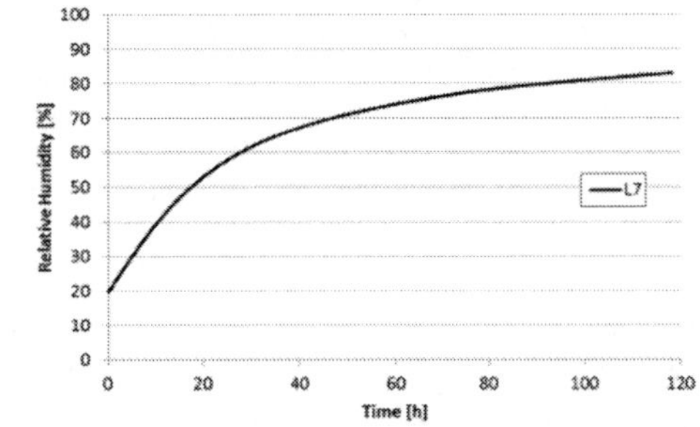

Figure 5: Moisture ingress monitored by a sensor located close to the modules edge for L7

After around 100 h of damp-heat exposure the EVA inside the laminate L7 is moisture saturated which was already stated by F. Galliano et.al. [3]. Next, all samples were stored under damp heat conditions for 300 h. Figure **6** displays the results after the accelerated aging procedure.

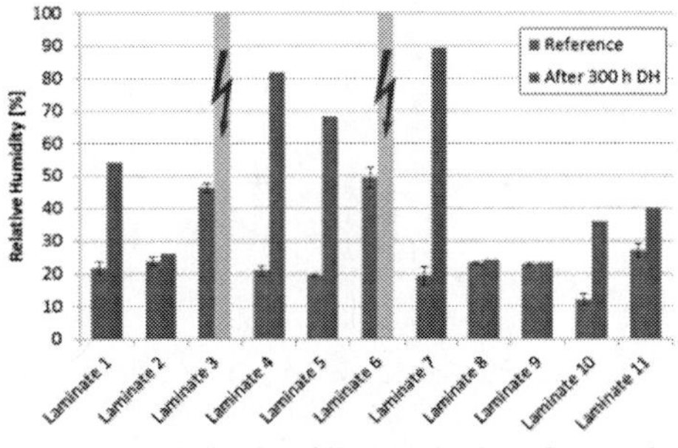

Figure 6: Relative humidity monitoring after 300 h of accelerated aging (85°C, 85% r.H., damp heat)

Samples L3 and L6 (direct lamination on the façade panel) strongly delaminated after around 100 h of testing. Here the sensors were partially laid open which resulted in incorrect measurement results. Hence these laminates were excluded from the results. Laminates L2, L8 & L9 performed best regarding their resistance against moisture ingress. For sample L2 a good performance was estimated due to the setup, comprising two glasses combined with a butyl edge sealing. Samples L8 and L9 should be in the range of laminate L10 (comparable setup), but resulted in almost no moisture increase, according to their slightly deeper positioning (distance from the edge ~5 mm more than other sensors) towards the center of the module. Hence this must be considered by comparing the laminates. Also laminates L4, L5 and L11 are very similar in their composition (choice of components), but showed different results. The variation of L4 can be explained by the choice of a different front sheet, and hence different water vapor transmission rate, compared to the others. L5 and L11 have almost the same composition, but they were differently manufactured (adaption of the manufacturing process parameters) which could explain the increased resistance against moisture ingress of sample L11.

4 SUMMARY AND OUTLOOK

The moisture ingress analysis described in this work was an important step for evaluating BIPV elements developed within the research project PV@Façade. Here it was essential to gather information about the different material compositions and their behavior due to increased stress conditions, to provide feedback for the project partners. As expected, the glass-glass setup with edge sealing (L2) performed best in regards to moisture resistance. For the non-glass modules, samples using TPO as encapsulation material showed good results. Laminates L8 & L9 also resulted in very low moisture absorption, but here it must be considered, that the distance of the sensor to the module edge is bigger than for the other samples. Hence, it is necessary that the sensors are put in correlation with the distance to obtain comparable results. For this, laminates with a set of stepwise arranged sensors are planned. One module (laminate L1) was already equipped with a set of six sensors (see Figure 7, Figure 8). The sensors were monitoring the relative humidity content during accelerated aging described in subchapter 3.2.

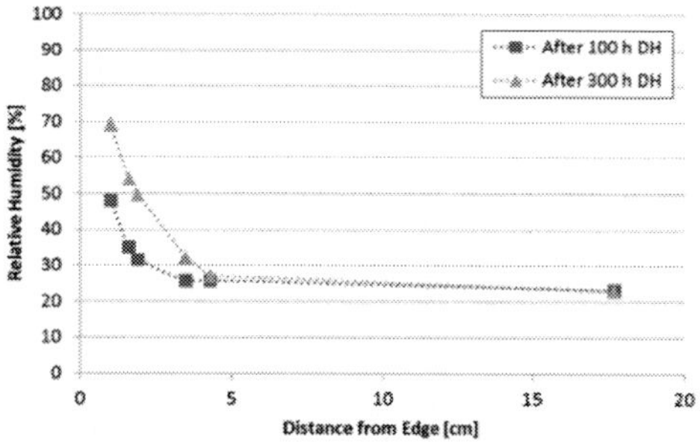

Figure 7: Moisture content in correlation to the sensor's distance from the module edge

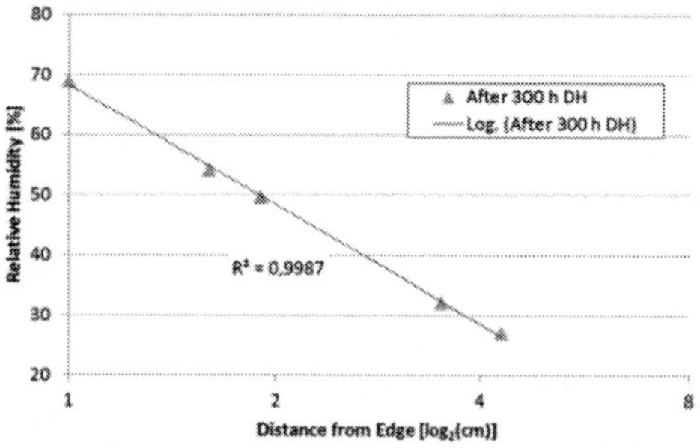

Figure 8: Logarithmic representation of Figure 7

Future work will deal with a more complex analysis of the moisture ingress for four selected material compositions being based on results from this first test series. Hence, for each module a set of sensors will be utilized to obtain distance-dependent humidity results.

5 ACKOWLEDGEMENT

The project PV@Façade was funded by the Austrian Climate and Energy Fund and conducted under the "Energy Mission Austria" program (Austrian Research Promotion Agency).

6 REFERENCES

[1] M.D. Kempe, "Control of moisture ingress into photovoltaic modules", 31st IEEE Photovoltaics Specialists Conference and Exhibition, Lake Buena Vista, Florida, 2005
doi: 10.1109/PVSC.2005.1488180

[2] M. Jankovec, G. Matic, E. Annigoni, F. Galliano, H.-Y. Li, L.-E. Perret-Aebi, F. Sculati-Meillaud, C. Ballif, M. Topic, "In-Situ Monitoring of Moisture Ingress in PV Modules with Different Encapsulants", 32nd European Photovoltaic Solar Energy Conference and Exhibition, Munich, 2016

[3] F. Galliano, E. Annigoni, H.Y. Li, F. Sculati-Meillaud, M. Jankovec, Ch. Ballif, L.-E. Perret-Aebi, "In-situ monitoring of moisture ingress in PV modules", 13. Nationale Photovoltaik-Tagung, Switzerland, 2015

EXPERIMENTAL CHARACTERIZATION AND NUMERICAL SIMULATION OF HUMIDITY-INDUCED DAMAGE IN PV CELLS

Mariacristina Gagliardi, Irene Berardone, Marco Paggi
IMT School for Advanced Studies Lucca
Piazza San Francesco, 19, 55100, Lucca, Italy

ABSTRACT: The prediction of photovoltaic module performances is an important topic for energy industry. Standard tests prescribed by IEC regulation give a picture of degradation under accelerated conditions. While results at the end of the IEC tests are widely reported, only few data show the degradation evolution with time. To provide additional details, we performed a thermal cyclic test on a PV mini-module, acquiring electroluminescence images after fixed times. Experimental results show the formation of electrically inactive zones and black spots due to moisture accumulation during time. The finite element analysis of reactions and diffusion phenomena in EVA confirms the effect of water accumulation as cell damage trigger. Results provide an accurate analysis of module degradation, under wide temperature inversions and humidity, and enable a thorough knowledge of damage evolution.
Keywords: Degradation, durability, electroluminescence, modelling

1 INTRODUCTION

The copolymer poly(ethylene-*co*-vinyl acetate) (EVA) is the most common encapsulant in photovoltaic (PV) module manufacture. EVA protects silicon cells against environmental agents, providing electrical insulation to metallic grid lines, and ensuring the layer adhesion. Weathering induces thermo-photo-oxidative phenomena in EVA, due to the prolonged UV exposure and absorption of contaminant from the external environment. High working temperatures (up to 85°C) accelerate the encapsulant degradation. The EVA copolymer chain contains two repeating units, ethylene (ET) and vinyl acetate (VAc). ET moieties degradation produces unsaturations (double bonds between carbon atoms – C=C–) and carbonyl bonds (>C=O) after oxidation, while VAc decomposes giving acetic acid (HAc) (Fig. 1). EVA degradation leads to the optical properties deterioration, which appears as discoloration of sunlight-exposed areas. HAc molecules dissolve in the moisture absorbed from the external environment. The contributory oxygen absorption induces an acidic and aerobic environment. This chemical environment generates optimal conditions for metallic grid line oxidation [1]. Starting from our recent study [2], we propose a finite element framework for the comprehensive simulation of EVA degradation, production of acetic acid, and corrosion of metallic fingers. The framework provides the solution of a reaction-diffusion model.

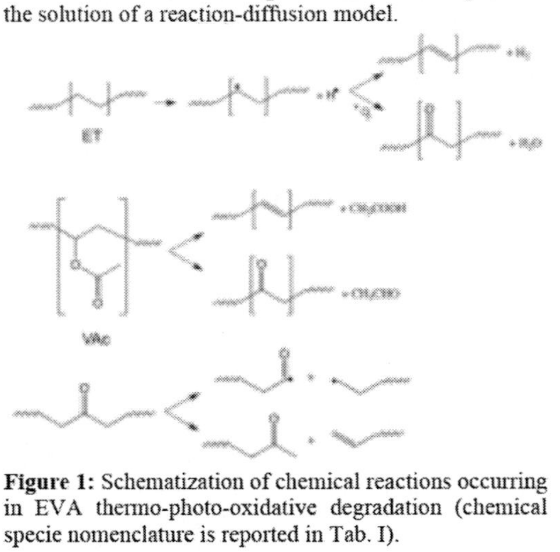

Figure 1: Schematization of chemical reactions occurring in EVA thermo-photo-oxidative degradation (chemical specie nomenclature is reported in Tab. I).

2 MATERIALS AND METHODS

The developed finite element model computes the solution of a reaction-diffusion system, which comprises coupled ordinary differential equations and partial differential equations. The proposed model enables the quantification of the chemical species produced after degradation, as well as the computation of concentration profiles, involving diffusion of small molecules, like oxygen and moisture, absorbed from the external environment, and HAc, formed upon VAc deacetylation). The computational model was implemented with MATLAB R2017a.

Chemical reactions on EVA considered were:

$$ET \xrightarrow{k_1} R^\bullet + H^\bullet, \tag{1a}$$

$$R^\bullet \xrightarrow{k_2} U^{(R2)} + H^\bullet, \tag{1b}$$

$$R^\bullet + O_2 \xrightarrow{k_3} C_b^{(R3)} + H_2O, \tag{1c}$$

$$VAc \xrightarrow{k_4} C_b^{(R4)} + CH_3CHO, \tag{1d}$$

$$VAc \xrightarrow{k_5} U^{(R5)} + HAc, \tag{1e}$$

$$C_b \xrightarrow{k_6} C_b^\bullet + R_t^\bullet, \tag{1f}$$

$$C_b \xrightarrow{k_7} C_{bt} + U_t. \tag{1g}$$

For the nomenclature of chemical species, see Tab. I. In the set of Reactions (1), k_i are the kinetic constants, which are dependent on the temperature, according to the Arrhenius law. The following set of equation constituted the reaction-diffusion system:

$$\frac{\mathrm{d}[ET]}{\mathrm{d}t} = -k_1[ET], \tag{2a}$$

$$\frac{\mathrm{d}[VAc]}{\mathrm{d}t} = -(k_3 + k_4)[VAc], \tag{2b}$$

$$\frac{\mathrm{d}[R^\bullet]}{\mathrm{d}t} = k_1[ET] - k_2[R^\bullet] - k_3[R^\bullet][O_2], \tag{2c}$$

$$\frac{\mathrm{d}[U]}{dt} = k_2[R^\bullet] + k_5[VAc], \tag{2d}$$

$$\frac{\mathrm{d}[C_b]}{\mathrm{d}t} = k_3[R^\bullet][O_2] + k_4[VAc] - (k_6 + k_7)[C_b], \tag{2e}$$

$$\frac{\mathrm{d}[C_b^\bullet]}{\mathrm{d}t} = k_6[C_b], \tag{2f}$$

$$\frac{\mathrm{d}[C_{bt}]}{\mathrm{d}t} = k_7[C_b], \tag{2g}$$

$$\frac{\partial[CH_3COOH]}{\partial t} - D_5\Delta[CH_3COOH] = -k_5[VAc], \tag{2h}$$

$$\frac{\partial[O_2]}{\partial t} - D_7\Delta[O_2] = -k_3[R^\bullet][O_2], \tag{2i}$$

$$\frac{\partial[H_2O]}{\partial t} - D_8\Delta[H_2O] = k_3[R^\bullet][O_2]. \tag{2j}$$

where D_i are the diffusion coefficients of chemical species, and are function of the temperature:

$$D_i = D_i^0 exp\left(-\frac{E_d}{RT}\right). \quad (3)$$

The model also considered temperature variation during time. Temperature profile was described by the Fourier's law:

$$\rho c\frac{\partial T}{\partial t} - \kappa\Delta T = Q_T(t), \quad (4)$$

where ρ, κ and c are, respectively, density, thermal conductivity and specific heat capacity of EVA. Q_T is the heat generated by chemical reactions, given by:

$$Q_T = \sum_{i=1}^{7} Q_i. \quad (5)$$

where Q_i were evaluated as:

$$Q_{ET} = \Delta H_{R1}k_1(T)[ET], \quad (6a)$$

$$Q_{R^\bullet} = (\Delta H_{R2}k_2(T) + \Delta H_{R3}k_3(T)[O_2])[R^\bullet], \quad (6b)$$

$$Q_{VAc} = (\Delta H_{R4}k_4(T) + \Delta H_{R5}k_5(T))[VAc], \quad (6c)$$

$$Q_{C_b} = (\Delta H_{R6}k_6(T) + \Delta H_{R7}k_7(T))[C_b]. \quad (6d)$$

After EVA degradation was computed, we considered the oxidation of silver gridlines. Gridline corrosion in PV modules is an important factor of power-loss [3]. Corrosive substances, such as the HAc produced by deacetylation, have reduced diffusion in the EVA layer in the absence of water. The high hydrophilicity of EVA, which promotes moisture absorption, enhances the oxidative action of HAc [1,4]. In the presence of water, HAc leads to the following equilibrium dissociation reaction:

$$CH_3COOH + H_2O \overset{K_a}{\rightleftharpoons} CH_3COO^- + H_3O^+. \quad (7)$$

The prediction of $[H_3O^+]$ over time is necessary to understand corrosion phenomena. Acetic acid molecules dissociate in the presence of water, and the equilibrium between molecules and ions depends on the dissociation constant K_a:

$$K_a = \frac{c_{CH_3COO^-}c_{H_3O^+}}{c_{CH_3COOH}}. \quad (8)$$

where $c_{CH3COO-}$ and c_{H3O+} are concentrations in water of the acetate and hydronium ions, while $c_{CH3COOH}$ is the concentration of non-dissociated acetic acid. K_a varies with the temperature [5] but we assumed that it was constant in the analyzed temperature range.

Table I: Nomenclature of chemical species

Chemical structure	Nomenclature	Definition
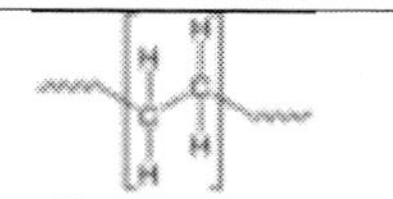	ET	Ethylene units
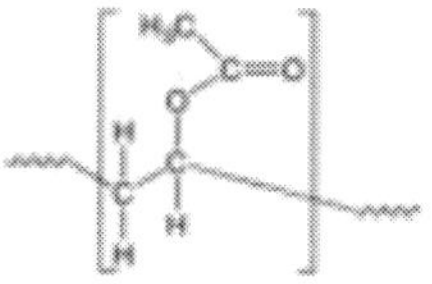	VAc	Vinyl acetate units
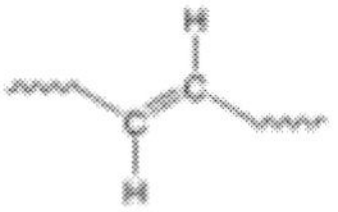	U	Unsaturated C=C bond in the main chain (vinylene groups)
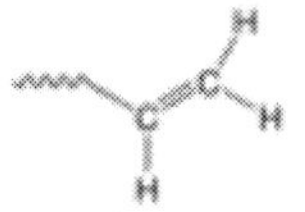	U_t	Unsaturated C=C bond in terminal position (vinyl groups)
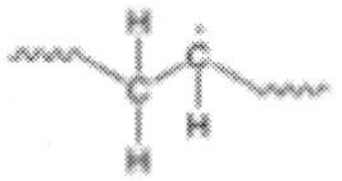	$R^\bullet$	Radical with one unpaired electron in the main chain
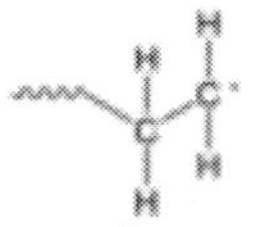	$R_t^\bullet$	Radical with one unpaired electron on a terminal C atom
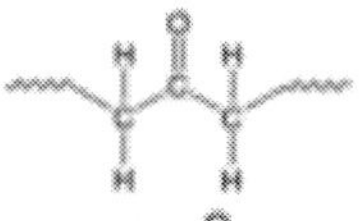	C_b	Carbonyl bond C=O in the main chain
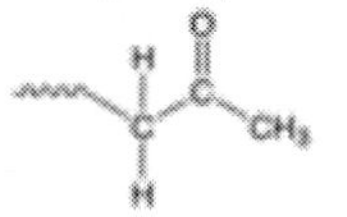	C_{bt}	Carbonyl bond C=O close to a terminal C atom due to back-biting isomerization
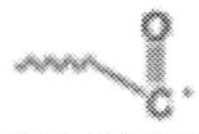	$C_{br}^\bullet$	Carbonyl bond C=O with radical on a terminal C atom
CH_3COOH	HAc	Acetic acid

According to the stoichiometry of Reaction (7), the value of $c_{CH3COO-}$ is equal to c_{H3O+}. Indicating with x the cumulative concentration of dissociated HAc molecules at the current time t, Eq. (8) becomes:

$$K_a = \frac{x^2}{c_{CH_3COO^-}|_{t-\Delta t} - x}. \tag{9}$$

Rearranging Eq. (9) we obtain:

$$x = c_{CH_3COO^-} = c_{H_3O^+} = \\ = \frac{-K_a \pm \sqrt{K_a^2 + 4c_{HAc}|_{t-\Delta t} K_a}}{2}. \tag{10}$$

Once ion, moisture and oxygen concentrations are known, silver oxidation can be computed. Silver (Ag) is a noble metal with high resistance to oxidation. This metal undergoes *dissolution corrosion*, characterized by the formation of cations Ag^+ that pass in solution under specific conditions of pH and oxygen concentration. The Ag dissolution reaction is given by [6,7]:

$$2Ag_{(s)} + \frac{1}{2}O_2 + 2H_3O^+ \rightarrow 2Ag^+ + 3H_2O. \tag{11}$$

In Reaction (11), silver is present in its solid form (Ag) and in the oxidized state (Ag^+). The oxidation-reduction of Ag can be splitted in two semi-reactions:

$$\begin{cases} 2Ag_{(s)} \rightarrow 2Ag^+ + 2e^- & oxidation \\ \frac{1}{2}O_2 + 2H_3O^+ + 2e^- \rightarrow 3H_2O & reduction \end{cases} \tag{12}$$

where e^- are the transferred electron moles. Ag oxidation occurs only if the free Gibbs energy ΔG of the overall process is negative. For electrochemical corrosion phenomena, the ΔG is defined as:

$$\Delta G = -nF(E^{eq}_{red} - E^{eq}_{ox}) \tag{13}$$

where n is the number of e^- moles involved in the process, F is the Faraday constant (96500 C mol^{-1}), E^{eq}_{red} and E^{eq}_{ox} are the equilibrium half-cell potentials. Nernst equations for semi-reactions (12) are:

$$\begin{cases} E^{eq}_{Ag^+/Ag} = E^0_{Ag^+/Ag} + \frac{RT}{2F}\ln\left(c^2_{Ag^+}\right) & ox \\ E^{eq}_{O_2/H_2O} = E^0_{O_2/H_2O} + \frac{RT}{2F}\ln\left(c^{\frac{1}{2}}_{O_2} \cdot c^2_{H_3O^+}\right) & red \end{cases} \tag{14}$$

In Eq. (14), $E^0_{Ag+/Ag}$ (0.8 V) and $E^0_{O2/H2O}$ (1.23 V) are the standard half-cell potential, c_i are the concentrations of species involved. The metal-ion concentration is often unknown, thus the conventional value of 10^{-6} mol m^{-3} is usually assumed [8].

Thermodynamic feasibility of Ag oxidation depends on oxygen and hydronium concentrations, which are both function of time and temperature. If Ag oxidation does not occur, the equilibrium between metal and its cation is stable, and the amount of solid is constant with time. If oxidation occurs, the amount of solid tends to vanish with time. Evaluation of thermodynamic oxidation feasibility can predict local conditions, in terms of pH and oxygen presence, inducing grid line corrosion. If free Gibbs energy is positive, Ag oxidation can be excluded.

If corrosion occurs, a different chemical equilibrium governs Ag consumption. The presence of acetate ions may lead to the formation of silver acetate salt, CH_3COOAg, according to the following equilibrium reaction:

$$Ag^+ + CH_3COO^- \rightleftharpoons CH_3COOAg; \tag{15}$$

This salt has a solubility product in water given by:

$$K_{ps} = c_{CH_3COO^-} \cdot c_{Ag^+}. \tag{16}$$

Amounts of Ag^+ and CH_3COO^- ions in water, indicated as s, depend on K_{ps}, resulting:

$$s = c_{CH_3COO^-,max} = c_{Ag^+,max} = \sqrt{K_{ps}}. \tag{17}$$

If $c_{CH3COO-}$ and c_{Ag+} are higher than s, silver acetate precipitates in its solid form, perturbing the equilibrium between Ag and Ag^+. Silver acetate precipitation induces c_{Ag+} decrease. The amount of precipitated silver acetate salt gives a quantitative prediction of the corrosion rate.

For the numerical analysis, we considered three cases: (i) homogeneous cell, (ii) cracked cell and (iii) cracked cell with encapsulant inhomogeneities. Geometries and meshes used for the finite element analysis are reported in Fig. 2. Details on the mathematical procedure used to solve the problem are reported in [2].

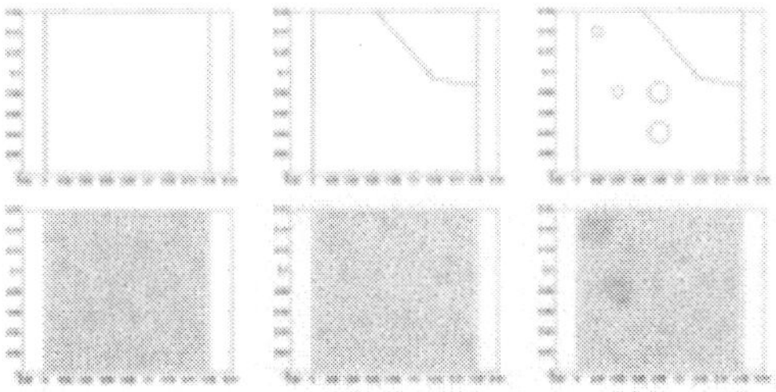

Figure 2: Geometries (top) and meshes (bottom) used to study three cases: homogeneous cell (left), cracked cell (middle) and cracked cell with EVA inhomogeneities (right).

Computational results were compared to experimental aging tests. The experiments were performed in climatic chamber (Angelantoni Challenge 250) at Politecnico di Torino [9]. Tests reproduced extreme cyclic thermo-hygrometric conditions, according to the IEC 61215 standards. Accelerated tests were monitored by electroluminescence (EL) technique. EL images were acquired every 40 cycles of accelerated tests, and provided quantitative data on electrical damage and cracking evolutions.

3 RESULTS AND DISCUSSION

Computational results provide concentration profiles of the chemical species involved in EVA degradation, moisture absorption and distribution, HAc dissolution and Ag oxidation.

EVA shows a homogeneous degradation over the Si cell with time. The decrease of VAc concentration decrease with the cycle no. (Fig. 4).

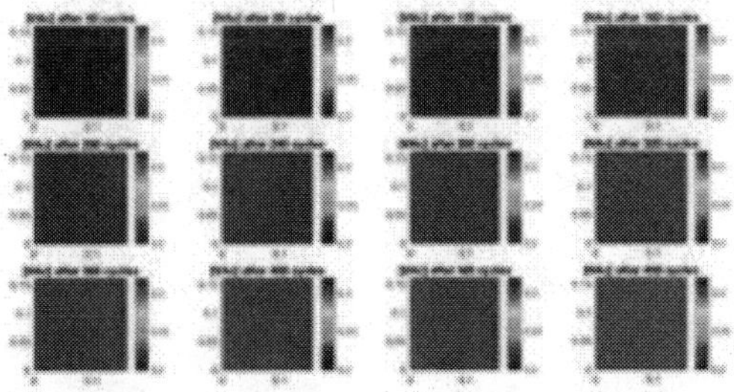

Figure 4: VAc concentrations after different cycle no.

Moisture concentration maps obtained from the analysis of the cracked cell (Fig. 5) show the effect of the crack on water accumulation, which can lead to the electrical disconnection of the interested area.

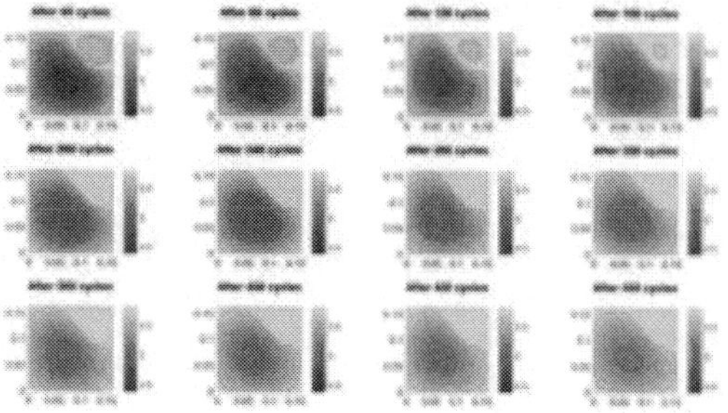

Figure 5: Numerically simulated moisture concentration maps in cracked cell after different cycle no.

Dark spot formation is clearly visible in maps obtained from the difference between results reported in Fig. 5 and those related to cracked cells without encapsulant inhomogeneities (Fig. 6).

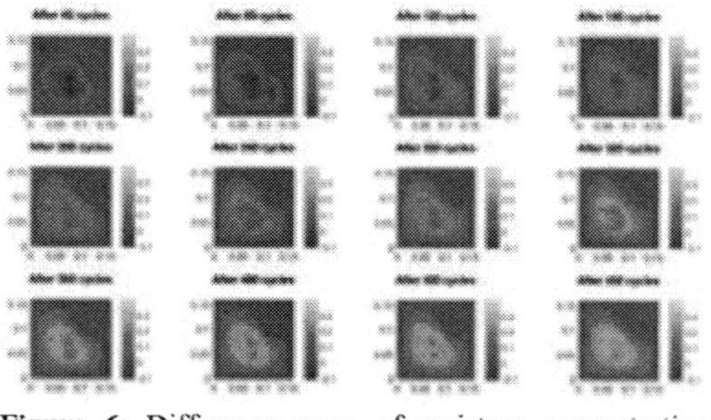

Figure 6: Difference maps of moisture concentration between cracked cell with and without encapsulant inhomogeneities after different cycle no.

EL images show the evolution of signals towards dimmer tones. A wide area of the mini-module central cell becomes electrically isolated after 400 cycles of climatic chamber. The enhanced degradation and the loss of electrical connection were presumably triggered by the pre-existing crack (Fig. 7). After additional cyclic tests, such damage propagated.

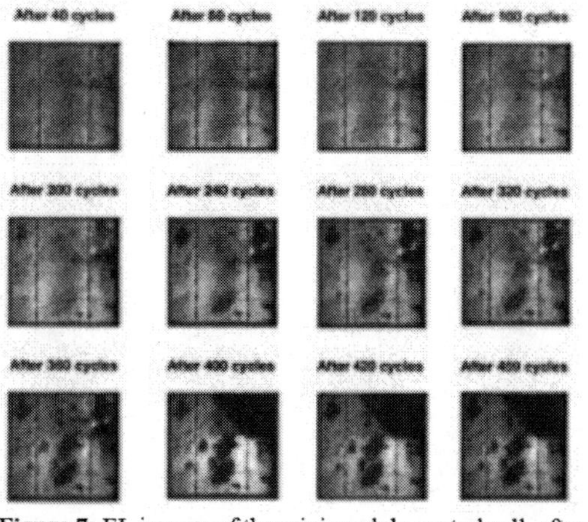

Figure 7: EL images of the mini-module central cell, after fixed cycle no., used to monitor damage evolution (wavelength range: 600-2300 nm).

pH values and the kinetic of silver acetate formation in three different points of the PV module (Fig. 7) highlight a gradual acidification of the environment, which stabilizes around 3.5. Values at the beginning of the analysis are higher in the portion over the crack due to the higher moisture concentration in this zone. Silver acetate concentration increases over time, following the VAc deacetylation rate.

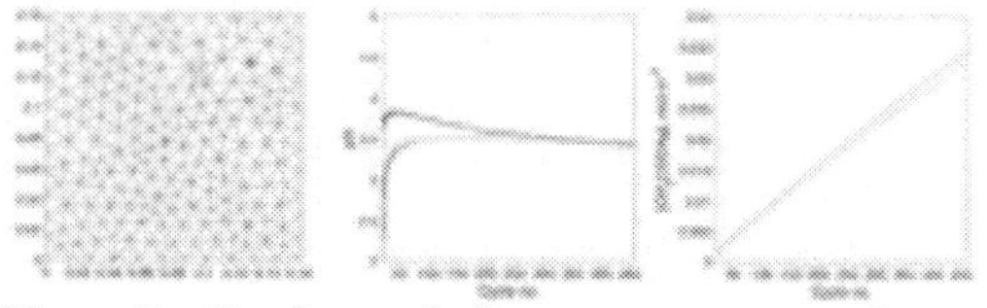

Figure 7: pH values and silver acetate concentration *vs.* cycle no. in three different points of the PV cell, indicated by colored dots.

4 CONCLUSIONS

The combination of computational and experimental results demonstrated a strict correlation between the appearance of dark spots, moisture absorption and diffusion within EVA, acidification and chemical specie distributions.

The proposed numerical scheme provided an efficient tool for the prediction of moisture content, enhanced by the presence of a crack, and the evolution of dimmer areas in the EL images. Obtained predictions can quantify thermo-photo-oxidative degradation of EVA and corrosion of metallic fingers.

The proposed numerical scheme can be considered a successful study to evaluate durability and performances of PV modules.

ACKNOWLEDGMENT

This research was funded by the European Research Council under the European Union's Seventh Framework Programme (FP/2007–2013)/ERC Grant Agreement No. 306622 (ERC Starting Grant Multi-field and Multiscale Computational Approach to Design and Durability of PhotoVoltaic Modules - CA2PVM).

REFERENCES

[1] M. Kempe, G. Jorgensen, K. Terwilliger, T. McMahon, C. Kennedy, T. Borek, Solar Energy Materials & Solar Cells, 91 (2007) 315.
[2] M. Gagliardi, P. Lenarda, M. Paggi, Solar Energy Materials & Solar Cells, 164 (2017) 93.
[3] O. Hasan, A.F.M. Arif, Solar Energy Materials & Solar Cells, 122 (2014) 75.
[4] B. Ketola, A. Norris, 25th European Photovoltaic Solar Energy Conference and Exhibition / 5th World Conference on Photovoltaic Energy Conversion, 6-10 September 2010, Valencia, Spain.
[5] J. Kendall, Chemical Society Transactions, 101 (1912) 1275.
[6] J. Liu, R. Hurt, Environmental Science & Technology, 44 (2010) 2169.
[7] B. Molleman, T. Hiemstra, Langmuir, 31 (2015) 361.
[8] E. Stansbury, R. Buchanan, ASM International (2000) 53.
[9] I. Berardone, PhD dissertation, Politecnico di Torino (2016).

CHANGE IN OUTPUT POWER BY LIGHT SOAKING FOR HIGH EFFICIENCY CRYSTALLINE SILICON PHOTOVOLTAIC MODULES

Ritsuko Sato[1], Sungwoo Choi[1], Yasuo Chiba[1], Tetsuyuki Ishii[2], and Atsushi Masuda[1]
[1]National Institute of Advanced Industrial Science and Technology (AIST),
807-1 Shuku-machi, Tosu, Saga 841-0052, Japan
[2]Central Research Institute of Electric Power Industry (CRIEPI),
2-6-1 Nagasaka, Yokosuka, Kanagawa 240-0196, Japan

ABSTRACT: The effects of the light-induced degradation for photovoltaic (PV) modules of n-type monocrystalline silicon heterojunction (SHJ) and p-type monocrystalline silicon passivated emitter and rear cell (PERC) were investigated. These module types were installed at National Institute of Advanced Industrial Science and Technology (AIST), then comparison between the outdoor measurement of each array and indoor measurement of the modules was carried out. In the case of SHJ, there was almost no difference in the change in normalized maximum power between short-circuit and open-circuit conditions, then no significant decrease in performance ratio (PR) occurred. On the other hand, PR of PERC was decreased with only a few amount of irradiation by light soaking at outdoor exposure.
Keywords: photovoltaic module, silicon heterojunction (SHJ), passivated emitter and rear cell (PERC), light induced degradation (LID), stabilization, performance ratio (PR)

1 INTRODUCTION

Recently, various types of high efficiency crystalline silicon (Si) photovoltaic (PV) modules are commercialized. However, in case of these high efficiency crystalline Si PV modules, it is found that the PV modules are degraded by light soaking, so-called light induced degradation (LID). Lee et al. reported the LID caused by the influence of boron-oxygen complex for p-type monocrystalline Si passivated emitter and rear cell (PERC) module [1]. On the other hand, there is also a report that the boron-oxygen complex is not the origin of LID [2]. Therefore, since the origin of LID has not been clarified yet, the metastability of output power for PV modules should be discussed.

As high efficiency crystalline Si PV module, Si heterojunction (SHJ) PV modules are also popular in addition to PERC. It is essential to investigate the amount of power generation for these PV modules. Therefore, it is important to continuously characterize the behavior of the PV modules and estimate the amount of power generation during outdoor exposure.

In this report, the light soaking for PERC and SHJ PV modules at the initial installation stage was studied. Although a simple method of the stabilization for PV modules has been already determined in the IEC 61215-2 [3], the details of pre-conditioning method for stabilization before the initial measurement for the PV modules were examined in this study.

2 EXPERIMENTAL PROCEDURE

The PV modules evaluated in this experiment were 2 types of SHJ and PERC, named SHJ a, SHJ b, PERC a, and PERC b, respectively. There are two modules for each PV type, totally 8 modules were examined. Change in output power by light soaking for commercialized these PV modules were investigated by repeating outdoor exposure and indoor measurements. These PV modules of four types were exposed outdoors under the condition of open-circuit or short-circuit, as shown in Table 1. Cumulative solar irradiation was recorded by mounted pyranometer (secondary standard) with the same angle of 26 deg as the test modules. The back side temperature of the PV module was recorded by type T thermocouple. Measurement duration was 37 days in autumn of Japan. Figure 1 shows the photograph of the exposure test.

Table 1: Circuit condition and number of tested PV modules.

	Short-circuit	Open-circuit
SHJ a	1	1
SHJ b	1	1
PERC a	1	1
PERC b	1	1

Figure 1: Photograph of tested PV modules.

The indoor measurement for these PV modules were carried out under standard test condition (25°C, 1SUN, AM1.5) using a pulsed solar simulator. Parameters of PV module, maximum power (P_{max}), short-circuit current (I_{sc}), open-circuit voltage (V_{oc}), and fill factor (FF), were obtained from current-voltage (I-V) curve data, and compared among each module.

3 RESULTS AND DISCUSSION

3.1 Indoor measurement for SHJ PV modules

Figure 2 shows the changes in P_{max} for SHJ a normalized by the initial value before outdoor exposure, as a function of the cumulative solar irradiation. P_{max} of SHJ a was immediatly decreased after installation, then P_{max} was increased by light soaking. P_{max} difference of approximately 0.5% between short and open circuit conditions at cumulative irradiation of 2.7 kWh/m^2 was similar to that at 122.0 kWh/m^2. It is found that the difference by the circuit condition at the initial deteriotation is almost kept even if cumulative solar irradiation is increased.

Figure 3 shows the changes in normalized P_{max} for SHJ b as a function of the cumulative solar irradiation. In case of SHJ b, P_{max} hardly changes at cumulative irradiation of 2.7 kWh/m^2. This point is different from SHJ a case. The amount of change in normalized P_{max} for SHJ b under open and short circuit conditions was +0.9% and +0.6%, at cumulative irradiation of 122.0 kWh/m^2, respectively. From the analysis of the I-V characteristics, it was found that the changes in P_{max} for SHJ a and b were mainly caused by change in I_{sc} and V_{oc}. The behavior of normalized P_{max} for SHJ b in the outdoor exposure period, however, was almost similar to that for SHJ a. The analysis of the characteristics in detail is under studied.

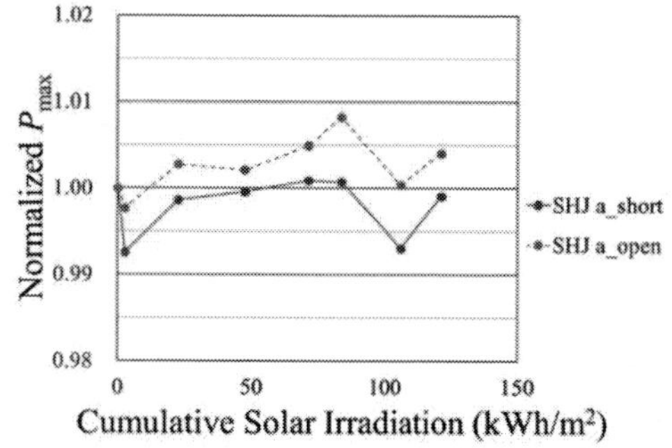

Figure 2: Change in normalized P_{max} for SHJ a module as a function of cumulative solar irradiation.

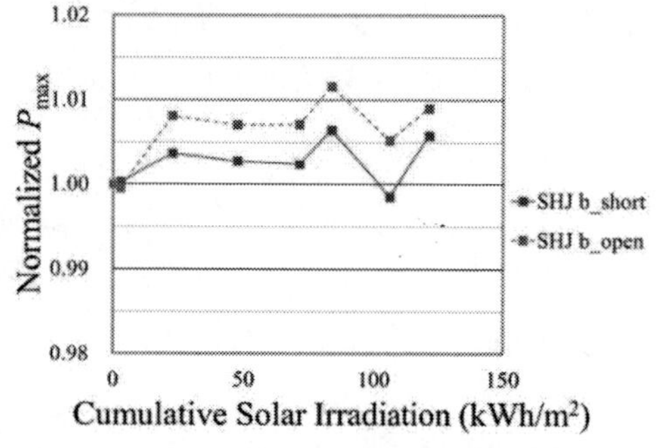

Figure 3: Change in normalized P_{max} for SHJ b module as a function of cumulative solar irradiation.

3.2 Indoor measurement for PERC PV modules

In Fig. 4, change in normalized P_{max} for PERC a module is shown as a function of cumulative solar irradiation. In case of PERC a, the P_{max} was decreased by light soaking in both short and open circuit conditions. P_{max} in each case was about 98.5-99.0% of the initial value at cumulative solar irradiation from 22.7 kWh/m^2 to 122.0 kWh/m^2. From the analyses of the indoor measurement for PERC a under short-circuit condition, it was found that V_{oc} hardly changes and the decrease in P_{max} is caused by the decrease in both I_{sc} and FF. On the other hand, PERC a under open-circuit condition presented the decrease in P_{max} mainly caused by the decrease in I_{sc}.

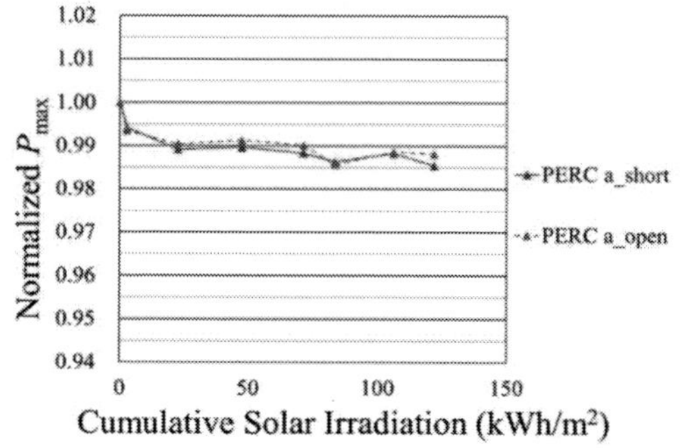

Figure 4: Change in normalized P_{max} for PERC a module as a function of cumulative solar irradiation.

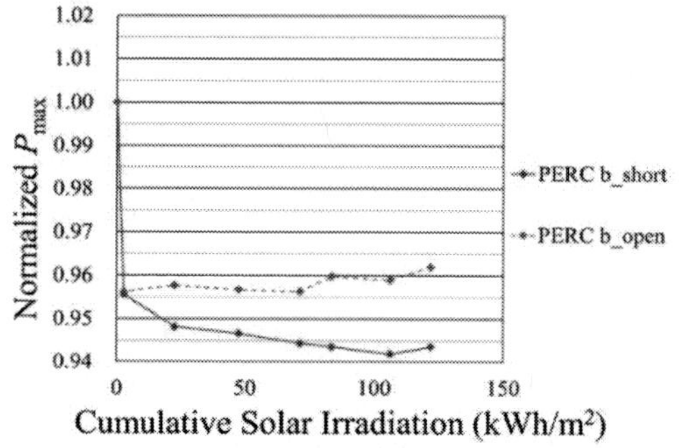

Figure 5: Change in normalized P_{max} for PERC b module as a function of cumulative solar irradiation.

Figure 5 shows the change in normalized P_{max} for PERC b. The remarkable change of -4.4% was observed for PERC b at cumulative solar irradiation of only 2.7 kWh/m^2 as shown in Fig. 5. Normalized P_{max} for PERC b with short and open circuit conditions gradually decreased and silightly increased, respetively, during light soaking with cumulative solar irradiation of 122.0 kWh/m^2. From the analyses of the I-V characteristics, it was found that I_{sc}, V_{oc} and FF were decreased in case of short-circuit condition, on the other hand, V_{oc} and FF were slightly increased in open-circuit condition. From these results in PERC a and b, it is considered that the behavior of output change may depend on manufacturers and also on circuit-conditions during light soaking.

3.3 Outdoor measurement for PV arrays

Various types of PV arrays have been installed at Kyushu Center, National Institute of Advanced Industrial Science and Technology (AIST), which are connected to grids through the power conditioner with the high-frequency insulated transformer type [4, 5]. The arrays composed of the same PV modules, SHJ a, SHJ b, PERC a and PERC b, were also installed. The direct current circuit of the PV modules was ungrounded. *I-V* curves were measured with an interval of 10 min. These PV arrays were operated by electronic maximum power point tracker (MPPT) under the outdoor exposure condition.

In order to compare the indoor measurement with the outdoor measurement for these arreys, weekly performance ratio (PR) was calculated from outdoor measurement data. After installation, in both cases of SHJ and PERC, PRs were gradually decreased, then became approximately 0.9 or less by outdoor exposure for 5 weeks. The PR was gradually increased by outdoor exposure between 9 and 27 weeks. The PR became 0.99 for SHJ a, 0.99 for SHJ b, 0.95 for PERC a and 0.92 for PERC b by outdoor exposure for 27 weeks. As far as the PR was measured for 27 weeks, PR for SHJ a decreased and PR for PERC increased compared to the PR by outdoor exposure after 1 week. However, since the PR in summer between June and August may be lowered by the influence of high temperature, the exact temperature coefficient for each module should be investigated.

4 CONCLUSIONS

We investigated the change in output power by light soaking for high efficiency crystalline Si PV modules; commercially available n-type monocrystalline SHJ and p-type monocrystalline PERC. In the modules used in this study, output change depended on circuit conditions during light soaking. In addition, the behavior of LID in PERC was greatly different depending on manufactures. No recovery was observed by light soaking for one PERC module with small LID. On the other hand, small recovery was observed by light soaking under open-circuit condition for the other PERC module with large LID, however, also in such PERC module output power gradually decreased by light soaking under short-circuit condition even after the initial LID. For SHJ modules less LID was observed and larger output power was obtained under open-circuit condition than those under short-circuit condition.

ACKNOWLEDGEMENT

This study is supported by the New Energy and Industrial Technology Development Organization (NEDO) under the Ministry of Economy, Trade and Industry (METI).

REFERENCES

[1] K. Lee et al., "Natural recovery from LID: Regeneration under field conditions?", *Proc. 31st European Photovoltaic Solar Energy Conf. Exhib., 2015*, pp. 1835-1837.

[2] M. Chang et al., "Light induced degradation of p-mono PERC from ingot, cell, module, to system", *Proc. 32nd European Photovoltaic Solar Energy Conf. Exhib., 2016*, pp. 1854-1857.

[3] IEC 61215-2 Ed. 1.0, 2016.

[4] T. Ishii and A. Masuda, "Annual degradation rates of recent crystalline silicon photovoltaic modules", *Progress in Photovoltaics: Research and Applications, 2017*; DOI: 10.1002/pip. 2903.

[5] S. Choi et al., "Time-dependent changes in copper indium gallium (di)selenide and cadmium telluride photovoltaic modules due to outdoor exposure", *Japanese Journal of Applied Physics* ***56***, 08MD06 (2017).

INDOOR MEASUREMENT OF ANGLE RESOLVED LIGHT ABSORPTION BY ANTIREFLECTIVE GLASS IN SOLAR PANELS

Mekbib W. Amdemeskel[1], Gisele A. dos Reis Benatto[1], Nicholas Riedel[1], Beniamino Iandolo[2], Rasmus S. Davidsen[2], Ole Hansen[2], Peter B. Poulsen[1], Sune Thorsteinsson[1], Anders Thorseth[1], Carsten Dam-Hansen[1]
[1] Department of Photonics Engineering, Technical University of Denmark, Frederiksborgvej 399, 4000 Roskilde, Denmark, mekwub@fotonik.dtu.dk
[2] Department of Micro- and Nanotechnology, Technical University of Denmark, Ørsteds Pl., 2800 Kongens Lyngby, Denmark

ABSTRACT: In this work, we present measurements of angle resolved light absorption of antireflective (AR) glass of PV samples, performed indoors using a collimated high radiance broadband light source. This indoor method proved to be viable and offered a significant simplification compared to outdoor measurements with trackers. The experimental results showed optical responses that are stable and suitable for indoor characterization of solar cells. We find the characteristic optical response of six different antireflective glasses, and based on such measurements, we perform PVsyst simulations and present the monthly DC energy production estimates across four distinct latitudinal locations with six different glass types. The results indicated that the AR glasses present different optical effects at the angles intervals between 0 – 45° and 60 – 90° and that the Diffuse AR glass can improve monthly yields by as much as 2% relatively to a bare cell.

Keywords: Antireflective glass, Angle of Incidence, Incidence Angle Modifier, Simulation.

1 INTRODUCTION

A solar cell output under working conditions is affected by several factors, such as the temperature of the cell, the intensity and incidence angle of the sunlight, which vary between different geographical locations and throughout the duration of a day in the year. In particular, the effect of the angle of incidence (AOI) on the optical properties of the cell – including the AR glass - is considerable for AOI larger than 45° and needs to be taken into account when assessing performance of solar cells. For instance, the effect of optical characteristics of various front glasses in PV modules was recently studied and the annual specific energy yields computed by Herrmann et al.[1]. Martin et.al investigated the angular reflection losses as function of tilt angle of solar cells at 79 sites under real conditions [2]. The effect of the presence of dust on the current output of the PV modules was also studied by exposing the front surface glass of the modules to dust [3]. Furthermore, the Centre for Renewable Energy Systems Technology in the UK has run a project to study optical loss reduction in PV modules with textured and AR coatings glasses, both via indoor and long term outdoor exposure [4]. Measurements of light transmissions on structured glasses using a solar simulator indicated that the current output increased with angle of incidence by increasing light transmission and decreasing temperature as compared to flat glasses [5]. In our work, we characterize six different glass types packaged into coupons using standard c-Si wafers. A bare c-Si cell without glass is included for reference. Our method is carried out by measuring short circuit current as function of incidence angle. In particular, we have built a set-up that allows automated, reliable measurements of the short circuit current as function of incidence angle with collimated, simulated sunlight. This indoor set-up has several advantages over measuring the AOI effect outdoors on a tracker: (i) the measurements are not compromised by the effects of atmospheric diffuse or ground reflected light; (ii) the spectrum and intensity of the light source remain constant throughout the test; and (iii) the rotation stage can move easily and accurately between -90° and + 90° AOI. Furthermore, indoor measurements using light sources such as flashlights are highly affected by the divergence of the beam. In contrast, the light source used here provides bright illumination across the UV-VIS-NIR range together with high spatial and power stability. In addition, the light source is well collimated by collection optics to give a stable and reliable power measurement, with max. 0.1° beam divergence with a beam diameter of 38mm. We have calculated the maximum divergence by measuring the beam diameter at different distances along the beam propagation direction using Equation 1.

$$\theta = 2\tan^{-1}(\frac{D_2 - D_1}{2l}) \qquad (1)$$

Where θ is the divergence angle, D_2 is the beam diameter at distance "2" and D_2 the beam diameter at distance "1".

In this work, we normalize the relative short circuit current to a cosine response thereby isolating the optical effect of the glass-air interface. This form of data presentation is frequently described as the "incidence angle modifier" (IAM) and is used in PV modelling programs such as PVsyst. In addition, we perform PVsyst simulations based on the measured IAM data and share monthly DC energy production estimates with six different glass types across four distinct latitudinal locations.

2 EXPERIMENTAL METHOD

Angle resolved absorption measurements of PV modules with AR glass were performed using the set-up sketched in Figure 1. The following components were used: a laser driven light source (LDLS) EQ-99FXC (Energetiq Technology, Inc.) with spectral emission between 190 and 2100 nm, an off-axis parabolic mirror for light collimation, a UV filter, a sample holder stepper motor (Thorlabs - NR360S) and a LabVIEW controlled short circuit current (I_{sc}) measuring transducer. The UV filter was positioned immediately after the light source in order to remove the UV-C part. The lab environment was kept at a temperature of 21°C using an air conditioning system.

The AOI (= θ) was varied by rotating the solar cells using a stepper motor around a vertical axis. The AOI are indicated with respect to normal incidence. The electrical transducer together with the LabVIEW controller was used to measure the I_{sc} for AOI between 90° to -90° with a 5° step size. Before starting all the measurements, we have aligned the samples normal to the collimated beam of diameter 38mm. For AOI larger than ± 75°, not all the light beam hits the solar cell. Therefore, we have computed an area, i.e. power correction factor, for the partial beam spot to recalculate the actual reduced power reaching the cell from the total power (full spot area). This area correction for large angles is similar to the standard cosine correction for measurements setups where the spot size is larger than the cell. In order to have track of the setup reproducibility, a stable reference cell was characterized before every new measurement. This setup has proven to be a viable method for determining the IAM for the study of optical loses [6]. Importantly, repeatability on the AOI was very high with a maximum relative standard deviation of 0.75%.

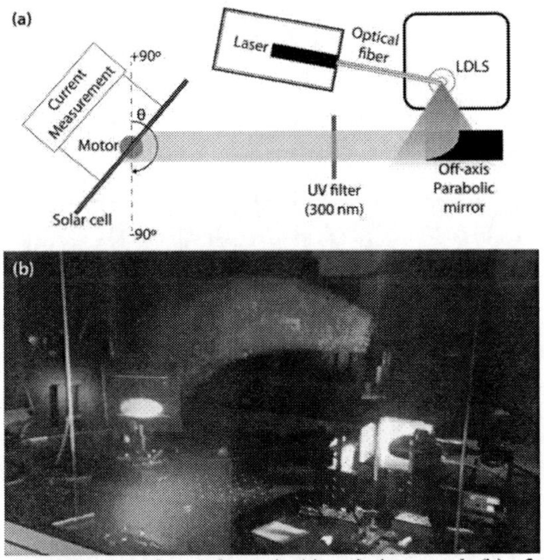

Figure 1: Top view Schematic (a) and photograph (b) of the setup for measuring angle resolved absorption.

Samples with the following commercially available AR glasses were used for the measurement, as well as a Bare Cell for comparison: Smooth Glass, Diffuse Glass, Structured Glass 1, Structured Glass 2, Structured Glass 3, and Structured Glass 4; where Structured Glass 1 had the smallest AR structure, 2 and 3 had intermediate structuring sizes, and Structured Glass 4 had the biggest (deep pyramidal structure). All seven samples are packaged with the same c-Si wafer type (multicrystalline Si 156 x 156 mm), with the 6 encapsulated samples using standard EVA encapsulation) and have the same quantum efficiency (QE).

The electroluminescence (EL) images of all samples were taken in order to evaluate the samples integrity. A modified DSLR Nikon camera, from Brightspot, was used for the images acquisition.

3 RESULTS AND DISCUSSION

We show in Figure 2 the photographs (top) and EL images (bottom) of the seven different samples studied in this work. The EL images indicated that the samples had good integrity and present no cracks or shunts, with the only exception of the Prismatic glass sample. As the AOI experiment required a small cell area, the cracked region was avoided, as well as the bus bars, shown in the EL images.

3.1 Angle of Incidence

We measured short circuit current versus angle of incidence (I_{sc}-AOI, or I_{sc}-θ) for the different PV samples. Once we corrected I_{sc} by the area factor, the area at which the beam misses the solar cell for angles greater than ±75, we normalized it with respect to the value at the normal angle of incidence, i.e. for AOI = 0°. Since we used a collimated light source, we neglected the diffuse component. It is worth mentioning that, for this setup, the normalized and area corrected current is in fact the same as the Incident Angle Modifier (IAM) used in e.g. PVsyst. We characterized seven solar cells, and the IAM plots for these samples are shown in Figure 3. The graphs look very similar, overlapping each other. Therefore, Figure 4 shows the difference of the IAM values relative to the Bare Cell. The Structured Glass 4 sample displayed the highest output for IAM between 80° and 85°, followed by the samples Diffuse Glass, Structured Glass 1, Structured Glass 3, Structured Glass 2 and Smooth Glass respectively, with that last sample underperforming the Bare Cell sample in this angle interval. For the angles between 70° to 0°, Structured Glass 4 was the only sample that present IAM lower than the Bare Cell from 0° until 45°. Results for the sample Diffuse Glass were very close to Structured Glass 3 between 0° and 45°, while Structured Glass 1 and 2 are the better performing samples in this angle interval.

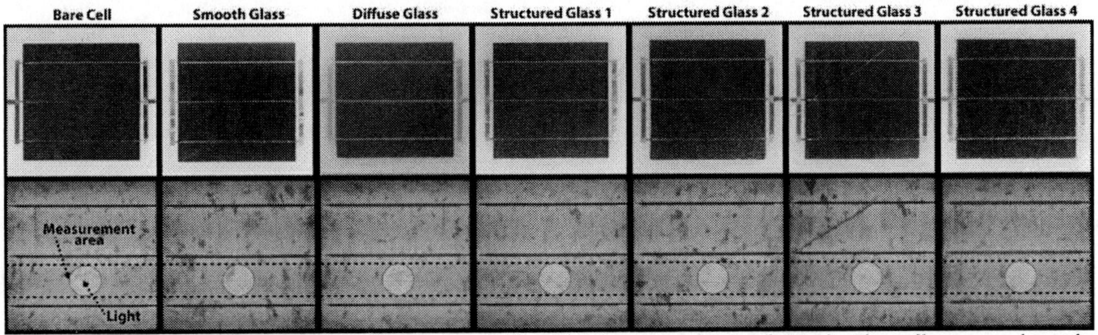

Figure 2: Photograph showing solar cells (top) and Electroluminescence image (bottom). The yellow spot show the approximate area illuminated by the LDLS at AOI = 0°.

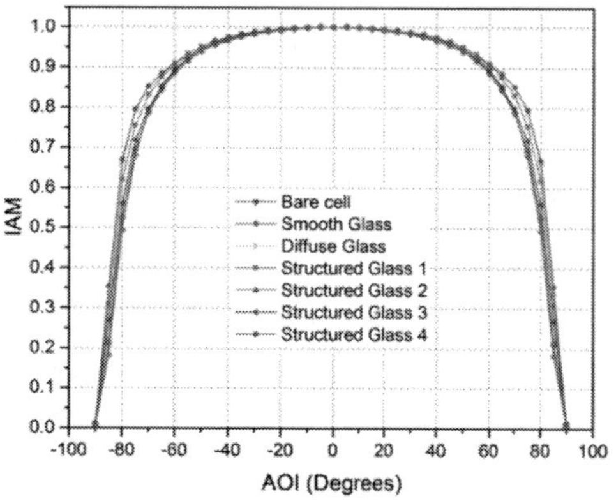

Figure 3: IAM-AOI plot for the Bare Cell sample and for the cells with different AR glass.

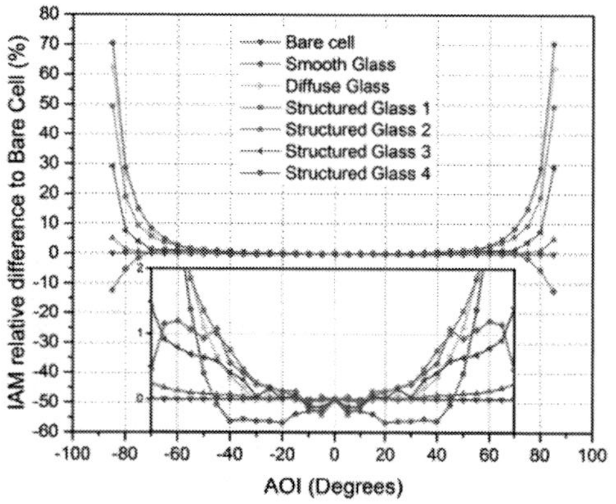

Figure 4: Difference of IAM values of the solar cells with different AR glass relative to the Bare Cell. The inset present the same graph with smaller y-scale for better view of the details.

3.3 Simulations

The measured IAM data shown in Figure 3 were used to create unique PV module files in PVsyst (.pan files). The seven unique modules were used to simulate the energy production of a 10 kW_{peak} grid-tied system in four locations. The locations were selected so as to span a variety of latitudes. The locations modelled are: Copenhagen, Denmark (55.7°N, 12.4°S), Rome, Italy (41.8°N, 12.6°E), Miami, United States (25.9°N, 80.3°W) and Nairobi, Kenya (1.3°S, 36.8°E). With the exception of the glass surface, the PV module bill of materials (BOM) was kept constant in all simulations. The PV balance of system (BOS) was also kept constant in all simulations. A fixed-tilt rack was used and the tilt relative to a horizontal plane was equivalent to the location's latitude. The monthly DC energy estimates relative to the Bare Cell are shown in Figure 5.

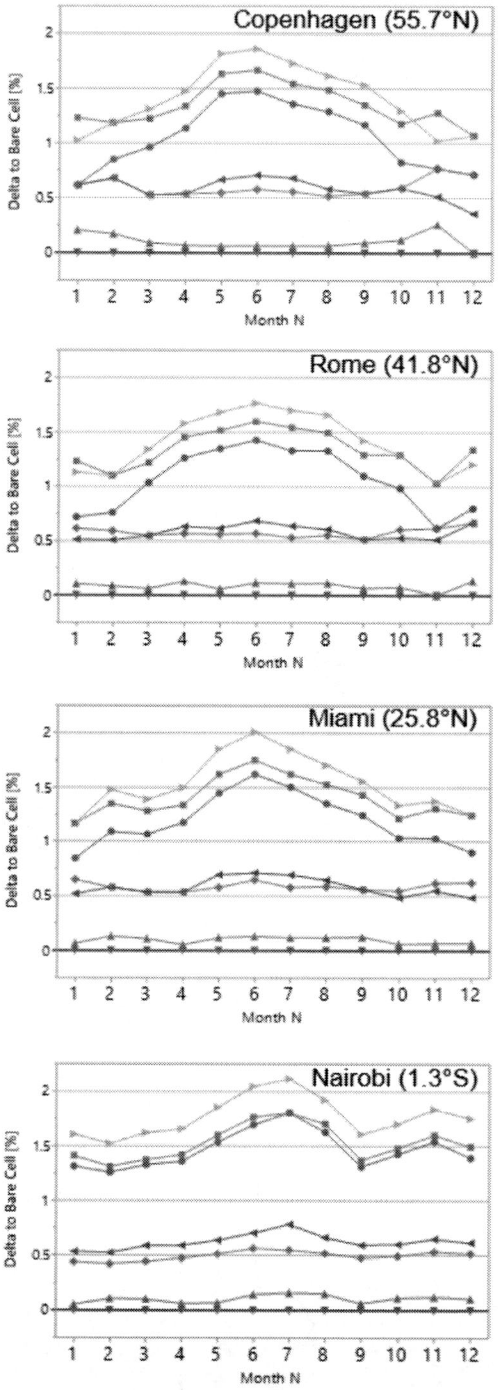

Figure 5: Summary of monthly DC energy production estimates when the seven glass types are used across four locations. Legend can be seen in Figure 4.

The PV system using Diffuse Glass is the top performer across all locations. On the summer solstice, the Diffuse Glass outperforms the Bare Cell by 1.8 to 2.2%. The Diffuse Glass sample outperformed Structured Glass 2 sample by as much as 1.8 – 2.0% depending on location. This result is not surprising given that Diffuse Glass sample IAM was higher than almost all samples between 60 – 85° AOI. The only sample that the Diffuse Glass did not outperform was Structured Glass 4, which underperformed even the Bare Cell from 0 – 45° AOI. The second highest performing PV system is the one using Structured Glass 1 followed by the system using Structured Glass 4 sample. This result is explained by the relatively low IAM performance of Structured Glass 4 at angles of 10 – 25° (Figure 4 inset) in spite of having high IAM performance between 60 – 85°.

The PV systems both in this simulation and typically in the field are oriented such that the AOI is minimized mid-day when solar irradiance is highest. For example, the worst case AOI at solar noon at the Copenhagen site is 25° (Winter Solstice). At AOI less than 25°, there is less than 0.5% difference in the IAM response between all glass types, as shown in Figure 4. Therefore, all the glass types provide nearly the same energy yield during the most irradiance-rich periods. Since the tilt angle of each modelled PV system was optimized for the respective latitude, the relative energy gains shown in Figure 5 are essentially the same irrespective of location.

4 CONCLUSIONS

In this work, we presented an indoor study on the effect of the angle of incidences on the short circuit current of solar cells with and without AR glass, using a broadband, collimated light source. The established setup allowed automated and repeatable measurements of short circuit current for AOI between +90° and -90°, viable for determining the IAM for the study of optical loses. The results indicated that the different AR glasses present diverse optical effects from angles intervals between 0 – 45° and 60 – 90°.

PVsyst simulations showed that Diffuse Glass sample can improve monthly yields by as much as 2% relative to Structured Glass 2 sample. The simulations assume a constant soiling loss for all glasses and locations, which likely will not hold true in the field. However, the intent was to study the effect on optical gains using various glass-air interfaces.

Based on the PVsyst simulations, we consider the setup presented a valuable tool for indoor measurements of the IAM i.e. the angular performance on solar cells and mini modules. Future work includes a Round Robin between other laboratories with AOI cell testing indoors and outdoors, for a comprehensive setup validation; and the modelling of different glass types on BIPV systems, where the installed tilt angle does not allow for receiving the optimal amount of solar irradiance.

6 REFERENCES

[1] W. Herrmann, L. Rimmelspacher, and M. Reuter, "Optical characteristics of PV module front glasses," pp. 4809–4814, 1801.

[2] N. Martín and J. M. Ruiz, "Annual angular reflection losses in PV modules," *Prog. Photovoltaics Res. Appl.*, vol. 13, no. 1, pp. 75–84, 2005.

[3] B. P. Rand, J. Genoe, P. Heremans, and J. Poortmans, "Solar Cells Utilizing Small Molecular Weight Organic Semiconductors," *Prog. Photovolt Res. Appl.*, vol. 15, no. February 2013, pp. 659–676, 2007.

[4] A. N. S. R. Williams, T. R. Betts, R. Gottschalg, D. Neumann, M. O. Prast, "Evaluating the Outdoor Performance of PV Modules with Different Glass Textures," *26th Eur. Photovolt. Sol. Energy Conf. Exhib.*, pp. 3152–3156, 2011.

[5] M. Duell *et al.*, "Impact of Structured Glass on Light Transmission, Temperature and Power of PV Modules," *25th Eur. Photovolt. Sol. Energy Conf. Exhib. / 5th World Conf. Photovolt. Energy Conversion, 6-10 Sept. 2010, Val. Spain*, no. Figure 2, pp. 3867–3872, 2010.

[6] M. W. Amdemeskel *et al.*, "Indoor Measurement of Angle Resolved Light Absorption by Black Silicon," *IEEE 44th Photovolt. Spec. Conf.*, 2017.

EFFECT OF ACCELERATED AGING ON THE PHOTOVOLTAIC ENCAPSULATION/GLASS INTERFACE

Akhilesh Dadaniya and Naresh Varma Datla
Department of Mechanical Engineering, Indian Institute of Technology Delhi -110016, India
a.dadaniya@yahoo.com, datla@mech.iitd.ac.in

ABSTRACT: Durability in photovoltaic (PV) solar cells are difficult to maintain as materials, and their interfaces degrade upon prolonged exposure to high temperature and humidity. Delamination is the major cause of degradation of the solar cell. Delamination aggravates due to thermal cycle loads and exposure to humid environments. In this study, failure resistance of the glass encapsulation interface was studied using water diffusion experiments, water diffusion simulation and peel experiments at an accelerated aging environment. Water diffusion studies were done using gravimetric tests. Water diffusion simulation was performed to determine water mass concentration at the encapsulation/glass interface. Failure resistance of the interface was determined from peel tests. Tests on water diffusion showed simple Fickian behaviour for water absorption, while the peel tests showed a significant decrease in peel strength with aging time.
Keywords: Delamination, Degradation, Environment effect.

1 INTRODUCTION

Photovoltaic (PV) module is a multilayer structure that contains cell and packaging layers, which are exposed to the outdoor environment for the prolonged duration of time. Different types of failures lower the mechanical and electrical performance of the photovoltaic solar cell. These failures include delamination, discolouration, corrosion, breaking, and hot spots [1]. Most of the packaging materials such as encapsulation and back-sheets are polymeric materials. Direct contact of the back sheet with outdoor environment leads to degradation of the back sheet due to humidity and temperature. This continuously reduces the electrical and mechanical reliability due to deleterious effects of humidity and temperature. Delamination is the major cause of degradation of the solar cell. Delamination of the back-sheet and encapsulation leads to penetration of water and other harmful elements within the module that leads to further degradation. The factors that lead to delamination failures include thermal cycling loading and humid environment [2]. Upon environmental aging, PV modules undergo water absorption and desorption that lead to both recoverable and irrecoverable strength loss. Critical fracture energy decreases in the corrosive environment because of the weakening of chemical bond due to chemical reaction. Temperature accelerates the chemical reaction between water and polymer that degrades photovoltaic solar cell structure in humid environments. This study aims to find the degradation in interfacial strength for encapsulation glass interface under accelerated environment condition.

2 MATERIAL AND METHOD

2.1 Material

The three PV module commercial materials considered in this study are the encapsulation layer, back-sheet, and float glass. Encapsulation material used is ethylene vinyl acetate (EVA). Back-sheet is a laminate of polyvinyl fluoride (PVF), polyethylene terephthalate (PET) and primer layer. Back-sheet and encapsulation are laminated with low iron contained float glass by using vacuum laminator. Figure 3 shows schematic of the PV module laminated specimen used in water diffusion simulation and peel test. Figure 1 shows the methodology which followed for encapsulation/glass interface durability testing.

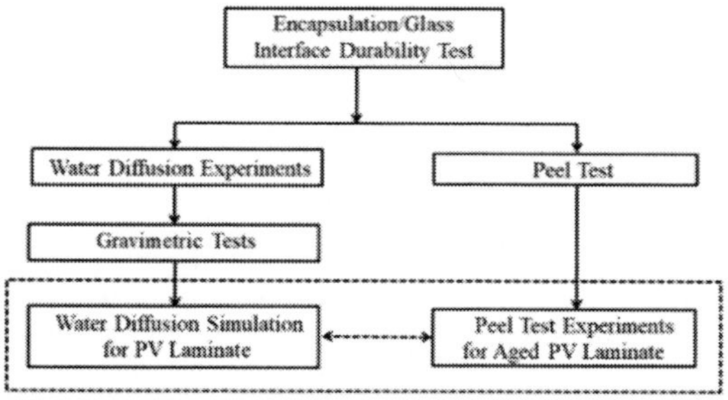

Figure 1: Flow Chart of the methodology.

2.2 Experiments

2.2.1 Gravimetric Test

Water diffusion behaviour in the commercial EVA and back-sheet were determined through gravimetric tests by following the ASTM standard [3]. 55 × 55 mm^2 bulk wafers of EVA and back-sheet were prepared with thicknesses of 0.70 and 0.36 mm, respectively. A constant humidity level of 78% RH was controlled by placing specimens in an airtight box containing a saturated solution of potassium chloride. Required temperature level was achieved by placing the airtight box in the temperature controlled oven. Continous monitoring of relative humidity and temperatures was done using capacitance sensor. Wafer weight during water absorption was periodically measured using a 0.01 mg resolution microbalance (GR-202, AND).

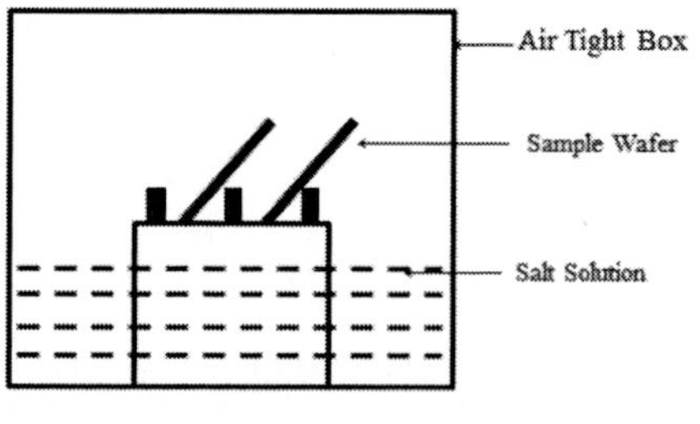

Figure 2: Humidity controlled air tight box.

2.2.2 Peel Test

Figure 4 shows the peel test setup used to determine the peel strength. Peeling was done under displacement control at a peel rate of 10 mm/min and a peel angle about 90° for a strip about 10 mm width. Peel tests were done both before and after aging. A pre-crack within the glass encapsulation interface was introduced to ensure crack propagation at that interface.

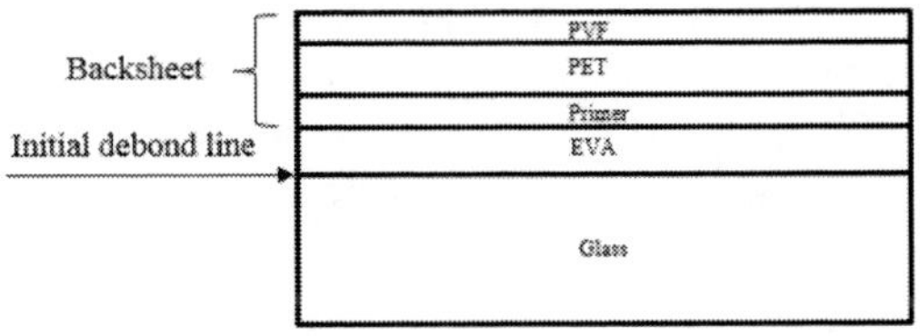

Figure 3: Specimen configuration for water diffusion simulation and peel test.

Figure 4: 90^0 peel test setup.

3 WATER DIFFUSION SIMULATIONS

Fick's second law of diffusion was implemented in a 2-D finite element model by using Abacus software [4]. The 2-D model for water diffusion simulation is shown in figure 5. Diffusion parameters from gravimetric tests were used. Normalized concentration approach was used to eliminate concentration discontinuity at the interface. Variation of water mass concentration over time was estimated at the encapsulation/glass interface for the aging environment of 85^0C/78%RH.

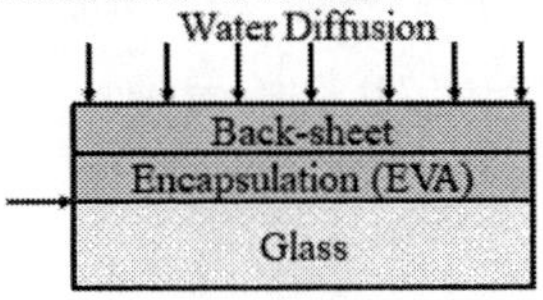

Figure 5: 2-D finite element model for determining water concentration at the interface.

4 RESULTS

4.1 Gravimetric Tests

Figure 6 and 7 shows the water mass uptake through a gravimetric test for back-sheet and encapsulation, respectively, at different temperatures and a fixed relative humidity of 78% RH. It can be seen that the water absorption followed a simple Fickian behaviour for both back-sheet and encapsulation. Diffusion coefficients obtained through the gravimetric test of back-sheet show good agreement with Arrhenius equation, as shown in figure 8. Extrapolation was done for diffusion coefficient of encapsulation at 85^0C through Arrhenius equation. Table 1 summarizes water diffusion properties obtained through a gravimetric technique for back-sheet and encapsulation at different temperature and fixed relative humidity of 78% RH.

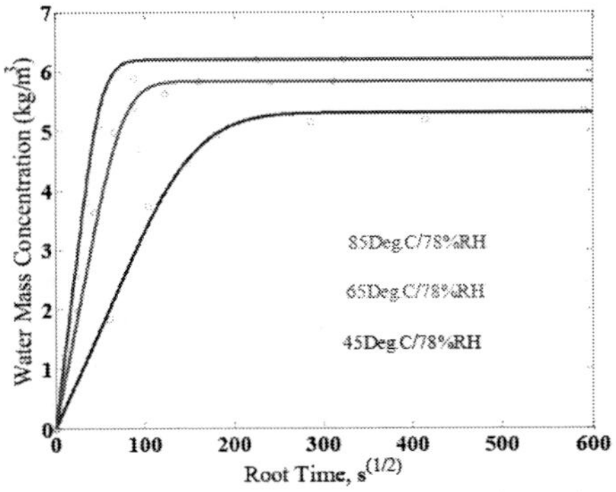

Figure 6: Water mass uptake for back-sheet through gravimetric technique.

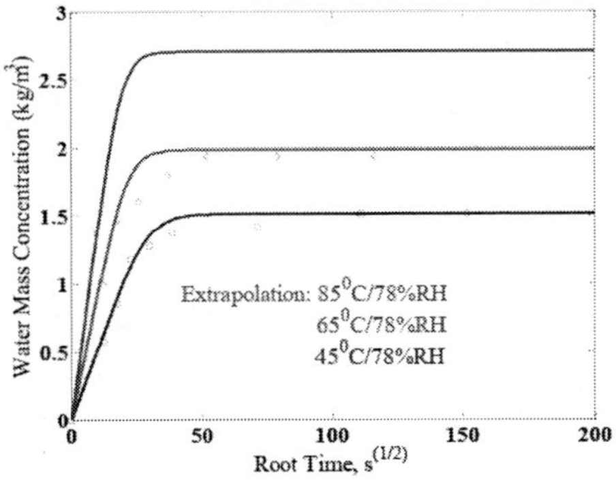

Figure 7: Water mass uptake for encapsulation through gravimetric technique.

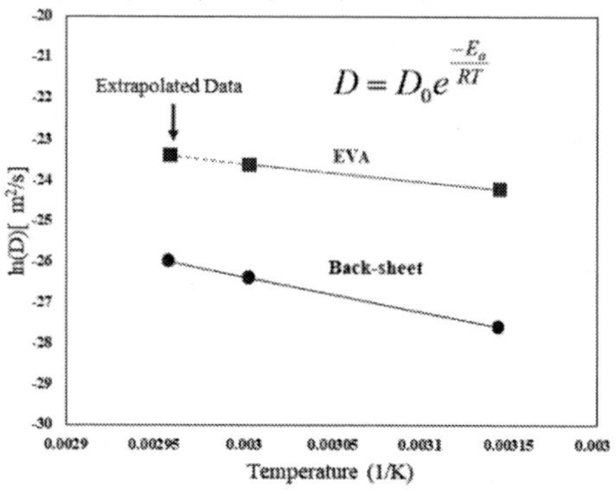

Figure 8: Temperature-dependent diffusion coefficient for back-sheet and encapsulation.

Table 1: Water diffusion properties of back-sheet and encapsulation.

Water Diffusion Properties				
RH: 78±3 %	Backsheet		Encapsulation (EVA)	
Temperature (°C)	Diffusion Coefficient (m²/s)X10⁻¹²	Solubility (kg/m³.Pa) X10⁻⁴	Diffusion Coefficient (m²/s) X10⁻¹²	Solubility (kg/m³.Pa) X10⁻⁴
45±2	1.03	6.8	30	1.9
65±2	3.4	3.7	55	1.2
85±2	6	1.3	**94**	0.6

4.2 Water Diffusion Simulation

The normalised mass concentration of water at the interface is shown in figure 9. It can be seen that the water concentration reached saturation within few hours, that can be related to the aggressive aging environment of 85°C/78%RH.

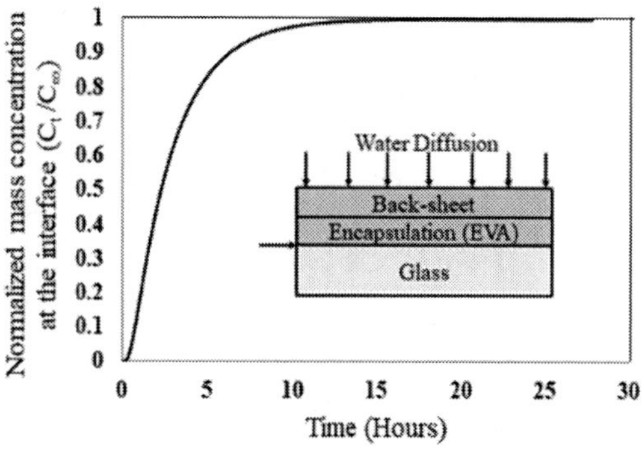

Figure 9: Normalized water mass concentration at the encapsulation/glass interface at 85°C/78%RH.

4.3 Peel Test

The measured peel strength for an unaged specimen tested at a peel angle of 90° is 68 N/cm. However, when aged at 85°C and 78% RH for about 60 days, the peel strength decreased to 24 N/cm, as shown in figure 10. The observed decrease in peel strength to about one-third of the initial strength is plausible because of the aggressive aging environment used in this study. The crack path was initially at the encapsulation/glass interface. However, as aging time reached 45 days the crack path shifted to the primer layer/PET interface.

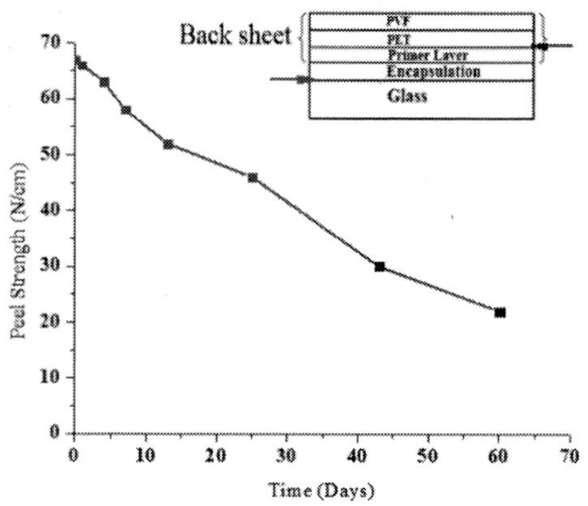

Figure 10: Peel strength degradation with aging time at 85°C/78%RH.

5. CONCLUSION

The effects of accelerated aging were studied on the water diffusion behaviour of EVA and back-sheet as well as the peel strength at the encapsulation/glass interface. Water absorption in the back-sheet and encapsulation showed simple Fickian behaviour. Water diffusion simulations showed that the encapsulation/glass interface reached saturation within hours because of the aggressive aging environment. The peel strength decreased significantly for a relatively short aging time. The crack path remained completely interfacial between the glass and EVA layer at realtively shorter aging times but later shifted to the primer layer/PET interface for prolonged aging times.

6. REFERENCES

[1]. Munoz, M. A., Alonso-Garcia, M. C., Vela, N., & Chenlo, F. (2011). Early degradation of silicon PV modules and guaranty conditions. *Solar Energy*, *85*(9), 2264-2274.

[2]. Jorgensen, G. J., Terwilliger, K. M., DelCueto, J. A., Glick, S. H., Kempe, M. D., Pankow, J. W. & McMahon, T. J. (2006). Moisture transport, adhesion, and corrosion protection of PV module packaging materials. *Solar Energy Materials and Solar Cells*, *90*(16), 2739-2775.

[3]. ASTM D570-98(2010)e1, Standard test method for water absorption of plastics, ASTM International, West Conshohocken, PA, 2010.

[4]. Crank, J. (1979). The mathematics of diffusion. Oxford University Press.

STUDY ON POTENTIAL-INDUCED DEGRADATION AND RECOVERY OF BIFACIAL N-TYPE SINGLE CRYSTALLINE Si PHOTOVOLTAIC MODULES

Mohammad Aminul Islam[1], Takuya Oshima[1], Hidenari Nakahama[2], Yasuaki Ishikawa[1]
[1]Nara Inst. Sci. & Tech., 8916-5, Takayama-cho, Ikoma, Nara 630-0192, Japan
[2]Nisshinbo Mechatronics Inc., Miai-cho, Okazaki, Aichi, 444-8560, Japan.
*Corresponding email: aminulislam@ms.naist.jp, yishikawa@ms.naist.jp

ABSTRACT: Potential induced degradation (PID) and corresponding recovery were studied on n-type single crystalline Si bifacial photovoltaic modules by applying -1000 V and +1000 V, respectively with environmental condition of 85 °C/85% RH. PID acceleration test time was increased step by step and three different time ranges were chosen for three different single cell modules. The module characteristics were investigated by means of electroluminescence (EL) imaging technique, lock-in-thermal (LIT) imaging technique, microwave photo-conductive decay (uPCD) measurement, and dark and light current-voltage measurement. It has been found that the surface polarization occurred very quickly on these modules, whereas the recovery depends on the time range of PID.
Keywords: PID; n-Si module; bifacial, EL; LIT; saturation current; carrier lifetime

INTRODUCTION

PID gains a considerable attention recently due to the significant negative impact on the performance as well as the reliability of photovoltaic (PV) systems [1, 2]. In contrast, PID in PV modules based on n-type crystalline Si solar cells have been less studied than p-type Si PV modules until today. Swanson et al. [3] firstly reported in 2005 for solar modules containing interdigitated back contact (IBC) cells; have been typically investigated by using high voltage stress with opposite polarity, and the negative surface polarization was observed. By the polarization effect, negative charges are left on the front surface of the n-type Si cell, leading to a degradation of the passivation at the front side of the n-doped basis, enhancing front surface recombination between electrons and holes at the surface [3]. Later on, Halm et al. [4] and Naumann et al. [5] reported PID in an n-type IBC solar cell. They concluded that the degradation of the front surface passivation, which enhancing surface charge recombination is the principal cause of the power drop of the module and this PID is completely reversible, repeatable and does not cause cell shunting. In addition, Hara et al. reported the influence of surface layer (n or p) of the n-type single-crystalline front junction (FJ) (not back contact) Si solar cell on PID by applying negative voltage stress [6]. They reported 17% of power reduction by applying negative voltage for 10 minutes only which is completely recovered after applying the same amount of positive voltage across the electrode. However, a detailed time-dependent analysis of PID phenomena and its recovery in n-base Si solar cells for high humid condition and long stress time (up to 300 hours) is still missing yet to date. In this study, we are reporting PID phenomena occurred in PV modules based on commercial n-type bifacial single crystalline Si (cs-Si) single cell modules for 1- 300 hours range of lab acceleration test. The recovery of the corresponding modules was also conveyed in this study and seems that this type of modules is almost 100% PID recoverable until 10 hrs. of PID.

This study investigated three different single cell modules, with emphasize given to the diode pn characteristics and minority carrier dynamics, including the complementary analysis of the light I-V, EL image and LIT image analysis. However, more detailed study is still demanding to fully understand the PID mechanism in n-type bifacial Si PV modules.

METHODOLOGY

The commercial bifacial n-type single-crystalline Si solar cells of front emitter was used as shown in Fig. 1(left).

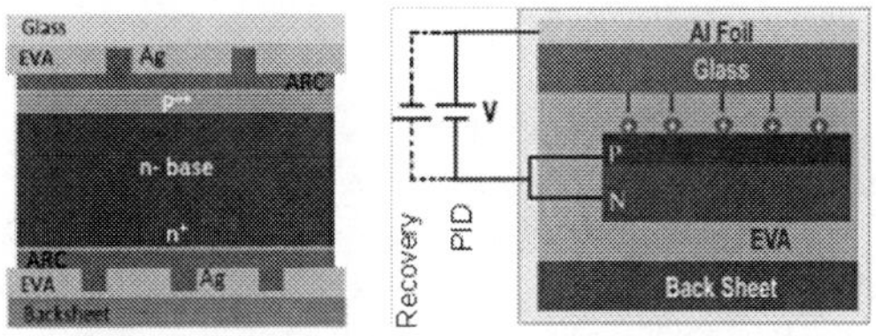

Fig.1 (left) Module structure of the single cell bifacial n-Si solar cells (back side is covered by the backsheet, (right) setup of PID and recovery acceleration test

The PV module components (glass/EVA/c-Si cell/EVA/backsheet) were laminated under vacuum condition by using a temperature of 130 °C for15min. PID has been provoked by biasing of "-1000 V" with respect to the module's glass frame in a climate chamber at environmental condition (85°C/85% of RH). PID recovery also studied using +1000V with respect to the module surface at the same environmental conditions. Three different modules are subjected by three different time frame as shown in Table 1. In order to ensure the bias voltage homogeneity over the whole cell surface, an Al plate of 0.1 mm thickness has been strictly attached to the entire cover glass using a glass plate of 1 cm over the plate.

Table1 Time frame for the PID acceleration test and recovery

	PID Time (hr.) (-1000 V)	PID recovery time (hr.) (+1000 V)	Environmental conditions
Module 1	0-10	1	85 °C,
Module 2	0-100	1-5	85% of RH
Module 3	0-300	1-10	

The EL image has been taken in forward bias with a current of 40 mA/cm^2 by using the InGaAs camera (model: Xeva-2.5-320) at the resolution of 320 x 256 pixels with a 30 μm pixel pitch. The local EL intensity has been measured using Hokawo 2.1 image analysis software. The carrier lifetime has been directly measured using the microwave photoconductive decay (μ-PCD) system (WT-1000B, Semilab Co. Ltd). A Matsusada PKT 80-50 Source Measurement Unit was used to record all I-V curves.

RESULTS AND DISCUSSION

Fig. 2 shows the EL image of the Modules in PID and recovery mode, and also, shows corresponding relative EL intensity. As it is seen that the module 1 under 10 hrs. of PID, decreases around 8% of EL signal, whereas the losses are increased with the increase of PID time. EL images of the modules reduced EL intensity over the whole surface that indicated the non-radiative recombination in the solar cell has increased uniformly during the PID stress. Under the recovery test, EL signal of the module 1 and 2 has been recovered 98%, signifying that the power loss on these modules occurred from the loss of front surface polarization which recovered very quickly. However, the EL signal of module 3 is recovered only around 90% in the longest time (10 hrs.) of recovery. Some point-like defects are observed in the EL image of the module and these defects remain in the recovered mode too. These dark color spots are signposted that the Ag fingers are diluted on those sites due to the moisture ingress which are not recoverable. On the other hand bus bar corrosion also observed in the module 3 as seen in Fig. 2 (c).

Fig. 3 shows the LIT images of the modules. The shallow hotspots are observed to spread over the surface demonstrating that the carrier recombination has dominated on the surface of the PID modules. It is seen that the 10 hrs. PID module is completely recoverable, whereas 100 hrs. and 300 hrs. of PID modules are not. In the case of long time of PID stress, ions may be damaging the SiNx layer, and/or emitter layer is contaminated by mobile ions [3] which is being remained unchanged under the recovery test. Also, the Ag finger dilution is confirmed from the LIT image of 300 hrs. of PID module as seen in EL images.

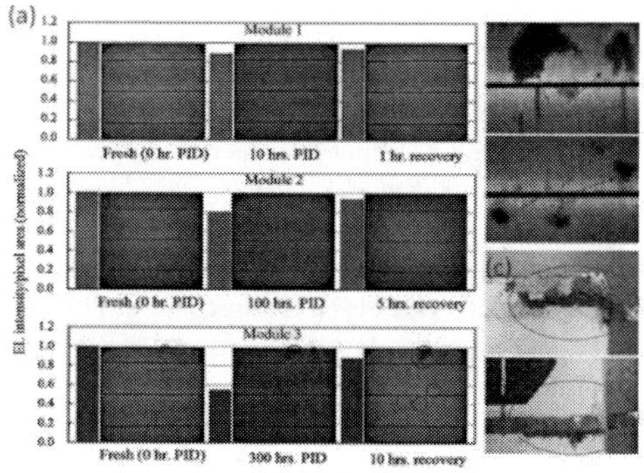

Fig. 2 (a) EL images of the modules (corresponding EL image intensity/pixel area shown with respect to the image of fresh mode), (b) enlarged EL images of the dark spots found on the module 3, and (c) photo of the delaminated part occurred on the module 3

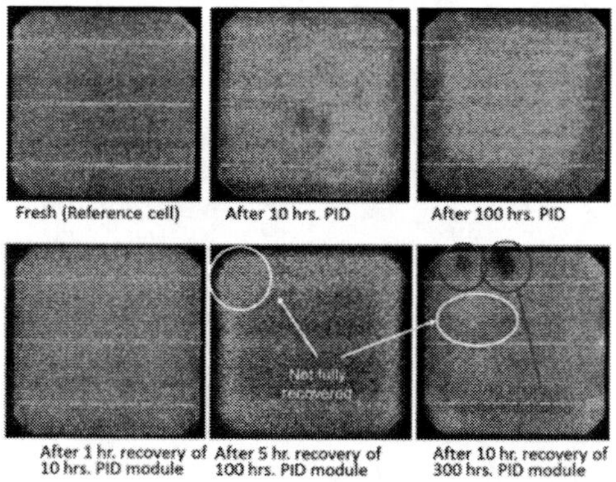

Fig. 3 LIT images for Module 1, 2 and 3 in PID and recovery mode

The obvious difference in the variation of carrier lifetime with respect to the laser power has been found for PID and fresh and recovery modules as shown in Fig. 4. The PID modules show lower carrier lifetime, which has no change under the increase of incident laser power indicating the domination of high surface recombination velocity due to the high surface killer defect density on those modules. The results show that the surface defects become saturated within 10 hrs. of PID and recovery depends on the PID test time. The module 3, in which PID was provoked for 300 hrs., the carrier lifetime became highest in 5 hrs., but not equal to the fresh mode. This also indicates that the module is still affected by the surface contamination. However, further increase of positive voltage bias time, carrier lifetime was decreased again. This phenomenon may be occurred due to the positive voltage bias affect the rare side of the PID modules.

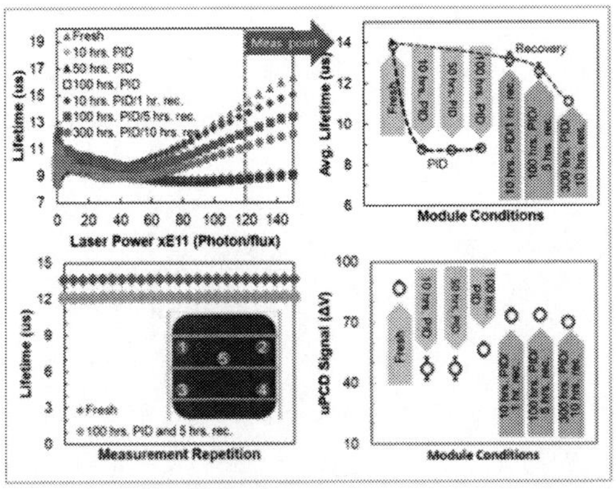

Fig. 4 (top-left) change of carrier lifetime with respect to the laser power irradiation, (top-right) average lifetime for the modules, (bottom-left) consistency of the measurement with the measurement points and (top-right) signal intensity for the modules with respect to their conditions, respectively

The dark IV characteristics also found quite consistent with the EL and effective carrier lifetime of degraded and recovery mode of the modules. To assess the degradation and recovery more clearly, the dark IV curve was fitted using the following two diode equation and saturation current of the first and second diodes I_{01} and I_{02}, shunt resistance Rsh, and series resistance Rs, ideality factors of the first and second diodes n_1 and n_2, respectively, were determined.

$$I(V) = I_{01}\left[exp\left(\frac{q(V-IR_s)}{n_1kT}\right) - 1\right] + I_{02}\left[exp\left(\frac{q(V-IR_s)}{n_2kT}\right) - 1\right] + \frac{V-IR_s}{R_{sh}}$$

Where, q is the elementary charge, k the Boltzmann constant, and T the absolute temperature (300K). I_{01}, I_{02}, R_{sh}, and R_s were positive, and n_2 was limited to greater than one. The two diode fitting was performed using the OriginPro2015J considering n_1=1. Fig. 5 shows the dark I-V curves and dependence of n_2, I_{01}, and I_{02} on the PID stress duration. The modules series resistance remains almost same for all the modules in all ranges of PID acceleration time, but shunt resistance is continuing to respond to the PID and recovery mode as like as the other parameters. The increase of diode saturation current indicates recombination is enhanced in the emitter and/or the base in the solar cells [7]. However, the increase of I_{02} and n_2 is saturated after 100 hrs. of PID, maybe the surface defects become saturated at this stage. It is seen that the dark characteristics of the PID modules of 10 hrs. and 100 hrs. could be returned to the states of fresh mode, however, 300 hrs. of PID still show somewhat higher saturation current compared to the fresh module after the recovery test. The results of DIV are well matched with the other experimental results as discussed in the earlier sections.

In table 2, it is seen that the efficiency loss of the modules is due to the loss of Isc and Voc. The similar results are also reported previously by others [6, 7]. The loss of Voc and Jsc is occurring due to the increase of surface recombination of the minority carriers. The module of 10 hrs. and 100 hrs. of PID has recovered almost fully, whereas 300 hrs. of PID module recovered only 88%. The all results are indicating that the module is degraded more rapidly than single crystalline p-Si modules and there is a degradation limit for n-type single crystalline Si solar cells and/or modules from where it could be recovered. Certainly, more studies are required to confirm the limit.

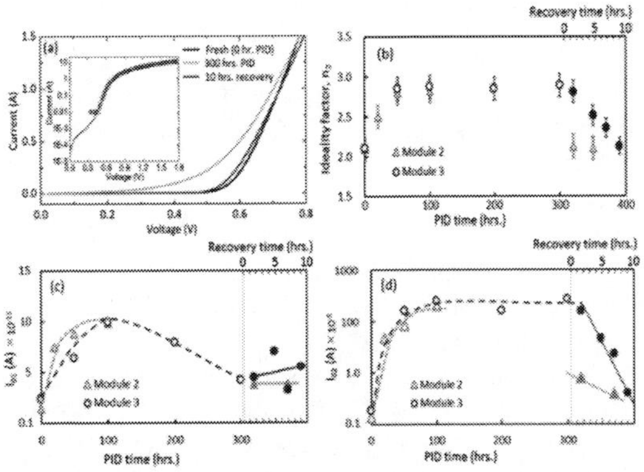

Fig. 5 Dark I-V characteristics of Module 2 and Module 3 in PID and recovery mode, (a) dark I-V curve of Module 3 (inset: fitted curve of the module in fresh mode), (b) the variation of ideality factor n_2 with respect to the PID and recovery time, and (c) and (d) the variation of dark saturation current with respect to the PID and recovery time

Table 2 Monitoring the loss and recovery of electrical parameters through light current-voltage measurement

PID/recovery (hrs.)	Fresh/-	10/-	100/-	10/1	100/5	300/10
Isc (%)	100	92	93	101	101	96
Voc (%)	100	93	93	100	95	96
FF (%)	100	98	96	98	98	95
η (%)	100	83	82	99	94	88
P_m (%)	100	83	83	99	99	88

CONCLUSION

We studied the PID mechanism and its recovery for single cell modules of front emitter n-type bifacial sc-Si solar cells by applying a negative and a positive voltage stress respectively. EL and LIT images reveal a uniform increase in non-radiative recombination due to the shallow hotspot from the surface polarization effect. This PID effect until 10 hrs. is completely reversible, and does not cause cell shunting. However, long time PID stress could be the cause of surface damage which is not recoverable. The uPCD carrier dynamics in the degraded and fresh/recovery states show that the module efficiency, as well as Voc and Isc losses, are related to the high surface recombination velocity. These facts suggested that the power and/or efficiency drop initially in n-type modules is related to the degradation of front surface passivation. After a long time of PID stress, the module suffers from additional causes, such as front surface damage or emitter contamination, Ag finger dilution by moisture ingress and busbar corrosion. The results suggest that there is a degradation limit for n-type single crystalline Si modules from where it could be recovered. Detailed characterization is needed to identify the root cause of the degradation and recovery mechanism we observed.

Acknowledgement
This work was supported by The New Energy and Industrial Technology Development Organization (NEDO), JAPAN.

References

[1] J. Berghold, O. Frank, H. Hoehne, S. Pingel, B. Richardson, M. Winkler, 25th EUPVSEC, Valencia, Spain, 2010, pp. 3753-3759.

[2] Mohammad Aminul Islam, Kazuki Noguchi, Hidenari Nakahama, Yasuaki Ishikawa, 43rd 1EEE Photovoltaic Specialists Conference (PVSC), 2016, pp. 885-888.

[3] R. Swanson, M. Cudzinovic, D. DeCeuster, V. Desai , Jörn Jürgens , N. Kaminar, W. Mulligan, L. Rodrigues-Barbarosa, D. Rose, D. Smith, A. Terao, and K. Wilson, In Technical Digests of the 15th International Photovoltaic Science & Engineering Conference (PVSEC- 15), Shanghai, China, 2005, pp. 410–411.

[4] Andreas Halm, Andreas Schneider, Valentin D. Mihailetchi, Lejo J. Koduvelikulathu, Lacramiora M. Popescu, Giuseppe Galbiati, Haifeng Chu, Radovan Kopecek, Energy Procedia 77 (2015) 356 – 363.

[5] V. Naumann, T. Geppert, S. Groβer, D. Wichmann, H.-J. Krokoszinski, M. Werner, C. Hagendorf, Energy Procedia 55 (2014) 498–503.

[6] Kohjiro Haraa, Kinichi Ogawa, Yusuke Okabayashi, Hiroyuki Matsuzaki, Atsushi Masuda, Solar Energy Materials & Solar Cells 166 (2017) 132–139.

[7] Seira Yamaguchi , Atsushi Masuda, Keisuke Ohdaira, Solar Energy Materials & Solar Cells 151(2016)113–119.

COMPARISON OF THE ELECTRICAL POWER ESTIMATED BY DIFFERENT MATHEMATICAL MODELS WITH THE ONE PRODUCED BY THE SOLAR PHOTOVOLTAIC GENERATORS CONNECTED TO THE ELECTRIC GRID AT ENSA-SAFI

Lahcen Boukhattem*[1], Nour-eddine Id omar[1], Fahd Oudrhiri Hassani[1], Mohamed Akhsassi[2], Aziz Ouknnou[3]

[1] LMPEQ, National School of Applied Sciences, Safi, Cady Ayyad University, Marrakesh
[2] LPSCM, Faculty of Science Semlalia, Cadi Ayyad University, Marrakesh
[3] Advanced Electrical Systems Control Team, ENSEM Casablanca
l.boukhattem@uca.ma

ABSTRACT: This work deals with the estimation of the electrical power produced by the solar photovoltaic generators of three technologies connected to the electric grid: Monocrystalline, Polycrystalline and Amorphous installed on the roof of the National School of Applied Sciences of Safi (ENSAS). The electrical power of the studied generators is 6 kWc. Different mathematical models estimating the electrical power were tested with climatic data measured in-situ during several days. The results showed that the electric power of each technology could be predicted well by a model among the studied ones with satisfying precision. The evaluation of different electrical parameters of solar panels permits the reconstitution of the characteristic I-V and comparison with the one given by the panel manufacturer under variable climatic conditions.
Keywords: Photovoltaic module, Electric power, Electric parameters, Simulation.

1 INTRODUCTION

Photovoltaic is a method of generating electrical power by converting solar irradiance into direct current electricity using semiconductors. Modeling is the first step in analyzing behavior and characteristics of PV panel in virtual environment. For perfect analysis of its performance, an accurate model of PV system is important considering its low power density and poor efficiency. Also, a reliable and accurate model of PV panel will help in predicting energy production from power plant under varying environmental conditions.

Several models have been proposed in the literature to deal with the electric power and characteristics I-V of PV panels. The proposed models are governed by either of the two aspects as follows. First aspect does the analytical study of PV panel by modeling it through electrical equivalent circuit [1-3] or through mathematical equations in order to obtain standard I–V and P–V characteristics of this PV panel. The second aspect covers the modeling of PV panel based on its characteristics under different environmental conditions. These conditions include the variation of solar irradiance, temperature and shading effect. An experimental verification is required to establish the accuracy of such models.

The PV panel model based on electrical equivalent circuit aspect was done by E. Saloux et al. and R. Chenni et al. [1-2]. They thoroughly analyzed one diode model and its practical verification. The two diodes model and associated mathematical formulation was described by Ishaque et al. and B.C. K. Babu et al. [3-4]. They concluded that the two diodes model is more accurate and presents a model whose I–V characteristics is closer to the practical behavior of PV cell. But the two diode model involves complex and more computational effort as compared to the one diode model. Mathematical representation of solar cell is based on Shockley diode equation [5-6]. This model represents the non-linear characteristics of PV cell with parameters such as shunt and series resistance, reverse saturation current and diode ideality factor. Shockley diode equation is an exponential expression.

In this paper, different mathematical models estimating the generation output of real PV power plant were tested with climatic data measured in-situ during several days. The evaluation of different electrical parameters of solar panels allows the reconstitution of the characteristic I-V and comparison with the one given by the panel manufacturer in the variable climatic conditions.

2 PV MICRO-PLANT OF SAFI SITE

3 grid-connected PV micro-plants, based on 3 different PV technologies Monocrystalline (SW 255 Mono), Polycrystalline (SW 255 Poly) and Amorphous (NT-155AX), of 2 kWp each, were installed on the roof of ENSAS-Safi (Figure 1-b). PV electricity generated is completely absorbed by the internal network of our institution. A weather Kit was also installed next to PV generators to monitor solar radiation, wind speed, ambient temperature and panel temperatures (Figure 1-a).

(a)

(b)

Figure 1: (a) PV micro-plants and (b) weather Kit.

3. RESULTS AND DISCUSSION

Among the models presented in the literature in order to estimate the electric power of the studied micro PV power plant are presented in Figure 2 [1-3]and [6-10]. The non-linear characteristics of PV panel were based on Shockley diode equation [5-6].

3.1 Electric power

Figures 2 (a, b and c) illustrate the variation of the electric power produced by the three studied technologies and the ones calculated using different models indicated in Figure 2, during 4 days (from 01/05/2017 to 04/05/2017). One can notice that the calculated electric power mach the experimental results well. Thus, the Villalva model [7] estimates well the electric power produced by the monocristalline technology with a maximum error of 14.42 % compared to the two other technologies. As for the poly-crystalline technology, it was well predicted by the Jones and Underwood models [8] with maximum error that doesn't exceed 11.19 % during the first day. The mathematical model that estimates the electric power of amorphous technology was the Saloux [1] model with a maximum error of 11.96 %.

3.2. Characteristic I-V

Figures 3 (a, b and c) present the comparison between the calculated characteristic I-V and the one of the panel concerning three technologies. This comparison was done for different solar irradiance values varying from 200 to 1000W/m^2. It can be noticed that the calculated characteristic by different models mentioned in Figures 3 (a, b and c) coincide with the one of the panel in the region where the module behaves as a generator of current. However, a significant difference was observed around the bend in the I-V characteristics curve. This is due to the electrical parameters of the studied panels.

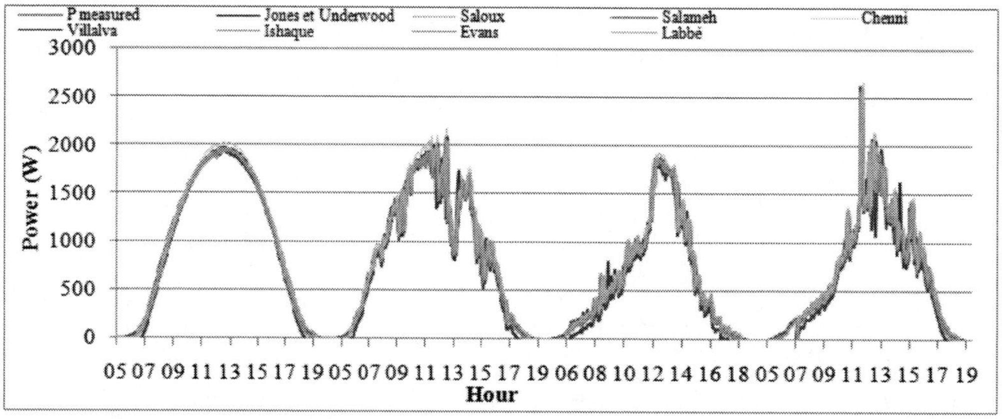

(a)

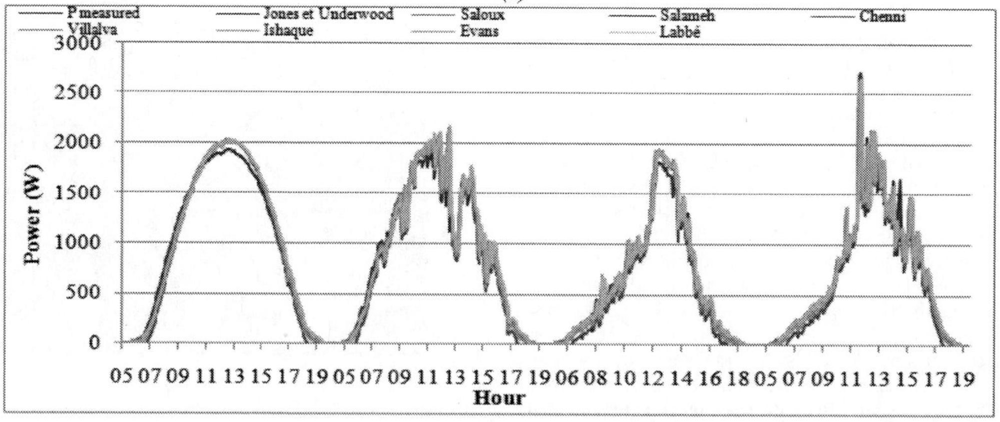

(b)

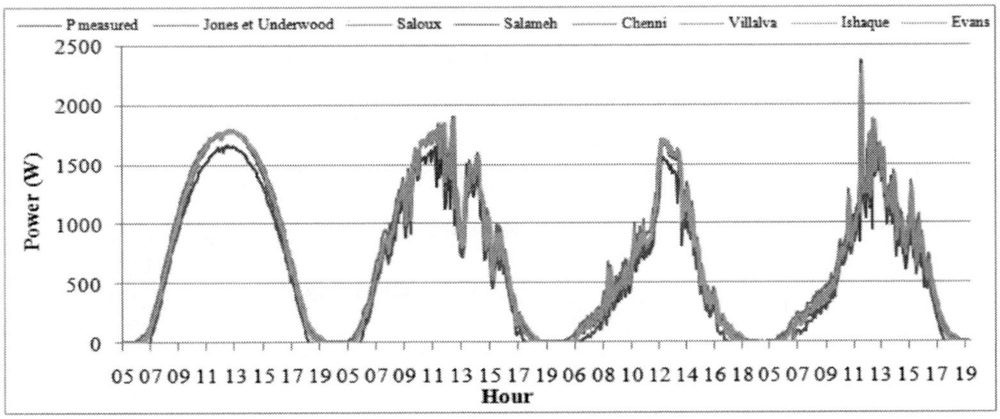

(c)

Figure 2: Measured and simulated electric power for technologies: (a) Monocrystalline, (b) Polycrystalline and (c) Amorphous.

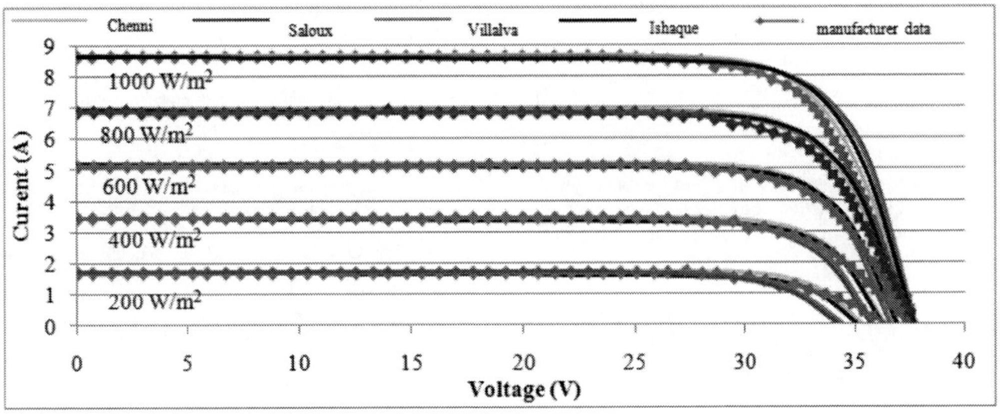

(a)

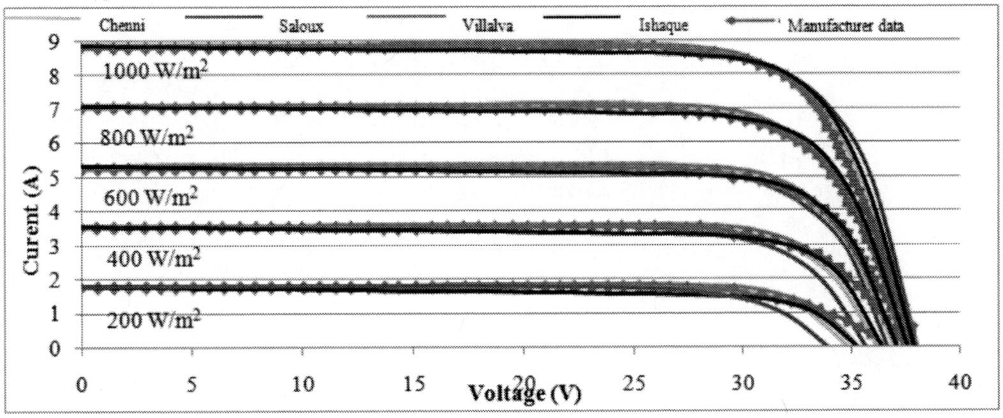

(b)

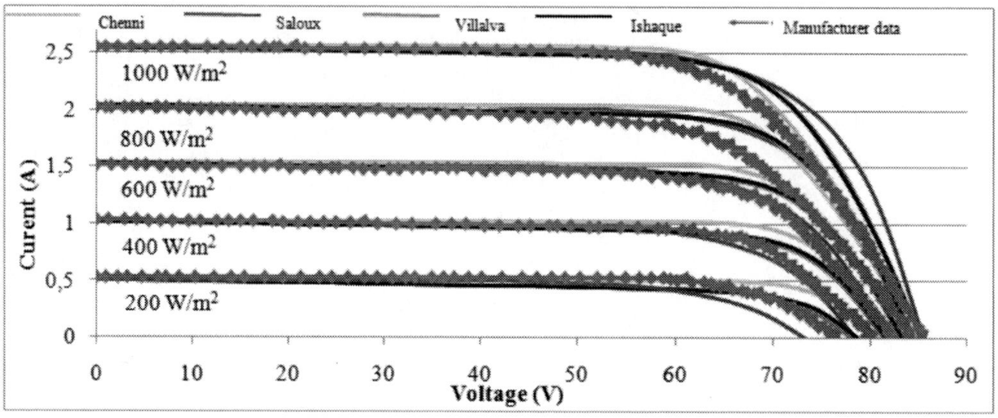

(c)

Figure 3: Characteristic I-V obtained by different models compared to the ones of 3 technologies PV for different irradiation values: (a) Monocrystalline, (b) Polycrystalline and (c) Amorphous.

4 CONCLUSION

In this work, different mathematical models estimating the electric power were compared to the measured ones concerning 3 technologies: Monocrystalline, Polycrystalline, and Amorphous. The results showed that the electric power of each technology could be predicted by a model with satisfying precision. Regarding the characteristic I-V, the result indicated that the calculated characteristics coincide with the ones of different technologies in the region where the panel behaves as a generator of electric current.

References

[1] E. Saloux and A. Teyssedou, "Explicit model of photovoltaic panels to determine voltages and currents at the maximum power point," Sol. Energy, vol. 85, pp. 713–722, 2011.

[2] R. Chenni, M. Makhlouf, T. Kerbache, and A. Bouzid, "A detailed modeling method for photovoltaic cells," Energy, vol. 32, no. 9, pp. 1724–1730, 2007.

[3] K. Ishaque, Z. Salam, and H. Taheri, "Simple , fast and accurate two-diode model for photovoltaic modules," Sol. Energy Mater. Sol. Cells, vol. 95, no. 2, pp. 586–594, 2011.

[4] B. C. Babu and S. Gurjar, "A novel simplified two-diode model of photovoltaic (PV) module," IEEE Journal of Photovoltaics, vol. 4, no. 4. pp. 1156–1161, 2014.

[5] J. C. D. Manning, "Development of a photovoltaic array model for use in power-electronics simulation studies," vol. 146, no. 2, 1999.

[6] D. L. Evans and L. W. Florschuetz, "Cost studies on terrestrial photo- voltaic power systems with sunlight concentration.," Sol. Energy, vol. 19, pp. 255–262, 1977.

[7] M. G. Villalva, J. R. Gazoli, and E. R. Filho, "Comprehensive Approach to Modeling and Simulation of Photovoltaic Arrays," vol. 24, no. 5, pp. 1198–1208, 2009.

[8] A. D. Jones and C. P. Underwood, "A modelling method for building-integrated photovoltaic power supply," Build. Serv. Eng. Res. Technol., vol. 23, no. 3, pp. 167–177, 2002.

[9] J. LABBÉ, "l'hydrogène électrolytique comme moyen de stockage d'électricité pour systèmes photovoltaïques isoles," 2007.

[10] B . Borowy and Z. M. Salameh, "Methodology for Optimally Sizing the Combination of a Battery Bank and PV Array," Energy Convers., vol. 11, no. 2, pp. 367–375, 1996.

ADVANCED PV MODULE HOT SPOT CHARACTERISATION

Stefan Wendlandt[1,2], Lasse Süthoff[1], Sarah Berendes[1], Janine Teubner,
Lars Podlowski[1], Stefan Krauter[1,2] and Paul Grunow[1]

[1] PI Photovoltaik-Institut Berlin AG, Wrangelstr. 100, D-10997 Berlin, Germany
Phone: +49 (30) 814 52 64 0, Fax: +49 (30) 814 52 64 101, e-mail: wendlandt@pi-berlin.com
[2] University of Paderborn, Faculty of Electrical Engineering, Informatics and Mathematics, Department for Electrical Energy Technology & Sustainable Energy Concepts, Pohlweg 55, D-33098 Paderborn, Germany, e-mail: krauter@nek.upb.de

ABSTRACT:

Photovoltaic modules are the key component of any solar power plants. Their long-term performance and availability are responsible for the electrical yield and thus for the profitability of the power plant. To minimise the failure risk, extensive module quality investigations should be carried out already during their planning phase. One of a major module stress factor is the solar cell overheating when partly shaded, or if broken cells or inhomogeneous module soiling occurs. At these conditions, the local cell temperature exceeds the normal operating module temperature which can lead to a significantly reduction of the long-term stability. A standard hot spot test is part of the PV module product certification. Unfortunately, this testing procedure takes a quite long time and the cell leakage current is used for determining the worst case cell. Therefor only some cells are thermal stressed. These are the motivation to develop an advanced hot spot method called rapid thermal hot spot method for PV modules. This paper will describe the technical set-up, the simulation models and the comparison between measurement and simulation.

Keywords: shadowing, hot spot, module reliability

1 INTRODUCTION

The shadowing effect of PV modules is well known [1][2][3]. There are several models for the IV curve in reverse bias [4][5][6]. Also potential root-causes for hot spots are well-investigated [7][8][9]. However, different research groups are still on the design and optimization of their solar cells, PV production equipment or PV modules to minimise the risk of a hot spot event. This would also minimize the risks of a faster degradation and safety issues.

A standard hot spot test is part of the PV module product certification [10][11][12]. Unfortunately the established procedure takes quite a long time and only some cells are thermal stressed under shadowed conditions. Therefore, a new method has been developed which allows to classify the thermal hot spot risk of each and every solar cell in a module [13]. Furthermore, the new method is much faster which allows to check the hot spot temperature of a much higher number of modules within the same time compared to the standard method. The rapid thermal hot spot method can be used in industry and in institutes to evaluate a new cell and module designs, a new cell process or to determine the thermal stability of new polymer materials (like encapsulation or back sheet materials foils). But it is also interesting for EPC's for statistical module testing as one aspect of quality insurance measures. The advanced PV module hot spot characterization consists out of an experimental and a simulative part (Figure 1). The simulative part is similar to the process presented from Geisemeyer *et al.* [14].

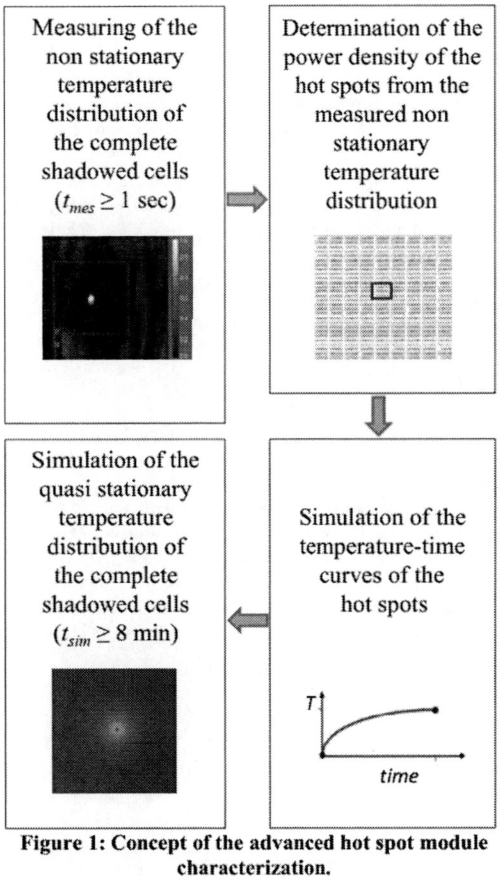

Figure 1: Concept of the advanced hot spot module characterization.

The main advantage of the rapid thermal hot spot method over IEC standard hot spot test (IEC 61215) is that it is a) faster and b) the real hot spot cell temperature is used for the valuation rather than the leakage current. This is an important change because the leakage current of the solar cells is not a clear indicator for the hot spot temperature. Figure 2 shows a very nice example for this fact [15]. However, the real cell temperature is the relevant parameter because it can lead to damages of the solar panel, e.g. by melting encapsulates and burning back sheets foils in extreme cases [16].

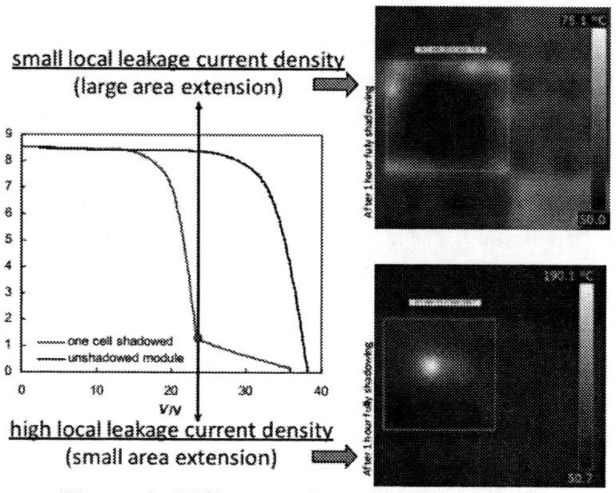

Figure 2: Different solar cell temperature distributions on cells with the same leakage current.

The advanced PV module hot spot characterization method allows the determination of the thermal stability of a large number of modules within a short time. Furthermore, the models is able to compute the quasi stationary temperature and thus gives a better understanding of the hot spot phenomena. The results will be used for a c-Si specific IR pattern catalogue for both field and for laboratory testing as well as for process of standardization.

2 EXPERIMENTAL AND MODELLING

2.1 Set-up

In the laboratory experiment the modules are homogenously irradiated and in an automated sequence each single solar cell is completely covered for a certain period of time. During that period of shadowing, the module in connected with an *MPPT* who operates the module in single module *MPP* or in string *MPP*. At the beginning and at the end of the cell shadowing an infrared image of the shadowed solar cell from the backside is taken. Figure 3 shows the experimental sequence.

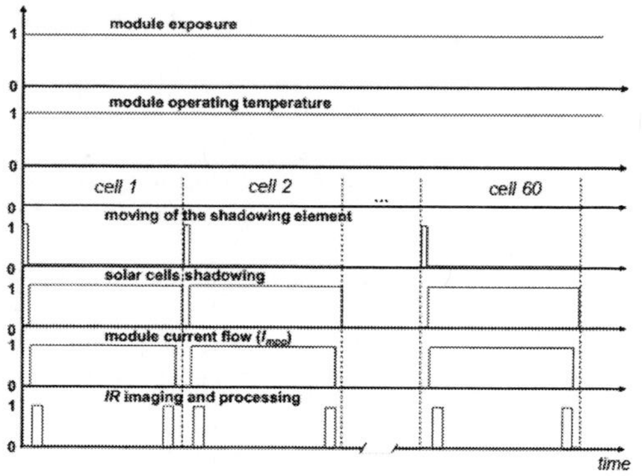

Figure 3: Experimental sequence of the advanced hot spot characterization for PV modules.

Figure 4 shows the back side view of the technical set-up. The vertical traversing axis with the fixed IR camera is visible. The camera is mechanically coupled to the shading element at the module front side and both units can be moved in sync in x and y direction along the module area.

Figure 4: Technical set-up (view from the back side) of the advanced hot spot characterization.

The technical details of the advanced hot spot set-up are presented in Table I.

Table I: Technical details of the advanced hot spot set-up.

Property	Value
Lighting unit:	
Irradiance	750…1000 W/m²
Homogeneity	Class C
Spectrum	AM 1.5_G (Class C)
IR camera system:	
Resolution	320 x 256 pixels
Thermal sensitivity	< 0.05°C @ +30°C
Image frequency	60 Hz

Temperature conditions:

Module temperature	≤ 65°C
Average ambiance temperature	≤ 40°C

Outdoor investigations have shown that these conditions for the module during the test are in good correspondence to the situation at a desert location.

The models to compute the quasi stationary temperature distribution of the shadowed solar cell are represent at next.

2.2 Theoretical Modellings

Due to the short shadowing period (typical: 15 sec) per cell the maximal hot spot temperature is not reached. Therefore, in a second step, the quasi stationary temperature distribution of the cells will be simulate with numerical methods based. These simulations carried out by finite difference method (*FDM*) and finite volume method (*FVM*). The following Figure 5 shows the properties of the applied 2D FDM model.

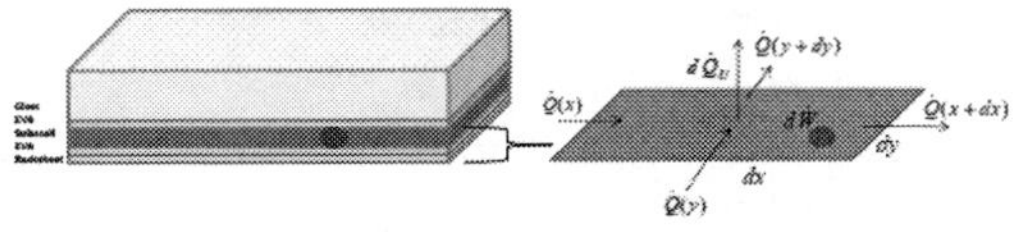

$$\frac{\partial \vartheta}{\partial t} = \frac{\dot{W}}{\rho c_p} - a\left(\frac{\partial^2 \vartheta}{\partial x^2} + \frac{\partial^2 \vartheta}{\partial y^2}\right) - \frac{\alpha}{z \rho c_p}\left(\vartheta_{xy} - \vartheta_u\right) \quad (1)$$

Figure 5: Physical and theoretical model of 2D FDM.

The applied 3D FVM model is shown in Figure 6.

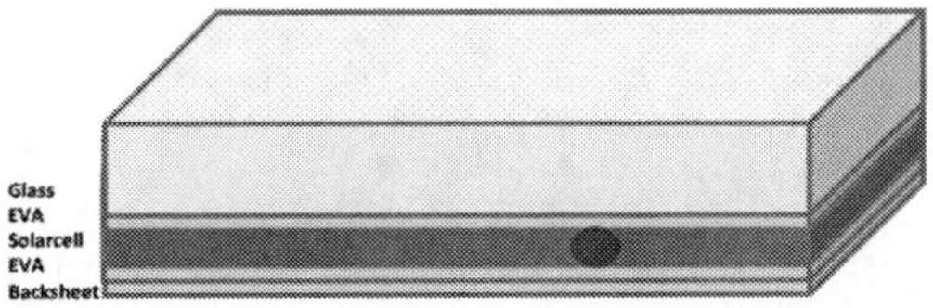

$$\frac{\partial \vartheta}{\partial t} = \frac{\dot{W}}{\rho c_p} - \frac{1}{\rho c_p}\nabla(\partial U) - \frac{\lambda}{\rho c_p}\left(\nabla^2 \vartheta\right) \quad (2)$$

Figure 6: Physical and theoretical model of 3D FVM.

The different simulation properties of the models are compared in Table II.

Table II: Model properties

Property	FDM-Model	FVM-Model
Dimension	2D	3D
Layers in z-dimension	1	n
Boundary Conditions	adiabatic	
Grid structure (x,y,z)	108, 108, 0	156, 157, 87
Software	MATLAB	OpenFOAM

After presenting the technical set-up and models to measure and simulate the hot spot temperature in the next chapter the results are displayed.

2 RESULTS

Table III shows the four most common failure modes which can be found for shaded solar cells. We can find very characteristic error pattern for every failure mod. This would basically allow an automated failure analysis from the IR image, where in a first step every identified hot spot is assigned to one of the failure modes as a conceptual analysis of IR images.

Table III: Typical hot spot pattern of shadowed solar cells.

IR-Image	Description / Reason
	Hot spot at the cell edge, Poor or no edge insulation, Solar cell process-induced
	Hot spot on the metallization area, To high / to low "Firing temperature" or to long / to short "Firing Time" of the contact paste, Solar cell process-induced
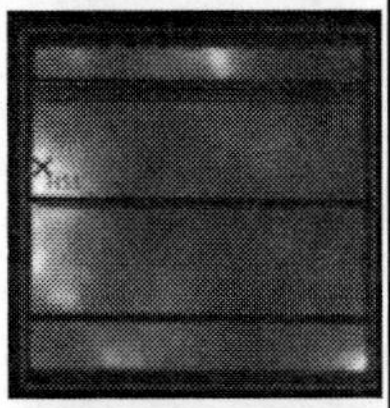	Wide spread hot spots, High net doping (high rate of impurities), Raw material-induced
	Micro crack, Displacement of the cell regions (p- / n-region), Caused by aging or transport, In case of mechanical stress can the hot spot active or not-active

In a second step the image gets quantitatively analysed. The requirements for the simulation of the stationary temperature distribution of a solar cell are the identification of the hot spots and the determination of their power loss density. That has been done by iterating until the simulated short term temperature distribution is equal to the measured ones. The expected stationary temperature distribution is calculated via finite difference methods (FDM) and finite volume methods (FVM).

Figure 7 presents the measured and simulated temperature distributions of sample 1 (solar cell with one major hot spot).

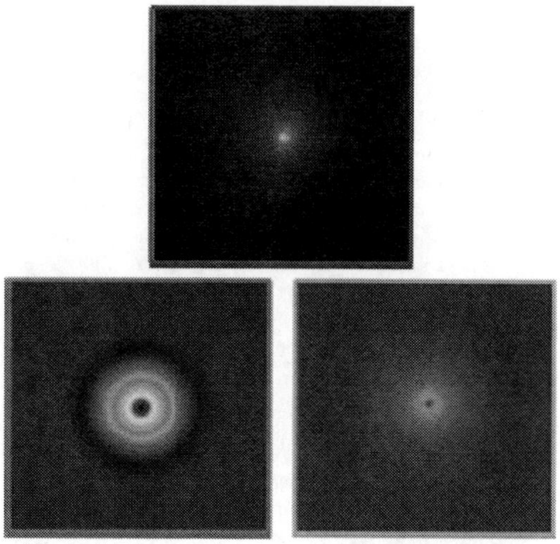

Figure 7: Measured and simulated quasi stationary temperature distributions of sample 1 (solar cell with one major hot spot). Red framed: measurement, blue framed: FDM 2D, green framed: FVM 3D.

The measured quasi stationary temperature distribution could be well reproduction with the models. The geometry for the simulated hot spot is more circular then it is in real. A reason for it could the heat conduction of the cell connector. In the next Figure the temperature-time diagram of the hot spot is displayed.

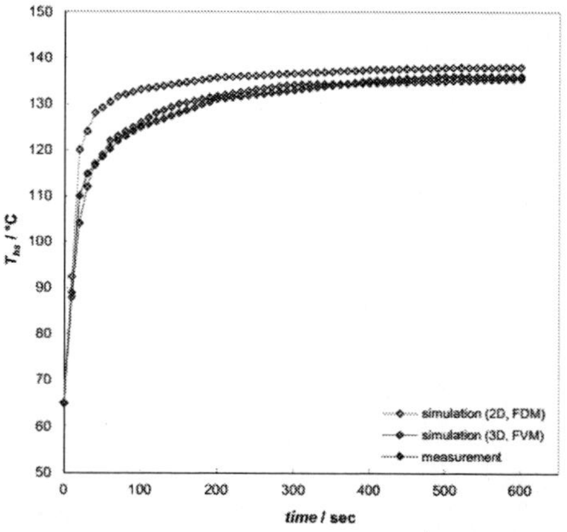

Figure 8: Measured and simulated temperature-time diagram of sample 1 (solar cell with one hot spot).

The curve shape could more exact simulated with the 3D FVM. However, both methods nearly show the trend of the measured temperature. The early high temperature slope of the 2D-Simulation can be explained by the missing third dimension.

At next a sample two with major two hot spots were investigated.

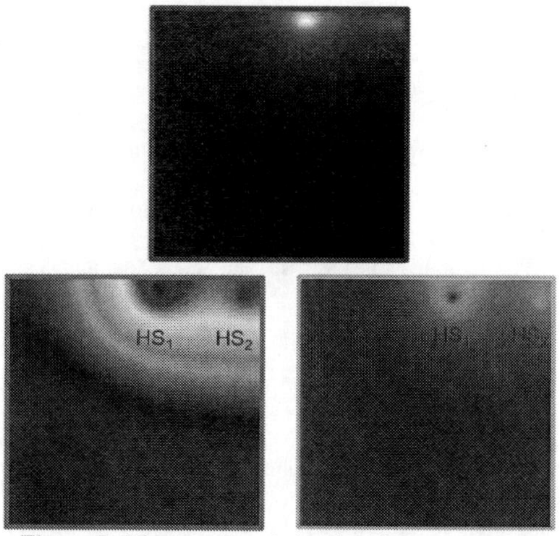

Figure 9: Measured and simulated quasi stationary temperature distributions of sample two (solar cell with two major hot spots). Red framed: measurement, blue framed: FDM 2D, green framed: FVM 3D.

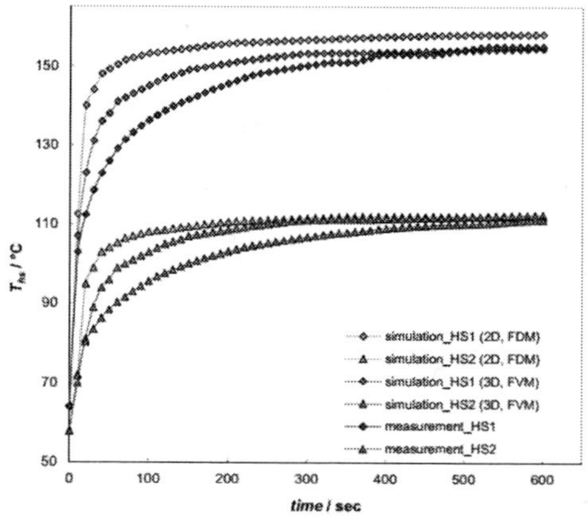

Figure 10: Measured and simulated temperature-time diagram of sample two (solar cell with two hot spots).

The temperature-time plots in Figure 10 clearly show that in both models the temperature rises faster then it has been measured. This can be explanid with deviations in the values of the heat conductivity. Conceptually it seems that the stationary temperature it can predicted more accurately by using the FVM modell.

Finally, the models where also applied at a solar with three major hot spot's which are very close to each other. Figure 8 displays the measured and simulated quasi stationary temperature distributions.

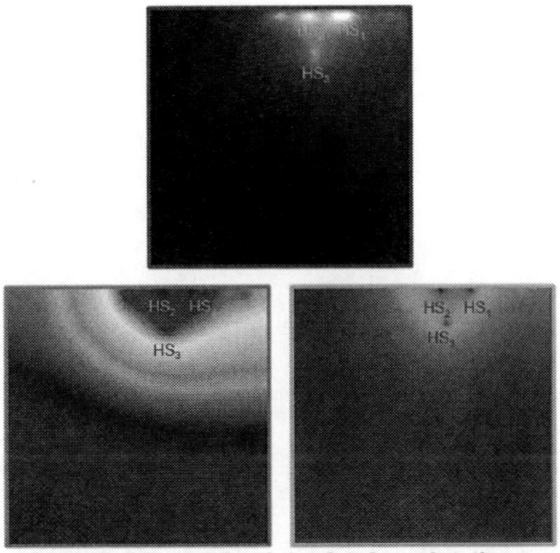

Figure 11: Measured and simulated quasi stationary temperature distributions of sample three (solar cell with three major hot spots). Red framed: measurement, blue framed: *FDM* 2D, green framed: *FVM* 3D.

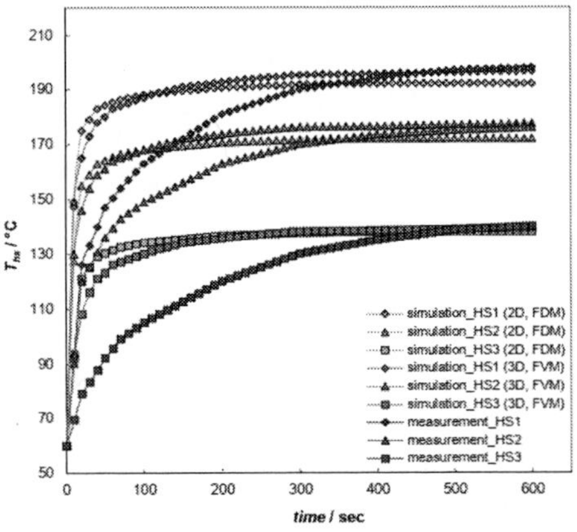

Figure 12: Measured and simulated temperature-time diagram of sample three (solar cell with three hot spots).

Figure 11 and 12 show again that the quasi stationary temperature (distribution) can well simulated but the slopes are not.

The coefficient of determination R^2 has been applied to estimate the models. The results are shown in the following table.

Table III: Coefficient of determination R^2

Sample	2D-FDM-Model	3D-FVM-Model
Sample 1	0.94	0.95
Sample 2		
HS_1	0.82	0.94
HS_2	0.80	0.95
mean	*0.81*	*0.94*
Sample 3		
HS_1	0.65	0.77
HS_2	0.73	0.84
HS_3	0.54	0.73
mean	*0.64*	*0.78*

Table III displays that coefficient of determination increases with the number (of nearby) hot spots. The can be described with the difficulty of simulation of superposition of temperature waves. This leads to simulation deviations in the range of the curve curvature.

4 SUMMARY AND CONCLUSIONS

A new procedure has been developed to determine the hot spot risk of PV modules. This is necessary because the standard hot spot test considers only the integral leakage current as the criterion for a high hot spot risk. Unfortunately in reality these cells do not necessarily have the highest temperature. Furthermore, the new method is much faster which allows to check the hot spot temperature of a much higher number of modules within the same time compared to the standard method.

The rapid thermal hot spot method consists out of two parts which is an experimental IR imaging followed by a simulation based on a theoretical model. In the experiment every solar cell gets covered for a defined time and an IR image of the cell gets taken before and after shading. The temperature distribution over the cell area before and after shading is the input for the simulation to compute the quasi stationary temperature distribution of every respective solar cell.

Two theoretical thermal models have been applied to simulate the quasi stationary temperature distribution of the shaded solar cell. The finite difference method (FDM) which operated in two-dimensional mode and the finite volume method (FVM) as a 3D model

The comparison of the simulated to the measured temperatures show a very good agreement for the stationary maximum temperature. The results are better with a lower number of hot spot's because the superposition of temperature fields per hot spot lead to a higher error. The results of the 3D FVM modelling conceptually do match better with the measured values compared to the 2D FDM results.

This paper demonstrates that the rapid thermal hot spot method is a very powerful tool for an advanced analysis of the hot-spot behaviour of solar cells in a solar panel. The rapid thermal hot spot method not only is better suited than the current standard test to identify the real "risky" cells but also it is much faster and allows for a higher number of tests in the same time. This makes it specifically interesting for quality assurance measures in module production as it allows for a higher sampling rate with more relevant results.

The method can also be used for the hot spot standard test especially the experimental part to select the worst case cells.

5 ACKNOWLEDGEMENTS

The authors would like to thank the whole Photovoltaik Institut Berlin team for its support.

6 REFERENCES

[1] D. Giaffreda *et al.*, "*Local Shunting in Multicrystalline Silicon Solar Cells: Distributed Electrical Simulations and Experiments*", IEEE Journal of Photovoltaics, Jan. 2014, p. 40 - 47

[2] S. Wendlandt *et al.*, "*Comparison of Hot Spot Investigations of Photovoltaic Modules at Laboratory and Field Conditions*", 29th European Photovoltaic Solar Energy Conference and Exhibition, 22nd – 26th September 2014, Amsterdam, Netherlands

[3] S. Janke *et al.*, "*Comparison of hot-spot endurance tests: temperature behavior of bare vs. encapsulated crystalline silicon solar cells*", 29th European Photovoltaic Solar Energy Conference and Exhibition, 22nd – 26th September 2014, Amsterdam, Netherlands

[4] M.C. Alonso-García *et al.*, "*Analysis and modelling the reverse characteristic of photovoltaic cells*", Solar Energy Materials & Solar Cells 90 (2006), p. 1105-1120

[5] J. W. Bishop, "*Computer simulation of the effects of electrical mismatches in photovoltaic cell interconnection circuits*", Solar Cells 25 (1988), p. 73 - 89

[6] V. Quasching, R. Hanitsch, Sol. Energy 56 (6) (1996) 513

[7] W. Herrmann *et al.*, "*Hot spot investigations on PV modules - new concepts for a test standard and consequences for module design with respect to bypass diodes*", 26th IEEE Photovoltaic Specialists Conference, Anaheim (1997), p. 1129-1132

[8] U. Jahn, "*PV module reliability issues including testing and certification*", IEA PVPS Workshop Frankfurt, 2012

[9] S. Wendlandt *et al.*,"*Comparison of Hot Spot Investigations of Photovoltaic Modules at Laboratory and Field Conditions*", 29th European Photovoltaic Solar Energy Conference and Exhibition, 22nd – 26th September 2014, Amsterdam, Netherlands

[10] O. Breitenstein *et al.*, "*Understanding junction breakdown in multicrystalline solar cells*", Proceedings 20th Workshop on Crystalline Silicon Solar Cells & Modules, p. 1 - 13 NREL, Breckenridge, USA (2010)

[11] W. Kwapil *et al.*, "*Investigations on the pre-breakdown of multi-crystalline silicon solar cells*", 23rd European Photovoltaic Solar Energy Conference, 1 - 5 September 2008, Valencia, Spain

[12] K. G. McKay und K. McAfee, „Avalanche Breakdown in Silicon", Phys. Rev. 94, 877–884 (1954)

[13] S. Wendlandt *et al.*, "*The Temperature as the Real Hot Spot Risk Factor at PV-Modules*", 27th European Photovoltaic Solar Energy Conference and Exhibition, 24th – 28th September 2012, Frankfurt, Germany

[14] I. Geisemeyer *et al.*, "*Fast Hot Spot Evaluation*", 29th European Photovoltaic Solar Energy Conference and Exhibition, 22nd – 26th September 2014, Amsterdam, Netherlands

[15] L. Podlowski *et al.*, "*Hot-spot measurements on crystalline silicon solar cells with different reversecurrent characteristics*", 31st European Photovoltaic Solar Energy Conference and Exhibition, 14th – 18th September 2015, Hamburg, Germany

[16] S. Wendlandt *et al.*, "*Thermal Stress Analysis at Encapsulation and Backsheet Materials for PV-Modules*", 31st European Photovoltaic Solar Energy Conference and Exhibition, 14th – 18th September 2015, Hamburg, Germany

SHADING AND HOT SPOT PERFORMANCE OF SHINGLED CELL ARRAY MODULE

Huaming Zhou, Lisong Zhou
Flex

ABSTRACT: As a disruptive module assembly technology, the shingled cell array modules(SAM) have gained more interest as they offer >10% higher module power and better module reliability than conventional ribbon interconnection based modules. Two kinds of mono PERC shingled array module (SAM) samples have been prepared and tested with rated power more than 325W under standard test conditions. The distinct shading and hot spot performance of these SAM samples are reported in the paper for the first time. In addition, the SAM demonstrated a significant reliability benefit of a lower hot spot temperature than conventional module.
Keywords: shingled cell array module, mono PERC, hot spot, shading,

1 INTRODUCTION

Most common solar modules are using metal ribbons for cell interconnection, and they have been proven reliable with low manufacturing cost. However, it suffers from high shading and resistive loss therefore resulting in lower module efficiency. In addition, it limits the usage of very thin wafers due to the nature of high stress in soldering process. Shingled cell Array Modules (SAM), however, eliminate the cell bus-bar shading by overlapping the cell bus-bar interconnection to the rear side of cell strips, and eliminate the cell to cell spacing within a string. As the result, it achieves extremely high ratio of active generation area over total module area [1]. In addition, the SAM can reduce the resistive loss by series interconnecting 1/4 - 1/6 cell strips in the string. And electrical conductive adhesives (ECA), used for interconnection between cell strips, could significantly reduce the stress of the interconnection therefore improve the module reliability performance [2]. With all the performance benefits, there is no report of their shading and hot spot performance in literature. In this paper, we have tested the shading and hot spot performance and reported the result for the first time for the shingled cell array modules with and without by-pass diodes.

2 SAMPLE PREPARATION

Different from conventional ribbon interconnection, the solar cells are first cut into 5 strips. On each strip, one bus-bar is located on one edge of front surface and the other bus-bar on the opposite edge of rear surface. 34 strips are interconnected using ECA in the shingled format, i.e., the bottom bus-bar on 2nd strip directly overlaps the top bus-bar of 1st strip, the connection repeats itself to have all 34 strips connected in one full string. After that, 10 full string array are laid up in parallel and ribbon foils are connected to the ends of each string and a circuit is formed. Finally, a similar backend process as used in the conventional glass backsheet module construction is used here, i.e., lamination, framing, Jbox, potting, and testing. In the shingled module, there are many connection ways to form various circuits. The SAM circuit in this paper can be considered as two Series-Parallel (SP) units linked in series as defined in IEC 61251, i.e., a parallel connection (P) of blocks, where each block consists of a series connection (S) of a certain number of cells, as shown in Fig. 1. Two module samples were prepared for this study: sample #1 has no by-pass diodes and sample #2 has two by-pass diodes each connected to a SP unit in parallel. All samples passed EL and I-V test with rated power more than 325W under standard test conditions. Fig. 2a shows the picture of finished sample and Fig. 2b represents the I-V curve from the flash test.

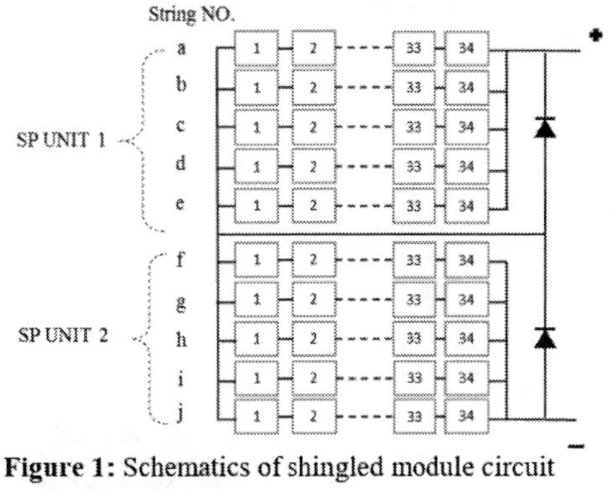

Figure 1: Schematics of shingled module circuit

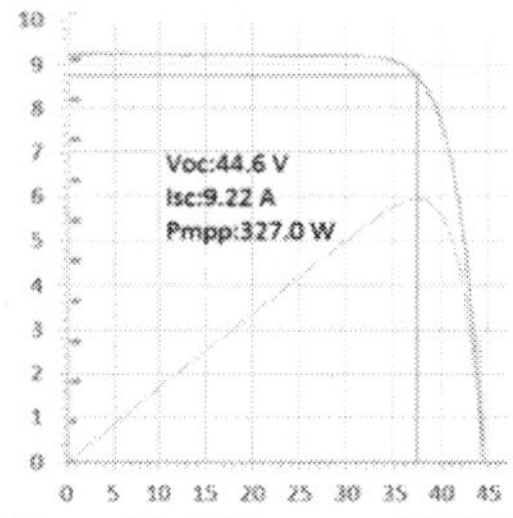

Figure 2a: Picture of SAM **Figure 2b:** I-V curve of SAM

3 EXPERIMENT AND RESULTS

3.1 I-V performance under shading

A black board is used to shade the cell strips (from 1 strip to multiple strips, respectively) in one string while the module is flashed by Pasan flasher, as shown in Fig. 3. And Fig. 4 and Fig. 5 show the I-V curves of sample #1 (without by-pass diode) and sample #2 (with by-pass diodes) under various shading conditions, respectively.

For sample #1 without the by-pass diode, when only one strip is fully shaded, the I-V curve shows a kink around 15V bias as in Fig.4(b). When module bias is <15V, the one shaded strip is reversely broken down and conducting the full string current from the other 67 un-shaded strips, since its reverse bias voltage is greater than the cell breakdown voltage which is around 30V for this cell. When module bias is > 15V, the reverse voltage on the shaded strip is less than the cell breakdown voltage and that string becomes an open circuit, therefore, limits the overall current to the sum of remaining 4 strings. As shown in Fig. 4c and 4d, when more than 2 strips in the same string are shaded, the reverse bias voltage on each shaded cell is less than the cell breakdown voltage, as a result, the shaded string becomes an open circuit and the module current is limited by the sum of remaining 4 strings. It is worth pointing out that the module could suffer from hot spot effect when the cell strip is broken down reversely, as it would happen without by-pass diodes.

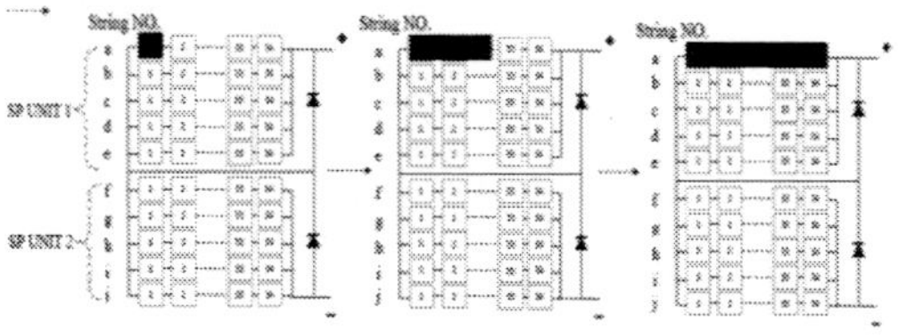

Figure 3: Shading sequence from 1 strip to 34 strips within one string

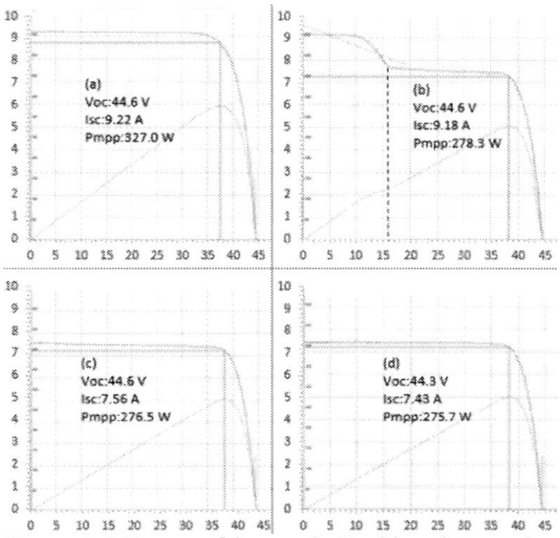

Figure 4: I-V curve of the sample #1 with various number of strips shaded in a string: (a) unshaded, (b) 1 strip, (c) 2 strips, (d) 34 strips

For sample #2 with by-pass diodes built in, we have a very different I-V characteristic, as shown in Fig. 5. First of all, I-V curves are almost identical if one or more than one strips within one string are shaded, and the kink voltage happens at around 20V. When module bias is less than around 20V (roughly half of module open circuit voltage), the reverse bias caused by the shaded strip drives the by-pass diode into conduction and by-passes the whole string with the shaded strip. The current of SP unit 2 limited by the shaded string will go through the by-pass diode. When module bias is above 20V, the by-pass diode parallel with shaded string is reversely biased, the string with any number of shaded strips becomes open circuit. And the module current is limited by the sum of unshaded strings. In contrast to sample #1, there are no cell strips operating under reverse breakdown region which effectively prevents from hot spot damage.

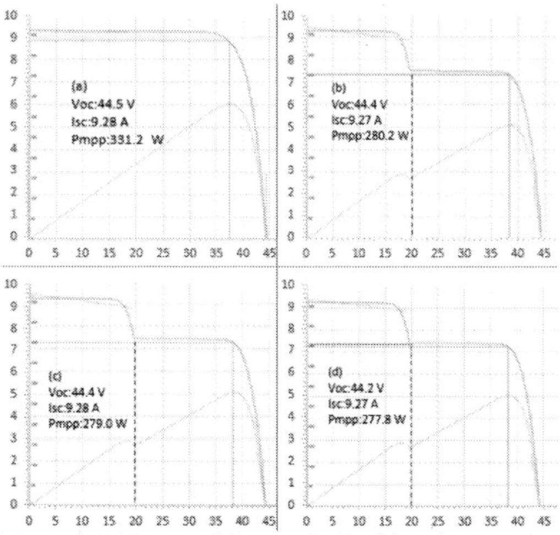

Figure 5: I-V curve of the sample #2 with various number of strips shaded in a string: (a) unshaded, (b) 1 strip, (c) 2 strips, (d) 34 strips

3.2 Current of by-pass diode under shading

It's quite interesting to understand how the by-pass diode functions as more strings are shaded in the SAM samples. The SAM sample is put into a chamber with a class BBB steady-state solar simulator. The real value of the irradiation density ranges from 800W/m^2 to 860W/m^2. Total current of the module (I_{sc}) and the current flowing through the by-pass diode (I_{diode}) are tested using two current meters with the resolution of 0.001A. The SAM sample is shaded with various shading area shown in Fig. 6. All samples were tested in short circuit at the chamber temperature of about 60°C.

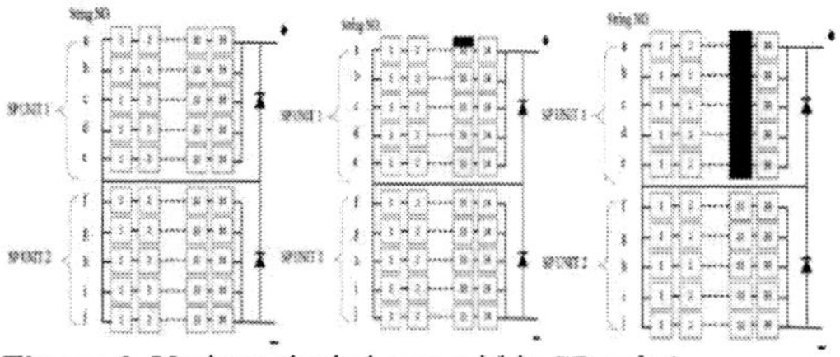

Figure 6: Various shaded area within SP unit 1

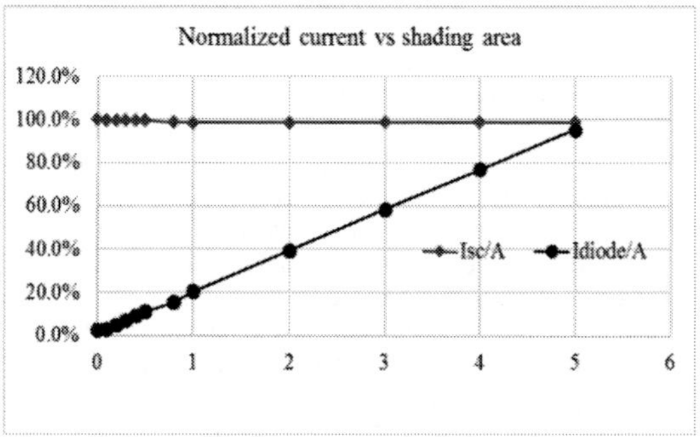

Figure 7: Normalized current vs shaded area

Fig. 7 shows the I_{diode} increases with the shaded area linearly while the I_{sc} is constant. In another word, the current limited by the shaded area will go through the by-pass diode which is quite consistent in the principle to conventional module [3]. In addition, if the same areas are shaded in both SP units separately, no I_{diode} is detected and the by-pass doesn't work.

3.3 Hot spot performance under shading

A class BBB steady-state solar simulator and a IR scanner (Fluke Ti400) were used to study hot spot performance of these SAM samples. All samples were tested in short circuit connection at an ambient temperature of about 10°C. A thermal couple and fan are used to control the average temperature of the back sheet within the range from 40°C to 60°C. Fig. 8 shows the temperature map of sample #1 after 1.5-hour exposure with no shading. The slight non-uniformity in temperature is due partially to the non-uniform irradiation intensity of the steady-state solar simulator, range from 930W/m^2 to 1070W/m^2, and partially to non-uniform cell current distribution. After that, one strip in sample #1 was shaded using a black board, and Fig. 9 shows the temperature map after 1.5-hour stabilization. The average backsheet temperature is at 49°C, however, the shaded area reaches a maximum temperature of 133.7°C after 1.5-hour exposure, as marked in the red circle.

Figure 8: The temperature map of sample #1 with no shading

Figure 9: The temperature map of sample #1 under shading (circled area) and temperature reading of 133.7℃

For sample #2, one strip in each string of the SP unit 1 is shaded by a black board. Fig. 10 shows the temperature map after 1.5-hr exposure. The temperature within SP unit 1 is uniformly < 50°C, as marked by red block in the Fig. 10. We know normally the cell will turn hotter under partial shading than the full shading condition. For sample #2, we have tested the hot spot temperature even at partial shading condition to exaggerate the result. Figure 11 shows the temperature map with one ¼ strip shaded in one string and one ½ strip shaded in another string. The average temperature is 54°C after 1.5-hr stabilization, the max temperature at the shaded locations is < 95°C. This hot spot temperature is well within safe temperature range under the normal operating conditions.

Figure 10: One strip in each string of SP unit 1 is shaded in sample #2

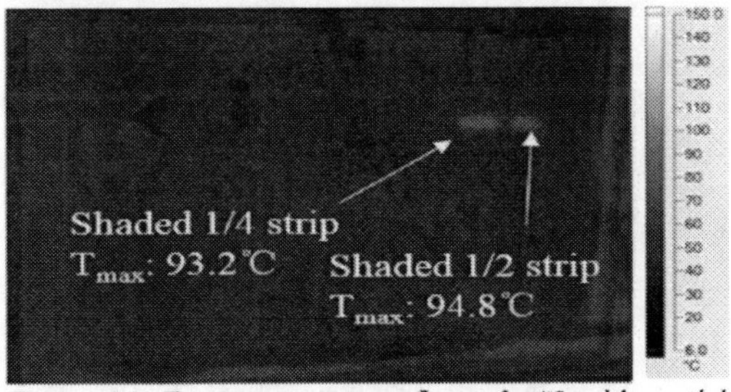

Figure 11: Temperature map of sample #2 with partial strip shading, the max temperature is about 95°C

3.4 Hot spot comparison

A SAM module and a 72-cell mono PERC module were put on the ground mount rack next to each other and the rack has a tilt angle of 45°. The same raw materials, such as cells, EVA and backsheet, are shared between SAM and mono PERC samples. In each shading condition, the hot spot temperature of these two modules was scanned simultaneously by the IR scanner (Fluke Ti400) after 15-minutes stabilization time. All samples were tested in short circuit connection at an ambient temperature range from 27°C to 40°C. Figure 12 shows the comparison of hot spot temperature between these two samples under three different shading area conditions: A:156.75mm×49mm, B:156.75mm×79mm, C: 156.75mm×118mm. obviously, a significant improvement (>30°C) of hot spot temperature is consistently observed for SAM modules.

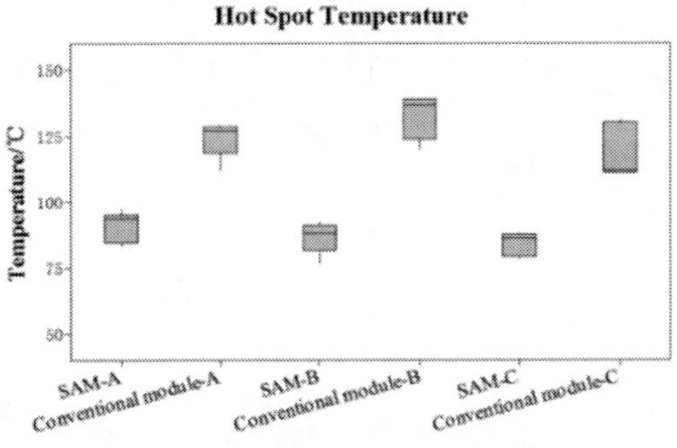

Figure 12: Hot spot temperature comparison between SAM and convention module

It is interesting to understand the trend of hot spot temperature with shaded area in the SAM samples. A SAM module was put into a chamber with a class BBB steady-state solar simulator. The real value of the irradiation ranges from 800W/m^2 to 860W/m^2. Several thermal sensors were glued on the spot of backsheet directly behind the shaded cell strip in the position as described in fig 13 inset. The shaded area is increased from 0% to 100% of a strip, and temperature data were recorded at each sensor location after a 15-minutes stabilization time. Figure 13 summarizes the temperature trend with various shading condition. It is obvious that the un-shaded partial area shows much higher hot spot temperature than the shaded partial area within the same strip. It also suggests that the hot spot may cause the most damage when it is half-way shaded as hot spot temperature is highest.

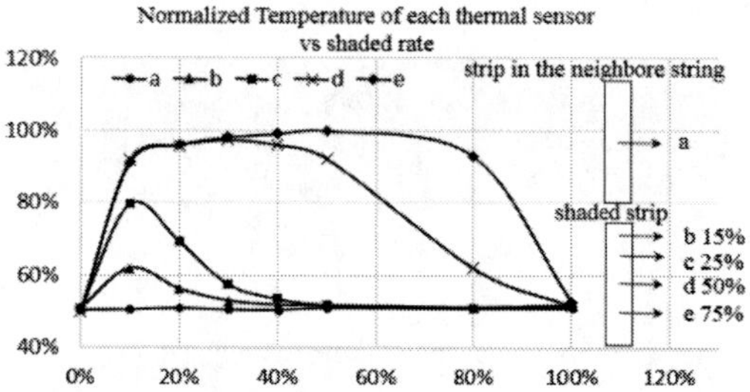

Figure 13: Hot spot temperature trend with various shading conditions at different locations

4 SUMMARY

We have for the first time reported the shading and hot spot performance for shingled array modules with and without by-pass diodes. The test result suggests serious hot spot damage risks for modules without by-pass diodes, and the by-pass diodes approach can effectively reduce the hot spot damage by conducting current through the diode under shading condition. In addition, we observed a >30°C lower hot spot temperature in SAM than conventional ribbon interconnected modules, and this can contribute to a significant reliability improvement to the conventional module.

5 REFERENCE

[1] J. Zhao, et al, Progress in Photovoltaics Research and Applications, vol 5, 269-276 (1997)

[2] G. Beaucarne, 6th Workshop on Metallization and Interconnection of crystalline silicon solar cells, 2016

[3] Stefan Wendlandt, et al, Operating Principle of Shadowed C-Si Solar Cell In PV Module, DOI: 10.18086/swc.2011.14.18 ISES Solar World Congress 2011.

TAPE INTERCONNECTION FOR SILICON SOLAR CELLS WITH EXTENDED LONG-TERM STABILITY

Jonas Buddgård, Torgny Lagerstedt, Andrew Machirant

JB EcoTech AB

Lärktunet 13, SE-181 48 Lidingö, Sweden

ABSTRACT: The purpose of this work is to reduce production cost for conventional photovoltaic modules by reducing the amount of silver used in the grid on the photovoltaic cell and simplify the method of interconnection of individual PV cells. A novel method of applying bus bars and interconnecting individual cells into a module is proposed. The new method have the possibility to reduces the amount of silver by 25%. Furthermore, the interconnection of individual cells to a module is simplified, thus making the production less costly and more robust. Finally, soldering can be completed without the use of lead and at a lower temperature (150-160°C) which enables the use of cells featuring advanced temperature sensitive architecture like HIT cells.

It is very important to ensure a long lifetime for all PV modules. For the proposed new interconnection method we believe that the most important tests to ensure a long lifetime are the UV-exposure test and the thermal cycling test. To ensure that the proposed interconnection tape do not discolor during the PV module´s lifetime this work will therefor test for extended UV exposure of 100kWh/m². This work will also expose the modules for extended thermal cycle test of up to 600 cycles. Because of the lower process temperature with this novel method the modules should be more robust to thermal cycles. A cross-section analysis with SEM of the interconnection between the tape and the grid fingers will also be conducted and shown in this work. Finally, electro luminescent measurements of the tested modules will be disclosed.

Keywords: Long term stability; Energy conversion; Silicon; silver; module production; cost reduction

1 INTRODUCTION

The purpose of this work is to reduce production cost for conventional photovoltaic modules by reducing the amount of silver used in the grid on the photovoltaic cell and simplify the method of interconnection of individual PV cells. One of the contributing cost factors for photovoltaic modules is the silver used in the grid on the PV cells. The grid consists of closely spaced narrow fingers and three to five wide bus bars perpendicular to the fingers. A novel method of applying busbars and interconnecting individual cells into a module is proposed. The new method holds the possibility to reduces the amount of silver by 25%. Furthermore, the interconnection of individual cells to a module is simplified and the process temperature is decreased significantly, from 240 to 160 degrees C, thus making the production less costly and more robust. All 60 cells in one PV module can be connected in a single step in the lamination process. Finally, soldering can be completed without the use of lead and at a lower temperature which enables the use of new high efficiency cells with delicate p/n junctions. The decreased process temperature also decreases the risk for cracking during soldering and makes the module more robust to thermal changes.

2. EXPERIMENT

PV cells are electrically interconnected in series to form a PV module. In this work, we are testing mini modules consisting of 6 cells, two rows each with 3 cells for thermal cycling tests and 2 cells modules for extended UV-test and electro luminescent measurements. These modules were produced with the proposed novel interconnection method and are tested for extended thermal cycling tests according to IEC 61215. We will disclose the test results and discuss it in detail to describe the difference between this low temperature interconnection and normal soldering when it comes to robustness to thermal cycles.

2.1 The proposed method for module production

The proposed tape method, enables a simplification of the production process and a reduction of cost of material. The tape act as a carrier for tabbing ribbons. In the production process, the tape with tabbing wires is placed on the top glass and the first EVA sheet. Tape with tabbing wires for a complete row or a complete module are laid out. After this a row (or rows) of cells is placed on the tape with tabbing wire. A row of cells is now placed on the previously placed tabbing wires. The active side, the topside, of the cell is here contacting the tabbing wire on the tape. The process continuous with a second layer of tape with tabbing wires which are placed on the back side of the cells. After this a second EVA sheet is placed on the cells and tabbing wires. This is followed by a back sheet. The module is finalized in the laminator step. When the increased temperature and high pressure on the tape and cells form a solid interconnection between cells. A schematic of the building of modules with the tape method is shown in figure 2.

The tape with tabbing wires is designed in such a way that an interconnection in series between cells is obtained in the lamination step, see figure 2.

By using the tape, the two steps, Stringing and lamination is merged into one step. The production process and production equipment is simplified and the required factory area is reduced. Also, the amount of silver can be reduced.

In the soldering method used today the cells are heated to of about 240 ^{0}C to solder the tabbing wires to the busbars. The temperature raise takes a few seconds and this fast increase from room temperature to 240 ^{0}C adds a risk of cracking of the thin cells. The soldering with the tape method is gentle and the risk of cracked cells is reduced because the soldering takes place in the lamination in which the increase from room temperature to 160 ^{0}C takes of about 10 minutes. Here it shall be noted that there is a trend towards using thinner cells which are sensitive to temperature gradients.

2.2 Description of tape with tabbing wire

The tape with tabbing wire is applied on both sides of the cell. The tabbing wires are cut and interrupted in such a manner that the individual cells are interconnected in series in the lamination process. A single string of tabbing wires on a tape is shown in figure 1. However, the tabbing can also be carried out by integrating the tape onto the EVA sheets. This is shown in figure 1.1.

Figure 1. Single string of tape. and. EVA sheet prepared with tape

3. DISCUSSION

The main reason for having the silver busbar on the PV cell in the traditional way of building modules is to increase strength in the bond between the cell and the tabbing wire. The high strength in the bond is needed because the row of 10 cells soldered together is moved from the soldering station and placed on top of the EVA sheet and top glass. Also, note that a turning of the row is needed. This strength is not needed when using the tape method. Because in the tape method, the cells are directly placed on the EVA sheet and top glass and the interconnection of cells is formed in the lamination process in which all the parts of a module are laminated into one unit. In this way, the process temperature for connecting the PV cells does not have to be above 160 degree C. This means that the Delta T, from the highest temperature the PV module experience during its lifetime to the lowest for the thermal cycling test or in real life out in the sun, is decreased dramatically. By approximately 80 degrees C (240C -160C). Finally, this gives a longer lifetime for the PV module and a lower price for the electricity it will produce during its lifetime.

4. RESULTS

The thermal cycling test was carried out at RISE - Research Institutes of Sweden and the UV- and cross-section analysis at Fraunhofer ISE.

Thermal cycling 600 cycles

The test was carried out by RISE according to the IEC 61215 standard. The thermal cycling test showed very small degradation even after 600 cycles. We tested 3 modules and the worst result was 2,3% and the best was only 1,1% degradation of maximum power after 600 cycles. See figure 4.1 below.

Tabell 2. Resultat initialt

Objekt	I_{sc} [A]	I_{mp} [A]	V_{oc} [V]	V_{mp} [V]	P_{mp} [W]	FF [%]	Cell eff, [%]	Module eff, [%]
160329 #01	8,74	8,12	3,81	2,96	24,06	72,22	16,48	11,88
160329 #02	8,76	8,19	3,81	3,05	24,95	74,80	17,09	12,32
160329 #03	8,70	8,10	3,81	2,96	24,01	72,35	16,44	11,85

Tabell 5. Resultat 600 cykler

Objekt	I_{sc} [A]	I_{mp} [A]	V_{oc} [V]	V_{mp} [V]	P_{mp} [W]	FF [%]	Cell eff, [%]	Module eff, [%]
160329 #01	8,72	8,09	3,80	2,94	23,79 (-1,1%)	71,84	16,30	11,74
160329 #02	8,74	8,13	3,80	3,00	24,37 (-2,3%)	73,45	16,70	12,03
160329 #03	8,73	8,00	3,80	2,93	23,44 (-2,3%)	70,78	16,06	11,57

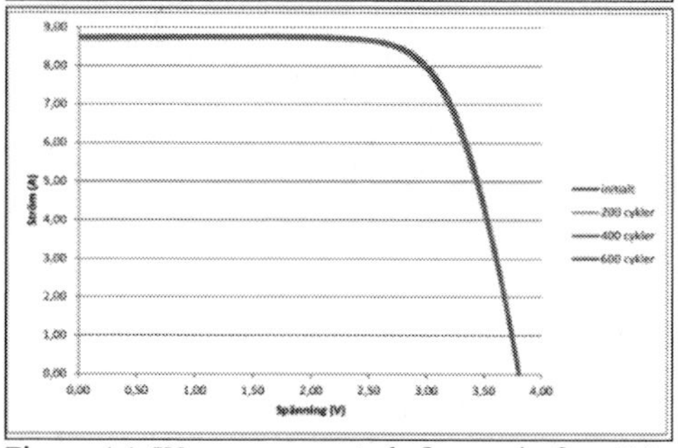

Figure 4.1 IV measurements before and after 600 cycles of thermal cycling -40 to + 85 C.

Extended UV test 100 kWh / m² with electroluminescence (EL) characterization

In figure 4.2 below the EL pictures before and after at 8 A is shown. No cell cracks or local interconnection issues visible, the dark areas at the lower and upper cell show contact finger interruptions which are a metallization issue. There is now visible degradation in EL after the 100 kWh of UV exposure.

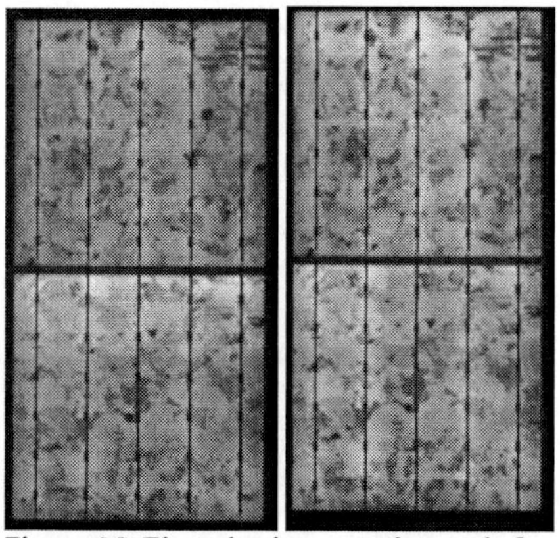

Figure 4.2 Electroluminescent pictures before UV expositor 100 kWh/m² to the left and after the test to the right.

Cross-section analysis with SEM

To determine the contact between the tape and the cell we carried out cross section analysis with the help from Fraunhofer ISE. The conclusion from this analysis was that maybe the process temperature was a little low. The tested module was manufactured at 150° C. In the picture 4.3 below you can see cavities at the metallization edges and there is almost now intermetallic phase growth at the boundary layer between solder and metallization.

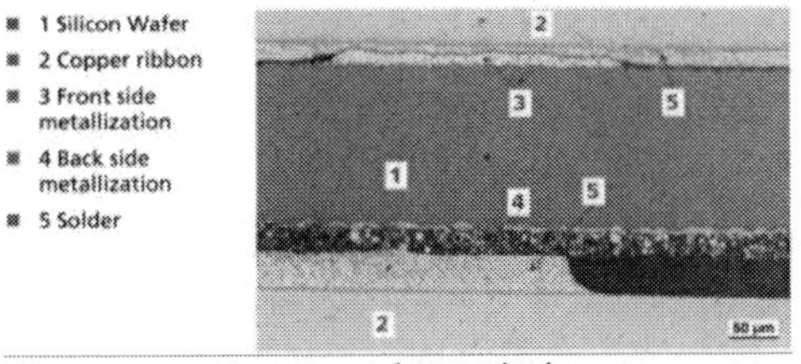

Figure 4.3 SEM cross-section analysis.

5. CONCLUSION

A novel tape method, of applying bus bars and interconnecting individual cells into modules reduces the amount of silver in the module. Also, the tape method enables simplification of the production process. The TC 600 and the 100kWh UV test was successful but the cross-section analysis indicated that the process temperature could be increased from the tested 150° C to maybe 160° C for a better connection.

6. TAPE PICTURES

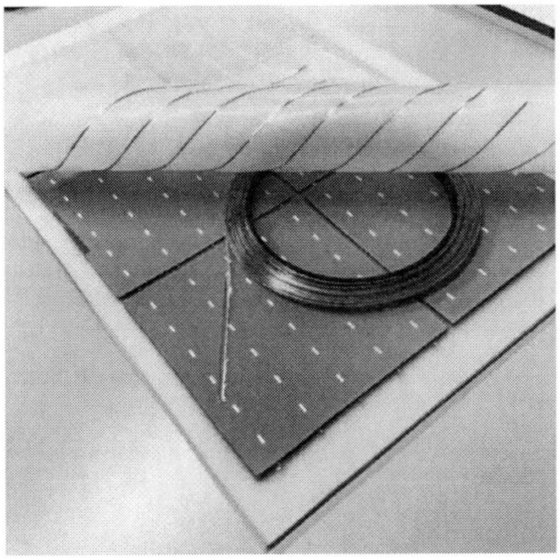

7. REFERENCES

[1] CONDUCTIVE GLUING AS INTERCONNECTION TECHNIQUE TOWARDS SOLAR CELLS WITHOUT FRONT BUSBARS AND REAR PADS
Andreas Schneider, Rudolf Harney, Severin Aulehla, Simon Hummel, Engelbert Lemp, Simon Koch* and Kevin Schröder*
International Solar Energy Research Center (ISC), Rudolf-Diesel-Str. 15, 78467 Konstanz, Germany

[2] A new interconnecting material for solar cells with high-potential on busbar-less solar cells
Daichi Mori, Akifumi Higuchi, Dexerials Corporation.

[3] Agroui K, Koll B, Collins G, Salama M, Hadj Arab A, Belghachi A, Doulache N, Khemici MW. Characterization of encapsulant materials for photovoltaic solar energy conversion, *Proc. SPIE 7048, Reliability of Photovoltaic Cells, Modules, Components, and Systems, 70480G (September 10, 2008); doi:*10.1117/12.794012

[4] Sachs EM, Serdy J, Gabor, AM, van Mierlo F, Booz T. *Light-Capturing Interconnect Wire for 2% Module Power Gain.* Proceedings of the 24th EU-PVSEC, *23 Sept 2009, Hamburg*

RELIABILITY ANALYSIS OF PHOTOVOLTAIC MODULES BY CONTACT STATES BETWEEN INTERCONNECTOR RIBBON AND AG ELECTRODE.

In-ae Kim, Yongki Min*, Ju-ho Choi, Chung-hyun Kim, Eun-ju Lee, Dong-seop Kim
Solar R&D Team,Solar Energy Business General, Shinsung E&G Co.,Ltd.
108-73, Hansam-ro, Daeso-myeon, Eumseong-gun, Chungcheongbuk-do, 27676, Korea

ABSTRACT: The warranty period of a PV module depends on the manufacturer and the minimum power performance is in accordance with the standard test conditions(IEC61215). But does not represent the installation performance overtime and installation. Generally, the PV module output is guaranteed to be at least 25years at 80% output, but maybe exceeded at power levels lower of 40-50years, depending on the installation location or climate. Therefore, in the soldering process, which has the great influence on the durability and output PV module, the soldering criterion is established through the Peel off test for the reliability and output degradation of the PV module are checked through the Humidity-Freeze test.

Keywords: Photovoltaic Module, Peel off test, Soldering, Tabbing, Humidity freeze

1 INTRODUCTION

Recently, Many companies of PV industry are developing high efficiency of solar cells and modules, and they are trying low cost of manufactured solar cells and modules.

Commonly used method for getting low cost and high efficiency is increasing number of bus-bar electrode line on the solar cell. The other method of reducing cost is decreasing amount of Ag paste. The more decreases number of electrode on a solar cell is, the narrower width of electrode are. For this reason, many companies of PV module have difficulties in applying a large number of interconnector solar cell, and thinner electrode cell to the modules.

Especially, many problems occur in the tabbing process, due to tabbing condition. The tabbing process is a eutectic process that connects a cell and a interconnector. However, micro-cracks can occur due to the different coefficient of thermal expansion of the cell and interconnector during soldering process. Because of the different coefficient of thermal expansion, the contraction and expansion of the heat is repeated, resulting in a crack. As a results, they can eventually affect the power, CTM loss and reliability of the photovoltaic modules [1,2,4].

To comprehend these influences on life time of PV modules, three states modules which are over-soldering, well soldering and non-soldering samples were manufactured by changing tabbing conditions of temperature and solder time. Several reliability test according to IEC 61215 were performed and analyzed [3].

2 SAMPLE PREPARATION

2.1 Soldering condition test

In order to established Soldering condition, we are manufactured three sample tabbing cell of solder state : Non-soldering, Soldering, Over-soldering. For reproducibility of the manufactured samples, 50ea samples were prepared for each soldering condition. The samples were confirmed the solder state by peel off test and Reproducibility was secured under each soldering condition.

2.2 Manufactured Module

First of all, to test reliability of module depending on soldering conditions, produce each conditional string.

Each soldering condition is set up through 180° Peel off test. The condition of Non-soldering is under 2N/mm of adhesiveness, whereas over 2N/mm in Soldering condition. The condition of Over-soldering is over 2N/mm and set according to the state of surface after peel off test. In case of Over-soldering, it is defined as a phenomenon where Cold solder joint occurs or the cell is rip out by interconnector during peel off test.

To confirm the condition of soldering, we measured surface of the cell with SEM & EDX after peel off test, and we proceed to modularize through lamination process after producing each soldering conditional string.

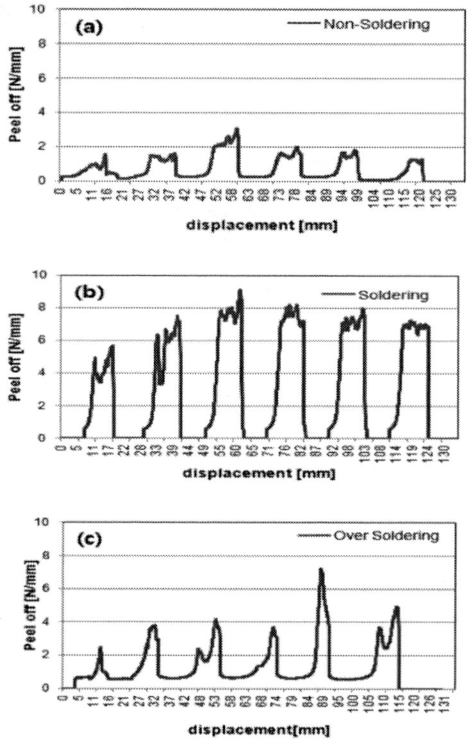

Fig 1. Peel off test of each soldering condition
(a)Non-soldering, (b)Soldering and (c)Over-soldering

2.3 Reliability test

Generally, reliability test according to IEC 61215 is performed such as Thermal cycle (TC), Damp heat (DH), Mechanical load test (MLT).

Especially, confirmed solder joint degradation and cell crack by TC, HF test. Therefore, we tested the TC 300, HF 40 in exceeded by the reliability test standard TC 200 and HF 10 in the IEC 61215.

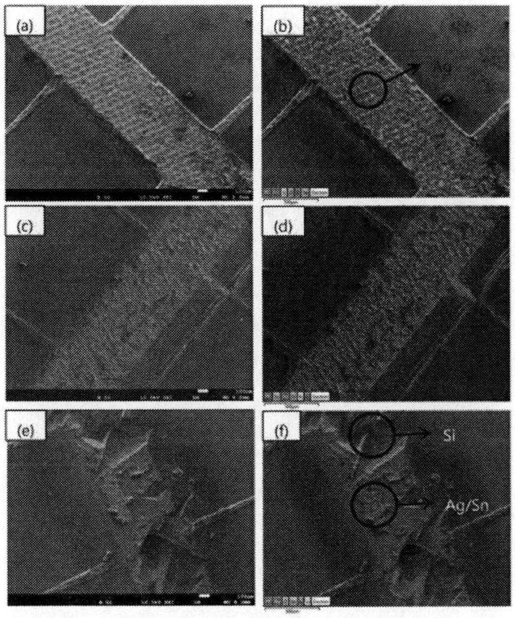

Fig 2. SEM & EDX image of the cell surface after 180° peel off test : (a),(b) Non soldering, (c),(d) Soldering and (e),(f) Over soldering

3 RESULT

The PV modules manufactured for each soldering condition are tested HF 10/20/30/40. At the end of HF 10 cycle, the output of the module and the Electronic luminescence (EL) were measured to confirm the power degradation and cell crack, and observed to change in the junction between cell and interconnector.

3.1 Power and EL

We measured the output and EL of the modules after HF test. The output of the module for each soldering condition dropped sharply after each HF 10/20/30 test.

As the result, each soldering conditions were not effect output of module.

However Vmp and Rsh have different results. As the HF test progressed, the Vmp of the Non & Over soldering modules gradually decreased and the Rsh also doubled compared to the soldering module.

This can also be confirmed by the EL test. The EL image of the module measured after HF 10 termination shows little cell crack than the initial module, but checking the EL image of the module after HF 20, it can be confirmed that the cell crack has increased considerably. Due to the different thermal expansion coefficients of interconnector and cell, it is expected that cracks will occur at the joint because they cannot withstand a large temperature difference of + 80℃ ~ - 40℃.

3.2 Cross section

We checked the junction of the interconnector and cell of each module by differentiating soldering condition.

It was difficult to identify the cross-section of the module by making sample with AR coated tempered glass. But, according to the soldering condition, the bonding area between electrode of the cell and solder was confirmed, and it was confirmed that the Intermatallic compound (IMC) between Ag and Sn/Pb was formed through the thickness of solder.

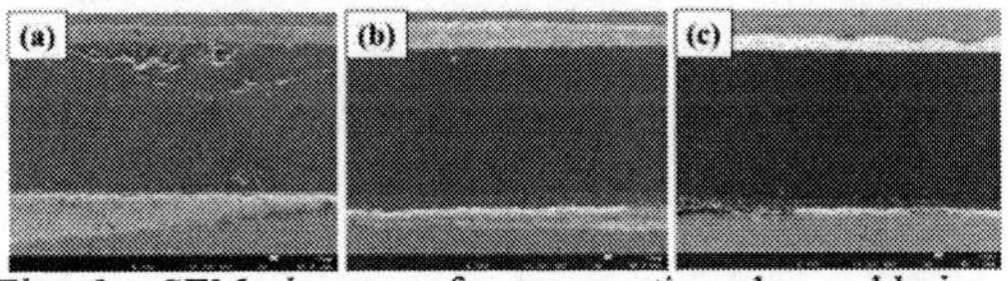

Fig 3. SEM image of cross-section by soldering condition : (a) Non-soldering, (b) Soldering and (c) Over-soldering

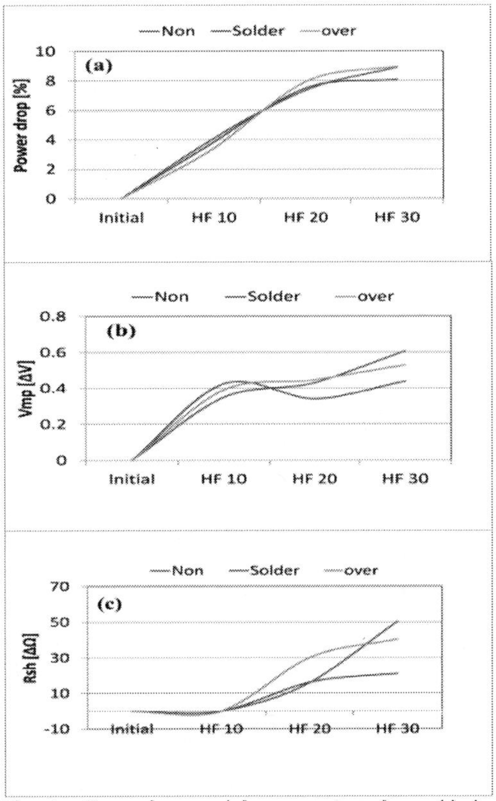

Fig 4. Comparing module parameters by soldering condition : (a) Output, (b) Vmp and (c) Rsh

4 CONCLUSION

In order to establish the soldering conditions, we manufactured the module under the conditions of Non, Well, Over soldering between the interconnector and the cell, and confirmed the reliability through HF test.

In this experiment, the soldering condition between

cells and interconnector may not have a significant problem in output drop, and it is necessary to carry out more various test than TC, HF, MLT in order to accurately expect to module reliability and life.

It is not known what factors are causing the durability of the module to failure, because modules are manufactured by sealing various materials (Glass, Cell, EVA, Backsheet, Al frame, etc.).

So, additional test is expected to be conducted to test the long-term reliability of the Photovoltaic module.

5 Reference

[1] R. Lathrop, K. Pfuke , Proceedings 26th European Union Photovoltaic Solar Energy Conference (2011).

[2] M. Vazquez, I. Roy-stolle, Progress in Photovoltaics. (2008) 10.1002.

[3] A. Zielnik, Validating Photovoltaic Module Durability Tests (2013).

[4] S. J, Kim, J. Y, Choi, etc, KSES, Vol. 31 (2011) 303.

ASSESSING THE IMPACT OF BROKEN AND DEFECTIVE INTERCONNECTION RIBBONS ON THE ELECTRICAL PERFORMANCE OF CRYSTALLINE SILICON PHOTOVOLTAIC MODULES

Eleonora Annigoni[1], Alessandro Virtuani[1], Fanny Sculati-Meillaud[1], Antonin Faes[2], Matthieu Despeisse[2], Christophe Ballif[1,2]
[1] École Polytechnique Fédérale de Lausanne (EPFL), Institute of Microengineering (IMT), Photovoltaics and Thin Film Electronics Laboratory (PVLAB), Rue de la Maladière 71b, 2000 Neuchâtel, Switzerland
[2] CSEM, PV-center, Jaquet-Droz 1, 2000 Neuchâtel, Switzerland

ABSTRACT: Fatigue of electrical interconnections is a frequently observed failure mode for photovoltaic (PV) modules. Fatigue is mainly due to the cells displacement that can occur due to temperature changes or mechanical loads to which the module is exposed like wind pressure or snow load. Other factors such as presence of acid in the encapsulant also affect the reliability of interconnections. In this work, we assess how the module performance is affected when one or more ribbons are cut or disconnected between cells. For this purpose, we manufacture a module composed of eighteen mono crystalline silicon cells connected in series by means of three interconnection ribbons and encapsulated in a glass/backsheet construction. An ethylene tetrafluoroethylene (ETFE) backsheet is employed as it can be easily opened without the risk of damaging the adjacent cells and allows cutting the selected ribbons. We then cut one or more ribbons, measuring the change in the module's electrical parameters. We observe that the power loss strongly depends on the ribbon's position with respect to both the cell (external vs central ribbon) and the string (string start or between two cells). For instance, if the external ribbon is disconnected from every cell, the maximum power (P_{max}) variation is -10.24%. If, instead, the disconnections are in the central ribbon the P_{max} changes by only -5.91%. The power variation is -35.09% when two ribbons are disconnected at every cell.
Keywords: Ribbons, Interconnections, Degradation, PV Module, Reliability

1 INTRODUCTION

Fatigue of interconnectors is one of the main failure modes for photovoltaic (PV) modules [1]. It is a consequence of the cells displacement that can occur due to temperature changes or mechanical loads to which the module is exposed. Other factors affecting the reliability of interconnections are related to moisture ingress in the sandwich and to presence of acids due to the photo-thermal decomposition of the encapsulant. Fatigue accumulation can result in degradation of the solder joints or in broken ribbons, as frequently reported from field inspections (see e.g. Ref. [2] and [3]).

Reliability of ribbons has gained more and more attention in the last few years. In Ref. [4] the authors investigate the fatigue behavior of copper ribbons resulting from temperature cycles, and obtain a relationship between the temperature cycle's amplitude and the number of cycles to failure (i.e. to ribbon's breakage). Finite-Elements Method (FEM) simulations are employed in [5] to reproduce the cell displacement and the stress distribution within the module materials resulting from thermal variations. In [6], Bosco proposes a climate-dependent number of thermal cycles (according to IEC 61215) as a stress test for modules in order to obtain an equivalent amount of solder fatigue damage as simulated by FEM. In [7], the authors simulate with FEM models the plastic strain energy accumulated by the ribbons during cyclic mechanical loading due to wind.

However, to our knowledge, there is no experimental study directly assessing the actual impact of broken interconnections on the electrical parameters degradation. In this work, we quantify the actual impact on module's performance when one or more ribbons connecting adjacent cells are disconnected.

2 INTERCONNECTION FAILURES

Thermal variations generate stresses within the module's structure because of the different coefficient of thermal expansion of the materials used to realize the sandwich. In particular, thermal variations can cause a displacement of solar cells in the module, as can similarly occur when the module is exposed to mechanical loads (due to wind or snow loads).

The fatigue mechanism in interconnect ribbons takes different forms. On the one hand, the stress induced by thermo-mechanical loads on the interface between the ribbons and the cell causes fatigue on the solder joints, which can, in the long-term, turn into cracks and ultimately failure (i.e. partial or complete detachment) of the solder joints. On the other hand, ribbons can break as result of the accumulated fatigue due to the cell displacement. One should also mention that solder bond failures potentially lead to hot spots [8]. In addition to these degradation modes, the presence of moisture and acids in the encapsulant (e.g. the acetic acid generated by the photo-thermal degradation of EVA) can result in the corrosion of the metal connections (i.e. the copper core and the tin coating of the ribbons, as well as of the solder bonds or the cell fingers) [9].

Failures in the internal circuitry were classified as having a major effect on module power in a recent extensive review of PV field reliability [10]. The experimental work presented in this paper aims at investigating how the module performance is affected when one or more ribbons are broken causing a disconnection in the electrical circuit.

3 EXPERIMENTAL WORK

For the experimental work, we manufacture at PVLAB a module with eighteen mono crystalline silicon (c-Si) solar cells with three busbars, as depicted in Figure 1. The cells are connected in series by means of commercially-available 1.70 x 0.17 mm copper ribbons coated with $Sn_{62}Pb_{36}Ag_2$. The encapsulation is done with a glass/backsheet construction with ethylene-vinyl acetate

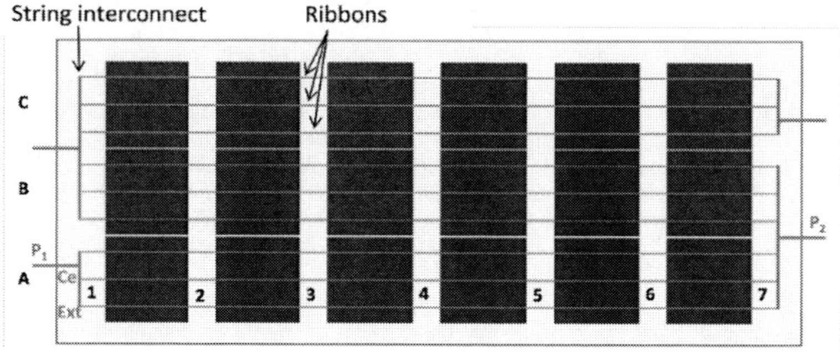

Figure 1: Sketch of PVLAB's module used in this work: 18 mono c-Si cells with 3 busbars are connected in series and encapsulated in a glass/backsheet construction. The 50 µm thick ETFE backsheet can be easily opened to cut the ribbons. Each line of six cells (A, B, and C) is considered as a separate string and used to analyze a specific configuration of disconnected ribbons. The IV and EL measurements are made at string level: string A, for instance, is measured between points P_1 and P_2. For each string we distinguish between the central ribbon (Ce) and an external one (Ext). The points or Positions where the ribbons are cut are denoted by numbers (1,…,7).

as encapsulant material. As backsheet we employ a 50 µm thick ethylene tetrafluoroethylene (ETFE) sheet. Such thin layer can be easily removed allowing us to cut the selected ribbon without the risk of damaging the adjacent cells.

After an initial characterization of the module, we cut the interconnection ribbons from one or more cells using an iterative process. When needed, disconnected ribbons can be repaired by soldering before proceeding with the following step. In order to facilitate the experimental work we consider the module as three separate strings (A, B and C in Figure 1) composed of six cells each. Each string is employed to analyze a specific configuration of the disconnected ribbons. For each cell, we distinguish between the central ribbon (Ce) and one of the two external ribbons (Ext), its side being irrelevant out of symmetry reasons. The locations, or Positions, where the ribbons are cut are labeled by progressive numbers (1 to 7, as shown in Figure 1).

The impact on the electrical performance is analyzed at string level: the string in consideration is characterized initially and after each step by means of current-voltage (IV) curve measurements using a LED-halogen based sun simulator that replicates the AM1.5 spectrum. In the case of string A, for instance, the IV measurements are performed between points P_1 and P_2 in Figure 1. The string is also characterized at each step by means of electroluminescence (EL), where a dc current (corresponding to the module' short-circuit current) is supplied in order to stimulate radiative recombination of carriers and consequent light emission. EL images allow observing the path that the light-generated carriers follow in presence of disconnected cells.

3.1 One disconnected ribbon per cell

We compare here the impact on the module's electrical performance of broken external (Ext.) vs central (Ce.) ribbon. We use string A to analyze the first configuration and string C for the second. In Figure 2 we plot the change in fill factor (FF) and in maximum power (P_{max}) of the full string after removing progressively one ribbon at a time. Results show that the loss in P_{max} follows, as expected, the same trend as the loss in FF. We observe a more pronounced degradation in the string performance when the external ribbon, rather than the central ribbon, is disconnected. Such difference is reflected in the EL images (Figure 3), where in the first case (top image) a larger part of the cell is dark, i.e. not contributing to carrier collection, than in the second case (bottom image). As a general observation, denoting by *l* the distance between the central ribbon and an external one, in the first case the maximum distance for the light-generated carriers from the external part of the cell to the closest connected ribbon is $l+l/2$. In the second case, instead, the maximum distance that carriers have to cover (from the center of the cell) through the metallic fingers to the closest connected ribbon is only $l/2$.

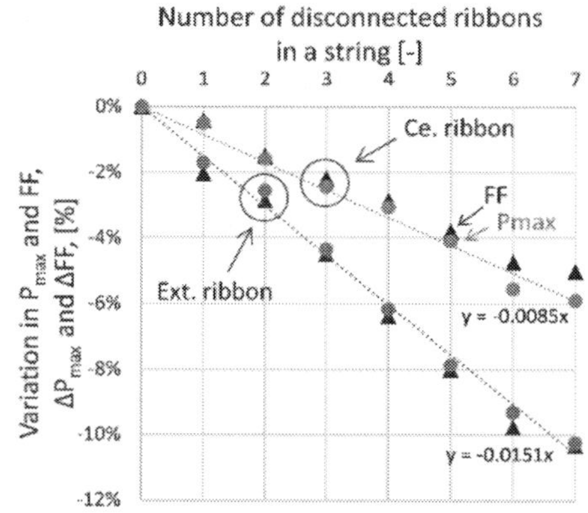

Figure 2: Change in FF and in P_{max} with respect to the initial values (all ribbons connected) as function of the number of cumulative cuts. For instance, two disconnected points means that the ribbon was disconnected in both Positions 1 and 2 (see Figure 1). Dotted lines are linear fits of the P_{max} values.

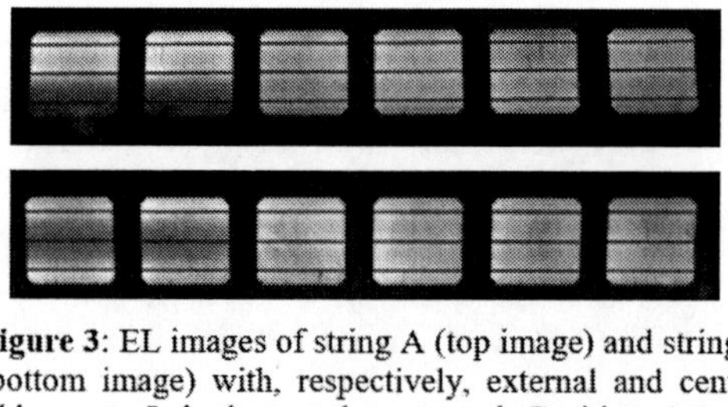

Figure 3: EL images of string A (top image) and string C (bottom image) with, respectively, external and central ribbon cuts. In both cases the cuts are in Positions 1 and 2.

3.2 Two disconnected ribbons per cell

After having disconnected the external ribbon for all cells in a string, we continue removing a second ribbon (the central one) from each cell. This is done iteratively for each cell, as shown in Figure 4. In this case, each cut in the central ribbon reduces the power by 3.88%, while if the external ribbon is still connected (as in case Ce. in Figure 2) the power reduction was only 0.85% (slope obtained if fitting the points in Figure 2 with a line).

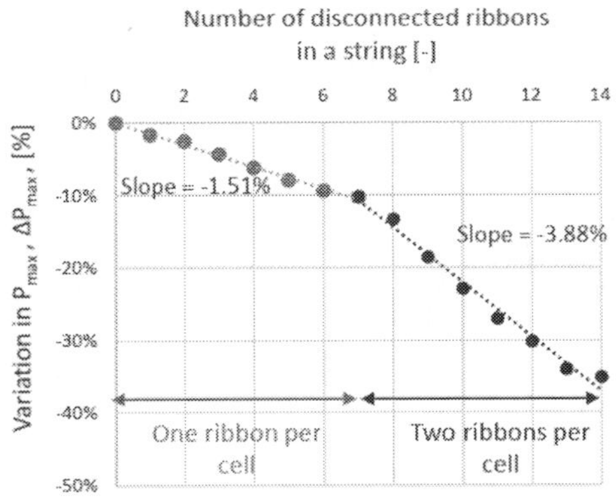

Figure 4: Change in P_{max} with respect to the initial values (i.e. all ribbons connected) of the Pmax of string A (composed of 6 cells) as function of the number of cumulative cuts. First the Ext. ribbon is iteratively cut until all the seven points (1 to 7) are disconnected (i.e. one ribbon per cell). Then a second ribbon per cell (the Ce. one) is iteratively removed (8 to 14), one cell at a time.

The relative variation of the series resistance (R_s) for the discussed configurations are shown in Figure 5. Such values refer to the series resistance of the full string in consideration (and not to that of a single solar cell). They were obtained from the one-sun IV curve by means of a linear fit in proximity of the open-circuit voltage (V_{oc}), which provides the slope 1/R. This value is not the exact absolute value of Rs, but it is related to it. Fitting the IV curve with the one-diode equation, indeed, resulted to be an inaccurate method to extract the R_s. This can be related to what observed by Bowden in [11]: when only part of the cell is affected by a high R_s (inhomogeneous distribution) the effects of this high R_s on the IV curve may be erroneously attributed to other electrical parameters of the curve instead, for instance to a low shunt resistance. Our case, with the disconnections affecting only some specific parts of the cells, falls within those described by Bowden and would therefore require specific measuring procedures and a dedicated R_s extraction method. Nonetheless, here we are mainly interested in observing a trend for the variation of R_s, rather than measuring absolute values, and a linear fit in proximity of V_{oc} gives us more robust and reproducible results.

We observe in Figure 5 a higher increase in R_s when disconnections are in the external ribbon than in the central one, which justifies the corresponding trend for the loss in FF and the P_{max}, seen in Figure 2. When the external ribbon is already completely disconnected for all cells in a string, each cut in the central ribbon increases the R_s by around 13%, with an impact on Rs three times higher compared to when the external ribbon is connected.

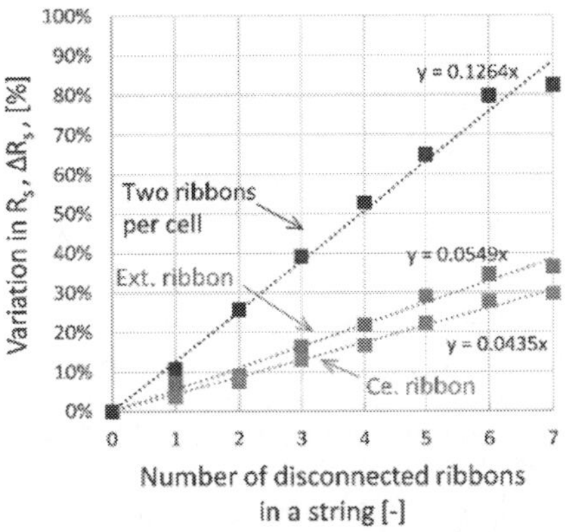

Figure 5: Change in string's series resistance (R_s) as a function of the number of cumulative cuts. Orange and green points refer to the configurations of Figure 2 (one disconnected ribbon per cell: Ext. ribbon versus Ce. ribbon). The blue points are for the case in which a second ribbon is iteratively cut (corresponding to the configurations in points 8 to 14 in Figure 4). Dotted lines are linear fits.

4 CONCLUSIONS

We quantify experimentally the effect of broken interconnection ribbons on the power performance of a PV module. Strings of six cells each were considered in order to analyze different configurations of disconnected ribbons. Cells have three busbars and are serially connected one another.

We observe that the power loss strongly depends on the ribbon's position with respect to the cell (external vs central ribbon) as well as the string (string start or between two cells). If the external ribbon is disconnected from every cell, the power variation for the string with respect to the initial case (i.e. all ribbons connected) is -10.24%. If, instead, the disconnections are in the central ribbon the power variation is only -5.91%. Finally, power changes by -35.09% when two ribbons are disconnected at every cell.

Complete rupture of interconnection ribbons is a particularly harsh event and not very frequent in real operating conditions. More often problems with interconnections rather involve a partial de-soldering of the ribbon from the busbar or from the string interconnect, or corrosion of the metal connections. However, here we consider the extreme case of ribbon breakage and present a first quantitative assessment of the effects of such failure mode on module's power performance.

5 ACKNOWLEDGMENTS

We gratefully acknowledge the financial support of EOS Holding. We are grateful to Jonathan Champliaud and Jacques Levrat for fruitful discussions, and to Joël Sunier for support in the experimental work.

REFERENCES

[1] IEA, "Review of Failures of Photovoltaic Modules - Report IEA-PVPS T13-01:2014," 2014.

[2] K. Kato, ""PVRessQ!" - A Research Activity on Reliability of PV System from an user's viewpoint in Japan," in *SPIE*, 2011.

[3] D. DeGraaf, R. Lacerda, Z. Campeau and Sunpower Corp, "Degradation Mechanisms in Si Module Technologies Observed in the Field; Their Analysis and Statistics," in *NREL 2011 Photovoltaic Module Reliability Workshop*, Golden, Colorado, 2011.

[4] R. Meier, S. Kraemer, S. Wiese, K.-J. Wolter and J. Bagdahn, "Reliability of copper-ribbons in PV modules under thermo-mechanical loading," 2010.

[5] U. Eitner, "Thermomechanics of Photovoltaic Modules," *PhD Thesis, Martin-Luther-Universität Halle-Wittenberg*, 2011.

[6] N. Bosco, T. J. Silverman and S. Kurtz, "Climate specific thermomechanical fatigue of flat plate photovoltaic module solder joints," *Microelectronics Reliability*, no. 62, pp. 124-129, 2016.

[7] M. Pander, S. Dietrich, R. Meier and M. Ebert, "Fatigue Analysis of Solar Cell Interconnectors Due to Cyclic Mechanical Loading," in *EU PVSEC*, Munich, 2016.

[8] J. Muñoz , E. Lorenzo, F. Martinez-Moreno, L. Marroyo and M. Garcia, "An investigation into hot spots in two large grid-connected PV plants," *Progress in Photovoltaics: Research and applications*, vol. 16, no. 8, pp. 693-701, 2008.

[9] IEA, "Assessment of Photovoltaic Module Failures in the Field," Report IEA-PVPS T13-09:2017.

[10] D. C. Jordan, T. Silverman, J., J. H. Wohlgemuth, S. R. Kurtz and K. T. VanSant, "Photovoltaic failure and degradation modes," *Progress in Photovoltaics: Research and Applications*, vol. 25, no. 4, pp. 318-326, 2017.

[11] S. Bowden and A. Rohatgi, "Rapid and Accurate Determination of Series Resistance and Fill Factor Losses in Industrial Silicon Solar Cells," in *17th European Photovoltaic Solar Energy Conference*, Munich (Germany), 2001.

[12] K. Kato, "PVRessQ! PV Module Failures Observed in the Field," Golden (US), 2012.

ADVANCES IN THE DEVELOPMENT OF A NOVEL MODULE DESIGN BASED ON ELECTRICAL CONDUCTIVE ADHESIVE GLUE FOR CONTACTING HIGHLY EFFICIENT N-TYPE SOLAR CELLS WITH PVD AL BACK CONTACT

*Enrique Cabrera[1], Andreas Schneider[1], Thomas Buck[1], Zih-Wei Peng[1], Thomas Fischer[2], Radovan Kopecek[1]
[1] International Solar Energy Research Center -ISC- Konstanz, Rudolf-Diesel-Str. 15, 78467 Konstanz, Germany
[2] Teamtechnik Maschinen und Anlagen GmbH, Planckstrasse 40, 71691 Freiberg, Germany
*Corresponding author: Phone: +49 7531 3618356; Fax: +49 7531 3618311; Email: enrique.cabrera@isc-konstanz.de

ABSTRACT: As to now no cost-effective approaches for the application of physical vapor deposition of an Aluminum (PVD-Al) layer onto a c-Si solar cell and subsequent module integration has been successfully developed at industrial production level. In the present study, n-type Si solar cells based on a promising industrial process for PVD-Al contacts were used to investigate its integration into the module and evaluate modules long-term stability. The interconnection technique based on electrical conductive adhesives (ECA) was mechanically evaluated by peeling tests and accelerated climatic indoor tests (TC200, DH1000). Peeling test results and cell to module (CTM) losses show that the ECA used is very well able to contact the PVD-Al back coming along with a contact resistance comparable to soldered contacts. No damages after lamination were detected by means of electroluminescence (EL). Preliminary results show losses in Pmpp lower than 1% in average after 1000 hrs of damp heat (DH) test. Contrary after 200 cycles of thermo-cycling (TC) Pmpp losses are in average higher than 5%. Even though no cracks and metallization damages were observed, galvanic corrosion between the ECA and the Al layer was detected after climatic testing which seems to be responsible for the higher losses. Corrosion was suppressed by the use of glass on front and rear side and POE as encapsulant, which provided a better water barrier against humidity.

Keywords: n-type cell, Al contact by PVD, PV Module, Accelerated Environmental Indoor Test

1 INTRODUCTION

High efficiency solar cells exhibiting a full area evaporated Al rear side are currently being developed by many research institutes and cell suppliers world-wide. One reason for this is the very high cells efficiency potential of > 21% [1]. Key are on one hand the reduction and simplification of high efficiency cell process steps to keep costs (hence $/Wp) low and on other hand the development of a reliable module interconnection technique. To keep the cell and module process simple it is desirable to process cells with full Al rear contact. For this new module concepts are required allowing for a direct interconnection of the Al layer and ribbon material. The standard industrial module concept is not applicable since soldering cannot directly be applied on Al surfaces.

Alternative interconnection technologies to soldering are required. Due to the lead-ban in electronic industry, an attractive technique is the use of ECA which offers a low temperature interconnection technique [2]. The application of ECA further reduce thermal mechanical stress during bonding hence decrease the micro crack generation rate, which are not always be detectable. Nevertheless several process step modifications are required to realize a high production throughput at low costs and guarantee a lifetime of 25, or even 30 years, with typical degradation rates of 0.3–0.5%/year for STC power valid for modules based on crystalline cells.

In the presented study, we show the feasibility of an alternative cell interconnection by means of ECA for cells with full Al rear contact. An analysis of CTM losses and climatic indoors tests (DH1000, TC200) were performed to prove the longevity of this approach.

2 EXPERIMENT

Modules based on 6 inch n-type Cz-Si PERC cells, so called MoSoN [3] with PVD-Al back contacts in combination with a 3.2 mm or 2.0 mm glass as front cover, POE or EVA as encapsulant, ribbons connected by means of ECA and a transparent backsheet (TBS) or 2.0 mm glass as rear cover were fabricated. In addition, a soldered reference group based on 6 inch p-type Cz-Si cells with Al/Ag pads using the same module materials was processed. The ribbon soldering process on the front and rear side was performed manually and automatically by a stringer machine using the same type of ribbon. The pressure and temperature for the lamination process was the same for all 4 groups, containing 4 to 6 modules per group, see Table I.

Table I: Experimental setup

Group	Interconnection	Encapsulant	Cover [front/rear]
MoSoN 1	ECA	POE	2 mm glass/ 2 mm glass
MoSoN 2	ECA	EVA	3.2 mm glass /TBS
Ref. 1	Soldering	POE	2 mm glass/ 2 mm glass
Ref 2	Soldering	EVA	3.2 mm glass/ TBS

For characterization, EL measurements were performed to control the mechanical and electrical integrity of the contact between the individual parts after lamination. Additionally, EL measurements accompanied the environmental indoor tests, applying the same current and exposure time which allowed to obtain comparable EL pictures. IV data was measured by a class AAA solar simulator with an illumination area of 22×22 cm^2 at STC conditions before (on cell level) and after lamination. This procedure allowed to determine the CTM losses and to quantify the environmental indoor test results.

3 RESULTS

3.1 Cell contacting and peel strength

ECA was deposited onto the MoSoN PVD AL rear contacts by using an automatic ECA dispensing system [4]. Subsequently, 2.5 mm wide and super soft ribbons with Sn60Pb40 coating were used as cell connector (dimension: 1.5 x 0.2 mm^2 - Rp0.2 = 50 MPa), ribbon was initially stretched. Finally, cell samples with ribbon and ECA were placed in a muffle furnace for curing at 150 °C for 10 minutes.

Table II shows the average peel strength of ribbon placed on the rear side, based on 2 samples. The peeling test allows to investigate the mechanical adhesion between the cells rear and the ribbon. The test was carried out at an angle of 180 ° in respect to the pulling direction of the ribbons and at a speed of 500 mm per minute. The peel force was measured in N/mm.

A very homogeneous peeling behavior over the ribbon length was observed for MoSoN cells during the peeling test, reaching 1.2 N/mm in average, while the reference soldered group exceeded 2.25 N/mm respectively.

Table II: Peeling test results

Cell	Interconnection	Ribbon location	Force [N/mm]
MoSoN	ECA	Left	1.32±0.20
MoSoN	ECA	Center	1.24±0.22
MoSoN	ECA	Right	1.03±0.45
Reference	Soldering	Left	2.25±1.30
Reference	Soldering	Center	2.32±1.23
Reference	Soldering	Right	2.19±1.36

3.2 Cell to Modules (CTM) losses

To investigate the electrical losses of the rear side connection after lamination CTM losses were evaluated. Figure 1 plots the CTM losses (y-axis) versus the 4 investigated groups (x-axis). The values correspond to the mean and standard deviation of IV data for 6 modules per group. High losses in Isc are noticeable for the MoSoN modules compared to the references. These high losses indicate that on one hand the front reflection of the cell is extremely low for the PERC cell type and thus hardly any encapsulation gains are achieved. Furthermore, the IQE of the PERC cells show a high sensitivity in the short wavelength range which is blocked by the standard EVA (no low UV cut-off material was used so far).

Lower FF and Pmpp losses for the glass-TBS configuration compared to the glass-glass configuration were found. The latter one contains solar cells with higher conversion efficiency, which provide higher sensitivity to CTM losses (solar cells with higher absolute FF).

In addition, no cracks and metallization damage were observed by EL after lamination for the MoSoN groups. This clearly shows that the chosen ECA is very well able to contact the aluminum back side with a small contact resistance.

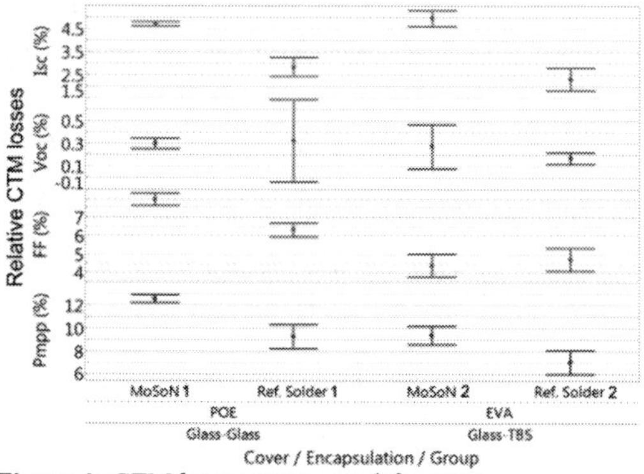

Figure 1: CTM losses versus module group

3.3 Climatic indoor tests

Accelerated environmental indoor tests were performed to prove the reliability of this alternative interconnection technique.

3.3 Thermo-cycling (TC) test

In order to investigate the effect of thermal stress on the module material compound, the climatic chamber was operated between a temperature range of -40°C and +85°C. Figure 2 shows the relative Isc, Voc, FF and Pmpp losses (y-axis) at initial state, after 50 cycles, after 100 cycles and after 200 cycles (x-axis). For this investigation only groups MoSoN 2 and Ref. solder 2 based on glass-TBS with EVA were used. The values correspond to mean value plus standard deviation for 6 modules per group.

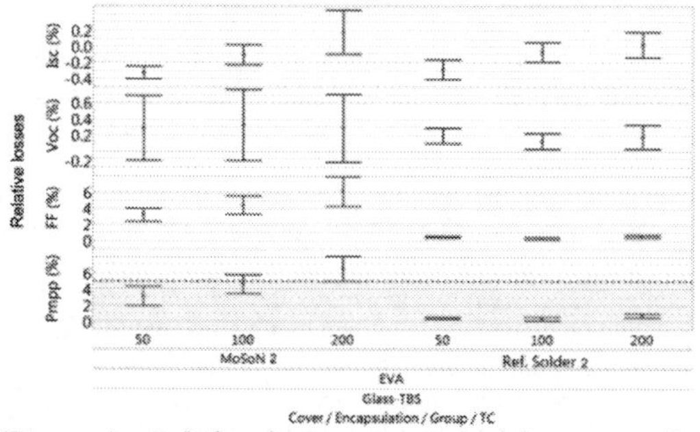

Figure 2: Relative losses versus module groups for TC200.

After TC50, a 3% FF and Pmpp loss for MoSoN 2 cell modules and almost no losses for the soldered reference groups were observed. Almost similar low losses in Isc and Voc for both groups were experienced. After TC 100, losses of the reference group remain at low level whereas the losses in FF and Pmpp of MoSoN cell module group increase to 5%. When increasing the number of cycles to 200, the losses in FF and Pmpp reach more than 5% as a result of the extreme thermo-mechanical impact of the test, while no significant change for the soldered references were detected. Thus, the reference samples pass the TC200 test while the MoSoN 2 samples fail. Figure 3 shows the EL pictures before and after TC 200 for the MoSoN 2 group. A very homogeneous degradation can be observed. In addition, Figure 4 shows a picture of the cell rear side of the MoSoN 2 cell after TC200. It appears that the losses detected on MoSoN 2 modules are due to galvanic corrosion occurring at the contact area between ribbon coating, ECA and cells Al layer. Since the galvanic potential of Al (-1.66 eV) is much lower than the potential of Ag (0.8 eV), which is contained inside the ECA, this reaction seems plausible.

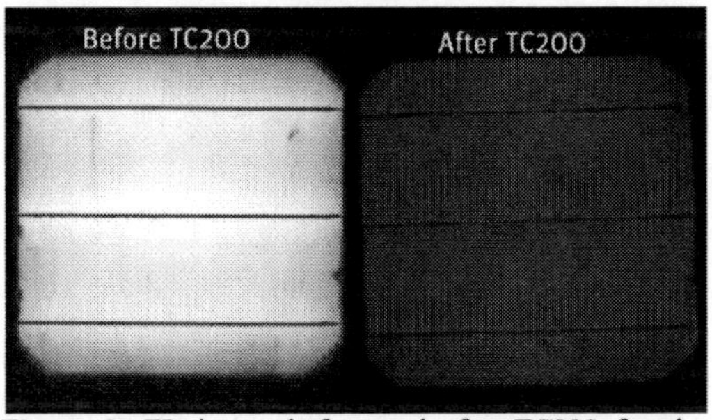

Figure 3: EL image before and after TC200 for the MoSoN 2 group.

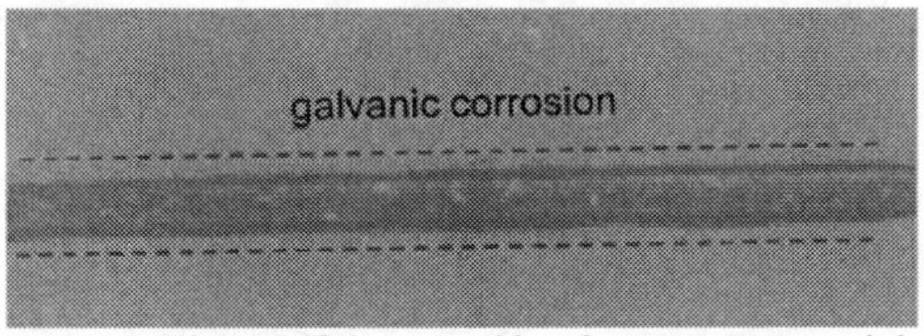

Figure 4: Picture of the rear side of a MoSoN 2 module after TC200. Galvanic corrosion between the ECA and Al contact was detected.

3.3.2 Damp heat (DH) test

For the damp heat test the climatic chamber was operated at a temperature of 85°C and a relative humidity of 85 %RH according to IEC 61215. In Figure 5 the relative losses in Isc, Voc, FF and Pmpp (y-axis) versus the time exposure in hrs of the damp heat testing (x-axis) are plotted. The values correspond to mean values plus standard deviation for 4 modules per group

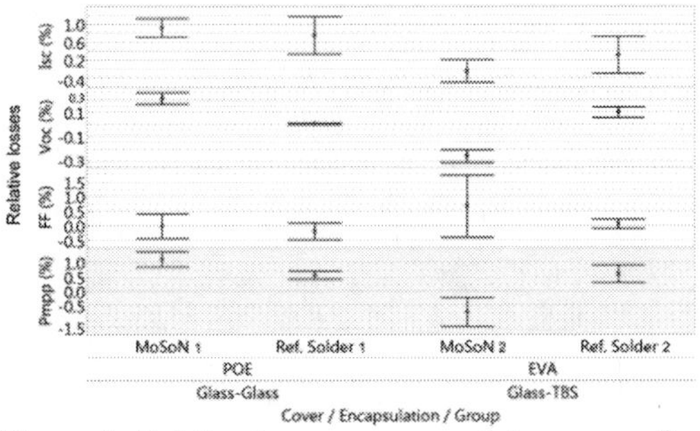

Figure 5: Relative losses versus module groups after DH1000.

From the beginning of the test up to 1000 hrs of DH exposure no significant power losses are determined, staying below 1% for all investigated group. Thus, the modules pass the DH1000 test according to IEC 61215 showing a power degradation below 5 % after 1000 hours. Figure 6 shows an EL image of the module before and after DH1000. It can be observed that the cells electrical integrity is fully given after the test. Furthermore, no delamination was detected. However, on the MoSoN 2 cell module group with glass-TBS and EVA configuration, see Figure 7(a), galvanic corrosion was detected which was also after TC200 observed and eliminated by the use of a glass-glass configuration and POE as encapsulant (MoSoN 1 group), see Figure 7(b). This material configuration offers a very stable composition against humidity penetration.

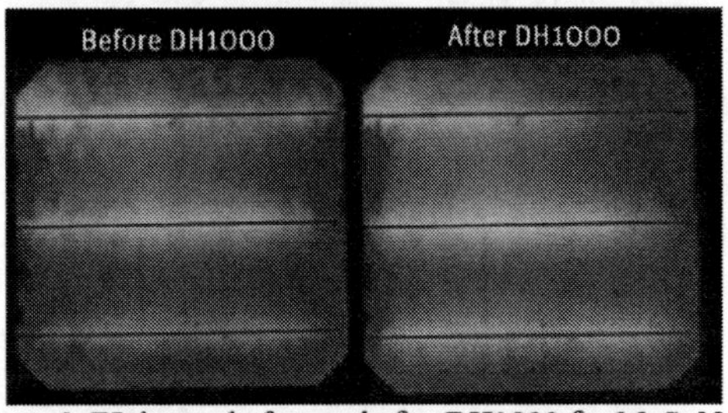

Figure 6: EL image before and after DH1000 for MoSoN 2 group.

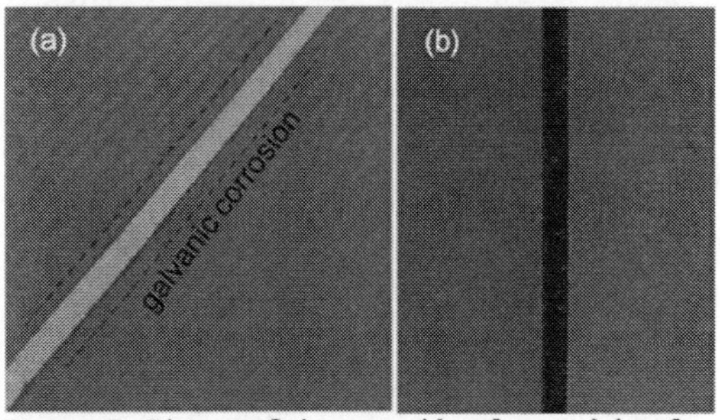

Figure 7: Picture of the rear side of a module after DH10000. (a) MoSoN 2: Galvanic corrosion between the ECA and Al contact was detected with the glass-TBS configuration. (b) MoSoN 1: no corrosion for module with glass-glass and POE configuration.

4 CONCLUSION

In the present study, the potential and reliability of a module interconnection technique based on MoSoN solar cells with PVD-Al back contacted by means of ECA were investigated. Preliminary results of electrically contacted MoSoN cells indicate that ECA can successfully contact the Al rear side reaching a low contact resistance and successfully passing the high humidity and temperature DH 1000 test. However, the TC 200 test showed larger than 5% losses in Pmpp. One reason is galvanic corrosion observed underneath the ribbons. The same corrosion effect was detected after DH1000 on modules with a glass-TBS and EVA material composition. Changing the materials to glass-glass and POE no galvanic corrosion was observed.

5 ACKNOWLEDGEMENTS

This work was funded by the BMWi within the IdeAl (Industrielles PVD-Al für hocheffiziente kristalline Si-Solarzellen und Module) project under contract number 02E2-41V7142.

6 REFERENCES

[1] Bernhard Klöter et al., "Current status of high-efficiency Q.ANTUM technology at Hanwha QCELLS", 39th IEEE Photovoltaic Specialists Conference (PVSC), 1246-1248 (2013). DOI: 10.1109/PVSC.2013.6744366.

[2] A. Schneider et al., "Comprehensive study of material dependency for silver based conductive glues", Energy Procedia 55, 509-518 (2014).DOI: 10.1016/j.egypro.2014.08.017.

[3] Zih-Wei Peng et al., "Investigation of Laser Contact Opening Pitch Affecting the Voids Formation on n-Type Rear Emitter Silicon Solar Cells", will be presented at 7th International Conference on Crystalline Silicon Photovoltaics, Freiburg, Germany, April 3-5 (2017).

[4] T. Meßmer et al., "Low cost semi automated assembly unit for small size back contact modules and low cost interconnection approach", Energy Procedia 98, 98-105 (2016). DOI: 10.1016/j.egypro.2016.10.085.

ADVANCES IN THE DEVELOPMENT OF A NOVEL MODULE CONCEPT BASED ON CONDUCTIVE STRUCTURES IN THE ENCAPSULATION MATERIAL FOR CONTACTING HIGHLY EFFICIENT N-TYPE BACK-CONTACT SOLAR CELLS

*Enrique Cabrera[1], Andreas Schneider[1], Jorge Rabanal[1], Diego Thaller[1], Lejo J. Koduvelikulathu[1], Benjamin Pérez[2], Blas Puerto[3], Paula Sánchez-Friera[3], Tina Brown[4], David Greenhill[4], Román Merino[2], Rene Camblor[3], Robin Pittson[4], Andreas Halm[1]
[1]International Solar Energy Research Center - ISC- Konstanz, Rudolf-Diesel-Str. 15, 78467 Konstanz, Germany
[2] Specialized Technology Resources –STRE, Parque Tec. de Asturias, prla. 36, 33428 – Llanera, Asturias, Spain
[3] Fundación PRODINTEC, Parque Científico Tec. de Gijon. Av. Jardin Botanico 1345. 33203 Gijon, Asturias, Spain
[4] Gwent Electronic Materials, Monmouth House, Mamhilad Park Pontypool, NP4 0HZ, United Kingdom
*Corresponding author: Phone: +49 7531 3618356; Fax: +49 7531 3618311; Email: enrique.cabrera@isc-konstanz.de

ABSTRACT: The new functional encapsulation layer which is presented in this study, known as conduction path on encapsulant (CPOE), has the potential to interconnect high efficiency IBC solar cells. In our study, single cell demonstrator modules were assembled reaching an efficiency of 19.3% with relative cell to module (CTM) losses below 2%. However, this concept so far requires reduction in the cell area (e.g. one-eighth cell size module design) since significant losses in FF were observed due to the high resistive losses in the CPOE. Accelerated environmental indoor tests reveal that the module based on CPOE technology passes the precondition ultra-violet (UV) degradation and damp heat (DH) exposure test for 1250 hrs. without any signs of degradation in the module packaging materials or electrically power losses. However, thermo-cycling for 200 cycles reveal relative Pmpp losses higher than 5%. Additionally, power degradation after applying a PID test at +1000V at 85°C was observed. Finally, outdoor energy yield studies of the CPOE modules compared with Zebra and p-type cell modules with soldered ribbons as references reveal higher energy yield (Ya) and performance ratio (PR) of CPOE and Zebra reference than p-type full cell module. In addition during cloudy days, CPOE modules show better energy performance when compared to Zebra reference modules.

Keywords: Bifacial IBC Cell, Alternative Module Interconnection, Accelerated Environmental Indoor Test, PID, Energy Yield Studies

1 INTRODUCTION

High efficiency IBC cells are currently being developed by many research institutes and cell suppliers world-wide. One reason for this is the very high cell efficiency potential of >24% and the bifacial gain of some of these technologies [1]. To keep industrial processing as simple as possible it is desirable to process rear contact cells with busbars for interconnection. This implies the application of an alternative module concept since standard soldering would lead to excessive cell warping due to the large number of busbars deposited only on one cell side. Despite efforts in process automation, thin wafer handling optimization and the use of low-temperature solder alloys, cell interconnection by soldering is still a very critical step in module manufacturing which often jeopardizes modules' longevity [2].

Alternative interconnection schemes for rear contact cells have been developed and presented such as those proposed by Eurotron and Formula E. Contrary to existing approaches, the scope of the InGrid project focuses on track formation by conductive ink printing directly onto the encapsulant foil, called CPOE, potentially leading to a very low cost process solution. Even though EVA substrates are well-known encapsulation product, we employ polyolefin elastomers (POE) due to their improved optical and electrical properties and to circumvent some of the disadvantages of EVA (mainly acetic acid evolution and moisture absorption). As the conductive ink material is directly placed on the encapsulant, lower lamination temperatures than for EVA are required.

In the presented study, we show the potential of the CPOE technology based on a conductive layer embedded onto the encapsulation material to interconnect solar cells of highest efficiency. Preliminary results of this novel module interconnection technique were performed and analyzed in terms of CTM loss studies, outdoor exposure and aging tests such as DH1250, TC200, UV15 and PID tests to prove the longevity of this approach.

2 EXPERIMENT

Modules based on n-type Cz-Si bifacial IBC cells, so called Zebra cells [3], reaching 22% peak efficiency with a bifaciality factor of 75% in combination with a 3.2 mm solar glass as front cover, POE as encapsulant, bussing ribbons with a Sn60Pb40 coating and 2.5 mm width or the application of an aluminium film (to collect current flowing through the cell), CPOE layer and transparent backsheet (TBS) or glass as rear cover were fabricated. In addition, low temperature solder paste (LTSP) between the cell and the CPOE was applied as well as adhesive strips (Scotch tape between the rear side cell and bussing ribbons)

The CPOE has as the function to grant improved functionality of the traditional rear side encapsulation layer. This is achieved by printing a metal interconnection circuit directly on top of the POE layer. The metal inks are based on silver nano particles in the bulk resistivity range of 1-10 mΩ/sq with a curing process at around 130°C. The screen-printing allows higher track thicknesses (up to 150µm) compared to other technologies (e.g. Ink-jet). The process speeds can reach up to 100 m/min with this technique, and it is possible to perform printing and sintering in the same production platform.

Table I: Module design for CTM losses analysis.

Design	Zebra size	LTSP	TAPE	additional POE	Bussing	Cover front - rear	Lamination optimization
(1)	1/2	YES	NO	YES	Al film	Glass-TBS	NO
(2)	1/2	NO	YES	NO	Al film	Glass-TBS	NO
(3)	1/2	YES	YES	YES	Al film	Glass-TBS	NO
(4)	1/2	NO	YES	NO	Ribbon	Glass-Glass	YES
(5)	1/2	YES	YES	NO	Ribbon	Glass-Glass	YES
(6)	1/8	YES	YES	NO	Ribbon	Glass-Glass	YES
(7)	1/8	YES	YES	NO	Ribbon	ARC Glass-ARC Glass	YES
(8)	1/8	NO	YES	NO	Ribbon	Glass-Glass	YES

For module construction the most promising CPOE metal formulation with a height of 50μm was used in terms of lowest CPOE track line resistance and contact resistance to the Zebra´s busbars. The Zebra cell features a three dimensional metallization scheme on the rear side consisting of an insulating layer to bypass every second of the parallel running contacting fingers for either the base or the emitter contact and busbars for current collection printed perpendicularly to the fingers on top of the insulation layer. Two different sizes of the zebra cell with an area of 11947.5 mm^2 (1/2 of a 6" n-type cell) and of 3042 mm^2 (which corresponds to 1/8 of a 6" mc-Si) were employed. Glasses with anti reflective coating (ARC) were also incorporated in some groups. The cell was encapsulated with POE toward the front glass. The rear contact cell was contacted and encapsulated with the CPOE, which connects the bussing ribbons or the aluminum film. In addition, LTSP between the cell and the CPOE was applied as well as adhesive strips in order to stabilize the contact between the cell and the CPOE during lamination. Thus, 8 different module designs (or groups) were fabricated to investigate the CTM losses. A matrix of the different groups is shown in Table I.

One sample (5th module concept) out of 2 to 4 modules was selected to carry out DH1250, UV15 degradation, TC200, PID and outdoor energy yield studies. As reference for the energy yield studies one module with a full size Zebra cell with soldered ribbons and a full size p-type cell module were assembled. The reference Zebra cell module was processed using the same materials and lamination recipe as used for the CPOE modules.

For characterization, EL measurements were performed to control the integrity of the electrical and mechanical contact between the module components after lamination. Additionally EL measurement accompanied the environmental indoor and outdoor tests, applying the same current and exposure time. IV characteristics were measured using a class AAA solar simulator with an illumination area of 22×22 cm^2 at STC conditions before and after lamination to determine the cell to module losses (CTM) and to quantify the environmental indoor and outdoor test results. In addition, spectral response measurements were carried out before and after PID testing.

3 ASSEMBLY PROCESS DEVELOPMENT

Eight different module designs were fabricated to investigate the interconnection quality between the CPOE and the cell through CTM loss studies, see Table I. The first module type consists of LTSP and an additional POE layer on the rear side which reduces the pressure above the CPOE. The LTSP was applied between the cell and the CPOE by dispensing. Glass and transparent backsheet were used as front and rear covers respectively. Figure 1(a) shows a picture after lamination of the module rear side. Significantly high number of air bubbles due to the use of the LTSP and the additional POE layer were observed. It seems that the LTSP damages the CPOE layer during lamination. Nevertheless, this approach results in an electrical interconnection between the CPOE and the cell without any shunting as shown in the EL images in Figure 1(b) (almost the same pFF values for module and cell were measured). However, high electrical losses are observed mainly due to increased series resistances. With this module concept a FF of 35% (relative CTM losses in FF of 55%) and η of 9% (relative CTM losses in η of 57%) were reached (see Figure 2).

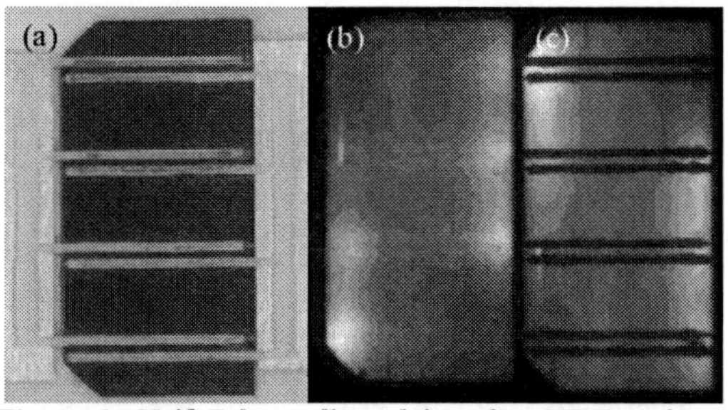

Figure 1: Half Zebra cell module using LTSP and an additional POE layer. (a) Picture of the module rear side after lamination. EL images of the front (b) and rear (c) side of the module.

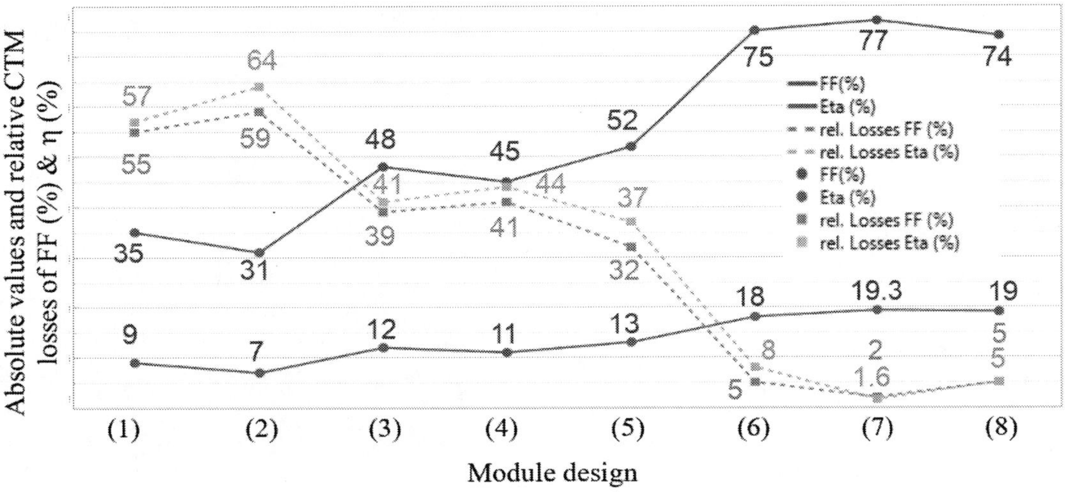

Figure 2: 8 different module designs showing the absolute values in FF (%) and η (%) with their respective relative CTM losses plotted in dotted curves.

The second module design avoids the use of LTSP and the additional POE layer by means of a Scotch tape layer above the CPOE metal layer. This concept provides a more homogeneous CPOE layer after lamination hence reducing the amount of bubbles and the CPOE damage. Even though no shunting was detected an electrical contact was reached resulting in a lower absolute FF of 31% and a reduced η of 7% if compared with the previous module concept. The relative CTM losses in FF and η were 59% and 64% respectively, see Figure 2. The reason for the larger losses can be explained by the direct mechanical connection between cell and CPOE, which stands for higher electrical losses due to the higher resistances losses in the CPOE. Additionally to be note is that the use of LTSP increases the FF by 4%.

The third module concept corresponds to the combination of the two last mentioned designs. This includes avoiding the excessive LTSP and the adhesive tape and without employing the additional POE layer on the rear side. By this the FF increases up to 48% and the efficiency η up to 12% with relative CTM losses in FF of 39% and in η of 41%, see Figure 2. Nonetheless, the high series resistance as well as the large number of bubbles after lamination remains.

By further optimization of the lamination parameters, the use of glass on the back side of the module and by avoiding the use of LTSP (fourth module concept), a FF of 45% and η of 11% with relative CTM losses in FF of 41% and in η of 44% were achieved, see Figure 2. Even better IV performance was achieved with the third module concept compared to the fourth one. No bubble concentration was observed for the fourth module concept. The use of glass on the rear side and optimal lamination parameters provides a better alignment between the CPOE and the cell and avoids bubbles after the lamination process.

The fifth module concept consists in the incorporation of LTSP in the fourth module design described above. The FF increases up to 52% and η up to 13% by reducing the relative CTM losses in FF by 32% and in η by 37%. The mechanical and electrical contacts are much more homogeneous for this module concept if compared to the previous module concepts.

In the previous five module concepts demonstrated, the achieved high electrical losses are a limitation of the CPOE concept. The conductivity of CPOE based on Ag is 15 times less if compared to the conductivity of soldered or glued standard Cu ribbons. Minimizing the resistive interconnection losses is crucial for successful module integration. For this in the following contacting techniques for module assembly, the cell area was reduced from 11948 mm^2 to 3042 mm^2. Thus, in an one-eighth-size solar cell module, the generating current is reduced to a quarter, which significantly reduce the series resistance losses to a sixteenth compared to half size cell design. The CPOE design was also changed. The contact length between the CPOE and bussing ribbons was increased from 40 mm to 156 mm in order to reduce the ohmic losses significantly. Also the height of the metal ink contained in the CPOE was increased from 50 µm to 150 µm.

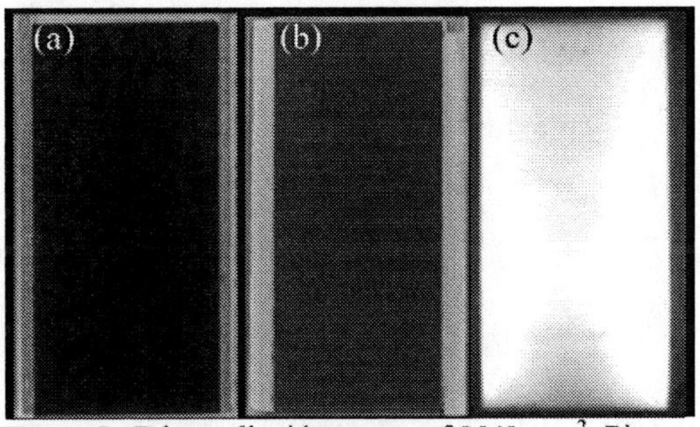

Figure 3: Zebra cell with an area of 3042 mm^2. Picture of (a) front and (b) rear module side. (c) EL image of the module front side after lamination.

The sixth module design includes the changes described above which results in a significant improvement of module performance. FF of 75% and η of 18% with relative CTM losses in FF of 5% and in η of 8% were achieved, see Figure 2. The cell is electrically interconnected, no major damages in electrical or mechanical integrity occur, as well as no bubbles are visible after lamination, see Figure 3. Applying ARC glass and reaching a better alignment between CPOE and cell after lamination (seventh module design), the FF further increases to 77% and η to 19.3% with relative CTM losses in FF of 2% and in η of 1.6%. Finally, the eighth module design again avoids the use of LTSP, shows a FF of 74% and an η of 19% with relative CTM losses in FF and η both of 5%. This lower result in FF and Pmpp confirms that the use of LTSP improves the FF by 3 to 4% and achieves a better electrical and mechanical contact between cell and CPOE.

4 MODULE QUALIFICATION TESTS

To prove the reliability of the CPOE technology as an alternative interconnection technique, the fifth module concept as described in the previous section was used for several qualification tests. This concept is based on glass for front side and rear side module protection, POE as encapsulant, CPOE interconnected half Zebra cells and bussing ribbons. Adhesive strips were applied between the rear cell and bussing ribbons in order to reinforce the components during the lamination process. The following results are based on averaging IV data of 2 to 4 modules, which were used to perform environmental indoor tests (DH, TC, UV and PID) and outdoor energy yield studies.

4.1 UV degradation test

In order to identify module materials which are able to withstand high doses of UV radiation, the UV chamber was operated at 60°C and UV-A+B doses of 15 kWh/m² (between wavelength of 280 and 385 nm and at least 5 kWh/m² between wavelength of 280 and 320 nm) according to IEC 61215. Figure 4 shows the relative losses in FF and Pmpp (y-axis) before and after UV preconditioning with 15 kWh/m² doses (lower x-axis) for front and rear side (upper x-axis). The values correspond to mean value of 3 modules. No significant power losses were measured after UV exposure (~0.5%). Additionally no evidence of degradation of the module packaging materials such as browning or discoloration was observed. In Figure 5 EL images show no changes for samples before and after the UV test.

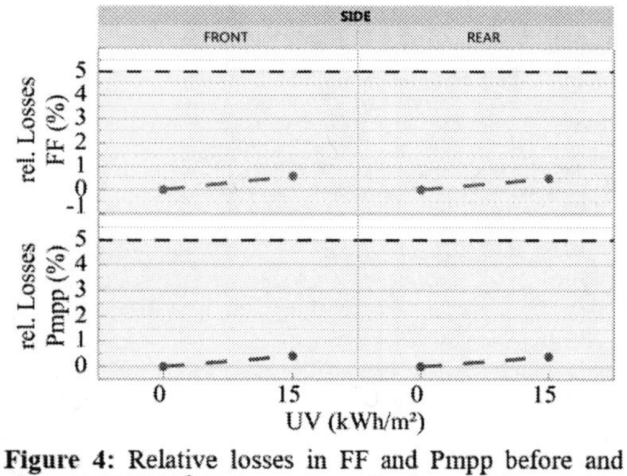

Figure 4: Relative losses in FF and Pmpp before and after 15 kWh/m² UV exposure for front and rear side. The blue dotted curves are a guide for the eye.

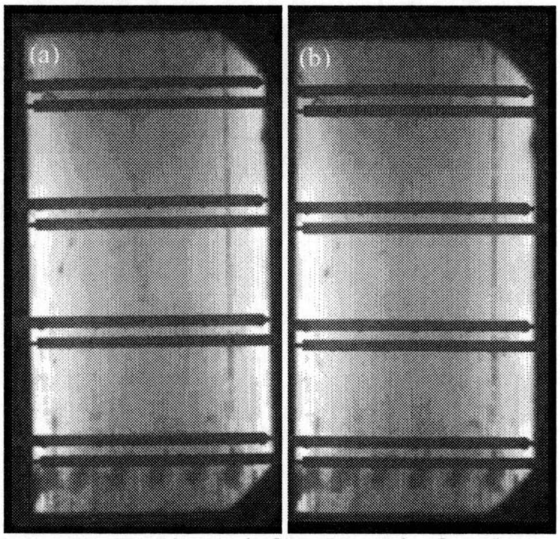

Figure 5: EL picture before (a) and after (b) UV degradation test (15 kWh/m²). No significant changes were observed.

4.2 Damp heat (DH) test

For the damp heat test, the climatic chamber was operated at a temperature of 85°C and relative humidity of 85 %RH. In Figure 6 the relative losses in FF and Pmpp versus the time exposure in hours for the front and rear side of the modules are plotted. The values correspond to mean value of 2 modules. From the beginning of the test up to 500 hrs., no significant power losses are measured (below 0.5%). After DH1250 long-term exposure a moderate degrading effect caused by the high humidity and temperature was observed. Relative losses of FF achieve approx. 2% and Pmpp of approx. 2.5%. Both values stay well below the threshold of 5% marked in the yellow area. Thus, the modules pass the DH1000 test according to IEC 61215 standard showing power degradation below 5 % after 1250 hours. Figure 7 shows an EL image of the rear side of a module before the test, after 500 hrs. and 1250 hrs. It can be observed that the cells electrical integrity is given. No delamination, corrosion and discoloration were detected.

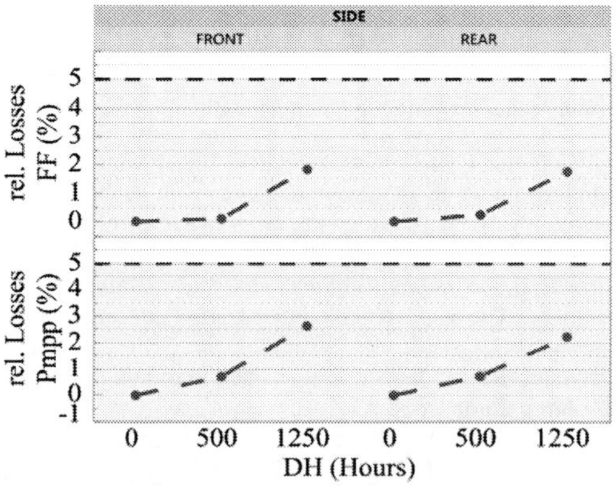

Figure 6: Relative losses of FF and Pmpp for DH before, after 500 hrs. and after 1250 hrs. for front and rear side. Blue dotted curves are a guide for the eye. The test is passed showing relative losses below 5%.

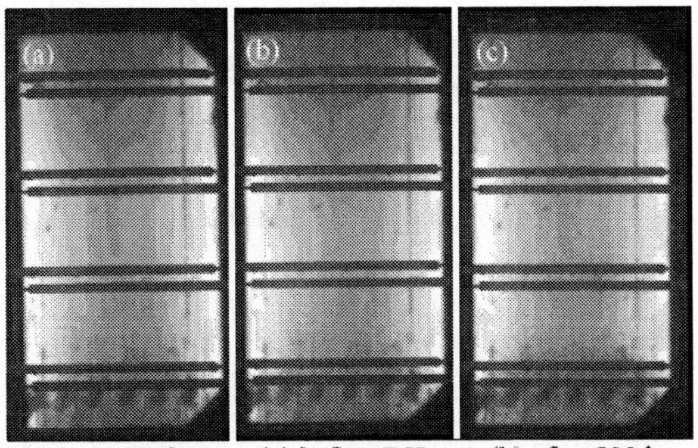

Figure 7: EL images (a) before DH test, (b) after 500 hrs. and (c) after 1250 hrs. of DH testing. No significant changes were observed.

4.3 Thermo-cycling (TC) test

In order to investigate the effect of thermo-mechanical stress on the module material compound, the climatic chamber was operated between a temperature range of -40 °C and +85 °C for 200 cycles (each cycle of 6 hrs.). Figure 8 shows the relative FF and Pmpp losses (y-axis) for the front and rear side (upper x-axis) at an initial state, after 50 cycles and after 200 cycles (lower x-axis). The values correspond to mean value of 4 modules. After TC50, larger than 5% relative FF and Pmpp losses are observed, which increase once exceeding TC200. However, no cracks underneath the CPOE and no metallization damages due to finger interruptions after TC 200 were found by EL analysis, see Figure 9. Thus, the modules do not pass the TC test since relative power losses exceed 5%. The relative FF losses were not due to shunting (pFF remains the same before and after TC200), but due to an increase of series resistance. This can be explained by possible contact damages between the CPOE and cell as a result of the mechanical stress during the TC test.

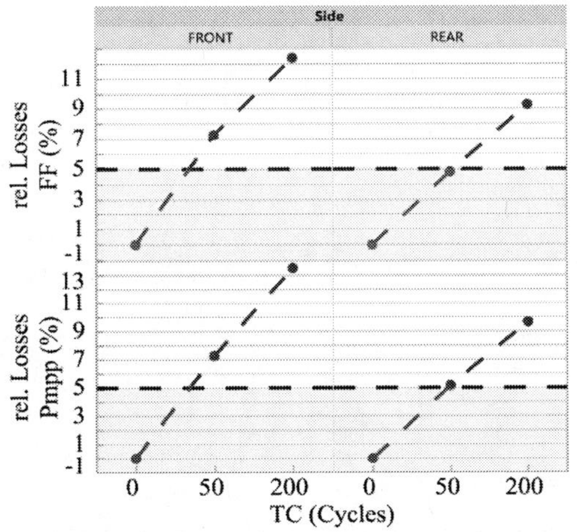

Figure 8: Relative losses of FF and Pmpp for failed TC test: before, after 50 and after 200 cycles for both sides. Blue dotted curves are a guide for the eye.

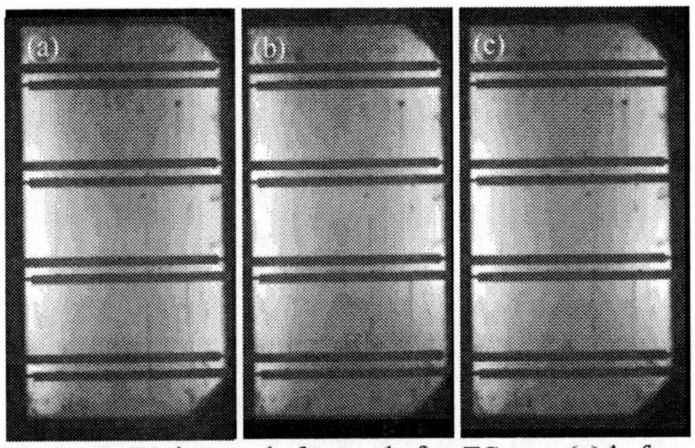

Figure 9: EL images before and after TC test: (a) before the test, (b) after TC50 and (c) after TC200.

4.4 Potential induce degradation (PID) test

PID stress tests of modules at system voltages of 1000 V and temperatures of 85 °C were carried out. The modules were wrapped with Al foil for electrical connection and placed with their sunny side on a metal plate, which was connected to one polarity of the voltage supply. The other lead of the module was connected to the other polarity of the voltage supply. A metal block was used to ensure the contact between the Al foil and the module. 3 different tests at 85°C were performed: (1) Applying -1000 V on the metal plate for 60 hrs. (2) +1000 V for 66 hrs. and (3) +1000 V for 75 hrs.

Figure 10 shows the relative short circuit current (Isc), open circuit voltage (Voc), FF and Pmpp losses (y-axis) versus the exposure time in hours (lower x-axis) for the front and rear side (upper x-axis). Additionally the voltage and polarity of the power supply are given (lower x-axis). The values correspond to mean value of 2 modules. From the beginning of the test up to 60 hrs. with -1000 V connected to the metal plate, a slight improvement in Isc and Voc of approx. 5% were detected, showing a 30% gain in FF and 40% gain in Pmpp. No significant change in EL images and external quantum efficiency (EQE) curves were observed, see Figure 11 and 12. It appears that the electrical contact between the CPOE and the cell improves during the 60 hrs. at 85°C, also stated in FF. Reversing the polarity to +1000 V on the metal plate high relative losses in Isc, Voc and Pmpp were measured after 66 hrs. The relative FF losses also slightly decrease, but the drop is smaller than the gain achieved before. Apart from the uniform degradation which can be observed by the EL analysis (compare Figure 11(f) and 11(g)), inactive cell areas (marked with red dotted squares in Figure 11(b) and 11(c)) were observed. The degradation is in agreement with the EQE drop observed in Figure 12. Both effects were also detected after additional 75hrs. and -1000 V applied on the metal plate. The root cause for these losses can eventually be the accumulated charges on the front of the cell, causing a surface passivation breakdown as it is reported in previous studies and which has been observed to be reversible with reversed bias [4].

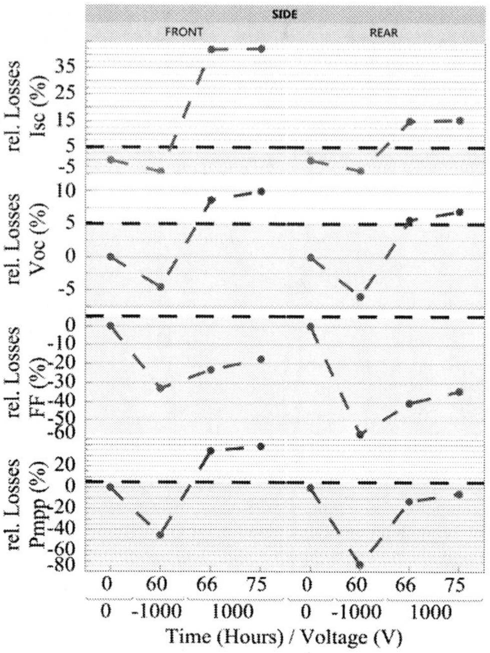

Figure 10: Relative losses of Isc, Voc, FF and Pmpp during PID testing before, after 60 hrs. and -1000 V, after 66 hrs. and 1000 V and after 75 hrs. and 1000 V. The blue dotted curves are a guide for the eye. The test failed showing relative losses larger than 5%.

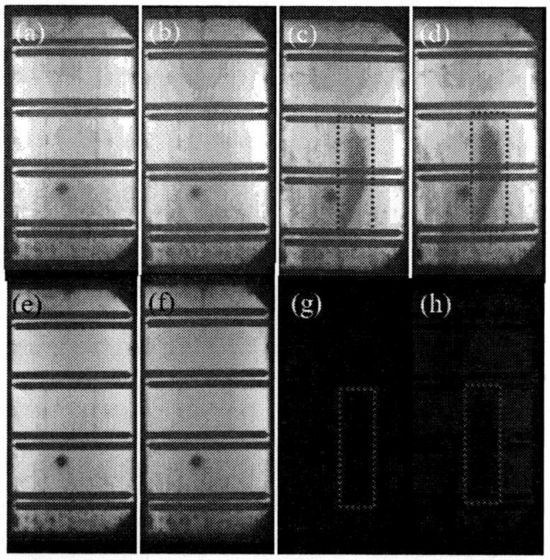

Figure 11: El pictures during PID testing: (a) and (e) before the test, (b) and (f) after 60 hrs. and -1000 V, (c) and (g) after 66 hrs. and 1000 V, (d) and (h) after 75 hrs. and 1000 V.

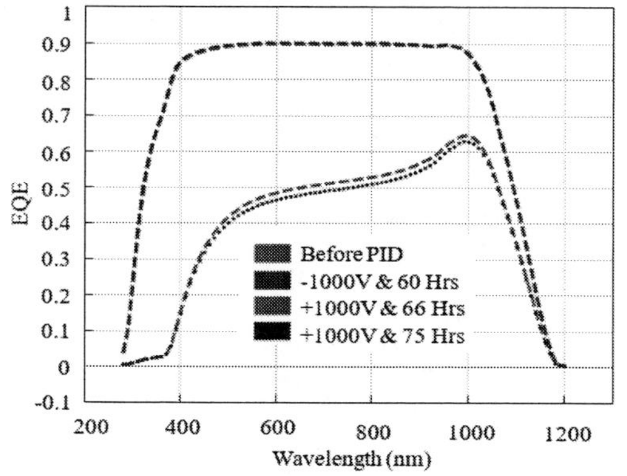

Figure 12: EQE curves before PID testing (red dotted curve), after 60 hrs. and -1000 V (blue dotted curve), after 66 hrs. and 1000 V (green dotted curve) and after 75 hrs. and 1000 V (black dotted curve).

4.5 Energy yield studies

Figure 13 shows the 6 modules installed on the rooftop at ISC Konstanz with an inclination of 30° facing south at coordinates 47°40'17.8"N 9°09'07.0"E and 0.5 m height above the ground. Underneath the modules white backsheet was placed to increase the ground reflection (in average an albedo of 31.2% was reached). The global irradiance (Gsky) was measured with a pyranometer located close to the modules (indicated as number 1 in Figure 13) and the module temperature (TrefMod) taken with a PT-100 sensor inside the modules (number 3 in Figure 13). Ambient temperature Ta was taken with a PT-100 sensor inside a white plastic housing indicated as number 2 in Figure 13. All sensors were mounted on the same plane of the modules.

The IV curve of each module was recorded every 5 min. for 100 days in the period between May and August 2017. Two CPOE modules were mounted, indicated with number 4 and 7 in Figure 13. Additionally one Zebra full cell module and one p-type full cell module with soldered ribbons as reference were installed, indicated with number 5 and 6 in Figure 13 respectively. Number 8 in Figure 13 corresponds to a module mounted with its sunny side down in order to measure the current generated by the rear albedo for rear irradiance calculation (Grear).

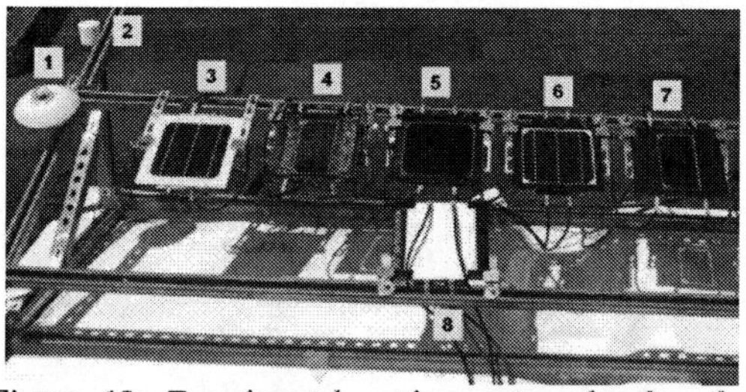

Figure 13: Experimental outdoor setup showing the pyranometer (1), temperature sensors (2 and 3), CPOE modules (4 and 7), Zebra cell module (5), p-type cell module (6) and one module for albedo current measurements (8).

Figure 14 plots the irradiance and temperature data registered in the measured time period. Smooth curves show the average for the global irradiance (Gsky), rear irradiance (Grear), reference module temperature (TrefMod) and the environmental temperature (Ta). The average values for the exposure time of 100 days are Gsky= 404.5 W/m^2, Grear= 125.8 W/m^2, TrefMod= 35.8 °C and Ta= 25.2 °C.

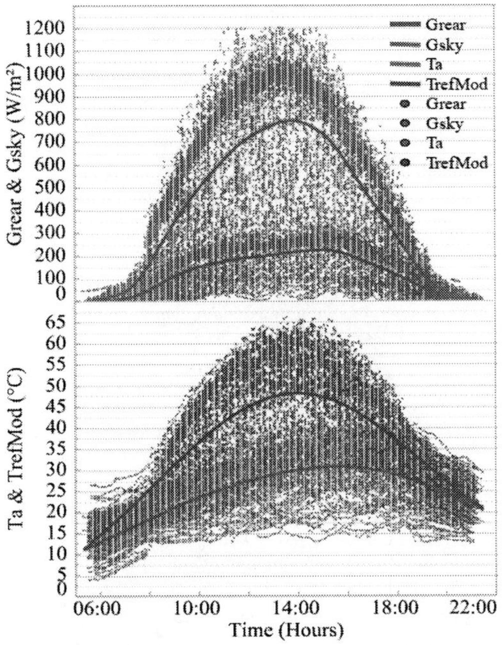

Figure 14: Grear, Gsky, Ta, TrefMod values over a day in hours.

Figure 15 plots the energy yield (Ya) and the performance ratio (PR) over the days of measurements of the CPOE and references modules. The Ya in Wh/Wp corresponds to the ratio between the energy (integration of the module output power over the time) and the Pmpp at STC conditions. While the PR is the ratio between the Ya and the ratio between the irradiance measured by the pyranometer (Gsky) and 1000 W/m^2. In the case of a bifacial module the Grear were added to the Gsky.

The average in Ya and PR for Zebra reference are 6.42 Wh/Wp and 108%, for CPOE modules 6.59 Wh/Wp and 114% and for p-type reference 4.98 Wh/Wp and 81.9%. Since CPOE and Zebra reference modules are bifacial, higher Ya and PR were achieved than for the p-type full cell module. This significant difference is due the bifaciality gain, meaning the energy contribution of the rear side in the Ya and PR calculation. CPOE modules show better energy performance than Zebra reference modules due to the cloudy days, which affect the average values (see dotted black circles in Figure 15). Excluding these days, the median in Ya and PR for Zebra reference are 7.03 Wh/Wp and 106%, for CPOE modules 6.98 Wh/Wp and 107% and for p-type reference 5.5 Wh/Wp and 84.17%. Thus, similar results are reached by the CPOE and Zebra reference module. In addition, a significant drop in Ya and PR from day 96 on for the p-type references (see dotted red circles in Figure 15) can be observed. The reason is that bussing ribbons (placed outside the module) were cut due to the high wind velocity registered at day 96.

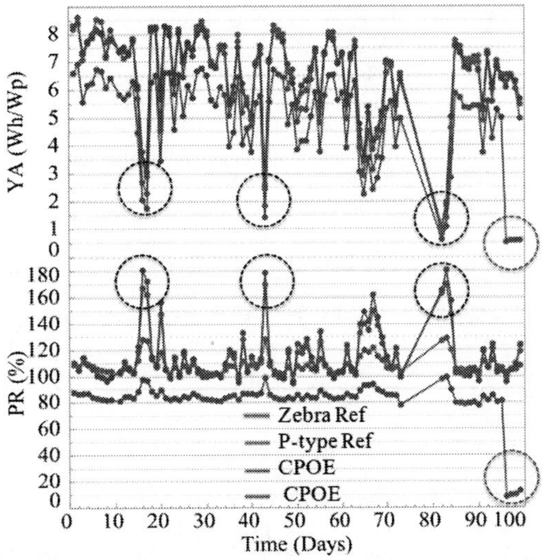

Figure16: Ya and PR versus the measured days for CPOE (red and purple curve), Zebra reference (blue curve) and p-type cell reference (green curve).

5 CONCLUSION

In the present study, the potential and reliability of the novel CPOE module interconnection technique which is based on a conductive layer embedded onto the encapsulation material for interconnecting IBC solar cells were investigated.

Modules based on 8 different contacting techniques were designed and evaluated in terms of CTM loss analysis. This includes the use of two POE layers on the back, the use of transparent backsheet or glass as rear cover, incorporation of adhesive strips between the rear cell and bussing ribbons, reducing the cell area, changing the CPOE metal contact scheme, optimization of the lamination process and adding LTSP between the cell and CPOE. For all samples an electrical contact was achieved without any local shunting. By application of certain improvements to these module designs, the air bubble concentration, detected after the lamination process, was eliminated. However, high losses in FF due to the high resistance losses caused by the contact between the CPOE and cell were observed for all half cell module concepts. In order to reduce the larger electrical losses, the cell area was reduced (from 11948 mm^2 down to 3042 mm^2) and the CPOE metallization scheme changed (increasing the metal ink height from 50 µm to 150 µm and the contact length between the CPOE and bussing ribbons from 40 mm to 156 mm). With these improvements CTM losses below 2% were achieved with a conversion efficiency of 19.5%. The limitation of the CPOE comes mainly from its poor conductivity, which was partly compensated by the use of LTSP, increasing the FF by 4%. Another current drawback to be mentioned is the alignment problem between the CPOE and cell. Glass-glass module configurations and the use of adhesive strips (Scotch tapes) between the rear side and the bussing ribbons were subject of the study in order to reach the same mechanical properties given by soldering or gluing, thus stabilizing the contact between the cells and the bussing ribbons with CPOE layer during lamination.

The CPOE modules were subject to accelerated environmental indoor climatic testing in order to prove their longevity. Samples passed the UV degradation test showing no evidence of transparency losses for the module packaging materials and any significant electrical power losses. The same result was achieved during DH test, achieving relative power losses less than 5% after 1250 hrs. However, TC200 failed with power losses larger than 5%. Already after TC50 relative losses of FF and Pmpp larger than 5% were observed. PID testing showed power degradation due to Isc and Voc losses after applying +1000 V at 85°C. The root cause for these losses can eventually be the accumulated charges on the front of the cell, causing a surface passivation breakdown. Degradation issues were detected by EL analysis. Outdoor energy yield studies on the CPOE modules compared with Zebra and p-type cell modules with soldered ribbons were performed. As CPOE and Zebra reference modules are bifacial, higher Ya and PR were achieved than for the p-type full cell module. This difference is mainly due the bifaciality gain, meaning the energy contribution of the rear side in the Ya and PR calculation. CPOE modules generally show a better energy performance than Zebra reference modules during cloudy days, which provide lower irradiance and lower electrical losses than the Zebra reference as a consequence.

6 ACKNOWLEDGEMENTS

This work was funded inside the SOLAR-ERA.NET programm within the InGrid project (Anpassung hocheffizienter n-Typ Rückkontaktsolarzellen für neuartige Verkapselungsmaterialien mit leitfähigen Strukturen) under contract number 0325821.

7 REFERENCES

[1] M. A. Green et al., ''Solar cell efficiency tables (version 48)'', Progress in Photovoltaics: Research and Applications 24 (7), 905-913 (2016). DOI: 10.1002/pip.2788.

[2] A. Schneider et al., ''Conductive Film Application to Replace the Soldering Process and Allow for Busbar-Less Cell Interconnection", 29th EUPVSEC, Hamburg, Germany, 158-160 (2014).

[3] G. Galbiati et al., "Large Area Back-Contact Back-Junction Solar Cells with Efficiency Exceeding 21%", 38th IEEE Photovoltaic Specialist Conference, Austin, USA. 1-4 (2012). DOI: 10.1109/PVSC-Vol2.2012.6656700.

[4] A. Halm et al., "Potential-induced Degradation for Encapsulated n-type IBC Solar Cells with Front Floating Emitter", 5th International Conference on Silicon Photovoltaics, Energy Procedia, 2015, 77, 356–363.

EVA and backsheet inspection for solar module

#Hsin-Hsin Hsieh[1], Yu-Hsien Lee[1], Yu-Tai Li[1], En-Yun Wang[1], Hung-Sun Wu[1], Ni-Hsin Tung[1]
[1]Center for Measurement Standards, Industrial Technology Research Institute, Taiwan

Ethyl vinyl acetate (EVA) and backsheet are key materials to the high reliability and performance of solar module. The complicated polymer compositions of EVA and backsheet are sensitive to humidity and UV light, thus resulting in degradation. The IEC standards concerning the polymer materials weatherability and durability usually take few months for testing. To reduce the testing duration, EVA and backsheet inspection is carried out to evaluate the applicability for solar module. The EVA inspection including 5 test methods in accordance with IEC 62788 standard, accelerated aging tests, and lamination conditions. The tests samples could be in the forms of cured status or raw materials. This may facilitate the solar manufacturers as incoming quality control or screening. In this research, 5 EVA and 5 backsheet samples are prepared for test. The results show a significant differentiation of the qualified materials from the failed samples. This paper delivers the 5 testing methods for EVA and backsheet materials, and their testing results are also revealed for discussion.
keywords：Ethyl vinyl acetate (EVA); backsheet; inspection;

1. Introduction

The weatherability and durability of EVA and backsheet materials are crucial for solar module. According to the literatures, the complicated polymer compositions are usually sensitive to moisture and UV light, thus resulting in degradation [1-3]. Common failure modes concerning polymers are snail trails (discoloration), yellowing, and leakage current degradation [4-6]. It is reported that the degradation of encapsulants is due to the photo-oxidation of EVA, resulting in either deacetylation (formation of corrosive acetic acid) or polyene oxidation [7]. To detect the degradation of solar module in infant status, IEC 61215 also reveal several testing methods that the most suggested are damping heat and UV exposure [8]. The high humidity and dosage of UV is to simulate and accelerate the rigorous weather condition while field application. According to the standard IEC 61215, the damping heat and UV exposure test take a long time for 1000 hr (>42 days) and inefficient UV exposure, yet most module products can pass the criteria easily. It indicates that the current testing methods may not fulfill the requirements while module applied in the field.

Recently, the reliability of solar materials has been highly valued. And the testing methods have been developed severely and strictly to monitor the failure of material; for examples, IEC 62804 technical specification for potential induced degradation [9], IEC 62892-1 for differentiate performance in multiple climates and applications [10], IEC 62788-1-2 for the volume resistivity of encapsulants and other polymeric materials [11], etc.

Apart from IEC standards, ASTM is generally referred for materials tests as indicators specified for photovoltaics. To approach the real PV field application, the apparatus and measurements are adjusted to imitate the long-tern application. For example, the accelerated UV test and chemical stability are well specified to test the polymer materials with standard procedure [12-13].

On the other hand, the manufacturing recipes and windows are the important issues due to the large diversity of massive production. The more knowledge about thermal stability and chemical reaction, the more feasible for manufacturers to optimize the production process and may eventually approach the target of cost-down and high reliability. Taking into account of quality and cost, the solar manufacturers need to evaluate the materials more precisely and efficiently. Thus, the requirement of screening qualified polymer materials for solar module is highly focused. In this paper, 5 tests are delivered to evaluate the feasibility of polymeric materials for solar module.

2. Testing methods

Inspection for EVA and backsheet materials contains 5 tests which are 2 chemical performance and 3 reliability tests.

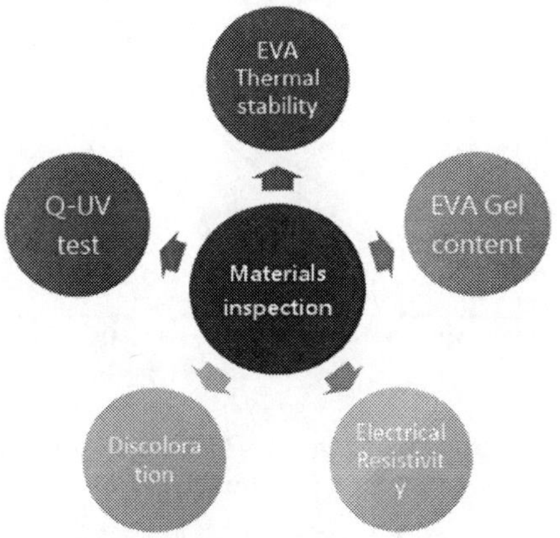

Fig 1 Materials inspections contains 5 tests.

2.1. EVA thermal stability

EVA thermal stability has been studied by Differential scanning calorimetry (DSC, TA instrument DSC Q2000). The preparation of the uncured EVA sample should be cleaned and sliced in a small piece of 5 mg. The heating condition is set from -40 °C to 220 °C with a heating rate of 40 °C/min. And then cool to room temperature. The second heating process is repeated for checking the curing fraction of the cured EVA. The typical thermal stability for EVA is shown in Figs. 2(a-b). In Fig. 2(a), DSC scans the glass transition reaction around -35 °C to -10 °C, followed by melting from 40 °C to 90 °C. And curing exothermic peak is detected from around 150 °C to 186 °C, which might be highly dependent on the chemical composition and content. When cooling down the sample, the thermal profile keeps linear due to full reaction of the material. At second heating process, a similar glass transition remains while the curing peak disappears. This is to check the molecular have almost transformed into stable cured status. On the other hand, assume the onset point is non-cure status, and the curing reaction stops at the offset of the peak. The curing fraction can be calculated according to the curing exothermic reaction. Fig. 2(b) features the curing reaction and duration of EVA materials. The thermal stability is key information to define or optimize the curing reaction during lamination process.

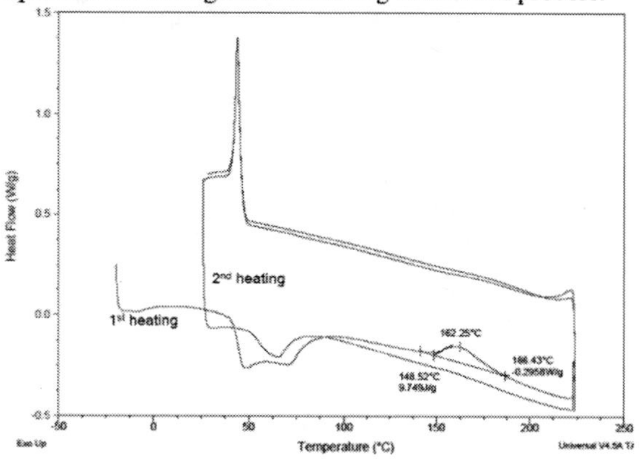

(a)

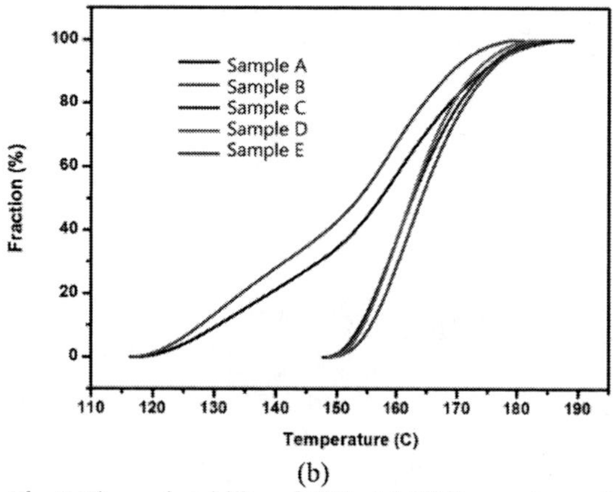

(b)

Fig. 2 Thermal stability of EVA: (a) DSC scanning and (b) curing fraction of EVA.

2.2. EVA Gel Content

Gel content method is a common way to evaluate the fresh or post-laminated EVA materials. The insoluble fraction produced in EVA plastics by cross-linking can be determined by extracting with toluene. EVA sample is sliced into small piece and combine with the toluene reagent.

The lamination process was generally set at (140~150) °C; however, the uniformity of the heating plate and temperature control is inevitable during massive production. To understand the EVA curing condition which is affected by the temperature, the samples prepared at different temperature were also prepared for gel content measurement. Fig. 3(a) illustrates the gel content of 5 EVAs curing at different lamination temperature. In general, most samples have gel content > 80 %. And the change of gel content is less than 10 % during 145 °C to 155 °C. Furthermore, it is also noted that some EVA samples is very sensitive to the temperature as depicted in Fig. 3 (b), for example, Sample B has a gel content variation of 23.5 %, in which the gel content ranges from 59.1 % to 82.6 % at temperature from 145 °C to 155 °C.

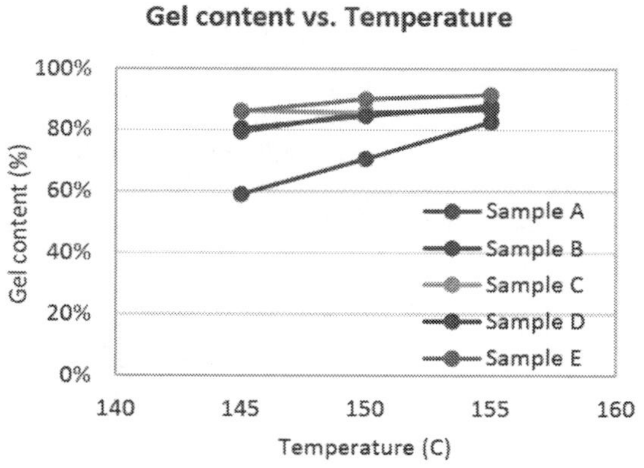

(a)

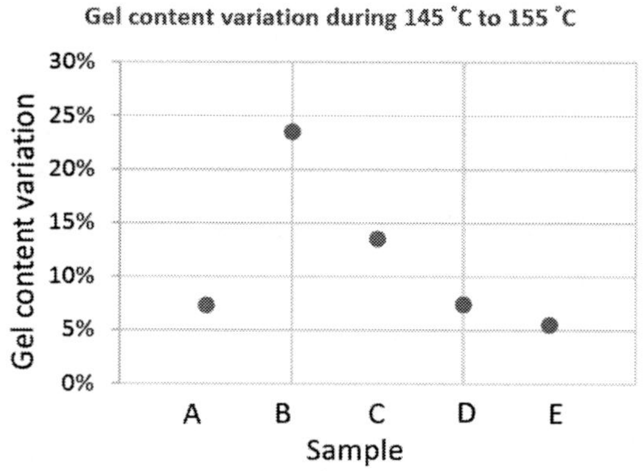

(b)

Fig. 3 Gel content of EVA samples at different lamination temperature.

2.3. Electrical resistivity

Volume resistivity of polymeric PV encapsulant and backsheet is measured according to IEC 62788-1-2 (Encapsulants - Measurement of resistivity of photovoltaic encapsulation and backsheet materials) [11]. The testing setup, sequence and example of measured current during the test are shown in Fig 4. The test sample, which has preconditioned under desiccated atmosphere for more than 48 h, is placed between the electrodes, and applied periodic voltage ±1000 V_{DC} with cycle time 1 min. The output current is simultaneously recorded until 5 cycles are completed. The current at the end of the cycles are taken into account in order to minimize noise originated from the background. After the weighted average of current is calculated, the final measured result could be derived in accordance with the general definition of volume resistivity.

Fig. 5 concludes the volume resistance of 20

individual PV encapsulants or backsheets measured by this specified method. The resistance ranges from 10^{14} Ω·cm to 10^{17} Ω·cm, which plays an important role of PV module reliability. Researches proved that encapsulants with greater resistance could help module products degrade less and sustain longer against potential induced degradation (PID). Thus the inspection of volume resistivity for PV encaplsulants will be essential in production line, especially when 1500 V systems become the mainstream in the field.

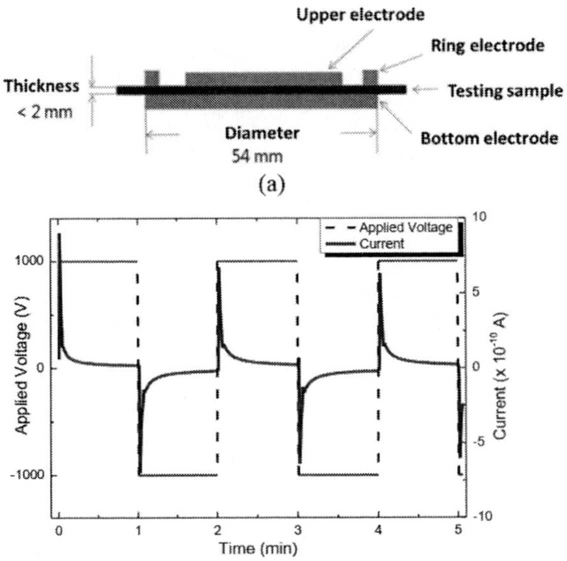

Fig. 4 Volume resistivity (a) testing electrode apparatus [11] and (b) example data

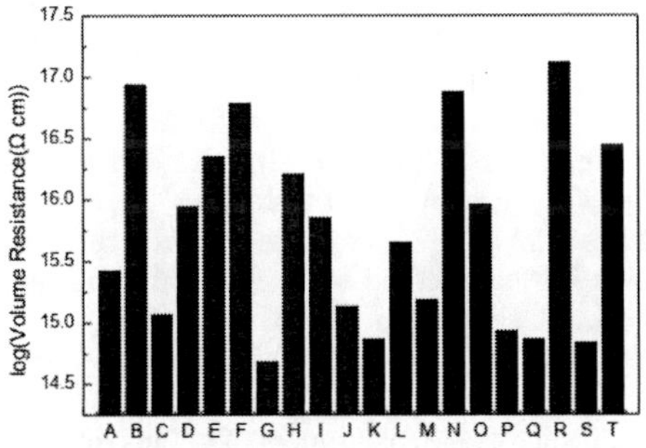

Fig. 5 Volume resistance of encapsulant and backsheet samples

2.4. Discoloration

The discoloration of encapsulant materials is also known as snail trail effect. Though there is so far no specific testing standard or criteria for evaluating the discoloration of PV module materials, some testing approach with chemical solution has been developed and provoked. It is reported that the discoloration of solar fingers is caused by the reaction between the solar metallization and degraded EVA chemicals which is triggered by moisture, temperature and electrical field [14]. And the Ag-polymer interface takes place which leads to the formation of particles containing Ag, P and S migrate within the encapsulation foil, resulting in the formation of the Ag-containing nanoparticles on the encapsulants surface. The Ag-containing nanoparticles features with a high optical absorption at wavelength of 425 nm. A typical absorption spectrum of polymeric EVA encapsulant and backsheet are shown in Figs. 6. The absorption spectrum at 425 nm of an EVA encapsulant material is less than the criteria, suggesting the chemical formula is stable during the accelerated discoloration test (Fig. 6(a)). On the other hand, high absorption signals at 425 nm are detected, indicating the high sensitivity of Ag-polymer reaction on both cell and air side of backsheet.

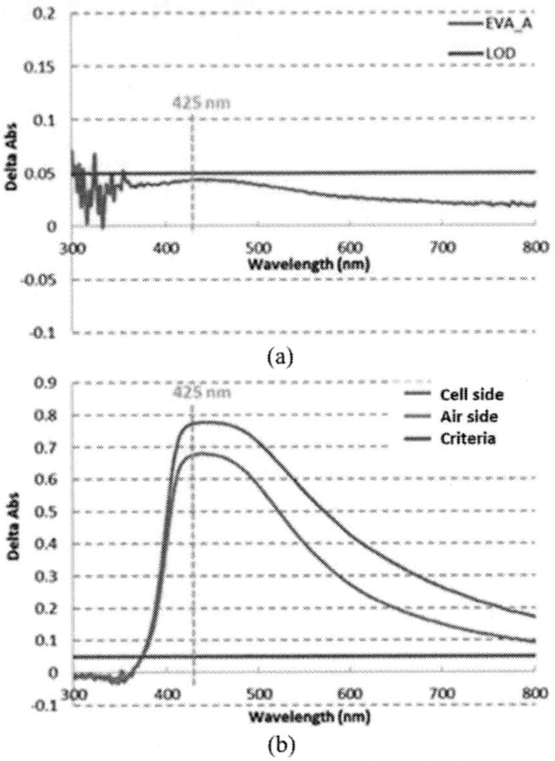

Fig. 6 Absorption spectrum of polymeric (a) backsheet and (b) EVA encapsulant.

2.5. Q-UV test

The Q-UV test of solar module is referred from ASTM G154 standard (Standard Practice for Operating Fluorescent Ultraviolet (UV) Lamp Apparatus for Exposure of Nonmetallic Materials) [12]. The aim of this test is to degrade the polymeric PV encapsulants with a high dosage and long duration. The apparatus of Q-UV test is shown in Fig. 7. The optical measurement before and after the testing is done for comparison. Figs. 8 illustrates typical EVA transmittance and backsheet reflection. It is shown that some encapsulant and backsheet degraded with the UV test.

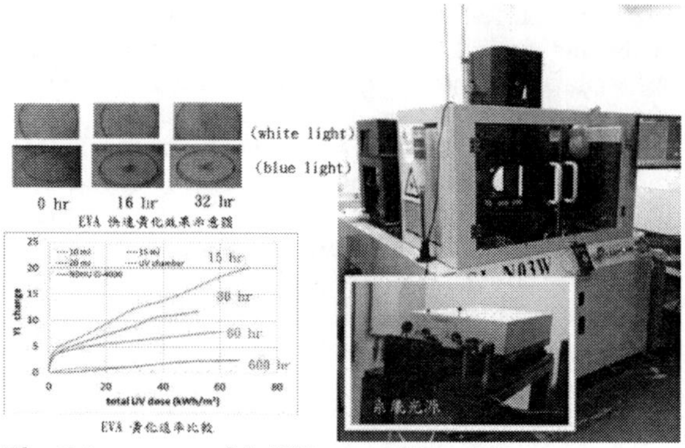

Fig. 7 Apparatus of Q-UV test and its testing results.

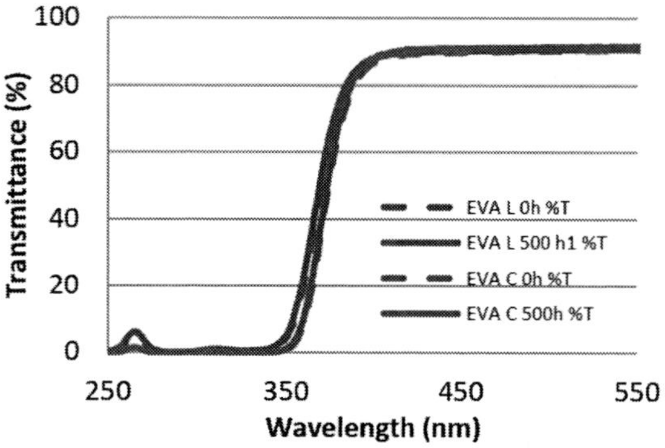

(a)

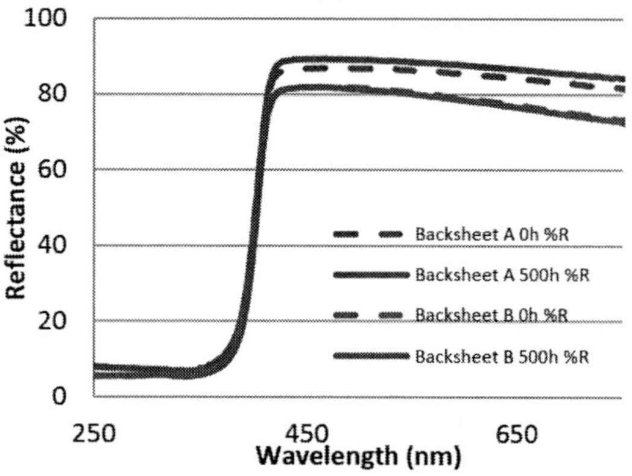

(b)

Fig. 8 Optical measurement: (a) EVA transmittance, (b) backsheet reflectance.

3. Conclusions

This paper illustrates 5 test methods in accordance with IEC 62788 standard, accelerated aging tests, and lamination conditions. It contains 2 chemical performance (EVA thermal stability, EVA gel content, and electricity resistivity) and 3 reliability tests (discoloration and Q-UV test). With these tests, it may facilitate the solar manufacturers as incoming quality control or materials screening. According to the testing results, it is useful for the manufacturers or suppliers to understand the advantages of the materials with good stability or performance. Yet, the cons are also revealed for satisfaction.

References

[1] W. H. Holley, S. C. Agro, J. P. Galica, L. A. Thoma, R. S. Yorgensen, M. Ezrin, P. Klemchuk, and G. Lavigne, "Investigation into the causes of browning in EVA encapsulated flat plate PV modules," in Conference Record of the 24th IEEE Photovoltaic Specialists Conference, Waikoloa, HI, 1994, vol. 1, pp. 893–896.

[2] D. Berman, S. Biryukov, and D. Faiman, "EVA laminate browning after 5 years in a grid-connected, mirror-assisted, photovoltaic system in the Negev desert: effect on module efficiency," Solar Energy Materials and Solar Cells, vol. 36, no. 4, pp. 421–432, Apr. 1995.

[3] K. Thaworn, P. Buahom, and S. Areerat, "Effects of organic peroxides on the curing behavior of EVA encapsulant resin", Open Journal of Polymer Chemistry, Vol. 2, pp. 77-85, 2012.

[4] Review of Failures of Photovoltaic Modules, International energy agency, Report IEA-PVPS T13-01:2014.

[5] F. J., Pern, A. W. Czanderna, K. A. Emery, and R. G. Dhere, Weathering degradation of EVA encapsulant and the effect of its yellowing on solar cell efficiency. IEEE (1991)

[6] H. Nagel, M. Glatthaar, and S. W. Glunz, Quantitative assessment of the local leakage current in PV modules for degradation prediction. 31st EUPVSEC (2015).

[7] A.W. Czanderna and F. J. Pern, "Encapsulant of PV modules using ethylene vinyl acetate copolymer as a pottant: A critical review" Sol. Energ. Mat. Sol. C. vol. 43, no. 2, pp.101-1018, 1996.

[8] IEC, 2005. Crystalline silicon terrestrial photovoltaic (PV) modules-Design qualification and type approval, IEC 61215-9 Ed. 2. (2005)

[9] IEC 62892-1. Testing of PV modules to differentiate performance in multiple climates and application-Part 1 Requirements for testing.

[10] IEC 62804 TS. Crystalline silicon photovoltaic modules-test method for the detection of potential-induced degradation. (2015)

[11] IEC 62788-1-2 Measurement procedures for materials used in photovoltaic modules - Part 1-2: Encapsulants - Measurement of volume resistivity of photovoltaic encapsulants and other polymeric materials, 2016.

[12] ASTM G154, Standard Practice for Operating Fluorescent Ultraviolet (UV) Lamp Apparatus for Exposure of Nonmetallic Materials.

[13] ASTM D3418-15, Standard Test Method for Transition Temperatures and Enthalpies of Fusion and Crystallization of Polymers by Differential Scanning Calorimetry.

[14] S. Meyer, S. Ritchter, S. Timmel, M. Glaser, M. Werner, S. Swatek, and C. Hagendorf, Energy Procedia, 38 (2013) 498-505.

IMPROVED ACCELERATED DURABILITY TESTING AND COMPARISON TO FIELD DEGRADATION

William Gambogi[1], Thomas Felder[1], Steven MacMaster[1], Kaushik Roy Choudhury[1], Bao-Ling Yu[1], Lucie Garreau-Iles[2], Hongjie Hu[3] and T.-John Trout[1]
(1) DuPont, Wilmington, DE, USA; (2) Du Pont de Nemours International S.A., Geneva, Switzerland; (3) DuPont (China) Research & Development and Management Co., Ltd., Shanghai, P.R.C.

ABSTRACT: It is important that accelerated test methodologies applied to photovoltaic modules address the key stresses applied to a photovoltaic module dominant in the field and reproduce the degradation observed in the field. Typically, durability is assessed based on extended exposure to the module qualification conditions but these stress conditions are intended to assess resistance to early life failures and not long-term durability. Multiple stress test protocols have been developed for crystalline silicon based photovoltaic modules that include temperature, humidity, thermal cycling, humidity freeze, direct UV from the front, UV from the back, and internal applied voltage. At elevated levels of stress, faster degradation is achieved and can be compared to field degradation. Test conditions must be carefully considered to be sure that the degradation mechanisms match the field. Sequential tests have been developed at DuPont and results are compared to field inspections and assessment of changes in materials properties observed in the field. Comparison of the performance of different materials and insights into these changes are applied. We use non-destructive testing including current-voltage, electroluminescent imaging and thermal imaging to assess performance. We use destructive testing of fielded modules to better understand the root cause of failures and the relationship to materials changes observed in accelerated testing and fielded modules.

Keywords: materials, module, durability, reliability, testing

1 INTRODUCTION

Long term durability and reliability of photovoltaic module performance is a key consideration on the financial viability of this technology. In early work on the subject, the Jet Propulsion Laboratory included stress testing of candidate module structures and component materials testing to identify issues in module design that might result in early life failures. The JPL study identified durable module designs and identified highly durable polymer backsheets [1].

This work led to IEC qualification protocol 61215 [2] that has been effectively applied to identify design flaws and additional test methods have been adopted to address new failure mechanisms (potential induced degradation (PID), for example), current IEC protocols are often confused as a method to assess long term durability. Frequently development groups will apply extended duration of the qualification conditions to assess long term durability despite recognition within PV reliability groups that this approach is ineffective [3].

A more effective approach applies multiple stress conditions, either sequentially or simultaneously, to better assess a broader set of stress conditions and the synergistic effects of the multiple stress conditions in the outdoor environment. Accelerated conditions can be developed based on expected outdoor stress levels for temperature, humidity, thermo-mechanical stress associated with expansion and contraction, and UV exposure. Careful consideration of environmental stresses are key to understanding outdoor performance and we have reported previously on this approach [4]. Other PV reliability groups have taken or are taking a similar approach (Fraunhofer, NREL, others) where outdoor stresses are simulated by sequential or simultaneous stress exposure. The limitation for this approach includes the long duration of the exposure and the need to correlate results with the outdoor environment for material sets which have only recently been developed or recently introduced into the market. In this paper, we will discuss approaches to shorten the exposure time and to better understand the results of this testing by comparison to the field.

2 MULTI-STRESS TESTING

2.1 Previous Results

In the past, we have reported on the use of sequential stress conditions where damp heat (DH), ultra-violet (UV) light exposure, and thermal cycling (TC) have been used in sequence to assess PV modules performance [4]. In this case, we have applied 1000 hours of damp heat exposure as an appropriate upper limit based on assessment of hydrolysis damage in PET core layers of backsheets from the field. This approach has been further supported by assessment of hydrolysis damage modelling by NREL [5]. We have also applied UV exposure to the back side of the module to address the albedo exposure typically seen by PV modules in the field which can vary depending on the surrounding ground cover and installation. Finally, we have also adopted the TC stress conditions used in IEC qualification but we have alternated it with other stresses to better match field exposures. We have applied these methods to mini-modules fabricated with differing materials sets including different backsheets types and assessed the relationship on backsheet failure in these tests to failure in the field.

2.2 Approaches to Reduce Exposure Time

Earlier work on applying sequential test methods centered on the following sequence:

DH1000/UVA1000/n*(TC200/UV1000)

Where DH1000 is 1000 hours of damp heat (85°C/85%RH), UV1000 is 1000 hours of UVA irradiation (70°C black panel temperature (BPT), 1.2 W/m^2-nm @340nm using a UVA fluorescent bulb) and TC200 is 200 cycles of thermal cycling (-40°C/85°C) as per the IEC 61215 standard. Mini-modules were characterized at each stress interval. The final stress sequence pair (TC200/UVA1000) was applied where n=1,

2 and 3. With n=3, the total exposure lasted over 9 months.

To reduce this 9 month test duration, we considered two known stress characteristics that influence exposure: temperature and irradiance. The initial work was conducted at 70°C BPT. In UVA exposures, the sample temperature is close to the black panel temperature and 70°C BPT is the upper limit of the equipment readily available. By moving to a xenon exposure in an industrial weatherometer for this part of the exposure sequence, higher BPT levels could be used. The xenon lamp source also had the advantage of running at higher irradiance levels (120 W/m^2 vs 65 W/m^2 (300nm-400nm)) allowing us to shorten the exposure time for the UV step. The overall exposure conditions for UV irradiance was reduced slightly to achieve an equivalent 20 years of outdoor exposure with an albedo level of ~ 12%. With the higher irradiance available from the xenon lamp based equipment the UV exposure step was reduced from 1000 hours UVA to 542 hours for xenon. As in the previous work, the UV exposure was applied to the back side (backsheet side) of the module.

3 RESULTS FROM NEW SEQUENTIAL TEST

A faster sequential test protocol was applied to single-cell mini-modules with differing backsheets using commercial 156mm silicon cells and commercial materials including framing, caulking, EVA encapsulant and 3.2mm solar glass. As in previous studies, the performance including visual inspection was assessed after each step of the sequential process. The backsheet types testing included Tedlar® PVF based backsheets (two-sided TPT and single-sided TPE). PVDF based backsheet, PET-based backsheets and PA-based backsheets. The polyamide (PA) based backsheets were particularly important as this backsheet has been found to fail in the field in a relatively short time (3 to 5 years) as will be discussed in more detail in the field section of this paper.

The test sequence that we have described previously [4] is summarized in Figure 1 below.

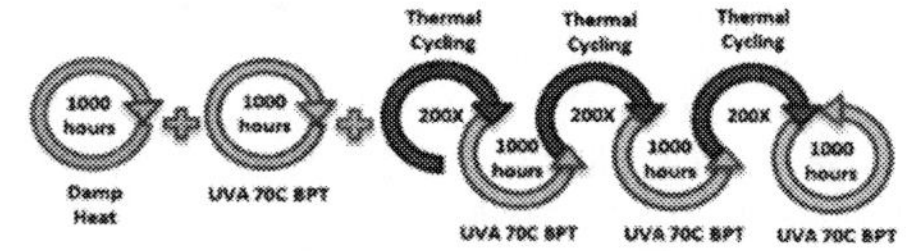

Figure 1. Previous Module Accelerated Sequential Test

The new sequence reported here shortens the test duration from 9 months to 6 months by increasing the intensity and temperature during UV exposure and reducing the duration of the final UV exposure. The shortened test is shown schematically in Figure 2 below.

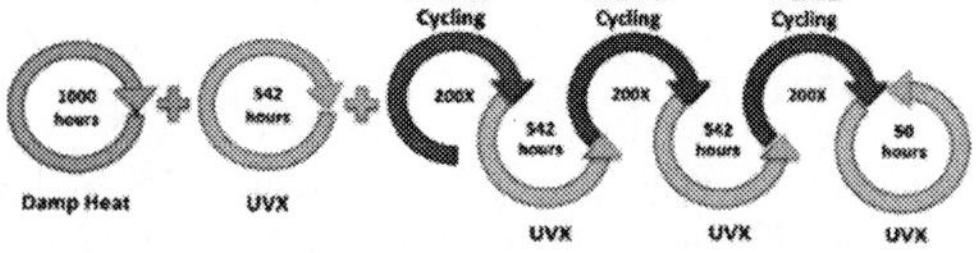

Figure 2. Faster Module Accelerated Sequential Test

In this shortened test sequence, we have replaced the 1000h UVA exposure (~65 kWh/m^2 300-400nm) with a xenon exposure at 120 W/m^2 for 542 hours leading to nearly equal UV dosage compared to 1000 hours UVA in the first three UV exposure steps. The final exposure duration was further reduced to 50 hours. The sample temperature has also been increased. For a UVA fluorescent source, the sample temperature is about the same as the black panel temperature (BPT) which was 70°C in the original work. The faster sequential test using xenon is run at about 90°C BPT, with a sample temperature of ~80°C is estimated. The higher exposure temperature is expected to accelerate the test.

3.1 Results of New Sequential Stress Protocol

The new test sequence has been used to evaluate several backsheets in single-cell mini-modules. The mini-modules were built with 156mm silicon cells and standard materials and processing including tabbing and stringing, framing and mounting the junction box. Tabbing wires were standard 0.16mm thickness. The backsheet color is measured and visual appearance of the modules noted after each step of the sequential test. All modules are photographed and defects are documented by zoom microscope. Table I summarizes the results. Color b* values are shown and backsheet cracking is denoted in red.

Table I: Summary of faster MAST testing

Sequence:	DH/UVX/ 3*(TC/UVX*)
Exposure (h)	4522
Exposure (mo)	6.2
PVDF1	0.6
PVDF2	0.5
PVDF3	0.3
PET1	3.1
PA1	0.2
TPT	0.3
TPE	0.1

cracked
yellow (delta b* > 2)
no yellowing or cracking

The results of this shortened sequential test match those of the previous 9- month sequential test.

3.2 Influence of busbar thickness

To investigate the impact of busbar thickness on susceptibility for cracking, we applied thermal cycling stress (-40°C<==> 85°C) to minimodules using three PVDF-based backsheets and two Tedlar® PVF-based backsheets. We used common industrial ribbon thicknesses of 0.25 and 0.27mm and we inspected minimodules after each cycle looking for any cracking. In this test, no damp heat or UV exposure preceded the thermal cycling stress. We found cracking in three commercial PVDF-based backsheets starting at 300 cycles of thermal cycling with all three PVDF-based backsheets failing after 600 cycles. An example of cracking observed over the busbar region is shown in Figure 3.

Figure 3. Cracking of PVDF-based backsheet after 600 thermal cycles.

3.3 Analysis of Mechanical Properties of PVDF Film

The susceptibility of PVDF-based backsheets to cracking can be understood by assessing the mechanical properties of PVDF in the machine direction (MD) and transverse direction (TD) of the film. We have studied eleven commercial PVDF film manufacturers and the results are shown in Figures 4 and 5.

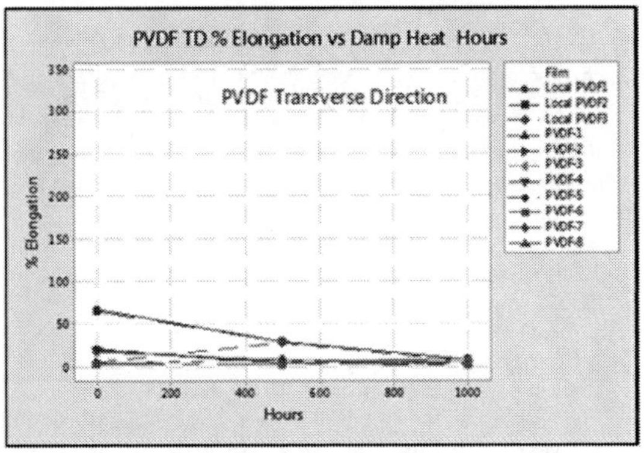

Figure 4. TD Elongation of PVDF films after exposure to damp heat (85°C, 85%RH). Nearly complete loss of elongation is observed.

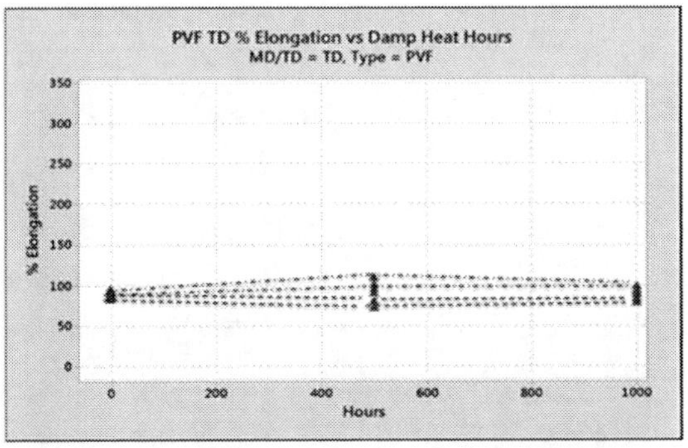

Figure 5. TD Elongation of PVF film after exposure to damp heat (85°C, 85%RH). There is little change in mechanical properties.

4 PERFORMANCE IN THE FIELD

4.1 Correlation to the Field

To validate the results from sequential testing, we have developed a global fielded module program in which we inspect modules over a wide range of time in the field and in many different climates with an emphasis on materials and defect analysis. As part of this program, we perform an assessment of the field including a visual inspection of modules and an identification of the backsheet using IR spectroscopy.

Common failure mechanisms for backsheets cited in several studies [6] include cracking, yellowing and delamination. It is generally understood the cracking is associated with a loss in the mechanical properties of the backsheet due to exposure to outdoor exposure and mechanical stresses on these materials resulting in cracking. Yellowing has also been associated with chemical changes and degradation of the backsheet materials and has also been associated with loss in mechanical properties. [7] It has also been found that cracking appears preferentially in areas of high localized mechanical stress including over busbar ribbons, between silicon cells where backsheets deform in lamination. Areas near junction boxes and near the frame also appear to be locations where cracking occurs, possibly due to cuts near the edge that propagate from these edges.

Two examples from the field demonstrate this tendency and show similar characteristics to failures observed in sequential stress testing described in the previous sections on accelerated testing.

First, we will discuss the case of polyamide-based backsheets failing in the field in less than 5 years. Figure 7 shows a module from China that has failed in the field after less than five years. Notice the backsheet has cracked over each busbar ribbon in the module along the entire length of the ribbon. We have also observed this damage in the area between silicon cells. Evidence suggests that the failure starts at a low level and progresses with time to include more modules and over a greater area of the module. For this backsheet, we have also observed delamination of the backsheet near the frame leading to leakage paths that compromise the operation and safety of modules in the field.

We have also presented in previous papers cracking in PVDF-based backsheets. We have associated this field failure with poor mechanical properties of the outer PVDF layer

4.2 PA Backsheets Cracking in the Field

As part of our global fielded module inspection effort, we have recently inspected and documented cracking of polyamide (PA) based backsheets. We had tested this backsheet material under component testing and found a significant yellowing and loss in elongation compared to its initial performance. From this component testing, we concluded that this backsheet would be more susceptible to cracking. We found this backsheet in the field and found surface cracking after one year as shown in Figure 6.

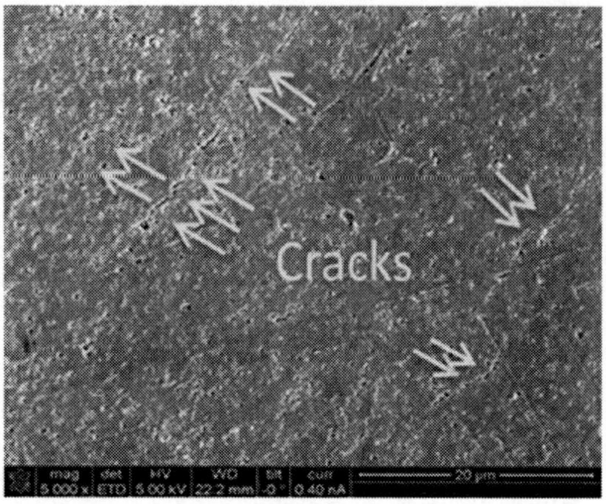

Figure 6. Surface cracking of polyamide (PA) based backsheets after one year in the field

Subsequent inspection of the field with this polyamide-based backsheet showed significant cracking preferentially in areas of high localized mechanical stress including the areas between the cells where the backsheet is deformed and in the area above the busbar ribbon where the backsheet is also locally deformed, as shown in Figure 7.

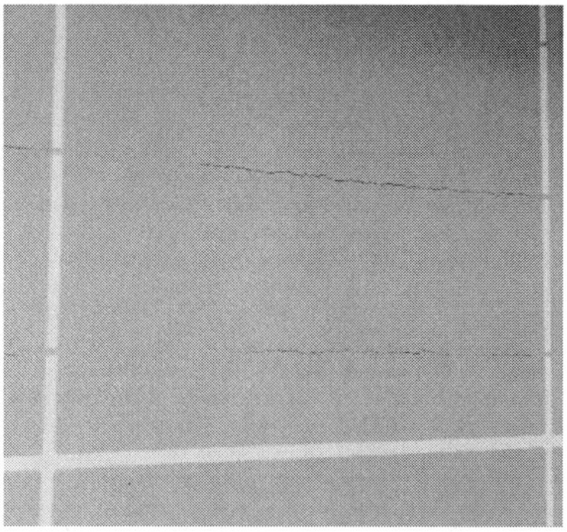

Figure 7. PA-based backsheet cracking after <5 years in the field. Cracking is observed along busbar ribbons.

4.3 PET Backsheets Yellowing and Cracking in the Field

We have previously reported on PET backsheets failing in the field by yellowing and cracking. [7] A recent field inspection found PET-based backsheet damage after 9 years in an Arizona field. 15 – 20% of the modules had cracked backsheets and the array showed 80% power loss. One of the cracked backsheets is shown in Figure 8. An example of yellowed PET backsheets is shown in figure 9.

Figure 8. Yellowing and cracking of PET backsheet in a 9 year old field in Arizona, USA

Figure 9. Comparison of replacement module color with original modules with PET backsheets, 5 years in the field in China

4.4 PVDF Backsheets Cracking and Delaminating in the Field

As reported previously [4], we have also observed cracking and delamination of PVDF-based backsheets. In Figure 10, we show examples from a recent inspection of a 4-year old field in Canada in which more than 60% of the modules have backsheets with severe cracking and delamination. Delamination of the outer layer compromises the electrical insulation of the module and exposes the core layer, typically PET, which is further degraded by environmental exposure leading to still greater risk for the insulation.

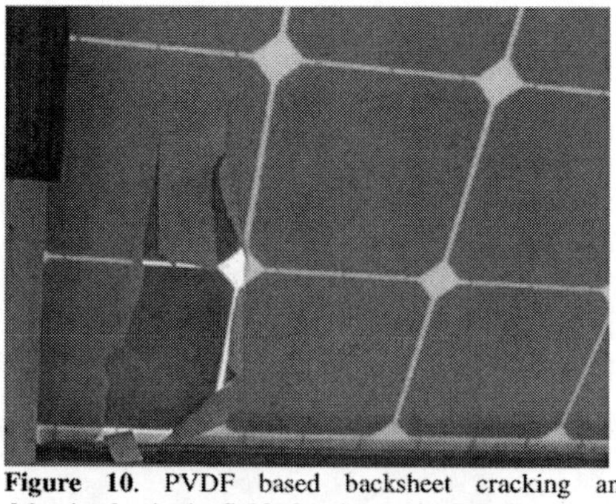

Figure 10. PVDF based backsheet cracking and delaminating in the field after 4 years in Canada

As with many defects, the observable defect starts at a low level and increases in frequency and severity with time. An example of a late stage failure in this field is shown in Figure 11 where the outer PVDF layer has completely delaminated from the backsheet exposing the underlying core PET layer.

Figure 11. Outer layer PVDF film completely delaminated from the core PET layer of the backsheet in less than 4 years in Canada.

5 CONCLUSIONS

Higher irradiance and higher sample temperature have been an effective approach to laboratory tests of backsheet durability. This approach was used to develop a shortened sequential 6-month duration test by replacing UVA exposure steps with xenon UV exposure. Results from this shortened sequential test match those for the previously developed longer 9-month sequential test. Cracking was observed in PVDF and PA based backsheets after exposure. PVF-based backsheets showed no signs of cracking, consistent with the stable mechanical properties well known and demonstrated for this material.

We continue to pursue test sequences which reduce the total exposure time while maintaining consistency with the field. Higher intensity UV exposure sources such as metal halide and higher sample temperature are under consideration.

6 REFERENCES

[1] E. Cuddihy, C. Coulbert, A. Gupta, and R. Liang, "Flat-Plate Solar Array Project Final Report" DOE/JPL-1012-125, JPL publication 86-31 (1986)

[2] IEC 61215 "Crystalline silicon terrestrial photovoltaic modules- design qualification and type approval" http://www.iec.ch/ (2016)

[3] S. Kurtz, J. Wohlgemuth, M. Kempe, N. Bosco, P. Hacke, D. Jordan, D. C. Miller, T. J. Silverman, N. Phillips, T. Earnest, R. Romero, "Photovoltaic Module Qualification Plus Testing", DOE/GO-102013-4327 (2013)

[4] T. Felder, W. Gambogi, K. Stika, B-L Yu, A. Bradley, H, Hu, L. Garreau-Iles, T.-J. Trout "Sequential accelerated tests: Improving the correlation of accelerated tests to module performance in the field", Proc. SPIE, (2016)

[5] M. Kempe, "Understanding the Temperature and Humidity Environment Inside a PV Module", NREL/PR-5200-58375 (2013)

[6] "Review of Failure of Photovoltaic Modules", International Energy Agency Photovoltaic Power Systems Programme, IEA-PVPS T13-01 (2014)

[7] W. Gambogi, T. Felder, T.-J. Trout, H. Hu, L. Garreau-Illes "Accelerated Reliability and Durability Testing for Photovoltaic Materials", SNEC, 2016

Concept of optimized encapsulant composition for PV module reliability under different climatic conditions

A. Mihaljevic[1*], G.Oreski[1], G. Eder[2], Y. Voronko[2], W.Mühleisen[3], L. Neumaier[3], C. Hirschl[3], R. Ebner[4], G. Pinter[5]
[1]PCCL Polymer Competence Center Leoben, Roseggerstraße 12, A-8700 Leoben,
Corresponding author: E: antonia.mihaljevic@pccl.at, T: +43 3842 42962-51
[2]OFI Österreichisches Forschungsinstitut für Chemie und Technik, Arsenal 213, Franz-Grill-Straße 5, 1030 Wien
[3]CTR Carinthian Tech Research AG, Europastraße 12, 9524 Villach
[4] AIT Austrian Institute of Technology GmbH, Donau City Str. 1, 1220 Vienna
[5]University of Leoben, Institute of Material Science and Testing of Plastics, Otto Glöckl Straße 2, 8700 Leoben

Abstract

Even though ethylene vinyl-acetate (EVA) is for more than 30 years employed as an encapsulating material it has some major drawbacks like formation of acetic acid during aging and the peroxide-initiated chemical crosslinking during module lamination. In several studies, the origin of EVA degradation has been linked to its composition and interactions at its interfaces with other PV module components (additives formulation, formation and permeation of acetic acid and residual peroxides). However, PV module degradation was found not only dependent on external stress factors like UV, temperature and humidity but also on internal stress factors like material composition (e.g. additives) and material combinations (e.g. encapsulant-backsheet combination). So far only few studies have looked on influence of internal factors on degradation of PV modules. The main purpose of this work was to understand the role of the encapsulant composition on the degradation behavior of PV modules due to interactions at interfaces of PV components under different stress conditions. Once the mechanisms behind PV module degradation are understood better, it is possible to optimize PV modules for different operating conditions.

Keywords: *EVA, encapsulation, degradation, climatic conditions, UV/Vis/NIR spectroscopy*

Introduction

PV manufacturer's warranty of 25 years of PV module mostly is not met due to different failure modes. In fact, the degradation rates of PV modules were found to be around 0.8 % per year. [1] Degradation was found dependent not only on external stress factors (UV, temperature and humidity level), but also on internal factors like materials combinations and interactions at their interfaces. Still, influence of different climatic conditions and material composition and combinations is underrated. In several studies, it was shown that material combinations could promote degradation of PV modules. However, up to now every module has pretty much the same composition regarding the polymeric materials (especially encapsulation) despite different operating conditions.

Even though ethylene vinyl-acetate (EVA) is still the most employed encapsulating material in PV industry, it has some major drawbacks like peroxide induced crosslinking and formation of acetic acid during aging by Norrish I and II mechanisms. Acetic acid is known as an auto catalyst and linked to several failure mechanisms (discoloration, PID effect, corrosion of metallization, delamination, etc.). It was shown that retention of acetic acid at interfaces of EVA and backsheet can promote degradation of EVA. [2] This means that "breathable" backsheets, i.e. backsheets with high acetic acid transmission rate (AATR) should be used in combination with EVA. On the other hand, peroxide-initiated chemical crosslinking during module lamination results in not only time-consuming and discontinuous lamination process, but the peroxides together with high additives content are also linked to the origin of EVA degradation. [3]

In order to increase reliability of PV modules it is necessary to adapt the PV modules regarding materials and material combinations. To achieve this aim, it is necessary to understand the role of the encapsulant composition on the degradation behavior of PV modules due to interactions at interfaces of PV components under different stress conditions. Once when these mechanisms are understood, it is possible to optimize PV modules for different operating (climatic) conditions.

In this work 6-cell modules (used for electrical characterization) and corresponding 1-cell modules (destructive methods of analysis) with different encapsulants were laminated. Different behavior after aging was correlated with type of the encapsulant and the (non)-favorable encapsulant/backsheet combination.

Experimental

Three types of encapsulants were used (see table 1). Next to the standard EVA, the alternatives: thermoplastic polyolefin TPO and polyolefin elastomer POE were used. In order to investigate influence of crosslinking time, modules with low crosslinked EVA and POE were produced as well. The backsheet was PET laminate.

Table1. Description of encapsulants

Name	*Formation of acetic acid*	*Crosslinking*
EVA	Yes	Yes
TPO	No	No
POE	No	Yes

For every composition, 4 single-cell and 12 6-cells modules were made. One module was used as a reference, while each of the rest was used for particular accelerated artificial aging test (damp heat, DH; Sun simulation, and sequential UV/DH test). In this paper only results of sun simulation and damp heat test will be presented. The parameters of the aging tests are listed in table 2.

Table 2. Accelerated aging parameters

Aging test	**Parameters**	**t [h]**
Sun simulation [300-2500nm]	50°C, 40% RH, 1000W/m^2	1000h, 2000h
Damp heat (DH)	85°C, 85% RH	1000h, 2000h

The changes in optical properties of modules with varying encapsulants were measured via UV/Vis/NIR spectrometer (Perkin Elmer) with integrating sphere in hemispherical reflection mode. After the aging, the 1-cell modules were delaminated in order to investigate degradation of extracted encapsulants via FTIR-ATR spectroscopy.

Preliminary results and discussion

In Figure 1. the exemplary UV/Vis/NIR reflexion spectra of three encapsulants above the backsheet are shown. Integrating the area under the curves in the blue region, the reflectance loss [%] was calculated (see Figure 2.) After the 2000h of aging the indication of influence of material combination and aging type on degradation of PV module was obtained.

After the damp heat test all of the encapsulants showed decrease of reflectance (yellowing) at the encapsulant/backsheet interface (see Figure 1 and Figure 2). After the sun simulation test, the decrease was lower or even an increase was observed (in case of POE), which was expected since sunlight triggers the reduction of chromophoric oxidation products. However, it should be kept in mind that the sun simulation test is done only until 1000h so far. The yellowing at the encapsulant/cell interface was lower after the damp heat test when compared to the encapsulant/backsheet interface. The reason is probably lowered diffusion of the moisture towards the center of the module. Yellowing above the cell after the sun simulation test is however slightly higher for EVA and TPO.

It can be seen that EVA showed very similar yellowing at both interfaces indicating that the yellowing is not consequence of a material combination. On the other hand, TPO showed very high yellowing after the damp heat test at the encapsulant/backsheet interface, but lower yellowing at the encapsulant/cell interface. This result indicates strong interactions of the TPO encapsulant with the PET laminate after the tested conditions (in the dark). By comparison, the yellowing of TPO at encapsulant/cell interface is only slightly higher after the sun simulation tests than after the damp heat test.

Among all three encapsulant types, the standard crosslinked POE so far proved to be the most stable encapsulant regarding yellowing.

A reduction of crosslinking time of EVA (EVA*) and POE (POE*) resulted in lower degradation at the encapsulant/cell interface (see Figure 3.) However, the reflectance loss at the encapsulant/backsheet laminate was found to be higher than for the standard crosslinked samples.

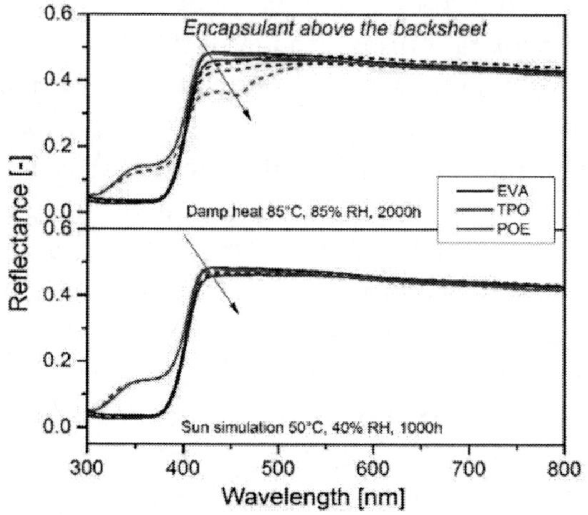

Figure 1. The UV/Vis/NIR spectra of reflection of encapsulants above the backsheet upon the agin

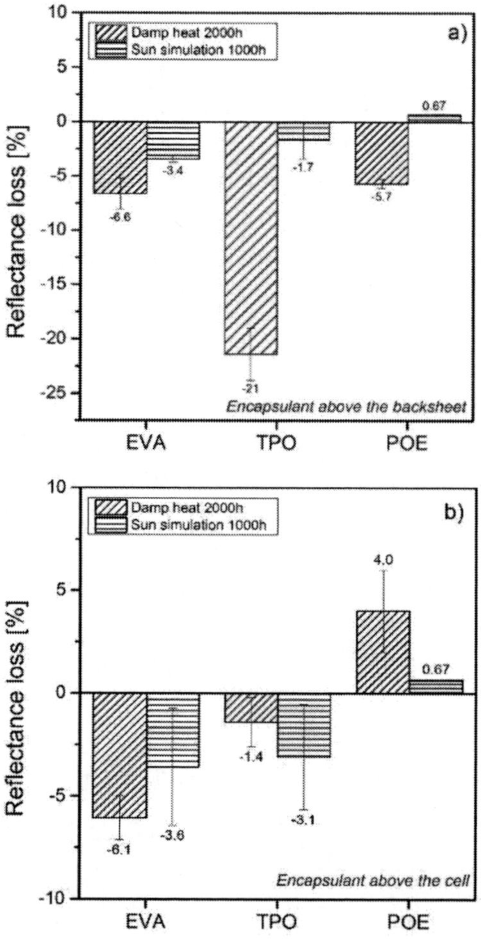

Figure 2. Calculated reflectance loss after damp heat and sun simulation tests at the

encapsulant/backsheet interface (a) and encapsulant/cell interface (b)

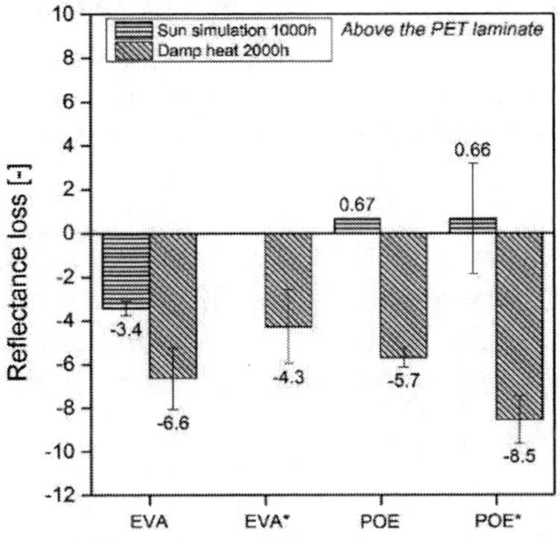

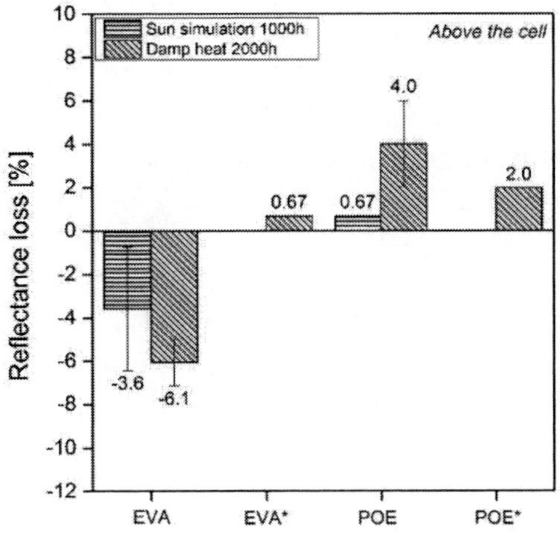

Figure 3. Calculated reflectance loss after damp heat and sun simulation tests at the encapsulant/backsheet and encapsulant/cell interface for the reduced crosslinking of EVA and POE

In Figure 4a-c. the FTIR-ATR spectra of extracted layers of encapsulants are represented. It can be seen that in all of the encapsulants certain degree of additive diffusion occurred. The spectra of the EVA after the aging did not significantly change except the formation of a broad peak around 1565 cm^{-1}, probably originating from the light stabilizer migration to the interface. However, the oxidation of TPO resulted in formation of lactones and ketones in different amounts depending on the aging conditions. At the two different interfaces (encapsulant/cell and encapsulant/backsheet) the degradation pattern was found to be the same. The aging of POE resulted in the formation of hydroxyl groups. Also diffusion of additives like for the other materials was observed.

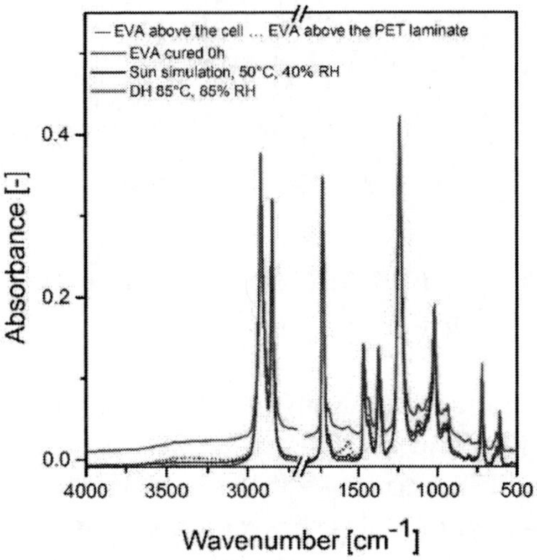

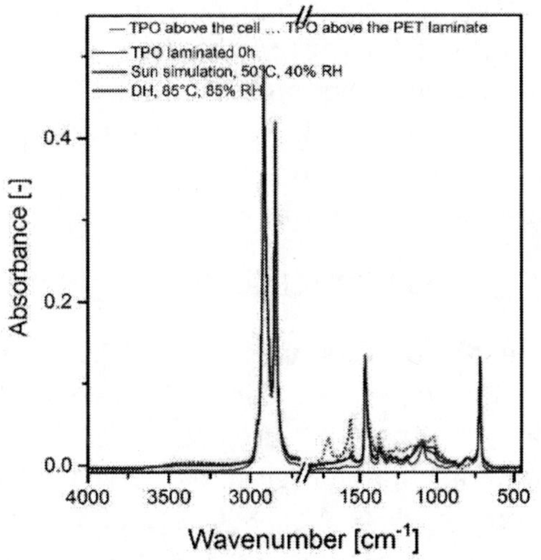

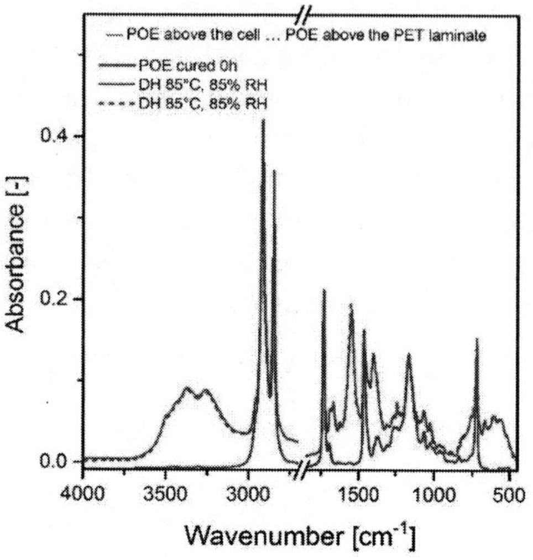

Figure 4. FTIR-ATR spectra of extracted encapsulants

Even-though the certain diffusion of the additives and reflectance loss occurred upon aging, it has to be kept in mind that none of the tested modules showed power loss so far. Nevertheless, the obtained results indicated (non)-favorable combinations of materials due to the interactions at the interfaces (in tested conditions and period of time). Reduced crosslinking of EVA and POE initially resulted in better behavior upon aging at the encapsulant/cell interface.

ACKNOWLEDGMENTS

This research work was performed at the Polymer Competence Center Leoben (PCCL) within the project "Infinity" (Energieforschungsprogramm 2015 - Leitprojekte, FFG No. 850414, Klima- und Energiefonds) in cooperation with the Chair of Materials Science and Testing of Plastics at the University of Leoben. The PCCL is funded by the Austrian Government and the State Governments of Styria and Upper Austria.

References:

[1] Köntges, M. et al: "Performance and reliability of photovoltaic systems-Subtask 3.2: Review of failures of photovoltaic modules", IEA PVPS Task 13, External final draft report, 2013.

[2] Oreski, G., Mihaljevic, A., Voronko, Y., Eder, G.C.: "Acetic acid permeation through photovoltaic backsheets: Influence of the composition on the permeation rate", Polymer Testing 60, 2017.

[3] Beinert, A., Peike, C., Dürr, I., Kempe, M.D., Reiter, G., Weiß, K.A.: "The influence of the additive composition on degradation induced changes in poly(ethylene-co-vinyl acetate) during photochemical aging", in 29th European PV solar Energy Conference and Exhibition, 22-26 September 2014, Amsterdam, The Netherlands

LONG TERM STABILITY TEST AND ANALYSIS OF AMORPHOUS SILICON GLASS-GLASS MODULES

U. Banik, N. Reininghaus, M. Vehse and C. Agert
DLR Institute of Networked Energy Systems
Carl-von-Ossietzky-Str. 15, 26129 Oldenburg, Germany

ABSTRACT: This paper focuses on analysis of damp heat tests performed on small 10cm x 10cm amorphous silicon glass on glass mini modules. The modules were prepared in-house in superstrate configuration with monolithic interconnections. Combination of encapsulants like EVA, Silicone, PVB with or without edge sealants like PIB and Silicone were tested during the experiment. The modules underwent 1000 hours in 85°C and at 85% RH as per IEC 61646 regulations. Performance and visual analysis of these a-Si:H modules were carried out at specific intervals during the test. Additionally, sensor modules using capacitive sensors and cobalt-chloride stripes modules were tested under DH separately to understand and correlate moisture ingress. This paper reports the best combination of materials that successfully withstands the test along with a comparison of moisture content measured inside various sensor modules. A brief analysis of the reasons of failure of other modules is also provided.
Keywords: thin film, damp heat, humidity sensor, degradation effect, encapsulant, edge sealant.

1 INTRODUCTION

Modern day solar panels are sold by manufacturers guaranteeing warranty of over 20 years. However, not enough statistical data are available on how the earliest installed modules fared after 20+ years in field[1]. Potential induced degradation (PID) is an undesirable phenomenon seen in PV modules that adversely impacts the output and reliability of the modules long after they have been installed onsite. Some causes for PID can be traced to corrosions from temperature and humidity exposure, system voltage, irradiance, manufacturing materials and techniques etc.[2]. Experiments showed that effects of degradation can result in reduction of shunt resistances, fill factor and open circuit voltage[2].

Instead of waiting for 20 years, another way to test or qualify modules for typical environmental usage is to put them through accelerated ageing tests in accordance to the IEC 61646 standard established by the International Electrochemical Commission. The so-called damp heat test (DH) is one of the environmental tests that a module must undergo to meet this standard. The intention of this is to check the exposure of modules to humidity by putting them in a chamber whose temperature is regulated to 85°C ± 2°C and relative humidity of 85% ± 5% for 1000 hours[1]. Moisture ingress in modules has been identified as a major concern in modern PV modules and the DH test is responsible for the most number of failures as tested in some labs[3]. It can weaken interfacial adhesive bonds, resulting in delamination and an increased number of ingress paths, resulting in a loss of passivation, electrochemical corrosion, and ultimately device failure [4].

Finite element method (FEM) simulation studies based on actual measurements showed that damp heat tests based on IEC cause comparable or even higher water uptake than a module in ambient outdoor exposure would[5]. Studies have showed that moisture ingress varies significantly based on the materials used for encapsulation [6][7]. Degradation of modules with EVA (ethylene-vinyl acetate) encapsulant has been previously studied with concluding evidence that formation of acetic acid within is detrimental[8]. Various humidity sensors have also been studied to quantify moisture penetration and compare water vapor transmission rates[9], [10]. Innovative methods to quantify moisture ingress like using low doped zinc oxide (ZnO) [11], calcium layers[12], [13] and cobalt chloride stripes[14] have also been investigated with success. Damp heat tests have also been carried out on thin-film modules to understand degradation processes and effects [16–18], but not many studies had concentrated on amorphous silicon modules.

The aim of this paper is to investigate the direct relationship between performance degradation and moisture ingress as well as to assess the encapsulating properties of various encapsulants and edge sealants in a-Si:H modules (see Table I). We used 3 types of samples that simultaneously underwent DH test according to IEC 61646: Working modules without additional sensors were used to measure the actual degradation of modules in terms of power output, dummy modules including a capacitive sensor were used to quantify the amount of moisture ingress causing the degradation, and another set of dummy modules containing several cobalt chloride stripes was used for a qualitative assessment to determine the origin of moisture ingress.

2 EXPERIMENTAL

A list of encapsulants and edge sealants used for our experiments, which are also commercially available, is given in Table I.

Material	Usage	Referred as
Liquid Silicone Elastomer	Encapsulant	Silicone
Ethylene-vinyl acetate sheet	Encapsulant	EVA
Polyvinyl butyral sheet	Encapsulant	PVB
Polyisobutylene tape	Edge sealant	PIB
Thixotropic adhesive silicone sealant	Edge sealant	Silicone S

Table I List of components used for encapsulation of modules in the experiment.

2.1 a-Si:H working modules

Deposition of a-Si:H pin layers were done on factory made TCO substrates (AZO on glass) using the PECVD reactors of a cluster tool type CS400PS by Von Ardenne GmbH. P1, P2 and P3 laser scribes for monolithic interconnections were done using a 3D-Micromac laser system. For the back-contact we used a combination of in-house AZO (100nm) and silver (70nm) deposited by sputtering and electron beam deposition respectively. Sputtering was done in the cluster tool described above, whereas the electron beam deposition was performed in a Dreva Lab 450 by Vakuumtechnik Dresden GmbH. All modules were edge isolated 5mm from the edge using the laser scribes so that no active layer comes in contact with moisture directly.

The connectors or bus bars out of the modules were soldered on top of the silver surface using an ultrasonic soldering iron (MBR electronic USS 9210). A sketch of the cross-sectional configuration of the active modules is shown in Figure 1 whereas a photograph of the completed module before encapsulation is shown in Figure 2. Lamination of the modules with encapsulant and edge sealant materials was done in a P. Energy L094A laminator under 1034 mbar pressure and variable temperature depending on the material.

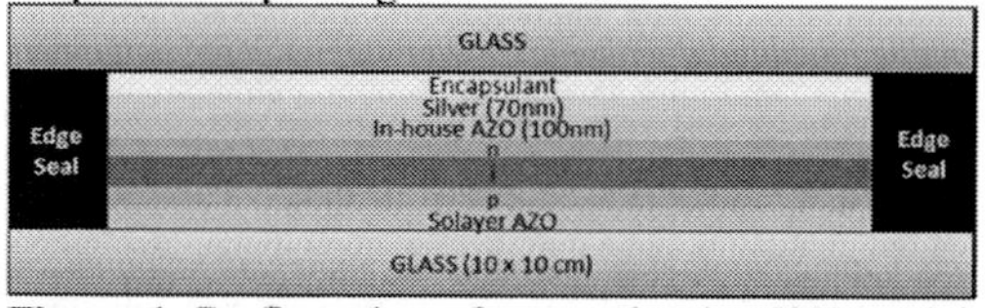

Figure 1 Configuration of a completed a-Si:H module structure. Laser scribes are not represented here.

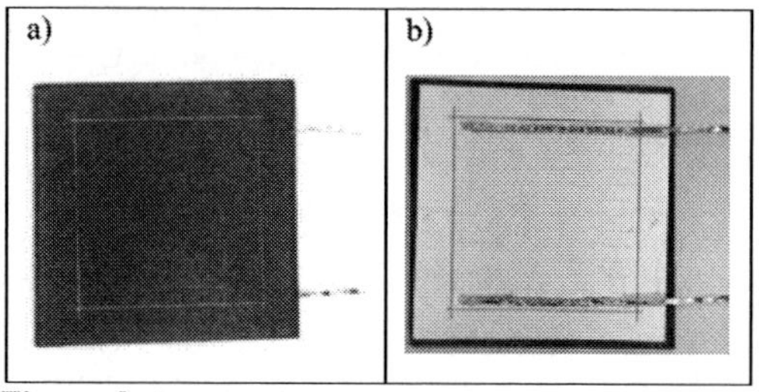

Figure 2 Completed module before being encapsulated, a) module as seen from the front side b) back side view with silver being the white opaque layer.

Only for the modules with Silicone S, we used the back-cover glass of 9cm x 9cm. The sealant was applied after laminating with the encapsulant. This allowed us to apply the sealant with a manual caulking gun onto the empty spaces (~1cm) left due to the difference in sizes of the two glasses. The same lamination technique was used to manufacture the cobalt chloride sensor modules explained in the Section 2.3

2.2 Capacitive sensor modules

We used two types of capacitive sensor, namely MK33-W and P14 Rapid, both from Innovative Sensor technology (IST). Before putting the sensor into the modules, they were soldered to tin and silver alloy wires and calibrated. Measurements were taken at ambient conditions as well as inside DH conditions at 85%RH and 85°C. As the MK33-W could be used with a higher operating temperature of 190°C[18], they were used for the EVA modules, which required higher temperature during lamination. For all other modules, the P14 was used, as in their case the curing temperature was below 150°C[19]. An example of our capacitive modules is shown in Figure 3

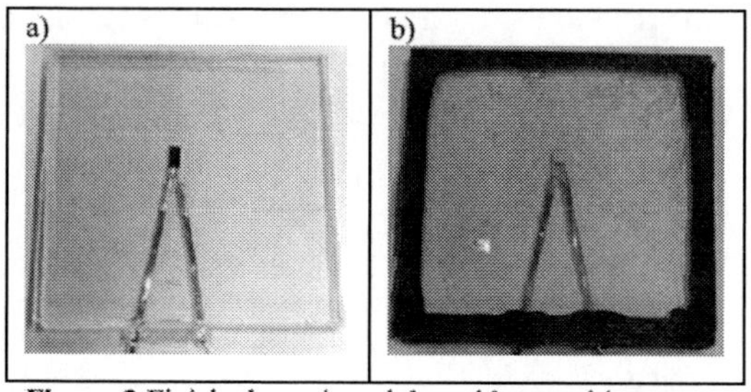

Figure 3 Finished quasi modules with capacitive sensors. a) EVA with MK33-W sensor and b) Silicone+PIB with P14-Rapid sensor.

Figure 4 shows capacitance versus humidity for both sensors measured at 25°C showing a good linearity in both cases. When compared to measurements taken at 85°C (not shown here) the sensors were more stable and had a better linear response at 25°C. Therefore, all capacitive measurements during the DH period for all samples were done by taking the sample out of the chamber and letting it cool down to room temperature.

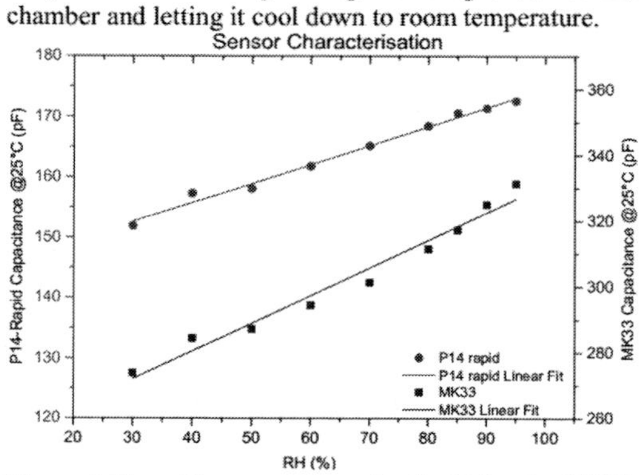

Figure 4 Capacitive measurements of a P14 and a MK33 type sensor at varying humidity and at constant temperature of 25°C

2.3 Cobalt chloride sensor modules

Cobalt chloride stripes were heated in the oven for about 1 minute at 80°C before encapsulating them with glass to remove residual moisture absorbed by the stripes from the environment. Photos of some of the prepared $CoCl_2$ modules are shown in Figure 5.

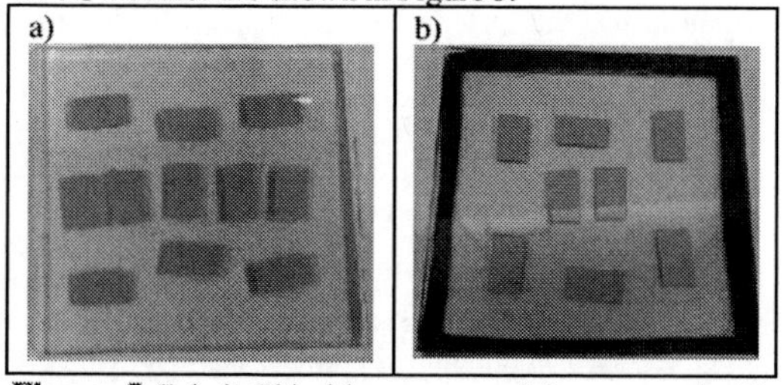

Figure 5 Cobalt Chloride sensor modules prepared with a) EVA + Silicone S and b) EVA+PIB

The $CoCl_2$ stripes turn from blue to pink when exposed to moisture. This means that they only allow a qualitative detection or a very rough quantitative determination of moisture ingress. Furthermore, due to various probable chemical reactions of the $CoCl_2$ with the encapsulants as well as the fact that different encapsulants could change the coloration in different ways, the conversion of color from blue to pink was not uniform over various samples. This made the comparison of humidity ingress even more difficult and indirect.

In an attempt to still quantify the humidity change using the $CoCl_2$ stripes, we determined a parameter to understand the change of spectrum of the $CoCl_2$ stripes which we named as R factor. It is the ratio between the reflectivity at 690nm to the reflectivity at 440nm, which describes how pink the stripe is. In particular, 690 nm corresponds to an absorption band related to the tetrahedral complex of $CoCl_2$ [20]. For each sample the result was then normalized to the value measured before degradation. This way the relative increase of *R* could be monitored, where an increase in R corresponds to an increase in humidity.

3 RESULTS AND DISCUSSION

Comparison of performance of a:Si-H modules made from various encapsulants are provided in Figure 6. The power at maximum power point (P_{mpp}) for the modules at each time interval was normalized to its P_{mpp} values at 0h to compare the degradation.

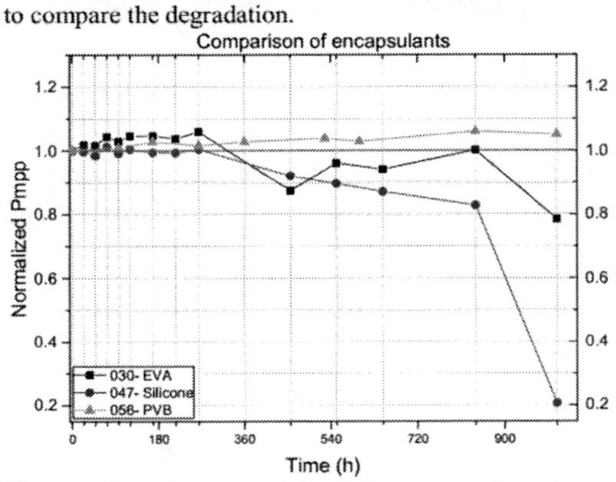

Figure 6 Performance degradation comparison for a-Si:H modules made from various encapsulants after 1008H

It can be seen from the graph that PVB as an encapsulant performed far better than silicone or EVA. Both EVA and silicone modules had visual degradation marks and failed the DH test as they lost more than 10% of the maximum power. To find out the reason behind such degradation we prepared capacitive humidity sensor modules with PVB and EVA as encapsulants. Data for liquid silicone are not shown here, as encapsulating the capacitive sensor led to the creation of air bubbles, leading to unreliable results. The findings from the sensor modules are provided in Figure 7. The results show that the rise in moisture level under DH conditions for the EVA module was far more than with PVB. This moisture ingress and saturation had a detrimental effect on the performance.

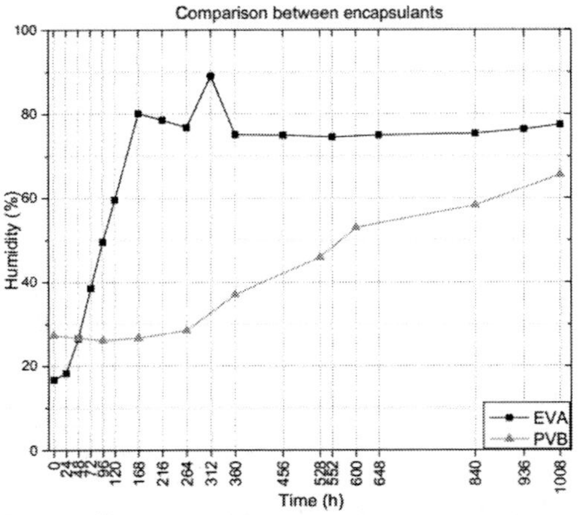

Figure 7 Change of humidity inside EVA and PVB sensor module during DH experiment.

To compare the difference in moisture at specific points of the module and to understand the path of moisture ingress we made cobalt chloride sensor modules for EVA, Silicone and PVB. Figure 8 shows for each encapsulant a comparison of the normalized R factor between the center and the edge of the modules.

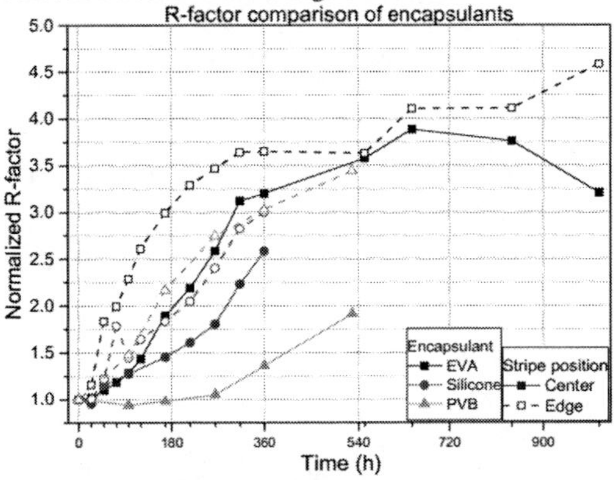

Figure 8 R factor comparison of cobalt chloride modules with various encapsulants

From this graph, we can see that for all encapsulants the edge stripes of the modules were always exposed to higher humidity compared to the center stripes. This clearly shows that an edge sealing component is required to stop moisture penetration from edges. Furthermore, we again observe that PVB leads to a much lower moisture ingress rate compared with silicone and EVA, suggesting it is the best of the three encapsulants.

Figure 9 shows a comparison of the normalized R factor for modules without an edge sealant and with PIB and Silicone S edge sealants, respectively. EVA was used as the encapsulant in all cases. It helps in comparing the effect of edge sealants on moisture ingress.

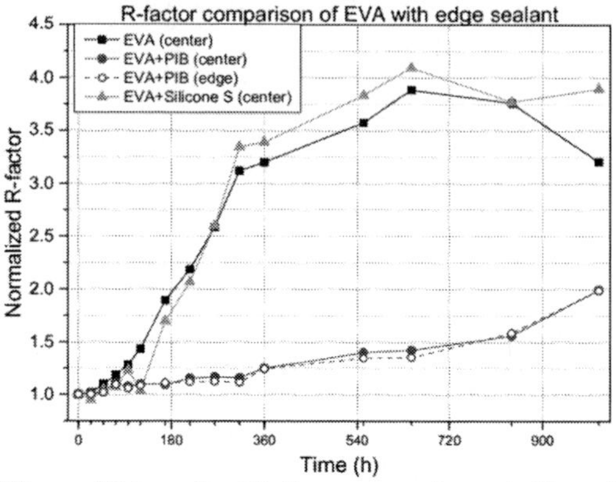

Figure 9 Normalized R factor comparison of effect of PIB and Silicone S edge sealants

The graph clearly shows that PIB blocked moisture significantly better than Silicone S. The Silicone S module almost reached the same values as the module without a sealant, meaning that the silicone sealant is not effective in blocking moisture ingress. In case of the PIB sealant, R factor values for the edge and center were almost the same, which meant that there was no significant moisture ingress from the sides anymore.

Finally, having established PIB as an effective edge sealant, we can now compare the actual moisture blocking capabilities of the encapsulants. Accordingly, Figure 10 shows a comparison of the three encapsulants in terms of relative humidity determined from capacitive sensor modules. Here, PIB was used as the edge sealant in all samples.

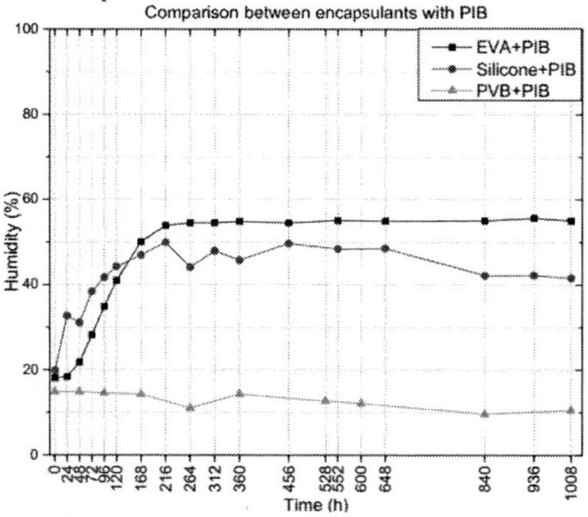

Figure 10 % Humidity comparison of capacitive modules prepared from various encapsulants and PIB

The experimental results show clearly that PVB+PIB was the best performer where the humidity did not rise. Rate of rise in moisture recorded inside the EVA+PIB was slower than Silicone+PIB but eventually the former reached a higher saturated humidity value. While with EVA and Silicone the final humidity was much higher than with PVB, all samples show values below 60%. As the PVB module without any edge sealant reached similar values without degrading (see Figure 6 and Figure 7), we decided to carry out performance tests on complete working modules for all three encapsulants.

Figure 11 shows a visual comparison of these modules after 1008h of DH test. EVA showed significant visual degradation whether used with PIB, silicone or without edge sealant. Significant cloudiness of the EVA was also observed, but degradation of the optical properties of the encapsulant would not hamper module performance in superstrate configuration. However, degradation of the back silver layer was evident, as can be seen in Figure 11 (c) and (d). The probable causes for such degradation could be due to formation of acetic acid[21][22]. Thus, EVA was excluded from the performance evaluation. No such degradation was seen with the other encapsulants.

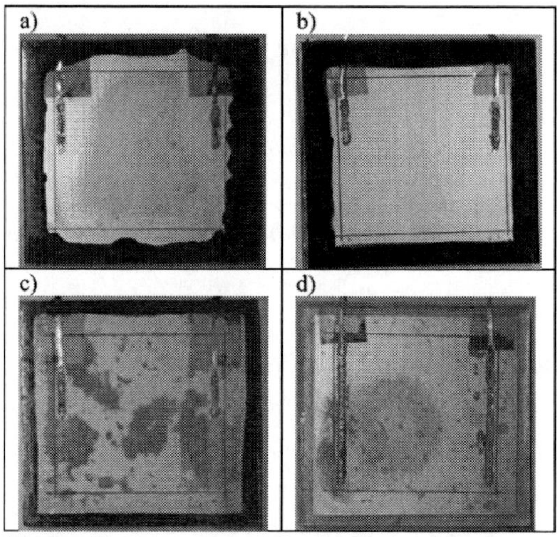

Figure 11 Comparison of visual degradation (back side) for a) Silicone+PIB b) PVB+PIB c) EVA+PIB d) EVA after 1008H or DH test

Performance analysis results of PVB and Silicone with PIB edge sealant are shown in Figure 12.

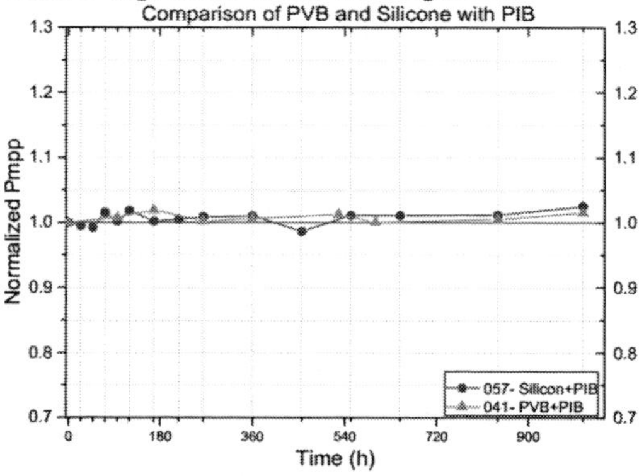

Figure 12 Performance degradation comparison for PVB+PIB and Silicone+PIB

Both these combinations passed the DH test based on the IEC 61646 regulations. Even after 1008H of DH test they did not show significant performance degradation or visual degradation patches. The electrical performance results were also well supported by our humidity sensor experiments.

4 CONCLUSION

The preliminary study of R factor gave us a better understanding of the moisture ingress rate in the modules with $CoCl_2$ as a sensor by measuring the change of color from blue to pink. $CoCl_2$ stripes can successfully provide location specific measurements unlike capacitive sensors. The results from the cobalt chloride sensors were in line with the capacitive sensor and shows potential for future research.

Among the a-Si:H modules tested we found that using EVA with PIB could cause moisture trapping in modules and could accelerate formation of acetic acid causing more degradation. Based on current knowledge, silicone or PVB encapsulant does not produce any threat to formation of acidic components and hence produces best results with PIB after 1008h of DH. Electrical performance analysis suggests very little degradation for these combinations after 1000h in DH. In comparison to Silicone S, PIB showed impressive resistance to moisture ingress with any encapsulant. This was supported well by our sensor measurements and is in agreement with other studies from the literature[23][13]. As an encapsulant PVB was far the best even when used without an edge sealant, whereas both EVA and Silicone failed.

5 ACKNOWLEDGEMENTS

The authors gratefully acknowledge the financial support for this work provided by INTERREG-Programme Deutschland-Nederland co-financed by the resources of the European Regional Development Fund (ERDF), the Ministry of Economy, Energy, Industry, SME and craft of NRW (MWEIMH NRW), the Ministerie van Economische Zaken, the Lower Saxony state chancellery and the Provinzen Gelderland en Overijssel. The authors would also like to recognize and thank the efforts put by Volker Steenhoff who helped in review process, Martin Kellerman who assisted with depositions and Ortwin Siepmann who helped us with laser scribes.

6 REFERENCES

[1] R. Arndt and R. Puto, "Basic understanding of IEC standard testing for photovoltaic panels," ... *[2012]. Http//Tuvamerica. Com/Services/Photovoltaics/* ..., no. 978, pp. 1–15, 2010.

[2] Advanced Energy Industries, "Understanding Potential Induced Degradation," *White Pap. - AE*, 2013.

[3] S. a. Sala, M. Campaniello, and a. Bailini, "Experimental study of polymers as encapsulating materials for photovoltaic modules," *2009 Eur. Microelectron. Packag. Conf.*, pp. 1–7, 2009.

[4] G. J. Jorgensen *et al.*, "Moisture transport, adhesion, and corrosion protection of PV module packaging materials," *Sol. Energy Mater. Sol. Cells*, vol. 90, no. 16, pp. 2739–2775, 2006.

[5] P. Hülsmann and K. A. Weiss, "Simulation of water ingress into PV-modules: IEC-testing versus outdoor exposure," *Sol. Energy*, vol. 115, pp. 347–353, 2015.

[6] P. Hülsmann, M. Heck, and M. Köhl, "Simulation of Water Vapor Ingress into PV-Modules under Different Climatic Conditions," *J. Mater.*, vol. 2013, pp. 1–7, 2013.

[7] M. Kempe, "Overview of scientific issues involved in selection of polymers for PV applications," *Conf. Rec. IEEE Photovolt. Spec. Conf.*, pp. 000085–000090, 2011.

[8] C. Peike *et al.*, "Origin of damp-heat induced cell degradation," *Sol. Energy Mater. Sol. Cells*, vol. 116, pp. 49–54, 2013.

[9] M. Jankovec *et al.*, "Long term stability of humidity sensors , laminated in EVA encapsulant," pp. 1–6, 2000.

[10] B. Lalaguna, P. Sánchez-Friera, F. Ropero, J. F. Gil, and J. Alonso, "Comparison of Moisture Ingress in Pv Modules With Different Backsheets Using Humidity Sensors," *Pv-Sec*, no. September, p. 3, 2008.

[11] C. B. S.Pélisset, R.Théron, M.Barnéoud-Raéis, L.-E.Perret-Aebi, M.Benkhaira, "Monitoring Water Vapour Penetration Using a Contactless Technique," *Pv-Sec*, no. September, pp. 3561–3564, 2009.

[12] M. O. Reese, A. a. Dameron, and M. D. Kempe, "Quantitative calcium resistivity based method for accurate and scalable water vapor transmission rate measurement," *Rev. Sci. Instrum.*, vol. 82, no. 8, 2011.

[13] M. D. Kempe, a. a. Dameron, T. J. Moricone, and M. O. Reese, "Evaluation and modeling of edge-seal materials for photovoltaic applications," *Conf. Rec. IEEE Photovolt. Spec. Conf.*, no. February, pp. 256–261, 2010.

[14] M. Miyashita, S. Kawai, A. Masuda, and S. Plant, "Measuring Method of Moisture Penetration Into Photovoltaic Modules," vol. 3, pp. 3–4, 2011.

[15] D. W. Lee *et al.*, "Effects of ZnO:Al films on CIGS PV modules degraded under accelerated damp heat," *Sol. Energy Mater. Sol. Cells*, vol. 105, pp. 15–20, 2012.

[16] M. D. Kempe, K. M. Terwilliger, and D. Tarrant, "Stress induced degradation modes in CIGS mini-modules," *Conf. Rec. IEEE Photovolt. Spec. Conf.*, 2008.

[17] R. Knecht, M. Knipper, I. Riedel, J. Parisi, and B. Sample, "damp -heat exposure on the electronic properties of Cu (In , Ga)(S , Se) 2 thin film solar cells," p. 2, 2009.

[18] Innovative Sensor technology, "MK33-W Capacitive Humidity Sensor Optimal for oil measurement applications MK33-W Capacitive Humidity Sensor Optimal for oil measurement applications." [Online]. Available: https://www.ist-ag.com/sites/default/files/DHMK33-W_E.pdf. [Accessed: 05-Apr-2017].

[19] Innovative Sensor technology, "P14 Rapid Capacitive Humidity Sensor Optimal for weather balloons and radio sondes P14 Rapid Capacitive Humidity Sensor Optimal for weather balloons and radio." [Online]. Available: https://www.ist-ag.com/sites/default/files/DHP14-Rapid_E.pdf.

[20] H. Morita, M. Miyashita, and A. Masuda, "Moisture ingress rate and route into c-Si PV modules," *Proc. Photovolt. Modul. Reliab. Work. 2014*, p. 601, 2014.

[21] C. Peike *et al.*, "Origin of damp-heat induced cell degradation," *Sol. Energy Mater. Sol. Cells*, vol. 116, pp. 49–54, 2013.

[22] J. Pern, "Module Encapsulation Materials, Processing and Testing (Presentation)," 2008.

[23] J. Pern, "Module Encapsulation Materials, Processing and Testing (Presentation)," *APP Int. PV Reliab. Work.*, pp. 1–33, 2008.

MODELING AND SIMULATION OF NON-UNIFORM ENCAPSULANT DISCOLORATION EFFECT IN CRYSTALLINE-SILICON PHOTOVOLTAIC MODULES

Humaid Mohammed*, **Archana Sinha, Rajesh Gupta**
Department of Energy Science and Engineering, Indian Institute of Technology Bombay, Mumbai- 400076, India
*Email: humaidmohd@iitb.ac.in, Phone: (+91-22) 2576 7837

ABSTRACT: Crystalline-silicon (c-Si) photovoltaic (PV) modules generally suffer from discoloration of encapsulation material after prolonged exposure to UV radiation under high temperature conditions, which ultimately affects the performance and reliability. In this work, distributed diode modelling approach of solar cell along with circuit simulation have been used to evaluate the effect of encapsulant discoloration area, discoloration intensity and number of weak cells on the performance of c-Si PV module. Results show that the discoloration primarily resulted in the reduction of short circuit current (I_{SC}) which further leads to loss in open circuit voltage (V_{OC}), fill factor (FF), maximum power (P_{Max}) and slight increase in series resistance (R_S) of module. P_{Max} loss has been quantified for yellow and brown discoloration for different discoloration areas. The variation in discoloration from yellow to brown, extending over 80% of the solar cell area leads to the P_{Max} loss in the range of 8-33%. Under the case of non-uniform discoloration, P_{Max} and mismatch power losses increased non-linearly with the number of weak cells and their discoloration extent. The presented approach can be used to study the impact of discoloration on the performance of PV module.

Keywords: c-Si, PV Module, Encapsulation, Modeling, Simulation

1 INTRODUCTION

Crystalline silicon (c-Si) PV module is a multi-layered structure, which comprises of front glass cover, solar cells sandwiched between two layers of encapsulant, back-sheet and aluminum frame. An encapsulant is a transparent polymeric material used as an adhesive in the structure of PV module to hold the solar cells in their position. These characteristics of encapsulant material enable it to perform vital functions such as structural support, optical transparency, electrical and physical isolation to the solar cells and different components from external environment. It has been observed that the front encapsulant layer starts to degrade when exposed to UV radiation, high temperature and humidity [1]. Recent studies [2]–[5] indicate that PV modules installed at hot and humid climatic locations are more prone to encapsulant discoloration. In most of the cases, the discoloration is visualized due to change in color of the encapsulation material from transparent to yellow or brown in the central region of solar cell.

Many studies have been conducted on the analysis and characterization of degraded encapsulant samples using different experimental methods, such as UV-visible and IR spectroscopy, thermo-gravimetric analysis, solvent extraction etc. [6]–[10] These methods are generally destructive and require degraded encapsulant sample from the module. Moreover, these methods only provide information about the chemical and structural changes that have occurred in the encapsulant. There are few studies where destructive methods have also been used to investigate the effect of encapsulant discoloration on electrical performance of solar cell [10] and modules [11]. Nondestructive characterization can also be done using imaging techniques such as dark lock-in thermography (DLIT) and electroluminescence (EL) for investigation of encapsulant discoloration in c-Si PV modules [12]. However, these methods only give qualitative information about the non-uniform discoloration and its effects. Performing a current-voltage (I-V) measurement on a PV module with visible discoloration is not sufficient to infer any conclusion about the effect of discoloration on module performance because it only shows the global effect of all the defects present in the module. Moreover, the weakest cell in the string of module will influence the performance and mismatch power loss of the module. Therefore, to study the effect of encapsulant discoloration in isolation, a distributed diode modeling and simulation approach has been used. The advantage of this approach is that it explicitly takes into account the spatial distribution of encapsulant discoloration and its effects. The aim of this work is to analyze the effect of non-uniform encapsulant discoloration on PV module's performance and quantification of power loss for different extents of discoloration using a distributed diode modeling and simulation approach.

2 METHODOLOGY

Distributed diode model of a solar cell is the electrical equivalent circuit with a three dimensional structure. This model utilizes the fact that current generation does not take place uniformly in a solar cell. The distributed diode model is constructed by dividing each solar cell of a module into a number of sub cells, as shown in Fig. 1 [13]. In this work, a typical solar cell was divided into 10000 sub cells so as to form a mesh of 100x100 nodes. Each sub cell has a representative node, which is modeled like a single diode model comprising of a current source, a diode and a shunt resistance connected in parallel.

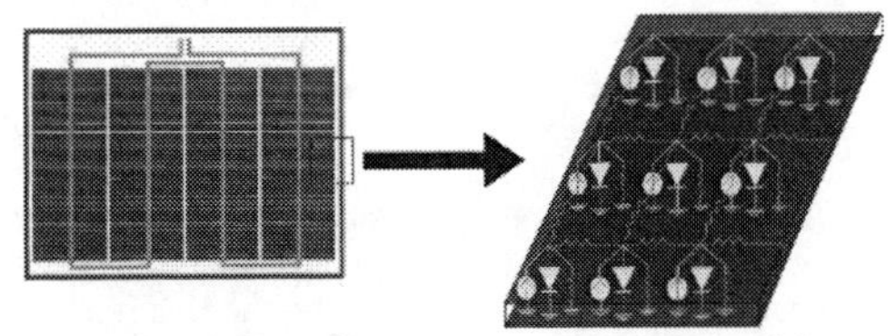

Figure 1: Distributed electrical model of PV module

All the nodes of a cell are interconnected with a series resistance (R_S) component and depending upon the location of node, the R_S component will be either, sheet resistance, finger or busbar resistance. Typical values of c-

Si solar cell parameters have been considered in the modelling [14]. The model has been simulated by PSpice circuit simulator [15]. Initially, simulation result of single solar cell has been cross-checked and latter 12 similar cells have been connected in series to model the module. One of the main traits of encapsulant discoloration is that the pattern appears in central portion of solar cell extending over the busbar as shown in Fig. 2. As a result, the transmission of light radiation incident on the solar cell decreases at the site of discoloration, which consequently reduces the current generation. The extent of discoloration can be characterized by the change in color of encapsulant from transparent to yellow or brown and the area over which the discoloration extends. In order to simulate different extents of discoloration; two parameters have been considered 1) fractional discoloration area (DA) of cell and 2) discoloration intensity (DI). The fractional DA of cell is defined as the ratio of the solar cell area affected by discoloration to the total cell area as given in eq. (1). DI is defined as the equivalent of percentage loss in incident radiation as given in eq. (2). In the present work, the equivalent of fractional DA of cell and DI are varied in the simulation to study various cases of discoloration.

$$\mathrm{DA} = \frac{Discolored\ area}{Total\ cell\ area} \tag{1}$$

$$\mathrm{DI} = \frac{Loss\ in\ radiation}{Incident\ radiation} x100 \tag{2}$$

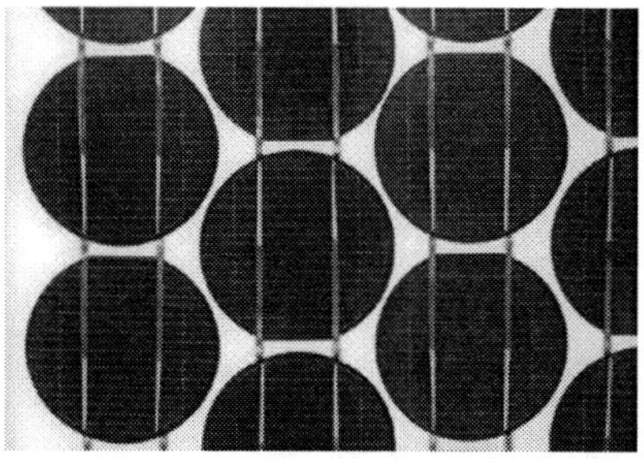

Figure 2: PV module having discoloration effect in the center of cells

In a PV module, the discoloration extent can either be uniform or non-uniform. The non-uniformity arises due to dissimilarity in either discoloration area (DA), discoloration intensity (DI) or both. The effect of encapsulant discoloration on PV module performance has been analyzed based on two factors. One factor is related to extent of discoloration which is the combined effect of DI and DA and the other factor is related to the number of cells in a module with most severe discoloration, which has been considered as weak cells. In case of uniform discoloration, weak cells will not be present, because all the cells would have the same DA and DI. Non-uniform discoloration also leads to inhomogeneity in performance of solar cells. As a result of this, each cell does not operate at their individual maximum power point. Therefore, the maximum power of the module comes out to be less than the sum of individual maximum power of cells. This difference has been termed as mismatch factor (*MF*), which is the ratio of module maximum power to the sum of the individual maximum power of cells as given in eq. (3)

$$MF = \frac{Maximum\ power\ of\ module}{\sum Maximum\ power\ of\ cell} \tag{3}$$

3 RESULTS AND DISCUSSION

In this section, simulation results obtained for change in electrical parameters of module under the impact of uniform and non-uniform discoloration are presented. The variation in P_{Max} loss and *MF* under different cases of discoloration have also been discussed.

3.1 Estimation of uniform discoloration impact on PV module performance

The impact of uniform encapsulant discoloration has been estimated by performing simulation to find the relative deterioration in performance of a module at two discoloration intensities: yellow and brown. Pern *et al.* performed the study on these two intensities of discoloration and found that yellow and brown discoloration leads to approximately 10% and 40% reduction in short circuit current (I_{SC}) respectively [10]. These two quantified levels of discoloration have been mainly considered as two reference points to estimate the P_{Max} reduction in a PV module for different fractional DA of cell. Two intermediate levels (corresponding to 20% and 30% reduction in I_{SC}) and two extreme levels beyond brown discoloration (corresponding to 50% and 60% reduction in I_{SC}) have also been considered to investigate the trend for wide variation in discoloration extents. For modeling DI of discoloration, radiation incident on solar cell has been reduced in the same proportion as per the reduction in I_{SC} given by Pern *et al.*[10].

The electrical parameters of module such as I_{SC}, open circuit voltage (V_{OC}), R_S, fill factor (*FF*) and P_{Max} have been obtained under the effect of uniform discoloration for varying DI and different fractional DA of cells. The relative change in these parameters has been estimated with respect to electrical parameters of a non-discolored PV module under STC in terms of ratio and plotted in Fig. 3 against DI.

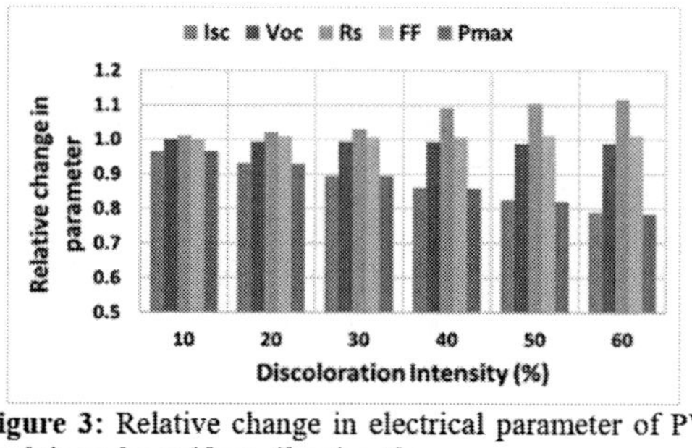

Figure 3: Relative change in electrical parameter of PV module under uniform discoloration

Fig. 3 shows that P_{Max} and I_{SC} are significantly affected by discoloration whereas other electrical parameters such as V_{OC}, R_S and *FF* are less affected. The decline in I_{SC} is due to reduction in incident radiation on solar cell at the site of encapsulant discoloration. Since, V_{OC} has logarithmic dependence on incident light radiation; therefore, for low levels of DI, the variation in V_{OC} is negligible. But as the DI increases, the decline in V_{OC} becomes noticeable due to increased voltage mismatch within a solar cell. The value of R_S increases with increase in DI because at the site of discoloration the voltage drop will be less which leads to increases in overall series resistance of module. However, the increase in R_S is not very large due to which its effect is not being reflected on FF. Descending trend of P_{Max} with DI is mainly due loss in I_{SC} with minor contribution from V_{OC}.

P_{Max} loss has been calculated for varying DI and plotted in Fig. 4 for different DA. It shows that for a constant DA, P_{Max} loss increases linearly with DI. Moreover, the P_{Max} loss increases at a faster rate for larger DA because of increased contribution from loss in I_{SC} and V_{OC}. The variation in DI from yellow to brown extending over 80% of the solar cell can lead to the P_{Max} losses in the range of 8 to 33%, which indicates that discoloration can result in significant deterioration of PV module performance. The results of uniform encapsulant discoloration is following the general trend of experimental results reported in literature [4]-[5]. Since P_{Max} directly transforms to efficiency, therefore increase in DI leads to reduction in efficiency of module.

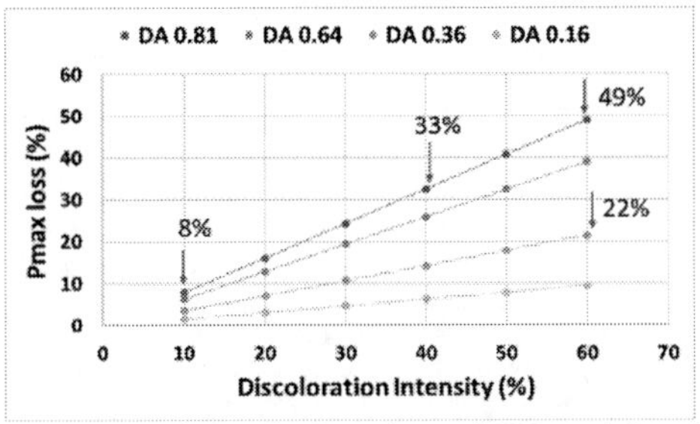

Figure 4: Variation in P_{Max} loss of PV module with discoloration intensity (DI) for different discoloration area (DA) under uniform discoloration.

3.2 Effect of weak cells on PV module performance

In case of non-uniform discoloration, module performance influenced by two factors, the percentage of weak cells in the module and their discoloration extent. The impact of weak cells on the PV module performance has been analyzed by simulating few cells in a module with relatively intense discoloration as compared to others, with each cell having same DA. In such a situation, module performance will be influenced by the cells with higher DI which will also limit the I_{SC} of the module and act as weak cell. By varying the number of weak cells in a module, while keeping their discoloration extents unaltered, electrical parameters such as I_{SC}, V_{OC}, R_S and FF and P_{Max} have been estimated. The relative change in these parameters has been estimated with respect to electrical parameters of a non-discolored PV module under STC in terms of ratio and plotted in Fig. 5 against percentage of weak cell. It shows that all the electrical parameters are not changing significantly. The value of I_{SC} is almost constant with increase in percentage of weak cells because the discoloration extent of the weak cell is unaltered and consequently the I_{SC} of module is limited by the performance of weak cell. Change in V_{OC} is negligible because the discoloration extent of weak cell is not resulting in significant loss of V_{OC} of weak cell. R_S increases slightly with the increment in percentage of weak cells due to reduced voltage drop at the site of discoloration in weak cells which leads to increase in overall series resistance of module. Although the DA is same throughout the cells of module, FF varies slightly due to shift in operating point of the module under current mismatch situation, as the number of weak cells in module increase. P_{Max} decreases with increase in percentage of weak cells mainly due to reduction from I_{SC}, with minor contributions from V_{OC} and FF.

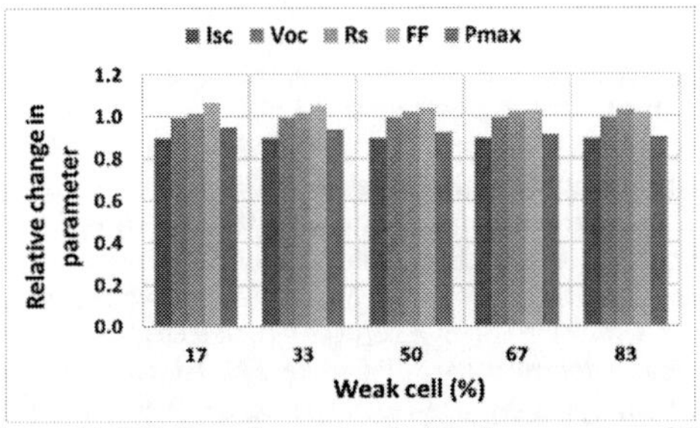

Figure 5: Relative change in electrical parameter of PV module with percentage of weak cell under non-uniform discoloration

For varying number of weak cells, P_{Max} loss has been calculated and plotted in Fig. 6 for different DA. It shows that P_{Max} loss increases almost linearly with increase in percentage of weak cells for small fractional DA of cells due to lesser amount of current mismatch. But, as the fractional DA of cell increases, the P_{Max} loss of module increases non-linearly with increase in percentage of weak cell, due to high current mismatch. This can also be comprehended from the fact that for small DA the variation in P_{Max} is quite less as compared to the variation in P_{Max} for larger fraction DA.

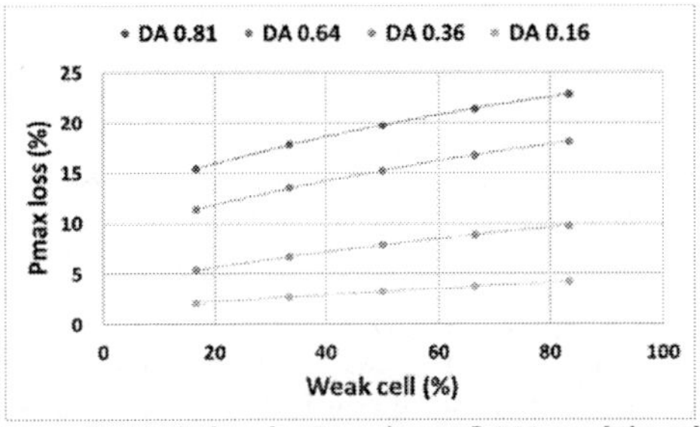

Figure 6: Variation in P_{Max} loss of PV module with percentage of weak cells for different discoloration area (DA) under non-uniform discoloration

MF has also been estimated simultaneously and results show that it does not vary significantly with increasing percentage of weak cells in a module. This is because MF is strongly influenced by the amount of current mismatch, and the discoloration extents of weak and healthy cells are kept unchanged. It has also been found that MF is lowest when number of weak cells is minimum, because in such a situation, disparity between maximum power point of weak cell and maximum power point of rest of the module is widest.

3.3 Effect of varying discoloration intensity of weak cell on PV module performance

The discoloration extent of weak cell is an important parameter as it determines the amount of current mismatch, which will influence the performance of PV module indirectly. The impact of DI of weak cells on the PV module performance has been analyzed by simulating cells with discoloration conditions similar to the previous case. By varying the DI of weak cells in a PV module, electrical parameters such as I_{SC}, V_{OC}, R_S, FF and P_{Max} are estimated.

The relative change in these parameters have been estimated with respect to electrical parameters of a non-discolored PV module under STC in terms of ratio and plotted in Fig.7 against DI of weak cell. It shows that I_{SC} and P_{Max} are majorly affected by the DI of weak cells. Other electrical parameters such as R_S and FF change slightly with increase in DI of weak cell, whereas the change in V_{OC} is almost negligible. It can be observed that I_{SC} of module declines corresponding to DI of weak cell. Change in V_{OC} is small because the discoloration extent of weak cell is not resulting in significant reduction. The value of R_S increases slightly because as the DI of weak cell increases the voltage drop at the site of discoloration reduces which leads to increase in overall series resistance of module. *FF* varies slightly due to shift in operating point of the module as the amount of current mismatch in module increases with increase in DI of weak cell. P_{Max} decreases consistently with increase in DI of weak cell due to major contributions from I_{SC} and minor contribution from V_{OC} and *FF*.

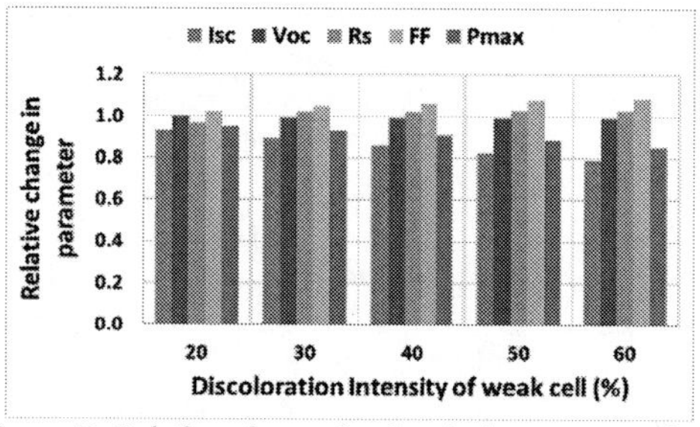

Figure 7: Relative change in electrical parameter of PV module with percentage discoloration intensity (DI) of weak cell under non-uniform discoloration

P_{Max} loss has been calculated corresponding to different DI and plotted in Fig. 8 for different DA. It shows that P_{Max} loss increases almost linearly with increase in DI of weak cells for small DA. But as the fractional DA of cell increases, the P_{Max} loss of module increases non-linearly with increase in DI of weak cells due to high current mismatch. It can also be explained on the on the basis of variation in P_{Max} which is less for small DA and more for large DA.

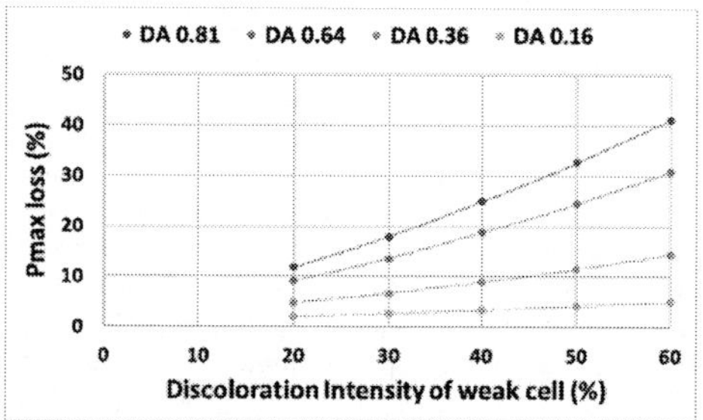

Figure 8: Variation in P_{Max} loss of PV module with discoloration intensity (DI) of weak cell for different discoloration area (DA) under non-uniform discoloration

MF has also been estimated simultaneously and plotted against increasing DI of weak cells for different fractional DA of cells in Fig. 9. Results show that the *MF* variation is insignificant for small fractional DA of cell but for larger fractional DA of cell MF varies non-linearly. This is because *MF* is strongly influenced by the amount of current mismatch. Therefore, the MF increases approximately linearly for small fractional DA of cell. It has been estimated that additional 25% of P_{Max} can be lost due to current mismatch under non-uniform encapsulant discoloration.

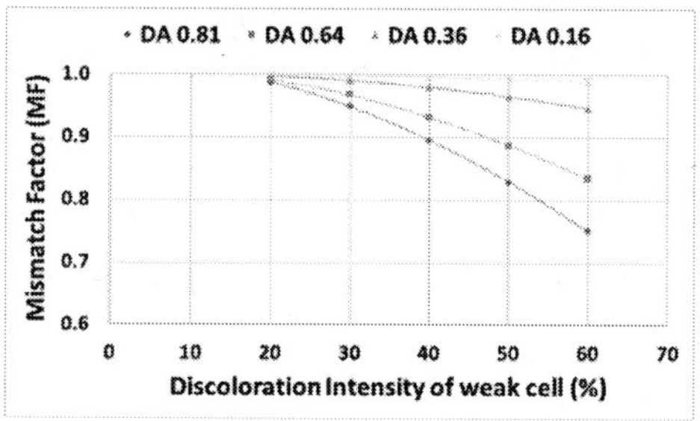

Figure 9: Variation in mismatch factor (*MF*) of PV module with discoloration intensity (DI) of weak cell under non-uiform discoloration

4 CONCLUSION

In this paper, the distributed diode model of solar cell has been extended to module in order to incorporate and analyze the effect of non-uniform encapsulant discoloration on PV module performance. Maximum power (P_{Max}) loss has been estimated for different fractional discoloration area (DA) of cell and different levels of discoloration intensity (DI). Under low discoloration extent, P_{Max} loss was mainly due to reduction in I_{SC}. However, under high discoloration extent, P_{Max} loss was contributed by the combination of I_{SC}, V_{OC} and *FF*. Results show that variation in DI from yellow to brown, extending over 80% of the solar cell leads to the P_{Max} loss in the range of 8 to 33%, which indicates that discoloration could have significant impact on output P_{Max}. In case of non-uniform discoloration, P_{Max} loss and mismatch factor (*MF*) has been analyzed by taking into consideration the discoloration extent and number of weak cells in a module. Results show that in this case P_{Max} loss increases non-linearly due to current mismatch and *MF* decreases significantly with discoloration extent of weak cells. The proposed approach can also be useful for estimating the P_{Max} loss in PV module suffering from different cases of discoloration.

5 REFERENCES

[1] A. W. Czanderna and F. J. Pern, Solar Energy Material and Solar Cells, Vol. 43, 1996, pp. 101–181.
[2] M. Köntges, S. Kurtz, C. E. Packard, U. Jahn, K. Berger, K. Kato, T. Friesen, H. Liu, and M. Van Iseghem, IEA-Photovoltaic Power Systems Program, 2014, pp. 1–140.
[3] O. S. Sastry, S. Saurabh, S. K. Shil, P. C. Pant, R. Kumar, A. Kumar, and B. Bandopadhyay, Solar Energy Material and Solar Cells, Vol. 94, 2010, pp. 1463–1468.
[4] A. Phinikarides, N. Kindyni, G. Makrides, and G. E. Georghiou, Renewable and Sustainable Energy Reviews, Vol. 40, 2014, pp. 143–152.
[5] D. C. Jordan, J. H. Wohlgemuth, and S. R. Kurtz, 27th European Photovoltaic Solar Energy Conference and Exhibition, 2012, pp. 3118–3124.
[6] T. Shioda, N. Maruko, A. Suzuki, K. Nishihara, H. Zenkoh, 26th European Photovoltaic Solar Energy Conference and Exhibition, 2011, pp. 4527–4530, 2011.

[7] E. Wanga, H. E. Yangb, J. Yena, S. Chia, C. Wanga, Energy Procedia, Vol. 33, 2013, pp. 256–264.
[8] F. J. Pern and A. W. Czanderna, Solar Energy Material and Solar Cells, Vol. 25, 1992, pp. 3–23.
[9] F. J. Pern, Solar Energy Material and Solar Cells, Vol. 41-42, 1996, pp. 587–615.
[10] F. J. Pern, A. W. Czanderna, K. A. Emery, and R. G. Dhere, The Conference Record of the Twenty-Second IEEE Photovoltaic Specialists Conference, 1991, pp. 557–561.
[11] N. C. Park, J. S. Jeong, B. J. Kang, and D. H. Kim, Microelectronics Reliability, Vol. 53, 2013, pp. 1818–1822.
[12] A. Sinha, O. S. Sastry, and R. Gupta, Solar Energy Materials and Solar Cells, Vol. 155, 2016, pp. 234–242.
[13] R. Gupta, P. Somasundaran, and D. K. Nandi, Applied Mechanics and Materials, Vol. 110-116, 2012, pp. 2453-2457
[14] J. Bengoechea, M. Ezquer, E. Zugasti, M. J. Rodríguez, A. R. Lagunas, 27th European Photovoltaic Solar Energy Conference and Exhibition, 2012, pp. 1242–1245.
[15] A. Zekry, A. Y. Al-Mazroo, IEEE Transactions on Electron Devices, Vol. 43, 1996, pp. 691-700

CORRELATION OF DEGREE OF EVA CROSSLINKING WITH FORMATION AND DISCHARGE OF ACETIC ACID IN PV MODULES

Jiang Zhu, Daniel Montiel-Chicharro, Tom Betts, and Ralph Gottschalg
Centre for Renewable Energy Systems Technology (CREST), Wolfson School of Mechanical, Electrical and Manufacturing Engineering, Loughborough University
Leicestershire, United Kingdom, LE11 3TU, Tel: +44(0)1509 635313, E-mail: J.Zhu@lboro.ac.uk

ABSTRACT: Ethylene vinyl-acetate (EVA) encapsulated crystalline silicon PV modules see encapsulant related degradation such as the hydrolysis of EVA which leads to generation of acetic acid that corrodes cells and cell metallisation. The formation and discharge of acetic acid in PV modules during an extended damp-heat test are studied in this work in dependence of the EVA crosslinking degree. To achieve different degrees of EVA encapsulation, mini-modules were laminated under different curing temperatures. The thermos-gravimetric analysis (TGA) is used to estimate the vinyl acetate (VA) content of the EVA before and after damp-heat ageing, from which the formation of acetic acid is evaluated. The net accumulation of acetic acid within modules is correlated to corrosion induced power losses. Results show mini-modules with highly cross-linked EVA form less acetic acid under damp-heat stresses, however, accumulated the highest amount of acetic acid leading to the most severe corrosion. Therefore, highly cross-linked EVA is not favoured in terms of long-term degradation due to DH stresses as it may trap the generated acetic acid within module

Keywords: EVA, vinyl acetate, VA, acetic acid, damp-heat, stress testing, lamination, curing, TGA

1 INTRODUCTION

PV modules rely on their packaging to provide durability. Typical c-Si modules consist of a number of interconnected solar cells encapsulated by two layers of EVA. This is then, in the majority of cases, sandwiched between a front glass layer and a back polymeric sheet. The outer layers of the front glass layer and the backsheet minimise encapsulant related degradations such as encapsulant decomposition (e.g. hydrolysis) and interconnect and cell surface corrosion [1-5]. However, when modules are exposed to humid environments, moisture can permeate through backsheet and interact with vinyl acetate (VA) leading to hydrolysis of EVA, where the acetic acid is formed as a by-product. The generated acetic acid can either diffuse out from the backsheet or accumulate within the module. Different types of EVA and backsheet will influence the diffusion behaviour of acetic acid, see e.g. [6].

Lamination of PV modules is typically with a flat-bed laminator where the lamination temperature, pressure and duration are controlled carefully to ensure good encapsulation quality. Variation in any of the factors may lead to changes in chemical reaction rates and alterations of phase transition of polymers. The lamination process can influence the crosslinking degree of EVA [7], which will be shown in this paper has a significant impact on net accumulation of the generated acetic acid. This will then influence the module degradation behaviour, e.g. corrosion, and thus affect the performance of PV modules [8-9].

A number of mini-modules were laminated in-house and subjected to an extended damp-heat (DH) stress testing. All materials used for the laminated mini-modules are commercially available in the PV market. Lamination temperature was set between 125°C and 150°C in order to obtain samples with optimal and sub-optimal crosslinking degrees of EVA. The EVA crosslinking was characterised by gel content measurements. The module power performance and degradation was characterised by I-V measurements during the course of DH test. The DH test is designed to drive humidity into module package. The humidity causes hydrolysis and acidification of EVA and ultimately corrodes busbars, fingers and cell-surfaces [10-11]. This may be influenced by the chemical structure of the EVA. As an example, curing temperature during lamination influences the EVA crosslinking as well as the stability of the materials. Thermo-gravimetric analysis (TGA) was carried out for the unaged and aged EVA samples, respectively, from which the VA content of the unaged and aged EVA can be estimated. This is then used for evaluation of the acetic acid formation rate. Based on the power and material measurements, correlations between the VA content and the formation and discharge of acetic acid were established.

2 EXPERIMENTAL SETUP

2.1 Sample preparation

Structure and specifications of the laminated single cell mini-modules used in this work are shown in Figure 1. The laminates consist of a glass front layer (2.9mm thick low iron float, un-tempered glass), two layers of encapsulant (460μm thick fast cure EVA) and a backsheet layer (tri-layer insulating polymer consisting PET/PET/primer layer). A 1.8W multi crystalline Si cell is laminated in between the two EVA layers. Lamination temperatures were set to different levels at 125°C, 135°C, 145°C, and 150°C, while the curing time was fixed at 10 minutes (These samples are referred to in the following as L125, L135, L145 and L150, respectively). Laminator was pre-heated to 120°C before placing samples on the flat bed. Vacuum time was 4 minutes and followed by 10 minutes curing under 1 Bar. Samples were taken out immediately after curing and left to cool in ambient environment.

At each lamination condition, four frameless samples were fabricated for DH testing. A large distance left to the boundaries of the module ensures that edge diffusion does not change the experiment.

2.2 Extended stress testing

The mini-modules were placed in an environmental chamber (see Figure 2) and subjected to extended damp-heat testing at 85°C and 85% relative humidity (RH) for a total exposure of 11000 hours. It might not be realistic for

PV modules to experience 11000 hours DH equivalent stress in the field, but this overstressing allows an investigation of the acetic acid formation and discharge behaviours as well as the corresponding degradation mechanism, i.e. the corrosion of cells and metallisation.

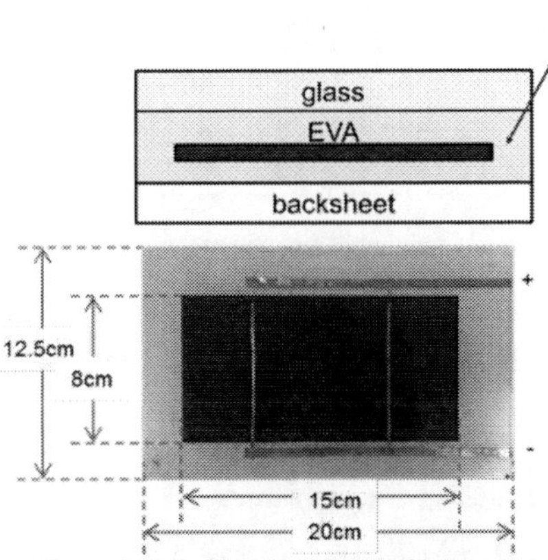

Figure 1: Specifications of the mc-Si mini-module laminated at CREST.

Figure 2: Testing samples in an environmental chamber.

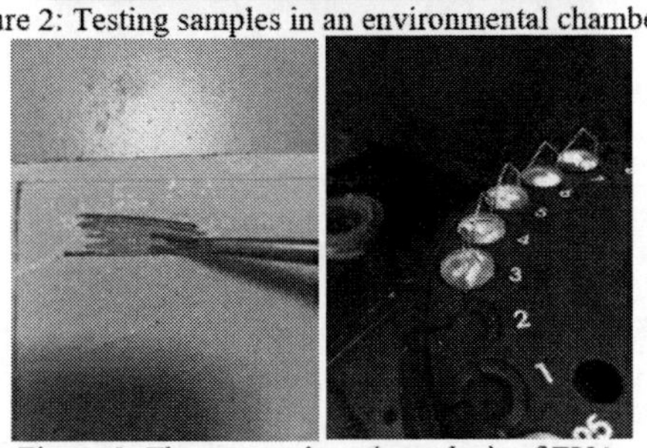

Figure 3: Thermogravimetric analysis of EVA.

The prolonged DH test is designed to drive humidity into EVA from module backsheet. Once the moisture inside the module permeates into EVA it will form acetic acid which will attack the electric circuitry and, in the saturated case, free water may corrode cells as well. This will cause power reduction due to losses of passivation, electrochemical corrosion and ultimately lead to module failure.

Measurements of gel content were taken for EVA cured at different temperatures before ageing by the solvent extraction method. Module I-V characteristics were measured indoors using the Pasan IIIb solar simulator at the standard testing condition (STC).

2.3 Thermogravimetric analysis of EVA

The dependence of thermal degradation behaviour on VA content of EVA copolymers can be studied by TGA. As shown in Figure 3, each TGA measurement uses about 5mg EVA sample extracted from the unaged or aged mini-modules. The sample is left in ambient environment for a few hours before measuring. This assumes no generated acetic acid remains in the tested sample of EVA. Then the sample is heated from 30°C to 600°C at the heating rate of 10°C per minute while the weight loss is monitored. It has shown the maximum rate of weight loss at the stage of evolving acetic acid exhibits a linear relationship with VA content and is independent of molecular weight and heating rate [12]. When temperature is at 350-380°C, the chemical reaction of EVA evolving acetic acid is expressed below:

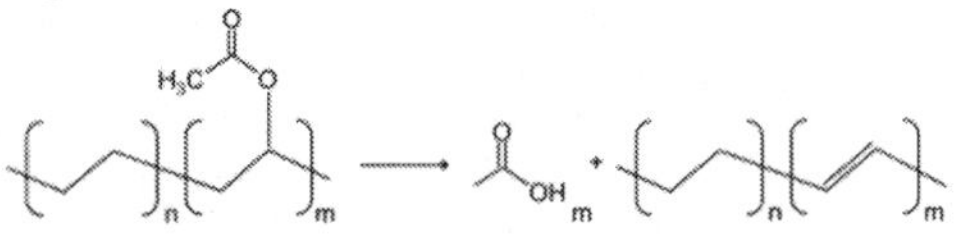

where n and m are the numbers of ethylene monomers and vinyl acetate monomers, respectively.

3 EFFECTS OF EVA CROSSLINKING ON WATER PERMEATION AND HYDROLYSIS

The lamination temperature has a significant impact on curing and crosslinking degree of EVA. The curing rate of EVA increases with the increasing temperature. If the temperature is too high, curing may occur too quickly without completely removing the gases within the module structure and the gases generated during crosslinking. If the temperature is too low, curing rate may become too slow and cannot achieve a good curing state during lamination. The crosslinking also converts a thermoplastic material into a network format thermosetting material so that the material will not flow under heat.

Figure 4 plots the averaged values of gel contents for EVA samples L125, L135, L145 and L150. The L125 shows gel content less than 60%. Samples of L135, L145 and L150 show a linear increase in gel contents from 72% to 85%. It has been reported in [13], for the EVA studied in this work, samples with lower degrees of crosslinking (gel content <60%) show water diffusion rate of about 35 $g/m^2/day$ under 23°C and 1 bar. The diffusion rate reduces with the increasing crosslinking degree and reaches about 10 $g/m^2/day$ for samples of gel content >80%. Higher curing temperature ensures good crosslinking of EVA, which has better resistivity to moisture ingress.

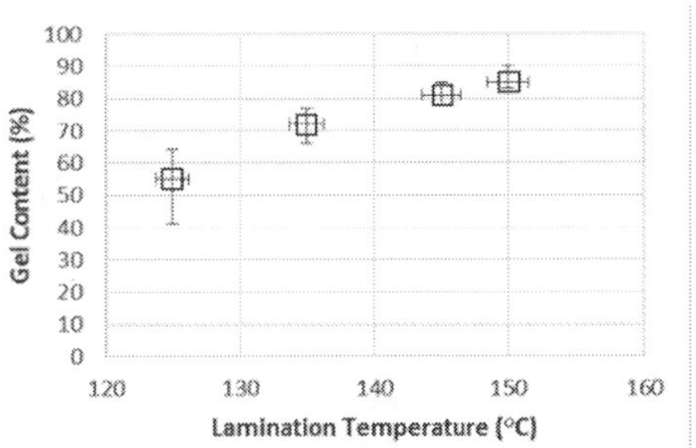

Figure 4: Averaged gel content of EVA versus lamination temperature.

Hydrolysis of EVA is a typical degradation mechanism due to DH stresses. The chemical reaction is expressed below in Figure 5.

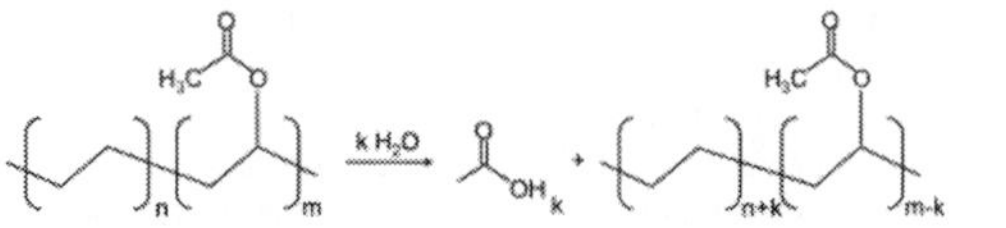

Figure 5: Chemical hydrolysis reaction of EVA

In Figure 5, k is the number of acetic acid generated due to hydrolysis during the DH test, n and m are the original numbers of ethylene monomers and vinyl acetate monomers, respectively.

The acetic acid is formed in the presence of water when the VA content of EVA is decomposed. Using TGA for the unaged and aged EVA samples, the differences in weight loss curves were observed as shown in Figure 6. According to the method in [12], the VA content can be estimated based on the TGA measurements and the averaged values are summarised in Table I for mini-modules L125, L135, L145 and L150. By comparing the VA contents before and after DH ageing, it is possible to estimate the amounts of VA content of EVA decomposed to produce acetic acid, which is equivalent to the amounts of acetic acid was formed. L125 samples generated more acetic acid than any other as 31.3% of VA was decomposed. L150 samples had the lowest generation with only 20.8% of VA hydrolysed during the 11000 hours of DH test. This indicates that less cross-linked EVA tends to produce more acetic acid than higher cross-linked EVA.

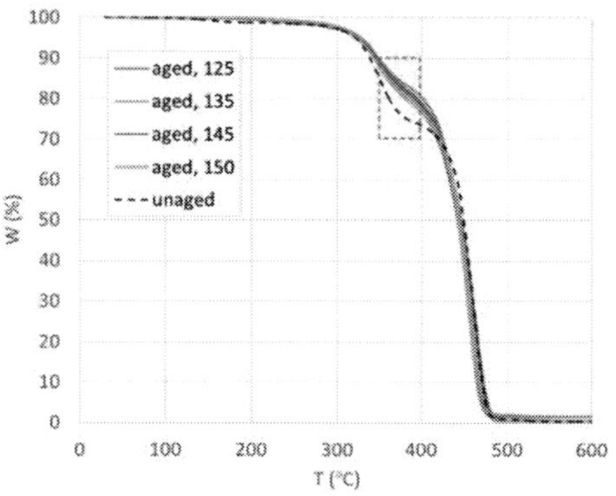

(a)

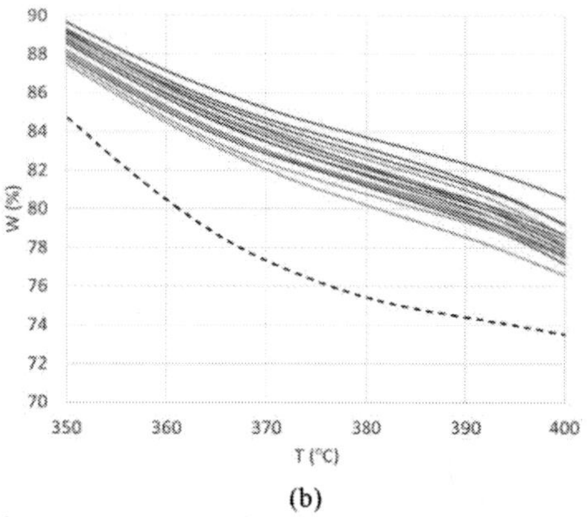

(b)

Figure 6: TGA curves for unaged and aged EVA samples extracted from mini-modules L125, L135, L145 and L150.

Tabel I: Evaluation of consumption of VA and formation of acetic acid.

	ΔW% - Unaged EVA sample	VA% - before DH ageing	ΔW% - Aged EVA sample	VA% - after DH ageing	Percentage of VA hydrolysed
L125	24	34.4	16.5	23.7	31.3%
L135	24	34.4	17.3	24.8	27.9%
L145	24	34.4	18.2	26.1	24.2%
L150	24	34.4	19	27.3	20.8%

4 EFFECTS OF EVA CROSSLINKING ON POWER DEGRADATION AND CORROSION

The mini-module power degradations due to DH stresses are reported in this section and their dependencies on EVA crosslinking degree are discussed.

Visual inspection saw all samples suffering from the same type of degradation mode, i.e. corrosion of busbars, fingers, cell surface and yellowing of EVA (see Figure 7), which are the expected degradation mechanisms in DH test.

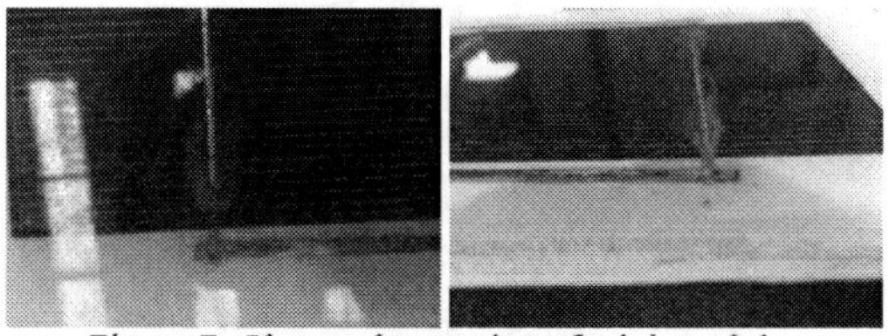

Figure 7: Observed corrosion of mini-modules.

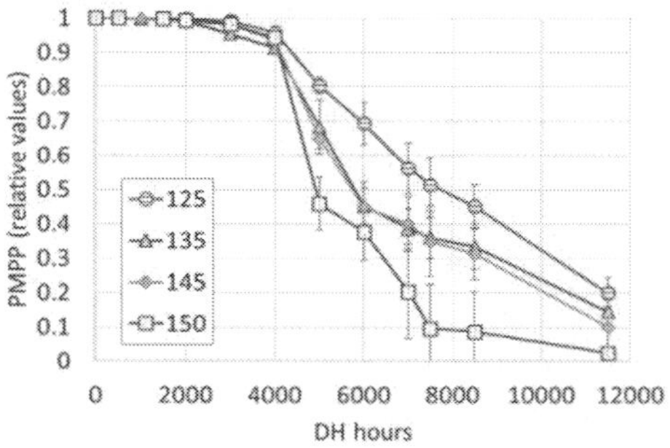

Figure 8: P_{MPP} degradation over time.

The averaged module power (P_{MPP}) of mini-modules L125, L135, L145 and L150 over DH testing time is plotted in Figure 7. Clear difference in the pattern of power degradation is observed for mini-modules with higher and lower cross-linked EVA encapsulation. Most of the samples started to degrade after 4000 hours in DH test. L125 samples saw the slowest deterioration, whereas L150 samples saw the fastest degradation in power. Assuming that the degradation is solely due to corrosion, which is correlated to acetic acid, the accumulation of acetic acid within the module is lower in L125 samples and higher in L150 samples. This would imply that lower cross-linked EVA tends to discharge more acetic acid than higher cross-linked EVA.

5 CONCLUSIONS

An extended DH stress test was carried out for EVA encapsulated mini-modules, whose EVA encapsulant exhibits different degrees of crosslinking with gel contents between 50% and 90%. The L150 samples with the highest crosslinking degree of EVA encapsulant producing the least acetic acid, however, showed the most severe corrosion and power degradation during the extended DH test.

The formation of acetic acid under DH stresses is evaluated by analysing the VA content in the EVA through TGA measurements for unaged and aged EVA samples extracted from the mini-modules. The formation rate appeared to be of positive dependence on the water diffusion behaviour in EVA as the amount of VA content hydrolysed increases with the increasing water diffusion rate in EVA.

The generated acetic acid either diffused out through backsheet or accumulated within module at e.g. EVA/cell interface and eventually the concentration saturated after about 4000-5000 hours DH test for this type of mini-modules, where corrosion led to rapid degradation in power. The EVA crosslinking degree appeared to be of negative dependence on acetic acid discharge rate. The L125 samples with the most acetic acid produced during the DH test had the least amount of acetic acid accumulation, which means acetic acid discharge rate for the L125 samples is the highest. Highly cross-linked EVA is not favoured in terms of long-term degradation due to DH stresses as it may trap the generated acetic acid within module leading to severer corrosion effect.

9 REFERENCES

[1] M.A. Quintana, D.L. King, T.J. McMahon, C.R. Osterwald, Commonly observed degradation in field-aged photovoltaic modules, *Proceedings of 29th IEEE PVSC,* 2002, New Orleans, USA.

[2] Oreski G, Wallner GM. Delamination behaviour of multi-layer films for PV encapsulation. *Solar Energy Materials and Solar Cells* 2005; 89: 139–151.

[3] Jorgensen GJ, Terwilliger KM, DelCueto JA, Glick SH, Kempe MD, Pankow JW, Pern FJ, McMahon TJ. Moisture transport, adhesion, and corrosion protection of PV module packaging materials. *Solar Energy Materials and Solar Cells* 2006; 90: 2739–2775.

[4] Jorgensen G, Terwilliger K, Glick S, Pern J, McMahon T. Material testing for PV module encapsulation, NREL/CP-520-33578, May 2003.

[5] Van Dyk EE, Chamel JB, Gxasheka AR. Investigation of delamination in an edge-defined film-fed growth photovoltaic module. *Solar Energy Materials and Solar Cells* 2005; 88: 403–411.

G. Oreski, A. Mihaljevic, Y. Voronko, G. Eder, Acetic Acid Permeation through PV-Backsheets: Dependence of the Composition on the Permeation Rate, *Proceedings of 32nd EU PVSEC,* 2016, Munich, Germany.

J. Zhu, D. Montiel-Chicharro, T. Betts, R. Gottschalg, Development of adhesive and cohesive failures in EVA-backsheet structures in environmental testing, *Proceedings of 32nd EU PVSEC,* Munich, Germany, 2016.

[8] A. W. Czanderna and F. J. Pern, Encapsulation of PV modules using ethylene vinyl acetate copolymer as a pottant: a critical review, *Solar Energy Materials and Solar cells*, vol. 43, no. 2, pp. 101–108, 1996.

[9] D. Wu, J. Zhu, D. Montiel-Chicharro, T. R. Betts, and R. Gottschalg, Influences of different lamination conditions on the reliability of encapsulation of PV modules, *Proceedings of 29th EU PVSEC,* 2014, Amsterdam, Netherlands.

[10] Kempe, M. D., Jorgensen, G. J., Terwilliger, K. M., McMahon, T. J., Kennedy, C. E., & Borek, T. T. (2007). Acetic acid production and glass transition concerns with ethylene-vinyl acetate used in photovoltaic devices. *Solar energy materials and solar cells*, 91(4), 315-329.

[11] M.D. Kempe, Modeling Rates of Moisture Ingress into Photovoltaic Modules, *Solar energy materials and solar cells*, 90, 2006, 2720-2738.

[12] Z. Wenwei, Z. Xiaoguang, Y. Li, Z. Yuefang, S. Jiazhen, Determination of the vinyl acetate content in ethylene-vinyl acetate copolymers by thermogravimetric analysis, *Polymer*, 35 (1994), p. 3348

[13] J. Zhu, C. Ng, D. MontielChicharro, T. R. Betts and R. Gottschalg, Water Vapour Transmission Rate of Ethylene Vinyl Acetate (EVA) Encapsulant with different Curing Levels, *Proceedings of 13th Photovoltaic Science, Application and Technology Conference*, Bangor, UK, 2017.

EFFECT OF THE FRAME SEALING ON THE FUNCTIONALITY OF A PHOTOVOLTAIC MODULE

Vanek Jiri, Jandova Kristyna, Sturm Martin, Josef Hylsky, David Strachala
Department of Electrical and Electronic Technology, Faculty of Electrical Engineering and Communication
Brno University of Technology, Technicka 10, 616 00 Brno, Czech Republic

ABSTRACT: This article deals with the effect of the quality of moisture isolation of the frame of a solar module on the overall functionality of the module. The main goal is to investigate the causes of frame damage, peeling of the frame seal and decrease in the photovoltaic module performance. The problems were examined using various diagnostic methods. The tested photovoltaic modules have been exposed to simulated climate-like environmental conditions and the mentioned defect has been detected. In this work the results of the investigated failure problem are presented. Measured values obtained from several types of measurements are compared and evaluated.
Keywords: defect analysis, humidity of insulation, photovoltaic, frame, destruction

1 INTRODUCTION

Within the functionality control of a photovoltaic system there has been found a large number of photovoltaic modules with a visually damaged frame and damaged bottom edge of the active area of the photovoltaic module. These modules were removed from the photovoltaic power plant and their condition and functionality has been analyzed. Within the measurement there has been revealed a serious defect effect on the overall power output of the photovoltaic modules.

The main aim of this study is to investigate, using different investigative techniques, the causes of deformation of the frame with subsequent destruction and damage of the solar module and the demonstration of this phenomenon due to peeling of the frame seal or its poor design. The tested photovoltaic modules have been exposed to similar simulated climatic influences as in the outdoor environment to determine what damage this defect can cause. Values obtained from several types of measurements are compared and evaluated

2 CONSTRUCTION OF THE PHOTOVOLTAIC MODULE.

The very construction of the photovoltaic module is very important in building a system that resists to influences and weather whims. In manufacturing, the individual module layers are stacked on each other in such a way as to prevent the penetration of water and dust particles into the cells. The top windscreen is hardened for greater durability and serves as a support substrate. Silicon cells are placed between two layers of an EVA film (ethylene vinyl acetate), heat-sealed and added to a glass and a back cover. The whole system is then placed in a frame, mostly aluminum, to increase resistance of the module against mechanical stress (Figure 1). Since the EVA material is highly hydrophilic [2], it is necessary to finally seal the entire system and the aluminum frame with a silicone seal that prevents the penetration of moisture into the structure of the photovoltaic module and the frame (Figure 2).

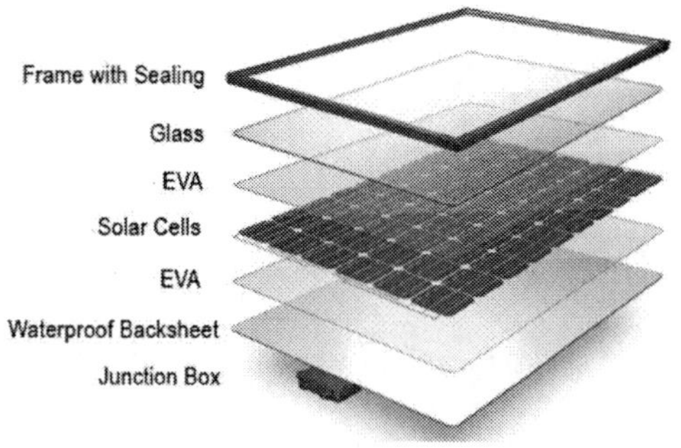

Figure 1: A module with crystalline silicon cells and EVA foil (EVA - Ethylene vinyl-acetate)

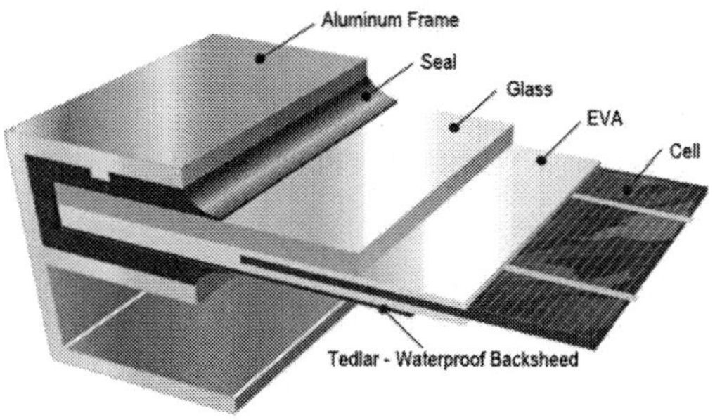

Figure 2: A detail of module frame seal

3 STATE OF THE STRUCTURE OF THE PHOTOVOLTAIC MODULE

During the analyses of the photovoltaic systems there was a case at a photovoltaic power plant where there was noted destruction of the frame of several photovoltaic modules (Figure 3, 4). These frames of the photovoltaic module were torn and deformed on the underside. The double wall of the frame profile was torn apart and curved.

Figure 3: Detail of the damage to the photovoltaic module - the plates of the frame are separated.

Figure 4: Next detail of the damage to the photovoltaic module - the plates of the frame are separated.

This mode of deformation implied either the application of an external force parallel to the frame plates, which could deform the structure of the frame, or the internal compressive force acting from the inside, and the plates separated from one another. After examining the power plant environment, we excluded an external force and suggested that the bursting of the frame was caused by the pressure generated by the freezing water in the winter period, accumulated in the space between the frames of the frame profile. To verify this theory, we decided to conduct a thorough analysis of the damaged modules using VA characteristics, electroluminescence and visual inspection.

4 EXPERIMENT

The damaged modules were first removed from the photovoltaic system and then the analysis process started. First, the modules were subjected to a solar simulator test to determine their current state of performance and other electrical properties. These were the Solar Swiss AG panels, type SSM-180/24M. From the solar simulator results it was evident that there was a significant performance decrease compared to the declared output by the manufacturer (Table I). Table II shows an example of fifth of the tested modules (sample 3.1).

Table I.: Values supplied by the manufacturer

Output	Pmpp	180 W
Voltage	Umpp	36 V
Proud	Impp	5 A
Open circuit voltage	Uoc	44.9 V
Short circuit current	Isc	5.52 A

Table II: Experimental data from the damaged modules

Sample	01032415	01032412	01025004	01032193	01032162
Pmpp [W]	162.6	162.4	161.7	165.4	163.9
Ump p [V]	35.4	35.4	35.0	35.4	34.9
Impp [A]	4.6	4.6	4.6	4.7	4.7
Isc [A]	5.061	5.116	5.145	5.220	5.261
Uoc [V]	43.612	43.604	43.680	43.766	43.517

The measured values for this module are distinctly smaller than the values stated by the manufacturer on the label. To assess the state of damage of the individual photovoltaic cells, the analysis was supplemented by detection of electroluminescent radiation.

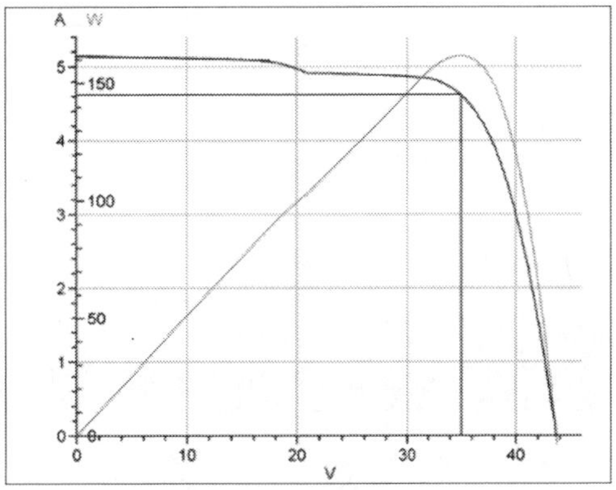

Figure 5: Example of the voltage-current characteristics of the most-damaged module, 01025004

Figure 6: Selected solar module (sample 01025004) investigated by the method of electroluminescent radiation.

This analysis shows significant damage to the left lower cell and significant decrease in cell performance in the second column from the right. The overall decrease in the solar module performance is related to these defects. The decrease in power in the second string is so significant that it can be analyzed on the voltage current characteristic as a sudden drop in current in the point near to 20 V due to the synergy of the bypassing diode.

Subsequently, a visual check of the module was performed and a poor seal between the photovoltaic module and the frame was found (see Figure 7). Water penetrating into the frame, and then into the module, caused a drop in overall power. Changing climatic conditions has resulted in freezing of accumulated water and subsequent deformation and damage to the frame as well as to the photovoltaic cells.

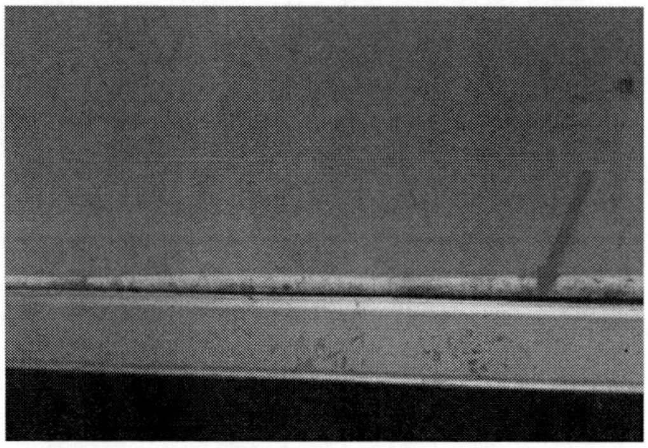

Figure 6: Detail of the insufficient seal of the photovoltaic module. The sealing putty does not adhere to the frame surface.

Figure 6: Detail of the insufficient seal in the corner of the photovoltaic module. The sealing putty does not adhere to the frame surface.

It was postulated that water infiltration has caused a drop in the performance of the whole module. This theory was further verified on a test module that was exposed to flowing water in the same way as it would be in a free installation outside the laboratory conditions. In places where the seal was missing or badly made, the water penetrated into the frame, flowed through these places, or accumulated there.

5 CONCLUSION

A link has been established between the decrease in solar module power and the deformation of the protective frame. This deformation had to be caused by a drop in temperature and a change in the volume of accumulated water in the interior of the protective frame due to poorly sealed seals. This deformation also caused further moisture penetration not only into the frame area, but also into the structure of the solar module with subsequent power drops.

ACKNOWLEDGEMENT

This research work has been carried out in the Centre for Research and Utilization of Renewable Energy (CVVOZE). Authors gratefully acknowledge financial support from the Ministry of Education, Youth and Sports of the Czech Republic under NPU I program (project No. LO1210).

REFERENCES

[1] Váry, Michal - Perný, Milan - Šály, Vladimír - Firický, Eduard: Monitoring of Grid Tied Photovoltaic System Efficiency in Operating Conditions. In: Electrotehnica, Electronica, Automatica. - ISSN 1582-5175. - Vol. 60, nr. 2 (2012), s. 9-12

[2] Bechnik B., Nejpoužívanější pojmy ve fotovoltaice Source: http://oze.tzb-info.cz/fotovoltaika/11772-nejpouzivanejsi-pojmy-ve-fotovoltaice

[3] JANDOVÁ, K.; VANĚK, J.; BAŘINKA, R. Detekce defektů v solárních článcích pomocí nedestruktivních metod. In 3. Česká fotovoltaická konference, Sborník příspěvků. Rožnov p.R.: 2008. s. 146-150. ISBN: 978-80-254-3528- 1.

[4] Jandová, K.; Diagnostické metody plošného rozložení defektů solárních článků, UETE FEEC VUTBR, dizertační práce, 2008

[5] TAUŠ P., TAUŠOVA M., Economical analysis of PV power plants according installed performance, Acta Montanistica Slovaca, 14 (2009), pp. 92–97

[6] DOSTÁL Z., M. ĎULÍK, M. KOŠČOVÁ, AND P. ŠIMON, Monitoring of degradation of photovoltaic panels using infrared method. Acta Montanistica Slovaca, 15(SPEC.ISSUE 2), pp. 111-115. (2011)

ANALYSES OF PHOTOVOLTAIC MODULES INFLUENCED BY VOLCANIC ASHES AT KAGOSHIMA IN JAPAN

Yasuo Chiba[1], Ritsuko Sato[1], Tadashi Hirayama[2], Shuma Kawabata[2], Yukio Yoshimura[3] and Atsushi Masuda[1]
[1]National Institute of Advanced Industrial Science and Technology, 807-1 Shuku-Machi, Tosu, Saga 841-0052, Japan
[2]Kagoshima University, 1-21-40 Korimoto, Kagoshima 890-0065, Japan
[3]Kagoshima Prefectural Institute of Industrial Technology, 1445-1 Hayatocho-Oda, Kirishima, Kagoshima 899-5105, Japan

ABSTRACT: We investigated the photovoltaic (PV) module located at Kagoshima Prefectural Institute of Industrial Technology (KIT) near Sakurajima volcano and exposed outdoor for a long time. The cell condition of the PV module is almost favorable, however, the PV module glass was influenced by the volcanic ashes from Sakurajima. From secondary ion mass spectrometry (SIMS) analyses, general soil elements were also detected. These results suggest that not only volcanic ashes but also general soil contaminants may chemically affect the PV module components in the future.
Keywords: Photovoltaics, Volcanic Ash, PV Module,

1 INTRODUCTION

In Japan, it is not seldom to see photovoltaic (PV) modules not only in towns but also in wide fields and mountains, which are installed at customer house and large-scale power generation plant. The installation costs of PV systems rapidly decrease, then the spread of the PV system installation is accelerating rapidly. One of the problems of these PV systems is, however, that nobody knows the lifetime of PV systems.

The other problem of the PV systems installed outdoors is the reduction of the amount of power generation due to mainly soil contamination on the surface of the PV modules. Recently, the problems of soil contamination in solar PV systems and plants were reported from all over the world [1-4]. In these reports, the fluctuation in the amount of power generation by soil is pointed out, however, there are few reports on the chemical influence of the soil contamination on the degradation of PV modules. Presently, we have investigated the influence of soils and volcanic ashes on four kinds of PV modules installed at Kagoshima Prefectural Institute of Industrial Technology (KIT), where is located at 31.7 °N and 130.7 °E in Japan (climate classification: Cfa).

Figure 1: Installed PV system in KIT.

KIT has a plenty of multicrystalline silicon PV modules exposed to the outdoors, as shown in Fig. 1. These PV modules are exposed for about 27 years. By analyzing these modules, it is possible to obtain knowledge about degradation which is expected to occur in the future. In this study, we examined the characteristics of the PV module, and further analyzed the glass which is component parts of the PV module. Kagoshima near Sakurajima volcano is a rare case in the world as the city is close to the volcano. KIT is located at about 20 km distance for north from Sakurajima.

2 EXPERIMENTAL PROCEDURE

Current-voltage (*I-V*) characteristics of the PV modules were estimated by using pulsed solar simulator at standard test conditions (STC; 25 °C, AM1.5G, 1SUN). Also the PV module was characterized by electroluminescence (EL). Furthermore, the PV module glass was analyzed by ion chromatography (IC), inductively coupled plasma mass spectrometry (ICP-MS) and secondary ion mass spectrometry (SIMS).

3 RESULTS AND DISCUSSION

As shown in Fig. 1, it is found that the color of encapsulant material of the PV module is brownish. Discoloration is considered to be one of the examples of deterioration by long term exposure. Table I shows *I-V* characteristics of the PV module. These data were the average values of the evaluated module measured by five times. The obtained maximum power (P_{max}) of the evaluated module was 46.66 W. The nameplate output power of the PV module was 58.7 W. The calculated value of the output rate against the nameplate was about 79.5%. Therefore, the annual degradation rate of the PV module was estimated to be about 0.76%/year.

Table I: The *I-V* characteristics of the evaluated module.

	I_{sc} (A)	V_{oc} (V)	FF	Eff. (%)	P_{max} (W)
Average (five times)	2.53	25.10	0.735	8.50	46.66

Also we characterized the PV modules by EL. From the EL images, although there are a few dark parts in the cells, it is suggested that the cell condition in the PV modules are roughly not poor.

After *I-V* and EL measurements, we analyzed the glass of evaluated module. In order to investigate the adhesion materials on the glass firstly, the measurement of IC was carried out. The results of the quantitative analysis of the anions are shown in Table II. From these results of IC analyses, the components of F^-, Cl^-, NO_3^- and SO_4^{2-} as an anion were detected. These components were consistent with the water-soluble components of volcanic ashes of Sakurajima reported by Sakamoto [5]. Also, the components of Na^+, NH_4^+, K^+, Mg^{2+} and Ca^{2+} as a cation were detected, as shown in Table III.

Table II: Quantitative analysis of anions of the adhension materials on the glass by IC.

Component	Detected amount (ng/cm^2)
F^-	2
Cl^-	5
NO_3^-	8
SO_4^{2-}	50

Table III: Quantitative analysis of cations of the adhension materials on the glass by IC.

Component	Detected amount (ng/cm^2)
Na^+	20
NH_4^+	10
K^+	1
Mg^{2+}	3
Ca^{2+}	6

Then, quantitative analysis of metallic elements of the adhesion materials on the glass by ICP-MS was carried out, as shown in Table IV. It was found that the metallic elements are detected with wide range from light element to heavy element. In particular, Cu, Zn and Pb were detected, which are consistent with the water-soluble heavy metal elements in Sakurajima volcanic ashes reported by Sakamoto [6].

Table IV: Quantitative analysis of main detection component of the adhension metallic elements on the glass by ICP-MS.

Component	Detected amount (ng/cm^2)
Na	30
Mg	4
Al	30
K	6
Ca	20
Ti	1
Mn	0.2
Fe	10
Cu	0.09
Zn	30
Pb	2
Si	92

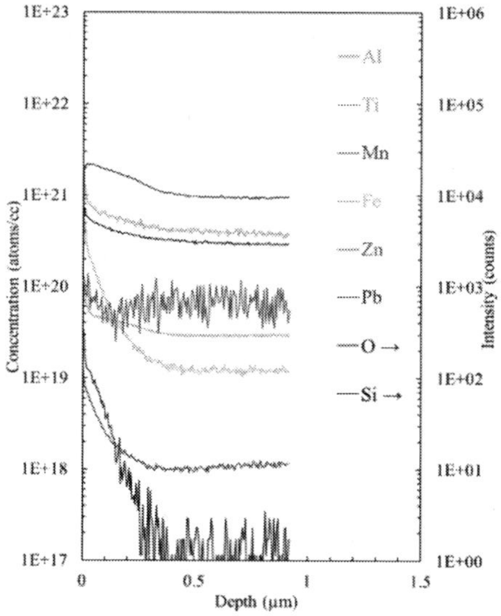

(a)

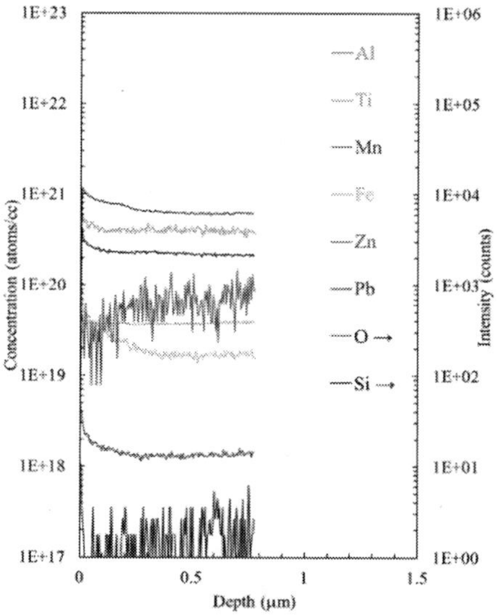

(b)

Figure 2: Excerpt of SIMS profiles on the PV glass surfaces of (a) the light receiving surface and (b) the encapsulant material surface.

Furthermore, SIMS analyses of both the light receiving surface and the encapsulant material surface of this glass were carried out. Figures 2(a) and (b) show the excerpt of SIMS profiles for the positive ions on the light receiving surface and the encapsulant material surface, respectively. It is found that the diffusion of the elements on the light receiving surface is different from that on the encapsulant material surface. Especially, Zn and Pb elements were diffused from the light receiving surface of the glass, and these elements were detected by ICP-MS as the water-soluble heavy metal elements [6]. Therefore, it is suggested that volcanic ashes may affect the module components in the long term. From Figs. 2(a) and (b), it is found that for example Al, Ti, Mn and Fe, which are also contained in general soil, are also diffused from the light receiving surface of the glass. These results suggest that not only volcanic ashes but also general soil contaminants may chemically affect the PV module glass in the future.

4 CONCLUSIONS

The PV modules exposed at KIT near Sakurajima volcano for a long time were characterized by *I-V*, EL, IC, ICP-MS and SIMS measurements. No abnormality was observed in the cells for these PV modules. The performance was maintained about 80 % against the nameplate. The ions and metallic elements based on the volcanic ashes from Sakurajima were detected by the measurements of IC and ICP-MS. It was found from SIMS analyses that the glass of PV modules is influenced by not only volcanic ashes but also general soil elements. As a result, it is considered that the PV modules may be chemically affected by the various contaminations in the future.

ACKNOWLEDGEMENT

This study was in part supported by the Japan Electrical Manufactures' Association (JEMA).

REFERENCES

[1] A. Massi Pavan *et al.*, Sol. Energy **85**, 1128–1136 (2011).
[2] M. García *et al.*, Prog. Photovolt.: Res. Appl. **19**, 211–217 (2011).
[3] F. A. Mejia and K. Kleissl, Sol. Energy **95**, 357–363 (2013).
[4] M. Vivar *et al.*, Sol. Energy **84**, 1327–1335 (2010).
[5] H. Sakamoto, Nature of Kagoshima **39**, 177-189 (2013) [in Japanese].
[6] H. Sakamoto, Nature of Kagoshima **40**, 237-246 (2014) [in Japanese].

DETERMINATION OF DEGRADATION RATES FOR PV MODULES AND PV GENERATORS APPLYING VARIOUS METHODS

Dirk Stellbogen, and Peter Lechner
Zentrum fuer Sonnenenergie- und Wasserstoff-Forschung Baden-Wuerttemberg (ZSW)
Meitnerstrasse 1, D-70563 Stuttgart, Germany
dirk.stellbogen@zsw-bw.de

ABSTRACT: The proper determination of degradation rates is a key issue in the technical and economic performance assessment of PV installations. Due to the considerable variations in the meteorological operating conditions of fielded PV devices over time, sophisticated analysis methods are required to derive meaningful degradation figures which are usually in the order of several per mill per year.

The study evaluated different procedures for the determination of performance change rates by applying them to multi-annual highly time resolved monitoring data of PV modules and PV generators from the ZSW test field Widderstall. A variation analysis has been conducted determining the linear trends of various sets of PR data derived with different irradiance filtering bands and of the respective medians over various intervals. Thereby, not only the change rates but also their uncertainty budgets have been computed.

The results show that regression analysis of high resolution performance data filtered to an irradiance band close to 1000 W/m² determines degradation rates with sufficiently low uncertainty levels and is much more reliable than trend analysis of aggregated median values.

Keywords: Performance, Reliability, Field testing, Degradation

1 INTRODUCTION

The proper determination of degradation rates is a key issue when accessing the actual and future technical and economic performance of PV installations. However, uncertainties are imposed onto this task by the variation of radiation and weather conditions in real field operation. This variation can be quite significant in comparison to degradation rates, which are usually observed to be in the order of several per mill per year. Consequently, the development and verification of sophisticated analysis methods is required in order to derive meaningful degradation figures.

ZSW has collected on its PV test field Widderstall long term monitoring data of both individual PV modules and PV generators of c-Si and thin-film technologies with high time resolution (1min or less). This data pool is used for an evaluation of different methods for the determination of degradation trends.

Several procedures for performance trend analysis of PV systems and devices are based on integral performance figures such as energy delivery or performance ratio over certain time periods. However, these metrics can be biased by seasonal variations of the distribution of meteorological operating conditions.

Other methods make use of filtering to select PV performance data recorded within a limited range of operating conditions (e.g., irradiance level, module temperature) and analyse a subset of the original operating data (at the available time resolution) to determine performance trends.

The present study evaluates these methods applying different selection criteria and selection ranges for the filtering methods.

While commonly applied methods for the derivation of degradation rates of PV devices are often lacking the determination of the uncertainty level, a special focus has been put on the assessment of the validity and significance of the investigated procedures with respect to their relative uncertainties.

2 APPROACH

The degradation rate study has analysed multi-annual high resolution data sets from monitoring of poly-c-Si and a-Si PV generators and CIGS type PV modules, which have been exposed and operated on the test stands at the ZSW test site Widderstall in Southern Germany (9.72°E, 48.53°N, 750 m altitude), see Figure 1 and Figure 2 [1, 2].

Figure 1: Detail of the PV generator test stand at the ZSW site Widderstall

Figure 2: PV module test facility at the ZSW site Widderstall

One-minute average Performance Ratios have been used for the evaluation of the PV generators, whereas instantaneous IV-curve scans at 10 sec intervals and their vertex parameters (Pmpp, Isc, Voc, FF) have been available for the PV modules.

From each of the full monitoring data sets, those records have been selected of which the measured irradiance value was in a certain band close to the referenced irradiance of 1000 W/m². For a sensitivity analysis, the width of this irradiance band has been varied from ±5 % to ±0.5 %; this means that filter bands from 1000 W/m² ±50 W/m² to 1000 W/m² ±5 W/m² have been applied.

The performance values have been linearly scaled to the reference irradiance (if applicable) and corrected for the temperature effect to the nominal temperature of 25°C.

A linear regression has then been performed on the resulting sets of scattered data to determine the respective change rate of the performance parameter. Alternatively, median values of the data have been determined for periods ranging from 2 days to 30 days and subsequently, a linear regression trend was computed for the median values.

3 RESULTS

3.1 PV Generators

The development of performance data of the poly-c-Si PV generator and the a-Si PV generator over a period of more than eight years is displayed in Figure 3 and Figure 4, respectively. The diagrams show the normalized and temperature corrected performance ratios filtered to irradiance bands of 1000 ±50 W/m² in the case of the poly-c-Si PV generator and of 1000 ±10 W/m² in the case of the a-Si PV generator. Additionally, the median values of the filtered data are plotted for periods of 5 days and 30 days, respectively.

A variation analysis has been conducted determining the linear trends of various sets of PR data derived with different irradiance filtering bands and of the respective medians over various intervals. Thereby, not only the change rates but also their uncertainty budgets at 95% confidence level have been computed [3].

The results of the regression procedure applied to the scattered PR data are shown in Figure 5. For the irradiance filter band of 1000 ±50 W/m², the method determines the degradation rates of the poly-c-Si and a-Si PV generators to 0.28 %/a and 0.49 %/a, respectively.

The calculated uncertainties are at a satisfactorily low level of ±0.025%/a for the poly-c-Si PV array and ±0.020%/a for the a-Si PV generator. From the graph, it can also be derived that no improvement of the regression quality is achieved if narrower irradiance bands are applied.

The sequence of median values can favourably visualize the annual and seasonal development of the performance parameter. However, the statistical significance of the changes rates is much lower for those derived from the medians than for the ones based on the sets of scattered data.

Figure 6 displays the results of the regression analysis on the medians of the PR data of the poly-c-Si and a-Si PV generators for various median periods at an irradiance filter band of 1000 W/m² ±20 W/m². The uncertainty budgets range for the poly-c-Si PV generator from about ±0.08 %/a at 5 day periods to ±0.12 %/a at medians of 30 days and for the a-Si PV generator from about ±0.15 %/a for medians of 5 days to ±0.45 %/a at 30 day periods, qualifying the derived change rates as unreliable.

Regarding the results for the PV generator with a-Si PV technology it can be concluded that the seasonal performance variations do not impair the linear regression of the scattered data but the median value analysis.

The evaluation also shows that the change rates derived from medians may be significantly different from the linear regression trends determined for the scattered data.

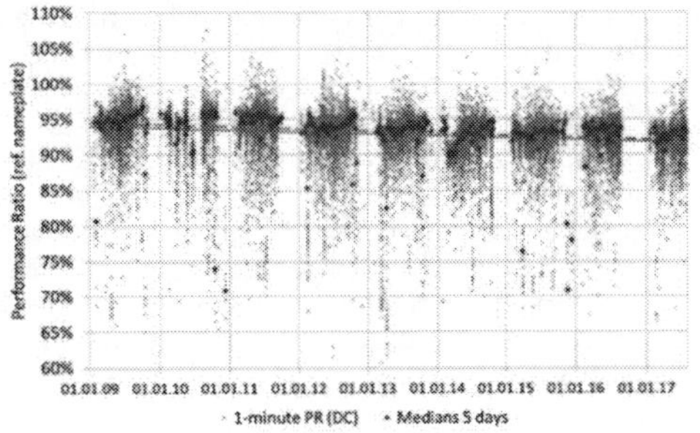

Figure 3: PV Generator poly-c-Si: DC Performance Ratio for records with irradiance of 1000 ±50 W/m², and 5-day medians thereof.

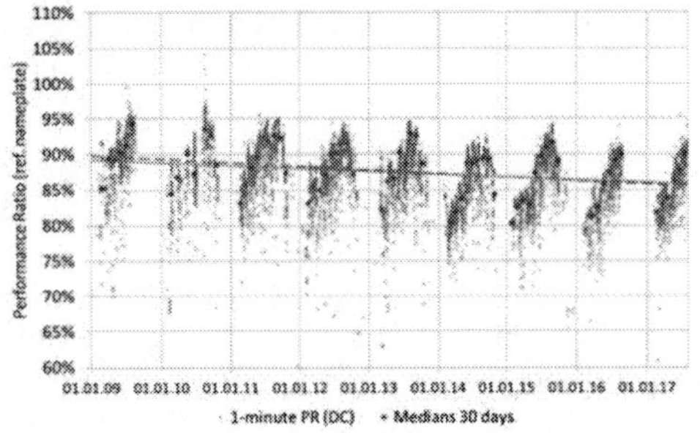

Figure 4: PV Generator a-Si: DC Performance Ratio for records with irradiance of 1000 W/m² ±10 W/m² and 30-day Medians

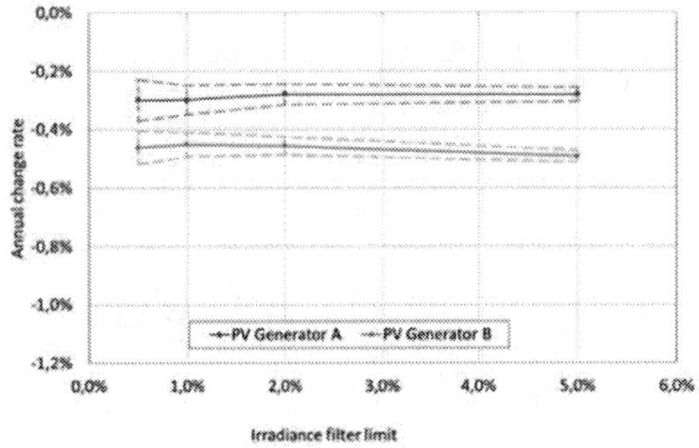

Figure 5: Results of scattered PR data regression for various irradiance filter bands: PV Generators poly-c-Si (A) and a-Si (B) degradation rates with respective 95 % confidence ranges

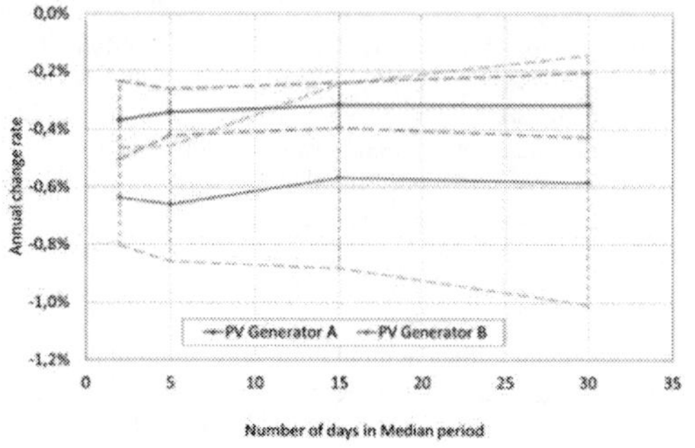

Figure 6: Results of median PR value regression for various median periods: PV Generators poly-c-Si (A) and a-Si (B) degradation rates with 95 % respective confidence ranges

3.2 PV Modules

The performance data of the PV modules individually sampled at 10s intervals exhibit an increased scatter which is due to short-term irradiance fluctuations in conjunction with the differing response times of the pyranometric irradiance sensor and the PV devices.

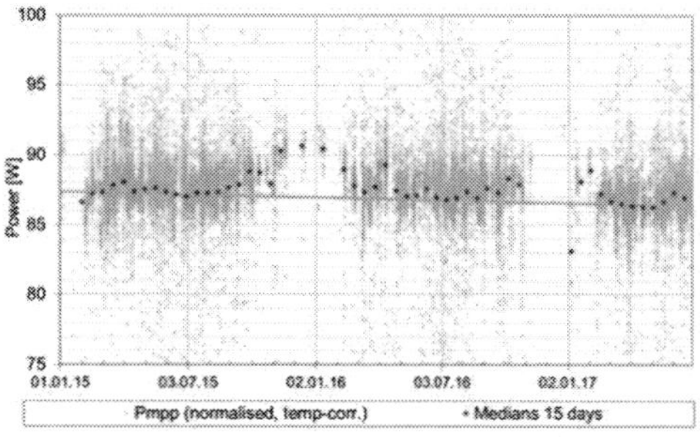

Figure 7: MPP power data for a CIGS PV module with measured irradiance at 1000 W/m² ±20 W/m² and 10 % stability filter, together with the median values for 15 day periods

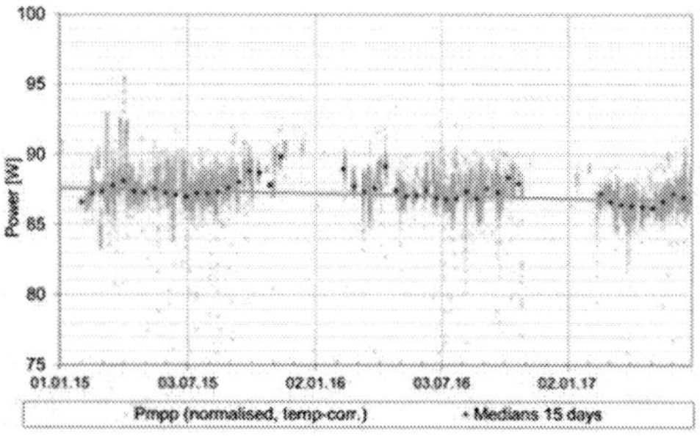

Figure 8: MPP power data for a CIGS PV module with measured irradiance at 1000 W/m² ±20 W/m² and 0.5 % stability filter, together with the median values for 15 day periods

In order to improve the trend analysis, an additional irradiance stability filter has been introduced, limiting the variation of the irradiance value of the respective record to its immediate ancestor and follower to a certain fraction of the reference irradiance. Figure 7 and Figure 8 show power performance data of a CIGS PV module over a period of 2.5 years with the measured irradiance in the band of 1000 W/m² ±10 W/m² and different stability filters of 10 % = ±100 W/m² and 0.5 % = ±5 W/m², respectively.

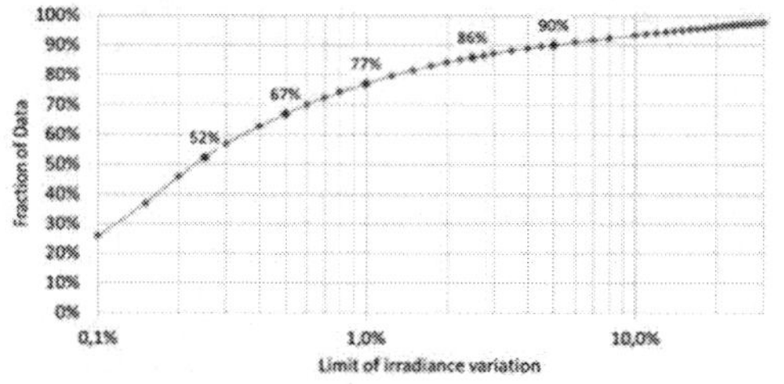

Figure 9: Fractional distribution of irradiance values in the range of 1000 W/m² ±20 W/m² with respect to variation limits in 10-sec data from the ZSW test site Widderstall (South Germany)

The application of the irradiance stability filter noticeably reduces the usable data fraction only for thresholds of about 1 % = ±10 W/m² and below (Figure 9).

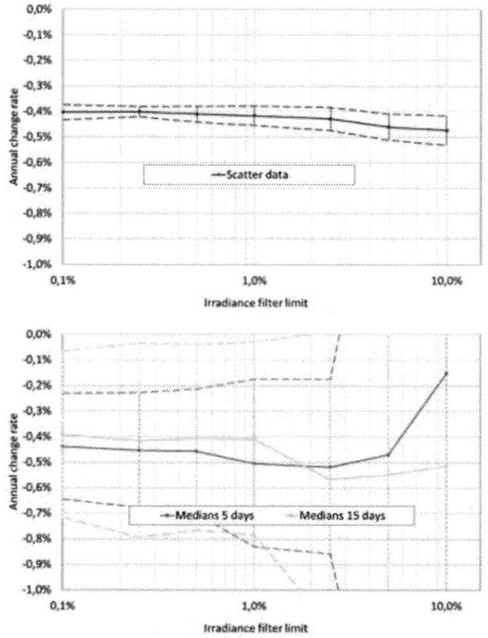

Figure 10: Degradation rates with 95 % confidence ranges versus the applied irradiance filter limit for a CIGS PV module; top: from regression of scattered data; bottom from median values of 5-day and 15-day periods, respectively.

The regression analysis of the power data from the CIGS PV module with various subsets applying different stability filters shows, that narrower limits yield an improvement of the significance of the determined degradation rates from a confidence level of ±0.06 %/a at 10 % stability filter to an uncertainty of only ±0.03 %/a at 0.5 % filter (Figure 10). For lower thresholds, the uncertainty increases again due to the strong reduction of the number of records.

The trend analysis of median values is unreliable for long aggregation periods of, e.g, 15 days as can be also seen in Figure 10. The uncertainty budget may even be larger than the derived degradation rate.

4 CONCLUSION

Regression analysis of high resolution performance data filtered to an irradiance band close to 1000 W/m^2 determines degradation rates with sufficiently low uncertainty levels and is much more reliable than trend analysis of aggregated median values.

ACKNOWLEDGEMENT

Parts of the work have been financially supported by the German BMWE within the projects FZ 0325715A and FZ 0324179, which is gratefully acknowledged

REFERENCES

[1] D. Stellbogen, H.-D. Mohring, A. Jagomägi, E. Mõttus, G. Friesen, D. Domine, F. Fabero, T. Betts, R. Gottschalg, T. Zdanowicz, M. Prorok, W. Herrmann, J.L. Martin, A.G. de Montgareuil, J. Merten, D. Faiman, Comparative outdoor characterisation of PV Modules across Europe, 25th European Photovoltaic Solar Energy Conference and Exhibition Proceedings, 2010

[2] D. Stellbogen, P. Lechner, H.-D. Mohring, Long term performance characterisation of PV generators, 28th European Photovoltaic Solar Energy Conference and Exhibition Proceedings, 2013

[3] J. Higbie, Uncertainty in the linear regression slope, Am. J. Phys. Vol.59, No.2, 184-185, 1991.

SOILING EFFECT ON PHOTOVOLTAIC MODULES' PERFORMANCE IN ARID ENVIRONMENT

Feras G. Alzubi and Abdulwahab T. Alasfour
Energy and Building Research Center,
Kuwait Institute for Scientific Research, Safat 13109, Kuwait
Phone: +965 2495 6291, Fax: +965 2498 9099
E-mail: feras.alzubi@gmail.com

ABSTRACT: The dust accumulation on photovoltaic systems is a detrimental agent as far as the reduction in the energy yield of the systems is concerned. The focus of this work is to accurately determine the soiling ratio of framed and frameless photovoltaic modules by monitoring their performance every ten minutes in an arid-desert environment. Moreover, we isolate the effect of the parameters and factors that can overestimate or underestimate the soiling ratio such as the light soaking degradation of modules, temperature coefficients deviations, and manufacturer mismatch of modules' efficiencies. All tested modules have shown lower efficiencies compared to manufacturer specifications after initial benchmarking. Furthermore, the relative efficiencies of tested modules decreased by 1 % after the light soaking process. The initial measured losses were incorporated with the field data of the monitored modules using an in-house built soiling-ratio model. The results showed that the reduction in output energy can reach 24 % due to dust accumulation over a time period of three months. On the other hand, frameless modules showed unstable behavior in terms of energy reduction.
Keywords: Soiling ratio, I-V characteristics, power output, STC measurement

1 INTRODUCTION

In most arid zones of the world, the presence of dust in addition to high temperature and low annual precipitation can enhance the accumulation of dust on photovoltaic (PV) systems. Dust deposition (soiling), which affects the light transmittance through a PV module can reduce its annual energy yield up to 34 % when tilted and 82 % when flat in desert environment [1]. Early soiling studies have focused on the drop of modules' performance due to dust only without consideration of the environmental and setup parameters [2]. Studies later on began to focus on the parameters that have a direct impact such as tilt angle and air pollution [3,4]. Different PV technologies have also been discussed as an important factor affecting the soiling severity level [5,6]. Simulations and induced soiling were additional approaches to predict the effect of dust on modules' performance [7,8,9,10]

Although most of the work done in this field clearly indicate the direct effect of soiling on modules, the techniques and methods used can represent high uncertainty. Gottschalg et al. had concluded that a deviation of 4.5 % can still be introduced in outdoor measurements of side-by-side PV systems [11]. In addition, further uncertainties are introduced to the measurements of soiling as a result of non-continuous monitoring of the PV modules reported by most of research groups. Moreover, the level of details and quality of measurements are advancing over time, which allow for a better energy loss determination by considering new parameters. Hence, further investigations with such parameters are needed. In addition, it becomes important to setup a scientific based approach to continuously monitor the performance of modules under real-life conditions outdoors.

The main objective of this work is to accurately measure the soiling ratio of mc-Si modules under outdoor conditions. This is achieved by isolating the parameters that can deviate the actual soiling loss such as light soaking degradation, manufacturer mismatch efficiencies, and temperature coefficients. In addition, the effect of the frame of module is studied. We ensure the robustness of outcome data by continuously measuring and tracking maximum power point and I-V characteristics in field.

2 EXPERIMENTAL SETUP AND MONITORING PROCEDURE

The testing setup consists of 10 mc-Si modules divided into two sets with five framed and five frameless

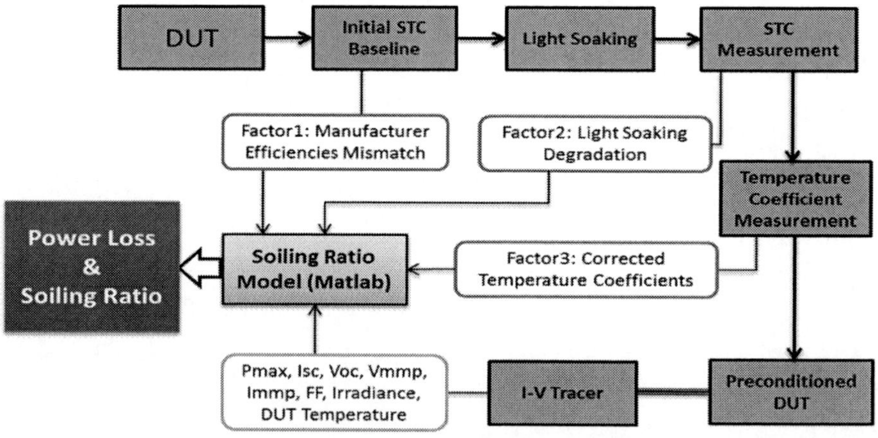

Figure 1: Schematic diagram of the experimental procedure and the work flow to calculate soiling ratio.

to consider the frame parameter effect. One of each set is kept clean, to be used as a reference module, and the rest are not cleaned (dusted). As an initial step prior to the outdoor installation of modules, performance base-lining was conducted indoor using Pasan 3B flash simulator to measure the deviations between the modules' actual performance and the manufacturer's specifications. This will accurately isolate the efficiencies mismatch factor from soiling effect by using the produced corrected STC values. Next, all modules are placed outside for two months for light soaking before applying a second base-lining under STC conditions to investigate the degradation in modules' efficiencies due to light soaking. Finally, all the PV modules are connected in field to a current - voltage (I-V) tracer measurement unit with a maximum power point tracker and a data logger unit to monitor the output during each 10 minutes sweep. The I-V characteristics at STC before and after light soaking in addition to the outdoor data from the I-V tracer are embedded in an in-house built soiling ratio model to accommodate for all factors including temperature coefficients to single out the power loss due to only soiling. Figure 1 below shows a schematic diagram of the followed methodology in this work.

3 DATA ANALYSIS

The studied 10 modules were baselined and characterized indoor under STC to investigate their PV characteristics' deviations from the claimed value by manufacturer. Figure 2 below shows the measured I-V characteristic parameters for framed and frameless modules, compared to the manufacturer specifications.

Both framed and frameless have shown similar deviation in the Voc values. However, for framed modules, deviations of Isc, PCE and MPP values were lower than frameless modules. Moreover, an average reduction of 12 % in MPP is observed for framed modules and 16 % for frameless modules. In addition, relative efficiency was less by an average of 1 % for framed modules and by 2 % for frameless modules. These losses are incorporated throughout the experiment to isolate the soiling effect from other environmental and setup parameters. Next, modules were exposed in field to sun light soaking for two months and re-baselined. Their relative efficiency values have shown a reduction of 1-2% from original claimed value. Finally, temperature coefficients measured and the outcomes values are embedded in the soiling model to accommodate the temperature effect correction in addition to the other losses.

4 RESULTS

After the modules preconditioned and factors affecting the calculation of soiling ratio were investigated, the modules were placed in field and being characterized by I-V tracer with a sweep every 10 minutes for testing the effect of the dust on their power outputs; Figure 3 below shows the dust accumulation over three months on an investigated module installed outdoor.

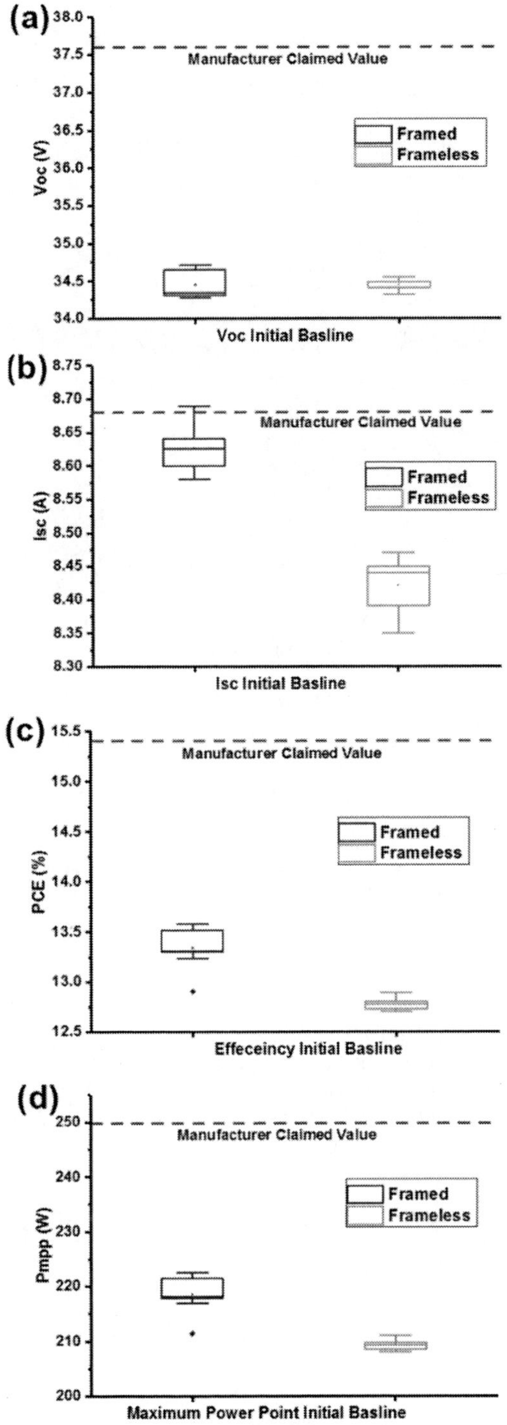

Figure 2: The STC measurement of the I-V characteristics of tested frameless and framed modules with respect to the manufacturer claimed values. Shown parameters are: (a) Open circuit voltage (Voc), (b) Short circuit current (Isc), (c) Efficiency (PCE), and (d) Maximum power point (MPP).

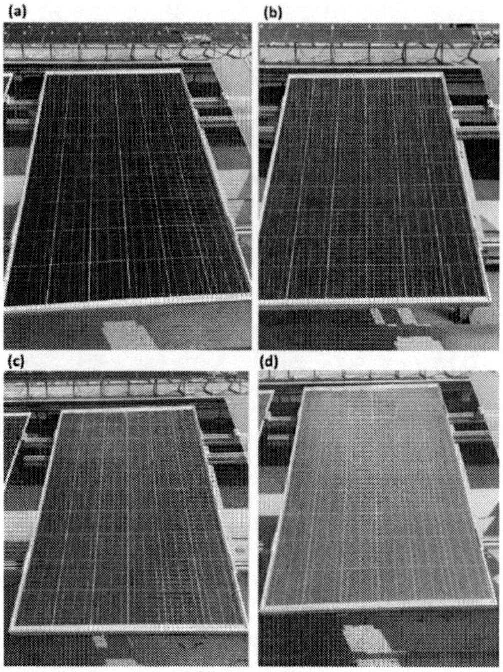

Figure 3: Outdoor tested PV module showing the dust accumulation (a) At the day of installation, (b) After one month, (c) After two months, and (d) After three months.

Both framed and frameless mc-Si modules were characterized under outdoor conditions. Loss in power or soiling ratio (SR) is calculated in the soiling model by using the values of corrected maximum power (P_{max}) when irradiance is over 900 W/m^2 using the following formula:

$$SR(\%) = \left(1 - \frac{P_{max(dusty)}}{P_{max(cleaned)}}\right) \times 100 \qquad (1)$$

Figure 4 below shows the loss due to soiling only calculated based on the P_{max} over three months period of time (may 16th to August 8th). The SR of the multi c-Si is (0.27 % /day) with a cumulative SR over three months of 24.1 %.

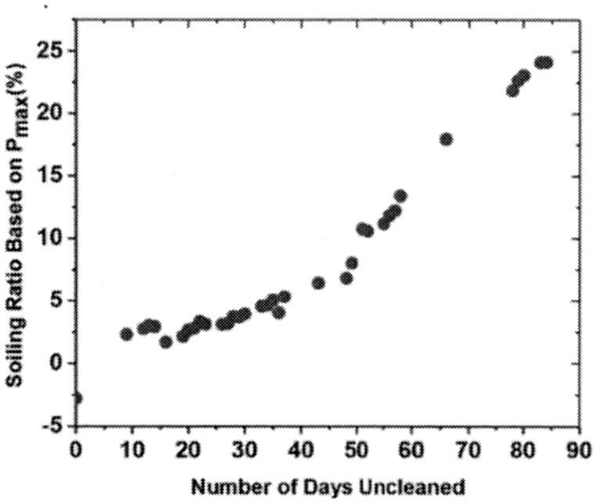

Figure 4: Soiling ratio based on the loss in the output power of framed modules for 90 consecutive days without cleaning.

At the beginning of the experiment, framed modules were visually found to increase the dust deposition greatly due to the existence of frame edges that can help the accumulation. In contrast, the frameless module structure allows dust to be washed out of the edge. As a result, output power loss is expected to be lower in case of frameless modules. However, similar trend as in Figure 4 of soiling ratio was extracted for frameless modules. This finding supports that all modules have similar spectral response and technology and soiling should affect the absorption and light scattering similarly in all modules. Further investigation of the power output values of frameless modules have shown an unstable behavior and fluctuation manner in the values. This behavior is mainly a result of having no frames, which allowed the humidity to have a greater effect on the Si-interlayers in the cells by penetrating through the unframed edges of the module. Figure 5 shows the behavior of the soiling ratio of dusty framed modules versus dusty frameless modules. During the first month in the testing site, values of soiling ratio are more scattered with wider range of 5 % compared for frameless modules to the framed modules with a range of 2.5 %. In general, frameless modules have shown higher soiling ratio values. We believe that these values are overestimated and can be attributed to the extra loss added due to humidity which was not accounted for in the soiling model.

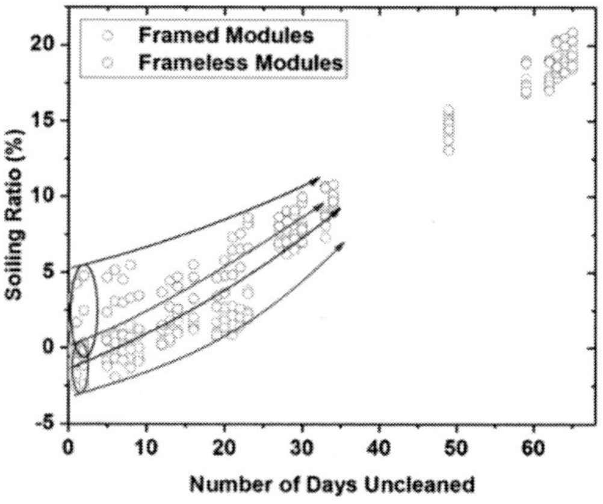

Figure 5: Soiling ratio based on maximum power loss of framed modules (red) versus frameless modules (Blue)

Figure 6 shows variations in the values of I-V characteristics (V_{oc} and I_{sc}) due to dust accumulation for clean modules versus dusty ones at the installation day versus after two months in field. Also the figure shows variation during the day from 10 am to 4 pm. For the module at the day of installation, V_{oc} values showed a dependency with a range of variation of 1 volt on the day time i.e. irradiance as can be seen in Figure 6 (a). This behavior was not observed later in the experiment. In the same figure by comparing clean to dusted modules after at day 60, we can see that the dust effect on V_{oc} value is very slight to negligible about 0.1 V. On the other hand, I_{sc} values in Figure 6 (b) are highly affected by the dust accumulation. As a result, soiling ratio based on loss in power is mainly due to the reduction in the I_{sc} values. This agrees with the fact that V_{oc} is a material property and should not be affected by dust accumulated on the surface of the module. Moreover, the dust affects the performance of the module mainly by partially screening light from reaching the cells. This leads to a reduction in the light absorption. Hence the amount of generated carriers (electrons and holes) is reduced and accordingly current is reduced.

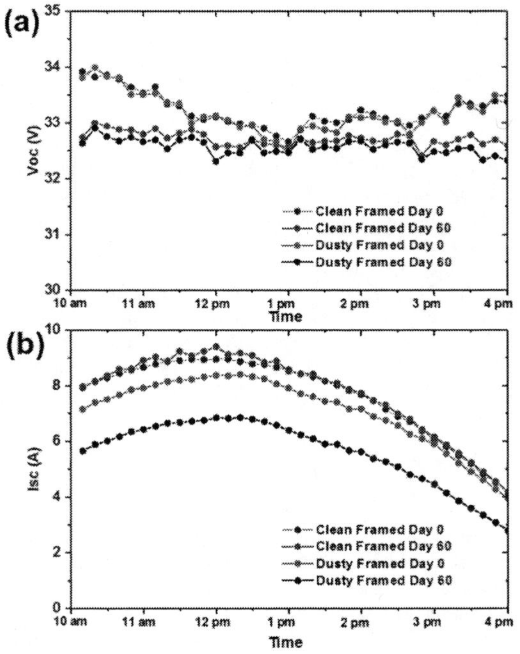

Figure 6: Values of Isc and Voc during the day for cleaned and dusted framed modules at the first day of installation and at day 60.

In our study, we also investigated the PV module performance in different seasons in the arid-desert area of Kuwait. The acquired results (not presented in this paper) show that in April, a clear drop in the output power of modules occurs as results of dust and rain combination in the region. In the hot and very dry summer, a lower power reduction is observed because the lower adhesion between the dust particles and surfaces. The maximum soiling ratio obtained 34 % occurs at the end of summer and beginning of winter where the humidity plays major role in power reduction in addition to the mud particles due to rain-dust combination. On the other side, in mid to end of winter, rain helps clean modules and output power increases again.

5 CONCLUSION

Through this study, we were able by this study to determine the effect of soiling on the performance of mc-Si PV modules. Soiling ratio based on power loss 24 % was observed for framed PV modules. This reduction is mainly due to the reduction in I_{sc} values due to dust accumulation. Frameless modules have shown similar trend of soiling ratio with unstable behavior of power outputs. Temperature coefficients, the manufacturer efficiencies mismatch, and the light soaking degradation factors were investigated and quantified to ensure that the reduction in the performance of the tested modules is only due to soiling. This allows for better understanding of the soiling problem in the way of presenting the best set of solutions. The findings of this work are being used to further explore the proper cleaning methods that can be used to clean the PV system while comparing the cost associated with different cleaning methods to the gain in energy.

6 ACKNOWLEDGMENT

The presented work is part of project EA048C that is financially supported by the Kuwait Foundation for the Advancement of Sciences (KFAS). Facility and equipment support was provided by the Kuwait Institute for Scientific Research (KISR) under Renewable Energy program.

7 REFERENCES

[1] H. Qasem, T.R. Betts, H. AlBusairi, R. Gottschalg, "Measurements of dust induced performance losses on Micromorph Photovoltaic Modules in Kuwait." in Proceedings of 27th European Photovoltaic Solar Energy Conference and Exhibition (27th EU PVSEC), 2012, pp 3010-3014.

[2] R. Hammond, D. Srinivasan, A. Harris, K. Whitfield and J. Wohlgemuth, "Effects of soiling on PV module and radiometer performance," *Photovoltaic Specialists Conference, 1997. , Conference Record of the Twenty-Sixth IEEE*, pp. 1121-1124, 1997.

[3] E. Al-Soleimani, S. Farhangi and M. S. Zabihi, "The effect of tilt angle, air pollution on performance of photovoltaic systems in Tehran," *Renewable Energy*, vol. 24, pp. 459-468, 11, 2001.

[4] M. Sadok and A. Mehdaoui, "Outdoor testing of photovoltaic arrays in the Saharan region," *Renewable Energy*, vol. 33, pp. 2516-2524, 12, 2008

[5] H. Pang, J. Close and K. Lam, *Study on Effect of Urban Pollution to Performance of Commercial Copper Indium Diselenide Modules.* New York; 345 E 47th ST, New York, NY 10017 USA: IEEE, 2006.

[6] H. AlBusairi and H. Möller. Performance Evaluation of CdTe PV Modules Under Natural Outdoor Conditions in Kuwait. 2010.

[7] M. García, L. Marroyo, E. Lorenzo and M. Pérez, "Soiling and other optical losses in solar-tracking PV plants in navarra," *Prog Photovoltaics Res Appl*, vol. 19, pp. 211-217, 2011.

[8] J. K. Kaldellis and M. Kapsali, "Simulating the dust effect on the energy performance of photovoltaic generators based on experimental measurements," *Energy*, vol. 36, pp. 5154-5161, 8, 2011.

[9] C. Farid C., "Numerical modelling of wind and dust patterns around a full-scale paraboloidal solar dish," *Renewable Energy*, vol. 39, pp. 356-366, 3, 2012.

[10] H. Qasem, T. R. Betts and R. Gottschalg, " Soiling Correction Model for Long Term Energy Prediction in Photovoltaic Modules," in 38th IEEE Photovoltaic Specialists Conference, Austin, TX., 2012.

[11] R. Gottschalg and T.R. Betts. Robustness of Energy Yield Measurements of Photovoltaic Modules. 2011.

ADVANCED METHOD FOR DETERMINING SOILING LOSSES ON PV MODULES IN DESERT CLIMATE

David Dassler[1,2,*], Stephanie Malik[1], Akshayaa Pandiyan[1], Aboubakr Benazzouz[3], Badr Ikken[3], Matthias Ebert[1]

[1] Fraunhofer Center for Silicon Photovoltaics CSP, Otto-Eissfeldt-Str. 12, 06120 Halle (Saale), Germany
[2] Anhalt University of Applied Sciences, Faculty EMW, Bernburger Str. 55, 06366 Koethen, Germany
[3] Institut de Recherche en Énergie Solaire et Énergies Nouvelles IRESEN, 16, Rue Amir Sidi Mohamed, Souissi, Rabat, Morocco
*=corresponding author: phone +49 345 5589-5214 | David.Dassler@csp.fraunhofer.de

ABSTRACT: Dust deposition on the surface of PV modules (soiling) leads to significant losses in energy yield particularly in desert climates, where high daily yield losses due to soiling can be observed. However, its estimation is still an open challenge in PV modeling. The intensity of soiling losses strongly depends on its environmental conditions. The aim of this paper is the development of a method to determine the soiling losses and to classify the soiling state of PV modules in desert climate. Therefore, the paper combines the results from indoor and outdoor measurements on PV modules with two types of front glasses. The outdoor parameters were measured in the research platforms of photovoltaic modules at "Green Energy Park" by IRESEN in Morocco for several months. During our investigation, different cleaning scenarios were implemented for different modules in both groups. The application of a multi-dimensional classification and clustering as well as parameter correlation enabled the determination of the impact of environmental data on module soiling on a daily basis. Thus, for each cluster its corresponding averaged soiling loss can be calculated. In terms of module current, the impact of the soiling is between the three classes 11 to 44 mA at high irradiation conditions.

Keywords: soiling, soiling losses, antireflective coating, modeling, clustering

1 INTRODUCTION

The photovoltaic industries are in a common pursuit of increased conversion efficiency for PV modules and have reached a great technological progress over the past years. However, irradiation losses due to external causes, like soiling and shading, remain as the prominent source to a PV module's energy yield cutback. Consequently, effort and investment to alleviate them becomes essential especially in cases of maintaining a large scale PV plant. This can be simplified by identifying and analysing the key influencing parameters with respect to external losses, and developing a method to estimate them based on the former.

Dust, dirt, sea salt, pollen, and snow originating from air pollution due to mining activities, construction, agricultural activities and other natural occurrences can lead to accumulation of particulate matter on the surface of solar panels. This is termed as *soiling*, and is one of the most significant external detriments for PV yield especially on dry or arid areas. This paper focuses on soiling caused by dust deposition in these areas. Nevertheless, the shown approach is applicable for several kind of soiling as well.

Soiling leads to reductions in current of the PV module, which consequently affects the power and energy yield. These losses are called *soiling losses* in this paper.

Soiling can be removed only through cleaning events which can occur naturally (strong precipitation) or manually (maintenance cleaning). Past researches on soiling have focused on desert or dry regions which are prone to large aerosols loading in the air, but consistently occur with large solar resources too. A recent study [1] by Mejia *et al.* reported a 7.4% loss in efficiency for a PV panel operated over an average of 145 day during drought in California, which is an order of magnitude higher than losses due to cell degradation. Similar case studies have supported the significant impacts of soiling losses towards a panel's energy yield for various regions, however efficient methods of quantitatively predicting the soiling is still a challenging task for researchers in PV modelling. The most influencing parameters in the magnitude of soiling losses are the dust properties (for instance, shape and size), pollution, moisture, humidity, frequency and intensity of rainfall, wind speed and the installation configuration of PV panels [2, 3]. In a study by Haeberlin *et al.* [4], using a 60 kW site, demonstrated the reversible effects of soiling losses before and after it was manually cleaned. Additionally, a recent work in Italy [5] reported a performance cutback of 6.9% for one site and 1.1% for another site not far from each other due to 8 week of soiling. It can be, therefore, interpreted that large variability of soiling can occur even over short distances.

In this work, the prime objective is to develop a method to determine soiling losses and to classify the soiling state of PV modules in desert climates, which can then be adapted for other climates based on outdoor measurements.

Pairs of modules with different front glasses were classed into different groups of cleaning cycles. Therefore, obtaining insights based on module and outdoor measurements with respect to different applied cleaning intervals. This enables the application of a multi-dimensional classification and clustering as well as parameter correlation to determine the impact of environmental data on module soiling.

The outdoor measurements over several months were performed in the outdoor research platforms of photovoltaic modules at "Green Energy Park" (GEP) in Ben Guerir, implemented in a partnership with Fraunhofer CSP. GEP is a solar testing, research and training platform, located in the green town of Ben Guerir, developed by the IRESEN with the support of the Ministry of Energy, Mines, Water and Environment and the OCP Group [6] to develop and provide the access to research infrastructures and expertise at optimum costs. IRESEN (Research Institute for Solar Energy and New Energies) was established in 2011 in order to lead and promote R&D in Morocco in the field of renewable energy.

2 METHODS

In order to identify soiling losses, the approach of the paper extends from outdoor measurements to extensive data analysis. Figure 1 structures the approach of this paper.

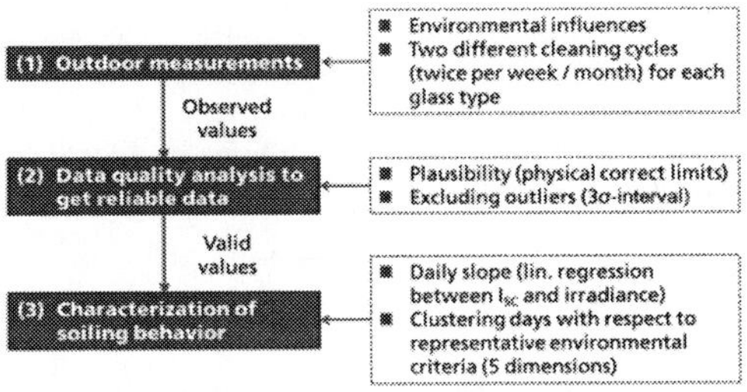

Figure 1: Flowchart of the investigations

In the first part, PV modules were characterized at STC and tested under outdoor conditions with different cleaning cycles in a desert environment. This data includes detailed information about the state of natural soiling of the modules and relevant environmental influences over the time.

The gathered data were then filtered in the second step to achieve validity for further data evaluation. Finally, the soiling losses were determined based on filtered data, considering the mean ratio between irradiance and module current. The current is only slightly temperature dependent and is directly affected through transmission losses caused by dust deposition on the module glass. Furthermore, the days were clustered due to different influencing environmental conditions. Thereby the determined soiling losses can be characterized by cluster information.

These methods are described in detail in the following chapters.

2.1 OUTDOOR MEASUREMENTS IN MOROCCO

The PV modules were built at Fraunhofer CSP Module Technology Center with two types of front glasses: with and without antireflective coating (ARC/non-ARC). Beside the glass, same module materials (solar cells, encapsulation, backsheet) were used.

Previous to outdoor exposure in Morocco, the performance of each module was characterized indoor at Fraunhofer CSP under standard test conditions (STC) with a class AAA module flasher (at 1000 W/m² with an AM1.5 spectrum and module temperature of 25 °C). Table 1 details the indoor characterization results at STC.

Table 1: Indoor power measurements at STC, measured with a sun simulator using class AAA flasher

Sample	Type	V_{OC} [V]	I_{SC} [A]	P_{MPP} [W]	FF [%]
Mod_1	ARC	37.1	8.9	245.2	73.9
Mod_2	ARC	37.1	8.9	245.7	74.6
Mod_3	Non-ARC	37.1	8.8	243.8	74.5
Mod_4	Non-ARC	37.1	8.8	244.1	74.6

The outdoor measurements over several months were performed at GEP in Morocco (see Picture 1). The module measurements (IV-curves and module temperature) were measured synchronously in a time interval of 10 s and between the measurements, every module is operated at maximum power point (MPP). The weather data, however, was measured every 60 s and dust particle concentration were measured with an interval of 10 min.

The measuring device for IV-curves has a current-full scale range (FSR) of 12 A with a resolution of 1 mA. The FSR of the voltage is 259 V with a resolution of 10 mV. The weather data acquisition system includes a pyranometer in the module plane. Wind direction and speed as well as ambient temperature, are logged using different devices.

Different cleaning approaches were applied for each glass type: twice per week and twice per month. This results in a "clean" and a "soiled" group containing both glass types.

Picture 1: Outdoor measuring set-up and installed PV modules in Green Energy Park of IRESEN, Ben Guerir, Morocco

The data was synchronized and filtered for further evaluation. The measured data have different intervals for various acquisition systems. The weather data and the dust particle concentration were assumed as constant for the subsequent 60 s and 10 min interval, respectively. The data was processed for plausibility by imposing physical limits and 3-sigma limits [7]. This ensures that the quality of the analysing data used for characterizing soiling losses is least affected by noise and outliers.

This paper considers only the time period from April 2017 to July 2017 (75 days).

2.2 DETERMINATION OF SOILING LOSSES

In order to determine soiling losses on a module level, the linear relationship between short circuit current (I_{SC}) and irradiance (G) was used (see Figure 2). By performing linear regression for a specific time period, representative average relationship can be approximated. Due to the fact that the current is zero without any irradiation, this paper assumes the linear regression model ($\widehat{I_{SC}}$) without intercept term, see Eq. 1.

$$\widehat{I_{SC}} = s_d\, G \qquad \text{(Eq. 1)}$$

The resulting slope of the regression coefficient (s_d) characterizes the quantitative conversion of irradiation energy into electrical current. In sense of this objective, the slope s_d determines the soiling state of the PV module.

The slope can be applied variably with respect to time period and PV module samples. In this case, the slope is determined per day and per module to have a valid linear regression process. The slope s_d can be calculated as shown in Eq. 2 for any day (d) and current (I_{SC}).

$$s_d = \frac{\sum_{t\in d} G_t\, I_t}{\sum_{t\in d} G_t^2} \qquad \text{(Eq. 2)}$$

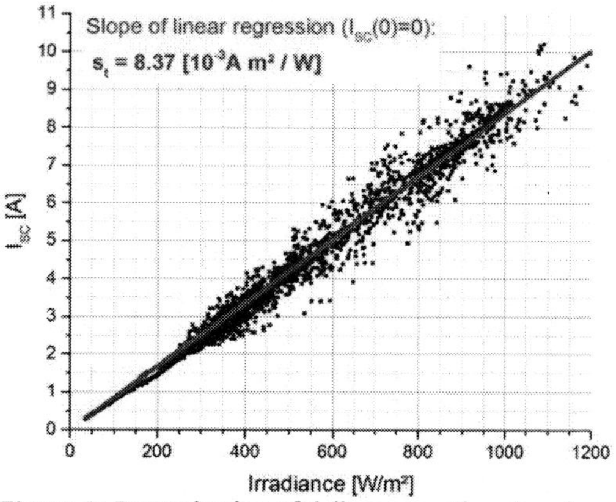

Figure 2: Determination of daily mean ratio using slope of linear regression: Eg. relationship between I_{SC} and irradiance for one module on 09.06.2017

Changes in slope from one day to the other can be interpreted as changes in the soiling state of the module (cleaning or soiling effect). This approach assumes no degradation or malfunction of the PV module, measuring set up or the irradiance sensor (regular cleaning of sensor was performed).

The daily differences in slopes ("*slope difference*" Δs_d) were calculated between one day and the day before (see Eq. 3). If Δs_d is positive, a cleaning event has happened (natural/manual) and if it is negative, the soiling has increased.

$$\Delta s_d = s_d - s_{d-1} \qquad \text{(Eq. 3)}$$

To be able to interpret results referring to this new introduced parameter (Δs_d), it is important to take the corresponding uncertainty into account. This uncertainty determination is based on the parameters, I_{SC} and G, where the standard uncertainty in the measuring process of both parameters is regarded for every component. In accordance with the "Guide to the Expression of Uncertainty in Measurement" (GUM) [8], the combined uncertainty (u_C) can be calculated along the measuring chain of I_{SC} and G. By assuming a coverage factor (c), an interval having a specific level of confidence is produced and the expanded combined uncertainty (U) per parameter can be determined.

The uncertainty determination of I_{SC} and G is based on an equipment documentary [9, 10]. The standard uncertainties of every component involved in I_{SC} measurements are treated as uncorrelated. Whereas, the standard uncertainties referring to the measurements of G are considered as correlated. This worst case scenario applied for G, as described in [11], results in higher uncertainty than the assumed uncorrelated uncertainties. For example, at 200 W/m² the difference between the uncertainty approaches 55.8%.
Subsequently u_C of I_{SC} and G was propagated through to determine $U\Delta s_d$ per Eq. 4.

$$U(\Delta s_d) = c\sqrt{u_c^2(s_d) + u_c^2(s_{d-1})} \qquad \text{(Eq. 4)}$$

With $c = 2$ (assumption of a normal distribution), $I = I_{SC}$ and

$$u_c^2(s_d) = \sum_{k\in d}\left[\frac{(I_k \sum_{t\in d} G_t^2 - 2G_k \sum_{t\in d} G_t I_t)^2}{(\sum_{t\in d} G_t^2)^4} u^2(G_k)\right] + \cdots$$

$$\sum_{k\in d} \frac{G_k^2}{(\sum_{t\in d} G_t^2)^2}\, u^2(I_k)$$

2.3 CLASSIFICATION

To solve the challenging task to classify environmental parameters with their special daily profile and their possible interaction between each other, the k-mean clustering method, popularly known for cluster analysis in data mining, was applied. k-mean clustering means separating N observations for n parameters into k clusters, in which each observation belongs to the cluster with the nearest distance (absolute difference) to cluster center. These cluster centers were determined iteratively, where in each step the observations were reassigned to that center with the nearest distance. Subsequently, the mean (centroid) for each cluster defines a new cluster center for the consecutive step.

The set of observations to be clustered includes 75 days ($N = 75$). An appropriate cluster solution containing three or four clusters can be chosen to avoid very small cluster division with less than 10 days per cluster. This work demonstrates 3 clusters estimation ($k = 3$). The used parameters for clustering ($n = 5$), aggregated per day (including night values) as sum or mean value, are listed inTable 2.

The quality of the standard k-mean method depends also on the randomly chosen initial cluster centers, before first iteration. To ensure a reliable minimum sum of distances within each cluster, a large number of repetitions (~ 10000) of the k-mean clustering has been realized.

Table 2: Characteristic parameters for clustering

Parameter	Aggregation per day
Irradiance	Irradiance yield sum [kWh/m²]
Dust particle (air)	PM total sum [10³ µg/m³]
Relative humidity	RH mean [%]
Ambient temperature	T_{amb} mean [°C]
Wind speed	WS mean [m/s]

3 RESULTS

3.1 SOILING LOSSES

Following the approach described in chapter 2.2, the resulting slopes per day for each module in the considered period are displayed in Figure 3. This also includes manual cleaning events for comparison. The soiling behavior per module can be seen as the decreasing slopes, and different applied cleaning treatments can be deciphered from the abruptly increasing slope.

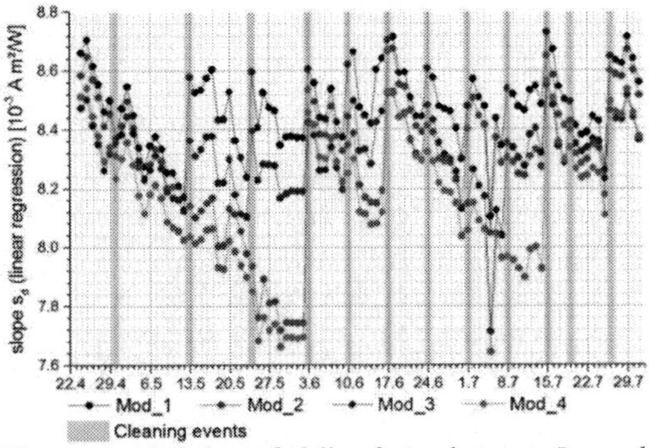

Figure 3: Variation of daily slopes between I_{SC} and irradiance for four modules with different cleaning cycles

An average soiling rate can be calculated by linear regression of the decreasing slopes (normalized to the cleaned state of the module) until the next cleaning event. Table 3 shows the averaged soiling rate for the four modules. Due to changes in cleaning cycles and samples, this average soiling rate was determined individually per module, following their soiling behavoir between two cleaning events. Nevertheless, module 1 & 3 was cleaned more frequently than module 2 & 4. Examining the results, module 1 & 3 show a higher soiling rate than the more soiled samples (see Table 3). The soiling rate of the cleaned modules lies near to each other and this respectively stands as well for the group of soiled modules. On the contrary, a strong effect between both glass types can't be seen.

Table 3: Maximum and average soiling rate [%/day] for each module

	Soiling rate [%/day]	
Module	Mean	Max (abs.)
Mod_1	-0.27	-0.57
Mod_2	-0.23	-0.36
Mod_3	-0.31	-1.16
Mod_4	-0.22	-0.34

3.2 CLASSIFICATION OF ENVIRONMENTAL DATA

The results of the 5-dimensional clustering applied on daily aggregated environmental parameters (listed in Table **2**), are shown in Figure 4. Both graphs are based on the same data and clustering process, in which the colors represent the clusters. The difference between them is due to the chosen environmental parameters and their dependency among each other in their corresponding cluster.

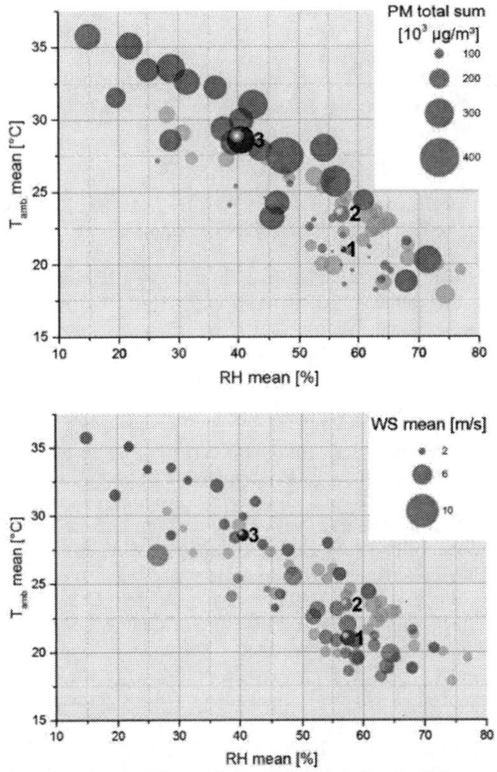

Figure 4: Results of 5 dimensional clustering of environmental data into 3 clusters based on daily values (colours refer to corresponding cluster)

It can be seen from the Figure 4 that:

- Cluster 1 is concentrated at high RH and moderate temperature.
- Cluster 3 is wide spread referring to RH and T_{amb}.
- The amount of cumulated air dust rises with rising T_{amb} and decreasing RH.
- The influence of wind speed doesn't show a clear tendency.

The ranges between the different clusters, regarding minimum, maximum and mean values are listed in Table 4. Referring to these values, the clusters are overlapped with each other for several parameters, because of the multi-dimensional relationship. The clearest distinction between the clusters in shown in the accumulated dust (PM total), whose values do not overlap.

Table 4: Characterization of clusters: minimum, maximum and mean values for each parameter and cluster

Parameter	Cluster 1	Cluster 2	Cluster 3
Irradiance [kWh/m²]	5.5 – 8 (Ø 7.1)	1 – 7.8 (Ø 6.6)	4 – 7.4 (Ø 6.6)
PM total [10³ µg/m³]	10.7 – 100.7 (Ø 59.1)	108.6 – 198.7 (Ø 155.5)	206.8 – 381 (Ø 262.7)
RH [%]	26.5 – 68 (Ø 55.1)	28.1 – 76.9 (Ø 55.5)	14.9 – 71.4 (Ø 40.9)
T_{amb} [°C]	18.2 – 27.1 (Ø 21.6)	17.9 – 30.4 (Ø 23.7)	18.8 – 35.7 (Ø 28.6)
WS [m/s]	2.2 – 6.5 (Ø 4.1)	2.2 – 5.2 (Ø 3.5)	2.5 – 4.8 (Ø 3.4)

Based on this clustering results, each further day (past/future) can be assigned to one of the three clusters if these five parameters are known for this day.

3.3 COMBINATION OF SOILING LOSSES AND CLASSIFIED ENVIRONMENTAL DATA

In the last step the classified days can be combined with the determined Δs_d, between one day to the other, of every module. The manual cleaning events have to be excluded in order to interpret the results only due to environmental impact. These results are shown in Figure 5 and in Figure 6.

In every cluster the soiling effects ($\Delta s_d < 0$) have a main emphasis, but natural cleaning effects ($\Delta s_d > 0$) occurre as well. A different soiling behavior with respect to the cluster can be seen. Cluster 2 shows the highest soiling impact followed by cluster 3 and 1. A different soiling behavior between the cleaned and soiled modules, as well in comparison with both glass types can be seen in Figure 6. In this figure, the mean Δs_d per module, per cluster with its corresponding expanded combined uncertainty ($U(\Delta s_d)$) is shown which is determined as described in chapter 2.2 to interpret the significance of the results. The relative mean of $U(\Delta s_d)$ is approximately 5.4% referring to all modules and clusters. The soiling behavior of the more frequently cleaned modules 1 & 3 (ARC and n-ARC respectively) in every cluster is worse than their corresponding soiled sample. This underlines the results of Table 3. Comparing the glass types with each other, the results are not that explicit as it was in reference with the three clusters (Figure 6). In cluster 1 the n-ARC modules 3 & 4 show an advantage in contrary to cluster 2 and 3.

Table 5: Deviation in I_{SC} determined per cluster by using mean value of slope differences (one day to the other) at different irradiances

Irradiance [W/m²]	Deviation in short circuit current [mA]		
	Cluster 1	Cluster 2	Cluster 3
1000	-11.3	-43.7	-26.5
800	-9.0	-35.0	-21.2
500	-5.6	-21.9	-13.3
250	-2.8	-10.9	-6.6

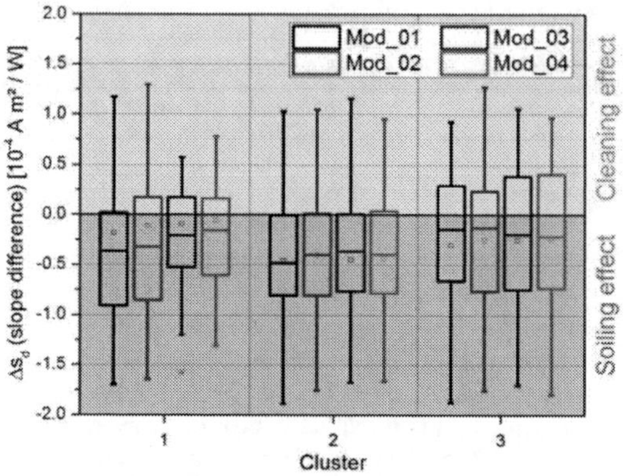

Figure 5: Slope difference Δs_d between two consecutive days; box plots (extreme values, quartile (25, 50, 75%) and mean) for each module and each cluster

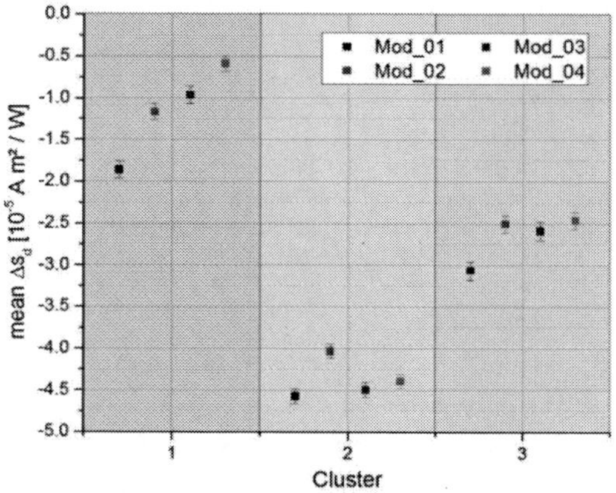

Figure 6: Mean of slope difference Δs_d between two consecutive days per module and cluster

Which deviation in short circuit current corresponds to the determined Δs_d Table 5 tries an answer on that, by assuming different irradiance levels.

4 DISCUSSION

Deciphering the effects of soiling based on the type of chosen glass, cleaning cycles and weather parameters isn't trivial. The proposed approach tries to put them into relation with each other. The results show a different soiling behavior between the cleaned and soiled modules as well as in comparing both glass types in the different clusters.

It is also relevant to notice that the mean values of slope differences of every module from all clusters are found in 'Soiling effect' quadrant, implying the decreasing slopes for all modules. The manual cleaning events are excluded in order to interpret the results only due to environmental impact; this could have resulted in higher slope differences. The soiling behavior of the more frequently cleaned modules 1 & 3 is for both modules types in every cluster worse than their corresponding soiled sample, which underlines a hypothesis that a cleaned module surface is particularly more prone to soiling effects than an already soiled layer of particulate matter to attract more dust and dirt until a saturated soiled state is reached.

Another interesting point to highlight is the higher slope differences (day-to-day) of cluster 2 than these of cluster 3, despite the higher concentration of dust in cluster 3. Even though cluster 2 has a less dust concentration than cluster 3, the mean humidity in cluster 2 is higher (Table 4). Against that result it is assumed that an increasing of the density of the aerosol particles suspended in the atmosphere induces more sedimentation of dust particles on the module surface. This is an illustration of the intricate relationship between the various environmental parameters that influence soiling and the importance of the proposed method, as clustering various environmental parameters and studying their soiling effects towards energy loss. Following this, the key influencing parameters can be determined more precisely, which enables a step towards the modelling of soiling effects of PV modules.

However, the end estimation of soiling losses are derived as a function of environmental parameters. Therefore, this work has an added advantage of being versatile to any location, given that the environment conditions are known. Comparing based on precision, this method has an good accuracy to differentiate between module properties as well as the soiling state of the modules.

5 CONCLUSION

The highly resolved outdoor measurements for PV modules with different front glasses and several environmental parameters, measured by high sensitive equipment in desert conditions, represents particular relevance. Through the different applied cleaning cycles, the development of soiling losses was investigated. The shown innovative application of multi-dimensional classification and clustering methods as well as parameter correlation to a large amount of outdoor data of different module samples enabled to quantify the relation between their soiling loss behavior and environmental factors, which demonstrates an advanced characterization method to determine and classify the soiling state of PV modules based on local climate conditions.

In the view of soiling loss estimation, the next step to improvise the proposed method would be to integrate the slope difference into clustering. This can provide information of the PV module soiling behavior directly to the environmental parameters. It would also be significant to study the influences per day and per night more in detail and to find aggregated environmental parameters referred to special time frames or conditions.

6 ACKNOWLEDGMENT

This work was partly supported by the Federal Ministry for Economic Affairs and Energy (BMWi) within the project "Fidelitas" (funding no. 0325735C) and by the Federal Ministry of Research and Education (BMBF) within the project "Wuestenmodule" (funding no. 03FH014IX4).

7 REFERENCES

[1] F. A. Mejia, J. Kleissl, Soiling losses for solar photovoltaic systems in California, Solar Energy Volume 95, September 2013, Pages 357-363

[2] A. Sayyah et al.; Yield loss of photovoltaic panels caused by depositions; Solar Energy, Sept. 2014

[3] T. Sarver et al.; A comprehensive review of the impact of dust on the use of solar energy: History, investigations, results, literature, and mitigation apparoaches; Renewable and Sustainable Energy Reviews 22, 2013, p. 698-733

[4] H. Haeberlin, J.D. Graf, 1998. Gradual reduction of PV generator yield due to pollution. In: 2nd World Conference on Photovoltaic Solar Energy Conversion.

[5] A.M. Pavan, A. Mellit, D. De Pieri; The effect of soiling on energy production for large-scale photovoltaic plants, Solar Energy, 85 (2011), pp. 1128-1136

[6] Malik, S.; Ebert, M.; Naimi Z.; Ikken, B.; Development of outdoor research platforms and tests on photovoltaic modules in the green energy park in Morocco, in Proc. of 1st Africa Photovoltaic Solar Energy Conference and Exhibition Africa PVSEC 2014, Durban; WIP Renewable Energies; 19 – 23

[7] Malik, S.; et al, "Outdoor data evaluation of half-/full-cell-modules with regard to measurement uncertainties and the application of statistical methods", 29th European Photovoltaic Solar Energy Conference and Exhibition, P. 3269-3273, 2014, Amsterdam

[8] ISO/IEC Guide 98-3 2008 (GUM: 1995)

[9] J. Kirchhof, B. Schulz: "Dokumentation ISET-mpp meter – Messunsicherheiten bei Modul-Messungen". Fraunhofer IWES, Kassel 2010

[10] Hukseflux Thermal Sensors, User Manual SR11_manual_v1608

[11] M.B. Strobel et al, Uncertainty in Photovoltaic performance parameters – dependence on location and material, Solar energy Materials & Solar Cells 93 (2009) 1124-1128

ADVANCED COATING FOR SOLAR CELL MODULE PROTECTION

Zhavnerko G.K., Shiripov V.Ya., Savich V.A., Sergeev O.V.*
Izovac Technologies Ltd., Bogdanovicha 155, Minsk, Belarus
*- DLR Institute of Networked Energy Systems,
Carl-von-Ossietzky-Str. 15, 26129 Oldenburg, Germany

ABSTRACT: Equipment and technology for anti-smudge (AS) -coating "Defensiz" fabrication has been developed. Fabricated AS coatings passed through environmental tests such as saline mist, thermal shock, abrasive sand test without changing the optical characteristics and wettability of the glass. A set of AS-protected and non-protected single 6-inch silicon heterojunction (SHJ) one solar cell modules have been assembled with corresponding electronics and software for data collection. They were monitored by operation at outdoor conditions. It has been proved that the power long-term generated by single solar cell modules protected with Defensiz coating is around 3-5% higher in comparison the modules without the coating. Finally, our investigations demonstrate how enhanced wear and optical characteristics of the cover glass may improve the module performance.
Keywords: Heterojunction silicon solar cell, module protection, anti-smudge coating

1 INTRODUCTION

Solar cell (SC) modules often operate in a polluted environment and at extreme conditions, where the cover glass surface is constantly contaminated by various substances. Soiling losses refer to decrease of generated power due to the shadowing with dirt, dust and other particles that cover the surface of the PV module. Another problem is snowfall or ice condensation, when SC module has to be cleaned. The average annual soiling power losses are around 5.2%, but at summer season are much higher [1].

Anti-Smudge (AS) "easy clean" protective coating on module's cover glass can be an efficient solution against soiling. It creates a layer of hydrophobic material to prevent grease from sticking to the surface and allows easy removal of dirt from the glass surface. The main requirements to AS coatings – it should provide a certain contact angle between liquid drop and the substrate and it should protect the substrate from scratches.

Nowadays there are several technological approaches like physical vapor deposition in vacuum or dip and spray coating from liquid phase for fabrication of AS coatings on glass. Hence, all of them are not efficient either in the sense of the achieved quality, thickness, transparence and durability of fabricated coatings or in economic and environmental aspects.

We proposed our innovative ultra-thin-film technology (GT-technology) [2] for cover glass protection with AS coatings. GT-technology is unique easy-scalable industrial technique [1] for vacuum-free room-temperature fabrication of many different functional monolayers of high uniformity. This technology allows modifying the substrates of various shapes, dimensions, material combinations and surface morphologies. Moreover, it is also compatible to the Roll-to-Roll (R2R) production of different coatings on the foils.

We already have successfully adopted GT-technology to following applications in modern PV devices [3,4], also for flexible PV:

- improved light trapping or complementary layers;
- conductive transparent layers ;
- antireflective and down-conversion layers;
- ultra-thin lamination and encapsulation coatings

Recently we have developed unique high-efficient AS coating "Defensiz" based on specially modified solution with perfluoroester (PFPE). "Defensiz" is transparent, extremely wear-resistant and hydro- and oleofobic coating, providing thus protective and easy-clean features.

That is why our idea is to grant the module cover glass surface with anti-abrasive and environmental protection properties by coating it with ultra-thin "Defensiz" using our innovative GT-technique. At this low thickness (3-5 nm only) "Defensiz" shows no any light absorption losses and production and material cost are negligible. "Defensiz" will reduce soiling effects due to low adhesion of contaminants to the glass surface, and will maintain the high transparency of the cover glass by efficient preventing of the matting and scratches.

There is no any comparable technology on mass-production market yet. Thus the pressing task is to introduce the technology in the production of PV modules.

In our study we investigate protection properties and their stability of GT-fabricated "Defensiz" coatings on glass substrates. Then, we compare long-term performance of small PV modules based on silicon heterojunction (SHJ) solar cells with and without "Defensiz" at outdoor conditions to prove its advantages.

Finally, we compare different technologies for AS-coatings fabrication and make recommendations of industrial application of "Defensiz" and GT-technology for efficient PV module protection.

2 EXPERIMENTAL PART

2.1 GT equipment and technology for glass surface modification

In our developed GT technology the film transfer from water interface to the surface is similar to Langmuir- Blodgett method, but based on roll principle of monolayer condensation [2] and almost horizontal deposition of PFPE monolayer by specially designed facility – small and middle range GT-coater for deposition on 60x40 cm^2 substrates or onto the foils (Fig. 1).

The deposition takes place at room temperature and air conditions with subsequent drying. The film thickness is only few nanometers, and useful material consumption is around 95-98%. There is no need in any vacuum or high temperature processes and dangerous or harmful reagents like in solvent treatment. That is why our technology is very environmental friendly and cost efficient.

Figure 1: GT-coater side view

2.2 Climatic tests

Abrasion, saline mist, thermo shock, and abrasive sand tests were carried out on modified glass substrates to confirm excellent performance of our coating.

First, we compared the stability of "Defensiz" coatings fabricated on glass substrates with different technologies: PVD, spray and GT. Abrasion test with stainless steel wool at of 1 kg load to 1 cm^2 wearing area was performed at air conditions. The hydrophobic properties were proved with contact angle measurements.

After that, the samples GT-coated with "Defensiz" were incubated in the climate chamber (MC-71 Tabai Espec Corporation, Japan) at 85 °C for 30 minutes and at -60 °C for thermal shock test to study stability of the coating.

Then two GT-"Defensiz" samples were treated in salt spray (SSD-450 chamber, Germany) with periodic spraying brine during 3 days for saline must test. The pH value was in the range 6.5 - 7.2 [at (20 ± 2) ° C]. The concentration of the brine was (5 ± 1)% by weight. The temperature in the chamber was set to 27 ± 2 ° C.

Finally, the samples were blown for 2 hours with the dried dust mixture containing 70% quartz sand, 15% chalk and 15% of kaolin in an amount of about 0.1% of effective volume of the chamber (dust chamber KTP-800, Russia) at a rate of 10-15 m/s for abrasive sand test. The particles size was about 0.8-1 microns and they passed through a sieve with a mesh number 0224 without the rest.

2.3 Experimental single module assembling

Our own SHJ solar cells with dimensions of 156x156 mm^2 were used in this work. Four single SC modules were assembled to monitor generated current (I, mA), voltage (V, mV) and power (P, W) every minute. The modues P1 and P2 were covered with GT-"Defensiz" protected glasses, P3 and P4 – with non-protected glasses. The obtained data were steadily saved into the database that averaged values for each hour of operation. A set of solar modules has been mounted in front of windows on Izovac building and the resulted data of generated electricity were then recorded and analyzed.

3 RESULTS AND DISSCUSSION

The results of climatic tests have proved that GT coater with supplement technology can be very useful for cover glass protection with DefensIz coating as extremely scratch-resistant, transparent, easy-cleaning (oleophobic) layer. Protected glass remained hydrophobic (105-108 degree for water contact angle) after more than 10000 cycles of abrasive wearing (Fig.2) what is at least twice as good as for competitive technologies.

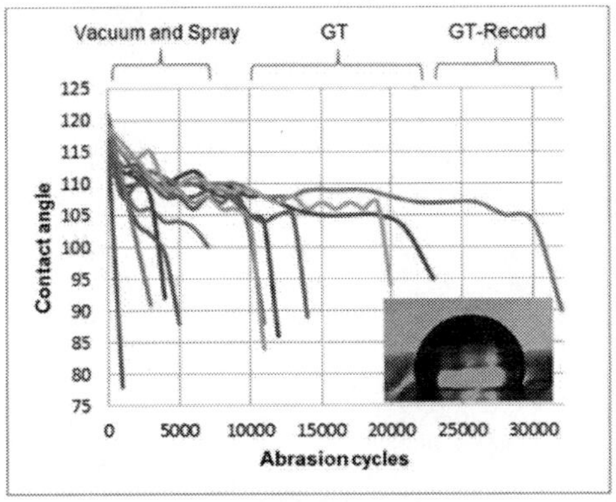

Figure 2: Abrasion stability of "Defensiz" coating fabricated with different techniques.

The protected glasses remained hydrophobic (Table 1) and transparent (Fig. 3) after additional environmental tests.

Table I: Parameters of glass substrates with AS coating prior and after environmental tests

Test	CA_{init}	CA_{final}	CA_{dif}
Saline mist	116.9	114.5	2.4
Thermal shock	116.7	115.6	1.1
Abrasive sand test	116.9	114.8	2.1

As it can be seen in Fig. 3, the spectra of glass substrates remained unchanged after climatic tests that testify excellent protective properties of AS film.

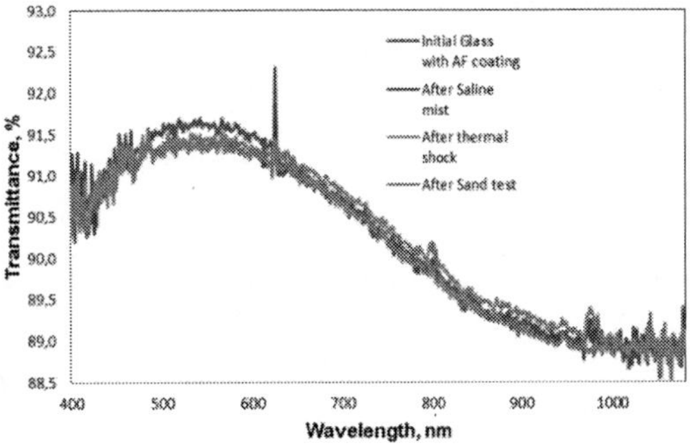

Figure 3: Transmittance of non-treated glass in comparison with modified samples after climatic tests

Then it was important to see the influence of the coating on power output, generated by single solar modules at outdoor condition for long timeIn general, each single module element has unique characteristics and it was necessary to normalize measured characteristics for statistical comparison. That is why, firstly, the generated energy was summarized within a day period and subsequent total power was divided on the values averaged for the first day to correct compare energy that was generated each from four modules. A tendency in variations of SC energy generated for a month observation period is shown in Fig. 4.

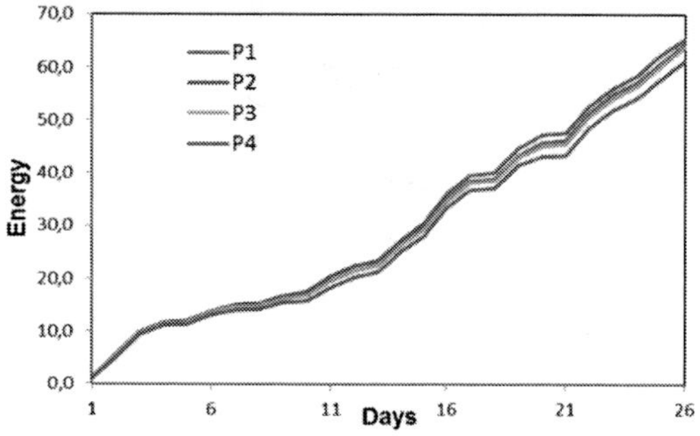

Figure 4: The amount of normalized energy generated by SC single modules for the first month of observation

As it can be seen, the normalized energy, generated by P1 and P2 elements with AS coating, exceeded the P3 and P4 elements without such a coating. Thus, we can conclude, AS coating improves long-term performance of SHJ solar cells by efficient preventing of transmission losses in the cover glass due to contaminations and surface damages.

2.4 Comparison of technologies

There are two principal methods of wear-resistant AS film formation: (i) by spray with an appropriate PFPE reagent or (ii) PFPE evaporation in vacuum. The detailed comparison of key performance indicators for GEN-5 glass size is given in Tab.2

Table 2: Parameters of glass substrates with AS coating prior and after environmental tests

	Spray (GEN-5)	Vacuum in-line machine	GT (R2R) deposition
Occupied area, m^2	16	100	10
Deposition area, m^2	1.43	0.96	0.36
Productivity, m^2/h	73	2,8	30
Wear resistance, cycles of abrasion	<1000	5000-7000	8000-12000
Useful material consumption, %	30%	~40%	>95%
Uniformity	Low on this scale	Low on this scale	High / Size independent
Scalability	Limited	limited	unlimited
Capital Expenditure (CAPEX)	$$	$$$$$	$
Energy consumption	Low	High	Very low
Maintenance cost	Low	High	Very low
Automatization	Yes	Yes	Yes
Environmental load	High	Huge	Low, Green technology

GT-technology is able to fabricate highly uniform nanolayer with superior deposition control and the ability to manipulate coating characteristics on the large scale substrates. The wearing stability of GT-coatings is significantly higher than for differently fabricated coatings.

In comparison with (i) and (ii) methods, where about 20 % to 30 % of very expensive material are only efficiently used for layer formation, our method allows nearly 100 % of useful material utilization.

With outstanding higher quality, productivity and environmental friendliness at much lower CAPEX, energy consumption, maintenance costs and area needed in front of alternative technologies and materials, the GT-technology makes "Defensiz" coating highly promising for efficient long-term protection of solar modules for PV industry.

Izovac Ltd. is ready for construction of industrial line for large size GT-modification (glass up to Jumbo size) to form scratch-resistant anti-smudge "Defensiz" layer on the glass.

4 CONCLUSIONS

Izovac Ltd has developed efficient GT-technology for anti-smudge coating "Defensiz". The coatings passed successfully through aggressive tests such as wearing abrasion, saline mist, thermal shock, abrasive sand test without changing the optical characteristics and wettability, keeping transparency, water-repellent and oleophobic properties of the cover glass used.

Our research demonstrated how improved wear and optical characteristics of the cover glass (the current power output) may improve the module performance. At outdoor conditions the long-term generated power by solar cell modules protected with "Defensiz" coating is around 3-5% higher than for the non-coated.

Comparison of key performance indicators for different technologies shows clear advantages of GT-technology in front of alternative methods.

That is why suggest to use our novel facilities and innovative GT-technology for efficient protection of module cover glasses in PV-Industry, that provides better module durability and grants the surface with "easy clean" properties.

REFERENCES

[1] M.R. Maghami, H.Hizam, C.Gomes, M.A.Rdzi. Renewable and Sustainable energy: Reviews, 59 (2016) 1307.

[2] G.K. Zhavnerko, V.Ya. Shiripiov. Device for formation of nanostructured coatings on solid surfaces. Taiwanese Patent No.: I542532. 2016.

[3] G.K. Zhavnerko, O.V. Sergeev, A. Neumüller, V.Ya. Shiripiov, Proceedings 31st European Photovoltaic Solar Energy Conference and Exhibition. 2015, Hannover. 220.

[4] G.K. Zhavnerko, O.V. Sergeev, V.Ya. Shiripiov, Proceedings 32st European Photovoltaic Solar Energy Conference and Exhibition. 2016, Munich. 822.

INVESTIGATING HOT SPOTS PERFORMANCE OF PV MODULE USING HALVED MULTI-CRYSTALLINE SILICON CELLS

Jingna Jiang, Jianxiong Ni*, Dandan Rong, Yabin Li, Yi He, Chao Ma, Jinchao Shi, Feng Li, Dawei Liu, Dengyuan Song
State Key Laboratory of Photovoltaic Materials and Technology,
State Key Laboratory of Energy Photovoltaic Technology
Yingli Group Company Limited, 3399 North Chaoyang Avenue, Boading, China 071051
*Jianxiong.ni@yingli.com

ABSTRACT: Using halved size cells in photovoltaic modules is an efficient way to reduce cell to module loss and increase the module power.In this work, we show the difference between halved size and full-size multicrystalline silicon solar modules from the two aspects. Firstly, we shows the electrical parameters difference between halved cell and full size PV modules in the same material combination. In the module with 144 halved cells, it is found that the power increasing of halved cell PV modules is above 9 W compared with full size cell modules. And the Imp of halved cell PV modules rises to 1.16%, the Isc of the module increases to about 1.32%. Secondly, the parameters differences are discussed in different shading conditions and different shading cell areas for halved size cells PV modules. The results show that halved size cells PV modules have better performance under different cell shading conditions than full size cell module. The Vmax changes to 2/3 of before while one cell is under 90% shading areas for halved size cells module.

Keywords: PV module; halved cell; power loss; optical gains; shading

1 INTRODUCTION

In the past few years, a great achievement was made in the photovoltaic industry in the world. And the market has increasing demand for the higher module efficiency. Reducing the cell to module losses are a concern for many manufacturers One method to improve PV module power is to produce modules using halved silicon wafer solar cells. A half-cut cell is a standard cell that has been split into two equal pieces to produce smaller cells, by using half-cut cells, the current per cell is also reduced by half. Therefore, the amount of internal losses in a half-cut module is 1/4 of a full sized cell module, and the power output is increased.

Recently, REC, BPSolar, Mitsubishi, Solarworld and Bosch offered solar modules with increased power due to half size solar cells. Jens Schneider and his partners investigate and find the electrical and mechanical properties of half size cell deteriorate. And the power increasing of halved cell PV modules is above 5% compared with full size cell modules. Hamed Hanifi investigate that the simulation results are in excellent agreement with the measurement results. Half-cell modules increase module power by 1.48% in Fill Factor due to reduced electrical losses in cell connectors and enhance about 3% of module current due to the optical gains from cells pacing. Parallel sub-strings alow the module to save up to 50% of the string's power under partial shading conditions. Stephanie Malik investigates the half-size cell and a full-size cell module power yield in outdoor. The half-size cell module shows a better behavior a higher irradiance levels for different ambient temperature classes. [1-5]

In this work, we investigate the difference between halved size and full-size multicrystalline silicon solar modules from the two aspects. Firstly, we shows the electrical parameters difference between halved cell and full size PV modules in the same material combination. Secondly, the parameters differences are discussed in different shading conditions and different shading cell areas for halved size cells PV modules.

2 SAMPLES PREPARARION AND EXPERIMENTS

2.1 Circuit design

Figure 1 shows the circuit design of halved cell module with 144 halved cells in 12 strings of 12 cells in the following experiments.

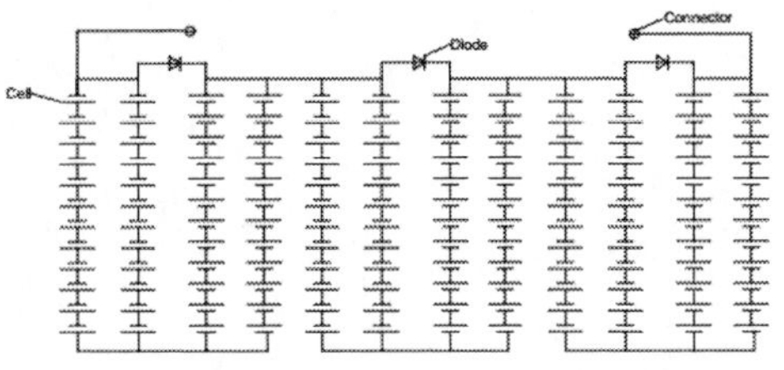

Figure 1: Circuit design of halved cell module

2.2 Samples and measurement setup

The reference modules with 72 full-cells in 6 strings of 12 cells and half-cell modules with 144 halved cells in 12 strings of 12 cells. Cell sizes are standard 156*156 mm multi-crystalline solar cells as full-cell layout and laser cut half-cells. The tab width is 1.0mm and the thickness is 0.25mm. Module manufacturing follows the standard material assembly order with EVA encapsulation and glass on front, backsheet on back side. Samples informantion and characteristics are shown in Table 1.

The initial characterization was performed at standard test condition (1000w/m^2 irradiance with AM1.5 spectrum and the cell temperature nomalized to 25 ℃). The IV measurement is PASAN solar simulator manufactured by Ophelms of German.

Table I: Samples types

Samples Number	Module Type
a	72 full size cell module
b	144 havled size cell module

2.3 Experiment contents

Firstly, we compare the electrical parameters difference between halved cell and full size PV modules in the same material combination.

Secondly, we discusses the electrical parameters difference in different shading conditions and different shading cell areas for halved size cells PV modules.

Figure 2 shows the different shading conditions for full cell module and halved cell module. The shadow area refers to the cell is in shading conditons.

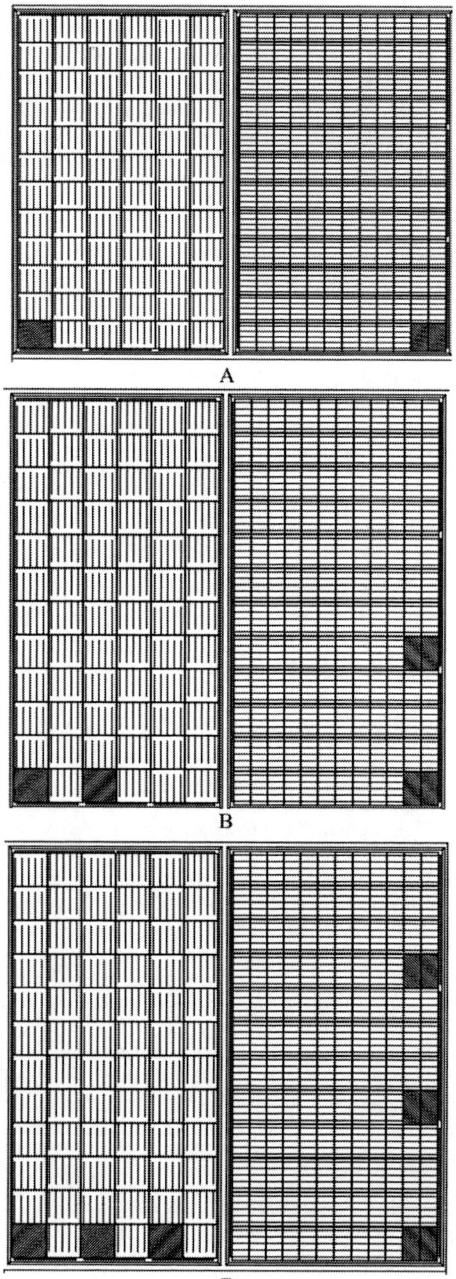

Fig 2. Different shading conditions for full cell module and halved cell module

Figure 3 shows the different shading areas for halved cell module. In this experiment, we discusses the electrical parameters difference in different shading cell areas for halved size cells PV modules.

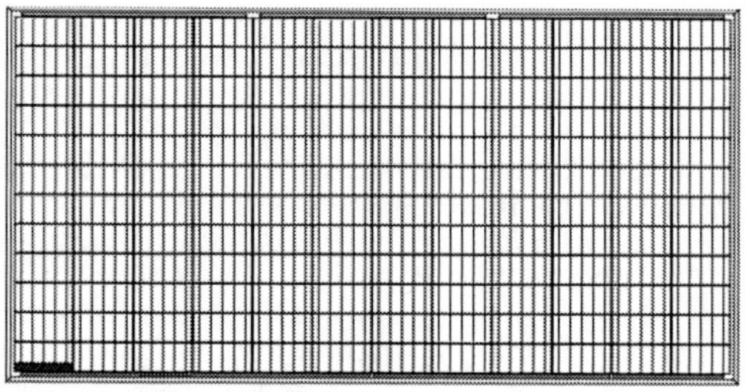

Fig 3. Different shading areas for halved cell module

3 RESULTS AND DISCUSSION

3.1 Parameters analysis

As shown in Table 2, comparing the halved cell module and full-size module, the power increasing of halved cell PV module is above 9 W, the Imp of halved cell PV module rises to 1.16%, the Isc of the module increases to about 1.32%. This is due to the higher optical gains and reduced electrical losses for the halved cell PV module. By cutting a full cell into two pieces the produced current by each pieces is halved as well. Thereby the internal loss in module is only 1/4 of a module made by full size cells so that the power output is increasing.

Table 2: Parameters of different types of module samples

Module Type	V_{oc}/V	I_{sc}/A	V_{pm}/V	I_{pm}/A	Power/ W
72 full size cell module	45.54	9.07	37.37	8.60	321.45
144 havled size cell module	45.50	9.18	37.97	8.70	330.17

3.2 Cell shading analysis

In shading condition A, a string of cells for full size cell module and two parallel sub-strings of cells for halved size cell module protected by a diode are influenced because one cell is in shading. In shading condition B, cells for full size cell module and halved size cell module protected by two dioded are influenced because two cells are in shading. In shading condition C, cells for full size cell module and halved size cell module protected by two dioded are influenced because three cells are in shading. As shown in Table 3, the results show that halved size cell PV module have better performance under different cell shading conditions than full size cell module.

Table 3. Power parameters in different shading conditions

Module Type	Parameters	Without shading	A	B
Full size cell module	I_m//A	8.52	8.51	8.49
	V_m//V	37.46	24.54	11.63
	P_{max}/P	319.33	208.83	98.72
	Power percent(%)	--	65.40%	30.90%
Half size cell module	I_m/A	8.62	8.61	4.60
	V_m/V	38.03	24.88	40.41
	P_{max}/P	327.60	214.08	185.73
	Power percent (%)	--	65.30%	56.70%

As shown in Table 4, the results show the parameters changes of halved size cell PV module under different cell shading areas. The Vmax changes to 2/3 of before while one cell is under 90% shading areas for halved size cells modle.

Table 4. Power parameters in different shading areas

	0%	50%	60%	70%
P_{max}	326.88	273.97	258.81	244.96
V_{max}	38.06	40.05	40.32	40.48
I_{max}	8.59	6.84	6.42	6.05

	80%	90%	95%	100%
P_{max}	226.85	213.76	213.65	213.71
V_{max}	40.70	24.92	24.90	24.90
I_{max}	5.57	8.58	8.58	8.58

4 CONCLUSIONS

The results show that halved size cell PV module have 9w power gian than full size cell module.And halved size cell module have better performance under different cell shading conditions than full size cell module. The Vmax changes to 2/3 of before while one cell is under 90% shading areas for halved size cells module.

5 ACKNOWLEDGEMENTS

The authors have received funding from the National High Technology Research and Development Program of China (863 Program) (Project No. 2015AA050301)

6 REFERENCES

[1] Jens Schneider, Stephan Schoenfelder, Sascha Dietrich, et al. Solar module with half size solar cells[C]//29th European Photovoltaic Solar Energy Conference and Exhibition, Amsterdam, Netherlands, September, 2014: 185-189.

[2] Hamed Hanifi, David Dassler, Jens Schneider, et al. Reduced shading effect on half-cell modules-measurement and simulation[C]//30th European Photovoltaic Solar Energy Conference and Exhibition, Hamburg, Germany, September, 2015.

[3] Stephanie Malik, David Dassler, Jes Frbel, et al. Outdoor Data Evalution of half/full-cell Modules with Regard to Measurement uncertainties and the application of statistical methods[C]. 29th European Photovoltaic Solar Energy Conference and Exhibition, 2015: 3269-3273.

[4] Siyu Guo, Jens Schneider, Fei Lu, et al. Investigation of the short-circuit current increase for PV modules using halved silicon wafer solar cells[J]. Solar Energy Materials & Solar Cells, 2015, 133: 240-247.

[5] Robert Witteck, Henning Schulte-Huxel, Hendrik Holst, et al. Optimizing the solar cell front side metalklization and the cell interconnection for high module power output[J]. Energy Procedia, 2016, 92: 531-839

ENERGY PERFORMANCE IMPROVEMENT AND THERMAL OPERATION OF CRYSTALLINE SILICON PHOTOVOLTAIC MODULES DESIGNED WITH INNOVATIVE PACKAGING COMPONENTS

George Makrides*[1], Marios Theristis[1], James Bratcher[2], Jeff Pratt[2] and George E. Georghiou[1]
[1]FOSS Research Centre for Sustainable Energy, Photovoltaic Technology Laboratory,
Department of Electrical and Computer Engineering, University of Cyprus
75 Kallipoleos Avenue, P.O. Box 20537, Nicosia, 1678, Cyprus
[2]Honeywell International, 101 Columbia Road, Morristown, NJ 07962, USA
*corresponding author George Makrides (phone: +357 22 894397, Fax +357 22 895370, makrides.georgios@ucy.ac.cy),
Marios Theristis (theristis.marios@ucy.ac.cy), James Bratcher (james.bratcher@honeywell.com), Jeff Pratt
(jeff.pratt@honeywell.com), George E. Georghiou (geg@ucy.ac.cy)

ABSTRACT: Energy performance improvements of photovoltaic (PV) modules can be achieved by using innovative packaging materials, specifically designed to decrease the operating temperature and increase power output. This is important, especially since the aesthetic appeal of black-coloured PV modules is becoming a global market driving feature. The scope of this work is to analyse the performance of different crystalline silicon (c-Si) PV modules manufactured with different packaging components (backsheet types) capable of reducing the operating cell temperatures and in parallel, to assess their thermal behaviour and reliability under real operating conditions in a warm climate. The comparative analysis of the thermal behaviour, based on acquired cell temperature measurements over a five-year period, verified that backsheets can be designed to influence the cell temperature and enhance performance by operating at lower temperatures in some cases exceeding 10 °C. The systems with the backsheets that retained lower operating temperatures produced higher annual energy yield, under the warm conditions exposed due to reduced thermal losses of approximately 1 %. In particular, the annual energy yield results showed that the systems equipped with the white control and black colour thermal management backsheet produced consistently the highest annual energy yield over the evaluation period. Finally, the results of the indoor and outdoor degradation rate analysis showed that, over the five-year period, there was no significant difference in the estimated degradation rate amongst the installed systems, since the results are within the uncertainty range.
Keywords: backsheet, c-Si, degradation, energy performance, photovoltaic, thermal performance.

1 INTRODUCTION

The geometry and thermophysical properties of the photovoltaic (PV) packaging materials, and especially the backsheet, have a direct impact on the thermal behaviour of PV modules. As such, optimum selection of materials can significantly reduce the operating temperature of the modules under real outdoor conditions [1-3] and may further influence the reliability and the degradation rate [4].

The global drive for lowering the cost of electricity (cost per Watt peak) from PV can be achieved by improving the outdoor performance of PV modules. In particular, for crystalline-Silicon (c-Si) PV modules performance improvements can be achieved by introducing new innovative materials, especially backsheets with heat dissipating capabilities. It is imperative though that PV modules with new packaging assemblies and materials are thoroughly tested not only for performance improvements but also for reliability. The reliability of backsheet material is currently tested under accelerated conditions according to the IEC 61215 qualification standard [5]. Furthermore, a number of backsheet components are evaluated, in terms of their suitability for PV module packaging applications, using indoor exposure reliability tests [6-7]. While indoor approaches may yield accurate results for infant mortality failures, it is generally accepted that in-depth information on the longer-term performance aspects can only be obtained from outdoor field studies.

The scope of this work is to present the results obtained from the outdoor testing of c-Si modules fabricated with different innovative packaging components. More specifically, four types of mono-crystalline Silicon (mono-c Si) PV modules with identical solar cells but different backsheet materials were installed and monitored outdoors since July 2012, at the outdoor test facility of the University of Cyprus (UCY) in Nicosia, Cyprus. The modules were installed to form grid-connected systems with an approximate nominal capacity of 1.2 kW_p, and were connected to identical inverters (SMA, SB1200) in order to exclude the influence of different maximum power point tracking (MPPT) methods. As such, a comparison between the performance and reliability of the systems could be directly performed.

2 EXPERIMENTAL SETUP

The outdoor test facility of the UCY includes a fixed plane infrastructure for outdoor performance assessments at both the module and system level. The PV systems with different backsheet materials were mounted in landscape arrangement on aluminium mountings, at the optimum plane-of-array (POA) angle for Cyprus of 27.5°. In order to enable a comparative performance analysis, the modules were installed side-by-side and therefore, they were exposed to identical conditions. The grid-connected PV systems were installed at location 5 to 9 (abbreviation ucy1-05 – ucy1-09) of the test infrastructure (see Fig. 1). Each PV system comprised of a PV array with five mono-c-Si PV modules, each of 245 ~ 252 W_p rated power, as indicated by the manufacturer's datasheets. The modules of each system were connected in series to form a string, at the input of each inverter.

Figure 1: PV systems with different backsheet material installed at the outdoor test facility of the University of Cyprus [8].

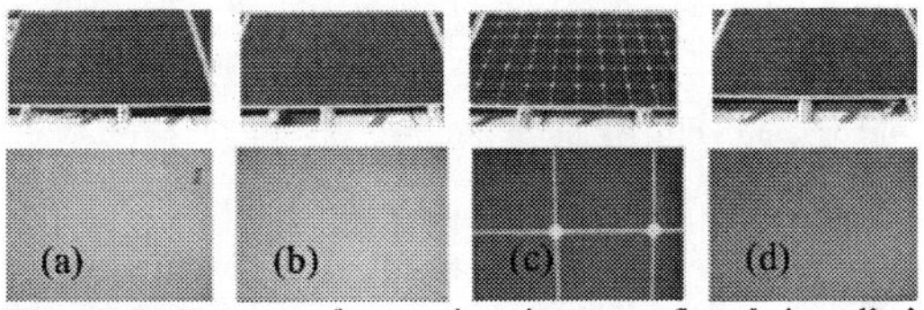

Figure 2: Front- and rear-view images of each installed technology where: (a) ucy1-05 black thermal management backsheet, (b) ucy1-06 FPE black control backsheet, (c) ucy1-07 FPE white control backsheet and (d) ucy1-09 FPF black control backsheet.

The front- and rear-view images of each installed technology show clearly the different backsheets used to carry out this investigation (see Fig. 2). Moreover, the technical specifications of the four installed systems and the polymer films used as backsheet materials (polyethylene (PE), polyethylene terephthalate (PET) and fluoropolymer) in the different configurations (Fluoropolymer/PET/EVA (FPE) and Fluoropolymer /PET/Fluoropolymer (FPF)) are summarized in Table I.

The performance of the different PV systems and the prevailing meteorological conditions were monitored at a resolution of one second and stored as fifteen-minute average data-sets. Specifically, the meteorological measurements include the incident global irradiance at the POA (G_I), wind direction (W_a), wind speed (W_s) and ambient temperature (T_{amb}). The PV system's operational measurements include maximum power DC current (I_{mp}), voltage (V_{mp}) and power (P_{mp}), as measured at the output of each PV system (DC side).

The installed data acquisition equipment, sensors and associated accuracy are summarized in Table II. Periodic calibrations and inspections of the installed sensors were performed to ensure quality measurements and reveal any deviations from the real measurements.

In addition, K-type thermocouple sensors were embedded in each module for the acquisition of cell temperature measurements in order to assess the capability of the backsheet material type to reduce cell temperatures and hence provide performance advantages, especially at locations with warm climates.

Table I: PV system technical characteristics. All modules were manufactured by GSS Gebäude Solarsysteme GmbH [8].

System 1 (ucy1-05) Black thermal management backsheet	
Model	ZRE 250GEF-60M6+F
Backsheet material type	Heat reflective cell-side / clear PET core / PE white air-side
Backsheet arrangement	Heat reflective black cell-side / white air side
Cell technology	Mono-c-Si
Maximum Power	250 W (±3.5 %)
Short circuit Current	8.45 A
Open circuit Voltage	36.90 V
Maximum Power Current	8.30 A
Maximum Power Voltage	30.10 V
System 2 (ucy1-06) FPE black control backsheet	
Model	ZRE 248GEF-60M6+F
Backsheet material type	PE black cell-side / clear PET core / Typical fluoropolymer white air-side
Backsheet arrangement	Standard black cell-side/white air-side
Cell technology	Mono-c-Si
Maximum Power	248 W (±3.5 %)
Short circuit Current	8.40 A
Open circuit Voltage	36.80 V
Maximum Power Current	8.25 A
Maximum Power Voltage	30.10 V
System 4 (ucy1-07) FPE white control backsheet	
Model	ZRE 252GEF-60M6+F
Backsheet material type	PE white cell-side / clear PET core / Typical fluoropolymer white air-side
Backsheet arrangement	Standard white cell-side/white air-side
Cell technology	Mono-c-Si
Maximum Power	252 W (±3.5 %)
Short circuit Current	8.50 A
Open circuit Voltage	37.10 V
Maximum Power Current	8.35 A
Maximum Power Voltage	30.20 V
System 4 (ucy1-09) FPF black control backsheet	
Model	ZRE 245GEF-60M6+F
Backsheet material type	Fluoropolymer black cell-side / clear PET core / Typical fluoropolymer black air-side
Backsheet arrangement	Standard black cell-side/black air-side
Cell technology	Mono-c-Si
Maximum Power	245 W (±3.5 %)
Short circuit Current	8.40 A
Open circuit Voltage	36.70 V
Maximum Power Current	8.13 A
Maximum Power Voltage	30.00 V

Table II: Summary of installed data acquisition equipment and sensors including manufacturer, type and accuracy information [8].

Equipment/Sensor	Manufacturer and model	Accuracy
Data acquisition	Delphin TopMessage	±0.01 %
Pyranometer 1	Kipp & Zonen CMP11	±2 %
Pyranometer 2	Hukseflux SR11	±5 %
Ambient temperature	Rotronic HC2S3	±0.1 °C at 25 °C
Relative humidity	Rotronic HC2S3	±0.8 % at 25 °C
Wind speed	NRG 40C	± 0.14 m/s at 10 m/s ± 0.45 m/s at 10 m/s
Wind direction	NRG 200P	±1 %
Cell temperature	K-type thermocouple	± 0.75 %
DC Voltage sensor	Muller Ziegler Ugt	±0.5 %
DC Current sensor	Muller Ziegler Igt	±0.5 %
AC Power	Muller Ziegler EZW	±1 %

2.1 Uncertainty analysis

The main sources of uncertainty in the outdoor performance evaluations due to the instrumentation are presented in this section. In particular, the global irradiance was measured using two thermopile pyranometers (pyranometer 1 and 2) in the spectral range of 285 - 2800 nm with a ±2% and ±5% expected daily uncertainty, respectively. In addition, the response time of the installed pyranometers is < 5 s (for 95% response) which is longer than the PV module's response time and the data logging interval (of 1 s) [9].

Furthermore, parameters including MPP voltage and current were obtained using an active sensor board and were then connected to the data logger's analogue channels. Both DC voltage and current were measured with an accuracy of ±0.5%.

The AC power and AC energy yield were measured using an AC meter which is primarily associated with a metering measurement uncertainty of ±1% and an additional ±1% uncertainty that accounts for differences caused by the inverters (efficiency and maximum power point tracking accuracy). DC power and energy were also calculated from the measurements of DC voltage and current and are therefore associated with an uncertainty of ±1%. It should also be noted that the measured annual DC and AC energy yield of each system were further normalised to the nominal standard test conditions (STC) power obtained using a class A+ solar simulator at UCY.

3 METHODOLOGY

As mentioned in Section 2, the systems were installed side-by-side and were exposed to the same prevailing climatic conditions. The different backsheet material packaging configurations of the PV modules are summarized in Table III.

Table III: Backsheet configuration of the different PV systems.

System	Backsheet material type	Description
ucy1-05	Heat reflective cell-side / Clear PET core / PE white air-side backsheet	Black thermal management backsheet (heat reflective on cell-side/white on air side)
ucy1-06	PE black cell-side / Clear PET core / Typical fluoropolymer white air-side backsheet	FPE black control backsheet (black on cell-side/white on air side)
ucy1-07	PE white cell-side / Clear PET core / Typical fluoropolymer white air-side backsheet	FPE white control backsheet (white on both cell and air side)
ucy1-09	Fluoropolymer black cell-side / Clear PET core / Typical fluoropolymer black air-side backsheet	FPF black control backsheet (black on both cell and air side)

3.1 Thermal behaviour

The effect of the different backsheets on the operating cell temperature of each module was assessed by acquiring measurements from thermocouple sensors that were embedded at three different locations of the PV modules (see Fig. 3). More specifically, the thermocouple sensors were embedded to the lower-, middle- and higher-positions on each PV module. For each system the temperature measurements were compared and statistically analysed in order to assess the thermal performance of each backsheet material.

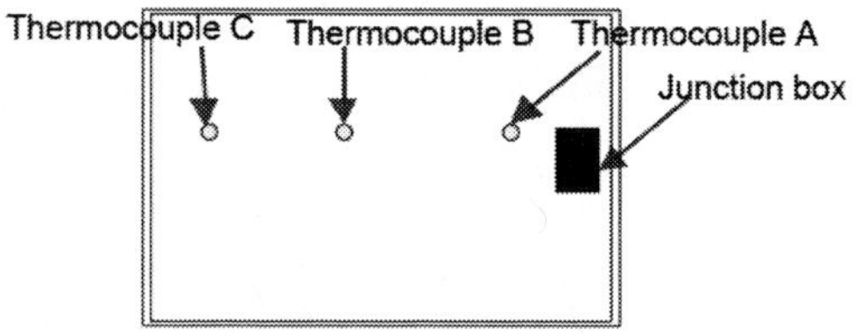

Figure 3: Schematic showing the embedded thermocouples inside each PV module.

3.2 Outdoor performance

The outdoor performance of the systems was assessed based on the acquired data-sets and calculated annual DC energy yield (E_{DC}) and performance ratio (R_P). In particular, the fifteen-minute average data-sets of V_{mp}, I_{mp} and G_I, were used for the analysis. The DC side P_{mp} was calculated from the V_{mp} and I_{mp} data. The raw data-sets were verified and corrected for outage periods, shading and other system faults.

The acquired data-sets were subsequently used to calculate the E_{DC} and R_P, at the DC side of the systems normalised to the nominal STC power obtained using a class A+ solar simulator at the UCY (before the outdoor installation of the modules) over the five-year period in order to yield results on the extent to which backsheets can influence the energy yield of PV modules.

3.3 Degradation and reliability

In order to calculate the degradation rate (R_D) of the different backsheet systems and further assess their reliability to the exposed outdoor conditions an indoor and outdoor analysis was conducted.

Standardized indoor test procedures were initially performed to determine the electrical characteristics of the modules under test at standard test conditions (STC) according to IEC 61215 Ed. 2 Procedure 10.2 [5]. In particular, two modules of each system were regularly dismounted and flash tested indoors for maximum power determination at STC. As such, the power loss was calculated by comparing the STC power after five years of exposure to the corresponding STC power obtained before the modules were installed outdoors. Electroluminescence (EL) imaging was also performed in order to identify and visually observe failures and fatigues due to the continuous outdoor exposure and also the thermal cycles.

Additionally, the annual degradation rate, R_D, of the systems was calculated using the statistical techniques of Ordinary Least Squares (OLS) [10], Classical Seasonal Decomposition (CSD) [10] and the Year-on-Year (YOY) [11] comparative technique. For all the systems the monthly R_P time-series was constructed over the five-year period. Additionally, it is important to note that for non-spectrally corrected pyranometer measurements that exhibit seasonal spectral variations, several complete cycles (typically 3 - 5 years) need to be completed in order to obtain reasonably accurate degradation rates [12]. The goal of the statistical analysis performed was to calculate the trend of the PV performance time series and translate the slope of the trend into the annual degradation rate, in units of %/year. Conversely, the comparative YoY technique performs linear comparisons on subsequent data-sets (e.g. monthly R_P) in order to extract the annual degradation rate.

4 RESULTS

The comparison of the thermal performance of each type of PV system over the five-year evaluation period, verified that module packaging can significantly influence the operating temperatures with differences exceeding the 10 °C. More specifically, over the evaluation period the maximum difference between the cell temperature measurements was 11 °C. In addition, the comparative thermal performance analysis during a typical clear-sky day, demonstrated that the measured operating cell temperatures of the black colour thermal management backsheet system (System 1) was comparable to that equipped with the FPE white control backsheet (System 3), while the highest cell operating temperatures were exhibited by the black control backsheet system (System 4) (see Fig. 4a). The maximum daily recorded cell temperatures of the installed systems as obtained from the embedded thermocouples clearly demonstrated that over a typical period of a week in summer, the highest operating temperatures were exhibited by System 4 with the standard black cell-side/black air-side backsheet while Systems 1 and 3, operated at similar lower temperatures (see Fig. 4b).

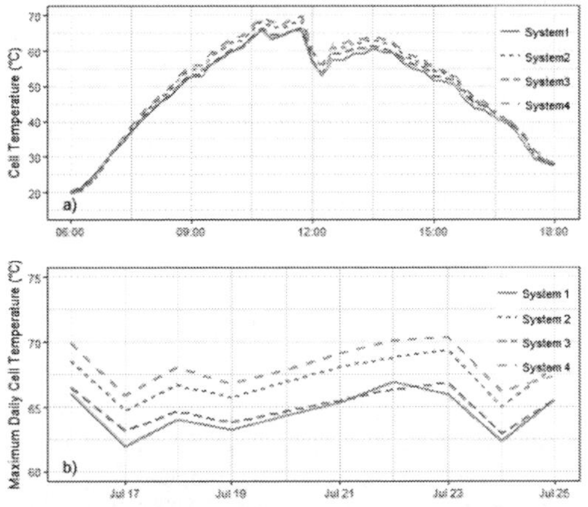

Figure 4: Thermal behaviour of the PV modules with different backsheet materials: a) operating cell temperatures during a typical clear-sky day and b) maximum daily cell temperatures for a typical summer week.

Useful information on the thermal behaviour of the different backsheet PV systems was also obtained through the comparative plots of cell temperature and the density distribution analysis performed over a period of a year (see Fig. 5). More specifically, the comparison of the cell temperature for systems 1 and 4, showed that for operating cell temperature above 25 °C, the thermal management system (System 1) operates at lower temperatures compared to the FPF black control backsheet system (System 4) (see Fig. 5a). The cell temperature density distribution plots of all the systems over a period of a year, further exhibited that the black colour thermal management and FPE white control backsheet systems operated at lower temperatures compared to the FPE and FPF black control backsheet systems which exhibited higher cell temperature density distribution at elevated temperatures (see Fig. 5b).

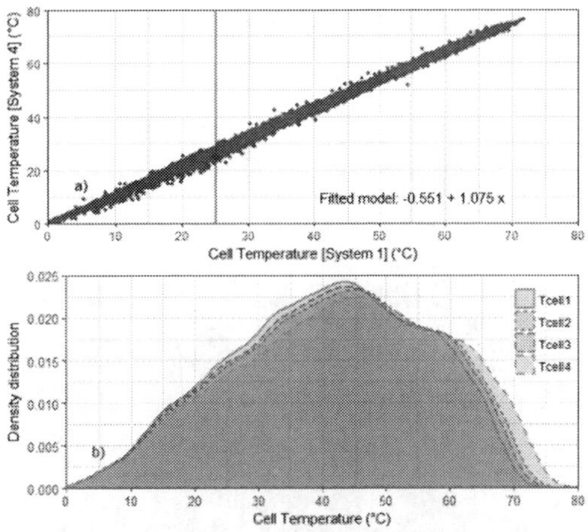

Figure 5: Operating cell temperature a) comparative analysis of System 1 against System 4 and b) density plot of all systems over a period of a year.

The full range of variation and density of the exhibited cell temperatures was also investigated by analysing the violin plots of each system over a period of a year (see Fig. 6).

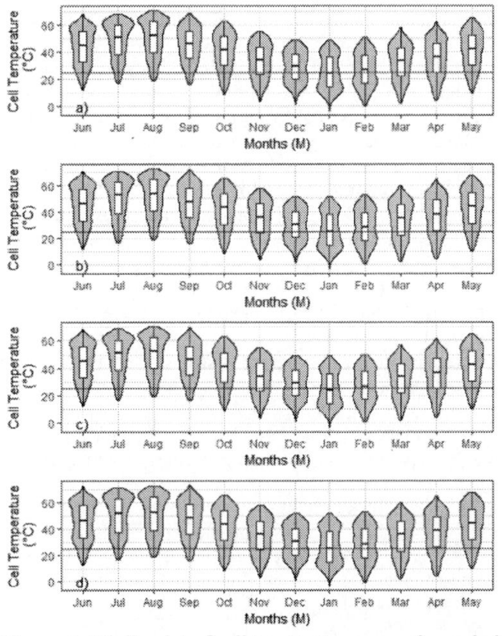

Figure 6: Violin plot of cell temperatures over the period of a year for system a) System 1, b) System 2, c) System 3 and d) System 4. The 25 °C cell temperature line is used as a reference.

The results obtained from the statistical analysis showed that throughout the period of a year, 78 % of the operation time, system ucy1-05 exhibited cell temperatures (15-minute average) 3 °C lower than the average cell temperatures of system ucy1-06.

The annual DC energy yield normalized to the STC rated power (UCY flasher results), demonstrated that System 3 (white control ucy1-07) and System 1 (black colour thermal management backsheet ucy1-05), produced the highest annual energy yield consistently for all years of operation (see Fig. 7).

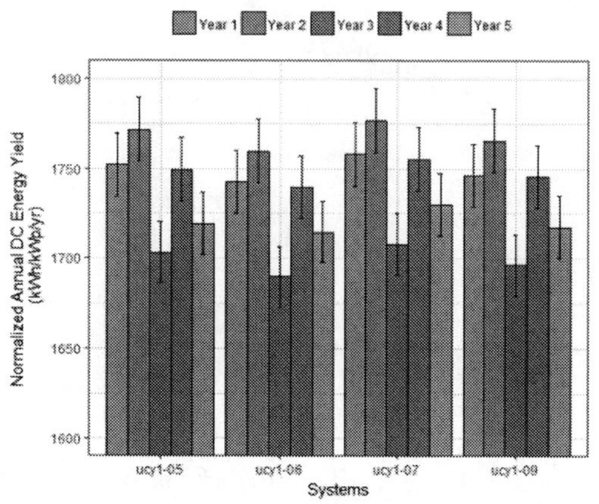

Figure 7: Annual DC energy yield, normalised to UCY flasher results for the installed systems, over the five-year evaluation period.

The power loss comparison of the STC power prior to outdoor installation to the results obtained after five years of exposure (final test on month 60) is summarised in Table IV. The obtained degradation rate results for the two modules of each system did not show any significant differences between the different backsheet material modules since all the modules under test exhibited similar progressive power loss. It is important to note that since the power loss results are comparatively similar and considering that the indoor tests are affected by the flasher's uncertainty (approximately ±3% for P_{mp}), the results may be obscured by this uncertainty. Additionally, the visual inspection and EL images obtained after five years of exposure were compared to the initial images (prior to outdoor installation). The comparison did not show any failures apart from some non-isolated cracks in some of the modules, especially for system ucy1-09 (which indicated that cracks within the modules do not cause significant power loss).

Table IV: Power loss comparison as obtained before the modules were installed outdoors (initial test) and after five years of exposure (final test on month 60) using the method described in IEC 61215 Ed. 2 Procedure 10.2 [5].

System	Average Power Loss Percentage Error (PE) for two modules of each system (%)
System 1 (ucy1-05)	1.59
System 2 (ucy1-06)	1.72
System 3 (ucy1-07)	1.16
System 4 (ucy1-09)	1.35

Finally, using the statistical techniques of OLS, CSD and YOY the annual degradation rate, R_D, of the systems was calculated over the five-year period from the outdoor acquired data-sets. As an example the results obtained by applying OLS to the monthly DC R_p time series of System 1 (ucy1-05) are depicted in Fig. 8.

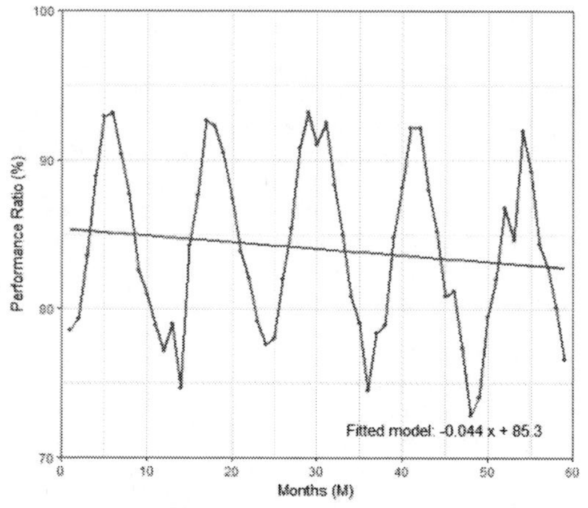

Figure 8: Monthly DC R_p of System 1 (ucy1-05) and linear fit applied using OLS, over the five-year evaluation period.

The annual degradation rates of all systems based on the outdoor analysis, summarized in Table V, showed that over the five-year period there was no significant difference in the calculated degradation rates amongst the installed systems, since the results of all techniques are similar and within the range of the standard error (s.e.).

Table V: Annual degradation rate of each system, over the five-year evaluation period by applying the OLS, CSD and YOY techniques.

System	Annual degradation rate (%/yr)		
	OLS	CSD	YOY
System 1 (ucy1-05)	0.52 ± 0.53	0.58 ± 0.10	0.31 ± 0.63
System 2 (ucy1-06)	0.50 ± 0.53	0.58 ± 0.09	0.31 ± 0.56
System 3 (ucy1-07)	0.50 ± 0.53	0.59 ± 0.09	0.28 ± 0.57
System 4 (ucy1-09)	0.51 ± 0.54	0.58 ± 0.10	0.26 ± 0.60

Both the indoor and outdoor degradation rate investigations showed that over the five-year period there was no significant difference in the calculated degradation rate results. In addition, the annual degradation rate results of the YOY technique were closely resembling the results of the indoor power loss comparison over the five-year period.

5 CONCLUSIONS

The performance and thermal analysis conducted over a five-year period for PV modules with different backsheet materials, clearly showed that energy performance improvements can be achieved by reducing the thermal loss using innovative packaging materials, specifically designed to decrease the operating temperatures. The results demonstrated that backsheets can influence the peak operating temperature of c-Si modules, in some cases, exceeding 10 °C. In this respect, the comparative analysis of cell temperature showed that system ucy1-05 equipped with a black thermal management backsheet operated systematically at lower cell temperatures compared to systems ucy1-06 FPE black control and ucy1-09 FPF black control backsheet, especially during relatively clear-sky days.

Accordingly, the annual energy yield produced from each system showed that systems ucy1-07 (white control system) and ucy1-05 (black colour thermal management backsheet system) produced the highest annual energy yield during all the years of operation.

Finally, the indoor and outdoor degradation rate and reliability analysis performed on all systems, showed that over the five-year period there was no significant difference in the calculated degradation rate results. The visual inspection and EL imaging did not show any significant failures apart from some non-isolated cracks identified for some modules especially in system ucy1-09 verifying that no significant reliability issues (failures) occurred over the five-year evaluation period.

6 ACKNOWLEDGEMENT

This work was funded through the IPERMON project which is co-financed by the SOLAR-ERA.NET Transnational Calls of the European Union under the grant agreement number SOLARERANET3-147. The authors would also like to acknowledge the financial support from the Cyprus Research Promotion Foundation.

7 REFERENCES

[1] J.D. Sandstorm, A method for predicting solar cell current-voltage curve characteristics as a function of incident solar intensity and cell temperature, *National Aeronautics and Space Administration, Technical Report 32-1142*, 1967.

[2] C.R. Osterwald, T. Glatfelter and J. Burdick, Comparison of the temperature coefficients of the basic I-V parameters for various types of solar cells, *in: Proceedings of the 19th IEEE Photovoltaic Specialists Conference*, 1987, pp. 188-193.

[3] G. Makrides, B. Zinsser, G.E. Georghiou, M. Schubert and J.H. Werner, Outdoor efficiency of different photovoltaic systems installed in Cyprus and Germany, *in: Proceedings of the 33rd IEEE Photovoltaic Specialists Conference*, 2008, pp. 1-6.

[4] D.C. Jordan and S.R. Kurtz, Photovoltaic degradation rates - An analytical review. *Progress in Photovoltaics: Research and Applications,* 2013; **21**: 12-29, DOI: 10.1002/pip.1182.

[5] IEC 61215:2005, Crystalline silicon terrestrial photovoltaic (PV) modules – Design qualification and type approval, 2nd Edition, 2005.

[6] G.J. Jorgensen, K.M. Terwilliger, M.D. Kempe and T.J. McMahon, Testing of packaging materials for improved PV module reliability, *in: Proceedings of the 31st IEEE Photovoltaic Specialists Conference*, 2005, pp. 499-502.

[7] W.J. Gambogi, J.G. Kopchick, T.C. Felder, S.W. MacMaster, A. Bradley, B. Hamzavytehrany, E.E.L. Kathmann, K.M. Stika, T.J. Trout, L.G.A. Garreau-Iles and T. Sample, Backsheet and module durability and performance and comparison of accelerated testing to long term fielded modules, *in: Proceedings of the 28th European Photovoltaic Solar Energy Conference and Exhibition*, 2013, pp. 2846-2850.

[8] G. Makrides, J. Bratcher, J. Pratt and G.E. Georghiou, Innovative packaging components for reduced cell operating temperatures improved power output and reliability in crystalline silicon photovoltaic modules, *in: Proceedings of the 28th European Photovoltaic Solar Energy Conference and Exhibition*, 2013, pp. 3064-3068.

[9] G. Makrides, B. Zinsser, M. Schubert and G.E. Georghiou, Energy yield prediction errors and uncertainties of different photovoltaic models, *Progress in Photovoltaics: Research and Applications*, 2013; **21**: 500-516, DOI: 10.1002/pip.1218.

[10] S. Makridakis, S. Wheelwright, and R. Hyndman, Forecasting: methods and applications, 3rd Ed. New York: John Wiley & Sons, 1998.

[11] D.C. Jordan, M.G. Deceglie and S.R. Kurtz, PV degradation methodology comparison - A basis for a standard, *in: Proceedings of the 43rd IEEE Photovoltaic Specialists Conference*, 2016, pp. 273-278.

[12] C.R. Osterwald, J. Adelstein, J.A. Cueto, B. Kroposki, D. Trudell, T. Moriarty, Comparison of degradation rates of individual modules held at maximum power, *in: Proceedings of the 4th World Conference on Photovoltaic Energy Conference*. 2006, pp. 2085–2088.

NEW METHOD OF SILICON PHOTOVOLTAIC PANEL FAULT DETECTION USING IMPEDANCE SPECTROSCOPY

Joanna Symonowicz[1,*], Nicholas Riedel[1,a], Sune Thorsteinsson[1,b], D. Sera[2,c], and Peter Behrensdorff Poulsen[1,d]

[1] Department of Photonics Engineering, Technical University of Denmark
Frederiksborgvej 399, 4000 Roskilde, Denmark
[2] Department of Energy Technology, Aalborg University, 9220 Aalborg, Denmark
[a] nrie@fotonik.dtu.dk, [b] sunth@fotonik.dtu.dk, [c] des@et.aau.dk, [d] ppou@fotonik.dtu.dk
* jokas@fotonik.dtu.dk / +48 517 611 149

ABSTRACT: The aim of our project is to develop a new method for photovoltaic (PV) panel fault detection based on analysing its impedance spectra (IS). Although this technique was successful in assessing the state of degradation of fuel cells and batteries [1, 2], it has never been applied to PV cells on a wide scale. In this paper, we show that, unlike current-voltage (I-V) tests, the IS method is capable of early detection of changes in PV panel parameters due to microcracks and potential-induced degradation (PID). Although our measurements are only successful under dark conditions, the results are similar for both laboratory environment and for outdoor tests in various weather conditions. A fully developed IS technique, accounting for all kinds of most common PV panel degradation types, would surpass the existing PV fault detection methods then it comes to cost and accuracy [3,4].

Keywords: c-Si, Degradation, Impedance, Experimental Methods

1 INTRODUCTION

The IS technique is based on applying a small sinusoidal signal (E) to a measured device over a span of a few decades of frequency (ω). The current response (I) to a sinusoidal potential (E) is a sinusoid at the same frequency but shifted in phase. An expression analogous to Ohm's Law allows us to calculate the complex system impedance [5-7] (see (1)).

$$(1)\quad Z = \frac{E}{I} = \frac{E_0 sin(\omega t)}{I_0 sin(\omega t+\varphi)} = Z_0 \frac{sin(\omega t)}{sin(\omega t+\varphi)}$$

Furthermore, with Euler's relationship, it is possible to express impedance as a complex function [6] (see (2)).

$$(2)\quad Z_0 \frac{sin(\omega t)}{sin(\omega t+\varphi)} = Z_0 \exp(i\varphi) = Z_0(cos\varphi + i sin\varphi)$$

A solar cell's behaviour can be described by an equivalent electrical circuit [8-11]. Such a circuit consists of common electrical elements that have a basis in the physical electrochemistry of the system. Each element contributes to the overall impedance response. Using a fitting method on measured data, it is possible to extract parameters of the equivalent circuit's individual parts [3,4, 8-11].

In this work we use the simplified dynamical equivalent model for a PV cell [11-13] which is presented in fig. 1 along with the corresponding impedance equation and its Nyquist plot. Although the model is derived for dark IS characterization, it should be a good approximation also for light characterization [3, 11-15].

Any changes in the equivalent electrical circuit parameters ought to indicate distortion in a PV module's performance [3,9]. The first use of AC measurements in fault detection was an investigation on how hot spots affect PV panel capacitance [16]. Nex, it has been proven that potential-induced degradation (PID) in c-Si PV panels results in a significant decrease in R_P and an increase in C_P values [3,9]. No other forms of degradation has been described by the IS method yet.

The focus of our experiment is PV module microcracks. This is a very common fault of PV panels, often seen in the field, which can easily be inflicted by a bad installation or any mechanical action that affects the integrity of the solar cells. After inflicting microcracks on silicon PV panels, we would like to spot changes in their R_P, R_s and C_P parameters in both indoor and outdoor experiments [17]. This would verify that the IS analysis is an accurate tool for field fault detection.

$$Z_{PV} = \left[R_s + \frac{R_p}{(\omega R_p C_p)^2 + 1}\right] - i\left[\frac{\omega R_p{}^2 C_p}{(\omega R_p C_p)^2 + 1}\right]$$

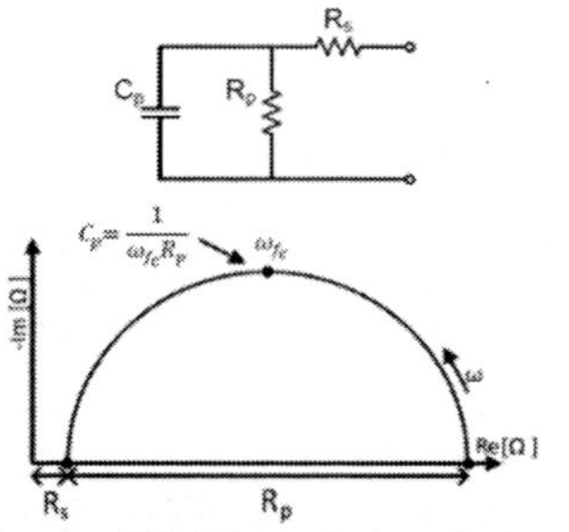

Figure 1: The simplified dynamical equivalent model for a PV cell - complex impedance equation Z_{PV} (top), its equivalent circuit model (middle), and its Nyquist plot visualizing how the parameters can be extracted by the fitting routine (bottom) [3,4, 8-11].

2 EXPERIMENTAL DETAILS

2.1 Stresses on modules

In our experiment we are considering ten silicon PV modules (see fig. 2). Three of them serve as a reference. Six panels, called the uCrack group, underwent mechanical stresses at the Department of Energy Technology at Aalborg University. The total weight used for the degradation was 400 kg (sand bags). The overall pressure put on panels was, thus, 3.9 kPa. The weights were distributed uniformly twice on panels' front surface and once on its back. The one remaining module (PID_05) received a PID test.

The degradation quantification was carried out by the I–V (current-voltage) and electroluminescence (EL) characterization. I–V measurements were performed in light using a Spi-Sun 5600SLP sun simulator. The EL photos were taken with a Photonic Science short wave infrared (SWIR) camera in a dark chamber with the PV module forward biased at 90% standard test condition (STC) short circuit current (I_{sc}). The result of the stresses was a P_{loss} of 4% in five of PV modules (uC_05 to uC_09) and a P_{loss} of 30% in the uC_04 panel. PID_05 did not display any performance reduction (P_{loss} of 0%). Figure 2 includes a visual image of the test modules along with their corresponding EL images and P_{loss} values.

Figure 2: Field station for the I-V, IS and weather conditions measurements built at DTU, Risø Campus.

2.2 Indoor measurements

All ten panels were tested once again at Danmarks Tekniske Universitet (DTU) Institut for Fotonik. We performed multi-irradiation I-V investigations using EcoSun 10L Ecoprogetti to recheck the P_{loss} values of 0%, 4%, and 30%. We then used a Gamry Reference 3000 Potentiostat [18] (AC=0,01V) for the IS tests in dark conditions at T=25°C. The IS test on irradiated samples ($G \neq 0 W/m^2$) was impossible due to equipment overloads [18]. The IS data was fit to the Z_{PV} equation using the Levenberg-Marquadt method.

2.3 Outdoor measurements

The outdoor measurements required the construction of a field station (see fig. 2). The block scheme of the station is presented in figure 3. Test samples are grid-tied via a string inverter when not measured. Each module has a DC-DC power optimizer. The modules are individually taken out of the energy production circuit and into the measurement circuit using relays. I-V and IS are measured on an individual panel before switching the next panel from energy production to measurement. The parameter extraction routine based on a simple RC circuit fitting method [3] runs automatically after the whole day of data collection. The IS curves are obtained by Emazys Z100 PV Analyzer (AC = 2V) [19] while I-V tests are performed by a Chroma single quadrant load.

The field station has been operating since November 2016 in various weather conditions. It possesses two pyrometers and a thermocouple on the back sheet of each module. Back sheet temperatures varied from -10 °C up to 60 °C while an irradiation was 1 to 1200 W/m^2.

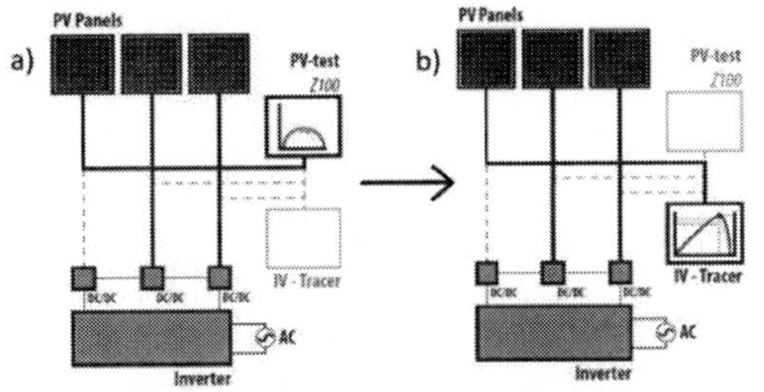

Figure 3: The block diagram of the field station. IS (a) followed by I-V (b) are measured on one panel at the time.

3 RESULTS

3.1 I-V measurements

Figure 4 shows the comparison between the DTU indoor and outdoor I-V measurements for similar irradiation conditions (200-800 W/m^2) and temperature of about 25°C. The values of maximum power points (Pmp) for our tests (EcoSun indoors and Chroma single quadrant load outdoors) correspond to the P_{loss} values detected by Spi-Sun 5600SLP sun simulator during the panel damage preparation.

Indoor measurements are usually higher by circa 5W. This may be due to variations in temperature throughout the outdoor measurements. The deviation can also be caused by different equipment sets used for I-V detection as well as long wires implemented in the field station. Since the deviation is constant, we assume that outdoor I-V measurements reflect indoor ones and that the field station operates correctly. This information is essential to ensure the good performance of the innovative IS tests.

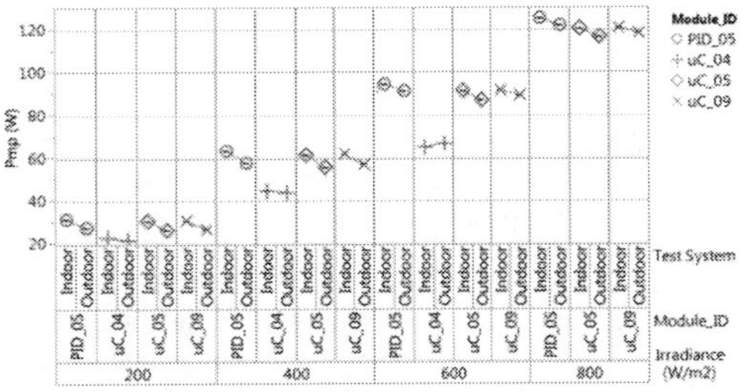

Figure 4: The comparison between indoor and outdoor I-V measurements at multi irradiance (200-800 W/m^2). The Pmp values of individual modules are very similar for indoor and outdoor test and confirm the correct functioning of our field station.

Figure 5 demonstrates that traditional I-V method fails in early detection of changes in AC circuit parameters due to microcracks and PID. The PID module ought to experience rise in C_p [3]. The uCrack group should exhibit lower R_p values and higher R_s values than the Control group [16]. Such performance was only detected for the uCrack panel with 30% P_{loss} which is a large scope of damage. It is well-known that I-V curve is insufficient to detect important parameters, such as interfacial diffusion, transition capacitance followed by resistive and capacitive components of p-p+ interface of a silicon solar cell [20]. Since these mechanisms regulate the values of AC circuit parameters, the extracted R_s, R_p, C_p are not exact [20].

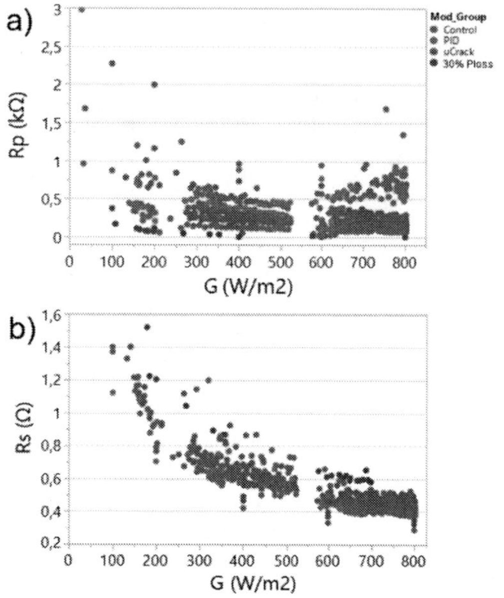

Figure 5: R_p (a) and R_s (b) extracted for indoor and outdoor I-V tests. The traditional I-V method is incapable of detecting changes in AC circuit parameters unless PID and microcrack damages are prominent.

3.2 IS Measurements

So far, our IS method has been only successful for the measurements in dark conditions (fig. 6a,b). Once $G \neq 0 W/m^2$ the IS curves are the same for all degradation groups and, thus, the values of extracted parameters do not differ, either (see fig. 6c). The plateau observed at high ImZ values for $G \neq 0 W/m^2$ (fig. 6c) is due to a limited frequency resolution of Emazys Z100 PV Analyzer [19], which goes down only to 100 Hz. Currently, we are updating the test bed so it can reach frequencies down to 1 Hz.

Similar shapes of IS curves for all degradation groups at high G might be caused by limitations of our equipment. However, it is equally possible that the changes due to microcracks are dimed during solar cell operation. This explains the failure of multi-irradiance IV technique. Nonetheless, the IS tests performed in total darkness seem to be the best option for two reasons. First, most of the previous research involving extraction of parameters from this AC circuit has been performed in total darkness [3, 8-12]. Hence, the model has been proven reliable numerous times for PV panels under no illumination. Second, once the PV modules performance is only tested at night, no power production time is wasted during the day.

From November 2016 until September 2017 the field station was operating only during the day. Hence, no dark condition data was collected during this period. In September 2017 we decided to take the night time data into account. That is why indoor data (fig. 7) was extracted in December 2016 while the outdoor data in fig. 8 was extracted in September 2017. The dissimilarities in their values are due to divergence in equipment and weather conditions. Nevertheless, in this work the focus is on how the parameters differ between module stress groups (uCrack, PID, and Control) and that information is clearly indicated in our data.

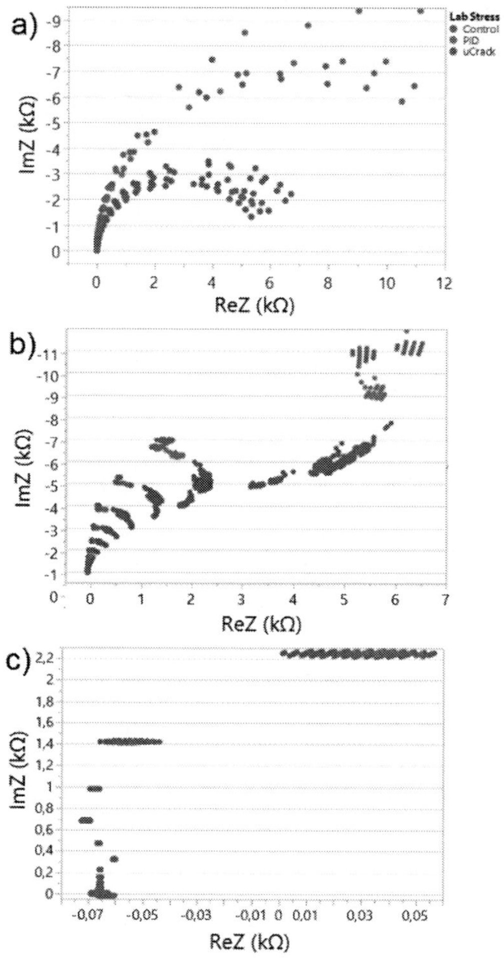

Figure 6: Overlay of IS spectra Nyquist plots at G=0 W/m^2 for indoor (a) and outdoor tests (b), and at G=700-800 W/m^2 for outdoor tests (c). The IS curves for individual degradation groups are distinguishable only at G=0 W/m^2.

For both indoor and outdoor tests at $G = 0 W/m^2$ we were able to observe that R_p values of control modules are higher than uCrack group (fig. 7a, 8a). The lower R_p values illustrate a bigger leakage current in PV cells [3,9] and explain the 4-30% P_{loss} we saw in the I-V data. The PID module exhibits a much smaller R_p value after 9 months of field station performance (fig. 8a). Simultaneously, its C_p and R_s values have increased. These are the symptoms of PID described in [3,9]. It is possible that the damage was minor at first (during indoor tests) and grew during PV panel functioning in altering weather conditions. C_p remains the same for uCrack and Control groups. The divergence of C_p behaviour for PID and microcracks (fig. 7c, 8c) is very convenient as it shows the proficiency of the IS method to distinguish between various kinds of faults.

Behaviour of R_s is problematic (fig. 7b, 8b). For indoor tests, uC_04 module with 30% P_{loss} exhibits 2,5 times higher R_s value than the rest of panels for indoor tests (fig. 7b). The remaining modules from uCrack group have the same R_s values as Control group. This is not the case for outdoor measurements were all uCrack

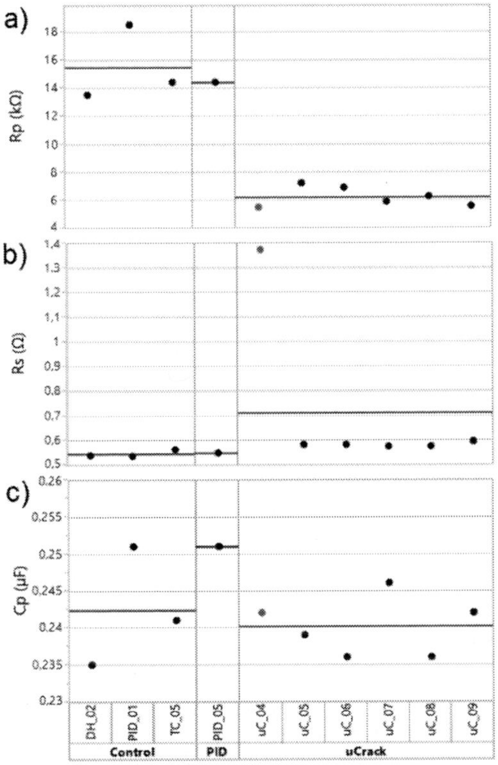

Figure 7: R_p (a), R_s (b), and C_p (c) parameters extracted from IS curves for indoor measurements in dark conditions at T=25°C.

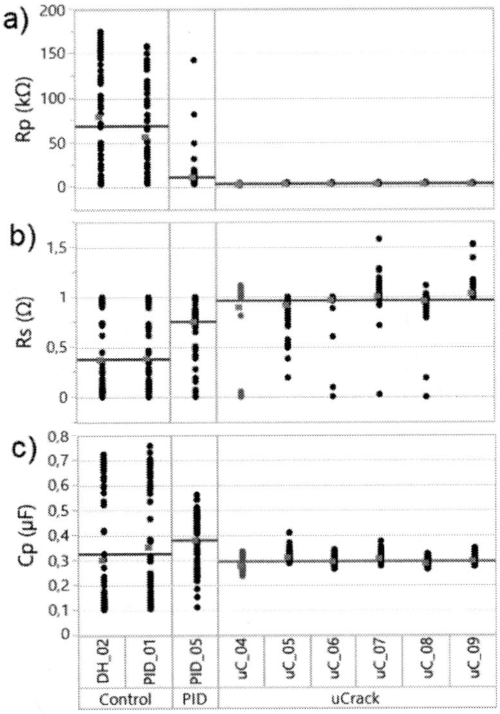

Figure 8: R_p (a), R_s (b), and C_p (c) parameters extracted from IS curves for field station tests in dark conditions at T=25°C. Blue dots indicate the mean values of distributions.

modules display increase in R_s by x2,5. In general, the upsurge in R_s indicates less current flow through the PV junction [3,9]. This might be due to mechanical surface damages [16]. We made sure that the Pmp values for individual modules are the same in September 2017 as in December 2016 using the outdoor I-V data. It is possible that signal AC=0,01 V of indoor potentiostat is not enough to expose higher R_s for minor microcracks [3]. Accordingly, indoor tests must be repeated for AC=2V.

We must bear in mind that the changes of R_s, R_p and C_p may account for various PV panel faults [14-15]. Therefore, at this point we are only able to detect a deviation in module performance which should be further investigated by a plant manager with I-V measurements. Nonetheless, once IS spectra are scrupulously tested for all the most common PV defects, it might be possible to specify a kind of fault based only on the IS parameters change rate.

4 CONCULSIONS

So far, we were able to detect alterations in PV panel parameters due their faults only in the dark conditions. However, the implantation of IS method exclusively at night does not interfere with solar power production, which is a huge advantage. We proved that the parameters are extracted more accurately than in case of I-V tests.

Our results confirm previous studies on how PID changes AC circuit parameters values [3]. The alterations in R_p and R_s after infliction of microcracks also agree with theory [16]. The latter results are different from the ones caused by PID. Therefore, the IS method has the potential to distinguish between various types of faults.

The observed differences indicate that the IS method has the potential to be a successful tool for solar panel fault detection. However, at the moment, the identified changes in R_p, R_s, and C_p only indicate a possible PV module performance failure, which should be further investigated by other detection techniques. In the future, we hope to be able to assess the P_{loss} value only through the field IS measurements. In order to achieve our goal, we first need to perform more indoor tests based on various defect types as a reference for the future outdoor data.

5 ACKNOWLEDGEMENTS

The work on the present study was funded by the Energy Technology Development and Demonstration Program (EUDP) through the project *New technology for localization and characterization of faults in solar panels*. We would also like to thank Matei-Ion Oprea from the Department of Energy Technology at Aalborg University for laying the foundations for our work.

6 REFERENCES

[1] D. Stroe et al., IECON Proceedings (Industrial Electron. Conf.), (2013) 6734–6739.
[2] I.F. Grigoras, Master thesis (Aalborg Univ.), (2013).

[3] M.I. Oprea, Master thesis (Aalborg Univ.), (2013).
[4] R.A. Kumar, M.S. Suresh, J. Nagaraju, Review of Scientific Instruments, Vol. 72 (2001) 3422-3426.
[5] E. Barsoukov, J.R. Macdonald, Wiley Interscience Publications, (2005).
[6] A.J. Bard, L.R. Faulkner, Wiley Interscience Publications, (2000).
[7] J.R. Scully, D.C. Silverman, M.W. Kendig, ASTM, (1993).
[8] R.A. Kumar, M.S. Suresh, J. Nagaraju, Solar Energy Materials and Solar Cells, Vol. 60 (2000) 155-166.
[9] M.I. Oprea et al., Proceedings of the IEEE 43rd PVSC, (2016).
[10] R.L. Mueller, M.T. Wallace, P. Iles, Proceedings of the IEEE 24th PVSC, Vol. 2 (1994), pp. 2034-2037.
[11] D. Chenvidhya, K. Kirtikara, C. Jivacate, Solar Energy Materials and Solar Cells, Vol. 86 (2005) 243-251.
[12] D. Chenvidhya, K. Kirtikara, C. Jivacate, Solar Energy Materials and Solar Cells, Vol. 80 (2003) 459-464.
[13] I. Mora-Sero, G. Garcia-Belmonte, P.P. Boix, M. Vazquez, J. Bisquert, Energy Environ. Sci., Vol. 2 (2009) 678.
[14] D.C. Jordan and S.R. Kurtz, Prog. Photovoltaics Res. Appl., Vol. 21 (2013) 12-29.
[15] M. Quintana, D.L. King, T.J. McMahon, and C.R. Osterwald, Proceedings of the IEEE 29th PVSC, (2002) 1436–1439.
[16] K.A. Kim, G.-S. Seo, B.-H. Cho, Fellow, P.T. Krein, IEEE Trans. On Power Electr., Vol. 31(2) (2016).
[17] J.I. van Mölken, U.A. Yusufoğlu, A. Safiei, H. Windgassen, R. Khandelwal, T.M. Pletzer, H. Kurza, Energy Procedia, Vol. 27 (2012) 167 – 172 .
[18] www.gamry.com/potentiostats/reference-3000/, [22.08.2017].
[19] emazys.com/z100-pv-analyzer, [22.08.2017].
[20] P. Yadav, K. Pandey, V. Bhatt, M. Kumar, J. Kim, Renewable and Sustainable Energy Reviews, Vol. 76 (2017) 1562–1578.

LONG-TERM PERFORMANCE OF PV MICRO-INVERTERS

Stefan Krauter and Jörg Bendfeld
University of Paderborn, Electrical Energy Technology – Sustainable Energy Concepts
Pohlweg 55, D-33098 Paderborn, Germany

ABSTRACT: To compare efficiency and yield of most micro-inverters available on the market during 2014-2017, an in- and outdoor test field at the University of Paderborn have been set up. The inverters have been fed by identical and calibrated crystalline silicon PV modules of 215 W_p. To monitor accurately AC power output and yield, each of the micro-inverters has been equipped with a calibrated precision electricity meter. For micro-inverters requiring control units for grid-feeding that has been acquired also. The test field is equipped with pyranometers in the module plane. The comparison includes energy yield over one to two years and efficiency-load characteristics. The weighted conversion efficiency according to EU and CEC standards has been measured and compiled. After two years of outdoor operation in Paderborn, all efficiency measurements have been repeated: None of the inverters has shown any degradation beyond the measurement accuracy of 0.1% of the precision power meter used (LMG 670 by Zimmer).

Keywords: Inverter, Yield, Small Grid-connected PV Systems, Power Conditioning

1 INTRODUCTION

Micro-inverters are inverters that are connected principally to a single PV module (sometimes to two modules, as indicated in the tables), so each module-inverter combination acts as an independent power plant. The micro-inverter consists of a maximum power point tracker (MPPT), the DC-AC inverter, and an islanding protection unit (see e.g., [1]). To achieve higher power outputs, several module-inverter combinations are interconnected in parallel on the AC output side. This configuration offers various advantages: Easier planning and easier installation, easy up- and downscaling of the power plant, including extensions or repair that could be carried out during power plant operation. Effect of shadowing is very limited, and due to low system voltages, potential induced degradation (PID) does not occur. Logistics is simplified. An excellent overview of the development and the advantages of micro-inverters has been compiled by H. Oldenkamp [2].

However, costs of power plants based on micro inverters are about 20% higher. Some of the inverters cannot be operated by themselves and require a control unit (often combined with a remote shutdown option and a monitoring system), or a protective device for grid interfacing (depending on national regulations), thus adding extra costs. Also, conversion efficiency may not be as high as for central inverters. Due to smart master-slave concepts, centralized solutions with multiple but relatively large inverters may offer higher yields under weak light conditions. [3] is giving a performance comparison of a micro inverter, a power-optimizer and a central inverter. Unfortunately, the only micro-inverter used in that comparison has been one with a quite poor performance (according to our comparisons [4] and [5]), and therefore the comparison results conceivably do not favor the category "micro-inverter" in general.

2 INDOOR TESTS

Due to the reproducible test conditions in the indoor lab, the inverters have been examined individually with predefined and controlled input data. Input has been a PV module simulator with data set according to the modules used in the outdoor test. The principal output data recorded has been the delivered AC power of the inverters. Besides input power, output is also a function of input voltage. If input voltage is getting too low, the inverters even stop operating. The following examinations are based on the possible range of input data (including voltage) given the specific PV module also used for the outdoor investigation. After two years of outdoor operation in Paderborn (see chapter 3), all efficiency measurements have been repeated: None of the inverters has shown any degradation beyond the measurement accuracy of 0.1% of the precision power meter used (LMG 670 by Zimmer).

2.1 Peak, EU- and CEC conversion efficiency

Peak efficiency is often reached close to the maximum load of the inverter. Peak efficiency in data sheets is not a helpful value since most of the time the inverters operate in the range of 20% to 30% of their rated power. Consequently, an adequately weighted efficiency is a more adequate measure. One type of weighted efficiency is the so-called "European Efficiency" η_{Euro} – it is calculated according to:

$$\eta_{Euro} = 0.03 \cdot \eta_{5\%} + 0.06 \cdot \eta_{10\%} + 0.13 \cdot \eta_{20\%} + 0.1 \cdot \eta_{30\%} + 0.48 \cdot \eta_{50\%} + 0.2 \cdot \eta_{100\%} \quad (1)$$

The other is the "CEC efficiency" by the California Energy Commission (CEC). CEC efficiency is computed as an average value of DC-AC conversion efficiencies at six pre-defined outputs between 10% and 100% of the rated power (with an emphasis on higher irradiance levels) is determined by:

$$\eta_{CEC} = 0.04 \cdot \eta_{10\%} + 0.05 \cdot \eta_{20\%} + 0.12 \cdot \eta_{30\%} + 0.21 \cdot \eta_{50\%} + 0.53 \cdot \eta_{75\%} + 0.05 \cdot \eta_{100\%} \quad (2)$$

For the European Efficiency the weighting factors are lower for the high relative power values.

2.2 Test arrangement and results

Ten different output power values from zero up to the maximum operational power of the PV modules have been considered (they are roughly identical for all seven micro inverters). The values used for the single input inverters (adjusted by controlling the current) are: 10 W, 25 W, 50 W, 75 W, 100 W, 125 W, 150 W, 175 W, 200 W, and 225 W. The averaging time for the power meter used (Zimmer LMG 670) was set to 500 ms. One micro-inverter showed an unstable AC power output for values below 100 W, for that device the averaging time had to be extended to two seconds.

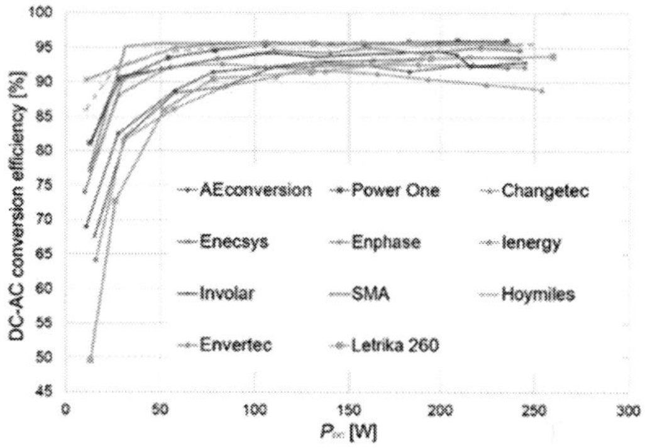

Figure 1: Measured DC-AC conversion efficiencies as a function of DC input power for eleven micro-inverters with single PV module inputs.

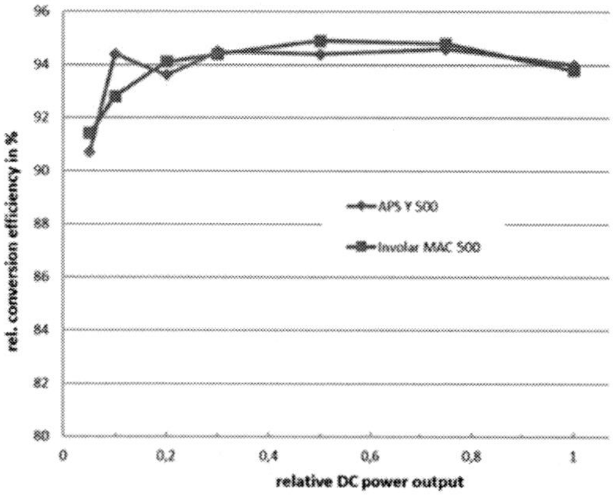

Figure 2: Measured DC-AC conversion efficiencies as a function of relative DC input power for two inverters that have inputs for two PV modules (2 · 250 W).

For each power-step 25 values have been recorded. The measured DC-AC conversion efficiencies of all inverters are shown in Fig. 1 and Fig. 2. Based on those measurements, the European (EU) efficiencies and the CEC efficiencies for the micro inverters have been calculated according to (1) and (2), eleven micro-inverters are designed for single modules, two inverters have inputs for 2 modules: Involar MAC 500 and the APS Y 500.

The ranking considering the European conversion efficiency is shown in Table I: AEconversion/Aptronic and Enecsys have the same EU-efficiency, thus sharing rank number ten.

Table I: Ranking of all tested micro-inverters by "European conversion efficiency", according to (1), same types and models as in Table III.

Rank	Manufacturer	European efficiency	Relative eff. vs #1
1	SMA	95.4 %	100.0 %
2	Enphase	95.2 %	99.8 %
3	Hoymiles	95.0%	99.5 %
4	ABB/Power One	94.6 %	99.2 %
5	Involar MAC 500	94.3 %	98.8 %
6	APS Y 500	94.1 %	98.6 %
7	Envertech	93.2 %	97.7 %
8	Involar MAC 250	92.7 %	97.2 %
9	Changetec	90.9 %	95.3 %
10	AEconversion	90.3 %	94.7 %
10	Enecsys	90.3 %	94.7 %
11	Ienergy	89.9 %	94.3 %
12	Letrika 260	88.7 %	93.0 %

The ranking for the CEC-conversion-efficiency is given in Table II, here the Letrika 260 and the Ienergy achieved the same results, therefore they are sharing rank number ten.

Table II: Ranking of all micro-inverters by CEC conversion efficiency, according to (2), same types and models as in Table III.

Rank	Manufacturer	CEC efficiency	Relative eff. vs #1
1	Enphase	95.6 %	100.0 %
2	ABB/Power One	95.5 %	99.9 %
3	Hoymiles	95.4 %	99.8 %
4	SMA	95.1 %	99.5 %
5	Involar MAC 500	94.6 %	99.0 %
6	APS Y 500	94.5 %	98.9 %
7	Envertech	94.1%	98.4 %
8	Involar MAC 250	93.9 %	98.2 %
9	Enecsys	92.0 %	96.3 %
10	Letrica 260	91.5 %	95.7 %
10	Ienergy	91.5 %	95.7 %
11	AEconversion	91.2 %	95.5 %
12	Changetec	90.9 %	95.1 %

3 OUTDOOR TESTS

3.1 Test arrangement and measurements

For the PV outdoor test lab (see Fig. 3) installed on the roof of the University of Paderborn (51.707° N, 8.771° E), a specific test system comprising of eight photovoltaic modules with eight micro-inverters has been employed. To examine thirteen inverters, five of them have been interchanged.Ten different output power values from zero up to the maximum operational power of the PV modules have been considered (they are roughly identical for all seven micro inverters). The values used for the single input inverters (adjusted by controlling the current) are: 10 W, 25 W, 50 W, 75 W, 100 W, 125 W, 150 W, 175 W, 200 W, and 225 W. The averaging time for the power meter used (Zimmer LMG 670) was set to 500 ms. One micro inverter showed an unstable AC power output for values below 100 W, for that device the averaging time had to be extended to two seconds.

Figure 3: PV outdoor laboratory at the University of Paderborn for comparison of micro-inverters using eight equal, calibrated PV modules.

Each PV module (QC-CO2 by Hanwa Q-cells) consists of 60 solar cells. In stated plant, these equal and calibrated modules are the input for each micro inverter. The goal of this investigation is to analyze the performance of the inverters in the lab and under real operating conditions comparing their energy yield simultaneously at the same climatic conditions and solar irradiance. The climatic conditions have been monitored during the whole test period. The meteorological monitoring equipment consists of two calibrated pyranometers in the plane of the module (CMP 21 and SP2 lite by Kipp & Zonen), a 3-D ultrasonic anemometer (by Thies), and a thermo-hygro sensor (by Thies).

Fig. 4 shows the electrical measurement setup. Each micro inverter has been directly connected to a calibrated electrical energy meter with a S0-interface. All S0-interfaces have been connected to an efficiency recording data server.

Fig. 5 shows an example of the collected data for single day (data recording every 5 min). Due to the high resolution in time, power and energy output, some specific characteristics could be observed: E.g., starting behaviors, MPP tracking (accuracy and speed), dropouts and performance.

Fig. 6 shows an example of the collected data for single day during spring of the year 2015. It is visible that some inverters have difficulties to follow the quick change of irradiance levels during that day (e.g. IEnergy, and Hoymiles – see also Fig. 7).

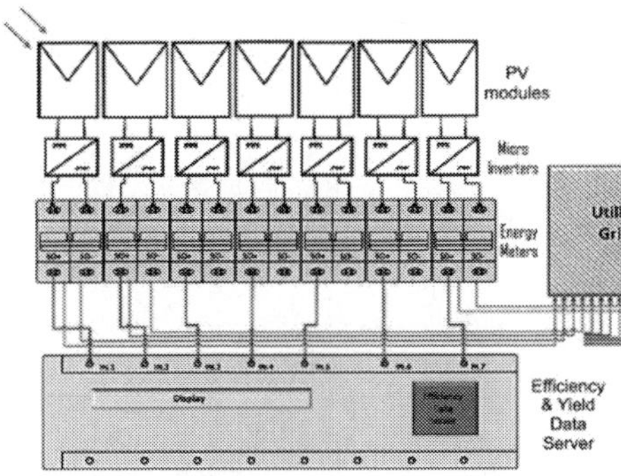

Figure 4: Scheme of the outdoor test setup for the micro-inverters (MI) investigated.

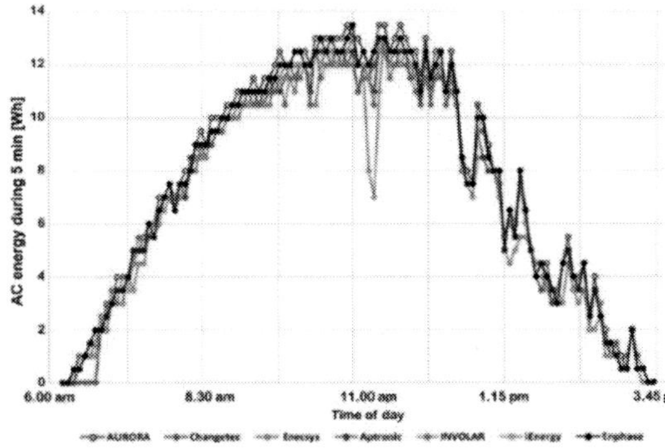

Figure 5: Actual recorded data of AC energy generation (over an integral of 5 min) for seven micro-inverters during the day of 10/31/2013.

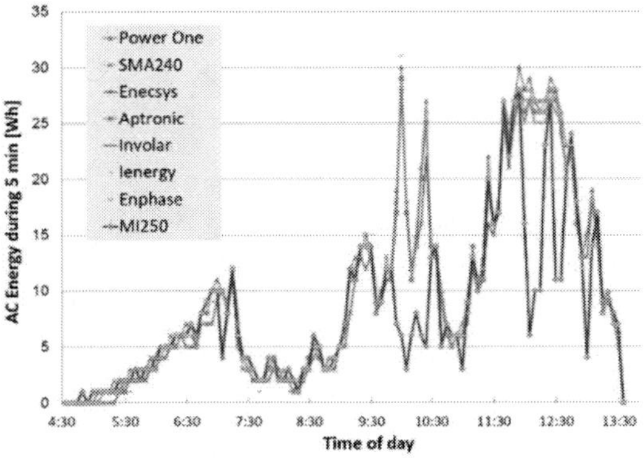

Figure 6: Actual recorded data of AC energy generation of seven micro-inverters (over an integral of 5 min) during the day of 4/6/2015 with changing irradiance. Some inverters have not been able to follow a quick increase of irradiance.

A ranking list according to the total AC energy yield of the micro inverters during the common operation period is shown in Table III (July 2015 until June 2016, (Envertech 6/1/2016 until 6/15/2016 only) relative energy yield vs. former #1 [4]. To make future progress in yield improvements more transparent, that #1 reference will be kept (as 100%). Eleven micro-inverters are designed for single modules; two new inverters have inputs for 2 modules: Involar MAC 500 and the APS Y 500. The

Letrika 260 has been at the outdoor test laboratory for insufficient time to achieve qualified yield results, but it seems to stay in the lower average of the range of inverters examined. It is equipped to perform active anti-islanding directly, so it does not need any additional device as a grid interface - even in countries that have require protection devices according to IEC 62116.

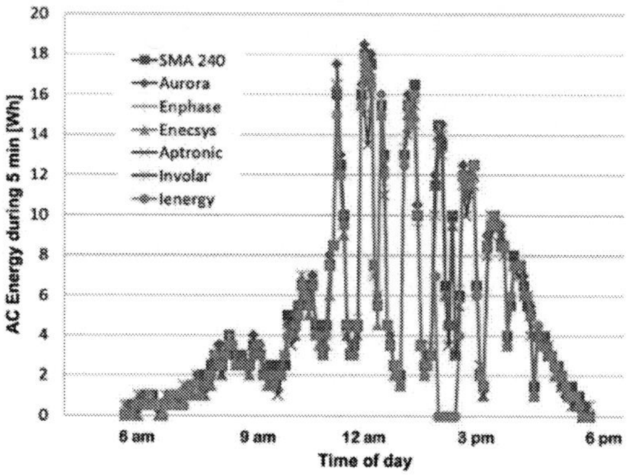

Figure 7: Actual recorded AC-energy generated within a 5 min interval during a day with changing irradiance levels. One inverter adapts quickly (blue line: ABB/Aurora), the other not (red line: Hoymiles MI 250).

Table III: Ranking of all micro-inverters by measured AC energy yield (in Paderborn 2015–2017), referenced to best performer of last examination [4] [5]. Update on 20/09/2017.

Rank	Manufacturer	Type, Model	rel. yield vs. ABB (Ex-#1) [4]
1	Involar	MAC 500	100.7 %
2	APSystems	Y 500	100.5 %
2	ABB/Power One/Aurora	Micro-0.25-i	100.0 %
4	SMA	Sunny Boy 240	95.5 %
5	Enphase	M 215	95.9 %
6	Involar	MAC 250	93.8 %
7	IEnergy	GT 260	93.4 %
8	Envertech	EVT 248	92.1 %
9	Lektrika	SMI 260	91.5%
10	AEconversion /Aptronic	INV 250-45	90.3 %
11	Enecsys	SMI-S-240W	88.7 %
12	Hoymiles	MI 250	78.4 %
13	Changetec	ELV 300-25	75.6 %

4 REPEATED INDOOR TESTS & YIELDS

4.1 Repeated efficiency measurements

After two years of outdoor operation in Paderborn, all efficiency measurements have been repeated, as shown ins chapter 2. None of the inverters has shown any degradation beyond the measurement accuracy of 0.1% of the precision power meter used (LMG 670 by Zimmer); Figures 1 and 2 are still valid.

4.2 Changes in relative AC energy yield

The yield ranking as shown in former publications [4] changed slightly, Table IV shows the relative AC yields of all eight microinverters measured during 2014–2015. While the conversion efficiencies and the subsequent EU- and CEC-efficiencies did not change beyond the measurement tolerance (as shown above), the yield differences can be explained by the changing weather conditions, with different fluctuation patterns of irradiance levels. Degradation of the MPPT capability seems unlikely, while the software did not change and the control hardware should not suffer from degradation.

Table IV: Former ranking of eight micro-inverters by measured AC energy yield (in Paderborn 2014–15) [4]

Rank	Manufacturer	Type, model	rel. yield vs. #1
1	Power One / Aurora / ABB	Micro-0.25-i	100.0 %
2	SMA	Sunny Boy 240	96.9 %
3	Involar	MAC 250	95.5 %
4	Enphase	M 215	95.0 %
5	AEconversion / Aptronic	INV 250-45	93.7 %
6	IEnergy	GT 260	93.1 %
7	Enecsys	SMI-S-240W	91.1 %
8	Changetec	ELV 300-25	75.6 %

5 RESULTS, DISCUSSION AND CONCLUSION

5.1 Results

Some of the results could be expected by taking a close look at the efficiency curves of Fig. 5. For example: If highest performance occurs at relatively low input power (and the corresponding low irradiance levels), the ranking for the European Efficiency (which puts a higher weight on lower irradiance levels) is higher, while the CEC ranking is worse due to the higher weighting of high irradiance levels (this is the case for the Changetec inverter). The best in the CEC ranking has been the Enphase inverter, which was second in European Efficiency ranking. For the SMA inverters the ranking positions have been the other way round: SMA is optimized for European irradiance conditions, Enphase for Californian conditions. While the conversion efficiency rating of the new micro-inverters with two inputs, Involar MAC 500 and the APS Y 500, has not been exceptional (ranks 5-8), they achieved rank 1 and 2 at the energy yield ranking. The layout of your paper should have exactly the same format as these notes.

5.2 Correlation of yield with conversion efficiency

Some of the rankings by yield (as shown in Table III, but also in Table IV) cannot be explained by the curves of Fig. 1 and 2 and the European or the CEC efficiency. We observed significant differences of the MPPT devices in their capabilities to adjust to an increase of irradiance quickly (reduction of internal load resistance). Best in this discipline has been the Power One / Aurora inverter, which has not been the best in the static conversion efficiency measurements, but in energy yield. The device MI 250 showed an extreme deviation of its efficiency rating from the yield results. An accurate but slow MPPT algorithm that had difficulties to follow changing irradiance conditions has possibly caused this effect.

Fig. 8 shows the relative energy yield as a function of European efficiency (which is often an important criterion for the purchase decision by a potential customer). As can be observed, there is just a quite vague correlation between those two values with high deviations. For the example indicated by the numbers 1,2,3 at Fig. 8: A customer wants to buy an efficient inverter and he usually takes the datasheet values (or even actual measurements) of the European conversion efficiency as an indicator. According to that, he would prefer Inverter (1), offering a slightly EU-efficiency: 95% instead of 94.6% (2), resp. 94.5% (3).

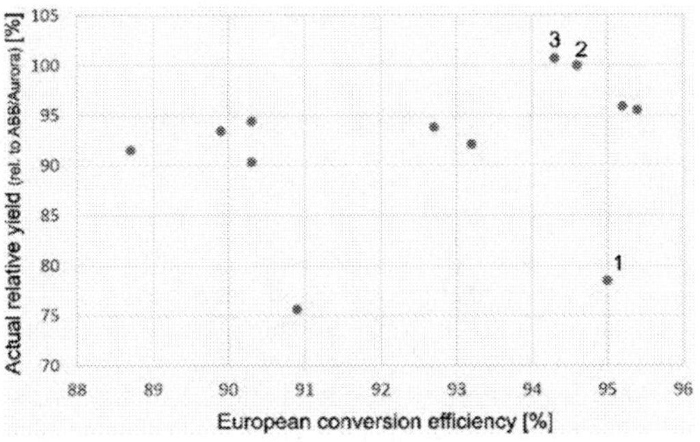

Figure 8: Electrical energy yield as a function of European efficiency: The user cannot expect a higher yield if he purchased an inverter with a higher European efficiency – the yield can be significantly lower, see also [5] for former measurements.

After purchase and during operation he will find out that his inverter will have a relative AC yield of 78.4% only, compared to 100% (2) or even 100.5% (3). This is unsatisfying; therefore, the actual on-site AC yield values (also considering the dynamic behavior of the inverters) should be the major indicator for a potential customer and should be given in the datasheet.

5.3 Conclusion

For the time being, no standard is able to overcome the constraints by the ranking via weighted efficiencies, so the on-site measurement of electrical energy yield is the best method to achieve reliable results that are helpful and relevant for the customer – or at least for customers with similar climatic conditions.

None of the inverters has shown any degradation after 2 years beyond the measurement accuracy of 0.1% of the precision power meter used (LMG 670 by Zimmer), but that might be different for harsher conditions.'

5.4 Outlook

To overcome the considerable deviation between yield and efficiency measurement results, a dynamic test will be deployed to find a better indicator for the relative yield to be expected by the different inverters. A statistics of the frequency of irradiance changes, its amplitudes, and the velocities of changes will be translated into a test procedure.

While the lab tests have been made at laboratory temperature conditions (20°–22°C) and the yield measurements at actual operating temperature for moderate Paderborn conditions (which did not cause a detectable degradation after 2 years), additional efficiency tests at elevated temperatures (e.g. 40°C and 60°C) will be carried out (see also [6]).

6 REFERENCES

[1] W-F. Lai, S-M. Chen, T-J. Liang, K-W. Lee, A. Ioinovici, "Design and Implementation of Grid Connection Photovoltaic Inverter" in *IEEE Energy Conversion Congress and Exposition (ECCE),* 2012, pp. 2426–2432.

[2] H. Oldenkamp, I.J. de Jong, "The return of the AC-module inverter" in *24th European Photovoltaic Solar Energy Conference (EUPVSEC)*, 2009, pp. 3101–3104.

[3] D. Stellbogen; P. Lechner; M. Senger, "Field and laboratory performance characterisation of microinverter and Power optimizer systems" in *32nd European Photovoltaic Solar Energy Conference (EUPVSEC),* 2016, pp. 1654–1659.

[4] S. Krauter; J. Bendfeld, "Cost, performance, and yield comparison of eight different micro-inverters", in *42nd IEEE Photovoltaic Specialist Conference (PVSC)*, 2015.

[5] S. Krauter; J. Bendfeld, "Deviations of results for energy yield from efficiency rankings of micro-inverters", in *32nd European Photovoltaic Solar Energy Conference (EUPVSEC)*, 2016, pp. 1508–1511.

[6] M. A. Hossain; Y. Xu; T. J. Peshek; L. Ji; A. R. Abramson; R. H. French: "Microinverter Thermal Performance in the Real-World: Measurements and Modeling", in *PLOS ONE* DOI:10.1371/journal.pone.0131279 July 6, 2015.

DEVELOPMENT OF A CONTROLLER-HARDWARE-IN-THE-LOOP (CHIL) TOOLBOX APPLIED FOR PRE-CERTIFICATION SERVICES FOR GRID-CONNECTED PV INVERTERS ACCORDING TO THE STATE-OF-THE-ART BDEW RL GUIDELINE AND FGW TR3 STANDARD

G.Lauss[1], D. Majstorovic[2], F. Leimgruber[1], O. Gagrica[2], R.Bründlinger[1], I. Morar[2], Z. Miletic[1], N. Fischer Celanovic[2];
[1]Austrian Institute of Technology, Giefinggasse 2, 1210 Vienna, Austria
[2]Typhoon HIL, Technoparkstrasse 1, 8005 Zürich, Switzerland
e-mail: georg.lauss@ait.ac.at

ABSTRACT: Real-time simulation methods for interconnection requirements of grid connected inverters are increasingly in the focus of distributed generation (DG) research. While laboratory testing methods are the predominant method for the verification of safety and quality related features of grid-connected generation units in the past, load flow modelling verification methods have been integrated in state-of-the-art standardisation frameworks recently. The next step is comprised in real-time simulation methodologies applied for compliance testing of entire power electronic systems integrated in distribution network systems. The controller hardware-in-the-loop (CHIL) simulation is an appropriate real-time simulation methodology that combines numerical simulations with software modelling approaches and classical hardware testing in laboratories. Control boards represent the hardware device directly connected to the power electronic periphery, which is entirely simulated in a real-time simulation environment. Hereby, input signals from voltage and current measurements and output signals for power system control are exchanged in real-time. Thanks to this setup the testing the true behaviour of the entire generation unit can be emulated precisely supported by the developed pre-certification toolbox, as the main outcome of this work. The toolbox offers possibilities of high level automation of test sequences representing a crucial benefit for an improved testing process. In addition, different parameter settings can be run in parallel on a CHIL platform, hence the total testing time can be shortened due to individual acceleration features. With the pre-standardisation toolbox for CHIL simulations a shorter time to market is achieved compared to high power laboratory testing.

Keywords: pre-certification, grid-connected PV inverters, real-time simulation, software modelling, inverter control, standardisation and compliance testing;

1 INTRODUCTION

Large-scale integration of renewable energies such as solar leads to an increasing level of distributed power generation. Each generation plant is obliged to comply with above-mentioned power quality and safety standards. The focus of this work is set to the development of a novel toolbox which supports real-time simulation testing with the CHIL methodology for grid-connected generation units. Methods of testing according to given standards [1]-[8] applying for grid connected inverters are increasingly in the focus of distributed generation (DG) research. While various principles have been developed over the last decades related to real-time simulation techniques, nowadays controller hardware-in-the-loop (CHIL) simulation [9] is becoming increasingly important for pre-certification services. The CHIL technique involves closed-loop simulation of two or more subsystems, at least one of which is a physical device and another a software-only subsystem executed on a digital real-time simulator (DRTS) platform. Hardware under test (HUT) is interfaced to the DRTS using analog or digital communication, which is sampled deterministically by the DRTS. The HUT is generally an embedded controller or other device with input or output using signal-level voltages of less than 50 V. This technique offers the ability to interface with physical device controllers in the state in which they will finally be implemented. In so doing, non-idealities introduced by embedding control logic on the target platform and physical signals are included in the overall closed-loop simulation. Compared with power hardware-in-the-loop simulation, no power amplification stage whatsoever is integrated in the CHIL methodology.

In the past, laboratories such as the SmartEST laboratory located at the Austrian Institute of Technology provided exclusive infrastructure for high power component testing (Figure 1). Thanks to the CHIL methodology, hardware testing facilities can be emulated in real-time environment platforms and testing of high-power components can be facilitated with respect to flexibility in system complexity, power and voltage scalability, and automated testing. This contribution gives statements based on reproducible laboratory test measurements and comparative CHIL test results based on a toolbox applied to pre-certification testing according to the BDEW and FGW TR3 standard. The development of a toolbox applicable to pre-certification testing is an important feature in combination with CHIL simulations and renews important prospects for upcoming concepts of fully standardisation testing.

Figure 1: View of the AIT SmartEST (Smart Electricity Systems and Technologies) high-power laboratory for component testing up to a rated power of 1 MVA.

Following this introduction, a thorough background of the idea of the toolbox and the design principles of the implementation are given in Section 2. The complete structure of the toolbox is introduced in Section 3 and various features are exemplarily presented in this section. In Section 4, the performance of the toolbox is validated in CHIL test results and a comparison to hardware testing results is done.

2 IDEA AND DESIGN OF THE DESIGNING TOOL

The driving motivation for the CHIL toolbox is to provide an end-to-end environment to support controller development from sketching test schematics through interactive and automated controller application up to standardized pre-certification testing and reporting.

Subsequently, the basic components "Schematic editor", "HIL Control Panel" and "Test Suite" from the Typhoon HIL Control Center are used in conjunction to provide the mentioned transition from test setup to manual testing and fully automated testing and reporting at the end.

One of the most challenging stages in the product development of a grid-connected converter is the certification according to the grid connection standard of particular countries representing target markets. The test iteration between the engineering teams and accreditation laboratories is a lengthy and costly process.

The standard tests can be very detailed, sometimes involving hundreds of test cases. With the constant struggle to reduce the time to market, the industry often employs test automation methods and technologies. Before the pre-certification was even considered, the Typhoon CHIL toolchain already had industry-proven test automation capabilities used in unit testing, regression testing, etc. It was natural to extend the application to standardized testing. Standards like BDEW, UL1741, etc., were implemented within a specialized test automation and precertification tool called "Test Suite".

The CHIL approach creates an opportunity to develop the pre-certification capacities in-house and reduce the amount of iterations with an accredited laboratory while testing the controller with laboratory-grade fidelity.

A C-HIL project workflow using Typhoon HIL toolchain can be described in following points:

- creating a converter and a testbed model using Schematic Editor
- compiling the model for execution on Typhoon HIL real time devices
- configuring the interface between HUT and HIL device in terms of I/O signal assignments and scaling
- validation of basic controller functionality through real-time simulation using manual and semi-automated test sequences using HIL SCADA tool

Without test automation, the controller validation, for the most part, requires constant supervision and input of a CHIL toolchain user. However, a more advanced user can use the "Script Editor" tool to write test scripts that will fully replace the manual actions and enable unsupervised, batch testing with automatic report generation.

The implementation and validation of test procedures, including stimulus generation, results analysis and reporting, is time consuming and requires programming and domain specific knowledge. Additional challenge poses the inclusion of standard compliance in test script writing. Standard test procedures and requirements need to be studied and interpreted before the script writing can take place.

The Test Suite project workflow can be described in following points:

- Test setup using test configuration wizard
- Individual or batch test execution
- Automatic report generation

The main structure of the Test Suite is shown in Figure 2. On the left side of the screen is the test library containing currently implemented standard tests. Since the grid connection standards are not fixed, but evolve with time (corrections, updates, etc.), Test Suite provides test version control. By double-clicking on any of the tests, the test configuration wizard opens. A major advantage when using Test Suite tool is that each standard test procedure has its own configuration wizard. The wizard guides the user in such a way that no detailed knowledge of standard test procedure is necessary to actually setup and run the test. The BDEW test scripts have been implemented by Typhoon HIL and have been validated by AIT Austrian Institute of Technology.

The initial setup steps involve: model binding (verification that the standard test facility emulating components are present in the test circuit); input of HUT operating parameters; test scope modification ("pre-certification mode" for full scope or "development mode" for testing specific test cases, operating points, etc.). Final step is the definition of user actions after starting, during, and before stopping the simulation. Less user actions provide higher level of test automation, but sometimes they are needed, by the very design of HUT (i.e. setting power references or machine states manually, before test script provides the stimulus).

Once initially configured, the test is stored in a project and occupies the main screen showing its latest run status (passed, failed, aborted, etc.). It is possible to manage the pre-certification for several standards under one project (i.e. if the converter is targeted at several different markets). Each test is accompanied with help documentation and linked to an automatically generated report based on the latest test run. As mentioned before, the standard test can involve hundreds of test cases. The size of the report can grow over several hundreds of pages. Therefore, the generation of automated reports with high quality formatting is another major time-saving feature of Test Suite tool.

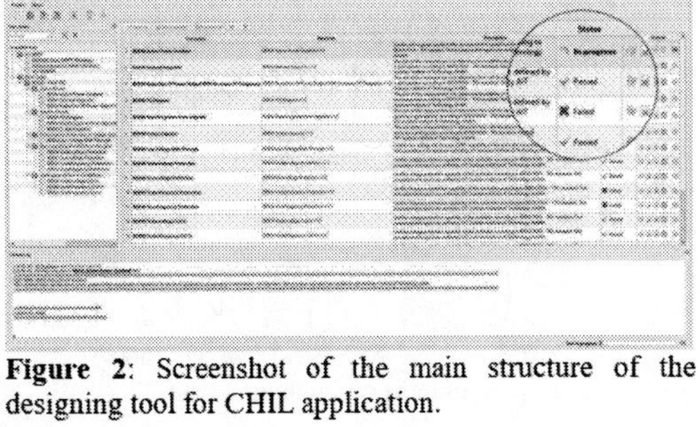

Figure 2: Screenshot of the main structure of the designing tool for CHIL application.

3 TOOLBOX STRUCURE AND FEATURES

This section discussed the main architecture of the implemented tool designed to perform high-fidelity CHIL simulation tests. Besides highlighting the structure in principle, the focus is set on multiple implemented features of the tool which enable the tests of discrete items as well as automated test sequences as a whole.

3.1 Main Structure

Issues relating to the integrated use of information and communication technology (ICT) for remote control are covered with the architecture of the CHIL testing platform. External signals acting as analog or digital input and output signals for embedded communication protocols which are emulated and implemented in the real-time simulator. This approach fulfils state-of-the-art high-level control functions and remote controllability as required in the standards, as standardized communication approaches, interfaces, and protocols are necessary for the usage of components in smart grid developments.

In the following, the most important components for the CHIL test setup are presented providing an embedded platform with the complete test setup and fully automated testing and reporting at the end. In general, the complete power electronic periphery is modelled in software and a state-of-the-art DRTS provides a suitable platform for this process. In contrast to that, the control board is the hardware device which is connected to the interfacing environment consisting of communication signals and control information signals.

Exemplarily, a power electronic circuit is discussed to explain the CHIL signal interface more in detail. Figure 2 highlights the schematics of a power generation source PV1, followed by the DC input stage, the symmetric DC-link stage (DC+, DCn, DC-), and the T-type electronic power bridge with phases A, B, and C as the generalized generator output stage.

For this particular case, the device under test (DUT) can be seen as a customized control board which controlling this grid-connected PV inverting system by means of given input and output signals. Input signals for the boosting stage of the DUT are comprised in voltage measurements of the DC link V_DC3, while output signals are voltage and current measurements as well as PWM signals for the gate driver of the DC the boost control stage. Analogously, input signals for the DC/AC bridge are comprised in voltage and current measurements of the phases A, B, and C, while output signals are comprised in PWM firing signals for the T-type switching devices responsible for the current control.

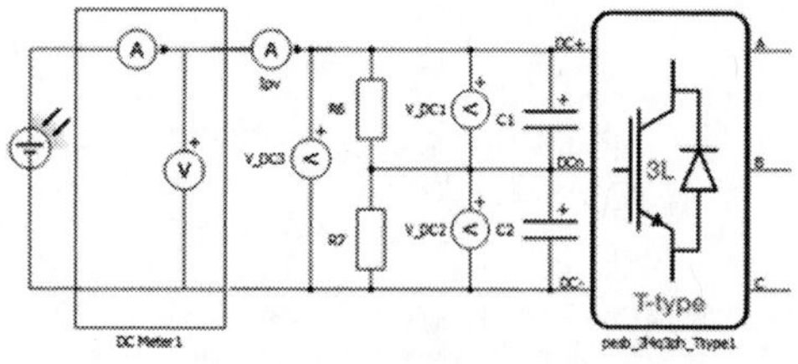

Figure 3: Schematic view of BDEW LVRT test setup with PV source, converter and grid components selectable from a prebuilt component library.

3.2 Implemented Features

There are numerous tests included in the standards classified with required test setups, data recording, and reporting structure. In this contribution, some test cases will be highlighted exemplarily to demonstrate the function of the developed toolbox.

The entire spectrum of the toolbox implementation is comprised of following subtests according to FGW TR3, Rev 23: 4.1 General specifications, 4.2 Active power provision, 4.3 Reactive power provision, 4.4 Voltage quality, 4.5 Disconnecting the PGU from the grid, 4.6 Verification of cut-in conditions, and 4.7 Performance in the case of voltage dips. As an example, the CHIL test for item BDEW TR3, 4.7, generally known as the low-voltage ride-through (LVRT), is set up by

- specifying schematics representing the test setup with PV simulator, HUT and grid components as built by the Schematic GUI editor
- specifying the parametrization of the mentioned components within schematics (grid simulator phase layout and frequency, PV system size and STC settings, and I/O channel setup).
- setting grid voltage configuration and rated power of the DUT
- choosing a test configuration and selecting faults types, voltage dip magnitude and active power reference setting (100%, 25%, 0%)

This procedure accommodates for individual test setup requirements such as I/O channel setup, DUT parametrization as well as an iterative development approach. Depending on the test configuration chosen, an estimation of the test sequence runtime is given as feedback in order to better schedule the test run. The development mode lets the user speed-up the test run by limiting the test types to a specific number of phases or duration times (Figure 4). As an example, only tests with

- 2-phase faults
- voltage dips of <0.05 and 0.20 – 0.25 p.u with respect to the initial voltage
- active power reference of both 100% and 25%

are executed if this is the current focus of the test. The estimated runtime of the subset of selected tests in this case is 7 minutes and 20 seconds. The actual test run produces a log consisting of simulation events including data capturing, fault simulation, and custom code execution. Furthermore, diagnostics of HUT behaviour and test run progress are highlighted during the test. At the end of the test run the user is notified about pass/fail results and has the option to open the detailed report in ISO/IEC OpenDocument format (OpenOffice and MS Word compatible). The detailed report contains high-level test results required by the BDEW standard such as the k factor calculation, rise and settle times for reactive current provision and diagrams of fault behaviour. A summary at the beginning of each test case states the cause of a pass/fail result and links to sections illustrating results. This allows fast and targeted error investigation.

The test sequence setup is stored in a project and can be rerun at a later time using the saved test configuration. A project can consist of more than one report type what means that reports according to UL1741 or EN 50530 can be added to the automated test sequence. The toolbox provides a Python API which can be used to design variants of test procedures for local testing and specific feature development testing.

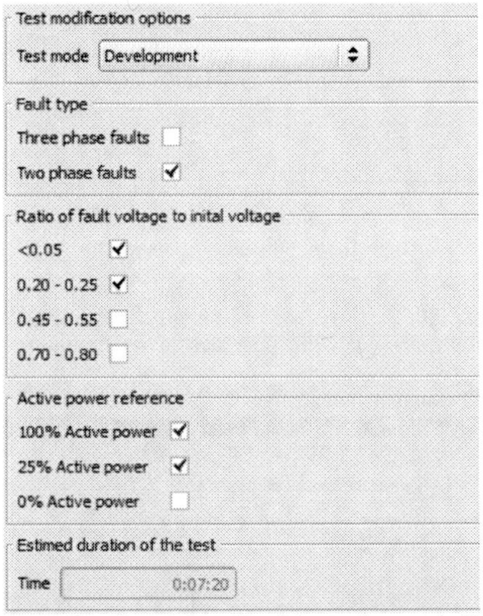

Figure 4: Development mode of the toolbox prebuilt component library with exemplary test subsetting.

In addition, fully automated test protocols are generated for each single test or the entity of the selected test sequence. Resulting reports are programmed in such a way that their appearance is identical to the required reporting format according to the selected standard.

4 EXPERIMENTS AND RESULTS

In this section, results of a test case for a CHIL setup are introduced. Experimental results of a control board connected to a CHIL environment, as mentioned above, are presented and particular focus is set to the application of the tool for LVRT testing. In the following section, the subsequent analysis of the results by means of automated generation of test reports including pass/fail criteria as well as other important notes is presented. Hereby, the automated report generation is implemented according to the rules and requirements stated in the relevant standard.

4.1 Experimental Results of the Low voltage ride through (LVRT) Test Scenario

During a grid voltage dip, the essential requirement from an inverter is not to trip, but to remain connected and support the voltage with reactive power (to "ride through" the event). After the fault has cleared the inverter should go back to normal operation and restore the active power level prior to the fault. Passing the LVRT test according to BDEW standard includes clearly specified pass/fail criteria and calculations. The RMS voltages and currents are calculated in the half-period running window. Symmetrical component calculations are used because the test must be able to evaluate the inverter response under unbalanced grid conditions.

As mentioned in Section 3.2, a specific test scenario is comprised in the LVRT test. This test configuration is specified and parametrized within the CHIL simulation setup and exemplary results are shown for one single LVRT scenario in the following. The dynamics of an LVRT event can be observed in Figure 5. Using HIL-embedded oscilloscope the voltages and currents are captured before, during and after the voltage dip.

Figure 6 and Figure 7 highlight the waveforms of 3-phaseline-to-neutral voltages and currents, respectively. They are automatically generated by Test Suite script and are shown in this section as they appear in generated reports regularly.

In Figure 6, the 3-phase line-to-neutral voltages U_L1 (blue), U_L2 (red), and U_L3 (green) are shown with a zoomed-in view using a time window with duration of 200 ms centred to the point of time at fault start. Hereby, the transients of the line voltages can be analysed in-depth and they can be compared to laboratory tests of hardware generation units in order to verify the CHIL methodology.

Figure 7 highlights the 3-phase line currents I_L1 (blue), I_L2 (red), and I_L3 (green) with the same time period settings as configured for Figure 5. Analogously, the transient behaviour of line currents at the point of time for the incipient fault start can be analysed and compared to results with hardware testing scenarios in order to get statements on the accuracy of the test.

The time step for this CHIL setup is set to 1 μs which provides high-fidelity for transient signals of the LVRT scenario. In general, the given results are compared to offline simulation and hardware experiments in order to evaluate the accuracy of the CHIL simulation. A comparison can be performed analysing the THDI or THDI, as well as an identification of the rise times of the apparent current signal.

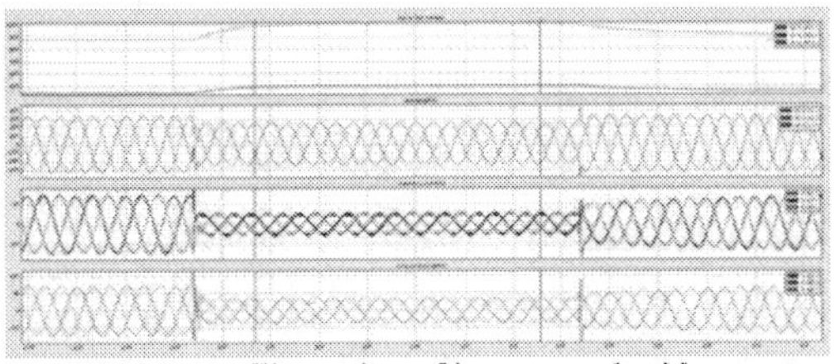

Figure 5: Scope illustration of inverter and grid currents and voltages at fault start and end.

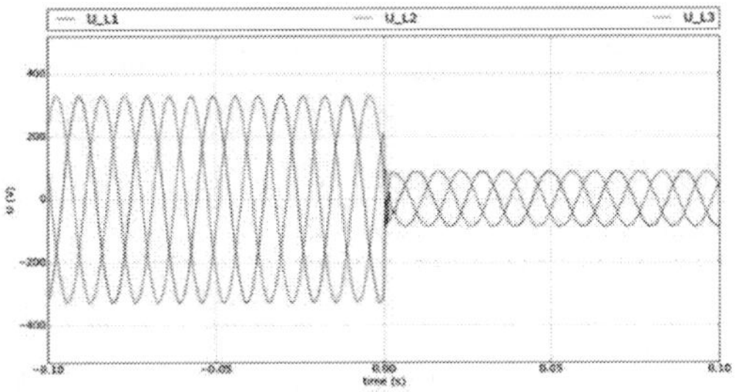

Figure 6: Waveforms of 3-phase line-to-neutral voltages at fault start.

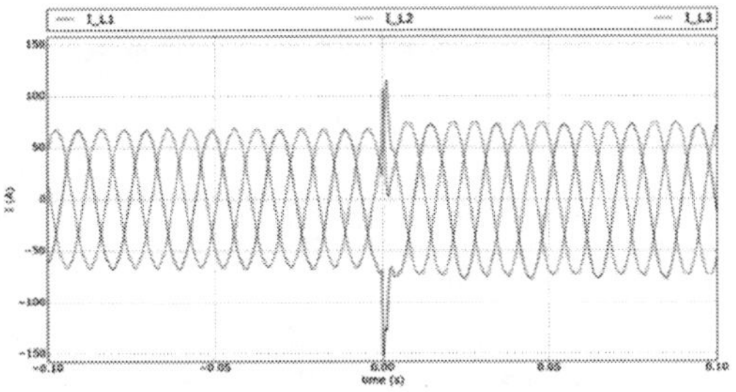

Figure 7: Waveforms of 3-phase line currents fault at start.

4.2 Automated Report Generation of Test Results

Subsequent automated generation of reports allows the user to visualize the given results in detail and one table highlighting important figures related to short circuit currents. This feature is an important support for engineers as test scenario information are processed by this tool automatically and as data and figures are provided which are directly related to discrimination criteria according to the given standard.

One of the LVRT test requirements is that positive sequence inverter currents remain within specified limits during the fault. Figure 8 illustrates resulting normalized value of the positive sequence apparent current Ib_1_per_norm marked with the blue solid line. The upper limit specified according to the applied standard (1.2 p.u.) is marked with by the red solid line and the lower limit (0.9 p.u.) is marked by the green solid line. A time period of 100 ms before and after the fault start is documented in order to be able to evaluate the rise time of the apparent current signal according to the given requirements.

In this particular case the inverter control passed the test run successfully. The summary of the test result is captured by a table, shown in Figure 9, which is directly taken from the automated test report.

In Figure 10 an example of inverter failing to provide specified levels of reactive current during the fault is highlighted. The apparent current does not reach the required value but shows incorrect behaviour.

Figure 11 presents the corresponding table which is automatically generated to provide more details. In this particular case, the rise and settling time are denominated with the expression "nan" (not available), as the reactive current does not even reach the evaluation level. The K-factor is equally denominated with the expression "nan" in case the calculated set point of the reactive current is equal or greater than 1.00 p.u. with respect to the nominal current of the inverter.

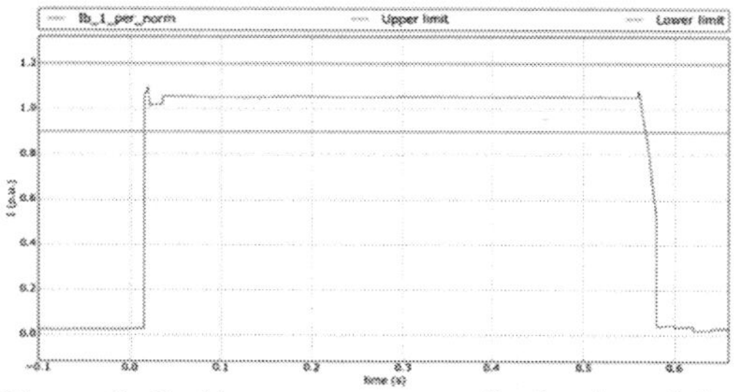

Figure 8: Positive sequence normalized value of the apparent current with upper and lower limitations; pass scenario.

Point in time	Description	Measured currents	Values
First half period after beginning of fault injection	Normalized rms values	I_L1 I_L2 I_L3	1.023 p.u. 1.037 p.u. 1.033 p.u.
First half period after beginning of fault injection	Peak value	i_L1 i_L2 i_L3	73 A 183 A 157 A
150 ms after beginning of fault injection	Positive sequence rms value Negative sequence rms value	NA	0.882 p.u. 0.071 p.u.
End of fault injection - 20 ms	Positive sequence rms value Negative sequence rms value	NA	1.063 p.u. 0.026 p.u.

Figure 9: Automated report generation excerpt of the set point and the measured value of reactive current infeed during fault, rise time, settling time and k-factor, pass scenario.

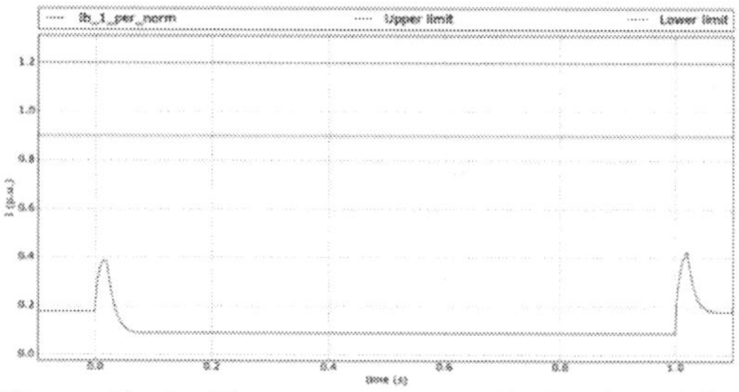

Figure 10: Positive sequence normalized value of the apparent current with upper and lower limitations; fail scenario.

Description	Value
Reactive current infeed actual value before fault (positive sequence)	0.19 p.u.
Reactive current infeed set point during fault (positive sequence)	1.0 p.u.
Reactive current infeed actual value during fault (positive sequence)	0.14 p.u.
Rise time	nan
Settling time	nan
Calculated K - factor	nan

Figure 11: Automated report generation excerpt of the set point and the measured value of reactive current infeed during fault, rise time, settling time and k-factor, fail scenario.

5 CONCLUSIONS

The work of this paper discussed the benefits of CHIL simulation and explains the introduced implementation of a toolbox that can be used by PV inverter manufacturer for precertification testing of their grid-connected inverters. The flexibility and carefully implemented features in the toolbox enables the user to customize the testing procedure before going in the full development mode. Another benefit is the fully automated test report which is of significant value for the user as it is compliant with the reporting requirements of the selected standard.

A comparison of the obtained results from the CHIL toolbox platform and laboratory tests with the physical PV inverters operated at rated power shows that the achievable accuracy within the CHIL platform is very high. In principle, an in depth-model of the power electronic circuit and its peripherals allows the CHIL simulation to produce the identical results as a laboratory test setup. This fact is an encouraging result which can push advanced testing methods more into the focus of rapid prototyping and opens possibilities for future consideration and significance in major standardisation frameworks.

The CHIL toolbox enables the iterative development of grid-connected PV inverters. Design verification testing happens in a closed-loop HIL environment where changes to controller logic and behaviour can be interactively verified for a given setup. Automated report generation according to relevant standards completes the end-to-end testing from controller code through HIL simulation up to compliant reports. The achieved degree of automation enables reproducible, scalable and transparent development for fast development cycles and time savings. Conducting pre-certification simulated testing upfront can cut costs on laboratory and staff later in the product development process.

6 References

[1] Ministry of Justice, "Regulation for ancilliary services by wind turbines (Regulation for ancilliary services - SDLWindV)" Berlin, Germany, 2009.

[2] BDEW Bundesverband der Energie- und Wasserwirtschaft e. V., "Technical Directive for Medium-Voltage Generation Units," Berlin, Germany, 2008.

[3] FGW e.V. Fördergesellschaft Windenergie, "Technical Guidelines for Power Generating Units and Farms Part 3 - Determination of electrical characteristics of power generating units and systems connected to MV, HV and EHV grids," FGW e.V. Fördergesellschaft Windenergie, Berlin, Guideline 23, May 2013.

[4] UL1741: "Inverters, Converters, Controllers and Interconnection System Equipment for Use With Distributed Energy Resources", January 2010.

[5] IEC 62116 Ed 1.0: Test procedure of islanding prevention measures for utility-interconnected photovoltaic inverters, September 2008.

[6] IEC 62116 Ed.2 Test procedure of islanding prevention measures for utility-interconnected photovoltaic inverters, March 2013.

[7] "TOR D4 – Parallelbetrieb von Erzeugungsanlagen mit Verteilernetzen", e-control GmbH, September 2013.

[8] CEI 0-21, "Reference technical rules for the connection of active and passive users to the LV electrical Utilities", November 2011.

[9] G. Lauss, M. O. Faruque, K. Schoder et al., "Characteristics and Design of Power Hardware-in-the-Loop Simulations for Electrical Power Systems," IEEE Transactions on Industrial Electronics, vol. 63, no. 1, pp. 406–417, Jan. 2016.

DLMS smart meter reading application for PV-micro-grids

M. AIT BENALI[1], A. OUTZOURHIT[1],
[1]Laboratory of Nanomaterials for Energy and the Environment (LN2E) - Cadi Ayyad University, Faculty of Sciences-semlalia Marrakech Morocco
simo.aitbenali@gmail.com
aoutzour@uca.ma

ABSTRACT: Smart-grid and smart micro-grids technologies can significantly enhance the integration of renewable energies and alleviate the effects of their intermittence. This mostly due to integration of modern communication technologies allowing an intelligent and flexible control of power flow in the grid. One of the pillars of smart grids/micro-grids is the smart meter. The standard communication between smart meters and the monitoring system is the Device Language Message Specification (DLMS) protocol. In this work, we have developed an application based on Gurux DLMS library to communicate with a smart meter and read the various measured parameters as function of time: active power, reactive power, power factor, apparent power, current and voltage.
This application was developed in order to monitor the energy consumption in a micro-grid installed in the TAHALA village in TAFRAOUT region of Morocco. The Graphical user interface of the application as well as the evolution of the active power obtained from a virtual meter are shown below. The application will be subsequently tested on DLMS smart meter.

Keywords: Micro-grid, PV plants, smart meters, metering platform, monitoring, DLMS protocol.

1 TAHALAGRID

The project TAHALAGRID falls within the context of the elaboration of the Sustainable Development goals and minimizing the effects of climate change. It is the first smart micro-grid network in Morocco.
The components of our smart micro-grid are described below with their characteristics:

1.1 Photovoltaic fields:

Two 8.25 kWp plants and one 1.5 kWp plants were installed in the TAHALA vllage to supply various public services of this site. They are connected to the local grid in the self-consumption mode without feeding excess energy into the grid.
The characteristics of PV panels (Elite Solar) constituting the photovoltaic field are:

Table I: The characteristics of PV panels

	Values
Nominal Power	250 Wc
Short –circuit curent, Isc	8.71A
Open circuit voltage; Voc	36.3V
Curent at maximum power	8.17A
Voltage at Maximum Power	30.6V

Figure 1 shows the two 8.25 kWp PV fields insataled on the roof of the women's club and the nursery of the village.

Figure 1: Two 8.25 kWp PV fields

The 1.5 kWp plants, shown in figure 2, is located in the old school,

Figure 2: PV field on the old school

1.2 Inverters:

The two 8.25 kWp plants were connected to two 12 kW Fronius inverters (Fig 3), while the 1.5kWp PV filed is connected to a 2 kW Aurora inverter (Fig 4).

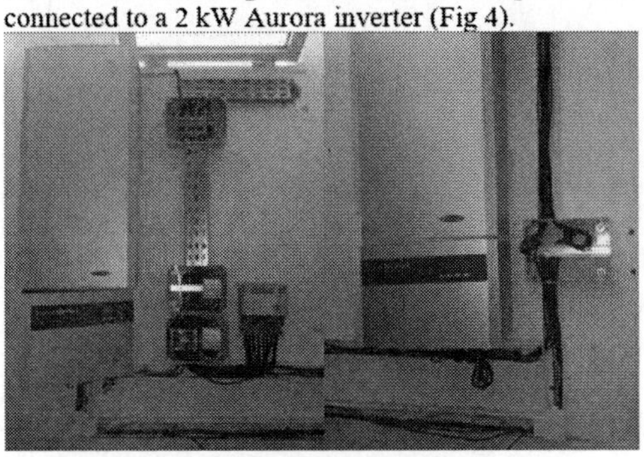

Figure 3: Fronius IG Plus inverters

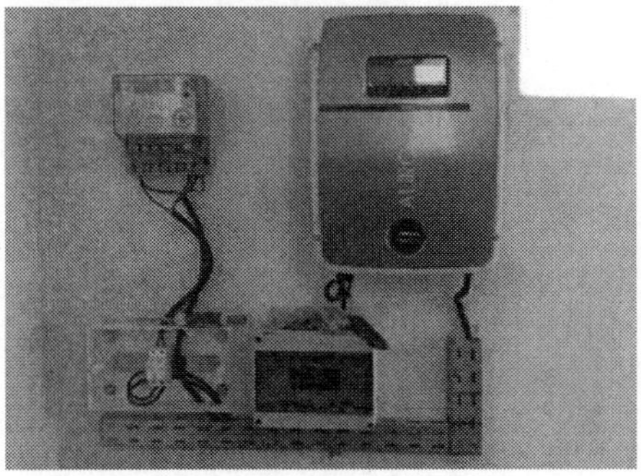

Figure 4: Aurora inverter.

The characteristics of the inverters are given by the tables II and III :

Table II : Characteristics of the Fronuis inverters

	Values
MPP range	230 – 500V
Maximum input voltage	600 V
Maximum input curent	55.6 A
Maximum input short circuit curent	83.4 A
Nominal power	12 kW
Nominal output voltage	400/230 V
Nominal output curent	17.4 A (AC)

Table III : Characteristics of the Aurora inverter

	Values
MPP range	200 – 470V
Maximum input voltage	520 V
Maximum input curent	12.5 A
Maximum input short circuit curent	15 A
Nominal power	2000 W
Nominal output voltage	400/230 V
Nominal output curent	10.5 A

1.3 Smart Meters:

The monitoring of the consumption of the various public facilities (sites) is performed by the EDMI smart meters (Fig 5).

Figure 5: EDMI Smart Meter.

A total of 4 smart meters were installed in the village. These smart meters can communicate with automatic meter reading systems using the standard communication the Device Language Message Specification (DLMS) protocol

2 DLMS Protocol :

The DLMS/COSEM protocol specifies an interface model and communication protocols for data exchange with metering equipment, each information in the smart meter is modeled by an object.

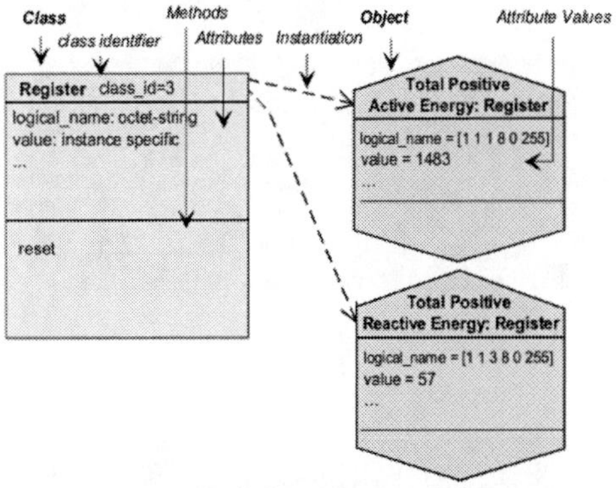

Figure 6: Smart meter informations modeling.

An object is a collection of attributes and methods. Attributes represent the characteristics of an object and methods allows examining or modifying the values of the attributes.
Objects can be referenced in two different ways:

- Using logical names (LN referencing): the object is referenced by its logical_name.

- Using short names (SN referencing): This kind of referencing is intended for use in simple devices. In this case, each object is identified with a 13-bit integer.

3 Programming of DLMS smart meter reading application :

In this work, we have developed an application based on the Gurux DLMS library [1] to communicate with a smart meter and read the various measured parameters as a function of time: active power, reactive power, power factor, apparent power, current and voltage. The Gurux DLMS library is written in Java and three types of communication protocols are available to the meters: Serial, TCP/IP and GSM. The steps to establish communication with a smart meter using the GuruxDLMS library are as follows:

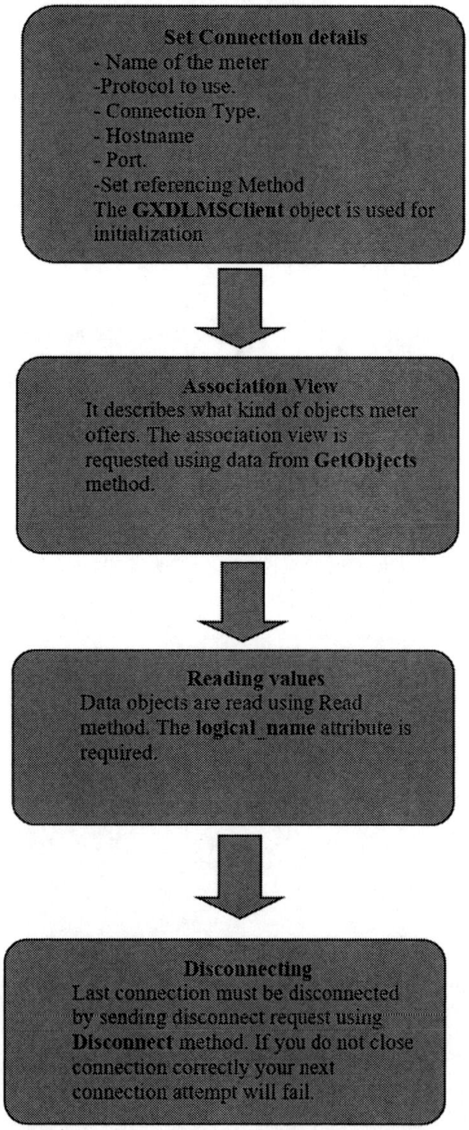

4 Preliminary Results

The Graphical user interface of the application as well as the evolution of the active power obtained from a virtual meter are shown below (from Fig 7 to 10) . The application will be subsequently tested on DLMS smart meter.

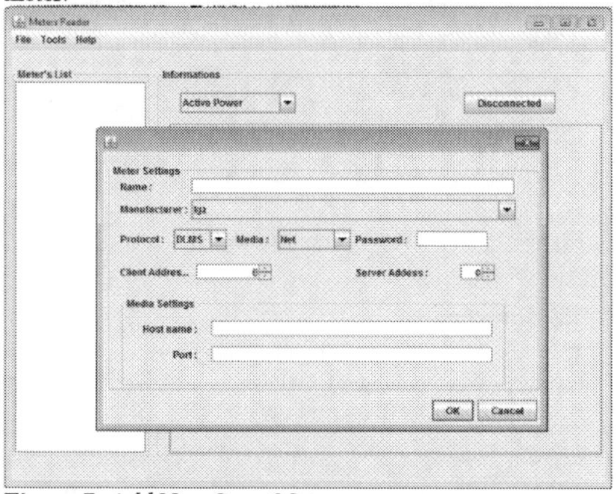

Figure 7: Add New SmartMeter.

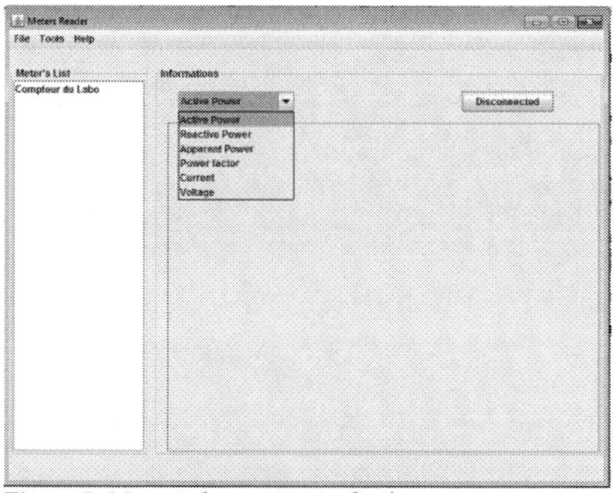

Figure 8: Measured parameters selection

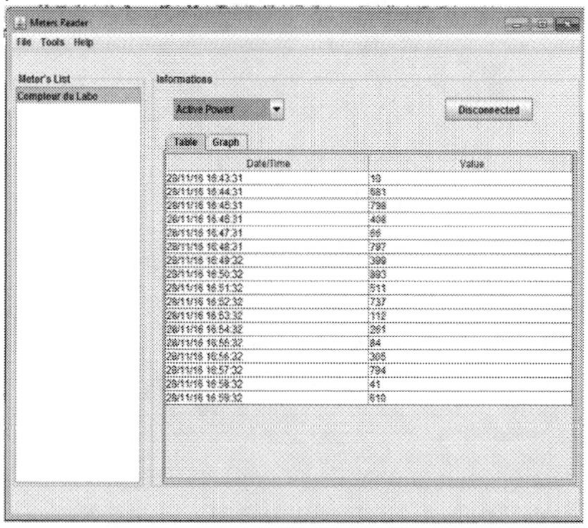

Figure 9: Table of active power data.

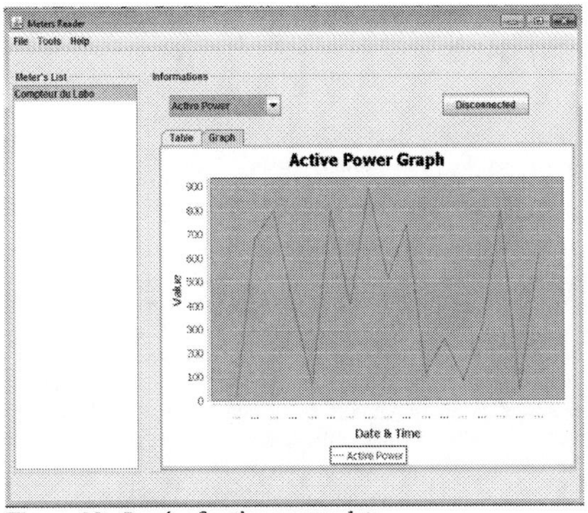

Figure 10: Graph of active power data.

5 CONCLUSION

An application was developed in order to monitor the energy consumption in a micro-grid installed in the TAHALA village Morocco. The micro-grid is feed by a 18 kWp photovoltaic plant in the self-consumption mode. The PV plant supplies electricity to the public services in the village and the smart meters are used to monitor the consumption of each service. The Graphical user interface of the application as well as the evolution of the active power obtained from a virtual meter are shown above. The application will be subsequently tested on DLMS smart meter.

6 REFERENCES

[1] Gurux link : http://www.gurux.fi/Gurux.DLMS

ACKNOWLEDGMENT

This project is funded by IRESEN in the framework of the InnoPV/Innowind2014 program

INTRODUCTION OF AN ADVANCED METHOD FOR TESTING OF BATTERY CHARGE CONTROLLERS FOR OFF-GRID PV SYSTEMS

Arash Khatibi [1], Jörg Bendfeld [1], Wolfgang Bermpohl [1], Stefan Krauter [2]
[1] Sustainable Energy Concepts, Paderborn University, Paderborn, Germany
[2] Kompetenzzentrum für nachhaltige Energietechnik (KET), Paderborn University, Paderborn, Germany
Email: arash.khatibi@uni-paderborn.de

ABSTRACT: With the fast-growing energy demands of today's world, the Sun as a massive clean energy source can compensate the predictable energy shortage in near future. Since solar irradiance is only available in some hours of the day, efficiency of a solar energy system is of high importance. In this matter, the advantage of considerable amount of energy that could be achieved if the sun could radiate all day long or there was no cloud shadowing, cannot be ignored. Since this situation is unreal, in solar off-grid systems a storage system and most commonly a battery system is considered. Besides, installing a Maximum Power Point Tracking (MPPT) allows the system to produce higher power in the low irradiation hours of the day or wintertime in comparison to systems without MPPT. This work focuses on the most usual approach for backup storage of an off-grid solar system, which consists of a battery and a battery charge controller. In this study, a test process will be introduced to be able to analyze these devices' performance and efficiency. Although this has been done before for PV inverters combined with Maximum Power Point Tracking (MPPT), it requires a different approach to implement efficiency analysis on charge controllers. Furthermore, by testing and comparing of various MPPT devices with different tracking algorithms advantages and disadvantages of tracking algorithms will also be shown.
Keywords: Solar Off-grid System, Battery Storage and Control, MPPT, Testing, Efficiency

1. INTRODUCTION

Permanent and non-stop rapid raise in energy demands of the world explains the importance of renewable energy systems (such as solar) and their advantages. Moreover, highly produced amount of CO_2 (Carbon Dioxide) disrupts the natural heat exchange between the Earth and space. This has led to global warming which makes huge problems in today's world such as extreme climate changes, rise of the open waters level and significant changes in the ecosystems. Considering solar energy potential, its availability, cleanness and limitless existence this energy source is a good choice for providing energy to locations far from a stable grid. Efficiency improvement in this area is an important matter of interest, which leads to comparable or even lower solar energy prices to conventional fossil fuels. As it is obvious, sun is only available for some hours of the day. Furthermore, in many places local climate conditions such as cloudy, rainy and snowy days do not let the sun irradiance reach the earth's surface. Considering an electrical storage system provides the possibility of storing excess produced solar electricity to use it later at nights or in case of low irradiance.

Meanwhile, a charge controller is responsible for making decisions of the system to operate in battery charging or discharging states. In other words, based on the amount of available produced energy, it decides if the battery can be charged simultaneously with load feeding or it should feed the load due to lack of enough irradiated energy.

In this study, an advanced method for testing of battery charge controllers with and without MPPT is designed and will be introduced. Since the efficiency and performance of MPPT as well as battery charge controller is tested, an irradiance profile has to be designed to be implemented to the system. This profile should cover all possible radiation conditions, no matter where on the globe this system is working. For this matter an irradiance simulator based on desired irradiance amounts has been used, therefore the test profile is designed to indicate the irradiance amount for each moment. For testing, this profile is used as input to the simulator that feeds the PV panel. Later on in [1], four different battery charge controllers as three of them with MPPT and one without MPPT will be tested and the results will be analyzed and visualized in diagrams to compare devices' performance and their efficiency in all tested conditions.

2. MAXIMUM POWER POINT TRACKING

Maximum amount of power produced in a photovoltaic panel depends on irradiance, panel temperature and connected load [2]. Irradiance variations lead to varying panel operating current, however temperature variations affect panel operating voltage [2].

By Maximum Power Point Tracking the charge controller detects the PV panel output voltage and then compares it to the voltage of the battery [3]. Afterwards it determines the maximum suitable power to be taken from the panel to charge the battery and converts panel working voltage to reach maximum charging current [3]. Efficiency of 93 to 97% can be expected in most modern MPP trackers [3]. This function can lead to a 20-45% power gain in winter time and 10-15% in summertime in comparison with systems with no MPPT function [3]. It is ideal to have an individual MPPT for each panel or small cluster of panels to minimize the risk of partial shading [2]. By this each panel can function at its peak efficiency [2].

Standard Test Conditions (STC) are uniform test conditions for comparing different PV modules which are defined as solar irradiance of 1000 W/m^2, Air Mass of 1.5 and module temperature of 25 °C [4]. Fig. 1 shows I-V characteristics of the PV panel used in this study without an MPP tracking device, which is backed up once with a 12V and once with a 24V battery at STC. Obviously maximum working power (without MPPT) is equal to the area under the diagram of I-V characteristics. As it can be seen the colored part shows the amount of power that can be taken when having a 24 V battery while the hashed part is the power when a 12V battery is used. As shown, these two powers are not as high as maximum power point (see Fig. 2) that can be achieved from the PV panel. In other words, the battery voltage limits the maximum possible

power to be taken from the PV module. In Fig. 2, I-V characteristics of the same module, but this time with MPPT, is indicated. The battery voltage could be 12V or 24V since MPPT is capable of changing the panel voltage in the best way to maximize the power taken.
This proves that MPP tracking is essential in systems with different battery and module voltages. On the other hand, if MPPT is not available in the system, more power could be gained by choosing battery voltages equal or close to PV panel voltage.

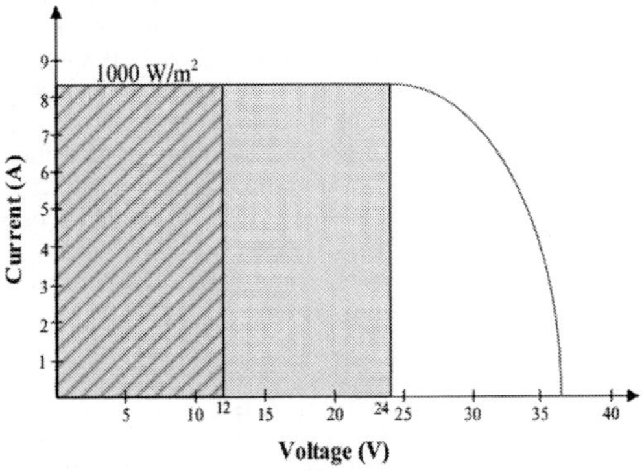

Fig. 1: Power limits for 12 and 24 volts battery in systems without MPPT at STC

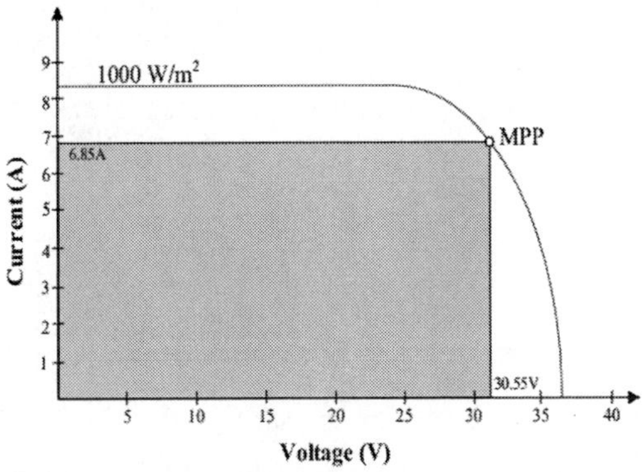

Fig. 2 - Maximum power in the same system with MPPT

3. STORAGE AND CHARGE CONTROLLER

Since the nature of solar energy does not allow the system to produce non-stop and continuous sufficient power and in addition, there are situations that electricity demand occurs at different times other than the production, the energy needs to be stored [4]. Consequently a storage including a battery and a suitable charge and discharge controller needs to be considered for the system [4]. Considering different investigations on characteristics of lead-acid batteries, the gel electrolyte lead acid battery was chosen to be used in the test process [5]. The chosen battery for this study is a 12V with the capacity of 65Ah. This gel battery is suitable as a buffer for irregular charging and discharging processes such as solar-powered applications [6].

The charge controller uses maximum available current from the source to charge the battery until it reaches its final charge voltage [7]. Depending on the battery status and behavior different charging stages (i.e. normal charging, boost charging and compensation charging) are used [7]. Since this behavior is different for each type of battery, the controller must be able to consider various cases and battery types [7]. Another necessity is to be able to prevent the battery from deep discharging [7]. Normally this is done by determining the State of Charge (SOC) of the battery or measuring its voltage. SOC and battery voltage can indicate the remaining energy of the battery. By knowing the exact remaining capacity, the load output will be switched off to prevent deeper discharge of the battery in case of battery discharge below a predefined SOC (with SOC controlling chargers) or battery voltage (with voltage controlling chargers) [7]. This highlights the importance of precise determination of SOC for optimized charge control. Normally SOC and battery voltage thresholds for deep discharging are defined by the manufacturer and cannot be adjusted [7].

4. TEST SETUP

Since in this test the efficiency of charge controllers with and without MPPT is the target, a test setup according to Fig. 3 is used. A computer enters desired irradiance amounts as power inputs to PV module every 0.5 seconds. As it can be seen the power, voltage and current measurements are carried out in points 1 to 4 for battery, load, PV module and simulator respectively. Later on in this script battery voltage, load power, PV power and simulator power are referred to measurements in these points.

Normally in solar off-grid systems an inverter is needed to convert DC from solar panels and also the storage system to the AC needed for the load [4]. However, since in this study only performance and efficiency of battery charge controllers with MPPT is desired, a DC load is used to omit efficiency and performance effects of the inverter.

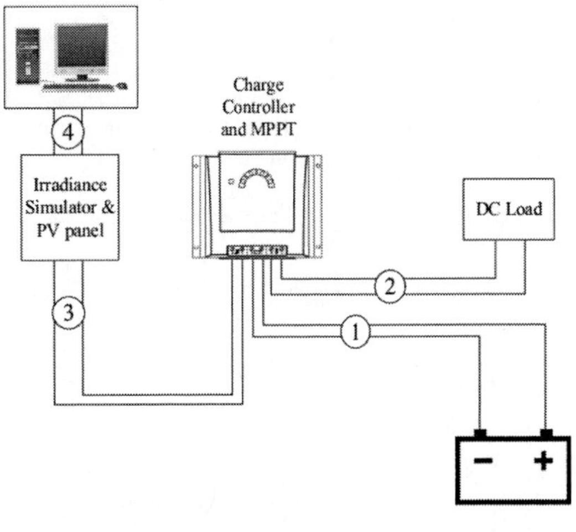

Fig. 3: Test setup with power, voltage and current measurement points for battery, Load, PV panel and simulator respectively (points 1 to 4)

5. TEST CONDITIONS

In the European standard for performance

characterization of PV inverters, EN 50530, the approach and methodology for the assessment of conversion and Maximum Power Point tracking efficiency is introduced [8]. This standard mainly addresses the tests for PV inverters capable of tracking MPP. The tests are done for steady state and dynamic states of the devices [8]. The test process in this standard has been modified to come up with a new test profile to be implemented on charge controllers with MPPT. Different test steps are brought later on to get implemented on devices.

EN50530 defines requirements for the PV simulator used in the test setup as "current source emulating the static and dynamic behavior of a PV array, in particular the current voltage characteristic" [8]. The standard clearly specifies the I-V curve models to be used by the PV simulator. The PV module used in this study's test is from Q.CELLS (QC-C02) with characteristics as in Table I in STC, which is compatible with the simulator characteristics defined by the EN50530.

Table I: Tested PV module performance at STC [9]

Nominal Power(+/- 2.5W)	P_{max}	W	210
Short Circuit Current	I_{SC}	A	8.00
Open Circuit Voltage	V_{OC}	V	35.60
Current at Maximum Power	I_{mpp}	A	7.40
Voltage at Maximum Power	V_{mpp}	V	28.40
Temperature Coefficient of I_{SC}	α	%/K	+0.06
Temperature Coefficient of V_{OC}	β	%/K	-0.36
Temperature Coefficient of P_{max}	ϒ	%/K	-0.43

5.1 Test conditions for steady state tests

In this study steady state conditions are modified in a way that stepwise irradiance is fed to the system which represents sudden increase in irradiance starting from zero to a percentage of STC irradiance [8], and then keeping the amount so that the tracker reaches its steady state. Table II shows the steps and duration of each for the steady state testing process. These steps are used to obtain and compare the response speed of the trackers under test after some fast changes in irradiance.

Table II: Test steps for steady state conditions

Step no.	Input irradiance [W/m^2] [8]	Duration [s]
1	50	30
2	100	30
3	200	30
4	300	30
5	500	30
6	750	30
7	1000	30
8	750	30
9	500	30
10	300	30
11	200	30
12	100	30
13	50	30

5.2 Test conditions for dynamic state tests

According to standard EN50530 for the dynamic tests on chosen devices, several ramp sequences of irradiance levels with different gradients should be considered and fed into the module [8]. The gradients of the ramps will rise from 0.5 W/m^2 per second to 100 W/m^2 per second for two different irradiance levels, levels of low to medium (100 to 500 W/m^2) and medium to high (300 to 1000 W/m^2) [8] (see Fig. 4). In the beginning of the test pattern and after each increasing or decreasing ramp a settling time is considered for the MPP tracker to stabilize, which are also defined in the aforementioned standard [8]. In addition, a dynamic test sequence for very low irradiance levels is considered which targets the performance of the tracker during start-up and shut-down periods [8]. These ramp tests will show the behavior of the trackers in different conditions that could happen in reality, indicating which tracking algorithm yields to a higher efficiency. Table III, Table IV and Table V consist of considered ramp tests for dynamic states and their conditions in order of applying to the setup system.

Table III: Ramp tests for low to medium irradiance levels [10]

Irradiance Variation 10% to 50% of G_{STC} (low to medium irradiance)		
Repetition	*Slope [W/m^2 per second]*	*Mode durations (in seconds) Rise/Settle/Fall/Settle*
1	0.5	800/10/800/10
1	1	400/10/400/10
1	2	200/10/200/10
1	3	133/10/133/10
1	5	80/10/80/10
1	7	57/10/57/10
1	10	40/10/40/10
1	14	29/10/29/10
1	20	20/10/20/10
1	30	13/10/13/10
1	50	8/10/8/10

Table IV: Ramp tests for medium to high and very low irradiance levels [10]

Irradiance Variation 30% to 100% of G_{STC} (medium to high irradiance)		
Repetition	*Slope [W/m^2 per second]*	*Mode durations (in seconds) Rise/Settle/Fall/Settle*
1	10	70/10/70/10
1	14	50/10/50/10
1	20	35/10/35/10
1	30	23/10/23/10
1	50	14/10/14/10
1	100	7/10/7/10

Table V: Ramp tests for very low irradiance levels [8]

Irradiance Variation 0.2% to 10% of G_{STC} (very low irradiance)		
Repetition	*Slope [W/m^2 per second]*	*Mode durations (in seconds) Rise/Settle/Fall/Settle*
1	2	49/10/49/10

In [10] the number each slope is getting implemented, differs from the tests considered in this work. This was due to test time limitations in this experiment such that the tests were done once for each ramp.

6. TEST PROFILE

The designed profile according to Table II, Table III, Table IV and Table V can be seen in Fig. 4. Firstly, the tests for irradiances starting from 100 to 500 W/m^2 and with slopes from 0.5 to 50 W/m^2 per second are implemented. Then the tests for irradiances between 300 and 1000 W/m^2 with gradients from 10 to 100 W/m^2 per second are executed. Afterwards the test for very low levels of irradiance between 2 to 100 W/m^2 with change rate of 2 W/m^2 per second is performed. Finally, steady state tests occur which look like stairs in the figure (from test time around 4450 seconds to the end).

When the setup is connected and ready according to Fig. 3, the irradiance profile is started through the simulator. The test profile as a whole takes about 82 minutes to run for each battery charger. Meantime a high quality power meter is measuring current, voltage and power of the components in 4 channels and saving them at a resolution of 2 Hz. Channels are connected to battery, load, PV module and irradiance simulator respectively.

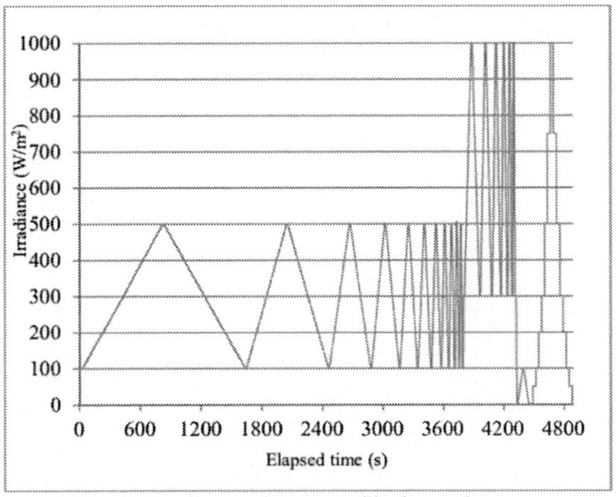

Fig. 4: Irradiance change profile for testing

7. EFFICIENCY CALCULATION

Dividing output power to input power of the charge controllers simply gives the efficiency in each measured moment. An important point to consider is that through the testing process, battery working state will vary regularly between charging and discharging phases. This makes efficiency calculation challenging, since in the charging phase the battery power is on the output side of the controller. On the other hand, in the discharge state, the battery is on the input side, since it is feeding the load through the charge controller. Following equations conclude efficiency calculations in both cases:

Battery in charging state
$$\eta = \frac{P_{Load} + P_{Battery}}{P_{PV}} \quad (1)$$

Battery in discharging state
$$\eta = \frac{P_{Load}}{P_{PV} + P_{Battery}} \quad (2)$$

After the tests, for each device efficiency is calculated every half a second (0.5 s) and according to aforementioned equations (1) and (2). As mentioned before, four different charge controllers with and without MPPT function are tested with this method, which led to many interesting results. Moreover, their performances have been analyzed in detail and compared which can be seen in the follow-up paper of this study in [1]. Distribution of parameters such as simulator power, PV power, load power and efficiency for each device are brought simultaneously to simplify understanding and following of devices' performance. Moreover, efficiency diagrams in some specific ramps of the irradiance profile (Fig. 4) are illustrated in details in [1].

8. CONCLUSION

This study focused on the backup storage of solar off-grid systems by introducing a new and advanced method for testing of battery charge controllers to control battery function, load feeding and maximum power point tracking. For this goal these devices had to be tested under different situations that could happen in real life, regardless of the installation location. A test circuit has been used to examine these devices under different circumstances. Since only the performance of battery charge controllers is desired, an inverter was omitted and a DC load was used. For testing, an irradiance test profile was needed to be implemented on the considered PV module. This profile is designed and modified based on standard EN50530 and is divided into two different states of steady state and dynamic tests. Dynamic tests are divided into low to medium, medium to high and very low irradiance level tests as well. In next step an irradiance simulator was used to enter the designed profile to the module. After measuring power, current and voltage in every 0.5 seconds of test time, these data needed to be analyzed and compared for different devices which is done in [1].

As it is seen in section "2", while designing a PV system if the components' voltages are not equal, a significant amount of power could be lost (more than half of the power in test with a device without MPPT) [1]. In other words, the battery voltage limit is forced on the PV module. This causes the system to work in the range of the battery voltage resulting in lower power production far away from the MPP. For compensating this power loss, a smart MPP tracker is the solution. MPPT is capable of changing the PV panel working voltage to get as close as possible to its V_{mpp}. By considering an MPPT in the system, the balance between battery voltage and PV panel voltage will be achieved, which results in higher energy production for feeding the loads and charging the storage (see I/V curves in system without MPPT in Fig. 1 in comparison to system with MPPT in Fig. 2).

Finally, it is worthwhile to point out an explanation about the used load for this testing method. As it will be seen in test results in [1], the load profile is different for each device under-test. This is because of manually changing the load (according to the produced power) to protect the battery from deep discharging and overcharging. This has to be done in case of the lack of a controllable load for test. To continue this study it is suggested to design a load profile based on the designed irradiance profile in this work (Fig. 4). For this matter battery state of charge, amount of produced power and accordingly required amount of load need to be calculated for every 0.5 seconds. Afterwards the experiment has to be repeated by implementing not only the irradiance profile but also the load profile. This provides the possibility to have the exact test conditions for all devices under test, therefore the comparison of the performance and quality of charge controllers with Maximum Power Point Tracking and their algorithms from various manufacturers could be fulfilled.

9. REFERENCES

[1] A. Khatibi, J. Bendfeld, W. Bermpohl und S. Krauter, „Testing and Analysis of Battery Charge Controllers for Off-grid PV Systems," in *EU PVSEC*, Amsterdam, 2017.

[2] M. Rosu-Hamzescu und S. Oprea, *Practical Guide to Implementing Solar Panel MPPT Algorithms*, Microchip Technology Inc., 2013.

[3] „ALL ABOUT MAXIMUM POWER POINT TRACKING (MPPT) SOLAR CHARGE CONTROLLERS," Northern Wind & Arizona Sun, [Online]. Available: http://www.solar-electric.com/mppt-solar-charge-controllers.html. [Zugriff am 27 10 2015].

[4] S. C. W. Krauter, Solar Electric Power Generation - Photovoltaic Energy Systems, 1 Hrsg., Springer-Verlag Berlin Heidelberg, 2006.

[5] A. Khatibi, „Analysis and Testing of Battery Charge Controllers with Maximum Power Point Tracking: Performance and Efficiency," Master's thesis, Dept. EIM-NEK, Paderborn Univ., Paderborn, Germany, 2015.

[6] Keckeisen Akkumulatoren, [Online]. Available: http://www.intact-batterien.de/en/gel-power.html. [Zugriff am 28 10 2015].

[7] Steca Elektronik, *Solar charge controller PR 2020 Operating instructions,* Steca Elektronik.

[8] R. Bründlinger, N. Henze, H. Häberlin, B. Burger, A. Bergmann und F. Baumgartner, „prEN 50530 - The new european standard for performance characterisation of PV inverters," in *European Photovoltaic Solar Energy Conference and Exhibition (EU PVSEC) <24, 2009>*, Hamburg, 2009.

[9] *Q-Cells solar module QC-C02,* Bitterfeld-Wolfen: Q.CELLS SYSTEMS.

[10] H. Haeberlin und P. Schaerf, „New Procedure for Measuring Dynamic MPP-Tracking Efficiency at Grid-Connected PV Inverters," in *24th European Photovoltaic Solar Energy Conference, 21-25 September 2009*, Hamburg, Germany, 2009.

[11] K. ViktorRudolf, „Financial analysis of utility scale photovoltaic plants with battery," *Energy Policy,* Bd. 63, Nr. December 2013, pp. 139-146, 2013.

[12] „Battery Maintenance," Trojan Battery Company, [Online]. Available: http://www.trojanbattery.com/tech-support/battery-maintenance/. [Zugriff am 21 11 2015].

[13] „Battery and Energy Technologies," Electropaedia, [Online]. Available: http://www.mpoweruk.com/life.htm. [Zugriff am 27 10 2015].

[14] „Solar Battery Care, Maintenance and Safety," SolarTown, 07 10 2012. [Online]. Available: http://www.solartown.com/learning/solar-panels/solar-battery-care-maintenance-and-safety-dont-touch-the-terminals/. [Zugriff am 28 10 2015].

[15] R. F. Coelho und D. C. Martins, „An Optimized Maximum Power Point Tracking Method Based on PV Surface Temperature Measurement," in *Sustainable Energy - Recent Studies,* A. Gebremedhin, Hrsg., InTech, 2012, pp. 89-114.

[16] D. P. Hohm und M. E. Ropp, „Comparative study of maximum power point tracking algorithms," *Progress in Photovoltaics: Research and Applications,* Bd. 11, Nr. 1, pp. 47-62, 22 11 2002.

TESTING AND ANALYSIS OF BATTERY CHARGE CONTROLLERS FOR OFF-GRID PV SYSTEMS

Arash Khatibi [1], Jörg Bendfeld [1], Wolfgang Bermpohl [1], Stefan Krauter [2]
[1] Sustainable Energy Concepts, Paderborn University, Paderborn, Germany
[2] Kompetenzzentrum für nachhaltige Energietechnik (KET), Paderborn University, Paderborn, Germany
Email: arash.khatibi@uni-paderborn.de

ABSTRACT: This study focuses on the backup storage of solar off-grid systems by analyzing four battery charge controllers designed to control battery function, load feeding and Maximum Power Point Tracking (MPPT). Although MPPT was so far implemented alongside inverters, recently some companies have introduced smart battery charge controllers combined with MPPT. A new test method is defined and implemented on these charge controllers to verify their functionality, performance and efficiency. This method consists of an irradiance profile implemented by an irradiance simulator to an off-grid solar system with a battery storage. Performance and efficiency of these devices through various irradiance situations are analyzed and compared which led to interesting results.
Keywords: Solar Off-grid System, Battery Charger, MPPT, Efficiency, Performance

1. INTRODUCTION

With the growing energy demands in today's world, rising energy costs is inevitable. In this situation, developing more efficient electricity production systems is highly beneficial. Furthermore, in the near future a highly growing demand for off-grid or grid connected solar systems is predictable. In this matter, having efficient and reliable system components helps a solar system to function efficiently to generate and deliver produced power to the loads. These facts highlight the importance of solar systems efficiency and their respective energy storage systems.

Considering a backup storage makes the system more reliable during times of low generation. Furthermore, a charge controller, capable of controlling charge/discharge process, assures the correct function of the system. These devices are expected to be able to store produced energy in batteries in the times of excess production and provide the needed amount of electricity power in the time of low generation. A premium battery charge controller should be able to protect the battery from deep discharging and excessive charging which is an important factor in maximizing their lifetime.

This paper focuses on efficiency and performance of battery charge controllers alongside MPPT. For this goal, first four devices to be tested are introduced. Next an off-grid solar system is set up to run the test and measure desired parameters to analyze the performance of these devices. In the next step, considered test profile is explained which is designed and modified in [1]. Finally, test results for one of the devices with MPPT are shown and they will be compared with another device without MPPT.

2. UNDER-TEST BATTERY CHARGE CONTROLLERS

For this experiment, four different battery charge controllers with various characteristics are chosen in a way that three of them (devices A, B and C) are combined with MPPT function and device D is without MPPT. Devices are chosen in various PV voltage controlling abilities and can handle different battery types including Lead-Acid (flooded, Gel and AGM) batteries. Battery charge controllers must be able to prevent the battery to discharge deeply which requires the knowledge of SOC. By knowing the SOC, the charge controller stops feeding the load to prevent deep discharge. Normally SOC and battery voltage thresholds for deep discharging or overcharging of connected battery are defined by the charge controller's manufacturer and cannot be adjusted [2]. However, charge controllers such as device B in this work also let the user customize them based on their needs. In Table I all four under-test devices are introduced briefly.

Table I: Introduction of under-test devices [2] [3] [4] [5]

	Device A	Device B	Device C	Device D
Manufacturer country	China	China	Germany	Germany
Connectable battery voltage	12-24 V	12-24 V	12-24 V	12-24 V
Maximum PV voltage	70 V	150 V	100 V	47 V
Battery type	Lead-Acid (flooded, gel, AGM)	Lead-Acid (flooded, gel, AGM, User-defined)	Lead-Acid (flooded, gel, AGM)	Lead-Acid (flooded, gel, AGM)
Type	Indoor	Indoor	Indoor	Indoor
Maximum claimed efficiency	97 %	Up to 98 %	> 98 %	-
Maximum charging current	20 A	40 A	20 A	20 A
MPPT function	Yes	Yes	Yes	No

3. TEST SETUP

An off-grid solar system is set up to run the test and analyze the performance of these devices based on the results. This system (Fig. 1) consists of an irradiance simulator to enter desired irradiance amounts, chosen charge controller, a gel electrolyte lead acid battery, an adjustable load and a high quality power meter. The chosen battery for this study is a 12V with the capacity of

65Ah. This gel battery is suitable as a buffer for irregular charging and discharging processes such as solar-powered applications [6]. Since only the performance and efficiency of battery charge controllers and MPPT is desired, an inverter was omitted in the test setup and consequently an adjustable DC load was used. The PV panel used in this experiment is a Q.CELLS (QC-C02) with nominal power of 210 Watts, short circuit current of 8 A and open circuit voltage of 35.6 V [7]. An irradiance simulator based on desired irradiance amounts has been used; therefore, a test profile is designed to indicate the irradiance amount for each moment. This profile should cover all possible radiation conditions, no matter where on the globe this system is working. The irradiance profile used is shown in Fig. 2 and covers various test conditions in dynamic state and steady state tests. Design and detailed explanation of this profile is brought in [1].

In the process of testing, this profile is used as input to the irradiance simulator. In parallel, a very precise power meter measures power, current and voltage of all system components (battery, load, PV module and irradiance simulator) with a frequency of 2 Hz at points 1, 2, 3 and 4 in Fig. 1.

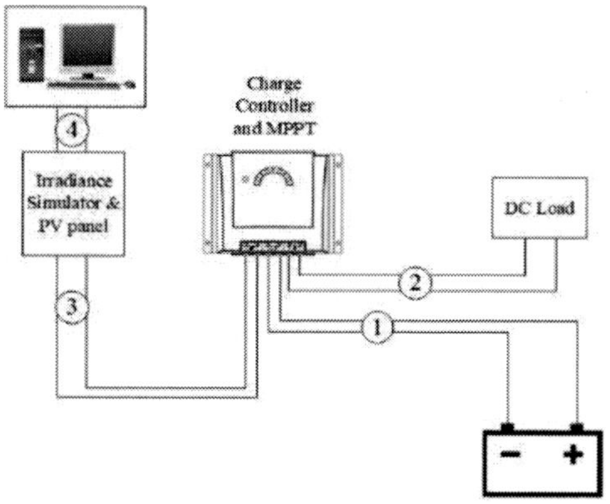

Fig. 1: Test setup with power, voltage and current measurement points for battery, Load, PV panel and simulator respectively (points 1 to 4)

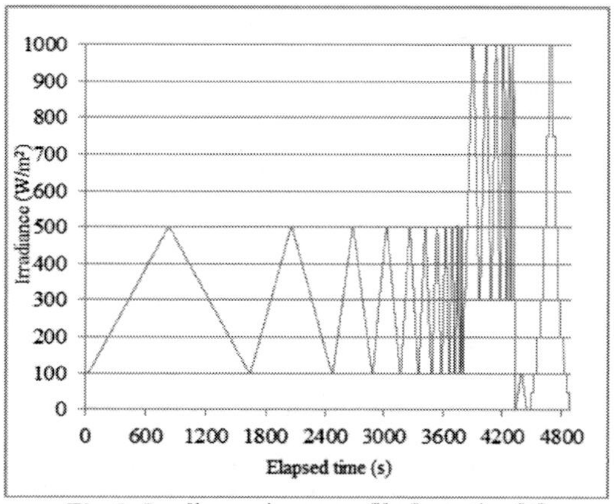

Fig. 2: Irradiance change profile for testing [1]

4. TEST RESULTS

After running the whole profile through the system and measuring the required parameters, efficiency needs to be calculated for every moment of the test and devices' performance will be analyzed.

4.1 Efficiency calculation

Some challenges must be overcome in efficiency calculation such as that the battery varies between charge and discharge mode. This also varies the components from being on the input and output sides of charge controllers which is important for efficiency calculation.

Battery in charging state $$\eta = \frac{P_{Load} + P_{Battery}}{P_{PV}} \quad (1)$$

Battery in discharging state $$\eta = \frac{P_{Load}}{P_{PV} + P_{Battery}} \quad (2)$$

In this paper, results of only device C and its comparison with device D (without MPPT) are presented. However, the other devices' results could be found in [8].

4.2 Device C

From now on, the term "simulator power" is the test profile power amount that is implemented to the system by the simulator. Furthermore, "PV power" is the amount that is controlled and modified by MPPT function of charge controller.

The battery SOC was 100% in the beginning of the test, which is the same for all the other devices. In the process of testing, device C changes the voltage of the PV in order to reach the maximum power point of the module. Depending on PV power, battery state varies between charge and discharge modes. The manufacturer claims that the maximum efficiency for this device is more than 98% [5].

In Fig. 3 four diagrams are displayed right after each other. This is required to follow what is happening in every state or time of the test. The first diagram of Fig. 3 shows the power or irradiance profile implemented by the simulator to device C which is slightly different from Fig. 2. This is due to losses in cables, connections, etc.

In the second diagram of Fig. 3, PV power and battery voltage during the test time are shown. In the beginning, device C starts charging the battery, since the load is very low (see third diagram of Fig. 3). When the load increases, the battery starts to discharge that makes it possible for the controller to feed the load. However, with more produced power, the controller feeds the load by the PV power and not by battery any more. The device C through the test profile does this logic well enough. There are some situations in which the PV power is small and even decreasing in the diagram (beginning of steady state tests), but the battery is still charging. This is due to no or a very small amount of load.

The third diagram of Fig. 3 is the diagram of power amounts that are fed to the load by this device. This ohmic load was changed manually during all the tests to avoid deep-discharging, over-charging and damage to the battery; therefore, it differs from device to device.

Last diagram of Fig. 3 demonstrates the efficiency through the power ramps and changes during the test profile. This diagram proves that efficiency stability is not outstanding since even in low steep tests it varies significantly. In some short data points, the efficiency reaches about 97% but considering less varying efficiency, this charge controller performs around 94% efficiency. However, low efficiencies around 83% are observed as well. It should be mentioned that because of low irradiance and small load needs in some cases, the efficiency is close to zero.

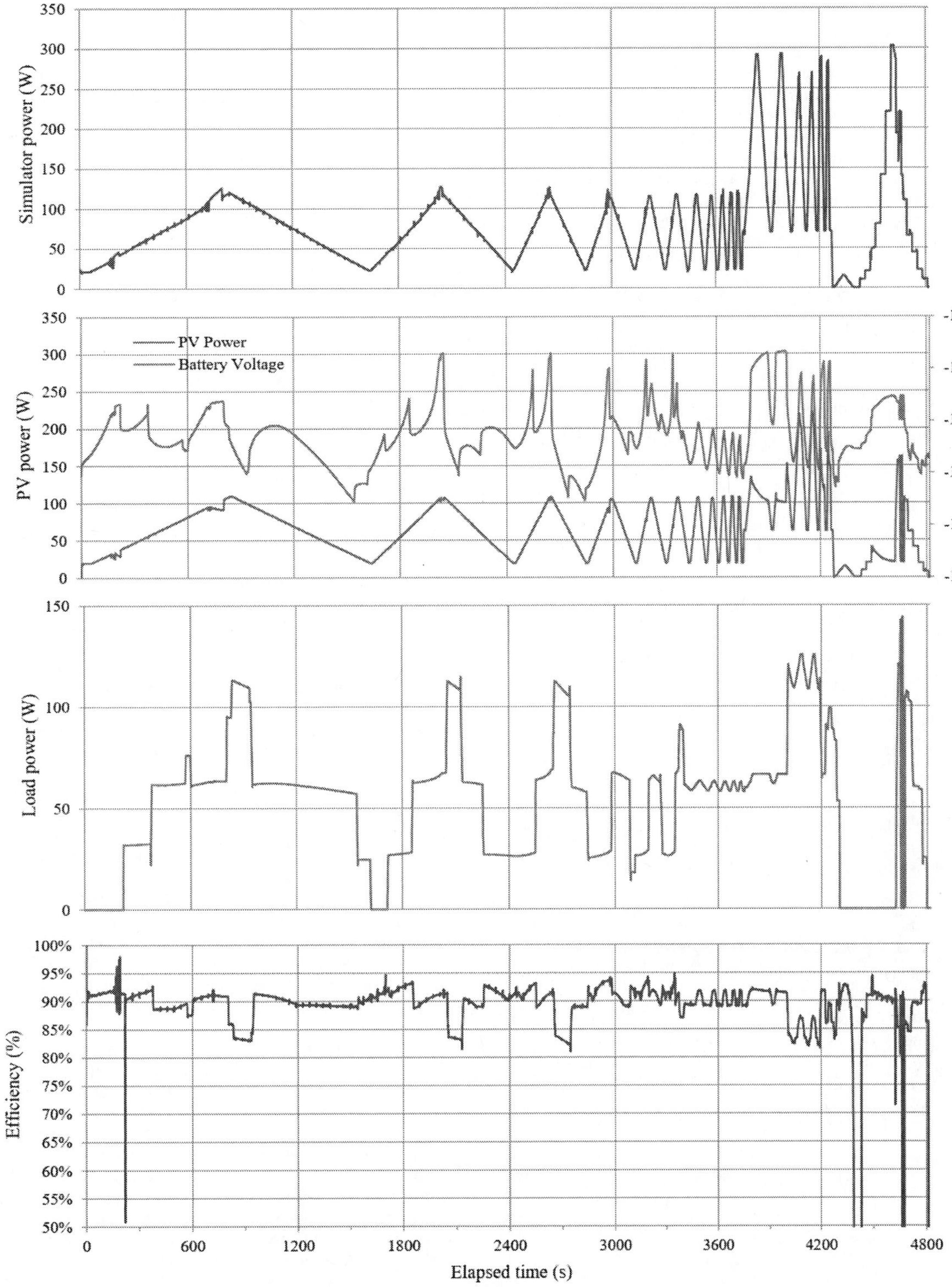

Fig. 3: Simulator power, PV power, battery voltage, load power and efficiency through the test (device C)

Fig. 4 illustrates the changes of PV power with respect to simulator power in the test process. In this figure all data points out of linear curve indicate situations where battery was full and load was low. In these conditions MPPT lowered the amount of delivered power in comparison with simulator power to compensate the extra produced power. Device C changes the module power in a cleaner way and with fewer spikes in comparison with other devices [8]. Furthermore, it takes and delivers relatively more power from the simulator to the battery and load. Like the other controllers with MPPT, device C changes the voltage to reach or get away from the maximum power point depending on battery and load needs.

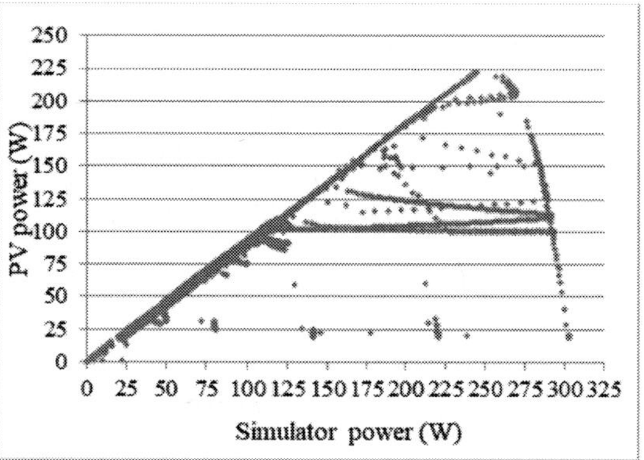

Fig. 4: PV power output to Simulator power (device C)

The efficiency diagram with respect to module power for device C is illustrated in Fig. 5. Distributed efficiency by power amounts in this device rises much faster than device B in this test [8]. This means that device C in general has higher efficiency while working in lower power levels. On the other hand, efficiency for maximum PV power is not higher than 87% which may not be expected from a premium charge controller. Thereby it can be said that maximum stable efficiency is around 93% and minimum efficiency that can be considered stable is almost 81%.

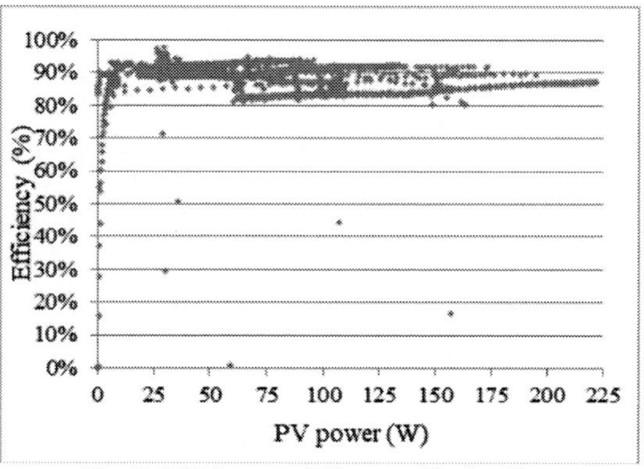

Fig. 5: Efficiency to PV power (device C)

5. COMPARISON TO DEVICE WITHOUT MPPT *(Device D)*

Device D provides higher stable efficiency percentage in comparison with other tested devices. In low to medium irradiance levels it reaches and stays around 97%, however later in medium to high power, very low power and steady state tests, it falls to some extent that is still higher or equal to the other devices analyzed.

Fig. 6 indicates the rate of module power changes with respect to simulator power. As it is expected the diagram is almost linear. This proves that there is no power tracking algorithm, since the input irradiance power variations are unaffected in PV output.

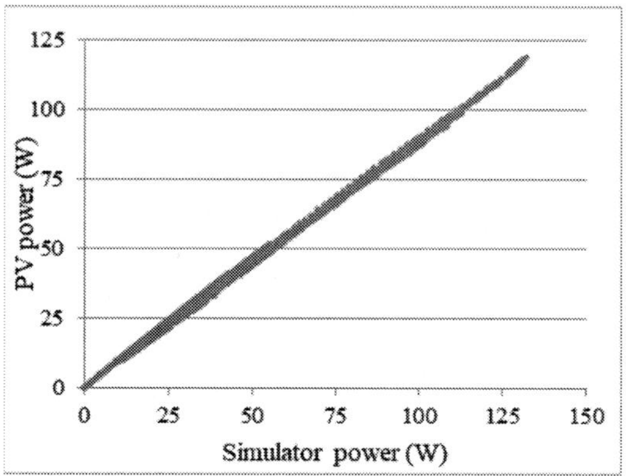

Fig. 6: PV power output to Simulator power (device D)

In Fig. 7 efficiency in different module power amounts is visualized. In the beginning of the diagram, the increase rate of the efficiency is much faster than other tested products, meaning that it works better in very low irradiance levels and it can raise efficiency significantly just with a small increase in input power. Additionally, for maximum input power amounts, device D has about 95% of efficiency which is quite interesting. On the other hand, in some points it works around 80 to 85%, which happened in other devices too.

Fig. 7 shows that device D has an outstanding efficiency in comparison with other tested charge controllers, however unlike other devices it does not have a converter circuit and this makes it not suitable for all systems. It means if all components of the grid or system match each other in the aspect of characteristics, it seems that device D is a really good choice even though it does not track the MPP. On the other hand, for systems with components with different characteristics, MPP tracking devices could be a better choice since they can change the voltage of the module in order to reach MPP in various situations.

In other words as it can be seen in Fig. 6 and Fig. 7, this device does not accept and tolerate powers more than 132 watts in simulator and 118 watts in PV output. Since the tested PV is a 210 Watt module, the output power is significantly lower than expected which is due to the battery voltage. As device D is only a charge controller without maximum power point tracking, it is unable to change the voltage of the module. This means the 12 Volt battery limits the module voltage to a maximum around 14 volts which leads to powers much lower than MPP. One solution could be to use batteries with voltages closer to the PV voltage (like 24 or 48V) or to choose smaller modules with lower voltages.

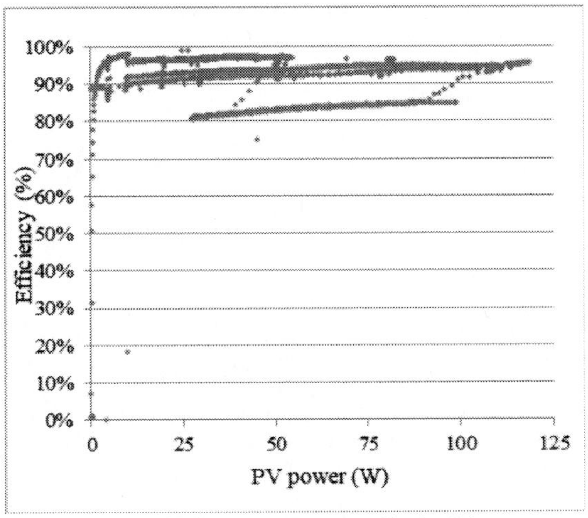

Fig. 7: Efficiency to PV power (device D)

6. COMPARISON

Fig. 8 contains efficiency changes for all four devices through the test process. As it can be seen and considering efficiency only, device D has the highest efficiency among all in most cases. However, as said before, it does not deliver enough power due to the lack of MPP tracking and working around battery voltage.

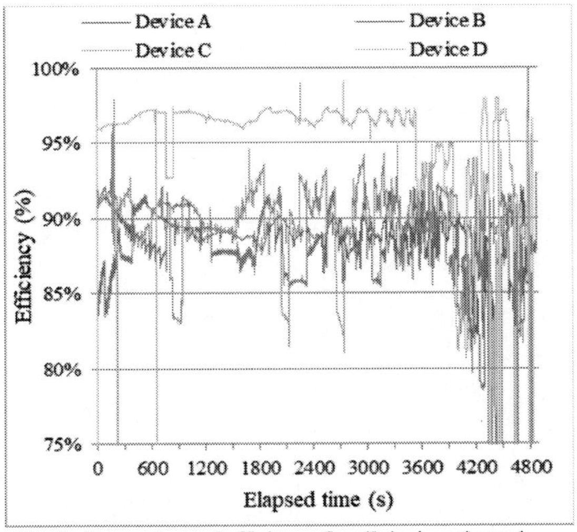

Fig. 8: Comparison of efficiency for all devices through test process

The diagram of the comparison of all devices in efficiency with respect to their working module power is visible in Fig. 9. Here also it is visible that device D delivers the power most efficiently in comparison with other devices, however it can only deliver half the full produced power. This is due to the lack of MPPT which leads the system to work in battery voltage range. Additionally, it has faster rise speed for very low powers, making it a better choice for low irradiance situations.

As can be seen in Fig. 9, device C is more efficient in high module powers compared to devices A and B. Choosing between device A and device B is not easy since both have high and low states in comparison.

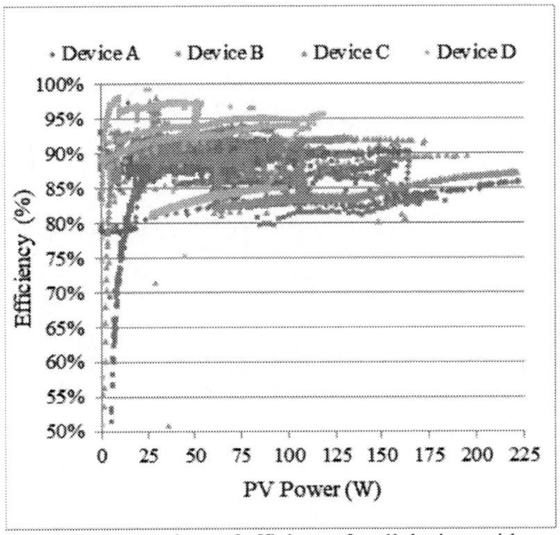

Fig. 9: Comparison of efficiency for all devices with respect to PV power

6.1 Devices' efficiency analysis in detail *(Device C)*

Since comparison diagrams are messy and few information can be extracted from them, in this section the efficiency analysis is shown in more detailed diagrams. For this, some individual irradiance slopes (from test profile) are considered and magnified as follows.

Fig. 10 and Fig. 11 illustrate efficiency changes during 2nd and 3rd slopes of the low-to-medium irradiance test for device C. Each diagram is divided into increasing ramp (increasing irradiance) and decreasing ramp (decreasing irradiance). As it can be seen, the diagrams have step-like shapes with increasing efficiency higher than decreasing in most of the data points.

Fig. 12 and Fig. 13 show 7th and 8th slopes of low-to-medium irradiance test with an increase/decrease rate of 10 and 14 W/m^2 per second respectively. It is shown that device C delivers the same amount of efficiency for data points in both increasing and decreasing slopes when irradiance change rate is high (7th and 8th slopes). This is the same for the rest of low-to-medium irradiance tests (9th to 11th slopes) when irradiance change rate is even higher.

Fig. 14 is the efficiency changes divided for irradiance increase and decrease parts of very low irradiance test with the rate of 2 W/m^2 per second. Fig. 15 shows the load power and battery voltage status during the time of the very low irradiance test for both increasing and decreasing slopes of irradiance. Load existence of about 83 watts in the beginning of the test (see third diagram of Fig. 3), explains the battery discharge and available efficiency during this period of increasing part in Fig. 14. Later on, when load is zero, battery starts to recharge. In the decreasing slope, efficiency moves to zero since there is no load to be fed and also battery state of charge stays the same.

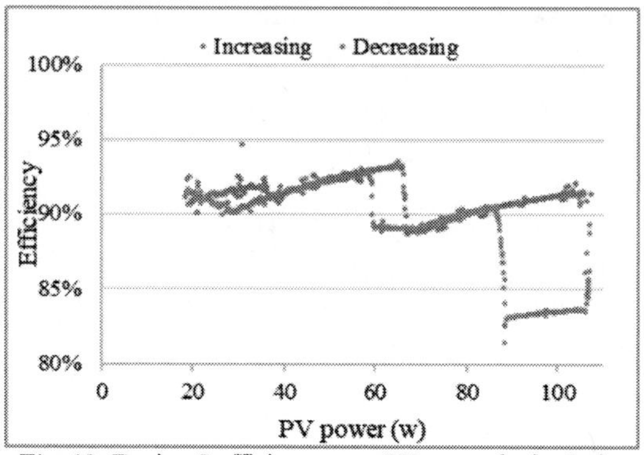

Fig. 10: Device C efficiency w.r.t PV power in the 2nd ramp of low-to-medium test (1 W/m² per second change rate)

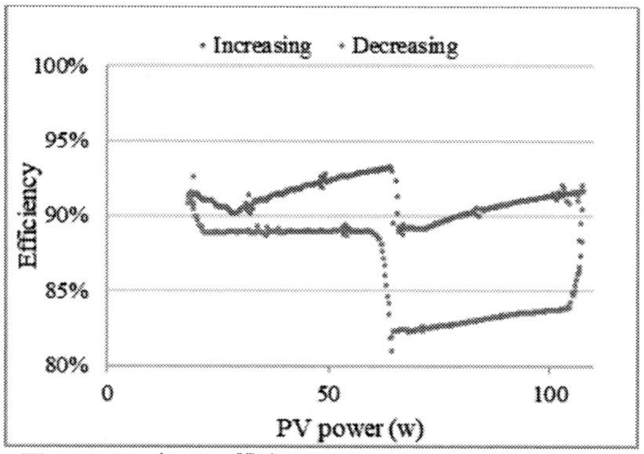

Fig. 11: Device C efficiency w.r.t PV power in the 3rd ramp of low-to-medium test (2 W/m² per second change rate)

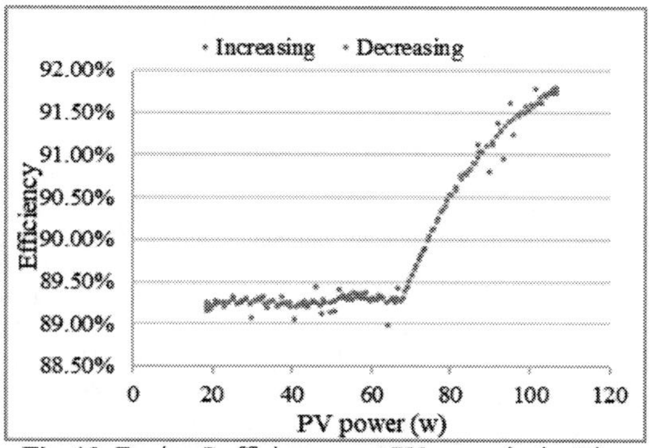

Fig. 12: Device C efficiency w.r.t PV power in the 7th ramp of low-to-medium test (10 W/m² per second change rate)

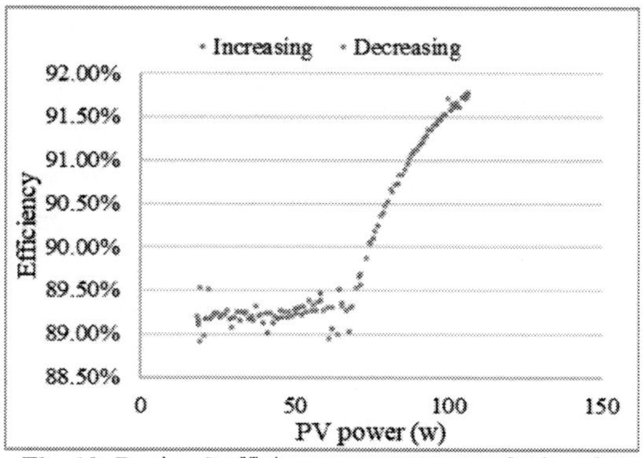

Fig. 13: Device C efficiency w.r.t PV power in the 8th ramp of low-to-medium test (14 W/m² per second change rate)

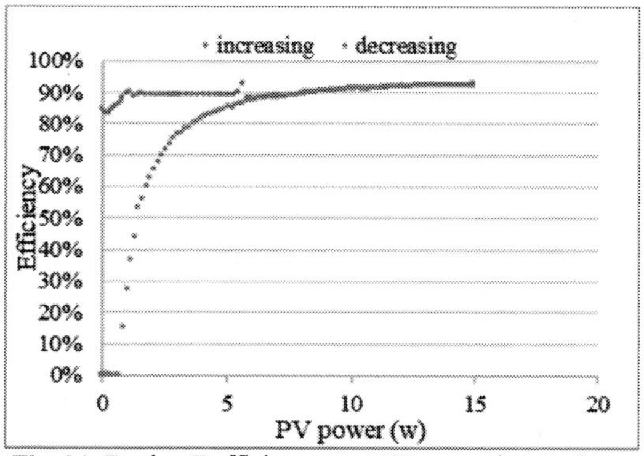

Fig. 14: Device C efficiency w.r.t PV power in very low irradiance test (2 W/m² per second change rate)

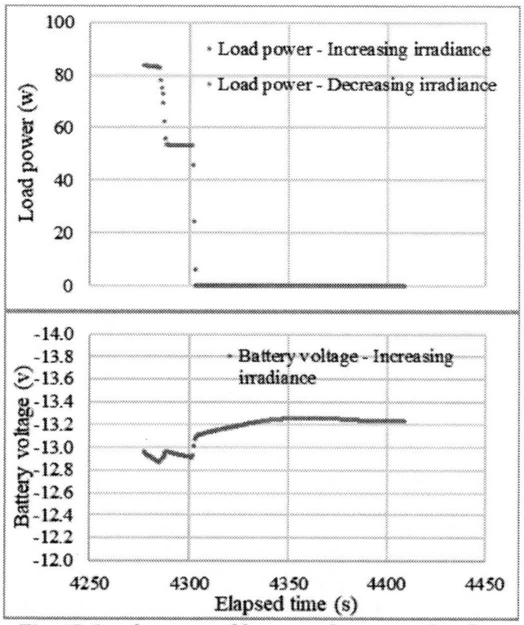

Fig. 15: Load power and battery voltage w.r.t time for the very low irradiance test (device C)

As it is mentioned before during the test profile, load had to be changed according to different states of irradiance. For example, when the battery is full, increasing the load can compensate extra produced energy. On the other hand, when the battery charge is low, decreasing the load lets the controller charge the battery and protect it from deep discharge.

7. CONCLUSION

The purpose of this experiment was to implement a new and advanced testing method on different battery charge controllers in order to analyze their performance and efficiency during various solar irradiance situations. This was done on four different devices and some important results are shown here. All available results in details are available in [8].

In this paper it has been shown that three devices with MPPT (devices A, B and C) deliver significantly higher power to the battery and load despite the fact that the difference between battery (12V) and panel voltage (30.55V as V_{mpp}) is considerable, albeit these devices have different efficiencies during test time. In the current study device D is the only device without MPP tracking, hence as it is shown, it does not deliver the expected amount of power that the panel is designed for. However, in its own power range, it shows the highest efficiency among tested devices. As a result, it is proved that while designing a solar system, all components should be in the same voltage range to prevent unnecessary power loss. Nevertheless, when it is not possible, considering an MPPT is essential. In this study it can be seen that three devices with MPPT are suitable for considered under test off-grid solar system (12V battery), although if all components are in the same voltage range and MPPT is not essential, device D could be a good choice considering its high efficiency.

The complete results for all four tested devices are available in [8]. Finally, an important point to consider is that the load for each experiment is manually adjusted. Therefore, the load diagram is unique for each device, which is because of lack of a controllable programmable load. For future work, a load profile could be designed based on the already designed irradiance profile in this work. For this matter, battery SOC and amount of produced power and required amount of load should be calculated for every 0.5 seconds. The test could be repeated one more time by implementing both irradiance profile (through irradiance simulator) and load profile (through a controllable load) for each device which results to exactly equal test conditions for all devices.

8. REFERENCES

[1] A. Khatibi, J. Bendfeld, W. Bermpohl and S. Krauter, "Introduction of an Advanced Method for Testing of Battery Charge Controllers for Off-grid PV Systems," in *EU PVSEC*, Amsterdam, 2017.

[2] Steca Elektronik, *Solar charge controller PR 2020 Operating instructions,* Steca Elektronik.

[3] *MPPT2024Z installation and operation manual,* 1.0 ed.

[4] *LCSC-M Series MPPT Solar Charge Controller User Manual,* 3.1 ed., LC Power.

[5] Steca Elektronik, *Solarix MPPT 2010 Installation and operating instructions,* Steca Elektronik.

[6] Keckeisen Akkumulatoren, [Online]. Available: http://www.intact-batterien.de/en/gel-power.html. [Accessed 28 10 2015].

[7] *Q-Cells solar module QC-C02,* Bitterfeld-Wolfen: Q.CELLS SYSTEMS.

[8] A. Khatibi, "Analysis and Testing of Battery Charge Controllers with Maximum Power Point Tracking: Performance and Efficiency," Master's thesis, Dept. EIM-NEK, Paderborn Univ., Paderborn, Germany, 2015.

[9] S. C. W. Krauter, Solar Electric Power Generation - Photovoltaic Energy Systems, 1 ed., Springer-Verlag Berlin Heidelberg, 2006.

[10] M. Rosu-Hamzescu and S. Oprea, *Practical Guide to Implementing Solar Panel MPPT Algorithms,* Microchip Technology Inc., 2013.

[11] R. Bründlinger, N. Henze, H. Häberlin, B. Burger, A. Bergmann and F. Baumgartner, "prEN 50530 - The new european standard for performance characterisation of PV inverters," in *European Photovoltaic Solar Energy Conference and Exhibition (EU PVSEC) <24, 2009>*, Hamburg, 2009.

[12] H. Haeberlin and P. Schaerf, "New Procedure for Measuring Dynamic MPP-Tracking Efficiency at Grid-Connected PV Inverters," in *24th European Photovoltaic Solar Energy Conference, 21-25 September 2009*, Hamburg, Germany, 2009.

[13] K. ViktorRudolf, "Financial analysis of utility scale photovoltaic plants with battery," *Energy Policy,* vol. 63, no. December 2013, pp. 139-146, 2013.

[14] R. F. Coelho and D. C. Martins, "An Optimized Maximum Power Point Tracking Method Based on PV Surface Temperature Measurement," in *Sustainable Energy - Recent Studies*, A. Gebremedhin, Ed., InTech, 2012, pp. 89-114.

[15] D. P. Hohm and M. E. Ropp, "Comparative study of maximum power point tracking algorithms," *Progress in Photovoltaics: Research and Applications,* vol. 11, no. 1, pp. 47-62, 22 11 2002.

DC-coupled buck-boost charge controller for utility scale photovoltaic plants

P.Burski, A.Neumann, R.Merz
University of Applied Science Karlsruhe, Moltkestraße 30, 76133 Karlsruhe
* Corresponding author: pawel.burski@hs-karlsruhe.de

ABSTRACT: Utility or industrial scale PV-systems provide lowest costs of energy. In order to provide 24h operation batteries are able to back up power during the day and provide electrical power during night. State of the art technologies use bidirectional battery inverters to charge and discharge the battery. In order to save costs and gain efficiency we present a non-inverting buck-boost charge controller operation at variable switching frequencies without interleaved buck-boost operation in order to minimize switching losses and to allow design of battery- and PV-system independent from each other.

1 INTRODUCTION

Large utility scale photovoltaic (PV) systems achieve electricity generation cost $K_{PV} < 5$ ct/kWh dependent on funding and PV plant location [1]. Thousands of solar modules with peak power $250\ W_P < P_{M,STC} < 350\ W_P$ [2] [3] according to standard test conditions (STC) convert radiated power of the sun into direct current (DC) and inverters feed in power P_{PV} of the PV-generator to the grid in form of alternating current (AC) using transformers and transfer stations. To provide electrical power in case of sundown it is possible to integrate batteries to the PV-System. Unfortunately, the DC-voltage V_{PV} of the PV-System depends on temperature and irradiation and the voltage of the battery storage V_{BATT} depends on state of charge (SoC) and its power $P_{BATT} = V_{BATT} I_{BATT}$ of operation.

The PV systems limited valid isolation voltage V_{ISO} depends on applied module type $V_{ISO} = 1000$ V or $V_{ISO} = 1500$ V. Increasing PV current I_{PV} decreases the PV voltage $V_{PV} < V_{PV,OC}$ below open circuit voltage $V_{PV,OC}$. At short-circuit conditions the string provides maximum current $I_{PV,SC}$ but no power $P_{PV} = V_{PV} I_{PV} = 0$. The maximum power point tracker (MPPT) of the inverter obtains the optimal operating voltage V_{MPP}.

Figure 1a) compares a typical PV-system with AC coupled battery to a DC-coupled system of figure 1b) in order to convert the variable voltage V_{PV} of the PV-system to the varying battery voltage V_{BATT}. Hence typical battery consists of a number of lithium-ion cells, for example based on nickel-manganese-cobalt (NMC) technology. Discharged, at state of charge SoC = 0%, the NMC-cells provide a minimum voltage $V_{CELL}(\mathrm{SoC} = 0\,\%) \approx 2{,}7$ V and completely charged at SoC = 100% $V_{CELL}(\mathrm{SoC} = 100\,\%) \approx 4{,}2$ V [4]. Battery systems with typical isolation voltage $V_{BATT,ISO} = 1000$ V connect up to $N_{CELL} = V_{BATT,ISO}(V_{CELL}(\mathrm{SoC} = 100\,\%))^{-1} = 238$ cells in series. Commercially available systems for instance combine 12 or 13 battery modules in series, each containing multiple lithium-ion-cells. The battery voltage range $V_{BATT}(\mathrm{SoC})$ results in $V_{BATT,MIN} \leq V_{BATT}(\mathrm{SoC}) \leq V_{BATT,MAX}$ depending on the amount of battery modules. With minimum voltage $V_{BATT,MIN} = V_{BATT,MIN12} = 520$ V and maximum voltage $V_{BATT,MAX} = V_{BATT,MAX12} = 805$ V for a battery containing 12 modules or minimum voltage $V_{BATT,MIN} = V_{BATT,MIN13} = 565$ V and maximum voltage $V_{BATT,MAX} = V_{BATT,MAX13} = 870$ V for a battery with 13 modules, each with limited state of charge $10\,\% \leq \mathrm{SoC} \leq 90\,\%$ [5].

To operate the PV-system and the battery at voltages V_{PV} and V_{BATT} independent from each other, the AC coupling of fig. 1a) applies one or more central inverters with integrated

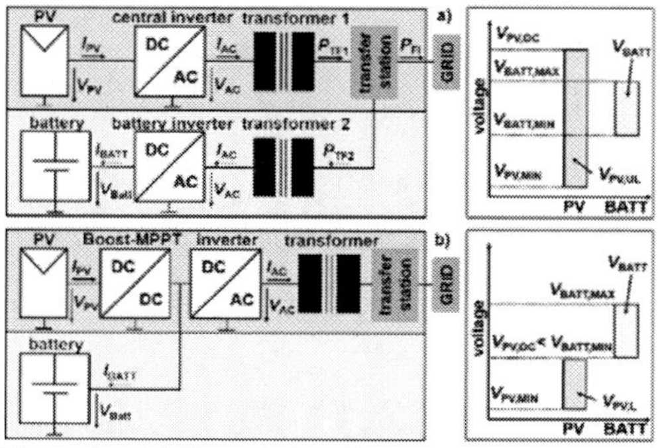

Figure 1: a) AC-coupled PV-battery system. A central inverter converts DC current I_{PV} from the PV-system to AC current I_{AC}. Battery inverter converts likewise current from the battery I_{BATT}. Transfer stations feed transformed power $P_{TF2} - P_{TF1} = P_{FI}$ from the transformers electrically isolated into the grid. b) DC-coupled PV-battery system. Boost-MPPT converter charges the battery with PV-voltage $V_{PV} < V_{BATT}$ resulting in limited voltage range $V_{PV,L}$.

MPPTs and transformers. The inverters operate the PV-system at its maximum power point (MPP) and convert the voltage V_{PV} to AC-Voltage V_{AC} resulting in AC-power $P_{AC} = V_{AC} I_{AC}$ and transformed power P_{TF1} of the transfer station for feed in power P_{FI} or power P_{TF2} for charging batteries. The battery inverter, in combination with a transformer connected to the transfer station, charges or discharges the battery with battery power $P_{BATT} = V_{BATT} I_{BATT}$ corresponding to transformed power $P_{TF2} = P_{TF1} - P_{FI}$. Charging and discharging the battery with power P_{BATT} according to the PV-power $P_{PV} = V_{PV} I_{PV}$ of the PV-system achieves a demand matched feed in power to the low-voltage or medium-voltage-network (LVN or MVN).

The separate transformation of DC-power of the battery to AC-power allows for independent PV- and battery system design and unlimited PV-voltage range $V_{PV,UL}$. During charging the battery, power dissipation at the battery-inverter, PV-inverter as well as at the transformers result in power losses. During discharging the battery inverter and transformer cause further energy losses. Therefore, the required three voltage-conversions cause increasing power losses and reduce efficiency $\eta_{AC} < 1$. Due to tolerances and cloud-drift central MPPT additionally result in mismatch-losses and decreases energy-yield of the system [6]. In case of failure or defect of central inverters higher loss of earnings occur compared to application of string-inverters and availability of the system decreases [7].

Figure 1b) illustrates the integration of the battery on the DC side of the PV-inverter. The lack of a separate battery inverter and a second transformer potentially increases system efficiency since less devices and voltage conversion steps are required. The PV-power charges the battery using a charge controller. Unfortunately existing charge controllers utilized in such PV systems operate as boost converters separating the voltage ranges of battery- and PV-voltage. The separation of the voltage ranges restricts the design of the PV-system and exacerbates planning of the whole system. Commercially available decentralized boost-converters with integrated MPPT (boost-MPPT) provide decentral MPPT resulting in increased energy yield and integrate the battery on the DC-side of the PV-system. In case of malfunction of single boost-MPPT's the remaining boost-converters continue operation and reduce loss of earnings compared to AC-coupling. Currently the assortment of such boost-MPPT products is very limited on the market. Available devices, such as the SBB16 from Helios-Systems offer a MPP-voltage-range of $400\ V \leq V_{PV} \leq 820\ V$ or the Sinusstrom Viper a MPP-voltage-range of $300\ V \leq V_{PV} \leq 800\ V$ of the PV-system designated as decentralized MPPTs supplying central inverters and increasing energy yield in systems without batteries [7][8][9]. Maximum PV-voltage $V_{PV,MAX} = V_{PV,OC} < V_{BATT,MIN}$ with respect to minimum battery voltage $V_{BATT,MIN}$ results in a limited voltage range $V_{PV,L}$ by $V_{PV,MIN} < V_{PV,OC} < V_{BATT,MIN}$ with $V_{PV,MIN} = 400\ V$ in case of the SBB16 Boost-MPPT or $V_{PV,MIN} = 300\ V$ in case of the Sinusstrom Viper. Low maximum input voltage $V_{PV} < V_{BATT,MIN}$ leads to disadvantage utilization of the boost-MPPT's. Due to operation of the PV-system at maximum power point voltage $V_{PV,MPP} < V_{PV,OC}$ and corresponding MPP current $I_{PV,MPP} < I_{PV,SC}$ the maximum power of boost-MPPT's $P_{BOOST,MAX} = V_{PV,MPP} I_{PV,MPP}$ is lower than manufacturer's specifications related to utilization of the entire voltage range $V_{PV,MPP} < 820\ V$ at the SBB16 or $V_{PV,MPP} < 800\ V$ at the Sinusstrom Viper. At constant device costs K_{LE} the power related specific cost $k_{Boost} = K_{LE}(P_{BOOST,MAX})^{-1}$ of the device increases at operation as battery charge controller with reduced power $P_{BOOST,MAX}$ compared to maximum power $P_{TOT,SBB16} = 25{,}6\ kW$ of each SBB16 MPPT or $P_{TOT,Viper} = 15\ kW$ of Sinusstrom Viper operated as decentral MPPT with central inverter without battery.

2 BUCK / BOOST CHARGE CONTROLLER

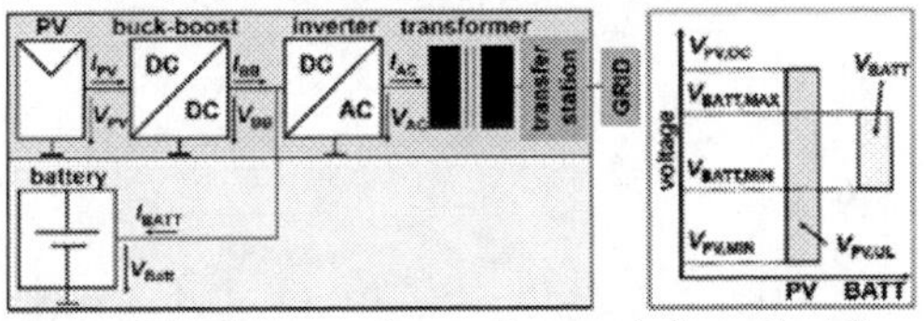

Figure 2: buck-boost charges battery independent from minimum and maximum PV- and battery voltage achieving unlimited PV-voltage range $V_{PV,UL}$. PV-system design same as AC-coupling but with higher efficiency and cost reduction compared to prior solutions.

Figure 2 introduces the application of a non-inverting buck-boost converter as charge controller with integrated MPPT (buck-boost-MPPT). Buck-boost-MPPT's are capable of charging batteries at all voltage ratios. Thus eliminating the restriction of PV voltage range with regard to the SoC dependent battery voltage $V_{BATT}(SoC)$. Replacing the boost-MPPT's of DC-coupled storage systems in figure 1b) with buck-boost-MPPT's provides unlimited PV-voltage range $V_{PV,UL}$. Therefore, the buck-boost-MPPT allows for independent PV-system design in same way as applications using central inverters or decentral string inverters without battery [7]. Additionally, consideration of temperature- und irradiation dependent PV-voltage according to battery voltage V_{BATT} is not necessary [10]. Battery implementation only depends on output voltage range $V_{BB,MIN} < V_{BATT}(SoC) < V_{BB,MAX}$ of the buck-boost-MPPT.

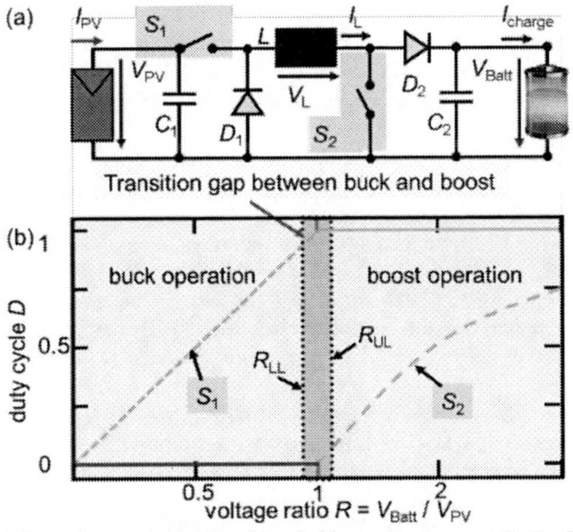

Figure 3: a) schematic of buck / boost converter. C_1 S_1, D_1 and L form Buck converter while S_1 switches with duty cycle D_{S1}. L, S_2, D_2, C_2 form boost converter while S_1 continuously is switched ON and S_2 switches with D_{S2}. b) Duty cycles D_{S1} and D_{S2} to adjust voltage ratio R for MPPT. Non-ideal switches with $D_{S1} < 1$ and $D_{S2} > 0$ limit the voltage ratio $R \neq 1$ and result in transition gap TG and prohibit continuous operation.

Figure 3a) explains the topology of a non-inverting buck-boost-converter, which is a combination of a buck and a boost converter. The power switch S_1, the diode D_1 and the inductor L in combination with the buffer capacitor C_1 form a buck converter. The power switch S_2, the diode D_2 and the inductor L in combination with the buffer capacitor C_2 result in boost converter topology. At voltage $V_{PV} > V_{BATT}$ the charge controller operates as a buck converter. Therefore the power switch S_1 sequentially switches ON during time $T_{ON,S1}$ and switches OFF during time $T_{OFF,S1}$. Usually the topology applies a constant pulse width modulated (PWM) switching frequency $f_{Switch1} = 1/T_{Switch1} = (T_{ON,S1} + T_{OFF,S1})$. The duty cycle of the PWM $D_{S1} = V_{BATT}/V_{PV} = T_{ON,S1} f_{Switch}$ adjusts current I_L of the inductor L to control the output current $I_{BATT} = 0.5(I_{L,Min} + I_{L,MAX})$ of the buck converter. The voltage $V_L(T_{ON,S1}) = V_{PV} - V_{BATT}$ during time $T_{ON,S1}$ increases current $I_L(t) = V_L(T_{ON,S1})L^{-1}t + I_{L,MIN}$. The switch S_1 is turned OFF during $T_{OFF,S1} = (f_{Switch})^{-1} - T_{ON,S1}$, the Diode D_1 takes over the current $I_L = I_{L,MAX} = I_{L,MIN} + V_L L^{-1} T_{ON,S1}$ while voltage $V_L(T_{OFF,S1}) = -V_{BATT}$ reduces inductor current $I_L = I_{L,MAX} + V_L(T_{OFF,S1})L^{-1}T_{OFF,S1}$.

At voltage $V_{PV} < V_{BATT}$ the charge controller operates as a boost converter switching S_2 sequentially ON during time $T_{ON,S2}$ and switching OFF during $T_{OFF,S2}$. Adjusting duty cycle $D_{S2} = 1 - (V_{BATT}/V_{PV})^{-1} = T_{ON,S2} f_{Switch}$ varies voltage ratio V_{BATT}/V_{PV} and input current $I_{PV} = 0.5(I_{L,MIN} + I_{L,MAX})$. The voltage $V_L(T_{ON,S2}) = V_{PV}$ during time $T_{ON,S2}$ increases current $I_L(t) = V_L(T_{ON,S2})L^{-1}t + I_{L,MIN}$. The switch S_2 is turned OFF during $T_{OFF,S2} = (f_{Switch})^{-1} - T_{ON,S2}$, the Diode D_2 takes over the current $I_L = I_{L,MAX} = I_{L,MIN} + V_L L^{-1} T_{ON,S2}$ while voltage $V_L(T_{OFF,S2}) = V_{PV} - V_{BATT}$ reduces inductor current $I_L = I_{L,MAX} + V_L(T_{ON,S2})L^{-1}T_{OFF,S2} = I_{L,MIN}$.

The duty cycle controls the current I_L and provides adjustable voltage ratio V_{BATT} / V_{PV} in both buck and boost operation. The voltage V_{BATT} is temporary approximated as constant and varies very slowly according to its state of charge (SoC). The duty cycles D_{S1} and D_{S2} adjust the input voltage V_{PV} to operate the PV-system at its MPP.

Figure 3b) outlines the corresponding duty cycle values D_{S1} and D_{S2} to adjust the voltage ratio V_{BATT} / V_{PV}. The duty cycles result to $D_{S1} < 1$ and $D_{S2} = 0$ while $V_{PV} > V_{BATT}$ according to a standard buck converter. The duty cycles result to $D_{S1} = 1$ and $D_{S2} > 0$ for $V_{PV} < V_{BATT}$ according to a standard boost converter. The rise-time t_R to switch the non-ideal switches S_1 or S_2 ON and fall-time t_F switching them OFF again limit the possible duty cycles for periodically switching to $(t_F+t_R)f_{Switch} = D_{MIN} < D_{S1}, D_{S2} < 1-(t_F+t_R)f_{Switch} = D_{MAX}$. The switch S_1 is continuously turned ON while boosting resulting in $D_{S1} = 1$ without impact of rise- t_R and fall-time t_F. The limitation of the duty cycles D_{S1} and D_{S2} result a transition gap $TG = V_{BATT}(1+D_{MIN}) - V_{BATT} D_{MAX}$ centered to the voltage ratio $V_{BATT} / V_{PV} = 1$ where the charge controller needs change from buck- to boost-operation. To prevent the transition gap commonly used buck-boost converters use an interleaved buck-boost operation or switch simultaneously S_1 and S_2 for straightforward implementation, independent of the voltage ratio V_{BATT} / V_{PV} at the same time. Unfortunately, interleaved operation as well as simultaneously switching S_1 and S_2 increase power losses during switching the S_1 and S_2.

3 BUCK-BOOST-TRANSITION

Continuous maximum power point tracking as well as preventing additional switching losses from interleaved buck boost operation or simultaneously switching, requires the MPPT to set $\Delta V_{MPPT} > TG$. In order to operate the PV-System as close as possible to its MPP-voltage $V_{PV,MPP} - \Delta V_{MPPT} \leq V_{PV} \leq V_{PV,MPP} + \Delta V_{MPPT}$ the voltage steps $\Delta V_{MPPT} > TG$ and therefore the transition gap needs to be as low as possible.

Usually the voltage modulation ΔV_{MPP} uses current mode controllers to adjust the average current $I_{L,RMS} = (I_{L,MAX} - I_{L,MIN})/2 + I_{MIN}$ of the current $I_L(t)$.

Analyzing the switching frequency

$$f_{Switch1} = \frac{V_{BATT}\left(1 - \frac{V_{BATT}}{V_{PV}}\right)}{L\Delta I_L}$$

operating the charge controller during buck operation or during boost operation

$$f_{Switch2} = \frac{V_{PV}\left(1 - \frac{V_{PV}}{V_{BATT}}\right)}{L\Delta I_L}$$

indicates an increasing switching frequency $f_{Switch1}$ and $f_{Switch2}$ by decreasing amplitude ripple current $\Delta I_L = I_{L,MAX} - I_{L,MIN}$ [11]. For $V_{PV} = V_{BATT}$ the switching frequency $f_{Switch1} = 0$ Hz as well as $f_{Switch2} = 0$ Hz becomes zero.

Figure 4 a) introduces a current mode controller for buck operation and b) for boost operation setting a constant current ripple $\Delta I_L = I_{L,MAX} - I_{L,MIN}$. For buck operation, if $I_L(t) \geq I_{L,SP} + 0.5\ \Delta L\ I_{L,MAX}$, ST_1 turns SW1 OFF and the voltage $V_L = -V_{BATT}$ of the inductor L lowers the current $I_L(t)$. When $I_L(t)$ drops to $I_L \leq I_{L,MAX} - \Delta I_L$, ST_1 turns S_1 ON again and $I_{L,MIN} + \Delta I_L/2 = I_{L,SP}$ match the set point current $I_{L,SP}$. For boost operation, S_1 needs to be turned ON and the Schmitt-Trigger ST_2 of fig. 4b) turns ON S_2 if I_L drops to $I_{L,MIN}$ and the voltage $V_L = V_{PV}$ increases the current to $I_{L,MAX}$. When $I_L = I_{L,MAX}$, ST_2 turns S_2 OFF again and the voltage $V_L = V_{PV} - V_{BATT}$ lowers I_L again. During S_1 is switched ON and S_2 is OFF, the slope $dI_L/dt = V_L L^{-1}$ of increase or decrease of I_L slows down with decreasing inductor voltage $V_L = (V_{PV} - V_{BATT})$ and phases out for $V_{PV} = V_{BATT}$ and the current will stay constant.

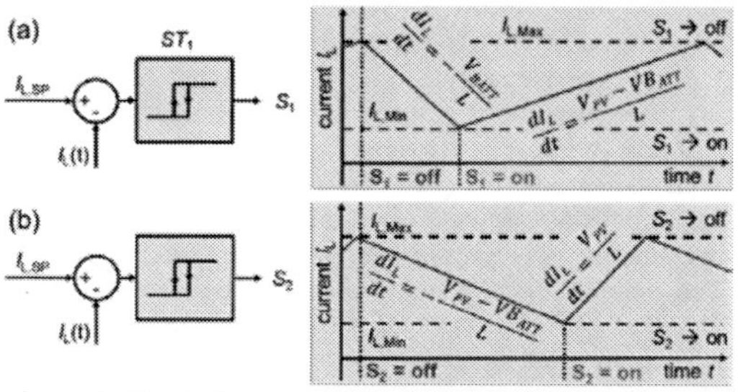

Figure 4: illustrates current mode controller with constant current ripple $\Delta I_L = I_{L,MAX} - I_{L,MIN}$. a) for buck operation ST_1 turns S_1 OFF if $I_L(t) \geq I_{L,MAX}$ and turns ON if $I_L(t) \leq I_{L,MAX}$. b) for boost operation ST_2 turns S_2 OFF if $I_L(t) \geq I_{L,MAX}$ and turns ON if $I_L(t) \leq I_{L,MAX}$.

Figure 5 analyses the frequency f_{IL} of I_L with respect to V_{PV} for different voltages $V_{BATT1} = 520$ V, $V_{BATT2} = 700$ V and $V_{BATT3} = 805$ V and constant current ripple $\Delta I_L = 10$ A. For $V_{PV} > V_{BATT}$ buck operation modulates the frequency $f_{IL} = f_{Switch1}$ of the ripple current ΔI_L by switching S_1. During boost operation $f_{IL} = f_{Switch2}$ matches switching frequency of S_2. As expected for $V_{PV} = V_{BATT}$ the current I_L stays constant and switching frequency $f_{IL} = 0$ Hz stops controlling the current I_L.

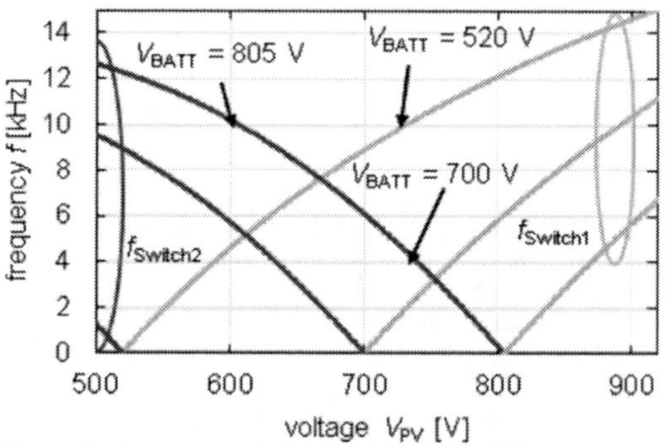

Figure 5: Frequency f of I_L over V_{PV} for different V_{BATT} and constant $\Delta I_L = 10$ A. Buck operation with $f_{IL} = f_{Switch1}$ for $V_{PV} > V_{BATT}$ and boost operation with $f_{IL} = f_{Switch2}$ for $V_{PV} < V_{BATT}$. For $V_{PV} = V_{BATT}$ I_L stays constant and $f_{IL} = 0$ Hz stops controlling I_L.

Figure 6 varies the ripple current ΔI_L to modify the frequency f_{IL} and therefore $f_{Switch1}$ and $f_{Switch2}$ during buck and boost operation. Reducing the ripple current, results in an increase of f_{IL}. As mentioned before, periodically MPPT with a variation of PV-voltage by voltage steps ΔV_{MPPT} defines a minimum frequency $f_{IL,Min}$. As illustrated in fig. 6, for V_{BATT} = 520 V decreasing the ripple as much as possible current also allows to operate the PV-system at voltages $V_{BATT} - TG/2 < V_{PV} < V_{BATT} + TG/2$ as close as possible to battery voltage V_{BATT}. Due to the transition gaps $TG_A < TG_B < TG_C$ decrease during ripple current ΔI_L decreases and accurate MPPT with $V_{PV} \approx V_{PV,MPP}$ becomes possible and prevents interleaved switching of S_1 and S_2.

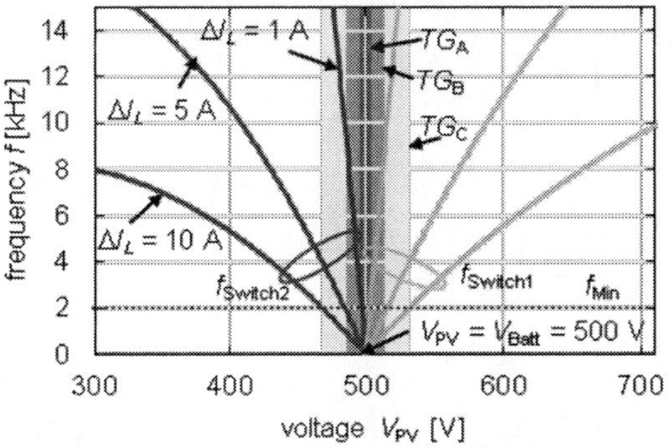

Figure 6: Varying ΔI_L modifies f_{IL} and therefore $f_{Switch1}$ and $f_{Switch2}$. Reducing ΔI_L increases f_{IL}. Periodically varying V_{PV} with MPPT by voltage steps ΔV_{MPPT} defines $f_{IL,Min}$. Due to $TG_A < TG_B < TG_C$ with decreasing ΔI_L accurate MPPT with $V_{PV} \approx V_{PV,MPP}$ becomes possible without interleaved switching.

Figure 7 describes a combined controller design for buck- and boost-operation capable to adjust the current ripple ΔI_L and to prevent interleaved switching of S_1 and S_2 using Schmitt-Trigger with fixed trigger levels. The MPPT varies the inductor current and sets a set point current $I_{L,SP}$. To use fixed switching levels with ST_1 or ST_2 to turn ON or OFF the switches S_1 or S_2, the controller uses the difference $I_{ERR} = I_{L,SP} - I_L(t)$ of set point current $I_{L,SP}$ and measured current $I_L(t)$. To adjust the current ripple ΔI_L a variable amplification A (A = 1...10) provides the modified current $I_{L,ST} = A\,(I_L(t) - I_{L,SP})$. To prevent simultaneously switching, a constant shift current I_{Shift} lowers $I_{L,ST}$ to $I_{Buck} = I_{L,ST} - I_{Shift}$ and the Schmitt-Trigger ST_1 turns OFF S_1 when $I_L(t) \geq I_{L,SP} + (\Delta I_L/2 + I_{Shift})$. In the same way a similar shift current I_{Shift} increases $I_{L,ST}$ and ST_2 switches OFF S_2 when $I_L(t) \geq I_{L,SP} + (\Delta I_L/2 - I_{Shift})$. Therefore, in case the current $I_L(t)$ increases and both switches S_1 and S_2 are switched ON, S_2 will switch OFF. If the current $I_L(t)$ continuous to increase ST_1 turns OFF S_2 at $I_L(t) = I_{L,SP} + (\Delta I_L/2 + I_{Shift})$ and the non-inverting buck-boost converter operates in buck mode. In case the current $I_L(t)$ decreases and both S_1 and S_2 are OFF, ST_1 first turns on S_1. Only in case S_1 is switched ON and, due to $V_{PV} < V_{BATT}$, the current $I_L(t)$ still decreases, ST_2 turns on S_2 and allows for boost operation.

Unfortunately, the shift current I_{Shift}, results a persistent mismatch $I_{L,RMS} = I_{L,SP} + I_{Shift}$ in case of buck operation and $I_{L,RMS} = I_{L,SP} - I_{Shift}$ and an additionally attached PID-controller needs to phase out the mismatch.

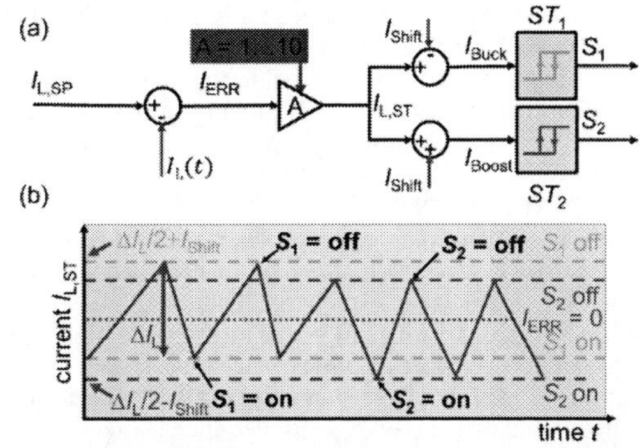

Figure 7: Controller to adjust $I_{L,RMS}$ to $I_{L,SP}$. Setting amplification keeps $I_{L,ST} = A\,I_{ERR}$ constant and reduces ripple ΔI_L. b) I_{Shift} added to $I_{L,ST}$ prevents interleaved switching of S_1 and S_2 but results in an mismatch of $I_{L,RMS} \neq I_{L,SP}$ and a PID controller needs to correct.

4 EXPERIMENTAL RESULTS

Figure 8a) and 8b) presents first experimental measurements of the current $I_{L,RMS}(t)$ controlled to the set point current $I_{L,SP}$ represented by the dashed line during MPPT with MPP-voltage V_{MPP} = 500 V and emulated battery voltage V_{BATT} = 500 V. Figure 8a) displays the current I_L(t) during the buck-boost converter changes from boost to buck operation and the voltage $V_{PV} = V_{BATT}$ keeps the current I_L(t) constant. Due to the MPPT reduces the set point current $I_{L,SP}$ in order to increase V_{PV} the buck-boost converter starts to operate in buck mode. To start buck operation the set point current I_L,SP needs to drop to $I_{L,SP} \leq I_L(t) + \Delta I_L(2A_1)^{-1} + I_{Shift}$ and trigger the Schmitt-Trigger ST_1 to switch OFF S_1 and results transition gap TG_{A1}. In order to increase the current $I_{L,RMS}$ the MPPT increases the set point current $I_{L,SP}$ again and the non-inverting buck-boost charge controller changes to boost operation again. Depending on the actual current I_L(t) the time to pass the transition gap TG_{A1} may vary. Figure 8 b) increases the amplification $A_2 > A_1$ and the time to pass the transition gap decreases. After passing the transition gap TG_{A2}, fig. 8b illustrates the current mismatch $I_{L,RMS} = I_{L,SP} - 2\,I_{Shift}$ due to I_{Shift} to prevent simultaneous switching of S_1 and S_2. In order to prevent static mismatch a PID controller controls $I_{L,RMS}$ to $I_{L,SP}$.

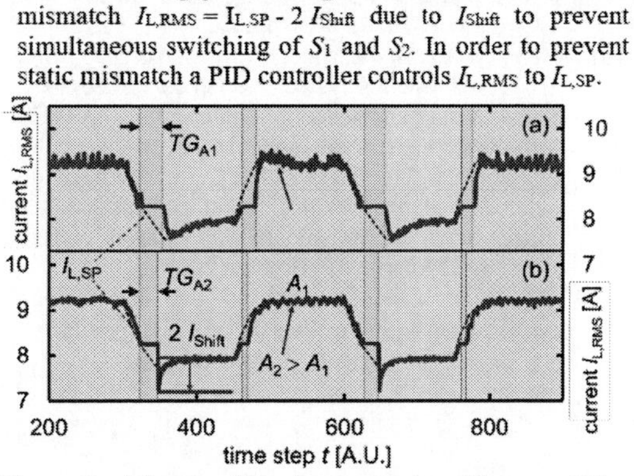

Figure 8: a) $I_{L,RMS}$ with gain A_1 sets transition gap TG_{A1} during transition from boost to buck operation. b) $A_2 > A_1$ results $TG_{A2} < TG_{A1}$. Switching from buck to boost initially results current shift $I_{L,RMS} = 2\,I_{Shift}$. A PID control fades out error and matches I_L to $I_{L,SP}$.

Therefore, for optimized MPPT the design of the controller allows for continuous transition from boost to buck operation and prevents interleaved or simultaneous switching of S_1 and S_2. The variable ripple current allows to adjust the transition gap TG to match voltage steps ΔV_{MPPT} as well as the frequency f_{MPPT} of the MPPT.

5 CONCLUSION

We presented a non-inverting buck-boost charge controller for DC-coupled high voltage battery storage. The non-inverting buck-boost charge controller allows for design of the PV-strings independent from the high voltage battery. In order to prevent electrical power losses, the novel control of the non-inverting buck boost converter operates the charge controller either in buck mode or in boost mode. In addition the non-inverting buck-boost charge controller reduces its switching frequency for $V_{BATT}/V_{PV} \approx 1$ as low as the frequency f_{MPPT} of voltage variation of the MPPT and prevents significant power losses during switching the power switches.

6 REFERENCES

[1] Johannes N. Mayer et al., Current and Future, Cost of Photovoltaics, edited by Mara Marthe Kleiner (Agora Energiewende, Germany, 2015)

[2] Aleo, datasheet S_19 gen2, www.bosch-solarenergy.de, (aleo Solar AG, Prenzlau, 03.2013)

[3] SolarWorld, datasheet SW340 – 350 XL MONO, www.solarworld.com, (10.2016)

[4] M. Ecker et al., Calendar and cycle life study of Li(NiMnCo)O2-based 18650 lithium-ion batteries, Journal of Power Sources Vol. 248 (2014), p. 839-851

[5] ads-Tec, datasheet StoraXe SRS0112/SRS0120, www.ads-tec.de, (ads-tec GmbH, Nürtingen, 06.2016)

[6] A.M. Pavan, S.Castellan, S. Quaia, S. Roitti, G.Sulligoi, Power Electronic Conditioning Systems for Industrial Photovoltaic Fields: Centralized or String Inverters?, (Clean Electrical Power, 2007. ICCEP '07. International Conference)

[7] S. Shimura, R. Herrero, M. K. Zuffo, J. A. B. Grimoni, Production costs estimation in photovoltaic power plants using reliability, Solar Energy Vol. 133 (Elsevier 2016), pp 294–304

[8] Helios Systems, datasheet String Booster Box SBB 16-10, www.helios-systems.net, (V1.0, 10.2016)

[9] Sinusstrom, datasheet Sinusstrom VIPER, http://www.sinusstrom.com/, (10.2016)

[10] E. Skoplaki, J.A. Palyvos, On the temperature dependence of photovoltaic module, electrical performance: A review of efficiency/power correlations, Solar Energy 83 (Elsevier 2008)

[11] U.Schlienz, Leistungselektronik und ihre Peripherie, (Vieweg & Sohn, Wiesbaden 2003) p. 16ff

APPROACH TO DETERMINE THE IMPACT OF COSMIC RAYS ON PV SYSTEMS

Martin Halwachs, M. Schwark, K.A. Berger, R. Ebner
[1]AIT Austrian Institute of Technology GmbH, Center for Energy, Business Unit Photovoltaic Systems, Vienna, Austria,
T: +43 (0) 50550-6406 , M: +43 (0) 664 8157806, Martin.Halwachs@ait.ac.at

ABSTRACT: The impact of free secondary particles, created in interactions of cosmic rays with earth's atmosphere, on semiconductor devices have been studied since the 1980s. Neutrons and Pions were found to be able to destroy semiconductor devices, hitting them while the devices is blocking high voltages. Usually this effect has been studied for high power applications. Within the project INFINITY various climate- and location-related effects are studied for their impact on PV system components in order to improve their future reliability. Due to dependency on device temperature, DC voltage and neutron flux, which changes mostly in altitude, the failure of inverts due to cosmic rays has been studied. In order to be able to predict failure rates of inverters, measurements and models from literature were mixed to provide a combined model, being highly related on local climate and device characteristics, which can be derived from specific measurements.
Keywords: Modelling, Radiation Damage, PV System, Inverter

1 INTRODUCTION

The failure rate of semiconductor devices, due to secondary particles, originating from cosmic rays has been studied since the 1980s [1]. Ziegler et al. described the effect of particles hitting memory chips and creating data failures. They stated the identified particles to be Neutrons and Pions, induced by the interaction of cosmic rays with earth's atmosphere. Models and simulations have been developed, based on failure rate experiments with different blocking voltages and fitted data from specific devices [2-4].

Neutrons as secondary particles from cosmic rays, may hit semiconductor devices. These particles can destroy devices, if they are currently blocking high voltage like diodes, transistors and IGBTs. The destruction is dependent on a neutron hitting the depletion region and high blocking voltage. These particles may then interact with silicon atoms, exciting them, followed by a disintegration to light ions. The moderation of these light particles, with energies between 10 and 100MeV, creates a highly concentrated electron-hole plasma. This plasma may stabilize in the presence of high blocking voltage between the electrodes, evolving into a thin channel, shortcutting the electrodes. The current is then dependent on the channel thickness, and the blocking voltage. High current can further effect high temperatures by Joule heating. If the resulting temperatures are higher than the melting point of silicon, these temperatures destroy the device [5].

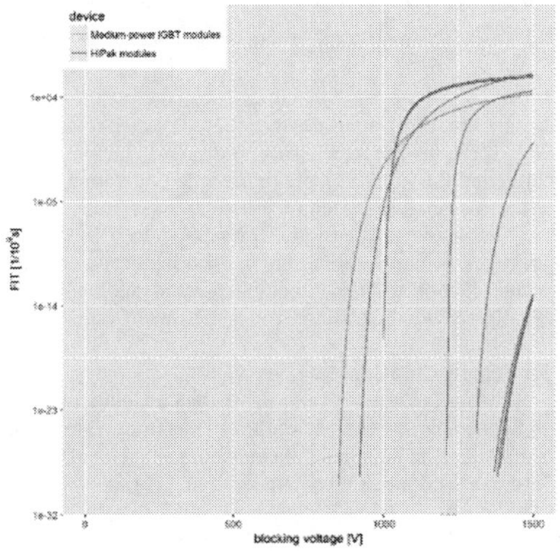

Figure 1: Plot of the model from [6], at sea level and 25°C, representing the device and blocking voltage dependent part. Only devices with low C_1 are shown.

Within the project INFINTY (started in 2015), PV system components are analyzed in the context of various environmental impacts in different climate zones. One main component in PV systems is the inverter. It utilizes power switches with huge chip area and highest isolation voltages of up to 1000V to 1500V. Thus it is studied, if they may suffer from cosmic ray impact as well, causing time dependent inverter failures.

Destruction of semiconductor switches due to cosmic rays, has already been investigated for high power electronics like in the model for several power switches by ABB [6,7]:

$$FIT(V,T,h) = C_3 \cdot \exp\left(\frac{C_2}{C_1 - V}\right) \cdot \exp\left(\frac{25 - T}{47.6}\right) \cdot \exp\left(\frac{1 - \left(1 - \frac{h}{44300}\right)^{5.26}}{0.143}\right)$$

Where V is the blocked DC voltage, T is the device temperature and h is the altitude against sea level. The coefficients $C_1$to C_3 are device dependent, stated within the document for different IGBTs and IGCTs. The model is valid in voltage ranges being large, compared to the pole C_1. Stated C_1 for IGCTs were higher than 2000V, for IGBTs the lowest started at about 800V. Thus these were not sufficient for modelling inverter failure, as the voltage range of PV inverters was not represented reasonably (see Fig. 1). So to model inverter failures a specific inverter type with its power switches was analyzed in detail.

2 MODEL

A model is presented to estimate failure rates of inverters due to cosmic rays. This model combines measurements, monitoring data and models from literature.

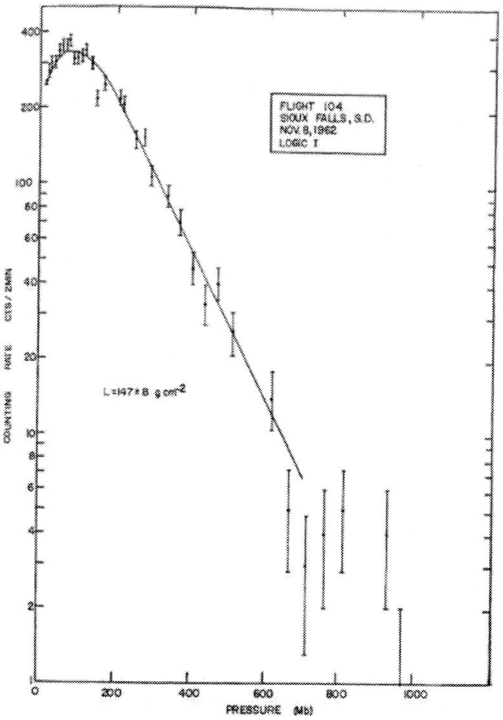

Figure 2: Measured data from Mendell and Korff [8 (Fig. 4.)], representing Neutron counting rate versus atmospheric pressure.

2.1 Height profile

The number of particles per height was gained from Mendell and Korff [8] (see Fig. 2) having fitted an exponential curve with constants a and b relative to sea level:

$$\frac{N(p[mbar])}{N(1000mbar)} = \exp(b \cdot (p - 1000)),$$

neglecting the impact of Pions due to less density in the atmosphere (100 times less at sea level) [9].

To transform atmospheric pressure in the Troposphere into height, the standard atmosphere was used, allowing for heights ranging up to 11 km [10].

$$p(h[km])\,[mbar] = 10^3 \cdot \left(1 - \frac{6.5 \cdot h}{288.15}\right)^{5.256}$$

This resulted in a factor representing particle flux relative to height in [km]:

$$f(h[km]) = \exp\left(b \cdot 10^3 \cdot \left(\left(1 - \frac{6.5 \cdot h}{288.15}\right)^{5.256} - 1\right)\right)$$

The linear regression resulted in

$$b = -0.0066724\, mbar^{-1}.$$

2.2 IGBT failures in time

The influence of neutron impact to semiconductor devices is highly dependent on its layout and structural composition and varies even between the same device types. Therefore the failures in Time (FIT) had to be measured to characterize devices [11]. The (FIT) corresponds to failed devices in 10^9 device-hours. It was modeled proportional to the blocking voltage by an exponential function (see Fig. 3):

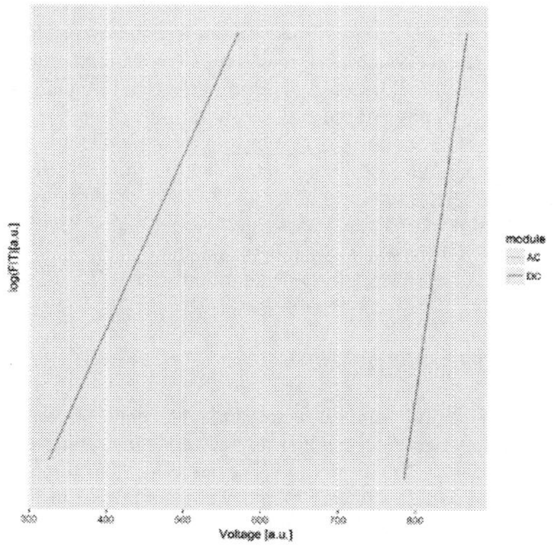

Figure 3: The graph shows the measured FIT having been different for the specific DC and AC modules. The units for FIT had to be censored.

$$FIT(U[V]) \left[\frac{1}{10^9 h}\right] = a \cdot \exp(b \cdot U)$$

For this research an inverter type with IGBT switches was chosen. For DC conversion and DC/AC conversion there were different IGBT modules used. This resulted in

$$FIT_{DC}(U_{in}) \left[\frac{1}{10^9 h}\right] = a_{DC} \cdot \exp(b_{DC} \cdot U_{in})$$

and

$$FIT_{AC}\left(\frac{2}{3} U_{in}\right) \left[\frac{1}{10^9 h}\right] = a_{AC} \cdot \exp\left(b_{AC} \cdot \frac{2}{3} U_{in}\right)$$

where V_{in} is the PV System DC voltage on the inverter input. Due to different devices the coefficients varied for the different devices. The additional factor of 2/3 converted the input voltage to the usual operating voltage of the AC module, based on practical experience.

The above regressions correspond to whole IGBT packages, usually containing all diodes necessary for a proper layout. If using manufacturer data, the FIT are usually provided in failures per 10^9 hours of usage.

2.3 Temperature dependence

The failure rate was also dependent on temperature, by higher temperatures providing more resistance to form a fine channel in between the electrodes [3]. The effect was described and modeled by Bauer et al. [12] as

$$\frac{FIT(T[K])}{FIT(300K)} = \exp\left(\frac{300 - T}{47.6}\right)$$

The same approach has obviously been used in the model used by ABB, shown above. In a first approximation the inverter inner temperature was used. As the device temperature is usually higher than this, this approach resulted in a slight overestimation of the effect. Therefore using the device temperature, if available, would be better in case of sensor readout from the blocking semiconductors.

2.4 Inverter Characteristics

In order to adapt to properly inverter failures, the model had to be adapted to the inverter characteristics:

- The actual fraction of semiconductors within the corresponding IGBT module, currently blocking the input voltage. This is assumed to be 50% all time with $r_{block} = 0.5$
- The total nominal input power per inverter P_{nom}

2.5 Input parameters

The above formalism is related on known sensor data over time. Ideally at least one full year of data has to be available with $T(t)$ being the inverter temperature or, even better, separate device temperatures as average readout over time and $U(t)$ being the DC input voltage over time.

2.6 Final Model

Assembling above pieces together the final model turned out to be

$$FIT\left[\frac{1}{W\ year}\right] = r_{block}\frac{f(h)}{P_{nom}}\ FIT_{yearly}$$

where the yearly AC and DC IGBT failure rate is

$$FIT_{yearly}\left[\frac{1}{year}\right] = 10^{-9}\sum_{i=0}^{n}\left(FIT_{DC}(U_{in}(t)) + FIT_{AC}\left(\frac{2}{3}U_{in}(t)\right)\cdot\exp\left(\frac{300-T(t)}{47.6}\right)\cdot\frac{dt}{3600}\right).$$

The factor 3600 translates time steps dt in seconds to hours.

3 PRELIMINARY RESULTS

To showcase the above model, temperature and input voltage data had to be used. This data was obtained by using inverter sensor readouts, to read voltage and temperature of one year. Data from several inverters on sea level at different locations in Austria were used. Fig. 4 shows the height dependence to demonstrate the model explained above.

Presently those height profiles, assume the monitored PV system being placed at different altitudes, while retaining climatic conditions. Though this is not an optimal solution, it can be assumed, that for higher altitude, the irradiance rises and temperatures fall, which will result in less beneficial conditions for the power switches. Therefore the demonstrated results show a minimum failure rate at the moment.

The current state of the model in Fig. 4 implies that an installation on Mount Everest, would have 0.15 to 0.5 inverts fail in a 1 MW PV system within 1 year. This means in a 10 MW system 1 to 5 inverters will fail due to cosmic rays in 1 year.

4 SUMMARY & OUTLOOK

A model to estimate device failure rates for inverters

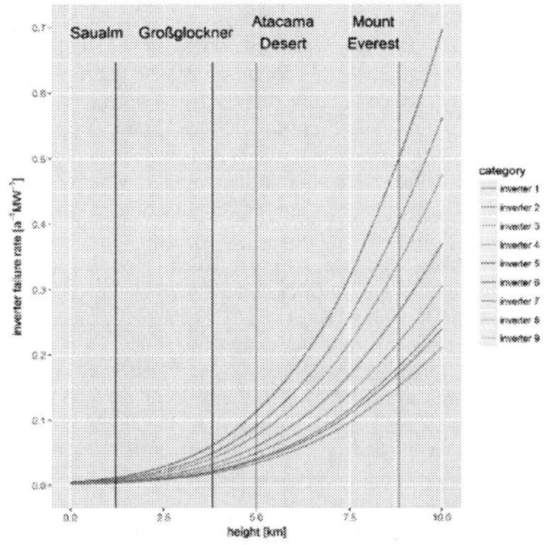

Figure 4: Plot of the developed model using inverter data from several inverters in Austria at about sea level, using the fitted altitude relation, without applying corrections to the climate in different heights. For height reference several mountain tips were added by vertical cuts.

was shown. This model consisted of several parts using neutron density flux at different altitudes, temperature dependencies from literature, device characteristics from the inverter data sheet and measured FIT from the power switch used. This model may be adapted to different inverter types, by exchanging input parameters in accordance with the device of interest.

The current state of the model in Fig. 4 implies that on the highest possible installation by now (the Mount Everest), a ratio 0.15 to 0.5 inverts fail in a 1 MW PV system within 1 year. This means in a 10 MW system 1 to 5 inverters will fail due to cosmic rays in 1 year.

Additional work has to be done in order to compare different climatic condition's impact to device failure rate and correct for varying climatic conditions with altitude. To validate the model, spontaneous device failure rates have to be used together with the device location of operation.

5 ACKNOWLEDGEMENT

This work was conducted as part of the Austrian "Energy Research Program" project INFINITY. Funded by the Austrian Climate and Energy Fund and the Austrian Research Promotion Agency (FFG), both are gratefully acknowledged

6 REFERENCES

[1] J. F. Ziegler *et al.*, "IBM experiments in soft fails in computer electronics (1978–1994)," *IBM journal of research and development*, vol. 40, no. 1, pp. 3–18, 1996.

[2] H. R. Zeller, "COSMIC RAY INDUCED FAILURES IN HIGH POWER SEMICONDUCTOR DEVICES," *Solid-State Electronics*, vol. 38, no. 12, pp. 2041–2046, Mar. 1995.

[3] W. Kaindl, "Modellierung höhenstrahlungsinduzierter Ausfälle in

Halbleiterleistungsbauelementen," Technischen Universität München, München, 2005.

[4] F. D. Bauer, "Accurate analytical modelling of cosmic ray induced failure rates of power semiconductor devices," *Solid-State Electronics*, vol. 53, no. 6, pp. 584–589, Jun. 2009.

[5] C. Weiß, "Höhenstrahlungsresistenz von Silizium-Hochleistungsbauelementen," Technischen Universität München, München, 2014.

[6] ABB, "Failure rates of IGBT modules due to cosmic rays," Application note 5SYA 2042-06.

[7] ABB, "Failure rates of IGCTs due to cosmic rays," ABB, Application note 5SYA 2046-03.

[8] R. B. Mendell and S. A. Korff, "Fast-Neutron Flux in the Atmosphere," *J. Geophys. Res.*, vol. 68, no. 19, pp. 5487–5495, Oct. 1963.

[9] J. F. Ziegler, "Terrestrial cosmic rays," IBM journal of research and development, vol. 40, no. 1, pp. 19–39, 1996.

[10] R. Stull, Practical meteorology an algebra-based survey of atomspheric science. Vancouver: UBC, 2015.

[11] "Test Procedure for the Measurement of Terrestrial Cosmic Ray Induced Destructive Effects in Power Semiconductor Devices." JEDEC Publication No. 151, Dec-2015.

[12] F. D. Bauer, "Calculation of cosmic ray limited maximum DC blocking voltages of high voltage silicon PIN diodes," Solid-State Electronics, vol. 52, no. 7, pp. 1052–1057, Jul. 2008.

ECO-SOLAR FACTORY: ESTABLISHMENT OF PAN INDUSTRIAL MATERIAL RE-USE OPPORTUNITIES

K. Wambach[1], I. Fechner[1], M.P. Bellmann[2], G.S. Park[3], J. Denafas[4], F. Buchholz[5], F. Madon[6], G. Noja[7], B. Ehlen[8], R. Roligheten [9], P. Romero[10], A. Bollar[11]

[1] bifa Umweltinstitut GmbH, Am Mittleren Moos 46, 86167 Augsburg, Germany;
phone: +49 821 7000-198, fax -100, e-mail: kwambach@bifa.de
[2] SINTEF Materials and Chemistry, Trondheim, Norway
[3]Norsun AS, Oslo, Norway
[4]UAB Soli Tek R&D, Vilnius, Lithuania
[5]International Solar Energy Research Center - ISC Konstanz, Konstanz, Germany
[6]Apollon Solar, Lyon, France
[7]Garbo Srl, Cerano, Italy
[8]Boukje.com Consulting BV, EM Bleiswijk, The Netherlands
[9] Steuler Solar Technology AS, Porsgrunn, Norway
[10]Aimen Laser Application Centre, Porrino, Spain
[11]Ingesea Automation SL, Elgoibar, Spain

ABSTRACT:
The European H2020-FOF-13-2015 Eco-Solar project, that was started in October 2015 envisions to increase resource and energy efficiency over the entire photovoltaic value chain, while simultaneously maximising recycling and remanufacturing possibilities, by introducing design for recovery, repair and reuse, and collaborating for improvements in waste reduction. Reusing materials and reducing the consumption of raw materials will make solar cell panels both cheaper and greener. A pan-industrial approach is used to create an optimized valorisation of by-products, recycled and recovered wastes and mitigate the environmental impacts of PV production and recycling. The best possible utilisation with respect to the eco-efficiency assessment results is identified and novel solutions for poorly recovered and reused materials today will be presented

Keywords: crystalline silicon, eco-efficiency analysis, life cycle assessment, life cycle costing, living lab

1 INTRODUCTION

Eco-Solar is a 3 years H2020 project, which involves 11 partners from 7 member states. Details can be found on the project website: http://www.ecosolar.eu.com.

The target is to identify reuse and recovery routes for the main products and the by-products within the solar and if this will not be possible in other industries. The major material streams are assessed with respect to reuse, remanufacturing, recyclability and utilization of the output for different applications to optimize the eco-efficiency utilizing the Living Lab Methodology. The use of the output may take place in completely different industrial branches. Potential customers and users are invited to test the materials and provide feedback on the experience made. This is done via a living lab, project webpage, trade fairs and conferences.

Eco-Solar aims to develop concepts to improving reuse of resources and components and increasing resource efficiency over the whole value chain. First results on the application of the Living Lab methodology to achieve a better recycling, utilisation and valorisation of by-products and wastes along the PV value chain are presented in 6 examples with a special focus on materials that are valuable or are present in large quantities.

A pan-industrial approach is used to create an optimized valorisation of by-products, recycled and recovered wastes and mitigate the environmental impacts of PV production and recycling. The best possible utilisation with respect to the eco-efficiency assessment results is identified and novel solutions for poorly recovered and reused materials today will be presented

2 AIM AND APPROACH USED

The whole value chain starting with the productions of Czrochalski crystals or multicrystalline Si ingots to the module production was analyzed for by-products and wastes, their quantities, potential values and environmental impacts. First results of the intermediate environmental assessment are presented in a separate paper [1]. Examples of targeted wastes and by-products selected so far are:

1. ingot production: silica, quartz and silicon nitride crucibles Quartz or silica crucibles are used only once in ingot melting as a current standard practice. Silica crucibles contribute up to ~30% of the conversion cost from Si-feedstock to the as-grown ingot. The used crucibles are usually landfilled, there is hardly any utilisation. Potential utilisations in other industries are under research
2. silicon, silver: silicon cut-offs, broken wafers and solar cells from the production steps like ingot cutting, wafer cutting, solar cell and module manufacturing and end of life treatment, damaged/defective modules, etc. The focus is set on silicon and silver from solar cell processes and recovery because the recovery of silicon is state of the art and there are worldwide specialized companies recycling the material as secondary feedstock or if this is not possible in ceramic or metal alloying applications. Here new potential applications of secondary silicon are targeted.
3. diamond wire from wafer cutting: The wires become thinner and thinner, after cutting hardly any diamonds are present. Several types of wires are used on the market. New solutions of the recycling of the diamond wire are examined.
4. glass sheets and cullet: Glass is usually recycled, only

heavily contaminated fines might be landfilled. Major uses are remelting of the glass in flat glass manufacturing, container glass or glass foam and glass fiber production depending on the quality. Processes to achieve a better quality and new applications of the cullet are investigated.
5. Cu interconnectors from modules: The yield and quality of the Cu interconnectors from recycling processes are examined and improved.
6. Al-frames: The quality and purity of the frames in different recycling processes are assessed to allow a better reutilization.

The following pictures give an overview of the mentioned materials and their potential re-uses.

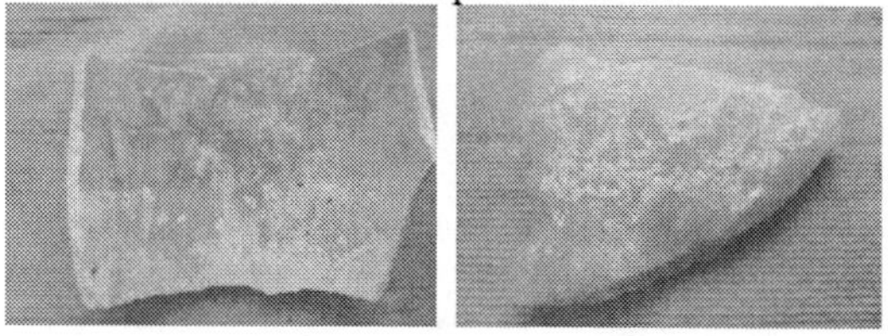

Picture 1 and 2: Silica (left)/quartz (right) crucibles ceramic/glass industry, horticulture, building industry, production of abrasives, ferrosilicon production, additives for dispersions/polymers

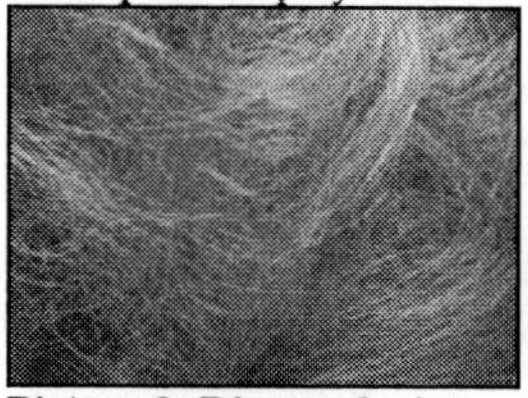

Picture 3: Diamond wire steel industry, diamond industry, metal recycling

Picture 4: Copper wire noble metal recycling, electronics industry

Picture 5 and 6: Retrieved (left)/chemically treated (right) solar cell metallurgy, noble metal recycling, solar cell producer

Picture 7: Wafer sawing slurry possibly containing diamonds metallurgy, metal recycling, organic chemistry (additives), production of abrasives, sawing wire industry, tool industry

Picture 8: Hot zone graphite: additives for ceramic formulas/ dispersions/ polymers/batteries, foundry/steal industry, automobile industry

Picture 9 and 10: Glass sheet (left)/cullet (right) glass recycling, building industry

Picture 11: Al-frame metal industry/ electronics industry

PRELIMINARY RESULTS AND CONCLUSION

Evaluation of crucible re-use opportunities

The silica and quartz crucibles used in the silicon ingot production are currently landfilled in despite of their sufficient quality with regard to possible re-use opportunities. Especially, the horticulture, the production of ferrosilicon, additives for dispersions/polymers and cement/concrete are suitable possibilities.

Table 1: Considerations about the reuse of different crucibles

used quartz and silicon nitride crucibles	***description***
result	high pure silica and silicon nitride crucibles are currently landfilled after use; secondary applications shall be developed for the material
unique selling point	very high purity in comparison to other similar materials
product market size	global market size ~ 50,000 t/a
product positioning	industrial application: ceramic/glass industry, horticulture, building industry, Si-production
requirements (need for authorisations, compliance to standards, norms, etc.)	material has to be modified in order to fulfil the needs of customers
competitors/ incumbents	raw material suppliers
foreseen product price	about 100 €/t
external experts/partners to be involved	silicon smelters, ceramic/glass industry, horticulture, building industry
sources of financing foreseen	savings from landfill costs

Table 2: Analysis of trace elements in a quartz crucible (vitreous $SiO_2 \geq 99.99\%$)[1]

	Al	*15*
Trace elements in ppm	Ca	0.6
	Cr	<0.8
	Cu	<0.3
	Fe	0.5
	K	0.6
	Li	0.7
	Mg	0.06
	Na	0.9
	Ni	<0.03
	Ti	1.3
	OH	30

It can be assumed that similar purity can be achieved with used quartz and silica crucibles after the recycling because the material was in contact with solar grade silicon only. Furthermore, there may still be some silicon residues on both types of crucibles and concerning the quartz crucibles it is obvious that the material went through a phase transformation to β-Cristobalite.
The horticulture is one thinkable re-use of the quartz crucible pieces. As an example, these pieces can be used as a filling material for the so-called gabiones.

Picture 12: Gabione bench filled with used quartz crucibles pieces

Laboratory test results of recycling potential

ICP-OES analytical results of a diamond wire (new and used)

Diamond wire is used for wafer cutting in the semiconductor and solar industry. The diamonds are fixed on the wire with a nickel coating or an organic coating. The project partner Norsun provided samples for the tests. The properties of the used diamond wire are determined by ICP-OES and REM-EDX.

Table 3: Diamond wire metal composition (example, new and used wire, ICP-OES)

element	*New, average value of mass [%]*	*Used, average value of mass [%]*
Cr	1.13	1.17
Fe	83.5	87.3
Ni	15.1	11.5
others	< 0.1	< 0.1

The used diamond wire is expected to be suitable for the reuse in the steel industry. There are experiments in progress to determine a cost-effective treatment in order to remove and to extract the diamonds from the used wire.

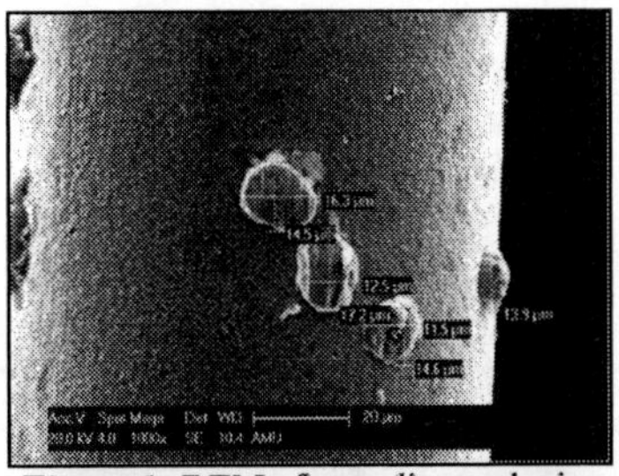

Figure 1: REM of new diamond wire

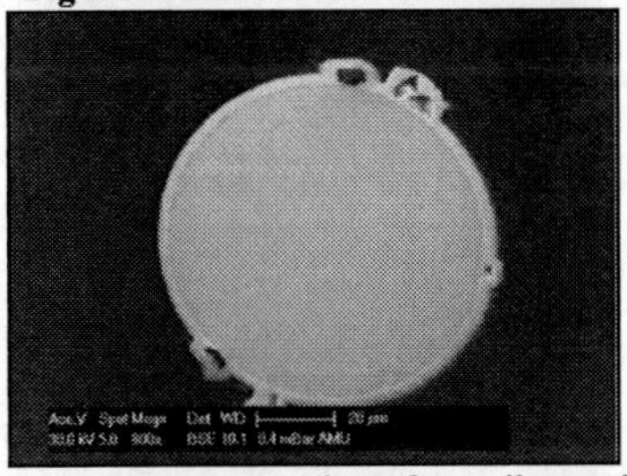

Figure 2: REM cross section of new diamond wire

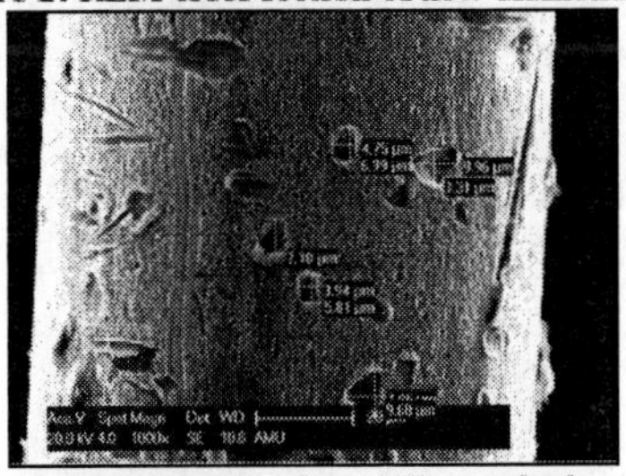

Figure 3: REM of used diamond wire

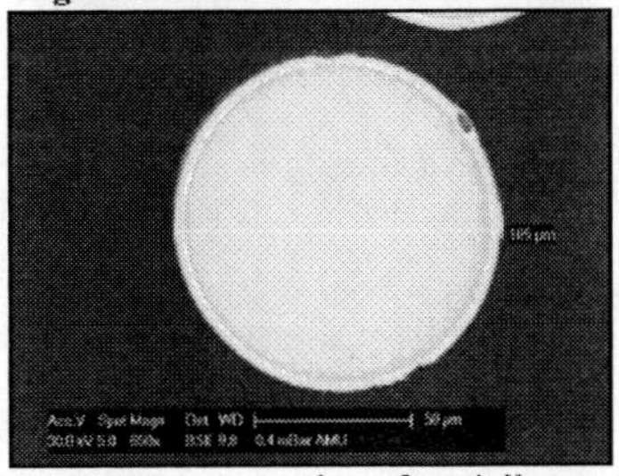

Figure 4: REM cross section of used diamond wire

Analysis of Used Copper Wire

The purity of the used copper wire is 99.93%. Other elements are below the limit of determination and therefore not detectable. The copper wire is analysed by ICP-OES. Copper is the dominant interconnector material but its market share will decrease in the next years due to alternative cell interconnector materials like conductive adhesives which shows lower thermal stresses. These thermal stresses, due to the soldering process, are an issue concerning a direct re-use of extracted copper wires in a new solar module. This material does not fulfil the quality demanded. Concerning the NICE Module this has to be checked.

[1] Source: product data sheet from Saint-Gobain Quartz given by SINTEF

ICP-OES analytical results of an example solar cell composition provided by ISC

Further investigations concerning the solar cells are planned. It is desired to replace the silver on a large scale by a more cost-effective metal like copper. This replacement still demands a lot of research and development and is expected to be introduced step by step in the next decades. By the time being, silver remains the most widely used metallization material in the solar cell production.

Table 4: Metallic components on a solar cell (example)

element	*average value of mass [mg/cell]*
Tl	6.4
Cd	not detectable
Ag	205.8
Al	934.7
Cu	0.4
Pb	4.4

Wafer sawing slurry and diamonds in the slurry

The used slurry from the wafer sawing process contains several components (e.g. diamonds, steel, additives) which may be still valuable if recoverable. In addition, a recovery or even a replacement of metalworking fluids additives is considered as worth for further research. In despite of the re-use opportunities, these additives are potentially toxic for human health and the environment, with long lasting effects to some extent.

Moreover, it can be assumed that most of the diamonds from the wire might be torn out during the sawing process and remain in the slurry. Therefore bifa is examining the slurry in order to find out if a subsequent treatment of the disposed slurry is feasible and cost-effective.

Example: Apollon Solar NICE Module, amount of possible secondary raw materials (Cu, Si, Ag, ...)

The waste regulation assessment is based on the data of the NICE Module (BOM, composition, PIB, etc.) delivered by Apollon Solar as well as on the disassembly experiment and the subsequent separation of the compounds at bifa. The definition of appropriate waste streams is the main goal for further considerations concerning recovery/recycling routes and reuse opportunities. Therefore, the quality of the extracted compounds is being examined. Furthermore, the waste streams have to be analysed with respect to hazardous waste criteria. Referring to the state of art modules, some module compounds can be seen as hazardous waste, for instance Cd-containing bus bars or polymers. Therefore, an analysis is taking place in order to confirm the indication that the NICE Module compounds are Cd-free. Identifying the share of the several compounds and allocating the waste key numbers to them is fundamental for the assessment. The definition of the different waste streams is based on the waste regulation. The experiments carried out confirm the assumption of a damage-free disassembly, nevertheless, an automatic disassembly process is crucial in order to compete with new raw materials. The recovery and re-use of intact components seem to be possible. The repair of recovered components such as defective solar cells and subsequent re-use are thinkable but the viability strongly depends on the current cost of new components in comparison with the recovery costs. In addition, the quality requirements will have to be met if a direct re-use is targeted. It is planned to take a closer look on this as a next step.

The compounds of the NICE Module are already allocated to different waste streams. The assessment of each waste stream has been started. It is expected that the new European waste regulation will not be an issue inhibiting reuse and recovery options.

Table 5: NICE PV Module compounds and possible recycling ways (non-exhaustive)

module compound	*material*	*mass [%]*	*possible use of output*
glass	soda-lime glass	30-65	glass industry ceramic industry horticulture building industry
composite material	Acrylate PU	5-10	thermal treatment separation
rear side covering	polyester Al steel glass	0-10 (without glass)	thermal treatment (polymers) plastic recycling metal industry glass industry
frame	Al steel PU PC	0-20	thermal treatment (polymers) plastic recycling metal industry
junction box	ABS, PC, PPO, PET	0-5	thermal treatment plastic recycling
wire	Cu polyolefines synthetic rubber	ca. 1	metal industry wire recycler electronics industry/ recycling
sealant	silicone polyacrylate PE – foam PU	0-10	thermal treatment
filler materials	Al_2O_3, TiO_2, $CaCO_3$, SiO_2, C	ca. 1	fillers mineral additive
solar cells	Si, Ti, Ag, Sn, Pb, Cu, Ni, Pd	5-10	metallurgy solar cell production
electrical connector	Cu, Sn, Pb, Al, Ag	ca. 1	metal industry electronics industry

Hot zone graphite: first customer feedback results

The hot zone graphite is currently used in order to insulate the crucibles during the production of solar grade silicon. As well as the crucibles, the hot zone graphite is used once and landfilled afterwards.

bifa has conducted a survey aiming to find out the requirements and needs of suitable customers of this material. These customers are producers of carbon electrodes, graphite film, polymers (carbon as conductivity additive) as well as members of the foundry and steal industry. As a result, most of these potential customers are interested in the hot zone graphite under certain conditions. A considerable condition is that the hot zone graphite pieces have to be ground to a specific particle size with regard to application. As a consequence, the hot zone graphite has to be treated in advance. The viability of the pre-treatment is being assessed at bifa. Additionally, some prepared samples have been sent to potential customers and results are expected in the near future.

Recovery of solar panel compounds: comparison of current mechanical treatment results

The experiments verify the feasibility of damage-free disassembly of a NICE Module. However, the development of a competitive automatic disassembly to decompose an end-of-life NICE Module into its different components is necessary.

Table 6: Comparison of mechanical treatments

no.	*kind of mechanical treatment*	*recovered materials*	*specific potential value [€/t]*
1	laminated glass recycling plant, incineration in municipal waste incineration plant	Al, glass, Cu	122
2	laminated glass recycling plant, incineration of mono	Si, Al, Ag, glass, Cu	316
3	laminated glass recycling plant, incineration	Solar Si, Al, Ag, glass, Cu	476

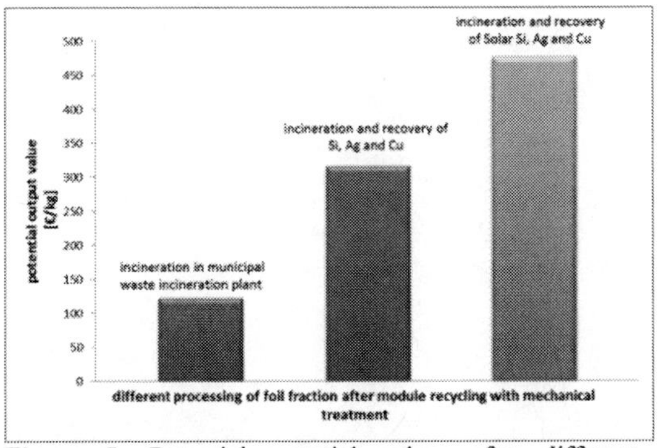

Figure 5: Potential material values after different recycling processes of PV modules

The comparison of mechanical treatments indicates that highly pure outputs of the recycling treatment will increase significantly the value of the secondary raw materials obtained. Moreover, the re-use of secondary raw materials within the solar industry can decrease the cost of solar panels making this kind of renewable energy source even more attractive. In addition, the establishment of a cost-effective treatment may have a positive impact on the eco-efficiency, too.

Table 7: Considerations about the recovery of solar panel compounds

solar panel components	*description*
target	using the state of art solar panel recycling systems in order to generate commercial benefits and to increase the eco-efficiency
result	recovery of solar panel compounds: high yield and high purity desired; secondary applications shall be easily enabled
unique selling point	equal quality for a lower price in comparison to raw materials
product market size	global market size [2] about 43,000 Mt/a
product positioning	industrial application: metal industry, metal recycling, metallurgy, solar cell/panel producer, glass industry, building industry, Si-production
requirements	material has to be modified in order to fulfil the needs of customers
competitors/ incumbents	raw material suppliers, recycling companies
foreseen product price	about 350 €/t
external experts/partners to be involved	laminated glass recycling plant owner, silicon smelters, metal industry, metal recycler, metallurgy, solar cell/panel producer, glass industry, building industry, Si-production
sources of financing foreseen	proceeds from the separated and purified solar panel compounds, using an established laminated glass recycling plant, savings from landfill costs

The question whether there is a subsequent treatment of the disposed wafer cutting slurry is still a challenge bifa deals with. The expectation has to be considered that the throughput is assumed to decline remarkably in the near future. Hence it follows, that the amount of possibly recovered material decreases significantly. Thus the conditions of a recovery or re-use route become a demanding task. Further research and input from project partners are necessary.

[2] IRENA and IEA-PVPS (2016), "End-of-Life Management: Solar Photovoltaic Panels," International Renewable Energy Agency and International Energy Agency Photovoltaic Power Systems

Table 8: Considerations about the recovery of wafer sawing slurry containing diamonds

wafer sawing slurry containing diamonds	***description***
target	using the state of art extraction recycling systems in order to generate commercial benefits and to increase the eco-efficiency
result	recovery of fluid, Si and diamonds: high yield and high purity desired; secondary applications shall be easily enabled
unique selling point	equal quality for a lower price in comparison to raw materials
product market size	global market size about >100 kt/a
product positioning	metallurgy, metal recycling, organic chemistry concerning additives, production of abrasives, sawing wire industry, tool industry
requirements	slurry has to be treated in order to extract the valuables
competitors/ incumbents	raw material suppliers, recycling companies
foreseen product price	<10 €/kg
external experts/partners to be involved	silicon smelters, metal industry, metal recycler, metallurgy, tool industry, Si-production
sources of financing foreseen	proceeds from the separated and purified slurry, using an established extraction technology, savings from landfill costs

Europe wants to reduce its needs for raw materials and raise the level of recycling of resources in the solar power industry. Our target is that after the successful completion of this project the greenhouse gas emissions from solar panel manufacturing will be reduced by 25 to 30 per cent and the waste generated will be decreased by 10% minimum.

REFERENCES

[1] K. Wambach, R. Peche, M. Seitz, M.P. Bellmann, G.S. Park, J. Denafas, F. Buchholz, R. Einhaus, G. Noja, B. Ehlen, R. Roligheten, P. Romero, A. Bollar, ECO-SOLAR FACTORY: ENVIRONMENTAL IMPACT OPTIMISATION OF PV PRODUCTION, EU PVSEC 2017, 33rd European Photovoltaic Solar Energy Conference and Exhibition, 25 – 29 September 2017, Amsterdam, The Netherlands

[2] DIRECTIVE 2008/98/EC OF THE EUROPEAN PARLIAMENT AND OF THE COUNCIL of 19 November 2008 on waste and repealing certain Directives

[3] REGULATION (EC) No 1272/2008 OF THE EUROPEAN PARLIAMENT AND OF THE COUNCIL of 16 December 2008 on classification, labelling and packaging of substances and mixtures, amending and repealing Directives 67/548/EEC and 1999/45/EC, and amending Regulation (EC) No 1907/2006

[4] IEA, 2014. Technology Roadmap: Solar Photovoltaic Energy. available at: http://www.iea.org/publications/freepublications/publication/technology-roadmap-solar-photovoltaic-energy---2014-edition.html.

[5] DIRECTIVE 2012/19/EU OF THE EUROPEAN PARLIAMENT AND OF THE COUNCIL of 4 July 2012 on waste electrical and electronic equipment (WEEE)

[6] Sander, K. e. a. (2007): Study on the development of a take back and recovery system for photovoltaic modules BMU grant No 03MAP092.

[7] European Chemicals Agency (ECHA): https://www.echa.europa.eu/de/web/guest/information-on-chemicals/registered-substances, 3 March 2017

ACKNOWLEDGEMENT

This project has received funding from the European Union's Horizon 2020 research and innovation programme under grant agreement No. 679692.

SUSTAINABLE RECYCLING OF WAFER-SILICON SOLAR MODULES

M. Tao,[a] W.-H. Huang,[b] and J. Schichtel[b]
[a] School of Electrical, Computer and Energy Engineering and [b] School for Engineering of Matter, Transport and Energy, Arizona State University, Tempe, Arizona 85287, USA
[a] Phone: +1 480 965 9845 Email: meng.tao@asu.edu

ABSTRACT: A major obstacle to solar module recycling is the cost associated with collecting and processing end-of-life modules. Our group recently reported a multi-step recycling process to breakdown wafer-Si modules and recover all the valuable, bulky and toxic materials including solar-grade Si, Ag, Al frame, glass, Cu, Sn and Pb. This paper presents a study on the financial and environmental sustainability of our recycling process. At today's prices for Ag and solar-grade Si, this process can produce at least $15/module from the recovered materials and potentially enable a profitable recycling business for Si modules without any government support. The risk in recycling revenue is future Ag-free modules and lower solar-grade Si prices, but additional revenues can be generated by recovering the glass unbroken and clean, resulting in $17.50/module in revenue. A network for collecting end-of-life modules is proposed based on the current distribution network for solar modules to contain the collection cost. To avoid HF wastes from the recycling process, alternatives need to be found for the fluoropolymer backsheet and SiN_x layer in Si modules.
Keywords: Recycling, Silicon

1 INTRODUCTION

The International Renewable Energy Agency (IRENA) estimates that end-of-life solar modules will appear in large quantities by the early 2030's and by 2050, they will total 78 million tonnes [1]. Among the various module technologies on the market, wafer-Si modules have always been the dominant one with a ~90% market share [2]. Thus, sustainable recycling of Si modules is a necessity for a sustainable solar industry. A major obstacle to module recycling is the cost associated with collecting and processing end-of-life modules. In Europe, this cost is partially offset by a small fee collected on each new module installed through legislative measures [3]. With the large number of new modules being installed and the small number of end-of-life modules to be recycled today, questions remain about the long-term prospects of this funding mechanism.

Table I: Composition of wafer-Si modules [4]

Material	Content (weight %)
Glass	74
Al	10%
Polymers	6.5%
Si	3%
Pb	0.1%
Cu	0.6%
Ag	0.006%

An alternative to subsidies is to maximize the financial returns from recycling. The most valuable materials in Si modules are solar-grade Si and Ag (Table I), which along with the bulky materials including Al and glass, can generate at least $15/module at today's prices for solar-grade Si and Ag. Our group recently demonstrated a technology to recover solar-grade Si, Ag, Cu, Sn, as well as toxic Pb from Si modules [5]. This paper extends our work and focuses on the sustainability, both financial and environmental, of our proposed recycling process for Si modules.

2 OUR RECYCLING PROCESS

Figure 1 presents our proposed process to break down wafer-Si modules and recover various materials. It involves two steps: module recycling and cell recycling. In the first step, the junction box and Al frame are mechanically removed for recycling. The polymers are then burned off in a furnace to separate the cells from the glass [6,7]. The exothermic reaction of burning polymers serves as a heat source for the furnace [7], reducing the energy input for the furnace. After the thermal process, the glass, broken or intact, is recycled for new glass and what is left is a string of interconnected cells.

In cell recycling, interconnected cells are first immersed into a leaching solution of concentrated HNO_3 to dissolve four metals, Ag, Pb, Cu and Sn. These metals are then recovered from the leaching solution one by one through a technology called sequential electrowinning [5]. The metal recovery rates are currently 74% for Ag, 83% for Cu and over 99% for Pb. With further efforts, a recovery rate of over 95% is expected for all metals. The recovered metals are at least 99% pure and can be sold as raw materials to the solar industry. The remaining cells are dipped into a solution of HF to remove the SiN_x layer and Al back electrode. Finally, the cells are submerged into a solution of NaOH to remove the emitter and back-surface field and to recover the base as solar-grade Si. By monitoring the sheet resistance of the Si wafer, the precise moments for complete removal of emitter and back-surface field can be observed [5], which maximizes the amount of solar-grade Si recovered. The recovery rate of Si is currently 91%.

Although our recycling process is landfill free, it generates chemical wastes which must be taken into consideration. There are three waste solutions, NaOH, HNO_3 and HF. NaOH and HNO_3 can neutralize each other to form a solution of $NaNO_3$, which is a fertilizer [8]. There is a well-established practice to treat HF wastes, i.e. $Ca(OH)_2$ is added to HF and the CaF_2 sludge is either used to regenerate HF or goes to landfill. There are also two gaseous exhausts from the recycling process. One is the exhaust from the polymer-burning furnace, which contains fluorine. The other is the exhaust from metal dissolution in HNO_3, which contains NO and/or

NO_2. Scrubbers are needed to trap F in water as HF and to trap NO and NO_2 as HNO_3. These acids can be used in the recycling process.

Further efforts for the recycling process include [5]: 1) improve the recovery rates to over 95% for metals, Ag, Cu and Sn; 2) replace the Pt anode with a lower-cost anode in sequential electrowinning and 3) develop specialized commercial equipment for Si module recycling.

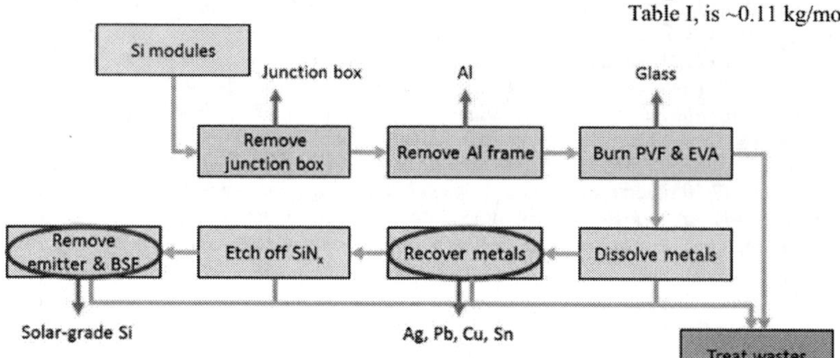

Figure 1: Our proposed process for recycling wafer-Si modules

3 FINANCIAL ANALYSIS

Our work demonstrates the technical feasibility of breaking down Si modules and recovering all the valuable, bulky and toxic materials. A fundamental question about our value-added process is whether it can pay for itself. Here a financial analysis is presented on the revenues our process can generate and also on the major cost contributors.

A typical 60-cell Si module weighs 18–18.5 kg. Each cell is a Si wafer of 15.6 cm × 15.6 cm × 180 μm, adding to ~0.61 kg of Si per module. If 90% of the Si is recovered as solar-grade Si, it is worth \$8.96/module at \$16.26/kg for solar-grade poly-Si or \$7.47/module at \$13.55/kg for 2nd grade poly-Si as of January 31, 2017 [9]. Each cell now uses ~115 mg of Ag, i.e. each 60-cell module contains ~7 g of Ag. If 98% of the Ag is recovered, it is another \$4.18/module at \$17.52/oz for Ag (January 31, 2017). As a result, recoverable valuable materials in a typical Si module add to \$11.65–13.14 depending on the quality of the recovered Si.

The Ag price has been fluctuating between \$14/oz and \$46/oz since 2010 [10]. If it goes back to \$46/oz, the recovered Ag would be worth \$10.97/module and the total value from recovered Si and Ag would be \$18.44–19.93/module. On the other hand, if the Ag price goes down to \$14/oz, the recovered valuable materials would be \$10.81–12.30/module. There is uncertainty in recycling revenue from ~\$11 to ~\$20 largely due to Ag.

Table II: Revenue from our recycling process

Material	Recovery %	Value	% Total
Glass	100%	\$0.43	2.8
Al	100%	\$2.49	16.5
Polymers	100%	n/a	0
Si	90%	\$7.47–8.96	49.6
Pb	99%	n/a	0
Cu	95%	\$0.50	3.3
Ag	98%	\$3.34–10.97	27.7

Additional revenues come from bulky materials including Al frame, glass and junction box. At 10% of the module weight (Table I), the Al frame is ~1.8 kg. The price of scrap Al extrusion is \$0.59–0.65/lb [11] or \$1.30–1.43/kg, translating into \$2.49/module on average from the 100% recoverable Al frame. The glass weighs ~13.5 kg. As the price of scrap glass is \$0.0145/lb or \$0.0320/kg in the U.S., the 100% recoverable glass brings in \$0.43/module. The amount of Cu, based on Table I, is ~0.11 kg/module. With the price of scrap Cu at \$2.07–2.36/lb [11] or \$4.56–5.20/kg and 95% recovery, the recovered Cu adds ~\$0.50/module.

The total revenue a typical 60-cell Si module can generate is thus \$14.25–23.35, as listed in Table II. At today's price for Ag and assuming 2nd grade poly Si, the revenue is \$15.07/module, with ~50% of the revenue from Si, ~28% from Ag and ~17% from Al. Polymers are recovered as a heat source but do not bring in revenues. The tiny amounts of Pb and Sn are neglected in the analysis.

As a metric ton (tonne) contains 55 Si modules, the revenue on a per weight basis from our recycling process is \$829/tonne. This number is much higher than the analysis by IRENA, which predicts \$15 billion in revenue from recycling 78 million tonnes of solar modules [1]. Our process would generate ~\$60 billion from 78 million tonnes of solar modules, as it recovers all the valuable materials in their pure, high-value forms. Without Si and Ag, Si module recycling generates only \$3.42/module or \$188/tonne.

The cost of sending municipal solid wastes to landfills is on average \$50/tonne in the U.S. [12]. Our process turns the negative \$50/tonne to a positive \$829/tonne. It is a net gain of \$879/tonne.

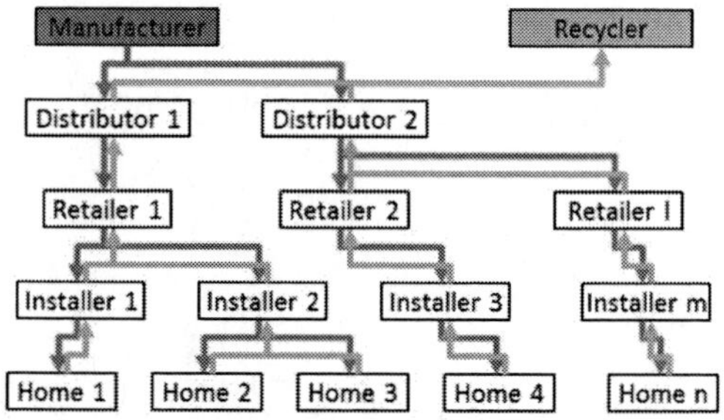

Figure 2: A network for module collection based on the current distribution network

The cost for recycling Si modules is difficult to estimate at this time. For a profitable recycling business, a cost below \$12/module is needed for a gross margin of 20%. A major contributor to the cost of module recycling is the cost to collect and transport solar modules which are scattered around in small quantities. To contain the collection cost, we propose to utilize the current distribution network of solar modules in the reverse order as a collection network (Fig. 2). Installers go to homes to perform repair and pick up end-of-life modules, which are shipped to retailers, then to distributors and finally to recyclers who operate centralized recycling facilities and generate revenues. To finance the network, each party in the network receives monetary compensation from the next party in the value chain.

The transportation cost is another contributor. Rail is \$0.045/tonne-mile in the U.S. [13]. Assuming a 500-mile average distance to centralized recycling facilities, the transportation cost by rail is \$23.32/tonne or \$0.37/module. Barge is ~25% cheaper than rail, but truck is ~6× more expansive than rail. Future recycling facilities must have access to railways.

4 FUTURE PROSPECTS

One thing certain about the future is that the amount of end-of-life modules to be recycled will grow exponentially over the next several decades. Since 2000, the annual installation of solar modules has increased by ~250×, and the economies of scale will help profitability in module recycling.

A potential risk in profitable recycling of Si modules is the elimination of Ag in Si cells [14]. Ag is expansive and the availability of Ag may limit terawatt-scale deployment of Si modules. Without Ag, the revenue from a typical 60-cell Si module would be down by ~28% (Table II) to \$10.89–12.38/module. The cost of recycling, therefore, has to be reduced to below \$9/module. If the price of Si goes down to \$10/kg, the recovered Si would be worth \$5.51/module and the revenue for all the recovered materials down to \$8.93/module.

One way to increase the revenue from module recycling is to recover the glass intact. The price of solar glass is currently ~\$5.50/m². This translates into \$8.57/module for the glass and makes glass as valuable as Si, if it is recovered unbroken and clean. Advanced equipment is needed to gently separate the junction box and Al frame from the glass, burn off the polymers and clean the glass. Table III shows the revenues under various scenarios.

Table III: Future revenue from Si module recycling

Variation	Value/module	Change
Baseline	\$15.07	
Ag-free cells	\$10.89	−\$4.18
Si at \$10/kg	\$8.93	−\$1.96
Glass intact	\$17.50	+\$8.57

Environmental sustainability and regulatory compliance also need some discussions. While the choices of NaOH and HNO_3 for our recycling process result in a safe, useful waste solution of $NaNO_3$, the HF wastes from cell recycling and from burning fluoropolymers create an environmental and regulatory problem. There are practically no processes generating HF wastes that have been approved in recent years in Europe.

To eliminate HF wastes from Si module recycling, an F-free backsheet has to be developed to replace the current polyvinyl fluoride based backsheet, without negatively affecting module reliability and lifetime. At the cell level, there is no common chemical other than HF that can remove SiN_x. A replacement for SiN_x needs to be found for Si cells to avoid HF in the recycling process. The replacement needs to have the right bandgap (over 3.5 eV), refractive index (over 2.0), positive fixed charge and hopefully a high content of hydrogen. It is noted that billions of Si modules with SiN_x have been produced and the production will continue for years to come, i.e. we have to deal with HF wastes for these billions of Si modules.

5 CONCLUSIONS

A multi-step recycling process for wafer-Si modules has been demonstrated to recover all the valuable, bulky and toxic materials. The financial and environmental sustainability of the recycling process is examined. At today's price for Ag, this recycling process can produce at least \$15/module from the recovered materials. The risk in recycling revenue is future Ag-free modules and lower solar-grade Si prices, but additional revenues can be generated by recovering the glass unbroken and clean, resulting in \$17.50/module in revenue. To avoid HF wastes from the recycling process, alternatives need to be found for the fluoropolymer backsheet and SiN_x layer in Si modules.

6 ACKNOLEDGEMENT

Financial support for this project is provided in part by the U.S. National Science Foundation under grant no. DMR-1306542.

7 REFERENCES

[1] M. Smith, A. Miller, Proceedings 17th European Photovoltaic Solar Energy Conference, Vol. I (2002) 903.

[2] K.W. Boer, Solar Cells 16 (1996) 591.

[1] International Renewable Energy Agency, End-of-Life Management: Solar Photovoltaic Panels, 2016.

[2] Fraunhofer Institute for Solar Energy Systems, Photovoltaics Report, 2016.

[3] European Union, Waste Electrical and Electronic Equipment Directive, http://www.solarwaste.eu.

[4] Bine Information Service, Recycling Photovoltaic Modules, 2010.

[5] W.-H. Huang, W.J. Shin, L. Wang, W.-C. Sun, and M. Tao, Solar Energy 144 (2017) 22.

[6] J.R. Bohland and I.I. Anisimov, 26th IEEE Photovoltaic Specialist Conference, (1997) 1173.

[7] L. Frisson, K. Lieten, T. Bruton, K. Declercq, J. Szlufcik, H. de Moor, M. Goris, A. Benali, and A. Aceves, 16th European Photovoltaic Solar Energy Conference, (2000) 2160.

[8] U.S. Department of Agriculture, Supplemental Technical Report for Sodium Nitrate (Crops), 2011.

[9] PVinsights, PV Poly Silicon Weekly Spot Price, http://pvinsights.com/index.php.

[10] InvestmentMine, Historical Silver Prices and Price Chart, http://www.infomine.com/investment/metal-prices/silver/all/.

[11] ScrapRegister, United States Scrap Metal Prices, http://www.scrapregister.com/scrap-prices/united-states/260/.

[12] U.S. Environmental Protection Agency, Municipal Solid Waste Landfills, 2014.

[13] U.S. Department of Transportation, Average Freight Revenue per Ton-Mile, https://www.rita.dot.gov/bts/sites/rita.dot.gov.bts/files/publications/national_transportation_statistics/html/table_03_21.html.
[14] M. Tao, Terawatt Solar Photovoltaics – Roadblocks and Opportunities, Springer, London, 2014.

REMELTING AND PURIFICATOIN OF Si-KERF FOR PV-WAFERS

M. Syvertsen[1], T. Halvorsen[2], K. Mørk[2], A. Nordmark[1], T. Kaden[3], A. Ulyashin[4]

[1]SINTEF, Material and Chemistry, Trondheim, Norway,
[2]ReSiTec AS, Setesdalsveien 110, NO-4617 Kristiansand, Norway,
[3]Fraunhofer THM, Am St.-Niclas-Schacht 13, 09599 Freiberg (Sachsen), Germany
[4]SINTEF, Material and Chemistry, Oslo, Norway

ABSTRACT: Kerf from slurry based wafer cutting has undergone several refining steps and remelted into ingots for PV-application. The ingots has been wafered and characterized and show promising quality.

1 INTRODUCTION

The CABRISS project is a European collaboration between 16 partners from 9 countries; 6 SMEs, 5 Industry and 5 RTO (Research Organization) [1]. The main vision of the project is to develop a circular economy mainly for the photovoltaic, but also for other industries such as electronic or metallurgy. It will consist in the implementation of recycling technologies to recover In, Ag and Si for the sustainable PV technology and others applications.

During production of silicon wafers in the step between silicon ingots to wafers, about 40 – 50 % of the material is lost due to the cutting technique. Due to the cutting technique, this silicon is in the form of particles with size from 1 – 100 µm. In addition, for slurry based cutting, the silicon particles are mixed with SiC-particles and PEG (Polyethylene glycol). The silicon in the slurry has to be separated from the SiC-particles and PEG. This is done in several steps, each step giving more and more pure silicon material. Finally, the material should be pure enough for directional solidification and production of new PV-wafers.

2 EXPERIMENTS

2.1 Input materials

The material used in the experiments presented here was delivered by ReSiTec. The material originates from cutting of silicon bricks into wafers by wire cutting. ReSiTec purify and dries the powder. The purity specification of the powder provided by ReSiTec is given in Table 1 and a sieve curve is shown in Figure 1.

Table 1: Purity specification of the tested material from ReSiTec.

Element	%
Si	> 99.7
Fe	≤ 0.04
Al	≤ 0.08
Ca	≤ 0.006
Cr, Cl, Ti	≤ 0.003

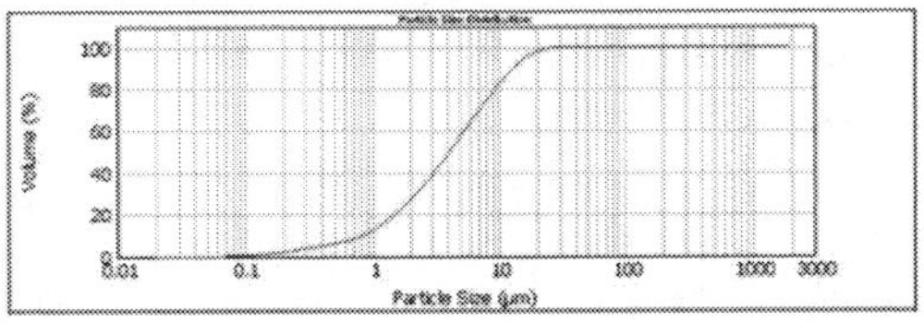

Figure 1: Size distribution of the silicon powder from Re-SiTec.

The silicon kerf was recycled in several step at an industrial scale. The kerf was removed from a highly diluted water slurry of app 0,5% solids. The silicon kerf was passivated to avoid degradation by oxidation. Contamination was to some extent removed by wet processing steps followed by drying. The recovery process is described in further detail [5].

2.2 First solidification in a Crystalox DS 250 furnace

The first melting tests were done by charging 8.9 kg of powder into a quartz crucible and performing a standard melting and crystallisation in the Crystalox DS250 furnace at SINTEF [2]. Normally 12 kg is used for these crystallisation.

Due to the nature of fine powder, it was not possible to charge such amount in the crucible. The idea was that solid inclusion will be pushed ahead of the solidification front during crystallisation. This principle has earlier been studied and verified [3,4].

A photo of the final ingot is shown in Figure 2. Visually, the produced ingot seemed successful; the bottom and sides of the ingot looked like solid silicon while a lump of porous oxides is concentrated on top.

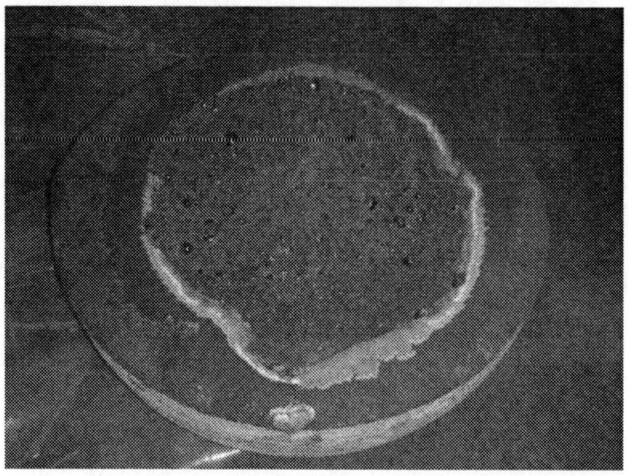

Figure 2: Photo of the first crystallisation test of the provided powder from ReSiTec.

The ingot was cut with the intension of wafering a brick. A cut plane of the ingot is shown in Figure 3. After cutting of the ingot it became evident that a single directional solidification is not sufficient for separation of silicon and the inclusions.

Figure 3: Cut plane of the first ingot. In addition to the particles on top, there are also large amounts of oxides embedded inside the ingot.

2.3 Separation of silicon and inclusions in semi-closed crucible

From the first two experiments, it was decided to do a separate inclusion removal step by melting and solidification. A schematic of this setup is shown in Figure 4. The crucible and lid are made of graphite. Argon gas is continuously flushing the chamber above the charge to minimize oxidation during melting. After melting, the silicon metal was poured into a silica crucible for solidification. As seen later, most of the inclusions were separated in this operation. Probably due to sticking between inclusions and the crucible wall.

Five batches of powder were melted and separated in this melting and tests in order to produce silicon for later crystallisation runs in the Crystalox furnace. The material yield in these melting tests were between 85 and 90 %. (Ratio of silicon tapped and powder fed into the crucible.)

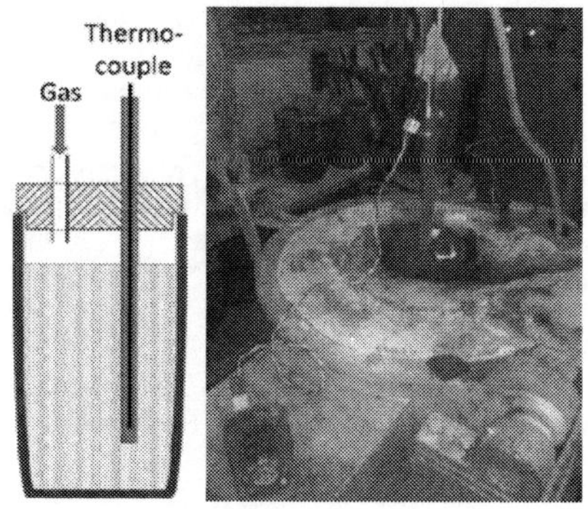

Figure 4: Schematic of the melting and separation setup on the left hand side and a photo of the experimental setup at the right. The inner diameter of the crucible is 150 mm and about 300 mm inner height.

After refining, solidification, and cooling, the material was etched to remove oxides from the surfaces.

2.4 Directional solidification of refined material

Two ingots were produced in the Crystalox DS 250 furace using refined material. One with 10% of refined material, and one with 100% refined material. Photos of these two ingots are shown in Figure 5 and Figure 6.

Figure 5: Photo of the ingot with 10 % refined material and 90 % poly-Si.

Figure 6: Photo of the ingot with 100 % refined material.

2.5 Cutting and analysis of cast material

The 10 % blended ingot even showed indications of crystal structure on top of the ingot. Also the ingot with 100 % refined material did not show much inclusions on top compared to the first ones. A brick was cut from each ingots for wafering. Photos of side cuts are shown in Figure 7 and Figure 8.

Figure 7: Side cut of the ingot with 10 % refined material.

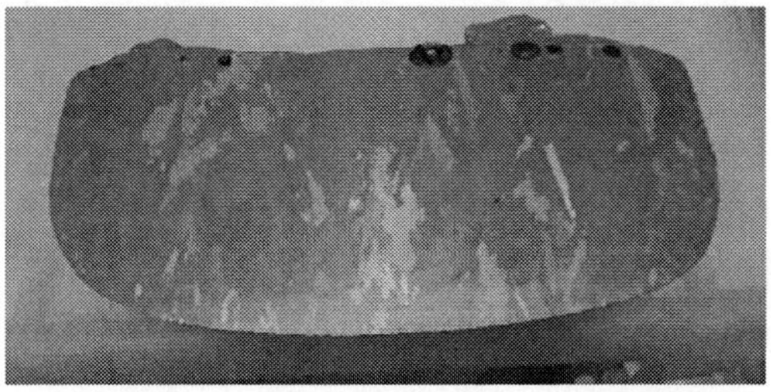

Figure 8: Side cut of the ingot with 100 % refined material

The side cuts indicates that the ingot with 10 % refined material (Figure 7) had a planar solidification front, whereas the ingot with 100 % refined material (Figure 8) lost the planar solidification front. The latter due to too high amount of impurities in the material.

Samples have been taken from both start-material (before refining) and three different heights in both the 10 % and 100 % ingots and analysed by ICP-MS at NTNU. The results are given in Table 2. The results are somewhat inconclusive.

Table 2: ICP-MS analysis of material before refining and after directional solidification in Crystalox DS250 furnace. All values are given in ppm by mass.

	B	P	Al	Fe	Ca	Co	Na	K
Powder	0.3	9.6	3.4	2.8	46	0.02	9.8	18
10 % top	1.9	0.6	570	2	20	192	9.3	2
10 % mid	1.4	0	160	6.4	20	139	2	0
10 % bot	1.7	0	770	5.0	9	145	14	3
100 % top	10	4.4	840	5200	11	127	4.1	0
100 % mid	10	4.7	670	6600	6	111	1.6	0
100 % bot	8	4.8	260	3800	2	93	0	0

The most likely reason for the strange ICP-MS measurements are contamination from the graphite crucible used for melting and refining.

2.6 Resistivity and lifetime measurements

The bricks cut from the ingots with 10 % and 100 % refined material were sent to Fraunhofer THM for IR imaging, resistivity and lifetime mappining, and wafering. Even for the 10 % ingot, there were no IR transmission detected. Probably due to (still) too high level of impurities.

The resistivity measurements showed 0.1 – 0.2 Ωcm for the 10 % ingot (Figure 9) and 0.2 – 0.3 Ωcm for the 100 % ingot. This is somewhat lower than expected. For the 10% ingot, it was added boron to the 90 % poly-silicon so that the resistivity should have been about 1 Ωcm. This indicated that there are elements present in the refined material acting as dopants. This must be taken into consideration in later work.

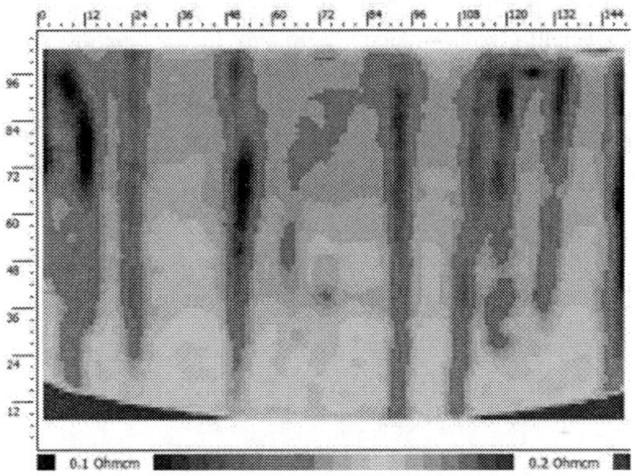

Figure 9: Resistivity mapping of a side of a square brick of the ingot with 10 % refined material.

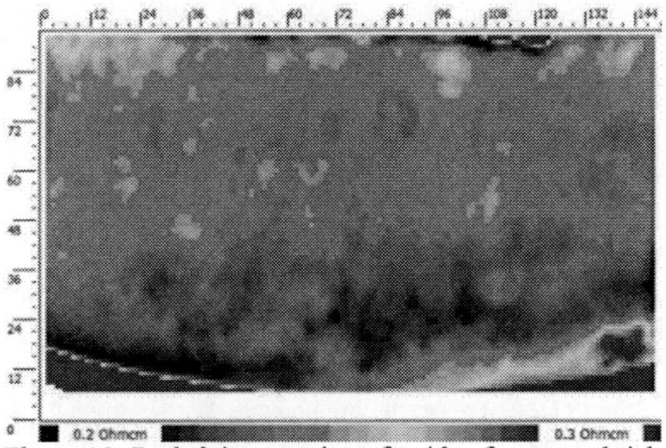

Figure 10: Resistivity mapping of a side of a square brick of the ingot with 100 % refined material.

The resistivity maps in Figure 9 and Figure 10 shows that the 10 % ingot has decreasing resistivity towards the top as expected for directional solidification, whereas the 100 % ingot did not obtain directional solidification.

Figure 11 shows a carrier lifetime mapping of the brick from the ingot with 10 % refined material. The maximum lifetime was less than 1 µs.

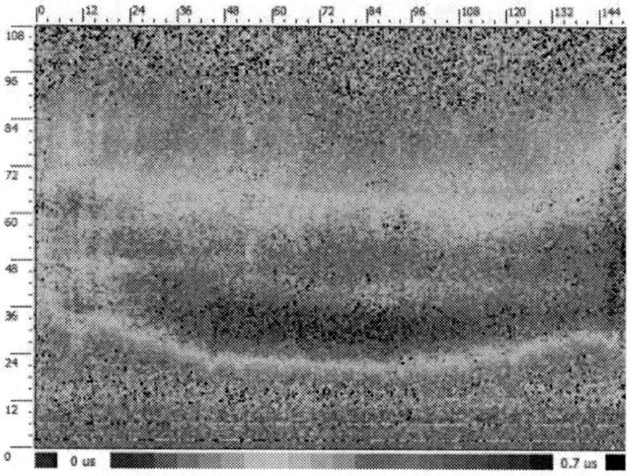

Figure 10: Lifetime mapping of the side of the 10 % ingot.

For the ingot with 100 % refined material, no lifetime was detectable. An explanation for this is that in this ingot, inclusions must be present in high concentrations.

3 CONCLUSIONS

Based on the experiments done so far, the following conclusions can be drawn:

- Refining of the powder by melting and separation of inclusions must be done if is should be used as feedstock for PV-material.
- By melting the powder in a graphite crucible and pouring it into an other crucible, the majority of inclusions will remain in the first crucible by sticking to the crucible walls.
- The refining process shown here gave 85 – 90 % material yield.
- With 10 % refined material, the ingot was still directionally solidified, whereas with 100 % refined material, the ingot was not.
- The quality of the material from the presented refining method is not high enough for ingots with 100 % refined material to be used as PV-material.

4 ACKNOWLEDGEMENTS

The project has received funding from the European Union's H2020 research and innovation programme under grant agreement N° 641972 with the consortium: CEA, SINTEF, IMEC, LOSER CHEMIE, SOLITEK, PYROGENESIS, RHP TECHNOLOGY, RESITEC, TU VIENNA, FERROATLANTICA I&D, SUNPLUGGED, THM FRAUNHOFER, PROJECT KOMPETENZ, PV CYCLE, INKRON, and ECM GREENTECH.

5 REFERENCES

[1] http://www.spire2030.eu/cabriss/
[2] Ø. Mjøs: *Directional solidification of silicon for solar cells.* Dr. Ing. Thesis NTNU 2006:109
[3] A.K. Søland: *Silicon for Solar Cells.* Dr. ing. Thesis NTNU 2004:140
[4] A. Çiftja: *Solar silicon refining; Inclusions, settling, filtration, wetting.* PhD thesis NTNU 2009:103
[5] T. Halvorsen, M.Moen, K.Mørk, D. Grosset-Bourbange, P. Rivat, H. Hajjaji, F. Coustier: Cabriss: Recycling of si-kerf, EU PVSEC 5EO.1.5, 2017.

PHYSICAL DELAMINATION OF PV-MODULES IN LESS THAN ONE SECOND

Harald Gross, Michael Rudolf Heuschkel
FLAXRES GmbH
01307 Dresden, Germany

ABSTRACT:
Economic recycling of photovoltaic modules presumes the separation of small amounts of precious materials from large volume, less valuable materials like glass in a cost-effective way. This suggests the delamination of a module into its basic components at an early stage of the recycling process. After removal of the junction box by mechanical means we propose a process that compartmentalizes the whole module in less than one second by a high-intensity and low-energy light source. Thermal simulations in this paper support experimental results for thin-film as well as wafer-based modules.

Keywords: Recycling, Rapid Thermal Processing

1 INTRODUCTION

Photovoltaic production-, module transportation-, installation- and early losses in combination with end-of-life modules generate waste volumes that are projected to reach several hundred thousand tons annually in the coming decade [1, 2]. Although statutory recycling quotas of 85 %, e.g. in the European Union can be fulfilled, the economic feasibility of any recycling process to date is questionable in respect to the truest sense of the word "recycling": feeding back the recovered materials into production for manufacturing of new photovoltaic modules.

Typically, the junction box, and if applicable the metallic frame of a module, is removed in a first process step by mechanical means for recycling. To recoup valuable materials like silver, indium, gallium or toxic materials as cadmium, led in an economic way, it appears to be requisite to separate these materials in a second process step from polymers and glass due to the dominance by weight and volume of the latter two. If the second process step can be performed without excessive amounts of energy or chemicals, the subsequent processes to refine the residual materials is most likely possible in a cost-effective manner. The obvious reason to enable an economic recycling of the residual materials is then their small mass and volume after the second process step.

We propose a chemical-free, low-energy, portable and fast method to separate glass, polymers and residual materials of a module from each other in a second process step by utilizing high intensity light.

2 RECYCLING PROCESS

2.1 Equipment

We have developed a high-intensity and low-energy light source that enables illumination of the whole photovoltaic module surface at once. Light absorbing layers in all kinds of modules can be heated from room temperature to temperatures between 400 and 1000 degrees Celsius in less than one second by this low-energy source. After illumination of the module, it takes only a second for the temperature of the light absorbing layer to drop back close to room temperature.

All experiments on photovoltaic modules have been performed with a laboratory equipment that facilitates sample sizes up to 10 cm x 6 cm so far. Although the exposure field of the laboratory equipment is rather small, many years of experience with such light sources at production scale and exposure fields of more than one square meter is proof for scalability. This experience allows a straight forward assembly of a light source with an exposure field of 2 m x 1 m that is tailored to recycling of photovoltaic modules.

2.2 Wafer-based Module Recycling

Polymers like ethylene-vinyl-acetate lose their adhesion irretrievably on both sides of the heated silicon in wafer-based photovoltaic modules during the light exposure. Depending on illumination period and light intensity, the bus bars can be unsoldered at the same time, whereas the contact fingers on the silicon surface remain unaltered. During the process, small amounts of gas are released that support the separation of the backside glass or backside polymer like polyvinyl-fluoride as well as the frontside glass. Therefore, the composite materials are delaminated from each other into four parts: front side glass with adjacent polymer layer, silicon wafers (including contact fingers), bus bars and finally back sheet polymer layers or back side glass with adjacent polymer layer.

The polymer layer that is directly attached to a glass sheet can be removed by another light exposure at different parameter settings. Although the second exposure requires additional energy, it enables the processing of modules with shattered front glasses, because the polymer layer keeps the glass pieces together during the first exposure step. For modules with a metallic frame it is even possible to retain the frame for mechanical stability reasons until after the second exposure step.

Figure 1 shows an example of a module piece with a size of approximately 4 cm x 5 cm after delamination into five components by two light exposures as described before. The tempered front side glass was already shattered before it was sawed out of a module before the experiment. For picture taking, the front side glass as well as the silicon wafer has been attached to a sticky tape right after each exposure step to keep them in one piece. The polymer composite on the backside consists of four different polymer layers in total that may be separated in subsequent process steps.

It is possible to treat various wafer thicknesses and polymer types with the same exposure parameters, which indicates a conveniently large process window for production.

Figure 1: Delamination of a module piece into five components after a second light exposure step.

2.3 Thin-film Module Recycling

Because the light absorbing layer of thin-film modules shows an approximately two orders of magnitude lower heat capacity, less energy is required to heat it up from room temperature to several hundred degrees Celsius. Typically, a polymer layer is laminated on one side of the light absorbing film that loses adhesion irreversibly during an exposure step pursuant to wafer-based modules.

Light absorbing layers and further films like the contact electrodes are directly deposited onto glass sheets during production of modules. The glass exhibits completely different mechanical parameters compared to polymers. This allows for separation of the films from the glass or each other by excessive, but controlled thermo-mechanical stress according to by rapid heating induced temperature gradients rather than by cracking chemical bonds at elevated temperatures. It is possible to enforce the temperature gradient by variation of duration and intensity of the light exposure. In this manner, the height of the mechanical stress or the height of the temperature can be emphasized.

In a first light exposure step, we have separated the frontside- from the backside glass at the interface of polymer and the light absorbing layer. Edge sealing materials like silicones or butyl rubbers behave similar to any other polymer. Utilizing a second exposure step employing a different parameter set, we stripped the light absorbing layer from the bottom electrode layer. Thereafter a third exposure step is performed to remove the bottom electrode layer including the bus bars at the edges of the glass sheet. Both layers generate particle flakes that can be collected separately. Thermo-mechanical simulations show, and small particle flakes are evidence for, excessive thermo-mechanical stress rather than evaporation of the films during the stripping process.

In reverse conclusion, films never reached evaporation temperatures while their exposure. By means of a final exposure step, the polymer layer was delaminated from the other glass sheet, which is the same in wafer-based modules. The resulting fractions of several light exposures are two clear glass sheets of different qualities due to their iron content, bus bars, flakes from the light absorbing film, flakes from the electrode layer, one polymer layer and edge sealing material. It remains to be seen, whether it is cost-effective to separate two different kinds of flakes from each other or to reduce the number of illuminations by one and strip both thin films at the same time. There is little difference between CdTe-, CIGS- or thin film Si-modules regarding their dissociation by a high-intensity light source since their thermo-mechanical properties are similar.

The next two figures show the same sample of a backside glass from a CIGS-module after separation from the frontside glass. For better visibility only a fraction of the sample has been illuminated through a shadow mask (dashed line). During the second exposure step, the light absorbing film has been stripped as can be seen in figure 2. The resulting CIGS-flakes at the right-hand side of the figure include also the top electrode layer, a transparent conductive oxide like aluminum doped zinc oxide.

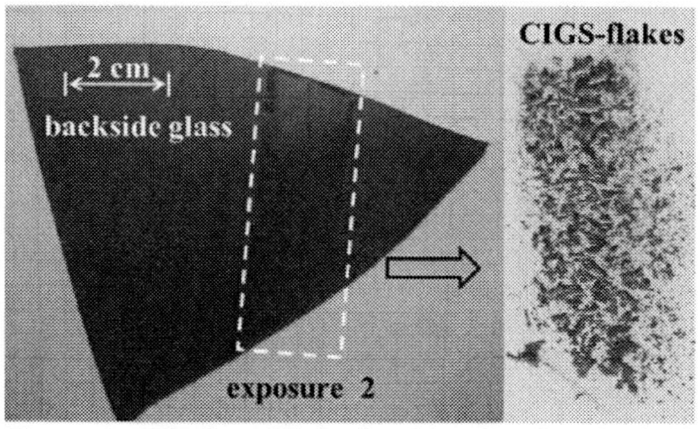

Figure 2: Stripping of light absorbing film from backside glass by exposure step 2 and resulting CIGS-flakes

To remove the bottom electrode layer that consist not only of a metallic film like molybdenum, but also e.g. sodium diffusion barriers, the process parameters are changed compared to exposure step 2. Figure 3 shows a clear window in the backside glass after the third exposure step.

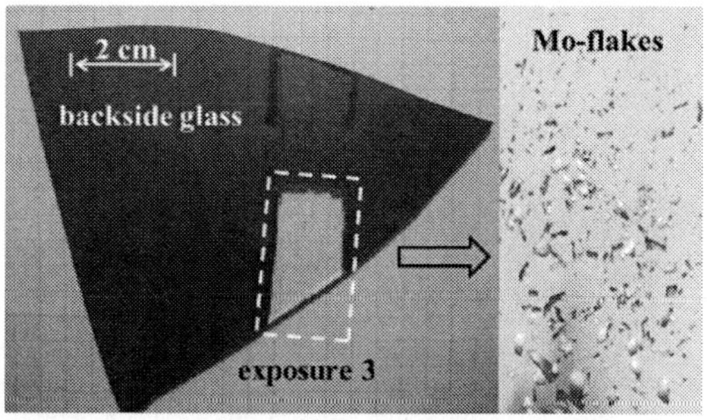

Figure 3: Stripping of bottom contact electrode from backside glass by exposure step 3 and resulting Mo-flakes

In case of CdTe-modules, the third exposure step can be performed at process parameters that detach not only the deposited films on the glass, but also the first few micrometers of the glass itself by inducing micro-cracks in the glass surface. This ensures a complete removal of toxic materials like cadmium.

Although thin-film silicon modules are not manufactured any more, the amount of waste of end-of-life modules is not negligible. Fortunately, this type of module can be disassembled in a similar manner like other thin-film modules by simple adaption of the illumination parameters.

3 THERMAL SIMULATIONS

There is no commercial equipment available that provides the measurement of a temperature profile at micrometer scale in combination with high speeds. The difficulty is not the rapid change of temperatures in the microsecond range, but the intrinsic averaging of the temperature over a relatively large volume. Furthermore, the interesting layers and interfaces, respectively, are buried inside a module in our case that restrains the emission of e.g. infrared light. Last, but not least, during an exposure step by our light source, electromagnetic waves interfere with measurement signals and trigger false readings. Therefore, we rely on thermal, mechanical and other simulations for a better understanding and prediction of the process.

Simulation tools in general have been proven to be powerful and the results are close to reality in cases where the underlaying models are chosen carefully. We have many years of experience with thermal simulations of thin films and have demonstrated their accuracy by comparing them with experimental results.

3.1. Thin-film Module Simulations

The following figures are examples to provide an insight into our recycling process utilizing a high-intensity light source. We especially want to show that temperatures of several hundred degrees Celsius inside thin films or even silicon wafers can be reached by low heating energies on glass substrates that stay close to room temperature during the whole process.

High film temperatures at low substrate temperatures utilizing low heating energies would be contradictory for a convection furnace. Furthermore, the previously described mechanical stress due to large temperature gradients is not only many orders of magnitude lower in a convection furnace, but also the furnace can only heat the module from the outside rather than solely the buried light absorbing layer.

Figure 4 shows a typical temperature behavior of a three micrometers thick light absorbing film over time, if heated by our light source for 0.5 milliseconds from room temperature to 600 degrees Celsius. In a parallel plane with a distance of five micrometers to the light absorbing layer, the maximum temperature is significantly lower. Since glass is a better heat conductor in respect to polymers in general, its maximum temperature is higher, too.

For a module with a size of e.g. 150 cm x 60 cm = 9000 cm^2 the required light energy is then 2.3 J/cm^2 x 9000 cm^2 = 20.7 kJ = 5.75E-3 kWh to reach 600 degrees Celsius. The conversion of electricity into light and then into heat in the light absorbing film implies a significant loss of energy. Nevertheless, the required electricity stays way below 0.1 kWh to dismantle the module into the fractions as mentioned before for thin-film modules.

It takes approximately one second until temperature equilibrium is achieved in the glass sheets after illumination. The final temperature of the module fractions is less than three degrees Celsius above the initial temperature and has neither a significant effect on the equipment design, production throughput nor the handling of the glass sheets afterwards.

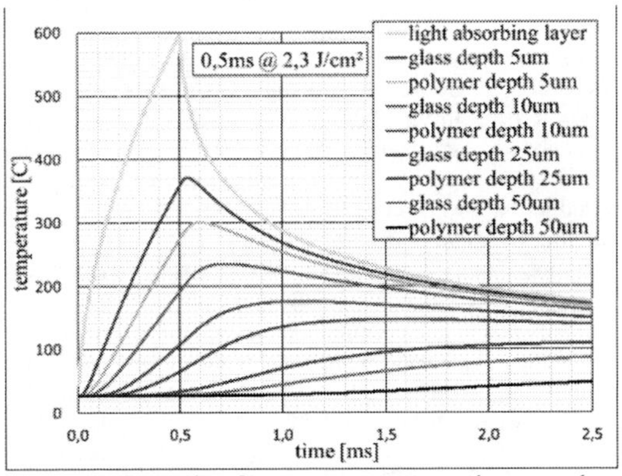

Figure 4: Example of temperature propagation over time for the light absorbing layer and parallel planes thereof at a selected distance.

Figure 5 depicts the temperature profile at the point of time of 0.5 milliseconds when the light exposure ends. At a distance of only 25 micrometers to the light absorbing layer, the temperature in the glass on one side or the polymer on the other side is significantly below 100 degrees Celsius. The temperature of the light absorbing layer reaches its maximum and drops after this point of time. Any other location of the composite will reach its maximum temperature at a later point in time, e.g. the glass plane with a distance of 25 micrometers at 1.5 milliseconds (see figure 4). However, in a plane at a distance of more than 50 micrometers the temperature will never rise above 100 degrees Celsius.

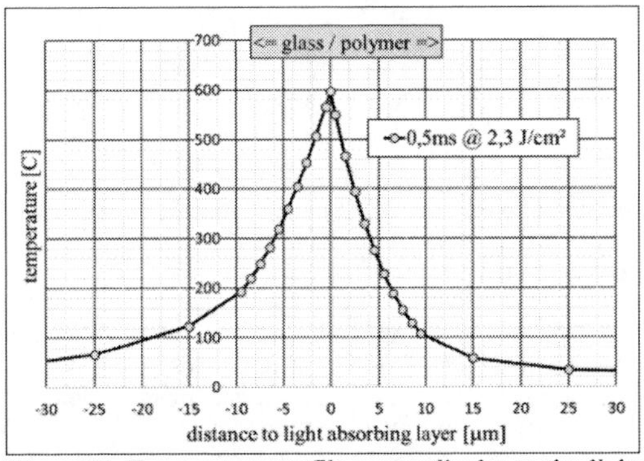

Figure 5: Temperature profile perpendicular to the light absorbing layer at maximum temperature of the latter.

To simplify the simulation model, neither changes of the thermal properties like conductive heat transfer, nor the formation of gases like acetic acid have been considered. Since these changes evolve at or close to maximum temperatures and at subsequent points of time, the simulations help to set parameters like illumination time and intensity to reach desired temperature gradients or maxima until the changes take effect. In other words, the simulations provide a good image of the actual process until material properties change significantly and fail beyond that point.

3.2. Wafer-based Module Simulations

As mentioned before, the wafer-based modules require more energy to heat the light absorbing layer to the same temperatures as for thin-film modules according to very different heat capacities. But silicon wafers are laminated to polymer foils on their front- and backside that exhibit a four times lower heat conduction coefficient compared to glass in thin-film modules. In this respect, the lower heat conduction coefficient of the polymer compensates for the higher heat capacity of the silicon. In contrast to the heat conduction coefficient of the polymer, the coefficient of silicon is about 600 times higher. If the illumination duration is sufficiently long and the exposure of the silicon wafer is performed only on the frontside, the temperature of its backside is the same in good approximation.

Figure 6 shows depth profiles of various maximum wafer temperatures. On the contrary to thin-film modules, the curves are mirror-symmetric in respect to the axis of ordinate due to the same material on both sides of the wafer and the high thermal conductivity of silicon in first order approximation. Five data points at close quarters on the axis of ordinate for each curve show a negligible temperature difference between the front- and the backside of the silicon wafer. The temperature difference increases with rising temperature and is reciprocally proportional with exposure time.

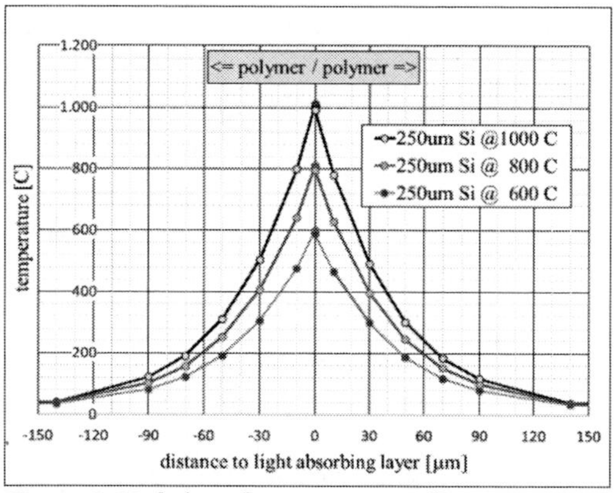

Figure 6: Variation of temperature profile perpendicular to the surface of a 250 micrometers thick silicon wafer at maximum temperature of the latter.

To reach the maximum temperatures of silicon in figure 6, the required energy is in all cases roughly 20 % higher compared to heating of silicon alone without any transfer of heat energy to its surroundings. Losses by heat radiation are emphasized at higher temperatures, especially at 1000 degrees Celsius according to the Stephan-Boltzmann-Law (Q=aT^4). However, such high temperatures occur only for a small fraction of the entire period of heating.

Experiments suggest much lower heat losses by conduction and radiation compared to the previously mentioned 20 %. This may be due to the failure of the simulation model in the case where small amounts of gas is generated on the surface of the polymer in combination with delamination effects.

4 SUMMARY AND OUTLOOK

We have evaluated a rapid process to compartmentalize composite materials like photovoltaic modules without chemicals by utilizing a large area and high-intensity light source at low energy consumption. This process allows for fractioning of an entire module into a clear frontside glass, and if applicable a clear backside glass, polymers, bus bars and light absorbing materials like silicon wafers or CdTe-films in less than one second. Thin-film modules can be treated in a similar way as wafer-based modules by changing exposure time and light intensity.

Thermal simulations have shown that temperatures of several hundred degrees Celsius inside thin films or even silicon wafers can be reached by low heating energies on glass substrates that stay close to room temperature during the whole process.

Taking the thermal mass of bus bars into account is an ongoing topic in simulations to get a better understanding of and to optimize the recycling process.

The construction of a pilot line, dedicated to dismantling modules with a size up to 2 m x 1 m is in progress. Its light source is going to enable the exposure of an entire photovoltaic module at once. Since the equipment fits into a standard sea container, it can easily be transported to any production site for solar modules, photovoltaic power stations or places for end-of-life module collection to minimize transportation routes, especially for glass.

5 REFERENCES

[1] IRENA IEA-PVPS Report: T12-06:2016 "END-OF-LIFE MANAGEMENT – Solar Photovoltaic Panels"

[2] GTM Research, Global Solar Demand Monitor Q1 2017

PV PRODUCTION FORECASTING MODEL BASED ON ARTIFICIAL NEURAL NETWORKS (ANN)

Spyros Theocharides*[1], George Makrides[1], Venizelos Venizelou[1], Paris Kaimakis[2],
Andreas Kyprianou[1] and George E. Georghiou[1]
[1]FOSS Research Centre for Sustainable Energy, Photovoltaic Technology Laboratory,
Department of Electrical and Computer Engineering, University of Cyprus
75 Kallipoleos Avenue, P.O. Box 20537, Nicosia, 1678, Cyprus
[2]University of Central Lancashire Cyprus, University Avenue 12-14, Pyla 7080, Cyprus
*corresponding author Spyros Theocharides (phone: +357 22 894398, Fax +357 22 895370,
theocharidis.spyros@ucy.ac.cy),
Venizelos Venizelou (venizelou.venizelos@ucy.ac.cy), George Makrides (makrides.georgios@ucy.ac.cy),
Paris Kaimakis (pkaimakis@uclan.ac.uk), Andreas Kyprianou (akyp@ucy.ac.cy), George E. Georghiou (geg@ucy.ac.cy)

ABSTRACT: Accurate photovoltaic (PV) production forecasting is necessary for the optimal integration of this technology into existing power systems and is important for both grid and plant operators as it allows energy management and flexibility. The purpose of this work is to present the methodology applied for predicting the power output of grid-connected PV systems, using a developed machine learning model based on artificial neural networks (ANNs) optimised according to the input parameters and network architecture. More specifically, a conventional neural network was designed, whose parameters were estimated by a feedforward supervised learning process. The comparative analysis between the different devised networks demonstrated that by increasing the input features and training in a random sample 70:15:15 % train, validation and test set approach with yearly data sets, improved the prediction accuracy. The performance assessment results obtained when the test set was applied to the optimal designed network, which comprised of 7 inputs and 12 hidden nodes, showed an average daily normalised root mean squared error (nRMSE) of approximately 0.76 %. Finally, by employing an ensemble meta-algorithm, developed based on the optimally identified network, the accuracy was further improved achieving 0.71 % nRMSE when validation was performed on the test set.
Keywords: artificial neural networks, forecasting, photovoltaic.

1 INTRODUCTION

A main challenge in the scope of ensuring large-scale deployment and sustainability of photovoltaic (PV) systems is to improve the accuracy of production forecasts, especially when there is such systems in high concentration. In particular, the high penetration levels and intermittent nature of the generated power due to the dependence of PV on weather conditions, pose power quality problems to grid operators. Accurate prediction of the day-ahead PV production is therefore an important feature that can assist utilities and plant operators in the direction of energy management through optimal operation, dispatchability and operation mode planning.

PV production forecasts have only been recently introduced into electricity system operation and most commonly adopted PV production forecasting approaches are based on time series analysis techniques [1]. In addition, parametric models for forecasting PV production have already been developed [2, 3] but their capacity to forecast the power output of PV systems is not an easy task since accurate knowledge of system characteristics and behaviour must be available. For this reason, a lot of work is currently undertaken for the development of more sophisticated forecasting algorithms that employ nonparametric models based on machine learning algorithms [4-6]. At the moment, as most approaches are not fully tested and verified on a large amount of field data and different technologies, there is yet no complete universal forecasting model and methodology to ensure accurate forecasts according to the technology and location. Additionally, in the scope of improving the accuracy of the power prediction of PV systems, adaptive methods that can capture system information and behaviour without the need of datasheet and installation parameters must be employed. This is very important especially since a lot of PV systems are de-centralized roof-top installations and knowledge of system information is not always readily available. Alongside, the systematic behaviour can only be captured by processing recent operational data sets. In this respect, artificial neural networks (ANNs) based on the classical approach of the feedforward neural network (FFNN) that are inspired by the way biological neurons work and pass information, are powerful machine learning algorithms that are very popular and proven in other fields for both prediction and classification purposes. The classical approach of the FFNN with an input, hidden and output layer of linear and non-linear activation functions can be viewed as a convenient way to predict the power output of PV systems. FFNN can be trained to develop relational weighted chains between internal nodes to overcome the limitations of traditional methods to solve complex problems, which can be modelled through a supervised learning technique based on historical data. Because of this chain of relationships, theoretically, multi-layered neural networks can be universal approximators and have tremendous potential to perform any linear and nonlinear mapping through a learning process based on historical time-series [7]. In addition, ANNs are very efficient for online training as they are capable of reflecting the information of new instances on a model very efficiently by just changing the weight values.

The scope of this work is to present the methodology to derive accurate day-ahead forecasts for PV systems, based on a nonparametric ANN model which was optimised according to the selection of input features and architectural parameters. Specifically, for this investigation PV operational and meteorological data sets were acquired over a period of a year for a grid-connected poly-crystalline Silicon (poly-c Si) PV system installed at the outdoor test facility of the University of Cyprus (UCY). The acquired data sets were utilized in order to train different designed ANNs by applying several combinations of training features while in parallel changing architectural parameters, in order to optimise the model and reduce the prediction accuracy. As such, a comparison between the performances of each devised ANN is directly performed.

2 EXPERIMENTAL SETUP

2.1. Outdoor test facility description

The outdoor test facility includes a fixed plane infrastructure for outdoor performance assessments at both the module and system level. The installed poly-c Si system was mounted in portrait arrangement on aluminium mountings, at the optimum annual energy yield plane-of-array (POA) angle for Cyprus of 27.5°.

Additionally, the PV system was connected to an advanced data-acquisition platform, used for the monitoring and storage of meteorological and PV system outdoor operational data. The platform comprises of meteorological and PV operational measuring sensors connected to a central data acquisition system. The performance of the system and the prevailing meteorological conditions were recorded according to the requirements set by IEC 61724 [8]. Specifically, the meteorological measurements include the incident global irradiance (G_I), relative humidity (RH), wind direction (W_a), wind speed (W_s), as well as ambient temperature (T_{amb}). The PV system operational measurements include maximum power current (I_{mp}), voltage (V_{mp}) and power (P_{mp}), as measured at the output of the PV system (dc side) [9, 10].

2.2 PV system description

The PV system comprises of five poly-c Si PV modules, rated at 235 W_p each as depicted in the manufacturer's datasheet. The modules of the system are connected in series to form a PV string at the input of a string inverter. The main technical specifications of the test PV system are summarised in Table I.

Table I: Installed PV system technical characteristics.

System 1	
Modules	5 × poly-c Si 235 W_p
System power (datasheet)	1175 W_p
Installation date	01/06/2015
Efficiency	14.40 %

3 METHODOLOGY

The methodology followed to develop an optimal nonparametric model based on ANNs for short-term forecasting included a network training phase which was effectively performed by applying a learning algorithm to a performance function. More specifically, to determine the systematic behaviour of the PV system and to design optimal ANN models a sequence of training and validation stages were performed varying the input parameters (features), training duration and samples used and architectural parameters of the devised networks. It is important to mention that this work focuses on improving the accuracy of derived machine learning algorithms (as the training and testing was performed on historical yearly data) that will subsequently operate on provided weather variable forecasts. The study does not, therefore, consider the Numerical Weather Prediction (NWP) associated errors in day-ahead PV production forecasts and only investigates the optimisation of the machine learning power prediction algorithms (see Fig. 1).

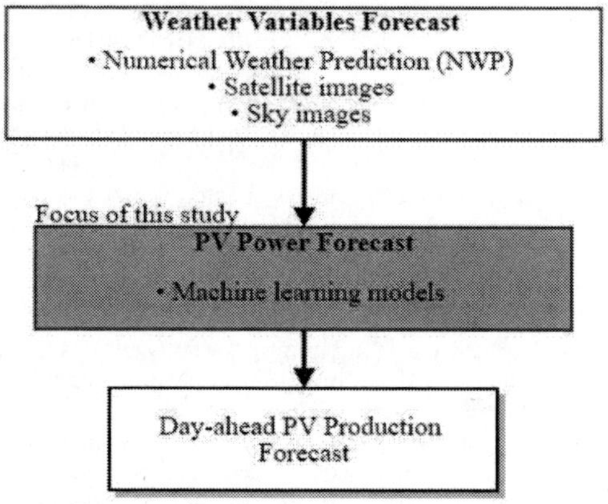

Figure 1: Flow-chart of a typical approach for generating short-term day-ahead PV production forecasts from NWP and machine learning models. The focus of this study is to improve the PV power prediction stage of the PV production forecasting process.

3.1 Neural network training and validation

The strength of ANNs is their ability to approximate non-linear functions through a supervised learning process (a set of input and output data is available). Training ANNs is performed by applying a learning algorithm to a cost function. More specifically, the training step in neural networks is formulated in terms of the minimisation of a loss function, which in our case comprises of a data and a regularisation term. The error term models the discrepancy between the produced output of the network and the desired output. This is measured utilising common indicators such as the mean square error (MSE). In a multi-layer network, minimising the error by training the network is achieved with the back-propagation (BP) algorithm, which is used to calculate the error contribution of each neuron after a batch of data by distributing the error back from the output through the network layers [11].

In addition, the regularisation term is used to prevent overfitting, by controlling the effective complexity of the neural network. The regularisation of the designed networks in this study was performed by adding a penalty equal to the L2-norm of the weights, in order to reduce the value of the weights by the same factor.

3.2 Model architecture

The acquired yearly data sets of the PV system were divided into three different sets in order to determine the optimal parameters for the neural network. The yearly data set was separated into the train, validation and test set with two different data split approaches. The first approach utilized successive samples that were split in a conventional 70:15:15 % portion. The second approach included samples taken randomly from the entire data set at a 70:15:15 % portion. The 70:15:15 % data split portion was selected in order to include as much systematic information for the training but also keep enough days for validation and testing. Although there are many different ways to split the data set (training/

testing set and cross validation), in this work we used data splitting that involves partitioning the data into an explicit training set used to prepare the model, the validation set used to tune the parameters of the model and the test set necessary to evaluate the model performance on unseen data. The training, validation and testing set, comprised of the model inputs which included the measurements of G_I, T_{amb}, RH, W_s, W_a and the calculated parameters of the Azimuth ($A\angle S$) and Elevation (A/S) Angle of the Sun, calculated using solar position algorithms [12] and used to address the angular response of the PV system. Finally the dc P_{mp} was the model output.

In order to achieve optimal prediction performance, different input feature combinations and hidden layer topologies (number of hidden units) were investigated while training the network models. The initially designed FFNN which was subsequently used in the tuning process comprised of two inputs (G_I, T_{amb}), four hidden nodes and an output layer (P_{mp}) (see Fig. 2). The optimal model parameters were identified by comparing the performance of the models created based on the training set, with the validation set. Re-iteration of the process (epochs) was performed until the results from the network became acceptable while avoiding over training the network, that may result in over fitting. Lastly, the tuned networks were then evaluated by feeding the test set input variables and assessing the actual measured power against the predicted power according to some performance assessment metrics.

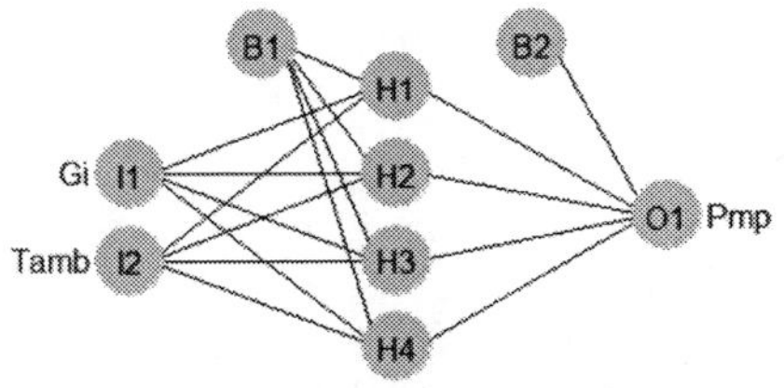

Figure 2: Network interconnection diagram (NID) of a FFNN ANN.

3.3 Model performance assessment

The performance accuracy was assessed based on several predefined metrics when the test set was applied to the designed networks. More specifically, the metrics used include the actual error (AE) which is the difference between the predicted and actual values, the absolute percentage error (APE) which measures the size of the error in percentile format, the mean bias error (MBE) which encapsulates the systematic tendency of a forecasting model to under or overestimate and the root mean squared error (RMSE) which provides more weight to the largest errors. All these metrics are stated as absolutes or else normalised by dividing by a reference value to facilitate comparisons. In this case for the RMSE the predictions were commonly normalised to the nominal power ($P_{nominal}$) of the system since this is a straightforward and easily accessible reference value.

$$nRMSE = \frac{100}{P_{nominal}} * \sqrt{\frac{1}{n} \times \sum_{i=1}^{n} (y_{i,actual} - y_{i,predicted})} \quad (1)$$

where $P_{nominal}$ is the nominal power of the respective PV system, n is the amount of all actual and predicted pairs, $y_{i,actual}$ and $y_{i,predicted}$ are the actual and predicted power points, respectively.

3.4 Ensemble learning

Ensemble methods are meta-algorithms commonly employed in order to combine single learners into one predictive model in order to improve the predictive force. Bootstrap Aggregation (Bagging) is a common method used to decrease the variance of the prediction by generating repetitive multisets for training, obtained using randomly drawn subsets from the train set of the original data. In this work, each trained model predicted the power when fed with the test set variables and the average of all learners was taken as the final result.

4 RESULTS

To establish an accurate output power prediction model for the PV system, it is initially necessary to analyse the importance of the variables on the output. For this purpose, the effects of each input parameter on the output power, was examined by calculating the Pearson Correlation Coefficient (PCC), which measures the direction and strength of the linear relationship between two variables. Table II summarizes the results from the PCC between the investigated features.

Table II: Pearson Correlation Coefficient between PV power output and environmental factors under typical weather conditions.

	Pearson Correlation Coefficient
	Maximum Power, P_{mp}
Incident global irradiance, G_I	0.9
Sun elevation angle, A/S	0.8
Relative humidity, RH	-0.7
Ambient temperature, T_{amb}	0.5
Wind direction, W_a	-0.3
Wind speed, W_s	0.2
Sun azimuth angle, $A\angle S$	-0.01

With the intention of selecting only the highly linearly correlated input parameters, we have selected the four highest positively and negatively correlated parameters. The parameters complying with this were the G_I, A/S, T_{amb} and RH. In order to implement an optimal model, different input parameter combinations at the input layer of each devised network were investigated comprising not only of the highly correlated parameters but also the remaining input parameters that have non-linear behaviour. However, the highly correlated input parameters were expected to achieve high accuracy and feature importance. In addition, the number of hidden units was set to an empirical value between the total of the number of inputs and output layer nodes. Table III tabulates the validation results based on the nRMSE of the input feature analysis to the ANN model.

Table III: Input feature parameters for ANN model training and the respective nRMSE applied to the test set.

Input features	Hidden Layer Neurons	Normalised RMSE (%)
G_I, T_{amb}	2	1.15
G_I, T_{amb}, AIS	3	1.10
G_I, T_{amb}, RH	3	1.18
G_I, T_{amb}, RH, AIS	4	1.10
G_I, T_{amb}, W_{alpha}, W_S	4	1.19
G_I, T_{amb}, W_S, AIS	4	1.10
G_I, T_{amb}, W_S, RH	4	1.12
G_I, T_{amb}, AzS, AIS, W_a, W_S, RH	7	0.87

The results clearly demonstrated that the optimal feature combination included all the input variables (G_I, T_{amb}, AzS, AIS, W_a, WS and RH), confirming that the use of more input parameters benefits the ANN model accuracy by learning both the linear and non-linear effects. Varying the number of input features resulted in nRMSE improvements from 1.15 % to 0.87 %. This is because increasing input features assist also in modelling non-linear effects, which is one of the strengths of neural networks.

Once the optimal input variable combination was identified the evaluation was performed again by comparing the validation results obtained when applying two different training sessions (training the network with the historical continuous and random sample data split at varying durations). Table IV summarizes the results for each training session implemented for continuous and randomly selected data while the test set, remained unaffected at 15 % of the entire data set. The results provide evidence that the duration of the training period affects the accuracy of the ANN model since improvements on the nRMSE were achieved for the same model when trained at 70 % of samples. Additionally, comparing the nRMSE results between continuous and randomly selected data indicated that randomly selected samples could yield better results compared to continuous samples.

Table IV: Test set validation nRMSE results of the training period and sampling type of the training data set.

Dataset length	Continuous data nRMSE (%)	Random data nRMSE (%)
30 %	1.30	0.99
40 %	1.28	0.97
50 %	0.98	0.91
60 %	0.95	0.85
70 %	0.90	0.80

In the scope of further improving the accuracy of the predictions of our optimal network, the effect of varying the number of hidden units on the prediction accuracy was also investigated. The results presented in Fig. 3, demonstrated that the nRMSE, decreased with the amount of hidden units in the case of the training set. In particular, although the training set was improving by increasing the number of hidden neurons, the nRMSE of the validation set showed a different behaviour when the number of hidden neurons was increased beyond a certain number. The results presented in the plot showed that the best accuracy for the validation set nRMSE was obtained when the selected number of hidden neurons was in the range of 12 to 16. In order, to minimise our error by decreasing the bias with more hidden units but at the same time not making our model too complicated and prone to overfitting the number of nodes was set to 12 in the hidden layer as an additional optimisation step.

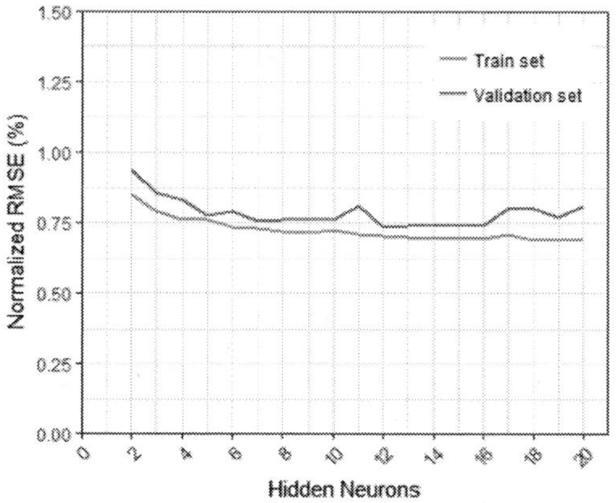

Figure 3: Plot of the daily nRMSE (%) for training and testing set for a single hidden layer.

At this stage, the neural network was designed by specifying the number of input variables and amount of hidden neurons. The optimal model therefore comprised of the input parameters of G_I, T_{amb}, AzS, AIS, W_a, WS and RH. The training set of the ANN model was developed by the 70 % randomly selected samples and a single hidden layer with 12 neurons established the final ANN model (see Fig. 4).

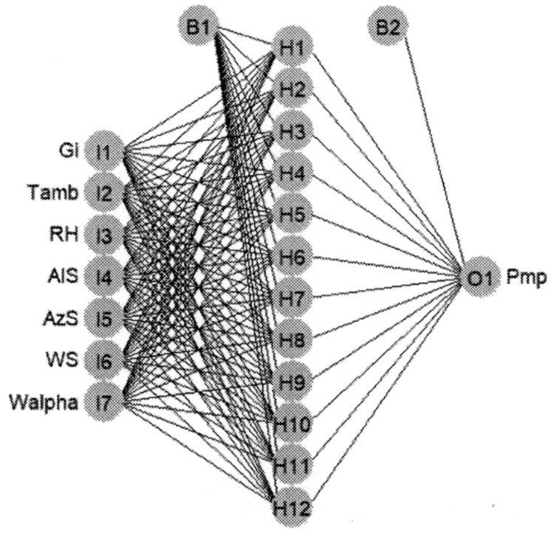

Figure 4: NID of the optimimally designed ANN. The thickness of the lines represents weight magnitude and the colour represents the weight sign (black = positive, grey = negative).

For the optimal designed network, the daily nRMSE showed that on average the nRMSE was 0.76 %, while some days were exhibiting accuracies close to 0.50 % (Fig. 5).

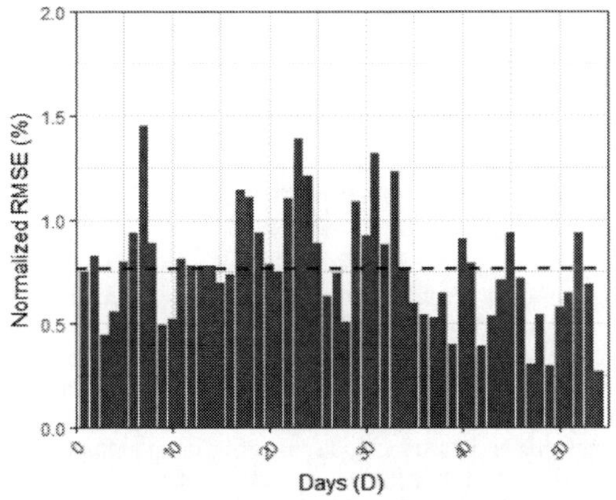

Figure 5: Daily nRMSE validation over the test set samples. The horizontal line presents the average daily nRMSE of 0.76 %.

Finally, in order to further lower our prediction error the common meta-algorithm technique of an ANN ensemble developed with the bagging approach was utilised, whereby 5 models of the same kind were trained with random samples of the training set and were subsequently used to provide predictions. The outcome used for validation purposes was the average of the 5 different models since each model predicted the power when fed with the test set variables. With this ensemble algorithm, the final achieved average daily nRMSE for the test set was approximately 0.71 %.

5 CONCLUSIONS

PV production forecasting based on machine learning algorithms such as ANNs can successfully be utilized to implement more agile forecasting procedures.

For the development of an optimal ANN model a procedure with subsequent stages of training and testing steps was followed, whereby the devised ANNs were assessed by tuning their input features and network architecture.

More specifically, a conventional FFNN was designed, whose parameters were estimated by a feedforward supervised learning process, trained using the BP technique and regularised by employing L2 regularisation. The final ANN included 7 input variables (G_I, T_{amb}, AzS, AlS, W_a, WS and RH), 12 hidden nodes and was trained with a random sample 70:15:15 % train, validation and test set approach. The prediction results obtained when the test set was applied to the optimal designed network, demonstrated that the average daily *nRMSE* was 0.76 %, while some days were exhibiting accuracies close to 0.50 %.

Finally, by employing an ANN ensemble meta-algorithm the prediction accuracy was further improved achieving an average daily *nRMSE* of 0.70 % when validation was performed on the test set.

6 ACKNOWLEDGEMENT

This work was funded through the INFORPV project which is co-financed by the SOLAR-ERA.NET Transnational Calls of the European Union under the grant agreement number SOLAR-ERA.NET/1215/02. The authors would also like to acknowledge the financial support from the Cyprus Research Promotion Foundation.

7 REFERENCES

[1] S. Pelland, J. Remund, J. Kleissl, T. Oozeki and K. D. Brabandere, Photovoltaic and Solar Forecasting: State of the Art, IEA, Technical Report 5-40, 2013.

[2] E. Lorenz, T. Scheidsteger, J. Hurka, D. Heinemann and C. Kurz, Regional PV power prediction for improved grid integration, *Progress in Photovoltaics: Research and Applications*, 2011; **19**: 757–771, doi:10.1002/pip.1033.

[3] S. Pelland, G. Galanis, and G. Kallos, Solar and photovoltaic forecasting through post-processing of the Global Environmental Multiscale numerical weather prediction model, *Progress in Photovoltaics: Research and Applications*, 2013; **21**: 284–296.

[4] I. Ashraf and A. Chandra, Artificial neural network based models for forecasting electricity generation of grid connected solar PV power plant, *International Journal of Global Energy*, 2004; **21**: 119-130.

[5] D. Caputo, F. Grimaccia, M. Mussetta and R. E Zich, Photovoltaic plants predictive model by means of ANN trained by a hybrid evolutionary algorithm, *in: Proceedings of the 2010 International Joint Conference on Neural Networks*, 2010, pp. 1-6.

[6] S. A. Kalogirou, Applications of artificial neural-networks for energy systems, Applied Energy, 2000; **67**: 17-35.

[7] M. P. Almeida, M. Muñoz, I. de la Parrab, O. Perpiñán and L. Narvarte, Comparative study of nonparametric and parametric PV models to Forecast AC Power Output of PV Plants, in: Proceedings of the 31st European Photovoltaic Solar Energy Conference and Exhibition, 2015, pp. 2230 - 2234.

[8] IEC 61724:1998, Photovoltaic system performance monitoring — Guidelines for measurement, data exchange and analysis, 1st Ed. Geneva, Switz. IEC, 1998.

[9] G. Makrides, B. Zinsser, M. Schubert and G. E. Georghiou, Energy yield prediction errors and uncertainties of different photovoltaic models, Progress in Photovoltaics: Research and Applications, 2013; **21**: 500-516.

[10] G. Makrides, B. Zinsser, M. Norton and G. E. Georghiou, M. Schubert, and J. H. Werner, Potential of photovoltaic systems in countries with high solar irradiation, Renewable and Sustainable Energy Reviews, 2010; **14**: 754-762.

[11] D. Rumelhart, G. Hinton and R. Williams, Learning representations by back-propagating errors, *Nature*, 1986; 323: 533–536, doi:10.1038/323533a0.

[12] R. Ibrahim and A. Afshin, Solar Position Algorithm for Solar Radiation Applications, *Solar Energy*, 2004; **6**: 577-589.

PREDICTIVE MAINTENANCE IN PHOTOVOLTAIC PLANTS WITH A BIG DATA APPROACH

Alessandro Betti †, Maria Luisa Lo Trovato ‡, Fabio Salvatore Leonardi ‡,
Giuseppe Leotta ‡, Fabrizio Ruffini † and Ciro Lanzetta †
† i-EM srl, via Aurelio Lampredi 45, Livorno (Italy)
‡ Enel Green Power SPA, Viale Regina Margherita 125, Rome (Italy)

ABSTRACT: This paper presents a novel and flexible solution for fault prediction based on data collected from SCADA system. Fault prediction is offered at two different levels based on a data-driven approach: (a) generic fault/status prediction and (b) specific fault class prediction, implemented by means of two different machine learning based modules built on an unsupervised clustering algorithm and a Pattern Recognition Neural Network, respectively. Model has been assessed on a park of six photovoltaic (PV) plants up to 10 MW and on more than one hundred inverter modules of three different technology brands. The results indicate that the proposed method is effective in (a) predicting incipient generic faults up to 7 days in advance with sensitivity up to 95% and (b) anticipating damage of specific fault classes with times ranging from few hours up to 7 days. The model is easily deployable for on-line monitoring of anomalies on new PV plants and technologies, requiring only the availability of historical SCADA and fault data, fault taxonomy and inverter electrical datasheet.
Keywords: Data Mining, Fault Prediction, Inverter Module, Key Performance Indicator, Lost Production

1 INTRODUCTION

The provision of a Preventive Maintenance strategy is emerging nowadays as an essential field to keep high technical and economic performances of solar PV plants over time [1]. Analytical monitoring systems have been installed therefore worldwide to timely detect possible malfunctions through the assessment of PV system performances [2-3].

However, high customization costs, the need of collecting a great number of physical variables and of a stable Internet connection on field generally limit their effectiveness, especially for farms in remote places with unreliable communication infrastructures. The lack of a predictive component in the maintenance strategy is also a hindrance to minimize downtime costs. In order to keep lower the implementation costs and model complexity, statistical methods based on Data Mining are recently emerging as a very promising approach both for fault prediction and early detection. However, while the Literature is mainly focusing on equipment level failures in wind farms [4-5], research for PV plants is still in an early stage [6].

The present paper describes an innovative and versatile solution for inverter level fault prediction based on a data-driven approach, already tested with remarkable performances on six PV plants of variable size up to 10 MW located in Romania and Greece and three different inverter technologies (Table 1). The proposed approach is easily portable on different plants and technologies and simplifies the update process to follow the PV plant evolution.

2 METHODOLOGY

The model is composed by two parallel modules both being capable of predicting incipient faults, but differing for the level of details provided and the remaining operational time after first indication of fault: a Supervision-Diagnosis Model (SDM) based on a Self-Organizing Map (SOM) predicting generic failures through a measure of deviations from normal operation, and a Short-Term Fault Prediction Model (FPM) based on artificial Neural Network (NN) addressing the prediction of specific fault classes. The main steps of the model workflow are described in the following.

Table 1: list of tested PV plants.

Plant Number	# of Inverter Modules	Inverter Manufacturer Number	Max Active Power (KW)	Plant Nominal Power (MW)
1	35	1	385	9.8
2	7	1	385	2.8
3	2	2	731.6	1.63
4	25	3	183.4	4.9
5	34	3	183.5	6.0
6	10	3	183.4	1.99

2.1 Data and Alarm Logbook Import

Historical data extracted from 5-min averaged SCADA data (Table 2) and inverter manufacturer electrical parameters for the on-site inverter technology are first collected to train the model. SCADA logbooks, as well as fault taxonomy, are also imported for offline performance assessment and normal training period selection in the case of the SDM, and for NN training for the FPM. Logbook import is achieved by matching fault classes listed in the fault taxonomy, which includes also fault severity, to the fault occurrences recorded in logbooks and discretizing them on the timestamp grid of SCADA data. In particular, a *k*-th fault is assigned to timestamp t_n if the following condition occurs:

$$t_{start,k} \leq t_n \leq t_{end,k}, \qquad (1)$$

where $t_{start,k}$ ($t_{end,k}$) are the initial (final) instant of fault event. Then SCADA data labelling is realized by

Table 2: electrical and environmental input SCADA data (GHI (GTI): Global Horizontal (Tilted) Irradiance).

DC Electrical tags	AC Electrical tags	Temperature tags	Irradiance tags
Current (I_{DC})	Current (I_{AC})	Internal (T_{int})	GTI
Voltage (V_{DC})	Voltage (V_{AC})	Panel (T_{mod})	GHI
Power (P_{DC})	Power (P_{AC})	Ambient (T_{amb})	

assigning fault codes (as integer numbers) to SCADA data. In the case of concurrent fault events, a prioritization rule is adopted considering only the most severe fault and, if necessary, the most frequent fault in that day.

2.2 Data Preprocessing

AC power (P_{AC}) depends primarily on the level of solar irradiance (GTI) and, secondarily, on ambient temperature (T_{amb}). A first-order regression of signals GTI and P_{AC} applied on training set samples allows to remove outliers corresponding to the furthest points from fitting: $|P_{AC} - GTI \cdot m + b| > thr \cdot (GTI \cdot m + b)$, where m and b are the slope and the intercept, respectively, computed by a least squares approach and thr is the threshold set by a trial and error process.

Further preprocessing steps include removal of days with a large amount of missing data, setting periodic tags to 0 in the nighttime, data range check and removal of unphysical plateaus.

2.3 Data Imputation

Missing instances in test set are imputed by means of a k-Nearest Neighbors (k-NN) algorithm using the training set as the reference historical dataset, selecting nearest neighbors according to the Euclidean distance and exploiting hyperbolic weights.

2.4 Feature Engineering: Data De-trending and Scaling

In order to remove the season-dependent variability from input data, a de-trending procedure has been applied by following customized approaches for each tag. T_{mod} training data have been de-trended by means of the least squares method to find the best line T_{fit} against T_{amb} and selecting only low GTI samples to remove the effect of panel heating due to sunlight:

$$T_{mod}^{detrend} = \left(T_{mod} - T_{fit}\right)/T_{fit}\ , \qquad (2)$$

where $T_{fit} = m_T \cdot T_{amb} + b_T$ is the fitting, m_T is the regression slope and b_T is the intercept. All the remaining input tags, but DC and AC voltages, have been de-trended according to a classical Moving Average (MA) smoothing to compute the trend component and applying an additive model for time series decomposition. In both cases also test instances have been de-trended by means of a moving window mask 1-day long for tracking of time-varying input patterns. Finally, input data normalization is performed to avoid unbalance between heterogeneous tags.

2.5 Data Processing: the Supervision Model (SDM)

SDM has been built by means of an unsupervised clustering approach based on a 20×20 Self-Organizing Map (SOM) algorithm (Figure 1) performing a non-

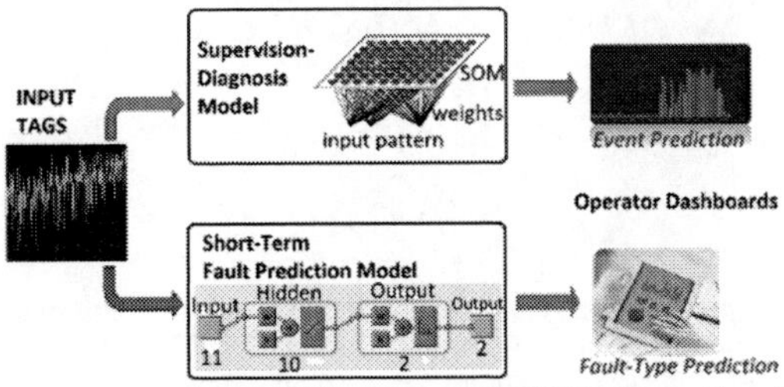

Figure 1: workflow of the Model SDM (SOM based) and FPM (NN based) from SCADA input tags to model outputs available in an operator dashboard.

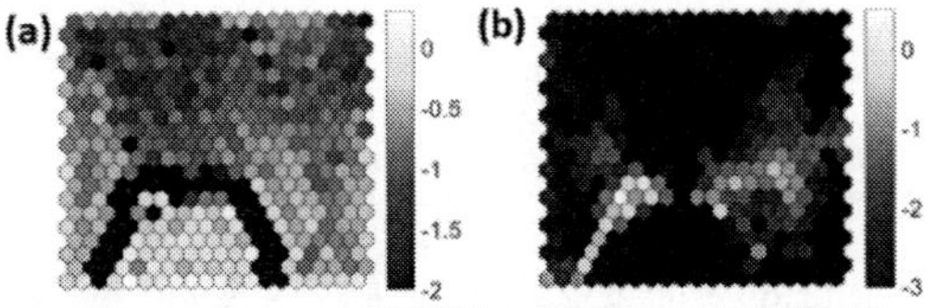

Figure 2: (a) $log(P_{ij}^{TRAIN})$ and (b) $log(P_{ij}^{TRAIN} - P_{ij}^{TEST-FULL})$ considering the full training and test phases. The larger the occupancy difference in (b), the smaller the contribution to KPI value (Eq. (3)).

linear mapping from a n-dimensional space (n=11) to a 2-dimensional space with an online weight update rule [7] and trained on a normal operation period, identified by a time interval, through a competitive learning process. SOM has the valuable property of preserving input topology: neighbor neurons in the SOM layer respond to similar input vectors. As a consequence, a change in the distribution of input instances, due for example to inverter malfunction, leads to a different data mapping in the output grid. A multivariate statistical process control analysis in the form of a control chart may be therefore built to detect this change in the patterns distribution.

A Key Performance Indicator (KPI) has been defined as in Eq. (3), measuring a process variation at generation unit level from the normal state towards abnormal operating conditions when a threshold is crossed:

$$KPI = \sum_{i,j} P_{ij}^{TEST} \cdot \frac{1-\left|P_{ij}^{TRAIN}-P_{ij}^{TEST}\right|}{1+\left|P_{ij}^{TRAIN}-P_{ij}^{TEST}\right|}\ , \qquad (3)$$

where sums run on the SOM cells, $P_{ij}^{TRAIN} = N_{ij}^{TRAIN}/\sum_{ij} N_{ij}^{TRAIN}$ is the normalized number of input patterns mapped into the cell (i,j) in the training phase and $P_{ij}^{TEST} = N_{ij}^{TEST}/\sum_{ij} N_{ij}^{TEST}$ is the normalized number of input patterns mapped into the cell (i,j) in a 24-hours test phase (Figure 2).

The ratio factor in Eq. (3) penalizes large deviations from the training operating conditions: indeed if $\forall\,(i,j)\ P_{ij}^{TEST} = P_{ij}^{TRAIN}$ then $KPI = 1$, and if $\left|P_{ij}^{TRAIN} - P_{ij}^{TEST}\right| \to 1$ ($P_{ij}^{TRAIN} = 0, P_{ij}^{TEST} = 1$, or viceversa), then $KPI \to 0$. From a physical point of view, Eq. (3) is a robust indicator detecting changes in the underlying non-linear dynamics of the generation unit from normal status, represented by $KPI = 1$.

To remove the seasonality pattern from Eq. (3) due to time-varying number of daytime instances, a KPI de-trending procedure has been followed by means of a linear regression of both signals to compute the KPI trend and selecting the de-trended component KPI/KPI_{trend} by assuming a multiplicative decomposition $KPI = KPI_{trend} \times KPI_{season} \times KPI_{cyclic} \times KPI_{residual}$. Once

Table 3: rules for switching ON of the 4 warning levels (w) of the SDM (d: day).

Warning Level (w)	Crossing of Threshold	KPI Derivative	Persistence
1	$3\sigma_{KPI^{TRAIN}}$	< 0	1 d
2	$3\sigma_{KPI^{TRAIN}}$	< 0	2 d of w 1
3	$5\sigma_{KPI^{TRAIN}}$	< 0	1 day
4	$5\sigma_{KPI^{TRAIN}}$	< 0	2 d of w 3

the de-trended KPI is available, a low-pass MA filter with a 4-weeks long window is applied. Finally, four warning levels of different severity are computed based on thresholds crossing and time persistence rules (Table 3). Model performances have been evaluated by means of usual classification metrics (accuracy, sensitivity and specificity) exploiting the alarm logbook knowledge and assigning a predictive connotation to sensitivity by assuming correctly predicted a fault event if in the last N days prior to fault a warning has been triggered by the SDM.

2.6 Data Processing: the Short-Term Fault Prediction Model (FPM)

Short-Term FPM is based on a 11-10-2 Pattern Recognition Feed-Forward Back-Propagation NN fed with 11 input tags (Table 2) and containing one hidden layer with 10 neurons and two output neurons (Figure 1) trained to recognize specific fault classes by means of a Bayesian regularization algorithm to prevent overfitting. NN architecture has been optimized building up an ensemble of statistical simulations to maximize classification metrics. Once the full dataset is labelled (Section 2.1) and preprocessed (Section 2.2), a data sampling step is followed, to compensate the large unbalance between the number of normal operation data of majority class and the one of low frequency failure data, causing a prediction bias affecting event classification.

Sample balancing is achieved by first collecting the number of fault instances $N_{fault,i}$ available for the i-th fault class and then assigning $2/3$ of them to training set, which is finally filled by randomly sampling normal instances up to the percentage required for almost balanced classes. Test set is instead built by means of the remaining $1/3 \times N_{fault,i}$ fault instances and randomly sampling normal instances up to the proportion necessary for representing the distribution of the labeled SCADA data. Normal and fault samples are sampled separately to avoid unbalanced fault distribution in training and test set.

Specifically, for prediction from current time t an event occurs (either faulty or normal) to the previous n hours a different training (test) set has been built by considering instances at time t $(t-n)$. In this manner, NN is trained to recognize events at time they occur (t) and tested at previous instants $(t-n)$. Due to the poor fault statistics available, missing data at time $t-n$ have been imputed by means of a k-NN algorithm. Model performances have been finally assessed by setting-up a Monte Carlo approach for each timestamp and averaging ensemble classification metrics.

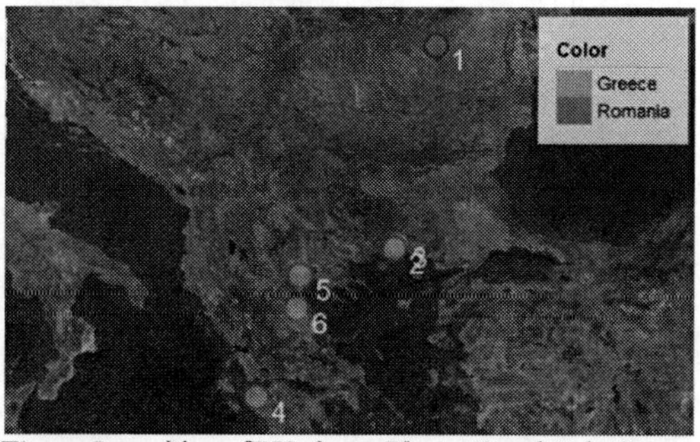

Figure 3: position of PV plants. Plants #2 and #3 in Greece are close to each other and their corresponding circles overlap in the figure.

3 RESULTS AND DISCUSSION

SDM has been tested on six different PV plants located in Romania and Greece (Figure 3) and corresponding, globally, to more than one hundred inverters of three well-known technology brands and typologies (inverter module, central inverter, master slave). The time period considered for offline performance assessment spans from 2014 to 2016, depending on data source availability.

In Figure 4 the KPI, warning levels, and fault occurrences percentage time series are shown for inverter A.3 of plant #1 in Romania with installed capacity of almost 9.8 MW and inverter technology #1. The two thresholds $3\sigma_{KPI^{TRAIN}}$ and $5\sigma_{KPI^{TRAIN}}$ are also represented by dashed and dotted black curves, respectively. According to alarm logbooks available, a series of thermal issue damages happened on different inverters in 2014-2015 which led to inverter replacements. In particular, a generalized failures occurred from 3rd to 28th of November 2014 to inverter A (AC Switch Open, severity 2/5) but it was recognized only later by the operator, with a severe lost production. As can be seen,

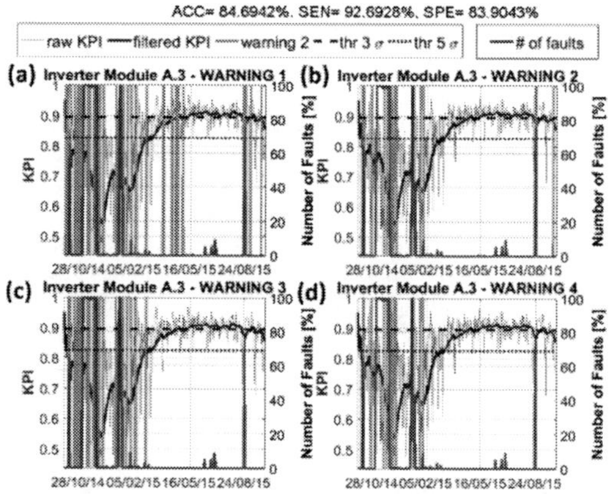

Figure 4: raw (sky-blue) and filtered (blue) KPIs, warning levels (green) and fault occurrences percentage (red, on right *side*) as a function of time for INV. A.3 of PV plant #1 (9.8 MW) and technology #1. Warnings 1 (a) to 4 (b) are shown from top left to bottom right for period October 2014 to August 2015.

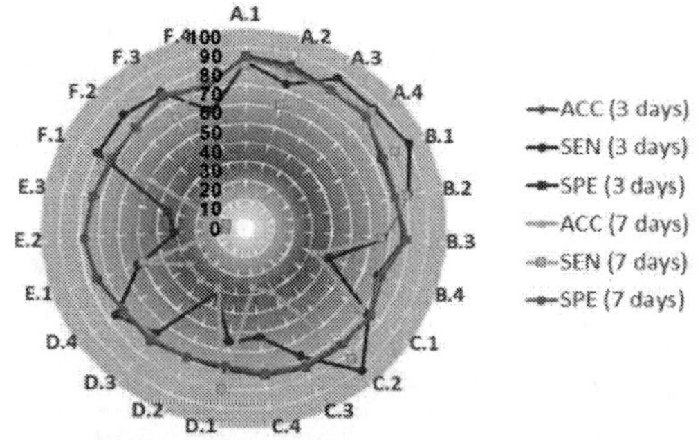

Figure 5: radar chart of classification metrics, expressed as a percentage 0-100% (ACC: accuracy, SEN: sensitivity, SPE: specificity), computed at 3 and 7 days prior of fault occurrences for inverter A-F of PV plant #1. Inverters G-H are neglected since no faults happened.

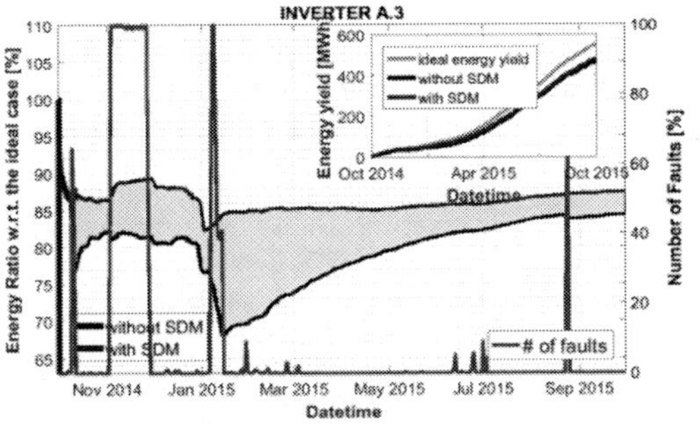

Figure 6: energy yield time series w.r.t. ideal case with (blue curve) or without (black) the predictive service: yellow area represents the energy gain enabling SDM. Fault occurrences percentage is also shown on the right (red). Inset: energy yield in the ideal case, as well as with or without SDM (INV. A.3 of Plant #1).

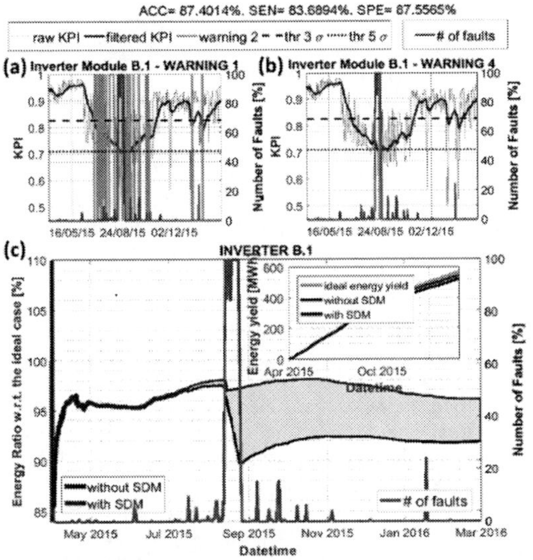

Figure 7: raw and filtered KPIs, as well as normalized fault occurrences and warning levels 1 (a) and 4 (b) for INV. B.1 of PV plant #2 (2.8 MW) and technology #1. (c): energy yield time series with respect to the ideal case with or without SDM enabled, as well as fault occurrences percentage as a function of time. Inset: energy yield in the ideal case, and with or without SDM.

Supervision Model well anticipates logbook fault events, with a clear correlation between the deep KPI degradation (with warning triggered up to level 4) and fault occurrences, even for events happening in 7-10 January 2015 (DC Ground Fault, i.e. high leakage current to ground on DC side, severity 1/5) and 23-24 August 2015 (DC Insulation Fault, i.e. overvoltage across DC capacitors, severity 2/5). Sensitivity roughly degrades from 93% to 84% anticipating faults from 0 up to 7 days, and with an overall accuracy of almost 85%.

Figure 5 shows a general overview of SDM performance on plant #1 at 3 and 7 days in advance with respect to fault occurrences. As can be seen, at 3 (7) days sensitivity (SEN) is larger than 60% for 17 (14) devices (on a total of 23), with a mean sensitivity of 72% (61%) at 3 (7) days. Accuracy (ACC) and specificity (SPE) are instead, on average, almost 80% for both the considered time horizons.

An estimate of the production gain achievable by means of a predictive service may be obtained by computing lost production as:

$$LP(t) = \int_o^t [P_{th}(t') - P_{AC}(t')]\, dt' , \qquad (4)$$

where $P_{th} = P_{nom}[1 - k_{pv}] \cdot GTI/GTI_{SC}$ is the theoretical power in normal condition [8], $GTI_{SC} = 1200\ W/m^2$ is the irradiance in standard conditions and $k_{pv} = (T_{cell} - 25)\frac{\gamma}{100}\ if\ T_{cell} \geq 25°C$ if $T_{cell} \geq 25°C$ and 0 otherwise. In Figure 6 the energy yield computed with respect to the ideal case ($P_{AC}(t) = P_{th}(t)\ \forall t$) is presented as a function of time for the case with (assuming $P_{AC} = P_{th}$ if a fault is correctly predicted) and without SDM enabled. As can be seen, an energy yield improvement (yellow area) up to 10-15% may be achieved with SDM.

In Figure 7a-b the KPI and warning levels 1 and 4, respectively, are illustrated as a function of time for inverter B.1 of PV plant #2 (2.8 MW) with installed technology #1. As may be observed, the strong fault spike occurred in August 2015 (red curve) due to an input over-current on DC side (severity 2/5) is well predicted by SDM, with a sensitivity larger than 95% and an overall accuracy above 80% event at 7 days in advance. The energy yield gain achievable when applying SDM is almost 6-7% (Figure 7c).

Some fault classes cannot be predicted due to their instantaneous nature. SDM, thanks to its parametric structure with respect to inverter technology and plant configuration and to an ad-hoc tuning of model parameters on each specific plant, guarantees early detection for these faults. In Figure 8 the case of plant #4 in Greece (4.9 MW) with installed inverter technology #3 is reported. As can be seen, in the second half of May 2016, SDM early detects a severe anomaly at inverter 3.5 due to an IGBT stack fault which led to inverter replacement. Due to the sharp KPI decrease below threshold $5\sigma_{KPI^{TRAIN}}$, warnings 1 and 3 are triggered almost instantaneously, whereas warnings 2 and 4 with a delay of almost 2 days.

Short-Term FPM model has been applied to three PV plants (#1, #2, #3) and two different inverter technologies (#1 for plants #1, #2 and #2 for plant #3). In Figure 9 the

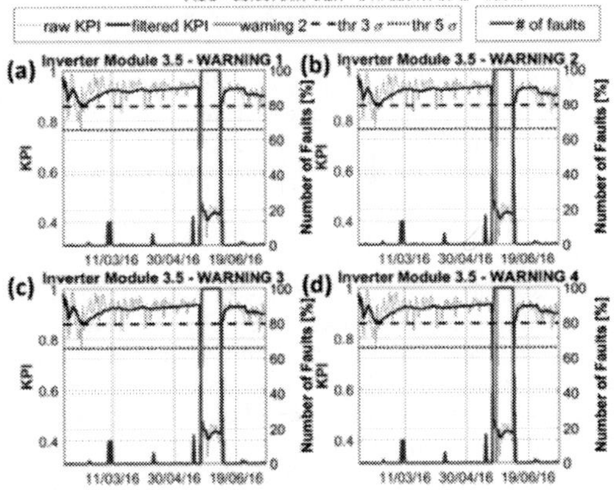

Figure 8: raw and filtered KPIs, as well as normalized fault occurrences and warning levels 1 (a) – 4 (d) for INV. 3.5 of PV plant #4 (4.9 MW) and technology #3.

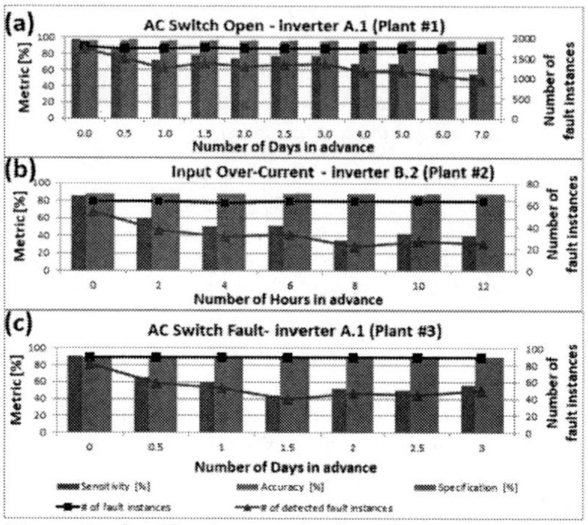

Figure 9: classification metrics (bar plot on the left) and number of faults and of detected faults (black and blue curves on the right, respectively) as a function of time in advance. (a): fault class AC Switch Open (plant #1); (b): fault class Input Over-Current (plant #2); (c): fault class AC Switch Fault (plant #3).

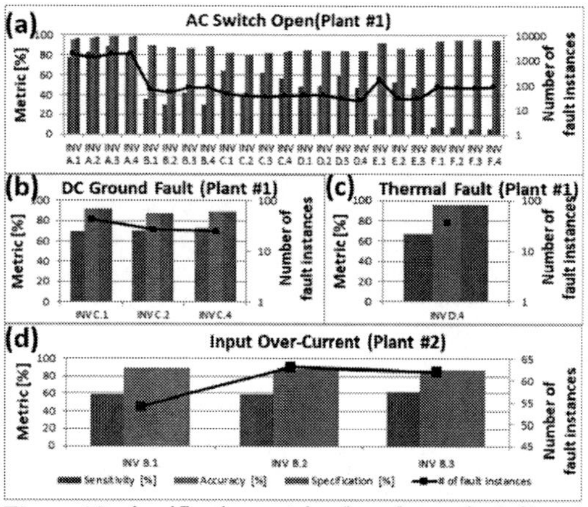

Figure 10: classification metrics (bar plot on the left) and number of fault instances (black curve on the right) as a function of inverter module for fault classes (a) AC Switch Open, (b) DC Ground Fault, (c) Thermal Fault (Low T_{amb}) of plant #1 computed at t-3Days, and (d) for class Input Over-Current of plant #2 calculated at t-2Hours.

classification metrics, as well as the number of detected faults, are shown as a function of the time prior to the fault occurrence for the highest frequency failure class of PV plants #1, #2 and #3 at fixed inverter module. As can be seen in Figure 9a, when thousands of occurrences are available, an outstanding prediction capability is achieved with sensitivity decreasing down to value of almost 50-60% seven days in advance. According to State of the Art [4], prediction performances degrade generally much faster on time horizon ranging from 1 hour to 12 hours for low frequency failures with less than 100 fault instances available (Figure 9b), with some exceptions depending on correlations among predictors and faults (Figure 9c).

In Figure 10 a global FPM performance overview on the inverter park is shown for classes AC Switch Open, DC Ground Fault and Thermal Fault (Low Ambient Temperature) of PV plant #1 computed 3 days prior of the failure, and for class Input Over-Current for PV plant #2 calculated 2 hours before the damage. In general, FPM is capable of detecting incipient faults for up to three or four fault classes for each inverter module. Figure 10a highlights however a strong correlation between FPM performances and failure data availability, with sensitivity (red bars) degrading dramatically from 80% (inverter A) to roughly 30-40% (inverter B-F), on average, when the number of fault instances (black curve) decreases from thousands to one hundred or less.

Accuracy (green bars) is still over 80% due to good negative samples classification performances. Figure 10b-d confirm previous conclusions: when few fault instances (roughly 30-40) are available, sensitivity is satisfactory generally on very short time horizons (2 hours in Figure 10d), but in some cases even larger (3 days in Figure 10b-c) depending on the strength of correlations among predictors and failure data.

4 FINAL REMARKS

An original methodology to predict inverter faults at two different levels of detail (prediction of a status/fault, prediction of a specific fault) has been presented and validated on SCADA data collected from 2014 to 2016 on a park of six PV plants up to 10 MW.

Results demonstrates that the proposed SDM effectively anticipates high frequency inverter failures up to almost 7 days in advance, with sensitivity up to 95% and specificity of almost 80%. SDM also guarantees early detection for unpredictable failures. FPM exhibits also excellent predictive capability for high frequency fault classes up to 7 days prior of the damage but, depending on fault statistics available, sensitivity may degrades also on horizons of few hours. Combination of these two prediction modules can therefore enable PV system operators to move from a traditional reactive maintenance activity towards a proactive maintenance strategy, improving decision-making process thanks to a complete information on the incoming failure before the fault occurs. The model is actually tested for on-line monitoring of anomalies in Romania and Greece and can be easily deployed on new PV plants thanks to the limited amount of information required.

Next steps may include the introduction of a deep-learning based automated system of fault detection in drone-based thermal images of PV modules [9] and integration of the predictive model in a smart solar monitoring software, including an intervention management system integrated with alarm handling and a business intelligence based reporting tool from intervention up to portfolio level [10].

5 ACKNOWLEDGMENTS

Authors thank Prof. Mauro Tucci and Prof. Emanuele Crisostomi from Destec Department of University of Pisa for fruitful discussions and helpful suggestions.

6 REFERENCES

[1] U. Jahn, M. Herz, E. Ndrio, D. Moser et al, Minimizing Technical Risks in Photovoltaic Projects, www.solarbankability.eu

[2] A. Woyte, M. Richter, D. Moser, N. Reich et al, Systems Report IEA-PVPS T13-03:2014 (March 2014)

[3] S. Stettler, P. Toggweiler, E. Wiemken, W. Heydenreich et al, 20th EUPVSEC (Barcelona, Spain 2005) 2490-2493
[4] A. Kusiak,W. Li, Renewable Energy 36 (2011) 16-23
[5] K. Kim, G. Parthasarathy, O. Uluyol, W. Foslien et al, Proceedings of 2011 Energy Sustainability Conference and Fuel Cell Conference (2011) 1-9
[6] F. A. Olivencia Polo, J. Ferrero Bermejo. J. F. Gomez Fernandez, A. Crespo Marquez, Renewable Energy 81 (2015) 227-238
[7] M. Tucci, M. Raugi, Neurocomputing 74 (2011) 1815-1822
[8] M. Fuentes, G. Nofuentes, J. Aguilera, D. L. Talavera, M. Castro, Solar Energy 81 (2007) 1396–1408.
[9] V. Jiri, R. Ilja, K. Jakub, P. Tomas, 32nd EUPVSEC (2016) 1931-1935
[10] White Paper: Beyond standard monitoring practice, www.3e.eu

SURVEY ON YIELD OF PV SYSTEMS IN GERMANY 2014 TO 2016

Henrik te Heesen*, Volker Herbort#
*Trier University of Applied Sciences, Environmental Campus Birkenfeld, Campusallee, 55768 Neubrücke (Nahe), Germany, h.teheesen@umwelt-campus.de
#Ulm University of Applied Sciences, Albert-Einstein-Allee 55, 89081 Ulm, Germany

ABSTRACT: More than one million small roof-top PV systems with a capacity up to 30 kWp have been installed within the last years in Germany. Previous publications have shown, that most PV systems do not reach the technical possible yield due to mistakes during the planning phase, the construction phase and mainly during the operation phase [1-7]. We want to derive the yield loss by these residential PV systems in Germany by analyzing public monitoring data from 22,000 PV systems. In conclusion, we find, that the annual yield loss in average is 5 to 10 percent and there is a nationwide annual revenue loss of €450m to €500m. The reference yield is derived by a data cleansing algorithm [8]. The actual yield of 50 percent of all PV systems in Germany is 5 to 20 percent smaller compared to the reference yield. Thus, a severe quality issue of PV Systems can be identified. Our publication addresses operators and planers of PV systems as well to make them aware of occurring technical issues of their PV system.

Keywords: Monitoring, Rooftop, System Performance

1 INTRODUCTION

Previous work has shown that more than 80 % of all PV systems in Germany are underperforming compared to their technical possibilities [1]. This observation can also be seen in other European countries [2, 3, 4, 5, 6, 7].
The first step to reduce large yield loss is to address PV system owners and operators about the yield they have to expect from their systems. This offers them opportunities to identify potential for improvement and optimization.
Figure 1 shows the distribution of PV systems in Germany with a capacity up to 30 kWp. In total, more than one million systems have been installed in this capacity class.

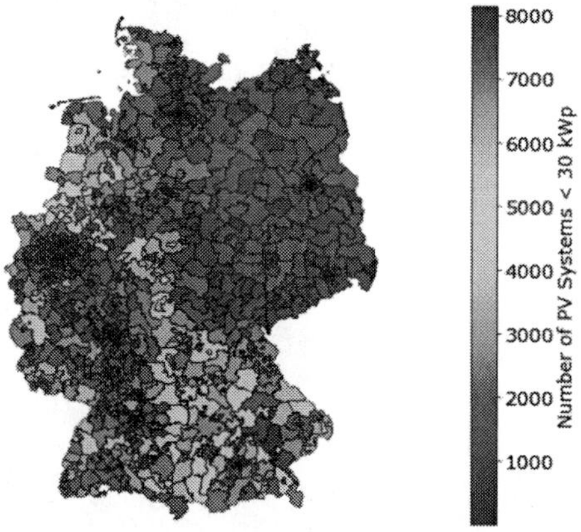

Figure 1: Distribution of installed PV systems with a capacity up to 30 kWp in Germany. Data source: Bundesnetzagentur.

2 METHODS

We use a modification of our algorithm to analyze the yield of photovoltaic systems [8]. The original algorithm considers the yield and the insolation as well. If the insolation data is neglected the results do not differ significantly from the previous one.
Thus, raw yield data and system information (location, capacity, inclination and orientation) are downloaded by screen scraping scripts from monitoring websites providing public data on a daily time interval. Implausible data are eliminated in a primarily step. Statistical values like median and quartiles are calculated for all systems within each one-digit postal code in Germany. PV systems with outlining values are neglected in the following steps. This data cleansing step is repeated once. Next, the yield of systems in each two-digit postal code and the neighboring two-digit postal code is cleaned. Finally, the remaining PV systems are used for further analysis.
The number of PV systems within the capacity classes between 0 and 30 kWp is shown in Figure 2. Most PV system have a capacity between 4 and 10 kWp and represent the group of small roof-top residential systems. The peak at 30 kWp relates to feed-in-tariff restrictions in former EEG (German Renewable Energy Law) versions.

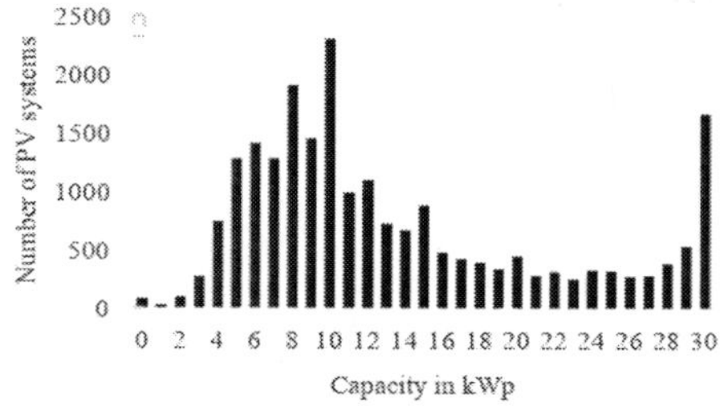

Figure 2: Distribution of the capacity of PV systems used for data analysis.

3 RESULTS AND DISCUSSION

The rated annual specific yield of small roof-top PV systems in Germany is shown in Figure 3. The yield values represent rated data which is the technical possible yield of well-operated PV systems with a southern orientation. A strong southeast-northwest gradient can be identified in all years. The highest yield of more than 1,150 kWh/kWp has been achieved in 2015 due to high irradiance during this year. In contrast, 2016 has been the worst year in comparison to 2014 and 2015.
The annual yield data corresponds to the monthly yield data shown in Figure 4. The boxplots represent the yield values of small roof-top PV systems in the northwestern part and the southeastern part of Germany. In 2014, the yield increases continuously during spring and summer from 90 kWh/kWp up to 130 kWh/kWp. There has been a long period with high irradiance from April to August

2015. The monthly specific yield in 2016 mostly stays below the corresponding months in 2014 and 2015. The third quartile values (upper limit of the boxes in the boxplot) represent the rated yield data for well-operated PV systems used in Figure 3.

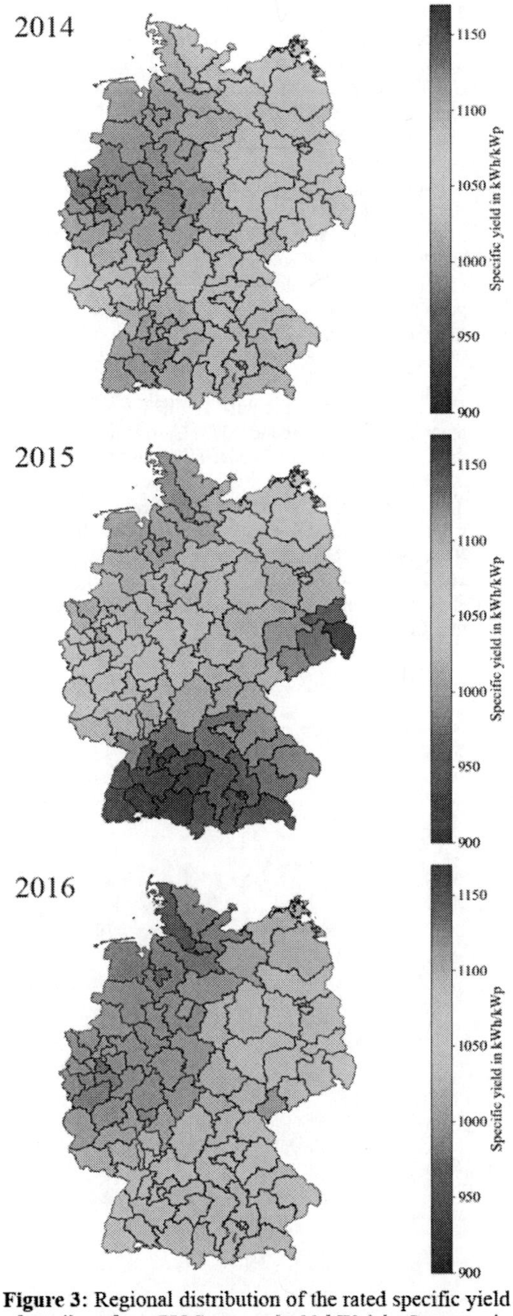

Figure 3: Regional distribution of the rated specific yield of small roof-top PV Systems (< 30 kWp) in Germany in 2014 to 2016. A strong southeast to northwest gradient can be seen.

The ratio of the mean specific yield of all PV systems aggregated over all two-digit postal code regions in Germany in comparison to the rated specific yield (3rd quartile) is shown in Figure 5. We used the daily yield information of the PV systems. The graph represents the 1st quartile (lower limit in orange), the median (transition orange to dark blue) and the 3rd quartile (upper limit in dark blue) of the daily mean specific yield data. PV systems without any yield data have been neglected. The spread of the ratio is increasing during winter time due to smaller yield values. The median ratio of the daily specific yield in Germany is increasing from January to June and reaches a ratio of 0.9. During autumn, the ratio decreases to 0.8. This means, that even in summer time residential PV systems in Germany show a yield loss of 5 to 15 % compared to the rated median value of well-operated PV systems.

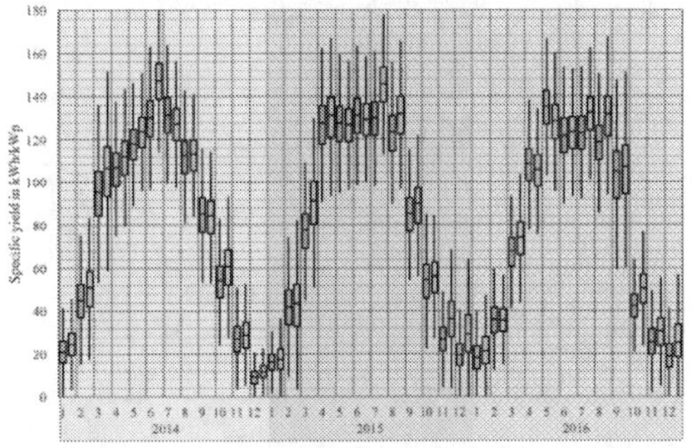

Figure 4: Boxplot of the monthly specific yield of PV systems in the northwestern (green) and the southeastern part (yellow) of Germany for the time period 2014 (light green), 2015 (light red) and 2016 (light blue).

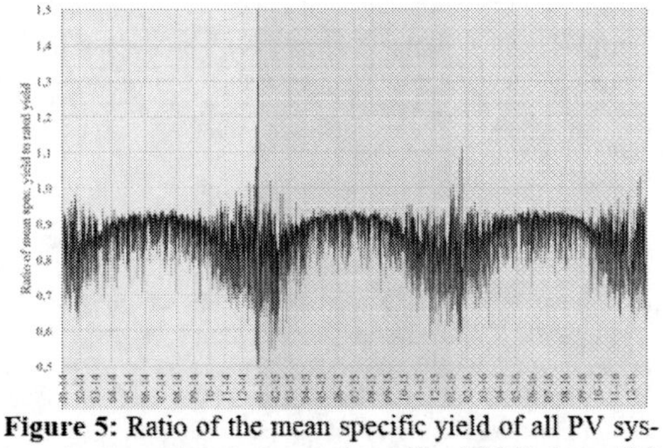

Figure 5: Ratio of the mean specific yield of all PV systems in comparison to the rated specific yield (3rd quartile) on a daily time interval. The lower limit in orange represents the 1st quartile of the mean yield data, the transition represents the median and the upper limit in dark blue represents the 3rd quartile of the yield data of the PV systems in the two-digit postal code regions.

Finally, we calculate the monetary losses due to yield losses based on the mean feed-in tariff in Germany [9]. We derive the daily yield loss for each two-digit postal code region in relation to the rated specific yield of each region. The daily yield losses are multiplied by the total installed capacity in each two-digit postal code region provided by the German Bundesnetzagentur. Summing up these values for all regions and all days gives the annual yield loss (see Figure 6 for the distribution of annual yield loss per region and Table 1 for annual data for Germany).

The yield losses on the map in Figure 6 correspond to the distribution of installed PV systems with a capacity < 30 kWp (see Figure 1). As there are less residential rooftop systems in the eastern part of Germany the losses are smaller as well. The largest yield losses can be identified in the postal code regions 84 (northeast of Munich) and 94 (at the border to Austria and the Czech Republic).

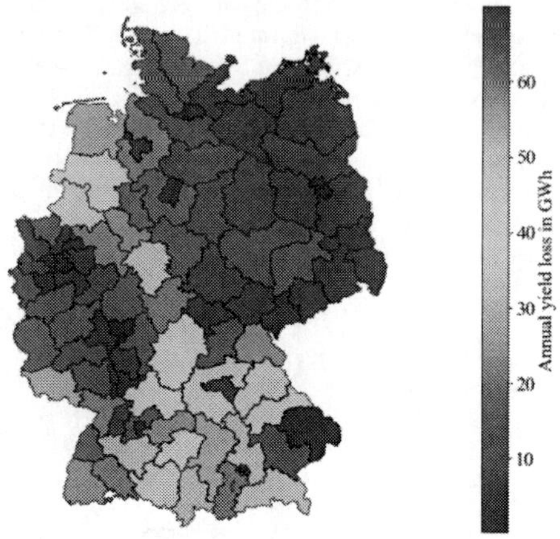

Figure 6: Annual yield loss in GWh for small roof-top PV systems for each two-digit postal code in Germany.

Table 1: Annual electric energy production of all PV systems in Germany, the calculated yield loss of small roof-top systems and the monetary losses based on the mean feed-in tariff [9].

Year	Electric energy production	Yield loss	Monetary loss in million €
2014	36.1 TWh	1.46 TWh	€460m
2015	38.7 TWh	1.59 TWh	€489m
2016	37.6 TWh	1.58 TWh	€460m

4 CONCLUSION

We analyzed daily yield information of small residential roof-top PV systems in Germany for 2014 to 2016. The data have been downloaded by screen scraping scripts from public PV monitoring websites. In total, we considered about 22,000 PV systems in our evaluation.
We calculated the rated specific yield for each two-digit postal code on a daily and a monthly interval. The comparison of this rated yield with the mean yield of PV systems shows a mean deviation of 5 to 10 % and leads to an annual loss of €450m to €500m. Reasons for this yield losses can be identified during the planning phase, the construction phase and the operation phase of a PV system. Mainly, the owner or operator needs to take more care about the PV systems to reduce the losses.
Next, we will analyze the daily data more in detail in order to derive conclusions about the quality of maintenance and the reaction time on incidents.

5 REFERENCES

[1] H. te Heesen, V. Herbort, R. Pfatischer, Survey on yield of PV systems in Germany 2013, in: 29th European Photovoltaic Solar Energy Conference and Exhibition, 2014, pp. 2198-2201.

[2] B. Kausika, W. Folkerts, G.J.H.M. van Sark, B. Siebenga, P. Hermans, A big data approach to the solar PV market: design and results of a pilot in the Netherlands, in: 29th European Photovoltaic Solar Energy Conference and Exhibition, 2014, pp. 4030-4033.

[3] J. Leloux, L. Narvarte, D. Trebosc, Performance analysis of 10,000 residential PV systems in France and Belgium, in: 26th European Photovoltaic Solar Energy Conference and Exhibition, 2001, pp. 3939-3946.

[4] E. Lorenz, J.W.H. Betcke, A. Drews, A.C. de Keizer, S. Stettler, M. Schneider, S. Bofinger, D. Heinemann, Intelligent Performance Check of PV System Operation Based on Satellite Data (PVSAT-2), final technical report. Tech. rep., Copernicus Institute, Utrecht University, 2004.

[5] T. Nordmann, L. Clavadetscher, W.G.J.H.M. van Sark, M. Green, Analysis of Long-term Performance of PV Systems, IEA PVPS Task 13, 2014. Subtask 1.

[6] M. Gromaire, H. te Heesen. Survey on Specific Yield of Photovoltaic Systems in France 2014. 30th European Photovoltaic Solar Energy Conference and Exhibition. pp. 2307 - 2309. Sept. 2015.

[7] O. Tsafarakis, W.G.J.H.M. van Sark. Analysis and Comparison of Performance of Photovoltaic Systems from the "Counting the Sun" Public Awareness Campaign on PV Performance. 31st European Photovoltaic Solar Energy Conference and Exhibition. pp. 2245-2248. June 2016.

[8] H. te Heesen, V. Herbort. Development of an Algorithm to Analyze the Yield of Photovoltaic Systems. Renewable Energy. Volume 87, Part 2, March 2016, 2016, Pages 1016–1022.

[9] BMWI. EEG in Zahlen: Vergütungen, Differenzkosten und EEG-Umlage 2000 bis 2017.

STATISTICAL ANALYISIS OF THE PERFROMANCE LOSS RATE OF PV PLANTS DISTRIBUTED IN A REGION: A REAL-CASE STUDY IN SOUTH TYROL

G. Belluardo[1*], M. Pierro[1,2], P. Ingenhoven[1], C. Cornaro[2], D. Moser[1]
[1]Institute for Renewable Energy, EURAC Research, 39100 Bolzano, Italy
[2]Department of Enterprise Engineering - University of Rome Tor Vergata, 00133 Roma (Italy)
*Phone: +39 0471 055626; Fax: +39 0471 055699; E-mail: giorgio.belluardo@eurac.edu

ABSTRACT: Information concerning the performance of photovoltaic (PV) plants in a region is of fundamental importance for many stakeholders of the sector. Nevertheless, often the lack of meteorological measurements – irradiance over all – prevent from correctly assessing the current health status of PV plants, i.e. the presence of ongoing mechanisms that gradually cause a performance degradation higher than expected and warranted. This study presents a methodology to infer statistical information on the performance loss rate (PLR) of a PV fleet in a region, when only few characteristics of PV plants (e.g. final yield, location, nominal power) are known. It makes use of satellite-retrieved values of global horizontal and diffuse horizontal irradiance, downloaded freely and only for an optimized number of representative locations. A fleet of 1899 PV plants located in South Tyrol (North-East of Italy) is used as a case study to validate the methodology, and results and possible applications are presented.
In general, the presented methodology looks promising in extracting statistical information from a fleet of PV plants distributed in a region, and in inferring the probabilistic performance of future plants installed in the same area. Issues concerning the quality of the input data, related especially to the irradiance and the local skyline, are identified and discussed for a further improvement of the methodology.
Keywords: system performance, performance degradation

1 INTRODUCTION

Nowadays, several PV actors are interested in collecting and processing data from several PV plants distributed in a region. For example, investment companies are interested in monitoring the health status of their portfolio and possibly increase their profitability through technical improvements and O&M cost-effective strategies; O&M companies strive to keep under control or increase the performance of their monitored plants to fulfil the performance conditions within the O&M contracts; last but not least, also utility companies are interest in now-casting and forecasting the PV-generated energy with a higher accuracy.

While the majority of non-residential plants are equipped with monitoring systems – from the basic to the more professional ones – that enable to immediately detect large faults, residential plants often lack of key measurements to detect the actual performance, irradiance on the modules' plane above all. In fact, by knowing the actual performance of a PV plant (we consider the Performance Ratio *PR*, as the most immediate and appropriate index to measure it), it is also possible to calculate its evolution with the time, i.e. its Performance Loss Rate *PLR*.

This paper presents a methodology that reconstructs the monthly series of PR when only the following information is available: monthly values of generated electricity, the location and the nominal power of the PV plant. This series is then used to calculate the Performance Loss Rate by applying one of the available statistical techniques.

The methodology is applied to a real-case study consisting of 1899 PV plants distributed in the South Tyrol region (North-East of Italy), to infer statistical information on the performance of the fleet, and its loss trend.

Finally, a probabilistic application is presented that enables PV designers to calculate the confidence interval of the performance of future PV plants by exploiting the above mentioned dataset.

2 METHODOLOGY

2.1 Data

The case study considers data from the net-metering of the local energy utility Edyna. In total, the monthly values of final yield *Y* of 1899 plants located in South Tyrol and corresponding to a total power of 64 MW are available. In addition, the nominal power and the coordinates of the plants are known.

The PV systems analysed are mainly distributed in the southwestern part of the region, around the cities of Meran and Bolzano, as shown in Figure 1. They are installed both in the valleys and in the plateaus, covering several altitudes whose frequency distribution is shown in the same figure.

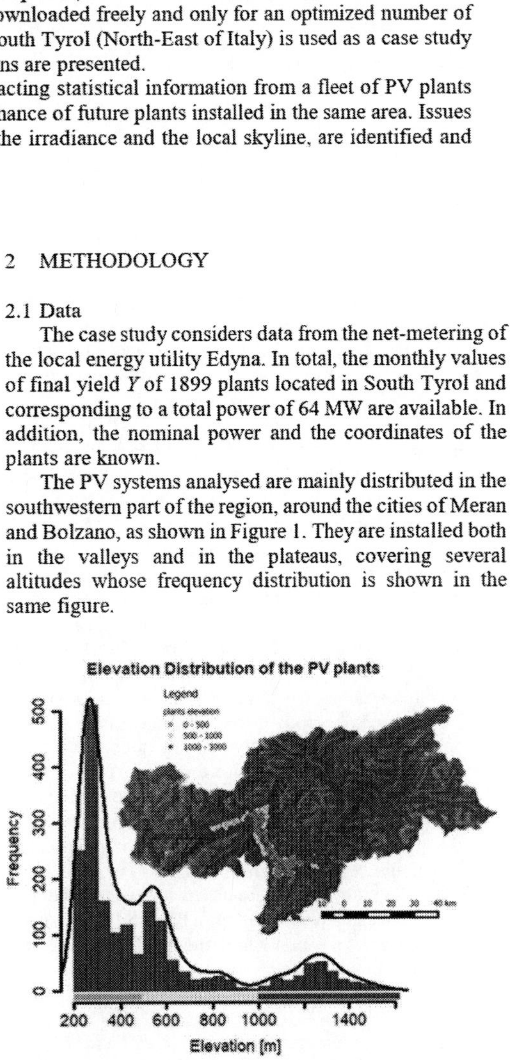

Figure 1: Locations of the 1899 PV plants analyzed in the South Tyrol region, divided per elevation classes, and frequency distribution of their elevation.

Figure 2 shows the distribution of the plants per power class. Almost half of the installed plants is residential (P < 10kW), while almost 9 out of 10 have a nominal power lower than 50 kW.

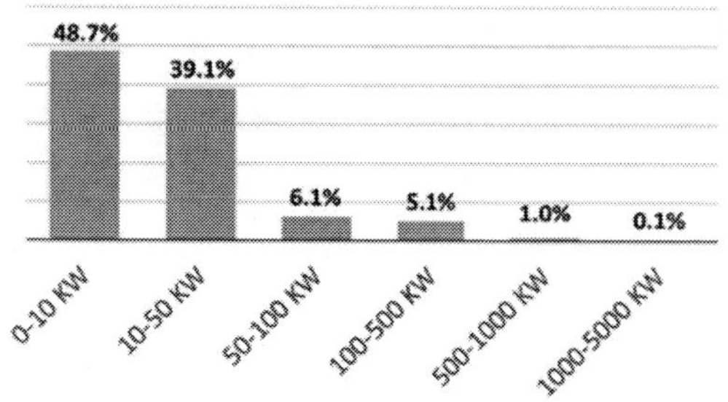

Figure 2: Percentage on total of PV plants, per power class.

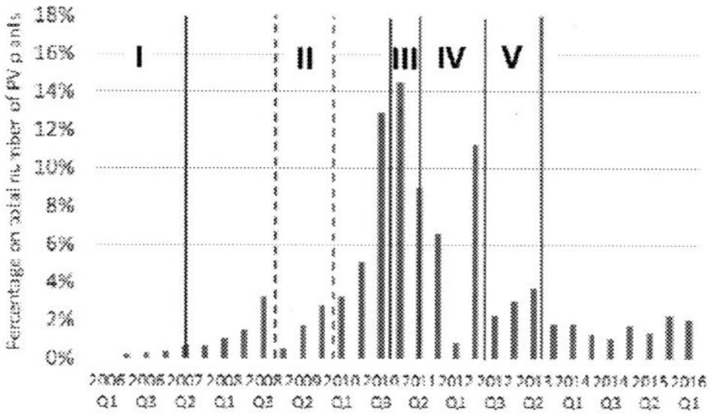

Figure 3: Percentage of PV plants installed, for four-month period of installation. Red lines corresponds to the limits of the several Italian incentives schemes "Conto Energia". Red-dashed lines show the limit of different incentives within the same Conto Energia.

In Figure 3 the percentage of PV plants installed for each four-month period of the year is displayed. The majority of the plants was installed between 2010 and 2011, following the trend registered in that period at national level [1]. Red lines correspond to the limits of validity of the feed-in incentives "Conto Energia" that ended in July 2013. Red dashed lines correspond to different incentives within the same Conto Energia. As expected, an increase in the installation volume occurred right before the end of each incentive period.

As presented in the next section, the methodology presented in this study retrieves the plane-of-array irradiance for each PV plant by using the global horizontal and diffuse horizontal irradiance for some representative locations. The irradiance data are freely downloaded from the Copernicus Atmosphere Monitoring Service (CAMS) [2] through the SoDa project [3], and correspond to satellite images with spatial resolution of 4-5 km (at 45° of latitude). They are then interpolated and combined with the local horizons downloaded from PVGIS [4] in order to adapt to the exact position and to the local morphology of each plant of the PV fleet.

2.2 Procedure

Part of the procedure is described in detail in [5]. Here the main steps are summarized.

The first step consists in a preliminary filtering of the data, in order to remove all the PV plants with an anomalous trend in the monthly-series of final yield (e.g. due to a change in the installed nominal power). After this step, data of 1750 plants remain valid.

With the next step, the available PV plants are grouped in small ensembles of close systems by using a clustering K-mean algorithm. A total of 50 centroids are therefore identified. This number is considered a good balance between avoiding high fragmentation into too many clusters, and the complex morphology of the territory. The identified centroids are considered as representative locations, for which irradiance data is downloaded as described in the previous section.

The next step consists in the minimization of a cost function $J(\beta,\gamma,\langle PR\rangle)$ using a gradient descent algorithm, in order to determine the tilt angle β, the orientation angle γ, and the mean monthly performance ratio $\langle PR\rangle$ of each PV system. The cost function considers the monthly measured values of final yield $Y(t)$ (where t is the generic month, starting from 0), and includes also the reference yield $Y_r(\beta,\gamma,t)$, defined as a function of the irradiation on the plane of the array. The latter is in turn calculated with a transposition model using the irradiance data of the corresponding cluster from CAMS and the horizon downloaded from PVGIS.

In the next step, the monthly values of Performance Ratio are retrieved from the ratio of the measured yield and the reference yield:

$$PR(t) = \frac{Y(t)}{Y_r(\beta,\gamma,t)} \qquad (1)$$

The Performance Loss Rate is then calculated using the *locally weighted scatterplot smoothing technique* (STL). As described in [6], the STL technique allows to decompose the original series of monthly PR into three components:

$$PR(t) = PR_{trend}(t) + PR_{seas}(t) + PR_{residual}(t) \qquad (2)$$

where PR_{trend} is the long-term trend of the time series, PR_{seas} is the seasonality and $PR_{residual}$ is the remainder, or noise. The absolute PLR is derived as the slope of the linear fit on PR_{trend}, while PR_0, i.e. the retrieved initial value of PR, corresponds to the intercept:

$$PR_{trend}(t) \cong PR_0 + PRL * \frac{t}{12} \qquad (3)$$

Figure 4 shows the time-series of monthly retrieved PR for the 1750 available plants, calculated according to Equation 1, as well as the decomposition into its trend and seasonality with STL (the residual component is omitted).

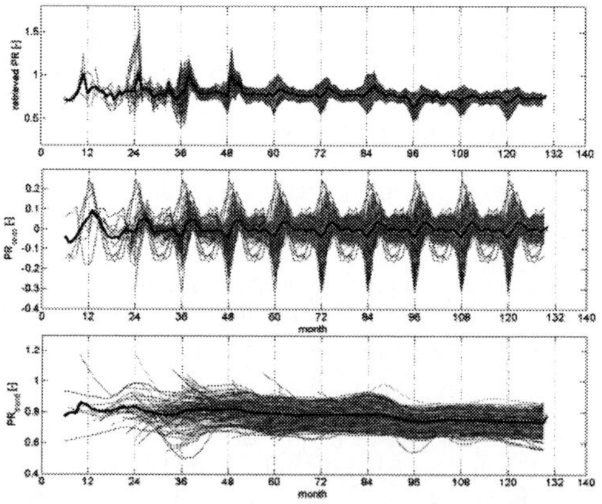

Figure 4: (top) Monthly PR series calculated for the 1750 plants; (center) monthly series of PR_{trend} and (bottom) monthly series of PR_{seas} after the application of the STL technique. Black lines show the average quantities.

2.3 Accuracy of the methodology

In a previous study [5], a validation of the methodology has been carried out by applying it to a benchmark plant not included in the PV fleet, and comparing the results with the real plant characteristics (tilt and orientation angles) and measured performance (final yield and performance ratio). Although the methodology itself is promising, several aspects concerning the quality of the input information have been identified that affect its performance at most:

- the uncertainty due to the quality of the insolation source (especially during winter);
- the complexity of the regional morphology (accuracy in accounting for the far-shading) and the impossibility to account for the shading due to close obstacles;
- the presence, within the datasets, of electricity production values of PV plants that are composed of several parts, differently tilted and/or oriented. These records result as if it was generated by a plant with a single orientation.

Future work will therefore focus on an improvement of the quality of input data. As for the insolation, we will focus on the use of more accurate satellite-retrieved datasets, or site adaptation techniques (combination of long-term satellite data and short-term series of measured data). Furthermore, the horizon of each PV plant will be extracted from the Digital Terrain Model available for South Tyrol with a resolution of 1 m x 1 m, in place of the actual 1 km x 1 km resolution of the horizon downloaded from PVGIS.

Minor changes in the cost function, especially in the Lagrangian constrain (see [5]) will also positively affect the computation of the tilt and orientation angles of the PV plants.

For all these reasons, in the next sections a possible application of the methodology will be introduced, and some results presented, with the only aim to show the potential of the methodology, and not to draw appropriate conclusions.

3 RESULTS AND APPLICATIONS

3.1 General analysis of PLR trends on the PV fleet

A first statistical analysis is performed on the PLR of the PV plants. In order to be more robust, first of all this analysis considers only 854 plants, for which a complete series of monthly values of measured yield is available.

A second filter is applied with the purpose of getting rid of monthly values of PR corresponding to a low-quality estimation of the monthly insolation from CAMS dataset. More specifically, each value of retrieved yield that exceeds the corresponding value of measured yield for more than x% (in absolute value) is filtered out. The retrieved yield is calculated as follows:

$$Y_{retr}(t) = PR(t) * Y_r(t) = \\ = \left(PR_0 + PLR * \frac{t}{12}\right) * Y_r(t) \quad (4)$$

The STL technique is therefore applied to the filtered PR series and new values of PLR are retrieved.

Figure 5 shows the probability density function of the 854 PV plants when no filter on the retrieved yield is applied, and when it is applied with threshold limits of 20%, 15% and 10%. It is evident that the standard deviation of the distributions, i.e. the variability of the PLR values, decreases with the application of more stringent filters, such as 10%. At the same time, the average value of PLR also slightly decreases: -1.17%/year for the unfiltered series, -1.12%/year for the 20% filter, -1.06%/year for the 15% filter, and -0.98%/year for the 10% filter.

Further analysis on the PLR values obtained with the 10% filter does not highlight clear trends in the distribution of PLR when clustering the data for nominal power, elevation of the plants, or year of installation.

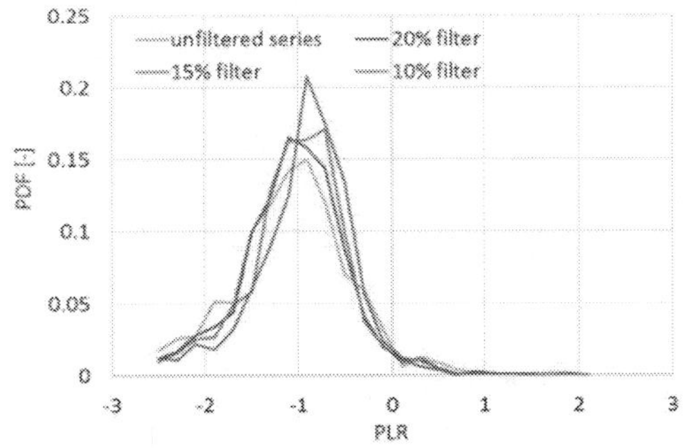

Figure 5: Probability density function of the PLR of the 854 PV plants with complete series of monthly PR values. The different colors correspond to the distribution of PLR values re-calculated after filtering out single monthly values of retrieved yield (as Equation 4) exceeding the corresponding measured yield of more than x% in absolute value.

3.2 Statistical analysis on a subset of PV plants

The methodology presented in this work allows, provided that high quality inputs are available, to retrieve information on the PR series of a consistent number of PV plants.

Another possible application is the extraction of information only from a subset of PV plants with specific characteristics (e.g. tilt and orientation angles) to be used in predicting the confidence interval of the PR series of a future PV plant with similar characteristics, and that will be installed in the same region. This paragraph explains the probabilistic approach to infer this kind of information, using an example from the real case.

From the PV fleet of 1750 plants, only 26 plants are selected, having an orientation angle between -10° and +10° (where 0° is South), and a tilt angle between 20° and 40°. One of this is kept out of the analysis, and used as test plant. For these plants, a RMSE of 5.7% between retrieved and measured yields is found. Figure 6 shows the probability density function of the PLR and the PR_0 of the 25 selected plants. In particular, in this case the mean value of PLR distribution is -0.93%. As discussed in section 2.3, this and the next results cannot draw to appropriate conclusions because of the several issues in the quality of the inputs that negatively affect the quality of results.

As for the seasonality of the PR-series, Figure 7 shows the probability density function of PR_{seas} of the 25 selected plants, for each month of the year. It is worth to note that the PDF corresponding to the winter months have a higher standard deviation, i.e. they show a higher variability amongst plants. This may be in part explained by the higher near- and far-shading effects during this period of the year.

According to Equation 2, the probability density

function of the PR, for a specific month, is given by a combination of the probability density functions of PLR, PR_0 and PR_{seas}.

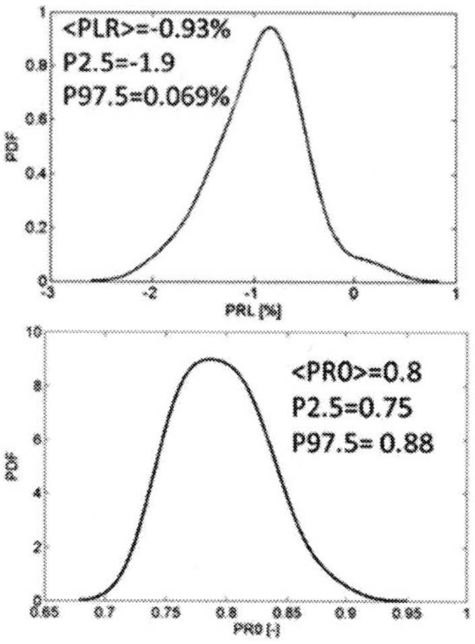

Figure 6: Example of probability density function of (top) the Performance Loss Rate *PLR* and (bottom) the initial Performance Ratio PR_0, for the 25 selected PV plants.

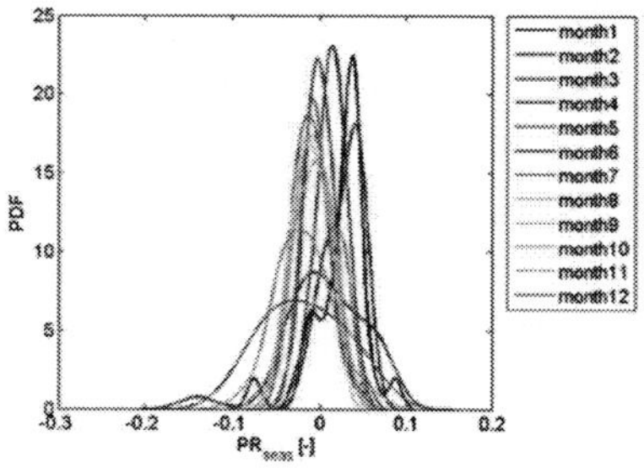

Figure 7: Example of probability density function of the seasonality component of the monthly PR-series, PR_{seas}, for each month of the year.

Therefore, for each month the percentile *p* of the PR of the selected plants $PR(t)_p$, i.e. the value of PR with probability p% that this value is higher than reality, is given by:

$$Prob\big(PR(t) \le PR(t)_p\big) = p \tag{5}$$

with

$$PR(t) = PR_{seas}(t) + \left(PR_0 + PLR(t) * \frac{t}{12}\right) \tag{6}$$

In the same way, also the normalized percentile *p* of the PR of the selected plants, for a specific month, can be defined as:

$$nPR(t)_p = \frac{PR(t)_p}{\langle PR(t) \rangle} \tag{7}$$

where $<PR(t)>$ is the mean value of the PR of the selected plants at month *t*.

In Figure 8 the percentiles of the PR for the first 12 months of operation of the 25 selected plants are shown.

Finally, the percentile for the month *t* can be calculated for a future PV plant using the following formula:

$$PR(t)_{p,project} = PR(t_{1,12})_{project} * nPR(t)_p \tag{8}$$

where $nPR(t)_p$ is given by Equation 7, and therefore derives from the probabilistic analysis on the subset of plants selected from the PV fleet, while $PR(t_{1,12})_{project}$ is the PR at one of the twelve months of a typical year of the test plant. The latter, an example of which is shown in Figure 9, can be calculated during the design phase with one of the available software.

This way it is possible to associate confidence intervals to the predicted performance that correspond to the probability of having, in reality, different degradations and seasonality.

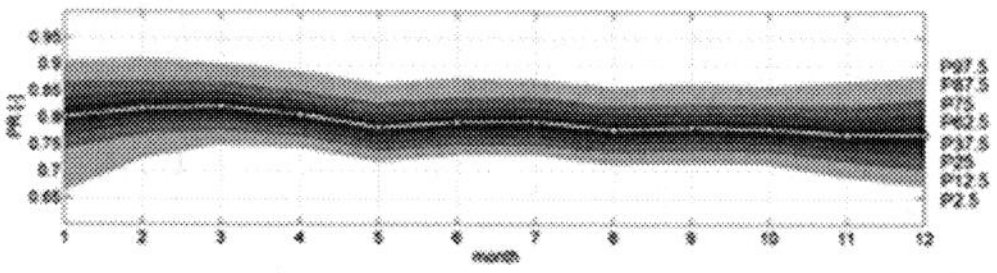

Figure 8: Example of confidence intervals of the PR for the 25 selected plants, for the first year of operation. For example, the interval with 95% confidence is given by the difference between the upper limit of p97.5 and the lower limit of p2.5. The white line corresponds to p50.

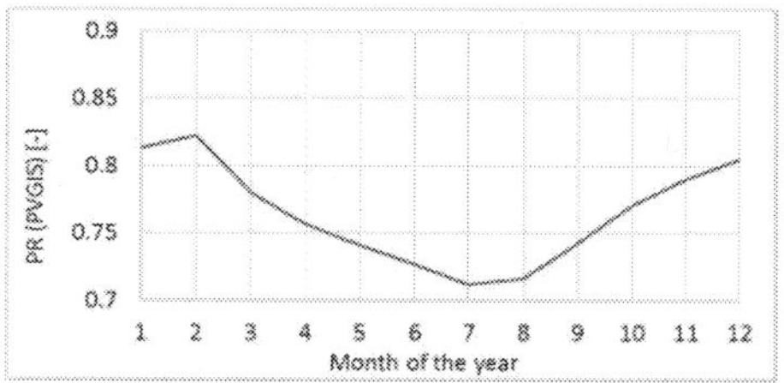

Figure 9: Example of typical year of monthly PR series, calculated for the test plant with the software PV GIS.

Figure 10 shows the confidence interval of the test plant, obtained by applying Equation 7, in terms of both PR and final yield. The red-dotted line corresponds to the monthly values of measured yield and to the monthly values of PR derived with Equation 1. For example, if we focus on the 50th percentile (p50), the predicted degradation after 25 years of operation is around 23%.

4 CONCLUSIONS

This paper concerns the statistical analysis of the performance loss rate of PV plants installed in a region. This analysis is based on values of monthly performance ratio that are retrieved with a methodology that uses a limited set of available information (measured yield, position and nominal power), satellite-retrieved irradiance and local skyline from PVGIS.

The methodology is applied on a real case study of 1899 PV plants installed in the South Tyrol region (North-East of Italy). Although promising, the results are affected by the quality of the input data, irradiance and sites' skyline accounting for near- and far-shading above all. Future work will focus on the increase of the accuracy of the methodology, via the following actions:

- use of more accurate satellite-retrieved datasets,

or site adaptation techniques (combination of long-term satellite data and short-term series of measured data) to increase the accuracy of input irradiance;

- use of sky-lines extracted from the Digital Terrain Model of South Tyrol with resolution 1 m (instead of 1 km resolution of PVGIS).

For these reasons, the results obtained for the real-case study cannot draw to definitive conclusions, but they will be confirmed or rejected after applying the above-mentioned actions to the input data.

A statistical analysis on the distribution of PLR on the entire PV fleet is first of all performed by using appropriate filters on the quality of the retrieved PR series. Results show that by applying a stringent filter, an average PLR close to -1%/year is obtained from a statistic of 854 plants. No particular relation in the distribution of PLR is highlighted when clustering per power, elevation of year of installation classes.

An example of possible probabilistic application of the methodology is finally presented and applied to a subset of 25 PV plants with tilt angle around 30° and orientation close to South, to compute the PR and yield of a future PV plant. The confidence interval of the PR and of the yield of a generic test plant is reconstructed for its entire lifetime. This might be used as a tool to assess the bankability of a future PV project bases on a probabilistic analysis of plants with similar characteristics (tilt and orientation angles) installed in the same region.

5 REFERENCES

[1] A. Jäger-Waldau. PV Status Report 2016, EUR 28159 EN; doi:10.2790/682995.

[2] Copernicus Atmosphere Monitoring Service (CAMS) Radiation Service, 2017. [Online]. Available: http://www.soda-pro.com/web-services/radiation/cams-radiation-service [Accessed: 18-Feb-2017].

[3] Solar Radiation Data (SoDa), 2017. [Online]. Available: http://www.soda-is.com [Accessed: 18-Feb-2017].

[4] Photovoltaic Geographical Information System (PVGIS), 2017. [Online]. Available: http://re.jrc.ec.europa.eu/pvgis/ [Accessed: 18-Feb-2017].

[5] M. Pierro, G. Belluardo, P. Ingenhoven, C. Cornaro, D. Moser. Inferring the Performance Ratio of PV systems distributed in a region: a real-case study in South Tyrol. To be published in Proc. 44th IEEE Photovoltaic Specialists Conference, Washington, 2017.

[6] P. Ingenhoven, G. Belluardo, D. Moser. Comparison of statistical and deterministic smoothing methods to reduce the uncertainty of performance loss rate estimates. Submitted to IEEE Journal of Photovoltaics.

[7] ITRPV. International technology roadmap for photovoltaic - 2016 results, March 2017.

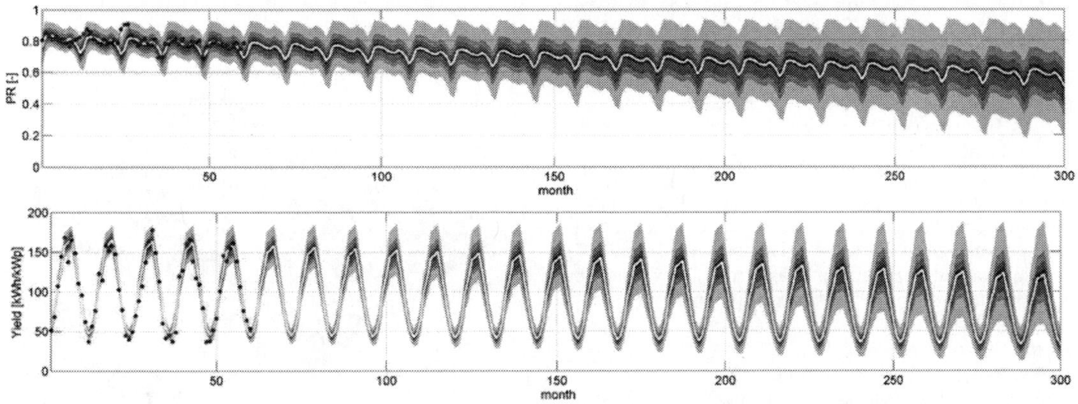

Figure 10: Example of confidence intervals of the predicted PR and yield of the test plant, projected on 25 years. Red points and lines correspond to the measured yield and the PR retrieved with the methodology presented in Section 2.2.

A 368-kWp Grid-connected PV System: Known and Hidden Losses

Georgi H. Yordanov[1,2*], Frank Smolders[3], André Olaerts[4], Gert Verbeek[1,2], Kris Baert[1,2], and Johan Driesen[1,2]
[1]KU Leuven, ESAT-ELECTA, Kasteelpark Arenberg 10, 3001 Leuven-Heverlee, Belgium
[2]EnergyVille, Thor Park 8310, 3600 Genk, Belgium; kris.baert@energyville.be
[3]KU Leuven, IIW, Kleinhoefstraat 4, 2440 Geel, Belgium; smolders.frank@gmail.com
[4]Affluent Energy, Winterslagstraat 224, 3600 Genk, Belgium; andre.olaerts@telenet.be

ABSTRACT: We present a method for real-time analysis and overall evaluation of some avoidable losses in large grid-connected PV systems. Plane-of-array (PoA) irradiance for each orientation is not required. For the 368.5-kWp flat-roof PV system at EnergyVille with modules facing South, East, West, and North (tilted at 10° from horizontal), we estimated 8.9% 'avoidable' losses of which 2.1% due to inverter downtime including breakdown and another 6.8% shading and mismatch losses between April 22, 2016 and April 21, 2017. String undersizing losses were identified for one of the inverters in the form of systematic underperformance by about 8%. Each sub-array's performance is compared to that of a non-shaded reference with the same orientation(s). We term as 'ideal' the operating conditions of the references. Plots of the relative array performance reveal the evolution of shading and other non-common-mode losses in the course of each day. Clear-sky comparisons are ideal for studying the shading loss evolution in the course of the day during different parts of the year, but also overcast days can reveal inverter or PV array issues at low input powers. Global horizontal irradiance (GHI) logging is therefore a minimum requirement. The method can be integrated in commercial PV-system monitoring software.
Keywords: Monitoring, loss analysis, system performance, visualization, software

1 THE ART OF PV SYSTEM MONITORING

The IEC standard 61724 for PV-system performance monitoring requires that PoA irradiance and back-of-module temperatures are measured and logged for all the different PV array orientations [1-4]. Examples from the literature include e.g. 8 different tilt angles on a curved roof and 20+ sensors to monitor unshaded as well as shaded PoA irradiance [5]. Performance Ratio (PR) analysis is done ideally for each orientation [6]. Over the past two decades, the PR of new grid-connected systems has steadily increased, averaging 85% more recently, with reported peak values as high as 96% [6]. However, detailed monitoring solutions can be costly, moreover some commercial PV-system monitoring tools only allow for logging one horizontal and one tilted irradiance sensor as in our case. In addition, the string inverters in some PV systems may differ significantly in terms of power rating, and absolute performance comparisons even between sub-arrays with the same orientation are more difficult to grasp. The new standard IEC 61724-1 specifies three system monitoring classes: class C offering basic system performance assessment (which is our case); class B offering in addition system loss analysis and documentation of a performance guarantee; and class A also offering extras such as precise degradation measurement and grid interaction assessment [2].

We found a method for real-time performance and non-common-mode loss analysis giving intuitive and easy to read plots [7]. It can reveal hidden systematic losses that are not seen in look-perfect daily AC power profiles. Our approach can be integrated in commercial monitoring tools to enable early detection of avoidable losses such as design flaws and inverter faults, as well as to quantify and visualize the evolution of unavoidable shading losses.

2 FLAT-ROOF PV SYSTEM AT ENERGYVILLE

A 368.5-kWp PV system (338 kVA AC rating) was installed on the flat roof of the EnergyVille-1 building in Genk, Belgium in early 2016. The system began operation on 9 April but performance logging started later on 21 April, and therefore we analyze data recorded between 22 April 2016 and 21 April 2017. Figs. 1-3 show 3D, top, and aerial+ close-up views of the building and PV arrays. The arrays with different colors in Figs. 1 and 2 correspond to 24 different inverters, with three different power ratings.

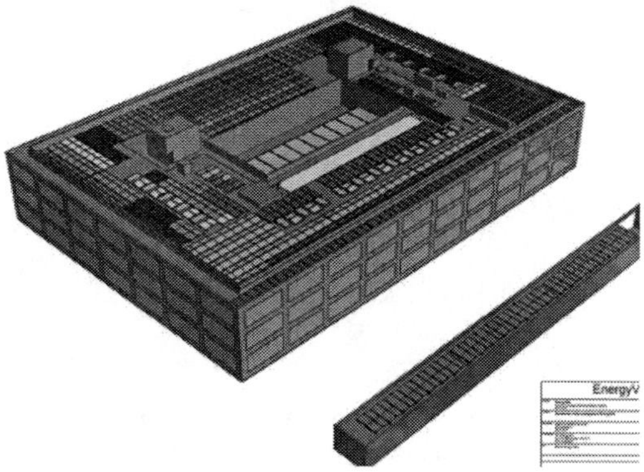

Figure 1: 3D drawing of the rooftop PV arrays.

2.1 PV array and roof structures layout

The PV arrays consist of 1068 modules SunPower X21-345 (345 Wp) made with Maxeon Gen III IBC cells with integral bypass diodes (96 per module). The tilt angle is 10° from horizontal. There are four azimuthal orientations: South, East, West and North. Each sub-array has equal number of modules with two opposite orientations, either South+North or East+West, and the strings with a given orientation are connected to one of the two maximum-power-point trackers (MPPT) of the corresponding inverter.

*Corresponding author; E-mail: ghy@kuleuven.be, petaflops@gmail.com; Tel. (office): +32 1637 9213.

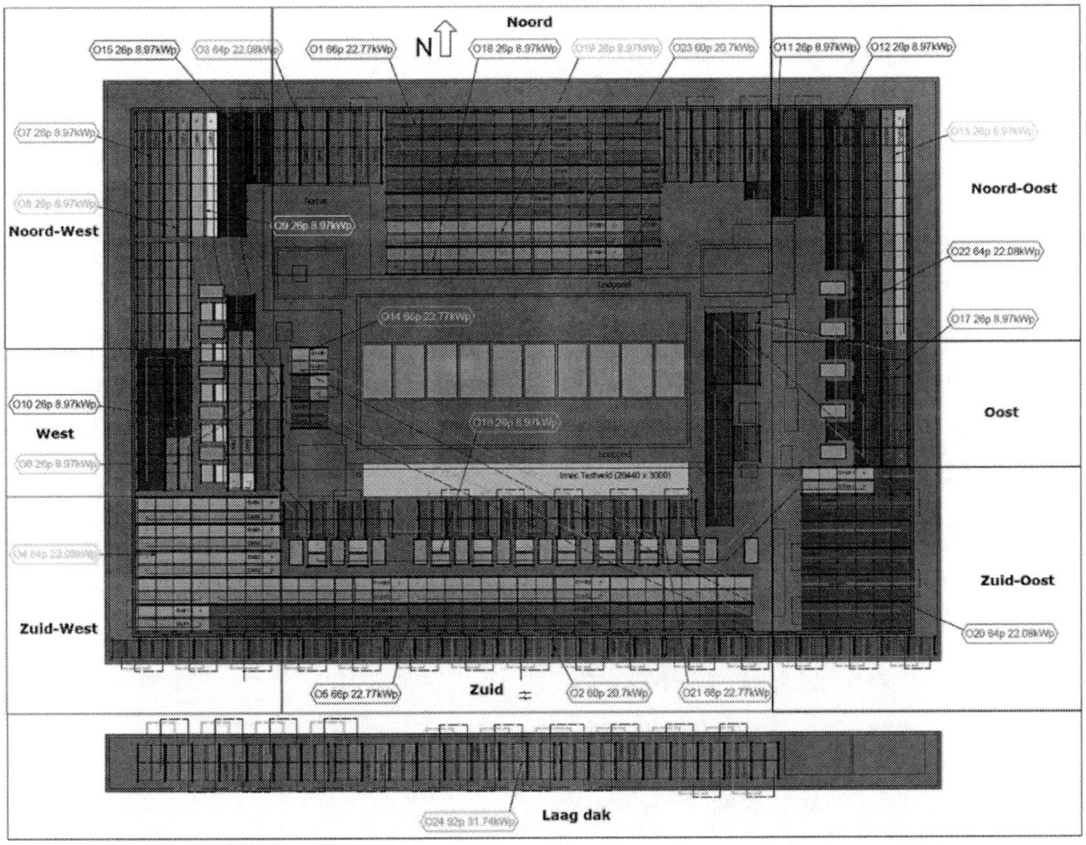

Figure 2: PV arrays layout. Different colors correspond to different inverters. Drawing courtesy of Affluent Energy BVBA.

Figure 3: Drone photo of EnergyVille-1 with the rooftop PV arrays. Inset: West view from the higher roof.

The system's no-exceed DC voltage is 1000 V for all inverters, but the optimal voltage ranges of the MPPTs depend on the inverter rating. The strings therefore have between 9 and 13 modules each. Up to five strings enter each MPPT input of an inverter, and the number of modules in a string differs for inverters of the same power rating. The roof area covered with PV modules is slightly over 40%. The PV arrays' geometries were chosen depending on the multiple structures on the roof such as air conditioning components, which cause shading. Other notable shading sources include the elevator and staircase risers; the fence (at some distance along the big roof's perimeter), and the inner yard's wall. Lunch tables and benches for the personnel are placed to the North of the staircase riser (the Western one) and colleagues may shadow the nearest PV modules unintentionally in good weather, as can any visitors or maintenance personnel walking on the roof. Also, there are safety ropes running along the PV array which is outside of the fence on the big roof and the one on the smaller, lower roof. The ropes and their poles cause some degree of shading as well. The neighboring trees and the IncubaThor building to the South-East are low-risers and only cause shading during clear-sky sunrises and sunsets.

The chosen closed-rack configuration (modules shield each other from strong winds) prevents the structure which is non-penetrating (fixed to the roof by gravity alone, relying on ballast weights) from being blown away and destroyed during severe storms.

Besides the shading, some PV modules see specular reflections from the glossy lids of the multiple light ducts on the roof, as well as from several South-facing research PV modules at 35° from horizontal, South of the inner wall, and their supporting structure (in the area denoted with "Imec Testveld" in Fig. 2; not present yet in Fig. 3 at the photos' shooting times).

2.2 Inverter configuration

The PV system relies on 24 three-phase inverters from the Aurora line by ABB-PowerOne. Thirteen are TRIO-8.5-TL-OUTD rated 8.5 kVA; another ten are TRIO-20.0-TL-OUTD rated 20 kVA; and the one beneath the lower roof is TRIO-27.5-TL-OUTD rated 27.6 kVA. Each has two independent MPPTs allowing connection of PV strings with different lengths and orientations. The PV arrays corresponding to each inverter are labeled in Fig. 2 with the inverter's number (e.g. O1, O2, ..., O24); the number of modules; and their total power rating. Note that the last two of the modules in PV array #24 are not connected to the inverter, so the accurate PV rating is *31.05 kWp*. The latter value was used for the calculations in the present work, and not the value given in Fig. 2. The MPPT voltage ranges of each inverter type are 320-800 V, 440-800 V, and 500-800 V, respectively. All have RS485 communication interfaces. The system's AC power rating is 338.1 kVA and the PV power rating is 368.5 kWp, an average undersizing ratio of about 92%. In terms of the maximum PV power acceptable by each inverter, the mismatches range between about 97% and 110% (depending on the expected degree of shading), averaging at 100.6%.

Some of the inverters failed either just after installation or during the first logged year of operation. These experiences are described later in section 2.4. In addition, buggy firmware of the small (8.5-kVA) inverters caused frequent morning start-up failures requiring manual PV-array disconnect-reconnects. Manual start-ups happened ideally ca. 9 a.m. on business days, but could be delayed until the afternoons, depending on authorized personnel availability. This is not done during weekends and on holidays, meaning that significant downtime losses could be expected.

2.3 Environmentals, logging, and online monitoring

The PV system comes with an ABB weather station VSN800 measuring (since 23 May 2016) the global horizontal irradiance (GHI), ambient air temperature, and wind speed and direction; and a data logger VSN700. All sensors and inverters communicate with the logger via the RS485 bus. The logger sends operational and weather data to ABB's online portfolio manager AuroraVision that are then available for download as 5-min, 15-min, hourly, etc. averages. The 1-minute averages option is listed but not actually available. The online monitoring tool shows fault flags and sends emails in case of inverter failures, but these come with great delays in the order of many hours. Regular on-site monitoring is therefore essential.

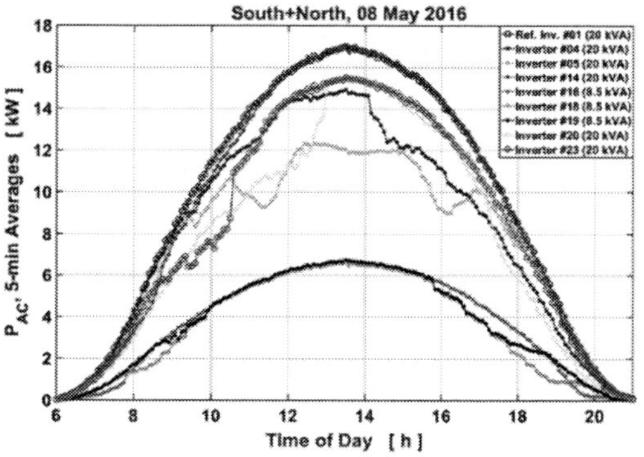

Figure 4: Example with daily AC power profiles of the inverters with South+North facing PV arrays on a clear-sky day.

The big inverters (20 kVA and 27.6 kVA) log more operational parameters than the small ones (8.5 kVA). In fact, the small inverters log the DC power at only one of two MPPT inputs. Furthermore, certain operational parameters on the DC side communicated by all inverters were found either unreliable or missing (empty columns), as well as some environmentals. We found that at least a half of the logged data were useless. To compare all the sub-systems' performance, we therefore analyze only their AC powers. It is indeed possible, by looking at the PV arrays and shadow sources geometries in Fig. 2, to attribute degraded performance to shading in different parts of a sunny day (see Fig. 4). However, an unshaded PV array may underperform systematically by e.g. flat 8% even on a clear-sky day and still have a near-perfect, bell-shaped AC power profile. Our relative performance analysis reveals such hidden losses, and gives a measure for comparing inverters with very different power ratings. The power plots displayed by the online monitoring tool cannot tell if such losses are or were present.

2.4 First logged year: key events

Unfortunately, replacement of an inverter results in wipeout of all previous data from the broken device in AuroraVision. A few such events occurred in the first logged year of system operation. Inverter #9 (8.5 kVA) did not work since installation and was replaced on 12 May

2016. On 28 September, the PV disconnect switch of inverter #13 (also 8.5 kVA) broke; the inverter was replaced on 10 November. Later on, there was a power outage in the region on 30 November caused by a sub-station fault, and as a consequence inverter #21 (20 kVA) suffered an internal short-circuit between phases L1 and L2. The inverter was repaired under warranty on 17 March 2017. These delays, along with the unpredictable morning start-up failures of mostly the small inverters, resulted in higher than anticipated overall downtime losses during the 1st year of operation. In the context of the present work, only the data set of inverter #21 suffered a nearly 2-month gap between 1 October and 30 November 2016, while we had a full data record prior to that which formed the basis of the work done in [7]. Luckily, the system's energy yields were calculated for those two months prior to the old inverter data wipeout, and this allowed the old inverter's monthly yields calculation. Since future inverter failures are not excluded, it is imperative to copy its data from AuroraVision before the replacement and wipeout.

3 NON-COMMON-MODE LOSS DETECTION

The method described in the present work was first presented in [7] while less than half a year of data was available from the PV system. For the period from 21 April to 30 September 2016, we estimated about 6% losses due to shading and another 1% due to inverter downtime. Since then, a higher rate of inverter downtime was observed, while the shading patterns varied in the course of the year. Later in section 4, we give the energy yield and loss values for the considered one-year period. In the following, we explain the concept of relative performance analysis using reference PV arrays with 'ideal' placement.

3.1 Relative performance plots

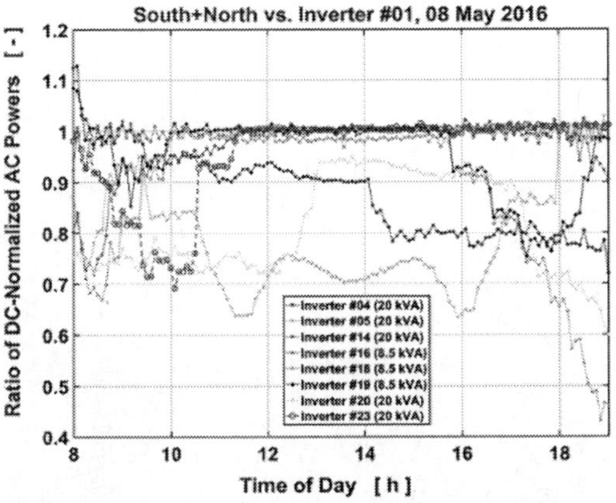

Figure 5: Relative performances corresponding to Fig. 4.

We chose as reference inverters #1 (in the northern part of the higher roof) for South+North oriented PV arrays and #2 (outside the southern fence) for East+West oriented arrays. We term as 'ideal' their operating conditions because they suffer from almost no shading. We assume that all the PV arrays are affected by the same degree of soiling and temperature-related losses at any given part of the day. Then, for each of the other 22 sub-systems, we compare the normalized AC output (versus the rated PV power) to that of the reference inverter with the same array orientations. When both sub-systems operate under 'identical' conditions, their relative performance curve is at unity; see Fig. 5.

3.2 Clear-sky reference days

Data from clear-sky days are most suitable for sub-system performance evaluation. We use the GHI data set to pick several representative days [7]. Before 23 May 2017 when weather data were not yet available, we look for nice bell-shaped daily DC power profiles from the two reference inverters. In fact, in [7] inverter #21 was used as an East+West reference. However, the corresponding PV array is somewhat affected by the nearby light ducts as well as by the long shadows of the fence and other objects to the East and West on the higher roof. Moreover, that inverter broke down before end-2016. Therefore, the sub-system with inverter #2 was chosen as a better reference for the present work.

Some of the relative performance curves in Fig. 5 exceed unity in the morning and evening hours. This is because at these times of the day even the reference PV array suffers from long shadows of distant objects. Yet this has a negligibly small effect on the accuracy of the loss accounting described later in the paper because the corresponding absolute AC power *differences* from the reference sub-system are very low (see Fig. 4).

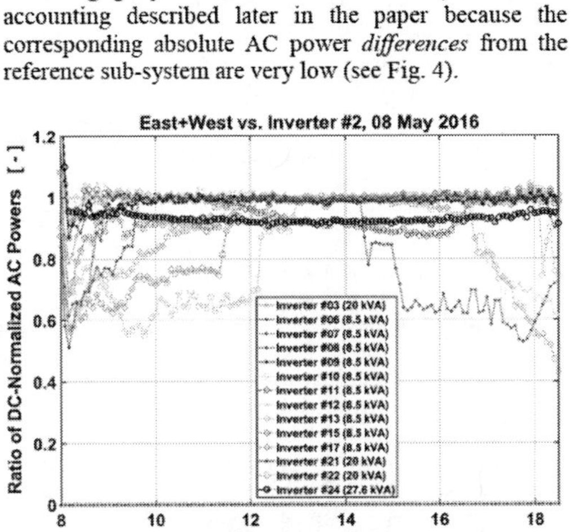

Figure 6: Performance of each East+West vs. Inverter #2.

Most of the losses emphasized by the relative performance plots were attributed to the varying shading patterns from other roof infrastructure. This was not the case of the PV array on the lower roof connected to inverter #24 beneath (27.6 kVA); see Fig. 6. Despite showing a nice bell-shaped AC power curve, its relative performance curve was systematically below unity by about 8%. This behavior continued throughout the summer. Array design analysis revealed that this was mainly due to short strings (5 strings with 9 modules connected to each of the two MPPT inputs) leading to MPP voltages below the 500-V lower bound for that inverter. We considered the shading effect of the safety rope and its poles (between the two rows of PV array #24) as too small compared to that of the MPPT losses.

Also, note that in Fig. 6 inverters #9 and #10 were down that whole day.

3.3 Performance on overcast days

Relative performance plots for overcast days can reveal

other performance issues when the inverters have operated below 20-25% of their rated powers. For example, note the curve of inverter #17 in Fig. 7 which underperforms by about 20% on the average. Such plots are quite 'messy' because the reference sub-system loses some superiority with respect to the others. Note that the dips seen in most curves shortly after 1 and 2 p.m. were due to GHI drops down to and below 50 W/m² [7]. All the 8.5-kVA inverters have one string with 13 modules connected to each MPPT inputs, but one string of inverter #17 is too close to the elevator riser and sees less diffuse light on overcast days. On clear-sky days without shading, that sub-system performs as good as the reference; see Fig. 6.

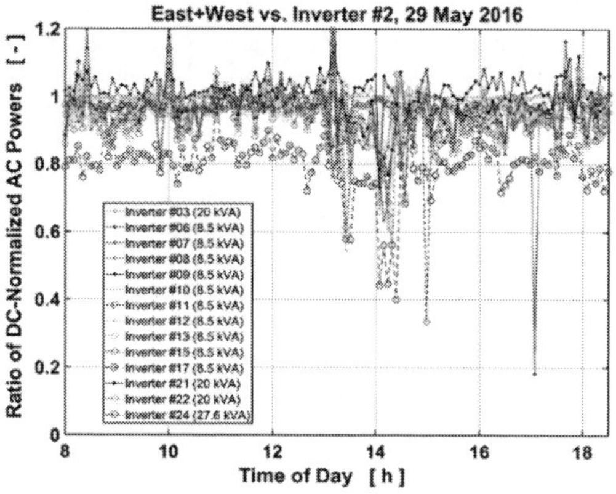

Figure 7: Performance of each East+West on overcast day.

3.4 Loss accounting and differentiation

We accounted for the downtime and other non-common-mode losses of each sub-system in the following manner. First, the 'ideal' (hypothetical) AC power (5-min average) of each sub-system is calculated for each time stamp from the reference sub-system power by re-scaling with the ratio of their rated PV powers. Then, the corresponding power loss is obtained by multiplying the ideal power with the relative underperformance (or overperformance) as defined in section 3.1. The power loss is then integrated over time to obtain the absolute energy lost to downtime, shading and other non-common causes. Its ratio to the ideal energy yield (obtained by integrating the ideal power over time) gives the relative energy loss for the sub-system.

We differentiate and quantify the downtime energy losses by integrating only the power loss when an inverter's AC power was zero (as in Fig. 6 for inverters #9 and #10). The difference to the total energy loss is then attributed to shading and other non-common factors with respect to the (largely) unshaded reference. Only for the big inverter #21 was the calculation somewhat different for the months of October and November 2016 as discussed above in section 2.4.

The above calculations can be done for an arbitrary period of time – one full year, monthly, quarterly, etc. by limiting the range of indices (time stamps) over which integration is done. Monthly loss analysis can be useful in revealing the evolution of shading and other non-common losses in the course of a year. The observed variations may vary from year to year due to interannual variability of the direct solar irradiation share in the annual total irradiation.

Soiling losses can be evaluated by cleaning the reference PV arrays and comparing another same-orientation array's relative performances on two clear-sky days shortly before and after the maintenance operation. Alternatively, all but the reference arrays can be cleaned first. This has not been tested so far for our system.

4 RESULTS AND DISCUSSION

4.1 Downtime, shading, and other non-common losses

Table 1 lists the relative energy losses for each but the reference sub-systems during the considered one-year period (rounded values; sums may differ from the totals).

Table I: Relative total, downtime, and other non-common-mode energy losses (mostly due to shading) per sub-system.

Inverter	kVA	Total [%]	Down [%]	Rest [%]
1 (S+N)	20	0.0		
2 (E+W)	20	0.0		
3 (E+W)	20	6.25	0.002	6.25
4 (S+N)	20	7.1	0.006	7.1
5 (S+N)	20	16.5	1.5	15.0
6 (E+W)	8.5	5.8	4.2	1.6
7 (E+W)	8.5	13.7	2.9	10.8
8 (E+W)	8.5	4.6	0.12	4.5
9 (E+W)	8.5	13.3	9.8	3.5
10 (E+W)	8.5	9.7	2.0	7.7
11 (E+W)	8.5	14.4	1.6	12.8
12 (E+W)	8.5	8.0	0.66	7.36
13 (E+W)	8.5	18.55	8.6	9.96
14 (S+N)	20	1.9	0.02	1.9
15 (E+W)	8.5	9.2	1.7	7.5
16 (S+N)	8.5	5.35	1.5	3.8
17 (E+W)	8.5	14.4	2.1	12.4
18 (S+N)	8.5	21.6	8.2	13.4
19 (S+N)	8.5	7.9	1.2	6.7
20 (S+N)	20	12.5	0.5	12.0
21 (E+W)	20	19.9	13.8	6.1
22 (E+W)	20	8.9	0.05	8.8
23 (S+N)	20	4.6	0.05	4.6
24 (E+W)	27.6	8.4	0.6	7.8
ALL	**338.1**	**8.9**	**2.1**	**6.8**

Figure 8: Frost cover on the modules in winter. Those closer to the viewer face South.

Notably, the downtime losses of the three inverters mentioned in section 2.4 and that of inverter # 18 are too high, in the order of or exceeding 10%. This was partly due to long delays in completing the service requests by the maker and partly due to the flaws in the firmware of the 8.5-kVA devices. Figs. 5 and 6 explain well the other losses as coming mainly from shading. This includes frost and snow on the PV modules in winter that were left for self-cleaning (melting by sunlight); see Fig. 8.

4.2 Final and specific energy yields

The system's energy yield during the one-year term considered was 294.7 MWh, or 799.9 kWh/kWp. The ideal (without any shading or downtime) energy yield from the system would have been 323.3 MWh, or 878.4 kWh/kWp. Both the actual and ideal yields are rather low for the present location. This is attributed to the non-optimal tilt angle of all modules, and the use of North-facing sub-arrays. However, the energy yield over one year for a commercial rooftop PV system of such a compact size is considerable, thanks to the high module density and efficiency. Cost efficiency and self-consumption analysis are outside the scope of the present work, partly because the PV modules and the inverters were acquired a few years before the actual installation when the market prices were much higher.

5 CONCLUSION

We presented a method for non-common-mode loss analysis in large grid-connected PV systems. It requires non-shaded reference PV arrays for each orientation type and reliable logging of AC power averages. Global horizontal irradiance logging is useful for identifying of clear-sky days that are very suitable for studying the shading effects and other losses' evolution. The method allows for real-time evaluation of such losses present in the different sub-systems and long-term accounting of the associated energy losses. Downtime losses can be easily distinguished from the others. The method has a potential to quantify the degree of soiling losses immediately prior to cleaning maintenance of the PV arrays. However, this was not tested in the present work.

We analyzed the first year of operational data available from a 368.5-kWp flat-roof system located in Genk, Belgium. The system features a rather low tilt angle of 10° from horizontal, four azimuthal orientations including North, and suffers significant shading from multiple elements of the building. With our method, we estimated 2.1% losses due to inverters downtime and 6.8% other non-common-mode losses, mainly due to shading. The specific energy yield was rather low due to the low tilt angle and many shading sources, but the total energy yield was considerable thanks to high module density and highly efficient PV technology.

Our method gives nice visualization of the above losses present in different parts of a large PV system and can be integrated in commercial performance monitoring tools.

ACKNOWLEDGEMENTS

The present work was partly funded by the European Community via the EFRO-SALK project nr. 936 "Sustainable Energy for Smart Cities." The authors thank the Faculty manager of EnergyVille, Jan Gordebeke; the ICT specialist Veronica Lucero Ortega; as well as Sara Colapinto from the front desk, for their kind support on many occasions. Georgi and Frank would also like to thank our colleagues Bieke Demaeghdt, Paulien Martens, and Nathalie Belmans for their kind attitude and support.

REFERENCES

[1] IEC Standard 61724 "Photovoltaic system performance monitoring – Guidelines for measurement, data exchange and analysis," 1st Ed., 1998.

[2] IEC Standard 61724-1 "Photovoltaic system performance – Part 1: Monitoring," 2017.

[3] IEC Technical Specification 61724-2 "Photovoltaic system performance – Part 2: Capacity evaluation method," 1st Ed., 2016.

[4] IEC Technical Specification 61724-3 "Photovoltaic system performance – Part 3: Energy evaluation method," 1st Ed., 2016.

[5] S. Wittkopf et al., "Analytical performance monitoring of a 142.5 kWp grid-connected rooftop BIPV system in Singapore," Renewable Energy 47 (2012), 9–20.

[6] A. Woyte et al., "Monitoring of photovoltaic systems: Good practices and systematic analysis," Proceedings 28th European Photovoltaic Solar Energy Conference (2013), 3686–3694.

[7] F. Smolders, "Schaduw -en uitvalverliezen in een 368,5-kWp netgekoppelde PV-installatie," Master's thesis (in Flemish language), KU Leuven, Geel, 2017.

COMPARISON OF SOILING SENSITIVITY OF THE PERFORMANCE OF MONOCRYSTALLINE AND AMORPHOUS PHOTOVOLTAIC SYSTEMS IN BENGUERIR, MOROCCO

H. ZITOUNI1, 2*, A. ALAMI MERROUNI1, M. REGRAGUI2, B. IKKEN1 and A. BENNOUNA3
1. Institut de Recherche en Energie Solaire et Energies Nouvelles (IRESEN); 16, rue Amir SidiMohamed Souissi - Rabat, Morocco Tel: +212 (0) 537 682 236 | Fax: +212 (0) 537 688 852 ;
Email: *houssinzitouni10@gmail.com, alami@iresen.org, ikken@iresen.org
2. Faculté des Sciences, 4 Avenue Ibn Battouta B.P. 1014 RP, Rabat, Morocco Tel: + 212 (0) 5 37 77 18 34/35/38, Fax: + 212 (0) 5 37 77 42 61 ;
Email: regrag@gmail.com
3. Cadi Ayyad University, Faculty of Science, Physics Department, BP: 2390, Marrakech, Morocco;
Email: benamin56@gmail.com

Abstract:
The dust and soiling accumulation on photovoltaic (PV) modules surface presents a big challenge for researchers, plants operators and O&M crews, especially in desert and arid regions. Indeed, in such areas PV plants performances are strongly affected by soiling that increases the cleaning costs, thus, the levelized electricity cost (LCOE).
The objective of this paper is to compare the performance of two different photovoltaic technologies (monocrystalline and amorphous) and identify the behaviour and the sensitivity of each one to soiling. In fact, Monocrystalline solar panels have the highest efficiency rates since they are made out of the highest-grade silicon and live the longest, but, they are the most expensive and if the solar panel is partially covered with shade, dirt or snow, the entire circuit can break down. Besides, the amorphous silicon panels are less sensible to the high temperatures and shading.
Results show that, the energy drop by 6-7% for the mono-crystalline and 8-9% for the amorphous.
Keywords: Dust effect, energy drop, performance ratio, soiling rate, photovoltaic, and power output.

1. INTRODUCTION:

The use of renewable energy sources and particularly solar can be a strategic solution for the countries with lack of fossil sources, to solve the supply problems [1, 2]. Photovoltaic is among the technologies enabling to supply electricity by a direct conversion of solar energy through the photovoltaic effect. However, one of the constraints of this technology in arid and semi-arid regions is the dust deposition and accumulation on the modules surface. This phenomenon dramatically affects the optical, the electrical performance as well as the lifetime of the PV modules especially in desert areas [3].
In Saudi Arabia [4] an experimental study showed that the output power of PV modules decreased by 20% after an exposition period of six months without cleaning. Besides, some studies [5-7] showed that also the PV performances are highly affected by dust. In Athens, Greece, Kaldellis et al. studied [6] the effect of dust deposition on the performance of the PV modules using three different dust samples: red earth, limestone and coal ash as well as natural dust sample.
A study in Egypt showed that the energy produced by a dusty module are 25 and 35% lower compared to a clean ones after a period of three months and one year of exposure, respectively [8, 9]. In Israel Boykiw [10] measured the effect of dust on a set of mono-crystalline panels electrical output in one week and also in one month and compared them with the clean panels. She founds that dust can cause 5–6% decrease in panel efficiency after only one week of exposition. In Nigeria, Sanusi [11] tested the effect of dust on Amorphous Silicon PV in the dusty season, and founds that for 70 days without cleaning there was a 20% reduction in solar absorption.
In order to better understanding and quantifying the impact of soiling on PV modules under Moroccan climate, the Research Institute for Solar Energy and New Energies (IRESEN) put in place the Green Energy Park (GEP) research platform [12]. GEP disposes of an outdoor PV test platform consisting different PV systems and modules from different technologies. Those systems are exposed to understand the environmental effect on their performances.
The objective of this study is the evaluation of the soiling's effect on the performances on two PV systems with different technologies: polycrystalline and amorphous. Each system is composed of two strings and they have a nominal capacity of 2160 Wc and 11040 Wc for the amorphous and the monocrystalline respectively.
Results show that for an exposition period of one month we measured an energy drop of 6-7% for the monocrystalline and 8-9% for the amorphous.

METHODOLOGY AND EXPERIENCE:

In this study two photovoltaic systems (monocrystalline silicon (m-Si) and amorphous silicon (a-Si)) (Figure: 1) were used. Those systems are grid connected and they are equipped with a monitoring system for the electrical data collection. More technical data are grouped in tables 1 and 2.

Figure 1- Photovoltaic systems used in the study

Type of technology	Amorphous	monocrystalline
The nominal power	135 W_c	240 W_c

The open circuit voltage Voc	61,3 V	37,3 V
The short circuit current Isc	3,41 A	8,3 A
The maximum voltage V_{mpp}	47 V	30,1 V
The maximum courant I_{mpp}	2,88 A	7,9 A

Table 1- The characteristics of the panels

2. RESULTS AND DISCUSSION:

2.1. Evaluation of the average daily energy of the two systems:

Figure 2 and Figure 3, present the energy produced by the dusted and cleaned strings for both technologies. As it can be seen, the difference in the energy production is obvious for both technologies. The energy of the dusted strings decreases with respect to the cleaned strings. The difference in productivity becomes more remarkable from may the 14th, that is to say after ten days of exposure under the actual conditions without cleaning. This significant decrease can be explained by the accumulation of dust on the surface of the un-cleaned module. In fact, dust induces generally non-uniform shading on the surface of the PV modules, so that the photovoltaic cells no longer receive the same intensity of sunshine. It is also found that the productions of the two systems are proportional to the irradiation since they vary in the same way but with a difference due to soiling.

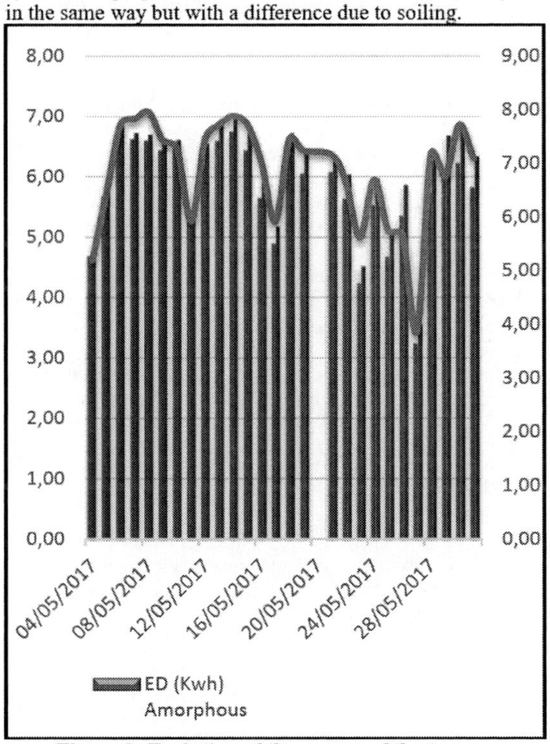

Figure 2- Evolution of the energy of the two amorphous strings as a function of the irradiation

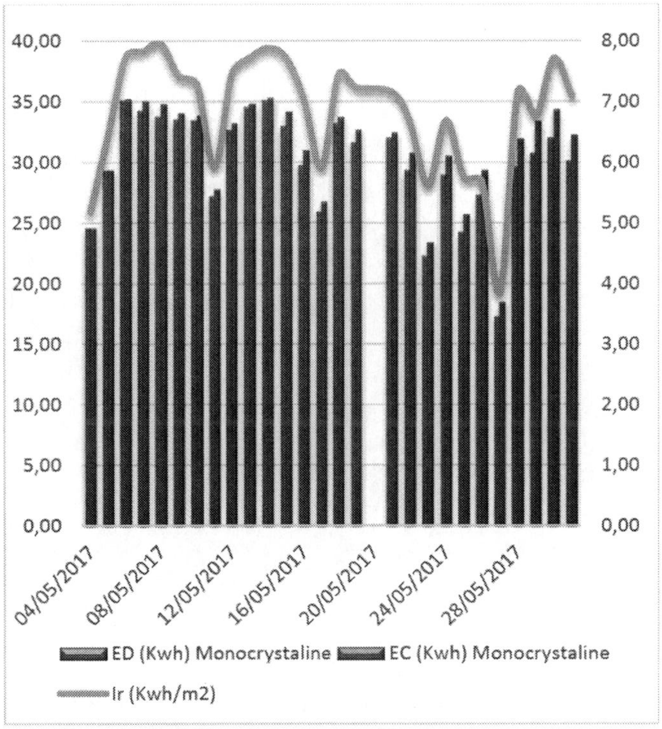

Figure 3- Evolution of the energy of the two monocrystalline strings as a function of the irradiation

2.2. Energy gains in Watt:

From Figure 4 and Figure 5, we notice in some cases that the two strings have the same productivity; which leads to a disturbance in the energy gain, it varies throughout the study period but it is noted that there is a kind of proportionality between this variation and the humidity.

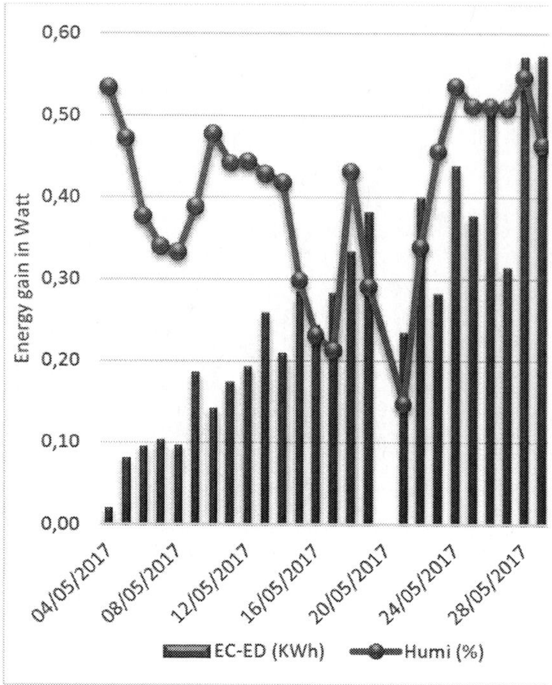

Figure 4- The amorphous energy gains in Watt as a function of the humidity

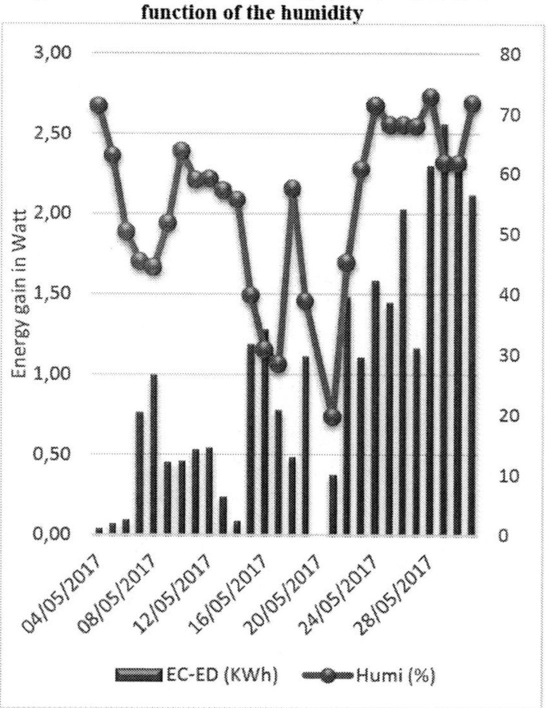

Figure 5- The monocrystalline energy gains as a function of the humidity

2.3. The soiling ratio:

For the systems, it was not possible to work with the Isc. For this reason, the soiling rate was calculated using the Energy of the cleaned and soiled strings.

From Figure 6 we notice that the soiling rate varies in a progressive way for both systems, but, it may be remarked that for the amorphous system the soiling rate varies in a greater degree than the monocrystalline system.

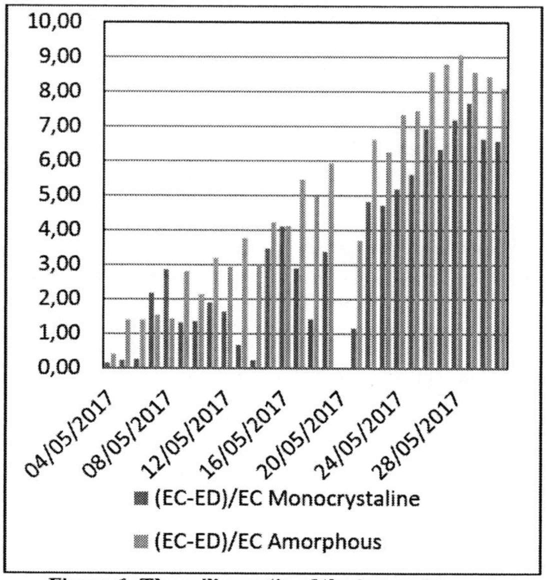

Figure 6- The soiling ratio of the two systems

3. CONCLUSION:

The objective of this paper is to compare the performance of two different photovoltaic technologies (monocrystalline and amorphous) and identify the opportunities and the challenges of each such as their sensitivity to soiling and hot climate. It has been found, that the energy of the un-cleaned strings degrades down to 8.05% (for the amorphous), and 6.58% (for the monocrystalline) and it has been found, that the progression of the soiling rate is linear for both systems, but, it more progressive for the amorphous system than the monocrystalline. From our results we can conclude that the amorphous technologies are more sensible to the soiling than the monocrystalline technologies.

REFERENCES:

[1] https://solarenergylocal.com/states/indiana/morocco/
[2] O. O.Ogbomo, E. H.Amalu,N. N. Ekere, et P. O. Olagbegi, « A review of photovoltaic module technologies for increased performance in tropical climate», Renew. Sustain. Energy Rev., nov.2016.
[3] Rajput, D. S. and Sudhakar, K., (2013), Effect of Dust on the Performance of Solar PV Panel, International Journal of Chem Tech Research 5 (2), pp. 1083-1086.
[4] M.J. Adinoyi, S.A.M. Said, Effect of dust accumulation on the power outputs of solar photovoltaic modules, Renew. Energy 60 (2013) 633e636.
[5] J.K. Kaldellis, M. Kapsali, Simulating the dust effect on the energy performance of photovoltaic generators

based on experimental measurements, Energy 36 (2011) 5154e5161.
[6] J.K. Kaldellis, P. Fragos, M. Kapsali, Systematic experimental study of the pollution deposition impact on the energy yield of photovoltaic installations, Renew. Energy 36 (2011) 2717e2724.
[7] Z.A. Darwish, H.A. Kazem, K. Sopian, M.A. Al-Goul, Effect of dust pollutant type on photovoltaic performance, Renew. Sustain. Energy Rev. 41 (2015) 735e744.
[8] E. Klugmann-Radziemska, Degradation of electrical performance of a crystalline photovoltaic module due to dust deposition in northern Poland, Renew. Energy 78 (2015) 418e426.
[8] Makrides G, Zinsser B, Norton M, Georghiou GE. Performance of photovoltaics under actual operating conditions; 2012.
[9] Ibrahim M, Zinsser B, El-Sherif H, Hamouda E, Makrides G, Georghiou GE, et al. Advanced Photovoltaic test park in Egypt for investigating the performance of different module and cell technologies. In: Proceedings of the 24th sympo- sium photovoltaic solar energy. Staffelstien, Germany; 2009.
[10] Boykiw E. The effect of settling dust in the Arava valley on the performance of solar photovoltaic panels. Department of Environmental Science Allegheny College Meadville, Pennsylvania (April 2011); 2011.
[11] Sanusi Y. The performance of amorphous silicon PV system under Harmattan dust conditions in a tropical area. Pac J Sci Technol 2012:13.
[12] http://www.greenenergypark.ma/
[13] T. Sarver, A. Al-Qaraghuli, et L. L. Kazmerski, « A comprehensive review of the impact of dust on the use of solar energy: History, investigations, results, literature, and mitigation approaches », Renewable and Sustainable Energy Reviews, vol. 22, p. 698 - 733, juin 2013.

This is a pre-peer review version of the paper submitted to the journal Progress in Photovoltaics.
It has been selected by the Executive Committee of the EU PVSEC 2017 for submission to Progress in Photovoltaics.

MACHINE LEARNING PV SYSTEM PERFORMANCE ANALYZER

Sandy Rodrigues[1,3], Helena Geirinhas Ramos[3], and F. Morgado-Dias[1,2]
[1]Madeira Interactive Technologies Institute, [2]Universidade da Madeira, [3]Instituto de Telecomunicações from the Instituto Superior Técnico of the University of Lisbon, Portugal.

ABSTRACT: Machine Learning Techniques (MLT) require large amounts of historical data needed for training the prediction models that is not always available. Since the Photovoltaic system behavior is nonlinear due to the unpredictable nature of the weather conditions throughout the year, MLT training require annual historical data to create the prediction model. The Photovoltaic system production meters store up to 3 months of historical power values. This information served as motivation to research different types of MLTs, in search of one that would accurately predict the daily solar AC energy values of a Photovoltaic system in any season of the year regardless of its location based on the available 3-month historical data used for training the prediction model. Therefore, five MLTs were compared which are the Regression Tree, Artificial Neural Network, Multi-Gene Genetic Programming, Gaussian Process and the Support Vector Machine for Regression. Results show that the Multi-Gene Genetic Programming presented the best prediction values throughout the four seasons of the year regardless of the Photovoltaic system location by only using 3 months of historical data for training the prediction model.
Keywords: Machine Learning, Rooftop, System Performance, Photovoltaic, PV System

1 INTRODUCTION

A Photovoltaic (PV) system performance analyzer is used to monitor the PV system output efficiency to ensure maximum output and maximize monetary gain on the PV system investment and consequently to shorten the payback time on the return of the investment (ROI). Model-based approaches have been mainly used in PV system performance monitoring and recently the data-driven approaches, which include Machine Learning Techniques (MLT), have been used to monitor the performance of PV systems, since they tend to deal better with the nonlinear behavior of the PV systems. However, both approaches are highly dependent on the available onsite historical data for comparison studies in performance analysis. For the case of PV system monitoring, annual onsite historical data is required due to the unpredictable nature of the weather conditions throughout the various seasons of the year that consequently produce nonlinear PV system solar production outputs. For this work, the onsite historical data includes daily solar irradiation values from generical local weather stations as the inputs and daily solar AC energy values as the outputs.

However, there are issues related to the gathering of the required onsite historical data namely that it is time consuming (wait a year to gather data), and that monitoring systems and pyranometers are costly. Another issue is related to the unavailability of the annual historical data that is unique to each PV system since the PV system behavior is correlated to the specific weather conditions of its location. In regard to solar irradiation historical data, an alternative to buying a pyranometer is to use solar irradiation measurements from a nearby weather station. Regarding the solar AC energy historical data, the PV system installation companies have access to three months of solar AC energy historical data stored by the onsite production meter. In regard to the unavailability of the unique historical data, research shows that a possible solution to solve this issue would be to apply Machine Learning Techniques to train prediction models and compensate for onsite historical data that is unavailable. There are a number of ways to predict PV solar AC energy output, however MLT have been mostly used in this type of prediction problem [1]–[4].

This information served as motivation to research the possibility of creating a MLT prediction model that does not require the unavailable onsite annual historical data from a given PV system and take advantage of the available (3 month) onsite historical data to analyses the performance of any PV system. Therefore, the aim of this work is to implement a PV system performance analyzer using a MLT to closely estimate the daily solar AC energy output of any PV system. In order for a PV performance analyzer to correctly detect anomalies, the predicted values have to be very accurate. In other words, the predicted values have to be closely correlated to the values of the measured data. The MLT prediction values should not be influenced by the distance to the weather station, should be indifferent to the PV system location and should not be influenced by the different seasons of the year. In other words, the projected prediction model is a general model that could be used as a "one-size-fits-all" prediction model which is able to estimate the solar production outputs of any PV system based on the historical data that is available which is then used to analyses the performance of a given PV system.

The main objectives of this work are to determine which MLT presents the best results when dealing with different seasons of the year (Spring, Summer, Autumn and Winter), and with PV systems in different locations. Achieving these main objectives is done by elaborating a MLT comparative study, and associating a rank score to each of the MLTs studied in this work.

There is a vast number of available MLTs, and recently Pedro Domingos [5] suggested that all MLT could be organized into five "tribes" or "schools" in which each one is represented by its own master algorithm, as presented in Table I [5]. This concept of the five MLT tribes narrows down the vast number of MLT and provides a MLT selection scheme that allows to cover all types of MLT in order to research the Machine Learning PV system monitoring topic.

Table I: The five MLT Tribe Concept Suggested by Pedro Domingos [5]

#	Tribe	Master Algorithm	Selected MLT for this work
1	Symbolists	Inverse Deduction / Induction	Regression Tree
2	Connectionists	Backpropagation	Artificial Neural Network (ANN)
3	Evolutionaries	Genetic Programming	Multi-Gene (MG) Genetic Programming
4	Bayesians	Probabilistic Inference	Gaussian Process
5	Analogizers	Kernel Machines	Support Vector Machine for Regression (SVR)

To achieve the aim of this work, a MLT Data-Driven approach is considered to analyses the errors of five MLTs namely the Regression Tree, the Artificial Neural Networks (ANN), the Multi-Gene (MG) Genetic Programming, the Gaussian Process and the Support Vector Machine for Regression (SVR), as mentioned in Table I. These five MLT estimates are then compared to the measured PV system solar AC energy production data. The Machine Learning models estimate the behavior of the system based on historical time series of inputs and targets and has no knowledge about the internal characteristics and processes of the PV system. The error analysis is based on the values of the root mean square error (RMSE), the normalized root mean square error (NRMSE). The error analysis of the MLTs was performed in two comparison experiments namely a seasonal comparison and a location comparison. The seasonal comparison considers the months of Spring (March, April and May), Summer (June, July, August), Autumn (September, October and November) and Winter (December, January and February). The location comparison considers three locations in the Island of Madeira. All together there are two experiments, seven scenarios and five MLTs.

Many research papers report the prediction/estimation of solar radiation values, but only recently PV solar output prediction/estimation research has taken place and is rising due to the growing number of PV system installations occurring all over the world. Since there is a significant amount of PV solar production being injected into the energy distribution grid, the energy distributor companies feel a need to have knowledge about the loads that are being injected into the grid in order to adjust their energy management. Only papers that consider PV output prediction/estimation models were included in the literature review, which is organized according to each of the five Machine Learning techniques used in this work.

Not much research on solar energy output predictions using Regression Trees has been done. Nonetheless, Zamo [6] conducted a comparison machine learning study that included the Regression Tree, the Support Vector Machines and other machine learning techniques that are not studied in this work. The results from this work show that the Support Vector Machine prediction values were better than those from the Regression Tree.

Artificial Neural Networks (ANN) are by far the Machine Learning Technique that is most used and/or compared to, in the PV output prediction research as an alternative to the Support Vector for Regression MLT. Many solar energy output prediction research papers conduct comparisons between ANN models and other machine learning models just as done by Hossain [7]. This work conducts a comparison study between machine learning algorithms such as the ANN, Support Vector Regression with radial kernel, and others that are not studied in this paper. The results show that the second-best prediction values were presented by the Support Vector Regression with radial kernel and the best predictions came from the ANN algorithm.

The Genetic Programming algorithm used in this work is part of the GPTIPS 2 toolbox [8]. This toolbox has been used by Aliesfahani [9] to implement a maximum power point tracker for a PV system based on a solar PV cell experiment and does not compare the results to other MLT. The toolboxes used for Genetic Programming and solar energy output prediction are the Brain Project toolbox by Russo [10] and the Fuzzy Rule toolbox by Prokop [11]. The Brain Project is the software tool used for the formal modelling of the data using Genetic Programming [10]. It compares the Genetic Programming model results to the ones done by Pedro and Coimbra [1] and found that their Genetic Programming results were worse than the ANN results. Prokop [11] used the Fuzzy Rule Genetic Programming model to compare to the ANN and SVR models. The results showed that the Genetic Programming models outperformed the ANN and SVR model results.

Not much research on solar energy output prediction using Gaussian Process has been done in the past. Only recently this topic has been explored by Huang [12]. This work carried out a comparison between solar energy prediction models such as the ANN model, the SVR model and the Gaussian Process model. The best results were presented by the Gaussian Process model.

The SVR model is very popular in the solar energy output prediction research field just like the ANN models. A model comparison study between an ANN backpropagation model and a SVR with radial kernel model was conducted by Yang [13]. The results show that the SVR model outperformed the ANN model. Another MLT comparative study was conducted by Junior [14], in which compares the SVR with the persistence method. The results show that the SVR model presented the best results for predicting daily power production.

This work contributes to further research in the PV system monitoring field by providing information about how

- Solar AC energy production behaves in different seasons of the year and in different locations;
- Different MLTs deal with known and unknown test data in the different seasons and locations of the Madeira Island.

The organization of the paper is as follows: section 2 provides information about the methodology of the experimental design of this paper as well as information about the historical data used to train and test the MLTs. Section 3 presents and discusses the results of the experimental study and finally section 4 concludes the main findings of the work.

This is a pre-peer review version of the paper submitted to the journal Progress in Photovoltaics.
It has been selected by the Executive Committee of the EU PVSEC 2017 for submission to Progress in Photovoltaics.

2 METHODOLOGY

This section describes the methodology that was used to determine which MLT would present the best results in predicting the daily solar AC energy values. This methodology describes the historical data and presents the experiment and scenario scheme. This is followed by the description of the raw data pre-processing method, Machine Learning Process, error analysis and finally the MLT parameter settings in Matlab 2016a.

The Statistics and Machine Learning Toolbox from Matlab version 2016a is the software used to train and test four of the five machine learning techniques studied in this work. While, the GPTIPS2 toolbox for Matlab is used to train and test the Multi-Gene (MG) Genetic Programming algorithm [8].

2.1 Scenarios for Machine Learning Technique Testing

This section describes the historical data as well as the experiment and scenario scheme used in this work.

The historical data of the PV systems used for training and testing the MLT estimate models were measured in the Island of Madeira (Portugal). This Island is an ideal location for solar energy analysis since there are many microclimates in a small amount of space making weather prediction a nonlinear and difficult task to carry through due to the orography features of the island that include beach and mountainous regions very close to each other as can be seen in Fig. 1.

The solar irradiation values (inputs) were acquired from one weather station of the IPMA institution (IPMA – *Instituto Português do Mar e da Atmosfera*) located in the city of Funchal and are daily measurements in kJ/m2. The solar AC energy values (observed outputs or targets) for this work were acquired from a PV system installation company called *Factor Energia*, in 15 minute power samples. Since the acquired solar irradiation values were only available in daily measurements, the solar power values were converted into solar AC energy values in kWh. Therefore, the proposed performance monitoring system analysis the PV system performance based on daily values.

Table II presents, information about the PV system installations in regard to the distance from the weather station, the altitude, and the PV system installed power expressed in direct current (DC). The PV system A installation is 440 meters away from the weather station, PV system B is 2.8 km and PV system C is 4.1 km. The PV system installations A and C have the same installed power of 4.14 kW, while PV system installation B has an installed power of 3.68 kW.

The input and target samples were measured in 2015 with a total of 361 daily samples from PV system A, a total of 260 daily samples from PV system B, and a total of 89 daily samples from PV system C. The training data that is used in this work came from the PV system A, which is closest to the weather station. The same weather station solar irradiation values are used in all scenarios of this work.

The Spring months are March, April and May, the Summer months are June, July and August, the Autumn months are September, October and November while the Winter months are December, January, and February.

Table II indicates the number of input and target (measured data) samples that are available in each PV system dataset for training and testing the MLTs.

The Training dataset is composed of: Spring dataset (92 samples), Summer dataset (89 samples), Autumn dataset (91 samples), and Winter dataset (89 samples) from PV system A.

The Testing dataset includes all the PV system A datasets used for Training the MLTs, referred to as "known data", as well as the PV systems B and C datasets of Spring, Autumn and Winter referred to as "unknown/unseen" data.

The PV system A, B and C Spring dataset from May all have 24 inputs and targets. The PV system C did not have available Summer, Autumn nor Winter historic data and PV system B did not have available Summer data. Therefore, the location experiment was only done in the Spring, Autumn and Winter seasons.

In order to determine which of the five MLTs presents the best results in PV solar AC output predictions two types of experiments were conducted with the first experiment carrying out four different scenarios and the second scenario carrying out three different scenarios. Since there are seven testing scenarios there are seven Test datasets, and each scenario uses five MLTs.

In this work, all five MLTs are trained by using the train dataset from PV system A which is the PV system that is closest to the weather station.

2.2 Machine Learning Process

This section describes the pre-processing technique applied to the raw data, the Machine Learning Process, and finally the error analysis.

Before training the Machine Learning prediction models, the raw historical data is pre-processed by normalizing both the inputs and targets with a pre-processing technique called Min-Max normalization which normalizes the values between 0 and 1. This pre-processing technique contributed to speeding up the Machine Learning model training time since the normalized values are small and lie between 0 and 1. It was

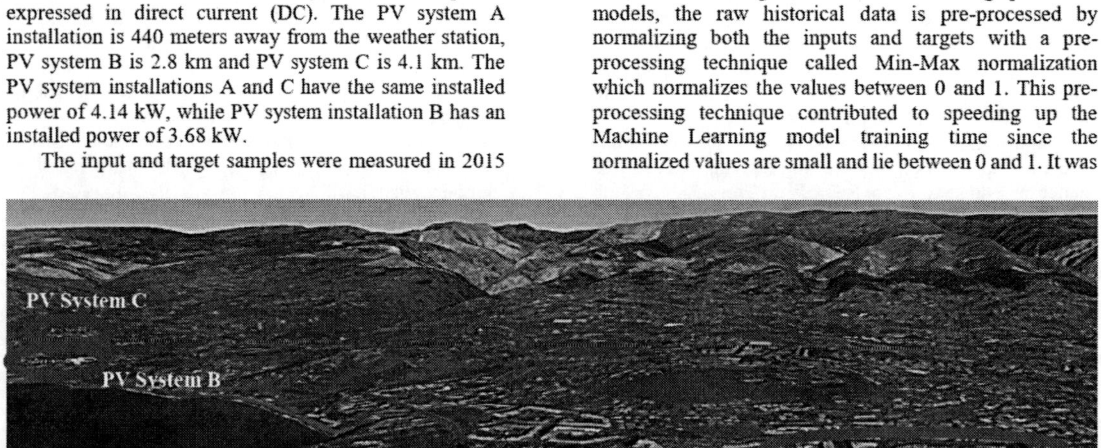

Figure 1: PV System Installation Locations and Weather Station Location in Funchal

This is a pre-peer review version of the paper submitted to the journal Progress in Photovoltaics.
It has been selected by the Executive Committee of the EU PVSEC 2017 for submission to Progress in Photovoltaics.

Table II: PV system installation location, installed power, and sample information

	PV system A	PV system B	PV system C
Distance to Weather Station	440 m	2.8 km	4.1 km
Altitude	14 m	77 m	121 m
PV System Installed Power	4.14 kW DC	3.68 kW DC	4.14 kW DC
Train Data	92 Spring Samples 89 Summer Samples 91 Autumn Samples 89 Winter Samples	--------------------------	--------------------------
Test Data	92 Spring Samples	--------------------------	--------------------------
	89 Summer Samples	--------------------------	--------------------------
	91 Autumn Samples	--------------------------	--------------------------
	89 Winter Samples	--------------------------	--------------------------
	24 Spring Samples (8 to 31 of May 2015)	24 Spring Samples (8 to 31 of May 2015)	24 Spring Samples (8 to 31 of May 2015)

assumed that all inputs were normalized with Min=1 kJ/m2 and Max=30000 kJ/m2 and all targets with Min=1 kWh and Max=30 kWh. The minimum and maximum values are based on the collected annual data and are used for generalizing the different PV system installed power installations (up to 4.14 KW) and for detecting outliers. Since this normalizing technique is used before training the models, it then has to be reversed and compared against the predicted output values coming from the machine learning models in order to obtain the estimated daily PV solar AC energy values in kWh.

In order for a Machine to execute a certain task, it needs to learn how to do the task by analyzing the relationships and trends from historical data. The Machine Learning Process for any MLT starts by training a Machine Learning (ML) model, then testing the ML model, to then comparing the measured data (targets) with the estimated data (outputs) that were generated by the ML model and finally performing an error analysis.

In order to compare the performance of the MLT, an error analysis is conducted by using two indicators namely the Root Mean Squared Error (RMSE) and the Normalized Root Mean Squared Error (NRMSE) after de-normalizing the data. The RMSE is easily interpreted due to having the same units as the targets and outputs (kWh) and computes the average error compared to the target (observed output) value (Equation 1). A large positive RMSE value represents a large deviation scale in the prediction values from the target values. The accepted RMSE value in this work, is a RMSE value lower that 5 kWh since this value represents a 2 € / day loss (5 x 0.40 €/kWh = 2 € / day loss) assuming that the grid injection price is 0.40 €/kWh (as practiced in Portugal). If this loss is an average daily loss over the course of 3 months, the total monetary loss is of 180€ (2€ x 90 days = 180€) which is quite a significant amount and is negatively reflected in increasing the investment payback time period.

The NRMSE provides the average percentage value of the error (Equation 2), in relation to the difference between the unnormalized minimum and maximum output values, which is 29 kWh (30 kWh–1 kWh = 29 kWh) assumed in this work.

$$RMSE = \sqrt{\frac{1}{n}\sum_{i=1}^{n}(Yt - Yp)^2} \qquad (1)$$

$$NRMSE = \frac{RMSE}{Y_{tmax} - Y_{tmin}} \qquad (2)$$

Where *Yt* is the observed targets, *Yp* is the predicted outputs and n is the number of observations.

The Machine Learning for Regression model parameter settings for this work are mainly the Matlab default Machine Learning model settings. Validation and cross-validation were not performed to any of the models in this work unless automatically done by the fitting command of the Matlab toolbox. The MLTs were only trained with simple/default parameter settings and then tested.

3 RESULTS AND DISCUSSION

In this section, the results of the two experiments and respective scenarios are presented in the tables and figures. First the results of the MLTs trained by the Spring dataset of PV system A of the experiments 1 and 2 (Seasonal Comparison and Location Comparison) are presented and discussed. Then the seasonal trends of the MLTs trained in the Spring season are presented and discussed. Finally, this is followed by the results and discussion of a Ranking Score that is associated to each of the seven scenarios.

3.1 Experiments 1 and 2 (Seasonal Comparison and Location Comparison) using the PV System A Spring dataset for training

In this section, the models were trained using only the Spring dataset (inputs and targets) of the PV system A. The error analysis results are presented in Table III. The highest RMSE value assumed to be acceptable in this work is 5 kWh due to the high monetary losses as explained in the error analysis description of the methodology section. The corresponding NRMSE value of RMSE = 5 kWh is approximately 17.2414 %, since the difference between the normalizing minimum and maximum values is 29 kWh.

Here, the MLT that consistently best predicts (green) the daily PV solar AC energy values in all the scenarios, except for scenario 1, is the Multi-Gene (MG) Genetic Programming with very low RMSE values that range between 1.1271 kWh and 1.8577 kWh in experiment 1 and between 0.6408 kWh and 1.0065 kWh in experiment 2. However, this MG Genetic Programming MLT does not perform as well when dealing with predicting the PV energy values of the season used for training (Spring scenario 1). For scenario 1 the MLT that best predicts the solar AC energy values is the Regression Tree MLT and presents a very low RMSE value of 1.5228 kWh.

The MLTs that perform second best (grey) in scenarios 1 and 2 is the ANN with low RMSE values. The MLTs that perform the second best in scenarios 3 and 4 is the SVR MLT with acceptable RMSE values. The second best MLTs for the location comparison experiment is the Regression Tree algorithm and presents very low RMSE values.

The third best (orange) MLT prediction values for scenario 1 is the MG Genetic Programming algorithm with a low RMSE value of 2.0960 kWh. For scenarios 2 and 3 is the Gaussian Process MLTs with acceptable RMSE values. As for scenario 4, the Regression Tree MLT presents the third best results, however the RMSE of over 5 kWh is an unacceptable prediction value. For scenarios 5, 6 and 7, the ANN presented the third best prediction values with very low RMSE values.

The fourth best (blue) MLT prediction values present acceptable RMSE values except for the Winter scenario 4. The Gaussian Process MLT presents a low RMSE value in scenario 1. For scenarios 2 and 3 the Regression Tree MLT presents the fourth best prediction values with acceptable RMSE values. In scenarios 5, 6 and 7 of experiment 2, the fourth best MLT prediction values are presented by the Gaussian Process algorithm.

Finally, the worst prediction values (white) belong to the SVR MLT in scenarios 1 and 2, from experiment 1 and scenarios 5, 6 and 7 from experiment 2. For scenarios 3 and 4, the worst prediction values belong to the ANN MLT.

Scenario 4 presents the worst prediction values when the MLTs are trained by the Spring dataset of the PV system A, except when the MG Genetic Programming and SVR MLTs are used. When these two latter MLTs are used, the RMSE values are within the acceptable prediction values assumed in this work.

The RMSE values increase successively from scenario 1 to 4 when using the Regression Tree and Gaussian Process MLTs. This does not take place with the other MLTs in experiment 1.

The ANN MLT does not deal well with the Autumn and Winter seasons, nor do the Regression Tree and Gaussian Process MLTs in Winter. This is an indication of how well these MLTs deal with new unknown data that have different variations to those that take place in the Spring season.

The MG Genetic Programming MLT presents the most unique prediction values since the Winter RMSE values are much better than the Spring and Autumn RMSE values. When the MG Genetic Programming MLT is trained by the Spring dataset from the PV system A, the Summer scenario presents the best prediction values with a very low RMSE value of 1.1271 kWh.

The PV systems of scenarios 5 and 7 from experiment 2 have an installed power of 4.14 kW, while the PV system in scenario 6 has a lower installed power of 3.68 kW. It is interesting to see that the PV systems with the same installed power (scenarios 5 and 7) present very similar RMSE values even though they are located 4 km away from each other with a difference in altitude of more than 100 meters which indicates that the weather conditions should be different (due to the orography of the Island and consequent microclimates) in both locations and consequently affecting the solar AC energy output of the PV systems. On the other hand, the PV system with the lower installed power (scenario 6) presents the highest RMSE values of the 3 scenarios in experiment 2. The similarity between the RMSE values of scenarios 5 and 7 might be due to the fact that both PV system installations have the same PV installed DC power of 4.14 kW which informs us that the model trains the behavior of the PV solar AC energy production pattern and is sensitive to the behavior patterns of the PV systems based on their installed power.

Overall the prediction values of the Spring season are better than the rest of the seasons except for the MG Genetic Programming MLT. This indicates that all the MLTs except for the MG Genetic Programming MLT deal badly with unknown data, while the MG Genetic Programming MLT deals best with unknown data (Summer, Autumn and Winter). Generally, MLTs deal better with known data then unseen or unknown data, as shown in Table III. However, the results in Table III show that the MG Genetic Programming does the exact opposite with the prediction values of the known data (Spring) presenting worse results than the ones from the unknown data (Summer, Autumn and Winter) results. This may be due to the fact that MG Genetic Programming uses symbolic regression as opposed to classical regression. While the classical regression algorithm requires a pre-specified model structure, the symbolic regression algorithm is not provided with a particular model nor with prior assumptions. Therefore, the symbolic regression algorithm infers the model from the data and attempts to discover the model structures as well as the model parameters by finding the fundamental relationships of the dataset and allowing the patterns in the data itself reveal the appropriate models. Thus, the symbolic regression algorithm is not affected by unknown gaps in the domain knowledge, since it does not require a specific model in

Table III: Error Analysis Results of Experiments 1 and 2 by training with the Spring dataset

Tribe	MLT	Error Analysis	Experiment 1 – Seasonal Comparison				Experiment 2 – Location Comparison		
			Scenario 1 Spring	Scenario 2 Summer	Scenario 3 Autumn	Scenario 4 Winter	Scenario 5 PV system A	Scenario 6 PV system B	Scenario 7 PV system C
1	Regression Tree (RT)	RMSE	1.5228 kWh	2.6709 kWh	3.5693 kWh	5.9739 kWh	1.2620 kWh	1.9410 kWh	1.3722 kWh
		NRMSE	5.2511 %	9.2101 %	12.3079 %	20.5997 %	4.3516 %	6.6931 %	4.7319 %
2	ANN	RMSE	1.8162 kWh	2.5327 kWh	25.7125 kWh	15.3871 kWh	1.6367 kWh	2.2001 kWh	1.5401 kWh
		NRMSE	6.2626 %	8.7334 %	88.6638 %	53.0590 %	5.6437 %	7.5864 %	5.3105 %
3	MG Genetic Programming	RMSE	2.0960 kWh	1.1271 kWh	1.8577 kWh	1.4988 kWh	0.6408 kWh	1.0065 kWh	0.7149 kWh
		NRMSE	7.2274 %	3.8864 %	6.4058 %	5.1683 %	2.2097 %	3.4706 %	2.4653 %
4	Gaussian Process	RMSE	2.1128 kWh	2.5532 kWh	3.4595 kWh	6.0252 kWh	2.1985 kWh	2.7621 kWh	2.0089 kWh
		NRMSE	7.2853 %	8.8043 %	11.9294 %	20.7767 %	7.5809 %	9.5245 %	6.9271 %
5	SVR	RMSE	2.4515 kWh	3.3579 kWh	3.0377 kWh	5.0739 kWh	2.5274 kWh	3.0424 kWh	2.1229 kWh
		NRMSE	8.4536 %	11.5790 %	10.4749 %	17.4963 %	8.7152 %	10.4911 %	7.3203 %

This is a pre-peer review version of the paper submitted to the journal Progress in Photovoltaics.
It has been selected by the Executive Committee of the EU PVSEC 2017 for submission to Progress in Photovoltaics.

order to be specified. The MG Genetic Programming MLT has its origins in Genetic Programming and is therefore composed of multiple independent trees named genes, in which their outputs are optimally linearly combined by means of the mean square error (MSE) of the resulting expression measured on the training data using the least squares method. Therefore, the MG Genetic Programming MLT employs a form of multiple linear regression to produce generalized linear models that create models with a small error and closely correlated to the measured data.

3.1.1. Seasonal Trends of the MLTs trained by the Spring dataset of the PV system A.

The seasonal trends of the MLTs trained by the Spring dataset of PV system A are visually illustrated in this section. This illustration provides a better understanding of the prediction value variation of the different MLTs throughout the different seasons of the year.

The bold blue line in all graphs below represents the measured data (targets) of the solar production output of the PV systems. The MLT estimates are very close to the measured data in the Spring season since they were trained by the Spring dataset. However, only in the month of April are the prediction values closely correlated to the measured values (Fig. 2).

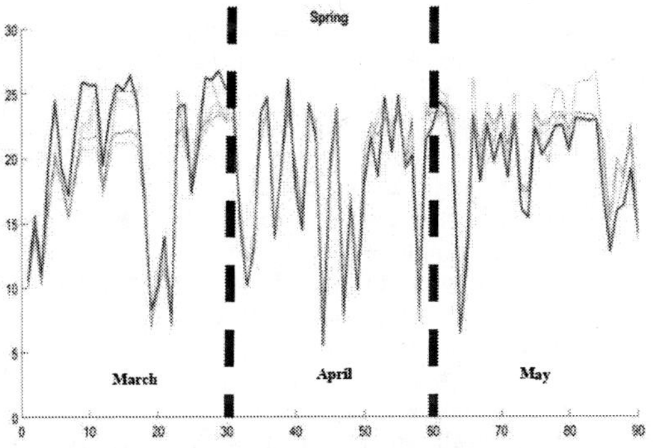

Figure 2: Estimation values of the Spring dataset

In the Summer season, the MLTs tend to overestimate the solar production in all three months (Fig. 3).

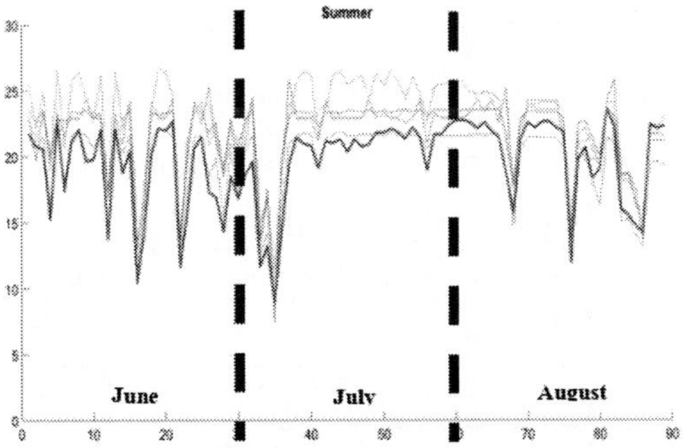

Figure 3: Estimation values of the Summer dataset

In the Autumn months, there are outliers created by the ANN MLT that was also identified in Table III. The ANN MLT does not know how to react when presented with very low energy outputs (unknown test data), since these were not introduced in the Spring database (known test data). It is curious to see that all the other MLTs know how to react when dealing with very low energy values or in other words unseen/unknown data. Here in the Autumn months, there is a tendency for the MLT to underestimate the solar production, however September presents a close correlation between the measured data and the MLT estimates (Fig. 4). This indicates that there could be a similar trend between the months of April and September which needs further research.

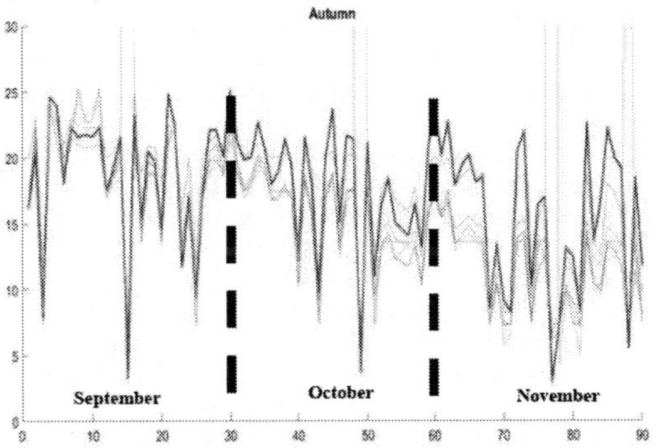

Figure 4: Estimation values of the Autumn dataset

Regarding the Winter months, the predicted values are highly underestimated, particularly in the month of December (Fig. 5). Here, it is clear to see how the MG Genetic Programming predicted values (purple) closely correlate with the measured values, throughout the months of Winter.

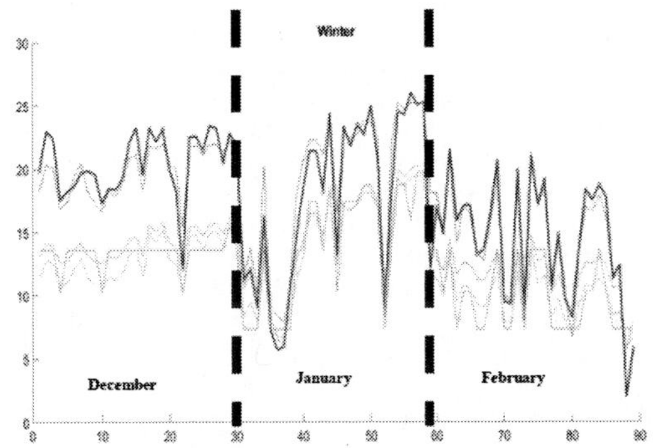

Figure 5: Estimation values of the Winter dataset

The closely correlated prediction values take place in the month of April and late September. The overestimated prediction values take place from the month of May right through to the mid-late September. While the prediction values are underestimated from the month of October right through to the month of April. The switch between the underestimated results with the overestimated results takes place in the month of April while the overestimated results switch with the underestimated results in mid-late September. Therefore, the trend of the prediction values of the Spring and Summer seasons are overestimated in relation to the measured data, while the prediction values of the Autumn and Winter seasons are underestimated. Overall, the predicted values do not closely correlate with the measured data in all the months of the year, except for the months of April and mid-late September.

This is a pre-peer review version of the paper submitted to the journal Progress in Photovoltaics.
It has been selected by the Executive Committee of the EU PVSEC 2017 for submission to Progress in Photovoltaics.

3.1.2. Machine Learning Technique Ranking Score.

This section presents a ranking score associated to each of the five MLT tribes within the seven scenarios of the two experiments carried out in this work. Each experiment has a scenario that uses known test data (scenarios 1 and 5) and various scenarios that use unknown test data (scenarios 2, 3, 4, 6, and 7). The solar AC energy values of the Summer months (scenario 2) are very similar to those of the Spring months (scenarios 1 and 5).

Table IV presents the ranking score of the MLT associated to each experimental scenario which is divided into known and unknown test data.

The Regression Tree MLT deals best with the known test data (scenario 1) of experiment 1, while the MG Genetic Programming MLT deals best with the known test data (scenario 5) of experiment 2. The MG Genetic Programming MLT also deals best with all the unknown test data scenarios (scenarios 2, 3, 4, 6, and 7) from both experiments 1 and 2.

The second place in the MLT ranking score belongs to the ANN, SVR and Regression Tree MLTs. In experiment 1, the ANN MLT deals well with the known (scenario 1) test data and unknown (scenario 2) but similar to the known test data which is the Summer season. The SVR MLT deals well with the unknown (scenarios 3 and 4) test data of experiment 2. Finally, the Regression Tree MLT deals well with the known (scenario 5) and unknown (scenarios 6 and 7) test data from experiment 2.

The Gaussian Process ranks third and fourth place throughout the scenarios.

The Regression Tree presents its worst results in fourth place in scenarios 2 and 3. Overall the Regression Tree presents better results in the Winter scenario than in the Summer and Autumn scenarios.

Interestingly the fifth place of the MLT ranking score of experiment 1 presents the same MLTs as the second place MLT ranking score, however the scenario positions of these two MLTs are inverted. In other words, the SVR MLT deals now better with the known (scenario 1) and unknown (scenario 2) but similar test data, while the ANN MLT deal better with the unknown (scenarios 3 and 4) test data of experiment 1.

4 CONCLUSIONS

The ideal PV performance analyser would have a prediction model that is not influenced by the different seasons of the year nor the distance between the PV system installation and the weather station. The challenge of this work was to determine whether the 3 month daily historical data of a given PV system installation could be used for another in a different location and/or for the same PV system installation but in a different season of the year, and attempt to accomplish the one model fits all approach.

Five Machine Learning Techniques were tested in two experiments (seasonal comparison and location comparison) with a total of seven scenarios, as explained in the methodology section. The five MLT are the Regression Tree, Artificial Neural Network (ANN), Multi-Gene (MG) Genetic Programming, Gaussian Process and the Support Vector Machine for Regression (SVR) MLTs. The GPTIPS2 toolbox for Matlab was used to implement the MG Genetic Programming MLT in this work, while all the other MLTs were implemented by using the Matlab 2016a toolbox.

In conclusion, it is possible to implement a "one-size-fits-all" prediction model for roof-top PV systems of up to 4.14 kW of DC installed power that does not need more than 3 months of daily historical data and that is not influenced by the seasons of the year nor the PV system installation location. This is possible by training the MG Genetic Programming MLT with at least 3 months (or 92 samples) of daily historical data (solar irradiation input and solar AC energy output) from any season of the year. Regarding the location of the PV system, the prediction values are very similar for the PV systems of the same installed power in different locations, however if the MG Genetic Programming MLT is trained with a database of a 4.14 kW PV system, the prediction values will not be as accurate for PV systems of a lower installed power in different locations. Even though PV system C is 4.1 km away from the weather station, all the MLTs provide accurate prediction values.

The MG Genetic Programming MLT presents very low RMSE values which range between 1.1271 kWh and 1.8577 kWh in experiment 1 and between 0.6408 kWh and 1.0065 kWh in experiment 2. However, this MG Genetic Programming MLT does not perform as well when dealing with predicting the PV energy values of the season by

Table IV: Ranking Score of the Machine Learning Techniques trained by the Spring season in each scenario

Rank	Experiment 1 - Seasonal Comparison				Experiment 2- Location Comparison		
	Scenario 1	Scenario 2	Scenario 3	Scenario 4	Scenario 5	Scenario 6	Scenario 7
	Spring	Summer	Autumn	Winter	PV system A	PV system B	PV system C
1°	Regression Tree	MG Genetic Programming	MG Genetic Programming	MG Genetic Programming	MG Genetic Programming	MG Genetic Programming	MG Genetic Programming
2°	ANN	ANN	SVR	SVR	Regression Tree	Regression Tree	Regression Tree
3°	MG Genetic Programming	Gaussian Process	Gaussian Process	Regression Tree	ANN	ANN	ANN
4°	Gaussian Process	Regression Tree	Regression Tree	Gaussian Process	Gaussian Process	Gaussian Process	Gaussian Process
5°	SVR	SVR	ANN	ANN	SVR	SVR	SVR
	Known data	Unknown data			Known data	Unknown data	

which it was trained with, but still presents very good prediction values (under 5 kWh). Overall, the MG Genetic Programming MLT deals best when tested by known and unknown/unseen data, since it presents prediction values that are very closely correlated to the measured data. The occurrence of these good results may be due to the fact that the MG Genetic Programming MLT uses the symbolic regression. This MLT is composed of multiple genes with optimally linearly combined outputs by applying the MSE to the outputs of the training data by using the least squares method. Therefore, the MG Genetic Programming MLT produces generalized linear models that create models with small errors and that are closely correlated to the measured data by using a form of multiple linear regression.

Therefore, the aim of this work is achieved by using the MG Genetic Programming MLT prediction model to implement an accurate PV performance analyzer trained by the available 3 month historic database that is stored in the production meter connected to the PV system. This PV performance analyzer accurately predicts the solar AC energy outputs of a PV system on every season of the year and in any location of the Madeira Island.

The overall trend of the prediction values of the Spring and Summer seasons are overestimated in relation to the measured data, while the prediction values of the Autumn and Winter seasons are underestimated. In general, the predicted values do not closely correlate with the measured data in all the months of the year, except for the months of April and mid-late September.

The first place MLT ranking score belongs to the Regression Tree and MG Genetic Programming MLTs. The Regression Tree deals best with the known test data (scenario 1) of experiment 1, while the MG Genetic Programming MLT deals best with the known test data (scenario 5) of experiment 2 as well as the unknown test data (scenarios 2, 3, 4, 6 and 7) of both experiments 1 and 2.

A PV system analyzer using MLT can be applied in real world applications. This Analyzer can be used by the PV system installation companies to offer new services to customers, since it has been proven to work with roof-top PV system in any season of the year and in any location of the Madeira Island. The services would include PV system performance analysis for PV systems that have at least 3 months of weather and solar AC energy historical data, for PV systems that do not have any available historical data, and for PV systems that have only the solar AC energy historical data but do not have the weather historical data. This new service can be used in real-time which allows for real-time problem solving by implementing an alarm that would go off in the event of poor PV output performance.

ACKNOWLEDGMENTS

Acknowledgments to the PV system installation company *Factor Energia* for providing the historical data from the three PV systems located in the Island of Madeira, Portugal. Acknowledgments also to the Meteorological institution IPMA (IPMA – *Instituto Português do Mar e da Atmosfera*), for providing the daily solar irradiation historical data. Acknowledgments also to the support from the Portuguese Foundation for Science and Technology for their support through Projeto Estratégico LA 9 - UID/EEA/50009/2013. The support from ARDITI – *Agência Regional para o Desenvolvimento da Tecnologia e Inovação* under the scope of the Project M1420-09-5369-FSE-000001 – PhD Studentship, is also gratefully appreciated.

REFERENCES

[1] H. T. C. Pedro and C. F. M. Coimbra, "Assessment of forecasting techniques for solar power production with no exogenous inputs," *Solar Energy*, vol. 86, no. 7, pp. 2017–2028, 2012.

[2] R. H. Inman, H. T. C. Pedro, and C. F. M. Coimbra, "Solar forecasting methods for renewable energy integration," *Progress in Energy and Combustion Science*, vol. 39, no. 6, pp. 535–576, 2013.

[3] J. Antonanzas, N. Osorio, *et al*, "Review of photovoltaic power forecasting," *Solar Energy*, vol. 136, pp. 78–111, 2016.

[4] Muhammad Qamar Raza, Mithulananthan Nadarajah, and Chandima Ekanayake, "On recent advances in PV output power forecast," *Solar Energy*, vol. 136, no. 136, pp. 125–144, 2016.

[5] P. Domingos, *The Master Algorithm: How the Quest for the Ultimate Learning Machine Will Remake Our World*. Basic Books, 2015.

[6] M. Zamo, O. Mestre, *et al*, "A benchmark of statistical regression methods for short-term forecasting of photovoltaic electricity production, part I: Deterministic forecast of hourly production," *Solar Energy*, vol. 105, pp. 792–803, 2014.

[7] R. Hossain, A. Maung, *et al*, "Hybrid prediction method for solar power using different computational intelligence algorithms," *Smart Grid and Renewable Energy*, vol. 4, no. February, pp. 76–87, 2013.

[8] D. Searson, D. Leahy, and M. Willis, "GPTIPS: An open source genetic programming toolbox for multigene symbolic regression," *Proceedings of the International of the MultiConference of Engineers and Computer Scientists*, vol. I, pp. 17–20, 2010.

[9] S. Aliesfahani and M. Shahbazian, "Maximum Power Point Tracking for Solar Photovoltaic System Using Genetic Programming Toolbox for Identification of Physical System," *Journal of Automation and control*, vol. 3, no. 1, pp. 25–28, 2015.

[10] M. Russo, G. Leotta, *et al*, "Genetic programming for photovoltaic plant output forecasting," *Solar Energy*, vol. 105, pp. 264–273, 2014.

[11] L. Prokop, S. Misák, V. Snásel, *et al*, "Supervised Learning of Photovoltaic Power Plant Output Prediction Models," *Neural Network World*, vol. 23, no. 4, pp. 321–338, 2013.

[12] C. Huang, A. Bensoussan, *et al*, "Improvement in artificial neural network-based estimation of grid connected photovoltaic power output," *Renewable Energy*, vol. 97, pp. 838–848, 2016.

[13] H.-T. Yang, C. Huang, *et al*, "A Hybrid Method for One-Day Ahead Hourly Forecasting of PV Power Output," *TRANSACTIONS ON SUSTAINABLE ENERGY*, vol. 5, no. 3, pp. 526–531, 2014.

[14] J. G. da S. F. Junior, T. Oozeki, *et al*, "Use of support vector regression and numerically predicted cloudiness to forecast power output of a photovoltaic power plant in Kitakyushu, Japan," *Prog. Photovolt: Res. Appl.*, vol. 20.7, pp. 874–882, 2012.

EFFECT OF PID ON ENERGY CONVERSION EFFICIENCY OF CRYSTALLINE SILICON PHOTOVOLTAIC POWER PLANT

Hong Yang[1] , He Wang[1], Xuefang Jiang[1], Chuanke Chen[1], Jipeng Chang[1], Junjun Zhang[2], Jingsheng Huang[2]
[1] School of Science, Xi'an Jiaotong University, Xi'an 710049, People's Republic of China
[2] China electric power research institute, Nanjing, 210003, People's Republic of China

ABSTRACT: The potential-induced degradation (PID) contributes to detrimental influence on the performance of crystalline silicon solar modules in the field photovoltaic power plant. PID decreases the energy conversion efficiency of solar modules significantly and hazards the reliability of photovoltaic system. This paper focuses on the degradation and power loss resulting from PID in the PV power plant under comprehensive operating conditions. And important location distribution law of PID in strings and modules was revealed for the first time by using thermography and electroluminescence (EL) imaging. The degradation law of module's output power and open circuit voltage (Voc) are presented by measuring current-voltage (*I-V*) and power-voltage (*P-V*) curves characteristic. The energy conversion efficiency of PV system was reduced tremendously led by PID. The degradation of PV modules in maximum output power is as high as 53%. The yield loss of an array affected by PID is about 40%. High temperature and relative humidity have proved to aggravate PID. The results of this work are considered as a basement to further study the mechanism of PID and the effect of PID on energy conversion efficiency of PV system in the field.
Keywords: Crystalline solar modules; Potential-induced degradation; Yield loss

1 INTRODUCTION

PV module is a key system component that converts solar radiation directly to electricity. The durability and the energy conversion of the PV modules are extremely critical for the total power output of the system, the cost effectiveness of the PV system and the commercial success of photovoltaic energy [1-3]. However, PID decreases the energy conversion efficiency of solar modules significantly and hazards the reliability of photovoltaic system. The degradation has been investigated in the publications [4-8]. In grid-connected photovoltaic systems, the individual modules are usually connected serially to strings so as to increase system voltage. Partial solar cells incorporated into PV modules in this architecture can be exposed to a voltage bias of several hundreds of volts. Because the insulation performance between the Al frame and the active layer is not perfect, these high voltages would cause leakage current between the solar cells and the module frames through the front glass and encapsulation. This current is rather slight but not negligible and has fatal impacts on the long-term system performance. This so-called potential-induced degradation can lead to remarkable power loss.

PID of p-type silicon solar cells is mainly characterized by a significant reduction of the parallel resistance Rsh which causes a drop of the fill factor and the power output. [8–9]. Sodium stemming from the soda lime glass plays a critical role for the leakage currents pathway [10-11]. Recently, the impact of sodium ions reaching solar cells was investigated: it is showed that large concentrations of sodium within the ARC and accumulation of those species on the ARC–silicon interface of PID-affected cells. Sodium decorated stacking faults are identified as the primary cause of potential-induced degradation [12].

The PV system is exposed to comprehensive field conditions for a long time. It is extremely important to investigate the effect of PID on energy conversion efficiency of PV system and the relationship between comprehensive field conditions and PID. The practical investigations of the PID in the PV power plant have been done in our work in the past years. We have measured the performance of PV plant affected by PID and achieved a mass of data based on actual operating of PV power station. The degradation regularity of PID-affected modules and the impacts on practical performance of PV system in the field have been detected. The degradation of PV modules in maximum output power is as high as 53%. The yield loss of an array affected by PID is about 40%. High temperature and relative humidity have proved to aggravate PID. The results of this work are considered as a basement to further study the mechanism of PID and the effect of PID on energy conversion efficiency of PV system in the field.

2 EXPERIMENTAL DETAILS

The PV power plant affected by PID is located in Dingbian County of YuLin, a small city in the northwest china in Shaanxi province at 37.6 degrees north latitude and 107.59 degrees east longitude which has been operating for one year. The capacity of the plant is 50MWp.

The installation connection information of this PV power plant is showed in Fig.1. The PV modules have been mounted on the flat field towards south with a 36° inclination, which is the best compromise between maximum solar radiation, the greatest available surface and minimum costs. The plant is divided into fifty PV units and every PV unit is configured with two 500kWp inverters. There are three main parts in PV plant such as strings, string boxes and inverters. In this PV plant, seventeen modules are interconnected serially into a string and the electric currents of sixteen strings connected in parallel gather into one string box. Each inverter connects with ten string boxes.

The grounded module string is characterized that the Al frame of every module is connected with ground but no pole of the string is grounded. The circuit diagram of grounded module strings is showed in Fig. 2 below.

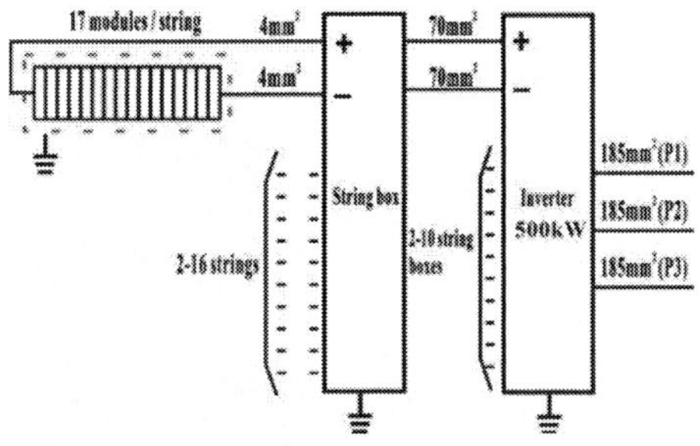

Fig. 1 Installation connection information of this PV power plant

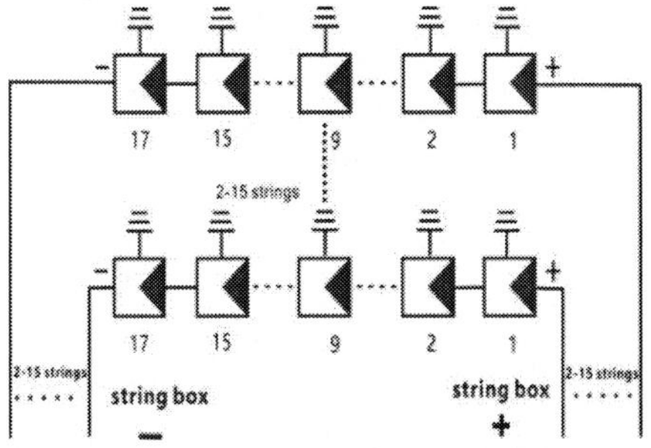

Fig. 2 The circuit diagram of grounded module strings without grounding pole

All the modules installed in this PV power plant are made of mono-crystalline silicon manufactured. Every module consists of 72 solar cells (125mm×125mm) connected in series. The main features of the PV modules are summarized in table 2.

The test conditions of the experiment in this PV power plant are about 27°C average ambient temperature, more than 800W/m^2 solar irradiation and 50% relative humid. Thermography images demonstrating temperature distribution pattern of modules and the location of PID had been taken in the field under operation by an infrared camera which has a resolution of 320×240 pixels and a spectral range of 7.5–13μm (FLIR Systems, Inc. T420). The sunlight exposure dose were measured by YFJ-10 solar irradiation meter, the operation temperature of module was measured by thermocouple. In addition, the *I-V* and *P-V* characteristics of PID-affected and sound modules dismounted from array were measured by using solar simulator (PASAN Sunsim 3C) under standard test conditions (STC: 1kW/m^2 irradiance, 25 °C module temperature and AM1.5 global spectrum). The electroluminescence (EL) images of the dismounted PV modules were measured by an EL measurement system (ITES Co., Ltd.) to verify PID effect. The yield of a PV array was acquired from inverter control panel (NB500kE).

3 RESULTS AND DISCUSSION

3.1 Potential of cells towards frame and Voc of PID module

In grounded module string, the potential of cells means that the potential of the cells embedded in the negative pole of modules towards frame when the modules are operating at maximum power point. The potential of cells after new installation and Voc of modules after one-year operation in grounded module string is showed in Fig.3.

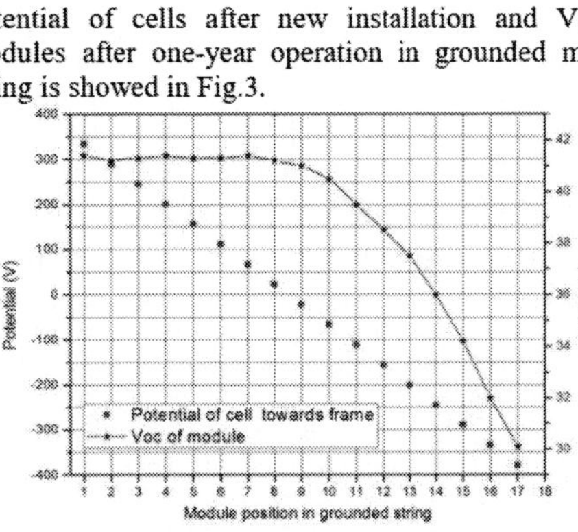

Fig. 3 The potential of cells after new installation and Voc of modules after one-year operation in grounded module string.

The cell's potential towards frames is positive from first to eighth module. On the contrary, that potential is negative from ninth to seventeenth. From the Fig.3, we find that the potential of cell in terms of metallic frame is decreasing continuously from module 1 to module 17 modules in grounded module string. In addition, the Voc from module 1 to module 8 is unchanged about 41.2V and the corresponding cell's potential is positive. The Voc of other modules from module 9 to module 17 is falling off constantly where the potential of cells towards frame is negative. These results clearly indicate that PID could be generated in the standard Si module in grounded module strings of PV power plant resulting from negative system voltage stress with significant degradation in the PV performance. Applying negative voltage towards the Si solar cell is one of the essential factors causing PID and the degradation is becoming more severe when the negative voltage is higher.

3.2 Characteristic of PID-affected modules

The compared *I-V* and *P-V* curves of PID-affected module and PID-free module are given out in Fig.4. The black lines represent the *I-V* and *P-V* curves of modules which are not affected by PID. And the red-star lines show the *I-V* and *P-V* curves of module which has the worst performance led by PID, as module 17 in grounded module string.

We find that the maximum power output (Pmax) of modules from the first to eighth have slight degradation which agreed with the common degradation law of PV modules about 3% resulting from LID and other factors after one-year operation in grounded module strings. However, the degradation of Voc and Pmax of modules from ninth to the seventeenth becomes more and more serious. The seventeenth module in grounded module string has the most severe power loss about 56%.

This is a new way that using thermography imaging to identify the localization and degradation level of PID-affected silicon solar modules in PV systems under operation which is considered to be a fast and reliable tool [14-15]. The thermography images of the PID-free module (left), module 1 in string, and PID-affected module (right), module 17 in string, are showed in Fig.6. From the two graphs in Fig.5, some results can be found that the temperature of cells in PID-free module exhibits a homogeneous temperature distribution. There are no cells of elevated temperature from module 1. However, in the PID-affected module, the temperature pattern is different from that in PID-free module. There are the elevated temperature cells in module 17. The temperature of edge cells is higher than the middle cells. Owing to PID effect, the performance of the edge cells in modules lags for the other cells'. A smaller portion of the irradiated power is converted to electrical energy. Thus, PID-affected cells exhibit a higher temperature than sound ones. This result agreed with other authors' work [6-8]. With PID influence, the shunt resistance (Rsh) of cells decreases dramatically and consequently the dark current of cells increase remarkably under shining. The middle-bottom areas in both thermography images have the highest temperature spot, because where the junction box built in.

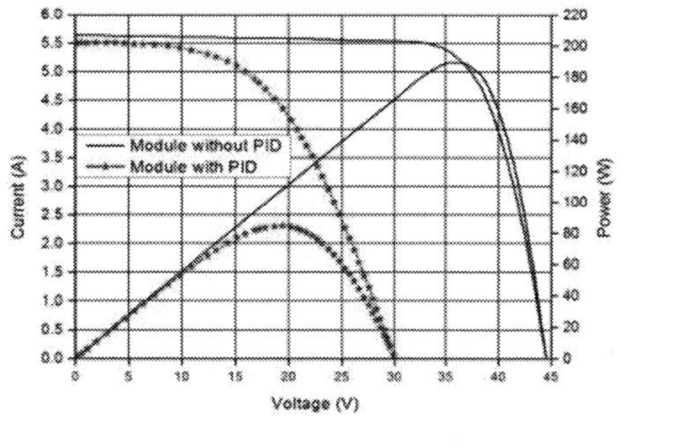

Fig.4 Thermography and electroluminescence images of PID-affected module in grounded module string

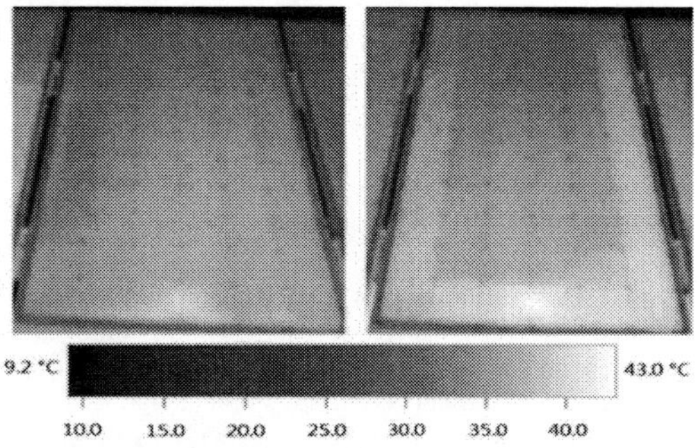

Fig.5 Thermographic images of monocrystalline silicon solar modules under operation, the temperature scale is in °C. PID-affected module contains cells on elevated temperature. The PID-free module (left) and PID-affected module (right) have different temperature distribution pattern.

Fig.6 gives out the EL images of the two modules. From Fig.6, we find that the cells in PID-free module have the same brightness. By contrast, the surrounding cells in PID-affected module are darker than the cells in middle. Because of PID effect, the appearance of PID-affected solar cells is dark in EL images caused by a higher rate of band to band recombination. The darker the appearance of cell is, the more serious the degradation affected by PID is. Comparing with the thermography images, the same distribution of PID-affected cells is showed in the electroluminescence images. And an interesting finding is that the appearance of cells adjacent to the Al frame is darker than the middle cells in PID-affected solar module. It is likely as we propose that the electric field adjacent to the Al frame is more intensive than middle area.

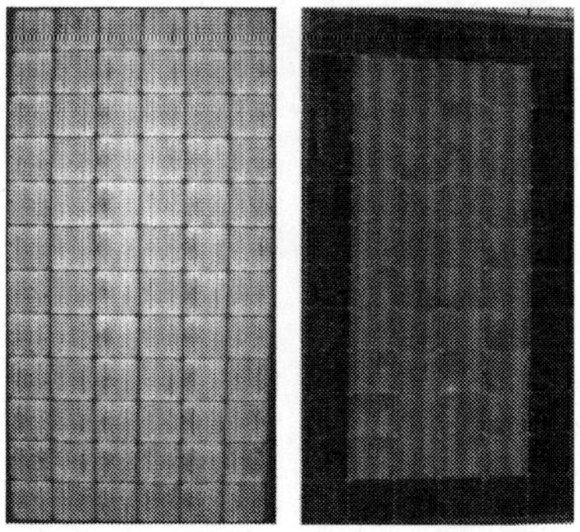

Fig.6 EL images of PID-free (left) and PID-affected (right) silicon solar modules dismounted from field which have different brightness distribution pattern. PID-affected module contains darkened cells.

3.4 Yield of PV array affected by PID in different temperature and RH conditions

Two 500 kWp PV arrays, one is affected and another is not affected by PID, had been mearsured consecutively about the yield for a few sunny days in July, 2016. In Fig.7, the bar chart shows the comparison of the yield per day between the two PV arrays and the line chart suggests the loss rate of yield in the PV array damaged by PID under various top ambient temperature conditions. From Fig.7, we find that the yield per day of the PV array affected by PID is less than that in common PV array. And this loss rate is more severe with the ambient temperature increasing.

Fig.8 shows the difference of yield per day between the two PV arrays and the loss rate of yield in the PV array affected by PID in different weather conditions: sunny day, cloudy day and shower day. And the three kinds of weather are corresponding to different relative humidity (RH): 20-40%, 40-60% and 60-90%. The three days have the same top ambient temperature: about 30°C. Similarly, the yield per day of the PV array affected by PID is less than that in common PV array. And the loss rate is becoming severe with the relative humidity increasing.

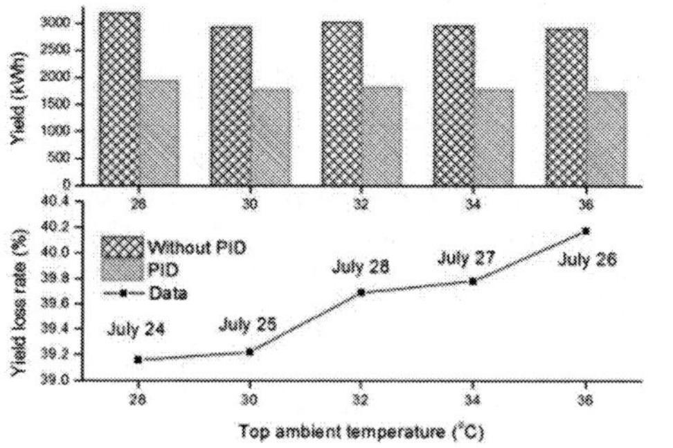

Fig.7 Comparison of yield per day between the two 500 kWp PV arrays (bar chart) and the loss rate of yield in the PV array damaged by PID (line chart) under various top ambient temperatures conditions for a few sunny days in July, 2016.

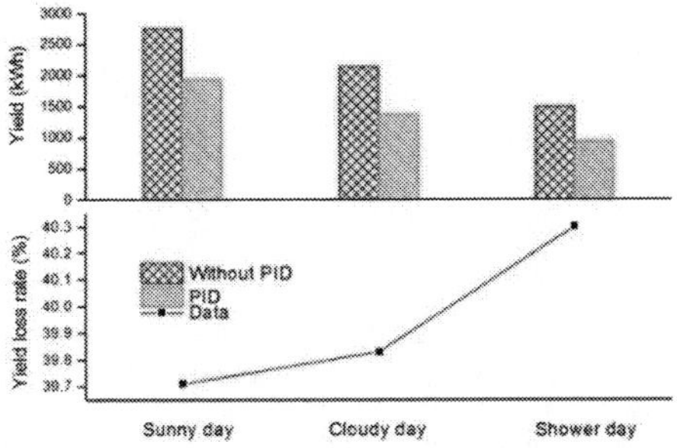

Fig.8. Difference of yield per day between the two 500 kWp PV arrays (bar chart) and the loss rate of yield in the PV array affected by PID (line chart) in sunny day, cloudy day and shower day in July characterized by different relative humidity.

4 CONCLUSIONS

The modules with power loss led by PID totally locate at latter half of the grounded module string. PID decreases the energy conversion efficiency of solar modules significantly and hazards the reliability of photovoltaic system. And this loss is becoming more serious while the module is more close to the end of grounded module string because the corresponding potential of cells in terms of metallic frame is increasing negatively. From a first visual comparison of thermographic images and EL images, there is a good agreement between the two images that seem to exhibit a similar degradation pattern. So PID mainly occurs on the surrounding areas of grounded modules. According to this degradation pattern, we propose that the possible cause which results in PID within the surrounding solar cells of grounded modules is electricity field concentration on surrounding areas of modules applying by metallic frame when cells are exposed to high minus system voltage stress. The end module in grounded module string has the worst impact by PID up to 53% power loss. The yield loss of PV array by PID is approximately up to 40%. High temperature and high relative humidity have proved to aggravate PID and further deteriorate the energy conversion efficiency of PV system.

ACKNOWLEDGEMENTS

The authors would like to thank the support of the National Key Technology R&D Program (Grant No. 2015BAA09B01) and 2016 Grid state science and technology program (NY71-16-027). This study was also supported by the Natural Science Foundation of China (Grant No.61274050 and 61376067).

REFERENCES

[1] A. Ndiaye, et al. Degradations of Silicon Photovoltaic Modules: A Literature Review. Sol. Energy 2013; 96: 140–151.

[2] Hong Yang , et al. Performance deterioration of p-type single crystalline silicon solar modules affected by potential induced degradation in photovoltaic power plant . Microelectronics Reliability, 2017, 72(1)：18-23.

[3] E. Dunlop, D. Halton. The Performance of Crystalline Silicon Photovoltaic Solar Modules after 22 Years of Continuous Outdoor Exposure. Prog. Photovolt: Res. Appl 2006;14: 53–64.

[4] S. Pingel, O. Frank, M. Winkler, S. Oaryan, T. Geipel, H. Hoehne and J. Berghold. Potential induced degradation of solar cells and panels. In Proceedings of the 35th IEEE PVSC, Honolulu, HI, USA, 2010; 2817–22.

[5] M. Schütze, M. Junghänel, M.B. Koentopp, S. Cwikla, S. Friedrich, J.W. Müller, and P. Wawer. Laboratory study of potential induced degradation of silicon photovoltaic modules. In Proceedings of the 37th IEEE PVSC, Seattle 2011; 821– 26.

[6] J. A. del Cueto and T.J.McMahon. Analysis of leakage currents in photovoltaic modules under high-voltage bias in the field. Prog. Photovolt: Res. Appl. 2002; 10:15-18.

[7] R. Swanson, et al. The surface polarization effect in high efficiency silicon solar cells. In Proceedings of the 15th IEEE PVSC Kissimmee, FL, USA 2005; 410–411

[8] M. Schütze, et al. Laboratory study of potential induced degradation of silicon photovoltaic modules. In Proceedings of the 37th IEEE PVSC, Seattle, 2011; 821–826.

[9] M. Koentopp, M. Krober and C. Taubitz. Toward a PID Test Standard: Understanding and Modeling of Laboratory Tests and Field Progression. IEEE Journal of Photovoltaics 2016; 6: 252-257.

[10] J. Bauer, et al. On the mechanism of potential-induced degradation in crystalline silicon solar cells. Physica Status Solidi RRL 2012; 6 (8): 331–33.

[11] V. Naumann, et al. The role of stacking faults for the formation of shunts during potential-induced degradation of crystalline Si solar cells. Physica Status Solidi RRL 2013; 7: 315–318.

[12] V. Naumann, et al. Explanation of potential-induced degradation of the shunting type by Na decoration of stacking faults in Si solar cells. Sol. Energy Mater. Sol. Cells 2014; 120: 383–89.

[13] P. Hacke, M. Kempe, K. Terwilliger, S. Glick, N. Call, S. Johnston, and S. Kurtz. Characterization of Multicrystalline Silicon Modules with System Bias Voltage Applied in Damp Heat. In Proceedings of the 25th EUPVSEC, Valencia, Spain 2010; 3760–65.

[14] C. Buerhop-Lutz, et al. Quality control of PV-modules in the field using a remote-controlled drone with an infrared camera. In Proceedings of the 27th EUPVSEC Frankfurt, Germany 2012; 3370–3373.

[15] T. Kaden, K. Lammers, H. Möller, Power loss prognosis from thermographic images of PID affected silicon solar modules, Sol. Energy Mater. Sol. Cells 2015; 142: 24–28.

QUANTITATIVE STUDY OF POTENTIAL INDUCED DEGRADATION OF A ROOF-TOP PV-INSTALLATION WITH IR-IMAGING

Claudia Buerhop[1], Tobias Pickel[1], Frank W. Fecher[1], Cornelia Zetzmann[2], Jens Hauch[1], Christian Camus[1], Christoph J. Brabec[1,3]
[1]ZAE Bayern e.V., D-91058 Erlangen, Immerwahrstraße 2,
[2]Rauschert, D-96332 Pressig, Bahnhofstraße 1,
[3]i-MEET, FAU Erlangen-Nürnberg, 91058 Erlangen, Martensstraße 7

ABSTRACT: Potential induced degradation (PID) causes severe damage and financial losses even in modern PV-installations. In Germany, approximately 19% of PV-installations suffer from PID and resulting power loss. This paper focuses on the impact of PID in real installations. The analysis focuses exemplarily on a 314 kWp PV-system in the Atlantic coastal climate. Infrared (IR)-imaging is used for identifying PID without operation interruption. Historic electric performance monitoring data for several years are analyzed on system and string level. Additionally, punctually measured module string current-voltage (I-V)-curves are considered. The data sets are combined for understanding the PID behavior of this PV plant. The number of PID affected cells within a string varies strongly between 1 to 22% with the string position on the building complex. A PID-resistance which correlates to the defined PID-defect-ration is presented. Finally, indicators for the reliable determination of PID is given.
Keywords: potential induced degradation (PID), IR-imaging, PID criteria, power loss, PID-resistance

1 INTRODUCTION

Potential induced degradation (PID) of photovoltaic (PV) modules has been a defect type becoming more and more relevant during the last years. PID causes severe damage and financial losses even in modern PV-installations. In Germany, approximately 19% of PV-installations suffer from PID [1]. This paper focuses on the understanding of PID generation in real installations.

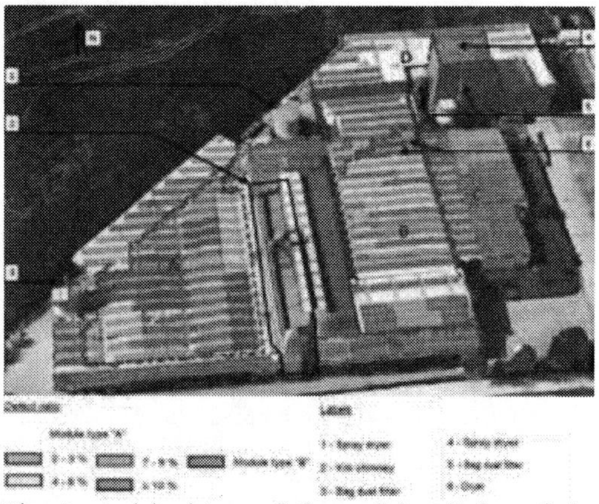

Fig. 1: Aerial picture of the roof installation of the investigated PV-plant on an industrial building. Positions of outlets emitting heat or dust are marked by the numbers 1 to 6. A, B, C, and D define different types of roof areas. The color code indicates the defect ratio of the strings [2]. (Imagery © 2017 Google, Map data © 2017 Google)

In this paper, we investigate a 314 kWp PV-plant on an industrial roof-top installation in Portugal exhibiting severe PID damage. Evaluation of the monitoring data of this PV-plant was challenging due to partial shading by various roof installations (chimneys and other roof outlets) resulting in inhomogeneously irradiated modules within the module string. Additionally, the diefferent roof areas exhibited varying orientation and strings were connected in a complex manner. (cf. fig 1). The annual energy yield for the last three years of operation was rather constant and gave no indication for accelerated degradation of the PV-plant [3]. Nonetheless, the monitoring data revealed degradation upon detailed analysis.

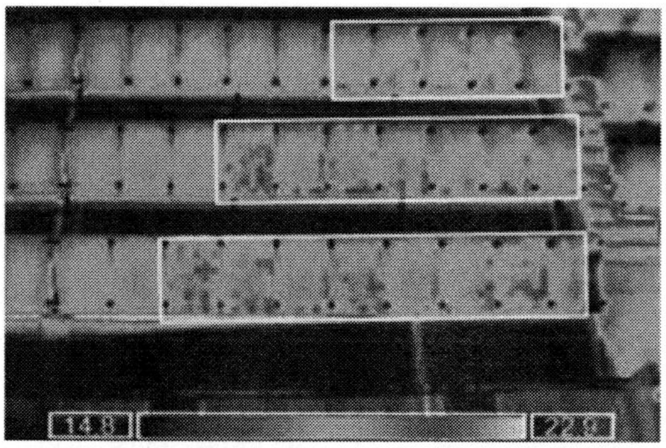

Fig. 2: aIR-image of the roof-top installation showing many PID-affected modules marked with frames

Aerial infrared (aIR) inspection has the property that it may be used for the direct visualization and localization of the PID-degraded modules [4-7], see fig. 2.

Besides monitoring data of the maximum power point (MPP) at string level and IR-imaging, electro-luminescence (EL) imaging, current-voltage (*I-V*) measurements (on module and module string level) and weather data completed our investigations. By these data, the performance ratios of the PV-plant and the module strings were evaluated and the distribution of PID-affected modules within a string as well as the locations of PID-cells within modules were analyzed.

Previous results [2, 3, 6] published on this PV plant are extended here by introducing a new parameter for characterizing PID.

2 EXPERIMENTAL PROCEDURE

A 314 kWp PV-plant installed in 2013 and located in the Atlantic coastal climate was examined. It consists of 1280 poly-crystalline PV-modules of P_{nom} = 245 Wp, 60 strings, and 15 inverters. Each string electrically connects 15 to 24 modules. The nominal open circuit and maximum power point (mpp) voltage are V_{oc} = 37.1 V and V_{mpp} = 30.3 V. For a string with 20 modules the voltages are 890.4 V and 727.2 V, respectively. We use the following nomenclature to address certain strings at specific inverters: "I1 S2" refers to string 2 of inverter 1.

The inspections were carried out with a measurement system consisting of an octocopter drone equipped with an IR- and VIS-camera and various sensors. The octocopter is remote-controlled and has a GPS-navigation system, which allows the manipulation of the x, y, z-coordinates. A GoPro Hero3+ is used as visual (VIS)-camera and an Optris PI 450 as IR-camera. The IR-inspection was carried out at suitable weather conditions, i.e. solar irradiance E = 920 - 950 W/m², ambient temperature T_{amb} = 28 - 29°C, and clear sky. This inspection procedure including preparation and set-up of the measurement system lasted approx. 45 minutes.

For the outdoor EL-measurements the system LumiSolarOutdoor Line-Powered from greateyes was used. It consists of a 10 kW power supply and a camera GE 2048 2048 DD NIR with cooled CCD-Sensor. The EL-imaging was performed at the mounted modules. EL-images of three complete strings, i.e. 72 PV-modules, were taken at night within 4 hours.

Additionally, *IV*-curves of all strings (60 strings) and all modules of one inverter #12 (in total 72 modules) were measured. Therefore, the IV-Curve-Tracer PVPM 1000CX with a Silicon reference cell (SOZ-03) was used. For practical purposes, the temperature of one module in the center was measured representatively for all modules using a PT100 temperature sensor. The measurement of the strings took 1 hour, the measurement of the individual modules lasted about 3 hours.

Furthermore, there was one weather station installed at the site. The monitoring data (string power, irradiance, temperatures) since February 2014 were analyzed in detail.

3 RESULTS AND DISCUSSION

3.1 Evaluating IR-images for local phenomena

PID-affected modules can be visualized by their typical temperature pattern using IR-imaging [6, 7]. The affected cells show a slightly increased temperature of 1-3 K with respect to the defect-free neighboring cells. A variety of patterns of differently affected PID-modules are marked in Fig. 3. EL-images show the same patterns as the IR-images [6].

Fig. 3: IR-overview showing several module rows of roof area B; homogeneous temperature distribution indicates defect-free modules whereas inhomogeneous temperature distributions mark defective modules. Typical temperature patterns of PID-affected modules show cells with slightly increased temperatures, E = 900 – 950 W/m², T_{amb} = 27°C, v_{wind} = 2 m/s, I12 S3

IR-imaging is one suitable tool to visualize the actual status of PID. For this PV-plant, the number of PID-affected cells does not increase monotonically with approaching the electrically negative poled string end (n = 21), see Fig. 9. Starting with module position 11, the number of PID-affected cells per module grows. The number does not increase steadily. Scattering to lower numbers is frequently observed. The last module does not necessarily exhibit the highest number of PID-affected cells, compare string 4/2 (module at penultimate position 20 has 47 PID-cells and last cell at position 21 only 5 PID-cells). This behavior is shown for four typical strings in fig. 4. Averaging across 24 strings results a continuous increase of PID-affected cells towards the negative string end, as expected.

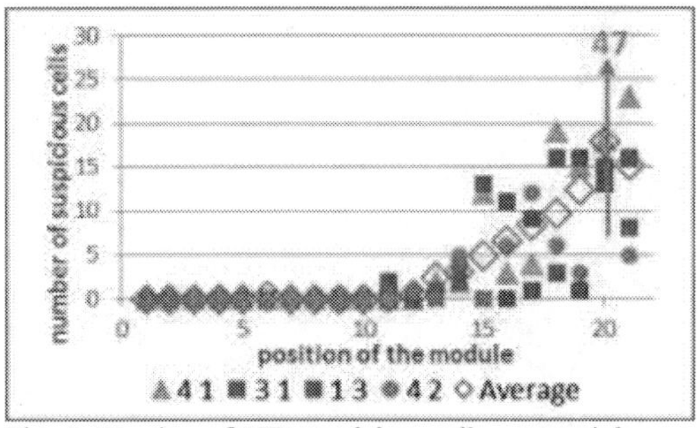

Fig. 4: Number of PID-suspicious cells per module as a function of the module position within a string of 21 modules based on IR-images. (Module 20 of string 4/2 has 47 suspicious cells.)

As indicated by the numbers in the Fig. 3, the modules of one string are not arranged in a straight line. As a consequence, modules close to building edges are cooler than those located in the roof center. This inhomogeneous temperature distribution influences the electrical parameters and PID-evolution locally. Cooler modules at the edges show an apparently higher V_{oc} (see Fig. 5) and high P_{mpp} because of the correction for STC based on the high temperature of the reference module with the sensor.

PID-development at such cooler locations is less pronounced as the findings reveal. On the other hand, PID-affected modules are found more often close to regions with elevated temperatures. Furthermore, it is more likely that the modules have many-PID affected cells.

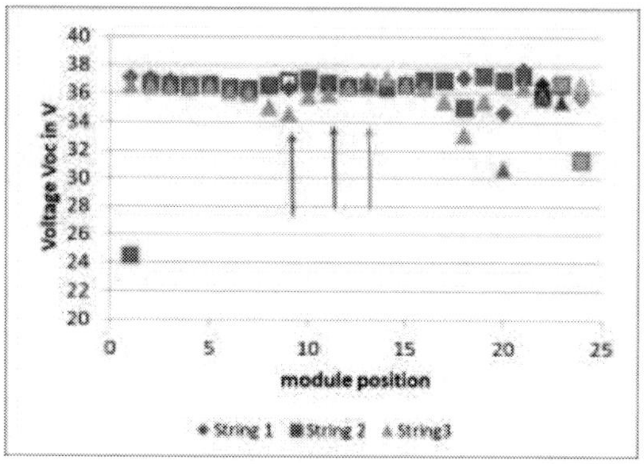

Fig. 5: Measured V_{oc} data corrected to STC for all modules of three strings of one inverter I12. Unfilled symbols indicate modules at edges. Changes of the main colors mark modules with different charge codes, which are predominantly found at the string ends. Module 1 of string 2 has a bypassed substring which causes the reduction of the voltage by one third.

Surprisingly, the voltage and power drop towards the end of the string (higher module position) is not as steadily as expected. Checking the batch codes of the individual modules has revealed, that it varies within one string. In this PV-plant, especially at the string end, modules with differing codes have been identified. An outstanding example is module 19 in Fig. 3, which is less PID-affected than its adjacent modules. Thus, the batch, to which module 19 belongs, seems to be more PID-resistant. This behaviour may be explained by changing material properties, different batches with differing codes and series numbers (however, further evidence is not available/accessible). This variation of module properties is most likely responsible for the scattering of electrical properties and PID-evolution at the modules at the string end.

3.2 Evaluating IR-images for extracting module power

IR-imaging provides an easy method to identify PID-affected cells and to determine a defect ratio or PID ratio k_m, which describes the number of PID-affected cells with an increased temperature within a module m

$$k_m = \frac{n_m}{N_m}$$

with n_m being the number of PID-affected cells within one module and N_m the number of all cells within a module, here $N_m = 60$.

Power measurements on module level for three strings reveal that the module power decreased linearly with increasing number of PID-affected cells in a module. This fact allows a fairly simple estimation of the module power on basis of IR-images [6].

Since the module power decreases linearly with increasing k_m, the resulting power P_{est} can be estimated, as follows

$$P_{est} = P_0(1 - k_m).$$

with the nominal power P_0.

Fig. 6 exhibits the comparison of the measured and the calculated power data for modules of three strings (72 modules). Temperature differences indicating different PID evolution stages of certain cells is not considered in this simple approach. However, the relative deviation between the measured power P_{meas} and P_{est} is roughly 5%. It is obvious that the modules from different batches match this approach according to their number of PID-affected cells, too.

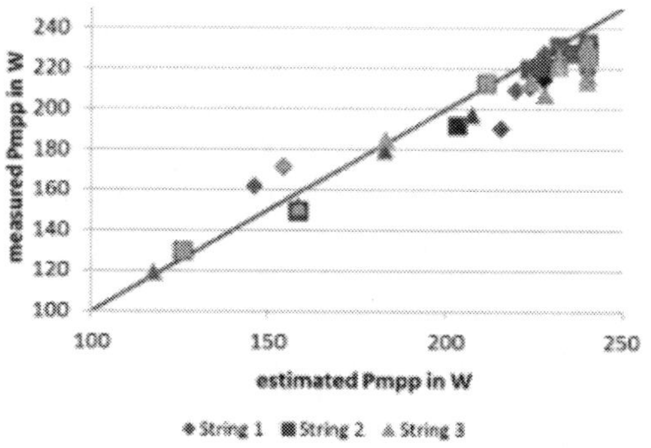

Fig. 6: Comparison of measured and calculated module power of PID-affected modules based on evaluating IR-images

3.3 Analyzing IR-images in combination with monitoring data

However, more interesting is the output of the serial-connected modules in strings. Yield data from the monitoring system are not directly useful in this case because of varying weather conditions.

Therefore, we calculated the performance ratio PR considering the irradiance conditions. Averaging over a certain time period, e. g. one month, a mean PR was calculated and related to the defect ratio or PID ratio k on string-level.

With the monitoring data the electric performance of the PV-plant can be determined for the historical operation period. The highest resolution available is on string level. This allows a space-resolved statement concerning the performance. For comparison of different days with varying solar irradiance the performance ratio PR is an appropriate approach. Then, temporal changes and degradation rates can be recognized and quantified. The performance ratio is determined as:

$$PR = \frac{energy\ production}{expected\ energy} = \frac{\sum \frac{P(t)}{E(t)} \Delta t}{\sum \Delta t} \cdot \frac{E_0}{P_0}$$

With irradiance $E_0 = 1000$ W/m² and $\Delta t = 5$ min., for $P \geq 10$ W and irradiance $E \geq 60$ W/m².

Furthermore, the defect ratio on string level k is determined basing on IR-images

$$k = \frac{n}{N}$$

with n being the number of PID-affected cells in one string and N the number of all cells within a string.

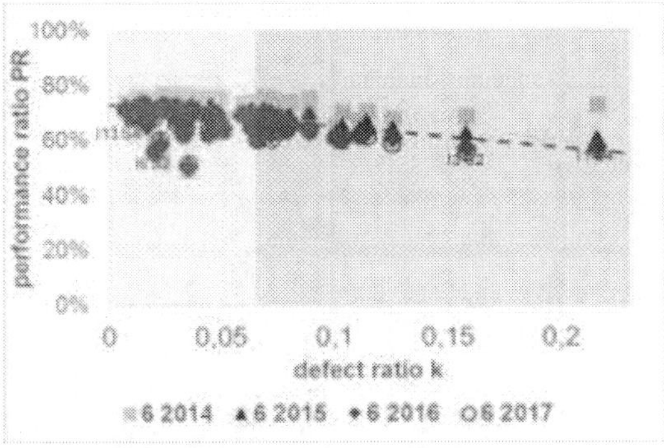

Fig. 7: Performance ratio *PR* versus PID-ratio *k* averaged for one month, here June, for various years. *k* bases on the data of June 2016.

PR decreases linearly with increasing PID-ratio, see Fig. 7. Monitoring data as well as IV-measurements of the module strings reveal this relationship [2]. Defect ratios up to 23% are observed. This high value is calculated for a module string installed on top of the spray dryer (marked 4 in fig. 1) where temperatures of 10 K higher than at other roof locations are frequently present. Defect ratios up to 10% are quite often detected. Their distribution across the roof-top installation is visualized in

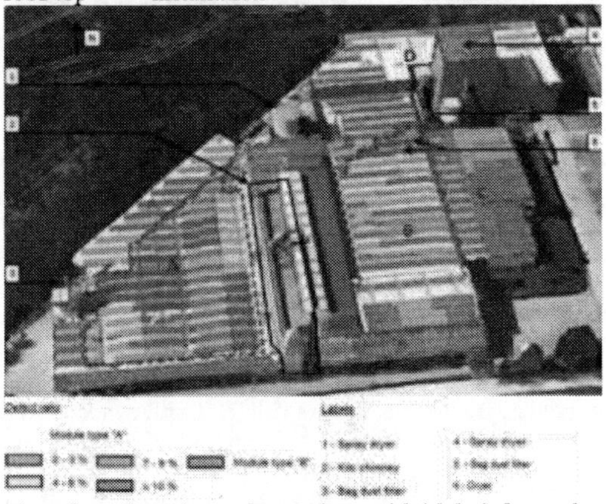

Fig. 1fig. 1 by color codes. Strings with high defect ratios are mostly located near heat and dust emitting as well as shading installations.

The reduction of *PR* during operation time is shown in Fig. 7. Although *k* is not known for past years (before the study started), the continuous loss of *PR* is obvious. The average *PR* loss rate for the investigated PV-plant is determined as about 3.65% per year [2] assuming a linear behaviour. However, fig.7 depicts clearly that the PR-loss between in the early years, between 2014 and 2015, was much larger than between the following years. Especially, for high defect ratios this temporal evolution is obvious.

3.4 Monitoring data analysis for calculating a (characteristic) PID resistance value

Monitoring data give evidence for electrical data of the module strings and the inverters. However, evidence for the appearance and the stage of development of PID-affected cells is not easily accessible.

Considering only data accessible by a monitoring system a characteristic parameter for PID-evolution should be defined. A rather simple approach calculating the conductance did not give a correlation to the defect ratio, as suggested by [8]. Therefore, a PID-specific resistance is calculated. Analyzing the IV-curves of PID-affected modules of different evolution state shows that the fill factor decreases, V_{oc} decreases, parallel and series resistances change, see Fig. 8. With the defect ratio the mpp changes as well as the calculated PID-resistance *R*(PID) [9]. With increasing *k* the resistance decreases.

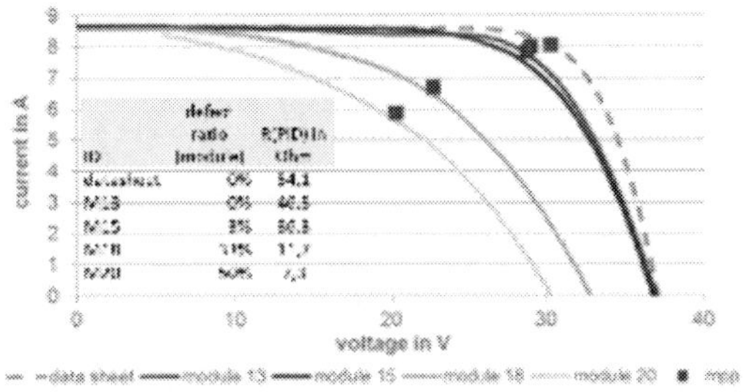

Fig. 8: Measured IV-curves for modules with different PID-ratio of string I12S3. With an increasing number the defects the PID-ratio increases as well.

Fig. 9 exhibits *R*(PID) versus PID-affected cells for all modules of three strings. From the 24 modules per string between 7 and 10 modules show typical PID features in IR-imaging. The correlation of *R*(PID) with respect to the PID-affected cells can be described by an exponential function. With increasing PID-evolution the resistance decreases continuously and approaches zero, if all cells will be affected. One module of string 2 with 20 affected cells, in Fig. 9 marked by a frame, has no PID but a bypassed substring.

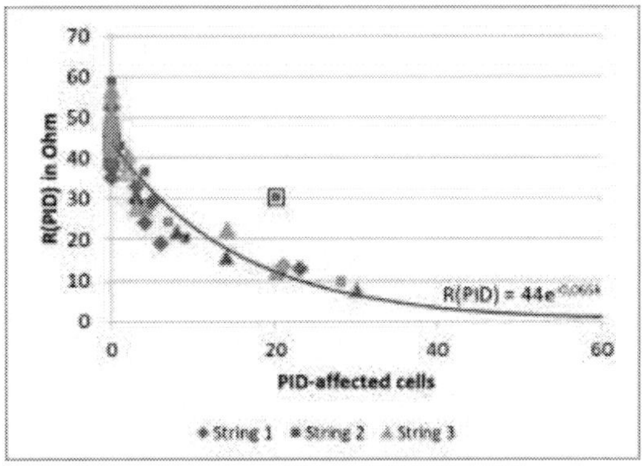

Fig. 9: Measured IV-curves for modules of three string of inverter 12 with different PID-ratios or different number of PID-affected cells. 60 PID-affected cells correspond to $k = 4.76\%$. The frame mark the module with the bypassed substring.

Transferring this knowledge to the analysis of string data yields Fig. 10. The exponential correlation between *R*(PID) and *k* can be observed. The *R*(PID) string values for low PID ratios *k* are higher than for modules. This may be due to a maximum power point of the string, which does not match the mpp current and voltage of individual modules. Further investigations will be done to identify further influencing parameters. The embedded monitoring data of four strings of one inverter fit well into the chart.

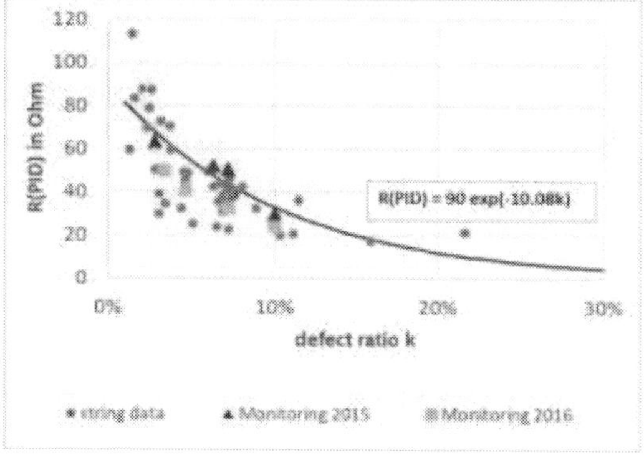

Fig. 10: Calculated string PID-resistance for measured data as well as for selected monitoring data versus their defect ratio *k*. The data are fitted using an exponential function.

3.5 Identifying PID-typical patterns using IR-imaging

The investigations of several PV-plants showed that a cell temperature increase of 1-2 K of certain cells is not a sufficient criterion. Fig. 12 exhibits an interesting example, where the IR-images suggest PID at the first glance. Therefore, IR- and EL-imaging as well as IV-curves of these modules were measured in the lab. Fig. 12 shows that local heat sources are visible at the busbars of many cells in the IR-image. The corresponding cells in the EL-image appear black because no current flows through the cell. The IV-curves are comparable to the ones of PID-affected modules. This defect is probably caused by soldering of the busbars. With the appearance similar to PID we call this „*pseudo*-PID".

In Fig. 12 the PID-distribution pattern of the cells within the modules of the PID-affected PV-plant is illustrated. For Fig. 12 143 PID-affected modules were evaluated using IR-images, which represents about 37% of the PID-modules and 11% of all modules in the PV-plant. Most affected cells concentrate at the module edges, particularly at the lower frame edge. In Columns 1 and 6 high average PID-rates of 19% and 16% are counted, whereas in the center columns lower values between 11% and 15% are found. More pronounced is the high clustering at the lower edge (row J) an average of 35% for PID-cell appearance. Within the rows A to I 10% to 15% are measured. This distribution differs strongly from failure patterns due to mechanical loading, e. g. transport and rough handling [10] as well as short-circuited cells fig 11.(right). Then, most fractured cells would be observed more in the center due to high bending radius.

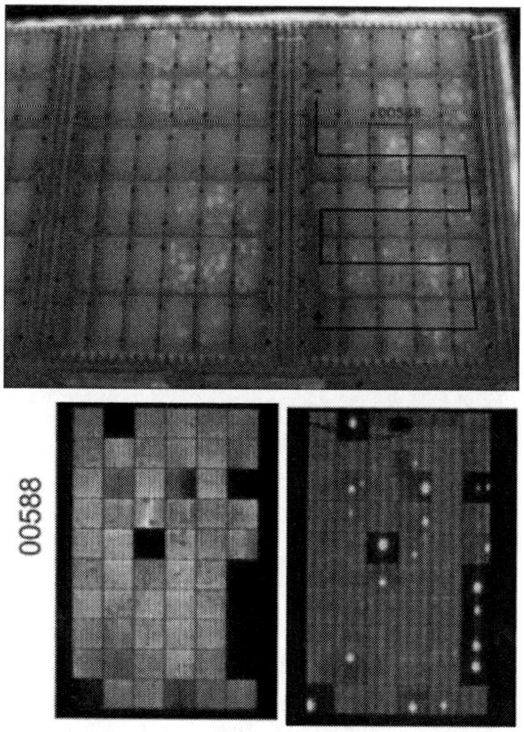

Fig. 11: "Pseudo-PID"-pattern visible in aIR-image at operating conditions without being PID (top), IR-image from the rear side at electrical bias of module 00588 showing localized heat sources at the busbar typical for soldering failure (bottom, right), EL-image (bottom, left).

According to this experience certain indicators can be formulated for PID: 1) PID-affected modules concentrate at one string end and are not distributed arbitrarily along the string, 2) PID is not detectable shortly after commissioning, 3) PID does not affect 60% of the modules, reasonable are less than half of the modules, 4) there should be a trend that the number of affected cells increases towards one string end, and 5) within a module the distribution of PID-affected cells is higher close to the frame, see fig 11.

	1	2	3	4	5	6
A	10%	14%	13%	10%	9%	3%
B	11%	10%	11%	6%	11%	13%
C	16%	8%	11%	12%	9%	15%
D	17%	6%	8%	13%	8%	21%
E	17%	7%	10%	9%	6%	17%
F	17%	3%	8%	12%	9%	15%
G	21%	7%	11%	10%	10%	20%
H	26%	8%	11%	9%	10%	15%
I	21%	10%	17%	13%	15%	13%
J	38%	34%	45%	33%	34%	30%

	1	2	3	4	5	6
A	5%	5%	0%	5%	0%	14%
B	0%	5%	14%	14%	14%	19%
C	5%	38%	5%	14%	0%	5%
D	14%	14%	5%	14%	5%	24%
E	29%	14%	10%	10%	19%	10%
F	5%	43%	24%	24%	19%	10%
G	24%	10%	10%	19%	5%	14%
H	14%	19%	0%	38%	10%	14%
I	0%	14%	19%	5%	29%	14%
J	33%	5%	10%	19%	43%	0%

Fig. 12: Average occurrence of a slightly heated cell in PV-modules based on IR-images, left: 143 PID-affected modules, right: 20 "pseudo-PID" modules with short-circuited cells

4 CONCLUSION

Fast aIR-imaging provides information for evaluating PID-affected PV-installations directly and quantitatively in terms of module power and performance ratio. More interestingly, the degraded module strings are not distributed homogeneously over the PV-plant, but are clustered. The reason for this clustering seems to be caused by different "micro-climates" (in terms of local operating conditions) on the roof which were localized in as well as different fabrication batches. Considering as well monitoring data the local development of PID can be predicted for installed modules for the first time. The distribution of PID-modules and PID-cells by IR-imaging give further hints for the formation of PID.

5 ACKNOWLEDGEMENT

We gratefully thank the German Federal Ministry for Economic Affairs and Energy (BMWi) for financial funding of this project.

6 REFERENCES

[1] M. Fuhs; PV magazine März (2016) 52.
[2] C. Buerhop, et al.; IET Renewable Power Generation (2017).
[3] C. Buerhop-Lutz, et al., SPIE 10370 Optics and Photonics, Optics + Photonics for Sustainable Energy (2017).
[4] J. A. Tsanakas, et al.; Renewable and Sustainable Energy Reviews 62 (2016) 695.
[5] C. Buerhop, et al., 32nd EU-PVSEC (2016) 2021.
[6] C. Buerhop, et al., SPIE Optics and Photonics, Optics + Photonics for Sustainable Energy (2016) 99380J.
[7] T. Kaden, et al., 29th EU-PVSEC (2014) 2994.
[8] T. Kilper, et al., 30th EU-PVSEC (2015) 1679.
[9] C. Buerhop-Lutz, et al.; in preparation.
[10] S. Koch, et al., 32nd EU PVSEC (2016) 1736 .

SCIENTIFIC INVESTIGATION OF A PV GENERATOR AFTER HAIL

W. Mühleisen (1), C. Hirschl (1), G. C. Eder (2), Y. Voronko (2),
M. Spielberger (3), H. Sonnleitner (4)
(1) CTR Carinthian Tech Research AG, Europastr.12, 9524 Villach/St.Magdalen, Austria
(Author and Contact: wolfgang.muehleisen@ctr.at; T: +43 4242 56300 238)
(2) OFI Austrian Research Institute for Chemistry and Technology, 1030 Vienna, Austria, (Gabriele.Eder@ofi.at)
(3) PVSV e.U., 9334 Guttaring, Austria, (markus.spielberger@pvsv.at)
(4) ENcome Energy Performance GmbH, 9020 Klagenfurt, Austria, (Horst.Sonnleitner@en-come.com)

ABSTRACT: A power plant in the south of Austria built with multi-crystalline PV modules was investigated after a hail storm. Standard measurement techniques like thermography, electroluminescence and ultraviolet fluorescence were used. It was found out that glass broken modules were not only responsible for a loss in power and yield. The remaining good-looking modules had invisible to the eyes cell damages causing the loss. Because of these impressive findings, the insurance decided subsequently to change the most affected bad working modules.
Keywords: PV Module, Inspection, Operation and Maintenance,

1 INTRODUCTION

The increase of weather disasters in the last years enhances the importance and relevance of a good operation and maintenance strategy for PV power plants. As an example, a PV installation in the south of Austria was investigated after it had been exposed to hail impact. This 20 kWp plant consisted of four separate generators of which two were orientated to the south, one to the east and one to the west. Due to the hail, 4 of the 80 crystalline standard modules, installed on a flat roof under an angle of 12 degrees, were affected by broken glass and hence needed to be replaced. However, even after this module change the monitored power plant showed a yield loss. This fact lead to a scientific investigation of the separate strings (Fig. 1) by using thermography (TG), outdoor electroluminescence (EL) and ultraviolet fluorescence (UV-F).

Figure 1: Overview of the plant

The complete analysis showed that there is a need to reconsider maintenance strategies and insurance cases. This is necessary due to the fact that in the described case only PV modules with broken glass (Fig. 2) have been replaced without considering possible damages of adjoining modules.

The impact of the hail on the strings mostly depended on its angle of incidence. To find a damaged string, a PV plant is usually controlled visually, by using the monitoring portal or by a service team with a digital handheld multimeter. Only in very rare cases more scientific inspection tools like thermography or electroluminescence are used.

Figure 2: Visually broken PV module due to hail

2 MEASUREMENTS AND RESULTS

In the beginning, the plant owner provided the monitoring data from the inverter web portal (Fig. 3). The month overview for July 2015 shows two circumstances that have changed from before to after the hail event. The first discrepancy is the changed gap in energy yield for inverter 1 (south plant) to inverter 2 (east-west plant). The second thing is that the energy yield level dropped after the hail event from 60 kWh to 45 kWh. Even after the module change because of broken module glasses, no improvement is displayed in the monitoring data.

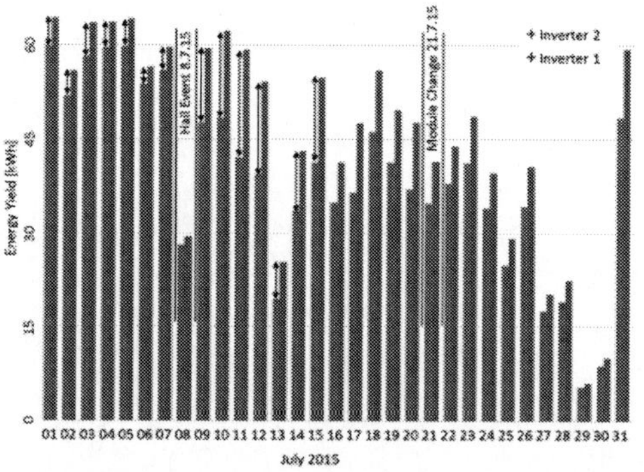

Figure 3: Web portal data of the plant with two inverters

This specific case shows that a plant owner/operator is helpless without using innovative and scientific relevant technologies to find the error.

2.1 *Visual inspection*

The visual inspection, which is important to find broken, shaded or discolored modules, showed only good-looking panels (Figs. 4, 5). The performed visual inspection methods are based on existing procedure sheets [3].

Figure 4: Picture of the west generator

Figure 5: 2nd Picture of the west generator

2.2 *Thermography inspection*

The first choice for many PV plant inspectors is thermography (Figs. 6, 7) because failures are often indicated by inhomogeneous heat distributions [1, 2].

Figure 6: Thermography image of the west generator

Figure 7: 2nd Thermography image of the west generator

In the Figures 6 and 7 a temperature span of 15 K was used to check the plant for discrepancies. The inspection with the thermography camera showed no significant damage, which would be indicated by relatively hotter or colder cells. The heat distributions look normal at a first glance, apart from some small differences at one connection box and for separate solar cells in the modules. But even the minor details are in the focus and therefore further inspection was done by using twilight EL and UV-F.

2.3 *Electroluminescence inspection*

Electroluminescence is a standard inspection technology which is mostly used in combination with a power supply [4, 5, 6]. Fed in current in forward direction into a PV module or a string, transforms the PV modules to big light emitting diodes. Broken solar cells can be seen with the EL camera as black areas in the picture (Fig. 8).

Figure 8: EL image of the west generator

Good working solar cells shine homogeneous. This is the case for the changed module which is now at the position instead of a glass broken one (Fig. 9). All around the healthy panel, there are the remaining cell cracked modules which were not changed because of the non-broken glass.

Figure 9: 2nd EL-image of the west generator

2.4 *Ultraviolet-fluorescence inspection*

The fluorescence method is the latest used one in PV plant inspection. Due to the fact that the encapsulation material is fluorescent, UV LED lamps (365 nm) in combination with a visual light filter can be used at twilight conditions or with a tent [7, 8].

In Fig. 10, there is an UV-F image of a hail damaged module without glass breakage. There are five cells with a milky to cloudy appearance over the whole cell area. These cells are in good condition. But many cells show dark tracks and a spider web structure. This is the indication for many cell cracks due to the hail.

Figure 10: UV-F image of the west generator

In opposition, Fig. 11 shows a new panel on the plant which has been replaced since the hail storm. The fluorescence is well advanced and the panel makes a healthy impression because of no dark tracks.

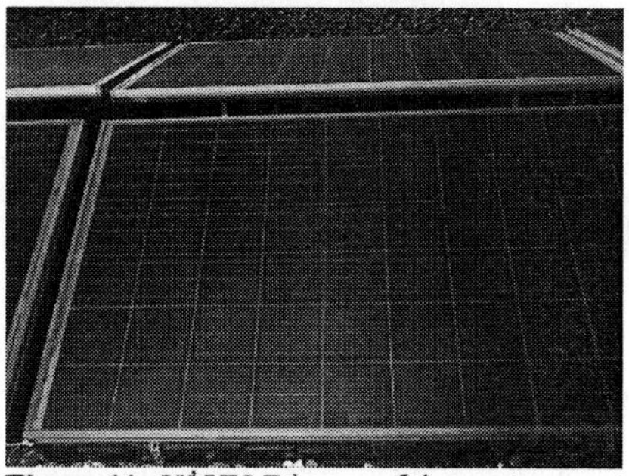

Figure 11: 2nd UV-F image of the west generator

2.5 *Current-Voltage inspection*

Two panels out of the most affected west generator were disassembled and measured again at the manufacturer. It turned out that both modules had a power loss of 35 % and 29 %.

5 SUMMARY AND CONCLUSION

By using the adapted standard laboratory equipment, electroluminescence and UV fluorescence outdoors, new findings and explanations of the condition of a hail affected plant in the south of Austria could be made.
Both methods showed many defects like solar cells with many fragments and micro cracks, which were not easily found by thermography. A great help was the web based monitoring portal indicating a power loss after the hail storm without an improvement even after the change of glass broken modules.
The visual inspection resulted in three glass-broken modules for the west generator and one broken module for the south generator. After the replacement of the visually damaged modules, the monitoring portal still showed a yield loss compared to the yield before the hail. The I-V-characteristic curve of two random modules showed a power loss in the range of about 30 % for the hail affected modules. A similar loss in the region of one quarter of the whole installed power was found by doing evaluation in the monitoring portal.
It can be concluded that without the usage of laboratory inspection tools like electroluminescence and ultraviolet-fluorescence, PV strings, damaged by hail, would have remained undetected with standard installer methods. This implies that PV plant owners suffer unexpected yield losses if only glass-broken modules are changed. In this example only four modules were changed even though 68 modules would have to be replaced. Thus, it is better to do a check for PV generators after a hail storm. In the case of found damages relating to hail, measurements are necessary to have enough documentary evidence claiming at the insurance.

6 ACKNOWLEDGEMENTS

This work was conducted as part of the Austrian "Energy Research Program" project INFINITY, using results from the predecessor project InnoModu. Funded by the Austrian Climate and Energy Fund and the Austrian Research Promotion Agency (FFG), are both gratefully acknowledged.

7 REFERENCES

[1] Auer, R. ,Jahn, U.,Buerhop-Lutz, C. ,Vordermayer ,C. ,Wotruba ,G. ,Zehner ,M. & Niess ,M., "Infrared analysis of PV Modules for Improving Quality", 22nd PVSEC 2007 – Milano, 2519-2522.

[2] Buerhop, C., Jahn, U., Hoyer, U., Lerche, B. & Wittmann,S. (2007), „Überprüfung der Qualität von Photovoltaik-Modulen mittels Infrarot-Aufnahmen, Korrelation von Infrarot-Messung und elektrischer Messung.", ZAE Bayern e.V. http://www.sev-bayern.de/content/downloads/IR-Handbuch.pdf

[3] Packard, C; Wohlgemuth, J; Kurtz, S; "Development of a Visual Inspection Data Collection Tool for Evaluation of Fielded PV Module Condition", Technical Report NREL/TP-5200-56154 August 2012, https://www.nrel.gov/docs/fy12osti/56154.pdf

[4] Mertens, K.; Stegemann, T.; Stöppel, T.: LowCost EL -Erstellung von Elektrolumineszenzbildern mit einer modifizierten Standard - Spiegelreflexkamera, 27. Symp. Photovoltaische Solarenergie, 2012

[5] Mertens, K., Pascual Gonzales, D., Diehl, M.: „LowCost – Outdoor - EL: Kostengünstige umfassende Vorort - Qualitätsanalyse von Solarmodulen", 30. Symp. Photovoltaische Solarenergie, 2015

[6] Mertens, K., Kösters, H., Diehl, M: „Low-Cost-Outdoor-EL: Cost-Efficient Extensive on-Site Quality Analysis of Solar Modules", Proceedings of 31st EU PVSEC, Hamburg, 2015

[7] Schlothauer JC, Jungwirth S, Röder B, Köhl M: Fluorescence imaging a powerful tool for the investigation of polymer degradation in PV-Modules. Photovoltaics International 2010: 149–54.

[8] Köntges M, Kajari-Schröder S, Kunze I: Cells Cracks Measured by UV Fluorescence in the Field. 27th EUPVSEC 2012; 2012: 3033–40.

The development of cell thickness reduction of crystalline solar cells in PV modules and its impacts on large PV power plants

Edwin Cunow
LSPV Consulting, Am Weidegrund 10, D-82194 Groebenzell, Germany
Phone: +49 (0)8142 6 52 59 00, e-mail: info@LSPV-Consulting.com

ABSTRACT

In last decades PV modules on basis of silicon solar cells have made an enormous progress regarding efficiency and production technology as well as in cost reduction. That can be deduced i.e. from world-wide market development and dissemination. This paper is calling to mind the development of wafer and solar cell thickness from 1980 up to today and provides an outlook to 2020 expectations. Further practical impacts are highlighted, for instance what a further increase of thickness of wafers and cells in production up to appliances in the field may have. Finally resulting effects on large scale ground mounted power plants are discussed. In contrary to PV roof-top installations PV modules for ground-mounted plants are experiencing very often different routs of transportation and installation experience linked with higher mechanical loads. Taken together it can be a challenge in regard of investor's expectation on quality. A list of effects due to decreasing cell thickness and a guideline for a reactive behaviour is given.
Keywords: c-Si PV Modules, Large grid-connected PV Systems, Performance, Quality

1 INTRODUCTION

A report on status and development of very large scale PV power plant was e.g. published by Keiichi Komoto et.al. in "Energy from the Desert – state of the art and into future" [1]. Understandable that the huge number the of applied crystalline silicon PV modules in multi Megawatt power plants having modules with even thinner solar cells require a particular attention, since an increase of module failure rate may result in disproportional high economic losses. An overview of common failure causes of PV modules has been given in "Review of Failures of Photovoltaic Modules" [2].

The present contribution is based on own evaluations and experience of a lot of onsite-inspections of PV plants. The presented results are based on inspection in the course of TDD services 2013 up to 2016 and on inspected 53 plants in the range of 1 MWp up to 6 MWp. Remarkable numbers of appearances of cell abnormalities have been inspiring and have given motivation for the production of this paper.

2. DEVELOPMENT OF WAFER AND SOLAR CELL THICKNESS BASED ON CRYSTALLINE SILICON

The real development of dimensions and resulting power of solar cells in the period 1980 to 2016 shows a steady increase of cell dimensions but at the same time also a reduction of wafer and cell thickness, as listed in Tab. 1. Together with an increase of power, efficiencies increased too (figures below are from [3]).

Year (ca.)	active cell area [cm²]	Raw cell / wafer dimensions / Ingot	Raw cell thickness (ca. µm)	Power / cell (ca. Wp)
1980	78,5	ø 100 mm Wafer	450	0.80
1987	98	quasi 100x100 mm², 5" Ingot	350	1.40
1987	104	quasi 103x103 mm², 5.3" Ingot	350	1.50
1994	155	quasi 125x125 mm², 6" Ingot	280	2.55
1998	177	6" Ingot	250	3.10
2006	220	quasi 150x150 mm², 8" Ingot	200	4.10
2011	241	156x156 mm², 8" Ingot	180	4.60
2016	241	156x156 mm², 8" Ingot	180 - 170	5.00

Tab. 1: Wafer / cell data regarding dimensions, area, thickness and power versus application year

The graph depicted in figure 1 below obviously exemplifies a steady reduction of cell thickness (red curve) in spite of opposing steady increase of cell area (green curve). In the same period, the cell efficiency (violet curve) doubled and due to threefold of cell area the cell power sixfold (blue curve).

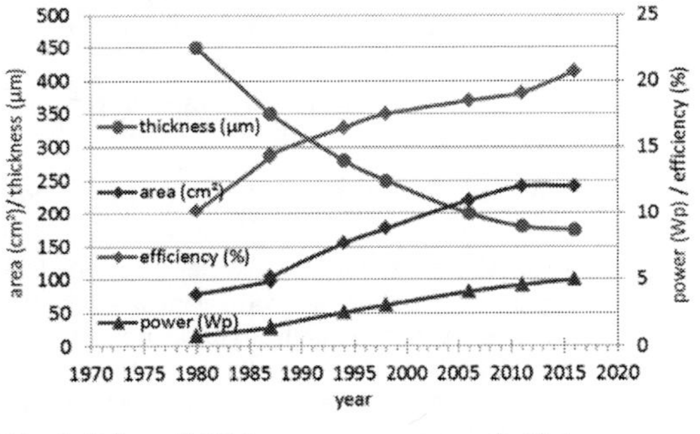

Fig. 1: Solar cell thickness, area, power and efficiency versus year of production

Up to 1992 wafer thickness of mono-crystalline silicon solar cells was about 320 µm and cell thickness after etching was about 270 µm. At this time micro flaws did not appear and micro cracks were almost unknown. Justified customer's complained files of this kind were unknown, also snail trails. Files were normally limited to delaminations and defect bypass diodes (mostly Shottky diodes).

2.1 Increase of sensitiveness in production

Micro cracks in silicon solar cells of solar modules have been remarkably noticed up from 2005 as the wafer thickness was reduced to around 200 µm and below and the cell thickness dropped down to about 180 µm and less. Up to then the cell-interconnection technology was performed by approx. 2 mm wide and up to 150 µm thick tin coated copper ribbons (cross section 0.3 mm²). Micro cracks, when identified, often appeared at solder joints. This was a reason for considering ribbons to become more narrow or thinner, ending in lower cross sections.

For future module technology with thin cell thickness the cell-connection technology is estimated be the decisive factor for preventing or minimising cracks and micro cracks.

2.2 Prognosis and Trends in cell manufacturing

An example of forecasted wafer thickness was published in the ITRPV Report 2016 [4]. There a gradual but steady wafer thickness reduction of 5 – 10 µm per year is assumed. But for the "limit of cell thickness of future module technology" a precipitous decline up from 2016 to 2018 is expected. Such a reduction of applicable cell dimensions to thinner cells is presuming a further advanced interconnection and module technology and should therefore merely be understood as bottom line of cell thickness development.

According to ITRPV Report 2016 [4] the current trend of wafer thickness looks as follows:

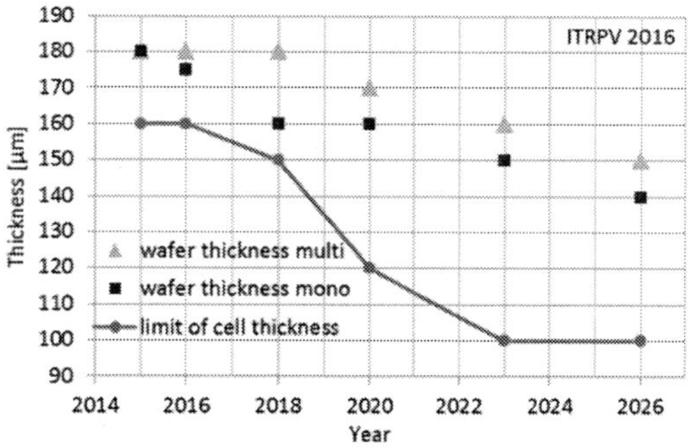

Fig. 2: Trend of wafer thickness according to ITRPV Report 2016 [4]

In the ITRPV Report 2016 [5] following comment is mentioned:

"It is important to note that the up- and-coming interconnection technologies will need to be compatible with the ever-thinner wafers that will be used in the future. In this respect, low-temperature approaches using conductive adhesives or wire-based connections have an inherent advantage due to lower thermal stresses associated with them" [5]

After an early reduction rated of ≥ 10 µm per year the current prognosis of a gradual reduction of wafer thickness was reduced to about 5 – 10 µm per year in the course of updated years forecast. Forecasted wafer thickness rose according to an evaluation of ITRPV prognosis from 2013, 2014 and 2015 as shown in figure 3.

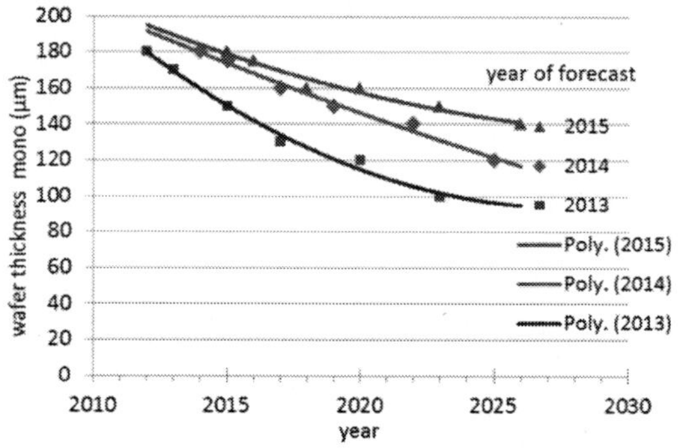

Fig. 3: Wafer thickness forecast according to ITRPV prognosis from 2013, 2014 and 2015

3 CELL INTERCONNECTIONS AND TRENDS OF DEVELOPMENT

Interconnection technologies with conductive adhesives have been evaluated in detail and even in common qualification tests have been successfully tried and tested. Nevertheless they have not been applied to a larger extent. Probably the adhesive joints as well as their bonding and electrical conductivity and the high quality requirements on guaranties have not been satisfactorily fulfilled.

Metallic ribbons with so far considerable cross sections of e.g. 2 mm x 0.15 mm or 0.3 mm² appear as mechanically too stiff. Hence interconnections with e.g. ø 0.3 mm wires (cross section of 0.07 mm²) and therewith less than ¼ fold mechanical loads are applied as shown in figure 5. However, an automated interconnection technique requires special new production equipment and referring investments.

Fig. 4: Multi-wire interconnection

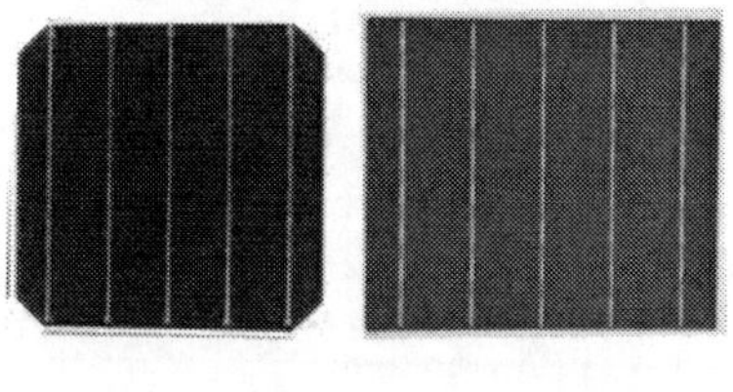

Fig. 5: 5-busbar-interconnection

For achieving a compromise between mechanical stability and sufficient flexibility in recent times, some small ribbons were applied. In this concern the 5-busbar-technology as shown in figure 5 is an example.

4. MODULES WITH DEFECTS IN SOLAR CELLS IN LARGE-SCALE POWER PLANTS AND VERIFICATIONS

Essential basis for an assessment of the condition of a PV plant is the plant performance. On one hand performance is expressed by its efficiency, represented by the performance ratio PR, on the other hand it is assured by a qualified monitoring system including qualified metrological and electrical sensoring. The plant performance as itself however presents an integral quantity without going into details on any reason of any performance losses.

Following graph in figure 6 shows a statistic of effects, which been collected from site inspections of plants with cumulated 53 Megawatts done from 2013 to 2016. All plants are been commissioned between 2010 – 2013 and are in operation. Figures are the result of mainly visual random samplings of about 30% of PV modules in PV arrays. Defects mainly emerged as visual module abnormalities caused e.g. by broken cells, hot spots, discolorations, delaminations and other module failures.

The graphic in figure 6 shows, that 26 out of 53 inspected plants revealed nearly zero cell defects, in 4 plants a percentage of about 1.5 % cell defects were seen, in 2 plants the percentage of cell defects was about 3,5 %, and so on.

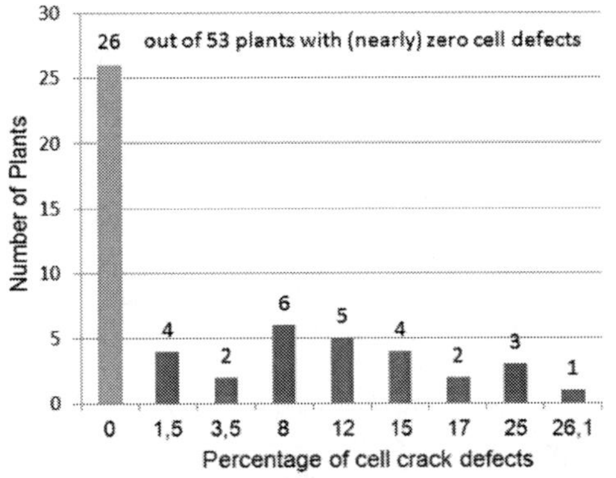

Fig. 6: Distribution of cell defect affected PV power plants

Module defects implicate increased maintenance efforts (costs for identification strings with referring modules, change of modules) and – depending on the kind of defects – partly remarkable yield losses.

Concerning existing plants and defective PV modules only a Status Quo can be stated. Accordingly it is advisable for the owner to decide at an early stage how to proceed with defective modules. A procedure could e.g. generally be a change of the product, or in some circumstances also a temporary stay of the modules in the plant with simultaneously monitoring referring strings in the framework of plant's SCADA information and in the course of maintenance routines.

Thus it is necessary to calculate and take precautious measures like e.g. provisioning of a suitable bulk of spare modules. Experienced investors know about this necessity. All the more it is important to early know any technical module defects or risks. A profound estimation of a number of spare modules is not easy. It usually is orientated on an optimistic or pessimistic view of module failures in the future. Often a true assessment of spare modules can be made after some years of operation and experience with the applied product "PV module".

For new erected plants incoming goods inspection for PV modules may become increasingly important. Likewise the commissioning phase of the DC parts can be an important action regarding identification of module abnormalities like e.g. snail trails and hot spots etc. Among others a thorough visual inspection may also deliver first information on the status of modules applied.

In plants being already for some years in operation the more it is important having qualified maintenance personnel which can identify and report on cell respectively module abnormalities and analyse the defects.

4.1 Electroluminescence technique, visual and IR examination on site

Electroluminescence: Hence for analysis in detail electroluminescence technique was developed. This technique originally was applied for quality assurance measures in photovoltaics, especially for analyses of defects in solar cell production. Objects of investigations are e.g. interruptions of contact lines, or doping inhomogeneities resulting in hot spots or cracks and micro cracks. In the course of the transition of solar cells to ever thinner wafers with increasing fracture behaviour during cell and module production it has developed to an enormous important and sound instrument.

The electroluminescence technique is allowing quick and cost-effective checks in production. Today mobile devices are able to check already installed PV modules in power plants for assessing the extent of losses.

The electroluminescence technique bases on a thermographic method. At a thermographic measurement a high resolution digital IR camera can display different warmed up cells or module areas. The display is able to visualise defects which are not visible with naked eyes. Concerning defect areas at or in PV modules beside shaded parts or soiling a lot of other failures can be visualised, like e.g. in the electroluminescence image in figure 7 below is indicated.

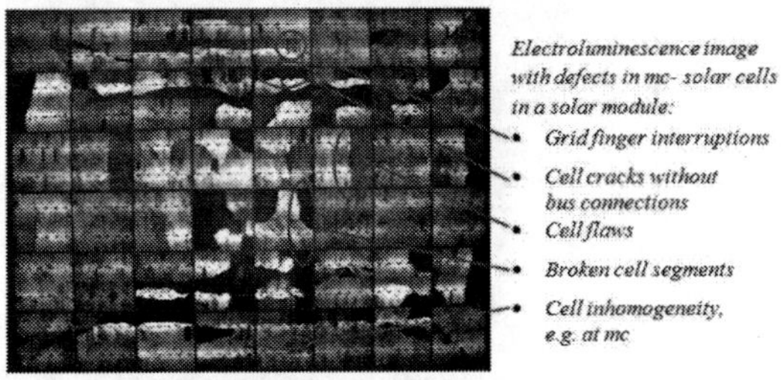

Fig. 7: Electroluminescence image

Visual examinations: Another measure for detecting cracks and micro cracks is a visual examination on site. It commonly is executed as random sampling. Indicators of cracks and micro cracks are typical discolorations of the metallisation. Since such discolorations are along the crack lines and darker as the cells such defects are commonly called *snail trails*. An example is shown in figure 8.

Not all cracks can visually be identified e.g. by snail trail appearances. Sometimes a magnifying glass is necessary since cell embedding material (back sheet and EVA foils) is still blocking humidity ingress which causes local delamination effects.

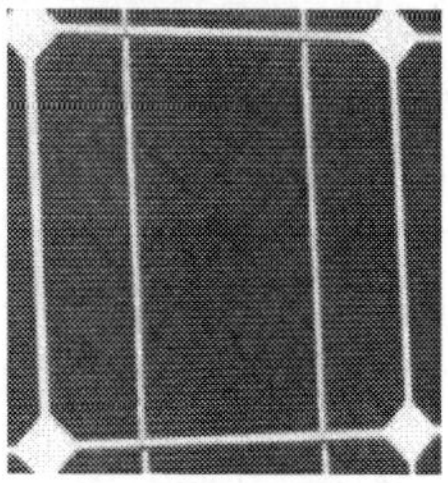

Fig. 8: „Snail trails“. Discoloration of a solar cell at the front side due to humidity passing through from the module rear side to the front side, hot spot at crack of solder joint

Discolorations emerge by moisture penetration through module rear side at cracks and micro cracks. The moisture ingress meets the metallisation on the front side and causes discolorations. Snail trails are therefore an indication of damages of solar cells in PV modules.

Fig. 9: Hot spot at crack of solder joint

Hot spots often can easily be visually detected due to their appearance as burned locations as shown in figure 9.

IR camera: By means of an IR camera on site thermal abnormalities in large installations of modules can easily be detected as shown in figure 10. A validation requires a minimum irradiation of 600 W/m² at module level.

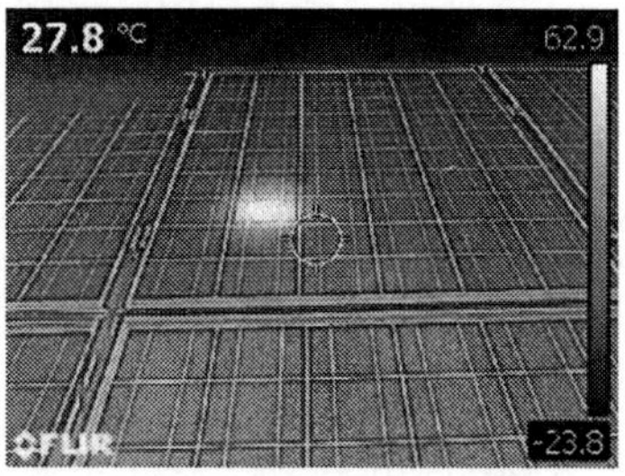

Fig. 10: Hot spot / overheated cell identified by IR camera

Advantageous is also that IR- tests can be done during normal operation of the plants. Experienced reviewer may detect e.g. hot spots, defective bypass diodes etc. without problems. In March 2016 a draft of a norm was published dealing with recommendations of usage of IR cameras in PV [6].

5 CONCLUSION: EFFECTS OF DECREASING CELL THICKNESS AND REACTIVE BEHAVIOUR

Consequences of decreasing thickness of solar cells could induce following risks:

1) for producers of solar cells and PV module
2) during transportation and assembly on site
3) during operation
4) for investors

Another momentum having influence on durable mechanical stability of ever thinner produced cells and cell interconnections is resulting from the fact of thickness reduction of module glass. In the course of more than ten years thickness of standard module glasses decreased from 4 mm to 3.2 mm. This development was mainly caused as one part of cost reduction measures. As a result temporarily appearing mechanical oscillations and short-time appearing deflections during transportation and wind pressures after mounting in the field are less damped. Consequences are short-term mechanical stress on cells and cell interconnections and thus inducing cell crazes and cell cracks rather likely.

1) A risk of decreasing cell thickness of solar cell and module manufacturing is an increased probability of micro crazes and micro cracks (MC):
   - in production of wafers and cells, including stringing process
   - as undetected pre-existing defects in wafers already
   - during handling of cells

Measures for keeping control at measuring in production lead to higher efforts, costs, and risks of an increase of cell rejection as well as a risk of undetected crazes.

2) Further risks concerning appearance of cell defects may emerge at transportation and module assembly, e.g.
   - after flash test during internal transportation, packaging and interim storage of modules
   - during transportation to the customer (from factory → to building site):
     - in the factory when loading module pallets with forklifts in containers and on the trucks
     - by loading module pallets with forklifts into containers and transport to building site
     - by inappropriate unloading module pallets by forklifts (figure 11) and transport over uneven building sites to mounting places

Fig. 11: Forklift and module transport

- during assembly by workers on support structures

  As guideline could serve the international standard IEC 62759-1:2015-07: Transportation testing of photovoltaic (PV) modules [7]

3) During the „operation“ phase micro cracks (MC) as risk events may appear and be identified as follows:
   - MC / cell cracks in the first years of operation may disclose small material defects by thermal stress and dynamic overstraining caused e.g. by short strong wind effects
   - During the years after a lot of temperature cycles micro crazes could be scaled up
   - MC as reason for hot-spots
   - If applicable verification whether MC or cell cracks have e.g. caused snail trails and could have been failures of the module producer

4) Beside risks caused by module producers investors may face risks also concerning cell and module failures. Additional measures from an EPC (Engineering, Procurement and Construction) as well as from an investor are required for keeping the risks as low as possible respectively to minimise them. Measures are e.g.:
   - Increased carefulness for any module handling on site
   - Increased efforts (and cost) concerning quality assurance on site by usage of measuring technique, e.g. EL random sampling by external service company
   - Targeted instructions and supervising activities of assembly personnel / temporary staff, creation of a catalog with possible failures as guide for visual failure identification, and last but not least and a matter of cost: employment of skilled workforce
   - During commissioning accurate IR examination of PV arrays for detection of hot-spots and other failures
   - Increased O&M efforts and maintenance by means of e.g. special attention and monitoring of formation of identified cell abnormalities, additional measuring of string values, eventually correction of spare part stock (risk of cost increase)

Depending on the results of analysis of identified defects / failures the investor may impose a warranty claim, which may be a difficult procedure.

SUMMARY

In the past years technical optimisation and progress in many and new production-steps, reduction of material volumes and as result of steady increase of cell and module efficiency product prices respectively performance price relations in EURO / Wp have become even more favourable.

The product „crystalline silicon PV module“ with cell thicknesses of much less than 200 µm is without doubts increasingly prone to any kind of mechanical stress.

According to the development of a further decrease of cell thickness, even more sensitivity and even increased efforts concerning handling of PV modules at all levels of production, transportation and assembly are inevitable.

Last but not least the last link in the „module value chain“ are investors respectively plant owners. They are expecting high quality products securing them a nearly trouble-free plant performance and operation with low level OPEX.

Acknowledgement

During investigations for this paper regarding cell issues valuable information have been given by Adolf Münzer. He spent all his professional life of over 30 years in photovoltaic and worked in well-known PV companies. He was active especially in solar cell research and module development and still remains to be always open to feedback from field experience.

REFERENCES

[1] *Very large scale PV power: state of the art and into future,* Keiichi Komoto, Christian Breyer, Edwin Cunow, Karim Megherbi, David Faiman, Peter van Fleuten, Routledge 2013, IEA PVPS Task 8

[2] *Review of Failures of Photovoltaic Modules, Photovoltaic Power systems*, Report IEA-PVPS T13-01:2014

[3] *Photovoltaik Technologie*, K.A. Münzer, book-on-demand, ISBN 978-3-86460-273-3

[4] *International Technology Roadmap for Photovoltaic (ITRPV), Seventh Edition,* October 2016, 2015 results including maturity reports, in co-operation with VDMA

[5] *International Technology Roadmap for Photovoltaic (ITRPV) 2015 Results including maturity report*, Seventh Edition, October 2016, in co-operation with VDMA

[6] VDE V 0126-23-3:2016-03, *Zerstörungsfreie Prüfung - Infrarot Thermografie von Photovoltaik-Modulen und Anlagen im Außenbereich (Non-destructing testing – Outdoor infrared thermography of photovoltaic modules and plants)*

[7] IEC 62759-1:2015-07: *Transportation testing of photovoltaic (PV) modules - Part 1: Transportation and shipping of PV module stacks* (IEC 82/729/CD:2012)

IMPLEMENTATION OF AN ACCURATE MEASUREMENT PROCEDURE TO DETERMINATE MAXIMUM POWER OF MODULES AT STANDARD TEST CONDITIONS IN THE FIELD THROUGH CORRELATION WITH MEASUREMENTS CARRIED OUT IN LABORATORY

*L. Pérez, J.A. Florez, M. Martínez, E. Alvarez, F. Domínguez, G. Castillo, A. Inza, H. Silva, R. Gómez, M. Fernández, V. Parra and A. Velasco.
Enertis Solar, S.L., C/Gomera 10,1°A. 28703, Spain. www.enertis.es
Phone : +34 91 651 70 21 Fax : +34 91 652 63 77, email : leonardo.perez@enertis.es

*Author for correspondence

ABSTRACT: A quality control procedure for modules has been implemented in a large-scale PV plant in Jordan. The procedure is easy to perform, uses portable equipment and has a relatively low uncertainty yet allowing a proper characterization of a PV module. This procedure is based on a methodology developed years ago by the Instituto de Energía Solar (IES) in Madrid – Spain. The key factor is using a module of the same technology and same Bill of Materials as the modules to be tested. This module must be calibrated in an accredited laboratory. This module will be called Reference Module (RM). On the field, both the RM and the *Device Under Test* (DUT) must be measured quasi-simultaneously so they both will be measured in the same irradiance and temperature conditions. Since the RM has the same spectral and thermal response as the DUT, a correlation between them could be used to determine the *Laboratory Equivalent Maximum Power* (LEMP) of the DUT. The simplicity of the procedure allows for rhythms of measurements up to 100 modules per hour, and in the right conditions, yields uncertainties values around 3%.
Keywords: Maximum power on site, Characterization, PV Quality control

1. INTRODUCTION

The performance of photovoltaic (PV) modules is not always as expected. Besides problems derived from improper manufacturing processes, incidents could happen in the transport of modules to the construction site that affect the performance of the PV module. Therefore, a quality control inspection is required upon arrival, measuring the maximum power produced by the module after its reception.

Furthermore, several works analyzing the testing of large amounts of modules have been published in recent years [1, 2]. In these publications, important differences between the peak power measured in an independent laboratory and the peak power declared by the manufacturer are reported. For this reason, a good quality control of PV modules is necessary in every PV projects to reach the estimated energy production and to control the degradation of PV modules during the lifespan of the PV plant.

As a standard in the PV industry, modules must be rated at Standard Test Conditions (STC), namely, normal irradiance of 1000 W/m^2, spectral distribution AM 1.5G and cell temperature of 25°C. The environmental conditions on the field generally differ greatly from STC values, so, in order to determine the maximum power at STC of PV modules under real sunlight on the field, the use of extrapolation equations is required. Since these equations perform a major adjustment, the result of the calculation usually carries a high uncertainty and low accuracy. Additionally, since the solar cell has a highly selective spectral response, the spectral composition of the incident irradiance has a great effect on the power produced by the module. This spectral composition varies greatly with environmental changes, like geographical location, season, weather and time of day.

Therefore, when it is necessary to test the maximum power of PV modules at STC on the field and with bounded uncertainty, it is usually done by using mobile laboratories [3, 4], which utilize solar simulators to perform the measurement in conditions of controlled irradiance and temperature. However, it is not always feasible to deploy a mobile laboratory on every site, especially when the project is built in emerging markets, where the photovoltaic industry is growing and the entities able to do this kind of tests are scarce.

This paper presents a testing procedure to obtain maximum power determination on the field with a relatively low uncertainty to achieve an appropriate characterization of a PV module. The procedure has been implemented in a PV plant located in Jordan, where a sample of two hundred modules were measured in a day.

2. MAXIMUM POWER DETERMINATION PROCEDURE

The procedure proposed in this document is based on the correlation of measurements between a module previously measured in a controlled laboratory environment, that would be the *Reference Module (RM)*, and the measures of the *Device Under Test (DUT)* on the field. The relation between these measurements has been proved to correlate by the Instituto de Energía Solar (IES). [5, 6]

The implementation of the proposed procedure in a large-scale plant allowed the test of up to 100 modules per hour, provided that the required irradiance is present during the testing. Since it is done using portable and easy-to-move equipment, the location of the plant does not entail any operative problem.

A detailed description is presented in two phases:

2.1 **Phase 1**. Creation of *Reference Modules* (RM).

The selection of the *RM* should be done according to IEC 60904-2:2015 [7]. The *RM* are created by precise and accurate calibration in suitable accredited laboratories. This precise measurement must be performed in accordance with IEC 60904-1:2016 [8]. For this kind of test the measurement uncertainty for the maximum power value should not be higher than ±2%. In order to carry out accurate measurements, the spectral response of the module should be also measured [9]. In addition, the number of *RM* required will depend on the size of the sample to be measured. Finally, the use of equivalent Bill of Materials as that of the *DUT* is recommended.

An accurate determination of the maximum power and I/V characteristics of each *RM* is crucial for the accurate execution of the procedure. The tests that must be performed for the creation of a *RM* are:

- Preconditioning of PV modules. The modules should be exposed to outdoor conditions prior to precision measurement (at least 60 kWh/m^2 according to IEC 61215-2:2016 [10])
- Visual inspection
- EL image for the inspection of cells.
- Measurement to determine the I/V characteristic and electrical parameters at STC, including spectral mismatch correction.

2.2 **Phase 2**. Maximum power determination on the field.

2.2.1 Previous steps and verification of *RM*.

The *RM* samples should be shipped to the site using proper packing and taking every precaution to avoid damages. In order to verify the good condition of the modules upon arriving on site is highly recommended to perform an electroluminescence test of the *RM's*. The use of damaged *RM* would yield inaccurate values throughout the entire test.

Additionally, to validate the procedure under the specific conditions of the site a cross measurement of the *RM* should be performed, where in each set of measurements, one of the modules is used as the *RM* of that set and its laboratory value is used to calculate the maximum power of the rest of the modules. The deviation of the calculated values on site should not differ more that 1% from the values obtained in the laboratory.

Table I: Fundamental requirements for the successful implementation of the proposed procedure

Installation of the modules	The *RM* should be installed in the same structure as the *DUT*.
Time	Tests should be performed in 4 hours around solar midday
Environmental conditions	Cloudless day. Required G > 600 W/m^2 Wind speed ≤ 2 m/s
Coplanarity	The *RM* and the *DUT* test should be in the same plane.
Temperature	The *RM* and the DUT should be under the same thermal conditions

Figure 1: Example of location of the reference module surrounded by the devices to be tested.

2.2.2 Maximum power determination.

Nearly simultaneous measurement of maximum power of both the *RM* and the *DUT* should be taken. In order to increase the efficiency of the procedure, measurements taken within one minute of each other will be considered quasi-simultaneous, which allows to measure the *RM* once and then measure up to five *DUT*. The variations in irradiance should not be greater than 1% between the measurement of the DUT and the RM.

Then to obtain the ***Laboratory Equivalent Maximum Power (LEMP)*** of the *DUT*, equation (1) is used:

$$LEMP = P_{mod}field * \frac{P^*_{ref}lab}{P_{ref}field} \quad (1)$$

Where:

- P_{mod} *field*: maximum power value of the *DUT* measured in the field.
- P^*_{ref} *lab*: maximum power value of the *RM* measured in the laboratory.
- P_{ref}*field*: maximum power value for the *RM* measured in the field.

Figure 2: Testing of the modules on site.

The use of a RM identical to the DUT allows to eliminate the uncertainties associated with the required STC values on the field, based on the assumption that the uncertainties associated with the extrapolation (spectrum and cell temperature effects) and the uncertainties associated with the field sensors (irradiance and temperature) are the same in both modules.

3 RESULTS OF THE IMPLEMENTATION IN A PV PLANT IN JORDAN

This procedure has been implemented to perform quality control tests in an 11MWp plant located in Jordan. The project required measurements with a limited uncertainty, and its location made it inviable to get a Mobile Laboratory to the site. The sample selected to be tested was comprised of 200 modules (amount estimated by using a level of inspection General I as per standard ISO 2859-1 [11]), with a nameplate power of 255 Wp, which had already been in operation approximately 6 months.

For this site, five modules were selected to be RM, named from A to E. These modules were sent to Enertis' laboratory in Madrid where they were measured accordingly to be made *RMs*, then sent back to the site. Four measurements were made to each module, and the average is presented as the *Laboratory Average* in the tables below.

To validate the procedure in the desertic conditions and to ensure that the modules did not present any damage, a cross measurement of the modules was performed, three sets of measurements were performed to each PV module. The average of these calculations is presented as the LEMP Average, and the result of these tests is presented in the following tables. It is notable that the highest deviation between laboratory and field values is of 0.95%.

Table II: Power calculation using module A as a reference module.

REFERENCE: MODULE A			
Module	***LEMP* Average [W]**	**Laboratory Average [W]**	**Dev [%]**
B	251.20	252.70	-0.59
C	254.37	254.80	-0.17
D	253.14	253.56	-0.16
E	256.29	255.37	0.36

Table III: Power calculation using module B as a reference module.

REFERENCE: MODULE B			
Module	***LEMP* Average [W]**	**Laboratory Average [W]**	**Dev [%]**
A	255.48	253.96	0.60
C	255.89	254.80	0.43
D	254.66	253.56	0.43
E	257.82	255.37	0.96

Table IV: Power calculation using module C as a reference module.

REFERENCE: MODULE C			
Module	***LEMP* Average [W]**	**Laboratory Average [W]**	**Dev [%]**
A	254.39	253.96	0.17
B	251.62	252.70	-0.43
D	253.57	253.56	0.00
E	256.72	255.37	0.53

Table V: Power calculation using module D as a reference module.

REFERENCE: MODULE D			
Module	***LEMP* Average [W]**	**Laboratory Average [W]**	**Dev [%]**
A	254.38	253.96	0.16
B	251.61	252.70	-0.43
C	254.79	254.80	0.00
E	256.71	255.37	0.53

Table VI: Power calculation using module E as a reference module

REFERENCE: MODULE E			
Module	***LEMP* Average [W]**	**Laboratory Average [W]**	**Dev [%]**
A	253.05	253.96	-0.36
B	250.30	252.70	-0.95
C	253.46	254.80	-0.53
D	252.24	253.56	-0.52

Once the modules and the procedure were verified, the procedure was used to test the 200 modules that comprised the selected sample. The modules were measured in batches of 50, each batch in a different structure.

The conditions of the site were ideal to perform the test, the irradiance was stable, with variations below 1% between measurements, the wind velocity was below 2m/s thus the variation in the temperature of the modules between measurements were negligible. With this procedure, it was possible to test one hundred modules per hour, making it possible to test up to four hundred modules per day.

The expected degradation in Si-C modules for the first year in operation, considering LID effects, is between 0.5% and 3% according to several studies [12]. The result obtained with the implementation of the proposed procedure fall well within this range as it can be observed in Table VII and Table VIII.

Table VII: Average results of the implementation of the proposed method sorted by structure.

Structure	***LEMP* Average [W]**	**Flash Average [W]**	**Deviation [%]**
1	257.16	258.73	-0.61
2	254.39	258.32	-1.52
3	255.99	259.30	-1.28
4	254.34	259.40	-1.95

Table VIII: Total average results of the implementation of the proposed method in a 11MWp plant.

***LEMP* Average [W]**	**Flash Average [W]**	**Deviation [%]**
255.47	258.94	-1.34

The percent deviation of each DUT with its flash value is presented in Figure 2. The greater deviation seen in the final part of the chart is due to the fact that the flash value of this modules was higher than the mean, probably hinting problems during the flash determination.

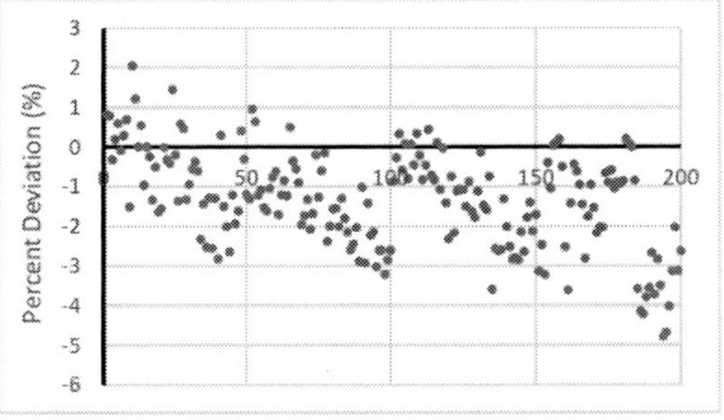

Figure 2: Deviation (%) of each measured module with its flash value.

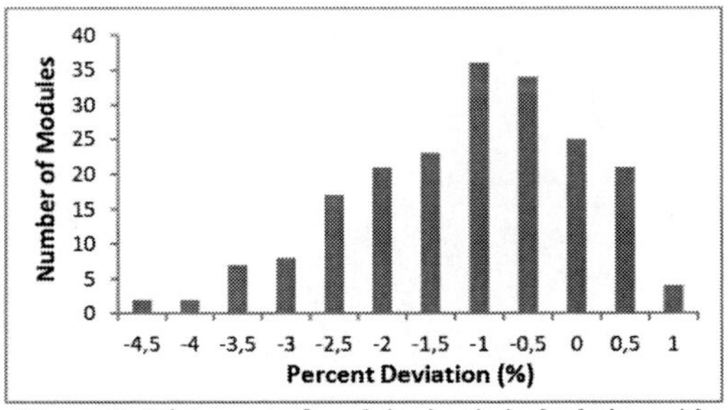

Figure 3: Histogram of modules by their deviation with flash values.

5 UNCERTAINTIES

The uncertainty sources that remains associated with the measurement procedure are:

- Uncertainties associated with the calibration of the *Reference Module.*
- Uncertainties associated with the I/V curve measurement equipment on site.
- Uncertainties derived from small environmental changes between measurements of the RM and the DUT.

Replacing the following variables in equation (1):

$P = LEMP$

$x = Pmod, field$

$y = P^{*}ref, lab$

$z = Pref, field$

Ecuation (1) becomes:

$$P = x \cdot \frac{y}{z} \quad (2)$$

According with the Law of Error Propagation, the uncertainty is calculated as:

$$u(P) = \sqrt{\left(\frac{\partial P}{\partial x} \cdot \delta x\right)^2 + \left(\frac{\partial P}{\partial y} \cdot \delta y\right)^2 + \left(\frac{\partial P}{\partial z} \cdot \delta z\right)^2} \quad (3)$$

$$u(P) = \sqrt{\left(\frac{y}{z} \cdot \delta x\right)^2 + \left(\frac{x}{z} \cdot \delta y\right)^2 + \left(\frac{x \cdot y}{z^2} \cdot \delta z\right)^2} \quad (4)$$

The uncertainty of each value could be determined as:

$$P = V \cdot I \quad (5)$$

$$P = \sqrt{\left(\frac{\partial P}{\partial V} \cdot \delta V\right)^2 + \left(\frac{\partial P}{\partial I} \cdot \delta I\right)^2} \quad (6)$$

$$u(P) = \sqrt{(I \cdot \delta V)^2 + (V \cdot \delta I)^2} \quad (7)$$

Using this equations, and always taking into acount the boundaries of needed irradiance, and the limits in the variation of irradiance and temperature, the maximum expanded uncertainty is around 3.5% with coverage factor K = 2 and coverage probability of approximately 95%.

6 CONCLUSIONS

The growing development of PV projects in remote areas calls for module quality control procedures that have a limited uncertainty, but also are able to be deployed in these remote and hard to access areas. In this paper, an easy-to-implement measurement procedure been presented, to determine the maximum power of PV modules in the field from the correlation with carefully calibrated reference modules and using portable equipment, with a limited measurement uncertainty.

This procedure has been implemented in an 11 MWp PV plant located in Jordan, a desert location with conditions of high irradiation and high temperatures. The method was validated in these conditions, by cross measuring five modules previously characterized in a laboratory (reference modules). The results after the implementation of the procedure showed that variations between the maximum power measured in the laboratory and the measurement in the field were below 1%. This demonstrates the robustness and validity of the procedure in such desert conditions. It is important to note that, given the simplicity of the procedure, rhythms of measurement up to one hundred modules per hour have been reached. With this pace of measurements an 11MWp plant, for which a sample of 200 modules was selected, was measured in only one day.

Finally, if the procedure is to be repeated in successive years to carry out annual plant degradation control, it is highly recommended to use the same reference modules used during the first measurement campaign and must be stored without being exposed to light or high temperatures. In this way, the validity of the relative measures is preserved, considering that the characterization of the reference module was performed with enough rigor (low uncertainty) to guarantee absolute values in the following years.

7 REFERENCES

[1] J. Coello, R. Carreño, M. Carames, J. L. Galindo. "Results of module testing in Enertis Solar during the quality control of 19 mw and 13 mw pv power plants in Spain". Proceedings of 24 th EUPVSEC (2009).

[2] M. Alonso-Abella, J.P. Silva, F. Fabero, N. Vela and F. Chenlo. "PV module peak power: comparison between manufacturer data and laboratory results", Proceedings of 23rd EUPVSEC (2008) 2837.

[3] L. Pérez, J. Coello, F. Domínguez, M. Navarrete. Development and implementation of a mobile laboratory - PV mobile lab quality to ensure the quality of photovoltaic modules "on site". 29th European Photovoltaic Solar Energy Conference, París (2013) pp 3500.

[4] Coello, J., Pérez, L., Domínguez, F., Navarrete, M., 2014. On-site quality control of photovoltaic modules with the PV MOBILE LAB. Energy Procedia 57 (2014), pp 89–98.

[5] Caamaño, E., Lorenzo, E., Zilles, R., 1999. Quality Control of Wide Collections of PV Modules: Lessons Learned from the IES experience. Prog. Photovoltaics Res. Appl. 7, pp 137–149.

[6] E. Lorenzo, R. Moretón, F. Martínez, R. Zilles. 2008. Retratos de la Conexión Fotovoltaica a la Red (XII). Era Solar Número 146 Año XXVI. (2008)

[7] IEC 60904-2. Photovoltaic devices - Part 2: Requirements for photovoltaic reference devices

[8] IEC 60904-1. Photovoltaic devices - Part 1: Measurement of photovoltaic current - voltage characteristics.

[9] IEC 60904-3. Photovoltaic devices - Part 3: Measurement principles for terrestrial photovoltaic (PV) solar devices with reference spectral irradiance data.

[10] IEC 61215-2:2016. Terrestrial photovoltaic (PV) modules - Design qualification and type approval - Part 2: Test procedures

[11] ISO 2859-1:2012. Sampling procedures for inspection by attributes -- Part 1: Sampling schemes indexed by acceptance quality limit (AQL) for lot-by-lot inspection

[12] B. Sopori, P. Basnyat, S. Devayajanam, S. Shet, V. Mehta, J. Binns, and J. Appel. Understanding Light-Induced Degradation of c-Si Solar Cells. IEEE Photovoltaic Specialists Conference, Texas (2012)

8 ACKNOWLEDGEMENTS

This investigation has been financed by the Spanish MINECO/CDTI project, ref. IDI-20151194, the Spanish MINECO project, ref. ENE2014-56069-C4-4-R and the Junta de Castilla y León project, ref. CP16/58.

DATA ANALYSIS FOR EFFECTIVE MONITORING OF PARTIALLY SHADED PHOTOVOLTAIC SYSTEMS.

Odysseas Tsafarakis[1], Kostas Sinapis[2], Wilfried G.J.H.M. van Sark[3]
[1,3]Utrecht University, Copernicus institute of Sustainable Development, Heidelberglaan 2, 3584 CS, Utrecht, The Netherlands
[2]Solar Energy Application Centre, HTC 21, 5656 AE, Eindhoven, The Netherlands
[1,2,3]E: O.Tsafarakis@uu.nl, sinapis@ecn.nl, W.G.J.H.M.vanSark@uu.nl

ABSTRACT: The purpose of this study is the development of an algorithmic tool to automate the process of analyzing monitoring data of partially shaded PV systems. The approach is to compare long-term and high resolution yield data of a partially shaded Photovoltaic (PV) system (investigated PV) with the yield data of an unshaded PV system or the tilted irradiance (reference PV system) and automatically detect the energy loss due to the expected shadow, caused by any surrounding obstacles and distinct it from any additional energy loss due to other malfunctions.
Keywords: see enclosed list of keywords

I. INTRODUCTION

The majority of the installed PV systems in the Netherlands are relatively small-scale installations, with an average capacity of 3.5 kWp[1]. About 70% of them is placed on residential rooftops[2], where different objects might obstruct the irradiance reaching the PV module and affects its energy output. Such small-scale, low-cost PV installations are mostly monitored through relatively simple, low-cost, data loggers of AC power, without any further equipment for solar measurements, i.e., pyranometers.

The ultimate aim of our research is to find an effective manner that allows monitoring large numbers of such installations in an automated manner. The proposed method offers the possibility to monitor partly shaded systems by excluding the effects of constant and unavoidable shadow from the performance evaluation and to detect faults that might occur during their operation.

II SCIENTIFIC INNOVATION

The innovation of the proposed method is its independence from the use of irradiance data. The performance evaluation is not based on the estimation of the energy generation/production of the studied PV systems and simulation of the losses due to shading (with the use of irradiance data and models) and its comparison with the real production[3]–[5] or the comparison of the production with the tilted irradiance, through the calculation of the performance ratio[5], [6]. It is based on the performance comparison and the detection of repeated or non-repeated errors/malfunctions between neighboring PV systems with similar characteristics (tilt and orientation).

III. METHODOLOGY

The basic principle of the proposed method/algorithm, is to study the daily persistence of any occurred error between the investigated PV and the reference PV system and identify these daily repeated errors, characterize them as shadows and create the **"shadow profile"** of the studied PV, *i.e. a table with the starting and ending dates of the shadow(s) during the year and the expected starting and ending times of the shadow(s) during each day.* Any difference between the yield of the investigated PV and the reference PV system, outside of the limits of these shadow profile time intervals is characterized as a malfunction and within these limits as shadow. An example of the shadow profile is presented in table 1. In this case, 5 different shadow periods are forming the shadow profile of the studied PV system.

During the analysis, the fluctuation of the Mean Absolute Error (MAE) between the production of the systems is studied ($|P_{reference} - P_{shaded}|$). Every timestamp with MAE higher than the normal is characterized as error and it is studied, in order to be characterized as shadow or malfunction. In order to set a threshold between the normal and the not normal operation, the iterative method Ran.Sa.C. (Random Sample Consensus) [7] is used.

Due to the nature of earth's orbit around the sun, the time interval of each shadow effect is varying from day to day. Additionally, due to the nature of the diffuse irradiance, the shadow is not present on cloudy days, where 5 days in row without any effect of shadow could appear. These facts are making the shading detection more difficult than a simple pattern recognition problem since different groups of hours and dates are forming the shadow profile. The clustering method D.B.S.C.A.N.[8] (Density-based Spatial Clustering of Applications with Noise) is used in order to identify shadows during different periods.

Thereafter the continuity of all the shaded periods of the "shadow profile" table is studied and presented in a "Date vs Time" plot, which demonstrates the change of the shadow effect during a long time studied period. We denote this as the **"Shadow story"**. An example of the shadow story is presented in figure I.

IV. RESULTS

As mentioned above, an example a partially shaded PV system is compared with a non-shaded PV system and their **"shadow window"**, i.e. the shadow periods of the system during the year (table I) and its **Shadow Story** (figure 1), i.e. the day vs time plot, are presented. The green marks in figure 1 indicate the measurements without a presence of error between the studied (partially shaded) and the reference (non-shaded) PV system. A shading obstacle was placed from August 10 to November 11 and impacts the PV system's performance during the morning hours. Since this error is repeated for long period, it is identified by the algorithm as shadow (black – higher than threshold difference – and blue – lower than threshold difference – marks) and not as malfunction. The shadow from the neighboring rooftop also affects the system in the late afternoon is also recognized as shadow.

N.	Start day	End day	Start time *(HH:MM)*	End time *(HH:MM)*	Cause
1	Aug 8	Oct 13	9:05	12:05	Pole
2	Oct 18	Nov 11	9:15	13:55	Pole
3	Jun 1	Jun 29	18:40	19:40	Roof
4	Jul 16	Aug 22	18:50	19:40	Roof
5	Aug 28	Sep 20	19:25	19:40	Roof

Table I: shadow windows example, where every row is a different shadow

Major remarks on the plot:

- At June 12 for external reasons, an obstacle was placed in front of the system, from 11:35 to 14:30. It is clearly within the hours were the pole is affecting the system (when it is placed). However, the algorithm recognized this incident as malfunction, since it is the only performance reduction, which occurred in that period and it is not in the table 1. A similar incident (roof maintenance) occurred for some days between October 26 and November 6. These dates are in shadow profile (shadow 2 in table I), however bigger part of the error was outside of the time intervals of the shadow profile, thus they marked as red.
- As depicted in figure 1, the shadow is shifting from August to November and impacts the system in later hours, due to the fact that the day is becoming shorter. In this case the role of the date clustering through DBSCAN is important, since the shadow period is separated in two smaller periods (shadows 1 & 2, Table I) and each one has different start and end time.

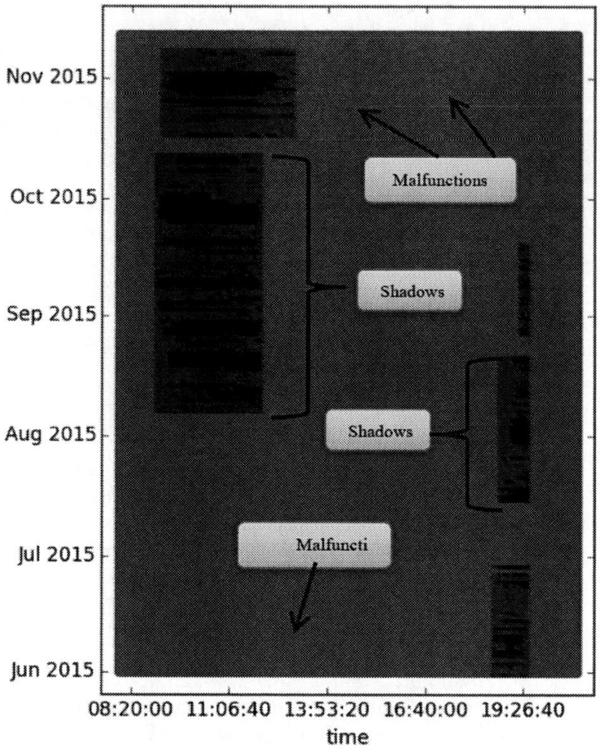

Figure 1: Shadow Story

V. FURTHER DISCUSSION OF THE RESOULTS

As depicted in figure 1, within the time intervals of the shadow profile (Blue points) there is a large numbers of measurements without error. It can be assumed that the reason of this fact is the high defuse irradiance, which, in contrast with the direct, doesn't obstructed by the shade causing object. This assumption is proved in this paragraph, where the blue points of figure I are further studied.

In Figure II, the plot of figure I is replotted with one major difference. The blue points of figure I are further studied and colored depending on the solar irradiance and specifically on the ratio of the diffuse horizontal irradiance (DHI) to the global horizontal irradiance (GHI) or R_{DHI}.

As depicted in figure II, in the majority of the points, R_{DHI} is higher than 0.5 (lighter blues) i.e. the DHI is higher than the direct normal irradiance (DNI). Further analysis of the blue points is presented in the scatterplot of figure III. In this scatter plot the three red lines are the 3 quantiles and the green line is the percentile where the R_{DHI} is 0.5. As it is clear in figure III, the 87% of the sample is above 0.5 thus the DHI is higher than the DNI, the irradiance which causes the shadow.

However it is very interesting that the upper right shadow (September, 18:00 to 19:00) of the plots I & II has still only dark blue dots in figure II. This fact is explained in figure IV, where the blue points are studied again and recolored. In this figure the blue dots are colored depending on the global tilted irradiance (GTI) and there are colored yellow if GTI < 100 W/m^2. Under this filter, all the blue marks of the specific profile are colored yellow, thus the irradiance is very low, in order to create any shadow effects.

VI. SUMMARY OF THE WORK

The proposed method is offering a monitoring solution to small residential PV systems, without the use of high-cost monitoring equipment which is completely cost ineffective

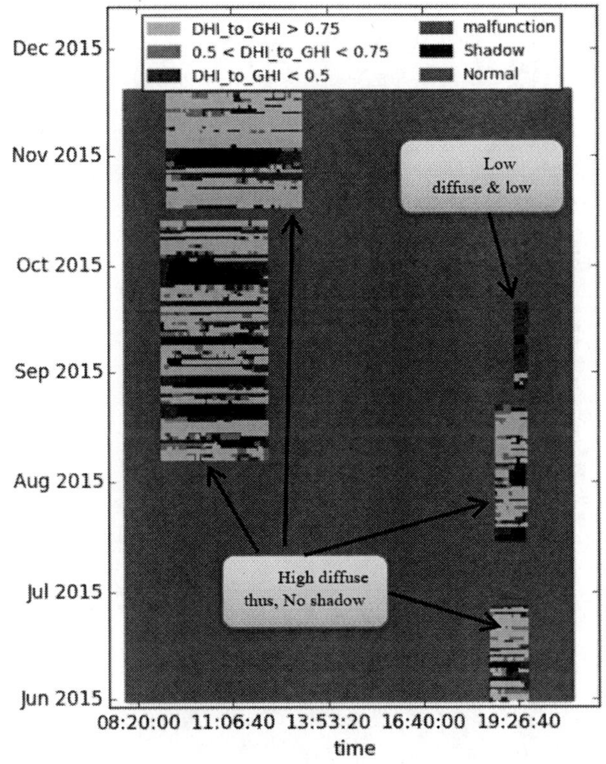

Figure II: Shadow story with measurements without fault inside shadow profiled colored depending on the ratio DHI/GHI

for this kind of installations. It only uses the PAC of two neighboring PV systems, with similar tilt and orientation and not necessary similar capacity (since the production can be normalized by divided with the W_{peak} of the system). The developed algorithm automatically isolates the timestamps with the non-normal operation and studies their concentration during the hours of the day and their continuity during the studied period and provides the owner of the system with the shadow story, a plot which demonstrates the effect of the shadow on his PV system during the year

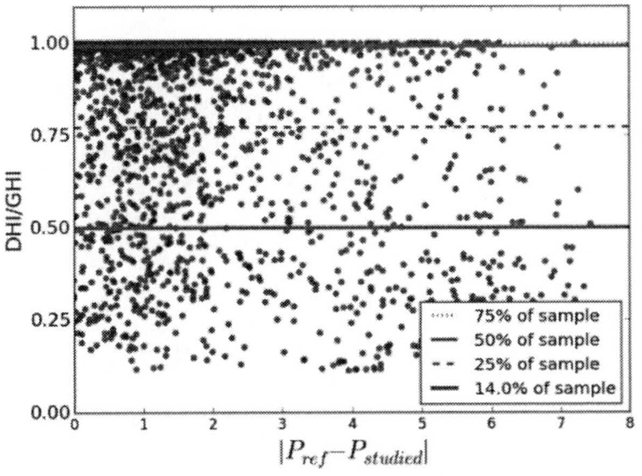

Figure III – Further analysis of the non fault measurments of shadow profiles

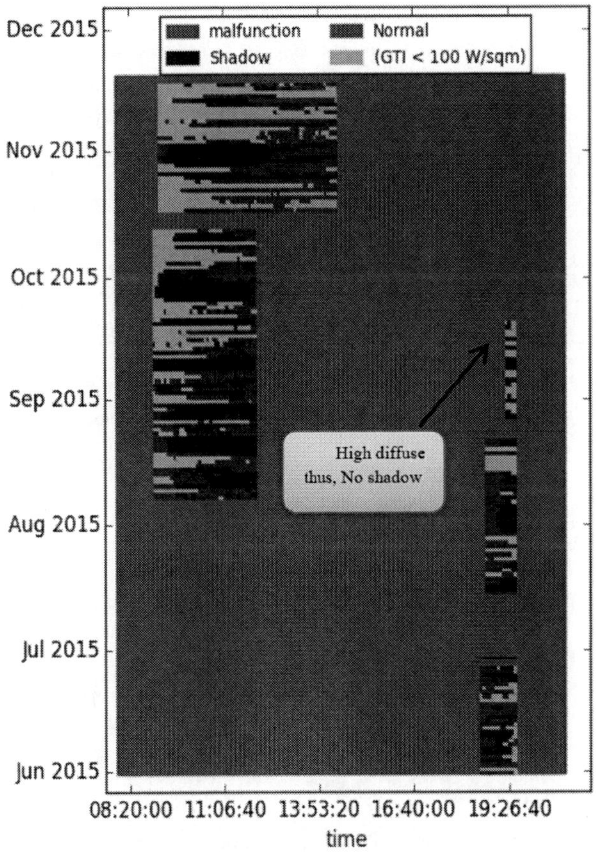

Figure IV: Shadow story with unshaded measurements inside shadow profile colored depending on the GTI

VII. REFERENCES

[1] W. van Sark, S. Hart, M. de Jong, P. de Rijk, P. Moraitis, B. B. Kausika, and H. van der Velde, "'Counting the Sun' – a Dutch Public Awareness Campaign on Pv Performance," *29th Eur. Photovolt. Sol. Energy Conf. Exhib.*, vol. 2014, pp. 3545–3548, 2014.

[2] Centraal Bureau voor de Statistiek (CBS), "Bijgeplaatst vermogen zonnestroom bijgesteld." [Online]. Available: https://www.cbs.nl/nl-nl/achtergrond/2015/51/bijgeplaatst-vermogen-zonnestroom-bijgesteld.

[3] A. Drews, A. C. de Keizer, H. G. Beyer, E. Lorenz, J. Betcke, W. G. J. H. M. van Sark, W. Heydenreich, E. Wiemken, S. Stettler, P. Toggweiler, S. Bofinger, M. Schneider, G. Heilscher, and D. Heinemann, "Monitoring and remote failure detection of grid-connected PV systems based on satellite observations," *Sol. Energy*, vol. 81, no. 4, pp. 548–564, 2007.

[4] S. Silvestre, A. Chouder, and E. Karatepe, "Automatic fault detection in grid connected PV systems," vol. 94, pp. 119–127, 2013.

[5] A. Woyte, M. Richter, D. Moser, M. Green, S. Mau, and H. G. Beyer, *Analytical Monitoring of Grid-connected Photovoltaic Systems*, vol. 13, no. 2. 2014.

[6] R. Eke and A. Senturk, "Monitoring the performance of single and triple junction amorphous silicon modules in two building integrated photovoltaic (BIPV) installations," *Appl. Energy*, vol. 109, pp. 154–162, 2013.

[7] M. a Fischler and R. C. Bolles, "Random Sample Consensus: A Paradigm for Model Fitting with Applicatlons to Image Analysis and Automated Cartography," *Commun. ACM*, vol. 24, no. 6, pp. 381–395, 1981.

[8] M. Ester, H. P. Kriegel, J. Sander, and X. Xu, "A Density-Based Algorithm for Discovering Clusters in Large Spatial Databases with Noise," *Proc. 2nd Int. Conf. Knowl. Discov. Data Min.*, pp. 226–231, 1996.

EFFECTS OF COMBINING SHADING ANALYSIS AND THE UNIQUE I-V CHARACTERISTICS OF THE PV MODULE

Rafael Herrero, Roberto Silva Simplicio, Cesar Biasi de Moura, Arthur Alves Miyazaki, Marcelo Knorich Zuffo
Escola Politécnica da Universidade de São Paulo
Av. Prof. Luciano Gualberto Trav.3 N.158. CEP: 05508-900. São Paulo - SP – Brazil e-mail: rherrero@lsi.usp.br

ABSTRACT: This study proposes a method for minimizing losses of partial shading, non-optimal orientation and tilt angle in PV carports. To accomplish this objective, a Matlab algorithm is present to study the effects of combining the incident irradiation analysis on PV module-level with a PV module electrical parameter in a 452.76 kWp parking-integrated PV base study. Results shows that the total energy output of sorted PV string reached a gain of 0.011% compared with unsorted method associated to irradiation. Even though the method results in a low gain and the manufacturing process has being improved, this investigation indicates that an accurate PV planning appears necessary for better performance of PV carports subject to partial shading conditions.
Keywords: Module Manufacturing, PV Module, Building Integrated PV (BIPV), Shading, Mismatch losses.

1 INTRODUCTION

Studies reveal that the global photovoltaic industry continues to grow exponentially and it is expected to reach between 396 and 540 GWp in the next 3 years [1].

Thanks to photovoltaic (PV) initiatives in Brazil since 2011 [2] when only 235 kWp PV systems were grid-connected [3], after 6 years of governmental financial support the cumulative installed PV capacity increased to 276.2 MWp [4].

From the total capacity installed, approximately 850 kWp are PV carports [5] distributed in nine facilities around the country as shown in Fig. 1.

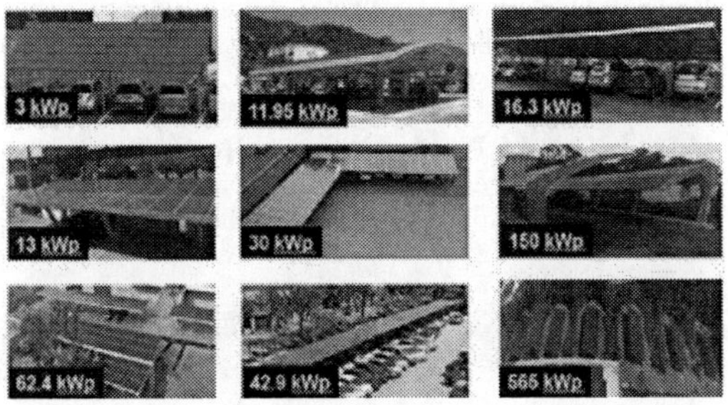

Figure 1: PV carports in Brazil

The collected data analysis shows that none of the facilities has both optimal orientation and tilt angle but, in one case, the system is correctly oriented to the North and, in another case, the tilt angle follows the local latitude. Three other facilities have low-tilt angle lower than 10°, which increases the dust accumulation and tends to difficult the self-cleaning [6].

This scenario reveals that previous establishment of parking layout orientation and limitations of metallic structure may cause non-optimum azimuthal and tilt angles, affecting the productivity (yield) and the performance rate (PR) of large PV systems assembled in this urban spaces.

The performance of this type of facilities could be worse when is subject to partial shading conditions caused by surrounding obstructions [7, 8], increasing mismatch losses due to partial shading.

As the yield and PR of PV plants play a fundamental role in the determination of grid parity [9], especially for PV carports, whose the Balance Of System (BOS) has a high cost due to the weight metal structure [10], a novel accurate method of PV module allocation appears necessary.

Considering that PV carports are also affected by PV losses as the ground mounted PV systems, a string configuration using the mismatch loss attributed to manufacturing tolerances associated with PV module-level irradiation pattern, reveals as an alternative to minimize the effects of partial shading and not optimal orientation and tilt angle.

Even with the improvements of PV module technology manufacturing process [11] and classification according to its maximum power based on a standard test condition (STC), the PV module power tolerances is still around ±2% and ±3% [12] and cannot be easily overcome, adding an inevitable dispersion between the nameplate and the I-V curve obtained in solar simulators.

Widely known as module quality loss by PV simulation softwares, the value of mismatch loss is usually assigned by PV designer based on the datasheet or it is assumed a statistical distribution. This occurs because currently it is not possible to import all flashtests data directly in simulators database. The mismatch loss tolerance is caused by manufacturing process, which occurs when PV modules with different I–V characteristics are interconnected [13].

According to [14], the current–voltage (I-V) curve of each PV module in a solar plant are similar, but non-identical to the main electrical parameters such as maximum power (Pmp), maximum power voltage (Vmp), maximum power current (Imp), open circuit voltage (Voc), short circuit current (Isc) and the fill factor (FF).

Over the years, several researchers [15, 16, 17, 18], studied the mismatch losses of PV cells and modules individually or interconnected, in order to evaluate the factors that affect the electrical parameters through mathematical models or experimental evidences.

A few years later, [19] obtained experimentally variations of 10% between the nominal power and catalog using PV modules from the same manufacturer and with the same electrical characteristics.

In the experiment of [20] were used 192 PV modules under five insolation levels between 100 and 800 W/m^2 and temperatures between 35°C to 55 °C, the results showed a mean maximum rated power of 14% lower compared with classification presented in the PV modules nameplate.

Using a Monte Carlo technique, [21] developed a statistical analysis tool to calculate the losses of mismatch in PV systems and requirements for modules acceptance.

One hundred of PV modules from different manufactures were analyzed by [22], indicating an improvement in the quality control compared to previous years with average values of ±3.5 %. However, many of them exceeded the power tolerance of -5%.

Using the [16] model, [23] analyzed the effects of different levels of shading in the performance of PV modules using Matlab. They found a power reduction of 29.64% when only one cell is completely shaded in a module with 36 cells associated in series.

In another study of mismatch loss, [24] presented another Matlab model and experimental results about the effects of mismatch in the association of PV modules. The total power loss obtained was approximately of 10%.

A statistical study of [25], presents different techniques used to reduce the electrical mismatch losses. The study concluded that a Isc selection is more effective than other parameters.

Using a Monte Carlo technique, [26] developed a study for selecting PV modules for two batches of different manufacturers. They found that the classification from the Imp resulted in a more consistent electrical mismatch reductions compared with Vmp and Pmp parameters. Besides, they concluded with a superficial economic analysis indicating a slightly gain in the selection of modules.

According to the study of [27] regarding the impact of mismatch loss in a PV module series association with different nominal powers, the mismatch losses will be below 1% if the number of PV modules with different powers is limited to three in a same string and the difference between them is 5 Wp.

Considering that previous studies has not addressed an association method to allocate the higher performance PV modules in the higher irradiation place at PV system, this study presents a method and the effects of combining an irradiation analysis and a PV module electrical parameter, against mismatch losses in large-scale PV plants subject to partial shading conditions.

To address this investigation, a irradiation analysis simulation was conducted for a 452.76 kWp PV carport case study, considering the distance and the measurements of obstructions in the surroundings.

The simulation was performed by using three-dimensional (3D) models developed in SKETCHUP software and a PV module-level simulation using ECOTECT software.

The input parameters for ECOTECT were the 3D model exported from SKETCHUP as 3DS file and Meteonorm 6.1 data from São Paulo.

After the simulation, it was identified the irradiation profile in order to establish arrays with minimum irradiation variation between them.

To fill these arrays, the proposed Matlab algorithm carried out a flashtest analysis from the entire PV module batch to indicate the electrical parameter that might be relevant for PV module sorting.

Subsequently, a comparative analysis was conducted considering the energy production of the PV string with and without PV module classification.

2 CASE STUDY

Fig. 2 presents the 3D model of two canopies of 132 meters length each (section A and B) totaling 452.76 kWp under 23°32′38′′S;46°4354′′W coordinates and the 12 buildings with 30–60 m of height. The two sections are 64° Northwest-oriented and 6° tilted over the horizontal.

From 1848 polycrystalline silicon technology PV modules foreseen to the entire PV carport, only 412 were produced in 2014 by Tecnometal – DYA Solar PV Brazilian manufacturer, with nominal peak power of 245 Wp model SV-245D20, 60 PV cells, 3 by-pass diodes, 1.62 m² of area and 15.61 % of efficiency.

Figure 2: 3D Model with surrounding buildings

2.1 Irradiation analysis

To obtain the irradiation profile for each PV module, an annual and month average daily global irradiation analysis per PV module was performed by ECOTECT.

After 109 hours of simulation for 1848 PV modules, the total irradiance loss over the year caused by buildings and mutual shading was 1.7 %.

The annual average daily global incident irradiation per PV module on section A and B can be observed in Fig. 3, where the range of colors from light yellow (3929.0 Wh/m²) to dark blue (3894.0 Wh/m²) indicates the variation among them.

When the PV module is fully irradiated all over the year it is indicated between light and dark yellow colors (3929.0 - 3918.5 Wh/m²), when it is partially irradiated, between light and dark red colors (3915.0 - 3908.0 Wh/m²) and when is less irradiated, between purple and dark blue colors (3904.5 - 3894.0 Wh/m²).

Figure 3: Annual average daily global incident irradiation per PV module on section A and B

2.2 Minimizing the mismatch losses attributed to string non-uniform irradiance

Considering the annual average daily global incident irradiation per PV module and avoiding the interconnection of PV modules subject to regions of non-uniform irradiance. Fig. 4 presents the PV array configuration proposed in this study.

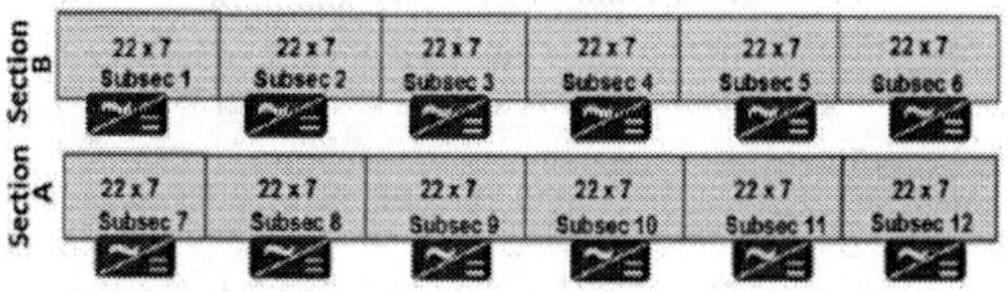

Figure 4: 12 PV arrays of 22x7 series-parallel configuration connected to a 40 kVA inverter.

After the array configuration, it is possible to place the higher performance PV modules in the higher irradiation places at PV carport. To identify these modules, a characterization of the flashtests indicated the electrical parameter relevant to use in the selection and PV strings arrangement.

2.3 Flash tests analysis

For the I–V analysis, an electrical characterization of 412 PV module flashtests was carried out, using a pulsed solar simulator model Spi-Sun SimulatorTM 4600SLP at Tecnometal manufacture. They were measured at the same test conditions in order to minimize variation in the absolute values of the parameters.

In Fig. 5, it is possible to see the block diagram of the algorithm, where the inputs and outputs to the main block are presented.

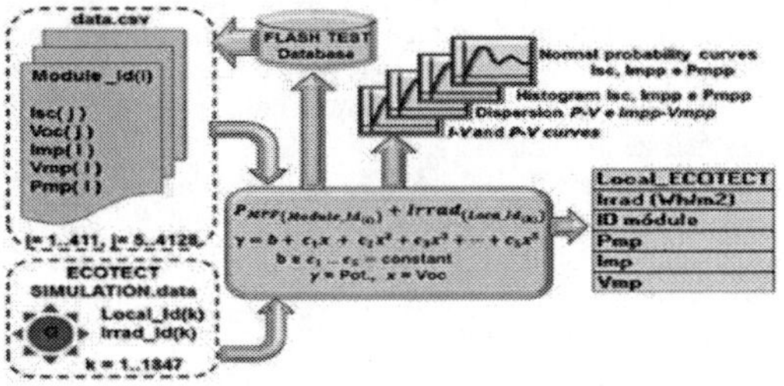

Figure 5: Block diagram of the algorithm proposed

In the first block (data.csv) of the algorithm, the electrical parameters (Pmp, Vmp, Imp, Voc and Isc) and the identification (Module_Id) of each PV module are collected and the database is loaded. Indices i = 1..412 and j = 5..4128 refers to the numbers of flashtests and the number of useful lines in each one to be parsed, respectively.

In the second block (ECOTECT SIMULATION.data) the irradiation analysis data file is accessed and the results of each PV module are loaded, considering the unique ID location (Local_id) and annual average daily global irradiance value (Irrad), where indices k =1..1848 refers to the number of PV modules position in the facility.

Regarding the output blocks, firstly, the values of Isc, Imp, Pmp, Voc and Vmp from the 412 PV modules flashtest are extracted and a statistical analysis is performed using the Data Statistics tool from Matlab R2015a.

Lastly, all strings were adjusted based on the lowest Isc value of the string through the algorithm. Remembering that, theoretically, in a series association, the product between the sum of Voc of each PV module and the smaller Isc are responsible for limiting the Pmp of the series association.

For this adjustment, it was necessary to use the following 5-degree polynomial equation in each module P-V curve of string, in order to identify the new Pmp with the Voc and lower Isc for each PV module of the string. Where Pmp: Maximum power point of PV module obtained in flash test (Wp), Voc: Open circuit voltage of PV module obtained in flash test (V).

$$Pmp = b + c^1Voc + c^2Voc^2 + c^3Voc^3 + c^4Voc^4 + c^5Voc^5$$

Fig. 6 shows the I–V (red color) and P–V (blue color) of the samples and Table I presents the statistics and percent variation of the electrical parameters to compare with the nominal characteristics presented in the manufacture catalog as: Pmp = 245.0 (±5 W), Imp = 8.06 A, Vmp = 37.41 V, Voc = 30.40 V and Isc = 8.65 A.

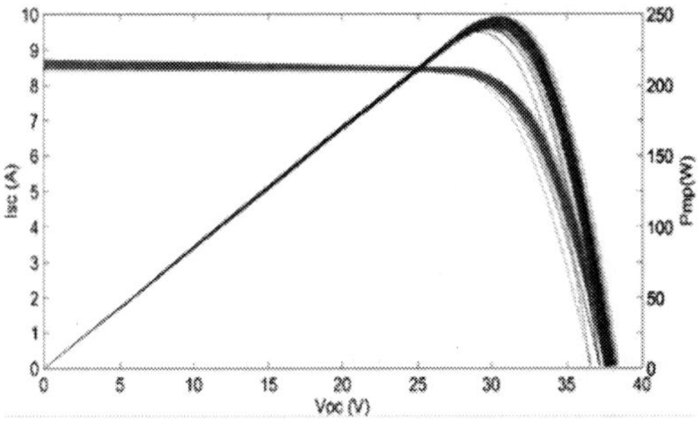

Figure 6: I – V and P – V plots from the 412 PV modules

Table I: Descriptive statistics of the main parameters of the 412 PV modules tested

	Pm (W)	Imp (A)	Isc (A)	Voc (V)	Vmp (V)
Min.	237.0	7.891	8.454	36.910	29.370
Max.	248.0	8.190	8.723	38.150	30.430
Median	242.5	8.040	8.588	37.530	29.900
Mode	237.0	7.891	8.454	36.910	29.370
Std.	7.778	0.2116	0.1899	0.6173	0.5267
Range	11.0	0.2993	0.2685	1.235	1.053

The statistical analysis reveals that in the batch of 412 PV modules, 225 (54.6%) presents Isc from 8.550 to 8.620 A, 165 (40.0%) between 8.640 and 8.720 A and 22 (5.34%) from 8.450 to 8.530 A.

The dispersion of Isc values are displayed in Fig. 7. It is possible observed the low variation between the highest (8.723 A) and the lowest (8.454 A) values of the Isc, corresponding to a difference of 269 mA, a median value of 8.588 A and low standard deviation of 0.1899.

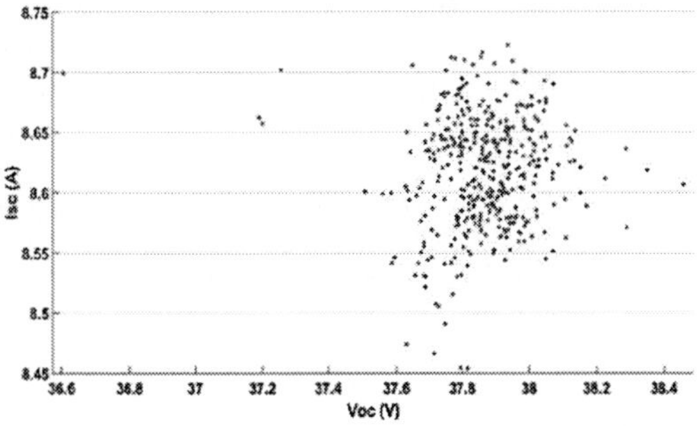

Figure 7: Dispersion of Isc values

Regarding the Imp analysis, was observed that 212 (51.46%) PV modules presented Imp from 8.100 and 8.180 A, 193 (46.84%) between 8.000 and 8.080 A and 7 (1.69%) from 7.890 to 7.980 A.

As can be seen in the Fig. 8, it was found a low variation between the highest (8.190 A) and the lowest (7.890 A) values of the Imp, which corresponding to a variation range of 299 mA, a median value of 8.040 A and a low standard deviation of 0.2116.

In a similar way that occurred in Isc analysis, it is possible to conclude that Imp also presented a low variation range compared with the nominal value, but not as lower as Isc.

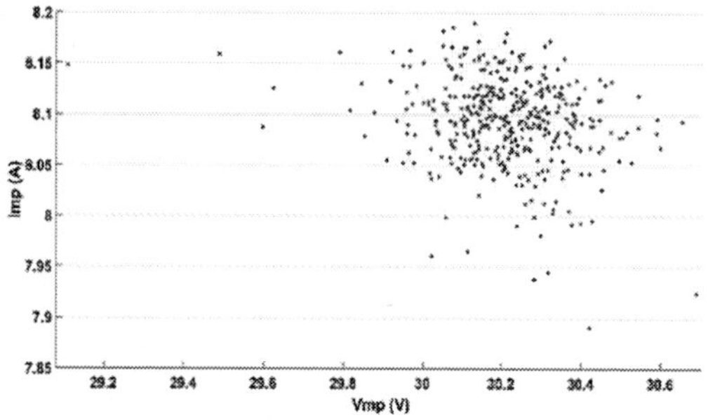

Figure 8: Dispersion of Imp values

After the statistical analysis of Pmp, it was noticed a relevant variation range when compared to Imp and Isc results.

Fig. 9 depicts the dispersion of Pmp, where it is possible to observe the highest (248.0 Wp) and lowest (237.0 Wp) values of the Pmp, corresponding to a variation range of 11 Wp.

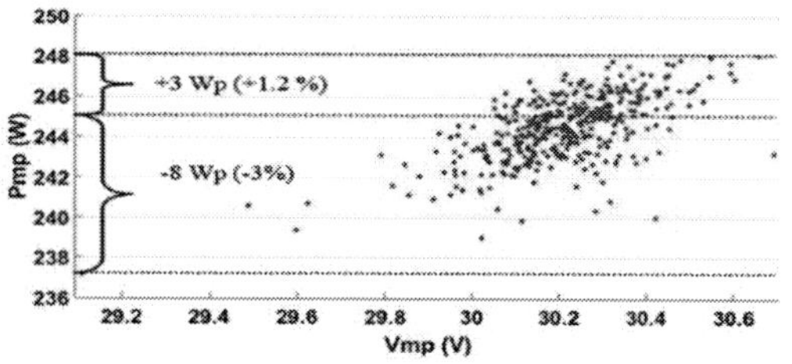

Figure 9: Dispersion of Pmp values

It is worth noting that the values of the positive nominal power tolerances are within the +5 Wp (+2%) defined by the manufacturer. However, the negative power tolerance exceed the -5 Wp (-2%), reaching -8 Wp (-3%) and consequently, decreasing the median value of the whole batch to 242.5 Wp obtained by the statistical analysis, which not corresponding to the PV module nameplate.

To validate the number of PV module out of the range of the negative power tolerance, Fig. 10 presents the power range divided in 11 intervals, showing the number of occurrences for each of them.

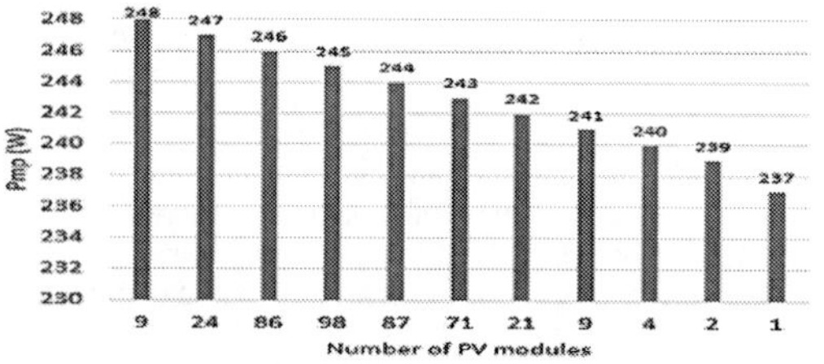

Figure 10: Distribution of the effective STC power of the considered 412 PV modules

It is possible to observe that only 98 (23.78%) PV modules from 412 analyzed, matched the 245 Wp defined by the datasheet, while 314 (76.21%) had presented electrical characteristics dissimilar.

Even with the Pmp divergence obtained in the flashtest analysis, it is notice that only 3 PV modules from the batch exceed the power tolerance of -5% specify by the manufacture catalog, which should be rated as 240 Wp.

Based on the observed Pmp electrical variation in the flashtest analysis, it is possible to conclude preliminarily that Pmp values is adequate to be combine with incident irradiation per PV module.

Next, the irradiation in a definied area of PV carport facility and flashtest analysis were associated to validate the chosen parameter.

3 RESULTS AND DISCUSSION

From the 412 PV modules available for this investigation, 396 have been analyzed at STC, considering 18 PV strings with 22 PV modules each.

For the first (1st) arrange was considered a descending sorting method by Pmp parameter. Next, a random iteration function based on a discrete uniform distribution using all the 396 flashtests was performed 10 times (2nd - 11th). The total nominal power of each string are summarize in Table II.

Considering the applied descending sorting method, as expected, the total nominal power obtained for the first (1st) sorted PV string reached the highest value of 5441 Wp, due to the fact that 22 highest performance PV modules are concentrated in this PV string. In the another hand, the eighteenth (18th) sorted PV string presents the lowest nominal power with 5295 Wp, indicating that the lower performance PV modules are concentrated on it.

Table II: Nominal power comparative from sorted and unsorted PV Strings

String	Sorted (Wp)	Unsorted (Wp)									
N	1	2	3	4	5	6	7	8	9	10	11
1	5441	5373	5388	5371	5386	5376	5397	5378	5370	5381	5368
2	5423	5381	5390	5369	5389	5376	5380	5374	5371	5386	5384
3	5416	5397	5368	5378	5382	5376	5373	5378	5373	5383	5379
4	5410	5382	5386	5378	5372	5389	5370	5369	5379	5365	5367
5	5403	5384	5364	5373	5388	5374	5372	5382	5384	5373	5388
6	5398	5363	5381	5376	5373	5370	5383	5379	5387	5381	5381
7	5393	5366	5372	5383	5363	5376	5382	5380	5383	5378	5378
8	5389	5376	5380	5372	5381	5366	5367	5398	5386	5369	5368
9	5382	5370	5383	5364	5370	5373	5365	5375	5376	5381	5385
10	5377	5382	5375	5392	5373	5379	5381	5374	5376	5370	5370
11	5373	5383	5373	5386	5379	5384	5388	5379	5377	5374	5373
12	5368	5377	5370	5393	5381	5385	5377	5380	5371	5379	5365
13	5361	5382	5363	5371	5367	5389	5381	5368	5374	5380	5378
14	5354	5386	5372	5385	5375	5378	5381	5385	5392	5374	5380
15	5349	5370	5387	5384	5380	5381	5380	5382	5384	5366	5383
16	5343	5370	5387	5381	5376	5391	5375	5369	5384	5379	5386
17	5333	5380	5392	5381	5382	5366	5364	5368	5370	5376	5374
18	5295	5377	5365	5363	5382	5370	5384	5381	5363	5384	5392

In Fig. 11 it is possible to see the average nominal power of PV modules for each PV string using the sorting method and unsorted, where for the 1st PV string the 22 higher performance PV modules has an average nominal power of 247.30 Wp and for 18th PV string an average of 240.70 Wp.

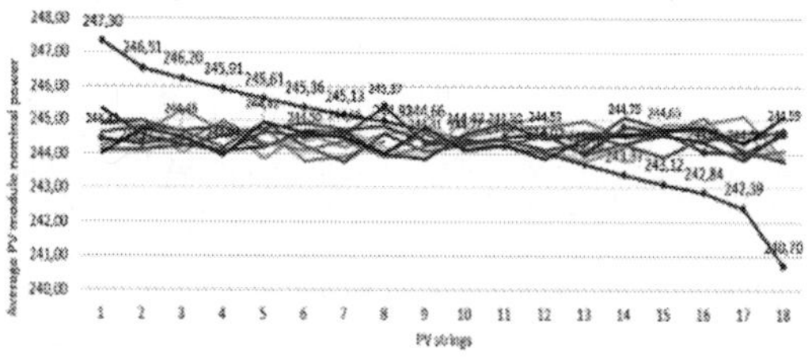

Figure 12: Comparative of average nominal power PV module for sorted and unsorted PV strings.

On the unsorted method is notice a mix of higher and low performance PV modules, which reduces the average nominal power of PV module of string.

At this point, it should be highlighted an average decreased power of 0.16% per sorting PV string. In terms of percentage, a total power loss of 2.7% between the first and the last sorted PV string, resulting in a nominal power variation of 146 Wp.

Considering this reduction, Fig. 12 presents the nominal power decreasing behavior as denoted with red marker circles, when the PV sorting method does not bring a nominal power gain, occurring in the eighth (8th) PV string.

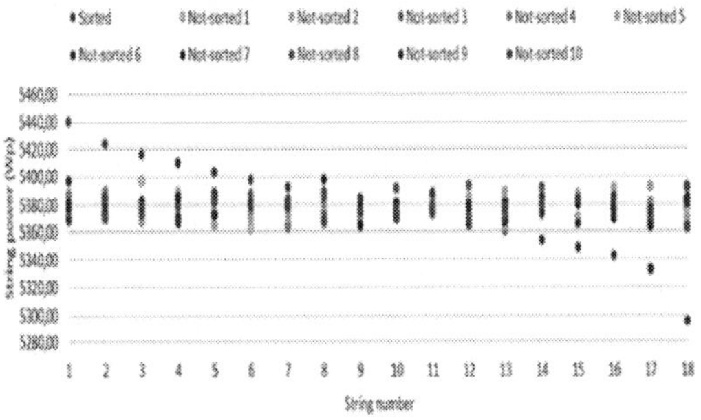

Figure 12: Sorted and unsorted total power PV string comparative

Even with the reduction observed for each sorted PV string, the sum of total nominal power from all sorted PV string is 0.009% to 0.011% higher than the 10 unsorted rounds as observed in Fig. 13.

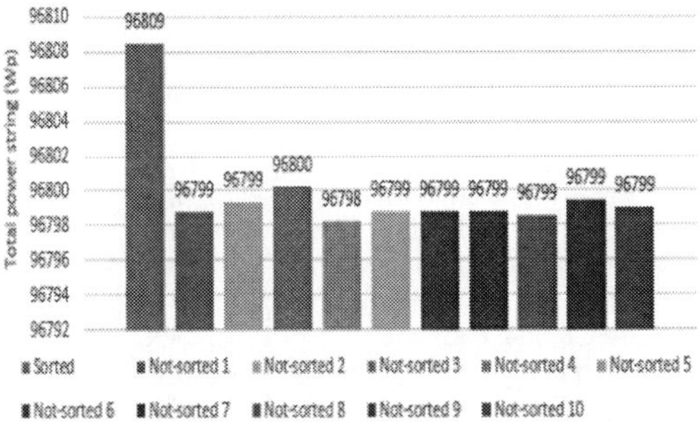

Figure 13: Total nominal power for sorted and unsorted PV strings.

Despite the PV sorting method by Pmp resulted in a low nominal power gain, it did not take into account the second part of the method proposed, which brings a PV module positioning based on the association between the high performance PV modules and incident irradiation.

In order to evaluate this association, the first step was calculating the efficiency of each PV string, by dividing the nominal power of each PV string and the total area (35.57 m^2) of 22 PV modules as shown in Table III.

It is possible to observe the highest value of 15.29% in the first (1st) and the lowest of 14.89% in eighteenth (18th) sorted PV string. However, in average the efficiency for all PV strings reaches 15.12%.

Table III: Efficiency for each PV string sorted and unsorted

String	Sorted (Eff%)	Unsorted (Eff%)									
N°	1	2	3	4	5	6	7	8	9	10	11
1	15,29%	15,10%	15,15%	15,10%	15,14%	15,11%	15,17%	15,12%	15,09%	15,13%	15,09%
2	15,24%	15,13%	15,15%	15,09%	15,15%	15,11%	15,12%	15,11%	15,10%	15,14%	15,14%
3	15,23%	15,17%	15,09%	15,12%	15,13%	15,11%	15,10%	15,12%	15,10%	15,13%	15,12%
4	15,21%	15,13%	15,14%	15,12%	15,10%	15,15%	15,10%	15,09%	15,12%	15,14%	15,09%
5	15,19%	15,14%	15,08%	15,10%	15,15%	15,11%	15,10%	15,13%	15,13%	15,10%	15,15%
6	15,17%	15,08%	15,13%	15,11%	15,10%	15,09%	15,13%	15,12%	15,14%	15,13%	15,13%
7	15,16%	15,08%	15,10%	15,13%	15,07%	15,11%	15,13%	15,12%	15,13%	15,12%	15,12%
8	15,15%	15,11%	15,12%	15,10%	15,13%	15,08%	15,09%	15,17%	15,14%	15,09%	15,09%
9	15,13%	15,10%	15,13%	15,08%	15,09%	15,10%	15,08%	15,11%	15,11%	15,13%	15,14%
10	15,12%	15,13%	15,11%	15,16%	15,11%	15,12%	15,13%	15,11%	15,11%	15,10%	15,09%
11	15,10%	15,13%	15,10%	15,14%	15,12%	15,13%	15,14%	15,12%	15,11%	15,11%	15,10%
12	15,09%	15,12%	15,10%	15,16%	15,13%	15,14%	15,12%	15,12%	15,10%	15,12%	15,08%
13	15,07%	15,13%	15,08%	15,10%	15,09%	15,15%	15,13%	15,09%	15,11%	15,12%	15,12%
14	15,05%	15,14%	15,10%	15,14%	15,11%	15,12%	15,13%	15,14%	15,16%	15,11%	15,12%
15	15,04%	15,09%	15,14%	15,13%	15,12%	15,12%	15,12%	15,13%	15,13%	15,08%	15,13%
16	15,02%	15,10%	15,14%	15,13%	15,11%	15,16%	15,11%	15,09%	15,13%	15,12%	15,14%
17	14,99%	15,12%	15,16%	15,13%	15,13%	15,08%	15,08%	15,09%	15,09%	15,11%	15,11%
18	14,89%	15,12%	15,08%	15,08%	15,13%	15,10%	15,14%	15,13%	15,08%	15,14%	15,16%

Next, the eighteen PV strings sorted and unsorted were positioned to fill the subsection 1, 2 and 3 from section B (Fig. 13) with uniform and non-uniform irradiation, in order to evaluate the average daily energy production for the arrangements.

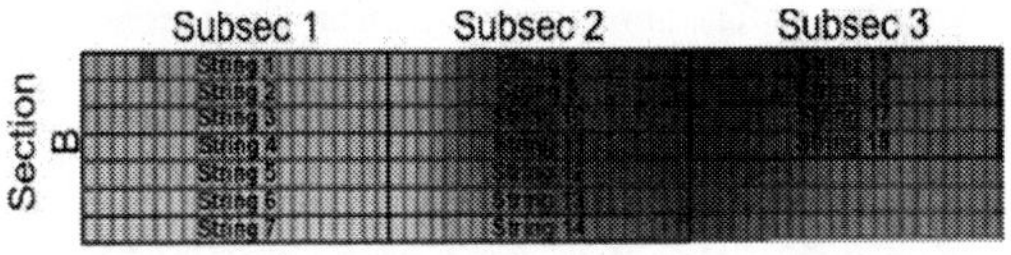

Figure 13: PV string configuration for subsection 1, 2 and 3 in Section B

The average daily incident irradiation and the daily energy production per PV string considering the sorted and unsorted method can be observed in Fig. 14. The range of colors from light yellow (higher values) to dark blue (lower values) indicates variation among them.

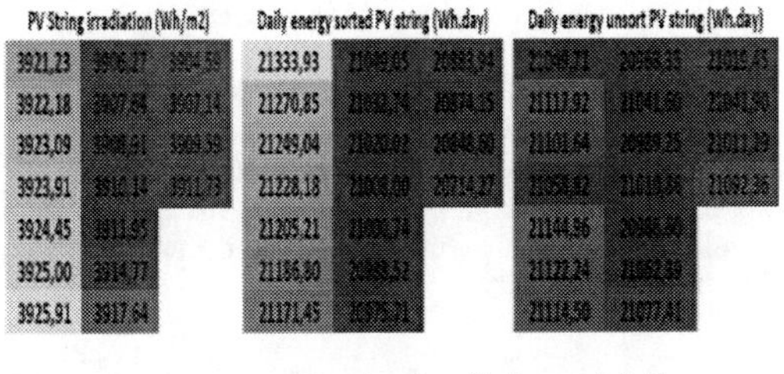

Figure 14: Average PV string irradiation and daily energy production for sorted and unsorted method.

The daily irradiation in each PV string was defined by the average daily incident irradiation in 22 PV module, according to our PV module-level ECOTECT simulation.

To obtain the daily energy production in each PV string, the product between the total PV string area of 35.57 m^2 (22 PV modules of 1.62 m^2 each), the PV string average daily irradiation and the PV string efficiency (Table III) was calculated.

These calculations shows the variation between the daily energy for both methods, where the daily energy production for sorted PV string ranges between 21333.93 and 20714.27 Wh and the average for unsorted PV strings between 21114.96 and 20968.35 Wh, representing a variation of 0.89% for the highest values and 1.21% for the lowest.

Regarding the total output from each PV strings, Fig.

15 presents a comparative between the sorted and unsorted PV strings, where the PV sorted associated to irradiation reaches a gain of 0.011%.

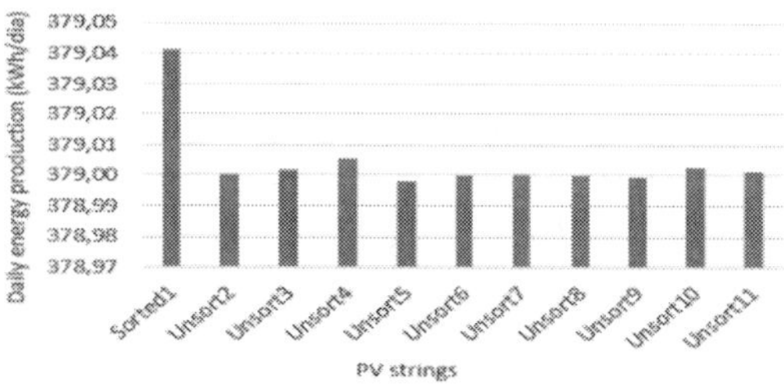

Figure 15: Total daily energy output for sorted and unsorted association.

4 CONCLUSION

The aim of this study was to investigate the effects of combining shading analysis and a PV module electrical parameter against mismatch losses in urban large-scale PV facilities

For this purpose, an accurate shading simulation based on an average daily global incident irradiation was carried out, demonstrating the PV module-level shading pattern in contrast to the conventional simulations. Besides a Matlab algorithm for flashtest parameter extractor and analyzer was developed.

The simulation results show that the method of sorting PV modules according to the Pmp electrical parameter in standard conditions presents a nominal energy gain of 0.011% when compared with unsorted method.

An interesting conclusion is that even the PV module having a Pmp manufacturing dispersion when compared to Imp and Isc as observed in the statistical analysis, the sorting method does not have a significant impact on the PV string energy output.

Although the method proposed has not obtained a relevant gain in terms of mismatch loss attributed to PV module manufacturing tolerances, the method could be applied for other urban solar plants with higher shading conditions than -1.7%. Or using Isc and Imp parameters for the same study case.

For these previously analysis, it is very important to understand and predict the shading behavior, in order to advance the PV string/array configuration and extract the maximum possible power from the PV system.

Considering that during the real conditions the PV module never operate at a solar irradiance of 1.000 W/m^2 and at a cell temperature of 25°C, for future work a third canopy (section C) composed of 395 PV modules is under construction parallel to the other two canopies (section A and B) that will provide the field data for fine tuning of this method.

Acknowledgements

The authors want to acknowledge the financial support by the Companhia Energética de São Paulo (CESP) and four other utility companies through the R&D public proposal from Brazilian government energy agency (ANEEL); the sponsors of Laboratory of Interdisciplinary Integrated Technologies (LSI-TEC);

References

[1] SolarPowerEurope. Global market outlook for solar power 2015–2019. Technical report, Solar Power Europe, (2015).

[2] ANEEL. Resolução normativa n. 482/2012, December (2012).

[3] IEA-PVPS. Trends in photovoltaic applications. survey report of selected iea countries between 1992 and 2011. Technical report, IEA, (2012).

[4] ANEEL. Registro de micro e mini geradores distribuídos, December (2016).

[5] R. Herrero, E. Melo, S. Shimura, C. B. De Moura, T. Costa, R. Simplicio, J. A. B. Grimonni, M. K. Zuffo. Comparing energy yield simulation in grid-connected 450 kwp parking-integrated photovoltaics-case study: Villa lobos project in São Paulo, Brazil. Proceedings 29th European Photovoltaic Solar Energy conference, (2014).

[6] A Nobre, R Malhotra, HT Chee, C Reise, K Kiefer, R Ruther, T Reindl. Degradation analysis of photovoltaic systems in a tropical environment. Proceedings 28th European Photovoltaic Solar Energy Conference & Exhibition, pages 3673–3677, (2013).

[7] A. Woyte, J. Nijs, R. Belmans. Partial shadowing of photovoltaic arrays with different system configurations: literature review and field test results. Solar Energy, 74(3):217–233, mar (2003).

[8] C. Zomer, A. Nobre, P. Cassatella, T. Reindl, R. Ruther. The balance between aesthetics and performance in building-integrated photovoltaics in the tropics. Progress in Photovoltaics: Research and Application, 22(7):744–756, oct (2013).

[9] A. M. Pavan, A Tessarolo, N. Barbini, A. Mellit, and V. Lughi. The effect of manufacturing mismatch on energy production for large-scale photovoltaic plants. Solar Energy, 117:282–289, (2015).

[10] L. Serrano-Luján, R. Garca-Valverde, N. Espinosa, M. Socorro Garca-Cascales, J. M. Sánchez-Lozano, A. Urbina. Environmental benefits of parking-integrated photovoltaics: a 222 kWp experience. Progress in Photovoltaics: Research and Application, 23(2):253–264, (2013).

[11] M. A. Green, K. Emery, Y. Hishikawa, W.Warta, E. D. Dunlop. Solar cell efficiency tables (version 47). Progress in Photovoltaics: Research and Applications, 24(NREL/JA-5J00-65643), (2016).

[12] H. Saha, G. Bhattacharya, and D. Mukherjee. Mismatch losses in series combinations of silicon solar cell modules. Solar Cells, 25(2):143–153, nov (1988).

[13] F. P. Gasparin, A. J. Bühler, G. A. Rampinelli, A. Krenzinger. Statistical analysis of i–v curve parameters from photovoltaic modules. Solar Energy, 131:30–38, (2016).

[14] Heinrich Häberlin. Photovoltaics. Wiley-Blackwell, feb (2012).

[15] L. L. Bucciarelli. Power loss in photovoltaic arrays due to mismatch in cell characteristics. Solar Energy, 23(4):277–288, (1979).

[16] J. Feldman, S. Singer, A. Braunsten. Solar cell interconnections and the shadow problem. Solar Energy, 26(5):419–428, (1981).

[17] C. Gonzalez. Photovoltaic array loss mechanisms. Solar Cells, 18(3-4):373–382, sep (1986).

[18] J.W. Bishop. Computer simulation of the effects of electrical mismatches in photovoltaic cell interconnection circuits. Solar Cells, 25(1):73–89, oct (1988).

[19] R Zilles, E Lorenzo. An analytical model for mismatch losses in pv arrays. International journal of solar energy, 13(2):121–133, (1992).

[20] C. E. Chamberlin, P. Lehman, J. Zoellick, G. Pauletto. Effects of mismatch losses in photovoltaic arrays. Solar Energy, 54(3):165–171, (1995).

[21] F. Iannone, G. Noviello, A. Sarno. Monte Carlo techniques to analyse the electrical mismatch losses in large-scale photovoltaic generators. Solar Energy, 62(2):85–92, (1998).

[22] ISE CalLab Fraunhofer. Annual report: Achievements and Results, (2004).

[23] S. Silvestre and A. Chouder. Effects of shadowing on photovoltaic module performance. Progress in Photovoltaics: Research and Application, 16(2):141–149, (2008).

[24] A. Chouder and S. Silvestre. Analysis model of mismatch power losses in PV systems. Journal of Solar Energy Engineering, 131(2):024504, (2009).

[25] P. Bakas, A. Marinopoulos, B. Stridh. Impact of pv module mismatch on the pv array energy yield and comparison of module, string and central mppt. 38th IEEE Photovoltaic Specialists Conference, pages 1393–1398, (2012).

[26] J. Webber and E. Riley. Mismatch loss reduction in photovoltaic arrays as a result of sorting photovoltaic modules by max-power parameters. Renewable Energy, (2013).

[27] A. M. Pavan, A. Mellit, D. D. Pieri, V. Lughi. A study on the mismatch effect due to the use of different photovoltaic modules classes in large-scale solar parks. Progress in Photovoltaics: Research and Application, 22(3):332–345, (2012).

OUTDOOR FIELD PERFORMANCE FROM BIFACIAL PHOTOVOLTAIC MODULES AND SYSTEMS

Joshua S. Stein[1], Daniel Riley[1], Matthew Lave[1], Chris Deline[2], Fatima Toor[3], Clifford Hansen[1]
[1]Sandia National Laboratories, Albuquerque, NM, USA;
[2]National Renewable Energy Laboratory, Golden, CO, USA
[3]University of Iowa, Iowa City, IA, USA

ABSTRACT: Bifacial PV modules and systems deliver more energy than equivalent monofacial modules in the same orientation. However, bifacial performance models are not yet mature enough to predict bifacial gains for all system configurations. Field performance data is needed at a number of different spatial scales in order to improve and validate these models. This paper reports on a number of bifacial field installations intended for this purpose.
Keywords: Bifacial; PV Array; PV Module; PV System; Photovoltaic; Performance

1 INTRODUCTION

Bifacial photovoltaic (PV) cells, modules, and systems offer a rapid pathway to significantly decreased levelized cost of energy compared with conventional monofacial PV modules. Unlike increasing cell efficiency, which takes many years to bring laboratory innovations to the production line, bifacial PV technology is available today but is underutilized. One major barrier to broader use of bifacial PV modules and systems is a lack of knowledge and experience with system designs that take advantage of the specific features of bifacial cells. Bifacial system performance cannot yet be predicted with confidence using current PV performance modeling applications because these tools typically assume that PV modules are illuminated on only one side. However, recent updates to software such as PVsyst are including the capability to model bifacial PV systems in limited configurations.

Analytic and empirical studies have shown that use of bifacial modules can potentially increase system yield achieved by at least 10% over a fixed latitude tilt monofacial array, and increased yield can be much higher under certain conditions. The bifacial benefit appears to increases with tilt angle, module height above ground surface, reflectivity of the ground, and other factors that influence the total amount of light reaching both sides of the PV cells. However, the sensitivity to these parameters is complex and as system size and ground coverage ratio increases, bifacial gains suffer as the array increasingly covers the ground with shadows and less light and more spatially vairable light is available to the back of the modules.

In order to better understand the factors that affect bifacial PV system performance Sandia National Laboratories (Sandia), the National Renewable Energy Laboratory (NREL) and the University of Iowa (UIowa) have teamed on a three-year research project aimed at better understanding the actual performance potential of bifacial PV systems.

The project aims to achieve the following three objectives:

1. Obtain field performance data from bifacial modules, strings, and arrays in a variety of orientations and environments.
2. Develop and standardize bifacial module rating methodology
3. Develop and validate bifacial performance models that can be used to inform bifacial array designs.

This paper describes current results obtained from the first objective. More results from this project are available [1-7].

2 BIFACIAL FIELD TESTING

Sandia has built a number of field testbeds using several types of bifacial PV modules to obtain performance data in a number of different configurations. In most of these set-ups we have included reference monofacial modules of the same size as comparisons. The following bifacial testbeds have been developed and are discussed here:

- Multiple modules, individually monitored on microinverters at five different orientations at three different climate sites (16 bifacial modules per site)
- Four vertical modules, individually monitored on microinverters in Turku Finland (60° N).
- String-level performance at four different fixed-tilt angles
- Bifacial string performance on single axis trackers
- Bifacial string performance on two dual-axis trackers

In this paper we are reporting on three different bifacial module technologies including:

- Prism Solar (co-diffused n-type mono silicon) 60-cell, glass-glass frameless
- SolarWorld (p-type mono-PERC) 72-cell, glass-clear backsheet, framed
- Sunpreme (HIT) 60 cell glass-glass, frameless

Each of these module types have different performance characteristics and production methods and costs. The n-type Si cells used by Prism Solar are more expensive to produce than p-type, but have low degradation rates, high bifacial ratios and experience little to no light induced degradation (LID). The p-type mono-PERC cells used by SolarWorld are currently less expensive to produce than n-type or HIT cells but have lower bifacial ratios and suffer from LID, which may recover over time in the field [8]. The HIT cells used by Sunpreme are much more expensive to produce but have high efficiency and bifacial ratios and a lower temperature coefficient. Differences between the modules are that both Prism Solar and Sunpreme use a glass-glass, frameless construction, while SolarWorld bifacial modules use a glass-clear backsheet with a frame, which is compatible with standard mounting systems.

2.1 Multiple modules, individually monitored on microinverters at five different orientations at three different climate sites.

Three nearly identical test systems are comprised of 16 bifacial and 16 monofacial modules divided into five different configurations that vary tilt and azimuth angles as well as ground albedo. Fig 1 shows the installed system in New Mexico (NM). Table I describes the five different configurations at this site. Nearly identical copies of this system are installed in Las Vegas, Nevada and Williston, Vermont. The differences are in the albedo between the ground surfaces at each site and tilted modules in Vermont are all at 30°. These systems are part of a project at the Regional Test Centers being performed with Prism Solar.

An additional module-scale bifacial system is deployed at the University of Applied Sciences in Turku, Finland (60°N) to test the performance of vertical, E-W facing bifacial modules from Prism Solar at high latitudes (Fig 2).

Figure 1: Prism Solar test array in New Mexico.

Table I: Orientation and ground surface of Prism Solar test system in New Mexico.

Label	Orientation		Ground Surface
	Tilt	Azimuth	
S15Wht*	15°	180° (South)	White gravel
W15Wht*	15°	270° (West)	White gravel
S30Nat	30°	180° (South)	Natural
S90	90°	180° (South)	Natural
W90	90°	270° (West)	Natural

*30° tilt in Vermont

Figure 2: Vertical bifacial array in Turku, Finland.

2.2 String-level performance at different fixed tilt angles

This system is focused on discovering how bifacial modules behave in series strings at different tilts. Sandia built four rows of racking, each at a different tilt angle (45°, 35°, 25°, 15°). Each row has two strings of eight modules (one bifacial and one monofacial). Bifacial modules are from Sunpreme (15° and 35°) and Prism Solar (25° and 45°). Fig 3 shows that bifacial and monofacial modules are interleaved to avoid any east or west bias in the light behind the rows. However, in the rows with lower tilt angles, there is some partial shading of the bifacial modules from the thicker adjacent monofacial modules.

Figure 3: String-level bifacial and monofacial test array in New Mexico.

2.3 String-level performance on single axis trackers

Sandia has also installed two rows of single axis trackers designed to hold four strings of bifacial modules. While this array is still under construction, front and back side irradiance data is available during tracker operation and is used estimate performance.

Figure 4: Single axis tracker with bifacial module strings in New Mexico.

2.4 Bifacial string performance on a two-axis tracker

As part of the Regional Test Center program, two dual-axis trackers from AllEarth Renewables have been installed in Williston, Vermont, each holding two strings (one of bifacial modules and one of monofacial modules) (Fig 5).

Figure 5: Two axis trackers in Vermont with bifacial and monofacial strings on each.

3 PERFORMANCE RESULTS

All of the test beds described above have been collecting data and some initial results are shared below.

Instantaneous bifacial gain at time t, $BG_i(t)$ is defined here as:

$$BG_i(t) = 100\% \times \left(\frac{P_{bifacial}(t)/P0_{bifacial}}{P_{monofacial}(t)/P0_{monofacial}} - 1\right) \quad (1)$$

where $P_{bifacial}$ and $P_{monofacial}$ are measured power values and $P0_{bifaical}$ and $P0_{monofacial}$ are front side power ratings measured on a flash tester at STC with the back of the bifacial module covered with an opaque material. An integrated bifacial gain in energy, BG_E (for example, one month) can be calculated as:

$$BG_E = 100\% \times \left(\frac{\sum_{1\,month} P_{bifacial} \;/\; P0_{bifacial}}{\sum_{1\,month} P_{monofacial} \;/\; P0_{monofacial}} - 1\right) \quad (2)$$

When performance from monofacial reference modules is not available, we compute a potential bifacial gain (BG_P) from measured front and back side irradiance as:

$$BG_P = 100\% \times R_b\left(\frac{G_f + G_r}{G_f} - 1\right) \quad (3)$$

where R_b is the bifacial ratio (quotient of backside to frontside flash rating) and G_f and G_r are the plane-of-array irradiance on the front and rear sides, respectively.

3.1 Single module monitoring on microinverters at five different orientations at three different climate sites.

As of September 2017, the Prism Solar RTC arrays have been operating for 19 months in New Mexico, 8 months in Nevada, and 4 months in Vermont. Initial results comparing total energy produced and bifacial gains in energy are shown in Figs 6, 9, and 10. Fig. 7 shows the avarage power and bifacial gains in New Mexico as a function of solar time and Fig 8. compares these quantities by month.

The main conclusions from these figures are that bifacial modules in these arrays significantly outperform the monofacial references. Bifacial gains vary between 15% and over 100%. The highest gains are associated with orientations that are not optimal for monofacial (e.g., W90). S15Wht and W15Wht arrays in New Mexico (albedo = 0.55) produced as much as or greater amount of energy than the monofacial at near latitude-tilt (S30Nat), which proves that low tilt bifacial over a high albedo surface can match or exceed latitude-tilt monofacial.

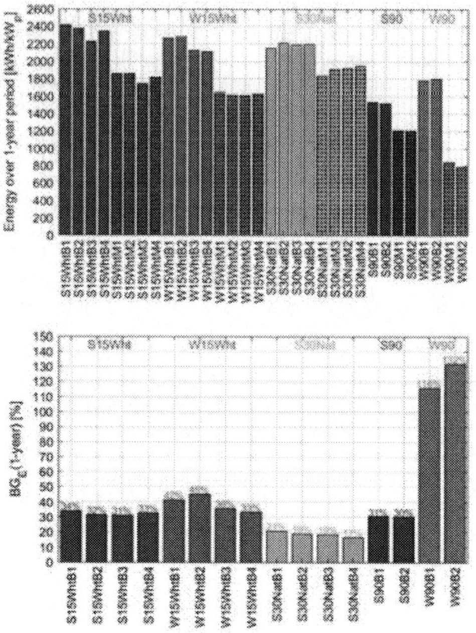

Figure 6: Chart showing total energy and bifacial gain from each module in the Prism Solar RTC array in New Mexico for the first 12 months of study.

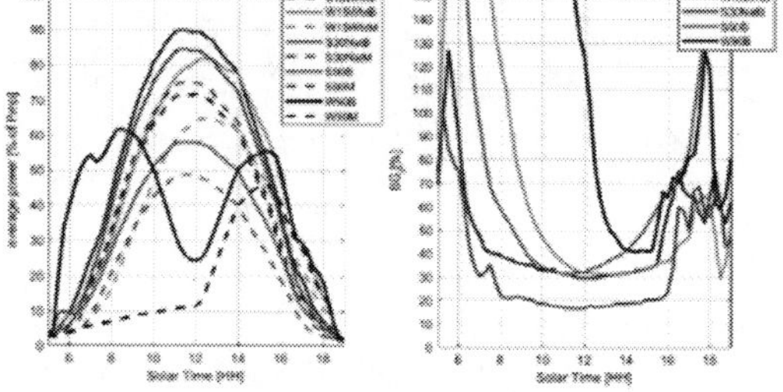

Figure 7: Average power (left) and bifacial gain (right) for Prism Solar RTC array in New Mexico for the first 12 months of study.

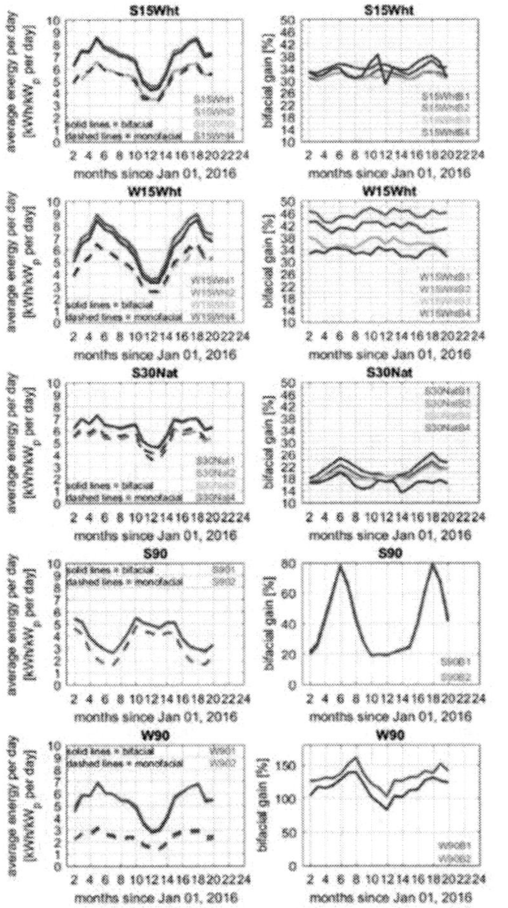

Figure 8: Monthly average power (left) and monthly bifacial gain (right) for Prism Solar RTC arrays in New Mexico.

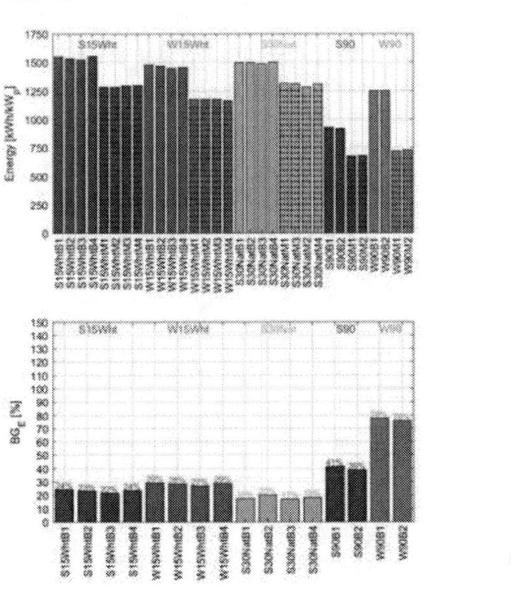

Figure 9: Chart showing total energy and bifacial gain from each module in the Prism Solar RTC array in Nevada for the first 8 months of study.

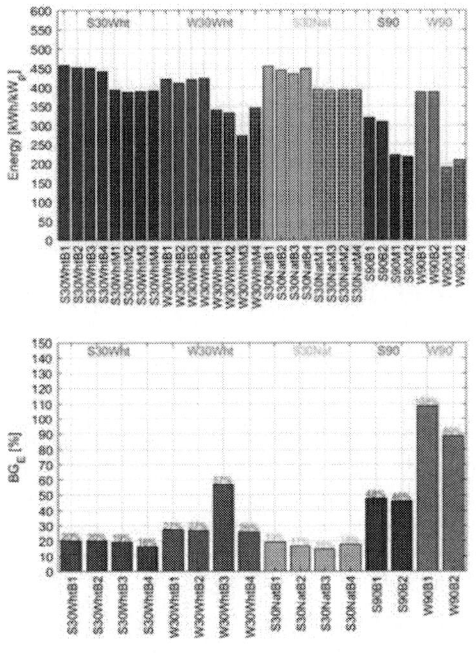

Figure 10: Chart showing total energy and bifacial gain from each module in the Prism Solar RTC array in Vermont for the first 4 months of study.

Initial results from the vertical bifacial system in Finland are compared with a monofacial reference system with an azimuth of 220° (~SW) and tilt of 43° in Fig 11. For the three days shown, the bifacial modules produced 37% more energy (normalized to front side rating). Power output from the vertical bifacial array occurs earlier and produces power for longer time periods than for the monofacial array.

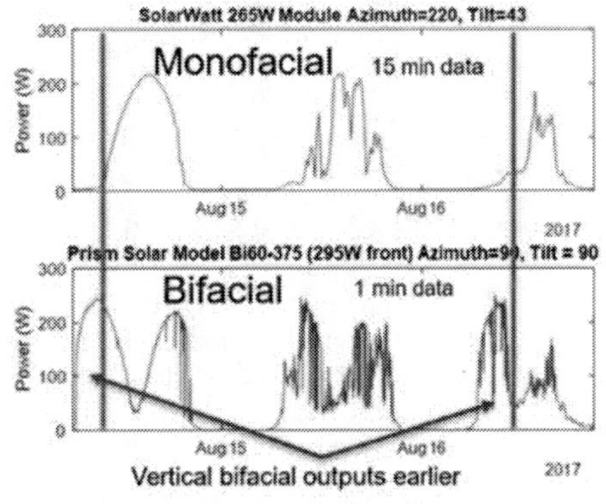

Figure 11: Three-day comparison of output from monofacial and vertical bifacial modules in Turku, Finland.

3.2 String-level performance at different fixed tilt angles

The performance difference between strings of bifacial and monofacial modules in New Mexico is compared in Figs. 12-13. Fig 12 compares the instantaneous bifacial gains plotted against sun azimuth. This plot shows how bifacial gains tend to be greater for S-facing bifacial arrays at the beginning and end of the day when the sun angles are higher. An exception occurs for the 15° and 25° tilted arrays due to partial shading caused by modules of differing thicknesses during these times. Fig 13 compares the energy produced from each string. It is notable that the energy produced appears to be inversely proportional to BG_E. An exception is that the BG_E for Sunpreme 15° and Prism 25° are nearly equal. This result may not be representative due to the partial shading in the lower titled arrays. Also worth noting is that the Sunpreme and Prism Solar modules use different cell technologies (HIT and n-type c-Si, respectively) and have different temperature coefficients and performance characteristics that cause some of the differences that are not related to the modules' bifaciality.

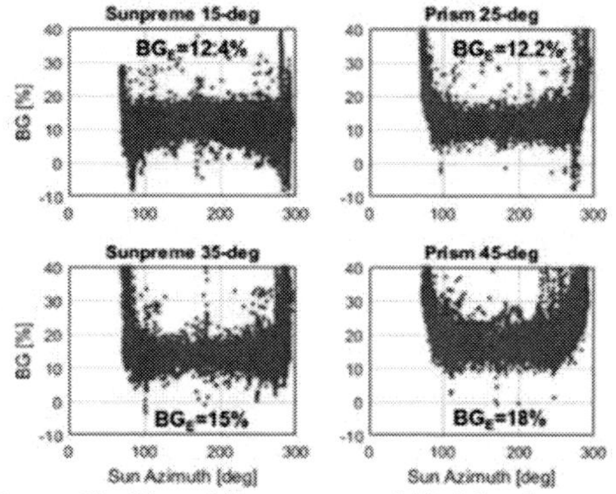

Figure 12: Plot of instantaneous bifacial gains vs. sun azimuth from strings of Prism Solar and Sunpreme modules from June 1 to Aug 31, 2017 in New Mexico (albedo = ~0.25).

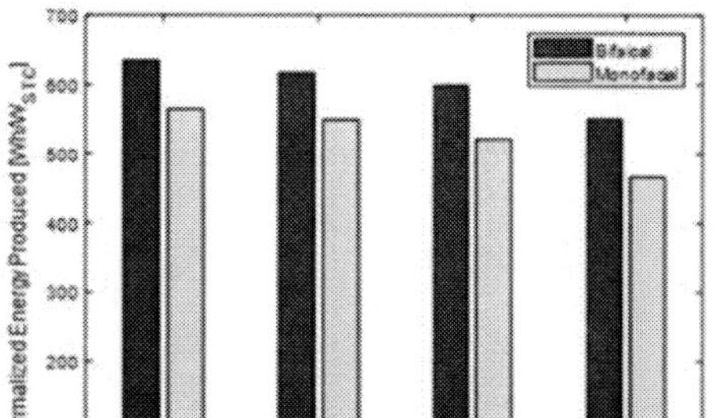

Figure 13: Normalized energy produced from bifacial and monofacial strings for each tilt angle in New Mexico.

3.3 String-level performance on single axis trackers

Because there are currently only two strings of bifacial modules installed on the single axis trackers in New Mexico, it is not possible to report bifacial gains for this array. Instead, we use the measured front and back side irradiance to estimate the potential bifacial gain (BG_P) (assuming the bifacial ratio =1) for each tracker. It should be noted that these trackers are controlled by light sensors which sometimes cause the trackers to face the wrong direction relative to the sum position. The data shown in Fig. 14 is calculated only when the tracker is within +/- 5° if its correct position. The estimated BG_P for these trackers is approximately 10%, but would be lower in the case of $R_b <$ 1.

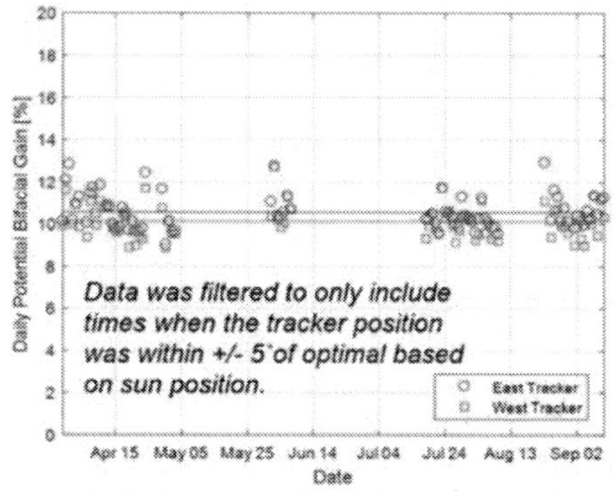

Figure 14: Daily potential bifacial gains for single axis trackers in New Mexico (albedo = ~0.25).

3.4 Bifacial string performance on two-axis trackers

Bifacial gains in energy were measured on two different two axis trackers each with a different bifacial module type. The first tracker used 60-cell Prism Solar bifacial modules with a R_b = ~0.93. The second used 72-cell SolarWorld framed bifacial modules with a lower R_b = ~0.62. Monofacial modules on both trackers are from SolarWorld. Both trackers have significant obstructions behind the bifacial modules that results in backside irradiance being partially blocked. Figs. 15 and 16 show the mean daily BG_E for the Prism Solar and SolarWorld strings, respectively. The Prism string BG_E is ~11% and greater than the SolarWorld (~6%) due to the difference in R_b. Using the measured irradiance and R_b for each module type, we calculated an average BG_P of 15.6% (Prism) and 9.5% (SolarWorld), which suggests that the obstructions cause approximately 4-5 percentage points of reduction in BG_E for each system. Both module types are perforing similarly when the differences in their bifacial ratos (R_b) are considered.

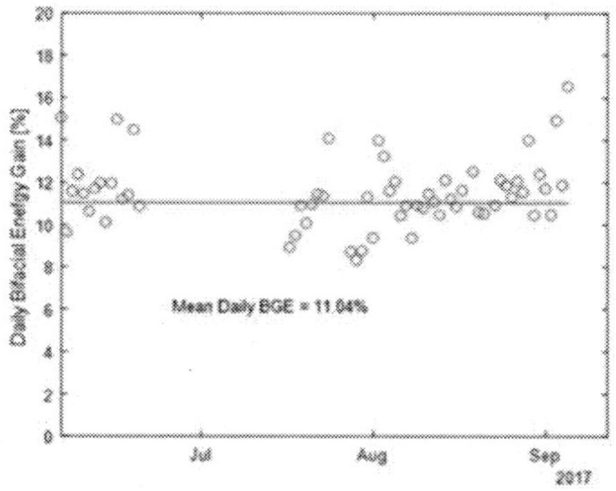

Figure 15: Daily bifacial gain in energy for a string of Prism Solar modules on a two axis tracker in Vermont (albedo = ~0.09).

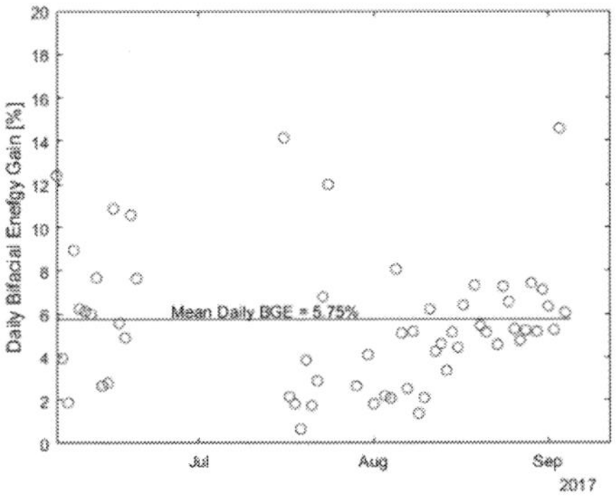

Figure 16: Daily bifacial gain in energy for a string of SolarWorld modules on a two axis tracker in Vermont (albedo = ~0.09).

4 CONCLUSIONS

Field data from a collection of bifacial PV systems in New Mexico, Nevada, Vermont, and Finland has been analyzed and compared with similar monofacial arrays and measured front and back irradiance. The following conclusions can be drawn from this analysis.

- Bifacial performance always exceeds monofacial performance when module output is normalized for front side STC rating and the rear side receives some amount of light.
- Bifacial gains increase as the orientation of the front side of the array (tilt and azimuth) deviates from the optimal orientation for monofacial.
- However, total energy production of tilted bifacial systems appears to be maximized at the same orientation as for monofacial modules. One exception is E-W bifacial vertical modules, which can outperform optimally oriented monofacial modules, especially with enhanced albedo. Other exceptions may exist, especially in cases with enhanced albedo.
- Bifacial gains for single bifacial modules and small systems are significantly higher than for larger systems. This is because a larger fraction of modules is at the edges of smaller systems and therefore more rear side irradiance is available. Bifacial gains for large multi row systems are expected to be significantly lower.
- Bifacial module performance benefits from module-scale MPPT. Rear-side irradiance varies significantly in space throughout the array leading to current mismatch in series connected modules.
- Bifacial gain of isolated modules and small arrays improves as the array height increases. This is because the module's view of the ground increases and light from more distant (unshaded) surfaces is available to the rear side. This is especially true for lower sun angles when shadows from modules high off the ground appear further away from the array. This is likely one of the reasons that the bifacial performance on the 2-axis trackers in VT was so high despite significant back side obstructions from the tracker supports.
- Bifacial performance is very sensitive to enhanced albedo of the ground surface.
- Vertical E-W bifacial modules produce energy earlier and later in the day than S-facing arrays. Such an output power profile may better match demand for electricity and could be a beneficial design under time of use rates.
- One must be careful when comparing different bifacial modules as they are not all alike. The bifacial ratio (flash rating of the back at STC divided by the front) can differ significantly between modules from different companies. Cell and module characteristics, module front side rating, bifacial ratio, temperature coefficient, and price all have to be considered when choosing the best bifacial module for a given project.

6 ACKNOWLEGEMENTS

Sandia National Laboratories is a multimission laboratory managed and operated by National Technology and Engineering Solutions of Sandia LLC, a wholly owned subsidiary of Honeywell International Inc. for the U.S. Department of Energy's National Nuclear Security Administration under contract DE-NA0003525.

7 REFERENCES

[1] Deline, C., S. MacAlpine, B. Marion, F. Toor, A. Asgharzadeh and J. S. Stein (2017). "Assessment of Bifacial Photovoltaic Module Power Rating Methodologies – Inside and Out." Journal of Photovoltaics 7(2): 575-580.

[2] Hansen, C. W., R. Gooding, N. Guay, D. M. Riley, J. Kallickal, D. Ellibee, A. Asgharzadeh, B. Marion, F. Toor and J. S. Stein (2017). A Detailed Model of Rear-Side Irradiance for Bifacial PV Modules. 44th IEEE Photovoltaic Specialists Conference. Washington DC. SAND2017-6554 C.

[3] Marion, B., S. MacAlpine, C. Deline, A. Asgharzadeh, F. Toor, D. Riley, J. Stein and C. Hansen (2017). A Practical Irradiance Model for Bifacial PV Modules. 44th IEEE Photovoltaic Specialists Conference. Washington DC. SAND2017-7000C.

[4] Riley, D., C. Hansen, J. Stein, M. Lave, J. K. B. Marion and F. Toor (2017). A Performance Model for Bifacial PV Modules. 44th IEEE Photovoltaic Specialists Conference. Washington, DC. SAND2017-6586C.

[5] Shishavan, A. A., T. M. Lubenow, J. Sink, B. Marion, C. Deline, C. Hansen, J. Stein and F. Toor (2017). Analysis of the Impact of Installation Parameters and System Size on Bifacial Gain and Energy Yield of PV Systems. 44th IEEE Photovoltaic Specialists Conference. Washington, DC. SAND2017-7001C.

[6] Stein, J. S., L. Burnham and M. Lave (2017). One Year Performance Results for the Prism Solar Installation at the New Mexico Regional Test Center: Field Data from February 15, 2016 - February 14, 2017. Albuquerque, NM, Sandia National Laboratories. SAND2017-5872.

[7] Stein, J. S., D. Riley, M. Lave, C. Deline and F. Toor (2017). Outdoor Field Performance from Bifacial Photovoltaic Modules and Systems. 44th IEEE Photovoltaic Specialists Conference. Washington, DC. SAND2017-6464C.

[8] Nicolai, M., M. Zanuccoli, M. Galiazzo, M. Bertazzo, E. Sangiorgi and C. Fiegna (2016). "Simulation Study of Light-induced, Current-induced Degradation and

Recovery on PERC Solar Cells." Energy Procedia 92(Supplement C): 153-159.

PERFORMANCE ESTIMATION OF BIFACIAL PV MODULES: A SIMULATION APPROACH THROUGH BOTH PHYSICAL AND SEMI-EMPIRICAL MATH MODELS AND ITS VALIDATION USING REAL BIFACIAL PLANT DATA.

M. Catena, I. Cascone, M. Carbone
ENEL GREEN POWER S.p.A.
Via Della Bufalotta 255 - 00139 Rome - ITALY
Phone: +39 095 287 7061 - E-mail: mario.catena@enel.com

ABSTRACT: Bifacial PV modules seem to be the most promising technology in the field of photovoltaic generation. The simple and innovative idea to build a module with two active faces seems easy and straightforward but now engineers have to face a new challenge: the determination of the effective energetic yield of both faces.
This paper will cover a computation method using both physical and semi-empirical math models, implemented in a custom developed software tool, to determine the total irradiance that hits the modules, on both sides, forecast the energy production and finally validate the achieved results with the experimental data of an in-operation 570 kWp bifacial PV plant.
This paper will show also an effective approach to find the optimal geometrical PV field layout for bifacial application as well the integration of the presented computation methods with existing PV simulation software packages.

Keywords: Bifacial, Design, Software, Simulation, Solar Radiation

1 INTRODUCTION

Bifacial PV modules can boost the performance of PV plants by 25-30% (even more using reflecting solutions) depending on their electrical behavior, mounting structures and terrain characteristics.

Therefore, computing performance of bifacial PV plants requires some attention due to many factors that can influence the overall production.

This paper will show computing methods as well as custom implemented software tools designed to be useful during the design and/or simulation process.

To determine accurately the amount of radiant energy that can be transformed into electricity by a bifacial PV field the following steps are necessary:

- Determination of soil albedo;
- Estimation of rear irradiance;
- Estimation of total irradiance that hits the module.

2 DETERMINATION OF SOIL ALBEDO

Albedo is a measure for *reflectance* or optical brightness (Latin *albedo*, "whiteness"). It is dimensionless and measured on a scale from zero (corresponding to a black body that absorbs all incident radiation) to one (corresponding to a white body that reflects all incident radiation).

Bifacial module production is strongly affected by terrain reflectance. Radiant flux reflected by terrain hits the rear active side of the bifacial PV module increasing its overall performance.

The percentage of reflected irradiance from the terrain has been always a challenge for designers of bifacial PV module-based plants since the irradiance that hits the backside of the module (and leads to a slight increase of total energy yield) is primarily dependent upon the characteristics of the terrain.

The conventional method of measuring the reflective property of the terrain is to install an instrument that is commonly called albedo-meter. It is composed of two identical pyranometers, one imposed over the other. Measurement is done using the definition of albedo: the ratio between reflected radiation and global radiation, i.e. the ratio of the two pyranometers' output. These instruments are expensive and need a fixed installation and a medium to long data acquisition campaign.

2.1 Albedo measuring – A software solution

As a workaround solution instead of using an albedo-meter, we developed a software tool, RefCalc, which is able to access the reflectance information that is stored within the pixels of a digital image.

By retrieving and manipulating such information with a sophisticated spectrum-derivative algorithm, it is possible to measure the reflectance using a simple digital image that can be taken quickly using a standard digital camera or even a cellphone.

The big advantage is that there is no long acquisition campaign, and no heavy and expensive equipment. It is as easy as taking a terrain snapshot.

Of course, there are some required specifications on how to take the picture.

The following image is an example of a terrain picture used in conjunction with RefCalc:

Figure 1: RefCalc Sample Picture

The bottom part of the picture is a white matte sheet of paper, although if available, a sheet of high reflectance material like Spectralon® is preferable. This area is used for reference purposes and acts as a baseline to have a relative measurement despite the irradiance conditions.

Any camera will work, whether a cell phone or SLR compact device, and resolution does not play a significant role due to the analysis algorithm. Nonetheless, a minimum resolution of 5 MP is preferable.

Pictures must be taken following some minimum requirements and rules:

- Ideally, photos should be taken between 11:00 and 13:00 (11 am – 1 pm) local time. In those hours, daylight conditions and sun height are optimal despite the irradiance.
- Avoid taking pictures in presence of clouds and other weather conditions since they can interfere with albedo measurements.
- Hold the camera at about 1 meter, perpendicular to the terrain;
- Do not cast a shadow on the terrain;
- Avoid using the camera flash.

Below in Fig. 2 is a screenshot of the software computing the reflectance of the previously seen soil picture.

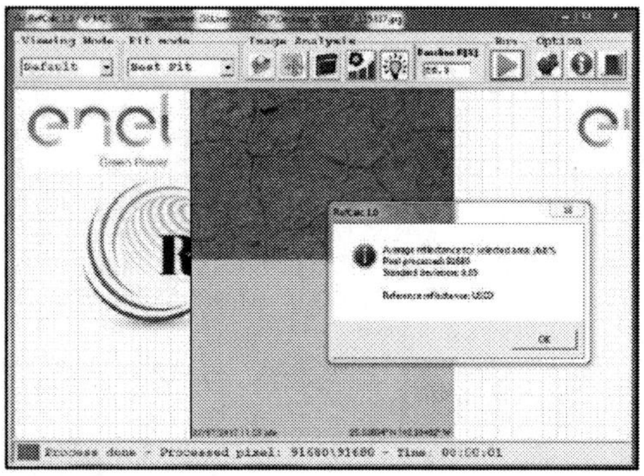

Figure 2: RefCalc screenshot showing reflectance computing of the soil picture

2.2 RefCalc – The algorithm

The Red, Green, Blue (RGB) digital values coming from a digital device, such as a smartphone or a digital camera, can be represented in the form of linear algebra as:

$$T = S \cdot I \cdot R$$

Where T is the matrix of RGB values, S is the spectral sensitivity matrix of the device (obtained by literature or knowing the device characteristics, i.e. if CMOS or CCD), I is the overall intensity (obtained via photo-histogram technique) and R is the matrix of reflectance that we want to obtain.

Matrix R that contains the spectral reflectance is represented with just the above three principal components through a matrix reduction. Therefore, the reflectance of the input spectrum R_S is computed as the product of the principal component matrix and the matrix of coefficients.

$$R_S = k \cdot B \cdot c$$

Matrix B is obtained from R with a reducing and normalizing Principal Component Analysis (PCA) algorithm, the term c is the matrix coefficient and finally k is a constant used to normalize the reconstructed spectrum, and in RefCalc represents the reference reflectance value of the media (the white area visible in the picture).

Finally averaging the R_S matrix, we will find the parameter we need: the terrain reflectance.

The software (see Fig. 3) will output the average soil reflectance as well as the standard deviation of the calculation that gives a good estimation of the uniformity of the soil, i.e. presence of stones, debris, etc.

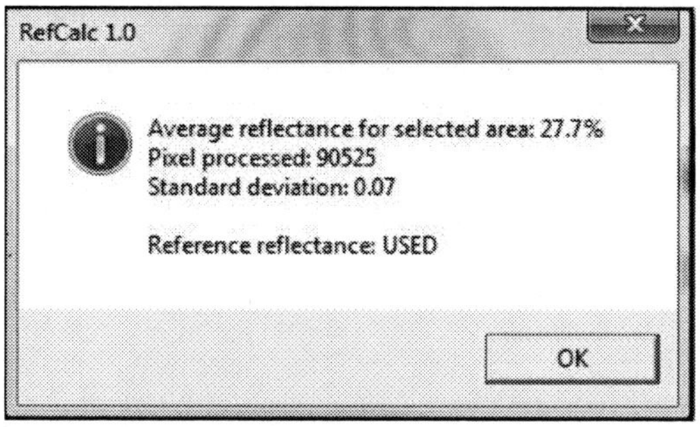

Figure 3: RefCalc output of simulation

3 ESTIMATION OF REAR IRRADIANCE

We have now obtained the terrain albedo but in the real world, a PV plant shaded system, the terrain reflection is influenced by the PV plant layout and typology (tracker, fixed differently tilted and/or oriented).

3.1 Method of estimation

There are several approaches in literature for representing the bifacial behavior and simulating its expected energy yield; some of them are too complex and many other methodologies are not so robust.

A good compromise between the many methods is what is proposed by Shoukry et al.[1] which specifies that the relevant parameters that should be chosen for modelling the bifacial PV module are: ground albedo; geographical location which include the sun's position; module height and tilt; global and diffuse irradiation; and module disposition, such as the row distance.

The total irradiance that reaches the back side of the module is the sum of the reflected irradiance from the two regions of the ground, the shadow region area and un-shadowed region. While the diffuse irradiance for the shadow area region is the only component that is reflected by the ground because the direct irradiance is blocked by the module, for the un-shadowed region area both direct and diffuse GHI components are reflected by the ground.

Therefore, according to Shoukry et al.[1], the reflected irradiance that strikes in the backside of the module is given as:

$$I_{refl\ back} = \rho \cdot GHI \cdot F_{Ansh \to Am} + \rho \cdot DHI \cdot F_{Ash \to Am}$$

Where ρ is the ground albedo of the site, GHI is the global irradiance at the horizontal position of the PV module and DHI is the diffuse component of the global irradiation.

FAnsh→Am and FAsh→Am are respectively the view factor between rear side of the module and the un-shadow area and the view factor between rear side of the module and the shadow area.

The total amount of the reflected irradiance captured by the rear side of the PV module can be described as follows:

$$I_{\text{refl back}} = \rho_n^{nsh} \cdot GHI + \rho_n^{sh} \cdot DHI \quad (1)$$

As noticed, we define ρ_n as *net albedo*, the reflectivity of soil that takes into account the effective shadowed and un-shadowed isotropic radiation reflection (i.e. assuming radiation has the same intensity regardless of the direction of emission) as actually viewed by the rear side of the modules mounting structures. This leads to the following *net albedo* expressions for both shadowed and un-shadowed areas.

The *net albedo* of un-shadow area is expressed as:

$$\rho_n^{nsh} = \rho \cdot F_{Ansh \to Am}$$

While the *net albedo* of the shadowed area is expressed as:

$$\rho_n^{sh} = \rho \cdot F_{Ash \to Am}$$

3.2 Determination of shadowed and unshadowed areas

The view factor is a geometric quantity (adimensional) that has its origin in the heat transfer formulation; we choose this method to determine the shadowed and unshadowed areas since it works with radiative transfer.

The view factor describes the fraction amount of the radiation that is exchanged between two generic surfaces; it mainly depends on the relative orientation to each other and the distance between them. The main assumption is that the surfaces are ideal diffuse reflectors and are independent of other surface properties and temperature (geometric property).

The analytical formulation for the calculation of the view factor between any orientation of surfaces A_1 and A_2 is the following:

$$F_{A1 \to A2} = \frac{1}{A_1} \int_{A1} \int_{A2} \frac{\cos\theta_1 \cdot \cos\theta_1}{\pi\, r^2} dA_1\, dA_2$$

Where r is the distance between the differential areas dA1 and dA2, A1 and A2 are respectively the angles between the normal vectors of the surfaces and the line that connects the differential areas (vector r).

Most of the common combinations of surface disposition can be found in manuals by means of tables (Howell [2]). One of the most important properties of the view factor is the unity rule, which means that the sum of all view factors from a given surface in an enclosure is equal to one, because the same amount of radiation emitted by a surface must be absorbed:

$$\sum_{j=1}^{n} F_{S_i \to S_j} = 1$$

The chosen method for the evaluation of the view factor is what Dobrowlsky et Al. [2] proposed, where in order to solve complex geometries, such as the ones found on module tracking system mounting structures, the total surface area is divided in triangles, see Fig. 5, generated by means of the Delaunay algorithm.

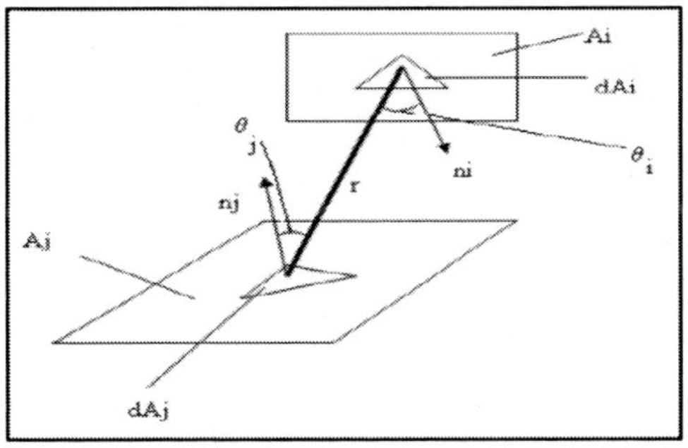

Figure 5: Delaunay algorithm approach as a solution of the view factor.

The view factor calculation can be re-written and approximated as:

$$F_{ij} = \frac{1}{A_i} \sum_{i=1}^{n} \sum_{j=1}^{m} \frac{\cos\theta_i \cdot \cos\theta_j}{\pi \cdot r^2} dA_i\, dA_j$$

Where r and ϑ_i and ϑ_j are given by:

$$\cos\theta_i = \frac{\ell_i(x_j - x_i) + m_i(y_j - y_i) + n_i(z_j - Z_i)}{r}$$

$$\cos\theta_i = \frac{\ell_j(x_i - x_j) + m_j(y_i - y_j) + n_j(z_i - Z_j)}{r}$$

Where $\mathbf{n_i}$ and $\mathbf{n_j}$ are respectively the versors of the two triangles represented in the Figure 5 that are composed with the following elements:

$$\vec{n}_i \equiv (\ell_i \quad m_i \quad n_i)$$

$$\vec{n}_j \equiv (\ell_j \quad m_j \quad n_j)$$

The principal advantage of Dobrowlsky et Al. [4] is related to the fact that it is the most comprehensive and accurate procedure and it can easily implemented in any programming language.

This method makes it possible to find the best PV mounting structure layout, which will maximize the energy that hits the rear side of the installed modules indeed maximizing the $\rho \cdot GHI \cdot F_{Ansh \to Am}$ factor.

4 ESTIMATION OF OVERALL IRRADIANCE

Once the shadow/unshadow ratio and the *net albedo* (starting from the original albedo value computed by RefCalc) are calculated, the next step is perform the calculation of equation (1).

To carry out the result, it is necessary to know the global horizontal irradiance GHI and the value of its main components: the direct irradiance, DNI and the diffuse irradiance DHI.

The relationship between those components, where Z is the solar zenith angle, is:

$$GHI = DHI + DNI \cdot \cos(Z)$$

This computation is more accurately managed using well known astronomical equations and a clear sky model, like the one proposed on the Sandia National Laboratory report [3], fitted to measured data (GHI radiation data provided by a satellite database or ground measured). Therefore, starting from a **T**ypical **M**eteorological **Y**ear (TMY) irradiance data file or even from averaged monthly data, it is possible to build a yearly profile of irradiance for both sides of the module. If there is no availability of hourly irradiance data, it is still possible to create an irradiance profile using solar distribution models.

The chosen model, based on the work of Collares-Pereira and Rabl and modified by Gueymard (CPRG) [4], offers a robust yet accurate way to predict the irradiance hourly trend. At the end of the process, we are able to obtain a yearly hour distribution of the irradiance received by both the front face and by the rear side of the module.

The **G**lobal **R**ear **I**rradiance (GRI) is the radiant flux received by the backside of the bifacial PV module, and **G**lobal **F**ront **I**rradiance (GFI) is the radiant flux received by the front face. The sum of these two components, defined as GBI, (**G**lobal **B**ifacial **I**rradiance, represents the total amount of solar energy received by the bifacial PV module.

Since any bifacial module does not convert the energy with the same efficiency on both its faces, we multiply the GRI for the BFF term, (**BiF**aciality **F**actor) i.e. the ratio of the power output at standard test condition (STC) from the rear side upon the power output at STC from the front side.

The global bifacial irradiance can now be written as:

$$GBI = GFI + GRI \cdot BFF$$

The complete method of calculation, bifacial layout view factors and irradiance computing is implemented in another custom-developed software tool, see Fig.4. SunEnergy:

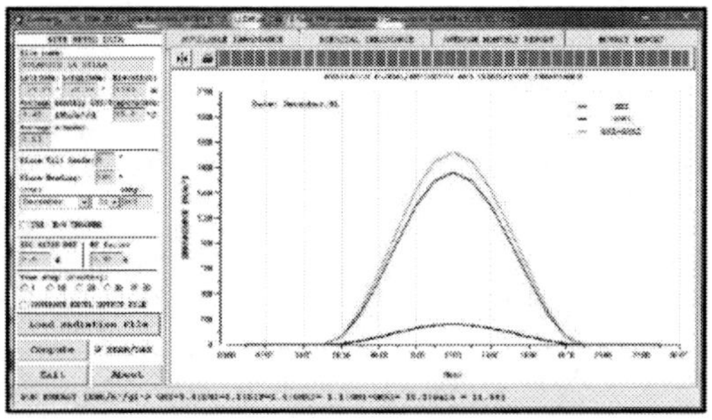

Figure 4: SunEnergy screenshot

This computed composite irradiance is used to determine the energy produced by a bifacial module given its STC electrical characteristics. The software produces an output in standard CSV format that can be imported by other design tools and used as irradiance descriptor file.

5 OVERALL IRRADIADIANCE COMPARISON

To validate the accuracy of the method and of the software developed, a comparison between the simulated output data and the real measured data coming from the SCADA of a real bifacial PV plant owned by EGP was performed.

5.1 Brief description of La Silla PV Plant

On September 26, 2015, EGP inaugurated its 1.7MW photovoltaic (PV) plant at the European Southern Observatory's (ESO), La Silla observatory in northern Chile.

The La Silla PV plant combines bifacial and smart modules with conventional modules for side-by-side testing. The smart modules have a microchip that allows each panel to deliver electricity regardless of any malfunctions affecting other panels, while the bifacial modules capture solar energy from both sides of the PV panel. Indeed, La Silla features a 570 kWp on a single axis E-W tracker field of bifacial MegaCell PV modules.

Data collected from installed sensors are using for model validation of the computed overall irradiance and produced energy

This has lead also to the determination of the net albedo i.e. the reflectance viewed by the rear side of the modules on tracker that mostly takes into account terrain reflectivity and tracker layout (size, pitch, elevation)

5.2 Validation of the model

Since we are validating data that are predicted by means of models versus actual measured data, the method of RMSE is more suitable for comparison and determination of the accuracy of the whole procedure.

The Root Mean Square Error is defined as:

$$RMSE = \left\{\left[\sum\left(r_{i,calc} - r_{i,meas}\right)^2\right]/n\right\}^{1/2}$$

RMSE provides information on the performance of the model used by allowing a term-by-term comparison of the actual deviation between the calculated value and the measured one. Table 1 shows the comparison between the model versus the measured data in both form of absolute error and RMSE deviation.

	AVERAGE DAILY DATA SET	
	GFI [W/m²]	GRI [kW/m²]
MEASURED	15560	1812
SIMULATED	15561	1825
SIM/MES [%]	0.01	0.71
RMSE	0.13	1.95

Table 1: comparison between measured vs. simulated data

The graph in Fig.5 shows the trend of measured irradiance versus the computed one. Both the back and rear irradiance are analyzed.

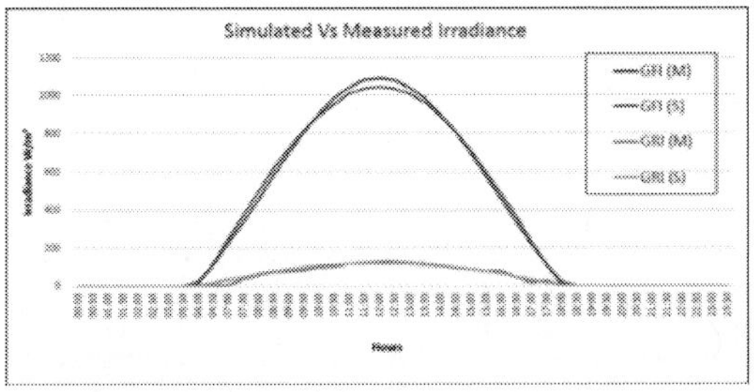

Figure 5: La Silla, simulated vs. measured irradiance trend.

The average daily data sets are data acquisition periods chosen taking into account data integrity, i.e. months when production and irradiance where congruent.

6 INTEGRATION IN OTHER SIMULATION TOOLS

We integrate the output of calculations made with the methods described in the previous paragraphs into PvSyst®, a well know PV design/simulation tool.

Basically, PvSyst needs to work a descriptor file of the PV module, a meteorological file (that contains the hourly annual irradiance and temperature values) and the PV Plant layout.

Therefore, given as input the irradiance output file generated by SunEnergy, the characteristics of the employed PV module (along with the electrical connection layout), and typology of mounting structure, it is possible to simulate the behavior of the bifacial PV part of the plant.

Knowing the measured monthly feed in-grid energy production, the result of the simulation made with PvSyst® are in Table 2:

	DATA SET 1		DATA SET 2		DATA SET 3	
	MW/h	EOQ	MW/h	EOQ	MW/h	EOQ
MEAS.	121.5	213.1	154.0	270.0	166.7	292.4
SIM.	125.3	219.7	154.3	270.6	167.6	293.9
RMSE	5.2		1.85		2.7	
E [%]	3.0		0.2		0.5	

Table 2: comparison between measured energy vs. simulated with PvSyst®.

The table shows both the AC energy production and the Equivalent Operating Hours (EOQ). The comparison is made between data coming from the plant acquisition system (AC side) and the output of the PvSyst simulation.

As in the previous chapter, data sets represent period of congruent data acquisition, checked for integrity and chosen between full data acquisition months.

7 CONCLUSIONS

The methods and software described in this paper have been successfully validated using a comparison with a real PV plant. The results obtained have proven that the approaches are valid and accurate.

Moreover, the integration with existing software offers a great feasibility of use.

Future development will include the development of a full featured custom design tool for bifacial PV plant.

8 REFERENCES AND FURTHER READING

[1] I.Shoukry, J. Libal, R. Kopecek, E. Wefringhaus, J. Werner, 6th International Conference on Silicon Photovoltaics, SiliconPV 2016, Energy Procedia 92 (2016) 600 – 608.

[2] L. Dobrowlsky de Carvalho Augusto, B. Giacomet and N. Mendes, NUMERICAL METHOD FOR CALCULATING VIEW FACTOR BETWEEN TWO SURFACES, Mechanic Engineering Graduate Program, Thermal Systems Laboratory, Pontifical University Catholic of the Paraná R. Imaculada Conceição, 1155 Curitiba-PR, 80215-901-Brazil, Proceeding: Building Simulation 2007.

[3] Matthew J. Reno, Clifford W. Hansen, Joshua S. Stein. "Global Horizontal Irradiance Clear Sky Models: Implementation and Analysis" - Sandia National Laboratories (2012)

[4] C. Gueymard. "Prediction and performance assessment of mean hourly global radiation", Solar Energy Vol. 68, No. 3, pp. 285–303, 2000.

[5] Howell, John R. A catalog of radiation configuration factors. McGraw-Hill Book Company, 1982.

[6] Gross, U., K. Spindler, and E. Hahne. "Shapefactor-equations for radiation heat transfer between plane rectangular surfaces of arbitrary position and size with parallel boundaries." Letters in heat and mass transfer 8.3 (1981): 219-227.

[7] Ville Heikkinen, Reiner Lenz, Tuija Jetsu et Al. "Evaluation and unification of some methods for estimating reflectance spectra from RGB images", J. Opt. Soc. Am. A/Vol. 25, No. 10/October 2008.

[8] Ufuk Alper Yusufoglua, Tae Hun Leea, Tobias Markus Pletzera et Al. "Simulation of energy production by bifacial modules with revision of ground reflection", Energy Procedia 55 (2014) 389 – 395

SIMULATION RESOLUTION OF PV SYSTEM PARTIAL SHADING STUDIES

Kari Lappalainen and Seppo Valkealahti
Tampere University of Technology, Laboratory of Electrical Energy Engineering
P.O. Box 692, FI-33101 Tampere, Finland
E-mail: kari.lappalainen@tut.fi, seppo.valkealahti@tut.fi

ABSTRACT: Many research groups have studied the operation of partially shaded photovoltaic (PV) systems during the past decades, and the well-known one-diode model of a PV cell has been mainly used to model the electrical behaviour of the PV systems. However, the effects of data sampling resolution have obtained little attention and even some meaningless results appear in the literature caused by unsuitable sampling resolution.

In this paper, the effects of the length of a simulation time step, the resolutions of irradiance and *I–U* curves and the size of a basic simulation unit on the accuracy of the simulations of PV system operation under partial shading conditions were studied. The operation of PV systems was simulated by using the one-diode model of a PV cell. The study aimed to find out adequate resolution for the simulation parameters to minimise the computing time while retaining sufficient accuracy of the simulations. The study was based on the apparent velocity and other characteristics of 7880 irradiance transitions caused by moving clouds and identified in measured irradiance data. The results show that the accuracy of 1 W/m^2 is sufficient for the irradiance and, accordingly, 1000 points for the *I–U* curves for PV string and array level partial shading studies.
Keywords: Modelling, PV system, Shading, Simulation

1 INTRODUCTION

Photovoltaic (PV) systems are prone to irradiance fluctuations caused by overpassing cloud shadows that are the main cause of fluctuating PV power production. Further, partial shading caused by moving clouds leads to mismatch losses of PV systems, which are the difference between the sum of the global maximum power point (MPP) powers of individual PV modules, as if they were operating separately, and the global MPP power of the whole PV system. Partial shading can also lead to failures in MPP tracking causing extra losses.

Solar radiation variability and irradiance transitions caused by the edges of moving cloud shadows have been studied in several papers, e.g. in [1]–[5]. A comprehensive study of the apparent velocity of shadow edges caused by moving clouds has been presented in [5]. When a cloud shadow is covering a PV array, the apparent speed of the shadow edge, i.e., the component of shadow speed normal to that shadow edge, actually defines how fast the PV array is becoming shaded. Thus, the apparent velocity of a linear shadow edge is a vital quantity in any analysis of the effects of overpassing cloud shadows on the operation of small PV systems and the PV arrays of large PV power plants. Still, the used assumption of linearity for the shadow edge might not be valid for large PV power plants as a whole [5].

The mismatch losses of PV generators caused by partial shading have been studied in several papers, especially during the past years, e.g. in [6]–[12]. In these studies, PV modules are mainly modelled using the well-known one-diode model of a PV cell. In [13], a model to study the effects of partial shading on PV array characteristics has been presented. During the past years, several studies have also been presented regarding the output power variation of PV systems, e.g. [14]–[16].

This paper presents a study on the effects of the length of a simulation time step, the resolutions of irradiance and *I–U* curves and the size of a basic simulation unit on the accuracy of the simulations of PV system operation under partial shading conditions. The studied PV systems were a string of 20 series connected PV modules and an array of 10 parallel strings of 20 modules. The study was based on the apparent velocity and other characteristics of 7880 irradiance transitions identified in measured irradiance data and was conducted using a mathematical model of irradiance transitions [3] and an experimentally verified MATLAB Simulink model of a PV module that is based on the well-known one-diode model of a PV cell. The aim of this study was to define adequate resolution for the simulation parameters to minimise the computing time while retaining sufficient accuracy of the results.

2 METHODS AND DATA

2.1 Simulation model for the PV modules

The electrical operation of the PV modules was modelled by an experimentally verified MATLAB Simulink model, based on the model presented in [17]. This model is based on the well-known one-diode model of a PV cell that provides the following relationship between the current and the voltage of the cell:

$$I = I_{ph} - I_0\left(e^{\frac{U+R_s I}{AU_T}} - 1\right) - \frac{U + R_s I}{R_{sh}}, \qquad (1)$$

where I is the current, I_{ph} the light-generated current, I_o the dark saturation current, U the voltage, R_s the series resistance, A the ideality factor, U_T the thermal voltage and R_{sh} the shunt resistance of the PV cell [18].

The simulation model for a PV module was obtained by scaling the parameter values used in the model of a PV cell by the number of PV cells in the PV module N_s. The thermal voltage of the PV module can be expressed as $U_T = N_s kT/q$, where k is the Boltzmann constant, T the temperature of the cell and q the elementary charge. Bypass diodes of the PV module were modelled using Eq. (1) by assuming that the light-generated current I_{ph} is zero and the shunt resistance R_{sh} is infinite. The dark saturation current $I_{o,\,bypass}$, the series resistance $R_{s,\,bypass}$ and the ideality factor A_{bypass} of the bypass diodes were determined by means of curve fitting to a measured *I–U* curve of a Schottky diode. The temperature of the bypass diodes was assumed to be constant and the same as the module temperature.

The characteristics of the simulation model were fitted to the characteristics of the NAPS NP190GKg PV modules used in the solar PV power station research plant of Tampere University of Technology (TUT). The NP190GKg module is composed of 54 series-connected polycrystalline silicon PV cells and three bypass diodes, each connected in anti-parallel with 18 PV cells. The electrical characteristics of the module, given by the manufacturer, in standard test conditions (STC) are presented in Table I. The parameter values of the simulation model for the PV modules and the bypass diodes are presented in Table II. The used simulation model, naturally, contains simplifications and assumptions. Moreover, the results of the simulations could slightly change if different PV modules were used as a reference. However, the basic behaviour would not change because the electrical characteristics of crystalline silicon PV modules do not differ substantially.

Table I: The electrical characteristics of the NAPS NP190GKg PV module for short-circuit (SC), open-circuit (OC) and MPP in STC.

Variables	PCC
$I_{SC,\,STC}$	8.02 A
$U_{OC,\,STC}$	33.1 V
$P_{MPP,\,STC}$	190 W
$I_{MPP,\,STC}$	7.33 A
$U_{MPP,\,STC}$	25.9 V

Table II: The parameter values of the simulation model for the PV modules and the bypass diodes.

Variables	PCC
A	1.30
R_s	0.329 Ω
R_{sh}	188 Ω
A_{bypass}	1.50
$R_{s,\,bypass}$	0.02 Ω
$I_{o,\,bypass}$	3.20 μA

2.2 Irradiance transitions and the shading of a PV array

The irradiance transitions caused by the edges of moving cloud shadows were modelled using the mathematical model presented in [3]

$$G(t) = \frac{G_{us} - G_s}{1 + e^{(t-t_0)/b}} + G_s, \tag{2}$$

where G is the irradiance, t the time, G_{us} the irradiance of an unshaded situation and G_s the irradiance under full shading. Parameter b is related to the steepness of the irradiance transition and the sign of it defines whether the transition is a fall or a rise, i.e., a decreasing or an increasing transition. Parameter t_0 adjusts the transition time defining the midpoint of the transition. In [3], the operation of the model has been validated with around 40 000 irradiance transitions identified in measured data. The average root-mean-square deviation between the curve fits and the measured transitions was around 12 W/m^2. The shading strength (SS), i.e., attenuation of irradiance due to shading, of an irradiance transition can be expressed as

$$\mathrm{SS} = \frac{G_{us} - G_s}{G_{us}}. \tag{3}$$

By utilising Eq. (2), irradiance transitions can be defined by using four variables: SS, parameter b and apparent speed and direction of movement, which have no correlation with each other [5], [19]. The duration of an irradiance transition was calculated by multiplying b by the experimentally obtained regression coefficient of 7.67 [19], and the parameter t_0 was then calculated from the duration.

The data used in this study was obtained from the TUT solar PV power station research plant located in Tampere, Finland. A total of 7880 irradiance transitions, consisting of 3834 falls and 4046 rises, were identified in 4 months (May–August 2013) of data, measured by irradiance sensors S2, S5 and S6 (see [20]) by using the method presented in [3]. The sensors were photodiode-based SP Lite2 pyranometers (Kipp & Zonen) mounted at a tilt angle of 45° from the horizontal plane and oriented nearly due south. In the identification of the irradiance transitions, the 40% limit of minimum acknowledged SS was used because it has been presented in [3] that moving shadows with lower SS have no significant effect on the operation of PV strings.

The apparent velocities of the identified shadow edges were calculated using the time lags between the shading of the three irradiance sensors using the method presented in [5]. The average values of the SS, the absolute value of b, and the apparent speed of the identified irradiance transitions were 59.2%, 1.91 s, and 8.66 m/s, respectively. The average length of the identified transitions calculated from b and the apparent speed was 112 m. For the studied PV arrays, a shading situation caused by an overpassing irradiance fall is symmetrical to a situation caused by a similar overpassing irradiance rise. Thus, the absolute values of parameter b for the identified irradiance transitions were used in the simulations.

In the simulations, a shadow edge was assumed to be linear across a PV array and the apparent velocity of the shadow edge was assumed to be constant. A simulation period started when a shadow edge moved over the first PV module of the array and ended when the shadow edge had moved across the array, i.e., when all the modules of the array were again uniformly shaded. In order to simplify the computation, the PV array was chosen to be under the constant STC irradiance of 1000 W/m^2 before each irradiance fall and the temperature of the PV modules was chosen as the constant STC temperature of 25 °C. During fast irradiance transitions, the changes of PV module temperatures are small having only a negligible effect on the operation of the modules.

The studied PV systems were a string of 20 series connected PV modules and an array of 10 parallel strings of 20 modules. In the studied PV systems, the series-connected PV modules were placed in straight strings. There were no gaps between the series-connected PV modules, and the distance between the adjacent strings was 2.0 m. The east-west orientation of the PV arrays was used in the simulations, i.e., the PV strings were placed from east to west. The PV modules were mounted at a tilt angle of 45° from the horizontal plane. The nominal powers of the studied PV string and array, under STC, were 3.8 and 38 kWp, respectively.

The mismatch losses of a PV system were calculated as the difference between the sum of the maximum powers of individual PV modules of the system, as if they were operating separately, and the global MPP power of the system. Relative mismatch losses were calculated with respect to the sum of the maximum powers of the modules. The rate of change in the power of the PV system was calculated with respect to the nominal power of the system.

3 RESULTS

In this study, the effects of four simulation parameters on the accuracy of PV system simulations under partial shading conditions were studied. The studied parameters were: the size of a basic simulation unit, the length of a simulation time step and the resolutions of irradiance and *I–U* curves. The simulations were conducted on a PV module and submodule level. The submodule is a series-connection of 18 PV cells protected by a bypass diode. It is the smallest element in a PV plant producing its own electrical characteristics. The studied PV modules contained three submodules. The used simulation time steps were 0.1, 0.5 and 1.0 s and the irradiance at the centre of each simulation unit with accuracies of 0.1, 1.0 and 10 W/m^2 was used for the whole simulation unit during a time step. The *I–U* curves of the basic simulation units were calculated with 100, 1000 and 10 000 points. The effects of the simulation parameters were studied by changing the studied parameter while keeping the other parameters unchanged in their reference values. A submodule as the basic simulation unit, the simulation time step of 0.1 s, the irradiance accuracy of 0.1 W/m^2 and the *I–U* curve accuracy of 10 000 points were used as a reference case, i.e., the most accurate case, to compare the results of the other cases. All the studied simulation cases are presented in Table III.

The effects of the four simulation parameters on the accuracy of the produced energy, power variation and mismatch losses of the PV string and array were studied during all the identified irradiance transitions. The produced energy and the average rate of change in the power of the studied PV string during all the identified irradiance transitions are presented in Table IV for all the studied simulation cases. Moreover, the differences with respect to the reference case are presented. The errors in the overall produced energy and the average rate of change in power caused by the decreased resolution of the simulations were small. Especially, the changes in the accuracy of irradiance had negligible effects on the accuracy of the produced energy. The difference with respect to the reference case was over 1% in the produced energy when the simulation time step of 1.0 s was used (case "1.0 s"), and in the average rate of change in power when the simulation time step of 0.5 or 1.0 s was used (cases "0.5 s" and "1.0 s").

The relative mismatch losses during all the identified irradiance transitions and the largest relative mismatch losses during a single irradiance transition for the studied PV string are presented in Table V for all the studied simulation cases. The difference with respect to the reference case was over 1% in the relative mismatch losses during all the identified irradiance transitions when the simulation time step of 1.0 s was used and when only 100 points were used for the *I–U* curves (case "100 points"). The difference with respect to the reference case was over 1% in the largest relative mismatch losses during a single irradiance transition when the simulation time step of 0.5 or 1.0 s was used. The difference with respect to the reference case was the largest in all the four studied quantities when the simulation time step of 1.0 s was used.

The produced energy and the average rate of change in the power of the studied PV array during all the identified irradiance transitions are presented in Table VI for all the studied simulation cases. Accordingly, the relative mismatch losses during all the identified irradiance transitions and the largest relative mismatch losses during a single irradiance transition are presented in Table VII. Again, the errors in the studied quantities caused by the decreased simulation resolution were

Table III: Studied simulation cases.

Case	Basic simulation unit	Time step (s)	Accuracy of irradiance (W/m^2)	Number of points in *I–U* curves
"Reference"	Submodule	0.1	0.1	10 000
"Module"	Module	0.1	0.1	10 000
"0.5 s"	Submodule	0.5	0.1	10 000
"1.0 s"	Submodule	1.0	0.1	10 000
"1 W/m^2"	Submodule	0.1	1.0	10 000
"10 W/m^2"	Submodule	0.1	10	10 000
"1000 points"	Submodule	0.1	0.1	1 000
"100 points"	Submodule	0.1	0.1	100
"1 W/m^2; 1000 points "	Submodule	0.1	1.0	1 000

Table IV: Produced energy and average rate of change in power of the studied PV string during all the identified irradiance transitions for all the studied simulation cases. The differences were calculated with respect to case "Reference".

Case	Produced energy (kWh)	Difference in produced energy (%)	Average rate of change in power (%/s)	Difference in average rate of change in power (%)
"Reference"	102.45		3.23	
"Module"	102.13	0.314	3.24	0.315
"0.5 s"	103.07	0.608	3.26	1.055
"1.0 s"	104.27	1.780	3.30	2.255
"1 W/m^2"	102.45	0.001	3.23	0.001
"10 W/m^2"	102.44	0.005	3.23	0.092
"1000 points"	102.45	0.001	3.23	0.001
"100 points"	102.34	0.102	3.23	0.104
"1 W/m^2; 1000 points "	102.45	0.002	3.23	0.001

Table V: Relative mismatch losses of the studied PV string during all the identified irradiance transitions for all the studied simulation cases. The differences were calculated with respect to case "Reference".

Case	Relative mis-match losses (%)	Difference in relative mismatch losses (%)	Largest relative mismatch losses during an irradiance transition (%)	Difference in largest relative mismatch losses (%)
"Reference"	2.99		19.18	
"Module"	3.01	0.614	19.18	0.011
"0.5 s"	2.97	0.609	18.63	2.864
"1.0 s"	2.93	1.817	18.41	4.023
"1 W/m^2"	2.99	0.000	19.18	0.010
"10 W/m^2"	3.00	0.487	19.29	0.571
"1000 points"	2.99	0.017	19.18	0.010
"100 points"	3.03	1.494	19.29	0.552
"1 W/m^2; 1000 points "	2.99	0.018	19.18	0.020

Table VI: Produced energy and average rate of change in power of the studied PV array during all the identified irradiance transitions for all the studied simulation cases. The differences were calculated with respect to case "Reference".

Case	Produced energy (kWh)	Difference in produced energy (%)	Average rate of change in power (%/s)	Difference in average rate of change in power (%)
"Reference"	1192.87		2.77	
"Module"	1189.65	0.270	2.78	0.269
"0.5 s"	1199.16	0.527	2.80	0.943
"1.0 s"	1211.24	1.540	2.83	2.102
"1 W/m^2"	1192.85	0.001	2.77	0.001
"10 W/m^2"	1192.91	0.003	2.77	0.091
"1000 points"	1192.85	0.002	2.77	0.001
"100 points"	1191.19	0.141	2.77	0.104
"1 W/m^2; 1000 points "	1192.83	0.003	2.77	0.001

Table VII: Relative mismatch losses of the studied PV array during all the identified irradiance transitions for all the studied simulation cases. The differences were calculated with respect to case "Reference".

Case	Relative mis-match losses (%)	Difference in relative mismatch losses (%)	Largest relative mismatch losses during an irradiance transition (%)	Difference in largest relative mismatch losses (%)
"Reference"	2.88		20.63	
"Module"	2.90	0.608	20.85	1.082
"0.5 s"	2.87	0.525	20.62	0.039
"1.0 s"	2.84	1.536	20.61	0.088
"1 W/m^2"	2.88	0.001	20.62	0.048
"10 W/m^2"	2.90	0.489	20.77	0.680
"1000 points"	2.89	0.035	20.63	0.004
"100 points"	2.97	2.985	20.64	0.064
"1 W/m^2; 1000 points "	2.89	0.037	20.62	0.042

small. The difference with respect to the reference case was over 1% in the produced energy, in the average rate of change in power and in the relative mismatch losses during all the identified irradiance transitions when the simulation time step of 1.0 s was used. The difference with respect to the reference case was over 1% also in the relative mismatch losses during all the identified irradiance transitions when only 100 points were used for the *I–U* curves and in the largest relative mismatch losses during a single irradiance transition when a PV module was used as the basic simulation unit.

The errors in the average rate of change in power caused by the decreased resolution of the simulations for the PV array were very close to those for the PV string. Moreover, the errors in the overall produced energy and the relative mismatch losses for the PV array were largely similar than for the PV string. However, the resolution of *I–U* curves has a larger effect on the relative mismatch losses on the PV array level than on the PV string level. The difference in the relative mismatch losses with respect to the reference case when only 1000 or 100 points were used for the *I–U* curves was about twice as large for the PV array as for the PV string. The largest differences in the errors caused by the decreased simulation resolution between the PV string and array occurred in the largest relative mismatch losses during a single irradiance transition, as was anticipated since it was the only studied quantity that was not computed from all the identified irradiance transitions but from a single transition.

The use of long time steps overestimated the produced energy and the average rate of change in power, and underestimated the relative mismatch losses, compared to the reference case. The relatively large differences between the use of 0.1 and 1.0 s time steps were anticipated since irradiance transitions caused by moving clouds can be very fast, over 300 W/m^2 in 0.1 s [3].

The results show that the irradiance accuracy of 1.0 W/m^2 and the number of *I–U* curve points of 1000 are sufficient for PV string and array level partial shading studies. The combined effect of these two factors on the accuracy of the produced energy, power variation and relative mismatch losses was negligible.

4 CONCLUSIONS

In this paper, the effects of the length of a simulation time step, the resolutions of irradiance and *I–U* curves and the size of a basic simulation unit on the accuracy of the simulations of PV system operation under partial shading conditions caused by moving clouds were studied. The effects of the simulation parameters were studied by changing the studied parameter while keeping the other parameters unchanged in their reference values. A submodule as the basic simulation unit, the simulation time step of 0.1 s, the irradiance accuracy of 0.1 W/m^2 and the *I–U* curve accuracy of 10 000 points were used as a reference case to compare the results of the other simulation cases. The study aimed to find out adequate resolution for the simulation parameters to minimise the computing time while retaining sufficient accuracy of the simulations.

The study was based on the apparent velocity and other characteristics of 7880 irradiance transitions identified in measured irradiance data and was conducted using a mathematical model of irradiance transitions and an experimentally verified MATLAB Simulink model of a PV module that is based on the well-known one-diode model of a PV cell. The studied PV systems were a PV string of 20 modules and a PV array of 10 parallel strings of 20 modules. The studied quantities were the produced energy, the average rate of change in power, the relative mismatch losses during all the identified irradiance transitions and the largest relative mismatch losses during a single irradiance transition.

The errors in the studied quantities caused by the decreased resolution of the simulation parameters for the PV array were close to those for the PV string. The largest errors were caused by the use of long 0.5 and 1.0 s simulation time steps instead of the reference value of 0.1 s, as was anticipated since irradiance transitions caused by moving clouds can be very fast. The use of a long time step overestimated the produced energy and the average rate of change in power, and underestimated the relative mismatch losses, compared to the reference case. The use of a submodule as a basic simulation unit provided better accuracy than the use of the whole PV module. The results show that the irradiance accuracy of 1.0 W/m^2 and the number of *I–U* curve points of 1000 are sufficient for PV string and PV array level partial shading studies.

5 REFERENCES

[1] T. Tomson, G. Tamm, Solar Energy, 80 (2006) 600.
[2] R. Perez, S. Kivalov, J. Schlemmer, K. Hemker, T. Hoff, Solar Energy, 85 (2011) 1343.
[3] K. Lappalainen, S. Valkealahti, Solar Energy, 112 (2015) 55.
[4] M. Lave, M.J. Reno, R.J. Broderick, Solar Energy, 118 (2015) 327.
[5] K. Lappalainen, S. Valkealahti, Solar Energy, 138 (2016) 47.
[6] D. Picault, B. Raison, S. Bacha, J. De La Casa, J. Aguilera, Solar Energy, 84 (2010) 1301.
[7] L.F.L. Villa, D. Picault, B. Raison, S. Bacha, A. Labonne, IEEE Journal of Photovoltaics, 2 (2012) 154.
[8] M.Z. Shams El-Dein, M. Kazerani, M.M.A. Salama, IEEE Transactions on Sustainable Energy, 4 (2013) 145.
[9] S.R. Potnuru, D. Pattabiraman, S.I. Ganesan, N. Chilakapati, Renewable Energy, 78 (2015) 264.
[10] N. Rakesh, T.V. Madhavaram, Frontiers in Energy, 10 (2016) 227.
[11] S. Vijayalekshmy, G.R. Bindu, S. Rama Iyer, Solar Energy, 135 (2016) 92.
[12] K. Lappalainen, S. Valkealahti, Solar Energy, 144 (2017) 548.
[13] H. Patel, V. Agarwal, IEEE Transactions on Energy Conversion, 23 (2008) 302.
[14] K. Lappalainen, S. Valkealahti, Applied Energy, 190 (2017) 902.
[15] R. van Haaren, M. Morjaria, V. Fthenakis, Progress in Photovoltaics: Research and Applications, 22 (2014) 548.
[16] R. Yan, T.K. Saha, P. Meredith, A. Ananth, M.I. Hossain, IET Renewable Power Generation, 10 (2016) 1229.
[17] M.G. Villalva, J.R. Gazoli, E.R. Filho, IEEE Transactions on Power electronics, 24 (2009) 1198.
[18] S.R. Wenham, M.A. Green, M.E. Watt, R. Corkish, Applied Photovoltaics 2nd ed., Earthscan, London 2007.
[19] K. Lappalainen, S. Valkealahti, Proceedings of 32nd European Photovoltaic Solar Energy Conference, Vol. I (2016) 1485.
[20] D. Torres Lobera, A. Mäki, J. Huusari, K. Lappalainen, T. Suntio, S. Valkealahti, International Journal of Photoenergy, 2013 (2013) Article ID 837310.

LA SILLA PV PLANT AS A UTILITY-SCALE SIDE-BY-SIDE TEST FOR INNOVATIVE MODULES TECHNOLOGIES

Agnese Di Stefano[1], Giuseppe Leotta[1], Fabrizio Bizzarri[1]
[1] Enel Green Power SpA,
Contact : agnese.distefano@enel.com , Zona Industriale Passo Martino, 95121 – Catania, Italy

ABSTRACT:
La Silla plant is a 1.7 MW innovative PV power plant installed in 2016 by Enel Green Power (EGP) in Chile, where two cutting-edge solar module technologies (bifacial and smart modules) are tested and compared with traditional ones at utility scale level.
The PV plant has the double purpose to test innovative technologies in real operating conditions of utility-scale PV plants, and meanwhile to produce energy and have revenues from this production. The side-by-side testing of the smart and bifacial modules with conventional panels allows comparing their performance with that of the conventional panels in the same site. The analysis of the data of each subfield showed that bifacial subfield performs better than the standard one with a gain of about 12%, while the smart subfield at the year one shows only a gain of about 1%.
The test confirms the potentialities of the solutions, that can be further investigated both for installation in EGP PV plants and for R&D development of 3SUN factory.
Keywords: bifacial, optimizers, side-by-side

1 INTRODUCTION

In the frame of finding solutions aimed to improve performances of PV systems, one technological sector to be focused deals with innovative module technologies. Many tests have been carried out in laboratory in order to compare the different modules technologies, but not many tests are performed at utility scale level, so the global behavior and the effect of the interaction between the strings is more difficult to be evaluated respect to the performances in a single string. Enel Green Power (EGP) installed a 1.7MW PV plant named La Silla (Chile) with the purpose to test innovative solutions at utility scale level, where two cutting-edge solar modules (bifacial and smart modules) are tested and compared with traditional ones.

The two innovative modules technologies have been chosen for tests, considering two aspects that can be exploited for the improvement of the performances of PV plants: one is minimize mismatch effects and maximize the performance of PV strings, the other is take advantage of the ground albedo for producing extra-energy from the back of the modules.

The aim of the utility scale test is to characterize two innovative technologies of high potential for future application and LCOE reduction of EGP PV plants, as well as for possible developments of EGP modules factory, including not only the effects of the single modules but also the interaction of the strings in a utility scale field.

The test is carried out as a side-by-side test comparison of the three subfields (standard, bifacial and smart modules). In particular, the testing protocol is divided in the following phases: 1) Calculation of the daily Performance Ratio and comparison of the performances of the subfields during the observation period; 2) Distribution of the PR of the single strings in the subfield and analysis of the variability of the string in each subfield. Moreover other tests have been carried out to monitor the tracking errors and their impact on the performances of the field.

The test allows monitoring the long period behavior of the technologies and detecting the variations in the gain due to the seasonality.

The scientific relevance is to test at utility scale level two new types of panel technology that have serious potential for future use, in order to see how much the levelised cost of electricity (LCOE) can be reduced at EGP's PV facilities.

2 TECHNOLOGIES DESCRIPTION

3.1 Maxim integrated optimizers

Smart modules with Maxim integrated optimizers consists in modules in which three optimizers are embedded in the modules, optimizing the performances of each cell string and thus reducing the mismatch effects.

The origins of mismatch could be several: systematic shading, irradiation mismatch (clouds, reflected light, soiling), thermal mismatch, different ageing of the modules, Potential Induced Degradation, installation mistakes and technical failures. All these reasons make the modules perform differently and the current of the strings is limited by the worst performing one [1].

Maxim optimizers are a solution feasible with utility scale PV plants because it does not require installation of additional components and it allows extracting the maximum current from each module [2].

The smart modules replace the bypass diodes in the module with solar cell optimisers, which enable maximum power plant tracking (MPPT) and direct current (DC) optimisation on each cell string inside the modules, enabling them to produce the maximum amount of energy possible regardless of shading issues and mismatch. This is done without the complexity of DC optimisers and with a simplified system design that also increases module reliability.

The key benefits of the smart module are an increased energy yield, better ground coverage and improvement of land use (10%-20% tighter inter-row spacing at the same energy production per panel), increased reliability (minimised power degradation over the panel and elimination of hot spots), and the lack of need for any additional components in the PV system (otherwise known as balance of system).

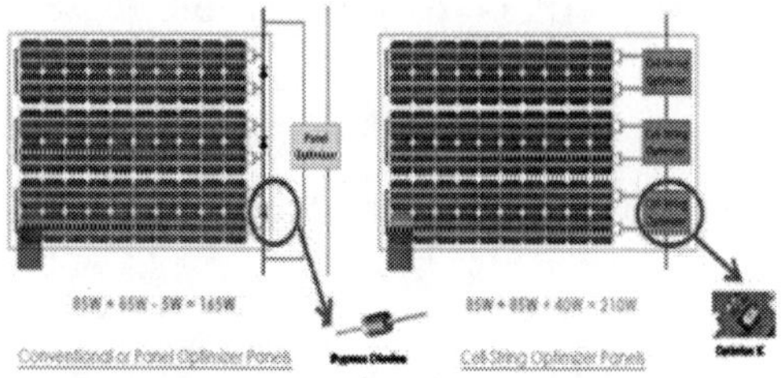

Figure 1: Smart modules architecture

3.1 Megacell bifacial modules

Bifacial solar cells enable the absorption of light also by the cell's rear side, hence increasing the energy yield of a bifacial module, compared to the energy yield of a monofacial module installed under the same conditions,

With their ability to additionally absorb light by their rear side, therefore increasing the energy yield, compared to standard modules, bifacial modules promise to further reduce the levelized cost of energy (LCOE) of photovoltaic modules. A parameter influencing the production is the albedo that is a reflection coefficient, and has a value of less than one. The albedo of a surface is the ratio of radiation reflected from the surface to the incident radiation. Its dimensionless nature lets it be expressed as a percentage and is measured on a scale from zero (no reflection) of a perfectly black surface to 1 for perfect reflection of a white surface.

The bifacial panels under test are highly efficient monocrystalline N-type photovoltaic modules that have long-term stability thanks to the technology's design and high performance. Their main benefits include a higher energy yield thanks to use of diffuse and reflected light on the rear side of the module, lower power degradation via use of highly efficient N-type silicon cells instead of the traditional P-type cells, and the same energy yield of a traditional PV plant using less land, leading to significant savings in labour and infrastructure building and as a result reductions in cost per KWh.

In order to determine the LCOE of bifacial modules and therewith their profitability, it is necessary to evaluate the exact gain in energy production, both with simulation tools and with experimental data on field (Fig. 2).

Figure 2: Bifacial modules under tests in Lab

The installation in La Silla plant allows to have a global information about the production of bifacial modules in a utility scale field, that is also influenced by the interaction between the rows and the shading of the ground.

3 EXPERIMENTAL SET UP AND TEST PROTOCOL

3.1 Experimental set up

The PV plant is located at the La Silla Observatory, which is run by the European Southern Observatory (ESO), Europe's pre-eminent intergovernmental astronomy organization (Fig. 3), and is the first observatory in Chile to be supplied by an innovative PV plant. The plant is divided in three sub-fields for side-by-side testing:

I) 575 kW crystalline standard modules as the reference field;

II) 575 kW smart modules with solar cell optimisers, with the aim of increasing the production by minimising the mismatch effect and mutual shading losses;

III) 575 kW of bifacial modules, with the aim of increasing energy production thanks to each panel being able to capture solar radiation on both of its sides.

La Silla Project is allowing to test these technologies in the same operating conditions of a utility-scale PV plant while producing energy for the observatory.

Figure 3: La Silla Observatory

In Fig. 4 the layout of the plant is showed in which three different subfields are located in the same location. In Fig. 4 and Fig. 5 the tracker structures with smart and bifacial modules are reported, showing also the different positioning of the modules on the structures, optimized for each technology.

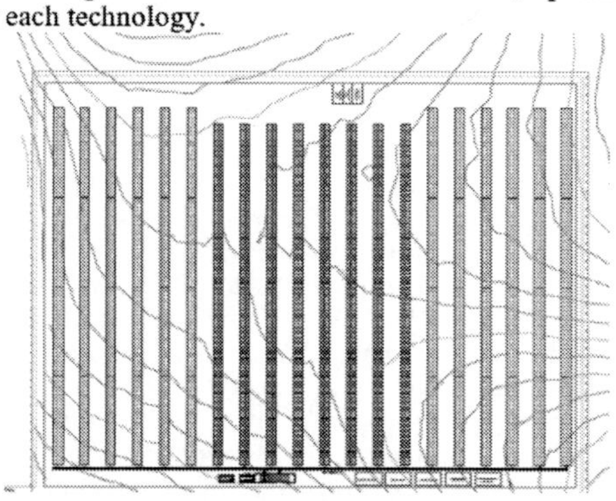

Figure 4 Plant layout of the experimental plant used for side-by-side tests

(a)

(b)

Figure 5: Smart modules (a) and bifacial modules (b) installed at La Silla plant

3.2 Test procedure

The production data are downloaded from a web platform and elaborated by EGP team. In particular two analysis are performed at inverter level and at string level:

1) Comparison of the three subfields:

Analysis of inverter data: Daily comparison and monthly average of PR data of the three subfields (IEC 61724)

Calculation of daily and monthly average gain of the innovative technologies respect to the standard

$$PR = \frac{Y_f}{Y_r} = \frac{\frac{E_{SYS}}{P_{nom}}}{\frac{E_{irr}}{I_{ref}}} \quad \text{Eq. 1}$$

$$Gain = \frac{PR_{inn} - PR_{std}}{PR_{std}}$$

Where E_{SYS} is the average energy measured for the subfield, P_{nom} is the nominal power of the subfield, E_{irr} is the measured irradiance and I_{ref} is the reference irradiance of 1000W/m2.

The gain is calculated as percentage difference between the PR of the innovative field and the PR of the standard field.

2) Analysis of each subfield:

Analysis of string data: Distribution of string PR within the same subfield to measure the variability of the strings performances

Moreover a daily map of the tracking errors is processed in order to detect the anomalies and correlate it with the underperformances of the field.

4 EXPERIMENTAL RESULTS

The performances of the technologies are monitored during the different months of operation in order to verify the stability and the long-term behavior of the technologies.

In Fig, 6 the comparison of the Performance Ratio of the three subfields is reported.

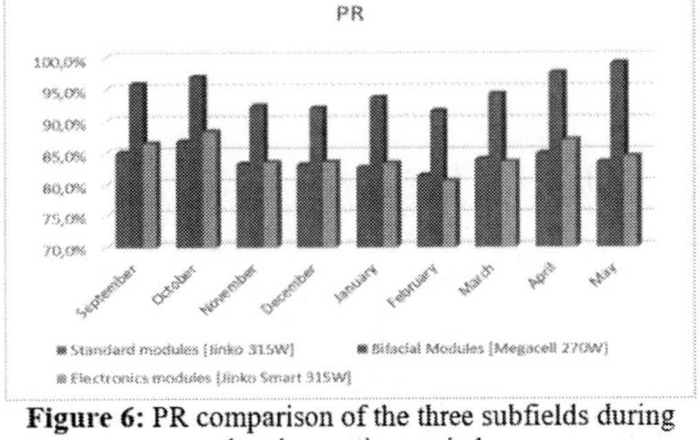

Figure 6: PR comparison of the three subfields during the observation period

The average PR values of the three subfield calculated during the observation period are reported in the following:

$PR_{m\_standard}$= 84.1%

$PR_{m\_bifacial}$= 94.8%

PR_{m_smart}= 85.2%

In order to have an immediate picture of the advantages of the technologies, the energy gain of the two innovative solutions respect to the standard is reported in Fig.7.

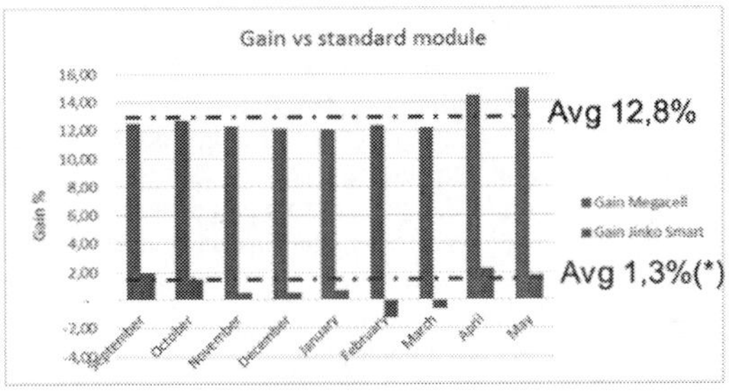

Figure 7: Energy gain of bifacial and smart fields with respect to standard field during the observation period

The bifacial technology shows a gain of about the 12,8%, while the smart modules performs slightly better than standard with a gain of about 1,3%. The gain of the smart modules has been calculated leaving the months of February and March out from the calculation, because the underperformances are related to tracking problems (as detailed below in the analysis of tracking errors).

The increase in the performances of smart modules is very low, but it has to be considered that the optimizers manufacturers claim to be able to recover the losses during all the lifetime of the plant, when the modules in the strings degrade in different ways and the mismatch losses increase [3]. So in this scenario the effectiveness of smart modules should be verified during the years while the system degrades.

After a macro analysis, the data of each subfield have been analysed comparing the performances of the strings. The distribution of the Performance Ratio of the strings is reported in Fig. 8, in which all the daily PR of each string during the observation period is considered in the distribution.

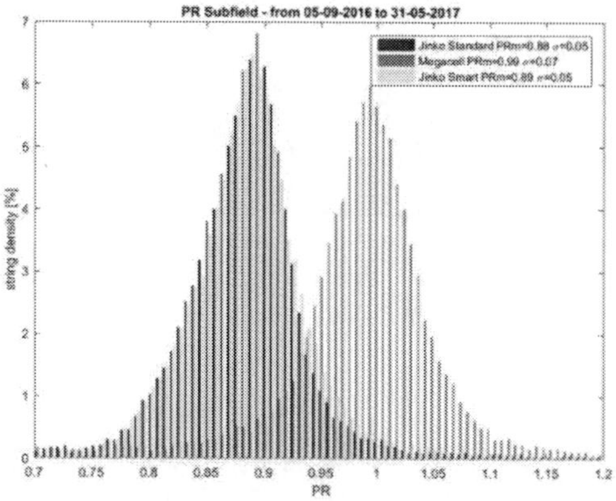

Figure 8: Distribution of PR of the strings within each subfield

The graph is characterized by a distribution peaked around average PR and a standard deviation of 5% for standard and smart field and 7% for bifacial field, that is indication of the variability of the performances of the strings.

After this analysis, a dedicated study regarding the tracking errors has been carried out.

The parameter under analysis is the tracking error, calculated as the difference between the ideal tracking angle (the one that guarantees the maximum energy production being perpendicular to the solar position) and the real position of the tracker. This parameter is summed for having a representative value for each tracker and for each day, as reported in Eq.2.

$$\text{error} = \sum_{time} |roll - target\_roll| \qquad \text{Eq.2}$$

In Fig. 9 the operation of the tracker in a normal working condition (with backtracking algorithm) and in stopped position is showed. The cyan line represent the tracking error, so in Fig.9a it is almost zero (where the values of roll and target roll are almost the same), while in Fig.9b it has significant values because the roll (always zero since the tracker is stopped) does not follow the target roll.

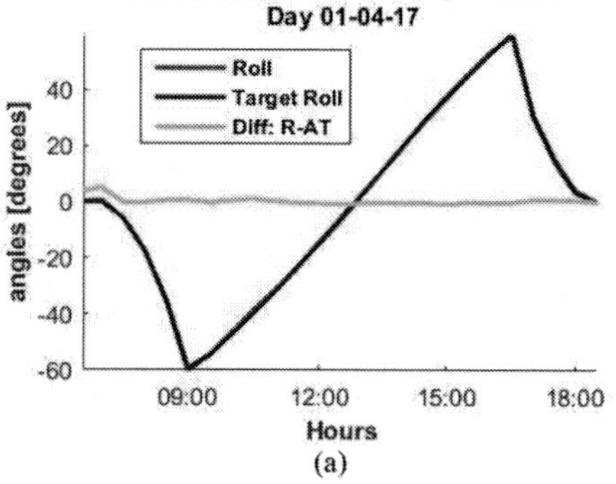

(a)

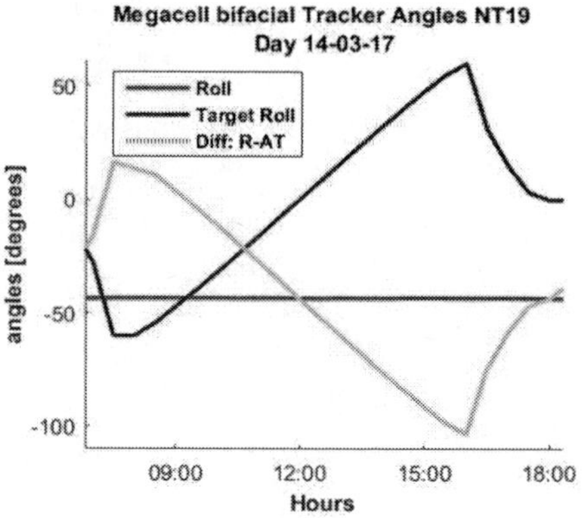

(b)

Figure 9: Tracking errors during the day in a condition of normal functioning (a) and tracker stopped (b).

In Fig.10 the daily maps of tracking errors are reported, in which the summed value as reported in Eq.2 is calculated for each tracker and for each day.

In particular, in Fig. 10 the maps of the smart subfield reffered to the may of March are reported in which a significant underperformance is detected from the data. This graph confirms that a tracker has been stopped for a significant part of the period, producing a loss of energy of about 2% in the smart subfield.

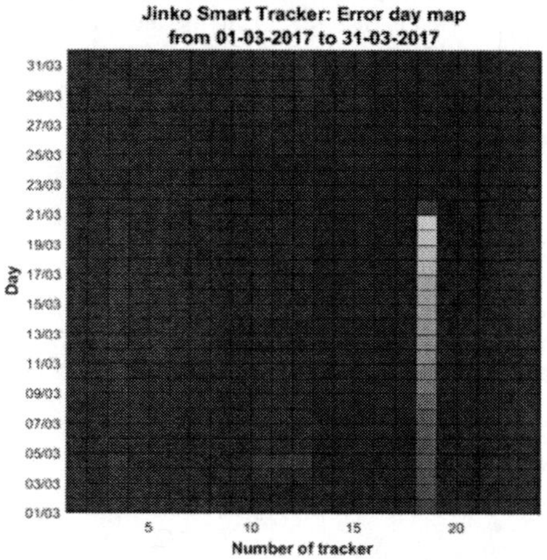

Figure 10: Tracking errors map of the smart subfield for the month of March. The tracking error is showed for each tracker and for each day of the month.

From this analysis, it can be also noticed that, apart for the cases in which some tracker is stopped and not working, the small tracking errors detected during the normal operation of the plant have a low impact on the global performances of the field. So in case underperformances are detected and the trackers are working, these problems should ascribed to other factors.

5 ECONOMICAL EVALUATIONS

The potential convenience of the optimized solutions has been evaluated building a business case in which both the modules technologies are installed on La Silla PV plant and comparing it with the economics of the standard

subfield.
Three scenario have been compared evaluating the economical parameters of LCOE and IRR, assuming the modules prices and all CAPEX information related to 2015, when all the purchase contracts have been defined for this plant. In addition, the assumptions of energy gain refer to the analysis made before the experimentation that should be matched with the first results coming from the data onfield.
The results of the economical analysis are reported in Table 1:

	Smart modules	Bifacial modules
ΔLCOE	-3.3% (G=3%) -4.7% (G=5%)	-5.3% (G=12%) -7.2% (G=15%)
ΔIRR	4.6% (G=3%) 7.1% (G=5%)	5.7% (G=12%) 9.1% (G=15%)

Table 1: Analysis of LCOE and IRR the two scenario with innovative modules technologies with respect to the standard case

The variation on LCOE and IRR of smart and bifacial fields with respect to the standard case (crystalline modules) is reported, assuming different values of gain. Since the extra-cost (respect to crystalline modules, assumed according to the offers obtained for this plant) of smart modules is much lower than bifacial modules, the threshold of energy gain needed to allow a reduction of LCOE is lower too: while for smart modules a gain of about 2,% is sufficient, for bifacial modules a gain of about 9% is needed.
The installation of smart modules gives an advantage in terms of LCOE reduction between 3.3% and 4.7% respectively assuming gain values of 3% and 5%. These values of gain are higher than the ones measured on field. Considering the gain of about 1% measured on field during the first months of operation of the plant, the LCOE results similar to the standard solution with no significant reduction, so from this evaluation the technology would not result convenient for utility scale installations in which there are not significant sources of mismatch (different orientations, shading, soiling and so on).
For what deals with bifacial modules, assuming a gain between 12% and 15% with respect to the standard crystalline case, the economical improvement in terms of LCOE reduction goes from 5.3% to 7.2%. In this case the assumptions are similar to the results obtained experimentally (gain of about 12%), confirming the potentiality of development of this technology.

6 CONCLUSION

The installation of La Silla PV plant has allowed to make a global evaluation of two different innovative PV modules technologies directly on field, considering all the factors that impact on a utility scale field.
The results of the analysis of the data collected during the test period showed that smart modules have about 1% of energy gain respect to standard crystalline modules, while bifacial modules show a gain of about 12%.
Despite both the technologies allow a certain performance increase with respect to standard technology, the smart modules do not seem at the moment to have a sufficient gain to allow a reduction of LCOE, while the results on bifacial modules confirm their potentiality.
Anyway, this consideration cannot be considered as a definitive result, because the advantage of distributed electronics should be evidenced after some year of operation of the plant, when the modules start to degrade in different ways and the system performances degradation results higher than the modules performance degradation. Moreover, tests on smart field removing backtracking algorithm should be also performed, in the conditions in which smart modules should compensate underperformances due to shading.
So the analyzed solutions are interesting solutions to be further investigated during the lifetime of the plant, also in the perspective of a future cost reduction in which the extra-cost with respect to standard crystalline modules should be reduced.
Summarizing, the test of innovative solutions directly on field on utility scale PV plants has been an opportunity to evaluate the technologies in a not controlled environment, which is different from laboratory, and in which different factors influence the performances of the strings as it happens in the normal operation of the plant.

7 REFERENCES

[1] B. Prasad Koirala, B. Sahan, N. Henze, "Study on MPP Mismatch Losses in Photovoltaic Applications" 24th European Photovoltaic Solar Energy Conference and Exhibition, Hamburg/Germany, 21.-25. September 2009

[2] Deline, C.; Marion, B.; Granata, J.; Gonzalez, S. (January 2011). A Performance and Economic Analysis of Distributed Power Electronics in Photovoltaic Systems. Golden, CO: National Renewable Energy Laboratory.Http://www.nrel.gov/docs/fy11osti/50003.pdf Accessed February 16, 2012.

[3] "Field Experience: DegradationRates, Lifetime and Failures", DJordan, NREL PVMRC 2016

Validation Study of Solar PV Energy Simulation Tools and Methodologies

Aspinall, Megan; Cahill, Maria; Noirault, Emma.
Prevailing Ltd
Tower House, Fairfax St, Bristol BS1 3BN

ABSTRACT: The global solar PV market has grown rapidly in over the last decade, with a significant increase in the volume of installations since 2010. Investment continues apace but as the industry matures and secondary market competition increases, margins are becoming tighter. The imperative for accurate energy yield prediction is evident.

Modeling methods used to forecast energy generation prior to project construction have evolved throughout this period. As the number of operational utility scale projects grows, actual production data can be used to assess the validity of current methods and standard assumptions.

The purpose of this validation study is to determine the accuracy of current pre-construction prediction methodology and identify areas for refinement.

The study indicates that there is a tendency for current pre-construction methods to be slightly conservative and that current pre-construction modeling methods may be lagging technological advancements. However, the variance between pre-construction energy modeling and operational assessments for projects constructed in recent years is increasing.

Keywords: System Performance, PV System, Performance, Modelling, Evaluation

1 INTRODUCTION

The global solar PV market has grown rapidly over the last decade, with a significant increase in the volume of installations since 2010. Investment continues apace but as the industry matures and secondary market competition increases, margins are becoming slimmer. The imperative for accurate energy yield prediction is evident.

Modeling methods used to forecast energy production prior to project construction have evolved throughout this period. As the number of operational utility scale projects grows, actual production data can be used to assess the validity of current methods and standard assumptions.

The purpose of this validation study is to determine the accuracy of current pre-construction prediction methodology and identify areas for refinement.

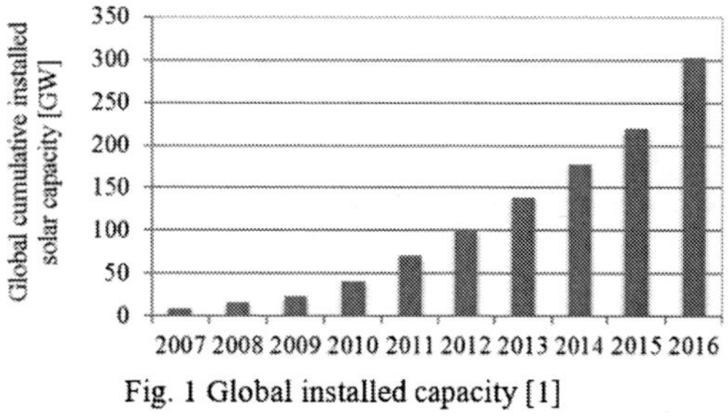

Fig. 1 Global installed capacity [1]

2 METHODOLOGY

2.1 Site selection

10 utility-scale solar PV projects were selected for the validation analysis. To ensure the sample group was diverse and representative, projects were selected with:

- Varying installed capacities, ranging from 1 MWp to 15 MWp;
- Differing equipment suppliers;
- Varying age range: 1 year to 5 years; and
- A UK geographical spread, as illustrated in Figure 2.

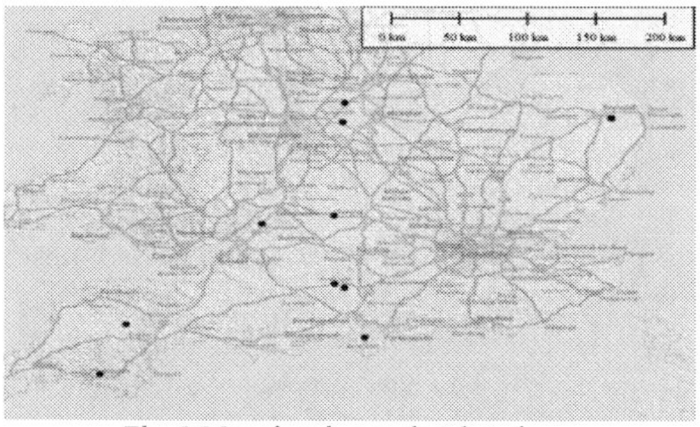

Fig. 2 Map showing project locations

2.2 Pre-construction energy yield assessment

Prevailing completed financial-grade pre-construction energy yield modeling for all 10 projects utilising:

- Satellite based resource inputs (Solargis, PVGIS and Meteonorm were considered as long-term reference sources);
- Project design details available publicly through planning portals;
- Satellite imagery and open source terrain data; and
- Equipment datasheets publicly available from suppliers.

Modeling was completed using PVsyst simulation software with industry standard equipment performance and loss assumptions.

Prevailing has developed a number of internal analysis tools that are used to refine the simulation software, including:

- A detailed shading tool, which interfaces with the PVSyst shading module combining feature coordinates and the terrain profile; and
- A soiling assessment tool that utilises rainfall data, cleaning regimes and module tilt angle to derive a precise soiling loss factor.

2.3 Operational energy yield assessment

Prevailing completed an operational energy assessment for all 10 sites, based on monthly operational production data obtained from public sources (e.g. ROC database).

In order to apply an appropriate long-term adjustment to the period of operational data available for each project, suitable data sources including Solargis and MERRA were considered. A degradation assumption of 0.5% pa was used to back-calculate energy yield in the first year.

2.4 Comparison of results

In the case of each project, the results of the pre-construction energy yield prediction and long term operational energy assessment were compared and the variance analysed.

3 RESULTS

As illustrated in Figure 3, some high-level conclusions can be drawn:

- The majority of pre-construction and operational analyses agree to within 5% of one another.
- Projects tend to out-perform their pre-construction estimates by 1% on average.
- There is a risk that 10% of projects may materially under-perform relative to their preconstruction estimate.

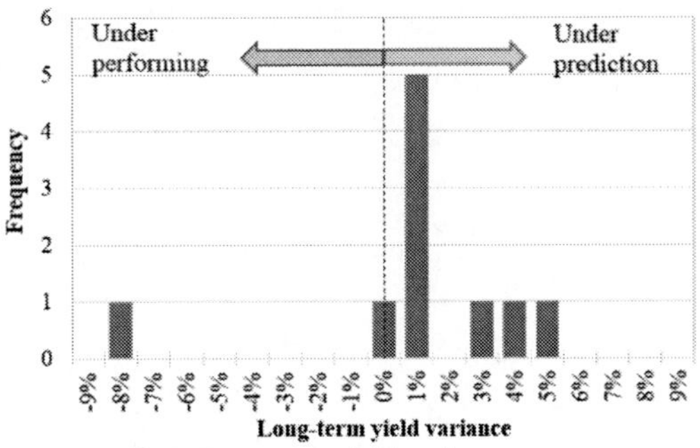

Fig 3. Long-term yield variance

4 DISCUSSION

In order to determine the cause of under-prediction, Prevailing considered a number of factors such as:

- Size
- Location
- Equipment manufacturer
- Age

Figure 4 shows the relationship between the variance in yield forecast and project age (the outlying project has been excluded). From the sample analysed, the variance is shown to be diverging with project age.

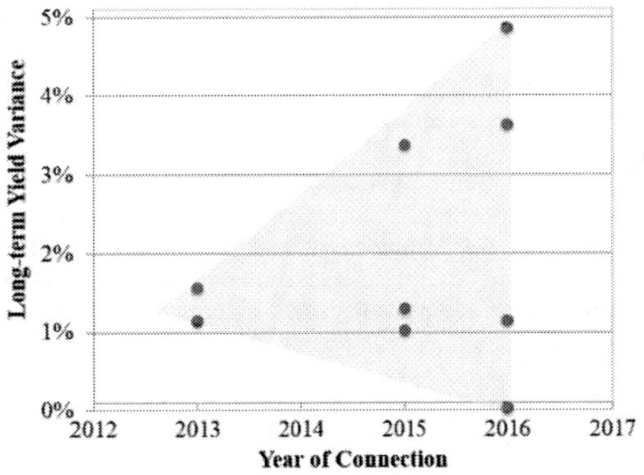

Fig. 4 Long-term yield variance against project age

4.1 Why are recent projects showing diverging results?

The diverging results observed in newer projects could be a result of less operational data being available for consideration in the operational energy assessment. Shorter time periods are more sensitive to short-term events.

Alternatively, it is possible that a higher degree of variance is seen with recently manufactured equipment. The conclusion here is that obtaining project specific equipment test data results in more accurate yield forecasts.

4.2 What are the likely causes of the 1% under-prediction bias?

The observed under-prediction bias is thought to be the result of:

- Conservative equipment performance parameters, including efficiency under varying irradiance and temperature conditions;
- Conservative loss assumptions, such as light induced degradation (LID), reflectivity (IAM factor), partial shading effects; and
- Conservative availability assumption (99% applied).

In undertaking pre-construction energy modeling, Prevailing engages with the equipment suppliers to obtain up-to-date testing records and detailed information on recent equipment advancements. Where robust information is provided, Prevailing seeks to make appropriate adjustments to pre-construction methods to limit any conservative bias.

4.3 What factors typically lead to an under-performing project?

Based on the study undertaken, approximately 1 in 10 projects are materially under-performing. This observation is in line with recent transactional experience.

Typical causes of under-performance include:

- A material grid outage, including a network upgrade or fault;
- A serial equipment defect, such as inverter software issue; and
- 'Scenario mismatch', where the as-built installation varies significantly from the modeled pre-construction design.

In a transaction scenario, these risks can be managed through comprehensive technical due diligence, including reviews of the equipment design and grid connection infrastructure. In some cases, the application of more optimistic future availability assumptions may be appropriate thereby reducing the variance.

5 CONCLUSIONS

The study indicates that there is a tendency for current pre-construction methods to be slightly conservative and that current pre-construction modeling methods may be lagging technological advancements. However, the variance between pre-construction energy modeling and operational assessments for projects constructed in recent years is increasing.

In the absence of any methodology refinement, a higher degree of variance may be expected with recently manufactured equipment.

Prevailing recommends detailed consideration of project specific factors to ensure pre-construction modelling results are as relevant and accurate as possible.

Prevailing intends to expand this validation study to include projects in high irradiance environments (e.g. Australia and USA) in the coming months.

6 REFERENCES

[1] World Energy Council, ENERGY RESOURCES SOLAR, accessible: https://www.worldenergy.org/data/resources/resource/solar/

This is a pre-peer review version of the paper submitted to the journal Progress in Photovoltaics.
It has been selected by the Executive Committee of the EU PVSEC 2017 for submission to Progress in Photovoltaics.

MANAGING TECHNICAL RISKS IN PV INVESTMENTS – HOW TO QUANTIFY THE IMPACT OF RISK MITIGATION MEASURES FOR DIFFERENT PV PROJECT PHASES?

Ulrike Jahn[1], Magnus Herz[1], David Moser[2], Giorgio Belluardo[2], Mauricio Richter[3]

[1]TÜV Rheinland Energy GmbH, D-51101 Cologne (Germany)
Tel. +49-221-806-2232, Fax: +49-221-806-1350, E-mail: ulrike.jahn@de.tuv.com

[2]Institute for Renewable Energy - EURAC Research, Bolzano (Italy)
[3] 3Esa, Quai à la Chaux 6, 1000 Brussels (Belgium)

ABSTRACT: Project investment has been and still is a primary financial factor in enabling sustainable growth in photovoltaic (PV) installations. When assessing the investment-worthiness of a PV project, different financial stakeholders such as investors, lenders and insurers will evaluate the impact and probability of investment risks differently depending on their investment goals. Similarly, risk mitigation measures implemented are subject to the investment perspective. The aim of this work is to evaluate the effectiveness of mitigation measures into the framework for the assessment of the economic impact of technical risks. The methodology is based on statistical analysis and can be applied to a single PV plant or to a large portfolio of PV plants in the same market segment. The technical risks during operation and maintenance were ranked based on the so-called Cost Priority Number (CPN), which was introduced in "Identification of Technical Risks in the Photovoltaic Value Chain and Quantification of the Economic Impact" [1]. The methodology has been further developed for the evaluation and effectiveness of the identified mitigation measures. Risk Mitigation Factors (RMF) are introduced which quantify the reduction of the economic impact of technical risks, achieved via the reduction of failure occurrence or reduction of costs for fixing the failures. The cost-benefit analysis can then include the combination of various mitigation measures and derive the best strategy depending on market segment and PV plant typology.

Keywords: Bankability, technical risks, operation & maintenance, risk mitigation measures

1 INTRODUCTION

In the project report "Technical Risks in PV Projects" [2], technical risks were identified and categorised for components (Modules, inverters, mounting structure, connection and distribution boxes, cabling, potential equalization and grounding, lightning and protection system, weather station, communication and monitoring, transformer station, infrastructure and environmental influence, storage system, miscellaneous) and phases (Product testing, PV plant planning/development, Installation/transportation, Operation/maintenance, Decommissioning) of the value chain of a PV project into a so called Risk Matrix. The technical risks were broadly divided into risks to which one can assign an uncertainty to the initial yield assessment [13] and risks to which one can assign a Cost Priority Number (CPN). This CPN can be described as a cost based risk priority number as frequently used in failure mode effect analysis. Basically, the CPN consists of frequency and the economic impact of a failure. The main outcome of the report were then further analysed and published in [1]. While failures arising from technical risks belonging to the first group have an impact on the overall uncertainty of the energy yield, failures with a CPN have a direct impact on the annual cost of running a PV plant. The latter are caused by economic losses due to downtime (reduction in the utilisation factor) and component repair or substitution (Operational Expenditures, OPEX) and it is given in €/kWp or €/kWp/year. The CPN methodology was thus developed to assess the economic impact of technical risks occurring during the operation and maintenance (O&M) phase of a PV project. The analysis presented in the report "Technical Risks in PV Projects" [2] was linked to a failure database over a portfolio of more than 700 PV plants, 440 MWp, ~2,000,000 modules, ~12,000 inverters, mounting structures, combiner boxes, cables and transformers for a total of ~2.4 million components. Although the database already includes PV plants from various market segments and countries, the comparison of the CPNs for various technical risks (e.g. quantifiable failures during O&M) was carried out on an annual basis for all plants. This is a shortcoming of the database and not of the methodology. The overall methodology was created to allow the estimation of the economic impact of failures on the levelised cost of electricity (LCOE) and on business models of PV projects. Not all the technical risks fit into the two groups previously described: some risks can be defined as precursor of a risk propagating along the value chain or have an indirect impact on the CPN of various risks in terms of occurrence, time to detect, time to repair, etc. Thus, the overall technical risk framework in the Solar Bankability project has been developed to determine the economic impact of a failure, but also to be able to assess the effectiveness of mitigation measures.

In a PV project, costs for correction of defects may increase exponentially with a factor of 10 by each step along the value chain from the product idea to the handover to the customer [3]. Defect prevention instead of defect correction should thus be considered as a first mitigation option with an effective risk management strategy during system design and planning. The reduction in occurrence of failures during the planning phase has in fact a direct positive consequence in terms of reduction in occurrence of failures during the operational phase, resulting in a lower CPN. Mitigation measures as defect correction will also have a cost. Therefore, the balance between the increased capital expenditure during planning must be countered by an effective decrease of operational (monetary) losses caused by downtime, component replacement or repair. It is important to this extent to analyse how risks propagate from one step of the value chain to the next; this allows us to identify mitigation measures and to understand if, for some specific failures, an effective mitigation measure is already in place.

This is a pre-peer review version of the paper submitted to the journal Progress in Photovoltaics.
It has been selected by the Executive Committee of the EU PVSEC 2017 for submission to Progress in Photovoltaics.

The cost of mitigation measures needs to be included in a cost benefit analysis, which has to consider the expectations of the stakeholders that are involved in a PV project [4]. Investors are seeking for long defect warranty periods, performance guarantees, reasonable low CAPEX and OPEX, high long-term plant performance and lifetime (ideally above the initial prediction). Banks have requirements similar to those of the investors which are looking for projects with a 10-15 year financing period and PV plant performance which can also be slightly below prediction. Insurers try to limit their liability to failures with an external root cause based on PV plants, which meet technical market standards and are maintained on a regular basis. On the contrary, EPC contractors will look for short defect warranty periods, minimum of additional guarantees and warranties, high sale price with low OPEX showing a very different time horizon compared to the investors.

As a consequence of the different needs between the key actors, O&M operators are in a difficult position to manage all these conflicting requirements for a long period of time. The best condition for O&M operators is in fact in the presence of long defect warranty period and low sale price to allow for higher OPEX. Recent trends in the PV market have put a lot of pressure on the O&M price which is reported to be as low as 8 Euros/kWp/year in Germany in 2016 [4]. A large share of these costs is labour intensive, for example site keeping and inspection, preventive maintenance, monitoring and reporting. It is therefore of extreme importance to identify what O&M scope is obligatory versus what is optional and the required reaction time depending on the severity of the failure by assessing the cost of various mitigation options during the operational phase which can be part of an effective O&M strategy.

Mitigation measures must be identified along the PV value chain and assigned to various technical risks. Typical mitigation measures during the design phase are linked to the component selection such as standardised products, products with known track record, O&M friendly design, like accessibility of the site, state of the art design of the monitoring system, LCOE optimised design, for instance tracker versus fixed tilt, central versus string inverter, quality check of solar resource data. Mitigation during the transportation and installations are linked to the supply chain management, for example well organised logistics and quality assurance during transportation, quality assurance (e.g. predefined acceptance procedures) and grid connection such as knowledge of grid code [5]. These mitigation measures reduce the uncertainty of the overall energy yield, increase the initial energy yield and reduce the cost of O&M during the operational phase (e.g. faster replacement of components, lower cost of site maintenance, lower occurrence and severity of defect, etc.).

Mitigation measures during the O&M phase are linked to maintenance, for instance preventive maintenance, visual inspection and spare parts management, monitoring and data quality (e.g. state of the art measurement equipment and software, performance evaluation, predictive monitoring), outsourcing (e.g. in-sourcing can reduce costs and dependency from suppliers) and remote monitoring like video surveillance or defined workflow to reduce replacement time. These mitigation measures directly affect the CPN of failures occurring during the operational phase by reducing the time to detect defects or the time to repair/substitute defects.

Compared to many other power generating technologies, PV plants have reduced maintenance and service requirements. However, a continuous O&M programme is essential to optimise energy yield and maximise the lifetime and viability of the entire plant and its individual components. Many aspects of O&M practices are interrelated and significantly affect the performance of all the components in the generation chain and project lifecycle. The PV technical risks were defined in the Project Report "Technical Risks in PV Projects" [2] in terms of downtime, production performance, operational costs and time to complete the required activities. It is important that risk ownership is also considered to better understand which key actor is responsible for the action of mitigating the risk. These risks can then be turned in opportunities to meet or even exceed the expectations of the developers and owners in terms of return on the investment. In particular, suitable planning, supervision and quality assurance actions are critical at all stages of a PV project in order to minimise the risk of damages and outages, optimise the use of warranties, avoid non-optimal use of resources and ultimately optimise the overall performance of the PV plant.

The scientific PV community has thoroughly investigated some specific failures and drawn recommendations on how to mitigate the economic impact for, e.g. soiling [6, 7, 8], grid integration [9] or potential induced degradation (PID) [10]. General recommendations on the mitigation measures to reduce the impact of technical risks are also found in more general publications given by companies active in the field as EPC contractors, consultants, and O&M operators [11, 12]. Some failures can be prevented or mitigated through specific actions at different project phases (e.g. for PID, a different encapsulant or glass during product manufacturing phase, a PID box in case of reversible PID during the operation/maintenance phase); others can be prevented or mitigated through a more generic action. For example, the monitoring of performance or visual inspection can be considered as generic mitigation measures that can have a positive impact on the reduction of the CPN of many failures. In practice, it is important to understand how mitigation measures can be considered as a whole to be able to calculate their impact and thus assess their effectiveness. It is not the aim of this paper to provide a set of specific mitigation measures for each technical risk as this would entail failure-specific cost-benefit analysis. It is the objective of this paper is to create a framework of well-defined mitigation measures, which have an impact on the global CPN (given as sum of CPNs of all technical risks). The cost-benefit analysis can then include the combination of various mitigation measures and derive the best strategy depending on market segment and plant typology. In addition to this, it is important to assess in the CPN analysis who bears the cost and the risk to derive considerations not only on the overall economic impact of the technical risks, but also on cost and risk ownership.

2 METHODOLOGY

Besides the technical risks associated with uncertainties during project planning phase [13], the second group of risks has a direct economic impact during operation. These risks were already identified and evaluated in the Project Report "Technical Risks in PV

This is a pre-peer review version of the paper submitted to the journal Progress in Photovoltaics.
It has been selected by the Executive Committee of the EU PVSEC 2017 for submission to Progress in Photovoltaics.

Projects" [2]. Failures which have a direct impact on the annual cost of running a PV plant caused by economic losses due to downtime (utilisation factor) and component repair or substitution (OPEX) are described by a CPN. The CPN is given in €/ kWp or €/ kWp/year. The methodology of quantification was also introduced in [1].

The developed method is now applied on the evaluation of the effectiveness of the identified mitigation measures. To this extent, Risk Mitigation Factors (RMF) are introduced which quantify the reduction in the economic impact of technical risks (i.e. reduce the occurrence of the failure, repair of existing component, substitution by spare component, substitution by new component). Therefore, three new parameters are defined as risk mitigation factors. Depending on the category of mitigation, the impact is:

(α) mitigation of number of failures
(β) mitigation of time to detection
(γ) mitigation of time to repair.

The quantitative impact is given by a number between 0 - 1. If a mitigation measure has no impact on a certain failure, the related risk reduction factor is 1. If the risk parameter is completely eliminated after the mitigation, the risk mitigation factor becomes 0. This means that with α of 0.5 the number of failures can be reduced to 50 % and with β of 0.5 the time to detection can be reduced to 50 % of the time without mitigation.

Thus, the mitigated number of failures is calculated as follows:

$$n_{fail,mit} = \alpha \cdot n_{fail} \quad (1)$$

The mitigated time to detection is calculated as follows:

$$t_{td,mit} = \beta \cdot t_{td} \quad (2)$$

and the mitigated time to repair/substitution is calculated as follows:

$$t_{tr/ts,mit} = \gamma \cdot t_{tr/ts} \quad (3)$$

When several mitigation measures with impact on the same risk mitigation factor are considered, the individual risk mitigation factors are multiplied with each other and the risk is further reduced. The risk mitigation factor α, respectively for β and γ, is then calculated as follows:

$$\alpha = \alpha_1 \cdot \alpha_2 \cdot \alpha_3 \ldots = \prod_{i=1}^{n} \alpha_i \quad (4)$$

Implementing the new mitigation factors into the CPN model leads to the following mitigated downtime costs (€):

$$C_{down,mit} = C_{down,td,mit} + C_{down,tr/ts,mit} + C_{down,fix,mit} \quad (5)$$

- mitigated downtime costs until the failure is detected (t_d, time to detection):

$$C_{down,td,mit} = \alpha \cdot n_{fail}/n_{comp} \cdot \beta \cdot t_{td}/t_{ref} \cdot PL \cdot M \cdot PPA \cdot Y = \alpha \cdot \beta \cdot C_{down,td} \quad (6)$$

- mitigated downtime costs from detection until before repair/substitution ($t_{tr/ts}$, *time to repair, time to substitution*):

$$C_{down,tr/ts,mit} = \alpha \cdot n_{fail}/n_{comp} \cdot \gamma \cdot t_{tr/ts}/t_{ref} \cdot PL \cdot M \cdot PPA \cdot Y = \alpha \cdot \gamma \cdot C_{down,tr/ts} \quad (7)$$

and mitigated downtime costs during repair/substitution (t_{fix}, time needed to repair or substitute a failed component):

$$C_{down,fix,mit} = \alpha \cdot n_{fail}/n_{comp} \cdot t_{fix}/t_{ref} \cdot M \cdot PPA \cdot Y = \alpha \cdot C_{down,fix} \quad (8)$$

where PL is the performance loss (or power loss), expressed in fraction (or %). Therefore PL=1 for failures causing total downtime. M is a multiplier to consider failures that cause problems at higher component level (e.g. 1 module takes down the whole array caused by theft of module).

- PPA is the missing income from power purchasing agreements.
- n_{fail} is the number of components affected by a specific failure.
- n_{comp} is the number of components in affected plants.
- t_{ref} is the reference time interval. For an annual analysis it defined as 8760 h

In addition to downtime costs, the costs for fixing the failures is the second main parameter. It considers the costs of repair and substitution of parts ($C_{rep/sub}$), costs of transportation(C_{trans}), labour costs (C_{lab}) and also the costs of detection and mitigation depending on the applied mitigation measures ($C_{det/mit}$). It is calculated as follows:

$$C_{fix,mit} = \alpha \cdot n_{fail} \cdot [C_{det/mit} + C_{rep/sub} + C_{trans} + (C_{lab} * t_{fix})] = \alpha \cdot C_{fix} \quad (9)$$

We have not identified mitigation measures that mitigate and reduce time needed to repair or substitute a failed component t_{fix}.

The calculation of the mitigated CPN is then given by:

$$CPN_{mit} = C_{down,mit} + C_{fix,mit} \quad (10).$$

Mitigation measures must be identified along the value chain and assigned to various technical failures. Some failures can be prevented or mitigated through specific actions at different project phases. For example, for potential induced degradation (PID), the mitigation measure could be using different encapsulant or glass during the product manufacturing phase, or installing PID boxes during the operation/maintenance phase (for reversible PID) [10]..

Others can be prevented or mitigated through more generic mitigation measures that can have a positive impact on the reduction of the CPN of many failures, e.g. visual inspection or performance monitoring. In summary it is important to understand how mitigation measures can be considered as a whole to be able to calculate their impact and thus assess their effectiveness.

The reported methodology allows the calculation of the impact of several mitigation measures with different level of details. To show its potentiality and the type of insight that can be gained, in this work we present the results from the analysis of the technical risks identified in

This is a pre-peer review version of the paper submitted to the journal Progress in Photovoltaics.
It has been selected by the Executive Committee of the EU PVSEC 2017 for submission to Progress in Photovoltaics.

[2], putting forward eight generic mitigation measures for PV technical risk management. They are categorised into two main categories. Preventive measures are applied before the failure occurs to prevent it from happening. The mitigation measures under this category are component testing, design review + construction monitoring, and EPC qualification. These measures can be implemented during the early phases of PV project lifecycle and are likely to increase the CAPEX. Corrective measures are mitigation measures that aim to reduce higher losses and costs, if the failure has already occurred. The costs are mostly related to the OPEX due to the implementation during the operation and maintenance phase.

The identified mitigation measures are here defined:

- **Component testing** of important plant components such as PV modules or inverters. The testing can be performed by the manufacturer in the factory, or independent testing at certified laboratory, or on-site at the PV plant;
- **Design review and construction monitoring** serve to catch issues caused by bad PV plant design and poor PV construction workmanship;
- **EPC qualification** focuses on ensuring the competencies of the field workers, e.g., by requiring certain technical qualification prerequisites or regular training of the field workers;
- Implementing **advanced monitoring** system for early detection and diagnosis of faults;
- Use of **basic monitoring** system to monitor plant level alarms and notifications;
- **Advanced inspection** (e.g. using infrared or electroluminescence camera) to detect defects not usually visible by with the naked eyes;
- **Visual inspection** to establish any visible changes in PV plant components;
- **Spare part management** to minimize the costs of downtime and repair/substitutions of components.

3 RESULTS AND DISCUSSIONS

The economic impact of various mitigation measures is here reviewed and quantitatively evaluated based on the described methodology. In addition, the measures with the highest cost-benefit ratio are identified in order to facilitate the selection of suitable measures and their combinations.

The following eight of the already introduced mitigation measures are assessed. The pre-selection was made after research of the most common measures as defined in the gap analysis published in [14] and a representation of measures influencing the three parameters described n_{fail}, t_{td} and t_{tr} / t_{ts}.

The implementation of the measures is performed based on a reference scenario, limited to the utility-scale segment of the database described in [2]. The downtime is based on a so-called "never detected" scenario without mitigation measures (the never detected scenario is defined with a 12 months lead-time to detection), here referred as LOSS scenario. The FIX scenario instead is dedicated to the cost of fixing the failure which is a combination of downtime from detection to repair/substitution and the cost of repair/substitution.

The total CPN is calculated from the sum of the individual CPN of all failures. The new CPN is the sum of the mitigated CPN including the costs $Cost_{mit}$ incurred for the mitigation measures. The benefit of the combination of the measures resulting from the difference between total CPN and new CPN is then given by:

$$CPN_{new} = CPN_{mit} + Cost_{mit} \tag{11}$$

The applied costs of the mitigation measures are defined in Table 1 where three different cost scenarios are introduced: medium cost scenario (1), low cost scenario (2) and high cost scenario (3). These scenarios mark the lower and upper range of the applied costs for the new CPN calculations.

Table 1: List of most significant measures, their mitigation factors, affected parameters, mitigation measures with medium, low and high cost scenarios. The annual value (in brackets) is based on a lifetime of 20 years' operation.

Mitigation Measure	Risk Mitigation Factor	Affected Parameter	Defined costs Scenario 1 (medium costs)	Defined costs Scenario 2 (low costs)	Defined costs Scenario 3 (high costs)
Component testing – PV modules	α	number of failures	3 €/kWp (0.15 €/kWp/year)	1 €/kWp (0.05 €/kWp/year)	10 €/kWp (0.5 €/kWp/year)
Design review + construction monitoring	α	number of failures	20 €/kWp (1 €/kWp/year)	10 €/kWp (0.5 €/kWp/year)	40 €/kWp (2 €/kWp/year)
Qualification of EPC	α	number of failures	3 €/kWp (0.15 €/kWp/year)	1 €/kWp (0.05 €/kWp/year)	10 €/kWp (0.5 €/kWp/year)
Advanced monitoring system	β	time to detection	2 €/kWp/year	1 €/kWp/year	3 €/kWp/year
Basic monitoring system	β	time to detection	0.5 €/kWp/year	0 €/kWp/year	1 €/kWp/year
Advanced inspection	β	time to detection	2 €/kWp/year	1 €/kWp/year	3 €/kWp/year
Visual inspection	β	time to detection	1 €/kWp/year	0.5 €/kWp/year	2 €/kWp/year
Spare part management	γ	Time to repair /substitution	10 €/kWp (0.5 €/kWp/year)	2 €/kWp (0.1 €/kWp/year)	20 €/kWp (1 €/kWp/year)

The influence of the mitigation measures was ranked by the risk mitigation factor for each failure and each of the eight mitigation measures in the following four impact classes defined in Table 2. Failures were initially identified and prioritised per component and project phase by a panel of experts composed by the project participants and thus included in the risk matrix. The final classification was carried out through a CPN analysis of the database of collected failures as reported in [2].

Table 2: Definition of impact classes with respect to risk mitigation factor (RMF)

Impact Classes	RMF (α,β,γ)
High	99.5 %
Medium	50.0 %
Low	25.0 %
No	0.0 %

Figures 1 and Figure 2 show two examples of risk mitigation benefits in terms of CPN for the top 10 risks of PV modules and inverters, respectively. For these two examples the cost of the mitigation measure was evenly divided over each specific failure. This means that when a measure mitigates 20 failures the partial cost of the mitigation measure for this specific failure is only one twentieth of the total cost. The CPN without any mitigation measures (blue bars) is compared to the CPN resulting from the best combination of mitigation measures (red bars) for each of the top 10 failures. When the CPN related to the mitigation measures is nearly zero, the benefit is the highest. This is the case for risks, which can easily be reduced by preventive measures e.g. bad installation, PID and defective back sheet for module risks (see Fig. 1) and wrong installation and grounding faults for inverter risks (see Fig. 2). A right combination of mitigation measures can be very effective taking a failure specific CPN down to nearly zero.

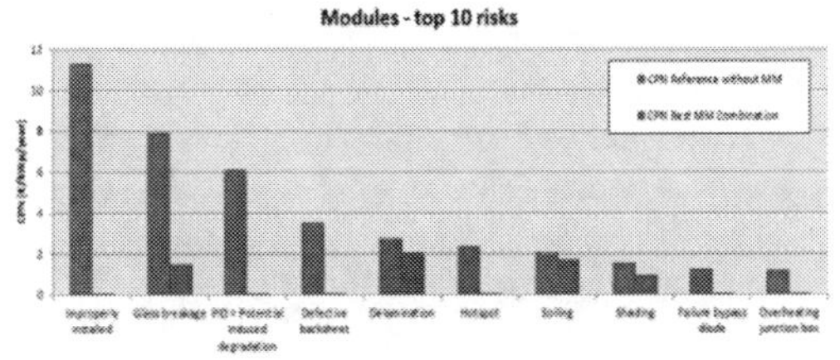

Figure 1: Top 10 risks of PV modules with and without mitigation measures in CPN. CPN is given in €/kWp/year

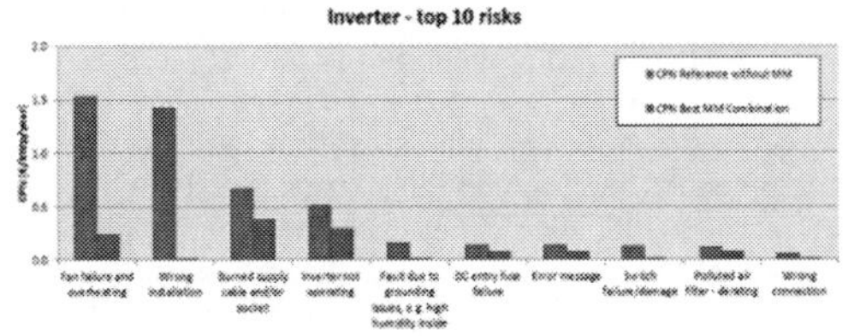

Figure 2: Top 10 risks of inverter with and without mitigation measures in CPN

3.1 Mitigation measures applied to the FIXING scenario

The costs related to FIXING the failures result from the sum of the costs of repair/substitution, the costs of detection, the costs of transportation, and the cost of labour as well as the downtime from detection to repair. Figure 3 shows the results of calculating the FIXING costs for selected failures when applying the eight mitigation measures listed in Table 1. The selection of failures was based on experts' panels and include the top 20 PV module failures, top 20 inverter failures, failures of mounting structure, combiner boxes, cabling as well as failures of transformer station as listed in [15].
The total CPN without mitigation measures is 104.75 €/ kWp/year for the defined FIXING scenarios, defined as FIX Reference. The CPNnew for all combinations of mitigation measures are shown in Fig. 3 below. In total there exist 8 MMs: their full binary combinatory corresponds to 28=256 combinations (0: #00000000: no MM, 256: #11111111: all the 8 MMs). The total CPN without mitigation measures is given by the green area at 104.75 €/kWp/year. This can be considered as the worst situation one can find in a portfolio of PV plants where no mitigation measures are applied. Applying specific mitigation measures (e.g. combinations 192 – 255 as given in [15]) it is possible to significantly reduce the risks and to obtain values of CPNnew in the order of 15 to 20 €/kWp/year (see Fig. 3). The index of combinations can be found in [15].

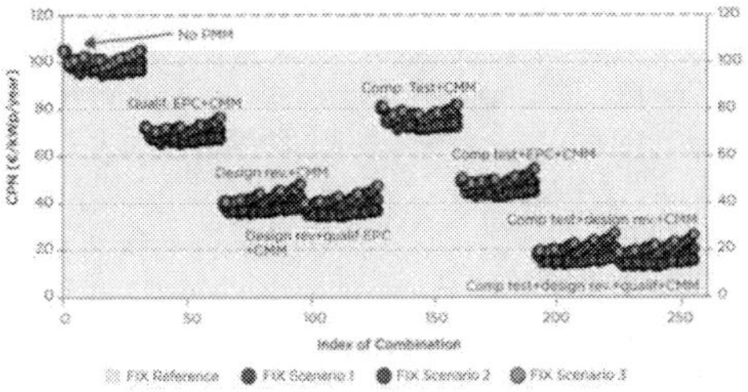

Figure 3: New CPN results of mitigation measure combinations for different FIX cost scenarios compared to CPN without MM – Index of mitigation measure combinations is given in [15].

The CPN_{new} analysis above shows that for 99 % of all mitigation measure combinations, the scenarios will result in economic benefit by reducing the CPN_{new} to values lower than the reference (104.75 €/ kWp/year). Savings up to 90 €/ kWp/year appear possible for the best combinations of selected mitigation measures. Furthermore, we can conclude that in general, mitigation measures which reduce the failure occurrence have the highest impact due to the related reduction in substitution costs. Preventive mitigation measures (PMM) have the highest impact on CPN_{new}, for instance, qualification of EPC will bring down CPN_{new} to 75 €/ kWp/year, and design review will further reduce the CPN_{new} to 40 €/ kWp/year. In fact, the highest savings can be achieved by applying all three preventive measures (component testing + design review + qualification of EPC). On the other hand, corrective mitigation measures (CMM) such as e.g. basic and advanced monitoring and visual and advanced inspection appear to have less impact on CPN_{new} for PV plants which are poorly designed, present low quality components and in presence of unqualified EPC contractors. In case of high quality PV projects, the CMMs play a much more important role leading to a possible cost reduction of up to 3 €/kWp/year.

Table 3 shows the corresponding savings of those mitigation measures for the best combinations (Rank 1 to 10) and for the three cost scenarios. The savings ranging between 85 and 91 €/ kWp/year are given by the difference between the total CPN (104.75 €/ kWp/year) and the new CPN. The savings for low cost scenario (2) are slightly higher than for the medium (1) and high cost (3) scenario (see Table 3).

Table 3: Best combinations of mitigation measures for medium (1), low (2) and high (3) cost scenarios and their savings in CPN (last column)

		[illegible]	[illegible]	[illegible]	[illegible]	[illegible]	[illegible]	[illegible]	[illegible]	[illegible]
[illegible]	[illegible]	[illegible]	[illegible]	[illegible]	[illegible]	[illegible]	[illegible]	[illegible]	[illegible]	[illegible]
	[illegible]	[illegible]	[illegible]	[illegible]	[illegible]	[illegible]	[illegible]	[illegible]	[illegible]	[illegible]
	[illegible]	[illegible]	[illegible]	[illegible]	[illegible]	[illegible]	[illegible]	[illegible]	[illegible]	[illegible]
	[illegible]	[illegible]	[illegible]	[illegible]	[illegible]	[illegible]	[illegible]	[illegible]	[illegible]	[illegible]
	[illegible]	[illegible]	[illegible]	[illegible]	[illegible]	[illegible]	[illegible]	[illegible]	[illegible]	[illegible]
	[illegible]	[illegible]	[illegible]	[illegible]	[illegible]	[illegible]	[illegible]	[illegible]	[illegible]	[illegible]
	[illegible]	[illegible]	[illegible]	[illegible]	[illegible]	[illegible]	[illegible]	[illegible]	[illegible]	[illegible]
	[illegible]	[illegible]	[illegible]	[illegible]	[illegible]	[illegible]	[illegible]	[illegible]	[illegible]	[illegible]
	[illegible]	[illegible]	[illegible]	[illegible]	[illegible]	[illegible]	[illegible]	[illegible]	[illegible]	[illegible]
	[illegible]	[illegible]	[illegible]	[illegible]	[illegible]	[illegible]	[illegible]	[illegible]	[illegible]	[illegible]
[illegible]	[illegible]	[illegible]	[illegible]	[illegible]	[illegible]	[illegible]	[illegible]	[illegible]	[illegible]	[illegible]
	[illegible]	[illegible]	[illegible]	[illegible]	[illegible]	[illegible]	[illegible]	[illegible]	[illegible]	[illegible]
	[illegible]	[illegible]	[illegible]	[illegible]	[illegible]	[illegible]	[illegible]	[illegible]	[illegible]	[illegible]
	[illegible]	[illegible]	[illegible]	[illegible]	[illegible]	[illegible]	[illegible]	[illegible]	[illegible]	[illegible]
	[illegible]	[illegible]	[illegible]	[illegible]	[illegible]	[illegible]	[illegible]	[illegible]	[illegible]	[illegible]
	[illegible]	[illegible]	[illegible]	[illegible]	[illegible]	[illegible]	[illegible]	[illegible]	[illegible]	[illegible]
	[illegible]	[illegible]	[illegible]	[illegible]	[illegible]	[illegible]	[illegible]	[illegible]	[illegible]	[illegible]
	[illegible]	[illegible]	[illegible]	[illegible]	[illegible]	[illegible]	[illegible]	[illegible]	[illegible]	[illegible]
	[illegible]	[illegible]	[illegible]	[illegible]	[illegible]	[illegible]	[illegible]	[illegible]	[illegible]	[illegible]
	[illegible]	[illegible]	[illegible]	[illegible]	[illegible]	[illegible]	[illegible]	[illegible]	[illegible]	[illegible]
[illegible]	[illegible]	[illegible]	[illegible]	[illegible]	[illegible]	[illegible]	[illegible]	[illegible]	[illegible]	[illegible]
	[illegible]	[illegible]	[illegible]	[illegible]	[illegible]	[illegible]	[illegible]	[illegible]	[illegible]	[illegible]
	[illegible]	[illegible]	[illegible]	[illegible]	[illegible]	[illegible]	[illegible]	[illegible]	[illegible]	[illegible]
	[illegible]	[illegible]	[illegible]	[illegible]	[illegible]	[illegible]	[illegible]	[illegible]	[illegible]	[illegible]
	[illegible]	[illegible]	[illegible]	[illegible]	[illegible]	[illegible]	[illegible]	[illegible]	[illegible]	[illegible]
	[illegible]	[illegible]	[illegible]	[illegible]	[illegible]	[illegible]	[illegible]	[illegible]	[illegible]	[illegible]
	[illegible]	[illegible]	[illegible]	[illegible]	[illegible]	[illegible]	[illegible]	[illegible]	[illegible]	[illegible]
	[illegible]	[illegible]	[illegible]	[illegible]	[illegible]	[illegible]	[illegible]	[illegible]	[illegible]	[illegible]
	[illegible]	[illegible]	[illegible]	[illegible]	[illegible]	[illegible]	[illegible]	[illegible]	[illegible]	[illegible]
	[illegible]	[illegible]	[illegible]	[illegible]	[illegible]	[illegible]	[illegible]	[illegible]	[illegible]	[illegible]

This is a pre-peer review version of the paper submitted to the journal Progress in Photovoltaics.
It has been selected by the Executive Committee of the EU PVSEC 2017 for submission to Progress in Photovoltaics.

3.2 Mitigation measures applied to the LOSS scenario

The LOSS scenario is focused on the economic losses due to downtime. Figure 4 and Figure 5 show the results of calculating the LOSS costs for the same failures as in Figure 3 when applying the eight mitigation measures listed in Table 1. For the calculation of the LOSS costs due to downtime, it is important to consider the missing income of feed-in tariffs or the missing income from Power Purchasing Agreement (PPA) or the missing savings in terms of Retail Cost of Electricity for PV plants on roofs. Here, we consider two values for PPAs, a high PPA (0.25 €/kWh) and lower PPA (0.10 €/kWh) for the three different cost LOSS scenarios of large-scale PV plants.

The total CPN without mitigation measures is 13.5 €/ kWp/year for the defined LOSS scenarios, also known as "never detected", defined as LOSS Reference based on high PPA value (0.25 €/ kWh). The CPN_{new} values for all 256 combinations of mitigation measures are shown in Figure 4 below. The total CPN without mitigation measures is given by the green area at 13.5 €/ kWp/year. Applying specific mitigation measures allows the significant reduction of the losses and obtain values of CPN_{new} in the order of 4 to 8 €/ kWp/year for the low cost scenario (2) and between 4 and 10 €/ kWp/year for the medium cost scenario (1) as shown in Figure 4.

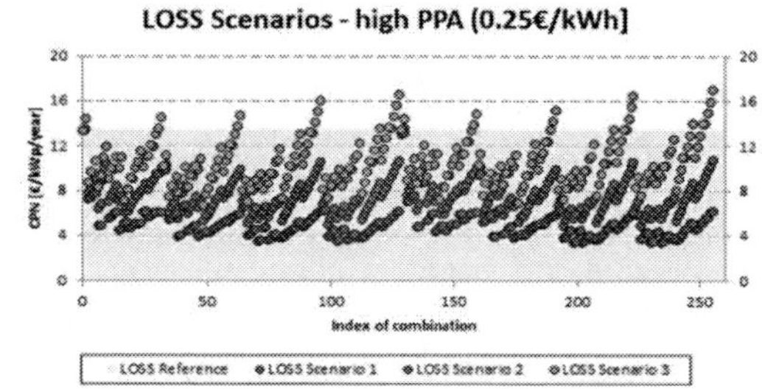

Figure 4: New CPN results of mitigation measure combinations of different LOSS scenarios based on high PPA compared to CPN without mitigation measures

Table 4: Combination of mitigation measures for the defined LOSS scenarios

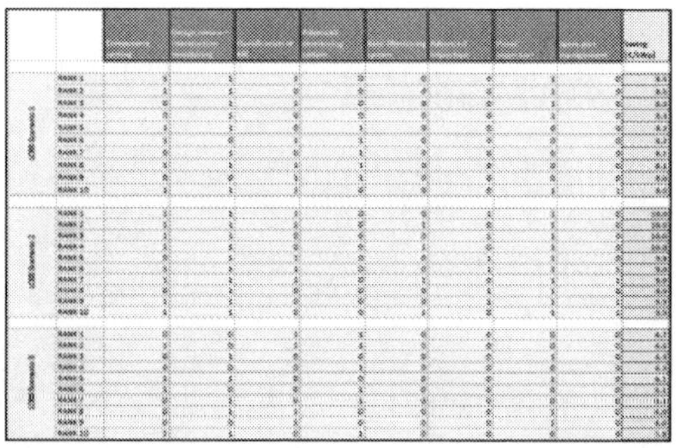

Table 4 shows the corresponding savings of the mitigation measures for the best combinations (Rank 1 to 10) and for the three LOSS cost scenarios. The savings ranging between 6 and 10 €/ kWp/year are given by the difference between the total CPN without mitigation measures (13.5 €/ kWp/year) and the new CPN (Figure 4). The savings for low cost scenario (2) are higher than for the medium (1) and high cost (3) scenario (see Table 4).

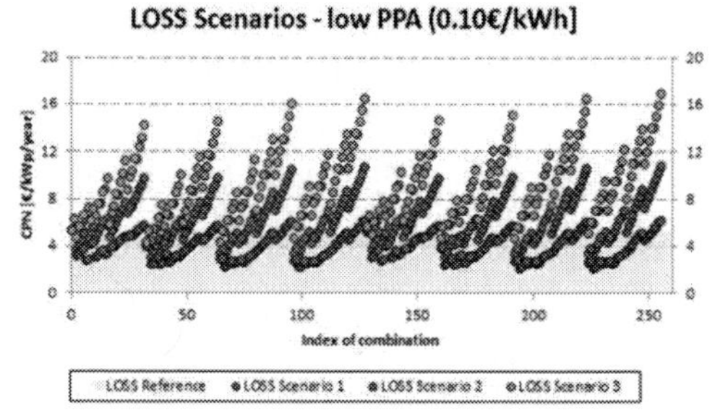

Figure 5: New CPN results of mitigation measure combinations of different LOSS scenarios based on Low PPA compared to CPN without mitigation measures

A combination of all mitigation measures is not recommended and lead to higher costs. E.g. three times the reference LOSS for scenario 3 and combination 255 (Figure 5). With low PPA it is more difficult to apply mitigation measures, which are economic beneficial. Best savings can only be obtained for low cost LOSS scenario 2 by applying the three preventive measures (component testing + design review + qualification of EPC). The savings may reach 3.2 €/ kWp/year for the best combinations of selected mitigation measures.

In summary: If a plant has to be designed, preventive measures are highly recommended. The application of corrective measures for a well performing PV plant where preventive measures had been implemented will yield in an additional benefit.

If preventive mitigation measures were not considered in the plant design phase, the operation of the PV plant is more risky and the corrective measures will allow to find all the technical issues and problems, not only for reasons to increase the yield, but also to avoid safety issues.

4 CONCLUSIONS

The overall methodology presented in this paper allows the estimation of the economic impact of failures on the levelized cost of electricity (LCOE) and on business models of PV projects and has been developed to determine the economic impact of technical risks, but also to be able to assess the effectiveness of mitigation measures.

There is a good reason that highlights the importance of assessing the impact of mitigation measures. It is the experience "Factor of 10" which means that in design phase, mistakes are costly, and the longer it takes to discover a problem, the more costly it becomes. According to David M. Anderson [16], it costs 10 times more to find and repair a defect at the next stage of the plant, and then it costs 10 times more at each subsequent stage of the project. The methodology presented in the paper allows for the inclusion of several mitigation measures assessing their effectiveness in terms of details and costs. To show the potential of the methodology and give an insight on the impact of mitigation measures, in this paper we have identified and categorized mitigation measures in:

This is a pre-peer review version of the paper submitted to the journal Progress in Photovoltaics.
It has been selected by the Executive Committee of the EU PVSEC 2017 for submission to Progress in Photovoltaics.

- Category 1 (before) represents all the preventive measures, which are applied before the failure occurs in order to prevent it from happening. The costs are mostly related to the CAPEX as the implementation is included in the initial investment costs;

- Category 2 (after) represents the corrective measures, which reduce higher losses and costs, if the failure has already occurred. The costs are mostly related to the OPEX due to the later implementation during the operation and maintenance phase.

Compared to many other power generating technologies, PV plants have reduced maintenance and service requirements. However, a continuous O&M programme is essential to optimise energy yield and maximise the lifetime and viability of the entire plant and its individual components. Many aspects of O&M practices are interrelated and significantly affect the performance of all the components in the generation chain and project lifecycle. Mitigation measures under Category 1 and Category 2 have an effect on the overall CPN as defined in the Project Report "Technical Risks in PV Projects" [2] in terms of downtime, production performance, operational costs and time to complete the required activities.

It is important that risk ownership is also considered to better understand which key actor is responsible for the action of mitigating the risk. These risks can then be turned into opportunities to meet or even exceed the expectations of the developers and owners in terms of return on the investment. In particular, suitable planning, supervision and quality assurance actions are critical at all stages of a PV project in order to minimise the risk of damages and outages, optimise the use of warranties, avoid non-optimal use of resources and ultimately optimise the overall performance of the PV plant.

The cost of mitigation measures was included in a cost benefit analysis, which has to consider the expectations of the stakeholders that are involved in a PV project [4]. Investors are seeking for long defect warranty periods, performance guarantees, reasonable low CAPEX and OPEX, high long-term plant performance and lifetime (ideally above the initial prediction). Banks have requirements similar to those of the investors which are looking for projects with a 10 - 15 - year financing period and PV plant performance which can also be slightly below prediction. Insurers try to limit their liability to failures with an external root cause based on PV plants, which meet technical market standards and are maintained on a regular basis. On the contrary, EPC contractors will look for short defect warranty periods, minimum of additional guarantees and warranties, high sale price with low OPEX showing a very different time horizon compared to the investors.

Mitigation measures with an impact on the overall CPN were identified as: component testing, design review + construction monitoring, qualification of EPC, basic and advanced monitoring system, visual and advanced inspection, and spare part management. The total CPN without mitigation measures was found to be 104.75 €/kWp/year for the defined FIX scenario, defined as FIX Reference. The impact of mitigation measures massively reduces this figure and in the report we have shown that it is possible to significantly reduce the risks and to obtain values of CPN in the order of 15 to 20 €/kWp/year. We can thus conclude that projects were preventive mitigation measures are successfully apply already present a low CPN value. Corrective mitigation measures can further reduce it by further 3 €/kWp/year and thus play an important role in terms of competitive edge during the operational phase of PV projects. PV projects of poor quality present high value of CPN and once in operation, the impact of corrective mitigation measures is only limited.

The value includes the economic impacts of the mitigation measures, their cost and the economic impact of the identified technical risks for all components after mitigation. Depending on when the failure occurs, the ownership of the risk (and consequently, cost) will vary between the involved actors, i.e. PV plant owner, investor, EPC contractor, insurance company, O&M operator. It is thus important as a next step to be able to assign the risk to the relevant stakeholder along the lifetime of a PV project and to evaluate who will ultimately benefit in terms of cost reduction from mitigation measures, which are implemented.

As consequence of the different needs between the key actors, O&M operators are in a difficult position to manage all these conflicting requirements for a long period of time. The best condition for O&M operators is in fact in the presence of long defect warranty period and low sale price to allow for higher OPEX. Recent trends in the PV market have put a lot of pressure on the O&M price which is reported to be as low as 8 €/ kWp/year in Germany in 2016 [4]. A large share of these costs is labour intensive (i.e. site keeping and inspection, preventive maintenance, monitoring and reporting). It is therefore of extreme importance to identify which O&M scope is obligatory versus what is optional. Furthermore, it is highly important to know the required reaction time depending on the severity of the failure by assessing the cost of various mitigation options during the operational phase which can be part of an effective O&M strategy.

In the coming years, as the availability of measured data will exponentially increase, it will be important to build large databases with a potentially uniform method to increase the confidence level of the statistical analysis and thus reduce the perceived risks for PV investors, EPC and O&M contractors as well as other stakeholders. With the availability of these large databases, the necessary information (minimum requirement) can be filtered out to perform tailored analysis in a uniform way i.e. same granularity, same data, and same formulas. Our CPN methodology and approach attempt to provide such a benchmark.

5 ACKNOWLEDGEMENTS

The research leading to these results has received funding from the European Union's Horizon 2020 research and innovation programme under the grant agreement No 649997 (Solar Bankability).

This is a pre-peer review version of the paper submitted to the journal Progress in Photovoltaics.
It has been selected by the Executive Committee of the EU PVSEC 2017 for submission to Progress in Photovoltaics.

6 REFERENCES

[1] Moser, D., Del Buono, M., Jahn, U., Herz, M., Richter, M., De Brabandere, K., "Identification of technical risks in the photovoltaic value chain and quantification of the economic impact", 2017: doi:10.1002/pip.2857

[2] Moser, D., Del Buono, M., Bresciani, W., Veronese, E., Jahn, U., Herz, M., Janknecht, E., Ndrio, E., de Brabandere, K., Richter, M., "Technical Risks in PV Projects - Report on Technical Risks in PV Project Development and PV Plant Operation," Solar Bankability WP1 Deliverable D1.1 and WP2 Deliverable D2.1, Mar. 2016. http://www.solarbankability.org/results.html#c87 (accessed 21.09.2017).

[3] Klute, I., 2016. Technical Risk Assessment during the Planning and Construction of PV plants/solar parks. http://www.solarbankability.org/fileadmin/sites/www/files/documents/1st_public_event/2.2_Klute_Technical_Risk_Assessment_during_the_Planning_and_Construction_of_PV_plants_solar_parks.pdf (accessed 21.09.2017).

[4] Bächler, M., 2016. Technical Risk Management during O&M of PV Plants. http://www.solarbankability.org/fileadmin/sites/www/files/documents/1st_public_event/2.3_O_M_Risk_Assessment_eon.pdf (accessed 21.09.2017).

[5] Herzog, O., 2016. Minimizing Financial Risks throughout Design, Construction and O&M [WWW Document]. URLhttp://www.solarbankability.org/fileadmin/sites/www/files/documents/1st_public_event/4.1_Minimizing_Financial_Risk_throughout_Design__Construction__O_M-Oliver_Herzog.pdf (accessed 8.1.16).

[6] Stridh, B., 2012. Economical Benefit of Cleaning of Soiling and Snow Evaluated for PV Plants at Three Locations in Europe. Presented at the 27th European Photovoltaic Solar Energy Conference and Exhibition, Frankfurt, Germany, pp. 4027–4029.

[7] Mani, M., Pillai, R., 2010. Impact of dust on solar photovoltaic (PV) performance: Research status, challenges and recommendations. Renew. Sustain. Energy Rev. 14, 3124–3131. doi:10.1016/j.rser.2010.07.065.

[8] Qasem, H., 2013. Cost Optimization of cleaning of dust on photovoltaic modules, in: 28th EUPVSEC. Paris, France.

[9] Appen, J. von, Braun, M., Stetz, T., Diwold, K., Geibel, D., 2013. Time in the Sun: The Challenge of High PV Penetration in the German Electric Grid. IEEE Power Energy Mag. 11, 55–64. doi:10.1109/MPE.2012.2234407.

[10] Pingel, S., Frank, O., Winkler, M., Daryan, S., Geipel, T., Hoehne, H., Berghold, J., 2010. Potential Induced Degradation of solar cells and panels, in: Photovoltaic Specialists Conference (PVSC), 2010 35th IEEE. IEEE, pp. 2817–2822.

[11] Vendrell, I., Korsakul, R., Verojporn, S., Smithtinand, P., Inradesa, P., 2014. Solar PV Technical Guidelines for Financiers.

[12] Lowder, T., Mendelsohn, M., Speer, B., Hill, R., 2013. Continuing Developments in PV Risk Management: Strategies, Solutions, and Implications (No. NREL/TP-6A20-57143). National Renewable Energy Laboratory.

[13] Belluardo, G., Herz, M., Jahn, U., Richter, M., Moser, D., "Evaluation of uncertainty in PV project design: definition of scenarios and impact on energy yield predictions" in: Photovoltaic Specialists Conference (PVSC), 2017 44th IEEE.

[14] Tjengdrawira, C., Richter, M., "Review and Gap Analyses of Technical Assumptions in PV Electricity Cost - Report on Current Practices in How Technical Assumptions are Accounted in PV Investment Cost Calculation" (No. Solar Bankability WP3 Deliverable D3.1). Jul 2016. http://www.solarbankability.org/results.html#c118 (accessed 21.09.2017).

[15] Jahn U., Herz M., Ndiro E., Moser D., Belluardo G., Richter M., Tjengdrawira C., "Minimizing Technical Risks in Photovoltaic Projects - Recommendations for Minimizing Technical Risks of PV Project Development and PV Plant Operation," Solar Bankability WP1 Deliverable D1.2 and WP2 Deliverable D2.2, Jul. 2016. http://www.solarbankability.org/results.html#c124 (accessed 21.09.2017).

[16]Anderson, D.M., 2014. Design for Manufacturability: How to Use Concurrent Engineering to Rapidly Develop Low-Cost, High Quality Products for Lean Production. CRC Press.

REVIEW OF DIFFERENT SOFTWARE SOLUTIONS FOR THE HOLISTIC SIMULATION OF DISTRIBUTED HYBRID ENERGY SYSTEMS FOR THE COMMERCIAL ENERGY SUPPLY

L. Schmeling[a,b,c], P. Klement[b,*], T. Erfurth[c], J. Kästner[c], B. Hanke[b], K. von Maydell[b], C. Agert[b]
[a] Carl von Ossietzky University of Oldenburg, Germany
[b] DLR Institute of Networked Energy Systems, Germany
[c] KEHAG Energiehandel, Germany
[*] Corresponding author: Carl-von-Ossietzky-Straße 15, D-26129 Oldenburg.
Tel.: +49 441 999 06-226, Fax: +49 441 999 06-107 E-Mail: peter.klement@dlr.de

ABSTRACT: Innovative commercial energy supply solutions combine CHP plants with renewable energy sources like PV, solar thermal or heat pumps, which results, under careful planning, in a profitable, yet sustainable hybrid distributed energy supply scheme. To get to this optimal, ecologically beneficial as well as economically advantageous system design and operation strategy a holistic modelling and simulation of the energy supply scheme is inevitable. This paper designs and tests an objective method to help users in finding a suitable simulation tool for a specific distributed generation project. This is done by firstly defining requirements and ranking them based on their importance for the project, followed by a detailed software finding phase, and finally the evaluation of the identified software solutions based on the requirement catalogue. The method is exemplarily applied and a limited overview over existing software solutions is presented. All of this helps any kind of interested user to find the optimal simulation software for a specific hybrid distributed energy generation project.
Keywords: Hybrid Distributed Generation, Commercial Energy Supply, Simulation Software

1 INTRODUCTION

Today companies are offered a wide variety of different energy supply schemes to choose from based on their specific requirements. Especially innovative distributed energy supply schemes have become popular, as they are an efficient and reliable source of all kinds of energies [1, 2]. A special benefit of those systems is the interconnection of different energy forms, e.g. electricity and heat or heat and cold. This leads to an efficient usage of available energy which is not always possible in current, centralised supply schemes [1]. The holistic view on energy flows in companies is vital to come to an optimal energy supply concept. Especially combining electricity and heat/cold energy supply in some form might influence the overall efficiency and thus economic aspect of energy supply [3-9]. Today mostly combined heat and power (CHP) plants are used to achieve this target. By combining a CHP with a renewable energy form like PV or solar thermal a flexible and reliable yet cheap energy supply system is formed [3, 5-7, 10-12] which is called hybrid distributed supply scheme [13]. The exact decision on which kind of system a company should invest in, looking for the maximum profit and therefore minimal cost of energy supply, is therefore as difficult as never before.

Hybrid energy supply feature complex interconnection of different technologies, country-specific funding and legislation and are highly dependent on outer influences. This leads to a high number of different options on how to provide energy in an optimal matter, making it difficult for current simulation software to model such systems. Therefore specialised software has to be used in order to get reliable results. There exist a large number of different software solutions to help certain users in their distributed generation project [14, 15]. Choosing the right software for the right project is vital to achieve a correct and relevant result. In this paper a methodology is introduced to structure the decision process and an exemplary case study is presented showing a short overview over available software solutions.

2 CLASSICAL, DISTRIBUTED AND HYBRID ENERGY SUPPLY

A classical energy supply scheme as used in many industries today is based on a complete separation of heat and electricity supply. Electricity is purchased from the grid; heat is produced in some sort of fossil fuel fired boiler.

In contrast, **distributed energy supply** describes mainly electricity generation either as part of the distribution network or even in the customer site of the meter [1]. Common approaches are combined heat and power (CHP) plants, photovoltaic (PV) or wind turbines (WTG) [2].

By using more than one distributed energy generation technology, the setup formed can be classified as being a **hybrid distributed energy system** [13]. Those kinds of systems feature different kind of generation concepts to supply the same kind of energy. This gives the operator the opportunity to decide at every point in time, based on the current load, forecasts and market prices, which kind of energy generation or combination of those might be the optimal one to use. To get even more freedom and possibilities some sort of storage might be added. This could be every sort of battery storage but thermal storage might be especially helpful [16, 17]. The interconnection and especially the operating and control strategy of those hybrid systems is critical to get to a profitable design [5, 9, 18] . A careful planning of which equipment to invest in and how to operate it is crucial before erecting equipment at site. This does not only include technical aspects but as well the skilful utilization of market opportunities and governmental subsidies [19].

3 SIMULATION OF DISTRIBUTED HYBRID ENERGY SYSTEMS

Specialised simulation software is used in the planning process of distributed hybrid generation to ensure a reliable result and therefore an economically beneficial energy supply scheme.

3.1 Simulation of complex systems

In general simulation of any kind of system is necessary, when its behaviour cannot be analysed and explained by simple theoretical and analytic methods. To do so, a model of the real-life system is constructed. The development of this model over time is then called a simulation [20]. With this procedure the time-dependent evolution of a complex system can be predicted without the need of working on the real-life system and therefore having to invest a lot of money and risking to destroy it.

A system is called complex when it consists of a large number of components which interact in a way that is non-linear and difficult to predict [20]. It can be positioned somewhere between an ordered and a chaotic system. This characteristics lead to a number of properties and features like emergence, resilience, and controllability which need careful consideration [20].

3.2 Simulation of complex energy systems

It should become clear from the basics of complex system simulation that energy systems, and especially electricity and heat supply, are a paragon for the application of simulation tools. These systems consist of a vast number of different technologies and components that are interconnected and therefore, in one or the other way, influence each other in a non-linear way. This is not only on a technical level but information like weather data, economic market behaviour and social factors play a vital role in the exact behaviour of energy systems. For that reason energy system analysis in form of modelling and simulation is a key part in today's energy research [20] and many ready-made simulation software tools are available on the market with rather different approaches. Each of these has its own right to exist and covers a certain fraction of the modern energy supply.

4 HOW TO FIND SUITABLE SOFTWARE?

In order to find appropriate software for a specific project a general comparison scheme of available software is developed. The process is structured in four subsequent steps:

1. Formulate Requirement Catalogue
2. Ranking of Requirements
3. Search For Existing Software Solutions
4. Evaluate Requirement Catalogue

In a first step, **formulate requirement catalogue**, the project specific hybrid distributed generation scheme is analysed from a technical, ecologic and economic, but pure theoretical point of view based on relevant literature, customer requirement and expert opinion. The mentioned requirements are then formulated and grouped for the later evaluation. Useful main categories are:

1. Basics
2. Technical
3. Economic and Legal
4. Ecological
5. Output

These main categories are divided in several sub-categories containing the identified requirements and thereby forming a comprehensive requirement catalogue. During the later on software tests the catalogue is constantly extended based on useful features found in the tested software.

Not every identified requirement is equally important to come to a reliable result, making a **ranking of requirements** advisable. Specialised post-processing tools are added manually without great effort and are therefore not as important as for example the correct simulation of certain kind of generation technologies, which, in most cases, cannot be added or manipulated by the user. Requirements are therefore ranked in three different categories: critical, relevant and optional.

The next step is to **search for existing software solutions.** This is done by different methods. There exist some scientific publications dealing with a general overview [14, 15, 21], but due to constant development of software a specific internet research is advisable to get up-to-date data. Another possibility is to contact different people working in this field and asking for their experience. The more software can be found the better the following decision for a software is.

The last step, **evaluation of the requirement catalogue**, brings together the identified and ranked requirements and found software solutions. Optimally, the software is installed (most software developers offer free trial versions) and tested following the requirement catalogue. For upcoming questions it is helpful to read the software manual or to contact the developers. The single requirements are then ranked from requirement fulfilled, partially fulfilled and [redacted]. Counting the different missed requirements based on their relevance leads to a detailed overview over each and every software. The decision on which software to use can then be based in this objective evaluation of the requirement catalogue giving an objective view on the decision process.

5 CASE STUDY

To show the work flow of choosing suitable simulation software and to present a first overview over available software solutions an exemplary project is analysed. The project under investigation is defined as follows: The software is dedicated for an engineering office trying to sell hybrid distributed energy supply solutions to small- to mid-scale industrial customers in Germany. The main focus is on the early planning phase of such systems where the general outline of newly developed, innovative schemes should be tested.

In a first step the requirement catalogue is developed. Therefore a detailed literature research of technical, economic, legal and ecologic basics is followed by an intense discussion with the future users and decision makers. The final requirement catalogue can be seen in Table I. In the same step the users are asked for their priorities of the requirements which are depicted in Table I as well by the colour code defined beforehand.

Afterwards, several online forums, search engines, scientific publications, and experts were considered in naming suitable software. The software tools identified and analysed were:

- ***RETScreen Expert***: The software describes itself as a "Clean Energy Management Software" and is designed to enable professionals and decision-makers to estimate the technical and economic potential of a wide variety of different supply schemes.

• ***energyPLAN***: With this software energy flows on a regional or national level are monitored and forecasted to help taking long-term investment decisions and testing different legal support policies to come to a stable and reliable future energy supply system.

• ***HOMER Pro***: When planning micro-grid structures, especially off-grid, *HOMER* helps the user to optimize the system design and to get a good estimation about the system behaviour over its lifetime.

• ***energyPRO***: By combining a simple technical approach with a deep economic optimization *energyPRO* helps the user to not only plan the design of the energy supply scheme but as well help optimize its operation and interaction in every detail.

• ***Polysun***: When focusing on the final design of solar-based heat and electricity setups *Polysun* empowers the user to make a detailed analysis on how to interconnect the system on an electric and hydraulic level and on how to program the control units.

• ***TRNSYS***: *TRNSYS* is widely used in academia to investigate the technical aspect of energy supply. No other software offers such a detailed technical analysis. Unfortunately, economic aspects are only considered as a side-line.

• ***EINSTEIN***: Developed by an EU funded research project (EISAV/EIE/07/210/2007) *EINSTEIN* aims at monitoring the thermal energy flows in a technical production and trying to meet this demand in an optimal manner. Electricity is only simulated in really basic terms.

• ***COMPOSE***: Based on a big community fed component database *COMPOSE* is a flexible simulation tool which adds features like risk analysis and innovative components to the field of energy system simulation.

• ***INSEL***: *INSEL* is primarily aimed at the simulation of PV setups and treats other distribution generation plants only as a side-line. It is perfect to test different operating strategies of different components.

Every software found was tested in great detail and exemplary simulations were conducted. The scenarios tested consisted of CHP, PV, solar thermal as well as heat and battery storage supplying a standardized industrial electricity and heat demand. Special attention was given to the legal and funding requirements, which could apply to distributed generation, by placing the consumer under current German legislation und funding. This gives a good impression on how usable the software tools are for real life planning tasks as demanded by the company.

The findings were collected and evaluated by filling out the requirement catalogue as seen in Table I showing the great diversity of all approaches. The spectrum ranged from research aimed software like *TRNSYS* and *INSEL*, via decision-making, more political approaches found in *energyPLAN* and *RETScreen*, to a wide variety of tools with a high orientation to practical applications like *energyPRO* or *HOMER*. The resulting evaluation of the requirement catalogue can be seen in Table I as well and is summarized and evaluated in Table II by counting all missed requirements based on their importance.

Table II: Summary of the evaluated requirement catalogue based on the number of missed requirements per category.

	Missed requirements		
	critical (✓)	relevant (o)	optional (x)
RETScreen	5	3	1
energyPLAN	3	7	1
HOMER	3	4	2
energyPRO	0	3	2
Polysun	1	2	1
TRNSYS	9	8	2
EINSTEIN	9	7	2
COMPOSE	9	8	1
INSEL	5	7	2

For the depicted use case the evaluation would result in the following Top 3 software:

1. *energyPRO*
2. *Polysun*
3. *HOMER*

Based on this evaluation the decision makers should be in the position to make an objective decision on which software is best suitable and are able to justify this decision to others.

Nevertheless, this process should not completely rely on objective measures, but as well take personal preferences, usability and other points into consideration which are difficult to measure. This is especially advisable when the result of the evaluation is not distinct and certain compromises are necessary. This procedure shall never replace the personal contact and discussion with everyone involved but helps in objectifying and structuring a subjective decision problem.

6 CONCLUSION AND OUTLOOK

This paper highlights the difficulties in modelling and simulation hybrid distributed energy supply schemes and gives a broad overview on how to find appropriate software, how to compare them and which ready-made software solutions are currently found in the market. This helps researchers and planning offices searching for an appropriate simulation tool to get a founded decision on which software to choose and what points to consider during this process. In general, the decision on which software to use for which project is mostly down to finding the best compromise between technical and economic focus and between detailed and rather broad analysis.

In the end this helps to get a detailed understanding how future energy supply might look like to come to a more intelligent and sustainable energy supply in the commercial sector.

Table I: Exemplary requirement catalogue of the case study (Sec. 5) where the project specific requirements are ranked from critical (✓) over relevant (O) to optional (X) and the requirement fulfilment of every software is marked as being fulfilled (✓), partially fulfilled (O) or missed (X).

		RETScreen Version 6.0.7	*energyPLAN* Version 12	*HOMER* Version 3.7.6	*energyPRO* Version 4.4	*Polysun* Version 10.0	*TRNSYS* Version 17.02	*EINSTEIN* Version 2.2	*COMPOSE* Version 3.14.11	*INSEL* Version 8.2.1
1	Basics									
1.1	Development State (✓)	✓	✓	✓	✓	✓	✓	X	O	✓
1.2	Cost Price (O)	O (869$/yr)	✓ (Freeware)	O (500-1500$/yr)	O (3600-9000€)	O (1999-7300€)	O (5060-7830$)	✓ (Freeware)	✓ (Freeware)	O (1700€)
1.3	Service and Support (✓)	X	O	O	✓	✓	O	X	O	O
1.4	Programming Interface (O)	X	✓	O	O	O	✓	X	✓	✓
2.	Technical									
2.1	Generation									
2.1.1	Boiler (✓)	✓	✓	✓	✓	✓	✓	✓	✓	X
2.1.2	CHP (✓)	✓	✓	✓	✓	✓	✓	✓	✓	O (Only electric)
2.1.3	Solar Thermal (O)	✓	✓	X	✓	✓	✓	✓	O	✓
2.1.4	PV (✓)	✓	✓	✓	✓	✓	✓	X	O	✓
2.2	Storage									
2.2.1	Thermal (✓)	✓	✓	X	✓	✓	✓	X	✓	✓
2.2.2	Electric (O)	X	✓	✓	✓	✓	✓	X	✓	✓
2.3	Consumption									
2.3.1	Electricity (✓)	✓	✓	✓	✓	✓	✓	✓	✓	✓
2.3.2	Heat (✓)	✓	✓	✓	✓	✓	✓	✓	✓	✓
2.3.3	Process Heat (O)	✓	✓	X	✓	✓	✓	✓	✓	✓
2.3.4	Cooling (X)	✓	✓	X	✓	✓	✓	✓	✓	✓
2.4	Load profile generator (O)	✓	X	X	X	O (Households)	O (Buildings)	✓	X	X
2.5	Manufacturing database (O)	O (US market)	X	✓	X	✓	X	✓	O (Community based)	✓
2.6	Export of Electricity (✓)	✓	✓	✓	✓	✓	✓	✓	✓	✓
2.7	Export of Heat (O)	X	X	X	✓	✓	✓	X	✓	✓
2.8	Stepsize =15min (✓)	X	O (1h)	✓	✓	✓	✓	O (1h)	O (1h)	✓
2.9	Weather database (O)	✓	X	✓	✓	✓	X	X	X	✓
3.	Technical and Legal									
3.1	Costs									
3.1.1	Capital Related Costs (✓)	✓	✓	✓	✓	✓	X	✓	X	✓
3.1.2	Demand Related Costs (✓)	✓	✓	✓	✓	✓	X	✓	X	✓
3.1.3	Operation Related Costs (✓)	✓	✓	✓	✓	✓	X	✓	X	✓
3.1.4	Other Costs (✓)	✓	✓	✓	✓	✓	X	✓	X	✓
3.2	Proceeds									
3.2.1	Selling of electricity (✓)	✓	✓	✓	✓	✓	X	✓	X	✓
3.2.2	German Funding System (✓)	X	X	X	✓	✓	X	X	X	X
3.3	German Legislation (✓)	X	X	X	✓	✓	X	X	X	X
3.4	Sensitivity Analysis (O)	✓	X	✓	X	✓	X	X	X	X
3.5	Risk Analysis (X)	✓	X	✓	X	X	X	X	✓	X
4.	Ecologically									
4.1	Emission Balancing									
4.1.1	CO2 (✓)	✓	✓	✓	✓	X	X	X	X	X
4.1.2	Other (O)	✓	X	✓	✓	X	X	X	X	X
4.2	Life Cycle Assessment (X)	X	X	X	X	X	X	X	X	X
5.	Output									
5.1	Detailed Generation Profile (✓)	X	✓	✓	✓	✓	✓	✓	✓	✓
5.2	Cash Flow (✓)	✓	X	✓	✓	✓	X	X	X	X
5.3	Technical Summary (O)	✓	✓	✓	✓	✓	X	✓	X	X
5.4	Economic Summary (O)	✓	✓	✓	✓	✓	X	✓	X	X
5.5	Ecologic Summary (O)	✓	✓	✓	✓	✓	X	X	X	X
5.6	Customer Report (O)	✓	X	✓	✓	✓	X	X	X	X

REFERENCES

[1] T. Ackermann, G. Andersson, and L. Söder, "Distributed generation: a definition," *Electric Power Systems Research,* vol. 57, no. 3, pp. 195-204, 2001.

[2] N. Jenkins, J. B. Ekanayake, and G. Strbac, *Distributed generation*. London: The Institution of Engineering and Technology, 2010.

[3] A. De Nardi, "Energy supply optimization in indutrial areas. The case study of the Ponte Rosso industrial area in San Vito al Tagliamento (Italy)," Doctoral thesis, Università degli Studi di Udine, Udine, 2015.

[4] G. Abdollahi and H. Sayyaadi, "Application of the multi-objective optimization and risk analysis for the sizing of a residential small-scale CCHP system," *Energy and Buildings,* vol. 60, pp. 330-344, 2013.

[5] E. S. Barbieri, Y. J. Dai, M. Morini, M. Pinelli, P. R. Spina, P. Sun, and R. Z. Wang, "Optimal sizing of a multi-source energy plant for power heat and cooling generation," *Applied Thermal Engineering,* vol. 71, no. 2, pp. 736-750, 2014.

[6] M. Carvalho, M. A. Lozano, and L. M. Serra, "Multicriteria synthesis of trigeneration systems considering economic and environmental aspects," *Applied Energy,* vol. 91, no. 1, pp. 245-254, 2012.

[7] M. Di Somma, B. Yan, N. Bianco, G. Graditi, P. B. Luh, L. Mongibello, and V. Naso, "Operation optimization of a distributed energy system considering energy costs and exergy efficiency," *Energy Conversion and Management,* vol. 103, pp. 739-751, 2015.

[8] C. Marnay, M. Stadler, G. Cardoso, O. Megel, J. Lai, and A. Siddiqui, "The Added Economic and Environmental Value of Solar Thermal Systems in Microgrids with Combined Heat and Power," in *3rd International Conference on Solar Air-Conditioning*, Palermo, 2009.

[9] H. Ren, W. Zhou, K. i. Nakagami, W. Gao, and Q. Wu, "Multi-objective optimization for the operation of distributed energy systems considering economic and environmental aspects," *Applied Energy,* vol. 87, no. 12, pp. 3642-3651, 2010.

[10] H. Wang, W. Yin, E. Abdollahi, R. Lahdelma, and W. Jiao, "Modelling and optimization of CHP based district heating system with renewable energy production and energy storage," *Applied Energy,* vol. 159, pp. 401-421, 2015.

[11] A. Safaei, F. Freire, and C. H. Antunes, "A model for optimal energy planning of a commercial building integrating solar and cogeneration systems," *Energy,* vol. 61, pp. 211-223, 2013.

[12] L. R. Rodriguez, J. M. S. Lissen, J. S. Ramos, E. A. R. Jara, and S. A. Dominguez, "Analysis of the economic feasibility and reduction of a building's energy consumption and emissions when integrating hybrid solar thermal/PV/micro-CHP systems," *Applied Energy,* vol. 165, pp. 828-838, 2016.

[13] J. F. Manwell, "Hybrid energy systems," *Encyclopedia of Energy,* vol. 3, pp. 215-229, 2004.

[14] D. Connolly, H. Lund, B. V. Mathiesen, and M. Leahy, "A review of computer tools for analysing the integration of renewable energy into various energy systems," *Applied Energy,* vol. 87, no. 4, pp. 1059-1082, 2010.

[15] S. Sinha and S. S. Chandel, "Review of software tools for hybrid renewable energy systems," *Renewable and Sustainable Energy Reviews,* vol. 32, pp. 192-205, 2014.

[16] S. Schramm and M. Adam, "Storage in Solar Process Heat Applications," *Energy Procedia,* vol. 48, pp. 1202-1209, 2014.

[17] R. M. Dell and D. A. J. Rand, "Energy storage - a key technology for global energy sustainability," *Journal of Power Sources,* vol. 100, no. 1-2, pp. 2-17, 2001.

[18] D. Buoro, M. Casisi, A. De Nardi, P. Pinamonti, and M. Reini, "Multicriteria optimization of a distributed energy supply system for an industrial area," *Energy,* vol. 58, pp. 128-137, 2013.

[19] M. Casisi, A. De Nardi, P. Pinamonti, and M. Reini, "Effect of different economic support policies on the optimal synthesis and operation of a distributed energy supply system with renewable energy sources for an industrial area," *Energy Conversion and Management,* vol. 95, pp. 131-139, 2015.

[20] E. A. Kremers, *Modelling and simulation of electrical energy systems through a complex systems approach using agent-based models*. KIT Scientific Publishing, 2013.

[21] H. Lund, *Renewable energy systems: a smart energy systems approach to the choice and modeling of 100% renewable solutions*. Burlington: Elsevier Science, 2014.

LABORATORY OF HYBRID SYSTEMS AND MINI-GRIDS

Claudomiro Fábio de Oliveira Barbosa, Jorge Augusto Leal Correa, João Paulo Alves Veríssimo, Israel Hidai Lobato Lemos, Marcos André Barros Galhardo, Edinaldo José da Silva Pereira, João Tavares Pinho, Alex Renan Arrifano Manito
Grupo de Estudos e Desenvolvimento de Alternativas Energéticas, Faculdade de Engenharias Elétrica e Biomédica
Instituto de Tecnologia, Universidade Federal do Pará
Av. Augusto Corrêa nº 01, CEP 66.050-000 – Belém, PA, Brazil
Tel. / Fax +55 91 3201 7299, cfob@ufpa.br

ABSTRACT: This work presents the Laboratory of Hybrid Systems and Mini-grids, installed at the Federal University of Pará (UFPA) by the Group of Studies and Development of Energy Alternatives (GEDAE). The laboratory aims to promote the research in hybrid systems and mini-grids through practices of installation and commissioning of the associated technologies and, especially, through experiments and operational tests, with results that will serve as basis for evaluating the performance of such systems. The main components are described and information about mounting, installation and commissioning of the components is also given here briefly. The results of operational tests show the effectiveness of system operation.
Keywords: Hybrid Power System, PV System, Mini-grid

1 INTRODUCTION

The prioritary use of local renewable energy resources, such as solar and wind, in combination with diesel generators or any fuel obtained from other sources, known as Hybrid Power Systems (HPS) or Hybrid Systems (HS), presents a viable alternative for supplying decentralized areas and distributed generation by means of mini-grids (MG). Currently, many studies investigate the connection of these MG forming small interconnected grids, which can be connected to the conventional grid. A MG can be connected to a traditional distribution grid in the form of a smart grid, which can be designed to supply local energy needs, and/or provide ancillary services to the main grid. However, MGs still have technical barriers to combine a range of technologies, as well as barriers due to economic, political and regulatory issues, that strongly interfere in the electricity market. Numerous electrification projects based on HPS and isolated MGs have been developed in Brazil. These projects have been largely underused both from a scientific point of view and from the perspective of capacity building. This is mainly due to limited resources available, which in most cases include, only the funds for the purchase of equipment, their installation and commissioning, in addition to a short period of operation. One way to partially overcome this inefficiency associated with many pilot projects based on HPS and MGs is to take advantage of laboratory structures available for teaching and/or research. In order to contribute to the construction of a better knowledge about this type of application, the Group of Studies and Development of Energy Alternatives of the Federal University of Pará (GEDAE/UFPA) has been working on the development of a laboratory structure to assist in a wide range of analyses to be performed in a real system, implanted in its test area.

2 LABORATORY OF HYBRID SYSTEMS AND MINI-GRIDS

The laboratory implementation aims to promote the research in hybrid systems and mini-grids through practices of installation and commissioning of the associated technologies, and especially through experiments and operation tests, with results that will serve as basis for evaluating the performance of such systems. The laboratory also allows capacity building by offering courses, trainings, lectures, and technical visits with practical demonstrations for engineers, technicians, university professors, high school teachers and students, and certificate programs.

2.1 Structure

The Laboratory of Hybrid Systems and Mini-grids (LHSM) is currently composed of:

- Generation system

Subdivided into renewable generation from seven Photovoltaic Generators (PVGs), having a total nominal capacity of 14.125 kWp, and non-renewable generation represented by one Diesel Generator (DGen) of 12 kVA.

PVG-01 (Fig. 1) consists of 5 modules YL 185 P-23b/ YINGLI connected in series, resulting in: 117.5 V (Vmp); 147.5 V (Voc); 7.87 A (Imp); 8.45 A (Isc); 0.925 kWp.

Figure 1: PVG-01

PVG-02 and PVG-03 (Fig. 2) are identical, and each one consists of 5 modules SPR-220-BLK-U/SUNPOWER connected in series, resulting in: 205 V (Vmp); 243 V (Voc); 5.37 A (Imp); 5.85 A (Isc); 1.1 kWp.

Figure 2: PVG-02 and PVG-03

PVG-04 and PVG-05 (Fig. 3) are identical, and each one consists of 12 modules KD250GH-4FB2/KYOCERA. Regarding their electrical connections, both generators have 3 strings of 4 modules in series, resulting in: 119.2 V (Vmp); 147.6 V (Voc); 25.17 A (Imp); 27.27 A (Isc); 3 kWp.

Figure 3: PVG-04 and PVG-05

PVG-06 and PVG-07 (Fig. 4) are identical, and each one consists of 8 modules KD250GH-4FB2/KYOCERA connected in series, resulting in: 238.4 V (Vmp); 295.2 V (Voc); 8.39 A (Imp); 9.09 A (Isc); 2 kWp.

Figure 4: PVG-06 and PVG-07

DGen (Fig. 5) is manufactured by Leon Heimer, with soundproof cabin; emergency power (standby) of 12 kVA; continuous power (prime) of 10 kVA; three-phase with 220 V phase-phase and 127 V phase-neutral; power factor of 0.8; automatic and microprocessed command and control.

Figure 5: DGen

- Power conditioning system

Composed of two Clusters (Cluster A and Cluster B), both with three Bidirectional Inverters (BINVs), resulting in a total capacity of 13.5 kW; three Grid-tie Inverters (GINVs) totalizing 2.1 kW; two other GINVs totalizing 4 kW; and two charge controllers (DC/DC converters) of 3 kW each.

BINV-01A, BINV-02A, and BINV-03A (Cluster A) are bidirectional inverters (4548-US/SMA model) of 4.5 kW each, which form the mini-grid for energy distribution (provide appropriate voltage and frequency signals) to supply the electrical loads. They are used to invert the DC power to AC (supply typical loads) and to rectify the AC power to DC (charge the battery banks).

BINV-01B, BINV-02B, and BINV-03B (Cluster B) have the same characteristics and purposes of the inverters above mentioned, but have their connection point on the mini-grid geographically distant from the connection point of BINV-01A, BINV-02A, and BINV-03A. This particularity allows to separate the mini-grid into two parts, which can operate isolated from each other or interconnected, and also with both connected to the main electrical grid.

GINV-01, GINV-02, and GINV-03 are inverters (700U/SMA model) of 0.7 kW each; unlike the BINV's, they are unidirectional, converting the DC power from the PVG's to AC power to be fed to the mini-grid.

GINV-04 and GINV-05 have the same purposes of the grid-tie inverters above mentioned, but are 2000HF/SMA model, with a nominal capacity of 2 kW each.

Charge controllers (CC-01 and CC-02) are DC/DC conversion units (FM60-150VCC/OUTBACK POWER model) destined to manage and control the energy flow from PVG-04 and PVG-05 to the batteries, thus protecting them both from excessive charge and discharge.

Fig. 6a-d present the above mentioned power conditioning equipment.

Figure 6a: Cluster A, CC-01, and CC-02

Figure 6b: Cluster B

Figure 6c: GINV-01, GINV-02, and GINV-03

Figure 6d: GINV-04 and GINV-05

- Energy storage system:

Composed of two battery banks (BB-01 and BB-02).

BB-01 and BB-02 (Fig. 7) are identical, each composed of 24 2 V/1,000 Ah batteries (model 17TFS-1000/FULGURIS) connected in series, totalizing individual capacities of 48 kWh. These banks can be connected in parallel, duplicating the storage capacity to 96 kWh.

Figure 7: Battery Banks

- General Distribution and Control Boards (DCB) and Protective Devices (PD):

Components intended for the easy and secure connections between the systems' elements, the protection of users and equipment, and the safe execution of operational maneuvers. Among these devices, the following can be highlighted:

DCB-Cluster A, manufactured by SMA (Multicluster Box MCB-12U model), serves as an integrated AC distribution system for the BINVs-01A-03A, GINVs-01-03 and DGen, maintaining connection with the mini-grid, which, in this case, is seen by the DCB as the load itself.

DCB-Cluster B was designed and implemented by the technical staff of GEDAE to shelter the electrical connections, buses and protective devices from Cluster B. Analogously to the Multicluster Box, the DCB-Cluster B provides the connections between the BINVs-01B-03B, GINVs (future expansion), DGen and the loads (mini-grid).

String boxes from photovoltaic generators and switch-fuse disconnectors are used for connection and protection of the equipment, as well as for the safe execution of maneuvers to change the configuration of the systems, either manually or automatically (in development), or maintenance of the equipment.

- Power distribution mini-grid

A three-phase aerial mini-grid (Fig. 8) with about 1 km linear length, designed to simulate a distribution grid. In addition to the arrangement of the 7 meter high 18 concrete poles and the cabling with multiplex aluminum 35 mm^2 conductors (4 conductors; 3 phases and 1 neutral), contactors and auxiliary switches with remote actuation are positioned at strategically chosen points. Depending on the state of the switches (ON or OFF), different grid configurations can be emulated, such as long, short, ring or radial, as well as the possibility of having two separate mini-grids connected to Cluster A and Cluster B, respectively.

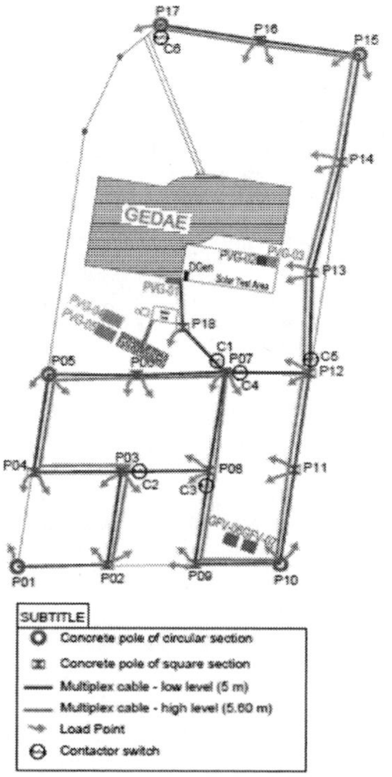

Figure 8: MG layout

• Electrical loads

Remotely controlled electrical loads of different types (resistive, mixed, and driving) are spread throughout the MG (in development). Furthermore, in the absence of power from the conventional distribution grid, GEDAE´s building presents a load for the laboratory.

• Monitoring system

The system's parameters can be monitored by the monitoring system (SM), which consists of data loggers, sensors, and communication and control units. For example, there is the Sunny SensorBox (together with the solar radiation sensor, module temperature sensor and anemometer it is responsible for acquiring meteorological data), and the Sunny WebBox (responsible for collecting and displaying operational information from the inverters). Other monitoring components are in the process of development and/or acquisition.

The systems' schematic with their main elements is presented in Fig. 9. In this schematic it can be noticed that there are two hybrid generation systems. One is formed by the PVGs-01-05, DGen, BINVs-01A-03A and the BBs-01-02. The other shares de same PVG-04, DGen and BB-01 from the first system and is also formed by BINVs-01B-03B. Other generation systems are the PVG-06 and PVG-07 with their respective inverters, GINV-04 and GINV-05, which are located in different points, introducing distributed generation (DG) in the mini-grid.

2.2 Installation and commissioning [1] and [2].

It is important to emphasize that during the assembly, installation and commissioning process of the laboratory components, the relevant standards and technical manuals of each component were followed.

The following is a summary of the installation and commissioning procedures for the main laboratory components.

In selecting the local for the installation of the generation systems, the following conditions were considered: area availability; inexistence of obstacles that could shadow the PVGs surfaces; no or little impact caused by the noise generated from the Dgen operation; and the proximity with the "Power Station" (PS) room. The exception to this last condition was the site selected for the installation of PVG-06 and PVG-07, for which the implementation of a photovoltaic generation in the form of DG along the mini-grid was sought.

The PS was chosen as the space to install the other equipment and accessories of the hybrid systems (bidirectional inverters, batteries, distribution boards, etc.), since it is wide, illuminated, ventilated and protected from direct climatic influences (such as rain and strong wind). These conditions are necessary for the correct operation of such equipment (Fig. 10). Moreover, this space guarantees the restrict access, as well as the safe movement of authorized people, aiming at the execution of the operation, tests and maintenance procedures. Once the locations were defined, the assembly and installation activities of the components began.

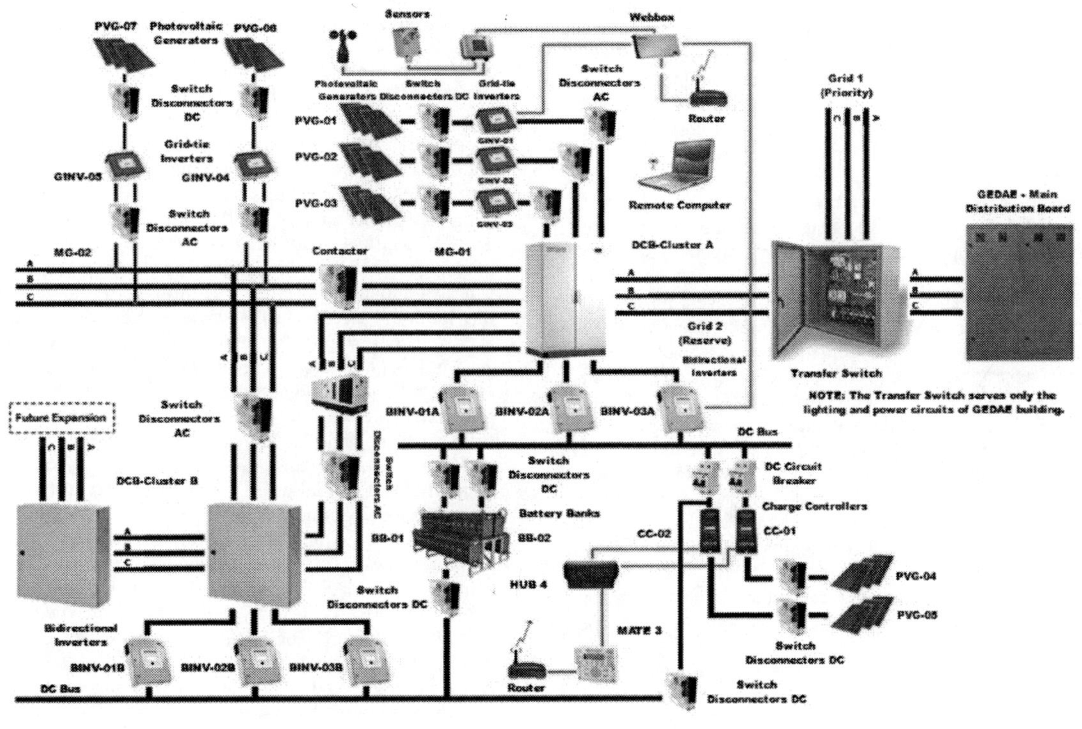

Figure 9: Schematic of the systems

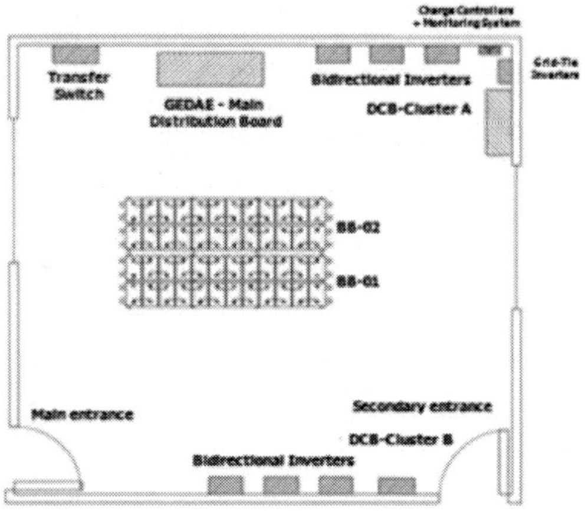

Figure 10: Layout of the main PS equipment

To support PVG-01, a structure strong enough to hold 5 modules was made, using aluminum profiles fixed to the wooden support beams of GEDAE's roof (rear of the building), projecting itself as a continuity of the roof. With this structure installed, the modules were positioned to the Geographic South, with a 10° inclination, fixed on the structure using rivets and were connected in series. To support PVG-02 and PVG-03 a structure using aluminum profiles was also used, except that it was not fixed to the roof. Its base was fixed directly to the floor of the test area of the building (area with adequate infrastructure for studies). After that, the PV modules of both generators were positioned to the Geographic North, with a 14° inclination, fixed on the structure using screws and nuts in stainless steel and connected in series (5 modules, each). Similar to the support of PVG-02 and PVG-03, for PVGs-04-07 structures composed of aluminum profiles were also used. Since the areas for installation of these generators presented several irregularities, such as uneven ground and flooded parts, it was necessary to build isolated reinforced concrete shallows to support the weight of the structures, modules and, in some cases, grid-tie inverters, and protection and connection boards. After these structures were positioned in their proper locations, the modules were fixed on them using bolts and nuts of stainless steel and positioned to the Geographic North, with the following inclinations: 12° for PVG-04 and PVG-05 and 11° for PVG-06 and PVG-07. Regarding the electrical connections between the modules, for PVG-06 and PVG-07 8 modules were connected in series. For PVG-04 and PVG-05, 3 strings were connected in parallel, each string consisting of 4 modules in series.

To ensure the proper functioning of DGen, 5/8" anti-vibration cushions were installed under its base and an extension was added on its escape pipe, so that the smoke would not affect anyone or any equipment near the DGen, such as PVG-01.

The BINVs and GINVs-01-03 were positioned on the PS walls using supports provided by the manufacturer. These supports were attached to the walls with screws and wall plugs. During this process, aiming for good ventilation, the minimal distance between equipment and surrounding obstacles was respected as described in the manuals and installation guides. GINV-04 and GINV-05 were positioned under their respective generators, in places sheltered from direct contact with water from rainfalls.

According to the charge controllers' manufacturer, they must be installed vertically and sheltered from severe weather, but there are no restrictions regarding minimal space from the top, bottom or the sides. However, the area destined to their installation is next to BINV-03A, which, according to the manufacturer, needs a distance of at least 30 cm from any surrounding obstacles on its sides. Following this instruction, the charge controllers were spaced 30 cm from BINV-03 and fixed directly on the wall using screws and wall plugs.

To accommodate the 48 batteries that form the two battery banks with 48 V nominal voltage, two shelves were mounted, each being capable of holding up to 24 batteries. Therefore, the batteries were arranged so that the positive terminal of one battery was close to the negative terminal of the subsequent one, making it easier to connect them in series (by means of cooper buses).

DCB-Cluster A and DCB-Cluster B were directly attached to the PS walls using screws and wall plugs. Since DCB-Cluster A does not have any kind of support in its bottom, a rigid rubber mat was used in order to protect its frame from direct contact with the floor.

Regarding the three-phase distribution mini-grid installation, a regular pattern for allocation of the concrete poles was sought, respecting the territory boundaries of GEDAE, as well as its unevenness. The poles were fixed on the ground inside holes of about 1.30 m depth and cross section of $0.36m^2$, obeying the technical standards of the local electric utility. For installing the mini-grid cabling, a frame and low-voltage pulley insulators (porcelain) were used.

In order to increase the physical length of the mini-grid, it was decided to use a second cabling level, which increased the tension on the poles, making it necessary to install tiers in some of them. The contactors and other accessories intended for changing the mini-grid configurations are in the final stage of assembly and installation. After the necessary tests are made, all the equipment in the mini-grid will be fully operational.

As electrical loads that emulate consumer units, boards containing resistive, mixed and driving loads that can be remotely controlled will be used. Prototypes of this loads are already in the stages of development and tests. It is noteworthy that other loads are already being used in laboratory tests, such as: 11 LED lamps of 20 W each (public lighting system – MG); a 3.6 kW programable electronic load; a 3 – 20 kW variable resistive load; and also the regular loads of GEDAE´s building (lighting and power circuits), which are connected to the mini-grid by a transfer switch that can operate manually or automatically.

With all the equipment positioned and fixed in their proper places, the interconnections between these elements were made through the use of cables. The sizing criteria for these cables and their conduits followed the ABNT standard (NBR-5410) and the specifications described on the installation manual of each equipment. It is important to note that all the structures and equipment, as well as the mini-grid, were properly grounded.

Regarding the commissioning process, it basically involved: visual inspection of the entire system; mechanical tests, applying some controlled forces to fixed equipment and conductors, trying to move or disconnect them, respectively; continuity tests to confirm the electrical continuity among all circuits extensions;

equipment start-ups and, when needed, parameterizations.

3 OPERATIONAL TESTS

The following results are from some tests that prove the operational effectiveness of the systems, as well as their performance.

- Test 1 – Photovoltaic system with battery storage and isolated operation.

DGen and PVGs-04-07 were switched off. For this test, the gray curve in Fig. 11 (obtained from the load variations during the test period) was used as the total demand profile. Fig. 11 also presents the average values obtained from the power delivered by the GINVs and the power delivered to the battery banks.

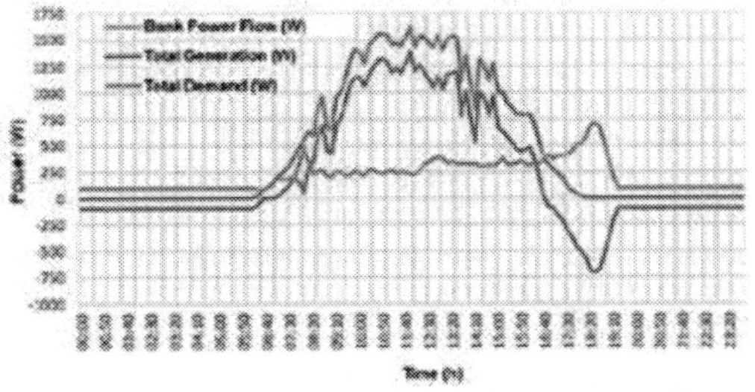

Figure 11: Behavior of the power delivered by the GINVs, loads demand, and power delivered to the battery banks (surplus)

In Fig. 11, the positive values of the AC power from Cluster A's BINVs, indicate that there is a generation surplus in this period and this energy is being stored in the battery banks. The negative values indicate that the demand is higher than the generation and, as a consequence, the energy is being drained from the battery banks to supply the loads.

For the energy to be stored in the battery banks during the period when the generation exceeds the demand, the BINVs drive the power flow towards the banks input. This operation strategy happens because the BINVs increase the voltage level of the DC bus, thus forcing the charging of the battery banks (Fig. 12). When the energy is demanded from the batteries, the BINVs regulate the voltage level on the DC bus, allowing the power flow to change its direction, going from the banks to the loads.

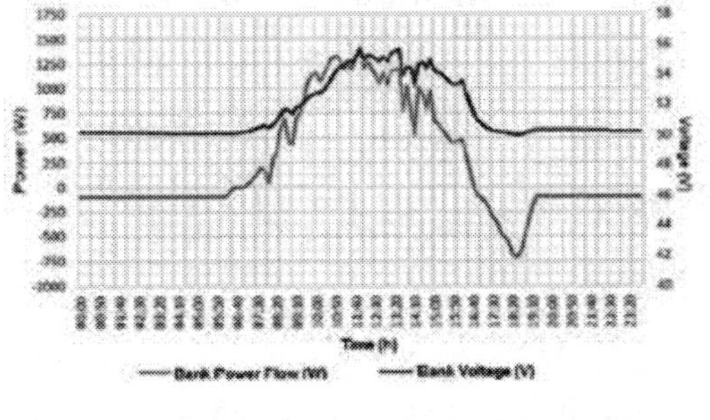

Figure 12: Relation of the voltage level from the battery banks according to the periods of charging and discharging

- Test 2 – Hybrid photovoltaic/diesel system with battery storage and isolated operation:

The total load value was maintained constant at 0.8 kW. In the first stage, the test initiated only considering PVGs-01-03 connected to the system and supplying the loads, having the surplus energy being stored into the battery banks. In the second stage, the DGen was included in the system. In the third stage, PVGs-01-03 were disconnected, simulating the absence of solar photovoltaic generation, letting only the DGen operating, supplying the loads and storing the surplus energy in the batteries. In the fourth and final stage, the DGen was turned off, leaving only the battery banks to supply the loads. The system's operation sequences, demonstrating its behavior with the connection and disconnection of these generators, is shown in Fig. 13.

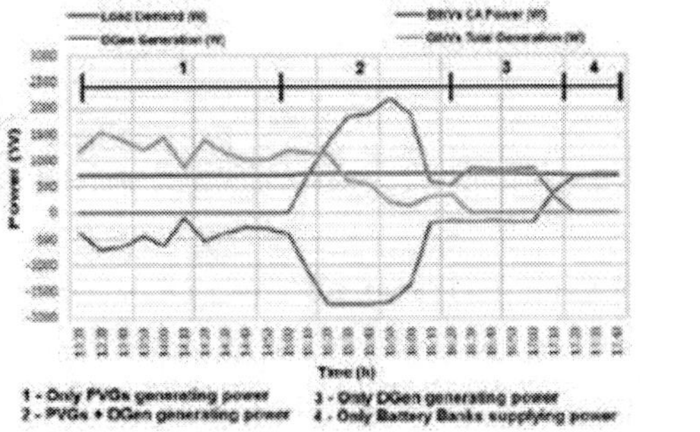

Figure 13: Operation sequences

From Fig. 13 it can be noticed that all the surplus generation, coming from both PVGs-01-03 and DGen, is stored in the battery banks. In addition, the system behaved in a stable way and, as soon as all generators stopped supplying energy (fourth stage), the BINVs (Cluster A) inverted the power flow and the loads started being supplied exclusively by the battery banks.

It is important to mention that, during the passage from the first stage to the second, when the DGen is turned on, the synchronization of the BINVs with the DGen occurs. This situation can be verified in Fig. 14a.

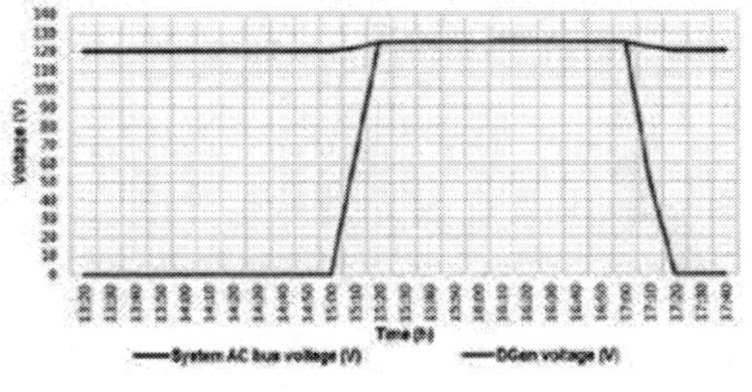

Figure 14a: Synchronism of the voltage level from the system with the voltage level from the DGen

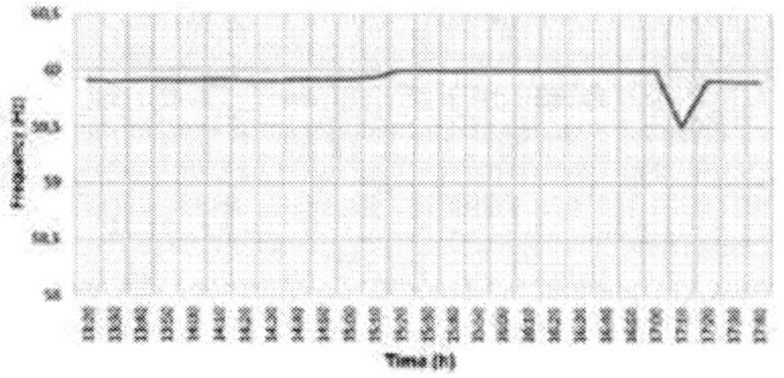

Figure 14b: Frequency of the system

From Fig. 14a it is observed that when the DGen is activated (approximately at 15 h), its connection with the AC bus is not made immediately. First, the BINVs identify that there is an active external source and, if their action is required (as it happened), they synchronize with the DGen. In this case, the nominal voltage level went from 120 V phase-phase to approximately 125 V phase-phase, and the frequency, which was oscillating between 59.9 Hz and 59.94 Hz, presented a better stability, being maintained at 60 Hz (Fig. 14b). After the DGen withdrawal, the system gradually returned to its initial state, presenting only a sudden short drop in the frequency value.

Concerning Cluster B, the tests made were similar to those obtained with the previous presented tests, as was expected.

It is worth pointing out that other tests are being performed to evaluate the operation and performance of the systems, such as the parallel operation of both clusters and the DG penetration of PVG-06 and PVG-07.

- Test 3 – Efficiency evaluation of CC-01 and CC-02:

In order to analyze the efficiency of the charge controllers, measurements were carried out every 10 minutes from 6:40 to 18:00 (start and end of the operation of the charge controllers, respectively), using an energy analyzer.

Fig. 15 presents the efficiency curve of the charge controllers versus normalized power, which is the result from the ratio of the output power to the nominal power. The graphic shows that their maximum efficiency occurs for a normalized power higher than 20 %. In this case, the efficiency of the charge controllers remained enclose to 98 %, which matches the value given in the manufacturer datasheet.

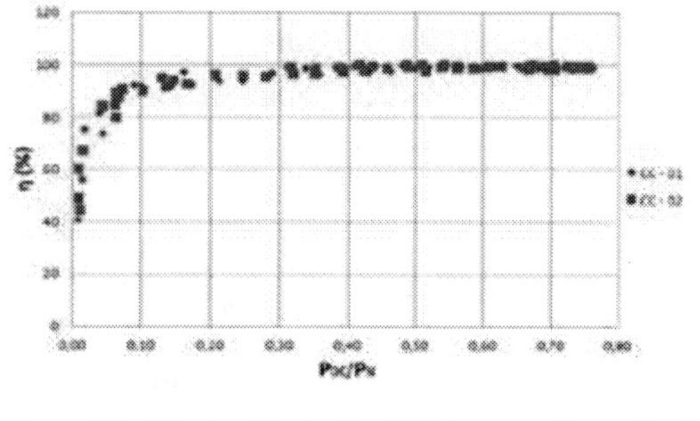

Figure 15: Charge Controller Efficiency x Normalized Power

It is worth mentioning that the normalized power values are directly proportional to the irradiance level on the plane of the photovoltaic generators. Thus, for higher levels of irradiance, the charge controllers will operate with higher efficiencies. The graphic in Fig. 16 shows the variation of the normalized power compared to the variation of the measured irradiance.

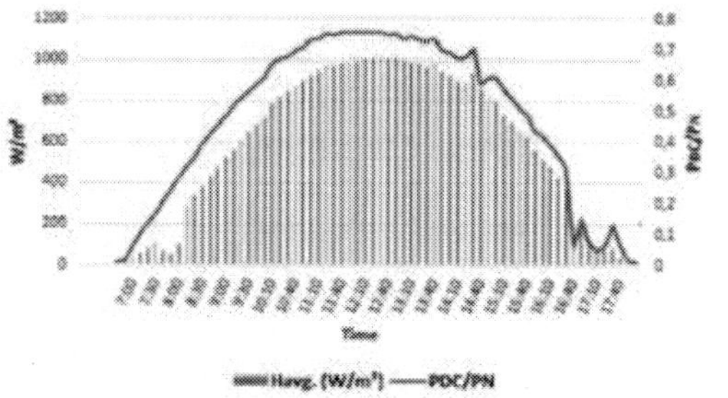

Figure 16: Irradiance and Normalized Power

Another important aspect in the analysis of the charge controller, is the fact that its DC input power depends on the point of the current-voltage (IxV) curve in which the photovoltaic generator is operating. That will vary depending on the ambient conditions to which the generator is subjected (basically irradiance and temperature). Fig. 17 shows the graphic of the irradiation compared to the modules′ temperature and the ambient temperature.

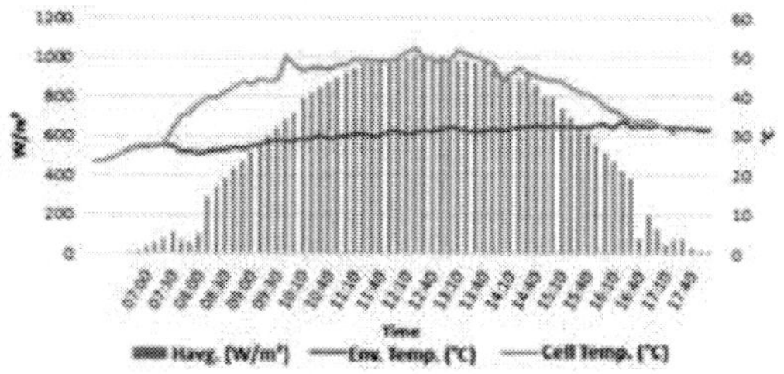

Figure 17: Irradiance and Temperatures

4 FINAL CONSIDERATIONS

With the available infrastructure, several situations and equipment technologies can be better evaluated, being the object of studies in several experimental researches related to the theme, contributing to the construction of knowledge about the real behavior of these technologies when subject to different configurations and environmental conditions. In Brazil, the knowledge and the application of the various operational situations mentioned here are still incipient and, in some cases, practically absent. Currently, the Agência Nacional de Energia Elétrica (ANEEL), through Brazilian Normative Resolution (RN) nº 493/2012, established the procedures and conditions to supply electricity by means of isolated micro-grids powered by intermittent sources, such as solar and wind. In addition, it is expected that the infrastructure will contribute to capacity building.

Acknowledgments

The authors would like to thank the National Council for Scientific and Technological Development (CNPq) for financial support to develop this research.

References

[1] C. F. O. Barbosa, P. H. A. Verissimo, J. T. Pinho, J. A. L. Correa, M. A. B. Galhardo. Laboratório de Sistemas Híbridos de Energia e Minirredes (Implementação e Ensaios Operacionais). VI Congresso Brasileiro de Energia Solar (2016).
[2] ABNT NBR-5410. 2004. Instalações Elétricas de Baixa Tensão.

FIELD TESTING OF PORTABLE LED FLASHER FOR NOMINAL POWER MEASUREMENTS OF PV-MODULES ON-SITE

Raphael Knecht[1]*, Franz Baumgartner[1], Fabian Carigiet[1], Christian Frei[2], Fritz Beglinger[2], Willem Zaaiman[3], Diego Pavanello[3], Michael Field[3], Roberto Galleano[3], Tony Sample[3]
[1]ZHAW, Zurich University of Applied Sciences, School of Engineering, IEFE
www.zhaw.ch/~bauf, Technikumstr. 9, CH-8401 Winterthur, Switzerland
*phone: +41589344576; email: kner@zhaw.ch
[2]Electrosuisse, Luppmenstrasse 1, CH-8320 Fehraltorf
[3]European Commission, Joint Research Centre, Directorate C, Energy, Transport and Climate, Energy Efficiency and Renewables Unit, Via E. Fermi 2749, 21027 Ispra (VA), Italy

ABSTRACT: Nominal power measurements of individual PV modules are needed to quantify the critical modules of PV plants offering lower energy production than expected. Today's state of the art procedure of shipping a small number of modules to a laboratory is time- and cost intensive and it bears the chance of accidental damage. The Portable LED Flasher (PLF) was developed to require no dismounting of the modules. The quality of the PLF was tested on three PV plants in Switzerland. Additionally, ten PV modules of each plant were measured in the certified indoor laboratory of SUPSI, resulting in a maximum deviation of 3% of the STC values. Furthermore, a round robin test on a single crystalline silicon reference module at 25°C was performed at the JRC's ESTI laboratory, the Swiss Mobile Flasher Bus and SUPSI resulting in a maximum deviation of the mean values below 1% compared to the PLF. A throughput of up to 150 modules or 500 modules respectively per day is expected and the total measurement costs are estimated to be about a tenth of the costs compared to an indoor laboratory. Module temperature measurement is crucial for a low total uncertainty. Thus, methods such as pre-shadowing of the module and approximation of cell temperature are the current focus of further improvement of the PLF measurement method.
Keywords: PV Module, Electrical Properties, Qualification and Testing, Cost reduction, Degradation

1 Motivation and Market Needs

The measurement of nominal power of PV modules is essential for the quality control of PV plants and its benefit depends highly on the costs for the measurement. Operators or owners of the plant have to apply economically feasible methods to prove insufficient STC power of module(s) for the purpose of further negotiations with the module manufacturer.

If these measurements are too expensive they may not be performed. The current state of the art procedure is to take a very small number of modules which will be measured indoors by a certified test laboratory. For these tests, the PV modules have to be dismounted from the PV plant and shipped to the laboratory, which is time- and cost-intensive and it bears the chance of accidental damage during transport or dismounting. Today, a required sample number can only be assessed if the measurement costs of each PV module are below the cost of a new PV module. This is not the case for the traditional approach of demounting a few samples and shipping them to a stationary, certified laboratory. The ZHAW developed an alternative approach by the use of the Portable LED Flasher (PLF). It is designed to measure PV modules directly on-site without the need to dismount them [1]. The PLF comprises a light engine with over 5000 LEDs to simulate the sun spectrum.

This paper reports the findings and experience of this new STC performance measurement method applied to a large set of PV modules on three different PV plants in Switzerland.

Table I: Four methods of nominal power measurements of PV modules and their dis-/advantages [1].

Measurement Method	Advantage	Disadvantage
Field Measurements under sunlight with reference cell 	• Inexpensive measurement device • No dismounting damage • Instantaneous result	• Uncertainty >5% • High fix costs at sub-optimal irradiance conditions • Temperature correction
Stationary measurement in certified laboratory 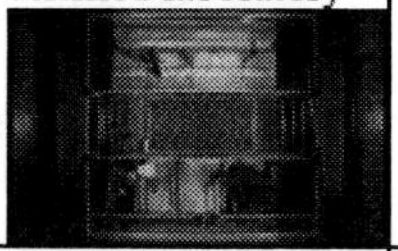	• Low uncertainty of 1.3–3.5%	• High costs • Dismounting, Transport • Low sample number
Field measurement with light engine in car 	• Uncertainty 3–4% • Instantaneous result	• Dismounting • Temperature correction
Field measurement with Portable LED Flasher 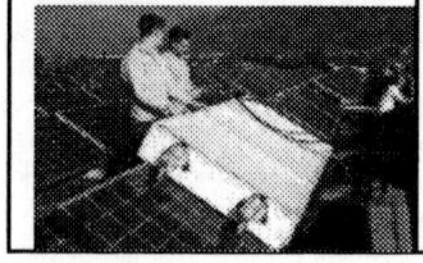	• Uncertainty 3–4% • Instantaneous result • No dismounting damage	• Temperature correction

2 SPECIFICATIONS OF IMPROVED PROTOTYPE

ZHAW invented the first functional model of the PLF in 2014 [2]. In a consecutive joint project, together with Electrosuisse an optimised device has been engineered. The optimised parameters and overall technical details of the new version of the PLF are shown in Table II. The major improvements compared to the first functional model comprise the weight reduction of the light engine of 40%, better spectral characteristics (Class A), higher measurement rates and flexibility due to the power autonomy (battery pack).

Table II: Technical specifications of the optimized Portable LED Flasher

Specification	**Value**
Outer dimension	2.0 m (with wheels: 2.6 m) × 1.1 m
Active optical area	2.0 m × 1.1 m
Transport dimension	1.1 m × 0.7 m × 0.6 m
Weight light engine	30 kg, with wheels: 38 kg
Spectrum	11 LED wavelengths, Class A
Illumination time	Typ. 130 ms – adjustable
Illuminance	20% higher than STC
Minimum theoretical measurement rate	10 Seconds
Power Supply	Lithium battery for 750 flashes
Temperature measurement	Either Platinum RTD or Thermocouple
Acquisition serial number	Barcode scanner
Handling and data storage	Tablet, instantaneous display of measurement result

The light engine of the PLF is divided into 18 equal LED boards comprising of 272 LEDs each. The intensity of the LEDs can be controlled individually for each sector containing approximately 32 LEDs. The spatial non-uniformity of the PLF has been determined using a crystalline reference cell and amount to 1.73%. According to IEC60904-9 the non-uniformity thus satisfies Class A.

983	993	1000	994	991	1007
982	989	1006	999	988	1000
977	1004	996	979	983	981
976	986	990	978	986	995
990	992	1001	990	987	982
985	997	1005	982	979	988
982	984	984	1003	983	989
979	986	981	990	987	991
986	999	981	1001	985	984
975	995	988	990	991	991
981	989	1005	1005	1007	1008
977	977	1008	1009	995	1009

Figure 1: Normalized irradiance of the PLF in W/m^2.

LEDs with central wavelengths of 430 nm to 940 nm are used to reflect the STC spectrum AM1.5G most precisely. The spectral mismatch is below the limit of Class A in all intervals of the spectrum.

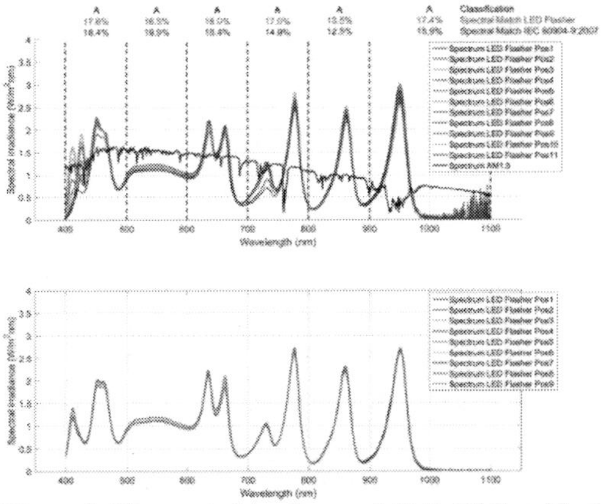

Figure 2: The spectral irradiance of AM1.5G (top, blue) and the measurements on the PLF with an Avantes AvaSpec-ULS2048 of ZHAW (top) in eleven different lateral positions in the intervals of IEC 60904-9. Bottom picture shows the spectral measurements performed with an Ocean Optics HR2000+ by ESTI in nine different lateral positions.

A major upgrade of the new version of the PLF is its flexibility regarding transport. Its light engine can be disassembled into three autonomously functional parts (See Figure 3). The possibility to transport the PLF with a station wagon has been proven in the field tests.

Figure 3: The light engine of the PLF disassembled and stacked ready for transport or storage.

3 ROUND ROBIN TEST WITH CERTIFIED LABORATORIES

In order to classify the uncertainty of the measurements with the PLF a round robin test with a reference module was performed. The same module was measured at the certified laboratory of SUPSI, at the certified laboratory of EU-JRC ESTI and with the Swiss Mobile Flasher Bus (SMFB) [3]. Furthermore, as part of the publicly funded project of Swiss Federal Bureau of Energy, samples of modules of three PV plants were measured by the PLF under outdoor conditions. Additionally, ten modules of each sample were shipped to the Swiss laboratory of SUPSI to compare with the results of the PLF.

To eliminate the uncertainty of temperature measurement, the PLF was installed in the indoor laboratories of ESTI during the round robin test. Immediately after measuring the reference module with the certified infrastructure of ESTI (Uncertainty P_{mpp} of 1.3%, k=2 [4]), the PLF performed the measurements on the same module. It could be shown that the difference of the averaged measurements with the PLF to ESTI, SUPSI and SMFB is below 1%, see Figure 4.

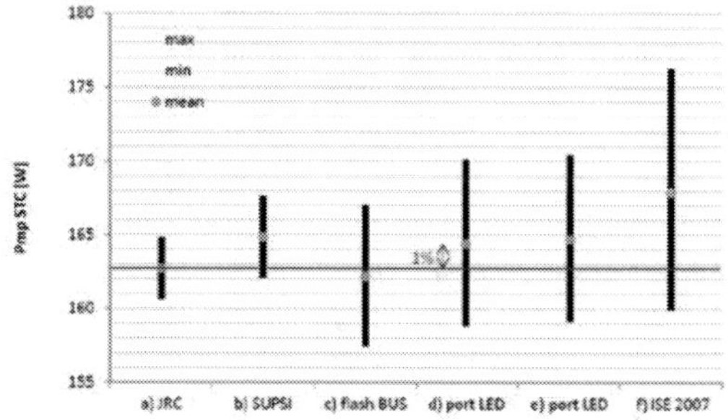

Figure 4: Results of the round robin test on the crystalline reference module (95% confidence interval). a) EU-JRC ESTI (combined uncertainty of 1.3%). b) SUPSI (combined uncertainty of 1.7%). c) SMFB. d) & e) PLF. f) ISE measurement in 2007

4 RESULTS OF FIELD MEASUREMENTS

The aforementioned field test of three PV plants lasted two consecutive days each. To estimate the uncertainty of the PLF under field conditions ten modules of each plant were measured by the certified laboratory of SUPSI. The measurement conditions of outdoor measurements entail module temperatures with mostly large discrepancies with respect to the STC of 25°C. The temperature measurement is crucial for the overall uncertainty. The maximum deviation of the individual measurement with the PLF is 3% compared to the measurements of the SUPSI laboratory. It has been observed that this deviation is lower in times of low solar irradiance, which can be explained by the lower temperature transients (see Chapter 5). This higher accuracy of nominal power measurement can also be achieved during night time measurements with the additional advantage of omitted yield losses to the system

Table III: Deviation of the outdoor measurement with the PLF to the indoor measurement in the certified laboratory of SUPSI. The values are averaged over ten PV modules per site.

PV Plant	Deviation of the PLF to SUPSI (k=2)
Die Werke, Wallisellen	(1.2 ± 1.81)%
AXA, Winterthur	(0.96 ± 0.74)%
LKW, Schaan FL	(1.51 ± 0.47)%

Figure 5: The PLF during the outdoor measurements at the PV Plant "Die Werke, Wallisellen".

The experiences during these first outdoor tests revealed throughput rates of up to 130 modules per day. Estimations expect a setup time of one hour and they promise a measurement throughput of at least 150 modules per day. The measurement costs per module are expected to be about a tenth compared to the costs of the common measurement in a certified laboratory.

The lion's share of the measurement time depends on the ease of handling since the actual electrical IV-measurement has a cadence of six measurements per minute. Thus, further investigations on how to reduce the time for roll on of the light engine, positioning on the module and temperature acquisition are being conducted by ZHAW.

The field measurements of the PV plants, after being in service since two, six and seven years respectively, revealed that no measured power of the modules were beyond its warranty boundaries. In Figure 6, the measured nominal power of the 180 modules are sorted in bins of 5 W power range. The nominal power according to the datasheet is 180 W. No measurement showed significant discrepancy compared to the lower power boundary of 164.7 W (95% of power warranty after 5 years and estimated uncertainty of PLF of 3.5% results in lower boundary of 91.5% which is equal to 164.7 W).

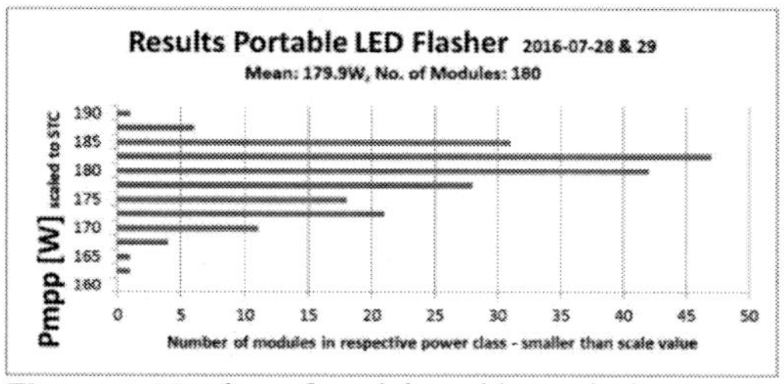

Figure 6: Number of modules with nominal power in classes of 5 W, as measured with the PLF on the PV plant "die werke, Wallisellen" on July 28th and 29th 2016.

With this new method it is possible to achieve a larger number of samples and thus, it enables the rearranging of PV plants into new strings with similar nominal power. This retrofit promises an increase in energy yield.

In order to identify modules with possible low nominal power compared to the datasheet value a thermal imaging of the plant can be conducted. The thermal imaging together with the visual inspection enables a pre-selection resulting in more significant samples of the whole array and plant.

5 IMPROVEMENTS OF TEMPERATURE MEASUREMENT AND UNCERTAINTY

The intended throughput of 150 modules per day requires a measurement time of about two minutes per module. In this short period the module will not reach thermal equilibrium with the ambient air and thus, it is difficult to extract the cell temperature within the PV module.

During the field testing an increased uncertainty has been observed during times of high solar irradiance. It has been found that this error was based on the uncertainty in the estimation of the module temperature due to the exponentially decreasing module temperature after shadowing of the module. The deviation between the measured temperature and the real cell temperature together with the inhomogeneous shadowing led to high uncertainties of the temperature corrected P_{mpp} value. Therefore, the ZHAW established an approach for improved cell temperature estimation leading to an improved temperature correction of P_{mpp}.

In order to determine the appropriate temperature acquisition method a 60-cell polycrystalline reference module was equipped with two PT100 RTDs. One of the RTDs was attached to the backsheet of the module and the other one to the front glass. Then, the electrical characteristics of the reference module were determined with the PLF at constant room temperature of 25±0.3 °C in the indoor lab of the ZHAW.

Table IV: Selected electrical data of the polycrystalline reference module as stated in the datasheet and as measured with the PLF at the ZHAW indoor laboratory (25±0.3 °C). *Estimated temperature coefficient, valid for this module and measurement setup

Electrical Quantity	Datasheet	Measured
P_{mpp} [W]	250.00	248.95
V_{oc} [V]	37.70	37.52
TC P_{mpp} [%/K]	-0.45	-0.53*
TC V_{oc} [%/K]	-0.33	-0.32*

Figure 7: Pictures of the outdoor measurement setup. The top picture shows the PLF on a neighbouring module while the reference module is exposed to the sunlight. The bottom picture depicts the setup during shading and measuring the reference module. On the right hand side the battery and the measurement electronics are shown.

The reference module was thereafter mounted to the PV array of the ZHAW PV outdoor lab and exposed to solar irradiance of around 800 W/m^2 (Figure 7 top) before being shadowed for half an hour by the PLF (Figure 7 bottom). During the period of shadowing the IV characteristic was measured in ten second intervals with the PLF. The aim of the measurement series was to find an optimised method to measure a constant temperature corrected P_{mpp} near its STC value over a large temperature range.

The decrease in measured temperature over time due to shadowing is depicted in Figure 8. The cell temperature was estimated using the measured V_{oc} value (Figure 10) and the temperature coefficient of V_{oc} (see Table IV). Consequently, the P_{mpp} was temperature corrected using the estimated cell temperature and the temperature coefficient of P_{mpp} (see Figure 11).

Newton's Law of Cooling fits the approximated cell temperature:

$$T(t) = T_{amb} + (T_{start} - T_{amb}) \times e^{\frac{-0.0063}{s} \times t}$$

The residuals of the curve fitting are shown in Figure 8 bottom.

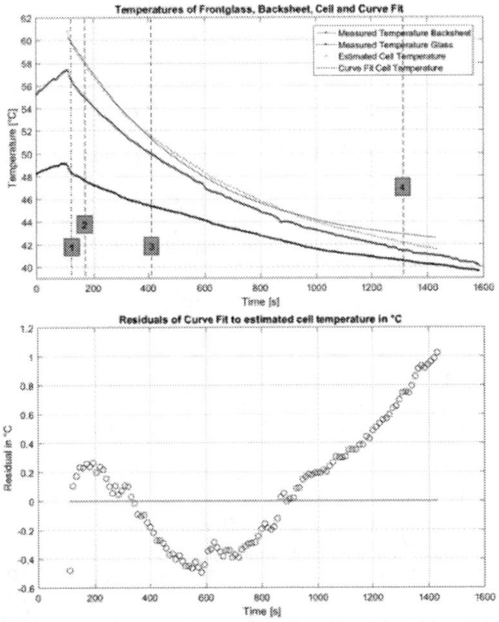

Figure 8: Top: Measurement results of two RTDs mounted on the backsheet (red line) and on the front glass (blue line) of the reference module. The cell temperature (cyan line) was estimated using the V_{oc} values and temperature coefficient of V_{oc}. Vertically dashed lines for 1) ten seconds, 2) one minute, 3) five minutes, and 4) 20 minutes after shadowing of the module. Bottom: Residuals (black circles) of the curve fit (top, black line) of the estimated cell temperature in degree celsius.

Figure 9 illustrates the problem of temperature measurement with a cross section of a PV module. The temperatures at the time steps 1-4 as indicated and explained in Figure 8 are shown as gradients over the layers of the PV module. The different thermal resistances and thicknesses of the different layers lead to temperature gradients with different slopes resulting in different temperatures at the front glass and at the backsheet. These temperatures differ from the cell temperature except in thermal equilibrium. During the first few seconds after shadowing, the difference in temperature between the cell and the glass/backsheet is relatively large (Figure 9, top red dashed line). This divergence decreases as the module approaches thermal equilibrium (Figure 9, bottom red dashed line).

Figure 8 shows that the estimated cell temperature is 4.4 °C higher than the measured backsheet temperature, which is caused by the thermal resistance of the encapsulation material and backsheet. This observation correlates to figures in literature [5]. Misinterpretation of the backsheet temperature for the actual cell temperature then could lead to an underestimation of P_{mpp} of up to 2.3% or 5.8 W.

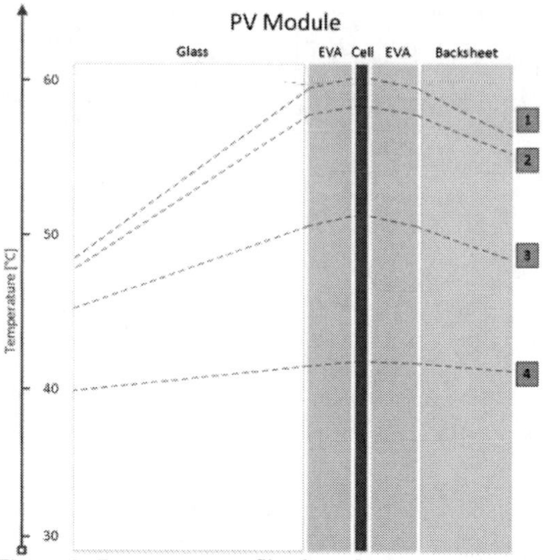

Figure 9: Temperature profile along the cross section of a PV module with the PV cell in the middle, encapsulated by EVA and front glass and backshee t (proportions in the lateral dimension are not to scale). Numbers 1 to 4 describe the lateral temperature characteristics after 10 seconds, 1, 5 and 20 minutes respectively.

The open-circuit voltage was corrected using the measured backsheet temperature (Figure 10, red) and the estimated cell temperature (Figure 10, cyan). It can be seen that the correction with the estimated cell temperature matches the STC value of V_{oc} as measured indoors at 25°C (Figure 10, black).

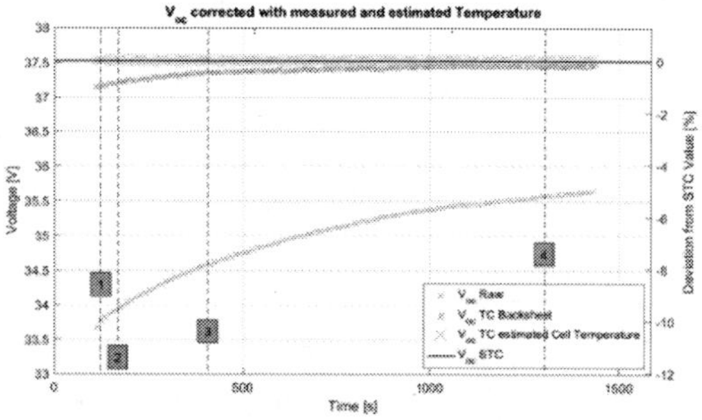

Figure 10: Raw data (green) and temperature corrected values of V_{oc} of the reference module during the same measurement as in Figure 8. The vertical lines indicate the temperatures after 10 s, 1, 5 and 20 min after shadowing of the PV module.

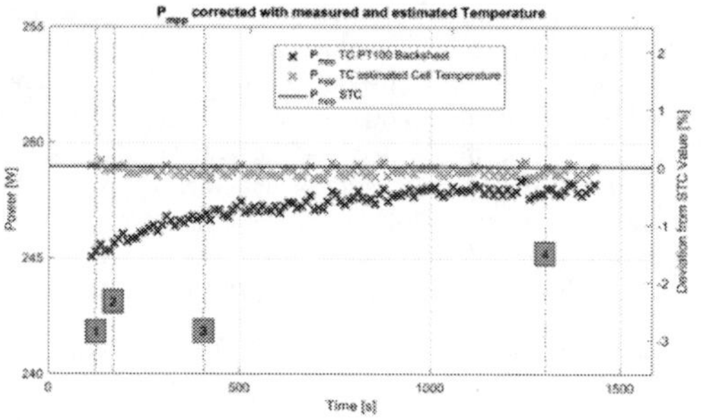

Figure 11: Temperature corrected values of P_{mpp} of the reference module during the same measurement as in Figure 8.

The maximum power was corrected using the measured backsheet temperature (Figure 11, red) and the estimated cell temperature (Figure 11, cyan). Again, the correction with the estimated cell temperature matches the STC value of P_{mpp} as measured indoors at 25°C (Figure 11, black) with very small deviation.

The RMSE of the P_{mpp} value estimated with the cell temperature is 0.1% while the RMSE of the P_{mpp} value corrected with the backsheet temperature is 0.83%. The uncertainty of the temperature measurement could be reduced considerably with this measurement setup resulting in a lower uncertainty of the nominal power measurement.

In order to alleviate the underestimation of P_{mpp} several measures are proposed:

I. Shadowing the next modules in line with a blanket ten minutes before the measurement in order to measure the electrical properties of the module nearer to thermal equilibrium. Homogenous shadowing leads to a homogenous temperature of the module.

II. Approximation of the cell temperature by measurement of front glass temperature and ambient temperature. To ensure a constant time delay between start of shadowing and measurement the measurement will be automatically started after a fixed time.

III. For large batches and lower uncertainty a calibration of temperature coefficients of the modules can be done on site.

As part of further research work, the ZHAW is establishing a tool to automatically determine the temperature coefficients of a module during a measurement series of roughly 15 minutes. This procedure is done on site of the PV plant and will contribute to the reduction of uncertainty in nominal power measurement with the PLF.

6 CONCLUSION

The new version of the Portable LED Flasher is superior compared to predominant nominal power measurement methods regarding the flexibility and costs. Due to elimination of demounting and shipping to a stationary laboratory, the costs for measurements can be reduced to one tenth. Thus, a higher number of modules for the same price can be evaluated which increases the significance of the sample and allow a more accurate assessment of the current economic value of the PV plant in question. The energy autonomy and mechanical structure of the light engine allow the transport to the site with a station wagon and measurements without mains connection for one day.

A round robin test confirmed the low uncertainty of <1% of the PLF in laboratory conditions (25°C). As a result of field tests the uncertainty was found to be below 3% in outdoor conditions. Further analysis of outdoor measurements led to a proposal of several measures that increase the accuracy of the critical temperature measurement. The temperature transient due to shadowing of the module by the light engine of the PLF can be alleviated by pre-shadowing the modules for at least 10 minutes before the measurement.

Furthermore, the actual cell temperature, which is relevant for the temperature compensation, can be estimated by measuring the ambient temperature and time between shadowing and IV-measurement. Successive research and development is done at the ZHAW to further reduce the uncertainty of the PLF.

REFERENCES

[1] F.P.Baumgartner, R.Knecht, F.Carigiet, C.Frei, and F.Beglinger, "Portabler LED Flasher - Resultate von Feldtests," presented at the 32nd PV-Symposium, Bad Staffelstein, 2017.

[2] D. Schär and F. P. Baumgartner, "PORTABLE LED FLASHER WITH IMPLEMENTED BYPASS DIODE TESTER," presented at the 29th European Photovoltaic Solar Energy Conference, Amsterdam, 2014.

[3] T. Achtnich, F. P. Baumgartner, N. Allet, M. Pezzotti, J. Haller, and B. Aeschbach, "SWISS MOBILE FLASHER BUS: PROGRESS AND NEW MEASUREMENT FEATURES," presented at the EUPVSEC, Valencia, 2010.

[4] ESTI, "PV Device Performance Sheet." 24-Aug-2016.

[5] Yang Yan *et al.*, "UNDERSTANDING THE UNCERTAINTIES IN THE MEASUREMENT OF TEMPERATURE COEFFICIENTS OF SI PV MODULES," presented at the 29th European Photovoltaic Solar Energy Conference and Exhibition, Amsterdam.

DYNAMIC IV ANALYSIS SYSTEM FOR DIAGNOSIS OF PV-MODULE STRINGS IN A LARGE SCALE PV-POWER PLANT

Author(s): Mark VERVAART (mark.vervaart@cea.fr, T. +33 (0)4 79 79 22 32), Sylvain LESPINATS (sylvain.lespinats@cea.fr), Franck AL SHAKARCHI (franck.alshakarchi@cea.fr)
Company / Institute(s): CEA, LITEN, Département des Technologies Solaires, F-73375 Le Bourget du Lac, France

ABSTRACT: Over a short period in the last decades, the world Photovoltaics market has been growing very fast (the installed module power has multiplied by a factor of 100). On the other hand, the production costs over the same period has been decreased by 75%. The performance monitoring of PV power plants, however, didn't follow the same development curve and the work on fault detection or performance monitoring didn't bring a significant number of commercial solutions so far. In the context of cost reduction of the electricity price, fault detection that doesn't need human intervention on site could be very fruitful.

The proposed system is capable of tracing the IV characteristics of a PV-module string in an active power plant, without interrupting the inverter, within between 10 and 100ms, depending on the array layout and module type. Hence limiting the impact on energy production. The IV curve is then interpreted by algorithms which detects an eventual problem. The IV analysis system communicates the state of the PV plant to the operator and indicates its performance level and the type of faults present in the module strings, in order to reduce the cost of maintenance and so exploitation costs.

This project has received support from the State Program "Investment for the Future" bearing the reference (ANR-10-ITE-0003) .

Keywords: Photovoltaic, PV Array, System Performance, Monitoring, Evaluation, Power plant

1 CONTEXT

In large PV power plants, monitoring often is done by communication with the inverters via an internet site of the manufacturer. By the time that an inverter significantly produced less energy than the comparable inverters at the same site, it might have been producing gradually less and less energy over a long period and therewith has been missing a high amount of energy.

Even when a power loss is detected, it might take a significant time to send a maintenance crew and find the problematic strings in the field.

Some systems have string monitoring, measuring the currents in running condition. That gives additional information, but not the type of problem involved. For identifying the type of problem, the complete IV curve should be traced.

The IV characteristics I(V) of a PV string allows to discriminate numerous possible defects inherent to PV-modules. It also allows an evaluation of the performance level of the modules and their aging.

These string characteristics are difficult to measure because one needs to stop the inverter and disconnect a string after another, therewith losing its energy during the evaluation period. Besides, this kind of analysis takes a lot of man power. On top of that, they are difficult to exploit because of the variability of the environmental conditions at the PV-site.

This innovation is based on the « dynamic » and « automated » character of the PV array analysis, which does need neither a human intervention, nor an energy production loss.

2 PRINCIPLE OF OPERATION

The analysis system consists of three parts

- A communicating hardware device that varies the working point on the PV module string and records the points of the I(V) characteristics of the string. The system is composed of a multiplexer unit with MOSFET switches, which puts the string to be analyzed on the IV tracer unit. This part is integrated in the junction box in the field.
- Measurements and file management by the local server (an industrial PC at the site). The string scans may be passed a couple of times per day, or on demand via a distant website on a tablet or smartphone.
- Data analysis algorithms and fault detection (done at the site by the same industrial PC). It processes the measurements and gives information (alerts) at the type of fault and the state of health of the power plant to the owner of the plant. The system uses fault recognizing algorithms that allows to indicate to the user if there is a string problem, and what type of problem. But it also will be able to follow the aging of the installation by following the characteristics over time.

2.1 The hardware of the analysis system

The industrial PC is based in a central place of the power plant. It gives order to multiple automates to scan a certain string number. The automate and multiplexers, placed in the junction boxes over the site, connect the right string to the IV tracer unit. The automate scans the IV curve, puts the string back to the array, and sends the results to the main PC by ethernet. The speed of the tracing is in between 10 and 100ms for 200 datapoints. This speed can be adapted for different module types and the physical distance of the different components.

For physical reasons, the multiplexer board is developed for groups of 4 string, because they have to be fitted in an additional combiner box that is added next to the usual combiner box. Multiple multiplexer boards can be mastered by one automate.

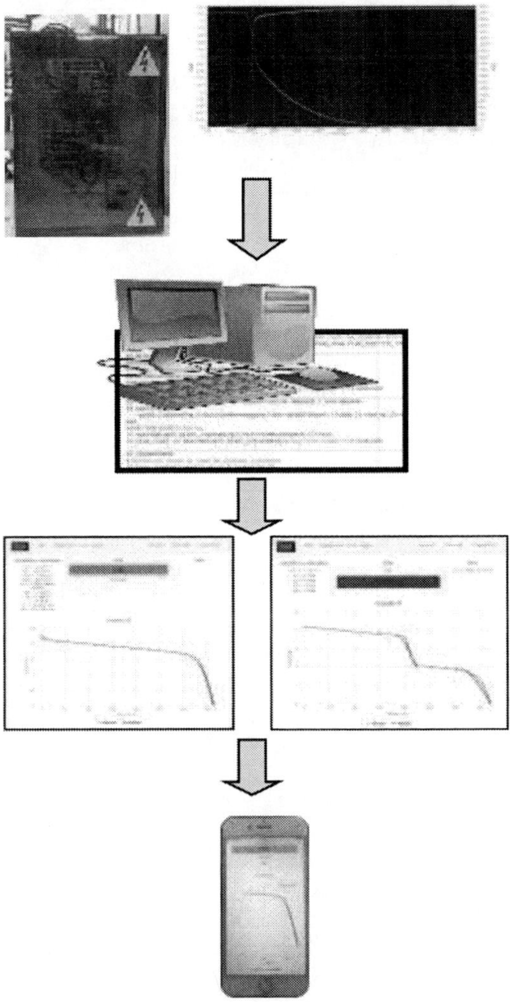

Figure 1: Multiplexer board and IV tracer in junction box. The automate sends the traced curves to the sever. After analyzing, eventual problems are being detected and a report is sent to an internet site or smartphone.

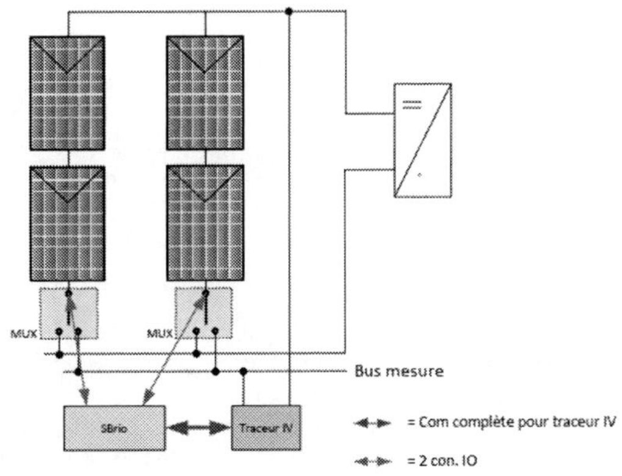

Figure 2: Main hardware overview

2.2 Data analysis algorithms and fault detection

After measurement, IV curves are post-treated so as to estimate several mains features. These features are provided by two different models. The first one is an empirical model. The second one is based on a physical 1-diode-equivalent-circuit model. The first model mainly allows estimating the degree of confidence to be given to the physical model. The second (physical) one mainly estimates the several features that allow discriminating the various types of defaults if exist. Figure 2 shows results reached with several typical examples of measurements through the two types of models.

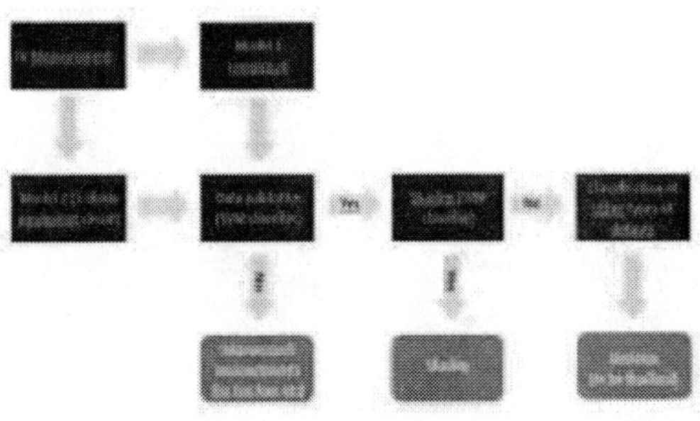

Figure 3 : Layout of the algorithms so as to perform a diagnosis from IV measurements.

An algorithm thereafter classifies curves from these features as presented in figure 3. Its ability to classify two types of defaults have currently been evaluated: series resistance faults and shading faults.
In order to evaluate shading default, semitransparent elements have been placed on a third of the modules of the string. Various types of semitransparent elements have been used so as to simulate more or less strong shades.

Table 1 shows the ability of our classification algorithms so as to detect shades:

	No shade	Shade
No shade detection	365 curves	0 curves
Shade detected	3 curves	12 curves

Yet, the statistical sensibility is 100% (every shades have been detected); the specificity is 99.2% which is also very high (only 3 false positive).
The ability of the algorithm have yet to be confirm from a substantial enough dataset, but preliminary tests tends to show that the increasing of series resistance may be efficiently detected. Simulations shows that various other types of defaults may be detected as well. However, these claims have to be confirmed on real measurements.

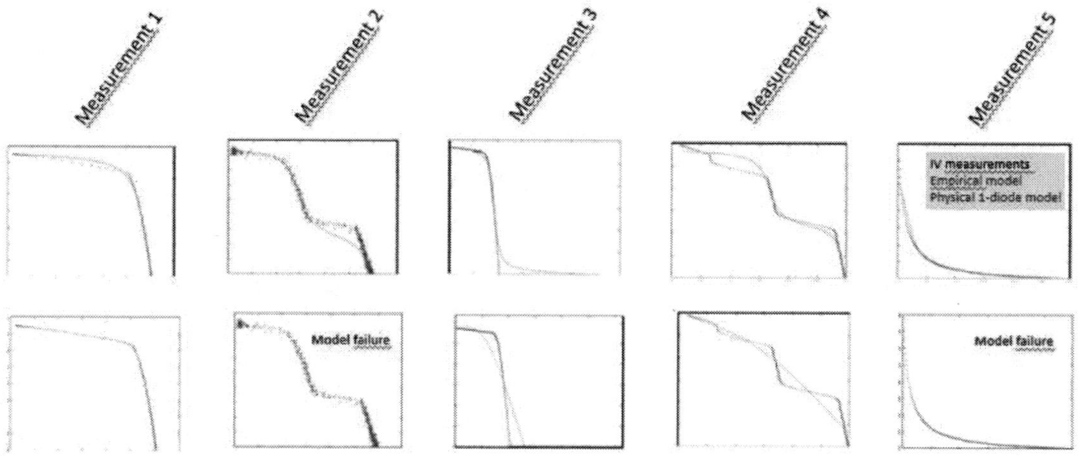

Figure 4 : data measurements versus the two models on 5 typical cases.

LUMINESCENCE IMAGING STRATEGIES FOR DRONE-BASED PV ARRAY INSPECTION

Gisele A. dos Reis Benatto[1], Nicholas Riedel[1], Claire Mantel[1], Sune Thorsteinsson[1], Peter B. Poulsen[1], Mingjun Chi[1], Anders Thorseth[1], Ole Bjarlin Jensen[1], Carsten Dam-Hansen[1], Søren Forchhammer[1], Kenn H. B. Frederiksen[2], Jan Vedde[3], Michael Petersen[4], Henrik Voss[5], Michael Messerschmidt[5], Harsh Parikh[6], Sergiu Spataru[6] and Dezso Sera[6]

[1]Department of Photonics Engineering, Technical University of Denmark, Frederiksborgvej 399, 4000, Roskilde, Denmark, garb@fotonik.dtu.dk
[2]Kenergy, Grønningen 43, 8700, Horsens, Denmark
[3]SiCon Silicon & PV consulting, J N Vinthersvej 5, 3460, Birkerød, Denmark
[4]Skive Kommune, Torvegade 10, 7800, Skive, Denmark
[5]Sky-Watch A/S, Østre Alle 6 Støvring, 9530, Nordjylland, Denmark
[6]Aalborg University, Pontoppidanstraede 101, 9220, Aalborg, Denmark

ABSTRACT: The goal of this work is to develop outdoor defect detection imaging and understand fully its challenges and limitations. The imaging is based on luminescence strategies that will be used for fast and accurate UAV-based inspection system for PV power plants. We studied electroluminescence (EL) acquisition under natural light conditions during several times of the day, under high sun irradiation, to unveil the sunlight noise characteristics towards an InGaAs detector. In order to bring more freedom to a drone-based inspection, we also show the preliminary results of a laser-line based photoluminescence (PL) strategy as a viable method for an outdoor module PL imaging system.
Keywords: Electroluminescence, Photoluminescence, Defects, Inspections

1 INTRODUCTION

Regular fault detection is highly important to ensure the expected return on investment (ROI) of both small and large-scale PV installations. Field experience shows that after 11-12 years of operation 2% or more of all PV panels fail, despite the fact that they are designed for lasting 25-30 years [1]. PV system inspections are even more necessary for older installations (installed before 2000), since modules manufactured before the 2000 are known to have more significant reliability issues [2]. Furthermore, a number of external factors such as transportation, wind and snow loads, and daily thermal stresses can cause critical damage to the PV panels, which are not always visually detectable. The frequency and detail level of an inspection is often limited by manpower and cost. Drone-based IR thermography surveys of solar plants are a reality today, and are expected to become an automated solar plant inspection method in the near future [3]–[5]. The accuracy of thermographic fault detection though, presents limitations, which can be surpassed when performed in combination to EL and/or PL imaging of the panels. In the laboratory, the combination of these defect detection techniques has already been tested [1], [6], however many challenges still need to be addressed in order to optimize, integrate and automatize the imaging system on a drone.

In this work, we present the results related to the development of the optimization of the outdoor luminescence techniques. First, we show the progress in the understanding and overcoming of the sunlight noise in EL imaging, using optical filters, lock-in signal and image processing. In the following, we present the first results using a laser-line approach under natural low light conditions, as a viable method for an outdoor module PL imaging, since it requires relative low optical power. The outdoor PL imaging enables a drone-based contactless inspection system that does not require hardware connections, as in the case for EL or PL using the sunlight as the light source.

2 METHODOLOGY

The luminescence emission peak for silicon-based solar cells at ambient temperature is centred at 1150 nm [7], near a water absorption band in the solar spectrum (AM 1.5) [8]. Two camera detectors are commonly used to detect such emission: i) cooled Si charge-coupled devices (CCD) ii) and short-wave infrared (SWIR) InGaAs sensors. Nevertheless, only the second allows a short enough integration times for non-stationary luminescence image acquisition; therefore, it is ideal for drone operation [9]. The InGaAs detector has a second advantage in that its quantum efficiency (QE) is not active in the most intense wavelengths of the solar spectrum, which is the main source of noise for daylight luminesce imaging. In the following described experiments, an InGaAs camera from Hamamatsu model C12741-03, and an OD>4.0 1150nm band-pass filter with 50nm FWHM was used. The sample was a mechanically stressed module consisting of 36 multicrystalline silicon 156 mm x 156 mm cells.

2.1 Electroluminescence

Figure 1a illustrates the outdoor EL setup which consists of a DC power supply unit (PSU), a PV module and an InGaAs camera connected to a modulating box. Such system synchronizes the image acquisition with an electrical forward bias applied by the DC power supply. Figure 1b shows the synchronizing circuit, driven by an Arduino logic controller, and Figure 1c shows the pulse width modulated (PWM) waveforms applied to the PV panel and camera. A PC connected to the camera was required for image data acquisition and settings control. The exposure time was established separately in the camera software.

One of the strategies to enhance the quality of the images obtained at high noise level during the day was sequential image acquisition. We acquired sequences of 100 images (50 under forward bias (EL signal) and 50 as background images for subtraction) at 6 Hz with 20 ms exposure time, while global horizontal irradiance (GHI) was observed through a fast response Si photodiode sensor in a weather station located few meters from the imaging position [10]. The complete image sequence took circa 17 seconds to acquire. Such strategy was used in order to estimate the effect of the sun irradiation on the imaging process and to better understand and overcome the noise characteristics towards the InGaAs detector. The iris aperture was f8 for all outdoor image acquisitions. A 25 mm fixed focus lens was used, and in order to photograph the whole panel, the camera was positioned at 2.6 meters from the solar module.

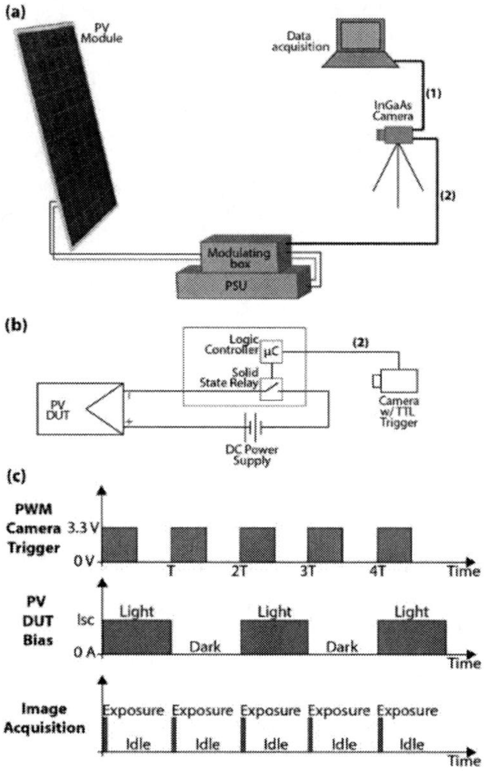

Figure 1: Outdoor EL imaging setup (a); correspondent basic trigger circuit that synchronizes the imaging and forward bias from a DC power supply (b); and PWM waveforms driven by the logic controller. Connection path (1) can be replaced with the frame grabber integrated to the drone CPU board, and (2) by radio connection to the modulating box.

As averaging several pictures is a very common method to denoising images, we averaged several EL images per scene to enhance image quality. The image processing included taking the difference between the average of light and dark images, linear scaling of the pixel values, automatic stretching of the initial dynamic range (the source images are 16 bits) to 8 bits for display and cropping of the whole image to the region of interest.

The signal to noise ratio averaged for the imaged scene (SNR_{50}) was calculated following the guidelines established in the EL technical standard (IEC TS 60904-13 [11]) using the equation (1), where noise level is obtained from the average absolute deviation of two EL images:

$$SNR_{50} = \frac{\sum_k \left(0.5\cdot\left(EL_1(k)+EL_2(k)\right)-BG(k)\right)}{\sum_k \left(|EL_1(k)-EL_2(k)|\cdot\sqrt{0.5}\cdot\left(\frac{2}{\pi}\right)^{-0.5}\right)} \quad (1)$$

In (1), EL_1 and EL_2 are EL images from the same scene, BG is the background image of the same scene, and all the operations are made pixel-by-pixel (k). We called "SNR_{50} of individual images" when the SNR_{50} is calculated with regular computation, i.e. using two single EL images and one single BG image. When SNR_{50} is calculated to averaged images, EL_1 and EL_2 correspond to two averaged images from 25 EL images (as we acquired 50 images with EL signal in each sequence), and BG correspond to the averaged image from 50 background images, all from the same scene/sequence. For this SNR_{50} value, we refer as "SNR_{50} of averaged images".

2.2 Photoluminescence

PL images under low natural light were acquired using an 800 nm laser diode at 6 W, while the camera and the laser were circa 0.7 meter from the sample. The laser beam coming from a fibre was shaped into a 4 mm thick line with <5% of optical power loss. The line beam was manually scanned over the area of interest. The sample consisted of the same module used for electroluminescence, with focus in the three cells from row 3, columns B, C and D (see Figure 2). The images were acquired with a 30 ms exposure time.

3 RESULTS AND DISCUSSION

3.1 Electroluminescence

The EL outdoor imaging experiments in this work were performed on a special testbed, designed for developing outdoor and drone based EL imaging. Here we used several strategies to perform the EL imaging at any time of the day and evaluate possible challenges. For guidance and qualitative comparison, Figure 2 shows the indoors EL image of the mechanically stressed module sample used in the work, which complies with the minimum SNR for EL laboratory measurements: $SNR_{50} \geq 45$. However, during the day such images are compromised by high and very often variable light noise levels, primarily from the sunlight itself. For such reasons, the signal to noise ratio requirement for outdoor EL is considerably lower: $SNR_{50} \geq 5$.

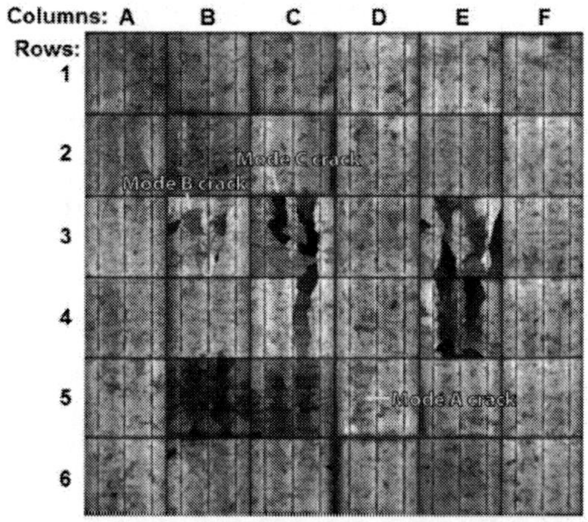

Figure 2: Indoors EL image of mechanically stressed PV module, showing mode A, B and C cracks. $SNR_{50} = 46.20$.

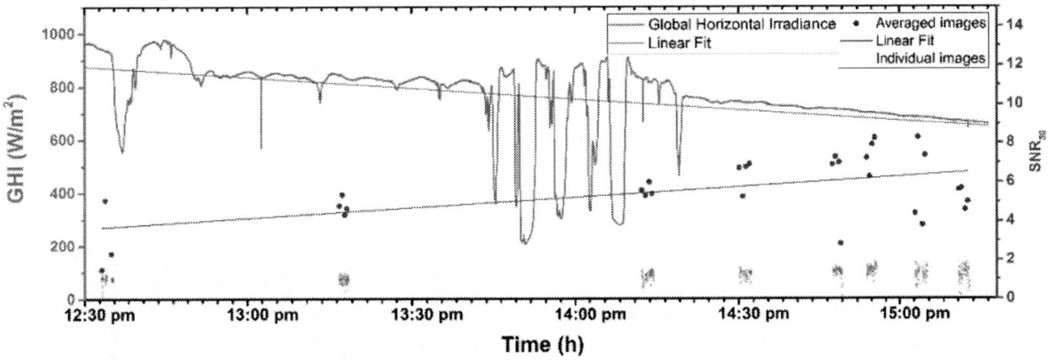

Figure 3: SNR_{50} of averaged (red dots) and individual (red crosses) EL images, from sequences acquired under the correspondent GHI (light blue line). The linear fit of the GHI (dark blue line) and the SNR_{50} of averaged images (dark red line) are also shown.

The sun intensity constantly varies, depending on the time of the day and cloud cover, while the InGaAs sensor is highly sensitive to variations in the light intensity. For this matter, we need to understand what the requirements are for EL imaging in a system including camera, EL signal and sun irradiation, in order to obtain the highest SNR. To obtain images as close as possible to the real field inspection condition, we performed the experiments in a bright, but partially cloudy day, with the panel was 75° tilted from the horizontal facing south. We acquired sequences of images in groups of at least three sequences.

In Figure 3, the SNR_{50} of the EL image sequences were plotted with correspondent global horizontal irradiation (GHI) in the moment of the image acquisition. In total, 31 image sequences were taken in a period of 3 hours. This measurement was performed under very high irradiation (~700-850 W/m^2) and as such, the iris aperture was very small (f8) to avoid saturation on the detector. Both SNR_{50} of individual (red crosses) and averaged (red dots) images were plotted in the graph of Figure 3. At first, we verified the substantial improvement in SNR_{50} of averaged images in relation to the individual images, with general improvement due to the denoising by averaging. At the same time, we observe that the average of SNR_{50} values of individual images remained constant along the GHI decrease during the day, while the linear fit SNR_{50} of the averaged images (dark red line) rose with very similar inclination as the linear fit (dark blue line) of the GHI throughout the day.

Figure 4 shows the zoomed in values for two periods of the graph from Figure 3, one with high variation of sun irradiation (Figure 4 top) and the other with high variation of SNR_{50} values of averaged images (Figure 4 bottom). The SNR_{50} of the individual images is not directly affected by the GHI variations. Figure 4 bottom show sequences taken in a relatively stable GHI period, however, the last sequence presents a high variation SNR_{50} of individual images, which may be caused by internal factors, such as the PSU electrical bias. At Figure 4 bottom, one of the SNR_{50} of the averaged images presented very low SNR_{50} value, which might be related to individual SNR_{50} variations between sets of images. One reason to this low value could be that the method chosen to denoise, the averaging, is not optimal due to varying acquisition conditions. For such cases, advanced denoising process could be applied.

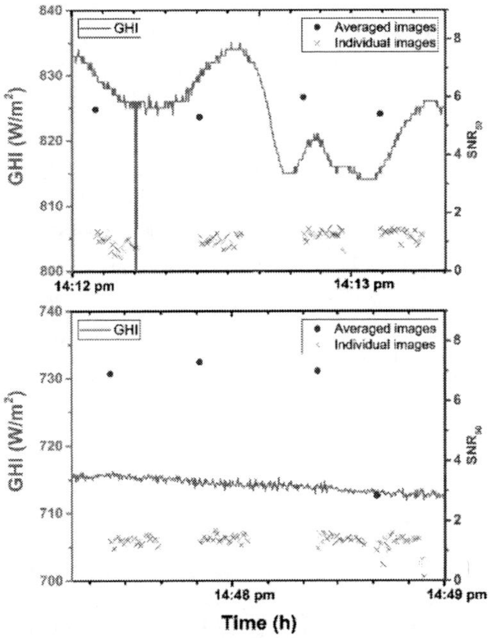

Figure 4: SNR_{50} of averaged (red dots) and individual (red crosses) EL images, from sequences acquired under the correspondent GHI (light blue line). Two sections zoomed in from Figure 3

The images corresponding to the lowest and the highest SNR_{50}, after the application of the image processing routine, are shown in Figure 5. Even though the lowest SNR_{50} does not comply with the minimum requirements of the standard ($SNR_{50} \geq 5$), the image presented similar qualitative information and is able show clearly mode C cracks present in the module (see Figure 2). Automatic defect detection methods such as EL intensity distributions (ELID) of mode C crack could be applied here, using the relative EL signal from the unbroken cells [12].

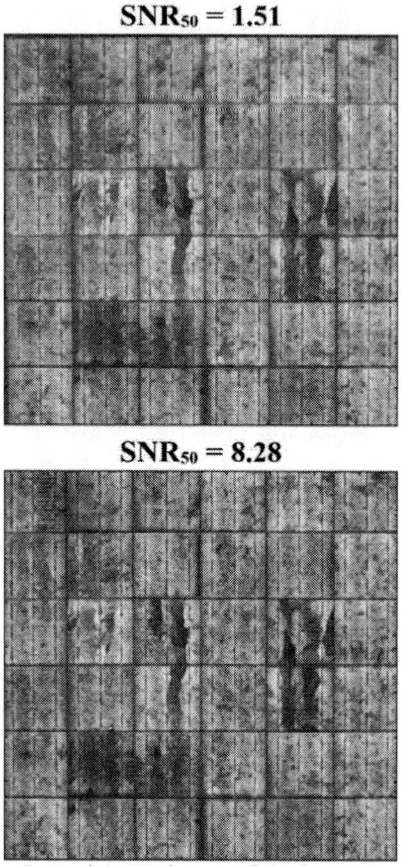

Figure 5: Outdoor EL images of the lowest (top) and highest (bottom) SNR_{50} obtained under high sun irradiation levels, after image processing, automatic cropping, and contrast correction.

3.2 Photoluminescence

Photoluminescence is a very challenging technique to be applied outdoors, due to the need for powerful light sources. At the same time, such technique could provide full freedom for a drone to execute luminescence imaging throughout strings in large-scale PV farms, independent of time demanding changes in electrical contacts. As a strategy toward PL outdoor imaging, we used a line shaped laser beam, with the aim to enhance the optical power per area and lower the required exposure time. Such PL strategy also brings up the possibility of using additional optical filtering (lenses and polarizers) which can be important alternatives for daylight luminescence inspection.

Figure 6 shows the comparison of EL and PL imaging using laser-line scan of cells B3, C3 and D3 of the module shown in Figure 2. PL emission induced by laser-line scan provided different information than EL. The broken and isolated regions of the cell (darker in EL) appear brighter than the unbroken or the still connected regions. As such regions have high series resistance, the probability of charge carriers generated by the laser to recombine radiatively is high, as they do not have proper electric connection to disperse [13], [14]. This method can be used either for complementary information to improve automatic assessment or for accurate contactless defect detection. Comparing the PL imaging induced by a homogeneous laser beam over a single cell illuminated area, the exposure time decreased from 700 ms to 30 ms from laser-area to laser-line beam respectively.

Figure 6: EL (top) and PL (bottom) PL image using laser-line scan of cells B3, C3 and D3 of the module from Figure 2.

4 DRONE INTEGRATION

For the integration of the EL setup into a drone, the connections 1 and 2 shown in Figure 1a need to be replaced with wireless connections. The modulation box is adapted to have radio communication to camera triggering. The synchronisation between imaging on the drone and the bias current needs to be accurate, therefore precise timestamping will be performed on an electronic board embedded on the drone. The final drone-based EL concept is shown in the top of Figure 7.

For non-stationary images, and therefore for drone applications, several images have to be acquired fast. At the current stage of this work, EL images for qualitative analysis can be acquired with eight EL images averaged and subtracted by eight background images. With a PSU capable to be as fast as the camera triggering for 20 ms exposure time, this amount of images can be obtained with a drone, equipped with an appropriate camera stabilization gimbal, keeping its position for 0.32 s. A faster camera could reduce the images acquisition time to 0.15 s. Moreover, further strategies can be used in terms of image processing in cases image displacements of few centimetres. Regarding the saturation of the detector due to high sun irradiation, which in this work needed to be controlled manually by the iris aperture, it will not be the case using an InGaAs camera adaptive to natural light variations.

As EL imaging is a technique that requires electrical contacting, it will be a time limiting factor for drone-based inspection, especially in large-scale solar plants. A PL imaging induced by a laser-line scan following the drone movement is a promising technique towards a contactless luminescence technique. As the laser-line concentrates the optical power in a small area, the total optical power acquired can comply with the weight and cooling requirements for a drone payload. An illustration of the scan orientation is shown in the bottom of Figure 7.

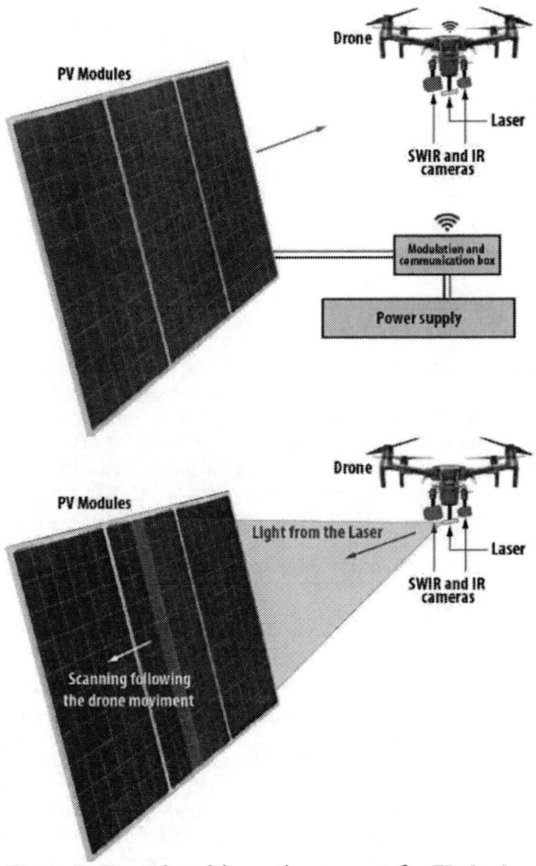

Figure 7: Drone-based inspection concept for EL (top) and PL laser-line scan (bottom).

5 CONCLUSIONS

In this work, EL imaging was performed successfully under high GHI, using the EL signal modulation and image processing. The averaging of the images from the modulation sequence lead to denoising by averaging, and the improvement of SNR_{50} was observed. We observed also that the SNR_{50} of the averaged images increases with linear decrease in GHI.

At the same time, the SNR_{50} of individual images was not directly affected by GHI variations, and SNR_{50} of averaged images was low when individual SNR_{50} presented high variation between sets of images, making denoising by averaging not optimal in such conditions. However, the lowest SNR_{50} image presented similar qualitative information and showing mode C cracks clearly.

The PL image using a laser-line scan excitation allowed covering a larger module area than a beam over an area, and at the same time lowered the exposure time for the technique from 700 to 30 ms.

The next steps of this work includes: forward bias and camera triggering carried out via wireless communication; controllable testbed with a moving camera to determine the limits of modulating EL image acquisition; scaling up laser-line scan for outdoor PL inspections.

6 ACKNOWLEDGEMENTS

The authors acknowledge the financial support from Innovation Fund Denmark for the project 6154-00012B DronEL – Fast and accurate inspection of large photovoltaic plants using aerial drone imaging.

7 REFERENCES

[1] M. Köntges *et al.*, "Review of Failures of Photovoltaic Modules," 2014.

[2] D. C. Jordan and S. R. Kurtz, "Photovoltaic degradation rates - An Analytical Review," *Prog. Photovoltaics Res. Appl.*, vol. 21, no. 1, pp. 12–29, 2013.

[3] S. Dotenco *et al.*, "Automatic detection and analysis of photovoltaic modules in aerial infrared imagery," *2016 IEEE Winter Conf. Appl. Comput. Vis.*, pp. 1–9, 2016.

[4] "The drones that inspect and maintain photovoltaic power plants," *sUAS News - The Business of Drones*. [Online]. Available: https://www.suasnews.com/2015/04/the-drones-that-inspect-and-maintain-photovoltaic-power-plants/. [Accessed: 25-Jan-2017].

[5] "Drones to cut plant inspection costs as South Africa eyes quality," *PV Insider*. [Online]. Available: http://analysis.pv-insider.com/drones-cut-plant-inspection-costs-south-africa-eyes-quality. [Accessed: 25-Jan-2017].

[6] S. Johnston and T. Silverman, "Photoluminescence and Electroluminescence Imaging Workstation," *NREL*, 2015.

[7] W. S. Yoo, K. Kang, G. Murai, and M. Yoshimoto, "Temperature Dependence of Photoluminescence Spectra from Crystalline Silicon," *ECS J. Solid State Sci. Technol.*, vol. 4, no. 12, pp. P456–P461, 2015.

[8] G. dos Reis Benatto *et al.*, "Development of outdoor luminescence imaging for drone-based PV array inspection," *IEEE Photovolt. Spec. Conf.*, 2017.

[9] J. Adams *et al.*, "Non-Stationary Outdoor EL-Measurements with a Fast and Highly Sensitive InGaAs Camera," *32nd Eur. Photovolt. Sol. Energy Conf. Exhib.*, 2016.

[10] DTU Fotonik, "DTU Risø PV Weather Data." [Online]. Available: http://ftnk-psolmaal.win.dtu.dk/B130_PV_Weather/index.html. [Accessed: 18-May-2017].

[11] I.E.C., "PNW/TS 82-901 Ed. 1.0 Photovoltaic devices - Part 13: Electroluminescence of photovoltaic modules (proposed future IEC TS 60904-13)." http://www.iec.ch/.

[12] S. Spataru, P. Hacke, and D. Sera, "Automatic Detection of Inactive Solar Cell Cracks in Electroluminescence Images," *44rd IEEE Photovolt. Spec. Conf.*, 2017.

[13] T. J. Silverman, M. G. Deceglie, K. Vansant, S. Johnston, and I. Repins, "Illuminated outdoor luminescence imaging of photovoltaic modules," in *IEEE Photovoltaic Specialists Conference*, 2017.

[14] I. Zafirovska, M. K. Juhl, J. W. Weber, O. Kunz, and T. Trupke, "Module inspection using line scanning photoluminescence imaging," *32nd Eur. Photovolt. Sol. Energy Conf. Exhib.*, 2016.

IMPLEMENTATION OF A FRIENDLY DAYLIGHT ELECTROLUMINESCENCE SYSTEM FOR THE INSPECTION OF SOLAR PV PANELS

O. Martínez[1], M. Guada[1], A. Moretón[1], S. Rodríguez-Conde[1], M.A. González[1], J. Jiménez[1], J. Pérez[2], M. Martínez[3], J.A. Florez[3], H. Silva[3], A. Velasco[3], L. Pérez[3], V. Parra[3]
[1]GdS-Optronlab Group, Dpto. Física de la Materia Condensada, Univ. de Valladolid
Edifico LUCIA, Paseo de Belén 19, 47011 Valladolid, Spain (oscar@fmc.uva.es)
[2]Dpto. Ingeniería Eléctrica, Escuela de Ingenieros Industriales, Univ. de Valladolid
Francisco Mendizábal 1, 47014 Valladolid, Spain
[3]Enertis Solar, SL.
Av. Bruselas 31, 1st floor, 28108 Alcobendas, Madrid, Spain (vicente.parra@enertis.es)

ABSTRACT: Electroluminescence (EL) is nowadays a well-recognized tool for the inspection of solar cells and panels, particularly interesting for the inspection of large plants. Routinely, when the modules are mounted at the PV site, EL measurements are carried out during nighttime and collected with Si CCDs (including conventional - modified - cameras), with good spatial resolutions and relatively low prices. However, daylight EL systems with InGaAs CCDs are promising tools for EL inspection, due to the possibility to analyse the solar panels during the day, which will be beneficial regarding the cost, practicability and module risk mitigation associated to the testing service. Also, in order to reduce production losses, the measurements should be conducted in a fast and reliable way. In this paper, we present a friendly daylight EL system using a cost-effective InGaAs CCD and appropriate filter setup. Good-quality images at high irradiations (above 1000 W/m^2) can be obtained in less than 20 s. Additionally, high-quality photoluminescence (PL) imaging can be registered by this friendly, cost-effective system.
Keywords: Electroluminescence, defects, Silicon, testing, daylight, on-site testing

1 INTRODUCTION

Electroluminescence imaging (ELi) is a well-known tool for the inspection of solar cells from the pioneer publication by Fuyuki and coworkers in 2005 [1]. The generalized use of Si CCDs, with resolutions of the order of 1 MPx or higher and relatively low cost, has allowed the use of ELi as a basic inspection tool for the control of the quality of modules and PV solar plants. The evolution of this technique in the last years has made of it one of the most prominent tools for the inspection of Si solar plants [2].

However, Si CCDs require strict dark conditions to perform the EL measurements, since any ambient light, with visible components, will be in competition with the small EL signal arising from Si solar cells. For this reason, EL measurements with Si CCDs were limited to be performed in a dark laboratory space (indoor EL), although, recently, outdoor EL become common using the Enertis' PV Mobile Lab tools or also by measuring during the night [3, 4].

The appearance in the last years of new CCDs with large efficiencies in the near IR has paved the way to performing EL measurements during the sunrise and nightfall, and has opened the door to perform daylight EL by using adequate filters and procedures [5].

The main objective of our work is to develop a cost-effective, optimized daylight friendly system, allowing EL images to be obtained with good resolution and optimized acquisition times.

2 ELECTROLUMINESCENCE

2.1 Basis

ELi is a relatively very easy characterization technique, requiring only to forward biassing the cell or the module and a CCD to collect the luminescence emitted by the recombination of the electron and holes (e-h) injected into the p-n junction of the solar cell.

ELi is usually based on Si-detector cameras, due to their resolutions and relatively low cost. The main drawback of Si CCDs for ELi inspection of Silicon solar cells is the inappropriate QE at the spectral range of the Silicon near band edge (NBE) emission, centered at ~1100 nm, Fig. 1. The QE of Si CCDs is high in the range 400 – 850 nm, but is very low beyond 1000 nm. However, in spite of the low QE of the CCD detector, it is possible to capture the luminescence emitted by the Si solar cells (alternatively one can use light excitation for Photoluminescence imaging, PLi). In a dark ambient, good EL images can be obtained with very short acquisition times, of the order of a few milliseconds. This makes ELi a very useful technique for identifiying defects related to faults of electrical connections and/or defects related to the Si material itself, in PV modules or plants, e.g., defects such as microcracks, those related to potential induced degradation (PID), etc., can be easily observed [6, 7].

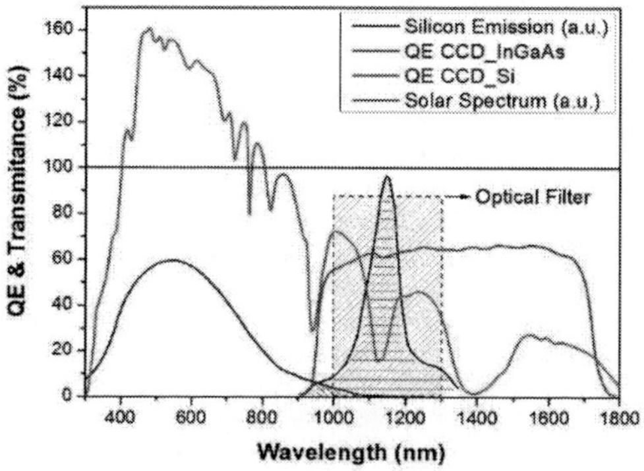

Figure 1: QE of Si and InGaAs CCDs, compared to the NBE emission of Silicon.

A dark condition is mandatory when using Si CCDs, in order to avoid any ambient light. This presents some drawbacks, such as the need to dismantle the panels in the case of measurements in a lab or in a mobile van (which forces EL to be a control tool for only a small portion of a PV plant) or the difficulties and inconveniences of performing the necessary electrical contacts during night; in particular, the manipulation of electrical interconnections by night becomes more complicated that by day, and also the measurements at night require large costs in workforce, security levels, etc.

2.2 Daylight EL

The appearance of IR CCDs such as InGaAs CCDs, with optimal QEs in the near IR region, open the possibility to perform EL measurements at not so strictly dark conditions. For instance, the NBE of Si at ~1100 nm lies on the maximum QE spectral range of InGaAs CCDs, as illustrated in Fig. 1. It is also beneficial the fact that the InGaAs CCD is blind to most of the visible spectrum. In spite of this, large amounts of IR ambient light will still arrive to the CCD, and compete with the comparatively low EL Si signal. For this reason, daylight EL is not straightforward, and the large parasitic ambient light that covers the EL signal must be rejected and filtered. This is especially critical for large background irradiation levels. For instance, at moderate irradiation levels of 700 W/m^2, signal to noise ratios (SNR) of 0.006 were recorded. Another critical point to consider is the significant fluctuation in the levels of the background light (Fig 2). Such large fluctuations in relatively short times make difficult the task to separate the EL signal and background from each other.

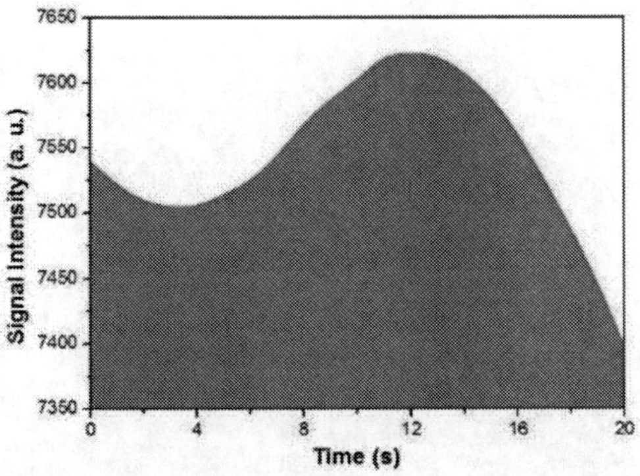

Figure 2: Fluctuations of the total signal detected by our cost-effective InGaAs CCD during 20 seconds in a partially sunny day (G = 700 W/m^2).

3 DEVELOPMENT OF A DAYLIGHT EL SYSTEM

3.1 Main components

Our approach to perform daylight EL measurements use three main components: an InGaAs CCD (hereinafter, CCD), a laptop and a power supply. Nowadays, high resolution InGaAs CCDs are still expensive. We have use a cost-effective InGaAs CCD (640 x 512 pixeles) as a proof of concept, demonstrating that our system allows, in spite of the resolution, to acquire high quality daylight EL images. The laptop will control the measurements; a proper interconnection between the three elements is mandatory. An appropriate software control was developed using LabVIEW, because of the advantages of this programming language for the control of different hardware components. In fact, one of the most important requirements of the CCD is the use of an USB connection, and the possibility of remote software control. In this same way, a precise control of the power supply is mandatory. Moreover, the uniqueness of our in-house developed software requires strict control of the times at which the power supply is turned on and the images are captured by the CCD. Aiming to this, an innovative forward biasing process has been developed.

3.2 Key aspects

The signal collected by the CCD will contain both the EL signal and the ambient noise, which needs to be filtered to extract the proper EL image. For instance, the use of adequate external filters can be beneficial for this purpose, such as band-pass filters around the NBE emission of Si (Fig. 1). However, software filtering methods are indispensable.

One of the main clues to perform good-quality EL measurements is to avoid direct light reflections from the panel, since this will increase the ambient noise substantially. For this reason, a proper orientation of the CCD with respect to the panel is very convenient.

Another key parameter to be controlled is the exposition time of the CCD. Large exposition times are beneficial, trying to collect as much EL signal coming from the Si panel as possible. However, the ambient light noise also grows linearly with the exposition time, Fig. 3. This should not necessarily be a problem if the filtering process is correctly developed. However, the increase of the exposition time has a limit, due to the saturation level of the CCD. This is especially critical for high irradiation conditions. In our approach, the selection of the exposition time is achieved automatically, in such a way that the saturation level of the CCD is never reached.

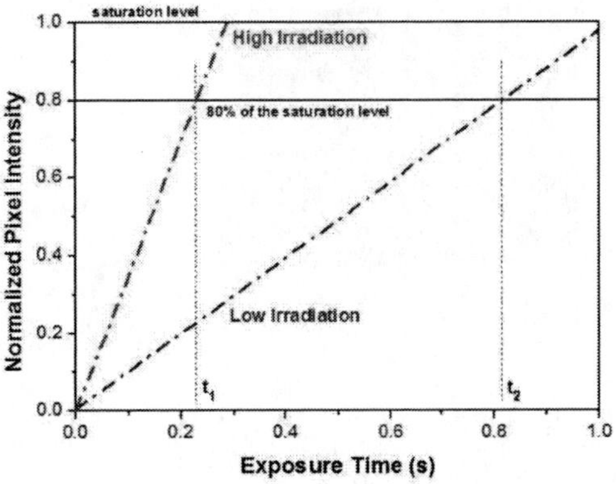

Figure 3: Increase of the total signal collected by the CCD as a function of the exposition time for two irradiation conditions. The saturation level of the CCD and a predetermined lower value of 80% of the CCD saturation level are indicated.

Two additional key points, which are always essential to conduct any modality of EL measurement, and especially relevant for large PV plant analysis, if one wants to carry as many EL measurements as possible, are both the way to polarize the panels (or strings) and the way to fix and control the CCD movements. These challenges should also be addressed for the case of daylight EL, with the advantage of being performed with solar light. On the contrary, some specific problems appear for the case of daylight EL, particularly for large irradiation conditions and hot days, such as those related to CCD heating, correct visualization in the screen of the laptop, etc.

3.3 Software development

Our software was developed in order to resolve the aforementioned problems, and to create a friendly and quick screening system. In particular, an automatic procedure was settled up to select the right exposition time avoiding CCD saturations. For this purpose, a live image of the modules is recorded, and the EL intensity histogram is evaluated. An automatic procedure allows selecting the largest exposition time avoiding the saturation of any CCD pixel.

Also critical, in certain cases, is the correct visualization of the EL images, which for sunny days can hardly be correctly visualized in the laptop. For this purpose, an automatic procedure has been developed to auto adjust the contrast of the images (Fig. 4). This allows, for instance, ensuring an optimum EL image, without the need to perform any post-procedure processing to improve the quality of the images.

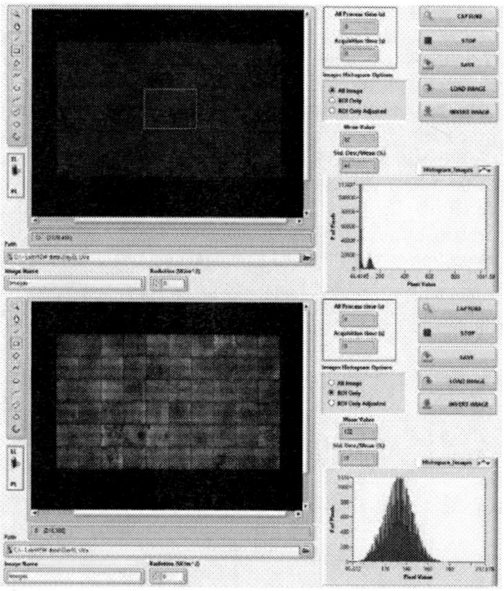

Figure 4: Automatic process developed to adjust the contrast of the EL images. Top: bare EL image; bottom: corrected EL image.

4 RESULTS

4.1 EL results

Fig. 5 shows the daylight EL image of a mc-Si test module obtained with our system on a sunny day at large irradiation conditions (G=850 W/m^2). As observed, the quality of the images is quite satisfactory, allowing to distinguish even small features and defects in the individual cells. The exposition time of the CCD was automatically fixed, being the total acquisition time (t_t) of this image 10 s. In order to assess the importance of the exposition time on the measurements, Fig. 5 also shows the EL image obtained on the same test module at very low-light conditions (G=70 W/m^2), without suitable exposition time adjustment (t_t=12 s). The saturation of some areas of the module leads to a much lower quality of the EL image.

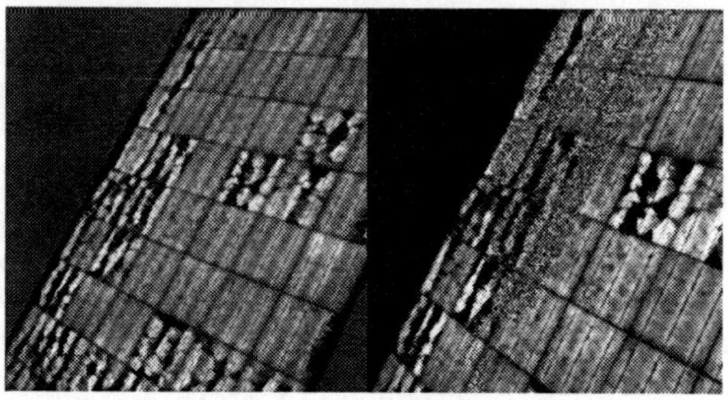

Figure 5: Right: daylight EL image on a mc-Si test module on a sunny day (G=850 W/m^2, t_t=10 s) by adjusting automatically the exposition time. Left: day light EL image on the same mc-Si test module at much lower irradiation conditions (G=70 W/m^2, t_t=12 s) with an incorrect exposition time.

As a proof of concept, this new tool has been tested on a PV rooftop (very low tilt angle) based on Cz-Si modules (Figs. 6 and 7). By biasing a complete string of 14 modules, the CCD was positioned in such a way to photograph six modules, as shown in Fig. 6. Despite the low focus angle, the quality of the EL image allows inactive-cell areas, finger interruptions and soldering issues be easily detected.

Figure 6: Top: Cz-Si PV rooftop (low tilt angle) surveyed with our system. Bottom: daylight EL image of a section of a 14 panels-string (G=650 W/m^2, t_t = 10 s).

As can be observed in Fig. 7, image resolution allows the minor and major EL defects such as metallic finger interruptions, cell breakage (dead regions), heterogeneous cell activity, soldering issues and cell inhomogeneities to be clearly detected and evaluated.

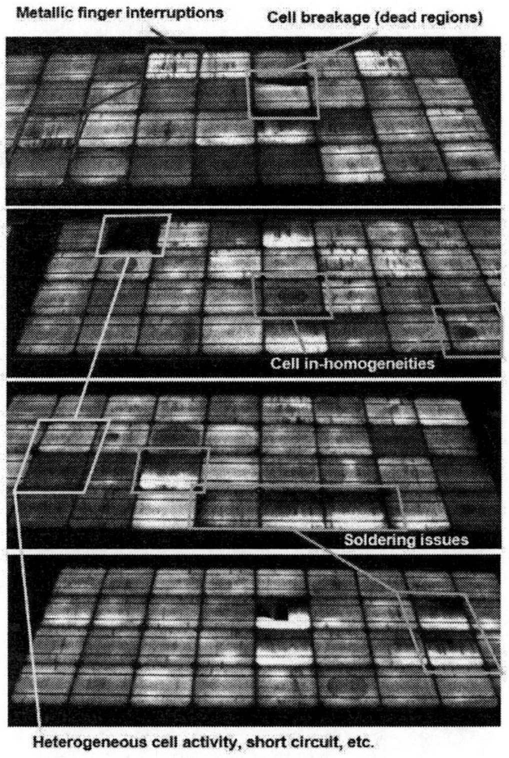

Figure 7: Daylight EL images of four individual panels of the first line of panels of the the Cz-Si PV rooftop tested. (650 W/m^2, t_t = 10 s for each image).

The present daylight EL approach has also been tested at high irradiation conditions. For instance, Fig. 8 shows a daylight EL image recorded on a cloudless, summer day in Madrid (27th of July 2017), at 13.00h (G=1050 W/m^2, t_t = 17 s). The image corresponds to a 60-cell, high-performance PERC module. The quality of the EL image (Fig. 8) is high enough to distinguish bulk defects, cell cracks, etc. on the individual solar cells.

4.1 PL results

The daylight EL system can be also adapted to allow PL imaging (PLi). In PLi, the panels do not need any external DC bias. In fact, the sunlight is used as the excitation source to create the corresponding e-h pairs, which the recombination generates the luminescence signal. This signal is also collected by the CCD, being conveniently filtered from the ambient noise by the above-mentioned software method.

Fig. 9 shows the PL image of the same previous PERC module. Again, the high-quality of the image allows the detection of major and minor defects in the panel. The advantages to obtain both the EL and PL images of a same module, or even the possibility to obtain only the PL images (without the need to use large power supplies to bias strings) makes of this method a very potential tool for new developments in the characterization and inspection of solar panels.

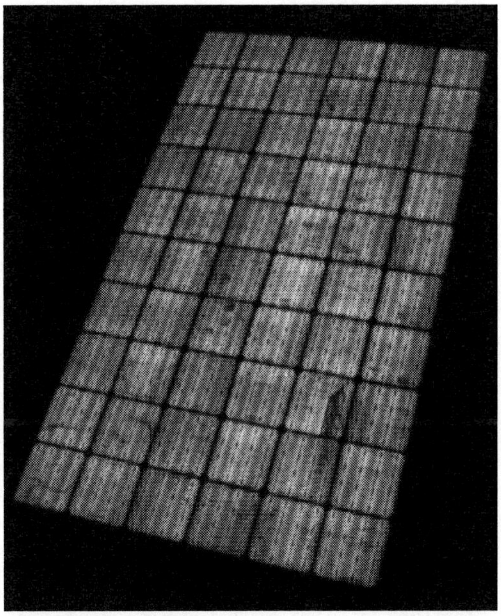

Figure 8: Daylight EL image of a high-performance PERC module (G=1050 W/m^2, t_t = 17 s).

Figure 9: Daylight PL image of a high-performance PERC module (G=1050 W/m^2, t_t = 17 s).

5 CONCLUSIONS

A friendly and cost-effective system for daylight EL testing has been developed. Good-quality and quick EL images of c-Si modules, even under high irradiance conditions, can be obtained. The system also allows recording PL signals, with no need to polarize modules.
The advantage of this tool is the easiness and high speed of the EL measurement process, given the current limitations of this valuable defect screening technique

when performing EL test on-site.

6 REFERENCES

[1] T. Fuyuki, H. Kondo, T. Yamazaki, Y. Takahashi, Y. Uraoka, "Photographic surveying of minority carrier diffusion length in polycrystalline silicon solar cells by electroluminescence", Appl. Phys. Lett. 86, 2005, 262108.

[2] O. Breitenstein, J. Bauer, K. Bothe, D. Hinken, J. Müller, W. Kwapil, M.C. Schubert, W. Warta, "Can Luminescence Imaging Replace Lock-in Thermography on Solar Cells?", IEEE J. of Photovolatics, 1 (2011) 159-157.

[3] S. Koch, T. Weber, C. Sobottka, A. Fladung, P. Clemens, J. Berghold, "Outdoor electroluminescence imaging of crystalline photovoltaic modules: comparative study between manual ground-level inspections and drone-based aerial surveys", 32th EU PVSEC, Munich 2016, pp. 1736-1740.

[4] M. Navarrete, L. Perez, F. Domínguez, G. Castillo, R. Gomez, M. Martinez, J. Coello, V. Parra, "On-Site Inspection of PV Modules Using an Internationally Accredited PV Mobile Lab: A Three-Year Experience Operating Worldwide", 31st EUPVSEC, Hamburg 2015, pp.1989-1991.

[5] L. Stoicescu, M. Reuter, J.H. Werner, "DaySy: Luminescence Imaging of PV Modules in Daylight; 29th EUPVSEC, Amsterdam 2014, pp. 2553-2554.

[6] M. Köntges, I. Kunze, S. Kajari-Schröder, X. Breitenmoser, B. Bjørneklett, "Quantifying the risk of power loss in PV modules due to micro cracks". 25th EUPVSEC, Valencia, 2010, pp. 3745–3752.

[7] T. Potthoff, K. Bothe, U. Eitner, D. Hinken, M. Kontges, "Detection of the voltage distribution in photovoltaic modules by electroluminescence image". Progress in Photovoltaics: Research and Applications 18, 2010, 100.

7 ACKNOWLEDGEMENTS

This investigation has been financed by the Spanish MINECO/CDTI project, ref. IDI-20151194, the Spanish MINECO project, ref. ENE2014-56069-C4-4-R and the Junta de Castilla y León project, ref. CP16/58.

DIGITAL PLANT LIFECYCLE RECORD – A NEW STANDARD FOR EFFICIENT PV O&M

Johannes Schmidt, Christian Bertsch-Engel
Institute for Applied Informatics (InfAI) / CEE Operations GmbH
Leipzig / Hamburg, Germany
schmidt@infai.org / bertsch@cee-group.de

ABSTRACT: Information management for PV power plants or renewable energy power plants in general is challenging. A holistic approach is required to ensure comprehensibility and interoperability of plant-related information. In November 2017, the German Institute for Standardization (DIN) is going to publish the new standard DIN 77005-1 specifying general requirements for plant lifecycle records (PLR). The application of this standard using a digital PLR provides significant opportunities for cost reductions in photovoltaics operation and maintenance (O&M) as well as asset management services.
Keywords: Information and Communication Technologies, Lifetime, Lifecycle Record, Standardization

1 INTRODUCTION

With system prices consistently getting cheaper, the proportion of operational expenditures (OPEX) gains significance in a levelized cost of electricity (LCOE) assessment of PV power plants. O&M and asset management services make up the largest part of a PV plant's OPEX structure. As a result, improving the efficiency of O&M and asset management services is a key pathway to ensure future LCOE reductions of PV. The crucial success factor for this pathway is the availability of profound and consistent knowledge concerning the power plant. Of particular interests are the plant structure, O&M processes, and any plant-related information including operational data. Standards and technical guidelines ensure a common understanding of this information, but as of today, no consistent and holistic information management approach exists. The new German standard DIN 77005-1 [1] for lifecycle records will address this gap. It provides general guidelines to manage plant-related information for technical objects.

In this paper, we give an overview of the upcoming draft of this standard including some remarks on its application for PV power plants. The remaining of this paper is structured as follows: Section 2 gives an overview about the current challenges in the lifecycle of a PV power plant. Section 3 introduces the Plant Lifecycle Record (PLR) approach including general normative foundations, its information model, and specific concepts like views and guidelines. Section 4 gives an overview about the standardization of the PLR with respect to DIN SPEC 91303 and the upcoming standard DIN 77005-1. Section 5 summarizes the application of the PLR for PV power plants and discusses upcoming standardization activities.

2 CHALLENGES

Parties involved in a PV power plant project (such as owners, EPC personnel or service providers) should be able to access plant-related information directly and appropriately to prepare, provision, and complete their specific tasks and services. In the wind energy domain, the management of a so-called Plant Lifecycle Record (PLR) is an integral component of the service portfolios of operators and manufacturers. This concept is borrowed from the space and aviation sector. For PV power plants however, no such general approach exists. There is no common understanding on the structure of such a PLR nor on its content or usage. Therefore, the quality of the comprising information is difficult to compare and the availability of information depends on the concrete responsible parties. Furthermore, the lack of common data formats and semantics hamper a close business-to-business integration, which is essential for automated and efficient process execution and cost reductions.

To our experiences, most of the parties spanning the entire lifecycle of technical assets and power plants are mainly dissatisfied with the quality and availability of information regarding the technical as well as operational documentation. The reasons for this are missing information from previous lifecycle phases, the lack of knowledge about completeness and consistency rules, or semantic and technical heterogeneity including data and document formats. The PLR approach addresses these challenges.

In a nutshell, the PLR contains the overall documentation of a technical object in chronological order including e.g. the technical documentation, the history of events and failures, the listing of all structural changes, as well as quality certificates and technical reports. To depict the problem scope, Figure 1 summarizes the lifecycle of a PV power plant focusing on the information flow. The lifecycle consist of a sequence of phases. Today, no common classification of that lifecycle exists. In other domains, like the chemical industry, these phases are standardized which facilitates the assignment of information related to these phases. In each phase, different stakeholders can be involved, cooperating within one or more phases or at the transition point of two phases (e.g. the commissioning of the PV plant). Some stakeholders are only involved in early phases; others participate in almost every phase. These stakeholders perform complex business processes that include business-to-business cooperation. Most of those tasks are knowledge-intensive, because a detailed and actual knowledge concerning the structure and state of the plant is needed. This information is mostly contained in documents and passed along the lifecycle. These documents relate to one another in different ways (e.g. "depends on", "based on", "relates to", or "defined by" relationships). In summary, this means every member of the value-added network for the PV power plant may lay down different requirements for the structure and overall content of the plant documentation, because they are responsible for different remits. To meet their specific information demands, the general interrelations between the actors, business processes, and information must be

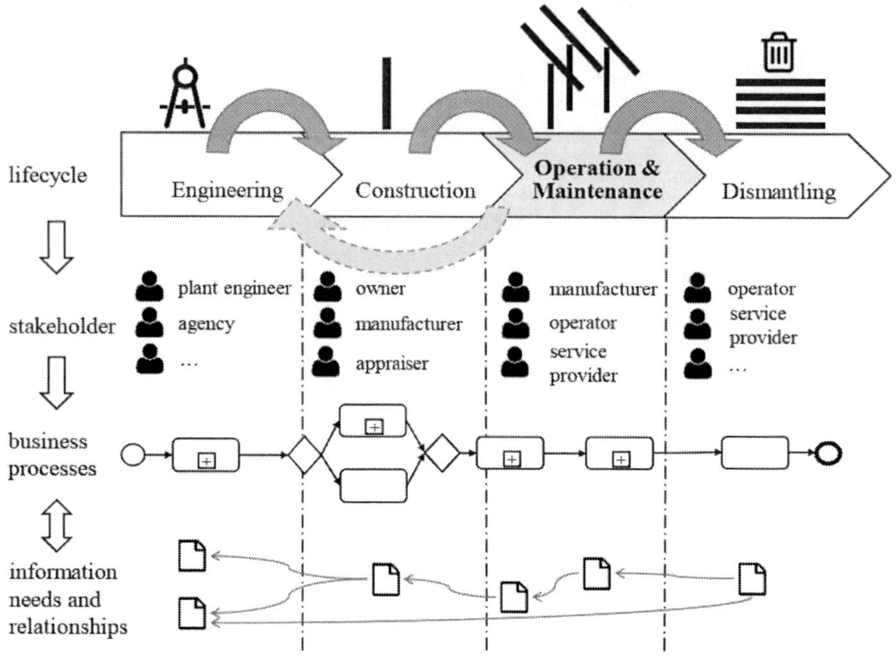

Figure 1: Lifecycle-oriented information management for PV power plants

known (as shown in Figure 1). However, today, there often are information losses at the transition point of two phases. One reason is that the information receiver is not aware of the information available in the preceding phases. On the other hand, information providers are not interested to pass every information, because of economic reasons or of corporate secrets.

Besides these challenges with regard to the availability of information, the proper handover of this information can be challenging, too. Today, the format and informative value of the delivered information is often negotiated individually. This results in different qualities as well as structures of documentations in practice. Furthermore, a return flow of information from e.g. the operation phase is desired to analyse the reliability of equipments or to support the development of successor models. However, no formal definition of such information flows exists. A structured and standardized approach for a PLR can help to improve these issues.

3 PLANT LIFECYCLE RECORD

3.1 Overview

To improve the availability and quality of plant-related information and provide a solid basis for analysis, reports, and decision support along the plant lifecycle, the sole introduction of a common information structure is not sufficient. Moreover, a holistic management approach is needed that addresses structural, semantic, and organizational aspects. Figure 2 shows a schematic visualization of a framework towards a systematic PLR. There are five successive layers that are defined as follows.

In every lifecycle phase, several legal and normative basics have to be met. Depending on the particular view, a PV power plant is a machine, a construction, a place of work, or an electrical installation. According to these aspects, individual standards and legal requirements have to be fulfilled in every lifecycle phase. The PLR shall help to manage these requirements and track the verifying documentation.

Building on the general legal and normative context, specific management approaches shall be installed that help to continuously improve the management of processes, technical assets [2], or quality aspects in general [3]. In some cases, national or international laws (e.g. environment management) require the application of management systems. The management layer of the framework provides general requirements for organizations to fulfil all necessary conditions.

The information layer deals with general requirements and approaches for structuring and managing plant-related information. To link information with the plant as well as its equipments and parts, a reference designation system shall be used to describe the hierarchical structure of the power plant. Furthermore, general standards for document management and structuring of documentations as well as information models for operational data shall be applied. The sub-layers document, plant structure, and data are extended by definitions and constraints defined in the information layer itself.

A PV power plant is designed for at least a lifetime of 20 years. To manage every lifetime information appropriately, methods and concepts are needed to build cluster of information with common reference entities and clear relevance. The previous information layer is responsible for the management of different types of information. At the linking and views layer, individual documents and data can be annotated to define views according to e.g. single stakeholder information needs or to link information to ensure efficient navigation within the PLR.

To adopt a PLR according to the specific type of plant or the specific need of the owner, guidelines and specifications can be applied. These specifications relate

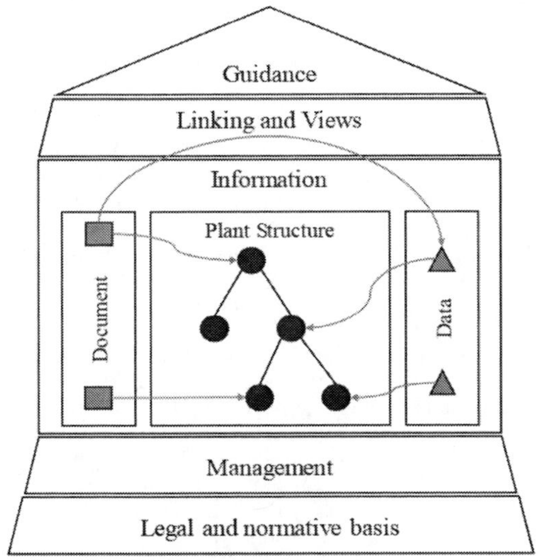

Figure 2: The PLR framework

to e.g. the definition of metadata or domain-specific interrelations of information types.

The following chapters explain in more detail the distinctive layers of the PLR approach.

2.2 Legal and Normative Basis

Throughout the entire lifecycle, a large number of legal and normative requirements and constraints must be observed. The manufacturer is responsible for state-of-the art design of the modules including the certification of equipments. The operator is responsible for the secure and economic operation of the power plant. The operation processes must always comply with the applicable laws and heed machine responsibilities or specific regulations for feed-in remuneration. Therefore, the operator has to prove the compliance with every operator obligations.

In this paper, the O&M phase is of most interest. It is the longest period in the plant lifecycle. The current state of the art should be regularly adapted, including definitions for technical services, maintenance processes [4], or general terms and definitions for maintenance and operation [5, 6].

2.3 Management

The safe and efficient operation of a PV power plant requires the implementation of, compliance with and continuous improvement of operational and management processes. These include, in the O&M's perspective, processes for maintenance, failure and accident management as well as processes for risk assessment, analysis and mitigation. The PLR enables the linking of management processes with individual components, plant parts or the entire plant.

Another aspect of this layer is the general management of the PLR and its information contained. A methodology is needed to open, access, update, deliver, integrate and close these records. A standardized methodology can ensure interoperability and comprehensibility of PLRs.

2.4 Information

The PLR provides common structures to manage plant-related information. The information layer in Figure 2 is divided into three sub-layers. The central layer is the plant structure. Every information must be assignable to specific elements of the PV plant. To ensure interoperability, a standardized reference designation system shall be used. Corresponding application guidelines are managed in the guidance layer.

RDS-PP® (Reference Designation System for Power Plants) is a modern reference designation system that defines a common language to designate systems and objects. It builds upon general structuring principles and reference designation rules described in IEC 81346 [7]. For wind turbines, an application guideline exists [8]. Today, such guidelines are missing for PV power plants, but a VGB PowerTech working group currently works on such a guideline for PV [9]. By using uniform, semantically defined and language-independent codes, each plant element can be uniquely referenced. Furthermore, information can be assigned to types of products or system elements to ensure durable and stable references. So, a differentiation between product, equipment, or system information is possible. The standard supports three aspects of a plant structure: the function designation, the product designation, and the location designation. These aspects are linked by transitions. A transition is used to assign e.g. a product class to a functional element. For example, a specific inverter can fulfil the function "Converting electric energy while retaining the energy type and changing the energy form". The application of structuring principles like RDS-PP® ensure semantic understandability of the plant structure and provides a starting point for information and data linking. A common understanding of information assignments is especially important, if different stakeholders will share plant-related information with each other.

Most plant-related information are structured as documents. IEC 61355-1 [10] describes a classification scheme and a designation system for documents. It is possible to distinguish different document types and define specific meta-data or management guidelines. The designation system for documents can be combined with reference designations.

Due to increasing digitalization, the importance of operational data increases. They form the basis for systematic failure analyses, anomaly detection or performance ratings. We recommend handling (selected) operational and master data in the PLR. Modern information models, like IEC 61850 [11], are object-oriented and specify the semantic of the data. Furthermore, these objects can be aligned with concepts of the PLR.

2.5 Linking and Views

The linking and views layer focuses on the relationships between the information in the PLR and their relevance for different project participants. These relations can be defined manually or are automatically derived from predefined specifications. Through the linkage of information, the overall value of the PLR increases.

Each stakeholder in a PV power plant project has its specific information requirements or provides specific information using particular terminologies and notations. For example, an employee of the maintenance team has a different information demand than an employee of the legal department or controlling division (see also [12]). In order to meet these requirements, the PLR supports five different views: the technical, process-related, legal, economic, and material view. It is possible to define custom views, if needed. Applying a view to information in the PLR means, that each stakeholder can annotate this information using specific taxonomies and classifications, providing specific contexts, or adding metrics. These annotations can help to index, query, interpret and validate the information.

2.6 Guidance

The information layer provides concepts to manage and link information in the PLR. The concrete definition of these relations and structures may be individually specified in the guidance layer. These specifications shall be based on standards.

Maintenance processes in general are well defined. DIN SPEC 91310 [13] considers maintenance of renewable power plants as a service. In theory, information models like IEC 61850 [11] ensure technical interoperability of operational data, but this standard is not common in practice. Finally, yet importantly, the efficient management of plant documentation is challenging. Only IEC 62446-1 defines documentation requirements for PV [14]. But, in general, there are many standards available that address document-related topics. In the guidance layer, one can specify, which standards are applied and which specific adoptions have to be observed.

2.7 Information model

To summarize the PLR framework, Figure 3 shows an information model that describes the most important concepts of the lifecycle record and their relations. To abstract from concrete data or documents, the concept of documented information is introduced, which is also used in management standards by the International Organization for Standardization (ISO). This concept provides meta-data, which apply to every specific type of information. To handle a related information, so called information sets are introduced. A lifecycle record contains these information sets and subordinate records. This results in a hierarchical recursive structure, which can be adapted flexibly and appropriately according to the complexity of the technical object. It is possible to manage a hierarchy of lifecycle records, that replicate the plant structure or one can define custom rules for this structure.

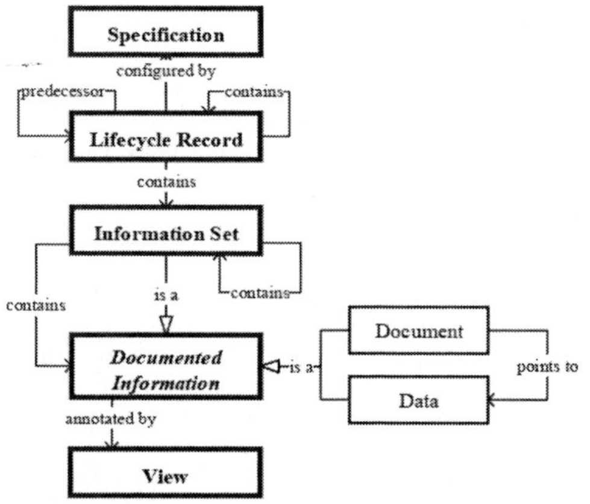

Figure 3: Except of the information model for a PLR

The concept of specification in Figure 3 represents other concepts, which are needed to define specifications for the PLR, like roles, laws, tasks, processes, the legal context and so on. This information forms the basis for view-based annotations.

3 STANDARDIZATION

In 2015, a DIN SPEC (technical specification) was released which includes a definition of the PLR term as well as a description of its common structure [5]. Based on this specification, a standardization working group started to work on a national standard for lifecycle records for technical objects. The scope of the standard covers facilities, power plants as well as machines or chemical plants. Representatives of PV operation companies were involved in this working group. In November 2017, the draft of the standard DIN 77005-1 will be available for public commenting.

To ensure a high applicability and compliance of the PLR, a wide range of standards are referenced and recommended. This includes standards for reference designation systems [7], document types [10], information models of operational data [11], the management of documents and records [17, 18], the linkage of information, or the usage of main documents [15, 16]. All these approaches are highly relevant for renewable energy power plants. The application of these standards result in an improved data exchange and reduces semantic inconsistencies. In total, a harmonized standard framework ensures improvements in business collaboration as well as reduction of process costs. Without these unification processes, operators of PV power plants will not be able to ensure an economic plant operation in the future. To realise these potentials, a specification for a digital PLR shall be developed to ensure interoperability between the information systems of the stakeholders. Using a common data exchange format, information losses and manual transfer of data can be prevented. Furthermore, the business-to-business integration is easier to manage and automate.

5 SUMMARY

Operators of PV power plants have to cope with several challenges including the management of and compliance with laws, economic objectives, the coordination of every involved party and many more. Today, no common approach is known for PV plants to manage plant-related information along the plant lifecycle. The upcoming standard DIN 77005-1 describes a holistic framework to introduce and manage lifecycle records for technical objects. It can be applied for the entire plant as well as for specific elements, components, or parts. However, this presupposes the application of basic standards for e.g. reference designation systems, document management or document types. In 2018, RDS-PP® application guidelines for PV power plants will be available. This standard can be easily combined with IEC 61355-1 to structure documentations using document types. To reduce redundant information sets, concepts like main documents are applicable, too.

The normative specifications in DIN 77005-1 are well suited for PV power plants. This standard helps every owner or operator to improve their processes according to a more traceable, automated, and structured approach. The normative part is detailed, but leaves space and flexibility to easily adopt the PLR for specific requirements whilst remaining interoperability and functionality for exchange.

REFERENCES

[1] DIN, *Lifecycle record of technical objects — Part 1: Terms and structure*, DIN 77005-1, 2017.

[2] ISO, *Asset management - Overview, principles and terminology*, ISO 55000, 2014.

[3] ISO, *Quality management systems - Requirements*, ISO 9001, 2015.

[4] CEN, *Maintenance process and associated indicators*, prEN 17007, 2016.

[5] DIN, *Fundamentals of maintenance*, DIN 31051.

[6] CEN, *Maintenance - Maintenance terminology*, prEN 13306, 2015.

[7] IEC, *Industrial systems, installations and equipment and industrial products - Structuring principles and reference designations - Part 1: Basic rules*, IEC 81346-1, 2009.

[8] VGB, *RDS-PP Application Guideline Part 32: Wind Power Plants*, VGB-S-823-32-2014-03-EN-DE, 2014.

[9] VGB, *RDS-PP Application Guideline Part 33: Photovoltaic Power Plants*, VGB-S-823-33, 2018.

[10] IEC, *Classification and designation of documents for plants, systems and equipment - Part 1: Rules and classification tables*, IEC 61355-1, 2008.

[11] IEC, *Communication networks and systems for power utility automation - ALL PARTS*, IEC 61850, 2017.

[12] C. Zinke, J. Schmidt, und A. Nareike, "Supporting Knowledge-Centered Business Collaboration for Wind Power Plants," in *IFIP Advances in Information and Communication Technology, Risks and Resilience of Collaborative Networks*, L. M. Camarinha-Matos, F. Bénaben, and W. Picard, Hrsg., Cham: Springer International Publishing, ISBN: 978-3-319-24140-1, 2015, DOI: 10.1007/978-3-319-24141-8_23, S. 255–262.

[13] DIN, *Classification of services for the technical management of renewable energy plants*, DIN SPEC 91310, 2014.

[14] IEC, *Photovoltaic (PV) systems - Requirements for testing, documentation and maintenance - Part 1: Grid connected systems - Documentation, commissioning tests and inspection*, IEC 62446-1, 2016.

[15] ISO International Organization for Standardization, *Technical product documentation - Use of main documents*, ISO 11005, Norm, 2010.

[16] IEC, *Structuring of technical information and documentation*, IEC 62023, 2012.

[17] ISO, *Information and documentation - Records management - Part 1: Concepts and principles*, ISO 15489-1, 2016.

[18] IEC, *Document management - Part 1: Principles and methods*, IEC 82045-1, 2001.

NEC2017 RAPID SHUTDOWN: USEFUL SAFETY FEATURE OR UNNECESSARY NUISANCE?

Urs Muntwyler, Daniel Gfeller, Jonas Wälten, Christian Renken, Martin Rutschi
Berne University of Applied Sciences (BFH), Engineering and Information Technology
Photovoltaic Laboratory (PV-Lab), Jlcoweg 1, CH-3400 Burgdorf, Switzerland
Phone +41 (0)34 426 68 11, urs.muntwyler@bfh.ch, www.pvtest.ch

ABSTRACT: The latest standard for safe installation of electrical wiring and equipment in the United States, the National Electrical Code 2017 (NEC2017 or NFPA 70), contains a requirement for a rapid shutdown functionality for photovoltaic installations on buildings. 30 seconds after activation of this feature, all voltages inside the installation must be reduced to non-hazardous levels. In practice this means, that the entire PV plant must be shut down on module level or an all-pole disconnection of all modules must be performed. The aim of this feature is the protection of the emergency responders in case of fire. Even though accidents caused by PV installations in firefighting operations are rare [1], the solar generator with its high voltage and the in principle non-existent ability for shutdown must be considered a potential danger. The protection of first responders is an important topic and should not be taken lightly by the PV industry. Under this point of view, the shutdown on module level is a very good solution and offers a maximum of safety. However, there is a risk that by solving this single problem, a whole bunch of new problems and even new sources of danger can be generated. In this contribution, pros and cons as well as the feasibility of such a shutdown feature are being discussed.
Keywords: Safety, System Performance, Rapid Shutdown, Fire Hazzard

1 NEC2017 RAPID SHUTDOWN

The National Electrical Code (NEC) is the most important standard for wiring and equipment of low voltage installations in the United States. Despite the term "National", the NEC is not a federal law. In fact, the jurisdiction of electrical installations lies with the individual states. Some states adopt the NEC directly, some adopt it in part and other states have their own regulations. But basically, the contents of the NEC can be considered a rough guideline for normative requirements of electrical installations in the USA. The NEC is released in a new edition every third year.

With the NEC2017, a new shutdown function for PV plants on buildings is introduced (NEC2017 690.12 Rapid Shutdown). This function is invoked by a manual switch, which must be placed on a well accessible location, namely near the buildings entrance. After activation, the rapid shutdown has a time limit of 30 seconds to bring the PV plant into a safe state. After this time, in the area of the solar generator, a maximum voltage of 80V between any two points of the PV plant or between any point of the PV plant and earth must no longer be exceeded. In every other part of the PV plant – e.g. at the connections to the inverter – the maximum voltage is 30V (see figure 1).

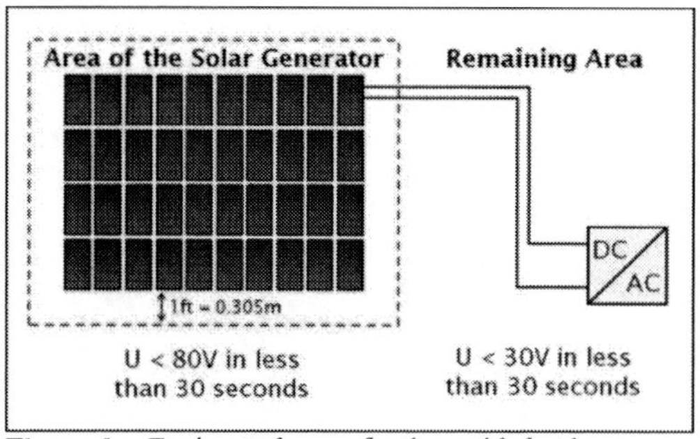

Figure 1: Designated areas for the rapid shutdown

De facto, this requirement can only be met by shutting down the PV array on module level or by disconnecting the strings between all modules. Moreover, power lines to the inverter must be disconnected as well and the inverter itself must guarantee that its input capacitors are sufficiently discharged inside the time limit. The aim of this feature is the protection of first responders, e.g. in case of fire.

2 PHOTOVOLTAICS AND FIREFIGHTERS

About six years ago, there has been a considerable increase in claims saying that PV plants pose a serious threat for firefighters. For example, there has been a report about a firefighter who had to be treated in a hospital after supposedly suffering an electrical shock from a PV plant. This was – to use the current buzzword – fake news. It is true that a firefighter has been injured by an electric shock. However, on the respective building, there was no PV installation at all, but rather a solar thermal installation. Obviously, the shock was caused by a regular low voltage line. Other reports saying that the fire brigade had to let buildings burn down because the use of water had been made impossible by PV installations on the roof have also been debunked hoax. False reports like these and the coverage by the media have caused a real hype that cost the PV industry a lot of its good image. But it did also lead to the PV industry making a realistic assessment of this issue. As a result, several investigations have been carried out [2]. Among other things, these investigations showed that sprinkling water on PV plants is harmless, if a safe minimum distance is being kept (according to DIN VDE 0132 five meters with full jet, one meter with water spray). A potential hazard is the solar generator, especially because it cannot simply be turned off. However, this is only a problem if the solar modules or the wiring are being touched directly and if there is at least one insulation fault. Moreover, the solar generator increases the roof load, which increases the danger of collapse.

In a building fire, highly toxic smoke and gases are formed. This smoke can reach temperature of 600°C and more and is so toxic that one lungful of it can be lethal. Moreover, the smoke worsens the view conditions which hinders the work of the firefighters. Finally, if temperature and concentration of smoke and gases reach a critical level, they can ignite. The result is a flameover with temperature of more than 1'000°C. Along with the high risk for persons, a flameover often leads to a fully developed fire which usually results in a total loss of the building. To prevent this, the fire brigade normally tries to ensure a good ventilation of the building, allowing the smoke to dissipate. With small buildings, this is normally done by opening or smashing the roof windows. Large buildings are often equipped with smoke flaps, which open automatically in case of fire alarm. In the USA, it is a common practice to use a chainsaw for cutting a hole in the roof. With the far more massive European construction methods this is not possible in most cases, simply because of the thick roof isolation.

To enable the ventilation, the firefighters often have to enter the roof. At this moment, a direct contact with the potentially dangerous solar generator is possible. Fire brigades are normally trained neither to touch nor to step onto the solar generator. However, the solar generator might be an obstacle for the firefighters (see figure 2). Moreover, the fire brigade must work under severe time pressure. Therefore, it can be assumed that there will be sometimes a direct contact with the solar generator, even if it is against the rules. In most cases this is not a problem as all components of the solar generator are insulated and are highly robust against mechanical damage (many solar modules can even be stepped on without damage). The situation looks different though if the solar generator has already been damaged by the fire. However, with a building fire so far advanced, there will hardly be any firefighter on the roof anymore because at this stage of the fire, it is too late for venting and the roof is already near collapsing.

Figure 2: In some cases, the direct contact with the PV plant is inevitable (image source: M. Rutschi, fire brigade of the city of Burgdorf)

3 SAFETY REQUIREMENTS

A shutdown feature such as the NEC 2017 rapid shutdown is a safety-related application, which in some cases might make the difference between life and death. With a PV plant, a near maintenance-free operation over a lifespan of 20 to 30 years is expected. Thus, the rapid shutdown must work properly even after years or even decades of inactivity. This poses extreme reliability requirements. In most cases, the shutdown feature will be realized with electronic circuits, which are being placed in close vicinity of the solar modules or even integrated into the module's junction box. The environmental conditions in this area are very harsh. For example, the temperature inside the junction box at partial shading (when the bypass diodes are active) can exceed 100°C. In the winter on the other hand, night temperatures below zero are possible. These extreme temperature cycles are being accompanied by humidity fluctuations, which from time to time can lead to condensation. Electronic circuits with the ability to withstand such conditions with a lifespan of 30 years or more must be carefully encapsulated and must not have any weaknesses in design or choice of components [1]. However, from a statistical point of view, an increase of complexity (which is inevitable with the shutdown feature) always has a negative impact on the failure rate, because every additional component is also a potential source of failure. To guarantee the required lifespan, careful failure rate analyses with empirical models and extensive investigations with methods for accelerated aging (e.g. temperature cycle tests in climate chambers) must be performed.

The shutdown feature has to work properly, even if the PV plant is already damaged (e.g. by the fire itself). Therefore, it must be designed according to the "fail safely"-principle. This means, that the system must be in a safe state in case of an error. In case of the NEC2017 rapid shutdown, the shutdown function must be activated if for example the connections between the manual switch (at the entrance of the building) and the devices for string disconnection (on the roof) are being interrupted. It is however very important, that the operational safety of the PV plant is not being corrupted by this. Because if the PV plant shuts down unnecessarily too often, the acceptance for the shutdown feature will be lost (as is the case with a fuse that always triggers and that at some point is simply being bypassed).

4 CONFORMITY WITH STANDARDS AND REGULATIONS

The normative requirements for the Rapid Shutdown are regulated in UL 1741. Even though the UL listing can additionally contain the title "High Reliability", the authors regard the existing requirements concerning safety and reliability to be insufficient for an application which can make the difference between life and death. However, other safety-related components of PV plants are already subject to strict standards. Solar modules for example are tested according to IEC 61215. There are also several normative requirements for the inverter. An example is the anti-islanding circuitry as it is described in VDE-AR-N 4105. This circuitry has to check the parameters of the power grid (voltage, frequency, impedance) continuously and, if limits are exceeded, disconnect the inverter from the grid. This function ensures that in case of a shutdown of the power grid there are no more PV plants injecting power into the grid, allowing maintenance work on the grid without the risk of an electrical shock. This is also a safety-relevant shutdown feature. Therefore, some of its requirements could be adopted to the NEC2017 rapid shutdown. This would mean that the rapid shutdown must be realised with double redundancy. To ensure the safety, always two separating contacts would have to be used in series. At least one of these contacts would need to be an electromechanical switch (not a semiconductor). If the rapid shutdown had the same safety requirements as the anti-islanding circuitry (and there is no reason why it shouldn't), the impact on the design would be extensive. Especially the electromechanical switch will be a problem, because it must be able to disconnect not only a single module, but the entire string (otherwise, a premature opening of a single switch will result in an electrical arc that would pose a serious fire hazard and a conflict with the fail-safe requirement). So, we are talking about a high-quality disconnection switch which is able to turn off a current of 10A at a DC voltage of up to 1'000V. And each solar module must be equipped with such a switch. With that, the NEC2017 rapid shutdown would become an intolerable cost factor.

If the firefighters have to manipulate on the solar generator (e.g. putting away some modules), they are working on an electrical installation. Aside from the fact that work like this should only be done by qualified electricians, there are some basic safety rules. In the German speaking countries, the following five rules apply (sometimes in slightly altered form):

1. Disconnecting
2. Preventing of unintentional restarting
3. Check for zero-voltage
4. Grounding and short-circuit
5. Covering of adjacent electrical circuits

Even though the work of firefighters is an emergency, their self-protection has priority. Therefore, the safety rules should be respected by the firefighters as well. Rule number one is covered by the NEC2017 rapid shutdown. Rule number two can easily be achieved by use of lockable switches. The problems start with rule number three. Firefighters are not generally qualified to perform electrical measurements in the field. And even if they were, the time pressure in an emergency and the unwieldy protection clothing would make reliable measurements nearly impossible. Rule number four would be even more complicated than rule number three. Rule number five should not be relevant in most cases. In principle, we have to assume, that the fire brigade has no means to check if the shutdown of the PV plant was successful. In this case, the plant has to be considered as "under voltage" and no manipulations on the solar generator can be performed. Therefore, the PV plant could just as well have been left turned on. The conclusion is that if basic safety rules are not being disregarded, the NEC2017 rapid shutdown does not pose any help to the firefighters.

5 POSSIBLE CREATION OF NEW PROBLEMS

A new feature like the NEC2017 rapid shutdown can always lead to new problems. Some of them have already be showed at the assessment of the safety required. From the author's point of view, there is a great risk. It can be assumed that the fire brigade would rely on the shutdown function. However, if a shutdown feature is present but not properly working, this could be disastrous, because in an emergency, even a failed attempt to shutdown would give the firefighters a false sense of safety. Automatically, this would lead to a careless handling with the solar generator which would still be under voltage. The risk of accidents would be enormous. It is therefore the author's conclusion, that if the reliable functionality of the NEC2017 rapid shutdown cannot be guaranteed, it would be better not to implement it at all.

We already mentioned that a safe disconnection of electrical contacts can only be realised with an electromechanical switch. This switch must be able to carry the string current over decades and in an emergency to turn of this current safely. If this switch fails – be it because of a too high contact resistance under normal operation or an unsuccessful disconnection in an emergency – an electric arc might arise, which represents a risk of fire. As there must be at least one such switch per module, even a medium sized PV plant will contain hundreds of switches. If only one switch fails, then the NEC2017 rapid shutdown, which should increase the safety, becomes a risk itself. Here also, the authors share the opinion that in case of doubt – if the quality cannot be guaranteed - it would be wiser not to implement the shutdown feature at all.

6 USE OF MICROINVERTERS AND POWER OPTIMIZERS

Next to classic topologies with string or central inverters, in the past few years many new PV plants with microinverters or power optimizers have been developed. With such systems, the solar modules are no longer connected in series directly. Instead, each module has its own DC/AC converter (microinverter) or DC/DC converter (power optimizer), which performs a separate maximum power point tracking on module level. Because of their topology, such devices are usually able to make their output voltage-free and they usually have means to communicate with each other or a central gateway. A shutdown feature like the NEC2017 rapid shutdown is therefore already implemented in such devices or can easily be upgraded (e.g. by a firmware update). With microinverters it is even enough to disconnect the devices from the power grid, which makes them turn off (as an additional safety though, the disconnected AC lines should be grounded). Still, some considerations concerning reliability and fail-safe behaviour must be done, before these devices are being entrusted with this safety-related function.

As far as the authors know, microinverters and power optimizers are currently the only cost-efficient systems to realize the NEC2017 rapid shutdown. It can be assumed that with the NEC2017 coming into effect, several manufacturers will bring their own products on the market. But until then, microinverters and power optimizers are de facto declared the market standard in those states where the NEC2017 is already in effect. Surely, manufacturers of such devices are glad about this development. However, speculating about in what way the publication of the rapid shutdown in the NEC2017 is attributed to lobbying shall not be part of this scientific publication. The authors would rather point out that microinverters and power optimizers do not only have benefits, but also disadvantages compared to classic PV topologies. The efficiency of such systems is usually lower but the costs are higher. Therefore, they are energetically and ecologically only viable in PV plants with strong partial shading. Moreover, they still have to prove their reliability in long-term field operation.

7 RAPID SHUTDOWN IN EUROPE?

In this section, we look into the topic, if a rapid shutdown like the one in NEC2017 would be reasonable and useful in Europe. Currently, most European PV plants are not even equipped with a string-level shutdown (better known as firefighter switch). To make an assessment about the benefits of such a feature, the differences to the USA have to be investigated. As we already mentioned, the common practice in the USA is to cut a hole in the roof using a chainsaw. This is possible, because many residential buildings in the USA have a wooden roof construction which is covered with roofing shingles. In Europe, most roofs are covered with tiles that cannot simply be cut with a chainsaw. Below the roof, there is usually a layer of stone wool or a similar isolation material which is also an obstacle. Therefore, this common practice from the USA is almost never applied in Europe. As a result, a PV plant does not pose an obstacle in this context. Still, there are some situations where the firefighters have to enter the roof. Even though the fire brigades are trained to steer clear of the solar generator, it would be beneficial if the PV plant could be simply brought into a safe state. It is however questionable, if this additional benefit is enough to justify the additional complexity and costs of the shutdown feature. Fact is, that because of the risk of collapsing, the roof is only entered, if the fire is not yet very far advanced. At this time, the solar generator is still intact and has no insulations faults. It becomes however critical, if for some reason the solar modules have to be put aside. But these are exceptional circumstances. Even though there are PV plants with a combined nominal power of around 100GW in Europe already, no heavy accidents of firefighters because of these PV plants have happened so far (as far as the authors know). The potential danger of a solar generator seems therefore be moderate. Thus, a rapid shutdown feature would be nice to have, but not a necessity.

8 ALTERNATIVE APPROACHES

In this section, we look into how the risk for the first responders could be avoided without the use of a rapid shutdown function. For this, we start with an assessment of how big the need for action actually is. As already mentioned, the authors have no knowledge of firefighters which have been seriously injured because of PV plants (however, there have already been firefighters suffering electrical shocks because of PV installations, which luckily did not result in significant injuries). But it is difficult to obtain reliable numbers on this issue. Furthermore, in most areas, PV plants on buildings are still an exception. But in the not too far future, various forecasts predict that a majority of all buildings will be equipped with a PV plant. Because of this, personal security – not limited to first responders only – is an important matter which should not be taken lightly by the PV industry. However, the threats coming from a PV plant are not any more dangerous than other threats which the firefighters must face (e.g. explosive gases, chemical products). The problems with PV plants are first that electricity cannot be registered with our senses from the outside and second that firefighters are mostly not qualified electricians and have only a rudimentary training in handling of electrical installations. It should not be neglected that electricity is also a danger that is present in the low voltage installations of nearly every building and therefore not a new invention (actually, AC installations are far more dangerous to people than the DC wiring of PV plants [2]). Only because the solar generator has a considerable size and is located on the roof, the probability for a direct contact is higher. In summary, the solar generator is a potential danger, but no reason for panic.

The dangers of PV installations (and other electrical installations) could be easily dealt with, if the firefighters would be given appropriate training and equipment. For this, firefighters should be taught basic electrotechnical skills as well as common grounding and protection technics. Their equipment should be complemented with gear and clothing protecting from electrical shocks. Possibly, the latter could be combined with the existing protection against heat and toxic gases. Furthermore, only non-conducting (or grounded) tools should be used whenever possible. This would of course require considerable investments and could therefore hardly be done all over the country these days. Of course, there would alternatively be the possibility of training and equipping only a handful of specialized firefighters. However, this would cause logistical problems, because the fire brigades would have to ensure that at every given time, at least one such specialist is available. Especially for a volunteer fire service, this would be very difficult. But as was already mentioned, it can be assumed that in the near future, PV plants on buildings will become standard. In such an environment, it should be carefully investigated which of the following is more useful and economic: To equip all PV plants with a shutdown feature (along with the risks and problems we discussed) or to train and equip the first responders so that they can handle the dangers posed by a PV plant. This would have the added benefit that the low voltage grid related accidents are also significantly reduced.

9 SUMMARY

In this contribution, it has been shown that a rapid shutdown feature as required by the NEC2017 would offer maximum safety. However, this does only apply if the feature is completely reliable. As the rapid shutdown is safety-related and aims to protect people, the requirements on this matter are very high. This is being made difficult because of rough environmental conditions and because the shutdown feature has to work properly even after decades of inactivity. Because the NEC2017 rapid shutdown is a new development, its minimum requirements are not yet regulated by normative standards. If the shutdown feature is present but not working, this may give the firefighters a false sense of safety and lead to even more dangerous situations. Furthermore, if the switchgear for the rapid shutdown is damaged or compromised, it is possible that electrical arcs occur, which in the worst case could cause a fire themselves. The shutdown functionality can be achieved easily by use of microinverters or power optimizers. However, the disadvantages of such systems compared to classic PV array topologies with string or central inverters must be considered. A possible alternative to the rapid shutdown feature would be an appropriate training and equipment of the fire brigades. Because the threats of PV systems are generally moderate, extensive upgrading of the fire brigades is presently neither useful nor necessary. However, this could change if in near future, PV plants on buildings are no longer an exception but a standard.

10 ACKNOWLEDGMENTS

This project is carried out in the frame of the Swiss Competence Centers for Energy Research on the Future Swiss Electrical Infrastructure (SCCER-FURIES) and with Swiss industry partners. We gratefully acknowledge financial support from the Swiss Commission for Technology and Innovation (CTI - SCCER program) and from Berner Fachhochschule BFH. We are also grateful for the financial support from Gebäudeversicherung Bern (GVB). Special thanks go to Mr. Martin Rutschi of the fire brigade of the city of Burgdorf, who, with his help and expert knowledge, made this paper possible.

11 REFERENCES

[1] C. Merz, G. Bettenwort, M Hopf, H. Knopf, J. Laschinski: “Safe PV Plants with Panel Level Electronics?”, 32nd European Photovoltaic Solar Energy Conference and Exhibition, Munich, Germany, June 2016.

[2] H. Häberlin, L. Borgna, P. Schärf: "PV and Fire Brigade Safety - No Panic, but Realistic Assessment of Danger and Possible Countermeasures", 26th European Photovoltaic Solar Energy Conference and Exhibition, Hamburg, Germany, September 2011.

ANALYSING THE EFFECT OF PV SYSTEM SIZE AND BATTERY STORAGE CAPACITY ON THE SELF-SUFFICIENCY DEGREE AND SELF-CONSUMPTION RATIO FOR DIFFERENT CONSUMERS

Manjunath Basappa Ayanna[1], Peter Klein[2], Stephen Koopman[3] & Tobias Bischof-Niemz[4],

[1]Energy Centre, Council for Scientific and Industrial Research, P O Box 395, Pretoria 0001, South Africa, +27128427285, MBasappaayanna@csir.co.za
[2]Energy Centre, CSIR, Pretoria, South Africa, PKlein@csir.co.za
[3]Energy Centre, CSIR, Pretoria, South Africa, SKoopman@csir.co.za
[4]Enertrag AG, Berlin, Germany, Tobias.bischof-niemz@enertrag.com

ABSTRACT: Small scale, distributed rooftop solar photovoltaic (PV) systems connected to the medium and low-voltage distribution networks have become cost competitive with end-consumer electricity tariffs in a number of countries. As a result, businesses and residential households are now increasingly self-generating electricity behind the meter. This paper studies the effect of the relative solar PV system size and battery storage capacity on the Self-Sufficiency Degree (SSD) and Self-Consumption Ratio (SCR) for different industrial, commercial and residential consumer profiles. The sizing of the solar PV system and battery storage is optimized to achieve a 100% degree of self-sufficiency, whilst maximizing the possible self-consumption ratio. The analysis excludes costs and therefore the optimization presented is based on minimising the size of the PV system and battery storage. Future work will consider the associated costs in the optimization.

Keywords: Self-Consumption Ratio, Self Sufficiency Degree, Battery storage, Optimisation

1 INTRODUCTION

As of the end of 2016 the global installed photovoltaic capacity has increased rapidly to a total of 303 GW [1], with many countries continuing to introduce competitive tendering to grant power purchase agreements. Although South Africa's power generation fleet remains predominantly coal-based, the country has an extensive solar resource with an average of more than 2,500 hours of sunshine per year and average direct solar radiation levels ranging between 4.5 and 6.5 kWh/m^2 per day [2]. Due to the large scale generation capacity expansion plan from the national electricity utility Eskom, South African customers have experienced a 300% increase in average electricity tariffs since 2008 [3]. The levelized cost of electricity under the Renewable Energy Independent Power Producer Program for solar PV generation has decreased from 3.65 ZAR/kWh to 0.62 ZAR/kWh in the last 5 years [4]. Therefore, due to increasing electricity prices and decreasing PV system costs there is growing interest from South African electricity consumers to install rooftop PV systems. Consumers will tend to self-consume their produced energy if the export tariff is lower than the import tariff. This article analyses the changes in Self-Sufficiency Degree (SSD) and Self-Consumption Ratio (SCR) with the increase in the size of PV system and battery storage, while highlighting the importance of optimal sizing.

2 CONCEPT OF SSD AND SCR

Figure 1 presents the balance of power flow between the solar PV generator, battery storage, grid and load. The power produced by the solar PV system (P_{PV}) is fed to the local load (P_{load}) and excess power is used to charge the battery storage ($P_{batt,c}$). When (P_{PV}) is insufficient to meet the load, the battery storage is discharged ($P_{batt,d}$) and the balance of power demand is imported from the grid ($P_{grid,i}$). When surplus power is available, such as in a case when the battery storage is fully charged; the power is exported to the grid ($P_{grid,e}$).

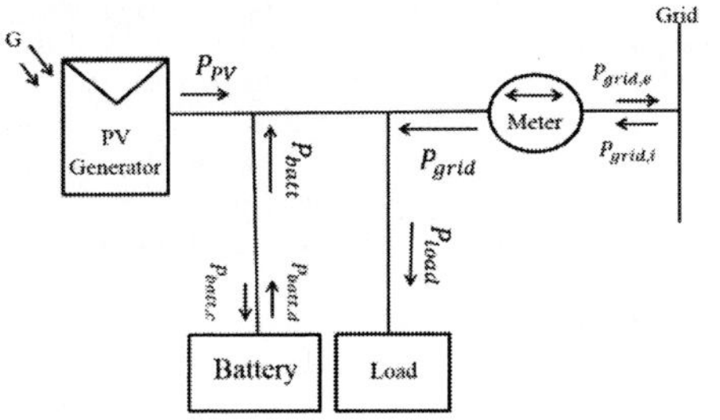

Fig. 1: Block diagram of a grid connected solar PV system with battery storage system

The balance of power in the system is given by Eq. (1). The power produced by a PV system (P_{PV}) and the demand by the local loads (P_{load}) are termed as fixed parameters as the generation from the PV system and the consumption by loads is known. The value of P_{batt} is considered as 0 for the PV system without battery storage throughout this analysis.

$$P_{PV} + P_{batt} - P_{load} + P_{grid} = 0 \quad (1)$$

Both P_{batt} and P_{grid} are assumed as variable parameters as the battery storage is either charged or discharged and the excess PV power is exported to the grid or required demand is imported from the grid.

Equation (2) defines a generic equation by including the charge/discharging states of the battery and import/export power from the grid. Table I defines the possible scenarios where the battery cannot be charged and discharged simultaneously, nor is power imported and exported from/to the grid at the same time.

$$P_{PV} + (P_{batt,d} - P_{batt,c}) - P_{load} + (P_{grid,i} - P_{grid,e}) = 0 \quad (2)$$

Table I: Possible scenarios

		Grid	
		Import	Export
Battery storage	Charge	$P_{grid,e} = 0$ $P_{batt,d} = 0$	$P_{grid,i} = 0$ $P_{batt,d} = 0$
	Discharge	$P_{grid,e} = 0$ $P_{batt,c} = 0$	$P_{grid,i} = 0$ $P_{batt,c} = 0$

From the data in Table 1, the following scenarios are derived as per the Eqs. (3) to (6).

$$P_{batt,c} = P_{PV} - P_{load} + P_{grid,i} \quad (3)$$

$$P_{batt,c} = P_{PV} - P_{load} - P_{grid,e} \quad (4)$$

$$P_{batt,d} = -P_{PV} + P_{load} - P_{grid,i} \quad (5)$$

$$P_{batt,d} = -P_{PV} + P_{load} + P_{grid,e} \quad (6)$$

In this work only Eqs. (4) and (5) are considered, while Eqs. (3) and (6) are reserved for further work with the financial assessment. Data for the PV generation and power demand (available in hourly values) was utilised to calculate the energy transfer in the system according to Eq. (8), over an analysis period of one year. It is important to take into consideration the round trip efficiency of the battery when considering the energy that can be discharged from battery storage system.

Therefore in energy terms:

$$E_{batt,d} = \eta_{batt} E_{batt,c} \quad (7)$$

$$E_{PV} + (E_{batt,d} - E_{batt,c}) - E_{load} + (E_{grid,i} - E_{grid,e}) = 0 \quad (8)$$

A time series graph for the period of one day is presented in Fig. 2, which depicts the energy demand supplied directly by PV (E_{PV-sc}), battery storage ($E_{batt,d}$) and grid ($E_{grid,i}$), and excess PV power charging the battery storage ($E_{batt,c}$). The excess PV energy feed into the grid ($E_{grid,e}$) only when the battery storage is full.

Two parameters are introduced to analyse the performance of the PV system with/without battery storage. Self-Consumption Ratio (SCR) is defined by Eq. (9) as the amount of solar PV energy consumed behind the meter as a percentage of the total energy produced by a solar PV plant (excludes export to grid).

$$SCR = \frac{E_{PV-sc} + E_{batt,d}}{E_{PV}} \quad (9)$$

The self-sufficiency degree (SSD) is defined by Eq. (10) as the proportion of the total energy demand met by the solar PV plant generation i.e. not imported from the grid.

$$SSD = \frac{E_{PV-sc} + E_{batt,d}}{E_{load}} \quad (10)$$

3 DEMAND PROFILE AND SOLAR GENERATION

Hourly energy demand profiles are required as an input into the model. A comparison of the energy demand profiles for different consumer that are considered in this paper is presented in Fig.3 for a week period. The solar PV profile constitutes hourly simulation results using PVSyst software for the geographical location of Pretoria in South Africa, for a fixed inclination installation.

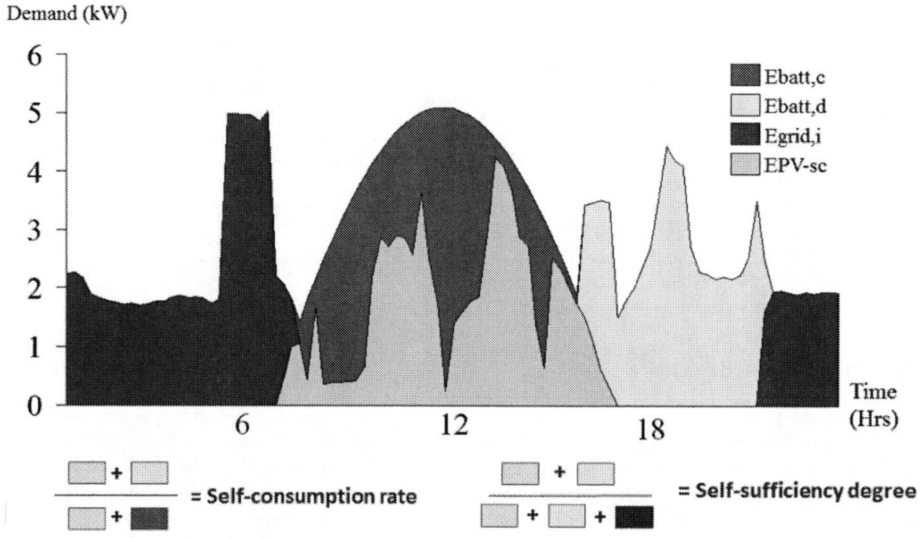

Figure 2: Demand & PV generation profile of a typical residential consumer without storage

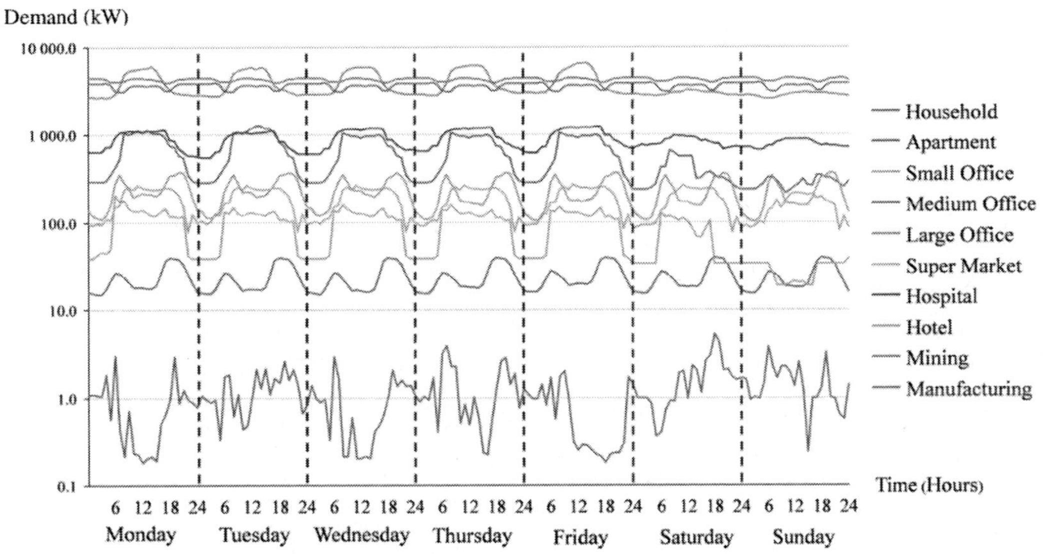

Figure 3: Typical energy demand profile of residential, commercial and industrial consumers.

The generic demand profiles include single family household, apartment, small, medium and large offices, hospital, supermarket, hotel, manufacturing and mining consumers. The demand from the single family household is measured and recorded on a real time basis on a social platform [5]. The demand profile for the medium sized apartment, small and medium offices, hospital, supermarket and hotel is derived from the real time measurement data obtained from [6]. The demand profile for both manufacturing and mining companies responsive to a Time of Use tariff is derived from [7].

3.1 Residential:

The peak demand from the residential consumer either a single family household or an apartment is in the morning and evening hours with mid-day hours having a lower demand.

3.2 Commercial:

The peak demand from commercial consumers, such as small, medium and large offices, supermarkets and hospitals on week days is during the day. The weekend (Saturday and Sunday) demand from these consumers lies around its baseload with little or no peak demand. Also, the peak demand in these consumers coincides with the solar PV system generation. A hotel has peak demand in the morning and evening hours with mid-day hours having lower demand similar to that of residential consumer. A supermarket has a uniform peak demand throughout the week.

3.3 Industrial:

The manufacturing and mining profiles have a consistent baseload throughout the year, considering the type of operational processes involved. According to Eskom, there are consumers responding to current peak signals i.e. Time-of-Use (ToU) responsive usage profile and are also those who do not respond to the current peak signals, and thus have uniform energy demand, mostly due to technical constraints within their day-to-day operations.

4 MODELLING APPROACH

In this study the PV system size and battery size are relative in size to the load. The relative PV system size is defined as total solar PV energy generated per total electricity consumption on an annual balance, while the relative battery size is defined as defined as total solar PV energy stored per total electricity consumption on an average daily demand. The SCR and SSD are influenced by the shape of the demand profile, the relative size of the PV system and the relative size of the battery storage.

The solar PV generation data is simulated for a period of one year and the data is overlaid on the energy demand data for the same period. A round trip battery efficiency of 90% is assumed. Equations (1) to (8), presented in Section 2 are solved for each consumer demand profile over a 1 year period. The use of annual data in this study is to absorb the seasonal variations in both the energy demand and solar PV generation. The SSD and SCR are calculated according to Eqs. (9) to (10), with reference to Fig. 2. The relative size of the solar PV system is varied up to a maximum size of 10000%. The relative size of the battery storage is determined against the average daily load and increased up to 400% capacity.

5 RESULTS

This section presents the results of the analysis. The effect of variation of the relative PV system size and relative battery size on the SCR and SSD values is presented. Finally the system is optimised to achieve 100% SSD, whilst maximising the possible SCR, for the lowest system and battery sizes.

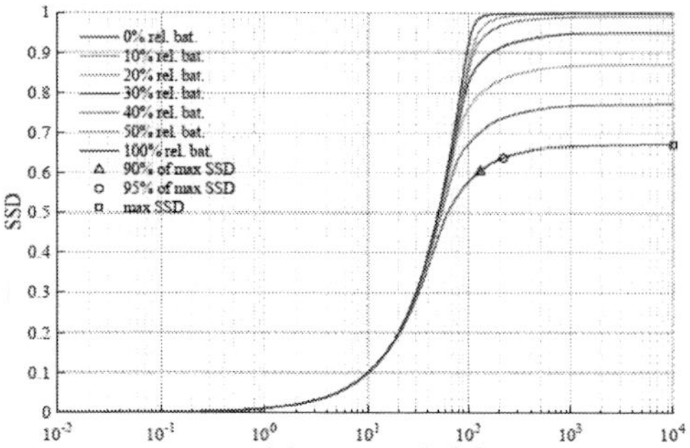

Fig 4: Increasing SSD with the increase in sizing of solar PV system and storage system

As shown in Fig. 4, increasing the relative size of the solar PV system, results in an increasing SSD that asymptotically reaches its maximum value, depending on the shape of the demand profile. The maximum SSD depends on how much load is aligned during non-solar PV-hours i.e. at night, early morning and evenings. Theoretically, the 100% SSD is only achieved if the demand profile matches exactly with the supply from the solar PV system (if no storage). The maximum SSD for each battery capacity is assumed to be the value achieved at 10000% relative size of the solar PV system. For the system without battery storage, 90% and 95% of the maximum SSD value were achieved at 125% and 202% of the relative size of the solar PV system (medium office profile). Thus, a 5% increase in SSD past 90% requires a significant 77% increase in the relative system size. The data in Table II presents the SSD value achieved at 90% of the maximum SSD and the corresponding SCR value for all demand profiles.

Table II: SSD and corresponding SCR values achieved at 90% of maximum SSD

	Rel. PV syst. size (%)	SSD (%)	SCR (%)
Household	205	46	22
Apartment	161	42	26
Small Office	139	57	41
Medium Office	125	60	48
Large Office	93	49	53
Super market	109	54	50
Hospital	106	49	47
Hotel	131	43	33
Manufacturing	85	40	47
Mining	89	42	47

The addition of battery capacity increases the maximum achievable SSD as energy can be shifted to non-solar PV-hours. Figure 5 shows how the maximum SSD (at 10000% relative system size) increases relative to the battery size for different consumer profiles. Conversely, as the relative size of the solar PV system increases, the value of SCR decreases. Naturally the maximum SCR of 100% is achieved with small solar PV systems where the generation is significantly lower than the demand. However, the high SCR comes at the expense of a lower SSD. By increasing the battery storage capacity, the range of high achievable SCR extends as a function of PV size.

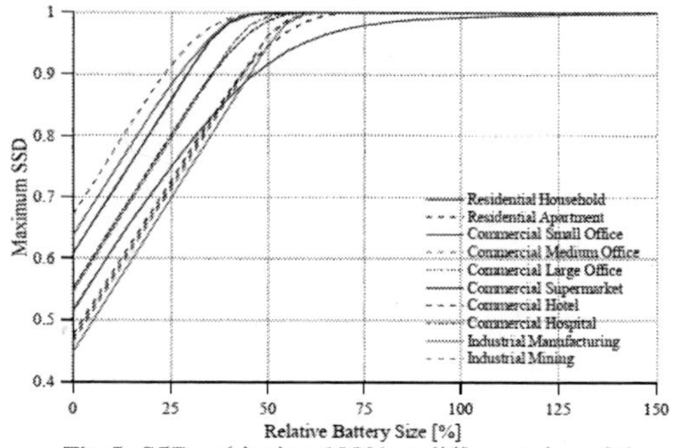

Fig 5: SSD achieving 100% at different size of the battery storage for different consumers

However, due to the round trip efficiency of the battery there is a small decrease in SCR from 100% with each increase in battery storage size. The graph in Fig. 6 depicts the behaviour of decreasing SCR with the percentage increase in relative size of the PV system and battery storage (medium office).

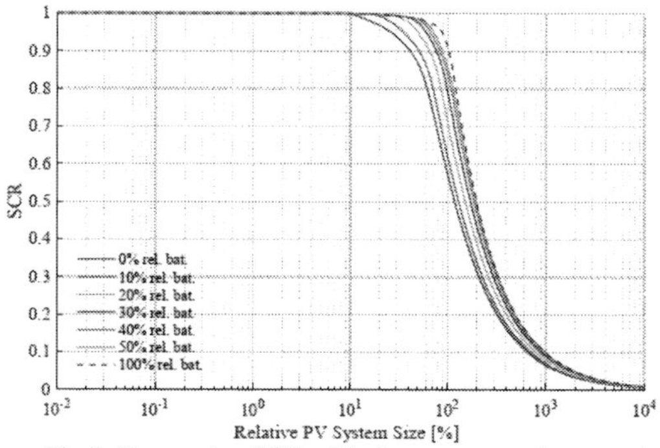

Fig 6: Decreasing SCR with the percentage increase in the size of solar PV system

The data in Table III presents the corresponding maximum possible SSD where 100% SCR is achieved with the increase in battery storage. The 100% SCR is achieved at different relative sizes of the solar PV system and battery storage depending on the type of energy demand profile. It should be noted that under these conditions the maximum possible SSD is equal to the relative PV system size.

Table III: Corresponding maximum possible SSD where 100% SCR achieved

	Relative Battery Storage Size			
	0%	10%	50%	100%
Household	2	8	8	8
Apartment	22	27	27	27
Small office	8	17	17	17
Medium office	11	24	26	26
Large office	22	31	32	32
Supermarket	36	44	45	45
Hospital	28	38	38	38
Hotel	21	29	29	29
Manufacturing	29	34	34	34
Mining	28	33	33	33

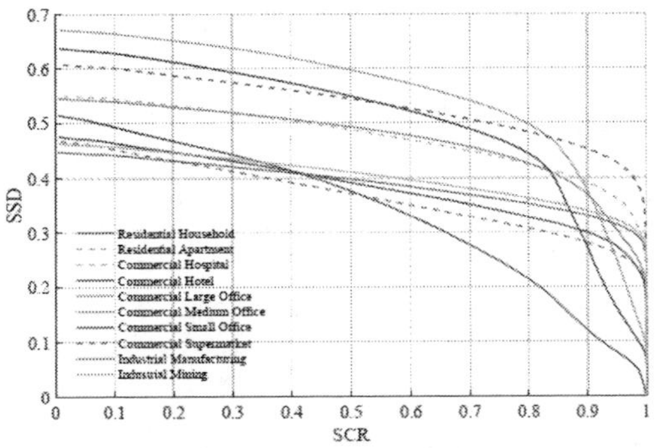

Fig. 7: Maximum achieved SSD and SCR with the increase in solar PV system size (without battery storage)

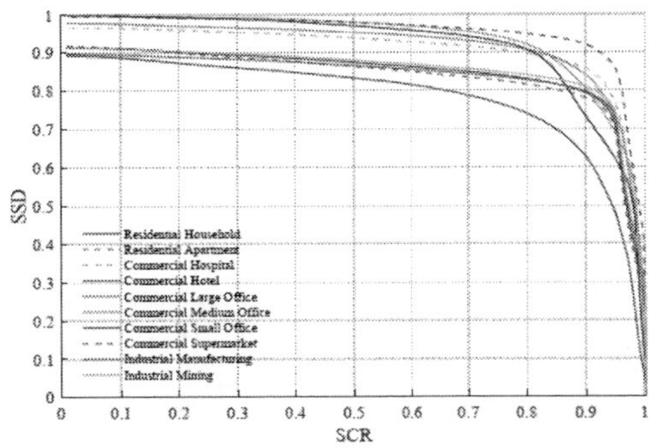

Fig. 8: Maximum achieved SSD and SCR with the increase in solar PV size (with 50% battery storage size)

In the PV system without storage, the SSD and SCR will attain their maximum at different solar PV system sizes depending on the type of demand profile. The set of achieved results of SSD and SCR is plotted on a two dimensional graph in Fig. 7 for different consumers to determine the ideal customer profile that attains high degree of self-sufficiency whilst simultaneously achieving a high self-consumption ratio. From the graph in Fig. 7, the ideal consumer demand profile for the PV integration is the one which achieved high SSD whilst simultaneously achieving the maximum SCR, for example a supermarket. This outcome is based on the energy-balance hypothesis and the decision is on the assumption that export tariff is lower than import tariff. If the export tariff is higher than the import tariff, then it would be the demand profile which gives a high SSD at a low SCR, probably the single family house-hold.

In the PV system with battery storage, the SSD and SCR will attain their new maximum at different PV system and battery storage size depending on the type of demand profile. The SSD reaches asymptotic 100% at certain PV system and battery storage size with a possible maximum SCR. The increase in SSD and SCR at relative 50% battery storage size with the increase in the relative size of the PV system can be seen in Fig. 8. The demand profile of Supermarket achieved 100% SSD with a possible maximum SCR at much lower PV system and battery storage size whereas the demand profile of household achieves similar results with higher PV system and battery storage size.

Depending on the investment, the size of PV system and battery storage can be optimized. A 100% SSD with maximum possible SCR with can be achieved with high battery storage and low PV system size or low battery storage and high PV system size. If the investment on the battery storage is expensive, one can invest on higher PV system size and achieve 100% SSD with a high SCR. This is the current market situation where the PV system is much cheaper and battery storage system is expensive. If the price of the battery storage goes down in the near future, one can then lower the PV system size and have high battery storage size still achieving 100% SSD. The graph in Fig. 9 represents the various combinations of relative size of the PV system and battery storage where 100% SSD is achieved (medium office profile). The region in grey indicates 100% SSD (2 decimal point accuracy) for any combination of PV system and battery storage size.

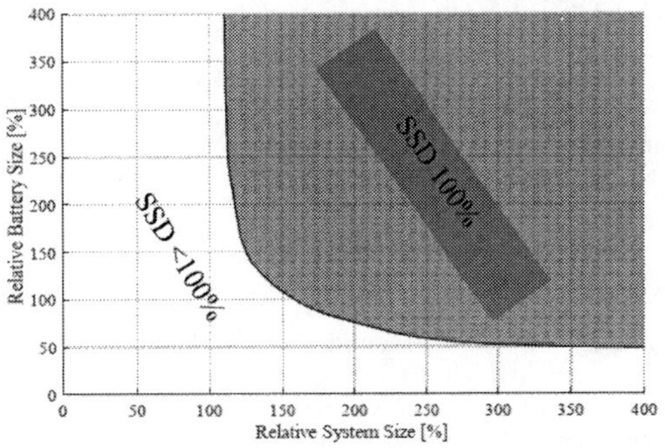

Fig 9: Multiple combinations of solar PV system and battery storage size where ≥100% SSD achieved

The marked square blocks in Fig. 10 represent the optimum combination where 100% SSD is achieved at minimum PV system and battery sizes. This optimisation does not take into account the different costs of the PV and battery systems as of yet. A high battery costs will shift the optimum point to the right of the curve, in order to reduce the battery size.

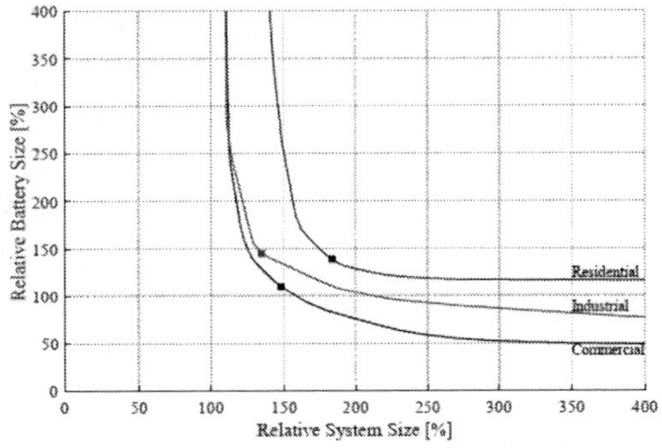

Fig 10: Different regions of 100% SSD depending on shape of the demand profile (marked points represent optimum combination)

The corresponding SCR values where 100% SSD achieved at equally minimum PV system and battery size for all the demand profiles are presented in Table IV.

Table IV: Corresponding SCR maximum where 100% SSD achieved at minimum PV system and battery sizes

	Rel. PV size (%)	Rel. Battery size (%)	SCR (%)
Household	184	138	54
Apartment	135	140	74
Small Office	153	129	70
Medium Office	148	110	69
Large Office	149	140	67
Super market	122	130	79
Hospital	146	115	71
Hotel	143	115	72
Manufacturing	135	145	74
Mining	138	140	72

6 CONCLUSION

This paper has presented the analysis of SSD and SCR for different demand profiles with varying PV generation and battery storage capacities. The SSD is a function of the alignment between the PV generation profile and the demand. For the PV system without battery storage, increasing the PV system size past a certain point does not significantly improve the SSD value. Therefore, if it is desirable to have an SSD value above 80%, battery storage must be considered. Conversely increasing the PV system size results in a decrease in the SCR value as excess energy generated cannot be shifted to the non-solar periods without a battery storage. In fact the shape of the demand profile itself has more of an effect on the SCR.

For the desirable case of full self-sufficiency where 100% SSD is achieved; a preliminary optimisation of the PV system and battery storage at their minimum capacities was presented. This did not take into account the costs of the PV system and battery storage. For different demand profiles, the corresponding SCR varied between 54-79%, with battery storage for 100% SSD.

The combination can be further optimized depending on investment costs by varying the size of the PV system and battery storage. If battery storage cost is high; then having a high PV system size and low battery storage size can achieve 100% SSD. If battery storage cost is low; then having a low PV system size and high battery storage size still can achieve 100% SSD. A higher SSD is desired to make an investment decision because installing a small PV system just to achieve 100% SCR is not justified. A high SCR is desirable if the import tariff is higher than the export tariff which makes solar PV a better business case.

7 OUTLOOK

This paper has focussed on understanding the trends in SSD and SCR for different demand profiles, PV system sizes and battery capacities. However, these results cannot be considered in isolation to investment costs and import/export electricity tariffs. Future work will build on the analysis presented here to carefully consider all financial aspects in order to determine optimum sizing for each demand profile while achieving 100% SSD and possible maximum SCR.

7 REFERENCES

[1] International Energy Agency (IEA). 2016 Snapshot of Global Photovoltaic Markets. (2016).

[2] Department of Energy of the Republic of South Africa. State of Renewable Energy in South Africa. (2015).

[3] Eskom. Historical average prices and increases. http://www.eskom.co.za/CustomerCare/TariffsAndCharges/Pages/Tariff_History.aspx. Accessed 18 Aug 2017

[4] M. Ngobeni. Stakeholder Consultation Workshop: Preparation of the Second Edition of the State of Renewable Energy in South Africa Report, Market Overview and Current Levels of Renewable Energy. (2016)

[5] Private Communication with Dr T. Bischof-Niemz. (2016)

[6] U.S. Department of Energy. OpenEI, Commercial and Residential Hourly Profiles for all TMY3 Locations in the United States. http://en.openei.org/doe-opendata/dataset/commercial-and-residential-hourly-load-profiles-for-all-tmy3-locations-in-the-united-states Accessed on 23 Feb 2016

[7] National Energy Regulator of the Republic of South Africa. Eskom's 2015/16 Time of Use (ToU) structural adjustment. (2014)

SYNTHESIZING RESIDENTIAL LOAD PROFILES USING BEHAVIOR SIMULATION

Noah Pflugradt, Urs Muntwyler
Berne University of Applied Sciences (BFH), Engineering and Information Technology
Photovoltaic Laboratory (PV-Lab), Jlcoweg 1, CH-3400 Burgdorf, Switzerland
Phone +41 (0)34 426 68 11, Noah.Pflugradt@bfh.ch

ABSTRACT: This research introduces a novel approach for synthesising residential electricity and water load profiles by performing a bottom-up simulation of the behaviour of the residents. For the simulation, the residents are modelled using independent software agents. The agents are using a psychological desire model to determine their actions. The model can simulate for example office workers, families, unemployment, or even factory shift workers. The implementation provides simulations of multi-family houses in detail and includes details such as illness periods, personal hobbies and device ownership. The detailed activity profiles are used to calculate load profiles. The software is available for free at www.loadprofilegenerator.ch.
Keywords:Demand-Side, Grid Integration, Modelling, Software, System Performance

1 INTRODUCTION

In the field of energy research and especially for renewable energy research load profiles are frequently needed. They are used both for sizing and for analysing systems in simulations. Applications include for example:

- testing the performance of battery or PV systems in a simulation and calculating the self-consumption
- verifying the performance characteristics of a given system in certain circumstances
- testing algorithms for energy management systems
- analysing low-voltage grids

Uses for load profiles can be divided into two areas: Sizing and analysis. For sizing systems, it is desirable to have average profiles that are representative of a wide range of customers. For analysing systems, on the other hand, it is helpful to have different profiles that reflect specific use cases to be able to evaluate the system performance in a wide range of conditions. Figure 1 shows an example of a measured profile like it might be used for an analysis and the averaged German standardized load profile H0.

This paper is focused on residential load profiles for electricity and water, primarily for system analysis. Due to the different user behaviour load profiles can vary wildly. An averaged profile across a couple of hundred households looks quite different than a measured profile from a single household and even between households there are large variations. A large family has a very different load profile than a working single.

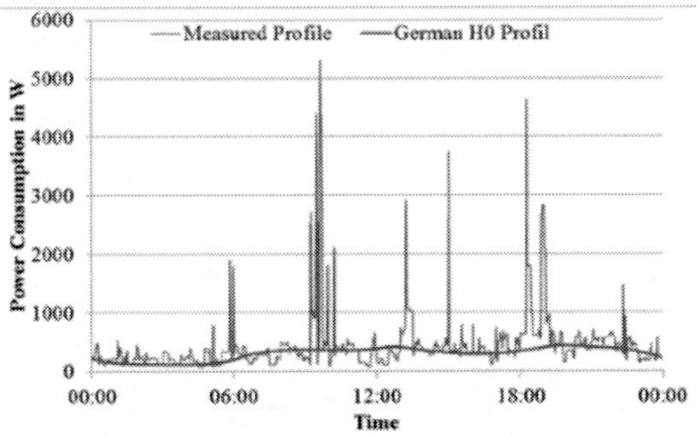

Figure 1: Measured residential load profile compared with the standardized German H0 profile

Frequently appropriate load profiles are not available, either due to privacy issues, because of funding issues or because the situation doesn't exist (yet).

Then the only solution is a load profile synthesis. This paper introduces a novel method for the synthesis using a behaviour simulation that models the residents of the house as independent, desire driven software agents. The method is implemented as a freely available software that can be downloaded at [1].

2 STATE OF THE ART

The problem of missing load profiles is not new, and there has been a lot of research into this, especially for electricity. One of the earliest papers on the topic was [2] from 1984. Most approaches use a probabilistic model which is based on the idea of rules such as "In the evening, 90% chance of turning on the TV". Of course, this simple approach is not sufficient to model all the intricacies of human behaviour. So, over the years researchers have used various refinements, such as combining multiple devices into activities and then calculating the probability profiles for the activities instead of the devices. Another improvement was using occupancy patterns to modulate the device probabilities, since very few devices get turned on if no one is home. Frequently sophisticated time-of-use studies are used to generate the probabilities for the activities, which yields very good results if the studies are available and the studies cover the situation that load profiles target [3]. One very interesting approach by [4] was to take the load profile from a substation and combine various statistics and device measurements to split the load curve apart into individual household load curves which were then used to analyse the flow in the low voltage grid. In [5] is a very comprehensive review of different load profile generators and their approaches.

3 METHODOLOGY

This section first introduces the basic model and then describes the various functions that were added to the basic model to improve the functionality.

3.1 Basic Model

The basic model is based on a model of human desires by the German psychologist D. Dörner [6]. He found that human decision-making is primarily driven by

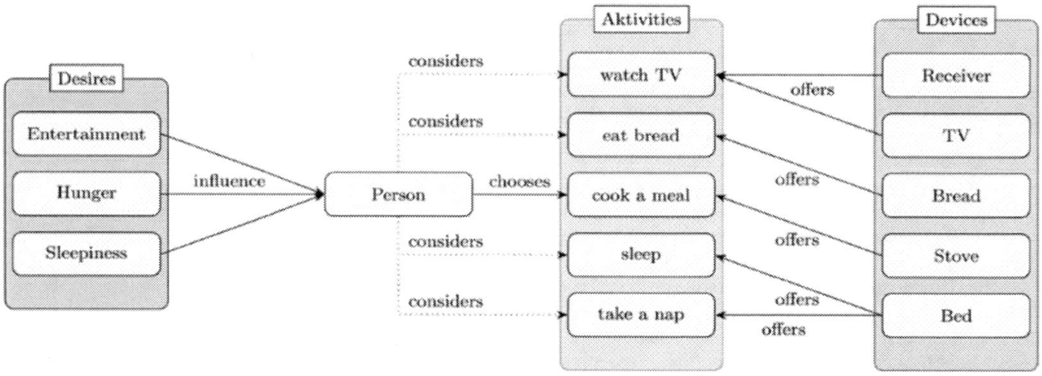

Figure 2: Basic idea behind the behaviour simulation

desires and thus the best way to model behaviour is to model the desires and activities separately. Then the decision-making process can be approximated as an optimisation problem that tries to reduce the unhappiness from the various desires to the lowest possible value. This model was used as the foundation for the behaviour based load profile generator described in this paper and is shown in Figure 2. Whenever a person finishes an activity, he/she will look at the currently available options and try to pick the best one. If, for example, the person is hungry, then that would be eating, if the person is tired, it would be sleep and so on. In parallel, as time passes every desire grows a small amount in every time step, though the amount depends on the desire.

For each activity, it needs to be defined beforehand which devices will be activated and which device load profiles will be used. Then the finished activity profile is used to combine the device load profiles into the finished household load profiles by simply adding them up.

3.2 Activity dependencies

Of course, not all activities can be executed at all times. The bLPG reproduces this by a sophisticated time limit model which can combine arbitrarily Boolean conditions, such as temperature, time, date, or other information from load profiles. Additionally, some activities depend on previous events. For example, if the dishwasher needs to be turned on depends on the number of dirty dishes, which are tracked in the bLPG using variables. Every cooking activity and every eaten meal produces a certain number of dirty dishes and as soon as the number of dirty dishes surpasses the minimum for the dishwasher, the activity “turn on the dishwasher” becomes available to the residents. The same applies for the laundry, the dryer and even the occasional bathroom dehumidifier.

Lighting consumes up to 15 % of the energy in a household [7]. The bLPG contains a lighting model that includes light devices in each room of the household which are turned on as needed. Experimentation and comparison with literature results from [4, 7] showed that using the global solar radiation with a limit of 50-75 W/m2 as a condition for turn on the lights yields realistic light patterns and total energy use.

3.3 Families

This basic model works quite well to generate realistic activity profiles for singles, as long as sufficient different activities are available for the person. For families, though the model needs to include interactions between family members. For example, the mother when the calls the children to eat dinner. Otherwise, the children would cook their food separately, which is in most cases not a very accurate depiction of the real world. To model this, the bLPG implements sub-activities. That means one person can start an activity and others can join later. For interrupting activities, it is possible to mark all activities as either interruptible or interrupting. One example of how this works is as follows: The mother starts to cook. Cooking has the sub-activity “eat a cooked meal” defined, which is marked as “interrupting”. That sub-activity becomes available to the other people in the household after 30 minutes of cooking. As soon as the activity is available every person in the household that is performing an interruptible activity such as watching TV, sleeping or doing the homework will stop or pause their current activity and join the meal. In this way, interactions can be modelled, which is crucial for realistic household activity profiles.

3.4 Device Switching

The bLPG uses measured device profiles to create the household profile. The device profiles are connected to each activity, since the activity determines how the device is used. But an important function is being able to switch out the devices in the household, for example to analyse the impact on the load profile of switching over to LED lighting from incandescent light bulbs. This sounds easy in theory, but the implementation is complicated, since one device might be used in multiple activities with multiple different load profiles. For example, a washing machine is used in the activity “wash at 30°C” and “wash at 60°C”, but with completely different device load profiles, both for water and electricity. So, switching out the washing machine needs to also switch the load profiles in multiple activities. To automate the switching process, so called “Device Actions” were defined that combine a device with a set of device load profiles which belong to that action, for example “run the 30°C cycle on the washing machine from manufacturer xxx”. Then device action groups were created that combine all exchangeable actions, for example “run a 30°C cycle”. Additionally, every device is sorted into a device category, such as “washing machine”. With these tools, it is possible to automatically switch out devices in households while ensuring that the correct device load profile is used in all cases.

3.5 Expansion

For many applications, entire houses, not just households are of interest. For this, the bLPG makes it possible to add multiple households together into a house and then add the typical house infrastructure, such as circulation pumps and other house infrastructure such as heating and cooling. Since the focus of the bLPG is electric and water load profiles from the user behaviour, the building simulation model is very simple. To provide a rough approximation of the heating and cooling demands for special use cases, a model based on the degree day/cooling hour method was implemented. This might be sufficient in many cases, but for detailed studies about the energy use of a heat pump, for example, a real building simulation should be used, such as Polysun, TRNSYS, Energy Plus or others.

3.6 Settlement Creation

Some studies such as low voltage analysis require large numbers of individual but different load profiles. Modelling each household manually takes 1-2 days. To model 1000 households this way is obviously not feasible, but using a small number of identical profiles is not possible either since then the concurrency of the activities would not reflect the real world, leading to very high spikes in energy consumption.

Therefore, the LPG includes a templating function. This function generates a template from existing household. The template contains rules such as "The mother in this households has between 1 and three hobbies". These templates can be edited and then used to generate new, similar but all individually different households.

Testing showed that one more component is very important. People's behaviour depends on some anchors that define the frame for all other activities. The typical anchors are the time of waking up and, where applicable, the time of work. Without these anchors the behaviour of the people in the simulation drifts all round the clock, much like it can be observed during vacations, especially for teenagers.

So, to make the random household generation yield good results every person has a living pattern such as "early rise office worker" defined and only gets assigned matching sleeping and working patterns. Then some different living pattern were defined. Similar patterns can then get varied randomly, for example by switching out the early rise office worker for the late rise office worker. By varying the living pattern for all the people in the household, it becomes possible to easily generate a large combination of very different households.

Built on top of this is a settlement template function. This function again uses rules to describe the combination of households that should be in the settlement. Such rules are for example "30% of the households should be families with one person working" or "80% of all households in this settlement should live in single-family houses with a photovoltaic panel".

These rules make it then possible to easily generate a large variety of different households that, when summed up, yield a characteristically smooth substation profile.

4 IMPLEMENTATION

The bLPG is implemented as a Windows program written in C#. It was developed as a PhD-Thesis [8] and has since then been continuously extended and improved. Right now, the entire implementation is about 50.000 lines of code. Figure 3 shows a screenshot of the program.

Figure 3: Screenshot of the bLPG

One important feature is the large number of charts and reports that are generated after the simulation. These make it very easy to check if any newly created household functions as intended, which speeds up the manual modelling process.

The software contains 60 predefined households which are based on statistical data, measurements and a small survey.

One side effect of the complex model is the long simulation time: About 0.5 to 1 minute per person per year. Generating all the reports and charts takes a multiple of that though so that a full household simulation for a year takes between 5 and 10 minutes. Of course, the reports can be turned off.

To make it easier to automate the process of generating a large number of households, a command line client is included that can be used to parallelise the simulation process on both the same computer and across a network, making it possible to generate even thousands of profiles within a reasonable time frame.

Currently, a Web API is in development to make it possible to include the LPG in other projects. The API will make it possible to request randomly generated load profiles for a template, which can be modified by changing certain traits on it.

5 VALIDATION AND RESULTS

One issue with load profiles is the difficulty in comparing the raw profile and thus checking if a profile is "correct". If for example, the evening peak inFigure 1 is 30 minutes later, that does not mean that the profile is "wrong" after all. It just means that the people in the household chose to eat later. Therefore calculating similarity for individual household profiles is not feasible. Instead, a more useful approach is twofold: Plausibility checks of the activities and comparison of a few metrics of the profiles.

5.1 Plausibility check

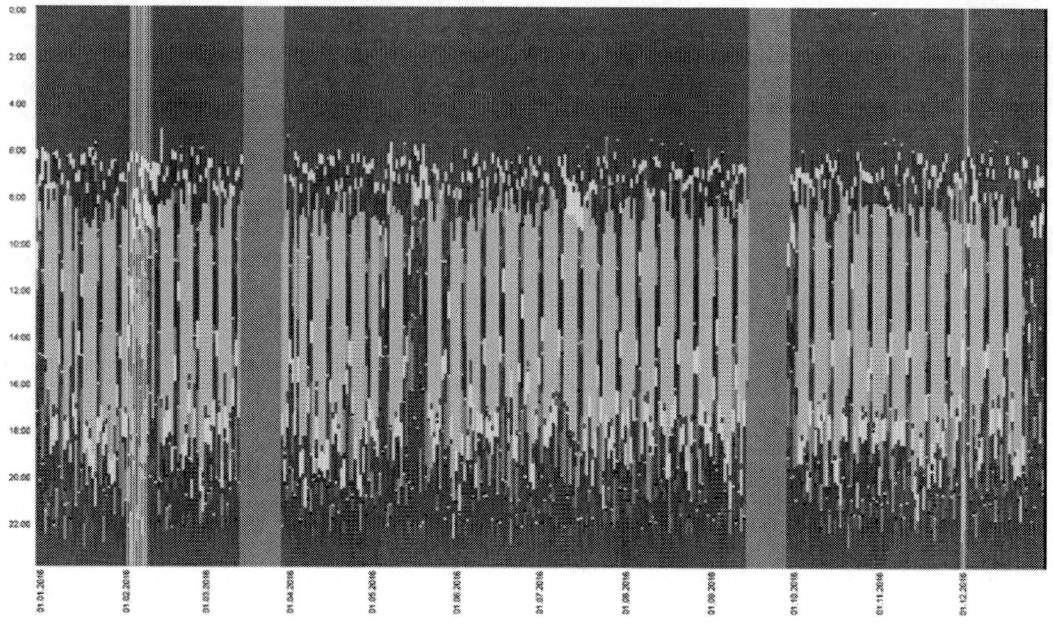

Figure 4: Example for the action carpet plots generated by the bLPG

Activity profiles are much easier to check for plausibility than load profiles. If the user of the bLPG is a domain expert and has a good understanding of the behaviour of the people in the household he/she is trying to model, then just reading the activity profile will give the user already a quick feedback about where the model needs to be improved. To accelerate the process the bLPG also generates visualisations of the activity profile such as the carpet plot shown in Figure 4. With this, it is possible to tell at a glance if the household is modelled correctly.

Another plausibility check is the energy consumption per device. Figure 5Figure 4 shows an example plot. Two things are visible: First, the energy consumption in the simulated load profiles is spread over many different devices, just like in reality. And second, any approach that tries to only model a few big consumers such as the washing machine, fridge and the dryer will not yield realistic profiles.

5.2 Validation Checks

The second part to check in the results are various metrics. The most obvious and the first to check is always the yearly energy/water consumption against the published country-wide averages. This has been done for the predefined households in Figure 6. For this first, all devices in all 60 households have been switched out for the most energy-saving devices, and then all simulations

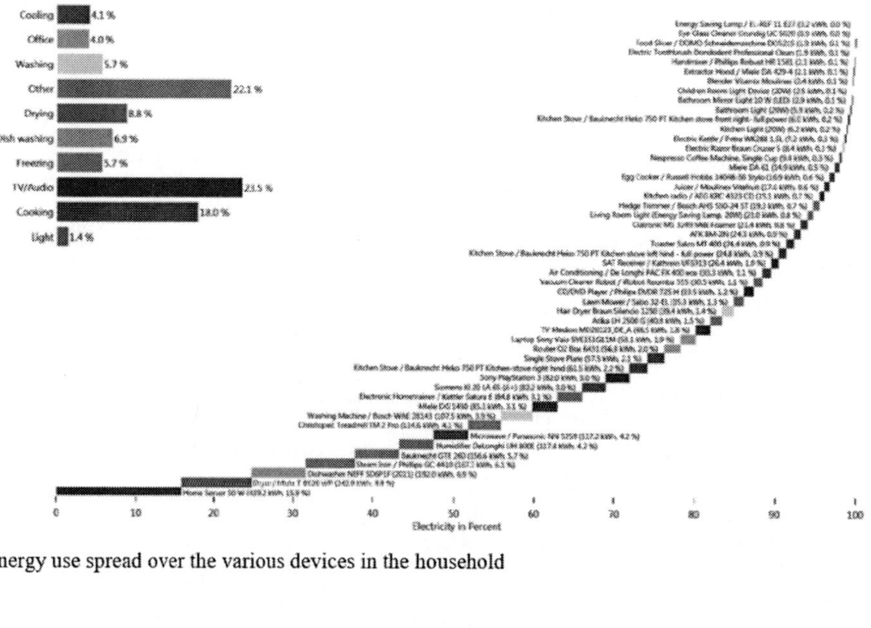

Figure 5: Energy use spread over the various devices in the household

were performed. Then, in a second step, all devices were switched to the maximum energy consumption devices, and the simulations were performed. The results were then compared with the average German Energy consumption from. The results show that the average is well in between the two extremes, which shows that the results are a good match.

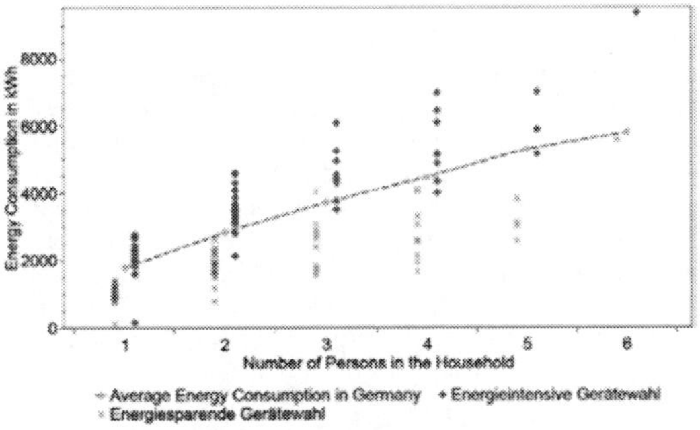

Figure 6: Comparison of the total energy consumption of each household with average German values

Another validation check is comparing the normalized duration curve of the simulated household with the normalized duration curves of real measurements. For this, a data set with 74 smart meter households has been used. The result is shown in Figure 7. The blue area is the reference area by the measured, smart meters. The coloured lines are the simulation results. The figure shows that the distribution of the load intensity is very similar to the measured profiles.

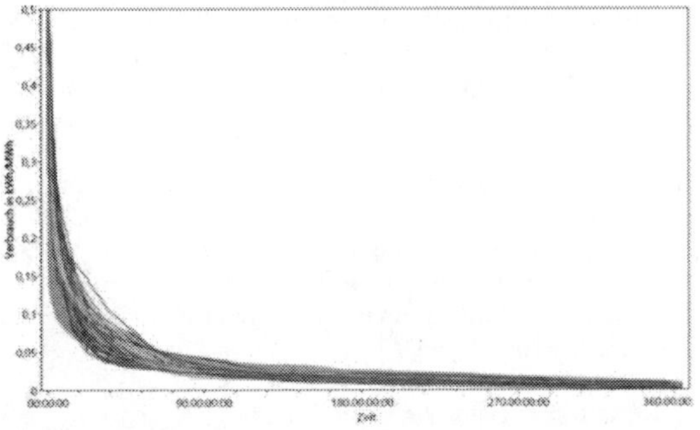

Figure 7: Validation of the duration load curves with measured data. The charts shows that the curves are very similar and thus the results of the bLPG are realistic.

5.3 Result

One interesting challenge is recreating the German standardised load profile H0 using the load profile generator. The challenge is recreating a settlement that is on average exactly like the settlement used to create the H0 profile. This requires among other things that the settlement needs to have a minimum of 300-500 households to yield a smooth enough curve, that the people all need to live in single-family houses. Otherwise, the base load will be too low. Another requirement is that the living pattern of the people has to vary considerably. Most importantly the settlement has to be mostly filled with Families, otherwise, the load curve won't be even remotely similar. The result is shown in Figure 8.

As is visible, the result is close, but still, has room for improvement. Development on that is ongoing.

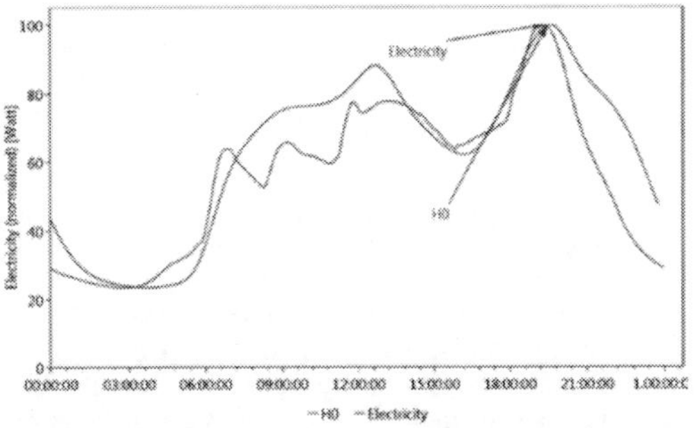

Figure 8: Comparison of the results of a sample settlement with the German standardized H0 profile

6 CONCLUSION

This paper introduced a novel approach for synthesizing residential load profiles. It described the model and a few selected validations and showed that the approach works very well for a wide set of problems. The bLPG is implemented as a freely available software that can be downloaded at [1] and has been used in various papers, thesis and projects over the years.

Acknowledgements

This study was carried out in the frame of the Swiss Centre for Competence in Energy Research on the Future Swiss Electrical Infrastructure (SCCER-FURIES) and with Swiss industry partners. We gratefully acknowledge financial support from the Swiss Commission for Technology and Innovation (CTI - SCCER program) and from Bern University of Applied Sciences BFH.

7 REFERENCES

8

1. Noah Pflugradt (2017) LoadProfileGenerator. http://www.loadprofilegenerator.de
2. Aigner DJ, Sorooshian C, Kerwin P (1984) Conditional demand analysis for estimating residential end-use load profiles. The Energy Journal 5(3): 81–97
3. Fischer D, Härtl A, Wille-Haussmann B (2015) Model for electric load profiles with high time resolution for German households. Energy and Buildings 92: 170–179
4. Stokes M (2005) Removing barriers to embedded generation: A fine-grained load model to support low voltage network performance analysis
5. Grandjean A, Adnot J, Binet G (2012) A review and an analysis of the residential electric load curve models. Renewable and Sustainable Energy Reviews 16(9): 6539–6565
6. Dörner D (2001) Bauplan für eine Seele
7. EnergieAgentur NRW (2013) Erhebung „Wo im Haushalt bleibt der Strom? Anteile, Verbrauchswerte und Kosten von 12 Verbrauchsbereichen in 1- bis 6-Personen-Haushalten
8. Pflugradt ND Modellierung von Wasser und Energieverbräuchen in Haushalten. Dissertation

EVALUATION OF THE EFFECT OF PV PLANT SIZE ON THE EFFICIENCY OF HOUSEHOLD LI-ION BATTERY STORAGE SYSTEMS

Nina Munzke, James Barry, Bernhard Schwarz, Felix Büchle
Karlsruhe Institute of Technology (KIT), INT-PCE
Hermann-von-Helmholtz-Platz 1, Gebäude 276
76344 Eggenstein-Leopoldshafen, Germany

ABSTRACT: The increase in electricity prices along with a decrease in the price of storage systems has led to a rapid expansion of the PV-battery storage system market in German households. In order to be economically viable, PV-storage systems must fulfil certain performance criteria. The overall performance and achievable self-sufficiency ratio of a PV-battery home storage system depends on (i) the efficiencies of the system components, (ii) the standby consumption, (iii) the reaction time of the storage system as well as (iv) the intelligence of the overall system control software. Several commercially available PV-battery systems have been analysed with respect to the different performance criteria. The current paper focuses on the efficiency of the battery itself as well as the overall system efficiency and their dependence on the PV plant size. It is shown that increasing the PV plant size leads to an increase in the overall system efficiency for all systems, while the battery efficiency varies slightly. However, the system efficiency ranking remains mostly constant, which means that the efficiency is a good measure of system performance.
Keywords: Batteries, Energy Performance, Performance, Storage, Efficiency

1 INTRODUCTION

Efficient and economic energy storage technologies are key elements for a sustainable future energy supply. Attractive and economical applications can be established through the successful implementation of long-lasting and cost-efficient storage systems for electrical energy.

Since electricity prices in Germany have increased in the last few years and storage prices are starting to decline, small German household applications will soon become economically feasible [1], [2]. There are several critically important aspects affecting commercial viability, such as the system design and dimensioning, the level of development of the system control software and the calendar and cycle life of the battery and electronic components.

An important question in this regard is to what extent the load profile and the size of the PV-system influence the performance of the battery energy storage system. In [3] it could already be shown that different load profiles with similar electricity consumption per year (3500 kWh and 4500 kWh) result in similar battery and system efficiencies, using a 3.5 kWp PV system as input. The current work therefore focusses on the question as to whether and to what extent the size of the PV plant influences the efficiency of the battery as well as the PV storage system as a whole.

2 METHODOLOGY

2.1 Input data and test procedure

The efficiency of the different power electronics components can either be determined using reference days (as described in [1]) or by measuring the individual power conversion pathways [2], [3]. In the same way the battery efficiency can be found using reference days or by means of charge/discharge tests (cycles), as detailed in [3]. At KIT the different test procedures were applied to various storage systems; the current work focuses on the evaluation of measurements using reference days.

These measurements require typical load and generation profiles as data inputs. For the tests described in this work, real PV data with a resolution of 1 Hz from the 1 MW solar-storage park at KIT's north campus were used. The data were recorded from an array with southerly orientation (0°) and an inclination angle of 30°, which corresponds to a typical house with south-facing pitched roof. The PV system sizes were chosen to be 3.5 kWp and 4.8 kWp, and 5 different days that correspond to the load reference days are used.

In order to generate appropriate and reproducible load profile data for a single family household (HH) the VDI 4655 standard [4] was used. The reference household upon which the results in this work are based has an annual electricity consumption of 4200 kWh and corresponds to a five-person household in the VDI 4655 classification. The load profile stems from region 12 of the VDI 4655 profiles, which matches the location where the PV data was sourced (Karlsruhe), and the data has a resolution of one minute.

The reference day measurement procedure allows one to calculate the average efficiency of both the battery (Eff_{BAT}) as well as the PV-storage system as a whole (Eff_{SYS}), for each household.

These quantities can be calculated for a synthetic year using Equation 1 and 2,

Equation 1: Battery efficiency in %

$$Eff_{BAT} = \frac{\sum_{\text{all ref. d}} (E_{\text{BAT dischg, DC, d}} * n_{\text{ref. d}})}{\sum_{\text{all ref. d}} (E_{\text{BAT chg, DC, d}} * n_{\text{ref. d}})} * 100\ \%$$

Equation 2: Total system efficiency, including PV inverter

$$Eff_{SYS} = \frac{\sum_{\text{all ref. d}} ((E_{\text{SC, d}} + E_{\text{FI, d}}) * n_{\text{ref. d}})}{\sum_{\text{all ref. d}} (E_{\text{PVS, DC, d}} * n_{\text{ref. d}})} * 100\ \%,$$

where $n_{\text{ref. d}}$ represents the number of reference days per type of day for region 12 (Karlsruhe) of the VDI 4655 classification. $E_{\text{BAT dischg, DC, d}}$ is the energy discharged during one reference day (measured on the DC side), $E_{\text{BAT chg, DC, d}}$ is the energy stored within the battery during one reference day (measured on the DC side), $E_{\text{SC, d}}$ is the energy directly consumed by the system operator per day, $E_{\text{FI, d}}$ is the energy fed into the grid per day and $E_{\text{PVS, DC, d}}$ is the energy yield of the PV plant per day.

In order to examine the dependence of conversion efficiencies on typical power flow profiles it is useful to calculate the so-called "residual power",

Equation 3: Residual power

$$P_{\text{residual}} = P_{\text{PV}} - P_{\text{load}},$$

This represents the power that would ideally have to be provided by the battery in order to cover the load, so that positive (negative) values mean that the battery should be charged (discharged). This was calculated for a whole year using real PV data (see description above) combined with a yearly load profile generated using the reference days as well as the frequency of each reference day.

2.2 Measurement apparatus

The storage systems are tested within a hardware-in-the-loop test environment, as detailed in the *Efficiency Guidelines for PV-storage Systems* [5]. Load and PV generators are used to emulate the households and PV system, respectively, and both are controlled by a LabView program, with a communication rate (for both data recording and setting of new targets) of 200 ms. The storage systems themselves decide whether to charge or discharge the battery, depending on the direction of the power flows with respect to the grid. The different powers are measured by sensors integrated within the test environment.

2.3 Storage systems under test

The evaluation is done for six different DC-coupled storage systems (cf. Table 1).

Table 1: Overview of storage systems under test

System	A	B	C	D	E	F
Useable capacity (kWh)	3.9	2.1	5.3	5.0	2.9	4.3
Converter nominal AC output power (kW)	5.0	3.7	4.6	5.0	3.0	6.0

The useable battery capacity is in this case defined as the capacity that is available to the system operator during normal operation. These values were determined using cycle tests according to [3] and represent the average of all cycles, i.e. the average value of the energy with which the battery was charged and discharged.

3 RESULTS

The efficiency of both the battery as well as the entire storage system over the course of a synthetic year can be determined from the reference day measurements using Equation 1 and Equation 2; the results for six storage systems for PV plant sizes of 3.5 kWp and 4.8 kWp are displayed in Figure 1 and Figure 2, respectively.

Table 2: Variation of the battery and efficiencies

	PV plant size	Min value for all systems (%)	Max value for all systems (%)
Battery efficiency	3.5 kWp	92.25	96.87
	4.8 kWp	92.56	98.21
System efficiency	3.5 kWp	79.54	94.15
	4.8 kWp	84.07	94.67

The variation of the battery as well as the system efficiency for the two different PV plant sizes is shown in Table 2. Note that the system efficiency in this case refers to the efficiency of the entire system, where losses due to direct use of PV power (both self-consumption and grid feed-in) have been taken into account (see Equation 2).

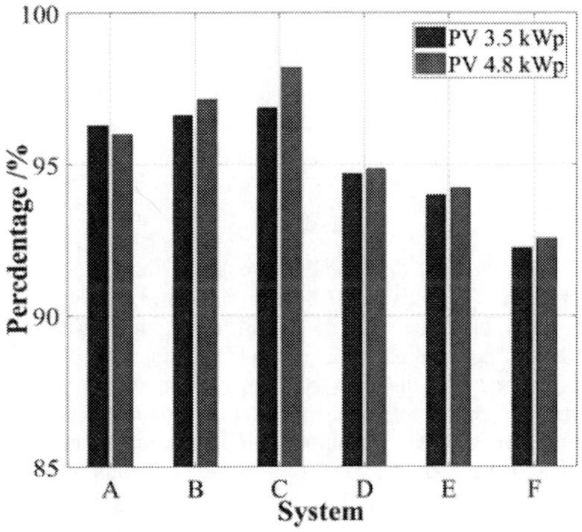

Figure 1: Comparison of battery efficiencies for the systems A to E, measured for a reference five-person household with an annual electricity consumption of 4200 kWh and a PV system size of 3.5 kWp and 4.8 kWp respectively.

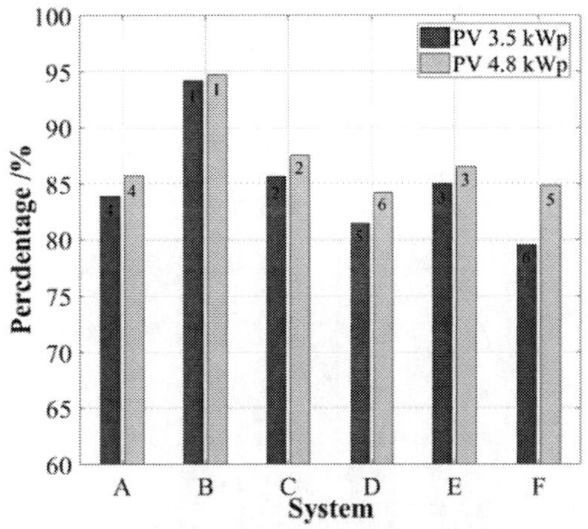

Figure 2: Comparison of total system efficiencies for the systems A to E, measured for a reference five-person household with an annual electricity consumption of 4200 kWh and a PV system size of 3.5 kWp and 4.8 kWp respectively.

The numbers within the bars in Figure 2 represent the rank of the different systems for both tests (3.5 kWp and 4.8 kWp), where 1 represents the highest efficiency. It is evident that in general the relative battery and system efficiencies of the six systems are independent of PV plant size, i.e., the ranking of the systems according to efficiency remains the same, with the exception of the rank of systems D and F in terms of system efficiency (see Figure 2). While the battery efficiency varies only slightly between the systems, the system efficiency increases for all systems when the systems are tested with a larger PV plant. This increase in efficiency is given explicitly in terms of percentage points in Table 3.

Table 3: System efficiency increase (in %) due to larger PV system size (from 3.5 kWp to 4.8 kWp), for storage systems A to E, for a reference five-person household with annual electricity consumption of 4200 kWh

System	A	B	C	D	E	F
Percentage increase in efficiency	1.8	0.5	1.9	2.8	1.5	5.3

4 DISCUSSION

The results described above show that the system efficiency slightly increases with a bigger PV plant. System efficiency is in general a function of energy losses that depend on several factors such as the efficiency of the battery, power conversion losses, standby consumption on both the AC and DC sides as well as additional consumption of peripheral devices such as sensors. An increase in efficiency means that the overall losses decrease in comparison to the PV energy yield.

Note that although maximum power point (MPP) tracking losses also play a role for some systems, they are excluded in this evaluation due to the fact that the maximum output power of some of the systems is smaller than 4.8 kW, so that no useful conclusions can be drawn by calculating the MPP tracking losses.

For both PV plant sizes the measurements show that power conversion losses account for between 59 % and 93 % of the total losses, depending on the system under consideration. Figure 3 shows the proportion of each type of losses for a PV plant of 3.5 kWp. The percentage of each loss only varies slightly (between 0 % and 4.5 %) when comparing the figures of the measurements with a PV plant of 3.5 kWp and 4. 8 kWp.

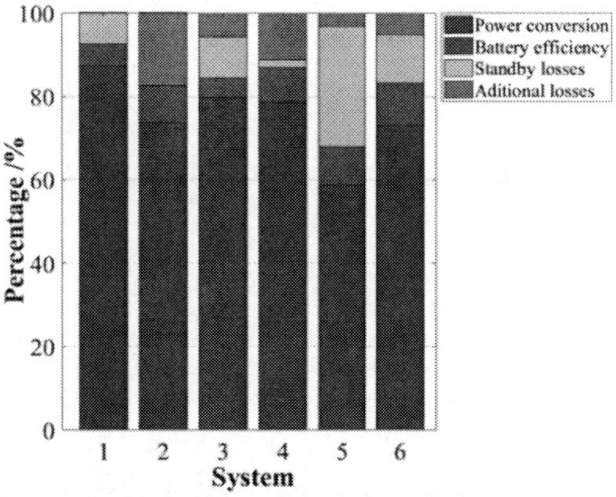

Figure 3: Comparison of the losses of system A to E, measured for the reference household with a PV system size of 3.5 kWp

The increase in system efficiency due to PV system size can be explained by closer examination of the distributions of both the charge and discharge powers of the battery as well as that of the grid feed-in power over the course of the year. To this end the residual power is calculated (see Equation 3), for PV plant sizes between 3.5 kWp and 6.5 kWp, for an entire year. The negative values of residual power, which represent the potential power which would ideally need to be discharged from the battery, are hardly influenced by the size of the PV plant (see Figure 4). Therefore it can be concluded that the PV plant size does not influence the battery discharge power.

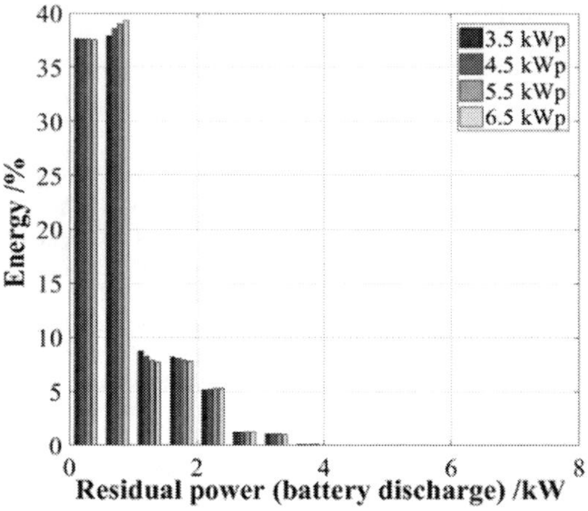

Figure 4: Comparison of the distribution of the residual power (negative power) over a whole year for a reference five-person household with an annual electricity consumption of 4200 kWh and a PV system size of between 3.5 kWp and 6.5 kWp

In comparison, Figure 5 shows the distribution of (positive) powers with which the battery would ideally need to be charged over the course of the year. In this case the distribution of energy at certain powers is largely influenced by the size of the PV plant. Since the energy that is not stored in the battery is fed into the grid, a larger PV plant also influences the amount of energy that is fed into the grid.

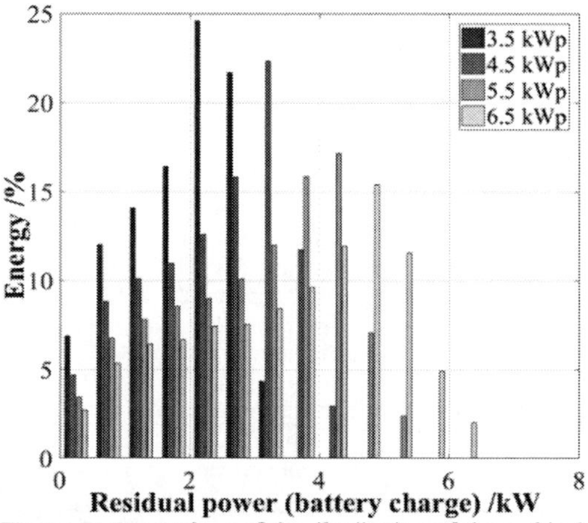

Figure 5: Comparison of the distribution of the residual power over a whole year for a reference five-person household with an annual electricity consumption of 4200 kWh and a PV system size between 3.5 kWp and 6.5 kWp

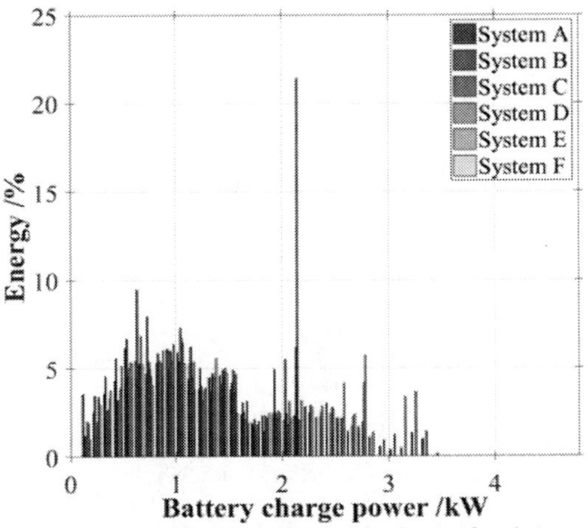

Figure 6: Distribution of energy flows for battery charging over a synthetic year for the systems A to E

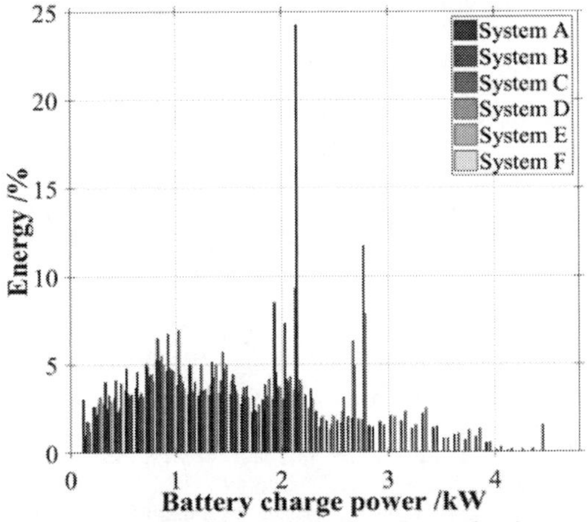

Figure 7: Distribution of energy flows for battery discharging over a synthetic year for the systems A to E

From the distributions calculated using Equation 3 and shown in Figure 4 and Figure 5 it can clearly be seen that a larger PV plant leads to the battery being charged with higher powers more often. These plots can be compared with actual measured powers over the course of the year, for the different storage systems and different PV plant sizes (see Figure 6 and Figure 7). The average charge powers of the different storage systems during the course of the year are shown in Table 4, and the difference in charge powers between a PV plant size of 3.5 kWp and 4.8 kWp ranges between 71 W and 124 W.

Using the efficiency curve for battery charging (PV2BAT) for the six systems as well as the distribution of the energy flows for battery discharging over a synthetic year the average efficiency (see Table 4) of each system for battery charging can be calculated. The efficiency curve PV2BAT is measured according to the efficiency guidelines [5]. In this case an increase in the PV plant size from 3.5 kWp to 4.8 kWp leads to an average efficiency increase for PV2BAT of between 0.6 % and 1.5 %, which is indeed one reason why the total system efficiency is proportional to the PV plant size.

Table 4: Comparison of the average efficiency of battery charging as well as the average battery charge power of systems A to E, measured for the reference household with a PV system size of 3.5 kWp and 4.8 kWp

	A	B	C	D	E	F
			Ø Eff. PV2BAT (%)			
3.5 kWp	84.3	94.3	87.9	85.5	90.1	83.2
4.8 kWp	85.6	94.9	88.2	86.6	90.8	84.7
			Ø Power (W)			
3.5 kWp	532	585	767	679	628	668
4.8 kWp	607	707	838	803	727	750

Using the average efficiency for battery charging as well as the energy stored in the battery over the course of the year, the input power of the battery charging device can be calculated. To determine the final AC energy that is either directly consumed by the system operator or fed into the grid, the battery input power needs to be subtracted from the produced PV power. By comparing the results of the measurements with 3.5 kWp and 4.8 kWp PV plants it can be shown that there is an increase in overall active AC energy output, i.e., the sum of the energy directly consumed and energy fed into the grid, of between 27.5 % and 42.7 % (see Table 5). This in turn results in less energy being stored in the battery and ultimately smaller total energy losses through the charge/discharge process. In addition, since the overall amount of energy with which the battery is charged over the course of the year increases by between 9 % and 18 %, the system spends less time in standby mode, so that a small decrease in standby consumption can be observed in some of the systems.

Table 5: Increase in energy that is either directly used or directly fed into the grid with a PV plant size of 4.8 kWp, for systems A to E

	A	B	C	D	E	F
Increase in energy directly used or fed into the grid (%)	42.7	35.5	40.3	46.5	27.5	49.4

The results above show that a larger PV plant will in the end influence the final AC power output of the DC/AC converter, since with more PV there is proportionally less energy transfer to and from the battery. Figure 8 and Figure 9 show the distributions of the converter's energy output over the course of a synthetic year, for PV plant sizes of 3.5 kWp and 4.8 kWp respectively. It can be seen that the AC energy output of the storage system shifts to higher powers, which results in an increase in the average power of between 165 W and 272 W (see Table 6).

Table 6: Comparison of the average AC power output of the converter of systems A to E, measured for a reference household with PV system sizes of 3.5 kWp and 4.8 kWp

	A	B	C	D	E	F
			Ø Power (W)			
3.5 kWp	532	585	767	679	628	668
4.8 kWp	607	707	838	803	727	750

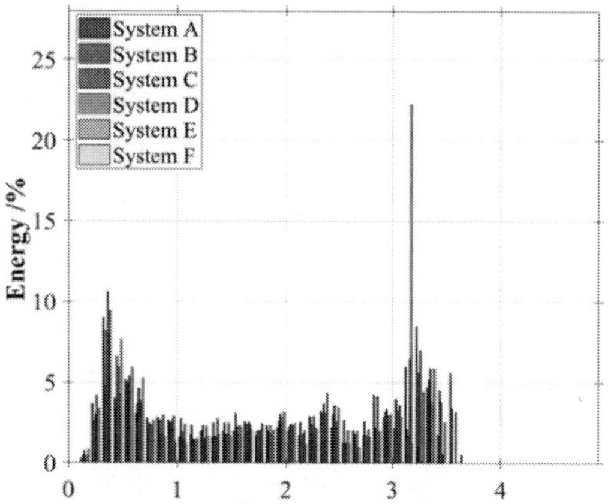

Figure 8: Distribution of AC energy output from the DC/AC converter over a synthetic year for the systems A to E for a reference five-person household with an annual electricity consumption of 4200 kWh and a PV system size of 3.5 kWp

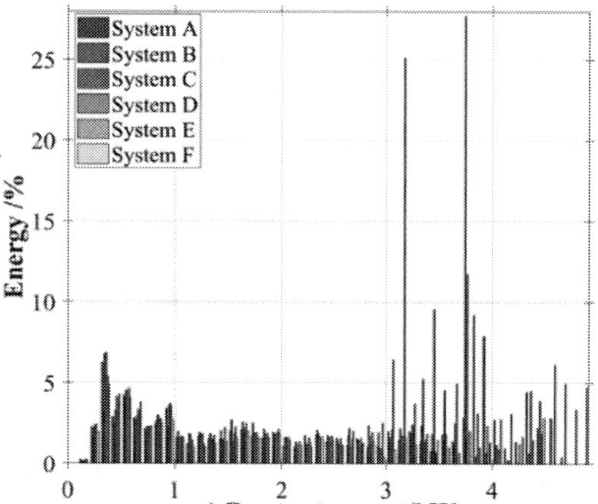

Figure 9: Same as Figure 9, for a PV system size of 4.8 kWp

Due to the fact that the AC output power is a sum of the power from battery discharging and PV power that is directly used by the system operator or fed into the grid, the efficiency of the converter cannot be determined by using the energy distributions and path efficiencies of the paths PV2AC or BAT2AC according to the efficiency guidelines [5]. To determine the increase in efficiency due to a large PV plant further evaluation is necessary. Nevertheless it can already be seen that the efficiency of DC/AC conversion most probably increases due to the fact that more energy is converted to AC at a higher power.

As described in the results section above, increasing the PV plant size leads to systems D and F changing their positions in the system efficiency ranking. It can be seen that system F has the largest converter size of all systems. The step up in its ranking is probably due to an over proportional increase of the converter efficiency of system F. This is mainly due to the fact that a 3.5 kWp PV plant is quite small for a converter which has a nominal output power of 6.0 kWp. Therefore the system efficiency of system F increases more than that of all the other system when it is operated with a larger PV plant.

5 CONCLUSION

With an increase in PV plant size, the overall system efficiency increases for all systems, while the battery efficiency varies slightly.

The increase in system efficiency is mainly due to an increase in average battery charge power and in the amount of energy that is directly used by the system operator or fed into the grid. The increase of the average battery charge power results in a small increase in the overall average efficiency of battery charging. The energy directly used by the system operator or fed into the grid increases by between 27.5 % and 42.7 %. This leads to higher system efficiencies because a smaller proportion of the energy is stored within the battery (fewer losses due to DC/DC conversion and battery efficiency) and due to the increase in converter efficiency. To determine the increase in efficiency of the converter further evaluations are necessary. In addition to the decrease in conversion losses the standby consumption also decreases.

As already discussed, increasing the PV plant size causes Systems D and F to change their position in the ranking of system efficiency. Nevertheless it can be concluded that a comparison of storage systems according to their battery efficiency and system efficiency is considered to be an appropriate method when taking into account that the ranking of the other four systems remains the same. A realistic comparison can be made if one simply makes sure that the PV plant size is selected appropriately.

ACKNOWLEDGEMENTS

The authors thank the German Federal Ministry for Economic Affairs and Energy (BMWi) and the Projektträger Jülich (PtJ) for support of the project “SafetyFirst” (grant no. 03ET6055A).

REFERENCES

[1] European Commission, “Database - Eurostat,” *eurostat*, 12-Jan-2016. [Online]. Available: http://ec.europa.eu/eurostat/data/database. [Accessed: 14-Jan-2016].

[2] O. Wollersheim and A. Gutsch, “Transparenz ist gefragt,” *PV-Mag.*, Jun. 2015.

[3] N. Munzke, B. Schwarz, F. Büchle, and J. Barry, “Lithium-Ionen Heimspeichersysteme: Performance auf dem Prüfstand,” presented at the 32. Symposium Photovoltaische Solarenergie, Kloster Banz Bad Staffelstein, 2017, vol. 32.

[4] VDI, “VDI-Richtlinie: VDI 4655 Referenzlastprofile von Ein- und Mehrfamilienhäusern für den Einsatz von KWK-Anlagen,” Beuth Verlag, 2008.

[5] “Effizienzleitfaden für PV-Speichersysteme.” BVES, BSW, Mar-2017.

GRID-RELIEVING EFFECTS OF PV BATTERY ENERGY STORAGE SYSTEMS WITH OPTIMIZED OPERATION STRATEGIES

Sebastian Zurmühlen[1,2,3], Georg Angenendt[1,2,3], Marcel Heinrich[1], Hendrik Axelsen[1,2,3], Dirk Uwe Sauer[1,2,3]
[1]RWTH Aachen University, Department of Electrochemical Energy Conversion and Storage Systems, Institute for Power Electronics and Electrical Drives (ISEA)
[2]RWTH Aachen University, Institute for Power Generation and Storage Systems (PGS), E.ON ERC
[3]Jülich Aachen Research Alliance, JARA-Energy
Jägerstr. 17-19, 52056 Aachen, Germany

ABSTRACT: The German distribution grids are facing a rising number of installations of private photovoltaic (PV) systems. Simultaneously, the number of installations of PV panels in combination with battery energy storage systems (PV BESS) is increasing. With appropriate operation strategies that consider grid limitations PV BESS have the potential to provide grid relief and thus reduce overall system cost. The effect of solar power generation units on the low voltage grid depends on the amount of installed generators, but also on the operation strategy of connected storage systems. The implemented operation strategy of the BESS determines the extent to which grid-relief can be achieved with storage systems.

A model of a distribution network based on field data is combined with the model of a PV BESS. The combined tool is used for power flow analysis. With this model of the whole distribution system, different penetration rates for PV systems as well as for PV BESS are simulated and the resulting influence on the grid is examined. Results show that a grid-oriented operation strategy can gain significant relief of the low voltage grid and therefore reduce otherwise necessary grid reinforcements in the near future. At the same time, these operation strategies do not lead to a significant decrease of the overall energy throughput of the PV BESS over the total lifetime of the system. This means the economics of such a system are only slightly influenced by the application of a grid-relieving strategy.

Keywords: Battery Storage and Control, Grid Integration, Grid Stability, Storage, Strategy

1 INTRODUCTION

The German power grid is facing a continuously rising number of photovoltaic power plants and therefore a growing share of fluctuating electricity production. This trend is expected to continue, due to goals set for the share of electricity from renewable energy sources by the federal government and falling prices of PV systems. The predominant share of the installed capacity of photovoltaic power systems is connected to the medium voltage distribution grid or the low voltage grid [1]. The fluctuating feed-in of the PV power plants leads to a reassessment of the capabilities of the grid and possibly to grid reinforcements

On the other hand, the number of battery energy storage systems (BESS) steadily increases, with 35,000 installed PV BESS in Germany by January 2016 [2]. The operation of these storage systems can heavily influence the effect that fluctuating PV power feed-in has on the low voltage grid. Numerous papers investigated the additional benefit of using more sophisticated operation strategies than the simple self-consumption maximization strategy which charges the battery as soon as negative residual load occurs, i.e. the production of the PV system surpasses the electrical consumption of the household. These strategies aim to reduce the maximum PV power feed-in to gain grid relief.

Moshövel et al. analyze the possible grid relief from reduced PV power feed-in by the use of different operation strategies for PV BESS [3]. They figure out that the implementation of forecast-based operation strategies offers increased grid relief over strategies to maximize self-consumption. To reduce PV power feed-in without a notable decrease of solar generated energy, they deploy two different forecast strategies, perfect forecast and persistence forecast.

Weniger et al. investigate forecast-based operation strategies as well. They introduce different strategies and compare their benefit [4].

Angenendt et al. show that forecast-based operation strategies, which limit the maximum state of charge of the PV BESS can enhance the battery lifetime and therefore decrease the levelized cost of electricity [5].

This paper investigates the influence of PV systems on the low voltage grid and how the installation of BESS with respective operation strategies changes how the grid is affected. The influence is determined by means of simulations, using a model of an existing urban low voltage grid with the simulated operation of a DC-coupled PV BESS based on [6].

The decentralized feed-in from photovoltaics can result in power flows from the low voltage grid to the superimposed transmission network in times of high solar production. For now, this does not drastically influence the low voltage grid, but this might change in the near future. With an ongoing trend of new installations for PV in the residential sector, the occurrence of situations in which the low voltage transformer will reach its capacity limit may increase. Especially low voltage grids with relatively long power cables and a high number of connected PV systems might reach the line's capacity limits in terms of power transfer capability and thermal stress.

The level of energy autarky that can be achieved by increasing the self-consumption of electricity from PV may interfere with a grid-relieving operation of the BESS. The trade-off between a maximization of self-consumption of solar energy and minimization of stress on the low voltage grid results from the non-concurrency of solar power generation and power consumption within a household. The ideal dimensioning of the rated PV power, the electrical storage capacity of the BESS, its rated power, and the power consumption of the household has been part of extensive research [7]. The next step is to ensure a grid-relieving effect of the BESS while simultaneously achieving a high rate of self-consumption by means of an operation strategy of the BESS that incorporates both aspects at the same time.

2 OPERATION STRATEGIES OF BATTERY STORAGE SYSTEMS AND GRID INFLUENCE

In recent studies two different kinds of operation strategies have been discussed. Operation strategies to reduce the PV feed-in power to gain grid relief [3], [4] and operation strategies which are used to enhance the battery lifetime in order to enhance the economics of PV BESS [5]. These strategies use forecast-based operation strategies to reduce the average SOC. A higher average SOC leads to higher calendric ageing as described in various publications, e.g. [8]. Therefore, an operating strategy is introduced which stores only the energy demand of the following night in order to reduce the average SOC, resulting in an enhanced battery lifetime and only a moderate reduction of the self-consumption rate. The resulting SOC curve is shown in the following graph.

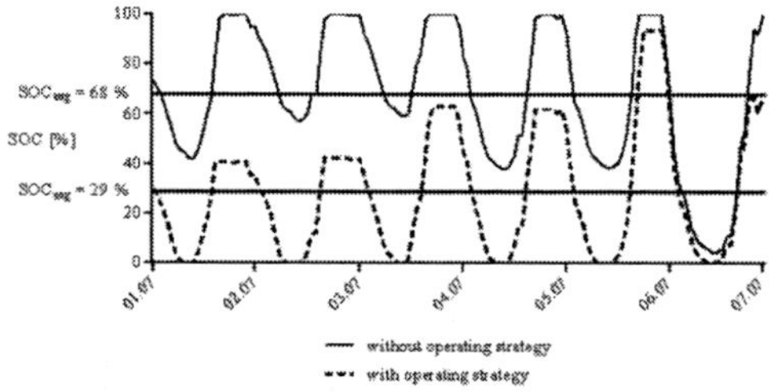

Figure 1: The state of charge of a PV BESS with and without lifetime enhancing operating strategy [5].

A combination of both types of strategy, enhancing battery lifetime while at the same time gaining grid relief, has been discussed in [9].

The presented approach uses the forecast-based lifetime enhancing operation strategy and extends it to take into account relevant network parameters of the low voltage grid. These parameters are the voltage level at the grid connection point, the grid frequency and the reactive power demand, i.e., the power factor cos(φ). The capability of the inverter system to act accordingly is defined in [10]. The principle is to store additional energy above the target SOC chosen by the forecast-based strategy to optimize the battery lifetime in order to support the grid and reduce the local voltage boost, if necessary. This strategy is called grid-oriented.

The relevant factors to determine grid influence of connected households are the voltage level at each grid connection point and the load on each electrical line of the network at any given time. We use the maximum value of the voltage variance ΔU and the maximum utilization of each line in relation to its power transmission capacity over a year as measures, as well as the number of times these two parameters exceed their limit. For the voltage, this means a violation of the $\Delta U \leq 3$ % limit and for the power line utilization the limit is set to 75 % of a line's transmission capacity.

The voltage increase at the end of a line with a high number of connected PV systems can quickly surpass the allowed voltage level of $\Delta U \leq 3$ %, thus resulting in a violation of the ±10 % U (400/230 V) limit [10], [11]. The operation strategy of the BESS is used to avoid a violation of these limits. The effectiveness is determined by power flow analysis.

3 MODELLING AND APPROACH

The presented approach combines a detailed model for households optionally including a PV panel and additional battery storage systems with a model for the low voltage grid and a power flow analysis.

Since not all households in the simulated grid have a PV system or a PV BESS installed, we use a set of varying load profiles to model the cumulative power demand in the analyzed low voltage grid. These load profiles were derived by using profiles for 32 typical domestic appliances and stochastically combining these into a household profile, also taking into account seasonal variations of usage [12]. Figure 2 illustrates the principal set-up of the different model components.

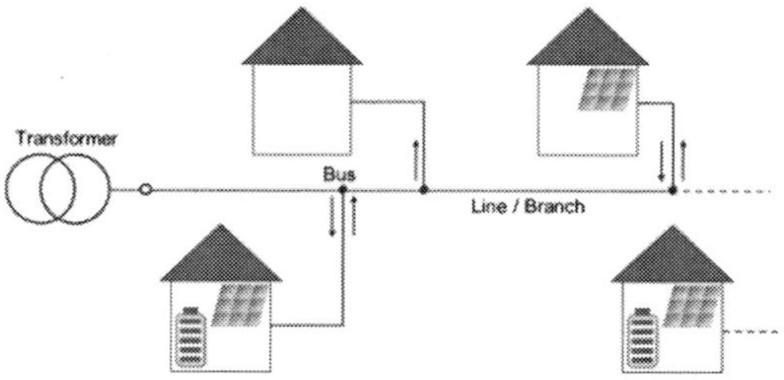

Figure 2: Model components overview – PV BESS and grid model.

The model for a household with a PV BESS is based on [6]. The system layout is shown in Figure 3 and consists of the PV panel, two DC/DC converters, one AC/DC converter, the battery energy storage system, the electrical load of the household and the grid connection. The energy management system (EMS) and the battery management system (BMS) are modelled as separate units as well. Basic operation prevents the BESS from charging from the grid, but only from the PV panel.

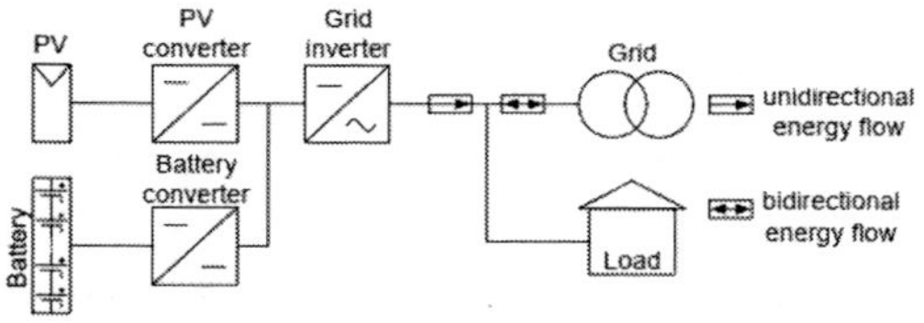

Figure 3: Model of the grid-connected, DC-coupled PV BESS [5], [6].

To determine the influence of PV battery storage systems on the low voltage grid, a quasi-stationary model has been developed to simulate the behavior of the PV BESS, combined with the repeatedly calculated power flow within the grid. The power flow calculation is based on [13] and has been adapted for an iterative simulation of the interlinked power system in combination with the PV BESS model.

The model simulates a large number of different PV battery storage systems which are connected to different busses within the power network and induce a drain of power or a power feed-in at each time step of the simulation. This is used to calculate the resulting power flow within the network topology and, thus, the voltage level at each node or bus in the system. For each time step, the solar production of all PV systems is simulated. This energy is added to the power consumption of each household for every time step, resulting in a residual load of the household. Depending on the existence of a storage unit, additional solar production can then be stored or fed into the grid. On the other hand, remaining positive residual load (thus a surplus of consumption over solar power generated) is either covered from energy stored in the battery system or taken from the grid. The decision, whether energy should be charged into the battery or discharged to cover the load within the household is made by the energy management system (EMS). The decision to charge or discharge the battery also depends on the distinct operation strategy, forecast data, and the actual state of the grid (e.g. frequency and voltage). The power flow calculation is conducted at an adjustable interval to update the voltage level at each node of the grid and to determine the load flow on each line.

The analyzed network is based on an existing urban low voltage grid in Germany. The modelled section has 124 nodes, 113 of which are households, 10 interconnections and the distribution substation. The grid has 128 power cable lines with a total length of 2.25 km. The layout of the grid is shown in Figure 5.

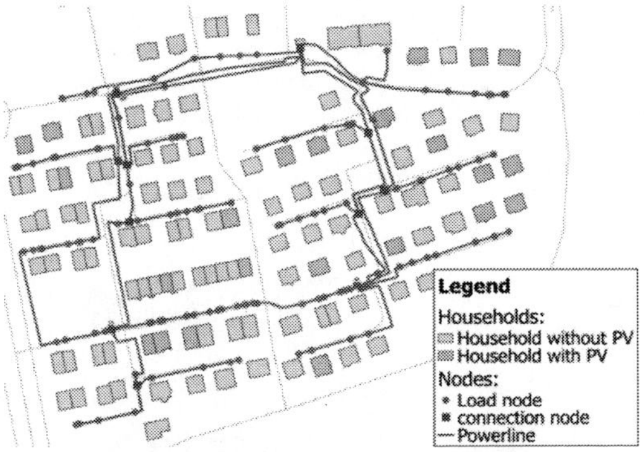

Figure 5: Layout of the modelled low voltage grid.

The results of the power flow analysis during simulation are fed into the EMS of the PV BESS model after each iteration step. The relevant parameters used to determine the grid-relieving operation are the actual voltage at the connection node of the BESS and the grid frequency. The EMS is able to adapt the charging strategy accordingly to reduce the power feed-in of the PV system and additionally, enable a feed-in of reactive power to decrease the voltage boost at line end [10].

4 RESULTS

Based on the aforementioned model different scenarios are analyzed. The results are evaluated based on the week with the highest PV feed-in. During this week the burden on the grid is expected to be the highest.

For the analysis, different scenarios are compared. First, a scenario without PV BESS is compared to scenarios with PV BESS. Different operation strategies for the PV BESS have been implemented. In the base scenario a strategy for maximization of self-consumption is used. This strategy is compared to grid-relieving strategies. The first grid-relieving operation strategy is the persistence forecast. This strategies aims to reduce the cut-off energy while at the same time reducing the PV feed-in peaks. Furthermore, this operation strategy aims to enhance the battery lifetime as well. The second grid-relieving strategy is the grid-oriented operation strategy which also takes into account grid parameters like the node voltage and transmission line utilization in order to reduce the influence of the PV systems on the grid. All three operation strategies are described in Chapter 2.

4.1 Effects of PV BESS operation strategies on the low voltage grid

A main indicator for the influence of PV systems on the low voltage grid is the effect of voltage boosts at the grid connection point caused by the power feed-in. If the voltage criterion is met, neither a voltage drop caused by the household load nor a voltage boost from feed-in results in a deviation from the standard grid voltage of more than three percent ($\Delta U \leq 3\,\%$) at a house's connection point. The analyzed grid has been designed to ensure this for the case of voltage fluctuations caused by electrical load. Through the impact of PV power feed-in on the other hand this limit can be well surpassed, if a certain amount of additional PV systems is installed. Figure 6 shows the voltage curve at the most critical node during a mid-summer day with a comparison for different installation rates of PV systems. In this scenario, we assume that there are no battery storage systems installed.

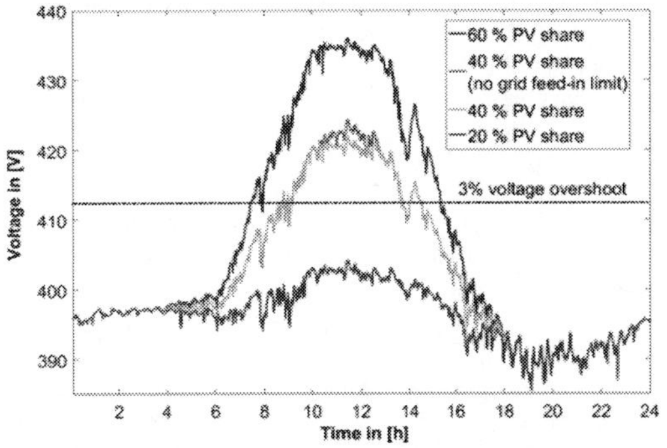

Figure 6: Comparison of node voltage for different share of installed PV systems.

In case of a 40 % share of installed PV systems the grid does already face a voltage increase above the allowed limit. A comparison of the orange and yellow curve shows that this is only slightly dampened by the 70 % limit for the maximum power feed-in by PV systems as it is part of the requirements for systems funded by the EEG. The difference between the voltage curve with and without this limit is almost negligible, as shown by the orange and yellow curves.

We evaluate the effectiveness of a BESS to act grid-relieving by its ability to reduce the voltage overshoot caused by PV feed-in and also by the reduction of stress put on the transmission lines in the grid. We use the 40 % PV share scenario, since it does already show situations in which the set limits are surpassed, but the assumed amount of installed PV systems is not unrealistic in the medium term. Figure 7 depicts the resulting voltage over the course of a day during summer for the same node of the grid as discussed before in Figure 6.

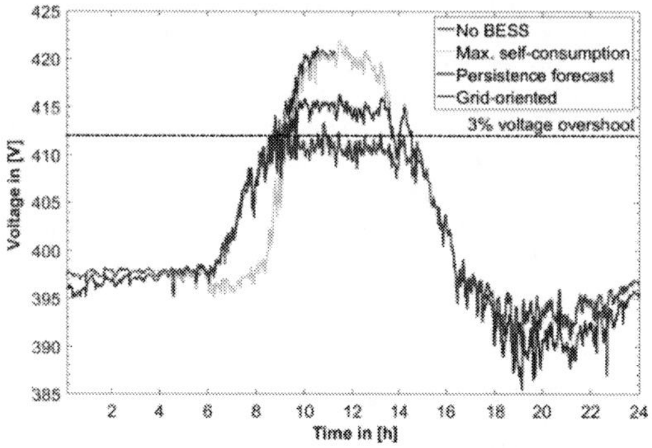

Figure 7: Comparison of node voltage for different PV BESS operation strategies.

It is clearly visible that the simple introduction of battery storage systems in the grid does not necessarily support the grid or help to reduce the occurrence of situations in which the grid is operated outside the allowed conditions. This is indicated by the overlapping yellow and red curves during midday. The battery systems which use a maximization of self-consumption charge as soon as additional power from the PV system is available and therefore become fully charged before the midday peak. Thus, they cannot help to reduce an overburden of the grid caused by massive feed-in from PV. If an operation strategy which uses a delayed charge is applied, like the forecast-based strategy to prolong battery lifetime, the battery is charged during noon, which reduces the power from PV fed into the grid. This strategy is illustrated by the green line in Figure 7. Still, in case of a significant share of households with photovoltaics, this does no longer suffice to ensure a grid operation within the set limits. This is caused by an accumulated voltage boost over the length of a transmission line. This may also cause unevenly distributed restrictions for households connected to the same line. The application of a voltage-based, grid-oriented strategy helps to avoid these instances and can ensure an operation within the allowed system bounds. This strategy, depicted by the blue curve, avoids a voltage overshoot at any node in the low voltage grid by storing additional power from the connected PV systems. Most of the time, the battery will have enough remaining capacity, since it has only been charged to the level necessary to provide for the demand of the following night. Only if the battery is already fully charged and the voltage boost still surpasses the three percent limit, the PV power will be cut off.

In addition to the voltage level at each node in the grid, the transmission limits of all power lines must be kept within the operating limits. Without any storage units providing grid relief, several lines of the simulated network surpass their allowed boundaries. This is shown in Figure 8. With the implementation of PV BESS with a grid-oriented operation strategy, all connected transmission lines are kept in the operating range. The same applies to the utilization of the technical assets of the distribution substation. If the transformer has been dimensioned to provide the maximum demand of all connected households, then a reasonable share of installed PV systems can lead to the transformer being overburdened during midday peaks. This means not only additional transmission capacities are needed to reinforce the grid, but the transformer has to be exchanged as well, which leads to additional costs without the use of BESS.

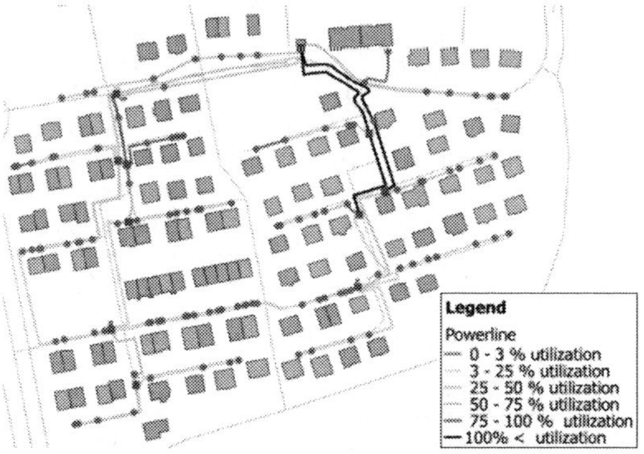

Figure 8: Transmission line utilization at 40 % PV share.

4.2 Influence of grid-reliving operation strategies on the economics of PV BESS

The evaluation of the economics of PV BESS using different operation strategies is done by the application of a detailed battery aging model based on [6].

As shown in the previous section the application of forecast-based operation strategies has a positive influence on the power grid. In this section the influence of the grid-supporting operation on the economy of a PV BESS is analyzed. The total energy throughput of a storage system is a good indicator for its profitability because this value shows how efficiently the BESS is used. The different operation strategies are evaluated using the 40 % PV share scenario.

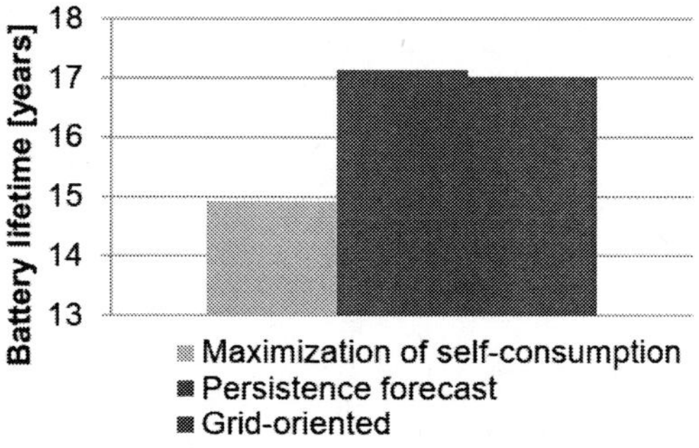

Figure 9: Average lifetime of the PV BESS.

As mentioned in Chapter 4, the persistence forecast operation strategy as well as the grid-oriented operation strategy use forecast-based algorithms to enhance the battery lifetime. The effect of these algorithms is depicted in Figure 9. The average lifetime of the installed PV BESS can be enhanced when the persistence forecast or the grid-oriented operation strategy are used in comparison to the simple maximization of self-consumption. The lifetime of the battery system using the persistence forecast operation strategy is slightly higher compared to the grid-oriented strategy because of the additional energy stored to relieve the grid. This leads to an increased average SOC at certain times a year with the grid-oriented strategy, which leads to slightly higher calendar aging of the battery system. This effect is rather negligible due to the few occurrences over the course of a year. The additional effect of cyclic aging is also insignificant in comparison.

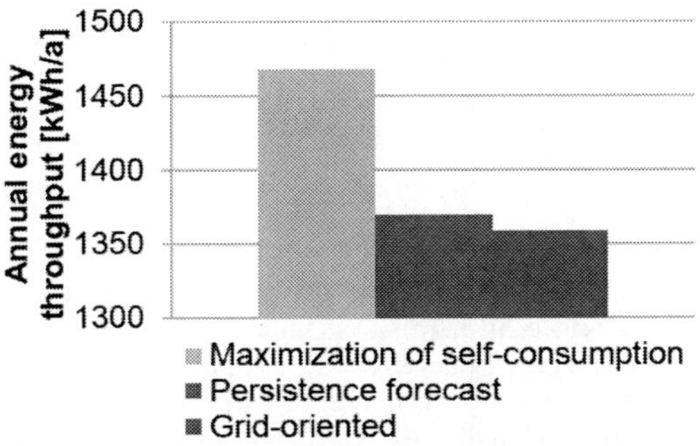

Figure 10: Annual energy throughput of the PV BESS.

Figure 10 illustrates the annual energy throughput of a PV BESS for the different operation strategies. If the maximization of self-consumption operation strategy is applied, the annual energy throughput is the highest. This strategy aims to increase the self-consumption by charging the battery as soon as positive residual power is available. Consequently, the energy throughput of the other strategies is lower, since charging is limited and also delayed with these strategies. Nevertheless, the annual throughput is only slightly lower because both strategies preferentially store cut-off energy in times of high power feed-in from the PV system.

The defining factor for the profitability of the BESS is the overall energy throughput over the total lifetime of the battery system. Since we do not assume any differences in the initial investment, the total yield from energy provided by the PV system depends on the effective lifetime of the BESS.

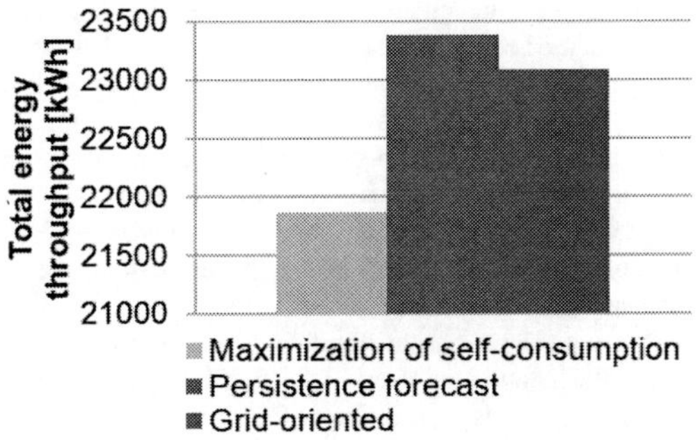

Figure 11: Total energy throughput of the PV BESS over lifetime.

To compare the energy throughput over the battery lifetime, the battery lifetime is multiplied with the annual throughput for each operation strategy. The results are depicted in Figure 11. The persistence forecast and the grid-oriented operation strategy have a distinctly higher energy throughput over the battery lifetime compared to the maximization of self-consumption. The total energy throughput only slightly differs between the persistence forecast and the grid-oriented operation strategy. This indicates that the grid-relieving operation of PV BESS does not influence the economy of such systems if these strategies are combined with lifetime enhancing algorithms. In summary, this paper shows that intelligent operation strategies of PV BESS can gain a significant grid relief without a negative influence on the economy.

5 ACKNOWLEDGEMENT

Part of this work took place and was funded under the project TILOS (Horizon 2020 Low Carbon Energy Local/small-scale storage LCE-08-2014). This project has received funding from the European Union's Horizon 2020 research and innovation programme under Grant Agreement No 646529.

Part of this study was conducted in the framework of the research project PV Home Storage Systems (PV-HOST) with the project number 0325477 supported by the German Federal Ministry for Economic Affairs and Energy (BMWi). The authors take full and sole responsibility for the content of this paper.

6 REFERENCES

[1] Harry Wirth, "Aktuelle Fakten zur Photovoltaik in Deutschland". Fraunhofer ISE, Freiburg, 2017.

[2] Kai-Philipp Kairies et al., "Wissenschaftliches Mess- und Evaluierungsprogramm Solarstromspeicher Jahresbericht 2016". Aachen, 2016.

[3] Janina Moshövel et al., "Analysis of the maximal possible grid relief from PV-peak-power impacts by using storage systems for increased self-consumption". Applied Energy, no. 137, pp. 567-575, August 2014.

[4] Johannes Weniger et al., "Integration of PV power and load forecasts into the operation of residential PV battery systems". Berlin, 2014.

[5] Georg Angenendt et al., "Enhancing Battery Lifetime in PV Battery Home Storage Systems using forecast based operating strategies". Energy Procedia, no. 99, pp. 80-88, 2016.

[6] Dirk Magnor et al., "Concept of a battery aging model for lithium-ion batteries considering the lifetime dependency on the operation strategy". 24th European Photovoltaic Solar Energy Conference, Hamburg, 2009.

[7] Dirk Magnor et al., "Optimization of PV Battery Systems Using Genetic Algorithms". Energy Procedia, no. 99, pp. 332-340, 2016.

[8] Madeleine Ecker et al., "Calendar and cycle life study of Li(NiMnCo)O2-based 18650 lithium-ion batteries". Journal of Power Sources, vol. 248, pp. 839-851, 2014.

[9] Georg Angenendt et al. "Reducing Cut-Off Energy of PV Home Storage Systems by Using Forecast Based Operation Strategies". 11th International Renewable Energy Storage Conference, Düsseldorf, 2017.

[10] VDE-AR-N 4105: 2017-07, "Erzeugungsanlagen am Niederspannungsnetz - Technische Mindestanforderungen für Anschluss und Parallelbetrieb von Erzeugungsanlagen am Niederspannungsnetz". 2017.

[11] Rolf Witzmann, "Empfehlung zur Richtlinie zum Anschluss von Erzeugungsanlagen an das Niederspannungsnetz". München, 2009.

[12] M. Bost et al., "Effekte von Eigenverbrauch und Netzparität bei der Photovoltaik". Berlin, Hamburg, 2011.

[13] R. D. Zimmermann et al., "Matpower: Steady-State Operations, Planning and Analysis Tools for Power Systems Research and Education". IEEE Transactions on Power Systems, Feb. 2011.

This is a pre-peer review version of the paper submitted to the journal Progress in Photovoltaics.
It has been selected by the Executive Committee of the EU PVSEC 2017 for submission to Progress in Photovoltaics.

Identifying risks, costs and lessons from ARENA-funded off-grid renewable energy projects in regional Australia

Bert Herteleer[1], Anthony Dobb[2], Olivia Boyd[2], Steven Rodgers[2], Lyndon Frearson[1]

[1]Ekistica, Desert Knowledge Precinct, Alice Springs, 0870, Northern Territory, Australia, bert.herteleer@ekistica.com.au
[2]Australian Renewable Energy Agency (ARENA), Level 2, 2 Phillip Law St NewActon, Canberra City, 2601, Australian Capital Territory, Australia

ABSTRACT

The Australian Renewable Energy Agency, ARENA, has provided grant funding to 18 off-grid and fringe-of-grid renewable energy projects under the Regional Australia's Renewables (RAR) program since 2013. This program was designed to address real and perceived risks associated with early-stage, pre-commercial renewable energy development, and provide a foundation of demonstration projects to enable the development of a competitive renewable energy sector. These projects range from low to high Renewable Energy Fractions at megawatt-scale in remote regions of Australia, and encompass a variety of sectors, such as mining, tourism, and remote communities. All projects use PV as a key technology, often supplemented by additional technologies. The experience from these projects shows that land acquisition, technical integration, stakeholder engagement, and access to finance are among the main reasons for project delivery delays. A qualitative assessment for the remoteness premium is given, based on a comparison of ARENA-funded on- and off-grid projects. This indicates that the structural barriers of governance, supply chains and finance need to be tackled further to lower soft costs. One of the key enablers for future lower renewable energy costs is ARENA's Knowledge Sharing model, through which the funding agency is recompensed by data and information that is provided to the market, and increases the impact of ARENA funding.

Keywords: Knowledge Sharing, structural barriers, off-grid, remoteness premium, risks.

1. INTRODUCTION

Globally, estimates for the number of people with limited or no access to reliable (on-grid) electricity is more than 1.2 billion [1], [2]. Predominantly, these people live in high-insolation areas, such as Sub-Saharan Africa and India [2]. In this context, the potential exists for a rapid cost-effective electrification with low(er) carbon dioxide emissions using photovoltaic (PV) power as the key low-emissions generating technology.

The sustained cost declines observed worldwide for grid-connected PV and reported on using learning curves [3], [4], [5], [6], together with high fossil-fuel powered generation costs in Remote and Regional Australia, have resulted in revived interest in the use of renewable technologies for electric power supply in these areas: approximately 2% of Australia's population lives in Remote and Regional Australia as classified by the Australian Statistical Geography Standard (ASGS) [7], yet are in locations with some of the highest renewable energy resource potential – in particular solar power.

The people and businesses in Remote and Regional Australia account for circa 6% of the country's annual electricity use, with a 5:1 ratio of energy use of mining to remote communities [8]. Diesel and natural-gas powered generators dominate the Australian off-grid electric power supply, with limited renewable power capacity, particularly within larger mines and communities. The gap between what is then identified on paper as potential for PV applications and what is installed in practice remains large. It was within this context of rapidly improving economics of solar and wind, and the expectation that these technologies were now cost competitive in many parts of regional Australia, that the Australian Renewable Energy Agency (ARENA) launched the Regional Australia's Renewables (RAR) program in 2013.

The RAR program was developed to stimulate renewable energy uptake in remote areas with two streams: one focused on industry such as mines and tourism facilities, and another focused on remote communities and islands. The industry-focused stream included funding for fringe-of-grid projects, which are connected to a centralised grid, but are located in weak areas of the network, such as at the end of very long transmission lines, where diesel back-up is often required to maintain power supply. Grant funding under the RAR program was designed to cover a part of the total project cost (the remainder being funded by equity and/or debt by the proponent), and was aimed at bridging the (perceived or apparent) risk gap and/or improve the financial viability of the project. A critical difference of the RAR program to similar funding in other countries, is ARENA'S Knowledge Sharing requirement, which obliges project proponents to provide ARENA with information. This includes:

- Standard project information (system size, technologies used, application purpose, expected renewable energy penetration and/or fossil fuel displacement, ...);
- Financial models, including the impact of the ARENA grant on project finances;
- Operational data post completion for a period of up to three years; and,
- Lessons learnt (Knowledge Sharing) reports for external distribution to the market.

Through this, ARENA receives a wealth of objective and verified information on projects that is otherwise difficult to obtain, which can help to answer key questions relating to the uptake of renewables in the off-grid and fringe-of-

This is a pre-peer review version of the paper submitted to the journal Progress in Photovoltaics.
It has been selected by the Executive Committee of the EU PVSEC 2017 for submission to Progress in Photovoltaics.

grid space: "What is the cost premium for remote locations (the 'remoteness premium') when compared to on-grid projects?", "What is the risk appetite by key stakeholders (users and financiers) for such projects?" and, "How do first-mover costs and integration challenges evolve over time?". Furthermore, the data provided allows ARENA to benchmark future investments, to ensure that its funding continues to stimulate innovation throughout the value chain. The intended goal of ARENA's Knowledge Sharing on RAR-funded projects is to accelerate the impact of its funding and increase awareness in stakeholders, so that future costs can be reduced over time and that risks can be identified, quantified and, where possible, managed or reduced.

The terms Renewable Energy Fraction (REF_{period}) and Renewable Power Fraction (RPF) are used in this article instead of renewable penetration, as they allow better clarity for communication regarding how renewable power and energy serve the load, and how both the generation and the load must be considered when designing, or discussing, power systems where a renewable generator is included. Ambiguity around the term renewable penetration is removed by describing characteristics using both the REF and the RPF metrics.

$$REF_{period} = \frac{E_{Ren}}{E_{load}} \ [\%] \qquad (1)$$

$$RPF = \frac{P_{Ren}}{P_{load}} \ [\%] \qquad (2)$$

In both Eq (1) and Eq (2), the subscript *Ren* is for renewable energy or power respectively, where the load is the consumption of electricity by the end user. Conversely, the FPF and FEF can be defined as the fossil-powered equivalents of the RPF and REF, respectively. The period for the REF is often clear from the context, although misunderstandings may arise by conflating daily with annual REF values, and not considering either the load or the generation's seasonality – a communication and lexicon issue that has appeared repeatedly with the projects in the RAR portfolio.

This article discusses the RAR program as a potential template for other funding bodies such as governments, multilateral development banks (MDBs) or non-governmental organisations (NGOs). Some of the key risks encountered throughout the RAR program and its funded projects will be discussed, together with lessons learnt by ARENA, the funding recipients, and stakeholders. The costs and project delivery models of the RAR-funded projects are used to illustrate the structural barriers [9] encountered in supplying renewable power in remote locations, such as Remote and Regional Australia. Lastly, the RAR program serves as a "sandbox" in which components of the structural barriers can be tackled or elucidated, to achieve high Renewable Energy Fractions on the grid (isolated or not), of which certain aspects will increasingly manifest themselves in the coming years [10].

2. RAR PROGRAM OVERVIEW

2.1 Historical context in Australia

The Australian Renewable Energy Agency became active on 1 July 2012 with the mandate to increase the uptake of renewable energy in Australia, systematically lower its costs, and support and stimulate innovation throughout the value chain, from early research to initial commercial deployment.

From the outset, the off-grid sector represented an opportunity to rapidly progress the commercial competitiveness of renewable energy in a niche market. The two years leading up to the launch of the RAR program in 2013 saw the price of crude oil climb to over US$100/barrel and hold steady [11]. In 2012/13 remote and regional areas of Australia had (and continue to have) some of the highest marginal costs of generation anywhere in Australia, due to reliance on diesel- and gas-powered generation. These combined cost fundamentals underpinned the assumption that renewable energy would likely be competitive with diesel on a pure cost basis, once de-risked via demonstration projects.

PV module costs in Australia dropped up to 80% between 2008 and 2012 [12]. While on-grid residential PV boomed in Australia in response to favourable government support schemes, uptake of PV in the off-grid market was sluggish, suggesting that PV costs were not the only driver for off-grid deployment. Figure 1 shows the magnitude of the PV capacity additions supported by ARENA under the RAR program compared to previous years, and suggests that, while small compared to the on-grid market (by the end of 2016, more than 5 GW of on-grid residential PV had been installed in Australia [13]), the off-grid renewable power sector is emerging [14], [15], [16] after a long period of limited growth.

The growing need for off-grid mines, communities, and tourist facilities to reduce their cost of power supply offered the promise that renewable energy solutions, once proven, would have room to replicate. ARENA estimates in 2012 placed the share of off-grid power demand at 6-10% of Australia's total power demand, with off-grid power demand expected to grow at 9% per year, on the back of the Australian mining investment boom. The opportunity for off-grid renewables was further supported by the significant impact that high oil prices and projected demand growth would have on state and federal government budgets. To provide a basic service to remote areas at a cost comparable to urban customers, state and federal governments provide subsidies for power supply to utilities and independent power producers (IPPs). The expectation of continued fuel subsidies further reinforced the idea for government entities that finding renewable powered solutions for remote communities would deliver long-term benefits.

This is a pre-peer review version of the paper submitted to the journal Progress in Photovoltaics.
It has been selected by the Executive Committee of the EU PVSEC 2017 for submission to Progress in Photovoltaics.

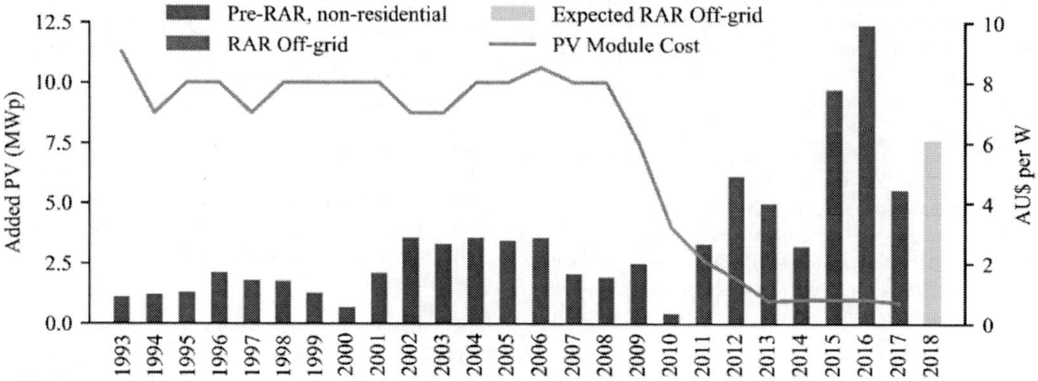

Figure 1: Annual PV capacity additions for off-grid, non-residential applications in Australia. Installation and cost data from 1993 to 2015 from *[25]*, *[12]*, RAR, and PV module cost data for 2016 onwards compiled by ARENA and Ekistica. The capacity additions for 2016 onwards are RAR (off-grid) only: other off-grid PV capacity additions in Australia are not shown here.

The vastness of Australia is well-known. Supply chains are drawn out over long distances and difficult terrain, complicating delivery of fuel and spares, and access to technical expertise. Moreover, large swathes of the country are regularly subject to the vagaries of the harsh weather conditions that can strongly constrain access, and impose severe strain on infrastructure. Australia has a country-specific definition for remoteness, which is based on distance from roads and population centres. This Remoteness Area classification is defined in [7], based on work by [17]. The off-grid projects in ARENA's RAR portfolio are in either Remote or Very Remote Australia (the highest on the remoteness scale). For context, reaching locations in Very Remote Australia often entails costly boat or air transport, or land travel from major Australian cities exceeding 500 km and up to 2000 km.

Attempting to answering a shared need of mines and remote communities for reliable, and less volatile costs for power supply, it was expected that a self-sustaining remote power supply market could develop, if renewable energy projects funded by ARENA's RAR program could shine a light on and tackle the structural barriers of governance, supply chains, and access to finance, through pilot projects serving mining and off-grid communities.

2.2 RAR program design

The fundamental premise of the design of the RAR program, and off-grid investment for ARENA, was that project economics, subsidy reduction potential, and strong forecast power demand growth constituted a powerful 'cost pull', but that an additional 'technology push' was needed to demonstrate to end-users that those theoretical cost savings could be realised without sacrificing power quality, reliability, and security. By building a suite of projects across community and industrial energy use, ARENA could retire risk perceptions surrounding the technology, and provide evidence of cost savings. The more forgiving cost environment of off-grid power supply meant that renewable energy (particularly solar PV) could be built and de-risked, at a time when utility scale PV was struggling to gain momentum on the Australian grid. Furthermore, by stimulating high Renewable Energy Fraction projects on isolated grids, it was (and is) possible to test renewable energy integration solutions in the off-grid "sandbox", with the findings then available for translation back on the main grid when needed.

The RAR program was designed to provide a total funding of up to AU$400 million (US$384M) for both off-grid projects and fringe-of-grid projects, with two separate funding streams focused on renewable energy supply for remote and regional communities (Community and Regional Renewable Energy or CARRE), and an industry funding stream, aimed at mining and tourism (Regional Australia's Renewables - Industry IRAR). The size of the RAR program funding to ARENA's total funding for projects at the time (AU$3,200M; US$3,072M) was designed to approximate the ratio of the remote and off-grid power demand to the total power demand in Australia.

To address the structural barriers to the commercial competitiveness of off-grid and fringe-of-grid renewable energy, the RAR program sought to tackle first mover costs and risks, and perceptions of the high cost and unreliability of renewable energy by funding a portfolio of demonstration projects. The program aimed to build 150 MW of total renewable energy capacity, including at least two utility-scale plants of $\geq$10 MW each, and to ensure that projects would be utilised for at least five years following commissioning. Sharing of project data and lessons learnt was added to the program, and included in the contractual obligations for successful applicants. By funding a suite of demonstration projects, ARENA also hoped to address strategic 'roadblocks', such as risks associated with the integration and management of renewable energy in a hybrid system, and increase skills and capacity associated with renewable hybrid systems in remote areas.

This is a pre-peer review version of the paper submitted to the journal Progress in Photovoltaics.
It has been selected by the Executive Committee of the EU PVSEC 2017 for submission to Progress in Photovoltaics.

ARENA's funding for the RAR program followed a stage-gate process that is still on-going:

1. Expression of Interest (EOI): Applicants were invited to submit a preliminary project proposal.
2. Full application: Following assessment of EOIs, shortlisted projects submitted a full project proposal to ARENA. This entailed a lengthier and more in-depth assessment process that involved external expert due diligence studies.
3. Funding Agreement: Once project proposals were agreed to, ARENA and the project proponents entered into funding negotiations, involving settling contractual obligations surrounding the funding amount to be paid to the proponent, project outcomes, and Knowledge Sharing.
4. Funds are disbursed to proponents upon successful completion of the agreed-upon milestones, subject to technical, legal, and financial due diligence.

To better enable communication among stakeholders and to tackle the lack of data capture standardisation, ARENA's RAR Data Specification [18] was developed. While the projects in the RAR portfolio will provide performance data and information to ARENA for subsequent distribution to the market, many more project configurations exist which are not yet captured in the RAR portfolio, and for which additional data will be beneficial.

3. RAR PROGRAM: APPLICATIONS, APPROVALS, AND RESULTS

ARENA received 72 applications for RAR funding in the initial application period (June – December 2013), indicating a high level of interest from off-grid and remote area consumers of electricity, and project developers. Of the 72 applications, 18 projects were approved for funding. Four of the approved projects (22%) are no longer active, with the funding made available by ARENA, but not disbursed. Figure 2 shows where the RAR-funded projects are located around Australia.

A description of the portfolio of active projects and the technologies used is given in Table 1. PV is the shared generating technology within the portfolio, although it generally needs to be supplemented by other technologies such as wind power or battery storage to achieve the design high annual Renewable Energy Fractions.

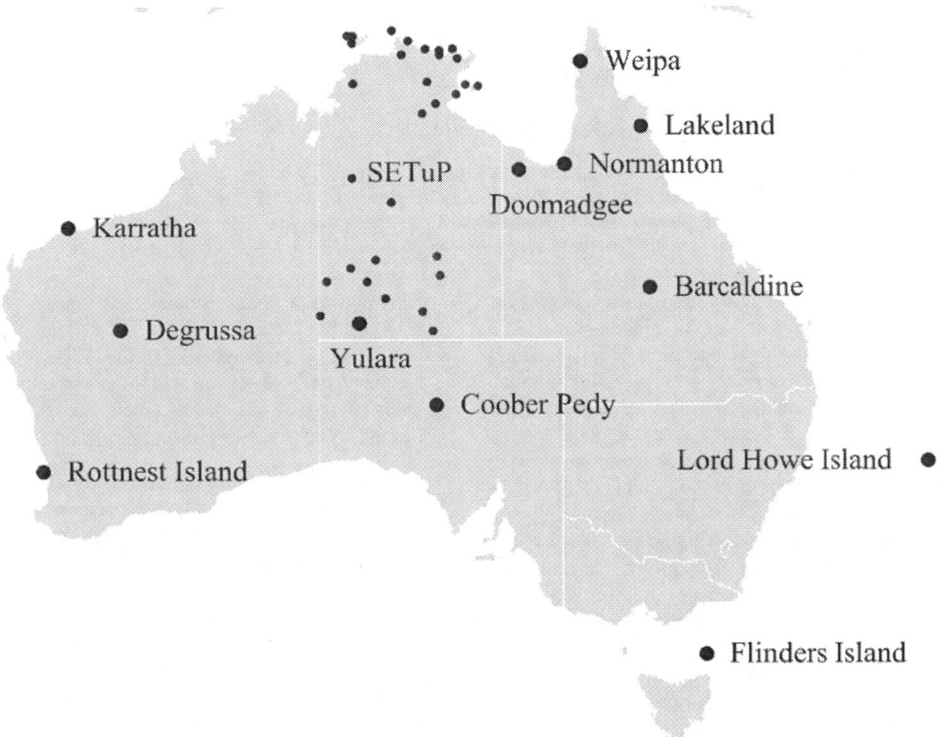

Figure 2: Locations of the ARENA-funded RAR projects around Australia. According to [7], much of the Northern Territory is classified as Very Remote Australia, with the locations of the SETuP-supported communities displayed in a different colour for easier identification.

This is a pre-peer review version of the paper submitted to the journal Progress in Photovoltaics.
It has been selected by the Executive Committee of the EU PVSEC 2017 for submission to Progress in Photovoltaics.

3.1 Costs and timing of RAR projects

To date, most active projects have been within 10% of their agreed-upon cost budgets. Figure 3 illustrates the scaled project costs, the design annual Renewable Energy Fraction, and the number of renewable technologies employed to reach the design annual REF values. The scaled project cost is determined as

$$P_c = \frac{\text{Total project cost (AU\$)}}{\text{Total RE net generating capacity } (W_{dc})} \left[\frac{\$}{W}\right] \tag{3}$$

where net renewable capacity is the DC (or DC-equivalent) nameplate net generating (i.e. solar and wind) capacity, and excludes storage.

3.2 Remoteness premium

With the data from the RAR projects' financial models at the time of funding by ARENA available, the PV-specific costs were extracted and compared to the ARENA-funded Large-Scale Solar (LSS) projects, which were granted funding in August 2016. The LSS projects are utility-scale, PV-only projects connected to either the National Electricity Market (NEM) or the South-Western Interconnected System (SWIS), and are in significantly less remote areas of Australia than the RAR-funded projects – more details about the LSS program can be found in [19]. Figure 4 shows the cost comparison of the RAR versus the LSS projects (AC system ratings used for both), which bring to light a few cost implications.

The effects of remoteness (transport costs, supply chain issues) is evident, with the highest PV-specific costs for the

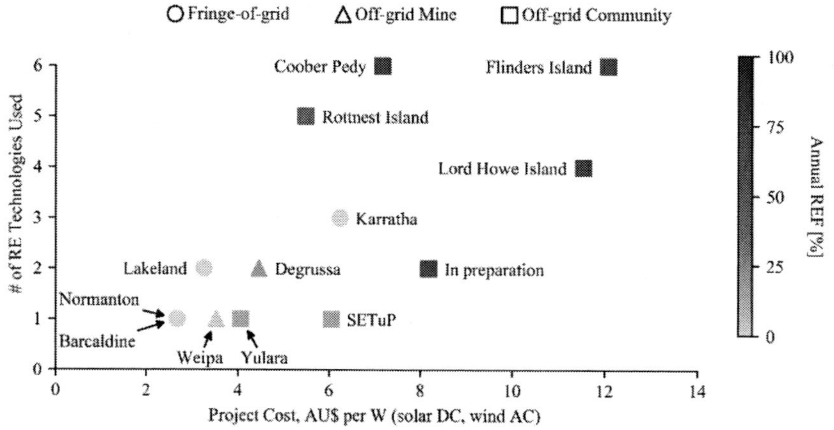

Figure 3: Comparison of the active ARENA RAR-funded projects, by project cost in AU$/W of net added renewable power capacity, annual Renewable Energy Fraction (REF), and number of renewable technologies employed. Costs shown are at the Final Investment Decision (FID) – ARENA Funding Agreement phase, and the dates for these vary between 2014 and 2016.

As can be observed from Figure 3, costs for projects vary strongly, with projects that aim for higher annual REF values generally showing higher (net renewable generating) costs. Similarly, more remote projects (particularly from a supply chain perspective) also are expected to have higher (transport) costs and a desire for higher REF values. Moreover, the associated technical complexity with integrating multiple technologies to reach high REF values (in general, more technologies are needed to reach higher annual REF values) also results in higher costs. Further cost differences relate to the timing of the projects: later projects were able to capture some benefits from lower global PV, wind and battery storage CAPEX costs.

Given the above, a more detailed look at the projects can give an indication as to what constitutes a "remote premium".

most remote locations. System scale impacts project costs to a similar extent, and this effect may reinforce or counteract the remoteness of the site. The projects with a later investment date were able to capture cost declines and industry learning, which lead to lower costs. The specifics, however, vary for each project and location.

For example, Degrussa, Coober Pedy and Yulara have similar PV-specific ($/W$_{ac}$) costs, yet have widely differing characteristics and system sizes. Degrussa is a remote mine with a single-axis tracking 10.56 MW$_p$ system, while Yulara at 1.83 MW$_p$ has multiple fixed-orientation arrays distributed on ground and rooftop, and Coober Pedy has one large 1.34 MW$_p$ fixed-orientation array, with the site's location having high integration costs. In the case of the islands in the RAR portfolio (Rottnest, Flinders and Lord Howe), the sub-megawatt size of the arrays and the remoteness work hand in hand to give PV-specific project costs that were significantly higher than Degrussa, Yulara or Coober Pedy.

This is a pre-peer review version of the paper submitted to the journal Progress in Photovoltaics.
It has been selected by the Executive Committee of the EU PVSEC 2017 for submission to Progress in Photovoltaics.

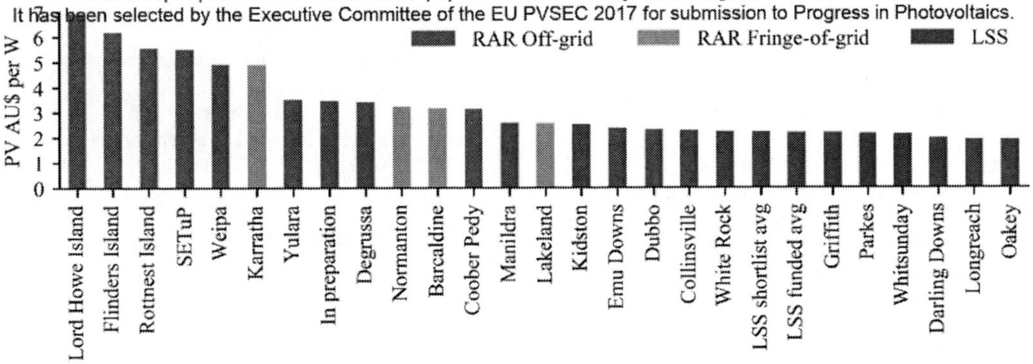

Figure 4: PV project component cost comparison of ARENA-funded projects at Final Investment Decision. RAR program FID dates were for 2013-2016, whereas Large Scale Solar (LSS) projects all received ARENA approval in 2016.

Similarly, the fringe-of-grid systems of Normanton, Barcaldine and Lakeland have the benefit of scale over Karratha and are less remote than the remainder of the RAR-funded off-grid projects, with commensurate lower costs. Lakeland has the lowest costs of the portfolio and the fringe-of-grid projects, due to its less remote location, and having the project refined several times prior to the final investment decision, thus enabling the project to capture improvements in the PV supply chain. Similar cost improvements are expected for the LSS projects for the final steps prior to construction: the values in Figure 4 were the basis for funding by ARENA to the projects, and for the projects to proceed to securing financing, and signing Engineering, Procurement and Construction (EPC) contracts. Nevertheless, contrasting the LSS projects to the RAR project portfolio, a scale and remoteness premium is clear. Given the many variables involved for the different projects, further study will be required to fully quantify the remoteness premium.

On the timing side, projects have – with few exceptions – not been able to complete within the planned, or expected, timeframe. The project duration is calculated from Full Application to Commercial Operation. For the projects yet to be completed, the expected completion date is calculated assuming that the project delivery plan is executed as planned with no knock-on delays. Figure 5 illustrates this and shows that when compared to the initially proposed timelines, all projects are on average 10 to 12 months late, and the shortest project completion is 20 months for the PV-only projects. On average, projects with more technologies to integrate take longer to complete.

However, the increased complexity of projects, evidenced by the number of technologies involved, has not led to the longest delays experienced in the portfolio, which are found among the PV-only projects. The main causes for delays in all projects were in obtaining financing, land use agreements, legal approvals, and to a minor extent, project execution delays.

These longer-than-expected project delivery schedules have a variety of causes, not all of which are a priori controllable or knowable, yet there are lessons learnt, and some risks bear further emphasis.

Land acquisition (or legal use thereof) for the project is vital, yet most, if not all, of the projects in the RAR portfolio did not have the luxury of *first* selecting and acquiring suitable land to *then* develop a project. The physical space around a remote location that is technically and economically feasible for renewable capacity additions to the existing electricity infrastructure is limited. This increases negotiation risks, as few alternative locations or counterparties exist, and many projects suffered significant delays.

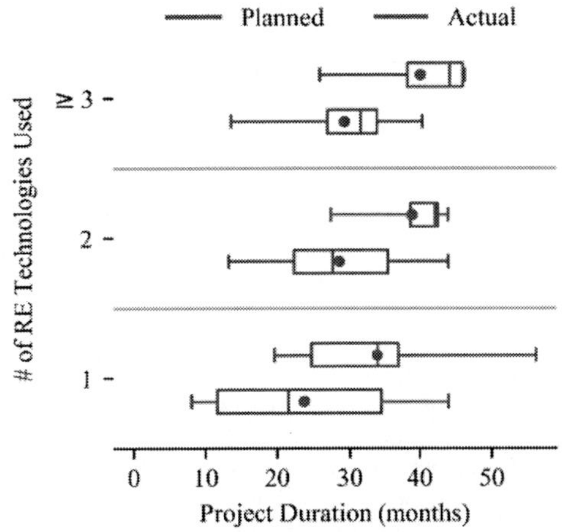

Figure 5: Box-plot comparison of RAR planned versus actual project durations to date, by number of renewable energy technologies employed. Each grouping has at least 4 (sub) projects. The dot and line indicate the mean and median, respectively.

The importance of stakeholder engagement and consultation is also higher in isolated grids, as there are limited to no opportunities to find more amenable counterparties if the project on the selected location is to proceed. In several projects, the need to educate co-decision makers, including local power station engineers who have had to change their operational paradigm, to board-level executives who had limited exposure to, and understanding of, the technologies proposed, caused delays beyond what had been anticipated.

This is a pre-peer review version of the paper submitted to the journal Progress in Photovoltaics.
It has been selected by the Executive Committee of the EU PVSEC 2017 for submission to Progress in Photovoltaics.

While efforts such as those employed at Yulara (geographically distributing PV around a location to reduce variability on the grid [20]) can be used to allow higher RPF and REF values on the grid with the same amount of spinning reserves, the nature of stand-alone grids is that they often are concentrating in operation: wind turbines and PV arrays can be dispersed only so far from the local load centre while remaining within economical limits, which thus lead to higher enabling technology needs and costs.

With extended project development timeframes, personnel change becomes a larger risk for all stakeholders. Similarly, significant electricity infrastructure upgrades tend to be few and far between, which carries the risk that the institutional memory of the local stakeholders may lapse. The experience of the RAR portfolio shows that the significant changes to the electricity infrastructure that these projects entail can and do take longer than the continued presence of decision-makers who changed throughout the course of the projects, causing a loss of the context of the project, the decision-making process and the incentives that drove these. This is compounded by the general lack of experience with renewables integration by local decision-makers, and the large variety in relevant regulations which often hamper the approvals and project delivery process. In some cases, project developers in the RAR portfolio needed to re-state the business case for a project to a new set of decision-makers, after having already committed to it. While this type of risk is common among many infrastructure projects, the long project lead times and relative unfamiliarity of stakeholders and decision-makers with renewable energy means that this risk is magnified for off-grid renewables.

The prolonged negotiations involved in the projects have also resulted in the risk of more stakeholders being involved, each with its own agenda and incentives. In one case, a project started with four main stakeholders. One of these, the state-owned electricity utility, was unbundled from a vertically integrated power company to three separate entities covering generation, distribution, and retail. This brought a new set of split incentives to the project. Where previously the loss of fossil fuel generation revenue due to the addition of end-user-owned, distributed PV could be compensated for internally by lower utility Operations & Maintenance (O&M) costs, unbundling of the utility unravelled the potential for internal transfer of cost savings. Including the financing aspects of the project, several more stakeholders were added between awarding the project funding by ARENA under the RAR program and project completion, bringing about negotiations which ballooned from 4 to more than 10 stakeholders at the same time, which pushed back the financial close of the project.

The need to adapt the project to local circumstances, such as for example educating the renewable energy project developers in mining safety regulations (which are additional to standard country-wide regulations), as well as educating the end-user about the technical capabilities of the renewable technology has also caused delays in completing the project within the anticipated timeframe. On a broader scale, equipment imported from overseas into Australia has occasionally failed to meet the local regulations, due to a lack of awareness or attention by foreign suppliers to the (slight) differences between Australian and international standards. The rectification work to achieve factory or site acceptance test certification then adds cost, and often creates knock-on delays for the project.

3.3 Lessons from the RAR portfolio

The key lesson (or rule) which is often forgotten, overlooked or underestimated by stakeholders for (renewable) electricity supply in remote locations, be they regulators, funding agencies, private sector participants or end users of electricity is that place, people and communication matter.

Place: communities, mining operations and tourist facilities are often in remote locations, far removed from supply of labour, spares, and institutional knowledge. Even if PV and other renewable hardware were to experience more aggressive cost declines than previously observed, this equipment still needs to be transported over long distances, installed with labour subject to remoteness or hardship premiums, to reach a location that is, by definition, difficult or costly to reach.

People: the expectations of the different stakeholders, some of which do not live in the remote environments discussed, need to be managed carefully, to ensure durable results for the end-users. For example, stakeholders sometimes expect that capital cost declines observed in other (often less remote) locations can be fully translated and passed on to the remote locations under study, while overlooking the integration and stakeholder engagement aspects.

Communication: Even if the stakeholders (e.g. ARENA and project proponents) have a solid understanding of the place and the people involved, there is still the need to communicate the concepts properly, and where needed, to spend time actively educating stakeholders to ensure that the magnitude and probability of risks are put into the appropriate context. Mitigating risk associated with end-user and stakeholder acceptance of the project is predicated upon good design and execution.

The risk appetite for a new technology to be added for power generation is often very low. The potential impact of a loss of revenue due to the risk of poor, or incomplete, integration of a renewable technology on the operations of a mine or a tourist resort is a multiple of the operational expenditure for the power and energy. While the absolute risk of failure in a remote site may not be materially different from an urban location with similar climatic conditions, the relative impact of such a failure on the local operations is much more severe, which then needs more operational guarantees to convince the local end-users to sign on for such changes. From feedback provided to ARENA, many developers of renewable power projects under-estimate the risk aversion that is present, over-estimate the capabilities of the technology being proposed, lack awareness of the technical limitations of the generation on-site, or are unable to fully articulate the relationship of risks to benefits, to convince the end-user to invest in the new technology.

The involvement of ARENA and other government departments also have led to higher due diligence, reporting and compliance requirements, which in some cases caused project delays. For some projects, the Key Performance Indicators (KPIs) to be reported on to satisfy ARENA's Knowledge Sharing requirements brought about prolonged negotiations. Further, as Knowledge Sharing was developed within ARENA during the RAR program's initial phase, some projects signed on to Knowledge Sharing requirements to answer questions that were no longer relevant by the project's commercial operation, which point to historically rigid requirements in the funding agreements. As the RAR program was among the first programs undertaken by ARENA, learnings such as these have since been incorporated into subsequent funding design, for example, for the LSS funding round [19].

While PV and similar renewable energy technologies have improved significantly in the past years, see [21], [22], [23], [24], they often lack all the characteristics that the traditional fossil-powered generators offer in one package or technology, such as dispatchability, (mechanical) inertia, rapid deployment, and all of these under the umbrella of financial viability. Various renewable and enabling technologies must be combined to achieve the same perception of reliability and control. For off-grid power supply, there are few cases where renewable technologies such as PV are accepted as a drop-in replacement or plug-and-play add-on to the incumbent technology. These situations are where enabling technologies ranging from low-load diesel generators to flywheels and batteries, to improved power station management and operation software can play a larger role for increased uptake of renewable power for isolated grids.

CONCLUSIONS

The continued experience and lessons from the projects funded by ARENA's Regional Australia's Renewables program point to issues for the integration of renewables which will increasingly manifest themselves in the coming years, for both on-grid and off-grid situations. A number of these issues are technical, and advances from either the on-grid or off-grid spheres can migrate to the other. On the other hand, with capital costs for renewable technologies decreasing over time, the relative importance of the structural barriers grows, as the ease of access to finance, the governance structures in place and the available supply chains will play an increasing larger role in project execution.

While the magnitude of the remoteness premium has not been quantified, it has been illustrated by comparing ARENA's RAR-funded projects to the LSS utility-scale projects in Australia. By definition, the remoteness premium is context-specific and dependent on a range of discrete variables such as scale, location, and quality of existing infrastructure, both electrical and transport. The impact of project scale and remoteness (or difficulty of access) have the potential to compound or counteract each other, depending on the project and locational configuration: large capacity additions in very remote locations may have similar specific costs to smaller systems in less remote areas, whereas lowest costs (approximating on-grid system costs) are achieved for large system sizes in less remote locations.

From the RAR portfolio project experience, attempting to reach higher Renewable Power Fractions and annual Renewable Energy Fractions is costlier and, on average, requires more time to complete, yet the longest project delays were experienced among projects that aimed for less ambitious annual REF values. From a technical perspective, once projects secured legal and financial approvals, project executions have been close to the contractual plans in time and expenses. The location- and project-specific delays experienced by the RAR projects point to areas where time and cost improvements can be achieved. The development and execution of the projects is contingent upon ensuring that stakeholder engagement and the particulars of the place are not forgotten.

ARENA's Knowledge Sharing requirements allow the RAR projects to be benchmarks for the future within Australia and the world. The Knowledge Sharing model – if properly designed and implemented prior to providing funding to projects – can significantly increase the impact of funding by governments, multilateral development banks and NGOs, and lower costs for future renewable energy projects.

ACKNOWLEDGEMENTS

The authors wish to thank Lachlan McLeod and George Dickeson for the graphical inputs, the projects team at ARENA for the insights and factchecking of data, and the project proponents for actively participating in Knowledge Sharing. Ekistica is ARENA's Knowledge Sharing partner and data handler under Work Order CEAP-WO-14/15-003.

REFERENCES

[1] OECD/IEA, "Energy access database," IEA Publishing, 2016. [Online]. Available: http://www.worldenergyoutlook.org/resources/energydevelopment/energyaccessdatabase/. [Accessed 23 June 2017].

[2] World Bank Group, "State of electricity access report 2017 (Vol. 2) : full report," 1 May 2017. [Online]. Available: http://documents.worldbank.org/curated/en/364571494517675149/full-report. [Accessed 25 June 2017].

[3] B. van der Zwaan and A. Rabl, "Prospects for PV: a learning curve analysis," *Solar Energy,* vol. 74, no. 1, pp. 19-31, 2003.

[4] G. F. Nemet, "Beyond the learning curve: factors influencing cost reductions in photovoltaics," *Energy Policy,* vol. 34, no. 17, pp. 3218-3232, 2006.

[5] Fraunhofer ISE, "Photovoltaics Report," 17 November 2016. [Online]. Available: https://www.ise.fraunhofer.de/content/dam/ise/de/documents/publications/studies/Photovoltaics-Report.pdf. [Accessed 23 June 2017].

[6] F. Kersten, R. Doll, A. Kux, D. M. Huljic, M. A. Görig, C. Breyer, J. W. Müller and P. Wawer, "PV Learning Curves: Past and Future Drivers of Cost Reduction," in *26th EU PVSEC*, Hamburg, 2011.

[7] Australian Bureau of Statistics, "Australian Statistical Geography Standard (ASGS): Volume 5 - Remoteness Structure, July 2011," Australian Bureau of Statistics, Canberra, 2013.

[8] AECOM, "Australia's Off-Grid Clean Energy Market Research Paper," 08 October 2014. [Online]. Available: https://arena.gov.au/assets/2014/12/ARENA_RAR-report-20141201.pdf. [Accessed 21 June 2017].

[9] M. Tuckwell and L. Frearson, "The Bushlight India Model – A Replicable, Scalable Model that Addresses the Structural Barriers to Remote Village Electrification," in *26th EU PVSEC*, Hamburg, 2011.

[10] B. Kroposki, B. Johnson, Y. Zhang, V. Gevorgian, P. Denholm, B.-M. Hodge and B. Hannegan, "Achieving a 100% Renewable Grid: Operating Electric Power Systems with Extremely High Levels of Variable Renewable Energy," *IEEE Power and Energy Magazine,* March-April 2017.

[11] IEA, "Key World Energy Statistics 2016," IEA, [Online]. Available: http://www.iea.org/publications/freepublications/publication/KeyWorld2016.pdf. [Accessed 23 June 2017].

[12] W. Johnston and R. Egan, "PV in Australia in 2015," 18 August 2016. [Online]. Available: http://www.iea-pvps.org/index.php?id=93&eID=dam_frontend_push&docID=3109. [Accessed 15 June 2017].

[13] APVI, "Australian PV market since April 2001," June 2017. [Online]. Available: http://pv-map.apvi.org.au/analyses. [Accessed 25 June 2017].

[14] IRENA, "Accelerating Off-grid Renewable Energy: Key Findings and Recommendations from IOREC 2016," Abu Dhabi, 2017.

[15] IRENA, "Innovation Outlook: Renewable Mini-grids," Abu Dhabi, 2016.

[16] K. Bunker, S. Doig, K. Hawley and J. Morris, "Renewable Microgrids: Profiles from Islands and Remote Communities Across the Globe," Rocky Mountain Institute and Carbon War Room, 2015.

[17] Hugo Centre for Migration and Population Research, University of Adelaide, "ARIA (Accessibility/Remoteness Index of Australia)," 3 December 2015. [Online]. Available: http://www.adelaide.edu.au/apmrc/research/projects/category/about_aria.html. [Accessed 19 June 2017].

[18] Ekistica, "Remote Area Renewables Knowledge Sharing: Data Specification to Benefit Financial and Operational Yields," 14 March 2017. [Online]. Available: https://arena.gov.au/assets/2017/05/ARENA-RAR-Knowledge-Sharing-Data-Specification-1.pdf. [Accessed 21 June 2017].

[19] B. Herteleer, A. Dobb, O. Boyd, S. Rodgers and L. Frearson, "ARENA's Large Scale Solar funding impact on utility-scale solar in Australia," in *Submitted to 33rd EU PVSEC*, Amsterdam, 2017.

[20] L. Frearson, P. Rodden, J. Backwell and M. Thwaites, "Investigating the Impact of Solar Radiation Variability on Grid Stability with Dispersed PV Generation," in *31st EU PVSEC*, Hamburg, 2015.

[21] B. Johnson, M. Sinha, N. Ainsworth, F. Dorfler and S. Dhople, "Synthesizing Virtual Oscillators to Control Islanded Inverters," *IEEE Transactions on Power Electronics,* vol. 31, no. 8, p. 6002–6015, November 2015.

[22] V. Gevorgian and Y. Zhang, "Wind Generation Participation in Power System Frequency Response," in *15th International Workshop on Large-Scale Integration of Wind Power into Power Systems as well as on Transmission Networks for Offshore Wind Power Plants*, Vienna, 2016.

[23] C. Loutan, P. Klauer, S. Chowdhury, S. Hall, M. Morjaria, V. Chadliev, N. Milam, C. Milan and V. Gevorgian, "Demonstration of Essential Reliability Services by a 300-MW Solar Photovoltaic Power Plant," 24 March 2017. [Online]. Available: http://www.nrel.gov/docs/fy17osti/67799.pdf. [Accessed 18 June 2017].

[24] G. Delille, B. Francois and G. Malarange, "Dynamic Frequency Control Support by Energy Storage to Reduce the Impact of Wind and Solar Generation on Isolated Power System's Inertia," *IEEE Transactions on Sustainable Energy,* vol. 3, no. 4, pp. 931-939, 2012.

[25] M. Watt and G. Watt, "PV in Australia in 2006," 30 April 2007. [Online]. Available: http://www.iea-pvps.org/index.php?id=93&eID=dam_frontend_push&docID=19. [Accessed 15 June 2017].

This is a pre-peer review version of the paper submitted to the journal Progress in Photovoltaics.
It has been selected by the Executive Committee of the EU PVSEC 2017 for submission to Progress in Photovoltaics.

Table 1: Portfolio of ARENA RAR-supported projects and technologies used

Project	Application	PV [MW_p]	Wind [MW]	Battery [MW /MWh]	Comment / Additional renewable technologies employed
Barcaldine	Fringe-of-grid	25	–	–	Stimulating utility-scale PV in remote region
Coober Pedy	Off-grid mining community	1.34	4.1	1 /0.5	2 x 0.85 MVA flywheel on fast-start diesel, 3 MW dynamic resistor for stability, integrating control system
DeGrussa	Off-grid mine	10.56	–	3.2 /1.81	PV 1-axis tracking, battery-aided high REF, battery peak output 6 MW
Doomadgee	Off-grid community	0.264 + 1	-	-	Increase of Renewable Energy Fraction by expanding PV on grid
Flinders Island	Off-grid community	0.25	0.9	0.75 /0.3	0.85 MVA flywheel on fast-start diesel and 1.5 MW dynamic resistor for stability, integrating control system
Karratha Airport	Fringe-of-grid	1.02	–	0.47 /0.32	Solar PV, batteries and cloud predictive technology for forecasting and ramp rate control, as per network operator requirements
Lakeland	Fringe-of-grid	13	–	1.4 /5.3	Battery for grid support and/or stand-alone operation (islanding distribution line from grid)
Lord Howe Island	Off-grid community	0.45	0.55 [1]	0.4 /0.4	Physical space limited for renewable energy technologies (small island, conservation & tourism measures)
In preparation	Off-grid community	0.51	–	3.75 /1.5	System upgrades to achieve high REF
Normanton	Fringe-of-grid	5.17	–	–	Demonstrator project: PV to aid network stability at the end of a very long (1000 km+) transmission line.
NT SETuP	Off-grid community	10	–	0.8 /1.6	30 communities, integrating PV with existing diesel. 1 site: high REF system, 1 MW_p PV & battery (diesel off possible), business transformation
Rottnest Island	Off-grid community	0.63	0.6	–	Demand management, efficiency measures, integrating control system for high Renewable Energy Fraction
Weipa	Off-grid mine	1.71	–	–	Demonstrator project: Solar PV used as diesel offset for mine
Yulara	Off-grid community	1.83	–	–	Distributed solar PV to mitigate variability in isolated grid [20]

[1] Lord Howe Island's application for wind turbine approvals has recently been denied, and the project proponents and ARENA are evaluating which changes are necessary for the project to continue.

IMPACT OF SELF-CONSUMPTION ON INTEGRATION OF PHOTOVOLTAICS IN MARTINIQUE: SIMULATION RESULTS FROM THE INSOLATIONS PROJECT

Author(s): Franck Bourry (franck.bourry@cea.fr ,+33 (0)4 79 79 21 57), F. Al Shakarchi (franck.alshakarchi@cea.fr +33(0)4 79 79 21 43), Nicolas. Martin (nicolas.martin2@cea.fr +33 (0)4 79 79 22 04)
Sarah Darivon (darivon.s@energie.mq, +33 (0)5 96 30 15 89), Laurent Bellemare
Company / Institute(s):
Univ. Grenoble Alpes, INES, F-73375 Le Bourget du Lac, France
CEA, LITEN, Solar Technologies Department, F-73375 Le Bourget du Lac, France,
Energy Agency of. Martinique, Imm. les Bosquets 2 - 26 ZI Petite Cocotte Champigny
F-97224 DUCOS, France

The large scale integration of photovoltaic (PV) generation may significantly impact the reliable operation of the power system, and particularly for insular grids. In this context, combining PV and storage systems can be considered as a solution to limit these impacts, relying on an increase of PV self-consumption. However, the large deployment of PV self-consumption systems also considerably affects the power system operation, from local solar home system operation to production mix at the territory level. It also modifies the economic flows between the different actors along the power system. In this context, the INSOLATIONS project has been carried out in order to technically and economically assess the impact of a large scale deployment of self-consumption solar home systems in Martinique island (French West Indies). An additional contribution from the project is the identification of the technical and financial required conditions for the deployment of PV self-consumption systems in Martinique.

This analysis has been done through simulations of individual systems and their aggregated impact on the global mix. Positive and negative externalities have been considered. The study is focused on self-consumption systems deployed at the residential level.

Keywords: Photovoltaic, Battery, self-consumption, residential, island, electricity mix, LCOE

1 SCIENTIFIC APPROACH

The main scientific innovation from this project relies in the proposition and in the application of a holistic approach covering all the necessary steps from self-consumption system simulation to economic evaluation at the territory level:

- The first step of the study consisted in generating PV production and residential consumption time series data from on-site measurements and characteristics from Martinique.

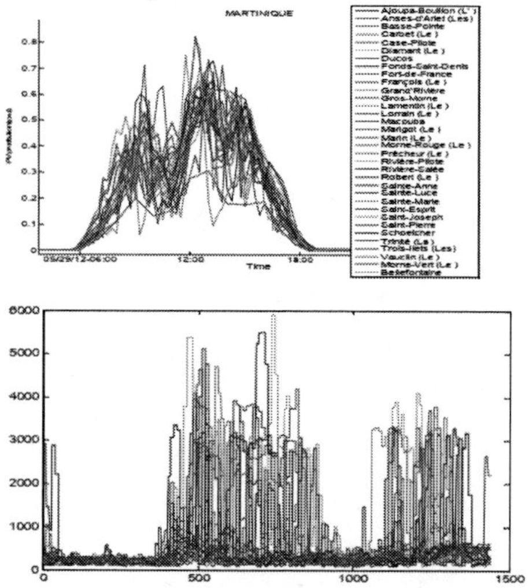

Figure 1 : PV generation and residential load data

- A set of technical requirements for the operation of PV self-consumption systems has been proposed. These requirements for the PV-storage system operation consist in (1) a limit of the PV variability so that no extra power reserve to compensate PV fluctuations is needed, (2) a compulsory injection to the grid during consumption peak time period to reduce expensive peak power generation, (3) a limit of the maximum power injected to the grid to reduce grid reinforcement and (4) a compulsory participation to ancillary services, such as primary frequency regulation.
- Based on PV and consumption data collection, and considering the proposed technical requirements, the operation of the self-consumption systems has been modeled and simulated. A wide range of configurations, covering the various configurations in Martinique, has been run :11 different load profiles covering 3, 6 and 9 kVA power subscription, 3 locations for the PV generation with 3 different PV sizing, 2 battery technologies (Lead-Acid and Lithium Ion) with 3 different battery sizing;

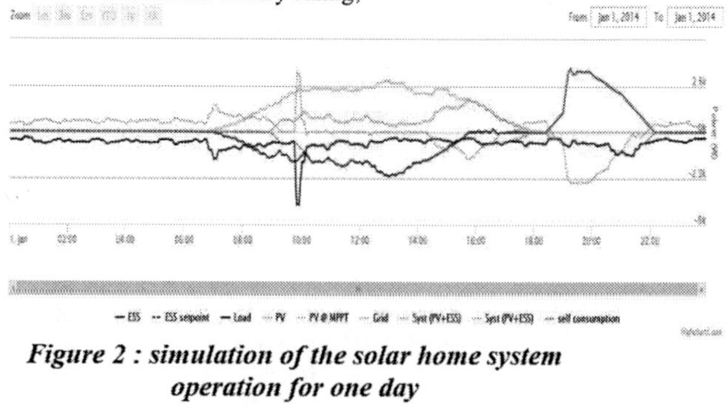

Figure 2 : simulation of the solar home system operation for one day

- Different scenarios for PV feed-in tariff, self-consumption tariff and electricity price from the grid have been proposed. These prices and tariffs are time-varying tariffs considering night time, base time and peak time. They have been used to compute the cash flow from the PV-storage system operation simulation. In addition, the initial installation, maintenance and operation costs have been modelled. All these elements led to an economic assessment from the end-user point of view: the Return On Interest (ROI) of each technical configuration has been computed, with a sensitivity analysis on the tariffs and loan rates.

- In addition to the analysis of local self-consumption solar home systems (SHS), the impacts of these SHS on the other actors from the power system have been identified. These impacts are illustrated through the following figure. They include the impact on the present Martinique electricity mix, the interactions with the grid operator (EDF SEI) and with economic actors such as installers and public authority.

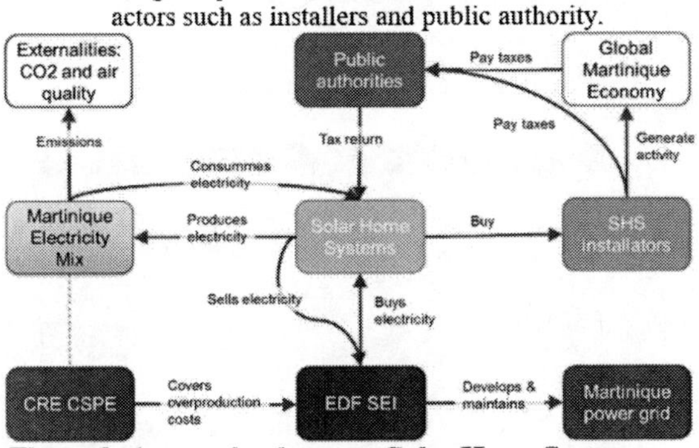

Figure 3 : interaction between Solar Home Systems and the actors of the power system in Martinique

This holistic analysis has led to the identification of the benefits from SHS deployment. These benefits are (1) the avoided fuel power generation and the reduction of its negative externalities, such as the reduction of the Green House Gases emissions and the improvement of air quality, (2) the cost reduction for grid development and maintenance, (3) the benefits from offering Uninterruptible Power Supply (UPS) and (4) the creation of local jobs. Finally, the coherence between these benefits from the whole power system point of view, and the costs related to the proposition of PV feed-in tariff and self-consumption tariff for the end-user, has been verified.

2 RESULTS AND CONCLUSIONS

The results from the study show that :

- A policy which targets the equipment of 25 % of the residential consumers with PV storage systems, is cost effective, both from the end-user and from the whole system point of view;

For the Martinique island, this represents 41 000 customers.

- The optimal PV size (kWp) and storage capacity (kWh) depend on the power subscription level (kVA); the ratio1 kWp per kVA for PV, and from 1 to 2 kWh per kVA for battery is the optimal sizing. With a deployment rate of 25 % among residential consumers in Martinique, this represents 220 MWp of installed PV and 440 MWh of battery storage systems.
- These conclusions are valid considering an oil cost of 80€/bl, with a relatively low consideration of externalities related to fossil fuel generation.
- Such policy does not affect neither the production units, nor the number of jobs from the existing system operator (EDF SEI).
- Such policy leads to the creation of 200 jobs.

Final requirements insist on the need for a global and open partnership in order to set up such ambitious policy.

Total costs in M€ considering 100% air quality cost			Fuel costs + 27% other variable cost		
			40€/bbl	80€/bbl	120€/bbl
			26.8	50.5	74.3
	15€/TCO2	3.70	43.0	66.7	[illegible]
	30€/TCO2	7.5	46.8	70.5	[illegible]
GHG costs	45€/TCO2	11.2	50.5	74.2	[illegible]

Total costs in M€ considering 50% air quality cost			Fuel costs + 27% other variable cost		
			40€/bbl	80€/bbl	120€/bbl
			26.8	50.5	74.3
	15€/TCO2	3.70	36.8	60.5	[illegible]
	30€/TCO2	7.5	40.6	64.3	[illegible]
GHG costs	45€/TCO2	11.2	44.3	68.0	[illegible]

Total costs in M€ considering no air quality cost			Fuel costs + 27% other variable cost		
			40€/bbl	80€/bbl	120€/bbl
			26.8	50.5	74.3
	15€/TCO2	3.70	30.5	54.2	[illegible]
	30€/TCO2	7.5	34.3	58.0	[illegible]
GHG costs	45€/TCO2	11.2	38.0	61.7	[illegible]

High and medium feasible
Medium feasible, not high
Close to medium feasibility (10% gap)
no feasibility

Figure 4 : Total costs in M€ for different scenarios

PVGIS VERSION 5: IMPROVEMENTS TO MODELS AND FEATURES

Thomas Huld[1], Irene Pinedo Pascua[1], Ana Gracia Amillo[1], Ruben Urraca[2], Ewan Dunlop[1]
[1]European Commission, Joint Research Centre, Viale Fermi, 2749, Ispra, Italy
[2]University of La Rioja Department of Mechanical Engineering, 26004, Logroño, Spain
Thomas.huld@ec.europa.eu, Irene.pinedo@ec.europa.eu, ana.gracia-amillo@ec.europa.eu, ruben.urraca@unirioja.es, ewan.dunlop@ec.europa.eu

ABSTRACT: We present the latest version of the PVGIS free online PV estimation tool, PVGIS 5. It includes significant improvements, not only to the data extended spatial coverage and increased length of the time series, but also to the accuracy of the estimation models implemented. The estimation of the PV energy yield now takes into account the effect of wind on the module temperature, which in turn affects its performance; and also the effect of the spectral distribution of the received irradiance. Instead of the long term averages available in previous versions, PVGIS 5 uses and provides the complete dataset of hourly values for several years and the corresponding statistical values. Thanks to the combination of satellite retrieved data and reanalysis values, the spatial coverage has been substantially increased, including now most of the Americas and northern Europe (above 65°N), while maintaining a high temporal resolution of hourly values and spatial resolution of few km. A new feature of PVGIS 5 is the TMY, whose data of solar radiation, ambient temperature, relative humidity among other variables, is suitable for studies on the energy performance on buildings.

Keywords: Solar radiation, Photovoltaics, PV yield estimation

1 INTRODUCTION

PVGIS is an EU project for performing research on the geospatial variation in solar radiation and PV system energy yield, as well as a free online tool for accessing solar radiation data and making PV performance estimates over large geographical regions.

The PVGIS online tool was first released around 2003 and was at the time the only freely available tool for estimating solar radiation and PV performance in Europe [1, 2]. Since then PVGIS has undergone a large number of modifications and improvements, increasing the geographical coverage and the capabilities of modelling. The last major update was implemented in 2014 with the introduction of solar radiation data for Asia [3].

This paper presents the latest version of PVGIS, PVGIS version 5. PVGIS 5 contains a large number of improvements in the calculation methods for solar radiation and PV energy yield estimates, and makes use of a number of new solar radiation data sources. The main new features are:

- Extension of data coverage to most of the Americas and parts of the Pacific, thanks to a collaboration with the National Renewable Energy Laboratory. In addition, new reanalysis products have been used to provide data coverage to Northern Europe.
- Improved calculation of solar irradiation and PV estimates using multi-year time series of solar irradiance and ancillary climatic parameters.
- Improved PV performance modelling taking into account the effects of wind speed and variations in the solar spectrum.
- Direct access to the hourly solar radiation and PV power values, as well as Typical Meteorological Year data for building energy performance studies.
- Options for accessing PVGIS as a web service, without using the web interface.
- A completely reconstructed user interface with improved user guidance and help features.

The graphical user interface to PVGIS 5 is shown in Figure 1. In the following we will present the new models, data, and features in more detail.

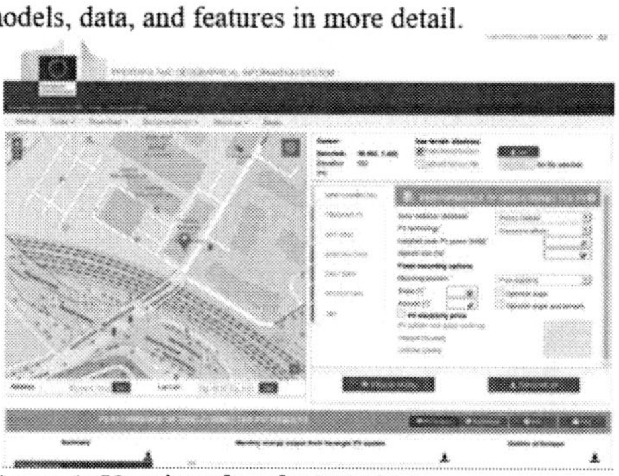

Figure 1: User interface for PVGIS 5.

2 NEW MODELLING FEATURES

PVGIS 5 introduces a number of changes to the calculation of the solar irradiation and the PV energy output.

2.1 Changes to the calculation of solar radiation

The main change regarding the calculation of solar radiation is that PVGIS now calculates the long-term averages of solar radiation using a full 10-years series of hourly solar radiation data. This is in contrast to the older versions that used long-term monthly averages of solar radiation data to reconstruct the daily profiles for a selected day in each month. The advantage of the new calculation method is that it can take into account diurnal patterns in the solar irradiance, which may also influence the calculation of the optimum slope and azimuth. For instance if a site tends to have more cloudy weather in the afternoon, the optimum azimuth will be more eastwards to catch the sun in the morning.

Performing the calculations with the hourly time series also makes it possible to provide new types of output, such as:

- Average irradiation for each month in the full time period, giving information about the variability in

the monthly irradiation.
- Download of the full hourly time series of solar irradiance.
- Providing information about the year-to-year variability of the solar resource, something that was not possible before in PVGIS.

2.2 Changes to the PV energy calculations

The calculation of PV energy production estimates now also uses the full 10-years time series of solar radiation data and other meteorological data. This solves the problem, first noted in [4] that solar irradiance and air temperature are correlated, often with temperatures being higher when there is sunshine than when it is cloudy. This correlation is lost when using long-term averages for the calculation, which can lead to overestimations of up to 2% of the PV energy output in some areas. With the new calculation method in PVGIS this will no longer be an issue.

In addition, PVGIS 5 introduces a number of improvements in the models for PV performance. The most important of these are:
- Calculation of the influence of variations in the solar spectrum on the PV energy output. The calculation is based on spectrally resolved solar radiation data calculated from satellite [5]. At the moment these data are available for Europe, Africa and Asia. Due to the very large amount of data needed for these calculations, the spectral correction is applied as a constant correction for each month for each PV technology. The spectral correction is performed for crystalline silicon and for Cadmium Telluride (CdTe).
- Calculation of the influence of wind on the module temperature and hence on the PV power. The model used is the 2-parameter model [6] with coefficients taken from [7]. The geospatial variation of this effect was studied by [8].
- We have introduced a new power model for CdTe modules, based on measurements performed at the ESTI Laboratory. The older model was based on measurements done outdoors where the measurements were subject to variations in the solar spectrum. With the new model it is possible to separate the effect of the spectrum from the effect of high temperature and low irradiance.

3 EXPANDED DATA COVERAGE

3.1 SARAH solar radiation database from METEOSAT Prime satellites

PVGIS version 4 contains two satellite-based solar radiation databases. Europe and Africa are covered by data from the CM SAF Operational Product [9, 10]. Data for the Middle East and Asia are from the CM SAF SARAH solar radiation data set [11, 4]. In PVGIS 4 only one of the two databases is available at any given location, and both are selected using the "PVGIS-CMSAF" option, thus the presence of two radiation data sets was invisible to the user.

PVGIS 5 has expanded the coverage of SARAH to Europe, Africa, and parts of South America. In these areas it is now possible to choose between the CM SAF operational product (still called PVGIS-CMSAF), and the SARAH data, in PVGIS called "PVGIS-SARAH". PVGIS-CMSAF contains hourly values from 2007 to 2016 with a spatial resolution of 1.5 arc-minutes, while PVGIS-SARAH has hourly values between 2005 and 2016 with a spatial resolution of 3 arc-minutes. A detailed validation of the two databases for Europe has been presented in [12].

3.2 Solar radiation data from NSRDB

The National Solar Radiation Database (NSRDB) has been developed by the National Renewable Energy Laboratory (NREL) [13], and data can be freely accessed at:

`http://nsrdb.nrel.gov`

Data from NSRDB have been made available for use by PVGIS. These data consist of global, diffuse and direct horizontal irradiance values for all land surfaces in the Americas between latitudes 60° N and 20° S, as well as for some island areas in the Pacific, most notably Hawaii. The spatial resolution is 2.4 arc-minutes. The temporal resolution of NSRDB is half-hourly; however, in PVGIS we use one value of solar radiation per hour. PVGIS uses data for the period 2005-2015.

3.3 Reanalysis solar radiation data

Geostationary satellites have a limited view of the earth, in particular it is not possible to use geostationary satellite data for areas too far from the equator. In PVGIS, the PVGIS-SARAH database is limited to 62°30'N (and lower latitudes in some areas), and NSRDB covers the Americas only as far north as 60° N. Northern Europe in particular has significant population living in areas north of those covered by the satellite-based data.

In order to provide data coverage also to high-latitude areas we have investigated using solar radiation data from climate reanalysis data sets. We have considered four data sets:
- **MERRA-2** MERRA and MERRA-2 have been developed by NASA's Global Modelling and Assimilation Office (https://gmao.gsfc.nasa.gov/reanalysis/MERRA/). Data have worldwide coverage at hourly time resolution and a spatial resolution of 0.5x0.625° lat/lon [14].
- **ERA-Interim** has been developed by the European Centre for Medium-range Weather Forecast (ECMWF, http://www.ecmwf.int). This data set has worldwide coverage at 3-hourly time resolution and a spatial resolution of 0.75° lat/lon [15].
- **ERA-5** This is the new reanalysis product from ECMWF, at the time of writing, the data period is limited to the years 2010-2016. Coverage is worldwide at hourly time resolution and a spatial resolution of 0.28° lat/lon [16].
- **COSMO-REA6** is a regional reanalysis product, covering Europe at hourly time resolution and a spatial resolution of about 6 km (we use 3' lat/lon in PVGIS). The data set is available at ftp-rea.dwd.de/pub/REA [17].

4 NEW DATA OUTPUT OPTIONS

The use of hourly solar radiation data makes it possible to provide new types of data output. The most important new features are:
- Output of hourly solar radiation and PV power values. Here, the user can request hourly values of solar radiation on an arbitrary plane (fixed or sun-

tracking) for one or more years. It is also possible to calculate the PV power output for each hour in the time period requested, using the PV performance models of PVGIS. This allows the user to calculate any desired statistical quantity for the data.

- Typical Meteorological Year (TMY) data. With this option it is possible to download TMY data for building energy performance studies. The data consist of one year of data, with each month selected as the most "typical" in a 10-years period [18]. The solar radiation data use the SARAH or NSRDB databases depending on the location, while the other meteorological quantities are taken from the ECMWF ERA-Interim reanalysis product (www.ecmwf.int). For the air temperature values we make use of the downscaling procedure of [19]. It should be noted that these data are NOT optimized for use in solar energy applications, since the construction method specified in [18] requires that equal weight is given to air temperature, relative humidity, and solar radiation, whereas TMY data for solar energy would need to have higher weight given to solar radiation.
- Monthly averages of solar radiation data. Earlier versions of PVGIS already had the option of getting monthly average data for solar radiation and temperature. However, in this case the values were long-term monthly averages calculated from many years of data. PVGIS 5 instead offers monthly average values for each year separately. This makes it easy to get an idea of the month-to-month variability in the solar resource.
- Year-to-year variability. In the PV performance calculations, the year-to-year variability (standard deviation) is now included.

5 VALIDATION OF SOLAR RADIATION DATA

The satellite-based solar radiation databases used in PVGIS have been validated by a number of authors in recent years. The operational CM SAF data product used in PVGIS as PVGIS-CMSAF has been evaluated by [9, 10]. The SARAH data have been validated in a number of publications [11, 3, 20]. In a recent paper [12] both the CM SAF operational product and the SARAH data set were validated against more than 300 ground stations in Europe.

Regarding the NSRDB data set, a validation study of the quality of the said data set is presented in [13].

All the satellite-based solar radiation products show very low overall bias (no general tendency to under- or overestimate solar irradiation), but there are stronger local deviations from measurements, with a typical standard deviation of the individual station bias values of 3-4%.

The two reanalysis-based databases have been released only very recently. A first attempt at validation has been presented in the contribution 6BV.3.4 to the present Conference [21], using a set of 293 stations in Europe. The results show that both ERA-5 and COSMO-REA6 generally have higher deviations from station measurements than the satellite-based products in the areas where these have data, but that both data sets are more accurate than previous reanalysis data sets such as ECMWF ERA-Interim or MERRA-2. Using as a global metric the mean of the absolute values of the individual station bias values, MAD(MBD), we find that SARAH has a relative MAD(MBD) of 3.0% while ERA-5 shows a value of 4.4% and COSMO 5.1% for the entire area. Looking only at the area north of 60° N (49 stations available), the three data sets all show MAD(MBD) around 5.5% (SARAH only contains data as far north as 65° N).

Figure 2 shows a comparison of validation results for Europe for the two chosen reanalysis data sets. For comparison, the validation results for the SARAH solar radiation data set are also shown.

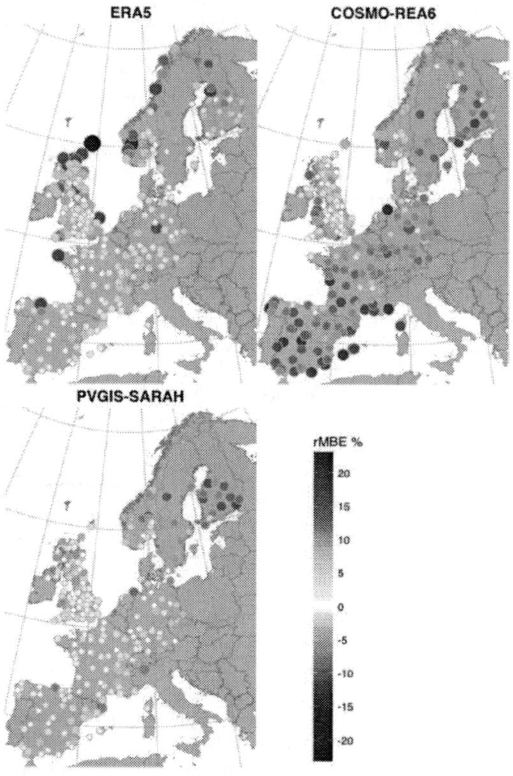

Figure 2: Annual relative mean bias deviation for each of the stations used for the validation in Europe, for ERA-5, COSMO-REA6 and SARAH.

6 COMPARISON OF RESULTS FROM DIFFERENT SOLAR RADIATION DATABASES

6.1 Comparison to the previous version of PVGIS

Figure 3 shows the relative difference between PVGIS 5 and PVGIS 4, in the annual average irradiation on an equator-facing plane at 20° inclination from horizontal. In PVGIS5, the PVGIS-SARAH database was used east of 40° E and the PVGIS-CMSAF database west of that longitude. In Africa there is a slight reduction in irradiation, caused by the change in the PVGIS-CMSAF database which now only uses data from 2007-2016, using the CM SAF "operational" solar radiation product, whereas the previous version of PVGIS used a weighted average of the CM SAF operational product and an earlier CM SAF solar radiation product based on Meteosat First Generation images Close to the border between the two databases there is a discontinuity caused by the change from one data set to another.

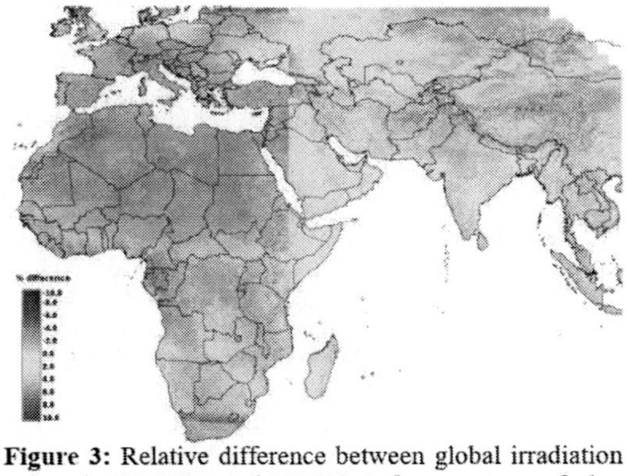

Figure 3: Relative difference between global irradiation values in PVGIS 5 and PVGIS 4, for an equator-facing plane inclined 20° from horizontal. For PVGIS 4, the PVGIS-CMSAF database was used, for PVGIS 5, PVGIS-CMSAF was used west of 40° E, while PVGIS-SARAH was used east of 40° E.

6.2 Comparison of solar radiation databases in PVGIS 5

In PVGIS 5 there are now two options for solar radiation data derived from the METEOSAT prime satellites located at 0° longitude: PVGIS-CMSAF, and PVGIS-SARAH. These data sets have been calculated using different algorithms, and although the geostationary satellite data used are the same, there are also differences in the aerosol data used for the calculations. It is therefore to be expected that there will be differences in the solar radiation estimates. Figure 4 shows the relative difference (in percentage points) between PVGIS-SARAH and PVGIS-CMSAF over Europe and Africa.

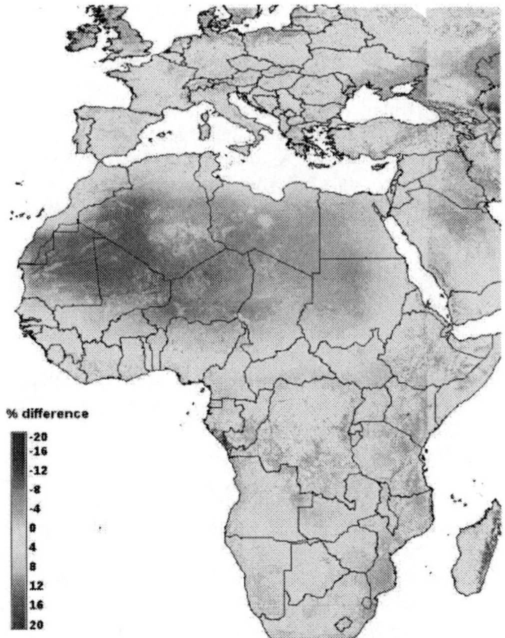

Figure 4: Relative difference (in percent) between the PVGIS-SARAH and PVGIS-CMSAF solar radiation database results, using the annual average global irradiation on an equator-facing plane inclined at 20°. from horizontal.

The most notable difference between SARAH and the PVGIS-CMSAF results is the higher values in SARAH for Northern Africa. This is mainly due to improved aerosol information in SARAH.

For Northern Europe we now have two new solar radiation databases derived from reanalysis data, COSMO-REA6 and ERA-5. Figure 5 shows the difference in annual average global horizontal irradiation between COSMO-REA6 and SARAH and ERA-5, respectively. The results show that COSMO-REA6 generally has lower estimates in much of Europe, with the exception of mountain areas in Northern Europe. In the areas of Northern Europe covered only by COSMO-REA6 and ERA-5, the two databases show rather similar results.

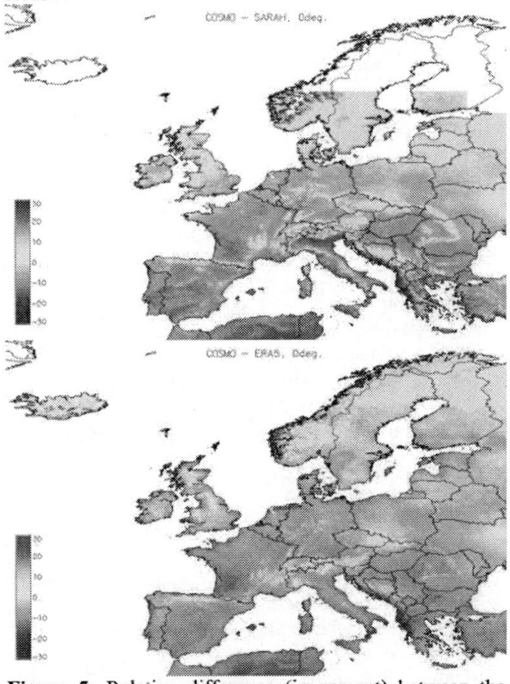

Figure 5: Relative difference (in percent) between the PVGIS-COSMO6 and PVGIS-SARAH (top), and PVGIS-ERA5 (bottom) solar radiation databases, using the annual average global horizontal irradiation.

Finally, figure 6 shows the difference between the COSMO database and the old PVGIS-CLASSIC data set which up to version 4 of PVGIS was the only data set available for Northern Europe, but which is no longer used in PVGIS 5.

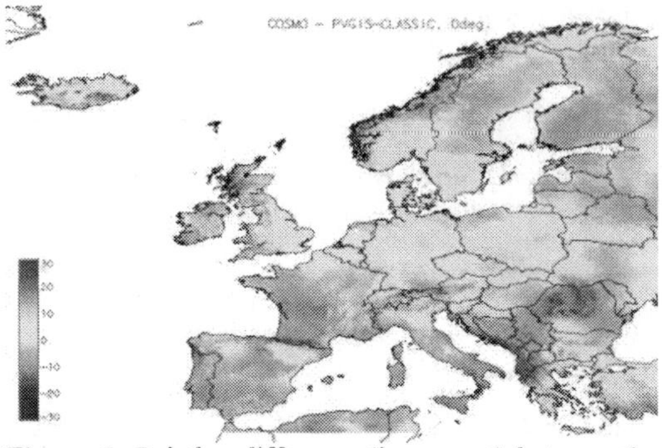

Figure 6: Relative difference (in percent) between the PVGIS-COSMO and the old PVGIS-CLASSIC data set used in PVGIS up to version 4, using the annual average global horizontal irradiation.

7 CONCLUSIONS

With the new version 5 of PVGIS we have introduced a number of improvements to the estimates of solar radiation and to the modelling of PV energy yield estimates. A number of new solar radiation data sets have made it possible to extend the geographical range of PVGIS to include most of the Americas. Two new data sets, based on reanalysis data, have enabled extending PVGIS spatial coverage in Europe to northern latitudes, which are not covered by satellite derived products.

Improvements to PV performance modelling include new models that take into account the effects of wind speed on module temperature and the influence of spectral variations. In addition, PVGIS now performs a calculation of PV energy output using a 10-years series of hourly solar radiation data, which makes it possible to take into account the correlation between solar radiation and air temperature.

The long term time series of hourly values provides a better description of the solar resource available at every location which in turn, allows an optimized design of the solar energy system, both connected to the grid (fixed system or sun-tracking) and off-grid systems. The TMY data can be applied to energy efficiency studies in buildings like those performed with the software EnergyPlus, whose data format is supported by the data that can be downloaded from PVGIS 5.

8 ACKNOWLEDGMENTS

We would like to thank the following organizations for access to the data that make PVGIS 5 possible:

- CM SAF for the satellite-based solar radiation data
- NREL for access to the NSRDB solar radiation data
- Deutscher Wetterdienst for making the COSMO-REA6 reanalysis data available
- ECMWF for producing the ERA-Interim and ERA-5 reanalysis data.

REFERENCES

[1] M. Šúri, J. Hofierka. A New GIS-based Solar Radiation Model and Its Application to Photovoltaic Assessments. *Transactions in GIS* 8, 89-96, (2004).

[2] M. Šúri, T. Huld, E. Dunlop. PV-GIS: a web-based solar radiation database for the calculation of PV potential in Europe. *International Journal of Sustainable Energy* 24, 55-67, (2005).

[3] A. Gracia Amillo, T. Huld, R. Müller. A new database of global and direct solar radiation using the eastern meteosat satellite, models and validation. *Remote Sensing* 6, 8165-8189, (2014).

[4] T. Huld, R. Gottschalg, H. Beyer, M. Topič. Mapping the performance of PV modules, effects of module type and data averaging. *Solar Energy* 84, 324-338, (2010).

[5] A. Gracia Amillo, T. Huld, P. Vourlioti, R. Müller, M. Norton. Application of Satellite-Based Spectrally-Resolved Solar Radiation Data to PV Performance Studies. *Energies* 8, 3455-3488, (2015).

[6] D. Faiman. Assessing the outdoor operating temperature of photovoltaic modules. *Progress in Photovoltaics, Research and Applications* 16, 307-315, (2008).

[7] M. Koehl, M. Heck, S. Wiesmeier, J. Wirth. Modeling of the nominal operating cell temperature based on outdoor weathering. *Solar Energy Material & Solar Cells* 95, 1638-1646 (2011).

[8] T. Huld, A.M. Gracia Amillo. Estimating PV module performance over large geographical regions: the role of irradiance, air temperature, wind speed and solar spectrum. *Energies* 8(6), 5159-5181, (2015).

[9] R.W. Müller, C. Matsoukas, A. Gratzki, H.D. Behr, R. Hollmann. The CM SAF operational scheme for the satellite based retrieval of solar surface irradiance – A LUT based eigenvector hybrid approach. *Remote Sensing of Environment* 113(5), 1012-1024, (2009).

[10] T. Huld, R. Müller, A. Gambardella. A new solar radiation database for estimating PV performance in Europe and Africa. *Solar Energy* 86, 1803-1815, (2012).

[11] R. Müller, T. Behrendt, A. Hammer, A. Kemper. A New Algorithm for the Satellite-Based Retrieval of Solar Surface Irradiance in Spectral Bands. *Remote Sensing* 4, 622-647, (2012).

[12] R. Urraca, A.M. Gracia Amillo, E. Koubli, T. Huld, J. Trentmann, A. Riihelä, A. Lindfors, D. Palmer, R. Gottschalg, F. Antonanzas-Torres. Extensive validation of CM SAF surface radiation products over Europe. *Remote Sensing of Environment* 199, 171-186, (2017).

[13] A. Habte, M. Sengupta, A. Lopez. Evaluation of the National Solar Radiation Database (NSRDB): 1998-2015, Technical Report, NREL/TP-5D00-67722, (2017). National Renewable Energy Laboratory, Golden, CO, U.S.A.

[14] R. Gelaro, W. McCarty, M.J. Suárez, R. Todling, A. Molod, L. Takacs, C.A. Randles, A. Darmenov, M.G. Bosilovich, R. Reichle, K. Wargan, L. Coy, R. Cullather, C. Draper, S. Akella, V. Buchard, A. Conaty, A.M. da Silva, W. Gu, G.K. Kim, R. Koster, R. Lucchesi, D. Merkova, J.E. Nielsen, G. Partyka, S. Pawson, W. Putman, M. Rienecker, S.D. Schubert, M. Sienkiewicz, B. Zhao. The Modern-Era Retrospective Analysis for Research and Applications, version 2 (MERRA-2). *Journal of Climate* 30, 5419–5454, (2017).

[15] D.P. Dee, S.M. Uppala, A.J. Simmons, P. Berrisford, P. Poli, S. Kobayashi, U. Andrae, M.A. Balmaseda, G. Balsamo, P. Bauer, P. Bechtold, A.C.M. Beljaars, L. van de Berg, J. Bidlot, N. d Bormann, C. Delsol,

R. Dragani, M. Fuentes, A.J. Geer, L. Haimberger, S.B. Healy, H. Hersbach, E.V. Hólm, L. d Isaksen, P. Kållberg, M. Köhler, M. Matricardi, A.P. McNally, B.M. Monge-Sanz, J.-J. Morcrette, B.-K. Park, C. Peubey, P. de Rosnay, C. Tavolato, J.-N. Thépaut, F. Vitart. The ERA-Interim reanalysis: configuration and performance of the data assimilation system. *Quarterly Journal of the Royal Meteorological Society* 137, 553-597, (2011).

[16] ERA5 data documentation. ECMWF. https://software.ecmwf.int/wiki/display/CKB/ERA5+data+documentation.

[17] C. Bollmeyer, J. Keller, C. Ohlwein, S. Wahl, S. Crewell, P. Friederichs, A. Hense, J. Keune, S. Kneifel, I. Pscheidt, S. Redl, S. Steinke. Towards a high-resolution regional reanalysis for the European CORDEX system. *Quarterly Journal of the Royal Meteorological Society* 141, 1–15, (2015).

[18] ISO 15927-4:2005. Hygrothermal performance of buildings – Calculation and presentation of climatic data – Part 4: Hourly data for assessing the annual energy use for heating and cooling. International Organization for Standardization.

[19] T. Huld, I. Pinedo Pascua. Spatial Downscaling of 2-meter air temperature using operational forecast data. *Energies* 8(4), 2381-2411, (2015).

[20] R. Müller, U. Pfeifroth, C. Träger-Chatterjee, J. Trentmann, R. Cremer. Digging the METEOSAT Treasure - 3 Decades of solar surface radiation. *Remote Sensing* 7, 8067-8101, (2015).

[21] T. Huld, R. Urraca, A.M. Gracia-Amillo, J. Trentmann. A global hourly solar radiation data set using satellite and reanalysis data. 6BV.3.4. 33rd European PV Solar Energy Conference and Exhibition. (2017).

COMBINE DEEP NEURAL NETWORK AND TREE BASED MACHINE LEARNING MODELS USING STACKED GENERALIZATION TO FORECAST HOURLY SOLAR IRRADIANCE IN TROPICAL REGIONS

Zibo Dong[a], Lu Zhao[a], Dazhi Yang[b], and Thomas Reindl[a]

[a]Solar Energy Research Institute of Singapore (SERIS), Singapore
[b]Singapore Institute of Manufacturing Technology (SIMTech), Singapore

ABSTRACT: Located on the Equator, Singapore has one of the most challenging climate for solar irradiance forecasting. The tropical rainforest climate in the area demonstrates high variability in solar irradiance due to the dynamic and unpredictable cloud formation. To provide a solid solution for hourly solar irradiance forecasting in the area, we design and implement the advanced stacked generalization to combine several state-of-the-art machine learning models: deep neural network, extreme gradient boosting, random forests and adaptive boosting in hourly solar irradiance forecasting for tropical regions. These models have been applied to various machine learning competitions and proved to be the top performers. By using stacked generalization, these individual machine learning models can be combined to improve the forecasting accuracy further. To appropriately design and implement these models in hourly solar irradiance forecasting, input features are carefully prepared and processed. After proper feature selection, the machine learning models are implemented and optimized specifically for our application. Then the models are combined together using stacked generalization to achieve the optimal forecasting accuracy. The final stacked model includes 10 extreme gradient boosting models, 5 deep neural network models, 3 random forests models and 2 adaptive boosting models to avoid overfitting. After 5-fold cross-validation using the training data from 2014-2015, not only the stacked model but also the individual models including the benchmark persistence model are tested on the 2016 data. The stacked model has the best performance among all the implemented models.
KEYWORDS: solar irradiance forecasting, deep neural network, machine learning, stacked generalization.

1 INTRODUCTION

To minimize the risk of destabilizing the electricity grid by a sizeable share of solar PV, it is indispensable to forecast the contribution of solar power ahead of time. For power system operators, it is important to understand the value of accurate solar energy forecasting on all time scales. The benefits of forecasting were ignored previously mainly because the solar PV penetration level is generally too low. With the rapid expansion of solar PV, particularly solar power installations, the value of forecasting must be examined by addressing the challenges in power system integration.

In the middle of the day, large instantaneous solar power generation is expected to correspond a large portion of the electricity demand. Depending on local irradiance conditions and the grid topology, it is challenging, and potentially possible, for power system operators to adjust power generation to meet the downward and upward net load ramping to compensate the solar power generation. Without proper power system management, solar power must be curtailed or the power system reliability is at risk, as already seen in Australia. The curtailment is an avoidable inefficiency in the grid. Accurate solar power forecasting can effectively reduce the inefficiencies through unit commitment and economic dispatch decisions.

Intra-hour forecasting usually makes use of statistical forecasting methods or sky imager analysis [1]. Day ahead forecasting (more than six hours) usually applies numerical weather prediction. For intra-day forecasting (one to six hours), machine learning techniques are commonly used [2-3]. In terms of performance, deep neural network and tree based models are most popular recently.

2 TREE BASED MACHINE LEARNING MODELS

Tree based models are widely applied for structured data. Recently, extreme gradient boosting, random forests and adaptive boosting have attracted a lot of attention due to their superior performance in various machine learning competitions. Tree based models are fundamentally based on a single decision tree, which partition the feature space into many different sub-spaces in a tree structure.

2.1 Single decision tree

Machine learning problem is generally to approximate the function

$$y = f(x_1, x_2, \dots, x_n)$$

where y is the target output value and $(x_1, x_2, \dots, x_n)$ is a set of input values.

A single decision tree partition the feature space Z into N different sub-regions $R_1, R_2, \dots, R_N$. The approximation function is modelled as

$$f(x) = \sum_{n=1}^{N} c_n I(x \in R_n)$$

where

$$I(x \in R_n) = \begin{cases} 1, & if\ x \in R_n \\ 0, & else \end{cases}$$

and the optimal choices for the parameter c_n for a given R_n is:

$$\hat{c}_n = \frac{1}{|R_n|} \sum_{x_i \in R_n} y_i$$

where $|R_n|$ is the number of sub-regions. This is derived following the standard linear regression theory for the common sum of squares loss function:

$$\mathrm{L} = \sum_i (y_i - f(x_i))^2$$

Ideally to minimize the loss function, the optimal partition $|R_n|$ should be identified. However, the number of possible partitions grows exponentially with the size of the feature space Z, which makes it computationally infeasible. Instead we can implement a greedy algorithm to find the subsequent splits of Z which minimize the loss function locally at each split.

First a pair of half-planes is defined, given a splitting variable j and a splitting point s:

$$R_1(j,s) = \{Z|Z_j \le s\}$$
$$R_2(j,s) = \{Z|Z_j > s\}$$

and optimize the loss function for j and s:

$$\min_{j,s}\left[\sum_{x_i \in R_1(j,s)} (y_i - \hat{c}_1)^2 + \sum_{x_i \in R_2(j,s)} (y_i - \hat{c}_2)^2\right]$$

The optimal choices for the $\hat{c}_1$ and $\hat{c}_2$ is the same as $\hat{c}_n$. The same splitting procedure is applied recursively on the two half-planes until the tree is grown to a certain size. The size of the tree decides the complexity of the model. A very large tree may lead to overfitting, while a small tree may not be able to capture the important dependencies in the data.

2.2 Random forests

A single decision tree suffers high variance as a highly non-linear classifier. This issue is addressed by random forests which establishes an ensemble of identically distributed single decision trees. The random forests contain N single decision trees. Each single decision tree is grown from a bootstrap sample from the training data. At each splitting node, the features are randomly picked to be considered for the best split. The final output of the random forests ensembles all the single decision trees. The approximation function is modelled as:

$$f(x) = \frac{1}{N}\sum_{i=1}^{N} T_i(x)$$

where T_i is the single decision tree and there are N single decision trees in the random forests. The linear averaging of the identically distributed decision trees will reduce the variance of the single decision tree.

2.3 Adaptive boosting

Adaptive boosting or adaptive boosted trees is also a type of ensemble trees. We can approximate $f(x)$ by summing a group of trees:

$$f(x) = \sum_{i=1}^{N} T_i(x)$$

The n-th tree $T(x)$ is grown based on a heavier weight on the error occurred in the previous n-1 trees. The n-th tree is added with a selected coefficient α to the previous n-1 trees to minimize the sum of the training error E:

$$E_{i,n} = \sum_{i=1}^{N} E[f_{n-1}(x_i) + \alpha_n T(x_i)]$$

where E() is error function.

2.4 Gradient boosting

Theoretically, extreme gradient boosting, or gradient boosted trees, is another type of ensemble trees. In this method, a new tree is growing on the error of the predecessor trees, which minimizes the loss function iteratively at each growing tree along the gradient descent direction in the solution space.

We can approximate $f(x)$ by summing a group of trees:

$$f(x) = \sum_{i=1}^{N} T_i(x)$$

The n-th tree is grown on the error quantity E calculated using the previous n-1 trees:

$$E_{i,n} = -\left.\frac{\partial L(y_i, f(x_i))}{\partial f(x_i)}\right|_{f=f_{n-1}}$$

This equation is the generalization of the minimization of the loss function iteratively towards the gradient descent direction in the solution space.

3 DEEP NEURAL NETWORK

Conventional neural network model consists of three layers: input, hidden and output, which is useful in simple applications. The structure is "shallow" because only one computation (hidden) layer is used to process the features. The concept of "deep" usually refers to more than one computation layer is used in the network structure. This allows us to process more complex input features because the hidden layer performs non-linear transformation computation of the previous hidden layer. Even though the deep neural network can learn much more complex functions, overfitting may occur when there are too many hidden layers.

In our experiment, a network structure of two hidden layers is chosen based on cross-validation results. The first hidden layer has 1000 neurons and the second hidden layer has 500 neurons. The hidden layers make use of ReLU activation function. We do not use dropout since it does not improve the result. Adam optimization method is used for the network optimization. We implement the neural network using Google's open source machine learning software library named TensorFlow.

4 STACKED GENERALIZATION

Stacked generalization is an advanced technique to combine a pool of machine learning models to reduce the generalization error. Usually a pool of first-level machine learning models is applied on the training data to generate predictions. Then a second-level machine learning model is used to combine the prediction of the first-level models.

In our application, 10 extreme gradient boosting models, 5 deep neural network models, 3 random forests models and 2 adaptive boosting models are implemented as first-level models. 5-fold cross validation is used to generate the predictions for the training data. Then an extreme gradient boosting model is used as the second-level model to combine the predictions from the first level. The overall structure of the proposed stacked generalization model is shown in Fig. 1.

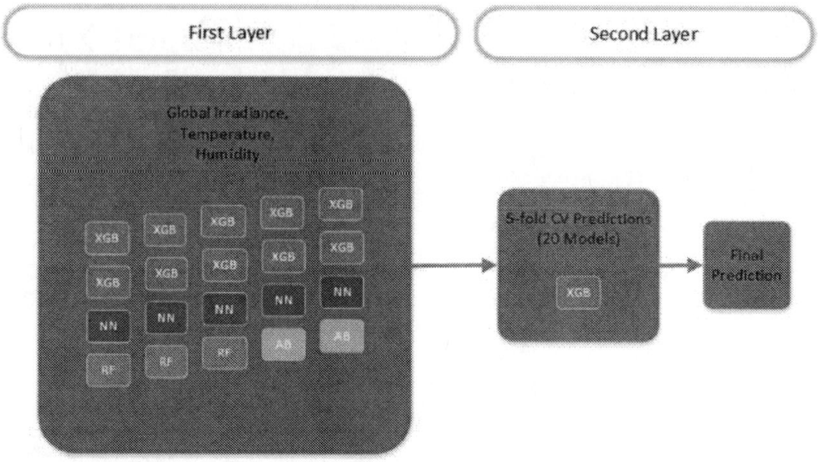

Figure 1: The overall structure of the stacked generalization model. (XGB: eXtreme Gradient Boosting; NN: Neural Network; RF: Random Forests; AB: Adaptive Boosting)

5 EXPERIMENT AND RESULTS

We compare the model performance using two data sets: single station forecast and all Singapore average forecast. Single station forecast is most relevant to the production forecast of a single photovoltaic power plant. Single station forecasting is very challenging in tropical regions like Singapore with high solar irradiance variability due to tropical rainforest climate. The all Singapore average forecast data set is the spatial average of 25 single monitoring stations across Singapore. This forecast is useful for the electricity grid to determine the overall PV generation in the country.

We use two years (2014, 2015) of historical one-hour average solar irradiance and environment monitoring data as training data. The training data includes both daytime and nighttime data. The 2016 monitoring data are used as test data. Only daytime hours (7am – 7pm) are included in the test data. 63 pre-processed features (Table I) are used to train the machine learning models: 48 hourly solar irradiance features, 6 hourly temperature features, 6 hourly humidity features and 3 time features.

Table I: Features

	Number of features
Irradiance	48
Temperature	6
Humidity	6
Time	3

Table II: Model performance comparison (nRMSE)

	Single station	All Singapore
Persistence	38.74%	30.77%
AB	29.41%	19.91%
RF	28.18%	17.94%
XGB	27.69%	16.67%
DNN	27.44%	15.31%
Stacking	26.91%	14.88%

6 ANALYSIS

Table II shows the average normalized root-mean-square error (nRMSE) comparison of hourly solar irradiance forecasting methods for the year of 2016. The first method is persistence which is used as a benchmark method. Due to the high solar irradiance variability in Singapore, the benchmark persistence model has 38.74% single station nRMSE and 30.77% all Singapore average nRMSE. The four individual machine learning models are adaptive boosting (AB), random forests (RF), extreme gradient boosting (XGB) and deep neural network (DNN). The AB model can reduce the nRMSE to 29.41% and 19.91%. The RF model can reduce the nRMSE to 28.18% and 17.94%. The XGB model can further reduce the nRMSE to 27.69% and 16.67%. The DNN model has the best performance among all the individual models. The DNN model reduces the nRMSE to 27.44% and 15.31%.

After combining all the individual models using stacked generalization as shown in Fig. 1, the final nRMSE obtained are 26.91% and 14.88%. Compared with the benchmark persistence model, the absolute nRMSE value reduction for the stacking model are 11.83% for single station and 15.89% for all Singapore average. Translating to improvement ratio, 30.54% and 51.64% performance improvements has been achieved by the stacked generalization method compared with persistence.

7 CONCLUSIONS

We designed and implement a stacked generalization model to forecast hourly solar irradiance for tropical regions with high solar irradiance variability. The proposed model consists of 10 extreme gradient boosting models, 5 deep neural network models, 3 random forests models and 2 adaptive boosting models. The performance of the proposed model has huge improvement compared with a benchmark persistence model. For single station forecast, the nRMSE has been reduced from 38.74% to 26.91%. For all Singapore average forecast, the nRMSE has been reduced from 30.77% to 14.88%. These accurate forecasts can help system operators and grid operators to optimize the PV integration to the power grid and cut costs on energy storage buffers.

REFERENCES

[1] Zibo Dong, Dazhi Yang, Thomas Reindl, Wilfred M. Walsh, Short-term solar irradiance forecasting using exponential smoothing state space model, Energy, Volume 55, 15 June 2013, Pages 1104-1113, ISSN 0360-5442, http://dx.doi.org/10.1016/j.energy.2013.04.027.

[2] Zibo Dong, Dazhi Yang, Thomas Reindl, Wilfred M. Walsh, Satellite image analysis and a hybrid ESSS/ANN model to forecast solar irradiance in the tropics, Energy Conversion and Management, Volume 79, March 2014, Pages 66-73, ISSN 0196-8904, http://dx.doi.org/10.1016/j.enconman.2013.11.043.

[3] Zibo Dong, Dazhi Yang, Thomas Reindl, Wilfred M. Walsh, A novel hybrid approach based on self-organizing maps, support vector regression and particle swarm optimization to forecast solar irradiance, Energy, Volume 82, 15 March 2015, Pages 570-577, ISSN 0360-5442, http://dx.doi.org/10.1016/j.energy.2015.01.066.

[4] Tianqi Chen and Carlos Guestrin, XGBoost: A scalable tree boosting system, 2016, arXiv:1603.02754v3[cs.LG]

TOWARD IMPROVED MODELING OF SPECTRAL SOLAR IRRADIANCE FOR SOLAR ENERGY APPLICATIONS

Yu Xie and Manajit Sengupta
Power Systems Engineering Center, National Renewable Energy Laboratory
15013 Denver W Pkwy, Golden, CO 80401

ABSTRACT: This study introduces the National Renewable Energy Laboratory's (NREL's) recent efforts to extend the capability of the Fast All-sky Radiation Model for Solar applications (FARMS) by computing spectral solar irradiances over both horizontal and inclined surfaces. A new model is developed by computing the optical thickness of the atmosphere using SMARTS2, a spectral irradiance model for clear-sky conditions. A comprehensive lookup table (LUT) of cloud bidirectional transmittance distribution functions (BTDFs) is precomputed for 2002 wavelength bands using an atmospheric radiative transfer model, libRadtran. The solar radiation transmitted through the atmosphere is given by considering all possible paths of photon transmission and the relevant scattering and absorption attenuation. Our results indicate that this new model has an accuracy that is similar to that of state-of-the-art radiative transfer models, but it is significantly more efficient.
Keywords: Solar Radiation, Optical Properties, PV System

1 INTRODUCTION

Radiative transfer models play a crucial role in assessing and forecasting solar resources for electricity conversion. NREL developed FARMS [1] to accurately and efficiently evaluate broadband solar irradiances for all-sky conditions over horizontal surfaces; however, the demands of solar energy research often exceed the capabilities of FARMS as well as other conventional radiative transfer models. For example, solar irradiances in numerous narrow-wavelength bands are not provided by FARMS, but they are particularly useful in solar cell research because of the spectral response of photovoltaic (PV) panels. Further, solar energy studies require solar irradiances over inclined surfaces because solar systems track the sun on multiple axes. Although numerous solar irradiance models can compute spectral or broadband solar irradiances over inclined surfaces, they are either designed for clear-sky conditions or use empirical models to simulate plane-of-array (POA) irradiances [2-4]. The purpose of this study is to extend the capability of FARMS by developing an all-sky model to efficiently provide physical solutions of spectral solar irradiances over both horizontal and inclined surfaces.

2 CLOUD BIDIRECTIONAL TRANSMITTANCE DISTRIBUTION FUNCTION

To develop a new irradiance model capable of rapidly simulating spectral irradiance on inclined surfaces, cloud bidirectional transmittance distribution functions (BTDFs) need to be combined with clear-sky absorption and scattering [5, 6]. Cloud BTDFs are functions of wavelength, cloud thermodynamic phase, solar zenith angle, viewing zenith angle, relative azimuth angle, cloud optical thickness, and effective particle size. Because the computation of BTDFs by libRadtran [7] is time-consuming, it is important to determine the appropriate spectral and spatial resolutions. Figure 1a compares the spectral resolutions of different radiative transfer models and BTDF databases. A clear-sky radiative transfer model, SMARTS2 [2], uses 2,002 wavelengths from 0.28–4.0 μm, accounting for the absorbing lines of trace gases in the atmosphere. Another spectral radiative transfer model, TMYSpec [3], considers wavelengths for every 10 nm from 0.3–1.8 μm; however, cloud BTDF is sensitive to wavelength because of the variation in the refractive indexes for water droplets and ice crystals which are more smoothly. Thus, fewer wavelengths can be used to compute cloud BTDFs to save computing time. We investigated BTDFs of water and ice clouds and found that 97 wavelengths (red dots in Fig.1a) from 0.2–4.0 μm provided the best balance between computing time and accuracy. The BTDF can be interpolated to 2,002 wavelengths and combined with SMARTS2 to compute all-sky spectral radiances.

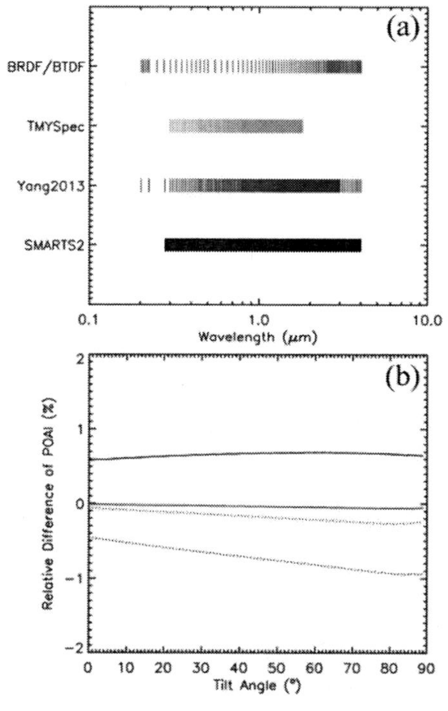

Figure 1: (a) Comparison of spectral resolutions; (b) POA irradiances computed from 50 viewing zenith angles and 36 relative azimuth angles (solid lines) and those from 25 viewing zenith angles and 18 relative azimuth angles (dashed lines) for water (red) and ice (blue) clouds.

To determine the resolution of viewing angles in the new radiative transfer model, we used libRadtran to compute cloudy-sky radiances of water and ice clouds for optical depth τ=5, effective particle size De=10 μm, solar zenith angle of 30°, and wavelength λ=0.6 μm. One hundred viewing zenith angles and 180 relative azimuth angles (for every 1°) were used in the computation. To save computing time, fewer viewing zenith angles and relative azimuth angles were also tested. Figure 1b compares the relative error of the POA irradiances from using 50 viewing zenith angles and 36 relative azimuth angles (solid lines) and 25 viewing zenith angles and 18 relative azimuth angles (dashed lines). POA irradiances from 100 viewing zenith angles and 180 relative azimuth angles were used as the ground truth. Compared to the solid lines in Fig.1b, 25 viewing zenith angles and 18 relative azimuth angles have less than 1% uncertainty.

2 FARMS WITH NARROWBAND IRRADIANCES OVER TILTED SURFACES (FARMS-NIT)

To compute the solar radiation transmitted through a cloudy atmosphere, we divide the atmosphere into three layers: a cloud/aerosol layer and the clear-sky layers above and below the cloud layer. For the clear-sky layers, the optical thicknesses in 2,002 wavelength bands are computed by a clear-sky model, SMARTS2 [2]. In each layer, a photon can either directly transmit through the layer or be scattered by air molecules or cloud particles. To derive surface radiation, we consider all possible photon paths that are listed in Table 1, where "D" denotes direct transmission and "S" denotes scattering within the layer. From Path 7, the direct solar radiation can be easily computed using the Beer-Bouguer-Lamber law [8] and the optical thicknesses of the clouds and the atmosphere. Following the Rayleigh scattering correction technique and [9], Rayleigh scattering is considered only once in the whole atmosphere. Thus, paths 1–4 and 6 are used in the computation of diffuse solar radiation. By solving the radiative transfer equation with a single-scattering event and combining the solution with cloud BTDFs from the precomputed LUT, we can efficiently compute the diffuse radiances for each photon path. The solar radiances at the land surface are given by the total values of the radiances from all photon paths. The irradiances for various wavelength bands can then be computed for both horizontal and inclined surfaces.

Table 1: Possible Paths of Photons Transmitted through the Atmosphere

Path	Upper Clear Layer	Cloud	Lower Clear Layer
1	S	D	D
2	D	S	D
3	D	D	S
4	D	S	S
5	S	D	S
6	S	S	D
7	D	D	D
8	S	S	S

3 RESULTS

Figure 2 shows the spectral POA irradiances computed using the FARMS-NIT for an atmosphere with a water cloud. The cloud optical thickness and effective particle size are 1 and 10 μm, respectively, with a solar zenith angle of 45°. The spectral POA irradiances in the 2,002 wavelength bands are simulated over two surfaces where the tilt angles, β, are 0° and 45°. The figure shows that the PV panel with 45° tilt receives significantly more solar radiation than the horizontal PV panel.

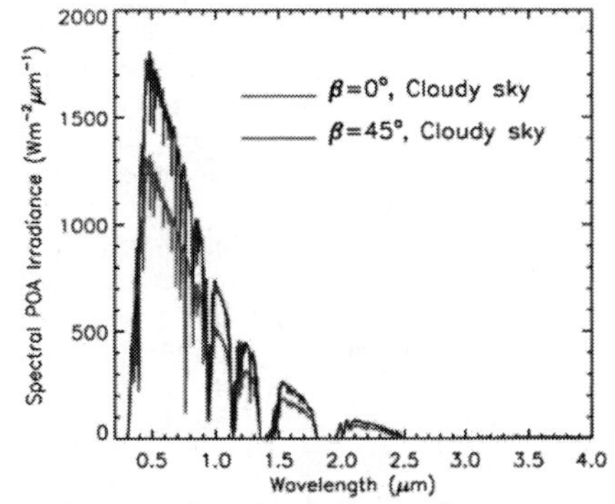

Figure 2: Spectral POA irradiances for an atmosphere with a water cloud. The solar zenith angle is 45°, and the surface albedo is 0. The cloud optical thickness and effective particle size are 1 and 10 μm, respectively.

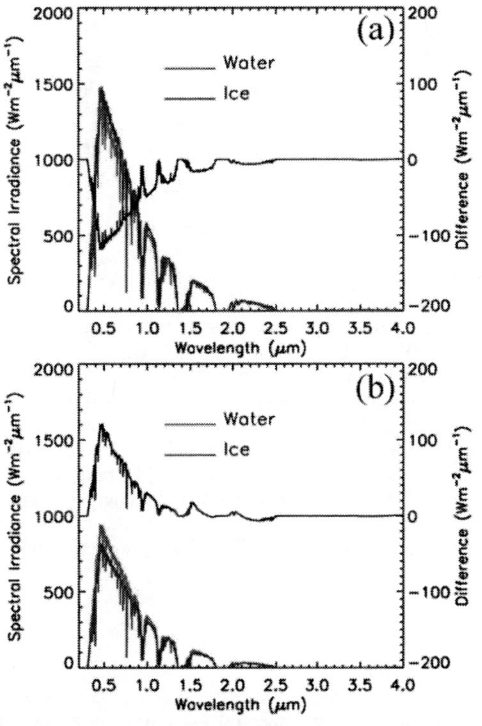

Figure 3: GHI simulated using FARMS-NIT when the solar zenith angle is 45° and surface albedo is 0.1. Clouds have optical thicknesses of (a) 1 and (b) 5 with an effective particle size of 10 μm. The black lines denote the difference in GHI between the water and ice clouds.

To understand the impact of ice and water cloud conditions on the simulations by FARMS-NIT, we computed cloudy-sky GHI when solar zenith angle is assumed to be 45°. The clouds are composed of water droplets or ice crystals with an effective particle size of 10 μm.

Figure 3 compares the GHI for optically thin and thick clouds. The black lines denote the difference in GHI between the water and ice clouds. For single-scattering events, ice crystals have more significant scattering than water droplets in the forward direction because of more diffraction. Ice crystals also have stronger backward scattering because of their nonspherical geometries and the relative internal and external reflections. Figure 3 shows that the GHI for ice clouds is higher than for water clouds when the clouds are thin due to the increased forward scattering by ice cloud particles; however, the forward scattering becomes less important for thick clouds when more solar radiation is reflected back to the space by ice crystals. As a result, the GHI for water clouds is higher than for ice clouds in Fig. 3b when an optical depth of 5 is considered.

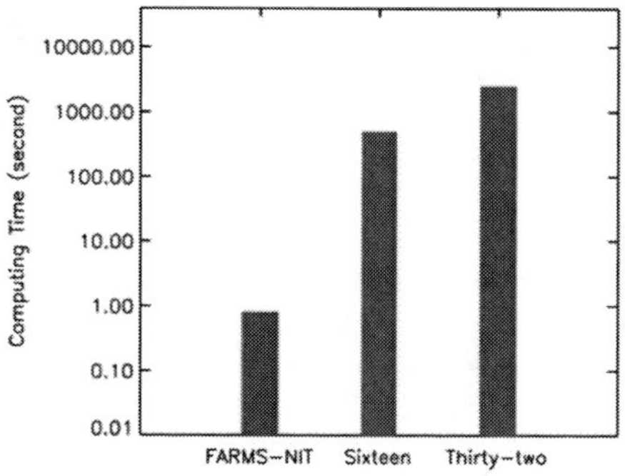

Figure 4: Computational time of FARMS-NIT and LibRadtran with 16 and 32 streams for computing spectral irradiances in 2,002 wavelengths.

Figure 4 shows the computational time for simulating spectral irradiances in 2,002 wavelengths. The cloudy-sky irradiances were computed using FARMS-NIT and a radiative transfer model for meteorology, LibRadtran. It is found that LibRadtran with 16- and 32-stream approximations consume 492 and 2,493 seconds for the computation, whereas FARMS-NIT uses only ~0.8 second. For clear-sky conditions, the computational time of FARMS-NIT can be further reduced, as fewer independent events need to be considered than for cloudy-sky conditions.

4 CONCLUSIONS

In this study, we developed a new model to efficiently compute spectral POA irradiances over both horizontal and inclined surfaces. The spectral solar radiances and irradiances were numerically derived by considering all possible photon paths in the atmosphere. Our results indicate that this new model can effectively demonstrate the spectral structure of cloudy-sky solar radiation. Note that a new clear-sky spectral model is also desired because the current models cannot precisely compute clear-sky radiances for the possible photon paths. The accuracy of the clear-sky and cloudy-sky models are estimated using state-of-the-art radiative transfer models and surface observations. Additionally, the performance of the spectral irradiance models depends on the computational capabilities. Further efforts are underway to significantly improve the computational efficiency of the new model by parameterizing the LUT of cloud BTDF or applying advanced search technology.

REFERENCES

[1] Y. Xie, M. Sengupta, and J. Dudhia, "A Fast All-sky Radiation Model for Solar applications (FARMS): Algorithm and performance evaluation," *Sol. Energy,* vol. 135, pp. 435-445, 2016.

[2] C. Gueymard, "SMARTS2: a simple model of the atmospheric radiative transfer of sunshine: algorithms and performance assessment," *Florida Solar Energy Center,* vol. Cocoa, FL, 1995.

[3] D. Mayeer, "Direct beam and hemispherical terrestrial solar spectral distributions derived from broadband hourly solar radiation data," *Sol. Energy,* vol. 86, pp. 2771-2782, 2012.

[4] R. Perez, P. Ineichen, K. Moore, M. Kmiecik, C. Chain, R. George, and F. Vignola, "A new operational model for satellite-derived irradiances: Description and validation," *Sol. Energy,* vol. 73, pp. 307-317, 2002.

[5] Y. Xie, "Study of ice cloud properties from synergetic use of satellite observations and modeling capabilities," Ph. D., Department of Atmospheric Sciences, Texas A&M University, College Station, TX, 2010.

[6] Y. Xie, P. Yang, G. W. Kattawar, P. Minnis, Y. X. Hu, and D. Wu, "Determination of ice cloud models using MODIS and MISR data," *Int. Remote Sens.,* vol. 33, pp. 4219-4253, 2012.

[7] B. Mayer and A. Kylling, "Technical note: The libRadtran software package for radiative transfer calculations: description and examples of use," *Atmos. Chem. Phys.,* vol. 5, pp. 1855-1877, Jul 26 2005.

[8] K. N. Liou, *An introduction to atmospheric radiation,* 2nd ed. Amsterdam; Boston: Academic Press, 2002.

[9] C. Wang, P. Yang, S. Nasiri, S. Platnick, B. Baum, A. Heidinger, and X. Liu, "A fast radiative transfer model for visible through shortwave infrared spectral reflectances in clear and cloudy atmospheres," *J. Quant. Specrtrosc. Radiat. Transfer* vol. 116, pp. 122-131, Feb 2013.

This is a pre-peer review version of the paper submitted to the journal Progress in Photovoltaics.
It has been selected by the Executive Committee of the EU PVSEC 2017 for submission to Progress in Photovoltaics.

VALIDATION OF AN ALL-SKY IMAGER BASED NOWCASTING SYSTEM FOR INDUSTRIAL PV PLANTS

Pascal Kuhn[1], Bijan Nouri[2], Stefan Wilbert[2], Christoph Prahl[2], Nora Kozonek[2], Thomas Schmidt[3], Zeyad Yasser[4], Lourdes Ramirez[5], Luis Zarzalejo[5], Angela Meyer[6], Laurent Vuilleumier[6], Detlev Heinemann[7], Philippe Blanc[7], Robert Pitz-Paal[8]

1 Researcher, M.Sc. Physics, German Aerospace Center (DLR), Institute of Solar Research, Plataforma de Almería, Ctra. de Senés s/n km 4, 04200 Tabernas, Spain, +34 950611743, Pascal.Kuhn@dlr.de

2 German Aerospace Center (DLR), Institute of Solar Research, Plataforma de Almería, Ctra. de Senés s/n, 04200 Tabernas, Spain. Bijan.Nouri@dlr.de, Stefan.Wilbert@dlr.de, Christoph.Prahl@dlr.de, Nora.Kozonek@dlr.de

3 CSP Services GmbH, Friedrich-Ebert-Ufer 30, 51143 Cologne, Germany. t.schmidt@cspservices.de

4 TSK FLAGSOL, Anna-Schneider-Steig 10, 50678 Cologne, Germany. zeyad.yasser@flagsol.de

5 CIEMAT. Energy Department, Renewable Energy Division, Av. Complutense, 40, 28040 Madrid, Spain. lourdes.ramirez@ciemat.es, lf.zarzalejo@ciemat.es

6 MeteoSwiss, Les Invuardes, 1530 Payerne, Switzerland. Angela.Meyer@protonmail.com, Laurent.Vuilleumier@meteoswiss.ch

7 Energy Meteorology Unit, Energy and Semiconductor Research Laboratory, Institute of Physics, Oldenburg University, 26111 Oldenburg, Germany. detlev.heinemann@uni-oldenburg.de

8 MINES ParisTech, PSL Research University, O. I. E. – Centre Observation, Impacts, Energy, CS 10207, F-06904, Sophia Antipolis CEDEX, France. philippe.blanc@mines-paristech.fr

9 German Aerospace Center (DLR), Institute of Solar Research, Linder Höhe, 51147 Cologne, Germany. Robert.Pitz-Paal@dlr.de

ABSTRACT: Due to the cloud induced variability of the solar resource, the growing contributions of photovoltaic plants to the overall power generation challenges the stability of electricity grids. To avoid blackouts, administrations started to define maximum negative ramp rates. Storages can be used to reduce the occurring ramps. Their required capacity, durability and costs can be optimized by nowcasting systems. Nowcasting systems utilize the input of upward-facing cameras to predict future irradiances. Previously, many nowcasting systems were developed and validated. However, these validations did not consider aggregation effects, which are present in industrial-size power plants. In this paper, we present the validation of nowcasted global horizontal irradiance (GHI) and direct normal irradiance (DNI) maps derived from an example system consisting of four all-sky cameras ('WobaS'). The WobaS system is operational at two solar energy research centers and at a commercial 50 MW solar power plant. Besides its validation on 30 days, the working principle is briefly explained. The forecasting deviations are investigated with a focus on temporal and spatial aggregation effects. The validation found that spatial and temporal aggregations significantly improve forecast accuracies: Spatial aggregation reduces the relative RMSE (GHI) from 30.9 % (one pixel of the GHI map of 25 m^2) to 23.5 % (4 km^2) on a day with variable conditions for one minute averages and a lead time of 15 min. Over 30 days of validation, a relative RMSE (GHI) of 20.4 % for the next 15-minutes is observed at pixel basis (25 m^2). Although the deviations of nowcasting systems strongly depend on the validation period and the specific weather conditions, the WobaS system is considered to be at least state-of-the-art.

Keywords: Photovoltaic, Plant Control, Grid Integration, Solar nowcasting, Nowcasting systems, All-sky imager

1 INTRODUCTION, AIM AND APPROACH

The solar resource shows variabilities, which influence the frequency, voltage and overall stability of electrical grids with high penetrations of solar power plants. Facing this challenge, legal limitations of the fluctuations of generated electricity (ramp rates) are already effective or being discussed [1]. Unsurprisingly, small island grids with high solar penetration had to rush ahead: Hawaii has defined a maximum negative ramp rate of 1 MW/min during certain times [1] and Puerto Rico has specified a maximum negative 1 min ramp rate of 10 % of the rated capacity [2]. For short-term periods between zero and 30 min, the origins of these ramp rates for photovoltaic (PV) plants are predominantly transient clouds. Forecasts covering this period are called nowcasts and can be achieved by all-sky imager (ASI) based systems [3-11]. By providing such forecasts, nowcasting systems can help the power plant to fulfill these ramp rate regulations. Moreover, they help to optimize the required size and the operations of combined PV-battery systems [12, 13]. Thus optimized operations potentially increase the lifespan of electrical storages and auxiliary devices [14].

The four-camera WobaS system, which is presented in this work, is installed at PSA (Plataforma Solar de Almería, Spain) [11]. WobaS stands for the research project 'Wolkenkamera-basierte Betriebsstrategien für Solarkraftwerke' – all-sky imager based operational strategies for solar power plant. Another WobaS system operates at the commercial solar power plant La Africana (near Córdoba, Spain) [15] and a third WobaS system is running at a solar research center of the University of Evora, Portugal [16]. WobaS systems use the inputs of four off-the-shelf surveillance cameras (Mobotix Q24 or Q25). Besides their low costs, these cameras were found to be easy to operate and reliable in harsh weather conditions. Synchronized by an NTP server, the cameras take images every 30 s with resolutions of 3 MP (Mobotix Q24) or 6 MP (Mobotix Q25). In these images, clouds are segmented using a 4-dimensional Clear Sky Library [11]. The 3-D positions and shapes of all visible clouds are determined via voxel-carving [17]. Voxel is an abbreviation for volume-element and voxel-carving is the process to derive 3-D objects out of multiple camera perspectives. This way, a voxel space filled with individual cloud objects is derived (Figure 1, left). Since the ground cameras cannot see on the top of the clouds (self-occlusion), the raw cloud voxels must be partially modelled.

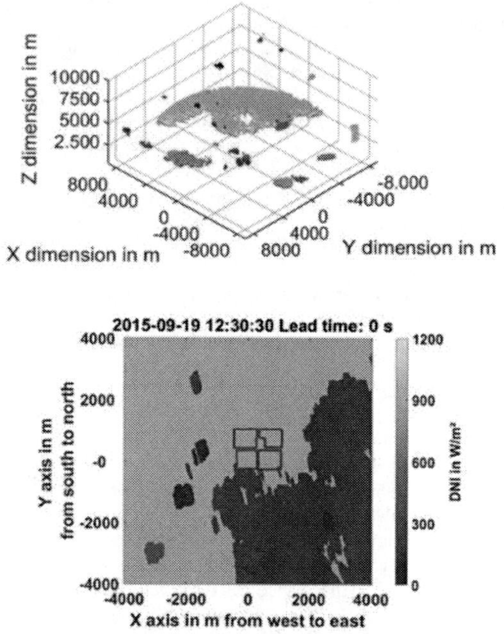

Figure 1: Several cloud objects in the voxel space (upper image) and corresponding nowcasted DNI map (image below) at the La Africana solar power plant.

By tracking 3-D cloud objects over multiple timestamps, cloud velocities and directions are derived and used to predict cloud movements. Future cloud positions are calculated for lead times up to 15 min ahead. With the sun position and the surface elevations known, the shadows on the ground are deduced [17]. If available, reference real-time irradiance measurements are used to determine cloud transmittances. With cloud transmittances derived, shadow maps are transformed into irradiance maps (see Figure 1, right). WobaS systems can use ground measurement stations for Direct Normal Irradiance (DNI) and Global Horizontal Irradiance (GHI). Without available ground measurements, modelled irradiance data can be used.

The validated WobaS system is located at PSA and has access to DNI, GHI and DHI (diffuse horizontal irradiance) measurements and is able to predict GHI, DNI and GTI (global tilted irradiance) maps. Spatial and temporal resolutions for these maps are 25 m^2, updated every 30 seconds for predictions up to 15 min ahead. The working principle of the WobaS system is illustrated in Figure 2.

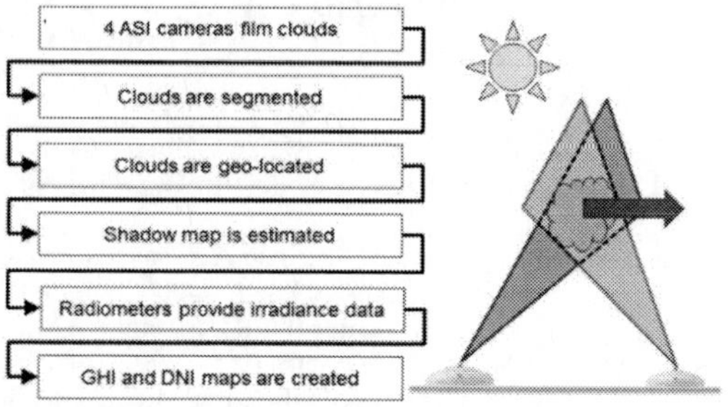

Figure 2: Working principle of the WobaS all-sky imager based nowcasting system. All-sky imagers are used to detect, 3-D geo-locate and track clouds to predict cloud shadows on the ground. This way, future irradiances in high spatial and temporal resolutions can be derived.

This publication is structured as follows: In section 2, the validation of a key sub-task of a nowcasting system, the cloud detection within all-sky images, is presented. The validation of nowcasted irradiance maps is presented in section 3. The focus of this validation lies on the investigation of spatial and temporal aggregation effects. Spatial aggregation effects are relevant for large solar plants, covering several km². Since the aggregated total electricity production of the PV plant is the most important parameter to forecast, the accuracies and validations of the average nowcasted irradiances over the complete plant area are of special interest. The validation of the nowcasted irradiance maps is conducted via pyranometer (GHI) and pyrheliometer (DNI) measurements (section 3.1) and with a reference shadow camera system (section 3.2). The conclusion is given in section 4.

2 VALIDATION OF CLOUD SEGMENTATION

Segmenting clouds in all-sky images is one key task of the nowcasting system and is surprisingly difficult. This originates from color and intensity dependencies within the images regarding pixel and sun positions as well as from artefacts due to saturation and glare effects in the circumsolar region. Many approaches, e.g. based on neural networks [18, 19] or ratios of the Red-Green-Blue (RBG) color channels [20, 21], are validated and discussed in the literature. The WobaS nowcasting system utilizes a 4-dimensional clear sky library (CSL) [11]. In the CSL, clear sky RGB values for every pixel depending on Sun Pixel Angle (SPA), the Pixel Zenith Angle (PZA), air mass (related to solar elevation for a given altitude) and Linke turbidity are stored. The two angles are illustrated in Figure 3 a: The PZA is the angle between the zenith above the camera and the pixel in the fisheye-projection. The SPA is the angular distance between the positions of the sun and the pixel.

The Linke turbidity is calculated from DNI or GHI measurements using the Linke turbidity model from [22]. Via the temporal variation of the Linke turbidity, shaded ground measurements are excluded using a Linke turbidity based method presented in [23, 24]. Afterwards, the current Linke turbidity is determined using a linearly time-weighted average of the most recent Linke turbidities derived from unshaded irradiance measurements.

This is a pre-peer review version of the paper submitted to the journal Progress in Photovoltaics.
It has been selected by the Executive Committee of the EU PVSEC 2017 for submission to Progress in Photovoltaics.

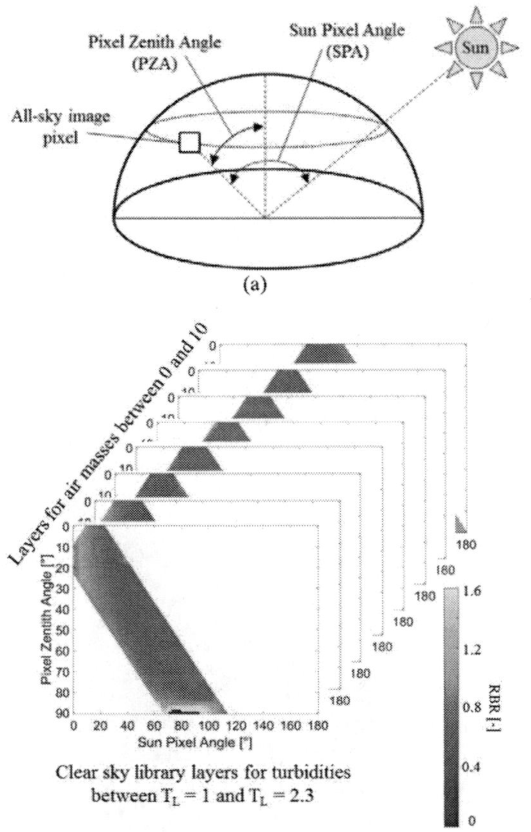

Figure 3: (a) Imaged semi-dome of the all-sky imager with relevant angles of the CSL marked. (b) Layers of the Clear Sky Library for a Linke turbidity range between TL = 1 and TL = 2.3 and the air masses between 0 and 10. In the CSL, a total of 80 layers per color channel for various Linke turbidities and air masses are stored.

The CSL consists of 80 layers per color channel. Each layer corresponds to a specific jointed air mass and Linke turbidity bin. Within a layer, the RBG values of images taken during clear sky conditions are stored relative to two angles (SPA, PZA, see Figure 3 a). In Figure 3 b, 8 layers of the Red-to-blue ratio (RBR) are depicted, which are calculated from the corresponding color channel layers.

If a pixel in the image under consideration is clouded, its RBG values deviate from the CSL and this way a cloud is detected. The detection is based on CSL color channel ratios (as depicted in Figure 3 b) or differences and situational thresholds. Figure 4 illustrates the approach in a simplified manner: From the raw image, ratios and difference of the color channels are calculated. Based on the current air mass and Linke turbidity, one CSL layer per color channel is selected. From these CSL layers, ratios (depicted: red-to-blue ratio, RBR) and differences are calculated and compared to the current image. Using multiple thresholds, cloud detection is then performed. The thresholds used for the segmentation depend on PSA, PZA and on the current weather situation (e.g. overcast, clear sky). The current weather situation is determined considering irradiance and illuminance fluctuations, detected movements and cloud coverages found in the previous segmentations.

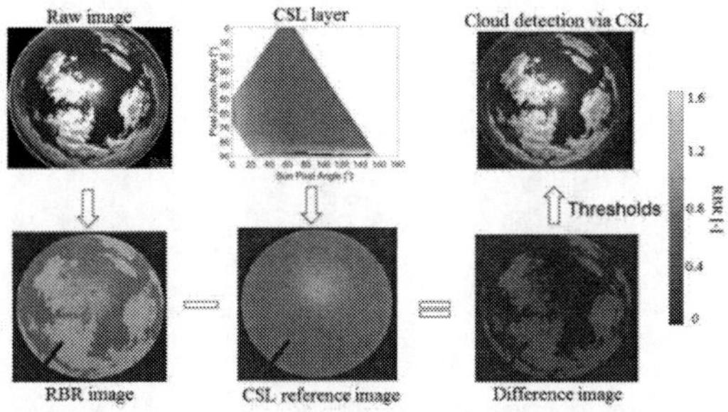

Figure 4: Working principle of the clear sky library (CSL) based cloud detection algorithm. Raw images are compared to clear sky reference images corresponding to similar Linke turbidities and air masses. One layer of the CSL is shown, showing the ratio of the corresponding red and blue CSL layers. The color bar, depicting the red-to-blue ratios (RBR), relates to the RBR image, the CSL layer, the CSL reference image and the Difference image.

The CSL based segmentation is validated with 530 manually segmented images. The approach is illustrated in Figure 5: all-sky images automatically segmented by the algorithm are pixel-wise compared to manually segmented references.

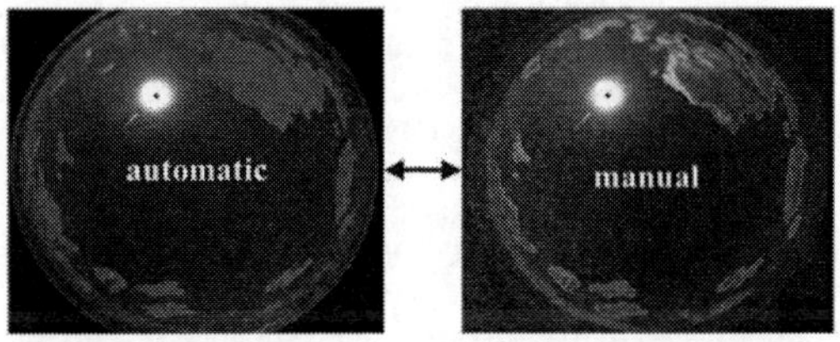

Figure 5: Validation approach of the CSL based cloud detection algorithm: All-sky images segmented by the algorithm ('automatic', left) are compared to 530 manually segmented images ('manual', right). In this example, some clouds at the very edge of the fish-eye projection are not correctly detected by the algorithm (filled red area in the left image).

In Figure 6, the results of the validation are presented for several weather situations divided into groups with high (above 30°) and low sun elevation angles (below 20°), hazy (Linke turbidity above 3.25) and less hazy conditions (Linke turbidity below 2.75) as well as different cloud coverages.

With one exception, the probability of correct detection is at 80 % or above, even exceeding 95 % for many situations. The probability of correct detection is defined as the percentage of pixels correctly detected as 'cloud' or as 'clear sky'. This probability of correct detection drops to 66 % for low solar elevations, high Linke turbidities and medium cloud coverages between 2.5 % and 80 %. Due to the low solar elevation and the high Linke turbidity, the saturation of the blue color of the sky is low and the sky in these circumstances appears whitish-gray. These situations are challenging, even for a

This is a pre-peer review version of the paper submitted to the journal Progress in Photovoltaics.
It has been selected by the Executive Committee of the EU PVSEC 2017 for submission to Progress in Photovoltaics.

human observer, as only partially illuminated clouds do not pose a strong contrast against the aerosol layers around them and can hardly be distinguished from them. The thin border between cloud and aerosols is discussed in [25].

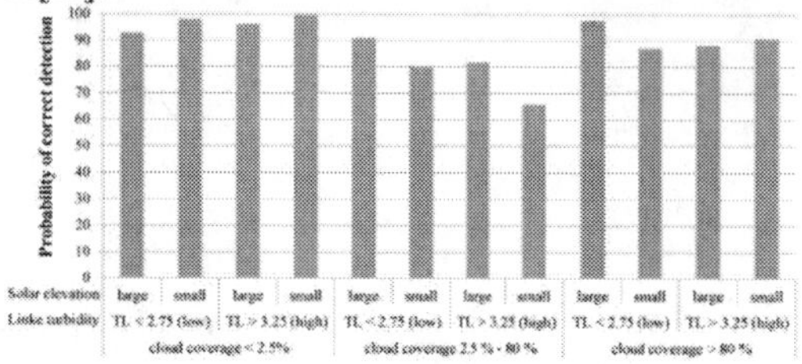

Figure 6: Validation results is visualized for several weather situations including large (above 30°) and small (below 20°) sun elevation angles, dusty and clear air as well as different cloud coverages.

For low sun elevations, the most relevant clouds near the sun are at the edge of the fisheye lens, therefore strongly distorted and are imaged from similar angles from all four WobaS cameras. Deriving 3-D shapes of such clouds is challenging and small pixel-wise errors in the cloud detection result in large deviations in the corresponding 3-D positions. Thus, for sun elevations below 15°, a persistence forecast based on recently measured irradiance data was found to be more effective. In the following validations, timestamps with such low solar elevations are excluded. Due to the rather low irradiance levels in these situations, this approach is considered acceptable, at least for a preliminary step. If low elevations were of special interest, an additional camera tracking the sun could be used.

3 VALIDATION OF NOWCASTED IRRADIANCE MAPS

In this section, we focus on the validation of DNI and GHI maps. These irradiance maps are the final output of the WobaS system. From DNI and GHI maps, GTI maps can be derived using regional-specific models [26-28]. In section 3.1, the irradiance maps are validated using three independent ground measurement stations on 30 days. These days are selected to reflect a wide range of irradiance variabilities and cloud heights. GHI, DNI and measured cloud heights of the 30 days validation period are depicted in Figure 7.

A shadow camera system [29], providing reference irradiance maps, is used for the validation in section 3.2. An auto-evaluation of the WobaS system was conducted on 30 days is briefly discussed in section 4. The validation setup at PSA is depicted in Figure 8. The 30 days of the validation are:

in 2015: September 8th - 11th, 15th, 18th, 19th, October 2nd, 4th, 8th - 11th, 15th, 17th, 18th, November 14th, 22nd, 24th - 29th; and

in 2016: May 11th, 14th, June 3rd, September 27th-28th and October 28th.

As discussed in the previous section, only timestamps with minimum solar elevations of 15° are considered in the validations.

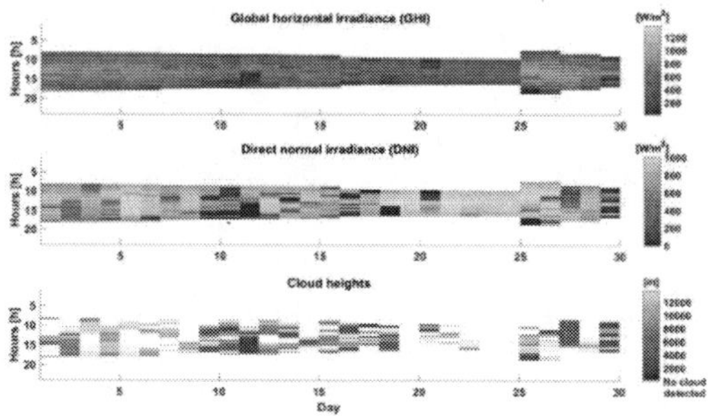

Figure 7. Overview of the 30 days validation period. During the validation period, a wide range of irradiance fluctuations and cloud heights (as measured by a ceilometer) are present. The pyranometer and the pyrheliometer which measured the irradiances depicted here is located at the southern-most X in Figure . The ceilometer for the cloud height measurements is at the position of the white triangle in Figure .

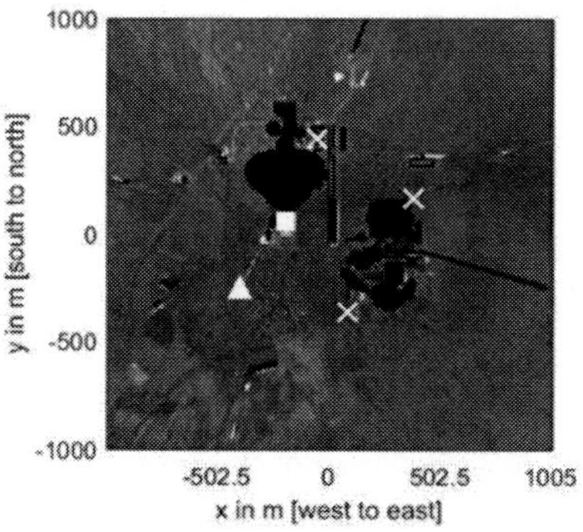

Figure 8: Orthoimage generated from the shadow camera system with ground measurements stations included: The square marks the position of the shadow camera system. The Xs mark the location of three groups comprising each a pyranometer, a pyrheliometer and an all-sky imager. The fourth all-sky imager and a ceilometer are located at the position of the triangle.

Bias, relative bias, mean absolute error (MAE), relative MAE, root mean square error (RMSE), relative RMSE, standard deviation (std) and relative std are calculated for each station (section 3.1) or each field size (section 3.2) and each day. The relative deviations are derived from the absolute deviations and the mean irradiance measured over the validation period. The formulas are defined in the following with oi as the irradiance values from the observations in the reference data set and pi as the predicted irradiance values from the nowcasted data set at timestamp index i. N is the total number of timestamps included in the evaluated validation interval (excluding low elevation angles). Om is set to the mean observed irradiance of the validation period.

The bias is the arithmetic average of the values of the differences:

$$bias = \frac{1}{N}\sum_{i=1}^{N} p_i - o_i$$

This is a pre-peer review version of the paper submitted to the journal Progress in Photovoltaics.
It has been selected by the Executive Committee of the EU PVSEC 2017 for submission to Progress in Photovoltaics.

The relative bias is given by:

$$relative\ \ bias = \frac{100\%}{O_m} bias$$

The Mean Absolute Error (MAE) is:

$$MAE = \frac{1}{N}\sum_{i=1}^{N} | p_i - o_i |$$

The relative MAE is defined as follows:

$$relative\ \ MAE = \frac{100\%}{O_m} MAE$$

The Root Mean Square Error (RMSE) is

$$RMSE = \left[\sum_{i=1}^{N} (p_i - o_i)^2 / N \right]^{1/2}$$

The relative RMSE is defined as follows:

$$relative\ \ RMSE = \frac{100\%}{O_m} RMSE$$

The standard deviation (std) of the fluctuations is:

$$std = \left[\frac{1}{N-1} \sum_{i=1}^{N} ((p_i - o_i) - bias)^2 \right]^{1/2}$$

The relative std is defined as follows:

$$relative\ \ std = \frac{100\%}{O_m} std$$

3.1 Comparing nowcasts with ground measurements stations

We look first at the root mean squared error (RMSE) and the mean absolute error (MAE) derived from three ground measurements stations in comparison to the corresponding pixels in the nowcasted irradiance maps of the WobaS system over the 30 days of validation.

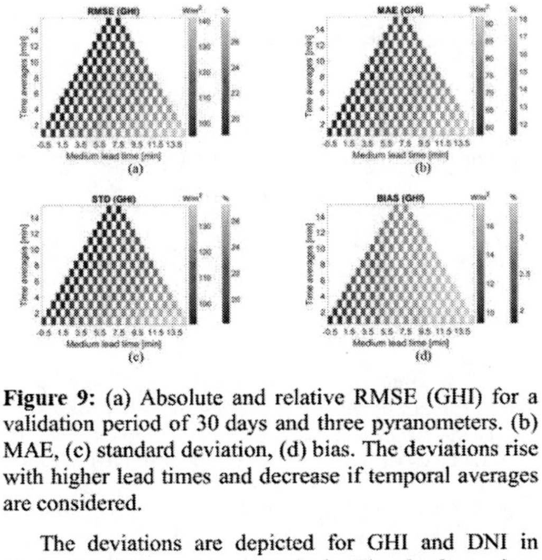

Figure 9: (a) Absolute and relative RMSE (GHI) for a validation period of 30 days and three pyranometers. (b) MAE, (c) standard deviation, (d) bias. The deviations rise with higher lead times and decrease if temporal averages are considered.

The deviations are depicted for GHI and DNI in Figure 9 and Figure 10, respectively. The absolute values are the average of the 90 daily stations-wise deviations (30 days and 3 stations). Each relative value is derived from the corresponding absolute values and the mean irradiance during the 30 days. On the y-axis, temporal averages from 1 min to 15 min are shown. The x-axis shows medium lead times from -0.5 min to 14.5 min.

Looking at Figure 9 a, both the absolute and the relative RMSE values for GHI are displayed. As expected from [30, 31], higher lead times show higher deviations for RMSE, MAE and standard deviation as the more distant future is more difficult to predict. Temporal aggregations reduce these deviations due to averaging effects. However, the behavior of the bias must be explained differently: Similar to the other deviations depicted in Figure 9, but on a lower level, the bias rises with increasing lead times and falls for higher temporal averages. Also, the bias found on 30 days is always positive (see Figure 9 d). All three effects hold for the depicted bias of 30 days. However, bias values for singular days within the 30 days show negative values and different behaviors. In general, there seems to be a minor tendency for positive biases within nowcasting systems (also observed in [30]) as usually more clouds remain undetected than sky clear pixels being incorrectly assigned to be clouded. This is especially true for large lead times where the relevant clouds might be outside the cameras' field of view. One reason for the underdetection is that the WobaS system uses its redundancy of four all-sky imagers to counter incorrect cloud detections: Only clouds seen by at least three of the four cameras are considered, the others are rejected. This potentially increases the bias further. With the bias ranging between 1.8 % and 3.4 %, these effects seem to be minor compared to the standard deviation and the RMSE. In Table 1, the minimum and maximum values are shown for the error metrics as the average daily deviations of 30 days.

Table 1: Minimum and maximum GHI deviations found as the average daily and station-wise deviations on 30 days. Compare with the graphs in Figure .

GHI metrics	Minimum value	Medium lead time	Temporal average
RMSE	18.7 % (95 W/m^2)	2.5 min	7 min
MAE	11.4 % (58 W/m^2)	1 min	4 min
Standard deviation	18.2 % (92 W/m^2)	2.5 min	7 min
Bias	1.8 % (9 W/m^2)	-0.5 min	1 min

GHI metrics	Maximum value	Medium lead time	Temporal average
RMSE	27.9 % (141 W/m^2)	14.5 min	1 min
MAE	18.1 % (92 W/m^2)	14.5 min	1 min
Standard deviation	27.2 % (138 W/m^2)	14.5 min	1 min
Bias	3.4 % (17 W/m^2)	14.5 min	1 min

This is a pre-peer review version of the paper submitted to the journal Progress in Photovoltaics.
It has been selected by the Executive Committee of the EU PVSEC 2017 for submission to Progress in Photovoltaics.

Table 2: Minimum and maximum DNI deviations found as the average daily and station-wise deviations on 30 days. Compare with the graphs in Figure .

DNI metrics	Minimum value	Medium lead time	Temporal average
RMSE	23.7 % (154 W/m^2)	2.5 min	7 min
MAE	13.6 % (89 W/m^2)	0.5 min	3 min
Standard deviation	23.0 % (150 W/m^2)	2.5 min	7 min
Bias	2.2 % (14 W/m^2)	-0.5 min	1 min

GHI metrics	Maximum value	Medium lead time	Temporal average
RMSE	35.5 % (232 W/m^2)	14.5 min	1 min
MAE	22.0 % (144 W/m^2)	14.5 min	1 min
Standard deviation	34.7 % (226 W/m^2)	14.5 min	1 min
Bias	4.6 % (30 W/m^2)	14.5 min	1 min

Figure 10 displays the error metrics found for the DNI predictions of the WobaS system similar to Figure 9. In general, the observed deviations for DNI are larger than the deviations found for the GHI predictions. This effect is caused by the DHI, which damps abrupt changes by scattering irradiance onto the areas shaded by clouds. Usually, inside the shadows of cumulus clouds, the DNI drops to zero and the GHI is equal to the DHI. Therefore, the relative changes of the DNI are larger than the relative changes of the GHI, which results in larger deviations for the same forecasts. Besides that, the behavior of the DNI deviations is similar to the GHI deviations: temporal aggregations reduce deviations whereas higher lead times result in larger deviations. Table 2 shows the maximum and minimum values of the DNI deviations.

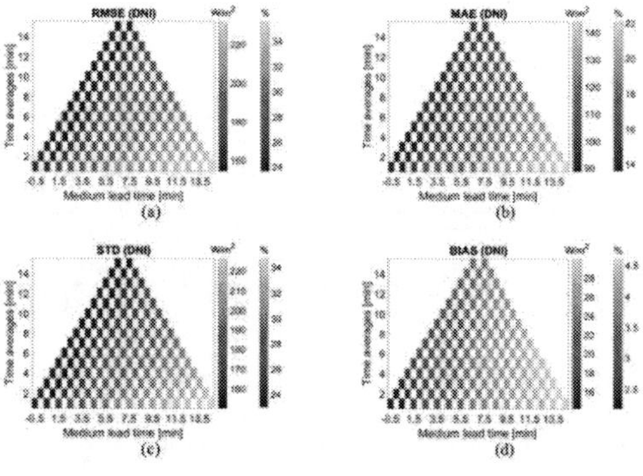

Figure 10: (a) Absolute and relative RMSE (DNI) for a validation period of 30 days and three pyrheliometers. (b) MAE, (c) standard deviation, (d) bias. The deviations rise with higher lead times and decrease if temporal averages are considered.

In Figure 11, temporal aggregation effects are shown differently by considering the deviations of the WobaS system for the next minutes ahead. For this so-called operator mode, the irradiances predicted for the positions of the ground measurements stations and the average of the next X min are compared to the aggregated measurements of these stations. The error metrics for GHI and DNI are depicted respectively in Figure 11 a and Figure 11 b. Due to temporal aggregations, the predictions for the next minute and the temporally aggregated predictions for the next 15 min ahead show similar deviations. These deviations are of interest if a back-up electricity generator of a solar plant requires time to be operational and the aggregated irradiance for the next e.g. 15 min must be predicted. The deviations displayed in Figure 11 correspond to the left-most value of each row in each graph of Figure 9 and Figure 10. Note the difference in the x-axis in these graphs: In Figure 9 and Figure 10, the medium lead times for various temporal averages are considered, whereas in Figure 11 the deviations for the next minutes ahead are displayed.

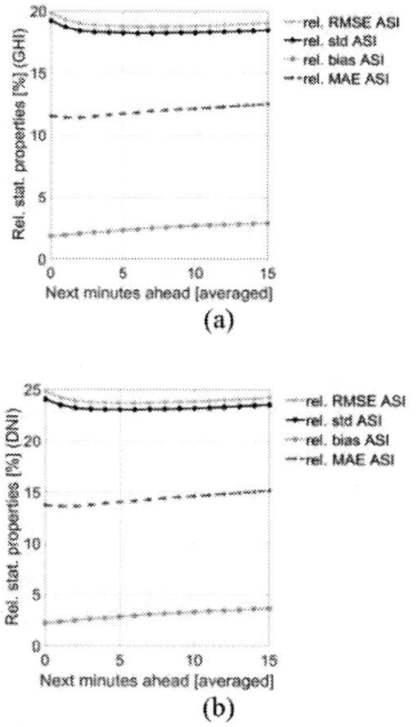

Figure 11: Error metrics for GHI (a) and DNI (b) using the operator mode. Three ground based reference stations are compared to corresponding pixels in the nowcasted irradiance maps for temporal averages up to 15 min. Depicted are the relative deviations of 30 days. Due to temporal averaging, the deviations for the next minute ahead and the deviations for the next 15 min ahead are found to be similar. This effect is of interest for plants with back-up generators.

In the graphs of Figure 11 ('ASI' – all-sky imager based WobaS system), the RMSE and MAE deviations found for 0 min ahead are slightly higher than the deviations found for 15 min ahead (RMSE (GHI): 19.8 % (0 min) to 19.0 % (15 min); RMSE (DNI): 24.8 % (0 min) to 24.2 % (15 min)). This is caused by interplay of two effects: With larger lead times, the predictions become less accurate, which can be seen in the graphs of Figure 9 and Figure 10. On the other hand, it is also visible in these graphs that with larger temporal aggregations, the deviations are reduced. In the operator modus, these two effects are combined and balance themselves: the increase of the deviations with larger lead times is compensated by the larger temporal aggregations.

The results found for the WobaS system are in alignment with publications on other all-sky imager based nowcasting systems: RMSE values between 155 W/m^2 and 250 W/m^2 for 10 min GHI forecasts are reported in [32] (validated on 15 days in Germany) and RMSE values for GHI forecasts up to 250 W/m^2 for 23 min lead time in [33] for a validation period of two month in Germany. A two all-sky imager based system validated on 9 days at the same location as the WobaS system (Spain) showed larger deviations for 1 minute averages of above 250 W/m^2 (RMSE, DNI) and similar deviations (RMSE, DNI: 31.6 %, lead time 15 min) for the operator mode [30]. A hybrid ASI/stochastic learning approach presented in [19] achieved RMSE values between 55 W/m^2 and 140 W/m^2 for DNI forecasts and a lead time of 10 min on a validation period of 6 months in California, USA. In [34], RMSE values for DNI forecasts between 169 W/m^2 (25 %) for 5 min lead time and 191 W/m^2 (28 %) for 15 min lead time as well as MAE values between 139 W/m^2 (22 %, 5 min) and 152 W/m^2 (24 %, 15 min) are found on a four week validation period in Taiwan.

Comparing observed deviations of nowcasting systems, which were validated on different days and in different weather situations, must be done with extreme precaution: the specific weather situations strongly influence the performance of nowcasting systems and the days included in the validation have a strong impact on the final error metrics. For the 30 days validation period used for the WobaS system, a wide range of different weather situations was chosen (Figure 7). Moreover, the weather above southern Spain with multiple cloud layers being frequently present might be more complex than the weather elsewhere. For instance, on the pacific coast of the USA, cumulus clouds at low altitudes seem to be predominantly present [35].

3.2 Comparing nowcasts with a shadow camera based reference system

At the Plataforma Solar de Almería, Spain, a unique and innovative shadow camera system is operational, which provides reference irradiance maps for the validation of nowcasts [10, 30]. The shadow camera system consists of six cameras taking photos from the top of an 87 m high tower. Out of the six photos, taken every 15 seconds, an orthoimage is calculated. From current and reference orthoimages as well as ground measurements of DNI and DHI, irradiance maps are derived. These irradiance maps have a spatial resolution of 25 m^2 over an area of 4 km^2. Comparing pixels of the derived irradiance maps to corresponding ground measurements, the shadow camera system shows RMSE (DNI) between 4.2 % and 16.7 % for 1 min averages. RMSE (GHI) deviations are below 10 %. Like the WobaS system, the shadow camera system benefits from spatial and temporal averaging with resulting spatially aggregated RMSE (GHI) between 3.3 % and 8.7 % and RMSE (DNI) between 4.3 % and 10 %. Figure 12 illustrates the approach and more details are given in [29].

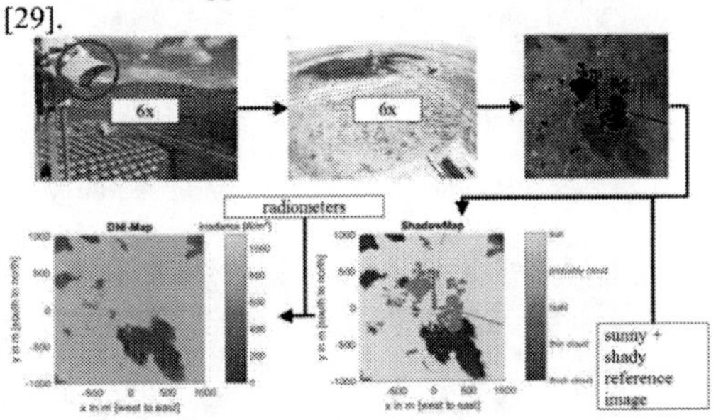

Figure 12: Working principle of the shadow camera system, providing reference irradiance maps for the WobaS system. Six downward-facing cameras acquire an image every 15 s. The set of six images are combined to an orthoimage. With reference orthoimages, taken during cloud-free and overcast conditions, available, shadows are detected and irradiance maps in high spatial (25 m^2) and temporal resolutions of 15 s are generated.

The effects of spatial aggregations are discussed by looking at one example day (19.9.2015). The irradiances of this day are depicted in Figure 13, showing a high level of variability.

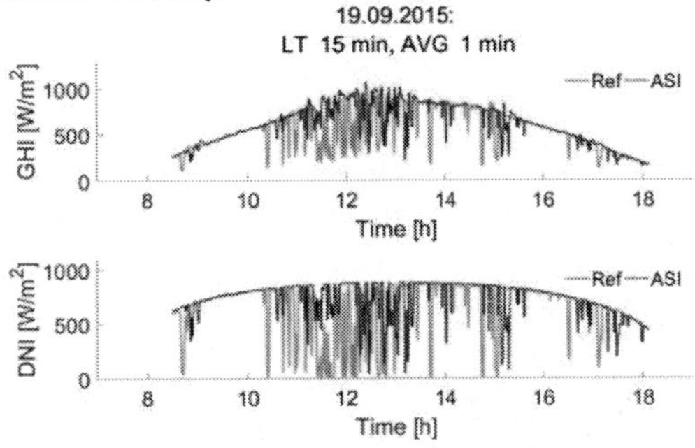

Figure 13: GHI and DNI ground measurements (green, "Ref"), compared to the WobaS forecasts of the pixel corresponding to the measurement station (blue, "ASI" – all-sky imager) for lead time 15 min ('LT') and 1 min time average ('AVG'). As can be seen, this example day shows a high variability.

Figure 14 displays the GHI deviations found for the WobaS system in comparison to the shadow camera system for 1 min temporal averages. Considered field sizes vary between 25 m^2 and 4 km^2 (x-axis). Lead times from 0 min to 15 min are depicted on the y-axis. The deviations increase with larger lead times. Due to spatial aggregation effects, the deviations decrease for larger field sizes. For instance, a RMSE value of 21.4 % (120 W/m^2) is found for lead time 0 min and a field size of 25 m^2, which shrinks to 13.0 % (73 W/m^2) for a field size of 4 km^2.

This is a pre-peer review version of the paper submitted to the journal Progress in Photovoltaics.
It has been selected by the Executive Committee of the EU PVSEC 2017 for submission to Progress in Photovoltaics.

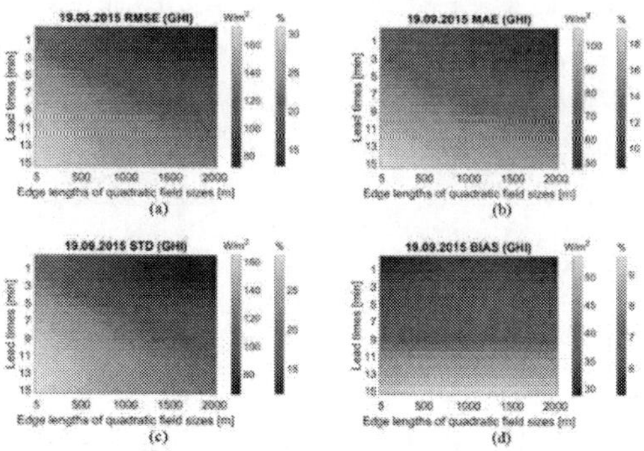

Figure 14: Absolute and relative error metrics for the 1-min GHI forecasted by the WobaS system in comparison to the shadow camera system for aggregated field sizes from 25 m^2 to 4 km^2 and lead times ranging from 0 min to 15 min.

The DNI deviations are depicted in Figure 15. Due to effects explained in the previous section, the deviations observed for the DNI are larger than the GHI deviations. In general, the DNI deviations show similar behavior in comparison to the GHI deviations. However, the DNI deviations drop more for larger field sizes. This effect can be explained with typical cloud sizes of this day: If clouds on that day have a usual diameter of 500-1000 m, then for field sizes above 1000 m the exact position of the cloud inside the considered field is not relevant for spatially aggregated irradiances. Prior to this critical size, small errors in the predictions of the shadow locations lead to large deviations as the shadow misses the field completely. This drop in the deviations is blurred out for the GHI due to the DHI reducing the offsets for small field sizes.

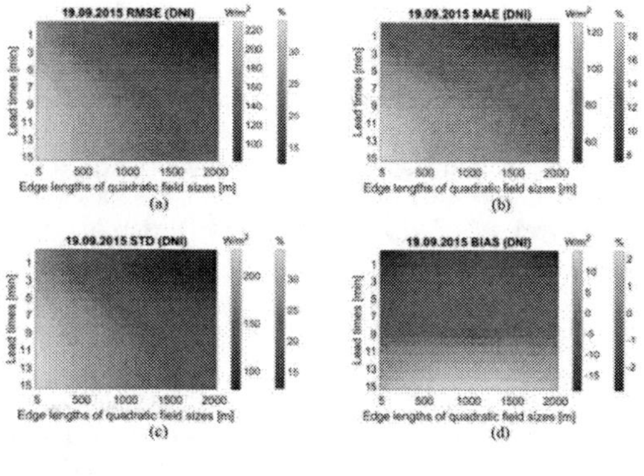

Figure 15: DNI deviations corresponding to Figure .

In the previous section, the operator mode was introduced. The operator mode considers deviations for the next minutes ahead. Due to temporal aggregation effects, the deviations found for the next minute ahead and the next 15 min ahead are similar if pixels in the nowcasted irradiance maps are compared to reference ground measurements (see Figure 11 and corresponding discussion). In Figure 16 (GHI) and Figure 17 (DNI), the operator mode is depicted considering different field sizes, lead times and deviations.

As an example, for field sizes of 4 km^2 and considering the next 4 min ahead, the observed relative RMSE values are 12.5 % for GHI (Figure 16 a). Without using the operator mode, the deviation observed for a lead time of 4 min and field sizes of 4 km^2 correspond to a RMSE of 15.0 % (GHI), as shown in Figure 14 a. If only one pixel is considered, temporal aggregation reduces the deviations from 24.9 % (RMSE, GHI, lead time 4 min) to 20.3 % (next 4 min ahead). For DNI deviations, temporal aggregations reduce the deviations found for one pixel from RMSE 30.6% (Figure 15 a, lead time 4 min) to RMSE 26.2 % (Figure 17 a, next 4 min ahead). For one pixel, temporal aggregations also reduce the DNI deviations from RMSE 34.8 % (lead time 15 min) to RMSE 27.6 % (next 15 min ahead). For field sizes of 4 km^2 and considering the next 15 min ahead, spatio-temporal aggregation reduces the RMSE to 14.2 % (Figure 17 a).

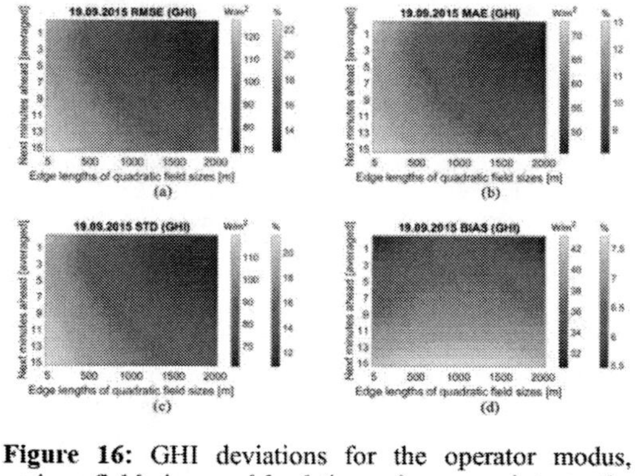

Figure 16: GHI deviations for the operator modus, various field sizes and lead times, in comparison to the shadow camera system.

As illustrated with one day considered as an example in this section, spatial aggregation effects significantly reduce forecasting errors. Since spatial aggregation effects are present in every industrial solar plant, these effects must be considered in the validations of nowcasting systems. For the total dispatched power of solar plants, this effect is discussed in [36].

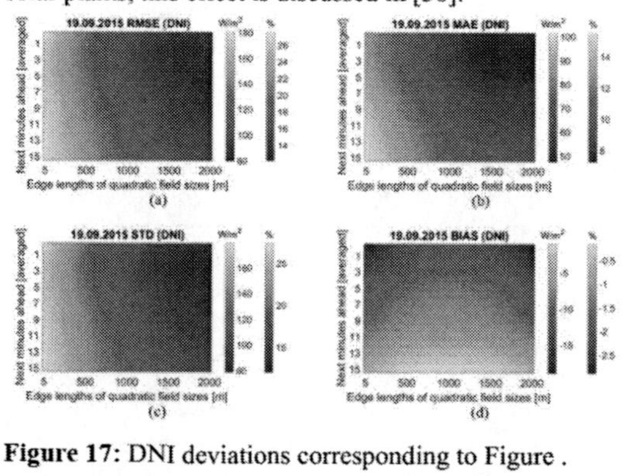

Figure 17: DNI deviations corresponding to Figure .

In the absence of a shadow camera system acting as a reference, auto-evaluations could be considered. For the whole validation period, such auto-evaluations of the WobaS system were performed. Auto-evaluations compare predicted irradiance maps for future timestamps with irradiance maps predicted on and for these timestamps. The WobaS system thus evaluates itself

This is a pre-peer review version of the paper submitted to the journal Progress in Photovoltaics.
It has been selected by the Executive Committee of the EU PVSEC 2017 for submission to Progress in Photovoltaics.

('auto-evaluation'). It was found that auto-evaluations of nowcasting systems should be treated with great precaution, especially regarding irradiance forecasts. In direct comparison, both the values of the deviations and their behavior as observed via auto-evaluations differ significantly from the deviations observed via reference measurements. Auto-evaluations could be used to test the robustness of a nowcasting system as well as the combined errors of inconsistent cloud detections, tracking errors and deviations in cloud height determinations. However, it is suggested to validate these sub-tasks separately with additional references to enable optimization. Validating sub-task separately leads to small improvements which in the end add up to a generally improved and robust nowcasting system.

4 CONCLUSION AND FUTURE WORK

In this publication, the WobaS nowcasting system was validated on 30 days representing various weather conditions with special focus on temporal and spatial aggregation effects. Temporal aggregation effects are relevant for defining storage requirements. Spatial aggregation effects are inherent in industrial size solar plants when the resulting energy production is under consideration. The deviations observed for the WobaS system for 1 min averages are coherent with other recently published nowcasting systems. However, directly comparing nowcasting systems using different data sets is delicate as the specifics of the meteorological situation during the selected days strongly influence the error metrics.

As key findings, both temporal and spatial aggregation significantly reduces forecasting errors. This was not studied in detail before for the temporal and spatial resolutions considered in this publication. Indeed, most nowcasting systems are validated based on few ground measurement stations, which yields error metrics that do not represent fitness-for-use and might not be relevant for industrial applications. The validation period should reflect all possible weather situations and their relative occurrence. In literature, due to technical constraints, often only few days are used for validation and this could be critical for the representability of the conclusions. Spatial aggregations are found to reduce RMSE (GHI) values from 21.4 % to 13.0 % for lead time 0 min and field sizes between 25 m^2 and 4 km^2. Temporal averaging reduces RMSE (GHI) deviations from 25.3 % (medium lead time 7.5 min, 1 min temporal average) to 19.0 % (medium lead time 7.5 min, 15 min temporal average). Furthermore, the effects of increased accuracies for larger temporal aggregations and increased deviations for larger lead times are found to be equivalent. Another key finding concerns auto-evaluations, which were found unfit for the validation of irradiance maps.

The WobaS system is currently operational at a commercial solar power plant. In the near future, nowcasting systems such as the WobaS system will contribute to handle ramp rate limitations. By providing nowcasts, the WobaS system helps to comply with such regulations, e.g. by reducing the required battery size. The conducted validation found the WobaS system to be reliable for operation in industry. Thus, with all-sky imager based nowcasting tools and relatively small batteries, the cloud induced variability of the solar resource will not lead to severe instabilities of electrical grids.

In future work, the benchmarking of further sub-tasks of the WobaS system will be presented. Future work will also conduct benchmarking of different nowcasting systems in comparison to the WobaS system. Moreover, there will be an on-going optimization of WobaS. This way, WobaS will be improved step-by-step.

ACKNOWLEDGEMENT

The research presented in this publication has received funding from the German Federal Ministry for Economic Affairs and Energy within the WobaS project, which enabled the development of the WobaS nowcasting system and the creation of reference irradiance maps from the shadow camera system. The initial development of the shadow camera system was partially financed by the European Union's Horizon 2020 programme (PreFlexMS, Grant Agreement no. 654984). The European Union's FP7 programme partially financed operations of the all-sky imagers and other ground measurements (DNICast project, Grant Agreement no. 608623).

REFERENCES

1. Crăciun, B.I., T. Kerekes, D. Séra, R. Teodorescu, and U.D. Annakkage, Power Ramp Limitation Capabilities of Large PV Power Plants With Active Power Reserves. IEEE Transactions on Sustainable Energy, 2017. 8(2): p. 573-581 DOI: http://dx.doi.org/10.1109/tste.2016.2612121.
2. Gevorgian, V. and S. Booth, Review of PREPA Technical Requirements for Interconnecting Wind and Solar Generation, NREL/TP ; 5 D 00-57089, 2013, available online: http://www.nrel.gov/docs/fy14osti/57089.pdf.
3. Chow, C.W., B. Urquhart, M. Lave, A. Dominguez, J. Kleissl, J. Shields, and B. Washom, Intra-hour forecasting with a total sky imager at the UC San Diego solar energy testbed. Solar Energy, 2011: p. 2881-2893 DOI: http://doi.org/10.1016/j.solener.2011.08.025.
4. Cheng, H.Y. and C.C. Yu. Solar irradiance now-casting with ramp-down event prediction via enhanced cloud detection and tracking. in 2016 IEEE International Conference on Multimedia and Expo (ICME). 2016. DOI: http://dx.doi.org/10.1109/icme.2016.7552863.
5. Ding, Y., X. Cheng, F. Cui, X. Zhu, and H. Zhou. Very-short term forecast of global horizontal irradiance based on ground-based sky imager and lifted condensation level calculation. in 2014 China International Conference on Electricity Distribution (CICED). 2014. DOI: http://dx.doi.org/10.1109/ciced.2014.6991849.
6. Yang, H., B. Kurtz, D. Nguyen, B. Urquhart, C.W. Chow, M. Ghonima, and J. Kleissl, Solar irradiance forecasting using a ground-based sky imager developed at UC San Diego. Solar Energy, 2014. 103: p. 502-524 DOI: http://dx.doi.org/10.1016/j.solener.2014.02.044.
7. Najera, Y., D.R. Reed, and W.M. Grady. Image processing methods for predicting the time of cloud

This is a pre-peer review version of the paper submitted to the journal Progress in Photovoltaics.
It has been selected by the Executive Committee of the EU PVSEC 2017 for submission to Progress in Photovoltaics.

shadow arrivals to photovoltaic systems. in 2011 37th IEEE Photovoltaic Specialists Conference. 2011. DOI: http://doi.org/10.1109/PVSC.2011.6185877.

8. Dev, S., F.M. Savoy, Y.H. Lee, and S. Winkler. Estimation of solar irradiance using ground-based whole sky imagers. in 2016 IEEE International Geoscience and Remote Sensing Symposium (IGARSS). 2016. DOI: http://dx.doi.org/10.1109/igarss.2016.7730887.
9. Sukič, P. and G. Štumberger, Intra-Minute Cloud Passing Forecasting Based on a Low Cost IoT Sensor—A Solution for Smoothing the Output Power of PV Power Plants. Sensors, 2017. 17(5): p. 1116 DOI: http://doi.org/10.3390/s17051116.
10. Blanc, P., P. Massip, A. Kazantzidis, P. Tzoumanikas, P. Kuhn, S. Wilbert, D. Schüler, and C. Prahl. Short-Term Forecasting of High Resolution Local DNI Maps with Multiple Fish-Eye Cameras in Stereoscopic Mode. in AIP conference proceedings (SolarPACES), accepted for publication. 2016.
11. Wilbert, S., B. Nouri, P. Kuhn, D. Schüler, C. Prahl, N. Kozonek, R. Pitz-Paal, T. Schmidt, N. Kilius, M. Schroedter-Homscheidt, and Z. Yasser. Wolkenkamera-basierte Kürzestfristvorhersage der Direktstrahlung. in 19. Kölner Sonnenkolloquium 2016. Köln, Germany.
12. Barnes, A.K., J.C. Balda, and A. Escobar-Mejía, A Semi-Markov Model for Control of Energy Storage in Utility Grids and Micrograms With PV Generation. IEEE Transactions on Sustainable Energy, 2015. 6(2): p. 546-556 DOI: http://dx.doi.org/10.1109/tste.2015.2393353.
13. Sani Hassan, A., L. Cipcigan, and N. Jenkins, Optimal battery storage operation for PV systems with tariff incentives. Applied Energy, 2017. 203: p. 422-441 DOI: http://doi.org/10.1016/j.apenergy.2017.06.043.
14. West, S.R., D. Rowe, S. Sayeef, and A. Berry, Short-term irradiance forecasting using skycams: Motivation and development. Solar Energy, 2014. 110: p. 188-207 DOI: http://dx.doi.org/10.1016/j.solener.2014.08.038.
15. Wilbert, S. and E. Reuschenbach. Solar researchers demonstrate new solar irradiance nowcasting system in a 50-megawatts solar power plant. 2017 [cited 2017 June 30]; Available from: http://www.dlr.de/sf/en/desktopdefault.aspx/tabid-10436/12676_read-48274/.
16. Aga, V. and C.F. Peruchena, PreFlexMS: Predictable Flexible Molten Salts Solar Power Plants. Impact, 2017. 2017(3): p. 58-60 DOI: http://dx.doi.org/10.21820/23987073.2017.3.58.
17. Oberländer, D., C. Prahl, S. Wilbert, S. Müller, B. Stanicki, and N. Hanrieder. Cloud shadow maps from whole sky imagers and voxel carving. in ICEM. 2015. Boulder.
18. Taravat, A., F.D. Frate, C. Cornaro, and S. Vergari, Neural Networks and Support Vector Machine Algorithms for Automatic Cloud Classification of Whole-Sky Ground-Based Images. IEEE Geoscience and Remote Sensing Letters, 2015. 12(3): p. 666-670 DOI: http://doi.org/10.1109/LGRS.2014.2356616.
19. Xia, M., W. Lu, J. Yang, Y. Ma, W. Yao, and Z. Zheng, A hybrid method based on extreme learning machine and k-nearest neighbor for cloud classification of ground-based visible cloud image. Neurocomputing, 2015. 160: p. 238-249 DOI: http://dx.doi.org/10.1016/j.neucom.2015.02.022.
20. Kazantzidis, A., P. Tzoumanikas, A. Bais, S. Fotopoulos, and G. Economou, Cloud detection and classification with the use of whole-sky ground-based images. Atmospheric Research, 2012: DOI: http://doi.org/10.1016/j.atmosres.2012.05.005.
21. Jayadevan, V.T., J.J. Rodriguez, and A.D. Cronin, A New Contrast-Enhancing Feature for Cloud Detection in Ground-Based Sky Images. Journal of Atmospheric and Oceanic Technology, 2015. 32(2): p. 209-219 DOI: http://doi.org/10.1175/JTECH-D-14-00053.1.
22. Ineichen, P. and R. Perez, A new airmass independent formulation for the Linke turbidity coefficient. Solar Energy, 2002. 73(3): p. 151-157 DOI: http://doi.org/10.1016/S0038-092X(02)00045-2.
23. Hanrieder, N., M. Sengupta, Y. Xie, S. Wilbert, and R. Pitz-Paal, Modeling beam attenuation in solar tower plants using common DNI measurements. Solar Energy, 2016. 129: p. 244-255 DOI: http://doi.org/10.1016/j.solener.2016.01.051.
24. Wilbert, S., S. Kleindiek, B. Nouri, N. Geuder, A. Habte, M. Schwandt, and F. Vignola, Uncertainty of rotating shadowband irradiometers and Si-pyranometers including the spectral irradiance error. AIP Conference Proceedings, 2016. 1734(1): p. 150009 DOI: http://dx.doi.org/10.1063/1.4949241.
25. Calbó, J., C.N. Long, J.-A. González, J. Augustine, and A. McComiskey, The thin border between cloud and aerosol: Sensitivity of several ground based observation techniques. Atmospheric Research: DOI: http://doi.org/10.1016/j.atmosres.2017.06.010.
26. Hay, J.E. and D.C. McKay, Estimating Solar Irradiance on Inclined Surfaces: A Review and Assessment of Methodologies. International Journal of Solar Energy, 1985. 3(4-5): p. 203-240 DOI: http://dx.doi.org/10.1080/01425918508914395
27. Yoshida, S., S. Ueno, N. Kataoka, H. Takakura, and T. Minemoto, Estimation of global tilted irradiance and output energy using meteorological data and performance of photovoltaic modules. Solar Energy, 2013. 93: p. 90-99 DOI: http://dx.doi.org/10.1016/j.solener.2013.04.001.
28. Gueymard, C.A. and J.A. Ruiz-Arias, Extensive worldwide validation and climate sensitivity analysis of direct irradiance predictions from 1-min global irradiance. Solar Energy, 2016. 128: p. 1-30 DOI: http://doi.org/10.1016/j.solener.2015.10.010.
29. Kuhn, P., S. Wilbert, C. Prahl, D. Schüler, T. Haase, T. Hirsch, M. Wittmann, R. Pitz-Paal, P. Blanc, L. Ramirez, L. Zarzalejo, A. Meyer, and L. Vuilleumier, Shadow camera system for the generation of solar irradiance maps. Solar Energy, accepted for publication, 2017.
30. Kuhn, P., S. Wilbert, C. Prahl, A. Kazantzidis, L. Ramírez, L. Zarzalejo, L. Vuilleumier, P. Blanc, and R. Pitz-Paal, Validation of nowcasted spatial DNI maps, in DNICast, Deliverable 4.1Validation of nowcasted spatial DNI maps, German Aerospace Center (DLR), http://www.dnicast-project.net/, available online:
31. Kuhn, P., S. Wilbert, D. Schüler, C. Prahl, T. Haase, L. Ramirez, L. Zarzalejo, A. Meyer, L. Vuilleumier, P. Blanc, J. Dubrana, A. Kazantzidis, M. Schroedter-Homscheidt, T. Hirsch, and R. Pitz-Paal. Validation

of Spatially Resolved All Sky Imager Derived DNI Nowcasts. in AIP conference proceedings (SolarPACES), accepted for publication. 2016.

32. Bernecker, D., C. Riess, E. Angelopoulou, and J. Hornegger, Continuous short-term irradiance forecasts using sky images. Solar Energy, 2014. 110: p. 303-315 DOI: http://dx.doi.org/10.1016/j.solener.2014.09.005.
33. Schmidt, T., J. Kalisch, E. Lorenz, and D. Heinemann, Evaluating the spatio-temporal performance of sky-imager-based solar irradiance analysis and forecasts. Atmos. Chem. Phys., 2016. 16(5): p. 3399-3412 DOI: https://www.doi.org/10.5194/acp-16-3399-2016.
34. Fu, C.-L. and H.-Y. Cheng, Predicting solar irradiance with all-sky image features via regression. Solar Energy, 2013. 97: p. 537-550 DOI: http://dx.doi.org/10.1016/j.solener.2013.09.016.
35. Wang, G., B. Kurtz, and J. Kleissl, Cloud base height from sky imager and cloud speed sensor. Solar Energy, 2016. 131: p. 208-221 DOI: http://dx.doi.org/10.1016/j.solener.2016.02.027.
36. Marcos, J., L. Marroyo, E. Lorenzo, D. Alvira, and E. Izco, Power output fluctuations in large scale PV plants: One year observations with one second resolution and a derived analytic model. Progress in Photovoltaics: Research and Applications, 2011. 19(2): p. 218-227 DOI: http://dx.doi.org/10.1002/pip.1016.

BIPV PRODUCTS OVERVIEW FOR SOLAR BUILDING SKIN (SUBTOPIC 6.3 / BUILDING, INFRASTRUCTURE AND LANDSCAPE APPLICATIONS)

P. Bonomo[1], I. Zanetti[1], F. Frontini[1]
M.N. van den Donker[2], F. Vossen[2], W. Folkerts[2]

[1]SUPSI, Institute for Applied Sustainability to the Built Environment (ISAAC), Lugano, Switzerland
[2]Solar Energy Application Centre (SEAC), Eindhoven, The Netherlands

mail to: pierluigi.bonomo@supsi.ch; phone +41 052 666 6320, fax +41 052 666 6349

ABSTRACT: In this paper, the authors report on a jointly conducted analysis of the current status of the BIPV market by giving a comprehensive overview on the available products, the current trends of technological innovation, the application ranges and a survey on the price levels. For this purpose a market survey was conducted among the various stakeholders in the BIPV sector, including manufacturers, installers and BIPV product suppliers. The survey focused on different application areas: roof and façade systems including both crystalline and thin film technologies. The market survey was conducted for the European market, with emphasis on The Netherlands and Switzerland. The most frequent products investigated were PV tiles and full roof solution (mainly for residential roofs). Although crystalline silicon technology was most commonly used, thin film technologies gained considerable share as well as façade systems. The data from the price survey showed that BIPV systems for the building skin can be executed at very similar costs than projects involving conventional building products.

Keywords: BIPV, building integration, multifunctional component, architecture, database

1 PREFACE

The BIPV market in Europe is in a transition. The past decades of a slowly emerging BIPV market have been characterized by the original dominant value of BIPV: a building that communicates an image of sustainability and innovation. As BIPV is more expensive than straightforward application of a PV system on a roof (BAPV), the appropriate question is always: what is the value that justifies this additional cost? In the past years this value in most cases was: image. The message that is communicated by the building: a message of sustainability, responsibility and even idealism. Payback time or return on investment have not been the major parameters in the decision process for applying BIPV.

This is changing now. In most European countries the new regulations on energy performance in buildings, derived from the EPDB 2010 [1], have been defined and the time for the regulations to become mandatory is very near. The energy performance regulations are now taking over as the main driving factor for the BIPV market. And that has huge consequences. Instead of BIPV finding its value in only being visible and supporting an image, the main value of BIPV products now becomes to be invisible. In other words: we do not want to see anymore the difference between a building component and a BIPV component both in aesthetical and construction terms.

Another dominant market factor in the new era is the fact that high-rise buildings (such as multifamily residential blocks) simply do not have enough roof area to meet the energy performance requirements, by using conventional PV modules. This leads the building designer naturally to the use of the façade for applying BIPV, which is a central part of the architectural concept.

This change in market driver has led to an intensified research and development aiming to create BIPV products that come in a variety of colours and sizes, while at the same time the manufacturing of the BIPV ideally should be as close as possible to existing building components, to its requirements and to the way of working in construction industry.

2 CLASSIFICATION OF BIPV SYSTEMS

The acronym BIPV technically refers to systems and concepts in which the photovoltaic element carries an additional building functionality, namely replacing an element of the building skin. The use of PV in the building sector opened up questions: re-imagine the envelope of solar buildings both in aesthetics and technology. The relationship between PV and construction has evolved in the years through gradual steps of innovation, resulting in a progressive technological transfer in the building technical elements, from addition of PV in conventional building elements, to the layering of the building skin with active cladding parts up to new integration concepts where PV is part of a unitary building envelope.

2.1 A basic concept: BIPV as a building component

Multifunctionality [2] refers to the functional or physical role of the PV modules in the building skin. In this definition, PV modules are considered to be building integrated if they represent a component of the building envelope providing a function as defined in the European Construction Product Regulation CPR (Construction Product Regulation n.305/2011) [3]. Thus the building performance of the BIPV module is required for the integrity of the building's functionality. Considering the growing market share of BIPV, the progress in standardization (EN 50583-2016, Parts 1 and 2) [4] and the tightening regulations with respect to CPR, we can state that a BIPV component is not merely an electrical device but it is becoming a mature building element as well.

2.2 Beyond functionalism, BIPV architecture.

BIPV also refers to the architectural concept: like any other building material, the potential of PV is its materiality and the opportunity to define the linguistic/morphological rules governing the signs, the structure and the composition of the architectural language. A large repertoire of examples is progressively showing the aesthetical innovation in contemporary architecture due to a solar design.

3 BIPV PRODUCTS DATABASE

The research methodology is founded on the analysis of a database collecting mainly products by manufacturers and installers based in Europe, focusing on systems commercially available in the Swiss and Dutch/Benelux markets. The SUPSI website www.bipv.ch and the SEAC-SUPSI BIPV status report of 2015 [5] served as basis for the products list. We removed products that were withdrawn from the market since 2015 and added new products that entered the market until today. This was executed through an internet web search, interviews with industries and trade fair visits. For more detailed definitions the complete report is available [6].

3.1 Product overview

Figure 1 shows the overview of the products grouped for BIPV categories. The most common product group is that of solar tiles (any size) immediately followed by the full roof solution. In 2015 this trend was already visible even though full roof systems where slightly more common than solar tiles. Products for rain-screen façades, where the photovoltaic module is used as a building cladding, and skylight/solar glazing follow with a significantly lower percentage. Products for roofs are much numerous compared to those for façades indicating that the roof market is currently wider than the façade market. It has to be remarked that products described as solar glazing for roofs are generally marketed also with a curtain wall variant.

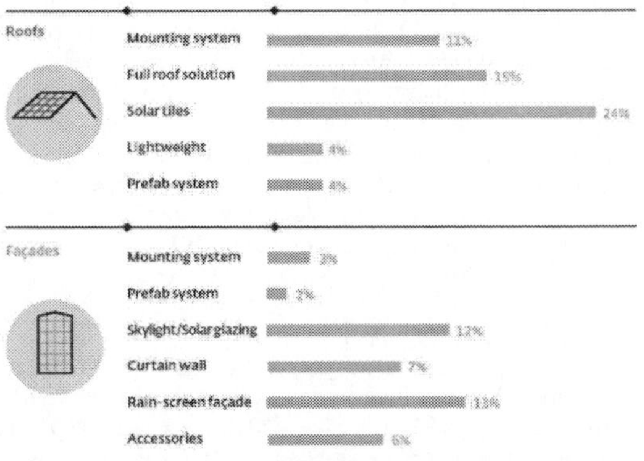

Figure 1: Results of the market survey presented in terms of the occurrence of product groups. Distribution of BIPV products according to different categories.

2.1 Used PV technology

The 8% of the BIPV products for roofs and 44% of the BIPV products for façades resulted to be using thin film technology. This is a very high technology share for thin film, when considering that only about 5% of all worldwide available PV module types are made using this technology [7]. We suggest the possible explanation for the success of thin film in BIPV, especially on façades, to be due to two key reasons: first of all the aesthetic appearance and secondly the low price/m^2, associated to the high relative substitutional costs when saving out conventional building materials.

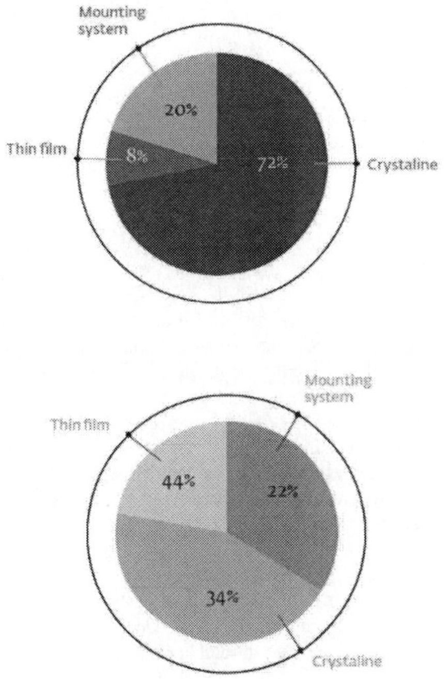

Figure 2: Pie charts of the technology used in the roof (top) and façade (bottom) BIPV application areas

During the data collection phase it was noticeable how products appeared and disappeared on the market in a consistent manner. About 35% of the companies listed in the current edition did not offer their products two years ago. Since the list only marginally grew (from 108 to 114 products) this means that also about 35% of the companies listed in the previous edition, stopped their activities or no longer offer BIPV products.

4 BIPV TECHNOLOGY TRENDS

The process of photovoltaic transfer in building ranges between new and tradition. In architecture, the replacement of an existing material with a new one is usually accompanied by the permanence of tradition looking both at architectural languages and technical systems. This slow process of innovation is linked both to new technology and new design models which can be related to morphological aspects, building skin image or construction and so on.

Regarding the building envelope, for example, the physical flexibility and lightness of thin film permit to incorporate PV in ultrathin and lightweight skins or membranes. The innovative trends such as Luminescent Solar Concentrators (LSC), new technologies (e.g. DSSC) or some treatments for glass are opening the scenario of a new language of photoactive technologies towards a colored or "invisible" PV and indoor PV [7].

The technological transfer of PV in the architecture field is changing building skin design approaches and opening new challenges. The industry makes available a plenty of products for building application: multi-functionality, cost-effectiveness, mass customization and other paradigms are ensuring a growing penetration on the real market. Thus, along with functional and construction aspects, definitely BIPV is today one of the new fundamentals for the contemporary innovation in architecture.

4.1 Prefab mounted BIPV

Using prefab elements brings a lots of advantages in building process: along with a better quality control of the construction component, it allows a fast building mounting and e.g., in case of renovation, lets people to stay inside their homes.

In some prefab systems the whole process of design, manufacturing and implementation is performed industrially.

These products are not only serially produced but are also more and more mass-customizable and, in case of BIPV cladding, it is possible to select: colour, stratigraphy (e.g. insulation thickness), dimensions, ratio transparency/opacity, types of interior and exterior layer, total thickness and fixing system. Consequently a successful integration of PV is possible thanks to advanced solutions where mass-production and flexibility are strategically optimized.

4.2 Coloured or patterned façades

The colouring of PV, namely its ability to be camouflaged or "designed", has been considered an essential requirement for market acceptance of PV façades in recent years.

In a BIPV glass façade, the conventional PV appearance can be camouflaged behind coloured patterns that completely dissimulate the original materiality of the PV cells. This involves a "shading" over the PV cells and a consequent reduction of the energy production that needs to be carefully optimized in order to obtain an energy efficient customization. Namely the challenge to optimally balance the aesthetical quality with the energy and electrical efficiency, reliability and safety is one of the drivers of innovation in the BIPV sector. Different customization techniques can be identified in the current developments [8].

Colouring and pattern of PV panels may occur by several different methods and treatments applied mainly on:

- the cells and backsheet
- the front glass
- an intermediate foil

Figure 3: A BIPV white module obtained through a laminated color foil with diffuse reflector (Solaxess).

4.3 Solar glazing

The combination of glass and photovoltaics seems to match well in terms of both aesthetics and functionality of the building skin and the "BIPV glass" market is expected to grow in the forthcoming years.

There are various technological routes to manufacture active solar glazing:

- Crystalline silicon glass-glass modules
- Crystalline silicon cell strips
- Thin film solar cell strips
- Thin film solar strips hidden by a lens system
- Luminescent solar concentrators
- Thin absorber layers

Figure 4: Thin film solar cell strips (Onyx).

4.4 Lightweight systems

Two techniques can be distinguished: modules based on thin-film technology and modules based on crystalline silicon technology. In general, thin film flexible panels have a smaller curvature radius and are more robust under continuous bending, flapping and moving conditions. They would therefore be a good choice for applications on tents, membranes, corrugated sheets, etc. Crystalline silicon flexible lightweight panels are higher efficient than their thin film counterparts, but less flexible and sometimes less good looking. They would therefore be a good choice for 'invisible' applications with a permanent fixture, such as flat roofs. New approaches are emerging to obtain such innovative applications [9].

5 BIPV PRICE SURVEY

A market survey was conducted through an on-line web form that was sent to producers, suppliers as well as installers of BIPV products. The market survey was conducted mainly for the Dutch/Benelux and Swiss markets.

The survey was sent out to about 120 contacts from which we collected 35 respondents. Roof solutions collected more feedback than façade products. This was probably because these kinds of systems have been on the market for a while and they are characterized by quite distinct cost levels associated with almost completely standardized mounting systems. If BIPV roof solutions are seeking to be as similar as possible to standard roofing applications, façade products are instead focusing more on tailored made building engineering solutions (rain-screen façade, curtain wall), especially in case of existing buildings, so that in most cases the price levels are still uncertain and less definable or, as mentioned before, cannot be really generalized from specific contexts.

Product group	# Participants in price survey
BAPV system and roof tiles	7
In-roof mounting system	4
Full roof solution	7
BIPV tiles	7
PV membranes	3
Metal panels	-
Solar glazing	-
Rain-screen façade	3
Curtain wall	-
Accessories	-
Solar glazing	4
Total	35

Table 1: Number of participants in the product survey with BIPV solutions for a roof or a façade.

In-roof mounting system, BIPV tiles and BIPV full roof solutions are already widespread available while BIPV metal panels, BIPV membranes, BIPV skylights and BIPV façades are still niche products for which it is difficult to rely in a consistent number of participants in the price survey.

The market of BIPV façade systems in Switzerland and Benelux, even though a relevant trend of increase is tangible, is still relatively small and there is a large cost variety depending on the building type, building skin application and on the specific context of intervention.

5.1 Costs: from LCOE to €/m^2

The reviewed price survey covered both BIPV and BAPV applications. The results on BAPV systems were included in order to have a benchmark and reference of the BIPV products but the main goal was to consider the prices of conventional building materials for further reference. The price study focused on end-user prices in the Benelux and the Swiss market, excluding VAT, thus obtaining an "average" European reference cost. The end-user prices are converted to €/m^2, which is the end-user PV system cost calculated over the area that the PV systems covers on the roof or façade. Using this unit of €/m^2 it is possible to directly compare various PV technologies to conventional building materials.

The ranges of cost are displayed in fig. 5. For façade systems is indicated the cost for providing only the cladding panel whilst for the roof is indicated the final cost. The grey bars represent the share of the costs that is avoided because the conventional building material (passive element) is replaced. The black bars represent the true additional costs that contractors need to spend in order to apply PV functionality in their building skin. The total cost results from the sum of both. The error bars show the full range (min-max) of all the different quotes.

Financially, a BIPV system becomes feasible in most of cases when BIPV replaces conventional building cladding elements (curtain wall, marble, etc.) and cost-effectiveness can be demonstrated taking into account the whole building envelope system. E.g. it is particularly interesting to use prefab elements or integrated solutions which allow to optimize not only the cost but also the timing and the complexity of the installation. In this case the technical and process costs can be evaluated together.

As mentioned before in a façade system (rain-screen façade, curtain wall, etc.) the BIPV cost is strongly affected by specific building factors that are the main drivers for cost-effectiveness and it cannot be easily generalized.

Moreover the cost of a BIPV unitized façade systems, which includes the supply of all the cladding elements, substructures, fixings, joints, connections, etc. (excluding insulation) as well as the needed building and electrical equipment (and usually also the installation of the system itself), cannot be compared with cases where the BIPV cladding is self-contained (it can be removed or replaced) from the rest of the façade system (typically, this is the case of a BIPV module that can be mounted as a façade cladding). In the second scenario the price is more similar to a conventional PV module.

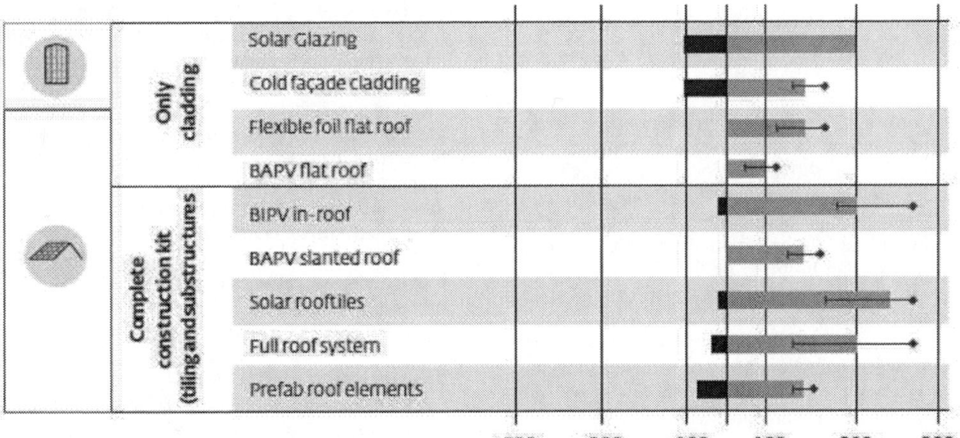

Figure 5: Cost of BIPV as resulted from the survey.

5.2 Cost of complete BIPV roof tiling construction

The graph refers to the final cost of a complete roof tiling construction, including mounting, transportation and other additional costs. This cost includes both the roof tiling and the mounting system (clamp, metal ducts, etc.). It shows a significant price range for the different conventional roofing materials.

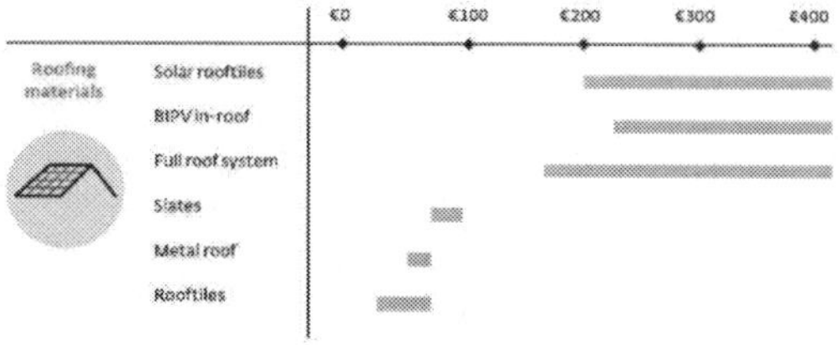

Figure 6: Cost of BIPV roof tiling compared to other roofing materials used in conventional pitched roof.

5.3 Cost of façade cladding

The prices for conventional façade applications were obtained using Swiss databases on building price information.

The next graph specifically refers to the cost of the cladding, namely the outer material layer that represents the exterior wall. The costs of the substructures, fixings and insulation are excluded. Only in case of curtain walls it is considered the cost of the whole system (including mounting and shading elements). Not to consider these variables could lead to confusing evaluations on costs.

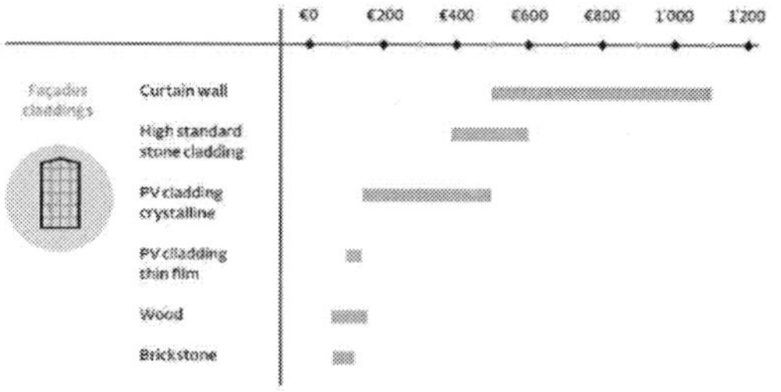

Figure 7: PV costs compared to other cladding materials used in the built environment as façade cladding materials.

Figure 7 provides only some reference ranges, which are not substitutive of an ad-hoc request of quotation for a specific building project. The cost per square meter is influenced by various factors such as: the global size of the project, the building typology and the window to wall ratio, the planned type of substructure (wood, aluminium, steel) which has a major influence on the durability of the installation itself, but also on the complexity of the technical details (the simpler, the cheaper), the size of the modules, the form of the façade (more or less articulated) from which it depends the possibility of using a modular system that is clearly simpler and usually lower priced compared to a custom made product. In case of transparent façades it could be necessary to plan an integrated shading system which could influence significantly the final cost of the BIPV installation.

In general the design, and consequently the required aesthetics, greatly affects the cost of the cladding layer. For example, the size and weight of the cladding element influences the characteristics of the substructure.

The market analysis indicates the following important conclusion: for façades a very interesting price point has been obtained, since BIPV systems are very comparable in price with conventional façade materials/systems and extra-costs to make it active are no higher than extra-costs for BIPV roofs! BIPV façades strengthen the promise of BIPV because these applications are cost-wise suitable as a substitute for the conventional solutions. This holds the promise of 'PV for free' in building contexts where façades are already quite expensive (unitized façade, curtain wall, rain-screen façade, etc.).

6 CONCLUSIONS

Compared to the previous BIPV Status Report, a greater product diversification is taking place even though the most common categories are still roof related (solar tiles and full roof solutions) indicating that this market is still significantly bigger than the façade market. Some of these new products appeared on the market also as a result of innovation projects supported at a national or European level.

Regarding the BIPV trends we expect that 'Invisible PV' will become a driving approach in the development of new products and that new technologies focusing on blending PV panels in buildings design will enter the market. Multi-functionality, cost effectiveness, mass customization and other paradigms are ensuring a growing penetration of the technology itself, but beyond functional and energy aspects: BIPV is slowly becoming part of the architectural concept!

Summarizing the considerations obtained through an online price survey we can affirm that standardized prices per square meter do not exist neither in building nor in BIPV, and suppliers are generally unwilling to share pricing information apart from very specific price quotations. Nonetheless, thanks to the participation of approx. 35 companies, we can conclude that BIPV is affordable and that the extra cost compared to a wide range of conventional building materials, especially on the high-end spectrum, is limited. However, there was no big price drop compared to 2 years ago. BIPV volumes remain low and no economy-of-scale effects are observed yet.

Overall, the BIPV sector is in a healthy shape. Many attractive products are available, reliable and offered at a competitive price. Good examples of aesthetically pleasing and affordable BIPV buildings can be found and are more and more entering the ordinary building stock. Normative approaches for product qualification are quite clear as well. It is time for the demand side to catch up and allow the suppliers of BIPV products to enlarge their market and realize economies-of-scale. We hope that this report helps to create awareness in the building and PV sector and positively add to the direly needed enlargement of the market.

7 ACKNOWLEDGMENTS

SUPSI would like to acknowledge the Swiss Federal Office of Energy which supports the website www.bipv.ch and the collaboration with Energieschweiz.

SEAC would like to acknowledge the Topsector Energy, the TKI Urban Energy and the Netherlands Enterprise Agency for financial support.

Finally, SEAC and SUPSI would like to acknowledge all companies that participated in the price and products surveys.

8 REFERENCES

[1] THE EUROPEAN PARLIAMENT AND THE COUNCIL OF THE EUROPEAN UNION, European Commission (2010): Directive 2010/31/EU of the European Parliament and of the Council of 19 May 2010 on the energy performance of buildings (recast). (2010) pp. 13–35.

[2] FROM BIPV TO MULTIFUNCTIONAL BUILDING COMPONENT, F. Frontini, A. Scognamiglio, G. Graditi, C. Polo Lopez, M. Pellegrino, 28th EU PVSEC 2013

[3] THE EUROPEAN PARLIAMENT AND THE COUNCIL OF THE EUROPEAN UNION: Regulation (EU) No 305/2011 of the European Parliament and of the council of 9 March 2011.

[4] EUROPEAN COMMITTEE FOR STANDARDS – ELECTRICAL, EN 50583-1:2016 Photovoltaics In Buildings - Part 1: BIPV Modules and Part 2: BIPV System

[5] BUILDING INTEGRATED PHOTOVOLTAICS REPORT 2015 - STATUS REPORT ON BIPV, F. Frontini, P. Bonomo, A. Chatzipanagi, G. Verberne, M. van den Donker, K. Sinapis, W. Folkerts, available online at www.bipv.ch/images/160112_BIPV_digitale.pdf

[6] BUILDING INTEGRATED PHOTOVOLTAICS STATUS REPORT 2017, F. Frontini, P. Bonomo, Isa Zanetti, G. Verberne, M. van den Donker, K. Sinapis, W. Folkerts,

[7] PHOTON MODULE DATABASE, available online at www.photon.info/photon_site_db_solarmodule_en.photon

[8] ACTIVE BIPV GLASS FAÇADES: CURRENT TRENDS OF INNOVATION, Saretta E., Bonomo P., Frontini F., , Glass Performance Days 2017, Tampere, Finland.

[9] 2017 IEEE PVSC-44 – Washington DC, Ultra-Lightweight PV module design for Building Integrated Photovoltaics, A Martins, V Chapuis, A Virtuani, et al

BUILDING-INTEGRATED PHOTOVOLTAICS (BIPV) OVER THE TIME – ILLUSTRATED BY MEANS OF EXAMPLES FROM COMPETITIONS

Gerd Becker, Fabian Flade, Roland Krippner, Bruno Schiebelsberger, Walter Weber
Bavarian Association for the Promotion of Solar Energy, Friedrich List Strasse 88, 81377 München, Germany
Phone: +49 (0) 89/278134-28, info@sev-bayern.de, www.sev-bayern.de

ABSTRACT: Starting in the year 2000, the Bavarian Association for the Promotion of Solar Energy (SeV Bayern) has organized the competitions "Building-Integrated Solar Technology". The aim of this award is to make exemplary solutions in qualitatively sophisticated architecture accessible to the greater public. Since the year 2000 seven competitions have been organized, with some 30 prize-winners. The awarded competition entries show the huge potential of coherent integrated solar technology (PV, but also solarthermal) as a part of advanced architectural and energy building concepts. The intension of the work is to illustrate how BIPV has changed within the last 17 years – by means of examples coming from the competitions.
Keywords: Building Integrated PV (BIPV), Façade, Solar Architecture

1 INTRODUCTION

The use of solar energy in and on buildings is a key aspect within energy efficient construction – for individual homes, industrial and administrative buildings, as well as for residential schemes.

Starting in the year 2000, the Bavarian Association for the Promotion of Solar Energy has organized seven competitions "Building-Integrated Solar Technology" in the years 2000, 2001, 2005, 2008, 2011, 2014, 2017 with some 30 prize-winners. The aim of the competitions was to make exemplary solutions in qualitatively sophisticated architecture accessible to the greater public. First, in 2000 and 2001, about a dozen entries included only contributions from Bavaria and in the end (2017) 123 entries came from all over the world.

The intension of the work is to illustrate how BIPV has changed within the last 17 years – by means of examples coming from the competitions [1]. For this purpose, three time periods with their competitions will be considered: Years 2000, 2001 and 2005, 2008 and 2011 and 2014 and 2017.

2 ASSESSMENT CRITERIA AND PRIZES

Within all competitions the following architectural, technical and other criteria had to be fulfilled:

- Architectural aspects (Integral conceptual design, integration of the solar technology in the building envelope, design quality and functionality)
- Innovativeness (energy and construction)
- Energy yield/energy efficiency
- Communicating the system to the general public

Typically in each competition two prizes 15.000 € plus 10.000 €) were awarded, additionally commendations and citations were given.

3 EXEMPLARY BIPV PROJECTS

Within the projects submitted, housing construction made up the largest share. However, there was also a significant share in projects in the non-residential construction sector. The other building types are largely distributed between residential, industrial/commercial, administration and municipal buildings. Some solarthermal systems are combined with BIPV.

3.1 Competitions of the years 2000, 2001 and 2005

The façade system (Figure 1) of a university building in Erlangen / Germany (built in 1999) consists of two subsystems, a solar awning and a one axis tracked solar jalousie. Both systems were designed to achieve a semitransparent effect through the arrangement of the cells and thus to maintain the visual relationship from the interior of the building to the outside. The nominal power of the movable shading gills is 7.7 kW, the one of the solar awning 22 kW. Final yields obtained lie in the range of 700 kWh/kW. The costs were high: 8,000 €/kW for the solar awning and 22,000 €/kW for the one axis tracked solar jalousie.

Figure 1: South orientation and movable shading gills at a building façade of a university building in Erlangen / Germany (2000/ Architect: Christof Präg, Erlangen)

At the office building (built 1999) of a law firm in Landshut / Germany (Figure 2) the sun protection for the whole building was solved in an innovative way. Both the 10 kW solar awning on the roof as well as the 16.8 kW solar shield are continiously tracking the sun. This configuration allows optimal sun protection of the building and high energy yields of the PV system. The costs were high as well: 20,000 €/kW for the solar awning and 22,000 €/kW for the one axis tracked solar jalousie. Final yields are about 950 kWh/kW.

Figure 2: Shading with tracking elements for an office building of a law firm in Landshut / Germany (1999 / Architects: HBH Architekten, München)

This low-energy house (Figure 3) - developed in its own initiative - shows the various possibilities of solar construction. A PV system together with a solar thermal system forms the weatherproof skin of the roof surface. The costs for the solar roof with a PV nominal power of 8.9 kW amounted to 6,000 €/kW. Final yields are about 800 kWh/kW.

Figure 3: Semi-detached house near Munich / Germany with a PV system together with a solar thermal system (1998 / Architect: Hans Kempter, Gräfelfing)

For the high-rise façade in Munich / Germany a combination of PV modules, printed glass and transparent areas is applied (Figure 4). The PV modules are arranged from the 3rd to the 11th floor up to a height of max. 47.0 m above road level. Location and arrangement guarantee shading free operation of the photovoltaic modules. The special area of the photovoltaic power plant of the high-rise façade includes the south façade as well as areas of the west facade of the so called "meeting tower" at the south end of the building. The PV nominal power is 10 kW, the final yield 600 kWh/kW.

Figure 4: PV modules are integrated in the southern and south western façade of a high rise building in Munich (1999/ PMP Architekten, München)

A car park with PV roof (Figure 5) proves that functional buildings can also be built in high-quality architecture. Here the integration of the photovoltaic modules into the roof of a car park is successfully carried out. This was made possible by modules that form a weatherproof roof skin in the roof. The modules are optimally ventilated from both sides so that the yield is high. The PV nominal power is 37.2 kW, the final yield 850 kWh/kW. The costs were roughly 8,000 €/kW.

Figure 5: PV roof for a car park in Munich (1999 / Architects: Ebe + Ebe, München)

High rise buildings are very well suited for PV systems, as the next example (Figure 6) illustrates. The integration of the front-high PV power plant in context of a redevelopment of two multi-family houses – each with nine floors – and totally 48 housing units in Freiburg / Germany has been successfully accomplished. In total PV modules of nominal power 51.2 kW with a surface of 230 m^2 have been installed. Final yields are roughly about 600 kWh/kW.

Figure 6: Multi-family house Freiburg with PV modules in the southern façade (1998 / rolf + hotz architekten, Freiburg)

3.2 Competitions of the years 2008 and 2011

The entire surface of the monopitch roof of Marché International Support Office in Kemptthal in Switzerland (Figure 7) is designed as an electricity generator. The layered execution of the glass-glass thin-film modules, which are relatively small in size and are anthracite-colored, has created an aesthetically pleasing roof surface that is exemplary in its elegance and establishes a benchmark. The materials used for the supporting structure and building skin combine, in a self-evident manner, standard timber construction methods, innovative Phase Change Materials (PCM) technology and photovoltaics.

Thin film modules of 44.6 kW form the roof. As an average the PV system produces 40,000 kWh in a year. This output electric energy covers the total demand including the heat pump.

Figure 7: Marché International Support Office in Kemptthal / Switzerland (2007 / Architect: Beat Kämpfen, Zürich) (Photo: Willy Kracher)

The new roof of "opusHouse" in Darmstadt / Germany has succeeded in visually incorporating solar thermal energy and photovoltaics along the building's street frontage in an exemplary manner (Figure 8). Solarthermal collectors and PV modules are colored to blend in with the neighboring roof surfaces. The different structural and thermal requirements (PV, thermal) are successfully resolved in the detailing of the eaves and ridge. The building possesses an 8.64 kW monocrystalline PV system and a solar thermal system on a double pitch roof to the east and to the west. The electric energy produced is 8,000 kWh in an average year plus 5,000 kWh on solarthermal base.

Figure 8: "opusHouse" roof with solarthermal and PV system (2007 / opus Architekten, Darmstadt) (Photo: Eibe Sönnecken)

For a new workshop - Hall design.s in Freising / Germany - for a wood-working company a PV system on a large roof has been installed, combined with other energy saving renewable technologies.

Figure 9 shows a subdued, elegantly designed building structure with a coherently zoned spatial arrangement which features a slightly sloping roof that points to the south and is covered by thin-film photovoltaic standard panels over the complete surface, to the south and the north. The PV system joins flush with the roof edge and requires no penetration, which evokes a "flat" picture of a "glimmering matte roof skin." A harmonious overall concept emerges through the color scheme matched to the building structure. The 1,035-m^2 PV generator produces approximately 70,000 kWh of electricity annually, which exceeds the workshop's requirements.

Figure 9: Hall design.s in Freising / Germany (2010 / Architects: Deppisch Architekten, Freising)

The refurbishment of a historic brewery "Mühlfeldbräu" in Bad Tölz / Germany shows how it is perfectly possible to transform historic tiled roofs into energy-generating roofs, see Figure 10. The central element is the fully glazed solar roof (280 m^2) of the brewery with an integrated system for light, air, heat, and electricity. The architects succeeded in designing a modular concept and structural design of a solar thermal system and PV generator with lighting elements, which produced a harmonious overall result that optimally demonstrates the many potentials of solar technology in existing buildings both in rural areas and in cities.

Figure 10: „Mühlfeldbräu" Bad Tölz / Germany (2009 / Lichtblau Architekten, München)

The next building (Figure 11 and 12) has won the solar decathlon 2007, which is a student's competition. The multifunctional combination of timber, as a sustainable construction material, with photovoltaics is particularly successful in the façades

This experimental building is equipped with great variety of technologies and concepts, such as vacuum insulation, thermoactive components with PCM, a heat pump and PV with a nominal power of 11.4 kW. It has been designed by a group of Darmstadt students (Professor Hegger) and realised as a prototype.

Figure 11: Solar Decathlon - multifunctional combination of timber with photovoltaics (2007 / contribution of the Darmstadt University of Technology)

The amorphous silicon modules are integrated within the timber louvres both design-wise and structurally (Figure 12), enabling both shadow free tracking as well as adjustable semi-transparency in the façade.

Figure 12: Amorphous silicon modules integrated within the timber louvres for shadow free tracking as well as adjustable semi-transparency in the façade

3.3 Competitions of the years 2014 and 2017

This chapter describes entries of the competitions 2014 and 2017. An important comment has to be made for the entries of the competition 2017. The deadline for the submission was July, 31, 2017. The meeting of the jury has not yet taken place. However, a few examples will be presented, without judgment and without prejudging the jury's decision.

The architects of the "Umweltarena" (Environmental arena) in Spreitenbach near Zurich / Switzerland refer to a "futuristic solar dress", whose structural and formal implementation provides an equally convincing solution when viewed both from a distance and close up. The southeast/northwest-oriented building succeeds in effectively activating the north-facing PV cells, enabling the roughly 5,400 m^2 of monocrystalline glass-glass modules to achieve a total installed output of 750 kW (Figure 13). The energy yield (2014: ≈ 640 kWh/kW) is in total more than double the building's own energy requirements. The costs for the 750-kW-PV systems including panels, inverters etc. amounted to 2,29 Mio SFr or roughly 3,400 €/kW.

Figure 13: 750-kW-Building Integrated PV System of the "Umweltarena" in Spreitenbach near Zurich / Switzerland (2014 / René Schmid Architekten, Zürich)

The Glattfelden single-family home (Figure 14). is a self-confident new-build scheme that is distinguished by its clearly structured timber facade and the energy roof, which is folded slightly inwards. Its roof form is reminiscent of the local saddle roofs and uses the different slopes to optimise the exposure of the PV modules and solar collectors

Using a standard mounting system, the architects have succeeded in combining the solar thermal and photovoltaic systems with 7.6 kW along with two skylights in an extremely elegant manner. Stainless steel solar absorbers and glass covers are matched in terms of their construction heights and the shingled arrangement also provides a slightly structured surface.

Figure 14: Glattfelden/Switzerland single-family home with PV and solarthermal systems in the roof (2013 / Architects: Mirlo Urbano Architekten, Zürich)

In the west façade of the SwissTech Convention Centre in Lausanne / Switzerland, 3-kW-dye-sensitized solar cells in various yellow, green and red hues are arranged in storey-high narrow strips across 300 m² in front of the glass facade. This project is the first time that the design possibilities offered by Grätzel cells (Grätzel was a Professor at EPFL) have been used on such a large scale anywhere in the world. These provide not only solar shading but also produce appealing lighting effects in the foyer.

Figure 15: SwissTech Convention Centre with dye-sensitized solar cells (2014 / Architect: Richter Dahl Rocha & Associés, Lausanne)

The District Town Hall in Freiburg/Germany is an energy - efficient building (Figure 16) in which PV contributes a significant share to the overall energy balance, which is heated and cooled with a groundwater-heat pump. To achieve the plus energy standard, PV modules were installed on the roof as well as in the façade and in the directions from northeast to northwest.

Figure 16: District Town Hall in Freiburg/Germany (2017 / Architect: Ingenhoven Architects, Düsseldorf)

Copenhagen International School CIS Nordhavn is a new school building for the Copenhagen International School (Figure 17). The school building's unique façades are covered with 12,000 solar panels, each individually angled to create a sequin-like effect (Figure 18). PV will supply approximately half of the school's annual electricity consumption. The solar cells cover a total area of 6,048 m², making it one of the largest building-integrated solar power plants in Denmark, estimated to produce 300 MWh per year.

Figure 17: Copenhagen International School in Copenhagen / Denmark (2017 / Architect: C.F. Møller Architects, Copenhagen) (Photos: Anders Mørk)

Figure 18: PV Cells, each individually angled to create a sequin-like effect

The multi-family house for nine families in Brütten / Switzerland (Figure 19) is the first energy-efficient building of its kind. It is completely independent of any external electric or thermal energy suppliers and has no grid connection: Off Grid! Its shell consists entirely of matte photovoltaic modules which deliver an annual electrical output of 90 - 105 MWh. About 70 % of this energy comes from the 512 m² roof-integrated 79.5 kW monocrystalline PV system. The remaining 30 % are produced by the 485 m² 47.0 kW facade installation of thin film modules. Heating is provided by a 28 kW water-to-water heat pump coupled to two 338 m-deep geothermal probes. The house has three energy storage facilities: a 192 kWh lithium iron phosphate battery for short-term storage; a power-to-gas installation with an electrolyser, a 120 m³ hydrogen tank (with a fuel cell) for long-term storage; and two 125-m³ hot water storage tanks in the ground.

Figure 19: Multi-family house in Brütten/Switzerland (2017 / Architect: René Schmid Architekten, Zürich)

Opened in January 2014, the new head offices for the Federation of Korean Industries (FKI Tower in Seoul / Korea) is a major new addition to the skyline of Seoul, Korea. The PV systems of the 50-story, 240-meter tower are facing southeast and southwest, they have a nominal power of 728 kW with an area of 5824 m².

The concept for FKI was developed by combining the client's request for a building that was contextual to the predominantly orthogonal architecture of the district and the city's requirement that all new large-scale commercial buildings generate a minimum of 5 % of their energy onsite. The use of building integrated photovoltaic panels (BIPV) was seen as an architecturally appealing way to meet the strict zoning requirement, while the optimization of the panels became a driving factor in developing the architectural expression.

Figure 20: FKI Tower in Seoul/Korea - BIPV details FKI Tower (2014 / Architect: Adrian Smith + Gordon Gill Architecture, Chicago)

4 CHANGES FROM THE YEAR 2000 TO TODAY

The example BIPV projects coming from the seven competitions show the huge potential of coherent integrated solar technology as a part of advanced architectural and energy technical building concepts. However, BIPV has developed, differences and trends can be detected between the first entries in the time from the year 2000 to now.

Within the first competitions (Competitions 2000 and 2001) only blue crystalline special modules have been applied, as the Façade of a university building in Erlangen (Figure 1), the office building of a law firm in Landshut (Figure 2), the semi-detached house near Munich with a PV system together with a solar thermal system (Figure 3), the high rise building in Munich (Figure 4) and the car park in Munich (Figure 5) illustrate.

Nowadays in an enhanced matter coloured and black modules are installed. The grid structure of the PV cells has deliberately made invisible, thus the modules appear like traditional façade materials. This is demonstrated by "Hall design.s" in Freising (Figure 9), Copenhagen International School (Figure 18) and the multi-family house in Brütten/Switzerland (Figure 19).

Concerning the orientation of the PV modules, tracking systems (Figure 1 and 2) can often be seen in the competitions of the years 2000 and 2008. Exemplary the façade of a university building in Erlangen (Figure 1), Shading with tracking elements for an office building of a law firm in Landshut (Figure 2) and Solar Decathlon - multifunctional combination of timber with photovoltaics (Figure 11 and 12) can be mentioned.

Nowadays fixed modules are facing many directions of the sky, as shown by the southeast/northwest-oriented 750-kW-Building Integrated PV System of the "Umweltarena" in Spreitenbach near Zurich (Figure 13), which succeeds in effectively activating the north-facing PV cells. At the District Town Hall in Freiburg/Germany (Figure 17) PV modules were installed in the façade and in the directions from northeast to northwest. At "Hall design.s" in Freising (Figure 9) PV modules are also facing north.

More and more BIPV systems are combined with modern energy saving technology, meanwhile this is a standard. At Marché International Support Office in Kemptthal / Switzerland (Figure 7) innovative Phase Change Materials (PCM) technology and photovoltaics are combined. The output electric energy of the PV system covers the total demand including the heat pump. The "Solar Decathlon" (Figure 11) building is equipped with great variety of technologies and concepts, such as vacuum insulation, thermoactive components with PCM and a heat pump. In 2007 this was an experimental building, nowadays such technologies are standard. The multi-family house in Brütten/Switzerland (Figure 20) is the first energy-efficient building of its kind.

Most of the BIPV systems of the first competitions were designed only for grid supply, then systems for self-consumption of the energy have been developed, today autonomous systems are possible. An excellent example for self-consumption of the PV energy is Marché International Support Office in Kemptthal / Switzerland (Figure 7). The output electric energy of the PV system covers the total demand including the heat pump. It is remarkable, that the multi-family house in Brütten/Switzerland (Figure 20) has no grid connection.

It is possible to place PV modules and solar thermal collectors side by side. The appropriate architectural solutions improved considerably over the time. At the semi-detached house near Munich with a PV system together with a solar thermal system (Figure 3) - built in 1999 - a rough solution is visible. The new roof of "opusHouse" in Darmstadt / Germany (Figure 8) has succeeded in visually incorporating solar thermal energy and photovoltaics, here architectural requirements have been taken into account. At "Mühlfeldbräu" in Bad Tölz (Figure 10) the architects succeeded in designing a modular concept and structural design of a solar thermal system and PV generator. Architectural requirements have also been successfully taken into account. For the Glattfelden/Switzerland single-family home with PV and solarthermal systems (Figure 14) a standard mounting system has been applied, the architects have succeeded in combining the solar thermal and photovoltaic systems in an extremely elegant manner.

Grätzel cells offer new design possibilities, however the energetic yield is small, as the SwissTech Convention Centre with dye-sensitized solar cells in various yellow, green and red hues shows (Figure 15).

No preference concerning thin-film or crystalline module technology could be detected. The costs for BIPV are going down. For the first projects (competition 2000) semi-detached house near Munich (Figure 3) and the car park in Munich (Figure 5) costs of 6,000 €/kW resp. 8,000 /kW have been achieved. The costs for the 750-kW-PV systems "Umweltarena" in Spreitenbach / Zurich (including panels, inverters etc.) amounted roughly to 3,400 €/kW.

It can also be noted that formerly mounting systems were custom-made products, today standardized mounting systems are applied, as the Glattfelden single-family home with PV and solarthermal systems (Figure 14) illustrates.

The high-rise buildings of the first competitions were equipped with only small PV systems, as illustrated by the high-rise façade in Munich, the PV system has a nominal power of only 10 kW (Figure 4). In contrary the FKI Tower in Seoul (Figure 20) can be seen. The PV systems of the 50-story, 240-meter tower have a nominal power of 728 kW with an area of 5824 m^2.

5 CONCLUSION

The architectural awards for building-integrated solar technology granted by the Bavarian Association for the Promotion of Solar Energy confirms the orientation on solar technology in a higher-level overall architectural concept. BIPV is developing, BIPV systems can easier be mounted, and the costs are going down. Today BIPV is a self-evident, unobtrusive application. More and more BIPV systems are combined with modern energy saving technology. Generally it can be stated, that more and more BIPV projects are nowadays solved with sophisticated architecture.

6 REFERENCE

[1] Roland Krippner (Hrsg.): Building Integrated Solar Technology; Architectural design with photovoltaics and solar thermal; Detail green books. Munich: DETAIL Business Information GmbH, June/2017)

DESIGN OF AN AUTONOMOUS SOLAR CHARGING STATION FOR E-BIKES

R.M.E. Valckenborg[1]*, R. Ghotge[2] and W. Folkerts[1]
[1] Solar Energy Application Centre (SEAC), High Tech Campus 21, NL-5656 AE Eindhoven, The Netherlands
[2] Carl von Ossietzky University of Oldenburg, Ammerländer Heerstraße 114, 26129 Oldenburg, Germany

*Corresponding author: valckenborg@seac.cc; Tel: +31 652803729

ABSTRACT: An autonomous solar charging station for e-Bikes has been developed. For an optimal design of all system components, both a prototype has been built, and a model has been created. Validation of the model has been done by taking measured irradiance and measured total electrical load profiles as input for a period of one month. The output in terms of PV yield, grid feed-in, SOC of battery bank, etc. has been compared with the real measured values of these parameters. This validation gives much confidence in the model, although care needs to be taken for low SOC values, at which a high accuracy of the model is mandatory to distinguish between 'able to power' and 'unmet load'. Full year round modeling results are performed by using TMY-data for irradiance, and an electrical load very nearby real measured load patterns including some seasonal effects. These simulation runs show a reliability of 94.9% for the current configuration with 7 e-Bikes and a well-equipped charging station. The system was found to have a relatively high degree of reliability (>90%) even when the system was used with twice the number of bikes it was designed for. Turning the heavy consuming advertisement screen completely off, a perfect reliability of 100% is found. This demonstrates that the designed system is quite flexible with the number of e-Bikes it can cater to and is far more sensitive to full-time auxiliary loads in the charging station.

Keywords: Battery Storage and Control, Optimum Sizing, Reliability, Stand-alone PV Systems, Storage

1 INTRODUCTION

A bicycle parking and charging station with a 5.52 kWp solar roof, was recently constructed in Eindhoven, the Netherlands (see Figure 1). The aim is to investigate the feasibility of stand-alone solar powered charging for electric bikes, so-called e-Bikes.

The primary goal of this research is to find the optimal system design for the autonomous operation of such a charging station. High autonomy of the system provides the flexibility of an off-grid installation. Even in a country with a high-quality grid, like the Netherlands, this could have the advantage of a better business case; the longer the distance between the preferred location of the station and the existing grid, the more economical benefit this off-grid installation gives. Also in areas with a lot of cabling and piping (gas, water, sewage, chemicals,…) in the ground, it is an advantage to build off-grid. Last, but not least, a significant amount of customers prefers to produce their own solar energy, instead of buying solar energy from somewhere else.

Figure 1: Prototype of the Solar Mobility Station installed in Eindhoven, the Netherlands.

For quantifying the autonomy of an off-grid system, many parameters exist [1]. We prefer to focus on the following Key Performance Indicators (KPI's) [2]:

- Unmet load, E_{unmet}: the sum of all electrical energy that the system was unable to deliver [kWh/year].
- Loss of Load Probability, *LoLP*, which is defined as the ratio of the unmet load and the full year load including that unmet load [Arno] $LoLP := E_{unmet} / E_{load,fullyear}$ [%]
- Reliability, $R := 1\text{-}LoLP$ [%]

2 EXPERIMENTAL SETUP AND RESULTS

2.1 Experimental Setup

The charging station is designed from an aesthetic point of view, with a ground floor layout of 6 m x 5.5 m and a roof under a 15° tilt. To keep the costs as low as possible, the project developers preferred the readily available 60-cells crystalline PV-panels with a standard dimension of 1.6 m x 1.0 m. However, it was impossible to get the solar roof densely packed with these PV-panels only. Therefore additional 48-cells crystalline PV-panels were also used in a way that is difficult to notice for passing viewers on street level. Each PV-field is connected to the 48V DC bus of the system (via an MPPT-controller); see Figure 2. The DC bus is connected to a battery bank of 4 series connected 12V lead acid batteries with a capacity of 220Ah and a combined voltage of 48V. Hence the nameplate capacity of the battery bank is 10.6 kWh. This type of batteries (AGM deep-cycle) has a maximum DOD of 80%, which implies that the usable capacity is 8.5 kWh.

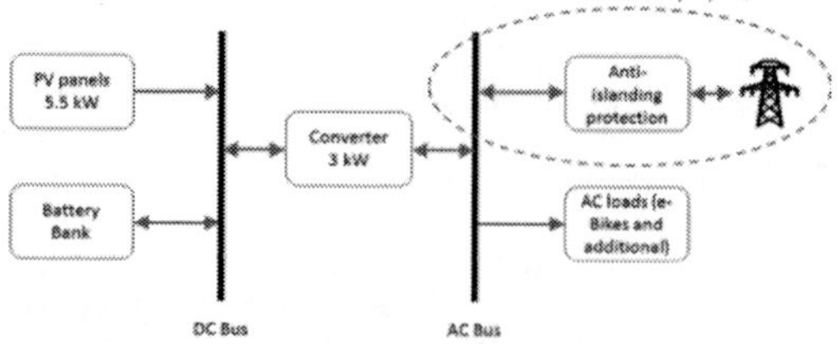

Figure 2: Electrical layout of the system

Because all the electrical applications in the charging station need different DC voltages (5V, 12V, 24V, 36V, ...), a design has been chosen in which all loads have their own normal charger on the AC bus. The electrical applications - and motivations to include them - are:

- e-bike chargers: the main function of the station is charging e-Bikes
- Screen: for advertising purposes
- LED lighting: only in night and when persons detected by a motion sensor
- Automatic sliding doors: for a secure shelter with a convenient entry
- Smart Card Reader: to allow only persons who are subscribed to the system
- Security camera: anti-theft mitigation
- WIFI-router: to make sure that the smart phones of the user always have internet
- Smart lock-by-app gateway: needed for communicating which user is allowed to open/close the lock of which e-Bike

The AC bus is connected to the grid with anti-islanding protection. The purpose of this connection is to be able to measure the unmet load. Please note: in a completely autonomous system, one can never measure how much kWh is missing or unmet, only how much time the load could not be served.

Between the DC and AC bus, an inverter charger (Victron Energy Multiplus 3000) is installed. This device can act as a charger when AC voltage is converted to DC voltage, e.g. for battery charging. It acts as an inverter when converting DC from PV or batteries into AC for the loads or the grid. The size of 3 kVA(=2.6 kW) seems rather low for a 5.5 kWp PV solar roof. However, one has to realize that we are not maximizing full year PV-yield. The sizing has been done making sure that in the very low irradiance days of winter, the PV-field is large enough to survive several of these days. The total AC-load that is needed to run the station is well covered by 2.6 kW.

2.2 Experimental Results

The measurement equipment of Victron Energy gives one-minute time series of all the electrical currents in Figure 2.

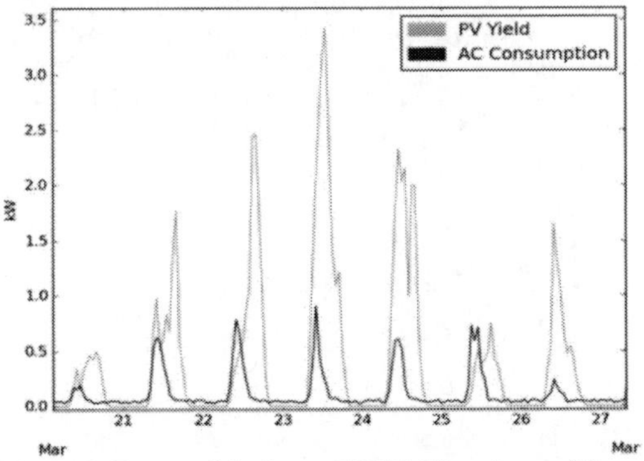

Figure 3: Power of the input (PV Yield) and one of the outputs (AC consumption) as a function of time for one week in March 2017.

Figure 3 shows the input electricity from the PV, and the output electricity of the e-Bike charging for an arbitrary week in the total test period. One can see that charging takes place mostly early morning. This is caused by the scenario of these e-Bike users: they were allowed to use the e-Bike for their commuting traffic, and they have no extra charger at home. When they arrive in the morning in the station, the e-Bike battery will be charged immediately. This coincides nicely with the pattern of the upcoming solar irradiance. This is beneficial for the life time of the battery bank in the station. However, one should note that other e-Bike scenarios, like the popular bike sharing principle, could give other charging patterns.

Although the experimental results are quite powerful, they also have limitations. First, it takes quite long to measure only one set of design parameters (battery capacity, number of e-Bikes, etc.). Because the autonomy is tested to its critical values only in winter, one needs a full year for each 'design'. Moreover, this prototype cannot easily be transported to another location or shifted towards another direction. All these aspects can be better investigated with a model.

3 MODEL METHODOLOGY AND RESULTS

3.1 Model Methodology

The electrical layout of Figure 2 also serves as the basic principle for the model, first described in [3]. All the energy generated by the PV system is used to power the e-Bikes and additional loads, charge the batteries or fed into the grid – each of which is individually measured. The major losses in the system are heating, battery storage losses, inverter and MPPT losses. Since the objective is to observe/use the system in stand-alone mode, the surplus energy sent to the grid may also be considered as a 'loss'.

Calculating each current effectively, is done in the software package HOMER [4]. Input in the model is irradiance of TMY-data of the location of installation, Eindhoven in the Netherlands [5]. The total AC load is the sum of a base load (based on real measurements) and a constructed e-Bike charging pattern load depending on the assumption (or measurements whenever available) of the e-Bike scenario.

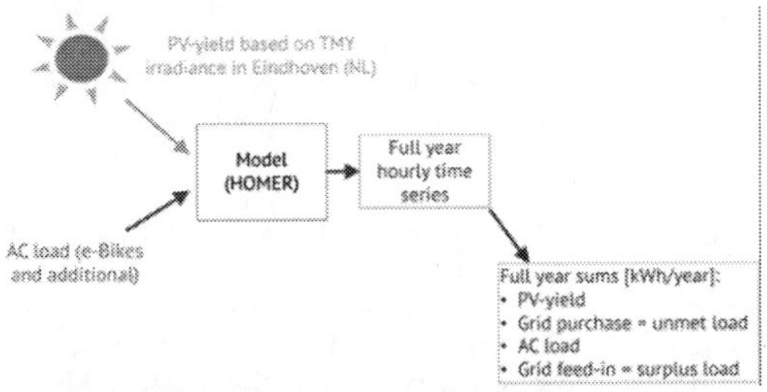

Figure 4: Sketch of modeling methodology.

The output of the model is a time series of 8760 hourly values of PV-yield, grid feed-in, grid purchase, SOC, and many more. The last step is summing up to full year values. See Figure 4 for a complete overview of the modeling methodology.

3.2 Modeling Results

The full year sums of the four most important parameters are plotted in Figure 5. One can see that the PV-yield is around 4500 kWh/year; off course not depending on the amount of e-Bikes in the station. A very small, but still visible, amount of unmet load can be seen. This gives a few days per year with just not enough electricity to fully charge the e-Bikes. Gauging if this is acceptable by the users will be monitored in the upcoming winter period (2017-2018). With increasing number of e-Bikes to charge, the unmet load becomes larger and most probably unacceptable in the case of 20 e-Bikes.

On the negative side of the y-axis, one can see the AC-loads and the grid feed-in. The latter is just surplus load that is of no interest because the prototype is not designed for maximum power production; moreover the kWh-sale value at the location of the prototype (premises of large institute) is just € 0.045 / kWh. Of course, this can be very different at other locations in the Netherlands!

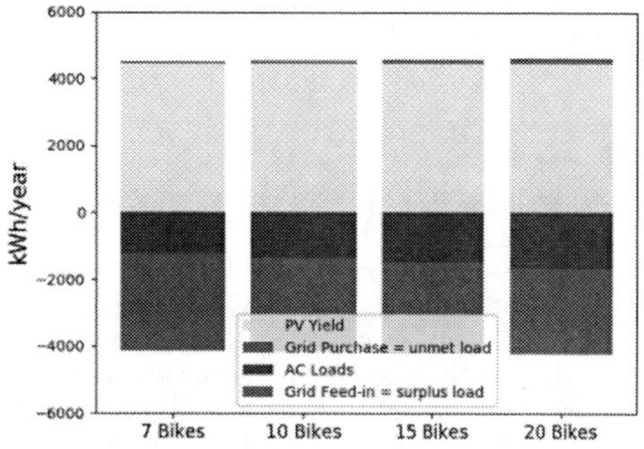

Figure 5: Full year sum results for the current design including the screen.

From Figure 5 the KPI's are calculated:

- unmet load, E_{unmet} = 62 kWh/year
- $LoLP$ = 5.1 %
- reliability R = 94.9 %

Zooming into the time series (not plotted in any Figure) looking for the unmet load, it is observed that 5 days in winter have a period of a couple of hours with unmet load. This is too much for the commercial proposition of a high-end product. But please note that the chosen type of screen (LCD) is a quite heavy user of kWh's. Several mitigation strategies can easily be implemented in a commercial follow-up, e.g.: alternative screen, or no screen at all, or turning screen off at low SOC of the battery bank.

The model has been of great value for understanding the interdependencies between the solar PV capacity, the battery capacity, the number of e-Bikes served by the station and the probability that the user will find sufficient loading capacity at the station when he arrives. Trusting and relying on the model becomes easier if the model is validated.

4 MODEL VALIDATION

4.1 Model Validation Setup

For the validation of the model, we make use again of the software package HOMER, but now in a different way. At the input side of the model, we have to consider:

- solar irradiance: G_{POA} [W/m^2] is measured minutely (and averaged into hourly values) with a sec. standard pyranometer at test facility SolarBEAT [6], less than 50 meters from the station.
- electrical load: the real total load (sum of all applications) is measured with a class 1 kWh-meter

These are fed into the program for the month of May 2017. At the output side of the model, we can monitor:

- PV-yield
- Grid purchase
- Grid feed-in
- AC loads
- SOC

all available at one hour time series.

The comparison of the measured and modeled version of a parameter shows the validation and the quality of the model.

4.2 Validation Results

In Figure 6, the comparison on daily PV yield is shown. The difference seems to be both on the over-estimation and under-estimation up till 2 kWh/m^2/day. Given the complexity of the full setup, this seems quite OK.

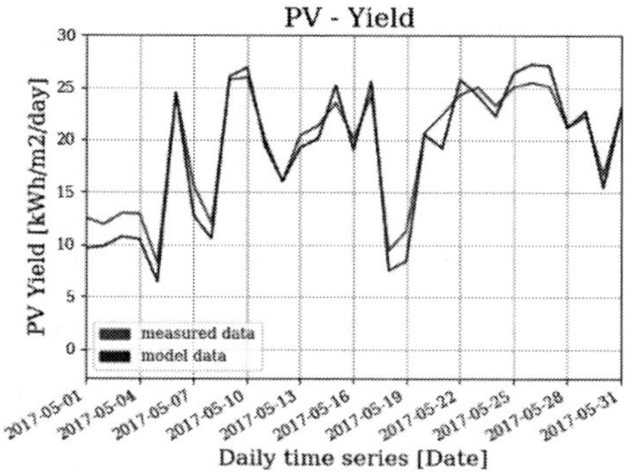

Figure 6: Daily PV Yield for May 2017. Measured yield in red and modeled yield in blue.

The most critical parameter is the SOC of the battery bank. Figure 7 shows the SOC for the same period and using the same colors for model and measurement.

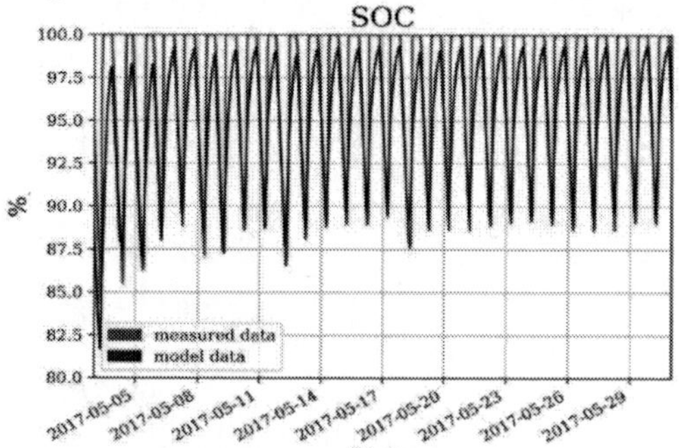

Figure 7: Timeseries of SOC for May 2017. Measured SOC in red and modeled VOC in blue.

The overall impression of Figure 7 looks quite good with deviations on the order of 2.5%. However, in the critical period of low irradiance persistent for a couple of short winter days, this 2.5% can make the difference between a fully charged e-Bike and a half full e-Bike, or even an empty e-Bike. Therefore, we will closely monitor the upcoming winter period 2017-2018.

5 OVERALL RESULTS

5.1 Current Design

With the current design parameters the KPI are presented for a situation of 7, 15, or 20 e-Bikes in Table 1.

Table 1: KPI's for current design.

Nr. of e-Bikes	Unmet load	LoLP [%]	R [%]
7	62 kWh/yr	5.1%	94.9%
15	116 kWh/yr	7.0%	92%
20	162 kWh/yr	9.6%	90%

Please note that the LoLP is by definition a critical parameter, because only the sum of electricity to the loads (≈ 1200 kWh/yr) is in the denomination, whereas the system could provide much more (≈ 4000 kWh/yr) in case of grid-connection. The 7 e-Bike case would give a questionable reliability, because it was observed that these 62 kWh are a sum of 5 days with unmet load for a couple of hours. So the 15 and 20 e-Bike cases with a reliability of 92% and 90% is too poor for a high-end product.

5.2 Increasing battery bank

Of course the reliability can be improved by increasing the battery bank. This has two major drawbacks. First and most important: the extra investment costs, because the batteries are the most expensive parts of the station. Secondly, the space occupied by the batteries in the current station comes from a neat design. Increasing that size gives a more bulky electricity cabinet, which could become too large to be accepted by the customer.

5.2 Scenario Screen Off

As mentioned before, the advertising screen consumes quite some electricity. Switching this screen completely off, shows very interesting results; see Table 2.

Table 2: KPI's for current design without screen.

Nr. of e-Bikes	Unmet load	LoLP [%]	R [%]
7	0 kWh/yr	0.0%	100.0%
15	7 kWh/yr	0.8%	99.1%
20	22 kWh/yr	2.2%	98%

Now, we see a completely 100.0% reliable solar charging station for the case of 7 e-Bikes. For the case of 15 e-Bikes a small unmet load of 7kWh/yr is seen, corresponding to a reliability of 99% which could be acceptable for the users.

5.3 Scalability

One has to realize that the latter needs already a dense packing, hence it is not meaning full to run the model for even more than 20 e-Bikes. On the other hand, less e-Bikes will not be viable from an economical point of view. For customers that would like solutions for more than 20 e-Bikes, a larger version of the station needs to be developed. In principle this would definitely be possible as all components are scalable, but goes beyond the scope of this paper.

6 CONCLUSIONS

The paper demonstrates the results obtained after investigation of real-time reliability and performance of a solar powered charging station. Based on the experience of the researchers, most, if not all of the issues faced over the course of the project were related to the e-bike systems, such as theft, bike security and bike range and bike range anxiety, rather than the energy availability of the solar charging station.

A model was designed using the HOMER software in order to apply the results of the real system to other scenarios, both in terms of system design as well as in terms of loading. This model was then validated against the measurements taken from the built system and found to have high degree of accuracy, giving acceptable results when matched with the live system.

Based on the model, the system was found to have a relatively high degree of reliability (>90%) even when the system was used with twice the number of bikes it was designed for. This increased to an extremely high value of >99% when the advertising screen was switched off. This demonstrates that the designed system is quite flexible with the number of e-Bikes it can cater to and is far more sensitive to full-time auxiliary loads in the charging station. Further options for increasing the energy reliability of the system are considered, in particularly the scalability of the designed system, which is important from a commercial perspective.

This study demonstrably shows the technical feasibility of an autonomous solar charging station for electric bikes, functioning with a high degree of reliability throughout the year, as well as the validity of the parallelly created model for estimating e-Bike loads for a given system.

7 REFERENCES

[1] M. Theristis, G.C. Bakos, and I.A. Papazoglou, 'Development of a reliability model for the estimation of the loss of load probability and cost for an off-grid PV system', 5BV1.31, EU-PVSEC (2012), pp. 4245-4248

[2] A.H.M. Smets, K. Jager, O. Isabella, R. van Swaaij, M. Zeman, 'Solar Energy', UIT Cambridge (28 January 2016)

[3] R. Ghotge, 'Estimation of Building Integrated Photovoltaic Yields and Energy Use for the Solar Charging of Electric bicycles', Master thesis, European Renewable Energy Master Programme, Faculty of Mathematics and Science, Carl von Ossietzky University of Oldenburg (2015).

[4] http://www.homerenergy.com/HOMER_pro.html

[5] Meteonorm 7.1, with Diffuse radiation model = Perez, Tilt generation model = Perez, and Radiation period = 1991-2010

[6] R.M.E. Valckenborg, A. de Vries, W. Folkerts, G.P.J. Verbong, 'The BIPV research facility SolarBEAT in the Netherlands', 6AV4.25, EU-PVSEC (2014), pp. 3707-3713

8 ACKNOWLEDGEMENT

This work is supported by 'Netherlands Enterprise Agency' (RVO) and the Dutch Topteam Energy via the project: 'Mobility Center' with grant number TKIZ TEID115004.

AN ARCHITECTURAL APPROACH FOR IMPROVING AESTHETICS OF PV

L.H. Slooff[1], J.A.M. van Roosmalen[1], L.A.G. Okel[1], T. de Vries[1], T. Minderhoud[2], G. Gijzen[2], T. Sepers[3], A. Versluis[4], F. Frumau[4], M. Rietbergen[5], L. Polinder[5], E.M.B. Heller[6], F. de Vries[6]
[1]ECN Solar Energy, P.O. Box 1, 1755 ZG Petten, The Netherlands
[2]UNStudio, PO Box 75381, 1070 AJ Amsterdam, The Netherlands
[3]TS Visuals, Schuit 14, 1724 VD Oudkarspel, The Netherlands
[4] Aldowa, Overschieseweg 32, 3044 EE Rotterdam, The Netherlands
[5]Design Innovation Group Nieuwegracht 1, 3512 LB Utrecht, The Netherlands
[6]Amsterdam University of Applied Sciences, Leeuwenburg, Weesperzijde 190, 1097 DZ Amsterdam, The Netherlands
Phone: +31 88 515 4314, Fax: +31 88 515 8214, E-mail: slooff@ecn.nl

ABSTRACT: In order to improve the social acceptance of photovoltaic modules, the choice in panel color and size should be enlarged. Although various approaches have been reported to change the appearance of PV modules, it often adds complexity to the manufacturing process. Here an approach is presented in which a design module can be manufactured on standard module lines, by adding a print interlayer to the module. First results are shown on small PV panels, including performance and stability tests. Also full size panels are shown with an aluminum back panel including mounting structures for easy mounting on roofs and facades. The results show that although there is a drop in conversion efficiency by applying a print, the overall drop is lower than expected based on the print coverage. The stability tests show promising results after thermal cycling, damp heat, UV degradation and outdoor exposure.
Keywords: Colored PV, design, photovoltaic, PV modules, performance

1 INTRODUCTION

The installed cumulative PV capacity has increased almost exponentially in last decade [1]. If this increase continues, PV will become much more visible in urban areas. Nowadays, the appearance of photovoltaic (PV) modules on buildings is not very attractive. Partly because the modules have a standard size that does not allow coverage of the full roof, partly because the panels are sensitive to shade and are thus only used in positions that are not affected by shade, but also because they have a completely different look compared to the surrounding surfaces. This often results in a patch-work like arrangement of visibly prominent panels as in Fig. 1. This aesthetically unappealing look will not favor large scale implementation of PV in urban areas. It needs better tuning of the aesthetics of the modules and more size flexibility.

Figure 1: Typical PV systems as installed on rooftops and facades.

2 COLORED PV MODULES

Various research groups and companies are developing modules with a different appearance compared to the traditional c-Si modules [2]. These concepts differ in the way the color is added to the module. Manz e.g. adds color by printing a ceramic layer on the front glass [3] Swiss Inco also uses the front glass but adds a multilayered coating to generate the color [4]. Others, like LOF e.g., vary the thickness of the anti-reflection coating of the cell, thereby tuning the color [5]. Other groups use multilayers on the cell [6]. There are also studies on nanoparticles on the cell that give the cell a nice color.[7] And there are several companies that insert a foil with a uniform or graphical print [8-11]. At ECN a procedure has been developed that adds the colored graphics at the inside of the module. The fabrication procedure is based on ECN's back contact foil technology [12-14] but can be used in combination with other module technologies as well. The modules can be manufactured without adding manufacturing complexity and thus cost.

3 DUTCH SOLAR DESIGN

In the back contact foil technology the back contact cells are interconnected via a conducting foil. This allows easy variation of the module size without the need for large changes in the module line. The appearance of the modules is easily changed by applying graphics at the inside of the module. The application of such colored modules typically lies in renovation of buildings in which the appearance of the buildings must not change and thus the modules have to mimic traditional building materials like bricks, stone, wood, marble and roof tiles. On the other hand they can be used in eye catching offices or company buildings for which more artistic designs might be used. The technology allows both approaches. In fact, the options are endless and will offer a new playing field for architects and designers.

As a start small brick modules have been made that look like a piece of brick wall, see Fig. 2. In this series quarter cells were aligned such that they form brick-like patterns. The coverage of the print was changed from 0% at the cell positions, left picture in Fig. 2 where the "cement" of the print was positioned between the cells, up to almost 45% in the most right picture of Fig. 2.

Figure 2: Typical PV systems as installed on rooftops and facades.

Of course, adding the color print does affect the power output of the panel, but the overall loss is much less than what is expected based on the coverage of the print. This is shown in Fig. 3, where the power loss of the panels is plotted as a function of the coverage of the print. The relatively low loss is due to the fact that the ink of the print is partly transparent. Fig. 3. shows also the results for a panel with the highest coverage with a structured front glass. As can be seen, the relative loss is comparable as with the flat glass. However, the appearance of the two panels is quite different.

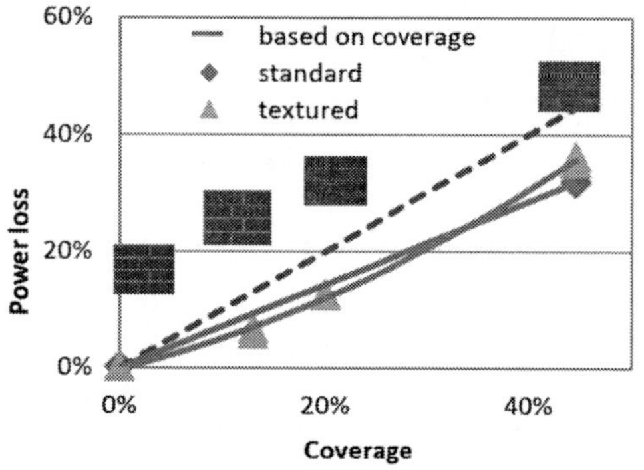

Figure 3: Power loss as a function of the print coverage. Also shown is a 1:1 relationship, showing that the actual measured loss is substantial lower than the coverage fraction.

Fig. 4 shows the panel with flat glass on the left and structured glass on the right. Clearly the left panel has much stronger reflections than the right panel in which the brick print remains clearly visible.

Figure 4: Panels with a brick design coverage of 45% with flat front glass on the left and structured front glass on the right.

As mentioned above, the ink of the print is partly transparent. Increasing the transparency will at some point start to affect the appearance of the panel. The transparency could still be improved without changing the appearance much. The output of the improved panel showed 7% less loss for the highest coverage of the print, resulting in an overall loss of about 25% for a coverage of 45%. Fig. 5 shows the two panels next to each other. There is only a very small difference, where the colors of the left panel are slightly lighter than of the panel on the right. The latter is the panel with the increased transparency of the ink.

Figure 5: Panels with a different ink transparency: left lower transparency than on the right.

Colored prints are built up from the basic colors: cyan, magenta, yellow and white. In a first step to be able to predict the performance of multi-colored prints, the performance of monochromic prints was measured, where both the coverage and dot size of the print was varied. Each panel consisted of 4 cells that could be measured individually. An example of the panels is shown in Fig. 6 for a dot size of 2 mm and different coverage fractions.

Current voltage measurements on these panels show that the current of the cells is hardly dependent on the dot size, but does depend on the coverage. This dependence differs per color. The results are given in Fig. 7. Cyan has the highest loss, followed by magenta, yellow and white.

Initial stability tests were performed on these panels. Figure 8 shows the result of outdoor testing, where the short circuit current is shown as a function of outdoor hours. Data are shown for irradiances between 450 and 550 W/m^2 for the cells behind the white and the cyan print with a dot size of 3 mm and coverage of 30%. The current is nicely stable for both white and cyan. The other colors are not monitored continuously so they cannot be shown. There might be a very small (0.7%) degradation observed after 900 hours, but this can also be due to dirt on the panels. Similar panels are also subjected to IEC 61215-2 thermal cycling, 2 panels, and damp heat tests, 2 other panels. Figure 9 shows the results for panels that had an additional Al panel laminated to the back side of the panel. The results show that after 1000 hrs damp heat testing (full IEC test) and 300 thermal cycling testing (1.5 times IEC test), the panels perform slightly better than at the beginning and also better than the reference that was kept in the dark at room temperature. The change is smaller after damp heat testing than after thermal cycling testing. Further inspection is needed to see where the differences arise from. As mentioned the panels shown above in the stability tests had a large size Al back panel laminated to the backsheet. This was done to be able to

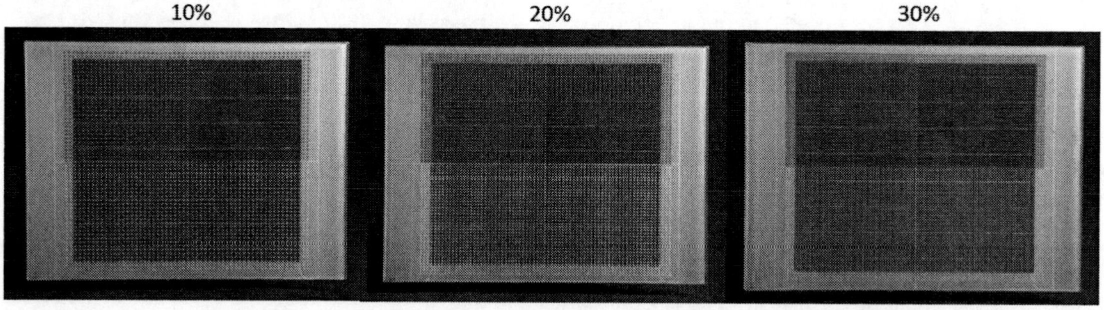

Figure 6: Panels with 4 cells where each cell has a single color print with a specific coverage and dot size.

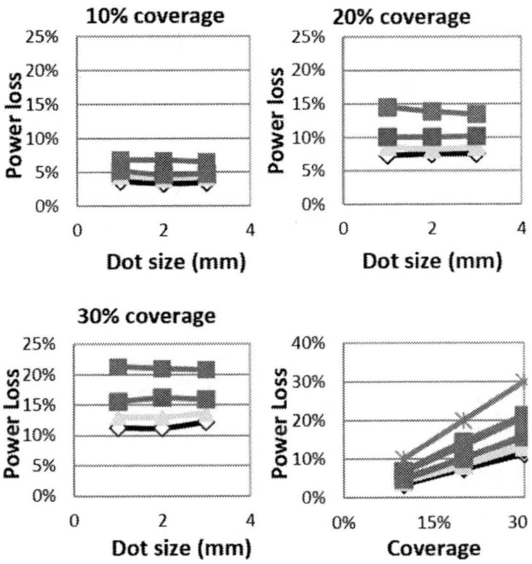

Figure 7: Power loss as function of dot size (1-3 mm) and coverage (10-30%) for the different colors: Cyan, magenta, yellow and white (indicated bij the open black triangles). The bottom right graph shows the average power loss if the 1-3 mm dot sizes as a function of coverage. Also shown in that graph is the 1:1 relation between power loss and coverage (purple dotted line)

use them as façade elements, where the Al back allows for easy mounting of the panels. This was continued in the fabrication of large size, 900x1200cm, façade elements with more artistic design prints. Figure 10 shows a series of these full size façade elements from the front side.

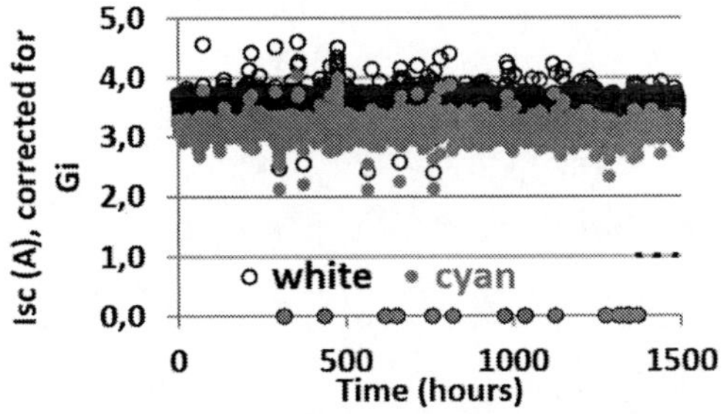

Figure 8: Outdoor stability test on laminates with white or cyan print. Data is selected for an irradiance between 450 and 550 W/m2.

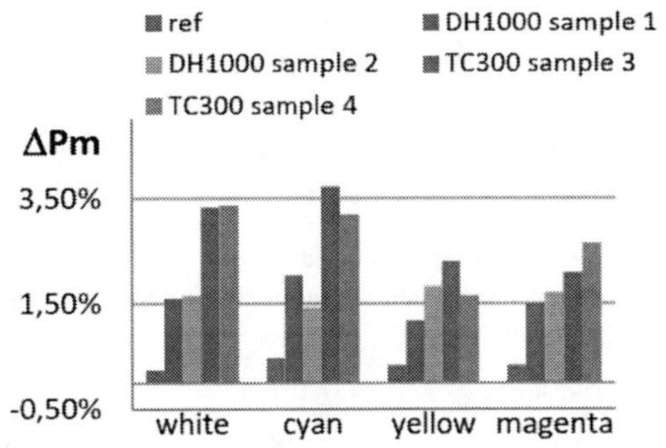

Figure 9: Relative change in power output after thermal cycling tests of 300 hours and damp heat test of 1000 hours. Each test contained two panels with 4 different colors. Also shown is a reference panel that was kept in dark at room temperature.

The output of these modules depends on the actual print, but is between 20 and 25% lower compared to a module without print. The rear of these modules have an Al plate with a special frame, see Fig. 11. This allows easy mounting on a façade.

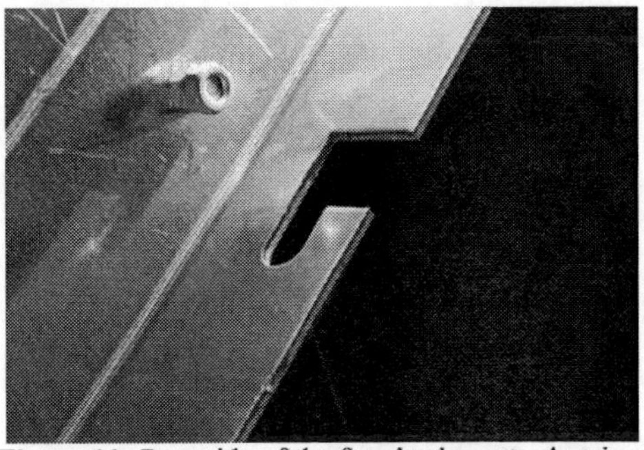

Figure 11: Rear side of the façade elements showing the Aluminum frame for mounting the modules to façades.

5 CONCLUSIONS

Multi-colored modules have been presented that combine a visually appealing appearance with an efficiency that is better than expected based on the coverage of the print. For a print coverage of 30% the output of the module is about 20% lower than the output of a module without print. The overall efficiency depends mainly on the coverage and the main colors that are used

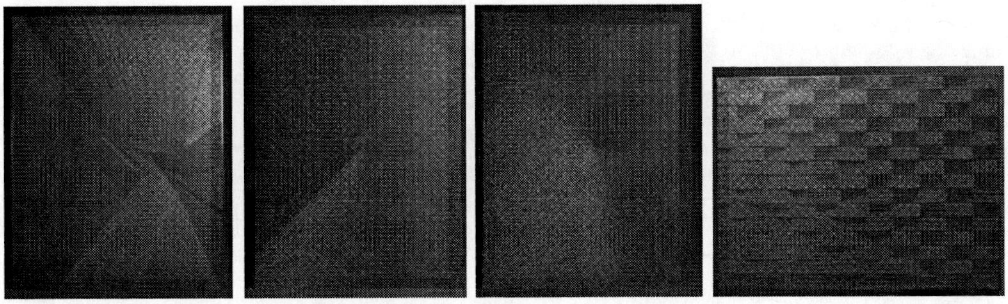

Figure 10: Façade elements with different full color prints.

but hardly on the size of the ink dots. By increasing the transparency of the print, the output could be increased by 7%. Stability tests have shown no sign of degradation after damp heat and thermal cycling tests comparable to the standard IEC test. Also outdoor tests and UV degradation tests show stable results. Large size façade element have been built with a mounting frame to facilitate application. Overall these panels are well suited for building applications and offer an aesthetic alternative to the standard modules.

6 ACKNOWLEDGMENTS

This work has been supported by the Dutch Government through the Topsector Energy Subsidies of the Ministry of Economic Affairs of the Netherlands.

REFERENCES

[1] N.M. Haegel, R.Margolis, T. Buonassisi, D. Feldman, A. Froitzheim, R. Garabedian, M. Green, S. Glunz, H-M. Henning, B. Holder, I. Kaizuka, B. Kroposki, K. Matsubara, S. Niki, K. Sakurai, R. A. Schindler, W. Tumas, E. R. Weber, G. Wilson, M. Woodhouse, S. Kurtz, Terawatt-scale photovoltaics: Trajectories and challenges , Science 356 (2017) 141.
[2] L.-E. Perret-Aebi, P. Heinstein, V. Chapuis, S. Pélisset, C. Roecker, A. Schüler, K. Lumsden, Y. Leterrier, J.-L. Scartezzini, J.-A. Manson, C. Balli, Proc. Cleantech for sustainable buildings 1 (2011) 761.
[3] https://www.manz.com/de/maerkte/solar/systemloesungen-fuer-gebaeudeintegrierte-photovoltaik-bipv/manz-cigs-solarmodule/
[4] Swiss Inso: http://www.swissinso.com/
[5] http://www.lofsolar.com/
[6] J.H.Selj, T.T.Mongstad, R.Søndenå, E.S.Marstein, Solar Energy Materials and Solar Cell, 95 (2011) 2576.
[7] V. Neder, S. L. Luxembourg, A. Polman, Appl. Phys. Lett. 111 (2017) 073902.
[8] https://www.sistinesolar.com/
[w]ww.kaleo-solar.ch/
[10] https://www.csem.com/
[11] H.C.A Rooms,. I. Barbu, Z.A.E.P.Vroon, R.Meertens, B.Vermeulen, Proc. SPIE 8110 Thin Film Solar Technology III (2011) 81100V.
[12] L. Hongfeng, Q. Zhe, C. Kaiyin, Z. Liyan, C. Xianzhi, L. Wei, G. Beaucarne, W. Peng, B. Chislea, Y. Yanghai, A. Zambova, K.M Broek, I.J. Bennett, J. Bakker, N. van Ommen, E. Frederikze, Photovoltaics International (PV-Tech) (2014) 24.
[13] G. Coletti, F. Ishimura, Y. Wu, E.E. Bende, G.J.M. Janssen, K. Hashimoto, Y. Watabe and B.B. Van Aken, Proc. EU PVSEC (2016) 715.
[14] K.M. Broek, I.J. Bennett, M.J.H. Kloos, W. Eerenstein, Energy Procedia 67 (2015) 175.

FLEXIBLE PNEUMATIC ACTUATOR FOR PV SOLAR TRACKING APPLICATIONS

Bratislav Svetozarevic, Johannes Hofer, Illias Hischier, Arno Schlueter
Architecture and Building Systems, Institute of Technology in Architecture,
Department of Architecture, ETH Zurich
Stefano-Franscini-Platz 1, 8093 Zurich, Switzerland

ABSTRACT: The global solar tracker market is currently growing at an unprecedented pace. The predominant application is in utility-scale ground-mounted photovoltaic (PV) installations, where the share of trackers has substantially increased over the past decade. However, the use of solar trackers in building integrated PV (BIPV) rooftops and facades is still limited. The main reason for this is the added weight and bulkiness of tracking mechanisms, as well as their mechanical complexity. To further expand the application range of solar trackers towards BIPV applications, as well as to offer a viable substitute for conventional solar trackers for ground-mounted PVs, in this work we present a solar tracker using a soft pneumatic actuator as the main tracking mechanism. This tracker is made of rubber, and therefore it is intrinsically compliant, i.e. robust to external disturbances, acting as a spring-damper system. This allows drastic reduction of tracker's weight and complexity. Moreover, flexible pneumatic actuators have a large power-to-weight ratio, they offer customization flexibility, and it is uncomplicated and economical to fabricate them on an industrial scale. In this paper, we give a brief summary of design, fabrication, characterization, and control of the flexible solar tracker. The focus of the paper is on long term measurements of the solar tracker's precision and its robustness to wind loads. Three thin-film BIPV options are evaluated: dynamic building facade, flat rooftop, and pitched rooftop. Finally, potentials of using flexible pneumatic solar tracker for conventional ground-mounted PV installations and concentrated applications are discussed.
Keywords: Tracking, Building Integrated PV (BIPV), Façade, Thin Film

1 INTRODUCTION

Compared to fixed photovoltaic (PV) installations, tracking the sun as it traverses the sky increases the direct normal irradiation (DNI) for 20% to 40%, depending on local weather conditions and the reflectivity of installation's surrounding [1, 2]. Besides the absolute gain over a course of a day, maximum solar irradiation gains from solar tracking occur in mornings and evenings, yielding more evenly distributed PV power output profiles. Therefore, more PV power capacity is available during on-peak hours, which do not necessarily coincide with the solar noon [3].

Even though solar trackers have existed for many years, it is only in the past decade that their installation has reached considerable numbers. Their predominant application is in large ground-mounted PV installations, where the share of trackers has increased from few percentages to 23% over the past decade [2, 4, 5]. The economic case for tracking systems is fully proven, and over 70% of U.S. ground-mount projects are now being installed with trackers. It is estimated that this trend will continue globally, with China and India experiencing the most significant growth [5].

As opposite to ground mounted installations, flat- and pitched-rooftop, which account for ~85% of all PV installations, lack conventional tracking options entirely [6]. The main reason for this is the added weight and bulkiness. Solar trackers traditionally consist of rigid electro- mechanical components, such as electromotors, gearboxes, bearings, rails, and cables [1]. In that setting, mechanical designs of trackers that can resist wind loadings are typically over-dimensioned, compared to the case of solar tracking without wind. The reason for this is in rigid coupling between the drive and the links in the mechanism, which is common for classical mechanics [7].

On the other hand, flexible mechanisms, also known as soft robots, have recently received considerable attention in the domain of robotics [8, 9]. They are completely made from elastic materials, such as rubber, which makes them intrinsically compliant, and, in that regard, inherently robust to external disturbances. Therefore, the tracking mechanism does not need to be over-designed, as the soft actuator acts as a spring-damper system and stabilizes disturbances coming from wind loads.

Further advantage of soft actuators compared to classical rigid mechanisms is their low weight. Among several motion technologies that exist for soft actuators, the pneumatic soft actuators (PSA) are particularly interesting, due to the cleanness and abundance of the operation fluid – compressed-air. PSAs have large power-to-weight ratio [10], which makes them particularly interesting for solar tracking applications where size and weight play crucial role, such as on rooftops and facades. Due to the recent commercialization of thin-film PV panels, and given the soft actuators, it is possible to integrate them in dynamic building facades [11].

Finally, the ease of fabrication of soft actuators, e.g. by casting of two-component silicone rubber [12], allows customization flexibility and quick prototyping. As casting is a common industrial process, the fabrication of flexible solar trackers on an industrial scale is possible [13].

The disadvantage of soft actuators compared to the classical actuators in general is in their speed of actuation and in a challenge to achieve precise control [9]. For slow applications, such as solar tracking, the speed of soft actuators is satisfactory [14]. Regarding precision, for flat-panel tracking applications, very high precision is not necessary, as $\pm$ 10° error in tracking still yields 98.5% of the output power of ideal tracking [1].

In this work, we present our recent development in the domain of soft, flexible actuators, and evaluate their usability for solar tracking applications, such as BIPV rooftops and facades, ground-mounted PVs, and concentrated PV applications. In Section 2 the design, fabrication, mechanical characterization, and control of

Figure 1: Adaptive Solar Facade with 50 flexible solar trackers, built in June 2015 at ETH Zurich Hönggerberg Campus, Switzerland [15]

the flexible tracker is presented. Long term measurements of the tracking precision and robustness to wind loads of the flexible solar tracker used in a dynamic BIPV facade are shown in Section 3. Potentials of flexible actuator for use on rooftops, ground-mounted PV installations, and concentrated PV is evaluated in Section 4. Conclusions and discussion are presented in Section 5.

2 PNEUMATIC ACTUATOR DESIGN, CHARACTERIZATION, AND CONTROL

As part of a novel dynamic building facade (Fig. 1) [11] a PV tracker with a pneumatic soft actuator has been developed (Fig. 2). The actuator consists of 3 bellow-like shaped chambers which are radially distributed around a central core. Rotation of the actuator is achieved through different thicknesses of the chamber walls, which determine the kinematics of the actuator. When pressure is applied to a chamber, it starts expanding first where the walls are the thinnest. In this design, the central core is of much larger thickness than the chamber walls, forming a motion restriction layer. Therefore, as pressure is applied to a chamber, the inner ribs, being the thinnest, tend to pop out and expand. The outer thicker ribs restrict the expansion of the chamber in radial direction. Vertical components of the pressure vectors act on the top and the bottom disc, rotating it around the central core. Three chambers radially distributed around the central core generate rotation of the top plate in 2 degrees-of-freedom with respect to the bottom plate [15]. This allows the actuator to perform as a dual-axis solar tracker, orienting a PV panel in azimuth and altitude angles.

The actuator has been experimentally characterized in terms of dynamic response [16], quasi-static pressure-deflection behavior [16, 15], modal analysis [15], and repeatability [16, 17]. Furthermore, a finite element method (FEM) simulation in Abaqus has been used to analyze stress-strain behavior. The results from simulations have been compared with the experimental data and used to update the actuator model. This model has then been used for the optimization of actuator's geometry [18].

A pneumatic control system for regulation of a single solar tracker has been introduced in [16]. It consists of five electro-pneumatic valves, a common chamber, a pressure sensor, and an inertial measurement unit (IMU) for measurements of the panel orientation. For the control of multi-panel installations, such as the Adaptive Solar

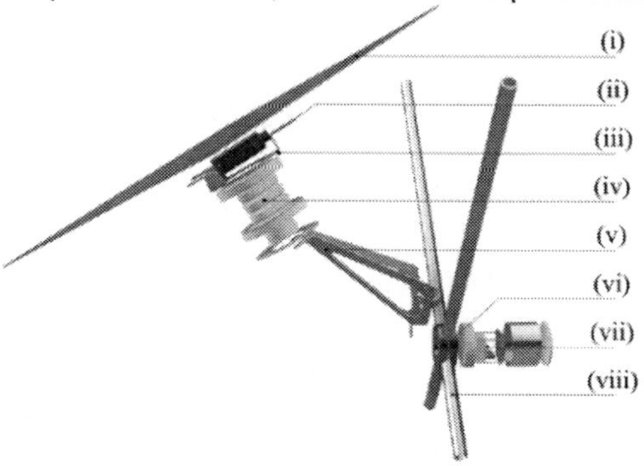

Figure 2: Flexible pneumatic solar tracker. a) Exploded view of one element of the Adaptive Solar Facade with flexible solar tracker. (i) thin-film PV panel, (ii) PV junction box, (iii) panel adapter, (iv) flexible solar actuator, (v) cantilever, (vi) junction element, (vii) integrated control module, (viii) rod net structure including pneumatic tubes and wiring, (ix) horizontal cross-section of the flexible solar tracker.

Facade, an integral control module (ICM) (Fig. 3) has been developed for the purpose of minimization of pneumatic and sensor lines [19]. Based on the above characterization, a feedback control algorithm has been proposed and demonstrated in [15].

3 EVALUATION OF ROBUSTNESS TO WIND LOADS

Due to the inherent compliance of the flexible solar tracker, it is necessary to characterize its behavior in wind. As described in Section 1, the flexible solar tracker behaves like a spring-damper system [15, 9]. Under wind loads, the tracker bends and alleviates the drag force acting on the panel. When wind load decreases, the panel retracts by itself.

Preliminary measurements have been conducted on the roof of HPZ building at ETH Zurich Hönggerberg Campus. A sun tracking experiment was performed with the IMU-based orientation feedback control [15]. A bang-bang control strategy with a predefined precision range was implemented. This control strategy was selected in order to minimize the effort of the control system. The tracker orientation was updated every 15 min. Maximum wind speed during the day was 33 km/h, while the average was 7 km/h. The statistics of the tilt angle in the steady state, i.e. after the transition between two consecutive tracking steps, is shown in Fig. 4a). The statistics is calculated for the last 5 min of each tracking step. For most of the steps, the tracker was within the predefined control precision range of $\pm 2.5°$. As expected, the tracker stays close to the lower bound of the control range, minimizing the control effort and energy. The figure also shows maximum deviations from the mean values due to wind, which are in the range of $\pm 3°$.

Modeled and measured PV power output performance are shown in Fig. 4b). The measurements correspond very well to the model. These results indicate that the flexible tracker can provide precise sun-tracking. The tracking precision can be adjusted in the control

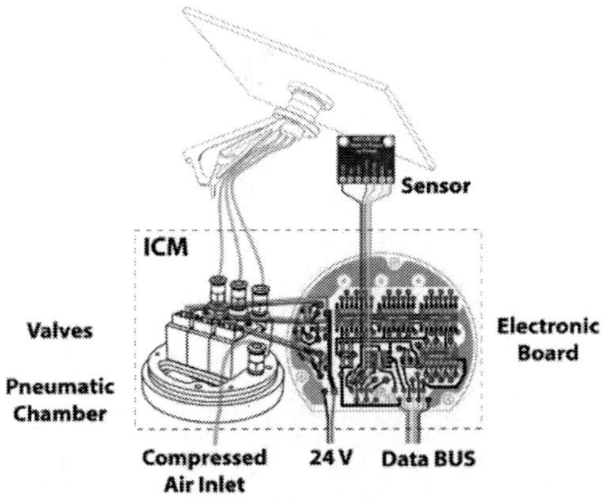

Figure 3: Integrated control module for flexible solar tracker.

system. More in depth analysis of the tracking precision over long term will be conducted during spring/summer time.

3 POTENTENTIALS FOR OTHER PV SOLAR TRACKING APPLICATIONS

Previous Section showed the viability of using flexible pneumatic solar tracker for BIPV facade applications, such as dynamic, modular solar facades [11]. The same actuator could be used for rooftop thin-film PV installations. For conventional crystalline PV panels, the actuator needs to be scaled up [18]. In-depth long term measurements are planned also for this type of installations, as well as for pitched rooftops. For concentrating solar application, e.g. concentrating PV (CPV) tracking accuracy is crucial. Depending on the desired solar concentration, tracking accuracies between ±0.1-2.0° are required. Preliminary measurements showed that it was possible to achieve this precision, which makes this tracker a viable option also for this domain.

4 CONCLUSIONS AND DISCUSSION

In this paper a flexible pneumatic solar tracker is presented. Its design, characterization, and control are briefly summarized from the previous works of authors. Measurements of the sun-tracking precision, robustness to wind loads, and PV power output for a thin-film PV panel integrated into building facade and rooftop, both flat and pitched, are presented. Measurements show that this flexible solar tracker is a viable option for dynamic BIPV facades and rooftops. For conventional crystalline PV panels, the actuator needs to be scaled up. More measurements are needed to confirm the validity for concentrated solar applications. Fabrication method has been already applied on an industrial scale.

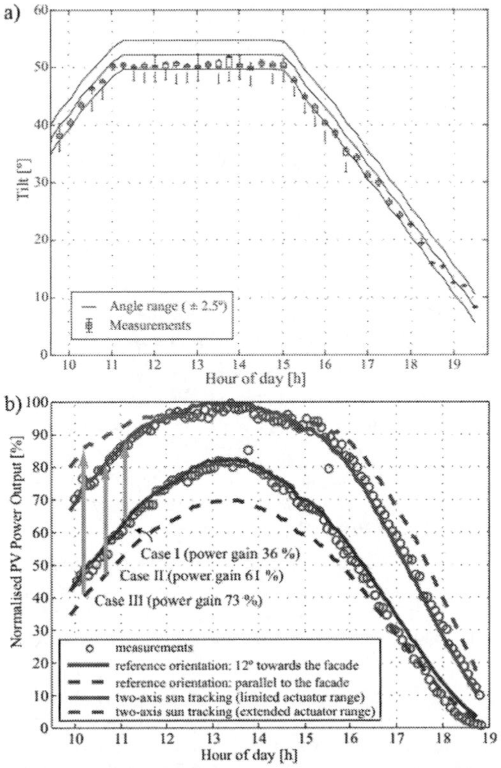

Figure 4: a) Statistics of the tilt angle during solar tracking over a course of a day with the precision in control algorithm set to ± 2.5°. b) Sun-tracking experiment: Modeled and measured power output of a single thin-film PV panel operating on the dual-axis flexible solar tracker

References:

[1] H. Mousazadeh, A. Keyhani, A. Javadi, H. Mobli, K. Abrinia, A. Sharifi, A review of principle and sun-tracking methods for maximizing solar systems output, Renewable and sustainable energy reviews 13 (8) (2009) 1800–1818.

[2] F. Baumgartner, A. Bu ̈chel, R. Bartholet, Solar wings: a new lightweight pv tracking system, in: 23rd European Photovoltaic Solar Energy Conference, Valencia, 2008, pp. 2790–2794.

[3] J. V. Paatero, P. D. Lund, Effects of large-scale photovoltaic power integration on electricity distribution networks, Renew- able Energy 32 (2) (2007) 216–234.

[4] B. Asiabanpour, Z. Almusaied, S. Aslan, M. Mitchell, E. Leake, H. Lee, J. Fuentes, K. Rainosek, N. Hawkes, A. Bland, Fixed versus sun tracking solar panels: an economic analysis, Clean Technologies and Environmental Policy (2016) 1–9.

[5] S. Moskowitz, The Global PV Tracker Landscape 2016: Prices, Forecasts, Market Shares and Vendor Profiles (2016).

[6] A. Lamoureux, K. Lee, M. Shlian, S. R. Forrest, M. Shtein, Dy- namic kirigami structures for integrated solar tracking, Nature communications 6 (8092) (2015) 1–6.

[7] A. Albu-Schaffer, O. Eiberger, M. Grebenstein, S. Haddadin, C. Ott, T. Wimbock, S. Wolf, G. Hirzinger, Soft

robotics, IEEE Robotics & Automation Magazine 15 (3) (2008) 20–30. C. Majidi, Soft robotics: a perspective - current trends and prospects for the future, Soft Robotics 1 (1) (2014) 5–11.

[9] D. Rus, M. T. Tolley, Design, fabrication and control of soft robots, Nature 521 (7553) (2015) 467–475.

[10] R. V. Martinez, C. R. Fish, X. Chen, G. M. Whitesides, Elas- tomeric origami: programmable paper-elastomer composites as pneumatic actuators, Advanced functional materials 22 (7) (2012) 1376–1384.

[11] Z. Nagy, B. Svetozarevic, P. Jayathissa, M. Begle, J. Hofer, G. Lydon, A. Willmann, A. Schlueter, The adaptive solar fa- cade: from concept to prototypes, Frontiers of Architectural Research 5 (2) (2016) 143–156.

[12] F. Ilievski, A. D. Mazzeo, R. F. Shepherd, X. Chen, G. M. Whitesides, Soft robotics for chemists, Angewandte Chemie 123 (8) (2011) 1930–1935.

[13] S. Kalpakjian, S. R. Schmid, K. V. Sekar, Manufacturing engi- neering and technology, Pearson Upper Saddle River, NJ, USA, 2014.

[14] B. Mosadegh, P. Polygerinos, C. Keplinger, S. Wennstedt, R. F. Shepherd, U. Gupta, J. Shim, K. Bertoldi, C. J. Walsh, G. M. Whitesides, Pneumatic networks for soft robotics that actuate rapidly, Advanced Functional Materials 24 (15) (2014) 2163– 2170.

[15] B. Svetozarevic, Z. Nagy, J. Hofer, D. Jacob, M. Begle, E. Chatzi, A. Schlueter, Soro-track: A two-axis soft robotic platform for solar tracking and building-integrated photovoltaic applications, in: Robotics and Automation (ICRA), 2016 IEEE International Conference on, IEEE, 2016, pp. 4945–4950.

[16] B. Svetozarevic, Z. Nagy, D. Rossi, A. Schlueter, Experimental characterization of a 2-dof soft robotic platform for architec- tural applications, in: Proceedings of the Robotics Science and Systems Conference, UC Berkeley, California, 2014.

[17] B. Svetozarevic, A. Schlueter, Z. Nagy, R. Shepherd, M. Begle, J. Hofer, An Adaptive, Energy-Harvesting Building Envelope Combining Soft Robotics and Thin-Film Photovoltaics, Nature Research, (Under review), 2017.

[18] B. Svetozarevic, M. Niffeler, M. Begle, Z. Nagy, A. Schlueter, Design Optimisation of Soft Pneumatic Actuator For Adaptive Solar Facade, Soft Robotics, (Under review), 2017.

[19] P.Jayathissa, S.Caranovic, M.Begle, B.Svetozarevic, J.Hofer, Z. Nagy, S. Arno, Structural and Architectural Integration of Adaptive Photovoltaic Modules, in: Proceedings of the Advanced Building Skins, 2016.

DETAILED MODELING OF BUILDING-INTEGRATED PHOTOVOLTAICS – FROM COMPONENT AND ENVIRONMENTAL DATA TO THE SYSTEM OUTPUT

J. Eisenlohr, H.R. Wilson, T. E. Kuhn
Fraunhofer Institute for Solar Energy Systems
Phone: +49 (0) 761 4588 5745 Email: johannes.eisenlohr@ise.fhg.de

ABSTRACT: Photovoltaic systems that are integrated into buildings face several challenges compared to free-standing PV plants. Besides their multi-functionality and a more complex construction process with the constraints given by a building and its facades and roof, BIPV systems are often geometrically very complex, leading to a more complicated electric circuit and effects like partial shading and hot spots. We addressed the complex electrical design with a detailed modeling approach that includes the relevant physical quantities to analyze the time-resolved yield of a BIPV system. This comprises the calculation of the time-resolved irradiance on each individual solar cell by ray-tracing, the temperature-dependent electric performance of each solar cell, the electrical interconnection including bypass and blocking diodes and the inverter behaviour. Beyond BIPV systems, our developed tools can also be beneficial for the planning of all PV systems, where partial shading occurs, e.g. for bifacial modules.
Keywords: Building Integrated PV (BIPV), System Performance, Façade, Shading, Simulation, Ray Tracing

1 INTRODUCTION

In this study we present a simulation tool suite developed at Fraunhofer ISE that allows for accurate design and yield calculation of complex building integrated photovoltaic (BIPV) systems. While for plant or rooftop configurations, simplified models at module level can be applied, geometrically and electrically complex BIPV systems require tools that are fully flexible with regard to the geometrical and electrical design of the system and a description at solar cell level.

The Fraunhofer ISE simulation tool suite was first presented and validated in [1] and further developed and applied to BIPV installations on the institute's main building [2] and a newly built laboratory building [3]. In these application cases, a complex geometry or a novel module concept required a detailed simulation. Still, the state of the art in widely-used simulation tools for photovoltaic systems does not account for detailed partial shading effects, as no ray-tracing routines for each time step are applied and irradiance values are not determined individually for each solar cell of the system. In many tools, methods using view factors or correction factors to calculate the irradiance on a tilted plane are applied, thus allowing for an accurate description only in very regular geometric configurations [4].

In detail, for the simulation of complex BIPV systems, several (additional) challenges arise:

- The geometry of the system is often very complex. The surroundings (other buildings etc.) and the geometry of the BIPV system itself, which can comprise several module orientations, need to be considered.
- The temperature of the solar cells depends on the installation situation and the actual irradiance that may differ from solar cell to solar cell (not only from module to module). Also hot spots due to high reverse bias can occur under partial shading conditions.
- The electrical interconnection can be very complex. Combinations of series and parallel connections as well as connections of modules with different sizes and different orientations may occur.
- Due to strongly varying irradiation conditions, the optimum inverter choice is more complex and cannot be roughly estimated based on standard test condition (STC) values.
- Feeding the power into different phases of the grid requires time-resolved balancing of the different phases due to regulations of the grid operator.

To address these challenges, the tool suite presented in section 2 has been developed. Briefly, it comprises the calculation of the time-dependent irradiance, temperature and IV characteristic of each solar cell in the system and derives the system's DC and AC yield. For such complex simulation methodologies, it is always crucial to get valid and detailed input data. Here, this consists of data about the environment (location, weather etc.), the building (topology, surroundings, materials etc.) as well as the BIPV products (thermal properties, electrical properties, interconnections etc.) and the inverters. It can be time-consuming and error-prone to collect all this data for a specific BIPV project in a specific planning phase. In the context of Building Information Modeling (BIM) all the required data can be included in a digital model of the building project and made accessible to planners (e.g. the façade planner or BIPV planner) at the time they need it in a non-proprietary, open format. This offers the possibility to automate certain simulation steps and to feed back results to the BIM model, which again can be made accessible to other partners. This is essential for BIPV systems, as they are multifunctional components in the building skin and different partners are involved in a planning and construction process. Using a digital model of the building, including 3D geometry and functionalities of the components, also allows for an application of the BIPV simulation tool suite in the (early) planning phase. With the BIPV simulation tool suite that can be used within the framework of BIM, we try to enable a physically detailed description of BIPV systems and thus to further facilitate the use of PV in building skins.

2 APPROACH: BIPV SIMULATION TOOL SUITE

The objective of all tools described in the following section is to calculate the detailed electrical and thermal behaviour of all solar cells at each time step. The tool suite comprises five major parts that are depicted in Figure 1 and described in detail in the following subsections 2.1-2.5.

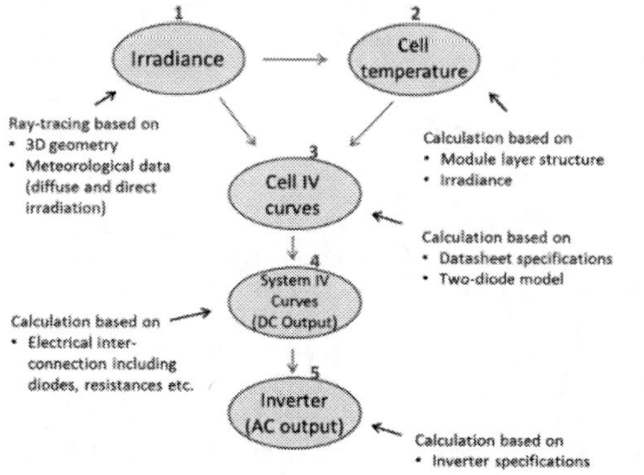

Figure 1: Schematic description of the five main steps for a detailed BIPV simulation. For each time step, the irradiance on each solar cell, the temperature of each solar cell, the cell IV characteristic for each temperature/irradiance combination, the system IV characteristic and the AC output are calculated.

2.1 Irradiance calculation

To calculate the time-dependent irradiance on each solar cell of the system, the ray-tracing tool RADIANCE [5] is used. There are two main inputs required for this step: the 3D geometry of the building and its surroundings including optical properties of the materials and meteorological data including diffuse and direct irradiation and temperature.

2.1.1 Geometrical input

The geometry and material definitions are needed in a RADIANCE-compatible format. As it is time-consuming to describe the building geometry correctly from scratch within RADIANCE, we use an in-house code that can translate a BIM model based on the industry foundation classes (IFC) into a RADIANCE-compatible geometrical and material description [6]. The IFC are an object-based file format with a data model providing the possibility to save the geometrical data and functionality of building projects. Proprietary BIM software packages provide import and export possibilities to IFC and thus, in principle, step 1 of the BIPV simulation can be started directly on the basis of a BIM model. However, the interfaces from proprietary BIM software tools to IFC are still insufficient, and, in practice, it is difficult to export a valid and error-free IFC model without major manual corrections to the originally exported result. This inadequate interface functionality still constrains an efficient combination of the BIM approach with our BIPV simulation tool suite in practice, but despite these challenges, the approach to start from a BIM model to generate the input data is promising and the most efficient solution up to date.

2.1.2 Meteorological input

The required meteorological data consist of diffuse and direct horizontal irradiance and ambient air temperature. Either these data can be measured directly on site, or, especially during the planning phase, reference data sets for the specific geographical location can be used. Starting with the direct and diffuse horizontal irradiance, the RADIANCE program gendaylit [7] is used to calculate the sky luminance distribution based on the Perez sky model [8].

2.1.3 Ray-tracing and post-processing

Using RADIANCE with the described input data, the direct and diffuse irradiance on each solar cell for each time step is calculated by backward ray-tracing. As the resulting irradiance will be compared to measurements under standard test conditions (STC) and normally incident irradiance to deduce the electrical behaviour (see 2.3 and 2.4), a correction for the angle of incidence has to be applied. Based on the module structure and its optical properties, the calculated irradiance is multiplied by an incidence angle modifier $F(\theta)$, resulting in an effective irradiance E_{eff} that can be compared to the 1000 W/m^2 under STC. Usually, we use the actual polar angle θ for the direct irradiance E_{dir} and an averaged polar angle of 60° for the diffuse irradiance E_{diff}:

$$E_{\text{eff}} = E_{\text{dir}} \cdot F(\theta) + E_{\text{diff}} \cdot F(\bar{\theta} = 60°) \qquad (1)$$

The angle-dependent incidence angle modifier $F(\theta)$ is a property specific to the BIPV module. It can be measured or simulated based on the module layer structure. Within this paper we treat it as a further input that is required to perform a BIPV simulation.

2.2 Temperature calculation

In order to calculate the cell temperature of each solar cell for each time step, differently detailed models can be applied: from a simple approach using the ambient air temperature and the irradiance level with a product-specific linear factor, through node models considering the module layer structure, up to detailed, spatially resolved thermal simulations. The presentation of different models is beyond the scope of this work; a detailed comparison can be found for example in [9].

2.3 Calculation of relevant cell IV characteristics

For all occurring combinations of irradiance and temperature, the cell IV characteristics are required. To obtain them, we usually deduce temperature-dependent parameters of the two-diode model from the data sheet specifications; for details see [1]. It is important to mention that the IV characteristic is not only needed in the "useful" quadrant (positive voltage and current, where power can be extracted), but also for negative voltages and negative current, as such operation points can occur in series and parallel connections when parts of the system are shaded. A detailed description how reverse bias conditions are modelled in the presented tool suite can be found in [1].

2.4 Calculation of the system IV characteristic (DC output)

Knowing the cell IV characteristics at a specific temperature/irradiance combination, the electrical interconnection and additionally the IV characteristics and location of blocking and bypass diodes, the system IV curve can be calculated based on Kirchhoff's circuit laws. By determining the maximum power point of the system IV curve, all cell operation points (current and voltage of each cell) can be extracted. In the planning phase, cells under reverse bias can be identified to avoid hot spots during partial shading situations. The electrical interconnection can be optimized with regard to yield maximization and system safety.

2.5 Calculation of the AC output

In most BIPV systems, an inverter converts the DC output calculated in step 4 to AC. Based on the data sheet parameters of the inverter, the inverter efficiency is calculated as a function of the input power and the input voltage using the model of Schmidt and Sauer [10]. Thus, the final AC output of the system is calculated and can be optimized by varying any of the inputs.

2.6 Post-processing of simulation results

Knowing all the time-resolved values of irradiance, temperature, current (DC and AC) and voltage (DC and AC), many quantities of interest can be deduced. Amongst others, typical performance indicators like performance ratio, specific yield in kWh/kW_{peak} or electrical mismatch of the chosen electric circuit plan can be extracted. When the system is divided into subsystems that feed into different phases of the grid, the time-resolved balance between these different systems can be evaluated. In addition, specific results of the individual steps can be used to optimize the system or to assess the multifunctional properties. For example, the results of the thermal simulation (step 2) can be used for consideration of the required additional thermal insulation of the building skin. Furthermore, the occurring operation points of the solar cells (step 4) can be used to optimize BIPV components for specific operating conditions instead of optimizing them for STC.

3 EXAMPLE PROJECTS

To demonstrate the versatility of the presented BIPV simulation tool suite, two example projects are presented in the following. The first is a house in Zürich with a very complex geometry and system design, while the second is an office building in Stuttgart, where BIPV modules covered with different screen-printing patterns have been installed on a south façade.

3.1 Residential house in Zürich

In 2016, a house in the Culmannstrasse in Zürich from the Art Nouveau period has been renovated with the goal of becoming a "plus-energy building" [11]. The roof and façade are very fragmented and a BIPV system consisting of 198 modules with 112 different sizes has been planned and installed. In addition to the various module sizes, 19 different module orientations required a detailed simulation to determine an electrical circuit design that allows a high yield and fail-safe operation. Figure 2 shows the building before and after installation of the BIPV system. Figure 3 shows a visualization of the object in RADIANCE after conversion from an IFC file and after adding the PV modules in the planning phase.

Figure 2: The building before and after installation of a complex BIPV system that was electrically designed and optimized with the presented BIPV tool suite.

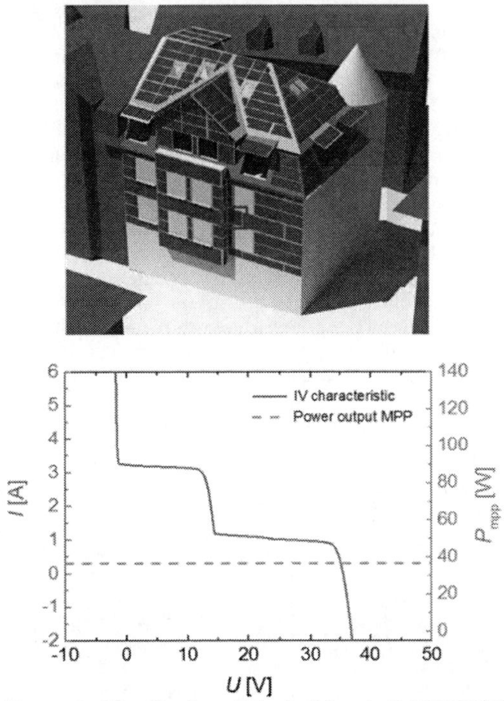

Figure 3: Visualization of the building in RADIANCE including the PV modules. For one module (highlighted with the red square), the IV curve and the maximum power point for the depicted irradiance situation (July 11th, 11am) is shown.

As an example for the simulation results, Figure 3 shows the irradiance situation at 11 a.m. on July 11th and the corresponding IV characteristic and MPP power for one selected module (within the red square). The partial shading of the PV module leads to the form of the IV curve with several local maximum power points. Knowledge of this behaviour for each cell and module for each time step is required to determine an optimum design of the complete system. The application of the BIPV simulation tool suite led to the following electrical design of the system: The system is divided into 14 subsystems with individual, small inverters. The subsystems comprise modules with different orientations, but by the use of appropriate series and parallel connections, the mismatch has been minimized during the simulations. The calculated mismatch losses for the

14 subsystems lie between 1.4 and 18.5 %. Overall, an annual yield of 14328 kWh has been predicted on the basis of a typical meteorological year for Zürich. After more than one year of operation, the output measured by the inverters indicates that approximately 90% of the predicted yield has been reached. Note that the comparison of simulated and measured values is based on different meteorological conditions (typical meteorological year for Zürich vs. actual conditions in 2016/17). Nevertheless, the successful operation of this complex system confirms the simulation methodology and demonstrates that very complex BIPV systems can be designed and optimized on the basis of the presented BIPV simulation tool suite.

3.2 Office building in Stuttgart

As a second example for the application of the BIPV simulation methodology, an office building in Stuttgart with BIPV modules on the south façade is shown in Figure 4 (left side). Modules covered with different translucent screen-printing patterns have been used to achieve a creative and architectural integration of PV into the facade. Figure 4 (right side) shows a visualization of the corresponding IFC model of this façade, which was the basis for ray-tracing calculations as presented in 2.1. The wooden fins of the façade cause shading mainly on the left and right sides of the modules early and late in the day. Therefore, it was decided to use modules consisting of three separately connected cell strings: one string with the left column of cells, one string with the right column of cells and one string with all the middle columns. Subsequently, for the complete façade, all left/right/middle cell strings were connected in series separately and connected to individual inverters or individual MPP tracking inputs of one inverter.

Figure 4: The south-oriented façade with BIPV modules with different screen-printed patterns between the vertical wooden fins. On the right is a visualization of the corresponding IFC model, where the modules are depicted in black.

The BIPV simulation tool suite enables detailed "what-if" analyses. For example, we investigated four different variations of the electrical interconnection design as depicted in Figure 5: first (A), a very simple approach with all cells within a module in series without any bypass diodes, and a series connection of all modules within one storey. Second (B), adding three bypass diodes to all modules – one for the left cell column, one for the right cell column and one for all middle cells. Third (C), the already described and realized approach with three separate strings within every module without any bypass diodes, and, fourth (D), three separate strings with additional bypass diodes as in the second variation. For the third and fourth variation, the module strings are not defined by the storey, but the left/right/middle module parts of all storeys are interconnected in series.

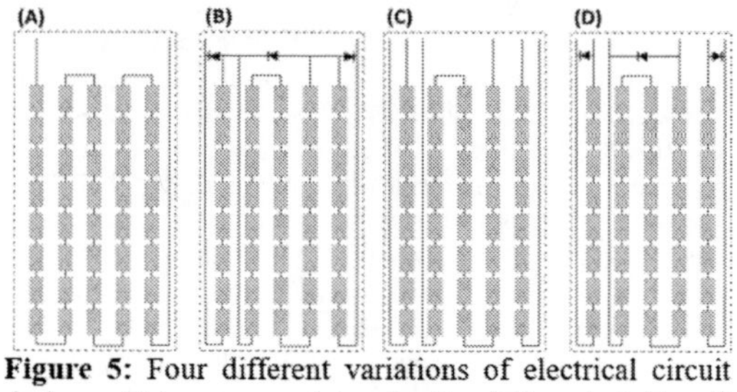

Figure 5: Four different variations of electrical circuit design within one module. As the left and right cell columns are shaded at certain times of a day, the options with bypass diodes (B), three separate strings (C) and a combination of both have been tested. Assuming similarly large strings (5 strings in total for the complete façade) for these four cases, Table I summarizes the resulting normalized DC output power.

Table I: Comparison of the normalized DC output power of the complete façade for different electrical interconnections within and between the modules.

	normalized DC output
One cell string (A)	1
One cell string with three diodes (B)	1.09
Three cell strings (C)	1.11
Three cell strings with three diodes (D)	1.13

Compared to the simple reference case with all cells within a module connected in series, adding bypass diodes at appropriate positions leads to an increase of the DC output power by 9 %, using three separate strings leads to an increase of 11 % and using three strings and diodes results in an increase of 13%.

To demonstrate the detailed effect of adding bypass diodes or using separate strings, Figure 6 shows a typical diurnal profile for the DC output of the complete façade. The four different variations of Table I can be clearly distinguished by different output levels in the morning and the afternoon/evening. At these times, the wooden fins cause shading of the left or right cell columns within the modules due to the oblique incidence angle of the direct solar irradiation. Around noon, when the direct irradiation is incident from the south, no partial shading occurs and the bypass diodes and/or multiple strings have no effect.

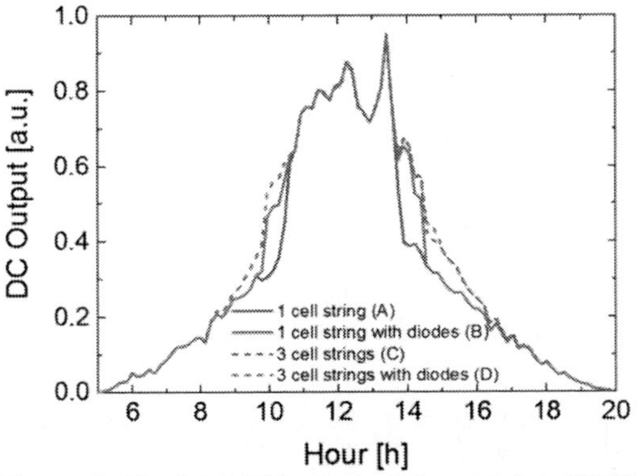

Figure 6: Simulated DC output of the complete BIPV system during one typical day. The four different curves correspond to the four different circuit designs from Figure5. The influence of partial shading in the morning and afternoon is clearly visible and decreases from (A) to (D).

4 CONCLUSIONS AND OUTLOOK

In this paper, we presented a tool suite for detailed simulations of building-integrated photovoltaic systems. Based on the 3D geometry of the building and its surroundings, the time-resolved irradiance on each solar cell position is calculated via ray-tracing using meteorological data for the corresponding geographical location. Subsequently, the temperature and the electric behaviour of each cell are modelled and the system's properties are determined considering parallel and series connections of cells and modules. Finally, the AC output is calculated for each time step based on the inverter efficiency. All these calculation methods are combined in a tool suite and programmed using Python and basic linux tools.

We are working on extracting the large amount of relevant input data - product data like module specifications as well as building and location-specific data – from digital models available in the context of building information modeling (BIM). Coupling such detailed simulations to BIM is a promising approach to facilitate the efficient use of innovative and complex products like BIPV in buildings.

The discussion of two example projects of complex BIPV systems has demonstrated the flexibility and applicability of the presented tool suite. A detailed understanding of these BIPV systems using our tools is possible and necessary to achieve an optimum yield and fail-safe operation.

ACKNOWLEDGEMENTS

The research leading to these results has received funding from the German Federal Ministry for Economic Affairs in the project "SolConPro" (project number 03ET1290A) and from the European Union within the "ConstructPV" project (Grant Agreement number 295981). The authors also thank Mr. Gallus Cadonau for commissioning the simulation project in Zürich.

BIBLIOGRAPHY

1. W. Sprenger, *Electricity yield simulation of complex BIPV systems* (Fraunhofer-Verlag, 2013).
2. W. Sprenger, H. R. Wilson, and T. E. Kuhn, "Electricity yield simulation for the building-integrated photovoltaic system installed in the main building roof of the Fraunhofer Institute for Solar Energy Systems ISE," Solar Energy **135**, 633–643 (2016).
3. H. R. Wilson, J. Eisenlohr, A. Armbruster, A. Spribille, M. Mittag, T. E. Kuhn, "Aufbau, Betrieb und simulationsgestützte Analyse einer hinterlüfteten BIPV-Fassade mit MWT-TPedge Moduelen," in *OTTI - 9. Forum Bauwerkintegrierte Photovoltaik (to be published 2017)* .
4. J. S. Stein and G. T. Klise, *Models used to assess the performance of photovoltaic systems*, SANDIA REPORT, 2009.
5. "Radiance - A validated Lighting Simulation Tool," https://www.radiance-online.org/.
6. M. Boudhaim, "A BIM-compatible Framework to Assess the Thermal and Optical Performance of Envelope Systems in Architecturally Special Buildings," in *30th International Conference on Efficiency, Cost, Optimisation, Simulation and Environmental Impact of Energy Systems (ECOS)* (2017).
7. "RADIANCE Manual Pages - gendaylit," https://www.radiance-online.org/learning/documentation/manual-pages/pdfs/gendaylit.pdf/view.
8. R. Perez, R. Seals, and J. Michalsky, "All-weather model for sky luminance distribution—Preliminary configuration and validation," Solar Energy **50**, 235–245 (1993).
9. Y. B. Assoa, L. Mongibello, A. Carr, B. Kubicek, M. Machado, J. Merten, S. Misara, F. Roca, W. Sprenger, M. Wagner, S. Zamini, T. Baenas, and P. Malbranche, "Thermal analysis of a BIPV system by various modelling approaches," Solar Energy **155**, 1289–1299 (2017).
10. H. Schmidt and D. U. Sauer, "Praxisgerechte Modellierung und Abschätzung von Wechselrichter-Wirkungsgraden," Sonnenenergie **4**, 43–47 (1996).
11. Swiss Solar Agency, "PEB-Dipolom: 118%-Jugendstil-PEB-MFH Culmannstr., 8006 Zürich," https://www.solaragentur.ch/node/624.

EFFECT OF MODULE ORIENTATION AND BATTERIES ON PERFORMANCE OF BUILDING INTEGRATED PHOTOVOLTAIC SYSTEMS.

Marco Lovati*[1,2],Rossano Albatici[2], Laura Maturi[1], David Moser[1]
[1]Institute for Renewable Energy, EURAC Research, Viale Druso 1, 39100 Bolzano (BZ), Italy
[2]Dipartimento di Ingegneria Civile, Ambientale e Meccanica (DICAM) – Università di Trento Via Mesiano, 77, 38123 Trento, 0461-282621
*Phone: +39 0471 055667; Fax: +39 0471 055699; E-mail: marco.lovati@eurac.edu

ABSTRACT: This paper explains the working principle of the free and open source software "POW" (Photovoltaic Optimization Ware), the tool works on the increase of self consumption of the load and on the balance of capacities between the components. The main innovation is providing the capacity of the BIPV system as an output instead of an input, this feature can better integrate the BIPV system in the building design process. The output of the tool is in fact the suggested capacity for each surface of the building (roof + facades) that should be installed to maximize the NPV (Net Present Value) of the system. Some results produced using the tool show that the annual cumulative irradiation cannot be the only parameter considered when installing a PV system in a building. It is shown that it can make sense to install part of the capacity in the least irradiated slope of a roof, thus to better fit the production curve to the demand curve and so increase self-consumption. Another example shows how part of a system can bē placed on the façade instead of the roof in the presence of electric heat pump as an heating system. This phenomenon persists also when combined with a prominent electric storage system. In general the orientation of the modules in terms of Azimuth loses importance when batteries are installed, but the orientation of the PV module in terms of tilt keeps holding primary importance, because it influences the seasonal production profile of the system.
Keywords: Building Integrated PV (BIPV), Building Integration, Economic Analysis, Software, Solar Architecture.

1 AIM OF THE TOOL

Photovoltaic Optimiztion Ware (POW) was developed within the H2020 4RinEU project and is intended to advice building designers about the correct PV and battery capacity to install and about which parts of the building envelope are the most useful for PV. The tool is especially significant in urban context or in case of PV installation in facades and other non-conventional surfaces. In urban context, shadows are common and affect the irradiation of parts of the envelope in different hours of the day during the year. In projects where the roof cannot be used for electricity production the facades can still be profitable and sharply reduce the electricity demand of the building. In such cases PV is often ignored as not believed to be economically competitive. This tool aims to increase the confidence in its use and to show how to make urban PV profitable. The ultimate aim is therefore to increase the prominence and capillarity of BIPV by including it in non-conventional areas of the buildings and of the city. Another desirable effect is to increase the quality of installations in terms of fitness to the energy needs of the building. The workflow controlling this tool is explained in the paper, a free and open source software tool was generated from the method in order to promote replicability.

1.1 Innovative features

An innovative aspect of this method is the ability to support complex geometries because of both the input type and the procedure used to calculate the irradiation. The irradiation is based on the RADIANCE powered DAYSIM software [1], which calculates it by means of ray-tracing [2]. This method presents some stochastic elements but proves much more versatile compared with purely geometric and deterministic methods. Aside of this, the main innovation presented by this method and tool is that the PV positions and capacity and battery capacity are not an input of the software but an output. Simulation tools, which greatly outnumber optimization tools in the field of building's physics [3], are used to predict the behavior of a specific system. Simulation software, in case of PV, can accurately model the electricity production and in some cases offer an economic outlook of the system.Nevertheless they require the position and capacity of the PV system as an input, therefore cannot fit into the design workflow. The optimization process, being similar to spontaneous processes in the architectural practice, could well integrate into the design workflow providing some basic rules for the design of BIPV systems in the early design stages [4].

2 APPROACH USED

The main task of the tool is to find, out of a complete cloud of points, the sub-system on which it makes sense to install a photovoltaic module (see Figure 1). Together with this information, comes the number of selected points and therefore the capacity of the PV system. Although not examined in this paper, the tool can optimize the battery capacity in relation with the PV capacity. To find the most economically rewarding PV system the method should take into account all the relevant aspects that can influence its performance and the profitability. Not all the inputs required by the simulation are known as some are strongly dependent on factors such as developments in the global economic outline or the aging of the photovoltaic material. For this reason, some inputs are values and others are lists, the lists will transform the result in a probabilistic curve where more values are possible.

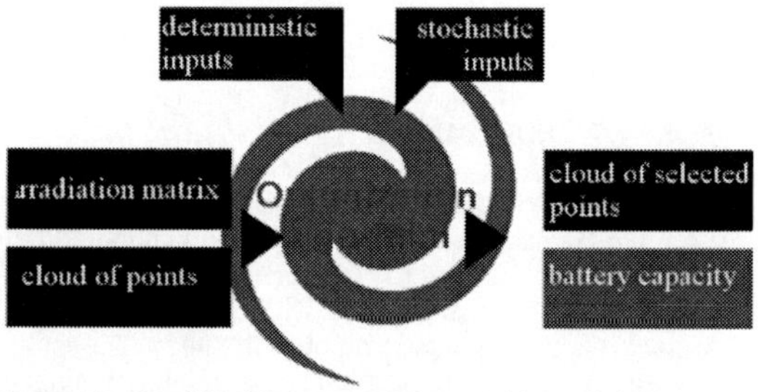

Figure 1: simplified scheme of the working of the POW tool, the battery capacity is represented in gray as it is not examined in the paper.

2.1 Inputs

The following table (see Table I) represents all the inputs required by POW

Table I: inputs of the simulation.

deterministic	stochastic
Cloud of points	
Irradiation matrix	
Weather file	PV degradation
Load vector	Load growth
Area of the modules **A**	Price el. growth
Efficiency of modules **η**	Price sold el. growth
N° of years for NPV **t**	Discount rate **i**
Price consumed el. $\mathbf{P_c}$	Maintenance [€/kWp] $\mathbf{C_t}$
Price sold el. $\mathbf{P_s}$	
PV cost[€/kWp] $\mathbf{C_0}$	
Battery cost[€/kWh] $\mathbf{C_b}$	

The **cloud of points** is a file containing all the tridimensional positions and directions where the irradiations are calculated, the direction associate to each points represents the plane over which the irradiation falls. The **irradiation matrix** contains a vector of hourly irradiations for each point of the cloud of points. The **weather file** is an *.epw (Energy Plus Weather[5]). The **load vector** [W] represents the hourly consumption of electricity of the building under exam. The **N° of years for NPV** represents the time horizon after which the Net Present Value (NPV) is calculated. **Price consumed el.** and **price sold el.** are the prices for the electricity bought or sold from-to the grid.

PV degradation refers to a loss of performance over the years by the photovoltaic system, it is expressed as a linear percentual degradation of the efficiency. All the **growth figures** are expressed in the form of a linear percentual growth of the initial value and represent variation of some parameters over the time of the NPV evaluation. **Discount rate** is a parameter to consider the cost of financing and the inflation rate.

2.2 Fitness function

The optimization algorithm needs a fitness function to evaluate the performance of the system. This function assumes a different value for each configuration of the system (i.e. capacity or positioning of the system). In the single target optimization implemented in POW the fitness function is represented by the NPV =

$$\sum_{t=0}^{n^{\circ}yr}\left(\frac{c \cdot P_c + s \cdot P_s - \omega_{PV} \cdot C_t}{(1+i)^t}\right) - \omega_{PV} \cdot C_0 - \omega_b \cdot C_b.$$

Equation 1: formula for the calculation of NPV

c is the production of electricity which is self-consumed in place, it depends on the weather, the capacity $\boldsymbol{\omega_{PV}}$ of the configuration, the load vector, the load growth, the battery capacity $\boldsymbol{\omega_b}$ and the PV degradation. **Pc** or **Ps** are the prices of the self-consumed or sold electricity, they change every year (dependence on **t**) according to the relative price growth. **s** is the production that is sold to the grid, it depends on the same variables as **c**. **Ct,i,C₀** and **Cb** can be found in Table I. The PV cost C_0 in the present version of POW is considered constant over the surface of the building. No difference is made in terms of cost [€/kWp] between PV in façade or on the roof. This aspect might be seen as an advantage for the PV on façade as it is expected to cost more due to premiums in aesthetics and mounting technology. In the façade integration, compared to the roof, PV is more likely to substitute another coating material, nevertheless it is unknown how to establish which comparable coating solution should be subtracted from the final price of the BIPV cladding to establish the real cost of PV.

The variables c and s are complementary parts of the power of the system described, in every hour of year (hoy), by the relation

$$c + s = A \cdot \eta \cdot PR \sum_{hoy=0}^{8760} \sum_{p=0}^{n} I_{hoy,p} \cdot c(T_p),$$

Equation 2: power of the system in every hour of the year

where **p** is one of a configuration of points $\{p_1, p_2, \ldots, p_n\}$. $\mathbf{I_{hoy,p}}$ is the irradiation over the point p in a specific hoy, **c(Tp)** is a reduction coefficient due to temperature losses and calculated with the linear relation as described in [6]. **A** and **η** are described in Table I, PR is a static performance ratio except the temperature effect which is calculated. The temperatures in each point and hoy is described by the relation $\mathbf{T_p} = T_{ambient} + k\ \mathbf{I_{hoy,p}}$, where k is the Ross coefficient as in [7]. A strong influence on the magnitude of c over s is exerted by the capacity of the battery, this is presently modelled in POW as an ideal storage (with efficiency of 1 and no losses overtime) which follows the behavior shown in Figure 2.

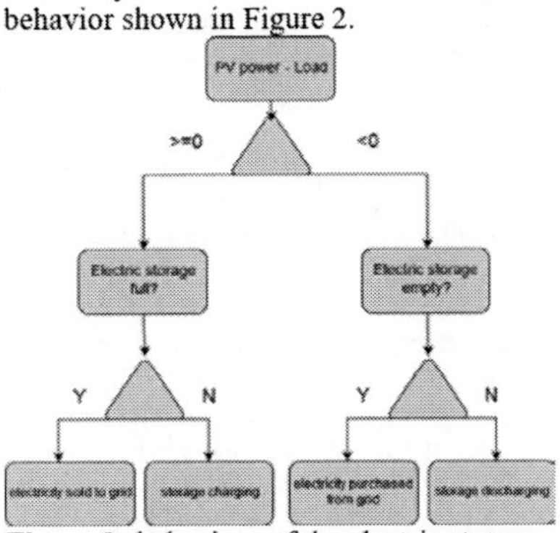

Figure 2: behaviour of the electric storage.

2.3 Optimization algorithm

The optimization algorithm is the procedure followed to obtain a well-performing, if not the best performing, configuration of the system. Presently POW allows for the use of a simple embedded algorithm, hence called One-By-One (OBO), or a single target genetic algorithm, hence called single target genetic (STG), based on the DEAP [8] library for Python. The main difference between the two optimization algorithms is that OBO has an additional (incremental) behavior and solves the problem one point at a time while STG directly finds the configuration in a single run. OBO has a completely empty starting point (0 kWp) and adds the modules one by one producing, at every iteration, a system of growing capacity. In every iteration OBO installs the most profitable module (i.e. the one that yields the highest NPV increase). At first, when the system capacity is low, all the electricity produced by the system is contemporarily consumed. The more modules are added, and the capacity grows, the more becomes crucial to match the load. Faced with the same problem STG would instead try various combination of points measuring the NPV of each complete system and be driven, generation by generation, towards a combination characterized by a high NPV. In both the algorithm the points are divided in groups according to their orientation (façade or roof-pitch) and each group is sorted from the most irradiated to the least, this is done because the orientation influences the hour of the year in which the collector produces electricity. Thanks to this organization of the points, OBO only has to select among the most irradiated modules of each group deciding which group can yield the highest NPV. STG also has a reduced number of combination as it should only find one percentage of utilization for each group instead of including every possible arrangement of points. If the battery is not part of the optimization, OBO really has an edge over STG as it can reach the best solution with a far smaller number of simulations. The aspect is apparent on the "Tower" building (Figure 3), among the example buildings examined. In this case study, which has 7 groups of 10 points each, OBO would have to perform (in the worst case) 7 simulations for each of the 70 points totaling at worst 490 simulations. On the same building STG would have to perform a good 10% of 10^7 simulations to have a good chance of finding the best combination. OBO is tough incapable of handling the optimization of a battery alongside the PV system, this because the battery has a sort of retro-active effect on the system. OBO can in fact install collectors in less irradiated groups in reason of a better contemporaneity between production and consumption. Whenever a battery is added, the production profile becomes less important compared to the sheer energy production rejecting the previous point selection criteria based on contemporaneity. In these situations OBO would have to run an entire optimization cycle for each battery capacity in the search space. A finer grain search space would then increase the number of simulations the most. In the former "Tower" building example the number of simulations would climb up to 490×10 if a grain of 10 steps is applied to the battery capacity parameter. On the other hand STG could give an excellent result with 70 simulations only (10% of 70 points × 10 batteries) provided that the points are divided in one group only, which, considering the battery, should not impair the result. The OBO algorithm has the advantage that saves the best PV module at each iteration, in this way the output is a list of positions for the PV collectors. If the whole list up to a chosen point is taken, it represents the best possible system of that specific capacity. Being that a collector area is assigned to every point, the number of points in a configuration is directly linked to the capacity of the system. This aspects of the OBO algorithm allows for the production of charts such as (Figure 6). In all the examples in this paper OBO was used, the main focus of the paper is in fact the positions and capacity of the PV system and the optimization of the battery

2.4 Assumptions for the examples

In the paper, three imaginary examples are examined in which the tool is used to find the best configuration of modules on the building skin to achieve a high NPV. These examples are not meant to be representative of any European building archetype, but are chosen to show a set of specific behavior of POW. The three examples will hence be called "Tower"(Figure 3), "Roof" (Figure 4) and "Block" (Figure 5).

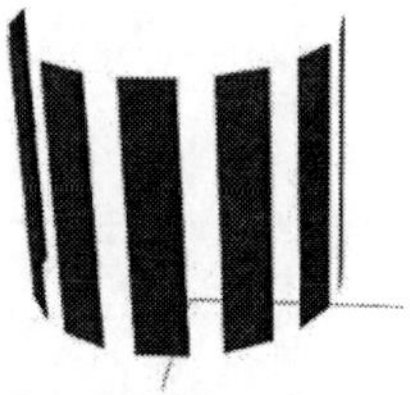

Figure 3: all possible PV positions for "Tower", seven sets of modules with varying azimuth angles.

Figure 4: all possible PV positions for "Roof", the real system would be over-dimensioned if completely covering the entire two slopes.

Figure 5: all possible PV positions for "Block", the roof and the opaque part of the southern façade are available for PV.

Despite different geometries and weather files, all the examples share the following set of inputs (Table II), these are imaginary inputs and do not represent real countries or systems.

Table II: common inputs for all the examples

Efficiency of modules η	0.15
PR (at STC)	0.82
N° of years for NPV **t**	25
Price consumed el. $\mathbf{P_c}$	0.2 €/kWh
Price sold el. $\mathbf{P_s}$	0 €/kWh*
PV cost $\mathbf{C_0}$	1800 €/kWp
PV degradation	[0.3,0.8]
Load growth	[0,2]
Price el. growth	[-1,-0.5,0,0.5,1]
Price sold el. Growth	[-1,-0.75,-0.5,-0.25,0]
Discount rate **i**	[0,1,2,3,4,5]
Mantainance $\mathbf{C_t}$	[18,36] €/kWp yr

*Can vary for the "Tower" building, see Table III

In the case of the "Tower" building, more scenarios have been explored to show how the positions in which it is possible to install PV rank differently. In the case of the "Tower" building the aim is not to find the best NPV producing system, but rather show how the modules rank differently by varying the scenario.

Table III: different scenarios for the "Tower" building

	Internal load	battery	Price sold el. **Ps**
I	low (1 mWh/yr)	no	0

II	low (1 mWh/yr)	10 kWh	0
III	medium (7 mWh/yr)	no	0.05

3 RESULTS

3.1 Position rank for the PV system "Tower"

The first example is not a real optimization, it only deals with the order in which the potential positions are ranked. As visible form Figure 3, which represents the seven groups of modules in "Tower" unrolled on a plane, the building has the capacity to host 70 PV modules (7 groups with 10 modules each, totaling ca. 15 kWp). A customer who wants to install 14 modules (a power of 3 kWp) needs to know which of the potential positions should be occupied, therefore needs the positions to be ranked. The easiest way of ranking the positions is by irradiation (Figure 6), but it is not always the most profitable one.

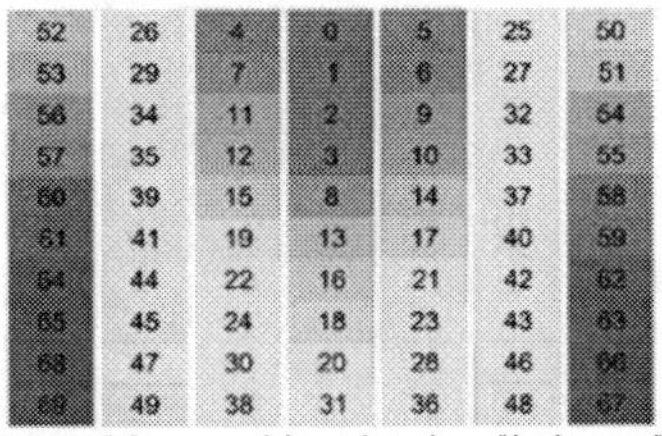

Figure 6: position ranking by irradiation, the scheme represents the seven groups of modules in "Tower" unrolled on a plane where left represent West and right represents East.

If the building is in the situation of *case I* from Table III, the best way of disposing 3 kWp is in fact in a horizontal array covering all the allowed groups of the building(see Figure 7).

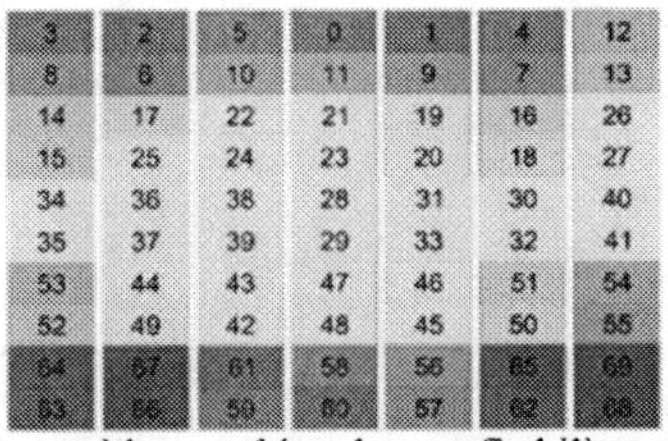

Figure 7: position ranking by profitability case I, the scheme represents the seven groups of modules in "Tower" unrolled on a plane where left represent West and right represents East.

If the building is in the situation of *case II*, the most irradiated position gain importance compared to *case I* because they allow to charge the battery during the central hours of the day. This ranking shows a distinct preference for the east facing modules(see Figure 8). This happens because after noon the battery is charged and the electricity production is not rewarding. On the other hand, in the morning it has already discharged encouraging a "demand matching" behaviour. This ranking could change if the charge/discharge algorithm of the battery is modified. The overall performance of the system will probably benefit from a "smarter" storage system.

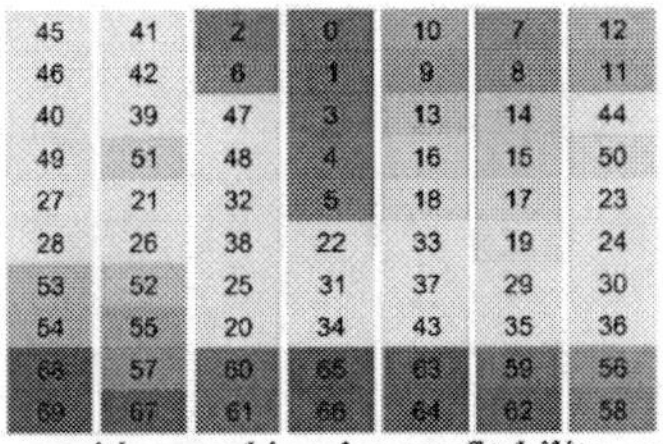

Figure 8: position ranking by profitability case II, the scheme represents the seven groups of modules in "Tower" unrolled on a plane where left represent West and right represents East.

If a higher electricity demand and a minimum valorization for the electricity sold to the grid is applied, such as in case III, the ranking pattern (Figure 9) becomes much more similar to the simple irradiation ranking of Figure 6. Installing a system of 3 kWp only one module differs between the ranking in Figure 6 and that in Figure 9.

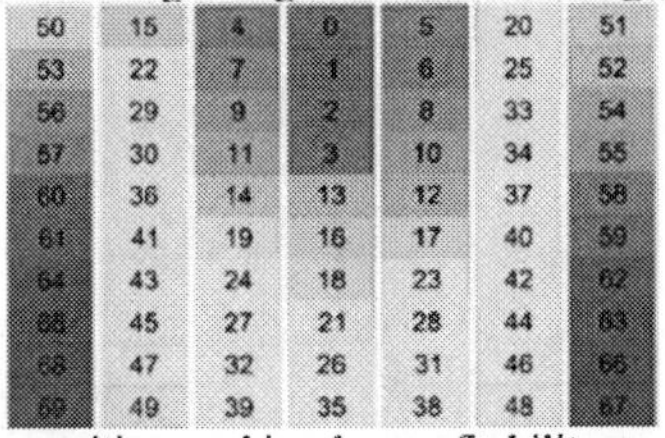

Figure 9: position ranking by profitability case III, the scheme represents the seven groups of modules in "Tower" unrolled on a plane where left represent West and right represents East.

3.2 optimized PV system "Roof"

The "Roof" example is composed by two surfaces lighted by two different annual cumulative irradiations (see Figure 11), in this example the battery is not considered. The electricity demand in the example is arbitrarily chosen and its annual cumulative value is of 36 MWh. The optimization, regarding PV capacity and positions only, is carried out using the OBO algorithm with median NPV as fitness function. The algorithm adds the PV collector one by one, and at every iteration it adds the one that better increases the total NPV of the system (see Figure 10). When OBO reaches the plateau it proceeds by installing the least damaging one (right side of the plateau in Figure 10). Despite being based on the median NPV each simulation produces a probabilistic curve of NPVs because the simulation is based on stochastic variables (Table III). In the grey area in Figure 11 all the possibilities between 25 and 75 percentile are represented, in the picture is visible that the plateau shifts toward highest capacities from the pessimistic scenario (lower bound of the grey area) to the optimistic one (upper bound). In this sense, the suggested capacity is 16 kWp and lies in the range (13,18) kWp. The thickness of the grey area is mainly due to the variability in the discount rate, when the optimization is run with a fixed discount rate the deviation of probability greatly reduces.

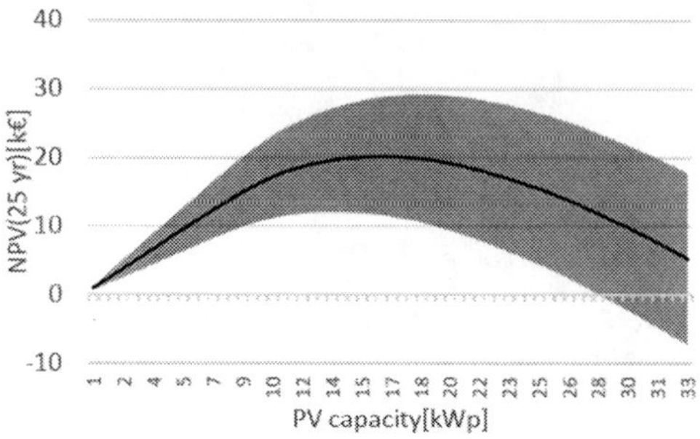

Figure 10: NPV vs installed capacity, the median value shows a plateau just above 16 kWp. The grey area represents all the possibilities between 25 and 75 percentile.

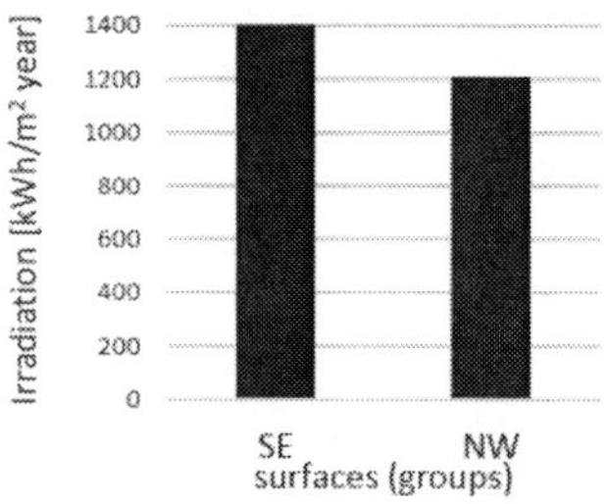

Figure 11: difference in irradiation between the South-East side of the roof and the North-West one.

For reasons of higher self-consumption, the tool suggests the installation of part of the capacity on the NW side of the roof (Figure 12) even if the SE slope alone could accommodate all the suggested capacity. In the optimistic scenario (i.e. a combination of stochastic variables that is favourable to PV) a higher PV capacity is suggested, the opposite happens in the case of a pessimistic scenario. The proportions of capacities over the two slopes cannot be known in pessimistic or optimistic scenarios. To know it, the optimization needs to be launched using the optimistic or pessimistic NPV as a fitness function.

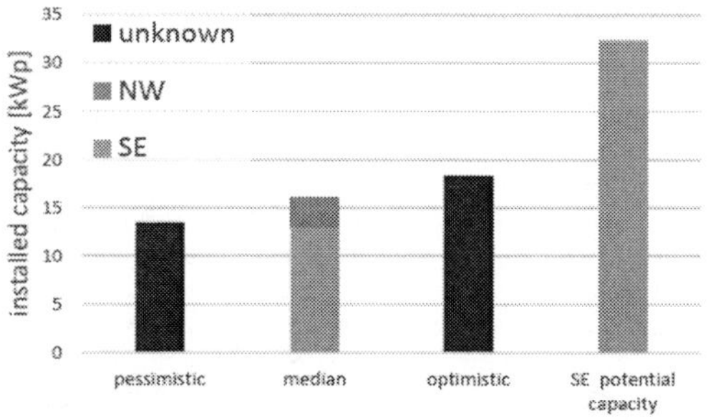

Figure 12: capacity to install as suggested by the optimization process.

3.3 optimized PV system "Block"

The third example is based on a project located near Modena, in the northern part of Italy, the building features a tilted roof (20° tilt) and an inclined façade(70° tilt). Both the surfaces are oriented south. Also in this case the irradiation is uneven (Figure 13), but less sharply than in the "Roof" example. This building is equipped with a battery of capacity 180 kWh, this capacity was chosen arbitrarily with the aim of observing the impact of a significant battery on the result of the optimization. The battery does not influence the initial cost of the system (i.e. the battery cost is assumed equal to 0). The electric load (totaling 182.4 MWh annually) is higher in winter as it covers the heating demand of the building. The result of the optimization shows that, if given the possibility, the software suggests to install a significant portion of the capacity (ca 27% of the total) on the façade of the building (see Figure 14). If the optimization is run without the possibility of using load matching (i.e. installing the capacity on the most irradiated spots first), a higher capacity is installed on the roof but no modules are installed on the façade. This is because the potential capacity on the roof exceeds the capacity required by the building+storage system (on an annual basis), the façade in this case is not even considered.

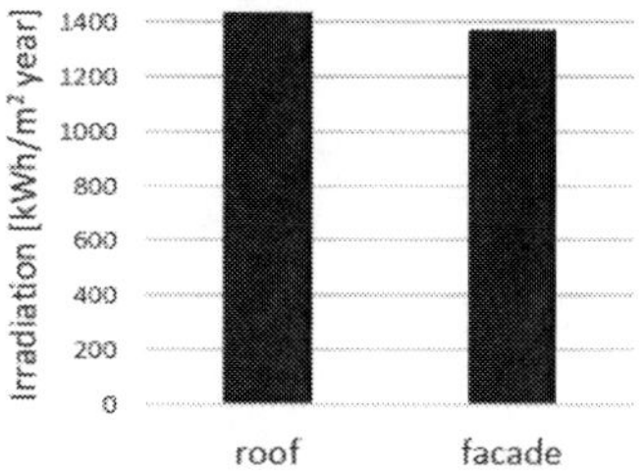

Figure 13: also in the "Block" building there are difference in terms of cumulative irradiation although less pronounced than in "Roof".

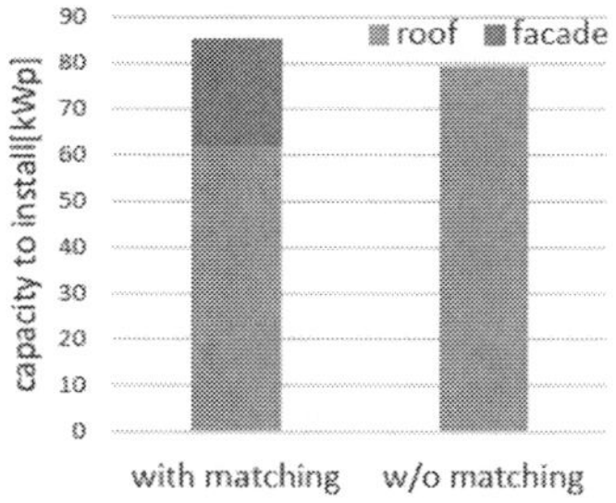

Figure 14: median (expected) output of two different optimizations, one allowing load matching, the other simply installing by irradiation rank.

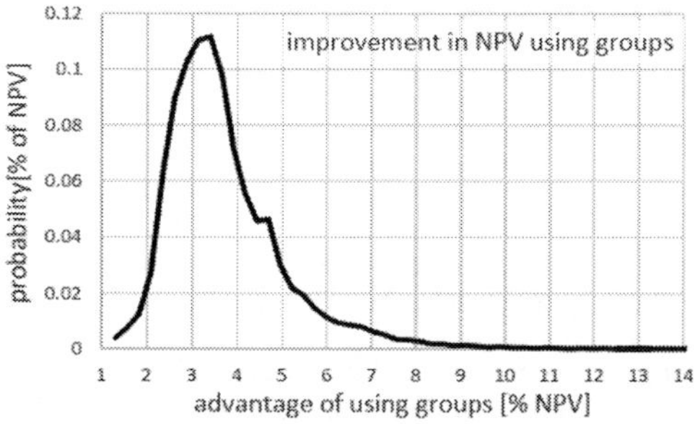

Figure 15: probability of improvement from the optimization w/o groups to the one with groups as % of NPV. Each combination of stochastic variables has a different % (all positive). the average is 3.86%.

The façade though, enjoying higher irradiations in winter when the daily demand is higher, proves more profitable than the roof. The optimization run with the use of groups increases the overall hosting capacity of the system (see Figure 14), and leads to an increased NPV (see Figure 15). The curve in Figure 15 represents the % of NPV improvement when optimizing using groups compared to an optimization without groups, it is visible that the % earning changes for each combination of stochastic variables and that in every combination considered the variation is positive. The average of the possible variation of NPV is +3.86%.

4 DISCUSSION AND CONCLUSION

The use of POW for the optimization of the PV system on different surface of buildings led to the first results. A better understanding of the energy phenomena of the PV system in relation to the building have been acquired, hopefully more case studies will uncover new things. Various improvements of the software and areas of development are noticeable from the first results. The changes in rankings of the Tower building shows that a ranking of positions in an energy system cannot be merely related to the irradiation, but the load and the electric storage should as well be taken into account. Figure 8 shows how the behavior of the storage (Figure 2) affects the result of the ranking. The current behavior of the electric storage praises East facing points even though they are slightly worst in terms of pure load matching in the absence of storage (see Figure 7). An improvement in the possibilities of the battery behavior would be much beneficial to the results of the simulation. The uncertainty in the load might be taken into account to increase the confidence and the reliability of the point rankings, this uncertainty should not be merely in terms of scale (which is already accounted for with different load growth) but also in terms of daily or seasonal shape of the curve. The effect of such uncertainty, which can at times advantage production in different sets of hoy, should give value to resilient configurations. An interesting experiment would be to run the optimization at larger scales. There the electric demand is better known and shows higher predictability. Furthermore, the absence of valorization for the electricity sold to the grid is a more realistic assumption when analyzing a large system. Such a system cannot in fact afford significant excess of production in specific hoy. The launch of the tool at higher complexity scales requires some preliminary effort in simplification of the model (i.e. some form of clustering of the irradiation points) to reduce runtime. With the present technology of the tool, a proportional fraction of a city scale aggregate demand curve can be applied to a single building according to the size of its electric demand, this might improve the overall performance and hosting capacity of the grid if applied consistently. The "Roof" example shows how the most remunerative configuration involves the allocation of part of the total capacity on the least irradiated surface, it shows nonetheless that in a different scenario the suggested capacity to install is different (see Figure 12). In order to know which proportion of the two capacities should be installed in a pessimistic or optimistic scenario the whole optimization should be launched setting different fitness function from the "median NPV". Aside from the discount rate, which is known to have a profound impact on the profitability of an investment, other aspects can vary to determine the overall NPV of a scenario. The combination of variables in a "pessimistic NPV" could vary based on the building geometry, weather file, capacity/load ratio or other aspects of the model (initial or depending on the configuration). A pessimistic NPV does not correspond to a precise combination of stochastic variables, but simply to an ever-changing unlucky set of variables that produces an NPV lower than the 75% of the possibilities. An interesting information would be the effect of "unlucky" scenarios on the proportions between the capacities in different facades. As a rule of thumb, a reduction of the least irradiated surface can be expected as effect of a higher maintenance cost, but the scale of the reduction is difficult to determine in advance. The notion that least irradiated surfaces tend to be excluded in presence of batteries cannot be accepted in case of different tilt angles, this angle influence in fact the seasonal production curve of a system and may benefit from the presence of storage systems. In the "Block" building the façade enjoys a lower annual cumulative irradiation compared with the roof and the building is equipped with an important battery (ca. 30% of the cumulative daily demand), despite these factors 27% of the total installed capacity is installed on the façade. The façade integrated modules, although annually less irradiated, get more irradiation in the cold months when the electric demand is higher. In this situation the presence of a battery is synergic and boosts the façade modules as it allows them to store their premium winter production. The "Block" building is a lucky example because it is equipped with an inclined façade. If the façade was vertical, at the latitude of Modena (Italy) where the building is simulated, its irradiation would have been lower than that of the roof in each month of the year. In that case there would have been no energetic or economic advantage in façade integration. It is good news that façade integration can make sense from the economic point of view also if part of the roof area is still available, it might tough be necessary to install the modules in a non-planar fashion. Non planar PV system exist [9], but they are now installed with angles for maximizing the annual cumulative production. It makes sense, from the energy and economic point of view, to install them more vertically, in a fashion similar to that of traditional wooden claddings.

5 AKNOWLEDGEMENTS

This project has received funding from the European Union's Horizon 2020 research and innovation programme under grant agreement No 723829.
The authors wish to thank Ing. Giulia Fratus and Doct.Ing. Graziano Salvalai from Politecnico di Milano for the precious cooperation.

6 REFERENCES

[1] C. F. Reinhart and S. Herkel, "The simulation of annual daylight illuminance distributions — a state-of-the-art comparison of six RADIANCE-based methods," *Energy Build.*, vol. 32, no. 2, pp. 167–187, Jul. 2000.

[2] G. Ward, "RADIANCE reference manual," *IEA SHC Task*, vol. 12, 1996.

[3] S. Attia, M. Hamdy, W. O'Brien, and S. Carlucci, "Assessing gaps and needs for integrating building performance optimization tools in net zero energy buildings design," *Energy Build.*, vol. 60, pp. 110–124, May 2013.

[4] M. Lovati, L. Maturi, J. Adami, and D. Moser, "Methodologies and tools for BIPV implementation in the early stage of the architectural design.," presented at the 12th conference on Advanced Building Skin, Bern, Switzerland, 2017.

[5] D. B. Crawley *et al.*, "EnergyPlus: creating a new-generation building energy simulation program," *Energy Build.*, vol. 33, no. 4, pp. 319–331, 2001.

[6] L. Maturi, G. Belluardo, D. Moser, and M. Del Buono, "BiPV System Performance and Efficiency Drops: Overview on PV Module Temperature Conditions of Different Module Types," *Energy Procedia*, vol. 48, pp. 1311–1319, 2014.

[7] E. Skoplaki and J. A. Palyvos, "Operating temperature of photovoltaic modules: A survey of pertinent correlations," *Renew. Energy*, vol. 34, no. 1, pp. 23–29, Jan. 2009.

[8] F.-A. Fortin, F.-M. D. Rainville, M.-A. Gardner, M. Parizeau, and C. Gagné, "DEAP: Evolutionary Algorithms Made Easy," *J. Mach. Learn. Res.*, vol. 13, pp. 2171–2175, Jul. 2012.

[9] R. Valckenborg, W. van der Wall, W. Folkerts, J. Hensen, and A. de Vries, "ZIGZAG STRUCTURE IN FAÇADE OPTIMIZES PV YIELD WHILE AESTHETICS ARE PRESERVED."

PVOPTI-RAY: OPTIMISATION OF REFLECTING MATERIALS AND PHOTOVOLTAIC YIELD IN AN URBAN CONTEXT

Michael Revesz [1,2], Sandro Oswald [2], Heidelinde Trimmel [2], Astrid Schneider [1], Philipp Weihs [2], Shokufeh Zamini [1], Martin Peyerl [3], Stefan Krispel [3]

[1] AIT Austrian Institute of Technology GmbH, Center for Energy, Photovoltaic Systems, Giefinggasse 2, 1210 Vienna, Austria
[2] University of Natural Resources and Life Sciences, Department of Water-Atmosphere-Environment, Institute of Meteorology, Gregor-Mendel-Straße 33, 1180 Vienna, Austria
[3] Smart Minerals GmbH

ABSTRACT: Within the project "PVOPTI-Ray", the effect of increasing the albedo of streets and building façades on the yield of building integrated photovoltaic (BIPV) systems, the comfort of pedestrians and the air temperature in the canyon were investigated. The aim was to find an optimised solution (with regard to the structure of an urban canyon and properties of surface materials on streets and buildings), where the yield of directly and indirectly integrated Photovoltaic (PV) is increased and the comfort of pedestrians is maintained within defined limits. Making use of reflected irradiation can improve the yield of BIPV systems on façades and increase their attractiveness inside cities. An existing software tool had to be improved regarding the ground reflected irradiance. Simulations show that increasing the ground albedo from 0.13 to 0.56 can increase the PV yield by around 25 % on building façades over one year. Measurements at increased ground albedo of 0.75 for a limited area of only ~20 m^2 in front of PV modules showed an increased PV yield by ~13 % during a clear sky day. The thermal comfort of pedestrians is not deteriorated by the installation of PV on building façades. However, larger albedo increases the heat stress in summer.

Keywords: Building Integrated PV (BIPV), Ground Albedo, Urban, PV Yield Modelling, Human Comfort

1 INTRODUCTION

While research in Photovoltaic (PV) and urban climate modelling is already quite advanced, the interaction of Building Integrated Photovoltaic (BIPV) in cities and urban climate has not yet received the required attention. In addition, city planners search for solutions to increase renewable energy production inside cities. Besides roofs, also building facades can be used for the installation of PV systems. Making use of reflected irradiation can improve the yield of BIPV systems on façades and make them a more attractive solution inside cities.

Within the project "PVOPTI-Ray", the effect of increasing the albedo of streets and building façades on the yield of BIPV systems and the thermal comfort of pedestrians was investigated (see the mutual influences in Fig. 1). The aim was to find an optimised solution (with regard to the structure of an urban canyon and properties of surface materials on streets and buildings), where the yield of directly and indirectly integrated PV is increased and the comfort of pedestrians is maintained within defined limits. The latter is very important in case of urban heat island effects, since additional heat stress on the human body should be avoided. While significantly larger yield could be achieved by using highly reflecting concrete for streets instead of dark asphalt, this gain could be at the cost of human comfort. Increased reflections of visible irradiation onto the human body could lead to both, thermal and visual stress [1, 2].

For this case study an improved model, with respect to the estimation of ground reflected irradiance, was required. Further, it was desired to use tools with comparably low computation time.

In the following, the main idea of the improved model and the results concerning the effect on PV yield for selected scenarios are presented. While the focus of this work is the PV yield on building facades, few results regarding the thermal stress on pedestrians will be presented in order to raise awareness of the implications.

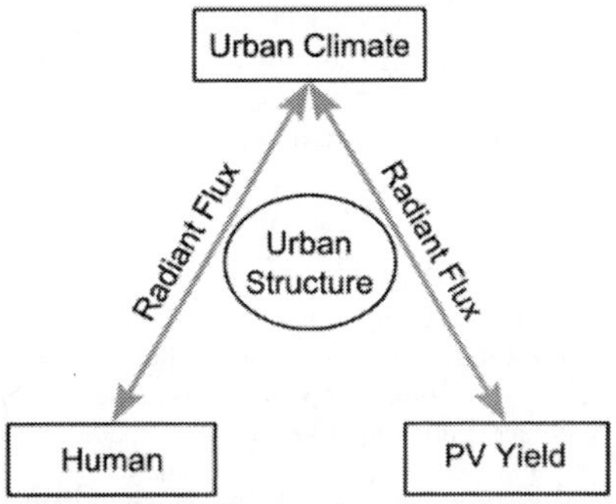

Figure 1: Mutual influences investigated in the project "PVOPTI-Ray".

2 MODEL ADVANCEMENT

In order to investigate and simulate the impact of ground reflected irradiance on the PV yield in an urban environment an adequate model is required. Raytracing methods might be quite accurate. However, computation time is high and therefore not suitable for simulating urban districts. Suitable options are GIS-based methods. GIS data is available for bigger cities and raster-based calculations can be performed rather fast [3].

For the estimation of PV yield on the building envelope, the open-source software SEBE (Solar Energy on Building Envelopes) was adopted [3]. SEBE is part of the simulation package UMEP (Urban Multi-scale Environmental Predictor), which is used for urban environmental simulations and uses digital surface models (DSMs) as input data for urban areas [4]. Since SEBE is used to estimate the solar energy on building surfaces, it was extended with a simple PV model, by Huld et al., for PV yield estimation [5, 6].

As it was the goal to estimate the change in PV yield on buildings in an urban canyon depending on the albedo of the ground surfaces, the estimation of ground reflected irradiation needed to be revised. The usual approach is very simplified (see Eq. (1)) and based on the assumption that the inclined surface receives ground reflected irradiance (G_{gr}) from a surface of infinite extent:

$$G_{gr} = G_0 \alpha \frac{1}{2}(1 - \cos(\beta)) \quad (1)$$

where G_0 is the global horizontal irradiance, α is the ground albedo and β is the receiving surface's inclination [7]. This method is justified for the reason that usually the albedo is rather low and on a surface with the usual tilt the last term in Eq. (1) is small as well. Thus, the contribution of ground reflections to the total irradiation is very small, more detailed models are not required and multiple reflections are neglected [8].

However, this does not hold for an urban canyon. For a vertical façade the last term of Eq. (1) becomes largest. On the other hand, opposite buildings block the view onto an infinite ground plane. Since it was intended to estimate the potential influence of different ground albedo on the yield of BIPV on facades inside an urban canyon, better models are required and can be justified.

The proposed improvement can be applied, because the simulation tool in use is working in a raster space. The new method for the ground reflected irradiance is to sum over visible ground sections as in Eq. (2):

$$G_{gr} = \frac{G_0}{p} \alpha \sum_{i}^{p} P_i \quad (2)$$

where p is the number of patches of the possibly visible hemisphere for a wall pixel and P_i is the wall-hemisphere patch from which direction ground pixels are visible. The latter represents a boolean value, 1 for visible to the wall pixel and 0 for invisible.

Nevertheless, even with this approach and the given tool, some simplifications have been required. For one, shadowed ground pixels contribute with reflectance of the full horizontal irradiance as if not being shadowed. On the other hand, reflections from opposite building walls are neglected. Further, it was not yet possible to implement the angular view factor between receiving wall pixel and reflecting ground pixel. Still, this approach is a better estimation compared to the use of Eq. (1).

3 URBAN SCENARIOS

For investigating the gain of PV yield on a building façade, the following scenarios and parameters were considered:

- Basic geometric urban structures:
  - Street/intersection, square.
  - Typical wall-height to street-width.
- Most common or promising PV-module types (crystalline Si, CI(G)S).
- Different materials for pavement and façades (distinguished by albedo and emissivity).

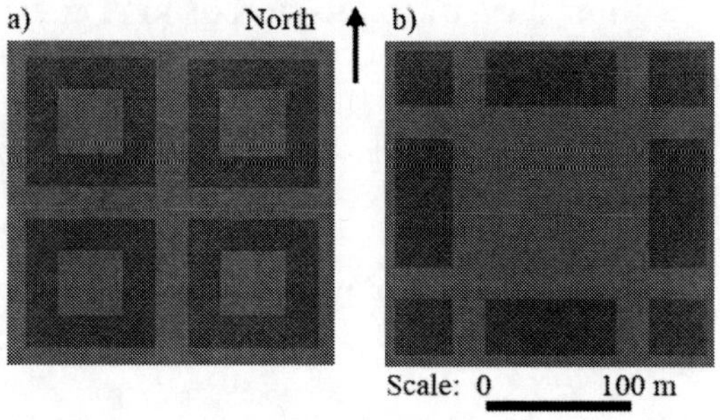

Figure 2: Land cover scheme of two scenarios, a) street canyon with intersection, b) buildings surrounding a small square. Grey: street, red: buildings. The street width is 20 m, building height 20 m.

The emissivity is not relevant for PV yield simulations, but used for the simulations related to urban climate. The scenarios were mainly based on typical urban canyons in Vienna. For each scenario different combinations of pavement and building materials with different properties (i.e. albedo and emissivity) were simulated and compared. Since it was not possible to implement ground areas with different values of albedo, the effect of vegetation as a ground surface is not simulated. Further, measurements at real sites are used for a comparison with the modelling results.

The two urban structures presented here represent a street canyon and an intersection at the centre (Fig. 2a) and buildings surrounding a smaller sized square (Fig. 2b). In both cases, a flat roof is assumed with a building height of 20 m. The street width is 20 m as well.

For each building scenario the ground albedo was set to 0.13 (typical for asphalt streets) and for comparison to 0.56 (possible with very bright concrete). The PV yield was estimated under those conditions for the PV technologies of crystalline silicon and CI(G)S. A hot clear sky day (09. June 2017) and a cloudy day (16. June 2017) were selected for the assessments. In addition, the PV yield over one year (1. September 2016 – 31. August 2017) was estimated. All weather data was measured in Vienna.

4 RESULTS

4.1 Model improvement

For the comparison of the original algorithm for calculating the irradiation on building walls with and the new algorithm to calculate the irradiation on building walls, the following situation is assumed: The two scenarios are A (street canyon, see Fig. 2a) and B (square, see Fig. 2b). The irradiance data used is arbitrarily chosen and for 1 hour around 10:00 CET at a clear sky day (09. June 2017) in June in Vienna. Table I shows the summary of estimated irradiations for a south and east facing wall. In all cases the average of the top 6 meters is taken.

For the two selected wall points, the irradiation predicted with the original algorithm is 7 – 9 % larger than with the new algorithm. Further, one can see that the irradiation estimated by the old algorithm is equal independent of the urban structure, i.e. street canyon or square. This shows that the area of ground, which is visible to each wall segment, is not taken into account. In addition, the irradiation on a wall decreases along its height as it is expected.

Table I: Comparison of the old and new algorithm for estimation of irradiation at selected wall points for 1 hour around 10:00 CET of a clear day in June.

Scenario	Wall orientation	old [kWh/m^2]	new [kWh/m^2]
B (square)	East	0.590	0.551
	South	0.390	0.359
A (street)	East	0.590	0.531
	South	0.390	0.340

Nevertheless, this improvement comes with an increase in computation time. The implementation of the PV model increased the computation time by a factor in the magnitude of the number of meteorological data points. In addition, the calculation time is increased by the better ground reflection algorithm.

4.2 PV yield increase

It is obvious that the PV yield increases if the ground reflected component is increased artificially, by altering the ground surface materials or even the facades to brighter colours. However, it is interesting to know quantitative by how much the PV yield could be increased in an urban canyon. In this chapter, this gain is presented for the two defined scenarios. Fig. 3 and Fig. 4 show the time series of the irradiance, which are used for the estimation of PV yield. Both represent the chosen days in Vienna, one cloudy day (16. June 2017) and one clear day (09. June 2017) respectively.

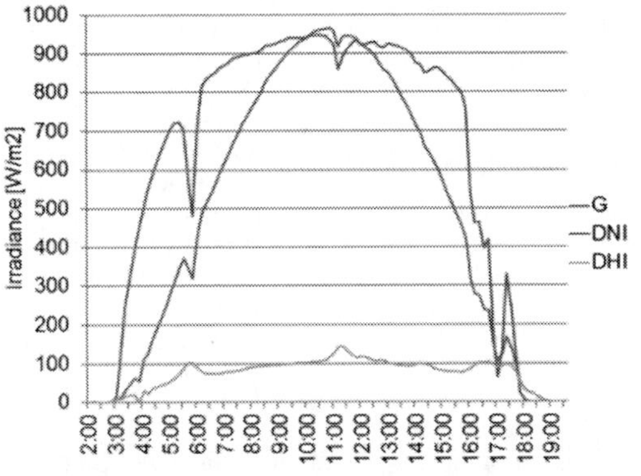

Figure 3: Irradiance profile in W/m^2 for a **clear** day used for estimations. Global horizontal irradiance (G) and direct normal irradiance (DNI) are measured values, diffuse horizontal irradiance (DHI) is estimated.

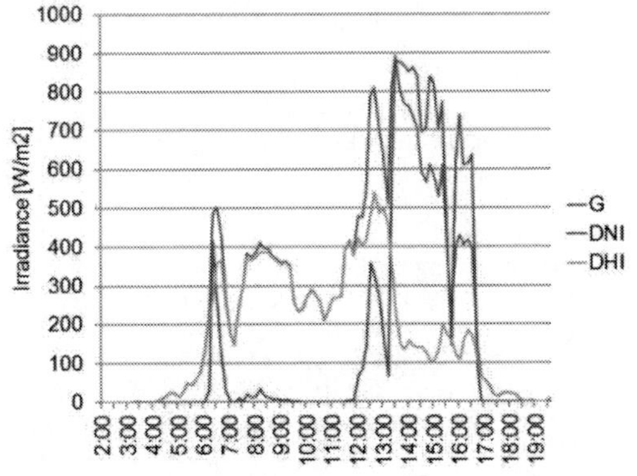

Figure 4: Irradiance profile in W/m^2 for a **cloudy** day used for estimations. Global horizontal irradiance (G) and direct normal irradiance (DNI) are measured values, diffuse horizontal irradiance (DHI) is estimated.

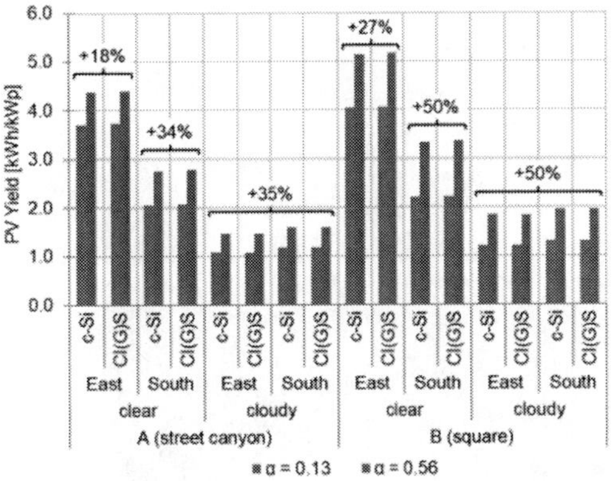

Figure 5: PV yield during one day in summer for each case of building scenario, ground albedo, weather condition and PV technology. Red: south facing wall, blue: east facing wall.

Table II: PV yield [kWh/kWp] on selected wall segments/orientations for 1 year of weather data in scenario B. Varied parameters: PV cell technology and albedo.

	α = 0.13		α = 0.56	
technology	**south**	**east**	**south**	**east**
c-Si	698	583	872	755
CI(G)S	706	594	879	765

During both selected days, the weather in the afternoon differs from the dominating weather condition. I.e. the afternoon of the clear sky day becomes cloudy and the cloudy day becomes scattered. Therefore, the results for the east facing wall represent the main weather condition of each day the best. In contrast, the south facing wall is exposed to a mixture of sky conditions and the corresponding results might be misleading in regards to the weather condition.

The results shown in Fig. 5 reveal the following findings: In general, it is of advantage, if a PV-façade installation is mounted on buildings at a square compared to at streets. In case of the chosen scenarios the much bigger ground surface of the square causes around double the gain compared to the street.

Further, during clear sky the gain by ground reflections is rather small compared to cloudy sky. In specific, the increase of ground albedo from 0.13 to 0.56 for a wall along a street results in a larger yield of around 18 % for the clear sky day. In contrast, for the cloudy sky condition the gain is even 36 %. This is clear as at clear sky the irradiance directly from the sun is much more dominant compared to the reflected components of irradiation. Despite the larger relative gain at cloudy sky, the absolute yield is very low. The south wall shows almost equal relative gains independent of the sky condition. This is because during each of the chosen days both weather conditions occur.

Simulations over one year of data show with increased albedo a gain of around 24 – 29 % for both considered PV technologies. For a south facing wall, the

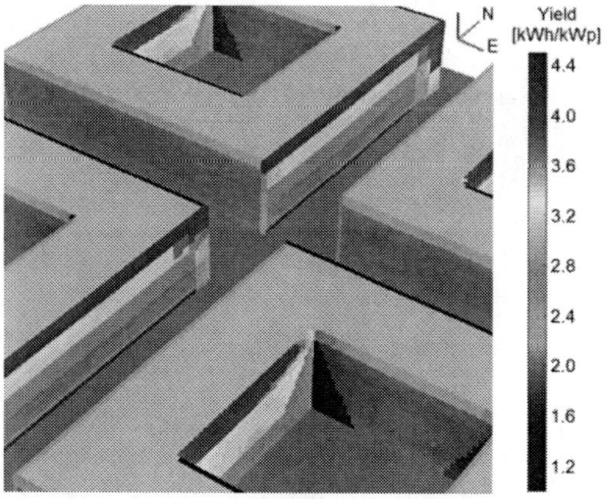

Figure 6: PV yield on façades, ground albedo is 0.13; simulated for the clear day, c-Si technology. The roof and ground are coloured in grey.

Figure 7: Gain of PV yield as a result of increasing the ground albedo from 0.13 to 0.56; simulated for the clear day, c-Si technology. The roof and ground are coloured in grey.

gain is slightly lower than for other orientations towards east or west. The absolute PV yield values for each case chosen are shown in Table II.

In case of the one-day course, the two PV cell technologies show very similar behaviour independent of the sky condition. Over a period of one year, CI(G)S technology has also no significant advantage over crystalline silicon technology when used as a façade integrated installation. Therefore, no technological preference for a PV installation in an urban environment on façades is given.

Fig. 6 shows the PV yield, in the case of a ground albedo of 0.13, on façades in an urban canyon. Fig. 7 shows the gain of yield as a result of increasing the albedo from 0.13 to 0.56. Nevertheless, for the decision where on a building a PV plant should be installed and where not, it remains important to assess not only the gain, but also the yield.

Measurements at a site which is similar to scenario A (street canyon) were taken. There the effect of a highly reflecting area in front of irradiance sensors and PV panels was tested. The street (asphalt) has an albedo of 0.13. For the measurement with higher albedo a white surface covering 29 m^2 (max. 9m x 3.7m), made of white painted panels, have been installed. The measured albedo was 0.75. For a clear sky day in August, the white area caused an increased yield of ~13 % during most of the day. Only in the morning and the afternoon the yield dropped due to shadowing. Given, that the measurement plane was facing to the south and the white surface is rather small compared to the total street surface (370 m^2 central, plus 200 m^2 on sides), the simulated results seem plausible.

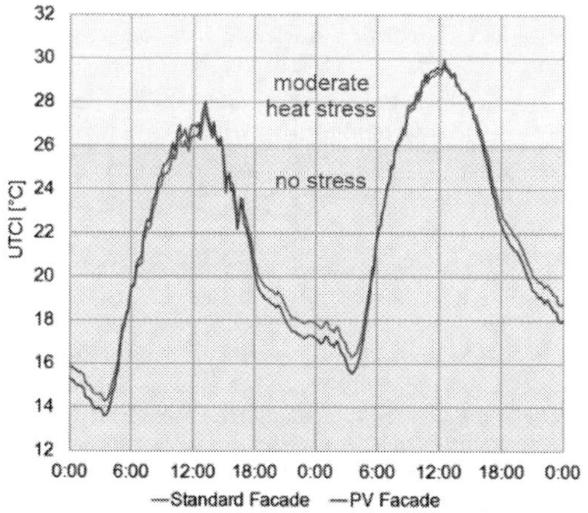

Figure 8: Thermal stress: UTCI for two consecutive hot summer days (07.-08. Aug. 2016, Vienna). Comparison of standard thermal insulated versus PV façade.

Concerning the optimization of surfaces inside an urban canyon regarding maximum PV yield, one can say: For the optimum scenario, all surfaces should be reflecting 100 %. However, the optimum is limited by the technical feasibility to produce surfaces as well as the human comfort. The latter concerns mainly the visual comfort due to glare, which was not assessed in this study. Another important criterion is the thermal comfort and the climatic effects.

4.3 Human comfort

One measure for the thermal comfort of humans is the so called Universal Thermal Climate Index (UTCI). This measure is influenced mainly by the air temperature, wind speed, humidity, the visible part of irradiation and the infrared portion of the irradiance. The UTCI is given in units of °C and representing the air temperature of a reference condition which causes the equivalent thermal stress [9].

Fig. 8 shows simulations, using TEB [10], and comparing building façades with thermal insulation as it is usual standard in Austria versus an insulated façade with BIPV installation. The two days shown are hot summer days (07. – 08. August 2016) in Vienna. While during the day the UTCI is slightly increased with PV installation, during night time the UTCI is around 1°C lower. This is mainly because of the relatively large emissivity of the PV module.

In contrast, increasing the ground albedo in an urban canyon has a larger effect in the thermal comfort of pedestrians (see Fig. 9). The higher amount of irradiance reflected onto the human body is increasing the UTCI, especially at clear sky condition in the direct sun. In the graph two times per day the UTCI changes immense: At around 6:00 in the morning, when a building shadow passes the test point and exposes it to the direct sun. And

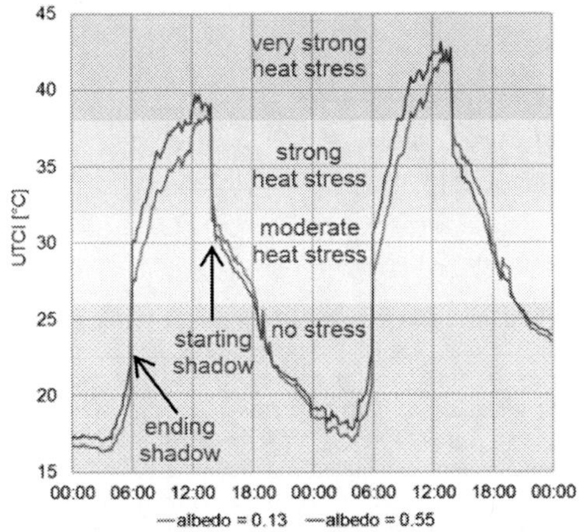

Figure 9: UTCI for two consecutive hot summer days (19.-20. June 2017, Vienna). Comparison of ground surface albedo of 0.13 versus 0.55.

later, at around 14:30, the test point lies in shadow again. However, for the following afternoon hours the UTCI for the darker ground surface is larger than for the brighter surface. The darker surface is heated up more and is

therefore emitting more infrared radiation. Once the darker surface has cooled down enough, the UTCI is again lower compared to the brighter surface.

It was not possible to assess the visual comfort for humans in the situation of high reflecting surfaces. However, it is expected that the visual stress will increase. Though, the extent of disturbance remains un-quantified.

5 CONCLUSIONS

It was found that it is technically possible and feasible to produce street surfaces with an albedo of up to 0.56. The effect of such higher reflecting surfaces in an urban canyon on the PV yield as well as the human comfort, compared to usual ground surfaces with an albedo of 0.13, was investigated.

Inside an urban canyon with a building-height to street-width ratio of 1:1 (20 m:20 m), which is typical for e.g. Vienna, the increase of ground albedo from 0.13 to 0.56 is expected to increase the PV yield by around 18 % for an east facing wall under clear sky condition. Over one year under typical Viennese weather conditions it was estimated, that the PV yield would be around 24 – 29 % larger for the same increase of reflectivity.

Therefore, it can be concluded, that BIPV on façades in an urban environment can significantly benefit from optimizing the albedo of surrounding surfaces. Further, it was concluded, that thinfilm PV technologies such as CI(G)S based ones have no benefit compared to crystalline silicon technologies under the given location and climatic condition.

Simulations of the human comfort showed, that the assessment of thermal stress on a pedestrian passing a BIPV installation and walking on a higher reflecting pavement is complex. While the thermal stress during a hot summer day is not much increased, different changes to the urban landscape have negative, but also positive effects. Many aspects, (e.g. albedo, thermal storage or shading), but also the climatic condition have to be considered for a well-made decision.

For a more accurate assessment of the PV yield it is suggested that the algorithms for estimating reflected irradiance will be further improved. E.g. the view-factor and the shaded ground surfaces should be considered in the model. Nevertheless, this case study shows that it is reasonable to install BIPV on façades in cities.

6 ACKNOWLEDGEMENT

The research leading to these results has received funding from the energy research programme (Stadt der Zukunft 3. Ausschreibung 2015) under grant agreement number 854691.

Many thanks to Fredrik Lindberg for his support with UMEP.

7 REFERENCES

[1] H. Lee, H. Mayer, D. Schindler, Importance of 3-D radiant flux densities for outdoor human thermal comfort on clear-sky summer days in Freiburg,Southwest Germany, Meteorologische Zeitschrift 23 **3** (2014) 315–330.

[2] R. Compagnon, J. Goyette-Pernot, Visual Comfort in Urban Spaces. In: RUROS – Rediscovering the Urban Realm and Open Spaces, Co-ordinated by CRES, Department of Buildings. http://alpha.cres.gr/ruros/ (2004)

[3] F. Lindberg, P. Jonsson, T. Honjo, et al., Solar energy on building envelopes – 3D modelling in a 2D environment, Sol. Energy 115 (2015) 369–378.

[4] F. Lindberg, C.S.B Grimmond, A. Gabey, et al., UMEP - An integrated tool for city-based climate services. Submitted to Environmental Modelling and Software. (2017)

[5] T. Huld, G. Friesen, A. Skoczek, et al., A power-rating model for crystalline silicon PV modules, Sol. Energy Mater. Sol. Cells 95 **12** (2011) 3359–3369.

[6] T. Huld, R., Gottschalg, H.G., Beyer, et al., Mapping the performance of PV modules, effects of module type and data averaging, Sol. Energy 84 **2** (2010) 324–338.

[7] H. Häberlin, Photovoltaik - Strom aus Sonnenlicht für Verbundnetz und Inselanlagen, VDE Verlag (2007).

[8] L. Eduardo, Energy Collected and Delivered to PV Modules, Handbook of Photovoltaic Science and Engineering, Wiley, 2nd Ed., pp. 984–1042.

[9] P. Bröde, D. Fiala, K. Blazejczyk, et al., Calculating UTCI Equivalent Temperatures, 13th Intl. Conf. on Environmental Ergonomics (2009)

[10] V. Masson, A physically-based scheme for the urban energy budget in atmospheric models, Bound-Layer Meteorol. 94 (2000) 357–397.

VALIDATION OF A FAÇADE PV POTENTIAL MODEL BASED ON LIDAR DATA

S. Freitas[1], M.C. Brito[1]
[1]Instituto Dom Luiz – Faculdade de Ciências, Universidade de Lisboa
Faculdade de Ciências da Universidade de Lisboa, Ed. C8 (8.3.33), Campo Grande, 1749-016 Lisboa, Portugal.
e-mails: srefreitas@fc.ul.pt, mcbrito@fc.ul.pt

ABSTRACT:
Cities have huge potential for PV applications, namely in the façades of buildings that feature high available surface area. Thanks to the advancements in computational technologies, full 3D calculations of solar radiation over Digital Surface Models of urban areas are now possible and the solar potential of façades can finally be quantified. The use of airborne LiDAR data has become more popular for this purpose, since it is an easy method of data acquisition for large areas. In this work, the solar potential model *SOL*, a LiDAR-based algorithm able to compute hourly solar irradiation for all the points of a DSM, was validated against electricity production records from a façade integrated PV system installed in the Solar XXI building, in Lisbon. Results point out that *SOL* estimations fairly agree with the measurements, featuring nMBE ranging from around 8% to 51%, nMAE from 20% to 58% and nRMSE from 29% to 81% - higher values for winter time. Most of these deviations are attributed to fine scale limitations on the rendering of shadows since the DSM only has a $1m^2$ resolution but results highlight the need for improvement of the diffuse radiation model, currently isotropic, which should portray more accurately the circumsolar and reflected components and the anisotropy characteristic of semi-cloudy sky conditions.

Keywords: solar potential, LiDAR, building façade, BIPV

1. INTRODUCTION

The solar potential of rooftops has been thoroughly explored in the literature and put into practice in the last two decades: the presence of PV panels and/or thermal collectors on rooftops of buildings has become very common and part of many modern cityscapes. Rooftops can be flat or pitched, which allows for the installation of solar panels in a way that the electricity yield is optimized. However, as cities grow vertically, the available area in rooftops that could host solar energy devices becomes scarcer. Centralized air conditioning units are also usually placed in the rooftops, occupying a large portion of free area. It is, therefore, evident that the vertical surfaces of buildings should not be overlooked in spite of their non-optimal inclination. Façades feature a larger available area than rooftops and face at least 4 different orientations, which makes them interesting for accomplishing better electricity demand-supply matches thanks to their off-peak production [1].

The study of solar potential of façades for electricity generation goes back almost two decades [2], with the Mataro public library [3] being one of the earliest examples of building integration PV, from 1998, but only recently it could be numerically quantified [1]. Thanks to the developments in computational techniques and technology, several models have been created to evaluate the solar resource and energy potential in the urban context. The most sophisticated algorithms are able to perform full 3D operations, essential for the assessment of solar irradiation in façades. One key element in these models consists on the digital representation of the buildings, and optionally trees and other urban equipment, which depends on the available data. Among other possibilities [4], vector footprints are usually used, after being extruded to the heights of the respective buildings, to create simple geometries that can later be enhanced with further details. This approach is mostly adopted when simulating solar potential with energy simulation plug-ins or scripted components within: i) Geographic Information Systems (GIS), such as gvSIG [5] or ArcGIS [6,7]; ii) Computer Aided Design (CAD), such as ImageModeler [8], Rhinoceros [9,10], Open Studio [11], PPF [12]; Design Builder [13]; or iii) dedicated urban simulation platforms/models such as CitySim [14] and SimStadt [15] with CityGML models [15–17].

Another option that has become popular in studies of large urban areas is the use of airborne LiDAR data [18]. The resulting 3D point cloud entails a large set of points with assigned x, y and z coordinates that can be used for energy simulation as it is or manipulated: resampled to a lower resolution, input to produce surfaces by Delaunay triangulation, etc. Several of the software mentioned previously can handle point cloud data sets, but some algorithms use the data to run solely/mostly on numerical computation programs such as MatLab [13,19–21]. One of these tools is the *SOL* algorithm [21], developed in the Faculty of Science, University of Lisbon, and the subject of the validation presented in this paper.

2. METODOLOGY

The validation of the model SOL was done using measured data from a PV system integrated in the south-facing façade of the Solar XXI building (Fig. 1, top), located in Lisbon (38.7723037 N, 9.177954 W). The PV system has a peak power of 12.16kW provided by 76 p-Si modules connected to 3 inverters. The modules are installed vertically with an air gap between them and the wall, to provide cooling or heating to the indoor spaces [22].

The experimental dataset comprises produced energy (Wh), power (W), continuous (DC) and alternating (AC) voltage and current grouped by inverter (Fig. 1, bottom), module and ambient temperature (°C) and irradiance on the vertical plane of the façade (W/m^2), all with a time step of 15min. Only the records associated to the months of June and November of 2012 and 2013 were provided for this study.

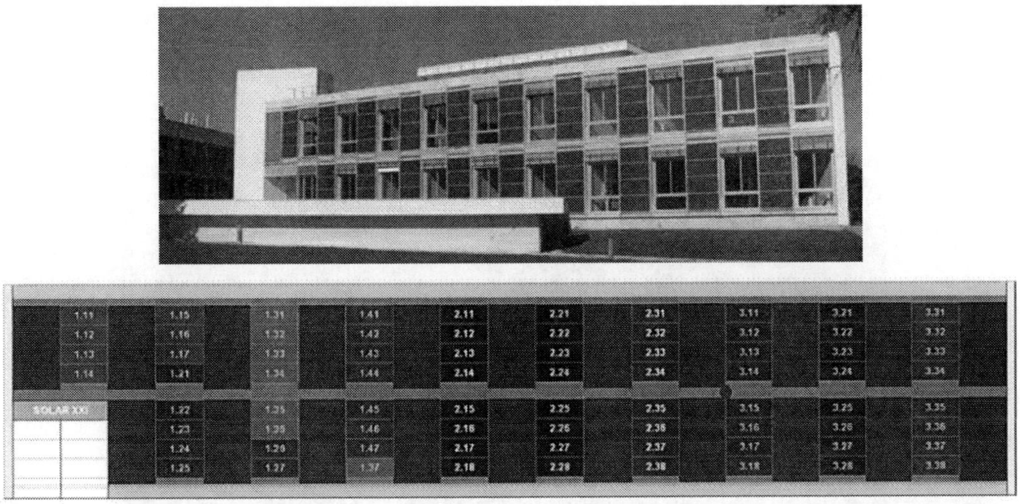

Fig. 1 – Façade integrated PV system in the Solar XXI building (top), distribution of modules in the façade and interconnections of the strings to the respective inverter (bottom) [22], and the location of the radiation sensor (red dot).

Hourly global and diffuse horizontal irradiation (Wh/m^2) measured on a nearby weather station were used as input to the model SOL to estimate the solar irradiation on a 1m^2 resolution Digital Surface Model (DSM) of the area. A sky vault with 400 divisions was used instead of the original 1081 to calculate diffuse irradiation, as discussed in [23]. Results were produced in an hourly time step excluding night time. Then, the façade points data were manually extracted (Fig. 2) and those that are in the approximate position of a module were assigned the respective number considering the arrangement in Figure 1.

Using the *SOL* output (solar radiation for each module) the hourly PV production of each module (PV_m) was estimated through the following expression [24]:

$$PV_m = P_{ref}\frac{Dir_m AL + Dif_m}{G_{ref}}\left[1 + \gamma\left(T_m - T_{ref}\right)\right] \quad (1)$$

where P_{ref}=160W is the nominal power of the PV module, Dir_m and Dif_m are respectively the direct and diffuse vertical irradiance estimated in the module m's position, and AL is a fraction to account for the angular losses due to the effect of the angle of incidence in the direct component of solar radiation, G_{ref}=1000W/m^2, T_m is the estimated module temperature, T_{ref}=25°C and γ is the power correction factor for PV module temperature.

One can set γ=0 if the effect of temperature is to be neglected and γ=-0.5% [25] when it is considered. The hourly temperature of the PV modules was calculated according to

$$T_m = T_a + \frac{NOCT - 20^oC}{800\ \mathrm{W/m^2}} GVI_m \quad (2)$$

where T_a is the ambient temperature and $NOCT$ is the nominal operating cell temperature, which was reported to be higher (=56°C [22]) than the manufacturer data (47°C [25]) due to the integration in the façade.

As for AL it can be estimated as per [26] through:

$$AL = \frac{1 - e^{-\frac{\cos\theta}{a_r}}}{1 - e^{-\frac{1}{a_r}}} \quad (3)$$

where θ is the hourly angle of incidence of the sun rays in the plane of the façade (output by the model *SOL*) and a_r=0.159 is an empirical angular losses coefficient for p-Si modules [26]. AL can be set to 1 if the angle of incidence effects are neglected.

The angle of incidence effect for the diffuse irradiation was disregarded since it is more significant when PV module tilt angles are small [27].

Considering the string arrangement in Figure 1, the hourly PV production by inverter (PV_i) was determined considering that a whole string (S) production is dominated by the module that produces less (Eq. 4). This conservative method seeks to portray the partial-shading effect in the strings.

$$PV_i = \eta_i \sum_{1}^{Ni} min\left(PV_{m,Si}\right) n_{Si} \quad (4)$$

where η_i=94% [22] is the inverter DC/AC efficiency, Ni is the total number of strings connected to inverter i, $PV_{m,Si}$ is the production of module m belonging to string S connected to inverter i and n_{Si} is the total number of modules in that string.

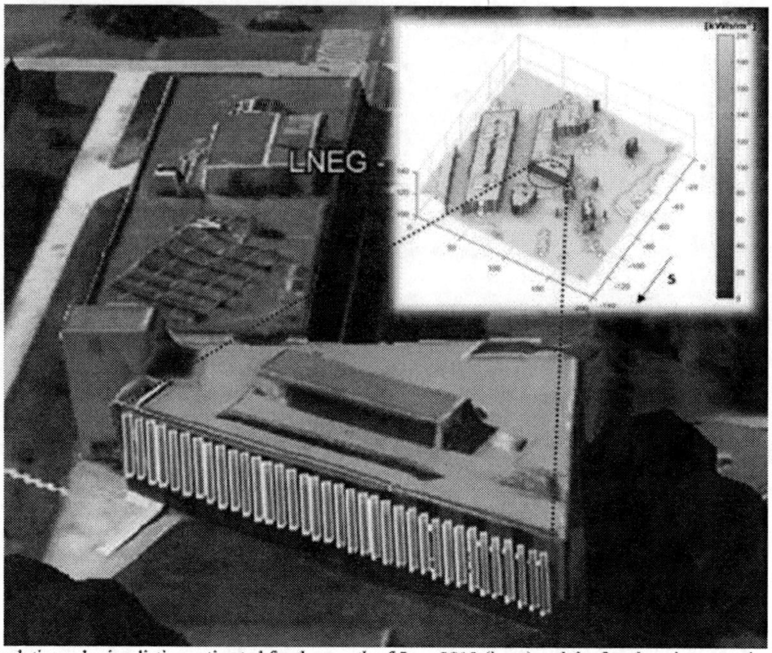

Fig. 2 – Cumulative solar irradiation estimated for the month of June 2012 (inset) and the façade points superimposed over a google earth view of the building.

The PV production estimated thought the model SOL and Equations 1-4 was performed using three different approximations: 1) without considering the temperature and incidence angle losses; 2) considering the effect of temperature but no incidence angle losses; and 3) with both. Results from the three approaches were compared against the measurements and performance errors calculated: the mean bias error (MBE, Eq. 5), which measures the systematic deviations; the mean absolute error (MAE, Eq. 6), which quantifies negative and positive deviations equally; and the root mean square error (RMSE, Eq. 7) that weights the larger deviations. To better understand the magnitude of these errors, they were normalized by the measured PV power.

$$nMBE = \frac{\frac{1}{H}\sum_{h=1}^{H}\left(PV_{i(h)} - PV_{i,meas(h)}\right)}{\frac{1}{H}\sum_{h=1}^{H} PV_{i,meas(h)}} \qquad (5)$$

$$nMAE = \frac{\frac{1}{H}\sum_{h=1}^{H}\left|PV_{i(h)} - PV_{i,meas(h)}\right|}{\frac{1}{H}\sum_{h=1}^{H} PV_{i,meas(h)}} \qquad (6)$$

$$nRMSE = \frac{\sqrt{\frac{1}{H}\sum_{h=1}^{H}\left(PV_{i(h)} - PV_{i,meas(h)}\right)^2}}{\frac{1}{H}\sum_{h=1}^{H} PV_{i,meas(h)}} \qquad (7)$$

H is the total number of hours and $PV_{i(h)}$ and $PV_{i,meas(h)}$ are respectively inverter i's estimated and measured output in hour h.

3. RESULTS

For the sake of readability, this section will only present and discuss the results for June and November of 2012. Results for 2013 are qualitatively similar.

In Figure 3 the estimated PV production is compared against the experimental measurements. It can be observed that electricity yields reach higher levels in the winter month, since lower sun heights allow the sunrays to reach the façade closer to the normal. However, the level of scattering is lower in the summer month, probably due to fewer cloud cover events. The model generally tends to overestimate PV generation in both months.

When temperature effects are considered (Fig. 3 middle), some points are drawn closer to the 1:1 relationship line, particularly those representing greater levels of production. Improvements of the same magnitude are visible when the angle of incidence losses are considered (Fig. 3, bottom), although they are barely visible for November due to the smaller angles of incidence.

Inverter 3's yields seem to reach lower levels than the other Inverters, which might be because of shadows cast by a group of trees located near the easternmost part of the façade.

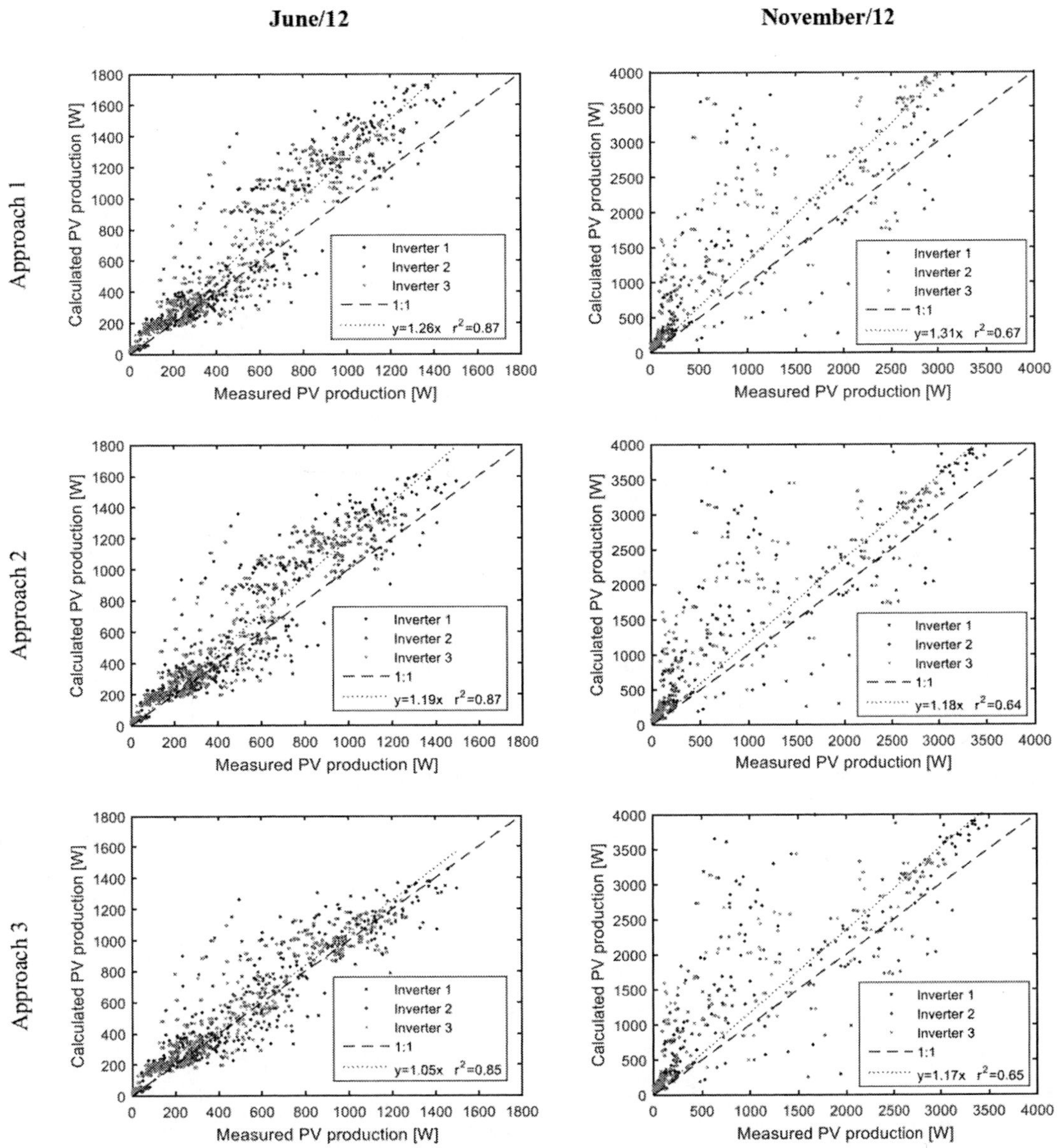

Fig. 3 – Comparison between the calculated and measured PV production by inverter, for June (left) and November (right) of 2012: Approach 1 (top); Approach 2 (middle); and Approach 3 (bottom).

The improvements from Approach 1 to Approach 3 can be analysed in more detail in Figures 4 and 5, which represent the hourly relative deviation. It is clearer that in June the model overestimates the production in the morning, around noon and early evening, especially for Inverter 3, whose strings are under partial shadow conditions more often than the others. The earlier and later hours of the day feature low levels of solar radiation which makes the modelling of PV production more challenging. Modelling temperature losses lead to slight improvements but the impact of incidence angle losses approach zero, for most of the boxes and median lines. This is not so visible for November, as the range of the vertical axis units is larger.

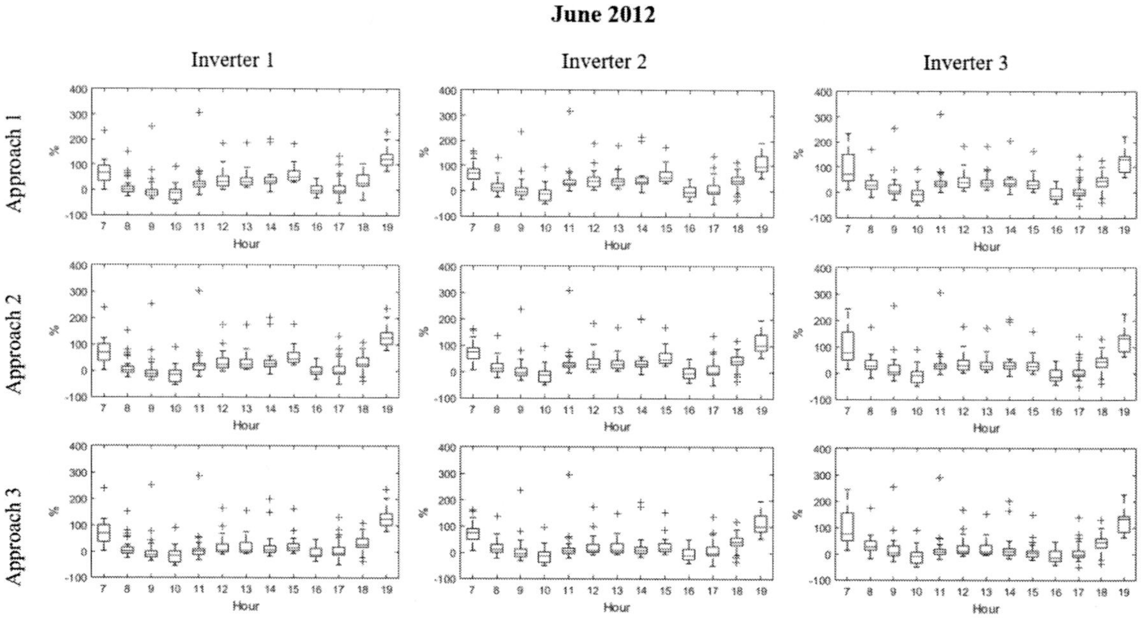

Fig. 4 – Boxplot representation of the hourly relative deviation in June/12, for the 3 Inverters (columns) and through the 3 Methods (rows). (The tops and bottoms of each "box" are the 25th and 75th percentiles, the red line in the middle of each box is the median, the black lines extending above and below each box are the extreme values and the red crosses are the outliers.)

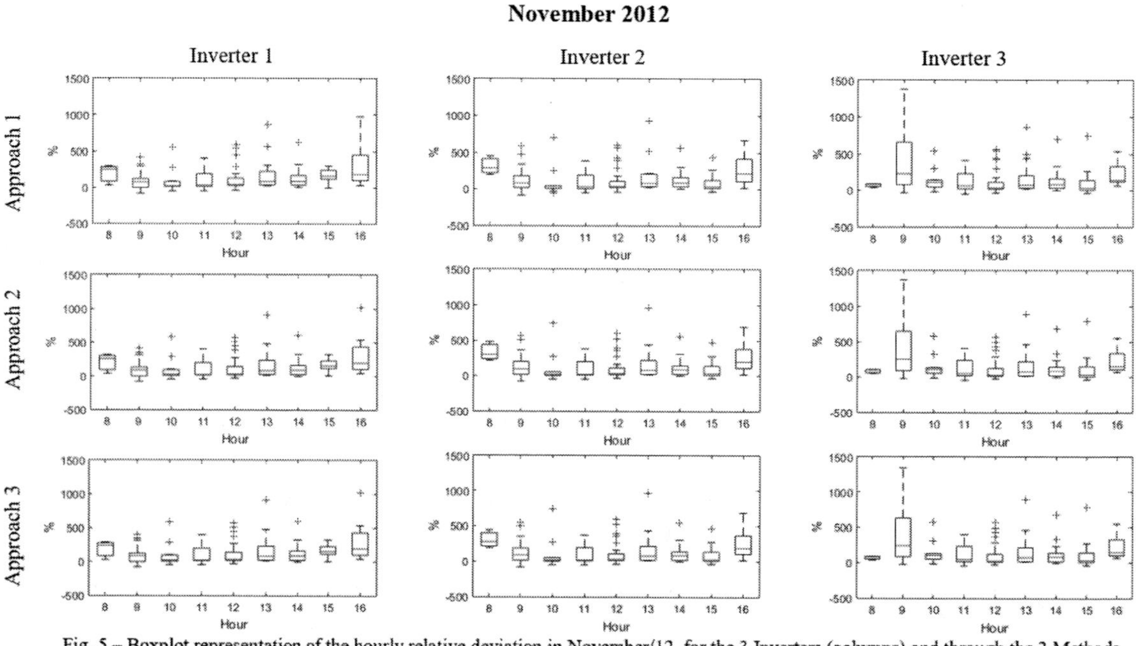

Fig. 5 – Boxplot representation of the hourly relative deviation in November/12, for the 3 Inverters (columns) and through the 3 Methods (rows). (The tops and bottoms of each "box" are the 25th and 75th percentiles, the red line in the middle of each box is the median, the black lines extending above and below each box are the extreme values and the red crosses are the outliers.)

To help understand in detail how the model performs in terms of the sky conditions and the sun's position relatively to the façade plane, the absolute deviation between estimated and measured PV production was plotted against the angle of incidence (Fig. 6, left) and against the diffuse fraction (Fig. 6, right). For higher incidence angles, the deviations are small, especially in June when the sunrays are almost parallel to the plane of the façade. This means that the correction described in Equation 3 is being effective at the hours when the sun is positioned in the easternmost and westernmost sides of the façade, around 9am and 4pm (Fig. 4 and 5), and when the sun is almost overhead in June, around 1pm (Fig. 4). On the other hand, it can be verified that the highest deviations in both months correspond to diffuse fractions between 30% and 80%, whilst smaller deviations are predominant for very high diffuse fractions (i.e. overcast days). This is not unexpected since the diffuse radiation model in *SOL* is isotropic, and thus describes better the isotropic conditions of overcast skies, whilst the anisotropy features of a clear sky day are missed (i.e. the proper contribution of the circumsolar radiation and the ground reflected component). However, since the sun is almost overhead at noon in June the high distribution of sky divisions in the zenith seems to portray the circumsolar contribution well, with smaller deviations being observed for diffuse fractions around 20%.

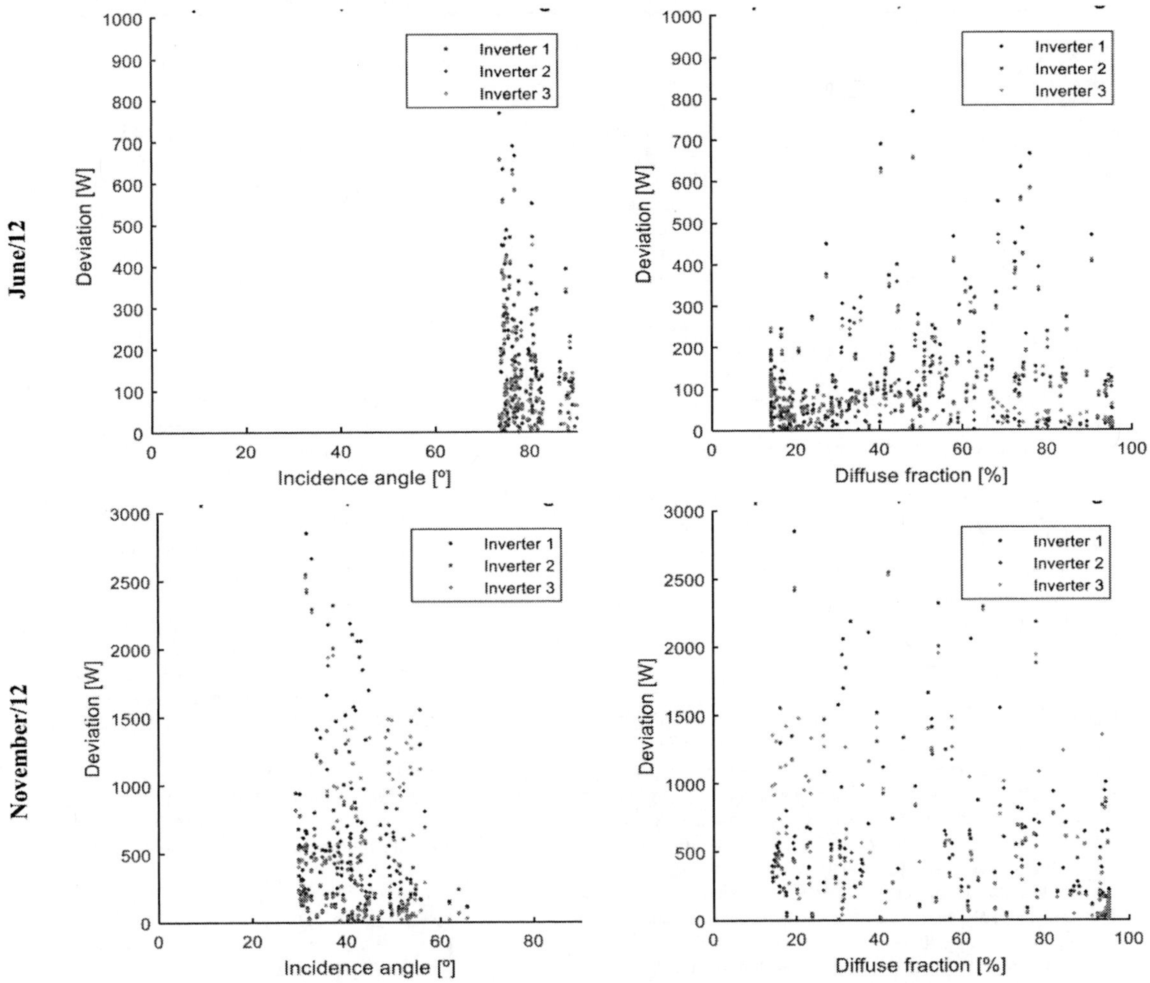

Fig. 6 – Absolute deviation between estimated and measured PV production plotted against the incidence angle (left) and diffuse fraction (right), for June and November of 2012, for Approach 3.

Table 1 – Normalized Mean Bias Error (nMBE), normalized Mean Absolute Error (nMAE) and normalized Root Mean Squared Error (nRMSE) by Inverter, for June and November of 2012, for Approaches 1, 2 and 3.

		nMBE [%]			nMAE [%]			nRMSE [%]		
		Inv 1	Inv2	Inv3	Inv 1	Inv2	Inv3	Inv 1	Inv2	Inv3
1	jun/12	25%	30%	29%	33%	36%	34%	45%	49%	48%
	nov/12	55%	42%	65%	64%	52%	70%	92%	74%	97%
2	jun/12	20%	24%	24%	28%	31%	30%	39%	42%	41%
	nov/12	43%	32%	52%	53%	43%	58%	77%	61%	82%
3	jun/12	7%	12%	12%	20%	20%	21%	29%	30%	30%
	nov/12	42%	31%	51%	52%	42%	58%	76%	61%	81%

Table 1 shows the errors. The consideration of temperature losses (from Approach 1 to 2) leads to an average reduction in the nMBE, for the 3 Inverters, of around less 5% in June and 11% in November. A similar improvement is observed for systematic errors, nMAE decreases 5% and 10%, respectively, and nRMSE 6% and 15%. Considering incidence angle losses (Approach 3) produces almost no changes in November (the errors differ only in 1%), but in June further reductions in the three errors are achieved (less 12%, 9% and 11%).

Considering the full model (Approach 3) the highest errors are registered for Inverter 3, namely in November, whereas Inverter 1 features the lowest errors in June and Inverter 2 the lowest errors in November. As discussed above, Inverter 3 yields are more overestimated due to morning partially shading of its strings, which might not be correctly modelled by *SOL* with a 1m^2 resolution grid. The errors for Inverter 1 and 2 are probably also associated to shading events, in particular due to a small group of trees southwest of the façade and a building to the west, which cast shadows all over the façade in the late afternoon [22]. Nevertheless, part of the error may also result from some outdated features in the DSM, which was created in 2006.

4. CONCLUSION

The 3D solar PV potential model *SOL* was validated against measurements from a system installed in a south-facing building façade in Lisbon. Hourly global and diffuse solar irradiation was calculated for a 1m^2 Digital Surface Model derived from airborne LiDAR data. The irradiation results were used to estimate the temperature of the PV modules and their electricity production, with and without taking into account angle of incidence losses. Estimations fairly agree with the measurements when the effects of temperature and incidence angle are considered. The normalized MBE for all the inverters that constitute the PV system ranges from 7% (31%) to 12% (51%), the nMAE from 20% (42%) to 21% (58%) and the nRMSE is around 30% (61% to 81%) in June (November).

Results highlight the need for finer spatial resolution of the input DSM, which should preferably be recently updated, and the upgrade of the diffuse radiation model in order to better describe the anisotropy of several possible sky conditions, such as [28] or [29].

5. ACKNOWLEDGMENTS

Part of this work was supported by MIT Portugal Program for Sustainable Energy Systems and FCT grant SFRH/BD/52363/2013, Fundação Calouste Gulbenkian 2014 Award and by Projects MITP-TB/CS/0026/2013 (SUSCITY), PTDC/EMS-ENE/4525/2014 (PVCITY), UID/GEO/50019/2013 - Instituto Dom Luiz.
The authors would like to thank LNEG for the experimental data used in this study.

6. REFERENCES

[1] Brito MC, Freitas S, Guimarães S, Catita C, Redweik P. The importance of facades for the solar PV potential of a Mediterranean city using LiDAR data. Renewable Energy 2017;111:85–94.

[2] Vartiainen E. Electricity benefits of daylighting and photovoltaics for various solar facade layouts in office buildings. Energy and Buildings 2001;33:113–20.

[3] Lloret A, Andreu J, Merten J, Chantant M, Servant JM, Aceves O, et al. The Mataro public library: a 53 kWp grid connected building with integrated PV-thermal multifunctional modules. Proceedings of the 13th European Photovoltaic Solar Energy Conference, Nice, France: 1995, p. 490–3.

[4] Freitas S, Catita C, Redweik P, Brito MC. Modelling solar potential in the urban environment: State-of-the-art review. Renewable and Sustainable Energy Reviews 2015;41:915–31.

[5] Esclapés J, Ferreiro I, Piera J, Teller J. A method to evaluate the adaptability of photovoltaic energy on urban façades. Solar Energy 2014;105:414–27.

[6] Liang J, Gong J, Zhou J, Ibrahim AN, Li M. An open-source 3D solar radiation model integrated with a 3D Geographic Information System. Environmental Modelling and Software 2015;64:94–101.

[7] Sarralde JJ, Quinn DJ, Wiesmann D, Steemers K. Solar energy and urban morphology: Scenarios for increasing the renewable energy potential of neighbourhoods in London. Renewable Energy 2015;73:10–7.

[8] Zhang W, Zhang Y, Li Z, Zheng Z, Zhang R, Chen J. A rapid evaluation method of existing building applied photovoltaic (BAPV) potential. Energy and Buildings 2017;135:39–49.
[9] Kanters J, Wall M, Dubois M-C. Development of a Façade Assessment and Design Tool for Solar Energy (FASSADES). Buildings 2014;4:43–59.
[10] Roudsari MS, Pak M. Ladybug: a parametric environmental plugin for grasshopper to help designers create an environmentally-conscious design. 13th IBPSA, Chambéry, France: 2013.
[11] Anton I, Tănase D. Informed Geometries. Parametric Modelling and Energy Analysis in Early Stages of Design. Energy Procedia 2016;85:9–16.
[12] Chatzipoulka C, Compagnon R, Nikolopoulou M. Urban geometry and solar availability on façades and ground of real urban forms: using London as a case study. Solar Energy 2016;138:53–66.
[13] Wegertseder P, Lund P, Mikkola J, García Alvarado R. Combining solar resource mapping and energy system integration methods for realistic valuation of urban solar energy potential. Solar Energy 2016;135:325–36.
[14] A.L. Martins T, Adolphe L, E.G. Bastos L. From solar constraints to urban design opportunities: Optimization of built form typologies in a Brazilian tropical city. Energy and Buildings 2014;76:43–56.
[15] Romero Rodríguez L, Duminil E, Sánchez Ramos J, Eicker U. Assessment of the photovoltaic potential at urban level based on 3D city models: A case study and new methodological approach. Solar Energy 2017;146:264–75.
[16] Wieland M, Wendel J. Computing Solar Radiation on CityGML Building Data. Proceedings of the 18th AGILE Conference on Geographic Information Science 2015:2–5.
[17] Bremer M, Mayr A, Wichmann V, Schmidtner K, Rutzinger M. A new multi-scale 3D-GIS-approach for the assessment and dissemination of solar income of digital city models. Computers, Environment and Urban Systems 2016;57:144–54.
[18] Tooke TR. Building energy modelling and mapping using airborne LiDAR. The University of British Columbia, 2014.
[19] Erdélyi R, Wang Y, Guo W, Hanna E, Colantuono G. Three-dimensional SOlar RAdiation Model (SORAM) and its application to 3-D urban planning. Solar Energy 2014;101:63–73.
[20] Lindberg F, Jonsson P, Honjo T, Wästberg D. Solar energy on building envelopes – 3D modelling in a 2D environment. Solar Energy 2015;115:369–78.
[21] Redweik P, Catita C, Brito M. Solar energy potential on roofs and facades in an urban landscape. Solar Energy 2013;97:332–41.
[22] Rodrigues C. Centrais solares fotovoltaicas ligadas à rede eléctrica. Provas de acesso à categoria de investigador auxiliar, INETI, 2008.
[23] Freitas S, Cristóvão A, Amaro e Silva R, Brito MC. Obstruction surveying methods for PV application in urban environments. Proceedings of the 32nd European Photovoltaic Solar Energy Conference and Exhibition, Munich, Germany: 2016, p. 2759–64.
[24] Marion B. A method for modeling the current-voltage curve of a PV module for outdoor conditions. Progress in Photovoltaics: Research and Applications 2002;10:205–14.
[25] BP_Solar. 160-Watt Multicrystalline Photovoltaic Module n.d.
[26] Martin N, Ruiz JM. Calculation of the PV modules angular losses under field conditions by means of an analytical model. Solar Energy Materials and Solar Cells 2001;70:25–38.
[27] Marion B. Numerical method for angle-of-incidence correction factors for diffuse radiation incident photovoltaic modules. Solar Energy 2017;147:344–8.
[28] Perez R, Ineichen P, Seals R, Michalsky J, Stewart R. Modeling daylight availability and irradiance components from direct and global irradiance. Solar Energy 1990;44:271–89.
[29] de Miguel A, Bilbao J, Aguiar R, Kambezidis H, Negro E. Diffuse solar irradiation model evaluation in the North Mediterranean Belt area. Solar Energy 2001;70:143–53.

PHOTOVOLTAIC SOLAR URBAN POWER PLANTS INTEGRATED IN URBAN FURNITURE ALLOWING FOR SOLAR COMMUNITIES WITHIN URBAN ENVIRONMENTS

Rodríguez San Segundo HJ*, Calo López A and de Vicente Suso C
The South Oracle, S.L.
Paseo de las Delicias, 1
41001 Sevilla (Spain)
Tel. +34 954 562 955, Fax +34 954 560 857
*Author for correspondence. E-mail: hugo@thesouthoracle.com

ABSTRACT: Community solar concepts are starting to make noise especially in certain US and European regions. In those concepts, any end-user can participate in either purchasing clean energy from a solar PV power plant and/or investing in a portion of such power plants. However, usually these power plants consist of conventional flat PV silicon modules, usually in the MW scale far away from the cities. We propose here to bring solar communities right to the front door of the end-users. With the aim of really socializing solar energy, a Photovoltaic Solar Urban (PSU) power plant is presented, made of very small, but widely spread solar generators especially designed to be attached to any urban furniture (for instance, streetlights) and generate right at the consumption points, avoiding transmission, conversion, tax costs of bringing the energy from far away (almost 40% of the total final price). And also using publicly, widely spread available, already existing and amortized supporting structures (urban furniture). To do so, a novel concept of mini solar module of 40 Wp has been developed and is currently being tested. First simulations on its performance show their worldwide profitability.
Keywords: Small grid-connected PV systems, Distributed system operators (DSO), PV market.

1 INTRODUCTION

Distributed generation has always been and is still one of the main advantages of photovoltaic (PV) solar energy over other sources of energy: it can be deployed almost anywhere and close to the consumption points. This notion was clearly present in the "dreamer" days of photovoltaics, and somewhat lost during the last decade, when PV growth has entered the world of big industries and big businesses.

However, distributed generation is slowly gaining ground again. Along the shared economy growth in the mobility and hosting sectors, many concepts for sharing electricity generation and consumption from renewable sources are starting to take off. For instance, community solar concepts are starting to make noise especially in certain US and European regions[1,2]. In those concepts, any end-user can participate in either purchasing clean energy from a solar PV power plant and/or participating in such power plants by buying a small portion of it. However, usually these power plants consist of conventional flat PV silicon modules, usually in the MW scale far away from the cities. It is but in cities where, as of today, 50% of world's population live, rising up to 70% in 2050[3].

Therefore, it is in cities where consumption is happening. In this paper, we propose to bring solar communities right to the front door of the end-users. With the aim of really socializing solar energy, a novel concept of solar power plant, the Photovoltaic Solar Urban (PSU) power plant, is presented. It is made of very small, but widely spread solar generators, which we have especially designed to be attached to any urban furniture (for instance, streetlights) and generate right at the consumption points, avoiding transmission, conversion, tax costs of bringing the energy from far away (almost 40% of the total final price of electricity in many markets[4]). And also using public, widely spread, available, already existing and amortized supporting structures (urban furniture).

Unlike the conventional solar streetlight concepts, which rely on new urban furniture and are mainly designed for self-consumption and autonomous operation[5,6,7]. PSU allows for municipalities to either become electricity producers themselves or rent out their otherwise useless and already amortized urban furniture. In any case, they turn them into sources of income. By allowing the people living in their municipalities to a.) either purchase clean energy from the PSU, and/or b.) involve themselves in investing in their own power plants, which they can see right outside their doors. This way, awareness and consciousness of the real power of solar energy is arisen, and solar energy is socialized in its true, original conception.

2 PHOTOVOLTAIC SOLAR URBAN (PSU) POWER PLANTS

2.1 Module concept

Such a power plant cannot rely on the conventional two-square-meter solar module concept, which is big, heavy and not easily adaptable to the many possibilities a city offers. In order to enter cities, solar has to be smaller, in order to be placed everywhere (i.e., ubiquitous), and that at affordable costs, so that it is accessible for everyone to invest in. Other type of solar modules is needed.

With all this in mind, we have developed and are starting to test a novel concept of mini solar module of slightly over 40 Wp[8]. Instead of glass, it uses a polymer as front cover and is therefore flexible. This unit can be used for many purposes and adapted to many features that are usually found in cities. For the PSU application, thus, the unit module consists of 10 156x156 mm solar cells in series connection. For such a small number of cells, no by-pass diode is necessary. The junction box is provided with a blocking diode, as standard in solar modules. The frameless structure allows the whole assembly to be under 2 mm thick, and to weigh one third the weight of a conventional module. Figure 1 shows a picture of the unitary PSU40 module.

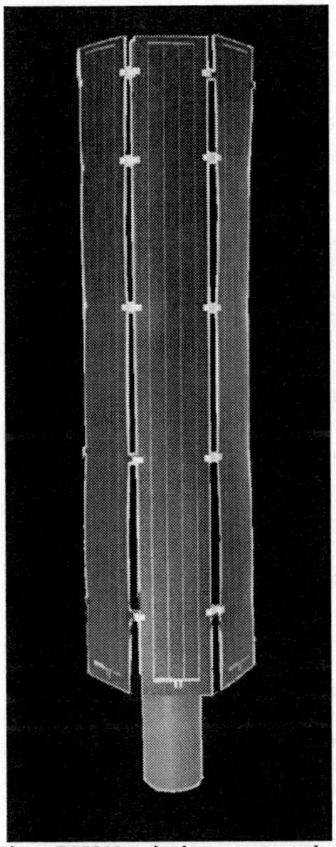

Figure 1: Three PSU40 units interconnected at 45°.

First series of PSU40 have been manufactured in a production line. Table I shows the average I/V characteristics of these first series, as well as their standard deviation, showing that production is consistent and these modules can be mass manufactured with low deviation. Both the industrial processing and the low variation from module to module makes IEC certification very feasible for this novel product. We foresee to do so within a year, so that bankability is not an issue.

Table I: I/V characteristics (average values) and corresponding standard deviations of the first series of industrially manufactured units of PSU40

	Isc (A)	Voc (V)	Vmp (V)	Imp (V)	Pmax (W)	FF (-)
Average	8,66	6,37	5,07	8,18	41,46	75,27
Std. deviation	0,09	0,01	0,02	0,09	0,52	0,85

2.2 PSU120 kit for streetlights

The PSU40 unit as presented in Figure 1 can be attached to basically any structure in cities: urban furniture of any kind, façades, rooftops, etc.

As a specific application, we are presenting in this paper the PSU kit to attach to the poles of existing streetlights. It consists of:

a.) Existing streetlights, which act as supporting structure. They are widely available in any city, and generally already paid for.
b.) 3 PSU40 units, attached to each other at 45° angles, and having Southeast-South-Southwest orientations. The whole kit is a unit of 120 Wp and therefore called PSU120.
c.) Stainless steel grips to attach the modules to one another and to the streetlight pole
d.) The 3 units are connected in parallel to a microinverter of 120 Wp, which is currently being developed for an optimum performance of PSU120 kits. For the current prototypes, commercial, non-optimized microinverters are being used.
e.) For every 70 to 100 PSU120 units, which are usually connected to an existing switchboard, one bidirectional meter and the corresponding electrical protections for grid feed-in are provided.
f.) Optionally, lighting control units can be introduced to control the lighting of the streetlights at night.

Figure 2 shows one single streetlight with all components of one PSU120 kit. The generated photocurrent is directly fed in to the low voltage grid.

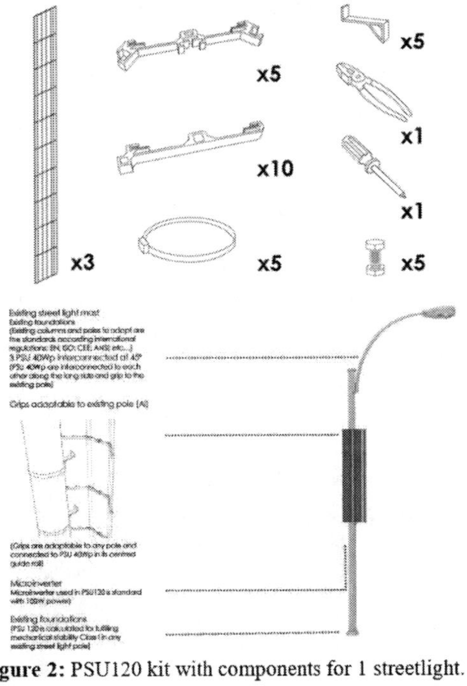

Figure 2: PSU120 kit with components for 1 streetlight.

2.3 Unitary PSU

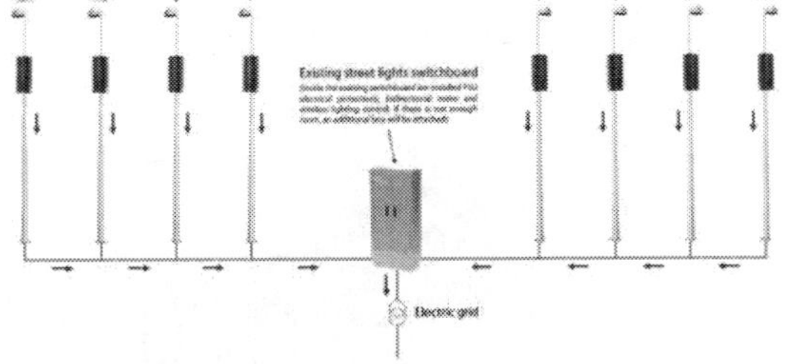

Figure 3: Unitary PSU consisting of 70 to 100 PSU120 kits, which are all connected to the existing switchboard that services the streetlights.

A unitary PSU could be as small as one PSU120 kit, therefore having a micro-plant, or as many micro-plants of 120 Wp each as wished. However, as stated, in order to optimize resources we will use the existing infrastructure already available for the streetlight configurations in cities. Nowadays, in most cities, every 70 to 100 streetlights are connected to one switchboard. We will use this switchboard to place the bidirectional meter and electrical protections, making a single Photovoltaic Solar Urban (PSU) power plant unit from 8.4 to 12 kWp. This is shown in Figure 3.

2.4 Building a PSU in a city

Figure 4 shows how the concept can be scaled up to any size in cities, simply by installing PSU120 kits on all suitable streetlights and further making unitary PSUs.

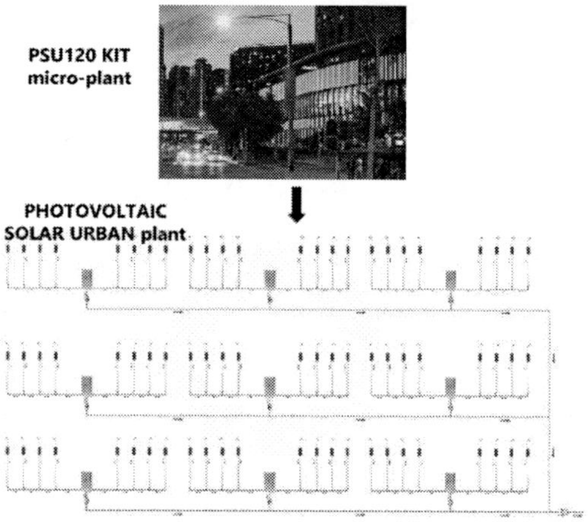

Figure 4: Building Photovoltaic Solar Urban (PSU) power plants in cities by using existing streetlight sets.

2.5 Mechanical resistance calculations

We are very aware that PSU have to withstand wind loads (the vertical positioning makes snow loads have no meaning). Therefore, we have carried out calculations for mechanical stability according to the international standards for mechanical resistance of streetlights, taking the strictest class (class I) into account. This class implies a terrain category I, i.e., the installation of streetlights along open seashores, lakes with exposed wind areas below 5 km in extension, or flat and open grounds with no obstacles. Wind speed is set at 120 km/h.

As seen in Figure 5, PSU120 kit withstands this category when placed in existing streetlights, eliminating any concern on that respect. The six ribs of each PSU40 and the gap between the panels make it possible for PSU120 to apply low mechanical load to the streetlight.

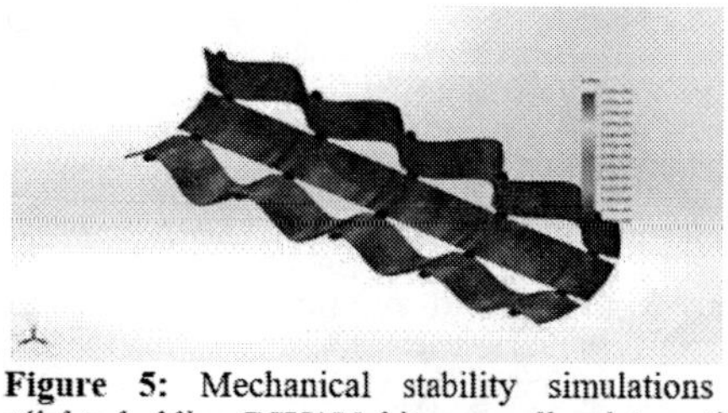

Figure 5: Mechanical stability simulations for streetlights holding PSU120 kits, regarding international standards for streetlights, class I.

3 PERFORMANCE SIMULATIONS

Regarding how PSU will perform, we have made simulations in numerous cities around the world. We have based these on numerous databases, both of solar irradiation and therefore production for vertically placed modules with SE/S/SW orientations, as well as databases for electricity prices around the world[9].

In the following, it is assumed that PSU operate for a lifetime of 25 years, a normal decrease in production of 0.5% per year, in an unsubsidized market and in a net-metering scheme, where every kilowatt hour (kWh) fed into the grid is paid at retail prices with a yearly escalation of 3%. It is also assumed that cities rent out their streetlights at 2€/unit annual rent.

Price for a PSU120 unit has been set at 200€, i.e., 1.67 €/Wp including installation, a very competitive price which sets Levelized Cost of Electricity (LCOE) in the region-dependent range of USD 127-221 per MWh. This is totally competitive with the USD 138-222 per MWh of residential rooftop PV, and quite close to the USD 88-193 per MWh of commercial rooftop PV.

Table II shows further results from calculations, obtaining that PSU are profitable for a wide range of latitudes around the world, with return on investment (ROI) from 6 to 15 years, depending on latitude. Counterintuitively, lower latitudes show a higher ROI, which is explained by the higher sun elevation, which penalizes vertically positioned modules.

Table II: Profitability in Return on Investment (ROI) of PSU in several cities.

Latitude	Location	PSU120 kWh/yr	Price ($/kWh)	Total income ($/yr)	ROI (yr)
24.714N	Riyadh	118,00	0,10	11,80	14,83
39.933N	Ankara	109,47	0,13	14,23	12,30
55.676N	Copenhagen	86,40	0,30	25,92	6,75
48.135N	Munich	84,00	0,30	25,20	6,94
41.903N	Rome	113,47	0,24	27,23	6,43
53.350N	Dublin	85,07	0,23	19,57	8,94
50.850N	Brussels	79,87	0,23	18,37	9,53
37.389N	Seville	123,60	0,22	27,19	6,44
40.417N	Madrid	122,53	0,22	26,96	6,49
43.263N	Bilbao	90,40	0,22	19,89	8,80
38.722N	Lisbon	115,73	0,22	25,46	6,87
51.507N	London	84,67	0,20	16,93	10,33
48.208N	Vienna	87,47	0,20	17,49	10,00
35.126N	Cyprus	120,40	0,19	22,88	7,65
52.370N	Amsterdam	81,20	0,19	15,43	11,34
59.329N	Stockholm	93,60	0,18	16,85	10,39
37.984N	Athens	116,13	0,17	19,74	8,86
48.857N	Paris	84,67	0,16	13,55	12,92

If we further assume that, due to shading and other effects, only one third to one half of the streetlights in a city are suitable of carrying PSU120 units, we can estimate the potential of a city to take PSU. A closer look at the case of Seville, for instance, where 30,000 streetlights are suitable, gives the following potential:

PSU power capacity:	3.6 MWp
PSU annual energy production:	3,700 MWh
PSU total energy production:	87,400 MWh
PSU total CO2 saved:	35,000 Ton
PSU net total income:	17,530,000 €
PSU IRR before taxes:	8 %
PSU IRR after taxes:	7%
PSU ROI:	6.4 years

Municipality incomes through rent: 2,187,555 €

This calculation can be made for all cities in Table II and others, showing that PSU is profitable. This is only taking into account streetlights, of which there are over 300 million available worldwide, equivalent to a 20 billion € market. But also other urban furniture, as well as façades and rooftops are widely available in cities. Therefore, the potential of PSU to introduce PV into cities is still greatly to explore.

4 CONCLUSIONS

A novel concept of solar photovoltaic plants has been presented: Photovoltaic Solar Urban (PSU) power plants. They are conceived to reach the highest consumption points: the cities, and to allow end-users to participate in solar communities from very low investments (200€), i.e., to really socialize solar energy. In order to do so, a novel design of a unitary module size of 40 Wp has been designed and manufactured. Three of these modules form a PSU120 kit, which can be easily attached around the pole of any existing streetlight. A unitary PSU takes advantage of the existing infrastructure for streetlights, and uses the switchboard shared by usually 70 to 100 streetlights to feed the generated energy into the grid. Mechanical stability calculations ensure that Class I, i.e., resistance to wind loads of 120 km/h, is met.

Profitability of PSU has been studied for several locations around the world, showing that Return on Investment (ROI) is between 6 and 15 years, for a 25 year lifetime.

5 ACKNOWLEDGEMENTS

The authors wish to acknowledge the helpful design and engineering support of Tomás Solano, Juan Solano and Alberto Maraña of Equs Arquitectos in developing the grips for the PSU120 kits, as well as providing mechanical stability calculations and renders of final solutions.

6 REFERENCES

[1] Farrel, J, "Community Solar Power: Obstacles and Opportunities", Ed. The New Rules Project, Sept. 2010.
[2] Coughlin, J, et al., "Guide to Community Solar: Utility, Private and Non-Profit Project Development", Publications of the UNLV, 1-2011.
[3] "World Population Prospects, The 2015 Revision". United Nations, New York, 2015.
[4] Green, R., "Electricity Transmission Pricing: an International Comparison", Utilities Policy, Vol. 6, Issue 3, 177-184, 1997.
[5] http://www.sepco-solarlighting.com/blog/3-types-of-solar-street-light-retrofit-systems
[6] http://www.bisol.com/uk/products/solar-street-lighting.html
[7] http://www.scotialight.com/
[8] Rodriguez San Segundo, HJ, Calo López, A and de Vicente Suso, C, "Photovoltaic Solar Urban Power Plant Composed of Generators Integrated in Urban Furniture, including Grid Feed-in", Spanish Patent Application ESP201601035, January 3rd, 2017.
[9] Thorough study by the authors from different sources: PVGIS and other solar irradiation databases, Public energy tenders, Public Electric Companies, Lazard and Eurostat.

SOLAR POTENTIAL ON COMMERCIAL TRUCKS: RESULTS OF AN IRRADIANCE MEASUREMENT CAMPAIGN ON 6 TRUCKS IN EUROPE AND USA

Ulrich Eitner[1]*, Matthieu Ebert[1], Tobias Zech[1], Cordula Schmid[2], Alliston Watts[2] and Martin Heinrich[1]
[1]Fraunhofer Institute for Solar Energy Systems ISE, Heidenhofstraße 2, 79110 Freiburg, Germany
[2]Fraunhofer Center for sustainable Energy Systems, Boston, USA
*Phone: +49 (0)761 4588 5825, ulrich.eitner@ise.fraunhofer.de

ABSTRACT: The use of photovoltaic modules on commercial trucks is considered to be a promising way to reduce fuel consumption and thus carbon dioxide emissions in the transportation sector [1, 2]. For refrigerating trucks, the energy demand (cooling energy) and availability (solar energy) are in close time correlation making the use of solar assisted cooling obvious. Several approaches and prototypes have been published since the early 90's [1–4], however, this technology has not experienced a commercial breakthrough yet. One reason for this stagnation is the uncertainty of payback time of such systems and reported technical drawbacks [5]. To overcome this reluctancy precise energy yield information is necessary to feed resilient return on investment calculations. In this work we present the result of a solar irradiance measurement campaign on 6 truck roofs operating in central Europe and north-eastern US. It is found that the best performing truck has the potential of a fuel saving of about 2113 L diesel/year and that the fuel saving potential strongly depends on the usage scenario of the truck. We find the operating conditions in the collected data to differ significantly from STC and typical PV application with 50% of the measured samples having an average daytime irradiance level below 200W/m² and 80% an average operating temperature below 25°C. In summary a cost competitive usage of photovoltaic systems on truck roofs seems possible but with tailored usage scenarios.
Keywords: Yield calculations, solar-assisted truck, irradiance measurement

1 INTRODUCTION

During the last decade production capacities for photovoltaics have drastically increased while the specific module prices (€/W_p) fell even stronger than the long-term trend (see Figure 1). This price decline opens opportunities for PV applications that seemed economically unreasonable in the past.

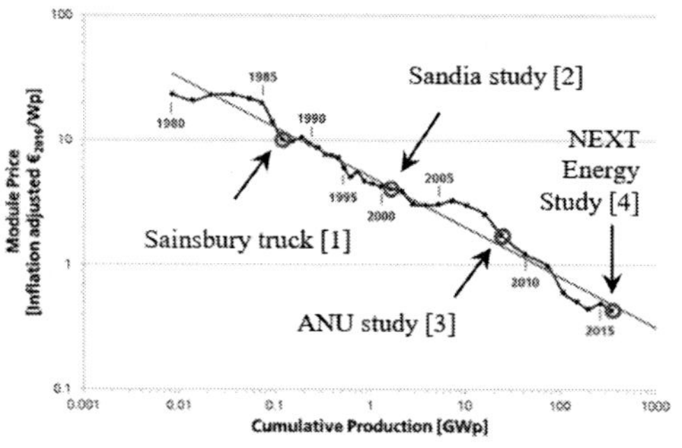

Figure 1: PV-learning curve taken from [5] and relevant studies regarding PV applications for trucks.

Especially for refrigerating trucks several studies were undertaken since the 1990s. The idea of solar assisted trucks, however, has not gained relevant attention in the market until now. In 1998, the concept of reducing diesel for a cooling truck by applying PV modules on the trailer roof has been demonstrated by the supermarket company Sainsbury and concluding a return-of-investment of approximately 15 years for the PV system including batteries [1]. The theoretical study of Sandia in 2001 for the US found only minor advantages for a solar system on trucks based on a module prize of 5\$/W_p, resulting reasonable economics after about 20 years [2]. A similar conclusion was published by ANU in 2009 for the application in Australia as diesel prize level then made the use of PV economically not attractive [3]. In a recent simulation study by Next Energy weather data are interpolated onto transportation routes in Germany, including the cooling effect on PV due to head wind on the road [4]. The calculations show a potential diesel reduction of up to 2000 L/year for crystalline silicon PV technology and up to 850 L/year for thin film PV. Based on the competitiveness of PV today, these studies are either outdated or require experimental verification.

Recently, several prototypes of solar assisted trucks have been reported to operate [6-10], showing the potential and interest in PV as a CO_2 reduction measure in the transportation sector.

The aim of this work is to assess and quantify the potential of harvesting solar energy during normal operation of trucks in the field. We therefore install small-size solar measurement setups on truck roofs and record the irradiance and temperature data over the period of up to one year.

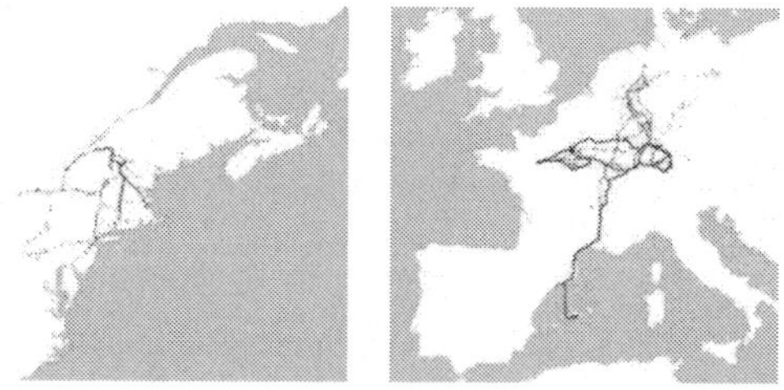

Figure 2: Measured transportation routes in north-eastern US (left) and central Europe (right) in this work.

2 METHOD

2.1 Measurement campaign

Six large cooling trucks (40 t) are equipped with a data logging system, irradiance and temperature sensors. An encapsulated solar cell with an integrated temperature sensor is used for measuring the irradiance. Two trucks are operating in North America, four others in Central Europe. The data is sampled in 1 min intervals and uploaded to a server once a day via GSM. The measurement campaign started in mid June 2016 and lasted until September 2017. Due to individual technical dropouts of the loggers some of the data could not be used. For some of the trucks the system was planned to be in operation for only a few months. Figure 3 shows the time intervals of data used for postprocessing for the 6 measured trucks.

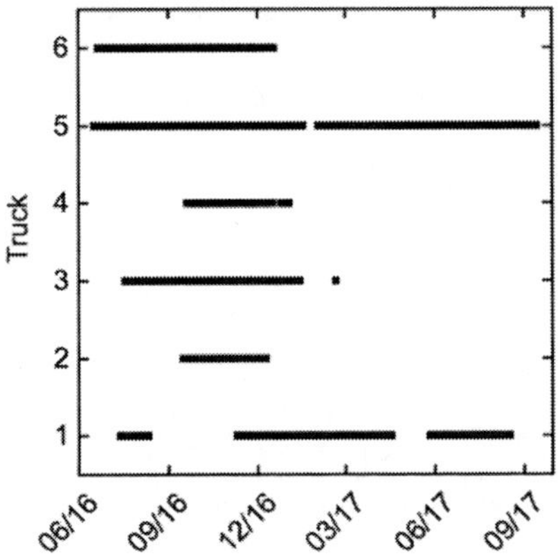

Figure 3: Measurement periods for 6 different trucks equipped with solar measurement equipment between June 2016 and September 2017.

2.2 Yield estimation model

To assess the energy yield a simple model is used. We use the short circuit current of the mini-modules laminated on the truck's roof and their cell temperatures. The sensor cells are measured under STC conditions before installation and a scaling factor for irradiance is calculated for each sensor. For the simple yield model a module efficiency of 18% is assumed. Factors like IAM or soiling are already included in the short circuit current measurement of the sensor cell and do not need to be considered in the model.

$$G = \frac{I_{sc}}{I_{sc,STC}} \cdot G_{STC} \cdot \left(1 - (T_{cell} - 25°C) \cdot \beta\right)$$

$$P_{\text{mod}} = G \cdot \eta_{\text{mod}} \cdot 1\left(- (T_{cell} - 25°C) \cdot \alpha\right)$$

$$E = \sum_{1}^{n} P_{\text{mod}} \cdot \Delta t$$

G is the irradiance in cell plane in W/m^2, I_{SC} the short circuit current in operation, $I_{SC,STC}$ the short circuit current measured at STC conditions before installation and T_{Cell} the temperature of the laminated cell. The temperature coefficients with respect to power and current are denoted as α and β. The energy yield E is calculated from the module power P_{mod} and the time stepwidth Δt.

2.3 Extrapolation of measured data

As the recording and filtering of the data does not cover a full year of operation for every truck due to measurement errors or demounting after a few months, an extrapolation is needed in order to generate results on a yearly basis. We therefore use stationary monthly irradiance datasets of locations near the transportation routes to fill the missing hourly data by applying the ratio of measured yield to location datasets for the recorded months.

3 RESULTS AND DISCUSSION

3.1 Annual yield of trucks

Figure 4 shows the measured yield data for each truck extrapolated to one year of operation. The total yield for a full trailer roof coverage of a 40t-vehicle ranges between 5297 kWh/a for truck 4 and 7395 kWh/a for truck 1. We split the total yield into 'yield moving' and 'yield stationary' in order to separate PV electricity generated during driving and parking. In fact, most of the yield is generated during parking (between 89 % and 65 %) while only 11 % to 35 % is harvested during driving.

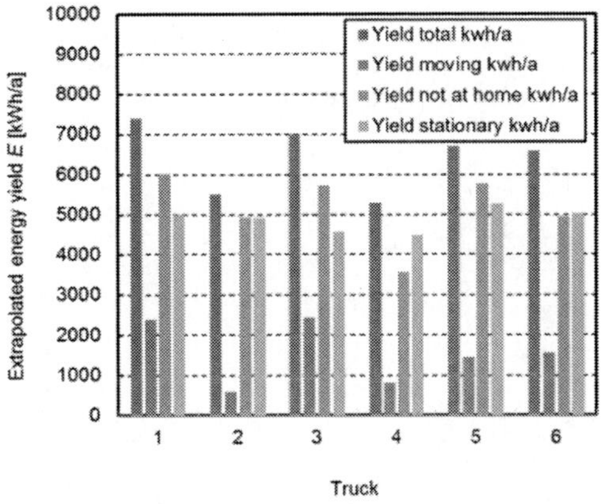

Figure 4: Yield data for trucks 1-6 assuming full roof coverage with PV modules. The data are extrapolated for one year as described in 2.3.

3.2 Irradiance, temperature and wind cooling

For the analysis of irradiance and temperature distributions we select values of truck 5 for one year (17.5.2016 – 16.5.2017). The temperature distribution is shown in Figure 5 ranging from -14.0°C to 66.6°C. The mean value is found at 22.6°C and the median is 19.9°C. Figure 5 shows the irradiance distribution, counting only daylight values with G above 10 W/m^2. Highest measured irradiances peak at 927 W/m^2 while the mean value is found at 277 W/m^2. 50% of the measured values are below 200 W/m^2.

The module power is directly influenced by its operating temperature, and decreases with increasing cell temperature. Further, the temperature of a PV module is depending on the air flow around it. The head wind speed of a moving truck is approximately around 22 m/s (80 km/h), which is a significant cooling wind stream for a PV module (in comparison NOCT windspeed is 1 m/s). The beneficial effect of this head wind cooling can be observed in the measured cell temperatures. In Figure 7 the cell temperature is plotted for the moving and the stationary time fractions of the measurement campaign. The linear regression curves clearly indicate that the head wind of a moving truck has a beneficial effect on the module temperature. For 800 W/m² irradiance the moving module is around 13°C cooler, which corresponds to approximately 5.5% relative efficiency increase.

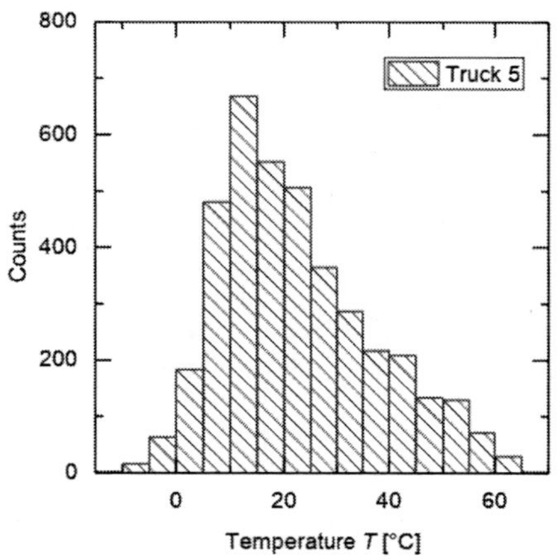

Figure 5: Temperature histogram of measured hourly values for truck 5 over one year in operation.

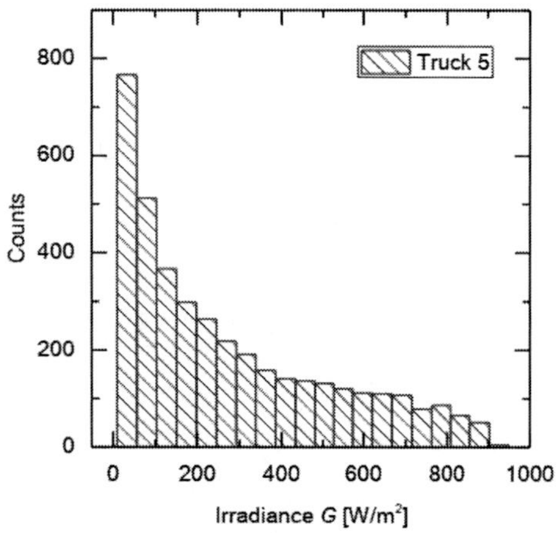

Figure 6: Irradiance histogram of measured hourly values for truck 5 over one year in operation.

Standard PV cells and modules are commonly optimized for Standard Testing Conditions (irradiance of 1000W/m², 0° incidence angle, 25°C temperature). On cell level this reflects in a certain metallization grid layout, optimizing the ratio between cell shading and ohmic losses in the metallization and bulk cell material. On module level the dimensioning of the cell interconnection ribbon and the optics of front glass and encapsulation are trimmed for an optimum at STC conditions. In the automotive application, such as an installation on truck roofs, our results show that STC conditions will hardly ever occur. The horizontal installation, the high air flow and other external influence factors cause operating conditions to differ significantly from STC.

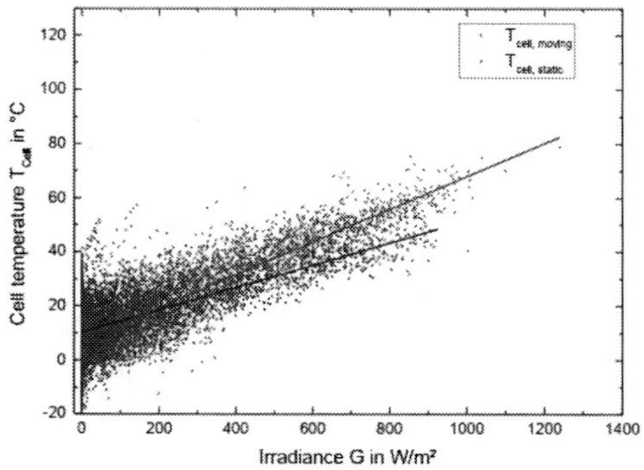

Figure 7: Measured temperatures of all trucks plotted vs irradiance. The blue values are recorded when the trucks were moving, the red values represent the status 'stationary'.

3.3 Potential for fuel savings

For the economic estimation, we translate the generated electricity into the same energy amount of diesel fuel. The amount of diesel V_{Diesel} (in l) is calculated from the PV energy yield E, assuming a utilization ratio of 100%, a diesel generator efficiency η_{Gen} of 35 % (for cooling) and a diesel energy density H_{Diesel} of 10 kWh/l,

$$V_{Diesel} = \frac{E}{\eta_{Gen} \cdot H_{Diesel}} .$$

The resulting diesel fuel reduction per year for a 40t trailer by using PV panels on the roof then sums up to 1513 L for truck 4 and 2113 L for truck 1, see data in Figure 4. The factor between these two transporting routes is thus 140%.

4 CONCLUSION AND OUTLOOK

From the collected and analyzed data we draw the following conclusions:

- the measured and extrapolated solar DC energy yield per year is between 7395 and 5297 kWh/a, in average over six trucks 6421 kWh/a.
- the energy harvested during driving is in average 23% of the total energy yield (min 11%, max 35%). This finding questions most of the assumptions used for yearly yield predictions on fuel savings from past studies [2–4]. It further leads to the need to differentiate by user group or usage scenario of the trucking company to calculate individual yield predictions. Important aspects are the use of PV electricity in stationary mode (truck loaded/unloaded, possibilities for gridfeeding at depot, consumption on board), the feeding into batteries of electric or hybrid vehicles and the actual energy need at certain times given by the refrigerated good.
- the temperatures and irradiance levels (analysed for truck 5) are significantly lower than standard PV

conditions. The average temperature during daylight operation is 22°C while the head wind effect is found to account for up to 13°C at high irradiance levels. Due to the horizontal positioning of the PV on truck roofs the irradiance levels are below STC conditions and average at 277 W/m^2.

Our experimental results for north-eastern US and central Europe are in the same range as previously published simulation results for Germany [4] and allow a conservative ROI calculation as it is based on real sample data. We find a significant fuel saving potential for commercial vehicles while the magnitude is, however, strongly depending on the usage scenarios (off-grid-time fraction) which are mainly part of the using profile.

We expect the future deployment of photovoltaic systems on commercial vehicles to be driven by four main factors:

- Transportation cost reduction by fuel saving [3]
- Noise emission reduction for deliveries in urban areas
- Legislative requirements for CO_2 and other pollutant emission reduction in transportation sector
- Ecological image gain

5 ACKNOWLEDGEMENT

The authors like to thank Nabil Mahmoud for his support during the measurement campaign. The companies Rolf Benzinger Spedition-Transporte GmbH, Dachser Group SE & Co. KG and Cabot Creamery are greatfully acknowledged for their valuable discussions and participation in the campaign.

6 REFERENCES

[1] A. S. Bahaj and P. James, "Economics of solar powered refrigeration transport applications," in Conference Record of the Twenty-Ninth IEEE Photovoltaic Specialists Conference 2002, pp. 1561–1564.

[2] David Bergeron - Sandia, "Solar Powered Refrigeration for Transport Application", 2001.

[3] B. Elliston and M. Dennis, "Feasibility of Solar-Assisted Refrigerated Transport in Australia," 2009.

[4] M. Kühnel, B. Hanke, S.Geißendörfer, K.v.Maydell, C.Agert, "Energy Forecast for mobile Photovoltaic Systems with Focus on Trucks for cooling Applications," Proc. EUPVSEC 2016, pp 2539 - 2542.

[5] Photovoltaics Report, Fraunhofer ISE, 2017

[6] Karsten Schäfer, "Auch Brummis lieben Strom", Photovoltaik, 2015.

[7] https://www.solaragentur.ch/sites/default/files/g-14-10-30_elektro-lkw_coop_solpreiskatenergieanl.pdf

[8] https://www.christoph-brand.com/deutschlands-erster-solarfoodtruck/

[9] https://cleantechnica.com/2016/08/22/the-solar-revolution-coming-soon-to-a-truck-near-you/

[10] https://www.energymatters.com.au/renewable-news/volvo-supertruck-solar-em5667/

BENEFITS FROM PV SYSTEM INTEGRATION WITH IRRIGATION AND DRAINAGE INFRASTRUCTURES: CASE STUDY FOR THESSALONIKI-IMATHIA-PELLA PLAIN IN GREECE

Nikolaos Chrysochoidis-Antsos, Christos Chrysochoidis
GOEV (Irrigation and Drainage Organization General) Thessaloniki (Greece)
43, 26th Octomvriou, 546 27, Thessaloniki (Greece)

ABSTRACT
Pumping stations assist the farming activities with irrigation and drainage. The pumping stations are powered by the Medium-Voltage grid and are operated through a network of open water canals and reservoirs. These infrastructures could be used to host PV systems that will feed the pumping stations with power. The annual seasonality of irrigation pumping stations matches well the seasonality of PV produced energy. A PV system design case study has been made to reflect the advantages in energy and cost under the net-metering scheme. Additionally, a number of benefits related to reduction to maintenance of infrastructures, efficient use of energy, reduction of evaporation losses and cost effectiveness are being presented within the study in this paper. Overall, PV system integration with irrigation and drainage infrastructures seems to have a positive effect on the agricultural economy.

Keywords: Rural Electrification, Water-pumping, Grid-Connected, PV systems

1 INTRODUCTION

Farming activities were and are important for bringing nutrition to humanity. Throughout ages it has been modernized with tools and technologies. A part of farming involves the water management. Water is essential for crops in order to grow and be ready for harvest. Irrigation and drainage are 2 main activities in modern agriculture that define the water management. This is facilitated with the use of pumping stations in order to "drive" water resources where they are needed. Pumps utilize electrical energy provided by the grid or in stand-alone operations. This electrical energy mainly comes from non-renewable resources and current global practices indicate the larger need to utilize renewable energy systems to provide the electricity needed. Moreover, energy efficiency has to increase even further while water consumption needs to be reduced in order to have a sustainable agricultural sector.
In order to modernize the agricultural activities as explained above an infrastructural change needs to be achieved with the minimum and less costly activities in order to ensure and reach the sustainable goals. One of these practices has been established in India, Gujarat with the Canal Solar project. Solar panels have been installed above the water canals in Gujarat in order to minimize the water evaporation and ensure electrification of rural areas [1].
This study uses as an inspiration this idea from India to apply it within Greece and specifically in Central Macedonia Region. The purpose is to provide affordable, cheap, green and sustainable energy to the pumping station infrastructure of Thessaloniki-Imathia-Pella plain, ensuring minimal infrastructural changes by retrofitting PV systems on the existing infrastructures of the irrigation and drainage works. The retrofitting is being done by utilizing the canal and reservoir networks nearby pumping stations, as well as the close access to Medium-Voltage grid. Thus economics of the system are improved significantly, while energy efficiency is increased annually.
PV generated energy is a mature system within the Greek energy infrastructures, which has in 2016 provided 7.05% of the annual electrical energy demand [2]. Moreover, maturity is observed in connections to Medium-Voltage grids with 45% of the PV systems being connected to it [2]. Additionally, PV system energy generation profiles match the seasonality of the pumping station operation. When marginally designing the PV system based on the energy demand and connecting with a net-metering scheme, economical benefits are observed for the farmers that pay for this electricity which provides them with the water for their crops.
The benefits are numerous and explained in this paper. These will benefit the 80.000 farmers that are paying for the water they use and the power needed for the pumping stations. A "helicopter view" study is presented for the plain infrastructures (57 irrigation, 10 drainage and 17 reservoirs "near pumping stations"). After that a case of an irrigation pumping station PV system retrofitting is presented.

2 CONCEPT

This chapter aims to
- introduce you to the concept of gound-mounted, roof-mounted and especially Canal PV and floating PV systems which seem to be favorable
- present infrastructural results of studying all the Thessaloniki-Imathia-Pella plain
- present case study results from an irrigation pumping station PV system retrofitted above a canal

2.1 What type of PV systems could be installed on pumping stations?

There are 4 types of PV system installations that could be facilitated next to pumping stations. These are the following:
- Ground-mounted PV systems
- Roof-mounted PV systems
- Canal-mounted PV systems
- Reservoir-floating PV systems

The **ground mounted PV systems** are the most typical utility scale PV systems that could be installed. However, the available area that belongs and is part of the irrigation and drainage infrastructures might be not enough in order to host all the power capacity needed to support this.

The **roof-mounted PV systems** solution is an interesting concept especially when trees are surrounding the pumping station and the modules being elevated can favor the obstacle free sun exposure. However, again available area is small and could potentially be not enough for supporting pumping station power demand.

The **canal-mounted PV systems** is a solution that could solve the area issue, as mentioned above for ground-mounted and roof-mounted, with a total network of 425 km [3]. The PV systems could be installed on the banks of a water canal. Metallic support structures have PV modules mounted on them on the designed orientation. Cabling is attached on the structure and depending on the string configuration they all meet at a central inverter point. After that point they are connected with the grid through a transformer. It has been installed in many projects worldwide. One of the cases is in India through the Canal Solar Project within the Gujarat region. The PV systems are installed for reduction of water evaporation and for rural electrification. The project had 2 pilots so far, one 1 MW and a 10MW [1] .It is of relevance to examine these cases as they are inspiring the concept of this project. There have also been some studies in the Mediterranean Sea region for the Mediterranean islands [4] as well as the river banks of a river in Spain [5]. These studies claim PV efficiency increase due to cooling of the PV modules and reduction of water evaporation.

Figure 1: Illustration of a Canal-Mounted PV System [1]

The **reservoir-floating PV systems** is a new innovation starting appearing worldwide and they start gaining industrial commercialization. Namely in 2016, 53MW have been installed on a worldwide basis, with a total cumulative capacity of 94 MW starting installations from 2013. Countries that lead this system are USA and Japan while some other countries follow with smaller demonstrations of this type of system [6].

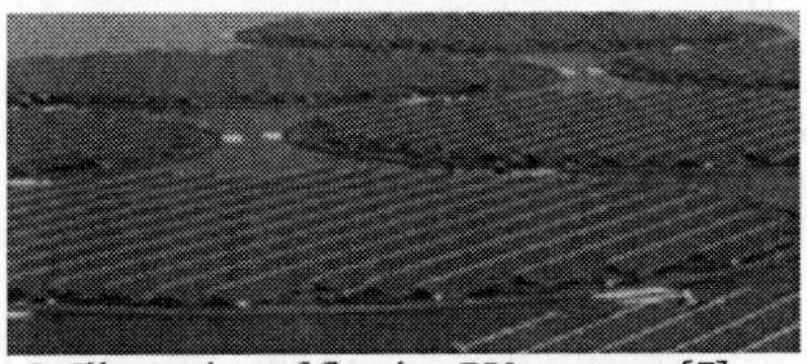

Figure 2: Illustration of floating PV-systems [7]

2.2 Case study for Thessaloniki-Imathia-Pella plain

The Thessaloniki-Imathia-Pella plain is being irrigated from 3 rivers (Aliakmonas, Axios and Loudias) with a total network of 425 km of canals. The water is being distributed to 58 irrigation pumping stations which send the water through underground pipeline networks or through gravity irrigation systems. The residual water is being pumped with the help of drainage pumping stations to the sea. These pumping stations are being powered through connection to the Medium-Voltage level grid networks and have a transformer that lowers voltage for the use in electric motors that support the pumping operation. PV systems could be installed to cover part of this electrical demand.

These types of PV systems could be retrofitted on the existing land reclamation infrastructures nearby pumping stations of the Thessaloniki-Imathia-Pella plain. It is worthwhile to have a first look on all the pumping stations of this plain on certain criteria that determine a successful retrofitting with PV systems and would help us filter out a common case in order to study. The data that are used for these criteria have been provided from balance sheets, GIS files (Geographical Information Systems) and telemetry data from the pumping stations of GOEV Thessaloniki (General Irrigation and Drainage Organization) and TOEV Krias Vrisi (Local Irrigation and Drainage Organization). The criteria are:

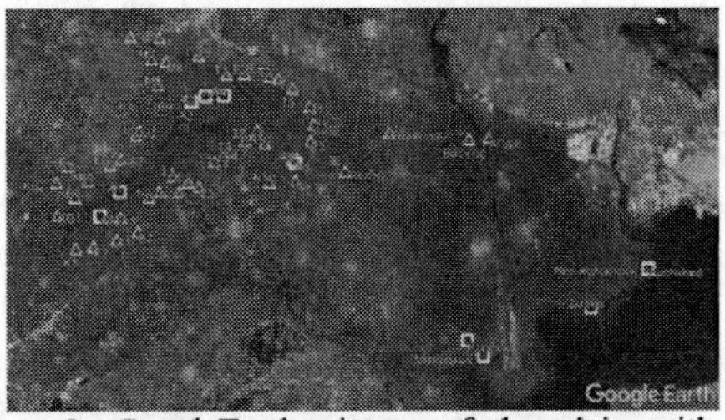

Figure 3: GoogleEarth picture of the plain with the irrigation pumping stations in triangle and the drainage in squares

Type of Station (Irrigation or Drainage)

Irrigation is the process of bringing water to the farming lands through the canal irrigation networks. The energy demand profile for powering the pumps of irrigation indicate seasonality in the case of Greece with most of the energy being consumed during the summer period, as the graph indicates below.

Drainage is a continuous and not a seasonal process of removing the residual water right after the irrigation process of crop fields.

The number of pumping stations is 80% for irrigation purposes while the rest 20% is for drainage. Therefore it makes sense to case study an irrigation pumping station.

Installed power of pumping station systems

The installed power of the system and the energy demand per station is an important factor that "drives" the sizing of the PV systems installed.

60% of the installed pumping power is within the 440-660kW range. Most of the pumping stations comprise of 110kW pumps connected to a low voltage network of 400 Volts that is transformed from a 20kV Medium-Voltage grid.

The PV system electricity will be fed to the pumps through the day but during the night the grid would be needed to support the operation or when PV electricity cannot cover the demand.

Availability for ground-mounted system

The average available land area for ground-mounted PV installation next to the pumping stations is 800 m^2, which could potentially host 100-130 kW of PV systems.

Availability for roof-mounted PV systems

The average pumping station rooftop area is 150 m^2, being able to host 20-25 kW of PV systems. A rooftop illustration can be seen in the figure 5.

Proximity to irrigation-drainage canals

Most of the pumping stations are close to canals at an average distance of 30-45 meters. Moreover, the canals have a long span, as picture indicates, which is important in order to have sufficient space to install the PV systems. A very small percentage of the canals have a curvature on their trajectory which could pose a difficulty on obtaining a constant orientation of panels when designing the support structures.

Orientation of the canal (3 standard domains)

As said before, canals have an orientation which should be taken into account when designing the support structure in order to ensure the optimal orientation. Therefore 3 arc domain categories have been proposed to categorize all systems. There are 12 canals facing North or South, 15 canals facing West or East and 23 facing North East or North West or South East or South West. It is of outermost importance to innovate and design support structures that could be retrofitted with respect to the orientation of each canal. Below in figure 4, an example of North East or South West canal and a South-North canal is presented.

Figure 4: Illustration of a pumping station and a North-East or South-West orientation canal and a South-North oriented canal

Proximity to reservoir

Another potential installation as said is with floating PV systems on top of the reservoirs of pumping stations. In total, 16 pumping stations have a reservoir in proximity of 30-70 meters. 3 are located next to drainage pumping stations and 13 are next to irrigation pumping stations. The area above the reservoirs is on an average 5500 m^2, which is more than enough to support a system ranging from 250-500kW.

Figure 5: Illustration of a pumping station with a reservoir

Proximity to medium-voltage transformer

An important factor to consider, for this retrofitting design, is the proximity to medium-voltage transformers. The PV systems are connected to inverter systems and after that they are connected to the transformer through the yoke of the switchboard of the pumping stations. In that way grid interconnection costs related to long cabling and extra transformers are avoided. The average distance found to the medium voltage transformer was found as 60 meters.

As a conclusion, ground-mounted and roof-mounted PV systems are excluded due to space limitations. In some cases of drainage pumping station there is available land and could be considered. The canal-mounted and reservoir floating configuration seem to be able to serve better most of the pumping stations. Reservoir floating though is less than the potential of canal-mounted for this point of study. Therefore the canal-mounted systems shall be studied. And the floating case is recommended to be studied in the future.

2.3 Case study of Kria Vrisi Irrigation Pumping Station

Since irrigation pumping stations are 80% of the pumping station systems, it makes sense to study this type of system. Regarding the irrigation pumping stations' installed power, 60% of those are within the installed power of 440kW-660kW. Therefore a representative case to be studied would be a 550kW pumping station.

From the local irrigation and drainage organization (TOEV) of Kria Vrisi, which operates 7 irrigation pumping stations, electrical demand profile data were collected for the years 2014-2017, as well as costs paid to the electricity provider.

By observing the annual plots of the electricity demand data, we could indicate the variation that the electricity consumption shows. These effects account to "drought" events in 2016 which explain the higher pumping energy consumed. The pumping energy demand reduces in 2017, thereby indicating the cyclic seasonality of drought events. Therefore the designed installed PV power capacity should be able to provide the average energy demand of these cyclic annual variations. A note shall be made that the graph below sums only data from January until August to indicate the annual variations. Also the data were only available for the period 2014-2017, however after interviewing personnel it was stated that this cyclic annual variation has similar pattern every 3 years, therefore these numbers are representative for this study.

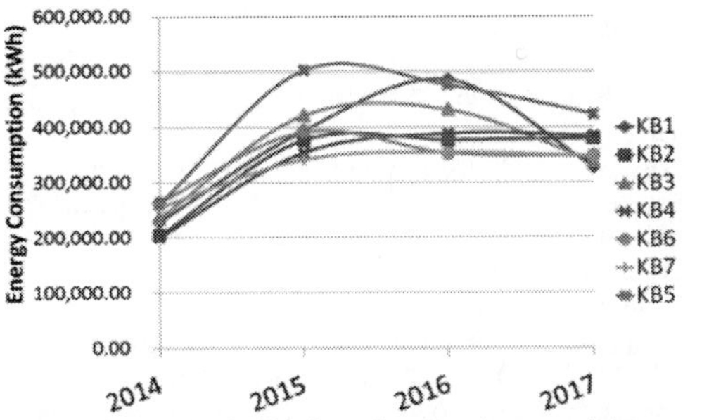

Figure 6: Graph indicating the Total Annual Energy demand for the 7 irrigation pumping stations of Kria Vrisi for the period 2014-2017 (note that 2017 demand is extrapolated)

Taking into account that land reclamation organization can apply for funds for self-producer with a net metering scheme, the PV system should be designed as such to match the limitations posed by the net metering. Net metering in Greece, indicates that energy meters are installed at the PV production and the grid consumption. In the end of a 3-year period, the difference between the energy consumed by the pumping station and the energy produced by the PV system indicates if you need to pay or not for electrical energy supplied by the grid. If this difference is positive, it means that there was more consumption than production and you pay the difference of those. If the difference is negative, it means that there was more production than consumption and you do not pay anything but as a disadvantage you do not get paid for the surplus of production. Having this in mind, the system should be designed in order to match the 3-year period energy demand.

Energy Analytics

1) Annual deviations

The figure 7 below indicates the variations in between the years 2014-2016. Therefore, data from 2015 year will be used which represent the average.

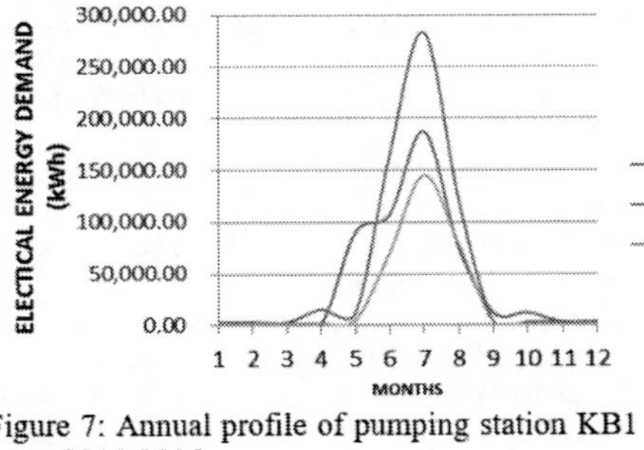

Figure 7: Annual profile of pumping station KB1 for the years 2014-2016

2) Electrical energy demand

The 2015 year data from the irrigation pumping station KB1 is the average of the above graph. The total annual electricity demand was 495 MWh. Therefore, this will be the design case for the PV system.

3) Explanation of modelling steps

For the modelling of the PV the following data have been used:

- local irradiance corrected for 33° optimized slope for Thessaloniki-Imathia-Pella plain (W/m^2) [8],
- local temperature profiles (°C) [8],
- corrections for canal top air temperature (%increase of efficiency/°C),
- a conservative factor of degradation of the 25 year lifetime of the system due to the humid environment (-0.5%/year) [9],
- a conservative estimate of 14% for other system losses (inverters, cabling losses due to the long canal length etc)

Modelling iterations of the PV system resulted in a 350 kW photovoltaic system installed on top of the canal next to the pumping station. The minimum spacing to ensure zero shading at noon on the day when the sun is at the lowest angle of the year (namely 22nd December at 12.00PM) should be 2.95 meters for a 1.5 meter solar module placed at the optimized slope of 33°. Taking into account the 7-m width of the canal neighboring this pumping station, the PV system installation will be along 500 meters of this canal.

Figure 8: GoogleMap picture of the pumping station used as a reference design

4) Results of simulations - Proportions and comparisons

Figure 8 below, indicates the annual profile of the pumping station electricity demand in blue while in green dashed line is the annual profile of the PV system production. It is clearly evident that PV overproduce during winter and autumn periods while in summer and some spring months the pumping station demand surpasses the PV system energy production. This is normal since the system has been marginally designed in order to produce the equivalent amount of the annual energy so that the net-metering scheme can apply economically.

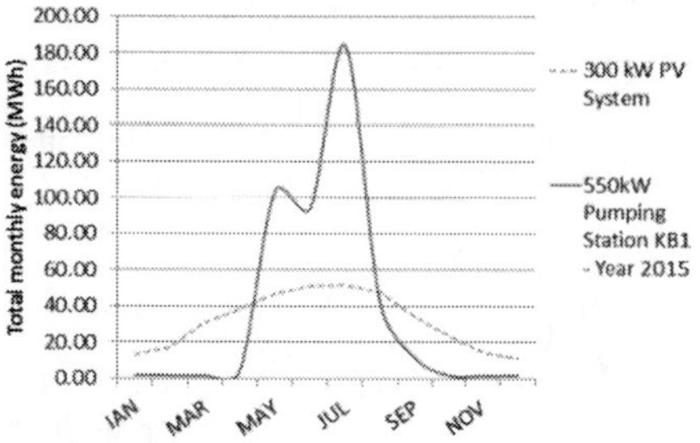

Figure 9: Annual PV production and pumping station electricity demand for 2015 for the KB1 station

The table below indicates that with the system designed PV system and grid electricity has an equal contribution to the total annual demand.

50% of the demand is covered by PV and especially during the daytime of the summer period, while grid is fulfilling the energy demand. One interesting conclusion could be that during the day the PV are relieving the power systems while grid covers major part of energy during the night when the electricity pricing is usually lower.

The annual PV electricity production is also divided into 2 equal parts where half of it is being fed into the grid, mainly from September until April where demand is lower. While the rest 50% is being fed during the daytime to the pumping stations from May until September.

Table I: Energy Analytics of the pumping stations demand and the PV production

	MWh	%
Total Annual Pumping Energy Demand	450	100
Energy covered from PV	242	54
Energy covered from Grid	206	46
Total Annual PV production	444	100
PV energy fed to pumps	242	54
PV energy fed to grid	202	46

5) System Economics

2016 prices

Data from 2016 indicate that for the Kria Vrisi division of pumping stations, the average price for the irrigation pumping stations electricity is 0.092 €/kWh. This is a special electricity price for rural (farming) activities and contains cost components of electrical energy, grid interconnection, carbon emission taxes, renewable energy development taxes and administrational costs. This price will be used to compare the PV system economics under the net-metering scheme.

The system has been designed marginally in order to meet the net-metering criteria, meaning that throughout a 3-year an energy balance assessment is made. In order to properly understand the concept 2 examples are given below [10].

Example 1 – Consumption exceeds Production

If during these 3 years 1000 MWh are being consumed from the pumps and 900 MWh have been produced for the given PV systems then the difference of 100 MWh has to be paid with the pricing of that time.

Example 2 – Production exceeds Consumption

If 1000 MWh have been consumed by the pumps while 1100 have been produced by the PV systems then no amount of money has to be paid from the distribution network operator to the producer of PV energy.

However due to the 3-year seasonality effect as explained above, the risk to experience these scenarios could be avoided.

Costs avoided due to retrofitting

Now that all energy modelling has been done and the system is able to deliver marginally the power to the pumping stations and the grid under the net metering legal scheme, an economic analysis is conducted taking into account capital expenditure (CAPEX), operational expenditure (OPEX) and costs related to replacement of component. The equation used is similar to the Levelized Cost of Electricity (LCOE) [11]. However, for the cost price the interest rate is not included.

Cost of Electricity by this system.

The cost of the system has several components as shown in the chart below. One major benefit of this retrofitting design is that some costs could be avoided due to the availability of infrastructures nearby. These costs relate to transformer, grid connection, minimal cabling distance to transformer and no requirement to rent land (publicly owned). However, due to the complex installation and mounting of solar panels above canals, costs are equalizing to the price of 1000 euros/kW installed [12]. These prices have been checked and validated through industrial interviews and references from well-known institutions [13].

The cost breakdown indicates that the support structures account for 20% of the system, since a special design of support structure should be realized, as seen in figure 10.

As a proposal, aluminum should be used for the profiles of the support structure for the following reasons:

- Is a lightweight material that can be easily transported, mounted and assembled
- Even though it is more expensive than steel structures it is a recyclable material with a high end-value
- Is a corrosion-resistant material and since the surrounding environment to be installed is humid, it is a favorable choice
- Lastly Greece is rich in bauxite deposits and produces aluminum

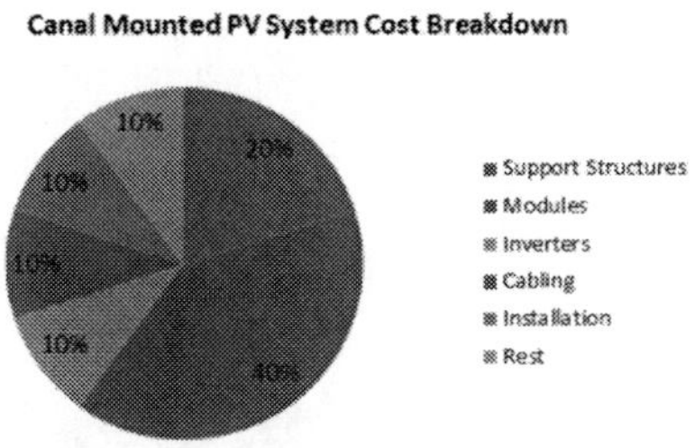

Figure 10: Canal-mounted PV system Cost Breakdown

Funding

This particular system being a part of public benefits could be either fully funded from governmental, regional or European funds or through private investments by having a bank loan.
The marginal cost of electricity for a 25 year lifetime operation taking into account all the degradation factors for the system would be 0.054 €/kWh following the net-metering scheme.
If the funds are interest-free (public investments) then the price is 0.054 €/kWh. If an interest of 4% is considering in the economical calculations [12], the loan could be paid off in a lifetime of 25 years with an electricity fixed price of 0.073 €/kWh. Practically this will be the price that will be reflected as a cost component the bill of water consumption of the farmers.
Comparing these values with the current price of 0.092 €/kWh, one can understand that through public investments the electricity price is 42% less and through private investments the price is 21% less.

The bottom line of this sub-chapter is that the farmer community in the end will not pay for electricity given the fact that this infrastructural development is done with public Greek or European funds. But the allocated cost for this electricity for the government of Greece or EU is 20-40% less than it is now. Therefore, it is a favorable system to realize.

3 BENEFITS

The benefits are analyzed within the following sections and results from observations, literature review, interviews with infrastructural companies and techno-economical modelling of the lifetime system's operation using the available data provided by GOEV as in chapter

3.1 Minimal Land Use

It is certainly a benefit not to waste the arable land with PV system installation but according to a recent study of Hellenic Statistical Service the current PV installation in Greece occupy only the 0.1% of it [14].
It is more a matter of cost for renting or buying land to install the systems rather than land usage, since the infrastructures belong to the public and the PV are retrofitted above canals or tanks, therefore increasing the value of this part of land.

3.2 Effective use of infrastructure

The infrastructures available could be categorized into structural and electrical. There is effective use of the infrastructures.

Considering the effective use of structural infrastructures, the following bullet points summarize the benefits:

- Less maintenance costs due to reduction of algae, moss and lichen accumulation on the bottom and side banks of the canals or the reservoirs, as seen in the pictures below. This reduction is due to shadowing of the covered in PV canal or reservoir. This reduction minimizes the frequency of cleaning the canals which reduces the cost of irrigation services. Additionally, having less algae means that the canals have less roughness which could reduce the operational loading of the pumps. Algae also can accumulate in the form of sludge near the pumping stations and could enter the pipeline network and thereby decreasing the lifetime of the pumps. Figure 11 indicates algae that is accumulated on the side banks of the irrigation canal. Also in figure 12, algae is accumulated in the reservoirs.

Figure 11: Illustration of the side banks of the open canal

Figure 12: Illustration of the algae accumulation in a reservoir next to a pumping station

- Reduction of water evaporation: According to some recent studies, water evaporation in open water channels in the Mediterranean region varies from 2mm/day in wet winter and 7 mm/day in dry summer. Another study for south-east Spain indicated that covering the irrigation canals could drastically reduce this water evaporation for the shadowed ranging from 50-80% reduction. While another more generic indicates 30% water savings [4].

Regarding the effective use of electrical infrastructures, the following bullet points summarize the benefits:

- No need for transformer upgrade or grid reinforcement or addition of transformer since already transformer(s) are present within the infrastructures, as in figure 13 below.

Figure 13: Photograph of the transformers next to a pumping station

- Transformer loading is decreased since part of electrical energy for the pumps is provided by the PV. Thereby the transformer overloads in a less frequent manner than before. Overloading causes heating which damages the insulation of transformer and thereby decreasing the lifetime of it [15].

- No modifications are needed on the existing electrical equipment. The PV modules are connected to the inverters and the inverter is connecting to the yoke of low voltage on the existing switchboard.

- The inverters have a low THD – Total Harmonic Distortion - thereby not overloading the transformer

- Some inverters offer the possibility to compensate reactive power during day and night, regardless of the PV operation. This process is essential, as the grid operator demand, exactly next to pumping stations which mainly consume induction loads.

3.3 Efficient electricity production

- Energy demand matches energy production (Annual profile charts for PV-Pumps)

The following figure 14 was obtained from the work of Iro Livaniou. This graph presents the daily pattern from May-September of an irrigation pumping station in Chania in the island of Crete in Greece [16]. The reason for using this graph is the lack of daily data for the Thessaloniki-Imathia-Pella plain, however the daily patterns for both Crete and Thessaloniki for the period of May-September would resemble alike.

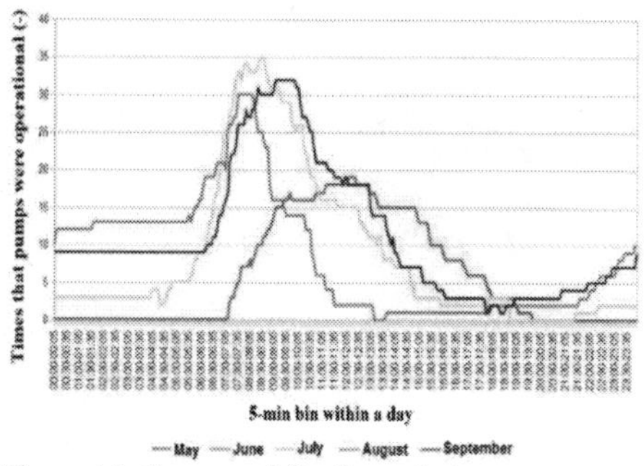

Figure 14: Summer daily times that pumps operate in an irrigation pumping station [16]

- Close proximity to the transformer (average of 30-60 meters) meaning that lower cabling losses until the point of grid interconnection. Moreover, close proximity to the electrical load of pumps motor meaning that the electrical energy is efficiently utilized without any transmission losses. As discussed in chapter 2, 50% of the pump energy would come from PV and the rest from the grid. Therefore, 50% of the energy arrives without any transmission losses.

- Natural cooling from the water-cooled air above canals or tanks can increase the PV system's efficiency with a maximum of 2.5% in the month of July as seen in the figure 15 below. In an annual basis 0.5-1.5% more energy is produced comparing to an equivalent ground-mounted PV system. For the case in India 2.5% more energy has been achieved in [17]. This increase in power output has been modelled with a cell temperature correction equation in this study.

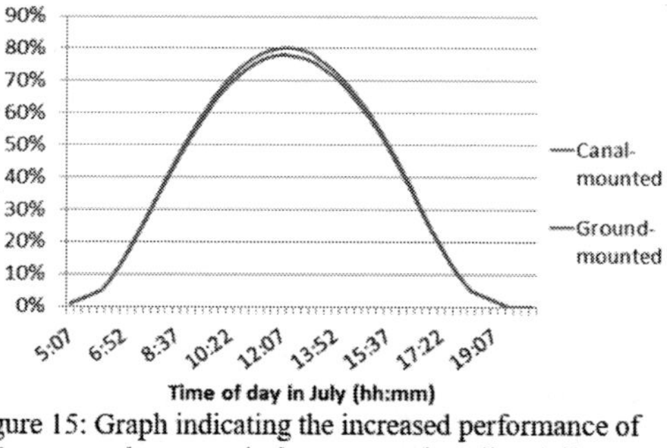

Figure 15: Graph indicating the increased performance of PV-systems above canals due to natural cooling (Pi/Pmax is indicating the proportion of power comparing to the rated power of the system)

- Minimum shadowing due to the flat landscape of the plain and low height crops surrounding the pumping station infrastructures (canals and reservoirs).

3.5 "Green" electricity production

- 60% of energy demand now comes from PV while the rest 40% comes from the grid, while 50% of PV is being fed to the grid. These results can be observed in an annual profile in figure 16. It is evident that PV provided most of energy in summer (nearly half of it).

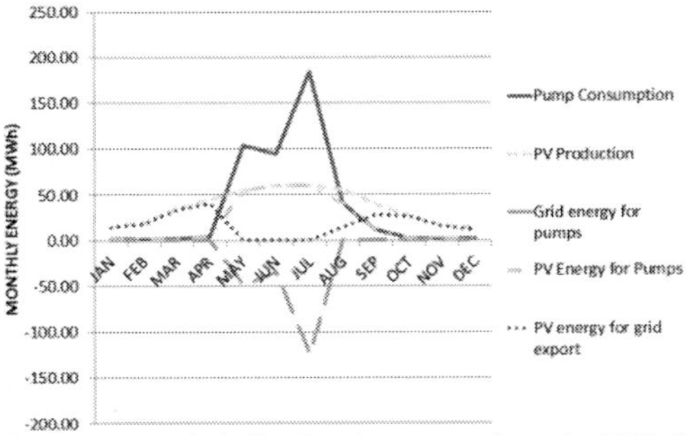

Figure 16: Graph indicating the energy flows in 2015 for the KB1 station

- Equivalent CO_2 reductions due to allocation which basically means that since 60% of demand is being covered by PV the rest 40% comes from grid electricity which uses the national energy mix. Currently the Greek electricity energy mix is carbon-based but looking to future projections it will lean upon a more and more renewable-based energy mix.

3.6 Cheap electricity production

The benefits of cheap electricity production are mentioned below and discussed in detail.

- The 3-year period net metering scheme allows for an economic and sustainable system integration, with a marginally and not over dimensioned system that can equally cover part of the energy demand and feed the grid with renewable energy

- A cost price of 0.052 €/kWh, which far below than what is now being paid 0.092 €/kWh from the farmers. This has been achieved since several cost components are avoided. Namely no land reclamation needed and no transformer is needed. Additionally, higher efficiency of the PV system proximity to the load leads to higher energy production which in turn reduces the system's size and therefore costs.

3.7 Probable Challenges

One important issue that should be considered is the need of preventive maintenance every 10 years for cleaning the open water canals or reservoirs. However, this type of maintenance is mainly due to concrete failure due to fauna accumulation or natural degradation, but the installation of PV systems that shadow the canal might prolong the maintenance procedures. Anyhow the system must be designed in order to let access of machinery for these types of maintenance works. Fortunately, the PV systems, in between their installation have gaps that would allow machinery to come through. In the worst case scenario the PV rack should be unmounted to allow the maintenance work.

Figure 17: Photographs from maintenance activities on the side banks of the open water canals

4 CONCLUSIONS

The 80.000 farmers on the Thessaloniki-Imathia-Pella plain of Greece could lower their costs of farming through PV system integration at the pumping stations that support the irrigation and drainage activities. This study indicated that the majority of the pumping stations are for irrigation with a system size range of 440-660 kW. The PV systems to be installed shall have a size that is designed in accordance with the net-metering scheme. Therefore it is marginally designed so as the production of PV electricity is equal with the pumping station electricity demand. The lowering of the cost is achieved due to the cost allocation with several infrastructures that exist within the pumping station area and the open canal networks. The PV systems are installed on top of canal or floating on reservoirs. A case study has been made and numerous benefits were addressed. Water evaporation can be reduced due to shading. Less maintenance will be needed for the electrical equipment of the pumping station and the canal networks since less algae will be accumulated on the canal. Also the transformer will be utilized less and therefore overloading events will not occur and the lifetime will increase. Finally, the daily and annual energy demand profile of irrigation match with the PV production profiles which in turn makes the system more efficient in energy utilization, by minimizing the energy absorption from the grid.

5 RECOMMENDATIONS

The following bullet points summarize the recommendations:

- The surplus of PV generated electricity during winter could be used for heating greenhouses producing vegetables and increasing the rural income
- The surplus of PV electrical energy and the existence of water means that through electrolysis system hydrogen fuels could be produced. This hydrogen fuel could be used to fuel up the futuristic farming machinery and vehicles (tractors, excavators) since hydrogen vehicles are gaining more and more momentum.
- Large potential for large-scale PV systems (425km of open water-channels) meaning that a potential of nearly 300 MW for PV system's installed on top of all canals.

- Hybrid energy system with PV and coastal small scale wind turbines for the drainage pumping stations located in the coastal region. However, a detailed study for the birds affected in Natura 2000 protected Delta of Axiou should be considered.
- As a recommendation these steps could be followed towards realizing such a project:
    - Detailed study for all the pumping stations in the Thessaloniki-Imathia-Pella plain
    - Construction of pilot plants in both configuration of canal-mounted and reservoir-floating
    - 1 year system operation and observation
    - Infrastructural development with PV integrated system for all the relevant pumping stations

6 REFERENCES

1. Rathore, S.S. *CANAL TOP SOLAR POWER PLANT AN INNOVATIVE INITIATIVE*. in *THE WORLD FUTURE ENERGY FORUM (WFEF) 2016 Presentation at International Conference on Affordable, Reliable and Sustainable Energy for the Future*. 2016. Beijing (China).
2. A. Kyritsis , D.V., N. Papanikolaou , S. Tselepis , C. Christodoulou , I. Gonos , S.A. Kalogirou *Evolution of PV systems in Greece and review of applicable solutions for higher penetration levels*. Renewable Energy, 2017. **109**: p. 487-499.
3. Goukos, D. *Management of irrigation wates in the plain of Thessaloniki*. Available from: http://docplayer.gr/2756080-Diaheirisi-ardeytikoy-neroy-stin-pediada-thessalonikis-dimitris-goykos-geoponos-eggeion-veltioseon-msc.html.
4. Ioannis Kougias , K.B., Arnulf Jäger-Waldau, Magda Moner-Girona, Fabio Monforti-Ferrario, Heinz Ossenbrink, Sándor Szabó, *The potential of water infrastructure to accommodate solar PV systems in Mediterranean islands*. Solar Energy, 2016. **136**: p. 174-182.
5. Antonio Colmenar-Santos , A.n.B.-E., Carlos de Palacio-Rodrı´guez, David Borge-Diez, *Water canal use for the implementation and efficiency optimization of photovoltaic facilities: Tajo-Segura transfer scenario*. Solar Energy, 2016. **126**: p. 168-194.
6. Holm, A. *Floating Solar Photovoltaics Gaining Ground*. 2017; Available from: https://www.nrel.gov/tech_deployment/state_local_governments/blog/floating-solar-photovoltaics-gaining-ground.
7. Sahu, A., N. Yadav, and K. Sudhakar, *Floating photovoltaic power plant: A review*. Renewable and Sustainable Energy Reviews, 2016. **66**: p. 815-824.
8. Commision, J.R.C.J.E. *Photovoltaic Geographical Information System - Interactive Maps*. Available from: http://re.jrc.ec.europa.eu/pvgis/apps4/pvest.php .
9. Mitsubishi, *MLU Series - Photovoltaic Modules*, Mitsubishi, Editor.
10. DEDHE), P.C.f.D.o.E.Δ.-. *PV as a selft-producer with a net-metering scheme*. 2017; Available from: https://www.deddie.gr/el/upiresies/fwtovoltaika-kai-alles-ape/fv-apo-autoparagwgous-me-energeiako-sumpsifismo-ne.
11. ENERGY, D.O.O.I. *Levelized Cost of Energy (LCOE)*. Available from: https://energy.gov/sites/prod/files/2015/08/f25/LCOE.pdf.
12. Lucie Tesniere, D.d.J., Paul Noothout, Sofia Boutsikoudi, Robert Brückmann, Flora Borek, Ivana Naydenova, Silvia Nicola, Boris Valach , Mak Dukan, Leo Jerkic, Marko Dabetic, *Mapping the cost of capital for wind and solar energy in South Eastern European Member States*. 2017, ECOFYS.
13. Wirth, D.H., *Recent Facts about Photovoltaics in Germany*. 2017, Fraunhofer.
14. HELAPCO, *Statistical Results of PV market in 2016*. 2016.
15. Adem Kaya, H.E.D., Mustafa Ekrem Burhan and Tugay Döner, *Transformer Aging Via Better Understanding The Lifetime Factors*. 2015.
16. Livaniou, I., *Viable energy management at pumping stations with utilization of solar energy*, in *Environmental Engineering - Lab of Renewable and Viable Energy Systems*. 2015, Polytechnical University of Crete: Chania.
17. Patil, M. *Canal-top solar power plants: One example of Narendra Modi's Gujarat Model that's working well*. 2017; Available from: http://www.firstpost.com/india/canal-top-solar-power-plants-one-example-of-narendra-modis-gujarat-model-thats-working-well-3225926.html.

PV ON LANDFILLS – A DUTCH CASE STUDY USING FLEXIBLE PV MODULES

K. Sinapis, M.N. van den Donker, W. Folkerts
Solar Energy Application Centre, High Tech Campus 21, Eindhoven, The Netherlands
Email: sinapis@seac.cc Phone: +31612339374

ABSTRACT: Un-doubtfully, there are great benefits from building PV systems on capped landfills and other waste containment sites. These projects revitalize otherwise unusable property at lower land costs. In this paper the authors describe a field test and case study for capping a Dutch landfill with a solar capping layer. Technical feasibility of the capping could be demonstrated in the field test. Scenarios for economic feasibility were calculated. Results show that the exposed geo-membrane solution is very promising in terms of performance, applicability and financial sustainability.

Keywords: Landfills, a-Si, Integrated system

1 INTRODUCTION

Spread over the Netherlands there are more than 4,000 landfills, with a total area of approximately 9,500 ha. Since 1980, only so-called IBC landfills (Isolate, Control and Monitor) are active. These landfills are covered by the Environmental Management Act and comply with strict environmental requirements. For example, pollution dusts should not leak from the landfill into the groundwater, and the landfill sites will be provided with both a bottom seal (at the beginning) and a top seal (when closed). These IBC landfills together comprise an area of approximately 1400 hectares, of which 850 hectares have to be sealed in the future. The decrease in the number of active landfills does not mean a decrease of the problem: landfill sites require perpetual care. The Environmental Management Act sets the province responsible for the perpetual after-care of landfills which have been landfilled after 1 September 1996. The closed landfills can be considered as 'lost space': because of the contents of the landfill, a new destination is only in very limited cases possible. Most closed landfills do not get a new destination, the scarce land can no longer be used for any other purpose.

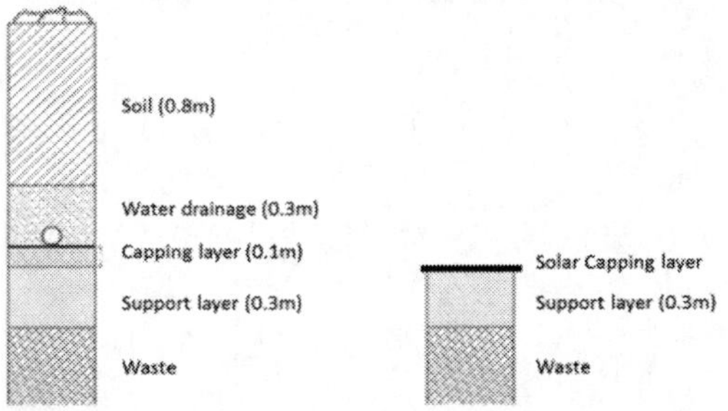

Figure 1: Traditional landfill sealing (left) compared to exposed geomembrane solar capping landfill sealing (right)

2 CONSIDERATIONS WHEN DESIGNING A PV SYSTEM ON A LANDFILL

The following constraints make the installation of solar PV on top of a landfill a challenge. For each landfill, its characteristics have to be evaluated and studied for feasibility. Finally the limited experience of PV in landfills is crucial since the first systems were installed relatively recently around 2010.

2.1 Challenges for installing solar PV on landfills

Despite the great potential of integrating PV on landfills, there are certain constraints that limit this application:

a) **Puncturing of the capping membrane**: The first priority for any landfill is that it should remain eternally covered. This is achieved using a high quality sealing of certified HDPE geomembrane and soil cover on top. The roots of grasses and other vegetation on the soil cover prevent the soil from washing away and exposing the HDPE geomembrane. Installing solar panels on top of the landfill causes the plants to die and the ground beneath it washes away. Additionally, the foundations and fastening frames for the panels can pierce the geomembrane film. This is a risk and hence the PV panels have to be mounted in such a way that it does not damage the HDPE geomembrane.

b) **Waste settlement**: Another common problem is that the surface of the landfill is not well defined. Depending on the type of waste material composition, the waste may decay at different rates throughout the landfill resulting in uneven surface of the landfill. This surface will be difficult to deal with when installing regular solar systems. The uneven surface causes stress on the mounting systems which may result in misalignment of arrays and hence decrease the energy production.

c) **Slope, orientation and stability**: Many landfills have steep slopes (15-30 degrees), and as the slope increases, the complexity of the PV system design increases, resulting in increased system costs. Installing PV arrays on steep slopes can lead to system design challenges associated with wind loads, soil erosion and foundation stability. These challenges often lead to increased system costs for traditionally anchored PV systems. The orientation (or azimuth) of the slope is also important. Generally the developers prefer south facing slopes to provide sufficient exposure to the sun over the course of the year. Slopes with orientations outside of 20-30 degree range of due south typically result in lower annual energy production from the PV system and will require additional design work and system layout modifications. The soil stability of the slope is another relevant engineering consideration. The soil has to withstand both the construction and operation of the PV system. The installation of the solar arrays on steep slopes can be challenging as the weight of the system places additional force on the slope and can lead to failure if the system is not properly designed.

2.2 An overview of PV integration methods on landfills

Various research and commercial projects have been conducted worldwide covering landfills with PV. In these projects, the following techniques were implemented to integrate PV on landfills.

a) **Solar panels mounted on the ground**: This technique has been used on majority of the projects worldwide. The landfill seal utilizes a HDPE foil and on top is covered with a thick soil layer. The PV system is mounted on top of the soil layer. For this technique, there are numerous examples worldwide. Figure 2 shows an example of the ground mounted PV system on the Canton landfill, Massachusetts. This technique, however, can only be applied to flat surfaces or surfaces without significant slope. If the slope is too steep, the soil will erode causing the HDPE foil to be exposed. Hence this technique is not suitable for landfills with steep slope.

Figure 2. Aerial view of 5.75 MW Ground-Mounted Solar System installed on the Canton landfill, Massachusetts

b) **Solar panels mounted top on top of an erosion protection layer**: To counter the problem of erosion of top soil layer, an additional erosion protection layer is applied. This additional erosion protection layer can be, for example, concrete or an anti-root geotextile. The advantage of this technique is that it ensure that rain water can easily drain from the surfaces thus ensuring that water does not drain into the landfill body. However the disadvantage is that there are additional costs for placing this erosion protection layer. An example of PV system using this technique is shown in Figure 3.

Figure 3. 3.8MWp PV system in Luneburg landfill

c) **PV integrated geomembrane**: PV integrated geomembranes are emerging PV technology solutions for landfill applications. A PV integrated geomembrane is a landfill cover, typically made of a TPO (Thermoplastic polyolefin) material, which can be used in place of a vegetative or final cover on the landfill. The PV integrated geomembrane provides good cover stability, reduces maintenance costs of the landfill cover, reduces erosion and infiltration of rainwater into the landfill body. A famous example for this technique is the PV integrated geomembrane in Hickory Ridge Landfill shown in Figure 4. Here, the thin film solar panels are glued on top of green colored TPO geomembrane covering the landfill. The geomembrane is made of high strength material for outdoor exposure. The geomembrane covered landfill slopes provide an ideal, clean and stable surface for the thin-film panels to be directly glued on top of it.

Figure 4. PV integrated geomembrane at the Hickory Ridge landfill in Georgia

3 FIELD TEST AND CASE STUDY

3.1 Field test

In this project the consortium has developed a landfill PV approach based on an exposed geomembrane solar capping. In figure 5 the installation of the solar capping can be seen, on an active landfill slope situated in Maasvlakte, the Netherlands.

Figure 5. Installation of the exposed geo-membrane solar capping phase I

The field test was installed in two phases, with two adhesion methods used.

Table 1. Summary of the two phases of the field test

Parameter	Flexible module v1	Flexible module v2
System rated size	1,1 KWp	3 KWp
PV module technology	a-Si single junction	a-Si single junction
Temperature coefficient	0,20%	0,20%
Inverter type	Fronius Symo 4.5 3-M	Fronius Symo 4.5 3-M
Inverter rated efficiency	97,2%	97,2%
Installation time	Oct-15	May-16
Number of modules	20	60

Precision shunts and voltage transducers were used to measure the DC currents and voltages respectively. DC power was then calculated using the measured voltage and current. AC power was measured with the help of AC power meters. Module temperatures were measured using thermocouples located at the back of the panels. In plane-irradiance was measured using a pyranometer installed in the same plane as that of the PV panels. Additional information such as the ambient temperature and wind speed was imported from the weather station "Hoek Van Holland" which is the closest located station to the site. Data acquisition was done by means of a data logger. A sampling interval of 5 minutes was chosen for all the DC parameters. The energy on the AC side was recorded every hour by AC power meters. Important performance metrics were calculated such as the Performance Ratio and specific yield.

3.2 Case study

The wealth of data aquired during the field test period was utilized to perform a case study simulation based large scale PV plant, covering the entire south slope of the landfill which is approximetely 7 hectares. To get an estimate about the usable area for the PV plant, a 3D model of the entire landfill was made with the help of SketchUp. The side view of the landfill model is shown below in Figure 8 with the south slope indicated by the white region. The roads are indicated by the brown region and the green region represents grass.

The total installed capacity assuming a-Si single junction solar modules is around 1.1MWp with an annual yield of 1110 MWh/year (PVSyst) or around 76% PR. The energy produced from the PV plant can be utilized for the operation of the waste and water treatment plants on the landfill site. There water purification unit is a water purification facility that runs continuously for 24 hours, 7 days a week consuming approximately 760 MWh per year. The Ashes Immobilization unit runs 9 hours daily except weekends consuming around 490 MWh per year.

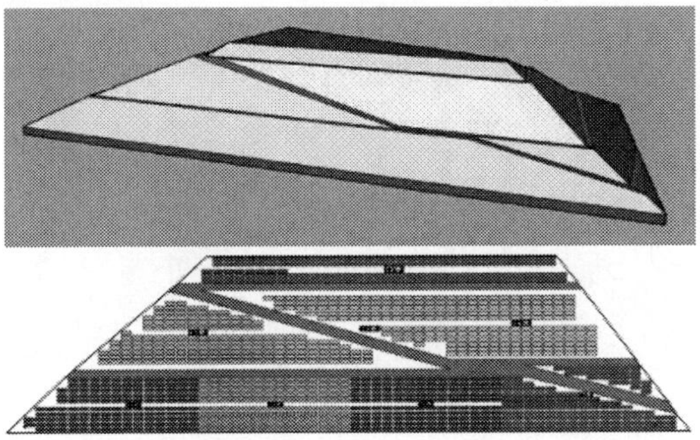

Figure 6. South slope of the landfill designed in SketchUp (top) and array layout including module groups and inverters (bottom)

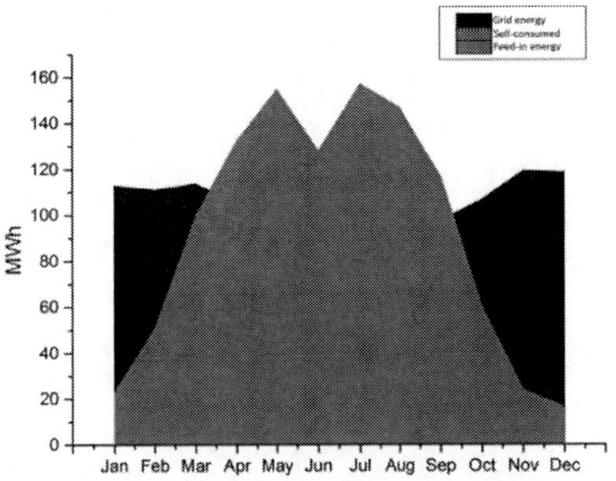

Figure 7. Comparison of monthly PV produced and consumed energy by the facilities of the landfill

Figure 7 shows the correlation of produced and consumed energy throughout the year for this layout. While the consumption is lower during summer months due to less water pumped for purification the PV production is higher. The excess energy can be then fed back to the grid. The produced energy during winter months cannot fulfill the energy needs of the water purification units and thus extra energy has to be supplied by the electricity grid.

3.3 Scenario modeling

Finally, we performed cost scenario modeling to explore the performance and economics of the landfill capping in various boundary conditions. Based on publicly available data and in-house calculations we estimated the investment costs required for the PV plant in the layout of section 3.2 to be above 1.50 €/Wp, which does not make it an attractive investment yet. We used scenario modeling to explore possibilities to get the costs down. Figure 8 shows the effect of the scenario calculations on projected PV plant capacity and investment costs.

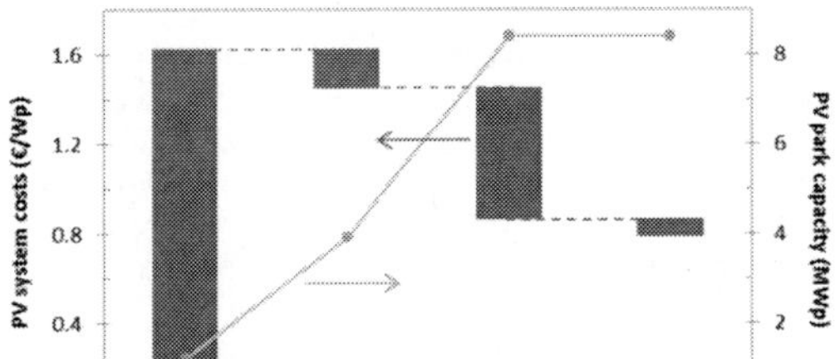

Figure 8. Scenario calculations for investment costs (left axis) and PV plant capacity (right axis) in the base case, and with scenarios of higher packing density, improved efficiency and lower module price.

(1) Higher packing density. As can be seen in Figure 7 the packing density of modules on the geomembrane was quite low. By using larger modules and packing them more closely together, the PV capacity could increase from 1.2 to 4 MWp.

(2) Higher efficiency. We used single junction amorphous silicon modules in the field test and case

study. By adopting a higher efficiency cell architecture the capacity of the PV plant could increase to 8 MWp. If the module manufacturing costs per m^2 remain similar a significant cost reduction could be reached as well.

(3) Lower module price. By increasing the market for flexible PV modules a further cost reduction of flexible modules towards the level of glass based modules can be foreseen.

4 CONCLUSION

We have investigated the use of a solar capping layer on Dutch landfills, and can draw the following main conclusions:

- Landfills can be a good opportunity for PV deployment. It is 'waste-space' without other possible uses and requires perpetual care that could be paid for by the revenue stream of the PV plant.
- There are many challenges in applying PV on land fills, such as settlement of the waste, erosion on the top layer and avoiding puncturing of the capping layer.
- A solar capping layer is a good option to overcome the challenges. On top of that it provides a potential cost advantage by making the landfill capping thinner. A solar capping layer consists of an exposed geomembrane with integrated PV modules.
- We have demonstrated the technical feasibility of a solar capping layer on a landfill in Maasvlakte, The Netherlands.
- The economics of the solar capping are promising, but also need further attention. It is a highly attractive investment if packing density can be increased, the efficiency of the flexible PV modules can be increased, and the cost of the flexible PV modules can be lowered.

ACKNOWLEDGEMENT

The authors like to acknowledge "PAS" project partners Hans Kolijn, Willem Boekholt (Renewi), Edward Hamers (HyET Solar), Hans Geusebroek and Tijs van Gisbergen (Sweco) and Carlo Scheerder (Genap). Many thanks go to the SEAC interns Lavanya Chandrashekar and Rico Berix that made good contributions to this research. Last but not least, SEAC would like to acknowledge the Topsector Energy, the TKI Urban Energy and the Netherlands Enterprise Agency for financial support.

RENEWABLE ENERGY HIGH PENETRATION SCENARIOS USING MULTI-NODES APPROACH: ANALYSIS FOR THE ITALIAN CASE

Matteo Giacomo Prina[1,2,*], Giampaolo Manzolini[2], David Moser[1], Wolfram Sparber[1]
[1]Institute for Renewable Energy, EURAC Research, Viale Druso 1, 39100 Bolzano (BZ), Italy
[2]Dipartimento di energia, Politecnico di Milano, Via Lambruschini, 4, 20156 Milano (MI), Italy
*Phone: +39 0471 055587; Fax: +39 0471 055699;
E-mail: MatteoGiacomo.Prina@eurac.edu
giampaolo.manzolini@polimi.it
david.moser@eurac.edu
wolfram.sparber@eurac.edu

ABSTRACT: The integration of high shares of variable renewable energy sources is becoming more and more challenging. The main options to achieve this task are the enlargement and reinforcement of the grid, the use of storage devices, and demand-side management. In this work, a multi-nodes modelling approach has been implemented to study the current situation of the Italian electric system. After evaluating the potentialities of variable renewable energy sources such as photovoltaic and wind power in terms of new installation and overall installed power, different scenarios have been developed. The final aim is to compare the single-node to the multi-nodes approach and to study the role of pumped hydro storage system as existing storage solution in Italy. At the increasing of the renewables generation, the results show the benefits in using the multi-nodes more complex methodology compared to the single-node one. Considering the overall potential of Italian variable renewable generation within the system, the single-node model presents an under-estimation of the curtailments of about 7% and thus an over-estimation of the percentage of renewables in the system equal to 6.6%. The already installed power of pumped hydro storage allows for a better integration of renewables and highlight its central role in increasing the percentage of renewable electricity consumption.

1 INTRODUCTION

This paper deals with the integration aspects of high share of variable renewable energy sources (VRES) into the electric grid. A model based on the Oemof python framework has been developed in order to perform a multi-nodes analysis on the Italian electricity sector [1, 2]. The model is characterized by hourly base dispatch optimization which means that the use of resources is optimized to satisfy the demand at least costs in terms of CO_2 emissions.

With this model the final aim is to assess the current and potential renewable energy sources (RES) electricity generation, the congestion problems and the degree of integration of RES into the grid. The purpose is also to identify the role of the already installed pumped hydro storage capacity and to quantify the importance of a multi-nodes versus a single-node modelling approach.

The novelty of the present paper is given by the multi-nodes modelling approach coupled to the hourly based dispatch optimization accounting for the national power system constraints: the variability of RES generation, the congestion problems between nodes and the role of storage.

This methodology is applied to the Italian case because of its morphological characteristics: there are significant differences in the variable generation by photovoltaic (PV) and wind power between areas, especially between north and south. The actual configuration of the Italian electric system consists in six main zones defined by the grid transmission capacity. The first critical obstacle concerns the geographical location of the different resources where the renewable energy generation and its potential is mainly placed in the south of Italy and the availability of pumped hydroelectric storage (PHS) is in the north.

The potential of PV and wind power has been estimated for the Italian territory. A multi-nodes model has been developed to assess eventual issues of the electric system configurations in the transition towards the installation of the whole variable renewable energy potential. Results show the importance of the multi-nodes approach compared to the single-node approach. The curtailments from renewables estimated through the single-node approach are largely under-estimated if compared to the values obtained through the multi-nodes approach. The results show also the fundamental role of the already installed pumped hydroelectric storage capacity to increase the renewables penetration in Italy. The possibility to install batteries or other types of electric storage is not considered in this paper, but it will be considered for future development of this work.

2 METHODOLOGY

High penetration of wind and solar sources is necessary to increase RES generation up to 100%, and, thus, issues related to the non-dispatchable power production must be faced. This paper assesses the integration of RES from a technical and economic point of view to regional scale, applying the model to the Italian case. For several nations and regions single node models have been implemented showing its capabilities together with the needs of widely RES based scenarios. Examples are the model elaborated for Germany by H. M. Henning et al. [3, 4] or other models elaborated through the EnergyPLAN software [5-10]. From a European perspective Pleßmann et al. [11] have developed some decarbonization pathways for European electricity supply until 2050. In this paper, a similar multi-nodes model is developed using the Open Energy System Modelling Framework (Oemof) which is a free, open source toolbox to assess energy supply systems. It is developed in Python and designed as a framework with a modular structure [1, 2]. Through this framework, for example, C. Möller et al. [12] have studied the technical potential of electrical storage technologies for a region in the northern part of Germany. In particular, the focus was the required storage capacity at different RES shares by using annual simulations in hourly time steps.

Firstly, the structure of the electric system with the corresponding transmission constraints has been analyzed

[13, 14]. Figure 1 shows the six main areas in which the Italian electric system is divided and the grid connection capacities between them.

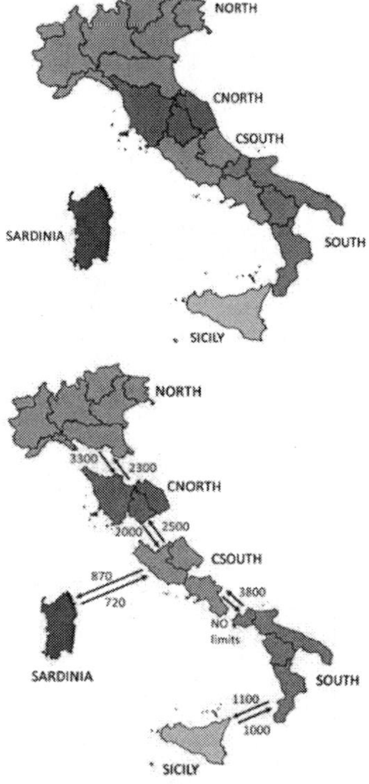

Figure 1: Division of the Italian territory into the different main areas of the electric system, identified by the transmission constraints of the grid [MW] [14].

The data set, on which this analysis is based, is made by data series with hourly resolution for the year 2014 [14]. The electricity production by hydro power has been normalized on the average equivalent hours of the previous five years For each node, the baseline of the current electric system has been created. Figure 2 shows the structure of the multi-nodes model created for the Italian electricity system.

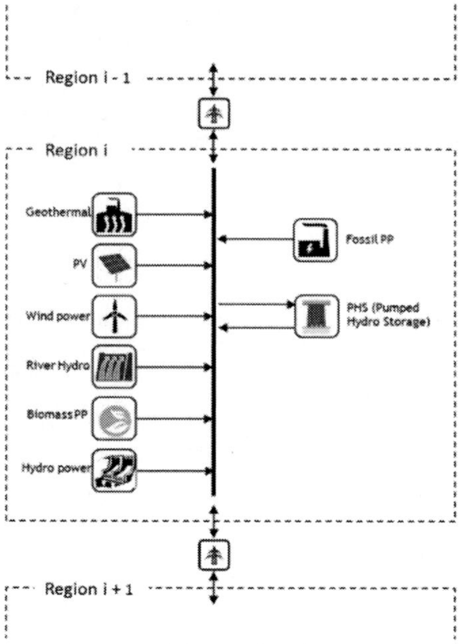

Figure 2: Structure of the multi-nodes model developed for the Italian electricity system.

Thus, the model takes into consideration not programmable RES like PV and wind power, RES with a fixed constant production like geothermal and biomass power plants, programmable RES like hydro power, programmable conventional power plants and pumped hydro storage.
The importance of a multi-nodes versus single-node approach can be easily identified by looking at the spatial characteristics of the Italian already installed renewable generation park. The distribution among the regions of the PV and wind installed power, and the capacity of pumped hydro storage is reported in Figure 3. This latter shows a fundamental and critical issue of the Italian electric system. While PV is homogeneously distributed, the majority of the wind installed power together with its potential is concentrated into the south of Italy in contrast to the available installed capacity of storage placed in the north.

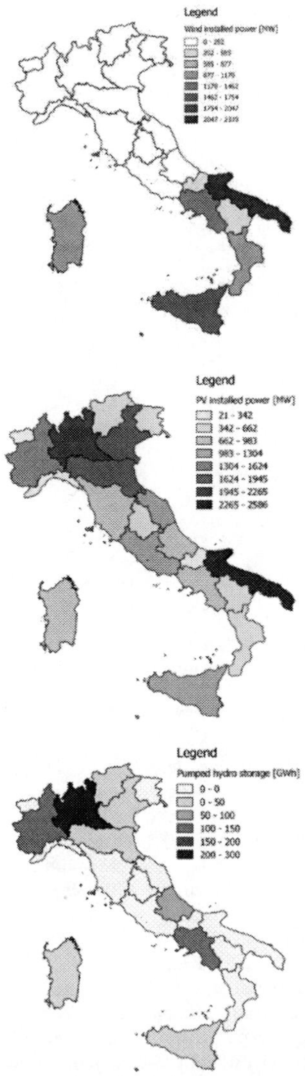

Figure 3: Distribution of the installed power of PV, wind and installed capacity of pumped hydro storage among the Italian regions [14, 15].

The maximum potential of rooftop PV and wind has been estimated for each region following the Re-shaping European project for wind [16] and some other studies for PVs [17, 18]. The overall potential is equal to about 120 GW and 49 GW for PV and wind power respectively, while the installed capacity in 2014 was 18.6 GW of PV and 8.7 GW of wind power. The distribution of these capacities in the Italian regions is based on different assumptions. For wind power, the assumption of the Re-shaping project is to replace existing turbines with larger ones, therefore the potential distribution reflects the current installed power distribution. A homogeneous distribution per dwelling is assumed for rooftop PV. Thus, the final distribution depends on the number of inhabitants per region and deviates from the current installed power distribution. Figure 4 summarizes the increase of installed PV and wind power for the six zones for the considered scenarios.

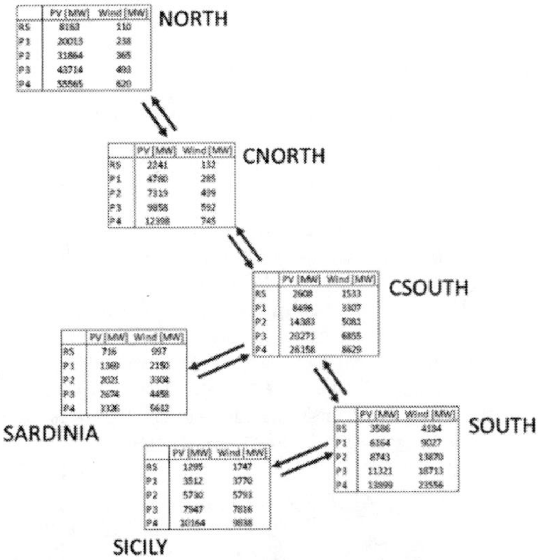

Figure 4: Installed power of PV and wind power divided per region for the reference scenario and four other scenarios with additional installed capacity.

The developed model is based on hourly base dispatch optimization which means that the use of the sources is optimized to supply the demand minimizing the CO_2 emissions. Therefore, the optimization starts to cover the demand for every hour in each node with the production from RES and then uses the power production from fossil fuel plants. If excess electricity production from RES occurs, the optimization evaluates the best option among the export of electricity to another node, storing electricity through pumped hydro storage if available or exporting to another node where there is availability of storage. The transmission losses that have been considered between two adjacent nodes are equal to 3% [1]. Figure 5 shows a diagram that summaries the main choices available within the dispatch optimization analysis.

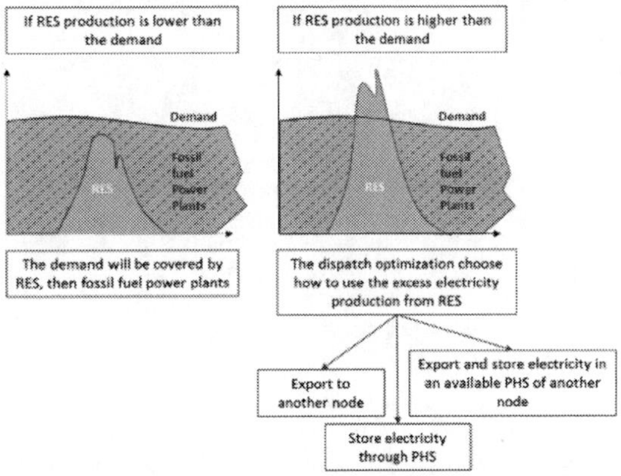

Figure 5: Diagram of the dispatch optimization operations.

The Oemof framework based on pyomo library [19-20] allows us to create and solve linear programming (LP) and mixed integer linear programming (MILP) optimization models specific to energy systems. It allows the flexible creation of the energy system model choosing between different components such as: source, sink, transformer and storage. Linear programming is used in this framework to produce a least-cost energy system where the costs can also be represented by the CO_2 emissions. Each component added to the energy system model brings a certain number of constraints to the model.

3 RESULTS

The developed baseline 2014 for Italy presents the following characteristics: an overall electricity consumption of 310.9 TWh and an average RES percentage of 33.4 %. Figure 6 shows the different RES production share on the overall electricity consumption of the zone. The highest value is in the South region characterized by large electricity production from PV and wind power. The overall minimum load of the whole Italy is about 19 GW with the maximum being 53 GW.

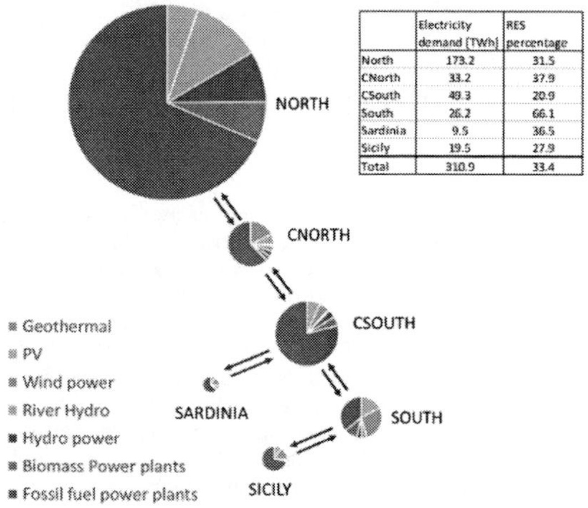

	Electricity demand [TWh]	RES percentage
North	173.2	31.5
CNorth	33.2	37.9
CSouth	49.3	20.9
South	26.2	66.1
Sardinia	9.5	36.5
Sicily	19.5	27.9
Total	310.9	33.4

Figure 6: Italian energy system baseline for the year 2014.

The multi-nodes approach has been inspected and compared to the single-node in order to quantify the differences linked to the single-node assumption.

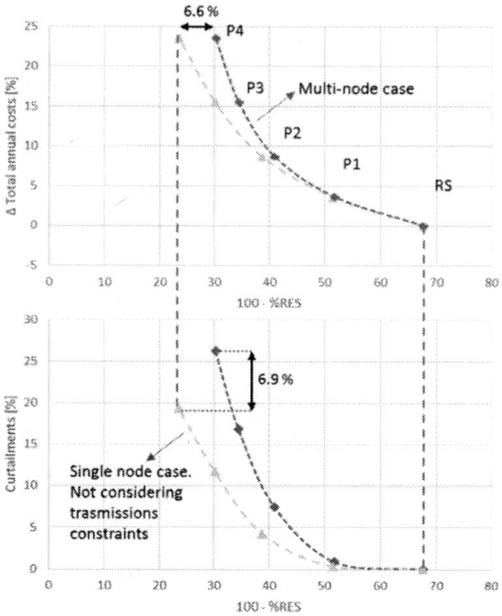

Figure 7: Mapping of the different configuration of the electric system increasing the capacity of PV and wind power. Comparison between the multi-nodes and single-node approach.

Figure 7 shows four different solutions with increasing capacity of PV and wind power starting from the reference scenario solution that refers to the Italian electricity system in 2014. The final solution P4 represents the configuration of the energy system where the maximum potential PV and wind power are exploited (i.e. 120 GW for PV and 49 GW for wind). The intermediate solutions, points P1, P2, P3 are characterized respectively by the 25%, 50% and 75% of the overall missed VRES potential (see figure 4). Figure 7 outlines the importance of the multi-node approach when large RES penetration is considered. While for modelling the reference scenario the use of single-node approach does not produce different results if compared to the multi-nodes, due to limited VRES installed power, it underestimates the curtailments for increasing RES capacities together with an over-estimation of the percentage of renewables in the system.

Figure 8 shows the hourly electricity production in the six main zones of Italy for scenario P3 in a summer week. Scenario P3 is characterized by PV and wind power equal to the 75% of their potential. This figure allows to highlight the over-generation problems, especially in south regions. The region R4 – SOUTH present peaks of over-generation higher than 10 GW while the transmission constraints of the grid from south to northern regions is in the order of 3-4 GW (see Figure 1).

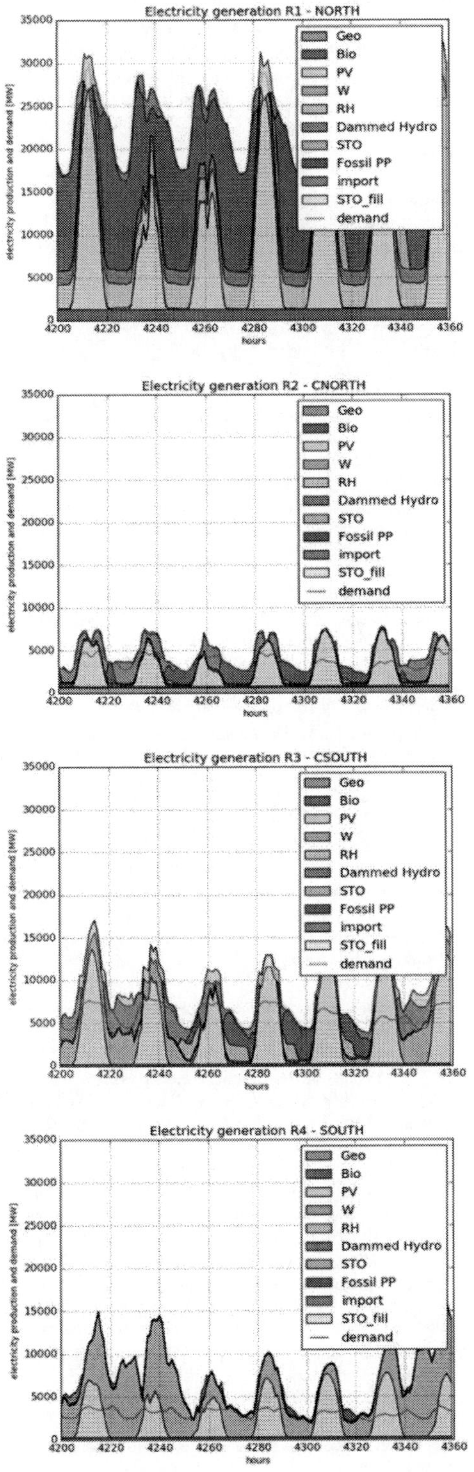

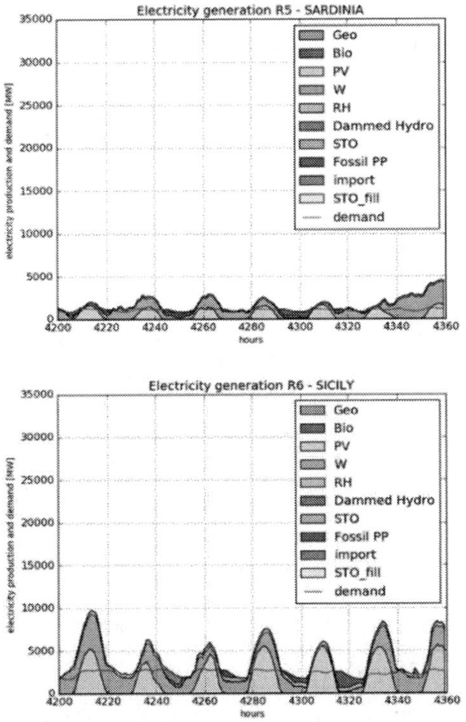

Figure 8: Hourly electricity production in the six main zones of Italy for scenario P3, where PV and wind power are increased to the 75% of their potential, in a week of summer.

In order to identify the future potential role of pumped hydro storage systems already installed over the Italian territory, other simulations has been run neglecting the pumped hydro storage capacity. Figure 9 shows the comparison between the actual case considering pumped hydro storage and the one not considering it. In this way, it is possible to estimate the future potential contribution of PHS. In the P4 scenario, with all the potential PV and wind power installed, the already existing pumped hydro storage capacity leads to an electricity consumption increase from renewables by 3.6%. The actual installed capacity of PHS permits to save curtailments of the grid equal to 4.7% of the total production from RES.

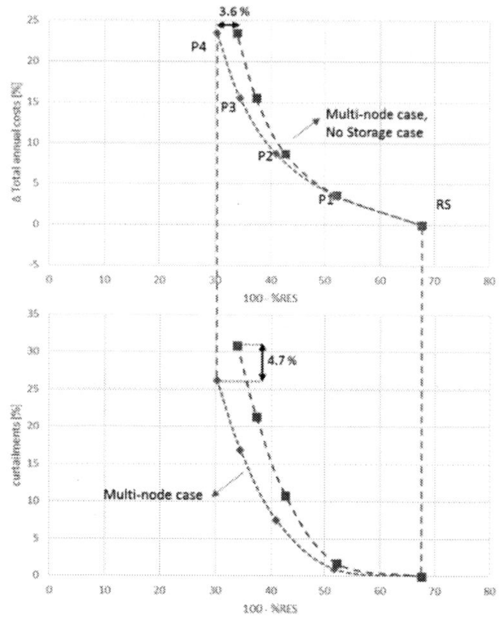

Figure 9: Mapping the different configuration of the electric system increasing the capacity of PV and wind power. Comparison with the case not considering pumped hydro storage.

4 CONCLUSIONS

A multi-nodes energy system model has been created to simulate one year with hourly time-step. The model has been applied to the Italian electricity system. The baseline for the Italian electricity system has been created starting from real data for the year 2014.

The aim of the paper is to investigate and quantify the importance of multi-nodes versus single-node approach and the role of the already installed storage at the increasing of variable renewable energy sources installed power.

The results have shown the importance of the multi-nodes versus single-node approach in energy system modeling. The multi-nodes approach has resulted particularly important for a territory like Italy where the majority of the installed power and especially the potential of variable renewable energy sources is placed in the south of Italy while the installed capacity of pumped hydro storage is placed mainly in the North. The simulation highlighted how the current transmission grid with an increase of variable RES is not sufficient to absorb the over-generation from RES without congestion problems and curtailments. The single-node modeling approach has shown not negligible differences with the multi-nodes approach. Indeed, the single-node model presents an under-estimation of the curtailments of about 7% and thus an over-estimation of the percentage of renewables in the system equal to 6.6%, considering the overall potential of VRES within the system.

Secondly, the potential benefits of pumped hydro storage in terms of integration of renewables have been inspected. The already installed capacity of pumped hydro storage has been considered and two different kind of simulations has been run with the multi-node approach: with and without this storage in order to evaluate its contribution at the increasing of RES. PHS has resulted to have a central role in the future integration of RES. For this reason, at the increasing of variable renewable energy sources, in the electricity market a mechanism to privilege pumped hydro storage systems has to be created in order to exploit the renewables integration potential role of these plants.

5 ACKNOWLEDGEMENTS

The authors wish to acknowledge the use of EURAC internal funding for the project "Regional Energy Modeling" and the availability and hospitability of the Oemof team at Reiner Lemoine Institut.

6 REFERENCES

[1] Oemof Developer Group. (2017, March 28). Open Energy Modelling Framework (oemof) - A modular open source framework to model energy supply systems. Version v0.1.4. Zenodo. http://doi.org/10.5281/zenodo.438676

[2] Hilpert, S.; Günther, S.; Kaldemeyer, C.; Krien, U.; Plessmann, G.; Wiese, F.; Wingenbach, C. Addressing Energy System Modelling Challenges: The Contribution of the Open Energy Modelling Framework (oemof). Preprints 2017, 2017020055 (doi: 10.20944/preprints201702.0055.v1).

[3] H.-M. Henning and A. Palzer, "A comprehensive model for the German electricity and heat sector in a future energy system with a dominant contribution from renewable energy technologies—Part I: Methodology," Renew. Sustain. Energy Rev., vol. 30, pp. 1003–1018, Feb. 2014.

[4] A. Palzer and H.-M. Henning, "A comprehensive model for the German electricity and heat sector in a future energy system with a dominant contribution from renewable energy technologies – Part II: Results," Renew. Sustain. Energy Rev., vol. 30, pp. 1019–1034, Feb. 2014.

[5] B. Ćosić, G. Krajačić, and N. Duić, "A 100% renewable energy system in the year 2050: The case of Macedonia," Energy, vol. 48, no. 1, pp. 80–87, Dec. 2012.

[6] D. Connolly, H. Lund, B. V. Mathiesen, and M. Leahy, "The first step towards a 100% renewable energy-system for Ireland," Appl. Energy, vol. 88, no. 2, pp. 502–507, Feb. 2011.

[7] L. Fernandes and P. Ferreira, "Renewable energy scenarios in the Portuguese electricity system," Energy, vol. 69, pp. 51–57, May 2014.

[8] H. Lund and B. V. Mathiesen, "Energy system analysis of 100% renewable energy systems—The case of Denmark in years 2030 and 2050," Energy, vol. 34, no. 5, pp. 524–531, May 2009.

[9] D. Connolly, H. Lund, B. V. Mathiesen, E. Pican, and M. Leahy, "The technical and economic implications of integrating fluctuating renewable energy using energy storage," Renew. Energy, vol. 43, pp. 47–60, Jul. 2012.

[10] D. Connolly, H. Lund, B. V. Mathiesen, and M. Leahy, "Modelling the existing Irish energy-system to identify future energy costs and the maximum wind penetration feasible," Energy, vol. 35, no. 5, pp. 2164–2173, May 2010.

[11] G. Pleßmann and P. Blechinger, "How to meet EU GHG emission reduction targets? A model based decarbonization pathway for Europe's electricity supply system until 2050," Energy Strategy Rev., vol. 15, pp. 19–32, Mar. 2017.

[12] C. Möller, K. Kuhnke, M. Reckzügel, H. J. Pfisterer, and S. Rosenberger, "Energy storage potential in the Northern German region Osnabruck-Steinfurt," in 2016 International Energy and Sustainability Conference (IESC), 2016, pp. 1–7.

[13] ENTSOE. < https://www.entsoe.eu/Pages/default.aspx>

[14] TERNA. < http://www.terna.it/en-gb/home.aspx>

[15] GSE. < http://www.gse.it>

[16] Reshaping european project < http://www.reshaping-res-policy.eu/>

[17] "DEPLOYMENT PATHWAYS FOR PHOTOVOLTAICS IN THE EU TOWARDS 2020: COMPARING ECONOMIC FACTORS WITH POLICIES AT MUNICIPAL LEVEL - EU Science Hub - European Commission," EU Science Hub, 21-Dec-2015.

[18] PV COSTS IN EUROPE 2014-2030, EU PV TECHNOLOGY PLATFORM. < http://www.etip-pv.eu/publications/other-publications/pv-costs.html/>

[19] Hart, William E., Carl Laird, Jean-Paul Watson, and David L. Woodruff. Pyomo – Optimization Modeling in Python. Vol. 67. Springer, 2012.

[20] Hart, William E., Jean-Paul Watson, and David L. Woodruff. "Pyomo: modeling and solving mathematical programs in Python." Mathematical Programming Computation 3, no. 3 (2011): 219-260.

DECENTRALIZED FUZZY-BASED VOLTAGE CONTROL FOR LV DISTRIBUTION SYSTEMS

Efraín Bernal A.[(a)], Maximiliano Bueno-López[(a),(b)] and Marta Molinas[(b)]
(a) Department of Automation Engineering, Universidad de La Salle, Cra. 2 No. 10-70 Bogotá-Colombia
(b) Department of Engineering Cybernetics, Norwegian University of Science and Technology, Trondheim-Norway

ABSTRACT: With the inclusion of Information and Communication Technology (ICT) components into the low-voltage (LV) distribution grid, some measurement data from smart meters are available for the control of the distribution networks with high penetration of photovoltaic systems (PV). This paper undertakes a decentralized voltage-control strategy for smart LV distribution networks, by using a Fuzzy Control in combination with the information collected from smart meters as source of information of the status of the system. Comparison studies for a smart LV distribution system to validate the method are presented, showing the robustness of proposed method compared with the traditional local approaches. To illustrate the applicability of the proposed method, a real-time digital simulator (RTDS) for smart distribution power grids is used. The analysis of the obtained results showed that the proposed method could to mitigate the negative impacts of the high PV penetration, as well as minimize the voltage rise and power losses
Keywords: Fuzzy logic, Photovoltaic system, Smart homes

1 INTRODUCTION

The increase of generation using renewables energies has created the need to make different adjustments in the conventional electrical network. The influx of distributed generation will pose new challenges for electrical power distribution. One of the most used sources is solar energy and as a result, residential rooftop-mounted solar photovoltaic (PV) panels are being installed at an increasing rate around the world [1]. This influence is stronger in some European countries, but in Latin America, the increase of generation with solar energy is still low. Latin American should be viewed as very promising for penetration of nonconventional renewable energy sources, with 8% by 2020 and 20–25% by 2025 in most countries, in countries like Uruguay and Chile, the percentage is higher. In Chile, Brazil, Colombia, and Panama some economic incentives are given, this is motivated by the pressing overall energy deficit and the uncertainty in hydroelectric generation [2]. Regulatory incentives have been delivered in other countries, but for example in Colombia, only in 2014 a law was created to regulate this type of generation [3]. While this occurs in Latin American, the total nominal power of photovoltaic (PV) installed in Germany is close to 38.5 TWh, according to the German Federal Network Agency, covering approximately 7.5 percent of Germany's net electricity consumption and distributed over 1.5 million power plants by the end of 2015 [4]. Indeed, it is planned to meet most of or all of the German energy demand with renewables by 2050, which means that a total of approximately 200 GW PV capacity has to be installed, (i.e., to reach this amount by 2050, something more than 4 GW PV must be installed annually [5].

Indeed, in the last years there has been more distributed generation units embedded with conventional and/or renewable resources into the German LV distribution networks; i.e., residential PV systems represent more than 250,000 solar arrays already connected to the LV distributed network of Bayernwerk AG [4], a regional network operator in Bavaria-Germany. As the penetration of residential and commercial PV increases into electrical distribution systems, problems such as voltage rise, overloading of network equipment, harmonic current emissions, network resonance, false islanding detection, and dc current injections are becoming more of an issue to be addressed carefully [5]. Reverse power flow or voltage increases may occur due to the increasing active power feed-in on the low and medium voltage levels, which make the connection of additional power generation plants more difficult or impossible. Furthermore, existing phase shifts and/or reactive power percentages in the grid reduce its transmission capacity and increase the accumulated transmission losses. One potential solution is the use of smart PV inverters as a promising inexpensive concept to solve the problems caused by PV penetration. Smart PV Inverters have the capability to supply inductive and capacitive reactive power to control the voltage at the connection point with the grid, and this ability of the smart PV inverters is limited only by their technical ratings. Its development and realization attract research efforts in a fairly large number of issues ranging from modeling [5], [6] to implementation [7], [8]. However, although in the last few years voltage control strategies are demonstrated to be able to solve voltage violation in LV distribution systems, international directives as the VDE-AR-N 4105 code do not include the possibility of implement other control strategies, whit which would allow to provide better solutions for this problem.

Different control strategies have been used to solve the voltage violation problem in LV distribution systems and Fuzzy Logic (FL) has appeared as an alternative solution. The Mamdani method is widely accepted in fuzzy expert systems for its ability to capture expert knowledge in fuzzy rules [8]. However, the Mamdani method would produce computational burden. On the other hand, the Takagi-Sugeno method improves the computational efficiency of the fuzzy inference, it also works well with adaptive, and optimization techniques, which makes it very suitable choice for control of a dynamic non-linear systems. It is necessary to tune the fuzzy logic system, which has been developed by adjusting the specific fuzzy sets and fuzzy rules to meet specified requirements desired.

At this aim, in this paper a fuzzy control strategy is proposed. It is based on reactive power control of smart PV inverters by using the data received from the smart meters. The proposed fuzzy formulation, which simultaneously minimizes the magnitude of the voltage rise and reduces the power losses, includes a function to evaluate the reactive power output of the inverters to improve the power quality in LV distribution network avoiding the use of an on-load tap-changing transformer (OLTC) in the substation. The remainder of this paper is organized as follows. Section 2 introduces the structure of the PV control strategies and Section 3 presents the proposed fuzzy decentralized control strategy (FDCS). The experimental setup and simulation results are illustrated in Section 4. Section 5 describes the conclusions drawn from the study carried out in this paper and suggests some guidelines for the future work.

2 REACTIVE POWER CONTROL STRATEGIES

Normally, in Low-Voltage distribution networks with PV penetration, the power flow goes from the LV distribution transformer to the load resulting in a decrease of the voltage across the power line. However, when the homes start to feed-in into the grid the active power generated by PV solar systems, a reverse power flow appears incurring excessive voltage increase, exceeding the allowable voltage limits. Fig. 1 shows an example of a LV distribution network with three residential homes equipped with PV solar systems where this voltage fluctuation problem is represented.

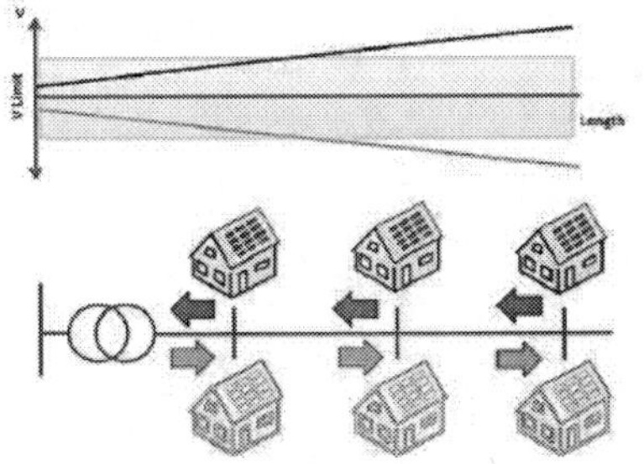

Figure 1: Low Voltage distribution network with PV penetration.

Germany is pioneered in the regulation and use of PV systems; they proposed the VDE-AR-N 4105 code to provide a reliable standard giving the possibility to residential PV systems to feed-in active power into the grid by using local reactive power controls to compensate the problems caused by power injection. This local control schemes (also known as droop-based regulation strategies) make autonomous control of the reactive power supply via characteristic curves. The fixed $cos(\varphi)$, $cos(\varphi)P$ and $Q(U)$ schemes based on droop characteristic are shown in Fig. 2.

The left part of the figure represents the droop-based regulation schemes, and the right part the proportion of related reactive power of the smart PV inverters. The local control strategie is based only in the value of each home without consider the effect on the neighbor home. Therefore, these schemes cannot guarantee optimal control of the smart LV distribution system.

In contrast, central control strategy aims for coordinated control of the complete LV distribution system by using the static and dynamic system information. The central control scheme can be described as a communication based control methodology that allows to optimize the LV distribution grid operation not only locally but also regionally with a common beneficial level for producers and consumers. An example of a central control scheme is presented in Fig. 3 [4].

Such throughput variations and packages loss can affect overall grid performance, by reducing the information flowing between the load meters and the central control station; that is why central voltage-control strategies have not been considered into the actual grid's control processes and standards.

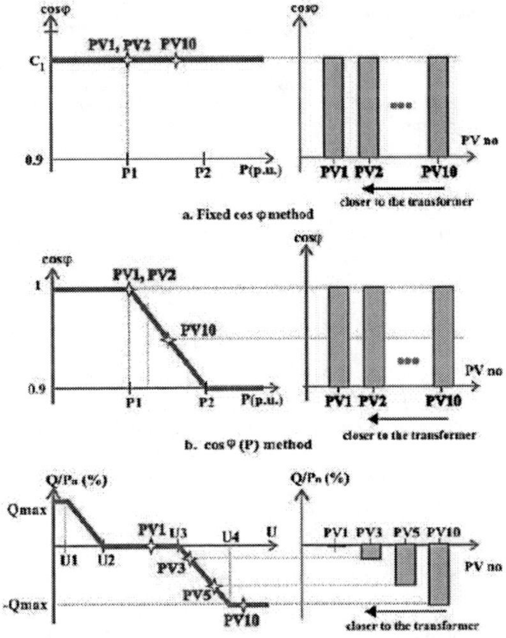

Figure 2: Standard local reactive power methods. a. Fixed $cos(\varphi)$. b. $cos(\varphi)P$. c. $Q(U)$.

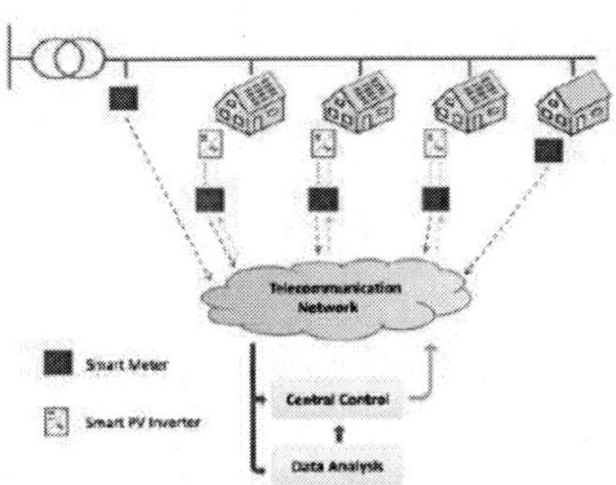

Figure 3: Example of a very well developed centralized voltage-control Strategy.

A current trend in academia and industry is to implement a media solution. Decentralized voltage control strategies are communication-based control strategies too, which interact with an intermediate level in the grid. This is an advantage as LV systems can be clustered into different cells using intelligent, controllable substations.

As can be seen in the Fig. 4, decentralized voltage

controllers influence voltage curves by adding a voltage step using on-load tap-changing transformers (OLTC). An example of a decentralized voltage control strategy is presented in Fig. 4.

The problem with OLTC transformers or voltage regulators is that, the amount of permissible voltage increase is limited if there is a load near the voltage regulator, a common case in LV distribution networks, therefore additional voltage regulators along the feeder may be necessary.

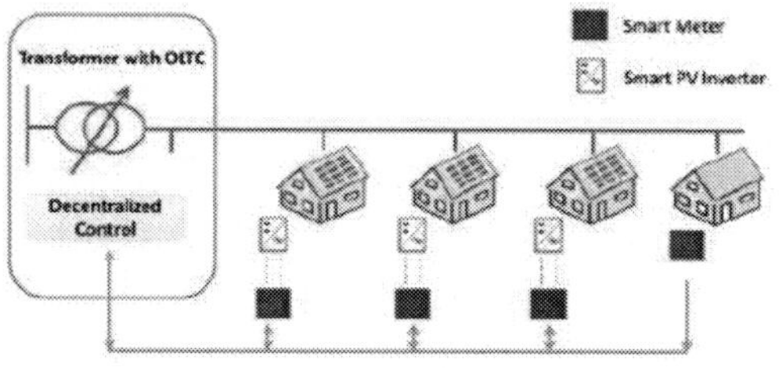

Figure 4: Principal design and components of a decentralized control voltage-control strategy.

3 PROPOSED FUZZY DECENTRALIZED CONTROL STRATEGY (FDCS)

To avoid the use of an OLTC transformer and to integrate the measurements of all the homes with PV systems into the control strategy, a Fuzzy Decentralized Control scheme is proposed and presented in Fig. 5. In Fig. 6 it is possible to observe with more detail the structure of the controller.

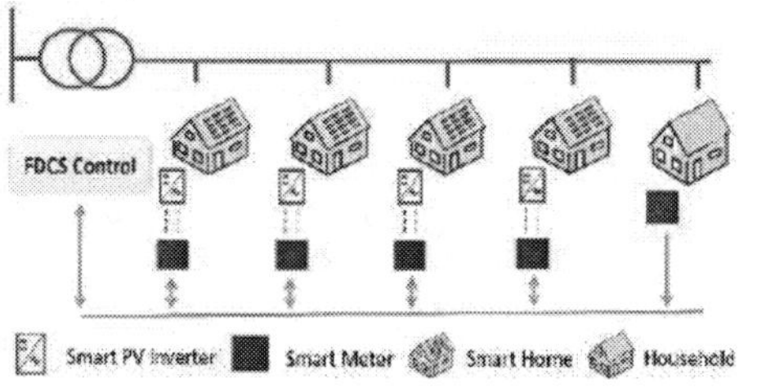

Figure 5: Conceptual scheme of the decentralized control strategy.

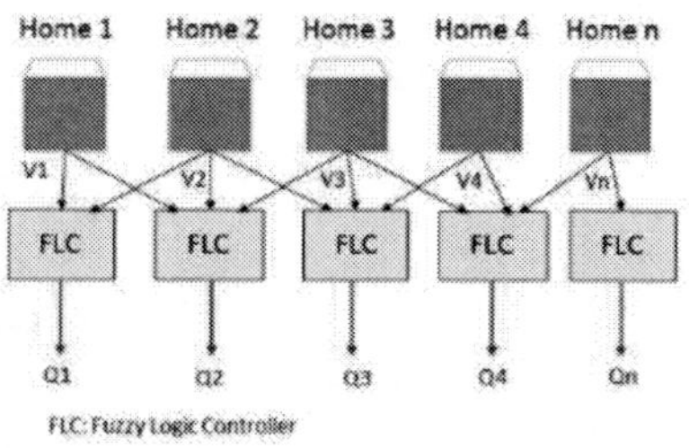

Figure 6: Fuzzy controller structure.

In a typical smart LV distribution grid, the smart PV inverters are controller by a local method as explained in the section before. However, the use of classical $Q(U)$ method in order to control the reactive power has various drawbacks in terms of voltage overshoots, switching frequency, even voltage instability when new smart homes are connected to the LV distribution system [2]. To mitigate this issues, a fuzzy control strategy is used for each smart PV inverter to control the reactive power applied.

For studying purposes a network topology in a low-voltage (0.4kV) distribution system is used. The main elements presents are: a distribution transformer connected to high voltage lines and supplying in a LV grid, transmission lines and loads which are represented by smart homes and non-smart homes. The model of the network is highly nonlinear.

The fuzzy controller has as inputs the errors $e(t)$ defined as the difference between the actual and the desired voltage. For these inputs, we use triangular and trapezoidal membership functions. The voltage in all smart homes is the principal value for the inputs.

The fuzzy controller for the smart home n has normally three inputs, $e_n(t)$ with $n = n-1, n, n+1$, as the voltage error for the evaluated smart home and its neighbours, for which triangular and trapezoidal membership functions are defined.

The input $e_n(t)$ is configured for five degrees with NB= Negative Big, NS= Negative Small, Z=Zero, PS= Positive Small and PB= Positive Big. Fig. 7 shows the membership functions for the inputs.

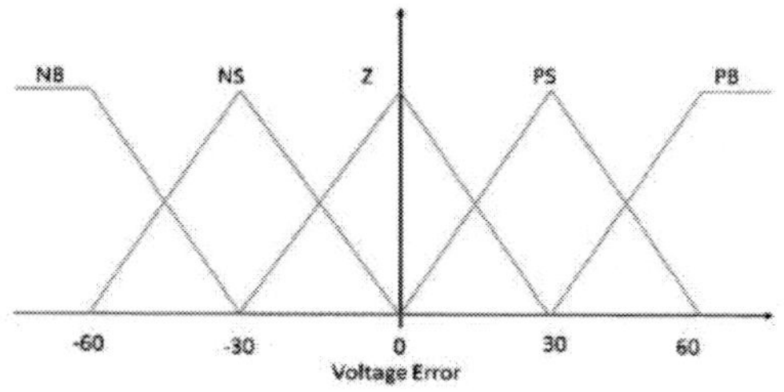

Figure 7: Fuzzy controller structure.

The output of the fuzzy controller (reactive power) has the next membership functions: QNB= Q Negative Big, QNS= Q Negative Small, Z=Zero, QPS= Q Positive Small and QPB= Q Positive Big. Fig. 8 shows the membership functions for the outputs.

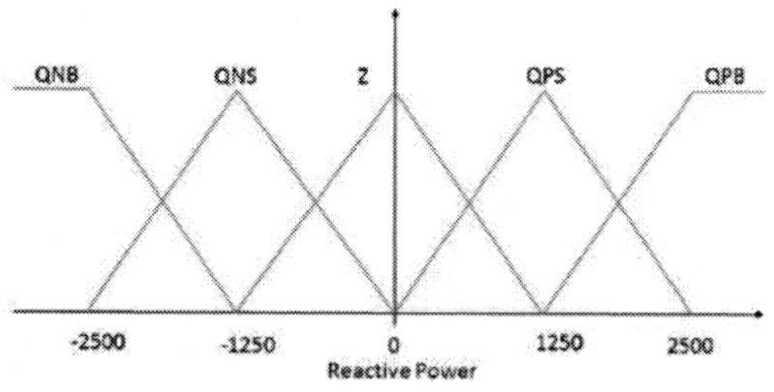

Figure 8: Fuzzy controller structure.

The fuzzy rules have the following form:

$$If\ e_{n-1}\ is\ X\ and\ e_n\ is\ Y\ and\ e_{n+1}\ is\ Z\ THEN\ Q_n\ is\ Qn,$$

Where X, Y and Z are the values of the voltage error for each home and Q_n is the reactive power handles by the $n - home$. We defined 125 fuzzy rules for the controller.

4 SIMULATION RESULTS

In order to validate and test the performance of proposed method, different solar profiles are simulated in a RTDS including the German smart home profile. In this case, a hardware in the loop real time simulator was constructed by using the xPC toolbox of MATLAB. The xPC operating principle is shown in Fig. 9.

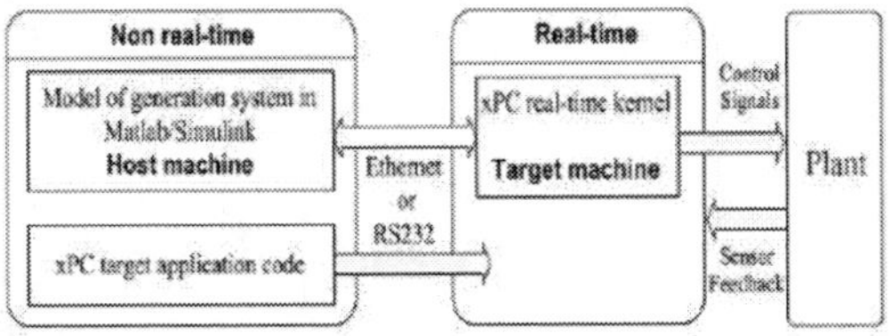

Figure 9: xPC architecture.

A smart LV distribution system is designed and embedded into the RTDS to test the proposed formulation. It is basically composed of a distribution transformer connected to high voltage lines, power lines and loads which are represented by typical households and smart homes, (i.e., a household is considered a smart home when it includes a smart meter, a smart PV inverter and some level of automation). In this case, the loads are defined as smart homes instead of passive loads to improve the simulation scenarios. The German smart home load profile used in this paper was defined in [1].

Fig. 10 shows how the developed MATLAB/Simulink SimPowerSystems model of the smart LV distribution system is embedded into the xPC target machine via the previously established network connection with the host machine. Once the model runs in the xPC target machine, the outputs of the selected smart homes are given as inputs to the voltage and current power amplifiers (i.e., the scaled three-phase voltages and currents at the connection point). Then, these amplified voltages and currents can be connected to commercial smart meters, (i.e., the maximal voltage value provided by the voltage power amplifier is ±440 Vac, a 40% over the nominal value 230 Vrms, and the maximal current value is 6A ac). More information about the hardware development for the real-time digital power simulator can be found in [4].
Fig. 11 exhibits the voltage behavior when all smart homes are injecting/absorbing active power to the grid for different day profiles without using the reactive power capability of the smart PV inverters. As expected, the voltage rise is superior to the desired limits (red and blue line), a typical case in a smart LV distribution grid with high PV penetration.
By using PV inverters with a local control $Q(U)$ method, the undervoltage and overvoltage are reduced due to reactive power injection and absorption of the smart PV inverters. However, despite the $Q(U)$ has in general terms a good behavior, some overvoltages can be observed (as shown in Fig. 11), especially in the smart home closer to the end of the line (smart Home 5). According with this, a LV grid expansion based on renewable energy systems or distributed generation would present strong issues on implementation and operation, since each smart inverter needs to be reconfigured for adjusting voltage levels adequately.

Figure 10: Experimental set-up for the evaluation of the proposed strategy in Universidad de La Salle, Colombia.

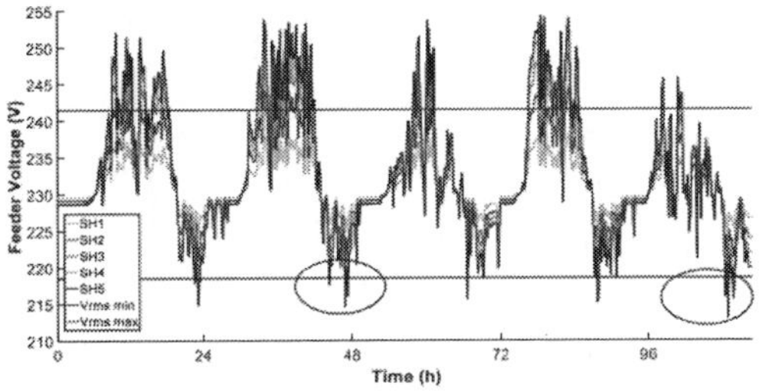

Figure 11: No-Control Method

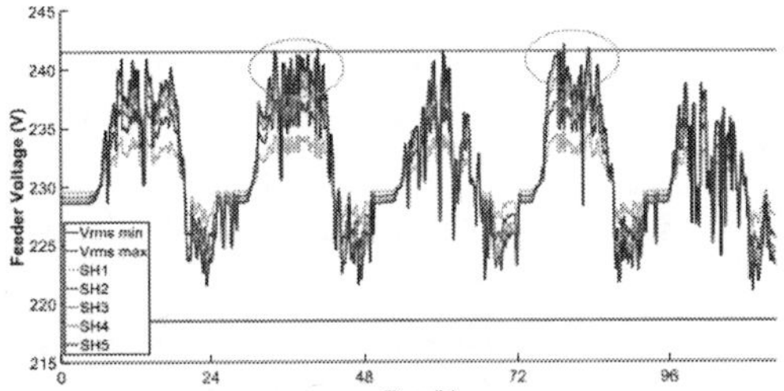

Figure 12: $Q(U)$ Method

On the other hand, the proposed FDCS maintains the voltage at the desired level irrespective of the load variations as shown in Fig. 13. In comparison to the $Q(U)$ method, the fuzzy controller allows the operating voltage to be within the set limits in all smart homes. The FDCS allows to apply higher levels of reactive power than the $Q(U)$ method according to the status of the voltage at the common connection point, as well as the voltage status of the neighbors, which is a great advantage in real operating conditions.
Finally, Fig. 14 provides a comparison of the power losses for selected different strategies. The difference in this quantity is greater as the PV active power injection increases. From these results, it can be seen that $Q(U)$ strategy reaches lower power loss in comparison to the reference case, (i.e., around 50%). Moreover, the proposed strategy has the lowest power loss, with a reduction of 72.3% with respect to the no-control method. In case of $Q(U)$ method, the losses are almost 54.7% higher than the proposed FDCS strategy. The

reason for this is because both strategies try to maintain the voltage profile closest to the nominal voltage magnitude value by injecting extra reactive power, but with the information from the smart meters of the neighbors the FDCS method inject additional reactive power in order to support the voltage control of the following smart PV inverter.

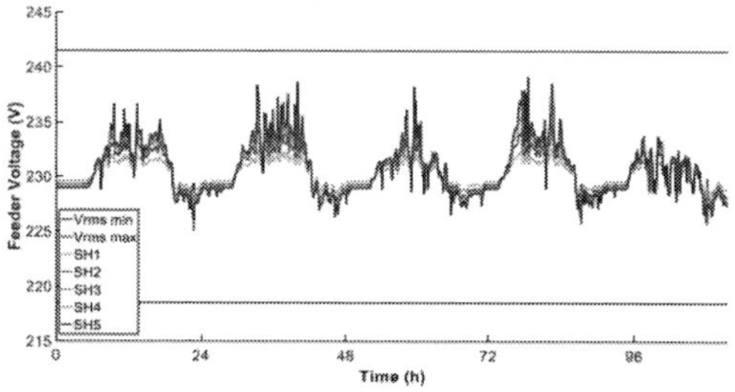

Figure 13: FDCS Method

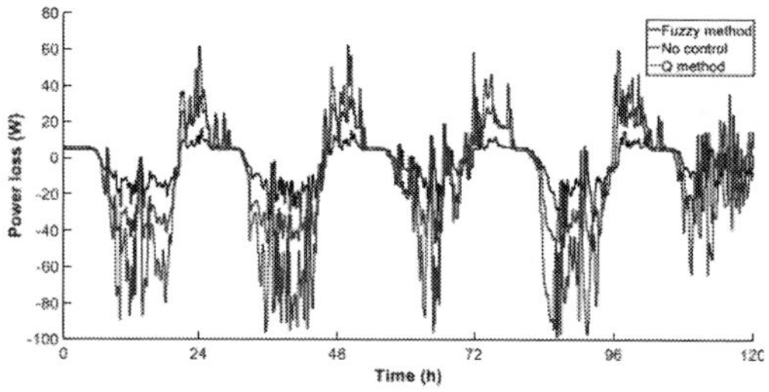

Figure 14: Comparison of the power losses for the different strategies

5 CONCLUSIONS

This work has developed a strategy based on fuzzy logic that improves the classical local control schemes like $Q(U)$ method, which are nowadays being used in PV inverters on German grid. The work done here is different and new on the basis that it uses three different voltage references, rather than one as the input for the fuzzy logic system to control the reactive power of the smart PV inverters, and without incurring in the use of an on-load tap-changing transformer (OLTC), as normally implemented, in decentralized control schemes.

The results showed that the proposed FDCS strategy improves the power quality, reduces the power losses of the system, as well as reduces the voltage fluctuations of the smart LV distribution system. The results are very promising as the voltage parameters are achieved fast and accurately, while the power loss is highly smoothed compared to typical local control strategies as proposed in the directive VDE-AR-N 4105.

6 ACKNOWLEDGEMET

This work was carried out during the tenure of an ERCIM 'Alain Bensoussan' Fellowship Programme.

7 REFERENCES

[1] J. D. Watson, N. R. Watson, D. Santos-Martin, A. R. Wood, S. Lemon and A. J. V. Miller, "Impact of solar photovoltaics on the low-voltage distribution network in New Zealand," *IET Generation, Transmission & Distribution*, vol. 10, no. 1, pp. 1-9, 2016.

[2] A. Vargas, O. R. Saavedra, M. E. Samper, S. Rivera and R. Rodriguez, "Latin American Energy Markets: Investment Opportunities in Nonconventional Renewables," *IEEE Power and Energy Magazine*, vol. 14, no. 5, pp. 38-47, Sept.-Oct. 2016.

[3] A. L. Gómez, C. A. Arredondo, M. A. Luna, S. Villegas and J. Hernández, "Regulating the integration of renewable energy in Colombia: Implications of Law 1715 of 2014," *2016 IEEE 43rd Photovoltaic Specialists Conference (PVSC)*, Portland, OR, 2016, pp. 3317-3321.

[4] S. Doring, " Einleitung. In Energieerzeugung nach Novellierung des EEG, 1–25. Springer, 2015.

[5] E. Bernal, Q. Li, and J. Xie, "A novel central voltage-control strategy for smart LV distribution networks", International Workshop on Data Analytics for Renewable Energy Integration, Springer, 2015, pp. 16–30.

[6] V. Calderaro, V. Galdi, A. Piccolo and G. Massa, "Optimal fuzzy controller for voltage control in distribution systems," *11th International Conference on Intelligent Systems Design and Applications*, Cordoba, Spain, 2011, pp. 1282-1287.

[7] X. Su, M. A. S. Masoum and P. J. Wolfs, "Optimal PV Inverter Reactive Power Control and Real Power Curtailment to Improve Performance of Unbalanced Four-Wire LV Distribution Networks," in *IEEE Transactions on Sustainable Energy*, vol. 5, no. 3, pp. 967-977, July 2014.

[8] G. Acampora, V. Loia and A. Vitiello, "Exploiting Timed Automata based Fuzzy Controllers for voltage regulation in Smart Grids," *2011 IEEE International Conference on Fuzzy Systems (FUZZ-IEEE 2011)*, Taipei, 2011, pp. 223-230.

[9] E. Bernal, M. Bueno-Lopez and F. Salazar, "Fuzzy-Based Reactive Power Control for Smart PV Inverters in LV Distribution Systems," *20th IFAC World Congress*, Toulouse, France, 2017, pp. 1-6.

[10] H. Mehrjerdi, S. Lefebvre, M. Saad and D. Asber, "A Decentralized Control of Partitioned Power Networks for Voltage Regulation and Prevention Against Disturbance Propagation," in *IEEE Transactions on Power Systems*, vol. 28, no. 2, pp. 1461-1469, May 2013.

[11] A. Giannitrapani, S. Paoletti, A. Vicino and D. Zarrilli, "Optimal allocation of energy storage systems for voltage control in LV distribution networks," in *IEEE Transactions on Smart Grid*, vol. PP, no. 99, pp. 1-12, August 2016.

[12] B. Perera, P. Ciufo and S. Perera, "Advanced point of common coupling voltage controllers for grid-connected solar photovoltaic (PV) systems," *Renewable Energy*, vol. 86, pp. 1037-1044, 2016.

[13] A. Cagnano, E. De Tuglie, "Centralized voltage control for distribution networks with embedded PV systems", *Renewable Energy*, vol. 76, pp. 173-185, 2015.

[14] M. Juamperez and G. Yang, "Voltage regulation in LV grids by coordinated volt-var control strategies", *Journal of Modern Power Systems and Clean Energy*, vol. 2, no. 4, pp. 319-328, Dec. 2014.

[15] J. D. Watson, N. R. Watson, D. Santos-Martin, A. R. Wood, S. Lemon and A. J. V. Miller, "Impact of solar photovoltaics on the low-voltage distribution network in New Zealand," *IET Generation, Transmission & Distribution*, vol. 10, no. 1, pp. 1-9, 2016.

IMPLEMENTATION OF CONTROL STRATEGIES FOR PV POWER RAMP-RATE LIMITATION USING ENERGY STORAGE: PROBLEMS AND SOLUTIONS ASSOCIATED WITH THE DIFFERENT BATTERY CHARGE/DISCHARGE POWERS

I. de la Parra[1]*, J. Marcos, M. Muñoz, M. García and L. Marroyo

Department of Electrical and Electronic Engineering – Institute of Smart Cities, Public University of Navarre (UPNa), Campus Arrosadía s/n, Los Pinos building, 31006 Pamplona, Spain
Institute of Smart Cities
inigo.delaparra@unavarra.es. +34 948 168 934. (*Author for correspondence)

ABSTRACT: Energy storage systems (ESS) have been widely proposed [1]–[10] as a solution for smoothing out PV power fluctuations and comply with new regulations [11]–[13] that limit the maximum fluctuation over a period of time, typically 1 minute. However, while implementing a ramp-rate control strategy in a real ESS, some problems linked with the maximum charge/discharge power permitted by the ESS arise and have so far not been appropriately studied. Herein, the problems of not taking into account this issue are shown, together with the solution proposed to solve it.

1. INTRODUCTION

Short-term variability of the PV power due to the intermittent nature of the solar resource together with the growing demand of installing new PV plants, pose one of the great challenges to those responsible for PV integration into electricity networks with conventional generators. While irradiance fluctuations can produce significant variations in the PV power output in a few seconds, the system operator response capacity is far away from this time scale, being 10 minutes the temporary window presented as the reaction capacity frontier for the grid operator. Therefore, any power fluctuation under this time scale might be a risk for the system stability. As PV output ramp-rates of the rated capacity can vary from 45-90%/min for multi-MW solar plants [14]–[18] being values high enough to compromise the grid stability in places with high PV power penetration levels, some transmission system operators (TSO) have issued new grid codes to address this matter, including new criteria to make it easier for the TSO to react appropriately against harmful power fluctuations, i.e., fluctuations with a time scale of less than 10 minutes [11]–[13]. Such criteria include setting power variation maximum ramps for the power being fed into the network by intermittent generation plants. In this way, for instance, the target specified in Mexico [12] varies from 1-5%/min. It is thereby ensured that the rest of the system, if provided with sufficient control capacity, may respond to any rapid power changes in the intermittent power generation plants. Given the fact that PV power fluctuations are greater than the restrictions imposed, some type of energy storage system (ESS) is required in order to comply with the same.

Different strategies have been proposed in the literature [1]–[10] to properly operate an ESS to comply with a maximum ramp rate limitation imposed by one of these new regulations. However, none of these papers have dealt with the problem that most of the commercially available ESS have different maximum charge and discharge powers [19] that can be insufficient to operate appropriately. The most common inconvenient is that, as the charge power is lower than the discharge power, there are some moments in which the ESS is not able to absorb all the energy required to comply with a ramp rate limitation when an upwards fluctuation takes place. In this work, we have studied this problem and proposed a solution to it. In addition, this solution has been evaluated with 5 seconds power measurements which were recorded at the output of 550 kW inverters at the 45.6 MW Moura (Portugal) PV plant for a maximum allowable ramp-rate value (r_{MAX}) of 2%/min.

2. EXPERIMENTAL DATA

The database used in this work is taken from Moura PV plant, which is located in Amareleja (South Portugal) PV plant. This plant occupies an area of 250 Ha and includes 2520 solar trackers with a rated output of 17.7 to 18.8 kWp, up to a total peak power of 45.6 MWp. The corresponding inverter power, Pn, is 38.5 MW and the ground cover ratio (GCR) is 0.162. The trackers are one-vertical axis models, with the receiving surface tilted 45° from the horizontal. Figure 1 shows the layout of Moura PV plant.

Figure 1: layout of the Moura PV plant.

The plant is divided into 70 units (Figure 2), each comprising 36 tracking systems connected to a 550 kW DC/AC inverter. The minimum and maximum distances between the units, are 220 meters and 2.5 km respectively. Thanks to extensive monitoring, 5 s synchronized records of the output power of all the inverters are available from May 2010.

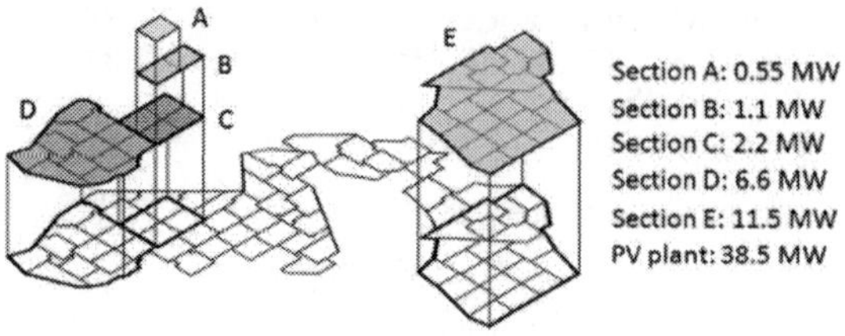

Figure 2: Field distribution of the Amareleja PV plant sections.

3 APPROACH

Different strategies have been presented in the literature to properly mitigate PV power fluctuations to a maximum ramp-rate restriction imposed by a certain regulation. In this way, one of the most used strategies is the classical ramp-rate control ($RR_{classical}$) [1], [20], [21]. In a recent study [8], we have proposed a new ramp-rate control strategy that halve the ESS requirements of the $RR_{classical}$. The control is based on the two PV plant production limits: the maximum PV plant power which occurs under clear sky conditions ($P_{PV,Max}(t)$) and the minimum PV plant power which occurs with complete cloud cover ($P_{PV,Min}(t)$). Therefore, as a function of the instantaneous PV power, it is then possible to obtain the state of charge (SOC) of the ESS needed to smooth out any potential fluctuations.

According to [8], battery requirements for both strategies are easily derived from the so called *worst fluctuation model* which is an effective method to calculate, for any PV plant size and maximum allowable ramp-rate (r_{MAX}), the maximum power and the minimum energy storage requirements alike. It is based on the worst fluctuation that can take place at a PV plant and it is a function of the shortest measurement for the PV plant perimeter and the maximum cloud speed.

According to this method, battery requirements for ramp-rate limitation are easily derived from the worst fluctuation observed at Amareleja PV plant (see Figure 3). This is properly described as an exponential decay from P_N to $0.1P_N$ (or an exponential increase from $0.1P_N$ to P_N) which means that the beam irradiance disappears and only remains the diffuse light, with a time constant, τ [s], which is empirically correlated with the shortest dimension of the perimeter of the PV plant, l [m], by Eq.(1) [9]:

$$\tau = \mathrm{a}\cdot l + \mathrm{b} \tag{1}$$

where a = 0.042 (s/m) and b = -0.5 s.

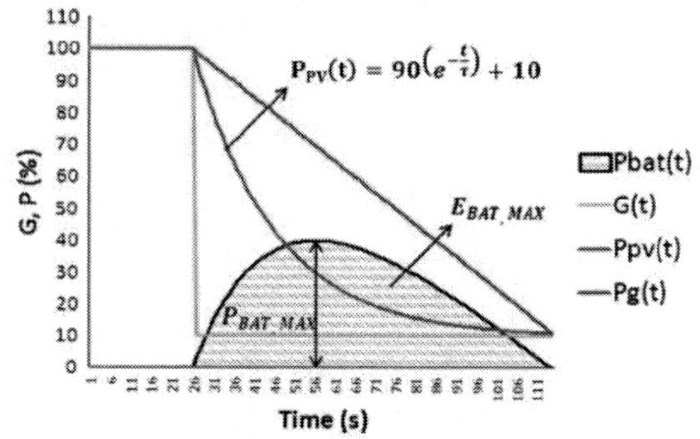

Figure 3: *Worst fluctuation model.* The blue line represents the generated PV power $P_{PV}(t)$ response to an irradiance fluctuation (orange line) whilst the red line is the power injected into the grid $P_G(t)$ with a ramp-rate control. The difference between $P_G(t)$ and $P_{PV}(t)$ is $P_{BAT}(t)$, the maximum difference corresponds to $P_{BAT,MAX}$ and the defined integral of $P_{BAT}(t)$ corresponds to the energy to be provided by the ESS, $E_{BAT,MAX}$. [9]

Battery requirements are easily derived from Figure 3. And are given by Eq.(2) [9]:

$$E_{BAT\_MAX} \approx \frac{0.9P_N}{3600}\left[\frac{90}{2\cdot r_{MAX}} - \tau\right] \tag{2}$$

Where P_N is expressed in [kW], r_{MAX} in [%/s], τ in [s] and $E_{BAT,MAX}$ in [kWh]. However, as the sign of the fluctuations is unknown, in the $RR_{classical}$ a double capacity battery is required to absorb both the upwards and downwards possible fluctuations. So the minimum battery required (C_{BAT}) is given by Eq.(3) [9]:

$$C_{BAT} = 2\cdot E_{BAT\_MAX} = \frac{1.8P_N}{3600}\left[\frac{90}{2\cdot r_{MAX}} - \tau\right] \tag{3}$$

And the required battery power is given by Eq.(4) [9]:

$$P_{BAT\_MAX}(t) = \frac{P^*}{100}\left[90 - \tau\cdot r_{MAX}(1+\ln\frac{90}{\tau\cdot r_{MAX}})\right] \tag{4}$$

On the other hand, [8] proposes the $RR_{clear\text{-}sky}$ strategy. It is proposed to implement a SOC control based on the actual power given by the PV plant and its production limits. Despite the variable nature of solar radiation, it can be taken advantage of the fact that its limits are well known. The instantaneous PV plant power generated, $P_{PV}(t)$, for specific values of irradiance, $G(t)$, and cell temperature, $T_c(t)$, can be easily estimated with a parametric model of the PV plant under consideration. It is possible to estimate at each moment, the PV plant production limits: the PV plant power under clear sky conditions, $P_{PV,Max}(t)$, and the PV plant power under totally cloudy sky conditions in which only the diffuse light reaches the PV arrays, $P_{PV,Min}(t)$. These powers represent both the maximum and minimum power that can occur at the PV plant at that moment in time. In this way, it is possible to calculate the maximum power variation that can take place, either positive or negative, from the instant power generated by the PV plant, $P_{PV}(t)$. So, as a function of the actual PV power, it is then possible to obtain the state of charge needed in order to either absorb or provide the necessary energy depending on the nature of the fluctuation, either upward or downward respectively. The logic of this SOC control can be found in [8]. Applying this control, again it is possible to operate with half the capacity required in Eq.(3), that is Eq.(5) [8].

$$C_{BAT} = \frac{0.9P^*}{3600}\left[\frac{90}{2\cdot r_{MAX}} - \tau\right] \quad (5)$$

While implementing any ramp rate control strategy in a real PV plant with an ESS some problems due to the different charge and discharge powers of the battery can arise. Particularly, as commercially available ESS usually have less power to charge than to discharge, it can be moments in which the ESS is not able to operate appropriately as it cannot absorb all the energy required to comply with a ramp rate limitation when an upwards fluctuation takes place. Figure 4 shows the problem over a very fluctuating day (15th February 2011). As can be seen in Figure 4 (b), when an upwards PV power fluctuation takes place (blue line) the limited PV power (dot red line) cannot comply with the maximum ramp-rate limitation. Moreover, Figure 4 (c) shows how the battery that has an asymmetric charge-discharge ratio (Pcharge/Pdischarge = = 0.5) (red line) cannot comply with the required charge battery power (black line).

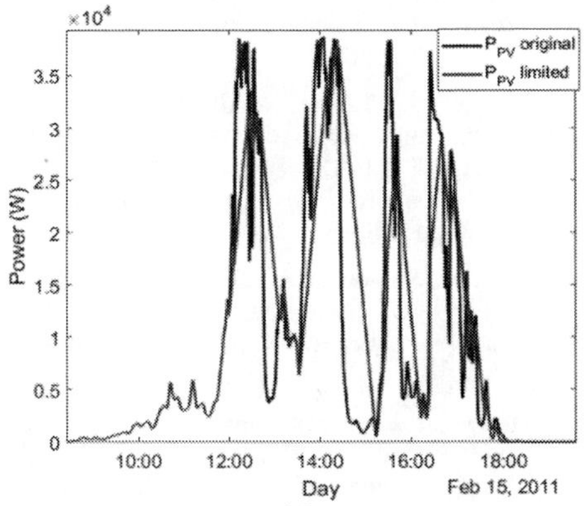

(a)

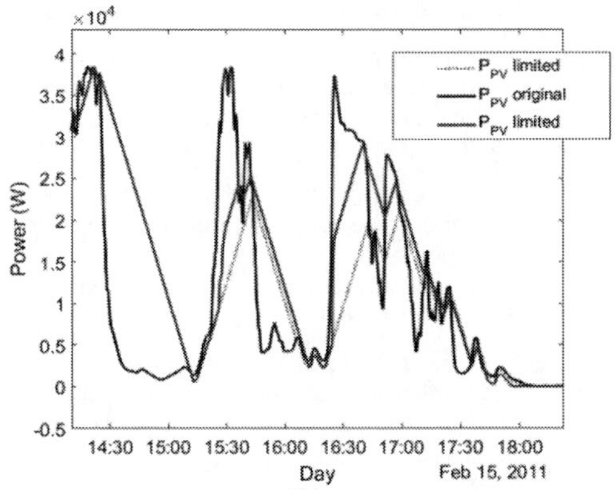

(b)

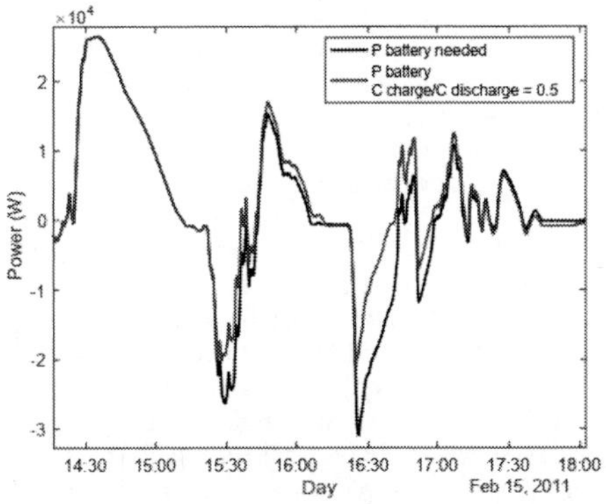

(c)

Figure 4: problem associated with not having enough battery charge power in the ESS during a very fluctuating day. (a) Evolution of the generated power the whole 15th February 2011 (b) Zoom where can be clearly seen the problem. (c) Battery power when the ESS has a symmetric charge-discharge power (black line) or power C charge/C discharge = 0.5 (red line).

4 PROPOSED SOLUTION

Herein, we have studied this problem and proposed a solution in which the maximum power that can be absorbed by the ESS is continuously calculating and, in case the ESS charge power is exceeded, the inverters will limit the power injected into the grid. Figure 5 shows the scheme control implemented for this technique. Although it is similar to the generic model used in [8], it has some modifications. At every step, the battery charge power that must be absorbed by the battery is checked that it does not exceed the maximum battery charge power capacity. In case it is exceeded, this power is subtracted from the original $P_{PV}(t)$ in order to be able to comply with the maximum battery charge power available. The way to subtract this power from the original PV power is by limiting the inverters, i.e., making the inverters work out of its maximum power point (MPP).

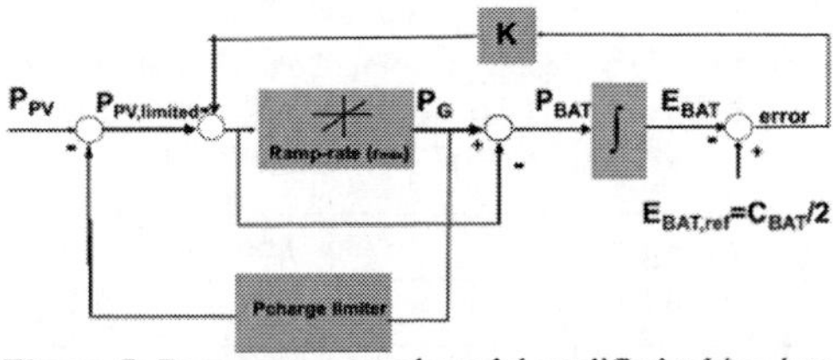

Figure 5: Ramp-rate control model modified taking into account the maximum battery charge power.

4 RESULTS

We have included and simulated this solution in both $RR_{classical}$ and $RR_{clear-sky}$ strategies over the course of a year (July 2010-June 2011) with data from Moura PV plant (5-s PV power production) and for a r_{MAX} = 2%/min.

Figure 6 shows the obtained results for a very fluctuating day (15th February 2012) for the $RR_{classical}$ strategy. It can be seen the original PV power (blue line), the limited PV power injected into the grid (red line), the power reference sent to the inverter (black line) and the modified PV power that takes into account the maximum available charge power of the battery (green line). In this case, the battery charge power that must be absorbed by the battery is verified that it does not exceed the maximum battery charge power capacity. When it is exceeded i.e. when the blue line surpasses the black line (around 15.30 and 16.30 in Figure 6), this power is subtracted from the original $P_{PV}(t)$ in order to be able to comply with the maximum power available. The way to subtract this power from the original PV power is by limiting the inverters, i.e., making the inverters work out of its maximum power point (MPP). In this way, this control is able to comply with the ramp rate limitation required at every moment with batteries with asymmetric charge-discharge powers. Of course, the difference between the original PV power (blue line) and the inverter power setpoint (black line) are energy losses associated with the inverters limitation.

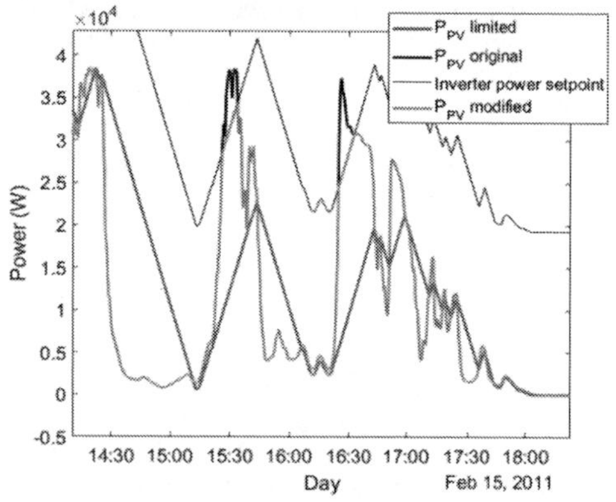

Figure 6: Evolution of the control proposed over a very fluctuating day (15th February 2017).

Furthermore, Table 1 shows the associated energy losses for different battery charge/discharge power ratios and for both $RR_{classical}$ and $RR_{clear-sky}$ strategies. The results show that these losses are insignificant, between 0.02% and 0.14% of the total production. In this way, this solution opens the door to study the viability of installing ESS with less charging power capacity depending on the cost of the battery per power unit and the energy losses.

Table 1: Energy losses based on the total production for both $RR_{classical}$ and $RR_{clear-sky}$ strategies and different charge-discharge battery power ratio

Pcharge/Pdischarge ratio (%)	Losses (%)	
	$RR_{classical}$	$RR_{clear-sky}$
33	0.14	0.08
50	0.03	0.02

5 CONCLUSIONS

Commercially available ESS usually have less power to charge than to discharge. As a result, it can be moments in which the ESS is not able to operate appropriately as it cannot absorb all the energy required to comply with a ramp rate limitation when an upwards fluctuation takes place. Herein, a solution in which the maximum power that can be absorbed by the ESS is continuously calculating. In case the ESS charge power is exceeded, the inverters will limit the power injected into the grid to the maximum that can be absorbed by the battery. This can be done by making the inverters to work in another point different from the MPP but entailing energy losses. However, these losses are insignificant, a fact that opens the door to study the viability of installing ESS with less charging power capacity depending on the cost of the battery per power unit and the energy losses.

6 AKNOWLEDGMENTS

The authors would like to thank ACCIONA for authorizing measurements at its PV plants and for the helpful collaboration of its staff. This work has been supported by the Spanish State Research Agency (AEI) and FEDER-UE under grants DPI2016-80641-R and DPI2016-80642-R.

7 REFERENCES

[1] N. Kakimoto, H. Satoh, S. Takayama, and K. Nakamura, "Ramp-Rate Control of Photovoltaic Generator With Electric Double-Layer Capacitor," *IEEE Trans. Energy Convers.*, vol. 24, no. 2, pp. 465–473, Jun. 2009.

[2] T. Kinjo, T. Senjyu, N. Urasaki, and H. Fujita, "Output Levelling of Renewable Energy by Electric Double-Layer Capacitor Applied for Energy," *IEEE Trans. Energy Convers.*, vol. 21, no. 1, pp. 221–227, 2006.

[3] T. Monai, I. Takano, and H. Nishikawa, "A Collaborative Operation Method Between New Energy-Type Dispersed Power Supply and EDLC," *IEEE Trans. Energy Convers.*, vol. 19, no. 3, pp. 590–598, 2004.

[4] T. D. Hund, S. Gonzalez, and K. Barrett, "Grid-Tied PV system energy smoothing," *2010 35th IEEE Photovolt. Spec. Conf.*, pp. 002762–002766, Jun. 2010.

[5] J. Traube *et al.*, "Mitigation of Solar Irradiance Intermittency in Photovoltaic Power Systems With Integrated Electric-Vehicle Charging Functionality," *IEEE Trans. Power Electron.*, vol. 28, no. 6, pp. 3058–3067, 2013.

[6] S. Rahman and K. S. Tam, "A feasibility study of photovoltaic fuel cell hybrid energy system," *IEEE Trans. Energy Convers.*, vol. 3, no. 1, pp. 50–55, 1988.

[7] J. Marcos, I. de la Parra, M. García, and L. Marroyo, "Control Strategies to Smooth Short-Term Power Fluctuations in Large Photovoltaic Plants Using Battery Storage Systems," *Energies*, vol. 7, no. 10, pp. 6593–6619, Oct. 2014.

[8] I. de la Parra, J. Marcos, M. García, and L. Marroyo, "Control strategies to use the minimum energy storage requirement for PV power ramp-rate control," *Sol. Energy*, vol. 111, pp. 332–343, Jan. 2015.

[9] J. Marcos, O. Storkël, L. Marroyo, M. Garcia, and E. Lorenzo, "Storage requirements for PV power ramp-rate control," *Sol. Energy*, vol. 99, pp. 28–35, Jan. 2014.

[10] I. de la Parra, J. Marcos, M. Garcia, and L. Marroyo, "Storage requirements for PV power ramp-rate control in a PV fleet," *Sol. Energy*, vol. 118, pp. 426–440, 2015.

[11] PREPA, "Puerto Rico Electric Power Authority Minimum Technical Requirements for Photovoltaic (PV) Generation Projects," 2012. [Online]. Available: http://www.fpsadvisorygroup.com/rso_request_for_quals/PREPA_Appendix_E_PV_Minimum_Technical Requirements.pdf.

[12] CRE, "Reglas generales de interconexión al sistema eléctrico nacional," 2014. [Online]. Available: http://www.cre.gob.mx/resolucion.aspx?id=7012.

[13] NERSA, "Grid connection code for renewable power plants (RPPs) connected to the electricity transmission system (TS) or the distribution system (DS) in South Africa," 2012. [Online]. Available: http://www.nersa.org.za/Admin/Document/Editor/file/Electricity/TechnicalStandards/South African Grid Code Requirements for Renewable Power Plants - Vesion 2 6.pdf.

[14] J. Jolmson, B. Schenkman, A. Ellis, J. Quiroz, and C. Lenox, "Initial Operating Experience of the 1.2-MW La Ola Photovoltaic System," *38th IEE Photovolt. Spec. Conf.*, vol. 2, no. 2012 IEEE 38th, pp. 1–6, 2012.

[15] J. Marcos, L. Marroyo, E. Lorenzo, D. Alvira, and E. Izco, "Power output fluctuations in large scale pv plants: One year observations with one second resolution and a derived analytic model," *Prog. Photovoltaics Res. Appl.*, vol. 19, no. 2, pp. 218–227, Mar. 2011.

[16] R. van Haaren, M. Morjaria, and V. Fthenakis, "Empirical assessment of short-term variability from utility-scale solar PV plants," *Prog. Photovoltaics Res. Appl.*, vol. 22, no. November 2012, pp. 548–559, 2014.

[17] T. Hansen, "Utility Solar Generation Valuation Methods, USDOE Solar America Initiative Progress Report," *Tucson Electr. Power, Tucson, AZ*, Aug. 2007.

[18] A. Mills, M. Ahlstrom, M. Brower, A. Ellis, and R. George, "Dark shadows," *IEEE Power and Energy Magazine*, pp. 33–41, 2011.

[19] SAFT SA, "Intensium Flex ®, Product brochure http://www.saftbatteries.com/battery-search/intensium%C2%AE-flex accesed January 2014," no. July, 2008.

[20] M. J. E. Alam, K. M. Muttaqi, and D. Sutanto, "A Novel Approach for Ramp-Rate Control of Solar PV Using Energy Storage to Mitigate Output Fluctuations Caused by Cloud Passing," *IEEE Trans. Energy Convers.*, pp. 1–12, 2014.

[21] L. N. Khanh, J. Seo, Y. Kim, and D. Won, "Power-Management Strategies for a Grid-Connected PV-FC Hybrid System," *IEE Trans. Power Deliv.*, vol. 25, no. 3, pp. 1874–1882, 2010.

OPTIMISATION OF THE LOAD FLOW CALCULATION METHOD IN ORDER TO PERFORM TECHNO-ECONOMIC ASSESSMENTS OF LOW-VOLTAGE DISTRIBUTION GRIDS

Fabian Carigiet[1]*, Franz Baumgartner[1], Petr Korba[1], Raphael Knecht[1], Michael Koller[2], Markus Niedrist[3]
[1] ZHAW, Zurich University of Applied Sciences, School of Engineering, Institute of Energy Systems and Fluid Engineering, Technikumstrasse 9, CH-8401 Winterthur, Switzerland, www.zhaw.ch\~bauf
*phone: +41 (0) 58 934 7292; e-mail: fabian.carigiet@zhaw.ch
[2] EKZ, Electric utility of the Canton Zurich, Switzerland, www.ekz.ch
[3] EKS, Electric utility of the Canton Schaffhausen, Switzerland, www.eks.ch

ABSTRACT: Decentralised power generation may lead to an inverse power flow compared to a centralised power supply system. Thus, voltage rises have to be limited at customer level with minimum extra costs during high PV power injection. Therefore, active and reactive power control of PV inverter will become more and more important because the additional grid operator's and end customers' investments could be nearly neglected. The decentralised control has to be triggered directly by the line voltage at the PV inverter ensuring an efficient use based on static characteristics, which are defined by the grid operator without the need of additional investment in IT infrastructure. The voltage dependent control of decentralised power generators is not implemented in the open source load flow calculation software Matpower. An elegant solution was found by integrating the dynamic change in active and reactive power directly into the load flow equations. This provides the basis for the techno-economic assessments which will be performed for different low-voltage distribution grid classes in Switzerland, Germany and Austria.
Keywords: Low-voltage distribution grid, photovoltaic, active power control, reactive power control, load flow calculation

1 INTRODUCTION

High injection rates of PV electricity in the distribution grid may increase the line voltage at the feed-in point of the inverter close to the limits defined by the grid code. Distribution system operators are responsible to comply with these voltage standards by planning regulations or retrofit the hardware in the grid.

In spring 2016, the three-year project "Cost effective smart grid solutions for the integration of renewable power sources into the low-voltage networks" (CEVSol) was started at the ZHAW in Switzerland. The Swiss electric utilities of the Canton of Zurich (EKZ) and Schaffhausen (EKS) are involved as industrial partners providing low-voltage distribution grid (LVDG) data, measurements and practical inputs. Further on, the ZAE Bayern and the Vorarlberger Kraftwerke (VKW) are associated partners for discussing and comparing the experience and best practice of LVDG analysis in Germany and Austria, respectively. Additionally, the VKW have already performed field tests regarding decentralised voltage dependent regulation of PV inverters [1].

The goal of this project is to classify the LVDG in different grid classes and to determine the most cost-effective solution for voltage stability within each class. There are various studies that classified their distribution grids using socio-techno-economic data. Therefore, model regions were extracted with the objective of extrapolating the needs of grid infrastructure of greater supply areas e.g. throughout Germany [2]–[5]. The different approaches can be taken into account partially as starting point for the CEVSol project. The methodology to be developed should provide the most cost-effective solution or a combination of solutions for the individual typical grid class. The goal is to maintain voltage stability by considering the corresponding strategy approach of the individual distribution system operator (DSO). The results should serve as a guideline for DSOs within their planning phase in order to assess the cost of the expected critical grid classes due to increasing PV capacity.

Classical grid reinforcement, active and reactive power control (APC and RPC), line voltage regulator, on-load tap changer, battery storage and load management are going to be analysed technically using load flow simulations. After that, the economic analyses will be performed resulting in a techno-economic assessment for each grid class. Possible grid classes are small/medium industries, city limits, urban settlements, village centres/peripheries and hamlets.

Two LVDGs representing a village with periphery and a hamlet are the first identified classes of LVDGs. During the technical analysis, it became clear that voltage dependent APC and RPC are powerful measures for maintaining voltage stability. Solar inverters will have a big influence in future LVDGs in terms of stabilising the voltage. Today, such products available on the market offer the possibility to control the reactive and active power directly by measuring the local line voltage. Thus, no external hardware will be needed.

The load flow simulations have been performed using Matpower, an open source toolbox for Matlab [6]. But this software could not handle voltage dependent control of decentralised power generators in a time-efficient manner. Therefore, an appropriate solution had to be developed.

2 APPROACH

An economic highly attractive solution would be to utilise the PV inverter as the control unit to stabilise the line voltage with a minimum use of additional IT infrastructure. This requires the implementation of some static characteristics for the APC and RPC as a function of the local line voltage during the installation of the inverter. Thus, each PV inverter needs static characteristic curves that describes its behaviour of P(V) and Q(V) described in Figure 3. According to these profiles, the inverter adjusts its active and reactive power injections depending on the measured line voltage without real time information from the grid operator. After the power adjustment, the line voltage will have changed, which will lead to another adjustment of power injections according to the defined characteristic curves. The APC and RPC will end as soon as the steady state voltage is reached. It is highly important to assess this effect of line voltage control within the grid code limits for different PV penetration levels on sunny days during the planning stage of the DSO. The standard load flow calculations cannot reflect this behaviour because the active and reactive power are constant input parameters of that simulation.

The most intuitive solution would be achieved by building an iterative process around the standard load flow calculation procedure shown in Figure 1. First of all, the load and generation profiles have to be initiated or updated. Then, it is proposed to solve the load flow calculation by the Newton-Raphson method. Once the calculation is converged, all node voltages are compared to the node voltages of the previous iteration. If the changes are greater than the allowed error value, the corresponding parameter for active and reactive power injection will be adapted and a further load flow calculation will be performed. This iteration is executed until steady state voltages levels are reached at each node. After that, the whole process has to be repeated for each time step.

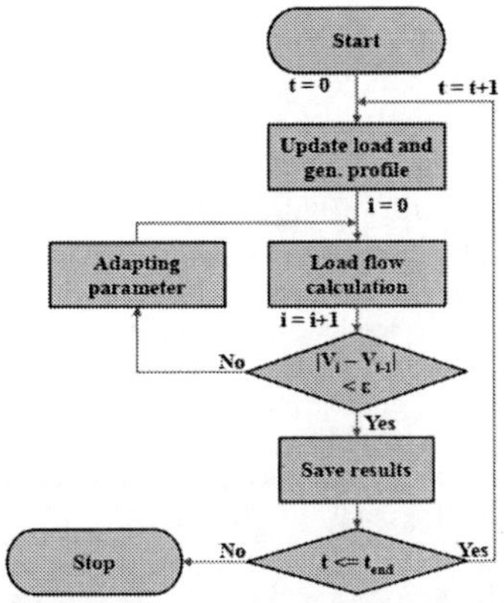

Figure 1: The iterative process adapts the parameter for APC and RPC and performs load flow calculation until the steady state voltages are reached at each node.

This procedure results in a too lengthy total calculating time. The more decentralised generators are controlled, the more complex it gets to define the loop that adapts the parameters for the APC and RPC and the longer it takes to converge to a stable solution. Therefore, the origin of the problems was identified and eliminated directly within the load flow calculation resulting in a slightly different mathematical formulation (see chapter 3) of the load flow equations. With that solution, only one load flow calculation has to be performed and the outputs are directly the steady state voltages.

The developed solution is applied in the analysis of the LVDG in Dettighofen, Germany. This village with periphery has a PV production share of 46% regarding the total electricity consumption. In 2014, the distribution grid was analysed and the effect of constant APC and RPC was investigated [7]. Therefore, the same grid data, PV production data and load data can be used to verify the implementation of the adapted load flow equations regarding the use of Q(V) as RPC and P(V) as APC.

3 ADAPTED LOAD FLOW EQUATIONS

The load flow equations of the active power g_h^p and the reactive power g_h^q balancing are written regarding the node h. The net active power injection p_h and reactive power injection q_h are the differences of generated power p_G and q_G and load power p_L and q_L, respectively [8].

$$g_h^p: \quad p_h = v_h \cdot \sum_{k=1}^{N} v_k (g_{hk} \cdot \cos\theta_{hk} + b_{hk} \cdot \sin\theta_{hk}) = p_G - p_L$$

$$g_h^q: \quad q_h = v_h \cdot \sum_{k=1}^{N} v_k (g_{hk} \cdot \sin\theta_{hk} - b_{hk} \cdot \cos\theta_{hk}) = q_G - q_L$$

The quantities g_{hk} and b_{hk} correspond to the conductance and susceptance of the line between the nodes h and k. The phase angle difference between those nodes is written as θ_{hk}. At each node, the power of a generator or a load is assumed to be constant during one time step within the load flow calculations. This assumption is no longer valid for the active and reactive power control or for voltage dependent demand side management.

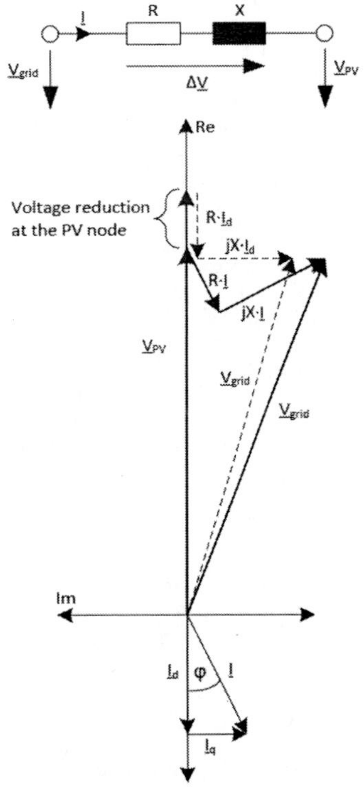

Figure 2: The phasor diagram of the line voltage (passive sign convention) shows the effect of under-excited reactive power injection at the PQ node. The grid voltage

V_{grid} is held constant at the end of a line and hence, the voltage V_{PV} is decreased compared to the case where only active power is injected.

Figure 2 shows the effect of under-excited reactive power injection at the end of the distribution line with the power input of the PV inverter at the line voltage label V_{PV} and without an additional load. The injection of reactive power reduces the line voltage V_{PV} as the basic principle of RPC during constant active power injection of the PV inverter. The supplied reactive power of the generator at a PQ node is an input value of the load flow equations and the reduced node voltage corresponds to their output value. The RPC will change the input value depending on the output value and the given characteristic curve Q(V). That means the angle φ is changed until the voltage V_{PV} is stable. This dynamic will be implemented in the load flow equations.

Therefore, the powers at the generator and load have to be considered as voltage dependent variables. Additionally, all nodes are treated as PQ nodes.

$$g_h^p = \underbrace{v_h \cdot \sum_{k=1}^{N} v_k (g_{hk} \cdot \cos\theta_{hk} + b_{hk} \cdot \sin\theta_{hk})}_{p_{bus}(\theta,v)} - p_G(v_h) + p_L(v_h) = 0$$

$$g_h^q = \underbrace{v_h \cdot \sum_{k=1}^{N} v_k (g_{hk} \cdot \sin\theta_{hk} - b_{hk} \cdot \cos\theta_{hk})}_{q_{bus}(\theta,v)} - q_G(v_h) + q_L(v_h) = 0$$

The Newton-Raphson method is often used to solve the non-linear load flow equations because it converges fast and efficient except for ill-conditioned cases [8]. The method is described using vector notation (bold variables) as follows:

$$\begin{bmatrix} \Delta\boldsymbol{\theta} \\ \Delta\boldsymbol{v} \end{bmatrix} = -J^{-1} \cdot \begin{bmatrix} \boldsymbol{g}^p \\ \boldsymbol{g}^q \end{bmatrix}$$

$$\begin{bmatrix} \boldsymbol{\theta}^{i+1} \\ \boldsymbol{v}^{i+1} \end{bmatrix} = \begin{bmatrix} \boldsymbol{\theta}^{i} \\ \boldsymbol{v}^{i} \end{bmatrix} + \begin{bmatrix} \Delta\boldsymbol{\theta} \\ \Delta\boldsymbol{v} \end{bmatrix}$$

The Jacobian matrix J contains the partial derivative of all load flow equations regarding all node voltages and phase angles. The red marked partial derivatives are added to the original matrix (black) due to the voltage dependency of the generators and loads.

$$J = \begin{bmatrix} \frac{\partial \boldsymbol{g}^p}{\partial \boldsymbol{\theta}} & \frac{\partial \boldsymbol{g}^p}{\partial \boldsymbol{v}} \\ \frac{\partial \boldsymbol{g}^q}{\partial \boldsymbol{\theta}} & \frac{\partial \boldsymbol{g}^q}{\partial \boldsymbol{v}} \end{bmatrix} = \begin{bmatrix} \frac{\partial \boldsymbol{p}_{bus}}{\partial \boldsymbol{\theta}} & \frac{\partial \boldsymbol{p}_{bus}}{\partial \boldsymbol{v}} + \frac{\partial p_L}{\partial v} - \frac{\partial p_G}{\partial v} \\ \frac{\partial \boldsymbol{q}_{bus}}{\partial \boldsymbol{\theta}} & \frac{\partial \boldsymbol{q}_{bus}}{\partial \boldsymbol{v}} + \frac{\partial q_L}{\partial v} - \frac{\partial q_G}{\partial v} \end{bmatrix}$$

The derivatives correspond to the slopes in the inverter profiles describing the behaviour of active and reactive power as shown in Figure 3.

Before the start of the simulation, each inverter profile has to be adjusted individually. The derivatives of the two solid lines are not continuous at the two edges (Figure 3). This may lead to a convergence problem of the Newton-Raphson method. In such cases, the profile can be approximated by an smooth logistic function (dashed lines) [9]. The behaviour of the reactive power is chosen in the following example so that half of the control range is exhausted at a voltage 1.03 pu. For the active power control curve, it has to be considered how the inverter is rated compared to the PV power plant. Here, it is assumed that the nominal power of the inverter corresponds to nominal power of the PV power plant. On a clear sky day in spring time, the nominal power of a PV module is never achieved due to the negative temperature coefficient resulting in a power decrease at higher module temperatures. This case is shown in the right lower part of Figure 6 and Figure 7, respectively. For this reason and for the purpose of verifying the APC, the active power control starts at the same voltage level as the reactive power control. Generally, the APC should start at higher voltage levels because it is economically preferable. The P(V) and Q(V) characteristics can be individually defined and the electric utilities should provide this information with their grid codes for installing and connecting a PV inverter to the distribution grid.

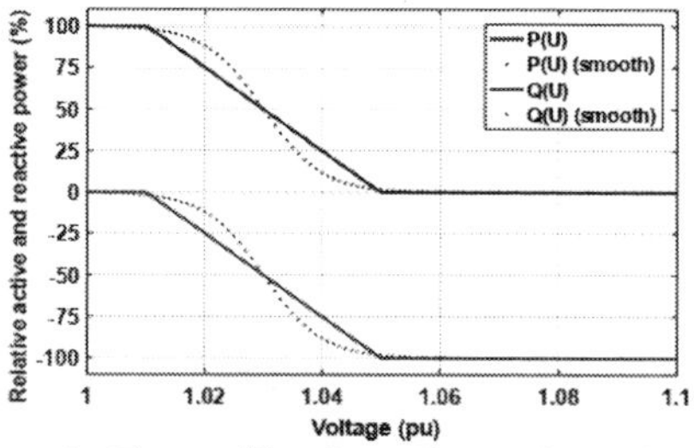

Figure 3: The profiles for active and reactive power control can be chosen individually for each decentralised generator within the load flow calculation. The negative percentages for the reactive power control represent the under-excited operation mode. The dashed lines are the alternative inverter profiles if the numerical method does not converge because the derivative of the solid curves is not continuous at the two edges.

4 LOW-VOLTAGE DISTRIBUTION GRID DATA USED FOR VERIFICATION

The implementation of the adapted load flow equations is verified by analysing the voltage dependent APC and RPC applied to the LVDG in Dettighofen, which is located in Baden-Württemberg, Germany. The same grid data, PV production data and load data is used as required for the case study in 2014 where the critical western part of that LVDG was analysed regarding different measures for voltage maintenance. This grid sector has a 400 kVA transformer and an installed PV capacity of 535.5 kWp [7]. Additionally, the distribution grid consists of 218 nodes, 228 lines and 33 PV power plants distributed over 25 different nodes. Figure 4 shows the single-pole schematic of the distribution grid with two long stubs connected to the distribution box VK6 in the south west with a total PV installation of 99.3 kWp. In this part of the grid sector, there are no major non-residential loads except some farmer houses with electric appliances.

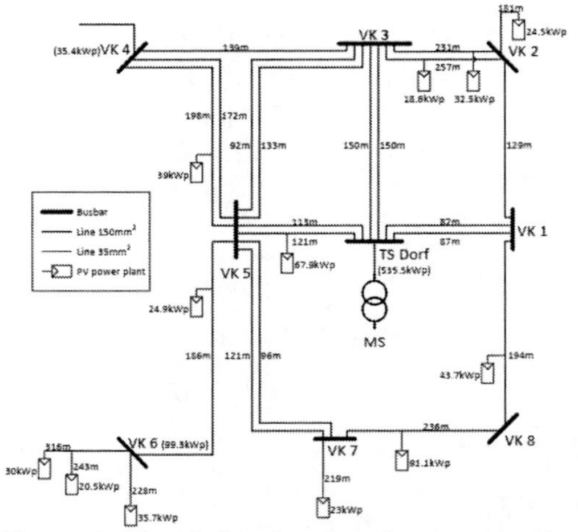

Figure 4: The single-pole schematic represents the critical western part of LVDG in Dettighofen, whereby only the largest PV plants are indicated. Two long stubs are connected to the distribution box VK6 carrying PV power with a nominal capacity of 99.3 kWp. In total, there are 33 PV power plant distributed over 25 nodes with an overall installed PV power of 535.5 kWp.

The simulations are performed for a clear sky day on Sunday 18th May assuming that all PV power plants are tilted by 30° and facing south. The generation profiles are modelled based on solar irradiance measurements [7]. The load profiles at each house connection are represented by the VDEW standard load profile H0 from E.ON Westfalen Weser (North Rhine-Westphalia) [10]. These profiles are scaled assuming a constant cosφ of 0.93 and an annual electricity consumption of 4500 kWh. The resolution of the used production and load data are 10 min intervals [7]. The identical APC and RPC profiles are assigned to each decentralised power generator according to Figure 3. As soon as the corresponding node voltage is higher than 1.01 pu, the active and reactive power will be decreased and increased according to Figure 5. The APC and RPC end when the node voltage reaches 1.05 pu.

5 RESULTS

First, the standard simulation is performed without any voltage control measures. The resulting node voltages are shown in Figure 6 as a snapshot of the situation at the time 14:00. The first numbers and the colour in the graphic indicate the relative voltage at the corresponding grid node. There are voltage rises of 5.0% and 5.6% relative to the nominal grid voltage at the ends of the two long stubs in the south western part of the LVDG, which is caused by decentralised feeding-in of the PV power plants with an overall nominal power of 99.3 kWp.

The second load flow simulation includes the voltage dependent APC and RPC together with the adapted load flow calculation. The voltages rises are limited to 2.9% and 3.1% according to results in Figure 7. The results of a previous study are equalised at the most critical nodes but there, the static comprehensive use of ACP and RPC were investigated [7]. In Figure 7, the first numbers represent again the relative voltages at the corresponding node. If there are PV power plants connected to that node, the percentage of active power curtailment (2nd number) is added followed by the percentage of utilisation of the reactive power control. It is evident that only the required quantity of active and reactive power is controlled. If only RPC is active, the voltage rises will be limited to 3.9% and 4.3%, respectively.

Figure 5 shows the simulated relative active and reactive power for the voltage dependent APC and RPC compared to the inverter profiles at the weakest node during the analysed day. The reactive power follows the given Q(V) profile. The active power has to be considered differently because the controlled active power is related to the nominal power of the PV power plant. The injected power does not reach the nominal PV power due to the power reduction at high module temperatures and in the morning and in the afternoon, where no perfect orientation to the sun is achieved. As a consequence, the simulated active power is lower or equal than the P(V) profile.

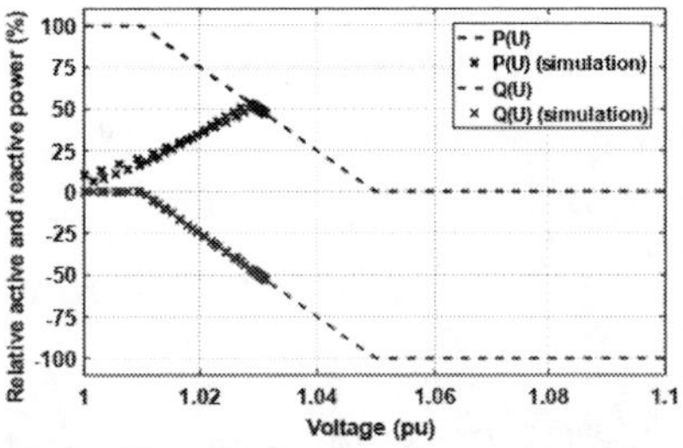

Figure 5: The simulated active and reactive power injection is compared to the given P(V) and Q(V) inverter characteristics during the course of the simulated day.

The more complex load flow calculation lead to losses of simulation speed performance over the analysed day as soon as the inverter control is activated. The detailed evaluation of the computation performances are shown in Table I. The computation time increases by a factor of 5 to 7 depending on the used inverter profile types. The reason is that there are higher amount of iterations per load flow calculation at the time at which the APC or RPC is used. The smooth inverter curve shows a better performance than the inverter profile with the edges according to Figure 3.

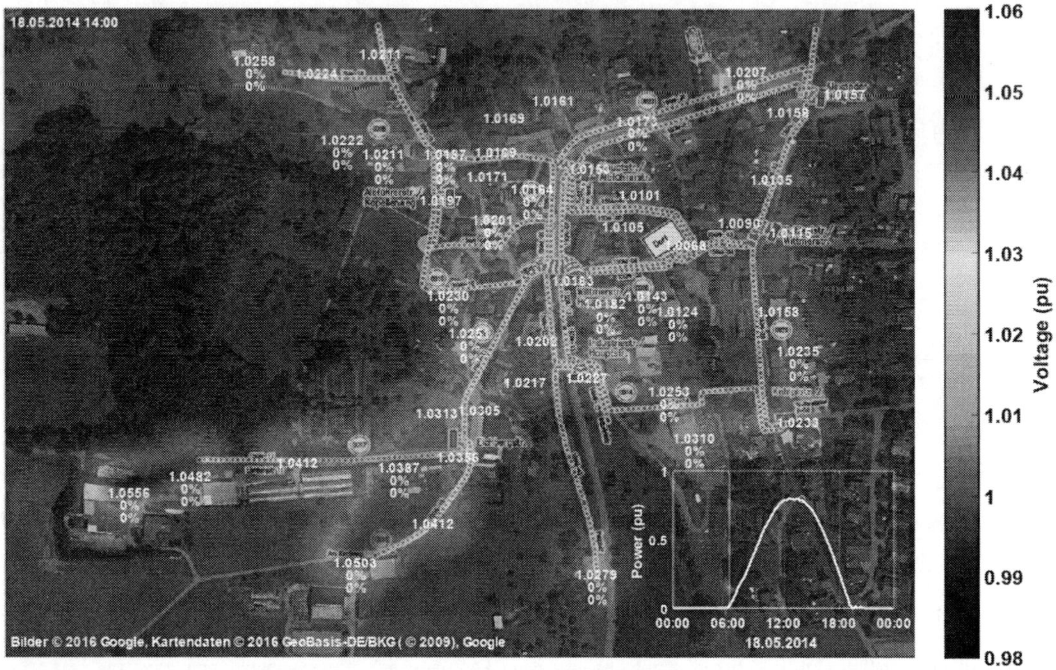

Figure 6: The mapping of the line voltage is given relative to the nominal voltage value at the south western part of the low-voltage distribution grid in Dettighofen, Germany, at 14:00 on a clear sky day in May. The number and colour in the graphic stand for the relative voltage at the corresponding grid node. Active and reactive power control are inactive. There are two long stubs with a total PV capacity of 99.3kWp, where the voltage rises reach their maximum of 5.0% and 5.6%, respectively, at 14:00 [11].

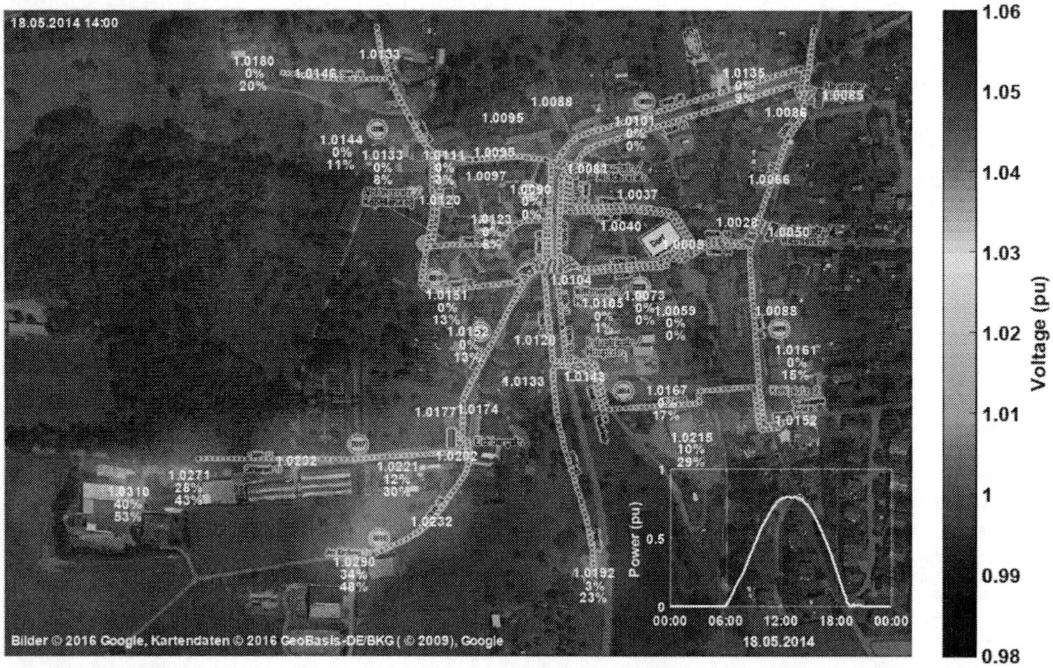

Figure 7: The voltage rise at the end of both stubs can be limited to 2.9% and 3.1%, respectively, as soon as the voltage dependent active and reactive power control is activated. The decentralised generators are only regulated where it is necessary according to the applied power flow calculation algorithm (Eq. in chapter 3) together with the inverter profiles in Figure 5. The results of a previous study can be achieved but there, the static comprehensive use of ACP and RPC were investigated [7]. The first number represents the voltage relative to the nominal voltage at that connection point. If there are PV power plants connected to that node, the percentage of active power curtailment (2nd number) is added followed by the degree of utilisation of the reactive power control [11].

Table I: The computation times and the number of iterations of the Newton-Raphson method are specified for the simulations (18.05.2014) without control and with APC/RPC according to the P(V) and Q(V) inverter characteristics described in Figure 3. The load flow calculations were performed by a Dell ultrabook with 16GB RAM and an Intel® Core™ i7-5600U CPU (2.6GHz).

Simulations	Standard w/o control	PQ control	PQ control (smooth)	Units
No of iterations of the Newton-Raphson method	2 to 3	2 to 14	2 to 17	#
Time per load flow calculation	4 to 30	14 to 81	9 to 63	ms
Total simulation time (incl. overhead)	0.78 (1.18)	5.16 (5.63)	3.60 (4.00)	s

6 CONCLUSION

The voltage dependent inverter control of active and reactive power according to the P(V) and Q(V) characteristics was successfully integrated into the open source load flow calculation software Matpower [6]. The developed and adapted solution can be applied to any other software tool. In the basic load flow equations, the power of a generator or a load is assumed to be constant during one time step. This is no longer valid for the voltage dependent APC and RPC or demand side management. This fact led to a modification of the load flow equation as well as the Jacobi matrix used in the Newton-Raphson method, which solves the non-linear load flow equations. Consequently, the simulation output is immediately the regulated voltages for each node of the distribution grid and time step. The user does not need to build complicated iterative processes around the complete load flow calculation. Different measures for voltage stability can be easily combined within the simulation allowing for a better techno-economical assessment.

In 2014, the LVDG with a PV share of 46% in Dettighofen, Germany, was analysed. Simulation results showed that the voltage rise could be limited to 3% by static centralised overall APC and RPC [7]. With the presented adaptation of the simulation method, roughly the same results could be achieved with targeted use of inverter control. Now, the two measures (APC and RPC) can be analysed more realistically with regards to technical as well as economical aspects.

The benefit of the local application of the APC and RPC is that each inverter feeds in reactive power individually if it is needed. This reduces the reactive power in the distribution grid and the amount of reactive power that is transferred to the next voltage level compared to the constant power factor approach. For example, the distribution system operator VKW in Vorarlberg, Austria, is already successfully applying the local P(V) and Q(V) control on approximately 2500 PV inverters [12].

The described simulation method allows different PQ strategies for each decentralised generator. Further on, the DSO can investigate where and which PQ-inverter profile should be used by the customers PV inverter. Thus, APC and RPC can be planned and deployed purposefully by the grid operator without further investments into information and communication technologies or hardware.

ACKNOWLEDGMENT

We thank the ZAE Bayern and the Vorarlberger Kraftwerke (VKW) for discussing and comparing the experience and best practice of LVDG analysis in Germany and Austria, respectively.

This project is funded by the Swiss Federal Office of Energy (SFOE) under the contract number SI/501370-01.

REFERENCES

[1] C. Elbs, R. Nenning, R. Pardatscher, and R. Witzmann, 'Einsatz der Q(U)-Regelung bei der Vorarlberger Energienetze GmbH', Vorarlberger Energienetze GmbH, Technische Universität München, 2014.

[2] L. Verheggen, S. Dierkes, H. Schuster, and A. Moser, 'Bewertung des Verteilungsnetzausbaus unter Berücksichtigung intelligenter Netztechnologien', presented at the 13. Symposium Energieinnovation, Graz, 2014.

[3] J. Büchner *et al.*, 'Moderne Verteilernetze für Deutschland (Verteilernetzstudie), Studie im Auftrag des Bundesministeriums für Wirtschaft und Energie (BMWi)', E-Bridge, IAEW, OFFIS, Forschungsprojekt 44/12, 2014.

[4] T. Eberl, K. Hufendiek, P. Wiest, and K. Rudion, 'Integrale Modellierung von Verteilnetzen und verteilter Erzeugung', ZfES Zentrum für Energieforschung Stuttgart, 29, 2014.

[5] M. Häni, 'Grundlagenbericht Stromnetze Thurgau mit Blick auf eine verstärkte dezentrale Stromproduktion', 2014.

[6] R. D. Zimmerman, C. E. Murillo-Sanchez, and R. J. Thomas, 'MATPOWER: Steady-State Operations, Planning, and Analysis Tools for Power Systems Research and Education', *IEEE Trans. Power Syst.*, vol. 26, no. 1, pp. 12–19, Feb. 2011.

[7] F. Carigiet, M. Niedrist, C. Scheuermann, and F. Baumgartner, 'Case study of a low-voltage distribution grid with high PV penetration in Germany and simulation analyses of cost-effective measures', *Prog. Photovolt. Res. Appl.*, 2015.

[8] F. Milano, *Power system modelling and scripting*. London: Springer, 2010.

[9] K. Turitsyn, P. Sulc, S. Backhaus, and M. Chertkov, 'Options for control of reactive power by distributed photovoltaic generators', *Proc. IEEE*, vol. 99, no. 6, pp. 1063–1073, 2011.

[10] 'Synthetisches Verfahren Strom | Westfalen Weser Netz', 2014. [Online]. Available: https://ww-

netz.com/synthetisches-verfahren-strom. [Accessed: 02-Nov-2016].

[11] F. Carigiet, F. Baumgartner, P. Korba, and V. Knazkins, 'Dynamische Wirk- und Blindleistungsregelung', *Bull. Fachz. Verbandsinformationen Von Electrosuisse VSE*, no. 2, 2017.

[12] C. Elbs, 'Erfahrung mit der Integration Erneuerbarer, insb. PV mit der Q(U)-Regelung', presented at the 32. Symposium Photovoltaische Solarenergie, Bad Staffelstein, 2017.

REDUCING THE GRID LOAD OF SOUTH AFRICAN OFFICE BUILDING BY IMPLEMENTATION OF ENERGY EFFICIENCY MEASURES AND INSTALLATION OF DEMAND OPTIMIZED PV

Benedikt Hanke[1*], Dorothee Peters[1], Meike Kühnel[1], Jéan Smit[2], Raimund Wiesmann[3], Ralf Saßmannshausen[3], Roland Hentschel[4], Karsten von Maydell[1], Carsten Agert[1]

[1]DLR Institute of Networked Energy Systems, Carl-von-Ossietzky-Str. 15, 26129 Oldenburg, Germany
[2]Buffalo City Municipality, Directorate Infrastructure Services, Department Electricity, Sherwood Ave, 5241 East London, Republic of South Africa
[3]Bundestechnologiezentrum für Elektro- und Informationstechnik e.V. BFE-Oldenburg, Donnerschweer Str. 184, 26123 Oldenburg, Germany
[4]Stadt Oldenburg, Industriestr. 1, 26121 Oldenburg, Germany
[*]corresponding author: benedikt.hanke@dlr.de

ABSTRACT

In East London, South Africa, a self-consumption optimized photovoltaic system has been erected at an office building operated by the municipality with a yearly energy demand of 50MWh per year after extensive energy efficiency measures (previously 84MWh/a). Remaining daily peak energy demands are attributed to heating and cooling of the offices correlating with the outside temperatures. Based on 2016 electricity monitoring, 70% of self-consumption have been achieved directly covering 45% of the electricity demand. On the balance sheet 64% of the buildings electricity demand is covered by the photovoltaic system. The peak power demand from the grid of 30kW has not been reduced by local renewable generation, but the times of demand greater 10kW have been reduced from 1485h/a to 277h/a by the photovoltaic system. The effects of a photovoltaic system on the annual load duration curve and power histogram of the building are discussed. Weekends with no office hours cover 29% of the time and are responsible for 6.2MWh/a or 63% of the feed-in energy into the grid from photovoltaics. The effects of self-consumption photovoltaic systems on the grid stability are discussed in the South African electricity grid and generation context.

Keywords: Small Grid-connected PV Systems, self-consumption, self-sufficiency

1 INTRODUCTION

The Beacon Bay Civic Center Offices in East London, South Africa – home of the Buffalo City Metropolitan Municipality (BCMM) electricity department – has been retrofitted with energy efficiency measures and renewable energies from photovoltaic to become an example for green retrofitting in the public building stock. The intention of this project was to identify the amount of energy savings possible within economic viability and to research the effect of local photovoltaic production on the grid load requirements.

With average annual wind speeds at 10m above ground of 4.5m/s and a solar radiation of 6200MJ/a (1720kWh/a) the area is well suited for the generation of electrical energy from wind and photovoltaic. Moderate climate conditions with temperatures between 7°C and 32°C and average daily sunshine durations of more than 6 hours throughout the year support the deployment of photovoltaic in the region. Located 32° 58' south of the equator, a wide range of panel inclinations is possible without losing significant amounts of energy (see Figure 1). This supports especially the possibility to install roof top solar systems.

The target of this publicly funded project (see section acknowledgements) was the reduction of the energy demand of the building by 25% using commercially viable measures and to generate at least 40% of the remaining energy demand by local renewable sources. This work focusses on the effect of the photovoltaic installation on the electricity grid demand of the location not reporting in detail on energy efficiency measures.

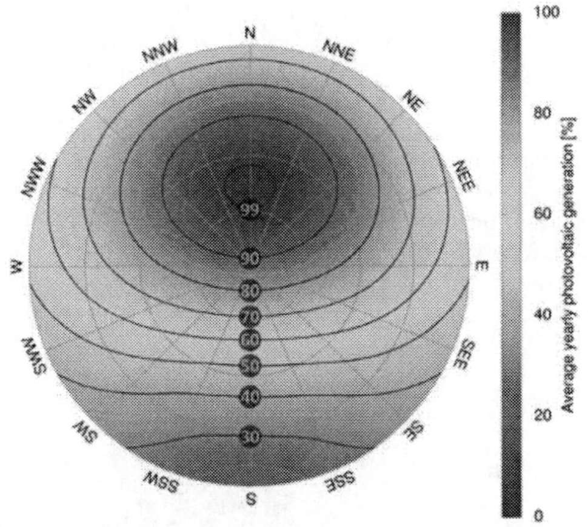

Figure 1 Average yearly photovoltaic generation potential hemisphere in East London, South Africa (S 32° 58.090' E 027° 56.320') based on solar radiation profiles from the *'meteonorm 7'* database.

2 BASELINE

Within the BCMM, the electricity demand of public infrastructure and buildings is not metered or billed. These demands are considered as losses to the grid and payed for by the budget of the electricity department. Metering was installed ahead of the project to estimate the electricity demand of the Beacon Bay Offices for planning of energy efficiency measures and for photovoltaic simulations.

2.1 Electricity Demand Estimation

The yearly electricity demand of the location was estimated 81.5MWh/a ± 2.5MWh/a based on 15 monthly meter readings from March 2014 until September 2015. Missing monthly values where replaced by worst case estimate (meter reading of June '15) to get a baseline for the energy efficiency and renewable energy generation goals. The monthly electric energy demand is given in Table 1 for reference. The average daily power curve is presented in Figure 2 (red curve, average load curve based on 232 days of measurement from Feb. – Oct. 2014). The profiles show a dayshift office building dominated energy demand profile that is ideal for high shares of directly consumed solar generation.

Table 1: Monthly electric energy demand of the Beacon Bay Office Building in East London 2014/2015.

Electr. Energy [kWh/month]	2014	2015	Average/ Estimate
January		7115	7115
February		7697	7697
March	8100	5401	6751
April	6437	5463	5950
May		6409	6409
June		8025	8025
July	6927	6020	6473
August	6009	5568	5789
September	5558	5568	5563
October			8025
November			8025
December	7983		7983

2.2 Energy Efficiency Measures

The proposed energy efficiency target to be achieved in the project is 25% electrical energy demand reduction, resulting in a saving target of 21MWh/a based on the estimated demand of the building. As main energy consumers, the ICT systems / digital infrastructure, air conditioning / heating, lighting and hot water preparation have been identified. The lighting and hot water preparation were identified as the most viable targets for energy savings, as they can be technically changed without interacting with the productivity, habits and the comfort of the office buildings users.

The fluorescent light tubes have been replaced by LED tubing by rewiring the existing light fittings. The conventional ballasts required to run fluorescent lights haven been removed from the electric circuit as they are not needed to run LED lights. This reduces the power demand of a single light point by 48W or 67%. Further rooms and locations that are irregularly used like kitchens, lavatories and conference rooms have been outfitted with motion detectors to reduce the operating hours of the lighting points. Introducing automated switching of the light in those locations allows for a light on demand operation without interfering with the comfort of the buildings users. Within the project 210 fluorescent tubes have been replaced and 30 motion detectors have been installed. The energy demand for lighting is reduced to an estimated 31% compared to the original demand saving roughly 20MWh per year. In air-conditioned rooms a second saving amounts from the reduction of electrical energy needed to light the rooms. As most of the electrical energy needed for lighting is dissipated as heat within the room, the energy needed to cool back this heat and keep the room at a constant temperature during times of air-conditioning is also saved.

The hot tap water for the kitchens has previously been prepared using 50l / 2kW_{el} hot water geysers. The hot water use per kitchen is estimated to 30l per working day (200 working days per year) at 60°C (temperature needed to prevent growing of legionella / germs). To heat 6000 liters of water per year from 10°C to 60°C requires 350kWh as an estimated yearly demand of single geyser. The heat losses of the geyser are specified to be 1.62kWh/day resulting in an additional yearly loss of 590kWh of electricity per geyser adding up to 940kWh. By replacing two of the geysers by a hot water solar system, up to 1.9MWh of electrical energy are saved per year on hot water preparation. The third water geyser is now assisted by a hybrid heat pump and air conditioning system, resulting in enhanced efficiency in the use of electric energy.

Based on measurements in 2016, the energy demand of the vicinity has been reduced by 31MWh ± 2.5MWh compared to 2014/15 by the described energy efficiency measures. The effect of the energy efficiency measures on the average daily load curve is presented in Figure 2 (orange area).

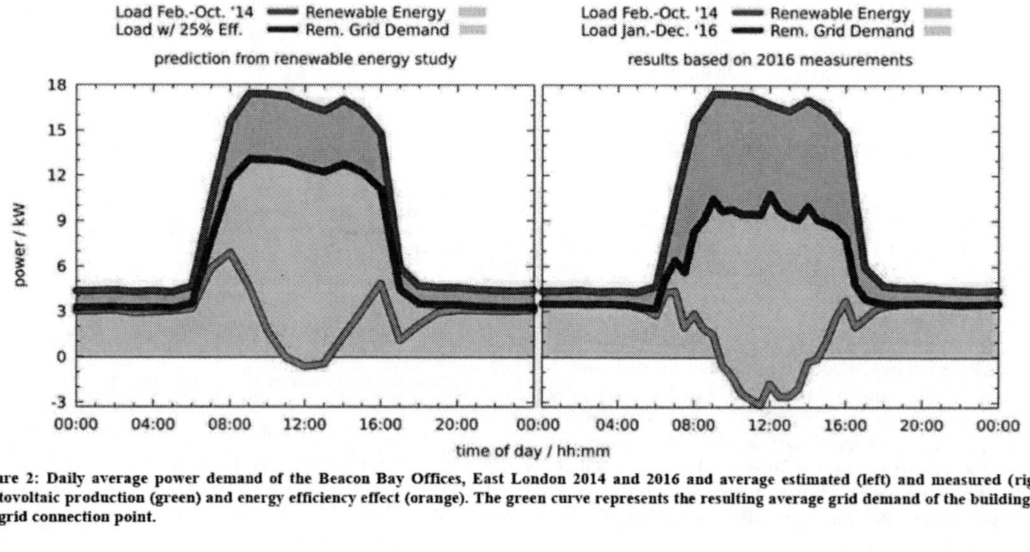

Figure 2: Daily average power demand of the Beacon Bay Offices, East London 2014 and 2016 and average estimated (left) and measured (right) photovoltaic production (green) and energy efficiency effect (orange). The green curve represents the resulting average grid demand of the buildings at the grid connection point.

2.3 Renewable Energy Generation

Based on a photovoltaic generation study using radiation data from *meteonorm 7* and the renewable energy simulation tool *insel*, a 16.5kW$_P$ system facing north with 33° tilt, or a 19.5kW$_P$ system facing east with 11° tilt both in combination with a 3.5kW$_P$ micro wind turbine with low cut in speeds was recommended to fulfill the given projects targets.

A 20kW$_P$ north facing grid connected photovoltaic system was installed on the roof of the public library that is located within the retrofitted office building complex. The installed 600W$_P$ micro wind turbine is solely feeding into the battery backup system for times of load shedding. The PV panels can be reconfigured automatically on the DC side to feed a battery charger during load shedding demand. The expected daily average energy profile based on simulation is shown in Figure 2 on the left hand side (green area).

3 RESULTS

The presented results of the effects on the electricity grid demand by introducing self-consumption optimized photovoltaic and energy efficiency measures are based on measurements from 2016 with 30 min resolution (366 days, 17 568 values). In 2016 50.3MWh of electrical energy have been consumed and 32.4MWh have been generated by the photovoltaic solar system. The direct self-consumption rate of photovoltaic energy was 70%, feeding 30% back into the grid. The self-sufficiency of the location was 64% on the balance sheet and 45% based on direct consumption. An overview of the yearly values is given in Table 2.

Table 2: Overview electricity demand and photovoltaic generation

	MWh	$\%_{2016}$	$\%_{PV}$
Demand '14/'15	83.8	167	259
Demand '16	50.3	100	155
PV generation	32.4	64	100
Self-Consumption	22.7	**45**	**70**
Feed-In	9.7	19	30
Grid Demand (balance sheet)	17.9	**36**	55
Grid Demand (absolute)	27.6	55	85

3.1 Daily Energy Demand

The daily electrical energy demand of the premisis is given in Figure 3 (bottom) in comparison to typical monthly temperatures of a long term average year taken from *'meteonorm 7'* database (top). The baseload of the buildings is roughly 50kWh/day. Peaks loads over 200kWh/day are associated with months of high temperature (air conditioning, Dec.-Mar.) and very low temperatures (electric heaters, Jun.-Aug.). During transition time (spring and autumn) typically between 50kWh and 150kWh are consumed per day.

The daily average power curves in Figure 2 (measurements 2016) clearly show the overachievement of the energy efficiency measures (orange area) in comparison to the planned achievements (left) from the energy-savings recommendations study. Therefor the average feed-in during midday is larger than expected. The characteristic morning and evening peaks at 8:00 and 16:00 can still be seen in the grid demand as originally expected but are now largely compensated by the bigger photovoltaic system and the greater effect of the energy efficiency measures.

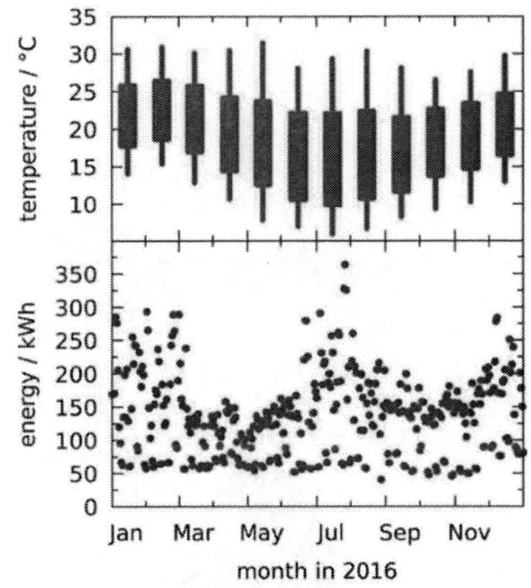

Figure 3: Daily energy demand in 2016 of the Beacon Bay Offices (bottom) and typical yearly temperature profile (top, source '*meteonorm 7*')

The exact knowledge of the demand side energy profile is essential for planning self-consumption dominated photovoltaic systems and easily leads to miscalculation on the viability of renewable energy generation systems. Additionally three demand peaks around typical work break hours become visible in the averaged profiles by reducing the energy demand of the buildings and its typical fluctuations.

3.2 Power Profiles

In Figure 4, Figure 5 and Figure 6 the 24h power profiles of all 366 days are plotted with maximum, median, average and minimum for the electricity demand of the buildings, the photovoltaic generation and the combined power profile at the grid connection point of the vicinity.

The minimum curve of the power demand profile (Figure 4) shows times of load shedding between 9:00 and 23:00 in multiple events during 2016. High power demands during the day are seldom compared to the typical demand. Especially during the night, only few events are responsible for the maximum curve of the mostly constant power demand during the night. See section on annual load duration curves below for a more detailed discussion of high power events.

The photovoltaic generation (Figure 5) exhibits a textbook-like generation profile over the year with a well defines peak which is not broadened by summer saving time measures.

The profile of the power demand / feed-in at the grid connection point (Figure 6) unfolds, that now even fewer events are responsible for the maximum power demand curve (red) with photovoltaic generation being synchronized in time with the demand and taking away a lot of the high power demand during the day, locally. The median and average curves are very well centered around net zero demand compared to their respective minimum and maximum values and reflect all properties of a well-balanced self-consumption optimized photovoltaic system.

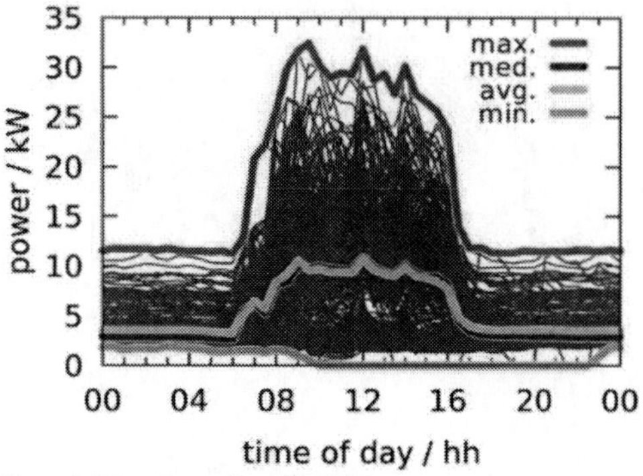

Figure 4: Power demand profiles 2016

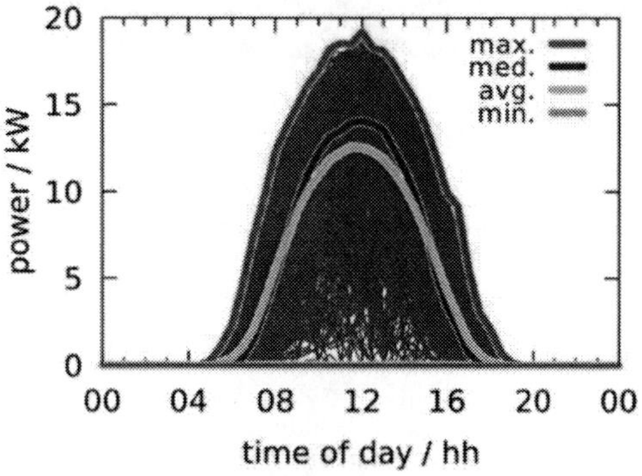

Figure 5: Photovoltaic generation profiles 2016

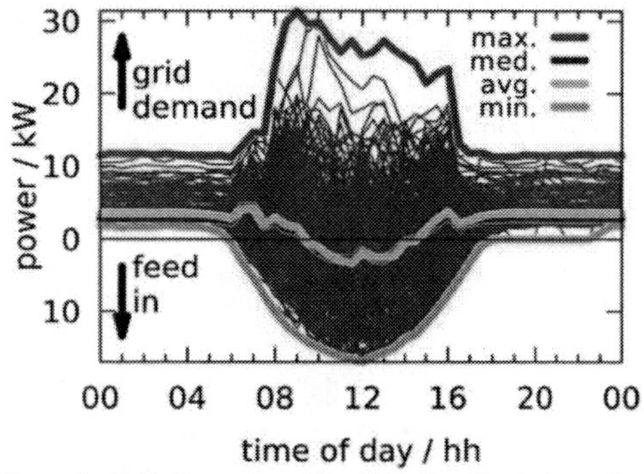

Figure 6: Grid demand profiles 2016 (combined demand and photovoltaic generation at grid connection point)

3.3 Annual Load Duration Curves

The annual load duration curves (aldc) and the histogram of the buildings electricity demand, photovoltaic generation and power demand at the grid connection point are shown in Figure 7 and Figure 8.

The aldc of the photovoltaic production (Figure 7, green) shows, that the PV system is active for 4308 hours per year with a linear power slope resulting in a mostly symmetric histogram (Figure 8, green) between 3-17kW; low powers being preferred and very large powers close to the nominal power rating of the photovoltaic system being stochastically under-represented the full histogram is skewed left peaking at 4476h of zero energy production.

The histogram of the buildings electricity demand is skewed right peaking at 3kW (Figure 8, blue). The peak represents the night time constant demand that is seen in the aldc of the demand starting at about 4500h (Figure 7). The demand for power in the power range of 10kW to 30kW is exponentially distributed over a summed duration of 1485h.

The resulting power demand of the combination of the photovoltaic generation and the demand of the buildings is a symmetric unimodal histogram peaking at 3kW (Figure 8, transparent orange area). The most hours reduced are located in the range above 10kW grid demand and above 10kW feed-in. The aldc of the grid connection points shows a remaining 277h with demand above 10kW and 1792h of feed in – less than half the productive time of the photovoltaic system.

The difference between the electricity demand of the building and the demand at the grid connection point is shown at the bottom of Figure 7, highlighting the effect of the photovoltaic system on the net electricity demand. Especially the peak power grid demand is not compensated by the photovoltaic system. As seen from the histogram (Figure 8) 2h with demands above 30kW and 15h above 20kW are still remaining. For about 4000h a year, the power demand is reduced. Additionally the feed-in into the grid comes to the delta curve starting at around 6000h.

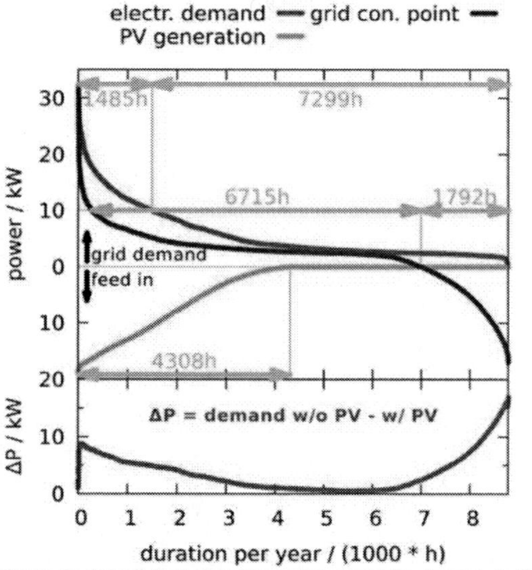

Figure 7: Annual load duration curves electricity demand (red), photovoltaic generation (green), Delta demand with PV and without PV (brown, bottom). The Delta curve is the red curve (demand without PV) minus the blue curve (demand at grid connection point) representing the effect on the aldc induced by the introduction of photovoltaic generation.

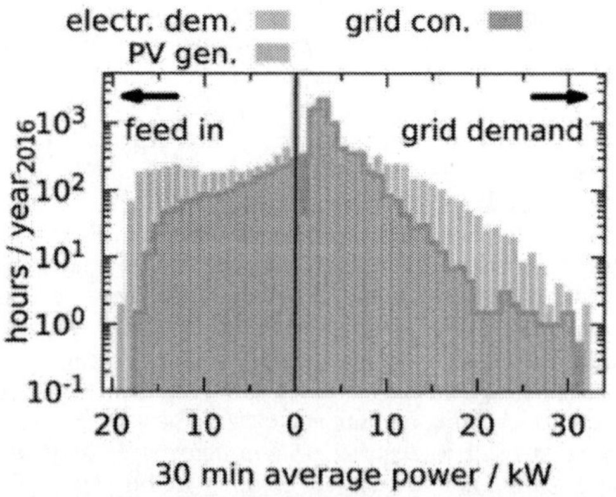

Figure 8: Histogram of the photovoltaic generation (green, feed-in side), building demand (blue, grid demand side) and demand at the grid connection point (transparent orange) with a bin size of 1kW.

3.4 Influence of Weekends and Weekdays

Roughly 29% or 2/7 days of the week account for weekends. In a building with worktime hours from 8:00 until 16:00, Monday through Friday (single shift, no weekends) and with a user dominated demand curve this has massive effects on the viability of a photovoltaic system. For ease of calculation, weekends are defined as Saturday and Sunday, weekdays as Monday through Friday.

The results for the yearly numbers are summarized in Table 3. 85% of the buildings electricity is consumed during the week, representing 70% of the time. Peak demands are also required only during the week. While the self-consumption hits 85% during the week, 70% of the electricity generated by the photovoltaic system is fed in during the weekends. During the week, the average grid demand curve is barely feeding in at all (Figure 9, orange curve), but the feed-in starts instantly during the weekends as the sun comes up due to the low base consumption of the buildings (Figure 10, orange curve).

Table 3: Grid demand and photovoltaic generation Mon-Fri compared to Sat-Sun

	Mon-Fri		Sat-Sun	
	MWh	$\%_{PV}$	MWh	$\%_{PV}$
Demand '16	42.8	182	7.5	84
PV generation	23.5	100	8.9	100
Self-Consumption	19.9	**85**	2.8	**30**
Feed-In	3.6	15	6.2	70
Grid Demand (balance sheet)	19.3	82	1.4	16
Grid Demand (absolute)	22.9	97	4.7	53

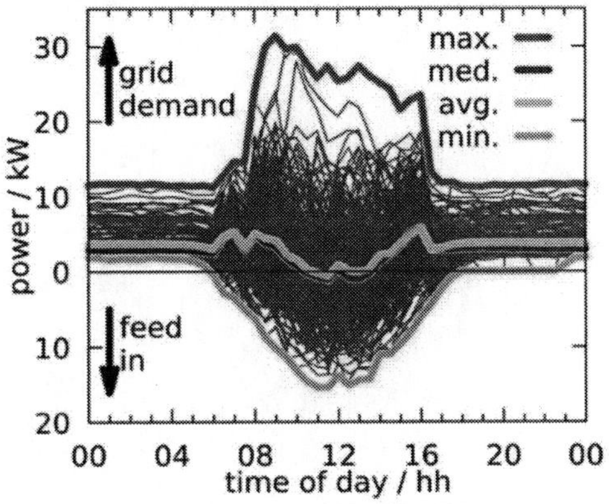

Figure 9: Grid demand during weekdays (Mon-Fri)

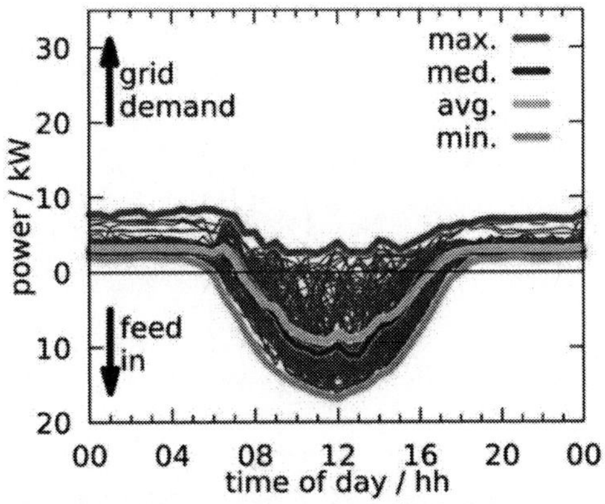

Figure 10: Grid demand during weekends (Sat-Sun)

4 DISCUSSION

A photovoltaic system optimized for self-consumption can significantly reduce the hours of high power demand (Figure 7 and Figure 8). The daily average electricity demand can be tuned to moderately fluctuate between feed in and demand (Figure 6). As the peak power demand is not reduced, the network resources required (transformers, cables) to connect the building to the grid cannot be reduced. Potentially distributed grid resources can be reduced due to a possible reduction of the concurrency factor of the peak load within the grid fragment of the building.

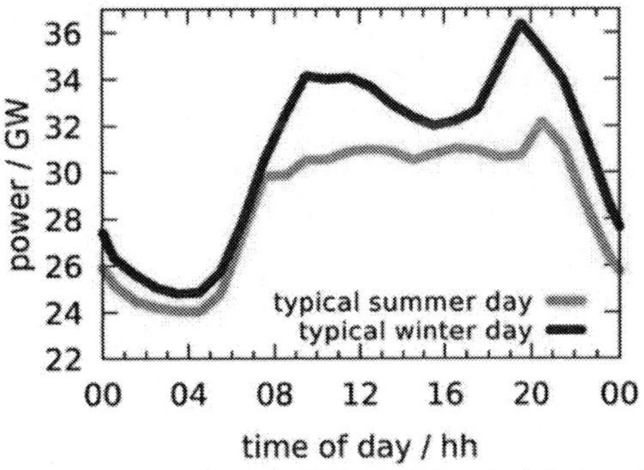

Figure 11: Load profile of the ESKOM grid for a typical winter day and a typical summer day [1], graphically reworked to improve readability in this paper

The grid demand for a typical winter and summer day in the ESKOM grid is given in Figure 11. The winter demand is higher than the summer demand due to heating application. The summer demand shows a table top mountain profile, while the winter demand has two distinct consumption peaks in the morning and in the evening. Surprisingly, the grid is more at strain during the summer than during the winter [2], as maintenance is conducted during summer time and no peaks are expected for calculating the planned capacity loss factor (PCLF). Additional problems in the power generation can then be problematic for the full day, while in winter they only effect short periods over the day. The reduction of peak demands during the summer time by self-consumption of locally generated electricity from photovoltaic systems might therefor have positive effects on the overall grid stability and reduces demand for high cost generation capacity (merit order).

While the times of high grid demand are reduced during the year, times of high feed in are introduced especially during the weekends (Figure 10). For feed in, regulatory action and grid codes are required for the photovoltaic system to keep the grid stable for high shares of photovoltaic generation in the grid.

5 CONCLUSIONS

One shift office buildings are very well suited for local photovoltaic electricity generation allowing for high shares of self-consumption (here: 70%) and self-sufficiency (here: 45% direct, 64% on the balance sheet). The peak power demand of the building is not reduced by the direct consumption of electricity from photovoltaics, but the hours of high power demand are significantly reduced. The feed in of electricity into the public grid is concentrated during the weekends. Energy efficiency is an essential prerequisite on the way to viable self-consumption photovoltaic systems.

This type of self-consumption optimized photovoltaic system stabilizes the public grid by reducing the times of high power demand especially during the summer. This reduces the risk for load shedding actions to be in place due to high maintenance activity during the summer. Additionally the table mountain top profile of the summer electricity demand increases the risk for blackouts if unexpected events occur, because the peak power reserve on standby during winter to cover the morning and evening demand is either shut down or on maintenance during this period of the year.

REFERENCES

[1] "Effective management of energy demand", ESKOM, April 2016, http://www.eskom.co.za/sites/idm/Business/Documents/EskomAdvisoryService14042016.pdf, page view 28.8.2017

[2] "Media presentation", ESKOM, Tshediso Matona, 15th Jan. 2015, https://www.teraco.co.za/content/Eskom_SystemStatusPresent.pdf, page view 28.8.2017

ACKNOWLEDGEMENTS

The authors greatly acknowledge the funding and support by the Federal Ministry for Economic Cooperation and Development of Germany under the SKEW / NaKoPa program and the Buffalo City Metropolitan Municipality. Further the City of Oldenburg and its department of international relations is deeply recognized for the organization of the project, co-funding and the application to the funding authority. The Buffalo City Metropolitan Municipality and its Department of Electricity is recognized for its political and organizational support and the co-funding of the project. The Federal State of Lower Saxony, Germany is acknowledged for providing the framework for the cooperation and for motivating and supporting the bilateral exchange.

The authors personally thank Darby Gounden of Buffalo City Municipality for her enthusiastic engagement with the project and the organization of workshops and her support in keeping to schedules.

Finally the authors acknowledge the paper size limitations for the conference proceedings of the EUPVSEC and regret the figure quality and jpeg artefacts.

AUTHOR INDEX

AUTHOR INDEX

AUTHOR INDEX

AUTHOR INDEX

AUTHOR INDEX

AUTHOR INDEX

AUTHOR INDEX

AUTHOR INDEX

AUTHOR INDEX

AUTHOR INDEX

AUTHOR INDEX

AUTHOR INDEX

AUTHOR INDEX

AUTHOR INDEX

AUTHOR INDEX

AUTHOR INDEX

AUTHOR INDEX

AUTHOR INDEX

AUTHOR INDEX

AUTHOR INDEX

AUTHOR INDEX

AUTHOR INDEX

AUTHOR INDEX

AUTHOR INDEX

33rd European Photovoltaic Solar Energy Conference and Exhibition (EU PVSEC 2017)

Amsterdam, The Netherlands
25-29 September 2017

Volume 4 of 4

Editors:

A. Smets
N. Taylor
P. Helm

ISBN: 978-1-5108-5397-3

Printed from e-media with permission by:

Curran Associates, Inc.
57 Morehouse Lane
Red Hook, NY 12571

Some format issues inherent in the e-media version may also appear in this print version.

Printed by Curran Associates, Inc. (2018)

For permission requests, please contact WIP – Renewable Energies
at the address below.

WIP – Renewable Energies
Sylvensteinstr. 2
81369 Munchen
Germany

Phone: +49 89 72012735
Fax: +49 89 72012791

wip@wip-munich.de

Additional copies of this publication are available from:

Curran Associates, Inc.
57 Morehouse Lane
Red Hook, NY 12571 USA
Phone: 845-758-0400
Fax: 845-758-2633
Email: curran@proceedings.com
Web: www.proceedings.com

33rd European Photovoltaic Solar Energy Conference and Exhibition

Proceedings of the International Conference held in Amsterdam, The Netherlands

25 September – 29 September 2017

Edited by:

A. SMETS
Delft University of Technology
Professor Solar Energy
Delft, The Netherlands

N. TAYLOR
European Commission Joint Research Centre
Ispra, Italy

P. HELM
WIP Renewable Energies
Munich, Germany

Coordination of the Technical Programme:

European Commission Joint Research Centre
Via E. Fermi 1
21020 Ispra (VA)
Italy

Institutional Support:

European Commission

UNESCO – United Nations Educational, Scientific and Cultural Organization, Natural Sciences Sector

WCRE – World Council for Renewable Energy

Supporting Organisations:

ARE – Alliance for Rural Electrification

EREF – European Renewable Energies Federation

ESA – European Space Agency

EUFORES –The European Forum for Renewable Energy Sources

EUREC – The Association of European Renewable Energy Research Centres

European Technology & Innovation Platform PV

IEA PVPS – International Energy Agency, Photovoltaic Power Systems Programme

REN21 – Renewable Energy Policy Network for the 21st Century

TKI Urban Energy - Topsector Energie

Institutional PV Industry Cooperation:

SolarPower Europe

SOLARUNITED

Supporting Associations:

AIE – European Association of Electrical Contractors

AREA – African Renewable Energy Alliance

CABA – Continental Automated Buildings Association

EPIC – European Photonics Industry Consortium

GENSED - Turkish Solar Energy Industry Association

MPIA - Malaysian Photovoltaic Industry Association

SASIA - Saudi Arabia Solar Industry Association

EU PVSEC realised by:

WIP
Sylvensteinstr. 2, 81369 München, Germany
Tel: +49 89 720 12 735, Fax: +49 89 720 12 791
Email: pv.conference@wip-munich.de
www.photovoltaic-conference.com
www.wip-munich.de

Proceedings produced and published by:

WIP
Sylvensteinstr. 2, 81369 München, Germany
Tel: +49 89 720 12 735, Fax: +49 89 720 12 791
Email: pv.conference@wip-munich.de
www.photovoltaic-conference.com
www.wip-munich.de

ISSN 2196-0992

33rd EUROPEAN PHOTOVOLTAIC SOLAR ENERGY CONFERENCE AND EXHIBITION
25 SEPTEMBER – 29 SEPTEMBER 2017
AMSTERDAM, THE NETHERLANDS

EU PVSEC COMMITTEES

P.A. Basore, PV Specialist Services, Solana Beach, USA
S. Bowden, Arizona State University, Tempe, USA
T.M. Bruton, Progress in Photovoltaics, Woking, United Kingdom
A. Danel, CEA, Le Bourget du Lac, France
B. Dauksher, Arizona State University, Tempe, USA
M. Debucquoy, imec, Leuven, Belgium
P. Delli Veneri, ENEA, Portici, Italy
T. Desrues, CEA, Le Bourget du Lac, France
D. Dimova-Malinovska, Bulgarian Academy of Sciences, Sofia, Bulgaria
S. Dubois, CEA, Le Bourget du Lac, France
T. Dullweber, ISFH, Emmerthal, Germany
P. Fath, RCT-Solutions, Konstanz, Germany
S. Gall, HZB, Berlin, Germany
I. Gordon, imec, Leuven, Belgium
J. Haschke, EPFL, Neuchâtel, Switzerland
K. Hesse, Wacker Chemie, Burghausen, Germany
J. Horzel, CSEM, Neuchâtel, Switzerland
J.-R. Huang, Motech Industries, Taoyuan, Taiwan
J. John, imec, Leuven, Belgium
D.-W. Kang, Cheongju University, Republic of Korea
S.D. Kim, San Jose, USA
A. Meeder, Geo-En Energy Technologies, Berlin, Germany
J. Meier, Meier Technologies, Berlingen, Switzerland
A. Metz, Alzenau, Germany
C. Molpeceres, UPM, Madrid, Spain
J.W. Müller, Hanwha Q CELLS, Bitterfeld-Wolfen, Germany
Y. Ohshita, Toyota Technological Institute, Nagoya, Japan
S. Peters, Hanwha Q CELLS, Bitterfeld-Wolfen, Germany
K. Petter, Hanwha Q CELLS, Bitterfeld-Wolfen, Germany
R. Preu, Fraunhofer ISE, Freiburg, Germany
J. Rentsch, Fraunhofer ISE, Freiburg, Germany
E. Sauar, Brighterlite, Oslo, Norway
H. Schlemm, Meyer Burger, Hohenstein-Ernstthal, Germany
J. Schmidt, ISFH, Emmerthal, Germany
M.C. Schubert, Fraunhofer ISE, Freiburg, Germany
B. Terheiden, University of Konstanz, Germany
M. Vetter, IPHT, Jena, Germany
G.P. Willeke, Fraunhofer ISE, Freiburg, Germany
D. Wood, Hemlock Semiconductor, Hemlock, USA
A. Yamada, Tokyo Institute of Technology, Japan
D.L. Young, NREL, Golden, USA
M. Zeman, Delft University of Technology, The Netherlands
J. Zhao, Meyer Burger, Hohenstein-Ernstthal, Germany

Topic 3 – Thin Film Photovoltaics

C.J. Brabec, University of Erlangen-Nuremberg, Germany
A.N. Tiwari, EMPA, Dübendorf, Switzerland
S.C. Veenstra, ECN, Eindhoven, The Netherlands
W. Wischmann, ZSW, Stuttgart, Germany

G. Bardizza, European Commission JRC, Ispra, Italy
R. Birkmire, University of Delaware, Newark, USA
D. Cheyns, imec, Leuven, Belgium
S. Cros, CEA, Le Bourget du Lac, France
T. Dalibor, AVANCIS, Torgau, Germany
B. Dimmler, Manz CIGS Technology, Schwäbisch Hall, Germany
K. Durose, Stephenson Institute for Renewable Energy, Liverpool, United Kingdom
S. Huang, UNSW Australia, Sydney, Australia
S.J.C. Irvine, Swansea University, St. Asaph, United Kingdom
C.-F. Lin, NTU, Taipei, Taiwan
M. Lira-Cantú, CIN2, Barcelona, Spain
M.Ch. Lux-Steiner, HZB, Berlin, Germany
M. Meuris, imec, Leuven, Belgium
J. Parisi, University of Oldenburg, Germany
R. Po, eni spa, Novara, Italy
M. Powalla, ZSW, Stuttgart, Germany
A. Redinger, University of Luxembourg, Luxembourg
A. Romeo, University of Verona, Italy
V. Sittinger, Fraunhofer IST, Braunschweig, Germany
A. Virtuani, EPFL, Neuchâtel, Switzerland

Topic 4 – Concentrator and Space Photovoltaics

G. Flamand, imec, Leuven, Belgium
C. Signorini, European Space Agency, Noordwijk, The Netherlands

I. Antón, UPM, Madrid, Spain
A.W. Bett, Fraunhofer ISE, Freiburg, Germany
R. Campesato, CESI, Milan, Italy
V. Khorenko, Azur Space, Heilbronn, Germany
R. King, Arizona State University, Tempe, USA
F. Rubio Berenguel, Soitec Solar, Freiburg, Germany
G. Siefer, Fraunhofer ISE, Freiburg, Germany
G. Timò, RSE, Piacenza, Italy

Topic 5 – Performance, Reliability and Sustainability of Photovoltaic Modules and Balance of System Components

M. de Wild-Scholten, SmartGreenScans, Groet, The Netherlands
R. Gottschalg, Loughborough University, United Kingdom
N.M. Pearsall, Northumbria University, Newcastle upon Tyne, United Kingdom
M. Perrin, CEA, Le Bourget du Lac, France
T. Sample, European Commission JRC, Ispra, Italy
K. Wambach, Wambach-Consulting, Aindling, Germany

K. Baert, KU Leuven, Heverlee, Belgium
G. Beaucarne, Dow Corning, Seneffe, Belgium
R. Einhaus, Apollon Solar, Lyon, France
G. Friesen, SUPSI, Canobbio, Switzerland
V. Fthenakis, Columbia University, New York, USA
W.J. Gambogi, DuPont, Wilmington, USA
L. Gomes de Freitas, Federal University of Uberlândia, Brazil
G. Graditi, ENEA, Portici, Italy
Y. Hishikawa, AIST, Tsukuba, Japan
U. Jahn, TÜV Rheinland Energy, Cologne, Germany
S. Krauter, University of Paderborn, Germany
A.R. Lagunas, CENER, Sarriguren-Navarra, Spain
C. Monokroussos, TÜV Rheinland, Shanghai, China
M.A. Muñoz-García, UPM, Madrid, Spain
H. Nagel, Fraunhofer ISE, Freiburg, Germany
H. Nussbaumer, ZHAW, Winterthur, Switzerland
S. Oktik, Sisecam, Istanbul, Turkey
G. Oreski, PCCL, Leoben, Austria
W. Palitzsch, Loser Chemie, Zwickau, Germany
J.N. Roy, IIT Kharagpur, India
R. Sinton, Sinton Instruments, Boulder, USA
M. Van Iseghem, EDF R&D, Moret-sur-Loing, France
E. Voroshazi, imec, Leuven, Belgium
A. Wade, First Solar, Mainz, Germany

Topic 6 – PV System Performance and Integration

F.P. Baumgartner, ZHAW, Winterthur, Switzerland
F. Bonemazzi, ENEL, Rome, Italy
F. Frontini, SUPSI, Canobbio, Switzerland

P. Lechner, ZSW, Stuttgart, Germany
O. Mayer, GE Global Research, Garching, Germany
F. Montanari, ENEL Green Power, Rome, Italy
C. Protogeropoulos, EEPS, Athens, Greece
A. Scognamiglio, ENEA, Portici, Italy
W.G.J.H.M. van Sark, Utrecht University, The Netherlands
I. Weiss, WIP - Renewable Energies, Munich, Germany

C. Alonso-Tristán, UBU, Burgos, Spain
J. Binder, ZSW, Stuttgart, Germany
P. Blanc, MINES ParisTech, Paris, France
P. Bonomo, SUPSI, Canobbio, Switzerland
F. Bourry, CEA, Le Bourget du Lac, France
I. Da Silva, Strathmore University, Nairobi, Kenya
B. Gaiddon, Hespul, Lyon, France
M. Grottke, WIP - Renewable Energies, Munich, Germany
I.B. Hagemann, Architekturbüro Hagemann, Aachen, Germany
C. Helmke, Clake Energy, Erfurt, Germany
T. Huld, European Commission JRC, Ispra, Italy
E. Kaijuka-Okwenje, Rural Electrification Agency, Kampala, Uganda
S. Koopman, CSIR, Pretoria, South Africa
P.A. Leicester, Loughborough University, United Kingdom
P. Malbranche, CEA, Le Bourget du Lac, France
G. Mütter, Alternative Energy Solutions, Vienna, Austria
C. Nyman, Soleco, Porvoo, Finland
K. Peter, ISC Konstanz, Germany
T. Reijenga, BEAR-iD, Gouda, The Netherlands
J. Remund, Meteotest, Bern, Switzerland
K.T. Roro, CSIR, Pretoria, South Africa
M. Schroedter-Homscheidt, German Aerospace Center, Wessling, Germany
M. Sengupta, NREL, Golden, USA
D. Stellbogen, ZSW, Stuttgart, Germany
J. Sutterlueti, Gantner Instruments, Schruns, Austria
S. Tselepis, Pikermi, Athens, Greece
X. Vallvé, Trama TecnoAmbiental, Barcelona, Spain
M. Zehner, Rosenheim University of Applied Sciences, Germany

Topic 7 – PV Economics, Markets and Policies

C. Breyer, Lappeenranta University of Technology, Finland
M. Getsiou, European Commission DG RTD, Brussels, Belgium
G. Masson, Becquerel Institute, Brussels, Belgium
T. Nordmann, TNC Consulting, Feldmeilen, Switzerland
S. Nowak, NET Nowak Energy & Technology, St. Ursen, Switzerland
E. Perezagua, Consultores de Energía Fotovoltaica, Madrid, Spain

S. Caneva, WIP - Renewable Energies, Munich, Germany
I. Kaizuka, RTS, Tokyo, Japan
D. Mayer, MINES ParisTech, Paris, France
P. Mazzucchelli, EUREC, Brussels, Belgium
S. Szabó, European Commission JRC, Ispra, Italy

SUBJECT INDEX

Topic Code

1 New Materials and Concepts for Photovoltaic Devices
2 Silicon Photovoltaics
3 Thin Film Photovoltaics
4 Concentrator and Space Photovoltaics
5 Performance, Reliability and Sustainability of Photovoltaic Modules and Balance of System Components
6 PV System Performance and Integration
7 PV Economics, Markets and Policies

Session Type

P = Plenary Session
O = Oral Session
V = Visual Session

Day Codes

A = Monday, 25 September 2017
B = Tuesday, 26 September 2017
C = Wednesday, 27 September 2017
D = Thursday, 28 September 2017
E = Friday, 29 September 2017

e.g. 3CP.1 ⇒ 3=Thin Film Photovoltaics, C=Wednesday, P=Plenary Session, 1=Session 1

TABLE OF CONTENT
of EU PVSEC 2017 Proceedings Papers

New Materials and Concepts for Photovoltaic Devices

Silicon Photovoltaics

Visual PRESENTATIONS 2AV.3 Heterojunction Solar Cells

Thin Film Photovoltaics

Plenary SESSION 3CP.1 Thin Film PV: Pushing the Limits with Breakthroughs in Industry and Research

Oral PRESENTATIONS 3AO.7 Optical Losses and TCO`s

Oral PRESENTATIONS 3AO.8 Module Stability and Characterisation

Oral PRESENTATIONS 3BO.9 Manufacturing and Performance Improvements

Oral PRESENTATIONS 3BO.11 Kesterites

Oral PRESENTATIONS 3BO.12 Organic Based PV

Oral PRESENTATIONS 3DO.7 Perovskite-Based Hybrid Tandems

Visual PRESENTATIONS 3CV.1 CI(G)S, CdTe and Related Thin Film Solar Cells and Modules (I)

Visual PRESENTATIONS 3DV.2 CI(G)S, CdTe and Related Thin Film Solar Cells and Modules (II) / Perovskite, Organic and Dye-Sensitised Devices

Concentrator and Space Photovoltaics

Oral PRESENTATIONS 4DO.4 III-V-Based Devices for Terrestrial and Space Applications

Visual PRESENTATIONS 4CV.4 III-V-Based Devices for Terrestrial and Space Applications

Performance, Reliability and Sustainability of Photovoltaic Modules and Balance of System Components

Plenary SESSION 5DP.1 Performance, Reliability and Sustainability of Photovoltaic Modules and Balance of System Components

Visual PRESENTATIONS 5DV.3 PV Module Performance and Reliability (II) / Inverters and Balance of System Components / Sustainability and Recycling

PV System Performance and Integration

Plenary SESSION 6DP.2 PV System Performance and Integration

Oral PRESENTATIONS 6BO.7 Advanced Field Performance Estimation

Oral PRESENTATIONS 6BO.8 Failure Modes and Degradation

Visual PRESENTATIONS 6BV.2 Design and Operation of PV Systems (II)

PV Economics, Markets and Policies

Oral PRESENTATIONS 7DO.8 Global PV Economics and Market Trends

Oral PRESENTATIONS 7DO.9 Innovative National PV Market Economics Business Cases

Oral PRESENTATIONS 7EO.3 PV-Related Policies, Strategies and Societal Issues

Visual PRESENTATIONS 7DV.1 PV Economics and Markets / PV-Related Policies, Strategies and Societal Issues

FOREWORD

The EU PVSEC 2017 in Amsterdam concluded with the strong message that the photovoltaic revolution is just getting started. Research and innovation across the sector, together with fast deployment and numerous application options, are paving the way to putting PV at the centre-stage of sustainable energy systems. This year 2,516 participants from 89 countries took part in this most inspiring platform. It provided a vital forum for information exchange in the field of PV solar energy research, demonstration and applications. In the exhibition 64 companies from all parts of the world welcomed visitors and presented their products and services.

The EU PVSEC 2017 Proceedings give a comprehensive overview of the photovoltaic solar sector, its current status and future prospects in science, research, innovation, development and deployment on more than 4,500 pages. Selection for inclusion in the conference was made by paper review experts and topic organisers (see the listing on pages III-V), to whom we express our sincere gratitude for their comprehensive review work and overall contribution to the success of the conference. In addition to the 619 submitted papers, the proceedings include 70 presentations (slides) shown during the plenary and oral presentations as well as 232 poster files of the visual presentations. In total this amounts to 921 publications.

The 2017 EU PVSEC Proceedings maintain our commitment to providing quick and open access to high quality scientific results. They constitute a powerful tool for targeted and quick information search and retrieval, enabling you to search by keyword, paper title and authors. We are sure that these features will help to simplify the use and exploit the full potential of this extensive source of information.

The Conference Proceedings are published as downloadable DVD file and are also fully accessible online. A DOI code (Digital Object Identifier) has been assigned to each paper. This ensures unequivocal and permanent identification and full citability. The papers can be viewed and downloaded in a full free open access from the EU PVSEC Proceedings website www.eupvsec-proceedings.com.

We are confident that these Proceedings will play an important role in providing a comprehensive overview of the current actors and activities in the global PV sector and that they will disseminate information on the state-of-the-art of technologies and applications. This can generate further research, add momentum to innovation and promote interest in PV worldwide.

We would like to cordially thank all authors and participants of the EU PVSEC 2017 for their contributions and look forward to welcoming you in 2018 at the 35th EU PVSEC 2018 in Brussels, Belgium.

The Editors

EU PVSEC 2017

25 - 29 September 2017

33rd European PV Solar Energy Conference and Exhibition

RAI Convention & Exhibition Centre • Amsterdam • The Netherlands

Selected Highlights

Nigel Taylor, Technical Programme Chair

.... with help from the session chairs and others

European Commission

Silicon

European Commission

2nd Fabrication Run of n-Type HP mc Solar Cells
IV Results

Fraunhofer ISE

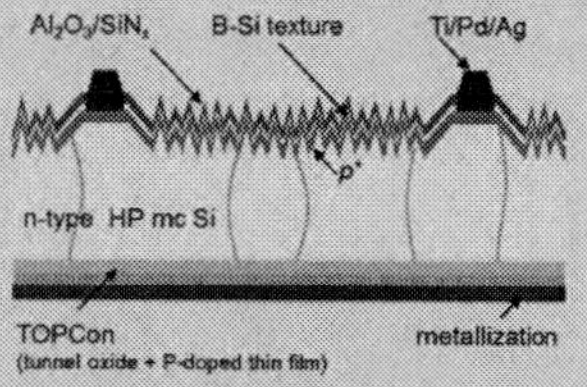

V_{oc} [mV]	J_{sc} [mA/cm²]	*FF* [%]	*pFF* [%]	η [%]
674.2	41.1	80.5	81.6	22.3*

Cell area 2×2 cm², aperture area (*) certified by Fraunhofer ISE CalLab

- 2nd cell batch
- Applying of „lessons learned" in cell fabrication
- Same material as previous solar cells

→ New record efficiency of 22.3% for a mc silicon solar cell

2DO.3.1 Approaching 22% Efficiency with Multicrystalline n-Type Silicon Solar Cells
J. Benick et al, Fraunhofer ISE, Freiburg, Germany

European Commission

Q.ANTUM technology on p-Cz substrates in mass production

Hanwha Q CELLS

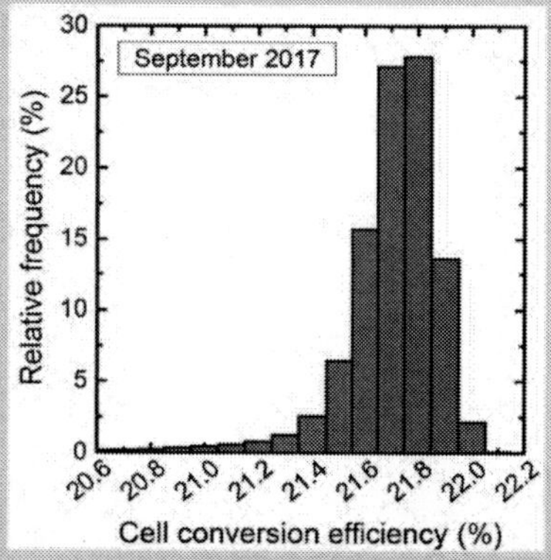

Platform change to Q.ANTUM[1] secures:

- Efficiency increase of +0.6%abs. per year
- Evolutionary development process[2] and upgrade of existing tool set
- Silver consumption: -50%
- Throughput increase
- **Avg. cell efficiency of 21.7% in production (max. >22%)**

2CO.9.6 Mass Production of Q.ANTUM Solar Cells and Modules on p-Type Cz Silicon Substrates
F. Fertig et al, Hanwha Q CELLS, Bitterfeld-Wolfen, Germany

European Commission

24.13% efficiency IBC cell

Trinasolar

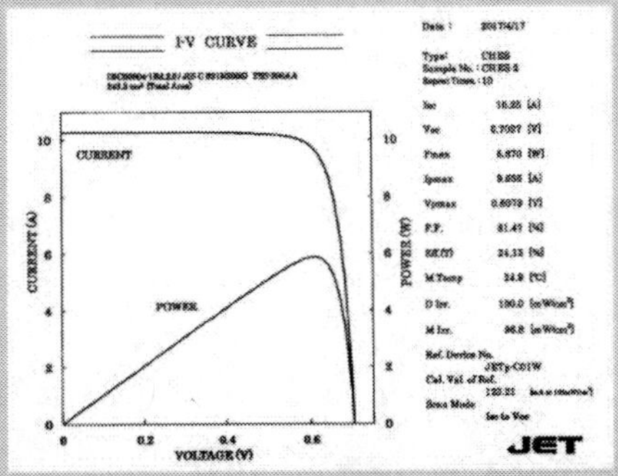

Compared to 23.5% cell (2016):

- 10 mV gain in V_{oc} by using selective emitter structure;
- J_{sc} is also improved by using less absorption in SiN_x;
- Higher FF is due to the reduced R_s, mainly from lower ρ_c on emitter.

Area	V_{oc}	I_{sc}	V_{mpp}	I_{mpp}	FF	Eff
[cm²]	[mV]	[A]	[mV]	[A]	[%]	[%]
244.9	702.7	10.25	607.9	9.66	81.47	24.13

* Independently measured at Japan Electrical Safety & Environment Technology Laboratories (JET)

2CO.11.5 Large-Area (6 Inch) Screen-Printed IBC Solar Cells with Efficiency Approaching 24% without Passivated Contacts

G. Xu et al, Trina Solar Energy, Changzhou, China

European Commission

KANEKA

Advantages of a-Si/c-Si heterojunction (HJ) back contact (BC) Si solar cell

a-Si/c-Si HJ : High V_{oc} and FF by excellent a-Si/c-Si interface

Back contact : High J_{sc} and FF by small optical and resistive loss

Si wafer-based: Reasonable production cost

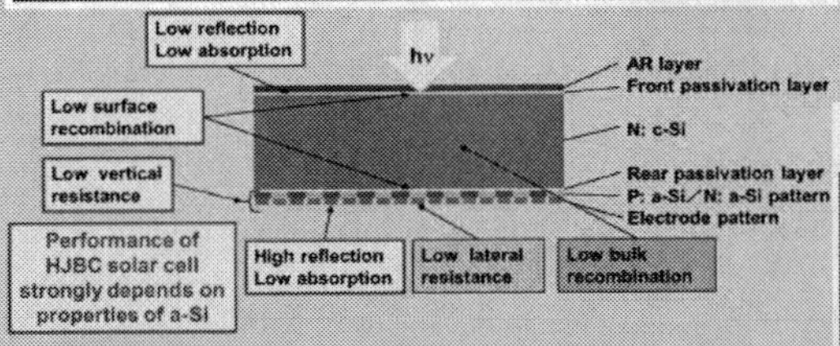

High efficiency c-Si solar cell; PV parameters

*da: designated area *ta: total area *ap: aperture area

	V_{OC} (V)	J_{SC} (mA/cm²)	FF (%)	Eff (%)	Area* (cm²)	Cell type
Panasonic	0.740	41.8	82.7	25.6	143.7 (da)	HJBC [1]
SunPower	0.737	41.3	82.7	25.2	153.5 (ta)	IBC [2]
Fraunhofer ISE	0.725	42.5	83.3	25.7	4.017(da)	TOPCon [3]
Kaneka	0.738	40.8	83.5	25.1	151.9 (ap)	HJ [4]
Kaneka	0.744	42.3	83.8	26.3	180.4 (da)	HJBC [5]
Kaneka	0.740	42.5	84.7	26.6	179.7 (da)	HJBC [6]
Kaneka	0.738	42.7	84.9	26.7	79.0 (da)	HJBC
Theoretical limit	0.753	43.6	89.2	29.3		100um-thick wafer [7]
Ratio (26.7%/limit)	0.980	0.979	0.952	0.911		

- High V_{oc}, J_{sc} and FF obtained by excellent PECVD passivation and structure design
- Challenge for improvement of FF is a key to achieve >27% efficiency

[1] K. Masuko et al., IEEE Journal of Photovoltaics, 4, 1433 (2014). [2] "SunPower 2015 Analyst Day" (2015). [3] A. Richter et al., Silicon PV proceedings (2017). [4] D. Adachi et al., Applied Physics Letters, 107, 233506 (2015). [5] K. Yoshikawa et al., Nat. Energy 2, 17032 (2017) [6] K. Yoshikawa et al., Silicon PV proceedings (2017) [7] A. Richter et al., IEEE J. Photovolt. 3, 1184 (2013).

2BP1.1 Record-Breaking Efficiency Back-Contact Heterojunction Crystalline Si Solar Cell and Module

K. Yamamoto et al, Kaneka, Osaka, Japan

J.L. Hernández, Kaneka, Westerlo-Oevel, Belgium

European Commission

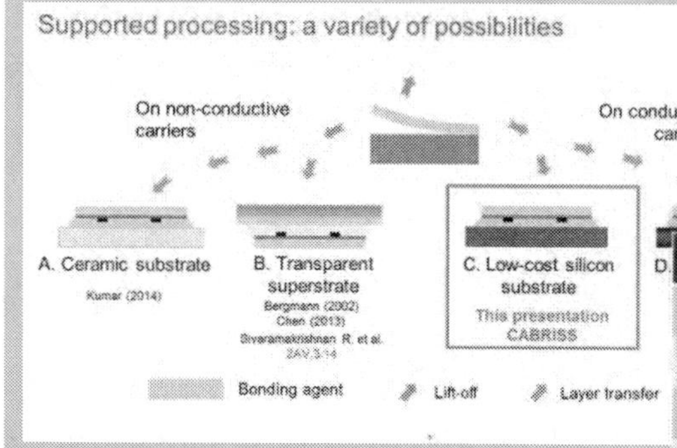

imec

TU WIEN · SINTEF · PyroGenesis · CABRISS · INKRON

Cell results

Material	Configuration		J_{sc} [mA/cm²]	V_{oc} [mV]	FF [%]	Eff. [%]	
Epifoil	Bonded to p⁺ Cz substrate	Best Avg.	33.2 32.8	681 676	69.2 69.1	15.7 15.3	*This work*
Thinned Cz	Bonded to p⁺ Cz substrate	Best Avg.	33.2 32.7	738 739	73.0 72.1	17.9 17.3	*This work*
Epifoil	Freestanding	Best Avg.	33.5 32.9	693 656	73.2 72.1	17.0 15.6	*Bearda et al. [2AV.3.26]*
Thinned Cz	Freestanding	Best	34.0 33.9	744 723	74.4 71.6	18.8 17.6	*Bearda et al. [2AV.3.26]*
Comparison with literature (cells bonded to conductive substrates)							
Epifoil (18 µm)	Bonded to steel	Best	34.5	632	77.2	16.8	*Amberwave/UNSW*
Slim-cut (25 µm)	Bonded to steel	Best	33.6	580	76.7	14.9	*Astrowatt*

2CO.12.2 Solar Cells on < 50µm Thick Epitaxial Foils Conductively Bonded to Low-Cost *Si* Carrier

H. Sivaramakrishnan Radhakrishnan et al, imec, Leuven, Belgium
N. Bednar & N. Adamovic, Vienna University of Technology, Austria
R. Roozeman & J. Heikkinen, INKRON, Espoo, Finland
A. Ulyashin & M. Syvertsen, SINTEF, Oslo, Norway

European Commission

Thin Film

European Commission

Key techniques implemented into the record sub(mini)modules

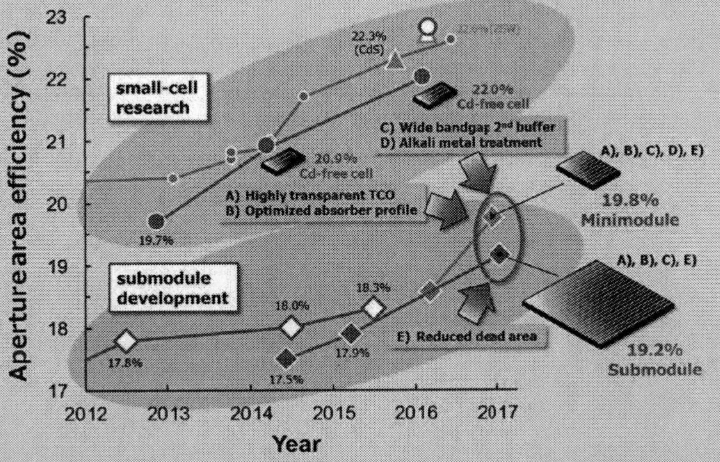

SOLAR FRONTIER

3CP.1.1 Present Status of Solar Frontier Cu(In,Ga)(Se,S)2 Record Efficiencies and Overall Progress

V. Bermudez et al, Solar Frontier, Atsugi, Japan

European Commission

HIGH EFFICIENCY MODULES

- Current in-house 30×30cm² champion module with 18.2% efficiency
- high Ga/high S absorber, wide bandgap buffer - highly Na doped In_xS_y:Na/i-$Zn_{0.85}Mg_{0.15}O$

AVANCIS CNBM

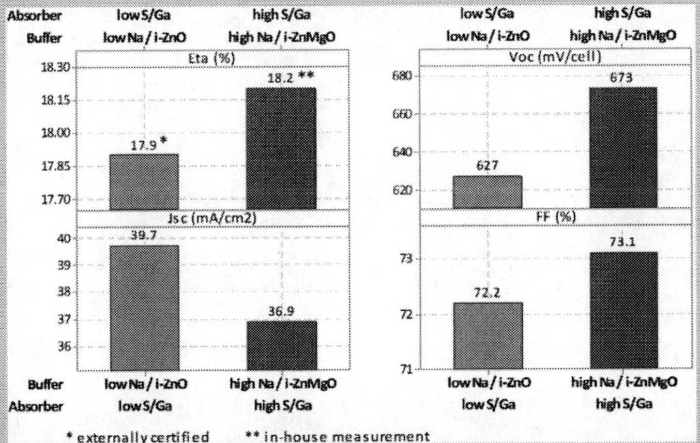

3BO.9.1 Wide Bandgap Sequential Absorber with Tunable Buffer Bandgap for CIGSSe Solar Modules at 18% Efficiency

R. Lechner et al, Avancis, Munich, Germany

European Commission

SOLIBRO Hi-Tech
A Hanergy Company

FULL SIZE: CHAMPION MODULE GEN 3

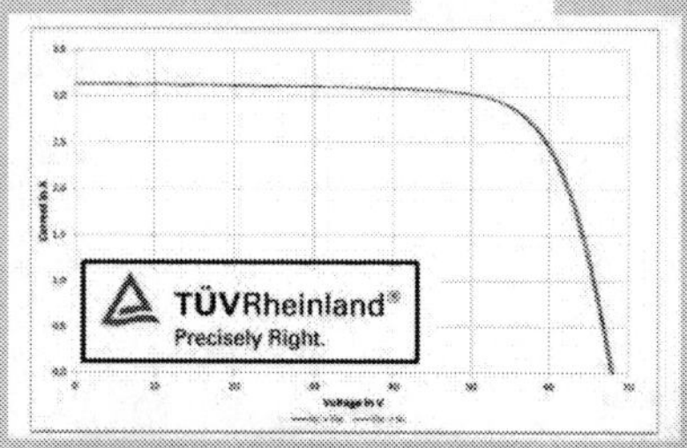

P_{mpp} [W]	159.4
V_{oc} [V]	67.9
I_{sc} [A]	3.12
FF [%]	75.2
η	17.0 %

- 17.9% Aperture Area Efficiency
- Externally verified
- Processed in SL2 production
- Module with future target process GEN3

17.0% @ 0.94m²
totale area efficiency
New world record

CELL AND MODULE DESIGN

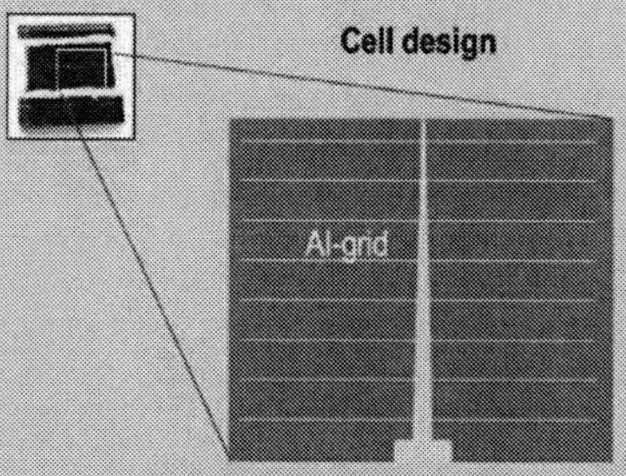

- Sheet resistance: 70 Ohms/sq
- Cell area: 10x10 mm²
- Dead area loss: ~ 3%
- Encapsulation: MgF_2

3CP.1.2 **17% Total Area Efficiency at Commercial Size CIGS Module**
P. Kratzert et al, Solibro Research, Uppsala, Sweden

Double Vision

Light Trapping

IV Results of III-V//Si Solar Cells

- Increase in J_{sc}
  - 11.6 → 12.7 mA/cm²
- Efficiency: 33.3%
- Record efficiency for 2T multi-junction solar cell

V_{oc} [V]	J_{sc} [mA/cm²]	*FF* [%]	η [%]
Gen 1			
3.046	11.7	87.5	31.3
Gen 2			
3.125	11.6	86.5	31.4
Gen 3			
3.127	**12.7**	83.8	**33.3**

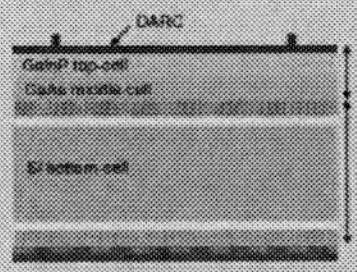

Surplus collection of carriers at 3rd terminal injected from top cell

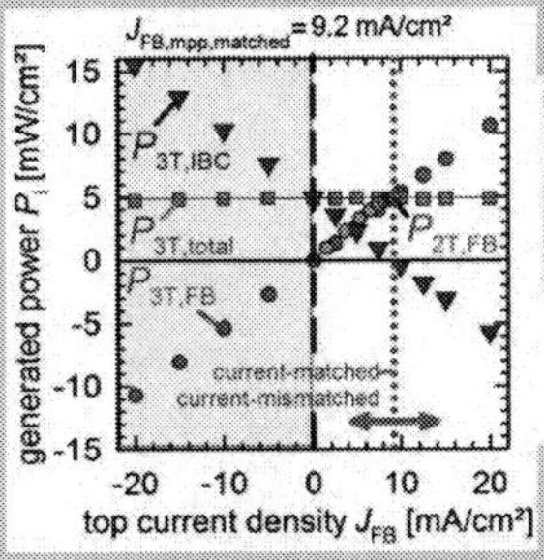

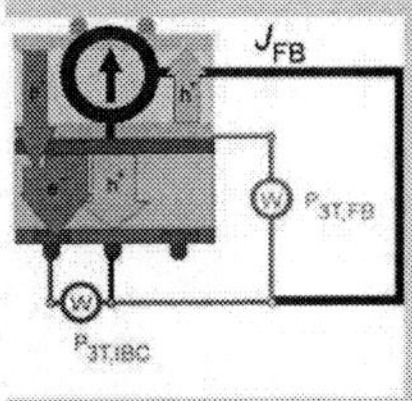

1AP.1.3 Monolithic III-V//Si Multi-Junction Solar Cell Exceeding an Efficiency of 31%
Fraunhofer ISE, Freiburg, Germany

1AP.1.2 : Maximum Power Extraction Enabled by Monolithic Tandems Using Interdigitated Back Contact Bottom Cells with Three Terminals
ISFH and NREL

European Commission

22.7% Perovskite-CIGS thin film tandem solar cells

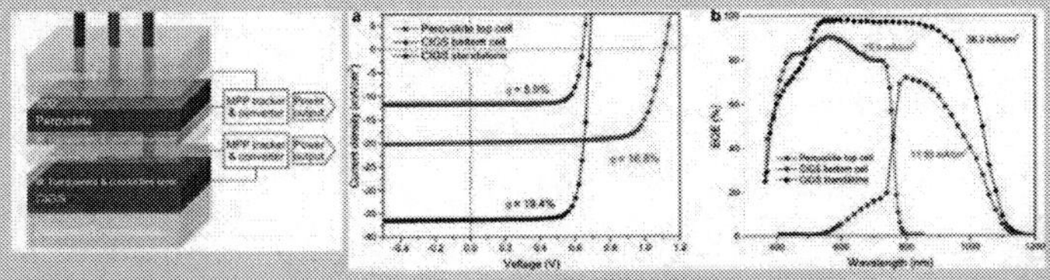

Solar Cells	V_{oc} (V)	J_{sc} (mA/cm²)	FF (%)	η (%)	MPP (%)	Cell area (cm²)	Efficiency gain
CIGS (stand alone)	0.684	36.4	78.2	19.4	19.4	0.213	
CIGS bottom cell	0.645	11.6	78.1	5.9	5.9	0.213	
Perovskite top cell	1.116	19.9	75.7	16.8	16.8	0.273	
Perovskite-CIGS 4-terminal tandem					22.7	0.213	3.3

Further improvement in Perovskite-CIGS thin film tandem solar cells via bandgap engineering, optical coupling etc.

Laboratory for Thin Films and Photovoltaics

13

3DO.7.4 Efficient and Stable NIR-Transparent Perovskite Solar Cells Prepared by Partial Ion Ex-Change Method for All-Thin-Film Tandem Applications
F. Fu, S. Pisoni, T. Feurer, A. Wäckerlin, S. Nishiwaki, A.N. Tiwari & S. Buecheler
EMPA, Dübendorf, Switzerland

European Commission

Performance

European Commission

~~MWp~~ MWh/MWp

Energy rating uncertainty

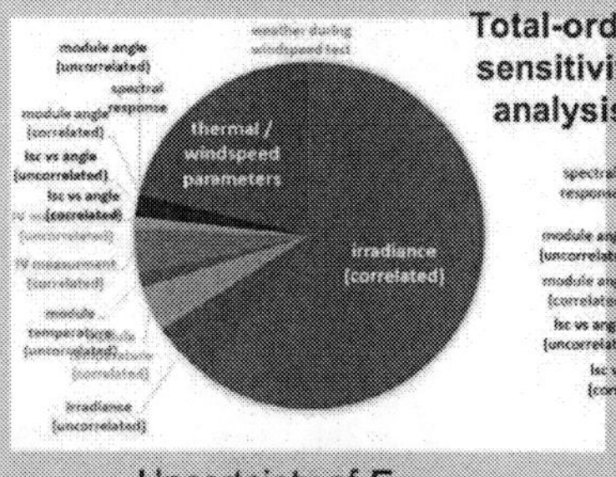

Uncertainty of $E_{mod,year}$
2.16 % (95% conf.)

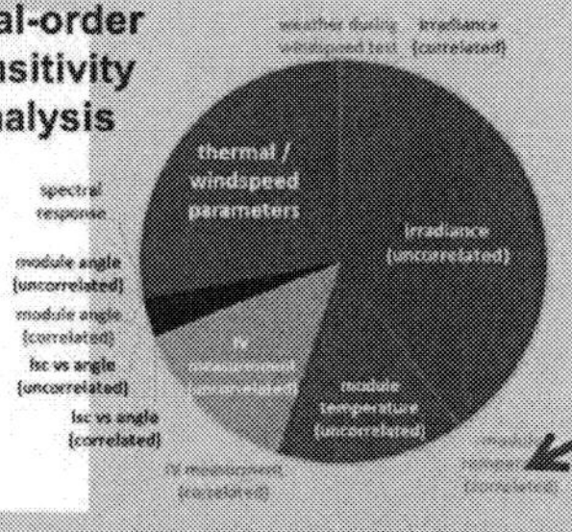

Uncertainty of ***CSER***
1.84 % (95% conf.)

IEC 61853 Energy rating standard

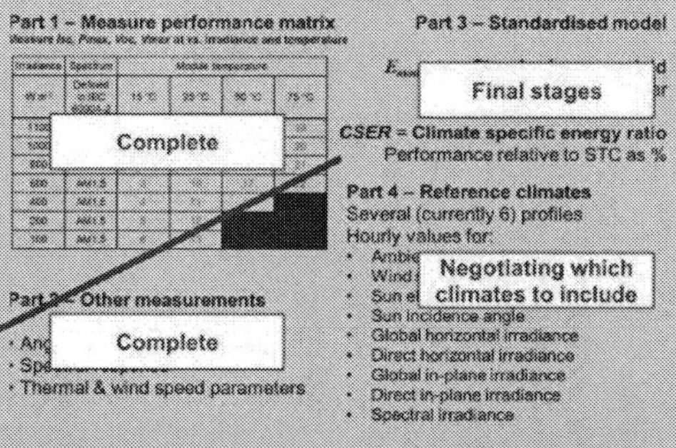

Simulation of 1000 modules, mixed types.
Uncertainty model based on survey of accredited test labs.

5CO.7.4 Progress in Energy Rating Standards: Accuracy and Optimisation
J.C. Blakesley, NPL, United Kingdom
T. Huld & H. Müllejans, European Commission JRC, Ispra, Italy

PhotoClass PV-Enerate

European Commission

Bifacial impact in the field

Bifacial Single Axis Tracker (NM)

- Daily Potential Bifacial Energy Gains were estimated from front and back irradiance data using reference cells.
- Potential gains increase when tracker is off-track, but yield decreases

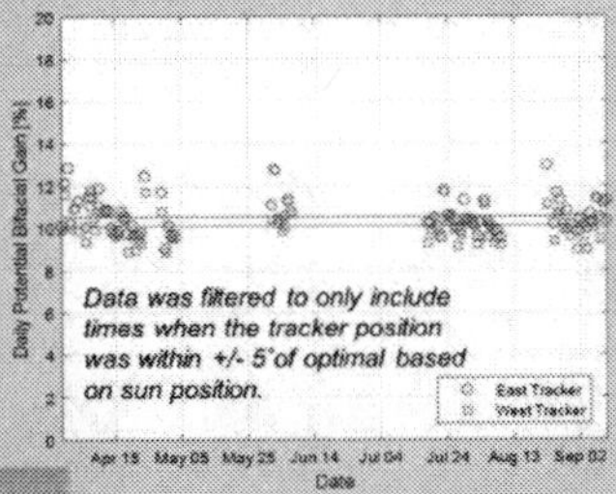

16

6CO.13.3 Outdoor Field Performance from Bifacial Photovoltaic Modules and Systems

J.S. Stein et al, Sandia National Laboratories, Albuquerque, United States
C. Deline, NREL, Golden, United States
F. Toor, University of Iowa, United States

European Commission

Light and elevated Temperature Induced Degradation

Piercing the Puzzle Together

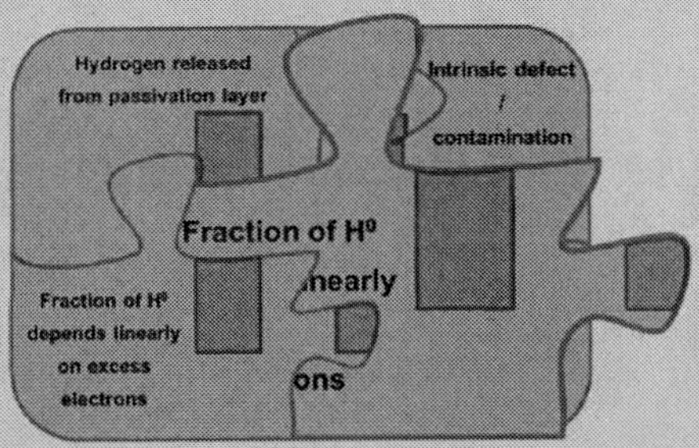

Model
Degradation

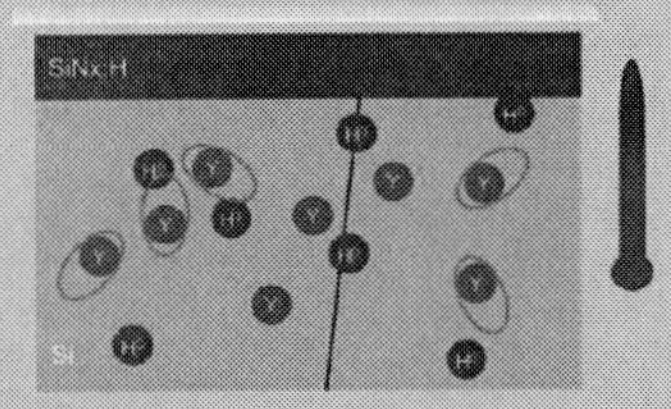

- LeTID observed in all p-type Si materials
- Defect activation only possible after fast firing of SiN_x:H-layer
- LeTID defect involves one fast-diffusing species
- Defect activated by inducing a change in charge state

→ Key to controlling LeTID: Manipulation of hydrogen

- Promising route for investigations: Float-zone silicon

2BP.1.4 Understanding Light-Induced Degradation in Multicrystalline Silicon: Possible Complex Formation Mechanisms

F. Schindler et al, Fraunhofer ISE, Freiburg, Germany

European Commission

Big data can help (1)

More than one million PV systems < 30 kWp have been installed in Germany.

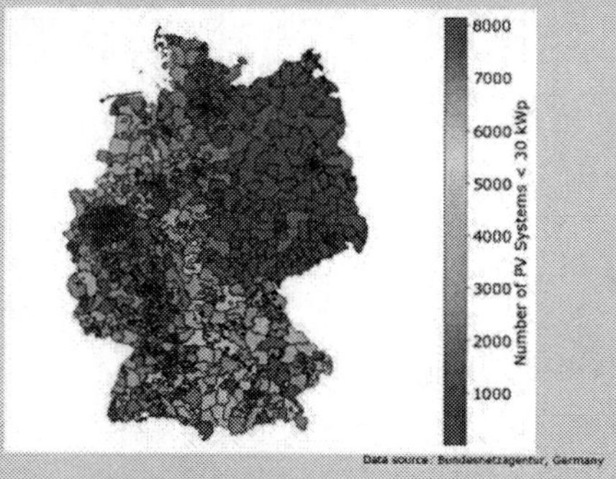

There is an annual loss of €450m to €500m by residential roof-top PV systems in Germany.

Year	Annual electric energy production in TWh	Annual yield loss in TWh	Monetary loss in million EUR
2014	36.1	1.46	€460m
2015	38.7	1.59	€489m
2016	37.6	1.58	€460m

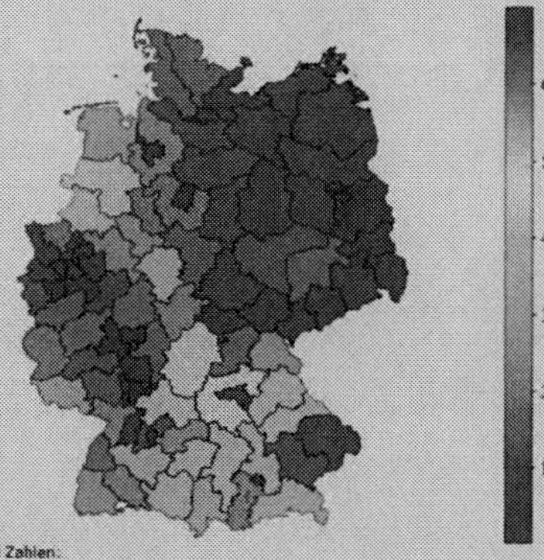

6BO.7.1 Survey on Yield of PV Systems in Germany 2014 to 2016

Source for mean Feed-In Tariff: BMWI, EEG in Zahlen: Vergütungen, Differenzkosten und EEG-Umlage 2000 bis 2017.

H. te Heesen & M. Rumpler
Trier University of Applied Science, Neubrücke, Germany
V. Herbort Ulm University of Applied Sciences, Germany

European Commission

Big data can help (2)

Model development: Model Architecture

i-EM

Two parallel approaches

- **Generic Fault Prediction (Supervision-Diagnosis Model)**: *predict inverter deviations from nominal behavior*
- **Specific Fault Prediction (Short-Term Fault Prediction Model)**: *predict a specific fault class*

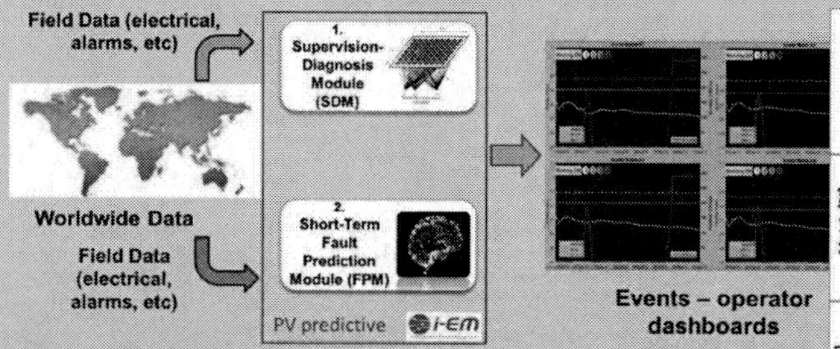

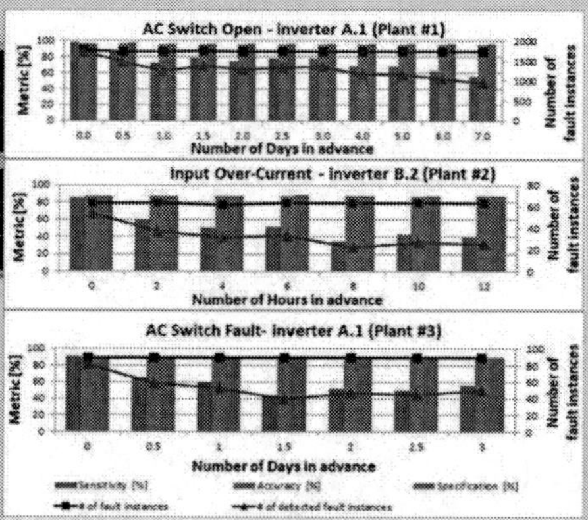

6DP.2.4 Predictive Maintenance in Photovoltaic Plants with a Big Data Approach

Alessandro Betti et al, i-EM (Livorno, ITALY), Enel Green Power (Rome, Italy)

European Commission

Non-destructive testing increasingly quantitative

Indication for PID

PID
- Homogeneous T increase
- Located at one string end
- Trend of PID-ratio increase towards the end
- < 50% of modules affected
- Incubation time
- PID-cells close the frame

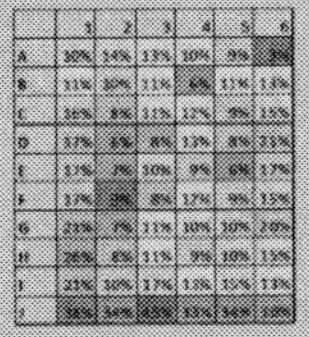

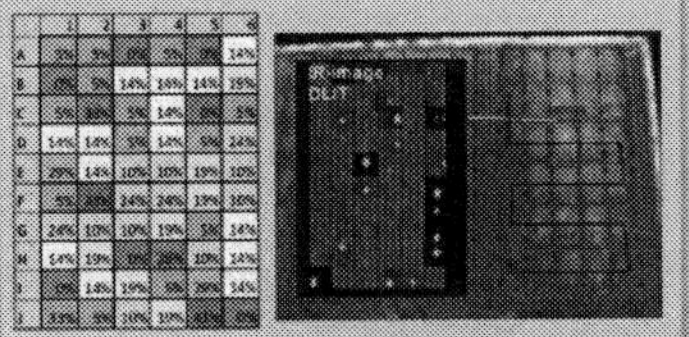

- Homogeneous T increase
- Arbitrarily distributed
- many modules may be affected
- Close to commissioning
- Affected cells everywhere

6BO.8.2 Quantitative Study of Potential Induced Degradation of a Roof-Top PV-Installation with IR-Imaging
C. Buerhop-Lutz et al, ZAE Bayern, Erlangen, Germany
C. Zetzmann Rauschert, Pressig, Germany
C. Brabec University of Erlangen-Nuremberg, Germany

European Commission

Which module is an insurance case?

6BO.8.3 Scientific Investigation of a PV Generator After Hail
W. Mühleisen et al, CTR, Villach, Austria
M. Spielberger PVSV, Guttaring, Austria
H. Sonnleitner ENcome, Klagenfurt, Austria
Y. Voronko OFI, Vienna, Austria

European Commission

Product Environmental Footprint Category Rules (PEFCR) for PV

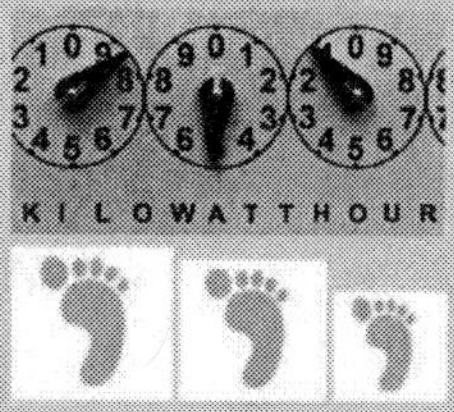

The PEFCR provides a harmonized set of rules to evaluate the environmental impacts of electricity produced from photovoltaic modules installed in the region of the EU28 and EFTA countries.

Circular Economy

Is it feasible to aggregate the complex results of a Product Environmental Footprint evaluation into a single score?

Communication Vehicles (CVs)

Performance label	Loyalty schemes	Campaigns targeting user behaviour
Performance label + QR code	Leaflets, catalogues, etc.	Marketing campaigns/ advertising
Performance improvement label	Instruction manuals	Declaration/ product passport
Barcode	Websites (producer, vendor)	On invoices
Pictogram, POS product advertisement	Websites (3rd party)	PEF external communication report
Consumer receipt – basket comparison	App based on performance	Performance tracking report.

Communication vehicles tested during the PEF Pilot Phase for the product category of PV electricity production

5DP.1.4 The Product Environmental Footprint (PEF) of Photovoltaic Modules – Lessons Learned from the Environmental Footprint Pilot Phase on the Way to a Single Market for Green Products in the European Union

A. Wade, First Solar, Mainz, Germany
P. Stolz & R. Frischknecht Treeze, Uster, Switzerland
G. Heath, NREL, Golden, United States
P. Sinha, First Solar, Tempe, United States

European Commission

Product Innovation

European Commission

BIPV / VIBV / IIPV

Solar module for automotive applications

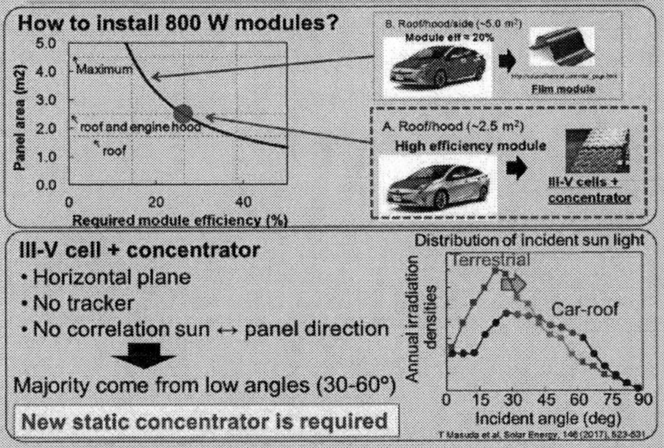

TOYOTA

- New static low concentrator was proposed for the automotive application.

- Prototype module was fabricated by using cutting tool for lens processing.

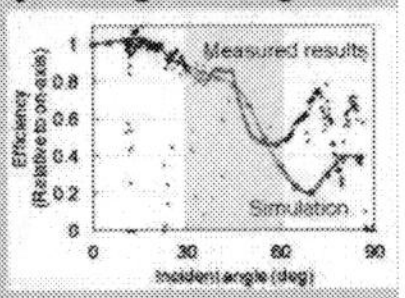

- Characteristics of lens is in good agreement with calculated results.

4DO.4.6 Solar Powered Vehicles with Static Concentrator Photovoltaics

T. Masuda et al, Toyota, Susono, Japan

K. Araki & M. Yamaguchi, TTI, Nagoya, Japan

European Commission

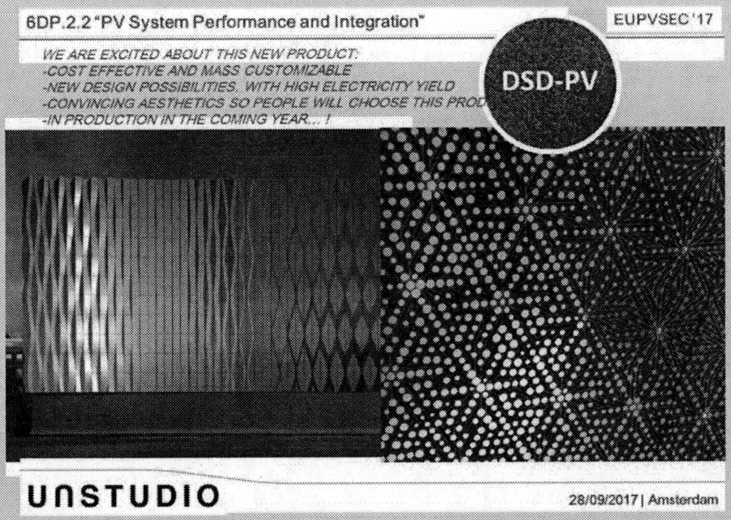

6DP.2.2 Aesthetics and Performance of PV

T. Minderhoud, UNStudio, Amsterdam, Netherlands

European Commission

Terrawatt

European Commission

Different Phases of VRE Integration – Towards Sector Coupling

Phase	Description	Country Examples
1	VRE capacity is not relevant at the all-system level	Most countries, incl. Mexico, Indonesia, South Africa
2	VRE capacity becomes noticeable to the system operator	Brazil, China, India, Sweden, Texas
3	Flexibility becomes relevant with greater swings in the supply/demand balance	Italy, Germany, Portugal, Spain, UK
4	Stability becomes relevant. VRE capacity covers nearly 100% of demand at certain times	Ireland, South Australia, Denmark
5	Structural surpluses emerge; electrification of other sectors becomes relevant	
6	Bridging seasonal deficit periods and supplying non-electricity applications; seasonal storage & synthetic fuels	

Electricity only accounts for around one fifth of total final energy demar
renewables will require multiplying their uses in buildings, indus

Acceptance
Landscapes
Clean Energy for All

EUPVSEC 2017, Opening, Paolo Frankl, IEA

European Commission

Open access proceedings

European Commission

Thanks!

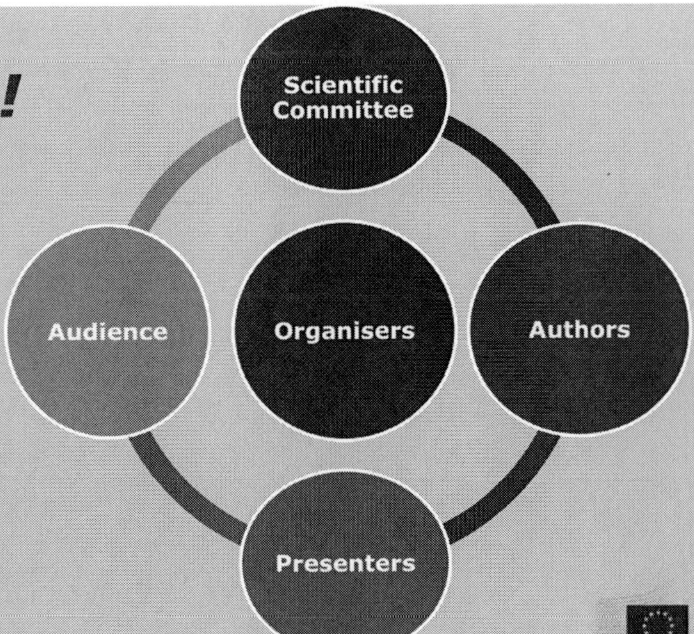

European Commission

LIST OF EXHIBITORS
(in alphabetical order)

3D-Micromac	DE
Abet Technologies	US
Advanced Silicon Group	US
ASYS	DE
Bentham Instruments	UK
BERGER Lichttechnik	DE
Berlin-Brandenburg Energy Network	DE
CSEM SA	CH
Delft University of Technology	NL
Dutch Energetics Management	NL
ECN	NL
Engineered Materials Systems	US
ETA Florence Renewable Energies	IT
EUREC	BE
European Commission, DG JRC	BE
FME	NL
greateyes	DE
h.a.l.m. elektronik	DE
Heraeus Photovoltaics	DE
Holland Solar	NL
HydroPV Technologies	NL
IEA PVPS	CH
InnoLas Solutions	DE
Ionx Cleaning Facilities	NL
ISC Konstanz	DE
IZOVAC Technologies	BY
Jonas & Redmann	DE
JRT Metallization Lines	DE
Kipp & Zonen	NL
KOPEL / KYOSHIN ELECTRIC	JP
Luvata Pori Oy	FI
Newport Spectra-Physics	DE
pv magazine group	DE
pv-tools	DE
Quantum Zurich	CH
RENA Technologies	DE
Semilab	HU
SINGULUS TECHNOLOGIES	DE
Sinton Instruments	US
Sisecam Flat Glass	TR
Solar Swiss Connect	CH
SOLARC Innovative Solarprodukte	DE
SOLARUNITED	DE
Solaxess SA	CH
Solibro	DE
Solliance	NL
Stäubli Electrical Connectors	CH
Sunprojects	NL
SUPSI ISAAC	CH
SVCS	CZ
Tempress Systems	NL
TFSC-Instrument	FR
TKI Urban Energy	NL
University of Ljubljana, LPVO	SI
Valentin Software	DE
Van der Valk Solar Systems	NL
Vela Solaris	CH
VITRONIC	DE
VON ARDENNE	DE
Wiley	US
WIP Renewable Energies	DE
ZSW Baden-Württemberg	DE

We thank the EU PVSEC 2017 Sponsors

BERGER Lichttechnik

LONGi Solar

RENA

SINGULUS

SHADING IMPACT ON 10 KWP ROOFTOP GRID CONNECTED PHOTOVOLTAIC SYSTEM

Roberto Silva Simplicio, Rafael Herrero Alonso, Cesar Biasi de Moura, Marcelo Knorich Zuffo
Laboratorio de Sistemas Integraveis da Escola Politecnica – Universidade de Sao Paulo
Av. Prof. Luciano Gualberto Trav.3 N.158. CEP: 05508-900. Sao Paulo - SP – Brazil.
E-mail: roberto.simplicio@lsitec.org.br

ABSTRACT: This paper describes the methodology for selecting an optimal configuration of photovoltaic modules in an urban environment, taking into account the shading caused by obstacles surrounding the site, a common scenario in distributed generation. The installation consists of approximately 10 kWp on a flat roof of a snack bar in Villa Lobos Park in Sao Paulo. The PVSyst V6.38 software quantifies the losses of the system due to shading effect. The analysis of twelve proposals of module layouts showed how these layouts affects the array yield. The best three layouts were chosen to make further studies. Subsequently the calculation of the return on investment for these layouts indicated an 11 years payback time in the worst case, even affected by surrounding obstacles.
Keywords: Building Integrated PV (BIPV); Shading; System Performance

1 INTRODUCTION

Building integrated photovoltaic, BIPV, has the advantage of reducing the losses during transmission and distribution since the generation is close to the consumer point. In countries with large territories, such as Brazil, the losses due to transmission and distribution, 15.1%, have a significant impact in the overall national energy balance [2]. Additionally, most of the Brazilian territory receives, in a year, more than 2,200 hours of solar irradiance making possible to generate up to 15 trillion MWh [1], equivalent to 28,000 times the total national electricity consumption for the year 2015 (522.8 TWh) [2]. This way, the PV energy source is a prominent alternative to the national electricity generation matrix, which, nowadays, has a great dependency on the hydroelectric sources as shown in Figure 1.

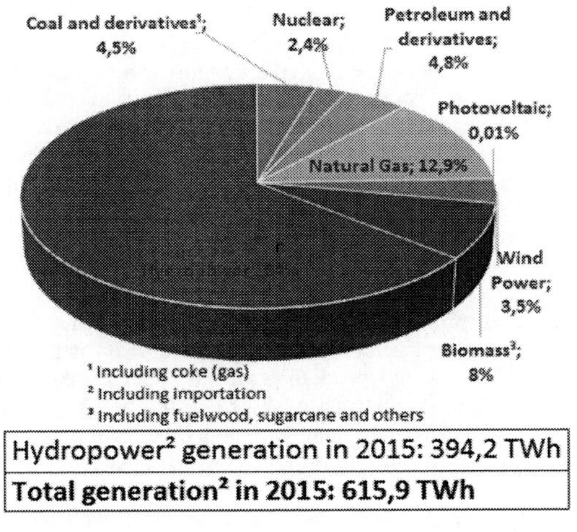

Figure 1: Brazilian electricity generation matrix in 2015 [2]

The Brazilian Energy Research Institute (Empresa de Pesquisa Energética - EPE) recently published a white paper estimating a national potential of 287 TWh/year of energy generated by PV modules installed on rooftops of buildings and houses. The correspondent peak power of 32 GWp would be able to generate 2 times the residential electricity consumption in 2013 [3]. The National Agency of Electric Energy (Agencia Nacional de Energia Eletrica - ANEEL) coordinated a public R&D project in 2013 that brought investors attention to distributed PV generation. The objective was to define the technical and commercial arrangements to include solar PV to the national energy matrix. This R&D project will provide the installation of 24.5 MWp in all Brazilian territory [4]. One of these projects, the "Villa Lobos Solar", will be able to install 547 kWp of photovoltaic in a public park, constituted by 446 kWp of BIPV in a parking lot, 53 kWp of solar trackers and 48 kWp in two micro generation. It also accomplishes a meteorological station with First Class precision instruments installed in the Electric Engineering building at Escola Politecnica of University of Sao Paulo.

ANEEL, in 2012, issued the standard 482/2012 that establishes the general technical requirements for connecting micro (less than 75 kW) and mini (75 to 5 MW) generation units to the national grid and the net metering rules for the energy injected into the grid [5].

Based on the described scenario, this paper has the objective of estimating the shading impact of one of these micro generation units installed in a snack bar, which is part of the "Villa Lobos Solar" project. The definition of the modules layout can minimizes the shading effect, thus raising the energy production. The shading analysis, as shown in this paper, is especially important to estimate the real energy output in urban areas, where the modules will be subject to shading of surrounding buildings, vegetation, chimneys and other obstacles.

2 REVIEW OF PV SYSTEMS AND METRICS

2.1 On-grid PV systems

There are two types of on-grid systems: distributed generation (on rooftop or facade), where the generation is close to the consumer; and centralized generation (solar farms or PV power plants), where the generation if far from consumer, requiring additional investments in transmission and distribution infrastructure. The advantages of distributed systems are the loss reduction; there is no need for larger areas (PV farms); the possibility of installation in any area; the mix in the energy matrix; postpone the additional costs in distribution and transmission lines infrastructure; and the property valuation [6].

2.2 Losses in PV systems

Although the focus of this study is to analyze the losses due to shading, it is important to point out that there are other sources of energy losses, such as optical, module technology/fabrication and electrical loss.

The optical losses happens as the consequence of total or partial shading of the modules. The position of the module in relation to the sun will cause effects like reflection, absorption, and spread of the sunrays before they hit the PV module surface [7].

PV module technology as well the manufacturing quality will directly affect how the module will respond to the combination of ambient temperature and solar irradiance level. It is an intrinsic of the manufacturing process of each PV module. [8].

The sources of electrical losses are the Joule effect of DC and AC cabling; losses in the inverter and transformer; and losses of system shutdown (either programmed or unplanned). Table I shows the main sources of PV system losses [14].

Table I – System losses [14]

	Typical	Minimum	Maximum
Module quality loss	0.95	0.88	0.96
Inverter and transformer	0.92	0.88	0.96
Mismatch	0.98	0.97	0.99
Diodes and connectors	0.995	0.99	0.997
Cables	0.98	0.97	0.99
Soiling	0.95	0.30	0.995
Availability	0.98	0.00	0.995
Shading	0.95	0.00	1.00

2.3 Shading effect on PV modules

Shading is very common on PV systems installed in urban environment by virtue of the existence of many different types of obstacles. The consequence of shading will depend on the type of PV modules, electrical configuration of the bypass diodes, string configuration, type of shading, among other factors. For partial shading, the production loss can be proportionally greater than the shaded area because of the current mismatch between modules connected in series and the voltage mismatch in parallel strings connected in the same array [7].

To minimize the impact of shading in the energy yield of a PV system, bypass diodes are integrated to the PV modules so that when they are directly biased, resulting from shading, they provide an alternative path for the electric current. In other words, shaded cell produces less current. Since it is in series with other cells, this string of cells imposes current to the shaded cell and it becomes reversely polarized, acting as a load to the system. If the voltage on the shaded cell becomes negative, the bypass diode becomes forward biased and the cell will operate close to V_{rev}, as shown in Fig 2 (a). The current versus voltage of a PV array will be the result of the combination of sub-strings connected in series and parallel with the optimal point of operation depending on the MPPT algorithm of the inverter. Fig. 2(b) shows two operating points of a PV array: [A] is the maximum global power point and at this point, the bypass diodes are directly polarized. In point [B], which is the local maximum power point, the bypass diodes are not directly polarized [9].

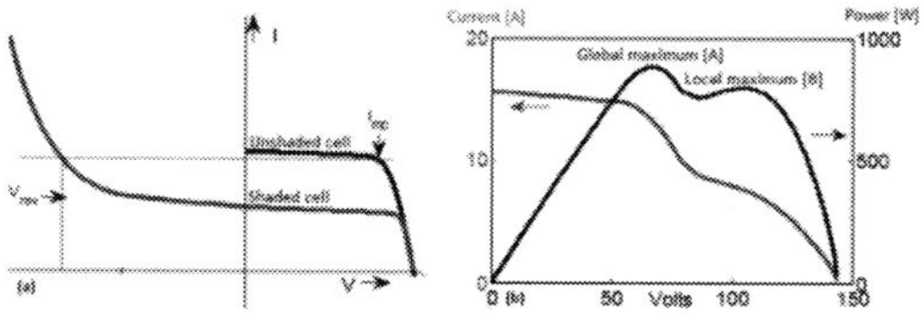

Figure 1 – Current versus voltage curve of a shaded cell (a); Current versus voltage and power versus voltage curve of an array (b) [9].

The effect of shading in microgeneration in urban environment is a subject of great interest. There are several papers in the literature that seeks the most important characteristics of a given location that makes a PV system installation economically viable. A methodology to quantify the available area for PV system installation for large populated areas such as cities, countries and even continents indicated more than six TWh of annual potential for the Cataluña in Spain [10]. A comparison of different PV module technologies and the impact in different locations taking into consideration the available area, solar irradiance profile and electric energy demand demonstrates the strategic advantages of PV installations in urban areas [11]. A methodology to estimate the PV energy generation potential of building integrated systems in Greece was show by [12]. The research about partial shading in systems with different configurations also draws much attention. The effects of local vegetation and other obstacles in different topologies: central inverter, distributed inverter and string inverter were subject of study [13].

2.4 Performance parameters of a PV system

The performance of a PV system is measured with well-defined parameters. They are useful to compare different systems and to evaluate the individual performance of each system. In this section, a brief description of the parameters used in this paper are presented. The time reference used by these parameters may be hours, days, months or years.

Final PV System Yield – Y_f. Y_f is the ratio between the energy delivered to the grid (E_{GRID}) and the rated system power P_0 and represents the number of hours that the system would need to work at nominal power to deliver the E_{GRID}. Units of Y_f are hours or kWh/kWp. Since this parameter is normalized by the rated power, it can be used to compare different systems sizes [15].

Reference Yield – Y_r. The Y_r is the solar incident energy delivered to the PV modules plane divided by the PV's reference irradiance G. It corresponds to the solar resource available in that location with the modules in a defined orientation, representing the number of hours that the system works at reference irradiance [15].

Figure 2 –3D model of the snack bar

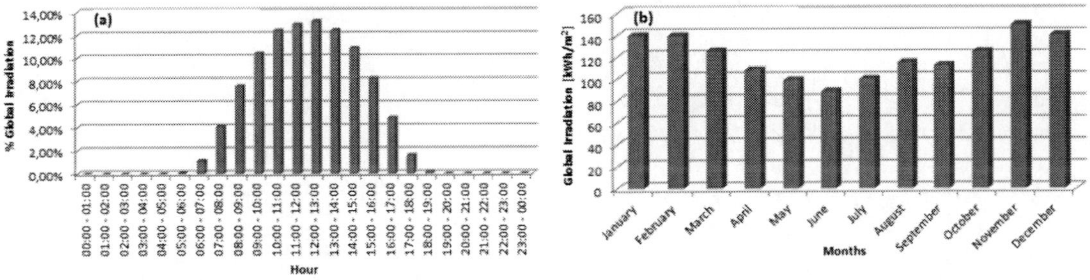

Figure 3 – hourly global irradiance distribution (a); global irradiance distribution for each month (b)

Performance Ratio – PR. PR is the Y_f divided by the Y_r. It is a dimensionless number that quantifies the total efficiency of the system, taking into consideration all possible losses of the system [15].

Array Yield – Y_a. Similar to Y_f, however using just the energy delivered by an array. It is the energy delivered by the array (E_{DC}) divided by the rated system power P_0. Y_a units are hours or kWh/kWp.

Collector Losses – L_C. This parameter identifies the systems with better use of the energy from incident irradiance. It is the difference between Y_r and Y_a and it can be expressed in hours or kWh/kWp. It includes all losses, from the incident irradiance to the inverter's input. In summary, it accounts for the optical, thermal and electrical losses in the DC side of the system.

Global Effective Irradiance – GlobEff. It is the global irradiance adjusted to the collector plane, accounting for the shading losses and incident angle modifier (IAM) losses. The effective irradiance unit is kWh/m².

3 CASE STUDIE

3.1 Methodology

The first step is to make a 3D model of the building and surrounding obstacles that may prevent the solar irradiance to reach the modules surface. The measurements were done during a visit to the location using measuring tape, digital photography and a known reference position. Based on these measurements, a 3D model was built using *PVSyst V6.3*, which was also used to make a one-year energy production simulation, using *Meteonorm 7.1* database for meteorological data.

Positioning, number of PV modules, elevation and azimuth angles and distance between rows of PV modules were the input parameters of this simulation. Among all possibilities of combinations of these parameters, twelve configurations with the lowest shading impact were chosen. After, the best three in terms of solar energy efficiency (highest GlobEff and lower L_C and Y_a) were analyzed in detail. Subsequently, the electrical series-parallel connection between PV modules that minimizes the electrical mismatch due to the shading pattern was designed. The idea is to place those PV modules that are exposed to similar average irradiance in the same string.

3.2 Project description

The PV modules will be installed on the rooftop of a snack bar building located at 23°33'S 46°38'W, with a total area of 120 m². The PV module series is SuryaVolt D SV-245D20 from *DYA Energia Solar* with 245 Wp of rated power. This PV module has 60 silicon polycrystalline cells with 15.2% efficiency. The Figure 3 shows the 3D model of all obstacles that may cause shading. The red color highlights the available space for installation.

3.3 Irradiance profile at building's location

From the simulation, 90% of the global irradiance for one year period is concentrated between 8 AM and 4 PM. Figure 4(a) shows the average hourly global irradiance distribution and Figure 4(b) shows the global irradiance distribution for each month.

Figure 5 shows the solar diagram and iso-shading curves for the area highlighted in Figure 3. From those curves, it is possible to identify the hours of the day with low and high shading impacts. The lowest is before 8:15 AM (iso-shading of 1%) between June and July. The highest is after 2 PM throughout the year.

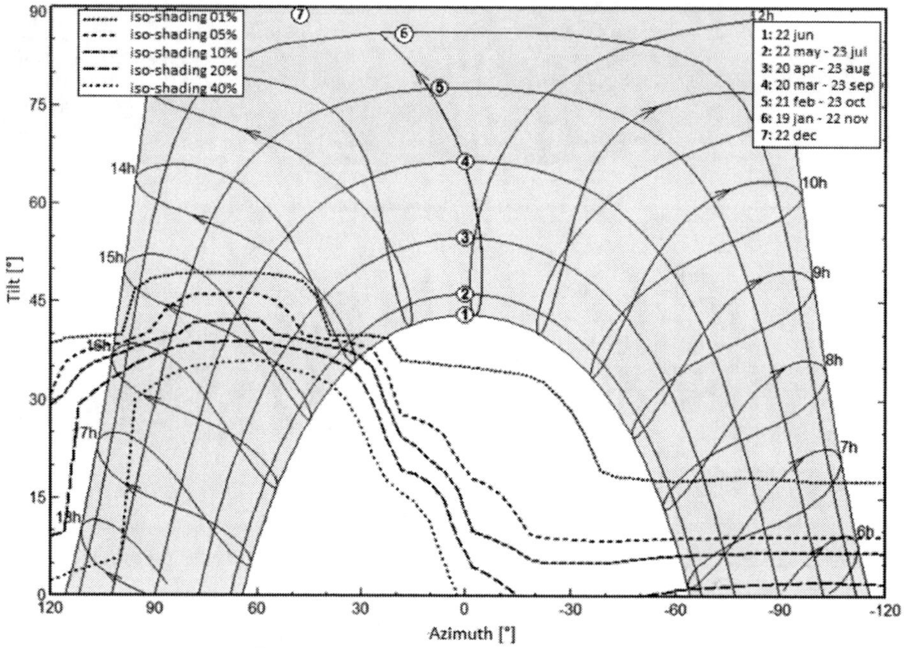

Figure 4 – Solar Diagram and iso-shading

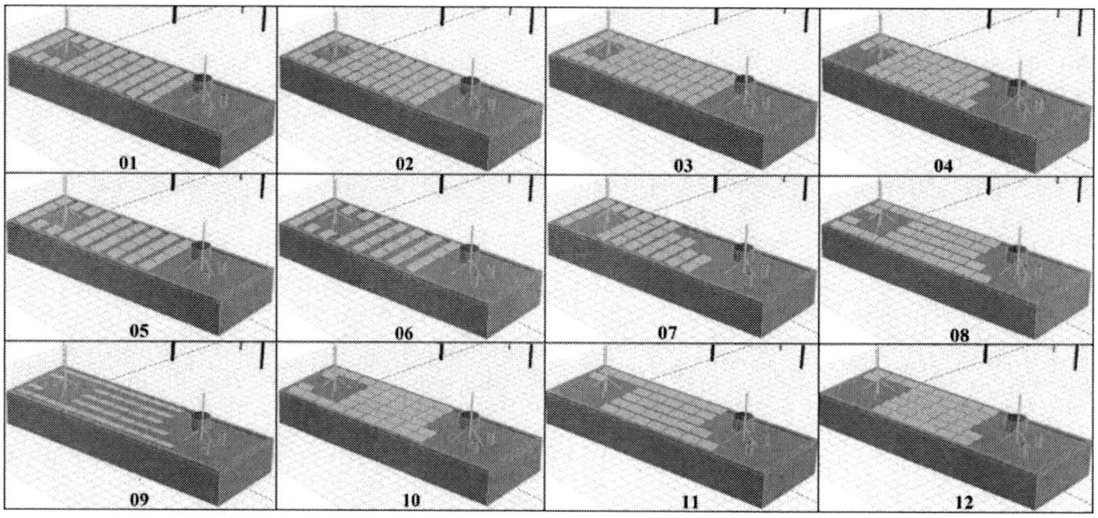

Figure 5 – PV modules layout proposals for the snack bar building

The shading from 2 PM to 4 PM blocks up to 19% of the global irradiance. For this reason, the layout was chosen in order to minimize losses during these hours in October through March since this is the time of the year with the highest global irradiance levels, as shown in Figure 4(b).

3.4 PV modules layout proposals

Several module layouts were simulated. Tilt angles varied from 10° to 30°, azimuth (north, east and west), distance between rows (from 1.0 to 2.0 m) and the number of modules. Simulations considered a time period of one year with the objective to find the layouts with the highest efficiency, as measured by L_C, Y_a and GlobEff.

The top twelve configurations in terms of performance are shown in Figure 6 and their characteristics are in Table II.

3.5 Preliminar results

Table III summarizes the simulation results of the twelve configurations shown in Figure 6. Configuration 9 presented the worst performance due to the azimuthal orientation (facing west). The main difference between configuration 01, 08 and 09, are the azimuthal angles: north for configuration 01, east for configuration 08 and west for configuration 09. The one with the best results, configuration 08, has the highest GlobEff and Y_a, and lower collector losses (L_C). Consequently, it has the highest PR.

Table II – Characteristics of the top twelve configurations

Configuration	modules	Total Power (kW)	Distance between rows (m)	Tilt	Azimuth
01	36	8,82	1,50	11°	9°
02	40	9,80	1,35	10°	9°
03	44	10,78	1,25	10°	9°
04	36	8,82	1,25	10°	9°
05	32	7,84	1,70	11°	9°
06	28	6,86	1,90	13°	9°
07	28	6,86	1,50	11°	9°
08	36	8,82	1,45	12°	-81°
09	36	8,82	1,45	10°	99°
10	40	9,80	1,15	10°	-81°
11	32	7,84	1,45	12°	-81°
12	36	8,82	1,15	10°	-81°

Table III – Simulation results

Configuration	modules	GlobEff (kWh/m²)	Ya (kWh/kWp)	L_C (kWh/kWp)	PR
01	36	1304	2,979	1,141	0,7038
02	40	1301	2,973	1,136	0,7050
03	44	1296	2,961	1,149	0,7024
04	36	1300	2,969	1,141	0,7032
05	32	1305	2,983	1,138	0,7037
06	28	1309	2,991	1,148	0,7012
07	28	1316	3,006	1,115	0,7079
08	36	1314	3,006	0,9375	0,7421
09	36	1209	2,762	1,137	0,6891
10	40	1291	2,954	0,9976	0,7284
11	32	1309	2,996	0,9477	0,7386
12	36	1286	2,941	1,0100	0,7245

Configurations 08, 10 and 11 have the lowest L_C. Because of the orientation of the building and the need of minimize the mutual shading between rows, the optimal tilt angle is different from the local latitude (23°). The azimuth angle of the top performers is 81°, almost directly facing east, instead of north, which is the theoretical optimal azimuth. This happens because this arrangement minimizes the mutual shading by requiring less number of rows. In conclusion, for this scenario, facing modules to the east results in better use of the solar resource. The PR values for configurations 08, 10 and 11 confirm their higher performance when compared with the other configurations.

3.6 PV system electrical connections

Once the shaded areas and the hours of the day with worst shading impact are determined, different electrical diagrams were proposed for the top three configurations, 08, 10 and 11, and another round of simulations was carried out. The simulations showed that the electrical connection with best performance is the one with two strings, both connected to a single inverter, in this case, Ingecon Sun 5 TL M from Ingeteam (5 kW). Using more strings would, in theory, increase the system's efficiency, however, would dramatically increase the cost of the system and this options was discarded.

3.7 Results

This section summarizes the results of the top three configurations. Table IV shows the shading losses for each configuration. The first line shows the losses due to the shading that prevents the sun's rays from reaching the surface of the modules. The non-uniformity of the sun's rays on module surface causes electrical mismatch losses between modules and strings, presented in the second line. As seen from the results, configurations 08 and 11 had the best performance.

Table V shows the main performance parameters for those configurations under analysis. Final Yield and Collector Losses, Y_f and L_C respectively, indicates that the top performer is configuration 08. Although this configuration does not produce more energy, it produces more energy per rated power. Configuration 08 has the highest PR and one of the lowest L_C among the twelve configurations analyzed.

Table IV – Losses due to shading

	Config. 08	Config. 10	Config. 11
No irradiation	5.5%	7.4%	5.8%
Electric mismatch	3.2%	4.8%	3.3%

Table V – Summary of the performance parameters

	Config 08	Config 10	Config 11
Y_r [kWh/m^2]	3.91	3.95	3.91
L_C [kWh/kWp]	1.009	1.130	1.022
Y_a [kWh/kWp]	2.90	2.82	2.89
Y_f [kWh/kWp]	2.82	2.75	2.80
PR [%]	72.2	69.5	71.8
E_{GRID} [MWh]	9.083	9.829	8.024

Table VI – Financial analysis

Configuration	Capital Investment	Energy production in 25 years	Savings in electricity bill in 25 years	Balance after 25 years	ROI
08	US$ 15,876.00	204,46 MWh	US$ 36,960.32	US$ 20,219.04	11 years
10	US$ 17,740.00	221,25 MWh	US$ 39,995.92	US$ 21,394.50	11 years
11	US$ 14,112.00	180,62 MWh	US$ 32,651.06	US$ 17,769.92	11 years

The financial analysis in Table VI considers 0.1857 US$/kWh as the reference for the cost of energy in July 2017 charged by the local energy company, AES Eletropaulo, including taxes. The analysis considered a period of 25 years and a linear efficiency degradation of the solar panels, where after this period the modules will be able to deliver 80% of the rated power according to the manufacturer's datasheet. The invested capital for similar project is US$ 1,800.00 per kWp according to [2] in 2015. With those inputs, it was possible to calculate the savings in the electricity bill for the life span for each system. Considering constant the cost of energy during the period of 25 years, a conservative approach, the results show a minimum break-even as 11 years and at least a saving of US$ 32,000.00 in energy bill during the whole life of the system.

4 CONCLUSION

Because of the recent events in feed-in policies in Brazil, a study focused on distributed generation attracts a lot of attention. The results show that the shading effect on small installations will be a common scenario in urban environmental due to the presence of objects around the installation. To mitigate this effect it is necessary to analyze each site during the preliminary project phase. As discussed in this article, the conclusion may be that facing the panels to north with the tilt angle equal to the latitude of the site it is not the optimal PV modules orientation.

Besides the limitations of the methodology presented in this paper and the simplified financial analysis, all analyzed systems had a breakeven of less than 25 years which proves that small PV systems aimed for residential applications are economic viable and demonstrate that this technology has a great potential in Brazil.

The incident irradiance losses caused by shading are the main cause of PR reduction, where 5.5% of the solar irradiance does not even reach the PV modules surface. Other important sources of losses such as electrical mismatch, Joule effect and high temperatures, also affect PR. From the different configurations analyzed in this paper it can be concluded that the distance between rows, which causes the mutual shading, must be taken in special consideration during the project phase.

The Project Villa Lobos is under implementation and expected to be in operation at the end of 2017. For a future work, real data from the actual location can be used for comparison and used to validate the methodology proposed here.

5 ACKNOWLEDGEMENTS

The authors want to acknowledge the financial support by the Companhia Energetica de São Paulo (CESP) and others four utility companies, through the R&D public proposal from Brazilian government energy agency/ANEEL; the sponsors of Laboratory of Interdisciplinary Integrated Technologies (LSI-TEC); the project "Desenvolvimento e instalacao piloto de geracao fotovoltaica para modelo estrategico de referencia tecnologica, regulatoria, economica e comercial, inserindo esta energia na matriz energetica nacional" (ANEEL code: PE-0061-0034/2011).

6 REFERENCES

[1] Rodrigues, D.; Matajs, R. Um banho de sol para o Brasil. São Lorenço da Serra: Instituto Vitae Civilis. 2005.

[2] EPE. Balanço Energético Nacional 2016 - Ano Base 2015. Brasília: EPE - Empresa de Pesquisa Energética MME - Ministério de Minas e Energia. 294 p. 2016.

[3] EPE. NOTA TÉCNICA DEA 19/14 - Inserção da geração fotovoltaica distribuída no brasil - condicionantes e impactos. Rio de Janeiro: EPE - Empresa de Pesquisa Energética. 2014.

[4] ANEEL. CHAMADA N°13/2011 Projeto estratégico: "Arranjos técnicos e comerciais para inserção da geração solar fotovoltaica na matriz energética brasileira". Brasília: ANEEL - Agência Nacional de Energia Elétrica. 2011.

[5] ANEEL, 2012. RESOLUÇÃO NORMATIVA N°482, DE 17 DE ABRIL DE 2012. [S.l.]: ANEEL - Agência Nacional de Energia Elétrica.

[6] Rüther, R. EDIFÍCIOS SOLARES FOTOVOLTAICOS. Florianópolis: LABSOLAR -

Laboratório de Energia Solar - Universidade Federal de Santa Catarina. 2004.

[7] DGS. Planning and installing photovoltaic systems: a guide for installers, architects and engineers. London: DGS - Deutsche Gesellschaft für Sonnenenergie. 2007.

[8] Luque, A.; Hegedus, S. Handbook of photovoltaic science and engineering. Second Edition. ed. [S.l.]: John Wiley & Sons. 2011.

[9] Deline, C. Characterizing shading losses on partially shaded pv systems. Albuquerque: NREL - National Renewable Energy Laboratory. 2010.

[10] Izquierdo, S.; Rodrigues, M.; Fueyo, N. A method for estimating the geographical distribution of the available roof surface area for large-scale photovoltaic energy-potencial evaluations. Solar Energy 82. pp. 929-939. 2008.

[11] Jardim, C. D. S. et al. The strategic siting and the roofing area requirements of building-integrated photovoltaic solar energy generators in urban areas in Brazil. Energy and Building 40. pp. 365-370. 2008..

[12] Loulas, N. M. et al, 2012. Photovoltaics in urban environment: A case study for typical apartment buidings in Greece. Renewable Energy 48. pp. 453-463. 2012.

[13] Woyte, A.; Nijs, J.; Belmans, R. Partial shadowing of photovoltaic arrays with different system configurations: literature review and field test results. Solar Energy 74, pp. 217-233. 2003.

[14] IEA. Analytical monitoring of grid-connected photovoltaic systems - good practices for monitoring and performance analysis - report IEA-PVPS T13-03: 2014. [S.l.]: IEA - International Energy Agency. 2014.

[15] NREL. Performance parameters for grid-connected pv systems. National Renewable Energy Laboratory. Lake Buena Vista, pp. 9. (NREL/CP-520-37358). 2005.

A ROBUST SLIDING MODE MPPT CONTROLLER APPLIED TO A STAD-ALONE PHOTOVOLTAIC SYSTEM

Hanane Yatimi, Elhassan Aroudam
Modeling and Simulation of Mechanical System Team, Physics Department, Faculty of Sciences, Abdelmalek Essaadi University.
Sebta Ave., Mhannech II BP 2121, 39002, Tetouan, Morocco.
yatimi.hanane@gmail.com, aroudam_hass@hotmail.com

ABSTRACT: The output characteristic of a photovoltaic (PV) module is nonlinear and changes with solar irradiance and cell's temperature. Thus, its maximum power point (MPP) is not constant. Therefore, in order to uphold the PV module operating at its MPP and then make full utilization of its output power, a Maximum Power Point Tracking (MPPT) technique is needed. This paper highlights the sliding mode control (SMC), in order to analyze, simulate and evaluate a standalone PV system under varying weather conditions. Simulation results show that the SMC provides good performance of the PV system under the weather changes in term of stability and robustness to solar irradiance and temperature variations. Through simulations, SMC can accurately track MPP of the PV system better than conventional technique can.
Keywords: PV System, Mono-crystalline, Boost Converter, MPPT, Sliding Mode Control (SMC)

1 INTRODUCTION

With the increasing demand for energy due to the indiscriminate use of electrical power by the people and industries, fossil energy reserves are becoming depleted; on the other hand the consumption of this kind of energy causes severe pollution and also endangering human health and natural life. Many renewable energy today are well developed, reliable and cost competitive compared with conventional fuel supplied generators. In particular, PV solar energy which has been widely utilized in many applications such as battery charging, water pumping, and satellite power systems, and so on, due to such advantages, it is clean, abundant and environmental friendly [1]. However, there are two main drawbacks of PV system, namely the high installation cost and the low conversion efficiency of PV modules. In fact, a PV module is a nonlinear power source, and its output power depends on weather conditions. To increase the efficiency of the PV system, the PV module must operate on its MPP, thus the need of using MPPT techniques. MPPTs play an important role in PV power systems because they maximize the output power from a PV system for a given set of conditions.

Many MPPT algorithms have been proposed in the literature. Most control schemes use the P&O algorithm [2], which is based on iterative algorithms, because it is easy to implement but the oscillation problem is unavoidable, Incremental conductance method [3] which requires complex control circuit, The fuzzy logic control search method (FLC) [5], and the neural network method [4]. In this paper, the sliding mode control is applied to track maximum power of PV system. The advantages of this control are various and important such as, high precision, good stability, invariance, robustness etc….[6]; It exhibits robustness properties with respect to both internal parameter uncertainties and external disturbances.

This paper is organized as follows. In Section 2, modeling of the overall PV system is developed and the proposed MPPT technique is explained. In Section 3, simulation results and discussions are presented to show the performance of sliding mode control. Finally, some conclusions are drawn.

2 PV SYSTEM DESCRIPTION

The studied system consists of a PV module, a boost converter, a MPPT stage, and a battery load, as shown in Fig. 1. T and G are the input signals of the system, and are respectively the solar cell temperature (in K) and the solar irradiance (in W/m²) on the surface of PV module.

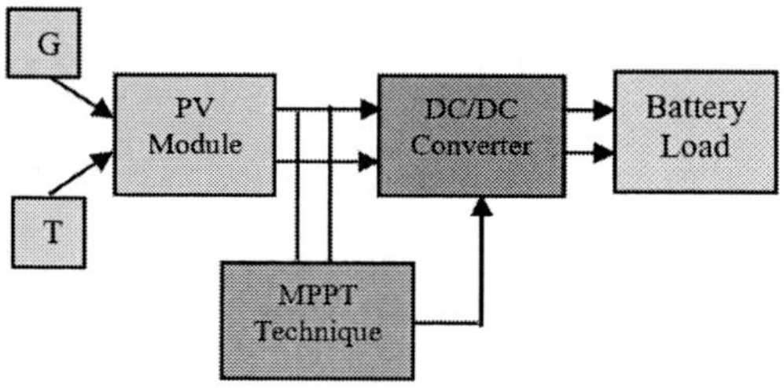

Figure 1: Block diagram of the stan-alone PV system.

2.1 PV module modeling

As the PV module is composed of group of cells, its model is based on that of a PV cell. The equivalent circuit of the PV cell is shown in Fig. 2 [7]. R_s and R_p are the series and parallel resistances which represent internal losses of the PV cell.

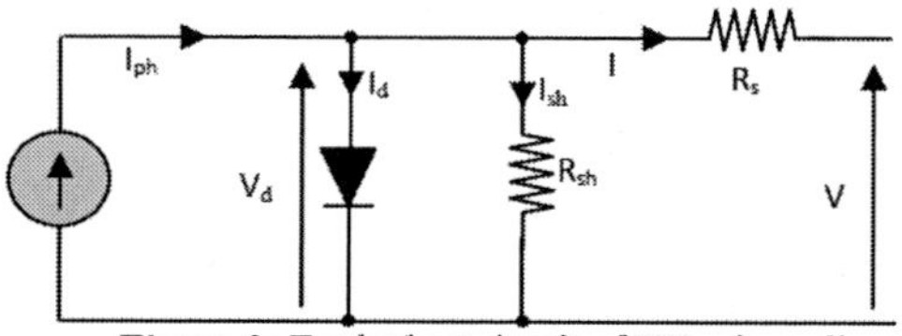

Figure 2: Equivalent circuit of PV solar cell.

The output current–voltage (I–V) characteristic of a solar PV cell is given in the Eq. 1 [7]:

$$I = I_{ph} - I_0\left[\exp\left(\left(\frac{q}{aKT}\right)(V+IR_S)\right)-1\right] - \left(\frac{V+IR_S}{R_{sh}}\right) \quad (1)$$

$$I_{ph} = \left[I_{SCR} + K_i\left(T-T_r\right)\right]\left(\frac{G}{1000}\right) \quad (2)$$

$$I_0 = I_{rs}\left[\frac{T}{T_r}\right]^3 \exp\left[\left(\frac{qE_{g_0}}{aK}\right)\left(\frac{1}{T_r}-\frac{1}{T}\right)\right] \quad (3)$$

I_{ph} : is the generated photocurrent (A); it is expressed by Eq.2. I_0: is the reverse saturation current of diode (A), it is given by Eq. 3. Where V, I, I_{scr}, I_{rs}, K_i, q, K, a, T_r, T, G, R_s and R_{sh} refer to PV cell output voltage and current, short-circuit current and saturation current at reference temperature, short-circuit temperature coefficient, electron charge ($1.60217*10^{-19}$ C), Boltzmann constant ($1.38*10^{-23}$ J/K), diode ideality factor, reference temperature (K), Temperature (K), solar irradiance (W/m2), Series and parallel resistances of PV cell (Ω), respectively. The PV module model is given by Eq. 4. Where Ns and Np are the series and the parallel solar cell per module, respectively.

$$I = N_p I_{ph} - N_p I_0\left[\exp\left(\left(\frac{q}{akT}\right)\left(\frac{V}{N_s}+\frac{IR_S}{N_p}\right)\right)-1\right] - \frac{N_p}{R_{sh}}\left(\frac{V}{N_s}+\frac{IR_s}{N_p}\right) \quad (4)$$

2.2 PV module characteristics

Fig. 3, [8] shows the behavior of a PV module simulation in accordance to solar irradiance variation (and constant temperature).

As can be seen in the curves of Fig. 3, the short-circuit current and the power increase are highly affected by the solar irradiance, while the effect on the open circuit voltage is quite low; so the more solar irradiance is high, the more the module generates power.

Fig. 4, [8] shows the behavior of a PV module simulation in accordance to temperature variation (and constant solar irradiance).

From the curves of Fig. 4, we can clearly detect that the temperature has a very important effect on the open circuit voltage and a no remarkable effect on both the short circuit current and the power of the module. The more temperature is low the more cell generates power.

These figures are using the real meteorological data locally measured at regular intervals (1 hour) [9] throughout the typical clear day of April in Northern Morocco (Tetouan). To conclude, the use of the MPPT techniques to extract the maximum available power at any changes is primordial.

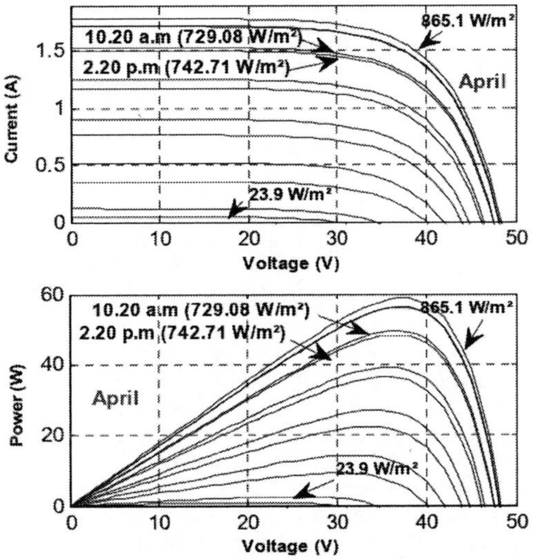

Figure 3: I-V and P-V characteristics at constant temperature and varying solar irradiance for the typical clear day of April-Tetouan/Morocco.

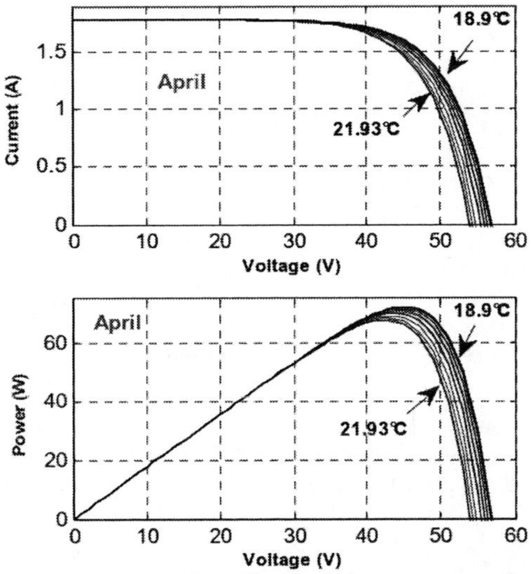

Figure 4: I-V and P-V Characteristics at constant solar irradiance and variable temperature for the typical clear day of one month of every season-Tetouan/Morocco.

2.3 Boost converter and battery model

A boost converter is a DC-to-DC converter steps up voltage. Its circuit is shown in Figure 5.

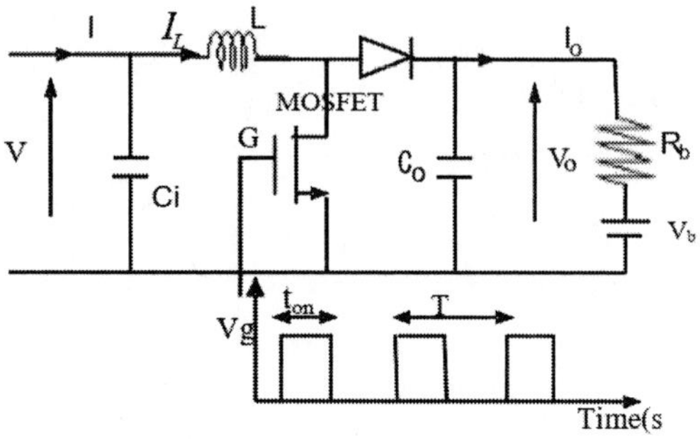

Figure 5: Circuit of boost converter and battery.

Its duty cycle is given by: $\alpha = t_{on}/T = t_{on} / (t_{on} + t_{off})$,Where T is the time period of the DC-DC converter, t_{on} and t_{off} are a positive variables. The battery is considered as a constant voltage Vb in series with a constant resistance Rb. By considering the mean values of the electric quantities over a chopping period, the state equations of the converter operating in continuous conduction mode is expressed by Eq. 5.

$$\begin{cases} C_i \dfrac{dV}{dt} = I - I_L \\ L\dfrac{dI_L}{dt} = V - (1-\alpha)(V_b + R_b I_o) \\ C_o R_b \dfrac{dI_o}{dt} = -I_o + (1-\alpha) I_L \end{cases} \tag{5}$$

Where I, V, I_L, I_o, V_o, L, C_i, C_o and α, refer to PV output current and voltage, inductor current, output current and voltage, inductance, input and output capacitor, duty cycle of the boost converter, respectively.

2.4 System model

By combining the different equations, the global mathematical model of the PV storage system, which is nonlinear, can be expressed as.

$$\begin{cases} \dot{x}_1 = \dfrac{I}{C_i} - \dfrac{1}{C_i} x_2 \\ \dot{x}_2 = \dfrac{1}{L} x_1 - \dfrac{\mu}{L} x_3 \\ \dot{x}_3 = \dfrac{V_b}{C_o R_b} + \dfrac{1}{C_o R_b} x_3 + \dfrac{\mu}{C_o} x_2 \end{cases} \tag{6}$$

With,

$$x_1 = V;\ x_2 = I_L;\ x_3 = V_o;\ u = 1-\alpha;\ y = \frac{\partial P}{\partial V}$$

Based on the power-voltage characteristic curves of the PV module (Figs.3-4), when the PV module is operating in its maximum power point $P=V.I$, we can get:

$$\frac{\partial P}{\partial V} = \frac{\partial (V.I)}{\partial V} = I + V\frac{\partial I}{\partial V} = 0 \tag{7}$$

The output system to be forced to zero in finite time is: $y=dP/dV$. The relative degree $r = 2$ of the system corresponds to the number of times the output y must be differentiated with respect to time before the input u appears explicitly in the resulting equations. We have the first and second time derivative of the controlled output y, after the second derivation of the output of the system we obtain the following expression:

$$\ddot{y} = f(x,t) + g(x,t)u \tag{8}$$

u is considered as the applied control law and α can be deduced from $u = 1-\alpha$ with the main objective of u is to steer the output to zero in finite time.

$$f(x,t) = \left(\frac{\partial^3 P}{\partial v^3}\right)\left(\frac{\partial V}{\partial t}\right)^2 + \frac{1}{C_i}\left(\frac{\partial^2 P}{\partial V^2}\right)\left[\left(\frac{\partial I}{\partial V}\right)\left(\frac{\partial V}{\partial t}\right) - \frac{V}{L}\right] \tag{9}$$

$$g(x,t) = \frac{1}{C_i}\left(\frac{\partial^2 P}{\partial V^2}\right)\frac{V_o}{L} \tag{10}$$

$$\frac{\partial^2 P}{\partial V^2} = 2\frac{\partial I}{\partial V} + V\frac{\partial^2 I}{\partial V^2} \tag{11}$$

$$\frac{\partial^3 P}{\partial V^3} = 3\frac{\partial^2 I}{\partial V^2} + V\frac{\partial^3 I}{\partial V^3} \tag{12}$$

3 PROPOSED MPPT TECHNIQUE

The suggested technique compels the system to run to its optimal operating point within variable temperature and solar irradiance. The main role of the MPPT technique is to modify the duty cycle of the boost converter until reaching the MPP.

The SMC has had great success over the last few years thanks to its simplicity of its implementation, accuracy, robustness and its strength against the uncertain systems and from external disturbances. It consists in moving the state trajectory of the system toward a predetermined surface called sliding or switching surface (SS) [10] and in maintaining it around this latter with an appropriate switching logic.

The design of SMC involves two tasks. The first one involves the design of a sliding surface S where the switching function $\sigma=0$, so that the sliding motion satisfies the design specifications. The second one is to design the discontinuous control such that the system enters the sliding mode σ and remains in it forever [11]. Therefore existence and reaching conditions are satisfied. These conditions are obtained from geometrical considerations: the deviation from the switching function σ and its time derivative $\dot{\sigma}$ should be opposite signs in the vicinity of a SS.These conditions are usually written more conveniently as $\sigma\dot{\sigma}<0$. By an appropriate choice of this surface, the output of our system is steered to zero in finite time, which assure the maximum power tracking and then improve the dynamic performance under rapidly varying conditions.

3.1 Choice of the sliding surface (SS)

The SS is chosen according to the output to be forced to zero in finite time and the relative degree of the system. The relative degree of the system r is defined to be the least positive integer i for which the derivative $y^{(i)}(t)$ is an explicit function of the control law $u(t)$ such that:

$$\frac{\partial y^{(r)}(t)}{\partial u}\neq 0 \text{ and } \frac{\partial y^{(i)}(t)}{\partial u}=0 \text{ for } i=0,\dots\dots,r-1 \quad (13)$$

The switching function can be selected as follows:

$$\sigma(t)=y^{(r-1)}(t)+\lambda_{r-2}y^{(r-2)}(t)+\dots+\lambda_0 y(t) \quad (14)$$

Where the coefficients $\lambda_0 \dots\dots \lambda_{r-2}$ are chosen so that the characteristic polynomial associated to $\sigma(t)$ have its roots strictly negative. Then, the output $y(t)$ tends to zero in a finite time when σ tends to zero in a finite time. In SS, when $\sigma(t)=0$. In our application r=*2* and the switching function:

$$\sigma=\lambda y+\dot{y} \quad (15)$$

$$\dot{\sigma}=\lambda\dot{y}+\ddot{y} \quad (16)$$

Where λ is a positive constant. Replacing $\ddot{y}$ by its expression (8), we obtain:

$$\dot{\sigma}=F+gu=(f+\lambda\dot{y})+gu \quad (17)$$

3.2 Determination of the control law of the SMC

The control law $u(t)$ forcing the output y to zero in finite time is composed of two terms u_{eq} and u_r. Where u_{eq} is the equivalent linear control term which makes the undisturbed nominal system state slide on the SS, and u_r is the term forcing the system to remain on the sliding surface in presence of disturbances and parameters variations. To determine these two terms, two states of the SS will be considered. When σ(t)=0, $u=u_{eq}$ is the control law obtained from the equivalent control method which is determined from the solution of equation $\dot{\sigma}(t)=0$ in Eq. (17), we obtain:

$$u=u_{eq}=-F/g \quad (18)$$

For the disturbed switching function σ(t)≠0, in order to demonstrate stability, the candidate Lyapunov function is adopted:

$$V=\frac{1}{2}\sigma(t)^2 \Rightarrow \dot{V}=\sigma\dot{\sigma} \quad (19)$$

By choosing:

$$\dot{\sigma}=-k\,sign(\sigma) \quad (20)$$

$$\dot{V}=-k\,\sigma sign(\sigma)=-k|\sigma|<0 \quad (21)$$

Where k is a positive parameter. From Eq. (17) and Eq. (20), we have:

$$-k.sign(\sigma)=F+g.u \quad (22)$$

$$u=u_{eq}+u_r=-\frac{F}{g}-\frac{k}{g}sign(\sigma) \quad (23)$$

4 RESULTS AND DISCUSSION

Simulation results are given with Matlab/Simulink to validate the proposed control strategy in rapid temperature and solar irradiance change. The Mono-crystalline SW 255 Mono PV module has been chosen to validate the model. Its key specifications under the STC are shown in Table 1. The parameters used in simulation are tabulated in table 2.

Table I: Electrical characteristics data of monocrystalline PV module SW255 at STC

Parameters	Values
Maximum power (P_{max})	255 W
Voltage at maximum power (V_{max})	30.6 V
Current at maximum power (I_{max})	8.43 A
Open circuit voltage (V_{oc})	38.7 V
Short circuit current (I_{sc})	9.05 A
Temperature coefficient of I_{sc}	0.044 %/K
Number of cells per module	60

Table II: Simulation parameters

PV module SW-255 Mono	DC-DC converter	Battery
R_s = 0.3058 Ω	L = 2.2.10^{-3} H	V_b = 48 V
R_p = 50 kΩ	C_i = 47.10^{-3} F	R_b = 2 Ω
A = 1.3352	C_o= 4.7.10^{-3} F	
I_{SCR} = 9.0501 A		
I_{rs} = 6.2217.10^{-8} A		
Ki = 4.4.10^{-4} A/K		
T_r = 298.15 K		

4.1 Theoretical results

For different values of solar irradiance G and temperature T, the computation of the theoretical optimum values of duty cycle α, PV voltage V, and PV power P, is assembled in table 3.

Table III: Theoretical results for given solar irradiance and temperature

G(W/m2) T(K)	800 298.15	1000 298.15	1000 308.15
I (A	6.7526	8.4301	8.3890
V (V)	30.635	30.6178	29.4301
P (W)	206.8647	258.1111	246.8896
A	0.4476	0.4633	0.4810

4.2 Simulation results and analysis

A simulation study was made to illustrate the response of the system to solar irradiance G and module temperature T variation. For this purpose, the irradiance G and module temperature T, which are initially at $800 W/m^2$ and 298.15K (25°C), are respectively switched at 3s and 6s to $1000 W/m^2$ and 308.15K (35°C), as shown in Fig. 6 and Fig. 7.

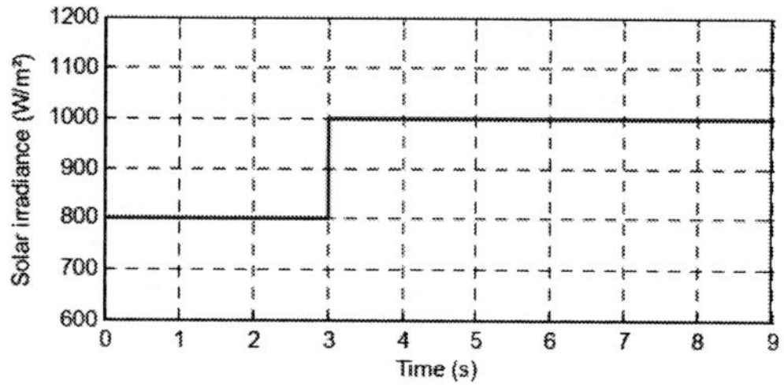

Figure 6: Change of solar irradiance.

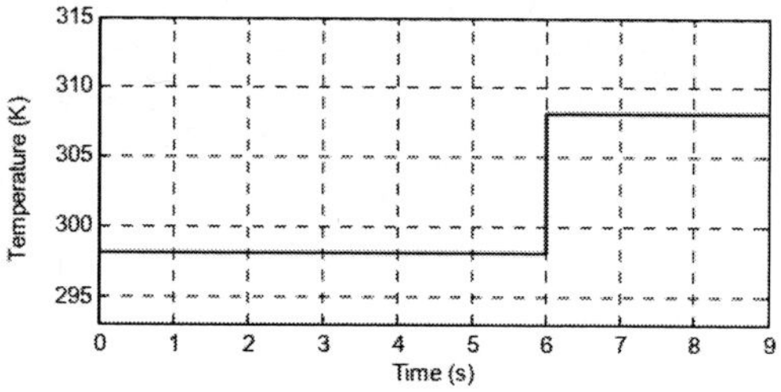

Figure 7: Change of temperature.

Fig. 8 and Fig. 9 show current-voltage and power-voltage characteristics and maximum power point (MPP) for different solar irradiance and temperature levels for PV module:

MPP1 (G=800W/m² and T=298.15K-25°C), (I_{MPP}=6.7526A, V_{MPP}=30.6350V, P_{MPP}=206.8647W).
MPP2 (G=1000W/m² and T=298.15K-25°C), (I_{MPP}=8.4301A, V_{MPP}=30.6178V, P_{MPP}=258.1111W).
MPP3 (G=1000W/m² and T=308.15K-35°C), (I_{MPP}=8.3890A, V_{MPP}=29.4301V, P_{MPP}=246.8896W).

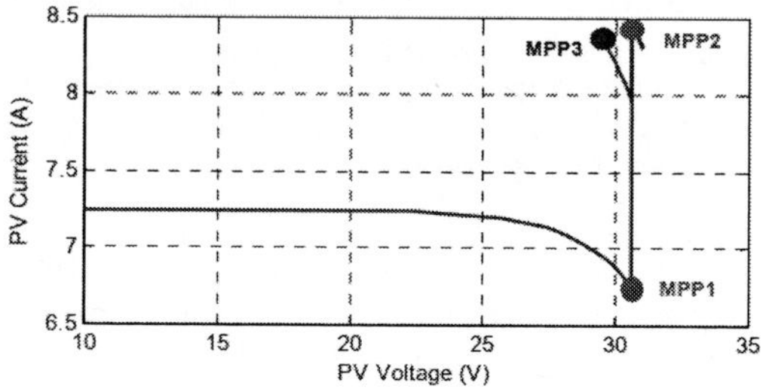

Figure 8: Characteristic curves current–voltage at different working irradiance and temperature levels.

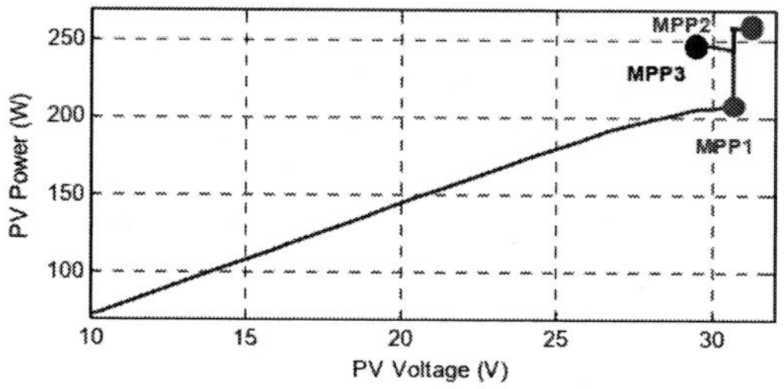

Figure 9: Characteristic curves power–voltage at different working irradiance and temperature levels.

It can be noted that the maximum power point is always achieved after a smooth transient response without oscillations. This way, the validation of the robustness of the proposed time sliding mode controller in presence of climatic variations is guaranteed.

Figs. 10-13 show respectively the good concordance between the current, the voltage, and the power of the PV module, the duty cycle control with the theoretical results.

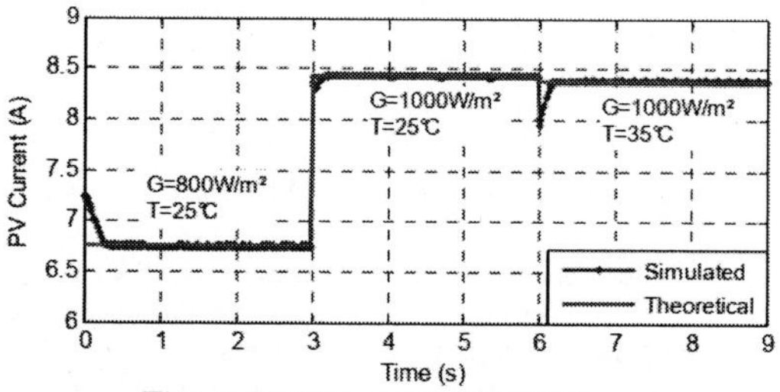

Figure 10: PV current with SMC.

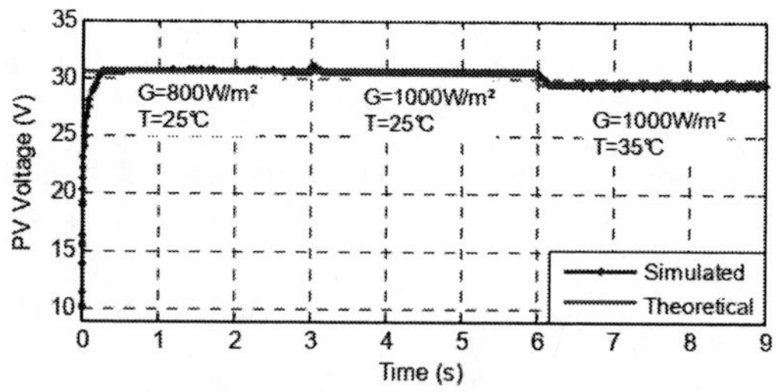

Figure 11: PV voltage with SMC.

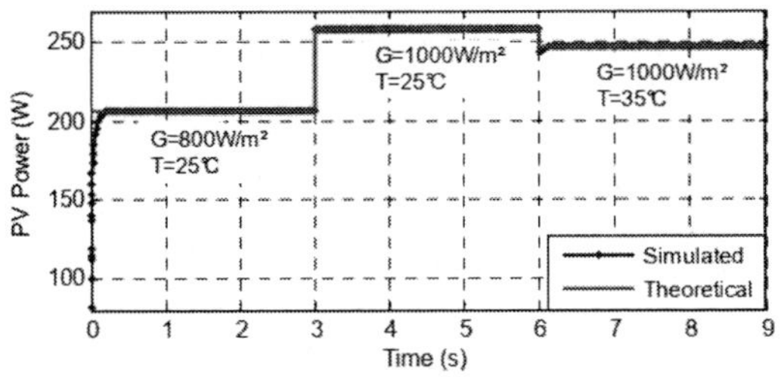

Figure 12: PV power with SMC.

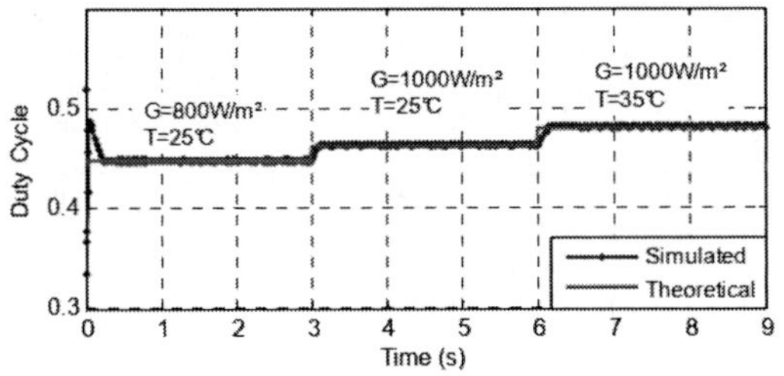

Figure 13: Duty Cycle (Control law α) with SMC.

Figs. 14-15 show the evolution of the controlled output y and the sliding surface. When the maximum power is reached, the sliding surface and the controlled output y=dP/dV converge to zero.

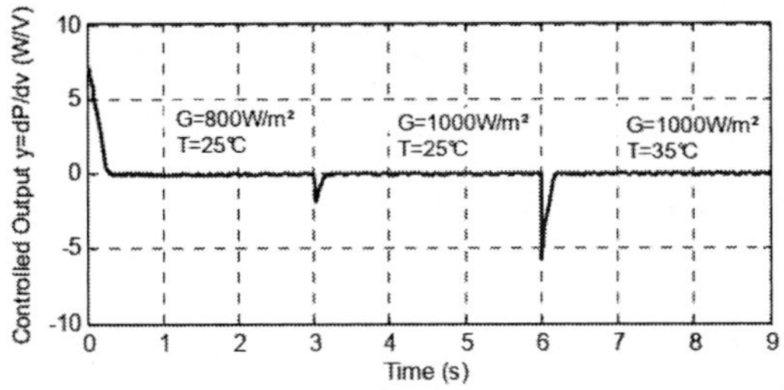

Figure 14: Controlled output (y=dP/dV).

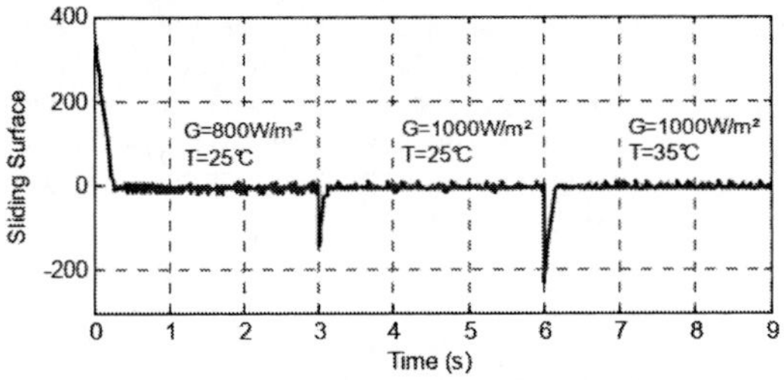

Figure 15: Sliding surface.

It is concluded from the simulations that, when the solar irradiance and temperature varies, the duty cycle of the boost converter α is judiciously adjusted to its desired value (Fig. 13), which forces the PV module voltage to follow its optimal value (Fig. 11). Consequently, the PV module power reached its maximal value (Fig. 13).

5 CONCLUSIONS

In this paper, modeling and simulation of a stand-alone PV system with battery storage were presented; numerical simulations were carried out for PV systems containing the SMC, under varying climatic conditions. The simulation analysis shows that SMC has shown good performance. Ensuring the robust stability property by the Lyapunov stability theorem, the numerical studies are demonstrated to verify the robust tracking performance (stability and accuracy) in both static and dynamic responses of the proposed method.

REFERENCES

[1] Fattori, F., Anglani, N., Muliere, G., Combining photovoltaic energy with electric vehicles, smart charging and vehicle-to-grid. Sol. Energy. 110 (2014) 438–451.
[2] Femia N et al, Optimization of perturb and observe maximum power point tracking method. IEEE Trans. Power Electron.. 20(4) (2005) 963–73.
[3] A. Zegaoui, M. Aillerie, P. Petit, J.P. Sawicki, A. Jaafar, C. Salame, and J.P. Charles, « Comparison of two common maximum power point trackers by simulating of PV generators," Energy Procedia, 6 (2011) pp. 678–687.
[4] A. Messai, A. Mellit, A. Guessoum, and S .A. Kalogirou, "Maximum power point tracking using a GA optimized fuzzy logic controller and its FPGA implementation," Solar Energy, 85 (2011) pp. 265–277.
[5] Lin, Whei-Min, Hong, Chih-Ming, Chen, and Chiung –Hsing, "Neural network- based MPPT control of a stand-alone hybrid power generation system," IEEE Trans. on Power Electron.,26(12) (2011) pp. 3571–3581.
[6] Chian-Song Chiu, Ya-Lun Ouyang, Chan-Yu Ku, Terminal sliding mode control for maximum power point tracking of photovoltaic power generation systems. Sol. Energy. 86 (2012) 2986–2995.
[7] Hanane Yatimi, Elhassan Aroudam, Assessment and control of a photovoltaic energy storage system based on the robust sliding mode MPPT controller. Sol. Energy. 139 (2016) 557–568.
[8] H. Yatimi and E. Aroudam. Evaluation of the Characteristics of the PV Module Considering Effects of Real Climatic Conditions, in: Sayigh Ali (Ed.), Renewable Energy in the Service of Mankind, Vol II, Springer International Publishing Switzerland, United Kingdom (2016) pp. 457-467.
[9] Aroudam El H.. Evaluation du Gisement solaire dans la région de Tétouan (Evaluation of solar field in the region of Tetouan), Ph.D. dissertation, Faculty of Sciences, Abdelmalek Essaadi Univ., Morocco, 1992.
[10] Slotine, J.J., "Adaptive Sliding controller synthesis for nonlinear systems," International Journal of Control, Vol 43. No. 6 (1986) pp. 1631-1651.
[11] Y. Levron and D. Shmilovitz, "Maximum Power Point Tracking Employing Sliding Mode Control," IEEE Trans. on Circuits and Syst., Vol. 60, No. 3, March (2013).

STUDY OF METHODS OF DEGRADATION ANALYSIS ON CRYSTALLINE SILICON PV MODULES IN A DISTRIBUTED MICROGENERATION UNIT LOCATED IN BRAZIL'S SOUTHEASTERN REGION

Aline M. Silva, Douglas B. Tsukamoto, Arthur C. Souza, Fernando C. Melo, Luiz C. G. Freitas and João B. V. Júnior
Núcleo de Pesquisa em Eletrônica de Potência (NUPEP)
Faculdade de Engenharia Elétrica (FEELT) - Universidade Federal de Uberlândia (UFU
Uberlândia, Minas Gerais, Brazil 38400-902
e-mails: alinefeelt@hotmail.com; fernando_cmelo@outlook.com; lcgfreitas@yahoo.com.br

ABSTRACT: This article presents a study aimed at understanding and analyzing the main forms of degradation that the PV modules of multicrystalline technology are subject to and their respective influence on the performance of electric power generation. For that, the forms of degradation found in the modules installed in the photovoltaic plant located in the premises of the Research Laboratory in Power Electronics of the Federal University of Uberlândia (NUPEP-FEELT-UFU) were identified and analyzed. Initially a visual inspection was performed and thermographic images were obtained and evaluated in order to identify Hot Spots and to associate them with the signs of degradation observed. Additionally, the I-V and P-V curves were measured in loco to obtain the main parameters used to measure the performance quality of the photovoltaic modules. For the validation of the results found, these were compared with the data obtained through a computational model, taking into account the same operational conditions, i.e., cell operating temperature and incident irradiation in the plane of the modules.
Keywords: Degradation; Photovoltaic Plant; Thermographic Images and I-V and P-V

1 INTRODUCTION

The degradation of the photovoltaic modules refers to the damages caused, generally, by chemical reactions and/or physical mechanisms, during the outdoor operation, causing gradual losses of its main electrical characteristics and material. According to [1], among the various types of degradation identified, the most frequently encountered are:

Discoloration: this is the degradation of the encapsulating material. One of the possible causes is the high temperatures [2], and its main effect is the loss of transmittance of the encapsulating material [3].

Delamination: consists of the loss of adhesion between the parts of the photovoltaic module, which can cause a greater reflection of light as well as the penetration of water inside the module structure [4].

Corrosion: process of deterioration of the metallic material, caused mainly by the humidity and generally occurs between the photovoltaic cell and the metallic frame, probably due to the failure of the seal.

Breaks and cracking: the photovoltaic cells are extremely fragile and, in this way, the breaks or fissures occur, in the majority of cases, when transporting, installing or maintaining the photovoltaic modules. This can facilitate the entry of moisture and the appearance of other forms of degradation.

Hot Spots: refers to the elevation of temperature in a particular site of the cell, usually caused by shading, dirt or by defective cells. In order to limit the reverse current caused by the shaded cells, the bypass diode is used in order to mitigate the increase in temperature [5].

According to [6], in addition to the main forms of degradation already mentioned, it is possible to mention snail track, which refers to the discoloration of the silver paste used for the gridlines on the cells. It should be emphasized that this form of degradation may not be confused with discoloration of the encapsulating material.

Other forms of degradation, according to [6] and [7], are LID (light-induced degradation) and PID (potential induced degradation). The LID refers to the small losses of power generation of the PV modules in the first moments of exposure to sunlight. The PID, in turn, occurs due to the not grounded installations operated at higher system voltages, thus, causing significant losses of energy on the system, usually under high relative humidity, longer rain or moisture condensation, being this form of degradation dependent on solar cell/module technology, of the location (relative humidity, rain and temperature) and type of PV system installation.

Based on [8] and [9], it is observed that at higher local values of the average temperature and relative humidity of the air the rate of degradation increases. In view of the above, it is worth noting that, in order to evaluate the operational state of the photovoltaic modules, besides visual inspection and thermographic analysis, it is necessary to know its main electrical characteristics.

Peak power is known by testing under standard test conditions (STC), i.e., ensuring an irradiance of 1000 W/m², cell temperature of 25 °C and A.M. = 1,5 (A.M. = Air Mass; This indication quantifies the thickness of the air layer, since the sensitivity of the photovoltaic cells changes according to the composition of the spectral light). From the curves I-V and P-V determine the main electrical parameters characteristic of the photovoltaic modules [10], namely: open circuit voltage (V_{OC}); short-circuit current (I_{SC}); maximum power voltage (V_{MPP}); maximum power current (I_{MPP}); maximum power point (P_{MPP}) and fill factor (FF), as shown in Fig. 1:

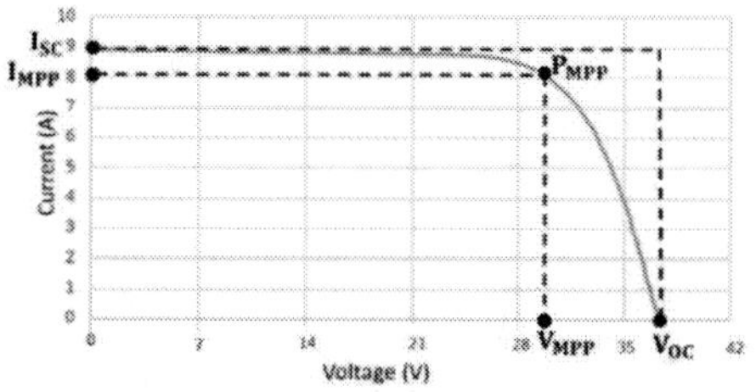

Figure 1: Representation of the I-V curve, under nominal test condition.

2 RESULTS

Fig. 2 shows the micro photovoltaic power plant used to carry out the proposed study, which consists of 48 modules of 5 different models, namely: 08 TPB156-60-P 235W from Sun Earth (string A), 08 AS-6P30-250 from

Amerisolar (string B), 16 CSI CS6P-255P from Canadian Solar (strings C and D), 08 KD140SX-UFBS and 08 KD135SX-UPU both from Kyocera (strings E and F, respectively), totaling a peak power Of 10.20 kWp. The 08 modules from Sun Earth feed an inverter 2 kW (FRONIUS IG 2000). The 16 modules of Canadian Solar along with the 08 of Amerisolar are feeding an inverter 5 kW (ABB PVI-5000-TL-OUTD). Kyocera modules, in turn, from 2015 onwards were not connected to commercial inverters, remaining available for prototype testing of solar inverters under development in the laboratory. However, they were duly examined. The main electrical characteristics in STC provided by the manufacturers are shown in Table I

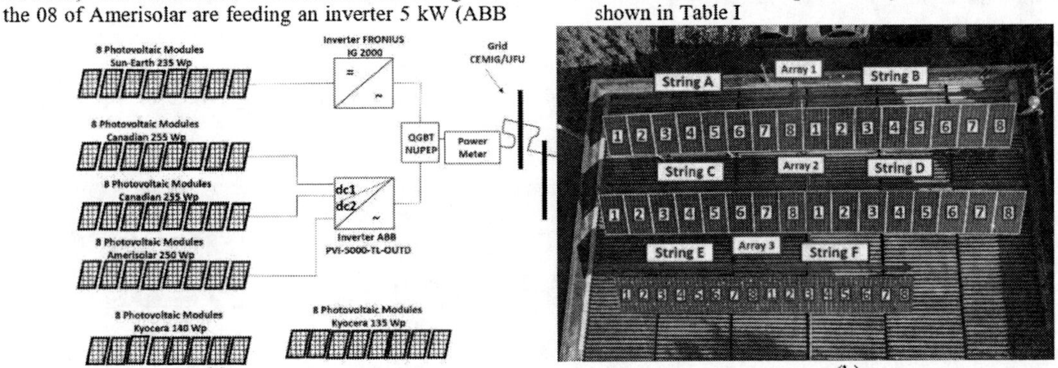

(a) (b)

Figure 2: Micro photovoltaic power plant installed (a) simplified single-line diagram and (b) mapping of photovoltaic modules.

Table I: Main electrical characteristics of photovoltaic modules (STC: T=25°C, G=1000 W/M², AM=1.5).

Parameter	Sun Earth TPB156-60-P 235W	Amerisolar AS-6P30-250W	Canadian CSI CS6P-255P	Kyocera KD135SX-UPU	Kyocera KD140SX-UFBS
Power, $P_{Máx}$	235 W	250 W	255 W	135 W	140 W
Voltage MPP, V_{MPP}	29,2 V	30,3 V	30,2 V	17,7 V	17,7 V
Current MPP, I_{MPP}	8,05 A	8,26 A	8,43 A	7,63 A	7,91 A
Open circuit voltage, V_{OC}	36,7 V	38 V	37,4 V	22,1 V	22,1 V
Short-circuit current, I_{SC}	8,47 A	8,75 A	9 A	8,37 A	8,68 A
Temperature Coefficient V_{OC}	-0,128 V/°C	-0,125 V/°C	- 0,127 V/°C	-0,08 V/°C	-0,08 V/°C
Temperature Coefficient $P_{Máx}$	-1.1 W/°C	-1,1 W/°C	-1,1 W/°C	-0,6 W/°C	-0,6 W/°C
Temperature Coefficient I_{SC}	0,00423 A/°C	0,00491 A/°C	0,00586 A/°C	0,00502 A/°C	0,00522 A/°C
Negative Power Tolerance, tol-	0%	0%	0%	5%	5%
Positive Power Tolerance, tol+	2,13%	3%	1,97%	5%	5%
Module Efficiency, η	14,4%	15,4%	15,85%	14,0%	14,0%

Initially, all the modules that comprise the system were photographed and mapped, as illustrated in Fig. 2(b), Then a visual inspection of the photovoltaic modules was performed for the detection of degradation signals and their respective thermographic images were obtained through the FLIR i5® infrared câmera.

By inspecting all installed PV modules, some visible forms of degradation have been identified only in the Kyocera modules (strings E and F). In Fig. 3 two photos are shown illustrating the discoloration of Kyocera KD135SX-UPU modules on the day of installation (Fig. 3(a)) and at the end of the third year of use (Fig. 3(b)), evidencing the blackening of the photovoltaic modules due to possible degradation of the encapsulant material, this was observed in all KD135SX-UPU modules. Fig. 3(c) refers to the image of the module most damaged by the snail track, located at position 6F (Fig.2 (b)), this fact was identified in three of the eight KD135SX-UPU modules from Kyocera (string F). In Fig. 4 (a) it is possible to observe a more severe degradation of the encapsulating material, located above the position where the junction box is installed, the discoloration at this location occurred in all modules of strings E (Kyocera-KD140SX-UFBS) And F (Kyocera-KD135SX-UPU).

Also with respect to the Kyocera modules, a severe degradation in two cells of the module located at position 8E (Fig.2 (b)) is shown in Fig. 3 (d). This occurred in February 2015, only 22 months after installation, due to a failure in the waterproofing of the junction box that permanently damaged the PV module. Currently module 8E is disconnected from the photovoltaic string.

Regarding the other modules, it should be noted that no visible signs of degradation have been identified, even in Sun Earth's TPB156-60-P 235W modules (string A) that have been in operation since August 2013.

In the sequence, thermographic images were obtained for all the modules of the installation in order to evaluate the thermal distribution and identify the existence of Hot Spots. The results found for the modules of the strings A (Sun Earth), B (Amerisolar), C (Canadian Solar) and D (Canadian Solar), they were deemed within normal, that is, without significant differences in the temperature distribution observed on the surface of the modules. However, in the PV modules of the strings E (Kyocera 140W) and F (Kyocera 135W), a noticeable increase in temperature was observed in the cells located immediately above the place where the junction box is installed, as evidenced in Figs. 4(b) and 4(c), could be attributed to this module an abnormal operating condition. This fact can be correlated to the existence of discoloration of the encapsulating material of said cells, and are characteristics of Hot Spots that may lead to the premature failure of said modules, in addition to a drop in system performance rate over years [10] and [11].

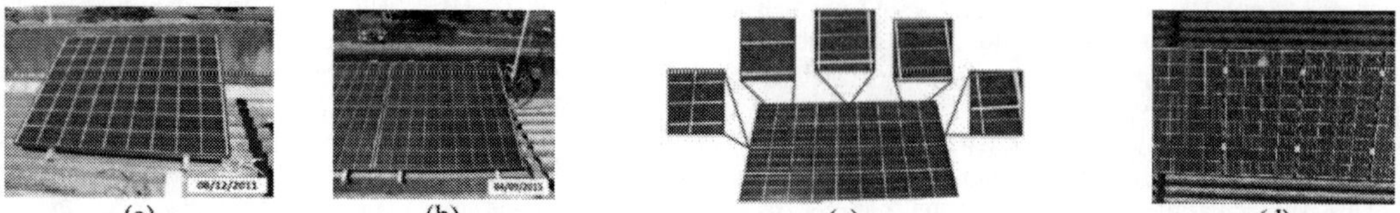

(a) (b) (c) (d)

Figure 3: (a) Image of KYOCERA modules KD135SX-UPU (array 3 - String F) showing the effect of the early discoloration of the encapsulating material in the entire photovoltaic module on the installation day (08/12/2011) and (b) after four years (04/09/2015), (c) module of the string F (6F-Kyocera) with the highest number of degraded cells and (d) module with cells damaged by junction box waterproofing failure (8E-Kyocera).

(a) (b) (c)

Figure 4: (a) Image of one of the KD135SX-UPU modules with highlight the discoloration observed in the cells installed on the junction box, (b) thermographic image of the complete module and (c) thermographic image with focus on the identified Hot Spot.

In sequence of the works, all identified modules, except 8E Kyocera that was disconnected from the system due to failure in waterproofing the junction box, were subjected to experimental tests aimed at obtaining curves IV and PV through the Solar IV instrument 400. The first procedure was the modules individual test. Later, after the connection of the strings, were obtained new curves for the same. For the validation of the practical results achieved, these were compared with the data obtained in computational model implemented in platform PVsyst®.

Preliminarily, it is possible to infer that the high operating temperatures that are submitted photovoltaic cells of the modules installed in Uberlândia-MG / Brazil, vary between 10 °C and 45 °C (7:00 a.m. to 10:00 a.m.) and between 45 °C and 65 °C (10:00 a.m. to 3:00 p.m.), in the months May and July, and between 15 °C and 45 °C (7:00 a.m. to 10:00 a.m.) and 45°C and 65 °C (10:00 a.m. to 3:00 p.m.) in the months August and October. The modules were analyzed separately, and then tests were performed for each photovoltaic string, both at a temperature between 45 °C and 65 °C. From the data found, by way of example, a result were selected for presentation in this work, namely:

String B - The results shown in Fig. 5 refer to module 4B (Amerisolar AS-6P30-250W). This presented greater discrepancy in relation to the expected maximum power of the 47 analyzed modules after the complete installation of the system. The maximum power point at operating conditions (OPC) was 178.63W (56 ° C and 927W / m²). In order to compare this result with the datasheet provided by the manufacturer, the metering equipment converts the electrical parameters obtained at the time of measurement (OPC) for values under standard test conditions (STC). In STC, the maximum power (PMPP) measured was 216,96W, while the nominal value supplied by the manufacturer is 250 Wp. Therefore, a discrepancy of -13,27%. The negative tolerance (tol-) reported by the manufacturer is equal to 0% and the positive tolerance (tol +) is equal to 3%. In this way, it is possible to infer that the module in question does NOT conform to what is presented in the manufacturer's datasheet. In PVsyst® ® the maximum power found was 203,34W (OPC), therefore, when comparing the maximum measured power (OPC) by Solar I-V 400 with the reference values obtained through the computational model, It is observed a negative discrepancy of -12.13% for PVsyst®. These results confirm the effectiveness of the measurements performed in loco, since the discrepancy found by the Solar I-V 400 instrument (-13.27%) remained close to those obtained in the computational model (-12.13%). Analogous procedure was adopted to obtain and analyze the I-V and P-V curves of the other modules of the photovoltaic system and the results are shown in Table II.

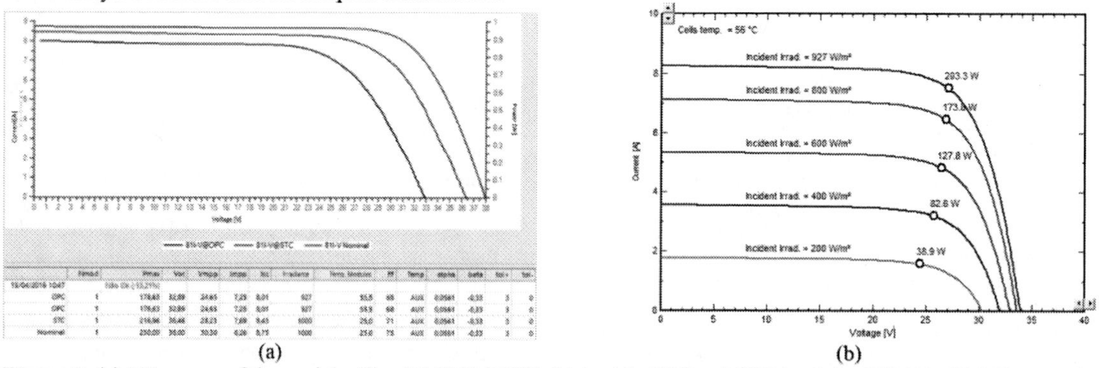

(a) (b)

Figure 5: (a) I-V curves of the module 4B - AS-6P30-250W obtained in OPC and STC by Solar I-V 400; (b) I-V curves in OPC by PVsyst®.

Table II: Results obtained for each module of the photovoltaic system analyzed, with values in STC and OPC obtained through Solar I-V and PVsyst®, respectively

Módulo	Parâmetros	PVsyst OPC	Solar I-V OPC	Discrepância (%)	Datasheet STC	Solar I-V STC	Discrepância (%)
1A	Temp. (°C)	47,8	47,8	-	25	25	-
	Irrad. (W/m²)	761	761	-	1000	1000	-
	Pmax (W)	158,6	144,92	- 8,62	235	209,95	- 10,66
2A	Temp. (°C)	46,7	46,7	-	25	25	-
	Irrad. (W/m²)	747	747	-	1000	1000	-
	Pmax (W)	156,4	145,04	-7,26	235	212,83	- 9,44
3A	Temp. (°C)	46,6	46,6	-	25	25	-
	Irrad. (W/m²)	748	748	-	1000	1000	-
	Pmax (W)	156,7	145,04	- 7,44	235	212,49	- 9,58
4A	Temp. (°C)	46,6	46,6	-	25	25	-
	Irrad. (W/m²)	751	751	-	1000	1000	-
	Pmax (W)	157,4	144,01	- 8,5	235	210,04	-10,62
5A	Temp. (°C)	46,6	46,6	-	25	25	-
	Irrad. (W/m²)	750	750	-	1000	1000	-
	Pmax (W)	157,2	145,91	- 7,18	235	213,08	- 9,33
6A	Temp. (°C)	46,5	46,5	-	25	25	-
	Irrad. (W/m²)	750	750	-	1000	1000	-
	Pmax (W)	157,2	143,93	- 8,44	235	210,16	- 10,57
7A	Temp. (°C)	46,7	46,7	-	25	25	-
	Irrad. (W/m²)	754	754	-	1000	1000	-
	Pmax (W)	158	143,44	- 9,21	235	108,61	- 11,23
8A	Temp. (°C)	46,9	46,9	-	25	25	-
	Irrad. (W/m²)	756	756	-	1000	1000	-
	Pmax (W)	158,2	144,41	- 8,71	235	209,52	- 10,84
1B	Temp. (°C)	54,8	54,8	-	25	25	-
	Irrad. (W/m²)	915	915	-	1000	1000	-
	Pmax (W)	201,2	188,86	- 6,13	250	231,52	- 7,39
2B	Temp. (°C)	55,3	55,3	-	25	25	-
	Irrad. (W/m²)	918	918	-	1000	1000	-
	Pmax (W)	201,4	184,33	- 8,47	250	226,02	- 9,59
3B	Temp. (°C)	55,5	55,5	-	25	25	-
	Irrad. (W/m²)	920	920	-	1000	1000	-
	Pmax (W)	201,7	182,7	- 9,41	250	223,44	- 10,62
4B	Temp. (°C)	55,5	55,5	-	25	25	-
	Irrad. (W/m²)	927	927	-	1000	1000	-
	Pmax (W)	203,3	178,63	- 12,13	250	216,96	- 13,21
5B	Temp. (°C)	55,4	55,4	-	25	25	-
	Irrad. (W/m²)	926	926	-	1000	1000	-
	Pmax (W)	203,1	185,1	- 8,86	250	224,61	- 10,16
6B	Temp. (°C)	55,3	55,3	-	25	25	-
	Irrad. (W/m²)	930	930	-	1000	1000	-
	Pmax (W)	204,2	184,44	- 9,67	250	223,31	- 10,68
7B	Temp. (°C)	54,8	54,8	-	25	25	-
	Irrad. (W/m²)	935	935	-	1000	1000	-
	Pmax (W)	205,8	192,03	- 6,69	250	230,13	- 7,95
8B	Temp. (°C)	54,4	54,4	-	25	25	-
	Irrad. (W/m²)	936	936	-	1000	1000	-
	Pmax (W)	206,4	188,92	-8,46	250	226,65	- 9,34
1C	Temp. (°C)	55,3	55,3	-	25	25	-
	Irrad. (W/m²)	959	959	-	1000	1000	-
	Pmax (W)	211,4	202,45	- 4,23	255	236,8	- 7,14
2C	Temp. (°C)	54,3	54,3	-	25	25	-
	Irrad. (W/m²)	962	962	-	1000	1000	-
	Pmax (W)	213,2	204,66	- 4,005	255	237,69	- 6,79
3C	Temp. (°C)	55,3	55,3	-	25	25	-
	Irrad. (W/m²)	963	963	-	1000	1000	-
	Pmax (W)	212,3	204,02	- 3,9	255	237,78	- 6,75
4C	Temp. (°C)	55,4	55,4	-	25	25	-
	Irrad. (W/m²)	965	965	-	1000	1000	-
	Pmax (W)	212,6	202,9	-4,56	255	236,03	- 7,44
5C	Temp. (°C)	55,6	55,6	-	25	25	-
	Irrad. (W/m²)	969	969	-	1000	1000	-
	Pmax (W)	213.2	204,32	-4,16	255	237,05	- 7,04
6C	Temp. (°C)	56,3	56,3	-	25	25	-
	Irrad. (W/m²)	978	978	-	1000	1000	-
	Pmax (W)	214,4	204,31	-4,7	255	235,38	- 7,69
7C	Temp. (°C)	56,3	56,3	-	25	25	-

	Irrad. (W/m²)	976	976	-	1000	1000	-
	Pmax (W)	213,9	209,76	-1,93	255	241,74	- 5,2
	Temp. (°C)	56,5	56,5	-	25	25	-
8C	Irrad. (W/m²)	978	978	-	1000	1000	-
	Pmax (W)	214.1	206,39	-3,6	255	237,68	- 6,79
	Temp. (°C)	54,2	54,2	-	25	25	-
1D	Irrad. (W/m²)	999	999	-	1000	1000	-
	Pmax (W)	221,3	211,63	-4,36	255	236,47	- 7,27
	Temp. (°C)	54,2	54,2	-	25	25	-
2D	Irrad. (W/m²)	1000	1000	-	1000	1000	-
	Pmax (W)	221,5	210,9	-4,78	255	235,66	- 7,58
	Temp. (°C)	53,9	53,9	-	25	25	-
3D	Irrad. (W/m²)	1003	1003	-	1000	1000	-
	Pmax (W)	222,5	212,2	-4,62	255	236,14	- 7,39
	Temp. (°C)	53,8	53,8	-	25	25	-
4D	Irrad. (W/m²)	1000	1000	-	1000	1000	-
	Pmax (W)	222	209,51	-5,62	255	234,23	- 8,14
	Temp. (°C)	54	54	-	25	25	-
5D	Irrad. (W/m²)	996	996	-	1000	1000	-
	Pmax (W)	220,9	212,2	-3,93	255	237,72	- 6,78
	Temp. (°C)	53,8	53,8	-	25	25	-
6D	Irrad. (W/m²)	998	998	-	1000	1000	-
	Pmax (W)	221,6	213,09	-3,84	255	237,98	- 6,67
	Temp. (°C)	54,3	54,3	-	25	25	-
7D	Irrad. (W/m²)	1001	1001	-	1000	1000	-
	Pmax (W)	221,6	208,05	-6,11	255	232,15	- 7,78
	Temp. (°C)	54,8	54,8	-	25	25	-
8D	Irrad. (W/m²)	1003	1003	-	1000	1000	-
	Pmax (W)	221,5	211,26	-4,62	255	235,89	- 7,49
	Temp. (°C)	48,9	48,9	-	25	25	-
1E	Irrad. (W/m²)	1024	1024	-	1000	1000	-
	Pmax (W)	127,3	118,87	- 6,62	140	129,07	- 7,81
	Temp. (°C)	50,1	50,1	-	25	25	-
2E	Irrad. (W/m²)	1024	1024	-	1000	1000	-
	Pmax (W)	126,5	118,47	- 6,34	140	129,64	- 7,4
	Temp. (°C)	50,5	50,5	-	25	25	-
3E	Irrad. (W/m²)	1026	1026	-	1000	1000	-
	Pmax (W)	126,5	117,57	- 7,05	140	128,08	- 8,52
	Temp. (°C)	51,2	51,2	-	25	25	-
4E	Irrad. (W/m²)	1022	1022	-	1000	1000	-
	Pmax (W)	125.5	118,22	- 5,8	140	129.88	- 7,22
5E	Temp. (°C)	52	52	-	25	25	-
	Irrad. (W/m²)	1022	1022	-	1000	1000	-
	Pmax (W)	124,9	117,86	- 5,63	140	130,07	- 7,09
	Temp. (°C)	52,3	52,3	-	25	25	-
6E	Irrad. (W/m²)	1022	1022	-	1000	1000	-
	Pmax (W)	124,7	116,32	- 6,72	140	128,55	- 8,18
	Temp. (°C)	52,7	52,7	-	25	25	-
7E	Irrad. (W/m²)	1024	1024	-	1000	1000	-
	Pmax (W)	124,7	115,94	-7,01	140	129,42	- 7,55
	Temp. (°C)	52,8	52,8	-	25	25	-
1F	Irrad. (W/m²)	1011	1011	-	1000	1000	-
	Pmax (W)	118,7	112,77	-4,99	135	125,83	- 6,79
	Temp. (°C)	53,3	53,3	-	25	25	-
2F	Irrad. (W/m²)	1009	1009	-	1000	1000	-
	Pmax (W)	118,1	113,05	-4,27	135	126,8	- 6,07
	Temp. (°C)	53,9	53,9	-	25	25	-
3F	Irrad. (W/m²)	1011	1011	-	1000	1000	-
	Pmax (W)	117,9	112,99	-4,16	135	126.65	- 6,19
	Temp. (°C)	54,3	54,3	-	25	25	-
4F	Irrad. (W/m²)	1009	1009	-	1000	1000	-
	Pmax (W)	117,6	112,05	-4,71	135	126,25	- 6,48
	Temp. (°C)	54,3	54,3	-	25	25	-
5F	Irrad. (W/m²)	995	995	-	1000	1000	-
	Pmax (W)	115,8	114,03	-1,52	135	130,21	- 3,55
	Temp. (°C)	54,1	54,1	-	25	25	-
6F	Irrad. (W/m²)	1016	1016	-	1000	1000	-
	Pmax (W)	118,4	106,61	-9,95	135	118,38	- 12,31
	Temp. (°C)	54,1	54,1	-	25	25	-
7F	Irrad. (W/m²)	1027	1027	-	1000	1000	-
	Pmax (W)	119,7	115,28	- 3,69	135	127,35	- 5,67

8F	Temp. (°C)	53,8	53,8	-	25	25	-
	Irrad. (W/m²)	1026	1026	-	1000	1000	-
	Pmax (W)	119,8	113,66	- 5,12	135	125,47	- 7,06

3 CONCLUSIONS

Through the thermography, were observed in all 15 modules of Kyocera the presence of hot spots in the cells located just above the junction box. In these cells, through a visual inspection, it was possible to observe an expressive discoloration, probably due to the high operating temperatures of the same ones. It was also observed in all 8 modules KD135SX-UPU a visible blackening of the same due to discoloration of the encapsulating material. In addition, It was observed a type of degradation denominated in the technical literature specialized by snail track. In the other PV modules, no signs of apparent degradation were found.

Analyzing the IV and PV curves obtained in loco, and the results obtained through the computational model analyzed in the platform PVsyst®, it is evident the reliability of the measurements taken with regard to obtaining the main electrical parameters used to determine the operational state of the installed photovoltaic modules.

Given the above, the analyzes performed for each module under operating temperatures around 45°C (10:00 a.m.) and 65 °C (3:00 p.m.), show that the 48 photovoltaic modules, except the Kyocera 5F, based on the data provided by manufacturers, presented underperform, indicating an average power degradation of -10.28% for Sun Earth (String A), -9.87% for Amerisolar (String B), -6.855% and -7.39% for Canadian Solar (Rows C and D), -7.68% and -6.76% for Kyocera (String E and F). The analyzes performed on each string also showed that the performances remained below the manufacturer's negative tolerance, with an average power degradation of -12.98% for Sun Earth (String A), -12.17% for Amerisolar (String B), -9.38% and -9.88% for Canadian Solar (String C and D), -8.36% and -9.78% for Kyocera (String E and F).

It is worth noting that despite the degradations present in some modules analyzed and the lower performance, these were not enough to invalidate them from their operation in the photovoltaic plant in question, only in the case of the Kyocera 8E module that was disconnected from the system due to the waterproofing of the box junction. In future works will be presented results for a longer time of operation of the analyzed modules, verifying the effect of the degradations and aging in the reduction of their performance.

4 ACKNOWLEDGMENTS

The authors would like to thank the financial support of CAPES, BRC ENERGIA LIMPA LTDA for the technical support and donation of the ABB PVI 5000-TL-OUTD-S inverter through the FEELT-47/15 Process, to FAPEMIG through the TEC-APQ Processes -01219-13, PPM-00565-13, PPM-00031-14, and CNPq through Processes 406845/2013-1, 304307/2013-0 and 301209/2010-3.

5 REFERENCES

[1] A. Ndiave, A. Charki, A. Kobi, C. M. F. Kebe, P. Ndiave, V. Sambou, "Degradations of Silicon Photovoltaic Modules: A Literature Review" in Solar Energy, Vol. 96, pp. 140–151, 2013.

[2] J. A. Tsanakas, L. Ha, C. Buerhop, "Faults and infrared thermographic diagnosis in operating c-Si photovoltaic modules: A review of research and future challenges", in Renewable and Sustainable Energy Reviews, Vol. 62, pp. 695-709, 2016.

[3] N. C. Park, J. S. Joeng, B. J. Kang, D. H. Kim, "The effect of encapsulant discoloration delamination on the electrical characteristics of photovoltaic module", in Microelectronics Reliability, Vol. 53, pp. 1818-1822, 2013.

[4] M. A. Munoz, M. C. Alonso-Garcia, V. Nives, F. Chenlo, "Early degradation of silicon PV modules and guaranty conditions", in Solar Energy, Vol. 85, pp. 2264-2274, 2011.

[5] H. Zheng, S. Li, R. Challoo, J. Proano, "Shading and bypass diode impacts to energy extraction of PV arrays under different converter configurations", Renewable Energy, Vol 68, pp. 58-66, 2014.

[6] M. Köntges, S. Kurtz, C. Packard, U. Jahn, K. A. Berger, K. Kato, T. Friesen, H. Liu, M. V. Iseghem, "Review of Failures of Photovoltaic Modules", in IEA International Energy Agency, Task 13, 2014.

[7] B. Jaeckel, M. Cosic, J. Arp, "Investigation of c-Si modules degradation and recovery effect under high potentials: CV-PID", in 2014 IEEE 40th Photovoltaic Specialist Conference (PVSC), pp. 937-942, 2014.

[8] N. Zlatov, I. Iliev, A. Terziev and V. Kamburova, "Influence of climatic data and degradation factor on the efficiency of the photovoltaic modules", in Research and Education in Mechatronics (REM), 2014 15th International Workshop on, pp. 1-5, 2014.

[9] V. Sharma, S. S. Chandel, "A novel study determining early life degradation of multi-crystalline-silicon photovoltaic modules observed in western Himalayan Indian climatic conditions", in Solar Energy, Vol. 134, pp. 32-44, 2016.

[10] F. P. Gasparin, A. J. Bühler, G. A. Rampinelli, A. Krenzinger, "Statistical analysis of I–V curve parameters from photovoltaic modules", in Solar Energy, Vol. 131, pp. 30-38, 2016.

[11] S. S. Chandel, M. Nagaraju Naik, V. Sharma, R. Chandel, "Degradation analysis of 28 year field exposed mono-c-Si photovoltaic modules of a direct coupled solar water pumping system in western Himalayan region of India", in Renewable Energy, Vol. 78, pp. 193-202, 2015.

OPERATIONAL FAULT-MODE DIFFERENTIATION IN A LARGE-SCALE PHOTOVOLTAIC POWER PLANT WITH FAULT-DIAGNOSTIC FUNCTION

Tohru Kohno[*†,***], Hideyuki Shitanishi[*], Masahito Toyosaki[*], Kenichi Gokita[*], Tomoharu Nakamura[*], Yuichi Nagayama[*], Koki Morikawa[**], and Mutsuko Hatano[***]

[*]Hitachi, Ltd. 1-280 Higashi-Koigakubo, Kokubunji, Tokyo 185-8601, Japan
E-mail: [†]toru.kono.cw@hitachi.com
[**]Tokyo Electric Power Company Holdings, Inc. 4-1 Egasaki, Tsurumi, Yokohama 230-8510, Japan
[***] Tokyo Institute of Technology 2-12-1, Ookayama, Meguro-ku, Tokyo 152-8550, Japan

ABSTRACT: Aiming to reduce a loss of long-term generation output for large-scale photovoltaic (PV) power plant, we developed a fault-detection architecture. In this work, we developed a method to identify cause of failure and degradation. This method is added to the existing fault-detection system and the effectiveness of proposed method was confirmed through the operation of large-scale PV power plants in service.
Keywords: Fault-detection architecture, Cause of failure and degradation

1 INTRODUCTION

Large-scale photovoltaic (PV) systems are promising as power plant utilizing renewable energy. The market of Large-scale PV has been growing dramatically in Japan since the introduction of a feed-in tariff scheme for renewable energy in July 2012. In 2017, the installed capacity of Japanese large-scale PV systems will reach 15 GW.

The key issues are long-term reliability of these power plants and sustaining steady operation without any faults or performance degradation. These faulty PV modules or performance degradation cause power loss of a plant. Therefore, it is important to detect faulty modules only when and where needed and grasp degradation at an early stage.

Concerning the fault diagnosis in PV systems, many kinds of techniques have been developed [1-4].We developed an architecture for detecting the number of faulty modules present in a PV array without adding extra sensors [5-7] and built the fault-detection systems in many PV power plants. The key feature of this architecture is that it calculates solar radiation and temperature from measured voltage and current of a PV inverter using the PV model based on semiconductor device physics.

In this work, we developed a method to identify cause of failure and degradation. A feature of this method is to estimate the fault mode from the amount of loss in two time zones of different solar radiation. This method is added to the existing fault-detection system using string monitors. The string monitor is equipped for every 84 modules. A fault-detection system equipped with this identification method was applied to Large-scale PV power plant. Its total output is 82 MW. Through the operation of this power plants in service from April 2014, the effectiveness of proposed method was confirmed.

2 IDENTIFICATION OF FAULT MODE

Figure 1 shows a sectional view of PV module. PV cells are connected in serial by interconnectors. Each interconnector is connected to the PV cell by soldering. PV cells are protected by filter between glass and back sheet. The junction boxes are fitted with bypass diodes.

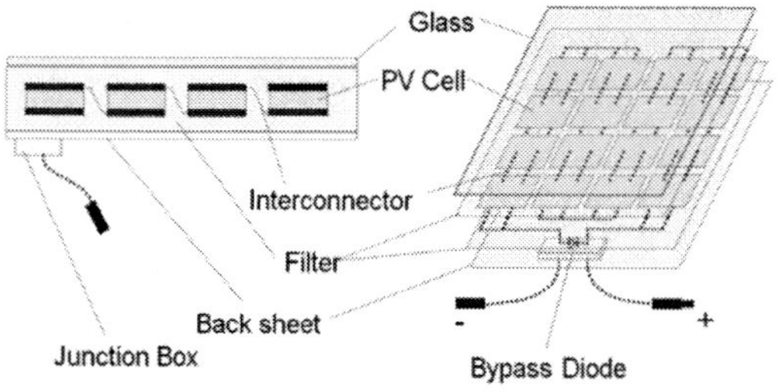

Figure 1: Sectional View of PV module

An equivalent circuit of PV module is shown in Fig.2. Faults and degradation of these components are detected by equivalent circuit parameters (series resistance, shunt resistance, bypass diode) of PV module, PV module's faults and degradation can be categorized into a couple of fault modes (solder peeling, cell degradation and broken wire) as shown in Table 1.

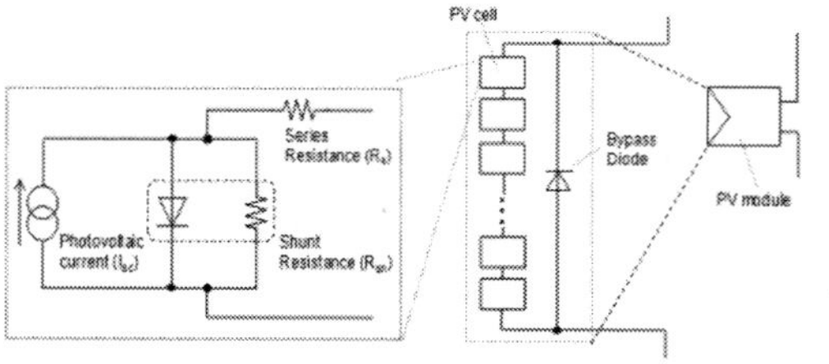

Figure 2: Equivalent circuit of PV module

Table 1: Correspondence between fault modes and parameters of equivalent circuit

Fault mode	Cause	Effect on equivalent circuit
Solder Peeling	Increased resistance in interconnector	Increase in series resistance(R_s)
Cell degradation	Development of PV cell's ohmic characteristic	Drop in shunt resistance(R_{sh})
Broken Wire	Bypass diode operates continuously	Bypass diode on

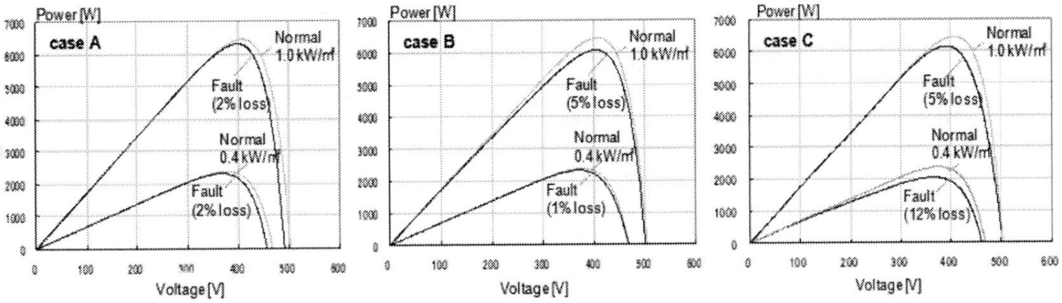

(a) Broken wires (b) Solder peeling (c) Cell degradation

Figure 3: PV string characteristics in each fault mode

If a fault mode is the broken wires, the bypass diode constantly operates. The power loss is dependent on the number of PV cells with broken wires and it has the same value regardless of the irradiation as shown in Fig. 3(a). In case of the solder peeling, series resistance should be increased. A voltage drop due to series resistance and current causes power loss that is proportional to current. Namely, power loss increases as the light irradiation increases as shown in Fig.3 (b). If a fault mode is the cell degradation, the shunt resistance should decrease and ratio of current leakage increases with decreasing the irradiation. Namely, power loss increases as the irradiation decrease as shown in Fig. 3(c).

3 LARGE-SCALE PV POWR PLANT APPLYING DEVELOPED METHOD

The configuration of the 337-kW PV system with string monitors is shown in Fig.4. This power plant has 244 PV systems. The string monitor is equipped for every 84 modules. The number of string-monitor channels is 12,200. There are 1220 Sting boxes equipped with string-monitors. The total number of PV modules is 341040 and the output of this power plant is 82 MW.

4 CALCULATION FLOW FOR IDENTIFYING FAULT MODE

The calculation flow to identify the fault mode is shown in Fig. 5. First, two time zones are selected. At these time zones, PV inverter is operating under maximum power point tracking. Secondly, at each time zone, irradiation p_a and temperature T_a are accurately calculated using operating voltage V'_{pmaxE} and operating current I'_{pmaxE} of PV inverter in equations (1) and (2) [5]. Next, the ideal power of each string or array is calculated at irradiation pa and temperature T_a. The power loss is calculated as ratio between measured actual power and ideal power. The power loss at the time zone that irradiation p_a is larger is defined as "loss 1". The power loss at the other time zone is defined as "loss 2". If the loss 1 and loss 2 are nearly the same, a process transitions to case A and number of bypass-diode operations is calculated in equation (3). If this number is larger than *num_th*, "broken wire" is judged to be most likely. If loss 1 is larger than loss 2, it transitions to case B and serial resistance is calculated in equation (4). If resistance is larger than threshold *Rs_th*, "solder peeling" is judged to be most likely. If loss 1 is smaller than loss 2, it transitions to case C. shunt resistance is calculated in equation (5). If shunt resistance is smaller than *Rsh_th*, "cell degradation" is judged to be most likely.

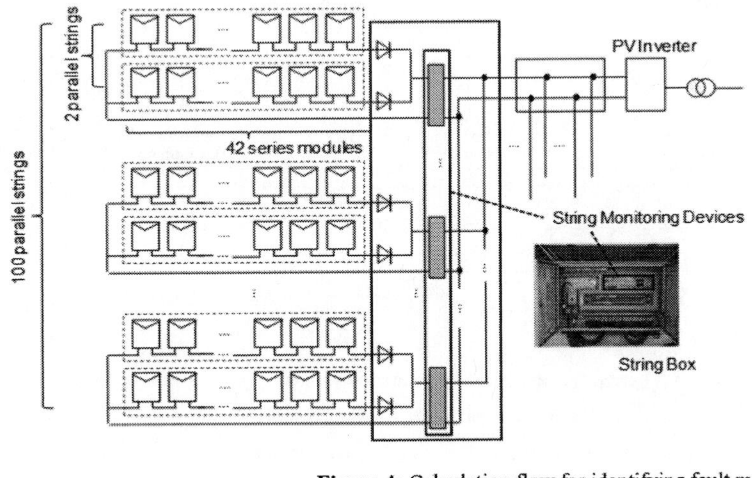

Total 82 MW

PV Module (240~245W)	341,040
PV Inverter (MPPT Function)	244
String Monitor - String Box - Monitor Channel	 1,220 12,200

Figure 4: Calculation flow for identifying fault mode

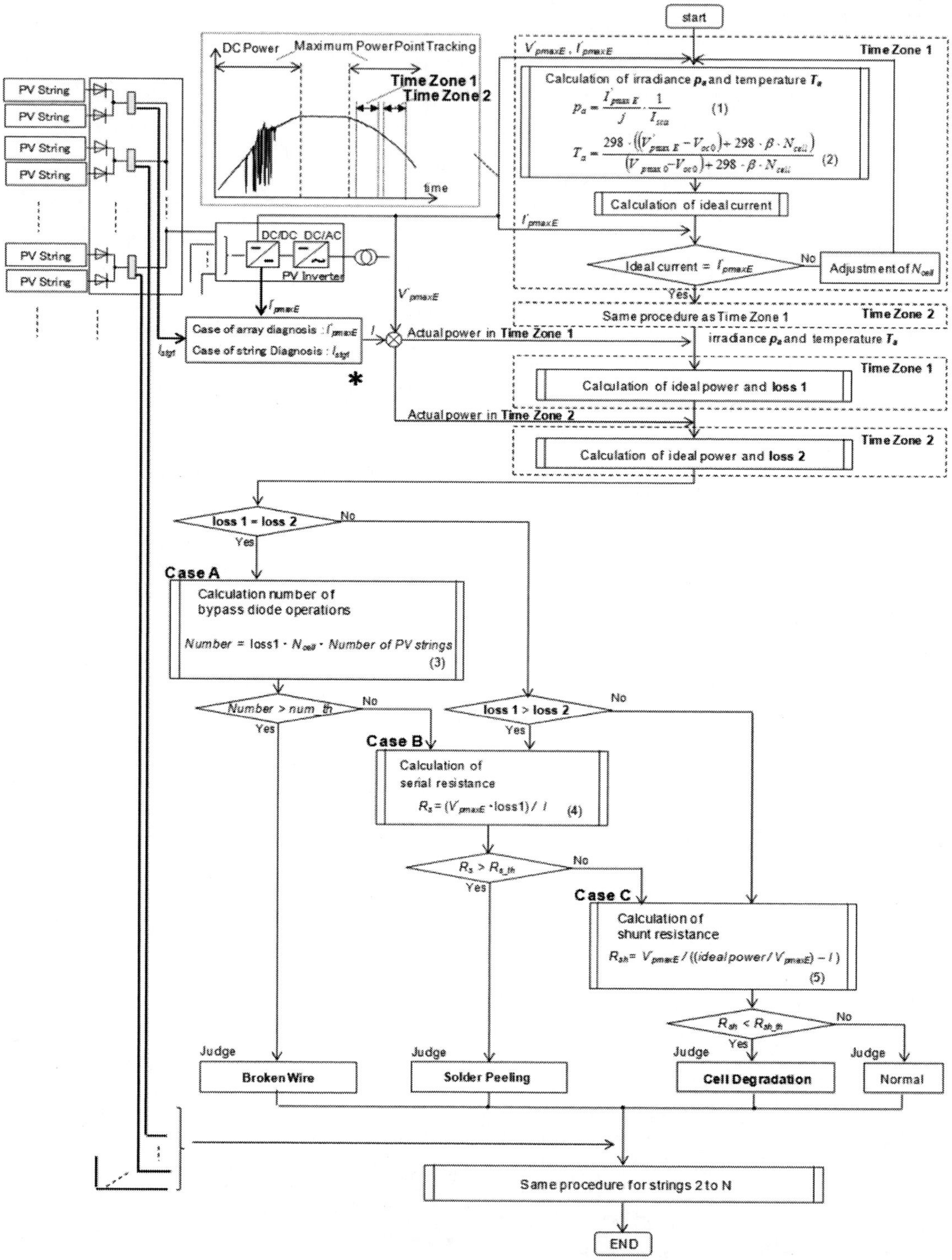

I_{sca}: Short-circuit current j: Predetermined constant V_{oc0}: Open-circuit voltage at 298 K

V_{pmax0}: Operating voltage at 298 K n_f: Diode Coefficient N_{cell}: Cell number of PV strings

k: Boltzmann constant β: Temperature coefficient of PV-cell's open-circuit voltage

Figure 5: Calculation flow for identifying fault mode

5 RESUTS

Figure 6 shows a diagnosis result for PV string including a PV module with glass breakage. The difference between actual current and other currents measured in string monitoring devices is apparently too low as shown in Fig. 6(a), it is thus difficult to judge whether this string has faulty module or not. On the other hand, the difference of actual power and ideal power that were obtained with procedure shown in Fig.5 is sufficient high to judge as shown in Fig. 6(b).

Actual fault detection results at the 82 MW power plant are shown in Table 2. The various faults (wire inserted to mimic "solder peeling", burnout in junction box, glass breakage, and shade of leaf) could be found in 105-hectares huge plant. These results show that the developed method can accurately determine "solder peeling" and "broken wires" even for a loss of only 0.7%. Moreover, at another power plant applied to array-diagnosis(* in Fig. 4), "cell degradation" for a loss of 5-6% was detected and void risking PV array developing into a more serious fault as like PID (Potential induced degradation).

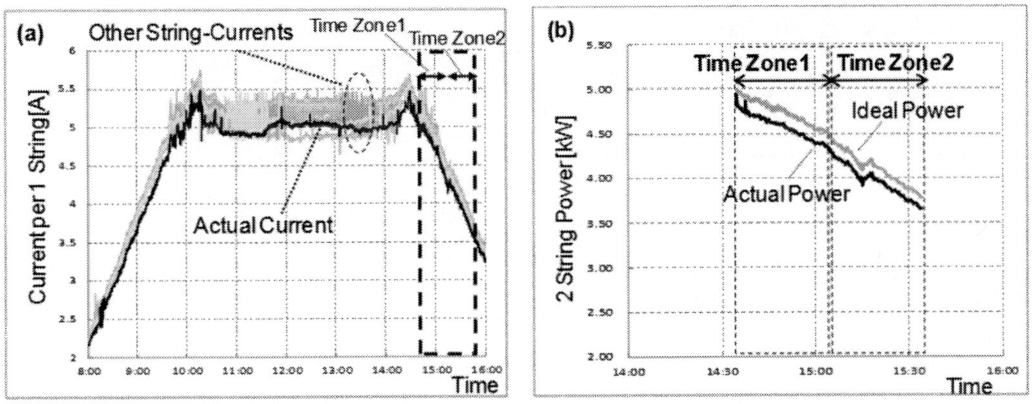

Figure 6: Diagnosis result for PV string including PV module with glass breakage

Table.2: Actual fault detection results at 82-MW power plant

Judge	Time Zone1 Ideal	Time Zone1 Actual	Time Zone2 Ideal	Time Zone2 Actual	Picture
Solder Peeling	**5.02** [kW] p_a 0.77 [kW/m²] T_a 23.7 [℃]	**4.94** [kW] **1.5%** loss	**4.69** [kW] p_a 0.69 [kW/m²] T_a 20.9 [℃]	**4.66** [kW] **0.6%** loss	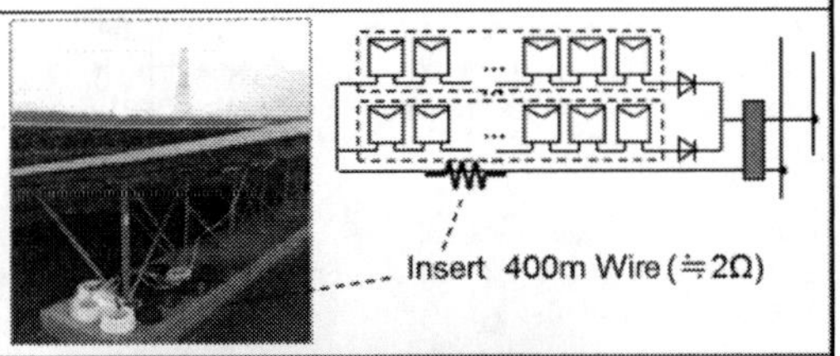
Broken Wire	**4.64** [kW] p_a 0.67 [kW/m²] T_a 44.3 [℃]	**4.44** [kW] **4.3%** loss	**2.97** [kW] p_a 0.43 [kW/m²] T_a 32.4 [℃]	**2.85** [kW] **4.1%** loss	 Short circuit of wiring caused by burnout in junction box
Broken Wire	**4.80** [kW] p_a 0.72 [kW/m²] T_a 20.4 [℃]	**4.71** [kW] **1.9%** loss	**4.69** [kW] p_a 0.69 [kW/m²] T_a 20.9 [℃]	**4.60** [kW] **1.9%** loss	
Broken Wire	**4.69** [kW] p_a 0.69 [kW/m²] T_a 30.9 [℃]	**4.66** [kW] **0.7%** loss	**4.18** [kW] p_a 0.61 [kW/m²] T_a 31.0 [℃]	**4.15** [kW] **0.7%** loss	

6 CONCLUSION

We developed a method to identify cause of failure and degradation and this method is added to the existing fault-detection system in 82 MW large-scale PV power plant. Through the operation of this large-scale PV power plants in service, we detected the various faults (wire inserted to mimic "solder peeling", burnout in junction box, glass breakage, and shade of leaf). Namely, the effectiveness of proposed method was confirmed. We will keep high-performance operation & maintenance service method by developed architecture in the future.

REFERENCES

[1] B.Cougo, et al, "A Photovoltaic System for Remote Fault Detection in Distribution Lines Using a Novel MPPT Algorithm", Industrial Electronics, 2007. ISIE 2007, pp 2397 - 2402

[2] T.Shimakage, *et al.*, "Development of Fault Detection System in PV System", INTELEC2013.

[3] K. Nishioka, *et al.*, "Evaluation of Output Performance of Various Photovoltaic Systems in the Hokuto Mega-Solar Project", INTELEC2010, 14.2.

[4] H.Koga et al, "A study Regarding Failure Detection for String Monitoring System in Photovoltaic – Field Trial in F Onomichi Photovoltaic Power Plant", IEEJ Transactions on Electronics, Information and Systems Vol.135 No.4, 2015, pp 396-402

[5] T.Kohno, K.Morikawa et al, "Experimental Verification of Fault-diagnosis Architecture in a Large-scale Photovoltaic Power Plant", 28th EU-PVSEC, proc, pp 3709-3713

[6] T.Kohno, K.Morikawa et al, "A Technique for Detecting Faults in a Photovoltaic Array", 27th EU-PVSEC, proc, pp 3610-3615

[7] T.Kohno, K.Morikawa et al, "High-Precision Method for Modeling I-V Characteristics of Photovoltaic Panel", 26th EU-PVSEC, proc, pp 3364-3368

INNOVATIVE SIMULATION TOOLS FOR AN EXHAUSTIVE AND SYNTHETIC CHARACTERIZATION OF THE SOLAR GLARE OCCURRENCES FOR THE DESIGN AND THE ADMINISTRATIVE INSTRUCTION OF LARGE-SCALE PHOTOVOLTAIC PLANTS

Christophe Vernay, Ana Maria Realpe, Daniel de Gabaï, Sébastien Pitaval
SOLAÏS
Immeuble Atlantis, 55 allée Pierre Ziller, 06560 Valbonne, France
cvernay@solais.fr

ABSTRACT: Photovoltaic (PV) is challenged to play a major role in the energy transition that must be conducted in the coming decade, through the development of large-scale PV plants. This emergence generates potential disturbances with the local environment starting with solar glare, i.e. the sun reflection on PV panels, which may jeopardize transportation safety. Some rare aviation administration already established restrictive policies for the construction of PV systems close to airports, such as the FAA in the United-States and the DGAC in France, but nothing yet for railways and motorways. These works, presented through a case study, aim at proposing to PV actors some innovative and generic simulation tools in order to address the glare problematic as part of their project development. The analysis of the glare is carried out in order to characterize the identified glare occurrences through an exhaustive set of physical quantities: occurring day and hour throughout the year, angle between reflected ray and user's field of view, etc. The overall analysis allows concluding on both the severity of the impact and the project compliance with the applicable requirements (e.g. aviation administration) while locating, within the finely-meshed PV plant, the impacting areas for which remediation must be considered.
Keywords: photovoltaic, glare, safety, simulation, characterization

1 INTRODUCTION

Photovoltaic (PV) is challenged to play a major role in the energy transition that must be conducted in the coming decade, through the development of large-scale PV plants, mostly ground-mounted, whose individual power now reaches hundreds of megawatt peak (MWp). As an illustration, Fig. 1 issued from PVresources.com, depicts the ranking of the 121 PV plants commissioned between 2010 and 2015 around the world, with an individual capacity exceeding 50 MWp. In 2010, only 6 plants were commissioned, with an average power of 73 MWp and a maximum of 97 MWp; their number reached 35 in 2015 with an average power of 151 MWp and a maximum of 575 MWp in California. On that period, Europe recorded 25 power plants with a maximum of 300 MWp in France.

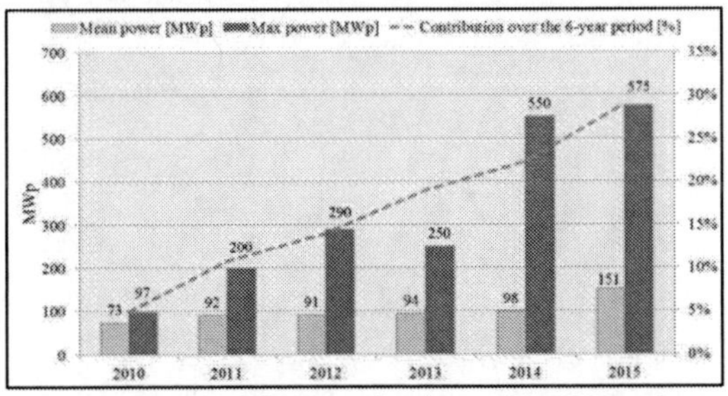

Figure 1: Breakdown of the 121 PV plants commissioned between 2010 and 2015 around the world, with an individual capacity exceeding 50 MWp.

The emergence of large-scale PV plants inevitably generates disturbances with the local environment, among which the sun reflection on the PV panels, also known as solar glare, whose consequences may vary from loss of visual comfort (e.g. residents) to the endangering of people in the case of transportation (e.g. airplanes, trains, cars). Besides maintaining good neighborly relations on the long term and preventing any legal action, PV developers shall consider the glare problematics early in the development process in order to successfully pass the administrative instruction, i.e. the building permit issuing process. Some high-level guidelines are already applicable depending on the country or the state. As an example, in Belmont, Massachusetts, a section of the law states that rooftop PV panels shall be placed and arranged such that glare shall not be directed onto adjacent buildings or properties [1]; other national laws indicate, for the same rooftop PV panels, a maximum time of exposure per day and per year. On the opposite side, complex policies were established by some aviation administrations for the construction of PV systems close to airports: the Federal Aviation Administration (FAA) in the United States or the Direction Générale de l'Aviation Civile (DGAC) in France, as detailed hereafter.

The FAA's opinion regarding the building of a specific PV generator is based on the assessment of the potential ocular hazard analyzed over the entire calendar year and defined as a function of the retinal irradiance and the subtended angle (size/distance) of the glare source [2]. Fig. 2 below defines three regions for the impact characterization: potential for permanent eye damage (retinal burn), potential for temporary after-image (flash blindness), and low potential for temporary after-image [3]. All potentials for glare in the airport traffic control tower (ATCT) cab is prohibited while only the low potentials for after-image are accepted regarding the final approach paths.

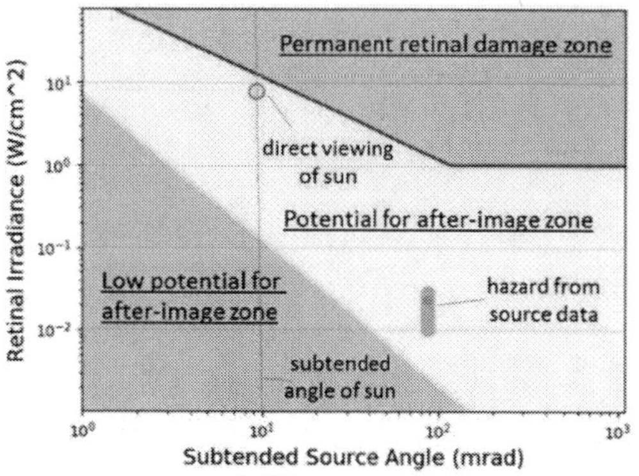

Figure 2: Impact of different retinal irradiances as a function of subtended source angle for short-term exposures.

In 2010, the French DGAC issued some rules for the impact characterization over control tower, the final approach path and the rolling of each runway, mostly based on the angle of view, i.e. the angle between the trajectory and the potential reflected ray, and the associated luminance ([4], [5]). The latter, expressed in candela per square meter (cd.m^{-2}), measures the luminous intensity per unit area of light travelling in a given direction; solar disk is about 16.10^{8} cd.m^{-2} at noon and 6.10^{5} cd.m^{-2} at sunrise [6]. Impacts over control tower are acceptable only when they are in the back of the air controller, whatever luminance is. Regarding the final approach and the rolling, Fig. 3 below describes three protection areas defined for each approach, namely A (around airport), B (around touchdown) and C (around runway): depending on the location of the PV project with respect to these three zones, the acceptance thresholds for the angle of view and the luminance vary (respectively ±30° or ±90° and 10^{4} or 2.10^{4} cd.m-2).

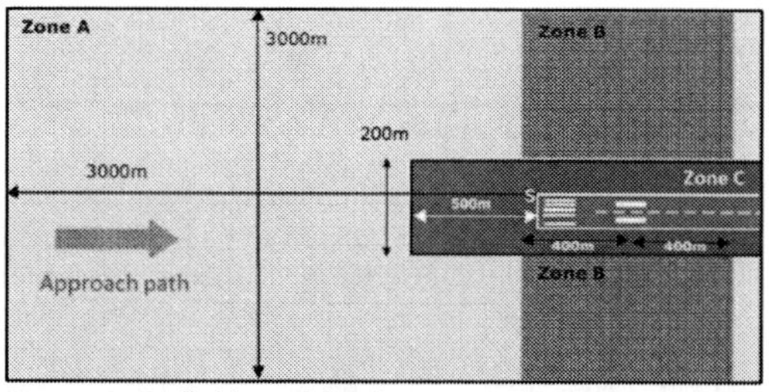

Figure 3: Definition of the three geographical zones A, B and C as per DGAC's requirements depending on the airport threshold (beginning/end of runway) and the touchdown point. The distances are different for a helicopter final approach and take off area (FATO) but the definition remains the same.

Finally, the DGAC systematically delivers a positive opinion to any project that implements glare-free PV panels, i.e. PV panels certifying a maximum luminance level of 10^{4} cd.m^{-2} for PV project in zone A or 2.10^{4} cd.m^{-2} for zone B.

Though quite different in the impact characterization, both the FAA and DGAC requirements are relevant references for the aviation administration outside the United States and France, but also for other administrative authorities that newly face PV projects close to sensible areas: networks of railways and motorways regarding security issues but also municipalities regarding urban constraints (e.g. co-visibility with historic monuments or neighbors). Considering the wide range of applications and targeted audience, the recent works carried out by SOLAÏS aim at proposing some innovative and generic simulation tools in order to address the glare problematic as part of the PV project development. The results are presented in this paper through a case study of a 1-axis tracking PV project close to a French airport, thus subject to the DGAC's requirements. Section 2 introduces the case study then presents the proposed modelling of the glare source (i.e. the PV panels) and the critical elements (e.g. final approach). Section 3 finally presents the identification and characterization of the glare impacts through a synthetic and visual approach for a better understanding of the results.

2 PROPOSED MODELLING

2.1 Introducing the case study

The case study considered in this paper was carried out by SOLAÏS in 2016 and consists in a ground-mounted PV plant using 1-axis tracking solution. It is located in Bretagne Region, south-west part of France, on the lands of the Bretagne Atlantique airport (also called Guiscriff - Scaër) whose main information is presented in Table I.

Table I: Information about Bretagne Atlantique airport

ICAO code	LFES	
Latitude	48° 3'4.96" N	
Longitude	3°39'55.74" W	
Landing strip	Paved, 1500 x 30 m	
1st magnitude orientation	24°	(QFU 02)
2nd magnitude orientation	204°	(QFU 20)

The 55-hectare footprint corresponds to the targeted cadastral parcels; it is divided into two areas, respectively 17 and 28 ha, located on both sides of the runway, as depicted in Fig. 4.

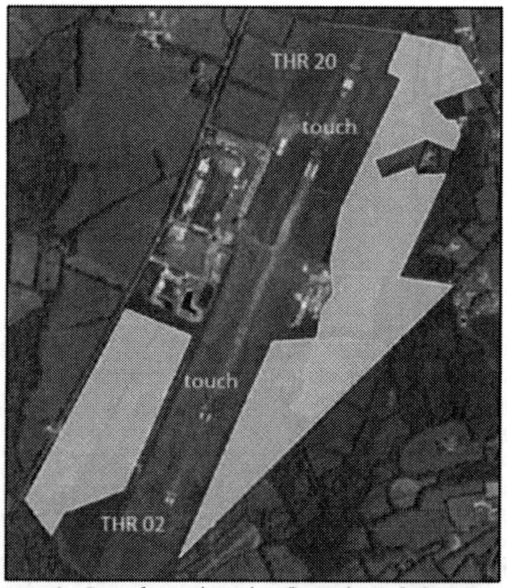

Figure 4: Superimposing the footprint of the ground-based PV plant (yellow) with the satellite image; thresholds (THR) and touchdowns are highlighted.

Considering the surface and the current state of PV panel efficiency, more than 40 MWp could be theoretically installed. A glare study must be conducted for the building permit examination in order to prove that the PV plant fulfils the DGAC's requirements.

2.2 PV plant mesh

A first challenge consists in defining a fine mesh of the PV plant in order to identify accurately the critical zones of the layout for which remediation should be potentially considered. For that purpose, the layout is firstly delimited by Jordan polygons, either concave or convex, which are defined as closed and simple, i.e. without intersection except the first and last points. An iterative process is then run in order to split each Jordan polygon into triangular facets whose maximum side length (L_{MAX}) can be changed down to ten meters. Fig. 5 below depicts the meshing for the case study using only two Jordan polygons (one per PV area), depending on L_{MAX} value.

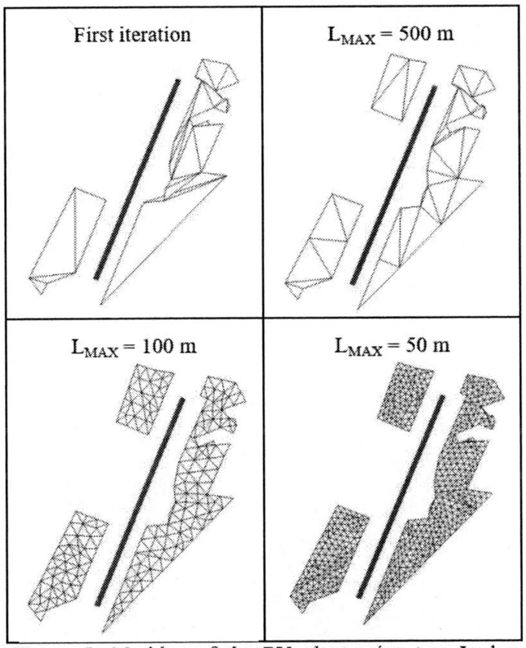

Figure 5: Meshing of the PV plant using two Jordan polygons depending on the maximum side length (L_{MAX}) of the triangular facets; runway is highlighted in purple.

It has to be noted that the size of the mesh, through L_{MAX}, doesn't have any influence on the overall results of the analysis because the latter only considers the intersection between a trajectory and a reflected ray for a given triangular facet as presented later in section 3. However, decreasing the mesh size allows refining the location of prohibited areas while increasing the computation time. This is all the more important when the PV plant layout intersects the protection areas defined by the French DGAC, cf. Fig. 6 for the QFU 20 approach.

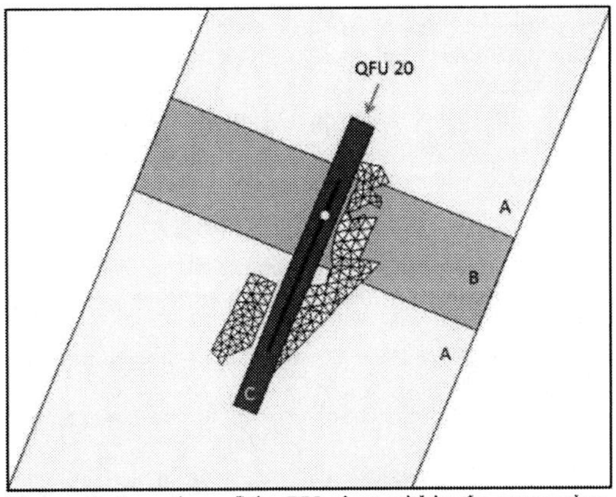

Figure 6: Location of the PV plant within the protection areas (A, B and C) as defined by the French DGAC.

2.3 PV panel modelling

The baseline considered in this study consists in classical crystalline PV cells encapsulated into a glass layer that protects them from the environment while allowing light energy to pass through. The disadvantage of this rigid outer layer is that part of the sun rays are reflected, thus generating glare, depending on the angle of incidence (AOI), i.e. the angle between the sun direction and the normal of the PV panel. The simplest approach to estimate the loss was developed in [7] and later adopted by the American Society of Heating, Refrigeration and Air Conditioning (ASHRAE, [8]) through the incident angle modifier (IAM) and the ASHRAE modelling defined by Eq. 1 and illustrated in Fig. 7. These profiles highlight the fact that PV panels behave as simple mirror for low incidences.

$$\text{(Eq. 1)} \quad \mathrm{IAM} = 1 - b_0\left(\frac{1}{\cos(\mathrm{AOI})} - 1\right)$$

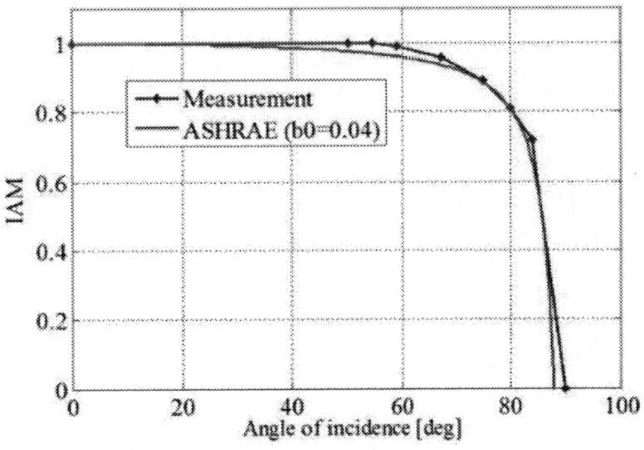

Figure 7: IAM profile of glass-protected PV panels based on ASHRAE modelling (b0=0.04) and measurements of a specific PV panel performed by Sandia National Laboratories.

In the present study case, the PV panels are mounted on a one-axis tracking system whose kinematic profile aims at maximizing the incident energy throughout the day. The rotation axis is horizontal and oriented north-south. A backtracking algorithm is implemented so as to remove the mutual shading, from one row to another one, at the sunrise and sunset. The tool allows processing the kinematic provided by any tracker manufacturer. However, in the early stage of a PV project development where manufacturer hasn't been selected yet, SOLAÏS proposes a generic tracking model depending on the "dihedral solar angle" defined as the dihedral angle between:

- The PV panel plane when the rotation angle is null;
- The plane containing the rotation axis and the sun direction.

Working with the dihedral solar angle allows working on a single profile, whatever the day is. Three different modes can then be considered, as depicted in Fig. 8:

1 Backtracking mode at sunrise and sunset to minimize the mutual shading;

2 Underperforming tracking due to the constraint on rotation angle;

3 Perfect 1-axis tracking around solar noon.

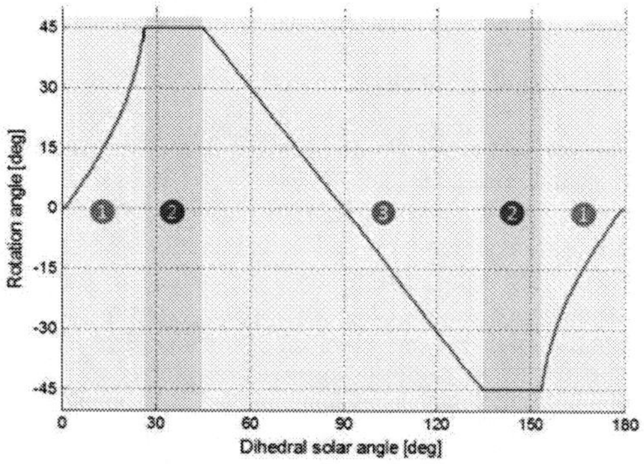

Figure 8: Daily kinematic profile of the 1-axis trackers depending on the dihedral solar angle, highlighting three different modes: backtracking (1), maximum rotation angle (2) and perfect 1-axis tracking (3); applicable whatever the day in the year.

Table II below presents the input parameters considered in the model where the PV panel thickness is used to slightly delay the rotation of the PV panels at the sunrise (symmetric for the sunset).

Table II: Input parameters for the tracking system of the case study

Distance between axes	4.5 m
Width of PV arrays	2.2 m
PV panel thickness	0.05 m
Azimuth of the axis	180° (north/south)
Inclination of the axis	0° (horizontal)
Maximum rotation angle	45°

3 CHARACTERIZING THE GLARE

A glare impact occurs when an incident sun ray, reflected on the surface of the PV panel, enters a user's visual field. A user can be either static (e.g. air controller, residents) or dynamic (e.g. pilot, train driver) and its localization is defined in the tool by using at least one segment; curved trajectories are then approximated using many segments. The impact identification consists in determining for each triplet {triangular facet, segment, sun position} the potential intersection between the segment and the infinite triangular prism derived from the reflection of the incident sun ray on the PV panels located in the triangular facet, as depicted in Fig.9.

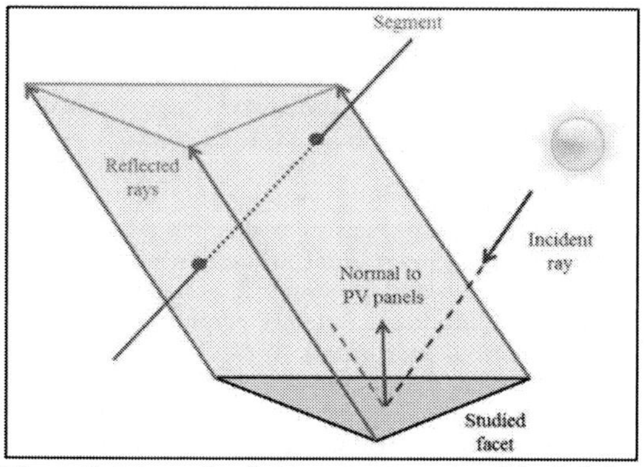

Figure 9: Intersection between a segment and the infinite triangular prism derived from the reflection of the incident sun ray on the PV panels located in the triangular facet.

The results for the specific case of the final approach coming from the north (i.e. QFU 20) are presented in two synthetic views. Fig. 10 first presents the geographical identification of the critical facets (in orange) for which at least one glare impact has been found. This 2D view provides useful information about the location of the critical areas for which glare has been identified (about 30 % of the total area). Regarding the development process, it allows PV developer balancing different trade-offs: abandoning the critical areas versus accepting another PV panel configuration with lower performance.

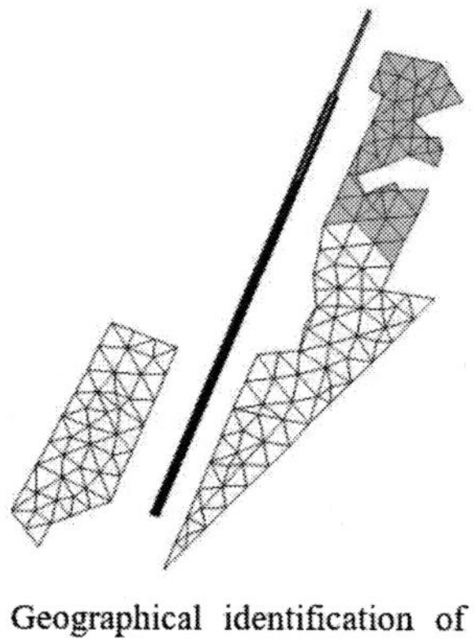

Figure 10: Geographical identification of the critical facets.

Then, a second chart characterizes the identified impacts throughout the year using true solar time (TST) on x-axis and day in the year on y-axis (Fig. 11); symmetric black curves corresponds with sunrise/sunset while topography may appear if present. The available physical quantities are the angle between the reflected ray and the user's visual field, the luminance of the reflected ray, the sun elevation, the angle between the reflected ray and the sun, the distance between the user (e.g. the pilot) and some critical locations (e.g. touchdown, railway crossing), etc. For this specific simulation, Fig. 11 show that a pilot approaching on QFU 20 is impacted by glare early in the morning (precise hour are available), during 2 to 34 minutes, between end of October and mid-February (accurate date are available), with an angle between user's field of view and reflected ray higher than 68°.

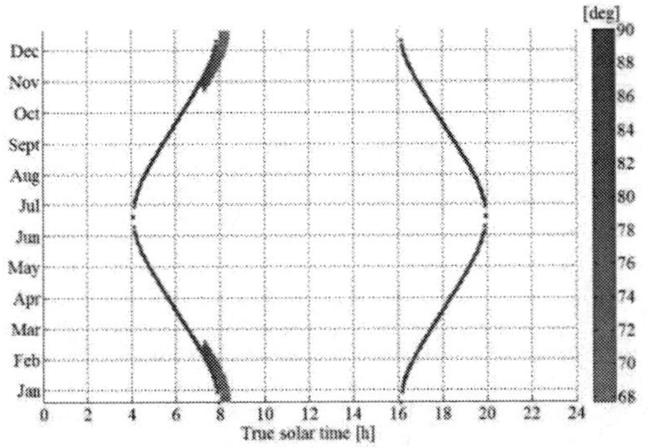

Figure 11: Characterization of the identified impacts through available physical quantities, here the angle between the reflected ray and the user's visual field is analyzed.

Finally, because the glare analysis is issued for people who may not have the adequate technical background (e.g. municipal councilor, administrative authority), another effort was made on the visual synthesis through the 3D envelop of the reflected rays embedded into Google Earth, as depicted in Fig. 12. This envelop, only applicable to fix structures, is defined by the direction of the reflected rays throughout the day for both summer and winter solstices; any trajectory that intersects this envelope is geometrically impacted by glare.

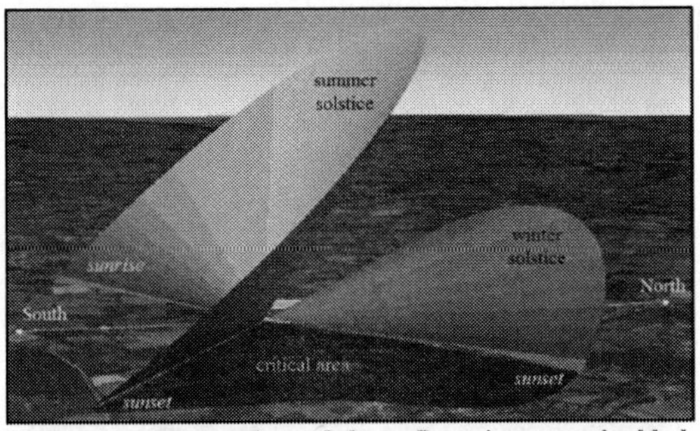

Figure 12: 3D envelop of the reflected rays embedded into Google Earth; applicable to fix structures only.

4 CONCLUSION

The tools proposed in these works are based on SOLAÏS' large experience regarding the glare studies requested since 2010 by the French aviation administration (DGAC) to the PV project developer. The latter now benefits from visual and synthetic results which are valuable for both the design and the administrative instruction of the project (building permit issuing process) and which ease the reading and the understanding of such a complex topic. The latest version of the tools finally intends to be as generic and adaptive as possible in order to address other national policies and other risk assessment (e.g. PV plants close to railways and motorways) while minimizing the study costs for the developers.

5 BIBLIOGRAPHY

[1] Colton, R., "Assessing rooftop solar PV glare in dense urban residential neighborhoods: determining whether and how much of a problem". Fisher, Sheehan and Colton Public Finance and General Economics, White paper, 2014.

[2] Aylward, A., Brecht-Clark, J., Farley, A., Hu, P., Ishihara, D., Johns, R., Lang, S., Partridge, E., Schmitt, R., Womack, K., "Interim policy, FAA review of solar energy system projects on federally obligated airports". FAA Register, Vol. 78, No. 205, October 2013.

[3] Ho C., Ghanbari, C., Diver, R., "Methodology to Assess Potential Glint and Glare Hazards From Concentrating Solar Power Plants: Analytical Models and Experimental Validation". ASME. J. Sol. Energy Eng. 2011;133(3):031021-031021-9. doi:10.1115/1.4004349.

[4] Verdier B., Théry P., "Dispositions relatives aux avis de la DGAC sur les projets d'installation de panneaux photovoltaïques à proximité des aérodromes". Note d'information technique de la DGAC. Révision 4 du 27/07/2011.

[5] Direction générale de l'aviation civile, "Gêne visuelle liée aux panneaux solaires implantés à proximité d'aérodromes". Rapport d'étude. August 2013.

[6] http://www.schorsch.com/en/kbase/glossary/

[7] Souka A.F., Safwat H.H., "Determination of the optimum orientations for the double exposure flat-plate collector and its reflections", Solar Energy vol .10, pp 170-174. 1966.

[8] ASHRAE Standard 93-2010, Methods for Testing to Determine the Thermal Performance of Solar Collectors.

PV-BATTERY AND DIESEL HYBRID SYSTEM FOR IRRIGATION OF A FARM IN PATAGONIA

Raphael Knecht[1]*, Franz Baumgartner[1]
[1]ZHAW, Zurich University of Applied Sciences, School of Engineering, IEFE
www.zhaw.ch/~bauf, Technikumstr. 9, CH-8401 Winterthur, Switzerland
*phone: +41589344576; email: kner@zhaw.ch

ABSTRACT: An existing off-grid irrigation system of a 12'000 ha farm in Patagonia today powered by electrical pumps and diesel gensets, will be extended by a PV plant and a battery system to improve ecological ratings of the products and reduce energy costs. An optimal photovoltaic plant and a battery energy storage system had to be designed. The compiled hourly demand profile served as input to a simulation model of a photovoltaic diesel battery hybrid system. With the given assumptions the PV array should be oriented to north at an inclination of 30°. The analysis of electricity cost indicated the optimal system size of 1800 kWp PV nominal power and 500 kWh of battery capacity. With this system 55.2% of the energy used for irrigation will be provided by the photovoltaic plant in the first year. The electricity cost amount to $0.136 per kWh electrical energy compared to the $0.432 per kWh electrical energy currently produced by the diesel-only system. Further analysis showed high dependency of the electricity cost on the time until connection to the utility grid and its electricity price and future diesel price.
Keywords: Off-Grid, Battery Storage and Control, Hybrid, Islanding, PV Pumping

1 INTRODUCTION

POMCO (Patagonia Organic Meat Company) owns a farm of 12'000 ha in Patagonia, Argentina where it develops fertile land along the Rio Negro river to cultivate different crops and cattle. These crops need artificial irrigation measures for optimal yield. The irrigation pumps have to be fed by means of off-grid solutions. POMCO installed a quantity of electrical water pumps which are connected to diesel generators, since the company expects a connection to the utility grid by 2022. These diesel generators are manually started when water pumping is required. With their goal to reduce diesel consumption and the runtime of the diesel engines POMCO aims to install a PV plant with battery storage hybrid system. The diesel generators will be centralized and a local grid will be installed in 2017. The PV plant, diesel generators and battery storage will go into operation in 2018 and will feed the irrigation pumps through this local grid. A central controller has to turn on/off the diesel generators according to power demand, State of Charge of the battery and solar irradiation.

The highest profit in terms of relation of investment costs and diesel savings is achieved by optimally dimensioning of the PV array and the battery. The main parameters which need to be determined are:

- Nominal DC-Power of PV Array [kWp]
- Azimuth and inclination of PV Array [°]
- Battery Capacity [kWh]
- Battery Inverter Power [kVA]

1.2 Approach

The software Polysun [1] was found to be best suited to simulate the system with the parameters mentioned before since it allows multiple PV arrays with different angles and azimuth. Furthermore it was found more straightforward to simulate multi-year scenario compared to the widely used software HOMER [2] or PVSyst [3].

After generating the demand profiles for the pumps a model of the hybrid system was created. Several simulations in Polysun and further analysis in Matlab provided a selection of system dimensions as a basis to decide upon the system design. Apart from technological feasibility also the economic viability was an important factor to optimize the system dimension.

While the hybrid energy generation means less diesel fuel consumption it also means relatively high investment costs. The cost of electrical energy as an average of the combination of diesel and PV electricity generation costs and the expected future price for energy bought from the utility served as the main indicator of system performance and economic viability.

2 SIMULATION PARAMETERS

In order to facilitate the simulation and due to constraints of the simulation software, model assumptions had to be made. The assumptions comprise the transformation of the demand profile as well as the simplification of the electrical performance of the diesel generators, battery inverter and control system.

2.1 Demand Profile

POMCO aims to develop new crop land year by year. With increasing crop land the demand for irrigation and thus electrical energy rises as well. Table I summarizes the planned peak power and the yearly energy demand for the next five years.

Table I: Planned peak power and yearly energy demand of the irrigation pumps for the next five years.

Year	Peak Power [kW]	Yearly Energy Demand [MWh]
2017	425	1253
2018	649	1842
2019	768	2384
2020	887	2987
2021	1024	3476

Due to lower temperatures and solar irradiation and thus lower evaporation in winter the irrigation demand is reduced from April to October. The energy demand from April to October makes up around 10% of the yearly energy demand (see Figure 1). The solar irradiance follows this demand profile perfectly (compare Figure 1 & Figure 2).

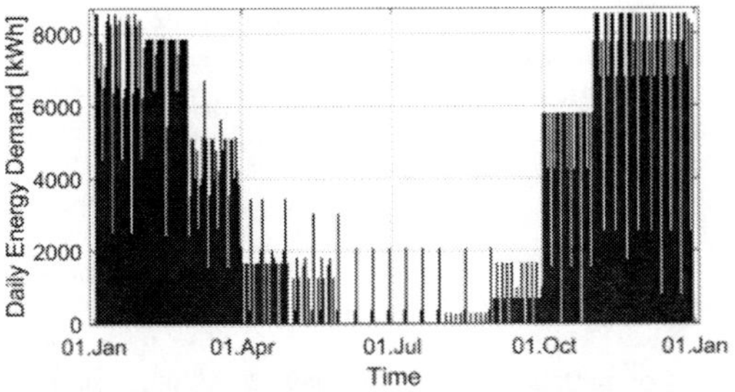

Figure 1: Daily sum of the energy demand of the irrigation pumps in kWh in 2017.

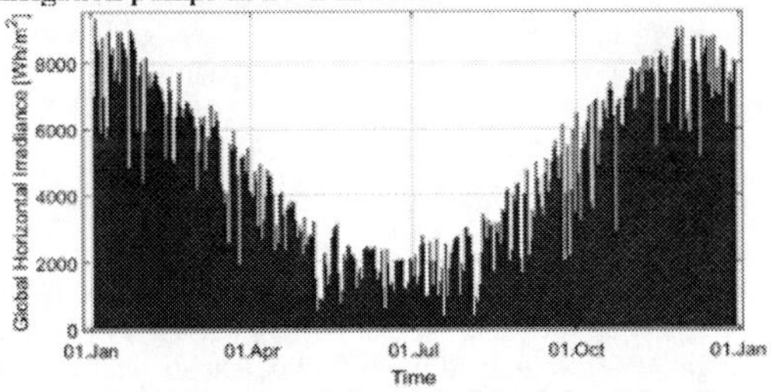

Figure 2: Daily sum of the global horizontal irradiance [Wh/m^2] as provided by the Meteonorm [4] profile of Neuquen (250 km North-West of the farm).

2.2 Simulation Model

The simulation software Polysun is a tool to primarily calculate energies of thermal and electrical components over a longer period (usually a whole year). Dynamic electrical phenomena and component transients in the duration range of a few miliseconds to a few minutes happening e.g. during the starting of a diesel generator or current harmonics are not the focus of the software. It also regards electrical components as purely resistive loads or sources. Consequently Polysun does not differentiate between apparent, reactive and real power. The power factor of the generators and pumps is not taken into account in the simulation. Polysun calculates only with active power. The demand profile is resolved in one-hour steps. This simplification leads to a neglection of peak power demands due to large inrush currents in the AC motors of the irrigation pumps.

The model behind the simulations in Polysun is illustrated in Figure 3.

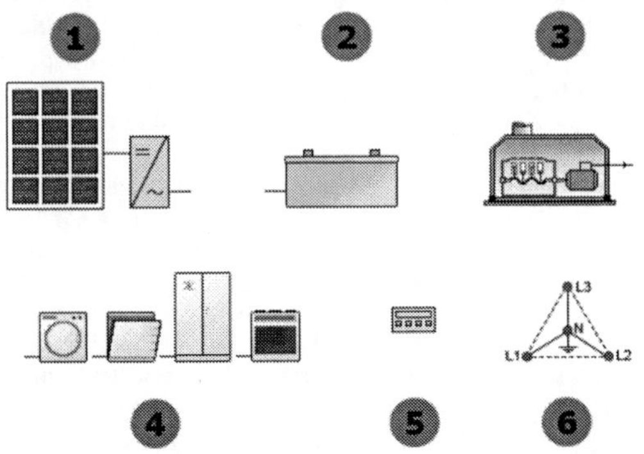

Figure 3: Schema of the model designed for the simulations in Polysun. Its components are: (1) PV Array and PV inverter, (2) Battery and Battery inverter, (3) Diesel Generator, (4) Load Profile, (5) programmable controller, (6) internal low-voltage grid.

The micro grid central controller (MGCC) is modelled by the programmable controller. A flow chart of the control logic is depicted in Figure 4. If the power demand of the irrigation pumps is lower than the power output of the PV array, the battery can be charged if not fully charged yet. On the other hand, if the power demand exceeds the current PV output, first the battery is discharged and after reaching a State of Charge (SOC) threshold of 33% the remaining power demand will be fed with the diesel gensets.

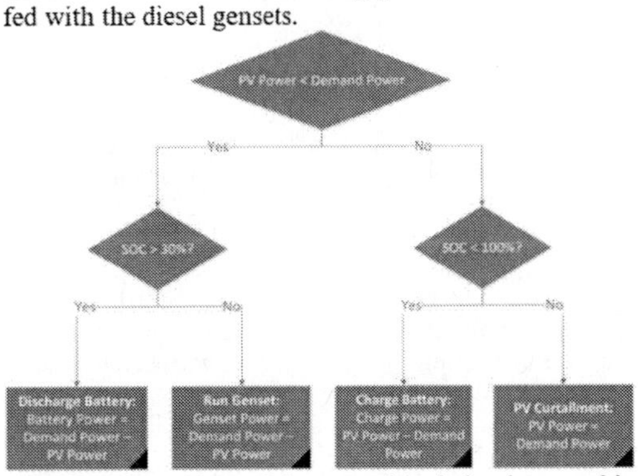

Figure 4: Flow chart of the control logic implemented in the Polysun model.

2.3 Electricity Cost

The common indicator for economic viability of renewable energy generation is the levelized cost of electricity (LCOE) [5][6]. Since the farm is expected to be connected to the utility grid, the LCOE does not represent the actual cost of electrical energy and the comparability to other renewable energy generation plants remains inaccurate. To get a more accurate approximation of the future electrical energy cost and also a more representative decision variable for investing, the weighted mean between LCOE and price of the energy supplied by the future utility grid was calculated:

$$ElectricityCost = \frac{I_{pv}+I_{batt}+I_{genset}+\sum_{t=1}^{20}\frac{M_t+F_t+E_{in,t}\times Pr_t}{(1+r)^t}}{\sum_{t=1}^{20}\frac{E_t}{(1+r)^t}}$$

Where:

I_{pv} : $CapEx\ of\ PV\ in\ the\ year\ t$

I_{batt} : $CapEx\ of\ Battery\ in\ the\ year\ t$

I_{genset}: $CapEx\ of\ Gensets\ in\ the\ year\ t$

M_t : $OpEx\ of\ PV, Battery\ and\ maintenance\ cost\ of\ Gensets\ in\ the\ year\ t$

F_t : Diesel fuel costs in the year t

$E_{in,t}$: Electrical energy fed into utility grid (negative) or supplied by the utility grid (positive)in the year t

Pr_t : Electrical energy price from utility in the year t

E_t : Total Electrical energy consumed in the year t

r : $Discount\ rate$

The following assumptions were made and were included in the calculation:

- Overall system life time of 20 years
- Replacement of battery after 10 years
- Replacement of Diesel Generators after 10 years
- Connection to utility grid in 2022
- Constant electrical energy price from utility (Pr_t) of $0.05 per kWh
- Constant diesel fuel costs (F_t) of $0.41 per kWh$_{el}$

- CapEx of $750.- per kWp for PV & BOS [7]
- CapEx of $1000.- per kWh for Battery, Battery inverter, MGCC, etc.
- CapEx of $370.- per kVA for Diesel Generators [8]
- OpEx of 2% of investment costs p.a. for PV and Battery
- Maintenance cost of 5% of investment costs p.a. for Diesel Gensets
- Constant discount rate (r) of 5% [9]

3 OPTIMAL SYSTEM DIMENSIONS

Applying the Formula in sub-section 2.3 reveals that the electrical energy cost for the diesel-only system without connection to the utility grid, amount to $0.432 per kWh. As mentioned in section 1 it seems economically viable to partly substitute the diesel-energy generation with PV energy, especially since the diesel price in the future is expected to rise. Consequently the most economically viable solution will hereinafter be identified.

3.1 Rough estimate of PV Power

It can be estimated that the optimal PV array size is for a yearly total PV production in the order of the yearly electricity demand. The theoretical nominal operating hours for a PV array inclined at 30° and oriented to north amount to 1680 h. Hence, for the purpose of a rough estimate the PV array size can be calculated:

$$W_{PV,yearly} = W_{demand,yearly} = 3476\ MWh$$

$$P_{n,PV} = \frac{W_{pumps,yearly}}{t_{nominal}} = \frac{3476\ MWh}{1680\ h} \approx 2000\ kWp$$

3.1. PV Nominal Power and Battery Capacity

As discussed in Section 1 the battery storage needs to be able to maintain the grid frequency in its specified boundaries by compensating for high ramp rates of the PV generator. These high ramp rates manifest mainly due to moving clouds and can only be limited in the upward direction by the PV inverter. The downward ramp rate thus has to be compensated by increasing the battery inverter output (see Figure 5: Time = 1s) to keep the grid frequency stable. For persistent coverage of the sky the diesel gensets have to be started (see Figure 5: Time = 11s) to make up for the PV power loss.

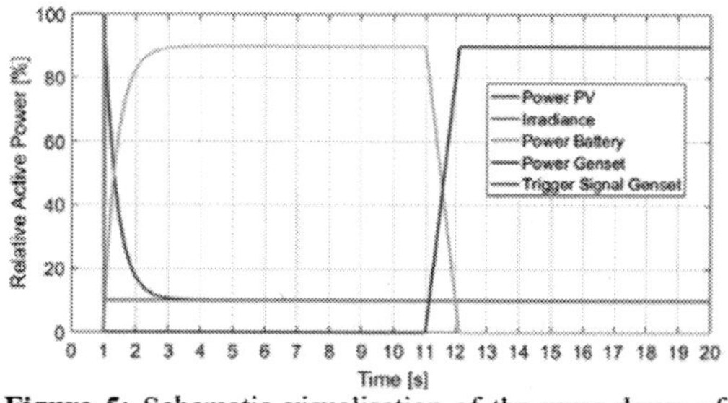

Figure 5: Schematic visualisation of the ramp-down of the PV generator (blue) from 100% active power to 10% active power (irradiance (red) decreases to 10% e.g. due to cloud covered sky) within two seconds. The battery (yellow) immediately has to compensate for the generation loss until the generator has started (purple).

The minimum required battery capacity depends on the maximum discharge power of the battery cells (see next paragraph) and on the spinning reserve to be maintained to cover for PV power fluctuations. This on the other hand depends on the delay time for starting up the diesel generators and the maximum power demand. It has to be assumed that the maximum spinning reserve equals the maximum power demand over 10 seconds, when a start time of the genset of 10 seconds is assumed [10].

Commercially available battery storage solutions feature maximum discharge rates of 0.5-3C [11]. The simulations were performed with batteries of maximum discharge rates of 1C and battery inverters of the same maximum power, hence the battery capacity required for the first year of operation must not be lower than

$$Capacity_{min} \geq \frac{P_{max}}{1C} = \frac{425kW}{1\frac{1}{h}} = 425kWh$$

The next bigger capacity value simulated is 500 kWh. It has to be noted that typical battery storage systems in this range are contained in 20 foot containers and feature fixed costs for the control system, switchgear, transformer, air conditioning and fire prevention. The battery system price thus is not proportional to the battery capacity as assumed in sub-section 2.3 due to fix costs of transport, engineering and installation. The electricity cost (see sub-section 2.3) for the simulated systems is depicted in Figure 6. The simulations performed comprise PV nominal values ranging from 100 kWp to 4000 kWp in 100 kWp steps and battery capacities ranging from 0 kWh to 4000 kWh in 100 kWh steps.

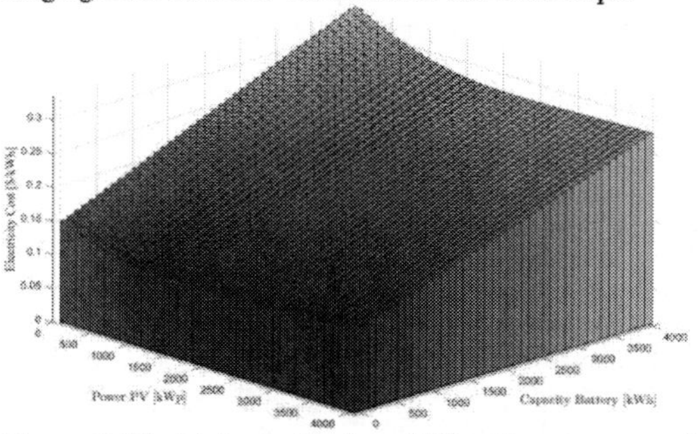

Figure 6: Electricity cost in $ per kWh at twenty years of amortisation. Each bar represents a simulated value of electricity price (Z-Axis) at specific PV nominal power (X-Axis) and battery capacity (Y-Axis). The chosen system size according to Table II is circled in red.

Figure 6 shows that the lowest electricity price of $0.117 is with a nominal PV power of 1700 kWp and without a battery. This system does not meet the goal of using the battery as a grid forming unit as required in Section 1 though. The lowest electricity price for systems with battery capacities ≥ 500 kWh of $0.136 is the system with 1800 kWp PV power and 500 kWh battery capacity. The PV array size is in the same order as the rough estimate of 2000 kWp in sub-section 3.1. Henceforth, these values will be chosen as optimal system sizes. Increasing the PV nominal power has a much lower impact on the electricity price than increasing the battery size.

An indicator of how much diesel is saved is the share of PV energy on the total energy demand as shown in Figure 7. The share of PV energy with the system as described above (1800 kWp, 500 kWh) in 2017 is expected to be 55.2%. Hence, roughly half of the energy needed for irrigation is produced from renewable PV energy. Due to increasing energy demand the PV share is expected to gradually decrease in the next few years if the PV array is not expanded (see Table II).

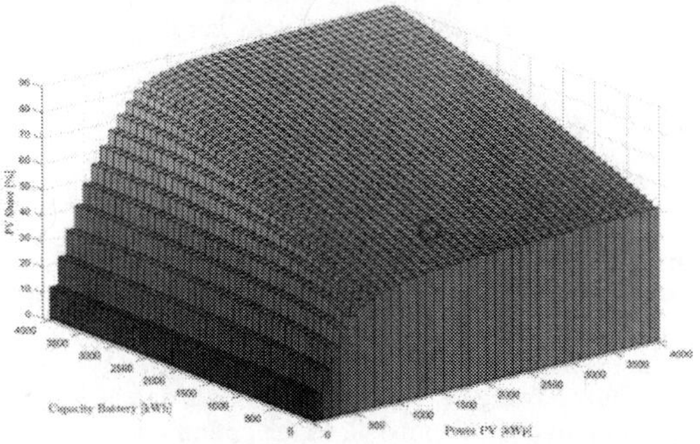

Figure 7: Share of PV energy on the total yearly electrical energy demand of 2017 in percent. Note that the X-/Y-axes are switched compared to Figure 6 for better readability. The chosen system size according to Table II is circled in red.

Summarized in Table II are the optimal system sizes and their corresponding performance indicators as discussed in this chapter.

Table II: Optimum system parameters, its electricity cost and PV Shares as explained in sub-section 3.1

Quantity	Value	Unit
PV Nominal Power	1800	kWp
Battery Capacity	500	kWh
Electricity Cost	0.136	$\frac{\$}{kWh}$
PV Share 2017	55.2	%
PV Share 2018	52.0	%
PV Share 2019	51.0	%
PV Share 2020	48.6	%
PV Share 2021	46.8	%

3.2 Sensitivity Analysis of Electricity Cost

Due to the assumptions made in sub-section 2.3 a sensitivity analysis on the electricity cost is performed. Figure 8 shows that the time until connection to the utility grid, the diesel fuel price and the energy price from the utility, have the highest impact on the electricity cost, whereas the CapEx and OpEx of PV plant, battery storage and gensets have a low impact. Since the electricity cost as charged by the utility ($0.05/kWh) is way below the LCOE, the averaged electricity cost rises for increasing time until connection to the utility grid. It has to be expected that the optimal system size changes with changing input values.

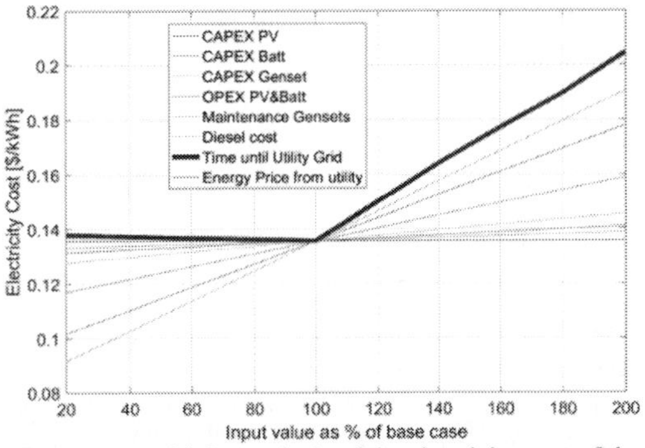

Figure 8: Sensitivity analysis of the electricity cost of the hybrid system (see Table 4.1) for its input values according to Equation 3.1 in percent deviation of its base value. The time until connection to the utility grid (ruby red) and the diesel energy price (turquoise) has the highest impact.

The dependency of the electricity cost on the time until connection to the utility grid is indicated in Figure 9. The electricity cost of the system discussed would increase to $0.299/kWh if the farm were connected to the utility grid only in 20 years (see Figure 9). The gradient of the electricity cost in direction of battery capacity in Figure 6 is much steeper than in Figure 9 since the battery does not contribute to the PV share of the total energy after connection to the utility grid.

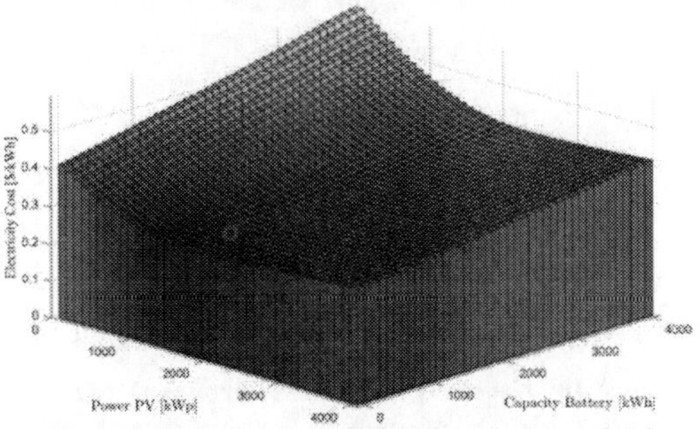

Figure 9: Levelized Cost of Electricity in $ per kWh at 20 years of amortisation without the connection to the utility grid. The chosen system size according to Table II is circled in red.

3.3 Battery Inverter Power

As discussed in the previous sub-section 3.1, as a worst case scenario it has to be assumed that all loads are simultaneously switched on when the PV-inverter output is zero. Only with this scenario considered a stable operation of the grid is possible. The discharge power of the battery inverter has to be higher than this peak power demand. The highest expected peak power according to Table I is 1024 kW. It has to be checked with the battery cell manufacturer if the resulting peak power discharge rate is compliant.

3.4 PV Array Inclination and Azimuth

The goal of optimizing the inclination and azimuth usually is to maximize the PV output over the year (Approach A). Since this PV plant currently is not utility grid-connected and hence cannot feed in excess electrical energy and in particular does not receive feed-in incentives, this is not the only solution.

Another approach is to maximise the fuel saving of the diesel generators (Approach B). Since the farm is expected to be connected to the public grid in a few years, it has to be assessed which approach is preferable. The simulation series which aimed to narrow down the optimum comprised 360° azimuth values from -180° (North) to +180° in steps of 30° and inclination values from 0° (horizontal) to 45° in 5° step values. The option for two PV arrays of identical nominal power but with different orientations was as well simulated. The nominal power of the PV array was chosen to be 1 MW since it is incidental to the optimization of the inclination and azimuth.

Figure 10 shows the relative theoretical PV yield over the first year as deviation from the maximum value of all inclinations and orientations in percent if the PV array were grid-connected i.e. all PV energy could directly be used or fed into the grid (Approach A). The theoretical yearly yield is maximised at -180° (North) for both arrays and elevated to 30°. The PV yield is not increased by using two differently oriented PV arrays (e.g. East-West orientation) and thus is not displayed in Figure 10.

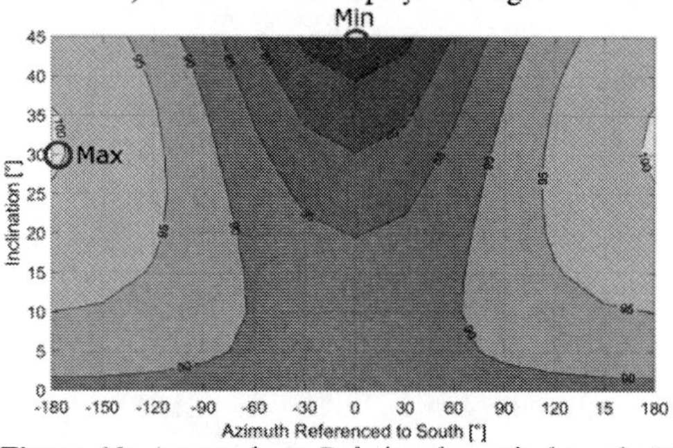

Figure 10: Approach A: Relative theoretical yearly PV yield (in % deviation from minimum value) plotted for each inclination from 5° to 45° (Y-Axis) and with azimuth values from -180° to +180° (X-Axis). The minimum value is at inclination 45° and azimuth 0° (South), the maximum value at inclination 30° and azimuth ±180° (North).

The value to maximise in Approach B is the fuel saving compared to the purely diesel generated electricity. Since this not only depends on the solar irradiation but also on the demand profile, the optimal inclination and orientation is different to Approach A. As a simplification only the demand values of 2017 were used. The error due to this simplification is marginal since the distribution of demand mostly remains the same in the future years while only the amplitude increases. Here, the maximum value was found with two differently oriented PV arrays, namely at an orientation of -90° and 90° respectively and at an elevation of 45°. Its magnitude is 7.2% higher than with the orientation of Approach A.

Since, in the scenario outlined by POMCO, the utility grid is expected to be connected to the farm in by 2022 and thus the PV plant is able to feed in excess energy for the major part of its life time, the Approach A has a lower electricity cost of $0.136/kWh compared to $0.141/kWh of Approach B. Approach B on the other hand has the advantage of more evenly distributed PV power and thus lower peak generation, which relieves the electrical grid. It can be expected that in the future the feed-in tariffs during peak PV production (e.g. lunchtime) are lower than during e.g. morning and evening hours.

4 CONCLUSION

The cleanest and most economical way to reduce fuel costs for the irrigation of the farm in Patagonia is to install a large-scale photovoltaic plant and a battery storage system. In order to maximise the fuel savings thorough analysis and simulations of the current and future demand of the irrigation were conducted.
For the following key data assumptions had to be made in order to simplify the model and the optimisation:

- Future demand in magnitude and time
- Future diesel price
- Time until connection to utility grid and its electricity price
- CapEx of PV system
- CapEx of battery

Due to the averaging of the electricity cost of off-grid generation in the first years with the cost of electricity of the utility the comparability with the LCOE of other hybrid systems is deceptive. Thus an average of LCOE and energy price from the utility grid was chosen as the main indicator for economic viability.

The optimal size of PV array and battery according to the estimations of sub-section 2.3 is 1800 kWp and 500 kWh. The gradient of the electricity cost in direction of higher PV nominal power is relatively low, thus the optimal size of the PV plant is much more flexible than the battery capacity. This system is expected to yield costs of $0.136 per kWh of electrical energy. In the first year 55.2% of the electrical energy will be provided by the PV plant whereas the rest needs to be covered by the diesel gensets. This PV-share on the total electrical energy demand will decrease marginally to 46.8% in the following five years due to increased energy demand.

It could be shown that the impact of delay in connection to the utility grid, volatility of the diesel price and the electricity cost of the utility have a high impact on the electricity cost, whereas the impact of changes in CapEx and OpEx is relatively low. Comparison of Figure 6 with Figure 9 highlights the dependency of the time until connection to the utility grid on the optimal battery size. Later examination of proposals for this hybrid system revealed that specific costs of especially the battery system decrease with increasing system size. This is due to the fixed costs of the containerized solution (Inverters, transformer, air-conditioning, fire prevention system, etc.) and the control as well as commissioning, engineering and economy of scale. Before commissioning of a system as discussed it is imperative to enquire concrete costs of a PV system and a battery in order to accurately describe the energy price.

Apart from technical feasibility and economical optimisation requirements to aesthetics, ease of installation and to the environment have to be respected during layout and dimensioning of the hybrid system. Two seperate PV arrays with different azimuth orientation, as proposed in Approach B of sub-section 3.4, possibly necessitate higher amount of labour input and land requirement. Since the farm is expected to be connected to the utility grid in five years, the optimal orientation is north with an inclination of 30°. The orientation of East-West with an inclination of 45° results in a <4% higher electricity cost and thus is equally interesting.

REFERENCES

[1] Vela Solaris AG, "Polysun," *Polysun Designer & Professional - Polysun Simulation Software*. [Online]. Available: http://www.velasolaris.com/english/product/polysun-designer-und-professional.html.

[2] Homer Energy, "Homer Energy," *HOMER Pro - Micrigrid Software for Designing Optimized Hybrid Microgrids*. [Online]. Available: http://www.homerenergy.com/HOMER_pro.html.

[3] PVsyst SA, "PVsyst," *PVsyst Photovoltaic Software*. [Online]. Available: http://www.pvsyst.com.

[4] Meteotest, "Meteonorm-Software," *Meteonorm*. [Online]. Available: http://www.meteonorm.com.

[5] Aishwarya S. Mundada, Kunal K. Shah, and J.M. Pearce, "Levelized cost of electricity for solar photovoltaic, battery and cogen hybrid systems," *Renewable and Sustainable Energy Reviews*, no. 57 (2016), pp. 692–703, 07-Jan-2016.

[6] K. Branker, M.J.M. Pathak, and J.M. Pearce, "A review of solar photovoltaic levelized cost of electricity," *Renewable and Sustainable Energy Reviews*, no. 15 (2011), pp. 4470–4482, 11-Sep-2011.

[7] National Renewable Energy Laboratory (NREL), "Distributed Generation Renewable Energy Estimate of Costs," Feb-2016. [Online]. Available: http://www.nrel.gov/analysis/tech_lcoe_re_cost_est.html. [Accessed: 05-Aug-2017].

[8] National Rural Electric Cooperative Association, "Costs of Utility Distributed Generators, 1-10 MW." Mar-2003.

[9] H. Al-Hammad, T. Becker, A. Bode, S. Gupta, and S. Kreibiehl, "Renewable Energy in Hybrid Mini-Grids and Isolated Grids: Economic Benefits and Business Cases," *Frankfurt School - UNEP Collaborating Centre for Climate and Sustainable Energy Finance*, Mai-2015.

[10] National Fire Protection Association (NFPA), "NFPA 110: Standard for Emergency and Standby Power Systems." 2016.

[11] Samsung SDI, "ESS Batteries by Samsung SDI." Mar-2017.

COMPARISON OF PERFORMANCE AND DEGRADATION OF DIFFERENT PV PLANT CONFIGURATIONS IN JOHANNESBURG, SOUTH AFRICA

T.J. Serameng[1, 2,*], K. T. Roro[3], E.E. van Dyk[2], J.L. Crozier[2], F.J. Vorster[2]
[1]ESKOM, Research Testing and Development, Private Bag X40175, Cleveland, 2022, South Africa
[2] Physics Department, Nelson Mandela Metropolitan University, P.O. Box 77000, Port Elizabeth, 6031, South Africa
[3]CSIR Energy Centre, Council for Scientific and Industrial Research, P O Box 395, Pretoria 0001, South Africa
[*]Author for correspondence: Phone: +27 11 629 5047, E-mail: SerameTJ@eskom.co.za

ABSTRACT: The main goal of this work is to investigate the impact of module technologies and mounting configurations on the performance of PV systems in central part of South Africa. In order to achieve this, a 400 kWp PV Solar plant has been installed and monitored at the ESKOM Research and Innovation Centre (ERIC) in Rosherville, Gauteng in the central part of South Africa (26°5'5" South, 28°58'1" East and 1625 m above sea level). The plant consists of multi-crystalline silicon (mc-Si) and copper indium gallium selenide (CIGS) thin film technologies. The plant comprises of two 25° fixed-tilt north facing arrays, two east-west single axis tracking arrays and two 10° fixed shed orientation (East and West) for both module technologies and one north-south single axis tracking for mc-Si. The analysis was conducted for one year, from January 2016 to December 2016. The evaluation was achieved by studying the operation of PV plant under different weather conditions and the impact of different configurations and technologies by analyzing mainly in terms of specific yields normalized per m^2 ($kWh/kWp/m^2$). The first part of the study investigates the impact on the overall performance of different PV technologies as a function of the module mounting configurations. The second part is ongoing and is aimed at gaining a better understanding of the degradation of PV-modules used by conducting standard indoor laboratory performance tests on modules.

Keywords: PV system, Performance, Analysis

1 INTRODUCTION

Photovoltaics (PV) are one of the key technology options to decarbonize how we generate energy which is the dominant contributor of the greenhouse gas emission [1]. Solar resources across the world are widely spread and abundant, hence, the energy from it cannot be monopolized by a single country. It is interesting to note that in contrast to conventional energy sources, PV and other renewable energies are the only ones to offer the prospect of a reduction rather than an increase in prices in future [2,3]. Solar PV systems can be deployed almost everywhere on this planet, since they are modular.

The optimization of solar PV electricity plant design and operation has direct effect on the operation and maintenance (O&M) costs. It is, therefore, clear that O&M costs play an important role for the economics of the PV installations. In order to capture more solar energy and optimize the electrical output from the PV system installers usually configure systems by adjusting the orientation and tilt of a system, or by using mechanisms that track the movement of the sun in the sky [7]. As a rule-of-thumb in the Southern Hemisphere, in South Africa for example, the annual energy production of a fixed system can be optimized by facing the module towards north and tilting at an angle equal to the local latitude. Several studies have been done to investigate the effects of orientation and inclination on the performance of PV systems in many parts of the world [8-18]. To the best of our knowledge, no study has been done that reported on the comparison of PV systems performance operating side-by side comprising different module technologies and mounting designs.

At a system level, integration of PV into a power system has many challenges. Among these, the matching between demand and supply is most common particularly when the penetration level of PV is high. This mismatch between supply and demand will cause significant power peaks. Sub-optimal inclination and azimuth angles of PV (e.g. East and West facing as opposed to the North facing) could be beneficial in this regard. Tröster and Schmidt [19] studied the impact of PV module orientation on grid operation by evaluating several mounting configurations for the city of Aachen in Germany. They found out that east-west facing systems have capacity factors not much lower than south facing systems, as well as power gradients and peak power outputs significantly lower, which has advantages for the electricity network and the integration of PV. The authors concluded that the optimal orientation depends on the priorities and the prevailing conditions in the specific location.

Due to the international drive for clean energy which has seen many countries participating in the renewable energy space and also due to the PV market growth globally and especially in South Africa over the past few years, it has become very important to compare, analyse and evaluate the performance of PV modules especially in actual South African climatic conditions. Performance monitoring of different module technologies and system configurations of Photovoltaic (PV) systems in South Africa are rare which result in few reports based on field results of PV systems installed and operated in South Africa. The ultimate goal of this work is to analyse the performance of PV systems, evaluate the energy output of various PV system configurations and module technologies in the South African climatic conditions. To achieve this, a 400 kWp PV Solar plant has been installed and monitored since January 2015 at the ESKOM Research and Innovation Centre (ERIC) in Rosherville, Gauteng, in the central part of South Africa. The plant consists of multi-crystalline silicon (mc-Si) and copper indium gallium selenide (CIGS) thin film technologies. The plant comprises of two 25° fixed-tilt north facing arrays, north-south single axis tracking, east-west single axis tracking and two 10° fixed shed East and West orientation for both module technologies. The analysis was conducted for duration of one year. The evaluation is achieved by studying the operation of the PV plant under prevailing weather conditions and the impact of different configurations and technologies on various systems overall energy output and other performance metrics.

2 ERIC SOLAR PV SYSTEM FACILITY

The ERIC Solar PV System facility is a 400kWp PV plant located at the Eskom Research and Innovation Centre (ERIC), Rosherville, South Africa. The site generally has a flat terrain and the facility occupies a total surface area of one hectare.

Figure 1: Aerial view of Eskom Solar PV facility in Rosherville, Johannesburg, South Africa

The PV plant utilizes ground mounted, multi-configuration PV systems comprising of crystalline and thin film technologies. It consists of four sections of 100 kWp of installed capacity each. Figure 1 shows an aerial photograph of the PV facility.

The various PV configurations are subject to the same environmental conditions in terms of temperature, radiation, humidity and wind speed. South Africa, in general, has an excellent solar resource providing a favourable condition for the development of solar PV projects. The facility has annual average global horizontal solar irradiation (GHI) of 2 400 kWh/m^2. The minimum and maximum monthly average temperature is estimated at 10.4°C and 21.7°C respectively. The long term monthly average wind speed is approximately 4 m/s. A detailed description of the facility can be found elsewhere [20].

The various PV configurations at the facility are listed in Table 1. In the table, crystalline refers to mc-Si and thin film to CIGS.

Table 1: ERIC PV system set-up, [20]

PV System Number	Orientation	Sub-orientation	Capacity (kWp)	Tilt Angle (°)
1	North facing fixed Crystalline (300 W mc-Si)		100.80	25
2	North facing fixed Thin film (145 W CIS)		102.66	25
3	Fixed E/W Crystalline	West Tilted Crystalline	33.60	10
		East Tilted Crystalline	33.60	10
4	Fixed E/W Thin film	West Tilted Thin film	17.11	10
		East Tilted Thin film	17.77	10
5	Tracking E/W Crystalline (Horizontal N/S axis)		33.60	-
6	Tracking E/W Thin film (Horizontal N/S axis)		33.60	-
7	Tracking N/S Crystalline (Horizontal E/W axis)		34.22	-

3 PERFORMANCE METRICS

The performance metrics used for the comparison are: final yield and solar gain.

3.1. Specific Yield

The specific yield was normalized ($Y_{f'}$) with respect to area and defined as:

$$Y_{f'} = \frac{E/P_{STC}}{A} \qquad (1)$$

where:

E is the total energy output of the PV system measured in kWh AC; P_{STC} is rated power of all modules evaluated at standard test conditions (STC: irradiance = 1 kW/m^2, cell temperature = 25°C, and air mass of 1.5); A is the array area in m^2.

3.2 Solar gain

The Solar gain of any configuration versus a North-facing fixed tilt is defined as:

$$S_{Gain} = \frac{S_x - S_{f,fixed}}{S_{f,fixed}} = \frac{S_x}{S_{f,fixed}} - 1 \qquad (2)$$

where:

S_x is the monthly solar insolation for any configuration;

$S_{f,fixed}$ is the monthly solar insolation a fixed-tilt (25°), North facing system.

4 PERFORMANCE RESULTS AND DISCUSSION

4.1 Analysis of the measured solar insolation

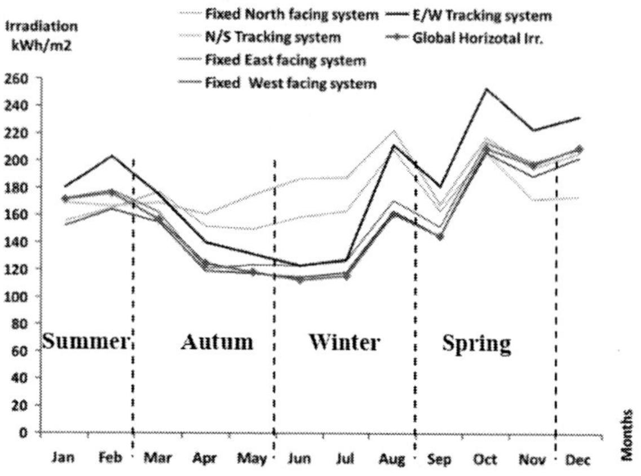

Figure 2: Monthly Solar insolation for the different configurations during the monitored period

Figure 2 shows the monthly variation of solar insolation for different configurations, with the insolation on the various array planes being as expected when compared to the GHI. The monthly global horizontal insolation varies from 112 kW/m^2 to 210 kW/m^2. Johannesburg, South Africa has highest GHI in the months of October and December, whereas the lowest during the months of May to July. It is interesting to note that the solar insolation for the month of September shows a decrease. This is most likely due to the high spring rainfall.

4.1.1 Monthly solar gain

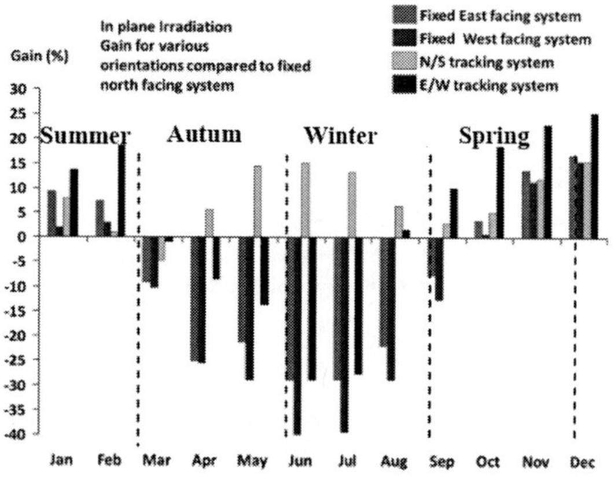

Figure 3: Monthly percentage change in irradiance for different configurations compared to fixed tilt.

Figure 3 depicts the monthly solar gain of different configurations compared to fixed optimum tilt system. It is evident from the figure that the north-south tracking system gained more solar compared to north facing fixed tilt configuration (except in March). The east-west tracking configuration gained in spring and summer, however, during autumn and winter the gain is negative. The east and west facing fixed systems only show a gain during the months from October to February. In other months, they show a loss.

The solar gain also shows expected trends, with overall annual gains for EW and NS tracked planes, respectively. The fixed East and West tilted configuration do not show an annual gain, but are included in the study to illustrate the utilization of such planes for deploying PV.

Figures 4, 5, 6 and 7 show the various typical daily irradiance profiles for the Solstices and Equinoxes in 2016. The figures illustrate the influence of the different configurations and orientations on the radiation incident on the plane of array. In general, these irradiance profiles are expected, but serve to illustrate the potential value of different configurations. The anomaly observed for spring and autumn around 15h00 is due to cloud cover that is typical for the Gauteng Highveld during these seasons.

The clear advantage of E-W tracking is observed for all seasons, with the customary broadening of the irradiance profile, except winter due to the low altitude angle of the sun. In addition, around solar noon the modules are horizontal and this low solar altitude compounds the lost irradiance. This is observed as the dip in irradiance around solar noon (Figure 6).

As expected, the N-S tracking should always yield a positive advantage in summer and winter, while at equinoxes there is no advantage over the fixed north facing configuration.

The fixed E and W sheds with a tilt angle of 10° also show profiles that may have an advantage under certain conditions. Figures 4 and 6 illustrate the expected offset in peak irradiance and that if systems with these configurations were to operate in tandem, there is a broadening of the irradiance profile without the use of tracking. In summer this advantage is clear, and the respective irradiance profiles are better than that of the fixed (latitude-tilted) north facing orientation. This is due to the low tilt and high altitude of the sun. This is, however, not the case in winter.

Figure 5 and Figure 7 show the typical spring and autumn equinox irradiation patterns.

From this brief analysis, the solar resource incident on the various fixed and tracked surfaces provides insight into potential PV energy output that can be expected. In addition, monitoring the solar resource also facilitates the validation of predicted energy yield from the plant, as well as optimizing plant modelling for Eskom.

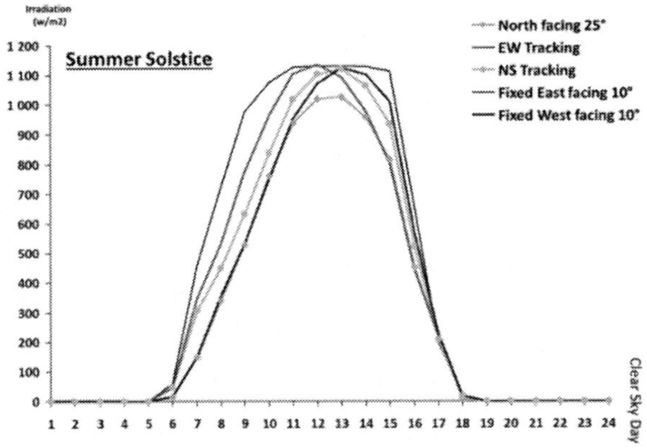

Figure 4: Clear sky day radiation received in summer

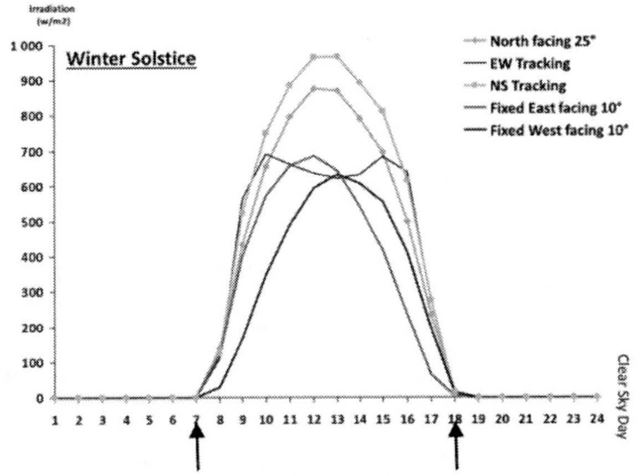

Figure 6: Clear sky day radiation received in winter

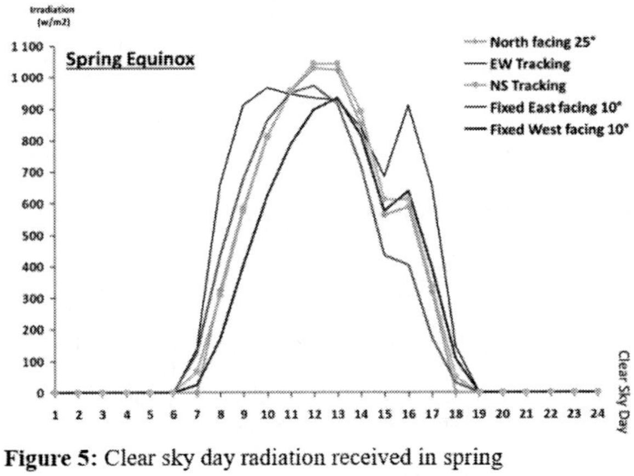

Figure 5: Clear sky day radiation received in spring

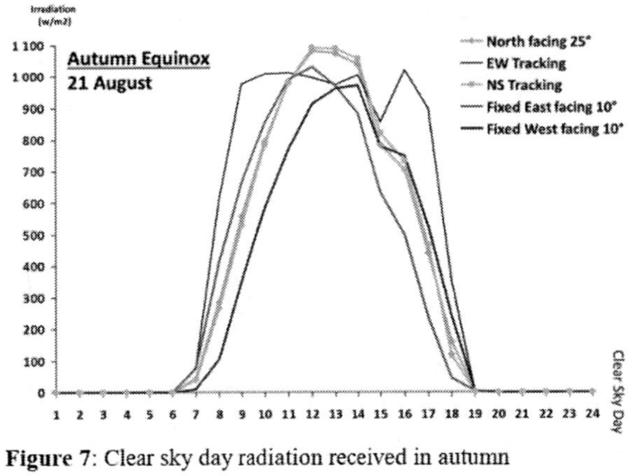

Figure 7: Clear sky day radiation received in autumn

4.2 Comparison of performance

When considering PV for different applications one of the critical factors that determine the suitability is the cost of energy. The life-time cost of a PV system is determined by its specific yield and the capital as well as the O&M cost. In order to make a suitable choice of technology and mounting configuration, a comparison was made using the normalized specific yield per m^2 (in kWh/kWp/m^2) in this study. This gives insight into possible structural costing.

4.2.1 Mounting configurations

Figure 8 depicts the monthly normalised yield values for various module mounting configurations used in this study. As expected the seasonal variations play a critical role. It is seen in the figure that during spring (Sept-Nov) and summer (Dec –Feb) months the tracking systems (east-west & north-south) performed better than the north facing fixed system. However, the north facing fixed system performed better than the east-west facing fixed system. In the autum (March- May) and winter (July-Aug) months, the north facing fixed system performed better than the east-west tracking and east-west fixed systems and lower than the north-south tracking system.

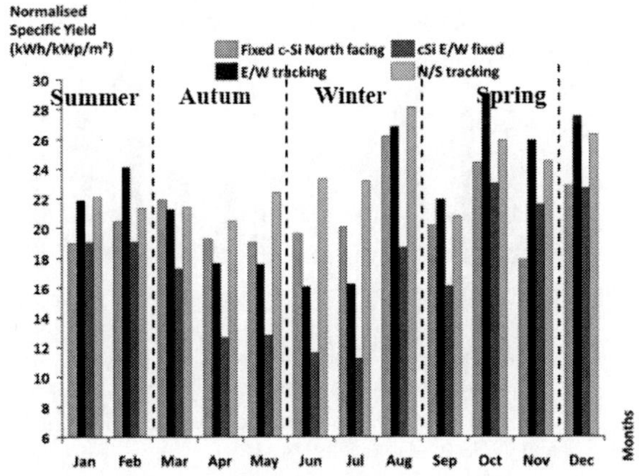

Figure 8: Monthly normalised yield of the different mounting configurations of mc-Si modules

4.2.2 Module technology

Figure 9 shows the monthly normalised yield of the PV systems that used mc-Si and CIGS thin film module technologies in a north facing fixed tilt configuration. It can be seen from the figure that throughout the year (except for November), the normalized specific yield values for the system that used crystaline modules is higher than the one that used the thin film. Normally, during the cost calculation of generating energy from PV, many elements add up to arrive at the total cost in their life time. One of the very important factors is the cost of modules. In some cases thin film modules may be cheaper per W_p than crystalline Si modules, however, thin film modules require larger areas in order to meet certain capacity requirements. Therefore, possible cost saved in procuring the modules is offset by higher structural and installation costs.

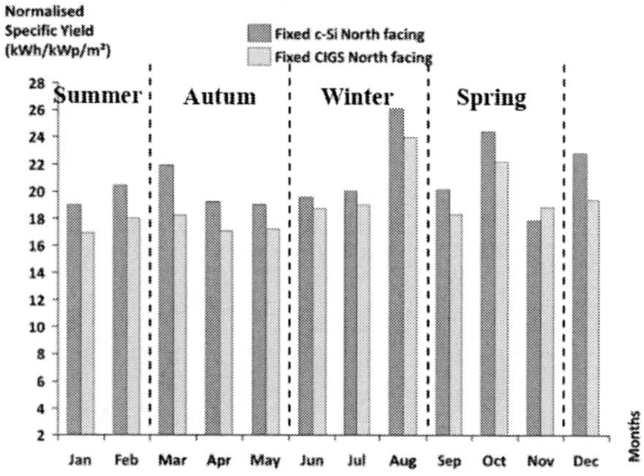

Figure 9: Comparison of monthly normalized specific yield of fixed mc-Si north facing and fixed CIGS north facing modules

5.2.3 Tracking

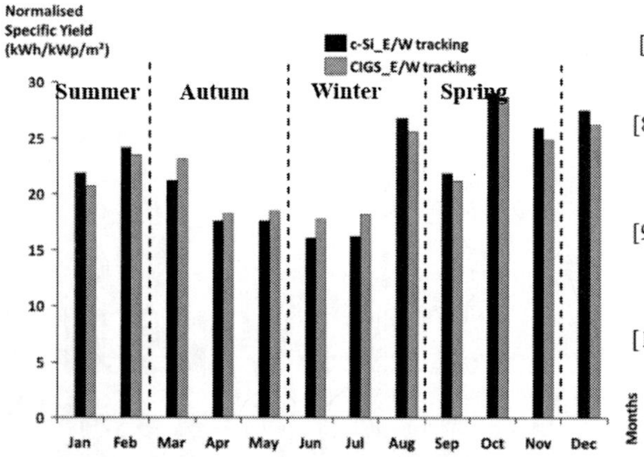

Figure 10: Monthly normalized specific yield of mc-Si and CIGS tracking modules

Figure 10 depicts the monthly normalized specific yield of the different PV technologies mounted on east-west tracking systems. It is shown in the figure that the tracking system that used the mc-Si performed better during the spring (Sep-Nov) and summer (Dec-Feb). However, during spring (March-May) and most of the winter months (June-July) the system that used the thin film technology performed better.

5 CONCLUSIONS

The performance of the 400 kW_p experimental PV plant installed at Rosherville, South Africa was evaluated for one year from January to December, 2016. The results that have been analysed thus far show the influence of the mounting configuration and technology type on the energy performance during a year. From the preliminary results it would appear that each technology type would require a different optimum mounting configuration that will depend if the energy yield must be optimised annually or monthly.
It is intended to find "best practices" in the overall PV system design and possible guidelines for future more extensive research and deployment of PV plants by ESKOM.

6 REFERENCES

[1] Malvoni, M., Leggieri, A., Maggiotto, G., Congedo, P. M., & De Giorgi, M. G. (2017). Long term performance, losses and efficiency analysis of a 960kW P photovoltaic system in the Mediterranean climate. Energy Conversion and Management, 145, 169-181.

[2]Arnulf Jäger-Waldau, PV status report 2016; EUR 28159 EN; doi:10.2790/682995

[3] World Energy Perspective – Variable Renewable Integration in Electricity Systems: How to get it right published by the World Energy Council 2016 with support from CESI Spa, Italy.

[4] IRENA (2016), 'Letting in the Light: How solar PV will revolutionise the electricity system,' Abu Dhabi.

[5] Snapshot of Global Photovoltaic Markets – IEA PVPS (2016)

[6] IRENA (2017), REthinking Energy 2017: Accelerating the global energy transformation. International Renewable Energy Agency, Abu Dhabi.

[7] https://www.eia.gov/todayinenergy/detail.php?id=18871

[8] J.D. Mondol, Y.G. Yohanis, B. Norton, The impact of array inclination and orientation on the performance of a grid-connected photovoltaic system, Renew Energy, 32 (1) (2007), pp. 118–140

[9] J.D. Mondol, Y.G. Yohanis, B. Norton, Optimising the economic viability of grid-connected photovoltaic systems, Appl Energy, 86 (7–8) (2009), pp. 985–999

[10] M. Hartner, A. Ortner, A. Hiesl, S. Nicoara Maximaler Ertrag vs. Kostenminimum: Der Einfluss der Ausrichtung von PV-Modulen auf den Marktwert und die Systemkosten Konferenz Energieinnovation/29. Symposium Photovoltaische Solarenergie Bad Staffelstein (2014) [English title: Maximum yield vs. minimum costs: the influence of the orientation of PV modules on the market value and the system cost]

[11] Waldmann, R. Bhandari ergleichende Analyse der technisch-wirtschaftlichen Bedingungen von PV-Anlagen Mit Süd- und Ost-West-Ausrichtung Z

Energiewirtschaft, 38 (2014), pp. 27–36 [English title: Comparative analysis of the technical and economic conditions of PV systems with south and east-west orientation]
[12] T. Huld, R. Gottschalg, H.G. Beyer, M. Topic, Mapping the performance of PV modules, effects of module type and data averaging, Sol Energy, 84 (2) (2010), pp. 324–338
[13] E. Tröster, J. Schmidt, Evaluating the impact of PV module orientation on grid operation, Energynautics GmbH, Darmstadt (2012), pp. 1–6
[14] M. Suri, T. Cebecauer, A. Skoczek, Solar electricity production from fixed-inclined and sun-tracking c-Si photovoltaic modules in South Africa, Proceedings of the 1st Southern African Solar Energy Conference (SASEC 2012), 21–23 May, Stellenbosch, South Africa. (2012)
[15] O. Asowata, J. Swart, C. Pienaar, Optimum tilt angles for photovoltaic panels during winter months in the Vaal Triangle, South Africa, Smart Grid Renew. Energy, 3 (2012), pp. 119–125
[16] T. Matshoge, A.B. Sebitosi, The mapping of maximum annual energy yield azimuth and tilt angles for photovoltaic installations at all locations in South Africa, J. Energy South. Afr., 21 (4) (2010), pp. 1–6
[17] W.G. Le Roux, Optimum tilt and azimuth angles for fixed solar collectors in South Africa using measured data, Renew Energy, 96 (A) (2016), pp. 603–612
[18] J. Kern, I. Harris, On the optimum tilt of a solar collector, Sol. Energy 17 (1975) 97-102.
[19] E. Tröster, J. Schmidt, Evaluating the impact of PV module orientation on grid operation, Energynautics GmbH, Darmstadt (2012), pp. 1–6
[20] Tshepo Serameng, Kittessa T. Roro, Manjunath Basappa Ayanna, Kumaresan Cunden, Senzo Myeni, Tobias Bischof-Niemz, E.E van Dyk, Stephen Koopman. EU PV SEC proceedings: June 2016.

ECONOMIC ANALYSIS OF PHOTOVOLTAIC POWER PLANT IN TURKEY

Abdullah Bugrahan Karaveli*[a], Bulent G. Akinoglu[abc] and Ugur Soytas[ad]
[a] *Earth System Science Program, Graduate School of Natural and Applied Sciences, Middle East Technical University, 06800, Ankara, Turkey*
[b] *Department of Physics, Faculty of Arts and Sciences, Middle East Technical University, 06800, Ankara, Turkey*
[c] *Solar Energy Research Center (GUNAM), Middle East Technical University, 06800, Ankara, Turkey.*
[d] *Department of Business Administration, Faculty of Economic and Administrative Sciences, Middle East Technical University, 06800 Ankara, Turkey*

*Corresponding author:
Tel. +90 532 510 7354, *E-mail address*: bugrahan.karaveli@metu.edu.tr;
[1] bulo@metu.edu.tr
[2] soytas@metu.edu.tr

ABSTRACT: Together with environmental and social contributions, its economic benefits owing to sharp decrease in costs of system components has increased preference of producing electricity from solar energy. Furthermore, strong financial support mechanisms such as feed-in-tariff scheme increase its economic feasibility and contribute its competitiveness against alternatives. Moreover, having high solar potential that is huge raw material soars the attractiveness of solar energy, *inter alia*. Consequently, accurate economical analysis of solar energy investments in countries having above mentioned features like Turkey gain utmost importance for the utilization of the potential in an appropriate manner. In that vein, this study examines the economical feasibility of a typical 1 MW PV PP in Turkey by defining various electricity pricing scenarios created in this study after designing a sample system by using own methodology authors of this article have been developed. Pursuant to results, 1 MW PV PP is able to earn around 2.7-6.3 M $ under the circumstances of the Turkish market. This amount pays the cost back in 3.5-6.6 years. Moreover, the LCOE of this system is 62 $/MWh for 8% discount rate and 50 $/MWh for 5%.
Keywords: Solar Energy, solar economy, photovoltaics, feed-in-tariff, day ahead market

1 INTRODUCTION

Turkey is a solar belt country [1] whose solar irradiation potential is 1311 kWh/m^2 in average according to SEPA (Solar Energy Potential Atlas) developed by the Ministry of Energy and Natural Resources of Turkey [2]. However going from north to the south, the potential multiplies around 1.5 fold [2]. Turkey's current PV installed power is 1361.4 MW according to the official results by TEIAS (Turkey Electricity Transmission Company) as of the end of June [3], while the cumulative CSP power is just 1.2 MW pursuant to EMRA (Energy Market Regulatory Authority) [4]. The total solar power corresponds to 1.7% of Turkey's total installed power that is 80343 MW [3]. Together with related legislation to subsidize the PV system installation both in a small roof top and utility scale ground-mounted, Turkey has feed-in-tariff scheme to support electricity production via PV and CSP systems as defined by Renewable Energy Law No. 5346.

Consequently, the investment environment to install PV systems is formed already in the country via legislation and incentives. However, it is obvious that the economical process and market variables are dynamic and needs some estimation that is not easy to be made concerning future legislative mechanisms and prices/supports. In that vein, this study questions the economic feasibility of PV systems installed in a location selected in Turkey according to estimation scenarios developed by the authors of this study for revenue analysis. In other words, this study aims to find out the economical feasibility of a photovoltaic investment in countries like Turkey and its competitiveness against its alternatives.

In order to attain the goal to do list is given as follows: A certain location is selected and photovoltaic system is simulated on that location. After defining the input (solar radiation), designing the system and determining the output (solar electricity), economical analysis is made by calculating all the related parameters to make correct analysis on its economy.

In detail, initially the solar radiation the location of interest is exposed which corresponds to the input of the system is calculated by using the methodology derived by the authors of this study. Then, the electricity production of the system that is the output is computed by utilizing the other methodology derived by the authors. In parallel with input and output calculations, the best design to maximize input and output done by selecting the best fitting technologies and the best design of the system.

After defining the output, we convert this output into economical asset under the revenue scenarios of the system written by the authors after defining the initial and life time cost of the system till dismantling. While defining these revenue scenarios, we have made estimations/projections about the Day Ahead Mechanism price, an electricity price reflected to the end-user, feed-in-tariff policies of the government and feed-in-tariff amounts in the future.

The methodologies used throughout the process mentioned above and derived by the authors will be used as the basis of a software to be created under the name of SOlAR TURnKEY by the authors of this study. The model used in solar irradiation calculations are described in detail in Ref [5-9]. The model for electricity calculation is the modified and innovated version of current models that are utilizing mainly 5-parameter model and efficiency and cell temperature relation. For the economy analysis this manuscript is the first and unique source that the methodology is described and presented.

Consequently, after making general introduction,

Section 2 explains the technical parameters and design steps of the system to be installed on the selected location. Section 3 is for the economical analysis of the system. Then, Section 4 compiles and discusses the results.

2 1 MW PV PP SYSTEM IN TURKEY

This part calculates all the related parameters to design and install the system by taking permission period into account depending on Turkish legislation.

2.1 Selection of a location

The location to install the system is selected as Develi town in Kayseri province where we have measured data to make the comparison of our calculations and where the second highest PV installed power is hosted after Konya province. Other reasons for this selection are the potential of the economy of the city and interest of local investors to the PV market.

2.2 Solar radiation

In order to be able to design the system in the appropriate manner and make reliable economic feasibility analysis of the PV, the initial information to know is the input of the system that is solar irradiation that a point of interest is exposed.

Consequently, in this part of the study we estimate total solar radiation falling on the tilted surface (on the module). The calculations are made in a monthly mean daily manner and for isotropic sky.

The total amount of the total solar irradiation falling on a horizontal surface for the selected location is 1786 kWh/m^2. Its diffuse component is 675 kWh/m^2 and beam component is 1111 kWh/m^2. With the optimum tilt angle pursuant to calculation made (31,47°), the total solar radiation on module is 1905 kWh/m^2.

2.4 Solar electricity

Other crucial step to make economic analysis is to calculate the output of the system, namely the electricity produced by the system. This calculation is also done in monthly mean daily manner and with respect to gross (total DC electricity production of PV modules) and net (net AC electricity supply of the system with all the losses) electricity calculation.

According to the estimations made annual net AC electricity production of the system is 1685213 kWh that is equal to 1.685 GWh. This amount decreases with a degradation rate year by year. This is also taken into consideration during revenue analysis.

2.5 System Design

The system will consist of PV modules (amount: 3125, peak power: 320 W, efficiency: 16.5%, polycrystalline), inverters (amount: 25 capacity: 40 kW), AC and DC cables, structural materials and remote monitoring and control system, soiling materials, hedges, transformer, buildings for inverters.

For the installation period, the field allocated for the system is graded, construction materials are mounted, required structures are constructed, PV modules are replaced, cable connections are made, inverters are replaced, and connection to the grid is made respectively.

Noting that the investor nets off the cost for the connection to the grid part with the distribution company according to the rules in Turkey.

In addition to the cost calculations the system is simulated and the distances between the rows are determined and the area required for the other structures like inverters are computed for the calculation of the area of the field required.

3 ECONOMICAL ANALYSIS

It is obvious that one of the most important parameters for the energy investments is the economical feasibility, affordability and profitability of the projects. Consequently, reliable and reasonable analysis of the economical aspects of the project is the utmost important step for the project to realize in a profitable and efficient way.

In order to check the economical feasibility, we initially determined the cost of the project with its fixed and variable components. Then we computed the LCOE (levelized cost of energy) so as to see its feasibility/competitiveness against alternatives. Then the revenue and profit analysis is made together with the calculation of payback time depending on the scenarios written by the authors of this study.

3.1 Cost and LCOE Analysis

The cost analysis of the PV system simulated includes all the expenditures starting from permission process till dismantling.

This project in parallel with related legislation is exempted from licensing process because of its being at the threshold/limit of 1 MW. However, that does not mean that the project does not require any permission. According to the related legislation in Turkey [10], for the projects with installed power below 1 MW are subject to the permission project by the related authority. In order to get this permission, the project owner/investor applies to the distribution company depending on the allocated capacity announced by the transmission authority, namely TEIAS.

This process includes some costs as application fee, expenditure for single line diagram etc. They all constitutes soft costs of the system together with project development process by EPC firm.

After permission is got, the side is graded and the system is installed with amount of equipment mentioned in system design part and finally it is connected to the grid. All of these expenses constitute fixed costs of the system.

According to the calculations we made, the system installation cost will be 1 million $.

Moreover, in order to maintain its duty there are some expenses for operational and maintenance purposes. So, they are called as variable costs of the system.

Depending on computations made within this study, the net present value of the variable costs throughout the lifetime of this PV Plant is 167376 $ with 8% discount rate.

Sum of all these costs together with land costs gives the total cost of the system when its net present value is calculated by abiding by reliable discount rate for Turkey that is 8% according to Central Bank of Turkey [11, 12].

By doing all these analyses /calculations below mentioned Table I is resulted / brought about.

Table I – Cost analysis of 1 MW PV PP

Equipment / Activity	Cost (NPV, 8% discount rate, $)
Land Cost	50000
Initial investment	1000000
Operation and Maintanance	167376
Dismantling and decomissioning	4970
Total	1222345

Moreover, LCOE corresponds to the cost of a unit of electricity production with respect to net present value of them. Then, it creates a chance to compare the system with its alternatives. It is calculated via the following Eq 1 [13, 14].

$$\mathrm{LCOE}_{PV} = \left. \sum_{t=0}^{N} I_t + \frac{O\&M_t}{(1+d)^t} \middle/ \sum_{t=0}^{N} \frac{E_t}{(1+d)^t} \right. \qquad (1)$$

where t is the year, N is the lifetime, I corresponds to initial cost, O&M corresponds to operational and maintenance cost, d is the discount rate and E is the electricity produced.

Hence, LCOE of the system is calculated as 62 $/MWh.

3.2 Revenue Scenarios and Analysis

After calculating the cost and the LCOE, the revenue of the system is calculated with respect to different scenarios. Before passing to the scenarios, there is a need to summarize related incentives and emphasize the distinctions of them. In Turkey, the PV systems are awarded with 13.3 $ cents/kWh during 10 years when they are permitted. This incentive can be maintained after the tenth year with Ministerial Council decision. If its equipment includes domestic contribution / local content, the amount increases till 20 $ cents/kWh for the first 5 years. The domestic contribution amounts are given/summarized in Table II.

Table II – Domestic contribution type and additive incentive

Domestic Contribution	Tariff ($cents/kWh)
Integration and structural mechanics	0.8
PV Modules	1.3
PV Cells	3.5
Invertor	0.6
Focusing equipment	0.5

Passing to the scenarios developed/created, we listed them as follows together with related/required explanations.

◑ Lifelong feed-in-tariff with constant amount (13.3$/kWh) (Option 1)

This option assumes that Ministerial Council has decided to carry on incentivizing the PV PP projects with fixed rate as it is done today. Moreover, there is no domestic contribution/local content in this option.

◑ Lifelong feed-in-tariff with variable value (Option 2)

Under this option, we assume that the FiT scheme is maintained as in Option 1 but with a decreasing amount. The rationale behind this idea is the sharp decrease in marginal cost of PV system components mainly PV modules. Hence, it is reasonable for the government to decide to decrease the amount of the support per unit electricity provided. Because of having different amount and time period to support the investment, the tender for Karapinar YEKA including the installation of 1000 MW PV system and construction of PV module manufacturing factory and establishment of R&D Center gives a hint to support this option. This option has some sub options as follows:

◑ 13.3 for first 10 years + 10.5 for last 20 years (Option 2.1)

According to Option 2.1, the support for solar electricity production is maintained by the government with 10.5 $ cents/kWh during 20 years after the first 10 years. 10.5 $ cents/kWh stems from the amount of the support for geothermal systems in Turkey.

◑ 13.3 for first 10 years + 7.3 for last 20 years (Option 2.2)

The similar policy with Option 2.1 has been followed in Option 2.2 but with different amount for the last 20 years. This amount is 7.3 $cents/kWh that equals to current FiT fee of wind and hydro electricity.

◑ 13.3 for first 10 years + 5 for last 20 years (Option 2.3)

At this time the FiT for the last 20 years becomes 5 $ cents/kWh.

◑ 13.3 for first 10 years + 2.5 for last 20 years (Option 2.4)

For Option 2.4, the FiT amount for the last 20 years corresponds to 2.5 $cents/kWh.

◑ Feed-in-tariff with domestic contribution (Option 3)

As mentioned previously, with the domestic contribution/local content in PV systems, there is a chance to increase the amount of support till 20 $ cents / kWh as shown in Table II. Nevertheless, the maximum tariff of domestic contribution suitable for the system simulated in this study is 6.2 $cents/kWh. The support for the domestic contribution is valid only for the first 5 years.

Moreover, as is obvious it is also supposed that the FiT is maintained during 30 years as it is now.

◑ 19.5 for first 5 years + 13.3 for last 20 years (Option 3.1)

In Option 3.1, it is assumed that the PV system is totally manufactured domestically. So it is awarded with the ceiling fee for the FiT that is 19.5 $cents/kWh.

◑ 14.1 for the first 5 years + 13.3 for the last 20 years (Option 3.2)

In this sub option it is postulated that only the integration and structural mechanics part of the system is manufactured within the territories of Turkey.

- Feed-in-tariff + Day Ahead Mechanism (DAM) (Option 4)

Option 4 consists of 10 year FiT as in action now then the Ministerial Council decides not to maintain it and repeals it. Consequently, solar electricity producers compete in Day Ahead Market Mechanism that is a mechanism in which the electricity, except the ones having special incentives, are sold/purchased. As of this point under the sub options of Option 4 we have made some estimation for the future price of the DAM.

Moreover, the PV system that we simulated in this study is able to produce electricity on when the Sun is available. When we check the hourly prices of the DAM, the total/average prices when the Sun shines is higher than the DAM price when there is no Sun [15]. Consequently, we have computed the average DAM price when there is Sun. Afterwards we have made estimations by taking that amount into account.

- 10 years 13.3 + 20 years current DAM w/o change (Option 4.1)

Under Option 4.1, DAM price is supposed to be fixed for the coming years since we have determined some clues paving way for these options to occur. These hints are concerning the pressure on the market price because of the share YEKDEM, that is, a support mechanism for renewable energy systems gets from the market. So the demand decrease pressures the price and does not let the increase of it. The monthly mean daily average price of the DAM when the Sun shines is 5 $cents/kWh. The stronger Dollar against Turkish Lira also supports this option to realize. The increase in prosumers who are producing/providing their self-consumption via solar systems on their roofs will also result in the decrease in demand side.

- 10 years 13.3 + 20 years current DAM with 1%/year increase (Option 4.2)

For this option with the increase in demand and electricity production costs, the DAM prices increase, the price of DAM is assumed to increase with annual %1 growth rate.

- 10 years 13.3 + 20 years current DAM with 1.5%/year increase (Option 4.3)

Like the previous sub option, in Option 4.3 the DAM price increases but with 1.5% slope in this time.

- Feed-in-tariff + Self Consumption with Different End User Electricity Prices (Opt. 5)

Like all the other options the first 10 years of this option is also allocated to FiT scheme. However, the producer decides to consume the electricity he/she produces by comparing end user electricity prices with DAM/FiT prices. Consequently, in order to be able to make related calculations for this option we need to make the reasonable and reliable estimations for the future end user electricity prices in Turkey. So as to realize this, we have taken the electricity prices of the last 5 years in Turkey and made the projections for the future.

- Feed-in-tariff + Self Consumption by household consumer (Option 5.1)

Depending on the estimations made the electricity price is assumed to increase with a rate that performs for the last years. Consequently, the amount starts with 12.5 $ cents/kWh and reaches up to 16.7 $ cents/kWh.

- Feed-in-tariff + Self Consumption for industrial purposes (Option 5.1)

Because of end-user electricity price differs depending on the purpose of consumption such as industrial purpose or household consumption, this sub option deals with industrial electricity use side. The end-user price for industrial use begins with 10.8 $ cents/kWh and reaches up to 14.4 $ cents/kWh.

Depending on all these options with their sub options we have reckoned related parameters and give the results in Table III in the Results section.

3.3 Payback Analysis

It is very significant for the investors to know the payback time of their investments according to the net present value. Accordingly, this part of the study determines the payback time of typical 1 MW PV PP investment in Turkey.

Pursuant to the scenarios mentioned, payback time differs since revenue differs. The payback time with respect to the scenario are summarized in Table III as well.

As seen from the calculations 8% of discount rate is accepted as referred from the Central Bank [11].

4 RESULTS, CONCLUSIONS AND DISCUSSION

4.1 Results

All the results from the calculations made are gathered under this subtitle so as to enable to check and analyze all the results. The analyses are given in Conclusion and Discussion part.

With the cost and revenue information mentioned till this part, the profitability and the payback time of the system in the NPV form is as given in Table III.

Table III: Economical analysis of the system with options

Option	Cost (Million $)	Revenue (Million $)	Profit (Million $)	Payback (years)
1	1.22	5.80	4.58	5.38
2.1		5.00	3.79	
2.2		4.10	2.88	
2.3		3.45	2.23	
2.4		2.74	1.52	
3.1		6.31	5.10	3.49
3.2		5.86	4.64	5.05
4.1		3.45	2.23	5.38
4.2		3.59	2.37	
4.3		3.66	2.45	
5.1		6.29	5.07	5.60
5.2		5.42	4.20	6.55

4.2 Conclusions and Discussion

The system is simulated to have 1 MW unlicensed ground-mounted PV system in Kayseri province located in Central Anatolia. The simulation consists of the technology selection and design of system equipments such as modules, BoS components and net input (solar radiation) and output (solar electricity) calculations by covering all the processes as of investment decision from permission to installation, grid connection, operation and dismantling.

After the system is simulated, its economic analysis is made. At the economic analysis part, the first step is to calculate the cost of the system including the costs of permission, initial cost, operation and maintenance cost till dismantling as net present value (NPV). Consequently, the NPV of the cost of this system is calculated as 1.22 million $.

Moreover, levelised cost of energy (LCOE) of the system is computed too whose result is acquired as 62 $/MWh.

As a result of revenue analysis, it is occurred that NPV of the revenue of this system is between 2.74-6.31 million $. Consequently, investment to solar electricity systems seems obviously feasible in any ways however the most feasible and profitable way is option 3.1 that includes 19.5 $cents/kWh feed-in-tariff with domestic contribution for the first 5 years and 13.3 $cents/kWh for the last 20 years.

Moreover, it's payback time seems to be between 3.49-6.55 years, differing depending on the options applied.

These results reveal that the PV PP installation in Turkey is highly profitable and economically and technically feasible way to meet electricity demand or to make electricity trade.

The location dealt within this manuscript is located close to the center of Turkey with respect to north to south line. That means if we go southern part of the country we can find locations having exposed to more solar radiation that means more input and more economical revenue.

Moreover, as obvious the discount rate of 8% is quite higher than expected in normal circumstances. Consequently, making the calculations with respect to 5% as done in Ref [13, 16] enhances the results and makes the PV PP more attractive. When we apply 5% discount rate, the NPV of the revenue becomes 2.82-6.49 million $ while payback time becomes 3.39-6.36 years. The LCOE occurs as 50 $/MWh.

The last but not the least, if feed-in-tariff decreases and electricity prices increases, self-consumption will be better option.

5 REFERENCES

[1] EPIA, Unlocking the sun belt potential of photovoltaics, October 2010

[2] DG for Renewable Energy of Turkey, Official web-site, Avaliable in: http://www.yegm.gov.tr/MyCalculator/Default.aspx. Last accessed: 05/09/2017.

[3] Turkish Electricity Transmission Company, Official web-site, Avaliable in: https://teias.gov.tr/sites/default/files/2017-07/kuruluguc.xls. Last accessed: 05/09/2017.

[4] Energy Market Regulatory Authority, Official web-site, Avaliable in: http://www.epdk.org.tr/TR/Dokumanlar/Elektrik/YayinlarRaporlar/AylikSektor. Last accessed: 05/09/2017.

[5] B. G. Akinoglu BG. Chapter 5: Recent advances in the relations between bright sunshine hours and solar irradiation. In: Badescu V., editor. 2008. Modeling solar radiation at the earth surface. Springer Verlag Berlin Heidelberg; p. 115-143

[6] A. B. Karaveli, B. G. Akinoglu. Comparison of quadratic and linear models on the estimation of diffuse solar radiation for location in Turkey, Proceedings of SEEP 2016: 9th international conference on sustainable energy and environmental protection (2016). p. 103-08. Avaliable in: https://drive.google.com/drive/folders/0ByjqGr0SgqwQUHM2cl9hZHFldFE

[7] A. B. Karaveli, B. G. Akinoglu. Estimation of diffuse component for two locations in Turkey, Proceedings of EuroSun 2016: Int. Conf. on Solar Energy for Buildings and Industry (2017). Avaliable in: http://proceedings.ises.org/paper/eurosun2016/eurosun2016-0145-Karaveli.pdf

[8] A. B. Karaveli, B. G. Akinoglu. Main article Development of new monthly global and diffuse solar irradiation estimation methodologies and comparisons. International Journal of Green Energy.

[9] A. B. Karaveli, B. G. Akinoglu. Comparisons and critical assessment of global and diffuse solar irradiation estimation methodologies. International Journal of Green Energy.

[10] Official Journal, 28783, 02.10.2013. Avaliable in (in Turkish): http://www.resmigazete.gov.tr/main.aspx?home=http://www.resmigazete.gov.tr/eskiler/2013/10/20131002.htm&main=http://www.resmigazete.gov.tr/eskiler/2013/10/20131002.htm. Last accessed: 06/09/2017.

[11] Central Bank of Turkey. Official web-site. Avaliable in: http://www.tcmb.gov.tr/wps/wcm/connect/tcmb+en/tcmb+en. Last accessed: 05/09/2017.

[12] Trading Economics. Official web-site. Avaliable in: https://tradingeconomics.com/turkey/interest-rate. Last accessed: 05/09/2017.

[13] A. B. Karaveli, U. Soytas, B. G. Akinoglu. Comparison of large scale solar PV (photovoltaic) and nuclear power plant investments in an emerging market. Energy, Vol. 84 (2015) 656-665.

[14] US Energy Information Administration, Official web-site, Avaliable in: https://www.eia.gov/conference/2013/pdf/presentations/namovicz.pdf. Last accessed: 05/09/2017.

[15] Energy Exchange Istanbul, Official web-site, Avaliable in: https://seffaflik.epias.com.tr/transparency/index.xhtml. Last accessed: 05/09/2017.

[16] A.N. Celik, T. Muneer, P. Clarke. Optimal Sizing and Life Cycle Assessment of Residential Photovoltaic Energy Systems With Battery Storage, Progress in Photovoltaics: Research and Application,Vol 16 (2007), 69-85.

Estimation of the final yield of Grid Connected PV Systems in the Eastern Africa - accuracy in dependence of input data sets and regional variation

Fabien Habyarimana*,#, Hans Georg Beyer+
Department of Physic, College of Science and Technology, University of Rwanda, BP3900, Kigali, Rwanda, +250783183932 habfabian@yahoo.fr
*Department of Engineering, University of Agder, BP 509, 4898Grimstad, Norway
+Department of Science and Technology, University of the Faroe Islands, 100 Torshavn, Faroe Islands, +298 292560
hans-georg.beyer@setur.fo

ABSTRACT: In Eastern Africa the growth of the share of PV-systems – both stand-alone and grid-connected - in ongoing. For support, this requires tools and data bases for the estimation of their energetic performance as expressed e.g. by the expected final yield. The accuracy of the respective data bases and simulation tools still is in need for a validation in this region.
Monitored data from a larger grid connected system in operation in Rwanda are used here to check, on one hand the accuracy of simulation tools for the estimation of the system yield and, on the other hand the reliability of available bases for meteorological data commonly used for the assessment of the expected yield. Bases with satellite derived irradiance data are hereby inspected. Bases on this validation example, the scheme (data base and simulation tool) tested is used to analyze the spread of the final yield of grid connected PV systems in Eastern Africa.

Keywords: final yield, Grid connected, PV-system, solar radiation, satellite-derived irradiance data, ambient temperature

1 CONTEXT

The International Renewable Energy Agency (IRENA) [6] states three suitable classes of global solar irradiance as follow suitable, highly suitable and excellent suitable for 1000-1500kWh/m2/year, 1500 -2500 kWh/m2/year and 2500 - 3000 kWh/m2/year respectively (fig.1).

In view of suitable areas for increased PV-installation, a scenario for increased PV capacity in Africa was analysed by [6]. Fig.2 gives the estimated annual PV generation for African regions, with Eastern Africa dominating.

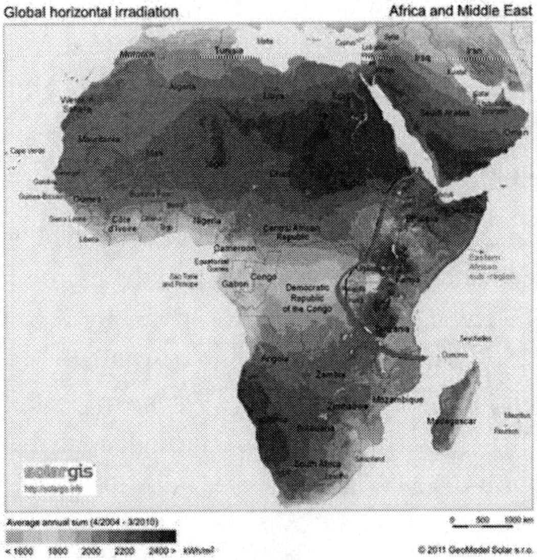

Figure 1: Solae Radiation Map for Africa sowing the annual sum of global irradiation on horizontal surfaces [7].

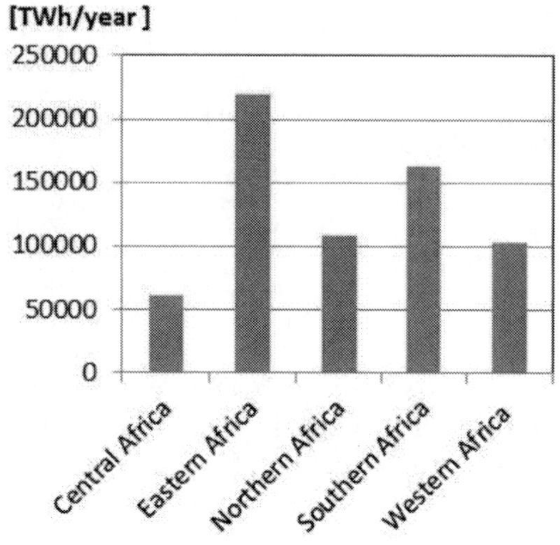

Figure 2: Potential annual PV generation estimated for African regions in TWh/year (from [6]).

Eastern Africa today is characterized by low electrification rates especially outside the metropolitan regions.

The central power grids are still mainly supplied by fossil fuel plants and - as second source - hydropower. However, besides an increasing contribution of PV to off-grid power supply (solar home systems, village power supply) the is a small, but increasing amount of grid connected PV–systems.

In 2016, there is an installed grid-tied PV-capacity of at least 9.2 MW in the region [9]. In Rwanda examples are given by e.g. the 250kWinst Kigali Solaire (installed 2007) and the 8.5 MWinst ASYV system (installed 2014) – see next section. Burundi contributes with 0.4 MWinst.

County	Populatio n without electricity millions	National electrific ation rate %	Urban electrificat ion rate %	Rural electrific ation rate %
Burundi	10	5	28	2
Djibouti	1	42	54	1
Eritrea	3	32	86	17
Ethiopia	73	25	85	10
Kenya	36	20	60	7
Rwanda	8	27	72	9
Somalia	9	15	33	4
South Suda	12	1	4	0
Sudan	24	40	67	26
Tanzania	36	30	57	18
Uganda	31	19	52	12

Table 1: Rates of access to electricity and total populations without access to electricity.

2 A SCHEMES FOR ESTIMATING THE YIELD OF GRID CONNECTED PV SYSTEMS IN EASTRN AFRICA

In Rwanda as in most Eastern Africa region, the density of radiometric ground stations is low. Satellite data should play a good contribution by offering global solar irradiance given that there is a good correlation with ground measurement data [1]. Together with information on ambient temperature they may serve as input time series for simulating the yield of grid connected Photovoltaic (PV) power systems.

Here the accuracy of a scheme for the estimation of the final yield of a system operated in Rwanda is set up. It is based on dedicated tool to model the systems power output (structure see [2]) based om hourly irradiance and temperature data [2]. Those data are generated from monthly information on irradiance and temperature taken from PVGIS [3] and RETScreen [4] respectively. For the generation hourly resolved data sets the WetSyn tool [5] is used.

This scheme is tested using measured data from a grid connected PV system at Agahozo Shalom Youth Village (ASYV) in Rwanda. The system has an installed power of 8.5MW. Monitored data on hourly power output, irradiance and ambient temperature are available. The test of the model accuracy is done in 3 steps: 1) test of the PV-modelling based on measured meteorological data, 2) modeling based on hourly data from Wetsyn initialized by measured monthly means of irradiance and temperature, and 3) modelling based on data generated from monthly information given by PVGIS and RETScreen.

3 QALITY OF YIELD ESTIMATIOM BASED ON ON-SITE METEOROLOGICAL DATA

Results for step 1) and step2) are shown in fig. 3. The scheme applied offers a reasonable basis for the estimation of the yield of systems, depending on the accuracy of the monthly meteorological information available.

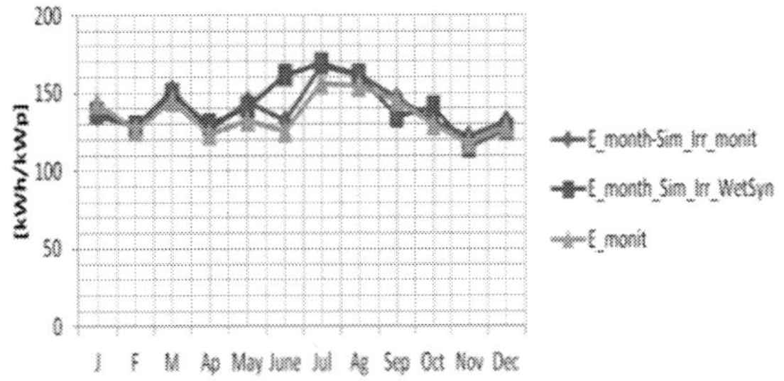

Figure 3: Monthly values of monitored AC yield (green) of the test system together with the simulated yield based on measured meteorological data (blue) and synthesized meteorological data the empirical monthly means (red) [deviation for the month of June is due to data inconsistencies].

3 QALITY OF YIELD ESTIMATIOM BASED ON SATELLITE DERIVED IRRADIANCE DATA

The quality of yield modelling when applying PVGIS irradiance data and ambient temperatures data from RETScreen is shown in fig.4. Shown are evolutions of the yield over the year. A bias in when using thr PVGIS data can be remarked. A similar deviation was reported in [1] for other locations in Rwanda. Using the simulation tools PVCrops –also using the PVGIS data – and PVSyst – using NASA-SSE satellite data – result in similar deviations.

This gives an impression on the current quality of the yield modelling. Thus,.with the current quality of the modelling – mainly limited by the accuracy of the information on irradiance – but differences in yield > 10% may be considered significant, when comparing the yield at different locations in this region.

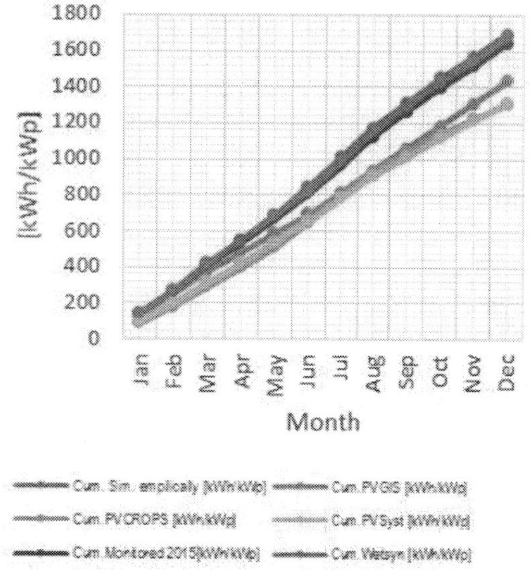

Figure 4: Evolution of the annual yield from measurement (dark blue) and various simulation approaches (the current scheme based on measured meteorology (light blue), synthetic sets based on measured monthly means (green), Synthetic sets based on PVGIS monthly averages (red) and the PVSYST – NASA-SSE (yellow) and the PVCrops-PVGIS (grey) tools).

4 SPREAD OF THE ESTIMATED YIELD FOR VARIOUS LOCATIONS IN EASTERN AFRICA

The PVGIS-REDScreen-Wetsyn based modelling scheme tested here is applied to estimate the annual yield at different locations in Eastern Africa. The evolution of the estimated annual irradiance sums for the sites: Kigali (Rwanda), Lodwar (Kenya), Soroti (Uganda), Dodoma (Tanzania), Gorowe (Somalia), Malakal (South Sudan), Bujumbura (Burundi) and Meleke (Ethiopia) are given in fig 5. For this set of sites, the annual irradiance sum shows differences of up to about 25%.

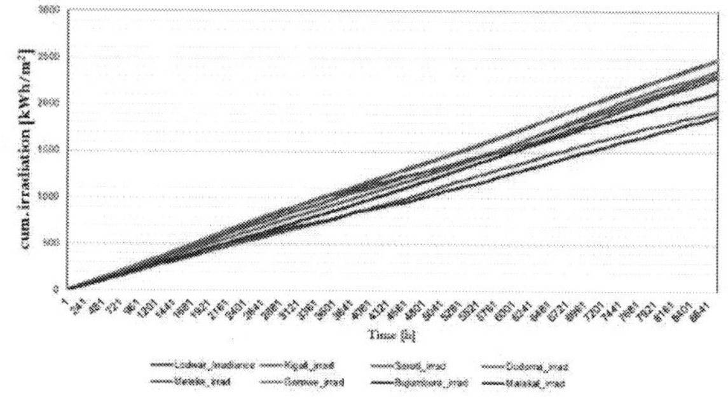

Figure 5: The evolution of the estimated annual irradiance sum for the East African sites inspected.

The comparison of the evolution of the annual yield shows – as could be expected – a similar pattern (see. fig. 6).

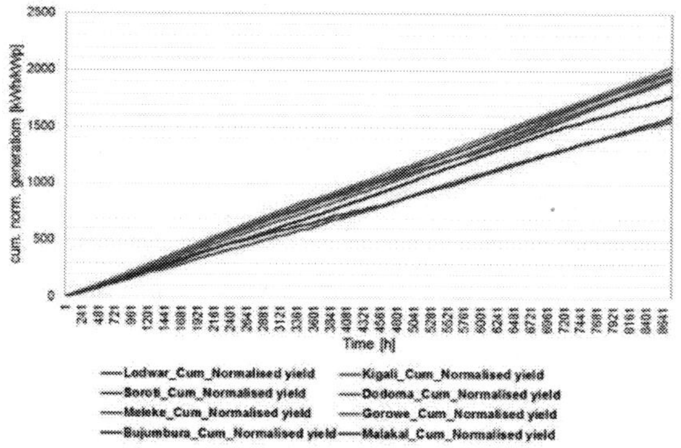

Figure 6: The evolution of the estimated annual yield for the East African sites inspected.

In view of the accuracy of the uncertainty of this yield estimation as discussed here, one may identify to main groups of sites with distinctly different yield characteristics deviating by ~25%. It may be remarked here, that these yield values are of almost twice the size of respective values of systems in central to northern Europe.

As supplementary information, from these simulations, information on requirements for an optimal sizing of system components – here the sizing of DC/AC inverter - can be gained. For this, the relevant information is the probability distribution of the PV-output as fraction of nominal power. Fig. 7 gives this distribution for the various sites. The differences in the occurrence of the higher output classes is remarkable. Inverter sizing must focus on good efficiencies for these classes. Comparison to inverter loading at European sites was shown in [13].

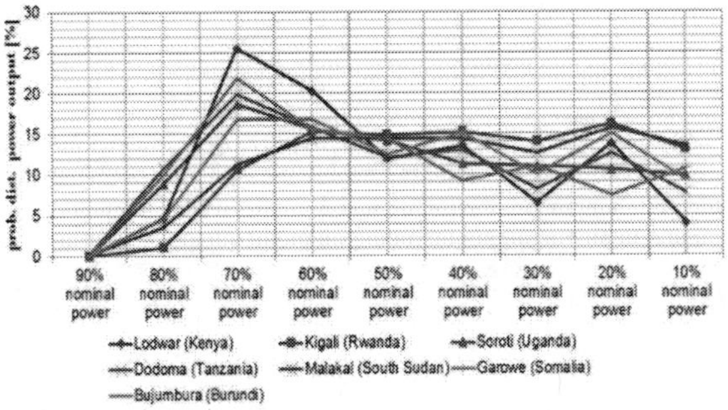

Figure 7: Probability distribution of estimated PV power output normalized to the rated power for the various location. There is a high spread in the occurrence of the elevated power classes.

5 COUNCLISION AND OUTLOOK

The accurate estimation of the final yield of grid connected PV-systems in Eastern Africa is up to now faced with the sparsity of data bases offering accurate information of the irradiance. It could be shown, that accurate information of the monthly radiation sum (or the mean monthly irradiance) would be sufficient, a tools to generate synthetic irradiance time series from that basis are at hand.

An inspection of the variability of the estimated yield for this region shows a spread of about 25% (well bigger than the uncertainty of the yield estimation). Together with finding of significantly different probability distributions of PV output classes this calls for a location specific sizing of the system components.

6 REFERENCES

[1] F. Habyarimana and H.G. Beyer, "Assessment of the quality of satellite derived irradiance data for Rwandabycomparison with data from a fleet of automated ground stations," in Proc. of the *ISES, Solar World Congress, Daegu,* Korea, 2015.

[2] F. Habyarimana and H.G.Beyer, "Validating the model for a 250 kW size grid connected PV-system in Rwanda based on sparse operational data.," in *31st European Photovoltaic Solar Energy Conference and Exhibition, 2015.*

[3] PVGIS, http://re.jrs.ec.europa.eu/PVgis/PV/, as of 2017 [Online].

[4] RETScreen, "RETSreen, International Clean Energy Decisive Support Centre: http: www.retsreen.net," Naatural resourcesCanada, as of 2017 [Online], available http://www.nrcan.gc.ca/energy/software-tools/7465

[5] M.Maier, et al. "Entwicklung und Programmierung eines Strahlungs- und Temperaturdatengenerators namens WetSyn in MS-EXCEL und JAVA", 17. Symp. Photovoltaische Solarenergue, Bad Staffelstein,, Germany, 2002

[6] The Royal Institute of Technology (KTH) and International Renewable Energy Agency (IRENA), "Estimating the Renewable Energy Potential in Africa ; A GIS-based approach" 2014.

[7] Africa, GeoSun, "Global horizontal irradiation Africa and Midl East," 2011., http://geosun.co.zo.
[8] "World energy out look 2016", IEA, 2016.[Online], available: WWW.world energy outlook.org.
[9] K. Kuran, "The Outlook for Utility-Scale and Commercial & Industrial Solar Projects in East Afric,", 2016., available: https://www-cdn.law.stanford.edu/wp-content/uploads/2017/01/East-Africa-Utility-and-CI-Solar-FINAL.docx.pdf
[10] PVSYST. As of 2017 [Online]. available: WWW.PVSYST.com.
[11] PVcrops. As of 2017 [Online]. available: WWW.PVCROPS.eu .
[12] I E A ; International Energy Agency, "African energy outlook, A fucus on energy prospectus in Sub-Saharan Africa," IEA, Paris, 2014.
[13] H.G.Beyer, F. Habyarimana, , Comparison of meteorological operation conditions governing the performance of grid connected pv-plants in tropical and moderate climate, EUPVSEC 2015, Amsterdan, the Netherlands, 2014

COMPUTATIONAL TOOL FOR THE MODELLING AND SIMULATION OF GRID-CONNECTED PHOTOVOLTAIC SOLAR SYSTEMS

Andreza Cardoso Ferreira, Luis Carlos Macedo Blasques, Marcos André Barros Galhardo, João Tavares Pinho, Marcelo Pinho Almeida
Grupo de Estudos e Desenvolvimento de Alternativas Energéticas, Faculdade de Engenharias Elétrica e Biomédica
Instituto de Tecnologia, Universidade Federal do Pará
Av. Augusto Corrêa nº 01, CEP 66050-000 – Belém, PA, Brazil
Tel. / Fax +55 91 3201 7299, andreza_car@hotmail.com

ABSTRACT: This paper describes the modeling of the electrical equipment and control mechanisms used in the power conditioning system of a grid-connected photovoltaic system (GCPVS), such as the photovoltaic generator, the DC/DC converter, the voltage inverter with PWM control, its output filter, as well as the converter controls and their functionalities, and the interconnection of the system with the distribution grid and load. In addition to the modeling of the grid-tie inverter, the main equipment of the power conditioning system, this paper describes the synchronization form of the photovoltaic (PV) generation with the electricity distribution grid, considering the power injection control method, the maximum power point tracking technique for maximizing the energy supplied by the photovoltaic generator, and the complete topology of the proposed PV system. The main objective is to provide a computational model capable of simulating the behavior of the GCPVS elements, varying the climatic conditions and the power demand on the AC side, presenting consistent results for different operating conditions, such as variations of climatic parameters (solar irradiance, wind speed, and ambient temperature) and load. In these cases, it is possible to evaluate the behavior of the system, supplying both the active and non-active power of the load, according to the need.

Keywords: Photovoltaic Solar Energy; Grid-Connected Photovoltaic Systems; Distributed Generation; Modelling; Operation Control

1 INTRODUCTION

The global trend of developing solutions as a way to compensate for the energy deficit drives the use of alternative sources integrated to the electrical grid, with the production of electricity near the place of consumption (distributed generation), the solar photovoltaic generation (PV) being one of great importance. Grid-connected photovoltaic systems (GCPVS) are used to complement conventional electricity generation, usually based on large hydropower plants or fossil fuels, and extensive transmission and distribution systems.

In some countries, consumers are encouraged to install GCPVSs to produce excess energy and they are financially remunerated for the exported electricity (feed-in tariffs). In other countries, such as Brazil [1], the monthly electricity bill is calculated based on the difference between consumption and electricity generation (net metering). Buildings with grid-connected photovoltaic systems that produce excess energy are not only consumers but also become producers of electricity [2, 3].

The development of computational systems that evaluate the operation of GCPVSs with good accuracy, being able to faithfully represent their actual behavior, is one of the subjects under study in order to spread the knowledge about the operation of GCPVSs. Thus, to perform simulations that provide reliable results, it is necessary to develop a system based on validated models.

The main objective of this paper is the development of a computational model that simulates, in details, the behavior of a grid-connected photovoltaic system. The global computational model consists of smaller models that represent each element of the system and can be an important aid tool for the evaluation and characterization of GCPVSs, being able to estimate the energy generation according to local climatic data such as solar irradiance, ambient temperature and wind speed, performance of the installation, and its impact on the distribution grid.

Besides that, the simulation of the GCPVS operation allows the evaluation of variables like active, non-active, and apparent power at the output of the PV system, at the load, and at the grid during the generation process. It can be used for both planning and analysis of the behavior of an already installed system.

2 MODELING THE GRID-CONNECTED PHOTOVOLTAIC SYSTEM

The GCPVSs have different classifications according to the arrangement of the inverters and the photovoltaic modules. For all topologies, the components that compose the GCPVS are practically the same: photovoltaic modules, inverters, cabling, mounting structures, protection devices, and bi-directional electricity meters.

The modeled grid-connected photovoltaic system is composed of a photovoltaic generator, a DC/DC boost converter (step-up converter) with Maximum Power Point Tracker (MPPT), a single-phase full-bridge inverter with PWM voltage control, a PLL (Phase-Locked Loop) synchronization circuit, an LCL filter, and the power and load distribution grid model, as shown in Figure 1.

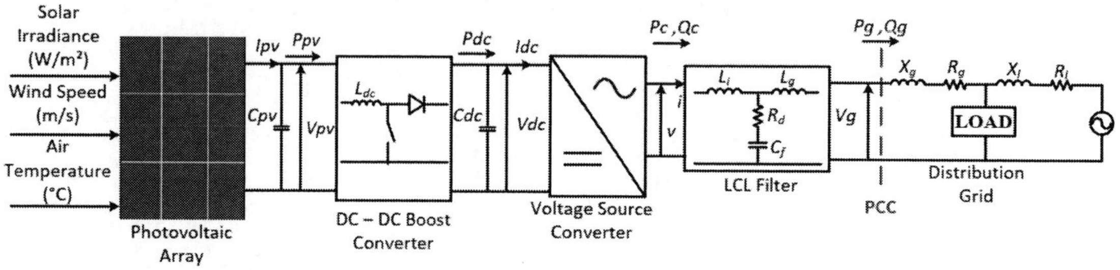

Figure 1. Grid-connected photovoltaic system with a two-stage converter [3]

The control strategy used is the direct control of the output voltage, where the inverter behaves as a voltage source to the grid. The voltage is synchronized with the grid using a PLL algorithm, which provides the frequency and the phase of the voltage at the point of common coupling (PCC) in order to ensure the current injection and operation of the inverter with unity power factor. In practice, as a function of the parameter settings, this power factor closely approximates the unity in the modeled system, but it is not exactly 1, remaining in the range of 0.99 most frequently during the operating time.

The frequency reference of the photovoltaic system is determined by the electricity distribution utility, and the GCPVS should always operate in synchronism with it, following the appropriate operating frequency range standard of each location. In Brazil, the nominal operating frequency is 60 Hz, and the tolerated range is from 59.9 to 60.1 Hz.

In the description of the modeling of each component of the GCPVS, the model of the photovoltaic generator is initially introduced. The produced energy varies according to the solar irradiance, ambient temperature and wind speed. Then, the models of the power conditioning components, the distribution grid, and the load are presented, as well as the synchronization stage with the grid, all implemented in the software *Matlab/Simulink*.

2.1 PV Solar Array

The photovoltaic cell is the basic element for the direct conversion of solar irradiance into electrical energy. The mathematical model of a diode is suitable to represent silicon PV cells, having a current source, I_{ph}, in parallel with the diode, one resistance in series, R_s, that represents internal losses, and a shunt resistance, R_{sh}, representing a current leakage , as shown in Figure 2 [4].

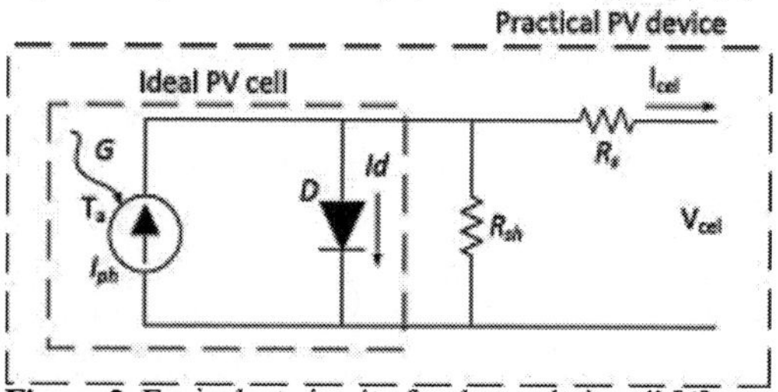

Figure 2. Equivalent circuit of a photovoltaic cell [4]

2.2 Power Conditioning System

For the developed system, a two-stage single-phase inverter was modeled, in string configuration, capable of extracting the maximum power from the PV system.

The two-stage single-phase voltage inverter consists of the integration between the booster stage (DC/DC converter) and the inverter itself (DC/AC converter), in a topology without galvanic isolation (transformer). The main advantage is the reduction on the number of components and weight, and the circuit simplification. The DC/DC converter also performs the maximum power point tracking (MPPT), while the inverter is responsible for controlling the current injected in the grid [8, 9]. Figure 3 shows a block diagram of the system, including the PV generator, the two stages of the power conditioning system mentioned, and the connection to the power grid.

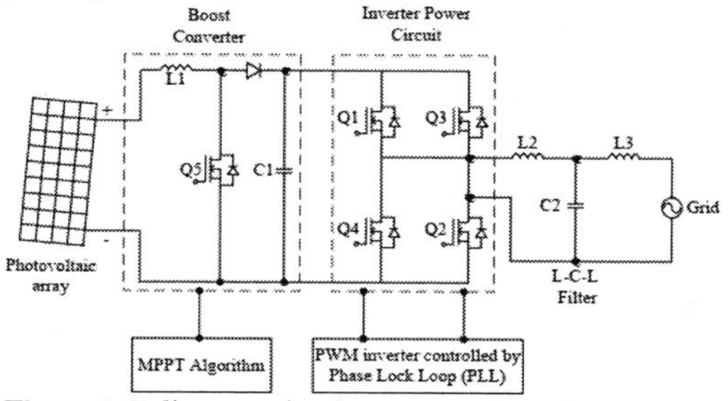

Figure 3. Full power circuit representation of the conventional converter [9]

The first stage of the energy processing (the first stage of the two-stage inverter) consists of the step-up converter (also called booster). It is the main element of the DC/DC stage, responsible for two important functions in the process of conditioning the energy generated by the PV system: the first concerns the maximum energy extraction provided by the generator, and the second is to raise the voltage level supplied by the system and to regulate the value of the DC voltage bus with the adequate characteristics for the inverter [10, 11]. A MPPT is usually integrated to the first stage, which is responsible for providing an increment in the energy generated by the system. Besides, the number of modules needed to form a string becomes more flexible, which makes the use of this converter indicated in many cases.

- DC/DC Boost Converter

The topology of the voltage step-up converter is not insulated and unidirectional, which is represented by and equivalent circuit with only four elements necessary for the conversion. With the inverter in operation, the output voltage of the PV generator is raised to an appropriate level (the optimal operating value). A capacitor is usually connected between the PV array and the step-up circuit to reduce the high frequency harmonic components [12,13].

- Maximum Power Point Tracker

In order to maximize the energy generated by the PV array, actions can be taken in order to get the system to operate at its maximum power point. The *Perturb & Observe* technique (P&O) is used to control the DC/DC converter shown in this paper, with the objective of changing the value of its work cycle.

With the use of the P&O technique, a reference variable (the output voltage of the PV module) is periodically increased or decreased, and the resulting power of the present cycle is compared with the power of the previous cycle, thus defining the direction of the disturbance. That is, a disturbance is introduced in a certain direction of the photovoltaic generator voltage, which causes a variation in the output power of the system. If the output power increases, the next disturbance continues in the same direction; otherwise, the disturbance is made in the opposite direction in the next iteration, resulting in an oscillation around the maximum power point of the PV generator. This method is considered a standard for maximizing the efficiency of a PV system and *Maximum Power Point Tracking* (MPPT) [14, 15].

- Single-phase full bridge voltage inverter

The second stage of power conditioning consists of the modeling of the single-phase voltage inverter. In addition to the basic function of converting direct current to alternating current, an inverter for the connection of photovoltaic systems to the grid must have characteristics such as high efficiency, power quality, reliability and safety [16].

The inverters can be classified according to the type of DC power supply, such as current source inverter (CSI) or voltage source inverter (VSI) [17]. The inverter model proposed in this work is a single-phase, bridged voltage source type, using IGBTs as switching devices, operating with the pulse width modulation (PWM) control strategy, which provides good control of waveform and output voltage value. The diodes associated with the switches guarantee a single direction for the current, necessary for the correct operation of the inverter [18].

The inverter modeled in this article uses Matlab's *SimPowerSystems* tool, which allows the simulation of complex elements without the need to idealize them, thus taking into account their nonlinearities and losses [19].

The operating principle of the voltage inverter is directly related to the modulation strategy, the PWM technique, used to open and close the semiconductor switches, and is the most used in the VSI inverters, as an efficient method of gain control and output voltage [20].

- PWM voltage control

For the inverter output voltage control using PWM modulation the semiconductors (switches) are controlled for switching in such a way that the output voltage of the inverter follows the reference voltage. An external voltage signal (corresponding to the grid voltage) is fed into a discrete single-phase PLL.

- Grid Synchronization

One of the fundamental conditions for connecting an inverter to the grid is the synchronism between the voltage produced by the inverter and the grid voltage. Synchronization techniques have an important function, providing the inverter with a reference signal in phase with the grid voltage, to avoid the possibility of damaging the inverter. To generate the synchronous reference signal, a *Phase Locked Loop* (PLL) can be used [22, 23].

Currently, the PLL algorithm has been the most common synchronization method to detect the phase angle of the grid voltage. This technique has a better rejection of harmonic components of the grid and other types of disturbances. It can be employed to provide the inverter with a unit power factor. Since the output current of the inverter that is injected into the grid must be synchronized with the grid voltage, synchronism plays an important role in the control of grid-connected inverters [24, 25].

The voltage control mesh model was developed to generate the reference voltage synchronization from the grid voltage through the phase angle generated by the PLL [19, 26].

The use of the PI controller is important to adjust the reference voltage, because its actuation guarantees both the delay compensation caused by the low-pass filter of the PLL, and the operation of the PV system without error in steady-state through its integral action. Therefore, the PI controller can ensure the good performance of the PLL. This method of parallelism is interesting since it allows the injection of power through the application of a traditional integral proportional control, being direct and intuitive [27].

- LCL Filter

The inverter output filter is of fundamental importance in the power conditioning system, since it has the objective of minimizing the amount of distortion, and limiting the high frequency harmonic components of the current injected in the grid, caused by the semiconductor switching [28, 29, 30].

2.2 Electrical distribution grid model

The distribution grid is usually represented at the PCC by a Thévenin equivalent circuit, in this case by a voltage source (V_s) of constant amplitude and frequency, with a series impedance (Z_s). In order to properly simulate the inverter connection to the grid, it is necessary that the inverter model react to possible changes in the grid behavior in the same way as it would in a real system [31].

2.3 Modeling of the demanded load

It is considered that the modeled GCPVS produces electric power to supply a local load, connected at the PCC, exchanging energy with the power grid. The GCPVS injects power into the grid when there is excess, i.e. when the generation is higher than the load demand, and there is the complement of energy from the grid when the generation is lower than the load demand. In order to verify the process of requesting active and non-active power from the grid and the inverter, a linear resistive (R) load and a non-linear load are considered.

The implemented nonlinear load model, shown in Figure 4, uses the Universal Bridge block, from the Sim Power Systems library, representing the uncontrolled single-phase bridge rectifier, whose RC components represent the type of load under analysis.

Among some examples of mixed loads, there are: fluorescent lamps, computers, televisions, electronics in general, etc.

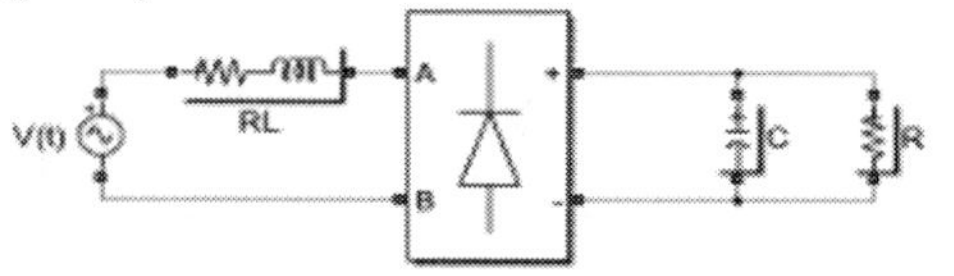

Figure 4. Non-linear load equivalent circuit

3 RESULTS

The modeling of the complete system is presented in Figure 5, composed of the PV generator, DC/DC Boost Converter with *Maximum Power Point Tracker* (MPPT), single-phase bridge inverter with PWM voltage control, synchronizing PLL circuit, LCL filter, power grid and load. The full model allows to perform different simulations, by varying the climatic parameters (solar irradiance, ambient temperature, and wind speed), as well as electrical parameters at the PCC.

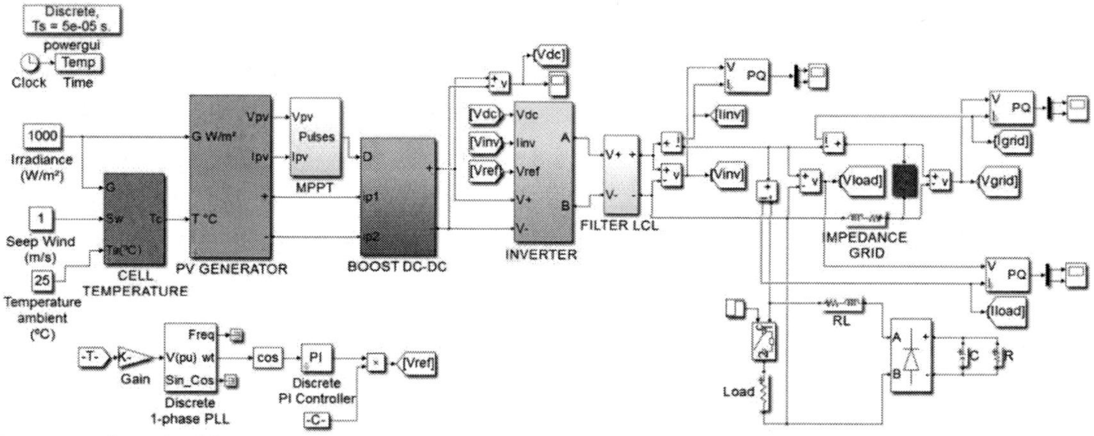

Figure 5: Schematic of the system

As an example of the complete model response, two case studies are simulated using the *Matlab/Simulink* software, the first with constant input parameters (solar irradiance, ambient temperature, and wind speed) and variation of the load demand, and the second with variable input and load parameters. Then, these cases are studied with the main purpose of analyzing the profile of the electrical parameters of interest.

In the case studies, a photovoltaic generator composed of ten polycrystalline silicon modules connected in series is used, resulting in a generator with 3,100 Wp and voltage of 363 V. The electrical characteristics of the used module are presented in Table I.

Table I. Electrical characteristics of the module. Kyocera KD245GH-4FB data sheet [32]

Parameters	Values
Maximum Power (Pmáx)	310 Wp
Maximum Power Voltage (Vmáx)	36,3 V
Maximum Power Current (Imáx)	8,53 A
Open Circuit Voltage (Voc)	45,6 A
Short-Circuit Current (Isc)	8,99 A
Efficiency	16 %
Voltage Temperature Coefficient	-0,32 %/°C
Current Temperature Coefficient	0,06 %/°C
Power Temperature Coefficient	-0,42 %/°C
NOCT	-40 a 85 °C
Number of cells per Module (Ns)	72

At the beginning of the first simulation a nonlinear load is used, with a resistance of 169 Ω and a capacitance of 100 mF. The rated voltage and rated frequency of the distribution grid considered for the simulation are 220 Vrms and 60 Hz, respectively. The theoretical active and non-active powers for this load are, respectively, 525.9 W and 18.83 VAr. After 10 s of the simulation, a second R load is added in parallel with the first, with a resistance of 12.4 Ω. The theoretical active power for this second load is 3,903.22 W. Table II presents the complete input parameters for this simulation.

Table II. Input parameters

Parameters	Load 1	Load 2
Solar irradiance (G)	1,000 W/m²	
Ambient temperature (Ta)	25 °C	
Wind speed (Vv)	1 m/s	
Active power demanded by the load (Pcd)	525.9 W	3,903.22 W
Non-active power demanded by the load	666.27 VAr	0 VAr
Load power factor (Fp)	0.62	

The power effectively supplied by the PV generator is 2,750 W (Figure 6), considering the input data in Table 2, the cell temperature calculation of the model used, and an open structure installation.

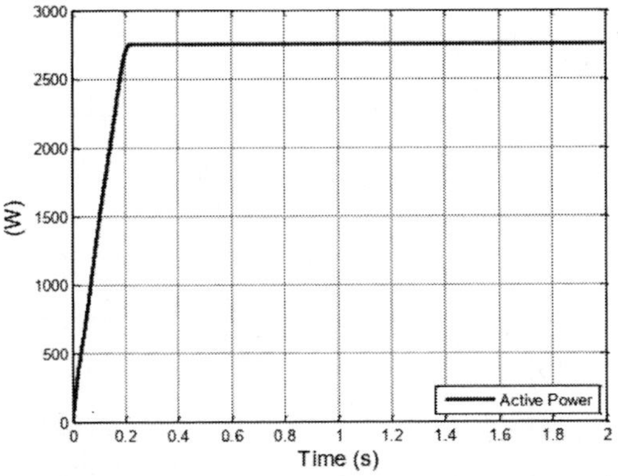

Figure 6. Active power provided by the photovoltaic generator

The voltage and current waveforms at the inverter output, at the nonlinear load, and at the distribution grid are shown in Figures 7 to 9. From Figure 9 it is possible to observe that the grid is absorbing active power from the PV system. When the load increases, it is possible to notice that the grid also provides active power.

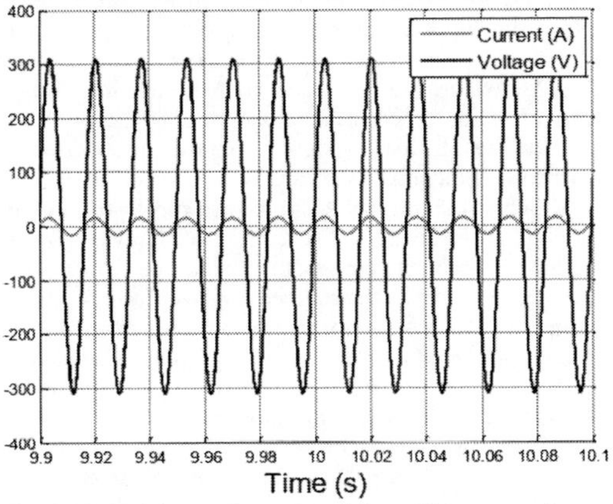

Figure 7. Voltage and current of the grid-connected inverter

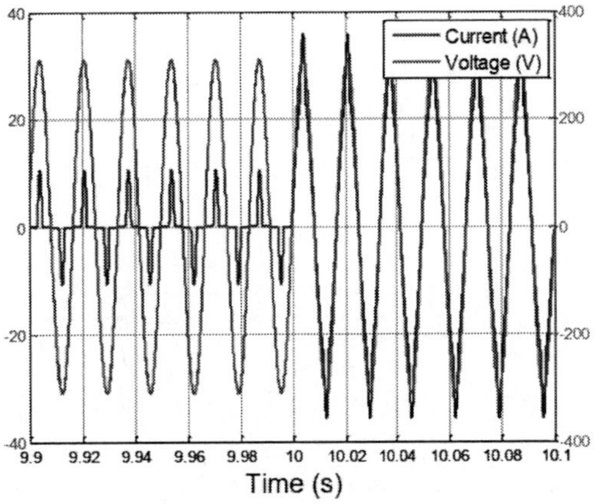

Figure. 8. RL load voltage and current

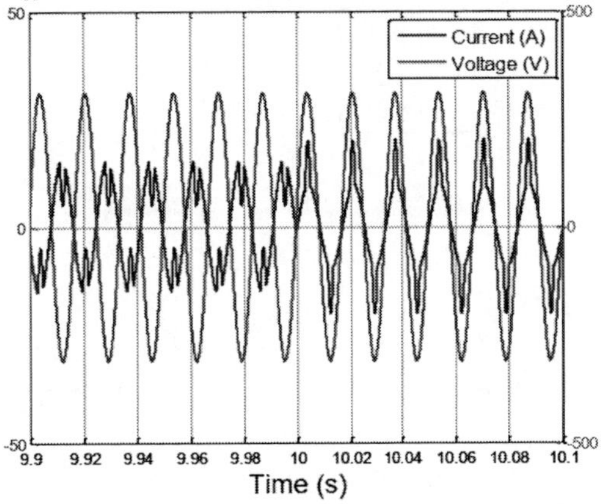

Figure. 9. Voltage and current of the distribution grid

The results for this simulation are presented in Table III, where it can be observed that, in the first moment, the photovoltaic system can supply the active power required by the load. However, when the load increases, the photovoltaic system can no longer fully supply the demand, being therefore necessary that the additional active power be supplied by the distribution grid. The graphs in Figures 10 and 11 confirm these comments. As for the non-active power, both with the first load present and with the addition of the second, the inverter supplies a small but insufficient portion to the load, being necessary that the grid provides the remainder, decreasing the power factor of the installation.

Table III. Results for each case

Parameters	Inverter	Load 1	Distribution grid (Load 1 only)	Load 2	Distribution grid (Load 1 + Load 2)
Voltage (*VRMS*)	219.9 V	219.9 V	220 V	219.9 V	220 V
Current (*IRMS*)	11.05 A	3.86 A	9.17 A	17.75 A	9.68 A
Active Power (*P*)	2,428 W	525.9 W	-1,902 W	3,903.22 W	2,001.12 W
Non-active Power (*N*)	95.95 VAr	666.27 VAr	672.53 VAr	0	728.5 VAr
Apparent Power (*S*)	2,429.90 VA	848.81 VA	2,017.4 VA	3,903.22 VA	2,129.6 VA
Power Factor (*Fp*)	0.999	0.62	0.942	1	0.94

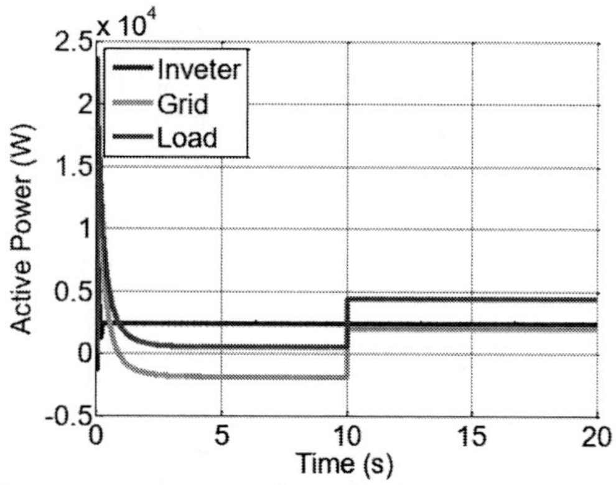

Figure 10. Active power for the simulated case

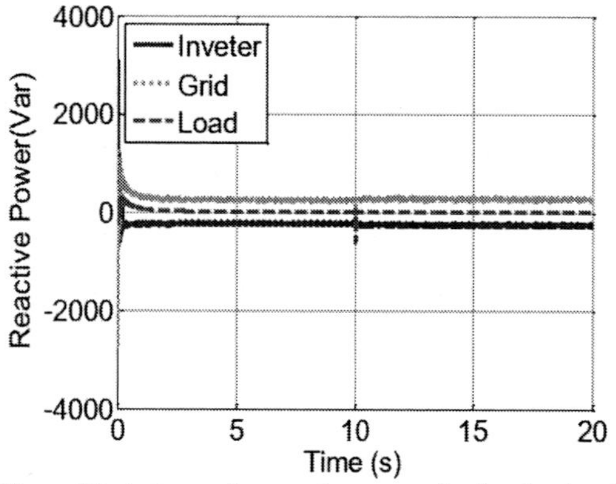

Figure 11. Active and non-active power for the simulated case

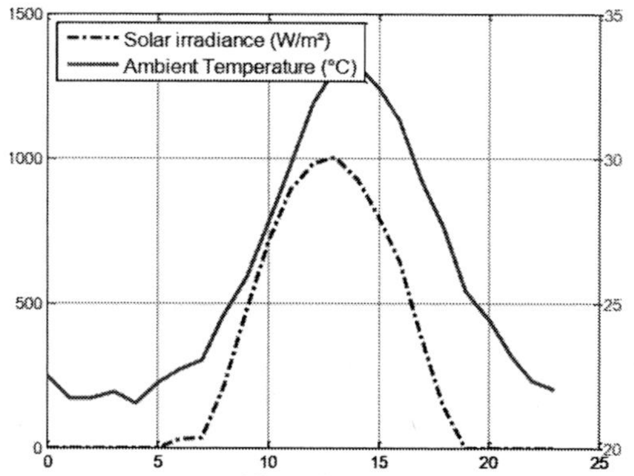

Figure 12. Hourly data of solar irradiance and ambient temperature

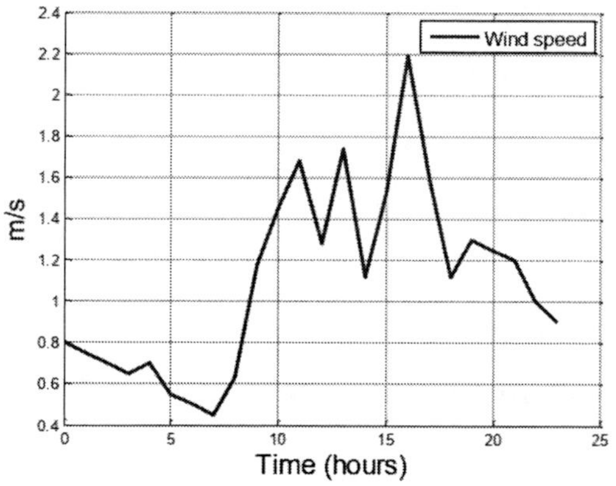

Figure 13. Hourly data of wind speed.

In the second case, hourly data of global irradiance (incident on a 10° inclined plane and oriented to the Geographical North), ambient temperature, and wind speed for the city of Belém, Pará, Brazil, are used, as shown in Figures 12 and 13 choosing the relevant climatic data on a clear sky day. The data are obtained from Meteonorm [33], a database that contains climatic parameters for different locations all over the world. In addition, the RL load is varied over the same period, in order to simulate real conditions.

Figure 14 shows the active power curve produced by the two-stage inverter, considering variations in solar irradiance, ambient temperature, wind speed, and load, and Figure 15 shows the respective power factors. For the simulated case, the maximum active power verified at the inverter output, at the peak moment of solar irradiance, is less than 2,500 W, which is expected due to the losses verified in the power conditioning process.

Based on these losses, the calculated system efficiency is about 91%. The daily active energy generated by the photovoltaic system is 16.78 kWh and the non-active is -2.26 kVArh, measured during the energy conversion process. The daily productivity, defined as the ratio of the active energy injected into the grid and the nominal power of the PV generator, is 5.41 kWh/kWp.

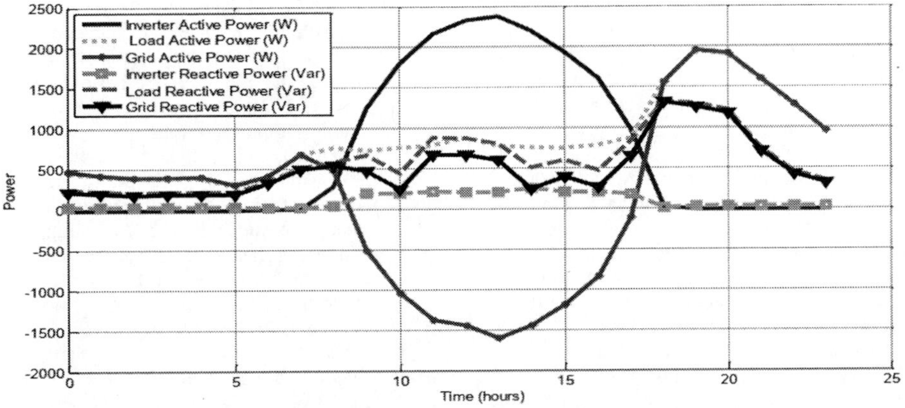

Figure 14. Active and non-active power demanded by the load, highlighting the GCPVS and grid contributions

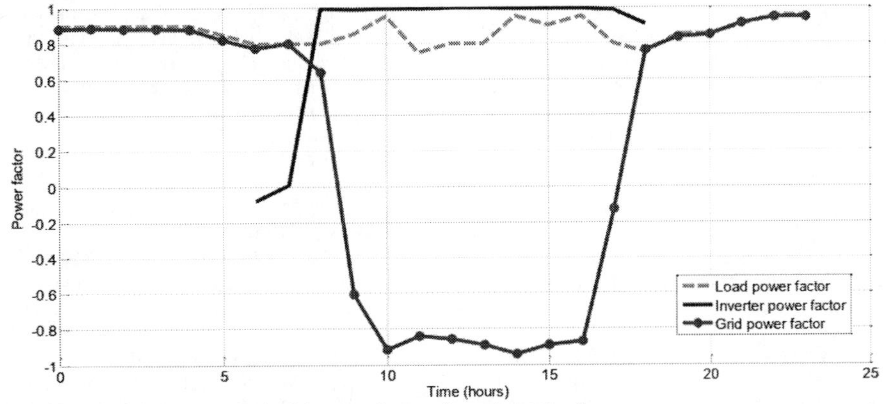

Figure 15. Variation of load, inverter, and grid power factors throughout the day

The simulated system operates as expected and through the simulations of the model employed in this work it is possible to visualize the characteristics of the dynamic operation of the GCPVS, such as the current waveform injected by the inverter as a function of the variations of incident irradiance and operating temperature of the photovoltaic cell, as well as the output voltage waveform and its possible variations with the load. Besides, the active and non-active power delivered by the inverter and the local distribution grid can be estimated.

4 CONCLUSIONS

In this paper a model of a grid-connected photovoltaic system (GCPVS) was presented. The model can be a useful tool for the evaluation of PV systems and as an important auxiliary tool to estimate the electricity production of similar GCPVSs. The computational model based on Matlab/Simulink included detailed representations of the GCPVS circuits and its controllers.

The simulations performed using the complete model proposed present consistent results, varying the climatic conditions and the power demand on the AC side.

Simulations for different load profiles were performed, but due to limited space only one case was chosen for analysis. In the simulated cases the values of effective voltage and current, and active and non-active powers at the output of the inverter, the load, and the distribution grid are close to the expected theoretical values, which makes the model reliable for use in other practical cases.

ACKNOWLEDGMENTS

The authors would like to thank the National Council for Scientific and Technological Development (CNPq) for financial support to develop this research.

REFERENCES

[1] Aneel – Agência Nacional de Energia Elétrica (Brasil). Resolução Normativa No. 482/2012. Disponível em: <http://www2.aneel.gov.br/cedoc/ren2012482.pdf>. Acesso em: 12/04/2017.

[2] Hernandez, J. A., Velasco, D., Trujillo, C. L. Analysis of the effect of the implementation of photovoltaic systems like option of distributed generation in Colombia. Renewable and Sustainable Energy Reviews. 15:2290–2298. 2011.

[3] Perera, B. K., Pulikanti, S. R., Ciufo, P., Perera, S. Simulation model of a grid-connected single-phase photovoltaic system in PSCAD/EMTDC," in POWERCON: IEEE International Conference on Power Systems Technology, pp. 1-6. 2012.

[4] Joshi, M., Vaidya, G.A. Modeling and Simulation of Single Phase Grid Connected Solar Photovoltaic System. Annual IEEE India Conference (INDICON). 2014.

[5] Tsai, H. L., Siang, C. S., Su, Y. J. Development of Generalized Photovoltaic Model Using MATLAB/SIMULINK, Proceedings of the World Congress on Engineering and Computer Science WCECS, October 22 - 24, San Francisco, USA. 2008.

[6] King, D. L., Boyson, W. E., Kratochvill, J. A. Photovoltaic Array Performance Model. Sandia Report SAND2004-3535, Sandia National Laboratories, California. 2004.

[7] Ayompe, L. M. Duffy, A. McCormack, S. Conlon, M. Measured performance of a 1.72 kW rooftop grid connected photovoltaic system in Ireland. Energy Conversion and Management. ISSN doi:10.1016/j.enconman.2010.08.007. 2010.

[8] Du, Y., Xiao, W., Hu, Y., Lu, D.D.C. Control Approach to Achieve Burst Mode Operation with DC-Link Voltage Protection in Single-Phase Two-Stage PV Inverters. Proc. IEEE Energy Conversion Congress and Exposition (ECCE), pp. 47–52. 2014.

[9] Chowdhury, A. S. K., Razzak, M. A. Single Phase Grid-Connected Photovoltaic Inverter for Residential Application with Maximum Power Point Tracking. Proceedings of the 2nd IEEE International Conference on Informatics, Electronics & Vision (ICIEV), May 17-18, Dhaka Bangladesh. 2013.

[10]Opiyo, N. Power Electronics for PV-Based Communal Grids. Smart Grid and Renewable Energy, 7, 67 – 82. 2016.

[11] Yang, W., Li, M. Design and simulation of single-phase photovoltaic grid-connected system based on quasi-sliding mode control. Control, Automation and Systems Engineering (CASE). 2011.

[12] Benkhelil, E., Gherbi, A. Modeling and simulation of grid-connected photovoltaic generation system. Revue des Energies Renouvelables SIENR12 Ghardaïa, pp. 295 – 306. 2012.

[13] Shirvastava, S., RatnaSaxena, A., Wadhwani, K. A. MPPT Based Optimization of Phtovoltaic System Using DC-DC Converter, International Journal of Hybrid Information Technology Vol.8, N°.10, pp.101-114. 2015.

[14] Sholapur, S., Mohan, K, R., Narsimhegowda, T, R. Boost Converter Topology for PV System with Perturb And Observe MPPT Algorithm. Journal of Electronics Engineering (IOSR-JEEE). 2014.

[15] Alsadi, S., Alsayid, B. Maximum Power Point Tracking Simulation for Photovoltaic Systems Using Perturb and Observe Algorithm. International Journal of Engineering and Innovative Technology – IJEIT, Volume 2. Issue 6. 2012.

[16] Dickson, A., Sitti, N, M., Donkor, D, O., Doe, N, P., Amakye5, E, A. Grid Connected Inverter with Unity Power Factor for Renewerable Energy (PV) Applications. IJISET - International Journal of Innovative Science, Engineering & Technology, Vol. 1 Issue 3. 2014.

[17] Mohammad Noor, S, Z. A review of Single-Phase Single Stage Inverter Topologies for Photovoltaic System. Proc. IEEE 4th Control and System Graduate Research Colloquium, Shah Alam, Malaysia, Vol. 19, N. 20, Aug. 2, pp. 69-75. 2013.

[18] Hanque, A. M., Sharma, S., Nagal, D. Simulation of Photovoltaic System Connected with Full Bridge Inverter Using Matlab/Simulink. IJRET: International Journal of Research in Engineering and Technology. Volume: 05, Issue: 01. 2016.

[19] Bratt, J. Grid Connected PV Inverters: Modeling and Simulation. Master of Science in Elecrical Enginneering, San Diego State University. 2011.

[20] Mishra, B., Kar, S. S. Design of an Improved PWM Inverter Using PI Controller. International Journal of Multidisciplinary Research and Development Volume 3; Issue 1; Page N°. 92-95. 2016.

[21] Chakraborty, S., Kroposki, B., Kramer, W. Advanced Power Electronic Interfaces for Distributed Energy Systems; Part 2: Modeling, Developement, and Experimental Evaluation of Advanced Control Functions for Single-Phase Utility-Connected Inverter. National Renewable Energy Laboratory, Golden, CO, Technical report TP-550-44313. 2008.

[22] Devi, P. B., Kumar, P. S. High Performace Based Single Phase Interleaved Flyback Grid Tied Inverter for Inverter for Solar Application. International Journal of Emerging Technology and Advanced Engineering. Volume 5, Issue 12. 2015.

[23] Vignesh, G., Kumarikala, N., Sankar, R., Radhika, D. Grid Tied PV Inverter With Intelligent Anti-Islanding Controland Load Sharing Tecnique. International Journal of Emerging Technology in Computer Science & Electronics (IJETCSE). Volume 13, Issue 1. 2015.

[24] Ciobotaru, M., Kerekes, T., Teodorescu, R., Bouscayrol, A. PV inverter simulation using MATLAB/Simulink graphical environment and PLECS blockset. 2006.

[25] Orłowska-Kowalska, T., Blaabjerg, F., Rodríguez, José., (Eds.). Advanced and Intelligent Control in Power Electronics and Drives, Studies in Computational Intelligence. E-Publishing Inc., Switzerland, pp. 56-61. 2014.

[26] Cai, W., Ren, H. Research on Grid-Connected Photovoltaic System Based on Improved Algorithm. Xi'an University of Science and Technology. 2012.

[27] Xavier L. S., Cupernito, A. F., Pereira, H. A. A Novel Adaptive Current Harmonic Detection Method Applied in Multifunctional Single – Phase Solar Inverters. Proc. IEEE 13th Brazilian Power Electronics Conference and 1st Southern Power Electronics Conference (COBEP/SPEC). 2015.

[28] Massawe, H.B. Grid Connected Photovoltaic Systems with SmartGrid functionality. Master of Science in Elecrical Power Enginneering, Norwegian University of Electric Power Engineering. 2013.

[29] Reznik, A., Simões, M. G., Al-Durra, A., Muyeem, S. M. LCL Filter Design and Performance Analysis for Grid-Interconnected Systems. Proc. IEEE Transactions on Industry Applications, Vol. 50, N°. 2. 2014.

[30] Kahlane, A.E.W.H., Hassaine, L., Kherchi, M. LCL Filter Design and Performance Analysis for Grid-Interconnected Systems. IEEE Transactions on Industry Applications, vol. 50, no. 2. 2014.

[31] Krenzinger, A., Prieb, C. W. M., Blauth, Y. B., Silvério, W. V., Vera, L. H., García, F. H., Simulação Computacional de Sistema Fotovoltaico Conectado à Rede. I CBENS - I Congresso Brasileiro de Energia Solar . 2007.

[32] Kyocera Solar, Datasheet – KD245GH-4FB. http://www.solarconvert.com.br/images/Produtos/KD245GH-4FB.pdf. Access in December 31th, 2016.

[33] METEONORM – Version 6.0 Meteotest, Bern, Switzerland. http://www.meteonorm.com/. Access in April 20th, 2016.

PV POWERED REVERSE OSMOSIS DESALINATION SYSTEM OPERATING AT VARIABLE PRESSURE CONDITIONS AND CONTROLLED BY A MULTI- AGENT DECENTRALIZED ENERGY MANAGEMENT SYSTEM

Christos – Spyridon Karavas[a], Konstantinos G. Arvanitis[a], George Kyriakarakos[a], Dimitrios D. Piromalis[b], and George Papadakis[a]*
[a]Agricultural University of Athens, Department of Natural Resources and Agricultural Engineering
[b]Technological Education Institute of Piraeus, Department of Automation
[a]75 Iera Odos street, GR 11855 Athens, Greece,
[b]250 P. Ralli&Thivon street, Egaleo 122 44 Athens, Greece
*Tel. +302105294209, Fax: +302105294032, email: gpap@aua.gr

ABSTRACT: Sea water reverse osmosis (SWRO) desalination units coupled with renewable energy systems (RES) such as photovoltaics and/or wind turbines is a very attractive proposition, especially in remote areas for covering potable water supply shortage. The intermittent energy production of RES, makes the integration of energy storage subsystems necessary, such as conventional battery banks, in order to ensure the shifting of energy. In small scale SWRO desalination systems, an energy buffer with adequate capacity is required to power the system for small time periods of time during the day. Moreover, SWRO desalination units show lower specific energy consumption when the units operate in variable load that in full load mode. Therefore, the utilization of an advanced energy management system for these systems is essential for their operation. This paper presents an autonomous SWRO desalination system powered by PV, including two alternative energy buffers; an electrical (hybrid capacitors) and a hydraulic (pressure vessels) for short term energy storage. A decentralized energy management system (DEMS) was designed and developed for the SWRO desalination system. The proposed EMS was based on multi agent system (MAS) and employed Fuzzy Cognitive Maps for its implementation. The experimental results, obtained from the application of the proposed MAS – DEMS, showed that the continuous and efficient operation of the desalination unit was ensured.
Keywords: PV System, Demand Side Energy Management System, Storage

1 INTRODUCTION

Water is one of the most important goods for human survival and development but the increase of the population and the climate change have led to one of most significant issues of the worldwide agenda, which is the water scarcity [1]. There are many areas in the world, such as islands and coastal regions, that have significant shortage in potable water supply [2] and more than two billion people do not have access to fresh water. An alternative and sustainable solution to this problem is the water desalination powered by RES [3]. Membrane desalination processes, such as electrodialysis and primarily reverse osmosis (RO), have more global acceptance today than the thermal desalination processes and have been implemented not only on water desalination but generally on water treatment [4]. RO is a process where the feed water enters a semi-permeable membrane under pressure, larger than the osmotic pressure, and the membrane produces fresh water and concentrate solution [5].

The continuously increase of the fossil fuels prices and their environmental impact as compared to the advantages of the renewable energy sources (RES), makes the combination of RES with desalination units a sustainable and economically viable solution [6]. In the past two decades, several desalination units powered by RES have been implemented such as RO desalination units using solar energy, wind energy, thermal energy or a combination of several renewable energy sources. Furthermore, taking into consideration that 97% of the earth's water comes from the oceans, sea water RO (SWRO) desalination is an attractive solution to the water scarcity [7]. SWRO desalination units have shown low specific energy consumption [8] and particularly when the SWRO desalination unit is coupled with energy recovery devices [9]. Moreover, excellent results have been attained when the SWRO desalination unit operates in variable load [10] rather than in full load conditions [11]. Finally, several research studies have dealt with the satisfaction of the water needs through the minimization of the water production cost [12] or through the detection of the optimal energy system from the techno-economic point of view that ensures the energy supply with high security [13].

The energy production by RES is variable as well as intermittent due to the fact that the electricity generation is subject to abrupt temporal and climate variations. This renders the embodiment of energy storage subsystems inevitable, in order to ensure the shifting of energy, by storing the excess energy and supplying it to the system when it is necessary. The traditional approach of storing electrical energy in renewable energy systems is the use of conventional battery banks, such as deep discharge lead acid batteries. In SWRO, an energy buffer with adequate capacity is required, in order to power the system for small time periods of time during the day, where there is no electrical production by RES, because of the continuous changes of the weather [14] and also to absorb rapid changes in power production. Hybrid capacitors and pressure vessels offer this possibility for energy supply as a substitute of batteries. Both can act as electrical and hydraulic energy storage devices, respectively.

As the system becomes more complex, apart from the need of energy storage, advanced energy management systems (EMS) should be incorporated, in order to manage and operate the SWRO desalination systems. In particular, computational intelligence techniques [15] have been exploited for the development of EMSs. Decentralized EMS (DEMS) based on multi-agent systems (MAS) provide distributed decision making and parallel computing. MAS emerge as a solution to decentralized coordination problems and have been used with success in many studies [16], demonstrating the beneficial application of MAS in energy systems [17, 18].

This paper presents an autonomous SWRO desalination system, managed by a MAS – DEMS, which was experimentally investigated at the Agricultural University of Athens, Greece. The autonomous SWRO desalination system consists of (a) a photovoltaic array of typical monocrystalline silicon panels connected of 1.7 kWp, (b) a SWRO desalination unit, which is equipped with a hydraulic energy recovery device of the Clark pump type, (c) two alternative energy storage units (an electrical and a hydraulic one) that are used for short term energy storage and for supplying the system during small periods of time i.e. feed the SWRO desalination system with the required energy (electrical or hydraulic), in order to operate the desalination unit about 10 - 20 minutes. In this way optimal use of the produced PV power is achieved while the SWRO unit operates at variable load (variable feed water pressure and variable feed flow) and (d) a MAS decentralized energy management system. The MAS – DEMS was designed and implemented for the efficient energy handling and maximization of water production of the autonomous SWRO desalination system. Each component of the SWRO desalination system is managed by an intelligent agent. The intelligent agent which manages the desalination units, it is also responsible to determine unit's operation point, corresponds to full or variable load using computational intelligent technique based on Fuzzy Cognitive Map theory.

2 SYSTEM CONFIGURATION

2.1 The PV system

The energy production source consists of a photovoltaic array (PV) of typical monocrystalline silicon panels connected of 1.69 kWp (Fig.1). Their main characteristics are presented in Table I.

Table I. Main characteristics of the PV modules

Type: Arco Solar	
Description	Characteristic value
Rated Power	47 W
Rated Voltage	14.2 V
Rated current	3.31 A
Open – circuit voltage	22 V
Short – circuit current	3.49 A

2.2 SWRO desalination unit

The SWRO desalination unit (Fig. 2) includes a rotary positive displacement pump, coupled with a DC electrical motor, a combined high pressure pump and energy recovery unit (Clark pump), a membrane vessel and hydraulic pipes.

Figure 1: The photovoltaic array

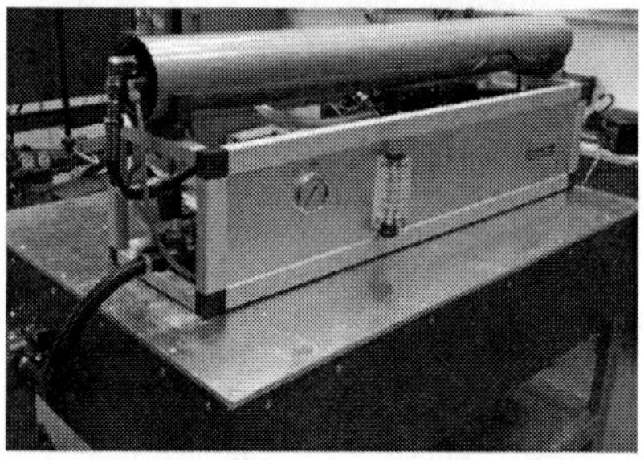

Figure 2: The SWRO desalination unit with the Clark pump

The main technical specifications of the SWRO desalination unit are shown in Table II. The feed water is a NaCl-H_2O solution which was prepared by dechlorinated tap water and its salt concentration was adjusted to 32000 ppm, which is the typical salinity of the seawater in Aegean Sea.

Table II: SWRO desalination unit main specifications

Membrane type	Filmtec SW 30 – 4040
Feed water pump type	Positive Displacement Recovery
Maximum pressure	17 bar
Maximum power consumption	750 W, 48 V DC
Energy recovery device	Clark pump
Maximum product flow rate	150 L/h ± 15%
Nominal Recovery ratio	14%
Rated pressure	55 bar
Rated feed pressure	11 bar

2.3 Energy storage systems

2.3.1 Electrical energy storage system

The electrical storage system includes a custom designed and manufactured array of hybrid ultra-capacitors modules (Fig. 3). Each ultra-capacitor has a nominal voltage of 2.3 VDC and the desired voltage of the system is 48 VDC. Each module includes 21 ultra-capacitors connected in series, allowing 48.3 VDC output and all modules are connected in parallel. The technical characteristics of a hybrid capacitor is shown in Table III.

Figure 3: The hybrid-ultra capacitor bank

Table III: Technical characteristics of hybrid capacitor

Parameter	Value
Type	IOXUS
Model	RHE2R3108SR
Capacity	1000 F
Rated voltage	2.3 V
Maximum Energy	0.6 Wh
Maximum power	2.1 kW/kg

2.3.2 Hydraulic energy storage system

The hydraulic pressure buffer (Fig. 4) includes three pressure vessels connected in parallel and feed the membranes of the desalination unit with sea water under pressure. Each pressure vessel has a nominal capacity of 200 L and it is able to feed the Clark pump with the appropriate pressure (from 6 to 11 bar) in order to operate the SWRO desalination unit. Each pressure vessel is equipped with a pressure switch which is connected with an electro valve for the safe control of the flow. The technical characteristics of the pressure vessels are shown in Table IV.

Table IV: Technical characteristics of pressure vessel

Parameter	Value
Type	Bladder pressure tank
Model	GT-U-200 PN 16 G 1 ¼ V
Capacity	200 L
Number of pressure tanks	3
Maximum Pressure	16 bar
Pre-charged pressure	4 bar
Material	Suitable for sea water

Figure 4: The hydraulic pressure buffer

3 MULTI AGENT SYSTEM

3.1 Intelligent Agent

An intelligent agent (IA) is an autonomous entity that is placed in an environment and is capable to monitor and acts upon its environment in order to achieve its design objectives [19, 20]. The intelligent agent processes the received information, makes comparisons with defined rules, investigates possible decisions and finally selects the optimum solution while it communicates with other agents in a specific environment. The agents have the ability to cooperate, coordinate and negotiate with each other making the interaction of the agent with its environment successfully. According to the literature [21, 22], an intelligent agent has four key features, namely are the autonomy, social ability, reactiveness and pro-activeness. Moreover, the agents are separated into five classes, according to their extent of perception, intelligence and capability. These classes are: (a) simple reflex agents, (b) model-based reflex agents, (c) goal-based agents, (d) utility-based agents, and (e) learning agents.

3.2 Fuzzy Cognitive Maps

The Fuzzy Cognitive Maps constitute an extension of cognitive maps that have been introduced by Kosko [23]. It belongs to the category of flexible computational methods (computational intelligence methods) and have been used in various technology fields such as to analyze electric circuits, to model and support control systems, to represent failure models and affect the analysis of a model system, etc. The FCM constitutes a modeling methodology based on human knowledge. The FCM incorporates that knowledge for modeling systems effectively, faces the ambiguity and uncertainty, and has the ability to adapt to unknown or changing environment. The FCM is suitable to handle available information from abstract knowledge. The construction of FCM is mainly based on experts who define the concepts of model nodes and causal links between nodes with weights. They describe the degree and type of the interaction between nodes [16].

3.3 Multi – agent system for the control of the SWRO desalination system

In order to control the SWRO desalination system under investigation, a multi-agent decentralized energy management system has been designed and realized. In the proposed system intelligent agents were modeled and installed. Each agent acts as an independent agent which control its own unit of the system. Every agent communicate and cooperate with other agents, thus forming an intelligent distributed computing system. The decisions concerning the energy production, the production and storage of potable water and the energy storage (electrical and hydraulic). Every agent receives information from sensors or through other agents. The agent, which control the desalination unit, determine the operation point of the particular controlled unit. The desalination unit operate in variable load, a fact that was utilized in the design of the algorithms. The intelligent agent makes decisions using the Fuzzy Cognitive Maps (FCM) theory.

The MAS of the SWRO desalination system under investigation consists of four intelligent agents:

(a) RES agent

The RES agent is a simple reflex agent. It is connected with the power sensors of the PV array and sets the operating point of the PV array at the maximum power point.

(b) Vessel agent

The Vessel agent is a simple reflex agent. This agent determines the point of electro valves (ON or OFF), setting as a goal the highest hydraulic energy storage.

(c) Hybrid capacitor agent

The hybrid capacitor agent is a simple reflex agent. This agent calculates the SOC (%) of the hybrid capacitor bank and the remaining energy capacity (Wh) in the hybrid capacitor bank at any time step.

(d) Desalination agent

The Desalination agent is a goal-based agent. The goal of Desalination agent is to produce the highest possible quantity of potable water. The desalination unit can operate in variable load and the agent determine unit's operation point with the help of an FCM.

4 EXPERIMENTAL INVESTIGATION

The SWRO Desalination system combined with Multi-Agent decentralized energy management system was experimentally tested. The experimental system was launched during a day with low solar radiation and variable cloudiness and it is presented in Fig. 5. The power consumption of the desalination unit, the water production and the feed pressure are presented in Fig. 6. It is seen in Fig. 6 that the feed water pressure to the membranes follows solar radiation intensity variations. It is observed that the desalination unit never stops operating and the hybrid short term storage ensures continuous system operation (continuous fresh water production) under sharp variations of solar radiation and the potable water production is guaranteed for the entire duration of the day. The total energy produced from the PV array during the system testing of the first day was 1274 Wh and the total potable water production was 173 L. The specific energy consumption, namely the energy consumption per unit volume of the produced fresh water was about 7.4 kWh/m^3.

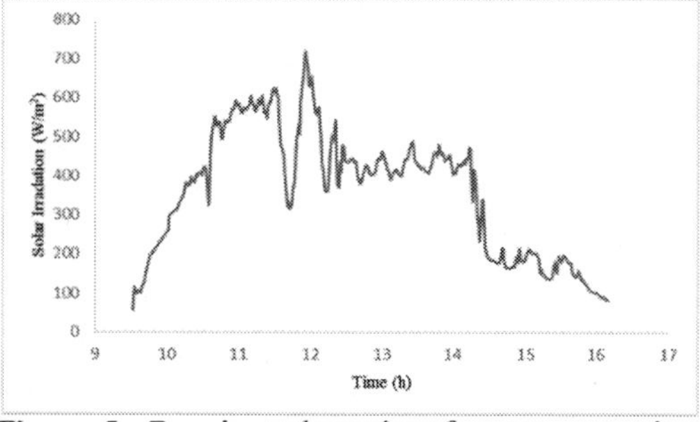

Figure 5: Experimental results of system operation during a cloudy day

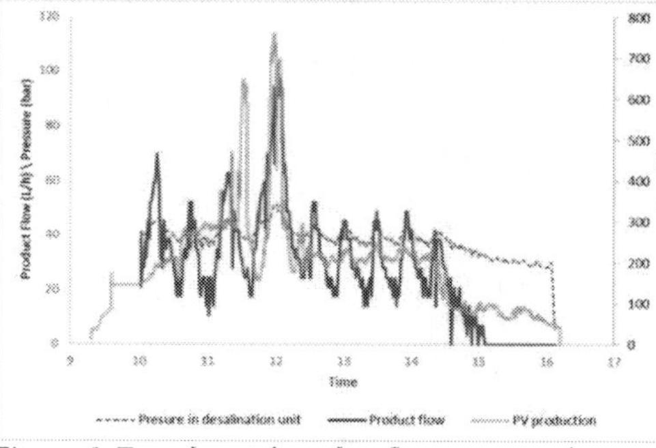

Figure 6: Experimental results of system operation during a cloudy day

5 RESULTS – CONCLUSIONS

This paper deals with a water production system relying on a reverse osmosis desalination unit, which is powered by solar energy and is operated in partial load mode. Two alternative energy storage components have been designed, incorporated and studied in the overall system instead of conventional energy storage units (batteries). The electric energy storage bank consisting of hybrid ultra-capacitors keeps the SWRO desalination system under operation for a period of 10 minutes. The hydraulic pressure buffer, in the form of a pressure vessel's array, is able to feed the desalination unit with sea water under pressure for about 20 minutes.

In the present work, the intelligent multi agent systems theory was investigated, and successfully applied for the decentralized energy management of the SWRO desalination system. The developed MAS consists of five agents that cooperate with each other, featuring distributed artificial intelligence in order to accomplish efficient control of the autonomous desalination system. The experimental system was launched with the help of the MAS – DEMS, during days with high solar radiation and during days with low solar radiation and variable cloudiness. The experimental results obtained from the application of the proposed energy system are encouraging and suggest that significant benefits related to the efficiency and safety of desalination units can be provided. The most significant benefit earned by using the decentralized architecture proposed in the paper is that the managed SWRO desalination system is operated at partial load and makes optimum use of the available power from the PV.

The continuous operation of the desalination unit is guaranteed as the desalination unit never stops producing potable water during the day and the MAS – DEMS gives the possibility to manage the generation, the storage systems and the desalination unit separately, keeping the electrical interchange between production and consumption stable and providing efficient operation with the use of Fuzzy Cognitive Map theory.

In conclusion, the combination of short term energy storage units along with an advanced energy management system is feasible and provides good results in terms of operation for autonomous PV-RO desalination units.

6 ACKNOWLEDGMENT

The present research work was conducted within the

research project "Development of Energy Management Systems for the Design and Control of Autonomous Hybrid Systems and Microgrids, Incorporating Computational Intelligence Techniques", funded by the Scholarship Foundation of Greece in the context of "RESEARCH PROJECTS FOR EXCELLENCE IKY / SIEMENS".

7 REFERENCES

[1] S. N. Gosling and N. W. Arnell, "A global assessment of the impact of climate change on water scarcity," Climatic Change, vol. 134, pp. 371-385, 2016.

[2] M. L. Parry, Climate change 2007-impacts, adaptation and vulnerability: Working group II contribution to the fourth assessment report of the IPCC vol. 4: Cambridge University Press, 2007.

[3] C. Fritzmann, J. Löwenberg, T. Wintgens, and T. Melin, "State-of-the-art of reverse osmosis desalination," Desalination, vol. 216, pp. 1-76, 2007.

[4] K. P. Lee, T. C. Arnot, and D. Mattia, "A review of reverse osmosis membrane materials for desalination—development to date and future potential," Journal of Membrane Science, vol. 370, pp. 1-22, 2011.

[5] L. F. Greenlee, D. F. Lawler, B. D. Freeman, B. Marrot, and P. Moulin, "Reverse osmosis desalination: water sources, technology, and today's challenges," Water research, vol. 43, pp. 2317-2348, 2009.

[6] N. Ghaffour, S. Lattemann, T. Missimer, K. C. Ng, S. Sinha, and G. Amy, "Renewable energy-driven innovative energy-efficient desalination technologies," Applied Energy, vol. 136, pp. 1155-1165, 2014.

[7] E. S. Mohamed and G. Papadakis, "Advances of Renewable Energy Powered Desalination," Handbook of Clean Energy Systems, 2015.

[8] E. S. Mohamed, G. Papadakis, E. Mathioulakis, and V. Belessiotis, "A direct coupled photovoltaic seawater reverse osmosis desalination system toward battery based systems — a technical and economical experimental comparative study," Desalination, vol. 221, pp. 17-22, 3/1/ 2008.

[9] E. S. Mohamed, G. Papadakis, E. Mathioulakis, and V. Belessiotis, "An experimental comparative study of the technical and economic performance of a small reverse osmosis desalination system equipped with an hydraulic energy recovery unit," Desalination, vol. 194, pp. 239-250, 2006.

[10] E. Dimitriou, E. S. Mohamed, C. Karavas, and G. Papadakis, "Experimental comparison of the performance of two reverse osmosis desalination units equipped with different energy recovery devices," Desalination and Water Treatment, vol. 55, pp. 3019-3026, 2015.

[11] E. S. Mohamed, G. Papadakis, E. Mathioulakis, and V. Belessiotis, "The effect of hydraulic energy recovery in a small sea water reverse osmosis desalination system; experimental and economical evaluation," Desalination, vol. 184, pp. 241-246, 11/1/ 2005.

[12] E. S. Mohamed and G. Papadakis, "Design, simulation and economic analysis of a stand-alone reverse osmosis desalination unit powered by wind turbines and photovoltaics," Desalination, vol. 164, pp. 87-97, 3/25/ 2004.

[13] Kyriakarakos, George, et al. "Design of a Fuzzy Cognitive Maps variable-load energy management system for autonomous PV-reverse osmosis desalination systems: A simulation survey." Applied Energy 187 (2017): 575-584.

[14] G. Kyriakarakos, D.D. Piromalis, K.G. Arvanitis, A.I. Dounis, G. Papadakis. On battery-less autonomous polygeneration microgrids: Investigation of the combined hybrid capacitors/hydrogen alternative. Energy Conversion and Management. 91 (2015) 405-15.

[15] G. Kyriakarakos, A.I. Dounis, K.G. Arvanitis, G. Papadakis. A fuzzy logic energy management system for polygeneration microgrids. Renewable Energy. 41 (2012) 315-27.

[16] C.-S. Karavas, G. Kyriakarakos, K.G. Arvanitis, G. Papadakis. A multi-agent decentralized energy management system based on distributed intelligence for the design and control of autonomous polygeneration microgrids. Energy Conversion and Management. 103 (2015) 166-79.

[17] S.D. McArthur, E.M. Davidson, V.M. Catterson, A.L. Dimeas, N.D. Hatziargyriou, F. Ponci, T. Funabashi. Multi-agent systems for power engineering applications—Part II: Technologies, standards, and tools for building multi-agent systems. IEEE Transactions on Power Systems. 22 (2007) 1753-9.

[18] N. Hatziargyriou, A. Dimeas, A. Tsikalakis. Centralized and decentralized control of microgrids. International Journal of Distributed Energy Resources. 1 (2005) 197-212.

[19] S. Russell, "Artificial Intelligence: A Modern Approach Author: Stuart Russell, Peter Norvig, Publisher: Prentice Hall Pa," 2009.

[20] M. N. Huhns, Distributed artificial intelligence vol. 1: Elsevier, 2012.

[21] Y. Shoham and K. Leyton-Brown, Multiagent systems: Algorithmic, game-theoretic, and logical foundations: Cambridge University Press, 2008.

[22] M. Wooldridge, An introduction to multiagent systems: John Wiley & Sons, 2009.

[23] Kosko B. Fuzzy cognitive maps. Int J Man-Mach Stud 1986;24:65–75.

A 360 KWP PV IRRIGATION SYSTEM TO A WATER POOL IN SPAIN

I.B. Carrêlo[1], R.H. Almeida[1,2], L.M. Carrasco[1], F. Martinez-Moreno[1], L. Narvarte[1]
[1] Instituto de Energía Solar - Universidad Politécnica de Madrid, 28031 Madrid, Spain
[2] Instituto Dom Luiz, Faculdade de Ciências, Universidade de Lisboa, 1749-016 Lisboa, Portugal
*Phone: +34 913365531; E-mail: isaac.barata@ies.upm.es

ABSTRACT: This paper describes a 360 kWp PV water pumping system to an elevated water pool in a borehole named San Cristóbal of the Irrigator's Community of Alto Vinalopó, Villena, Spain. This system is able to work along the year, being the maximum need to satisfy in July. The goal of this paper is to show the design, the implementation of the system and some first results by analysing the monitoring done since 1st of June until the end of August. During this period, 173780 m3 of water were pumped and it was working more than 800 hours thanks to its north-south horizontal axis tracker.
Keywords: Design, Monitoring, Stand-alone PV Systems, Water-Pumping

1 INTRODUCTION

The use of photovoltaic (PV) generators for water pumping is not new [1], but with regard to irrigation at large-scale, it is innovative. Nowadays, the state of the art relating to irrigation in agriculture is based on power supply by the national electrical grid or by using diesel generators [2].
However, the rise of the electricity price (in Spain, for example, around 1250%, since 2008) [3], as well as the disappearance of the fossil fuels has forced the irrigators' communities to seek for alternatives.
As a result of this, photovoltaic water pumping systems appeared as a different way for supplying the irrigators needs.
Hence, some regional agriculture associations are intrigue by the opportunity of integrating large stand-alone PV pumping systems in their systems [4].
This paper illustrates the design (Section 2), the implementation (Section 3) and first results (Section 4) of a stand-alone PV water pumping system installed in a borehole named San Cristóbal of the Irrigator's Community of Alto Vinalopó, Villena, Spain.

2 DESIGN

2.1 System configuration

The PV irrigation system consists of a submersible pump of 250 kW (Caprari - E12S55FUS/10A+ MAC 12340 /1C/DF/V-8) in a 400m-deep borehole that provides a daily peak water flow of 226 m^3/h powered by a 360 kW_p PV generator. .The water pumped during the day is accumulated in a water pool of 173000 m^3, elevated 12 meters from the ground. As the pump is installed at 300 m, and the dynamic level of water in the well at maximum water flow, Q_{max}, is 257 m. The total manometric head is 269 m.
To be able to irrigate more hours a day, the PV generator has a north-south horizontal axis tracker divided in 18 rows with 80 panels each one (2x40). This system includes a frequency converter of 355 kW (OMRON A1000 CIMR-AC4A0675AAA), able to transform the direct current (DC) generated by the PV modules to alternating current (AC) at a variable frequency, allowing the pump to work at different frequencies according to the available power. This system also includes an external Programmable Logic Controller (PLC) which plays an important role for the proper functioning of the system (See figure 1).
It is also installed a datalogger system, that not only saves and send by internet the status of each variable from the PLC and frequency converter (by Modbus - serial) but also allows the Irrigator's Community to know the real status of the system by seeing its touch screen.
Since there is no electrical grid connection, in order to fed all the auxiliary charges (monitoring and control systems), it was installed a stand-alone PV system with batteries. This system has a nominal power of 1.5 kW and a total capacity of 32.9 Wh (up to 6 days of autonomy).

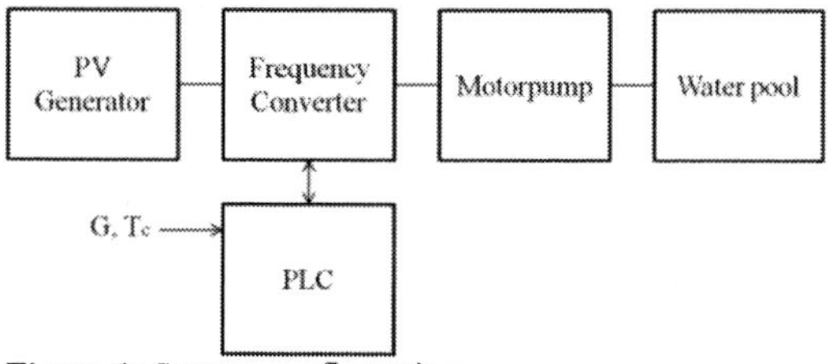

Figure 1: System configuration

2.2 System operating logic

Regarding the control of the system it is important to make a difference between normal and deviant working conditions.
Normally, the PLC is always estimating the available PV power, through the measurement of the irradiance and cell temperature that is possible to get from a reference module which is installed in the tracker, side by side to the rest of modules. Once this estimated PV power is equal or higher than a threshold that allows the pump to work, the PLC gives the run signal to the frequency converter and the system starts pumping water to the water pool.
From this moment, the PLC will execute a Maximum Power Point Tracking (MPPT) routine (which maximizes and improves the efficiency of the PV production), by estimating the set point voltage and sending it to the frequency converter in order to work in this condition.
However there are some deviant conditions that can occur and are expected in the logic of the system.
The first is the instability that occurs due to passing clouds. When the clouds pass over the PV generator, the irradiance reduces and for this reason the current

provided by the panels also decreases ($G\sim I_{DC}$). Hence, the V_{DC} falls down and the frequency converter may have an error [5] [6]. In this installation this does not occurs because there were developed algorithms to avoid this problem decreasing the effect of power intermittences due to clouds in the overall system performance.
There are four more abnormal conditions that are predicted in this system: the motor overheating, empty engine, leakage current and the measure of the isolation of the PV generator. In all of these situations there are installed sensors that give information by analog or digital signals to the PLC. Once the PLC receives these information and "knows" what is happening, sends a stop signal to the frequency converter and does not reset until the problem is solved. Through the screen the user may know what the problem is.

3 IMPLEMENTATION

In order to have a better visualization of the overall system under analysis, the following figures illustrate its main parts. In the figure 2, it is possible to see half of the PV generator oriented to East, in the morning.

Figure 2: PV Generator

The following figure is the reservoir to where the water is pumped. From this waterpool, the water falls by gravity to others reservoirs that the irrigator's community has. Our system has no control of the water that is in the waterpool, pumping all the water possible, since the frequency of the pump is variable (from 38 to 45.5 Hz).

Figure 3: Waterpool

The next figure presents the building where are all the electronic, control and monitoring devices. It is possible to see the stand-alone system installed in the roof.

Figure 4: Building with electronic, control and monitoring system.

Inside this building it is the frequency converter (figure 5). On the right side of the frequency converter, there is another frequency converter in charge of connecting an external fan in case of the temperature inside the box surpasses a specified level.

Figure 5: Frequency Converter

Also inside of the building, in a box, it is installed all the control system (marked in blue) and the monitoring system (marked in orange), that can be seen in the next figure.

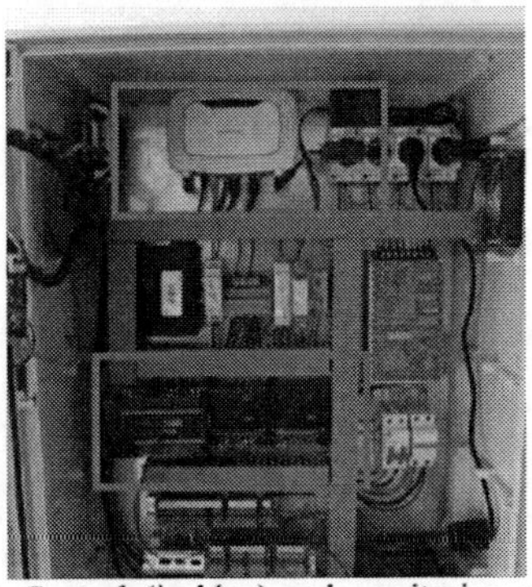

Figure 6: Control (in blue) and monitoring system (in orange)

4 FIRST RESULTS

Regarding the results presented in this paper, an analysis for 92 days (from 1st June to 31st August) is performed. Due to a monitoring problem in one of the days (the file was corrupted), only 91 days are taking into account.
In the next figure it is possible to verify the behavior of the pumping system for a clear sky day.

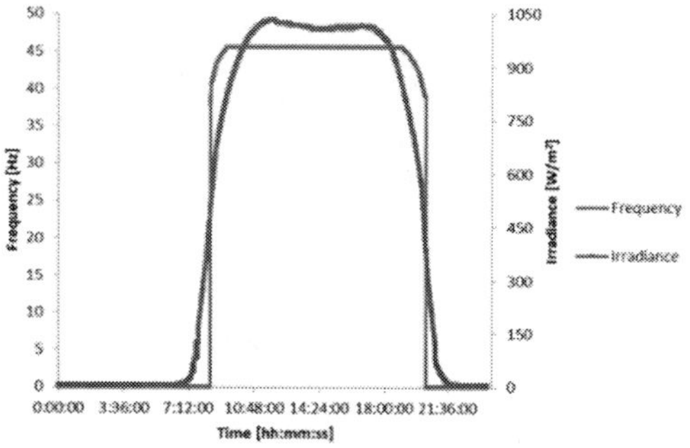

Figure 7: Pump Frequency related with the irradiance received by the PV generator (2nd June 2017).

As it was referred before, when the estimated power surpass a certain threshold, the PLC gives the run order to the frequency converter. This estimated power was adjusted *in situ* in order not only to verify that no error occurs when the system starts to pump but also to optimize the operation of the system, guaranteeing more pumping time. Regarding the maximum frequency, 45.5 Hz were applied in this system due to the dynamics of the well.
According to these 91 days, the volume of pumped water was 173780 m^3 during 823 hours. It is important to refer that if for any reason the pump stop, it will be in that condition for more 8 minutes, in order to empty the tubing. Taking into account the water pumped and the working hours, the average water flow rate is 211.2 m^3/h. The energy consumed in DC (before the frequency converter) is 203370 kWh and the AC energy is 199690 kWh.
Table I summarizes these monitoring results.

Table I: Monitoring results

Variable	Unit	Value
Pumped Water	[m^3]	173780
Irrigation Hours	[h]	823
Q_{avg}	[m^3/h]	211
AC Energy	[kWh]	199690
DC Energy	[kWh]	203370

From these values it is possible to get the frequency converter efficiency, 0.98 (Table II).
Besides the frequency converter efficiency, having the hydraulic energy, it is also possible to calculate the motorpump efficiency, 0.62 (Table II).
The performance ratio (PR) is one of the most important variables to do an evaluation of a PV system. It is expressed by the ratio between the AC energy yield and the theoretical energy production by the PV generator. In this paper, it is made a distinction between two electrical PR. The difference is in the denominator of the equation. While PR considers all the theoretical energy PV production, the $PR_{I\text{-}H}$ only considers the PV energy production when the system is pumping. For that reason $PR_{I\text{-}H}$ is greater than PR (0.75 and 0.64, respectively).

Table II: Values resulting from the monitoring system

Variable	Value
PR	0.64
$PR_{I\text{-}H}$	0.75
η_{MP}	0.62
η_{FC}	0.98

5 CONCLUSIONS

The results show that the demonstrator has been working in accordance to the initial expectations. More working time is needed to fully demonstrate the long-term viability of the solution but the first results have been showing a high reliability and performance – it can be mentioned that, on average in the 91 days, the system was able to pump nearly 2000 m^3 per day.

6 ACKNOWLEDGEMENT

This work has been possible thanks to the funding from the European Union's Horizon 2020 research and innovation programme in the project MArket uptake of an innovative irrigation Solution based on LOW WATer-ENergy consumption (MASLOWATEN), under grant agreement n°640771, as well as the financial support by MIT Portugal Program on Sustainable Energy Systems and the Portuguese Science and Technology Foundation (FCT), grant PD/BD/105851/2014 to Rita Hogan Almeida.

7 REFERENCES

[1] R. Enochian, "Solar-and Wind-Powered Irrigation Systems", United States Department of Agriculture, Economic Research Service, Agriculture Economic Report Number 482, 1982.
[2] J. Carroquino, R. Dufo-López, J. L. Barnal-Augustin, "Sizing of off-grid renewable energy systems for drip irrigation in Mediterranean crops," Renewable Energy, Vol 76 (2015) 566-574.
[3] Grant Agreement number 640771 – MASLOWATEN, European Commission
[4] J. Munoz, J. Carrillo, F. Martínez-Moreno, L. Carrasco and L. Narvarte, "Modeling and simulation of large PV pumping systems," in Proceedings of the 31th European Photovoltaic Solar Energy Conference and Exhibition, Hamburg, 2015.
[5] L. Narvarte, E. Lorenzo, M. Aandam, "Lessons from a PV pumping programme in south Morocco" in Progress in Photovoltaics Research and Applications 13, 261–270, 2005.
[6] J. Fernández-Ramos, L. Narvarte, F. Poza-Saura, "Improvement of photovoltaic pumping systems based on standard frequency converters by means of programmable logic controllers" in Solar Energy 84, 261–270, 2010.

A 160 KWP CONSTANT PRESSURE PV IRRIGATION SYSTEM IN SPAIN

I.B. Carrêlo[1], R.H. Almeida[1,2], F. Martinez-Moreno[1], L.M. Carrasco[1], L. Narvarte[1]
[1] Instituto de Energía Solar - Universidad Politécnica de Madrid, 28031 Madrid, Spain
[2] Instituto Dom Luiz, Faculdade de Ciências, Universidade de Lisboa, 1749-016 Lisboa, Portugal
*Phone: +34 913365531; E-mail: isaac.barata@ies.upm.es

ABSTRACT: This paper describes a 160 kW_p PV irrigation system at constant pressure installed in Valladolid, Spain. The objective of this paper is to present the design, as well as the implementation of the system. Moreover, monitoring data, from 22nd June to 31st August 2017, is analyzed and some first results are presented. During this period, the system was working, on average, almost 10 hours per day, which corresponds to 2085 m^3 of pumped water per day.

Keywords: Design, Monitoring, Stand-alone PV Systems, Water-Pumping

1 INTRODUCTION

Nowadays there is a paradigm-shift in the energy systems, the electricity price is increasing [1] and fossil fuels are depleting. These facts led to an investment in renewable energies. As it is known, irrigation for agriculture is a highly consuming activity of electricity and water [2]. Because of that, the world of irrigation requires innovative solutions [3], less water and energy dependent, to face up this problem. A H2020 project called MASLOWATEN [4] recommends solutions, and includes the installation of 5 real scale photovoltaic water pumping systems as a way to answer the irrigators' needs. This paper is about one of these PV water irrigation systems that is installed in Valladolid, Spain.

The objective of this paper is to show the design (Section 2), the installation (Section 3) and first results (Section 4) resulting by analyzing the monitoring data obtained since 22nd June 2017 to 31st August 2017.

2 DESIGN

2.1 System configuration

The previous irrigation system was composed by a diesel generator that fed two frequency converters which in turn fed two pumps. The first one is a submersible pump of 92 kW (Caprari - E10S50S/6C+MAC10125DS-8V), which pumps water from a 140m-deep borehole to an elevated reservoir of 1000 m^3 and the other is a 30 kW surface centrifugal pump (Caprari - MEC-AS2/80A+FELM 30KW 2P) in charge of pumping at a required constant pressure by an accurate performance of the pivot sprinkles. It was also already installed an irrigation automatism (Agronic 2500).

MASLOWATEN project has several premises. One of them is the concept of adapting the PV pumping system to the existent one. Accordingly, the PV system is integrated in the irrigation system with no changes is this last one.

The PV system includes a 160 kW_p PV generator divided into 8 North-South horizontal trackers, as well as three additional frequency converters (FCs), able to transform the direct current (DC) generated by the PV modules to alternating current (AC).

Two of these FCs fed the previous pumps, allowing both of them to work with PV energy. The third one feeds the pivots (working all the time at 50 Hz, simulating an electrical grid). A Programmable Logic Controller (PLC) was also added to the system in order to facilitate the communication between irrigation and energy parts of the system.

For a better understanding of the system, Figure 1 includes the main components of the PV irrigation system (in green), together with the previous system (in red). As it can be seen, the previous system is still operative. Accordingly, the PLC allows the user to select between the previous and the new PV system allowing the user to come back to it if necessary. This makes easier to get the confidence of the user in the new PV irrigation system.

As well, a datalogger system is integrated in the system in order to monitor the system operation.

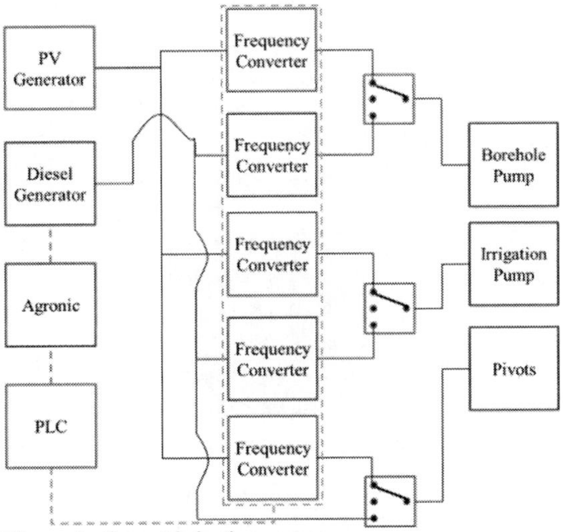

Figure 1: Current irrigation system

2.2 System operation logic

By definition this system is considered a stand-alone PV pumping system. However, since the diesel generator was kept, there were created 3 operation modes that the user can manually choose:

- PV mode, in which all the energy comes from the photovoltaic system.
- PV/Diesel mode, in which both the pivots and

the irrigation pump are fed by the PV system and the other pump is fed by the diesel generator.
- Diesel mode, in which the diesel generator fed the entire system (previous system).

Since the PV production is limited to the peak power installed as well as the weather conditions, the power is not the same along a day. So, it was necessary to create an operating logic control based on the available PV power, the water level in the tank and the end-user irrigation needs.

If there is no irrigation, only the borehole pump (BH pump) is allowed to work. Otherwise, both pumps can work.

Table I shows the system operation logic according to user's need to irrigate based on the available PV power.

Table I: Operation of the system when there is an irrigation program

<table>
<tr><th></th><th>Increase in PV power</th><th>Decrease in PV power</th></tr>
<tr><td></td><td></td><td>Irrigation Pump</td></tr>
<tr><td>30 kW</td><td colspan="2"></td></tr>
<tr><td></td><td colspan="2">Irrigation Pump</td></tr>
<tr><td>35 kW</td><td colspan="2"></td></tr>
<tr><td></td><td colspan="2">Irrigation Pump</td></tr>
<tr><td>36 kW</td><td colspan="2"></td></tr>
<tr><td></td><td colspan="2">Irrigation Pump</td></tr>
<tr><td>45 kW</td><td colspan="2"></td></tr>
<tr><td></td><td colspan="2">Irrigation Pump</td></tr>
<tr><td>51 kW</td><td colspan="2"></td></tr>
<tr><td></td><td>Irrigation Pump</td><td>Irrigation Pump + BH Pump</td></tr>
<tr><td>110 kW</td><td colspan="2"></td></tr>
<tr><td></td><td colspan="2">Irrigation Pump + BH Pump</td></tr>
</table>

As it can be seen a hysteresis logic was programed to avoid the continuous stop and start of the pumps regarding to small variations of the PV power.

Regarding to the water level control in the tank, there are installed 4 floating markers (L1, L2, L3 and L4) as it can be seen in the next figure.

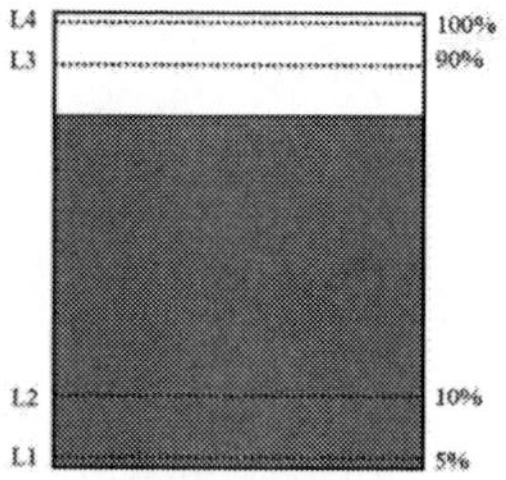

Figure 2: Water tank with the 4 levels

As well as in the case of the threshold power control, in order to avoid stop/start cycles of the pumps due to water tank levels, it is also used a hysteresis control. L1 and L2 are related to the irrigation pump, while L3 and L4 to the borehole one.

In what concerns L1 and L2, if an irrigation program starts with a level equal or higher than L2, the system will start to irrigate. If during the irrigation program the water level falls below L1, the pump will stop and a signal will be received by Agronic to conditionally stop the current irrigation program. A new start or a continuation of an irrigation program can only occur when the water level is equal or higher than L2.

Similarly, if the BH pump is working, this pump will stop automatically when water reaches L4. Then, this pump will only start again when water level is equal or lower than L3.

Besides the operating logic that exists, both frequency converters have their own program. The frequency converter from the borehole pump has a Maximum Power Point Tracking (MPPT) routine (which maximizes and improves the efficiency of the PV production). On the other side, the irrigation pump does a pressure control routine, in which from an analog signal of the actual pressure, by perturb and observe method reach the pressure setpoint. It is able to pump in three different pressure setpoints (3, 4 and 5 bar) according to the field sector that is being irrigated. This frequency converter is also able to send error signals to the PLC of high and low pressure level. Both frequency converters have passing clouds routine that allows the system to not destabilize with fast deviations on the irradiance that incises over the PV generator.

The FC of the pivots, besides the passing shadows algorithm, is parameterized to work continuously at 50 Hz, simulating the grid.

3 IMPLEMENTATION

In order to have a better picture of the overall system under analysis, the next figures show the real system. Figure 3 shows the PV generator and its integration on the field.

Figure 3: PV Generator

The following figures (Figure 4 and 5) show an inside view of the building with the 5 frequency converters (marked in green) and the monitoring and control system (marked in orange).

Figure 4 and 5: Inside view from the building with electronic, control and monitoring system

The following figures include the 1000 m^3 intermedium water tank, between the borehole pump and the irrigation system, and a pivot structure with the low pressure sprinklers.

Figure 6: Water tank

Figure 7: Pivots with low pressure sprinklers

4 FIRST RESULTS

The results presented in this section were obtained from data recorded between 22nd June and 31st August 2017. In this period, due to a monitoring problem in 4 days, we are taking into account 67 days.

The irrigation pump worked for 1225 hours and the borehole pump for 1283 hours, which means a total volume of water of 123921 m^3 and 128273 m^3, respectively. These values indicate that each pump worked nearly 20 hours per day, which means that the system worked not only with PV but also with the previous system (during nighttime).

It is interesting to mention these values in PV mode. The irrigation pump worked for 637 hours and the borehole pump for 580 hours, which means a total volume of water of 61751 m^3 and 77926 m^3, respectively.

It is possible to quantify the PV penetration occurred in this period. In table II, in addition to the previously mentioned data, it is made a distinction if penetration is done considering the irrigation and pumping hours or the volume of water provided by the photovoltaics and the diesel generator, 47% and 58%, respectively. This difference can be explained because when the system works in the diesel mode, the borehole pump frequency is limited to 42 Hz, because of the diesel generator capacity. Notwithstanding when the system is working in the PV mode it pumps most of the time at 50 Hz, pumping for that reason more water.

Table II: Measured data and first results

Variable	Unit	Value
Irrigation hours	[h]	1225
Pumped hours	[h]	1283
Total volume of irrigation water	[m^3]	123921
Total volume of pumped water	[m^3]	128273
PV hourly penetration	[%]	47
PV water flow penetration	[%]	58

Regarding to the performance ratio (PR) parameter, in this paper, it is made a distinction between the typical PR, 0.61, and an irrigation hours PR (PR_{I-H}), 0.68. This last one considered the irradiation only when the system is pumping.

Table III: Performance ratios

Performance Ratio	Value
PR	0.61
PR_{I-H}	0.68

Until here, only global values were presented. To better know the way the PV system works, the following figure represents a typical daily profile of irradiance and frequency of both pumps.

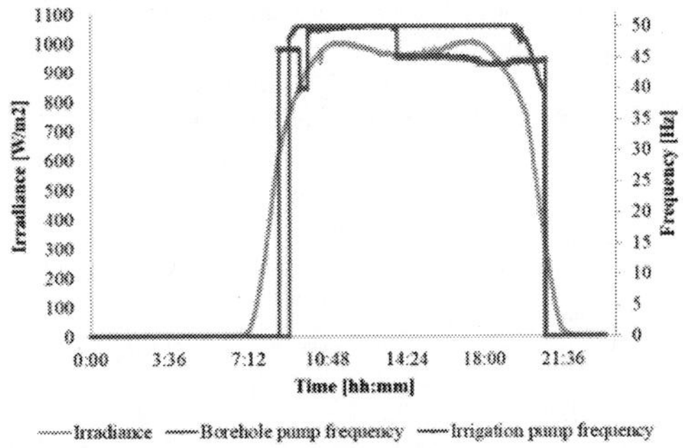

Figure 8: Pumps frequency related with the incident irradiance (11th July 2017)

Based on the irradiance profile, one can say that 11th July 2017 was a clear sky day (no clouds). During this day, the irrigation pump started to pump over 8:30 AM and the borehole pump 30 minutes after. Both pumps work for almost 12 hours fed by the PV system. During the day the irrigation pump changed the output frequency regarding to the change of field sectors that are irrigating and its correspondent pressure setpoint. On the other hand, once the tank was not full the borehole pump was able to pump almost all day at its maximum frequency (50 Hz).

5 CONCLUSIONS

The results show the performance of a real demonstrator with so high requirements of irrigation, around 20 hours per day. Considering this, the PV penetration was almost 60%. Nevertheless more working time is needed to show the long term viability of the system, but these first results show a high reliability of the system – on average it is pumping over 2000 m^3 a day.

6 ACKNOWLEDGEMENT

This work has been possible thanks to the funding from the European Union's Horizon 2020 research and innovation programme in the project MArket uptake of an innovative irrigation Solution based on LOW WATer-ENergy consumption (MASLOWATEN), under grant agreement n°640771, as well as the financial support by MIT Portugal Program on Sustainable Energy Systems and the Portuguese Science and Technology Foundation (FCT), grant PD/BD/105851/2014 to Rita Hogan Almeida.

7 REFERENCES

[1] R. Langarita, J. Sánchez Chóliz, C. Sarasa, R. Duarte, S. Jiménez, "Electricity costs in irrigated agriculture: A case study for an irrigation sheme in Spain", Renewable and Sustainable Energy Reviews, Vol 68 (2017), 1008-1019.
[2] T. Jackson, S. Khan, M. Hafeez, "A comparative analysis of water application and energy consumption at the irrigated field level", Agricultural Water Management, Vol 97 (2010), 1477-1485.
[3] R. Johansson, Y. Tsur, T. Roe, R. Doukkali, A. Dinar, "Pricing irrigation water: a review of theory and practice", Water Policy, Vol 4 (2002), 173-199.
[4] Grant Agreement number 640771 – MASLOWATEN, European Commission.

LARGE-SCALE HYBRID PV-GRID IRRIGATION SYSTEM

R.H. Almeida[1,2], I.B. Carrêlo[1], L.M. Carrasco[1], F. Martinez-Moreno[1], L. Narvarte[1]
[1] Instituto de Energía Solar - Universidad Politécnica de Madrid, 28031 Madrid, Spain
[2] Instituto Dom Luiz, Faculdade de Ciências, Universidade de Lisboa, 1749-016 Lisboa, Portugal
*Phone: +34 913365531; E-mail: rita.hogan@ies.upm.es

ABSTRACT: This paper describes a 120 kWp hybrid PV-grid irrigation system at constant pressure, which has been installed in a 233 ha farm devoted to the intensive cultivation of olive trees in Tamelelt, Morocco. The description includes important information regarding system configuration, working control and implementation. In addition, monitoring data between July 18 and August 31, 2017, is analyzed and first results presented. The system has been working according to the initial expectations since July 2016 and in the 45 days currently under analysis, the number of irrigation hours were 251, being 71% the PV penetration. The PR of the system is 0,33, while the PR during irrigation hours increases to 0,69.
Keywords: Hybrid, PV Pumping, Water-Pumping

1 INTRODUCTION

Agricultural irrigation is a highly consumer of electricity and water. Accordingly, MASLOWATEN, a H2020 project [1], proposes some innovative solutions, less water and energy dependent, as well as the installation of 5 real scale systems for different irrigation applications to show its long-term viability. Therefore, four photovoltaic (PV) systems were installed in the South of Europe (2 in Spain, 1 in Portugal and 1 in Italy) and one in Morocco. This paper will be about this last system.

Morocco has an electrical consumption for water productive applications of 587 GWh, the 20,3% of the total electricity consumption for productive activities with 7% of yearly increase. The electrical consumption for irrigation in this region is estimated to yield 2500 GWh/year. This leads to a current potential market of 1,5 GW of large PV pumps, which represents a potential market volume of 2250M€, with an expected increase in the range of 7%-11% per year.

Worldwide, the high cost of electricity is encouraging the reduction or even elimination of the grid consumption [2]. This leads to the increase in the use of photovoltaic solutions [3], [4].

This paper will present the design (Section 2), implementation (Section 3) and first results (Section 4) of a 120 kWp hybrid PV-grid system for drip irrigation installed in Morocco. Finally, some conclusions (Section 5) are summarised.

2 DESIGN

2.1 System configuration

The system was installed in Tamelelt, Morocco, in a 233 ha farm devoted to the intensive cultivation of olive trees. This farm belongs to ELAIA, and the annual irrigation needs goes from March to December every year, being the months with highest consumption May and June.

A previous system was already installed and included two centrifugal surface pumps of 45 kW for irrigation at constant pressure (4.2 bar), operating at 50 Hz driven by soft starters powered from the national grid.

The farm is divided in eight sectors and the irrigation is done in 4 different shifts, each one with two sectors. The area irrigated in each shift varies from 56 to 59,8 ha and the water flow from 368,3 to 393 m^3/h.

The design of the PV system was done taking into account the characteristics of the irrigation system already installed and simulations were done using SISIFO [5], 6].

The new and current system includes a PV generator of 120 kWp, electrically divided into two equals fields of 60 kWp, each one feeding one frequency converter of 55 kW (Omron 3G3RX-A4550) - which replaces the previous soft drivers - and the 45 kW pumps (Caprari – MEC-AS4/125C+ FELM 45KW 4P) .

It is important to mention that the 120 kWp are a combination of 480 PV modules of 250 Wp, organized in 24 strings of 20 modules. These PV modules are mounted in a North-South axis tracker, occupying only 0.15% of the total available area.

Now, each frequency converter is fed both by the PV generator and the grid. Accordingly, three operating modes are available: PV mode, hybrid PV-grid mode, and grid mode.

In addition to the previous components, the system also includes a Programmable Logic Controller (PLC) – which made the communication between photovoltaic and irrigation parts of the system, a monitoring system with a touch screen, an irrigation automatism (Agronic 4000, from Progres), a filtration system with automatic cleaning, a fertirrigation system with two auxiliary pumps and a pump filling water from the fertirrigation tanks. The first two components mentioned were added to the system with the installation of the PV components. All these auxiliary loads are fed from the grid.

Figure 1 includes the main components of the system.

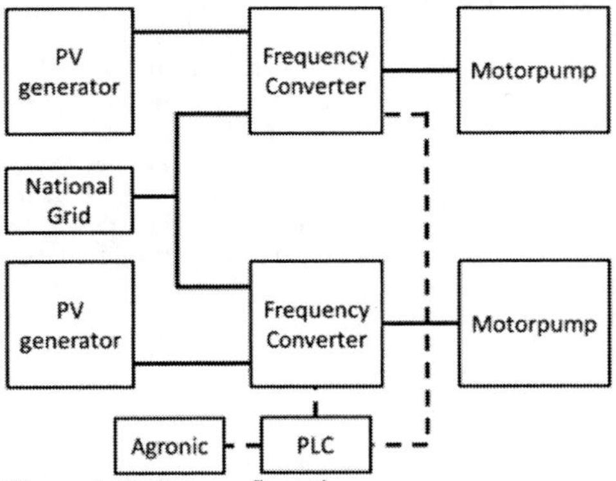

Figure 1: System configuration

2.2 System operating logic

The control of the system is done through the irrigation and photovoltaic controllers, Agronic and PLC accordingly. In normal operating conditions, the irrigation controller sends a signal to the PLC when there is a need to irrigate. At this time, the PLC sends a run signal to both frequency converters and the system starts to work.

The initial idea was that according to the end-user desire, the grid can be connected or disconnected to the frequency converters, i.e., there is a contactor in each frequency converter that can be opened and grid is not used. This can be very useful to avoid grid consumption in high price periods. In this case, if there is not enough PV power, the PLC will send a signal to the Agronic to temporally stop irrigation. Despite having this possibility, the end-user decided to be all the time connected to the grid, no matter the available PV Power.

The selection of the operating mode is done automatically according to the PV power available on each moment. This PV power is calculated in the PLC based on the measured data of irradiance and cell temperature of a calibrated solar cell installed in the PV tracker.

Accordingly, three operating modes are available:

- PV mode: if the PV power is enough to fulfill the required load, the power consumption from the grid is null;
- Hybrid PV-Grid mode: if the PV power is not enough, the grid will inject the correspondent lack of PV power to maintain the pressure in the irrigation system;
- Grid mode: if the PV power is null, the system is totally fed by the grid (like in the previous system, but now using the frequency converters instead of the soft starters).

The two frequency converters work in a master-slave mode, i.e., one of the frequency converters (the master) control the pressure in the irrigation system and sends a digital signal (representing the frequency) to the other frequency converter (the slave). It is important to note that a specific algorithm has been implemented to maximize the use of PV production.

It should also be mentioned that if there is no water on the water tank, if there is some pressure issue (pressure higher or lower than the expected one), or if one or both frequency converters have some type of mal-functioning, the PLC will send a signal to Agronic and irrigation program will be stopped.

3 IMPLEMENTATION

The PV system was installed in the field in July 2016. This means that it has been working in half of the last irrigation campaign (2016 one) and in the current one.

In order to have a better understanding of the system, the following figures include the olive tree plantation (Figure 2), the integration of the PV generator on the plantation (Figure 3), the engine room (Figure 4) and the set of pumps (Figure 5).

Figure 2: Olive tree plantation.

Figure 3: PV generator.

The engine room includes all the components that can be seen in the next figure plus the irrigation controller. The two frequency converters are in the electric boards on the ground (the master on the right and the slave on the left), while the upper board has the PLC inside and the touch screen on the door. On the left wall, the big box is the DC protection box and the small ones are related to the monitoring system.

Close to the engine room, there is the pumping room, where one can find all the additional components mentioned in Section 2. Figure 5 shows only the set of pumps.

Figure 4: Engine room.

Figure 5: Set of pumps.

4 FIRST RESULTS

The results presented in this section were obtained from data recorded between July 18 and August 31, 2017 (45 consecutive days).

During this period, the irrigation time was 251 hours, which means 92433 m^3 of pumped water and an average water flow of 368,5 m^3/h.

The PV system was able to produce 16342 kWh of DC energy, converted in 15383 kWh of AC energy. The frequency converter also registers the DC power in the DC bus, which is, in fact, the sum of the instantaneous power coming from the grid and coming from the PV generator. Since this last value is also being measured, the PV penetration, i.e., the percentage of power/energy that is coming from the PV generator is calculated. During the period under analysis this penetration is 71%, which also means that the equivalent quantity of water extracted with PV is 65609 m^3. Table I summarizes these results.

Table I: Measured data and first results

Variable	Value
Irrigation hours [h]	251
Volume of pumped water [m^3]	92433
Volume of pumped water from PV [m^3]	65609
Average flow [m^3/h]	368,5
DC energy from PV [kWh]	16342
AC energy from PV [kWh]	15383
PV penetration [%]	71

The previously mentioned measured data, plus irradiation (see Table II) and nominal peak power (120 kWp), also allows the calculation of the performance ratio (PR) of the system. In addition to the typical PR of a PV system, a specific one for irrigation applications is also presented. This last one only considered AC power and irradiance during irrigation hours. The typical PR in this time interval is 0,33, while the PR during irrigation hours (PR_{I-H}) is around the double, 0,69 (see Table III).

Table II: Irradiation

Irradiation	Value [kWh/m^2]
Total irradiation	385,4
Irradiation during irrigation hours	184,5

Table III: Performance ratios

Performance ratio	Value [ ]
PR	0,33
PR_{I-H}	0,69

Until this point, only global values were presented. To a better understanding of the system working along a specific day, two particular days are presented in Figures 6 and 7, representing the first one a clear day and the last one a cloudy day. Irradiance as well as the frequency of the master pump are plotted (it should be mentioned that the frequency of the slave pump has exactly the same behavior of the master one and for that reason it is not included in the figure).

It can be seen that, in both figures, the frequency remains constant no matter the variations in irradiance and, accordingly, in the available PV power. This is particularly noticeable in the case of the cloudy day (Figure 7).

Since drip irrigation systems have high requirements regarding both constant pressure and flow, the variability of the PV power along a day maybe a threat.

The main advantage of a hybrid PV-grid system like this one, with hybridization in the electric part, is its reliability due to its robustness to a variation in irradiance and consequently in PV power. In fact, the frequency converter will always receive their needed power, being the only difference the percentage that is coming from the PV and from the grid. Consequently, the irrigation part of the system will see no variation in pressure and flow due to clouds.

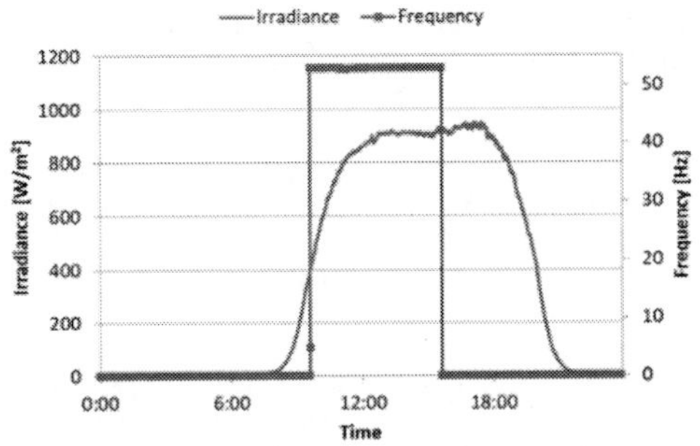

Figure 6: Irradiance and frequency of the master pump in a clear day.

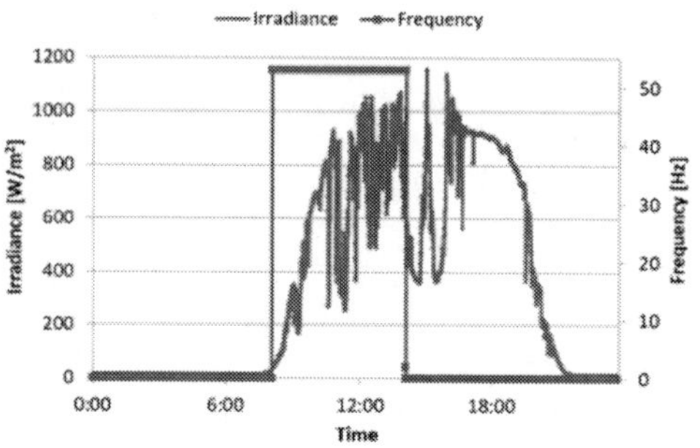

Figure 7: Irradiance and frequency of the master pump in a cloudy day.

Still keeping in mind the two particular days under analysis, it is worth mentioning that the PV penetration in the cloudy day is 62%, while in the clear day this value increases to 82%.

5 CONCLUSIONS

Considering the high requirements of pressure and flow of drip irrigation systems like the one under analysis, the PV-grid hybrid solution implemented guarantees a high system reliability.

During the 45 days under analysis, a 71% of PV penetration is achieved in the 251 hours of irrigation. This corresponds to a little more than 92000 m3 of pumped water and a PR during irrigation hours of 0,69.

6 ACKNOWLEDGEMENT

This work has been possible thanks to the funding from the European Union's Horizon 2020 research and innovation programme in the project MArket uptake of an innovative irrigation Solution based on LOW WATer-ENergy consumption (MASLOWATEN), under grant agreement n°640771, as well as the financial support by MIT Portugal Program on Sustainable Energy Systems and the Portuguese Science and Technology Foundation (FCT), grant PD/BD/105851/2014.

7 REFERENCES

[1] Grant Agreement number 640771 – MASLOWATEN, European Commission.
[2] R. Langarita, J. Sánchez Chóliz, C. Sarasa, R. Duarte, S. Jiménez, "Electricity costs in irrigated agriculture: A case study for an irrigation scheme in Spain," Renewable and Sustainable Energy Reviews, Vol 68 (2017), 1008-1019.
[3] L. Narvarte, E. Lorenzo, "Sustainability of PV water pumping programmes: 12-years of successful experience", Progress in Photovoltaics: Research and Applications, Vol 18 (2010) 291-298
[4] C. Gopal, M.Mohanraj, P.Chandramohan, P. Chandrasekar, "Renewable energy source water pumping systems – A literature review", Renewable Energy and Sustainable Energy Reviews, Vol 25 (2013) 351-370.
[5] J. Muñoz Cano, L. Marroyo, M. Collares-Pereira, N. Tyutyuyndzhiev, M. Conlon, A. El Moussaoui and B. Wilkin, "An open-source simulation tool of grid-connected PV systems," in Proceedings of the 28th European PV Solar Energy Conference and Exhibition, Paris (2013).
[6] J. Munoz, J. Carrillo, F. Martínez-Moreno, L. Carrasco and L. Narvarte, "Modeling and simulation of large PV pumping systems," in Proceedings of the 31th European Photovoltaic Solar Energy Conference and Exhibition, Hamburg (2015).

A 140 KW HYBRID PV-DIESEL PUMPING SYSTEM FOR CONSTANT-PRESSURE IRRIGATION

R.H. Almeida[1,2], I.B. Carrêlo[1], F. Martinez-Moreno[1], L.M. Carrasco[1], L. Narvarte[1]
[1] Instituto de Energía Solar - Universidad Politécnica de Madrid, 28031 Madrid, Spain
[2] Instituto Dom Luiz, Faculdade de Ciências, Universidade de Lisboa, 1749-016 Lisboa, Portugal
*Phone: +34 913365531; E-mail: rita.hogan@ies.upm.es

ABSTRACT: This paper describes a 140 kWp hybrid PV-diesel pumping demonstrator for constant pressure irrigation in a 200 ha olive tree plantation at Herdade de São Bernabé, Alter do Chão, Portugal. The PV system was installed in June 2016 and has been working since then. A previous system fed exclusively by a diesel generator was already implemented, and the PV part was added, maintaining unchanged all the irrigation part of the system. The new system configuration and control logic, as well as the implementation in the field, are detailed in this work. The first results obtained from monitoring data during June, July and August 2017 show that 125760 m3 were pumped (55179 m3 with photovoltaic) during 590 irrigation hours. The PR during irrigation hours is 0,42 and PV penetration 43%.

Keywords: Hybrid, PV Pumping, Water-Pumping

1 INTRODUCTION

The majority of agricultural irrigation systems are currently based on electricity from the grid of from diesel generators [1]. However, the increasing price of the energy from these sources [2], [3] boosted the development of alternative solutions using renewable energies, namely photovoltaic solar energy [4], [5].

These new solutions create new problems associated with the variability of the solar resource [1], both along the day and along the year, which leads to instability in the control and lower efficiencies, as well as problems in the hydraulic system [6] and in constant pressure irrigation plants. The system under study has robust algorithms in the standard frequency converters, programmable logic controllers and irrigation automatisms that minimized these problems.

Regarding these PV solutions, if the number of irrigation hours is higher than the number of sun hours, it is necessary to have a hybrid solution. This means that the cost of diesel or grid can decrease but cannot be eliminated. These systems can be hybridized in the electric or in the hydraulic part of the system [7]. In this particular case, a hydraulic hybridization is done.

This paper presents the design (Section 2), implementation (Section 3) and first results (Section 4) of a 140 kWp hybrid PV-diesel system for drip irrigation. Finally, some conclusions (Section 5) are summarised.

2 DESIGN

2.1 System configuration

The system was installed in a previously diesel powered drip irrigation system in Alter do Chão, Portugal. It is a 200 ha farm devoted to the super-intensive cultivation of olive trees. The annual irrigation needs of the farm goes from May to October every year.

As far as the irrigation part of the system is concerned, the farm is divided in 30 sectors, combined in 7 different shifts. The set point pressure is 5,7 bar and the working flow varies between 200 and 240 m^3/h.

The PV generator system and all its components were added to a previously installed irrigation system with a diesel generator, an irrigation automatism (Agronic 4000), a frequency converter (Diesel FC) and a soft-starter, as well as two centrifugal pumps (see Figure 1).

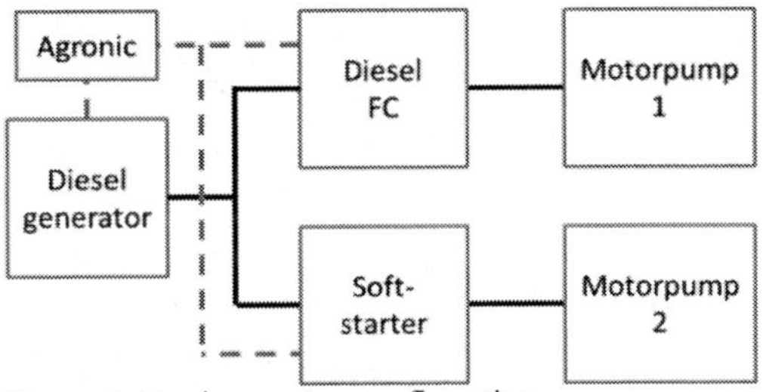

Figure 1: Previous system configuration

In order to partially substitute the diesel consumption, the current system (Figure 2) includes a 140 kWp PV generator using a North-South horizontal axis tracker. The PV generator feeds two 55 kW frequency converters (Omron 3G3RX-A4550). Each of this FC is electrically connected to a 45 kW centrifugal pump (Caprari MEC-MRS 100-2D 45 kW).

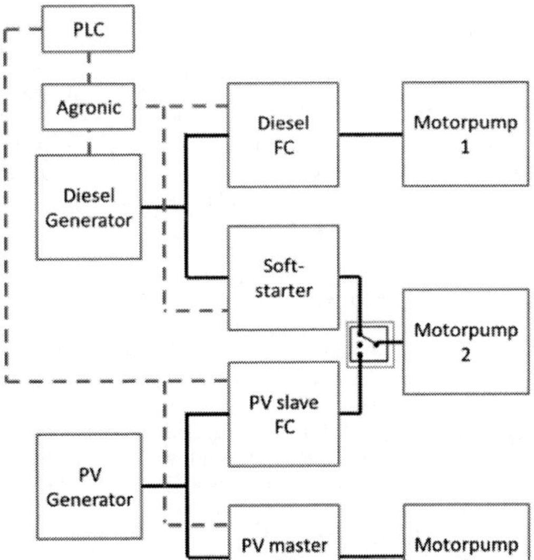

Figure 2: Current system configuration – the new components of the system are marked in orange, while the previous ones are marked in green.

As one can see, two additional FC (PV master FC and PV slave FC) were added but only one pump (motorpump 3). The pump previously fed by the soft-starter (motorpump 2) can now also be fed by the PV generator. A contactor was added before this pump to allow the change of energy source. The new components of the system are marked in orange, while the previous ones are marked in green.

Moreover, a programmable logic controller (PLC) was also added to the system, as well as a monitoring system. This PLC controls the three different operating modes that the system is currently able to work - PV, hybrid and diesel.

A 1 kWp stand-alone PV system with storage capacity is included to feed auxiliary loads, namely the PLC and the Agronic.

2.2 System operating logic

The control of the system is done by means of the PLC and the irrigation automatism. When the irrigation automatism sends a signal of irrigation to the PLC, this one calculates the instantaneous available PV power after checking the irradiance and cell temperature.

Table I includes the PV power thresholds (with hysteresis) for the three operating modes.

Table I: Power thresholds with hysteresis for the different operating modes - PV, hybrid and diesel

	Increase in PV power	Decrease in PV power
	PV	
105 kW		
	Hybrid	PV
95 kW		
	Hybrid	
55 kW		
	Diesel	Hybrid
45 kW		
	Diesel	

If the available PV power is enough to irrigate at the pressure set point, the PLC sends a run signal to both PV frequency converters. The PV master FC will regulate its output frequency according to the pressure set point value, while the PV slave FC will follow the frequency of the master through an analog signal.

Otherwise (i.e. if there is not enough power to work 100% with PV energy), the system can work in hybrid or diesel mode. In both of these last modes the diesel generator needs to be connected and for that reason the PLC sends a digital signal to the Agronic that will "inform" that it is necessary to connect the diesel generator. With the available PV power, PLC is also able to select between hybrid and diesel mode – if there is enough power for one pump to work with photovoltaics, the system will work in hybrid mode, if there is not it will work on diesel mode.

In the hybrid mode, the control of the pressure is done by the diesel frequency converter, while the PV master FC works in a maximum power point tracking (MPPT) routine.

In the diesel mode, the diesel FC controls the pressure and the soft-starter is also working.

It is important to note that the current system includes three pumps, although only two can work simultaneously (as in the previous system).

Table II allows an easy visualization of the components working and not working in each operating mode.

Table II: Status of the main components in each operating mode – PV, hybrid and diesel

Component	Mode		
	PV	Hybrid	Diesel
Diesel generator	OFF	ON	ON
Soft-starter	OFF	OFF	ON
Diesel FC	OFF	ON	ON
PV master FC	ON	ON	OFF
PV slave FC	ON	OFF	OFF

In order to increase the lifetime of the components of the system, the following protections are programmed (it is important to mention that these protections overlap the selection mode based on the available PV power):

- If the diesel generator receives 3 run signals in the same irrigation program due to, for example, passing clouds that reduce intermittently the PV power, the diesel generator will be working for at least one hour. This means that during this hour the system will work only on hybrid or diesel modes (according to the available PV power). This will protect the diesel motor against a high number of starts.
- If the soft-starter receives 3 run signals in the same irrigation program, it will be working for at least one hour. This means that the system will work on diesel mode during this time. This will protect against continuous transitions between PV and diesel modes, which can affect the irrigation system.

In both of the previous cases, once the timer finishes, the system will return to its normal operating conditions based on the available PV power.

Furthermore, an additional switch is included in the system. This switch allows the end-user to return to the previous system, i.e., if the end-user desires to use only the diesel generator, he/she only needs to activate this switch. Obviously, in this case, the available PV power will not influence the working of the system.

3 IMPLEMENTATION

This system has been working since June 2016. The following figures allow the reader to have a better visualization of the real system. Figure 3 shows part of the 140 kWp N-S axis PV generator, Figure 4 includes the set of pumps and the filter bench, while Figure 5 shows the majority of the electrically components of the system (irrigation and PV controllers, as well as frequency converters and soft-starter), as well as the stand-alone system. Finally, in Figure 6 the olive tree plantation is shown.

Figure 3: PV system generator

Figure 4: Set of pumps and filter bench

Figure 5: Engine room

Figure 6: Olive tree plantation

4 FIRST RESULTS

The first results presented are based on monitoring data recorded in the months of June, July and August 2017 (92 consecutive days).

The DC energy produced by the PV generator was 21019 kWh, which means 19866 kWh of AC energy. The water pumped with this energy was 55179 m^3, being the total pumped water 125760 m^3. The system was working during 590 hours with an average flow of 213 m^3/h. The PV penetration during this period was 43%. A summary of these values is presented in Table III.

Table III: Measured data and first results

Variable	Value
Irrigation hours [h]	590
Volume of pumped water [m^3]	125760
Volume of pumped water from PV [m^3]	55179
Average flow [m^3/h]	213
DC energy from PV [kWh]	21019
AC energy from PV [kWh]	19866
PV penetration [%]	43

Two performance ratios are calculated: the typical one and one that only considers the irrigation hours. Based on the measured irradiance presented in Table IV the typical PR is 0,16, while the PR during irrigation hours (PR_{I-H}) is 0,42 (see Table V).

Table IV: Irradiation

Irradiation	Value [kWh/m^2]
Total irradiation	895
Irradiation during irrigation hours	341

Table V: Performance ratios

Performance ratio	Value [ ]
PR	0,16
PR_{I-H}	0,42

Figure 7 includes the daily irradiance during three days in June 2017 and Table VI the number of irrigation hours in each operating mode in the same periods.

As it can be seen, June 2 and June 16 have similar irradiance profiles (typical of a non-cloudy day), while June 6 has a profile which represents a cloudy day until mid-afternoon. These profiles, if nothing else is taking into account should mean that PV penetration will be similar on June 2 and 16, being the lowest penetration on June 6. Although, seeing the values in Table VI this does not happen. June 2 is a great day regarding PV penetration (around 70%), while June 16 is not (around 20%). This happened because a hydraulic issue occurred during the day and there was the need to irrigate during nighttime. On the other side, even on June 6 (the cloudy day) the system was able to work 100% based on PV for 2 hours and in the hybrid mode during 3 hours. So, the user and the problems in the irrigation system influence dramatically the performance of the PV irrigation system independently of the quality of the PV plant.

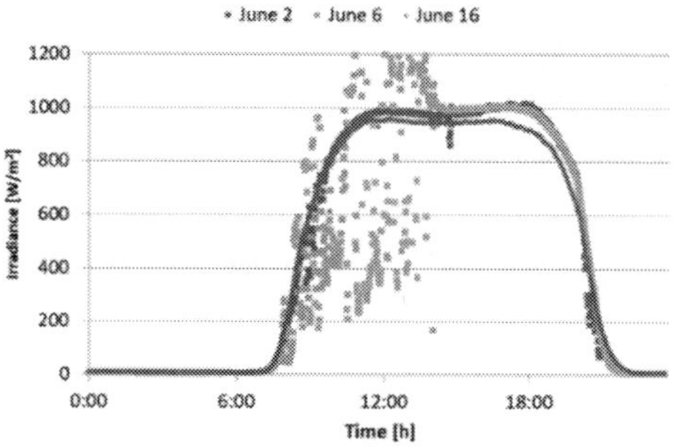

Figure 7: Irradiance daily profiles for three different days – June 2, June 6, and June 16, 2017

Table VI: Number of irrigation hours in three different days in each operating mode

Day	Mode			
	PV [h]	Hybrid [h]	Diesel [h]	Total [h]
June 2	10	1	4	15
June 6	5	1	9	16
June 16	2	3	12	17

5 CONCLUSIONS

This paper has presented the design and implementation of a 140 kWp PV-diesel pumping system installed in Alter do Chão, Portugal. The system has been working since June 2016 in accordance with the initial expectations of guaranteeing constant pressure and flow. As far as first results are concerned, 125760 m^3 were pumped during 590 irrigation hours between June and August 2017, being 55179 m^3 pumped with PV energy. The PR during irrigation hours is 0,42 and PV penetration 43%.

6 ACKNOWLEDGEMENT

This work has been possible thanks to the funding from the European Union's Horizon 2020 research and innovation programme in the project MArket uptake of an innovative irrigation Solution based on LOW WATer-ENergy consumption (MASLOWATEN), under grant agreement n°640771, as well as the financial support by MIT Portugal Program on Sustainable Energy Systems and the Portuguese Science and Technology Foundation (FCT), grant PD/BD/105851/2014.

7 REFERENCES

[1] J. Carroquino, R. Dufo-López, J. L. Barnal-Augustin, "Sizing of off-grid renewable energy systems for drip irrigation in Mediterranean crops," Renewable Energy, Vol 76 (2015) 566-574.

[2] R. Langarita, J. Sánchez Chóliz, C. Sarasa, R. Duarte, S. Jiménez, "Electricity costs in irrigated agriculture: A case study for an irrigation scheme in Spain," Renewable and Sustainable Energy Reviews, Vol 68 (2017), 1008-1019.

[3] M. Abu-Aligah, "Design of Photovoltaic Water Pumping Systems and Compare it with Diesel Powered Pump," Jordan Journal of Mechanical and Industrial Engineering, Vol 5 (2011) 272-280.

[4] P. E. Campana, et. al, "Suitable and optimal locations for implementing photovoltaic water pumping systems for grassland irrigation in China", Applied Energy, Vol 185 (2016) 1879-1889.

[5] L. Narvarte, E. Lorenzo, "Sustainability of PV water pumping programmes: 12-years of successful experience", Progress in Photovoltaics: Research and Applications, Vol 18 (2010) 291-298

[6] J. Fernández-Ramos, L. Narvarte Fernández, F. Poza-Saura, "Improvement of photovoltaic pumping systems based on standards frequency converters by means of programmable logic controllers", Solar Energy, Vol 84 (2010) 101-109.

[7] Technical specifications and quality control procedures for contractual frameworks, available at www.maslowaten.eu

A NEW METRIC FOR ASSESSING LOCAL MECHANICAL LOAD SCENARIOS FOR PV MODULES AT SPECIFIC LOCATIONS

Christian Camus[1*], Pascal Offermann[1], Martin Weissmann[2], Claudia Buerhop-Lutz[1], Jens Hauch[1] and Christoph J. Brabec[1,3]
[1] Bavarian Center for Applied Energy Research (ZAE Bayern), PV-Systems, Immerwahrstr. 2, 91058 Erlangen, Germany
[2] Meteorologisches Institut, Ludwig-Maximilians-Universität, Theresienstr. 37, 80333 München, Germany
[3] Materials for Electronics and Energy Technology (i-MEET), Friedrich-Alexander-Universität Erlangen-Nürnberg, Martensstraße 7, 91058 Erlangen, Germany
[*]Corresponding Author: christian.camus@zae-bayern.de

ABSTRACT: In this paper, a new metric for assessing the location-specific mechanical loads imposed by snow or wind gusts which a photovoltaic module can be expected to be exposed to at a given location is introduced. For demonstrating this general approach, two locations of moderate distance (~110km) in Northern Bavaria are compared. It is shown that even on moderate distances the expected loads differ by a factor of four or more highlighting the importance of location-specific requirements for PV modules. For assessing the loads, the historical weather data of the closest surrounding weather station is evaluated in order to deduce an expectation value for the site of interest. This data is analyzed with respect to the frequency and severity of the relevant weather events such that the annuality of a weather event of a given severity can be determined. Finally, the severity of an event corresponding to a single occurrence in 20 years, i.e. a standard product warranty, is chosen to represent the requirements of a location with respect to a certain parameter.
Keywords: Photovoltaic, Forecast, Reliability, Weather

1 INTRODUCTION

Nowadays, photovoltaic (PV) modules are usually manufactured according to certain IEC standards, which are meant to assure product safety and reliability. However, these standards usually test a certain set of conditions or scenarios, independently from the actual location at which a particular module is going to be installed in the end. Thus, a module will always overfulfil many local requirements whilst possibly failing one in which its properties might not be sufficient for a particular location finally leading to a module failure within its expected lifetime or even guarantee period.

Especially, in times of high price pressure on the manufacturer side it would be highly desirable to cut down requirements where they are too strict, whilst ensuring robustness where needed. As most PV modules are going to be installed at their installation time for their entire technical lifetime a customization to the local requirements in terms of climate parameters or safety demands would be possible and highly desirable. However, in order to do so a method is needed to assess the climatic conditions for any given location for deducing the respective requirements.

Among the climate conditions affecting module reliability one has to distinguish between factors which lead to a constant ageing of the module and the materials therein, such as UV dose temperature cycles and factors for which the most severe single event can be expected to be most relevant. Mechanical loads can be assumed to belong to the latter group with wind gusts and snow loads being the most critical ones for non-desert climates.

According to present data for differentiating the wind conditions for e.g. Germany only a very coarse site classification can be made. For instance, DIN EN 1991 1 4, which provides a wind zone map for Germany, categorizes entire Bavaria, Hesse, Baden-Wuerttemberg as well as parts of Rhineland-Palatinate and North Rhine-Westphalia as one single category. For the so-called Test Reference Years (TRY) published in [1], 15 weather regions are differentiated for Germany, which still constitute a rather coarse classification scheme, even though they can be modulated by factors accounting for elevation and urbanity. Furthermore, the TRYs only state hourly values, thereby automatically neglecting wind gusts which can be expected to constitute a major factor in mechanical load-induced PV module degradation.

Here, an approach for a site-specific assessment for expected loads is presented exemplarily for mechanical loads imposed by wind gusts and snow loads which can generally be transferred to any other parameters such as UV irradiation, temperature cycles etc., even though slight modifications might be needed. Thus, here meteorological snow and wind gust data will be used to deduce a long-term forecast for the site-specific stress exposure due to mechanical loads, whereas to date snow forecasts have been mainly discussed in terms of yield forecasting as snow covering prevents the modules from power generation [2,3].

2 GENERAL APPROACH

Currently, various studies focus on the determination of regional degradation rates by monitoring modules installed at various locations all over the world [4]. However, though yielding highly important data, these studies are limited by two facts: Firstly, they analyze the integral influence of the local weather, so the influence of a single parameter on the degradation cannot be deduced. Secondly, by nature these studies are very time-consuming as the modules degrade in real-time. Thus, it will take several years to be able to determine the regional lifetime differences for modules. This paper's approach aims at overcoming these limitations by using a standardized statistical analysis approach for assessing expected mechanical loads due to snow and wind. Thus, a 20-years-projection can be obtained right away and single parameters can be separated. This method then allows for deducing location or region-specific qualification-tests for any given location, which are tailored for assessing the suitability of a module for the respective location and can be conducted in the lab in an accelerated manner.

This study is based on the assumption that for mechanical loads, a module failure will be caused by the most severe load occurring during product lifetime rather than by the cumulative amount of its loading history. For other weather parameters, such as UV radiation or thermal cycles a similar approach could be chosen, but in that case the cumulative effect would have to be considered instead.

The new metric allows assessing the location-specific mechanical loads imposed by snow or wind gusts which a photovoltaic module can be expected to be exposed to at a given location by analyzing historical weather data of close-by weather stations of the German Weather Service (DWD) in order to deduce an expectation value for the station of interest [5]. This data is analyzed with respect to the frequency and severity of the relevant weather events such that the annuality of a weather event of a given severity can be determined. Finally, the severity of an event corresponding to *a single occurrence in 25 years*, i.e. a standard product warranty period, is chosen to represent the requirements of a location with respect to a certain parameter (e.g. snow load).

Here, two locations (Erlangen, elevation: 279m, and Arzberg, elevation: 468m) of moderate distance (~110km) in Northern Bavaria are compared. In order to deduce the expected wind gust and snow load data for both locations, the weather data of the close-by weather stations of the DWD (see Table I) was imported for further analysis. Here, DWD weather stations within a distance of 30km (for snow, up to 64km for wind) were taken into account. Generally, all available data was considered (here, starting 1950), even though data was not provided for all stations for the full period of time. If no data was available for some of the stations data was deduced based on the data of the remaining stations as long as at least data was available for three stations. The incompleteness of this data constitutes on of the main reasons for transferring the approach presented here into an automated method with is based on complete and consistent reanalysis data (see section 5).

Table I: Weather stations and respective distances to the sites of interest: a) Erlangen, b) Arzberg

a) Erlangen

Station (years of data)	Distance in km	applied for
Erlangen-Bruck (3)	2.2	Snow
Uttenreuth (35)	4.0	Snow
Nuremberg-Netzstall (41)	8.1	Snow
Möhrendorf (66)	8.6	Snow
Nuremberg-Buchenbühl (25)	10.0	Snow
Nuremberg (60)	23.2	Snow, Wind
Bamberg (44)	34.3	Wind
Roth (15.5)	40.1	Wind
Kümmersbruck (15.5)	64.8	Wind

b) Arzberg

Station	Distance in km	applied for
Waldsassen (9)	9.3	Snow
Selb (2)	15.6	Snow
Wunsiedel (10)	16.4	Snow, Wind
Bad Elster (66)	23.7	Snow
Tirschenreuth (43)	25.4	Snow
Fichtelberg (66; 20)	27.4	Snow, Wind
Hof (39)	36.1	Wind

The resulting weather data for both locations was calculated from the station data by applying the Inverse Distance Weighting (IDW) method [6]. Thereby, results of stations being closer to the site of interest than others are weighted according to their inverse distance to the respective site. The formula for calculating the interpolated data value $z(x)$ for point x (e.g. Erlangen) is given in (I), where d_i is the radial distance between station i (e.g. Uttenreuth) and point x, z_i is the measured value at station i and n is the number of stations used for calculation:

$$z(x) = \frac{\sum_{i=1}^{n}\left(\frac{1}{(d_i)^2} * z_i\right)}{\sum_{i=1}^{n}\frac{1}{(d_i)^2}} \qquad \text{(I)}$$

In (I), the radial distance di is calculated as:

$$d_i = \sqrt{(r_x - r_i)^2 + (h_x - h_i)^2} \qquad \text{(II)}$$

Here, r_x is the easting of point x, r_i is the easting of station i, h_x is the northing of point x and h_i is the northing of station i.

In the following sections the procedure for calculating the snow loads and wind gusts for both locations of interest is described and the respective results are presented.

3 SITE-SPECIFIC SNOW LOAD SCENARIOS

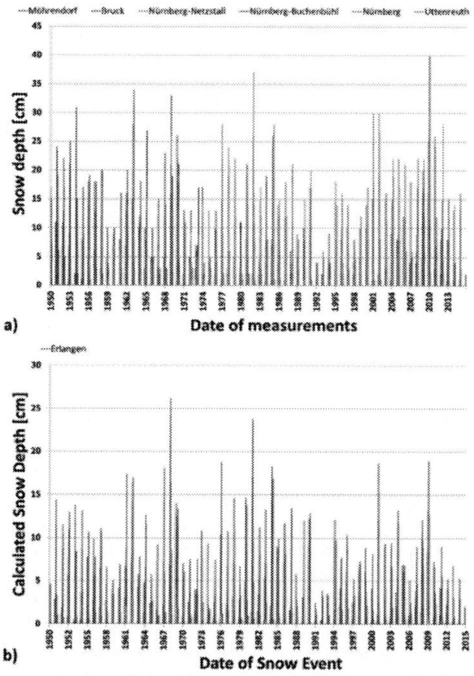

Figure 1: a) Historical snow depth data as derived from the weather station data around Erlangen; b) data derived for Erlangen by IDW from a)

For the deduction of snow load data, snow depth data

provided by the DWD weather stations was used [5]. These measurements are conducted manually at each station at 6 UTC on a reference area as long as this area is covered by at least 50% by snow [7].

If this prerequisite is fulfilled, these days will also be classified as "snow cover days". In Figure 1 a) the historical snow data for the relevant stations around Erlangen is displayed. In Figure 1 b), the derived data for Erlangen is shown which was obtained by applying the IDW method as described above. Data for Arzberg was obtained analogously. Subsequently, the data is analyzed for the frequency distribution of the snow depths in order to deduce the probability for a certain snow depth and of the resulting snow load. Figure 2 depicts the snow depths frequency distribution for both sites as deduced from the IDW data (traces labelled "unscaled"). However, due to the procedure described here, the actual number of snow cover days will be overestimated, as a non-zero snow depth will be computed as soon as one of the sites exhibits a non-zero snow-depth. For example, for Erlangen this procedure would yield an unrealistically high number of 70 snow cover days in total. In order to compensate for this effect these distributions were normalized to the average number of snow cover days that were deduced from the DWD's snow cover maps, which provide average numbers of snow cover days for Germany on a 1km² pixel raster [9].

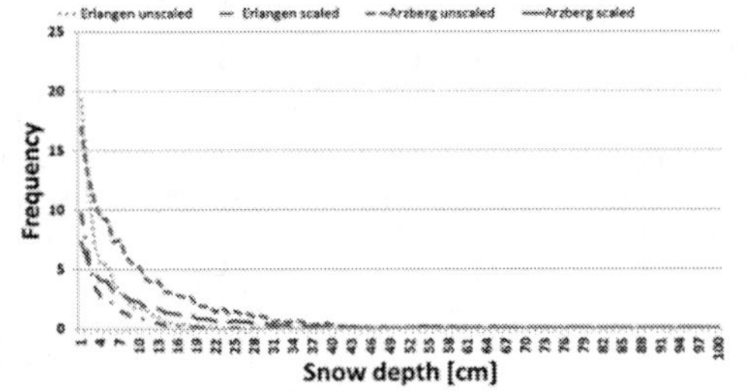

Figure 2: Annual snow depth frequency distribution as derived by the IDW Method from weather station data for Erlangen and Arzberg. Traces refer to data scaled or unscaled to the average number of snow days for each site, respectively.

As the snow coverage on the ground differs from the coverage on PV modules due to the tilted mounting and the elevated module temperature, the correlation between both kinds of snow coverage needs to be investigated. Therefore, webcam images recorded at the ZAE Bayern's PV test field in Arzberg were analyzed in order to deduce a coarse estimate for this correlation. Here, snow coverage on 4 standard crystalline silicon modules tilted by 30°C with an orientation of 33° divergent from south towards east was compared to snow coverage on a reference ground area during winter of 2015/2016 (starting on Nov. 24th with the commissioning of the webcam). This comparison yielded that the number of "module snow cover days" corresponds to 72% of the number of "ground snow cover days". Even though this comparison certainly only gives a very rough estimate of the dependency of both parameters as it is strongly influenced by the conditions of this particular site it can still be used for a first approximation within this analysis and will be used subsequently.

Assuming an average snow density of ρ_{snow} = 300kg/m³ [8], the frequencies deduced from Figure 2 can be converted into frequencies of snow loads on the ground. In order to deduce the mechanical load onto the module plane, the tilting angel of the modules also needs to be taken into account. In the following the tilt angel will be assumed as 30°.

Figure 3: Webcam image of the ZAE Bayern's test PV test site in Arzberg. The snow reference ground and module areas are marked by the red rectangles.

Thus, the normal mechanical load F_{snow} on a PV module caused by a snow load can be approximated from meteorological snow depth data for a given site as

$$F_{\text{snow}} \approx 0.72 \cdot \rho_{\text{snow}} \cdot d_{\text{snow}}^{\text{ground}} \cdot g \cdot \cos\alpha \qquad \text{(III)},$$

with the gravitational acceleration g = 9.81m/s², the ground snow depth $d_{\text{snow}}^{\text{ground}}$ and the module tilt angle α. Applying (III) to the frequency distributions of Figure 2 the corresponding snow loads distributions can be deduced. Subsequently, the annuality A of the snow loads can be calculated by inverting the annual frequency values. As the annuality states the period during which a certain event is statistically likely to occur this quantity can be used to assess certain locations with respect to the maximum snow load MSL_x (t), which is to be expected within a module lifetime t for a certain site x. For representing the latter, the module guarantee period is chosen, which is typically 25 years corresponding to an annuality of 25. Thus, the snow load annualities for Erlangen and Arzberg as shown in Figure 4 can be deduced.

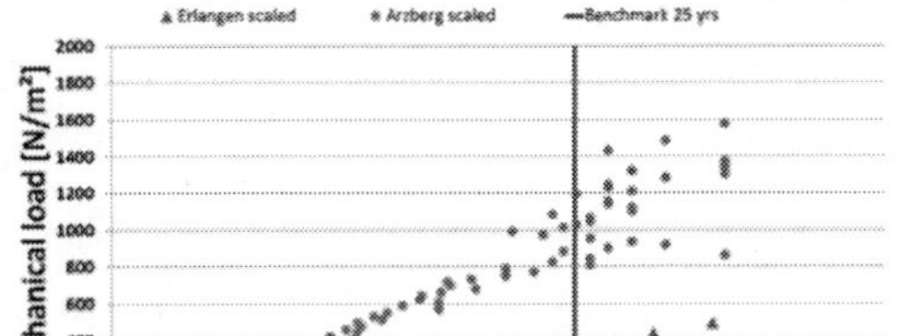

Figure 4: Annualities of mechanical snowloads as deduced for Erlangen and Arzberg for 30°-tilted modules. The red line marks the 25 year benchmark representing a standard PV module guarantee period.

From Figure 4, it becomes clear that very different snow loads, varying by roughly a factor of two, can be expected for Erlangen and Arzberg. The corresponding maximum snow loads, which are to be expected within the module guarantee period of 25 years can be deduced as $MSL_{Erlangen}(25)\approx 400N/m^2$ and $MSL_{Arzberg}(25)\approx 900N/m^2$ Both values correspond to rather moderate load values for which no significant module damages are to be expected [10]. Nevertheless, the maximum snow load quantity can be easily applied for comparing various sites with respect to the mechanical loads that can be expected to be imposed on a PV module at those locations. In particular, this value can be deduced, as described here, by the mere use of publically available weather data.

4 SITE-SPECIFIC WIND LOAD SCENARIOS

In principle, the same approach as it was applied for snow loads can be applied for wind loads. However, for wind, fewer weather stations can provide an appropriate data set than it was the case for snow. Due to the high volatility of wind it seems reasonable to assume that the wind speed of wind gusts is more important for assessing the effect of mechanical loads on the lifetime of PV modules than the daily or hourly average wind speeds. Therefore, in the following, wind gusts are discussed. Unfortunately, wind gust data are provided by much fewer weather stations than snow or averaged wind data. Therefore, the range for relevant stations had to be extended for the wind analysis to up to 64km around the site of interest in order to obtain wind gust data over a sufficiently long period of time (see Table I). Obviously, this will negatively affect the accuracy of the wind load estimation, which is one of the main reasons for extending our approach to Reanalysis data in chapter 5.

For the empirical estimation, the procedure was slightly modified compared to the previous snow analysis due to the high volatility of wind gusts in space and time. If wind gust data was averaged over several stations over time the heterogeneity of the values would be even higher than for snow so that the averaged value would not be meaningful anymore. Therefore, for each site, the wind speed distribution for each site was calculated. Firstly, maximum daily (i.e. wind gusts) and the average daily wind speeds measured at a given site were sorted in descending order with respect to their wind speed for every year of the evaluation period. Then, the average annual distribution was calculated by averaging over all highest wind speed values, all second highest wind speeds and so on.

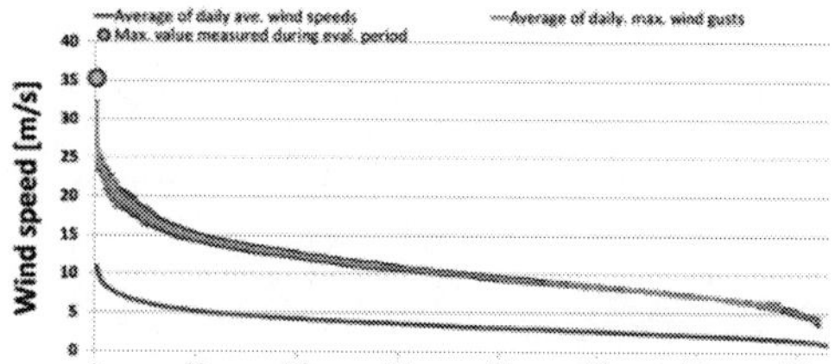

Figure 5: Wind speed values as measured for Wunsiedel. The average values of all years of the daily maximum wind speeds (wind gusts) and of the average wind speeds are displayed. The highest overall wind speed value of the data set is also shown. The error bars symbolize the standard deviation of the spread of the data.

Figure 5 exemplarily displays the average maximum and average daily wind speeds for Wunsiedel. Additionally, the highest wind speed measured in the entire data set is shown. Once these values have been computed, the resulting wind speed distributions for both sites of interest were calculated from the data of the sites listed in Table I by applying formulas (I) and (II). According to the formalism described in [11] the mechanical loads imposed by the wind gusts can be calculated. Figure 6 depicts the mechanical loads as a function of the wind speed assuming 30°-tilted modules and front-facing wind. Additionally, the average loads for the commonly used Beaufort classes are shown.

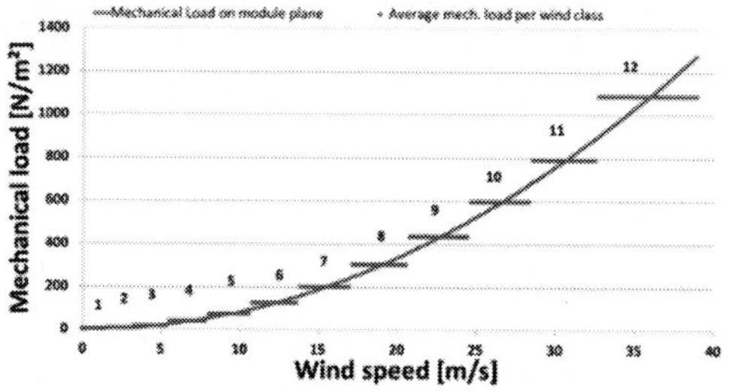

Figure 6: Mechanical loads as they are imposed on a 30°-tilted module for front-facing wind according to [11]. Additionally, the average loads imposed by wind of a given Beaufort class are shown.

The daily maximum wind gust speeds and the resulting mechanical loads based on [11] for an average year as calculated by applying formulas (I) and (II) are depicted in Figure 6.

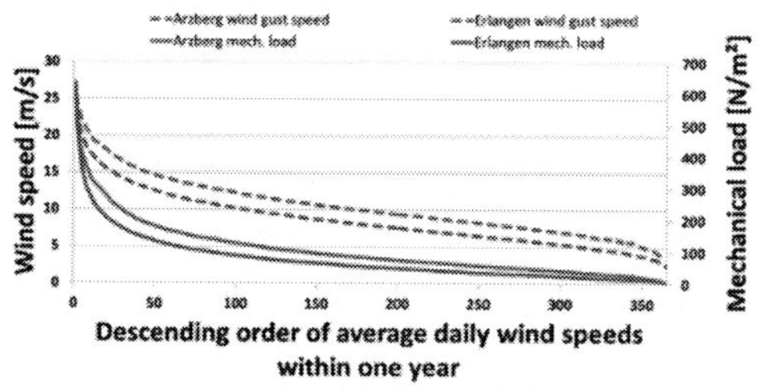

Figure 6: Daily maximum wind gust speeds and resulting mechanical loads for a 30°-tilted module and front-facing wind as calculated for Erlangen and Arzberg based on the IDW-method in descending order for an average year.

The wind data used in this analysis does not take the wind direction into account. It has been shown by Sander et al. [12] that front-facing wind bears by far the most severe risk of module damage. Thus, one may argue that statistically only ¼ of the wind events with a given mechanical load should actually impose a risk on the modules. On the other side many locations exhibit a preferred wind direction distribution due to the local ground topology (e.g. valleys or coastal regions). Therefore, we decided not to introduce any statistically motivated reduction of wind frequencies knowing that the resulting distribution will constitute an upper limit of the mechanical load situation of a PV module at a given location. It can be seen that the wind speeds and mechanical loads are roughly 25% higher for Arzberg than for Erlangen. Therefore, also for wind the mechanical loads can be expected to be significantly increased for Arzberg.

Even though Figure 6 provides an excellent overview of the wind situation at a given site for an average year, it cannot serve for deducing an expectation value for the maximum load to be expected within a typical module lifetime as a consequence of the averaging. Thus, an annuality as shown above for snow cannot be extracted for seldom but severely strong storm events. In order to calculate a metric comparable to MSL_x (t) all historical data needs to be taken into account. Therefore, the wind gust data of the sites listed in Table I were classified into Beaufort wind classes (0 = no wind, 12 = hurricane). Subsequently, the frequency of the occurrence of wind gusts of these classes was calculated for Erlangen and Arzberg by applying formulas (I) and (II), which were then converted to annualities of the corresponding mechanical loads by using the formalism described in [11]. Figure 7 shows the resulting annualities of the mechanical loads based on maximum daily wind gust values.

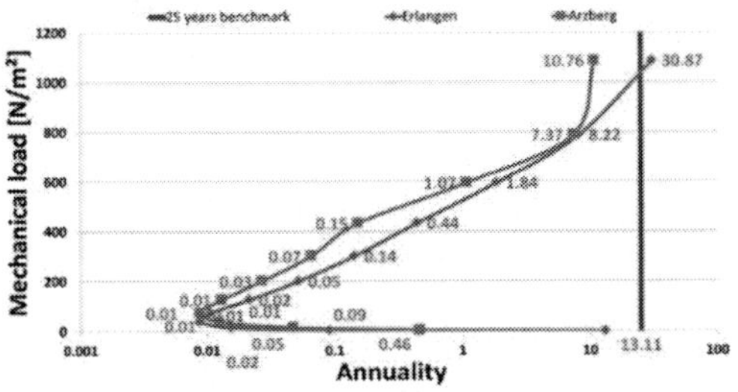

Figure 7: Annualities of mechanical loads for Erlangen and Arzberg as they are imposed on a 30°-tilted module for front-facing wind according to [11]. Wind gust data for the stations in Table I were classified according into Beaufort classes and values for Erlangen and Arzberg were computed using the IDW method. If no event of a given class was observed for a certain site this site was excluded from IDW calculation for this class.

Similarly to the snow analysis, it can be seen that a wind gust event of a certain Beaufort class can be expected much more frequently for Arzberg than for Erlangen. Here, the annualities for Arzberg roughly amount to one third to half of the Erlangen value. Thus, wind gust of a certain Beaufort class can be expected to occur two to three times more often in Arzberg compared to Erlangen. For wind, severe wind gusts of the highest Beaufort class 12 corresponding to mechanical loads of ~1100N/m² can be expected to occur roughly three times within a module lifetime of 25 years. For Erlangen, only up to one of such events would be expected. However, within the observation period now wind gusts above 35.3 m/s were measured. Thus the maximum mechanical load imposed by wind gusts $MWL_x(t)$ for a given site x and a period t would rather be stated as a pair of values: the maximum mechanical load value and its expected number of events during the period t. Hence, for Erlangen $MWL_{\mathrm{Erlangen}}(25)$≈(1100N/m², 2.5) is obtained and for Arzberg $MWL_{\mathrm{Arzberg}}(25)$≈(1100N/m², 0.8).

5 AUTOMATION BASED ON REANALYSIS DATA

The methodology described in the previous chapters allows for predicting the expected mechanical loads on a PV module for virtually any location based on historical weather data of surrounding weather stations. However, this methodology has several drawbacks with respect to its precision and practicability. Firstly, the method requires extensive manual data extraction and verification due to the heterogeneous structure and completeness of the data sets of the various stations. Secondly, the precision is highly dependent on the distribution of surrounding weather stations and local meteorological conditions. Therefore, the method, though proving the value of the concept, is not suited for automation allowing for a time-efficient evaluation of potential sites for a PV system. Here, meteorological reanalysis (RA) appears to be an appropriate way for increasing precision and automatability [13]. In principle, RE combines physical observations with a physical atmosphere model by a data assimilation approach. Thereby, a consistent set of weather data set is obtained which can be used for deducing for any location within the area covered by the respective RE system. For Europe, two main RE systems exist: ERA-Interim [13] and COSMO [14], of which COSMO provides a very high special and temporal resolution (6km, 1h; ERA-interim: 80km, 6h). In this chapter it shall be evaluated in how far RE data can be used for deducing weather data for a given location. In order to be able to access the accuracy of the results a locations of existing weather station are chosen. Here, only wind gust data is evaluated. However, in principle the approach could also be applied for snow data.

In the following, the procedure for accessing the applicability of RE data for the prediction of mechanical loads caused by wind gusts is explained. Generally, RE data always needs to be calibrated by real weather data. Therefore, weather station data as well as data obtained from ERA-interim and COSMO were fitted by a Generalized Extreme Value (GEV) distribution, which is determined by the parameters k, μ and σ given by the following formulas:

$$f(x) = \frac{1}{\sigma} \cdot \upsilon(x)^{k+1} \cdot e^{-\upsilon(x)} \qquad \text{(III)}$$

with:

$$\upsilon(x) = \left(1 + k \cdot \left(\frac{x-\mu}{\sigma}\right)\right)^{-1/k} \qquad \text{(IV)}$$

Here, k describes the shape of the distribution function, whereas σ describes its amplitude and μ its position on the x-scale. For the calibration of k, μ and σ the wind gust data of 17 DWD weather stations was fitted by a GEV as well as the COSMO and ERA-interim RE data of for the corresponding locations. Subsequently, correlations between k_{station}, μ_{station}, $\sigma_{\mathrm{station}}$ and k_{ERA}, μ_{ERA}, σ_{ERA} and k_{COSMO}, μ_{COSMO}, σ_{COSMO}, respectively, were determined by linear regression. With these correlations values for k, μ and σ can then be calculated for any location. In order to evaluate the precision of these deduced data sets a weather station (Dresden) was chosen, which was not part of the 17 stations used for calibration. Therefore, the RE data extracted from ERA-interim and COSMO was fitted by a GEV. The resulting k, μ, σ were then converted to k_{station}, μ_{station} and $\sigma_{\mathrm{station}}$ by applying the correlations that were deduced from the linear regressions for the respective RE system. Figure 8 displays the data and fits deduced from both RE systems as well as the real weather station data.

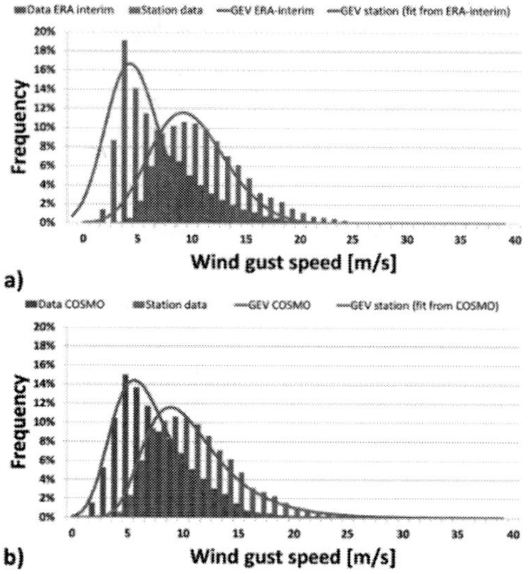

Figure 8: Data extracted from the Dresden weather station in comparison to data and fits deduced from ERA-interim (a) and COSMO (b) RE systems. The GEV distributions describing the station data were obtained by fitting the RE data and then converting it to "real" data by applying the correlations deduced for $k_{station}$, $\mu_{station}$ and $\sigma_{station}$ from the k, μ, σ of the GEV fitted to the respective RE data

It needs to be noted that the full blue lines in Figure 8 do not constitute a fit to the station data but purely result from the RE datasets. It can be seen that these curves describe the station data very well. Thus, it can be concluded that calibrated RE data is well suited for deducing weather data for any given location, thus, paving the ground for automizing the approach described in the previous chapters.

6 CONCLUSIONS

A methodology was presented that allows for assessing the expected mechanical loads a PV module would be exposed to at a specific location. This methodology is based on the evaluation of the data gathered by surrounding weather stations. Here, mechanical loads imposed by snow and wind loads have been analyzed. It could be shown that even for locations just separated by a moderate distance of just 110km the loads to be expected can vary by a factor of two or more. Furthermore, an approach for replacing the station data by data generated from RE data was validated and it could be demonstrated that, if calibrated properly, the data obtained from RE systems describes the station data very well. Our future will now focus on the development of a fully automated tool which allows the calculation of expected wind and snow loads for any (European) location directly from RE data.

7 ACKNOWLEDGEMENT

Funding of this work by the German Federal Ministry of Economic affairs in the framework of the project AQUAM (FKZ: 0325807) is gratefully acknowledged.

ECMWF is greatly acknowledged for providing ERA-Interim reanalysis data.

DWD is greatly acknowledged for providing COSMO reanalysis data and weather station data.

8 REFERENCES

[1] Bundesamt für Bauwesen und Raumordnung (BBR), Testreferenzjahre von Deutschland für mittlere, extreme und zukünftige Witterungsverhältnisse (2014), accessed: 30th of Aug. 2017: http://www.bbsr.bund.de/EnEVPortal/DE/Regelungen/Testreferenzjahre/Testreferenzjahre/03_ergebnisse.html?nn=437706,

[2] E. Lorenz, D. Heinemann, C. Kurz, Prog. Photovolt: Res. Appl., 20 (6) (2012), 760

[3] B. Marion, R. Schaefer, H. Caine, G. Sanchez, Solar Energy 97 (2013) 112

[4] M. Schweiger, 6th PV Performance Modeling and Monitoring Workshop, Freiburg, Germany (2016)

[5] German Weather Service (DWD, Deutscher Wetterdienst) website; accessed: 6th of Feb. 2017: https://www.dwd.de/DE/leistungen/cdcftprasterdaten/cdcftpraster.html,

[6] B. Rudolf, H. Hauschild, M. Reiss, U. Schneider, Meteorol. Zeitschrift, N.F. 1 (1) (1992)

[7] German Weather Service (DWD, Deutscher Wetterdienst) website; accessed: 31st of Aug. 2017: http://www.dwd.de/DE/service/lexikon/Functions/glossar.html?nn=103346&lv2=102248&lv3=102452

[8] O. Bruland, Å. Færevåg, I. Steinsland, G. E. Liston, K. Sand, Hydrology Research, 46(4) (2015) 494

[9] DWD, snow cover days data: ftp://ftp-cdc.dwd.de/pub/CDC/grids_germany/annual/snowcover_days/ accessed Feb. 2017

[10]C. Buerhop, S. Wirsching, S. Gehre, T. Pickel, T. Winkler, A. Bemm, J. Mergheim, C. Camus, J. Hauch, C. J. Brabec, 44th IEEE Photovoltaic Specialist Conference, Washington DC, USA (2017)

[11]DIN EN 1991-1-4/NA:2010-12, National Annex - Nationally determined parameters - Eurocode 1: Actions on structures - Part 1-4: General actions - Wind actions, Beuth (2010)

[12]M. Sander, S. Dietrich, M. Pander, M. Ebert, J. Bagdahn, Sol. En. Mat. Sol. Cells, 111 (2013) 82

[13]D. P. Dee, S. M. Uppala, A. J. Simmons, P. Berrisford, P. Poli, S. Kobayashi, U. Andrae, M. A. Balmaseda, G. Balsamo, P. Bauer, P. Bechtold, A. C. M. Beljaars, L. van de Berg, J. Bidlot, N. Bormann, C. Delsol, R. Dragani, M. Fuentes, A. J. Geer, L. Haimberger, S. B. Healy, H. Hersbach, E. V. Hólm, L. Isaksen, P. Kållberg, M. Köhler, M. Matricardi, A. P. McNally, B. M. Monge-Sanz , J.-J. Morcrette, B.-K. Park, C. Peubey, P. de Rosnay, C. Tavolato, J.-N. Thépaut and F. Vitart, Q. J. R. Meteorol. Soc. 137 (2011) 553–597

[14]C. Bollmeyer, BONNER METEOROLOGISCHE ABHANDLUNGEN, 69 (2015)

SYSTEM SIZING FOR RESIDENTIAL PV AND EES SYSTEMS

Tobias Melloh, Tristan Fehling, Gerhard Kleiss[1], Bernard Nacke
Institute for Electrotechnology (ETP), Leibniz Universität Hannover
Wilhelm-Busch-Str. 4, D-30167 Hannover, Germany
[1] corresponding author: kleiss@etp.uni-hannover.de

ABSTRACT: Looking at the market for residential Electrochemical Energy Storage (EES) systems in combination with PV systems (PV), a significant growth has been observed recently. For these systems a new purchase motivation of the residential customers can be observed: private customers are looking for autarky in order to avoid purchasing electricity from the grid. This is in contrast to preceding purchase motivations triggered by comparatively high feed-in tariffs leading to installation of larger grid-connected PV systems without storage on private households (having promised more financial benefit for the user in the past). This work determines levels of autarky for different sizes of both PV system (PV) and storage (EES). A simplified algorithm for system operation is applied to a standardized consumption profile that is calculated under consideration of 3 different typical European climatic conditions. Based on these results design rules are suggested leading to comparatively small system sizes as the economic optimum under the financial boundary conditions considered here.

KEYWORDS: PV, ESS, system sizing, autarky, cost, site-dependence

1. INTRODUCTION

Today, there are an increasing number of residential applications of photovoltaic (PV) solar energy systems in combination with an electrochemical storage system (EES). Since such systems are in most cases still connected to the electrical grid, many users are looking for a high level of autarky in their electricity supply in order to reduce cost of electricity purchase from the Distribution System Operator (DSO).

We may call this approach in the context of this text the "user's egocentric approach", as no or little awareness is given for the overall grid situation. In case all residential users would follow exactly the scheme proposed here, this would incur a serious burden for the DSO. Other approaches may deliver not only benefits for the individual customer but also for the utility operator. This is not discussed in this paper but will be discussed elsewhere.

The aim to minimize private electricity consumption directly creates a need for choosing the appropriate size of both PV generator and EES battery. Prices and economics of both PV and EES have seen considerable changings recently [1], [2]. This is why it is important to re-calculate models and derive new design rules for system sizing. The influence of climatic operation conditions can be expected and is studied in this work by comparing results from several climatic zones: temperate continental, Mediterranean and temperate coastal. The selected locations are taken from [3] and offer a good broad-band view for locations in Europe.

2. METHOD

2.1. Model of General System Set-Up and Operation

The PV-EES system investigated is based on the 4 following components:

- PV Generation (P)
- ESS Storage (S)
- Grid (G) (consumption or feed-in)
- Load requirement (L)

The scenario which has been studied has a flow of energy following the subsequent rationale:

(a) The actual load requirement L must always be met: in the first place (Priority 1a) through the PV system (P), in the second rank (Priority 1b) through the ESS (S), and in case of necessity the grid (G) as Priority 1c.

(b) Only if the load requirement is satisfied and if there is additional energy available from the PV system, the ESS (S) can be loaded by PV (Priority level 2).

(c) Only if the ESS is fully loaded under such conditions the residual power may be injected into the grid G (Priority level 3).

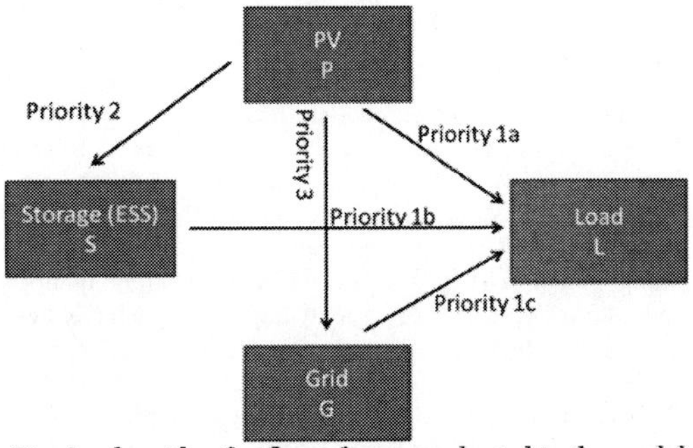

Fig. 1: algorithm for flow of energy adopted in the model.

Of course, there are many other strategies to run PV-EES combinations [4], [5], including the possibility to charge the EES directly from the grid under conditions of varying tariffs [6]. Another necessity to adapt the operation algorithm may arise due to regulatory requirements limiting the feed-in from the PV system into the grid [7]. The latter case was investigated in the frame of the current work, but results are not presented here, as the main trends turned out to be similar to what is presented here.

To calculate the load L, we use a Standard Load Profile (SLP) H0 as published by the German Distribution System Operators' association BDEW relevant for residential applications [8]. This SLP is given on quarter-hourly base and is expanded to an hourly base for purpose of calculation within this study. Fig. 2 shows the typical SLP yearly normalized to a yearly consumption of 4000 kWh as investigated in this work.

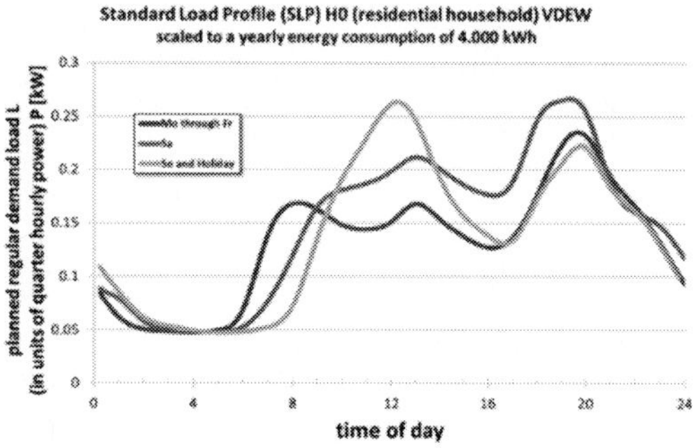

Fig. 2 BDEW load profile H0 for the month of January, scaled for a yearly consumption of 4'000 kWh

2.2. PV Specific Model

In order to simulate the PV-EES system, a module efficiency of $\eta_{mod} = 17\%$ is assumed. Likewise, a maximum inverter efficiency of $\eta_{inv} = 97\%$ is considered for calculation. Both assumptions may be considered a conservative approach. Temperature related losses are accounted on the base of calculating the power reduction by a temperature coefficient

$$\eta_{mod}(T) = \eta_{mod}(T_{STC}) \cdot [1 + (T_{mod} - T_{STC}) \cdot TC_\eta] \quad \text{(eqn. 2)}$$

with the temperature coefficient set to TC_η = -4'600 ppm/K. Operating temperatures T_{mod} have been calculated by FAIMAN's approach [9]

$$T_{mod} = T_{amb} + \frac{G_{POA}}{u_1 + u_2 \cdot v} \quad \text{(eqn. 3)}$$

with applicable coefficients u_1, u_2 for typical crystalline PV modules taken from [10]. There is no further consideration of the modules' relative performance with respect to irradiance and/or other impacts. Prior studies [11] have shown that an influence is existing, but it may be regarded small as within this study the total energy produced is the leading quantity (and corrections are of second degree). We have been considering the thermal correction (eqn. 1 and 2) as this will exclude any possible temperature related (AM versus PM) asymmetry. The model contains no provision for temperature-related de-rating of the inverter or the EES, as those components are deemed to be operated indoors within a controlled environment. Moreover, no degradation effects are considered as a time frame of 10 years is considered, corresponding to typical warranty periods for many residential EES systems promising lawless operation. In order to attain a site-dependent result, the meteorological data given per Table I have been used for computation.

Table I: sample data sets provided in the Committee Draft of IEC 61853-4 [3] for different climatic regions

Latitude	Longitude	Type *(example in country)*
38°N	3°15'W	Mediterranean *(Spain)*
48°N	18°E	temperate continental *(Slovakia)*
56°N	4°W	temperate coastal *(UK – Scotland)*

2.3. ESS parameters and example calculation

It is assumed that the EES is exclusively operated at state of charge (SOC) between 10% and 90% of the nominal capacity. Degradation effects are not accounted for. Results in this report are considered for a time-of-use of 10 years, typically corresponding to less than 10.000 loading / unloading cycles. Moreover, no thermal derating is considered, since it is assumed that the EES is placed in a cool and dry place of a residential home.

To calculate the results a time series for one year is determined (results are later multiplied by 10 to study the 10 years frame), where at every time step a decision of the energy flow is made on the base of the algorithm shown in Fig. 1.

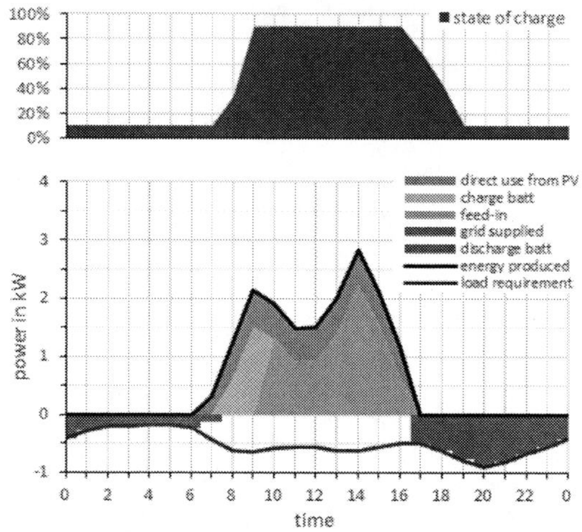

Fig. 3 typical result for a partially sunny winter day: temperate continental location, PV-size 5 kWp, ESS capacity 2 kWh

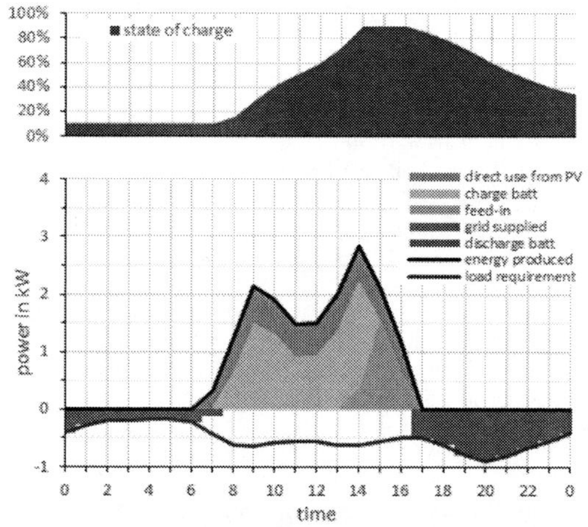

Fig. 4 typical result for a partially sunny winter day: temperate continental location, PV-size 5 kWp, ESS capacity 8 kWh

Fig. 3 and Fig. 4 show simulation results for different ESS sizes (2 kWh and 8 kWh) obtained for a partially sunny winter day at the temperate continental location, where the size of the PV system is 5 kWp. It is obvious that the smaller storage (in comparison to the larger storage) reaches faster its state of maximum charge and can contribute less to the load requirement: More "surplus" energy must be fed into the grid and the level of autarky (as defined in the next section) is smaller.

3. RESULTS

3.1. General

Autarky (self-sufficiency) a is defined as the normalized share of cumulated actual supply of a PV ESS system to the energy requirement (L) from either the PV system (P) or the ESS (S):

$$a = \frac{\sum_{1\,year} Load\ served\ in\ timestep\ by\ S\ or\ PV}{\sum_{1\,year} Load\ requirement} \qquad \text{(eqn. 3)}$$

If the autarky factor was 100% the energy requirement (L) would be met under any circumstances at any time by the PV ESS system ("Priority 1a" and "1b" of Fig.1) and no consumption of grid energy ("Priority 1c") would be necessary. We are deliberately not applying the term "self-consumption" in this work, since the definition is often not consistently defined in literature and it is often related to PV-systems without storage, only.

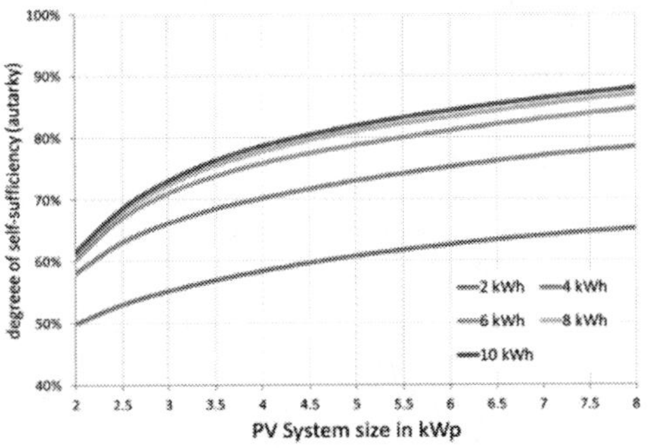

Fig. 5: degree of self-sufficiency (autarky) for a combined PV ESS system in dependence of the PV system size for several possible storage sizes. Assumed yearly consumption of the residential household: 4'000 kWh. PV, location: temperate continental

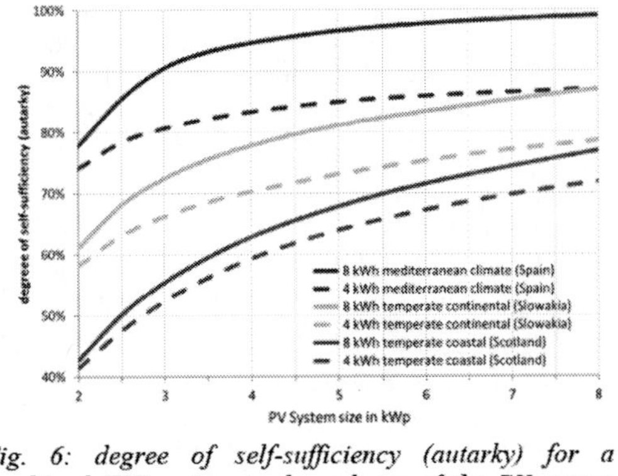

Fig. 6: degree of self-sufficiency (autarky) for a combined ESS system in dependence of the PV system size for 2 possible storage sizes at different locations. Assumed yearly consumption of the residential household 4'000 kWh

Fig 5. shows the degree of autarky for a 4'000 kWh-per-year-consumption household with PV and ESS, whereas various sizes of ESS and PV generator sizes for the temperate continental location are given. A certain saturation can be seen for both higher PV generator sizes and ESS sizes. In particular, there is only little additional contribution to the self-sufficiency factor once a threshold of 8 kWh of ESS size is exceeded.

It is instructive to look at the effect of differences between ESS's of capacities of 4 kWh and 8 kWh for the described total system and to compare the projected performance under different climatic environments. This is depicted in Fig. 6. It can be seen, that the degree of higher autarky (self-sufficiency) can be attained at a higher level in regions with more solar energy supply (in this case the Mediterranean region). Interestingly enough, the split between the 4 kWh and the 8 kWh ESS contribution is higher at the high-irradiance region compared to the other regions.

3.2. Economic consideration

The economic considerations are based on the following estimations:

Table II: price assumptions (i.e. cost for residential used) considered in this study.

Element		Assumed value
Specific Cost of PV	C_{PV}	1'000 EUR / kWp
Specific Cost of ESS	C_{ESS}	1'250 EUR / kWh
Cost of electricity	C_{grid}	0.28 EUR / kWh

It is assumed that a primary motivation for the residential customer is the intention to save energy expenditure by means of a combined PV ESS system. With the autarky a as given in Fig. 3 and 4 above, the private customer can avoid a yearly payment of $AYP = a \cdot C_{grid} \cdot total\ consumption$, which is in our example ($total\ consumption = 4'000$ kWh):

$$AYP = a \cdot 1'120\ € \qquad \text{(eqn. 3)}$$

In order to calculate economic results, it is assumed that the system has a life time of 10 years and that the savings AYP per year (eqn. 3) can be realized during every year on equal level. The net benefit is the total savings during 10 years ($AYP \cdot 10$), while depreciation and inflation are neglected .

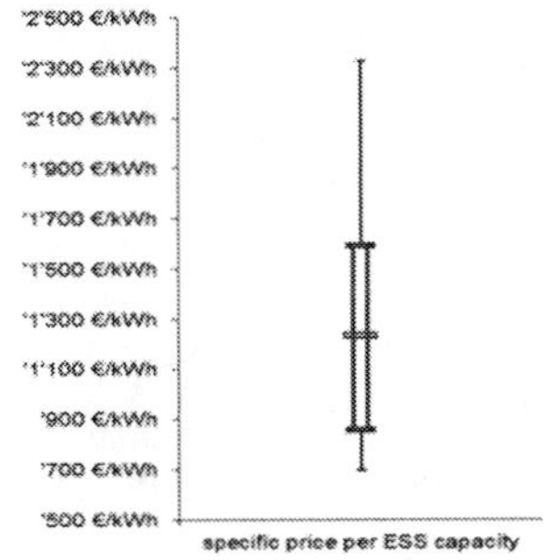

Fig. 7: Box Plot for specific end-customer prices researched in February 2017 for residential German ESS of size up to 10 kWh. A median of 1'250 EUR / kWh is used for further calculation.

3.3. Cost of Autarky

Fig. 8 shows possible maximum degrees of autarky at the different sites for combinations of the EES-PV system ranging between 1 kWp ... 10 kWp (PV) and 2 kWh ...8

kWh (EES). The maximum possible autarky a can be reached at high cost (large EES and large PV generator) at more than 99% for the Mediterranean location. Obviously these high levels of autarky cannot be easily met for the two temperate locations at reasonable effort due to the lower irradiation.

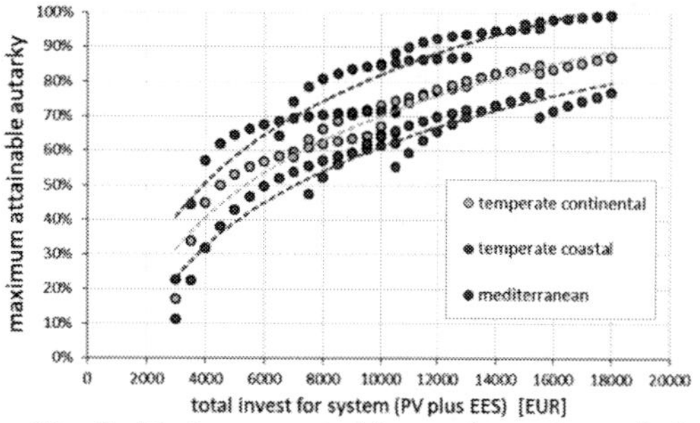

Fig. 8: Maximum attainable autarky in the studied PV-ESS System for the 3 locations. Results are shown for groups of systems comprising of 2 kWh, 4 kWh, 6 kWh and 8 kWh system at each location for different sizes of the PV system (between 1 kWp and max. 10 kWp). Dashed lines show trends for the different locations.

As there has been shown that the level of autarky can be chosen for each site within certain limits, it is interesting to study which level of autarky will deliver the best economic result for the residential user. Under the assumptions as given in the preceding section, it can be shown that for a 10 years period positive results will be achieved for the Mediterranean and the temperate continental location. However, due to the comparatively weak irradiation level, the chosen temperate coastal location (UK / Scotland) does not show a positive gross margin after 10 years operation yet. The results are summarized in Fig. 9.

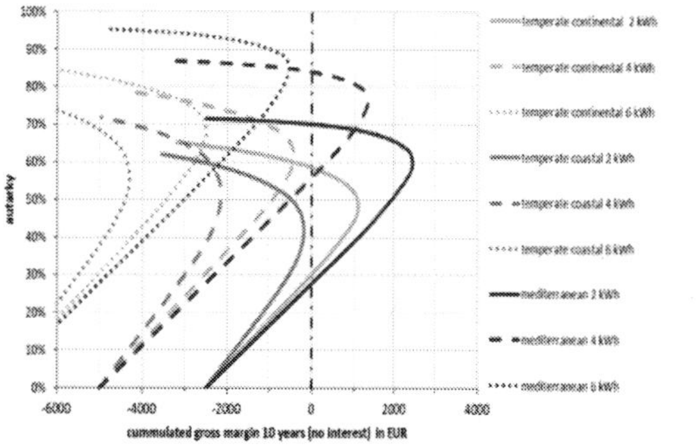

Fig. 9 level of autarky with respect to the extrapolated gross margin (AYP – invest) for different PV-EES configurations at the three climatic sites. Calculation for 10 years lifetime – no interest rate considered.

4. DISCUSSION AND CONCLUSION

A straight model to work out a design rule for PV EES systems under today's relevant boundary conditions (price / system sizing) is established for a residential system (private residential user). This leads to the conclusion that a comparatively small PV system with a comparatively small ESS will deliver under today's pricing situation favorable results, whereas the optimum level of autarky does not exceed 50% … 60%. It is a statement of major importance, that PV-ESS systems are deemed to be economically viable on a 10 years scale without any other subsidy and also without a "penalty" for self-consumption. Investigations for different climate zones within the EU show that a higher solar irradiation is contributing to the profitability of this scenario.

In conclusion, the economic optimum is found for comparatively small system sizes and only medium degrees of autarky. This finding does not necessarily reflect today's PV / ESS market, but has been described earlier [12] with application of different methods.

5. LITERATURE

[1] David Wedepohl (Bundesverband Solarwirtschaft), "Preise für solarstromspeicher halbiert," *Press Release 29.5.2017*.

[2] T. B. Randall, "World Energy Hits a Turning Point: Solar That's Cheaper Than Wind," *published online 2016-12-15*. [Online]. Available: https://www.bloomberg.com/news/articles/2016-12-15/world-energy-hits-a-turning-point-solar-that-s-cheaper-than-wind. [Accessed: 19-Sep-2017].

[3] IEC TC 82 WG 2, "CD IEC 61853-4 Ed. 1.0 Photovoltaic (PV) module performance testing and energy rating - Part 4: Standard reference climatic profiles," *82/1067/CD*, vol. 82/1067/CD, 2016.

[4] J. Weniger, T. Tjaden, and V. Quaschning, "Sizing of Residential PV Battery Systems," *Energy Procedia*, vol. 46, pp. 78–87, 2014.

[5] A. U. Schmiegel and A. Kleine, "Optimized Operation Strategies for PV Storages Systems Yield Limitations, Optimized Battery Configuration and the Benefit of a Perfect Forecast," *Energy Procedia*, vol. 46, pp. 104–113, 2014.

[6] A. Sani Hassan, L. Cipcigan, and N. Jenkins, "Optimal battery storage operation for PV systems with tariff incentives," *Appl. Energy*, vol. 203, pp. 422–441, 2017.

[7] J. Bergner, J. Weniger, T. Tjaden, and V. Quaschning, "Feed-in Power Limitation of Grid-Connected PV Battery Systems with Autonomous Forecast-Based Operation Strategies," in *29th European Photovoltaic Solar Energy Conference and Exhibition*, 2014, pp. 2363–2370.

[8] BDEW, "Standardlastprofile Strom." BDEW Bundesverband der Energie- und Wasserwirtschaft e.V., Berlin, 2002.

[9] D. Faiman, "Assessing the outdoor operating temperature of photovoltaic modules," *Prog. Photovoltaics*, vol. 16, no. 4, pp. 307–315, 2008.

[10] M. Koehl, M. Heck, S. Wiesmeier, and J. Wirth, "Modeling of the nominal operating cell temperature based on outdoor weathering," *Sol. Energy Mater. Sol. Cells*, vol. 95, no. 7, pp. 1638–1646, 2011.

[11] G. Kleiss, H. Schülbe, and B. Nacke, "Energy rrating of crystalline solar modules: investigation of uncertainties due to binning in mass production," in *Proc. 32nd EC-PVSEC*, pp. 2239 – 2242.

[12] G. Kleiss, "Methode zur dynamischen Auslegung eines PV Systems mit Elektrischem Energiespeicher," *Presentation*, August 2016, DOI: 10.13140/RG.2.2.35664.71680, 2016.

ENERGY PERFORMANCE OF A 1.2 MWP PHOTOVOLTAIC SYSTEM DISTRIBUTED OVER EIGHT BUILDINGS AT UTRECHT UNIVERSITY CAMPUS

Wilfried van Sark[1], Arjen de Waal[1], Jasper Uithol[2], Niekol Dols[2], Fréderique Houben[2], Richard Kuepers[2], Michiel Scherrenburg[3], Benno van Lith[4], Ferry Benjamin[5]

[1]Utrecht University, Copernicus Institute of Sustainable Development, Heidelberglaan 2, 3584 CS Utrecht, The Netherlands, E: w.g.j.h.m.vansark@uu.nl
[2]Utrecht University, Real Estate and Campus, Heidelberglaan 8, 3584 CS Utrecht, the Netherlands
[3]Utrecht University, Programme Team Sustainability, Heidelberglaan 8, 3584 CS Utrecht, the Netherlands
[4]BAM Bouw en Techniek - Regio Midden, Regulierenring 35, 3981 LA Bunnik, the Netherlands
[5]ProfiNRG, Leidsestraatweg 20, 3481 EX Harmelen, the Netherlands

ABSTRACT: A distributed PV system comprising of eight subsystems on separate buildings totaling an installed capacity of 1.2 MWp has been realized at buildings of the Utrecht University campus Utrecht Science Park. A detailed design process was followed taking into account the presence of surrounding buildings. This has lead to a system with modules oriented South, West, East, and even North. Most of the modules are installed at a tilt of 10 degrees. A full year of monitoring results is presented here and shows that the system is performing well, with an average performance ratio of 0.825, thus delivering more than 1 million kWh annually.
Keywords: Energy Performance, PV System, Design, Large Grid-connected PV systems, Education and Training

1 INTRODUCTION

Utrecht University aims to be CO_2-neutral by 2030 with an intermediate aim to generate half of its energy consumption in 2020 in a sustainable way, including thermal storage, the purchase of wind power and the use of solar energy. This has motivated the design and installation of 4,600 PV modules on several buildings in the Utrecht Science Park "De Uithof" in the period May-September 2016. The grid-connected system is designed such that it is predicted to generate one million kWh of electricity annually, which constitutes ~2% of the annual electricity demand. The installation of the PV systems contributes to a reduction of the CO_2 footprint. Moreover, it offers the potential to enhance research and education activities of the faculties involved.

In this contribution we will describe the performance of the system linking that to the specific design and layouts of the systems in detail.

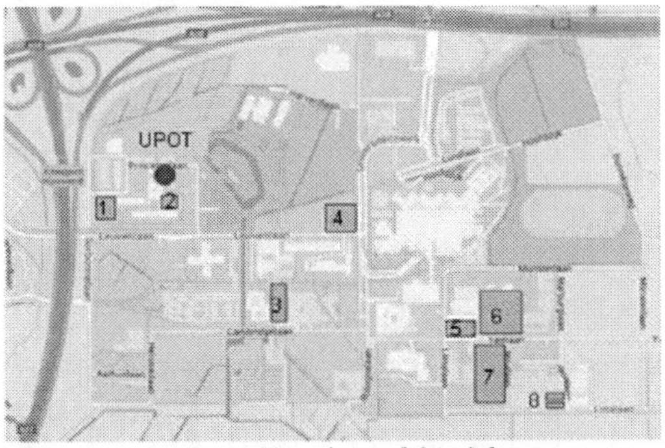

Figure 1: Overview of locations of the eight systems at the UU campus, spanning about 2 km^2. 1: Caroline Bleeker building, 2: Victor J. Koningsberger building, 3: Library, 4: David de Wied building, 5: Jeannette Donker-Voet building, 6: Willem C. Schimmel building, 7: Martinus G. de Bruin building, 8: Waterberging Diergeneeskunde/Jongveestal (Tolakker). The Utrecht Photovoltaic Outdoor Test facility (UPOT, red dot [1]) is used for measured irradiation.

2 SYSTEM DESIGN AND LAYOUT

The design of the system was started by identifying suitable buildings with flat roofs among the many University buildings located at the Utrecht Science Park. Two main criteria were used: 1) roof was strong enough to withstand the additional weight of the PV subsystem, and 2) the building itself would not be renovated nor demolished within the next 25 years. From the remaining selection the suitability with regard to shading from nearby buildings and/or obstacles on the roofs was determined. This has lead to the eight buildings as shown in Figure 1. Several PV systems consist of a combination of East and West oriented modules, others are South oriented, and one is North and South oriented (library building, #3 in Fig. 1), due to surrounding high-rise buildings, see also Fig. 2. This does influence annual yield, however with a tilt of 10 degrees, orientational effects can be considered to be small. Building #8 has a 30-degree tilted roof and is facing South.

Figure 2: Photograph (top) of University Library system, with detail (bottom, © Ivar Pel). The system consists of 860 modules of 270 Wp, and 7 inverters. Total capacity is 232.2 kWp. Half of the modules are oriented towards South and half towards North, at 10-degree tilt.

The installation of the various subsystems has taken place between May and September 2016, a video can be found here [2]. In this video UU's chair of the University Board stresses the importance of this system for UU research and education and how it supports UU's ambition regarding sustainability.

The layout of the system is optimized using separate inverters per orientation. In total 4,474 modules of 270 Wp and 126 of 265 Wp are installed and 37 inverters of different nominal AC capacity (12, 20, 23, and 30 kW). The average ratio of AC inverter capacity to PV DC capacity is 0.825. Details of the subsystems are listed in Table I.

Table I: Subsystem details. Inverter/PV ratio is defined as AC inverter nominal capacity divided by DC PV rated capacity. Building number refer to Fig. 1. Total DC PV rated capacity is 1218 kWp.

building	size (kWp)	orientation	inverter/PV ratio
1: Bleeker	132.3	W-E	0.81
2: Koningsberger	22.4	S	0.90
3: Library	232.2	N-S	0.85
4: Wied	121.1	W-E-S	0.80
5: Donker-Voet	46.4	W-E	0.73
6: Schimmel	233.3	W-E	0.85
7: Bruijn	285.7	W-E	0.81
8: Jongveestal	131.2	S	0.81

Monitoring of performance is executed using inverter-based monitoring at 5-minute time resolution, while calibrated kWh meters are installed per building that measure at 15-minute time resolution. The latter is required in order to receive SDE+ subsidy (Stimulation of Sustainable Energy Production [3]) per generated kWh. In addition, for every orientation and building pyranometers will be installed (in total 15), and temperature sensors will be mounted to ~50 modules. This allows to link the energy yield with solar irradiation and to determine the performance ratio for every system and orientation. A web-based platform is under development in which measured power, irradiance, and module temperatures are collected, stored and visualized. Irradiation data is measured at the Utrecht Photovoltaic Outdoor Test facility [1]. In this paper we will show monthly averaged performance data.

3 PERFORMANCE EVALUATION

3.1 Data analysis

The monthly averaged irradiation data measured at UPOT is depicted in Fig. 3. The summer of 2017 clearly peaked in May in the Netherlands. The 12-month (September 2016 – August 2017) global horizontal irradiation sum is 1035 kWh/m^2.

Figure 4 shows the generated amount of energy for all subsystems for all months, as taken from inverter readings. The maximum amount of energy is generated in May 2017, i.e., 161 MWh adding all subsystems, while a minimum of 15 MWh is found for December 2016. The total amount over the full period is 1007 MWh.

The specific energy yield in kWh/kWp is depicted in Figure 5. We conclude that most systems show similar specific energy yield, however, the Donker-Voet and the Bleeker system clearly do not, especially in the months

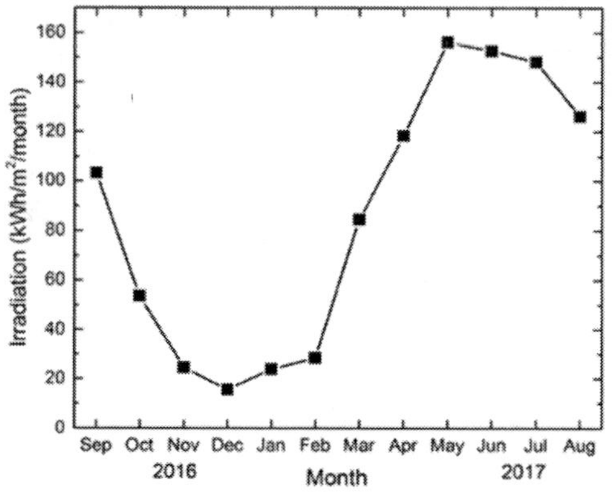

Figure 3: Monthly averaged global horizontal irradiation in kWh/m^2 for the full monitoring period, measured at UPOT.

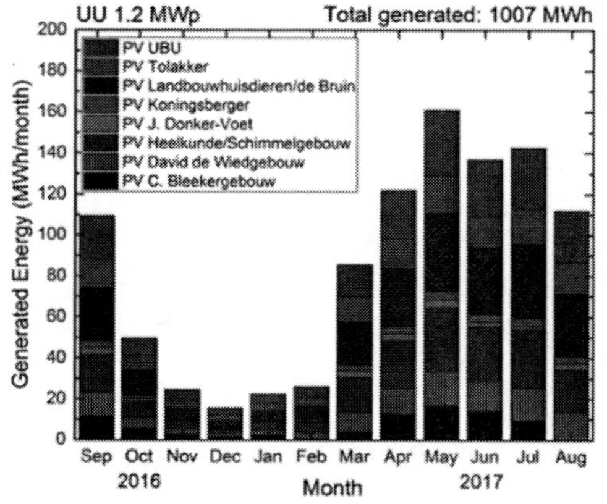

Figure 4: Monthly generated energy in MWh for the full monitoring period for all subsystems.

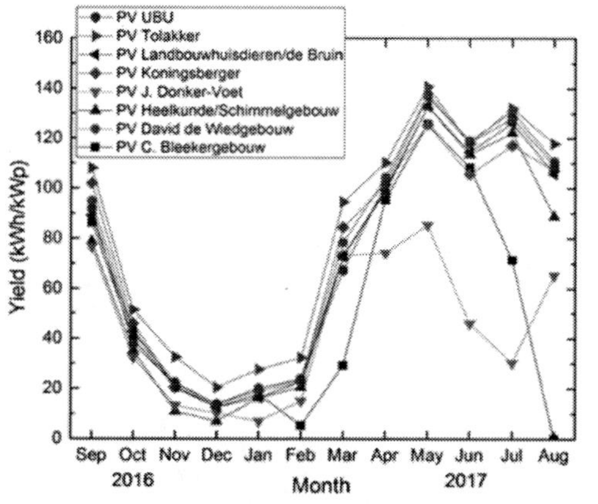

Figure 5: Monthly specific energy yield in kWh/kWp for the full monitoring period for all subsystems.

from February 2017 onwards. As the Jongveestal (Tolakker) system has a tilt of 30 degrees and is facing South, the yield is higher than other systems. The annual performance data, i.e., annual generated amount of energy, specific yield and performance ratio (*PR*) is summarized in Table II.

Table II: Subsystem annual generated energy, specific yield and performance ratio. Note, the Jongveestal is a 30-degree tilt system.

building	energy (MWh)	yield (kWh/kWp)	PR
1: Bleeker	81.0	613	0.592
2: Koningsberger	28.9	865	0.835
3: Library	202.1	871	0.841
4: Wied	110.7	899	0.868
5: Donker-Voet	24.6	529	0.511
6: Schimmel	186.1	798	0.771
7: Bruijn	244.6	856	0.827
8: Jongveestal	129.4	986	0.857

Interestingly, most systems show a good to very good performance ratio value of 0.8 or higher [4], demonstrating that the systems operate as designed. However, using the inverter data to calculate performance ratio may lead to the conclusion that two systems (Donker-Voet and Bleeker building) are not performing well, as is also obvious from looking at the annual yield. Inspection of the data on a daily level revealed that for some days, and even for the Bleeker building a full month, data from the inverter were not transmitted correctly. We can conclude this, and excluding system malfunction, as the data from the calibrated kWh meters for the Donker-Voet and Bleeker building showed that the monthly energy yields for these buildings are similar to the other buildings. If we use the energy measured from these calibrated kWh meters, an additional amount of 11.9 and 20.6 MWh for the Donker-Voet and Bleeker building system respectively can be added to the generated amount of energy of 1007 MWh. We thus find that the performance ratio of the complete system is PR = 0.825, which can be considered an excellent value taking into account the many high-rise buildings at the campus that cause shading losses.

3.2 Educational activities.

We have developed practical assignments for master students allowing them to get acquainted with analysis of various parts of the data, as part of the master course Photovoltaic Solar Energy Physics and Technology [5], in which we use a recent textbook [6]. Students were given irradiation data from UPOT (1-min time resolution) as well as inverter readings for four systems (5-min time resolution), differing in capacity but foremost in orientation. The sequence of assignments is designed such that students start analyzing PV systems that are optimized for yield (South oriented, 30 degree tilt), followed by more complex systems (mix of East and West orientations).

The following assignments were given to the students:

a) Koningsberger system: Make graphs of the time-series of this system. What is the maximum power generated? The PV/inverter ratio is about 1; could a smaller inverter capacity have been chosen smaller?
b) Show the variation of daily performance ratio of the Koningsberger system.
c) Jongveestal system: Make graphs of the time-series of this system. What is the maximum power generated for each inverter? Do you see effects of the different PV/inverter ratio values? Does the daily performance ratio differ per inverter?
d) Bleeker system: Make graphs of the time-series of this system. What is the maximum power generated? The PV/inverter ratio is about 1; could a smaller inverter capacity have been chosen smaller?
e) Wied system: Compare time-series of this system and explain differences in relation to the four inverters and the number of West, East, and South oriented panels. What is the maximum power generated? Why is the PV/inverter ratio lower than 1.
f) Show the variation of daily performance ratio for the West, East, and South oriented panels for the Wied system. Explain your results.

An example of a student's answer is shown in Figure 6, in which the student shows a daily power profile illustrating that the system combines East, West and South oriented modules. In fact, the Wied system combines four inverters and three orientations: inverter #1 and #2 serve a capacity of 30.51 kWp PV modules, inverter #3 37.36 kWp, and inverter #4 24.84 kWp, see Table III, with inverter AC capacity of 20, 20, 30, and 20 kW, respectively. The student realized that a combination of several orientations makes it possible to lower the inverter capacity below the DC PV capacity as maximum power per inverter will never be reached. However, around midday inverters clip shortly at 100 kWp.

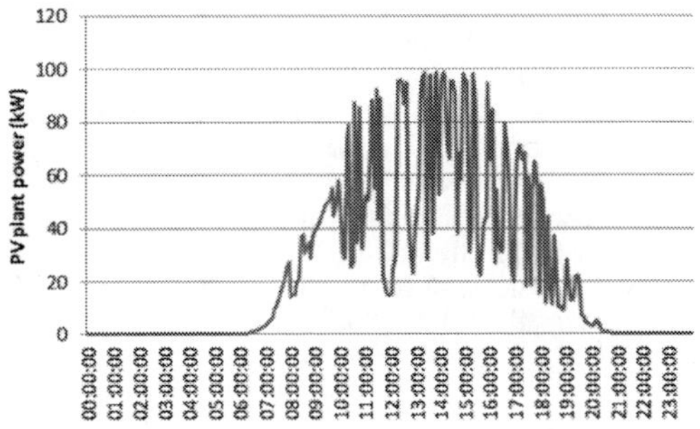

Figure 6: PV power generated on 28 April 2017 as a function of time for the Wied system.

In Figure 7 the student graphed the generated amount of energy for a period of four weeks roughly covering the month of May 2017. Variations are obviously due to variations in irradiance, while also differences in inverters can be discerned. The calculated average PR per inverter is given in Table III. The student concluded based on the data that "when the south-oriented PV capacity becomes higher (inverter #3), the performance ratio will be improved". However, the differences in PR values are not statistically significant. An MSc student will further investigate the performance of the system.

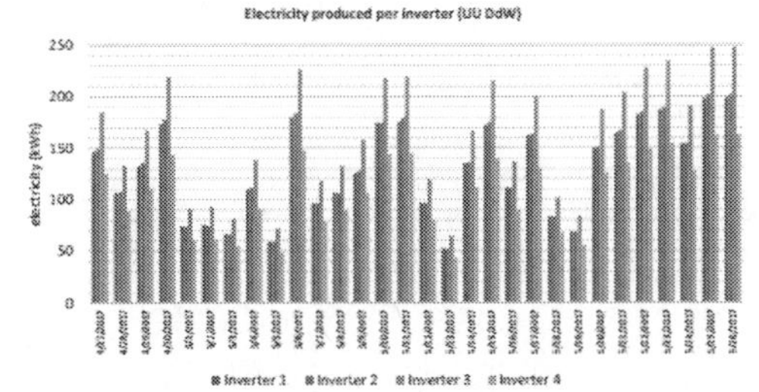

Figure 7: Daily energy yield in kWh for the Wied system showing differences per inverter for the period 27 April 2017 to 26 May 2017.

Table III: Inverter analysis for the Wied system.

inverter	PV capacity (kWp) W	E	S	*PR*
1	12.42	12.42	5.67	0.899
2	12.42	12.42	5.67	0.903
3	12.42	12.42	12.42	0.911
4	12.42	12.42	0	0.907

4 CONCLUSION

In this paper we have described the design and realization of a 1.2 MWp distributed PV system on eight different buildings at the Utrecht University campus Utrecht Science Park. A combination of East, West, South, and North oriented PV subsystems is shown to perform well, with annual yield of over 1000 MWh and an average performance ratio of 0.825. Detailed performance analysis is presently part of a Master level course on photovoltaics.

5 ACKNOWLEDGEMENTS

We would like to thank Utrecht University's governing board for its visionary decision to invest in a 1.2 MW PV system as part of UU's sustainability strategy. We gratefully acknowledge both UU's governing board in particular Prof. Pijpers as well as the dean of the Faculty of Geosciences Prof. Hoekstra for their generous financial support allowing us to perform our analysis. Finally, we thank all the students that have performed data analysis as part of the PSEPT Master course, as well as Panos Moraitis for assisting them.

6 REFERENCES

[1] W.G.J.H.M. van Sark, A. Louwen, A.C. de Waal, R.E.I. Schropp, *UPOT: the Utrecht Photovoltaic Outdoor Test facility,* Proceedings 27th European Photovoltaic Solar Energy Conference (2012) 3247-3249.

[2] Installation video: https://www.uu.nl/en/organisation/real-estate-and-campus/campus-usp-de-uithof/development/work-on-the-outdoor-areas/placing-solar-panels (last accessed 21 September 2017).

[3] https://english.rvo.nl/subsidies-programmes/sde (last accessed 21 September 2017).

[4] N.H. Reich, B. Mueller, A. Armbruster, W.G.J.H.M. van Sark, K. Kiefer, Ch. Reise, *Performance Ratio revisited: Are PR > 90% realistic?*, Progress in Photovoltaics 20 (2012) 717-726

[5] https://www.osiris.universiteitutrecht.nl/osistu_ospr/OnderwijsCatalogusZoekCursus.do (search for GEO4-2513, Photovoltaic Solar Energy Physics and Technology)

[6] A. Reinders, P. Verlinden, W. van Sark, A. Freundlich (Eds.), *Photovoltaic Solar Energy, From Fundamentals to Applications*, Wiley, 2017.

COMPARISON OF THE SIMULATED AND THE MEASURED OUTDOOR POWER GENERATION PERFORMANCE OF PV MODULE & SYSTEM

Jong-Keun Lim, Moonseok Kim, Sanil Yoon, Hyunshin Kang, JungWoo Yang, Jin-Hyung Ahn, Taejun Kim, and Seungwon Lee
Hyundai Heavy Industries Green Energy Co., Ltd.
17-10 Mabuk-ro 240beon-gil, Yongin 16891, South Korea

ABSTRACT: The performance ratio (PR) of PV individual module and PV system were predicted by the PV simulation tool, PVsyst, and the predicted PR was compared with the real PR which was measured with electrical output and weather data, which was collected under field conditions. In this study, the difference between the simulated PR and the measured PR could be minimized by optimizing the PVsyst loss parameter (including the shading loss, IAM loss, soling loss, quality loss, LID loss, mismatch loss, wiring loss, inverter loss), the meteorological data, PANfile etc. Using the on-site measured meteorological data and optimized input parameters, the difference between the simulated PR and measured PR was within 1 %.

Keywords: performance ratio, PVsyst, field conditions, loss parameter

1 Introduction

It is not easy to accurately predict the performance of the PV system under real field condition. The performance of PV system can be affected by various factors, such as meteorological data, specification of the solar module, design of PV system, losses and shadows etc. [1] [2]. We used PVsyst software, the well-known PV simulation tool to compare the simulation data with the real field date collected in our outdoor test bed. In this study, one of our objectives was to evaluate the difference between the simulated performance ratio (PR) by PVsyst and the measured PR of the PV system. In order to minimize the difference, we collected the real-time meteorological data including solar irradiation, ambient temperature, humidity, wind speed etc. Also, we used the most reasonable PVsyst loss parameters, which were obtained by direct measurement, certification, and production distribution.

In this study, we reduced the errors which can affect the simulated PR as low as possible and evaluated the accuracy of the predicted power generation of PVsyst.

2 Outdoor testbed setup

We established our outdoor test-bed to investigate the performance of the individual PV modules and PV system. Figure 1 shows the whole view and configuration of our outdoor test-bed to monitor the PV individual module and PV system. We used module optimizer (Tigo company 2ES50XL model) to check the performance of the individual modules with 295W mono-PERC structure. And the 2.36kW PV system is consisted of 8 series 295W mono-PERC modules with ABB inverter. The generated power at PV system is directly delivered to the grid. All data including I-V data of PV module and system, meteorological data (solar irradiation, ambient temperature, humidity, wind speed etc) are monitored via a web interface. This makes it possible to compare the real PR and simulated PR.

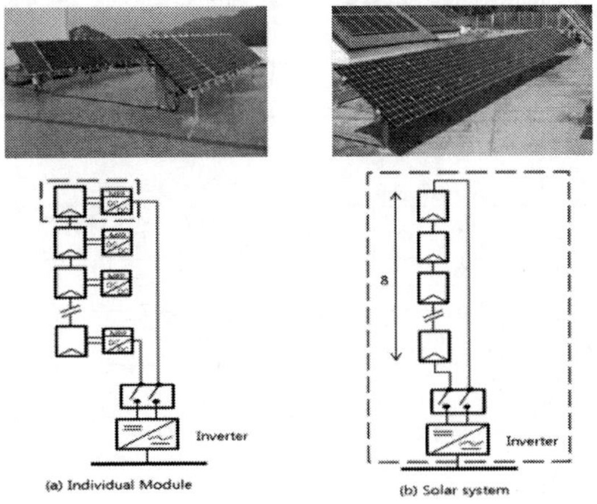

Figure 1: PV monitoring system (a) individual module (b) solar system

3. Procedure to evaluate the performance of PV module & system

After finishing the setup of our facility, we evaluated the performance of the 295W mono-PERL module and 2.36W PV system. We have to consider various factors to compare the predicted PR and the measured PR, which are depicted in Figure 2.

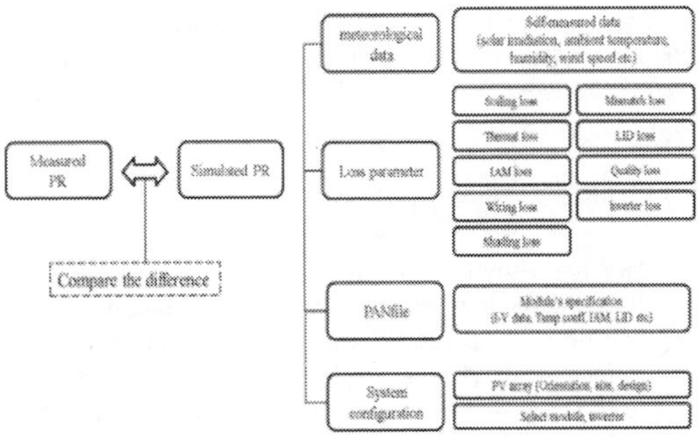

Figure 2: The procedure to compare simulated and measured PR

4. Optimization of PVsyst simulation

4.1 Meteorological data

The meteorological data directly affect the PR of PV module/sytem. PVsyst provides the meteorological data sources measured by government agencies or a variety of organizations. But these data can be different with on-site self-measured data by up to ±10%. [4] So, we used the on-site self-measured data including solar radiation (global horizontal, Plane of Array), temperature, wind spee, etc to minimize the error by the meteorological data.

4.2 Loss parameter

Table 1 shows the PVsyst loss parameter to be considered to calculate the performance ratio. In case of the individual module, array mismatch loss, ohmic wiring loss, and inverter loss are excluded from computing the simulated PR. In case of the PV system, all loss parameters are included. Also, we installed our modules at an area without shading to remove the error of the calcuated PR by shading loss.

Table 1. PVsyst loss parameter of individual module and solar system

<PVsyst loss parameter of individual module>

PVsyst loss parameter	Individual Module (295W)	
	Not optimized	Optimized
Shading loss	No shading area (0%)	
IAM loss	Ashrae model	Certified value
Soling loss	3%	1%
Module quality loss	3%	0.42%
LID loss	2%	1.6%
Array mismatch loss	Not considered	
Ohmic wiring loss	Not considered	
Inverter loss	Not considered	

< PVsyst loss parameter of solar system>

PVsyst loss parameter	Solar system (2.36kW)	
	Not optimized	Optimized
Shading loss	No shading area (0%)	
IAM loss	Ashrae model	Certified value
Soling loss	3%	1%
Module quality loss	3%	0.42%
LID loss	2%	1.6%
Array mismatch loss	1%	0%
Ohmic wiring loss	1%	1.3%
Inverter loss	Use ABB inverter model	

1) Mismatch loss

The mismatch loss default of PVsyst is 1%. But mismatch loss of test modules was almost 0% when it was calculated from the relative standard deviation of Voc, Isc of the installed modules.

2) LID loss

The LID can be occured due to the boron-oxygen (BO) defects that form during the excess carrier generation at field condition.[3] Our test modules made from mono-PERC solar cells were operated at MPPT mode during 4 weeks to check their light-induced degradation (LID). The average of LID was about 1.6%. The measured value of LID is lower about 0.4%p than PVsyst default value, which is shown in Figure 3.

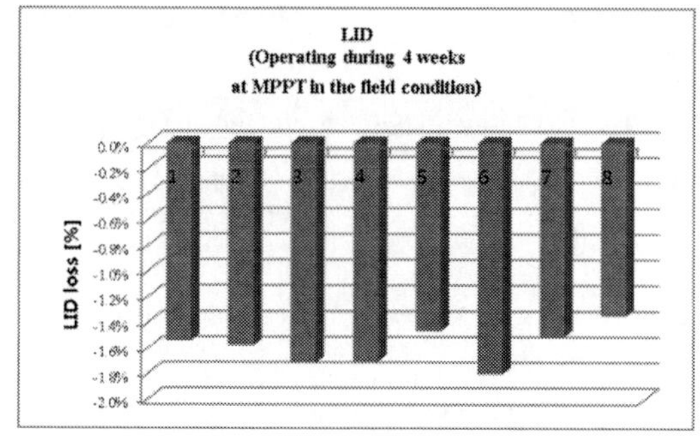

Figure 3: The measured LID loss

3) IAM loss

We have sent our modules to CFV solar test laboratory to determine their Angle of Incidence [AOI] profiles. The IAM default value of PVsyst is the value calculated from Ashrae model, which is a little lower than the measured value. To improve the accuracy of simulated PR, we have applied the measured IAM value. Fig 4 shows the measured IAM profiles by certificated CFV.

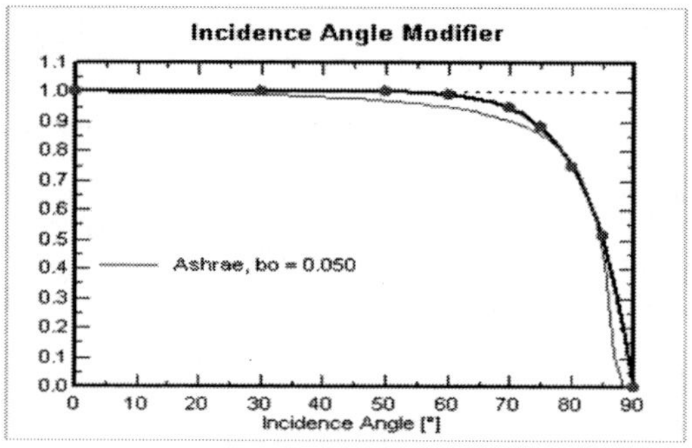

Figure 4: The measured IAM profile of our module (certificated by CFV)

4) Quality loss

The module quality loss is defined as the deviation of the average effective module efficiency with respect to manufacturer specifications. The default value of PVsyst is 3%, but the measured value of the quality loss is 0.42%. The power is mesaured at the solar simulator of our lab based on calibration module certificated by Fraunhofer ISE lab. When measuring the power, the accuracy of the measurement is not perfect. The measurement uncertainty of calibration module is ±2% at Pmax, and the measurement uncertainty of our solar simulator is approximately ± 1% at Pmax.

5) Ohmic wiring loss

The ohmic wiring loss default of PVsyst is 1%. But ohmic wiring loss of test modules is 1.3%, which is calculated value from wiring length, thickness of our PV systems. Table 2 shows the series resistance of wiring, which is applied to simulate the wiring loss in PVsyst.

Table 2. The calculated ohmic wiring loss

Wiring	Wire length [mm]	Wire thickness [mm^2]	Series Resistance [ohm]
Module –Moudle	16m/circuit	4 mm^2	75 Ω
Module-Inverter	120m/circuit	4 mm^2	564 Ω
Total	120m/circuit	4 mm^2	639 Ω

6) Soiling loss

Soiling loss can occur by dust, plant products, soot, salt, bird droppings etc. Also, this loss randomly varies by rain, wind, unpredictable events etc. So, it is very difficult to accurately measure the power loss by soiling. We checked the power of two modules, one which was cleaned everyday ("module without soling loss") and another left in field condition without man-made cleaning ("module with soling loss"). The soiling loss could be measured from the difference of the net power [kWh] generated from two modules during the experiment period. The measured soiling loss was about to 1%, which is lower than PVsyst soiling default value 3%.

7) Inverter loss

We used inverter (model:PVI-5000-TL-OUTD) manufactured by ABB company. When simulating the peformance of the PV system, we changed the characteristics of inverter file in the database of PVsyst according to the recent datasheets of our inverter's model. The simulated inverter loss value was about 4%. Fig 5 shows the inverter efficiency curve, which means the input/output power transfer function during normal operation.

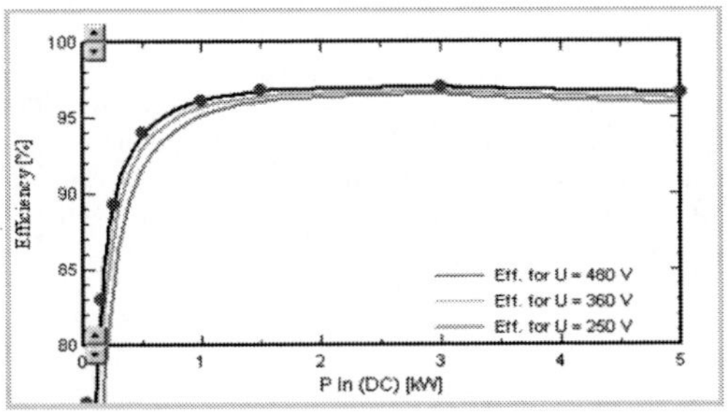

Figure 5: The used inverter's efficiency curve

4.3 PAN file

PVsyst has a PV module database to represent the characteristics of the modules provided by the manufacturers themselves. But, PVsyst cannot guarantee the accuracy and reliablity of the modules' parameters in the database. In order to make our module's PANfile more accurate, we sent our test modules to CFV solar test laboratory to obtain I-V characteristics, temperature coefficient, incident irradiance characteristics etc. The PANfile from CFV solar lab was a little different to the PANfile in the PV module database of PVsyst.

5. Results

In this experiment, we should consider a variety of factors to predict the performance of PV module/system. Although it is not easy to accurately simulate and measure the PR, it is proved that the difference between the simulated PR and the measured can be minimized when optimizing the meteorological data, the PVsyst parameter, PANfile etc. Especially, the PVsyst parameter loss was totally optimized by direct measurement, certification, and production distribution. In case of indvidual module, the PR difference between the simulated value and the measured value was improved to 0.3%p when the all loss paramter are optimized. In case of PV system, the the PR difference was improved to 0.5%p. In two cases, the difference of PR is within 1 percent, which is very accurate when considering that it is the predicted value. .

Table 3. Measured and simulated Performance Ratio (PR) of Individual module and solar system

	Loss parameter	Simulated PR (A)	Measured PR (B)	△PR (A-B)
Individual Module (295W)	Not Optimized	81.4 %	90.8 %	-9.4%$_p$
	Optimized	91.2 %		+0.3 %$_p$
Solar system (2.36kW)	Not Optimized	86.8 %	82.5 %	+4.3%$_p$
	Optimized	82.0 %		-0.5%$_p$

References

[1] Wilfried G.J.H.M. van Sark1, Atse Louwen1, Arjen C. de Waal1,2, Boudewijn Elsinga1, Ruud E.I. Schropp2, UPOT: THE UTRECHT PHOTOVOLTAIC OUTDOOR TEST FACILITY, 27th European Photovoltaic Solar Energy Conference and Exhibition

[2] N.H. Reich, B. Mueller, A. Armbruster, W.G.J.H.M. van Sark, K. Kiefer, and C. Reise, Performance ratio revisited: is PR > 90% realistic?, Progress in Photovoltaics: Research and Applications 20 (2012) 717-726

[3] Kyumin Lee, Moonseok Kim, Jong-Keun Lim, Jin-Hyung Ahn, Myung-Ick Hwang, and Eun-Chel Cho, natural recovery from lid: regeneration under field conditions, 31st European Photovoltaic Solar Energy Conference and Exhibition

[4] "Cautions for Interpreting the Results," NREL / rredc.nrel.gov/solar/codes_algs/PVWATTS/interp.html

THE PRACTICABILITY OF OUTDOOR MEASUREMENT METHODS FOR PHOTOVOLTAIC INSTALLATIONS

W. Mühleisen[1], L. Neumaier[1], C. Hirschl[1], M. Spielberger[2], H. Sonnleitner[3], Y. Voronko[4], G. C. Eder[4], R. Ebner[5]
[1] CTR Carinthian Tech Research AG, Europastr.12, 9524 Villach/St.Magdalen, Austria
(Author and Contact: wolfgang.muehleisen@ctr.at; T: +43 4242 56300 238)
[2] PVSV e.U., Unterer Markt 7, 9334 Guttaring, Austria, (markus.spielberger@PVsv.at)
[3] ENcome Energy Performance GmbH, Lakeside B08, A-9020 Klagenfurt, Austria, (horst.sonnleitner@en-come.com)
[4] OFI Austrian Research Institute for Chemistry and Technology, Arsenal Objekt 213, Franz-Grill-Str. 5, 1030 Vienna, Austria, (yuliya.voronko@ofi.at, gabriele.eder@ofi.at)
[5] AIT Austrian Institute of Technology, Giefinggasse 2, A-1020 Vienna, Austria (r.ebner@AIT.ac.at)

ABSTRACT: Different sized power plants with crystalline silicon PV modules were inspected directly on-site. For the inspection purpose, measurement equipment was used in different situations. First at plants with good infrastructure and accessible power sockets for powering whole PV strings, second with only a small battery powered system for powering separate modules. Characterization methods like thermography, electroluminescence and ultraviolet fluorescence were performed for field use with the goal of testing the practicability in a non-laboratory surrounding. The checked PV plants of different sizes were evaluated to accessibility and difficulties faced by the local conditions on-site. In the majority of cases, and despite of adverse conditions, the checks were successfully performed.

Keywords: PV Module, Performance, Inspection, Operation and Maintenance

1 INTRODUCTION

For about ten years now, it has been very common to inspect photovoltaic modules individually in laboratories with a flash test system, electroluminescence (EL) and thermography (TG). Some advantageous conditions of laboratories are constant temperature, absolute darkness and module holders for good positioning. Because of a global increase in installation capacity of about 300 GW at the end of 2016 [1], strong efforts were taken to allow for measurements not only in the laboratory but also directly in the field. Therefore, standard laboratory equipment was adapted for outdoor use with focus on handling and time efficiency. The requirements to plant inspections are different depending on purpose. Sometimes it is only of interest to rewrite a string plan for a clear documentation. Another aim could be to check many solar panels to find out what is wrong with the system.

Commercial and professional plant inspections are state of the art and available for plant owners. This includes, for example, drone flights with a thermography system or mobile laboratories in trailers equipped with standard technique like EL or a flash test system for measuring the current voltage curve (I-V-curve). Recently, also research institutes offer outdoor inspection on a large scale – for instance the ultraviolet-fluorescence (UV-F) method. Despite the mostly extensive offer, in many cases, only the inspection technique is favorized, which is under good control.

There are also some plant owners who are interested in an evaluation of their monitoring data and in a quick check of their small or big PV plant. For that it is not necessary to measure modules in a series but to check monitoring data and to use three standard inspection techniques like TG, EL and UV-F. The combination has supplied big advantages as these measurement techniques yield complementary results and some failures cannot be found with only one inspection tool alone.

Regardless of the chosen concept, every plant is different and in case of EL the needed access to the PV modules or strings can be more or less difficult. There are PV plants on flat roofs, which are easy to reach but without the possibility to power whole PV strings because of missing power sockets. In most cases there were no three phase power sockets available but 230 V one phase power sockets. For that the limitation in power is the fuse with 16 A so the usable power is available up to 3600 Watts. When driving a PV string with 18 modules with each 250 Wp theoretically 4500 Watts would be used. Due to the fuse limitation and due to non-ideal device efficiency only the half Isc can be used in practice reducing light intensity for EL measurement. One idea to solve this problem could be to create a connection from the grid but this needs an electrician at your side.

One alternative is to power separate modules with a DC-DC converter or a laboratory power source. Both devices could be powered with a battery system, which is more flexible in the field. However, it is only suited for separate modules because of limited power of the battery system.

2 SET-UP

Besides electrical measurements of the performance, three measurement methods are in the main focus of our work. First of all TG is used to detect common failures leading to inhomogeneous temperature distribution within the PV-module like hot spots or connection breaks while quickly providing an overview if all PV-modules within a plant are working correctly. However, this method needs bright sunny days and there exists a small number of module defects, which cannot be visualized by thermography but by outdoor EL and UV-F. The latter both techniques are used during the night, where EL works in combination with a power source to power whole PV strings for the detection of faulty modules, while a UV-Fluorescence lamp can be supplied by a battery and visualizes degradation effects in the encapsulation material. The fluorescence image-patterns of crystalline Si-modules are influenced by the permeation properties of the backsheet used and allows to visualize cell cracks and furthermore to distinguish micro cracks of silicon solar cells based on their age [9, 10].

2.1 Used Equipment

There are many inspection tool manufacturers on the market. The equipment used here is a standard technology and in a low-priced and medium-priced segment.

Thermography camera
For the measurement, a professional thermal imaging camera (Testo 890-2) with a resolution of 640 x 480 pixels and a thermal sensitivity in the range of 40 mK, was used. The Testo 890 is equipped with an additional 3.1 megapixel sensor for visual photo imaging. With every shot, thermal and visual pictures are taken simultaneously. Photovoltaic thermography was boosted by the group of Dr. Buerhop (Bavarian Center for Applied Energy Research) since 2007 [2, 3].

Electroluminescence camera
As Electroluminescence camera a modified digital single-lens reflex camera (Canon EOS 700D) was used. Its 18 megapixel APS-C CMOS sensor can reach the sensitivity of ISO 12800. A special infrared light suited lens with manual focus helps to capture clear images with short exposure times within seconds. This idea of an affordable modified camera was already introduced by the group of Prof. Mertens (Muenster University of Applied Sciences) in 2012 and has been enhanced since then [4, 5, 6].

Ultraviolet-fluorescence lamp
As ultraviolet-fluorescence lamp a self-made device was used in combination with a photo camera. Three battery powered and tunable LEDs (380nm, 100 W) which are powered by a Li-accumulator, were used for the inspection. Such powerful UV lamps are necessary to shoot good photographs for reports. Another torch was tested too (TROTEC 15F, 365 nm, 9 W) which is enough to inspect PV modules with bare eyes and protective goggles. The group of Prof. Roeder and Dr. Schlothauer (University of Berlin) made some experience with UV light for PV module usage in 2010 [7] and the group of Dr. Koentges (Institute for Solar Energy Research Hamelin) have refined the outdoor UV-inspection tool for PV modules since 2012 [8].

Power source string use
To drive a complete string with many PV modules it is necessary to use a high power source. The used DC power supply offers an output power of up to 5 kW with a provided voltage of up to 1000V and a current of up to 5 A (BK Precision - PVS10005).

Power source module use
To drive a separate PV module a 600 W power source (VOLTCRAFT HPS-16010) was used. The provided voltage/current range is 1 – 60 V/ 0 – 10 A.

Battery system 1: DC-AC-DC
For the field use a non-grid connected system was created by using a car battery, a 12 V to 230 V (300 W) inverter and a 600 W power source (Fig. 1). With this system one 60-cell standard module can be powered to measure EL. The total system efficiency for the chain 12 V_{DC} to 230 V_{AC} and 230 V_{AC} to 37 V_{DC} is at 65 % for a current of 3 A at module side.

Figure 1: Battery DC-AC-DC system

Battery system 2: DC-DC
Due to the efficiency purpose a second battery system was created with only a DC-DC step up converter from 12 V_{DC} to 37 V_{DC}. The total system efficiency is for this system at 88 % for a current of 3 A at module side (Fig. 2).

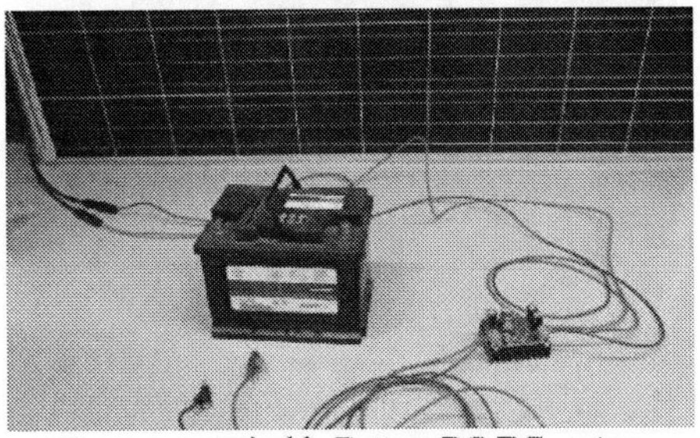

Figure 2: more practicable Battery DC-DC system

3 RESULTS

Exemplary the results of the investigations of three PV-plants, one in France and two in Austria, are shown in the following. Thermography was used at an irradiation of 800 W/m² and separate strings were found which were hotter in relation to adjoining ones (Fig. 3). Hotter strings are an indicator for non-distributed electrical energy with the effect of more heated panels. These strings were inactive because of blown fuses.

Figure 3: TG overview of a plant from a lift

Another failure was identified by TG at an inverter due to connection problems (Fig. 4).

Figure 4: TG - Inverter connection problem

Relative hotter cells at the module edge are often caused by dirt that operates like shading (Fig. 5). A cleaning step was suggested instead of the claimed module change due to "hot spots".

Figure 5: Hotter solar cells due to shading dirt

In Fig. 6 a wrong connected module was identified by TG. The module was hotter because it could not convert the light energy and was additionally powered by the string.

Figure 6: Wrong connected module

When using outdoor EL ambient light disturbs the measurement leading to poor resolution due to blooming of the EL signal which could cause misleading results (Fig. 7).

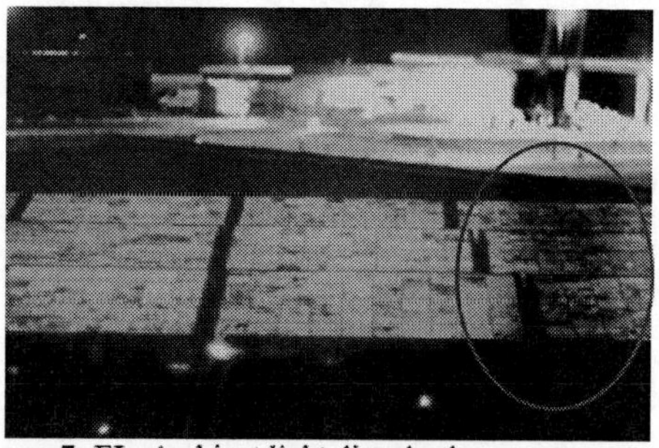

Figure 7: EL: Ambient light disturbs the measurement

Directly on a roof or with help of a cherry picker better overview images can be obtained, however, in most cases little space is available and, thus, the field of view is not optimal (Figs. 8, 9).

Figure 8: EL - Field of view is limited due to surroundings – no string overview possible

Figure 9: EL - View from a lift – overview too small

When using UV-F the image quality strongly depends on the level of darkness, scattered light from street lamps, traffic or even full moon can reduce the contrast and resolution of the UV-F-image drastically (Figs. 10, 11). The south side module without disturbance showed an impressively good fluorescence signal. Fractions in the cells could be visualized well.

Figure 10: UV-F - South side – good image

The fluorescence signal on a west side module was cross-faded by the scattered light from the street lamps as well as the full moon. Nevertheless, the detection of fractioned cells was still possible.

Figure 11: UV-F: West side – less significant image

4 CONCLUSIONS

It can be concluded, that thermography is well suited to give a general overview of a PV-plant and is capable of identifying inactive modules/strings or hot spot cells. The camera is additionally indispensable for the inspection of cables, connectors or inverters that may be a fire risk. The comfortable usage as handheld device is an advantage of this inspection tool and very practicable. However, special defects, such as e.g. fractioned cells caused by external mechanical impact like hail storms or rough handling can be characterized better with electroluminescence or ultraviolet-fluorescence. The usage of further equipment on a plant like a lift or a drone is an advantage for making an overview picture and very practicable. But inspection from the ground is also possible with the disadvantage of needing more time and not often reaching an ideal measuring angle.

The measurements with the used electroluminescence showed that ambient light is disturbing although the position of the PV modules was good and well accessible on the flat roof. A plant on a badly accessible roof was only checked with the help of a cherry picker with the disadvantage of only seeing the modules in function. The string powering is only limited possible because of the fuse. This lead to the situation that only the half I_{SC} or less of a module can be adjusted. Nevertheless, string powering saves time when measuring modules in series and helps to make an overview picture. However, measuring more than four to six modules side by side is not practicable because of losing detailed information. Therefore, the usage of a battery powered system with the possibility to connect and power separate modules is favorized in future.

The ultraviolet-fluorescence inspection tool is well suited to detect micro cracked cells and offers the possibility to make a differentiation between older and newer cracks. However, the fluorescence signal is weak and disturbing ambient light is a disadvantage to make clear pictures.

5 ACKNOWLEDGEMENTS

This work was conducted as part of the Austrian "Energy Research Program" project INFINITY. Funded by the Austrian Climate and Energy Fond and the Austrian Research Promotion Agency (FFG), both are gratefully acknowledged.

6 REFERENCES

[1] International Energy Agency (IEA), "Snapshot of Global Photovoltaic Markets 2016 ", http://www.ieapvps.org/fileadmin/dam/public/Press_Releases/Press_Release_T1_-_Snapshot.pdf

[2] Auer, R. ,Jahn, U.,Buerhop-Lutz, C. ,Vordermayer ,C. ,Wotruba ,G. ,Zehner ,M. & Niess ,M., "Infrared analysis of PV Modules for Improving Quality", 22nd PVSEC 2007 – Milano, 2519-2522.

[3] Buerhop, C., Jahn, U., Hoyer, U., Lerche, B. & Wittmann,S. (2007), „Überprüfung der Qualität von Photovoltaik-Modulen mittels Infrarot-Aufnahmen, Korrelation von Infrarot-Messung und elektrischer Messung.“, ZAE Bayern e.V. http://www.sev-bayern.de/content/downloads/IR-Handbuch.pdf

[4] Mertens, K.; Stegemann, T.; Stöppel, T.: LowCost EL -Erstellung von Elektrolumineszenzbildern mit einer modifizierten Standard - Spiegelreflexkamera, 27. Symp. Photovoltaische Solarenergie, 2012

[5] Mertens, K., Pascual Gonzales, D., Diehl, M.: „LowCost – Outdoor - EL: Kostengünstige umfassende Vorort - Qualitätsanalyse von Solarmodulen“, 30. Symp. Photovoltaische Solarenergie, 2015

[6] Mertens, K., Kösters, H., Diehl, M: „Low-Cost-Outdoor-EL: Cost-Efficient Extensive on-Site Quality Analysis of Solar Modules", Proceedings of 31st EU PVSEC, Hamburg, 2015

[7] Schlothauer JC, Jungwirth S, Röder B, Köhl M: Fluorescence imaging a powerful tool for the investigation of polymer degradation in PV-Modules. Photovoltaics International 2010: 149–54.

[8] Köntges M, Kajari-Schröder S, Kunze I: Cells Cracks Measured by UV Fluorescence in the Field. 27th European Photovoltaic Solar Energy Conference and Exhibition 2012; 2012: 3033–40

[9] G. C. Eder, Y. Voronko, B. Kubicek, K. Knöbl, „Fluorescence Spectroscopy and Imaging on aged polymeric PV Encapsulants", 20th ESOPS, Dresden, 09.2016; DOI: 10.13140/RG.2.2.20827.36648

[10] G. C. Eder, Y. Voronko, P. Grillberger, B. Kubicek, K. Knöbl, „UV-Fluorescence measurements as tool for the detection of degradation effects in PV-modules", European Weathering Symposium EWS, Vienna, 09.2017

THE USE OF LOGISTIC REGRESSION FOR EVALUATING CLIMATE-RELEVANT PV MODULE FAILURES

N. Vollert[1], L. Neumaier[1], W. Mühleisen[1], M. Halwachs[2], L. Maul[3], Y. Voronko[4], A. Mihaljevic[5], C. Hirschl[1]
[1] CTR Carinthian Tech Research AG, Europastr.12, 9524 Villach/St.Magdalen, Austria
[2] AIT Austrian Institute of Technology GmbH, Giefinggasse 2, 1210 Vienna, Austria
[3] Fachhochschule Technikum Wien, Giefinggasse 6, 1210 Vienna, Austria
[4] OFI Austrian Research Institute for Chemistry and Technology, Franz Grill Str. 5, 1030 Wien, Austria
[5] PCCL Polymer Competence Center Leoben GmbH, Roseggerstr. 12, 8700 Leoben, Austria
Corresponding author: E: natalie.vollert@ctr.at; T: +43 4242 56300 239
Co-Authors: E: lukas.neumaier@ctr.at, wolfgang.muehleisen@ctr.at, martin.halwachs@ait.ac.at, maul@technikum-wien.at, yuliya.voronko@ofi.at, antonia.mihaljevic@pccl.at, christina.hirschl@ctr.at

ABSTRACT:
When it comes to the reliability of PV modules it is important to guarantee a constant energy yield while at the same time achieving a low failure rate of the modules. Amongst the numerous environmental influences that increase the risk of failures in PV systems, it is assumed that the different climate zones play an important role. This works deals with analyzing the issue of deterioration of the energy yield of PV systems by using logistic regression models. It will be shown that the risk for such power losses is higher in climate zones that include extreme conditions, e.g. high temperatures and snow loads and hence, a clear connection between failures that result in power losses and the different climate zones can be made.

Keywords: failure analysis, climate influence, logistic regression, power loss

1 INTRODUCTION

It is well known that several environmental factors increase the risk of failures in PV plants. Therefore, it is the goal of the project INFINTY to determine which of these failures are climate-related and how climate zones affect the different components in more detail. This is an important information since the components can then be adapted to withstand the special circumstances in different environments, e.g. a module situated in a snow climate must be adapted to tolerate high loads. Thus, an INFINITY project-database has been created, based on an adopted failure survey sheet which evolves from the activities of IEA PVPS Task 13 [1].

2 DATA PREPARATION

Currently, the database is filled with a total of 1243 failure entries coming from 488 surveys based on different subsystems. Amongst them, there are also some errors that are not relevant for obtaining climate-related influences, e.g inverter software errors, which are excluded from the analysis yielding 1138 failures.
Each of these failure entries contains fundamental information needed to describe the corresponding PV plant and the failure itself like failure type, failed component, time of failure occurrence etc. However, not all of this information is available for each failure and therefore a filtering mechanism for removing all empty or unknown fields must be applied to the data before any analysis further reducing the sample size depending on the attributes of interest.
For classifying the climate zones, the Koeppen-Geiger climate classification [1] is used in this work. Basically it would consist of 5 main and several sub-classes, but due to a lack of data coverage over all sub-classes only the 5 main climate classes are used. Fig. 1 shows how the failures of the different PV components are distributed amongst the climate zones.

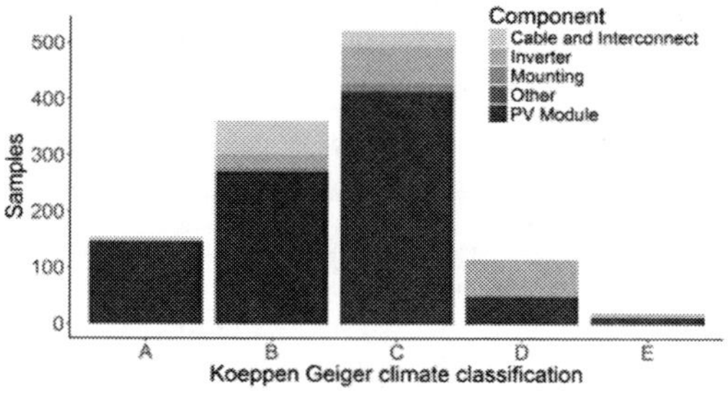

Figure 1 Distribution of failures of different components over the 5 main climate zones of the Koeppen Geiger climate scheme.

It can be easily seen from Fig. 1 that the majority of the failures is assigned to the PV module and hence, we will restrict all of our following analyses to failures of this component.

2.1 Power loss data

As one of the most important issues in PV plants is achieving constant energy yield, we take a closer look on a possible correlation between the power loss caused by each failure entry and its climate zone.
Actually, the power loss attribute is divided into 11 subclasses, i.e. 10 classes in 10% steps and one class for "no power loss". However, for the analysis we combine the 10 classes forming one class "power loss" class and just consider a two-class problem, i.e. "power loss" and "no power loss", again due to a lack of data coverage. Fig. 2 shows how the different climate zones are distributed within these two classes. Note that all failure entries with unknown power loss or unknown climate zone are already discarded.

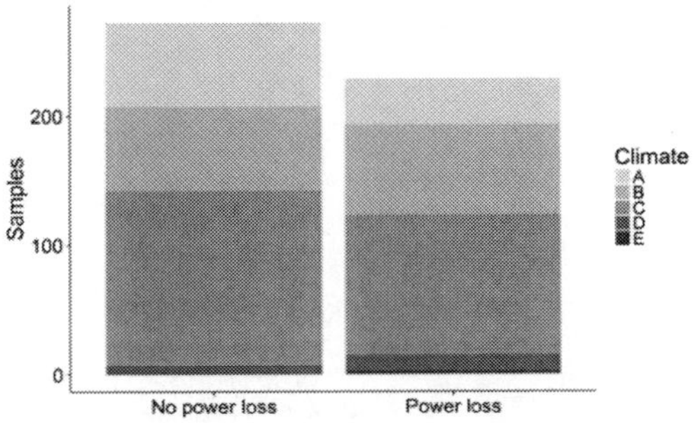

Figure 2 Distribution of the climate zones of the data within the two power loss classes.

Of course there are also some other influential variables that may be useful for describing the power loss, e.g. the material of the cell. Unfortunately, such attributes are often described as unknown in the surveys and incorporating them in the analysis would strongly reduce the sample size. However, despite the climate zone we also include the "time until failure", describing the difference between system installation date and failure occurence, as an input variable in our analysis as this seems to be also correlated to the power loss, see Fig. 3.

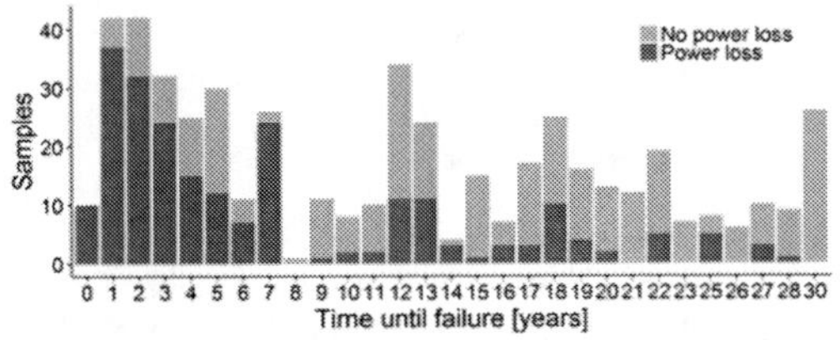

Figure 3 Distribution of the two power loss classes over different failure times in years.

Finally, by using the attributes "power loss", "climate zone" and "time until failure" and filtering out all unknown entries, the remaining total sample size reduces to 500 failure entries.

3 LOGISTIC REGRESSION APPROACH

Here, the power loss is adapted to be a binary response, i.e. it is distinguished between "no power loss" and the overall group "power loss", which is independent of the amount of the real power loss. A very flexible method for modeling such two-class categorical data is a binomial logistic regression approach [2]. Here, the log-odds transformations of class probabilities are modeled as linear functions in the input variables $\boldsymbol{x}$, which are "climate" and "time until failure" in our case. The model has the form

$$\log\frac{P(Y = \text{"power loss"} \mid \boldsymbol{x})}{P(Y = \text{"no power loss"} \mid \boldsymbol{x})} = \beta_0 + \boldsymbol{\beta}^T\boldsymbol{x}. \quad (1)$$

Generally, this model shows how different input variables influence the probability for a power loss, e.g. if a higher risk can be linked with certain climate zones. The formula for calculating this probability is given by

$$P(Y = \text{"power loss"} \mid \boldsymbol{x}) = \frac{\exp(\beta_0 + \boldsymbol{\beta}^T\boldsymbol{x})}{1 + \exp(\beta_0 + \boldsymbol{\beta}^T\boldsymbol{x})}. \quad (2)$$

As usual, the parameters $\boldsymbol{\beta}$ have to be estimated, which is typically done by maximum likelihood. However, as this estimation procedure often struggles with stability issues a Bayesian approach is used in this work. Here, the parameters are estimated as the means of there a-posteriori distributions which are explored by Markov Chain Monte Carlo (MCMC) methods [3].

4 RESULTS

The 500 samples have been split into a train data set of 301 samples for building the model and a test data set of 199 observations for testing it afterwards. Care has been taken to assure a regular distribution of the different climate zones in both data sets. The statistical software R [4] has been used to estimate the parameter values of β_0 and $\boldsymbol{\beta}$ and the probabilities for a power loss for the different climate zones and two different times of failure occurrence are summarized in Table I.

Table I Power loss probabilities according to eq. (2) for all climate zones. "*time=0*" and "*time=10*" refers to a instantaneous failure and an occurrence of the failure after 10 years, respectively.

P(Y="power loss")	time=0 [yrs]	time=10 [yrs]
climate A	0.84	0.57
climate B	0.76	0.45
climate C	0.77	0.46
climate D	0.92	0.74
climate E	0.79	0.49

According to Table I, the probabilities for a power loss when an instantaneous failure occurs are quite high for all climate zones. The reason for this is that such failures most often directly indicate some power loss due to the sort of failure, e.g. disconnected cables, and therefore, they are completely independent of the climate zone. When the time until failure increases, the results generally represent what can also be seen in Fig. 3: the later the failure occurs the lower the probability for a power loss. However, it can be also seen that climate zone A and D, referring to tropical and cold climate, respectively, still have a higher risk for a power loss, so PV systems situated in these climate zones can be identified as having the highest risk for a loss in energy yields. Note, that there are only 3 data points for climate zone E in our data set. Thereof, 2 are in the training set, where one yields a power loss and the other doesn't, and hence, it is clear that the estimated model parameter is very unreliable. In fact, it can be expected, that climate zone E actually shows a similar behavior as climate zones A and D, as these are the climate zones that contain extreme conditions.

Now the train data set is used to validate the model, see

Fig. 4 and classification errors are shown in Fig.5. The misclassification rate of about 24% could be improved by using more attributes of the PV system to describe the power loss, e.g. manufacturing date of the PV module (age). However, these attributes are often defined as "unknown" in the surveys and hence, introducing them as input variables would have reduced the sample size for modeling enormously. This, in turn, would also reduce the information about the influence of the climate zone, which is of main interest here.

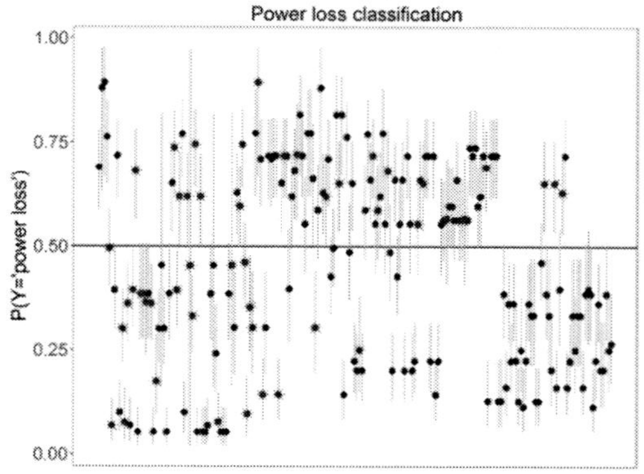

Figure 4 It is shown how the model classifies the test data. The horizontal red line refers to the border between the two groups, i.e. $\mathbf{P(Y = 'power\ loss') = 0.5}$. The vertcial grey lines refer to the 95% confidence intervals for each prediction and points equipped with red stars refer to points assigned to the wrong group.

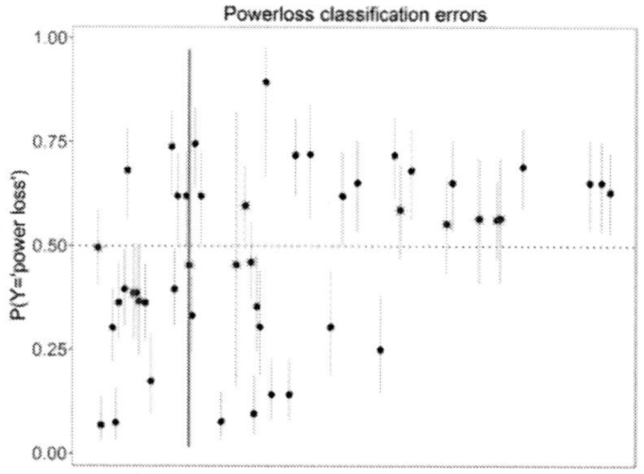

Figure 5 This time only the points that are misclassified are shown. Here, points equipped with a red star are samples for which the border is included in the corresponding confidence interval (grey vertical lines), i.e. for these samples it is clear that the assignment is uncertain. One confidence interval is especially noticeable and therefore shown by a red vertical line: it almost covers the whole range. This sample points refers to the one test point from climate zone E where a reliable classification cannot be made.

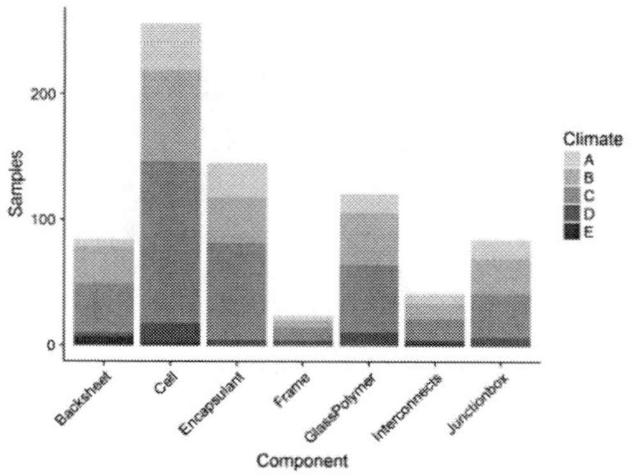

Figure 6 Distribution of the climate zones for the different components of the PV module. Most of the data refers to "Cell", so especially this component could be analyzed in more detail in upcoming work.

5 SUMMARY & OUTLOOK

The analysis in this work gives a first overview how different climate zones may influence failures of PV systems. This work especially deals with the influence of the climate on a possible power loss of the systems and climate zones A and D have been identified as being the ones making a power loss most likely. Unfortunately, there is a lack of failure data for climate zone E, which makes a reasonable statement about its influence impossible. This is a very informative result as these are exactly the climate zones with the most stressful conditions, indicating that the different climate zones are in fact an influential factor for PV module failures.

Future work may contain a more detailed analysis of these climate influences, e.g. how they affect the different components of the PV module. An overview about the distribution of the failures of the different components and there relation to the climate zones is shown in Fig. 6. Furthermore, a similar model describing the relation between different climates and inverter failures can be of high interest in upcoming studies.

6 ACKOWLEDGEMENT

This work was conducted as part of the Austrian "Energy Research Program" project INFINITY. Funded by the Austrian Climate and Energy Fund and the Austrian Research Promotion Agency (FFG), both are gratefully acknowledged.

7 REFERENCES

[1] M. Kottek, J. Grieser, C. Beck, B. Rudolf and F. Rubel; World Map of the Köppen-Geiger climate classification updated; Meteorologische Zeitschrift, Vol. 15, p. 259-263, 2006

[2] IEA-PVPS Task 13 – Performance and Reliability of Photovoltaic Systems (Photovoltaic Power Systems Program, Int. Energy Agency, FFG project number: 754374).

[3] T. Hastie, R. Tibshirani and J. Friedman; The Elements of Statistical Learning – Data Mining, Inference, and Prediction; Springer, 2009

[4] J. Gill; Bayesian Methods – A Social and Behavorial Sciences Approach; CRC Press, 2015

[5] R Development Core Team; R: A Language and Environment for Statistical Computing; R Foundation for Statistical Computing, Vienna, Austria

FIRE SAFETY OF PV MODULES AND BUILDINGS: OVERVIEWS, BOTTLENECKS AND HINTS

P. Bonomo[1]*, E. Saretta[1], F. Frontini[1], M. Caccivio[1], G. Bellenda[1], G. Manzini[2], P. Cancelliere[3]
[1] SUPSI-ISAAC, Campus Trevano, CH 6952, Canobbio, Switzerland
[2] RSE S.p.A., Milano, Italy
[3] Italian National Rescue and Services - DCPST, Rome, Italy
* mail to: pierluigi.bonomo@supsi.ch; phone: +41 (0)58 666 62 75

ABSTRACT: The enhancement of the fire safety of buildings with PV modules installed/integrated on roofs or facades is a topic that still concerns the PV industry and research communities due to the fact that the presence of PV modules can facilitate a fast spread of the flames. Considering that PV modules are spreading all over the world on buildings, there is the need to guarantee higher safety requirements both in order to avoid economic losses and to protect people who can be involved in case of fire. This is involving, on one hand, a continuous improvement of the international and national regulation frameworks and, on the other hand, new research activities for the evaluation of the PV module behavior in case of exposure to an external fire that are aimed at considering real conditions for PV installations in order to go beyond the actual framework and further improve the fire safety aspects.
Therefore, this work aims at presenting a clear understanding of the existing regulation framework (at the International, Italian and Swiss levels) and at resuming the main results of the current studies and activities of some institutions (RSE and Italian National Rescue and Services – DCPST – Italy, SUPSI-ISAAC – Switzerland).
Keywords: PV Modules, PV Systems, Fire Risk, Buildings, BIPV

1 INTRODUCTION

The use of photovoltaics (PV) in buildings have been subjected to a wide spread in recent years. This is due to the technology maturity of PV modules for producing the required electrical energy of the building and to the decrease of their price. In order to maintain the confidence into PV modules, there is nowadays the need to guarantee higher and higher PV safety requirements. Among these requirements, one is represented by fire safety. Indeed, the fire spread is more easy if there is a PV system on the roof or façade and, in some cases, there are also additional fire ignition events within the module (e.g. junction box overheating) which could be the cause of accidental fires. As a result, an improvement of fire safety is required in order to both protect people who can be involved in case of fire (e.g. building inhabitants, firefighters…) and avoid economic losses. For the latter, it is interesting to note that there are some insurance companies that are offering insurance services for the fire risk of PV systems relying on the safety requirements set in international and national standards. About that, recently, the updated version of the IEC 61730 introduced a new approach, including a new fire test for PV modules, and also some national guidelines are setting new requirements.

In a continuously changing legislative framework for fire safety of PV modules and systems, this paper is aimed at providing a clear understanding about the state-of-the-art of existing standards and regulations in order to provide some hints for the fire safety of PV systems in buildings. Therefore, in section 2 some real fire accidents involving PV systems in buildings are presented. Then, section 3 shows the standards framework at the international level for the fire safety of PV modules. Next, section 4 is focused on national regulations (Italy, Switzerland) and section 5 on some of the current research activities carried out by RSE – Ricerca sul Sistema Energetico / Research on Energy system company and Italian National Rescue and Services / DCPST - Central Directorate for Fire Prevention and Technical Safety. Moreover, section 6 provides the Swiss regulation framework for fire safety of BIPV and some considerations developed by SUPSI / ISAAC/ Swiss BIPV Competence Center.

Finally, section 7 tries to give some hints about the current situation of fire safety of building with PV systems by the point of view of regulations and standards and their possible developments.

2 REAL FIRE CASES AND MAIN SOURCES OF FIRE HAZARDS

According to a recent survey of the Italian National Fire Rescue and Service (CNVVF), around 2500 fire-related accidents have occurred in the nearly 550'000 PV systems currently installed in Italy. Figure 1 shows the numbers of PV fires in Italy from 2002 to 2015.

Some severe fires involving buildings with PV systems have lately occurred in many European countries and in the USA [1], [2]. Even though PV solar energy converters could not be the cause of ignition, in case of an outbreak of a fire in hosting buildings, they will intensify fire spread and trigger flame propagation mechanisms. Therefore, fire safety issues of a photovoltaic solar energy converter must be addressed both as a cause and as a victim of a fire.

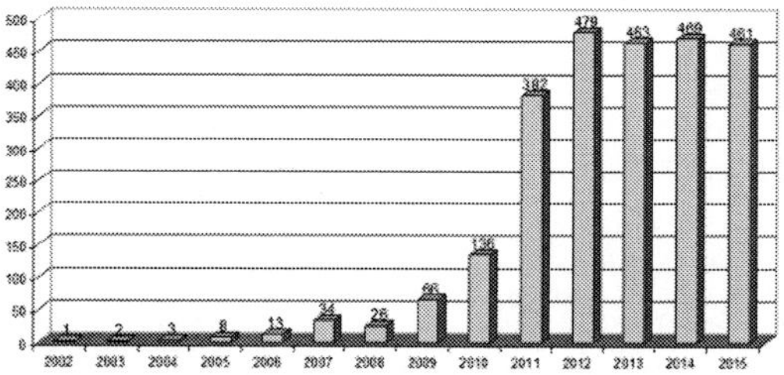

Figure 1. Italian PV fire related accidents 2002-2015

Some possible causes of fire arising from a building integrated array of PV modules, excluding power converters as a possible source, have been identified.

Generally speaking, the possible causes of ignition of an electrical fire within a photovoltaic array, sub-array or module can be divided into two main groups:

1. Localized over-heating,
2. Electric arc faults.

Excessive over-temperatures are a consequence of electric arc-faults. Table 1 shows a summary (not exhaustive) of the main sources of fire within a building integrated PV system is shown.

Table 1. Common fire causes within an array of PV modules

POTENTIAL FIRE SOURCE	Effects
Unprotected hot spots at module level	Localized over-heating
Faulty or undersized bypass diodes	
Poor electrical insulation of the PV circuit to under-construction and/or ground	Electric ARC fault – (parallel)
Poor or neglected wiring and insulation of wiring harness	
Poor electrical interconnections and bonding cell-to-cell (module level)	Electric ARC fault– (series) and/or localized over-heating
Faulty connectors (module-to-module)	
Faulty electrical connections inside the junction boxes	

3 PV FIRE SAFETY: INTERNATIONAL OVERVIEW

3.1 International

The safety standards applied at international level is IEC 61730-2:2016 Photovoltaic (PV) module safety qualification - Part 2: Requirements for testing. The standard was prepared by the IEC TC 82 "Solar Photovoltaic Energy Systems" (first edition 2004).

In the last few years, an increasing attention has been payed also to the development of a standard aimed to provide necessary requirements and testing procedures for protective devices to be installed in PV systems to reduce the risk of igniting an electrical fire: the future IEC 63027 is under development to define common requirements for arc fault detectors to be installed on PV circuits. This standard will allow to require arc fault detectors for PV systems, where reasonable, by installation standards.

Table 2. Fire hazard tests from IEC 61730:2016 (table 4).

Test	Title	Referenced standards	Based on 61215-2
MST 21	Temperature test	ANSI/UL 1703:2015	-
MST 22	Hot-spot endurance test	-	MQT 09
MST 23*	Fire Test	-	National/Local code
MST 24	Ignitability test	ISO 11925-2	-
MST 25	Bypass diode thermal test	-	MQT 18
MST 26	Reverse current overload test	ANSI/UL 1703:2015	

* Fire tests are locally regulated and typically only required for building integrated or building added products, typically to verify their ability to resist fire from external sources

Since at the international level fire rating classification of the PV module or panel has not been reached an agreement, the 2016 version of the 61730-2 standard states that the PV modules mounted in or on buildings should comply with national building and construction regulations and the related requirements. Table 2 summarizes the fire test for PV modules in the IEC 61730-2:2016.

If at a national level there are no requirements or local fire regulation, the standard offers a set of international and national standards that could be used to test and rate the PV module or panel. The listed standards are as following:

- **ISO 834-1** Fire-resistance tests - Elements of building construction - Part 1: General requirements;
- **ISO 834-3** Fire-resistance tests - Elements of building construction - Part 3: Commentary on test method and guide to the application of the outputs from the fire-resistance test;
- **ISO 5657:1997** Reaction to fire tests - Ignitability of building products using a radiant heat source;
- **ISO 13501-5** Fire classification of construction products and building elements. Classification using data from external fire exposure to roofs tests
- **ENV 1187-1 through 4**, Test methods for roof coverings under the influence of a thermal attack of burning brands and radiant heat;
- **ANSI/UL 790** Standard Test methods for Fire Tests of Roof Coverings;
- **ANSI/UL 1703:2015**, Standard for Flat-plate photovoltaic modules and panels.

Annex B of the IEC 61730-2:2016 specifies also how to use the different standards in order to perform a fire test and characterise a PV module.

The Annex highlights that fire resistance of PV modules installed on or over a building roof or facades is not only related to the flammability characteristic of the module itself but also the combination and configuration of roofing material as well as the mounting system of PV panels play important roles in the fire rating characterisation and PV module classification process.

3.2 USA

One of the most representative Code where specific requirements for PV systems have been requested is the National Electrical Code (NFPA 70 National Electric Code, 2011). As described in the previous section, for nearly two decades, the fire test and classification of PV modules has been performed according to fire exposure tests outlined in UL 1703 Flat-Plate PV Modules and Panels, using the ANSI/UL 790 test method. At first, this approach ignored the racking assembly's impact on the spread of flame. As described in the report from Solar America Board for Codes and Standards [3] it is of paramount importance to assess fire exposure of roof assemblies, considering the coupling of both the panel/module and the roof or facade substrate.

PV modules or panels intended for roof mounting, and designated for installation above, upon, or integral with a building roof structure having a surface consisting of a Class A, Class B, or Class C type roof covering material, shall be subjected to a spread-of-flame test as described in Tests for Fire Resistance of Roof Covering Materials, UL 790. At no time during or after the tests shall: "the flame spread beyond 6 ft (1.82 m) for Class A, 8 ft (2.4 m) for Class B, or 13 ft (3.9 m) for Class C rating

3.3 Europe

The current European version of the IEC international standard (EN 61730-2:2007 Photovoltaic (PV) module safety qualification - Part 2: Requirements for testing, EN 61730-2:2007/A1:2012) does not specify a mandatory fire rating procedure, since consensus on this issue has not been reached to date. Fire protection classification is left to national or regional codes.

In Europe these regulations are only to be found for roof integrated systems (CEN/TS 1187:2012 Test methods for external fire exposure to roofs) and BIPV (EN 50583-1:2016 Photovoltaics in buildings - Part 1: BIPV modules, EN 50583-2:2016 Photovoltaics in buildings - Part 2: BIPV systems). About roof systems even in that European classification system, four different test methods are taken into account:

a) test 1 – T1 with burning brands;
b) test 2 – T2 with burning brands and wind;
c) test 3 – T3 with burning brands, wind and supplementary radiant heat;
d) test 4 – T4 with two stages incorporating burning brands, wind and supplementary radiant heat.

About BIPV, the EN standards above are focused both on roof and façade modules. In addition, within CENELEC the document CLC/TR 50670:2016 "External fire exposure to roofs in combination with Photovoltaic (PV) arrays - Test method(s)" has been recently published (CLC/TC 82 "Solar Photovoltaic energy systems"). It provides some test methods for an initial assessment of fire performance of PV modules on roofs (BIPV - Building Integrated PV, is not covered).

4 PV FIRE SAFETY: NATIONAL REGULATIONS

In accordance with international standards, each country can develop its national regulation framework. In detail, in the following paragraphs, the Italian and Swiss regulation frameworks are described.

4.1 Italy

The Italian National Fire Rescue and Service has issued a national resolution to test and classifying the reaction to fire of PV modules and panels based on the Decree of the Ministry of the Interior issued on June 26, 1984 [7]. This technical regulation states that reaction-to-fire performance classifications for combustible materials are determined by using the results from a combination of the four tests listed below:

- **UNI 9176,** Preparation of the materials to reaction to fire test, D method;
- **UNI 8457,** Combustible products which can be hit by flames on one surface - Small flame test;
- **UNI 9174,** Reaction to fire of products subjected to a flame in the presence of radiant heat;
- **UNI 9177,** Classification of reaction to fire of combustible materials.

According to test results, combustible materials and products are divided into 5 categories, 1, 2, 3, 4 and 5, with 1 the best level of performance and 5 the worst (class 0 is for non-combustible materials).

The test standard UNI 8457 applies a little ignition flame to the lower border of the specimen under test for 30 seconds, observing and measuring the flame propagation speed, the damaged zone of the specimen, the post glowing time and the flaming droplet/particles.

The more severe standard UNI 9174 implies that the specimen under test is placed on an incombustible support and exposed to a radiant panel and hit by a flame to the lower side of the specimen. During the test, the heat emitted by the radial panel is set to 6,2 W/cm^2. During the radiant panel test, ignition velocity, flame propagation speed, damaged zone extension, post glowing time and flaming droplet/particles are recorded. Figure 2 details the possible combinations of the roof-top substrate with the fire rate of the PV module installed.

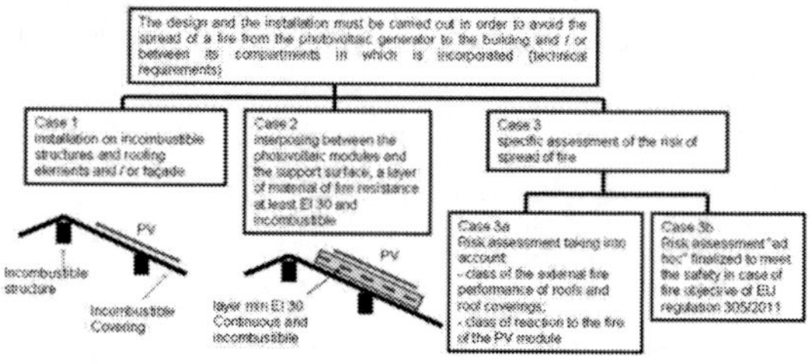

Figure 2: Rooftop and PV module possible couplings based on Fire ratings and Risk assessment.

4.2 Switzerland

In Switzerland, there is not a specific standard for testing and classifying the fire behaviour of PV and/or BIPV modules. However, the Swiss Association of Cantonal Fire Insurance Underwriters (VKF) has developed a Memorandum for Solar Systems where fire hazards are addressed. Among recommendations, the fundamental ones are:

- solar systems in buildings should not involve an unacceptable increase of the danger due to the formation and propagation of the fire, and
- the existing fire protection regulations for buildings should be fulfilled.

5 PV FIRE SAFETY RESEARCH ACTIVITIES

Together with a continuously improving regulation framework, also the need for new test methods can arise, involving research activities to develop new procedures. About that, a research program is currently carrying out by RSE S.p.A., Italian National Fire Rescue and Service, Politecnico di Milano and Istituto Giordano S.p.A. with the main purpose of developing a new step of test protocols about fire behavior of PV modules [5]. That program is mainly focused on testing PV module samples taking into account the influence had by: tilting Heat Release Rate (HRR) and size of ignition flame, sample exposure time to ignition flame and presence of sample initial deterioration. Furthermore, some tests are using the same arrangement of the sample and the burner of the TR 50670, while others test are performed by way of arrangements more similar to those of original European Specifications for building products (i.e. EN 13823:2010 "Reaction to fire tests for building products - Building products excluding floorings exposed to the thermal attack by a single burning item", EN ISO 11925-2:2010 "Reaction to fire tests - Ignitability of products subjected to direct impingement of flame - Part 2: Single-flame source test"), as represented in figure 3.

To perform this task, each one of those protocols is carried out in the same facilities (with some modifications) of European Horizontal Specifications for building products and they are using some of the same fire ratings variables (e.g. HRR – Heat Release Rate, FIGRA - FIre

Growth RAte, SMOGRA - SMOke Growth RAte, LFS - Lateral Flame Spread, THR - Total Heat Release, TSP - Total Smoke Production, post – combustion time, Flaming droplets, etc.

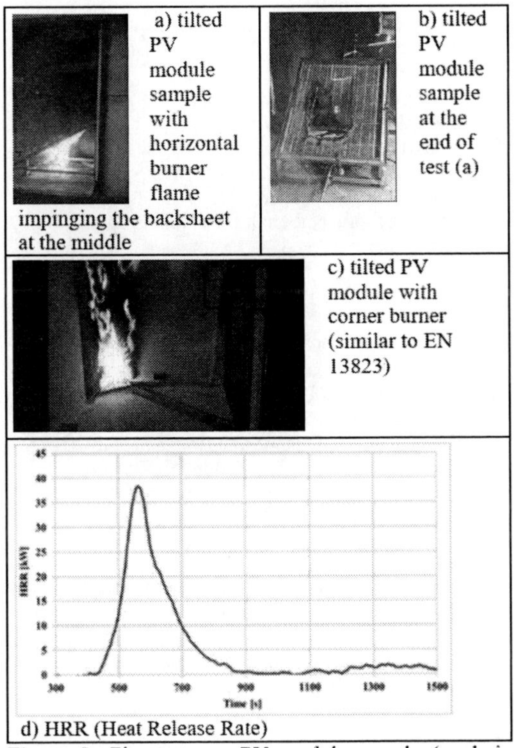

Figure 3: Fire tests on PV module sample (work in progress tests*). *About RSE S.p.A. activity; this work has been financed by the Research Fund for the Italian Electrical System under the Contract Agreement between RSE S.p.A. and the Ministry of Economic Development - General Directorate for Nuclear Energy, Renewable Energy and Energy Efficiency in compliance with the Decree of March 8, 2006.)

6 BIPV FIRE SAFETY IN SWITZERLAND

6.1 BIPV and fire safety regulations in buildings

Since 1st January 2017, the set of fire protection regulations has been updated by the Swiss Association of Cantonal Fire Insurance Underwriters (VKF). This legislative framework includes: the fire protection norm (main objectives and concepts), the fire protection directives (what should be done operatively), the explanatory notes (integrations), additional fire protection documents as well as some memorandums about specific building systems.

The memorandum about solar systems consists of several recommendations for preventing outbreak of fire, but the most important ones are: (i) the presence of solar systems in buildings should not involve an unacceptable increase of the danger due to the formation and propagation of the fire and (ii) the existing fire protection regulations should be fulfilled. In fact, the use of solar systems is subjected to the requirements of the Fire Protection Directive 14-15 "Use of Building Materials" for the fire reaction of the roofs and exterior walls.

Specifically, different reaction-to-fire requirements are set when BIPV modules are considered as the outer surface of the roof of as the outer cladding of an exterior walls, as shown in table 2, whereas classes are defined as:

- **RF1**: No reaction-to-fire
- **RF2**: Minimum reaction-to-fire
- **RF3**: Fire reaction allowed
- **RF4**: Reaction-to-fire not allowed
- **Materials with critical behavior (cr)**: Material producing inadmissible effects under fire due to their production of smoke and/or to the drop/fall of incandescent material fragments and/or to their corrosivity, etc… that need particular prescriptions.

Table 2: BIPV reaction-to-fire requirements

Building use	Class
BIPV as outer surface of roof	RF1 or RF3cr, when there is not a ventilated cavity and there is a RF1 substrate
BIPV as outer cladding of exterior walls (residential)	
Low height	RF3cr and RF1 substrate
Mid height	RF2cr (RF3cr, if recognized) and RF1 substrate
High height	RF1
BIPV as outer cladding of exterior walls (non-residential)	
Low height	RF3cr
Mid height	RF2cr
High height	RF1

Furthermore, BIPV systems wider than 1200m^2 should be coupled with a RF1 substrate or with combustible substrate, if there is a subdivision of the substrate in areas lower than 1200m^2 by means of RF1 materials (0.5m large).

It is interesting to note that the reaction-to-fire of a building element depends on building characteristics, such as the stratigraphy of the envelope and the geometry.

Moreover, Fire Protection Directive 13-15 "Building Materials and Constructive Parts" provides the correlation between the Swiss and the European reaction-to-fire classes.

With regard to the fire-resistance characteristics of BIPV systems, there are some recommendations when BIPV modules are installed near to REI walls but there is no evidence of other prescriptions in Swiss regulations. This is probably due to the fact that fire-resistance – defined as "the ability of a specimen to withstand fire or give protection from it for a period of time" (ISO 13943:2008) – is generally demanded to massive part of buildings such as load-bearing walls, beams and roof.

However, IEC standard provides some test method to assess fire-resistance of BIPV modules and systems on roof coverings when they are subjected to external fire sources but these tests do not include vertical facades and walls. In fact, the Directive 14-15 claims that a fire on the outer wall should not propagate to a height of two floors above the floor where the fire has developed.

Therefore, there is the need to evaluate the ability of BIPV modules used as façade claddings to withstand fire for a certain time period.

6.2 Further considerations about BIPV and fire

In order to support the breakthrough of BIPV in the market, it is necessary to deal also with the regulation framework. Indeed, BIPV – intended as a construction product capable to produce energy – should be able to fulfil both IEC and constructive requirements

In case of fire, the standard IEC61730 adopt pass/fail criteria but there is not a direct correlation between the results of the electro-technical test with the reaction-to-fire classes, required for BIPV as defined in the Swiss Fire Protection Directive 14-15 "Use of Building Materials".

On one hand, maintaining the twofold requirements about fire safety (related to the building and electro-technical perspectives) can be a conservative approach to reduce the risk. On the other hand, this can involve misunderstandings in defining the needs and the requirements, if a clear overview of the two domains (constructive and electro-technical) are not compared and harmonized.

7 CONCLUSIONS

Because of the events and the studies carried out so far, fire risk for photovoltaic installations is not negligible and must addressed to avoid economic losses and protect people who can be involved in case of fire.

This awareness has led to the arising of three main domains that can be further improved:

- **Testing methods and procedures**

The increasing number of PV related fire accidents was vital for the development of modern test methods and classification but more can be done in order to consider the real position of the PV (or BIPV) module in the building apart from the features of the materials composing the PV systems. Indeed, fire behavior of PV installed on or over a building roof or facades is not only related to the characteristic of module itself but also the combination and configuration of roofing material, building substrates as well as the mounting system of PV modules should be taken into account during the fire rating process.

- **Monitoring during operating conditions**

Since one of the main causes of fire is related to the cables and connectors, the maintenance of the PV and BIPV systems should be addressed in order to check the conservative state of the BOS.

Moreover, diagnostic and/or monitoring tools focused on the prevention of fire ignition should be taken into consideration, as well as safety devices capable to disconnect the PV system at the module level in case of fire. The latter is due to the fact that one of the main issue in case of fire is that the PV modules cannot be immediately disconnected, continuing to energize the source of fire.

- **Clear regulation framework**

To achieve a codification of construction, design and installation of PV and BIPV systems and their components, a process capable to involve various organizations (control authorities, standardization bodies, modules manufacturers, etc.) should be taken into serious consideration to increase the fire safety related to building installations.

In conclusion, this paper, listing, describing and discussing the current regulative frameworks, the test methods and procedures at the international and national levels (for Italy and Switzerland) points out some hints to overcome current issues with the main aim to increase the fire safety of PV and BIPV installations.

8 AKNOWLEDGMENTS

Italy

Thanks to the teams of FireLab (DENG – Politecnico di Milano) and Reaction to fire lab. of Istituto Giordano S.p.A.

Switzerland

This activity has been conducted as part of the Active Interfaces research project, financially supported by Swiss National Science Foundation (SNSF) within the framework of the National Research Program NRP70 "Energy Turnaround".

9 REFERENCES

[1] Marv Dargatz, DC Arc Faults and PV System Safety, December 2009 Jay Holman;

[2] NEC and Fire Safety in PV Systems, February 2011;

[3] Decree of the Ministry of the Interior June 26, 1984 "Classification of reaction to fire and approval of materials for the fire prevention" (in Italian);

[4] Fire Classification Rating Testing of Stand-Off Mounted Photovoltaic Modules and Systems, Solar America Board for Codes and Standards, www.solarabcs.org, by Larry Sherwood, Bob Backstorm, Dwayne Sloan, Christopher Flueckiger, Bill Brooks and Andrew Rosenthal, August 2013

[5] Cancelliere PG., Manzini G., Mazzaro M., A Review of the Photovoltaic Module and Panel Fire Tests, IFireSS 2017 – 2nd International Fire Safety Symposium, Naples, Italy, June 7-9, 2017

FLOATING PHOTOVOLTAIC MODULE TEMPERATURE OPERATION CHARACTERISTICS

Waithiru Charles Lawrence K.[1], Chang-Sub Won[1], Dong-Chan Kim[1], Kwang-wook Kim[1],
Bo-ram Kang[1], Gun-Hyun Lee[1], Ogeuk Kwon[2], Sumin Lee[2]
[1]Power Conversion Research Team, LSIS Corporation Korea
[2]Water Energy Research Center, K-water
[1]431-831 LSIS R&D Campus, LS-ro 116 beongil 40, Dongan-gu, Anyang-si
[2]125 1689beon-gil, Yuseong-daero, Yuseong-gu, Daejeon 34045
Telephone: +82-31-8090-7759 Fax: +82-31-8090-7788 Email: cswon@lsis.com

ABSTRACT: There has been a rapid reduction in the $/Wp prices of photovoltaic (solar PV) energy recently, resulting in near exponential deployments with an annual capacity of 200 GW expected by 2020. Achieving high efficiency is necessary for many solar manufacturers to break even. In addition, new innovative installation methods are emerging to complement the improvement of system performance. The floating PV (FPV) solar market space has emerged over the past decade as a method for utilizing the cool ambient environment of the FPV system near the water surface to boost the power output performance of the PV module and ultimately the yield of the PV system. PV module temperature, which is the most critical factor affecting efficiency, ultimately governs the effective performance of solar cells, module, and all semiconductor materials in general. We propose the first ever electrical efficiency equations (η_{c,FPV_1} and η_{c,FPV_2}) for an FPV module installed on water based on two new predictions of FPV temperature operation models (T_{m_1} and T_{m_2}), whose coefficients are derived from FPV site data with MATLAB. The theoretical prediction of module temperature shows respective errors of 2% and 4% when compared to the FPVM measured data.
Keywords: PV System, PV Module, Floating PV Systems (FPV), Floating PV Module (FPVM)

1 INTRODUCTION

A report published by IRENA in 2016 [1] shows that the global cumulative capacity of installed solar systems was 222 GW, with China, Germany, Japan, and USA installing 43 GW, 40 GW, 33 GW and 22 GW, respectively. In many markets, we see the growing conflict between environmentalists and solar enthusiast concerning installation land policy. A new and innovative installation method to cater to the installations of the future is necessary. A floating photovoltaic (PV) system is one such method that utilizes the cooling effect of water on its surface to improve the efficiency of the PV module and ultimately the performance of the PV system, with minimal interference with the marine environment.

Extensive studies on the efficiency, power, and temperature of the conventional PV system module have been carried out by Evans and Florschutez [2], Duffie and Beckman [3], and many others [4]. Considering the importance of device temperature in efficiency analysis, model 1 proposed in this paper correlates the temperature of the FPV module (FPVM) to the ambient temperature, solar radiation, and wind speed of the FPV environment. The influence of water temperature of the FPV installation is incorporated in model 2. When compared to the field data of a real FPVM, the average error of models 1 and 2 is 2% and 4%, respectively. The two temperature models are based on analysis of data obtained for 5 min from two FPV sites on Hapcheon lake in Korea. Important comparisons are performed with ten reference temperature models and the resulting findings are presented. The characteristic analysis of the FPV models shows resemblance to that of the models proposed by Lasnier and Ang 1990 [5] and Duffie and Beckmans 2006 [3]. Duffie and Beckmans predictions are thus preferred for size optimization, simulation and design of solar photovoltaics. Kurtz [6], Koehl [7] and Skoplaski[8] that include wind speed in temperature predictions are also included in analysis. A simple comparison of the temperature profiles of FPVMs with the conventional land- or rooftop-based modules shows that the mean value of the yearly PV module temperature of an FPV system is 21 °C, which is 4 °C below that of conventional PV modules, translating into 10% more kWh energy production by the FPV system.

The aforementioned research is important in analyzing the correlation between efficiency and temperature. Solar cells only convert a small amount of absorbed solar radiation into electrical energy with the remaining energy being dissipated as heat in the bulk region of the cell [9],[10]. A rise in the operation temperature of a solar cell and module reduces the band gap, thus slightly increasing the short circuit current of a solar cell for a given irradiance, but largely decreasing the open circuit voltage, resulting in lower fill factor and power output. The net effect results in a linear relation for the electrical efficiency (η_c) of a PV module as follows:

$$\eta_c = \eta_{Tref}\left[1 - \beta_{ref}\left(T_m - T_{ref}\right)\right] \quad (1)$$

where η_{Tref} and β_{ref} are the electrical efficiency and temperature coefficient of the PV module, respectively, with the values of 14.5% and 0.004 K-1, respectively, for the FPV analysis case. T_m and T_{ref} is the PV module operational temperature and reference temperature respectively.

Based on the two predictions of the temperature of an FPVM proposed herein, we propose two corresponding modifications to Eq. (1) based on the input parameters of T_m as follows:

$$\eta_{c,FPV_1} = \eta_{Tref}\left[1 - \beta_{ref}\left(2.0458 + 0.9458T_a + 0.0215G_T - 1.2376V_w - T_{ref}\right)\right] \quad (2)$$

$$\eta_{c,FPV_2} = \eta_{Tref}\left[1 - \beta_{ref}\left(1.8081 + 0.9282T_a + 0.021G_T - 1.2210V_w + 0.0246T_w - T_{ref}\right)\right] \quad (3)$$

where $\eta_{c,FPV}$, T_a, G_T, and V_w represent average values of efficiency, ambient temperature, solar irradiation, and wind speed of the FPVM. Eq. (3) includes an additional variable, i.e., water temperature (T_w).

As discussed later, Eq. (2) is shown as a graphical illustration in Fig. 7 demonstrating a reduction in the efficiency of FPVM by 0.058% per 1 °C increase in the temperature of the FPVM.

2 FLOATING PV SYSTEM

2.1 Site Information of Floating PV System

Fig. 1 shows the aerial views of Korea's first 100 kW and 500 kW FPV power stations. Based on the previous research on module reliability [11], a special anti-damp proof FPVM with a unique encapsulation [9] was certified and installed. A unique mooring system designed locally anchored the floating system on the dam floor, aligning the FPV system to the correct azimuth. A weather station was installed on the floating platform with radiation sensors, temperature sensors to monitor water temperature and the temperature of the FPV module, GPS positioning sensors, an anemometer to monitor wind speed, and a security camera for a visual view under severe weather situation such as typhoons. Data acquisition was based on IEC standard 61724[10] .A low-loss cable transmitted DC power from the FPV system to dry land where an electric room housing a PV inverter and monitoring computers were installed.

Figure 1: Aerial view of the 100 kW (Up) and 500 kW (Down) floating systems on Hapcheon dam, Korea

Table I: General FPV system information

Project Type	Test bed	Commercial	
		Floating PV	**Rooftop PV**
Site Name	Hapcheon Dam 100kW	Hapcheon Dam 500kW	Haman 1,000kW
Site coordinates	N 35.5° 33'36"	E 128° 02' 26"	N 35.5° 33'36"
Installation Capacity	100kW	500kW	1,000kW
Installation Date	2011 Nov.	2012 June	2012
Module Slope	33°	33°	30°
Module type	c-Silicon	c-Silicon	c-Silicon
Module#	417pcs	1,667pcs	4,000pcs
Mounting	Aluminum, Steel	Aluminum	Aluminum
Mounting type	Fixed	Fixed	Fixed
Water Depth	20 meters	40 meters	n/a

2.2 Power Outputs of Floating PV versus Rooftop-based System

In Table 2, we compare the floating systems of capacities 100 kW and 500 kW with a rooftop system of capacity 1,000 kW, located 60km southeast of the 100kW site. The output summary is given in Table 2.

Table II: General system performance and output

Project Type		Floating PV		Rooftop PV
		100kW	500kW	1000kW
Annual Output (kWh/year) Avg.		130,300	693,219	1,197,547
Monthly Output (kWh/year) Avg.		357	1,859	3,281
Normalized Power	kWh/KWp/year	500kW	1,000kW	1,198
	kWh/KWp/day	2012 June	2012	3.28

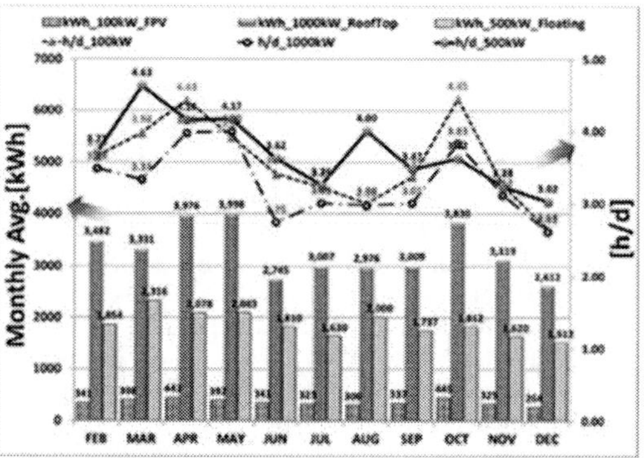

Figure 2: Average Monthly Power (kWh) output comparison for 2012 of 100 kW, 500 kW floating systems (Hapcheon dam) ; and 1000 kW rooftop system at Haman site (2013).

Table 2 shows yearly energy results of the three PV stations. As shown in Table 2, the y-axis (left) is the monthly average energy output. For example in April 2013, average monthly output from the three PV Systems was 443kWh, 2078kWh and 3976kWh for the 100kW, 500kW and 1000kWPV systems respectively. Multiplying respective monthly average but days in month, and summing monthly outputs gives 130.3MWh, 693.2MWh and 1,197.5MWh respective total yearly output.

For the 100kW FPV station, October and December are the best and worst performing month at 445kWh and

264kWh respectively, compared to the station's yearly average of 357kWh. Similarly for the 500kW FPV station, March and December are the best and worst performing month at 2,316kWh and 1,512kWh respectively, compared to the station's yearly average of 1,859kWh. Finally for the 1,000kW rooftop PV station, May and December are the best and worst performing month at 3,998Wh and 2,612kWh respectively, compared to the station's yearly average of 3,281kWh. Whereas the rooftop produces more power quantitatively, the FPV systems are more efficient in qualitative power delivery.

With reference to y-axis on the right, output energy is normalization to name plate peak power (kWp) with units hours per day (h/d). Table 2 shows an average of monthly values giving yearly normalized output as 3.58 h/d, 3.80 h/d, and 3.28 h/d, for the 100kW, 500kW and 1000kW sites respectively, as shown in Table 2. Analysis of the latter values proves the two FPV systems are outperforming the rooftop systems by 9% and 16%, warranting investigation into temperature performance.

3 FLOATING PV TEMPERATURE MODEL

In this section, we formulate a multiple linear equation for the dependent variable (y; FPV module operation temperature) using four independent linear variables x_1, x_2, x_3, and x_4 representing solar irradiance (G_T), ambient temperature (T_a), wind speed (V_w), and water temperature (T_w), respectively.

Table III: Multiple regression variables for FPVM temperature (symbol; TM)

Term	Predictor Variables	Symbol	Unit
x_1	Ambient temp.	T_a	°C
x_2	Irradiance	G_T	W/m²
x_3	Wind Speed	V_w	m/s
x_4	Water temp.	Tw	°C

The multiple linear equation is linear for unknown parametersβ_0- β_{k-1}, and is of the form given in Eq. (4).

$$y_i = \beta_0 + \beta_1 x_{i1} + \beta_2 x_{i2} + \cdots + \beta_{k-1} x_{ik-1} + \epsilon_i \quad (4)$$

for $i = 1(1)n$, where y_i is the predicted value of y, and assumes i^{th} independent error $\epsilon_i \sim N(0, \sigma^2)$ following a normal distribution with independent mean and variance squared. The matrix can be expressed as

$$Y = X\beta + \epsilon \quad (5)$$

where

$$Y = \begin{pmatrix} Y_1 \\ Y_2 \\ \vdots \\ Y_n \end{pmatrix},\ \beta = \begin{pmatrix} \beta_1 \\ \beta_2 \\ \vdots \\ \beta_n \end{pmatrix},\ \epsilon = \begin{pmatrix} \epsilon_1 \\ \epsilon_2 \\ \vdots \\ \epsilon_n \end{pmatrix},\ X = \begin{pmatrix} 1 & x_{11} & x_{12} & \cdots & x_{1k-1} \\ 1 & x_{21} & x_{22} & \cdots & x_{2k-1} \\ \vdots & \vdots & \vdots & \cdots & \vdots \\ 1 & x_{n1} & x_{n2} & \cdots & x_{nk-1} \end{pmatrix}$$

The multiple linear regression form is expressed in Eq. (5) with Y, β, ϵ, and X representing y observations, vector of parameters, error, and $n \times k$ matrix vectors, respectively. The goal is to estimate the model parameters.

The field data of floating PV are given in the forms of $Y_i, x_{i1}, x_{i2}, x_{i3}$, and x_{i4}, for T_m, T_a, G_T, V_w, and T_w , respectively.

We use the standard least-squares minimization to determine the aforementioned model parameters by minimizing the sum of squares of residuals (SS_{Res}) as shown in a matrix form in Eq. (6).

$$SS_{Res} = \Sigma_1^n e_i^2 \quad (6)$$

where $e = (Y - \hat{Y})$ and $\hat{Y} = \hat{\beta}_0 + \hat{\beta}_1 x_1 + \hat{\beta}_2 x_2 + \cdots + \hat{\beta}_{k-1} x_{k-1}$.

Substituting the former into Eq. (6) leads to the definition of SS_{Res} in terms of the unknown parameters in Eq. (7).

$$SS_{Res} = \Sigma_1^n (Y - \hat{Y})^2 = (Y - \hat{Y})'(Y - \hat{Y}) = Y'Y - 2\hat{\beta}'X'Y' + \hat{\beta}'X'X'\hat{\beta} \quad (7)$$

where equation is expanded using $\hat{Y} = X\hat{\beta}$. Integrating SS_{Res} with respect to $\hat{\beta}$ results in normal equations, which have to be solved for unknown equations in Equation (8). For easy computation, an alternative matrix equation is presented for solving the coefficients.

$$\frac{\partial SS_{Res}}{\partial \hat{\beta}} = \frac{\partial (Y'Y - 2\hat{\beta}'X'Y' + \hat{\beta}'X'X'\hat{\beta})}{\partial \hat{\beta}} \quad (8)$$

$$\hat{\beta} = (X'X)^{-1} + X'Y \quad (9)$$

where $X'$is the inverse X matrix of predictor variables listed on Table 3

Our X matrix contains more than 100,000 data points, as is plotted in Fig.3, and corresponds to FPV data collected every 5 min in 2013. The Y matrix corresponds to the measured module temperature. The coefficients of models 1 and models 2 in Equation4 for the FPVM suggested in this paper are expressed as follows

$$T_{m_1} = 2.0458 + 0.9458T_a + 0.0215G_T - 1.2376V_w \quad (10)$$

$$T_{m_2} = 1.8081 + 0.9282T_a + 0.021G_T - 1.2210V_w + 0.0246T_w \quad (11)$$

T_{m_1} and T_{m_2} explains the operation temperature behavior of the FPVM with seasonal variables T_a, G_T, V_w and T_w.

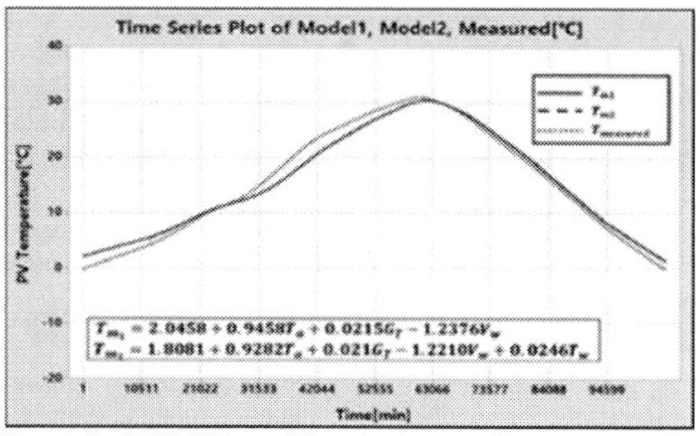

Figure 3: 2013 PV module temperature data; measured a nd modeled data (100 kW system)

Fig.3 below, is a time series plot of T_{m_1}, T_{m_2} and $T_{measured}$ for 2013. $T_{measured}$ is actual FPV measured data. From the graph, predicted PVM temperatures are almost always higher than measured except during second quarter (Q2) where $T_{measured} > T_{m_1}$, T_{m_2}. Coincidentally, wind speeds (T_W) are also low during same period, implying the dominance of T_W in the two models Equation 10 and Equation11.

Equation12 introduces the average error of FPV models, by comparing real to predicted values as shown. Calculations show T_{Error} ranging 2.06% and 4.40%, for respective FPV models;

$$T_{Error} = \int_{n=1}^{n=k} (T_{measured} - T_m) \quad (12)$$

where k are total data points. Inclusion of T_w in Eq.11 increases the error by 2% when compared to Eq.10.

4 IV. FPV MODEL COMPARISON

4.1 Comparison with Land-based PV System

In this section, we compare the FPV with an arbitrary 1,000 kW rooftop system whose information is given in Table 1. It is evident the FPV system produces a large portion of energy at lower temperatures [11].

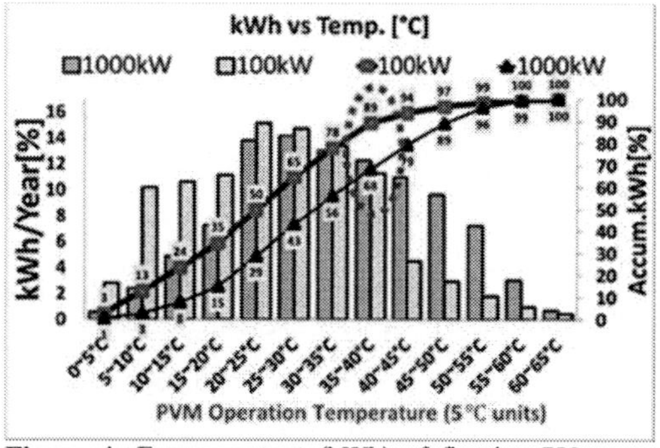

Figure 4: Energy output (kWh) of floating PV verses rooftop PV for various module temperatures

Fig. 4 illustrates the correlation between PV power output (kWh) and corresponding module temperature when respective power is produced. 100% of energy produced by the 100kW site is 130MWh, while energy produced by the 1000kW site is 1,197MWh per year. This two MWh outputs occur when PVM operation temperature varies periodically between the lowest (0°C) and the highest value (65°C). Using Minitab statistical software, yearly energy is sorted based on corresponding module temperature, as shown in Fig.4. For example, with reference to the 20~25°C range, 15% of 130MWh and 14% of 1,197MWh is produced by the two respective PV sites. With reference to y-axis (right) on the Fig.4, cumulative energy for respective temperature ranging from 0°C to 65°C is plotted against corresponding energy (kWh) increasing from 0% to 100% of total yearly output. Cumulatively, 89% of all energy produced by the FPV system, and 68% of all energy delivered by the rooftop system occurs when module temperatures of both systems are below 40°C, as indicated. Energy produced beyond the 25°C Standard Test Condition (STC) condition has the negative power loss effect due to loss of open circuit (V_{oc}) and fill factor (F.F.) [9]. As is evident, a larger percentage of cumulative energy of the FPV is produced at lower temperatures.

4.2 Comparison with Selected Temperature Models

A select group of PV temperature models [4] is presented in Table 4 for comparison. The models incorporate a reference state for example air temperature (T_a), and the corresponding values of relevant variables (G_T,V_w, etc.). Owing to the complexities involved, some authors presented explicit correlation in addition to implicit relations requiring iterations.

In Fig.5 below, models listed in Table 4 above are plotted against both ambient (T_a) temperature, and solar radiation (G_T).

Table IV: PV module models

Model	Empirical Models
Ross (1976)[10]	$T_c = T_a + kG_T$ where $k = \Delta(T_c - T_a)/\Delta G_T$
Rauschenbach (1980)[12]	$T_c = T_a + (G_T/G_{T,NOCT})(T_{c,NOCT} - T_{c,NOCT})(1 - \frac{n_m}{\gamma_a})$
Risser& Fuentes(1983) [13]	$T_c = 3.81 + 0.0282 \times G_T 1.31 \times T_a - 165V_w$
Schott (1985)[14]	$T_c = T_a + 0.028 \times G_T - 1$
Ross& Smokler (1986) [12]	$T_c = T_a + 0.035 \times G_T$
Mondol et al., ('05,'07)	$T_c = T_a + 0.031G_T$ $T_c = T_a + 0.031G_T - 0.058$
Lasnier & Ang (1990) [5]	$T_c = 30.006 + 0.0175(G_T - 300) + 1.14(T_a - 25)$
Servant (1985)[15]	$T_c = T_a + \alpha G_T(1 + \beta Ta)(1 - \gamma V_w)(1 - 1.053n_{m,ref})$
Duffie& Beckman [3]	$T_c = T_a + (G_T/G_{NOCT})(9.5/5.7 \times 3.8V_w)(T_{NOCT} - T_{a,NOCT})(1 - n_m)$
Koehl (2011) [7]	$T_c = T_a + G_T/(U_0 + U_1 \cdot V_w)$
Kurtz S (2009) [6]	$T_c = T_a + G_T \cdot e^{-3.473 - 0.0594 \cdot V_w}$

Skoplaki (2008) [8]	$T_c = T_a + (G_T/G_{NOCT}) \cdot (T_{NOCT} - T_{a,NOCT}) \cdot h_{w,NOCT}/h_w \cdot [1 - \eta_{STC}/\tau \cdot \alpha (1 - \beta_{STC} T_{STC})]$

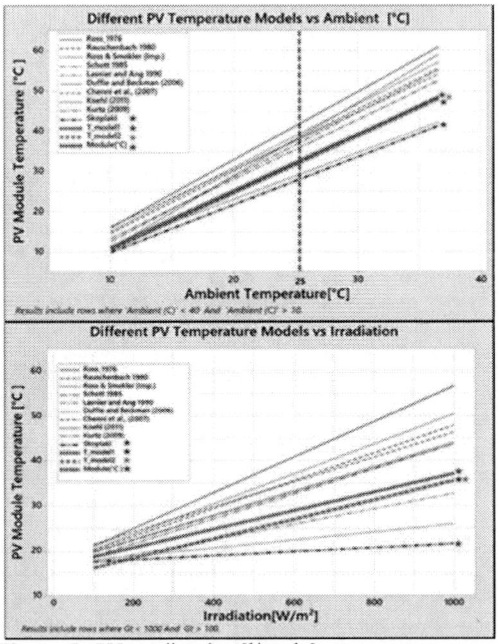

Figure 5: PV predicted cell/module temperature verses ambient temperature (Top) and irradiance (Bottom)

As can be seen, all models vary linearly with both T_a and G_T with varying model specific gradients. The implication is difference model interpret heat dissipation by the PV module differently when exposed to the elements. For example, Ross [12] and Rauschenbach [16] model display the highest PV operation temperatures when exposed to G_T at constant T_a. Koehl, Kurtz and Skoplaki's research incorporates wind (V_w) data in temperature prediction.

In Fig.5 (top), Duffie & Beckman and Skoplaki [8] record low temperatures with increasing T_a suggesting adequate heat dissipation by the modules due to incorporation of wind data. To the contrary, Ross and Rauschenbach show high temperatures near 60°C suggesting the PVM retains heat. Model 1 (T_{m1}) and Model2 (T_{m2}) plots are almost identical, and vary slightly with real PV module data, and operate at much lower temperatures than all other models.

In Fig.5 (bottom), Skoplaki has lowest operation temperature with increasing G_T, while Ross has highest temperature values because of PVM heat retention. Skoplaki model reacts very slowly to rising G_T due to quick heat dissipation by the V_w factor. A slight deviation from real temperature by Model 1 (T_{m1}) and Model2 (T_{m2}) is noted with increasing G_T.

Based on the two graphs in Fig.5, we conclude that our two FPV models operating temperatures are significantly lower than conventional PV module ranges.

4.3 Comparison of Models with Minitab Model

Minitab has well-defined algorithms that describe the change of any dependent variable y with the interaction between the respective independent variables x_i. Refer to Appendix 1 for graphs showing the interaction between independent variables. Minitab generates an equation that shows the interaction between the dependent variable (module temperature) and independent variables. The Minitab equation Eq. (12) shown below is highly accurate (0.1%) but incurs the risk of equation complexity due to over-fitting.

$$\text{Module} = -1.9034 + 1.12322\,x_1 + 0.028655\,x_2 - 0.6517\,x_3 - 0.09362\,x_4 - 0.001328\,x_1^2 - 0.000014\,x_2^2 + 0.08382\,x_3^2 - 0.000604\,x_1 \cdot x_2 - 0.031334\,x_1 \cdot x_3 + 0.001389\,x_1 \cdot x_4 - 0.000981\,x_2 \cdot x_3 + 0.000545\,x_2 \cdot x_4 + 0.039145\,x_3 \cdot x_4 \quad (12)$$

for x_1, x_2, x_3, x_4, representing T_a, G_T, V_w, T_w.

In Fig.6, four histograms compare the normal distribution of real FPV module temperature data to Model 1, Model 2, Minitab's predicted values.

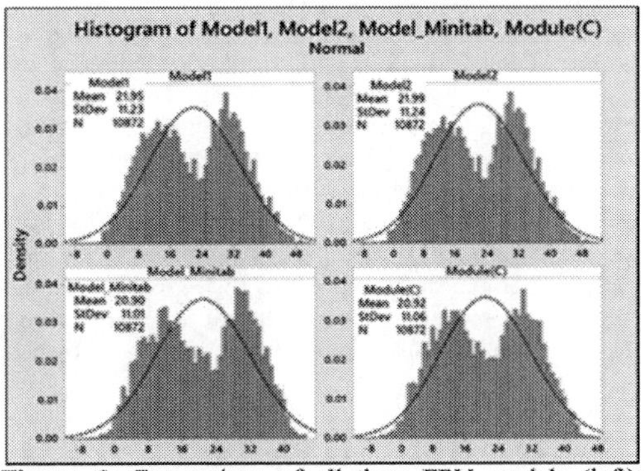

Figure 6: Comparison of all three FPV models (left); Histograms of temperature predictions; Minitab, model 1, model 2, and actual temperature (right)

The x-axis shows operational temperature from -10°C to over 50°C. Y-axis plots the density of respective temperature range throughout the year. All model distributions show a bimodal shape, with 2 peaks temperatures at 10°C and 30°C. The dotted red line shows the normal distribution curve of respective data.

Mean values are 21.95°C, 21.99°C, 20.90°C and 20.92°C for Model 1, Model 2, Minitab's and to real field data (Module(C)). When compared to real data, mean errors as 4.92%, 5.11%, 0.1% and 0.0% (base value) for the respectively values. A 0.1% error is indicative of the Minitab model's high accuracy in comparison to real measured data. The behavior of a plot of Equation12 is identical to $T_{measured}$ in Fig.3.

5 FPV MODEL EFFICIENCY AND POWER PREDICTION

Operating a PV system on the water surface has the added benefit of increasing conversion efficiency due to the cooling effect on water's surface.

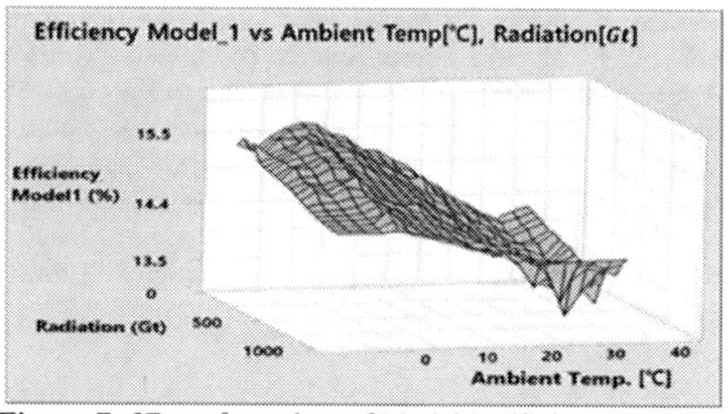

Figure 7: 3D surface plots of Model 1 efficiency/Ta/Gt

Fig. 7 is a 3D plot of FPV module efficiency/T_a/G_T . In the plot, a decrease in ambient temperature (T_a) has a positive effect of increasing efficiency between 1% ~ 2% points. The plot shows the importance T_a in defining PVM operation temperature and ultimately conversion efficiency. Radiation (G_T) plays a secondary role given the minimal impact on efficiency. It can be observed that, at higher radiation level G_T is varying more frequently, and this impacts power stability.

In summary, as observed from the FPV data, low ambient conditions are ideal for higher system efficiency and power performance as shown by seasonal variation in efficiency in Fig 8. In June through August when ambient temperature are high, PVM efficiency drops between 1~2% points. For a land based system, a more severe dip is expected. During fall and winter when temperature drops, we notice a step rise in efficiencies to mid-15% level. Based on graphical description on Fig.8, for FPV module temperature models 1, we predict that a 1 °C increase in T_m results in a 0.058% decrease in η_{c,FPV_1} , as shown for in Eq.13 .

$$\eta_{c,FPV_1} = 15.96 - 0.058T_{m_1} \tag{13}$$

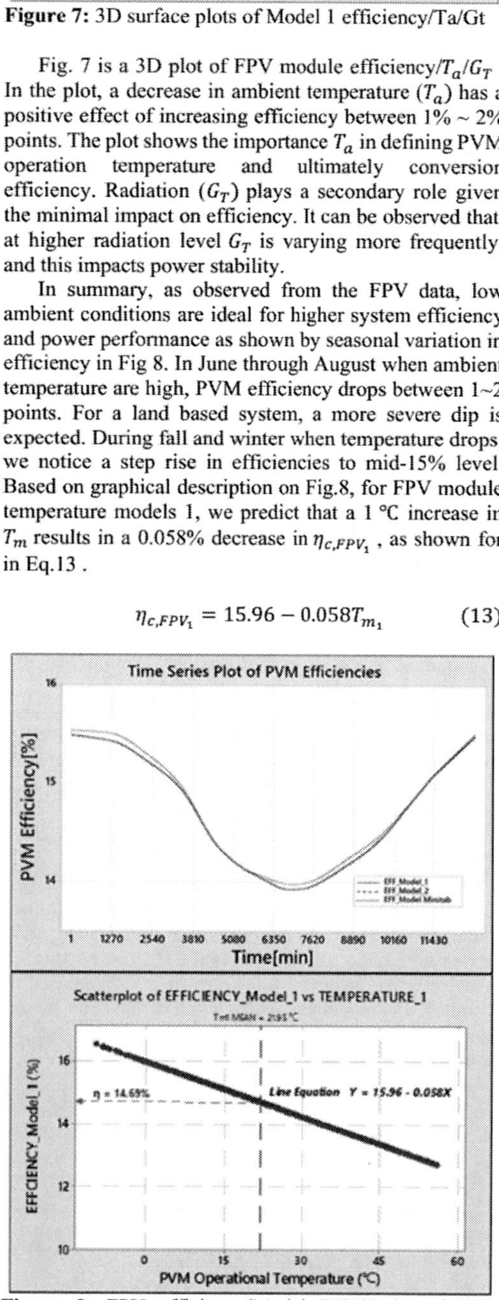

Figure 8: FPV efficiency/Model FPV(Tm) and FPV (100kW) versus efficiency Gt

Operational temperature has an important role in the energy conversion process [17]. From Fig 8, the electrical efficiency of the PV module depends linearly on the operation temperature as shown. Latifa's [18] has done important work (2014) comparing crystalline (c-Si) and amorphous silicon (a-Si) coefficients per °C. This work shows coefficient values for a-Si closely identical to FPV.

6 CONCLUSION

A floating PV system was installed in 2011, and its floating module temperature is analyzed in this paper. The theoretical prediction of module temperature shows an error in the range of 2%–4% when compared to the measured data. The performance ratios (PR) on both the AC and DC sides were analyzed, which showed that the floating PV system exhibits PR 10% better than that of a land-based system. The results suggest lower thermal losses associated with thermal heating of the FPV modules owing to cool temperature environment near the water surface. The analysis of the FPV module operation model showed the critical role played by low ambient conditions in boosting the operation efficiency of the PV system module.

Two prediction models of the FPV module temperature are suggested for the analysis of performance of the FPV module and system. Model 1 includes the effects the independent variables, i.e., ambient temperature (Ta), solar irradiance (Gt), and wind speeds (Vw). When compared to the measured data, the equation error of model 1 is 2%. Model 2 includes the three aforementioned independent variables in addition to water temperature (Tw). Although the error of model 2 increases slightly to 4%, the results are within the reasonable range of error.

Through this research, a correlation between the temperature of the operating environment and system efficiency is derived. Beyond solar irradiation of 100 W/m², the floating system records an ideal efficiency averaging more than 14.69% based on yearly mean PVM temperature of 21.95 °C . It was observed that approximately two-thirds (68%) of the annual yield was produced by the FPV system when the module temperature was less than 40 °C.

7 ACKNOWLEDGMENTS

This work was supported by the Korea Institute of Energy Technology Evaluation and Planning(KETEP) and the Ministry of Trade, Industry & Energy(MOTIE) of the Republic of Korea (No. 20153010012060).

8 REFERENCES

[1] IRENA, Solar PV in Africa: Costs and Markets, Chapter 3 Page 29

[2] Evans, D. L. "Simplified method for predicting photovoltaic array output." Solar energy 27.6 (1981): 555-560.

[3] Duffie, J. A., and W. A. Beckman. "Solar Engineering of Thermal Processes Wiley: Hoboken." (2006).

[4] Skoplaki, Elisa, and John A. Palyvos. "On the temperature dependence of photovoltaic module electrical performance: A review of efficiency/power correlations." Solar energy 83.5 (2009): 614-624.

[5] Lasnier, F., Ang T.G., 1990. Photovoltaic Engineering Handbook. Adam Hilger, New York, NY, p.80
[6] Koehl, Michael, et al. "Modeling of the nominal operating cell temperature based on outdoor weathering." Solar Energy Materials and Solar Cells 95.7 (2011): 1638-1646.
[7] Kurtz, Sarah, et al. "Evaluation of high-temperature exposure of rack-mounted photovoltaic modules." Photovoltaic Specialists Conference (PVSC), 2009 34th IEEE. IEEE, 2009.
[8] Skoplaki, Elisa, and John A. Palyvos. "On the temperature dependence of photovoltaic module electrical performance: A review of efficiency/power correlations." Solar energy 83.5 (2009): 614-624.
[9] Malik, A. Q., et al. "Influence of temperature on the performance of photovoltaic polycrystalline silicon module in the Bruneian climate." ASEAN Journal Science and Technology Development 26.2 (2010): 61-72.
[10] King, David L., Jay A. Kratochvil, and William E. Boyson. "Temperature coefficients for PV modules and arrays: measurement methods, difficulties, and results." Photovoltaic Specialists Conference, 1997., Conference Record of the Twenty-Sixth IEEE. IEEE, 1997.
[11] Malik, A. Q., et al. "Influence of temperature on the performance of photovoltaic polycrystalline silicon module in the Bruneian climate." ASEAN Journal Science and Technology Development 26.2 (2010): 61-72.
[12] Ross Jr, R. G. "Interface design considerations for terrestrial solar cell modules." 12th Photovoltaic Specialists Conference. 1976.
[13] Risser, V. V., and M. K. Fuentes. "Linear regression analysis of flat-plate photovoltaic system performance data." 5th Photovoltaic Solar Energy Conference. 1984.
[14] Schott, T. "Operation temperatures of pv modules: a theoretical and experimental approach." EC Photovoltaic solar energy conference. 6. 1985.
[15] Servant, Jean-Micheal. "Calculation of the cell temperature for photovoltaic modules from climatic data." Proceedings of the 9th biennial congress of ISES–Intersol. Vol. 85. 1985.
[16] Rauschenbach, H.S., 1980. Solar cell array design handbook. Van Nosstrand Reinhold, New York, pp:390-391.
[17] Martineac, Corina, et al. "Temperature Influence on Conversion Efficiency in the Case of Photovoltaic Cells." Development and Application Systems (2010): 69.
[18] Sabri, Latifa, and Mohammed Benzirar. "Effect of Ambient Conditions on Thermal Properties of Photovoltaic Cells: Crystalline and Amorphous Silicon." International Journal of Innovative Research in Science, Engineering and Technology (2014).

DAILY PERFORMANCE COMPARISON OF PV MODULES OPERATING IN MAUI

Severine Busquet, Jonathan Kobayashi, and Richard E. Rocheleau
Hawaii Natural Energy Institute, University of Hawaii
1680 East-West Rd, Honolulu, HI 96822, USA

ABSTRACT: The performance ratio (PR) has been used to compare PV modules based on their daily average efficiency but this single criterion is not enough to understand the performance variability between technologies and locations. A PV test platform, commissioned on Maui in February 2016, conducts side-by-side comparison of 10 different PV modules. A new energy rating analysis approach is used to present the first-year performance. PR is separated into current (IP) and voltage (VN) performances. The optical performance is also defined from the short-circuit current. Simple empirical models are used to predict these performance criteria as a function of the irradiation, the main environmental parameter. This method helps visualize the performance differences between PV modules and could lead to a better understanding of the performance variabilities between locations.
Keywords: energy rating, outdoor testing, data analysis and modelling

1 INTRODUCTION

Energy rating is typically conducted by calculating the performance ratio (PR), as defined in IEC 61724 [1]. PR is the operating performance of the PV module corresponding to the daily average efficiency relative to the datasheet specifications defined at standard test conditions (STC). Experimental results vary between modules and locations [2-4]. PR does not provide enough information to understand the variabilities because of the multiple parameters affecting this criterion. Complex modelling tools predicting the instantaneous production of the PV module can provide additional insights on energy rating if the module reference data set is available [5]. These models are also essential tools to diagnose potential operating conditions such as shading by comparison of the PV system power output with the output of a prediction model [6].
The Hawaii Natural Energy Institute (HNEI) of the University of Hawaii designed a PV test platform to test side-by-side 15 grid-connected PV systems and 10 PV modules. Energy rating is the first focus in analyzing the collected datasets. Our goal is to characterize the operation and performance of the PV modules/systems without the need of complex modelling. To address this, PR is separated into current (IP) and voltage (VN) performances. The optical performance is defined from the short-circuit current. These additional criteria help understand the performance differences between PV modules and identify the impacts of the environmental and operating conditions.

The analysis approach was previously introduced [7]. It was explained how to extract the impact of the environmental conditions from the operating conditions (soiling, shading) using an empirical model of the optical performance as a function of the irradiation. The focus of this document is to describe the impact of the irradiation on the PV performances showing the differences between PV technologies and module designs.

Test protocols and analysis method are introduced in section 2 and 3. Section 4 presents the first-year results that are discussed in section 5 before concluding.

2 TESTING

2.1 PV test platform

The new PV test platform was commissioned in February 2016 in Kihei, Maui. It consists of 15 grid-connected PV systems using 10 PV modules and 3 system architectures (microinverter; string inverter; string inverter with optimizers) for a total rated power of 22 kW. An additional PV module of each PV technology is tested with an IV tracer to collect detailed module performance. All PV modules are installed with an orientation of a 20° tilt and 197°N azimuth. The latitude, longitude, and altitude of the test site are 20.7°N, 156.4°W and 60 meters, respectively. The PV modules described in **Table I** comprise of 4 standard (S) p-type mono and polycrystalline, 3 high efficiency (H) n-type monocrystalline (including a HIT and a bifacial), and 3 thin-films including 2 CIGS (C) and 1 CdTe (D). Except for the CdTe modules that were donated by the PV company, all modules and auxiliaries were purchased (in 2015) by the contracted PV installer without informing the PV manufacturers of the assessment. This insures fair testing and unbiased evaluation of commercially available PV technologies.

Table I: Description of the PV modules

PV name	PV technology	P_{MP} [W]	η [%]
S1	Standard p-type polycrystalline	250	15.4
S2	Standard p-type polycrystalline	250	15.2
S3	Standard p-type polycrystalline	260	15.5
S4	Standard p-type monocrystalline	265	16.5
H1	High efficiency n-type monocrystalline with heterojunction intrinsic thin layer	240	19.0
H2	High efficiency n-type monocrystalline with rear contact	245	19.7
H3	High efficiency n-type monocrystalline, Bifacial Hybrid Cell Technology	300	18.2
C1	Copper indium gallium selenide (CIGS)	145	13.3
C2	CIGS	170	13.8
D1	Cadmium telluride (CdTe)	77.5	10.8

2.2 Monitoring

The Daystar multitracer is used to collect IV curves on a sample of each of the 10 PV modules in test. The modules operate at maximum power point during normal operation with I-V curves measured every 3 minutes. In the following analysis, there is no data provided for the module C2 due to overvoltage issues leading to an insufficient dataset. The problem was corrected during the first-year maintenance conducted in January 2017 and datasets for C2 will be available for the second year of monitoring.

A data acquisition system (DAS) collects data from the following sensors. The reference solar sensor is a secondary standard thermopile pyranometer (Kipp & Zonen, SMP21-A) that measures full spectrum global irradiance. A masked set of thermopile pyranometers (Delta-T Devices SPN1) provides the diffuse and global irradiance. A spectroradiometer (EKO, MS-700) monitors the light spectrum with wavelengths from 350 nm to 1050 nm. All solar sensors are coplanar with the PV modules to assess the solar resource they receive. A weather station (Gill MetPak Pro) collects additional parameters such as ambient temperature (AT), rainfall, wind speed and direction.

3 METHODOLOGY

3.1 Performance criteria

Energy rating requires evaluating the daily average performance of the PV modules. For each day, the average for all measured and computed data is calculated using data with angle-of-incidence (AOI) lower than 70° to limit error due to the directional response of the solar sensors. The irradiance, power and current are integrated over time to calculate the energy collected. PR is calculated (1) by dividing the PV energy production by the solar energy received by the PV modules, and both are normalized using the STC specifications. The novel approach was inspired from the loss factor model [8]. PR is separated (2) into current performance (IP) and normalized voltage (VN), also referred to as voltage performance. IP is calculated similarly to the PR except that PV current, rather than power, is integrated (3). The average operating voltage is divided by the STC maximum power point voltage of the PV module to obtain VN (4). I_{SC} is the maximum current that a PV module can produce under specific environmental conditions and is used to calculate (5) the short-circuit current performance (IP_{SC}) called optical performance in what follows.

$$(1)\ \mathrm{PR} = \frac{\int_{\Delta t} P_{PV}\cdot dt}{P_{MP,STC}} \times \frac{G_{STC}}{\int_{\Delta t} G\cdot dt},\ (2)\ \mathrm{PR} \approx \mathrm{IP} \times \mathrm{VN}\ ,$$

$$(3)\ \mathrm{IP} = \frac{\int_{\Delta t} I_{PV}\cdot dt}{I_{MP,STC}} \times \frac{G_{STC}}{\int_{\Delta t} G\cdot dt},\ (4)\ \mathrm{VN} = \frac{\overline{V_{PV}}}{V_{MP,STC}},$$

$$(5)\ \mathrm{IP}_{SC} = \frac{\int_{\Delta t} I_{SC}\cdot dt}{I_{SC,STC}} \times \frac{G_{STC}}{\int_{\Delta t} G\cdot dt}$$

where P_{PV} is the power of PV module [W]; G_{STC} is the STC irradiance [=1 kWm^{-2}]; Δt is the period of analysis [hour] (using hour to provide energy results directly into Wh); I_{PV} and V_{PV} are the operating current [A] and voltage [V] of the PV module; $P_{MP,STC}$, $I_{MP,STC}$ and $V_{MP,STC}$ are the power [W], current [A], and voltage [V] of the maximum power point at STC; I_{SC} and $I_{SC,STC}$ are the operating and STC short-circuit current [A].

3.2 Empirical models

To study the effect of the environmental conditions, data were selected to reduce the impact of the operating conditions (soiling, shading) as explained in [7]. From the 330 days available for the first-year analysis, half were excluded because of important operating condition impacts.

Figure 3 shows the performances (PR, VN, IP, IP_{SC}) of a PV module (H2) as a function of irradiation. The graphs indicate days with AT higher and lower than 27°C. The last plots are the predictions using the empirical models. Performance is generally linearly correlated to irradiation with different sensitivities on overcast and sunny days. The same modelling approach is used for all performance criteria defined above in (1), and (3) to (5). The empirical models use two linear functions (6) with a breaking point observed at 3 kWhm^{-2}. Fitting coefficients are calculated using least squares regressions. To characterize sunny days, the regression is completed selecting days with irradiation above 3 kWhm^{-2}. For overcast days, the fitting curve is calculated with days whose irradiation is below 3 kWhm^{-2} and constrained to intersect with the other linear fit at the breaking point.

(6) $Perf = O_{IRR} + C_{IRR} \times IRR$ calculated

(6-1) on overcast days (irradiation < 3 kWhm^{-2})

(6-2) on sunny days (irradiation ≥ 3 kWhm^{-2})

where $Perf$ is one of the PV performances defined in (1, 3-5); O_{IRR} and C_{IRR} are the offset and coefficient of the linear fit of $Perf$ versus irradiation.

4 RESULTS

This section presents the results obtained during the first year of operation from April 2016 to March 2017.

4.1 Environmental conditions

Yearly global irradiation is 5.4±1.5 kWhm^{-2} per day with 67% coming from the direct beam and 33% from the diffuse component. AT averages 28.1±1.9°C. Spectral energy estimated using the average photon energy (APE) [9] is 1.92±0.18 eV higher than the STC spectrum (APE_{STC}=1.88 eV). **Figure 1** plots the energy coming from the direct beam and diffuse component as a function of the global irradiation. Overcast days with irradiation below 3 kWhm^{-2} receive mostly diffuse light. Above 3 kWhm^{-2}, the energy received from the direct beam increases generally linearly with increasing irradiation while the diffuse component energy decreases.

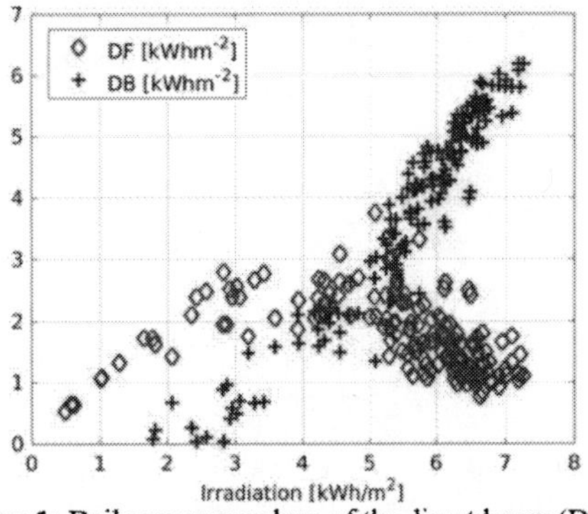

Figure 1: Daily average values of the direct beam (DB) and diffuse irradiation (DF) as a function of irradiation from April 2016 to March 2017.

4.2 Yearly performance

Figure 2 shows the yearly average performances calculated excluding days affected by important impact of the operating conditions. The difference between PR and the product of IP and VN is estimated within ±0.6% for all PV modules. PR is between 88% and 93% for most PV modules. Higher performance is calculated for D1 with PR at 98%. Most PV modules (crystalline and CdTe) have VN between 87% to 93%. The CIGS module C1 has higher VN (96%). IP averages between 97% and 102% for the crystalline modules. C1 has a low IP around 92%. D1 IP is high at 106%. Optical performance (IP_{SC}) vary from 99% to 104% for all modules except C1 at 92%. The low IP of C1 is related to a low IP_{SC}. However, the high IP of D1 is not associated with a high IP_{SC} (102%).

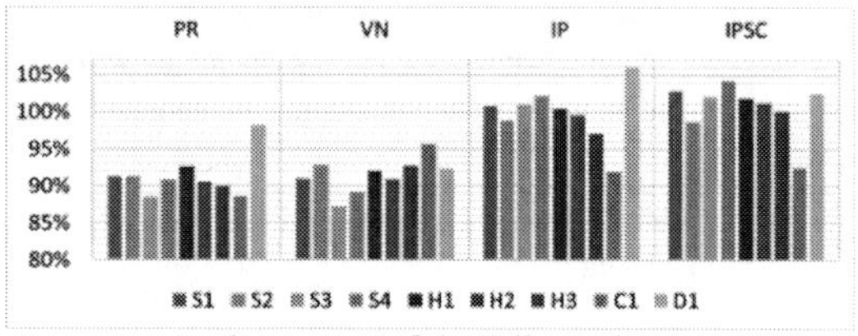

Figure 2: Yearly average of the daily performances (PR, VN, IP, IP_{SC}) of the PV modules from April 2016 to March 2017.

4.3 Impact of the irradiation

Figure 3 shows the impact of irradiation on the performance of a PV modules (H2). From top to bottom, the graphs display PR, VN, IP, and IP_{SC}. Each graph includes the calculated performance (distinguishing when AT is below and above 27°C) and the prediction. For all modules, we observe similar trends. PR generally decreases with increasing irradiation. VN increases with increasing irradiation below 3 $kWhm^{-2}$. On sunny days, VN decreases with increasing solar energy and displays distributed data with higher values corresponding essentially to days with low AT. IP and IP_{SC} decreases with increasing irradiation with different sensitivity in overcast or sunny days. There is a very low impact of AT on the IP and IP_{SC} which relates to the low temperature coefficient on I_{SC} and the high correlation between IP and IP_{SC} [7]. These plots suggest that irradiation is the key environmental parameter of the PV performances at the location. AT is another parameter affecting VN and PR on sunny days.

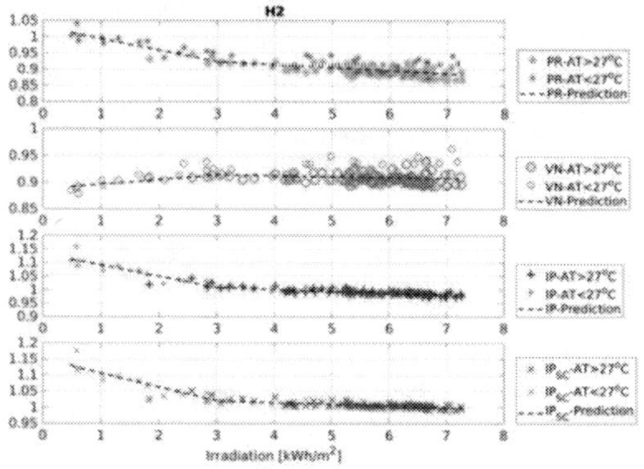

Figure 3: Daily performances of a PV module (H2) as a function of the irradiation. From top to bottom: PR, VN, IP, and IP_{SC}.

4.4 Comparison between PV technologies

Figure 4 compares the PV performances (PR, VN, and IP) of all PV modules using the empirical models. The last graph labelled error plots the difference between PR and the product of IP and VN.

All crystalline modules exhibit an IP with similar values and impact of the irradiation. As observed on the yearly average values (**Figure 2**), IP is higher for D1 and lower for C1 compared to the crystalline modules. On sunny days, the sensitivity of IP to the irradiation is the highest for H2, H3, and D1 and the lowest for C1. On cloudy days, D1 IP is the most improved with decreasing irradiation and C1 IP the least. On sunny days, from the most to least affected VN by increasing irradiation: The S modules, D1, the H modules, and C1. In cloudy conditions, the most decrease of VN with decreasing irradiation is observed on C1 and H3. The H modules tend to exhibit higher VN decrease than the S modules on overcast days.

PR is similar to the product of VN and IP on sunny days with an error within ±1%. The error is more important in overcast conditions and suggests that PR is more affected by IP than VN. PR is higher for D1 than for all other modules at all levels of irradiation. C1 is the least performer, competing with the crystalline modules only at high irradiation. In sunny conditions, H1 outperforms the other crystalline modules while S3 underperforms. H3 underperforms the other crystalline in overcast conditions.

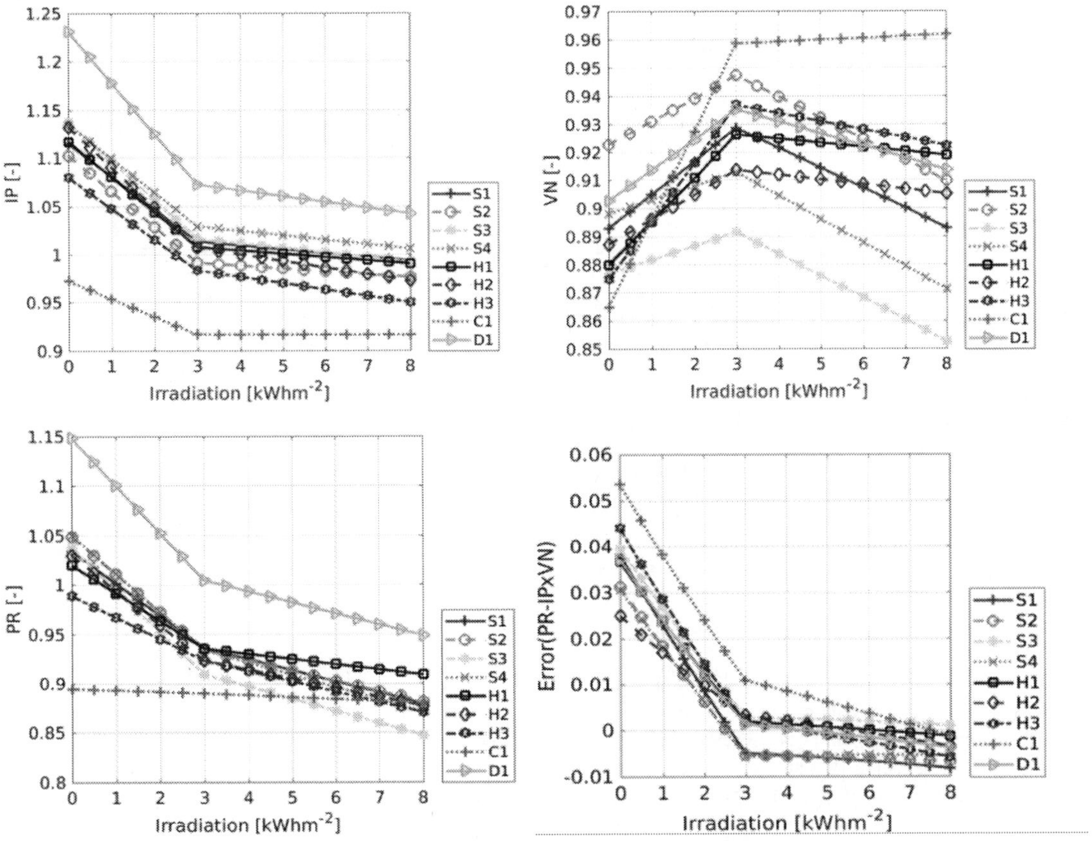

Figure 4: PV performances (IP, VN and PR) as a function of the irradiation for all PV modules using the empirical models. The last plot includes the error between PR and the product of IP and VN.

5 DISCUSSIONS

The yearly average PR of the PV modules depends on the current and voltage performances as it is the product of IP and VN at our mostly sunny location. The CdTe module exhibits superior IP while the C1 has an underperforming IP compared to the crystalline modules. In terms of VN, S3 underperforms and C1 overperforms the other modules. This results in PR values from 88% to 93% for most technologies with higher performance for the CdTe at 98%.

The daily performances are linearly correlated to the irradiation with a change in sensitivity observed at 3 kWhm^{-2}. This corresponds to the breaking point where direct beam irradiation is contributing to the daily solar energy. This is coherent with some modelling approach [10] where I_{SC} depends on different coefficients for direct beam and diffuse irradiance.

The impact of the irradiation on the module performances is observed dependent on the PV technologies and module designs. PV technology affects the thermal loss (1-VN) observed higher for the S modules than the H and thin film modules, relative to the temperature coefficients. H modules tend to have most sensitive IP in sunny conditions and higher VN decrease in overcast conditions but there is variation between the 3 n-type modules which have very different designs. CdTe and CIGS modules exhibit different sensitivity compared to the crystalline. The CdTe module has the most IP improvement in overcast conditions at the site. This could be related to its spectral sensitivity expected to be higher than the crystalline and the CIGS [2]. However, its optical performance does not exhibit specific enhancement as seen on the yearly average values (**Figure 2**). The tested CIGS shows low sensitivity to the irradiation on sunny days and has poor efficiency in low light conditions [7] affecting on both IP and VN.

From the first-year results, we conclude on the best PV technologies for specific irradiation environment. Within the crystalline modules, H1 would perform best for mostly sunny locations thanks to low thermal loss and low IP decrease with increasing irradiation. S3 is the least performing due to low VN. For mostly cloudy locations, the S modules would be a better choice than the H modules as they have lower voltage decrease. Despite a high VN, the CIGS C1 underperforms due to its low optical performance and poor low light efficiency. It should not be considered as representative of the CIGS technology. High VN was also monitored on the second CIGS whose PV system exhibited the highest PR at the site with an IP at 98% [7]. Results on the C2 module in test with the IV tracer will be available for the second year of operation. The CdTe is the best candidate for all solar conditions due to very high IP. Further investigations are needed to explain such high current performance.

The important advantage of this approach is to dissociate the parameters of the PV performance. It is expected that IP versus irradiation would mostly depend on the location (spectral energy) and orientation (angle-of-incidence) while VN would vary mostly with AT, wind conditions and mounting. This analysis method should help better understand the performance variabilities of PV modules when operating in various environments.

6 CONCLUSIONS

We have compared the daily performance of PV modules in test in Maui during the first year of operation. A novel analysis approach expands the number of performance criteria introducing the optical, current, and voltage performances in addition to the performance ratio. Using simple empirical models, these performances are studied as a function of the irradiation, the main environmental parameter. This method helps visualize the performance differences between PV technologies and module designs and could lead to a better understanding of the performance variabilities between locations.

7 AKNOWLEDGEMENTS

HNEI is grateful for the support of the United States Department of Energy (DE-EE0003507) and the Office Naval Research (N00014-10-10310) that funded the development, building and installation of the PV test platforms and HNEI DAS described in this document. Many thanks to Maui Economic and Development Board for hosting our test station. We also recognize the great support of our contractors including the PV installer HNU Energy and the electronic developer Grandalytic. Finally, we are thankful of First Solar for donating a set of their PV modules.

8 REFERENCES

[1] IEC 61724, Photovoltaic system monitoring - Guidelines for measurement, data exchange and analysis, Edition 1.0, 1998-04.

[2] Makrides G., Zinsser B., Norton M., Georghiou G.E., Performance of Photovoltaic under actual operating conditions, DOI: 10.5772/27386.

[3] Canete C., Carretero J., Sidrach-de-Cardone M., Energy performance of different photovoltaic module technologies under outdoor conditions, Energy (65), pp. 295-302, 2014.

[4] Del Cueto J.A., Comparison of energy production and performance from flat-plate photovoltaic module technologies deployed at fixed tilt, 29th IEEE PV Specialists Conference Proceedings, pp. 1523-1526, 2002.

[5] D. Dirnberger at al., PV module energy rating: opportunities and limitations, *Progress in Photovoltaics: Research and Applications* (23), pp. 1754-1770, 2015.

[6] The Performance of Photovoltaic (PV) Systems – Modelling, Measurement and Assessment, ISBN: 978-1-78242-354-6, Elsevier, 2017.

[7] Busquet S., Kobayashi J., Rocheleau R. E., Operation and performance assessment of grid-connected PV systems in operation in Maui, Hawaii, 44th IEEE Photovoltaic Specialists Conference, 2017.

[8] S. Sellner, J. Sutterluti, L.Schreier, S. Ransome, Advanced PV module performance characterization and validation using the novel Loss Factor Model, 38th IEEE PV Specialists Cenference Proocedings, pp. 2938-2943, 2012.

[9] Ishii T. et al., A simplified methodology for estimating solar spectral influence on photovoltaic energy yield using average photon energy, Energy Science and Engineering v. 1, pp. 18-26, 2013.

[10] King D.L., Boysin W.E., Kratochvill J.A., Photovoltaic Array Performance Model, Sandia Report, SAND2004-3535, 2004.

INVESTIGATING A POTENTIAL LINEAR MODEL FOR PREDICTION OF MONTHLY SNOW-INDUCED PRODUCTION LOSSES FOR ROOFTOP PV

Michiel van Noord[1], Mark Murphy[2], Torsten Berglund[1]

[1] Esam AB - Consultants for Sustainable Development, Kungsholmstorg 10, SE-112 21 Stockholm, Sweden
phone: +46 708-717 828; e-mail: michiel.vannoord@esam.se
[2] Department of Applied Physics and Electronics, Umeå University, Umeå, SwedenCompany / Institute(s)

ABSTRACT: This study investigates the potential of an easy applicable linear model of typical yield losses from PV rooftop systems due to snow coverage, by relating these losses to air temperature. The basis for this study is measurements conducted during two winter seasons, on three locations and a total of six PV systems. A yield prediction model based on "snow free" Performance Ratio is used to calculate daily and monthly snow losses. Correlation of monthly snow loss factors (% of expected yield) with average monthly air temperature is visually analyzed. Three temperature correlation zones are identified: A zone with practically no snow losses, a transition zone and a zone with almost full snow losses. A linear regression is used to model snow loss factors for the transition temperature zone. Comparison with historical data indicates that an improvement of the transition zone model is desirable. Addition of snowfall amounts and/or global irradiation into the model are proposed for future development.

Keywords: System Performance, Modelling, Snow, Rooftop

1 BACKGROUND

Annual snow-induced yield losses of 20% have been observed on rooftop PV systems in Northern Sweden (latitudes 60°N and higher, climate zone Dfb and Dfc) by the authors. With snow loss levels well above 5% it becomes relevant to predict snow losses for photovoltaic systems as input to decision processes amongst investors and PV system owners. For these purposes, it is sufficient to predict long-term averages for monthly losses, that are easily applied to simulation results of commercial software, resulting in average annual yield losses.

A number of studies on the effects of snow on PV yield and the underlying principles have been published historically [1, 2, 3, 4, 5]. From [6, 7] we know that the main influencing weather parameters for snow-induced yield effects are considered to be air temperature, global irradiation and amount of snowfall. Several studies [1, 5, 6] indicate that the effect of module inclination on snow effects is weak for inclination angles below 30°. This is also in line with observations in Southern Sweden [8].

The validity of previous studies for Northern Sweden and Scandinavia is unclear. Most previous work was conducted in North-America at typical latitudes of 37°N to 51°N, as compared to 60°N and higher for Northern Scandinavia. Also climate types may differ. Another important fact is the limited statistical basis of many studies, where most studies cover just one or two PV systems or locations during one winter season.

Models resulting from attempts to predict snow-induced yield losses are often complicated to use, due to demand for continuous hourly data and advanced multi-factor fitting [6] or site-specific geometrical analysis [7]. In order to meet the typical target group of consultants, property owners and property developers, this study investigates the potential of a rather simple and generic model based on publicly available meteorological data. Such an easily-applied model would increase the commercial application in trade-off for a higher uncertainty.

This article presents the first step to reach an "engineering-model" as described above.

2 METHOD

2.1 Measurements

During two Winter seasons (2014/2015 and 2015/2016) meteorological data and yield from six rooftop PV systems, distributed in pairs over three geographical locations has been collected. These locations or sites are: Umeå, Örnsköldsvik (or Övik) and Bleka. All PV systems in this study have a moderate inclination (14°–30°), which is typical for roofs on commercial buildings, and an azimuth between south-facing and west-facing. The geographical sites were located in Northern Sweden, with latitudes between 63° and 64°N. For more details, see Table I and Figure 1.

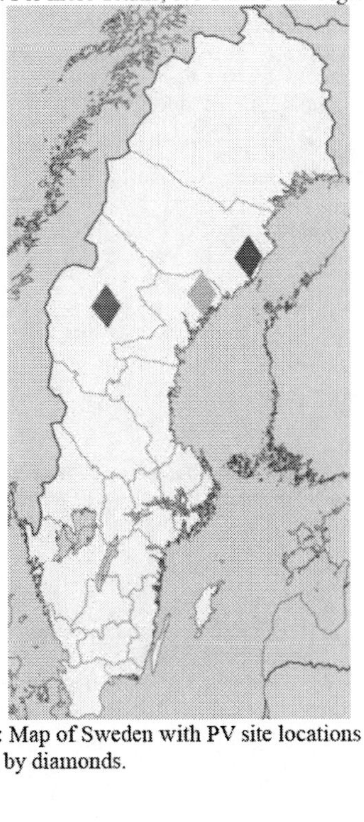

Figure 1: Map of Sweden with PV site locations indicated by diamonds.

Table I: Studied PV systems

PV system *(color indicates location in Fig. 1)*	Power (kWp)	Inclination	Azimuth
Umeå West	7.9	14	90
Umeå South	57	18	0
Övik West	35	20	80
Övik South	33	20	-10
Bleka Shed	20	21	30
Bleka Roof	10	30	30

For each day and PV array the amount of snow coverage for each module row was categorized in four classes, defining the snow coverage index *S* (see Table II). Arrays with multiple rows use one digit per row, starting with the top row, e.g. 033 for a three row array with the two bottom rows fully covered.

Table II: Snow coverage class definitions

Class	Amount of snow coverage
0	no coverage
1	up to 1/3 of module height covered some time of the day
2	day average of 2/3 of module height covered
3	module fully covered
9	ripe or ice on (part) of module

2.2 Calculation of snow-induced yield losses

Using only snow-free days (*S*=0) from each calendar month in both winter seasons, the "snow-free" monthly Performance Ratio is calculated for each month and PV system, using

$$(1)\ PR_{sf,m} = \frac{\sum_{i\in m_{sf}} E_{AC,i}}{\sum_{j\in m_{sf}} P_{STC}\left(\frac{G_{poa,j}}{G_{STC}}\right)}$$

where m_{sf} includes all days with *S*=0.

This "snow-free" Performance Ratio is then used to calculate monthly snow-induced yield losses:

$$(2)\ \hat{E}_{AC,sl,m} = PR_{sf,m} \cdot P_{STC} \cdot \frac{\sum_{i\in m_{sl}} G_{poa,i}}{G_{STC}}$$

where m_{sl} includes all days with $S \neq 0$.

The monthly snow loss factor, SLF_m, is then defined as:

$$(3)\ SLF_m = \frac{\hat{E}_{AC,sl,m}}{\hat{E}_{AC,sf,m}}$$

where $\hat{E}_{AC,sf,m}$ is the expected monthly "snow-free" yield:

$$(4)\ \hat{E}_{AC,sf,m} = \hat{E}_{AC,sl,m} + \sum_{i\in m_{sf}} E_{AC,i}$$

2.3 Visual analysis

As a final step snow loss factors are plotted against meteorological parameters for visual analysis.

3 RESULTS

3.1 Snow-induced yield losses and snow loss factors

Figure 2 shows the expected "snow-free" yield on a daily basis for PV system Övik South in December 2015. Color settings are used to illustrate the three different components of the "snow-free" yield: actual yield on days with and without snow cover and snow-induced yield losses. Global irradiation is plotted in the same figure to show feasibility of the PR-based calculation of "snow-free" yield.

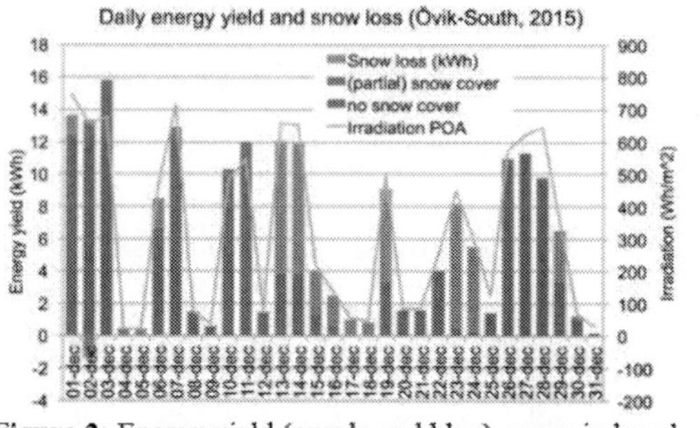

Figure 2: Energy yield (purple and blue), snow-induced yield loss (orange) and irradiation (green) for PV system Övik South in December 2015.

Looking at the resulting snow loss factors for all months and PV systems studied (Fig. 3) it can be observed that they cover the whole spectrum from 0% to 100%. Missing data for both Övik systems and Bleka roof in November 2014 is due to measurement or datalogger failures.

Comparing PV systems at the same site shows typically only small differences for Umeå and Övik, with three exceptions for the Umeå site. This is probably due to the difference in building height (1 story), which for the Umeå West system results in higher horizon and more shading. Also Bleka shows several months of substantial differences between the two PV systems snow loss factors. The system with PV mounted flat on the roof (Bleka roof) suffers from larger snow losses than the system with sheds (Bleka shed), despite higher inclination (30° resp. 21°).

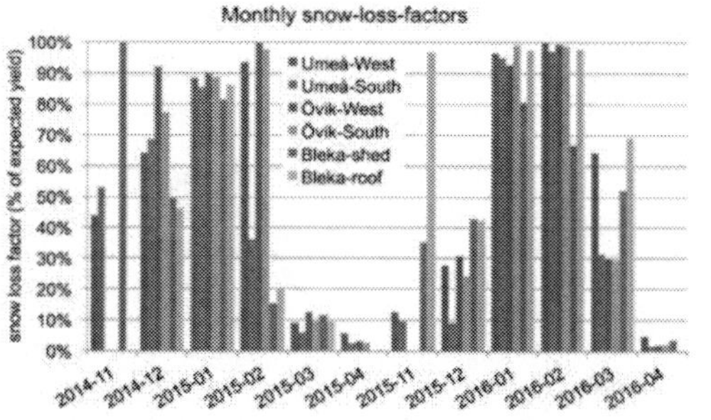

Figure 3: Snow loss factors for each month and PV system studied.

When considering snow loss factors it is important to be aware of typical annual irradiation distribution in Northern Sweden, which is illustrated by the example in Figure 4. It then becomes clear that low snow loss for March or April give rise to larger absolute snow losses than high snow loss factors in November, December or January.

Annual yield losses due to snow coverage are calculated from monthly snow loss factors and were found to be in the range of 1,5% to 10%. Winter season 2015/2016 showed higher annual snow losses for all PV systems, due to high snow

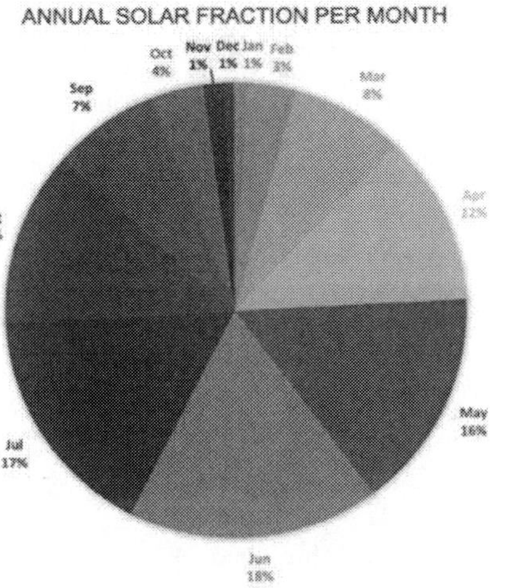

Figure 4: Distribution of global irradiation over the year for a south facing surface with 15° inclination in Umeå.

3.2 Snow loss factors dependency on air temperature

Analyzing the correlation with weather parameters showed that the monthly snow loss factors are highly dependent on monthly average air temperature. From a graphical analysis three temperature zones can be identified (see also Figure 5):

a. zero-snow-loss temperatures (blue triangles)
b. transition temperatures (pink squares)
c. near-full-snow-loss temperatures (purple diamonds)

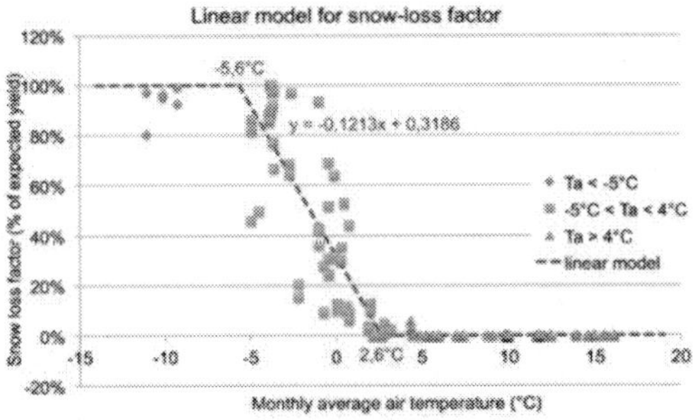

Figure 5: Snow loss factor vs monthly average air temperature.

Based on the three temperature zones, a multi-linear model is developed by least squares fitting on the transition temperature zone:

$$(5)\ \widetilde{SLF}_m = \begin{cases} 1, & T_a < -5.6°C \\ -0.12T_a + 0.32, & -5.6°C \le T_a \le 2.6°C \\ 0, & T_a > 2.6 \end{cases}$$

One implicit assumption that is included in the proposed model is that snow precipitation will always occur during months with negative temperatures. Our research shows that this assumption is clearly valid for Northern Sweden, but for other regions this has to be investigated.

3.2 Comparison of linear model with historical data

The model in (5) is applied to historical data from the SMHI meteo station in Umeå and compared with historical energy yield data from PV system Umeå West, for three winter seasons.

For the Umeå West case the snow free Performance Ratios ($PR_{sf,m}$) had to be adjusted for global horizontal irradiation to be able to use SMHI:s irradiation data:

$$(6)\ \dot{PR}_{sf,m} = PR_{sf,m}\frac{G_{poa}}{G_{hor}}$$

The result of this comparison is presented in Figure 6. The linear model shows to be in good correspondence with historical snow loss factors for the zero-snow-loss and near-full-snow-loss temperature zones. For the transition zone correspondence varies and shows significant deviations from historical data.

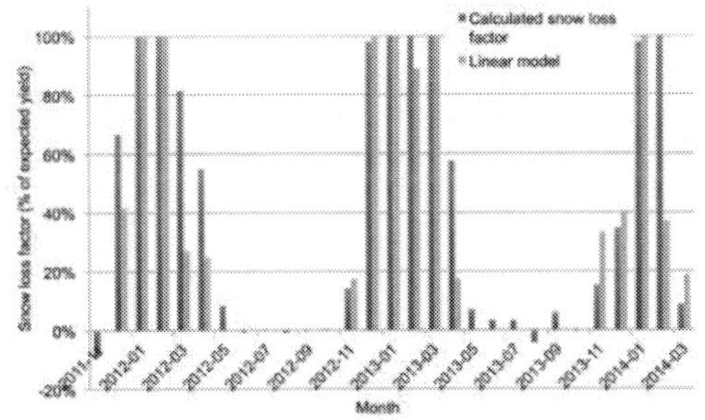

Figure 6: Comparison of the linear snow loss model with historical data for Umeå West PV system.

Due to the typical irradiation distribution the errors in snow loss factors for March or April are magnified when calculating annual or absolute snow losses.

At the same time, it is unclear if the Umeå West system can be considered representative for typical PV systems in Northern Sweden. As mentioned earlier this array suffers from significant shading in winter and a high horizon profile. Model deviations shown in Figure 6 for early and late winter months are of the same order as the differences between snow loss factors for Umeå West and Umeå South for the same months (see Fig. 3). Unfortunately historical production data from other PV systems in combination with irradiation and temperature data was either unavailable or unsuitable (due to historical cleaning of modules from snow coverage).

3.3 Other weather parameters to consider

In spite of some uncertainties in which conclusions should be drawn from the historical comparison above, it is highly probable that additional parameters have to be added to the snow loss factor model for the transition zone. The main candidates for these parameters are the amount of snow precipitation and global irradiation on the plane of the array. Adding global irradiation to the model is likely to make it more compatible with the Umeå West system. Adding snow precipitation is likely to make the model less dependent on geographical or climatic variations.

4 CONCLUSIONS

A snow-free Performance Ratio can be used to model system performance, given a sufficient number of snow-free days. If sufficient data is available, the resolution of Performance Ratio calculation should be increased to bi-weekly, or a moving average should be used, to increase precision.

This study shows that a rather generic approach to long-term average snow-loss prediction can be feasible.

The linear transition model for monthly snow-loss-factors should be used with caution, especially for PV arrays suffering from significant shadowing or high horizon angles. Nevertheless the model might be the best available method for a rough estimate of monthly (and annual) snow losses for rooftop PV in Northern Sweden (tilt angles of 14-30°).

Comparison with historic data was only possible for the Umeå-West system and it indicates that model precision for the transition temperature zone should be improved. However validation attempts on other, more typical PV systems are worth investigating.

Improvements of the model are likely to be achieved by adding plane of array irradiance or snow precipitation height to the transition model. Also expansion of the statistical data set (temporal and spatial broadening as well as system variety) is necessary in order to improve reliability of any model.

5 ACKNOWLEDGEMENTS

This work was supported financially by the Swedish Energy Agency, the SolEl-program for applied research on solar electricity and the region of Västerbotten. The authors would like to thank these organisations. Additional thanks goes to Bostaden, Landstinget Västernorrland and the inhabitants of Bleka for allowing us to study their PV systems. Finally, we would like to thank all members of the reference group for valuable inputs during the project: Jan Lindberg (LVN), Lena Ahlgren (Umeå Energi), Mikael Ivarsson (Bleka), Royne Söderström (Bostaden), Thomas Olofsson (Umeå Universitet), Tobias Boström (Tromsø Universitet), Tobias Walla (Energimyndigheten) och Tomas Landelius (SMHI)

6 REFERENCES

[1] Brench, B.L., *Snow-Covering Effects on the Power Output of Solar Photovoltaic Arrays*, Report C00-4094-61, (1979).

[2] Ross, M. M. D., *Snow and Ice Accumulation on Photovoltaic Arrays: An Assessment of the TN Conseil Passive Melting Technology*, report # EDRL 95-68 (TR), CANMET Energy Diversification Research Laboratory, September (1995).

[3] Becker et al, *An Approach to the Impact of Snow on the Yield of Grid Connected PV Systems*, Bavarian Association for the Promotion of Solar Energy, Munich, (2006)

[4] Sandell, M., *The effect of snowfall on the power output of photovoltaic solar panels in Halifax, Nova Scotia*, Honours Candidate Thesis, Dalhousie University, (2012).

[5] Andrews et al, *The Effects of Snowfall on Solar Photovoltaic Performance*, Solar Energy, June (2013), Vol.92, pp.84-97

[6] Andrews, R.W. and Pearce, J.M., *Prediction of Energy Effects on Photovoltaic Systems due to Snowfall Events*, 38th IEEE Photovoltaic Specialists Conference, June (2012).

[7] Townsend, T. and Powers, L., *Photovoltaics and Snow: An update from two winters of measurements in the SIERRA*, 37th IEEE Photovoltaic Specialists Conference, June (2011).

[8] Stridh, B., *Utvärdering av sveriges första MW-solcellspark, Mälardalens Högskola, January (2016).*

[9] Peel MC, Finlayson BL & McMahon TA, *Updated world map of the Köppen-Geiger climate classification*, Hydrol. Earth Syst. Sci., (2007), 11, 1633-1644

STATISTICAL ANALYSIS OF INFRARED-INSPECTIONS OF PV-PLANTS

Claudia Buerhop[1], Tobias Pickel[1], Hans Scheuerpflug[1], Jens Hauch[1], Christian Camus[1], Christoph J. Brabec[1,2]
[1]ZAE Bayern e.V., D-91058 Erlangen, Immerwahrstraße 2,
[2]i-MEET, FAU Erlangen-Nürnberg, 91058 Erlangen, Martensstraße 7

ABSTRACT: First statistical evaluation of IR-inspections of PV-plants reveals that 81% of the installed PV-plants show IR-abnormalities. More than 140 PV-plants with more than 200,000 PV-modules were inspected and evaluated statistically. Main IR-abnormalities or failures in modules and string installations are analyzed, respectively. The average failure rate for PV-modules is about 6% and for module strings approximately 1%. The differentiation based on the installation type reveals that small residential installations relatively seen show more defective modules than large field installations. A first estimation of the missed income by not detecting and maintaining major module and plant failures yield a missed income of about 100 Mio. EUR considering the PV-systems installed in Germany. IR-imaging is a valuable method to give fast and reliable information about the actual quality and failure rate in inspected PV-installations.

Keywords: IR-imaging, statistics, failure, performance.

1 INTRODUCTION

Quality control of PV-plants is of increasing interest for service and maintenance. Failures in PV-installations may cause a safety risk as well as a severe power reduction and poor performance in the future. Therefore, many studies have been carried out about the reliability of PV-modules in the past years. Also, there are various surveys of failure rates of PV-modules during operation [1-8].

Among the top ten module failure mechanisms are: connector failures and glass breakage as retrieved from field inspections [6]. The data presented suggest that 34% of the inspected modules have some type of failure. The IEA report [1] states that the most important failures in the field are junction box failure, glass breakage, defective cell interconnect, loose frame and delamination. De Graaff's [3] survey points out that 2% of the modules do not meet the manufacturer warranty. After a certain life time the most dominant module failures are delamination, cell fracture and discolouration [7]. Most investigations focus on the impact on the actual power output.

Fig. 1: aIR-images of two PV-plants showing different defect sites, left: industrial roof installation with numerous defective modules and two suspicious module strings, right: IR-overview of a field installation with two suspicious module strings

In order to check the PV-plant quality and locate and identify potential failures IR-inspections are a valuable tool. Drone-based measurement systems enhance the fast, mobile, flexible and non-destructive inspection under real operating conditions without operation interruption [9]. Electrically malfunctioning modules show up in the IR-image because of their elevated temperature [10]. This drone-based IR-inspection is called aIR-imaging and respectively aIR-images or IR-images. Typical IR-images of two different installations are presented in Fig. 1. Here, module and string failures can be distinguished. A module failure normally shows single sites (small spots, individual cells, cell arrays typically protected by a bypass diode, so called substring) of increased temperature within the module. A String failure is typically caused by a cabling problem of installed PV-modules, e. g. the connection to the inverter is interrupted. The modules within the string are at open circuit. No current flow is possible and so finally all absorbed solar irradiation is transferred to heat. Therefore all these modules show a homogeneously higher temperature.

The scope of this paper is to present an updated overview of the abnormalities in IR-inspections of PV-plants based on a statistical evaluation of the data collected in the last years. Thereby the evaluation is governed by the question *"what are typical occurrences for IR-findings and how are they distributed and influenced?"* A yield discussion concludes the paper.

2 EXPERIMENTAL PROCEDURE

The aIR-inspections were carried out with a measurement system consisting of a drone equipped with an IR- and VIS-camera and various sensors. The drone is an unmanned aerial vehicle (UAV). Here an octocopter is used which is equipped with a navigation system and various sensors.

During the last years a large number of PV-generators were investigated. They differ in cell technology (crystalline and thin film), in power output (kWp), in the number of modules, operation duration and installation location (industrial roof, residential roof, field installation, tracker) and geographic location. IR-inspections of approximately 140 PV-plants are evaluated statistically in this paper. Their peak power adds up to roughly 52 MWp with about 400,000 installed PV-modules. Mostly, PV-plants with crystalline silicon cell technology were analyzed. That means, 95% of the inspected PV-plants and roughly 40% of the investigated PV-modules are of crystalline technology.

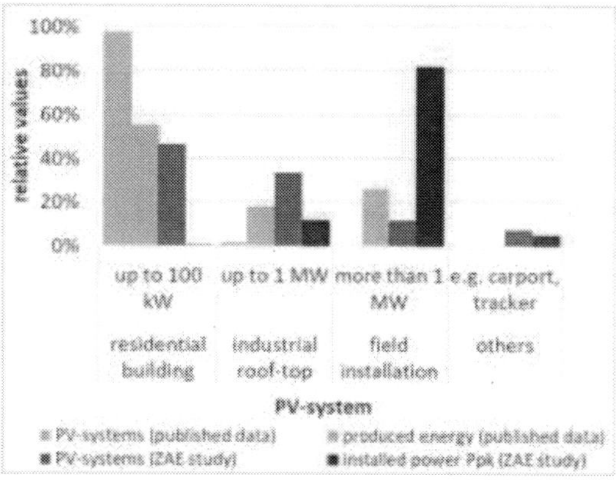

Fig. 2: Overview of distribution of PV-systems: firstly, published data for 2015 (1.5 Mio systems, 37 GWp installed) showing their number and produced energy according to the installed power bins [11] and secondly, evaluated PV-systems during the statistical ZAE-study of findings using aIR-inspection.

The percentage of residential house installations of up to 100 modules is approx. 45%, industrial roofs with 100 to 1,000 modules amount to 33% whereas distinct industrial installations (carports, and others) constitute approx. 8%, and the large-scale field PV-plants with more than 10,000 modules constitute roughly 12% of the inspected PV-plants. Fig. 2 illustrates that the number of inspected roof-top PV-systems on industrial building is fairly high as well as for large MW-PV-systems. However, a reasonable number of rather small PV-installations on residential and apartment buildings was analyzed. As Fig. 2 illustrates the large number of residential PV-systems produce less than half of the PV-energy.

The investigated PV-plants were equally distributed between the following cases

1. Service measurements, initiated by operators assuming power reduction based on lower than expected energy yields
2. Research projects, here the PV-plants performed within the expected power output range, but were still investigated periodically.

This study includes a high number of industrial roof-top and field PV-installations as well as small residential PV-systems.

The PV plants investigated included PV-plants with and without monitoring systems. In the case that monitoring (on string as well as on module level) was actually present we could determine the resulting yield at operating conditions.

The IR-images were evaluated regarding their temperature distributions. At suspicious sites the absolute temperature as well as the temperature difference to adjacent cells is determined. In parallel to the IR-images, visual images are recorded. The comparison of the IR- and the VIS-images allows the recognition of artefacts, as e. g. bird droppings, which do not constitute a permanent yield risk and were thus not considered in this study. Therefore, IR-features evaluated here, are indeed due to defective areas causing irregular heating in the modules and in the PV-generators.

3 RESULTS AND DISCUSSION

The percentage p_i of a PV-plant with a specific IR-abnormality of failure type i is given by:

$$p_i = \frac{n_i}{N} \qquad (1)$$

with n_i being the number of affected PV-plants with the specific failure type and N the total number of inspected PV-plants. Calculating the failure rate of a given PV-plant the equation is given as following:

$$q_i = \frac{m_i}{M} \quad \text{and} \quad q_i = \frac{s_i}{S} \qquad (2)$$

with m_i the number of affected PV-modules with the specific IR-abnormality of failure type i and M the total number of PV-modules in the PV-plant, and s_i the number of modules strings with elevated temperature and S the total number of strings in the PV-plant.

3.1 Defect classification

Based on their temperature distribution the IR results were assigned to failure modes identified by their characteristic IR-fingerprint as described in [10]. The occurrence of the specific IR-findings or failure modes in PV-plants differs strongly, see Fig. 3. Only 19% of the PV-plants exhibited no IR abnormalities. Surprisingly, almost 30% of the PV-plants show more than one complete *suspicious module string* (compare both IR-images in Fig. 1). Typical is the homogeneous temperature increase of about 5 -6 K. Possible reasons for the disconnection may be wrong installation or destruction by animals or natural degradation and deterioration.

Approximately 45% of the PV-plants have *bypassed substrings*. The temperature difference to adjacent substrings ranges from 3- 5 K.

Further detected module failures are *defective bypass diodes*, *cell fracture*, *defective solder joints* as well as *short-circuited cells*, *potential induced degradation and delamination* by their typical IR-patterns. While most failure types cause entirely heated cells areas, defective solder joints are recognized as a localized, point-shaped heat sources. Depending on the conditions defective solder joints and fractures can locally reach fairly high temperatures T > 100°C. In contrast, slightly increased temperatures are detected for short-circuited cells or PID [12]. The number of PV-plants with PID is rather small compared to the proposed severity of PID [13]. However, if PID is detected in a PV-plant, then roughly 7 to 40% of the modules are affected [12]. IR-imaging enables a first quality check of the number of the PID-affected modules as well as a quantitative estimation of the resulting power loss [12, 14]. Further analysis showed that standard monitoring data does not reveal the presence and impact of PID clearly and does not provide information about the location of these PID-affected modules [15].

Glass breakage was observed most frequently in thin-film modules [4] (percentages up to 16% were detected) as well as after recent strong storm events also for crystalline modules (approx. 1%).

If the temperature difference is negative, the temperature of the defective site is lower than that of the surrounding cells, there is delamination. For this failure up to 80% of the modules of a plant can be affected.

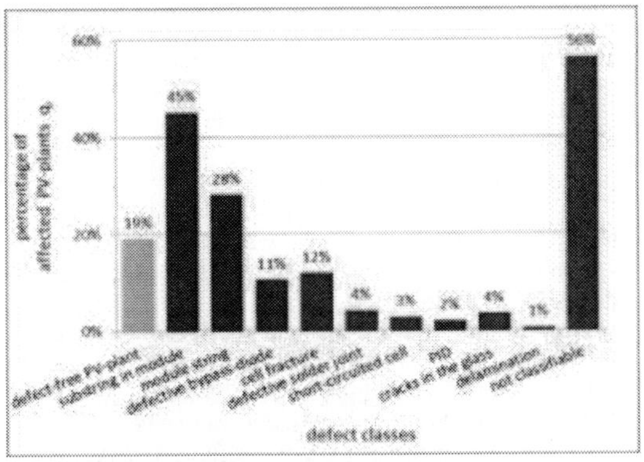

Fig. 3: Distribution of the most frequent findings in IR-images

Suspicious substrings and *module strings* are the features most frequently detected in PV-plants.

We found almost 20,000 IR-suspicious modules equalling 6% of all modules in the inspected PV-plants. The temperature profile of one disconnected module string is commonly seen in PV-plants. Apparently, 1.1% of the strings seem not to be properly connected to the inverter. Table 1 gives an overview of the occurrence of the most prominent failure types and their occurrence.

Table 1: Occurrence of different failure types with respect to the entity of the sampling and the frequency in affected PV-plants

	module failure[1]	substring failure	module string failure
all modules	5.8%	0.16%	
all strings			1.1%
Average in affected PV-systems	8.4%	2.3%	3.9%
Standard deviation	±12,86%	±4,51%	±5,8%

3.2 Operation period

Based on the inspected sampling in Fig. 4, there is no visible tendency of the failures with respect to operating period. The high module failure of 80% in Fig. 4 was observed for PV-plant with delaminated modules. The rate of PID-affected modules is larger for longer operation periods, 42% for 5.5 year of operation and, respectively, 35% for 3 years.

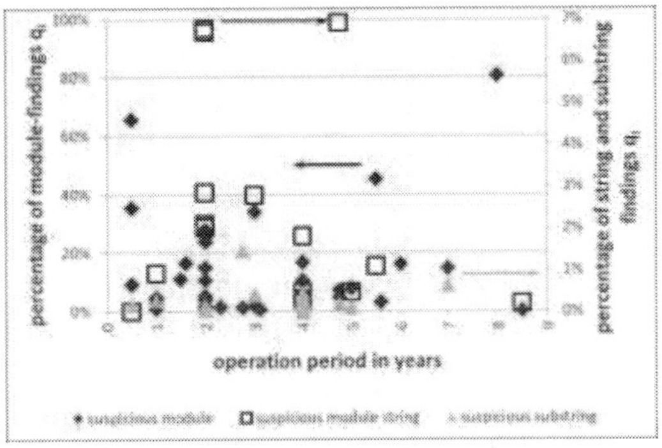

Fig. 4: Percentage of IR-findings (modules, substrings, and module strings) as a function of the operation time

Outstanding are the high percentages half a year after commissioning of two PV-systems, 38% and 65%. In these cases, faulty manufacture is obvious.

Disregarding failure types as delamination, PID, and fabrication defects, IR-suspicious module failures seem to be rather constant during first half of lifetime. Possible replacements of modules are not taken into account because unknown and badly documented.

3.3 PV-plant size

The correlation between the size of the PV-plant and the failure distribution is presented in Fig. 5, where the PV-plant is given as the number of installed PV-modules.

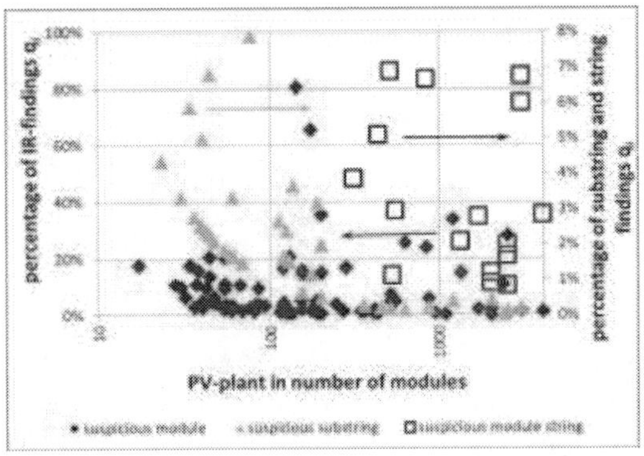

Fig. 5: Percentage of IR-findings (modules and strings) as a function of PV-plant size in number of installed PV-modules

The module failure rate (neglecting PID, delamination, and manufacturer faults) ranges up to 20% for PV-plants up to 200 modules. For large PV-plants (> 200 modules or > 1 MWp) it reaches even values up to 35%. The distribution of substring failure looks differently. Small installations up to 200 modules can reach substring failure rates of 8%. For larger ones a rather constant level of less than 0.5% is visible.

Small and medium-sized PV-plants (with few strings due to the limited number of modules) rarely show string failures, whereas it is quite often detected in larger PV-plants. The occurrence for string findings in large PV-plants (more than 300 modules) ranges between 0.5% and 7%. Due to the large number of strings, the probability to detect suspicious module strings is rather high.

[1] without substring failure

3.4 Installation type

The installation locations differ between large field installations, medium and large industrial roof installations and small and medium residential roof installations. Three IR-findings, suspicious module (except substring), suspicious substring, and suspicious module string are evaluated in detail. As Fig. 6 illustrates nearly all (> 80%) PV-plants of any type show IR-findings.

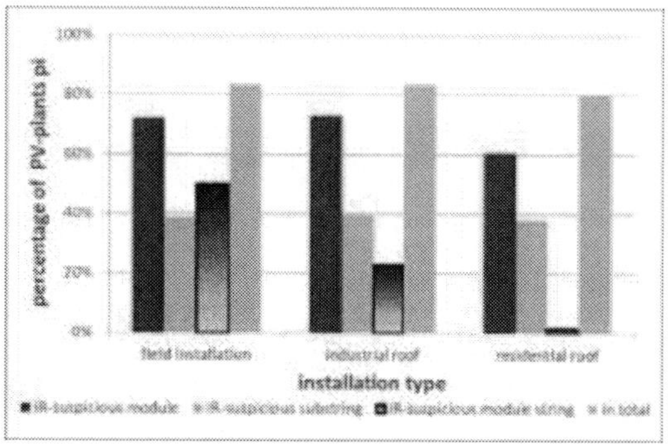

Fig. 6: Frequency of PV-plants with IR-findings for suspicious modules, substrings within modules and module strings for different installation locations showing a least one specific failure type

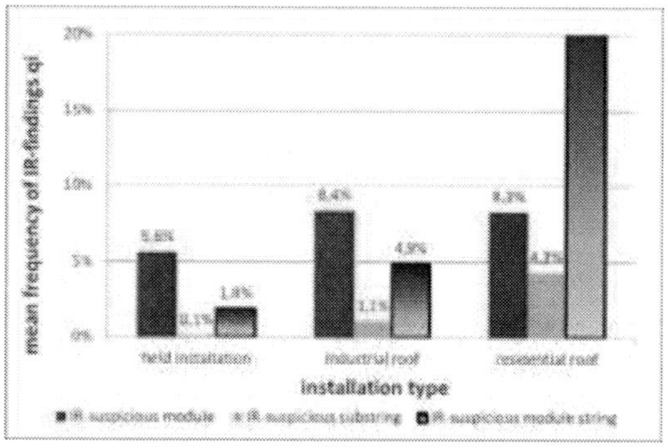

Fig. 7: Frequency of PV-plants with IR-findings for suspicious modules, substrings within modules and module strings for different installation locations. Note, string failure occurred only once for a single residential PV-system.

Module and substring abnormalities are assumed to be indicators for transport, installation and degradation problems. String findings usually indicating disconnected strings are interpreted as an indication of cabling problems due to installation or deterioration of all kinds.

3.5 Yield & financial loss analysis

In ordert to address the financial aspects a closer look on the yield loss is needed. Therefore, the analysis of the yield loss caused by certain error cases is necessary and will be discussed.

Almost 45% of the PV-plants suffer from bypassed module substrings. For modules with three bypass-diodes that means one third of the power is missing. The theoretical influence on the string output is studied exemplarily for different installation cases. 1) In a roof-top installation with module optimizer (e. g. Solaredge) two modules are connected and MPPT-tracked [16]. 2) PV-plants equipped with monitoring on module- and string-level (e. g. Sunsniffer) were analysed [17]. Here 18 modules are connected within a string. 3) Finally, most frequently found are standard string monitoring systems providing data on the string level. [2] describes a 3 MWp PV-plant studied in detail. Table 2 lists calculated and measured yield data. The yield ratio describes the relationship between the yield of the string with the defective substring module and the yield of a string with defect-free modules. Theoretically the string power loss is less than 2% according to this failure. The measured values show a much higher loss. Because the module optimizer only tracks two modules, the yield loss is rated to a string of 18 modules for better comparison. As the data show, the yield loss using module optimizers is the lowest but still higher than the theoretical value. In conclusion, bypassed substrings cause significant power and yield losses in operating conditions.

Table 2: Field data of the influence of 1 bypassed substring failure on the string performance for different string configurations

energy yield ratio		module optimizer	module-level string data	string data
	string length	2 modules	18 modules	20 modules
theoretical	relative power loss within one string	17%	1,85%	1,7%
	yield ratio of 1 resp. 2 modules	80 - 82%	96,60%	
	yield ratio for string with 18 resp. 20 modules		95,70%	93,40%
measured	yield loss ratio for string with 18 or 20 modules		4,30%	6,56%
	rated yield ratio to 18 modules	98%		
calculated	rated yield ratio loss for 18 modules	2%		

On the basis of the IR data the missing income due to undetected and not maintained failures may be estimated, compare Table 3. In Germany 40 GWp PV-power were installed in 2017. Assuming 200 Wp modules and strings with 20 modules, there are 200,000,000 modules and 10,000,000 strings each string with a power of 4,000 Wp. Here, no differentiation is made with respect to the installation type and plant size. Table 1 gives the failure rates for the main defects. Note that in this calculation, bypass-diode failures are added to the substring failures because they have the same impact on the power output. According to Table 1 720,000 strings are affected by substring and bypass diode failure and 110,000 strings are disconnected, as well as 11,600,000 modules. Further calculations focus on substring and string failure because their impact on the output is known. Table 2 shows that for substring failures the affected string produces 6.56% less power. This results in a power loss for strings and substrings in 628 MWp. Considering annual solar irradiance in Germany of 980 kWh/m² / 1 kW/m² gives the missed energy of 616 MWh. In order to calculate the missed income an average price 0.17 EUR/kWh is considered. This yields a missed income of 105 Mio. EUR.

Table 3: approximation of power loss and missing income due to mayor IR-detected defects

installed PV-power in Germany in W	40.000.000.000		
	module failure		strings (20 modules)
installed	200.000.000		10.000.000
power of 1 unit in W	200		4000
	module failure		string failure
	substring + bypass-diode	others	disconnected string
appearance acc. to sampling	0,36%	5,80%	1,10%
scaling to PV-systems in Germany, number of modules, resp. Strings	720.000	11.600.000	
affected strings	720.000		110.000
measured power loss, lit data	6,56%		100%
resulting power loss for the affected string in W	188.928.000		440.000.000
summed power loss for Germany in W	628.928.000		
missed energy for 1 year, irradiance 980 kWh/m2/1kWp/m² in kWh	616.349.440		
missed income (assumed pay 0.17 EUR/kWh) in EUR/year	104.779.405		

Not detected and not repaired major faults in PV-systems can cause significant income losses. The rough estimation yields a significant economic damage only for not maintained defective sites, like module substrings and unconnected strings, every year. Other faults like PID, if it occurs, then many modules are affected, are not taken into account yet.

4 CONCLUSION

aIR-imaging is a valuable tool for identifying failures in all kinds of PV-installations. Many module failures are distinguished. Since the evaluation reveals that most of the PV-plants show IR-abnormalities, IR-inspection is recommendable for quality inspections of large field PV-plants as well as inaccessible residential installations.

5 ACKNOWLEDGEMENT

We gratefully thank the German Federal Ministry for Economic Affairs and Energy (BMWi) for financial funding of this project.

6 REFERENCES

[1] M. Köntges (2014) IEA.
[2] M. Dalsass, et al., IEEE PVSC 43rd (2016).
[3] D. DeGraaff, et al., Photovoltaic Module Reliability Workshop (2011).
[4] C. Buerhop, et al., 32nd EU-PVSEC (2016) 2021.
[5] J. Wohlgemuth, et al., 31st EU-PVSEC (2015) 1517.
[6] U. Jahn, et al., solarbankability (2016).
[7] K. Schulze, et al., 28. Symposium Photovoltaische Solarenergie (2012).
[8] D. C. Jordan, et al.; Progress in Photovoltaics: Research and Applications 24 (2016) 978.
[9] C. Buerhop, et al., Third Southern African Solar Energy Conference SASEC (2015) 224.
[10] C. Buerhop, et al.; Solar Energy Materials and Solar Cells 107 (2012) 154.
[11] energymap.info, (2016) http://www.energymap.info/energieregionen/DE/105.html.
[12] C. Buerhop, et al., SPIE Optics and Photonics, Optics + Photonics for Sustainable Energy (2016) 99380J.
[13] M. Fuhs; PV magazine März (2016) 52.
[14] C. Buerhop, et al., (year) Institution of Engineering and Technology.
[15] C. Buerhop-Lutz, et al., SPIE 10370 Optics and Photonics, Optics + Photonics for Sustainable Energy (2017).
[16] C. Buerhop-Lutz, et al., 33rd EU PVSEC (2017).
[17] J. Teubner, et al.; Energy Procedia (2017).

REVERSE VOLTAGE SIMULATION OF CRYSTALLINE SILICON PV MODULE WITH DAMAGED BYPASS CIRCUIT

Naotaka Oka[1], Yasuhito Takahashi[1], Koji Fujiwara[1], Yoshiyuki Ishihara[1], and Shogo Nishikawa[2]
[1]Department of Electrical Engineering, Doshisha University
1-3 Tataramiyakodani, Kyotanabe, Kyoto, 610-0321, Japan
Phone: +81-774-65-6266, Fax: +81-774-65-6266
E-mail: ytakahashi@mail.doshisha.ac.jp (*Corresponding author: Yasuhito Takahashi)
[2]Department of Electrical Engineering, Nihon University
1−8−14 Kanda Surugadai, Chiyoda−ku, Tokyo, 101−8308, Japan

ABSTRACT: The authors first describe the modelling of a photovoltaic module (PV) using parameters derived from genetic algorithm and the experimental data of its current vs. voltage (*I-V*) characteristics. Then, the possibility of reverse voltage occurrence when changing the number of damaged bypass circuits in a PV module is examined. This is also investigated by varying the specifications of the PV module with damaged bypass circuits and power conditioning systems (PCSs) with a different maximum power point tracking (MPPT) method. In addition, the PV module usable for replacement is discussed taking account of the safety range of specifications of a PV module. As a result, the occurrence of reverse voltage does not change by the number of damaged bypass circuits in a PV module regardless of its specifications. Moreover, the safety range of the PV module for replacement varies by the operation methods of the PCSs; however an adequate value of short circuit current is 97 % of originally installed PV modules in all types of MPPT operations.
Keywords: Interchangeability, Safety, Reliability, Polycrystalline

1 INTRODUCTION

The main point of this study is to examine the variation of reverse voltage occurrence when varying the number of bypass circuits in a photovoltaic (PV) module for its safety evaluation. The demand of PV systems in Japan is increasing rapidly, which is reaching closed to 11 GW by 2015 [1]. Accordingly, the reports on PV accidents are also increasing. Some of them are caused by a reverse voltage.

Generally, protection devices such as bypass diodes and blocking diodes are installed in a PV system. However, these devices may deteriorate due to aging degradation or breaks down because most of the latest bypass diodes are Schottky diodes which have only quite low reverse voltage ratings [2-3]. Thus, there is a risk of accidents like emitting smoke and catching fire, when a PV system operates with damaged devices.

In our previous research, the authors have investigated on occurrence possibility of reverse voltage when all bypass circuits in a PV module are damaged [4]. However, in practice, there is a possibility that not all, but rather some bypass circuits are damaged in a PV module. In this paper, the possibility of reverse voltage occurrence when changing the number of damaged bypass circuits is examined. For examination, current vs. voltage (*I-V*) and power vs. voltage (*P-V*) characteristics of a polycrystalline silicon PV module were modelled. Next, maximum power point tracking (MPPT) methods of three types of power conditioning systems (PCSs) were modelled based on the specifications and experimental data. Then, the variation of reverse voltage occurrence when changing the number of damaged bypass circuits was investigated. Moreover, the safety performance of the PV system were also evaluated when varying the specifications of the PV module with damaged bypass circuit and PCSs. Lastly, the safety range of the PV module specification for replacement was discussed.

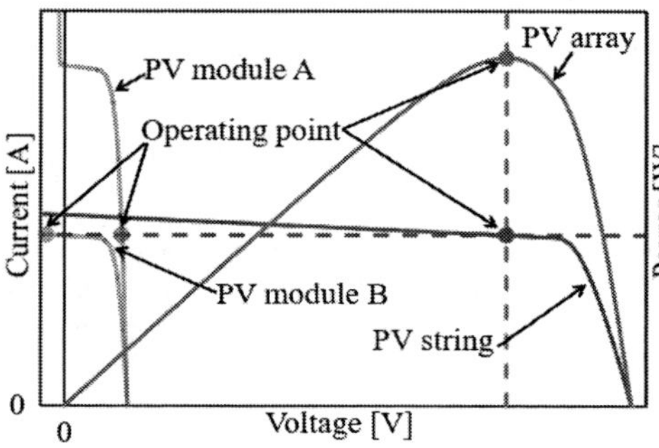

Figure 1: *I-V* characteristic of PV module A and B. Also, *I-V* characteristic of PV string and *P-V* characteristic of PV array where PV modules A and B with different I_{sc} are connected.

2 MECHANISM OF REVERSE VOLTAGE OCCURRENCE

In the case where a bypass diode is connected to a PV module, current flows through the bypass circuit when the PV module covered with the shadow to avoid reverse voltage. However, in the case where the bypass diode is not connected to the PV module, reverse voltage may occur when the PV module operates beyond short circuit current (I_{sc}). In detail below, the mechanism of reverse voltage occurrence is explained when the bypass diode is not connected to the PV module.

Fig. 1 shows *I-V* characteristic of PV module A and PV module B which have different I_{sc}. Also, *I-V* characteristic of the PV string and *P-V* characteristic of the PV array where PV modules A and B are connected. PV module A has bypass diodes and PV module B does not have them. Moreover, I_{sc}^{A} which is I_{sc} of PV module A is bigger than I_{sc}^{B} which is I_{sc} of PV module B. This PV string is consisting of eight PV module A and one PV module B. Also, the PV array has three strings. From Fig. 1, reverse voltage does not occur in PV module A which has bigger I_{sc}, because the operating current of the PV string is below I_{sc}^{A}. In contrast, reverse voltage occurs on PV module B, because the operating current is above I_{sc}^{B}. To sum up, when a PV string contains the PV module with low I_{sc} and bypass circuit is not attached to it, there is a possibility of reverse voltage occurrence.

3 MODELLING PV MODULE AND POWER CONDITIONING SYSTEMS

3.1 PV Module Modelling

The *I-V* characteristic of a polycrystalline silicon PV module during the light irradiation can be calculated with the following equation:

$$I = I_{ph} - I_o\left\{\exp\frac{q(V+R_sI)}{nmk(T+273.15)} - 1\right\} - \frac{V+R_sI}{R_{sh}} \quad (1)$$

Here, I is the output current of a PV module [A], V is its terminal voltage [V]. The quantities q, k, m, n and T denote the electronic charge (1.6 × 10^{-19} C), the Boltzmann constant (1.38 × 10^{-23} J/K), the number of serially-connected cells in a PV module, the diode ideality factor and the PV module temperature [°C], respectively. The *I-V* characteristic of a PV module is calculated by using four parameters: I_{ph} is the photoelectric current [A], I_o is the reverse saturation current [A], R_s the series resistance [Ω], and R_{sh} is the shunt resistance [Ω]. These parameters are defined as functions of irradiance and PV module temperature. They are derived from the experimental data based on a genetic algorithm. In addiation, the value of n is set as 1.5 in this paper.

Fig. 2 shows the comparison between the measurement and simulation values of *I-V* and *P-V* characteristics of the polycrystalline silicon PV module used in this study (measurement condition: 0.658 kW/m², 18.0 °C). The specifications of it are I_{sc} = 8.21 A, open circuit voltage (V_{oc}) = 25.61 V, and the maximum power (P_m) = 153.0 W, respectively. From Fig. 2, the measurement and simulation values agree well. Moreover, a difference of the maximum power point is -0.90 % (-0.87 W). The above results enable us to confirm sufficient reproducibility of *I-V* and *P-V* characteristics to examine the reverse voltage occurrence on a PV system.

3.2 Power Conditioning System Modelling

Fig. 3 shows the operation methods of three PCSs from companies A, B, and C. Two major differences are operation interval and voltage amplitude. Based on the experimental data, the operations of PCSs are modelled using C programming language.

To validate the modelling of operation methods of PCSs, the measurement and simulation value of the operating voltage were compared. Fig. 4 shows the PV array composition for the validation. It is composed of three PV strings in parallel and a PV string consists of nine serially-connected PV modules with three bypass diodes and three PV clusters in each of them (27 PV clusters with a bypass diode on each). Figs. 5 and 6 show the comparison of the simulation and measurement operation voltage value of three PCSs and each meteorological condition. From Fig. 5, the measurement and simulation values agree well. In addition, root mean square error (RMSE) and mean bias error (MBE) are used for confirming the reproducibility of MPPT operations of each PCS. As the value of RMSE and MBE become closer to zero, it means the estimated value is accurate. RMSE and MBE are calculated from the following equations:

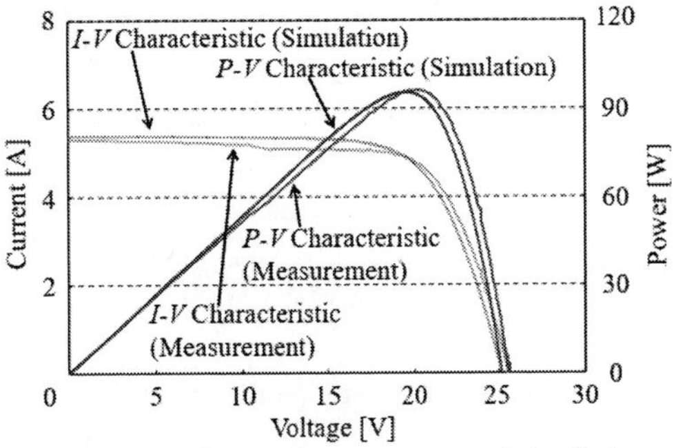

Figure 2: Comparison of measurement and simulation value of *I-V* and *P-V* characteristics of polycrystalline silicon PV module in this study.

PCS		A	B	C
Basic operation	Amplitude [V]	4.0	1.6	2.5
	Cycle [s]	8.0	4.0	0.2
Operation interval [s]		1.0	1.0	0.1
Basic operation		1 s, 4 V	1.6 V, 1 s	0.1 s, 2.5 V
Increasing operation		2 V, 1 s	0.8 V, 3.2 V, 2.4 V, 1 s	1 V, 3 V, 2.5 V, 0.1 s
Decreasing operation		2 V, 1 s	2.4 V, 0.8 V, 1 s, 3.2 V	3 V, 0.1 s, 1 V

Figure 3: Specification and operation method of PCSs used in examination.

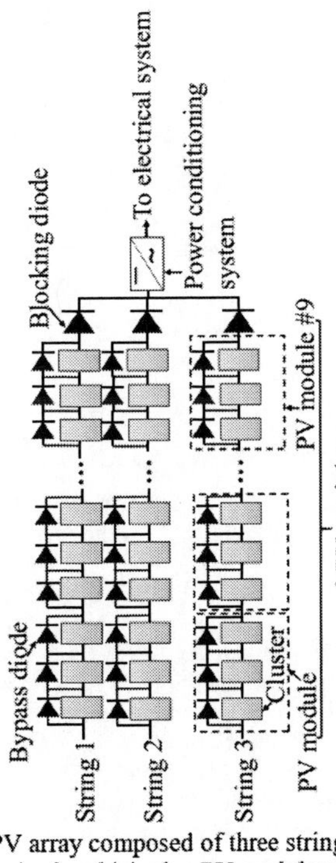

Figure 4: PV array composed of three strings with blocking diodes in which nine PV modules with bypass diodes are serially connected.

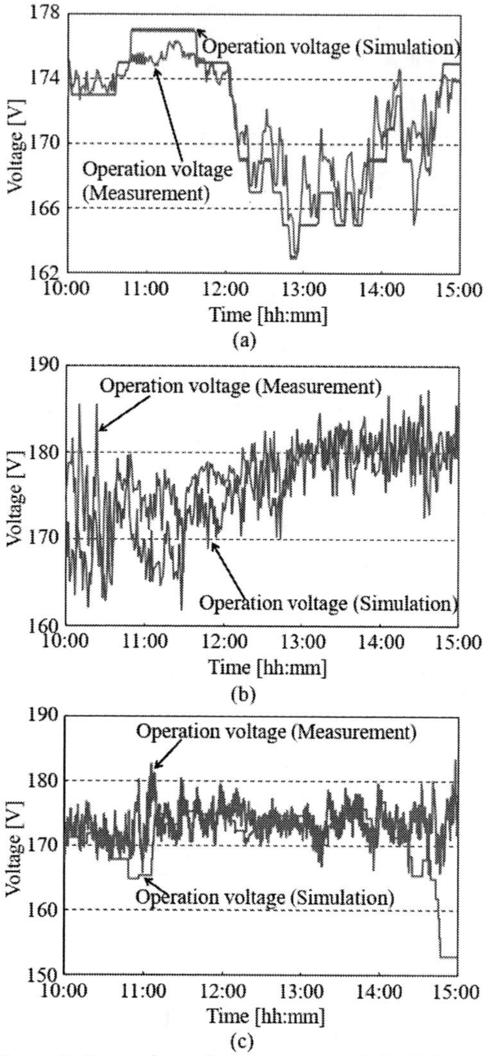

Figure 5: Comparison of measurement and simulation value of operation voltage of PV array: (a) PCS: A, (b) PCS: B, and (c) PCS: C.

$$\mathrm{RMSE} = \sqrt{\frac{1}{n}\sum_{i=1}^{n}\left(x_i - x_{\mathrm{true},i}\right)^2}\,, \tag{2}$$

$$\mathrm{MBE} = \frac{1}{n}\sum_{i=1}^{n}\left(x_i - x_{\mathrm{true},i}\right). \tag{3}$$

Here, n is the number of samples (n = 300), x_i is the estimation value [V], and $x_{\mathrm{true},i}$ is the measurement value [V], respectively.

Table I shows RMSE and MBE of three PCSs used in this study. From Table I, it can be confirmed that PCS A has high reproducibility. As for PCSs B and C, they do not have as high reproducibility as PCS A; however they have enough accuracy for estimation on reverse voltage. Moreover, the simulation value of operation voltage tends to be estimated low in all PCSs. Therefore, the above results enable us to confirm sufficient reproducibility of PCS operation to examine the reverse voltage occurrence on a PV system.

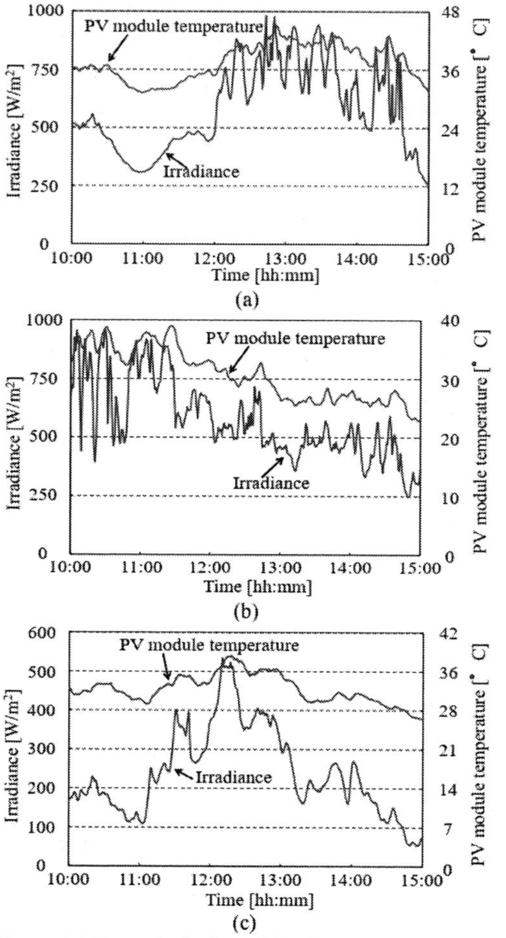

Figure 6: Meteorological condition for comparison on simulation and measurement value of operation voltage of PV array: (a) PCS: A, (b) PCS: B, and (c) PCS: C.

Table I: RMSE and MBE of PCSs (A, B, and C) used in this examination

Power condition system	A	B	C
RMSE [V]	1.73	5.09	6.05
MBE [V]	-0.59	-3.00	-2.45

Table II: Possibility of reverse voltage when changing number of damaged bypass circuit with PCS A, B, and C

PCS		A		B		C	
Number of damaged bypass circuits		1	3	1	3	1	3
k_i	0.50	100	99.9	94.2	91.9	100	100
	0.66	20.4	17.0	41.4	39.6	83.3	82.3
	0.75	0.3	0.3	12.9	10.4	45.1	44.6
	0.90	0	0	2.1	2.0	4.6	4.5
	0.95	0	0	0.2	0.2	0	0

4 DISCUSSION ON VARIATION OF REVERSE VOLTAGE OCCURRENCE

Examination on reverse voltage occurrence was

conducted using PV array shown in Fig. 4. To examine the reverse voltage occurrence, the operation voltage of PV module #9 was observed. Also, the meteorological data from 10:00 A.M. to 4:00 P.M. in June 8, 2013 taken at rooftop of Funabashi Campus, Nihon University (Funabashi, Chiba) was used (see Fig. 7). This day had a high average of irradiance and also the irradiance fluctuated frequently during the daytime. In other words, it is the day when reverse voltage is likely to occur. In this study, the ratio of I_{sc} of a PV module for replacement to an originally-installed PV module is defined as k_i. To simulate a PV module for replacement, k_i of PV module #9 was set as 0.50, 0.66, 0.75, 0.90, and 0.95. In addition, the number of damaged bypass circuits on the replacement PV module was set to one or three.

Table II shows the possibility of reverse voltage when changing number of damaged bypass circuits and k_i. From Table II, it can be confirmed that reverse voltage occurs more when only one bypass diode is damaged where k_i is between 0.66-0.90. However, the maximum difference of reverse voltage occurrence possibility is about 3 %. Thus, it can be determined that the number of bypass diodes does not affect the possibility of reverse voltage occurrence significantly. This result is obtained because k_i of the PV clusters without bypass diodes are as low as ones with bypass diodes. Hence, the reverse voltage occurs evenly at PV clusters with low k_i regardless of bypass diode existence.

In addition, a difference of the possibility of reverse voltage by PCS can also be confirmed from Table II. The reason is the magnitude of voltage fluctuation. Compared to PCS A, PCS B has a bigger value of voltage fluctuation when they are in increasing or decreasing operation. As a result, reverse voltage more likely to occur in PCS B than PCS A on the occasion where operation voltage changes by maximum power point tracking due to sudden change in irradiance. Therefore, the operation points of the PCSs are different even though they operates under the same meteorological condition.

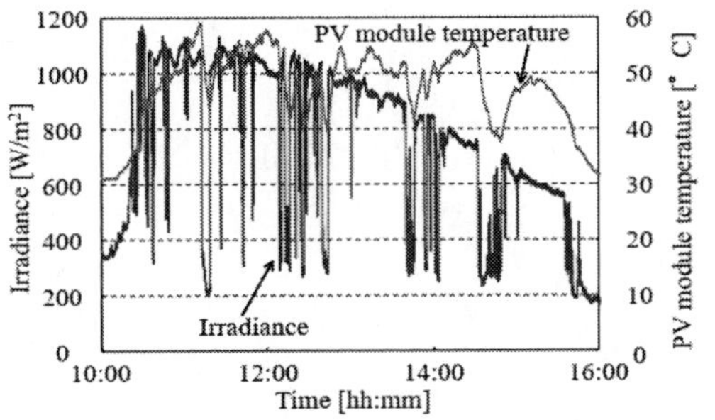

Figure 7: Meteorological condition for examination on reverse voltage occurrence (June 8, 2013, 10:00A.M. to 4:00 P.M.).

5 DISCUSSION ON SAFETY RANGE SPECIFICATION OF PV MODULE

Examination on determining the safety range of PV modules with one or three damaged bypass diode was conducted using the PV array shown in Fig. 4. The safety range refers to the area where reverse voltage does not occur. In other words, k_i of the replacement PV module in which reverse voltage does not occur can be determined. To examine the reverse voltage occurrence, the operation voltage of PV module #9 was observed. Also, the meteorological data shown in Fig. 7 was used. To simulate a PV module for replacement, k_i of PV module #9 was varied from 0.75 to 1.00 in steps of 0.01.

Fig. 8 shows the safety range of reverse voltage occurrence. From Fig. 8, the safety range of reverse voltage occurrence does not change whether the number of damaged bypass circuits is one or three. Moreover, if k_i is larger than 0.97, reverse voltage will not occur in all PCSs. As a result, it can be confirmed that I_{sc} over 97 % of the original PV module is required as the specification for replacement PV modules which prevents reverse voltage under the condition of having damaged bypass circuits on this system.

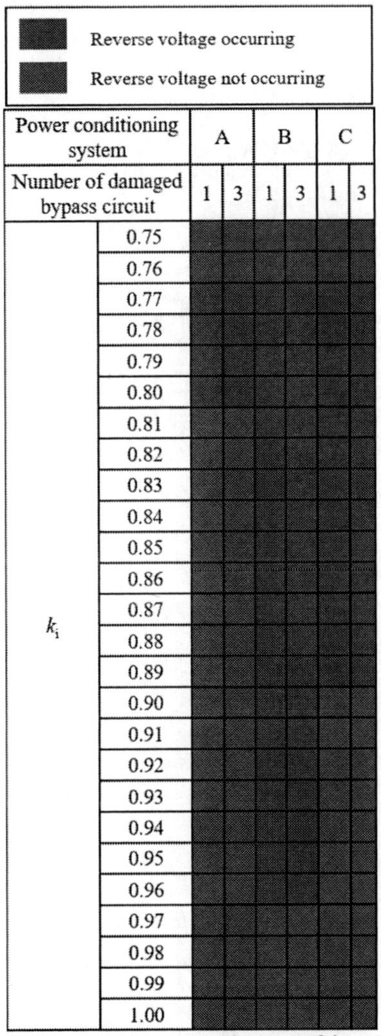

Figure 8: Reverse voltage occurring area of three PCSs when varying k_i from 0.75 to 1.00 when the number of damaged bypass circuit is one or three.

6 CONCLUSIONS

Examinations on the variation of reverse voltage

occurrence by the number of bypass circuits in a PV module were considered by simulation. The results of the examinations clearly demonstrate that the number of damaged bypass circuits does not affect the occurrence possibility of reverse voltage significantly. Moreover, the occurrence possibility of reverse voltage differs by the operation method of PCSs and I_{sc} of the PV module. In addition, the safety range of the specification for replacement PV modules is over 97 % of I_{sc} of the original PV module on this PV system.

For the future examination, the safety range of the specifications of PV modules for the replacement is needed to be obtained in different PV array constitution. This is because the possibility of reverse voltage differs by the number of serial connections in a PV string. Hence, it is necessary to evaluate the safety range of the replacement PV module on various PV array constitutions.

References

[1] H. Yamaya, T. Ohigashi, H. Matsukawa, I. Kaizuka, and O. Ikki, Proceedings 43rd Photovoltaic Specialists Conference, (2016), pp.3304-3308.
[2] H.Haeberlin, Proceedings 23rd European Photovoltaic Solar Energy Conference, (2008), pp.2934-2947.
[3] S. Oke, H. Sakai, I. Nanno, T. Hamada, M. Fujii, and N. Ishikura, Proceedings JSES/JWEA Joint Conference 2016, (2016), pp.285-288.
[4] K. Horie, Y. Takahashi, K. Fujiwara, Y. Ishihara, K. Takahashi, S. Nishikawa, and H. Kato, IEEJ Transactions on Power and Energy, vol. 134, no. 8, (2014), pp. 673-681 (in Japanese).

POTENTIAL INDUCED DEGRADATION OCCURRENCE IN PHOTOVOLTAIC POWER PLANT

Josef Hylský, Dávid Strachala, Jiří Vaněk, Jan Mucha
Brno University of Technology, Faculty of Electrical Engineering and Communication Technologies,
Department of Electrotechnology
Technicka 10, 616 00 Brno, Czech Republic

ABSTRACT: This paper discusses a photovoltaic power plant affected by Potentially Induced Degradation (PID). In the work is presented the initial analysis of PID detection through the thermal image camera and subsequent examination of the extent of degradation on individual PV modules placed in the negative part of string WR14. As an PID indicator was chosen maximum power point parameter and open circuit voltage parameter. From the measured values it was found that the extent of PID degradation is not directly proportional to the negative voltage potential or on the PV module location in the string.
Keywords: Photovoltaic power plant, photovoltaics modules, Potential Induced Degradation (PID)

1 INTRODUCTION

In recent years, PID (Potential Induced Degradation) has become more known for owners of photovoltaic power plants. This is due to more frequent articles in specialized journals [1], websites and because of the real cases of this degradation in photovoltaic power plants in the Czech Republic. PID was a few years ago taken as an uncertain debatable issue. Distributors, service companies and owners did not take it seriously. Today it is different. PID incidence is increasing and owners of PV plants experience a decrease in performance associated with this degradation.

The incidence of PID degradation is gradual and lasts for several years before it becomes fully observable - in extreme cases from 2 to 5 years. This degradation is accelerated by environmental conditions [2], such as humidity and heat. Therefore, the incubation time cannot be accurately determined and depends on the location and the conditions in which the PV power plant operates.

The reason for the degradation process begins with the design of PV plants with increasing performances, higher number of PV modules in series and higher potential between cells and grounded frame for security reasons. When using a floating potential (neither the positive or the negative pole of the string is grounded) a negative voltage potential arises at the negative pole, which causes the leakage currents from the PV cells passing through the sandwich structure of the PV module into the grounded frame. Conversely, the resulting negative electric field between the aluminum frame and the PV cells accelerates the movement of the positive sodium ions contained in the cover glass of the PV module that travel across the sandwich structure to the PV cells.

If the PV module is created of PV cells susceptible to PID degradation (they contain stacking fault in their active part) sodium ions reduce parallel resistence Rsh and PV module performance decreases. This would give a very brief description of the factors influencing PID degradation [3,4].

We can look at three levels on the issue of PID protection or healing. At the level of the whole photovoltaic systém, at the level of the photovoltaic module and at the level of the photovoltaic cell, see Figure 1.

Figure 1: a) Photovoltaic system, b) Photovoltaic module, c) Photovoltaic cell.

To understand the occurrence of PID degradation, the level of the photovoltaic cell is described, where is the physical nature of the power reduction. If the sodium ions get through the sandwich structure of the PV module up to the surface of the cell, they drift by an electric field through an anti-reflective layer of silicon nitride SiN_x, where they accumulate in silicon oxide SiO_x. The presence of stacking faults through the PN junction allows the sodium ion further penetrate into the PV cell resulting in short-circuiting the PN junction. As a result of the above-described process, the value of the parallel resistence decreases and consequently the power of the whole PV cell is lower.

Protection solution at this level is economically and technologically most acceptable because it does not interfere with the already functional PV system. It could be said that by modifying the technological process it is possible to modify the structure of the PV cell to be PID resistive.

PID protection and treatment at the levels of the PV module and the PV system is economically disadvantageous due to higher intervention and changes in their structure. At the level of the PV module is needed to find an alternative to Ethylene vinyl acetate film (EVA film) in a PV module sandwich structure. Alternative material must protect PV cells from external influences, ensure light transmission to cells and its resistance must be high enough to prevent the drifting of sodium ions to PV cells. Material with the above-described properties is, for example, Acit (partially neutralized polyethylene) [5]. This material would meet the requirements, on the other hand its price is so high that the PV modules would be very expensive and it would reduce their competitiveness.

Nowadays the biggest problems are the PV modules installed on PV plants, where the change of EVA foil or PID resistive PV cells is not possible. One of the solutions is swapping of PV modules from positive to negative string section and vice versa, which prevents an increase of PID degradation. Another solution is in grounding a negative pole using a galvanic isolated inverter. This eliminates the negative voltage potential. The third option to remove PID degradation is to design a PV plants in such a way that no high voltage potential occurs, it means small numbers of PV modules connected in series. In this protection, the outcome is not certain because of the undefined magnitude of the negative voltage potential at which PID degradation may occur. One of the most used options at the level of the PV system is the use of so-called PID doctors, which at a time when the PV power plant does not generate electricity (at night) will change the effect of the electric field and thus regenerate the affected PV modules. At these levels all listed options are costly and reduce the gain from the distributed power of the PV plant.

2 ANALYSIS OF AFFECTED PHOTOVOLTAIC POWER PLANT

A 2 MW PV power plant affected by PID degradation is equipped with Evergreen photovoltaic modules with output power ranging from 205 W to 210 W.

Table I: Parameters of module ES - A - 2O5-fa3 (Measured at STC).

P_{MPP} (W)	E_{ff} (%)	U_{MAX} (V)	I_{MAX} (A)	U_{OC} (V)	I_{SC} (A)
205	13.1	18.2	11.2	22.7	11.9

Table II: Parameters of module ES - A - 210-fa3 (Measured at STC).

P_{MPP} (W)	E_{ff} (%)	U_{MAX} (V)	I_{MAX} (A)	U_{OC} (V)	I_{SC} (A)
210	13.4	18.3	11.48	22.8	12.11

The number of PV modules in the individual strings is 32. Pv modules are connected in series so that the sum of their voltage does not exceed 600 V (maximum PV plant string voltage in the Czech Republic). At the output of each string is inverter that converts DC power into AC power. Since the inverter does not have a galvanic isolation (eg a transformer), it is not possible to apply the grounding of the negative pole of the string. For this reason, the grounding of the PV plant is designed as floating. This connection divides the string into a negative and a positive part. The most affected PV modules will occur at the most insensitive pole of string, which depends on the effect of high negative voltage potential according to the principle of PID. With gradual transition to the positive part, PID degradation should disappear.

For an analysis was chosen a WR14 string in which the highest power drop was measured by 0.5 kW. Initial diagnosis and confirmation of PID degradation was performed using a thermal imaging camera, see Figure 2. The images compare PV modules from the negative pole of the string with the PV module from the positive pole of the string. The images were taken from the back of the PV module to minimize reflection of light from the module glass.

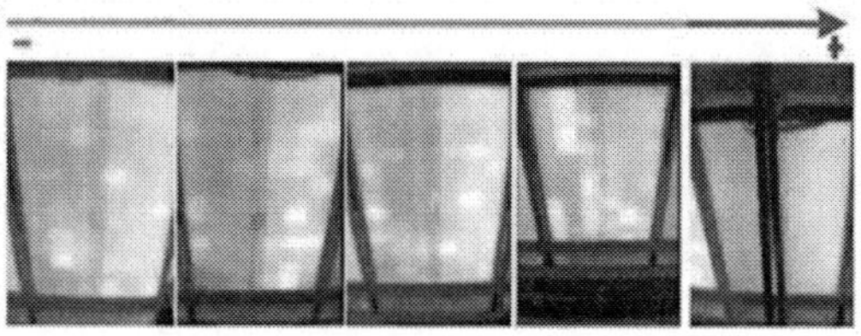

Figure 2: Pictures of selected PV modules using the thermal imaging camera take from the back side of the WR14 string.

As it can be seen from the pictures, degraded cells are in PV modules in the negative pole of the string. On the other hand, the PV module located in the positive part of the string is without defective cells, which confirms the occurrence of PID degradation.

To confirm the occurrence of PID, the PV modules were analyzed by further diagnostic measurements like VA characteristic and an electroluminescence to determine the extent of the degradation. Fig. 3 compares the maximum power of the PV modules in the negative pole with the reference value of the module from the positive pole of the string.

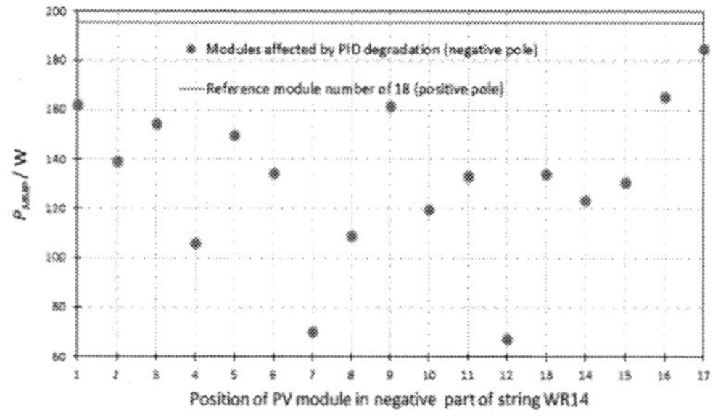

Figure 3: Comparison of dependent of maximum power on position of PV modules in negative part of string and of reference PV module from positive pole of string (position number 1 = The PV module with the largest negative voltage potential).

The maximum power parameter P_{MPP} has been selected as an indicator determining the degree of PID degradation. From the above-described text, the PID degradation decreases with the gradual reduction of the negative potential of the voltage, but this is not apparent from Fig. 3, where the P_{MPP} maximum power values do not increase in the negative pole. For example, positions 7 and 12 show a rapid decrease in maximum power in comparison with negative PV modules. This was verified by the electroluminescence measurement, where were measured the PV modules from the 3rd, 12th and 17th positions of the negative string and the reference module the positive part of the string, see Figure 4.

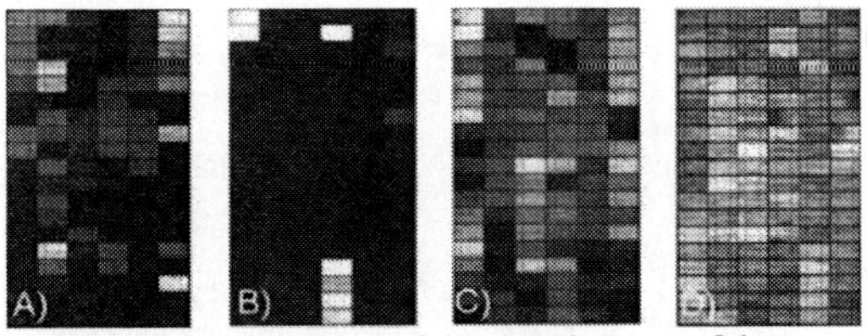

Figure 4: A) 3rd position of the negative part of the string, B) 12th position of the negativ part of string, C) 17th position of the negative part of the string and D) a reference sample of the positive part of the string.

Electroluminescence measurement with high accuracy reveal PID degraded PV cells and in this confirms that PV modules in negative voltage potential are more degraded. As shown in Fig. 4, PV module B) is much more degraded than PV module A), which is in a position with a higher negative voltage potential. No faulty PV cells were detected in PV module D, indicating that the positive voltage potential does not have influence on the PV modules.

The PID degradation extent was expressed as a percentage drop of the maximum P_{MPP} on the negative voltage potential of the individual PV modules, see Figure 5.

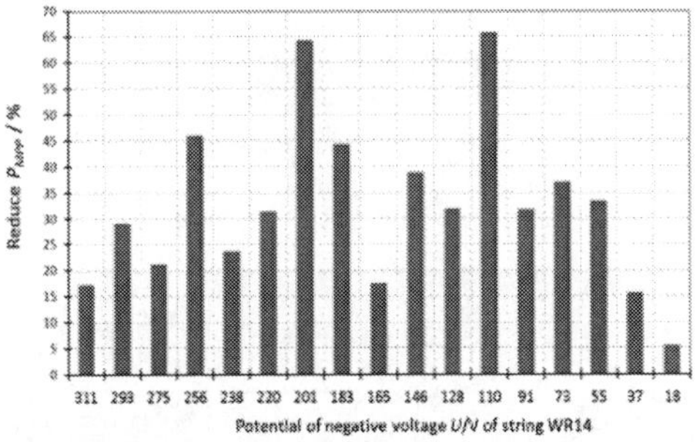

Figure 5: PV modules percentage drop of P_{MPP} output relative to the negative potential.

Figure 5 shows the largest reduction of the maximum power at positions 7 and 12. The PMPP drop is around around 65%.

3 CONCLUSION

This paper describes the diagnosis of a 2 MW PV plant that has been suspected of reducing power due to PID degradation. One of the most affected strings of the PV power plant was analyzed. Its maximum power output dropped by 0.5 KW. The PV modules were first measured by a thermal imaging camera. Thermographic images confirmed the occurrence of PID degradation. In order to determine the extent of PID degradation, all PV modules from the negative part of string were dismantled and the parameters of the affected PV modules (maximum power of the PV module - PMPP) were measured.

Measured values were compared with the parameters of the reference PV module uninstalled from the positive part of the string. Based on the diagnosis, it was found that the extent of PID degradation is not directly dependending on the magnitude of the negative voltage potential. With decreasing negative voltage potential, PID degradation on individual PV modules is not reduced. This can be attributed to the fact that PV cells themselves and their susceptibility play a major role in the extent of PID degradation in individual PV modules. How large the negative voltage potential is needed to produce PID degradation is the reason for further research in this area.

3.1 References

[1]W. Luo et al., Energy Environ. Sci, 10, 43-68 (2017)
[2] S. Pingel et al., in 2010 35th IEEE Photovoltaic Specialists Conference,, p. 002817-002822, IEEE (2010)
[3] V. Naumann, C. Hagendorf, S. Grosser, M. Werner, and J. Bagdahn, Energy Procedia, 27, 1-6 (2012)
[4] V. Naumann et al., Energy Procedia, 33, 76-83 (2013)
[5]J. Kapur, K. M. Stika, C. S. Westphal, J. L. Norwood, and B. Hamzavytehrany, IEEE Journal of Photovoltaics, 5, 219-223 (2015)

3.2 Acknowledgement

This research work has been carried out in the Centre for Research and Utilization of Renewable Energy (CVVOZE). Authors gratefully acknowledge financial support from the Ministry of Education, Youth and Sports of the Czech Republic under NPU I programme (project No. LO1210).

A WEB APPLICATION FOR YIELD OPTIMIZATION OF PHOTOVOLTAIC SYSTEMS

Henrik te Heesen, Martin Rumpler
Trier University of Applied Sciences, Environmental Campus Birkenfeld, Campusallee, 55768 Neubrücke (Nahe), Germany, h.teheesen@umwelt-campus.de

ABSTRACT: There are about 1.5 million photovoltaic systems in Germany. Most private owners cannot afford the commission of the operation and maintenance operate most of their systems on their own. Furthermore, monitoring systems are too complex or cannot be afforded by many system owners. This is one reason that about 80 percent of the photovoltaic systems in Germany do not generate the yield that is technically possible. Therefore, we present an innovative web-application that allows system owners and operators to monitor the actual status and profitability of their photovoltaic systems. Also, they can get in contact with a specialized company, which can repair or maintain the system in case of problems such as too small yield, damages or malfunctions. Our goal is the development of an application that is easy to use and designed for desktop computers as well as for mobile devices.

Keywords: Defects, Monitoring, Performance, Solar Home System

1 INTRODUCTION

In Germany, there are more than 1.5 million photovoltaic systems [1]. Most of these installations are operated by private owners who cannot afford to commission a maintenance company monitor the system regularly. Monitoring systems are usually still too costly or technically complex for many system owners. For this reason, about 80 percent of the photovoltaic systems in Germany do not generate the revenues that would be technically possible [2]. Our solution is an innovative web application that offers system owners the possibility to monitor the current state and the economic viability of their photovoltaic system and, in the event of problems, e. g. yield losses, technical faults or damage, to commission a specialist company to carry out maintenance or repair work. The main focus in the development of the web application is that the app should be as easy as possible to use and should be designed for use on a PC and mobile devices (smartphones and tablets).

2 SCOPE OF FUNCTIONS

The UCB PVapp can only be used as a registered user. Only an e-mail address a self-selected password must be given as a unique identifier and. After logging on, the following basic functions are available to the user:

2.1 ADD, CHANGE AND DELETE PV SYSTEMS

One or more PV systems can be monitored. When a system is installed, the following data must be provided: Nominal power of the PV system in kWp, orientation and inclination of the modules, the date of installation, and the postal code of the site.

2.2 ADD, CHANGE OR DELETE METER READINGS

The user can enter, change and delete meter readings of the energy yield for each PV installation. The date, the time, and the meter reading are recorded via an input mask. The user is reminded once a month to enter this data, because otherwise it is not possible to evaluate and monitor his system (see Figure 1).

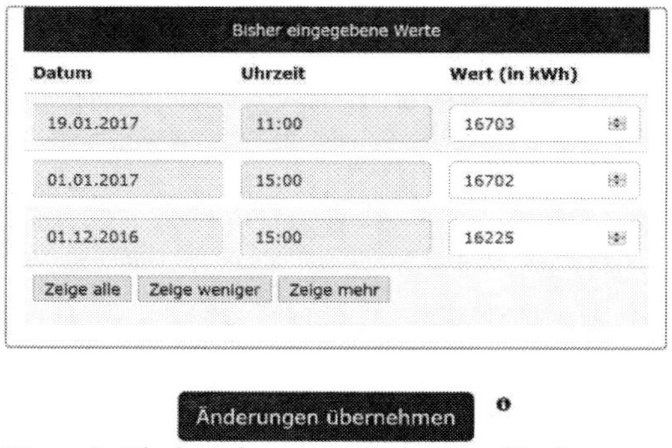

Figure 1: Display of meter readings entered by the user.

2.3 VISUALIZATION OF YIELD DATA

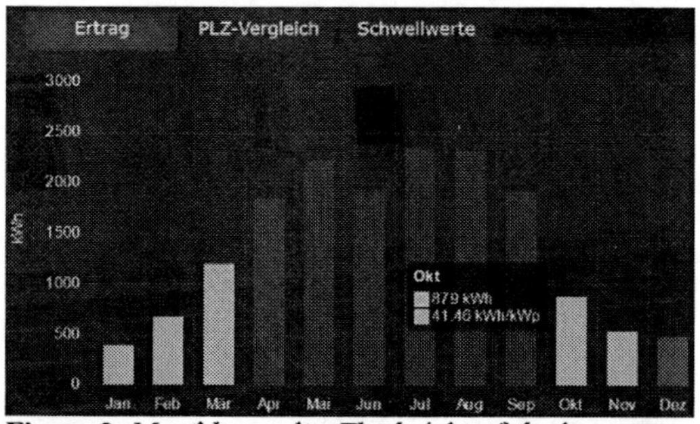

Figure 2: Monthly results. The height of the bars represents the monthly electricity yield of the PV system; the color reflects the assessment of the yield in comparison to reference systems.

The user can display his yield data for a specific calendar year. A rated-to-actual yield comparison is provided for each month based on the analysis of the yield data (see Figure 2). The bar color indicates how well the system performs compared to other systems in the same two-digit postal code region. Green means that the system belongs to the best 25 % of the systems, yellow means that the system is worse than the best 25 % but better than 50% of the remaining systems and red means that it is one of the worst 50 % of the systems.

In the future, our web application will provide additional useful functions. As a further option, a comparison with the PV system of another user will be possible (if the other user has agreed to publish his yield data). In this comparison, the performance of the two photovoltaic systems is normalized in order to obtain comparable data. There will also be a checklist to help system owners to identify possible causes if the PV system does not perform well. In case of technical problems, qualified electric companies are then proposed in the respective postcode region.

3 ANALYSIS OF YIELD DATA

The starting point of the analysis is the measurement data of PV systems, which have an internet-based remote monitoring system and whose data are publicly available. The raw data are initially anonymized so that no conclusions can be drawn about the individual system or the owner. Data processing and cleansing are then carried out [3]. The central component is a data logger which records and stores the measured data of the individual components and transmits them to a central server over the internet. In addition, configuration, location and orientation of each PV system and the inclination of the solar modules are included in the analysis. The measurement data are typically available in 5 or 15 minute intervals for each PV system. Daily data is used for the evaluation and is aggregated to monthly values.

The most important key figure for assessing a single PV system is its energy yield. The energy yield is the amount of energy produced in a given period and is measured via the feed-in meters of a system. Since the energy yield depends on many factors of a PV system (nominal power, location, orientation, inclination, components, configuration), a direct comparison of the energy yields of different systems is not possible. The specific yield can be used to compare PV systems in a region with a similar configuration (orientation, inclination, components).

The yield data for all systems located in a specific postal code region are processed. PV systems whose yields are outside certain limits (i.e., providing too low or implausible values) are neglected in the further data analysis. All yield data refer to PV installations with modules that are oriented towards the south and have an inclination of about 30 degrees.

The key figures relevant to the analysis for analyzing the statistical results are the median and the quartiles. The distance between the 1st and 3rd quartile is called the interquartile range (IQR). Since the algorithm is based solely on statistical methods, the results of the algorithm are significant for a sufficiently large number of PV systems in a region [3]. The safety probability is already 97% for a tolerated error of 5% from 50 evaluable systems in a postcode region.

The user can display the yield thresholds in the yield overview. The red, yellow and green bars indicate the monthly yields of the user's PV system (see Figure 3). The half-transparent areas in the background visualize the reference data. For this purpose, the median as well as the first and third quartile from the regional comparison are converted to the system performance of the respective user. The dark red area shows the median. Systems which have produced a smaller yield in the respective month are colored in bright red. The yellow area ends at the 3rd quartile. This reference value is additionally indicated by a black bar. Systems that generate larger yields are among the best 25% in the respective postal code region and are colored in light green.

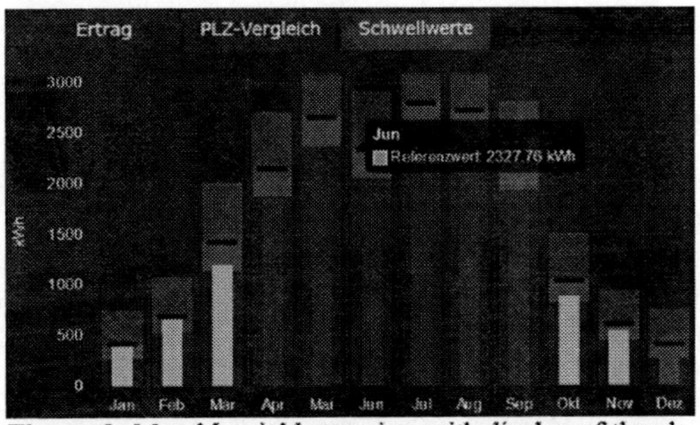

Figure 3: Monthly yield overview with display of thresholds. The display for advanced users provides a detailed comparison of the PV's own PV yield with expected reference yields.

4 REALIZATION OF UCB PVAPP

The UCB PVapp was realized as a single page web application with standard web technologies. The visualization of the UCB PVapp automatically adjusts itself to the size and resolution of the respective device (PC, Tablet, Smartphone). To realize this, we decided to use the CSS Framework Bootstrap. Bootstrap contains HTML and CSS-based design templates for typography, forms, buttons, tables, grid systems, navigation, and other interface elements as well as additional JavaScript extensions. We also use the JavaScript framework jQuery. This is a free JavaScript library, which provides functions for DOM navigation and manipulation. The Chart.js framework has been chosen for the visualization of the yield data, because it is one of the few available visualization frameworks that makes combinable diagrams possible (bar graph and area diagram in combination) and also offers a lot of design possibilities. In addition, the framework has a very extensive documentation. The server-side backend is implemented in PHP using the Symfony framework.

5 REFERENCES

[1] Bundesnetzagentur. Photovoltaic installations data submissions and EEG-supported feed-in tariffs. Available online: https://www.bundesnetzagentur.de

[2] H. te Heesen, V. Herbort, M. Rumpler. Studie zum Ertrag von Photovoltaikanlagen in Deutschland 2016. Hochschule Trier. March 2017.

[3] H. te Heesen, V. Herbort. Development of an Algorithm to Analyze the Yield of Photovoltaic Systems. Renewable Energy. Volume 87, Part 2, March 2016, 2016, Pages 1016–1022.

LABORATORY INFRASTRUCTURE FOR RESEARCH AND CAPACITY BUILDING ON ISOLATED AND GRID-CONNECTED SMART MICRO-GRIDS

Alex R. A. Manito[1]
Kauê J. F. N. C. de Souza[1]
André R. Mocelin[1]
Teddy A. F. Melendez[1]
João Tavares Pinho[2]
Roberto Zilles[1]

1 – Laboratório de Sistemas Fotovoltaicos, Universidade de São Paulo
2 – Grupo de Estudos e Desenvolvimento de Alternativas Energéticas, Universidade Federal do Pará

ABSTRACT: Despite the extensive discussion on the design of isolated systems using intermittent renewable sources, the integration of autonomous Rewable Energy Systems (RES) for joint operation in a smart micro-grid or connected to a utility main grid is a relatively new subject. The integration allows increased flexibility and modularity, which are important characteristics to improve the overall robustness of the system and tackle unpredicted adversities. This paper presents the implementation of the facility dedicated to the research on micro-grids at the Laboratório de Sistemas Fotovoltaicos (LSF) of the Universidade de São Paulo, in Brazil. The facility allows the simulation of several topologies and even energy tariff conditions, providing useful insights in the operation and control of an area with several interconnected micro-grids and connected to the main distribution grid. Moreover, the infrastructure is expected to contribute to capacity building and the development of national standards for connection and operation.

key-words: Autonomous Renewable Energy Generation Systems, mini-grids, smart grids, energy management, distributed generation, control strategies.

1 INTRODUCTION

The convenience of centralized power generation systems was greatly related to technological constraints, the wide use of site concentrated primary resources, economies of scale, and the way the energy markets were organized. In that topology, increases in demand were usually met only by increases in generation capacity and reinforcement of the transmission and distribution grids. However, this approach has difficulties in the matching between the demand and the generation, often requiring the existence of spinning reserves, generation and distribution capacity that are underused most of the time. Moreover, in recent years, issues like the depletion of the energy potential from certain types of generation, the trends of price increase of fossil fuels, the need to address environmental problems and climate changes have emerged. These issues need to be addressed while meeting the ever-increasing demand from all sectors and even the emergence of new demand niches such as electric vehicles.

In this context, renewable distributed generation near the consumers, possibility of storage and improved energy management in the form of mini and micro-grids could alleviate the problems of centralized generation by inserting new solutions and operating models that achieve adequate levels of service quality with lower environmental impacts. A mini/micro-grid presents itself to the main grid as a controlled unit that meets the local requirements of reliability and security, being capable of flexible operation, depending on the needs of the consumer/producer (prosumer) and the main grid. Among the benefits of this topology are: the reduction of feeder loading in regions with high load density, peak-shaving/valley-filling, loss reduction, postponement of investments for the expansion of the distribution system, and reduced environmental impacts. Although the first tests with micro-grids go back to the 1980s, only recently micro-grids in grid-connected mode of operation seem to be crossing the academic and laboratory borders and entering the market.

The current way to operate de grid cannot be fully transposed to the micro-grid topology, and adaptations need to be made so that the potential benefits can be harnessed. In this context, many studies have been carried out in the field of micro-grids, which address different relevant aspects, not only restricted to the technical levels, but also those related to regulation and market. For instance, [1] presents a work on the opportunities and challenges concerning the micro-grid concept. References [2] and [3] consist of a two-part review work on the important aspects of micro-grids in active network management; [4] also presents a review article on the implications to protection and stability due to the increasing level of distributed generation. [5] and [6] present control strategies for micro-grid and multi-micro-grid concepts. [7] presents a work regarding energy planning and the micro-grid concept using smart-grid mechanisms. [8], [9], and [10] present several examples of micro-grids being installed around the world.

Large-scale application of such topologies still has to overcome difficulties and barriers, so that all these benefits can be taken advantage of. The inertia and skepticism when adopting new technologies are not totally unjustified. The new equipment and the new operating viewpoints lack proof of their durability and effectiveness, both under normal and abnormal conditions, whereas well-established technologies already underwent a long learning process, which grant them high economies of scale and well-established human resources for operation and maintenance. Moreover, economic and regulatory barriers also need to be overcome, since they influence several stakeholders from different areas of the electricity market.

This paper presents the implementation of the facility dedicated to the research on micro-grids at the Laboratório de Sistemas Fotovoltaicos (LSF) of the Universidade de São Paulo, in Brazil. The facility allows the simulation of several topologies, and even energy tariff conditions, providing useful insights in the operation and control of an area with several micro-grids, acting as prosumers connected to the main distribution grid. Moreover, the infrastructure is expected to contribute to capacity building and the development of national standards for connection and operation.

2 THE INFRAESTRUCTURE OF LSF

The challenges associated with micro-grids are not limited to those of technical nature, as stated before. Some studies address the fact that the regulatory barriers are harder to be overcome [11] due to the consequent changes in the grid's mode of operation that micro-grids would produce. This kind of transitions face some inertia, which diminishes as the stakeholders gain trust in the new philosophy of operation and the equipment utilized.

Due to all uncertainties and unknowns in the operation of the new equipment, it is clear that the market for micro-grids cannot be created overnight. The regulations for the connection of distributed resources are still undergoing a learning phase in many countries, what demands extensive tests, research, pilot projects and gradual adoption, so that drawbacks that may not be anticipated can be handled without compromising the quality of the service.

The main objective of this facility is to emulate a smart grid with several producers, consumers, and prosumers connected to it, and evaluate the outcomes of several scenarios, including the emulation of a small energy market. Its Micro-Grid Central Controller (MGCC), which acts as the distribution system operator, must have the capability of exchanging information, and control each client connected to the Point of Common Coupling (PCC). The clients, on the other hand, are capable of granting ancillary services to the grid, such as demand shifting, and storage, which may contribute to flatten the demand curve, and also to power quality improvement.

The facility was designed to enable a high level of flexibility and scalability, which permits to reproduce many real cases in the operation of the whole system, and considers the benefits and drawbacks of many topologies, including different levels of non-dispatchable renewable energy penetration and storage capacity. Aside from the different topologies that the facility allows, it also permits the emulation of different scenarios of information availability, which greatly influences the risks involved with the operation of the micro-grid, permitting an evaluation of the amount and type of information that would be vital for the reliable operation of the system. Fig. 1 presents a schematic of the topology implemented at LSF.

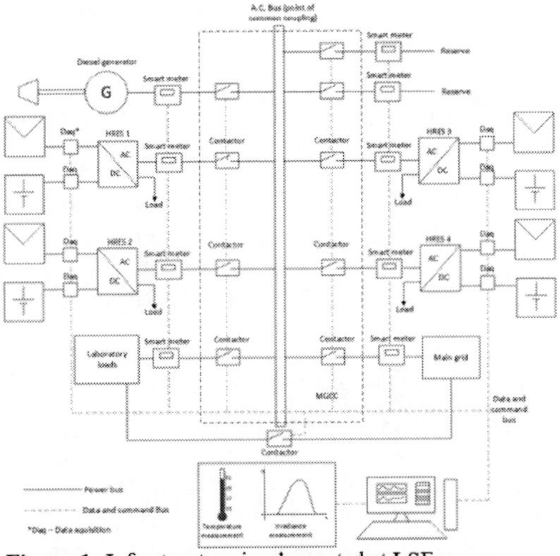

Figure 1: Infrastructure implemented at LSF.

The infrastructure of LSF is composed of four Hybrid Renewable Energy Systems (HRES), each with its respective loads, storage, distributed generation, and grid forming inverter. The four HRESs are connected to a PCC, where a diesel generator and the main grid are also connected. Moreover, two standby circuits were introduced to allow the connection of temporary components like loads or distributed generation.

2.1 Equipment installed in the laboratory

The facility assembled at LSF corresponds to four HRESs, as stated before, connected to a PCC controlled by a MGCC. HRES1 is composed of three SMA Sunny Island inverters of 5 kVA each, forming a three-phase system, a 23.5 kWh battery bank and PV systems coupled both in a.c. and d.c., with overall installed capacity of 6.71 kWp. HRES2 is composed of three Studer Xtender inverters of 6 kVA each, forming a three-phase system, a 19.2 kWh battery bank, and a 2.8 kWp PV system. HRES3 is a single-phase system composed of a 4 kVA Schneider inverter, a 4.8 kWh battery bank, and a 0.4 Wp PV system. HRES4 is a single-phase system composed of a 5 kVA SMA Sunny Island inverter, a 9.6 kWh battery bank, and a 1.28 kWp PV system. Aside from the HRESs, there is also a 40 kVA diesel generator, which may be connected to all systems. Table I summarizes the HRESs configurations.

Table I: Nominal capacities of each HRES.

	Photovoltaic generator	Storage capacity	Bi-directional inverters
HRES 1	6.71 kWp	23.5 kWh	15 kVA
HRES 2	2.8 kWp	19.2 kWh	18 kVA
HRES 3	1.28 kWp	9.6 kWh	5 kVA
HRES 4	0.4 kWp	4.8 kWh	4 kVA

Fig. 2 presents some of the equipment installed in LSF. The MGCC is implemented with a CompactRIO from National Instruments.

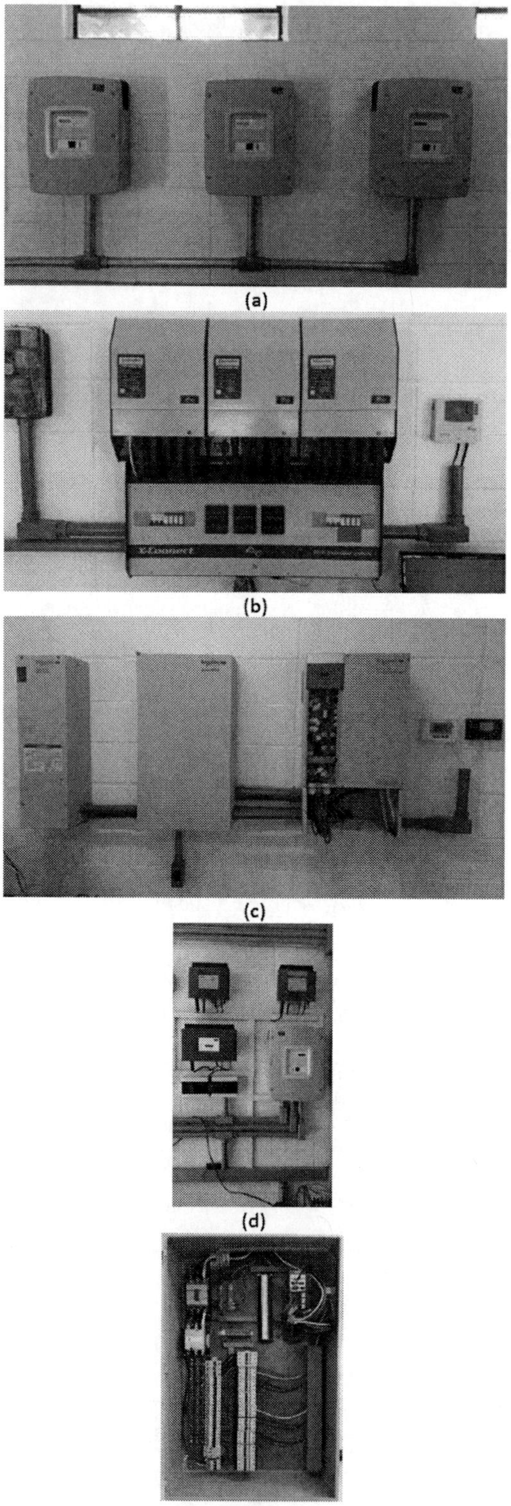

(a) (b) (c) (d) (e)

Figure 2: HRES installed in LSF. (a) SMA three-phase system. (b) Studer three-phase system. (c) Schneider single-phase system. (d) SMA single-phase system. (e) MGCC.

2.2. Monitoring interface

Monitoring allows improved analysis and response over the system. Due to the amount of data collected by all sensors, it was necessary to create an algorithm to quickly process and display the data from the sensors, otherwise the amount of data could be cumbersome to deal with, resulting in no actual benefit from the monitoring.

The measurements collected are visualized in a unified interface created in LabVIEW. Each HRES is monitored both at its AC and DC coupling. The electrical parameters, such as voltage, current, battery bank state of charge, among others, as well as weather data (such as temperature and irradiance), are acquired and serve as inputs for the operation algorithm programed in the MGCC. The interface allows the selection of parameters by equipment and the choice of parameters within the selected equipment. Fig. 3 presents the monitoring interface of the data acquired by the system.

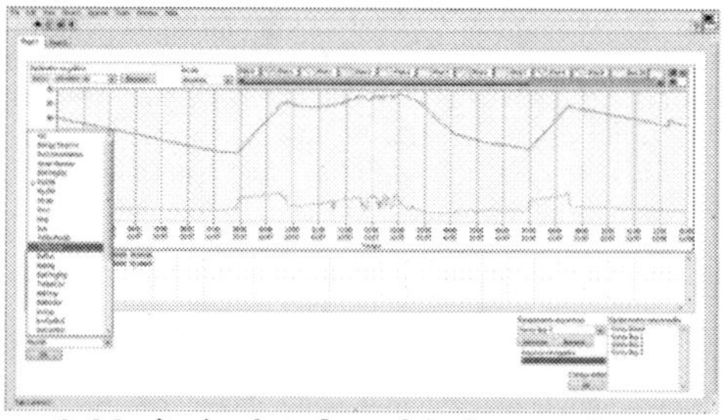

Figure 3: Monitoring interface of the integrated systems.

3 POSSIBLE TOPOLOGIES

The MGCC is responsible for coordinating the operation of all elements connected to the PCC, according to pre-defined operational characteristics and constraints, which can be programed in the MGCC control algorithm. The operation algorithm is responsible for making decisions regarding the best way to operate the system, based on the boundary conditions of a given moment, considering the resources available and the demand to be met.

Table II presents the possible topologies. The column loads presents the micro-grids (prosumers) that can be (but not necessarily will be) assigned as loads in a particular mode of operation. As presented in Table II, each HRES can be either a load or a generator to the entire smart grid. In each mode of operation, one of the prosumers needs to operate as a voltage source and the others need to operate as current sources or loads that will either inject or draw power from the PCC.

4 OPERATIONAL SITUATIONS

Fig. 4 presents the production of one of the PV systems over the course of 44 days. In the figure, it is possible to observe the stochastic nature of PV systems, with days of both abundant production and lack of production. In days of abundant production, the excess energy may be stored in an energy storage system (like batteries), however the storage capacity is still limited, which may lead to the situation presented in Fig. 5.

Table II – Possible configurations.

	Load				
power generator	Grid	HRES1	HRES2	HRES3	HRES4
Main grid		x	x	x	x
Diesel generator	x	x	x	x	x
HRES1	x		x	x	x
HRES2	x	x		x	x
HRES3	x	x	x		x
HRES4	x	x	x	x	
HRES1+Main grid			x	x	x
HRES2+Main grid		x		x	x
HRES3 + Main grid		x	x		x
HRES4 + Main grid		x	x	x	
HRES1 + Diesel gen.	x		x	x	x
HRES2 + Diesel gen.	x	x		x	x
HRES3 + Diesel gen.	x	x	x		x
HRES4 + Diesel gen.	x	x	x	x	
HRES1 + HRES2	x			x	x
HRES1 + HRES3	x		x		x
HRES1 + HRES4	x		x	x	
HRES2 + HRES3	x	x			x
HRES2 + HRES4	x	x		x	
HRES3 + HRES4	x	x	x		

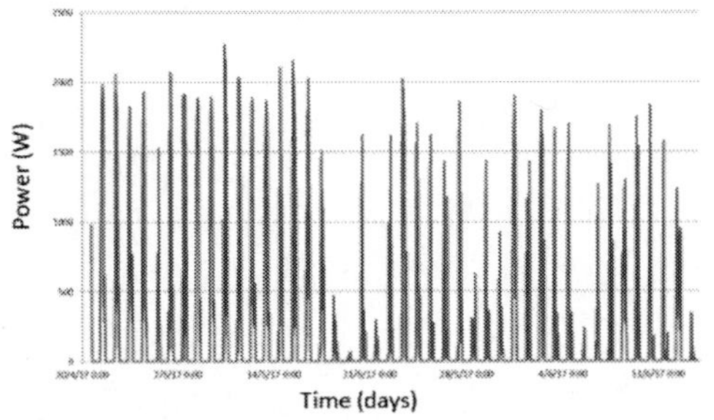

Figure 4: Production of the PV system connected to HRES1.

In Fig. 5 the state of charge of the battery bank (blue) and the frequency of the HRES1 (red) are depicted. It is possible to observe that the frequency of the system started to increase due to a rise in the state of charge of the batteries. This is done by the control logic of the grid forming inverter to disconnect distributed generations when the storage system is full. This presents a situation where the integration of several systems in a mult-micro-grid would be desirable. In this situation the micro-grid with production excess could negotiate this amount with the main grid or even with other micro-grid.

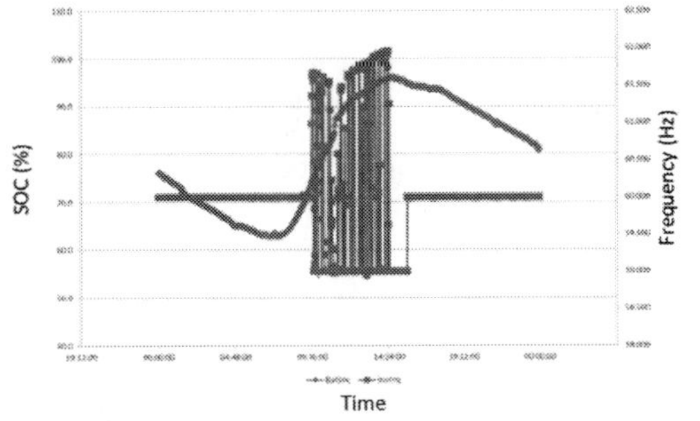

Figure 5: Situation of frequency rise to disconnect the DG of HRES1 due to energy excess.

Fig. 6 presents a situation in which the excess energy of HRES1 is injected into the grid in days of excess production. It is possible to observe that on somedays the bidirectional meter show negative power, which represents injection of energy into the grid and would avoid the disconnection of GD.

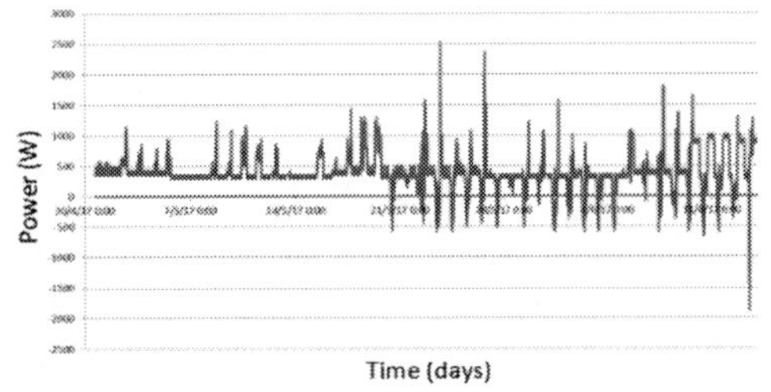

Figure 6: Demand/injection profile of the laboratory.

As stated before, one of the objectives is the simulation of energy markets and the interaction of the prosumers following and even placing price signals to negotiate energy. In the facility, it is possible to simulate the operation with different tariffs that could vary in a programmable period. For instance, it is possible to program tariffs that vary throughout a day and would reflect situations like periods of need or excess of energy. Moreover, the measurements in the PCC may be used as signals to trigger different tariffs that could be based on the need for ancillary services and even dynamically change the simulated tariffs scenario according to the needs of the smart-grid and prosumers in a given situation.

5 CONCLUSIONS

In the case of grid-connected operation, micro-grids are still presented as a relatively new subject that only now starts to gain the market. The adoption of this alternative to the electrical system still needs to transpose many barriers like those of technical and regulatory nature before the potential benefits can be fully appreciated. The paper presented the facilities dedicated to the research on micro-grids carried out in the Laboratório de Sistemas Fotovoltaicos (LSF) of the Universidade de São Paulo, in Brazil. The main objective of the facility assembled in LSF is to carry out tests that can help achieve a better understanding of this kind of systems.

Moreover, the possibility to simulate an environment where prosumers may interact actively with the grid by buying, selling and even placing prices over the energy that they would be willing to negotiate will provide useful insights over the operation of such environments.

6 REFERENCES

[1] Muhammad Irfan, Jamshed Iqdal, Adeel Iqdal, Zahid Iqbal, Raja Ali Riaz, Opportunities and challenges in the control of smart grids. Renewable and Sustainable Energy Reviews, 71 (2017), pp 652-674, DOI: 10.1016/j.rser.2016.12.

[2] Omid Palizban, Kimmo Kauhaniemi, Josep M. Guerrero. Microgrids in active network management – part II: System operation power quality and protection. Renewable and Sustainable Energy Reviews. 36 (2014), pp 440-451. DOI: 10.1016/j.rser.2014.04.048.

[3] Omid Palizban, Kimmo Kauhaniemi, Josep M. Guerrero. Microgrids in active network management – part I: Hierarchical control, energy storage, virtual power plants, and market participation. Renewable and Sustainable Energy Reviews, 36 (2014), pp 428-439. DOI: 10.1016/j.rser.2014.01.016.

[4] Prasenjit Basak, S. Chowdhury, S. Halder nee Dey, S. P. Chowdhury, A literature review on integration of distributed energy resources in the perspective of control, protection and stability of microgrids. Renewable and Sustainable Energy Reviews, 16 (2012), pp 5545-5556. DOI: 10.1016/j.rser.2012.05.043.

[5] J. Vasiljevska, J. A. Peça Lopes, M. A. Matos, Integrated micro-generation, load and energy storage control functionality under the micro-grid concept. Electric Power System Research, 95 (2013), 292-301. DOI: 10.1016/j.epsr.2012.09.014

[6] J. Vasiljevska, J. A. Peça Lopes, M. A. Matos, Evaluating the impacts of multi-microgrids concept using multicriteria decision aid . Electrical Power Systems Research, 91 (2012), pp 44-51. DOI: 10.1016/j.epsr.2012.04.013.

[7] Francesco Guzzi, Diana Neves, Carlos A. Silva. Intergation of smart grid mechanisms on microgrid energy modelling. Electric Power Systems Research, 95 (2013), pp 292-301. DOI: 10.1016/j.epsr.2012.09.014.

[8] N. W. A. Lidula, A. D. Rajapakse, Microgrids research: A review of experimental microgrids and test systems. Renewable and Sustainable Energy Reviews, 15 (2011), pp 186-202. DOI: 10.1016/j.rser.2010.09.041.

[9] Mariya Soshinskaya, Wina H. J. Crijns-Graus, Josep M. Guerrero, Juan C. Vasquez. Microgrids: Experiences barriers and success factors. Renewable and Sustainable Energy Reviews, 40 (2014), pp 659-672. DOI: 10.1016/j.rser.2014.07.198.

[10] Taha Selim Ustun, Cagil Ozansoy, Aladin Zayegh, Recent development in microgrids and example cases around the world – A review. Renewable and Sustainable Energy Reviews, 15 (2011), pp 4030-4041. DOI: 10.1016/j.rser.2011.07.033.

[11] Liang Tao, Dr. Christine Schwaegerl, Sankar Narayanan, Dr. Jian Hui Zhang. From laboratory microgrids to real markets – Challenges and opportunities. Power Electronics and ECCE Asia (ICPE & ECCE), 2011. DOI: 10.1109/ICPE.2011.5944600.

PERFORMANCE ANALYSIS OF THREE PV TECHNOLOGIES UNDER DIFFERENT OPERATING CONDITIONS

Ilham BAGHDADI, Ali EL YAAKOUBI and Adel ASSELMAN
Optic and Photonic Team, Faculty of sciences, Abdelmalek Essaaidi University, Tetouan 93002, Morocco

ABSTRACT: Recently, solar the most attractive renewable energy sources until present, due to the development and the advancement of the technologies. This source enables to generate electricity by the conversion of the solar radiation. In order to investigate the performances of a photovoltaic system installed on the roof of the Faculty of Science in Tetouan - Moroccco, a study and analysis is performed in this paper. The installation consists of three photovoltaic chains of three technologies (monocrystalline, polycrystalline and amorphous) connected to the grid. This work under study consists in the first part of presenting of the photovoltaic technologies and the monitoring station where the data has been collected. Secondly, an analysis of the system performances of the three technologies under different metrological conditions is given. The results demonstrate that the power output of the three technologies is very affected by the considered operating conditions.

Keywords: Grid connected-photovoltaic; solar radiation; monocrystalline; polycrystalline, amorphous.

1 Introduction

Electricity is now the easiest form of energy to exploit. But first it must be produced in high-power generating units before being transported and distributed to the consumer. This represents a heavy financial burden for the electrification of the isolated sites in our country.

PV solar energy has genuinely developed and has increasingly become as a promising energy option with many advantages such as abundance, absence of any pollution and availability in greater or lesser amounts at any point of the terrestrial globe. For the past decade, Morocco has implemented a national energy policy favorable to the development of renewable energy, with the aim of securing its energy supply in a context of strong growth in energy demand, in order to control the future costs of energy services compared with the upward trend in oil products prices and lastly to preserve the environment by mitigating greenhouse gas emissions. In this context there is a good opportunity for Morocco to develop the renewable energy sector in order to reduce its dependence on foreign countries and to meet its energy demand in a sustainable and autonomous way. Morocco has plenty of potential for renewable energy, including solar energy with 5.5 kWh / m^2 / day; with a sunshine of 3,000 hours per year.

The main objective of this work consists of estimating and investigating the performances of the three technologies in term of ac output power under operating condition, a sunny and cloudy day. The clear, cloudy and rainy day is the 25 Jun, 20 December and 18 November 2016, respectively. Various studies has been performed in literature [1]–[5], but this work introduce a new study with different and modern technologies and climatic conditions.

2 Material and models

Our installation consists of three photovoltaic chains of different technologies (monocrystalline, polycrystalline and amorphous), each chain injects 2kw into the network of the Faculty of Sciences of Tetouan under standard conditions. The capacity of this connected grid installation is 5,86Kwp. The installation is connected to the grid, by a monitoring station to supervise the production of the installation, and the meteorological conditions of the site. The data that we have used in this work are obtained from this station of this year. Fig.1 and 2 represent images of the installation and the metrological station, respectively. The properties of the boards used are listed in Table 1.

Fig.1 PV installation

Fig.2 Metrological station.

Trademark	SOLARWORLD		NEXPOWER
Model	SUNMODULE plus SW 255 poly	SUNMODULE plus SW 255 mono	XTREM+(NT-155AF)
Solar cell	Poly-crystalline	Mono-crystalline	Amorphous
Maximum power at STC (Pmax)	255 *Wp*	255 *Wp*	155*Wp*
Optimum orerating voltage (Vmp)	30.9*V*	31.4V	65.9V
Optimum operating current (Imp)	8.32A	8.15A	2.43A
Open circuit voltage (Voc)	38V	37.8V	85.5V
Short circuit current (Isc)	8.88A	8.66A	2.57A

Table 1. Main characteristic of the different modules.

3 Results and discussions

Fig.3, 4 and 5 show the AC output powers of the photovoltaic PV technologies for the clear, cloudy and rainy day, respectively. During the clear day the powers are higher and similar to solar irradiations compared to the cloudy and rainy day, where the fluctuations of solar irradiation conducted to fluctuations in the system's powers. As illustrate the figures, the mono-crystalline technology generates more power than the two others technologies

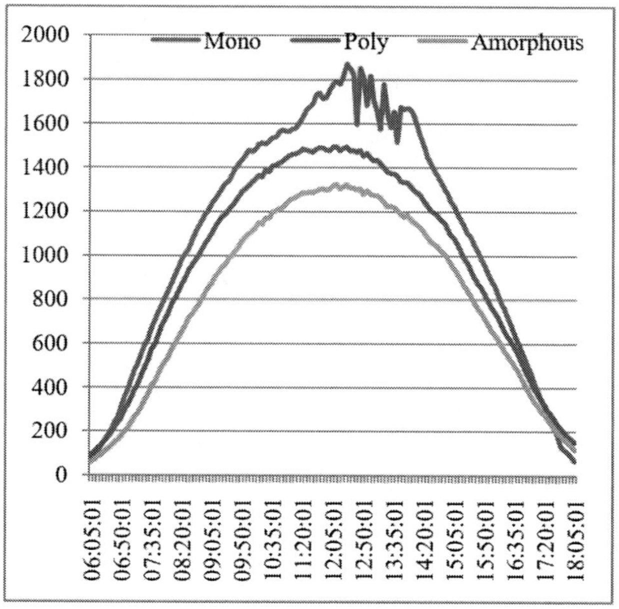

Fig.3 AC power output of the clear day.

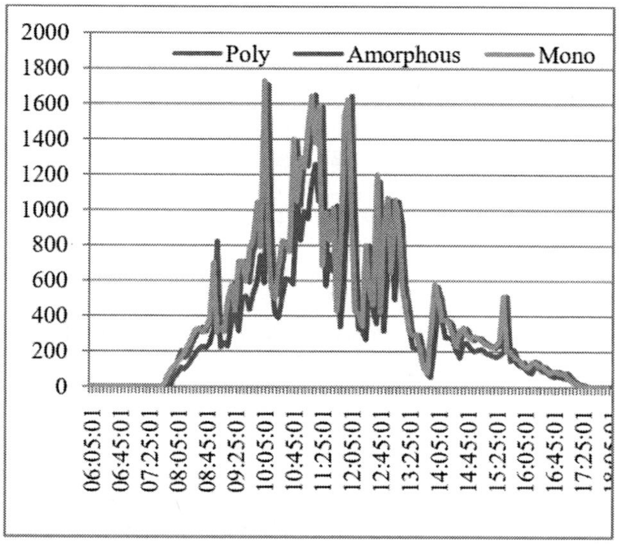

Fig.4 AC output power of the cloudy day.

Fig.6 shows daily variation of solar irradiance of the three days. The clear day is characterized by a maximum irradiation during the noon, this leads to an increase of the ambient temperature and the temperature of the modules. During the other two days oscillations are observed on the solar irradiation, especially for the cloudy and rainy day, and thus are produced on the power output of the system.

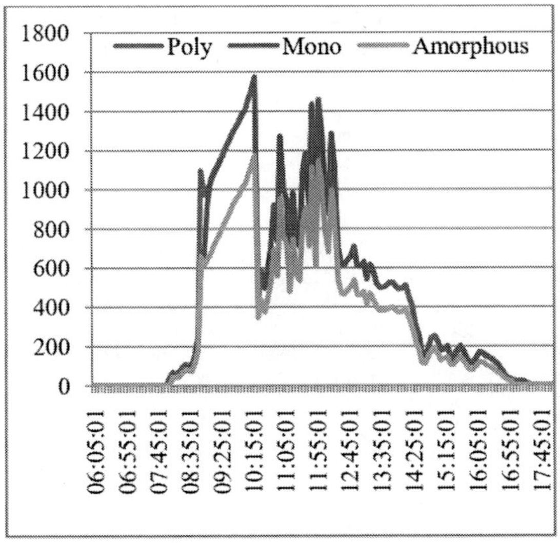

Fig.5 AC output power of the rainy day.

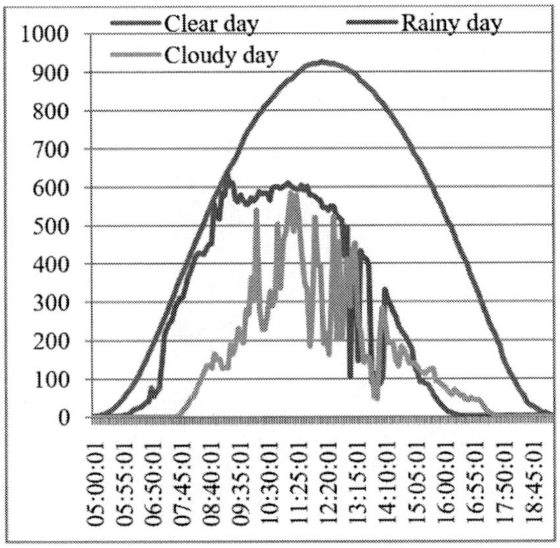

Fig.6 Solar irradiation of the three considered days.

4 Conclusion

This study presents an analysis of the AC output power from three PV technologies under different conditions. The experimental results demonstrates that the produced power is very affected under the operating conditions, and the mono-crystalline technology generates more power compared to the others technologies.

Acknowledgements

The authors acknowledge the Institut de Recherche en Energie Solaire et Energies Nouvelles (IRESEN) in Morocco.

References

[1] R. Dabou *et al.*, "Monitoring and performance analysis of grid connected photovoltaic under different climatic conditions in south Algeria," *Energy Convers. Manag.*, vol. 130, pp. 200–206, Dec. 2016.
[2] K. Attari, A. Elyaakoubi, and A. Asselman, "Performance analysis and investigation of a grid-connected photovoltaic installation in Morocco," *Energy Rep.*, vol. 2, pp. 261–266, Nov. 2016.
[3] D. D. Milosavljević, T. M. Pavlović, and D. S. Piršl, "Performance analysis of A grid-connected solar PV plant in Niš, republic of Serbia," *Renew. Sustain. Energy Rev.*, vol. 44, pp. 423–435, Apr. 2015.
[4] B. Shiva Kumar and K. Sudhakar, "Performance evaluation of 10 MW grid connected solar photovoltaic power plant in India," *Energy Rep.*, vol. 1, pp. 184–192, Nov. 2015.
[5] A. Al-Otaibi, A. Al-Qattan, F. Fairouz, and A. Al-Mulla, "Performance evaluation of photovoltaic systems on Kuwaiti schools' rooftop," *Energy Convers. Manag.*, vol. 95, pp. 110–119, May 2015.

VERY SHORT-TERM SOLAR IRRADIATION FORECASTING METHOD USING STATE ESTIMATION BASED ON KALMAN FILTERS FOR PV-DIESEL HYBRID SYSTEMS

A. Notholt
Reutlingen University
Alteburgstr. 150
72762 Reutlingen, Germany

ABSTRACT: The purpose of this article is to provide insight of a new simple forecasting method based on a state-estimation algorithm known as the Kalman Filter. While the accuracy of such algorithm is not comparable to state-of-the-art forecasting algorithms for PV-power production it does not require any internet connection, eyefish cameras or time intensive training. The algorithm was tested with several months of real high-resolution data with adequate results for the intended application: The minimization of the necessary spinning reserve on a PV-Diesel hybrid system to increase the solar fraction and reduce diesel consumption.

Keywords: Forecasting, Hybrid, Plant control

1 INTRODUCTION

An important objective in PV-Diesel hybrid systems is the maximization of PV Energy input in order to minimize the Diesel consumption. Nevertheless, especially in systems without energy storage, this objective is counteracted by the necessity of having enough spinning reserve to account for fast drops in PV produced for example by cloud passing.

While several research efforts focus on the very short term forecasting of this events in order to allow the system controller to start enough dispatchable generators to compensate for the power losses, these methods often require the use of satellite imaging or eyefish cameras on site. Since a considerable amount of systems may not have a stable internet connection and the investment in cameras may exceed the potential benefits in small systems, a new approach is presented.

This new approach utilizes a Kalman filter in order to estimate the minimum PV production at a given forecasting frame.

2. IMPORTANCE OF FORECASTING

Industrial PV-Diesel hybrid systems face often the important challenge of choosing between economic operation and security of supply. In these industrial systems, the renewable energy generation is usually not enough to supply the complete load requirements and thus in most cases the genset system must always run. In some advanced systems such as the Degrussa Mine Solar Power Project [1] may in some cases allow a complete takeover of solar and energy storage but these events and systems still represent a very small proportion of the operation time.

Since generators are for the most part of the day active, it is imperative to follow their requirements in order to maintain guarantees and prevent premature failure. The most important characteristic is the fact that gensets need to run with a minimum power output. Minimum loading of 30% for short periods of time and 50% of average loading is often required.

The minimum load requirements directly oppose to the actual objective of a PV-System to inject as much power into the grid as possible. Because of this a system controller will try to use as few generators as possible to allow for a higher PV power production.

PV is however stochastic and in some regions quite volatile. Figure 1 shows the typical ramp rates for a 5 MW system in south America [2]. It can be seen that ramp rates of up to 50% of the nominal PV power per minute are probable for 70 hours a year, assuming the typical duration of these events of 1 minute (see reference [2]), a total of 4200 events are likely to appear during one day.

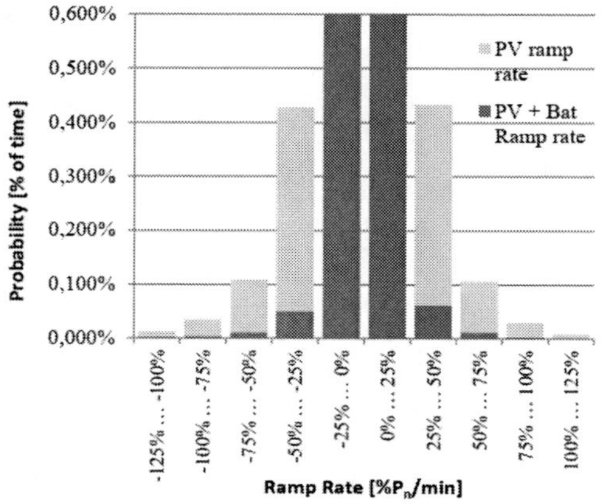

Figure 1: Rate of change of PV power output in the system of Cobija, Bolivia, 2016

Because of this, a system controller must maintain a sensible amount of spinning reserve in he system. The amount of reserve power is defined by the load but also by the probable loss of PV generation during the time in which the next generator is able to start. This is achieved by turning on enough gensets to cover the PV decrease.

This leads to the optimization deadlock: Larger PV power generation requires larger spinning reserve which in turn requires a larger minimum load thus throttling PV injection. Because of this it is very important for all Hybrid systems to assume a sensible PV decrease, which is as small as possible.

This can be achieved with complex forecasting or by estimating the state using state estimators like Kalman filters.

3 THE KALMAN FILTER

From the literature: "A Kalman filter is an optimal estimator, it infers parameters of interest from indirect, inaccurate and uncertain observations. It is recursive so that new measurements can be processed as they arrive. If all noise is Gaussian, the Kalman filter minimises the mean square error of the estimated parameters."[3]

The estimated state in this case is the solar irradiance, which then is input as basis for a PV calculation based on temperature and array size. Kalman filters use a model for the state estimation.

Figure 2 shows the block definition of the used algorithm. In this project, the model-based estimation has been replaced by the use of historical values and their gradients. Since most weather conditions have a certain persistency, this can be a suitable assumption. The model can be however complemented by a state estimation based on the sun and array position in order to achieve better results.

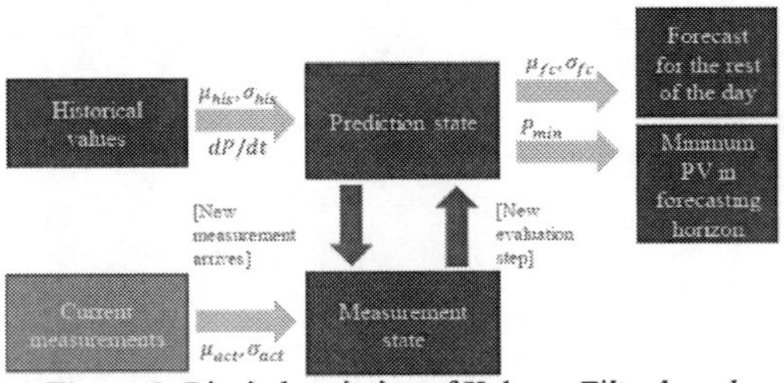

Figure 2: Block description of Kalman-Filter based Forecasting

Note that the algorithm computes both the minimum expected PV production on the forecasting horizon as well as the forecast for the rest of the day.

Since the algorithm uses statistical data, the forecasting algorithm calculates in timeslots or "bins" of a fixed width (T_b) which can be set up as required.

3.1 The algorithm in detail

The first assumption of the algorithm at the beginning o the day is that today is going to be the same as yesterday. With this information, sunrise and sunset values, sun duration and probable production can be calculated. Since the historical values are statistically evaluated, the average and standard deviation are available.

As the day passes by, new data is collected, providing average and standard deviation as well. The Kalman Filer is then used to estimate the next bins by using the following information:

- Day before absolute value [kW] (μ_{his}) and standard. Deviation (σ_{his})
- Current day absolute value [kW] (μ) and standard deviation (σ)
- Current day power gradient [kW/min] (δ)

For each forecasting bin the next average and standard deviation is calculated by:

$$\mu_{k+1} = \alpha \cdot \mu_k + \beta \cdot \mu_{his,k} + \gamma(\mu_k + \delta \cdot T_b)$$

$$\sigma_{k+1} = \sqrt{\sigma_k^2 + \sigma_{his,k}^2}$$

The subindex k denotes the actual bin and so $k+1$ denotes the next bin in the forecast. These formulae can then be used to forecast any number of bins in the future by always taking the previously calculated forecast and combining it with the historical value.

The weighted average coefficients α, β and γ are calculated based on the standard deviation thus allowing the most probable values (those with the smaller standard deviations) to have a greater weight. The coefficients as proposed in this paper are:

$$\alpha = \gamma = \frac{\sigma_{his,k}}{2\sigma_{his,k} + \sigma_k}$$

$$\beta = \frac{\sigma_k}{2\sigma_{his,k} + \sigma_k}$$

3. RESULTS

3.1 Data source

This algorithm has been confronted with 10 months of high resolution data from the island of St. Eustatuis: "SMA and its subsidiary SMA Sunbelt Energy GmbH commissioned a PV Hybrid System on the Caribbean island of St. Eustatius, to completely electrify the island with utility scale PV, battery storage and diesel generators. By integrating 1,9 MWp of PV power and 1 MW of batteries the local utility company is now reduces its fossil fuel consumption and save more than 800.000 liters of diesel fuel and 2,200 tons of CO_2 per year".[4]

3.2 Results

As shown in Figure 3, the measured PV production is over the forecasted minimum. Moreover, a very good correlation between the forecasted average and the actual production can be achieved. Since the algorithm assumes a certain amount of persistence, the forecast fails sometimes during the days where a change in weather is occurring. However, since the algorithm reacts on current data, the error is corrected very fast.

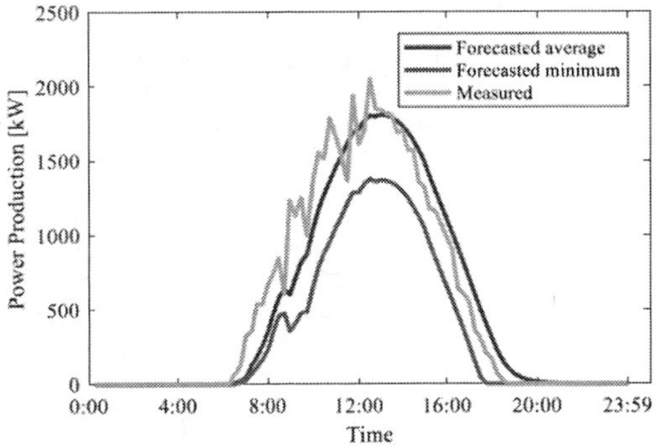

Figure 3: Results of measured and forecasted values

Table I shows the absolute results for the forecasts on 30m, 1h and 2h together with the absolute RMS error, which represents around 14% of the nominal output power. The results consider a safety factor of 0.5 standard deviations for the calculation of the minimum value or:

$$P_{min} = \mu_k - 0.5 \cdot \sigma_k$$

Table I: Forecasting errors depending on horizon and safety factor

Forecast Horizon (s)	Safety Factor (σ_k)	Absolute RMSe (kW)	Time out of Target (%)
1800	0.5	203.0	22.46
1800	1.0	261.2	16.01
3600	0.5	288.6	22.95
7200	0.5	415.9	25.43

During the evaluated period it can be observed that the forecasting algorithm is accurate in more than 75% of the time. The absolute RMS error is however small enough to assume a genset is able to cope with the excess loading (most gensets may be overloaded some 20% for a short period of time until the next genset starts).

This, however, means that in such systems, a small storage solution may be of great help in the optimization of the system. By increasing the safety factor, it is possible to reduce the error but it will diminish the targeted solar fraction.

It can also be observed that the forecasting horizon has a quadratic impact on the quality. While the forecasting of one hour increases the inaccuracy time by 2.2% more time, a second doubling of the forecasting frame yields an increase of 13% or six times larger.

4. CONCLUSIONS

Although Kalman filters are widely used for state estimation, its application in solar systems has not yet been documented. This study shows that this approach may be used for simple state estimations but does not yield the best results. Nevertheless, the application of this approach is still possible in PV-Hybrid systems to improve the use of PV against a purely static approach with good results.

Possible approaches for refining the algorithm includes:

- Inclusion of a simple geometric model to calculate the clear-sky irradiance and use this as an additional input.
- Separation of the clear sky irradiance and clear sky index components to improve overall efficiency of the algorithm

The Kalman Filter is a linear state estimation algorithm, which only provides good results if the system has normal distributions and its characteristic is linear. It is also possible to review non-linear state estimation algorithms that can provide better results while maintaining he same premise: Forecast with good confidence without costly equipment or internet connection.

5. ACKNOWLEDGEMENTS

The author would like to thank the companies SMA Solar Technology AG and SMA Sunbelt Energy GmbH for the support in the development of this algorithm. Please note that this information is the sole responsibility of the Author and no endorsement nor statement from the mentioned companies is implied.

3 REFERENCES

[1] Spreadborough, M. “Embracing Change, Driving Innovation – the DeGrussa Solar Project”, Presentation 24 May 2016

[2] Notholt, A., “Industrial hybrid systems with high PV penetration – performance analysis and key success factors”, 32nd EU PVSEC, Munich, Germany 2016

[3] Kleeman, L. “Understanding and Applying Kalman Filtering”, Monash University, Clayton.

[4] SMA Solar Technology AG, Press Release on the Sustainable power supply for carribean island St. Eustatius, 2016-05-24

QUICK AND EFFECTIVE PLANT EVALUATION USING DARK-IV STRING CURVES

K. Mertens, A. Arnds, M. Diehl*)
Münster University of Applied Sciences, Electrical Engineering, Testing Lab for Photovoltaics
Stegerwaldstraße 39, D-48565 Steinfurt, +49 2551/962-111, mertens@fh-muenster.de
*) photovoltaikbuero, Ternus und Diehl GbR, D-65428 Rüsselsheim, md@pvbuero.de

ABSTRACT: Most measurement techniques for **on-site evaluation** of PV plants show **different problems**: for peak power measurements and thermography **stable weather** and **minimum light intensities** are necessary. EL measurements at night offer detailed fault analysis but are very **time consuming.**
We propose a **method** that offers the possibility **to evaluate** a solar generator **very quickly** under **stable conditions**: the **dark-*IV* string curve method.** The method can reveal errors like **PID**, **defective** or **missing bypass diodes** as well as **defective connectors**. The method has the **potential** to become a **standard method to analyze PV plants.**

Keywords: dark-*IV* curves, dark-*IV* string curves, electroluminescence, outdoor EL, shunted bypass diodes, missing bypass diodes, defective connectors, series resistance, quick measurements, pvServe

1 INTRODUCTION

With the **price reduction** of solar modules, the **measurement** of modules **in testing labs** has become more and more **unattractive**. Instead, different **on-site measurement techniques** have been developed in the last years: **peak power measurement, thermography** and **LowCost-Outdoor-Electroluminescence** [1,2].

They offer the **possibility** to **detect faults** in the solar generator and to determine the quality of the plant. However, there are different **problems arising** with these methods. **Peak power measurements** can **only** be done **at** good and **stable weather**, which **reduces** the **possible measurement periods drastically**. The same holds for **thermography: stable weather** and **minimum light intensities** of 600 W/m^2 are necessary to assure correct measurement results. **EL measurements** at night offer **detailed fault analysis** [1,2] **but** are **very time consuming.**

We propose a **method** that offers the possibility to **evaluate the solar generator very quickly** by executing **dark-*IV* string curve measurements.**

2 PERFORMING DARK-*IV* STRING CURVE MEASUREMENTS

Up to now, **dark-*IV* curves** only **have been used** to evaluate **single cells** [3] or **single modules** [4]. Instead, **we propose** to **measure whole dark-*IV* string curves of a plant at night.** Figure 1 shows the principal setup. A high **voltage power supply** (0 – 1000 V, 0 – 5 A) is connected to the string under test. Then a rising **positive voltage** is applied to the string and the resulting current is measured. The **result** is the **dark-*IV* curve** of the string. In our case, we use the **special power supply *pvServe***, which can acquire the **curves** automatically, even without a laptop, **and stores them** on a memory card [5].

A measurement is performed in **only one second.** Therefore, all the strings of a plant can be measured very quickly. Then the **curves are evaluated** in order to **find out deviations** with respect to the "normal dark-*IV* curve".

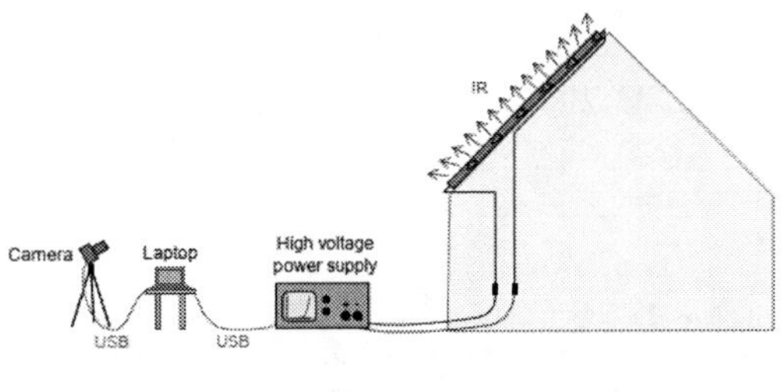

Figure 1: Principle measurement setup: the whole dark-*IV* string curve is measured automatically within one second and evaluated afterwards. The strings that appear to be faulty are then examined by outdoor-EL

3 DETECTING POTENTIAL INDUCED DEGRADATION (PID)

A well-known problem of solar modules is **potential induced degradation** (PID). It can be unambiguously detected with outdoor-EL measurements [6]. However, EL **measurements at night** take a **long time** because one has to find an **optimum perspective** to do the exposures and to get the desired picture quality. The **effort** can be **reduced drastically** by **detecting** the **"suspicious strings"** beforehand and then taking the EL photos of these strings only. **Fig. 2** shows an **example of a dark-*IV* measurement. Strings A to D** have **almost identical curves** whereas **strings E and F** show **strong deviations.** This is mainly due to the fact that the **PID** effect **reduces** the **shunt resistance** of the solar cells [6].

In analogy to the fill factor *FF* of the bright-*IV* string curve, we can define a "**dark-*IV* fill factor**" FF_D:

$$FF_{\mathrm{D}} = \left| \frac{V_{\mathrm{MPP}} \cdot (I_{\mathrm{Max}} - I_{\mathrm{MPP}})}{V_{\mathrm{Max}} \cdot I_{\mathrm{Max}}} \right| \qquad (1)$$

Here V_{Max} is the maximal applied voltage and I_{Max} is the current flowing at that voltage.

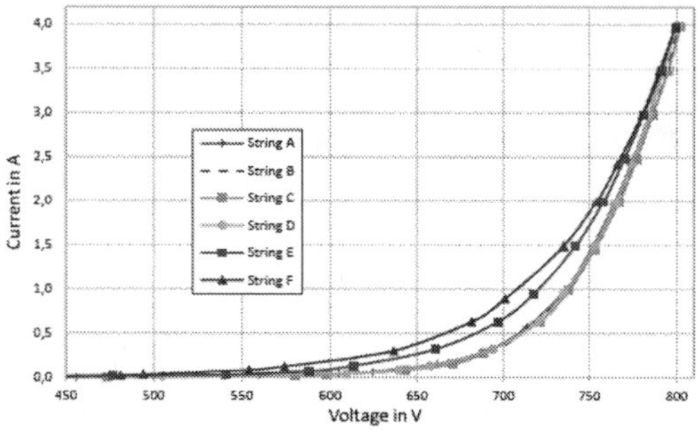

Figure 2: Dark-*IV* curves of different strings: strings E and F are suspicious with their small fill factors FF_D

The results for strings A to F are shown in Table 1: the "good strings" A to D all show dark-*IV* fill factors around 79.8 % whereas the two **strings affected by PID** lie **clearly below**.

Name	String A	String B	String C
FF_D	79.7%	79.8%	79.9%
Name	String D	String E	String F
FF_D	79.8%	76.1%	73.6%

Table 1: Results of calculating the dark-*IV* fill factor FF_D: the strings E and F are affected by PID and therefore show notedly reduced values

In Fig. 3, the EL pictures of the corresponding strings are depicted. Obviously, **only strings E and F are affected by PID as predicted** by the string curve measurement.

4 DETECTING DEFECTIVE BYPASS DIODES

Beside the possibility to pre-detect PID affected modules, the **dark-*IV* method** can **reveal other defects**. Not uncommon defects are **bypass diodes** that are **in short-circuit** state. In this case, the bypass diode short-circuits the corresponding cell substring of the solar module. This **fault can be easily detected** by dark-*IV* measurements. Fig. 4 gives an example. Here the dark-*IV* curve of **string G** shows a **clear reduction of** the string's **voltage**. The **voltage difference** ΔV at I_{Max} is about 22 V. This large drop can only be caused by **two defective bypass diodes** resulting from **2 x 20 cells** that are **shunted**.

The **proof** for this can be seen in Fig. 5: clearly visible **two modules** are **affected** by a shunted bypass diode each.

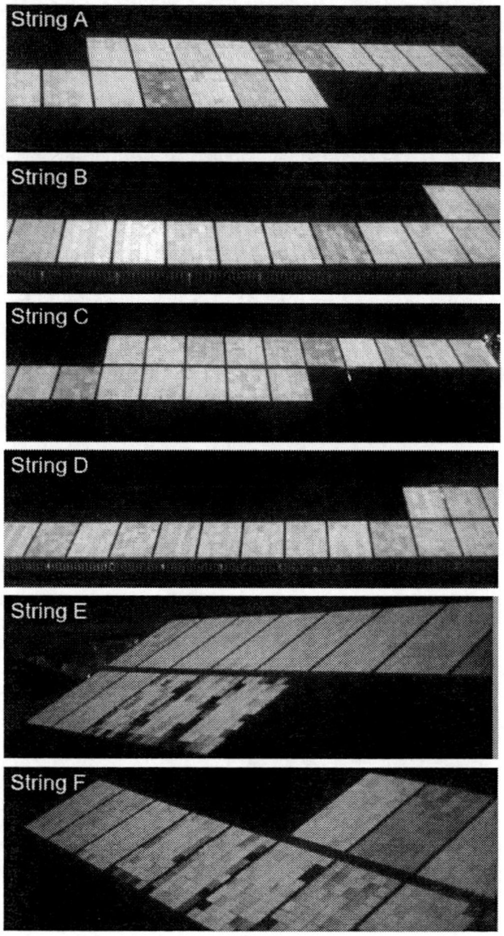

Figure 3: Dark-*IV* curves of different strings: strings E and F are suspicious with their small fill factors FF_D

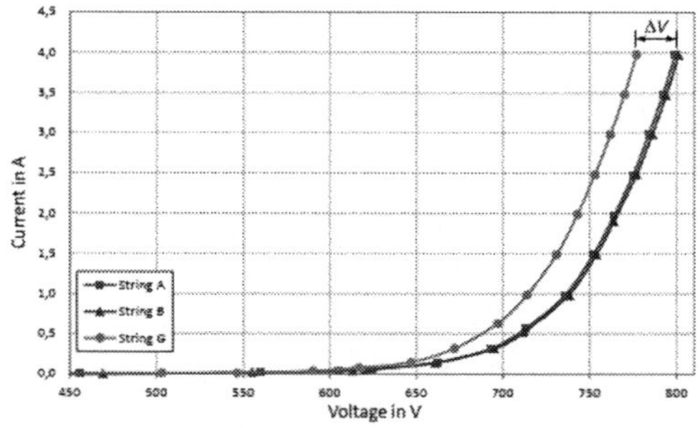

Figure 4: Dark-*IV* string curves with string G showing a considerable voltage drop

Figure 5: EL exposure of string G with two black stripes resulting out of two shunted bypass diodes

5 DETECTING MISSING BYPASS DIODES

Also **missing or blown bypass diodes** can be detected by the dark-*IV* curve method. This is done by **applying a negative voltage** to the string. In case of **no fault**, all the bypass diodes will show a **typical diode behavior**. If at least one **bypass diode** is **missing**, a current will only flow at high negative voltages where the **cells** under the missing bypass diode **come to an avalanche breakdown**. A **trick** to find out the **corresponding module** is to **press** e.g. **1 Ampere through the string**. As the cells in avalanche breakdown are **heated up**, the module can be found by looking on the module field with a **thermography camera**.

6 DETECTING DEFECTIVE CONNECTORS

Defective connectors can have **high series resistances** that lead to **hazards**. Using the dark-*IV* string curve method, the **curve of** the **defective string** will show a **rising voltage drop with rising current** with respect to the “good strings”. Therefore, this fault can also be detected easily.

7 CONCLUSION

The **dark-*IV* string curves method** is a **powerful tool** in evaluating PV plants. **Many different faults can be detected** easily and quick. As the **measurements** are done **at night**, the **ambient conditions** are **very stable** resulting in a **high repeatability** and **accuracy**.

We think that the **dark-*IV* string method** has the potential to **become a new standard method to analyze** PV plants.

LITERATURE

[1] Mertens, K., Kösters, H., Diehl, M: „Low-Cost-Outdoor-EL: Cost-Efficient Extensive on-Site Quality Analysis of Solar Modules“, Proceedings of 31st European Photovoltaic Solar Energy Conference and Exhibition, Hamburg, 2015

[2] Mertens, K.: Arnds, A., Behrens, G., Domnik, A., „LowCost-Outdoor-EL: Significant Improvements of the method“, Proceedings of 32nd European Photovoltaic Solar Energy Conference, Munich, 2016

[3] Aberle, A. et al., "New Method for Accurate Measurement of the Lumped Series Resistance of Solar Cells", 23rd IEEE PVSC, 1993, pp.133-139

[4] King, D.L. et al., "Dark current-voltage measurements on photovoltaic modules as a diagnostic or manufacturing tool", Photovoltaic Specialists Conference, 1997

[5] See website www.pvserve.de/en

[6] Mertens, K.: Photovoltaics – Fundamentals, Technology and Practice, 2nd edition, John Wiley & Sons, London, 2017

NOVEL SOILING DETECTION SYSTEM FOR SOLAR PANELS

Marc Korevaar[1], Joop Mes[1], Pramod Nepal[2], Gijs Snijders[1], Xander van Mechelen[1]

[1]Kipp & Zonen B.V., Delftechpark 36, 2628 XH Delft, the Netherlands
[2]Delft University of Technology, PVMD/EKL, Mekelweg 4, 2628 CD Delft, The Netherlands
Email: marc.korevaar@kippzonen.com, Telephone: 0031-152755242

ABSTRACT: Environmental factors such as wind, rain and back panel temperature all affect the PV plant performance ratio; but, in many environments, soiling of PV modules has the largest impact on the performance ratio. Cost effective soiling measurements within a solar plant can give reliable insights into the soiling behaviour over the whole production area, allow for an optimal cleaning schedule and trigger action after a dust storm.
Kipp & Zonen present a novel system for measuring the soiling on PV plants, the DustIQ. Using the Optical Soiling Monitoring technology the DustIQ measures the soiling without need for manual daily cleaning, moving parts or water. The system communicates via a Modbus® protocol through a two-wire RS-485 connection.
Different amounts and colours of soiling were characterised and a linear relationship between DustIQ signal and PV glass transmission loss was found. And different colours of dust result in a different linear slope. The effect of dust grain size was also investigated and no effect on the DustIQ signal and transmission loss was found. Furthermore, a rooftop PV monitoring station was built up where modules can be artificially soiled.
Keywords: Environmental effect, Monitoring, Optical losses, Performance, Plant Control

1 INTRODUCTION

In recent years, photovoltaic (PV) technology has emerged as one of the leading renewable energy technologies currently available. The PV plant performance is, aside from hardware, determined by three main parameters: irradiance, temperature and soiling. Irradiance and temperature are well understood and high quality measurements are available. However, for soiling there is still limited understanding and only basic measurements.
The extent of the effect of soiling on the PV plant performance is shown by the accumulation of soiling per day for different locations, shown in fig. 1, reproduced from previously published work [1]. Given as well is the average annual solar insolation for these locations showing their high potential for PV.

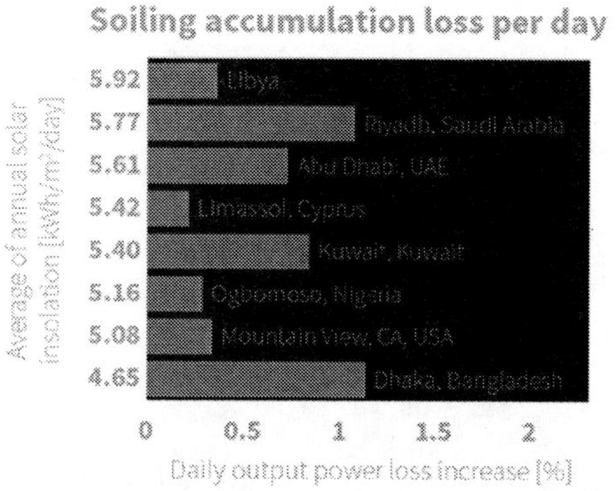

Figure 1: Soiling loss accumulation per day for different locations with average annual solar insolation given (recreated from [1]).

Previously published measurements of a reference cell at the Photovoltaic Laboratory of the University of Málaga in the south of Spain [2], are shown in fig. 2. The measurements show the accumulation of soiling on a not cleaned reference cell, measured using a shunt resistor.

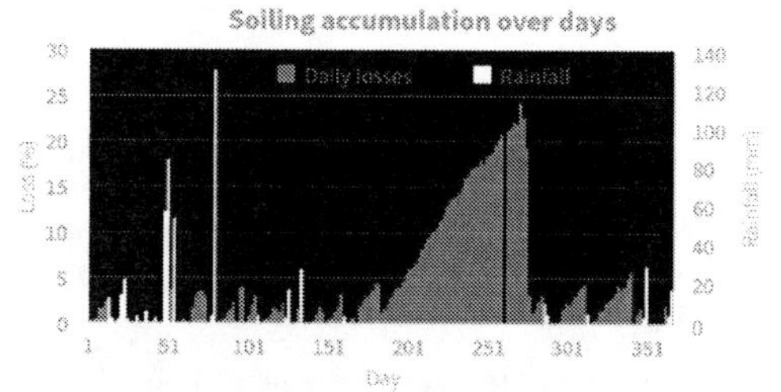

Figure 2: The loss of a reference cell over a year time with rainfall events plotted as well (recreated from [2]).

Natural rainfall events can clean the cell but in the summer months the lack of rain allows the loss to accumulate to over 20%.
The amount of soiling is dependent on the location and a map showing the world distribution, from previously published work [3], is shown in fig. 3. The darker colours indicate a higher level of dust than the lighter colours and the number represents the μg/m³ of PM10. Information on the size distribution of soiling on PV panels in Kuwait, from previously published work [4], is represented in table I.

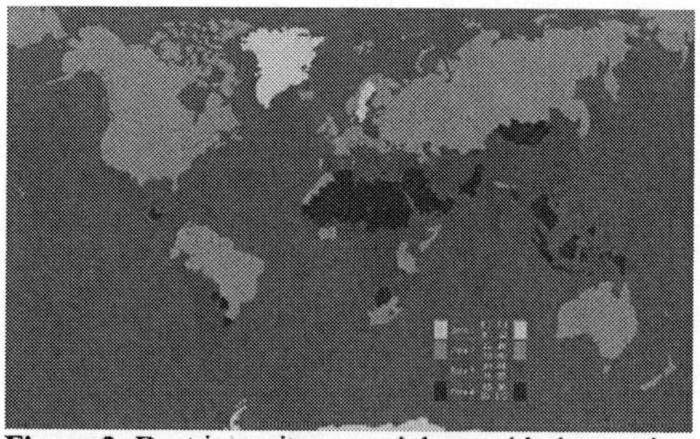

Figure 3: Dust intensity around the world, the number represents the μg/m³ PM10. (recreated from [3])

Table I: Dust grain distribution and grain type from [4].

D(μm)	% of the total sample	Grain type
1000-500	0.00	Coarse grained
500-250	0.00	Medium grained
250-125	0.82	Fine grained
125-63	4.78	Very fine grained
63-31	8.16	Coarse silt
31-16	16.47	Medium silt
16-8	23.82	Fine silt
8-4	20.19	Very fine silt
<4	25.75	Clay

From the abovementioned work by other groups we can see that the PV plant performance can be significantly influenced by module soiling. But when and where to clean the modules? To answer this question there is a need for accurate measurements of the amount of soiling on PV modules.

Existing systems all use the principle of a clean and not cleaned solar panel. It requires daily manual cleaning [5] or cleaning with moving parts/liquids.

This need for accurate measurements and the drawbacks of existing systems has been the inspiration for Kipp & Zonen to develop the DustIQ, a novel soiling sensor that avoids daily cleaning, moving parts/liquids, and provides multiple sensors across a plant for a similar budget. This information can be used by an O&M company to agree with stakeholders an optimal cleaning schedule and to trigger action after an event such as a dust or sand storm.

Here we will describe the measurement system, the DustIQ, and show the characterization of the system. Furthermore, we describe the rooftop PV measurement system for performing soiling tests.

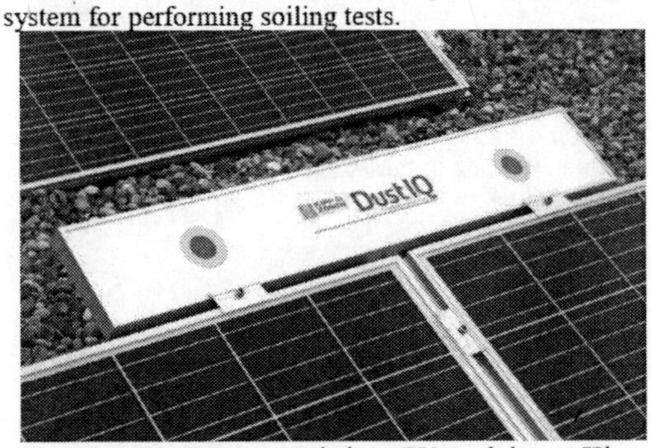

Figure 4: DustIQ mounted along PV modules on Kipp & Zonen roof in Delft.

2 SYSTEM & METHODS

2.1 System

For the accurate and cost effective detection of soiling on PV panels, Kipp & Zonen presents the new DustIQ, shown in fig. 4.

The system consists of glass plate, the same as in a PV panel, mounted next to the PV modules to ensure that the instrument follows the same soiling and cleaning patterns as the PV panels themselves. Below this glass plate there are two sensors that use the Optical Soiling Measurement (OSM) technology to measure the scattered light it receives from the soiling on top of its glass panel, shown in fig. 5. With an on-board calculation the transmission loss due to soiling is determined from the two OSM sensors. It can be shown that this transmission loss gives an equivalent loss in short circuit current and power. This loss in power is directly related to the Soiling Ratio (SR), which is the ratio of the irradiance measured with a cleaned PV panel to a soiled PV panel. The SR is also delivered by the DustIQ.

A unique feature is that the soiling measurement does not use the irradiance of the sun and is therefore independent of the sun position and sky conditions. The system communicates via a Modbus® protocol with all Kipp & Zonen smart instruments, with compatible Modbus®-enabled sensors from other manufacturers, and with suitable inverters and data acquisition systems. Connectivity is through a wired connection; two-wire RS-485. A back panel temperature sensor is included, to be mounted to the PV module.

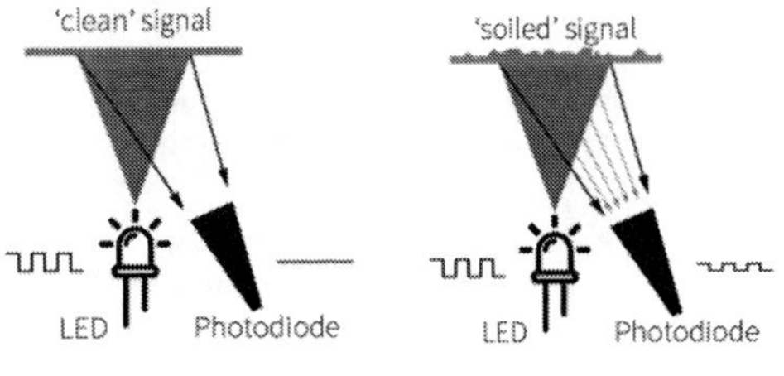

(a) **(b)**

Figure 5: diagram of DustIQ function with pulsed LED and photodiode with **(a)** clean glass and **(b)** soiled glass, giving rise to detection of scattered light.

2.2 Methods

The color of soiling can vary per location and influences the amount of scattered light measured with OSM and the corresponding transmission loss. An example of different dust colours is shown in fig. 6 (a). We provide a colour chart with which the customer can characterize the soiling colour and configure the DustIQ for that particular soiling, shown in fig. 6 (b).

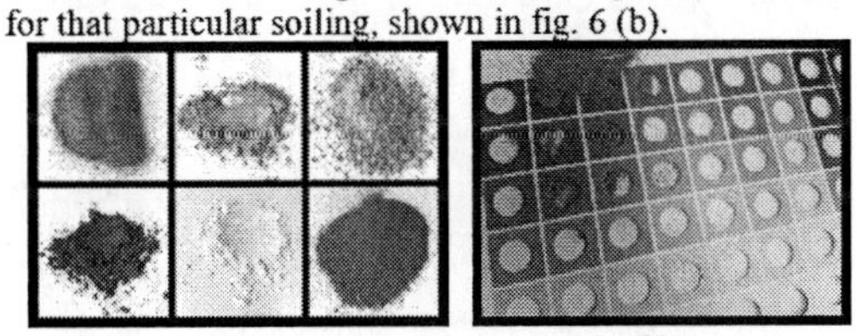

(a) **(b)**

Figure 6: **(a)** different colours of soiling dust and **(b)** colour chart for characterization.

In the near future we will expand the colour choice process by adding the possibility of a local dust calibration with an onboard sensor.

This sensor has the same optical and spectral properties as a PV module and allows for on-site calibration by relating the signal loss to the DustIQ signal due to the total ensemble of soiling types.

3 RESULTS

3.1 DustIQ characterization

The DustIQ was characterized for different soiling samples (as shown in fig. 6 (a)) with different amounts of soiling. The measured transmission loss of the PV glass and the corresponding DustIQ diode signal are plotted in fig. 7. The response between transmission loss and DustIQ signal is linear.

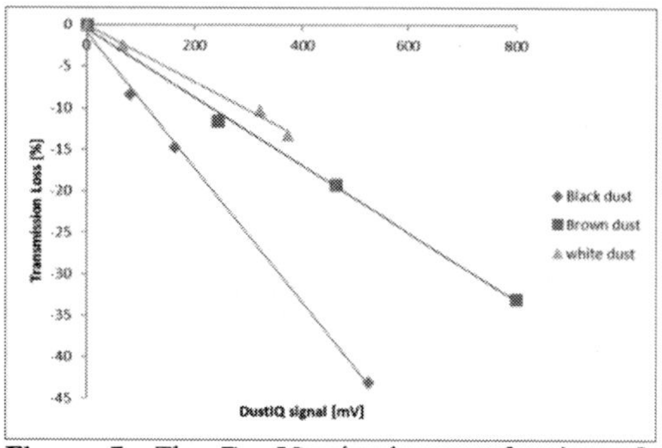

Figure 7: The DustIQ signal as a function of transmission loss for different colours of dust.

Furthermore, we investigated the effect of grain size on relation between DustIQ signal and transmission loss, and we found that there is no significant effect of grain size, as shown in fig. 8.

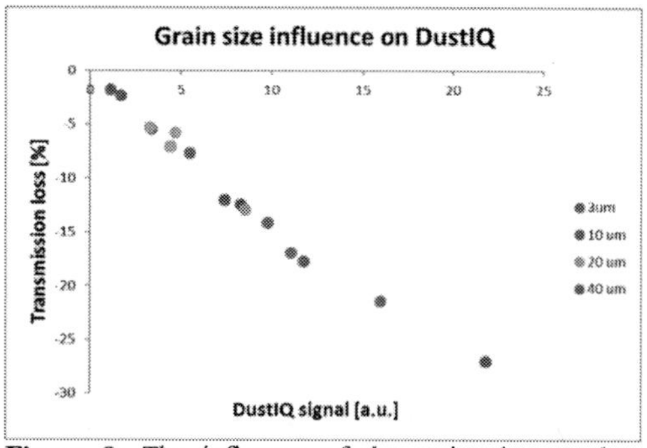

Figure 8: The influence of the grain size on the transmission loss.

3.2 Rooftop Soiling Ratio measurement

The amount of soiling is expressed in the Soiling Ratio (SR) as defined as the ratio in short circuit current, I_{sc}, or maximum power, P_{max}, between a soiled and a cleaned PV panel in the IEC 61724-1 [6]. We have artificially soiled PV panels at the Kipp & Zonen roof in Delft as shown in the photo in fig. 9 and measured IV curves and I_{sc}, the latter as a function of time.

Figure 9: Photo of roof PV setup with artificially soiled PV panels

The ratio of I_{sc}, obtained with the clean and soiled PV panel is plotted as a function of time in fig. 10. This SR is not constant over the day but is largest around solar noon, and gets smaller closer to sunset and sunrise, as is also reported in literature [2,7]. The SR delivered by the DustIQ can, with the help of post-processing, give this same daily variation if desired by the customer.

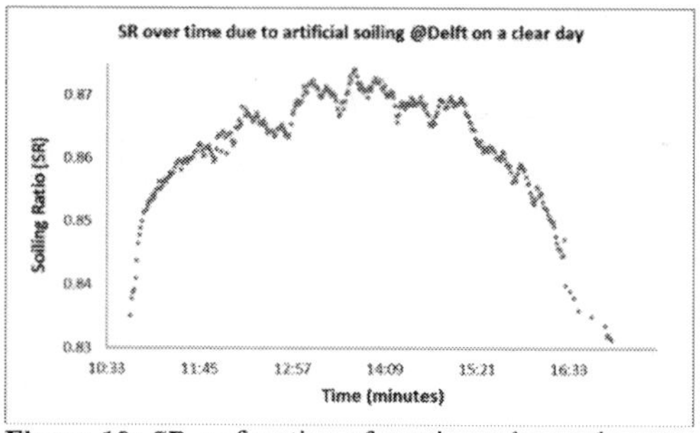

Figure 10: SR as function of varying solar angle over the course of a day with clear weather as determined from PV panels in fig. 9.

4 CONCLUSIONS

We demonstrated the relevance of soiling for PV plant performance and locations particularly prone to soiling. The new DustIQ, using the OSM technology, allows the PV plant owner to know the amount of soiling on the PV panels in real time, day and night.

The new soiling measurement technology signal is linear with transmission loss and is independent of grain size of soiling. The colour of the soiling influences the slope of the relation between transmission loss and DustIQ signal. We built a measurement setup to determine the SR of artificially soiled PV panels on the roof of the Kipp & Zonen building. In the near future we plan to do more outside tests also with the DustIQ on the roof in Delft (Netherlands) as well as at selected desert test locations.

5 ACKNOWLEDGEMENTS

We would like to thank Keith Wilson and Clive Lee for their contribution to this work.

6 REFERENCES

[1] A. Sayyah, M. Horenstein, and M. Mazumder, "Yield loss of photovoltaic panels caused by depositions", Solar Energy 107, 2014.

[2] J. Zorrilla-Casanova, M. Philiougine, J. Carretero, P. Bernaola, P. Carpena, L. Mora-Lopez, and M. Sidrach-de-Cardona, "Analysis of dust losses in photovoltaic modules", World Renewable Energy Congress, 2011.

[3] M. Maghami et al., "Power loss due to soiling on solar panel", Renewable and Sustainable Energy Reviews 59, 2016.

[4] H. Qasem, T. Betts, H. Mullejans, H. AlBusairi, and R. Gottschalg, "Dust-induced shading on photovoltaic modules", Prog. Photovolt: Res. Appl., 2014.

[5] M. Gostein, T. Duster, and C. Thuman, "Soiling Station Employing Module Power Measurements", Proc. IEEE PVSC, 2015.

[6] International Standard IEC 61724-1.

[7] M. Gostein, J. Caron, and B. Littmann, "Measuring Soiling Losses at Utility-scale PV Power Plants", IEEE Photovoltaic Specialists Conference, 2014.

IMPROVEMENTS OF PHOTOVOLTAIC SYSTEMS BY USING SOLAR TRACKING IN EQUATORIAL REGIONS

Vaca-Jiménez, S.[1,2], Morales, C.[1], Ordóñez, F.
[1]Faculty of Mechanical Engineering, Escuela Politécnica Nacional; Ladrón de Guevara E11-253, Quito – Ecuador
email: santiago.vaca@epn.edu.ec, carlos.morales01@epn.edu.ec, freddy.ordonez@epn.edu.ec
[2]Center for Energy and Environmental Sciences; Nijenborgh 6, 9747 AG Groningen, The Netherlands

ABSTRACT: Solar tracking may increase the profitable energy of PV solar systems, since the PV panels are oriented towards the sun during daylight hours. Thus, the radiation intercepted by them is higher. In this work, solar-tracking devices are assessed with the aim of quantifying the possible improvement in the energy harvested by using solar tracking in an equatorial region (Quito - Ecuador). For this, a series of simulations were made to compare the attainable power output of a PV system installed in a fixed position against the alternative of using different tracking systems (two-axes and one-axis). Additionally, the effect of nearby-, self- and horizon-shading were studied to identify which losses could potentially hinder the improvement of the energy harvested by a PV installation with solar tracking.
Results hinted that systems with one-axis tracking are comparable to two-axes tracking for this region, and that shading could represent power losses of around 30% for tracking devices in the case of nearby obstructions, especially when obstructions are present nearby in the East direction. Self-shading could represent a 50% of losses, and horizon-shading around 3%. These results could be used in the future to define strategies to optimize tracking systems in equatorial latitudes.
Keywords: Tracking, Modelling, Energy Performance, Shading, Simulation.

1 INTRODUCTION

Nowadays, PV energy systems are the fourth most installed renewable energy in the world, just behind bioenergy, hydropower and wind [1]. Nevertheless, the current efficiency of PV installations is relatively low, so any way to improve it is worthwhile.
Inside the feasible options to improve the overall efficiency of a PV installation, solar tracking has shown to increase the profitable energy, since the PV panels are oriented towards the sun during daylight hours, maximizing the energy harvested annually.

1.1 Solar tracking systems

In the literature we find that there are two categories of tracking systems, when considering the degrees of freedom (DOF) of the tracker: 1-axis trackers and 2-axes trackers (see Fig. 1) [2].

1.2 Literature review

There are many studies focused in the assessment of the profitability and behavior of tracking systems. The improvement of the energy harvested or the efficiency of the PV arrays have been by far the most studied. Studies have shown improvements in the 15-50% range in comparison to fixed (tilted) systems [3]. Between the most representative of them, there are the studies of [4] and [5], in which they found an improvement between 33% to 35% of solar tracking against the fixed arrays. However, these increases in the performance depend on many factors, like weather conditions [6].Also, many studies have focused in understanding the effects that losses could have in the overall performance of the system considering that tracking devices will consume more energy and resources. For instance, [7] found that the shadows casted by close objects could reduce the electricity output by 6%. Moreover, [8] hinted that shadows casted by obstacles located far from the installation (horizon) could also be representative in the performance.

1.3 Equatorial Latitudes

Considering than the sun's path in the equatorial latitudes is very different than in higher latitudes (refer to Fig. 2), and that this path creates different climatic conditions, it is interesting to see only few studies have focused on these latitudes. This leaves a great knowledge gap to fill, as most of the postulates and premises defined in most of the existing literature should not be applied in equatorial regions without considering the differences that exist between them.

To our understanding, the study made by [7] is the only comprehensive study that tries to compare two systems (one fixed and one tracking) in a latitude of 5°. They found an average improvement of 11% in the daily efficiency of the PV array.
Therefore, this work attempted to contribute to the existing knowledge, and fill the remaining gaps, using Quito, Ecuador (latitude 0°) as the case study.

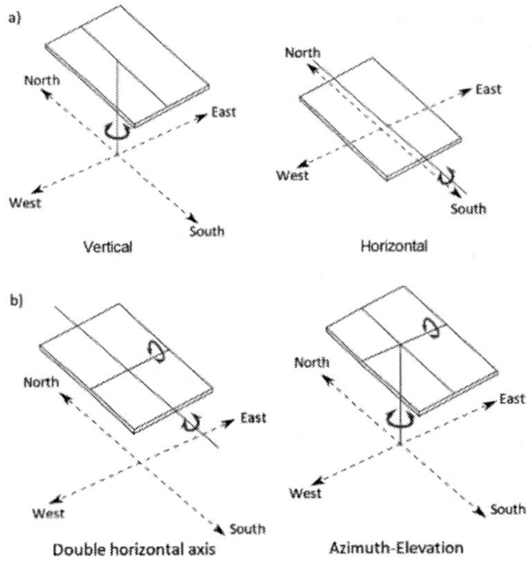

Figure 1: Solar trackers DOF classification. a) Types of 1-axis trackers, b) Types of 2-axes trackers.

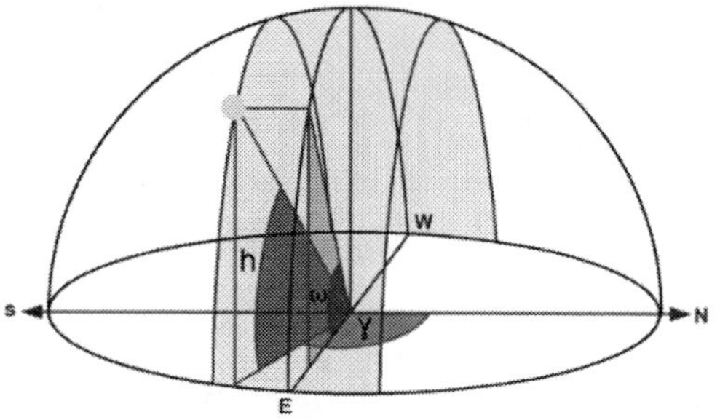

Figure 2: Annual sun's path for Quito, Ecuador (*0.16° S, 78.44° W*). Elevation angle (h), hour angle (ω) and azimuth angle (γ).

This study had two stages: in the first one, we compared the annual theoretical performance of tracking systems against a conventional fixed installation. Afterwards, we analyzed the possible effects of shading in the overall performance of these tracking systems.

2 METHODOLOGY

2.1 Description of the Model

The base part of the model resides in the description of the solar radiation into a surface that has a specific inclination and orientation. In this case, the global incident radiation (I_T) is the sum of the direct/beam (I_B), diffuse (I_{DT}) and reflected (I_R) components that arrive from the sun to the designated surface. This is described by equation 1.

$$I_T = I_B + I_{DT} + I_R \tag{1}$$

The direct component is defined by the angle of incidence (AOI), which is the angle formed between the incident direct radiation over the surface of the panel and the normal to the surface.

To calculate this component, we used equation 2.

$$I_B = E_b \cos AOI \tag{2}$$

$$AOI = \begin{cases} \pi & if\ a < -1 \\ 0 & if\ a > 1 \\ \arccos a & if -1 \le a \le 1 \end{cases} \tag{3}$$

$$a = \sin Z \cos(\gamma - \gamma_s) \sin \beta_s + \cos Z \cos \beta_s \tag{4}$$

Where, E_b is the global radiation over a horizontal surface, Z is Zenit angle, γ is the azimuth angle, γ_s and β_s are angles that describe the PV panel and that are defined by the tracking mechanism.

The diffuse component of the radiation is estimated by the HDKR model (as described in [7]), in which the sky is considered anisotropic. In contrast with other models available in the literature (like the isotropic, or the Perez model), the HDKR turns to be the most suitable considering the weather conditions of Quito (high radiation and clear skies), and the complexity of the calculations. The equations governing this component are:

$$I_{DT} = I_D \left[\frac{I_B{}^2}{I_T{}^2} + B \left(1 - \frac{I_B}{I_T}\right) \left(\frac{1 + \cos \beta_s}{2}\right) \right] \tag{5}$$

$$B = \left(1 + \sqrt{\frac{I_B}{I_T}} \sin^3 \frac{\beta_s}{2}\right) \tag{6}$$

$$I_D = I_{T(iso)} + I_{CS} \tag{7}$$

Where, $I_{T(iso)}$ is the isotropic model and I_{CS} is the circumsolar radiation.

Finally, the reflected component is defined mainly by the albedo (ρ) of the ground of the installation. The equation in this case is defined as follows,

$$I_R = \rho (E_b \cos Z + E_d) \left(\frac{1 - \cos \beta_s}{2}\right) \tag{8}$$

In this case, the albedo factor used was 0.2 that is the closest to urban settings according to [9], [10].

Three PV configurations (arrays) were used in this study to analyze the effect that solar tracking has on the overall performance of the system: 1) A fixed array of modules (horizontal), 2) 1-axis horizontal tracker, and 3) 2-axes double horizontal tracker. For the tracking systems, the tracking algorithm is defined in terms of a typical equatorial solar path, and based on the hourly position of the module (defined by the angles as Fig. 3 describes).

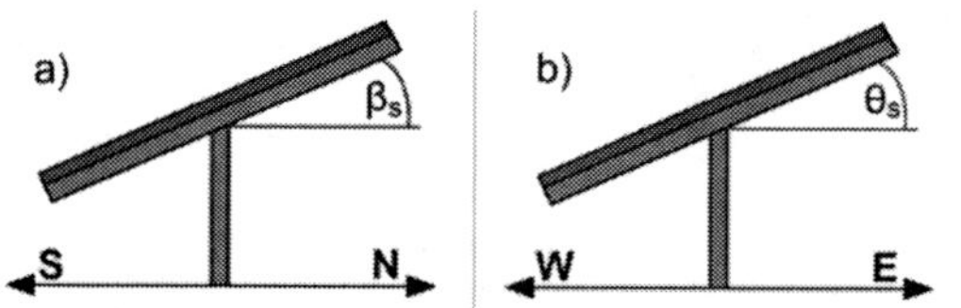

Figure 3: PV modules position. Tilt angle (β_s), and Rotation angle (θ_s).

The definition of the parameters of each configuration is presented in Table I.

Table I: Description of the parameters used for the modelled PV configurations.

Configuration	Parameters	
Fixed	Tilt (β_s)	0°
	Rotation (θ_s)	0°
	Azimuth (γ_s)	0°
1-axis horizontal tracker	Tilt (β_s)	0°
	Rotation (θ_s)	$\omega - 90°$
	Azimuth (γ_s)	0°
2-axes double horizontal tracker	Tilt (β_s)	90°-h
	Rotation (θ_s)	$\omega - 90°$
	Azimuth (γ_s)	0°

2.2 Simulation and Software

The software used for the simulation was the NREL's System Advisor Model (SAM) due to the availability of tools and ease of use. In specific, this study found of great aid the module destined to evaluate the shading effects in PV systems.

This software has been validated several times in different studies, but it is worth to note one of them that stands out. [11] found that SAM's estimation of the annual generation of a PV system has an error around the 3% compared to real on-site data.

The simulation considers the losses in the harvested energy due to the lack of cleanliness on the surface of the panels, but it does not consider the effect this could have in the mechanisms in charge of the tracking system. The idea of this paper is to analyze solely the PV system, and therefore, no other losses were considered in terms of the electricity grid. Moreover, this project is the initial part of a series of studies regarding PV installations with tracking systems, including a future experimental assessment of the configurations studied in this work.

2.3 Location

The case study considers a theoretical installation in Quito, Ecuador, on the premises of the Escuela Politecnica Nacional (*0.16° S, 78.44° W* and 2 804 m.a.s.l.).

2.4 Meteorological Data

In the proximities of the selected study site there are a few meteorological stations. Nonetheless, the equipment used to collect the solar radiation data is still very rudimental, and lack the required detail and accuracy to be used in this type of study (i.e. only recorded global radiation or daylight hours, etc.). In consequence, the meteorological data was obtained from the NREL's National Solar Radiation Database (NREL-NSRDB) because it contains a long log of historical measurements and it uses the physical model to define the radiation data. Additionally, this database was validated recently with on-site measurements [12].

We used the typical meteorological year (TMY) for the simulations. Therefore, the results obtained in this study correspond to a likely (and representative) year, not an historical one.

2.5 Modelling of losses due to shading by different types of obstacles.

The impact that shading could have in the overall performance of the PV installation with tracking mechanisms is significant. Thus, it is quite important to analyze the effect that most common obstacles can have on a tracking system.

In general terms, the losses due to shading are defined by the rate of power reduction (in percentage), defined in this paper as *S*.

The first types of obstacles studied in this paper are the modules themselves. When a PV array is configured with solar tracking, their movement in specific moments could cast shadows to adjacent arrays (refer to Fig. 4). These are defined as self-shading.

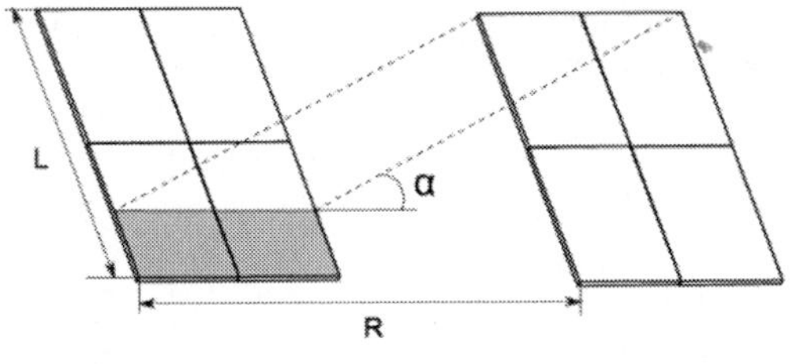

Figure 4: Shadows casted by adjacent rows of PV modules.

The relations used to model the shading effect of adjacent cells are described in equations 9 (diffuse component) and 10 (beam component):

$$S_D = \frac{I_{DT} - I_{DT}\left(\frac{2}{1+\cos\beta_s}\right)\left(1-\cos^2\frac{\alpha}{2}\right)\left(\frac{N_h-1}{N_h}\right)}{I_{DT}} \quad (9)$$

$$S_R = \frac{F1(I_B\cos Z)A + F2\left[I_D\left(\frac{2}{1+\cos\beta_s}\right)\right]A}{\rho\sin\frac{\beta_s}{2}\left[I_B\cos Z + I_D\left(\frac{2}{1+\cos\beta_s}\right)\right]} \quad (10)$$

$$F1 = \rho\sin\frac{\beta_s}{2} + \frac{\rho}{2}\left[1 + \frac{K}{L} - \sqrt{\frac{K^2}{L^2} - \frac{2K}{L}\cos(180-\beta_s) + 1}\right] \quad (11)$$

$$F2 = \rho\sin\frac{\beta_s}{2} + \frac{\rho}{2}\left[1 + \frac{R}{L} - \sqrt{\frac{R^2}{L^2} - \frac{2R}{L}\cos(180-\beta_s) + 1}\right] \quad (12)$$

$$A = \frac{(N_h - 1)}{N_h} \quad (13)$$

$$\alpha = \arctan\frac{L(\sin\beta_s)}{(R-L)(\cos\beta_s)} \quad (14)$$

$$K = R - L\left(\frac{\sin(180-\alpha-\beta_s)}{\sin\alpha}\right) \quad (15)$$

Where, *Nh* is the number of rows that conform the array in the East-West direction.

In this paper, self-shading was studied in terms of the constructive distribution of the arrays (how the modules are assembled together), and the separation between them, defined as Ground Cover Relation – GCR (Eq. 16). Three array configurations were studied: 1x4, 2x2 and 4x1 (see Fig. 5).

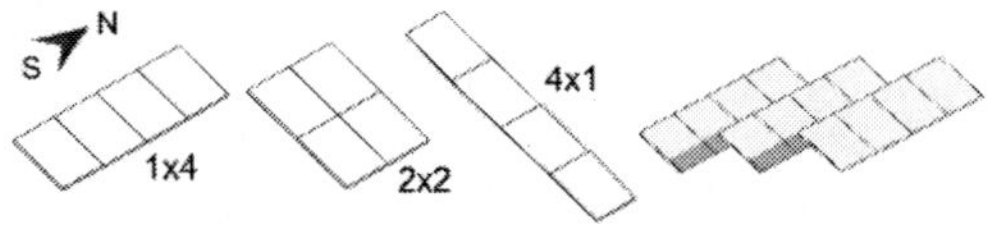

Figure 5: Shadows casted by adjacent rows of PV modules.

$$GCR = \frac{L}{R} \quad (16)$$

The chosen GCR for the study is 0.9.

The impact of the other types of shading were defined by the loss factor (*T*) that describes the percentage of the time that specific shadow has its effect in a day. Once the *T* is defined manually for all these obstacles, the affectation due to shading (*S*) is calculated with equation 17.

$$S = 1 - \frac{T}{100\%} \quad (17)$$

The second type of obstacles analyzed were the nearby buildings around the PV installation. In this case, we studied how the nearby obstacles casted shadows into the PV modules. To define the general effect of these shading, the study contemplated the existence of 8 obstacles in within the distance of 4 meters from the center of the PV array. We varied the height of those obstacles to generalize the results, as the shading effect will only be defined by its position with respect to the PV installation, and the angle that the highest point of the building describes in reference with the PV module (see Fig. 6).

Finally, the last type of shading structures studied are the far obstacles, or also defined as horizon obstacles. In this case, Quito is located between two massive mountain chains, which are surrounding the city. The city and the mountains extend in a North-South direction, so the horizon obstacles (in this case the mountains) are East and West of the city. The first step to analyze these losses was to measure the heights and positions of the obstacles and graph the profile of the horizon in the sun's path for Quito. For this, the tools available in Google Earth© were used, using as reference the location of the Escuela Politécnica Nacional.

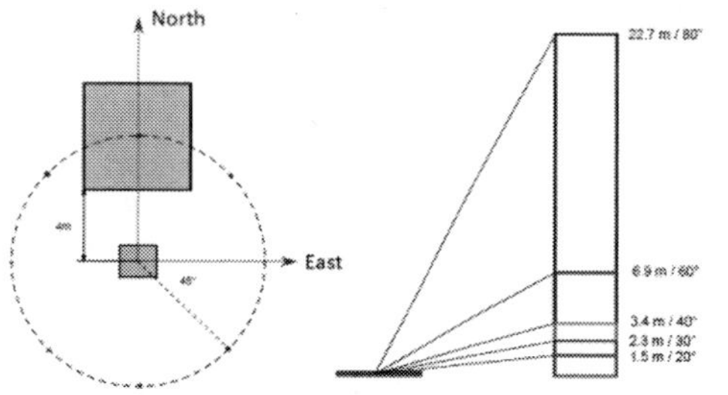

Figure 6: Shadows casted by nearby buildings, and their definition of spatial location and elevation angle.

As seen in Fig. 7, the mountains are in the path of the sun. Moreover, the west obstacle is high enough to obstruct the sun's path in the afternoon for almost an hour (from 17h00 to 18h00).

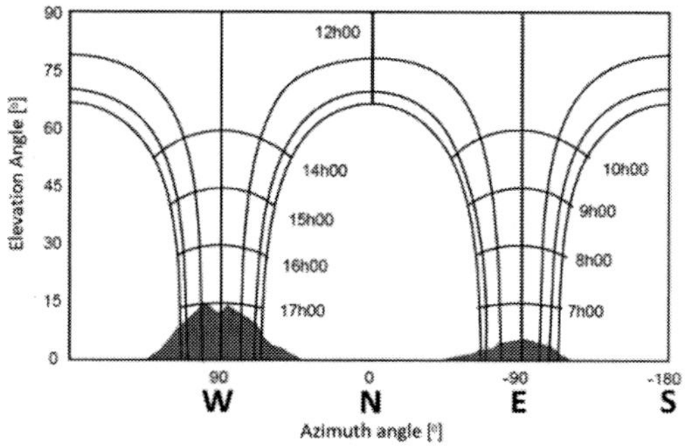

Figure 7: The horizon obstacles' profile graphed in the sun's path.

3 RESULTS

As defined in the previous section, three system configurations were modelled and studied using the meteorological data of Quito, Ecuador. This section will show the different results of the simulation.

3.1 Improvements of the energy harvested due to tracking

Table II. shows the theoretical improvement in the cumulative energy incident over a normalized 1 m^2 of PV installation, for the three configurations.

As expected, solar tracking gives an important advantage in comparison to a horizontal fixed position in the same location. On one hand, two-axes tracking reaches the highest improvement, with 34.62% more of the harvested electricity annually. On the other hand, one-axis tracking reaches improvements around the 30%. Thus, both improve the energy output significantly.

When comparing both tracking systems, the difference of their annual performance is low (approx. 4%). In the case of the monthly electricity output (see Fig. 8), it is possible to identify that even in the worst-case scenario (in the solstices), the difference between 1-axis and 2-axes trackers is not higher than 6.5%.

Table II: Monthly and Annual energy harvested for the three configurations of the PV systems [kWh/m^2].

Month	Fixed	1-axis	2-axes
January	166.51	215.70	226.26
February	140.33	177.79	180.61
March	146.85	174.58	174.80
April	144.46	185.82	187.71
May	154.69	192.85	200.33
June	151.54	194.39	207.00
July	172.19	235.89	249.30
August	178.55	236.65	242.26
September	167.57	215.60	216.00
October	166.93	204.19	205.97
November	150.33	189.50	196.14
December	149.00	181.43	190.63
Year (TMY)	1888.95	2404.39	2477.01

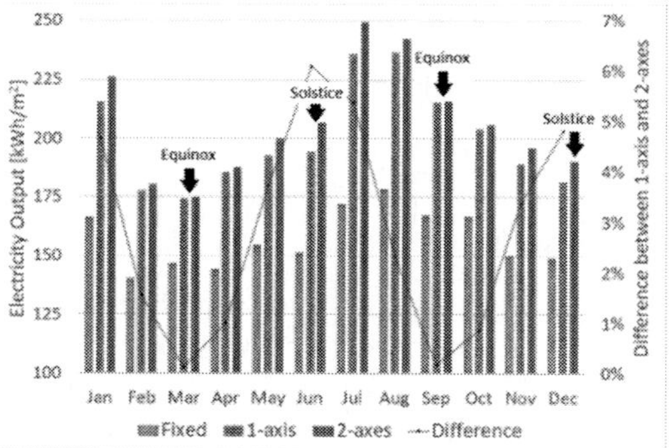

Figure 8: Monthly electricity output for the three PV systems configurations (bars). Difference of electricity output between 1-axis and 2-axes (line, expressed in percentage).

3.2 Shading impacts in the performance of the system

In terms of the arrays' configuration of the PV modules, the simulation defined that the best configuration to avoid excessive self-shading is the 4x1. This does not represent that all arrays should contain four modules aligned together, but the fact that the arrays should be defined in an East-West direction. Actually, the worst is the North-West configuration (refer to Fig.9).

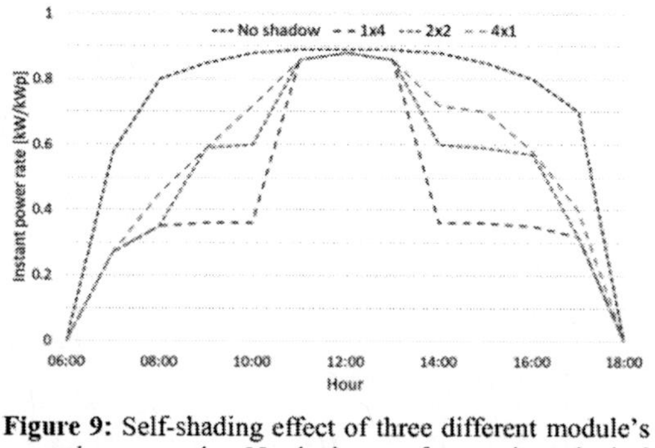

Figure 9: Self-shading effect of three different module's arrays in a sunny day. No shadows, refers to a hypothetical scenario in which the self-shading does not exist.

Furthermore, about the far obstacles (horizon), the simulation showed that these types of shadings do not affect significantly the performance of the PV with tracking systems. Since, both tracking systems only reduced their annual electricity output by 2.6% when these losses were considered (see Fig. 10). However, it is also worth to mention that the fixed configuration had the lower reduction in the performance.

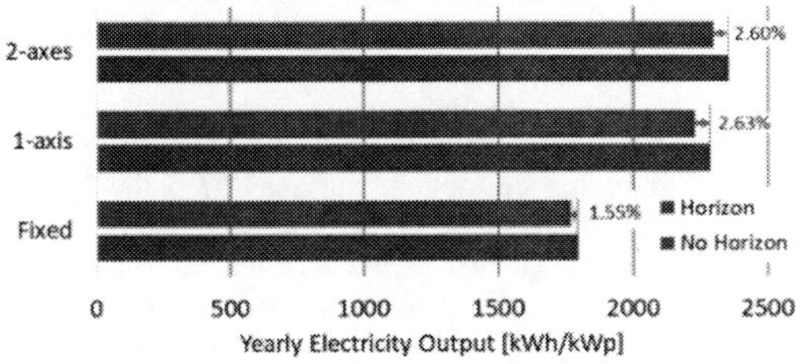

Figure 10: Effects of shadows casted by horizon obstacles in the performance of the three configurations of PV systems.

Finally, in the subject of the nearby obstacles, the results obtained suggest a significant impact on the annual performance from high buildings located East of the PV installation (refer to Fig. 11). Interestingly, this effect does not have the same impact from buildings located West of the installation. Similar trends are observed in the NE and SE in relation to the NW and SW counterparts. North and South obstacles cast an important shading into the installation, but according to our findings, there is not as great as in the case of the East of West obstacles.

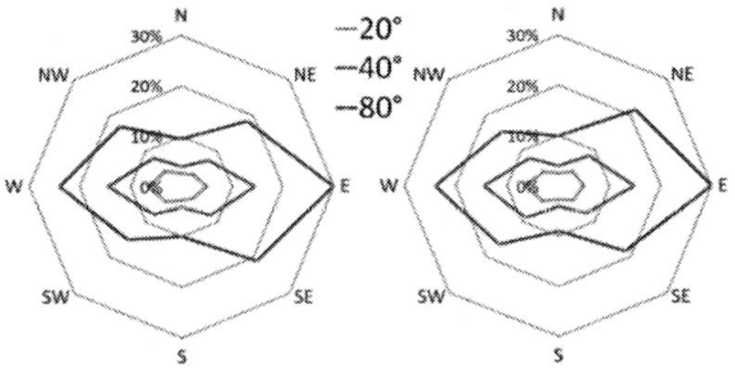

Figure 11: Shadows casted by nearby obstacles and the effects on the PV system defined by height (described by the angle) and position of the obstacle in relation to the PV installation.

4 DISCUSSION

Solar tracking systems in equatorial latitudes do represent a viable alternative to significantly improve the electricity output of a PV installation. In this matter, there are two main factors that affect the performance of solar tracking PV systems in these regions: 1) Yearly sun's path, 2) Meteorological conditions.

4.1 Sun's path

On one hand, the sun's path affects greatly the annual electricity output. Taking the point of view of an observer on the equatorial line, the sun's position in the sky defines a small difference in the elevation angle through the year (±23° in the solstices), but a constant rotation angle (180° all year long, from east to west) as these latitudes have around 12 hours of sunlight yearlong. This fact is the main reason why trackers in general harvest a significantly higher amount of energy annually, and especially why a horizontal 1-axis tracker has a similar annual power output than a double horizontal 2-axes tracker as they do not have to make great changes in their tilt angle to position themselves facing the sun. Therefore, the results obtained in this study contrast with the existing literature in the fact that 2-axes trackers are found to be more effective in increasing the electricity output than 1-axis trackers (for instance as described in [13]). According to our results, the difference between the energy harvested by the two tracking systems is lower than 7%.

Therefore, one can infer that for equatorial latitudes the 1-axis tracking system provides more benefits than the two-axes tracking systems as the latter implies more energy consumption and higher costs (as seen in [12]).

In a similar fashion, our results suggest that the tracking system should have an array of modules based on a 4x1 configuration, in which the adjacent modules are connected to each other in rows align with an East-West direction. This is because the shadows casted by the rows of this installation affect only the tilt angle and not the rotation one. Interestingly, most of the PV installations in Ecuador (not tracking) have a North-West configuration of arrays, which indicates that the results obtained in this study could have an important role in the future deployment of PV power plants, especially the ones using solar tracking to improve their performance.

These results have a global connotation as the sun's path is similar in all equatorial regions and depends only on the latitude. Thus, similar results could be observed in other equatorial latitudes despite the climate or specific geographical features.

4.2 Meteorological conditions

On the other hand, the meteorological conditions affect the specific location of the PV installation. In terms of solar radiation, the TMY of Quito describes a unique condition in which, historically, the irradiation early in the morning is higher than later in the afternoon because of a more abundant presence of clouds in the sky that block the direct component of the solar radiation. In consequence, the overall available radiation is lower than in the morning, which translates in a lower impact of the possible obstacles located in the west. This is the case for the horizon shadows, in which despite having a high mountain located west of the city (4 784 m.a.s.l.), the results suggest that this obstacle located in the horizon of the city does not represent a major obstruction to the possible energy harvested by a solar tracking device, even though it is situated in the annual path of the sun (as seen in Fig 6., Pichincha mountain nearly obstructs one hour of solar radiation from 17:00 to 18:00). The lower mountain in the east affect little to the performance, not because of its location, but due to its height. It was not studied the effect of a high mountain in the east side of the city, but it is expected in this case that the result could change meaningfully.

Accordingly, the high nearby obstacles located east of the installation have a greater impact on the system's performance than the ones located in the west, even though they have similar theoretical hours of sunlight. The results also indicate that the shadows casted by obstacles located North or South of the installation have a similar effect on the yearly output of the installation.

At the end, these results obtained for Quito could not

be extrapolated to other equatorial latitudes as the climatology of the city is different due to many geographic features. It is obvious that these results are going to change significantly if the study place is in a coastal area, or in a lower height highland, despite of being located in an equatorial latitude.

5 CONCLUSIONS

Solar tracking systems in equatorial latitudes do represent a viable alternative to significantly improve the electricity output of a PV installation. However, there are many aspects that should be analyzed before the implementation of the tracking system to avoid significant losses due to shadows and the configuration of the arrays of modules.

The 2-axes double horizontal tracker harvested a slightly higher amount of solar energy annually than the 1-axis tracker. Therefore, the latter was suggested to be the best option for a tracking system in an equatorial latitude as it represents the most cost-effective alternative.

In terms of the possible losses due to the shadows casted by different obstacles inside and outside the installation it was found that the arrays of modules should be defined in an east-west configuration to avoid excessive self-shading.

The results also suggest that the mountainous landscape in the horizon of Quito does not affect significantly the annual electricity output of a tracking system.

Finally, nearby obstacles could cast the most influential shadows that would reduce significantly the energy harvested by a solar tracking system. Thus, it is strongly suggested that the installation should avoid nearby and high obstacles, especially those located east or west of the PV modules.

6 ACKNOWLEDGEMENTS

The authors of this study would like to thank the Escuela Politécnica Nacional for the support and funding for this project (PIJ 15-07).

7 REFERENCES

[1] IEA, "World Energy Outlook 2015," 2015.

[2] D. F. Tamayo, "Diseño de un seguidor Solar Pasivo para uso con diversos paneles de captación de energía solar," Universidad Autonoma de Yucatan, 2011.

[3] S. S. Hegedus and A. Luque, *Handbook of photovoltaic science and engineering*. John Wiley {&} Sons, 2003.

[4] A. Coelho and R. Castro, "Modeling and validation of PV power output with solar tracking," in *2011 International Conference on Power Engineering, Energy and Electrical Drives*, 2011, pp. 1–6.

[5] H. O. Njoku, "Upper-limit solar photovoltaic power generation: Estimates for 2-axis tracking collectors in Nigeria," *Energy*, vol. 95, pp. 504–516, Jan. 2016.

[6] S. Heslop and I. MacGill, "Comparative analysis of the variability of fixed and tracking photovoltaic systems," *Sol. Energy*, vol. 107, pp. 351–364, Sep. 2014.

[7] R. G. Vieira, F. K. O. M. V. Guerra, M. R. B. G. Vale, and M. M. Araújo, "Comparative performance analysis between static solar panels and single-axis tracking system on a hot climate region near to the equator," *Renew. Sustain. Energy Rev.*, vol. 64, pp. 672–681, 2016.

[8] J. Ranalli, R. Vitagliano, M. Notaro, and D. J. Starling, "Sensitivity of shading calculations to horizon uncertainty," *Sol. Energy*, vol. 144, pp. 399–410, 2017.

[9] C. Demain, M. Journée, and C. Bertrand, "Evaluation of different models to estimate the global solar radiation on inclined surfaces," *Renew. Energy*, vol. 50, pp. 710–721, Feb. 2013.

[10] A. G. Touchaei, H. Akbari, and C. W. Tessum, "Effect of increasing urban albedo on meteorology and air quality of Montreal (Canada) – Episodic simulation of heat wave in 2005," *Atmos. Environ.*, vol. 132, pp. 188–206, May 2016.

[11] J. Freeman, J. Whitmore, L. Kaffine, N. Blair, and A. P. Dobos, "System Advisor Model: Flat Plate Photovoltaic Performance Modeling Validation Report," *Nrel/Tp-6a20-60204*, no. December, p. 129, 2013.

[12] C. E. Morales Cumba, "Estudio Energético en Sistemas Fotovoltaicos empleando seguimiento solar en regiones ecuatoriales," Escuela Politécnica Nacional, 2017.

[13] S. Deepthi, A. Ponni, R. Ranjitha, and R. Dhanabal, "Comparison of Efficiencies of Single-Axis Tracking System and Dual-Axis Tracking System with Fixed Mount," *Int. J. Eng. Sci. Innov. Technol.*, vol. 2, no. 2, pp. 425–430, 2013.

ADVANCED FAILURE DETECTION ALGORITHMS AND PERFORMANCE DECISION CLASSIFICATION FOR GRID-CONNECTED PV SYSTEMS

Andreas Livera[1], Alexander Phinikarides[1], George Makrides[1], Juergen Sutterlueti[2] and George E. Georghiou[1]
[1]FOSS Research Centre for Sustainable Energy, Photovoltaic Technology Laboratory,
Department of Electrical and Computer Engineering, University of Cyprus
75 Kallipoleos Avenue, P.O. Box 20537, Nicosia, 1678, Cyprus
[2]Gantner Instruments GmbH, Montafonerstraße 4, 6780 Schruns, Austria
(Telephone: +357 22 894398, Fax: +357 22 895370, email: livera.andreas @ucy.ac.cy)

ABSTRACT: The scope of this paper is to present the development of failure detection routines (FDRs) that will operate on acquired data sets of grid-connected PV systems and determine and classify accurately the exhibited failures. The developed FDRs comprise of a failure detection and a classification stage. Specifically, the implemented failure detection stage was based on a comparative algorithm that detected discrepancies between the measured and simulated electrical measurements (dc current, voltage and power of the array) of a test PV system. Furthermore, statistical algorithms for identifying outliers, anomalies and normal system operation limits were also used for failure detection. Accordingly, for each identified failure there was a subsequent decision stage which performed a classification process based on developed logic and decision trees. The decision trees were constructed with a supervised learning process, trained with continuous samples split in a 70:30 % train and test set approach of acquired data sets which included the feature patterns exhibited during normal and faulty operation. The results obtained by emulating three faults (shading, inverter and bypass diode failure), showed that the developed FDRs were capable of detecting accurately the faults upon their occurrence. Finally, when applying binary classifier to the test set which included several emulated failure patterns, the obtained results demonstrated that the classification accuracy was 98.7 % for the inverter failure, 95.3 % for bypass diode fault and 96.6 % for partial shading.
Keywords: decision trees, failure detection routines, outliers, photovoltaics.

1 INTRODUCTION

Over the last few years, research into photovoltaics (PV) has shifted from a race for the highest efficiency to the increase of the performance reliability in the field. A major part of current research activities focuses on the reliability of the installations and the guaranteed lifetime output. In this domain, the accurate identification of performance losses and failures in grid-connected PV systems is crucial for their further penetration in the energy mix [1]. PV performance losses and failures can occur during the operational lifetime of a PV system due to different factors such as maximum power point (MPP) tracking error, electrical disconnection, wiring losses, shading effects, and faulty equipment [2]. Such faults decrease the output power of the PV system and also degrade the module's properties.

In order to ensure reliability and high performance guaranteed output, most PV installations around the world are by default continuously monitored, either via the inverter or via proprietary monitoring hardware and software / dashboards. The vast majority of monitoring systems show a simplistic view of the energy yield and Performance Ratio (*PR*) and more rarely, the most sophisticated of these dashboards can provide an overview of the expected energy yield and a comparison to the measured energy yield and can signal the operator when the PV plant has failed. For the expected PV power production, different predictive models (based on parametric modelling and machine learning approaches) are used [3], [4]. On the other hand, the most common implementation of failure detection is performed at the inverter level. Even in the case that no monitoring system is set up, the inverter can be used to detect grid errors (voltage and frequency deviations) or array faults that might be dangerous, such as arc or ground faults and act to ensure the safety of the owner and the installation. With the current advances in the machine learning field and the rapid developments and innovations in data analysis algorithms, failure detection and classification of critical infrastructure has attracted a lot of attention at many different levels. In the case of PV plants however, the current state of the art within the PV industry for automated identification and classification of even the most common faults is simplistic or proprietary. Most published implementations are concerned with defining thresholds for each monitored parameter and comparing the measured value to the threshold in order to make a decision on the fault condition [2]. These thresholds can be either static (i.e. predefined) or dynamic (i.e. there is feedback from the stream of measured values). The advantages of these approaches include the simplicity of implementation whereas the disadvantages include inaccuracy and false prediction. Other implementations search for discrete changes in performance from individual sensors, instead of cross-correlating measurements from multiple sensors or taking into account the history of the measurement [5].

In this perspective, the implementation of fault analysis tools is essential to ensure the reliability and optimal performance of PV systems by quickly identifying and accurately quantifying the factors behind the various failure mechanisms. The ability to detect and diagnose potential failures at an early stage or before occurrence is also crucial to reduce costs associated with operation and maintenance (O&M) and system downtime.

The scope of this work is to propose a methodology for the development of failure detection routines (FDRs) that will operate on acquired data sets of grid-connected PV systems and determine accurately the exhibited failures. The developed routines were based on the comparison between the measured data with modelling prediction techniques as well as on outlier detection rules. Data driven-algorithms were also used in order to analyse the detected fault by conducting logic and decision trees to achieve reliable outcomes on the occurred fault.

2 METHODOLOGY

2.1 Experimental setup

At the outdoor PV Technology test site of the University of Cyprus (UCY), grid-connected PV systems of different technologies and approximately 1 kWp capacity each, were installed and commissioned in June 2006 [6]. The performance of each PV system and the prevailing meteorological conditions were recorded according to the requirements set by the IEC 61724 [7] and stored with the use of an advanced measurement platform.

The measurement platform comprises of solar irradiance (Kipp & Zonen CM21 pyranometer), temperature (PT 100) and electrical (potential divider, shunt resistor, NRZ) sensors connected to a central data acquisition system that stores data every second. The measurements were also averaged every one- and fifteen-minutes and stored in the database. Specifically, the meteorological measurements include the incident global irradiance at the POA (G_I), wind direction (W_a), wind speed (W_s) and ambient temperature (T_{am}). The PV system's operational measurements include the module temperature (T_m), maximum dc current (I_{mp}), voltage (V_{mp}) and power (P_{mp}), as measured at the output of each PV system (dc side).

The PV system used for the emulation of faults comprises of 5 poly-crystalline Silicon (poly-c Si) PV modules (see Fig. 1), each of nominal power 235 Wp as depicted by the manufacturer's datasheet. The system is connected in series to form one string at the input of a grid-connected inverter. The system is installed in an open-field mounting arrangement due to South, at the optimum annual energy yield angle for Cyprus of 27.5°.

Three emulated faults (inverter failure, partial shading and bypass diode fault) were introduced to the PV system and formed the basis for the training and testing phase of the developed routines.

Figure 1: Grid-connected PV system used for the emulation of failures.

2.2 Approach

The developed FDRs comprise of a failure detection and a classification stage. The failure detection stage was based on a comparative algorithm that detected discrepancies between the PV system measured and simulated electrical measurements (dc current of the array, voltage and power), obtained using empirical parametric models for each day. In parallel, the failures were also detected by comparing the operating points with normal operation limits set by defined thresholds (three-sigma limits) over a weekly time window and verified using statistical outliers detection techniques.

Accordingly, the failure classification stage was based on data-driven algorithms, which were used to analyse the detected fault through the application of developed logic and decision trees. The constructed logic and decision trees considered the affected parameters that are relevant to the fault pattern and thus required historical data sets in order to learn the systematic performance behaviour.

2.3 Failure detection – Comparative to empirical parametric method

The failure detection algorithm stage was based on the idea of comparing the measured against the simulated daily PV performance. More specifically, the failure detection stage encompassed a comparative algorithm that detected discrepancies between the PV system measured and simulated electrical measurements (current, voltage and power of the array) at the dc side, obtained using empirical parametric models for each point. A failure was identified when significant deviations between the measured and simulated parameter were detected, signifying a noticeable performance discrepancy. The detection technique used in this study, required as input features the acquired incident global irradiance at the POA, PV systems' specifications and module temperature.

In particular, the model used for the simulated array dc power production at MPP as part of the detection stage of the FDRs was based on a parametric approach according to the following equation [2].

$$P_{mp} = G_I(a_1 + a_2 G_I + a_3 \log(G_I)) \cdot (1 + \gamma(T_m - T_{STC})) \qquad (1)$$

where γ is the maximum power temperature coefficient (%/°C), T_{STC} is the Standard Test Conditions (STC) temperature of the cell (°C) and the regression coefficients a_1, a_2 and a_3 are calculated so that the model result is as close as possible to the monitored data.

Additionally, the empirical parametric model utilized to simulate the maximum voltage produced by a PV system is given by:

$$V_{mp} = V_{mp,STC} \cdot (1 + \gamma(T_m - T_{STC})) + (\beta_1 \log \frac{G_I}{G_{STC}}) + \beta_2 \qquad (2)$$

where $V_{mp.,STC}$ is the STC maximum voltage (V), G_{STC} is solar irradiance at STC and the regression coefficients β_1 and β_2 are calculated so that the model result is as close as possible to the monitored data.

Respectively, the empirical model to simulate the maximum current is defined as:

$$I_{mp} = \frac{G_I}{G_{STC}} \cdot I_{mp,STC} \cdot (1 + \gamma(T_m - T_{STC})) + a_0 \qquad (3)$$

where $I_{mp,STC}$ is the STC maximum current (A) and a_0 is the regression coefficient calculated to fit the model result to the monitored data.

2.4 Failure detection – Comparative mean 3σ limit method

In this approach failures were detected when normal operation limits set by defined thresholds were exceeded. The thresholds imposed were the three-sigma limits (a statistical calculation that refers to data within three standard deviations from the average) which are commonly used to set the upper and lower control limits in statistical quality control charts [2]. Control charts were used to establish and visualize the investigated operation and limits. Specifically, the normal operation limits of the PV monitoring system were calculated using the ratio between the measured and simulated electrical measurements (dc current, voltage and power of the array). The ±3 standard deviation (±3σ) interval was used to calculate the normal operation limits as follow:

$$\text{Lower limit} = \mu - 3\sigma \quad (4)$$

$$\text{Upper limit} = \mu + 3\sigma \quad (5)$$

where μ and σ are the average and standard deviation, respectively, of the values of the ratio between the measured and simulated dc current, voltage and power. The ratio was used to determine how close the measurements were to their simulated values, according to the PV production model; the closer the ratio was to 1, the closer the measured parameters were to the modelled values. An additional step taken to reduce high bias errors occurring at low irradiance levels was to filter the data sets at global irradiance levels over 50 W/m^2, because system performance is affected by sunlight levels [2].

2.5 Failure detection – Statistical approach

Statistical algorithms were used for the identification of global and local outliers and anomalies. More specifically, the Local Outlier Factor (LOF) algorithm was used for identifying density based local outliers [8]. With LOF, the local density of each point is compared with that of its k-Nearest Neighbours (k-NN) and if that point has a significantly lower density than of its neighbours (with an LOF value greater than one), the point is in a sparser region than its neighbours, which suggests to be an outlier.

In addition, the Bonferroni Outlier Test algorithm was used for outlier testing. This algorithm reports the Bonferroni p-value for the most extreme observation based on the linear regression model of the measured and simulated dc power of the array.

Moreover, the Seasonal Hybrid Extreme Studentized Deviates (S-H-ESD) algorithm was implemented for detecting anomalies in a time series data-set that follows an approximately normal distribution. The S-H-ESD algorithm is an extension of the Generalized Extreme Studentized Deviates (ESD) and can be used to detect both global and local anomalies by employing time series decomposition and using robust statistical metrics together with ESD [9].

2.6 Failure classification – Logic and decision tree

For the classification stage of the detected failure patterns, logic and decision trees were constructed in order to categorize the different detected fault features.

A logic tree is a simple top-down approach commonly employed to tackle a problem by breaking possible solutions down into parts. In this approach the detected condition was dissected into different sub components in order to yield more information about the fault condition and designate a fault condition.

Contrary, a decision tree is a flow-chart-like tree structure used to detect the faulted points from normal ones. A decision tree consists of a root and internal (decision) nodes. Starting from the root node, every instance will be split by the test at internal nodes and proceed down to the terminal nodes to classify either normal operation or fault condition [10]. For the decision trees implemented in this work, supervised learning based models were constructed with emulated faults for the classification stage. The process to construct such a decision tree, which can operate in a learning approach to classify fault conditions, was divided into four steps. First, meteorological and PV operational data were collected in both normal and emulated fault conditions. In the second step, the experimental data were carefully examined for missing or erroneous data. Then, specific feature labels were inserted to the data in order to assist the learning model phase for classification. The third step, was to train the decision tree model by using 70 % of randomly chosen data, while the remaining 30 % of the data were used to test the model at the final step.

The k-NN algorithm was used for object classification. For each data point, the k-NN algorithm finds the k closest observations (the ones with the smallest Euclidean distance to the data point under consideration), and then classifies the data point to the majority. To assess the performance of the classification model on the test set which included emulated failure patterns, confusion matrices of a binary classifier were used. The four outcomes of the binary classifier are defined as true positive (TP) (correct positive predictions), true negative (TN) (correct negative predictions), false positive (FP) (incorrect positive predictions) and false negative (FN) (incorrect negative predictions). The classification accuracy of the binary classifier is defined as the ratio of correct predictions to total predictions made and is calculated as follows:

$$Classification\ accuracy = \frac{TN + TP}{TN + TP + FN + FP} \quad (6)$$

3 RESULTS

3.1 Inverter failure

Inverter failure was emulated by switching off the inverter of the test system, while the measurement platform continued to function properly. Since the measured power production during the fault, which occurred during a clear sky day, was significantly lower than the simulated power production (beyond set threshold of < 1 % × simulated power), the FDRs detected the power loss and required further analysis in order to classify the fault (Fig. 2). Besides, the measured current was nearly zero during the fault occurrence, while the measured voltage was approximately equal to the system's open circuit voltage. All three electrical parameters were affected in this failure scenario.

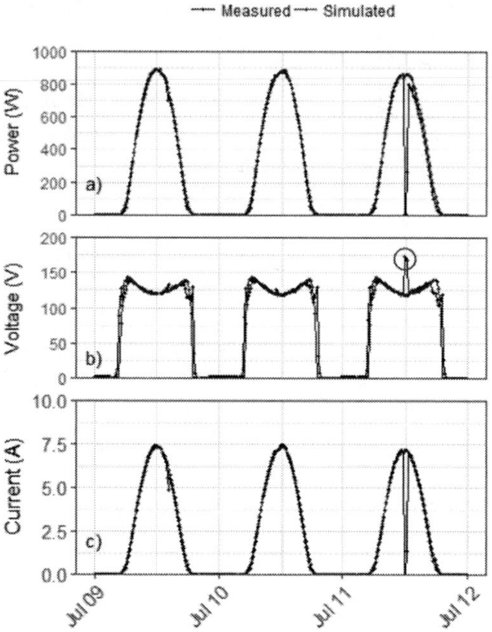

Figure 2: Measured and simulated electrical parameters of the test PV system. Inverter fault occurred on 11th of July.

Furthermore, the inverter fault was identified by the S-H-ESD algorithm when applied to the array voltage data sets and verified the occurrence of an anomaly. The algorithm successfully detected the anomaly on the 11th of July and the abnormal point due to the inverter fault is highlighted with a red circle in Fig. 2b.

As in the previous cases, the inverter fault was also detected by employing the comparative mean 3σ limit method and the results exhibited by the dc power ratio control chart (Fig. 3), clearly indicate the occurrence of the fault as an outlier exceeding the normal operational limits (3σ).

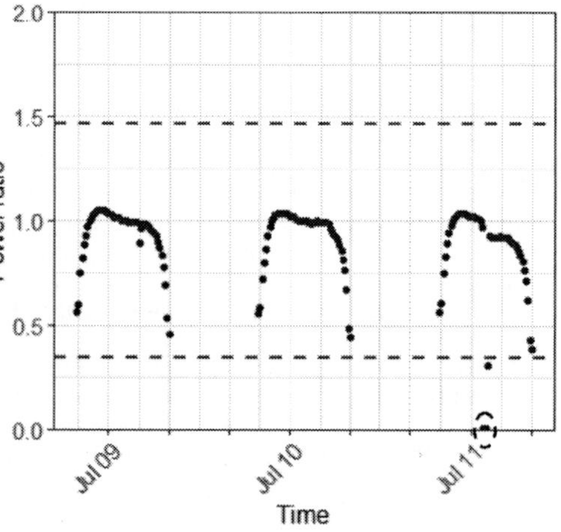

Figure 3: Power ratio control chart. The horizontal lines indicate the 3σ lower and upper limits.

For the classification stage of the fault, data-driven algorithms were developed based on logic and decision trees. The logic tree which was successful in detecting and classifying the inverter fault is shown in Fig. 4.

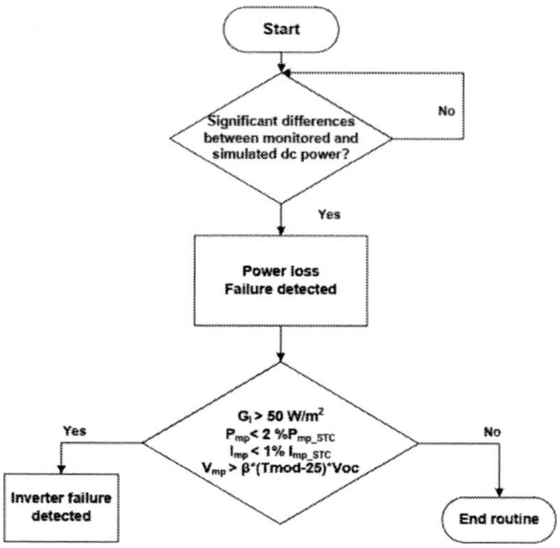

Figure 4: Logic tree for classifying inverter failure.

Similarly, the constructed decision tree included a root node, which defined a voltage comparison with a threshold value evaluated during the training process and then was split by comparing the electrical (P_{mp} and I_{mp}) and meteorological (G_I) measurements to three subsequent internal nodes. The final terminal nodes exhibited the classification of either normal or fault condition.

When applying a binary classifier to 700 instances of the test set (including three emulated inverter failures), the decision tree obtained results showed that the classification accuracy was 98.7 % when using the optimal k value for classification (k=5).

3.2 Bypass diode fault

The bypass diode failure was emulated by short-circuiting one of the three bypass diodes located at the junction box of a module of the test system. As depicted in Fig. 5, the measured power produced over the fault applied period, 12th of June until the 16th of June 2017 (four days), was lower than the simulated power production (outside the measurement uncertainty of ±2 % and beyond a set power threshold level calculated by the ratio of faulty to total diodes of the array multiplied by the nominal power of the system). In addition, for this particular period, the voltage produced by the investigated PV system was approximately 10 Volts (ratio of faulty to total diodes of the array multiplied by the nominal voltage of the system) lower than the simulated voltage and exhibiting the same pattern shift for all days (Fig. 5), while the difference between the measured and simulated current was negligible. For this failure scenario the electrical parameters affected were the power and voltage.

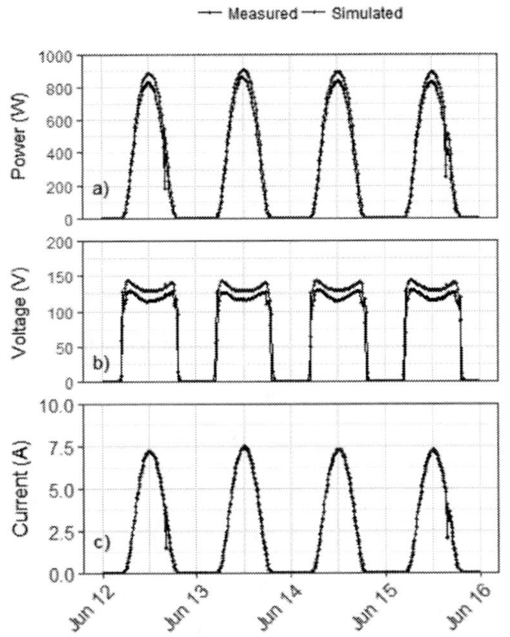

Figure 5: Measured and simulated electrical parameters of the test PV system. Bypass diode failure occurrence over four days.

For this particular failure, the steps of the developed logic tree that successfully detected and classified the bypass diode fault are presented in Fig. 6.

Similarly, the constructed decision tree performed a power comparison with a threshold value at the root node and then was split to compare the electrical (V_{mp} and I_{mp}) and metereological (G_I) measurements into three internal nodes. The final terminal nodes exhibited the classification of either normal or fault condition.

The results showed that for 900 instances and using the 5 closest observations for the classification stage, the accuracy when the decision tree was applied to the test set (including 4 days of emulated bypass diode failures) was 95.3 %.

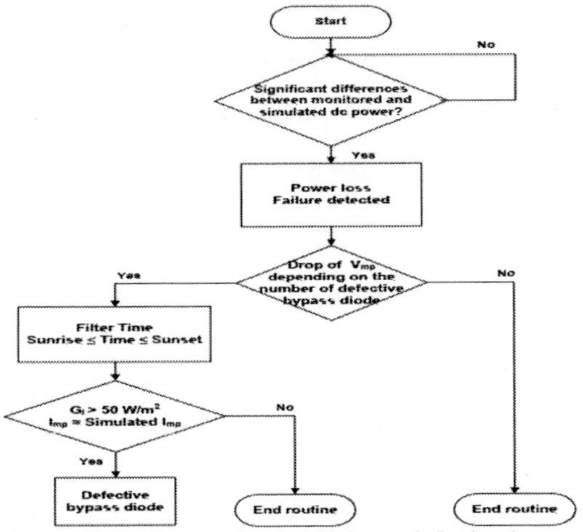

Figure 6: Logic tree for classifying a defective bypass diode failure.

3.3 Shading

Partial shading of the array was achieved by placing an object on the surface of a PV module of the array, for a specific time duration while the solar irradiance was relatively constant. As illustrated in Fig. 7 the measured power production on day 4 (10th of August) when the effect was emulated, was significantly lower than the simulated power production. In addition, during day 4, the measured current was significantly lower than the simulated one. Likewise, the measured voltage was significantly lower than the simulated maximum voltage produced by the PV system. All three electrical parameters were affected in this failure scenario.

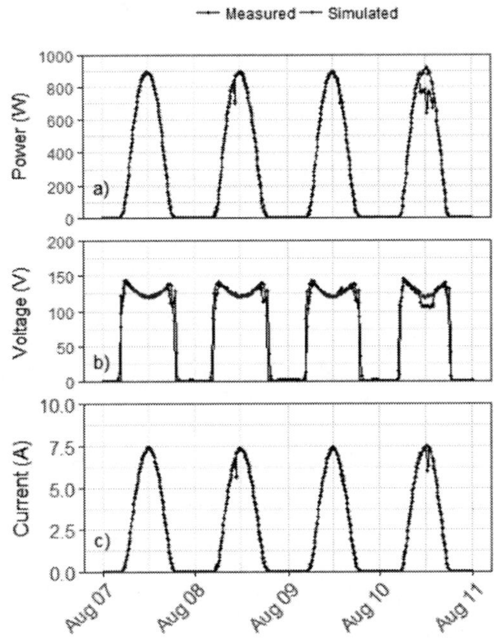

Figure 7: Measured and simulated dc power of the test PV system. Partial shading (one module partially shaded) occurrence during day 4.

Applying the Bonferroni Outlier Test algorithm to verify the occurrence of partial shading and based on the given data set, the output of the algorithm reported that the most extreme observation occurred on day 4 during midday.

For the classification stage, the developed logic tree is presented in Fig. 8.

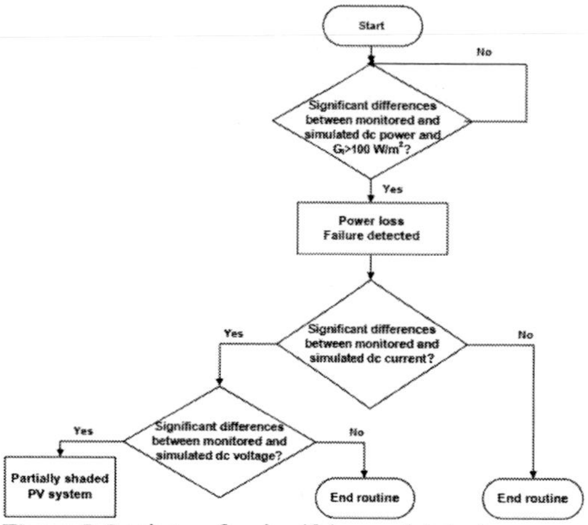

Figure 8: Logic tree for classifying partial shading.

Finally, the developed decision tree performed a power comparison with a threshold value at the root node and then was split into three internal nodes in order to test the electrical (V_{mp} and I_{mp}) and meteorological (G_I) measurements. The final terminal nodes exhibited the classification of either normal or fault condition.

The results showed that for 700 instances and using the optimal k value for classification (k=5), the classification accuracy on the test set (including two emulated faults) was 96.6 %.

4 CONCLUSIONS

PV performance losses and faults can occur during the plant operation due to different factors such as maximum power point tracking error, electrical disconnection, wiring losses, shading effects, dust and faulty equipment. Advanced PV monitoring systems encompassing FDRs can provide a clear indication whether the system is performing as expected and therefore safeguard the optimal performance and reliability.

The obtained results clearly demonstrate that the developed FDRs are capable of detecting specific failures associated with grid-connected PV systems. Moreover, the developed FDRs can classify the faults by applying well designed logic and constructed decision trees. In particular, to assess the performance of the developed routines, three emulated faults (inverter failure, bypass diode fault and shading) were introduced to a grid-connected PV system. The emulated faults were accurately detected by the detection stage and the experimental results showed good accuracy of fault classification stage on the test set.

Finally, the classification accuracy was 98.7 %, 95.3 % and 96.6 % for inverter failure, bypass diode fault and partial shading, respectively.

5 ACKNOWLEDGEMENTS

This work was funded through the IPERMON project which is co-financed by the SOLAR-ERA.NET Transnational Calls of the European Union under the grant agreement number SOLAR-ERA.NET/1214/08. The authors would also like to acknowledge the financial support from the Cyprus Research Promotion Foundation.

6 REFERENCES

[1] A. Woyte, M. Richter, D. Moser, M. Green, S. Mau, and H. G. Beyer, "Analytical Monitoring of Grid-connected Photovoltaic Systems - Good Practice for Monitoring and Performance Analysis.," *Tech. rep., IEA-PVPS T13-03: 2014.*

[2] R. Platon, J. Martel, N. Woodruff, and T. Y. Chau, "Online Fault Detection in PV Systems," *IEEE Trans. Sustain. Energy*, vol. 6, no. 4, pp. 1200–1207, 2015.

[3] R. Platon, S. Pelland, and Y. Poissant, "Modelling the Power Production of a Photovoltaic System: Comparison of Sugeno-Type Fuzzy Logic and PVSAT-2 Models," in *Europe Solar Conference (ISES)*, 2012.

[4] W. Chine, A. Mellit, A. M. Pavan, and S. A. Kalogirou, "Fault detection method for grid-connected photovoltaic plants," *Renew. Energy*, vol. 66, pp. 99–110, 2014.

[5] N. Gokmen, E. Karatepe, S. Silvestre, B. Celik, and P. Ortega, "An efficient fault diagnosis method for PV systems based on operating voltage-window," *Energy Convers. Manag.*, vol. 73, pp. 350–360, 2013.

[6] A. Phinikarides, G. Makrides, B. Zinsser, M. Schubert, and G. E. Georghiou, "Analysis of photovoltaic system performance time series: Seasonality and performance loss," *Renew. Energy*, vol. 77, pp. 51–63, 2015.

[7] IEC 61724:1998, "Photovoltaic system performance monitoring — Guidelines for measurement , data exchange and analysis," *1st ed. Geneva, Switz. IEC*, 1998.

[8] M. M. Breunig, H.-P. Kriegel, R. T. Ng, and J. Sander, "LOF: Identifying Density-Based Local Outliers," *Proc. 2000 Acm Sigmod Int. Conf. Manag. Data*, pp. 1–12, 2000.

[9] B. Rosner, "Percentage Points for a Generalized ESD Many-Outlier Procedure," *Technometrics*, vol. 25, no. 2, pp. 165–172, 1983.

[10] L. R. Zhao Y, Yang L, Lehman B, J. F. de Palma, Mosesian J, "Decision tree-based fault detection and classification in solar photovoltaic arrays," *27th Annu. IEEE Appl. Power Electron. Conf. Expo.*, 2012.

CHARACTERIZATION OF A STAND-ALONE PV COOLING/HEATING SYSTEM

Celena Lorenzo*, Luis Narvarte
Instituto de Energía Solar - Universidad Politécnica de Madrid, 28031 Madrid, Spain
*Phone: +34 91 336 55 31, mail: celena.lorenzo@ies.upm.es

ABSTRACT: Recently, the environmental objectives set by the European Union to reduce CO_2 emissions and to increase the renewable share, together with the reduction of PV prices since 2010, have motivated a great interest in PV heat pump systems for heating and cooling applications. This papers describes the initial characterization test and results of a stand-alone PV heat pump prototype installed in Madrid (Spain), using a water tank as thermal storage instead of batteries as electricity storage (a more expensive and technically complex solution). This initial test has the objective of characterizing the response of the converter that operates the compressor of the heat pump. Such converter was powered by a constant frequency input (the electric grid), in order to quantify the variable frequency output that the compressor demands. In future work, the converter will be powered by a PV array (which means a variable frequency input) and programmed to provide the required output without overcoming any limit value of the system. This way, the heat pump system will be operated completely stand-alone.

Keywords: characterization, heat pump, qualification and testing, refrigeration, stand-alone PV systems.

1 INTRODUCTION

In the last decades, the demand of reversible heating and cooling systems has increased importantly, together with the need of generating energy in a more efficient and sustainable way. According to the European Heat Pumps Agency, heat pumps market increased by 3.5% in terms of units sold in 2014 (which means an installed electrical power of 1.88 GW) [1] and by 12% in 2015 (which means an installed electrical power 2.11 GW) [2]. However, even if heat pumps are one of the most efficient technologies for conditioning applications, the overall power consumption in Europe in this sector keeps increasing every year, so bigger efforts must be dedicated to achieve the environmental objectives of the European Union for the year 2020 –i.e. reducing the CO_2 emissions by 20% compared to 1990, increasing the renewable share to 20% and improving the electrical efficiency to 20% [3].

Photovoltaic (PV) energy is one of the most promising renewable energy sources for powering heat pumps and reducing their environmental impact, for several reasons. First, the PV technology can be applied to big and centralized installations, as well as to small and decentralized ones, and its cost is almost independent from the system size and its application. Second, the PV generating cost is similar to the electricity sale price in applications which lack any energy storage system and even lower in large power systems (bids of 1.9 and 2.3 US cents/kWh were approved in China and Abu Dhabi in 2016 [4]), which makes it competitive with almost any other energy source.

Before the PV panels price started dropping in 2010, most research related to solar heat pumps focused on hybrid photovoltaic/thermal (PVT) systems that combined a PV module with a thermal collector on its back surface, which worked as the evaporator of the heat pump. This type of technology is still the dominant one, but the reduction of PV prices has motivated an increasing attention in only PV (OPV) heat pump systems –where the compressor is powered by the electricity generated by a PV array. The objective of this paper is to describe a OPV heat pump prototype installed in Madrid, intended for stand-alone operation with thermal storage, and to discuss the results of the initial experimental works. These have been performed grid-connected, in order to precisely characterize the variable-speed compresor before substituting the grid (which provides constant output frequency) with a PV array (with variable output frequency).

2 CURRENT STATE OF THE ART

Roughly, solar heat pumps can be divided into two categories -hybrid photovoltaic/thermal (PVT) or only photovoltaic (OPV) - that can present two configurations –grid-connected or stand-alone-. Figure 1 shows the schematic of a basic PVT (a) and OPV (b) heat pump system with both configurations. The first experimental study of a PVT heat pump was reported in 1997 and it was a grid-connected system [5], although stand-alone PVT systems have also been widely explored. The biggest inconvenience of this hybrid technology is the high investment and maintenance costs of solar collectors. However, the first work related to OPV heat pump systems was not published until 2010 [6]. Since then, research on this type of heat pumps has been abundant and has explored a wide variety of solutions. The majority of them consists in grid-connected OPV heat pumps for increasing the local self-consumption in buildings. This solution has the inconvenient of forcing the PV array to operate at the voltage of the grid -and not at its maximum power point- unless a back-up storage system is installed in between. The possibility of operating stand-alone OPV heat pumps has been explored only very recently. Reported studies are very few and all of them consider a Battery Energy Storage System (BESS) coupled with the PV array [7], [8]. Such BESS are expensive and might be prone to create technical and security problems.

In conclusion, research in the near future should focus on stand-alone OPV heat pumps that are able to operate without back-up BESS. This solution raises two main technical challenges: managing PV power intermittences due to cloud-passing, and solving day-night generation cycle.

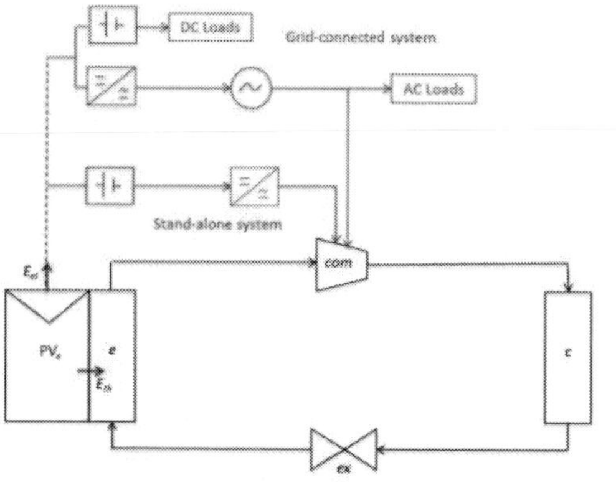

(a)

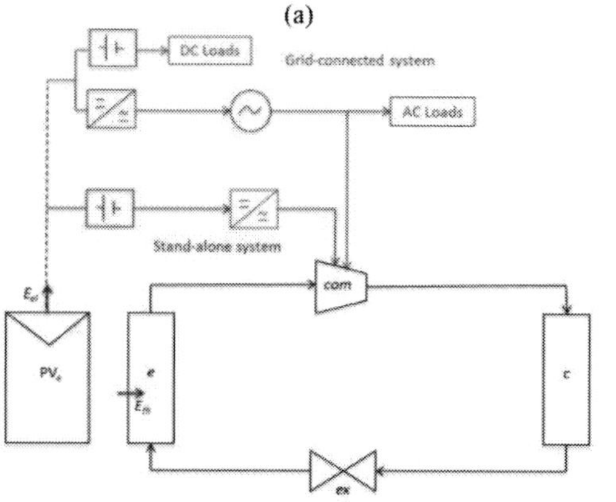

(b)

Figure 1: Schematic of a basic PVT (a) and OPV (b) heat pump system with grid-connected and stand-alone configurations.

3 EXPERIMENTAL SET

Figure 2 shows the schematic of the PV heat pump installation tested for this work, operating in cooling (a) and heating (b) modes. This installation mainly consists of the following elements:

- PV Heat Pump: a commercial heat pump that needs to be precisely characterized with the electric grid, in order to determine its operating parameters and limits, before powering it exclusively with a PV array.
- Main Water Tank: it simulates the room to be acclimatized, so it is cooled by the PV heat pump when operating in cooling mode and vice versa.
- Electric Resistance: in order to calculate how much thermal energy is extracted by the PV heat pump from the main water tank in the cooling operating mode, an electric resistance is turned on to keep the water temperature constant (see Figure 2.a). This way, thermal loads from the heat pump and the resistance are known to be equal.
- Air Blower: it is used in the heating operating mode for calculating how much thermal energy is provided by the PV heat pump to the main water tank (see Figure 2.b). Such thermal energy is equal to the one extracted by the air blower for keeping the water temperature constant.
- Test Chamber: it is an enclosed and thermally isolated space that simulates the ambient, so its temperature must be kept constant while running cooling or heating tests. The PV heat pump releases heat to the test chamber when operating in cooling mode, and retrieves it in heating mode.
- Fan Coil: it is used for maintaining a constant temperature inside the test chamber, compensating the thermal load of the PV heat pump (the fan coil extracts heat from the chamber when the PV heat pump is in cooling mode, and vice versa).
- Auxiliary Water Tank: heat extracted (or demanded) by the fan coil for regulating the temperature inside the test chamber is transmitted to (or obtained from) the auxiliary water tank.
- Auxiliary Heat Pump: it controls the temperature inside the auxiliary water tank, so the thermal load required by the fan coil can be obtained from water.

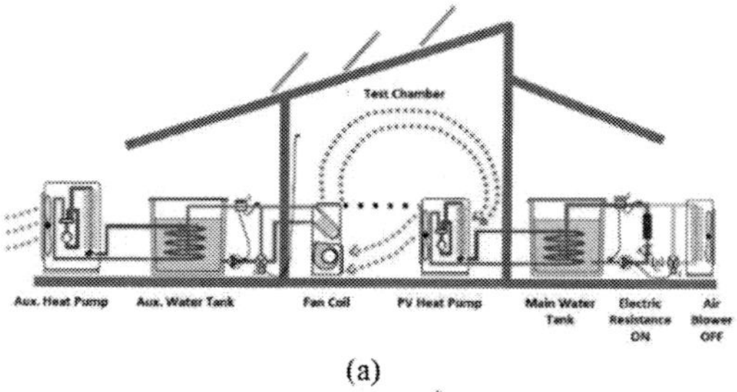

(a)

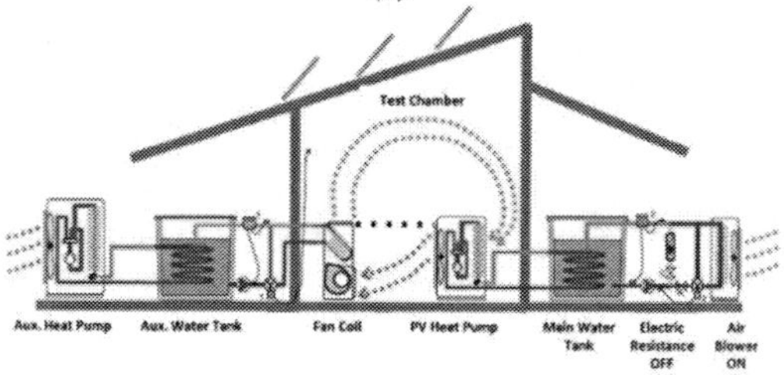

(b)

Figure 2: Schematic of the PV heat pump installation tested, operating in cooling (a) and heating (b) modes.

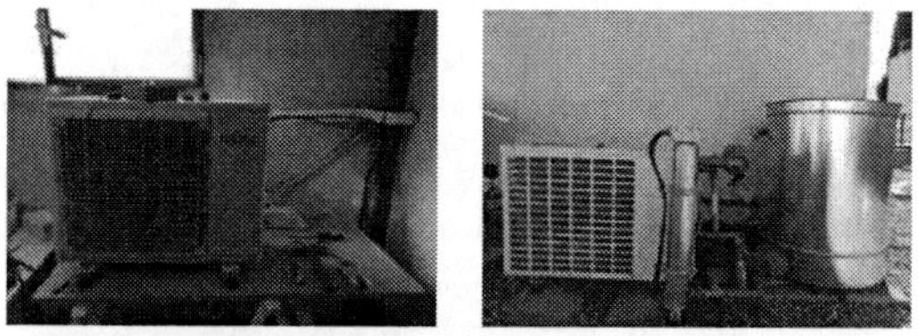

(a)

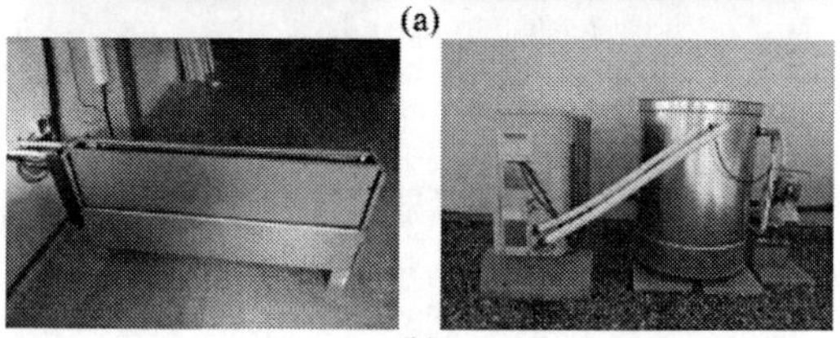

(b)

Figure 3: Images of the PV heat pump prototype installed in Madrid (Spain), composed of the tested system (a) and the auxiliary system (b).

Figure 3 shows real images of the prototype installed in Madrid (Spain). Figure 3.a shows the tested system, which includes the PV heat pumps, main water tank, electric resistance and air-blower, Figure 3.b shows the auxiliary system, including the fan coil, the auxiliary water tank and the auxiliary heat pump. The final objective of this research work is to efficiently operate this prototype, powering the PV heat pump exclusively with a PV array (stand-alone operation) and without any BESS. The auxiliary heat pump, the fain coil, the electric resistance and the air blower are powered by the grid.

4 TEST AND RESULTS

As mentioned before, the compressor of the PV heat pump needs to be characterized before powering it with a PV array, in order to determine its operation parameters and their limit values. This compressor is equipped with a commercial frequency converter that regulates its operating frequency depending on the thermal conditions set for the experiment. The objective of the test described in this paper is to analyze the response of this commercial converter with a constant frequency input (the electric grid in this case). Such converter will be substituted with another one, external to the heat pump, when the compressor is powered by a PV array. Hence, its response needs to be replicated in order to respect all the operation limits of the compressor, but with a variable-frequency input.

Initial tests have been carried out for the cooling mode and forcing a constant temperature of 20°C in the test chamber. Figure 4 shows the frequency output against voltage for one of the phases of the frequency converter. Both variables present a linear relationship with a slope very close to 1, although frequency is slightly smaller than the voltage. Figure 5 shows the current and the frequency output of the converter against ΔT, i.e. the difference between the temperature in the water tank and the temperature set as target (18°C for the cooling test). It can be observed that both variables (as well as the voltage, as a consequence of results showed in Figure 4) remain almost constant during the cooling process, with operating values close to 160 Hz, 5 A and 170 V, respectively. Moreover, it can be deduced that when the frequency increases, the current decreases proportionally and vice versa. Finally, Figure 6 shows active, reactive and total power output of the converter versus ΔT. Active and reactive powers have very similar values during the cooling process ($\cos\varphi \approx 0.7$), and they remain almost constant (total power varies between 800 and 900 W during most of the test). If comparing Figures 4 and 5, it can be observed that the power follows the current, varying proportionally and opposite to the frequency. In addition, these figures show that the heat pump system is started very fast (frequency, current and power grow from zero to the operation values almost instantly) and stops gradually, slowing down when $\Delta T \approx 2$°C.

5 FUTURE WORK

In order to completely characterize the heat pump prototype described in this paper, the same test performed for the cooling process will be repeated for the heating process (with a set temperature in the main water tank of 30°C). Additionally, both cooling and heating tests will be replicated for different temperatures in the test chamber, which simulates the ambient, for simulating a very warm summer and a very cold winter, respectively.
Once the response of the converter is precisely characterized with a constant frequency input, the compressor of the heat pump will be powered by a PV array, which provides a variable frequency input. Hence, the converter will have to be programmed for providing the output demanded by the compressor at every instant, without overcoming any limit value of the system. For operating the stand-alone system, there are two technical problems that need to be solved when programming the converter: managing the PV power intermittences due to cloud passing, and solving the day-night generation cycle by means of the thermal storage provided by the water tank.

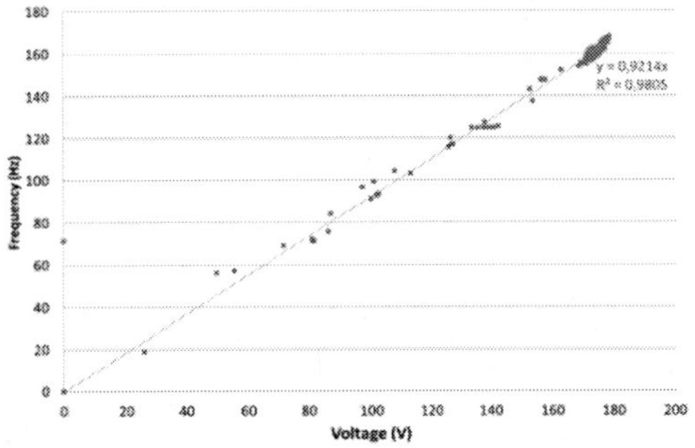

Figure 4: Output frequency versus voltage for one of the phases of the frequency converter.

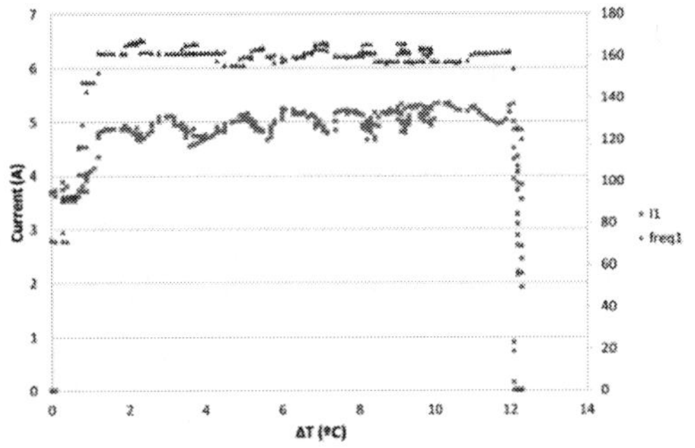

Figure 5: Output frequency and electric current versus ΔT for one of the phases of the frequency converter.

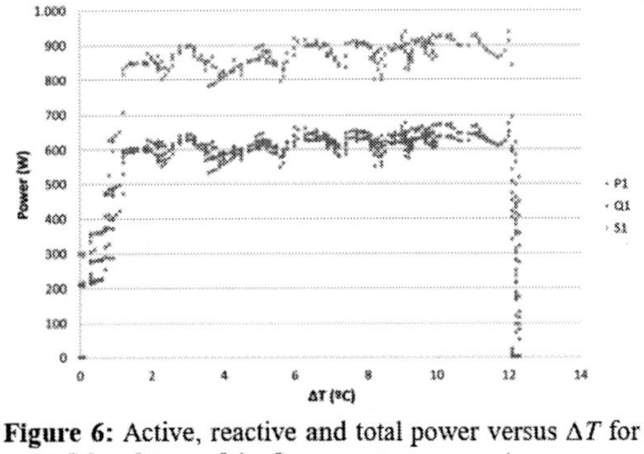

Figure 6: Active, reactive and total power versus ΔT for one of the phases of the frequency converter.

6 REFERENCES

[1] European Heat Pump Agency, *Heat Pump Market and Statistics Report 2015*, November 2015.

[2] European Heat Pump Agency, *Heat Pump Market and Statistics Report 2016*, November 2016.

[3] Oficina de Publicaciones de la Unión Europea, *Europa 2020: la estrategia europea de crecimiento*, Luxemburgo 2013.

[4] J. F. Weaber, *Solar power cost down 25% in five months*, Electrek September 2016, online: https://electrek.co/2016/09/26/solar-power-cost-down-25-in-five-months-theres-no-reason-why-the-cost-of-solar-will-ever-increase-again/

[5] S. Ito, N. Miura, J. Q. Wang, M. Nishikawa, *Heat Pump Using a Solar Collector With Photovoltaic Modules on the Surface*, Solar Energy, Vol. 112 (1997), pp. 147-151.

[6] Kei. Inoue, Y. Iwafune, *Operation of Heat Pump Water Heaters for Restriction of Photovoltaic Reverse Power Flow*, 2010 International Conference on Power System Technology, Hangzhou (2010).

[7] M. Spinnler, L. Hörth, F. Böing S. Wolf, T. Sattelmayer, *Novel heat pump system for highly transient PV operation*, 29th European Photovoltaic Solar Energy Conference and Exhibition, Amsterdam (2014).

[8] G. F. Abdullah, W. Saman, D. Whaley, M. Belusko, *Life cycle cost of standalone solar photovoltaic system powering evaporative cooler and heat pump water heater for Australian remote homes*, Energy Procedia, Vol. 91 (2016), pp. 681-691.

SOLAR FARM CLEANING ROBOT: ECO-FRIENDLY CLEANING OF SOLAR FARMS WITH REDUCED ENERGY AND WATER CONSUMPTION

K. Molnar[1], Z. Bilau[1], I. Bogár[1], M.P. Bellmann[2], B. Ryningen[2], W.R. Glomm[2], Shabnam Arbab[3]

[1] ProDSP Technologies, Budapest, Hungary
phone: +36 1 321 5181, e-mail: karoly.molnar@prodsp.hu
[2] SINTEF Materials and Chemistry, Trondheim, Norway
phone: +47 934 986 19, e-mail: martin.bellmann@sintef.no
[3] Department of Architecture and Technology, NTNU, Trondheim, Norway

ABSTRACT: The performance of a PV module will degrade due to fouling. Our primary aim is to supply the demands of the solar panel industry by providing them with a technology that reduces the use of water in solar panel cleaning. This way solar energy becomes a more efficient and cleaner energy source. The approach is to create an autonomous device, the Solar Farm Cleaning Robot, that is capable of cleaning solar panel farms in an eco-friendly, energy-efficient and cost-effective manner. The Solar Panel Cleaner Robot is a based on a proprietary cleaning roller that cleans surfaces easily in a wide zone with the assistance of a special microfiber textile. The operation does not require any water or energy supply hoses, and therefore, it is perfectly suitable for cleaning even at those places where electricity or water supply is not resolved. SINTEF has evaluated some parameters for the cleaning. A climate chamber for soiling and a test rig for the cleaning was build. PV panel glass plates was intentionally soiled in the climate chamber before different washing parameters were tested in the test rig. Light transmittance was measured before and after each cleaning.
Keywords: PV System, System performance

1 INTRODUCTION

Solar energy is one of the most environment-friendly energy sources known by humanity as of today, practically available in unlimited quantities. No pollutant is emitted directly during its utilization, the process of its transformation into electric current is environmentally friendly and nowadays experience shows that it is becoming more and more economical.

Economic performance of solar panels requires regular cleaning, the lack of which may result in a loss of 10-15 % or more in solar panel efficiency, depending on the extent of pollution. We have designed our product as a response to the demand revealed by our wide-scale market research and based on the information that has come to our knowledge from this industry sector. As we proceeded with our preliminary researches, it became obvious that most appliances available on the market for the cleaning of solar panels were wasteful in their usage of water, outdated and unable to meet expectations.

One of the pivotal points of solar panel cleaning is the quantity of water used for cleaning, from environmental- and economical aspects and from a feasibility-oriented point of view, as well. Since industrial solar panel parks harbor large surfaces of solar cells – to reach 1 MWp output, an area of the approximate size of a football field must be planted with solar cells –, the quantity of water necessary to clean a given unit of solar panel surface is becoming more and more a cardinal question nowadays with respect to the environmental impact and to the economic performance of the industry sector, as well as regarding its impact on society.

2 THE SOLAR FARM CLEANING ROBOT

2.1 Project goals

Most of the PV panels installed in Europe is installed in a humid temperate climatic zone with no dry seasons or with dry summers. Since Europe is densely populated and have high traffic- and industry density, frequent cleaning is assumed to be essential for PV modules installed most places in Europe.

Our primary aim is to supply the demands of the solar panel industry by providing them with a technology that reduces the use of water in solar panel cleaning. This way solar energy becomes a more efficient and cleaner energy source. The approach is to create an autonomous device, the Solar Farm Cleaning Robot, that is capable of cleaning solar panel farms in an eco-friendly, energy-efficient and cost-effective manner.

Figure 1: SFC Robot Mechanical Design

The Solar Panel Cleaner Robot is a based on a proprietary cleaning roller that cleans surfaces easily in a wide zone with the assistance of a special microfiber textile. The operation does not require any water or energy supply hoses, and therefore, it is perfectly suitable for cleaning even at those places where electricity or water supply is not resolved.

2.2 Project results

We have successfully built the prototype of the SFC Robot. The properties of the solution are as follows.

- Energy- and liquid provision is carried out from an inner source, in an autonomous manner.
- Movement of the instrument is assured by moving elements and by an engine.
- Cleaning is not realized by a watery washing process but by softening the pollu tion with an environment-friendly detergent solution and removing it with a wiping cylinder developed in-house, in a water-saving and efficient process.
- Cleaning is always done with clean, easily replaceable and washable wiping material, water consumption rate is rather favorable, approx. 0,015 l /m2.
- Cleaning is carried out quickly, efficiently and economically (0,5-1 MW / 8 hours).
- Thanks to its autonomy, the robot can also be used at locations that are hardly reachable or that have no water resource.
- The robot requires little space, it is easily transportable and it can be operated with low energy consumption.
- The robot is of simple construction and it can be fabricated economically.
- The robot works even on solar panel rows planted closely to each other (it moves along on the surface of the solar cells).

Figure 2: SFC Robot Prototype

3 PV PANEL CLEANING EXPERIMENTS

3.1 Experiments setup

SINTEF has evaluated some parameters for the cleaning. A climate chamber for soiling and a test rig for the cleaning was build. PV panel glass plates was intentionally soiled in the climate chamber before different washing parameters were tested in the test rig. Light transmittance was measured before and after each cleaning.

Micro fibre cloths have shown to be good alternative for water and detergent for dust removal and cleaning of hard surfaces indoors. The cloths should be used with damp to reduce the wear on the surfaces. A Climate chamber was built to be able to deposit relevant soiling layers on the PV glass plates. The light transmittance was measured after soiling and thereafter cleaning with different parameters was conducted in a test rig designed for this purpose. The light transmittance was thereafter measured again.

3.2 Results

First, different soiling layers were tested out. What is clear, however, is that the soiling reduces the light transmittance considerably and cleaning restores the transmittance almost completely in all the runs. The reference value for a clean glass plate with the antireflective coating is 94%.

Six heavy soiled glass plates were cleaned with varying load and constant tray speed and varying tray speed with constant load. Even for the smallest load (1,7 kg) a transmittance ratio of almost 90 % was achieved after cleaning. In addition the highest tray speed also gave a transmittance ratio of almost 90 % after cleaning.

Figure 3: Cleaning the artificially soiling

Since a high speed and minimum load is desirable, the last experiments were conducted with the smallest load (1,7 kg) and varying tray speeds. It was further decided to do the last tests on a little bit less soiled glass plate, since that would be closer to reality. Three layers of concrete (instead of six) was applied. The results are that even the fastest tray speed gave transmittance ratio of almost 90 %.

From these measurements it can be concluded that a tray speed of 50 cm/s and a load of 1,7 kg is sufficient to regain a transmission ratio of above 90 % (reference sample is 94 %) for heavy soiled glass plates. This is however only valid for a washed distance of less than 50 cm and for soiling that does not stick to the surface. If a rotating cloth is applied, it is assumed that these numbers will be valid over any distance and be limited by the speed of the rotation of the cloth and not the cleaned distance.

ACKNOWLEDGEMENT

This project received funding from Norway Grants.

norway grants

DESCRIPTIVE STATISTICS ON THE CLIMATE RELATED PERFORMANCE AND RELIABILITY ISSUES FROM GLOBAL PV INSTALLATIONS

Martin Halwachs[1], K.A. Berger[1], L. Maul[2], L. Neumaier[3], Y. Voronko[4], A. Mihaljevic[5], N. Vollert[3], W. Mühleisen[5], M. Schwark[1], R. Ebner[1], Ch. Hirschl[3]
[1]AIT Austrian Institute of Technology GmbH, Center for Energy, Business Unit Photovoltaic Systems, Vienna, Austria,
T: +43 (0) 50550-6406 , M: +43 (0) 664 8157806, Martin.Halwachs@ait.ac.at
[2]Fachhochschule Technikum Wien, Vienna, Austria,
[3]CTR Carinthian Tech Research, Smart Systems, Villach, Austria,
[4]OFI Österreichisches Forschungsinstitut für Chemie und Technik, Werkstoffanwendungen, Vienna, Austria
[5]PCCL Polymer Competence Center Leoben GmbH, Leoben, Austria

ABSTRACT: The Austrian national project INFINITY focuses on design improvements for PV system components operating in different climatic regions. This topic appeared increasingly often in the past Photovoltaic conferences, being a result from globalization of the PV industry and market, where the same components are mounted under different environmental stress. In this context the initial target of the project was to gather failure data from PV installations in various climate zones, mostly from literature and cooperation like the IEA PVPS Task 13. A standardized data input was achieved by using an extended format of the IEA PVPS Task13 failure survey, allowing to enter PV system and failure information in rich detail. The component's structure and failure structure was elaborated by combining input from reliability workshop proceedings, Task13 and experience in Photovoltaic reliability testing. This structure was further mapped to a database to collect and prepare the input for further statistical analysis. This paper presents the results from comparing different climate zones in the frequency of failures from 1970 until now. From this already some differences can be seen between the different climate zones.
Keywords: c-Si, Environmental Effect, Climate, Reliability

1 INTRODUCTION

In the past years several studies investigated the performance and reliability of PV power plants (e.g. [1-3]). The Austrian funded project INFINITY aims on design recommendations and possible improvements for PV system components operating in different climate zones. One task is the analysis of state of the art performance and reliability issues in different climate zones. Therefore reliability issues of PV installations were collected from literature and cooperation.

In order to differentiate locations in different climate zones the Köppen Geiger classification scheme was used (see Fig. 1, [4]). It allowed to specify each PV installation's climate zone with respect to main climate zone, precipitation and air temperature within a 0.5 degree latitude/longitude grid.

2 DATA COLLECTION

Based on the PV failure survey developed within the IEA PVPS Task 13 [5], reliability issues were collected. Starting with the components defined in the original survey: "PV module", "Inverter", "Cable and Interconnector", "Mounting and Other"; each **component**'s key elements (**parts**) were elaborated together with common associated failures (or degradations) as **samples**.

The developed extended surveys contain more detail on reported issues, while still collecting data in an anonymous way. Data from literature, PV system qualification within the project and cooperation like IEA PVPS Task 13 were structured into the format of the former mentioned surveys.

In order to allow the collection of surveys with variable component types (e.g. mono and poly crystalline Si PV modules in the same installation) such PV systems were separated in uniform units (**sub-systems**) of specific component types. This separation was necessary to clearly identify characteristics of component types and according failures. Different sub-systems within the same system were found to be affected by different failures [6].

Each survey was capable of defining failures for a single sub-system by specifying the components and failures found (see Fig. 2). In Total surveys for 488 sub-systems and 1243 different failure entries were created, collected and stored in an SQLite database.

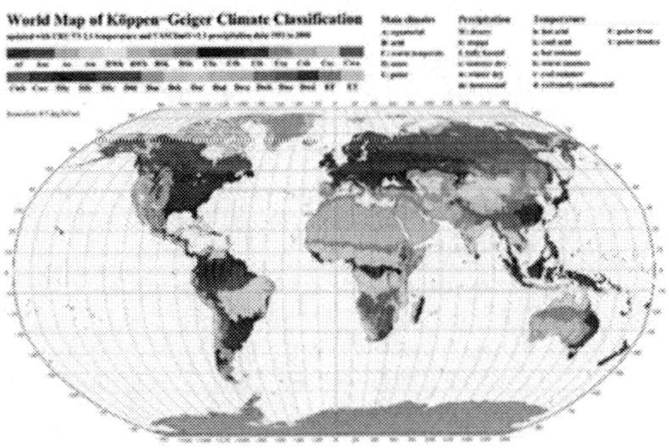

Figure 1: World map of Köppen-Geiger climate classification [4].

3 DATA FILTERING

In order to find relations between climate zones and failures and / or parts, the collected data was filtered for several aspects.

3.1 Climate relation

As all arising failures from various components were collected, failures without any climate relation at all were removed from the evaluations. This mostly affects software and connectivity issues of inverters.

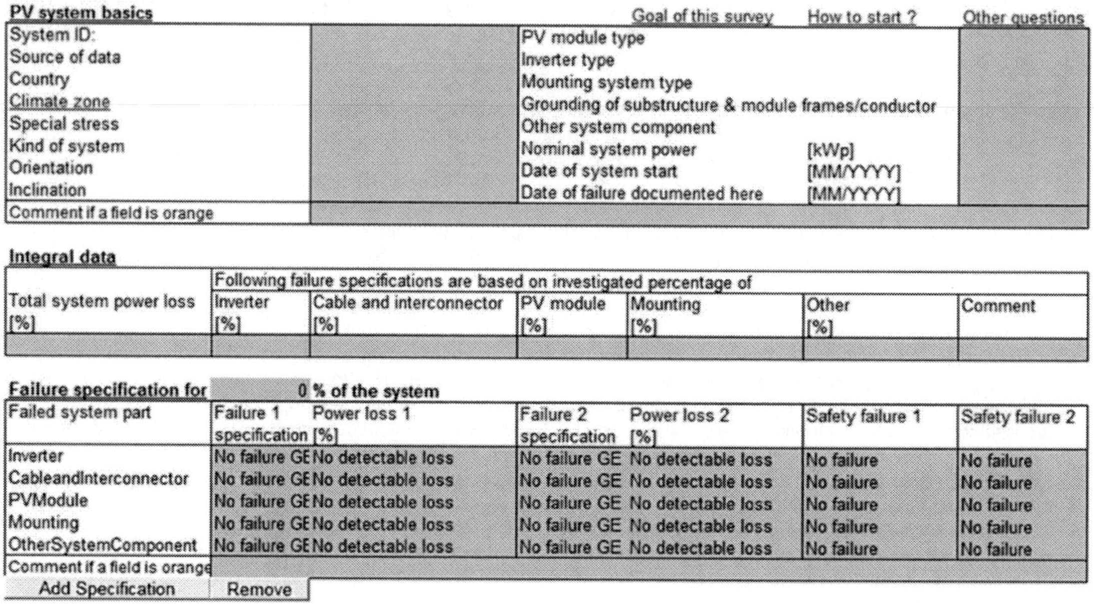

PV system basics — Goal of this survey — How to start ? — Other questions

System ID:		PV module type		
Source of data		Inverter type		
Country		Mounting system type		
Climate zone		Grounding of substructure & module frames/conductor		
Special stress		Other system component		
Kind of system		Nominal system power	[kWp]	
Orientation		Date of system start	[MM/YYYY]	
Inclination		Date of failure documented here	[MM/YYYY]	
Comment if a field is orange				

Integral data

Total system power loss [%]	Following failure specifications are based on investigated percentage of Inverter [%]	Cable and interconnector [%]	PV module [%]	Mounting [%]	Other [%]	Comment

Failure specification for 0 % of the system

Failed system part	Failure 1 specification	Power loss 1 [%]	Failure 2 specification	Power loss 2 [%]	Safety failure 1	Safety failure 2
Inverter	No failure GE	No detectable loss	No failure GE	No detectable loss	No failure	No failure
CableandInterconnector	No failure GE	No detectable loss	No failure GE	No detectable loss	No failure	No failure
PVModule	No failure GE	No detectable loss	No failure GE	No detectable loss	No failure	No failure
Mounting	No failure GE	No detectable loss	No failure GE	No detectable loss	No failure	No failure
OtherSystemComponent	No failure GE	No detectable loss	No failure GE	No detectable loss	No failure	No failure
Comment if a field is orange						

Add Specification — Remove

Figure 2: Extended failure survey, originally from IEA PVPS Task 13. This survey allowed to enter the system properties (on top) and the found failures with their occurrence and impact. Each component may be defined in more detail by a dropdown list.

3.2 Detailing general failures

The data collection was performed by several scientific partners. In order to not affect the data quality by different failure interpretation, the failures were entered as close to the stated failure in the original source as possible. This resulted in several failures in too little detail (e.g. PV module soiling).

Where in doubt:

- "Soiling" in all combinations (homogeneous / inhomogeneous from Biofilm or Dust) was interpreted as failure of the front glass
- "PID", "hot spot", "snail tracks", "cracked" and "electrical mismatch" were interpreted as solar cell failure.
- "Delamination over cells" was interpreted as front encapsulant failure
- "Foreign particle embedded" and "signs of moisture penetration" were interpreted as encapsulant failures and
- "Increased series resistance" was interpreted as interconnect failure.

These replacements left more than 60 failures where the failure could not be linked to a specific PV module part only: "Breaks", "burn marks", "discoloring", "corrosion" and "delamination" as well as 2 "lightning strikes".

3.3 crystalline silicon

Within the project INFINITY the focus was on mono and poly crystalline silicon technologies. Therefore all other technologies, including all unknown technologies were dropped from the evaluations

3.4 Detail in climate zones

The initial attempt was to use the whole Köppen Geiger specification, divided in all possible sub climate zones. But a histogram for the individual components showed some of these containing very little amount of samples (see Fig. 3).

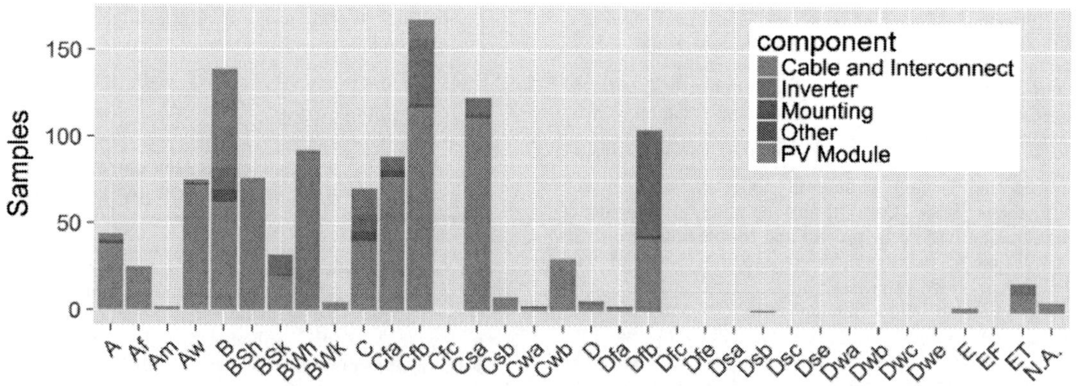

Figure 3: Histogram of failures by their Köppen Geiger climate zone. The color resolves for different components.

To achieve a better distribution of samples, climate zones were solely broken down to the main climate zones A (equatorial), B (arid), C (warm temperature), D (snow)

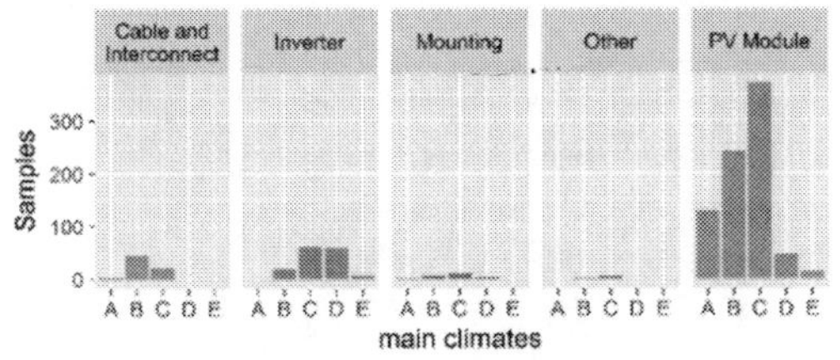

Figure 4: Distribution of failures for different components in main climate zones.

and E (polar), see Fig 4, removing samples, where no exact location was stated by the original source to derive the climate zone (N.A.).

4 RESULTS

After applying the above stated filters a total of 950 failure entries, 344 sub-systems in 45 countries (see Fig. 5) was left. But the distribution of samples in the various climate zones according to Fig. 4, solely shows a sufficient amount of PV module failures, to be discussed in detail. Therefore the following results will focus on the analysis of PV module failures.

Figure 5: The number of failure entries per country was marked from dark grey to blue worldwide (top) and in Europe (bottom). Most failures were collected from the US.

Table I and II show the total amount of PV module failures in different climate zones and the distribution of PV module technology. The amount of unknown PV module type relates to inverter failures, where the module technology does not matter for evaluation and may be some kind of crystalline silicon but was not stated by the original source.

Table I: Distribution of failures per climate zone, counting each failure type per individual sub-system

climate	A	B	C	D	E
failures	128	242	373	46	13

Table II: Number of Failures for each PV technology, counting each failure type per individual sub-systems

PV technology	Failures
mono cSi	407
multi cSi	268
x-Si	127

For data collection a tree structure for PV parts was developed splitting up into each part of the component. An example path is: PV module – encapsulant – front encapsulant. In the same manner failures were structured into groups like heat damage or physical damage. This structure allowed multiple parts to have multiple failures and vice versa. So visualizing solely distribution for parts in different climate zones, omits information on the various possible failures. The same happened, if solely showing failures without the according parts, accumulating to probably misleading information. For this analysis only other and unknown failures in an undefined PV module part were removed, leading to total 791 failures. Whereas unknown and other failures in certain parts and unspecified PV module parts with distinct failures were kept.

Fig. 6 shows histograms for the combination of part and failure having been found in each climate zone. The data was sorted descending for each climate zone separately. The three level tree structure of PV module parts was reduced to 2 levels, showing only the tip of the structure and not the whole root. The 2 level failure structure is fully visible with root and tip element separated by a "/".

Additionally this analysis was done for all collected data (installations starting at 1970) and for all sub-system installations after the year 2000. Viewing both side by side shall allow to identify which failures modes have already been sorted out and which failure modes are still relevant to be investigated.

From Fig. 6 a very interesting finding is the number of PID found most prominent in younger times in A and C climate, showing 6 and 10 sub-systems being affected.

Another finding is the amount of soiling in A climate environment, which sums totally to 8 sub-systems being affect by different type of soiling. Homogeneous soiling by dust is also still an issue in climate B, which is reasonable for B, thinking about desert and stepped environment.

Climate D on the other hand shows mostly snow and hail affected failures (possibly also interpreting cell cracks as a result from high load on the module) with climate C showing also a high amount of hail damage. Cell cracks and cell hot spots in climate D are obviously all found in sub-system installations after the year 2000.

Same can be found for defect bypass diodes and pottant discoloring in climate B, which is the only climate zone having this high amount of defect by-pass

Discoloring and delamination are spotted usually in older installations for A, while in B and C there is a

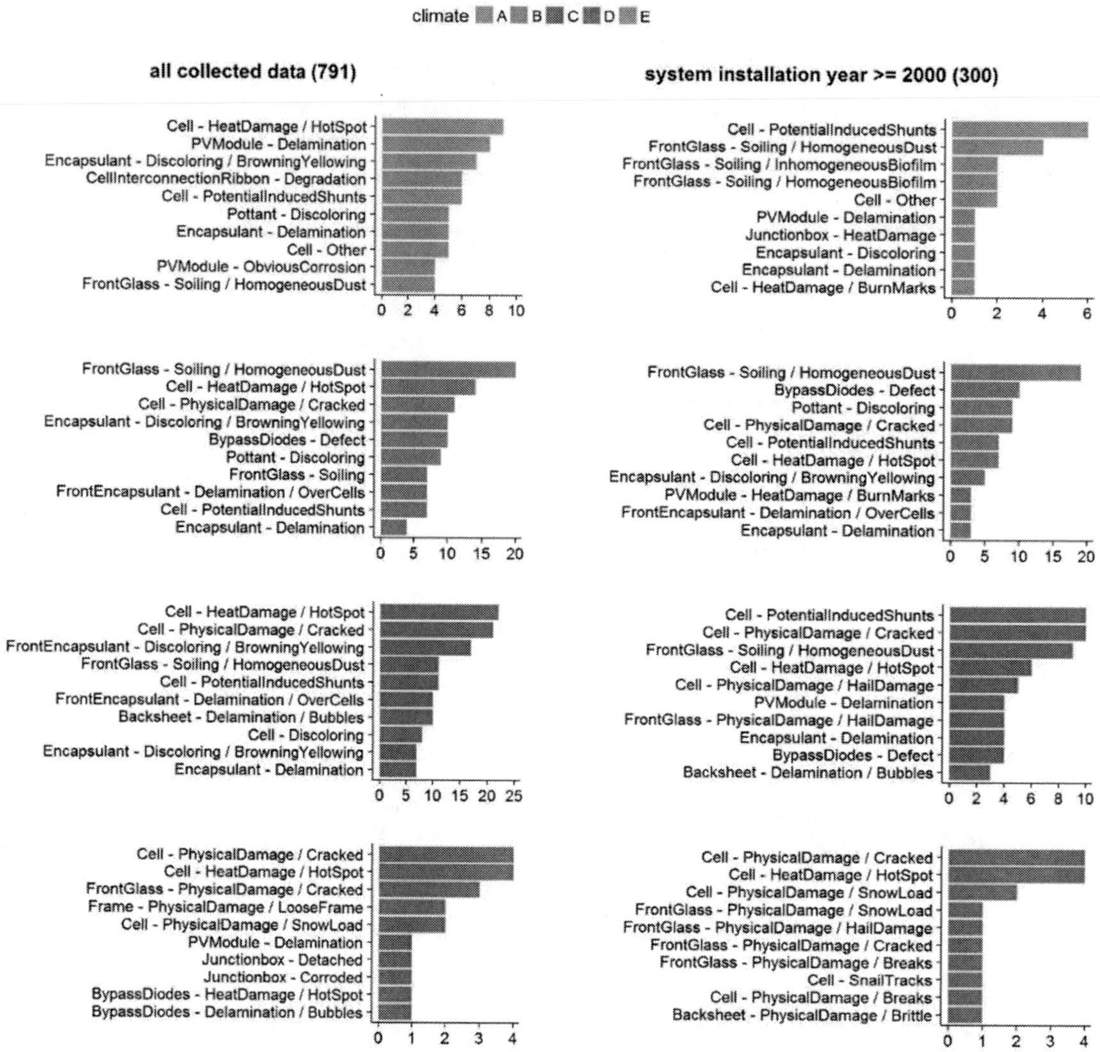

Figure 6: Top 10 number of samples per failure with corresponding part (some failures may occur for various parts) within each climate zone. Climate zone E was neglected due to too little samples. The left side shows distribution for all data collected, the right side represents all data from sub-system installations after the year 2000. Substructures are separated with a "/".

reasonable lower, but still high presence corresponding to a shorter time period for the development of these defects.

5 CONCLUSIONS

A descriptive analysis of PV module failures from the data collection in the Austrian national project INFINITY was presented. Data collection and filtering was described in order to allow for reproducible comparisons.

The collection of data resulted in 950 failure entries from 344 sub-systems in 45 countries all over the world. In order to identify climates for PV system locations the Köppen Geiger climate classification scheme was used, based on anonymous GPS data from PV installations or regions.

Data was collected mainly from climate zone C and poorly from climate zone E. The current basis of data allows to compare PV module failures between climate zones A to D. Several differences in the climate zones where found and shown.

At the current state of project INFINITY solely descriptive statistics on the collected data basis are shown. This data will be used to identify potentials to improve PV module parts to design PV modules for specific climates.

6 ACKNOWLEDGMENTS

This work was conducted as part of the Austrian "Energy Research Program" project INFINITY. Funded by the Austrian Climate and Energy Fund and the Austrian Research Promotion Agency (FFG), both are gratefully acknowledged

7 REFERENCES

[1] M. Alonso-Abella, F. Chenlo, A. Alonso, and D. Gonzàlez, "TOLEDO PV PLANT 1 MWp – 20 YEARS OF OPERATION," presented at the 29th European

Photovoltaic Solar Energy Conference and Exhibition, Amsterdam, The Netherlands, 2014, pp. 2728–2733.

[2] R. Dubey, et.al. (2014 Mar.). "All India survey of PV module degradation: 2013," Nat. Cent. Photovoltaic Res. Edu-cat., Mumbai, India, [Online]. Available: www.ncpre.iitb.ac.in/uploads/All_India_Survey_of_Photovoltaic_Module_Degradation_2013.pdf

[3] F. Bandou, A. Hadj Arab, M. S. Belkaid, P.-O. Logerais, O. Riou, and A. Charki, "Evaluation performance of photovoltaic modules after a long time operation in Saharan environment," International Journal of Hydrogen Energy, vol. 40, no. 39, pp. 13839–13848, Oct. 2015.

[4] M. Kottek, J. Grieser, C. Beck, B. Rudolf, and F. Rubel, "World Map of the Köppen-Geiger climate classification updated," Meteorologische Zeitschrift, vol. 15, no. 3, pp. 259–263, Jun. 2006.

[5] M. Köntges et al., "Assessment of Photovoltaic Module Failures in the Field," IEA PVPS Task 13, Subtask 3, IEA-PVPS T13-09:2017, May 2017.

[6] M. Alonso-Abella, F. Chenlo, A. Alonso, and D. Gonzàlez, "TOLEDO PV PLANT 1 MWp – 20 YEARS OF OPERATION," presented at the 29th European Photovoltaic Solar Energy Conference and Exhibition, Amsterdam, Niederlande, 2014, pp. 2728–2733.

PHSO: A GRAPHIC USER INTERFACE OPTIMIZER FOR THE SIZING DESIGN OF THE PV HYBRID SYSTEMS

Carlos D. Rodríguez-Gallegos[a,b]*, Oktoviano Gandhi[a], Thomas Reindl[a], S. K. Panda[b]
[a]Solar Energy Research Institute of Singapore (SERIS), National University of Singapore (NUS), Singapore 117574
[b]Department of Electrical and Computer Engineering, National University of Singapore (NUS), Singapore 117583
*Corresponding author. Email address: carlos.rodriguez@nus.edu.sg

ABSTRACT: Access to electrical energy is a basic and indispensable requirement to enhance people's quality of life. Nevertheless, the extension of main grids is, in general, not cost effective for regions located at far and remote locations (such as islands) or whose communities only consist of few members. In these cases, off-grid systems are installed instead, which predominantly use diesel generators (DGs) as power source. These micro-grids, however, have drawbacks with respect to their cost (high transportation cost of fuel and maintenance requirements) and environmental concerns (CO_2 emissions). To alleviate these burdens, the addition of photovoltaic solar panels (PV) and batteries (BAT) to these off-grid systems is proposed. This paper provides a method to obtain the optimal number of solar panels and batteries that could be added to the off-grid systems (currently operated by diesel generators) to reduce the total cost (investment and operations) of the system as much as possible. For this, a graphical user interface called PV Hybrid System Optimizer (PHSO) was created in MATLAB, from which the user can select relevant parameters regarding the installed DGs and the desired share of PV and BAT to be installed. Other input variables such as the load and weather profile, among others, can also be entered or selected from a library.

Keywords: PV hybrid system, sizing optimization, cost reduction, grid management

1 INTRODUCTION

The addition of solar panels (PV) and batteries (BAT) into off-grid systems initially operated only by diesel generators (DGs) to reduce the total cost and CO_2 emissions has taken place from a long time ago. This trend is likely to continue due to the continuous reduction of the price related to solar panels and batteries and the price increase of diesel in time.

As such, the ability to precisely design and define the number of PV and BAT to be installed for particular scenarios becomes essential. Poor designs can end in undersize and oversize of these elements. As a result, the savings will not be maximized as much as possible.

Different tools have been developed in an effort to optimize the sizing design of the hybrid systems. Based on many publications, HOMER [1] appears as a standard tool to perform these designs. Despite its popularity, this and many other tools lack several characteristics, such as the consideration of the:

- Performance degradation of the PV in time
- Yearly load variation
- Yearly variation of the diesel price
- Yearly variation of the battery price
- Spinning reserve condition, among others

These factors are important to generate more realistic models and to obtain more credible results. In this paper, a graphic user interface is developed in MATLAB, namely, the PV Hybrid System Optimizer (PHSO). This tool is built by considering the previous factors and others that will be explained in the following sections. Nevertheless, the current version of PHSO has its own limitations as it can only consider systems formed by DGs, PV and BAT.

The rest of the paper is arranged as follows: Section 2 deals with the mathematical modeling to represent the electrical performance and cost related to each device. In Section 3, an insight of the developed algorithm is provided while the case study is defined in Section 4. Their results are then presented and analyzed in Section 5. Finally, the conclusions of this work are presented in Section 6.

2 MATHEMATICAL MODELLING

The initial system is assumed to be composed only by DGs and the load. From here, solar panels and batteries can be added if desired. A schematic of the electrical connection is presented in Fig. 1. The main grid is AC. The solar panels and batteries are connected to this grid through inverters.

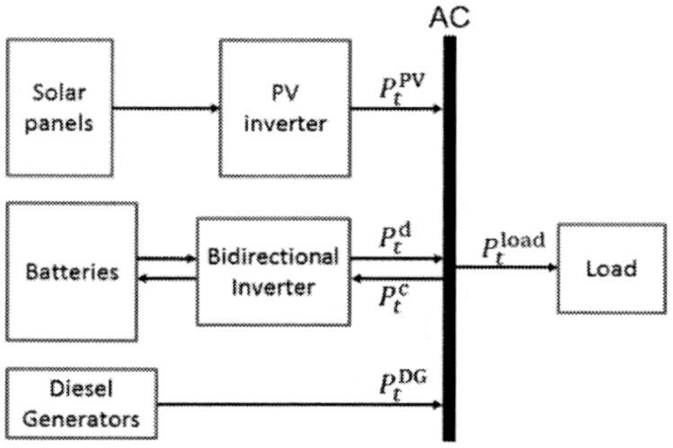

Figure 1: PV hybrid system design

In the following sub-sections, several variables will be described with a t sub index to represent their value for the tth hour.

2.1 Diesel generators

The diesel consumption, $\kappa_{j,t}^{diesel}$, of the jth DG at the tth hour is calculated based on its output power, $P_{j,t}^{DG}$, and nominal power, $P_j^{DG,nominal}$ [2]:

$$\kappa_{j,t}^{diesel} = a \cdot P_{j,t}^{DG} + b \cdot P_j^{DG,nominal} \quad (1)$$

where a and b are constants whose values depend on the generator quality.

Once the diesel consumption is estimated, its

associated cost can be calculated based on the diesel price. Other costs such as the O&M cost of the DGs, $c^{DG,OM}$, are also considered.

The CO_2 emissions are calculated by assuming that 2.6 kg of CO_2 are produced per liter of diesel [3].

Each generator should work within its minimum allowed power, $P_j^{DG,min}$, and nominal power. This constraint assures the proper operation of the DGs [4]:

$$P_j^{DG,min} \le P_{j,t}^{DG} \le P_j^{DG,nominal} \quad (2)$$

Furthermore, at least one DG must be operating at any time. This condition is applied in an effort to avoid grid stability issues due to the low inertia of the PV and BAT systems.

2.2 Solar panels

The total cost to buy and install the solar panels (c^{PV} and $c^{PV,ins}$, respectively) is defined as:

$$C^{PV}=\left(c^{PV}+c^{PV,ins}\right)\cdot P^{PV,STC}\cdot n^{PV} \quad (3)$$

The cost related to the acquisition of the PV inverters, $c^{PV,inv}$, with lifetime of $l^{PV,inv}$ is also considered together with the O&M cost of the PV system, $c^{PV,OM}$.

The maximum power generated by a solar array into the grid, P_t^{PV} (see Fig. 1), consisting of n^{PV} solar panels is estimated as [5-7]:

$$P_t^{PV}=P^{PV,STC}\cdot\frac{I_t}{1000}\cdot\left(1+\gamma\cdot\left(T_t^{cell}-25\right)\right)\cdot n^{PV}\cdot\beta_1\cdot\eta^{PV,inv} \quad (4)$$

where $P^{PV,STC}$ corresponds to the power of a single solar panel under standard test conditions, I_t is the solar irradiance, γ is the power temperature coefficient, T_t^{cell} is the temperature of the solar cells, β_1 corresponds to the PV yearly degradation power and $\eta^{PV,inv}$ is the PV inverter efficiency. To estimate T_t^{cell}, the following equation is employed [5]:

$$T_t^{cell}=T_t^{amb}+\frac{I_t}{800}\cdot\left(T^{NOCT}-20\right) \quad (5)$$

where T_t^{amb} and T^{NOCT} are the ambient temperature and the nominal operating cell temperature, respectively.

2.3 Batteries

Batteries with a capacity S are used. The acquisition cost of these batteries is defined as [8, 9]:

$$C^{BAT}=c^{BAT}\cdot S\cdot n^{BAT} \quad (6)$$

where c^{BAT} is the cost per unit energy of a battery and n^{BAT} is the total number of installed batteries.

The O&M cost of the batteries, $c^{BAT,OM}$, as well as the cost related to their bidirectional inverters, $c^{BAT,inv}$, with lifetime of $l^{BAT,inv}$ and efficiency $\eta^{BAT,inv}$ are also considered.

The employed approach to estimate the total energy E that a battery will discharge during its lifetime is the one proposed in [10].

The battery state of charge for the tth hour SOC_t is calculated based on its value from the previous hour SOC_{t-1} and the current charging and discharging power (P_t^c and P_t^d, respectively, see Fig. 1)[11]:

$$SOC_t=SOC_{t-1}+\frac{\eta^{BAT,inv}\cdot\eta^c\cdot P_t^c}{S}-\frac{P_t^d}{\eta^{BAT,inv}\cdot\eta^d\cdot S} \quad (7)$$

where η^c and η^d are the charging and discharging efficiency of the batteries, respectively.

To avoid severe degradation, their SOC_t values must be kept within a safety range [8, 9]:

$$SOC^{min}\le SOC_t\le SOC^{max} \quad (8)$$

Furthermore, P_t^c and P_t^d must not exceed the maximum charging and discharging power:

$$0\le P_t^c\le n^{BAT}\cdot P^{c,max}\cdot\eta^{BAT,inv} \quad (9)$$

$$0\le P_t^d\le\frac{n^{BAT}\cdot P^{d,max}}{\eta^{BAT,inv}} \quad (10)$$

where $P^{c,max}$ and $P^{d,max}$ stand for the maximum charging and discharging power of a single battery, respectively.

2.4 Spinning reserve

Even if the load demand is satisfied, it is necessary to ensure that the operating system is able to react fast enough in case there is a sudden increase in the load demand or a sudden decrease in the irradiance. For this, the system must have a reserve on the available power it can immediately provide. This is known as the spinning reserve, SR. The value of SR_t at the tth hour is calculated based on the spinning reserve provided by the k DGs which are currently in ON state, SR^{DG}. DGs which are in OFF state are not considered as some time is required to turn them on. Batteries are also able to contribute to the spinning reserve, SR^{BAT}, as shown:

$$SR_t=SR_t^{DG}+SR_t^{BAT} \quad (11)$$

On the one hand, SR_t^{DG} constitutes the available extra power that the DGs can still provide, if necessary:

$$SR_t^{DG}=\sum_{l=1}^{k}P_l^{DG,nominal}-P_{l,t}^{DG} \quad (12)$$

On the other hand, SR_t^{BAT} is the extra power that batteries can still discharge, if necessary:

$$SR_t^{BAT}=n^{BAT}\cdot P^{d,max}\cdot\eta^{BAT,inv}-P_t^d \quad (13)$$

Based on the previous equations, the condition for the spinning reserve can be defined with respect to the load demand, P_t^{load}, and the power produced by the PV system, P_t^{PV}:

$$SR_t\ge\alpha\cdot P_t^{load}+\beta\cdot P_t^{PV} \quad (14)$$

where α and β are the spinning reserve coefficients.

2.4 Load demand

At each time step, the load demand has to be fulfilled. This power can be provided by the PV, BAT or DGs. The power schedule of these devices is decided based on the algorithm described on the subsequent sections. The load demand condition can then be stated as:

$$\sum_{l=1}^{k} P_{l,t}^{\mathrm{DG}} + P_t^{\mathrm{c}} - P_t^{\mathrm{d}} + P_t^{\mathrm{PV}} \geq P_t^{\mathrm{load}} \quad (15)$$

3 ALGORITHM

The sizing algorithm applies a linear increase on the installation number of PV and BAT. For each combination, the performance of the simulated system is obtained giving the total cost, CO_2 emissions, and percentage of times the constraints were fulfilled, among others.

With respect to the dispatch strategy, the power generated from the DGs and BAT (PV always operates at maximum power point) at each time slot are obtained by implementing a rule-based approach that follows the subsequent criteria, arranged in priority order:

1. Fulfillment of the load demand
2. Fulfillment of the spinning reserve condition
3. The DGs should not operate below its minimum allowed power
4. Reduction of the operational cost (the diesel cost and battery replacement cost are compared)
5. Reduction of the CO_2 emissions

4 CASE STUDY

In order to validate the PHSO tool, a case study is evaluated for an off-grid system that is initially composed of two DGs of 50 kW each. This system is assumed to be located at the Southeast Asian region, in one of the Indonesian islands. This location is chosen due to the many islands solely relying on DGs for electricity. The system lifetime for this preliminary investigation is considered to be 25 years.

The accumulated load profile of an Indonesian island was measured during the year 2016. The normalized profile is presented in Fig. 2-a. This profile is applied for the current case study considering a peak value of 75 kW during the first year. A yearly load increase of 2% is assumed.

The hourly irradiance and ambient temperature profiles are obtained from the measurements performed by SERIS in Singapore (latitude of 1.3026, longitude of 103.7729, tilted angle of 10°and azimuth angle of 180°). Based on five years historical data (2012-2016), the typical meteorological year data of the solar irradiance and ambient temperature were calculated. Due to the proximity of Singapore to many Indonesian islands, a similar weather profile is expected. These profiles are shown in Fig. 2-b and Fig. 2-c.

Furthermore, the forecasted diesel price is estimated with respect to a reference case for future oil prices presented in [12] and by setting an average diesel price of 0.645 USD/l in Indonesia during 2016 [13]. The expected battery cost (for $LiFePO_4$ technology) is obtained from [14]. These are shown in Fig. 2-d.

The rest of the parameters to perform the sizing design are presented in Table I. These were obtained based on a literature review.

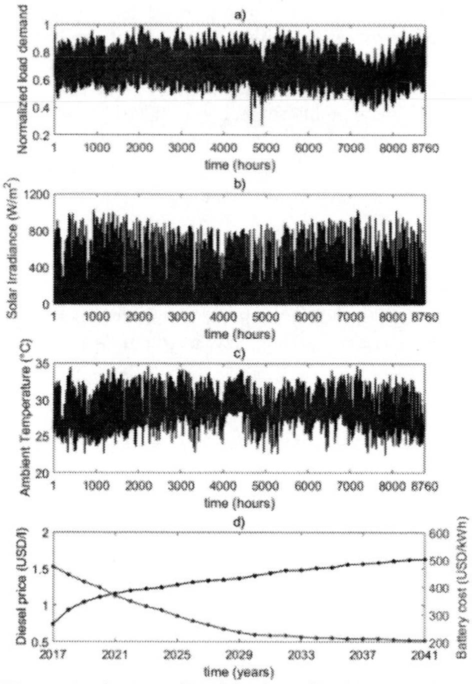

Figure 2: a) Normalized load profile; b) TMY data of the irradiance; c) TMY data of the ambient temperature; and d) forecasted diesel price and battery cost

Table I: Specifications for the system components

Diesel generators	
$P^{\mathrm{DG,nominal}}$	50 kW each
a	$2.46 \cdot 10^{-4}$ l/Wh
b	$8.4 \cdot 10^{-5}$ l/Wh
$P^{\mathrm{DG,min}}$	$0.3 \cdot P^{\mathrm{DG,nominal}}$
$c^{DG,OM}$	0.14 USD/year/kW
Solar panels (multicrystalline)	
$P^{\mathrm{PV,STC}}$	325 W
c^{PV}	0.341 USD/W_p
$c^{\mathrm{PV,ins}}$	0.45 USD/W_p
γ	-0.4 %/°C
T^{NOCT}	45.5°C
$c^{\mathrm{PV,OM}}$	7.1 USD/year/kW_p
β_1	0.8%/year
$l^{\mathrm{PV,inv}}$	9 years
$\eta^{\mathrm{PV,inv}}$	98%
Batteries ($LiFePO_4$)	
S	3840 Wh
SOC^{max}	95%
SOC^{min}	20%
$P^{\mathrm{c,max}}$, $P^{\mathrm{d,max}}$	3840 W
E	8832 kWh
η^{c}, η^{d}	96%
$c^{\mathrm{BAT,OM}}$	7 USD/year/kWh
$l^{\mathrm{BAT,inv}}$	9 years
$\eta^{\mathrm{BAT,inv}}$	94%
Spinning reserve	
α	10%
β	20%

For this case study, the number of installed solar panels will start with a minimum of zero reaching a maximum of 3000 with a step increase of 50. In a similar way, the initial number of installed batteries is set to zero reaching a maximum of 600 with a step increase of 20.

5 RESULTS

The obtained results are presented in Fig. 3. The accumulated costs of the analyzed system for the next 25 years are given in Fig. 3-a while the total CO_2 emissions are shown in Fig. 3-b. Furthermore, Fig. 3-c provides the percentage of times that the load was satisfied ($Load_c$), Eq. (15), while Fig. 3-d shows the percentage of times that the spinning reserve condition was fulfilled (SR_c), Eq. (14). These plots reveal a variety of results generated from the different designs. As such, with the current graphs, a particular design can be chosen considering the desired priorities (e.g. maximize the savings, minimize the CO_2 emissions, assure that loads are always satisfied, among others).

In the present investigation, the results that achieved the lowest cost will be analyzed in detail. The designs that generated the lowest cost for systems composed of DGs and PV (DG+PV) as well as systems composed of DGs, PV and BAT (DG+PV+BAT) are highlighted in Fig. 3 (systems composed of only DGs and BAT were not cost effective, as such these are not discussed).

A comparison on the performance of these systems with respect to the initial one, in which only DGs operate (DG), are presented in Table II. Here, it can be appreciated that by adding solar panels (DG+PV system), the total cost and CO_2 emissions were minimized. Furthermore, the values of $Load_c$ and SR_c were increased. If batteries are also considered (DG+PV+BAT), the results were further improved as now any extra power produced by the PV will not be wasted but will be stored on the batteries. Consequently, the $Load_c$ and SR_c values were also increased.

To have a better insight in the cost sharing due to the hybridization process, the cost contributions are presented in Fig. 4 for the DG+PV and DG+PV+BAT systems. For the DG+PV case, even though solar panels were added, the diesel consumption possess most of the cost share (95%) while the acquisition of the PV only accounts for 1% of this cost. The reason for this outcome can be attributed mainly to two factors:

1. In Section 2.1, a condition was stipulated indicating that at least one DG must be operating at any time. Therefore, diesel will always be consumed (even in the extreme cases when DGs do not need to produce any electrical power) due to the "$b \cdot P_j^{DG,nominal}$" term from Eq. (1), which is related to the internal losses of the DGs.
2. Solar panels cannot operate at night. Consequently, only the DGs can provide power at these periods, enhancing by this the diesel consumption.

As such, even if more solar panels were added, they would not be able to further reduce the diesel consumption (that is the reason why only 650 solar panels were assigned for this design, as shown in Table II).

When batteries are also included (DG+PV+BAT), Fig. 3-f shows that the diesel consumption still represents most of the cost share (57%). However, its share has been reduced with respect to the DG+PV case. Because batteries were introduced, more solar panels can be installed so that batteries cannot only operate at daytime but also at nighttime. Due to the high price of batteries, they account for 16% of the total costs. Nevertheless, the accumulated cost was reduced as now less diesel is consumed. Furthermore, even though a total number of 1800 solar panels were installed (see Table II), their acquisition cost only accounts for 4% of the total one.

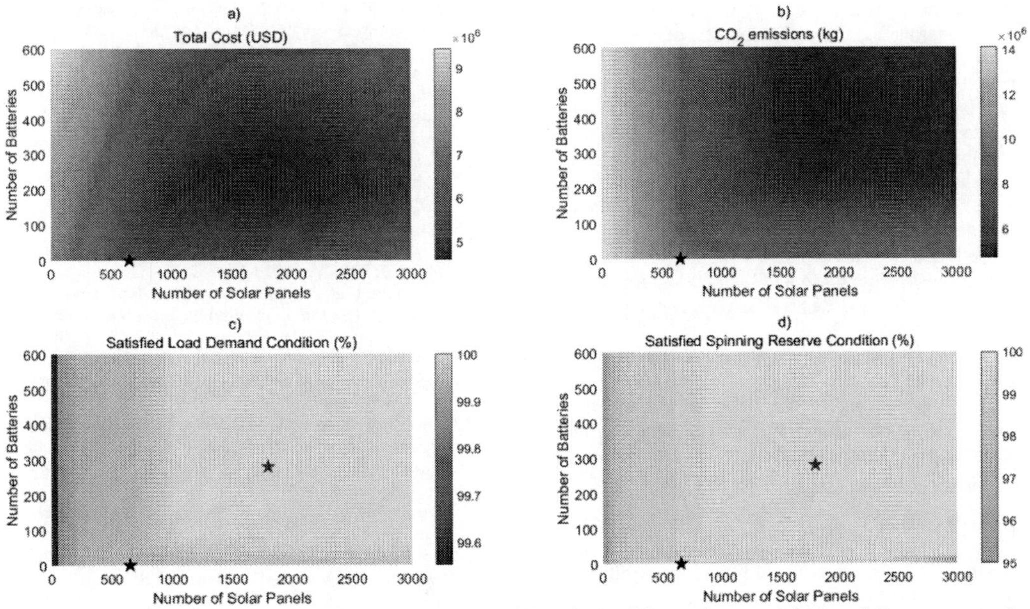

Figure 3: a) Total cost; b) CO_2 emissions; c) percentage of time the load demand was satisfied; and d) percentage of times the spinning reserve condition was satisfied. The black and red stars denote the designs which achieve the lowest cost for the DG+PV and DG+PV+BAT scenarios, respectively

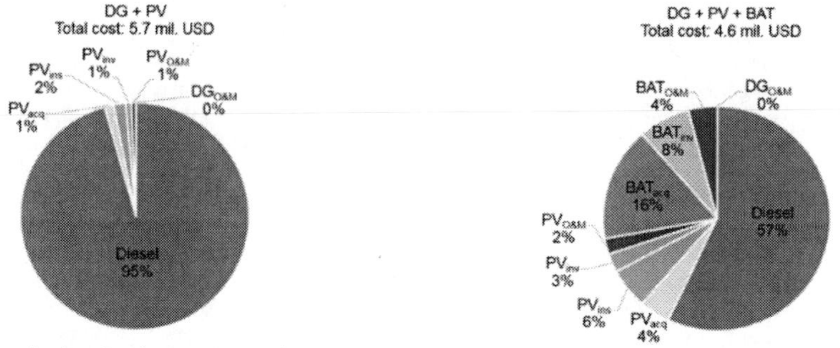

Figure 4: Cost sharing for the DG+PV and DG+PV+BAT systems. Here, the subscripts represent: *OM*, the operation and maintenance cost; *ins*, the installation cost; and *inv*, the inverter cost. As for example, PV_{acq} stands for the acquisition cost of the solar panels

Table II: Obtained results of selected designs

	DG	DG+PV	DG+PV+BAT
# PV	-	650	1800
# BAT	-	-	280
Cost (mil. USD)	7.2	5.7	4.6
CO_2 (mil. kg)	14.1	10.6	5.2
$Load_c$ (%)	99.552	99.947	99.999
SR_c (%)	95.49	99.32	99.97

6 CONCLUSIONS

In this paper, a graphical user interface called PV Hybrid System Optimizer (PHSO) was introduced as a tool to optimize the design of off-grid systems composed of diesel generators, solar panels and batteries.

With respect to other software which also aim to optimize this design, PHSO has an advantage as it performs a more detailed model of the analyzed system by considering factors such as: time dependence degradation of PV, battery cost in time, diesel price in time, spinning reserve condition, among others.

A case study in which two diesel generators were originally operating as the only sources of electrical energy (50 kW each) was raised. The results showed the expected outcome of the system with respect to the costs, CO_2 emissions and fulfillment of the constraints when solar panels and batteries were added. For the current scenarios, the addition of PV (DG+PV system) was cost effective as it was able to reduce the costs to a maximum of 5.7 mil. USD (a cost of 7.2 mil. USD was calculated when only the DGs are present). Furthermore, the addition of solar panels and batteries in parallel was able to enhance the maximum cost reduction (4.6 mil. USD) and improve the fulfillment of the load demand and spinning reserve conditions.

The PHSO tool can then be applied to different case studies to estimate the ideal design of off-grid systems avoiding the under sizing or oversizing of the system components. However, further work is still required to improve the dispatch strategy of these off-grid systems.

5 ACKNOWLEDGEMENTS

The authors of this paper would like to thank all the co-workers from the Solar Energy Research Institute of Singapore (SERIS). This work was supported by the Singapore's National Research Foundation (NRF) within the research project "NRF2013EWT-EIRP003-043".

7 REFERENCES

[1] HOMER Energy. (February, 2017). *Microgrid Decisions Made with Confidence*. Available: http://www.homerenergy.com/team.html
[2] D. Yamegueu *et al.*, "Experimental study of electricity generation by Solar PV/diesel hybrid systems without battery storage for off-grid areas," *Renewable Energy*, vol. 36, no. 6, pp. 1780-1787, 2011.
[3] U.S. Environmental Protection Agency, "Emission Facts: Average Carbon Dioxide Emissions Resulting from Gasoline and Diesel Fuel," ed, 2009.
[4] H. Nacfaire, *Wind-diesel and wind autonomous energy systems* (no. Book, Whole). New York; London; Elsevier Applied Science, 1989.
[5] E. Skoplaki and J. Palyvos, "On the temperature dependence of photovoltaic module electrical performance: A review of efficiency/power correlations," *Solar energy*, vol. 83, no. 5, pp. 614-624, 2009.
[6] O. Gandhi *et al.*, "Continuous Optimization of Reactive Power from PV and EV in Distribution System," in *IEEE Innovative Smart Grid Technologies - Asia (ISGT Asia)*, Melbourne, Australia, 2016, pp. 281-287.
[7] O. Gandhi *et al.*, "Competitiveness of Reactive Power Compensation using PV Inverter in Distribution System," in *7th IEEE Innovative Smart Grid Technologies (ISGT Europe)*, Torino, Italy, 2017.
[8] C. D. Rodríguez-Gallegos *et al.*, "Placement and Sizing Optimization for PV-Battery-Diesel Hybrid Systems," in *The 4th IEEE International Conference on Sustainable Energy Technologies (ICSET 2016)*, Hanoi, Vietnam, 2016, pp. 83-89.
[9] C. D. Rodríguez-Gallegos *et al.*, "Optimal PV and Storage Sizing for PV-Battery-Diesel Hybrid Systems," in *42nd Annual Conference of IEEE Industrial Electronics Society (IECON)*, 2016, pp. 3080-3086.
[10] T. Lambert *et al.*, "Micropower system modeling with HOMER," *Integration of alternative sources of energy*, vol. 1, no. 15, pp. 379-418, 2006.
[11] O. Gandhi *et al.*, "Economic and technical analysis of reactive power provision from distributed energy resources in microgrids," *Applied Energy*, 2017.
[12] U.S. Energy Information Administration. (2017, April). *Real Petroleum Crisis: Crude Oil: Brent Spot*. Available: https://www.eia.gov/outlooks/aeo/data/browser/#/?id=12-AEO2017&region=0-0&cases=ref2017~highprice~lowprice~highrt~lowrt&start=2015&end=2050&f=A&sourcekey=0
[13] GlobalPetrolPrices.com. (2017, April). *Diesel prices, liter*. Available: http://www.globalpetrolprices.com/data/
[14] P. Bronski *et al.*, "The economics of grid defection: When and where distributed solar generation plus storage competes with traditional utility service," *Rocky Mountain Institute*, 2014.

INFLUENCE OF SMALL DEFECTS ON THE PRODUCTION AND THE SAFETY OF PV PLANTS

Marcelo. P. Almeida[a], Alex. R. A. Manito[a], Gilberto Figueiredo[a], Roberto Zilles[a]
[a] Instituto de Energia e Ambiente – Universidade de São Paulo
Avenida Professor Luciano Gualberto 1289, Cidade Universitária, 05508-010 São Paulo, Brasil

E-mail: marcelopa@iee.usp.br, alex@iee.usp.br, figueiredo@iee.usp.br, zilles@iee.usp.br

ABSTRACT: Photovoltaic (PV) plants are continuously subjected to defects caused by various factors during their lifetime, and many of these defects are related to the PV generator, such as localized hotspots, decoloring, infiltration, cuts on the backsheet, shattered glass, defective by-pass diode, etc. Though most defects may create safety risks to both personnel and equipment and/or diminish system's performance, some have little or even no impact on performance and safety, and the actions that must be taken to eliminate them could be more expensive than the eventual production losses. These small defects should not be neglected, but sometimes monitoring, instead of replacing, is the most appropriate maintenance course of action, particularly for big PV installations.
Keywords: Defects, Performance, Safety, PV module

1 INTRODUCTION

It is common to seek for a problem-free photovoltaic (PV) plant, in which even the minor defects are intended to be corrected as soon as possible, mostly due to the fear of them becoming a serious problem to be dealt with in the future. Besides, some defects may create safety risks to both personnel and equipment, contributing to this general behavior.

When it comes to defects, safety and energy production are the most important issues to be considered, so an evaluation of their actual influence on system's performance and safety should be carried out in order to decide the best maintenance course of action.

This paper sheds some light on the thin line between precaution and exaggerated zeal by presenting an evaluation of the impacts of three frequent defects considering overall safety and performance.

2 SMALL DEFECTS

During their lifetime, PV modules may present defects that are caused by either manufacturing problems or environmental factors, such as temperature, humidity, UV irradiation and mechanical shock. Defects may cause module degradation, leading to power loss and, in more severe cases, electrical safety issues, especially when the defects reduce the insulation between active parts and the external surface of the module, including the metallic frame, increasing the risk of electrical faults and endangering both people and property with electrical shock and fire hazards [1].

Common defects are yellowing/browning, corrosion/oxidation, discoloration, delamination, bubbling, broken/cracked cells, damaged by-pass diodes, shattered glass. The literature has many good studies describing and analyzing PV modules' defects and their origins and consequences [2]–[9].

Visual inspection, IV-curve analysis, thermography and electroluminescence are the most common methods to spot specific defects of PV modules, including those that do not cause sensible power, current or voltage reduction.

Although hot spots are not considered to be a defect by definition, they are a common manifestation of defects and they can be easily identified by a field thermography.

Some defects, such as by-pass diode failure or PID, may have a significant impact on the modules production. Others, such as massive delamination, severe glass loss or extensive water vapor infiltration, impact on both safety and production.

However, there are some defects that have little or no effects on safety and production, especially when they are restricted to a single module of a large PV generator. These small defects should not be neglected, but it is important to differentiate them from those defects that could actually somehow compromise the PV plant.

3 FIELD EXAMPLES

Real field cases were analyzed in order to assess the influence of three small defects on the performance and safety of PV generators. Each defect was localized in a different generator, but they were all composed by the same 260 Wp PV module. Shattered front glass, damaged by-pass diode and perforated module were the analyzed problems.

The defective modules were subjected to infrared inspection and the whole generators to a performance evaluation. In addition, the insulation resistance of the defective modules was also measured. An operational performance evaluation was chosen instead of a simple IV curve, so the actual impact on system's performance could be verified. The insulation resistance measurement was carried out between the short-circuited poles of the module and its metallic frame under dry conditions.

3.1 Shattered front glass and defective by-pass diodes

Two of the real cases that are described in this paper were identified in PV generators installed in the Institute of Energy and Environment of University of São Paulo. Among the installations of the institute, there are three identical 1.56 kWp PV systems detailed in Fig. 1. For convenience, they will be named S1, S2 and S3.

Due to high pressure caused by inadequate fastening, the front glass of a module from system S1 has shattered, as shown in Fig. 2. The shattering profile is very irregular, so some cells are more affected then others. The infrared profile can be seen on Fig. 3, where the maximum difference between the coldest and the hottest cells is no larger than 15°C and the maximum cell temperature is around 50°C.

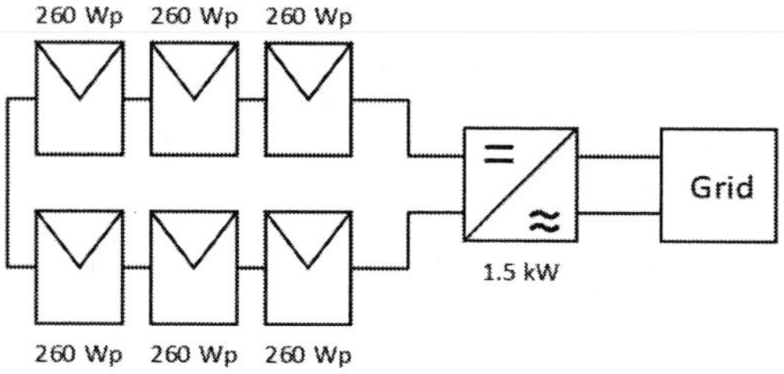

Figure 1: Diagram of PV systems S1, S2 and S3.

Figure 2: Module with shattered front glass from the generator of system S1.

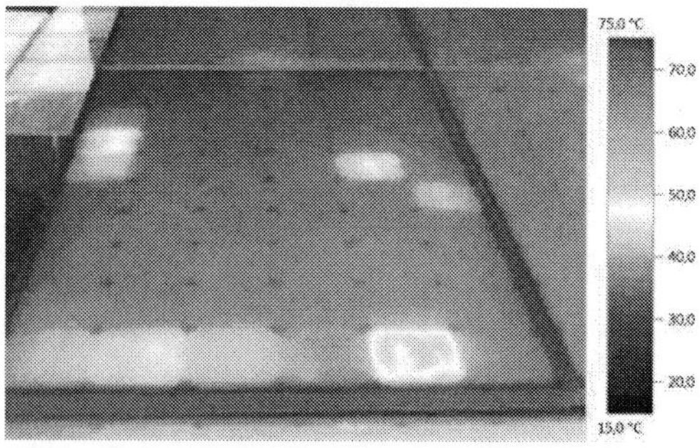

Figure 3: Infrared profile of the module with shattered glass from the PV generator of system S1.

On its turn, system S2 has a module with one of its three by-pass diodes in a defective condition that causes them to behave as an undesirable resistance. Therefore, this diode dissipates a lot of heat, but not enough to melt its weld or the connection box. Fig. 4 shows an infrared image of the connection box and the front surface of the module. On Fig. 4(a) it is possible to identify that the defective by-pass diode is working with a temperature above 220°C., and on Fig. 4(b) it is easy to see the affected area of the PV module with several hot spots.

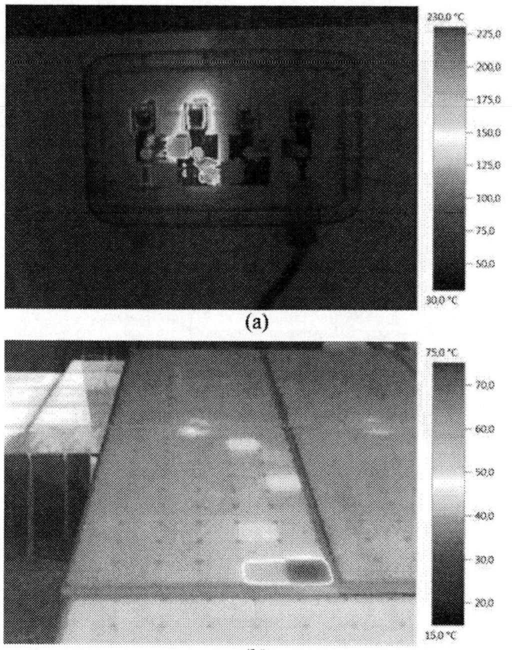

(a)

(b)

Figure 4: Infrared images of the (a) connection box and (b) the front surface of the module with defective by-pass diodes from the generator of system S2.

Finally, system S3 has no defective module, so it will be used as a reference for performance and insulation resistance measurements. The performance indicator chosen here was the energy production during the four consecutive days the measurements were collected.

System S1 and S2 presented a decrease of 10% and 6% in energy production, respectively. Although the module with defective by-pass diode has a worse performance on its own when compared with the module with shattered glass, when operating with other modules in series, the later limits the operation of the whole array. Therefore, the PV generator with the defective by-pass diode performs better, as the only impact of the defect is the reduction of the rated power of the generator in 1/3 of the rated power of a module.

Regarding the insulation resistance, the differences between the defective modules and the good ones are well within the uncertainty of the measuring device, so it is safe to say that these two particular defects have not compromised the modules insulation.

3.3 Perforated module

On 2015 a PV module from a 156 kWp PV plant located in the Institute of Energy and Environment of University of São Paulo was hit by a small rock when the nearby field was mown. The impact caused the glass to shatter and made a half inch diameter hole in the glass, exposing the PV cell. Fig. 5 shows the damage caused to the PV module and its infrared image, where a hot spot above 100°C can be easily identified.

The PV module has been working properly for over 2 years, despite the apparent seriousness of the damage. The PV system consists of 30 parallel strings of 20 PV modules each, and Fig. 6 presents a comparison between the DC operational current of each strings during a day. The red line is the DC current of the string with the damaged module, the orange lines are the DC currents of two other string which are connected in parallel in the same string box, and the grey lines are the DC current of the other strings from different string boxes (which are all connected in parallel inside the central inverter).

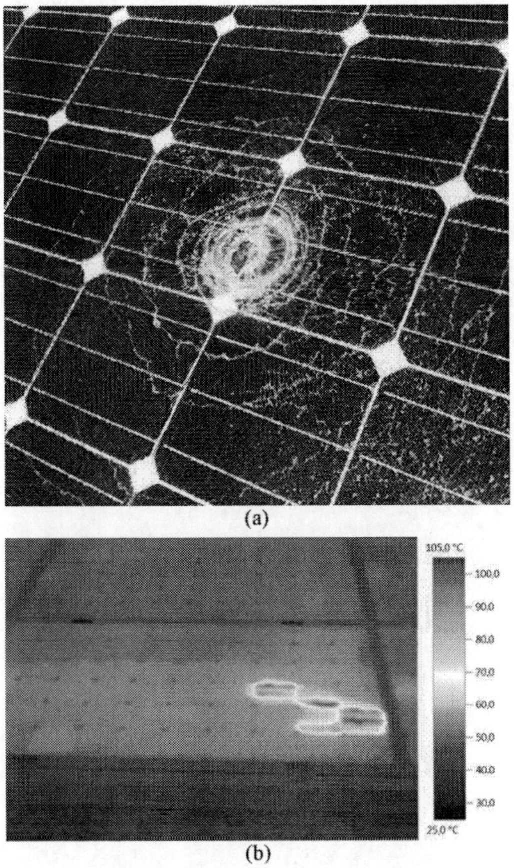

Figure 5: Damaged PV module and its infrared image.

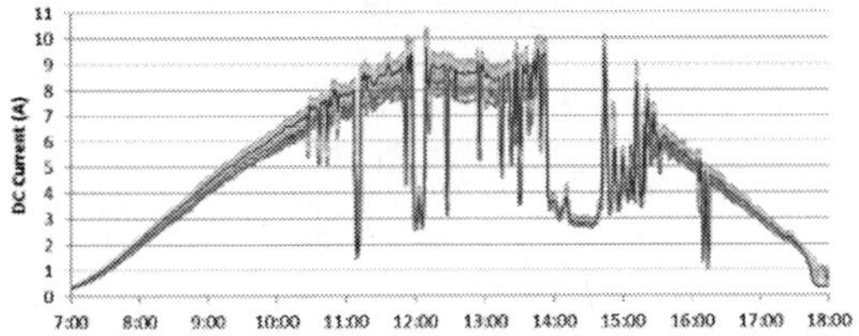

Figure 6: DC current of parallel strings, including the one with the damaged module (red line).

It is clear that the damage has not compromised the generation of the PV module, although intense hot spots have been developed. As no major change in insulation resistance was observed and there has been no dangerous degradation (fire hazard, for example), the module has been kept operational.

4 DISCUSSIONS

When dealing with small defects, it is important to have in mind which kind of PV application is under consideration. Distributed PV plants on rooftops, parking areas or facades are more likely to be touched by people and are closer to flammable parts. In this case, safety issues are extremely relevant. On the other hand, big centralized PV plants have restricted access and are usually ground mounted. Safety issues are obviously important in this case, but at a different level.

Therefore, these two different scenarios dictate how cautious one should be regarding the small defects, but one should always have in mind that undesirable insulation faults can be easily detected [1], [10].

Performance is also a relevant aspect, and the size of the PV generator is of great influence. In the examples of item 3.1, the defects caused small, but no negligible reduction of performance due to the relatively small PV generator. The defective by-pass diode, for example, short-circuited one third of a single module, which corresponds to almost 5% of the whole PV generator. If this defect was located in the 156 kWp plant described in item 3.2, the relative loss of peak power would be one hundred times smaller and no significant impact would be observed.

Small defects should not be underestimated or neglected. The main idea of this paper is to show that PV plants can withstand some kinds of small defects without compromising their safety and production.

This is especially true to larger PV plants, where the revenue reduction due to small defects may be inexpressive when compared to the replacement costs of these modules, which is a major O&M expense in actual PV plants [11]. Small defects may evolve into real problem, so maintenance routines must include the observation of all known small defects.

5 CONCLUSION

Small defects usually do not cause significant impacts, if any, on the production of the PV plant they belong to, so the main concern must be the safety of people, equipment and the surroundings. Therefore, small defects should be categorized according to its potential of inflicting damages rather than its visual appearance or supposed future influence, so those components with a low damaging potential could be only monitored, instead of being replaced.

This monitoring is definitely easier and cheaper than a fully replacement of a component. PV plants, especially the large ones, have maintenance procedures that could be adapted to include periodic inspection of those defective components with almost no increase in the costs. In those cases when the defect is not covered by the warranty given by the manufacturer of the equipment or the installer, replacements are even more difficult and expensive.

6 REFERENCES

[1] M. K. Alam, F. Khan, J. Johnson, and J. Flicker, "A Comprehensive Review of Catastrophic Faults in PV Arrays: Types, Detection, and Mitigation Techniques," *IEEE Journal of Photovoltaics*, vol. 5, no. 3, pp. 982–997, 2015.

[2] C. Ferrara and D. Philipp, "Why do PV modules fail?," *Energy Procedia*, vol. 15, pp. 379–387, 2012.

[3] D. Polverini, M. Field, E. Dunlop, and W.

Zaaiman, "Polycrystalline silicon PV modules performance and degradation over 20 years," *Progress in Photovoltaics: Research and Applications*, vol. 21, pp. 1004–1015, Apr. 2013.
[4] D. C. Jordan, T. J. Silverman, B. Sekulic, and S. R. Kurtz, "PV degradation curves: Non-linearities and failure modes," *Progress in Photovoltaics: Research and Applications*, vol. 25, pp. 583–591, 2017.
[5] J. A. Tsanakas, L. Ha, and C. Buerhop, "Faults and infrared thermographic diagnosis in operating c-Si photovoltaic modules: A review of research and future challenges," *Renewable and Sustainable Energy Reviews*, vol. 62, pp. 695–709, 2016.
[6] S. Djordjevic, D. Parlevliet, and P. Jennings, "Detectable faults on recently installed solar modules in Western Australia," *Renewable Energy*, vol. 67, pp. 215–221, 2014.
[7] A. Ndiaye, A. Charki, A. Kobi, C. M. F. Kébé, P. A. Ndiaye, and V. Sambou, "Degradations of silicon photovoltaic modules: A literature review," *Solar Energy*, vol. 96, pp. 140–151, 2013.
[8] M. A. Munoz, M. C. Alonso-García, N. Vela, and F. Chenlo, "Early degradation of silicon PV modules and guaranty conditions," *Solar Energy*, vol. 85, no. 9, pp. 2264–2274, 2011.
[9] IEA-PVPS-T13-01, "Review of Failures of Photovoltaic Modules," 2014.
[10] M. C. Falvo and S. Capparella, "Safety issues in PV systems: Design choices for a secure fault detection and for preventing fire risk," *Case Studies in Fire Safety*, vol. 3, pp. 1–16, 2015.
[11] A. Golnas, "PV System Reliability: An Operator's Perspective," *IEEE Journal of Photovoltaics*, vol. 3, no. 1, pp. 1–6, Jan. 2012.

A COMPARATIVE STUDY OF TWO MODELS FOR EVALUATING THE POWER OF PHOTOVOLTAIC MODULES IN A STANDALONE POWER PLANT

Amine El Fathi, Mohamed Akhsassi, Amin Bennouna and AbdelkaderOutzourhit*
LNEE- Department of physics, Faculty of sciences Semlalia, Cadi Ayyad University
BP: 2390, Marrakech Morocco

ABSTRACT: In this work, we will compare two models which are able to calculate the PV module power based on its temperature, in-plane irradiance and other physical parameters such as the temperature coefficient of power (TCP) and low light efficiency change (LLEC) for PV modules. In this regards, we will analyze the measured values of in-plane irradiation, cell temperature, power and batteries state of charge (SOC) of a standalone PV power plant installed in a remote area (Elkaria village)which feeds 16 households with electricity. We will also compare the output energy of the two models with experimental measurements.

Keywords: PV pant, nonlinear model, efficiency

*Email: aoutzour@uca.ma

1 INTRODUCTION

The power and the efficiency of photovoltaic (PV) modules or arrays need to be accurately determined in order to assess their electrical performances. There are several parameters affecting the PV module power. The first of these is the in plane irradiance of the module, G_g, and the second one is the PV module temperature, T_c.The latter parameter affects negatively the PV module efficiency. The temperature coefficient of power (β_{Pmp}) for crystalline silicon encapsulated solar cells is usually not measured but determined using a calculation procedure according to the IEC 60891 standard, and it is in the range of -0.52 to -0.37 %.°C^{-1} [1]. King et al. [2] showed that the efficiency of three PV modules technologies can decrease by up to 10 % for the highest module temperatures.The low light efficiency change (γ_{Pmp}) can be calculated by using the electrical parameters of PV modules which are usually measured by the manufacturers at Standard Test Conditions (STC): an irradiance level of 1000 W/m², a cell temperature of 25°C and an air mass AM= 1.5 spectrum. In fact, these conditions occur rarely on site since solar radiation of 1000W/m² makes it difficult to have a cell temperature of 25°C [4].

Several authors, such as, Evans et al.[3], Duffie & Beckman [4], Skoplaki et al. [5], Koehl et al. [6], Williams et al. [7] andDurisch et al[8]developed electrical models for PV systems taking into account the non-linearity effect on PV power/efficiency calculations.

In this work, two models for PV module power/efficiency and onsite experimental measurements will be used to assess the performance of standalone PV power plant operating under real conditions. The objective is to specify the most accurate model, whichcan be latter used in future works in order to predict the energy production of such systems.

2 DESCRIPTION OF THE PV POWER PLANT

2.1 Location

The PV power plant of Elkaria is installed in a coastal region in the middle of Morocco between the cities of Safi and Essaouira. This region has an interesting irradiance levels and an average annual wind speed between 7 and 8 m/s[9].

2.2 Power plant configuration

As described in details in a previous work [9], the PV plant consists of 32 (Sunpower SPR-225-WHT) panels with mono-crystalline silicon cells using the back contact technology. The peak power and efficiency of each panel are respectively 225 $W_p$and 18.1 %. Some pertinent characteristics of the panel are summarized in the table 1.

Table I: Pertinent parameters of the PV panels Measured at Standard Test Conditions (STC).

Solar cells monocrystalline	72 SunPower all back-contact
Efficiency	18.1%
Peak power	225 W
Rated voltage	41 V
Rated current	5.49 A
Open circuit voltage	48.5 V
Short circuit current	5.87 A

The panels were mounted at an optimum inclination angle of 35° for this location. The photovoltaic array is divided into four strings of 8 panels each mounted in series. As shown in Fig.1, each two independent strings are connected in parallel to a PV-inverter (Sunny Boy 3800, SMA, Germany) with a rated power of 3.8 kW. The two inverters are then connected to the local isolated grid than runs through the village (about 2.5 km long) powered by the grid forming unit. The later consists of two Sunny Island SI5040 bidirectional inverters (SMA, Germany) and a lead-acid battery bank (Hawker Pb TYS7/2AT 1101 Ah). The rated input DC voltage of the SI5040 is 48 V while its rated power is 5 kW. The DC sides of these inverters are connected to the battery bank which consists of a series combination of 24 batteries (2 V each) with a C_{100} capacity of 1100 Ah [5]. An inside view of the PV power plant is given in Fig.2. The grid control is performed by the Sunny Island (BI) battery inverter using the Droop Mode control[9].

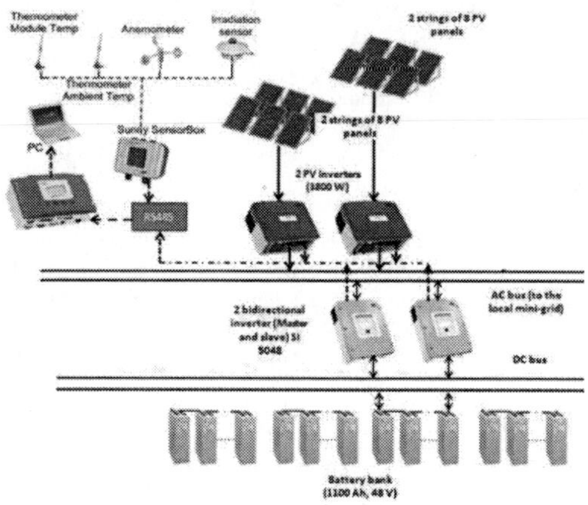

Fig.1. The block diagram of the PV-power plant

The Sunny Boy control data logger (SMA, Germany) was used to record the relevant parameters from the the Sunny Sensor Box (SMA, Germany) such as irradiance, wind speed, ambient and panel temperature. Other relevant parameters are also recorded to provide information about the delivered and consumed power, and AC/DC currents and voltages levels. The data is recorded and saved on a daily basis. The Data logger communicates with the various instruments through an RS485 cable and interface.

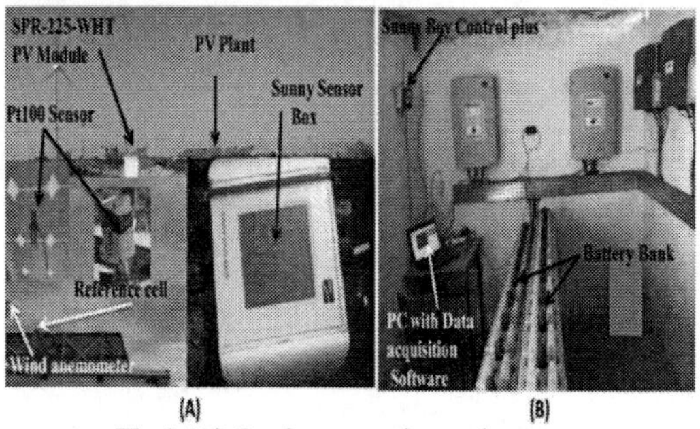

Fig.2.(A) Outdoor experimental setup, (B) Indoor experimental setup.

3 USED MODELS

In addition to the solar radiation, the power of a PV module depends on its temperature[3], [7], [10]–[12].

3.1 Non-linear model

The most known nonlinear model for photovoltaic module efficiency is given by the following equation:

$$P_{dc} = P_{STC}\frac{G_g}{G_0}\left[1 + \beta_{P_{mp}}\left(T_c - T_{ref}\right)\right]\left[1 + \boldsymbol{\gamma}_{\boldsymbol{P_{mp}}} Ln\left(\frac{G_g}{G_0}\right)\right]$$

Where P_{STC} and $\beta_{Pmp}$are the peak power of the PV plant and the temperature coefficient of maximum power, respectively, at Standard Test Conditions (3.6kWp for the two strings, and -0.39%/°C, respectively). T_{ref} and G_0 are the reference temperature and reference solar irradiance (25°C and 1000W/m², respectively). $\gamma_{Pmp}$is a dimensionless coefficient which is between 0.03 and 0.12 for single crystalline silicon [3], [7], [10]–[12].

3.2 Evans model

Another model used in the litterature similar to thenon linear oneis given by:

$$P_{dc} = P_{STC}\frac{G_g}{G_0}\left[1 + \beta_{P_{mp}}\left(T_c - T_{ref}\right) + \gamma Ln\left(\frac{G_g}{G_0}\right)\right]$$

Where γ is equal to 0.12 for single crystalline silicon[3].

4 RESULTS

In addition to the measured in-plane irradiance G_g, the module temperature T_c and the battery SOC levels, Fig.2, Fig.3, Fig.4 and Fig.5 show the measured (for the DC voltage and current of the inverter) and calculated DC power of the 3.6 kW_pfield for four typical days. These represent the most frequently encountered days since the operation of the PV power plant.

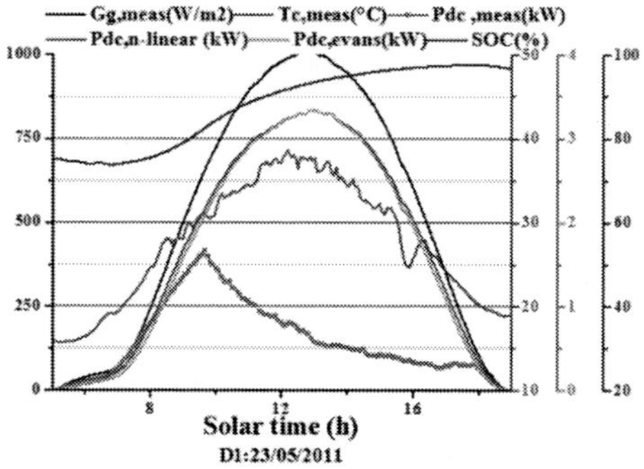

Fig.2. Measured and calculated DC power of a standalone PV plant (3.6 kWp), battery SOC, irradiance and cell temperature vs. solar time for a summer clear day.

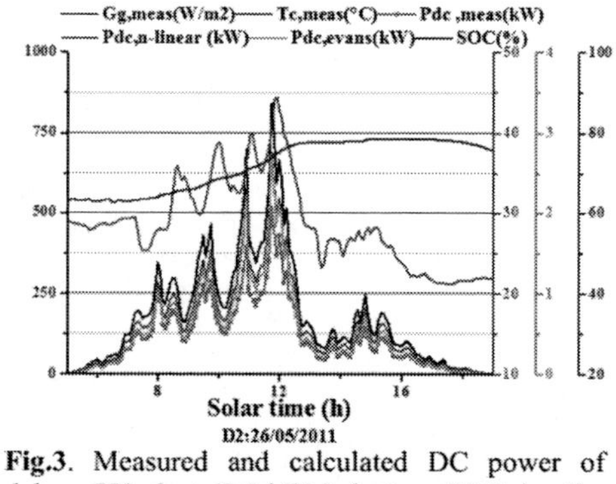

Fig.3. Measured and calculated DC power of a standalone PV plant (3.6 kWp), battery SOC, irradiance and cell temperature vs. solar time for a cloudy day.

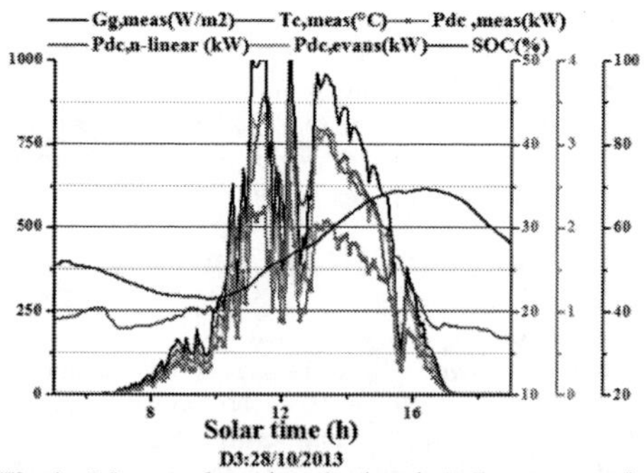

Fig.4. Measured and calculated DC power of a standalone PV plant (3.6 kWp), battery SOC, irradiance and cell temperature vs. solar time for a semi-cloudy day.

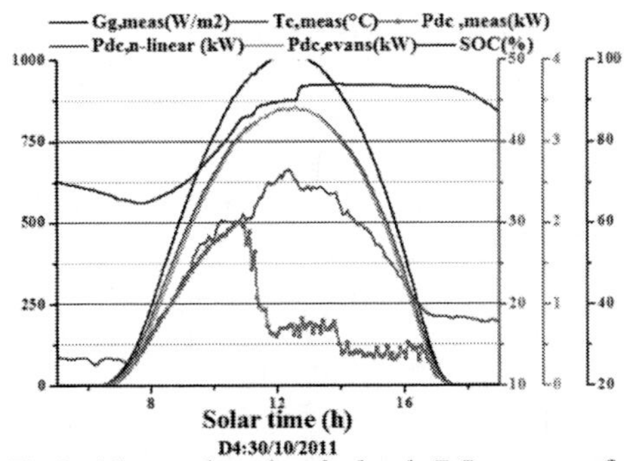

Fig.5. Measured and calculated DC power of a standalone PV plant (3.6 kWp), battery SOC, irradiance and cell temperature vs. solar time for a winter clear day.

In Table 2, the DC energy of one PV inverter is depicted and compared with the calculated DC energy by using the two models. The observed difference between measured and calculated DC energy especially for clear days is due to the way the SI is managing the power plant. Depending on the weather conditions, the energy demand, and the battery state of charge, the PV inverter reduces the power fed to the local grid when the battery is charged and the demand is low.

Table 2: Comparison of two models with experimental measurements DC output energy along four days.

Edc (kWh)/Day	D1(summer clear)	D2(cloudy)	D3(semi-cloudy)	D4(winter clear)
Measured	8.28	6.29	9.67	8.37
Non linear model	24.11	8.35	14.44	22.29
Evans model	23.37	7.38	13.76	21.72

4 CONCLUSION

The comparison of calculated DC power with the experimental data shows that:

- The nonlinear model gives the best predictions,
- The Evans model underestimates the power of the PV modules,
- The two models approach the experimental measurements when the SOC is low (no limitation of the PV production).

5 PERSPECTIVES

This work has the perspectives of:

- Measurement of the electrical power of the PV plant in full load (connected to the grid) to validate the two models,
- Extractionof the coefficients of temperature (TCP) and the coefficients of irradiance (LLEC).

ACKNOWLEDGMENT

This work is partially funded by Research Institute for Solar Energy and New Energies (IRESEN) in the framework of the InnoPV/Innowind2014 program.

6 References

[1] *Planning and installing photovoltiac systems : a guide for installers, architects, and engineers, Deutsche Gesellschaft fur Sonnenenergie (DGS)*, 5th ed. Berlin: Landesverband Berlin Brandenburg e.V, 2014.

[2] D. L. King, J. a Kratochvil, and W. E. Boyson, "Photovoltaic array performance model," *Online*, vol. 8, no. December, pp. 1–19, 2004.

[3] D. L. Evans, "Simplified method for predicting photovoltaic array output," *Sol. Energy*, vol. 27, no. 6, pp. 555–560, 1981.

[4] J. A. Duffie and W. A. Beckman, *Solar Engineering of Thermal Processes*, 3rd ed. 2006.

[5] E. Skoplaki, a. G. Boudouvis, and J. a. Palyvos, "A simple correlation for the operating temperature of photovoltaic modules of arbitrary mounting," *Sol. Energy Mater. Sol. Cells*, vol. 92, no. 11, pp. 1393–1402, 2008.

[6] M. Koehl, M. Heck, S. Wiesmeier, and J. Wirth, "Modeling of the nominal operating cell temperature based on outdoor weathering," *Sol. Energy Mater. Sol. Cells*, vol. 95, no. 7, pp. 1638–1646, 2011.

[7] and D. G. I. Williams, S. R., T.R. Betts, T. Helf, R. Gottschalg, H.G. Beyer, "Modelling Long-term Module Performance Based on Realistic Reporting Conditions with Consideration to Spectral Effects," in *In Proceedings of 3rd World Conference on Photovoltaic Energy Conversion*, 2003, pp. 1908–1911.

[8] W. Durisch, J. Urban, and G. Smestad, "CHARACTERISATION OF SOLAR CELLS AND MODULES UNDER ACTUAL OPERATING CONDITIONS," pp. 359–366, 1996.

[9] A. El Fathi, L. Nkhaili, A. Bennouna, and A. Outzourhit, "Performance parameters of a standalone PV plant," *Energy Convers. Manag.*, vol. 86, pp. 490–495, 2014.

[10] G. Notton, V. Lazarov, and L. Stoyanov, "Optimal sizing of a grid-connected PV system for various PV module technologies and inclinations, inverter efficiency characteristics and locations," *Renew. Energy*, vol. 35, no. 2, pp. 541–554, 2010.

[11] F. Fouladi, P. Henshaw, and D. S.-K. Ting, "Enhancing Smart Grid realisation with accurate prediction of photovoltaic performance based on weather forecast," *Int. J. Environ. Stud.*, vol. 70, no. 5, pp. 754–764, 2013.

[12] E. Kaplani and S. Kaplanis, "Thermal modelling and experimental assessment of the dependence of PV module temperature on wind velocity and direction, module orientation and inclination," *Sol. Energy*, vol. 107, pp. 443–460, 2014.

EXPERIMENTAL YIELD STUDY OF BIFACIAL PV MODULES IN NORDIC CONDITIONS

Elin Molin [ac*], Bengt Stridh [b], Andreas Molin [c,d], Ewa Wäckelgård [a]
a Dalarna University, Energy Technology, SE-791 88 Falun, Sweden
b Mälardalen University, Department of Energy, Building and Environment, P.O. Box 883, SE-721 23 Västerås, Sweden, and ABB AB, Corporate Research, SE-721 78 Västerås, Sweden.
c PPAM Solkraft AB, Corporate Research, Husarvägen 57, SE-590 72 Ljungsbro, Sweden,
d Linköping University, Division of Energy Systems, Department of Management and Engineering, SE-58183 Linköping, Sweden
* Corresponding author. E-mail: elin.molin@ppam.se, Tel. +46 (0)70 567 66 31

ABSTRACT: Bifacial solar modules offer new opportunities in terms of orientation. In Sweden, at northern latitudes, south orientated modules receive direct irradiance also on the backside. In summer, the sun rises in the north-east and sets in the north-west, having rather perfect conditions for vertical east and west orientated bifacial modules. This field study was conducted during June 2016 at two test sites on almost flat roof-tops in Linköping, Sweden. Test site 1 have monofacial and bifacial modules south-orientated with 40° tilt, which is optimal for yearly energy yield for monofacial. Test site 2 have monofacial optimally south-orientated modules and bifacial vertical east-west orientated modules. The specific yield for bifacial modules peaks at 8.1 kWh/kW$_p$ for one day in June 2016, which was 12% higher than for the monofacial modules. The bifacial energy gain at the test sites was in average 12 % higher on a monthly basis in June 2016 for both the bifacial systems although the albedo is only about 0.05 and the tilted bifacial modules have a self-shading effect due to low elevation above the roofs resulting in shading of the reflection area.

Keywords: Bifacial, HIT, monofacial, field study, Sweden.

1 BACKGROUND

The bifacial photovoltaic (PV) module is a flat plane solar module that produces electricity from both the frontside and the backside, in contrast to the monofacial module. This innovative approach gives a possibility for higher energy production per module area and thereby also a lower levelized cost of energy. Bifacial solar modules also offer new opportunities in terms of orientation. This work studies frameless glass-glass modules that can be used for solar shields, carports, greenhouses, visual barriers and noise barriers. This is the first reported field study in Sweden with bifacial solar modules. In Sweden, at 55°-69° northern latitudes south orientated modules will receive some direct irradiance also on the backside, which is beneficial for bifacial modules. Snow, with its higher albedo, can result in a higher yield in winter when modules can be mounted vertically and the snow do not cover the modules. Bifacial modules have earlier been studied both in simulations and in field studies but for Nordic conditions the studies are few. In Albuquerque, USA, the additional energy that bifacial PV arrays could generate under different conditions and orientations was quantified [1]. The bifacial gain over a six-month period for five test orientations ranged from 18% to 136%. A simulation study showed that a bifacial module mounted east-west needs less albedo when the latitude increases than a monofacial to still have a higher yield [2]. A comparison of tilt angle, elevation above ground and albedo influence on bifacial energy yield for south orientated modules were modeled and measured for one test conditions in New York and six test conditions in Arizona [3]. The yield increased as each different parameter of tilt angle, elevation and albedo increased. The study believed the HIT (Heterojunction with Intrinsic Thin-layer)-cell should be studied separately, since cell architecture and temperature coefficients substantially differs from the more traditional bifacial cells used in developing the model. Report [4] shows that the annual yield gain increases strongly with albedo of bifacial east-west modules over south-oriented monofacial modules. In a location like Amsterdam the gain is 10% at albedo 0.2 and 30% at albedo 0.5. The model has been developed through simulations and measurements. The studies in [5] examines a model to calculate the production on the backside of a south-orientated bifacial PV module. An effect pointed out was self-shading due to low elevation above ground resulting in shading of the reflection area.

2 AIM

The aim of this study is to:

- Quantify the additional energy yield that bifacial PV arrays could generate under different irradiance conditions and orientations using our two test sites as examples.
- Compare the yield in June 2016 of a 40° tilted south orientated (optimal for yearly energy yield) monofacial modules with bifacial HIT-modules installed in the same direction and a vertical mounted bifacial HIT-panel facing east and west.
- Study how cloudy and sunny days affects the energy yield in the chosen configurations.

3 METHODOLOGY

The bifacial modules used in this study are PPAM Transparium, using HIT-cells with a thin amorphous silicon layer on monocrystalline silicon substrate on the front- and backside as described in [6]. The bifacial modules are frameless glass-glass modules while the monofacial modules have a metallic frame. This field study was conducted during 2016 on two test sites on roof-tops in Linköping, Sweden. Test site 1 has monofacial and bifacial modules south-orientated with 40° tilt, which is optimal for yearly energy yield for monofacial.

Figure 1. Test site 1 (Ryd). Bifacial (back) and monofacial (front) modules in the same direction (south, 40° tilt). 12 meters between the installations.

Test site 2 has monofacial optimally south-orientated modules and bifacial vertical east-west orientated modules. The roofs are almost flat with tar paper surface, which is commonly used on such roofs in Sweden. Therefore, a low albedo of about 0.05 is expected [7].

Figure 2. Test site 2 (Byggvesta). With east-west orientated bifacial modules (left) and south orientated monofacial modules tilted 40° (right). 22 meters between the installations.

Table 1 shows the set-up for the installations. The modules are mounted in one string with one single-phase SMA Sunny Boy 3300 inverter with one maximum power point tracker per string. The measurements are recorded with SMA Sunny web boxes collecting 5-minute average values from the inverters and used sensors. None of the sites had shading from the horizon or nearby objects during the studied period in June 2016. All diagrams use central European summertime (CEST).

Table I: Installations set-up. Datasheet values [8][9]. Module power is P_{mp}. Bi = bifacial. Mo = monofacial.

Site 1	PV modules	
No	6	6
Power (W_p)	Bi 360	Mo 320
P_{mp} tolerance	+5/-3%	+3/-0%
P_{mp} coefficient (%/°C)	-0.28	-0.42
Site 1	PV system	
Power (W_p)	2160	1920
Azimuth (°)	0	0
Tilt (°)	40	40
Latitude/longitude	58.4 N 15.6 E	

Site 2	PV modules	
No	7	6
Power (W_p)	Bi 300	Mo 320
P_{mp} tolerance	+5/-3%	+3/-0%
P_{mp} coefficient (%/°C)	-0.28	-0.42
	PV system	
Power (W_p)	2100	1920
Azimuth (°)	-90/90	0
Tilt (°)	90	40
Latitude/longitude	58.4 N 15.6 E	

4 RESULT AND DISCUSSION

Monthly specific yield in June 2016 was 156 kWh/kW$_p$ for the bifacial system at site 1, 12 % higher than the 139 kWh/kW$_p$ for the monofacial system. For site 2 the monthly specific yield was 163 kWh/kW$_p$ for the bifacial system, 12 % higher than the 145 kWh/kW$_p$ for the monofacial system. The bifacial system giving the highest monthly specific yield was the vertical that due to the orientation can use direct irradiance for a much longer time in summer than the south oriented modules. Daily specific yield for June 2016 is shown in Fig. 3-4 and the yield for the first week of June is shown in detail in Fig. 5-6.

In general, the bifacial modules have a higher daily yield than monofacial modules during June 2016. The daily yield was highest for the bifacial modules for all days in June at site 1 and for 24 of the 30 days in June for site 2. The bifacial modules facing south at site 1 have in average during the month a 12% higher specific power at noon than the monofacial modules, see example in Fig 7. One effect of importance that can contribute to these results is the lower temperature coefficients of the bifacial compared to the monofacial modules. With a measured module temperature of up to 55°C during sunny days the power will decrease 12% for the monofacial modules but only 8% for the bifacial modules, compared to the name plate power. When using vertical bifacial modules, the daily yield can be lower than for south oriented monofacial modules if it is clear weather midday, but cloudy in early and late day.

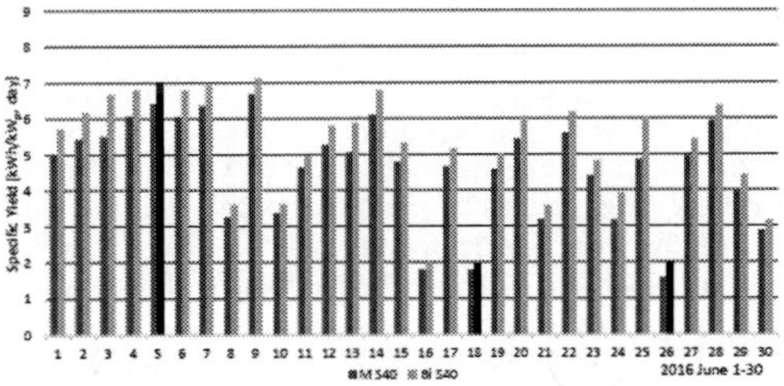

Figure 3. Site 1. Specific yield per day in June 2016. Days marked with red and black are shown in detail below.

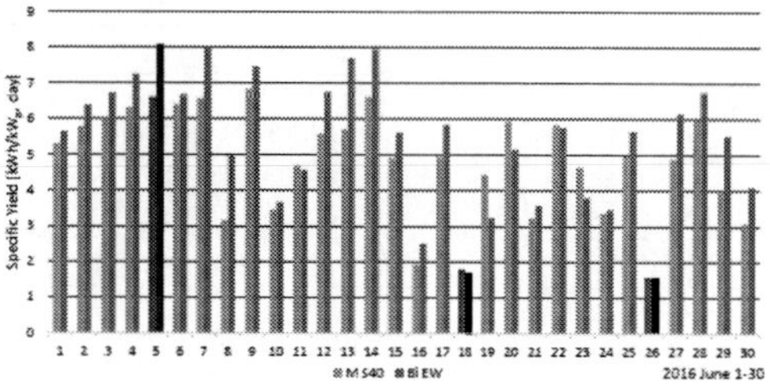

Figure 4. Site 2. Specific yield per day in June 2016. Days marked with red and black are shown in detail below.

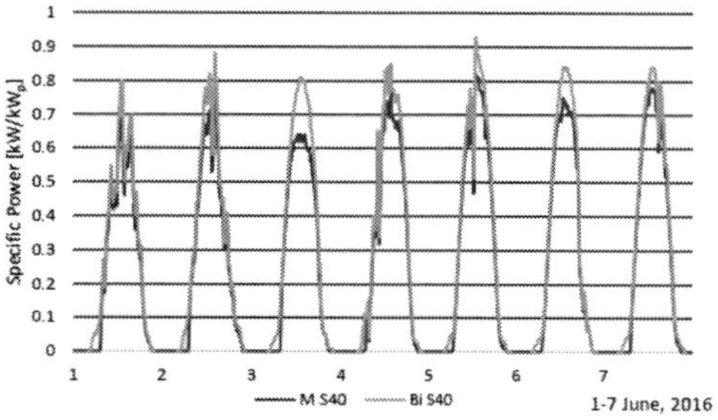

Figure 5. Site 1. Specific power, first week in June 2016.

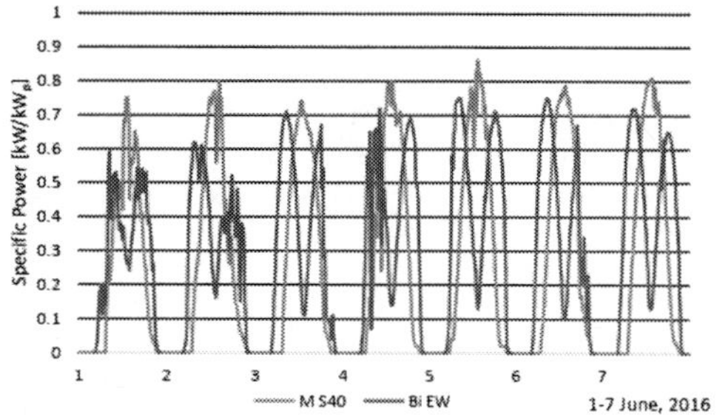

Figure 6. Site 2. Specific power, first week in June 2016.

The highest daily yield during June 2016 was noted for the vertical east-west bifacial modules at the clear day of June 5 with 8.1 kWh/kW_p, that was 12 % higher than for the monofacial south oriented modules, see Fig. 8. The east-west bifacial modules had peaks in specific power early and late day, instead of during midday. In Fig. 7, a clear gain of the bifacial modules in site 1 is seen. This can partially be explained by the temperature coefficient, but not solely, since bifacial modules in site 1 have a specific power 0.95 kW/kW_p while the monofacial modules have specific power is 0.82 kW/kW_p at noon, a difference of 14%. On a cloudy day like in Fig. 9 the difference is 6%. The south oriented bifacial modules with 40° gave 7.4 kWh/kW_p at the clear day of June 5, which was 12 % higher than the monofacial modules with the same orientation, see Fig. 8. In Sweden, also clear days have some clouds which can be seen in Fig. 7-8.

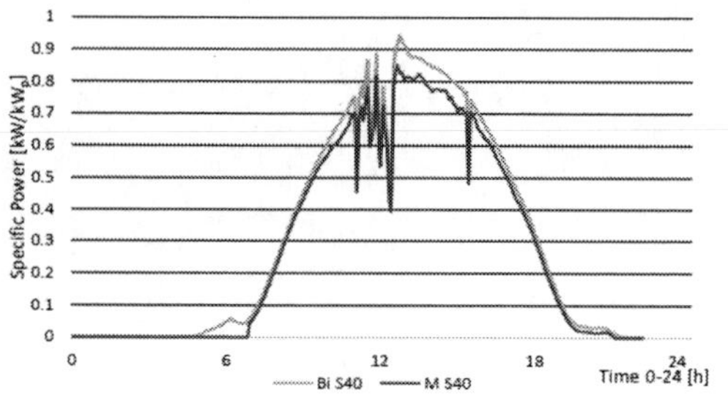

Figure 7. Site 1. Clear day, 5 June 2016.

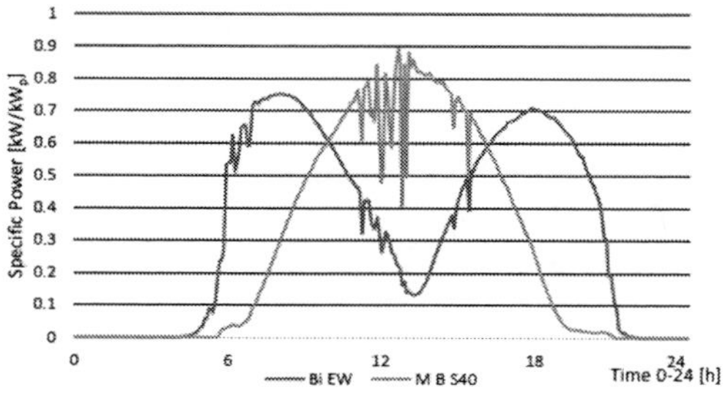

Figure 8. Site 2. Clear day, 5 June 2016.

The specific power for two cloudy days in June 2016 with lowest yield is shown in Fig. 9-12. Although the specific yield for the bifacial modules is lower the 26 June (3.5 kWh/kW_p) than 18 June (3.7 kWh/kW_p) the specific power is much higher sometimes under the day of 26 June, illustrating different cloud effects on the power for the two different days.

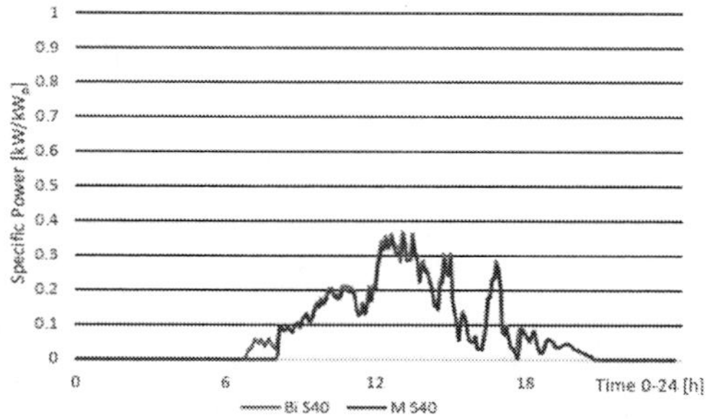

Figure 9. Site 1. Cloudy day 18 June 2016.

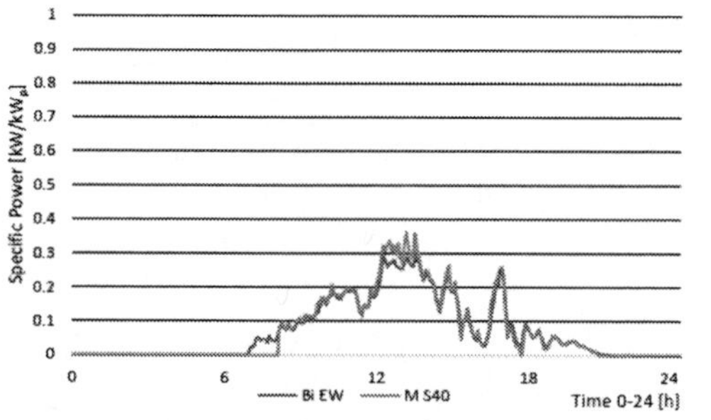

Figure 10. Site 2. Cloudy day, 18 June 2016.

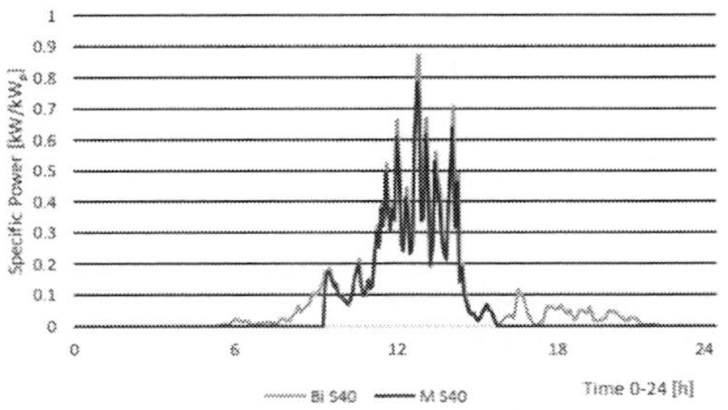

Figure 11. Site 1. Cloudy day, 26 June 2016.

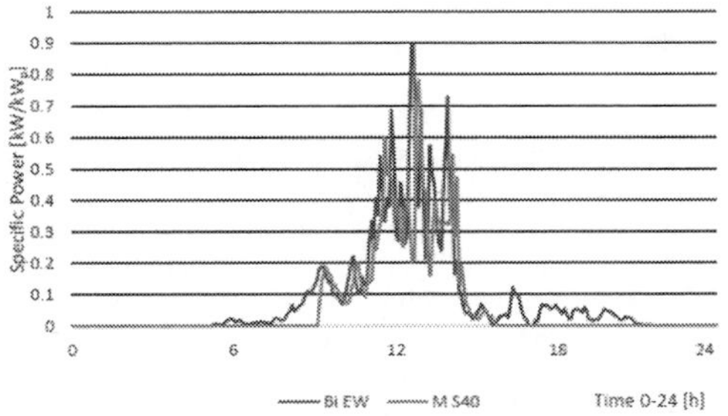

Figure 12. Site 2. Cloudy day, 26 June 2016.

In summer, the sun rises in the north-east and sets in the north-west, resulting in good conditions for the bifacial modules. In Linköping at northern latitude 58° the south orientated modules will receive some direct irradiance also on the backside of the modules in morning and in evening. This is a beneficial effect for the bifacial system in this study. Since the expected albedo 0.05 of the roof is so small the reflected irradiance should have a minor effect on these test sites.

To explore seasonal trends the average energy per day and the monthly bifacial gain in energy BGE (%) was calculated.

$$BG_E\,(1\text{ month}) = 100 * \left(\frac{\sum_{1\text{month}} \frac{P_{bifacial}(t)}{P_{mp_{bifacial}}}}{\sum_{1\text{ month}} \frac{P_{monofacial}(t)}{P_{mp_{monofacial}}}} - 1 \right)$$

The resulting BGE for June 2016 was 12% for the sites. This means that in June 2016 both south oriented bifacial modules with tilt 40° and vertical east-west bifacial modules gave 12% higher specific energy yield than monofacial modules with orientation optimized for highest yearly yield (south, 40° tilt). The gain is lower than in [1] because of the low albedo of the roof and in site 1, also the self-shading effects mentioned in [5] due to the low height of 0.3 m above the roof.

5 CONCLUSIOS

The bifacial energy gain in June 2016 is at the test sites 12 % on a monthly basis for both of the bifacial system configurations although the albedo is only about 0.05 and the tilted bifacial modules have a self-shading effect since the height on the plant is only 0.3 m above the roof. Therefore, it is likely that the higher yield for the bifacial modules is a result of a higher amount of direct and diffuse irradiance than for the monofacial modules, in combination with a lower temperature coefficient for the bifacial modules. Bifacial modules have a promising yield in summer compared to standard monofacial modules and can be used in vertical installations with east and west directions to get a more even production during a day, which can increase the self-consumption if the consumption is higher in early and late day, instead of during midday. To improve the bifacial gain, a surface with higher albedo than tar paper should be used. The one-month study of the two sites will be continuously measured for one year and be reported in a full-study to come. This will also include measurements of irradiance on both sides of the bifacial modules at the two sites.

6 ACKNOWLEDGEMENTS

The work has been carried out under the auspices of the industrial post-graduate school Reesbe, which is financed by the Knowledge Foundation (KK-stiftelsen). The authors thank Rickard Linderström of Byggvesta AB and Peter Alfredsson of Stångåstaden for invaluable support with giving us access to the project sites.

7 REFERENCES

[1] Lave, M., Stein, J.S., Burnham, L., "Performance results for the prism solar installation at the New Mexico regional test center: Filed Data from Februari 15-August 15, 2016."

[2] Guo, S., Walsh, T.M., Peters, M.," Vertically mounted bifacial photovoltaic modules: A global analysis" (2013) Energy, 61, pp. 447-454.

[3] J. E. Castillo-Aguilella and P. S. Hauser, "Multi-Variable Bifacial Photovoltaic Module Test Results and Best-Fit Annual Bifacial Energy Yield Model," in IEEE Access, vol. 4, no. , pp. 498-506, 2016.

[4] Janssen, G.J.M., Van Aken, B.B., Carr, A.J., Mewe, A.A., "Outdoor Performance of Bifacial Modules by Measurements and Modelling" (2015) Energy Procedia, 77, pp. 364-373.

[5] Yusufoglu, U.A., Pletzer, T.M., Koduvelikulathu, L.J., Comparotto, C., Kopecek, R., Kurz, H., "Analysis of the annual performance of bifacial modules and optimization methods."(2015) IEEE Journal of Photovoltaics, 5 (1), art. no. 6960821, pp. 320-328. Cited 1 time.

[6] K. Maki et al., "High-efficiency HIT solar cells with a very thin structure enabling a high Voc," 2011 37th IEEE Photovoltaic Specialists Conference, Seattle, WA, 2011, pp. 000057-000061.

[7] Mat Santamouris, "Environmental Design of Urban Buildings: An Integrated Approach", p112, ISBN-13: 978-1902916-42-2.

[8] PPAM Onyxium, datasheet.

[9] PPAM Transparium, datasheet.

EARLY DEGRADATION OF PHOTOVOLTAIC MODULES BASED ON n-TYPE SOLAR CELLS

Gilberto Figueiredo, Marcelo Almeida, Roberto Zilles
Universidade de São Paulo/Instituto de Energia e Ambiente
Av. Professor Luciano Gualberto, 1289. Sao Paulo/Brazil
Phone: +55 11 3091-2656, email: figueiredo@iee.usp.br

ABSTRACT: The work presents the investigation of in field early degradation of photovoltaic modules installed at a 156 kWp grid connected installation located in the south hemisphere. The operation started in 2014 and the generator is composed by n-type monocrystalline silicon solar cells. Recent measurements revealed severe loss of generation capability up to 16%, although a performance assessment performed in 2015 showed no evidence of power degradation by then. Field measurements revealed a homogenous degradation over the strings of the installation. Also, the pattern of electroluminescence images and isolation resistance measurements showed indication of strong ageing characteristics normally found on photovoltaic modules installed for a long time in field or on samples after Damp Heat (DH) test.
Keywords: Degradation, Electroluminescence, Performance Assessment, n-type

1 INTRODUCTION

Grid connected photovoltaic systems are well known for its fast installation. The decision process during the pre-project phase is extremely important to guarantee a good relation between price and quality of the products to be installed. Warranty conditions must protect the investor against any unexpected loss of performance related to the material. Initial surveillance and specific measurements are useful to detect early undesirable operation [1]. Also the interest on the degradation mechanisms is strongly influenced by the growth of the PV market worldwide, since it influences the uncertainty of the output power forecast and, consequently, the return of the investment.

Brazil is facing a rapid increase of grid connected photovoltaic (PV) installations after the publication of ANEEL's (Brazilian Agency of Electrical Energy) regulation for distributed generation up to 5 MW. In addition, since 2014 EPE (Brazilian Energetic Research Company) realized 3 specific PV auctions for installations up to 30 MW connected to the medium/high voltage transmission system. By the end of 2017 it is expected to be installed 1 GW_p of PV plants mostly in the northeast region. Recent estimations stated that 7 GW_p of PV will be connected to the Brazilian grid by the end of 2024.

The challenge of installing and operating bigger PV systems, besides the concern about the technical problems of such application, influenced ANEEL to publish, in 2012, an open call to the utilities to invest R&D budget on PV systems research involving universities and research groups. The first project to start its operation took place at the University of São Paulo (USP) and consists of 0.52 MWp installed capacity system divided into three different project conceptions: Building Integrated Photovoltaic System (BIPV), Building Applied Photovoltaic System (BAPV) and standard ground mounted photovoltaic system (PV Central).

The operation started in 2014 and by the end of its first year a performance assessment showed normal condition of the generator and inverter capability. As the installations are also used as training facilities, continuous surveillance was performed by the application of periodic predictive maintenance techniques. Measurement of modules instantaneous operational voltage indicated symptoms of degradation and further analysis were performed.

Degradation investigation showed propensity for PID with negative polarization, but also indication of strong aging since even the modules located at the positive pole of the strings are also presenting power loss. *IxV* curves of modules and strings were measured on field besides the measurements on laboratory: modules *IxV* curves and electroluminescence.

2 SYSTEM DESCRIPTION

The 156 kW_p PV Central installation is located at the Institute of Energy and Environment (IEE) and represents a portion of what could be a larger PV plant. The topology was chosen to investigate the particularities of a ground mounted single inverter PV system. Fig. 1 shows the the PV generator and inverter unit.

Figure 1: PV central overview

The system is basically composed of a 156 kWp PV generator (30 strings of 5.2 kW_p each), 140 kW grid connected inverter and a step up transformer (140 kVA) to inject the energy into the 13.8 kV primary distribution feeder. The PV modules are made from n type solar cells with nominal power of 260 W_p.

2.1 PV generator technology

All the PV generators are made of n-type solar cells: a recent generation of PV technology, more efficient and with advantages over the standard p-type based solar cells [2, 3, 4]:

- Less sensible to many common metallic impurities: Iron, for instance, is a concern in other processes which demands higher quality feedstock.
- Higher minority carrier diffusion lengths were also

observed for this type of technology.

- No Boron dopant: boron-oxygen is absent and the defects caused by such complexes will not be formed upon illumination and degrade the bulk lifetime of the material. As a result, Light Induced Degradation (LID) is not expected to occur.
- Possibility to create bifacial PV cells and modules: it is not an exclusivity of the n type solar cells. In practice it is easier to realize on this material compared to the traditional p-type cells with full Al back surface field, or PERC configuration cells with blanket rear metallization.

The challenges concerning the industrial processes to produce n-type solar cells are mainly related to the simultaneous formation of p-type emitter and Back-Surface-Field (BSF) and the passivation of the highly doped p-type regions. Another issue is to find a suitable metallization for the boron doped emitter [4]. Besides all the challenges, improvements on the process technology and techniques have made the n-type solar cells achieve 20% of conversion efficiency for a PV module [5].

PID has also been detected in standard n-type c-Si technology and it is suspected that the effect in this case is not caused by Na^+ diffusion, as in p-type solar cells, but by positive charges – mostly K^+ centers– trapped in the SiNx layer. This conclusion relies on the fact that PID of n-type PV modules occurs within 10 s and the degradation is strongly saturated within 120 s. Quicker degradation implies that PID of n-type PV modules is not associated with sodium drift but caused by positive charge accumulation in front passivation films. The trapped charges lead to an increasing concentration of the minority carrier in the p-layer which consequently increases the charge recombination rate on the surface [6].

Besides being also sensible to negative polarization, in this case, PID acts differently from the standard p-type solar cells. When affected by PID, n-type cells have its V_{oc} and I_{sc} diminished (which is coherent since PID increases the front surface recombination) while the Fill Factor (FF) is not affected. On possible explanation relies on the suspicious that Stacking Faults (SF), in the emitter of the n-type solar cells, are not active after PID and different kinds of degradation occur, such as surface recombination, inversion of the emitter and changes in the doping density [7].

2.2 On field measurements

A performance assessment was performed at the PV central 1 year and a half after the beginning of operation. It was measured 146.7 kW_p of installed power, which represents 6% deviation from the nominal power. It is important to notice that the instrumentation used in the measurements was installed at the dc side of the inverter so the results take into account the losses by PV module dispersion, dc cabling, MPPT and soiling since the generator was not cleaned from the start of its operation. Besides the manufacturer of the PV module stated that there was no LID on the technology, the different on power was not a concern by then due to the loss sources.

Fig. 2 presents the results of the performance assessment. The chart shows the relation between measured dc power (P_{dc} considering the effects of the losses by temperature) and the actual irradiance (H) measured with a calibrated PV modules of the same model. In order to charactezire the installed power at Standard Test Conditions (STC), it is only taken into account points above 700 W/m^2. The actual installed power of the PV generator os considered to be the angular coefficient of the fitting line which represents the relation between P_{dc} and H.

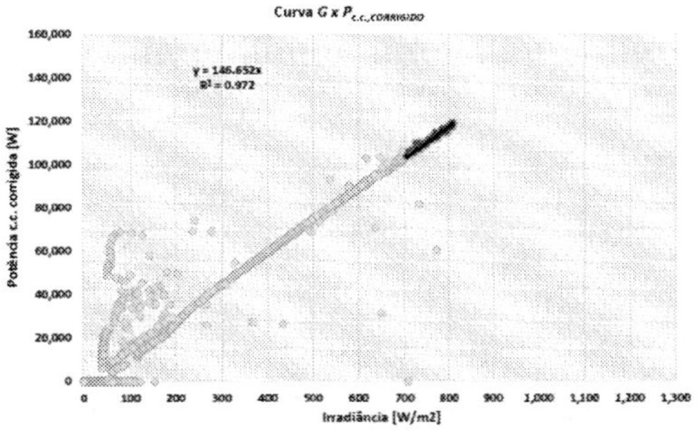

Figure 2: Performance assessment after the first year of operation (July 2015)

Fig. 3 presents the PxV curves of one string of the installation consisting of 20 series connected PV modules. The first measurement was performed in 2016 and showed 14% degradation compared to the nominal value. Then, the power loss in the beginning of 2017 was 25% of the nominal value and it is clear that the degradation process was still ongoing. Table I shows the measured power of the stringboxes of the installation which is composed by 3 PV strings each.

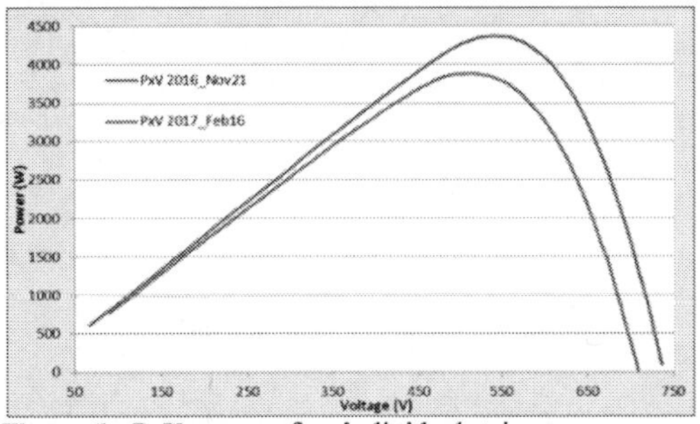

Figure 3: PxV curves of an individual string.

Table I: Measured PV power.

Stringbox	P_{dc} (kW$_p$)
01	13,7
02	13,0
03	13,01
04	12,01
05	13,4
06	12,7
07	13,4
08	12,8
09	13,3
10	13,3
Total	**130,62**

The soiling losses were neglected since the generator was clean in the night before the measurements presented on table I. The total measured STC power exhibited 16% decrease in 3 years when compared with the nominal intalled power. It is almost the warranty for the lifetime of the product (18% degradation) and twice the 10 years minimal rate output (8% degradation). Then some laboratory measurements were carried out for further investigation.

3 Laboratory measurements

Some samples were collected from the field for further investigation at the laboratory. A solar simulator was used for *IxV* curves measurements and a modified photo camera for the electroluminescence images of the PV modules to help on the diagnosis of the degradation sources

3.1 *IxV* curves

Back in the first quarter of 2015, a soiling investigation was performed to an installed module located at the negative pole of the PV generator. The measurements revealed normal operation and no degradation could be observed since the sample exhibited maximum power of 258.5 W_p. By the end of 2016, the same module was measured after 1 and a half year more in operation but then the STC power decreased to 221.2 W_p. (14.5% less than initial).

The module was then kept in the dark at the laboratory as a reference device for the degradation level of the affected individual modules. Nevertheless, the device degraded even more at ambient temperature and without sunlight exposure (167 W_p). This could be an indication of moisture ingress in the packing of the solar cells because low temperatures facilitate the dissemination of the water through the encapsulation until reaching the active parts. Fig. 4 shows the *IxV* curves of the mentioned PV module measured in the first two years of regular operation (dark blue and green) and then after being stored in the laboratory.

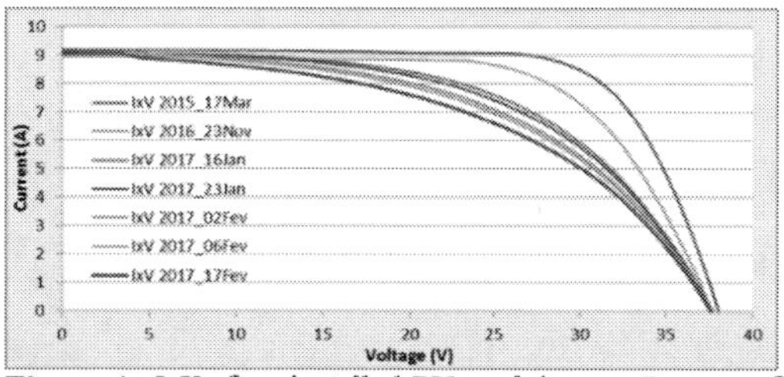

Figure 4: *IxV* of an installed PV module over 2 years of operation and 3 months stored.

Increase of series resistance (R_s) is easily recognizable by the evolution of the *IxV* trace over the years. Such particular characteristic can be attributed to moisture ingress since its dissemination might affect Anti-Reflexive Coating (ARC), the back sheet, the contacts and the PV cells Nevertheless, no visual evidences appeared so far (e.g.: delamination, AR deterioration, oxidation, corrosion, etc.) and IR inspection of the PV generators have not shown hot sports formation.

Small decrease of V_{oc} and the trace in the I_{sc} zone, which is compatible with a pattern of mismatch losses between cells, are also perceptible from the Fig. 4. Similar results can be found in [8] for modules biased during DH tests with the modules positively biased suffering stronger degradation.

3.2 Electroluminescence images

Electroluminescence is a useful technique mostly used by PV cells manufacturers to visualize defects of several sources, such as [9, 10]: micro-cracks, broken contacts, disconnected cells, finger failure, shunts of the cell, humidity corrosion, etc. It is also helpful at the PV system level [11] besides the need to disconnect the module and the strict conditions of illumination (depends on the device used to take the photographs) to perform the measurement.

Fig. 5 presents degraded modules electroluminescence images after 3 years of regular operation. Fig. 5(a) belongs to the sample presented in Fig. 4. Fig. 5(b) shows a module degraded 26% on power and located at the negative pole of a string. A less degraded module (9.5%) located at the positive pole of the same string is presented on Fig. 5(c).

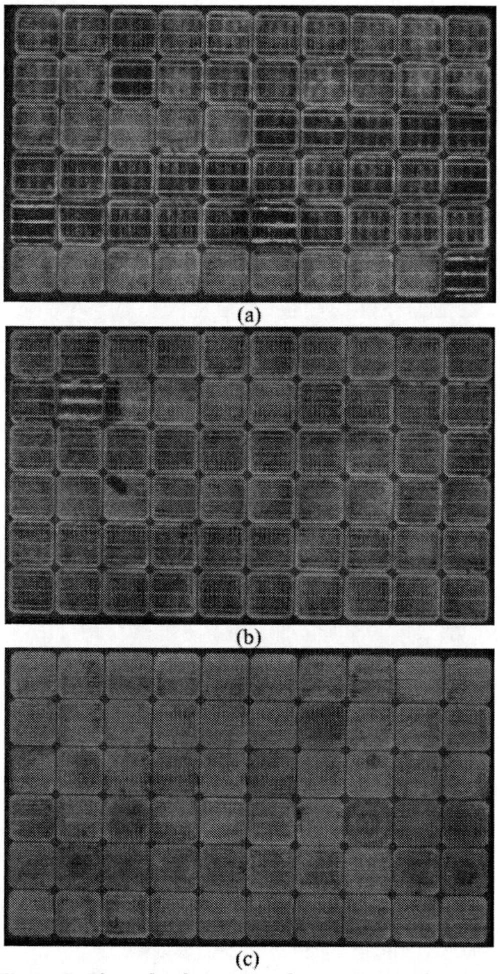

Figure 5: Electroluminescence of n-type PV modules after 3 years of operation.

The most degraded modules presented considerable dark regions and a brighter area close to the bus bar because of the current concentration, as can be seen on Figs. 5(a) and (b). Besides, both image patterns are very similar to samples submitted to a DH cycle [12]. It is clear now that the moisture ingress is the main source of the degradation process. Nevertheless, differences on power up to 5% were found between operating modules positively and negatively biased. Polarization can also affect this technology but is less severe than potential induced degradation of p-type PV modules and saturates at low degradation rates.

Fig. 5(c) shows a less degraded module with slighter and smaller dark areas. Swirl shaped regions in some cells are clearly perceptible and known as striation rings. They are correlated with distribution of oxygen and carbon impurities in the crystal leading to the formation of oxygen clusters that are associated with recombination active defects [13, 14]. Due to the curved growth interface during ingot growth these clusters are always ring or disc shaped and they are not a concern in such pattern presented on Fig. 5(c). Nevertheless, in other cases, it can also be responsible for a significant reduction of the solar cell efficiency.

4 CONCLUSIONS

PV technology is constantly presenting fast development on the manufacturing of PV cells and modules. A PV project is a challenge of balancing the relation between quality of product/services and the costs to achieve a feasible deal. As the possibilities are many, quality control of the products is extremely important. In addition, warranty conditions must protect the investor against the occurrence of unexpected early degradation and bad performance.

The importance of initial surveillance of a grid connected PV installation to detect unexpected initial degradation is presented on this work. Periodically measurement in the first years of operation: visual inspection, IR images and *IxV* curves on field are normally part of predictive maintenance routine and the measurements must be recorded for further analysis. In this paper, laboratory measurements with Electroluminescence images were also helpful to investigate the root causes of the degradation.

5 REFERENCES

[1] M. A. Muñoz et al., Solar Energy, ed. 85, 2011.

[2] M. Webber et al., Proceedings 24th EUPVSEC Hamburg, 2009.

[3] A. R. Burgers et al., Proceedings 25th EUPVSEC Valencia, 2010.

[4] A. R. Burgers et al., Proceedings 26th EUPVSEC Hamburg, 2011.

[5] A. R. Burgers et al., Proceedings 27th EUPVSEC Frankfurt, 2012.

[6] K. Hara et al., Solar Energy Materials and Solar Cells, ed. ,2015.

[7] S. Yamaguchi et al., Applied Physics Express, ed. , 2016

[8] K. Brecl et al., Journal of Solar Energy Engineering, ed. 139, 2017.

[9] D. M. Tsai et al., Solar Energy Materials and Solar Cells, ed. 99, 2012.

[10] J. A. vön Molken et al., Energy Procedia, ed. 27, 2012.

[11] A. Mansouri et al., Proceedings 27th EUPVSEC Frankfurt, 2012.

[12] W. F. F. Mohamed, PhD Thesis, Cairo University, 2014.

[13] P. Manshanden et al., Energy Procedia, ed. 92, 2016.

[14] A. Le Donne et al., Applied Physics Letters, ed. 109, 2016.

A Detailed Performance Model for Bifacial PV Modules

Clifford Hansen[1], Daniel Riley[1], Chris Deline[2], Fatima Toor[3], Joshua Stein[1]

[1]Sandia National Laboratories, Albuquerque, NM, 87109 USA
[2]National Renewable Energy Laboratory, Golden, CO, USA
[3]The University of Iowa, Iowa City, IA, 52242, USA

Abstract — **We present a detailed performance model for bifacial PV modules. Our model employs configuration factors to compute irradiance the rear surface of each cell in the module, estimates cell temperatures from irradiance and ambient temperature, and predicts of P_{MP}. Model performance is analyzed using measurements of bifacial module performance outdoors in Albuquerque, NM, USA.**

I. Introduction

Bifacial photovoltaic (PV) modules can accept light on both the front and rear surfaces. Currently, research efforts are underway to describe, test, rate, and model bifacial PV modules, as bifacial PV becomes a larger portion of the overall PV market. Sandia National Laboratories (Sandia), the National Renewable Energy Laboratory (NREL) and the University of Iowa are working together to provide the necessary data and analysis to better describe and predict bifacial PV performance to facilitate wider adoption throughout the PV market.

Our effort to develop a detailed performance models for bifacial PV modules seeks to create an accurate and computationally efficient model to predict the output of a bifacial PV system under any weather conditions and for a wide range of module orientations and system configurations, similar to the effort reported in [1]. Our effort is informed by extensive data describing measured performance of bifacial modules outdoors in a wide range of configurations [2]. Available performance models, e.g., [3], [4], typically estimate annual performance ratio or annual energy by means of correlations between module orientation and the *bifacial gain*, a ratio between power when both sides of the module receive illumination and power when only the front side is illuminated.

We test our model using data collected for bifacial modules mounted on a south-facing adjustable tilt rack (Fig. 1) with monofacial PV modules on the west half (left in figure) and bifacial PV modules on the east half. Reference cells measure irradiance along the middle of the rack: at the top and bottom of the front, and at the top, middle and bottom of the rear. Reference cells are calibrated outdoors against a primary reference cell (calibrated by NREL) to reduce variation among cells to less than 4 W/m^2 at irradiance of 1000 W/m^2.

II. Modeling Approach

The performance model for bifacial PV systems comprises a sequence of sub-models similar to the approach employed for modeling monofacial PV. Separation of irradiance into direct and diffuse components (if needed) is followed by translation to the module's plane using the same methods employed for monofacial modules. A configuration factor approach is used to estimate irradiance cell by cell on the module's back surface. Front and rear-side irradiance can be adjusted to account for reflection, shading, irradiance spectrum, and soiling. Cell temperature is estimated from irradiance and weather; cell temperature and irradiance are then used with calibrated cell models to estimate the IV curve for the module.

Fig. 1. Sandia's adjustable rack for measuring bifacial PV module performance.

A. Front and Rear Irradiance

The irradiance available for the front surface of bifacial PV modules is modeled in the same manner as for monofacial PV. Similar methods of transposing direct and diffuse irradiance onto a tilted surface may be used, and the site albedo can be used to estimate the amount of ground-reflected light present on the front surface.

Determining the irradiance on the rear side of a PV module requires new models. Rear side irradiance comprises primarily diffuse irradiance by way of reflections from the ground and other nearby surfaces, sky diffuse irradiance, with rare circumstances when direct irradiance illuminates the rear surface [5]. The module's orientation, proximity to nearby surfaces, and transparency [6] affect the area of ground and nearby surfaces and fraction of the sky visible from the module's rear surface. Rear-side irradiance models at different levels of detail are needed to support analyses from module and system design to annual energy production assessments [7].

Rear side irradiance can vary substantially depending on a module's position in the row, and rear irradiance may vary by more than 50 W/m^2 across a module's back surface [5]. A cell-level rear-side irradiance model [5] can estimate the irradiance mismatch across a single module and between modules in an array. An array-scale model [8] computes irradiance along a cross-section of a fixed-tilt array with regular rows; the computational simplicity makes this model suitable for energy production assessments for large systems. A similar approach can be used to model rear-surface irradiance for bifacial modules on horizontal east-west single axis trackers [9]. Ray-tracing models (e.g., [10]) provide detailed assessment of effects of module design details, such as cell to cell spacing and surface reflections, but with significant computational cost. Selecting the best method for determining rear irradiance may depend on the use case and requirements for computation time, accuracy, and resolution.

The detailed cell-level and the array-scale rear-surface irradiance models use view factors, also termed shape and view factors, configuration factors quantify the fraction of irradiance reflected from one surface that arrives at a receiving surface. View factor models [11], [3] calculate back surface irradiance E_2 (W/m^2) by

$$E_2 = \alpha \times G_1 \times F_{1\to 2} \qquad (1)$$

where G_1 is the total irradiance (W/m^2) on the reflecting area being considered (e.g., an area of the ground), α is the reflecting surface's albedo (unitless) and $F_{1\to 2}$ is the view factor (unitless) from the reflecting area to the receiving surface. The total irradiance on the back surface of a cell or module is the sum over all contributing surfaces. A view factor model implicitly assumes that all reflecting surfaces are Lambertian, i.e., irradiance is scattered isotropically.

Formally, view factors are calculated by integration (Eq. 2) using terms in the illustration. An emitting surface (dA1) reflects incident irradiance, part of which is incident on the receiving surface: the view factor $VF_{1\to 2}$ quantifies the fraction of irradiance emitted by dA1 that is received by dA2.

$$F_{1\to 2} = \frac{1}{A_1}\int_{A_1}\int_{A_2}\frac{\cos\theta_1\cos\theta_2}{\pi s^2}\,\mathrm{d}A_2\,\mathrm{d}A_1 \qquad (2)$$

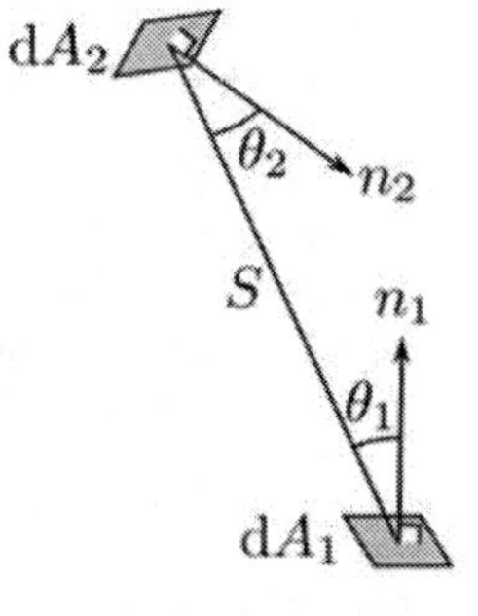

A rear surface irradiance model is assembled by specifying the set of reflecting surfaces and the irradiance incident on each surface. For example, irradiance on the sunlit ground is modeled by global horizontal irradiance (GHI) and irradiance on a shadowed area is modeled by diffuse horizontal irradiance (DHI), where each irradiance is reduced by the fraction of the hemispherical sky dome occluded by adjacent rows of modules.

Fig. 2 illustrates the modeled variation in rear-surface irradiance between cells at the edge of a row in the array (left side in figure) and cells toward the middle of a row (right side of figure). Cells at the left edge of the row receive significantly more rear-surface irradiance than cells in the middle of the row's modules, because edge cells 'see' sunlit ground to the left of the row. Cells at the lower and upper edges also receive more irradiance than middle cells because of sunlit areas in front of and behind the row's shadow. Sandia has developed an efficient method [5] for computing the required view factors; implemented in Matlab™; simulations with 150 time steps, 4 60-cell modules, and 10 nearby objects contributing shadows (modeling the adjustable rack shown in Fig. 1) were complete in about 15 seconds.

Calculation of view factors at a cell level permits modeling of rear-surface irradiance for arrays with subsets of modules in different configurations, e.g., a mix of southward facing, fixed tilt modules and vertical E-W facing modules. In concept, a cell-level irradiance model enables an array performance model to directly account for mismatch conditions among cells and modules, although explicit mismatch modeling is not done in our present work.

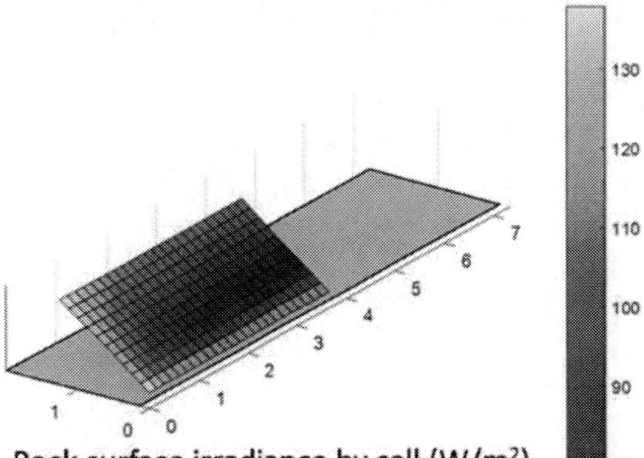

Fig. 2. Example calculation of cell-by-cell rear-surface irradiance, solar noon, assumed clear sky, single row of 8 modules in portrait orientation at fixed 35° tilt (left-most 4 modules shown).

B. *Reflection, Shading, Spectrum and Soiling*

Reflections. Reflections from a module's front surface are modeled as is done for monofacial modules (e.g., [12]): an empirically determined expression computes the fraction of the direct irradiance on the module's face is reflected away from the angle of incidence of the direct irradiance. Reflections from the rear surface are not explicitly represented; the effect of any reflections can accounted for in the empirical relationship between total incident irradiance (front surface + back surface) and the current produced by the module.

Shading. PV racking, junction boxes, or module wiring may shade parts of the rear-side of a bifacial PV module from irradiance sources. To discover the effects of rear-surface shading, we measured IV curves after placing cardboard strips in various locations and orientations behind a bifacial module (e.g., Fig 3, see [13] for experimental details). Comparison of IV curves with and without the obstructions showed that P_{MP} is generally reduced in proportion to the obstruction area, with the proportionality decreasing with increasing distance between the obstruction and the module. However, obstructions affected both I_{SC} and I_{MP} in a complicated manner depending on obstruction orientation; this is not surprising as obstructions across several cell strings will have a different effect on current than obstructions confined to a single cell string (e.g., compare Fig. 3 and 4).

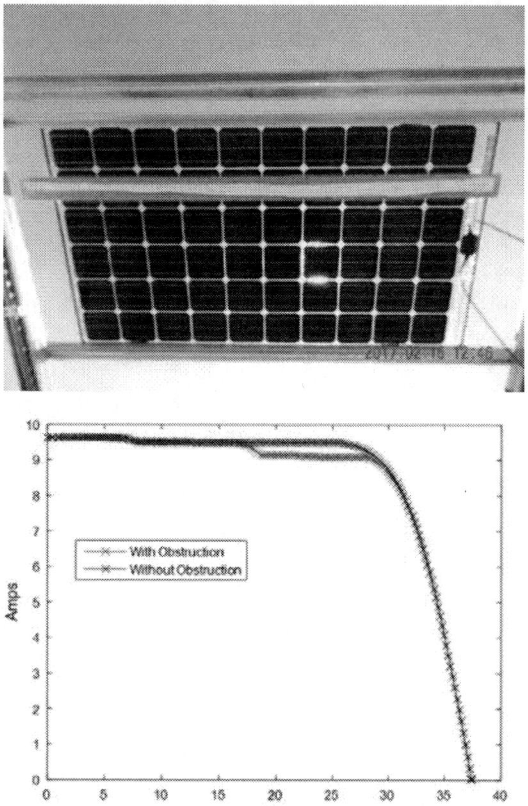

Fig. 3. Example of rear-surface irradiance obstruction experiment: obstruction (top) and measured IV curves (bottom).

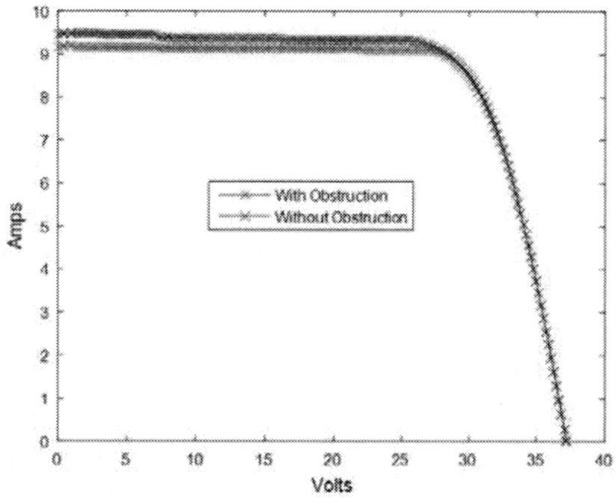

Fig. 4. IV curve with obstruction across cell strings rather than along a single string, as in Fig. 3.

The reduction of rear-surface irradiance due to obstructions can be modeled directly using ray-tracing methods. View factor methods appear suitable only for the simplest obstruction cases, e.g., a single rectangular obstruction with edges parallel to the module's edges. In either case, the computational effort to configure and complete the calculations appears impractical for power and energy simulations. As an alternative we propose the following simple model for the effect of obstructions on module P_{MP}:

$$P_{MPo} = P_{MP}\left(1 - k_{obs} \times A_{obs} / A_{mod}\right) \quad (3)$$

where P_{MPo} is the P_{MP} reduced by the effect of obstructions, A_{obs} and A_{mod} are the obstruction and module aperture areas, respectively, and k_{obs} is an empirical factor representing the effect of the distance between the obstructing objects and the module's rear surface. Module IV curves with obstructions (e.g. [13], Fig. 3) indicate values of k_{obs}=1.0 for obstructions placed against the module's rear surface, and k_{obs}=0.6 for obstructions placed ~5cm from the rear surface.

Spectrum. The spectrum of the incident light alters the module's current due to the spectral response of the module's cells. The effect of spectrum changes can be estimated with existing models, e.g., [14], [15]. Spectral irradiance on the rear side differs from spectral irradiance on the front side due to reflections from the ground and/or nearby objects. Here, we neglect the effect of spectrum of rear-surface irradiance, reasoning that its effect is less than the uncertainty in estimating the broadband rear-surface irradiance.

Soiling. We assume that the effects of soiling on bifacial PV modules can be represented by two factors applied to front and rear-surface irradiance, respectively. For bifacial modules at fixed tilts, we have observed little to no soil accumulation on the rear surfaces in Albuquerque, NM, USA. It stands to reason that vertical modules will accumulate soil equally on both faces, although we have no measurements available to quantify the accumulation of soil on vertical module surfaces.

C. Cell Temperature

Cell temperature can be modeled using an approach similar to that used for monofacial modules, e.g., [14]. We measured temperature with thermocouples attached in between cells on a bifacial module's rear-surface concurrent with ambient temperature and wind speed (at 3m above ground). The measurements verify that module temperature models in common use for monofacial modules also describe well the measured data for bifacial modules. For example, the module temperature model in [14]

$$T_{mod} = \left(E_f + E_r\right) \times \left[e^{a+b \cdot WS}\right] + T_{amb} \quad (4)$$

fits well to the collected data: T_{mod} is the average temperature across the rear surface of the module, T_{amb} and WS are the ambient air temperature and wind speed (m/s), respectively, E_f and E_r are the front and rear surface irradiance (W/m^2) respectively, and a and b are empirical constants determined from the fitting. For glass-cell-glass modules in open racking, values of a = –3.47 and b = –0.0594 are suggested [14]. Cell temperature T_C is obtained from Eq. 4 as

$$T_C = T_M + \Delta T\left(E_f + E_r\right)/1000 \quad (5)$$

where ΔT = 3 °C for open racking. We expect that other cell temperature models, e.g., [16] would be successful as well.

Fig. 5 shows the residuals when modeling a bifacial PV module temperature with Eq. 4. The blob of residual values lying below the linear tread generally corresponds with dynamic temperature conditions resulting from passing cloud shadows. Shading immediately reduces the irradiance and hence module temperature predicted by Eq. 4. In reality, module temperature decreases take place over time periods on the order of 10 minutes, thus, the steady-state model in Eq. 4 underpredicts module temperature after the transition from fully lit to shaded conditions. Similarly, module temperature is overpredicted after the transition from shaded to fully lit conditions.

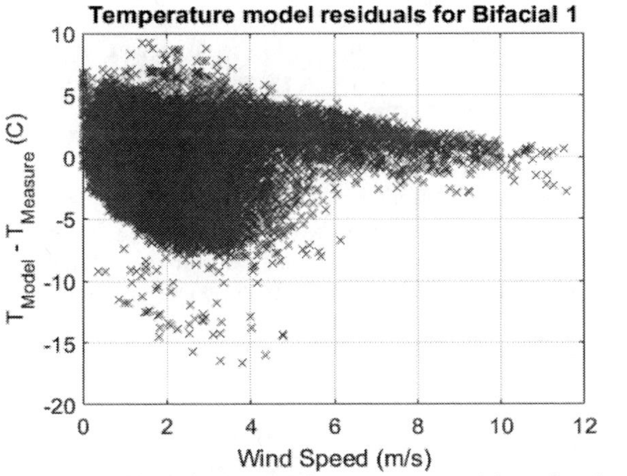

Fig. 5. Residuals for the module temperature model as a function of wind speed for a bifacial system.

Plots of the residuals for a co-located monofacial system exhibit nearly identical features and magnitudes: the RMSE of the model for the bifacial system is about 3.4 °C, while the RMSE for a co-located monofacial system is 3.1 °C. The similarities in model performance between bifacial and monofacial modules, and the physical reasoning behind Eq. 4 (see [14]), provide confidence that typical values for a and b determined for glass-glass monofacial modules should also be reasonable for bifacial modules.

D. Electrical Performance

We present a simple model for P_{MP} for bifacial PV modules akin to PVWatts [17] which can account for rear-surface shading by obstructions using Eq. 3. The model operates on cell average irradiance (front and rear surface); thus the effects of mismatch from cell-to-cell variation in irradiance are not explicit.

We estimate maximum power for a bifacial module, P_{MP} (W) from broadband front irradiance E_f (W/m^2), rear irradiance E_r (W/m^2), cell temperature T_C (°C), module bifacial ratio R_b (unitless), temperature coefficient for maximum power γ (1/°C) and the module power rating at STC conditions P_{MP0} (W).

$$\begin{aligned} P_{MP} &= P_{MP0}\left(1+\gamma\left(T_C - T_0\right)\right) \\ &\times\left(M \times E_f + R_b \times E_r\right)/E_0 \\ &\times\left(1 - k_{obs} \times A_{obs}/A_{mod}\right) \end{aligned} \tag{6}$$

The bifacial ratio R_b is the ratio of power produced by the rear side to the front side under identical illumination and temperature and P_{MP0} is determined by a front-side-only flash test at standard test conditions; these measurements are discussed in [6]. The term M in Eq. 6 represents the combined adjustment to front irradiance from spectrum, reflections and soiling.

III. Verification and Analysis

Fig. 6 compares the measured and modeled (using Eq. 6) maximum power for the top bifacial PV module on the adjustable rack, shown in Fig. 7. For the results in Fig. 6, the front irradiance E_f is the irradiance measured by a coplanar reference cell, and the rear irradiance E_r is estimated by averaging measurements from two reference cells adjacent to the module's western corners (indicated by green triangles in Fig. 7). For the test module P_{MP0} = 270 W and R_b = 0.93 as determined using IV curves measured separately for the front and rear. We set M = 1 because the reference cells have similar reflection and spectral response as the module under test. The mounting rack and connector wiring do not obstruct rear irradiance, thus A_o = 0. Except at high P_{MP}, the simple model predicts the module power to within 5W (about 2%), but at high P_{MP} the model overestimates power by as much as 20W (about 8%). The few points far below the 1:1 line likely result from temporary shadows affecting the front irradiance measurement, or bad data.

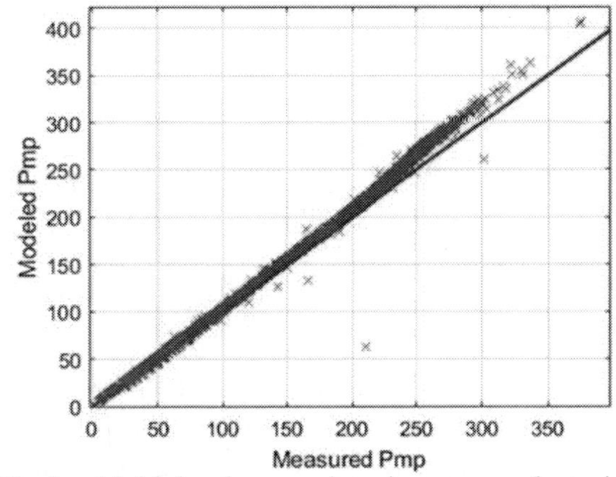

Fig. 6. Modeled and measured maximum power for a single bifacial PV module on 5-min intervals from 30 June to 25 July, 2016, in Albuquerque, NM. 1:1 line in red.

The overestimation at high P_{MP} could result from a combination of two causes:

1. Mismatch loss due to variation in rear-surface irradiance among cells.
2. Non-linear dependence of P_{MP} on irradiance and/or temperature.

The shadows on the ground cause variation among the rear-facing reference cells of up to 40 W/m² (top and middle cells indicated in Fig. 8). Similar variation is also present among the module's cells, so it is reasonable to anticipate a mismatch in current on the order of 3% (40W being roughly 3% of the total front+rear irradiance of around 1250 W/m²), with a corresponding reduction in the module power. The mismatch would be most prominent during clear skies periods at midday, when irradiance is high and shadows on the ground are well-defined (see Fig. 7). In addition, the departure from linear dependence of power on irradiance may be a property of the particular bifacial module under test, a possibility we intend to investigate further.

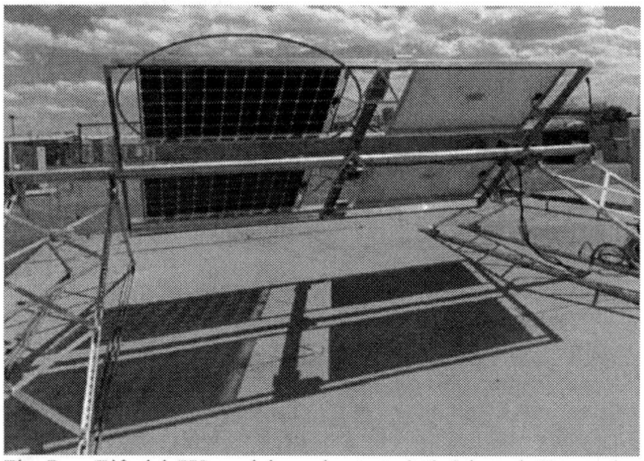

Fig. 7. Bifacial PV module under test circled in red. Top and middle rear-facing reference cells indicated by green triangles. View is to the south.

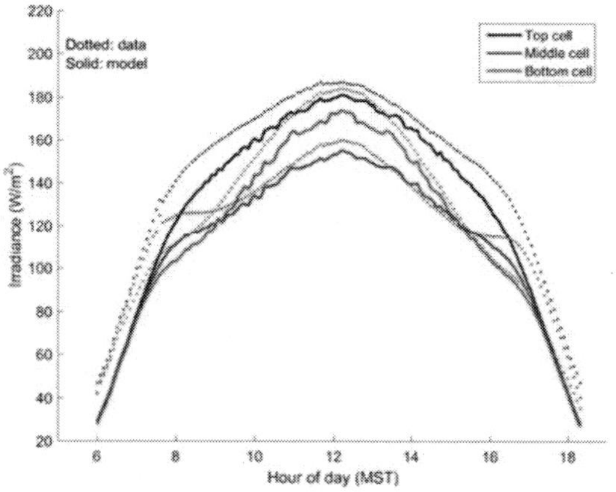

Fig. 8. Modeled and measured irradiance for rear-facing reference cells in the adjustable array: 14 July 2016, a clear day, array titled at 15° with center at 1.63m above ground.

With the above model discrepancies in mind, we compute P_{MP} using modeled rear-surface irradiance instead of the reference cell measurements. We use the cell-level rear-surface irradiance model described above (and detailed in [18]) to compute the cell-by-cell irradiance for the top-right bifacial module (6 × 10 cells in landscape orientation; Fig. 7). This model was previously verified [5] to be within 10% of measurements over a wide range of conditions using an array of rear-facing reference cells mounted on a module-sized plate (Fig. 11). Fig. 8 compares rear-surface irradiance model predictions with measurements for each of the three rear-facing reference cells. The model generally follows the trends in the data, although with a consistent underestimate of the measured values. Cell-by-cell modeling of the module (Fig. 9) shows a cell-to-cell variation as great as 50 W/m², or roughly 5% of total (front + back) irradiance of 1250 W/m² at midday.

Fig. 10 compares residuals for predicted P_{MP} using either measured or modeled rear-surface irradiance. The model overestimates P_{MP} by approximately 20 W (about 7%) at full sun conditions, with either irradiance source. Causes for the overestimate include mismatch losses not accounted for by the model, or module behavior, as previously discussed. Two bands are apparent in the residuals when using measured rear-surface irradiance at moderate P_{MP} levels. These bands result from the dynamic effect on ground-reflected irradiance of the passage of the sunlit and shadowed areas beneath the cell as the sun moves from east to west. The bands are not present when using the modeled rear-surface irradiance because of the module's larger aperture smooths these dynamics to a large extent.

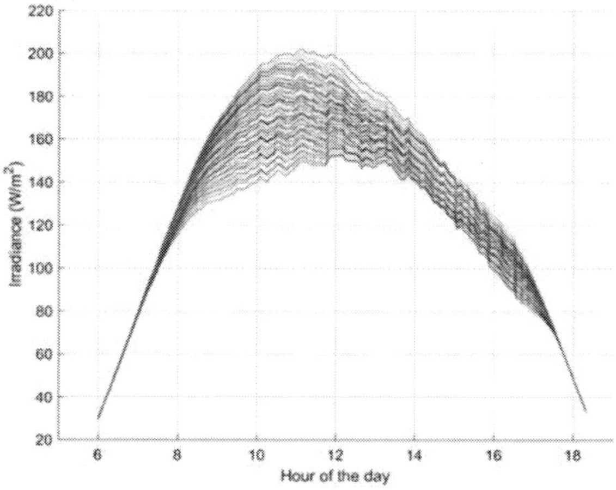

Fig. 9. Modeled rear irradiance by cell on the top right bifacial module in the adjustable array: 6 July 2016, a clear day, array titled at 15° with center at 1.63m above ground.

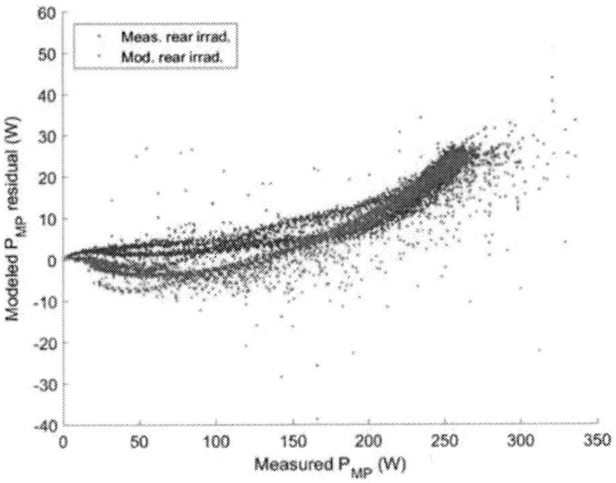

Fig. 10. Residual of modeled PMP using modeled and measured rear irradiance.

Fig. 11. Sandia's high spatial resolution rear irradiance module blank with ten reference cells measuring rear irradiance, used for validation of the irradiance model.

ACKNOWLEDGEMENTS

Sandia National Laboratories is a multi-mission laboratory managed and operated by National Technology & Engineering Solutions of Sandia, LLC., a wholly owned subsidiary of Honeywell International, Inc., for the U.S. Department of Energy's National Nuclear Security Administration under contract DE-NA0003525.

REFERENCES

1. Janssen, G.J.M., et al., *Outdoor Performance of Bifacial Modules by Measurements and Modelling.* Energy Procedia, 2015. **77**: p. 364-373.
2. Stein, J., C. Deline, and F. Toor, *Outdoor Field Performance from Bifacial Photovoltaic Modules and Systems*, in *44th IEEE Photovoltaic Specialist Conference*. 2017: Washington, DC.
3. Yusufoglu, U.A., et al., *Simulation of Energy Production by Bifacial Modules with Revision of Ground Reflection.* Energy Procedia, 2014. **55**: p. 389-395.
4. Wang, S., et al., *Bifacial Photovoltaic Systems Energy Yield Modelling.* Energy Procedia, 2015. **77**: p. 428-433.
5. Hansen, C.W., et al., *A Detailed Model of Rear-Side Irradiance for Bifacial PV Modules*, in *44th IEEE Photovoltaic Specialist Conference*. 2017: Washington, D.C., USA.
6. Deline, C., et al., *Evaluation and Field Assessment of Bifacial Photovoltaic Module Power Rating Methodologies*, in *43rd IEEE Photovoltaic Specialist Conference*. 2016: Portland, OR.
7. Hansen, C.W., et al., *Analyis of Irradiance Models for Bifacial PV Modules*, in *43rd IEEE Photovoltaic Specialist Conference*. 2016: Portland, OR.
8. Marion, B., et al., *A Practical Irradiance Model for Bifacial PV Modules*, in *44th IEEE Photovoltaic Specialist Conference*. 2017: Washington, DC.
9. Anoma, M., et al., *View Factor Model and Validation for Bifacial PV and Diffuse Shade on Single-Axis Trackers*, in *44th IEEE Photovoltaic Specialist Conference*. 2017: Washington D.C., USA.
10. Asgharzadeh, A., et al., *Analysis of the Impact of Installation Parameters and System Size on Bifacial Gain and Energy Yield of PV Systems*, in *44th IEEE Photovoltaic Specialist Conference*. 2017: Washington, DC.
11. Iqbal, M., *An Introduction to Solar Radiation*. 1983, Toronto: Academic Press Canada.
12. Martin, N. and J.M. Ruiz, *Calculation of the PV modules angular losses under field conditions by means of an analytical model.* Solar Energy, 2001. **70**: p. 25-38.
13. Riley, D., et al., *A Performance Model for Bifacial PV Modules*, in *44th IEEE Photovoltaic Specialist Conference*. 2017: Washington, DC.
14. King, D.L., E.E. Boyson, and J.A. Kratochvil, *Photovoltaic Array Performance Model.* 2004, Sandia National Laboratories: Albuquerque, NM.
15. Lee, M. and A. Panchula, *Combined Air Mass and Precipitable Water Spectral Correction for PV Modelling*, in *4th PV Performance Modeling and Monitoring Workshop*. 2015: Cologne, Germany.
16. Faiman, D., *Assessing the outdoor operating temperature of photovoltaic modules.* Progress in Photovoltaics: Research and Applications, 2008. **16**(4): p. 307-315.
17. Dobos, A., *PVWatts Version 5 Manual.* 2014, National Renewable Energy Laboratory.
18. Hansen, C.W. and D. Riley, *A Computationally Efficient Method for Detailed Modeling of Rear-Side Irradiance for Bifacial PV Modules*, in *44th IEEE Photovoltaic Specialist Conference*. 2017, IEEE: Washington, D.C., USA.

WHERE HAS ALL THE POWER GONE? A HEALTH CHECK OF ITALIAN SOLAR ELECTRICITY IN 2016

Alessandro Virtuani[1], Mauro Pravettoni[1], Luca Parini[1], Luca Morganti[1], Artur Skoczek[2], Juraj Betak[2], Marco Mussetta[3], and Stefano Marchionna[4]

[1] Officina del Sole (O'Sole), Milan, Italy
[2]Solargis, Bratislava, Slovakia
[3]Politecnico di Milano – Dipartimento di Energia, Milano Italy
[4]Ricerca sul Sistema Energetico (RSE), Milan, Italy
* Corresponding author: Tel. +39 333 7476 939, e-mail: alessandro.virtuani@yahoo.com

ABSTRACT: In 2015 Italy had a level of PV penetration >8%. However, as reported by professionals and the specialized press with sometimes alarmist tones, solar electricity generation in 2016 has been much lower compared to 2015. With final 2016 data now available, we give a closer look at the aggregate data and calculate some Key Performance Indicators (KPI) for the whole country and for each region to assess the health status of solar PV electricity in Italy twelve years after the start (2005) of the feed-in-tariff subsidy scheme. At country level, the 2016/2015 relative variation in final yield Y_f is - 5.4%. About 80% of this variation can be attributed to a lower insolation in 2016 compared to 2015 (-4.3%). The remaining losses are reflected in the –1.1% PR relative variation (2016/2015) and can be attributed to a mix of factors: degradation in performance (system/components), moderate use of O&M, etc. At country level, the average PR is below 70%. This low value is partly attributed to the fragmentation and small average size of PV systems in Italy (~ 28 kWp in 2016), to their non-optimal orientation/inclination and to presence of shading. Other reasons, linked to the dynamics and features of the national feed-in tariff incentive schemes are briefly recalled. We further note, that the alarmist tones circulated about the pronounced decline in generation of solar electricity in 2016, are not justified. On the contrary, 2015 has been a year with an availability of solar resources considerably higher than 2016 and of the long-term average. Nevertheless, still concerns about the actual (and long-term) performance of the Italian solar park as a whole exist.
Keywords: PV system, operation, performance, reliability

1 INTRODUCTION

With a share of 8.4% of electricity generation covered by solar electricity, in 2015 Italy has become the country with the highest penetration of photovoltaics (PV) into its electricity mix [1]. In 2016, Italy has added ~+370 MWp of capacity totaling an overall installed capacity of 19.3 GWp. During year 2016, however, professionals and the specialized press (see e.g. [2, 3]) has reported with quite alarmist tones a considerable decline in production of PV electricity generation in Italy. Provisional figures from Terna (Italian power transmission company) for 2016 were in fact reporting a strong decrease in the electricity production from PV (-10%, 2016 vs 2015 for the period January-September) [4]. A different analysis from GSE [5] covering the first eleven months of 2016 on a subset of plants (~7 GWp of size >55 kWp, corresponding to ~36% of the total) pointed out an year-on-year decline in PV production of -4.6% (2016 vs 2015 for the period January-November).

Now that the final figures for 2016 are confirmed, we analyze year-on-year variations in solar resources at country and a regional level and compute a set of Key Performance Indicators (KPIs) for the whole country and for each distinguished region using aggregate values. These include the Final Yield Y_f and the Performance Ratio PR. The assessment of solar resource for Italy was carried out using a satellite model developed by *Solargis*.

We finally discuss the factors leading to a potential underperformance of Italian solar PV in its entirety, and how these factors are strongly linked to the dynamics and features of the national feed-in tariff incentive schemes in place from 2005 to 2013.

2 SOURCES AND APPROACH

2.1 Input data

In this paper, we make use of official figures released by *Gestore Servizi Energetici* (GSE, the Italian public company in charge of managing subsidies to renewables) for statistics related to Italian solar PV [6, 7].

Country's electricity generation and consumption data are obtained by *Terna* (Italian transmission system operator TSO [8, 9]), and insolation data for 2015, 2016 and long-time averages (1994-2016) by *Solargis* [10, 11, 12].

The inputs to the Solargis model are multispectral satellite images from Meteosat satellite and atmospheric parameters like aerosols and water vapour. Spatial resolution of Meteosat data considered in the calculation scheme is approximately 3 km at sub-satellite point. Model results are served in 2 arc-minutes (app. 4x4 km) regular grid in WGS84 geographical coordinate system. For the purpose of this study, the spatial resolution of solar data products is enhanced to 15 arc-seconds (nominally 500 m) using digital terrain model comprising a combination of data from the Shuttle Radar Topography Mission. The primary temporal resolution of satellite data is 15 min. The atmospheric parameters are updated daily. The air temperature at 2m data is derived from the NOAA Climate Forecast System, with spatial resolution of approximately 20 km and temporal resolution of 1 hour.

For simplicity, in this work we use, for the whole country and for each region, the yearly sum of Global Tilted Irradiation (GTI) calculated at 25°-tilt. This, rather than using GTI calculated at optimal tilt, introduces a slight underestimation in the effective availability of solar resources (~1.5% and ~0.4%, respectively, for the

Northern and Southern part of the country) and, consequently, a moderate overestimation of the same magnitude in the computation of the PR.

For a south-facing installation, the optimal tilt would be 36°, 34° and 30°, respectively, for Milan, Rome and Palermo.

2.2 Key performance indicators

We make use of two KPIs, which are generally used at the solar plant/array level, and that, utilizing aggregate average values, we use to compute country and regional performance indicators. These are:

(1) the *final yield* Y_f(kWh/kWp): i.e. the annual net AC energy output E_a [kWh] of a system divided by the peak DC power P_{STC} [kW] of the installed PV array at standard test conditions (STC: 1000 W/m², 25°C, AM1.5):

$$Y_f = E_a / P_{STC} \qquad (1)$$

Y_f is a direct function of the cumulative irradiation (i.e. higher Y_f should correspond to sunnier regions) and, therefore, it generally depends on the latitude as well.

(2) the *performance ratio PR* (%) is an index illustrating the overall effect of losses on the system/array's rated output due to device temperature, incomplete utilization of the irradiance (e.g. due to shading or non-optimal orientation), and system component inefficiencies or failures [13]. It provides an indication of how a system effectively performs under real operating conditions, compared to how it would be operating if it were constantly exposed to standard laboratory (STC) ones.

$$PR = (E_a \times G_{STC}) / (H_a \times P_{STC}) \qquad (2)$$

with G_{STC} [kW/m²] the irradiance at STC (1000 W/m²), and H_a [kWh/m²/y] the yearly cumulative irradiation. In this work, for H_a we use the Global Tilted Insolation at 25° tilt (GTI-25°) for the whole country and for each separate region.

3 RESULTS

3.1 Deployment of solar electricity in Italy

Figure 1 shows the historical deployment of solar PV in Italy since the start of the national feed-in tariffs scheme (*Conto Energia*) in 2005 (see Table I as well). The figure shows cumulative and newly added PV capacity per year and the average size of PV plants, which is larger in Southern Italy (~35 kW), rather than in the central (~28 kW), and northern part of the country (~21 kW). In 2016, in fact 91% of the plants have a size ≤ 20kW, whereas 60% of the PV capacity corresponds to installation with a size > 200 kW. The *Conto Energia* incentive scheme ended in 2013 and has had a boom of new installations in year 2011, with a peak of newly added capacity in this year (~9.5 GWp).

At the end of year 2016, the weighted average age of the Italian solar PV park (considering 0.5 years increments) is of 5.2 years.

3.2 Solar resources in Italy

Figure 2 shows the GTI at 25° tilt calculated by *Solargis* for the whole country and for each region for years 2015 and 2016, together with the long-term average (1994-2016). These data clearly indicate that for nearly all regions (and for the whole country) the annual insolation has been remarkably higher in 2015 rather than in 2016, and that GTI values for 2016 are more aligned to the long-term average.

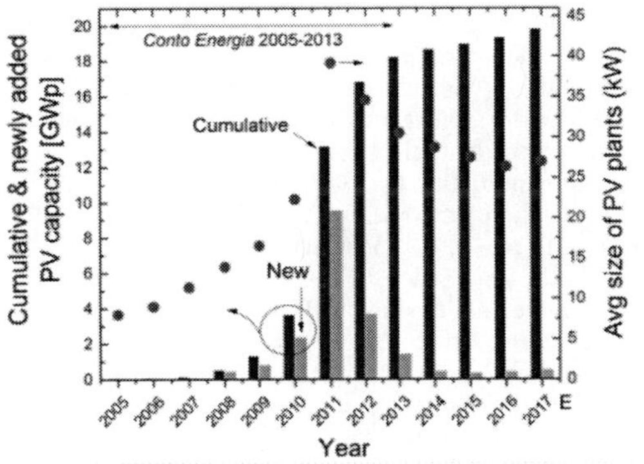

Fig. 1: average size of PV pants, cumulative and newly added PV capacity in Italy since 2005

On average, at country level 2016 has had -4.4% less GTI than 2015. More specifically, as can be observed in inter-annual variability map for Italy showing the relative difference in GTI (2016 vs 2015, see Fig. 3):

- In central and Northeast regions (e.g. Umbria, Abruzzo, Veneto, Friuli-Venezia Giulia) GTI in 2016 was on average -6% lower than in 2015. In some Southern regions and islands (Calabria, Sicilia, Sardegna) it was lower than -3%;
- Locally (particularly along the Apennines and in parts of Veneto) the drop was higher than -8% or even -10%;
- Only very rarely the year-on-year difference was positive (e.g. in Southern Calabria, along the coastal region of Toscana or in Piemonte).

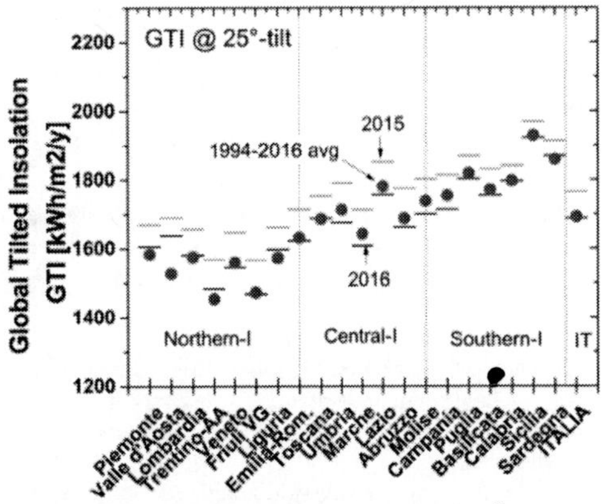

Fig. 2: global tilted irradiation (GTI, kWh/m²/y) at 25° tilt for all Italy and its regions in 2015 and 2016. The long-term GTI average (1994-2016, dots) is shown as well.

3.3 Key performance indicators

In Fig. 4 we show the calculated final yield Y_f and performance ratio PR for each region and for the whole country in years 2015 and 2016. The PR is calculated using GTI values at 25°-tilt, which, as previously observed in Section 2, leads to a slight overestimation of this indicator. To compute regional and national Y_f and PR we do not use the total PV capacity (GW_p) available at year's end, but we make use of what we call *adjusted PV capacity* (kW_p): i.e. the available PV capacity at the end of the year minus half of the newly added PV capacity in the specific year.

The use of this parameter should promote a smoother match between the newly-added capacity (MW) and the electricity generation (MWh) from the new installation in a given year. This adjustment works well under the assumption that the rate of new installations is constant over the year, which has been the case in years 2015 and 2016 (see [6, 7]), and which has presumably been the case for the years (after 2013) when new installations could not benefit from feed-in-tariff-like incentive schemes (*Conto Energia*). On the contrary, this newly-defined parameter can only partly mitigate this mismatch for the previous years with booming installations, when a rush to connect the new installation to the grid towards the end of the year has frequently been reported, and, therefore, no constant rate of new-added capacity over the year can be assumed.

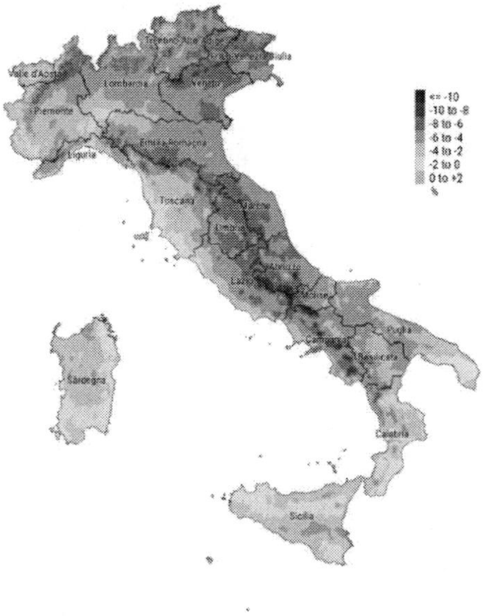

Fig. 3: inter-annual variability map of solar resources for Italy showing the relative difference between GTI ($kWh/m^2/y$) at 25° tilt in 2016 vs 2015 data.

Due to the larger insolation in 2015 compared to 2016, the Y_f has been understandably higher in 2015 for all regions (with the exception of Valle d'Aosta) and for the whole country.

Similarly, for most regions and for Italy the PR has been higher in 2015 rather than 2016. For this parameter, which should be uncorrelated to the annual cumulative irradiation, the trend is, however, less clear with some regions exhibiting a higher PR in 2016.

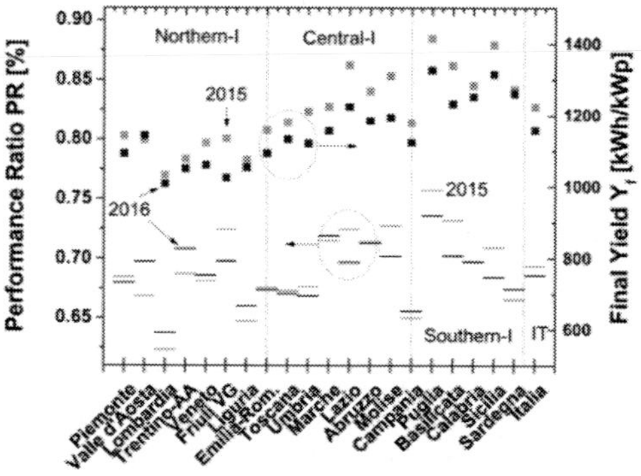

Fig. 4: performance ratio PR (lines) and final yield (squares) for years 2015 and 2016 for Italy and for each region.

4 DISCUSSION

From the long-term average GTI data in Fig. 2 (regions and country), we note that the insolation in 2016 has been more in line with the historical availability of solar resources than 2015, a remarkably sunny year. So that the previously mentioned alarmist tones about the decline in production of PV in Italy in 2016, are not justified.

On the other hand, from Fig. 4 and Table I, we observe that for the whole country between 2015 and 2016 the electricity generation from PV and the average final yield have decreased by -3.7% and -5.4%, respectively. About 80% of the decline in Y_f in 2016 with respect to 2015 can in fact be ascribed to the lower GTI (- 4.3%), but still the decline in GTI does not explain the full discrepancy. This is reflected in the lower country average PR with a -1.1% decline in 2016 compared to the previous year.

We further note that the country average PR is below 70% (68.6% in 2016), a value that is well below the 80-85% value exhibited by well-functioning and properly maintained PV plants. Furthermore, in Fig. 4, we observe that, at regional level, some outliers exist. Regions with the lowest PR are Lombardia, Liguria, and Campania. Puglia, the region with largest average size of solar plants (~58.8 kW in 2016), exhibits the highest PR.

If we consider the country average PR, this is a value reflecting the average performance of the entirety of solar PV installations (>732'000 in 2016) and simultaneous contributions from well and poorly designed/performing/maintained PV plants.

As the average size of PV plants in Italy is relatively small (~27 kW, see Figure 1), these numbers include contributions from a vast majority of small-scale PV plants that, more frequently than larger scale ones, may suffer from non-optimal orientation/inclination, shading, sub-optimal selection of components, poor design practices, malfunctioning, limited use of operation and maintenance (O&M) services.

As previously mentioned, in the booming years of PV in Italy (until 2012) fostered by the *Conto Energia* incentives, a rush in terminating installations by year's end has frequently been reported. This often at the expense of overall system quality.

In addition, there exist anecdotal evidence that the quality of some components (particularly modules and inverters) available in years of high demand in the Italian/European market is sometimes below standard, leading to a severe underperformance of the corresponding installations.

The combination of these reasons could explain most of the relatively low performance of Italian solar electricity when considered in its entirety.

We are not able, however, at this stage to assess the impact of the different contributions and to recognize if a accelerated long-term decrease in performance - beyond a physiological one - is taking place over the years. This would be the subject of a future work.

These evidences from the Italian experience about the correlation of overall PV quality with market dynamics and incentives could hopefully provide a constructive feedback to the several countries (particularly in sun-belt regions) that are about to design feed-in tariffs (or other incentive schemes) to promote a domestic diffusion of solar electricity. Overgenerous feed-in tariffs, combined with non-optimal timing and sharp year-end rate reduction mechanism will in fact generate distortion in the market, cause non-healthy temporal dynamics and considerably increase local market prices for solar PV; and, not ultimately, possibly cause the flow of often below standard components to the domestic markets.

Table I: figures of PV electricity generation, cumulative, newly-added and adjusted PV capacity, solar resources, country average final yield and performance ratio in Italy for years 2015 and 2016 (and relative variation).

Solar PV in Italy – all country	**2015**	**2016**	**2016/2015 relative variation**
PV Elec. Generation (GWh)	22'942	22'104 *	**-3.7%**
Newly added PV capacity (MW/y)	306.4	382.4	+24.8%
Cumulative PV capacity (MWp)	18'900.8	19'283.2	+2 %
Adjusted PV capacity (GWp)*	18'747.6	19'092.0	+1.8
Global Tilted Irradiation GTI at 25°-tilt (kWh/m2/y)	1'764.39	1'688.26	-4.3 %
Avg Final Yield (kWh/kWp) **	**1'223.7**	**1'157.8**	**-5.4%**
Avg Performance Ratio PR (%) **	**69.4**	**68.6**	**-1.1 %**

* The adjusted PV capacity takes into account only half of the newly added capacity per year.
**Rated to the adjusted PV capacity.

So that, a more "sustainable" diffusion of solar PV in a specific country should be backed by carefully tailored incentives schemes.

So that, a more "sustainable" diffusion of solar PV in a specific country should be backed by carefully tailored incentives schemes.

Finally, we conclude with two general remarks:

(1) In 2016, a national average PR of 75% – which is a reasonable target - would have led to a +9.3% increase in solar electricity generation (corresponding to +2'070 GWh);

(2) The GTI figure of Fig. 2 do not include error bars, which are significantly large and that would be reflected in the corresponding uncertainty of the calculated PR figures. This will be the subject of a subsequent work.

5 CONCLUSIONS

At country level, the 2016/2015 relative variation in final yield Y_f is - 5.4%. About 80% of this variation can be attributed to a lower insolation in 2016 compared to 2015 (2016/2015 GTI: -4.3%).

The remaining losses are reflected in the –1.1% PR relative variation (2016/2015) and can be attributed to a mix of factors: degradation in performance (system/components), moderate use of O&M, etc.

At country level, the average PR is below 70%. This low value is partly attributed to the fragmentation and small average size of PV systems in Italy (~ 28 kWp in 2016), to their non-optimal orientation and to presence of shading. Other reasons, linked to the dynamics and features of the national feed-in tariff incentive schemes in place from 2005 to 2013, are briefly recalled.

We further note, that the alarmist tones circulated among professionals and in the specialist press about the pronounced decline in generation of solar electricity in Italy in 2016, are not justified. On the contrary, 2015 has been a year with an availability of solar resources considerably higher than 2016 and of the long-term average. Nevertheless, still concerns about the actual (and long-term) performance of the Italian solar park as a whole exist.

Finally, by having a look at regional PR values, we notice that some outliers are present: regions with the lowest PR are Lombardia, Liguria, and Campania, the region with the highest PR is Puglia, the region with the plant's largest average size.

6 ACKNOLEDGMENTS

We gratefully acknowledge Luca Benedetti (GSE) for delivering 2016 data prior to publication.

REFERENCES

1 IEA Report: Trends 2016 In Photovoltaic Applications - Survey Report of Selected IEA Countries between 1992 and 2015, Report IEA PVPS T1-30:2016, ISBN 978-3-906042-45-9

2 AAVV "Calo di produzione: che sta succedendo al fotovoltaico italiano?", QualEnergia.it, www.qualenergia.it/articoli/20160808-calo-di-produzione-che-sta-succedendo-al-fotovoltaico-italiano;

3 AAVV, "Fotovoltaico 2016, in Italia produzione calata di oltre il 4%", www.rinnovabili.it, http://www.rinnovabili.it/energia/fotovoltaico/fotovoltaico-2016-italia-produzione/;

4 Terna report: *Rapporto Mensile sul Sistema Elettrico*, September 2016;

5 GSE Report: "La produzione fotovoltaica in Italia nel 2016 - Analisi preliminari" www.gse.it/it/Statistiche/RapportiStatistici/Pagine/default.aspx;

6 GSE Report "Solare Fotovoltaico - Rapporto Statistico 2015": *idem*;

7 GSE Report « Solare Fotovoltaico - Rapporto Statistico 2016 »: *idem*;

8 Terna Report: "Consumi 2015-2016", http://www.terna.it/it-it/sistemaelettrico/statisticheeprevisioni/datistatistici.aspx

9 Terna Report: "Produzione 2015-2016", *idem*;

10 Cebecauer T., Šúri M., Perez R., 2013. Semi-Empirical Satellite Models. In Kleissl J. (ed.) Solar Energy Forecasting and Resource Assessment, Academic press, pp. 21-48

11 Cebecauer T., Šúri M., Skoczek A., 2011. SolarGIS: Solar Data And Online Applications For PV Planning And Performance Assessment. 26th European Photovoltaics Solar Energy Conference, September 2011, Hamburg, Germany

12 Cebecauer T., Šúri M., Skoczek A., Fantík J., 2011. Analysis of PV Power Generation in the Czech Electricity Transmission System CEPS. 26th European Photovoltaics Solar Energy Conference, September 2011, Hamburg, Germany;

13 International Electro-technical Commission, IEC-61724:1998(E), I.S, Photovoltaic system performance monitoring – guidelines for measurement, data exchange and analysis. International Electrotechnical Commission;

PERFORMANCE AND RELIABILITY OF A PROFESSIONAL SMALL-ISLAND HYBRID PV-SYSTEM

Heinz A. Ossenbrink
Consultant
Bad Feilnbach (DE), email: ossenbrink11@gmail.com

ABSTRACT: This paper describes the seven years operational experience with a professional hybrid PV system which generates electricity for a resort on a small island in located in the Kingdom of Tonga, South Pacific. Performance and Reliability is the most important and strategic requirement for the successful operation of a premium segment resort. A secondary requirement is to reduce expenditure for diesel-fuel and the operating hours of the generator to extend its lifetime.
On-site, real-time data as well as the data stored by the control system have been analysed with the aim to review performance and reliability, as well as to optimise the overall operation. In addition, a detailed calculation of the cost structure has been performed, leading to a proposed extension of the system.
Keywords: Battery Storage and Control, Stand-alone PV Systems, Villages, Hybrid, Reliability

1 INTRODUCTION

Figure 1: View of Fafa Island

Fafa Island is a premiums holiday resort, the #1 of all comparable resorts in the Kingdom of Tonga. Reliably providing electricity and water supply is a strategic requirement for the successful operation of the resort.

Therefore, a PV Hybrid System was installed. It powers the resort consisting of 18 bungalows for an average of 30...40 guests, a restaurant with kitchen, the office, staff quarters and maintenance facilities. Moreover, the use of the PV-module surface contributes to about half of the rainwater collection area, reducing significantly the need for operating the sea-water desalination sub-system.

The PV System must meet two principle objectives:

- Provide reliable, high-quality and round-the-clock electricity services to the Fafa Island Resort
- Reduce expenditure for diesel-fuel and operating hours for generator to extend its lifetime

By meeting these objectives, the PV system gives additional benefits to the resort:

1. The module surface of about 120 m^2 is a perfect means to collect rainwater for the fresh water supply, adding yearly at least 70 m^3 of water to the existing facilities
2. the silent and clean generation of electricity contributes to the "Green" image of the resort
3. in the long-term, the PV-system provides lower electricity cost

Figure 2: Location of vthe PV System on Fafa (blue Arrow)

2 SYSTEM DESCRIPTION

2.1 Technical Data

The system had initially (June 2006) a nominal PV power of 15 kWp with a battery bank of flooded lead-acid (FLA) cells of nominal 96 kWh together with 3 inverters (SMA Sunny Boy) totalling 15 kW, supported by three others (SMA Sunny Island), also totalling 15 kW. These SMA Sunny Island inverters control the charge/discharge management of the battery banks, generate the 400 V three-phase AC grid (to which the Sunny-Boy grid-inverters feed) and manages the diesel generator operation as back-up supply.

As PV module prices continued to drop, and the electricity demand of the Fafa Island Resort was increasing, an extension of the PV system was commissioned on 15/01/2016. It comprised an additional 5.5 kWp PV array, an 5 kW 3-phase grid-feeding inverter (SMA-Sunny TriPower), and the replacement of the battery bank by sealed, valve regulated lead-acid cells (VRLA) of a nominal capacity (I_{10}) of 130 kWh.

2.2 Operation Mode

The load of the resort can be described in three daily cycles: from sunrise to sunset, the resort has a moderate electricity consumption, fairly well distributed through the day, with peak loads caused by laundry, and the noon operation of the kitchen. Around sunset, the demand caused by the kitchen and illumination can reach the maximum system power (25 kW). After dinner until sunrise, only base-load (3 kW) is observable, caused by a number of refrigeration services, with the walk-in cold store being the largest consumer (2.5 kW)

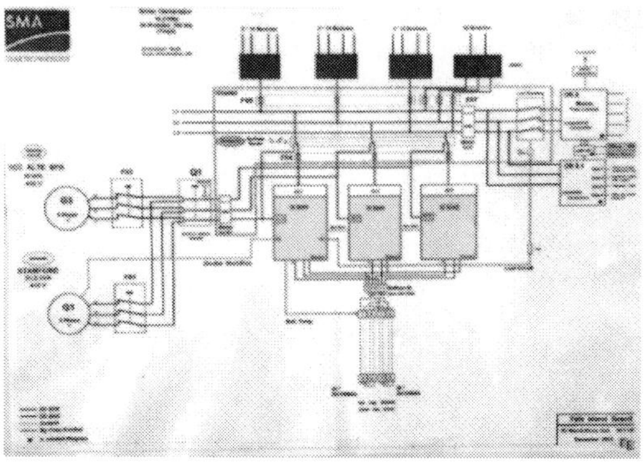

Figure 3: Electrical Layout of the System. The three "SunnyIsland" inverters (yellow) create a local grid and allow the whole system control and battery management.

3 ASSESSMENT

3.1 Data collection

The major data source are the data collected by the SMA Sunny Island inverters, which store about 50 system parameters in one-minute intervals on a memory card. In addition, on site real-time measurements have been performed using timed readings of the two electricity meters (one metering the energy from the diesel generator, the other the overall energy supplied to the resort), readings of the battery state-of-charge in 2-hour intervals, and single-cell voltage measurements of the battery bank. The on-site measurements allow a fairly good determination of battery capacity, PV performance, charge cycles and load profiles.

3.2 PV System

The measured operational system power corresponds to 78% of the nominal STC power of 20 kW_p, which is exactly the expected value for NOC (Nominal Operating Conditions, at 48 °C Module temperature, and 800 W/m^2).

It must be concluded that all 4 PV sub-arrays are exactly within their specified performance, neither degradation or partial module-failures can be deduced from this measurement or has been observed.

3.3 Battery Storage

As a Battery bank has been replaced end of 2015, a careful analysis has been performed of both the old and new battery system, by taking monitoring data, instantaneous single-cell measurement and review of full charge/discharge cycles. After one year of operation of the new battery bank, a capacity of slightly above 80% of the nameplate capacity (at 20°C) has been calculated. Considering the elevated average operational temperature of 32°C, this capacity value is with the design value. Performing a similar analysis for the old, replaced battery block, a remaining capacity of 65% ... 70% was calculated from the monitoring data. This is in contrast to the very rapid (less than one hour) discharge observed prior to the exchange. The reason most probably was a wrong setting of the nominal capacity in the control system, which subsequently assumed a 10 times smaller capacity in its battery management.

A single cell voltage measurement has been performed at no-load condition. From the bar graph to the right it is apparent, that the voltage of each 24 cells the two battery banks are very much in agreement. After the system analysis, major changes have been made to the control parameters of the battery management. For instance, there are neither automatic equalising nor full-charge cycles foreseen any more. In general, recommendations of the battery manufacturer appear too conservative, without providing reliably data on dependence of lifetime vs. control system setting.

Figure 4: Battery and control room

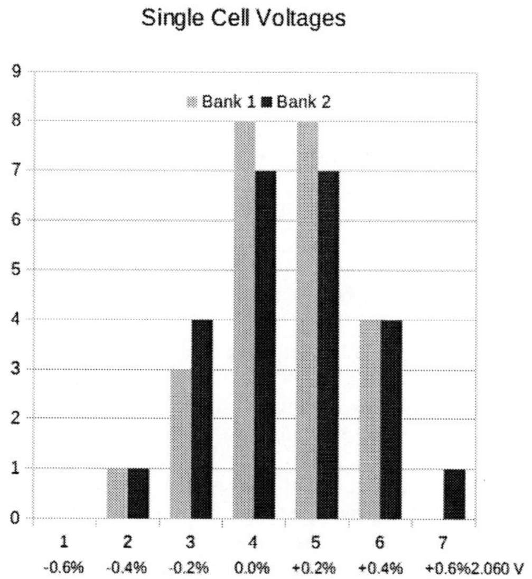

Figure 5.Result of the battery single-cell voltage measurements

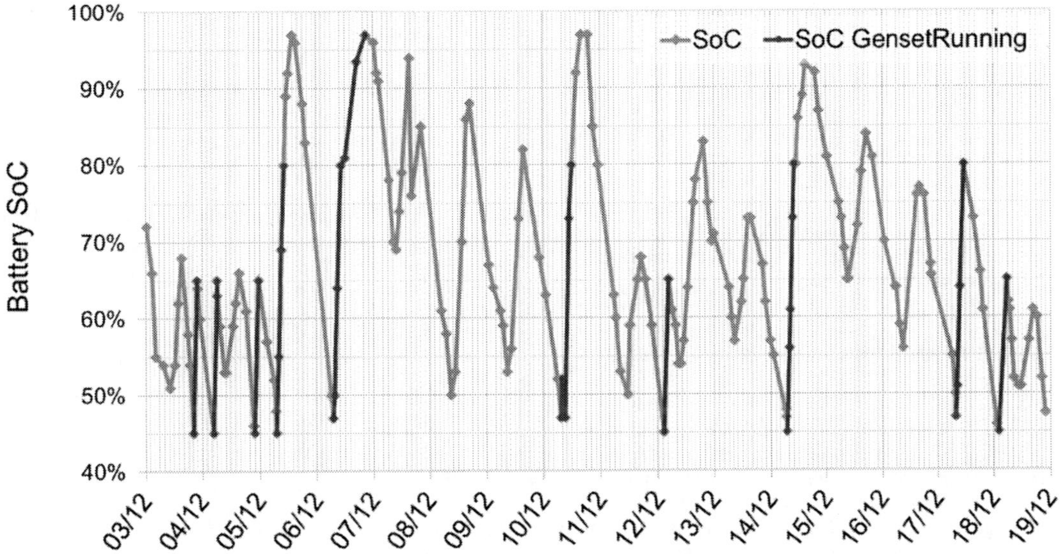

Figure 1 Measured State-of-Charge (SoC) of the battery system during two weeks of medium weather. The read lines indicate battery charge by the diesel generator. One can observe the low SoC threshold (set to 45%) resp. the high SoC threshold (set to 80%) which controls the back-up generator during the night (red lines). On 7th Dec the battery management system was enforcing an equalising phase (up to 95% SoC), even though it was a sunny day. Consequently, the PV system wasn't use to its full power. The management of the batteries has been changed there after, allowing deeper discharge, avoiding equalising cycles, and by switching off major loads during the night. The SoC decreases now much less during the nights, avoiding the start-up of the generator in most cases.

3.4 Diesel Generator

The diesel generator is controlled by the battery management system of the Sunny Island Inverters. Its operational performance in the past was below 30% conversion efficiency (consumption 0.4 l/kWh) due to long low power operation for equalising and full charge cycles. Diesel is delivered to the island in 200 l drums, the handling of which is time consuming, dangerous (spills!) and costly. Therefore, the battery management has been changed to reduce generator operating hours, and sacrificing an estimated battery lifetime.

3.5 Load Management

The control system is capable of giving different priorities to loads. This has been optimised to increase self-consumption and to provide more reliability. Moreover, the cold store is now automatically off-line between 22:00 and sunrise. On a yearly basis, about 60% of electrical energy is used for cooling appliances, and the load management was optimised to get as much electricity storage in cooling during the day, and switching off refrigerators during night time. Since all illumination has been changed to LED-bulbs, electricity consumption due to lighting has a share of only 3% ..5%.

4. BENEFITS AND COSTS

4.1 Additional Benefits

The surface area covered by the PV modules is about 160 m^2, and is used to collect rainwater to be stored in large concrete tanks. Yearly water delivery is about 200'000 l water, almost half of the yearly water consumption. Since its installation, the operation time of the desalination system which is used during extended dry periods, was reduced by 80%.

Part of the PV Array, rainwater is collected by the gutters and fed into concrete storage tanks.

4.2 Cost Structure

Taking the operational costs of the PV system, including the yearly costs of a 7 years battery operation, electricity on Fafa Island is provided for less 0.45 €/kWh over a 15-year financial lifetime. This needs to be compared with the equivalent cost of the diesel generator electricity of about 0.65 €/kWh, which includes maintenance, repair and depreciation. The PV system saves about 13'000 € per year on electricity costs. In addition, the water collection area reduces the operational time of the reverse-osmosis seawater desalination system by 750 hours and saves 3 MWh (3000 €).

4.3 PRICING

The Gen-set operation delivers about 30%...40% to the overall electricity needs of the resort. Hence the costs of the generator produced electricity is well known from the fuel-costs as well as maintenance and repair cost. Within the accounting scheme of the resort, the PV-electricity is accounted for at 90% of the generator costs, and yields a payback-time of about 15 years.

5 RESULTS AND CONCLUSIONS

As the control system allows the configuration of more than hundred parameters which can be optimised for a variety of priorities, the paper describes the major issues which determine reliability, performance and costs. Each of the subsystems (PV array, Inverter, Battery Storage, generator, loads) is first analysed separately and then further assessed in terms of system performance and costs. For instance, optimising battery performance does not lead necessarily to a better system performance. Transient effects (e.g. cloud-passage) can interfere with the load management and, when not properly addressed in the control system can disrupt the system. In optimising the parameters of the control system, the system reliability had the highest priority, whilst cost of the electric energy supplied had a lower priority. An important criterion for optimising the control system was to minimize the operational time of the generator for reducing yearly maintenance cost, whilst ensuring maximum efficiency.

The system showed a reliability of in average less than 1 hour of electricity loss per year, far better (99.99% availability) than any grid system in industrial countries. It survived heavy weather, thunderstorms and strong winds. Already seven years operation fully paid back the investment costs (~5 €/kW$_p$ in 2006). The system contributes to the environmental image of the resort and avoids extended use of the seawater desalination unit. The approximately 20 MWh of yearly electricity production corresponds well with the demand of many non-electrified villages in under-developed regions of the world. The economic assessment confirms that the 10-year energy cost is well below those of a conventional, diesel-generator electricity system. Based on the presented analysis, and because of the significant drop of PV-module prices in the last seven years, the system will be in future double its PV-capacity, avoiding the investment costs of a second diesel generator.

5 CONCLUSIONS

- The PV System is overall operating close to perfection and well within its design specifications
- The battery bank is performing within the specifications, exhibiting at least 80% of its initial capacity
- The electricity demand of the resort with about 100 kWh/day is high in relation to the PV system capacity (production 40..50 kWh/day, usable storage 50 kWh/day)
- It appeared to the author that much of concern and doubts about the batteries was inclining the operation and control strategies towards "caring" the batteries rather than minimizing generator running time
- As the diesel generator operates already 30 years, its remaining lifetime can be considerably prolonged by doubling the PV capacity.

DESIGN AND PERFORMANCE OF A REAL SCALE REFRIGERATED PHOTOVOLTAIC PLANT INSTALLED IN A HYDROELECTRIC PLANT

SILVA, Vinícius Oliveira da; GIMENES, André Luiz Veiga; RELVA, Stefania Gomes; UDAETA, Miguel Edgar Morales; GALVÃO, Luiz Claudio Ribeiro.
Energy Group of the Electric Energy and Automation Engineering Department of the Polytechnic Scholl of the University of São Paulo.
Avenida Luciano Gualberto, travessa 3, n° 158, Sala AN-16 do prédio Anexo, Zip code: 05508-010.
vinicius.oliveira.silva@usp.br

ABSTRACT: The aim of this work is to evaluate the impact of using a cooling system in a real scale Solar Photovoltaic Power Plant named cPVPP in the production of electricity in order to determine operating parameters and a model for this prototype system model. The methodology is based on the measurement, verification, and data analysis of temperature, and electricity generation of two strings, test (cooled) and comparison (uncooled). By analyzing the data has observed that, for the period between 09:00am and 5:30pm, the PV modules of the test string operate at temperatures below those of the PV modules of the comparison string. During the tests, in which the temperature of the PV modules of the comparison string operated above 55.0 °C, the average and the maximum temperatures recorded in the PV modules of the testing string lied below 37.0 °C, operating below the NOCT of the PV module. Regarding the generation of electricity, the test string generated 3.0 kWh/day more than the comparison string. Hence, the cooling system decreases the operating temperature of the PV modules, particularly during the maximum power generation period which is from 11am to 3pm. This leads to efficiency average gains of up to 5.9% in the generation of electricity, 10.3% in the power, and 5.3% in the PR and PF.
Keywords: Hybrid solar system, Photovoltaic/thermal, Cooling system, empirical M&V, Cooled PVPP.

1 INTRODUCTION

Brazil has enormous potential for exploitation of solar energy, since the average annual radiation varies between 4.200 a 6.700 kWh/m², values considerably higher than the average of European countries where the exploitation of this resource is much more recurrent. However, for achieving a more significant reduction of production costs within the photovoltaic chain in the country through economies of scale it is necessary to stimulate the further development of the market for solar energy [1].

The production of electricity and/or heat causes an increase in temperature in the equipment involved in the operation. To resolve problems due to excessive operating temperatures of the dynamic energy production equipment, in case of internal combustion engines, the radiator is used as cooling system and heat exchangers of the most diverse to hydraulic turbines. The PV module, electricity production equipment solid state, also suffer from the effects of increased temperature during operation, because it absorbs 80% of solar radiation, of which only 5-20% are converted into electricity the remaining energy is converted into heat [2].

This phenomenon further jeopardizes the yield of modules, because with the increase in module temperature silicon observes the power loss due to the voltage decrease [3], that is, the vast exposure to sunlight associated with high ambient temperature raises the temperature operation of the module directly impacting the efficiency and energy produced, this effect occurs with more intensity in tropical regions, low latitudes, with high incidence of solar radiation [4].

These characteristic is present in the region of the city of Ilha Solteira which features historical highs 42°C of temperature and 802 W/m² of solar irradiance [5]. To minimize this effect arises hybrid photovoltaic/thermal system (PVT) in which the generated heat is extracted from the PV module to the heat absorber system. Therefore, the PV/T system consolidates the production of electricity from PV modules and heat of solar concentrators. installed at the HPP in the region of the city of Ilha Solteira, Brazil, which is a tropical region, with low latitude (22°), high incidence of solar radiation and ambient temperature [6].

The PVPP is equipped with 344 PV modules rigid (p-Si), with a capacity of 290 W each, totaling 99,760 WP installed capacity and 623 m² of area. Besides that, it has ten solar inverters, six with 12.0 kW and four with 5.0 kW, both are coupled on three-phase transformer of 100 kVA. All energy is consumed in the auxiliary system of the HPP [6].

This study will enhance the operation of photovoltaic power with cooling system in tropical countries can improve performance of the panels facilitating the large-scale integration into the Brazilian market, since the country has great potential for generating and will provide the study complementary of electricity production between UFV cooled and HPP. In addition, it will propose an operating model for a Photovoltaic Solar Power Plant on a commercial scale.

Figure 1. Cooled Photovoltaic Power Plant (cPVPP).

2 METHODOLOGIES

The methodology is based on the measurement, verification, and data analysis of temperature, electricity generation, testing strings (cooled) and comparison strings (uncooled), see Figure 1, in the cPVPP prototype operation. The cooling equipment is named Modular Cooling Unit (MCU). The MCUs are installed on the bottom surface of the PV module (Figure 2).

Figure 2. MCU installed on the cPVPP.

Performance tests of the strings, cooled and uncooled on the PVPP, consist on measured of operating temperature of the PV Modules, input and output of coolant with the use of PT-100, the output voltage, current, power e energy data are measured by the inverter. All measured data is stored in datalogger and used to evaluate the performance of the PVPP in the production of electricity, for later to determinate the model of a cPVPP.

The cooling system operates as described follow (Figure 3):

i. Point *I* (Input line) is the input line of the cooling system, which captures HPP process water;
ii. Water circulates in the MCU throughout the day;
iii. The MCUs are connected in series of two, ie, the fluid circulates in an MCU and then passes through the second to be discarded;
iv. The water is discarded on Point O (output line) and goes downstream of the HPP.

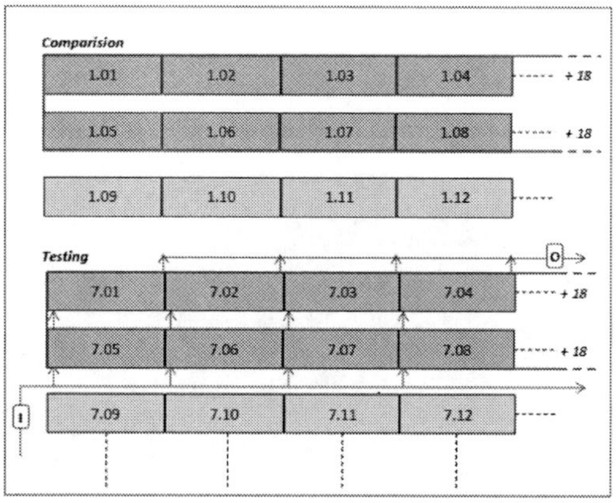

Figure 3. cPVPP's operation flowchart.

3 ANALYSIS AND RESULTS

It was confirmed that the temperature of the PV modules of the testing string operate at lower temperatures compared to the comparison string, as always operated at temperatures below 37.0 °C, or at least 8.0 °C below the minimum limit NOCT (Nominal Operating Cell Temperature) 47±2°C, and 8.4 °C below the maximum limit NOCT, while the comparison string of modules operated at temperatures up to 59.97 °C, 14.97 °C value above the minimum limit NOCT, at this very moment, the difference between the maximum and minimum operating temperature of the PV modules was 25.86 °C (Figure 4a).

To leave no doubt in the period comprising the 11:00am to 03:00pm in which higher operating temperature of the PV modules of both strings is the largest power generation and in turn are registered, the temperature of PV modules in string comparison is always above 45.0 °C while the test string not exceeded 36.23 °C (Figure 4a).

About the water flow, it was almost constant during the day. Its temperature variation occurred because of the water interruption in the rockfill system of the hydroelectric plant. The maximum temperature difference of the water between input and output of two MCU linked in series was 2.50 ° C (Figure 4b). This has occurred at the same time of maximum operating temperature of the photovoltaic modules.

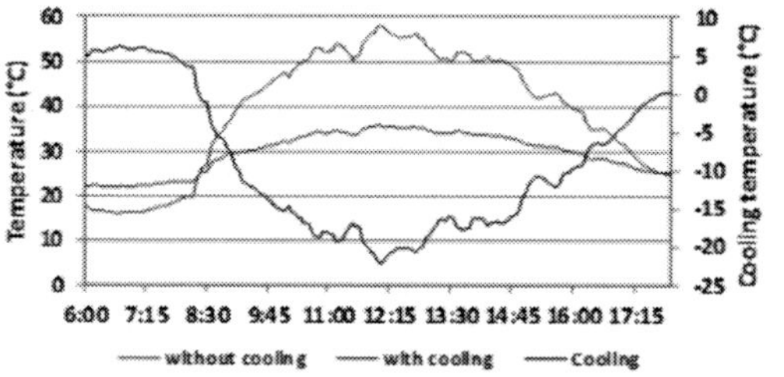

a

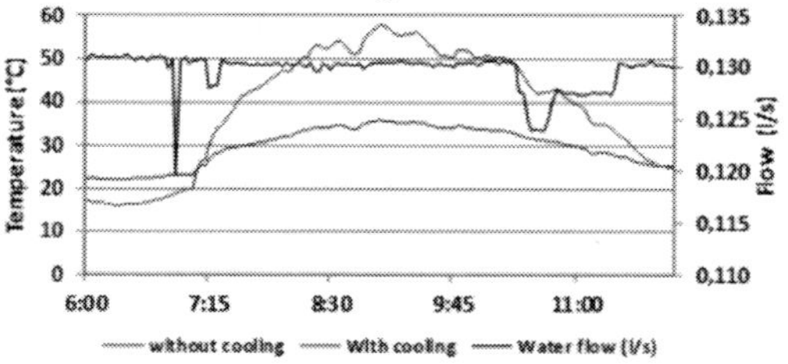

b

Figure 4. (a) Temperature difference between maximum and minimum temperature measured on PV modules and (b) Temperature of the PV module with and without cooling and water flow.

As for the generation of electricity, it was proven that the use of MCU in PV modules increases the efficiency of the electricity production of UFV, as the test string generated up to 3.0 kWh/day of energy more than the string comparison, a total of 11 kWh for examined period, with the increase of 4.0% in performance ratio (PR) and 10.3% in the power, a gain of 612 Wp (Figure 5a). It is equivalent more than 4.3 PV modules operating in those conditions this results are evident in the performance of index of yield (Y) , PR and capacity factor (CF), in which the testing string presented gain of 5.4% (Figure 5b), over the comparison string (Table 1).

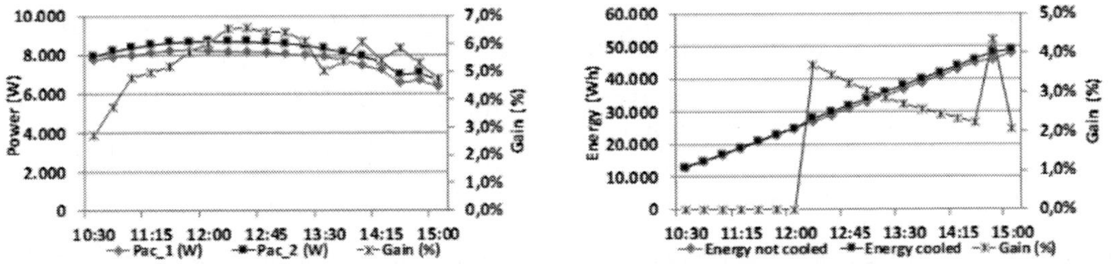

a *b*

Figure 5. (a) Curve of power and gain of the systems with (PAC_2) and without (PAC_1) cooled for a short period and (b) Power generation and gain of the system with and without cooled for a short period and gain of the system with and without cooled for a short period.

Table 1. Consolidated performance data.

String	Data	Unit	Period 1 (P1)	Period 2 (P2)	Period 3 (P3)	Period 4 (P4)
Comparison	Power's peak	(kWp)	8,873	8,901	8,698	8,873
	Maximum gain	(%)	8.7%	7.9%	10.3%	7.2%
	Power generation	(kWh/day)	54	56	57	56
	Day's gain	(%/day)	5.9	3.7	5.6	5.7
Comparison	Y	(kWh/kWp)	4,31	4,39	4,47	4,39
	PR	%	71,0	70,2	71,3	(1)
	CF	%	18,0	18,3	18,6	18,3
Testing	Y	(kWh/kWp)	4,39	4,62	4,62	4,55
	PR	%	72,3	73,9	73,8	(1)
	CF	%	18,3	19,3	19,3	18,3
Gain		%	1,82	5,36	3,51	3,57

4 CONCLUSIONS

The results obtained were conclusive and indicate that the test string in the periods of peak generation and temperature operated with temperature always lower than the NOCT while the comparison string always operated above, a difference greater than 25 ° C. In this moments, the power generation and test string power was always higher than the comparison string, proving that the cooling system reduces the operating temperature of the PV modules by increasing the energy generated in the UFV. Achieving 5.6% higher performance for the analyzed period.

Thus, the scale model architecture of a cPVPP is effective and possible to be reproduce in scale. If there is expansion of this type of system in conjunction with HPP it is possible to increase the bulk energy in a commercial HPP and store water in its dam.

ACKNOWLEDGEMENTS

To CAPES, for the scholarship. To CESP, for funding the P&D ANEEL PE-0061-0037/2012, which allowed for the conclusion of this article. To the team of researchers and collaborators that helped direct and indirectly to the accomplishment of this P&D ANEEL.

[1] E.B. Pereira, F.R. Martins, S.L. Abreu, R. Rüther.; Atlas brasileiro de energia solar (ABES), São José dos Campos: INPE, 2006

[2] van Helden W. G. J., van Zolingen R. J. Ch, Zondag H. A.; PV thermal systems: PV panels supplying renewable electricity and heat. PROGRESS IN PHOTOVOLTAICS: RESEARCH AND APPLICATIONS, Prog. Photovolt: Res. Appl. 2004; 12:415–426 (DOI: 10.1002/pip.559)

[3] Prieb, C. W. M.; Desenvolvimento de um Sistema de Ensaio de Módulos Fotovoltaicos; Dissertação para Obtenção do Título de Mestre em Engenharia, UFERGS, Junho de 2002.

[4] Silva, V. O.; Udaeta, M. E. M.; Gimenes, A. L. V.; Galvão, L. C. R. . Analysis and Evaluation of Cooling Systems for the Selection and Prototype Development for Solar Photovoltaic Power Plant - ILS. In: EU PVSEC 2014, Amsterdam, Netherland, 2014, DOI: 10.4229/EUPVSEC20142014-5BV.2.13.

[5] Unesp, 2013. Acessado em 12/08/2013. Site: http://clima.feis.unesp.br/recebe_formulario.php.

[6] Silva, V. O.: 'Estudo e Modelagem da arquitetura modular de uma Usina Solar Fotovoltaica Arrefecida com Protótipo de Verificação'. Master thesis, University of São Paulo, 2015

DEVELOPMENT OF ROBUST ALGORITHM FOR AUTONOMOUS SYSTEM HEALTH MONITORING OF ULTRA LARGE SCALE BASED SOLAR FARM

Sirshendu Arosh[1], Kuntal Ghosh[2], Surya Prakash[3], S.P. Duttagupta[4]
[1,3,4] Department of Electrical Engineering, IIT Bombay,
[2,4] Centre for Research in Nanotechnology, IIT Bombay,
Mumbai, India.
Email Id: {[1]sirshenduarosh, [3]suryaprakash, [4]sdgupta} @ ee.iitb.ac.in, [2]ghosh.kuntal@iitb.ac.in

ABSTRACT: System health monitoring of the ultra large scale solar Photovoltaic farm is an integral part to harvest steady and stable output energy. This requires adaption of an effective and innovative, operation and maintenance methodology for such largescale systems. Hence a camera integrated automated maneuverable unmanned aerial vehicle is proposed which can scan large scale solar farm without investing much time. Images from camera facilitate to analyze the degree of threat on the system so that corrective measures can be enabled precisely only for the affected SPV modules. This work mainly accounts for introducing novel analyzing method that incorporates a robust algorithm which can re-iterate the degree of threat in the system dynamics. This leads to reliability enhancement of such large scale SPV power system.

Keywords: Hotspot, Non-Uniform illumination (NUI), unmanned aerial vehicle (UAV), airborne early warning system (AEWS), LAB image, Kalman Filter (KF), Active contour, Monitoring, PV array, Thermal Processing, Tracking, Building integrated PV (BIPV).

1 INTRODUCTION

The concept of smart cities is emerging due to the drastically growing urban population. Mostly these smart cities meet its energy demands from the distributed large-scale renewables (solar, wind, biofuel, geo thermal etc.), installed in discrete places according to the availability. Among these sources solar energy found to be more suited due to proven integration process of solar PV modules with the buildings i.e. building integrated PV (BIPV) and easy installation at the rooftops [1]. These constellations of discrete SPV array cumulatively construct a large scale SPV farm. In addition, ultra large scale solar PV farms are also accelerating its presence in remote areas gradually by replacing the conventional grids. This phenomenon leads to appearance of solar PV technology more potent among the other renewable energy sources. But the reliability and performance monitoring pose a severe challenge for this technology. Non-uniform illumination (NUI) is the primary reason for disruption in power generation using solar PV modules. Sun movement, environmental mist and aerosol particles, high rise and bird drop develops NUI on solar PV modules [2]. This leads to a voltage drop between the affected SPV modules, which eventually cause a temporal degradation in generated power [3]. This defect has potential to generate hotspot in the affected modules and can create permanent damage to the affected modules if timely measurement is not adopted. Hence, a large scale SPV array requires continuous performance monitoring to detect any unwanted aforementioned hazards which can create NUI. Primarily it has been performed using sensor network [4]. But to manage such large scale SPV system need a huge number of sensors which additionally pose a problem while data communication, data fusion etc. Hence, as a preventive measure, an automated system health monitoring method is proposed in this article which could diagnosis the fault and its degree, to avoid any aggravation to the system. There are several literatures where scientists have explored the potential threats of the SPV system such as, Garcia et. al. [5] has detected hotspots and its associated effects in the power generation in a grid connected solar plant. Then Tyutyundzhiev et. al. [6] proposed the concept of multirotor UAV based SPV module is inspection by using both heat-map image and visual image. Quarter et.al. [7] and Grimaccia et. al. [8] has also highlighted the cost effectiveness and time saving method using light unmanned aerial vehicle. Leva et. al. [9] has investigated about the relation between the altitude and the defect detecting capability using aerial vehicle. Aghae et. al. [10] has proposed basic image processing based algorithm for finding out contrasts in solar cell using both IR and visual image. On the other hand, Arosh et. al. [11] and Ghosh et. al. [12] has tracked the dynamic non-uniform illumination (NUI) using optical flow techniques and Hungarian-Kalman filter respectively by optical camera to prevent damage in the rooftop SPV array.

This article attributes to development of a precise visual-heatmap camera input based robust algorithm for power optimization. This algorithm can also estimate the future positioning of the Non-uniform illumination (NUI) to function as an airborne early warning system (AEWS).

2 APPROACH

Primarily this work demonstrates the process of rapid fault diagnostics of discrete SPV modules in an ultra large scale SPV farm. Moreover, it leads to prediction of the degree of defects in the SPV modules in order to facilitate the corrective measures of those modules. This process requires deployment of an airborne platform (such as aerostat, drone) which carries the both visual as well as thermal imaging camera to monitor the NUI dynamics on the ground, terrestrial and building integrated SPV system. According to the region of interest (ROI) (means size of the farm) and field of view (FOV) of the camera, the aerial platform is deployed. In the beginning the thermal imaging camera maps the entire field within its FOV to locate the invariably illuminated SPV modules which can pose a potential threat to the stability of the system. The possible defects which can be associated with such ultra large scale solar PV farm are (i) Dynamic NUI due to cloud movement and sun movement (ii) Environmental mist and aerosols in the form of dust and bird droppings which prevent solar irradiance to the cells and (iii) crack on the SPV modules due to thermal expansion. All these problems lead to generation of hotspots which can be identified by thermal imaging camera. Then the visual camera is enabled to re-affirm the associated degree of defects (weather it is temporary or permanent) to the identified co-ordinates (by thermal imaging camera) and accordingly preventive measures are initiated.

Only the image frames are obtained from the visible camera which are transmitted to the base computer and processed further to diagnose the faults and finds its co-ordinates. In general, unwanted objects which disrupt the generated power have different color identifications; can be easily classified by color-segmentation method. In this article, CIE LAB color scale is used for color segmentation. The NUI is nothing but a change in light intensity at the background; which can be filtered out from a single frame. Then the NUI detection technique can be applied on the frames to detect and to track the movement of the NUI over frames.

This complete process is tabulated in the following algorithm.

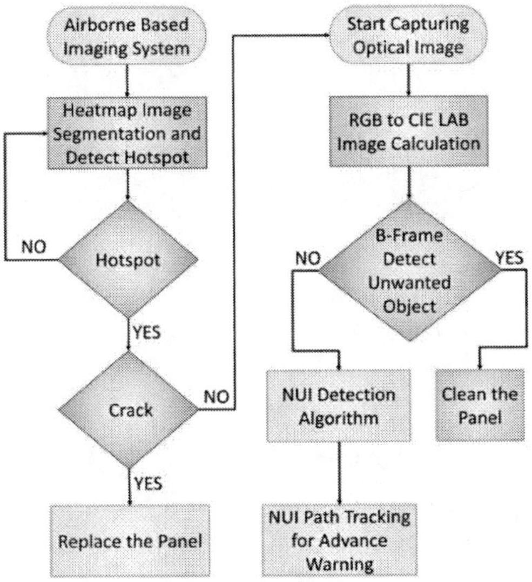

Figure 1: Overall system algorithm

3 SCIENTIFIC INNOVATION AND RELEVANCE

Primarily this research work aims to

(i) Reduce the operation & maintenance cost
(ii) Real time fault diagnosis of the SPV modules and its corresponding analysis.
(iii) Detecting the NUI on SPV array and tracking
(iv) Further to enhance optimized power generation

Following section gives brief description of each aforementioned points.

Integration of renewables mostly ultra large scale SPV farms to the grids is emerging rapidly to develop smart cities and to supply power in remote areas. But the recurring cost of operation and maintenance (O & M) for such system and degrading reliability are matter of concerns which shorten the MTTF (mean time to failure). Mostly dense wireless sensor network is deployed for O & M of such solar farm, which poses challenges in the fields of multi-sensor data fusion, network protocols etc. Hence alternatively this camera mounted on mobile aerial platforms is proposed in this work to cater deduction in the data processing time and to localize the defected SPV module. It has been found that this process is fast to identify the degree of threat to the system and enable a feedback corrective process to root out the threat from the system. This process provides more robust preventive solution and increases mean time between failure (MTBF) rates. Corrective measures can include feedback precision cooling across the faulty module and reconfiguration switching methodologies etc.

3.1 CIELAB image processing

CIELAB color space is an international standard such as sRGB, HSV, XYZ color space. CIELAB color space provide and near perfect approximation to human vision and it describes all possible visible color in three axes namely L as lightness, A as color opponents from green to red and B as blue to yellow. CIELAB shows more perpetual uniformity and it is appropriate to this work since the background i.e. solar panel has almost uniform color of blue and the most of the common obstacles are non-blue in nature color wise. CIELAB is a perfect choice for such differentiation between small color changes. The color components are denoted by L^*, a^*, b^*.

The CIIELAB color model is defined in terms of another color space named XYZ. XYZ color space is computed form gamma corrected sR'G'B' values to linear sRGB values, next those are converted using ITU-R Recommendation BT.709 using the D65% white point reference. Next XYZ values are converted to L^*, a^*, b^* [13-15].

$$L^* = 116\left(\frac{Y}{Y_n}\right)^{\frac{1}{3}} - 16, \qquad for\ \frac{Y}{Y_n} > 0.008856$$

$$L^* = 903.3\left(\frac{Y}{Y_n}\right), \qquad otherwise$$

$$a^* = 500\left(f\left(\frac{X}{X_n}\right) - f\left(\frac{Y}{Y_n}\right)\right)$$

$$b^* = 200\left(f\left(\frac{Y}{Y_n}\right) - f\left(\frac{Z}{Z_n}\right)\right),$$

$$where\quad f(t) = t^{\frac{1}{3}},\quad for\ t > 0.008865$$

$$f(t) = 7.787t + \frac{16}{166}$$

Where X_n, Y_n, Z_n are the tristimulus values of the reference white point specified using the White point parameter.

3.2 NUI boundary detection and NUI tracking [11]

Non-uniform illumination (NUI) may cause non-uniform current distribution in the PV cell array which can cause hotspots in the solar cell. The detection of NUI pattern is detected by the optical camera. Once the hotspot is detected by thermal camera, NUI detection algorithm is computed over the cell from optical camera and NUI boundary is detected. Multiple frames from optical camera is used for detection of moving boundary and predict the future position of the boundary. The algorithm [8] follows few steps

(i) NUI detection algorithm [11]:

NUI detection algorithm is applied for each frame in ROI. First normalized image pixels are taken and temporary dark map is computed using non-linear sigmoid function. Then, using optimal thresholding, NUI candidate is computed over the images.

(ii) Active contour Detection[11]:

NUI area can be deformed and irrigular in nature and it is tough to find out the area of NUI. To get uniform area, active contour method is used.

(iii) Boundary point extraction:

To detect the movement of the NUI, boundary points of the computed active contour is extracted. 5% of those computed points are taken for shake of fast computation compromising the accuracy.

(iv) Hungarian-Kalman estimation[11]:

The nature of the ROI can be arbitrary in shape and the boundary points can be overlapped during tracking. Kalman filter is used for tracking the points on the boundary assuming the points are moving linear speed with zero or unform accelaration. In this process, some points may added or some points may obliterate. To find out, best tracked-measured coupling, Hungarian algorithm is used during Kalman filter correction step to get higher confidance matched prrior points and postrior points

4 RESULTS AND DISCUSSION

Current study expedites an algorithm to diagnose faults on solar cell primarily due to NUI. This study is a calibrated response, which has been performed on single solar cell within a module. Figure 3 depicts the concerned experimental setup consists of SPV module, optical imaging system, thermal imaging system and processing and video capturing unit. The captured videos are segregated into the frames which have been cropped according to the region of interest (ROI). Next, the calibrated condition made of NUI and un-wanted object (dust) on SPV module is constructed, to check the viability of the algorithm in Fig. 4. Following figure exhibits the constructed hotspot region on those co-ordinates of the SPV, which has been identified at different altitudes by on-board thermal imaging camera of an unmanned aerial vehicle (UAV). This camera is employed to collect sequential frames from the recorded video. From those frames, the non-uniform heat-map images are obtained by thresholding the intensity. According to the figure 7, NUI detection algorithm (refer section 3.2) alone is unable to differentiate between shadow and dust on a SPV module. In order to segregate between NUI and dust particles, the results of CIELAB B-plane output is carried out in fig. 6.

The un-wanted objects are mostly of different colour than the background, utilizing this phenomenon B-frame clearly segments the dust from the background PV module.

As the NUI is a temporary phenomenon and changes co-ordinated according to the positioning of the sun and cloud, a tracking methodology is required to avert performance degradation. Multiple frames are used for this purpose which has been successfully monitored for a single PV module in figure 8 [11]. It is found that the frames detect NUI co-ordinates according to the time and spatial domain. Alternatively, figure 9 shows the output performance of active contour algorithm over an arbitrary shaped single NUI detected frame.

Next, the NUI tracking can be further improvised by implementing the Hungarian-Kalman filter algorithm as shown in fig. 10. Here the green points show the priori points before correction by Kalman filter and red points indicate the postiori points corrected with Hungarian optimization.

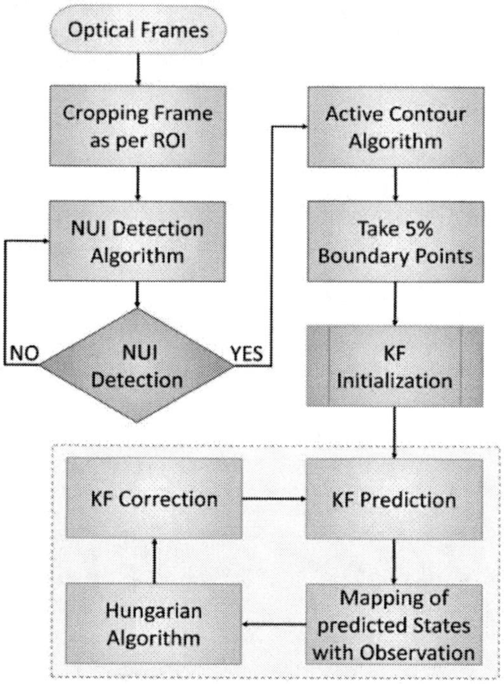

Figure 2: Dynamic arbitrary shaped NUI boundary detection using Hungarian Kalman filter [11]

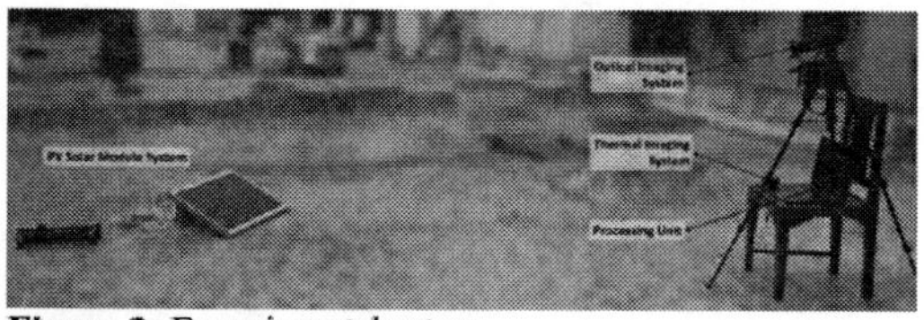

Figure 3: Experimental set-up

Figure 4: PV solar panel model frame for detection of NUI and un-wanted object

Figure 5: Heat map images shows hotspots on the solar cell from different distance.

Figure 6: Un-wanted object detection and contour detection

Figure 7: Non-uniform illumination detection

Figure 8: NUI detection and movement over PV module [12]

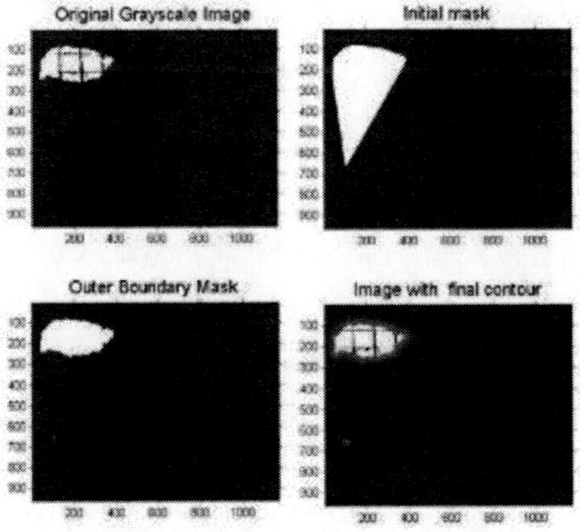

Figure 9: Active contour on single NUI pattern [11]

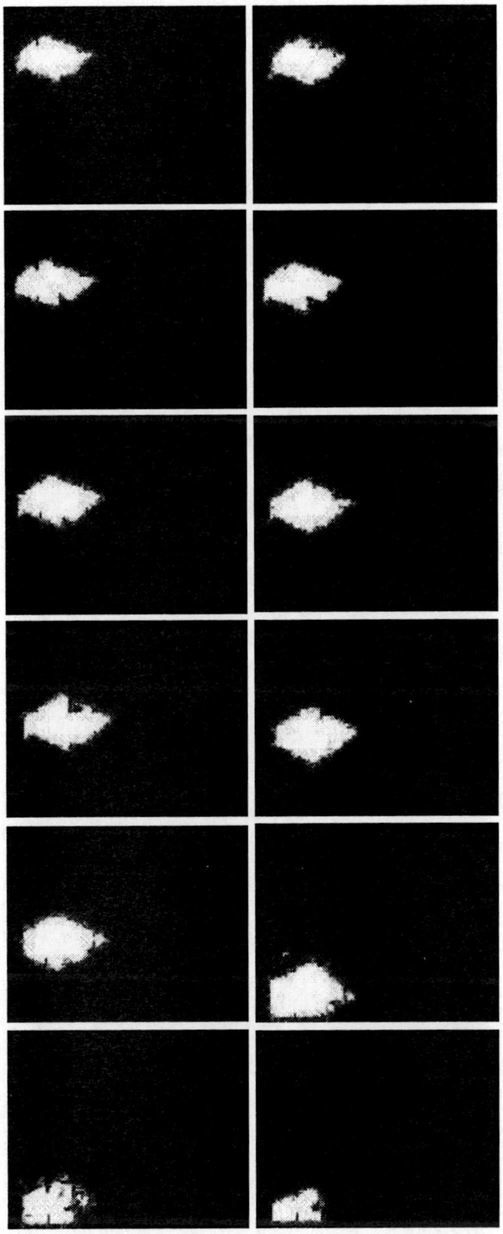

Figure 10: NUI tracking and position estimation using Hungarian-Kalman filter [11]

5 CONCLUSIONS

Effective operation and maintenance is an integral part of PV solar farm to obtain the maximum efficiency. Since the solar field is a huge area for surveillance, UAV and mobile aerial platforms [16,17] (aerostat, kite, airship etc.) mounted with imaging instruments(optical camera, thermal camera) are most prominent to monitor the entire field instead of using dense sensor network. Current work describes the different faults (NUI due to cloud, sun movement, high rise, dust etc.) which are potent to disrupt the optimized power generation on a ultra large scale solar PV farm and proposes an algorithm to differentiate the faults using image processing techniques in real time by mounted cameras in a calibrated scale. Further it also tracks the NUI coordinates to avert performance degradation using multiple frames and Hungarian-Kalman filter. On upon tracking a reconfiguration switching network can be enabled to prevent permanent degradation due to hotspots. Overall this study proposes a multiprone fault diagnosis and system health monitoring technique to generate maximum power and to increase the relaibility of ultra large scale solar PV farm sidelining the conventional sensor network.

5 REFERENCES

[1] Benemann, Joachim, et al. *"Building-integrated PV modules"* Solar Energy Materials and Solar Cells 67.1 (2001): 345-354.

[2] http://publications.naturalengland.org.uk/file/6000213410447360 (accessed 09/16/2017).

[3] Patnaik, B., et al. *"Reconfiguration strategy for optimization of solar photovoltaic array under non-uniform illumination conditions."* 37th IEEE Photovoltaic Specialists Conference (PVSC), 2011.

[4] Prieto, Miguel J., et al. *"Development of a wireless sensor network for individual monitoring of panels in a photovoltaic plant."* Sensors 14.2 (2014): 2379-2396.

[5] García, Miguel, et al. *"Observed degradation in photovoltaic plants affected by hot-spots."* Progress in Photovoltaics: Research and applications 22.12 (2014): 1292-1301.

[6] Tyutyundzhiev, Nicolay, et al. *"Advanced PV modules inspection using multirotor UAV."* 31st European Photovoltaic Solar Energy Conference and Exhibition, Hambug, 2015.

[7] Quater, Paolo Bellezza, et al. *"Light unmanned aerial vehicles (UAVs) for cooperative inspection of PV plants."* IEEE Journal of Photovoltaics 4.4 (2014): 1107-1113.

[8] Grimaccia, F., et al. *"Planning for PV plant performance monitoring by means of unmanned aerial systems (UAS)."* International Journal of Energy and Environmental Engineering 6.1 (2015): 47-54.

[9] Leva, S., M. Aghaei, and F. Grimaccia. *"PV power plant inspection by UAS: Correlation between altitude and detection of defects on PV modules."* 15th International Conference on Environment and Electrical Engineering (IEEE EEEIC), 2015.

[10] Aghaei, Mohammadreza, et al. *"Innovative automated control system for PV fields inspection and remote control."* IEEE Transactions on Industrial Electronics 62.11 (2015): 7287-7296.

[11] Arosh, S, et al. *"Dynamic Arbitrary Shape Boundary Detection and Tracking Using Hungarian Kalman Filter"* 2016 IEEE Industrial Electronics and Applications (IEACon), November 2016.

[12] Ghosh K., et al. *"A novel method to predict Non-uniform illumination pattern on a large-scale rooftop SPV array."* 2016 First International Conference on Sustainable Green Buildings and Communities (SGBC), December 2016.

[13] https://in.mathworks.com/help/vision/ref/colorspaceconversion.html#bqdn5yt-1(accessed 09/16/2017).

[14] Leon, Katherine, et al. *"Color measurement in L∗a∗b∗ units from RGB digital images."* Food research international 39.10 (2006): 1084-1091.

[15] Schwarz, Michael W., William B. Cowan, and John C. Beatty. *"An experimental comparison of RGB, YIQ, LAB, HSV, and opponent color models."* ACM Transactions on Graphics (TOG) 6.2 (1987): 123-158.

[16] Ghosh K., et al. *" Solar PV Integrated Lighter-Than-Air Platform (LTAP) for Airborne Power Generation under Variable Wind Gust."* 31st European Photovoltaic Solar Energy Conference and Exhibition, Hambug, 2015.

[17] Ghosh K., et al. *" Solar PV Integrated Lighter-Than-Air Platform (LTAP) for Airborne Power Generation."* 29th European Photovoltaic Solar Energy Conference and Exhibition, Amsterdam, 2014.

A QUANTITATIVE STUDY OF VARIABLE ORIENTATION MEHODS ENHANCING SOLAR POWER GENERATION ON TETHERED AEROSTATS

Sowmya Gupta[1], Siddhartha P. Duttagupta[2], Mira Mitra[3], Leena Vachhani[4]
[1]Centre for Research in Nanotechnology & Science, [2]Department of Electrical Engineering
[3]Department of Aerospace Engineering [4]Systems and Control Engineering
[1, 2, 3, 4]Indian Institute of Technology Bombay, Powai, Mumbai 400076 India.

ABSTRACT: Attitude stabilization of Solar Powered Lighter Than Air Platforms is required for stable solar power generation. The instability in longitudinal and lateral dynamics is primarily due to wind turbulence or internal disturbance of the system. Such instability leads to variations in the generated solar power. Attitude control is achieved by manipulating the tether tension of the actuated winches. Location of a confluence point and number of tether segments are key factors impacting the control of the solar powered aerostats. For maximum power generation, the system has to be rotated to optimum angles at regular intervals of time. In this work, experiments are conducted every hour during the day, for calculating the impact of rotation angles on the solar power generated on tethered aerostats. Based on the experiments, an optimum constant panel rotation angle of -30 degree in East-West direction, is suggested for maximum power generation. It is found experimentally that average power generated is 16.88% larger when panel is rotated once every two hours and additionally, it is 19.74% larger when panel is rotated once every hour compared to the average power generated from the constant angle of -30 degree (E-W direction) .

Keywords: Stand alone PV systems, Stability, Lighter than air systems, Solar powered tethered aerostat.

1 INTRODUCTION

Integration of flexible solar photovoltaic (SPV) module technologies are found to be most viable and provide to be an effective solution to harness solar power from mobile aerial platforms. These green energy power generation units can find applications as stand alone power source for remote area locations, a portable green energy source,surveillance of sensitive areas, mapping of the earth surface for resources and minerals . These systems, of aerodynamic shape , filled with a buoyant gas like helium/hydrogen can be tethered to ground for limited mobility. The advantage being, easy deploy-ability, no fuel consumption and the platform that can remain afloat for long duration of trials.

Such phenomenon requires continuous steady and stable positioning of SPV integrated aerial platform. Present work aims to quantify power generation at variable tilt angles in order to formulate the robust algorithm to constraint the positioning of the *Solar Powered Lighter Than Air Platform* (SPLTAP) within specified limits.

The SPLTAP experiences pitching, rolling and yawing oscillations in presence of the winds. The disturbance to the system are altitude dependent winds. These winds are superimposed by stochastic turbulent gusts which are characterized by scale lengths and spectra adapted from Von Korman model [1]. Among the oscillations of SPLTAP, combined pitching and rolling oscillations leads to dynamic tilt angles of the mounted SPV modules. This invariably leads to generation of variable incident solar flux (Watts/m^2) on those tilt angles. In this work first, the incident solar flux is measured at variable tilt angles. Further, power generation by SPV modules at variable tilt angles are measured and accordingly boundary limit is defined. This boundary limit further aids in developing the control algorithm for maximum solar power generation on SPLTAP.

2 SOLAR POWERED LIGHTER THAN AIR PLATFORM

2.1 System Layout

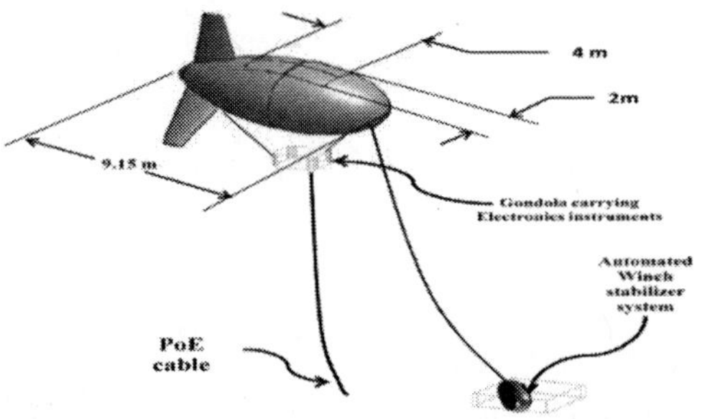

Fig. 1: System layout of Solar Powered Tethered Aerostat
Acknowledgement: K. Ghosh , Research Scholar, I.I.T Bombay Powai, Mumbai 400076, India

Fig. 1 depicts the system layout of SPLTAP. The system consists of flexible SPV modules mounted on the aerostat . The gondola carries electronic circuitry for data acquisition. With the dimensions of the aerostat shown in Fig. 1, this system is capable of generating 1.2 kW$_p$ solar power with each panel rating 300 W$_p$.

The system is stabilized in position and orientation by a plurality of tethers controlled by automated winches. Depending upon the degrees of freedom to be controlled, the system can be bi-tethered or tri-tethered.

2.2 Literature Survey

C. M. Lambert et al. ([2], [3]) has worked on position stabilization of tethered aerostats as a telescopic receiver C. Vermillion et al. [4] have presented methods for attitude and altitude stabilization of tethered . Load cells and IMU sensors are incorporated into the system for tether tension and tilt measurement of the system.

SPLTAP can be oriented at a desired orientation by setting commands for a predefined set-point [4] viz : *Pitch setpoint (θ_{sp}), Roll setpoint (Φ_{sp}) and Altitude setpoint (Z_{sp}).* The following equations govern the stabilization of SPLTAP derived from the works of C. Vermillion et al. [4].

$$U_{front,left} = u_{ave} + \Delta u_{forward/aft} + \Delta u_{left/right} \quad (1)$$

$$U_{front,right} = u_{ave} + \Delta u_{forward/aft} - \Delta u_{left/right} \quad (2)$$

$$U_{aft,left} = u_{ave} - \Delta u_{forward/aft} + \Delta u_{left/right} \quad (3)$$

$$U_{aft,right} = u_{ave} - \Delta u_{forward/aft} + \Delta u_{left/right} \quad (4)$$

Commands to the tethered aerostat governed by the above eq. (1)-(4). Pitch control and roll control is achieved by tracking with the aim of maximum power generation onboard.

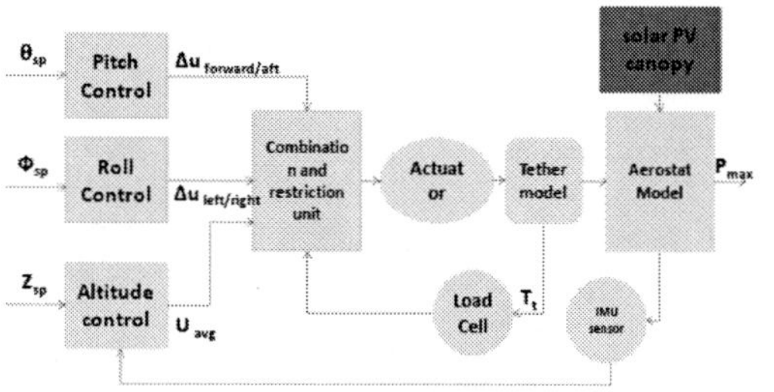

Fig. 2. Attitude and altitude stabilization of tethered aerostat

3 SOLAR POWER GENERATION UNDER WIND INDUCED DISTURBANCE

3.1 System Equations

Attitude destabilization of SPLTAP due to incident wind results in solar power fluctuations. In this section the impact of wind disturbance on the generated power is stated [5].

The incident solar irradiation (Watts/m^2) on the platform is given by the following equation :

$$I = I_b \cos\theta \quad (5)$$

The solar radiation incident on the platform is variable in nature and depends on the orientation of the system. The angle cos Θ is described by the following equation

$$\begin{aligned}\cos(\theta) = {} & \cos(\beta)\sin(\delta)\sin(\phi) \\ & + \cos(\delta)\cos(\Phi)\cos(\omega) \\ & - \cos(\delta)\sin(\omega)\sin(\beta)\sin(\gamma) \\ & + \sin(\beta)\cos(\gamma)\sin(\delta)\cos(\Phi) \\ & - \cos(\delta)\cos(\omega)\sin(\phi)\end{aligned} \quad (6)$$

Symbols meaning :

(I_b): Radiation Intensity on SPLTAP under no wind

(I) : Radiation intensity on LTAP under wind induced disturbance

(Θ): Angle between incident solar radiation and normal of SPV modules integrated on LTAP

(β): Aperture tilt angle,

(Φ): Latitude

(ω): Solar hour angle

(δ): Declination angle of sun

(Υ): Aperture azimuth angle

In the subsequent sections, simulations have being carried out to study the impact of variable tilts on the generated solar power .

3.2 Simulation Results

Results have been generated by simulating a a 64 Watts rated polycrystalline Si module for SPLTAP. The voltage is varied in steps from 0 to 16 volts and the corresponding current and power is calculated.

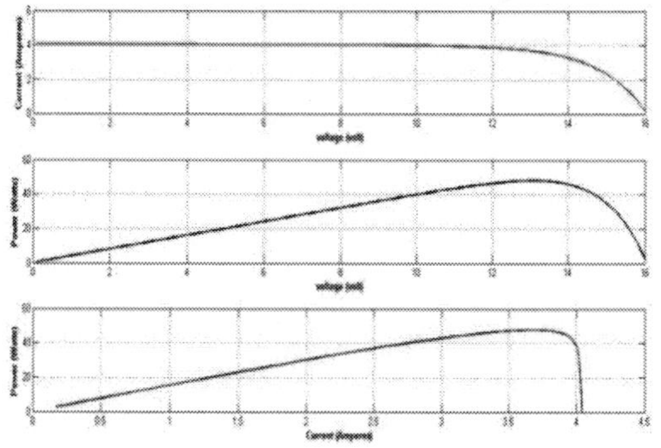

Fig 3. Solar power generation under no wind conditions

Results have been generated by simulating a a 64 Watts rated polycrystalline Si module for SPLTAP. The voltage is varied in steps from 0 volts to 16 volts and the corresponding current and power is calculated.

The figure below depicts the impact of the variable platform tilt on the generated power. The generated power is fluctuating in nature. The power variation is in the range of 30.5 Watts to 31.5 Watts.

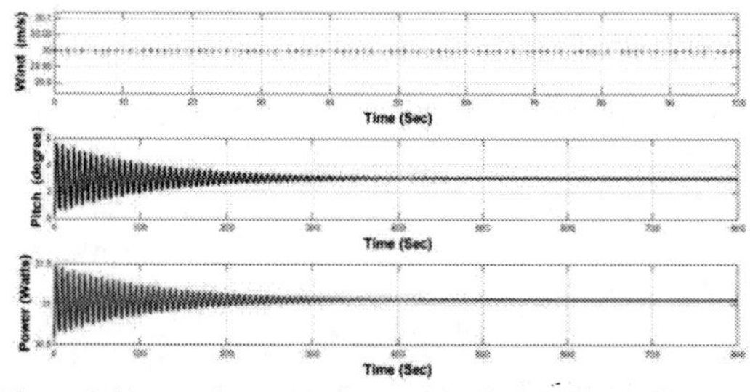

Fig 4. Solar power generation under wind induced tilt.

The generated oscillatory power cannot be used in conventional MPPT controllers like perturb and observe, hill climbing or constant voltage. Hence there is a need for stabilizing these power fluctuations of SPLTAP for load distribution.

For generating maximum power onboard, the platform needs to be tilted at an optimum direction by linearly varying tilt angles viz pitch (Θ), roll (Φ) and yaw (ψ).

Surface plots has been obtained to illustrate the variation of incident solar radiation when the pitch is varied linearly from -90^0 to $+90^0$ and the solar inclination angle is varied from 0^0 to 100^0.

Following equations calculate the solar power incident on the tilted platform :

$$S_{horizontal} = S_{incident} \sin\alpha \quad (7)$$

$$S_{module} = S_{incident} \sin(\alpha + \beta) \quad (8)$$

where

α :	Solar tilt angle
β :	Tilt angle of the platform
S_{module} :	Radiation incident on a tilted surface
$S_{horizontal}$:	Radiation on horizontal surface
$S_{incident}$:	Radiation perpendicular to the sun

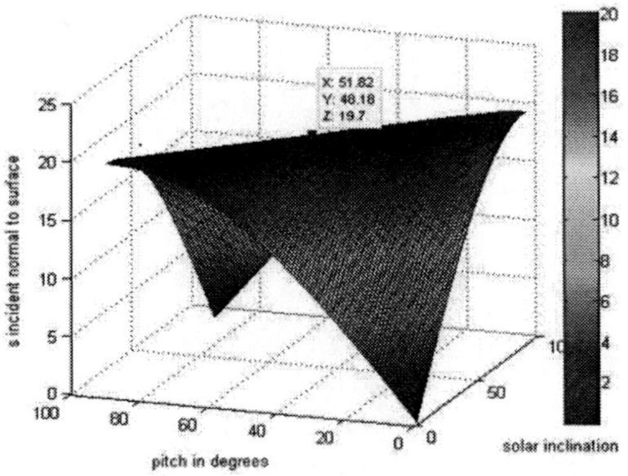

Fig 5. Solar power generation under linearly varying tilt for maximum power generation by tracking.

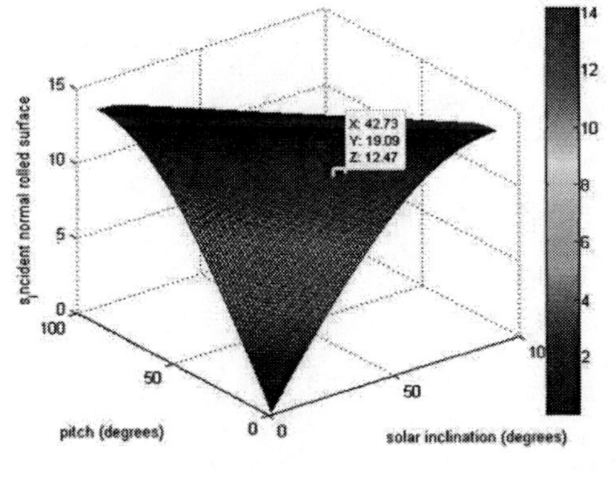

Fig 6. Solar power generation under linearly varying pitch and 45^0 roll surface.

Fig. 5 and 6 depicts the variation of incident solar radiation as a function of pitch , pitch plus 45^0 roll platform respectively. The magnitude plot of solar flux (S_{flux}Watts/m^2) differ drastically with each orientation coupling (viz: pitch , pitch+roll, pitch+roll+yaw dynamics) for the SPLTAP.A need to orient the SPLTAP in an optimum direction at which maximum solar energy can be harnessed is identified.

3.3 Experimental Results

Experiment test was conducted to study the impact of tilt on the generated power. A platform with 7 Watts rated solar panel is angularly varied from angles : -90^0 to $+90^0$ in 15^0 step variations. With the platform orientation in E-W direction , the tilt angle is varied on full scale for every hour and corresponding readings are recorded for the entire. Average power computations were carried out for each tilt angles and a tilt angle at which maximum average power is generated was identified.

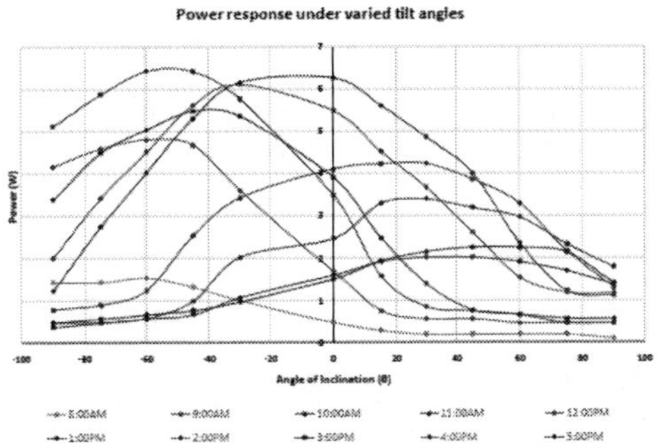

Fig. 7 Power response for a 7 watt rated SPV integrated platform under variable tilt.

Fig. 7 depicts the power computations for the solar platform under variable tilt angles. Fig. 8 depicts the optimum angle of the platform for maximum power generation.

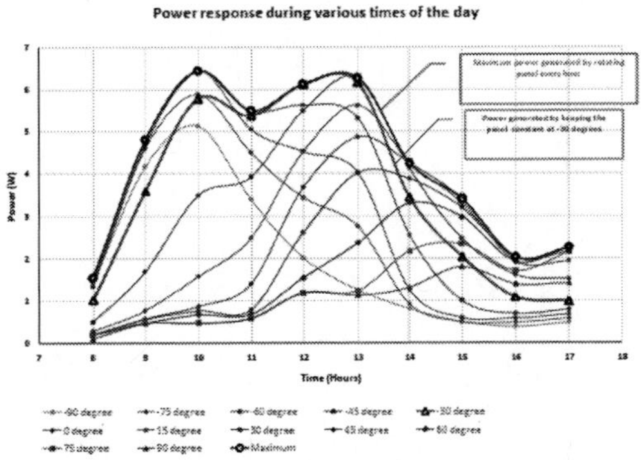

Fig. 8 Power response for a 7 watt rated SPV integrated LTAP depicting the maximum power generation.

By analyzing the experimental results, we can conclude that on a particular day, a tilt of *-30^0 (E-W)* direction with respect to the horizontal (0^0) generates maximum average power . However this is to be noted that the optimum tilt is variable in nature and will be different depending upon the seasons and latitude.

As compared to -30^0 orientation, generated power increases on changing the tilt on hourly basis by *19.74 %*,while an increase of *16.88 %* is reported on changing the tilt angles once in every two hours.

Thus there is a need to tilt the solar powered platform continuously by means of actuated tethers for maximum power generation.

4 CONCLUSIONS

a) The impact of wind on the power generating capability of SPLTAP has been studied.

b) Surface plots depicting the variation of tilt and roll angles of the platform on the incident solar radiation has been obtained.

c) Power response under variable tilt (-90^0 to $+90^0$) is generated by experiments and optimum tilt angle for maximum solar power generation has been calculated.

d) Experimental results suggest hourly tilt variations of SPLTAP for maximum power generation .

5 REFERANCES

[1] B. Etkin, *Dynamics of atmospheric flight*. Wiley, 1972.
[2] C. Lambert, M. Nahon, and D. Chalmers, "Implementation of an Aerostat Positioning System With Cable Control," *IEEE/ASME Trans. Mechatronics*, vol. 12, no. 1, pp. 32–40, Feb. 2007.
[3] C. Lambert and M. Nahon, "Stability Analysis of a Tethered Aerostat," *AIAA*, vol. 40, no. 4, pp. 705 715, 2003.
[4] A. D. G. Christopher R. Vermillion, Benjamin W. Glass, "Systems and methods for attitude control of tethered aerostats," 2012.
[5] C. S. Solanki, *Solar photovoltaics : fundamentals, technologies and applications*. .

ON THE CALCULATION OF THE STC POWER OF PV GENERATORS BY USING TYPICAL MONITORING SYSTEM DATA

M. Muñoz*, M. García, I. de la Parra, J. Marcos, L. Marroyo
Department of Electrical and Electronic Engineering
& Institute of Smart Cities
PUBLIC UNIVERSITY OF NAVARRE
Campus Arrosadía
31006 Pamplona, Spain
Phone: +34 948 168 934
Email: mikel.munoz@unavarra.es

ABSTRACT: The properly in-field characterization of the power at Standard Test Conditions, P_{STC}, of PV generators is becoming increasingly important in order to evaluate their performance and its evolution in time. Within the state of art, the P_{STC} characterization procedures of PV arrays are mainly based on I-V curve measurements or P_{DC} measurements performed by precision wattmeters. Those characterizations are usually carried out during discrete measurement campaigns, which does not allow a continuous tracking of the P_{STC} evolution. In this paper a new P_{STC} characterization procedure is proposed which is based on the DC power measurements performed by the own PV inverters connected to the PV arrays. This procedure enables an automatic and continuous calculation of the P_{STC}, which allows to observe its evolution and to detect possible anomalous trends, premature degradations, etc. The procedure has been validated in several PV generators of the large-scale Amareleja PV Plant (45.6 MWp). As showed in this paper, by using several day data and applying the adequate filters, a high accuracy in the P_{STC} calculation can be achieved, a similar accuracy to that obtained by using precision wattmeter measurements.

Keywords: STC power, PV arrays characterization, PV power, PV performance.

1. INTRODUCTION

The sizing of photovoltaic (PV) plants has been increasing from the very first systems in the 80's, going from around 100kWp up to several hundreds of MWp at present. Those plants are composed of several PV generators, being a PV generator the set of modules connected to the same PV inverter. The number and size of those PV generators make more difficult to evaluate the performance and efficiency of the whole system. Thus, a properly in-field characterization of the power at Standard Test Conditions, P_{STC}, of those generators is becoming increasingly important.

Within the state of art, different standards which discuss the photovoltaic systems characterization [1]–[4] and show some particular experience [5], [6] can be found. Regarding to the P_{STC} characterization, one procedure that provides excellent results was proposed by Martínez [7]. Such procedure is based on high accuracy DC power measurement by means of a wattmeter besides in-plane irradiance and module temperature measurements through refence modules. However, although the PV plants are each time more and better monitored, is not usual the use of precision wattmeters to record data at the different levels of the PV installation, mainly due to economical reasons.

In this paper, a new P_{STC} calculation procedure is proposed, based on the data provided by the PV inverters, instead of those from external wattmeters. This procedure enables an automatic and continuous calculation of the P_{STC}, which allows to observe its evolution and to detect possible anomalous trends, premature degradations, etc. The analysis, observations and conclusions, along with guidelines to gain in accuracy are shown at the present work.

2. EXPERIMENTAL SET UP

The Amareleja PV plant [8], property of Acciona Energía S.A, has been used for the implementation and validation of this procedure. Particularly, 8 of the PV generators installed in this PV plant (Fig. 2.1(a)) has been considered. Each PV generator has a peak power of 168kWp and is composed of 9 azimuthal trackers. One reference module is attached to one of the trackers in order to measure the operating conditions, i.e. in-plane irradiance, G, and module temperature, T_C (Fig. 2.1(b)). Each generator is connected to a 125 kW inverter manufactured by Ingeteam S.A. (Fig. 2.1(c)). Those inverters provide the DC power measurement of the PV generator, which is recorded by the SCADA of the PV plant as a 10 minutes average.

Fig. 2.1 Amareleja PV plant: (a) PV generators. (b) Tracker with reference module. (c) PV inverter.

3. THE P_{STC} CALCULATION PROCEDURE

The P_{STC} calculation procedure proposed is based on the regression method described in [7], which requires the use of a precision wattmeter in order to make available high temporal resolution measurements of the P_{DC} delivered by a PV generator during a clear day. The procedure here proposed uses only the P_{DC} data provided by the PV inverters, allowing a lower temporal resolution for the power and reference conditions measurements. Therefore, there is no need of including additional measuring equipment (wattmeter) to the PV generator. In order to get a good accuracy on the P_{STC} calculation, some restrictions (filters) should be applied to the data, which makes necessary to consider the data from several days. The details about the proposed method are described below.

- P_{STC} calculation:

Once a time series of $[P_{DC}, G, T_C]$ values is obtained, a way of deriving P_{STC} consists of, first, correcting the actual P_{DC} values to 25°C, by means of Equation (1):

$$P_{DC}(G, 25°C) = \frac{P_{DC}}{[1 + \gamma(T_C - 25)]} \quad (1)$$

And second, to obtain the best fit for Equation (2) [7]:

$$P_{DC}(G, 25°C) = \frac{P_{STC}}{1000} G \quad (2)$$

Better results can be obtained by adding a term to the previous model, in oreder to include the PV module efficiency variations due to the irradiance:

$$P_{DC}(G, 25°C) = \frac{P_{DC}}{[1 + \gamma(T_C - 25)]} \cdot \left[a + b * \frac{G}{1000} + c \cdot \ln\left(\frac{G}{1000}\right)\right] \quad (3)$$

Where a, b, c are empirical parameters that should be adjusted by means of experimental measurements.

- Accuracy of the PV inverter measurements:

In order to get a good accuracy in the P_{STC} calculation, the accuracy of the PV inverter measurements should be also tested. For this purpose, a precision wattmeter was installed on two different PV generators of the Amareleja PV plant. The differences between the measurements from the PV inverter and the wattmeter are shown in **Table 1**. For P_{DC} values near the nominal power of the PV inverter (125 kW), the mean and standard deviation of the differences are less than 1%.

Table 1 Average and standard deviation of the difference ($P_{WAT} - P_{PV\_INV}$).

PV Generator	Error		Type of day
	AVG [kW]	σ^2 [kW]	
GEN1	-0,23	0,39	Sunny
GEN2	-0,19	0,28	Sun and clouds

- Temporal resolution of data records:

The procedure described in [7] uses a 60-s sampling time but the minimum sampling time registered by the PV inverters at Amareleja PV plant is 10-min. Despite this, a carefully analysis carried out by using an external wattmeter showed that recording data every 10 minutes provides similar results than recording data every minute if some restrictions (discussed below) are taken into account. The same cannot be said if only one-hour data records are available, since this causes a noticeable precision loss.

- Data filtering:

One of the most critical points of the procedure is to decide which data are valid to include in the linear regression and which should be discarded. Martínez et al. [7] proposes several restrictions which could be grouped into three: values of in-plane irradiance greater than 800 W/m² are needed, the PV system must be shadow-free and, finally, the inverter must not be saturated. However, when more restrictions better results can be obtained. The following filters have been implemented:

- Elimination of data corresponding to some possible measurement problem:

$$P_{AC} < 0 \quad (4)$$
$$G < 3 \quad (5)$$
$$P_{DC}[G, 25°C] < 10 \quad (6)$$

- Data corresponding to low irradiance values are rejected:

$$G < 800W/m^2 \quad (7)$$

- Extreme irradiance values are also rejected in order to avoid the irradiance increase situations due to cloud enhancement.

$$G > 1050W/m^2 \quad (8)$$

- Elimination of data corresponding to situations with variable irradiance. For that purpose, it is observed whether the difference in radiation with the anterior and posterior moment has exceeded 5%, in absolute value. If this occurs, the data is removed

$$\frac{|G_t - G_{t-1}|}{G_t} 100 > 5 \quad (9)$$
$$\frac{|G_t - G_{t+1}|}{G_t} 100 > 5 \quad (10)$$

- Elimination of data corresponding to inverter saturation situations. Usually, PV field power installed is bigger than the nominal PV inverter. Therefore, the PV could produce more power than the inverter nominal power. Then, the inverters limit the output AC power to their nominal value. In order to ensure those data are not taken into account, a 0.95 coefficient is applied to the limiting power of the PV inverter.

$$P_{AC}(t) > 0.95 * P_{Lim}(t) \quad (11)$$

- Only data from clear days are considered. A cloudy day may imply instability. Thus, the theoretical clear day irradiance is calculated, G_{clear_day} [9]. If the measured irradiance is less than 95% of the corresponding to the clear day the data is eliminated.

$$G < G_{clear\_day} \cdot 0.95 \qquad (12)$$

- In order to avoid wrong values due to installation failures (like blown fuses, etc.), another filter is applied to the data. On the basis of the linear regression calculated for the previous day, a +5% upper limit and a -5% lower limit around that regression are created. Every point out of those limits is removed and a new regression is done.

$$a_{dow} = (P_{STC_{yesterday}} * (1 - 0.05)/1000) \qquad (13)$$

$$a_{up} = (P_{STC_{yesterday}} * (1 + 0.05)/1000) \qquad (14)$$

$$\frac{P_{DC(G,25)}}{G} > a_{up} \text{ ó } \frac{P_{DC(G,25)}}{G} < a_{dow} \qquad (15)$$

Once all those restrictions have been applied, the number of valid data is considerably reduced. Therefore, it necessary to consider the data of several days in order to obtain a good fit in the linear regression. The number of days needed depends on climatology, seasonality, etc. Therefore, in order to automate the process, it should be determined how many days must be selected. Different tests have been performed using data from 15 and 30 days. It has been observed that, although 15 days data offer good results, the use of 30 days data slightly increases the accuracy of the procedure. Fig. 3.1 shows the difference between the P_{STC} calculation using 15 (dotted line) and 30 (solid line) days for 2 of the PV generators (GEN1 (blue) and GEN2 (orange)) throughout one year.

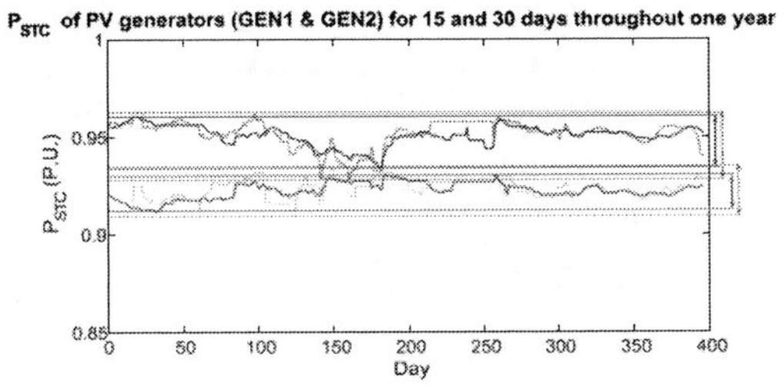

Fig. 3.1 P_{STC} of PV generators (GEN1 and GEN2) for 15 (black) and 30 (red) days throughout one year

Fig. 3.2 shows the proposed filters for the data of the 30 days prior to the calculation of P_{STC} of one PV generator (GEN2), and Fig. 3.3 shows the regression performed with those data.

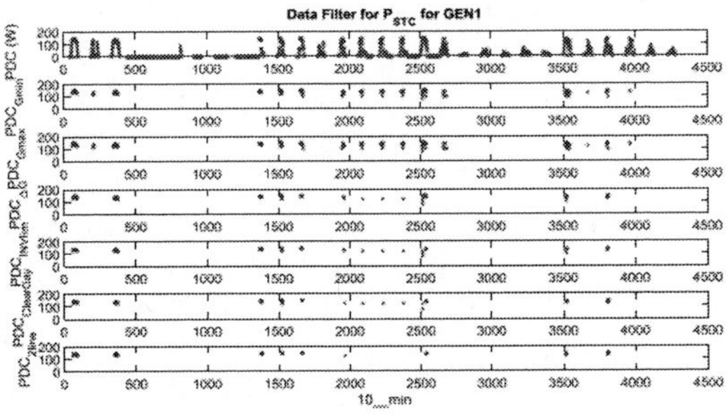

Fig. 3.2 Filter data process for GEN1 with 30 days

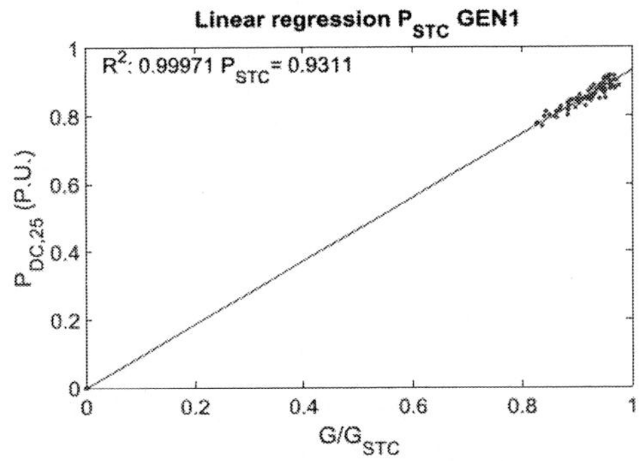

Fig. 3.3 Linear regression for GEN1

4. PROCEDURE VALIDATION

By using the previously described regression method, the P_{STC} of 8 PV generators of the Amareleja PV plant has been calculated daily along a whole year. The DC power output and operating conditions measured in those PV generators in the previous 30 days has been used as input data for the regressions. Fig. 4.1 shows P_{STC} calculation for those generators throughout one year and Table 2 shows the statistical results. As can be seen, the dispersion in the calculated values is around ±1%.

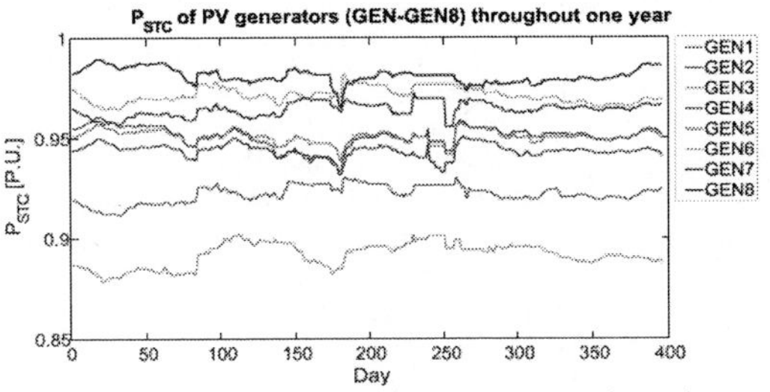

Fig. 4.1 P_{STC} calculation for GEN1-GEN8 throughout one year.

Table 2 Statistical results of 8 PV generators P_{SCT} calculations

GEN	P_{STC} AVG (P.U.)	P_{STC} σ^2 (P.U.)	GEN	P_{STC} AVG (P.U)	P_{STC} σ^2 (P.U.)
GEN1	0,970	0,0052	GEN5	0,955	0,0055
GEN2	0,924	0,0081	GEN6	0,893	0,0095
GEN3	0,976	0,0061	GEN7	0,966	0,0049
GEN4	0,951	0,0065	GEN8	0,954	0,0049

Likewise, the P_{STC} values obtained by the proposed method has been c ompared with the values obtained during a measurement campaign carried out during a clear day. In that measurement campaign, a precision wattmeter with 60-s sampling time configuration was used. Fig. 4.2 shows a histogram with P_{STC} values obtained by the proposed method and the red dotted line the P_{STC} calculated using a precision wattmeter for the GEN2.

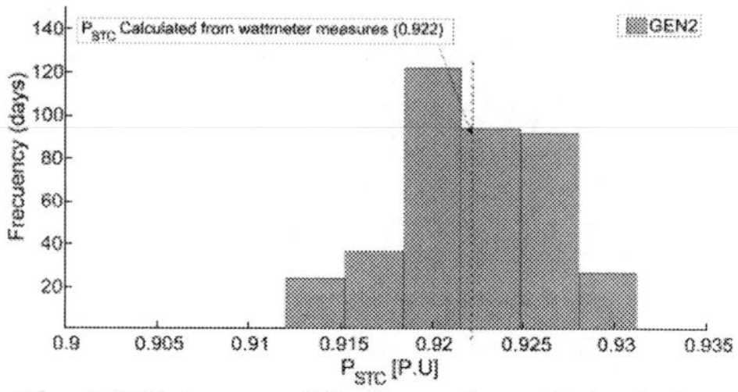

Fig. 4.2 Histogram of the P_{STC} values obtained along a year by the proposed method for the GEN2

The difference between the annual mean value of the P_{STC} (0.924) obtained through the proposed procedure and the wattmeter measurement (0.922) is around 0.2%.

5. CONCLUSIONS

In view of the experience obtained from the Amareleja PV plant, the PV inverters measurements can be used to obtain an accurate P_{STC} characterization of the PV generators of a PV plant. At this PV plant, the inverters accuracy is enough (attend to manufacture recommendations) and the sampling time provided by manufactured is sufficient. In any case, it would be interesting to check that PV inverters supplier for other manufactures also meet the commented requirements.

A procedure based on linear regression using the DC power measurements provided by the PV inverters and the in-plane irradiance and the module temperature registered by reference modules has been developed. Several filters are necessary to eliminate undesired data. The number of days required to obtain an accurate linear regression depends of climatology. At the Amareleja PV plant, 30 days are more than enough.

The dispersion in the obtained P_{STC} values is around ±1% and the difference with the values obtained by a precision wattmeter is around ±0.2%.

6. ACKNOWLEDGEMENTS

This work has been partially financed by the Seventh Framework Programme of the European Commission with the project PVCROPS (Photovoltaic Cost R€duction, Reliability, Operational Performance, Prediction and Simulation – Grant Agreement no: 308468) and by the Spanish Ministry of Economy and Competitiveness under Grant DPI2013-42853-R.

REFERENCES

[1] I. Electrotechnical Commission, "IEC Standard-61829. Crystalline Silicon Photovoltaic (PV) Array–On-site Measurements of I–V Characteristics." 1995.

[2] I. Electrotechnical Commission, "IEC Standard-60904-1. Photovoltaic Devices. Part 1: Measurements of Photovoltaic Current–Voltage Characteristics." 2006.

[3] I. Electrotechnical Commission., "IEC Standard-60891. Photovoltaic Devices. Procedures for Temperature and Irradiance Corrections to Measured I–V Characteristics." 2009.

[4] G. Blaesser and D. Munro, "Guidelines for the assessment of photovoltaic plants. Document C: Initial and periodic tests on photovoltaic plants. Joint Research Centre of the European Communities,Ispra Establishment," 1995.

[5] F. Spertino, A. Abete, and R. Napoli, "Experimental testing of grid-connected PV Systems with different power in order to assess the yearly energy production," in *European Photovoltaic Solar Energy Conference*, 2006, p. 2312–2315,.

[6] D. L. King, W. E. Boyson, and J. A. Kratochvill, "Photovoltaic Array Performance Model," *Sandia National Laboratories: Albuquerque, NM and Livermore, California*, no. December, 2004.

[7] F. Martínez-Moreno, E. Lorenzo, J. Muñoz, and R. Moretón, "On the testing of large PV arrays," pp. 100–105, 2011.

[8] Amareleja (ACCIONA), "Amareleja PV plant," 2008. [Online]. Available: http://www.acciona-energia.es/areas_actividad/solar_fotovoltaica/instalaciones/plantaamaraleja/planta-amaraleja.aspx.

[9] Rigollier, C., Bauer, O., Wald, L., 2000. On the clear sky model of the ESRA – European solar radiation atlas – with respect to the Heliosat method. Sol. Energy 68, 33–48.

HIGH-FIDELITY SOLAR POWER INCOME MODELING FOR SOLAR-ELECTRIC AIRCRAFT: DEVELOPMENT AND FLIGHT TEST BASED VERIFICATION

P. Oettershagen, R. Siegwart
Autonomous Systems Lab, Swiss Federal Institute of Technology Zurich (ETH Zurich)
Leonhardstrasse 21, 8092 Zurich
Switzerland

ABSTRACT: Solar power models are a crucial element of solar-powered aircraft design and performance analysis. During the conceptual design phase, their accuracy directly relates to the accuracy of the predicted performance metrics and thus the final design characteristics of the solar-powered aircraft. Likewise, during the operations phase of a solar-powered aircraft accurate solar power income models are required to predict and assess the solar power system performance. However, the existing literature on solar-powered aircraft uses highly simplified solar power models. This paper therefore, first, introduces a high-fidelity solar power model that takes into account the exact aircraft attitude, aircraft geometry, and physical effects such as temperature and the sun radiation's angle of incidence that influence the overall solar power system efficiency. Second, models that require a reduced set of input data and are thus more appropriate for the initial design phase of solar-powered aircraft are derived from the high-fidelity model. Third, the models are compared and verified against flight data from a 28-hour continuous day/night solar-powered flight. The results indicate that our high-fidelity model allows a prediction of the average solar power income with an error of less than 5% whereas previous models were only accurate to about 18%.
Keywords: Modeling, performance, solar aircraft, solar UAV

1 INTRODUCTION

1.1 Motivation

Solar-powered Unmanned Aerial Vehicles (UAVs) promise significantly increased flight endurance over conventional aircraft. This benefits applications such as large-scale disaster relief, border patrol or aerial inspection in remote areas [1]. Designing and testing solar-powered UAVs requires a solar power model which accurately predicts the amount of power P_{solar} collected by the aircraft's solar modules. More specifically, in the conceptual design phase, the solar power is used to predict the central aircraft performance parameters such as the flight endurance T_{endur} or, if perpetual flight is feasible, the excess time T_{exc} [2]. During the flight operations phase, the solar power model is integrated into path planning algorithms [3] to plan safe paths with sufficient solar power income and thus energetic margins. Inaccurate solar power models lead to sub-optimal solar aicraft designs or unsafe flight operations.

Figure 1: ETH Zurich's AtlantikSolar UAV, which holds the current world record in flight endurance for aircraft below 50kg mass, was designed with the solar power model presented in this paper. Here, it is operating during a 230km mission over the Arctic Ocean in July 2017.

This paper extends our previous research [4] and presents a high-fidelity solar power model for solar-eletric UAVs. The model was used to design ETH Zurich's AtlantikSolar UAV (Figure 1), a small-scale solar-powered aerial vehicle which has demonstrated an 81-hour continuous flight [2] that is the current *world record*[1] in flight endurance for all aircraft below 50kg total mass. This solar power modeling paper is released in conjunction with the aforementioned UAV design literature.

1.2 Contributions

To answer the need for higher-fidelity solar power income models during all stages of solar-powered UAV design, testing and operations, this paper contributes by

- Introducing the *Full Solar Power Model* (FM), a fully time-resolved high-fidelity solar power model that takes into account the exact aircraft attitude, aircraft geometry, and physical effects such as temperature and the sun radiation's angle of incidence that influence the overall solar power system efficiency.
- Deriving the *Conceptual Analysis Model* (CAM) and *Conceptual Design Model* (CDM) from the high-fidelity model. The two models are more appropriate for the initial design phase of solar-powered UAVs because they require less input data (e.g. certain technological aircraft parameters).
- Comparing and verifying all models against flight data from a 28-hour continuous solar-powered flight of the AtlantikSolar UAV. The systematic errors are retrieved for each model. The systematic errors can be used by the solar-powered UAV designer to correct the design estimates such that a more precise model-based performance prediction of solar-powered UAVs is possible even in the early design stages.

---

[1] A video of the flight is available at https://www.youtube.com/watch?v=8m4_NpTQn0E

The contributions of this report mean that a solar-powered UAV designer

- Can select the most appropriate model to use in a specific development stage (conceptual design, detailed flight-test based analysis or flight planning).
- Knows what physical effects (e.g. the angle of incidence dependence of the solar module efficiency or the solar module temperature) are driving the solar power system performance.
- Knows the prediction accuracy that can be expected from each of these models.

2 BACKGROUND

Stein [5] provides an excellent overview over the physical- and electrical-effects that should be included in accurate solar power models. However, the solar power models employed in the current solar-powered UAV design literature are more simple. For example, Shiau [6] considers only the average solar radiation over time. A second group of papers [7, 8, 9, 10] does model time effects that arise primarily due to the daily solar cycle. A common mathematical model for the instantaneously collected solar power also used in our previous work [10] is

$$P_{solar}^{nom} = I_{solar}(\varphi_{lat}, h, \delta, t, \vec{n}_{sm}) \cdot A_{sm} \cdot \eta_{sm} \cdot \eta_{mppt} \quad (1)$$

Here, $I_{solar}(\varphi_{lat}, h, \delta, t, \vec{n}_{sm})$ is the solar radiation on a unit $1\,m^2$ area that is modeled as described by Duffie [11]. It is a function of geographical latitude φ_{lat}, altitude h, current day-of-year δ, local time t and solar module normal vector $\vec{n}_{sm}$. For the conceptual design stage of a solar-powered UAV, the solar module area A_{sm}=const is mostly considered a horizontally-oriented area ($\vec{n}_{sm} = [0,0,-1]$ in an inertial aircraft-centered North-East-Down coordinate system) given that the exact orientation depends on the specific mission profile which is unknown until shortly before the flight operation. The solar module efficiency $\eta_{sm} = \eta_{sm}^{STC} \cdot \epsilon_{sm}^{cbr}$ only includes a wing camber efficiency reduction factor ϵ_{sm}^{cbr}, and the maximum power point tracker efficiency is η_{mppt}=const. While these assumptions are acceptable for the very early solar-powered UAV design stages, they do not yield sufficient accuracy for the later detailed design and analysis stages. As an example, consider that a solar-powered UAV shows less-than-expected power solar income during a flight test. The above model is not accurate enough to analyze the issue. Instead, we require a high-fidelity fully time-resolved solar-power model to detect, understand and mitigate these solar power system issues.

3 A HIGH-FIDELITY SOLAR POWER MODEL

This section introduces a fully time- and aircraft-state dependent solar power model — the *Full Solar Power Model* (FM). It allows to, first, more accurately predict aircraft performance in a specific mission (day-of-year δ, time of day t, latitude φ_{lat}, chosen flight path), and second, to detect, understand and mitigate solar power system anomalies. We also use the model to provide a categorization of physical effects that degrade solar module and thus UAV performance, resulting in *lessons learned* for future UAV designers.

3.1 Geometric Modelling

We first define an extended geometric representation to calculate the solar power income separately for each of the aircraft's non-horizontal solar modules. Note that we use the AtlantikSolar UAV (Figure 1 and Figure 2, with further details given in [2]) as an example for implementing the extended geometric model in this paper. For each of the N solar module-covered areas A_{sm}^i we define the orientation relative to the aircraft body frame B through the surface-normal vector

$$[\vec{n}_{sm}^i]^B = [\vec{n}_{sm}^i]^B (\Delta\phi_{dih}, \Delta\theta_{wing}, \Delta\theta_{A_{sm}^i}) \quad (2)$$

In lateral direction, the surface normal vector is determined by the wing dihedral angle $\Delta\phi_{dih}$. In longitudinal direction, due to the wing upper surface profile every A_{sm}^i is mounted at an additional solar cell pitch angle $\Delta\theta_{A_{sm}^i}$ with respect to the wing chord line. The relative pitch orientation between the aircraft longitudinal or x-axis (which is approximately aligned with the aircraft's Inertial Measurement Unit longitudinal axis) and the wing chord line is $\Delta\theta_{wing}$ and is added on top. Table 1 summarizes the numerical values chosen for AtlantikSolar. Using the aircraft attitude (Euler-angles roll ϕ, pitch θ and yaw ψ) we can then express the orientation of each A_{sm}^i in the inertial frame I through

$$[\vec{n}_{sm}^i]^I = R_B^I(\phi, \theta, \psi) \cdot [\vec{n}_{sm}^i]^B \quad (3)$$

where R_B^I is the transformation or direction cosine matrix from the aircraft body to the inertial reference frame. By rewriting Eq. (1) for all A_{sm}^i and performing the corresponding summation, Eq. (4) yields the total incoming solar power P_{solar}. One can furthermore separate the global solar irradiation I_{solar}^i into its direct component $I_{solar}^{i,dir}$ and diffuse component $I_{solar}^{i,diff}$ with corresponding efficiencies $\eta_{sm}^{i,dir}$ and $\eta_{sm}^{i,diff}$:

$$P_{solar} = \sum_i^N P_{solar}^i = \sum_i^N I_{solar}^i \cdot \eta_{sm}^i \cdot A_{sm}^i \cdot \eta_{mppt} \quad (4)$$

$$= \sum_i^N (I_{solar}^{i,dir}\eta_{sm}^{i,dir} + I_{solar}^{i,diff}\eta_{sm}^{i,diff}) \cdot A_{sm}^i \cdot \eta_{mppt} \quad (5)$$

The extended geometric model also allows to consider shading effects which can, for certain aircraft configurations, greatly degrade module performance at low sun inclination. It considers shading by the horizontal and vertical tail plane under a sun vector $\vec{r}_{sun}$ using a straightforward ray tracing method (Figure 5) and yields the shaded area per solar module $A_{sm,shaded}^i$. The method does allow a first qualitative assessment, but it does not provide a quantitative measure for the power loss. Calculating the actual power loss based on the model output is possible, but requires modeling the cell-cell and cell-diode interactions within the respective solar module and thus requires knowledge of the specific solar-cell and diode configuration on the airplane. While no generic model can therefore be given here, solar-powered UAV designers are encouraged to implement their own custom plugins to calculate the quantitative power loss.

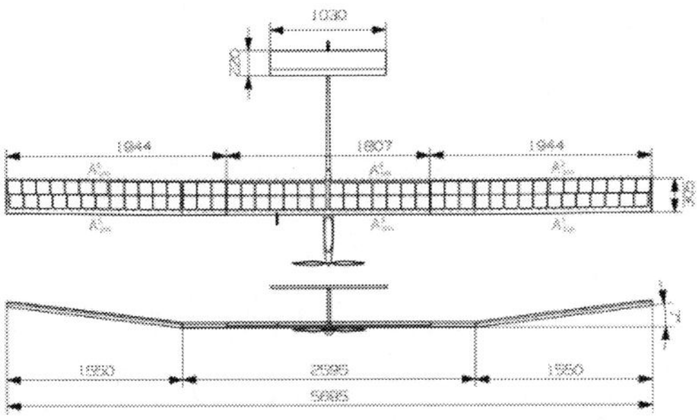

Figure 2: AtlantikSolar UAV airframe. Dimensions are given in mm. The solar module geometry of the surfaces A^i_{sm} is given in light blue.

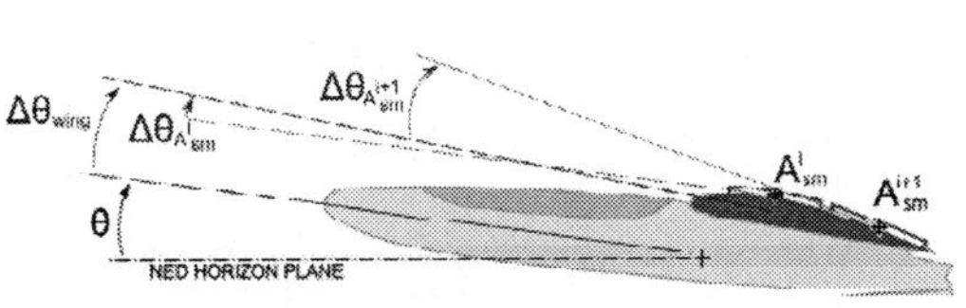

Figure 3: Geometric arrangement of solar module surfaces in the full solar power model.

3.2 Solar Module Efficiency Modeling

The standard test conditions (STC) under which solar modules provide their nominal efficiency $\eta_{\text{sm}}^{\text{nom}}$ are a total (direct + diffuse) irradiation $I_{\text{solar}}^{\text{STC}}$=1000 W/m², a temperature $T_{\text{sm}}^{\text{STC}}$=25°C and an angle of incidence $\gamma_{\text{sm}}^{\text{STC}}$=0°. Deviations from the STC or nominal conditions will cause efficiency changes that are modeled via the three respective component efficiency factors

$$\epsilon_{\text{sm}}^{I} = \frac{\eta_{\text{sm}}}{\eta_{\text{sm}}^{\text{STC}}\left(I_{\text{solar}} = I_{\text{solar}}^{\text{STC}}\right)}\Big|_{T,\gamma=\text{const}} \quad (6)$$

$$\epsilon_{\text{sm}}^{T} = \frac{\eta_{\text{sm}}}{\eta_{\text{sm}}^{\text{STC}}\left(T_{\text{sm}} = T_{\text{sm}}^{\text{STC}}\right)}\Big|_{I,\gamma=\text{const}} \quad (7)$$

$$\epsilon_{\text{sm}}^{\gamma} = \frac{\eta_{\text{sm}}}{\eta_{\text{sm}}^{\text{STC}}\left(\gamma_{\text{sm}} = \gamma_{\text{sm}}^{\text{STC}}\right)}\Big|_{I,T=\text{const}} \quad (8)$$

The exact correlation for $\epsilon_{\text{sm}}^{I}(I_{\text{solar}})$ is retrieved from Townsend [12] and is shown in Figure 4a). It mostly results in a relative decrease of η_{sm} because the total solar irradiation per unit area is usually lower than under standard test conditions. Second, solar module efficiency losses at higher temperatures are represented through the linear relationship

Table 1: Solar model geometric parameters for AtlantikSolar. Square brackets denote sets of variables, in this case for the different solar modules distributed along AtlantikSolar's wing.

Parameter	Value	Source
$\Delta\phi_{\text{dih}}$	7.0°	Aircraft specs
$\Delta\theta_{\text{wing}}$	5.7°	Aircraft specs
$\Delta\theta_{A^i_{\text{sm}}}$	$\Delta\theta_A[1,3,5] = -0.5°$, $\Delta\theta_A[2,4,6] = 9.4°$	Measured
A^i_{sm}	$A[1,3,4,6] = 0.184\text{m}^2$, $A[2,5] = 0.307\text{m}^2$	Aircraft specs

$$\epsilon_{\text{sm}}^{T} = 1 - c_l \cdot \left(T_{\text{sm}} - T_{\text{sm}}^{\text{STC}}\right) \quad (9)$$

where the loss factor c_l =0.3%/K, applicable to AtlantikSolar's SunPower E60 solar cells, is extracted from solar cell manufacturer data sheets [13]. The resulting correlation is shown in Figure 4b). The instantaneous solar module temperature is approximated using flight test data and the linear relationship

$$T_{\text{sm}} = T_{\text{amb}} + \Delta T_{\text{max}} \cdot \frac{P_{\text{solar}}}{P_{\text{solar}}^{\text{max}}} \quad (10)$$

Here, T_{amb} is the ambient temperature measured by the airplane, and $\Delta T_{\text{max}} \approx 12$°C is the temperature difference between solar module and ambient temperature that was measured in flight at cruise speed approximately at maximum insolation (i.e. for AtlantikSolar at $P_{\text{solar}}^{\text{max}}$ ≈265W). Third, the exact angle of incidence component efficiency $\epsilon_{\text{sm}}^{\gamma}(\gamma_{\text{sm}}^{i})$ for the AtlantikSolar UAV modules is taken from Townsend [12] and is given in Figure 4c). The data shows the expected relative loss of efficiency due to increased reflection that can be of significant importance at high angles of incidence. However, it should be noted that while Townsend [12] considers the same solar cell type, his study provides data for a glass front cover and not the specific foil front cover used on the AtlantikSolar UAV. The data thus has to be considered a first-order approximation. To determine the angle of incidence γ_{sm}^{i} for each solar module surface, we use the unit-vectors $\vec{n}_{\text{sm}}^{i}$ and $\vec{r}_{\text{sun}}$ from the previous section and define

$$\gamma_{\text{sm}}^{i} = \left|\arccos\left(\vec{n}_{\text{sm}}^{i} \cdot \vec{r}_{\text{sun}}\right)\right| \quad (11)$$

Together with the wing camber component efficiency $\epsilon_{\text{sm}}^{\text{cbr}}$ introduced before, we can now express (omitting the surface-index i for legibility) the overall solar module efficiency used in Eq. (4) separately for direct and diffuse radiation as

$$\begin{aligned}\eta_{\text{sm}}^{\text{dir}} &= \eta_{\text{sm}}^{\text{STC}} \cdot \epsilon_{\text{sm}}^{I} \cdot \epsilon_{\text{sm}}^{T} \cdot \epsilon_{\text{sm}}^{\gamma,\text{dir}} \cdot \epsilon_{\text{sm}}^{\text{cbr}} \\ \eta_{\text{sm}}^{\text{diff}} &= \eta_{\text{sm}}^{\text{STC}} \cdot \epsilon_{\text{sm}}^{I} \cdot \epsilon_{\text{sm}}^{T} \cdot \epsilon_{\text{sm}}^{\gamma,\text{diff}}\end{aligned} \quad (12)$$

Here, $\epsilon_{\text{sm}}^{\gamma,\text{diff}} = 0.83$ is a constant factor that approximates the respective component efficiency over the varying incidence angles in the diffuse radiation. It is retrieved by assuming a uniform distribution of the diffuse radiation $I_{\text{solar}}^{\text{diff}}$ over the sky's semi-sphere and by performing area-weighed averaging of $\epsilon_{\text{sm}}^{\gamma}(\gamma_{\text{sm}})$ over that area. If no separate calculation of direct- and diffuse-efficiencies is performed, then we use Eq. (12) and $\eta_{\text{sm}} = \eta_{\text{sm}}^{\text{dir}}$.

3.3 Maximum Power Point Tracker Efficiency Modeling

The maximum power point tracker efficiency η_{mppt} is a function of the solar power input P_{solar} to the MPPTs. The relationship was measured (Figure 4d) for the AtlantikSolar MPPT in a laboratory test setup. The MPPT performance agrees well with the data sheet values. It reaches efficiencies of $\eta_{\text{mppt}}^{\text{max}}$=97% but drops to 70% efficiency at low solar power income. However, during the major part of the power income range, the assumption η_{mppt}=95% used for the conceptual design phase is a good fit.

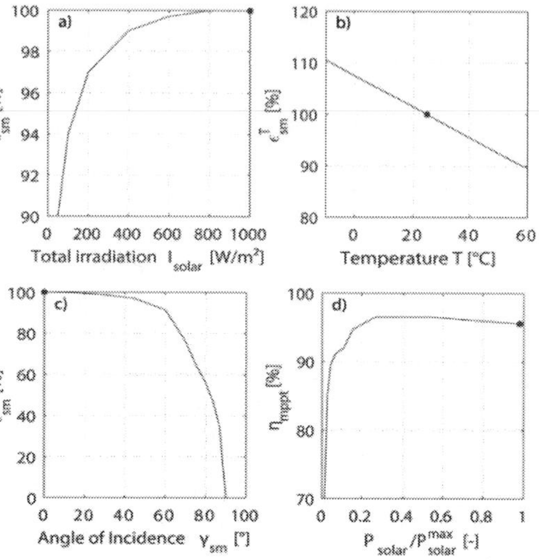

Figure 4: Component efficiencies used in the solar power model. Blue dots represent standard test conditions (STC).

4 DERIVATION OF SIMPLIFIED MODELS FOR CONCEPTUAL UAV DESIGN

The full solar power model (FM) of the previous section requires the current aircraft and thus solar module attitudes $\vec{n}^{i}_{sm}$, the irradiation level I_{solar} and the current solar module temperature T_{sm} as input parameters. However, quantities $\vec{n}^{i}_{sm}$ and T_{sm} are not available during the UAV conceptual design and analysis phase because the exact operating conditions (e.g. the chosen flight path) are unknown. Therefore, Table 2 introduces models that run with a less extensive set of input data. These are the

- *Conceptual Design Model* (CDM), which is and was used in the conceptual design of AtlantikSolar in [2]. It assumes a single, flat and always horizontal solar module surface, a constant MPPT efficiency η_{mppt} and a constant solar module efficiency η_{sm} that is only a function of a user specified constant temperature T_C.
- *Conceptual Analysis Model* (CAM), which adds the angle of incidence correlation of Figure 4c) to obtain a time-varying efficiency η_{sm}. It is used for a more detailed performance prediction once technical parameters such as the γ_{sm}-correlation are known, but the exact conditions such as the flight path, wind conditions and thus flight attitude are still unknown. This is often the case during the late conceptual design phase and the detailed design phase. In [2] the CAM is used to provide an accurate performance outlook after the design and flight verification has been completed.
- *Verification Model* (VM), which compensates for offsets in the required heading and roll angles (especially in a loitering mission) caused by horizontal winds. The VM approximates the aircraft geometry with a single flat surface with the time-varying and wind-dependent orientation of the central wing $\vec{n}^{ctr}_{sm}$. The VM is not used for actual solar-powered UAV design or analysis, but only used internally in this paper to assess the magnitude of the CDM/CAM modeling errors that are *not* caused by wind using 'wind-disturbed flight test data'.

Table 2: The solar power models proposed for different stages of UAV development. The table indicates whether certain properties or variables are assumed constant (subscript 'C') and if not, what their dependencies are.

Name Scope	CDM Design	CAM Analysis	VM Verification	FM Full model
Surfaces	Single	Single	Single	Multiple
Geometry	Flat	Flat	Flat	Realistic
Attitude	horiz.	horiz.	$f(\vec{n}^{ctr}_{sm})$	$f(\vec{n}^{i}_{sm})$
η_{mppt}	const	const	const	$f(P_{solar})$
η_{sm}	$f(T_C)$	$f(\gamma_{sm}(\vec{n}_C), T_C)$	$f(\gamma_{sm}(\vec{n}^{ctr}_{sm}), T_C)$	$f(\gamma^{i}_{sm}, I_{solar}, T_{sm})$

5 SOLAR POWER MODEL VERIFICATION

This report assesses the developed solar power models using flight data from the 28-hour solar-powered flight performed by AtlantikSolar AS-2 [4] in 2015. Note that this flight only includes a single day/night solar cycle. A verification that uses the more comprehensive dataset from the three day/night cycles of AtlantikSolar's 81-hour flight is performed (though in less detail) in [2].

5.1 Full Solar Power Model Verification

Figure 6 compares modeled and measured solar power system data for three segments (early- to late-morning) of the aforementioned 28-hour flight. To assess the accuracy of the developed solar power income model, we employ the standard definitions for the absolute average-, RMS- and maximum model error $\hat{E}_x$ (where a negative error represents an underestimate by the model). In addition, we define the respective relative error of quantity x as $\hat{e}_x = \hat{E}_x / x^{exp}_{avrg}$. The three time segments of the aforementioned flight exhibit characteristically different solar power system behavior. More specifically, during

- *Early-morning flight* (Figure 6, left) at t=[5.71h,5.79h] local solar time, the low sun elevation causes a significant variation of the solar irradiation levels with the aircraft yaw angle. Both ϵ^{γ}_{sm} and η_{mppt} therefore play a significant role, while ϵ^{I}_{sm} and ϵ^{T}_{sm} only have a minor effect. The resulting overall module efficiency η_{sm} is only 16-18% compared to the standard operating condition efficiency of η^{STC}_{sm}=23.7%. The model errors are $\hat{E}^{RMS}_{P_{solar}}$=8.54W, $\hat{e}^{RMS}_{P_{solar}}$=25.4%, $\hat{e}^{max}_{P_{solar}}$=43.4% and $\hat{e}^{avrg}_{P_{solar}}$=5.14% ($P^{model}_{solar}$=35.3W and P^{exp}_{solar}=33.6W). Two graphs help to explain this overestimation by the model: First, the recorded MPPT voltages (Figure 6, bottom) indicate that in low irradiation conditions (i.e. the solar panels face away from the sun), the employed MPPTs fail to track the solar panel maximum power point. This clearly indicates sub-optimal power system behavior that, especially for flights in low total radiation or low sun elevation conditions, needs improvement. Second, the power plots show small notches shortly before the model predicts maximum solar power income. Our geometric airplane model indicates (Figure 6, bottom) that shading of the center and right solar modules by the aircraft tail exactly at t = 5.71h can explain this. Figure 5 visualizes the output of the shading calculations in the full solar power model at this time.

- *Mid-morning flight* (Figure 6, center) at t=[8.02h,8.12h] the average solar power income has increased considerably to $P_{\text{solar}}^{\text{model}}$=140.5W and $P_{\text{solar}}^{\text{exp}}$=142.1W. The average error is only $\hat{e}_{P_{\text{solar}}}^{\text{avrg}}$=−1.1%, and the RMS and maximum errors have decreased to $\hat{E}_{P_{\text{solar}}}^{\text{RMS}}$ = 11.3W, $\hat{e}_{P_{\text{solar}}}^{\text{RMS}}$=7.98% and $\hat{e}_{P_{\text{solar}}}^{\text{max}}$=21.3%. Due to the higher sun elevation, the variations in $\epsilon_{\text{sm}}^{\gamma}$, η_{sm} and η_{mppt} are much less pronounced. Solar module shading and sub-optimal MPP-tracking are not observed anymore.
- *Late-morning flight* (Figure 6, right) at t=[9.52h,9.62h] local time we retrieve $P_{\text{solar}}^{\text{model}}$=204.7W and $P_{\text{solar}}^{\text{exp}}$=211.5W. With an estimated ϵ_{sm}^{T}=96%, the elevated temperature of T_{sm}=37°C at T_{amb}=27°C now noticeably reduces the solar module efficiency. Despite the approaching solar irradiation maximum, γ_{sm} remains between 31° and 49°, thus underlining the importance of considering an exact $\epsilon_{\text{sm}}^{\gamma}$ correlation (Figure 4) even at high sun elevation. The RMS and maximum errors are $\hat{E}_{P_{\text{solar}}}^{\text{RMS}}$=9.39W, , $\hat{e}_{P_{\text{solar}}}^{\text{RMS}}$=4.4% and $\hat{e}_{P_{\text{solar}}}^{\text{max}}$=9.9% respectively. The average error has increased in magnitude to $\hat{e}_{P_{\text{solar}}}^{\text{avrg}}$=−3.2%. The full solar power model thus tends to underestimate the incoming solar radiation for high irradiation conditions.

A comprehensive assessment over a full day (Figure 7) confirms the overestimation of P_{solar} during the early morning and the underestimation during the higher radiation conditions of the late morning. The plots clarify that solar module shading occurs from sunrise to around t=6.26h (or more generally until the sun elevation angle is above 19 degrees for this UAV geometry) and accounts for less than a 10% decrease in P_{solar} even during the most significant shading problems around t=5.71h. Although shading by AtlantikSolar*'s* T-tail is not significant over the full day in the presented conditions, aircraft designed for optimum performance at low sun elevation conditions could consider alternative tail configurations (e.g. an inverted T-tail) or optimized path planning to reduce the effects of shading. In contrast to shading, the imperfect maximum power point tracking under low light conditions is found to occur until 7.77h and thus plays an even more significant role. The discrepancies in the high-radiation conditions around noon are expected to lie in the first-order approximation for the $\epsilon_{\text{sm}}^{\gamma}$- and ϵ_{sm}^{T}-correlations.

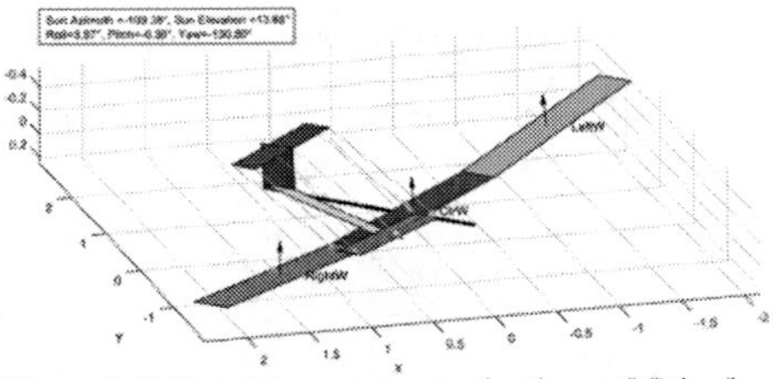

Figure 5: Full model geometry at solar time t=5.71h of AtlantikSolar's 28-hour flight. The black arrows are the surface normal vectors $\vec{n}_{\text{sm}}^{i}$ and the orange arrows are the sun rays $\vec{r}_{\text{sun}}$ which can be used to infer the angles of incidence γ_{sm}^{i}. The grey and dark-orange surfaces are the shadows of the rudder and elevator projected onto the solar-module covered wing.

Removing these deviations would require accurate measurements of $\epsilon_{\text{sm}}^{\gamma}$ with high-precision laboratory equipment such as solar flashers that was not available for this paper. Note that to exclude errors in the underlying solar irradiation I_{solar}, the model by Duffie [11] was compared to the *Sandia National Labs PV_LIB toolbox* model [5] and found to agree within 3% maximum error.

Overall, given all these uncertainties, the full model accuracy is satisfactory: The average deviation over the domain in which the MPPTs operate without limiting their output current t=[4.17h,9.72h] is $\hat{e}_{P_{\text{solar}}}^{\text{avrg}}$=-1.75% ($\hat{e}_{P_{\text{solar}}}^{\text{avrg}}$=97.9W and $P_{\text{solar}}^{\text{exp}}$=99.6W). The RMS error is, mostly due to the higher amplitude of the heading-dependent variations of P_{solar}, $\hat{E}_{P_{\text{solar}}}^{\text{RMS}}$=9.84W or $\hat{e}_{P_{\text{solar}}}^{\text{RMS}}$=9.88%. These model errors are relatively small compared to all other possible design errors (caused e.g. by wrong user inputs for subsystem masses as described in [2]) or external meteorological factors (e.g. clouds), and the full model can thus be considered accurate enough for solar-powered UAV design and performance analysis.

5.2 Comparison of Full- and Simplified Models

Figure 7 and Table 3 provide a detailed performance comparison of all models. As stated before, even the full model (FM) shows deviations from the flight test data. The findings for the other models are:

- *Verification Model*: Over the full day, the difference between verification and full model is small. The additional physical effects modeled in the FM thus do not have a significant influence for the chosen aircraft and power system. Specifically, η_{mppt} and η_{sm}^{I} mainly deviate from their nominal values in the morning, and the temperature dependence leads to a decreased P_{solar} during noon. Consequently — and this is an important finding — the conceptual design of solar-powered UAVs can be safely performed without these physical effects of the FM.
- *Conceptual Analysis Model*: The relative error $\hat{e}_{P_{\text{solar}}}^{\text{avrg}}$ differs by 7 percent points between VM and CAM. The only difference between these models is the inclusion of the current aircraft attitude through $\vec{n}_{\text{sm}}^{\text{ctr}}$. The deviation is caused by two effects: First, due the concave-downwards nature of $\epsilon_{\text{sm}}^{\gamma}(\gamma_{\text{sm}})$ in Figure 4c), oscillations of the *instantaneous* angle of incidence $\gamma_{\text{sm}}(t)$ around an *average* $\gamma_{\text{sm}}^{\text{avrg}}$=const result in a lower average P_{solar} than for $\gamma_{\text{sm}}(t)$ = const = $\gamma_{\text{sm}}^{\text{avrg}}$. Second, during the loitering turns, an easterly wind of 2m/s causes the solar panels to face away from the sun longer than towards the sun. Consequently, the CAM overestimates P_{solar} during the morning. This confirms that wind has a significant influence, and although wind data is usually not available in the conceptual design stage, its effects especially during the morning and evening need to be considered as early as possible.
- *Conceptual Design Model*: The large deviation between CAM and CDM is explained by the inclusion of the γ_{sm}-dependence of η_{sm} in the CAM. The respective component efficiency $\epsilon_{\text{sm}}^{\gamma}$ is only 20% around sunrise. Clearly, $\epsilon_{\text{sm}}^{\gamma}$ needs to be estimated and considered as early as possible during the conceptual analysis stage.

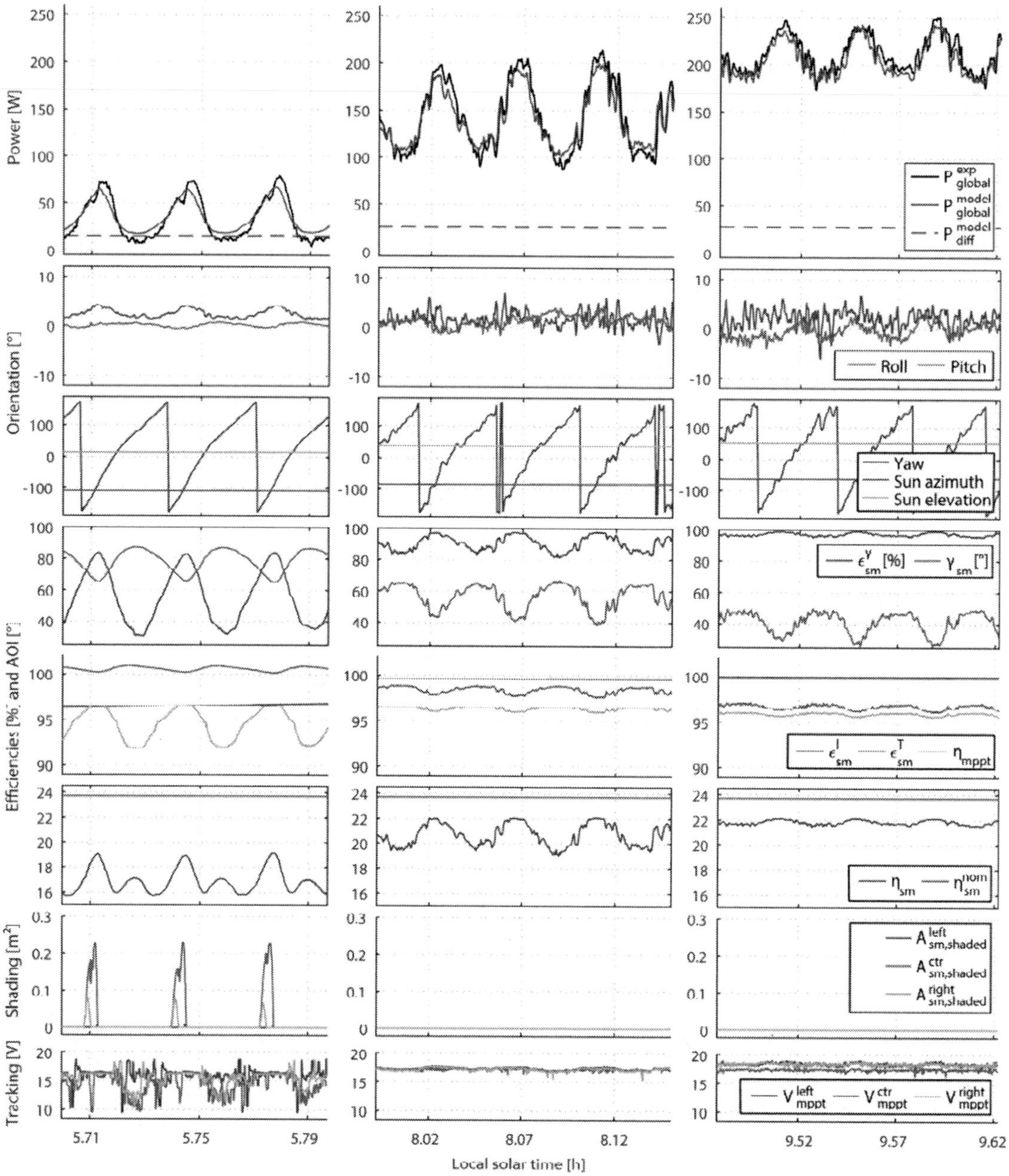

Figure 6: Comparison of results from our full solar power model and test flight data collected during the second day of a continuous 28-hour flight of AtlantikSolar AS-2 on July 1st 2015. The three patches show early-morning (left), mid-morning (center) and late-morning (right) flight. The graphs show experimental (P_{global}^{exp}) and modeled (P_{global}^{model}) global solar power income, measured aircraft orientation, modeled component and total efficiencies, modeled shaded solar module area and the measured maximum power point tracker voltage as an indicator of solar power system tracking behavior. The data is recorded at 2Hz and a two-sided moving average filter with a semi window length of 2 samples is applied. The model efficiency and angle data is plotted as averages over the six solar module sections of the UAV.

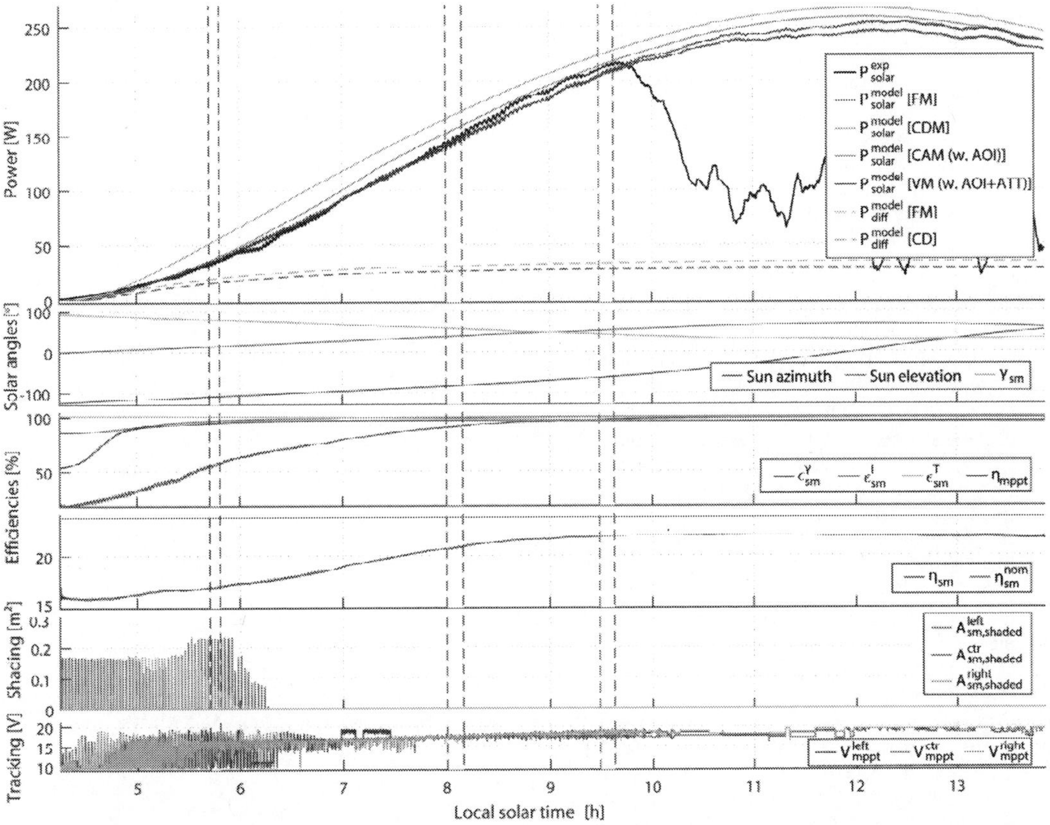

Figure 7: Comparison of the solar power models of Table 2 and flight data. The data is recorded at 2Hz and, except for the shading and tracking plots, a two-sided moving average filter with a semi window length of 800 samples is applied. The grey patches represent the detailed plots of Figure 6. The model efficiency and angles are averages over all UAV solar modules. The decrease in P^{exp}_{solar} at $t = 9.72$h is only because the batteries are nearly full and the MPPTs thus reduce their power output.

Overall, the comparison shows that while the full model allows accurate solar power income prediction, the less sophisticated conceptual design (CDM) and analysis models (CAM) show errors that are of significance during the conceptual design. The two most important factors that a designer of solar-powered UAVs should consider as early as possible to decrease these deviations are the angle of incidence correlation ϵ^{γ}_{sm} for η_{sm} and the related effects of changing aircraft attitude and horizontal winds. Both effects increase in importance when operating in low sun elevation angles.

Table 3: Estimation errors of the four solar power models. The RMS errors are calculated based on pre-filtered data (using the same filter as Figure 7), in order to allow a meaningful comparison between the attitude-dependent VM and FM models and the attitude-independent CD and CA models.

Name Scope	CDM Design	CAM Analysis	VM Verific.	FM Full model	Flight
P^{avrg}_{solar}	117.35 W	104.89 W	97.65 W	97.87 W	99.61 W
$\hat{e}^{avrg}_{P_{solar}}$	17.81 %	5.31 %	-1.96 %	-1.75 %	—
$\hat{E}^{RMS}_{P_{solar}}$	19.95 W	6.99 W	3.94 W	4.17 W	—
$\hat{e}^{RMS}_{P_{solar}}$	20.03 %	7.02 %	3.96 %	4.19 %	—

6 CONCLUSION AND LESSONS LEARNED

Of the models derived in this paper and verified via AtlantikSolar's 28-hour solar-powered flight, the Full Solar Power Model (FM) provided accurate solar power prediction results with an error in the average P_{solar} of -1.75%. The FM however requires input data that is not available during the conceptual design stage of solar-powered UAVs. The models derived for these purposes, the Conceptual Design Model (CDM) and Conceptual Analysis Model (CAM), have to make simplifications that result in an overestimation of the average P_{solar}. of 18% and 5% respectively. The lessons learned that can be used by other solar-powered UAV designers are

- The Conceptual Design Model (CDM) should only be used in the very early design stage of a solar-powered UAV. It needs to be taken into account that the solar power income and thus overall flight endurance of the designed UAV will be lower than predicted by the CDM.
- The Conceptual Analysis Model (CAM) should replace the CDM as soon as possible, even in the early conceptual design phase of a UAV. The CAM incorporates the angle of incidence sensitivity of the solar module efficiency η_{sm} and thereby reduces the

prediction error significantly, i.e. from 18% to 5% in the specific 28-hour flight presented in this paper.

- The Full Solar Power Model (FM) shall be used in the flight operations phase when the wind conditions are known and the flight path is planned. With a modeling error of approximately 2%, the FM is sufficiently accurate for a detailed performance analysis, i.e. it can also be used to verify that a solar power system is working optimally.

REFERENCES

[1] N. J. Colella e G. S. Wenneker, «Pathfinder: Developing a solar rechargeable aircraft,» *IEEE Potentials,* vol. 15, pp. 18-23, 1996.

[2] P. Oettershagen, A. Melzer, T. Mantel, K. Rudin, T. Stastny, B. Wawrzacz, T. Hinzmann, S. Leutenegger, K. Alexis e R. Siegwart, «Design of small hand-launched solar-powered UAVs: From concept study to a multi-day world endurance record flight,» *Journal of Field Robotics (JFR),* vol. 34, pp. 1352-1377, 10 2017.

[3] L. Wirth, P. Oettershagen, J. Ambuehl e R. Siegwart, «Meteorological Path Planning Using Dynamic Programming for a Solar-Powered UAV,» in *IEEE Aerospace Conference*, 2015.

[4] P. Oettershagen, A. Melzer, T. Mantel, K. Rudin, T. Stastny, B. Wawrzacz, T. Hinzmann, K. Alexis e R. Siegwart, «Perpetual flight with a small solar-powered UAV: Flight results, performance analysis and model validation,» in *IEEE Aerospace Conference*, 2016.

[5] J. S. Stein, «The photovoltaic performance modeling collaborative (PVPMC),» in *38th IEEE Photovoltaic Specialists Conference (PVSC)*, 2012.

[6] J.-K. Shiau, D.-M. Ma e C.-W. Chiu, «Optimal Sizing and Cruise Speed Determination for a Solar-Powered Airplane,» *Journal of Aircraft,* vol. 47, 2010.

[7] A. Noth, «Design of Solar Powered Airplanes for Continuous Flight,» 2008.

[8] S. Morton, L. Scharber e N. Papanikolopoulos, «Solar Powered Unmanned Aerial Vehicle for Continuous Flight: Conceptual Overview and Optimization,» in *IEEE International Conference on Robotics and Automation (ICRA)*, 2013.

[9] A. T. Klesh e P. T. Kabamba, «Solar-Powered Aircraft: Energy-Optimal Path Planning and Perpetual Endurance,» vol. 32, 2009.

[10] P. Oettershagen, A. Melzer, T. Mantel, K. Rudin, R. Lotz, D. Siebenmann, S. Leutenegger, K. Alexis e R. Siegwart, «A Solar-Powered Hand-Launchable UAV for Low-Altitude Multi-Day Continuous Flight,» in *IEEE International Conference on Robotics and Automation (ICRA)*, 2015.

[11] J. A. Duffie e W. A. Beckman, Solar Engineering of Thermal Processes, 3rd a cura di, Hoboken, New: Wiley, 2006.

[12] T. Townsend e J. Wilson, «SunPower Yield Report,» 2013.

[13] SunPower, *E60 Solar Cell,* 2013.

APP FOR DESIGNING RESIDENTIAL SOLAR PV SYSTEMS WITH DC-COUPLED BATTERY STORAGE

Ashley Gritzman, Toby Kurien, Tapiwa Chiwewe, Jeofrey Ditsela
IBM Research | Africa, South Africa Lab
41 Juta Street, Braamfontein, Johannesburg, 2000, South Africa
ashley.gritzman@za.ibm.com

ABSTRACT: In recent years, decreasing costs of solar PV have fuelled interest in the residential PV market. In contrast to North America and Europe which encourage grid-tie PV, in many African cities it is illegal to feed power back to the grid, and in the few cities where it is legal, the feed-in tariff is unattractive. The inability to feed excess energy back to the grid, coupled with grid instability and blackouts, has resulted in increased popularity of battery-based PV systems in African cities. However, homeowners embarking on this journey quickly discover that they are ill equipped to navigate the many options presented by suppliers. In this paper, we present a consumer-friendly app that empowers homeowners to design their own battery-based PV system. The app selects components from a live catalogue, which is based on the availability of equipment from nearby retailers, and optimises the system on price and performance. The app is accessible via a web app which can be viewed on a mobile, tablet, or desktop device.

Keywords: Solar PV, System Design Software, Battery Storage, DC-Coupled, Residential, Off-grid

1 INTRODUCTION

As the hype around solar PV continues to grow, homeowners find themselves asking when is the right time to jump on board. However, as they embark on this journey to a greener future, they quickly discover that they lack the basic knowledge required to get started [1], [2]. On first contact with a supplier, the homeowner may face questions including "How many kWh do you consume during winter?" or "What size inverter do you need?". Homeowners often struggle to answer even these basic questions, and as such, they are not able to make informed decisions on the options presented by suppliers.

Our goal is to empower consumers to design their own solar PV system, which in turn will enable them to obtain competitive quotes from suppliers. Suppliers will also find this app useful to quickly produce system designs and cost estimates for their customers.

The landscape of solar software already contains a number of useful tools, both free and commercial [3]. The majority of these tools simulate the energy output of a solar system using irradiance data at a particular location (e.g. PVWatts [4] and PVGIS [5]). Some tools assist with system sizing by calculating the required solar power and battery storage, but stop short of recommending specific components and configurations to use. Other software tools offer system configuration, whereby the user first selects the components to use (e.g. solar modules and inverter), and then the software designs the system configuration.

The objective of the app presented in this paper is to design the system without depending on the user to choose the components. Instead, the algorithm selects components from a catalogue of equipment available at the user's location, which may include stock from several nearby retailers. Since the app has been developed independent of equipment manufacturers, the algorithm is free to make component selections without bias towards a particular company or brand.

In contrast to North America and Europe which encourage grid-tie PV, in many African cities it is illegal to feed power back to the grid, and in the few cities where it is legal, the feed-in tariff unattractive. The inability to feed excess energy back to the grid, coupled with grid instability and blackouts, has resulted in increased popularity of battery-based PV systems in African cities.

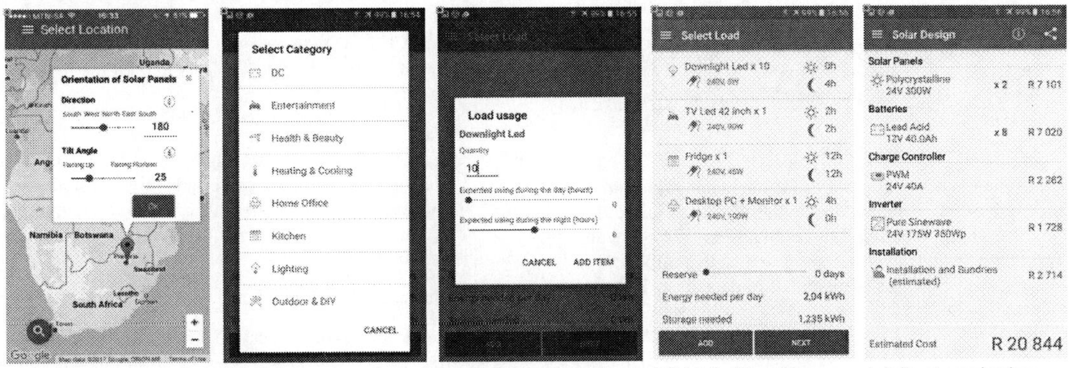

(a) Select location, direction & tilt (b) Categories of load items (c) Specify quantity, day & night hours (d) List of load items (e) System design and cost estimate

Figure 1: Screenshots from web application https://empowersolar-demo.mybluemix.net/

This paper presents an app for designing solar PV systems with DC-coupled battery storage. The app selects specific components from a catalogue (solar modules, charge controllers, batteries, inverters), and specifies the electrical configuration in terms of the number of units in series/parallel. The app ensures the electrical compatibility between all components by verifying that currents and voltages fall within the recommended operating ranges. The algorithm chooses between potentially thousands of options by optimising on price and performance. Since the algorithm uses components and pricing from suitable retailers, this app may form the basis of a price comparison engine.

Section 2 of this paper describes the sources of data used in the app, and Section 3 details the algorithm to design the DC-coupled PV system. The interfaces to access the app are discussed in Section 4, and Section 5 presents an example use case.

2 DATA SOURCES

2.1 Solar Radiation

We obtain solar radiation for a given tilt and orientation from PVGIS [5], which uses the CM-SAF solar radiation database [6], and inclined-plane models described in [7]. The main source of data for the CM-SAF dataset is the images of the Meteosat series of geostationary satellites above Europe and Africa [6].

2.2 Database of Appliances

Instead of asking the user to specify the load in terms of power and energy, we adopt a consumer-friendly approach that does not require technical know-how, whereby the user selects from a database of common household appliances. Table I shows a sample from the appliance database, which includes values for voltage, average power, peak power, and the type of inverter required. The appliances are group into the following eight categories:

1. DC	5. Home Office
2. Entertainment	6. Kitchen
3. Health & Beauty	7. Lighting
4. Heating & Cooling	8. Outdoor & DIY

Table I: Sample of appliances in the database

Id	Appliance	Voltage	Avg. Power	Peak Power	Inverter Type
1	fridge	240 AC	100	150	Pure
2	laptop	240 AC	30	80	Modified
3	led downlight	12 DC	5	5	None
4	led tv 42"	240 AC	90	90	Modified
5	oven	240 AC	2000	2500	Modified

2.3 Solar Component Catalogue

The catalogue contains a list of suppliers and their current stock of solar components including solar modules, batteries, charge controllers, and inverters. We manually capture the technical specifications of each component in the catalogue from the corresponding datasheet. It is also important to know what geographic areas the suppliers can service.

At present, the current catalogue only contains components from one supplier, which is an online retailer in South Africa. However, additional suppliers can be added in future, and interested suppliers are encouraged to contact us.

We envision that the catalogue will link to the supplier's inventory, and automatically update the availability and price of components.

3 ALGORITHM TO DESIGN SYSTEM

3.1 Solar Radiation

Given the site location in terms of latitude and longitude, as well as the tilt and orientation of the solar array, the app retrieves the average monthly solar radiation from PVGIS [5]. Subsequent calculations are based on the mean monthly solar radiation, which results in surplus energy during summer months, but in winter the load will need to be supplemented from an alternative source (e.g. grid or diesel generator).

3.2 Electrical Requirements of the Load

The user selects a list of appliances from the database, and specifies the quantity, and typical usage in terms of daytime hours and night-time hours. It is important to distinguish between daytime and night-time loads, as the energy for daytime loads can be used directly and does not need to be stored in the battery bank. The daytime and night-time energy requirements are calculated as follows:

$$E_{day} = \sum_i P_i Q_i t_{i-day} \qquad E_{night} = \sum_i P_i Q_i t_{i-night}$$

where P_i is the power of appliance i, Q_i is the quantity, t_{i-day} is the number of daytime hours, and $t_{i-night}$ is the number of night-time hours.

The number of daytime hours (t_{i-day}) and night-time hours ($t_{i-night}$) are specified on a range of 0 – 12h by the user. However, the number of sun hours of solar radiation is well below 12h per day, so the app limits the daytime hours to the number of sun hours, and adjusts the day/night energy usage accordingly. For example, if the user specifies that an appliance operates for 8h during the day in Johannesburg, which receives approximately 6 sun hours per day, then the app adjusts the energy usage to 6 daytime hours, and 2 night-time hours.

The peak power during the day PP_{day} is calculated by summing the power of the daytime loads, and similarly the peak power during night PP_{night} is calculated by summing the power of the night-time loads. The overall peak demand PP_{24h} is then the maximum of PP_{day} or PP_{night}, which is later used to determine the size of the inverter.

Regarding the type of inverter, if any appliance requires a pure sine wave inverter, then a pure sine wave inverter is preferred. Otherwise, the algorithm will choose the inverter by optimising on price, which will usually result in selecting a modified sine wave inverter. If no appliances require an inverter, then a DC system will be designed, without an inverter.

3.3 Selecting the Components

Since the geographic location of the installation site is known, the app selects suppliers from the component catalogue that can deliver to the specified location.

The diagram in Figure 2 shows the major components of a DC-coupled solar PV system, including solar modules, batteries, charge controllers, and inverters. Individual components are combined into groups to achieve the required electrical specification. For example, several modules are combined into a solar array to generate the solar energy required, or several inverters are combined into an inverter bank to achieve the required power output.

For each component, it is necessary to select the individual component from the catalogue, and to specify the electrical configuration in terms of number of units in series/parallel. In specifying the components and configurations, the objective is to design the most cost-effective system that meets the electrical requirements.

The algorithm designs three systems, one for each of the nominal voltages (12, 24, and 48V), and selects the final system on price.

Table II shows the efficiency parameters applicable to the calculations that follow. For each component, we consider the downstream energy losses. For example, in designing the battery bank, we consider only the efficiency of the inverter and cabling. In designing the solar array, we consider all the efficiency parameters in Table II. To keep the formulae that follow simple, the efficiency adjustments are not shown.

Table II: Efficiency of components

Component	Efficiency
Solar Module	0.9
Battery	Depends on battery selected: Lead acid 0.8 Lithium 0.94
Charge Controller	0.96
Inverter	Depends on inverter selected: Range 0.85 – 0.95
Cabling	0.98

Inverter Bank

Design of the inverter bank involves selecting the inverter from the catalogue, and specifying the configuration. The following factors are considered:

- Individual inverters must be compatible with the nominal system voltage (12, 24, or 48V).
- The minimum power of the inverter bank is determined by the peak power demand (PP_{24h}), with an additional buffer of 20% to cater for inrush currents when appliances are switched on.
- Individual inverters can be configured in parallel to achieve the required power rating, with a restriction of up to four inverters.
- The inverter type depends on whether a pure sine wave inverter is necessary.
- No inverter is required if all the appliances can operate from a DC power source.

Battery Bank

Design of the battery bank involves selecting the battery from the catalogue, and specifying the configuration. The following factors are considered:

- Required storage capacity is determined by day/night load characteristics, and reserve days specified:
  i. *Only day loads specified* – need storage to supply day load for 30 minutes, to cater for passing cloud cover
  $$E_{day} / h_{sun} * 0.5$$
  where h_{sun} is the number of sun hours
  ii. *Both day & night loads specified* – need storage to supply night energy requirements
  $$E_{night}$$
  iii. *Reserve days specified* – need storage to supply first night and additional reserve days
  $$E_{night} + E_{24h} * d_{reserve}$$
- The storage capacity must be increased to account for the battery depth of discharge (DoD): lead acid = 0.5; lithium = 0.8.
- Batteries are combined in series to obtain the nominal system voltage (12, 24, or 48V), and in parallel to achieve required storage capacity.

Solar Array

Design of the solar array involves selecting the solar module from the catalogue, and specifying the configuration. The following factors are considered:

- Array power is determined by the energy required over 24h, and the mean monthly solar radiation ($E_{24h}/E_{SR-mean}$).
- Solar modules are combined in series to obtain a voltage that is well matched to a PWM controller:

PWM Voltage	String Voltage Range
12V	17 – 20V
24V	34 – 40V
48V	68 – 80V

The current algorithm does not support maximum power point trackers (MPPTs).

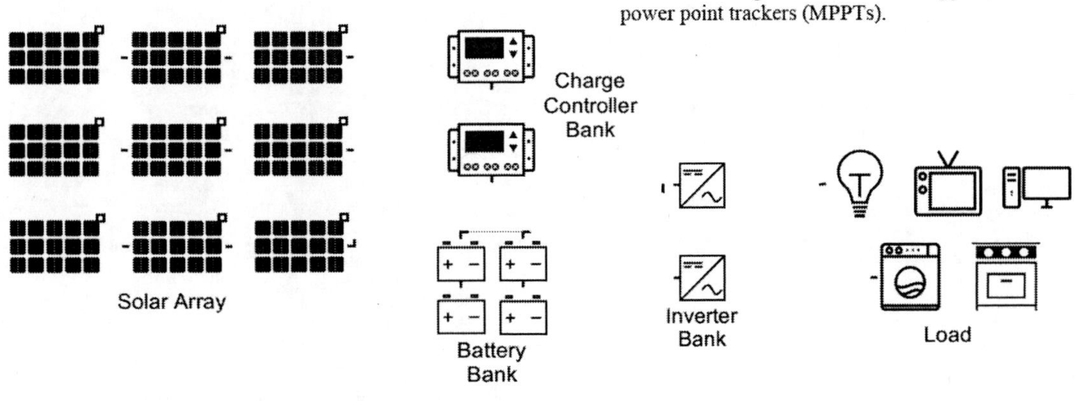

Figure 2: Diagram of solar PV system with DC-coupled battery storage

- Strings are combined in parallel to achieve required array power.

Charge Controller Bank

Design of the charge controller bank involves selecting the controller from the catalogue, and specifying the configuration. The current algorithm only supports PWM controllers, but support for MPPT controllers will be added in the future. The following factors are considered:

- Individual PWMs must be compatible with the solar array voltage (see above) and the nominal system voltage (12, 24, or 48V).
- The current rating of the PWM bank must be greater than the short-circuit current of the solar array. Individual PWMs can be configured in parallel to achieve the required current specification.
- Multiple strings can be connected to a single PWM, but the combined current through a single PWM cannot exceed the maximum current rating.

4 INTERFACES

We offer two interfaces to access the data and algorithms in the app, namely the web app and API.

4.1 API

The database of appliances, catalogue of solar components, and design algorithms are accessible via a RESTful API. Table III lists the API calls, and the OpenAPI (Swagger) specification can be accessed at http://solaralgorithm.mybluemix.net/api/v1/spec/.

Table III: RESTful API calls to the database of appliances, catalogue of solar components, and design algorithms (parameters not shown)

`GET solarpanels batteries chargecontrollers inverters`	Retrieve the catalogue of solar components.
`GET appliances`	Retrieve the database of appliances.
`GET solarrad`	Obtain the plane-of-array solar radiation for a specified location using PVGIS.
`GET elecreqs`	Calculate the electrical requirements from the list of appliances.
`GET solarsystem`	Design a solar PV system with battery storage to power a list of appliances. The design uses the solar components available in the catalogue, and specifies the size and quantity of solar modules, batteries, charge controllers, and inverters. The algorithm designs 12, 24, and 48V options, and selects the cheapest system that meets the requirements.

4.2 Web App

Figure 1 shows several screenshots from the web app. The user selects their location on a map, and builds up their load profile by selecting appliances and specifying the corresponding quantity, day hours, and night hours. The web app queries the API, and returns the system specifications including the solar modules, batteries, charge controllers, and inverters. The web app is responsive and can be viewed on a mobile, tablet, or desktop device.

5 EXAMPLE USE CASE

In this section, we apply the app to design solar systems for two load profiles: a residential home, and a small office. To demonstrate the effect of solar radiation on the resulting system design, we compare different locations spanning cities in South Africa, Africa, and Europe. For each location, the tilt of the solar array is set to the latitude of the location, and orientation is set to North for cities in the Southern Hemisphere, and South for cities in the Northern Hemisphere.

Table IV presents the list of appliances and usage characteristics of the *Residential* and *Office* load profiles. The *Residential* profile includes both daytime and night-time appliances, while the *Office* profile includes only daytime appliances. Recall that the number of *Day Hours* and *Night Hours* are specified on a range of 0 – 12h but the app adjusts the energy usage according to the number of sun hours at the location. Therefore, battery storage is required even in the case of the *Office* profile with only daytime appliances. In the example scenarios, no additional reserve capacity is requested, above what is required to supply the load for one full 24h period.

Energy 24h is the total daily energy required, and *Peak Power* is the maximum power demand to supply all the daytime loads simultaneously. Since the app uses the mean solar radiation in calculations, in summer months the system will produce surplus energy, and in winter months the load will need to be supplemented from an alternative source (e.g. grid or diesel generator).

Table IV: List of appliances and usage characteristics of the *Residential* and *Office* load profiles

Id	Appliance	Avg. Power	Quantity	Day Hours	Night Hours
RESIDENTIAL					
93	led downlights	5	40	0	3
62	fridge-freezer	100	1	12	12
10	tv	90	1	0	3
35	internet router	12	1	12	12
40	laptop	30	1	2	2
68	kettle	1800	1	0	0.2
78	washing machine	2300	1	1	0
47	dish washer	2300	1	1	0
69	microwave	800	1	0	0.2
70	oven/stove	2000	1	0	1
	Energy 24h	10.8kWh			
	Peak Power	5.032kW			
OFFICE					
120	desktop pc	200	4	8	0
40	laptop	30	6	8	0
35	internet router	12	1	8	0
85	flourescent tube 5ft	58	20	8	0
	Energy 24h	17.2kWh			
	Peak Power	2.152kW			

The catalogue of solar components available for the design comprises the following:

Solar modules	polycrystalline power {10, 30, 50, 100, 120, 150, 250, 300W}
Batteries	12V lead acid capacity {7.2, 12, 18, 40, 96, 102, 105, 230Ah}
PWM Controllers	current {10, 20, 30, 40A} voltage {12, 24, 48V}
Inverters	power {175, 300, 500, 700, 1000, 1500, 2000, 3000, 5000W} voltage {12, 24, 48V}

Table V presents the designs produced by the app for the *Residential* load profile, in each of the cities. *Solar Rad* is the mean solar radiation in the plane of the solar array. The system voltage is either 24 or 48V, and the size of the battery storage is similar across cities ranging from 13.8 to 14.7kWh. The main difference between the cities is the size of solar array which ranges from 2.7kWp in Upington to 6kWp in Amsterdam/Berlin.

Table VI presents the designs for the *Office* load profile, showing the battery bank in Upington of 5.5kWh is more than four times smaller than the battery bank in Amsterdam/Berlin at 24.5kWh. This large difference is explained as follows: of the 8 hours that the office is active in Upington, the sun is shining for 7 of them, and the solar energy flows directly from the array to the load, bypassing the batteries. In Amsterdam/Berlin however, the sun shines for only 3 of the 8 office hours, hence the battery bank is required to store the energy for the remaining 5 hours.

Another point worth noting is the size of the inverter bank in Cairo/Upington of 3kW (2 x 1.5kW), whereas in the other cities the inverter is 5kW. The peak power demand of the load does not change depending on the city, rather the different inverter is selected for compatibility with the system voltage. The 1.5kW inverter in Cairo/Upington is only compatible with 12 or 24V systems, not 48V systems which are required in the other cities.

Figure 3 presents the breakdown of the system cost in the different cities. For the *Residential* load profile, the total cost in Amsterdam/Berlin is $8600, which is 30% more than the cost in Johannesburg of $6600. The cost of the inverters and batteries is similar across all the cities, with the largest variation occurring in the cost of the solar arrays. The cost of the solar array is generally the largest cost component, except for Cairo/Upington where the batteries cost more than the solar array.

For the *Office* load profile, the differences between cities are even more extreme, e.g. the cost in Amsterdam/Berlin is almost 70% above that of Johannesburg. In this scenario, both the cost of the batteries and solar array exhibit high variability between the cities.

The analysis of the *Residential* and *Office* load profiles presented in this section is not intended to be comprehensive. Rather, these examples illustrate the type of comparisons and insights that can be obtained using the app.

6 CONCLUSION

This paper presents an app for designing battery-based solar PV systems, with the goal of empowering consumers to design their own systems. Instead of asking the user to specify the load in terms of power and energy, a consumer-friendly approach is employed whereby the user makes selections from a database of common household appliances. The final system is chosen based on price and performance considerations, and the design output includes components from the supplier catalogue, the electrical configuration, and cost estimates.

The app is accessible via an API, or web app https://empowersolar-demo.mybluemix.net/.

We invite suggestions on ways to improve the app, and any interested suppliers or retailers are encouraged to contact us.

7 REFERENCES

[1] E. Karakaya and P. Sriwannawit, "Barriers to the adoption of photovoltaic systems: The state of the art," *Renew. Sustain. Energy Rev.*, vol. 49, pp. 60–66, 2015.

[2] V. Rai, D. C. Reeves, and R. Margolis, "Overcoming barriers and uncertainties in the adoption of residential solar PV," *Renew. Energy*, vol. 89, pp. 498–505, 2016.

[3] Photovoltaic-software.com, "Photovoltaic software." [Online]. Available: http://www.photovoltaic-software.com/.

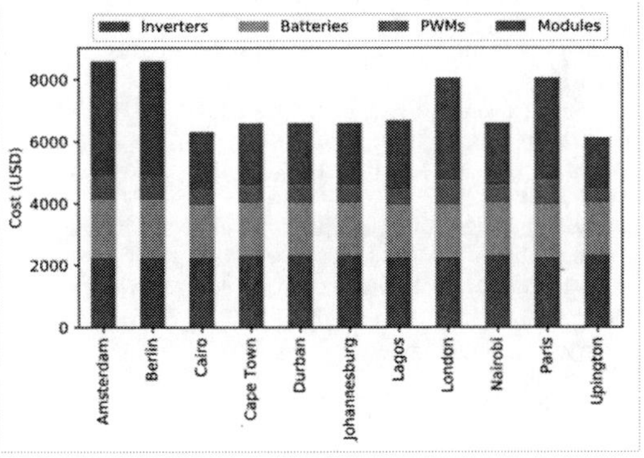

(a) Residential

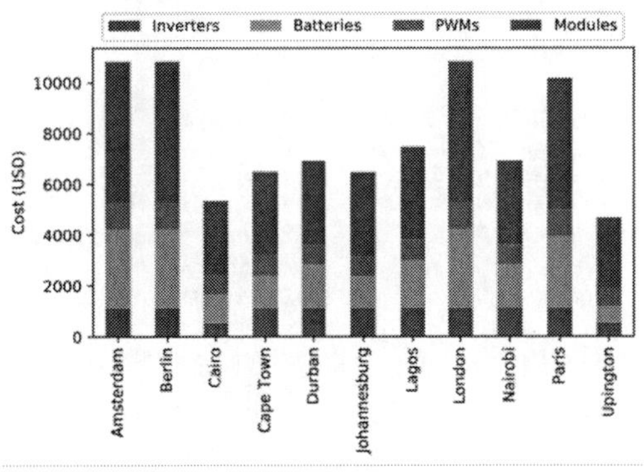

(b) Office

Figure 3: Breakdown of system cost in the different cities for *Residential* and *Office* load profiles

[4] NREL, "PVWatts Calculator." [Online]. Available: http://pvwatts.nrel.gov/.

[5] European Research Commission, "Photovoltaic Graphical Information Systems (PVGIS)." [Online]. Available: http://re.jrc.ec.europa.eu/pvgis/.

[6] T. Huld, R. Müller, and A. Gambardella, "A new solar radiation database for estimating PV performance in Europe and Africa," *Sol. Energy*, vol. 86, no. 6, pp. 1803–1815, 2012.

[7] M. Šúri, T. a. Huld, and E. D. Dunlop, "PV-GIS: a web-based solar radiation database for the calculation of PV potential in Europe," *Int. J. Sustain. Energy*, vol. 24, no. 2, pp. 55–67, 2005.

Table V: Configuration of solar system designed to supply *Residential Load*

City	Solar Rad	V	PV	Batteries	PMWs	Inverters
Amsterdam	3.35	48	6.0kWp, 20 x 300W (2 series by 10 strings)	14.7kWh, 12 x 102Ah (4 series by 3 parallel)	120A, 3 x 40A	10.0kW, 2 x 5.0kW
Berlin	3.34	48	6.0kWp, 20 x 300W (2 series by 10 strings)	14.7kWh, 12 x 102Ah (4 series by 3 parallel)	120A, 3 x 40A	10.0kW, 2 x 5.0kW
Cairo	6.47	48	3.0kWp, 10 x 300W (2 series by 5 strings)	13.8kWh, 12 x 96Ah (4 series by 3 parallel)	80A, 2 x 40A	10.0kW, 2 x 5.0kW
Cape Town	6.10	24	3.3kWp, 11 x 300W (1 series by 11 strings)	13.8kWh, 12 x 96Ah (2 series by 6 parallel)	120A, 3 x 40A	9.0kW, 3 x 3.0kW
Durban	5.69	24	3.3kWp, 11 x 300W (1 series by 11 strings)	13.8kWh, 12 x 96Ah (2 series by 6 parallel)	120A, 3 x 40A	9.0kW, 3 x 3.0kW
Johannesburg	6.21	24	3.3kWp, 11 x 300W (1 series by 11 strings)	13.8kWh, 12 x 96Ah (2 series by 6 parallel)	120A, 3 x 40A	9.0kW, 3 x 3.0kW
Lagos	5.25	48	3.6kWp, 12 x 300W (2 series by 6 strings)	13.8kWh, 12 x 96Ah (4 series by 3 parallel)	80A, 2 x 40A	10.0kW, 2 x 5.0kW
London	3.48	48	5.4kWp, 18 x 300W (2 series by 9 strings)	13.8kWh, 12 x 96Ah (4 series by 3 parallel)	120A, 3 x 40A	10.0kW, 2 x 5.0kW
Nairobi	5.71	24	3.3kWp, 11 x 300W (1 series by 11 strings)	13.8kWh, 12 x 96Ah (2 series by 6 parallel)	120A, 3 x 40A	9.0kW, 3 x 3.0kW
Paris	3.66	48	5.4kWp, 18 x 300W (2 series by 9 strings)	13.8kWh, 12 x 96Ah (4 series by 3 parallel)	120A, 3 x 40A	10.0kW, 2 x 5.0kW
Upington	7.00	24	2.7kWp, 9 x 300W (1 series by 9 strings)	13.8kWh, 12 x 96Ah (2 series by 6 parallel)	90A, 3 x 30A	9.0kW, 3 x 3.0kW

Table VI: Configuration of solar system designed to supply *Office Load*

City	Solar Rad	V	PV	Batteries	PMWs	Inverters
Amsterdam	3.35	48	9.0kWp, 30 x 300W (2 series by 15 strings)	24.5kWh, 20 x 102Ah (4 series by 5 parallel)	160A, 4 x 40A	5.0kW, 1 x 5.0kW
Berlin	3.34	48	9.0kWp, 30 x 300W (2 series by 15 strings)	24.5kWh, 20 x 102Ah (4 series by 5 parallel)	160A, 4 x 40A	5.0kW, 1 x 5.0kW
Cairo	6.47	24	4.8kWp, 16 x 300W (1 series by 16 strings)	9.2kWh, 8 x 96Ah (2 series by 4 parallel)	160A, 4 x 40A	3.0kW, 2 x 1.5kW
Cape Town	6.10	48	5.4kWp, 18 x 300W (2 series by 9 strings)	10.1kWh, 8 x 105Ah (4 series by 2 parallel)	120A, 3 x 40A	5.0kW, 1 x 5.0kW
Durban	5.69	48	5.4kWp, 18 x 300W (2 series by 9 strings)	13.8kWh, 12 x 96Ah (4 series by 3 parallel)	120A, 3 x 40A	5.0kW, 1 x 5.0kW
Johannesburg	6.21	48	5.4kWp, 18 x 300W (2 series by 9 strings)	9.8kWh, 8 x 102Ah (4 series by 2 parallel)	120A, 3 x 40A	5.0kW, 1 x 5.0kW
Lagos	5.25	48	6.0kWp, 20 x 300W (2 series by 10 strings)	14.7kWh, 12 x 102Ah (4 series by 3 parallel)	120A, 3 x 40A	5.0kW, 1 x 5.0kW
London	3.48	48	9.0kWp, 30 x 300W (2 series by 15 strings)	24.5kWh, 20 x 102Ah (4 series by 5 parallel)	160A, 4 x 40A	5.0kW, 1 x 5.0kW
Nairobi	5.71	48	5.4kWp, 18 x 300W (2 series by 9 strings)	13.8kWh, 12 x 96Ah (4 series by 3 parallel)	120A, 3 x 40A	5.0kW, 1 x 5.0kW
Paris	3.66	48	8.4kWp, 28 x 300W (2 series by 14 strings)	23.0kWh, 20 x 96Ah (4 series by 5 parallel)	160A, 4 x 40A	5.0kW, 1 x 5.0kW
Upington	7.00	24	4.5kWp, 15 x 300W (1 series by 15 strings)	5.5kWh, 2 x 230Ah (2 series by 1 parallel)	160A, 4 x 40A	3.0kW, 2 x 1.5kW

TECHNO-ECONOMIC ANALYSIS OF OFF-GRID PHOTOVOLTAIC LED ROAD LIGHTING SYSTEMS FOR TURKEY

Anıl Can Duman, Önder Güler
Istanbul Technical University, Energy Institute
Ayazaga Campus, 34469, Maslak, Istanbul, TURKEY
dumanan@itu.edu.tr, onder.guler@itu.edu.tr

ABSTRACT: In this study, techno-economic analysis of off-grid PV LED road lighting systems have been carried out for two provinces of Turkey. DIALux software is used for lighting calculations and HOMER software is used for energy system modeling and simulations. Calculations were made for two road lighting classes (M4, M5), two pole spacing options (maximum, minimum), five scenarios (current case and future projections for cases of 25% increase in electricity unit prices, 25% decrease in PV panel and battery costs, 50% decrease in PV panel and battery costs and 25% increase in electricity unit prices with 50% decrease in PV panel and battery costs) and for two provinces (Antalya and İstanbul). As a result of the study, it has been obtained that today off-grid PV LED road lighting systems cannot payback themselves both in Antalya and İstanbul. In the most favorable future scenario, payback periods can go below system lifetime and reach 14,23 and 21 years for Antalya and İstanbul respectively in M4 road lighting class. The results showed that, even though the cost of single pole system increases in maximum pole spacing case, the cost of the entire system reduces due to the installation of lesser amount of lighting poles.
Keywords: Stand-alone PV Systems, Sizing, Simulation, Economic Analysis

1 INTRODUCTION

Renewable energy systems have begun to be increasingly supported by governments, businesses and consumers in recent years as a more environmentally friendly and sustainable solution rather than conventional energy systems. Photovoltaic (PV) technology, which does not need fossil fuel, can provide a green and sustainable alternative to energy consumption needs by reducing CO_2 emissions. Structures of PV systems are suitable to meet energy consumptions of road lighting installations in off-grid applications. Today, off-grid PV LED road lighting systems which can only be feasibly used in rural areas where transportation is difficult and grid electricity does not reach is becoming attractive in the rest of the world due to the decrease in PV panel, battery and LED luminaire costs [1]. Developments in material technology provided PV costs which were 1.34 $/W in 2011 to reduce under 0.5 $/W today. It is estimated that prices will decrease further in the coming years [2] and recent developments in LED technology have made it possible to use lower-powered LED luminaires instead of higher-powered conventional luminaires. Reduction in the luminaire powers has made it available to make smaller PV panel and battery sizing which enables providing lighting criteria of the roads with off-grid PV systems more feasibly.

In this study, techno-economic analysis of off-grid PV LED road lighting systems have been carried out for two provinces of Turkey. Istanbul, financial and cultural center of Turkey at the north where irradiation and sunshine duration are low and Antalya, tourism center of Turkey at the south where irradiation and sunshine duration are high. DIALux software is used for road lighting calculations and HOMER software is used for energy system modeling and simulations. Calculations are made for two provinces (Antalya and İstanbul), two road lighting classes (M4 and M5), two pole spacing options (maximum and minimum) and five scenarios (in addition to current case, future projections for cases of 25% increase in electricity unit prices, 25% decrease in PV panel and battery costs, 50% decrease in PV panel and battery costs and 25% increase in electricity unit prices with 50% decrease in PV panel and battery costs)

2 SOLAR ENERGY POTENTIAL IN TURKEY

Turkey, with a surface area of 783,562 km^2, lies between 36° - 42° northern latitudes and 26° - 45° eastern longitudes. The country has a relatively high solar potential and has the highest sunshine period in Europe after Spain [3].

In the study carried out by the Electricity Affairs Survey Administration (EİE) which used solar radiation intensity and sunshine data for 1966-1982 of the General Directorate of State Meteorology Affairs (DMİ), the average annual sunshine period of Turkey was found as 2640 hours (daily total 7.2 hours) and the average total radiation intensity was calculated as 1311 kWh/m^2-year (total 3.6 kWh/m^2-day). Fig. 1 shows the solar potential map of Turkey [4]. Turkey has the potential to generate 376 billion kWh of electricity annually with solar energy due to being located within the zone between 40° north and 40° south latitudes which is called the "sun belt" [5]. By the end of September 2016, the installed power capacity of Turkey has reached 78072.2 MW of which 28.6% consists of natural gas + liquefied natural gas (LNG), followed by 24.5% hydraulic (dam), 22.2% coal (import + domestic), 10% hydraulic (river), 6.3% thermal (other), 6.7% wind, 0.9% geothermal and 0.8% solar energy. The 0.8% share of solar energy is equivalent to 660.2 MW installed power [6].

3 STRUCTURE OF OFF-GRID PV LED ROAD LIGHTING SYSTEMS

Off-grid PV LED road lighting systems consist of PV panel, battery group, charge controller and LED luminaire as shown in Fig. 2. PV panels charge the battery group by converting solar radiation into electricity, the charged battery group provides energy for the load during lighting hours and a charge regulator controls the output voltage of the panel and protects the battery from overcharges and discharges.

Figure 1: Solar potential map of Turkey

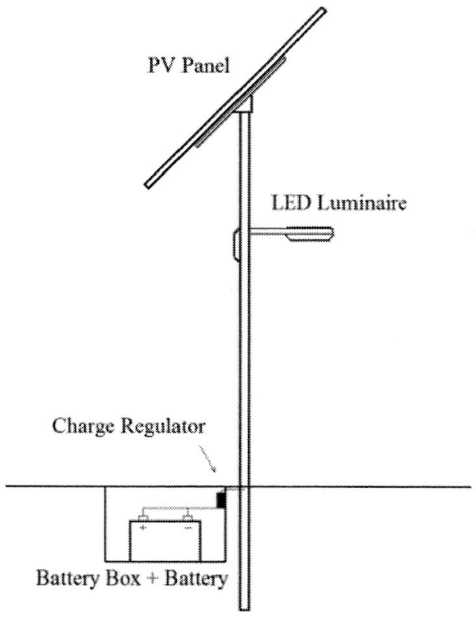

Figure 2: Structure of an off-grid PV LED road lighting system

4 ROAD LIGHTING

Road lighting is defined as illumination of intersections, squares, roads, inner and outer main traffic roads and covers most of the general lighting based energy consumption. According to 2014 Turkish Statistical Institute (TÜİK) data, 1.9% of 207,375 GWh electricity consumption in Turkey has been used for general lighting purposes [7]. General lighting consists of illumination of roads, streets, overpasses, underpasses, bridges, squares and pedestrian streets for general use of the public except for motorways and privately controlled access roads, and parking lots, gardens, historical sites open to the public for their free use and traffic signaling.

In Turkey, 20% of the total electricity energy is consumed for lighting. There is a potential of saving 40% of this amount by using energy efficient lighting systems [8].

Today, road lighting criteria are determined by the International Commission on Illumination (CIE) and the European Union Standards (CEN) [9,10]. With these criteria defined, each country chooses its preferences considering its geographical, economic, climatic and cultural situation. Turkish Electricity Distribution Co. (TEDAŞ) holds the responsibility for the installation and maintenance of approximately 5 million lighting poles located in cities and rural areas in Turkey [11]. The specifications taken into consideration during road lighting designs are "TEDAŞ Technical Specification for LED Light Sourced Road Lighting Luminaires" and "Procedures and Principles on the Usage of LED Luminaires in the General Lighting Scope" [12,13].

4.1 Road lighting calculations

This study is carried out for the M4 and M5 road lighting classes, where relatively lower-powered luminaires could be used since in off-grid PV road lighting systems an increase in luminaire power means an increase in sizing of PV panel, battery, charge regulator and lighting pole.

For the M4 and M5 road lighting classes, road lighting calculations were made through the left-sided lighting arrangement for a 2 lane road with the width of 7 meters. The LED luminaires guarantee protection class IP66 and the maintenance factor according to CIE 154-2003 [14] is set as 0.89. Road class has been chosen as R3. Observer age for calculations is set as 23. According to the LED Light Sourced Road Lighting Luminaires Technical Specification, the minimum luminous efficacy is 115 lm/W, the color temperatures of the LED packages used in the design of LED luminaires are 4000 K ± 5% and LED luminaires should operate for minimum 50,000 hours. Table I shows the road lighting criteria need to be followed for the selected road lighting classes. Technical information and light distribution curves of the luminaires used in road lighting calculations are given in Fig. 3.

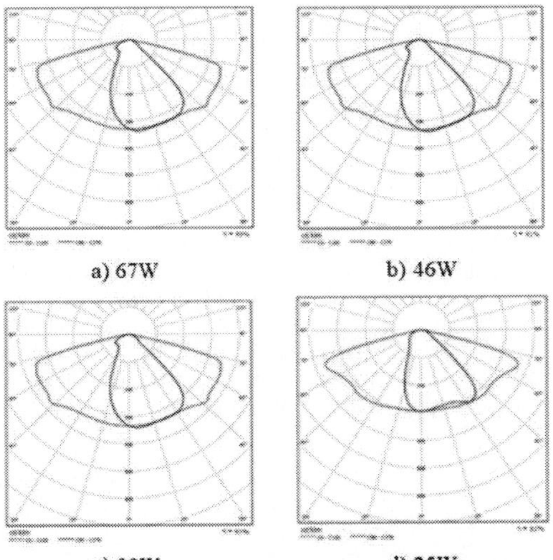

Figure 3: Light distribution curves of luminaires

According to the TEDAŞ LED Light Sourced Road Lighting Luminaire Technical Specification for the case of left-sided arrangement, minimum pole spacing to be provided for M5 and M4 road lighting classes is 28 m [12]. Thus, for road lighting calculations, search spaces were selected ranging between 28 m – 55 m for pole spacing in 1 m increments, between 7 m - 10 m for pole length in 0.5 m increments and between 0 m – 1.5 m for boom length in 0.5 m increments with 0° angle of inclination.

Table I: Road lighting criteria for selected road lighting classes

Lighting Class	L_{avg}(cd/m²)	U_o	U_l	TI (%)	SR
M4	≥0.75	≥0.4	≥0.5	≤15	≥0.5
M5	≥0.5	≥0.35	≥0.4	≤15	≥0.5

L_{avg}: average luminance ; U_o: overall uniformity ; U_l: longitudinal uniformity ; TI: threshold increment ; SR: surround ratio.

In the study, maximum and minimum pole spacing options for the selected road lighting classes have been examined.

In order to meet lighting criteria at maximum pole spacing, there is need for higher-powered PV panels, higher-capacity batteries and charge regulators to supply the load requirement of higher-powered luminaires, and thus more durable and higher lighting poles to carry all of these equipment. On the other hand, maximum pole spacing means using a minimum number of lighting poles which also enables a reduction in maintenance & operating costs.

On the contrary, in the case of minimum pole spacing, lower-powered PV panels, lower-capacity batteries and charge regulators are needed to supply the load requirement of lower-powered luminaires. By contrast, minimum pole spacing means using a maximum number of lighting poles and also an increase in maintenance & operating costs.

Road lighting calculations according to maximum and minimum pole spacing scenarios for M4 and M5 road lighting classes are shown in Table II.

Table II: Road lighting calculations for M4 and M5 road lighting classes

Lighting Class	M4		M5	
Luminaire Power (W)	67	39	46	25
Luminaire Luminous Flux (lm)	8280	4641	5642	2875
Luminous Efficacy of Luminaire (lm/W)	123.59	119	122.65	115
S (m) (Spacing)	47	28	53	28
h (m) (height)	10	8.5	10	8.5
k (m) (boom length)	0	0	1	0
L_{avg} (cd/m²)	0.81	0.82	0.50	0.52
U_o	0.49	0.52	0.41	0.52
U_l	0.51	0.83	0.41	0.83
TI (%)	12	10	12	9
SR	0.63	0.50	0.61	0.50

5 HOMER MODELING

In the study, modeling and simulation of the energy systems have been carried out in HOMER software. Monthly average solar radiation data is obtained from the National Aeronautics and Space Administration (NASA) Meteorology and Solar Energy Database. The project lifetime is set at 20 years. Real interest rate is calculated as 3%. Since the LED luminaires are supplied with DC voltage and the system is off-grid, the inverter is not modeled.

PV panel price is taken as 0.52 $/W through a market research over polycrystalline panels. Operation and maintenance costs of the PV panel is set as 7 $/yr. In the simulations, the power of the PV panel is searched in the range of 100-1000 W in 10 W increments. PV derating factor which includes the effect of dust on modules, high temperature, shading, snow cover, aging and cable losses is taken as 80% in the study which can vary between 70% and 90% in hot climates. The percentage of ground reflectance, which is the proportion of the radiation coming to the earth to reflected by the earth, is taken as 20%. PV lifetime is set as 20 years. Since PV lifetime is equal to the system lifetime, replacement cost and salvage value for PV are not included in the calculations.

12 V lead acid batteries are used in the modeling with allowed minimum state of charge (SoC) of 30% and round trip efficiency of 86%. According to different capacity requirements 9 battery options were proposed. Battery prices varied between $117 - $997.5 and nominal capacity varied between 33.3 Ah – 500 Ah. As the battery and panel are installed on the same pole, the operation & maintenance cost of the batteries are included in the operation & maintenance cost of the PV panels.

Calculations were made assuming that the road lighting would be performed between twilight and dawn. So, loads were out of operation during the daytime and civil twilight, which is the part of the day when the angle between the horizon and the sun is less than 6° and where the objects can still be easily selected and people can perform daily tasks without any requirement of artificial lighting. Daylight saving time has been taken into consideration. Table III shows the daily and monthly average lighting durations for Antalya and İstanbul.

Table III : Average lighting times and durations for Antalya and İstanbul

	Daily Average Lighting Time		Daily Average Lighting Duration		Monthly Average Lighting Duration	
	Ant	İst	Ant	İst	Ant	İst
Jan	06:41 17:33	06:56 17:31	13h 08m	13h 25m	407.13h	415.92h
Feb	06:20 18:03	06:31 18:06	12h 17m	12h 25m	343.93h	347.67h
Mar	05:41 18:31	05:46 18:40	11h 10m	11h 06m	346.17h	344.1h
Apr	05:55 20:00	05:56 20:13	09h 55m	09h 43m	297.5h	291.5h
May	05:19 20:28	05:14 20:48	08h 51m	08h 26m	274.35h	261.43h
Jun	05:07 20:49	04:58 21:12	08h 18m	07h 46m	249h	233h
Jul	05:21 20:45	05:13 21:06	08h 36m	08h 07m	266.6h	251.62h
Aug	05:48 20:14	05:46 20:30	09h 34m	09h 16m	296.57h	287.27h
Sep	06:13 19:30	06:17 19:40	10h 43m	10h 37m	321.5h	318.5h
Oct	06:40 18:45	06:49 18:49	11h 55m	12h 00m	369.42h	372h
Nov	06:09 17:15	06:22 17:15	12h 54m	13h 07m	387h	393.5h
Dec	06:35 17:11	06:52 17:08	13h 24m	13h 44m	402h	412h
				Total	3961.17h	3928.51h

Load profile of 25W and 46W luminaires for M5 road lighting class, and 39W and 67W luminaires for M4 road lighting class have been created in HOMER software. Daily, seasonal and yearly load profile of 25W luminarie is shown in Fig. 4 as sample.

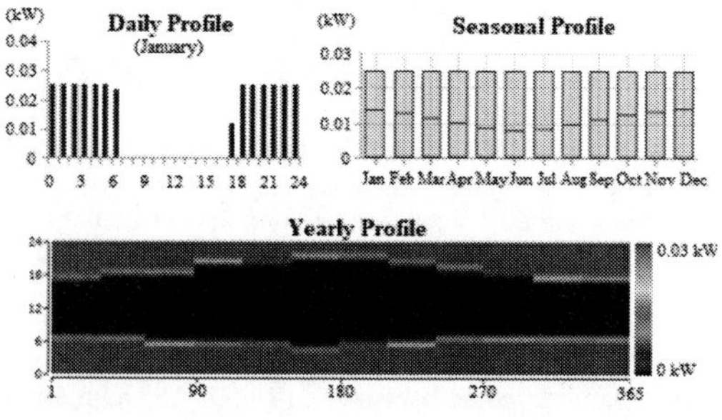

Figure 4 : Load profile of 25W luminaire in HOMER software as sample.

In the study, systems' contribution to the environmental sustainability is also considered that off-grid PV lighting systems allow reduction of CO_2 emissions to the atmosphere. The International Energy Agency (IEA) data has been used to determine the amount of CO_2 reduction. According to the data, the amount of CO_2 emission produced per kWh in Turkey is 490 gCO_2 / kWh [15].

6 CALCULATIONS

Antalya is located in the southwest of Turkey in the Mediterranean Region between 29° 20' – 32° 35' east longitudes and 36° 07' – 37° 29' north latitudes. İstanbul is located in the northwest of Turkey in the Marmara Region between 28° 01' and 29° 55' east longitudes and 41° 33' and 40° 28' north latitudes. The global radiation and clearness index data for Antalya and İstanbul are given in Fig. 5. The data were obtained from NASA's Surface Meteorology and Solar Energy Database.

Simulation results for the case of using left-sided lighting arrangement in the M4 and M5 road lighting classes according to the minimum and maximum pole spacing scenarios for Antalya and İstanbul are given in Table IV.

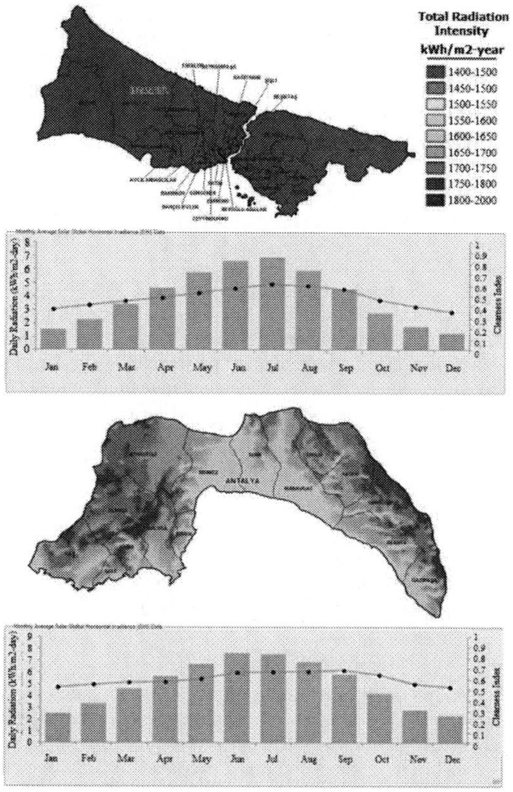

Figure 5: Global radiation and clearness index data for Antalya and İstanbul

6.1 Investment payback periods for Antalya and İstanbul

The payback periods of the energy system investments obtained from calculations for İstanbul and Antalya are given in Table V and VI. The electricity tariff for general lighting is taken as 0.128 $/kWh (0.37 TL/kWh) in the calculation of payback periods [16]. The results showed that under current condition in Turkey, payback periods of off-grid PV LED road lighting systems cannot go below 20 years which is the system lifetime. For this reason, in addition to current case, calculations were also made for future scenarios in which 1) electricity unit prices increases 25%, 2) PV and battery prices decrease 25%, 3) PV and battery prices decrease 50% and 4) PV and battery prices decrease %50 and electricity unit prices increases 25%.

Table IV: Simulation results for the M4 and M5 road lighting classes according to the minimum and maximum pole spacing scenarios for Antalya and İstanbul

Province	Antalya				İstanbul			
Road Lighting Class	M5		M4		M5		M4	
Pole Distance (m)	28	47	28	47	28	53	28	47
Luminaire Power (W)	25	46	39	67	25	46	39	67
Battery Capacity (V, Ah)	12V 100Ah	2 x 12V 100Ah	12V 166.6Ah	12V 250Ah	12V 100Ah	12V 250Ah	12V 166.6Ah	12V 416.6Ah
PV Panel Power (W)	170	280	240	480	490	680	710	890
PV Tilt Angle (°)	36.90				41.10			
Cost of Energy ($/kWh)	0.281	0.254	0.258	0.229	0.395	0.359	0.367	0.362
Battery + PV Initial Cost ($)	308.5	585.10	490.05	759.60	474.55	863.60	734.45	1316
Battery + PV Net Present Cost ($)	414.60	689.24	594.19	907.11	578.69	967.74	838.59	1420
Charge Regulator Cost ($)	75	75	75	100	100	100	125	125
Operation & Maintenance Cost ($)	104.4	104.14	104.14	104.14	104.4	104.14	104.14	104.14
PV Electricity Production (kWh/yr)	274	448	384	779	618	857	900	1132
PV Excess Electricity Production (kWh/yr)	159	236.3	204.5	471	504.5	647.4	722.1	828.2
Excess Electricity Production / Electricity Production (%)	58.1	52.8	53.3	60.5	81.6	75.6	80.3	73.1
Load Electricity Consumption (kWh/yr)	99	183	155	266	98	181	154	264
Unmet Electric Load (%)	0							
Autonomy (hr)	74.04	80.51	79.13	69.09	74.69	101.39	79.72	116.15
CO_2 Emission Reduction ($kgCO_2/yr$)	48.51	89.67	75.95	130.34	48.02	88.69	75.46	129.36

Table V: Investment payback periods for Antalya

Lighting Class	Luminaire Power (W)	Payback Period (years)	Energy System Net Present Cost ($)	Load Electricity Consumption (kWh/yr)
For the current case				
M5	25	38.64	489.6	99
	46	32.63	764.24	183
M4	39	33.73	669.19	155
	67	31.64	1007.11	266
For 25% increase in electricity unit prices				
M5	25	30.91	489.6	99
	46	26.10	764.24	183
M4	39	26.98	669.19	155
	67	23.66	1007.11	266
For 25% decrease in PV and battery costs				
M5	25	32.51	411.98	99
	46	26.38	617.96	183
M4	39	27.55	546.68	155
	67	23.68	806.37	266
For 50% decrease in PV and battery costs				
M5	25	26.38	334.37	99
	46	20.13	471.69	183
M4	39	21.37	424.17	155
	67	17.78	605.62	266
For 25% increase in electricity unit prices and 50% decrease in PV and battery costs				
M5	25	21.11	259.37	99
	46	16.11	396.69	183
M4	39	17.10	349.17	155
	67	14.23	505.62	266

Table VI: Investment payback periods for İstanbul

Lighting Class	Luminaire Power (W)	Payback Period (years)	Energy System Net Present Cost ($)	Load Electricity Consumption (kWh/yr)
For the current case				
M5	25	54.10	678.69	98
	46	46.08	1067.74	181
M4	39	48.88	963.59	154
	67	45.72	1545	264
For 25% increase in electricity unit prices				
M5	25	43.28	678.69	98
	46	36.87	1067.74	181
M4	39	39.10	963.59	154
	67	36.57	1545	264
For 25% decrease in PV and battery costs				
M5	25	44.65	560.05	98
	46	36.76	851.84	181
M4	39	39.57	779.98	154
	67	35.98	1216	264
For 50% decrease in PV and battery costs				
M5	25	35.18	441.42	98
	46	27.44	635.94	181
M4	39	30.25	596.37	154
	67	26.26	887.29	264
For 25% increase in electricity unit prices and 50% decrease in PV and battery costs				
M5	25	28.15	341.42	98
	46	21.96	535.94	181
M4	39	24.20	471.37	154
	67	21.00	762.29	264

Table VII: System installation costs per kilometer for Antalya and İstanbul according to the M4 and M5 road lighting classes

Province	Antalya				İstanbul			
Lighting Class	M4		M5		M4		M5	
Luminaire Power (W)	39	67	25	46	39	67	25	46
Pole Spacing (m)	28	47	28	53	28	47	28	53
Pole Length (m)	8.5	10	8.5	10	8.5	10	8.5	10
Battery + PV Net Present Cost ($)	594.19	907.11	414.60	689.24	838.59	1420	578.69	967.74
LED Luminaire Cost ($)	291	316.5	285	295.5	291	316.5	285	295.5
Charge Regulator Cost ($)	75	100	75	75	125	125	100	100
Galvanized Steel Polygon Lighting Pole Cost ($)	137.91	184.10	137.91	184.10	137.91	184.10	137.91	184.10
Boom Cost ($)	0	0	0	7.35	0	0	0	7.35
Pole Mounting Cost ($)	73.20	97.72	73.20	97.72	66.73	89.08	66.73	89.08
Cable Cost ($)	3.06	3.6	3.06	3.6	3.06	3.6	3.06	3.6
Cabling Cost ($)	1.02	1.2	1.02	1.2	1.02	1.2	1.02	1.2
Single Pole System Cost ($)	1175.38	1610.23	989.79	1353.71	1463.31	2139.48	1172.41	1648.57
Number of Poles per km	36	22	36	19	36	22	36	19
Installation Cost per km ($/km)	42313.68	35425.06	34632.44	25720.49	52679.16	47068.56	42206.76	31322.83
Single Pole System CoE ($/kWh)	0.258	0.229	0.281	0.254	0.367	0.362	0.395	0.359
Annual Operation Duration (hours)	3961.17				3928.51			
Annual Electricity Consumption per km (kWh)	5561.48	5838.76	3565.05	3462.06	5515.63	5790.62	3535.66	3433.52

6.2 System installation costs per kilometer for Antalya and İstanbul

In the study, system installation costs of off-grid PV LED road lighting systems for Antalya and İstanbul according to the M4 and M5 road lighting classes were calculated separately for both single pole system and entire system depending on number of poles to be built in 1 km in order to make a comparison between maximum and minimum pole spacing options. Calculation results are given in Table VII.

7 CONCLUSIONS

As a result of the study, it has been observed that off-grid PV road lighting system investments cannot payback themselves both in Antalya and İstanbul under current conditions and payback period of the systems cannot go below 20 years which is the system lifetime. For M4 road lighting class, in the current case, system payback period is 31.64 years for Antalya and 45.72 years for İstanbul. In the most favorable scenario (25% increase in electricity unit prices and 50% reduction in battery and PV prices), payback period can go below 20 years and reach 14.23 years for Antalya and 21 years for İstanbul. For M5 road lighting class, in the current case system payback period is 32.63 for Antalya and 46.08 years for İstanbul. In the most favorable scenario, payback periods go down to 16.11 for Antalya and 21.96 for İstanbul.

The results showed that, cost of energy (COE) which is the average cost per kWh to produce useful electricity energy by the systems, are lower in Antalya than in İstanbul. While the COE decreased as the luminaire powers increased in Antalya province, this situation continued to be the same in İstanbul province only to a certain extent. Since maximum PV panel capacity that can be mounted on a lighting pole is limited, in İstanbul when sunshine and solar radiation conditions became inadequate, systems than could not increase the limited PV capacity anymore increased battery capacity instead which are limitless but also more costly.

As a result of the study, it is found that, the maximum pole spacing should be preferred in an off-grid PV LED road lighting system design between the maximum and minimum pole spacing options. Even though the cost of the single pole system increases in maximum pole spacing case, the cost of the entire system is reduced due to the installation of lesser amount of lighting poles.

The study showed that, off-grid PV road lighting systems are not feasible under current conditions in Turkey. Nonetheless, the systems can be less costly than on-grid systems if they are installed in rural areas where the grid electricity is not reachable and new transmission lines are required to energize the road lighting systems. As the lighting criteria were aimed to be uninterruptedly met throughout the year, batteries and PV panels were sized according to the lowest radiation intensities and sunshine durations of winter months, which increased the costs. Most of the electricity produced couldn't be consumed.

Despite the unfeasible results, the systems have the added benefits of contributing to the reduction of CO_2 emissions, increasing awareness of environmental policies, contributing to the environmental images of institutions and organizations and supporting works and experiments in PV lighting.

REFERENCES

[1] Baurzhan S, Jenkins GP. Off-grid solar PV: is it an affordable or appropriate solution for rural electrification in Sub-Saharan African countries? (2006). Renew Sustain Energy Rev ;60:1405–18.
[2] Feldman, D., Barbose, G., Margolis, R., Bolinger, M., Chung, D., Fu, R., Seel, J., Davidson, C. & Wiser, R. (2015). Photovoltaic System Pricing Trends Historical, Recent, and Near-Term Projections, National Renewable Energy Laboratory Publications.
[3] Cengiz, M. S., Mamiş M. S. (2015) Thermal Solar Energy Use and Turkey Analysis.
[4] www.eie.gov.tr/MyCalculator/Default.aspx. Accessed 07.08.2015.
[5] Aksungur, K.M., Kurban, M., Filik, Ü.B. (2013) Türkiye'nin Farklı Bölgelerindeki Güneş Işınım Verilerinin Analizi ve Değerlendirilmesi, 5. Enerji Verimliliği ve Kalitesi Sempozyumu, Kocaeli.
[6] http://www.teias.gov.tr/yukdagitim/kuruluguc.xls. Accessed 30.09.2016.
[7] www.tuik.gov.tr/PreIstatistikTablo.do?istab_id=1579> Accessed 09.08.2015.
[8] www.lighting.philips.com/pwc_li/main/products/controls/assets/ankara%20-led-road-lighting-pilot-project-case-study.pdf. Accessed Aug. 2015.
[9] International Commission on Illumination (2010) "Lighting of Roads for Motor and Pedestrian Traffic," CIE Central Bureau CIE 115.
[10] European Committee for Standardization (2003). Road Lighting. Part 2: Performance Requirements, CEN, Brussels EN 13201-2.
[11] Güler, Ö. & Onaygil, S. (2007). Yol Aydınlatması Tesisatlarında Armatür Fotometrik Değerlerinin Önemi, IV. Ulusal Aydınlatma Sempozyumu, pg. 125-135, İzmir.
[12] TEDAŞ (2015). Technical Specification for LED Light Sourced Road Lighting Luminaires.
[13] Republic of Turkey The Ministry of Energy and Natural Resources (2015). Procedures and Principles on the Usage of LED Luminaires in the General Lighting Scope.
[14] CIE Pub. 154, (2003) The Maintenance of Outdoor Lighting Systems, International Commission on Illumination, Wien.
[15] ABB (2011). Turkey Energy Efficiency Report.
[16] http://www.tedas.gov.tr/#!tedas_tarifeler. Accessed April 2016.

SOLAR PHOTOVOLTAIC FOR SUSTAINABLE USE IN TRITURATION OIL OLIVE UNIT AND ENERGY EFFICIENCY IN COLD AND HYDRIC STORAGE.

S. Mounir[1], S.Ladouy[2], K.EL Harrouni[3], A.Khabbazi[4], Y.Maaloufa[5]
[1]National School of architecture Fez, 80 Avenue saint louis street Saada Fez.
[3]National School of architecture Rabat, Avenue Allal el FassiMadinat al Irfane B.P 6372 Rabat.
[2,4,5]LEME EST , University Mohammed V in Rabat , 227 Avenue Prince Heritier Sale Morocco.
Company / Institute(s)
soumia.mounir@gmail.com,sara.ladouy@gmail.com, Kelharrouni@gmail.com,abdelhamidkab@hotmail.com, y.maaloufa@gmail.com

ABSTRACT: The object of this work is the use of photovoltaic solar in sustainability, especially in agriculture so as to feed a trituration oil olive unit and to manage the excess of electricity for a refrigeration system to store growing vegetables, fruits and water pump to irrigate soil.
The study concerns a rural area in Ouazzane a city in the north of Morocco. The first step consists of studying the irradiation of this city using PVgis. The second step consists of sizing the photovoltaic installation necessary for the trituration oil olive unit which works just three months per year, the rest of the year , the energy produced by the PV system is stored as a thermal or hydric energy to be used to pump water for irrigation and refrigeration system necessary to store vegetables and fruits. The capacity of this PV solar system is 21 Kwp which assured by 150 PV panel. The preliminary results indicate an important gain in term of energy and CO2 gas emissions.
Keywords: PV solar, trituration oil olive unit, refrigeration system, pumping water.

1 DEADLINES AND DELIVERY INTRODUCTION

Industrialization of oil olive consumes a huge average of energy and hurts the environment by the big emissions of CO2. The use of Solar PV system on oil olive trituration unit will present an important gain in term of energy.

A state of the art was done such as the work of the Setiawan et al.[1]which study the development of solar water pumping system as a renewable energy to solve water supply problem in Purwodadi village when they used 32 solar panels to produce 3200Wp, the flow rate produced is about 0.4-0.9 L.s-1. The work of chandel et al.[2] which present a literature study on solar water pumping, evaluate the economic viability and propagation of this technology. The work of closas et Rap[3] where groundwater pumping system are studied using PV solar, they discuss how many policies and projects promoting financial and economic cost of these solutions and their limitations concerning the equipment related to solar and their impacts on the groundwater. Verma et Dondapanti[4]are treatedthe techno economic sizing analysis of PV solar system for domestic refrigerators, this work develops a solar PV system for the use in domestics refrigerator. PV solar system developed can generate 47085kWh/year.

The use of renewable energy in agriculture a huge sector of consumption of electricity has become a necessity to benefit freely from energy. For this purpose, authors studied the sustainability of solar photovoltaic system to feed anoil olive trituration unit working three months per year and to manage the excess of energy on a refrigeration system necessary to store vegetables, fruits and a pumping water system necessary for irrigation.

The site studied is Ouazzane a north city in Morocco. The trituration unit consumes 8.44 kW and produces 150 kg.h-1 of oil olive. First step of this scientific work is the description of the installation treated and the study of the Irradiation of Ouazzane city. The second step is the sizing of the solar installation, to determine the number of the Panels and the accumulator then the monitoring of this installation according to the different sequences of working. Finally, a study of the impact of this installation on the environment was studied to determine the footprint of carbon and the gain obtained in term of the CO2 gas emissions.

2 RESULTS AND DISCUSSIONS

2.1 Description of the PV solar installation

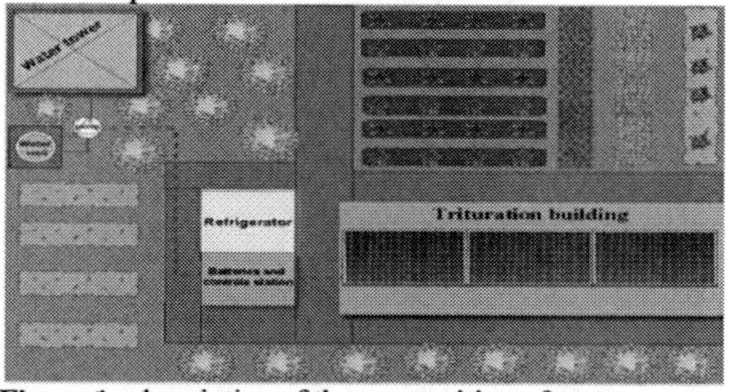

Figure 1 : description of the composition of our installation feeding by the PV solar panels.

The installation object of our study is composed of PV solar panels which feed electricity to a trituration oil olive unit. This unit machine works just three months per year and the rest of the energy produced is used for a refrigeration system and a water pump which fills out a water tower.

2.2 Study of the Irradiation of the site ouazzane

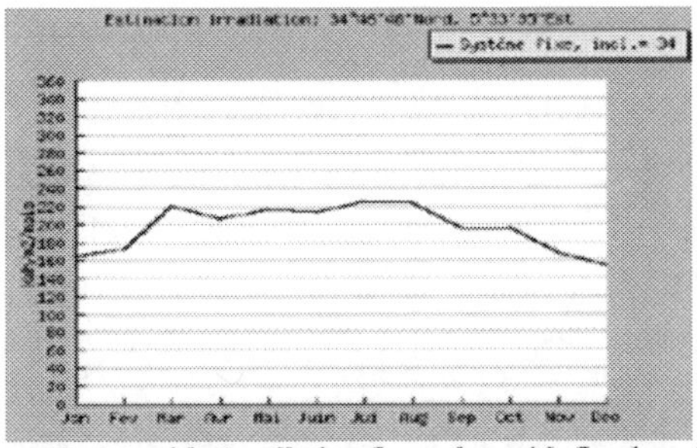

Figure 2 : Monthly Irradiation for a plan with fixed angle.

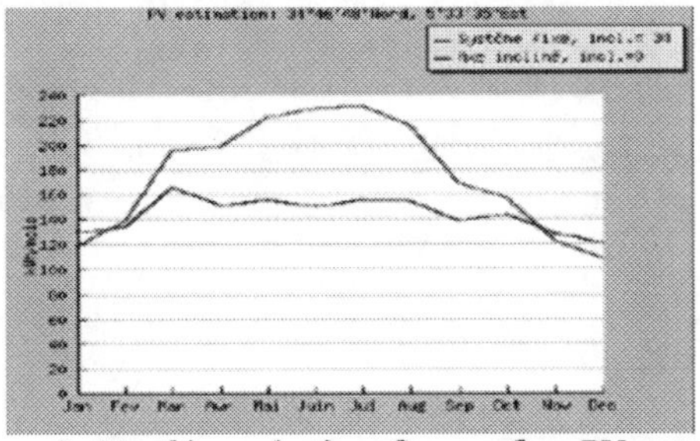

Figure 3 : Monthly production of energy for a PV system with a fixed angle.

The analyses of the figures indicate that in spring and summer, we have the maximum irradiation attending 6 kWh/m2. However, in the winter the average of monthly irradiation decreases.

2.3 Sizing of the PV solar installation

Our trituration oil olive unit works 10 hours per day,its power is 8.5Kw; it consumes a flow of water 80 $l.h^{-1}$.

The energy consumed

$$E_c = 8.5 \times 10 = \frac{85KWh}{day} \quad (1)$$

$$E_p = \frac{E_c}{K} = 130.77\ Kw \quad (2)$$

For the coefficient K we take 0.65 for the system of PV solar with battery

The average of Irradiation taken for Ouazzane is $6.32kWh.m^{-2}$

$$P_C = \frac{E_P}{I_r} = 21Kwp \quad (3)$$

Number of panel

The characteristic of the panel used is 1.42 m2 and of power capacity 140 W

$$N = \frac{P_c}{P_{cu}} = 150 \quad (4)$$

The capacity of accumulator

$$C = \frac{E_C \times N}{D \times U} \quad (5)$$

Where

N: number of the autonomy day we have taken N=3 day

D: Maximum admissible discharge, we take 0.8 for the battery of plumb

U: Battery voltage (v)

C= 13281 (Ah)

The system is working only for three months but during the rest of the year, the PV array produces electricity that needs to be managed in a way to cover the needs of the rural community. That's why half of the energy will be stored as hydric energy using water tower for irrigation purposes, the other half will be stored as thermal energy to ensure cold for vegetables produced by the region.

The PV ensures a $P_c = 21Kwp$

The tower will be dimensioned to use the half of this energy as amaximum of

$$P_{cTower} = 11KWh/_{day} \quad (6)$$

The potential of Hydraulic energy E_h (kWh/d) available per day is given by

$$E_h = E_{cTower} \times I \times R \times F = 11 \times 6 \times 0.5 \times 0.65 = 22.6kW \quad (7)$$

R: pump efficiency in the volumetric pump R=0.5

This hydric energy is the motor-pump system energy.

The Hydraulic energy E_h (kWh/d) required per day to supply a volume V of water (m3) at TDH is given by

$$E_h = \rho \times g \times V \times TDH \quad (8)$$

We take TDH=3m

$$V = \frac{E_h}{\rho \times g \times TDH} \quad (9)$$

The water tower volume is V=5m3

2.3 Monitoring of the PV installation

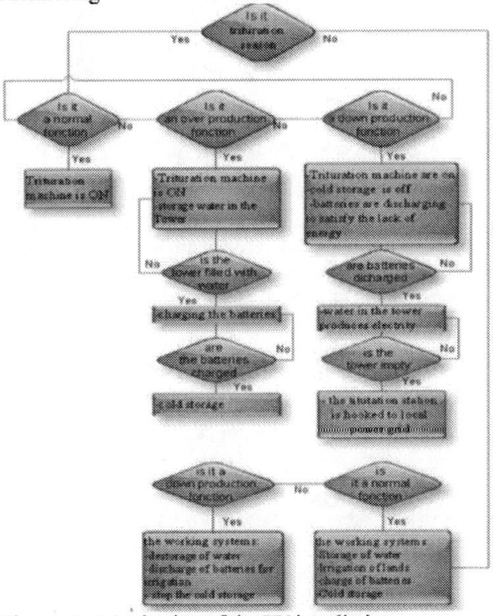

Figure 4 : Monitoring of the PV installation.

The schema shows the monitoring of the PV installation and their sequences when the trituration oil olive unit is producing oil and when it doesn't work.

The Organigram explains the function of the PV installation all over the year:

A cold storage consists of producing an important mass of ice that will be used in separate isolated rooms to store fruits, vegetables and fish;

During the down production function is stopped for two reasons:

During this period the amount of solar radiation is low and thus we maintain an ambient temperature so the system can ensure a longer autonomy;

During this period the demand for cold will be low since the fruits vegetables and fish are exposed to a lower ambient temperature.

2.3 Impact of the installation in the environment

A state of the art was studied on the impact of our PV installation studied on the emission of the CO2 was revealed that [5,6]if the modules used are made in Spain the carbon footprint is 37.3 gCO2/kWh for Mono-Si, equal to 3170 gCO2/day concerning our installation which consumes 85kWh/day.But, if the modules used are made in China the carbon footprint is 72.2 gCO2/kWh for Mono-Si, equal to 6137gCO2/day. However,[7]if the production of electricity was assured by coal the footprint carbon is 1060 gCO2/kWh equal, to 90100 gCO2/day or if electricity is assured by fuel the footprint carbon is 730 gCO2/kWh equal to 62050 gCO2/day if electricity is assured by the gas the footprint carbon is 418 gCO2/kWh equal to 35530 gCO2/day.

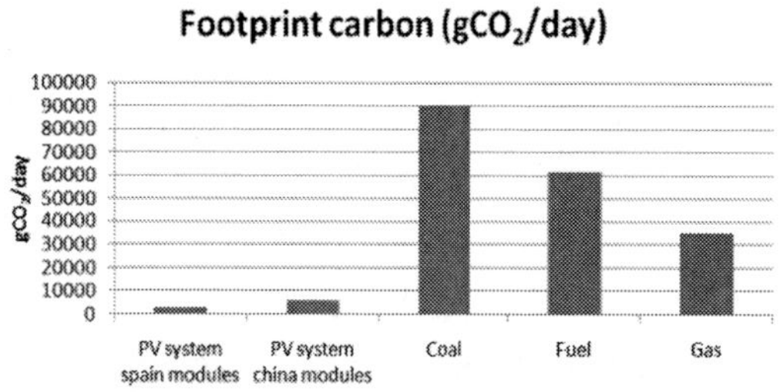

Figure 5: Footprint carbon according to the system of production of electricity.

To illustrate the gain obtained from using PV system for the trituration oil olive unit, authors presented the histogram 1 which shows that the modules made in Spain permit a gain of 91% comparing to a system using the gas to produce electricity and 96% compared to a system using coal to produce electricity. However, for the modules made in China, they permit a gain of 82.72% compared to a system using gas and a gain of 93.18 comparing to a system using coal.

3. CONCLUSION

This scientific work aims to study the sustainability of a PV system used to feed an oil olive trituration unit which works just three months per year and to manage the rest of free energy for a refrigeration system and for a water pump necessary to irrigate the soil. The study was done on a site in the north of Morocco Ouazzane city where authors have studied its average irradiation. The PV ensures a maximum of 21kWp. Half of the energy was managed to feed a tower sized to contain 5m3 of water and the half of energy of 11 kWh/day will be used in a refrigeration system necessary to freeze fruits and vegetables. Finally, a study of the impact of the PV system on the environment the results were proved that there is an important decrease ofCO2 emission and a gain attending 96% compared to other systems of producing electricity.

REFERENCES

[1] A.A. Setiawan, D.H. Purwanto, D.S. Pamuji, N. Huda, Development of a Solar Water Pumping System in Karsts Rural Area Tepus, Gunungkidul through Student Community Services, Energy Procedia. 47 (2014) 7–14. doi:10.1016/j.egypro.2014.01.190.

[2] S.S. Chandel, M. Nagaraju Naik, R. Chandel, Review of solar photovoltaic water pumping system technology for irrigation and community drinking water supplies, Renew. Sustain. Energy Rev. 49 (2015) 1084–1099. doi:10.1016/j.rser.2015.04.083.

[3] A. Closas, E. Rap, Solar-based groundwater pumping for irrigation: Sustainability, policies, and limitations, Energy Policy. 104 (2017) 33–37. doi:10.1016/j.enpol.2017.01.035.

[4] J. kishor Verma, R.S. Dondapati, Techno-economic Sizing Analysis of Solar PV System for Domestic Refrigerators, Energy Procedia. 109 (2017) 286–292. doi:10.1016/j.egypro.2017.03.068.

[5] D. Yue, F. You, S.B. Darling, Domestic and overseas manufacturing scenarios of silicon-based photovoltaics: Life cycle energy and environmental comparative analysis, Sol. Energy. 105 (2014) 669–678. doi:10.1016/j.solener.2014.04.008.

[6] J.M. Pearce, Optimizing greenhouse gas mitigation strategies to suppress energy cannibalism, (n.d.). http://inis.iaea.org/Search/search.aspx?orig_q=RN:47087370 (accessed July 19, 2017).

[7] Quel bilan carbone des sources de production d'électricité, Réseau Durable - Réflex. Sur Lénergie En Réseau. (2016). http://reseaudurable.com/bilan-carbone-production-electricite-france/ (accessed July 19, 2017).

PERFORMANCE PARAMETERS OF AN OFF-GRID BUILDING INTEGRATED PHOTOVOLTAIC SYSTEM IN SOUTH AFRICA

Edson L. Meyer[1], Carine L. Buma[2], Raymond Taziwa[3]
University of Fort Hare, Institute of Technology
1 Fort Hare Drive, 3rd Floor Chemistry Building, Alice 5700, South Africa
emeyer@ufh.ac.za[1], cbuma@ufh.ac.za[2], rtaziwa@ufh.ac.za[3]

ABSTRACT: Less than 4% of South Africa's energy is being generated from renewable energy sources. With South Africa's vast solar resource, a lot more can be done to nurture the dependence on renewable sources of energy such as solar energy. The use of off-grid photovoltaic systems is one major way of reaching out to the rural and isolated communities inaccessible to the utility grid. Monitoring the performance of such systems is the key to ensuring growth and development in the sector. The performance of a stand-alone photovoltaic system is not as straight forward as for a grid-connected system. The performance of a typical off-grid building integrated photovoltaic system was monitored for a period of seven months. The work demonstrated that the system performed better in winter than in the summer. It was observed that the average reference yield, array yield, final yield, array capture loss and system losses for the system were of 4.5, 2.5, 1.1, 1.9 and 1.4 h/d respectively. Average array, inverter and system efficiencies were 10%, 78% and 8% respectively. Also, the system recorded an average performance ratio of 27%.
Keywords: Off-grid system, building integrated photovoltaic system, performance ratio.

1 INTRODUCTION

The global demand for energy has skyrocketed greatly over the last few decades principally due to industrial development and population growth. South Africa is not left out in this massive increase in the demand for energy. The major concern is that over 90% of South Africa's energy is being generated from coal [1], with less than 4% coming from renewable sources. It is worth noting that this country is endowed with a very rich solar resource. On average, it receives annual 24-hours global solar irradiance levels of about 220 W/m^2 compared with about 150 W/m^2 for parts of the USA, and 100 W/m^2 for Europe and the United [2]. The advancement of the photovoltaic (PV) technology in the country is one major way of offsetting this gross dependence on unclean sources of energy. However, compared to other developed countries, South Africa's PV market is still in the early stages. One of the major ways of ensuring a brighter future for this technology in the country is through accurate and continuous performance monitoring of existing PV systems on the field.

Off-grid PV systems are one of the major applications of PV systems. This is very important in that it can be used to power locations inaccessible to the utility grid and for individuals wishing to get some degree of energy independence. This technology is equally very important to most remote communities in Africa where connecting to the national utility grid is still far-fetched.

An even more innovative advancement in the PV technology is the building integrated photovoltaic (BIPV) system. However, after over three decades of existence, this technology has successfully claimed just about 1% of the total PV installations worldwide. Very few of such systems have been reported in South Africa. Continuous research is required in this sector in order to promote its growth.

One of the most important quantities usually determined when evaluating the overall performance of a PV system is the performance ratio (PR). Several studies have demonstrated the continuous improvement in PV system PR over the years. It rose from 0.5 to 0.75 in the late 1980s and 0.65 to 0.7 in the 1990s, to more than 0.8 currently [3]. However, PR values as low as 0.35 are still common in stand-alone PV systems [4].

The PR being a dimensionless quantity that indicates the overall effect of losses on the rated output, does not represent the amount of energy produced because a system with low PR in a high solar resource location might produce more energy than a system with a high PR in a low solar resource location. Nonetheless, for any given system, location and time, if a change in component or design increases the performance ratio, the final yield increases accordingly. The performance ratio can identify the existence of a problem, but not the cause. To identify the cause of the existing problem, a research at the site is needed.

The performance analysis of stand-alone PV systems is not as straight forward as it is for grid connected systems. A poor performance does not necessarily mean that the system encounters technical problems but can also be the consequence of a bad matching between the production and the load [5]. The IEA PVPS task II experts collected operational data for a set of PV systems amongst which were over thirty stand-alone PV systems. The reason was to access their performance over time, and create a database. The results showed that the stand-alone systems presented a wide range of PR values which did not reflect the proper operation of the system from a technical view point. The typical range of PR values obtained was between 0.1 and 0.7. Over 80% of the PR was between 0.3 and 0.4 while just 10% ranged between the 0.6 and 0.7. An oversized stand-alone PV system usually faces frequent array disconnection in order to prevent the batteries from overcharging. This tendency has a negative effect on the value of the PR for the system.

The aim of this work is to evaluate the performance of a 3.8 kW_p stand-alone building integrated photovoltaic system that was installed at the Fort Hare Institute of Technology, South Africa. The results obtained from this study will create awareness of the potential of such a system being used to curb the problem of energy poverty and energy insecurity in various parts of the world, especially in less developed countries in the African continent, more particularly in remote rural communities.

2 SYSTEM DESCRIPTION

This off-grid BIPV system is located at the

University of Fort Hare, Alice, South Africa.32° 47' 0" South, 26° 50' 0" East. The system has a peak capacity of 3.8 kW. The SANYO HIT (Hetero-junction with Intrinsic Thin layer) PV modules used here have a high capacity to frame area ratio, with a conversion efficiency of 16.1%. These PV modules have been integrated to form part of the roof of a passive zero energy solar building. The twenty HIT PV modules occupy a total roof area of 24 m^2, at an inclination of 25°.

A Victron Energy MultiPlus5048 bidirectional inverter is used in this system which converts the DC energy produced by the panels to AC energy needed by the loads. The inverter has a nominal output power of 5 kVA at 25°C. There is also a FLEXmax 80 MPPT charge controller. This regulator ensures peak performance from the PV panels, by letting them deliver their maximum power at any given instance. The charge controller also protects the battery by controlling its charging current and voltage, thereby prolonging the battery life span. A C/100 900 Ah rated capacity flooded lead acid battery is used in this system to store energy during excess generation. The battery bank is a 48 V system.

Also, there is a data acquisition system composed of sensors, transducers and data loggers for measuring and recording electrical system parameters and the meteorological parameters affecting system performance.

Below is a schematic representation of the house containing the BIPV modules and the balance of system components.

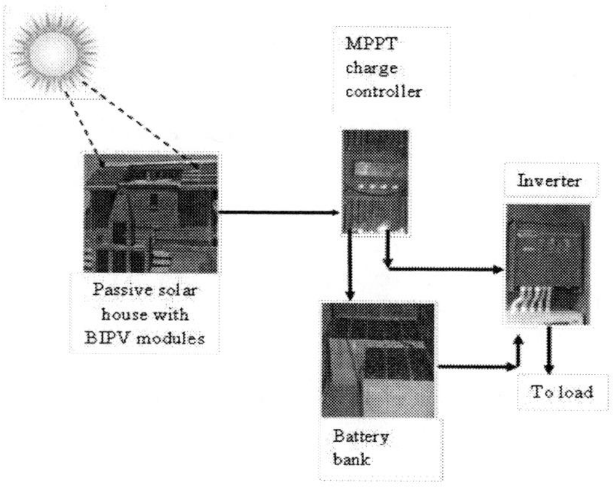

Figure 1: Schematic representation of the off-grid BIPV system components

3 METHODOLOGY

This system is made up of sensors and data loggers for measuring and recording both electrical and meteorological parameters of the system. The electrical parameters include the current and voltage flowing from and into each balance of system component. The meteorological parameters include; solar irradiance on the plane of array, ambient temperature, cell temperature, relative humidity, wind speed and direction. There are two data loggers used for measuring the electrical and meteorological parameters respectively. The electrical data is sampled every 30 seconds and averaged over five minute intervals. Likewise, the meteorological data is sampled every minute and averaged over thirty minute intervals.

Data was collected for a period of seven months; between February and August, 2017. This data was used to evaluate the performance of this off-grid BIPV system based on certain normalized system performance parameters. This was done based on the IS/IEC 61724, 1998 standards. These performance parameters are discussed below;

The yields have units of kWh/day/kW_p (or h/day) and indicate the amount of time during which the array would be required to operate at the array's rated output power, P_o to provide a particular quantity [6].

3.1 Reference yield (Y_R)

The reference yield is the theoretically possible energy output of the PV plant. It is defined as the anticipated output from the same system with nominal efficiency determined under STC of PV modules. In other words, the reference yield is the number of peak sun-hours. It is the ratio of the total in-plane solar radiation (H_{id}) [kWh/m^2], to the reference irradiance (1 kW/m^2) at standard test conditions.

$$Y_R = \frac{H_{id}}{G_{STC}} \quad (1)$$

3.2 Array yield (Y_A)

The array yield represents the number of hours per day that the array would need to operate at its nominal or peak power to contribute the same quantity of energy to the system as that actually measured in practice. It is the ratio of the daily energy output of the PV array (E_A) to the peak power of the installed PV array.

$$Y_A = \frac{E_A}{P_o} \quad (2)$$

3.3 Final yield (Y_F)

The final yield is the usable portion (used by the load, E_L) of the energy derived from the PV modules (E_{pv}). In a standalone system, the energy stored in the battery bank (E_{bat}) is equally taken into consideration. The final yield is obtained from;

$$Y_F = \frac{E_{PV}}{P_o} \quad (3)$$

$$E_{pv} = \frac{E_L}{1+\left(E_{bat}/E_A\right)} \quad (4)$$

3.4 Performance ratio (PR)

The performance ratio is one of the most important parameters to consider when evaluating the performance of a PV system. It is the ratio of the actual energy output to the theoretically possible energy output. It is independent of the orientation of the PV plant and the incident solar radiation on the PV plant. This is why the performance ratio can be used to compare PV plants at different locations all over the world, or for performance comparison of a particular system over the years. It is the ratio of utilizable AC electricity to the amount of energy which could be generated in case modules were operated under STC continuously and without any further losses in the system [7].

$$PR = \frac{Y_F}{Y_R} \quad (5)$$

3.5 Losses

The losses in a PV system may be caused by high

array temperatures, incomplete utilization of the irradiation, improper system sizing and system component inefficiencies or failures [8]. Energy losses indicate the amount of time during which the array would be required to operate at its nominal power in order to produce power for the losses. The losses are in the two parts: system losses (Ls) and array capture losses (Lc). The capture loss is the difference between the reference yield and the array yield. It is caused by incomplete utilization of the available radiation. The system loss is the difference between the array yield and final yield. It mainly comes from the inverter in converting the DC power to AC [9]. and from losses in the other balance of system components.

$$L_c = Y_R - Y_A \tag{6}$$

$$L_s = Y_A - Y_F \tag{7}$$

3.6 Efficiencies

The efficiencies involved in a PV system are in two major levels; at the level of the array and the whole system in general. The system efficiency embodies the efficiencies of the balance of system components and the PV array. Given that the major component where losses are incurred is the inverter, the system efficiency is simply the product of the array efficiency and the inverter efficiency.

$$\eta_{array} = \frac{E_A}{A_{area} \times G_{id}} \tag{8}$$

$$\eta_{sys} = \eta_A \times \eta_{inv} \tag{9}$$

4 RESULTS AND DISCUSSIONS

4.1 Load profile for the building powered by the off-grid BIPV system

During the analysis period, the house powered by this PV system was occupied by two adults and three kids.

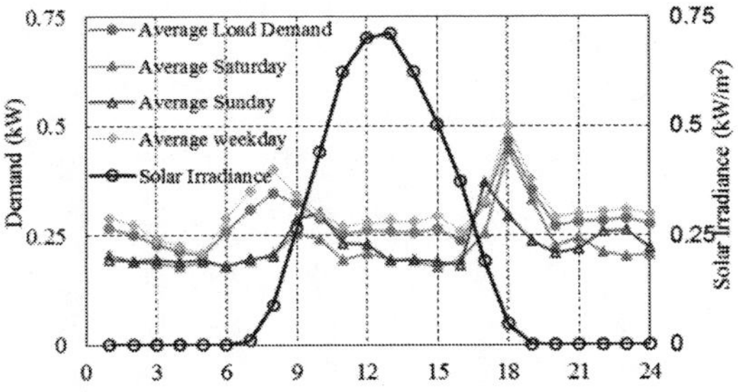

Figure 2: Load profile for the house powered by the off-grid BIPV system

Figure 2 above presents the load profile for an average weekday, Saturday and Sunday. An average demand profile for the whole period is also illustrated. Furthermore, the figure also shows the variation of solar irradiance at various hours of the day.

It can be observed that the consumption pattern for the weekend is slightly different from the weekday pattern. On average, the morning demand peak occurs at 7 AM (0.35 kW) while the evening demand peak occurs at 5 PM (0.45 kW). These peaks fall within the utility grid's peak demand periods (between 7 am and 9 am in the morning and between 5pm and 9pm in the evening). At night, from 7 PM to 3 AM, the average demand is 0.26 kW. The figure reveals that the highest consumption occurs during the working days or weekdays (Monday to Friday) with an average consumption of 0.31 kW, followed by the Sunday consumption of 0.23 kW and finally, a Saturday consumption of 0.22 kW. It can equally be observed that peak irradiation occurs at about midday when the load demand is least. Thus, this system is not being utilized at its full capacity. If this system was connected to the grid to feed in this surplus energy, a lot of energy and monetary savings will be generated. It is recommended that the occupants of this house make an effort to rearrange their consumption pattern by using energy intensive appliances at about midday when the PV generation is highest.

4.2 Solar radiation and ambient temperature variation

The performance of a PV system is influenced by so many factors amongst which are meteorological factors. The principal meteorological factors affecting the power output and efficiency of the system are ambient temperature and solar radiation. Figure 3 below shows the variation of solar radiation (average number of peak sun hours per day) and ambient temperature for the various months considered for this study.

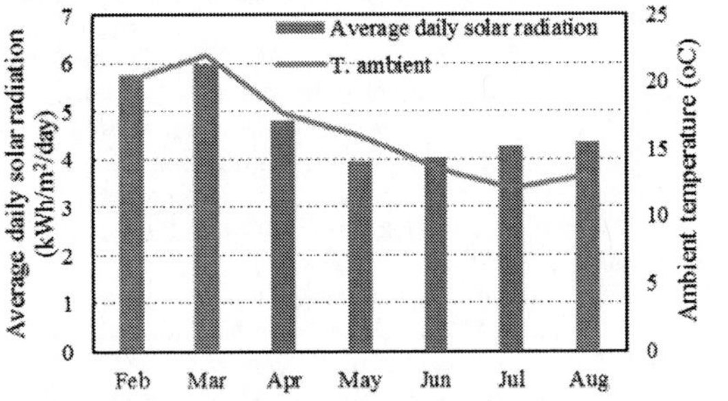

Figure 3: Average solar radiation and ambient temperature profiles for the analysis period.

It can be observed that the highest radiation occurs in March, with an average of 5.9 peak sun hours per day while the least radiation occurs in May, with an average daily radiation of 3.9 peak sun hours per day. The temperature variation follows a similar pattern. However, the highest average daily temperature is noticed in March (22.0°C) while the least is observed in July (12.1°C). From these profiles, the PV system is expected to have the best performance in July, given that it is the month with the least temperature and considerably high levels of solar radiation.

4.3 Yield factors for the system

As earlier mentioned, the yields indicate the amount of time during which the array would be required to operate at its rated output power to provide a particular quantity. Figure 4 below shows the relationship between the reference, array and final yields for this system for the various months considered for the study.

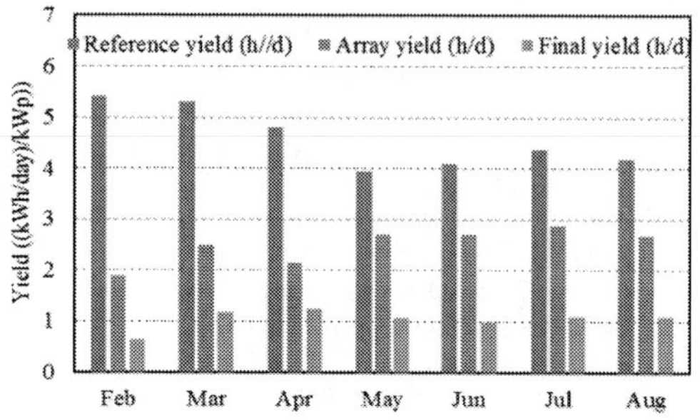

Figure 4: Variation of yield factors for this system

For each month, the reference yield is highest, followed by the array yield and then the final yield. The reference yield depicts the number of peak sun hours for a particular month. It represents the amount of energy which would be produced if the system operated at STC conditions and at the rated efficiency. This is highest in February, with a total number of 5.4 peak sun hours per day. The least value occurs in May (3.9 peak sun hours per day).

Given that this is an off-grid system with a battery back-up and a charge controller, only the amount of energy needed to charge the battery bank and to feed the loads can be used irrespective of the available energy from the sun. This partly explains the huge difference between the reference yield and the array yield. Moreover, the inefficiencies involved in the conversion of the available solar radiation to electricity also contributes to this difference. There is an average of 2.1 kWh per day of energy difference between the available solar radiation and the amount of energy tapped by the MPPT charge controller. This is termed the capture loss as illustrated on Figure 5 below.

Furthermore, of the energy that is tapped by the charge controller to feed the battery bank and the loads, part of it is lost before if finally gets to the loads. This is as a result of energy losses in the various system components such as the charge controller, the battery bank, the inverter and other wiring losses. This explains the gap between the array yield and the final yield in Figure 4 above. This difference is termed the system loss. It is equally depicted on Figure 5 below.

The system recorded an average reference yield, array yield and final yield of 4.5 h/d, 2.5 h/d and 1.1 h/d respectively.

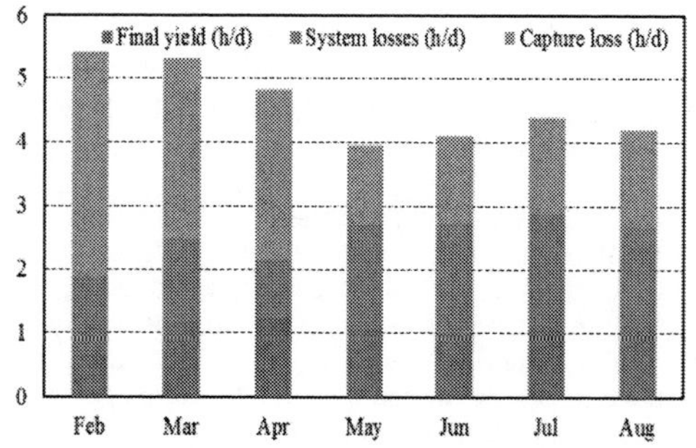

Figure 5: An illustration of the two major types of losses incurred in this system

Figure 5 above illustrates the major losses incurred in the system. A combination of the final yield, system loss and array losses results in the reference yield. So, the full length of each bar represents the amount of energy this system would have produced at standard test conditions and at the arrays rated efficiency, in the absence of losses.

Furthermore, it can be noticed that the system encounters a lot of capture losses. Given that this system has been in existence for about eight years, there are signs of degradation observed on the surface of some modules. This is principally module browning caused by encapsulant delamination. Previous studies on the modules reveals a drop in the short circuit current of the modules. This effect contributes greatly to the high capture losses. Another reason for the high capture losses is the huge difference between the load consumption and the PV generation potential. There is a lot of energy which is not tapped because the battery bank gets fully charged and the load is demanding very minimal energy when the PV is still at its peak generating capacity.

Additionally, it can be observed that most of the capture losses occur in the summer months of February March and April (an average of 3 h/d) when temperatures are still very high. During the winter months (between May and August), the capture loss is greatly reduced as temperatures have become quite low (an average of 1.4 h/d). High operating temperatures reduce the output voltage and thus the power output of the modules.

On average, the system recorded capture losses of 1.9 h/d and system losses of 1.4 h/d.

4.4 Performance ratio of the system

As earlier mentioned, performance ratio is a globally accepted parameter used to compare similar PV systems at various locations on the globe. A low value for performance ratio in an off-grid PV system does not necessarily imply poor system performance or technical problems. It is equally an indication of poor matching between the PV generator and the electrical appliances powered by the system. An oversized system incurs losses because the PV system is not being used at its full potential. Figure 6 below illustrates the changes in the performance ratio for this system at various months of the year. The figure also reveals the dependence of the performance ratio on temperature.

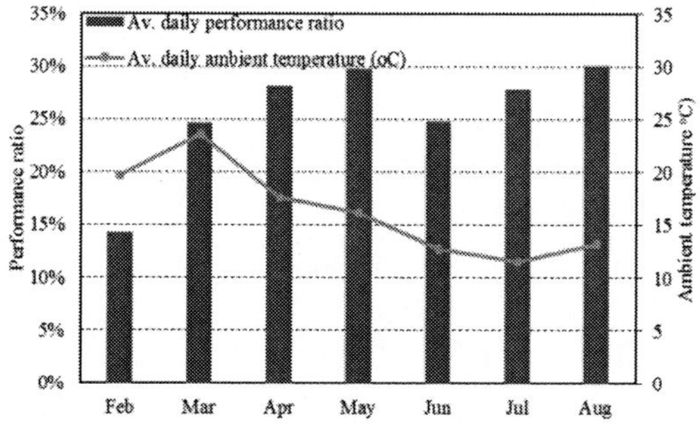

Figure 6: Variation of performance ratio and ambient temperature for various months of the year

As shown on Figure 6 above, this system shows a relatively low performance ratio compared to other stand-alone PV systems in different parts of the world. It however falls within the range of the PR values obtained by the IEA PVPS task II experts. They carried out a survey on about thirty off-grid PV systems and discovered that over 80% of the PR values ranged between 0.3 and 0.4 while just 10% ranged between 0.6 and 0.7. The months with the highest average daily performance ratio are May and August, while the performance ratio is least in February (a daily average of 14%). However, the highest temperature is in March and the least is in July. This system recorded an average performance ratio or 27%. It is well known that the performance of a PV system is adversely affected by high temperatures. The fact the month with the highest performance ratio does not exactly correspond to the month with the least ambient temperature justifies the fact that the performance ratio of this off-grid system is also affected by other factors, load matching being one of the key factors. Figure 7 below illustrates the correlation between the performance ratio for this system and the energy demand. It shows a linear relationship with a correlation coefficient of 0.63. This goes to justify the fact that the performance ratio for an off-grid system is highly dependent on its load consumption.

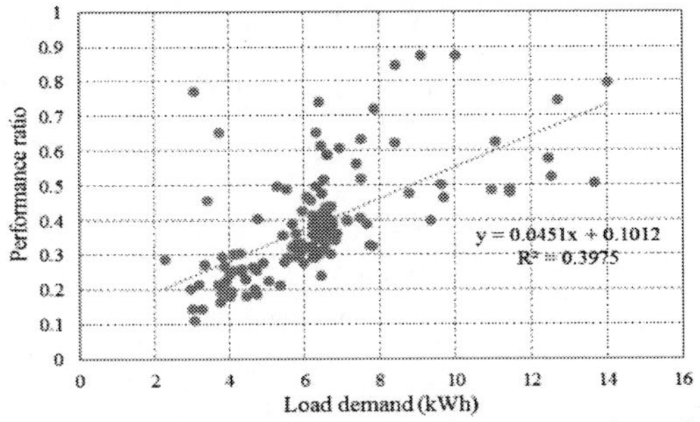

Figure 7: Correlation between performance ratio and energy demand

4.5 System efficiency

The overall efficiency of the system is equally an important parameter used in describing the performance of a PV system. It encompasses the efficiencies of the various balance of system components used in the system. The system efficiency shown on Figure 8 below is a product of the array efficiency and the inverter efficiency. The inverter efficiency was used because it is one of the major balance of system component where losses are incurred. The array efficiency accounts for the percentage of incident radiation converted to electricity by the modules.

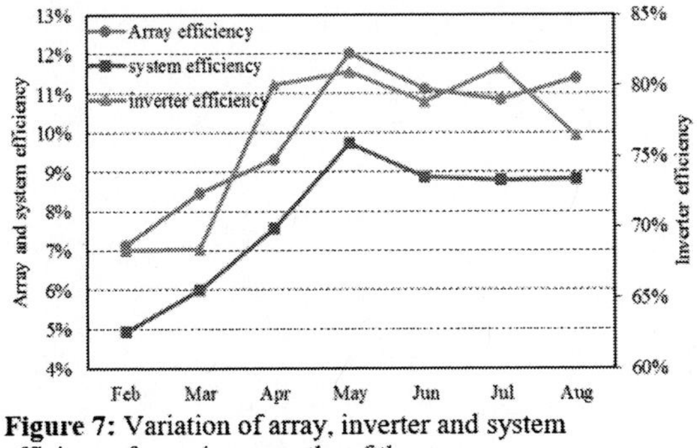

Figure 7: Variation of array, inverter and system efficiency for various months of the year

The least array, inverter and system efficiencies of 7%, 68% and 5% respectively occur in February (summer). Similarly, the highest array, inverter and system efficiencies of 12%, 81% and 10% respectively occur in May (winter).

Figure 8 below represents the correlation between array conversion efficiency and the operating array temperature.

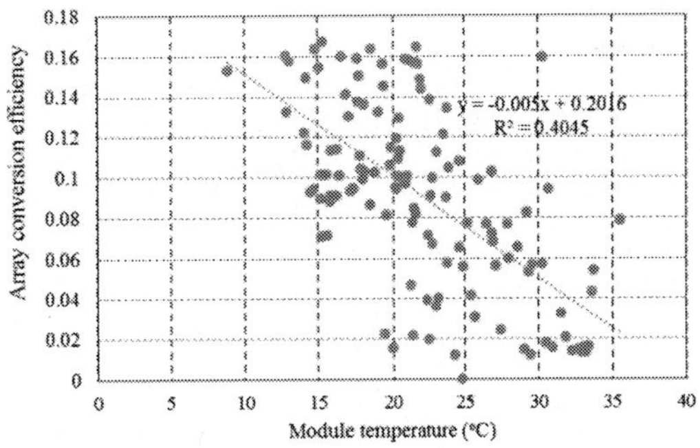

Figure 8: Correlation between array conversion efficiency and module temperature

Figure 8 above represents a negative linear relationship between the array efficiency and the operating array temperature with a correlation coefficient of -0.63.

4.6 Inverter efficiency

The inverter is one of those components whose efficiency is highly dependent on load demand. This is because being a stand-alone PV system, the input power into the inverter is determined by the demand from the loads. In the absence of any demand, little or no energy is fed to the inverter. Figure 9 below reveals the relationship between the inverter efficiency and the normalised input power into the inverter. The normalised input power was calculated as the inverter's input power as a percentage of its rated capacity.

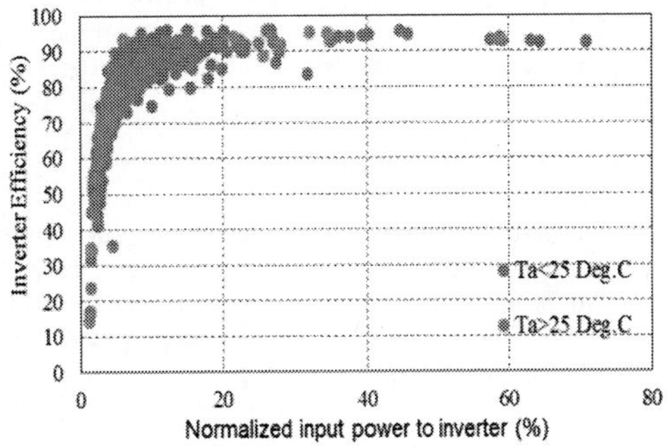

Figure 9: Variation of inverter efficiency with normalized inverter input power at various temperatures

Also, this calculation was done for ambient temperatures above 25°C and below 25°C given that the inverter is rated at 25°C. From Figure 9, it can be observed that the efficiency of the inverter increases with an increase in the normalized input power into the inverter up to when the inverter is operated at about 15% of its rated power. The efficiency then remains fairly constant and then starts dropping slightly when operated above 50% of its rated capacity. This low efficiency at low load can be associated with the switching losses at low input power. The input capacitance of the MOSFETS which are the switching devices in the inverter, consume some power whether there is output power from the inverter or not. Furthermore, at very high input power into the inverter, the inverter efficiency drops due to resistive losses which increase as output current increases. In addition, the temperature rise of the electronic devices in the inverter that accompany high output power may cause the inverter to reduce its output power and hence, its efficiency. This is called temperature derating. Also, it can be observed from Figure 9 that the higher inverter efficiency also corresponds to periods when ambient temperature is lower than 25°C, and the lower ambient temperatures highly contribute to the high efficiency values of the inverter.

5 CONCLUSIONS

It is recommended that off-gird PV systems be used only is locations inaccessible to the grid. In locations where the PV system is installed close to the utility grid, the system should be connected to the utility grid to feed in the surplus energy that is otherwise wasted due to poor system sizing, which is of course difficult to avoid.

Furthermore, the study shows that temperature is a very important parameter affecting the performance of a PV system. The system performed better in the winter months than in the summer months even though high irradiation values were recorded in the summer months. The performance ratio ranged from 14% in February to 30% in August, with a summer average of 22.3% and a winter average of 28.3%. Furthermore, the summer values for the array efficiency, inverter efficiency and system efficiency were 8%, 72% and 6.3% as opposed to the winter values of 11.25%, 79.25% and 9.25% respectively.

This system encounters a lot of array losses (1.9 h/d on average) principally due to the mismatch between the array capacity and the load consumption.

It is recommended that energy intensive appliances be used at about midday when PV generation is maximum, so as to minimize the wasted PV generation potential.

6 ACKNOWLEDGMENTS

The authors wish to acknowledge the Department of Science and Technology, the National Research Foundation and GMRDC for their financial support. Also, the technical support of Mr Melvin Adonis and Mr Vincent Schoemann is much appreciated.

7 REFERENCES

[1] D.V. Brits and R. Zietsman. The Feasibility of Building Integrated Photovoltaic Systems for Single Residential Buildings in the Western Cape, South Africa. Proceedings of the 5th built environment conference (2010).

[2] C.L Buma, E.L Meyer, R Taziwa, Energy Management in the balance of system components in a stand-alone building integrated photovoltaic system in Alice. Proceedings of the South African universities power engineering conference (2017).

[3] A. Woyte, M. Richter, D. Moser, S. Mau, N. Reich, U. Jahn, Monitoring of photovoltaic systems: good practices and systematic analysis. European photovoltaic solar energy conference proceedings. Paris, France (2013).

[4] C. Iskander and E. Scerri, Performance and cost evaluation of a stand-alone photovoltaic system in Malta. Renewable energy journal. Vol. 8 Issues 1–4, May–August (1996) 437-440.

[5] D. Mayer and M. Heidenreich, Performance analysis of stand-alone Photovoltaic systems from a rational use of energy point of view. International energy agency photovoltaic power system program. Austria (2003).

[6] IS/IEC 61724, Performance monitoring – guidelines for measurement, data exchange and analysis. Indian standards (1998).

[7] W.G. Van Sark, N.H. Reich, B. Muller, A review of PV performance ratio development. Proceedings of world renewable energy forum conference Vol. 6 (2012).

[8] W. Xinfang, L. Yongsheng, X. Juan, L.Wei, S. Xiaodong, D. Wenlong, Z. Chunjiang, Z. Yunbo, P. Lin, L. Jia, Monitoring the performance of the building attached photovoltaic (BAPV) system in Shanghai. Journal of energy and buildings. Vol. 88 (2015) 174-182.

[9] E. Kymakis, S. Kalykakis, T. M. Papazoglou, Performance analysis of a grid connected Photovoltaic Park on the island of Crete. Journal of energy conversion and management, Vol. 50 (2009) 433-438.

SOLAR RESOURCE FOR HIGH PENETRATION AND LARGE SCALE APPLICATIONS – A NEW JOINT TASK OF IEA PVPS AND IEA SOLARPACES

Jan Remund[1], Lourdes Ramirez[2], Stefan Wilbert[3], Philippe Blanc[4], Elke Lorenz[5], Carmen Köhler[6] and Dave Renné[7]
[1] Meteotest, Fabrikstrasse 14, CH-3012 Bern, Switzerland, jan.remund@meteotest.ch Tel. +41 31 307 26 26, Fax +41 31 307 26 10
[2] CIEMAT, Lourdes.Ramirez@ciemat.es
[3] German Aerospace Center (DLR), Stefan.Wilbert@dlr.de
[4] Mines Paristech, philippe.blanc@mines-paristech.fr
[5] Fraunhofer Institute for Solar Energy Systems ISE, elke.lorenz@ise.fraunhofer.de
[6] i-em, carmen.koehler@i-em.eu
[7] Clean Power Research, drenne@mac.com

ABSTRACT: A new joint Task of IEA Technological Collaboration Programmes PVPS (Photovoltaic Power Systems Program) and SolarPACES (Solar Power And Chemical Energy Systems) has started in July 2017 and lasts for three years. The PVPS Task 16 will be led by Jan Remund and the SolarPACES Task V by Lourdes Ramirez.
This paper shows the main topics of the work plan. The main goal is to lower uncertainty of resources and forecasts and therefore lower the costs of financing costs as well operation and grid integration costs. The Task has started in June 2017 and will last till May2020. To give an example of the work to be done the discussion about different forecast metrics is shown in this paper.

Keywords: science collaboration, solar radiation, forecasts, uncertainties

1 INTRODUCTION

A new joint Task of IEA Technological Collaboration Programmes PVPS (Photovoltaic Power Systems Program) and SolarPACES (Solar Power And Chemical Energy Systems) has started in July 2017 and last for three years. The Task will be led by Jan Remund (for IEA PVPS) and Lourdes Ramirez (for SolarPACES).

This paper shows the main topics of the work plan which are aligned with the current industry needs. The paper also explains and justifies the added value of the development of each topic.

Fifty-four organizations from 22 countries will take part in the new Task. Some of these partners were involved in the previous IEA Solar Heating and Cooling Task 46 but new partners like Vaisala, Fraunhofer ISE, Sunpower and Energinet (TSO of Denmark) are now included in order to fulfil a wide range of expertise. Countries – like Morocco, United Arab Emirates and Chile - are now new participating as well.

The Task is organized in four subtasks in order to cover all the identified objectives and will be led by experienced scientists:

1. Resources (Stefan Wilbert)
2. Bankability (Philippe Blanc)
3. Forecasts (Elke Lorenz, Carmen Köhler, Kristian Nielsen)
4. Dissemination (Dave Renne)

The Task offers a unique platform of exchange for private and public organizations as well as data providers and science organizations. A strong emphasis on additional collaboration is promoted. Benchmarking studies, data standardisation, description of state-of the art methodologies and dissemination are in the focus. Added values like lower resource uncertainties and higher forecast accuracies, will lead to lower implementation costs of PV and CSP projects.

To give some examples of the outcomes proposed:

- Definition of a benchmarking framework.
- Definition of standard methods for merging satellite, NWP and ground data.
- Enhanced models for description of short and long term variability.
- Definition of methods for regional solar power forecasting.

For more information we refer to the official IEA PVPS webpage: http://www.iea-pvps.org/index.php?id=389

The main work of the Task has started in summer 2017. Therefore no results are available yet. The first workshops and webinar will be given in spring 2018.

In the following chapter we show one example of one topic the team wants to work on: how to measure uncertainty and how to compare solar forecasts.

2 ERROR METRICS

Standard metrics of solar and PV production forecasts include mean bias error (mbe), mean absolute error (mae) and root mean squared error (rmse) [1]. rmse is mostly used in form of relative error normalised by average radiation or installed PV power (which results in lower values).

Alternative metrics have been proposed during the last year by Lorenz et al. [2], Frias-Paredes et al [3] or Vallance et al. [4].

The following Figure 1 based on [4] however shows, that the standard metric rmse isn't optimal for all cases.

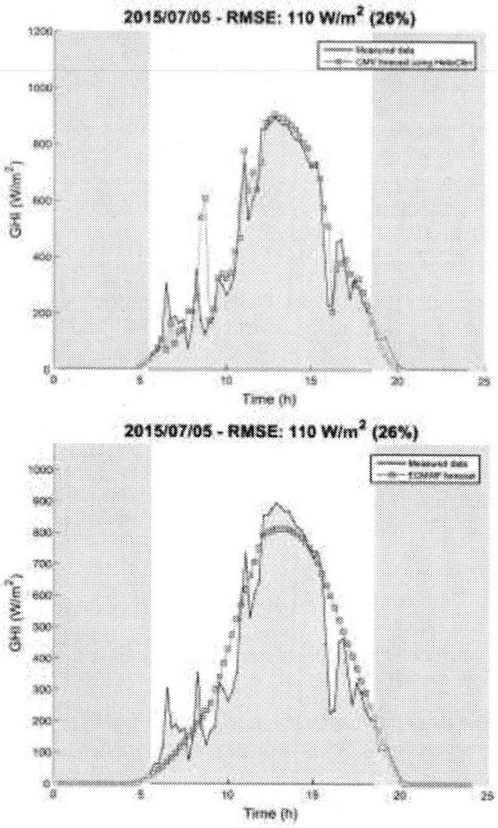

Figure 1: Two examples of forecasts showing the same rmse values.

Both forecasts in Figure 1 show the same rmse. The top forecast includes variations, but misses, under- or overestimates some peaks and valleys. This results in double penalty when using rmse. The bottom forecast is smoothed and doesn't include any variation aside those induced by the daily sun path.

This kind of conditions have been reported often in many papers [2] and favours temporally and spatially smoothed forecasts like those based on global weather prediction models and penalizes regional or local models.

3.1 Suggestions by Vallance et al.

Vallance et al [4] suggest to use a combination of different metrics to include all aspects of solar forecasts uncertainties and display them as spider graph. This graph includes ramp score, rmse, mae, bias and temporal distortion metrics based on Frias-Paredes et al. [3] (Figure 2). The temporal distortion metrics are the temporal distortion index (TDI) and the temporal distortion mix (TDM).

The disadvantage of this type of graph and also parameters like the temporal distortion is, that they are much more complex then mae or rmse, not understood by clients and also not easy to review.

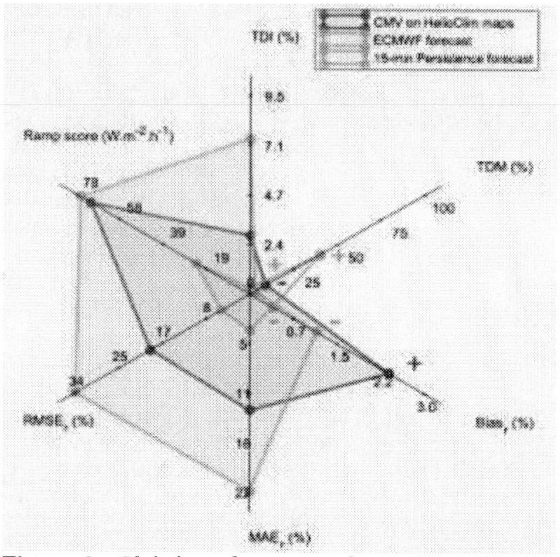

Figure 2: Global performance of 15-min persistence forecast, the cloud motion vector (CMV) method based on HelioClim maps and the 15-min interpolated IFS (ECMWF) forecast shown as spider graph [4].

3 GENERAL COMMENTS AND FUTURE WORKS

The final word (if ever) isn't spoken in this topic. Nevertheless a broader view and discussions on uncertainties is needed in future. This issue will handled in subtask 2 (bankability) of the Task 16.

Our message is that metrics has to be chosen depending on the usage of the forecasts and for instance, in the case of solar energy plants operation the information related to the expected variability and the probability distribution are crucial. In order to improve the forecast characterization for the plant operation the use of second order error metrics like KSI [5] will be prospected.

4 REFERENCES

[1] Hoff, T.E., Perez, R., Kleissl, J., Renne, D., Stein, J., 2013. Reporting of irradiance modeling relative prediction errors. Prog. Photovoltaics Res. Appl. 21, 1514– 1519. http://dx.doi.org/10.1002/pip.2225.

[2] Lorenz, E., Kühnert, J., Heinemann, D., Nielsen, K. P., Remund, J., & Müller, S. C. (2016). Comparison of global horizontal irradiance forecasts based on numerical weather prediction models with different spatio-temporal resolutions. Prog. Photovolt: Res. Appl. http://doi.org/10.1002/pip.2799

[3] Frías-Paredes, L., Mallor, F., León, T., Gastón-Romeo, M., 2016. Introducing the Temporal Distortion Index to perform a bidimensional analysis of renewable energy forecast. Energy 94, 180–194.

[4] Vallance, L., Charbonnier, B., Paul, N., Dubost, S., & Blanc, P. (2017). Towards a standardized procedure to assess solar forecast accuracy: A new ramp and time alignment metric. Solar Energy, 150, 408–422. http://doi.org/10.1016/j.solener.2017.04.064

[5] Espinar, B., Ramírez, L., Drews, A., Beyer, H.G., Zarzalejo, L.F., Polo, J., Martín, L. Analysis of different comparison parameters applied to solar radiation data from satellite and German radiometric stations, Sol. Energy. 83 (2009) 118–125. doi:10.1016/j.solener.2008.07.009.

A GLOBAL HOURLY SOLAR RADIATION DATA SET USING SATELLITE AND REANALYSIS DATA

Thomas Huld[1], Rubén Urraca[2], Ana Gracia Amillo[1], Jörg Trentmann[3]
[1]European Commission, Joint Research Centre, Viale Fermi, 2749, Ispra, Italy
[2]University of La Rioja, Department of Mechanical Engineering, 26004, Logroño, Spain
[3]Deutscher Wetterdienst, 135 Frankfurter Strase, 63067, Offenbach, Germany
Thomas.huld@ec.europa.eu, ruben.urraca@unirioja.es, ana.gracia-amillo@ec.europa.eu, joerg.trentmann@dwd.de

ABSTRACT: Datasets using images from geostationary satellites are extended widely used method to obtain hourly estimates of surface irradiance, but their coverage is limited to +/- 65° latitude, and even there they are not always freely available. Reanalysis products can be an alternative to generate a global hourly radiation dataset, but the quality of solar radiation data in the most common datasets (ERA-Interim and MERRA-2) is rather low. Looking for a product to complement satellite data, we evaluate the new ECMWF-ERA5 reanalysis (hourly, ~31 km) using 40 BSRN stations and two satellite products (SARAH and NSRDB).

The results show a significant improvement obtained with the new ERA5. The yearly bias (6 W/m^2) is reduced around 50% compared to MERRA-2 and ERA-Interim, and it is comparable to SARAH and NSRDB for inland regions with homogeneous terrain. Still, the weaknesses of ERA5 are the spatial grid, which is too coarse for coasts or mountains, and a poor modeling of clouds, which causes a high variation of the bias with cloudiness and leads to larger absolute errors than satellite products. Hence, we recommend using satellite products when available, but taking into account its limitations, ERA5 is a reasonable alternative to complement the coverage of satellite datasets.

Keywords: Solar radiation, reanalysis, satellite-based solar radiation estimates

1 INTRODUCTION

Over the last 10 years solar radiation estimates based on satellite data have emerged as the main source of solar radiation data for solar energy applications [1]. Using images from geostationary meteorological satellites, it is possible to obtain values for solar irradiance at ground level with a spatial resolution of a few km and time resolution of hourly or better. There exist now several freely available data sets of solar radiation values based on satellite data, supplying data for long time periods, often exceeding 20 years [2, 3, 4].

There are certain limitations to solar radiation estimates based on geostationary satellite data. Some areas near the poles of the earth are not seen at all by the geostationary satellites. In addition to this, the quality of the satellite-retrieved solar radiation estimates decreases at high latitudes due to the low viewing angle of the ground as seen from these satellites. For this reason, geostationary satellite data are not useful at latitudes greater than 60° -65° from the equator. Furthermore, at the present time there are still areas of the earth for which there are no freely available satellite-based solar radiation data available. This is the case for East Asia, Oceania and the Western part of the Pacific.

Polar orbiting satellites view the whole earth and may be used to estimate the solar radiation at ground level. However, the orbits of these satellites only bring them over a given point on the earth surface once or twice a day, so the time resolution of these data is daily or less. One example of such a data set is the CM SAF CLARA-A2 [4] dataset that provides daily average irradiance values with a spatial resolution of 0.25° lat/lon.

One alternative source of solar radiation data is the various reanalysis products based on numerical forecast models. These data have global coverage and generally high temporal resolution (1-3 hours). The disadvantage is that the global reanalysis products available at present have low spatial resolution (0.5°-1° latitude/longitude). In addition, it has been found that the most popular reanalysis products, such as MERRA and ERA-Interim, tend to have large bias in the solar radiation estimates due to a general underestimation of the cloud cover [5, 6]. The European Centre for Medium-range Weather Forecast (ECMWF) has recently started releasing the new ERA5 reanalysis product with higher spatial resolution (about 31 km) and hourly time resolution. At the time of writing, 7 years of data have been made available (2010-2016) with a longer time period to be released in 2018.

In this paper we will investigate whether reanalysis data, and especially the new ERA5, can be useful to provide solar radiation estimates for areas not currently covered by geostationary satellite data. The investigation will make use of ground-based measurements to assess the quality of each of the reanalysis data sets and compare these with solar radiation data derived from images of geostationary satellites.

2 SOLAR RADIATION DATA SOURCES

2.1 Satellite-based solar radiation data

Two data sets derived from geostationary satellite data have been used in this study:

- CM SAF SARAH [3] (www.cmsaf.eu) uses data from the European METEOSAT satellites to produce solar radiation estimates for Europe, Africa, most of Asia and parts of eastern South America. It uses a semi-empirical method to derive hourly solar irradiance values as described in [3, 7, 8], and the data set has been validated by a number of authors [3, 8, 9]. The data consist of hourly global and direct horizontal irradiance values (GHI and BHI, respectively) with a spatial resolution of about 5 km at nadir.
- The National Solar Radiation Database (NSRDB) has been developed by the National Renewable Energy Laboratory (http://www.nrel.gov). The latest NSRDB version is the Physical Solar Model (PSM) [10] and it

derives global, diffuse and direct irradiance from the U.S. GOES class of satellites [11]. The data cover the Americas and Pacific Islands from 60° N to 20° S, have a spatial resolution of about 4 km at nadir and cover the period 1998-2015 in half-hourly resolution (for the present study we have used one image per hour).

Combined, the satellite data sets derived from geostationary satellites have a geographical extension as shown in Figure 1.

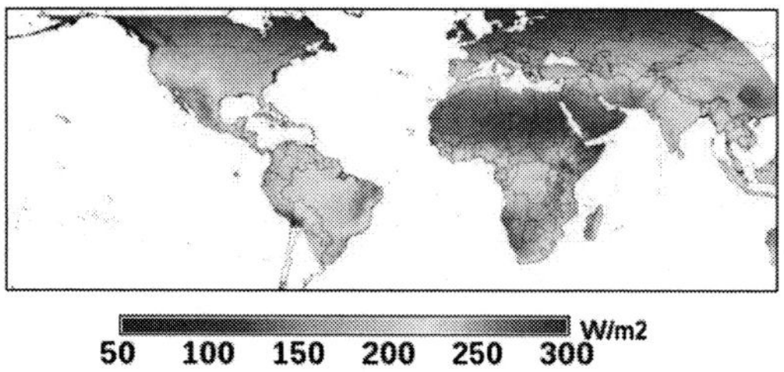

Figure 1: Average annual global horizontal irradiance estimated by the two satellite-based solar radiation data sets used in the present study: NSRDB (Americas) and SARAH (Eurasia, Africa, parts of South America).

2.2 Reanalysis data

Three reanalysis-based solar radiation data sets have been used here:

- ERA-Interim [12] is the current reanalysis dataset of the ECMWF (www.ecmwf.int). It contains values of global horizontal irradiance with a spatial resolution of 0.75°x0.75° (~81 km) and 3-hourly time resolution for the period 1979-present.
- MERRA-2 [13] has been released by NASA's GMAO and provides global horizontal irradiance values with a spatial resolution of 0.5x0.625° (~50 km) and hourly time resolution, from 1979-present
- ERA5 [14] is the new global reanalysis being developed by the ECMWF to replace ERA-Interim. It will cover the period 1950-present, but at the time of writing, only the data for the period 2010-2016 have been released. The spatial resolution is 0.2825°x0.2825° (~31 km), with hourly time resolution. Unlike MERRA-2 and ERA-interim, ERA5 provides both global horizontal and direct horizontal radiation values.

3 VALIDATION OF THE REANALYSIS AND SATELLITE-BASED DATA SETS

3.1 Ground-based measurements

For the validation of the different solar radiation data sets we have used data from the stations in the Baseline Surface Radiation Network (BSRN), which are well-known for their high standards of quality, and which have been extensively used for validation of solar radiation products. A total of 40 BSRN stations have been used for this study, employing only the years with at least 7500 hours of valid data in the period 2010-2014, and excluding stations in the Antarctic and Izaña.

3.2 Validation using the BSRN station measurements

The summary statistics in the BSRN stations reveal a substantial improvement in the estimation of surface irradiance by ERA5 compared to the other reanalysis datasets. ERA5 still mildly overestimates, but the overall bias is reduced to 6 W/m^2 and a mean absolute deviation of the individual biases equal to 8.2 W/m^2. ERA-Interim and MERRA-2 generally show high positive biases (MBE = 11.5 W/m^2 and MBE = 12.8 W/m^2 respectively). The quality of ERA5 worsens in places with high spatial variability of surface irradiance (such as coasts, islands and mountains), which indicates that the spatial grid of 31 km is still not adequate there. However, in inland stations over homogeneous terrain, the bias of ERA5 is similar to the one of satellite products, and in some cases even better (NSRDB – Brazil).

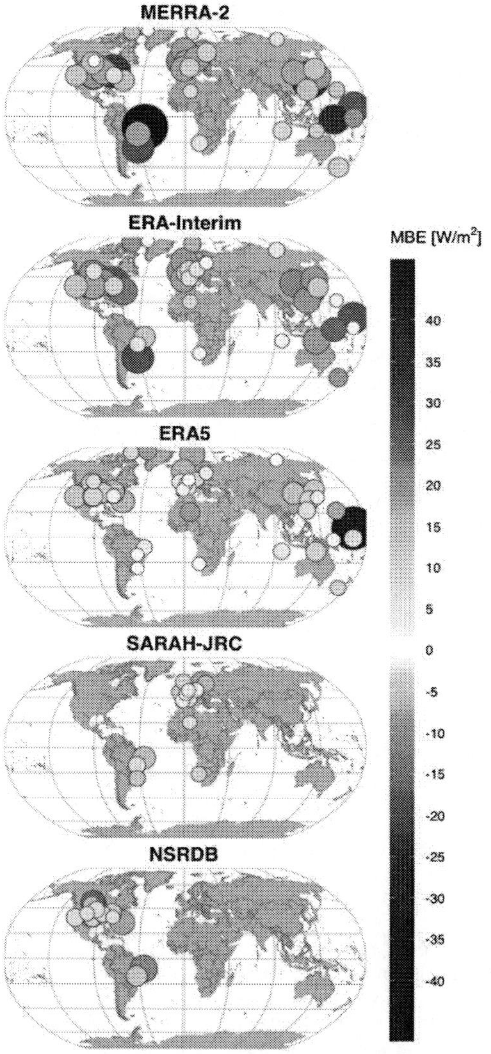

Figure 2: Mean bias error (MBE) of GHI for each of the three reanalysis datasets when validated against BSRN station data, as well as for the two satellite-based products. Red colors represent overestimation of the products (error = product – records).

Apart from the spatial resolution, the main limitation of

ERA5 compared to the satellite products is the prediction of the cloud pattern. The bias of ERA5 strongly varies with the degree of cloudiness of the atmosphere, overestimating under cloudy and intermediate conditions (KT < 0.6) and slightly underestimating under clear-skies. This pattern is less pronounced in ERA5 than in ERA-Interim and MERRA-2, leading to the moderate overall bias, but it still results in large absolute deviations in ERA5 compared to the satellite products.

4 COMPARISON OF REANALYSIS AND SATELLITE DATA SETS

The number of stations used for the validation is low compared to the very large areas that are covered by the solar radiation data sets. For this reason we have also made a direct comparison of the various reanalysis products against the satellite-based data sets in their respective areas of coverage.

4.1 Comparison of reanalysis data with NSRDB data

For the Americas we have compared the global horizontal radiation from the three reanalysis data sets against the values from the NSRDB data set. The results are given as the difference in the average irradiance, calculated over the period 2010-2015. The results are shown in Figure 3 for the three reanalysis data sets.

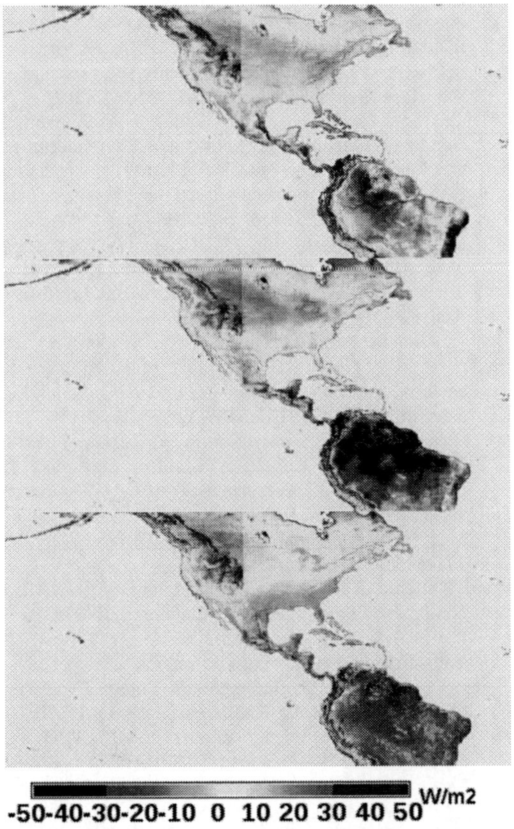

Figure 3: Difference in average irradiance between the three reanalysis data sets and the NSRDB solar radiation data set for the Americas. Values are given in W/m^2. Top: MERRA-2, Middle: ERA-Interim, Bottom: ERA5.

Note that the colour scale ranges from -30 to +30W/m^2, areas with differences outside this range are coloured in dark brown or black. Red colours represent overestimation of the reanalysis products with regard to the NSRDB data set.

In general, the results show that ERA5 is more similar in irradiance values to NSRDB than the other two reanalysis products. This is in line with the validation results that show lower absolute bias values for ERA5 compared to MERRA and ERA-interim. It should be noted that some of the high differences observed might be attributed to the quality of the NSRDB, such as the negative deviations in Brazil, a region in which the NSRDB obtained negative biases in the three BSRN stations. In this line, the discontinuity in the maps along the 105° W longitude line is also most likely due to differences in NSRDB when using data from the two sets of GOES satellites stationed at 75° W and 135° W.

4.2 Comparison of reanalysis data with SARAH data

We have performed a similar comparison with the satellite-based SARAH data. In this case, the comparison uses data from 2010-2016, again showing the difference in average global horizontal irradiance. The results are shown in Figure 4.

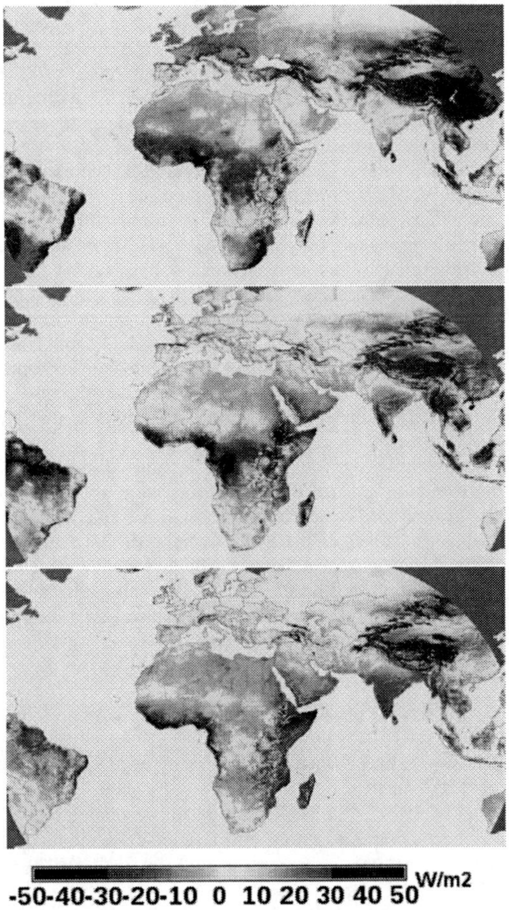

Figure 4: Difference in average irradiance between the three reanalysis data sets and the SARAH solar radiation data set for Eurasia and Africa. Values are given in

W/m^2. Top: MERRA-2, Middle: ERA-Interim, Bottom: ERA5. Note that the colour scale ranges from -30 to +30W/m^2, areas with differences outside this range are coloured in dark brown or black. Red colours represent overestimation of the reanalysis products with regard to the SARAH data set.

Again here, we see that ERA5 tends to have the lowest differences to the satellite-based data, with the main exception being that ERA5 shows much lower irradiance than SARAH in tropical West Africa, while MERRA-2 shows somewhat smaller differences in the area. All reanalysis data products show much higher irradiance in the Tibetan plateau, and much lower irradiance in tropical areas.

5 CONCLUSIONS

We have investigated the quality of global horizontal solar radiation data from three reanalysis data sets: MERRA-2, ERA-Interim and the new ERA5 products. Using the worldwide BSRN set of measurement stations as well as satellite-based solar radiation products, we have performed an initial validation of the three reanalysis data sets.

Results show that the annual mean bias varies strongly between station locations, but there are general patterns in the results. Both ERA-Interim and MERRA-2 tend to overestimate GHI, with overall MBE of 12.8 W/m^2 and 11.5 W/m^2 for MERRA-2 and ERA-Interim respectively. Hence, the use of these products to assess the surface irradiance should be generally avoided.

ERA5 shows less tendency to overestimate, with a MBD of 6.0 W/m^2 for all stations. Excluding coasts, islands and mountains, the bias of ERA5 is comparable to the satellite products NSRDB and SARAH. This was also corroborated in the comparison of the maps between both sources of data. Apart from its spatial resolution, the main limitation of ERA5 is the modeling of clouds, which makes the bias vary with the cloudiness and leads to larger absolute deviations than satellite products. However it would be reasonable to use the global horizontal irradiance values from ERA5 in areas that are not covered by the satellite data.

This investigation has only looked at the global horizontal irradiance. However, for many solar energy applications it is important to look also at the direct or diffuse radiation, both for sun-tracking or concentrating solar energy system, and for inclined-plane PV systems. We plan to extend the present study to also look at the diffuse/direct irradiance and the implications for the quality of solar energy yield estimates.

6 ACKNOWLEDGMENTS

We would like to thank the following organizations for access to the data:

- CM SAF for the satellite-based solar radiation data
- NREL for access to the NSRDB solar radiation data
- ECMWF for producing the ERA-Interim and ERA5 reanalysis data.
- NASA's Global Modeling and Assimilation Office (GMAO) for producing MERRA-2.
- BSRN for the high-quality ground records.

REFERENCES

[1] [1] J. Polo, S. Wilbert, J.A. Ruiz-Arias, R. Meyer, C. Gueymard, M. Súri, L. Martín, T. Mieslinger, P. Blanc, I. Grant, J. Boland, P. Ineichen, J. Remund, R. Escobar, A. Troccoli, M. Sengupta, K.P, Nielsen, D. Renne, N. Geuder, T. Cebecauer. Preliminary survey on site-adaptation techniques for satellite-derived and reanalysis solar radiation datasets. Solar Energy 132, 25–37, (2016).

[2] M. Sengupta, A. Habte, S. Kurtz, A. Dobos, S. Wilbert, E. Lorenz, T. Stoffel, D. Renné, C. Gueymard, D., Myers, S. Wilcox, P. Blanc, R. Perez. Best practices handbook for the collection and use of solar resource data for solar energy applications. NREL technical report, (2015).

[3] R. Müller, U. Pfeifroth, C. Träger-Chatterjee, J. Trentmann, R. Cremer. Digging the METEOSAT treasure - 3 decades of solar surface radiation. *Remote Sensing* 7, 8067–8101. (2015)

[4] K.G. Karlsson, K. Anttila, J. Trentmann, M. Stengel, J.F. Meirink, A. Devasthale, T. Hanschmann, S. Kothe, E. Jääskeläinen, J. Sedlar, N. Benas, G.J. van Zadelhoff, C. Schlundt, D. Stein, S. Finkensieper, N. Håkansson, R.Hollmann. CLARA-A2: The second edition of the CM SAF cloud and radiation data record from 34 years of global AVHRR data. *Atmospheric Chemistry and Physics* 17, 5809-5828, (2017)

[5] A. Boilley, L. Wald. Comparison between meteorological re-analyses from ERA Interim and MERRA and measurements of daily solar irradiation at surface. *Renewable Energy* 75, 135–43, (2015).

6] C. Träger-Chatterjee, R.W. Müller, J. Trentmann, J. Bendix. Evaluation of ERA-40 and ERA-interim re-analysis incoming surface shortwave radiation datasets with mesoscale remote sensing data. *Meteorologische Zeitschrift* 19, 631–640. (201)

[7] R. Müller, T. Behrendt, A. Hammer, A. Kemper. A New Algorithm for the Satellite-Based Retrieval of Solar Surface Irradiance in Spectral Bands. *Remote Sensing* 4, 622-647, (2012).

[8] A. Gracia Amillo, T. Huld, R. Müller. A new database of global and direct solar radiation using the eastern meteosat satellite, models and validation. *Remote Sensing* 6, 8165-8189, (2014).

[9] R. Urraca, A.M. Gracia Amillo, E. Koubli, T. Huld, J. Trentmann, A. Riihelä, A. Lindfors, D. Palmer, R. Gottschalg, F. Antonanzas-Torres. Extensive validation of CM SAF surface radiation products over Europe. *Remote Sensing of Environment* 199, 171-186, (2017).

[10] National Renewable Energy Laboratory (NREL). National Solar Radiation Database (NSRDB) - Physical Solar Model (PSM) version. (2017). URL: htttps://nsrdb.nrel.gov/current-version.

[11] A. Habte, M. Sengupta, A. Lopez. Evaluation of the National Solar Radiation Database (NSRDB): 1998-2015, Technical Report, NREL/TP-5D00-67722, (2017). National Renewable Energy Laboratory, Golden, CO, U.S.A.

[12] D.P. Dee, S.M. Uppala, A.J. Simmons, P. Berrisford, P. Poli, S. Kobayashi, U. Andrae, M.A. Balmaseda, G. Balsamo, P. Bauer, P. Bechtold, A.C.M. Beljaars, L. van de Berg, J. Bidlot, N. d Bormann, C. Delsol, R. Dragani, M. Fuentes, A.J. Geer, L. Haimberger, S.B. Healy, H. Hersbach, E.V. Hólm, L. d Isaksen, P.

Kållberg, M. Köhler, M. Matricardi, A.P. McNally, B.M. Monge-Sanz, J.-J. Morcrette, B.-K. Park, C. Peubey, P. de Rosnay, C. Tavolato, J.-N. Thépaut, F. Vitart. The ERA-Interim reanalysis: configuration and performance of the data assimilation system. *Quarterly Journal of the Royal Meteorological Society* 137, 553-597, (2011).
[13] R. Gelaro, W. McCarty, M.J. Suárez, R. Todling, A. Molod, L. Takacs, C.A. Randles, A. Darmenov, M.G. Bosilovich, R. Reichle, K. Wargan, L. Coy, R. Cullather, C. Draper, S. Akella, V. Buchard, A. Conaty, A.M. da Silva, W. Gu, G.K. Kim, R. Koster, R. Lucchesi, D. Merkova, J.E. Nielsen, G. Partyka, S. Pawson, W. Putman, M. Rienecker, S.D. Schubert, M. Sienkiewicz, B. Zhao. The Modern-Era Retrospective Analysis for Research and Applications, version 2 (MERRA-2). *Journal of Climate* 30, 5419–5454, (2017).
[14] ERA5 data documentation. ECMWF. https://software.ecmwf.int/wiki/display/CKB/ERA5+data+documentation.

COMPREHENSIVE ANALYSIS OF SOLARIMETRY ELEMENTS FOR PRIMAR ENERGY FORECASTING METHODOLOGIES RELATED TO PHOTOVOLTAIC POWER PLANTS

RELVA, Stefania Gomes; UDAETA, Miguel Edgar Morales; SILVA, Vinicius Oliveira; GIMENES, André Luiz Veiga; GALVÃO, Luiz Claudio Ribeiro.
Energy Group of the Electric Energy and Automation Engineering Department of the Polytechnic Scholl of the University of São Paulo.
Avenida Luciano Gualberto, travessa 3, n° 158, Sala AN-16 do prédio Anexo, Zip code: 05508-010.

ABSTRACT: The purpose of this study is to analyze the elements required for modeling primary solar energy for photovoltaic generation in large scale. Subjects related to irradiance data treatment algorithms, the irradiance estimates processes for agriculture and the probability analyses of atmosphere transmittance indexes are methodological elements. Data provided for two solarimetric stations: ESUPP and ESPEL are analyzed. Results showed that the partial annual daily average for resource availability in the countryside of São Paulo State (ESUPP location), is 5.9 kWh/m², and the annual daily average for resource availability in the metropolitan area of São Paulo City (ESPEL location) is 4.41 kWh/m²; the relation between the global horizontal, diffuse horizontal and direct vertical components varies between both solarimetric stations; ESPEL data presented trimodal characteristics for the probability density curves of the transmittance index for global irradiation, while ESUPP data reflect on the same curve, with bimodal characteristics. Thus, although solar primary energy modeling has tools that can be widely used, it requires specific inputs for each region under study. There is an unquestionable need for reliable radiometric data, measured in ground, for scientific, technologic and commercial advances to improve the use of solar resource as energy source.
Keywords: Solar Radiation; Data Quality Assessment; Primary Energy Modelling; Energy Performance.

1 INTRODUCTION

The efficient use of an energy resource in a power plant is considered of utmost importance within a lot of perspectives, e.g. energy efficiency, sustainable development, security of energy supply, optimization of the operation of the electric system - both generation and transmission. In this sense, the coefficient of the ratio "generated energy / resource availability" must be as large as possible. To optimize this coefficient, it is essential to know the behavior of the primary energy source along the geographic space and along the time. This demonstrates the importance of modeling the primary energy of intermittent renewable resources.

Currently there is an increase in installations of solarimetric stations around the world and in the use of data from these stations for energy studies and forecasting. Thus, an in-depth study of the elements inherent to solarimetry is important to understand the available and necessary inputs for the formation of models of predictability and characterization of primary solar energy for the use in photovoltaic power plants, optimizing the ratio "generated energy / resource availability".

Brazil has a huge potential for exploitation of solar energy, with a great irradiance flux per square meter. However, the first contracting of photovoltaic projects in energy auctions was only in 2014. Brazil still has not a national photovoltaic or solarimetric industry and has few installed radiometers and an increasing number of photovoltaic power plants. Therefore, very few Brazilians studies about irradiance data treatment algorithms, probability analyses of atmosphere transmittance indexes and irradiance flux forecasting models are found. So, this study analyses these international existing procedures and verifies if they are appropriate to Brazilian climate reality with focus on primary energy modeling for photovoltaic power plants in large scale. Hence this study contributes to international knowledge on specificities of energy resource behavior caused by different climate profile.

2 SOLAR PRIMARY ENERGY MODELING AND SPECIFICITIES

The primary energy modeling is the construction of the distribution profile of the source as a function of time and geographical space. The resource availability prediction models are categorized into three main strands: physical models, statistical models and empirical models [1]. The empirical models are those parameterized by other climatic variables, such as ambient temperature, relative humidity, cloudiness, among others that may guide the future forecast of energy availability. The physical models constitute a theoretical approach that seeks to define the equation of radiative transfer to describe mathematically the physical processes that occur in the atmosphere [1]. The statistical models are those that use the computation of probabilistic events from the historical series to define the probability of future energy. These methods can be used to forecast resource availability for the short, medium and long term.

Atmospheric transmittance is the main factor in determining the amount of radiation on the Earth's surface at any given time. However, the characterization of atmospheric transmittance goes far beyond the identification of cloudy conditions. The atmosphere is heterogeneous and difficult to define. This heterogeneous profile changes over the time and space, making the definition of a transmittance pattern even more difficult [2]. Thus, access to ground measurements of irradiance and others climatic variables is essential for understanding the behavior of the resource over the time and space.

Besides the importance of access ground measurements of irradiance, it is also necessary to ensure the quality of these data. Nowadays there are some reference procedures for irradiance data quality assessment which are briefly discussed in the next section.

2.1 Irradiance data quality assessment

Considering data of global horizontal irradiance (GHI) measured by a pyranometer, data of diffuse horizontal irradiance (DHI) measured by a shaded pyranometer under a tracking sphere and data of direct normal irradiance (DNI) measured by a pyrheliometer on a sun-following tracker, in a specific moment (t): the measurement of GHI performed by the pyranometer may be equivalent to the measurement of DHI plus the vertical component of DNI measurement performed by the pyrheliometer. This relation is shown in Equation 1:

$$GHI\ (t) = \ DHI(t) + \ DNI\ (t) \times \cos\theta_Z(t) \quad (1)$$

Where cos Θ_z is the cosine of zenith angle. This equation is one of the pillars of irradiance data quality assessment.

Still considering the same measurements at the same moment (t), it is possible to state that the physical maximum value of GHI may be the vertical component of extraterrestrial normal irradiance (DNI_0) at that moment (t) which has a zenith angle cos Θ_z, thus the relation for maximum value of GHI may be:

$$GHI(t) \ \leq \ DNI_0\ (t)\,.cos\ \theta_Z(t) \quad (2)$$

Following the same reasoning, the physical maximum value of DNI may be the extraterrestrial normal irradiance (DNI_0) at that moment (t):

$$DNI(t) \ \leq \ DNI_0\ (t) \quad (3)$$

From the moment the sun rises to the moment it sets, it is possible to check for natural clarity, even on days when the sun appears completely covered by clouds (at which point the direct irradiance tends to zero), so it is possible to conclude that diffuse horizontal irradiance (DHI) never obtains null values throughout the day and the diffuse irradiance is at most equal to the global horizontal irradiance:

$$\text{DHI}(t) \leq GHI(t) \ \leq \ DNI_0\ (t)\,.cos\ \theta_Z(t) \quad (4)$$

However, for few-minutes mean of irradiance measurement it is possible the occurrence of negative measurements and overirradiance conditions. The overirradiance phenomenon can result from high direct radiation plus high diffuse radiation due to nearby cloud scattering [3]. It is possible that the solar global irradiation at the surface can be higher than that at the top of atmosphere for a short time period, because of the diffusive effects of clouds that are not in the way of the solar beam [4]. The negative values are justified by some possible situations like sudden shadowing of the equipment, uncertainty rates inherent to any radiometer and sudden cooling of the ambient temperature, causing a difference between the temperature of the equipment domes [5].

The limits of acceptable values of irradiance data depend on the microclimate specificities, the averaging time of the historical series and the zenith angle because when solar elevation is small the equipment uncertainty and atmosphere interferences rise. The Baseline Radiation Network (BSRN) project [6] stablished maximum limits for irradiance data larger than extraterrestrial irradiance and negative maximum limits for irradiance data. The BSRN [6] also states: DHI/GH <1.05 for cos Θ_z < 75° and GHI > 50 W/m² and DHI/GH <1.10 for 75° < cos Θ_z < 93° and GHI > 50 W/m².

Considering the relation of Equation 1, BSRN states: GHI/GHI_c = 1± 8% for cos Θ_z < 75° and GHI > 50 W/m and GHI/GHI_c = 1± 15% for 75° < cos Θ_z < 93° and GHI > 50 W/m². Where GHIc corresponds to the calculated global horizontal irradiance values through the DHI and DNI measured values.

The Brazilian Environmental Data Organization System (SONDA) adopted the BSRN data quality assessment procedure, however, the Brazilian project assumes a broader limit for the relation of Equation 1, adopting that the relation GHI/GHI_c may be equal to 1 ± 10% instead of 8% for cos Θ_z < 75° and GHI > 50 W/m².

There is currently no definitive, ideal, or widely accepted procedure for optimal data quality assessment for irradiance measurements [7]. The BSRN, the National Renewable Energy Laboratory (NREL), the Royal Meteorological Institute of Belgium (RMIB) developed relevant studies and data quality assessment procedures that proves the efforts to ensure the quality of radiometric data.

Based on the condition that irradiance data behavior depends on: the different averaging time of the historical series, the zenith angle variation and the location climate specificities, the NREL data quality assessment procedure [8] considers the historical series of the solarimetric station for the determination of the quality criteria. For this, NREL procedure uses the transmittance index analysis considering different air mass intervals. RMIB data quality assessment procedures [9], besides maximum and minimum limits and relations based on Equation 1, also considers test to detect shadow obscuring the global sensor and persistence test (among others), but the procedures are applicable only in ten-mean or thirty-mean averaged data.

Throughout the solarimetric technological development, a series of stations was discontinued and, scales, equipment and technologies - considered as reference standard - were altered as can be verified in [10]. These changes are natural in any technology development process. Currently, although significant improvements have been achieved in the uncertainty levels of radiometric equipment, there are low levels of installation, registration and maintenance of these equipment. The World Meteorological Organization (WMO) and International Organization for Standardization (ISO) made relevant advances in radiometers measurement accuracy classification and analysis [11].

It is also important to notice that data quality assessment may assume that station has adequate maintenance and installation procedures, with calibrated equipment [8, 9]. So, data quality assessment may be carefully chosen, considering the historical series specificities. Thus, assuming a treated and reliable irradiance data, modeling procedures can be applied aiming the characterization of resource behavior. Based on ground measurements, empirical and statistical models are used to this end.

2.2 Empirical and statistical modeling process

The number of empirical models of irradiance estimation found in the literature is quite significant. Many of the global irradiation estimation models are developed by the agronomic sciences since global daily irradiation is an input in the reference evapotranspiration definition models. There is also a great number of models which define diffuse and direct irradiation based on measurements of global irradiation. These models are called "decomposition models" [12]. Many of them use as input only the data of global irradiance or global irradiation. Other portion of irradiance empirical estimation models aims the definition of solar radiation for inclined surfaces. These models, in general, use as input diffuse and direct irradiation data. Several of them can be found in [13].

The most common empirical estimation model of global irradiance is the Angströn-Prèscott equation which correlates global daily irradiation with daily hours of insolation and with daily horizontal extraterrestrial irradiation. In general, empirical models of global irradiation estimation equate how much of the extraterrestrial global irradiance does not reach the Earth's surface due to the atmospheric transmittance, so the global radiation estimation models can be generically represented by Equation 5:

$$GHI^{daily} = GHI_0^{daily} * k_t \quad (5)$$

Where GHI^{daily} is the daily global horizontal irradiation, k_t is the transmittance index and GHI_0^{daily} is the daily extraterrestral irradiation that reaches an Earth horizontal surface. The definition of transmitance index is the aim of the most part of the empirical models. Several of them use daily thermal amplitude, or daily insolation hours, correlating these variables with GHI^{daily} and GHI_0^{daily} through linear regression, determining empirical coefficients of correlation. With these empirical coefficients, futures GHI^{daily} values can be found based on existing data of others climatic variables. Several of these models can be found in [14, 15, 16, 17, 18].

The transmittance index is also used to modeling irradiance behavior through the construction of probability density distributions and cumulative distributions. These analysis is generally used aiming the characterization of atmospheric condition (cloudy sky and clear sky) as a function of air mass. The air mass index (AM) indicates a multiple of the solar radiation path into the atmosphere. There are several air mass equations in the literature, in this study, since data with zenith angle larger than 75° were excluded, the adopted equation is:

$$\mathrm{AM} = 1/\cos\Theta_z \quad (6)$$

Equation 7 is a simplification of the relation presented in [19].

While empirical models of irradiation estimation are used mainly for daily or monthly sums of irradiation, the statistical models of transmittance index values and irradiance values can be used for instantaneous data or for few-minutes averaging data. Thus, it is possible to identify the percentage of time in which the transmittance index indicates clear or cloudy sky conditions, as long as the historical series available is long enough to characterize a relevant period, such as one season or one year.

In the last decades, several authors found a bimodal character for probability density distributions of transmittance index [20, 21, 22, 23]. For some locations the bimodal character decreases as air mass increases [21, 23], for others, the bimodal character increases as air mass increases [20, 22]. The bimodal character indicates, at first, that there are only two cloud cover states: cloudy sky and clear sky. Intermediate values may occur when radiation passes at the fringe of thick clouds and when very thin clouds prevail [20]. Therefore, intermediate values of transmittance index do not only represent a transition state, but also a permanent state of thin clouds covering the sky.

Through the analysis of irradiance data of two Brazilian solarimetric stations, this study evaluates the irradiation behavior pattern applying some data quality assessment criteria, discussing linear correlation between climate and irradiation variables and producing probability density distributions of k_t for different air mass intervals.

3 METHODOLOGY AND AVAILABLE DATA

DNI, DHI and GHI one-minute mean data from two solarimetric station in Brazil are used. One of the solarimetric station (named ESPEL station) is located at São Paulo city (22.486° S, 52.955° W), a region characterized by high levels of atmospheric pollution, at an elevation of 756 meters above sea level. The second station (named ESUPP station) is in a hydropower plant, very close to the hydropower dam, in the countryside of state of São Paulo (22.486° S, 52.955° W), at an elevation of 261 meters above sea level. Both stations have sensors of wind speed and direction, ambient temperature, atmospheric pressure, relative humidity and precipitation. They also have tracker system with GPS and sunsensor coupled to ensure the right position of the pyrheliometer and of the shading sphere of the diffuse radiation pyranometer.

The ESUPP started recording data in the early evening of September 19th, 2015. On November 14th , 2015, the solar tracker was turned off. The equipment had to be sent to the manufacturer in Japan for maintenance, implying, until the closing of this study, the lack of measurements of direct normal and diffuse horizontal irradiances.

The ESPEL began recording data at the beginning of November 2014, however the data recording was stabilized only on December 23th, 2014. On January 30th, 2015, the pyranometer and pyrheliometer connection cables and wrapped around the tracker causing the cables breakage, because of this, the station remained unregistered until March 24th , 2015, when the problem was solved. However, in the process of reinstalling the equipment, the sunsensor has been installed upside down. This sensor was withdrawn on April 27th, so that the positioning of the solar tracker was, until May 20th, 2015, only in charge of the GPS system. As a result, the misalignments of the shading sphere and the pyrheliometer were visually perceived in early morning and at the end of the day during the period from March 24th to April 27th, 2015, as can be verified in the photographic record obtained at 6:50 am of March 26th, 2015, shown on Figure 1. On October 2th, 2015, there were problems with the connection of the datalogger resulting in the storage of data in the memory card until November 17th, 2015, the date on which the memory card stopped working, causing the lack of data recording until January 2th, 2016. The historical series assessed in this study ends on June 30th, 2016. The station sensors are cleaned and the alignment is checked weekly.

Figure 1: Photographic record of the misalignments of the shading sphere.

To ensure the data quality and to simplify the irradiance characterization, data corresponding to zenith angles equal to, or more than, 75° were excluded. The data quality assessment was performed considering the following conditions: (i) verification of missing record lines; (ii) verification of repeated record lines, (iii) verification of null data, (iv) verification of negative data, (v) verification of irradiance values higher than extraterrestrial irradiance, (vi) verification Equation 1 relationship.

Thus, the irradiance data from both solarimetric station were used to calculate the probability density distributions of k_t for different air mass intervals. The air mass intervals were chosen based on [22] and extraterrestrial irradiation and zenith angle for each minute of historical series were calculated based on [19]. The linear relation between irradiation data and others climate data is also analyzed, based on [24], through the calculation of Person's coefficient between daily means of global horizontal irradiation and daily means of ambient temperature, daily sum of precipitation and daily mean of relative humidity.

3 RESULTS AND DISCUTIONS

ESPEL data corresponds to period from March 24th, 2015 to June 30th, 2016, with lack of data from November 17th, 2015, to January 2th, 2016. Thus, considering the existing data, the principal data quality assessment results are: (i) some of the minutes over the historical series have not been recorded, and (ii) others were recorded twice; (iii) 14 null records of GHI and DHI were found in the middle of the day, indicating certainly a wrong data set; (iv) negative DNI data were found in all months of historical series, the month with the largest amount of negative DNI data was May 2016 (38.2% of DNI records for this month are negative values) and the month with the lowest amount of negative DNI data was April 2015 (38.2% of DNI records for this month are negative values), all negative values are very close to zero and were checked early in the day or late afternoon; (v) GHI values higher than extraterrestrial irradiance in a Earth horizontal surface (GHI_0) were found mainly in the winter (37,3% of GHI records of July 2015 are higher than GHI_0) because of this, the criterion of physically possible data established by BSRN [6] was also applied. 50 DHI data from April 2015 and 3 from May 2015 were identified as higher than the values established by BSRN, a value of the month of June 2016 and another in the month of July of 2016 of GHI were identified as higher than the values established by BSRN; however, all these data correspond to values very close to the BSRN limit and consistent with the values of previous and posterior measurements of the historical series. No data was deleted; (vi) for the varication of relationship on Equation 1, GHI/GHI_c was calculated, this coefficient is named in this study as k_{tc}, about 30% of the data presented GHI measurements of 12% to 13% higher than the respective DNI*cos Θ_Z + DHI values, while 49% of the data presented values of k_{tc} between 0.9 and 1.1, that is, within of the range acceptable by the Sonda Project. The parcel of data with k_{tc} between 1.12 and 1.13 is quite significant.

The first possible explanation for this would be a possible misalignment of the solar tracker, so that the DNI measurements were not recorded in their entirety, reducing the total of this component in relation to the real. However, due to the constant process of maintenance and monitoring of the station and exclusion of data with zenith angles higher than 75°, this possibility is unlikely. The second possible explanation, and more likely, would be the interference caused by pollution in the radiation flux in the region. Considering the Mie scattering theory, is indicated that particles with a diameter greater than the wavelength of the radiation, such as particulate matter, cause a significant scattering process of the radiation including frontal scattering, reducing the portion of direct radiation and increasing the amount of diffuse radiation, which may not uniformly affect the global radiation sensor and the diffuse radiation sensor, generating a decompensation of the relation.

GHI data period from ESUPP station was from September 20th, 2015 to May 31th, 2016; DNI and DHI data period used was from September 20th, 2015 to November 14th, 2015, that is a very restrict period of data. The principal data quality assessment results are: (i) some of the minutes over the historical series have not been recorded; (ii) no duplicate records found; (iii) 151 null records of GHI and DHI were found in the morning period indicating certainly a wrong set of data, representing less than 0.1% of the data sample with zenith angle lower to 75°; (iv) no negative data found; (v) GHI values higher than extraterrestrial irradiance in a Earth horizontal surface (GHI_0) were found mainly in the winter (May with 1.200 records), all data has been approved in the criterion of physically possible limits established by BSRN [6]; (vi) 0.6% of the sample from the period from September 20, 2015 to November 13, 2015 was not approved by the Sonda Project criterion.

Aiming at a better understanding of this difference of patterns related to the last data quality criterium between data from two solarimetric stations, relative frequency curves of k_{tc} were plotted for different air masses using ESPEL data. Graphics of Figure 2 shows the relative distribution for AM = 1.0 (which represents the air mass values between 1.0 and 1.2) and AM = 3.0 (which represents the air mass values between 2.8 and 3.5) calculated for each minute of historical series using Equation 6. The frequency of k_{tc} values greater than 1.1 increases as the air mass increases. For AM = 1.0, the number of acceptable values (between 0.9 and 1.1) is 52.8%, while for AM = 3.0 it is 38.8%. Although the amount of high values of ktc increases as the AM increases, the amount at AM = 1.0 is still quite significant.

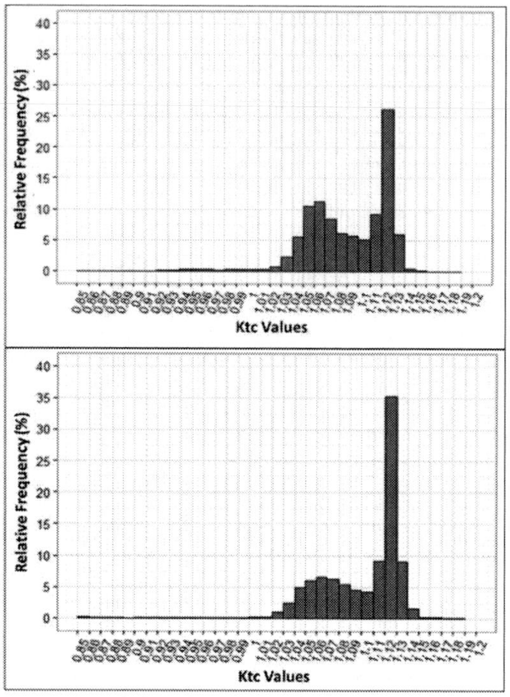

Figure 2: Relative distribution of k_{tc} for AM = 1.0 (top chart) and AM = 3.0 (bottom chart).

The same analysis was performed for ESUPP data. In this case, for AM = 1,0, the amount of acceptable values (between 0.9 and 1.1) is 99.5%, for and for AM = 3.0 it is 98.3%. That is, although there is a slight drop in the amount of acceptable values as the air mass increases, there is no, in this case, strong relationship between increasing air mass and increasing high k_{tc} values.

This analysis corroborates the hypothesis that pollution and high concentration of particulate matter in the atmosphere generates a significant change in the behavior of the resource.

Still considering the division of the data in air mass intervals, the GHI data of both station were used to k_t calculation and probability density distributions (PDF) of transmittance index were plotted. Graphs in the Figure 3 shows the PDF of k_t for AM = 1.0 and for AM = 3.0 for ESPEL data. Graphs in the Figure 4 shows the same for ESUPP data.

The PDFs were calculated for five air mass intervals: AM =1.0 (AM values between 1.0 and 1.2), AM =1.5 (AM values between 1.4 and 1.6), AM =2.0 (AM values between 1.8 and 2.2), AM =2.5 (AM values between 2.3 and 2.7) and AM =3.0 (AM values between 2.8 and 3.5).

Excepted for AM = 1.0, all the other PDFs of ESPEL data showed a trimodal profile. This indicates a greater occurrence of intermediate values of transmittance index than those found in the Literature. As the air mass increases, the first peak of distribution also increases indicating, as expected, more atmospheric interference on radiation.

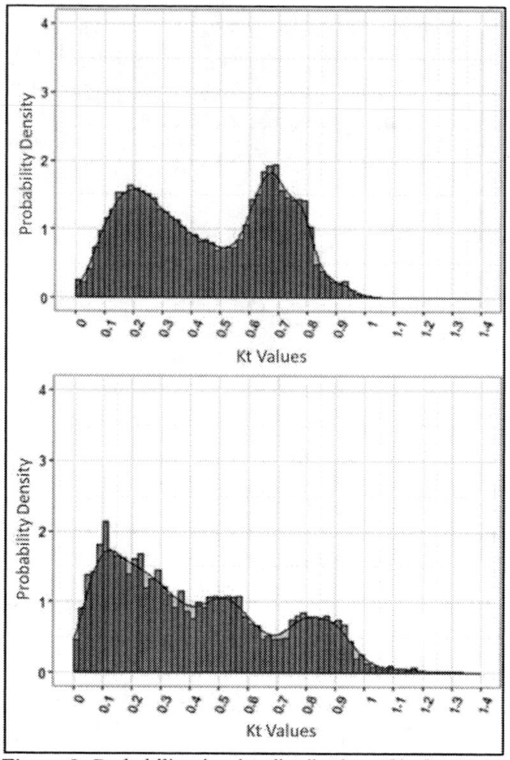

Figure 3: Probability density distribution of k_t for AM = 1.0 (top chart) and AM = 3.0 (bottom chart) for ESPEL data.

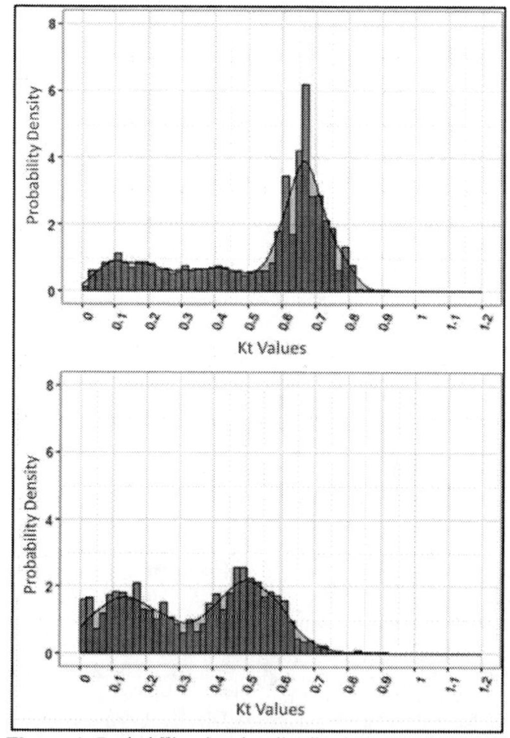

Figure 4: Probability density distribution of k_t for AM = 1.0 (top chart) and AM = 3.0 (bottom chart) for ESUPP data.

Data from ESUPP presents similar behavior than those found in the Literature, with the increase of the bimodal character as the air mass increases. Considering only the PDFs for AM= 1.0, ESUPP has a more defined mainly peak, indicating high occurrence of clear sky conditions, while ESPEL has a more homogenous distribution between cloudy and clear sky conditions.

From theses analysis, it is possible to highlight and endorse the possibility of huge interference of the pollution and high concentration of particulate matter in the irradiation behavior.

In terms of amount of available energy, daily means of global horizontal irradiation were calculated. The partial annual daily average for resource availability in the countryside of State of São Paulo is 5.9 kWh/m² (ESUPP region), and the annual daily average for resource availability in the metropolitan area of São Paulo City (ESPEL region) is 4.41 kWh/m². The daily means of GHI were used to examine the linear relation between this variable and others climatic data. Data from any of these variables were not considered for zenith angles greater than 75°. This premise assumes that since there are no positive values of radiation during the night, the consideration of the other climatic variables during the nocturnal period could indicate a false correlation between the variables.

For ESUPP data Pearson's coefficient results are: -0,3 for precipitation, 0,64 for ambient temperature and -0,67 for relative humidity. For ESPEL data, only the longer period with continuous measurements was used: from April 1, 2015 to November 16, 2015. ESPEL Pearson's coefficient results are: -0,34 for precipitation, 0,76 for ambient temperature and -0,79 for relative humidity. This means that, for both stations, no other analyzed climatic variable present a strong linear relation with daily irradiation, seeing that Pearson's coefficient between |0.5| and |0.8| indicates a moderate relation (direct relation if positive coefficient and inverse relation if negative), and values between |0.1| and |0.3| indicates week correlation [24].

The analysis of Pearson's coefficient produced in [24] indicates relationships between insolation and other climatic variables in existing hydroelectric dams. In these analysis, the strongest relationships were found between insolation and evaporation, insolation and relative humidity and insolation and cloudiness.

The largest number of empirical models of radiation estimation present in other studies [14, 16, 15, 17, 18] indicate the use mainly of insolation and temperature data and indicate that the more meteorological variables are considered, the better the results obtained in a simple linear regression model will be. Thus, it is evident that each location and even each season of the year for the same location may have a specific correlation pattern between climatic variables. These empirical models for global irradiation may be chosen according to climatologic available data and they can help in the process of formatting complete historical series.

Thus, the modeling of the solar resource for photovoltaic generation in scale must be done in two moments: (i) prediction of the availability of the resource in the long term (annual daily average) - important for the definition of generation potentials, firm energy calculation and economic feasibility of the projects and (ii) prediction of the availability of the resource in the short and very short term - important for the operation of photovoltaic plants, mainly those of centralized dispatch. About this second step it was verified that the data quality assessment procedures have a very strong relation with the analysis of temporal aspects of the resource. The constant application of quality procedures in data collected by an solarimetric station is the first step in the characterization of the behavior of the resource in that region during the hours of the day, the days of the months and the months of the year. Besides that, it is important to notice that the most part of empirical models for irradiation in an inclined surface is linked to diffuse and direct values, and, if these values do not fit in relationship of Equation 1, the performance of some empirical models can be compromised.

The zenith angle is the main factor in determining the temporal aspects of the resource, it is a metric for the estimation of uncertainties and, through the calculation of the air masses, it allows the understanding of the atmospheric interference in the analyzed data. The calculation of transmittance coefficients is widely used for characterization of the resource and for data quality assessment.

5 CONCLUSIONS

By analyzing data from two stations, it is possible to conclude that there is a huge difficult in the obtainment of complete and sequential historical series from ground measurement data in Brazil. At the present, the Brazilian supply of radiometers and other solarimetric instruments still faces obstacles, especially regarding installation, maintenance and operation. There are several obstacles related to the installation process of the equipment such as improper installation of cable lengths, favoring the screwing of the same ones in the solar tracker and causing the disruption of the installations. Another obstacle is the time required to correct the faults. Since the suppliers of the equipment do not yet have national facilities, the correction of problems in the equipment often requires the transport of the same ones out of the country, causing months' failures in the historical series. Thus, given the need for one year of continuous measurements of global horizontal irradiance data to accept projects from the futures energy auctions in Brazil, stakeholders should be concerned with the planning of the installation of these stations for periods of more than one year before the planned date of the insertion of the projects in the auctions process.

Results showed that the partial annual daily average for resource availability in the countryside of São Paulo State is 5.9 kWh/m², and the annual daily average for resource availability in the metropolitan area of São Paulo City is 4.41 kWh/m². Moreover, it was observed that the relation between the global horizontal, diffuse horizontal and direct vertical components varies between both solarimetric stations, such that a significant part of the ESPEL data shows global horizontal irradiance (GHI) between 12% and 13% above the respective diffuse horizontal irradiance (DHI) plus the vertical portion of the respective direct normal component ($DNI\cos\theta z$), while the same relation for the ESUPP data attend the $GHI{=}DHI{+}DNI\cos\theta z$ relation within the acceptable range of 10%, give or take. The most possible reason for this is the pollution interference since the solar station has weekly maintenance procedure. Data quality assessment procedures have a very strong relation with the analysis of temporal aspects of the resource. This is verified by means the measurement variations through k_t analysis, and the relations between global, diffuse and direct irradiances flux. Both processes are a benchmark for both short-term availability analysis and data quality analysis.

ESPEL data presented trimodal characteristics for the probability density curves of the transmittance index for global irradiation, while ESUPP data reflect on the same curve, with bimodal characteristics. Thus, although solar primary energy modeling has tools that can be widely used - such as data quality parameters and empirical and statistical models - it requires specific inputs for each region under study. The variability of the resource is totally linked to the climatic specificities of each region, and they need a close and precise look at the ground measurements to define appropriate procedures for the predictability of the resource and appropriate procedures for treatment of data.

In conclusion, there is an unquestionable need for reliable radiometric data, measured in ground, for scientific, technologic and commercial advances to improve the use of solar resource as energy source, considering the current scenario, which demands alternative sources for clean energy. It is also evident that there is still a long path to be tracked in order to consolidate a reliable, monitored and amplified network for solarimetric measurements in Brazil and in the world.

ACKNOWLEDGEMENTS

To CAPES, for the scholarship and PROAP financial support; to CESP, for funding the P&D ANEEL PD-0061-0043/2014, enabling the accomplishment of this article. We would also like to thank the team of researchers and collaborators who directly or indirectly influenced the P&D ANEEL.

REFERENCES

[1] E. J. WEBER. Tesis. (2011) Federal Unversity of Rio Grande do Sul, Porto Alegre.

[2] F. R. MARTINS et al.. Revista Brasileira do Ensino de Física 26 (2004) 145.

[3] K. YANG et al.. Agricultural and Forest Meteorology 150 (2010) 38.

[4] G. SHI et al. Journal of Applied Meteorology and Climatology 47 (2008) 1006.

[5] L. J. B. MCARTHUR (2005).Geneva: WCRP-121; WMO/TD-No. 1274.

[6] C. N. LONG, E. G. DUTTON, [Online]. Available: http://bsrn.awi.de/fileadmin/user_upload/bsrn.awi.de/Publications/BSRN_recommended_QC_tests_V2.pdf. [Access 2017].

[7] C. A. GUEYMARD, J. A. RUIZ-ARIAS. Solar Energy 128 (2016) 1.

[8] E. MAXWELL (1993). Colorado: National Renewable Energy Laboratory.

[9] M. JOURNÉE, C. BERTRAND. Solar Energy 85 (2011) 72.

[10] F.VIGNOLA et al, (2012). Boca Raton: Taylor & Fracis Group.

[11] World Meteorological Organization, (2012, Geneva: Chairperson, Publications Board.

[12] C. RAICHIJK, F. TADDEI. Avances en Energias Renovables y Medio Ambiente 16 (2012) 23.

[13] J. E. HAY, Renewable Energy 3 (1993) 373.

[14] F. CARVALHO et al. XVIICBA: Congresso Brasileiro de Agrometeorologia (2011).

[15] S. F. M. DANTAS. Monograph (2013). Federal University of the Semi-Arid Region

[16] A. D. L. CASA et al.. Revista de Investigaciones Agropecuarias 32 (2003) 45.

[17] V. J. SILVA et al.. Revista Ciência Rural 42 (2012) 276.

[18] K. OUALI, R. ALKAMA. Energy Procedia.50 (2014) 670.

[19] M. IQBAL (1983) Toronto: Academic Press, 1983.

[20] H. SUEHRCKE, P. G. MCCORMICK. Solar Energy 40 (1988) 413.

[21] A. SKARTVEIT, J. A. OLSETH. Solar Energy 49, (1992) 477.

[22] M. TOVAR et al..Solar Energy 62 (1998) 387.

[23] H. F. ASSUNÇÃO et al. Theoretical and Applied Climatology 75 (2003) 213.

[24] S. G. RELVA et al.. 5th International Youth Conference on Energy (IYCE) (2015).

BEST PRACTICES TO UNCERTAINTY ESTIMATION FOR THE NATIONAL SOLAR RADIATION DATABASE (NSRDB 1998–2015)

Aron Habte, Manajit Sengupta
National Renewable Energy Laboratory, Golden, CO, 80401, USA

ABSTRACT:

It is essential to apply a traceable and standard approach to determining the uncertainty of solar resource data. Solar resource data are used for all phases of solar energy conversion projects, from the conceptual phase to routine solar power plant operation, and to determine performance guarantees of solar energy conversion systems. These guarantees are based on the available solar resource data derived from measurement stations or modeled data sets, such as the National Solar Radiation Database (NSRDB). Therefore, quantifying the uncertainty of these data sets provides confidence to the financiers, developers, and site operators of solar energy conversion systems, and ultimately reduces deployment costs. In this study, we implemented the *Guide to the Expression of Uncertainty in Measurement* [1] to quantify the overall uncertainty of the NSRDB data. First, we start by quantifying measurement uncertainty, then we determine each uncertainty statistic of the NSRDB data, and finally we combine them using the root sum squared method. The statistics were derived by comparing the NSRDB data to the seven measurement stations from the National Oceanic and Atmospheric Administration's Surface Radiation Budget Network (SURFRAD), National Renewable Energy Laboratory's (NREL's) Solar Radiation Research Laboratory (SRRL), and the Atmospheric Radiation Measurement (ARM) program's Southern Great Plains (SGP) Central Facility, in Billings, Oklahoma. The evaluation was conducted for hourly values, daily totals, monthly mean daily totals, and annual mean monthly mean daily totals. Varying time averages assist in capturing the temporal uncertainty of the specific modeled solar resource data required for various phases of a solar energy project; some phases require higher temporal resolution than others. Overall, by including the uncertainty of measurements of solar radiation made at ground stations, bias, and root mean square error (RMSE), the NSRDB data demonstrated expanded hourly and annual uncertainties of 17%–29% and 5%–8% respectively.

Keywords: uncertainty, pyranometer, NSRDB, GUM

1 INTRODUCTION

Recognizing the importance of solar resource to industry and other stakeholders, the National Solar Radiation Database (NSRDB) has been updated over the years to meet demand and be useful for deployment of current technologies. Recently, the National Renewable Energy Laboratory (NREL) released a version of the NSRDB (1998–2015), which contains gridded solar irradiance—global horizontal irradiance (GHI), direct normal irradiance (DNI), and diffuse horizontal irradiance at a 4-km by 4-km spatial resolution and half-hourly temporal resolution covering 18 years. Details of the data set are available at https://nsrdb.nrel.gov. Additional details about the NSRDB are also described in [2], [3]. Understanding the impacts of atmospheric constituents on the solar resource and quantifying the uncertainty are essential to accurately designing utility-scale solar energy conversion systems. This study attempts to provide a comprehensive uncertainty estimate that demonstrates the performance characteristics of the NSRDB (1998–2015) when including all sources of uncertainty at various time-averaged periods, a protocol that is not often used in model evaluations. Further, to have a quality uncertainty analysis representing the spatial coverage of the model, it is advisable to obtain enough ground measurement locations that encompass various climatic regions. However, it is challenging to obtain long-term, high-quality ground measurements of solar radiation in various regions because ground measurement stations are expensive to commission and operate. For this study, nine good quality stations were chosen representing seven U.S. climatic regions (Figure 1).

2 METHOD

Determining the accuracy and uncertainty of satellite-derived data sets is based on a wide range of approaches and data analysis techniques. Some of these techniques use mean bias error (MBE) and root mean square error (RMSE) as statistical measures; however, to perform comprehensive uncertainty estimation, all known sources of uncertainties must be included [4], [5], [6], and one of these sources of uncertainty is the ground-based measurement uncertainty.

2.1 MEASUREMENT UNCERTAINTY

Many studies have reported that the uncertainty of ground-measurements from pyranometers could range from 3%–5% [7], [8], [9]. In this report, we considered the conservative value of 5% uncertainty for the ground-based measurement data (Figure 2).

Determining the uncertainty of the solar radiation measurement using pyranometers and/or pyrheliometers includes sources of uncertainties, such as the specifications of the instrument, calibration procedure, measurement conditions and maintenance, and environmental conditions. In this paper we follow the steps required to quantify a traceable uncertainty estimate using the *Guide to the Expression of Uncertainty in Measurement* (GUM) method.

2.2 NSRDB UNCERTAINTY STATISTICS

Statistical measures were used to investigate the performance of the NSRDB relative to ground-based measurements. These included:

MBE in units of irradiance (W/m^2) and percentage of reading (%) were calculated using the following equations:

$$MBE(W/m^2) = \frac{1}{n}\sum_{i=1}^{n}(x_i - x_{true})$$

$$MBE(\%) = \frac{\left(\frac{1}{n}\sum_{i=1}^{n} (x_i - x_{true})\right)}{\left(\frac{1}{n}\sum_{i=1}^{n} (x_{true})\right)} * 100$$

where n represents the total number of measurements at each time average, x_i represents values from the NSRDB (1998–2015) satellite-derived data, and x_{true} represents the concurrent ground-measured data for both GHI and DNI at a given site, which were evaluated separately. Similarly, RMSE values were computed using the ground measurements (x_{true}):

$$RMSE(W/m^2) = \sqrt{\left(\frac{1}{n}\sum_{i=1}^{n} (x_i - x_{true})^2\right)}$$

$$RMSE(\%) = \sqrt{\frac{\left(\frac{1}{n}\sum_{i=1}^{n} (x_i - x_{true})^2\right)}{\left(\frac{1}{n}\sum_{i=1}^{n} (x_{true})^2\right)}} * 100$$

These statistics provided insight into the performance of the modeled data set relative to the ground-based measurements [10], [3]. For instance, RMSE is useful in determining the error distribution [11]. MBE describes the direction (+/-) of the error or difference bias.

To understand and quantify the overall uncertainty of the modeled data set, the approach outlined in the GUM was used. The method applies RMSE, MBE, and the ground-based measurement uncertainties as sources of estimated combined uncertainty. This approach provides uncertainty estimation on a 95% confidence interval (U_{95}%) representing two standard deviations (coverage factor of $k = 1.96$) and assuming a Gaussian or normal distribution of the data. Details of this methodology are described in [7], [12], [13]:

$$U_{95}\% = k * \pm \sqrt{\left(\frac{U_{meas}}{k}\right)^2 + \left(\frac{MBE}{k}\right)^2 + \left(\frac{RMSE}{k}\right)^2}$$

where U_{meas} is the estimated uncertainty of the surface measurement (ground truth), and MBE and RMSE are derived from the model validation analyses.

3 RESULTS AND DISCUSSION

Figure 2 shows that the hourly data set demonstrated a higher uncertainty estimation of the NSRDB. As the time average changed from hourly to annually, the uncertainty decreased and approached the measurement uncertainty. The estimated uncertainty ranged from approximately 30% on an hourly basis to 5% annually. However, these ranges differ from one location to another because of multiple reasons, such as climatic and microclimate differences among regions and differences in instrumentation, maintenance, and operation of measurements of solar radiation made at ground stations.

The surface measurement uncertainty was left unchanged at 5% throughout the averaging period because the inherent uncertainties of the radiometers should remain the same throughout the averaging. The 5% uncertainty was derived using the method outlined in the GUM, which applies both the statistical and nonstatistical sources of uncertainty [7], [12]. Evaluating the random component of the ground measurement requires a statistic, such as standard deviation; however, the inherent nature of changing weather during the ground-based measurements cannot be considered a source of uncertainty to determine the standard deviation because it is not about the instrument that measures the changes. In future studies, statistical sources of uncertainties of measurements could be included by analyzing the time response and stability of the radiometer that measures the solar resource and the data acquisition system. We believe that these sources will not have a significant effect in reducing the estimated uncertainty of the ground-based data as the averaging time increases, but this needs further investigation.

Figure 3 illustrates the relative GHI and DNI biases and RMSE. The results show that the hourly-averaged satellite-derived data have a bias of approximately ±5% for GHI and less than +10% for DNI; however, RMSE is higher for the hourly averages: the GHI of the NSRDB demonstrated on average up to 20% RMSE when compared to the ground-based GHI measurements and up to 30% RMSE on average when compared to the ground-based DNI measurements.

On the other hand, the RMSE in the NSRDB (1998–2015) decreased significantly with the increase in the time average. This is partly because as averaging time increased, there is a cancellation of the random error. Further, the reason for the high RMSE in the hourly data set is that the NSRDB pixel represents a 4-km by 4-km area, whereas a ground-based station represents only a small area above the measuring station. The subpixel variability in clouds appears to contribute to the higher differences.

4 SUMMARY and CONCLUSION

A full characterization of uncertainty estimation provides the basis to assess the predicted output of planned solar conversion systems and is thus a key factor when determining the bankability of the project. The integrity of modeled data can be improved by applying a comprehensive description of the model uncertainty. As reported in the paper, this can be achieved using the GUM method. Proper uncertainty estimation provides confidence in the data; however, it is important to determine the uncertainty using a standard methodology that others can also use and will obtain identical results.

ACKNOWLEDGEMENTS

This work was supported by the U.S. Department of Energy under Contract No. DE-AC36-08-GO28308 with the National Renewable Energy Laboratory.

References

[1] Joint Committee for Guides in Metrology Working Group 1 (JCGM/WG 1). (2008). *Guide to the Expression of Uncertainty in Measurement.* Accessed February 08, 2017: www.bipm.org/utils/common/documents/jcgm/JCGM_100_2008_E.pdf

[2] Xie, Y., M. Sengupta, and J. Dudhia. 2016. "A Fast All-Sky Radiation Model for Solar Applications (FARMS): Algorithm and Performance Evaluation." *Solar Energy* 135: 435–445.

[3] Sengupta, M., A. Weekley, A. Habte, A. Lopez, C. Molling, and A. Heidinger. 2015. "Validation of the National Solar Radiation Database (NSRDB) (2005–2012): Preprint." Paper presented at the European PV Solar Energy Conference and Exhibition, Hamburg, Germany, September 14–18.

[4] Sengupta, M., A. Lopez, A. Habte, Y. Xie, A. Weekley, C. Molling, C. Gueymard, P. Edwards, and D. Getman. 2015b. "Release of the New National Solar Radiation Database (NSRDB)." Presented as an NSRDB webinar on December 8. https://nsrdb.nrel.gov/sites/default/files/NSRDB_webinar_presentation_small.pdf.

[5] Wilcox, S. M. 2012. *National Solar Radiation Database 1991–2010 Update: User's Manual* (Technical Report NREL/TP-5500-54824). Golden, CO: National Renewable Energy Laboratory. http://www.nrel.gov/docs/fy12osti/54824.pdf.

[6] Šúri, M., and T. Cebecauer. 2014. "Satellite-Based Solar Resource Data: Model Validation Statistics Versus User's Uncertainty." *Proceedings of the 43th ASES National Solar Conference.*

[7] Reda, I. 2011. *Method to Calculate Uncertainties in Measuring Shortwave Solar Irradiance Using Thermopile and Semiconductor Solar Radiometers* (Technical Report NREL/TP-3B10-52194). Golden, CO: National Renewable Energy Laboratory. http://www.nrel.gov/docs/fy11osti/52194.pdf.

[8] Myers, D.R., et al. 2002. "Recent Progress in Reducing the Uncertainty in and Improving Pyranometer Calibrations." *Journal of Solar Energy Engineering*124: 44–50.

[9] Habte, A., M. Sengupta, A. Andreas, S. Wilcox, and T. Stoffel. 2016. "Intercomparison of 51 Radiometers for Determining Global Horizontal Irradiance and Direct Normal Irradiance Measurements." *Solar Energy* 133: 372–393.

[10] Wilcox, S.M., and D. Myers. 2008. *Evaluation of Radiometers in Full-Time Use at the National Renewable Energy Laboratory Solar Radiation Research Laboratory* (Technical Report NREL/TP-550-44627). Golden, CO: National Renewable Energy Laboratory. http://www.nrel.gov/docs/fy09osti/44627.pdf.

[11] Chai, T., and R.R. Draxler. 2014. "Root Mean Square Error (RMSE) or Mean Absolute Error (MAE)?—Arguments Against Avoiding RMSE in the Literature." *Geoscientific Model Development* 7: 1,247–1,250.

[12] Habte, A., M. Sengupta, I. Reda, A. Andreas, and J. Konings. 2014. "Calibration and Measurement Uncertainty Estimation of Radiometric Data: Preprint." Paper presented at Solar 2014, San Francisco, California, July 6–10.

[13] Aron Habte, Manajit Sengupta, Anthony Lopez. 2017. Evaluation of the National Solar Radiation Database (NSRDB): 1998-2015. NREL/TP-5D00-67722. http://www.nrel.gov/docs/fy17osti/67722.pdf.

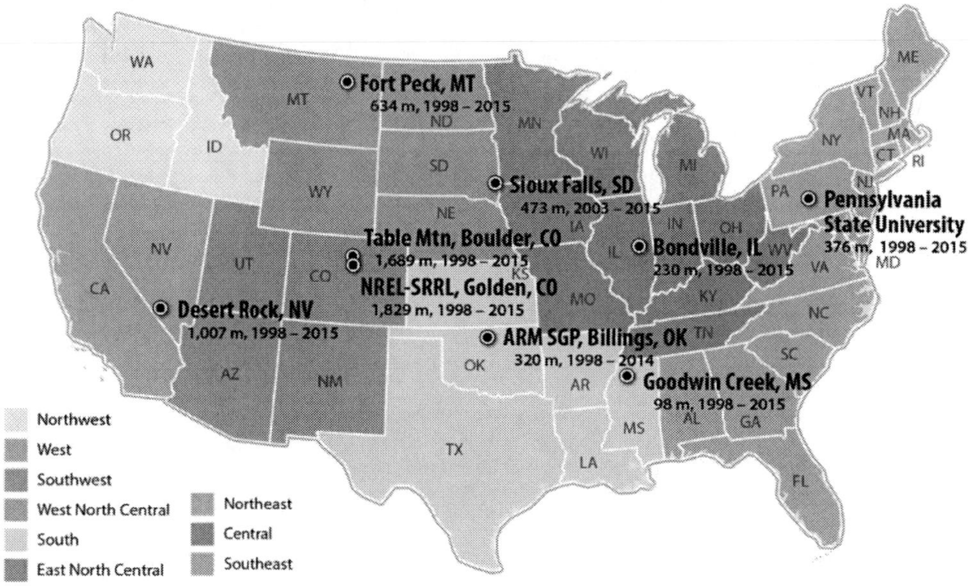

Figure 1: SURFRAD network, NREL's SRRL, and the ARM's SGP locations overlaid on the United States climatic regions

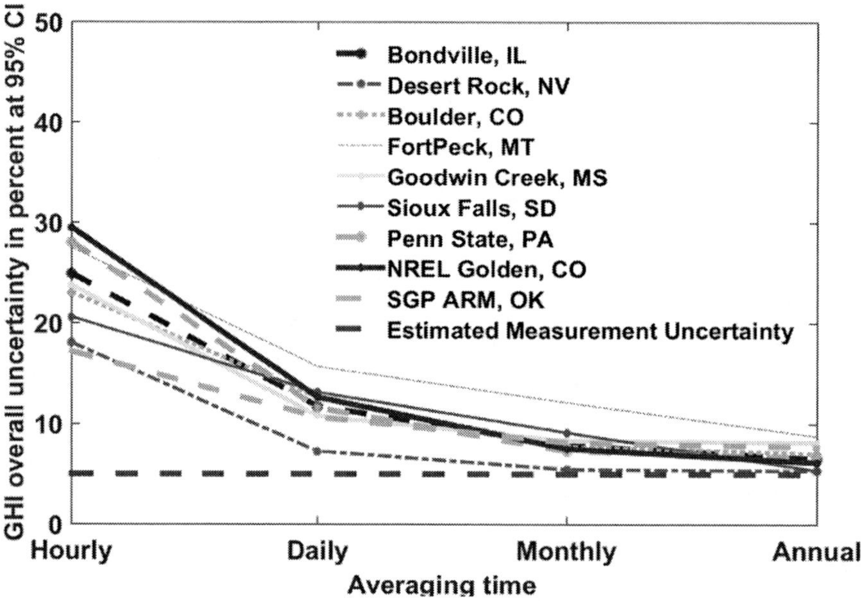

Figure 2: NSRDB uncertainty under various time averages

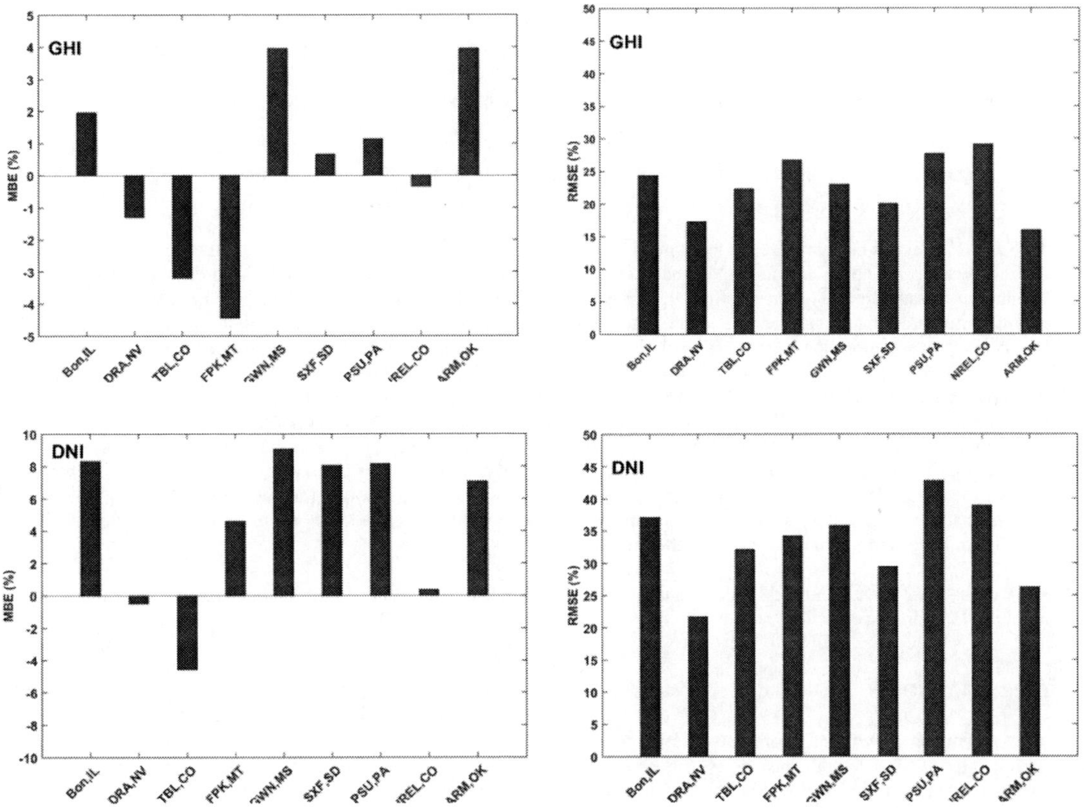

Figure 3: MBE percentage (left) and RMSE percentage (right) comparison results for both GHI (top) and DNI (bottom) between the NSRDB and ground-measured data for nine locations. Data obtained from the hourly data.

INTRA-DAY FORECASTS OF PV POWER WITH NUMERICAL WEATHER PREDICTION DATA AND MACHINE LEARNING IN KYUSHU, JAPAN

Joao Gari da Silva FONSECA Junior[a*], Fumichika UNO[b], Takashi OOZEKI[b], Kazuhiko OGIMOTO[a]

[a] Collaborative Research Center for Energy Engineering, Institute of Industrial Science (IIS), The University of Tokyo, 153-8505 Komaba 4-6-1, Meguro-ku, Tokyo, JAPAN. E-mail: jfonseca@iis.u-tokyo.ac.jp
[b] System and Applications Team, Research Center for Photovoltaic Technologies, National Institute of Advanced Industrial Science and Technology, AIST Tsukuba Central 2, 305-8568 Tsukuba, Ibaraki, JAPAN.

ABSTRACT: In this study we verify the typical levels of error of intra-day forecasts of regional PV power generation for the Kyushu island in Japan. To do that, we applied a method to yield continuous intra-day forecasts with a variable lead time (2h to 24h ahead of time). The method requires the use of numerical weather prediction, NWP data, recently measured data, and machine learning to model the relation between input and output variables. Forecast of hourly PV power generation were done for 3 years considering a scenario of high penetration of PV power in Kyushu. The PV power forecasts were based on insolation measurements and NWP data yielded by the meso-scale model of the Japan Meteorological Agency, JMA. The results show that the annual root mean square error normalized by the PV installed capacity, NRMSE, varied from 0.0297 GWh/GW_{peak} to 0.0318 GWh/GW_{peak} for hour-ahead forecasts (calculated with values from 5h to 20h of each day of the target period). The same error varied from 0.0753 GWh/GW_{peak} to 0.0820 GWh/GW_{peak} for 24 hours ahead forecasts. Furthermore, the results also show for how long the use of recent clearness index values, or its persistence, is effective to reduce forecast errors. For all 3 years of results, past clearness index use was effective for forecasts of up to 6 hours-ahead. For a lead time longer than that its use did not have any meaningful effect on the forecasts.

Keywords: PV Power, Intra-day Forecasting, Clearness Index Persistence, Numerical Weather Prediction, Machine Learning Application

1 INTRODUCTION

The continuous growth of photovoltaic, PV, power generation around the world has been followed by increasing demands regarding the accuracy of methods used to forecast PV power. In Japan, for example, better PV power forecasts became important with the beginning of the power market liberalization, and with PV installed capacity reaching 35 GW in 2016 [1]. This is an even more pressing matter in Kyushu island in the south of Japan. Kyushu has low interconnection capacity to other areas, 7.05 GW of installed capacity (of 17.1 GW authorized), and annual average power demand varying roughly from 7 GW to 18 GW.

With the objective to create a reference base with which improvements in PV power forecasts' accuracy in Japan can be assessed, we have evaluated the performance of a machine learning method used with numerical weather prediction data to yield forecasts of PV power and insolation in Japan. In these studies we considered only a day-ahead (with variable lead time) and hour-ahead forecasts and only 1 year of data [2][3].

Due to the sharp growth of PV in Kyushu, in this study our objective is to evaluate the typical accuracy of intra-day forecasts of regional PV power generation for this area in a scenario of high PV penetration. To do that, we made forecasts for Kyushu with 25 different lead-times; from 1 to 24 hours ahead of time and a day-ahead forecast (17h to 32h hours ahead). The error of each forecast then was evaluated. Moreover, to study the annual variability of the forecasts, they were done for a period of 3 years. Finally, we also studied the effect of the use of recent clearness index values as input data, on the accuracy of forecasts with different lead times.

2 FORECAST METHOD

To forecast PV power, first insolation was forecasted in each of the 8 prefectures of Kyushu island. After, considering each prefecture level of PV penetration, insolation was converted to PV power.

The forecasts were done using as input data numerical weather prediction, NWP, data, extraterrestrial insolation and the clearness index calculated with the most recent measured insolation regarding the release time of the forecast. The NWP data comes from meso-scale model developed by the JMA, which has a spatial resolution of 5km by 5 km, temporal resolution of 1 hour and a lead time varying from 0h to 39h ahead of time in its current version. Insolation forecasts done with this system are not published by the JMA and therefore were not used as input data. The output data were the insolation measured in each weather station of the JMA within Kyushu island and near by.

The data were used with a support vector regression algorithm to obtain forecast models. Each forecast model was trained with 840 hours of past data. This approach to forecast hour-ahead insolation was developed and presented previously [3]. The main difference here is that the most recent measured insolation and NWP data change according to the lead time as showed in Figure 1. In the case of the most recent clearness index 1-hour-ahead forecasts will use in their input a clearness index calculated with insolation measured 1 hour before the target hour, 2 hours-ahead forecasts will used the same variable measured 2 hours before the target hour and so on. The NWP data are updated 4 times a day (Japan Standard Time, JST, 0h, 6h, 12h and 18h); from each release time weather variables forecasts for the next 30 hours are used. Regarding the training data, an approach that selects it according to the release time of the NWP data was employed. For example, forecasts that used NWP data released at JST 0h, are done with a model trained only with NWP data released at JST 0h and so on. In addition, for each forecast model, input-output data available until the lead time of the forecast are used as training data. Thus, a 1 hour-ahead forecast for 14h, would be done at 13h, with a model trained with all input-output data available until 13h. This is possible because both the training and forecast procedures can be executed short time (in the order of a few minutes) in a desktop computer. Nonetheless, it should be noted that following this procedure, 1-hour-ahead forecasts would be done in slightly less than 1 hour ahead of time. Depending on the model and forecast target the lead-time oscillated around 55 minutes. For 2 hours ahead forecasts the lead time would be 1h55m and so on. For simplification purposes we considered the lead time as rounded hours.

Finally, the forecasts were done just for hours in which the extraterrestrial insolation was not zero to avoid calculation for night periods, which were also excluded from the error evaluations.

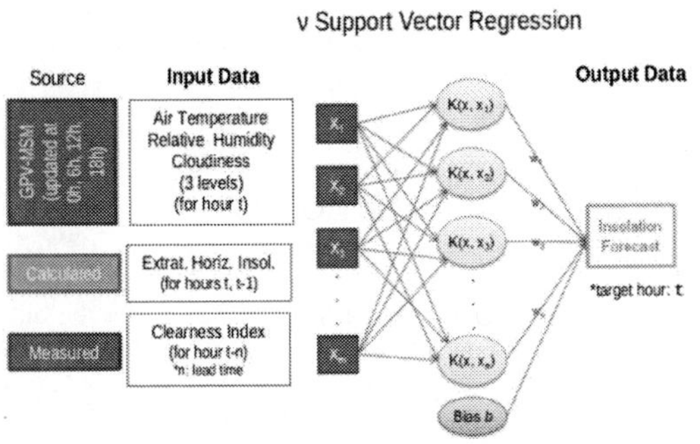

Figure 1: Method used to generate intra-day forecasts of Insolation for Kyushu.

Once insolation for each prefecture of Kyushu is forecasted for a given hour and lead time, the PV power generated in it was calculated using Eq. 1. In this formulation all non-linearities and losses involved in insolation to PV power are represented by a performance ratio factor. This calculation is done for each target hour and prefecture. After the PV power generation of all prefectures for a given target hours is calculated, the regional yield is obtained summing them as showed in Eq. 2.

$$P_{f,i}=P_{ins,i}\cdot\frac{H_{f,i}}{G_s}\cdot K \qquad (1)$$

$$P_{f,total}=\sum_{1}^{N} P_{f,i} \qquad (2)$$

In Eq. 1 $P_{f,i}$ is the PV power forecast, in kW, of a prefecture i, in a target hour. It is calculated for a given based on the insolation forecast of such prefecture $H_{f,i}$, in kW/m^2, and on the insolation in standard conditions G_s, regarded as 1 kW/m^2, and the PV installed capacity $P_{ins,i}$ in the prefecture i and the performance ratio K, regarded as 0.9 in this study. The values of P_{ins} is provided in section 3. In Eq. 2 $P_{f,total}$ is the total regional PV power forecast, in kW, for a given hour.

3 TARGET AREA DESCRIPTION

The forecasts' target was Kyushu island in the south of Japan. Hourly insolation data from 3 fiscal years (April 1st to March 31st of 2013, 2014, 2015) of 8 weather stations in and near Kyushu island were used to make the insolation forecasts of each prefecture in Kyushu. Moreover, the values of the NWP variables described in section 2 of the nearest point to each weather station was also retrieved to be used in the forecasts. The approximate locations of the weather stations used in the study are presented in Fig. 2.

With the exception of Fukuoka prefecture, each prefecture's insolation was forecasted based on the data available in the weather station within the prefecture as showed in Fig. 2. For Fukuoka prefecture's forecasts 2 weather stations were used due to their locations. One in Fukuoka prefecture and a second one in Shimonoseki city, which is in Yamaguchi prefecture in the border with Fukuoka prefecture.

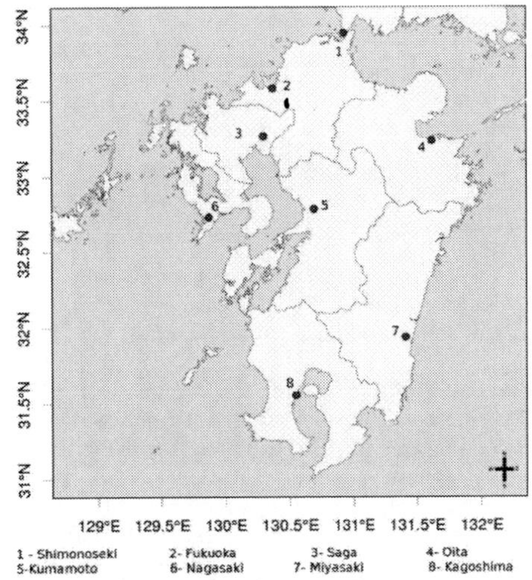

* Kyushu map built with the maptools R library and data publicly available from the GADM database.

Figure 2: Map of Kyushu area in Japan with the locations of weather stations target of the forecasts.

Regarding the PV penetration scenario, it was assumed an installed capacity of 15.695 GW spread differently per prefecture around Kyushu (as of 2017/3 there are 7.05 GW of PV installed in the region from a total of 17.1 GW of PV authorized [1]). This scenario was provided by RTS, a consulting company in Japan, and it is based on the expected installed capacity in Kyushu by 2030. The scenario chosen regards the PV installed capacity proportional to the power demand of each prefecture within Kyushu. Following this assumption, the PV ratio per prefecture in Table I was used.

Table I: PV Power Penetration scenario in Kyushu, 2030.

Location/ Prefecture	Installed Capacity (GW)	Region Ratio (%)
Kita Kyushu - Fukuoka	1.570	10
Fukuoka city - Fukuoka	3.538	22.54
Saga	1.070	6.83
Nagasaki	1.463	9.32
Kumamoto	2.169	13.82
Oita	1.954	12.45
Miyasaki	1.568	9.99
Kagoshima	2.363	15.05
Total	15.695	100

4 RESULTS

The results in Fig. 3 show the variation of 3 error parameters per year and per forecast horizon. Three kind of errors were calculated, the normalized root mean square error, NRMSE, normalized mean absolute error, NMAE, and the normalized mean bias, NMBIAS. The errors are normalized by the peak PV power generated in the period studied.

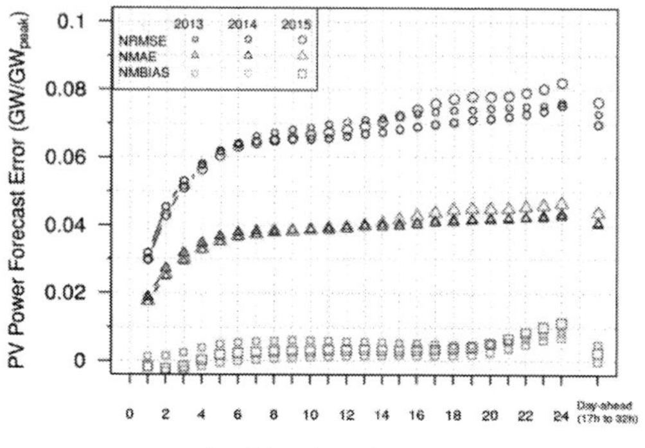

Figure 3: Regional PV forecast error for Kyushu according to the lead time and per year (2013 to 2015).

In the 3 years studied, the forecast NRMSE and NMAE presented smaller variation for shorter lead times than for longer ones. For example, for the 1 hour lead time the highest variation was 6.5% between 2014 and 2013. For a lead time of 18 hours the NRMSE variation reached 9% between 2014 and 2015. In general, for longer lead times 2015 presented forecasts with the highest errors, and for shorter lead times 2013 had forecasts with slightly higher biases.

Regarding the effect of the lead time on the forecast error, 24 hours-ahead forecasts had a NRMSE around 0.08 GW/GW_{peak}. For hour-ahead forecasts the same error was reduced to 0.03 GW/GW_{peak}. These results indicate an average reduction of the 62.5% of the error with the decrease of the lead time.

Moreover, a second trend was observed. From Fig. 3 one can see that the forecast errors decreased gradually from a lead time of 24 hours to one of 6 hours, dropping sharply after that, until a lead time of 1 hour. As the forecasts use as input recently measured clearness index values, we infer that the sharp decrease of the forecast errors is related with the level of effectiveness of using it in forecasts with short lead times. Whereas the persistence of the clearness index is strong for 1 hour, it should considerably lower for 6 hours. Thus, whereas for a lead time of 1 hour, the past hour clearness index value should be a good predictor of future insolation and PV power generation values, for a lead time of 6 hours it should be considerably less so. In fact, according to the results in Fig. 3, after 6 hours, there seems to be no benefit of using past clearness index as one of the input variables of the forecasts.

To confirm the effectiveness of the past clearness index in the forecast of regional PV power in Kyushu, we built a model that does not use it as one of the input variables and made forecasts with this model for 1 year, 2013. The results of the 3 error parameters of the forecasts done with both models (one with and another one without past clearness index) are in Fig. 4.

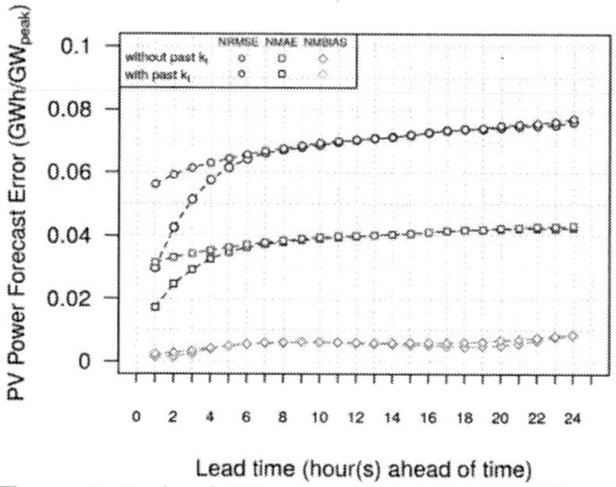

Figure 4: Regional PV power forecasts with different lead times for Kyushu in 2013 from a model using past clearness index as input and from another one not using it.

The results in Fig. 4 show that when just NWP data are used the forecast NRMSE dropped from 0.077 GW/GW_{peak} (for the 24 hour ahead lead time) to 0.056 GW/GW_{peak}, (for the hour ahead lead time), a reduction of 27% of the error with the lead time. This reduction represents the effect of updating the NWP data 4 times a day and using it to train the forecast model. On the other hand, when the clearness index was used the NRMSE dropped from 0.077 GW/GW_{peak} to 0.03 GW/GW_{peak} when the lead time changed from 24 hours to 1 hour; a reduction of 61% for 2013 data. For a lead time of 1 hour the reduction of the NRMSE was strong enough to reach the same level of the NMAE of the forecasts done without using past clearness index. Furthermore, after 6 hours there was no significant difference between both forecasts, which indicates that the assumption based on the results in Fig. 1 was correct. For Kyushu the use of past clearness index is only effective to reduce the NRMSE and NMAE up to a lead time of 6 hours. It is interesting to note also, that although the results in Fig. 4 show that the past clearness index was effective to reduce forecast errors, it had negligible effect to reduce the forecast biases.

To understand better the distribution of the forecast error according to the lead time, we plotted in Fig. 5 a box plot of the forecast errors of the 3 years studied. The 95th percentiles in Fig. 5 show that in general the 3 years had similar forecast errors. The year of 2015 had forecast errors with slightly higher positive bias than the forecasts of the other 2 years. Looking to the extreme forecast errors, for lead times of up to 13 hours, 2014 had the lowest positive extreme forecast errors. For longer lead times, 2015 and 2013 had forecasts with similar trends, yielding the highest positive errors. As noted also in Fig. 3 the bias does not vary strongly with the forecast horizon. Another, result that can be inferred from the results in Fig. 5 is the expected reduction of extreme errors with the forecast horizon. For negative forecast errors the extreme values decreased from near to -0.46 GW/GW_{peak} to -0.26 GW/GW_{peak} and -0.18 GW/GW_{peak} for 2013 and 2014, respectively. For 2015 the reduction was from .-0.40 GW/GW_{peak} to 0.23 GW/GW_{peak}. A reduction of at least 42.5 % in the worst case. For positive forecast errors, strong reductions were obtained and a reduction of extreme forecast errors of at least 70% was observed in the best case.

The 95th percentiles in Fig. 5 show that for short-term forecasts, with a lead time of up to 11 hours, the years of 2014 and 2015 has similar forecast errors. Looking at 2013, it had forecasts with a distribution slightly displaced to the side of positive errors. This characteristic is reflected on the higher positive bias of 2013 forecasts comparing with 2014 and 2015, noted in Fig. 3. For higher lead times, however, the forecast of all 3 years had similar distribution of forecast errors.

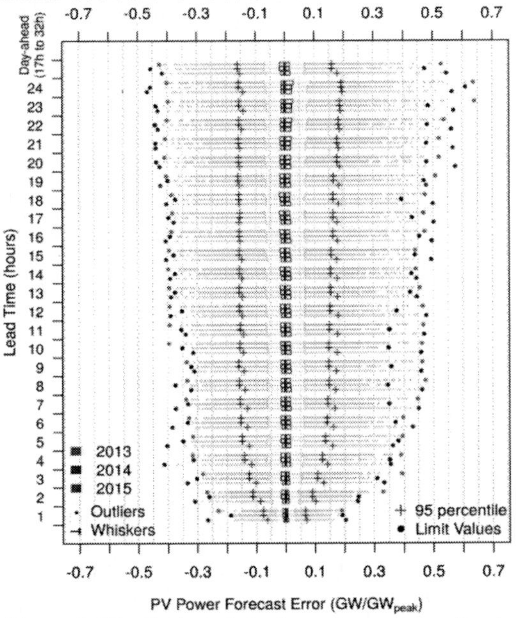

Figure 5: Distribution of the forecast error according to the lead time in Kyushu, from 2013 to 2015.

Finally, the whiskers in Fig. 5 indicate that most of the error distributions have low bias. The magnitude of the positive outliers is the cause of the positive biases noted in Fig. 5 and Fig. 3.

Another way to assess the expected error reduction with the lead time is presented in Fig. 6. In this figure, forecasts errors from forecasts with different lead times are plotted against a day-ahead forecasts for all 3 years of data.

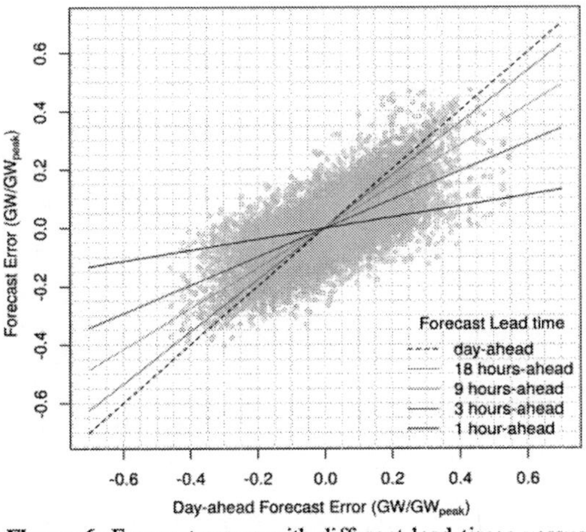

Figure 6: Forecast errors with different lead-times versus day-ahead forecast error of Kyushu, 3 years.

In this case, for Kyushu, it is clear that large forecast errors are considerably reduced with forecast lead time. For example, forecast errors that reached 0.4 GW/GW$_{peak}$ in day-ahead forecasts are reduced to average values of 0.35, 0.27, 0.18, 0.08 GW/GW$_{peak}$ when the lead-time is reduced to 18, 9, 3 and 1 hour ahead of time. Another important characteristic that Fig. 6 shows regards the distribution of forecast error reduction with the lead time. For high forecast error considerable error reduction are often reached when the lead time is shortened. However, for average to low forecast errors, many times a reduction of the forecast horizon did not reduce the forecast error. For example, day-ahead forecast errors of 0.05 GW/GW$_{peak}$ many times increased to values as high as 0.1 GW/GW$_{peak}$ when the lead time was reduced to 1 hour ahead of time. One factor that can explain this behavior is the use of clearness index persistence even in situations with unstable weather. Another one is related with the limitations of GPV-MSM model, from which cloudiness forecasts are retrieved. The model, as used, was updated at every 4 hours and it did not capture all cloudiness variation that may occur in shorter time scales.

In Fig. 7, the monthly variation of the NRMSE of forecasts with 3 lead times, in the 3 years studied is presented.

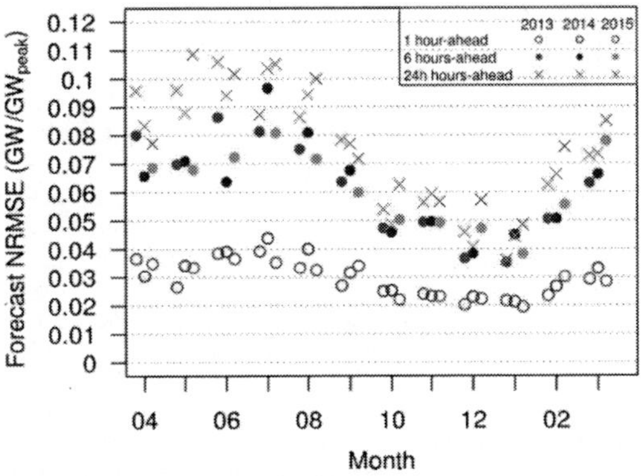

Figure 7: NRMSE variation per month in 2013, 2014, 2015, of forecasts done 1 hour, 6 hours and 24 hours ahead of time.

The results in Fig. 7 show that forecasts done 6 hours ahead of time have errors closer to the errors of 24 hours ahead forecasts than those of the 1 hour head forecasts. This is due to the vanishing effect of using past clearness index values in the forecasts. As the results in Fig. 4 show, clearness index persistence is weak after 6 hours, which makes the improvement rate of the forecast errors with the lead time considerably smaller than in the period between 1 hour and 6 hours ahead.

Another trend is also noted in Fig. 7, annual variation of the NRMSE of the forecasts in any given month seems to be small at 1 hour ahead and to become large for 6 hours ahead forecasts, and even larger for the 24 hour ahead forecasts. However, if one looks the percent variation, the difference between the errors of forecasts with different lead times in a given month, from one year to another were in the same order. For example, in April, the NRMSE forecasts of hour ahead forecasts changed from 0.03 GW/GW$_{peak}$ in 2013 to 0.039 GW/GW$_{peak}$ in 2014, a variation of 30%. The NRMSE of the 24 hours ahead forecasts in the same period varied from 0.098 GW/GW$_{peak}$ to 0.082 GW/GW$_{peak}$, a variation of 20%. In spite of that, not always the year with lowest forecast NRMSE in one lead time was the year with the lowest value in another lead time. The range of variation per lead time per year is presented in Table II.

Table II: Range of variation of the NRMSE of PV power forecasts with 3 different lead times, in 3 years.

Lead time	Highest Value (GW/GW_{peak})	Lowest Value (GW/GW_{peak})
24 hours ahead	0.11 (May, 2015)	0.036 (Jan 2013)
6 hours ahead	0.098 (Jul, 2015)	0.035 (Jan 2013)
1 hour ahead	0.043 (Jul, 2014)	0.019 (Jan, 2015)

5 CONCLUSION

In this study, we proposed and tested a simple approach to obtain forecasts of PV power generation with different lead times (from 1 hour-ahead to day-ahead) using as input data only NWP data and recently measured clearness index. The method was applied to data of 3 year of Kyushu island in Japan.

The results showed that, for Kyushu, day-ahead forecasts (lead time between 17 hours and 20 hours) annual NRMSE values from 0.070 to 0.077 can be obtained. Such values drop to values around 0.03 GW/GW_{peak} when the lead time decreases to 1 hour ahead. Considering that just two sources of data were used, those are good results particularly for the hour-ahead lead time.

Using the results of this study as starting point and a reference basis for short-term forecast errors in Kyushu area, in following studies, the use of satellite, PV power online measurements, and more advanced forecast models should be investigated as a potential source of improvement for PV power forecasts in Kyushu and in Japan.

Finally, the effect of the clearness index in the forecasts done in Kyushu was also isolated. The results showed that for the area studied, there was little to no benefit of using recently measured clearness index for forecasts with a lead time higher than 6 hours. To verify if such result is also valid for other regions is also an interesting topic to the development of short-term PV power forecast methods in Japan.

ACKNOWLEGMENT

This paper is based on results obtained from a project commissioned by the New Energy and Industrial Technology Development Organization (NEDO).

References

[1] METI, なっとく!再生可能エネルギー 各種データの公開 . [Online in Japanese], Available: http://www.fit.go.jp/statistics/public_sp.html, [Last accessed: 10-Feb-2017].

[2] J.G.S. Fonseca Jr., T. Oozeki, H. Ohtake, K. Ogimoto, 'Evaluating the Effect of the Forecast Horizon in the Error of Regional Forecasts of Insolation in Japan: the Kanto Region Case', in Proc. 31st European Photovoltaic Solar Energy Conference and Exhibition, Hamburg, Germany, 2015, pp. 2091–2095.

[3] J. G. S. Fonseca Jr., H. Ohtake, T. Oozeki, and K. Ogimoto, 'Local and Regional Hour-Ahead Forecasts of Solar Irradiance with Training Data Selection and Support Vector Regression', IEEJ Transactions on Power and Energy, vol. 136, no. 12, pp. 898–907, 2016.

STATISTICAL TECHNIQUES USED TO IMPROVE SOLAR RESOURCE ASSESSMENTS FOR PHOTOVOLTAIC PLANT APPLICATIONS

Clohessy C.M [1], van Dyk E.E [1], Sharp G.D.[1] & Hugo J[1]
Nelson Mandela University
Department of Statistics, Nelson Mandela University, University Way, Port Elizabeth, South Africa, 6001
chantelle.clohessy@mandela .ac.za
Tel: +27833711881

ABSTRACT: This study proposes statistical techniques that can be used to improve resource assessment methods in the solar energy domain. The use of these techniques are illustrated for a case study of a 1MW photovoltaic system proposed for the Nelson Mandela University Summerstrand South Campus in Port Elizabeth, South Africa. Multivariate profile analysis is used to compare ground-modelled to ground-measured solar resource data. The results provide strong support for the use of this technique for the assessment of solar resource data. Variance component models are used to model energy yield estimates of the proposed photovoltaic system. The variance components are simulated using Bayesian simulation techniques. Bayesian tolerance intervals are derived from the variance components and are used to determine what percentage of future energy yield values that fall within an interval with a certain probability. This method incorporates both the varaibility and uncertainity of the energy yield estimates.
Keywords: Photovoltaic, Solar Radiation, Solar Resource Assessment, Energy Performance, Bayesian Tolerance Intervals

1 INTRODUCTION

Estimation of the annual performance of a Photovoltaic (PV) system is essential for system developers. The annual performance indicates the expected levels of energy yield from a given system. Evidence from the literature shows that point estimates, graphs and probability of exceedance are the most commonly used measures to assess the energy output of a PV system.

The probability of exceedance is a statistical metric used to describe the probability of a given level of energy generation will be met or exceeded. A simple illustration would be the 90th percentile (P90), where 90% of the observations exceed some level while 10% fall below the threshold. Figure 1 illustrates the exceedance probability for the net energy production for a given year.

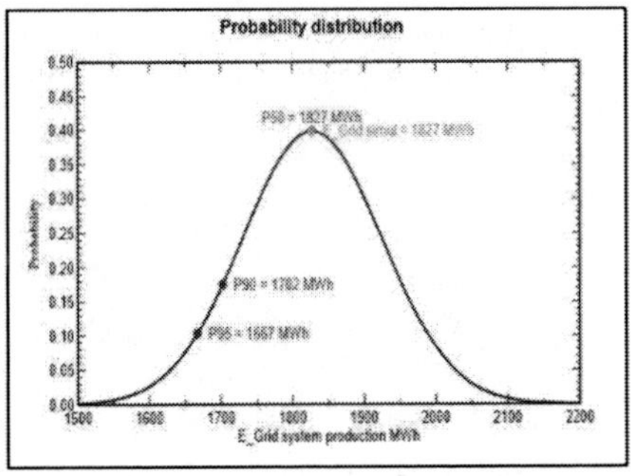

Figure 1: Yearly probability of exceedance for energy yield measurements

The disadvantage of using the probability of exceedance (point estimate) is that it is a single value and does not provide any information on the variability of the expected energy yield. This study proposes methods that can be used to assess the expected energy yield of a system with a certain degree of probability using the variability of the energy yield measurements. These methods are more informative and give the expected ranges of energy yield with regards to its limits and observations.

The use of these methods for PV assessments was found by providing the unique link between manufacturing processes and energy generation from a PV system. A manufacturing process is defined as a process in which raw materials are transformed into a final product [1], while energy generation from a PV system is a process in which energy from sunlight is transformed into electricity [2].

Wolfinger (1998) proposed the use of variance component models to assess variability in manufacturing processes. These models are used in quality control assessments as they can adequately handle multiple sources of varaibility. This study extends the work of Wolfinger (1998) and Hugo (2012) by proposing variance component models to assess energy yield of a PV system. Bayesian simulation methods are used to derive the tolerance limits and intervals for the energy yield of a given system.

2 DATA COLECTION

2.1 Solar Resource Data

The measured solar resource data was retrieved from South African University Radiometric Network (SAURAN) [5]. This data was collected at the Outdoor Research Facility (ORF) at the Nelson Mandela University in Port Elizabeth, South Africa for the years 2013, 2014 and 2015. The measured data collected included: global horizontal irradiation (W/m^2), diffuse irradiation (W/m^2), and temperature (^{o}C). These values were measured every 15 seconds. The average hourly values were calculated from the 15 second observations. The system used to collect the global horizontal irradiation and diffuse irradiation was a Kipp and Zonen CMP11. Temperature was collected via a Campbell Scientific CS215 sensor. The measurement instrumentation was positioned on the roof at the ORF. The source data was used to simulate energy yield estimates from a specific PV system.

The source data used in this study was assessed using multivariate profile analysis [2].

2.2 Energy Yield

Energy yield (MWh) in this study refers to the amount of potential energy that can be injected into the grid by a given PV system. Energy yield is a function of power output and time.

Daily energy yield estimates were calculated using a simulation programme, PVSyst [6] for a 1MW system. Table I summarises the specifications and the characteristics of the 1MW system.

Table I: Specifications for a 1MW PV System

Number of modules	In series: 15 In parallel: 212
Total number of modules	No. 3180 Unit nominal power: 315Wp
Array global power	Nominal (STC): 1002kWp At operating conditions: 901kWp (50°C)
Total Area	Module area: 6170m^2 Cell area: 5573m^2
Model Type	Sunny Central 500MV-11IT
Characteristics	Operating voltage: 450-820V Unit nominal power: 500kWac
Inverter	Number of inverters: 2 units Total power: 1000 kWac

3 METHODOLOGY

The measured resource data was assessed using a two-sample profile analysis and the daily energy yield output of the 1MW system was evaluated using tolerance intervals. Section 3 summaries the methodology used in both these analyses.

3.1 Two-sample Profile Analysis

The multivariate two-sample profile analysis was used to assess the measured resource data. This was done by comparing the measured data to simulated data from Meteonorm [7].

Daily average values were necessary in the evaluation for the two-sample profile analysis. A requirement for a two-sample profile analysis is that the number of daily average values for each month has to be the same. As such, 28 days were selected randomly for each month, to bring all the monthly values in line with February which has the fewest days.

Two tests were conducted on the profiles to determine whether the Meteonorm generated estimates were parallel and coincidental to the measured data. These two tests included the parallel test and the same level test [2].

The package *profileR* in R3.1.1 [8] was used for the two-sample profile analysis plots and tests.

3.2 Variance Component Model

The following one-way random effects model was used to model the daily energy yield estimates in a given season

$$y_{ij} = \mu + a_i + \varepsilon_{ij},$$

The value y_{ij} represents the jth ($j = 1,...,d$) energy yield measurement on a randomly selected month i ($i = 1,...,m$), μ is the common location parameter, a_i represents the ith month effect and ε_{ij} represents the error component. The random month effect (a_i) as well as the error component (ε_{ij}) are assumed to be independent and normally distributed as $a_i \sim N(0, \sigma_m^2)$ and $\varepsilon_{ij} \sim N(0, \sigma_\varepsilon^2)$. The variance σ_m^2 represents the variability between the different months in a given season, while σ_ε^2 represents the variability between the daily energy yield measurements. These variance components were used for the estimation of the different tolerance intervals.

3.3 Simulation of the Variance Components

The following steps were followed to simulate the variance components for the energy yield estimates:

- A chi-square distribution with v_1 degrees of freedom was used to simulate $\frac{v_1 m_1}{\sigma_\varepsilon^2}$. The variance component σ_ε^2 was then obtained by calculating $\sigma_\varepsilon^2 = \frac{v_1 m_1}{\chi_{v_1}^2}$.

  Where the value $v_1 m_1 = \sum_{i=1}^{m}\sum_{j=1}^{d}(y_{ij} - \bar{y}_{i.})^2$ represent the observed sum of squares between days (SSE).
- A chi-square distribution with v_2 degrees of freedom was used to simulate $\frac{v_2 m_2}{\sigma_\varepsilon^2 + d\sigma_m^2}$. Since $\sigma_\varepsilon^2 + d\sigma_m^2 = \frac{v_2 m_2}{\chi_{v_2}^2}$, the variance component σ_m^2 was then determined by calculating $\sigma_m^2 = \frac{(\sigma_\varepsilon^2 + d\sigma_m^2) - \sigma_\varepsilon^2}{d}$. Where the value $v_2 m_2 = h\sum_{i=1}^{d}(\bar{y}_{i.} - \bar{y}_{..})^2$ represents the observed sum of squares between months (SSA).
- The constraint $(\sigma_\varepsilon^2 + d\sigma_m^2) > \sigma_\varepsilon^2$ needed to be met to ensure that the simulated variance component σ_m^2 was positive.
- For this study the steps were repeated $\tilde{l} = 10000$ times and if sets of variance components did not meet the constraint they were disregarded.

3.4 The (0.90; 0.95) One-Sided Tolerance Limit

To construct the one-sided (α, β) tolerance limit for the balanced one-way random effects model the marginal posterior distribution of q must be estimated. The marginal posterior distribution of q represents the $(1-\alpha)^{th}$ quantile of the $N(\mu, \sigma_e^2 + \sigma_m^2)$ distribution. The following steps were followed for the construction of the lower one-sided (α, β) tolerance limit:

- The simulated pair of variance components σ_m^2 and σ_e^2 were substituted into the following equation for the lower one-sided limit (q_l):

$$q_l \mid \sigma_e^2, \sigma_a^2, y \sim N\left(\bar{y}_{..} - z_\alpha(\sigma_e^2 + \sigma_m^2)^{\frac{1}{2}}, \frac{\sigma_e^2 + d\sigma_m^2}{md}\right)$$

  where z_α represents the αth quantile of a standard normal distribution.
- Steps 1 was repeated $\tilde{l} = 10000$ times and two histograms were drawn. The histograms represented the estimated marginal posterior distributions of q_l given as $p(q_l \mid y)$.

3.4 Fixed-in-Advance Tolerance Interval

The following steps were followed for the construction of the fixed-in-advanced tolerance interval for the one-way random effects model:

- The simulated pair of variance components subjected to the constraint $(\sigma_e^2 + h\sigma_a^2) > \sigma_e^2$ and simulated target value μ (this value was also simulated from a normal distribution, see Clohessy (2017)) were used for the estimation of the fixed-in-advanced tolerance interval.
- The simulated pair of variance components, σ_m^2 and σ_e^2, as well as the target value μ were substituted into the formula for the content (c) of the interval level $[s; \infty]$ given by

$$c = 1 - \Phi\left[\frac{(s-\mu)}{(\sigma_e^2 + \sigma_a^2)^{\frac{1}{2}}}\right] = 1 - \int_{-\infty}^{\frac{(s-\mu)}{(\sigma_e^2+\sigma_a^2)^{\frac{1}{2}}}} \frac{1}{\sqrt{2\pi}} e^{-\frac{1}{2}z^2} dz$$

- Steps 1 and 2 were repeated $\tilde{l} = 10000$ times and a histogram was drawn. The histogram represented the posterior distribution of content c above the preselected limit s. The estimated energy yield fixed-in-advance tolerance intervals will be compared for different preselected lower specification limits s.

4 RESULTS

4.1 Two-sample profile analysis

To illustrate this assessment technique the graphical illustration for global horizontal irradiance is given in Figure 2. This is followed by the inferential results in Table II.

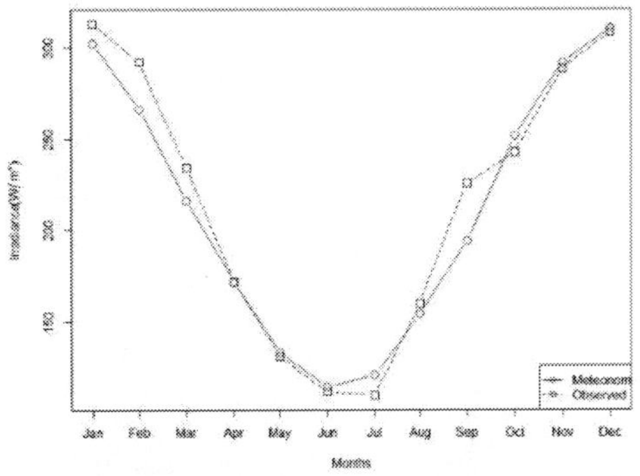

Figure 2: Global horizontal irradiance profile plot.

From Figure 2 the global irradiance horizontal estimates from Meteonorm are parallel, and at the same level as the 2013 measured data. The month of September has a marginally larger monthly average value for the measured data when compared to that of the software data. After further inspection of the measured data it was found that a number of days in this month had large hourly averaged values. This was a result of less cloud cover in this month due to low rainfall and high wind speeds

Table II: Inferential results for global horizontal radiation.

	F	p-value
H_0: Profiles are Parallel	1.198	0.317
H_0: Profiles are Same Level	1.596	0.212

The inferentical results support the interpretations of the visual observations from Figure 2. The null hypotheses are not rejected at a 5% significance level for the global horizontal irradiance profile. This lends support to the observation that the Meteonorm estimates and the 2013 source data are parallel and coincidental.

The two-sample profile analysis provides visual and statistical motivation for both the trend and shape of the measured data when compared to software generated data. The parallel test was used to compare the group profiles of the software to that of measured data. This was done to determine if the increase or decrease of the slopes from one average month to another was the same. The same level test provides statistically motivation on whether the monthly averages of Meteonorm were at the same level as the measured data. These test were used to assess all the solar resource data in this study. Refer to Clohessy (2017) for further details.

4.2 Results: One-sided tolerance limits

The (0.90, 0.95) lower one-sided tolerance limits for all four seasons are shown graphically in Figure 3.

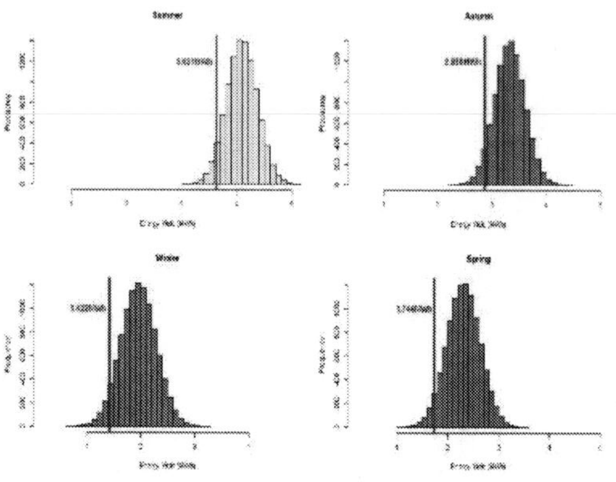

Figure 3: The one-sided (0.90; 0.95) tolerance limits.

The interpretation of the intervals are illustrated for the season of autumn. The lower limit for autumn indicates that 90% of future daily energy yield measurements are greater than 2.858MWh with 95% probability.

The results show that summer had the largest lower one-sided limit of 3.631MWh. This is expected as summer has higher daily energy yield measurements due to the high global horizontal irradiance values found in this season. Spring has the second smallest lower one-sided limit of 1.744MWh and winter has the smallest lowest one-sided limit of 1.422MWh. The small one-sided lower limit of spring would be due to the large variability of energy yield measurements found for this season. The small one-sided limit for winter would be due to the weather conditions experienced during this season.

These one-sided limits provide meaningful results and will be beneficial for the assessment of the feasibility of a PV system. These results can be compared to different PV systems and also these limits provide useful information on the lower bounds for the future energy yield measurements. The tolerance intervals for the lower limit includes the percentage of measurements that fall above a given value but also the likelihood of this event occurring. This interval is more informative than the single estimate from the probability of exceedance [2].

4.3 Results: Fixed-in-advance tolerance intervals

The Bayesian 100(0.95)% fixed-in-advance tolerance interval results are given in Table 3.

Table III: Fixed-in-advance tolerance intervals (%).

	3MWh	4MWh	5MWh
Summer	[99.6; 100.0]	[94.; 99.1]	[67.4; 86.6]
Autumn	[98.8; 99.9]	[82.4; 98.1]	[32.3; 76.8]
Winter	[82.2; 94.9]	[30.4; 56.6]	[2.3; 10.5]
Spring	[83.8; 99.9]	[26.6; 93]	[10.0; 41.7]

For illustration purposes, the interpretation of the Bayesian 100(0.95)% fixed-in-advance tolerance interval for autumn at a preselected limit of 4MWh is given. The fixed-in-advance tolerance interval for autumn was found to be [82.4; 98.6]. This indicates that for future months in the season of autumn, between 82.4% and 98.6% of energy yield measurements will fall within the interval $[4, \infty]$.

The other 3 seasons fixed-in-advance tolerance intervals are interpreted in a similar manner.

The fixed-in-advance tolerance intervals show that the intervals differ for the different preselected limits. For example; summer has an interval of [99.6; 100.0] for the preselected limit of 3MWh. This interval indicates that between 96.6% and 100% of future daily energy yield measurements fall within this interval with 95% probability. While for the preselected limit of 5MWh, the season of summer only has between 67.4% and 86.6% of future daily energy yields falling above the limit of 5MWh with 95% probability. The lower limit s could be used to assess the profitability of a solar energy system. For example, if different sites fixed-in-advance tolerance intervals are compared, with the lower limit indicating profitability, the most profitable system for a given site could be identified.

4 Conclusion

The advantages of using the methods proposed in this study include:

- The multivariate profile analysis provide visual and statistical motivation for the trend and shape of solar irradiation data.
- The tolerance intervals provide a unique statistical assessment method for energy yield measurements by incorporating both the variability and uncertainty of the estimates.
- These one-sided limits provide meaningful results, which will be beneficial for the assessment of the feasibility of a PV system. These results can be used to compare different PV systems by assessing the lower bounds of future energy yield measurements.
- The fixed-in-advance tolerance intervals provide a percentage interval of energy yield measurements that fall above a preselected limit, which is easy to interpret and estimate an associated probability of the event occurring in the future.

5.2 Acknowledgements

The project would not have been possible without the financial support of the National Research Foundation Thuthuka grant. This work was supported by the National Research Foundation (NRF) Thuthuka grant (grant number: TTK13060618606) and the Nelson Mandela Metropolitan University (NMMU) Research Capacity Development (RCD) Department (grant number: NA).

5.3 References

[1] R. Rajput. A textbook of manufacturing technology: Manufacturing processes. Firewall Media. (2007).

[2] C. Clohessy. Statistical Viability Assessment of a Photovoltaic System in the Presence of Data Uncertainty. (2017). PhD Thesis, Nelson Mandela Metropolitan University.

[3] M. Brooks, S. Clous , W. Nieker, P. Gauche, C. Leonard, M. Mouzouris, R. Meyer, N. Westhuizen, E. van dyk. & Vorster F. SAURAN: A new resource for solar radiometric data in Southern Africa. Journal of Energy in Southern Africa. (2015). Vol 25, 1-10.

[4] PVSYST. Photovoltaic and simulation program, Version 6.2.5. (www.pvsyst.com) [accessed November 2015].

[5] O. Bulut. & Desjardins C. profileR: Profile analysis of multivariate data in R. R package version 0.2-1. http://CRAN.R-project.org/package=profileR. (2013).
[6] J. Remund, S Müller, S Kunz. & C. Schilter. Meteonorm Version 7.0, Handbook. Part II: theory. METEONORM. Version 7 (www.meteonorm .com). [accessed June 2014].
[7] R. Wolfinger. Tolerance Intervals for Variance Component Models using Bayesian Simulation. Journal of Quality Technology. (1998). Vol 30, 18-32.
[8] J. Hugo J. Bayesian tolerance intervals for variance component models. (2012). PhD Thesis, University of the Free State.

A NEW APPROACH FOR REGIONAL PHOTOVOLTAIC POWER ESTIMATION AND FORECAST

Marco Pierro[a,e], Matteo De Felice[d], Enrico Maggioni[c], David Moser[e], Alessandro Perotto[c], Francesco Spada[c], Cristina Cornaro[a,b]

[a]Department of Enterprise Engineering, University of Rome Tor Vergata, Via del Politecnico 1, 00133 Rome, Italy, e-mail: cornaro@uniroma2.it, marco.pierro@gmail.com, frabucci@gmail.com
[b]CHOSE, University of Rome Tor Vergata, Via del Politecnico 1, 00133 Rome, Italy 7
[c]Ideam Srl, via Frova 34 Cinisello Balsamo, Italy, e-mail: alessandro.perotto, enrico.maggioni, francesco.spada@ideamweb.com
[d]Casaccia R.C., ENEA Climate Modelling Laboratory, Rome, Italy e-mail: matteo.defelice@enea.it
[e]EURAC Research, Viale Druso, 1, 39100 Bolzano, Italy e-mail: david.moser@eurac.edu

ABSTRACT: On regional scale, the estimation of the solar power generation from the environmental conditions and the solar power forecast is essential for Distribution System Operators, Transmission System Operator, energy traders, and aggregators. Indeed, high photovoltaic penetration results in a stochastic variability of the electric demand that could compromise the stability of the grid, increase the amount of energy reserve and the energy imbalance cost.

In this context, two upscaling methods and several models were developed and used for estimation and forecast of the photovoltaic distributed generation in a small area of Italy with high photovoltaic penetration. The upscaling methods are based on chains of data-driven models (k-mean clustering, principal component analysis and ensemble of neural networks) that use as inputs satellite derived irradiance and numerical weather prediction. These models allow a complete assessment of the accuracy of the regional power forecast at different horizons, from 0 hour ahead (power estimation) to 48 hours ahead. The power estimation model provided a RMSE of 3% of installed capacity. Intra-day forecast (from 1 to 4 hours) obtained a RMSE of 5% - 7% and a skill score with respect to the clear sky persistence from -8% to 33.6%. The one and two days ahead forecast achieved a RMSE of 7% and 7.5% and a skill score of 39.2% and 45.7%. The ensemble smoothing effect on cluster scale was also studied. It was proved that the RMSE reduction achieved on cluster scale due to the ensemble smoothing follows the same exponential law found for single systems. In this case the smoothing effect reduces the RMSE of day-ahead forecast of 12% with respect to the mean single cluster value. It was also shown that the skill score increases linearly with the size of the region with a rate of 0.1% for km.

Furthermore, a method to estimate the forecast error was also developed. It was based on an ensemble neural network model coupled with a probabilistic correction. It can provide a highly reliable computation of the prediction intervals.

Keywords: Regional photovoltaic generation forecast; upscaling forecast methods; ensemble smoothing effect; prediction intervals.

1 INTRODUCTION

The residual electric demand (net load), that should be covered by other non-variable energy sources, is the difference between the energy need (load) and the distributed photovoltaic (PV) generation. Therefore, high photovoltaic penetration (i.e. amount of load fulfilled by solar generation over one year) introduces into the net load a dependence on meteorological conditions.

For this reason regional photovoltaic power estimation and forecast is becoming essential for Distribution System Operators (DSO), Transmission System Operator (TSO), energy traders, and aggregators.

In particular the estimation (nowcast) of regional PV power generation from the environmental conditions is needed since in Italy the actual energy meters used by DSO do not allow a real time power monitoring of the distributed photovoltaic production. Thus, power estimation can be used for PV power supervision, real time control of residual load and energy reserve activation in case of deviation. Day-ahead and intra-day PV power forecast can be employed by DSO and TSO for transmission scheduling to reduce energy imbalance and related cost of penalties, residual load tracking, energy trading optimization, secondary energy reserve assessment.

On regional scale, ongoing research is focused on upscaling methods [1-5] that allow the forecast of the overall generation of aggregates of PV systems without using the prediction of the power output of each PV plant (bottom-up approach). Indeed, thanks to the "ensemble smoothing effect", the performance of the forecast of each PV system only slightly affects the performance of the regional prediction. For this reason, up-scaling methods can achieve similar accuracy of the bottom-up strategy but with a much lower computational and data handling efforts.

The "ensemble smoothing effect" is related to the correlation between the forecast errors of different PV plant generation [6-10]. This correlation decreases with the distance (or with the size of the area) so that the regional forecast accuracy is not the average of the accuracy of power prediction of each plant.

In this paper new upscaling methods for estimation and forecast are developed. They are based on chains of data-driven models (k-mean clustering, principal component analysis and ensemble of neural networks) that use as inputs satellite derived irradiance and numerical weather prediction. These methods allow the study of the ensemble smoothing behavior on cluster scale and a complete assessment of the accuracy of the regional power forecast at different horizons, from 0 hour ahead (power estimation) to 48 hours ahead. Moreover, the models requires very few input information that should be provided by users. Thus they could be easily adopted by forecast providers to deliver the regional PV power estimation and forecast to DSO, TSO for grid management and balancing issues.

Furthermore a parametric probabilistic method [6, 10] to compute prediction intervals of the day-ahead forecast was also developed. It is based on an ensemble of neural networks to predict the forecast errors and a probabilistic correction that takes into account the deviation of the probability density function (PDF) of the errors from the standard normal distribution.

The proposed method allows a very reliable computation of the prediction intervals so that the frequency of the observations that fall inside each interval is almost equal to the associate confidence level. Thus prediction intervals can be employed by DSO and TSO to reduce the energy reserve (by the use of 95th quantile of the forecast errors) [1, 2] but also by energy traders, and aggregator companies to estimate the probability of a specific PV generation bid on the energy market.

The upscaling methods were trained and tested on the real case study: the distributed solar generation of a small part of the South Tyrol region in the North of Italy, characterized by a PV penetration similar to the one achieved at national level.

2 DATA DESCRIPTION

2.1 Local Experimental Data

The case study is a small part of South Tyrol Region in the North of Italy. In this area (around 800 km^2), the only variable renewable energy is the photovoltaic generation due to 1985 distributed plants with a total installed capacity at the end of 2015 of 68.2 MWp over a total installed capacity in the entire South Tyrol of around 240 MWp). Figure 1 shows the location of the region and of the PV plants in the area of interest.

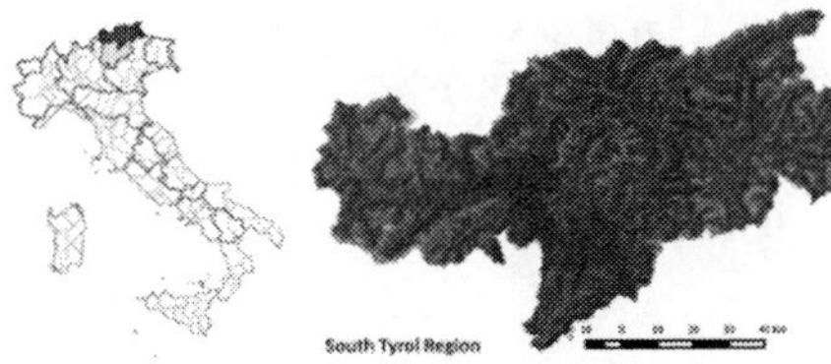

Figure 1: PV plants in the region of interest.

The PV power generation data (two years: 2014-2015) were provided by the local Distribution System Operator (Edyna) with a time resolution of 15 minutes.

These data were used to train and test power estimation and forecast models.

2.2 Satellite derived irradiance data

The satellite derived irradiance data used for power estimation come from the Geostationary radiative fluxes products, under Météo-France responsibility. They were obtained by OSI SAF SSI algorithm (Ocean and Sea Ice - Satellite Application Facility - Surface Solar Irradiance) applied to the satellite images provided by METEOSAT-10 (MSG-3) at 0° longitude, covering 60S-60N and 60W-60E, with a 0.05° horizontal resolution, and an hour granularity [3].

2.3 NWP data

The numerical weather predictions used for the day-ahead forecast were generated by the Weather Research and Forecasting (WRF–AWR 3.6.1) developed by NCAR (National Center of Atmospheric Research). The model is run operationally by the US National Weather Service and, being open source and easily portable, it is widely used around the world for research and weather forecasts [4]. Daily hindcasts were performed for the year 2014 and 2015. The model was initialized at 12 UTC, analyzing the 24 hour forecasts starting from the following 00 UTC, which is the typical procedure for the NWP solar day-ahead forecast. The model domain is centered over Italy with a horizontal resolution of 12 km, a higher resolution inner domain is nested centered on the region of interest, with a horizontal resolution of approximately 3 km. This horizontal resolution was necessary because of the complex orography of the region. The output was written every 20 minutes to achieve a better synchronization with the PV generation data. Details on WRF physics configuration can be found in [5].

3 METHODOLOGY

In this section, four different methodologies and models are described:

1. Upscaling methods and models for regional PV power estimation;
2. Upscaling methods and models for day-ahead forecast of the regional PV power (one and two days ahead);
3. A model for the intra-day forecast of the regional PV power (from 1 to 4 hours ahead);
4. A model able to provide the confidence intervals of the PV forecast;

The developed upscaling method consists on the application of spatial clustering of PV plants and then the use of satellite derived irradiance and numerical weather prediction (NWP) data (centered on each cluster centroid) as inputs for Ensemble of Artificial Neural Networks (ANNsE) that estimates or predict the regional PV power output (PO).

For the estimation and day-ahead forecast, two different methods were tested:

- Models output average: provides the regional result by the average of the power estimation/forecast of each cluster. This can be considered a cluster bottom-up strategy.
- Model inputs average: uses the model inputs calculated on each cluster to directly provide the regional power estimation/forecast. This approach is based on the spatial smoothing of the inputs features.

For intra-day forecast a further ANNsE model that makes use of past power estimation and one day ahead forecast (derived from the two previous models) was built.

To estimate the prediction intervals at different confidence levels, a model to predict the error of the day-ahead forecast was developed. It is based on ANNsE model to predict the forecast errors and a probabilistic correction that takes into account the deviation of the Probability Density Function (PDF) of the errors from the standard normal distribution.

All the ANNsE were trained on 2014 and tested on 2015.

3.1 Upscaling methods and models for estimation

The models output average consists of the estimation of the power yield of each cluster using an ANNsE based model (GNN). The estimation of the regional yield is then obtained by the average of the GNN output weighted with the PV capacity (P_n): $W_C = P_n(\text{cluster})/P_n(\text{regional})$. The GNN model inputs the sun azimuth and elevation to account the sun position and the satellite derived irradiance centered on the cluster centroid.

The model inputs average provides directly the regional estimation of power yield through a single ANNsE model (6GNN). The 6GNN uses the average clusters sun position (sun azimuth and elevation) and the satellite irradiance centered on each cluster centroid, as inputs.

Figure 2 shows the two different approaches and models used to provide the estimation of the power yield (PO/P_n).

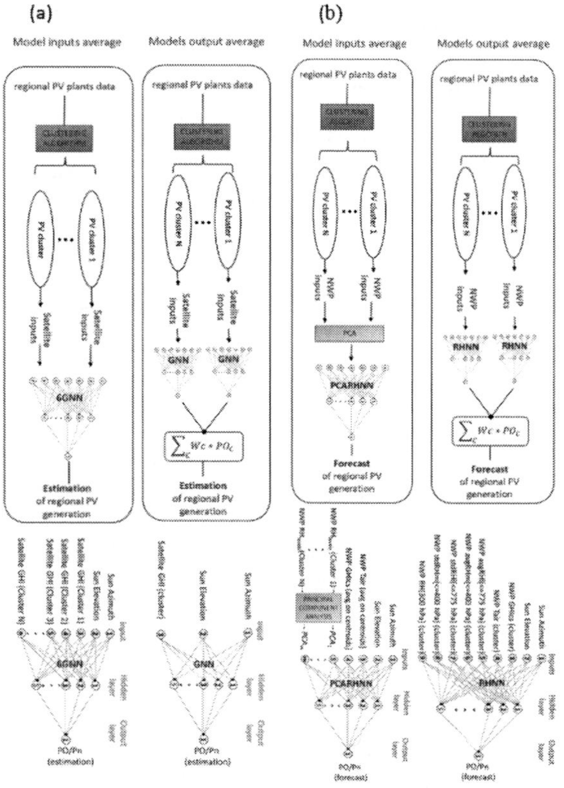

Figure 2: shows the two different upscaling approaches and models used to provide the estimation (a) and day-ahead forecast (b) of the regional power yield (PO/P_n).

3.2 Upscaling methods and models for day-ahead forecast.

Similarly to power estimation, the models output average approach consists of forecasting the yield of each cluster using an ANNsE based model (RHNN). The regional forecast is obtained by weighted average of the cluster forecasts. The model RHNN uses nine input features. The first four are the sun azimuth and elevation, the clear sky global horizontal irradiance (GHI_{cs}) and the 2 m temperature (T_{air}) predicted by WRF. The other five inputs are the average values and the standard deviation of the relative humidity predicted by WRF at different atmosphere levels corresponding to low clouds approximately below 2500 m, mid clouds between 2500 and 7500 m and high clouds up to 9000 m. The RHNN model is similar to the one developed in [5, 6] for the PV power forecast of a single plant.

The "model inputs average" provides directly the regional PV power forecast thought a single ANNsE model (PCARHNN). PCARHNN uses as inputs the sun azimuth and elevation, the clear sky irradiance (GHI_{cs}) and the 2 m temperature (T_{air}) predicted by WRF on each cluster centroid and averaged over all the clusters. The other input features of the ANNsE result from the principal component analysis (PCA) pre-processing of the relative humidity of 20 vertical atmospheric levels (RH_{level}) predicted by WRF and calculated on the cluster centroids.

Figure 2 shows the two different approaches and models used to provide the day-ahead forecast of the power yield.

It should be noted that all the ANNsE models (RHNN and PCARHNN) do not input the WRF prediction of the GHI. Indeed it was proved in [7] and [5] that the irradiance prediction provided by WRF radiation scheme should be post processed with a Model Output Statistics to reach a satisfactory accuracy level.

Details on the procedure to generate the artificial neural network ensemble can be found in [8].

3.3 Model for intra-day forecast

The intra-day forecast model essentially correct the day-ahead power prediction using previous power estimation. It is based on an ANNsE and for each hour it inputs the current and the past three yield estimation together with the day-ahead forecast of the next four hours. The model provides directly the forecast of regional power output from 1 to 4 hours ahead.

Figure 3 (a) shows the model used for the intra-day forecast of the PV power yield.

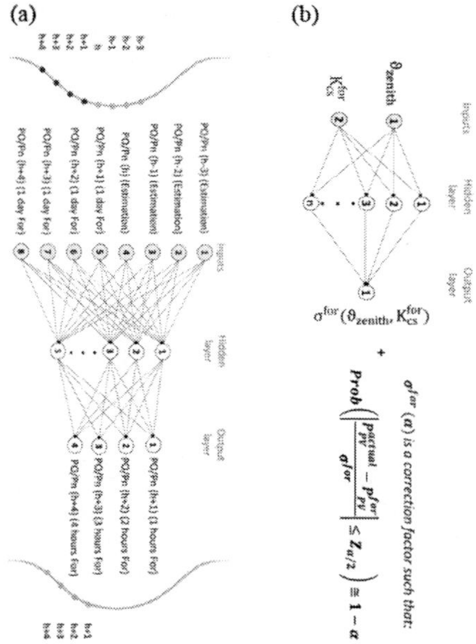

Figure 3: (a) model for intra-day forecast of PV normalized power output; (b) model for prediction of day-ahead forecast error.

3.4 Prediction of day-ahead forecast error

The prediction intervals could be calculated forecasting the standard deviation of the residuals (forecast errors).

The standard deviation of the forecast errors (σ^{for}) can be modelled as a function of the zenith angle (ϑ_{zenith}), the predicted clear sky index (K_{cs}^{for}) and the confidence level $(1-\alpha)$:

$$\sigma^{for} = \sigma^{for}(\vartheta_{zenith}, K_{cs}^{for})\,\sigma^{for}(\alpha) \qquad (1)$$

The term $\sigma^{for}(\vartheta_{zenith}, K_{cs}^{for})$ was calculated also using an ANNsE model. The second term $\sigma^{for}(\alpha)$ is a correction term that takes into account the deviation of the distribution of the forecast errors from the standard normal. The model is sketched in Figure 3 (b).

4 METRICS

According to solar forecast literature, the main metrics used to evaluate the models accuracy are reported in Table 1.

Table 1: accuracy metrics

Acronym and formulae
$e_h = (X_h^{for} - X_h^{obs})$
$RMSE = \sqrt{\frac{\sum_{h=1}^{n} e_h^2}{n}}$
$MAE = \frac{\sum_{h=1}^{n} \lvert e_h \rvert}{n}$
$MBE = \frac{\sum_{h=1}^{n} (e_h)}{n}$
$\text{Skill Score} = 100\left(\frac{RMSE(ref) - RMSE(for)}{RMSE(ref)}\right)\%$

Where
X_h^{obs} = Observed hourly PV power yield [kW/kWp]
X_h^{for} = Forecast hourly PV power yield [kW/kWp]
Pn = Capacity of the PV fleet [kWp]
n = number of sun hours in a year

The accuracy of the forecasts has been compared to the accuracy of clear sky persistence model used as reference [9] , [6].

The skill score is the metric that measures the performance improvement of the developed forecast model with respect to the reference (clear sky persistence).

5 RESULTS AND DISCUSSION

5.1 Power estimation and forecast (overall results)

The k-mean clustering aggregates all the PV plants in six areas corresponding approximately to six municipalities: Naturno, Tirolo-Merano, Lana, Nalles, Bolzano, Collalbo-Soprabolzano.

Figure 4 shows the spatial clustering of the PV systems in the controlled area.

Using the clustering results the two upscaling methods (the model inputs average and models output average) were applied.

For power estimation, the model inputs average obtained an RMSE of 3% while the models output average a RMSE of 3.2%. For the day-ahead forecast the two upscaling methods led to similar accuracy with an RMSE of 7.1%.

Thus for both the power estimation and the power day-ahead forecast, the performance of the simplest upscaling method (model inputs average) was considered to assess the accuracy of regional power forecast at different horizons, from 0 hour ahead (power estimation) to 48 hours ahead.

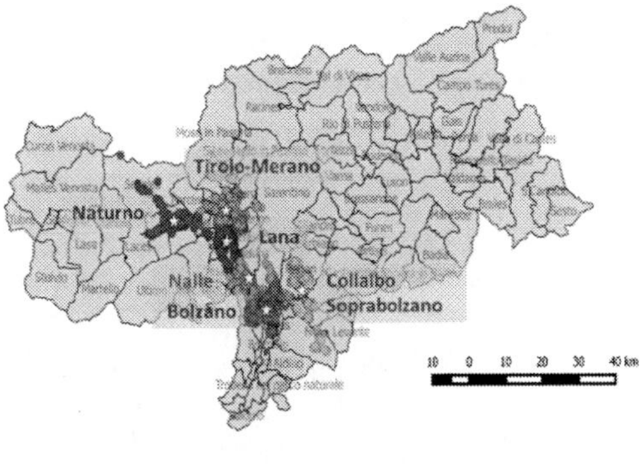

Figure 4: PV plants spatial clustering

Figure 5 shows the RMSE of regional PV power estimation and intra-day and day-ahead forecast vs forecast horizon, while in brackets the skill score (SS) with respect to the RMSE of the clear sky persistence is reported.

The power estimation upscaling method achieved to 3% of RMSE and 2% of MAE.

The intra-day forecast obtains a RMSE of 5% - 7% (SS from -8% to 34%) and a MAE of 3%-4%.

Day-ahead forecast achieves to an RMSE of 7% and 7.6% (SS from 39% to 45%) and MAE of 4%-5%.

For horizon longer than 4 hours the intra-day forecast provides a lower accuracy with respect to the day-ahead forecast. Thus satellite derived data can be used to correct NWP up to 4 hours ahead. Similar result was found in [10] as well as in other papers. On the contrary, the lower is the horizon the higher is the persistence accuracy. Thus in this case for 1 hour ahead the smart persistence provides a slightly better accuracy.

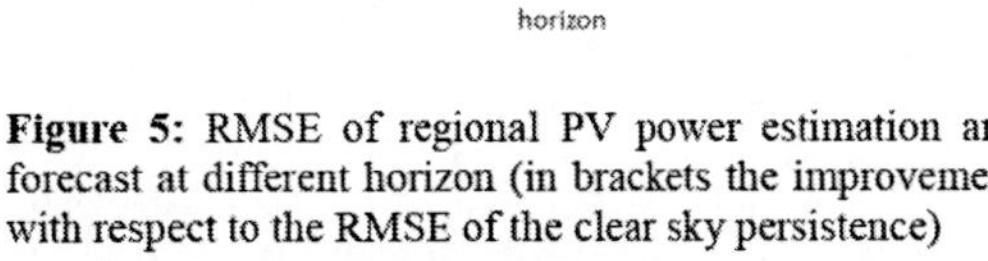

Figure 5: RMSE of regional PV power estimation and forecast at different horizon (in brackets the improvement with respect to the RMSE of the clear sky persistence)

5.2 Ensemble smoothing effect

The ensemble smoothing effect is related to the number of systems in the controlled area (N_{plants}), the PV capacity distribution (P_n^i with $i = 1{:}N_{plants}$) and the pair correlation of the error (ρ^{ij}). In particular, in [11] it was showed that the mean correlation of the forecasting error between two different site decreases exponentially with the distance as: $\rho^{ij} = e^{a\,(d_{ij})^b}$ so that the RMSE of the ensemble decreases with the size of the controlled area following the same law.

The models output average allows the study of the smoothing effect on cluster scale computing the RMSE of the forecast of different ensembles of adjacent clusters.

Figure 6 show the one day-ahead forecast error as a function of the maximum distance between the centroids of each ensemble of cluster. Each cluster behave like a single plant since the RMSE reduction due to the ensemble smoothing follows the same exponential law found in [11] for single systems.

Figure 6 also reports the skill score with respect the RMSE of the persistence model as a function of the maximum distance between the centroids of each ensemble of cluster. It can be observed that the skill score increase linearly of 0.1% per km. The skill score is usually calculated to compare the forecast accuracy achieved in different sites or in different years since it doesn't depend on the irradiance variability. In case of plants with similar characteristics it remains almost constant between different sites or years [12] and [6]. On the contrary, for the regional forecast the skill score increases linearly with the controlled area, thus the reduction of RMSE is achieved not only with the decrease of solar variability (RMSE of persistence model) but also with the improvement of forecast capability of the models due to the smoothing effect.

The fit of the mean RMSE (with coefficients of determination R2 of .985) allows the extrapolation of the RMSE on larger ensemble areas. For day-ahead power forecast a RMSE of 4% of Pn and a skill score of 60% could be achieved with distance around 200 km.

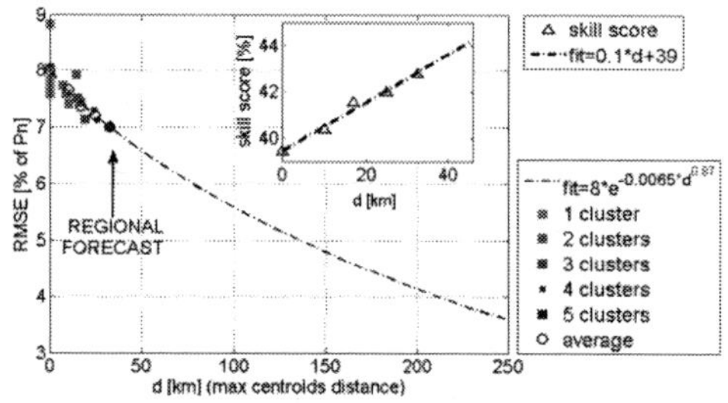

Figure 6: Ensemble smoothing effect of day-ahead forecast

This confirms that the de-correlation distance, for hourly values, is around to 100-200 km [13, 14]. Moreover the extrapolation proves that the obtained accuracy is in the "state of the art" range. Indeed, in [9] a RMSE of 4.1% - 4.3% of Pn with a skill score of 48% - 52.8% was obtained for distances around 500 and 750 km while in [15] a RMSE between 6% and 7% with a skill score between 50% and 60% was achieved for distances between 200 and 400 km.

5.3 Prediction intervals

Using a parametric probabilistic approach, the prediction intervals are usually estimated supposing a normal distribution of the forecast errors [16, 17], but this is a poor assumption (see figure 7).

Figure 7 reports the PDF of the error normalized with the standard deviation predicted with the ANNsE model (section 3.3): $Ne = e/\sigma^{for}(\vartheta_{zenith}, K_{cs}^{for})$.

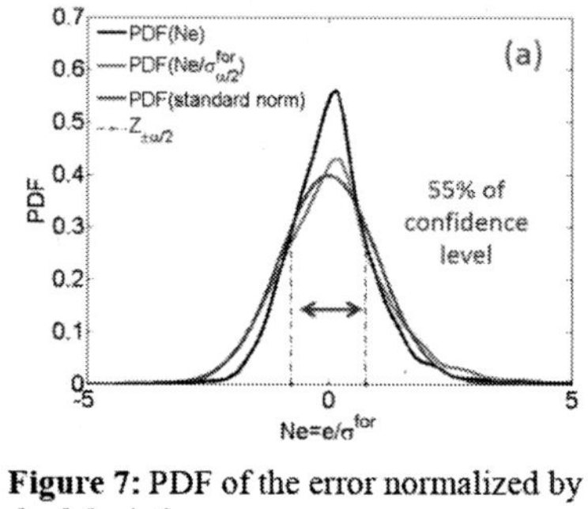

Figure 7: PDF of the error normalized by the predicted standard deviation.

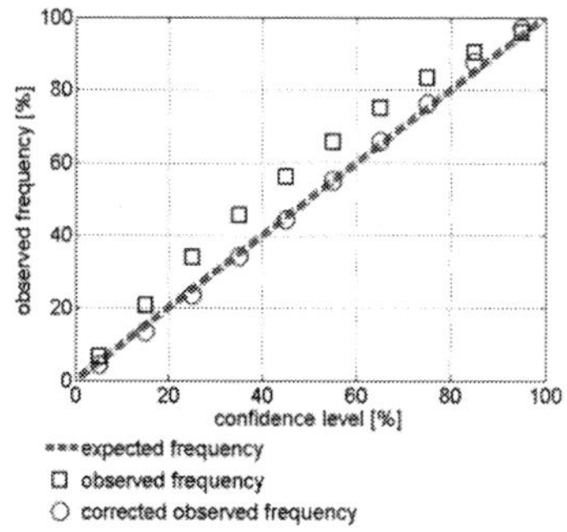

Figure 8: reliability plot.

It can be observed that the PDF of Ne (black curve) is far to be Gaussian.

Figure 8 shows the reliability plot i.e the frequency of observations that fall inside each interval vs the corresponding confidence level.

The Gaussian hypothesis provides a correct estimation of the 95th quantile but is not reliable for other confidence levels since the observed frequency (square dots) is greater than the expected one (dash line). Thus the ANNsE model provides an overestimation of the prediction intervals for almost all the quantiles.

Nevertheless, for each confidence level it was possible to predict a correction coefficient $(\sigma_{\alpha/2}^{for})$ so that:

$$P\left(\left|\frac{Ne}{\sigma_{\alpha/2}^{for}}\right| \leq Z_{\alpha/2}\right) \cong 1 - \alpha$$

In this way, the integral between $-Z_{\alpha/2}$ and $Z_{\alpha/2}$ of the corrected PDF$(\frac{Ne}{\sigma_{\alpha/2}^{for}})$ is now almost equal to the integral of the standard normal distribution and for all the confidence level $(1 - \alpha)$ a very reliable estimation of prediction intervals was obtained (circle dots in Figure 8).This good result was possible since the PDF of Ne of different years (2014 and 2015) are very similar, so that the corrections factors calculated on the PDF of the year 2014 are effective also for the year 2015.

Figure 9 reports the trend of the prediction interval for five days of February 2015. It can be observed that the extent of the interval is reduced when passing from overcast to clear sky days.

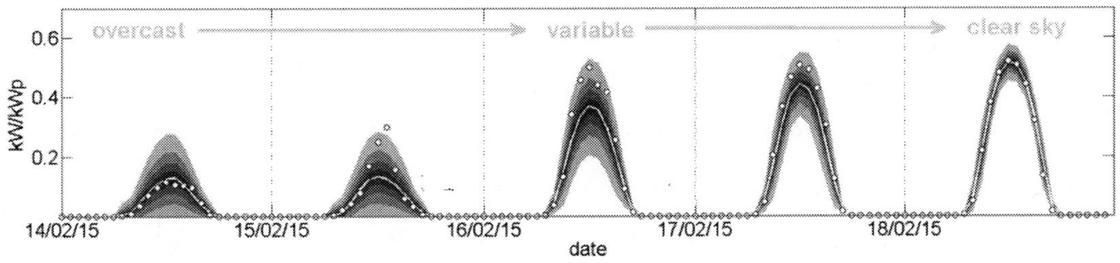

Figure 9: Example of prediction interval trend for five days of February 2015. The grey colors correspond to different confidence levels: 95%-75% -50%-25%; from the clearest grey showing the interval with 95% of confidence to the darker grey showing the interval with 25% of confidence. The dots represent the observed values while the white line is the forecast.

7 CONCLUSIONS

This study allows a complete assessment of the forecast accuracy of regional power forecast at different horizons, from 0 hour ahead (power estimation) to 48 hours ahead.

The developed forecast models were used to estimate and forecast the distributed generation of 1975 PV plants in a small part of the South Tyrol Region with an installed capacity at the end of 2015 of 67.2 MW_p. This region of around 800 km^2 has a complex orography and variable weather condition

The power estimation model achieved 3% of P_n in RMSE and 2% in MAE. Intra-day forecast (from 1 to 4 hours) obtained a RMSE of 5% - 7% and a MAE of 3%-4%. The skill score with respect to the accuracy of the clear sky persistence model was between -8% and 34% in RMSE. One and two days-ahead forecast provides a RMSE of 7% and 7.5% (with a skill score of 39% and 45%) and a MAE of 4% and 5%.

The proposed upscaling methods were also used to study the smoothing effect on cluster scale. This effect improves the accuracy of regional power forecast with respect to the mean clusters value. It reduces the RMSE of day-ahead forecast of 12%. It was proved that each cluster behave like a single plant since the RMSE reduction due to the ensemble smoothing follows the same exponential law found in [16] for single systems. Extrapolating the RMSE reduction on wider areas, the developed upscaling methods bring an accuracy comparable with the one found in [10] and [18] for areas greater than 200x200 km2. Thus the models could be easily adopted by forecast providers to deliver the regional PV power estimation and forecast to DSO, TSO for grid management and balancing issues.

The skill score increases linearly with the size of the region with a rate of 0.1% per km, thus the reduction of RMSE is achieved not only by the decrease of solar variability (RMSE of persistence model) but also with the improvement of forecast capability of the models due to the smoothing effect.

A model to estimate the forecast error was also developed. It was based on an ensemble neural network model that forecasts the standard deviation of the error coupled with a probabilistic correction that takes in to account deviations of the PDF of the error from the normal distribution. Indeed using a parametric probabilistic approach, the prediction intervals are usually estimated supposing a normal distribution of the forecast errors but this is a poor assumption. This hypothesis provides a correct estimation of the 95th quantile but is not reliable for other confidence levels. The proposed model allows a very reliable computation of the prediction intervals so that the frequency of the observations that fall inside each interval is almost equal to the associate confidence level. Thus prediction intervals can be employed not only to reduce the energy reserve (by the use of 95th quantile of the forecast errors) but also to estimate the probability of a specific solar energy bid on the energy markets.

8 ACKNOWLEDGEMENTS

The authors acknowledge the Distributor System Operator Edyna Srl, for providing the data and for co-funding the work. A special thanks to the technical director Andreas Bordonetti for the trust granted to Eurac and to its staff Tschager David, Bruno Fasoli and Danilo Pederiva for the useful discussions.

The authors acknowledge also Stiftung Südtiroler Sparkasse/Fondazione Cassa di Risparmio for co-funding the PhD of Marco Pierro.

REFERENCES

[1] . J. Zhang, . B. Hodge, S. Lu , H. Hamann e B. Lehman , «Baseline and target values for regional and point PV power forecasts: Toward improved solar forecasting,» *Solar Energy*, vol. 122, pp. 804-819, 2015.

[2] P.-J. Alet, «Photovoltaics merging with the active integrated grid: a white paper,» European Photovoltaic Technology Platform, 2015.

[3] EUMETSAT, «Operational Services Specitication,» EUMETSAT, 2017.

[4] W. Skamarock, J. Klemp, J. Dudhia, D. Gill e D. and Barker, «A description of the Advanced Research WRF version 3.,» NCAR Tech. Tech. rep., Note NCAR/TN-4751STR, 2008.

[5] M. Pierro, F. Bucci, M. De Felice, E. Maggioni, A. Perotto, F. Spada, D. Moser and C. Cornaro, "Multi-Model Ensemble for day ahead prediction of photovoltaic power generation," *Solar Energy*, vol. 134 , p. 132–146, 2016a.

[6] . M. Pierro, F. Bucci , M. De Felice , E. Maggioni , . A. Perotto, F. Spada, D. Moser e C. Cornaro, «Deterministic and Stochastic Approaches for Day-

Ahead Solar Power Forecasting,» *Journal of Solar Energy Engineering,* vol. 139 (2) , p. 021010, 2016b.

[7] M. Pierro , F. Bucci , C. Cornaro , . E. Maggioni, A. Perotto , M. Pravettoni e F. Spada , «Model output statistics cascade to improve day ahead solar irradiance forecast,» *Solar Energy Journal,* vol. 117, pp. 99-113, July 2015.

[8] C. Cornaro , M. Pierro e F. Bucci, «Master optimization process based on neural networks ensemble for 24-h solar irradiance forecast,» *Solar Energy,* vol. 111, p. 297–312, 2015.

[9] E. Lorenz , T. Scheidsteger, J. Hurk, D. Heinemann e C. Kurz, «Regional PV power prediction for improved grid integration,» *Progress in Photovoltaics: Research and Applications,* vol. 19 , p. 757–771, 2011.

[10] B. Wolff , J. Kühnert , E. Lorenz , O. Kramer e D. Heinemann , «Comparing support vector regression for PV power forecasting to a physical modeling approach using measurement, numerical weather prediction, and cloud motion data,» *Solar Energy,* vol. 135, pp. 197-208, 2016.

[11] . E. Lorenz, J. Hurka , D. Heinemann e H. G. Beyer , «Irradiance forecasting for the power prediction of grid connected photovoltaic systems,» *IEEE Journal of Selected Topics in Applied Earth Observations and Remote Sensing;,* vol. 1, n. 2, pp. 2-10, 2009.

[12] E. Lorenz, J. Remund, S. C. Muller, W. Traunmull, G. Steinmaurer, D. Pozo, J. Ruiz-Arias, V. L. Fanego, L. Ramirez e M. G. Romeo, «Benchmarking of different approaches to forecast solar irradiance,» 2009.

[13] . R. Perez, T. Hoff e . S. Kivalow, «Spatial and temporal characteristics of solar radiation variability,» in *International Solar Energy (ISES) World Congress*, Kassel, Germany, 2011.

[14] T. Hoff e R. Perez, «Modeling PV fleet output variability,» *Solar Energy,* vol. 86 , pp. 2177-2189, 2012 .

[15] J. S. Fonseca Jr, T. Oozeki, H. Ohtake, K. Shimose and T. K. Takashima, "Regional forecasts and smoothing effect of photovoltaic power generation in Japan: An approach with principal component analysis," *Renewable Energy,* no. 68, pp. 403-413, 2014.

[16] E. Lorenz, . J. Hurka, D. Heinemann e H. G. Beyer, «Irradiance Forecasting for the Power Prediction of Grid-Connected Photovoltaic Systems,» *IEEE* , vol. 2 (1), pp. 2-10, 2009.

[17] R. Marquez e C. Coimbra, «Forecasting of global and direct solar irradiance using stochastic learning methods , ground experiments and the NWS database,» *Solar Energy,* vol. 85 (5) , pp. vol: 85 (5) pp: 746-756, 2011.

[18] J. G. Fonseca, T. Oozeki, H. Ohtake, . K. Ichi Shimose, T. Takashima e K. Ogimoto, «Regional forecasts and smoothing effect of photovoltaic power generation in Japan: An approach with principal component analysis,» *Renewable Energy,* vol. 68, n. 403-413, 2014.

COULD A PHYSICAL MODEL BE USED TO FORECAST PV?

Owen T Huxley, Ciaran Allen, Jamie Taylor, Julian Briggs, Aldous M Everard, Alastair Buckley
Sheffield Solar, Department of Physics and Astronomy, The University of Sheffield
Hicks Building, Hounsfield Road, Sheffield, S3 7RH, UK

ABSTRACT: We investigate the use of a deterministic approach to modelling PV using a physical model and assess whether this approach is suitable for forecasting PV. We find that on an individual system basis for hourly data, the physical model is not appropriate for modelling PV because the errors inherent in a physical model are too large at this scale. However, on a national or regional scale, or for daily or weekly data, the physical model offers a fast,easy and viable method for operational forecasting of PV generation. The physical model achieved an r^2 of 0.77, a mean normalised RMSE of 36%, and a mean bias error of -0.002 kWh. When the data was aggregated spatially and temporally, the performance of the model was found to improve significantly with the scale of aggregation. The performance of the model under different sky conditions was analysed and the model was found to perform best under very cloudy or very clear conditions, although the model typically overestimates the PV generation for clearer conditions. A commercial irradiation forecast was analysed and compared with ground based observations, leading to an r^2 of 0.88 and a mean normalised RMSE of 34%. This error is comparable with the error in the PV model and the two errors combined are deemed to be use-prohibitive for the applications considered.
Keywords: Forecasting, Modelling, Photovoltaic

1 INTRODUCTION

PV deployment has consistently exceeded expectations [1]. Decreasing installation costs and governmental support through Feed-In Tariffs have made solar panels very popular. Around 900k PV systems are currently installed [2] in the UK and the energy they produce has a significant impact on power flows across the transmission network. The importance of solar is only set to increase with the system price still falling and its continued, albeit diminishing, support from local and national government. Therefore, forecasting PV generation is an ever more critical issue for the Transmission Network Operators (TNOs) and Distribution Network Operators (DNOs). Improving PV generation estimates will enhance the TNOs capability to balance demand with supply. PV generation is highly weather dependent and the UK climate, often cloudy and overcast, makes forecasting PV very challenging.

To utilise the full potential of any energy source, forecasting energy production is a key issue. For conventional energy production methods (coal, gas or nuclear) this is trivial. However, for renewable energy sources such as wind and solar - where the energy production is weather dependent - forecasting requires a good computer model. There are many types of computer model which could be used; statistical models, physical models and machine learning algorithms. Statistical models and machine learning algorithms are sometimes referred to as black box models. This is because the model is fed the input data and returns the modelled data using parameters through training. In contrast to this a physical model utilises a set of equations derived from physical principles between the input data and the output data.

We assess the suitability of a physical model for forecasting PV generation, both on an individual system basis and on a nationally aggregated level. We assess the inherent uncertainty in the physical model when applied to an historical dataset and then investigated the additional uncertainty associated with using a forecasted meteorological data set. We briefly examine how these uncertainties compare and might interact with each other when used together.

2 DATA SOURCES

2.1 Physical model

The physical model used here is the Global Solar Energy Estimator (GSEE) described in a 2016 paper published by Stefan Pfenninger & Ian Staffell [3]. The GSEE model is easily available online via an API called Renewables Ninja (RN) [4]. The API allows users to download PV generation data modelled by the GSEE methodology for the calendar year 2014. This generation data can be downloaded for calculations performed with two different irradiance datasets: MERRA [5] and SARAH [6]. This report uses the SARAH dataset because it offers a considerably higher spatial resolution than MERRA.

2.2 Measured Data

To validate the GSEE modelled generation, it was compared to measured PV generation data from 238 sites across the UK. The measured PV generation data came from two independent PV companies, hereafter known as dataset A and dataset B. These datasets provide monitored PV generation data, complete with system metadata which includes latitude, longitude, azimuth, elevation and total installed capacity (TIC). We filter data during the calendar year 2014 for each site in the two datasets and down-sample to hourly data since the SARAH dataset only provides hourly data.

2.3 Forecast and Measured Irradiance Data

To use a physical model for PV forecasting the input parameters, namely irradiation, temperature, and wind speed, must be forecasted. The analysis of the physical model in this paper was performed with the SARAH reanalysis dataset. This data was used for convenience because it was easily accessible through the RN API. However, we must also consider the additional error associated with forecasted meteorological parameters. Therefore, an historical irradiance forecast was compared with ground-based irradiance observations. The measured irradiance values came from the Midas dataset.

3 METHOD

3.1 Model validation

The accuracy of the GSEE physical model was investigated. Data was downloaded from RN for the 238 locations within the measured PV datasets. RN returns hourly power generation predictions timestamped at the start of the hour. The measured datasets are timestamped at the end of each hour. Therefore, the RN data was shifted forward by 1 hour. Figure 1 shows a time series comparing dataset A to modelled data for one site over a period of 5 days in September 2014.

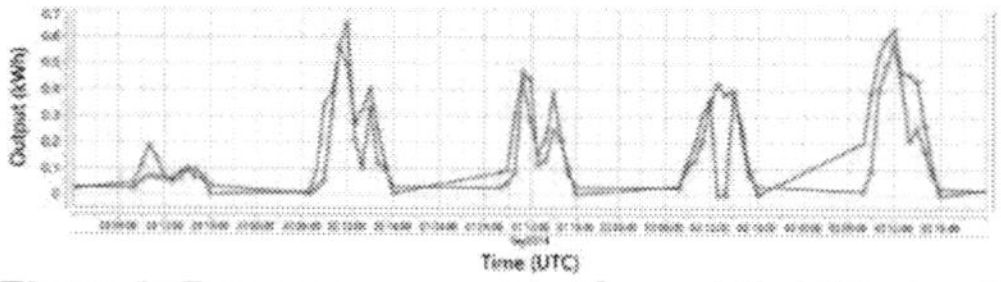

Figure 1: *Time series comparing the modelled (blue) and measured (red) data for a PV site in dataset A. The model performs well at predicting the general trend of output. However, it shows poor performance in the mornings and evenings and erroneous zero predictions around midday.*

We see that the model performs poorly at sunrise and sunset. To remove this large uncertainty in our dataset, we chose to limit the analysis between 10:00 and 14:00 GMT. Forecasting around midday is the highest priority for a typical end user, so excluding sunrise and sunset is acceptable. There are erroneous zero values around midday in both the modelled and measured data. Figure 1 shows an example of this in the modelled data. For the measured data, this is attributed to an error in measurement and these values are removed as they cannot be used for model validation. They are also omitted from the modelled data because this is an error unique to the RN physical model implementation. We are interested in assessing the suitability of physical models in general, as opposed to this one specific model therefore these anomalies were ignored. A bivariate plot of modelled versus measured data once the zero values and measurements outside of the hours of 10:00 to 14:00 are removed can be seen in Figure 2. The non-parametric density plot shows that most of the measurements are found at low irradiance.

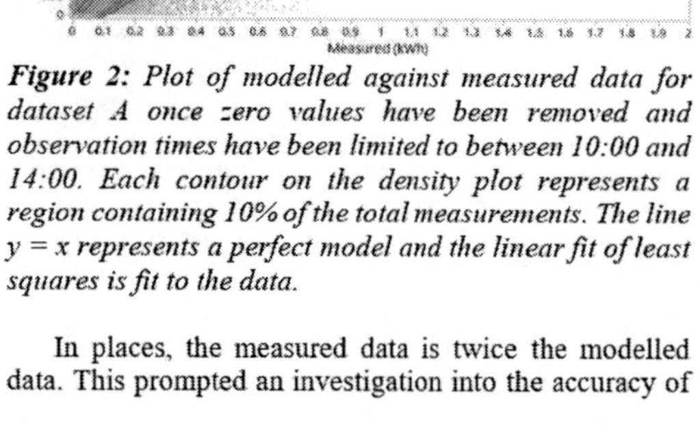

Figure 2: *Plot of modelled against measured data for dataset A once zero values have been removed and observation times have been limited to between 10:00 and 14:00. Each contour on the density plot represents a region containing 10% of the total measurements. The line $y = x$ represents a perfect model and the linear fit of least squares is fit to the data.*

In places, the measured data is twice the modelled data. This prompted an investigation into the accuracy of the measured data. A visual inspection of each PV site's measured vs modelled plot was undertaken and some sites were found to have bad measured data. We identify three classes of 'bad' sites; "cutting" of the generation profile caused by incorrectly sized inverters, faulty or under-performing sites – identified by many anomalously low measurements, and sites with incorrect metadata - such as incorrect TIC. Figure 3 shows examples of each of these 'bad' sites from dataset B. We find that 10.5% and 12% of sites in datasets A and B respectively are 'bad'.

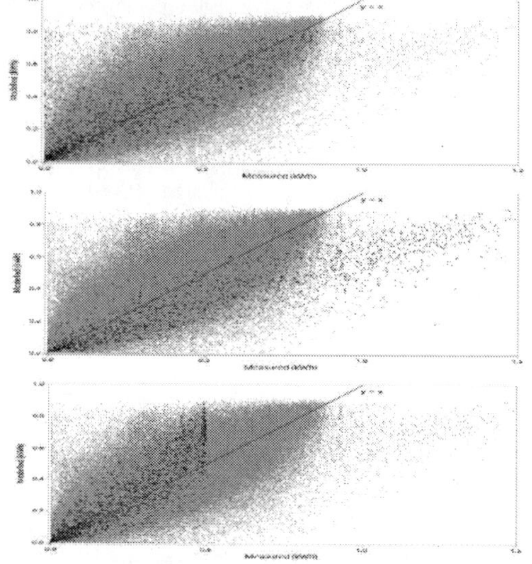

Figure 3: *Modelled against measured plots for 3 sites taken from dataset A which were identified as bad sites. The top figure shows an example of a faulty system. Here a portion of the data appears to have been measured incorrectly with many values close to zero. It is assumed that this is due to the measured data rather than the modelled data as it is only seen in a couple of sites. The middle figure shows an example of sites where the metadata provided is suspected to be incorrect. In each of these examples we see the measured generation is exactly half or exactly double what would be expected of a typical system. It is assumed here that the TIC value provided has been entered wrong. The bottom figure shows an example of cutting due to an incorrectly sized inverter. The measured data provided by the system appear to cut off above a certain power which depends on the level of cutting that affects the system.*

The final part of the validation process involved removing outlying data using the Tukey method. In the Tukey method an outlier is defined as any point that falls more than 1.5 times the interquartile range above or below the upper and lower quartiles respectively. We calculate the residuals using:

$$residual = measured - modelled \quad (1)$$

and calculate the normalised residual using:

$$normalised\ residual = \frac{measured - modelled}{measured} \quad (2)$$

Tukey outlier limits are then applied using these two parameters achieving two different results. Figure 4 shows two plots of modelled data against measured data after these two limits have been applied.

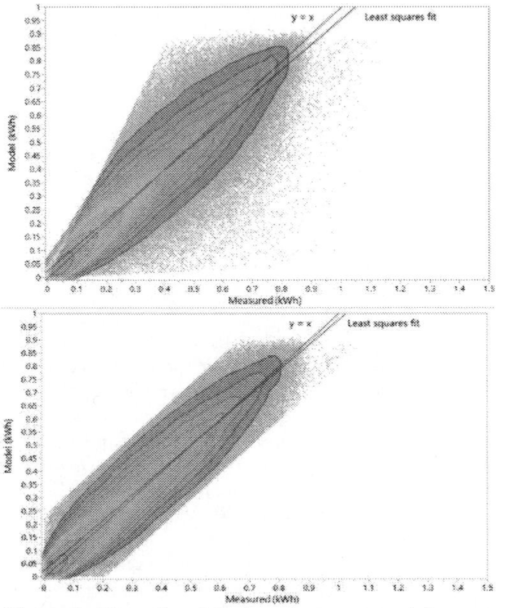

Figure 4: *Plots of modelled against measured data once times have been limited between 10:00 and 14:00, erroneous zero values have been removed and Tukey outlier limits have been applied to dataset 2. The top figure shows the data with limits applied to the normalised residuals while the bottom figure shows the data with limits applied to the absolute residuals.*

3.2 Temporal and spatial aggregation

An investigation into how the physical model performs when the predictions are aggregated both spatially and temporally was then undertaken. The aim was to assess the response of the physical model to binning data into larger regions and longer temporal resolutions. With the hypothesis that this would improve accuracy by decreasing the significance of poor data. Dataset A was the only dataset used for both aggregations as it was the only available dataset at the time of this analysis. To calculate the temporal aggregation, the modelled and measured PV generation data were grouped into hourly, daily, weekly and monthly time periods and then summed. To calculate the spatial aggregation, a region in the South of England with the most uniform distribution of sites was selected. This region and the grid can be seen in figure 5. Sites were aggregated by 4x4, 2x2, 1x1, 0.5x0.5 and 0.25x0.25 degree squares and finally a national aggregation of all sites was completed.

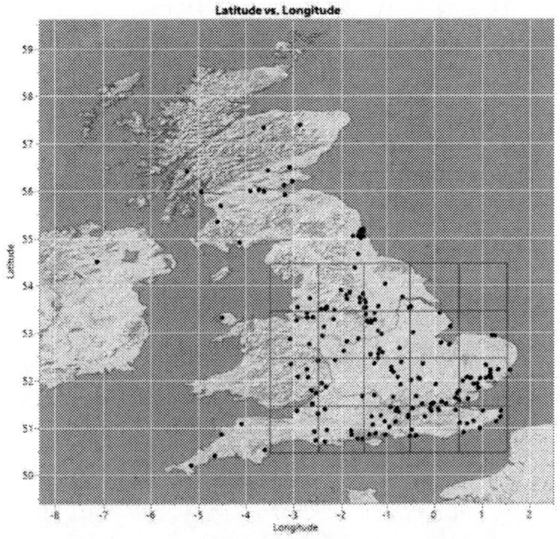

Figure 5: *Map showing the region over which the spatial aggregation was undertaken. The red 1x1 degree grid shows the regions where sites were binned. The shift of the grid 0.5 degrees in both latitude and longitude was undertaken to maximise the uniformity of the captured regions. When smaller binning of regions was undertaken, a finer grid over the same region was used. The national aggregation involved summing the measured and modelled data for every system in our dataset, even those outside of this grid region.*

3.3 Clearness index

Clearness index (K_t) is the ratio of top-of-atmosphere irradiance to global horizontal surface irradiance and is used as a parameter to describe how clear the atmosphere is i.e. how much cloud cover there is. In this investigation, the physical model's response to different clearness indices was tested to ascertain how well it performs under different sky conditions. We calculate the hourly clearness indices using:

$$K_t = \frac{H_0}{I_{global}} \qquad (3)$$

Where K_t, I_{global} and H_0 is the hourly clearness index, the global irradiance and the extra-atmospheric irradiation respectively, each integrated over an hour.

Extra-atmospheric irradiance is a measure of irradiance at the top of the atmosphere. Global irradiance is a measure of the total irradiance from both direct and diffuse (scattered) components at ground level. We calculate extra-atmospheric irradiance at 1-minute resolution for each site. The extra-atmospheric irradiance is then converted into the plane of the PV system which gives the global irradiance. Finally, the irradiance is averaged over each hour to give the irradiation. The clearness index is then calculated using equation [3].

3.4 Forecast Irradiance

To use a physical PV model as a PV forecast model the irradiance dataset needs to be forecasted. This forecast will have its own associated error which will add to the total uncertainty when combined with the physical model. To investigate the size of the error introduced by the forecasted irradiance, a forecasted irradiance dataset was compared to a measured irradiance dataset over 1 year for the year ending 30th June 2016. MIDAS [7] provided ground-based irradiation observations. The forecasted irradiance dataset is confidential. The residual between the two datasets was calculated from equation 1.

4 RESULTS

Validation of the physical model involved many successive data selection steps. Table 2 shows a summary of the fit statistics at each step for the two datasets.

Table 1: *Summary of statistics for validation of GSEE model at various stages of reduction of data. When the data is limited between 10:00 and 14:00 there is an increase in the r^2 value of the model's predictions. This is because r^2 is a measure of how well the model explains the variance in the data: low irradiance values inherently are more likely to have a low variance.*

	R Squared		**Mean Bias Error (kWp)**	
	A	B	A	B
No zeros	0.76	0.78	-0.005	-0.012
10am-2pm, no zeros	0.71	0.69	-0.005	-0.019
10am-2pm, no zeros, no bad sites	0.75	0.73	-0.011	-0.015
10am-2pm, no zeros, no bad sites, Tukey Residuals	0.83	0.86	-0.014	-0.013
10am-2pm, no zeros, no bad sites, Tukey Normalised Residuals	0.78	0.77	0.002	-0.004

The model estimates generation with an average r^2 value of 0.77 over the 2 datasets for the raw data. This decreases to an average of 0.70 when the data is limited from 10:00 to 14:00 and erroneous zeros are removed. Continuing with the removal of 'bad sites' and the application of the Tukey outlier limits to the residuals and normalised residuals, the average r^2 improves to 0.74, 0.845 and 0.775 respectively. Throughout these steps the mean bias error remains largely unchanged. After the final step, the model had a mean normalised RMSE of 36%

The suitability and effectiveness of these steps is discussed in full in section 5. The temporal and spatial aggregation investigations were performed on dataset A only. Figure 6 shows plots of the mean bias error and standard deviations of predictions made by the model at different aggregation levels both temporally and spatially. We see a similar significant improvement with both mean bias error and standard deviation when the scale of aggregation is increased. Both statistics improve as the level of temporal and spatial aggregation increases.

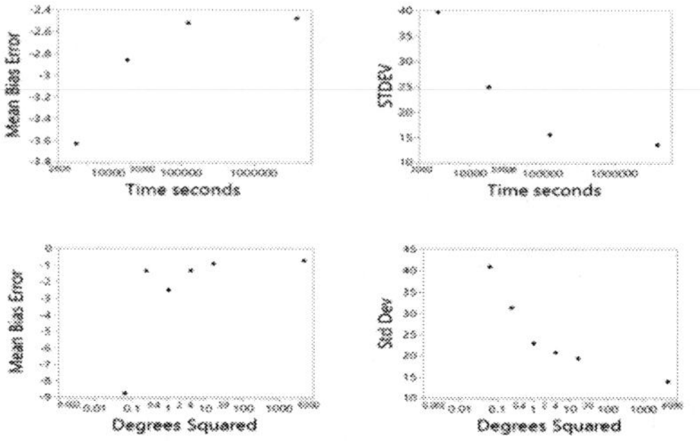

Figure 6: *Plots of mean bias error and standard deviation for different temporal aggregations are shown in figures 6a and 6b respectively. The same plots for spatial aggregation measured in degrees squared are shown in figures 6c and 6d respectively. Degrees squared is measured using the latitude and longitude bins. We see a clear logarithmic relationship between these statistics and the scale of aggregation in both tests.*

The clearness index investigation conducted in this study looked at the accuracy of the physical model under different sky conditions by calculating hourly clearness index to accompany the measured and modelled data. Figures 7a to 7d show plots of modelled vs measured data for dataset B grouped into clearness index values of kc ≤ 0.3, 0.3 < kc ≤ 0.6, 0.6 < kc ≤ 0.9 and 0.9 < kc . Figure 8 shows a summary of statistics of the model fit as the clearness index increases for both datasets. This analysis is conducted with a smaller clearness index bin size of 0.1 from values of 0 to 1.2. We see that both datasets respond in a comparable manner at different values of clearness. Under very overcast skies (low clearness index) and very clear skies (high clearness index), the r^2 of the model is highest as these conditions are more easily modelled than changeable and cloudy skies. Similarly, the mean bias error shows that the bias is small under overcast skies. In contrast, the bias becomes more negative under clearer skies.

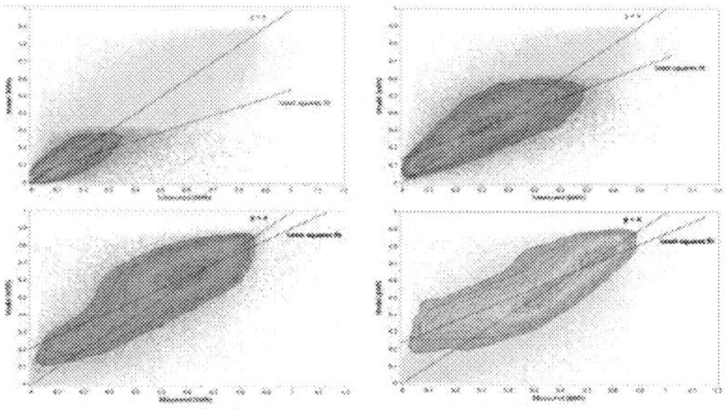

Figure 7: *Modelled vs measured data for dataset B separated into various levels of sky clearness. Figures 7a to 7d show plots for; $K_t \leq 0.3$, $0.3 < K_t \leq 0.6$, $0.6 < K_t \leq 0.9$ and $K_t \geq 0.9$ respectively. The points in dark grey, the non-par density plots and the line of fit are related to the data within the respective clearness index group. The light grey points show the overall dataset. The non-par density plots show 10% contours.*

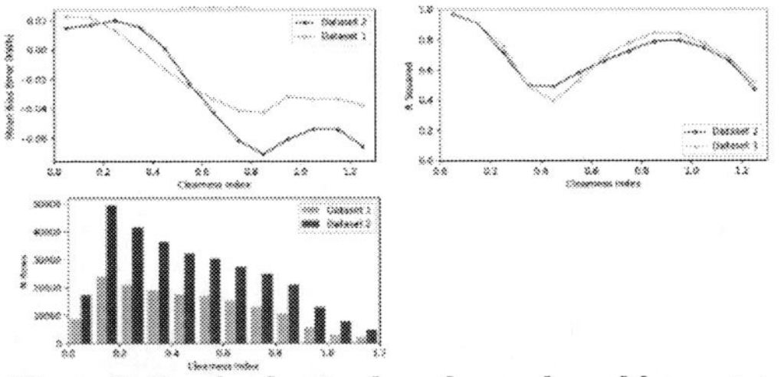

Figure 8: *Graphs showing how the number of data points (bottom left), the mean bias error (kWh) (top left) and the* r^2 *(top right) value varies for the GSEE model under different sky conditions. Data is binned into groups of size 0.1 kc and the statistics for both dataset A and dataset B are shown. The* r^2 *value responds how we might expect however the mean bias error appears to become more negative at high clearness values.*

The final investigation conducted in this study looked at the accuracy of a forecast irradiation dataset. A forecast irradiation dataset was compared to a measured irradiation dataset (MIDAS) for the calendar year ending 30th June 2016. There were 98 weather stations across he UK in the MIDAS dataset. Figure 9a shows a plot of the forecasted irradiation against the measured irradiation. The r^2 of this plot is 0.88. Figure 9b shows a histogram of the residuals. The mean of this histogram is -5.833 kWh. Therefore, the irradiation forecast also tends to overestimate, like the PV model. The irradiance forecast has a mean normalised RMSE of 36%, which is comparable to the error in the PV model.

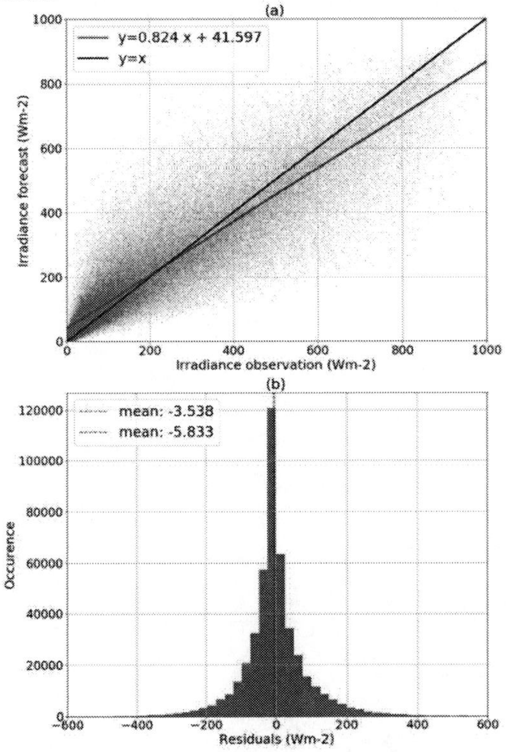

Figure 9: *(a) Scatter plot of irradiance forecast and observations (b) Histogram of residuals between irradiance forecasts and observations. Comparison uses 98 systems from across the UK for the year ending 30th June 2016. Observation data are extracted from MIDAS.*

5. DISCUSSION

Validation of the physical model shows that irradiation to PV power generation with a historical dataset can be performed reasonably well with a suitable outlier method employed. Initial calculations of the effectiveness of the model give an average r^2 value of 0.77, a mean normalised RMSE of 36%, and a small negative mean bias error. Referring to Table 1, we see that the r^2 value of the GSEE model fit gets worse when the data is limited between 10:00 and 14:00 and erroneous zero values are removed. This is because r^2 is a measure of the proportion of variance in the generation that is explained by the model. The removed values were mostly located at the origin. These values artificially improve the r^2 value of the model. Therefore, a reduction in the r^2 value is expected when they are removed.

Approximately 10% of the sampled PV sites have bad data. We define a site as having 'bad' data if it possesses one of the three characteristics shown in Figure 3. These bad sites were removed from the datasets. This results in the r^2 value of the model to be improved by about 0.04 for both datasets. This is important because we do not want to compare the model to incorrect measured data.

The last step of the validation process involved removal of outliers using the Tukey method. By applying the Tukey method to the normalised residuals, estimated generation values that are large overestimations are removed. This is due to the normalisation factor being the measured value as seen in equation [2]. This outlier removal method has the effect of ignoring outliers at large underestimated values and results in the bias error being reduced significantly.

The second method applies the Tukey outlier limits to the absolute residual values. We see that this improves the r^2 drastically. However, it also has the effect of streamlining the data and we suggest that an outlier method such as this will need to be carefully justified.

Throughout the validation process the mean bias error remains largely unchanged, small and negative. Hence we can conclude that the model tends to overestimate PV generation. This skew was not removed with the removal of bad sites. Therefore, it is hypothesised that this is a bias in the model.

The temporal and spatial aggregation investigation showed that as the scale of aggregation was increased, the bias on the model was reduced and the precision of the model was improved. This result is expected.

We see a general trend of both the modelled and measured generation power increasing as the clearness index increases, which is expected. Figure 9 also shows that there is a spur of low generation measured data that remains as the clearness index increases. The modelled generation values of the same data points can be seen to increase with clearness index and this causes the spur to move upwards. This results in a population of data points with a large negative residual at low measured PV generation. At higher clearness indices, the main component that prevents transmission of solar irradiance through the atmosphere becomes aerosol content [8]. As the model begins to skew upwards at these high clearness values, we suggest that the poor understanding of the effect of aerosols on global irradiance results in this bias. We expect that the model should perform best on very overcast days (low clearness index) and very clear days (high clearness index) since these are the conditions under which irradiance is most easily modelled. This result is seen in the r^2 plot in figure 8. The effect of aerosols is seen in the mean bias error plot where the error is low at low clearness index but becomes increasingly negative on clearer days where aerosol effects become more important.

Figure 9 shows the error in the forecast irradiation dataset. The r^2 of the plot in 9a is 0.88, and the mean normalised RMSE is 34%. This error is similar in magnitude as the error in the PV model. It is hard to know exactly what the error would be when the PV model is used with the forecast irradiance dataset because the error in the SARAH dataset has not been assessed. However, it is reasonable to assume that the error would propagate through from the irradiation data to the modelled PV data. There is likely more error in a forecasted irradiation dataset than a measured satellite based irradiation dataset so it can be assumed that the PV model performance would decrease. Figure 9b shows the distribution of the residuals. The mean of the histogram is 5.83. This suggests that the forecast irradiation data has similar negative bias to the PV model. However, here the bias is positive because the residuals have been calculated the opposite way around to they were for the PV model analysis.

5 CONCLUSION

It has been have shown that by applying an appropriate outlier removal technique a physical model can achieve a good r^2 of around 0.77. We also identify many sources of inaccuracy in the measured data in the form of 'bad' sites. A significant source of negative bias on the model predictions is identified to be poor modelling of how aerosols affect the transmission of irradiance. This is particularly apparent on clear sky days where the effect of aerosols has the most significant impact. The accuracy and precision improve significantly with temporal and spatial aggregation as expected. A forecasted irradiation dataset was shown to have a r^2 of 0.88, and a mean normalised RMSE of 34% This error is comparable to the error in the PV model which had an r^2 of 0.77 and a mean normalised RMSE of 36%. It is reasonable to expect that the irradiation dataset would increase the error in the output of the modelled PV data when the two are used in conjunction. The physical model has been shown to predict PV power relatively well using a historic dataset. The error is large on an individual system basis for hourly data and it is suggested here that this physical model is not suitable for PV forecasting on this scale in the UK. However, the model performance improves significantly when the data is aggregated spatially and temporally and could be used on a national scale successfully. The model's performance is best at high and low clearness and hence would suit a sunnier climate best.

6 REFERENCES

[1] Felix Creitzig. et al. "The underestimated potential of solar energy to mitigate climate change". *Nature Energy*

[2] Nick Jesson(2017). *Solar Voltaics Deployment* (31 Aug 2017) [Online]. Available: https://www.gov.uk/government/statistics/solar-photovoltaics-deployment

[3] Stefan Pfenninger. Iain Staffell. "Long-term patterns of European PV output using 30 years of validated hourly reanalysis and satellite data". *Energy*, 114, pp.

[4] Stefan Pfenninger. Iain Staffell. *Renewables Ninja*. Available: https://www.renewables.ninja/, Aug 11, 2017 [Nov 20, 2016]

[5] Rienecker. et al. "MERRA: NASA's Modern-Era Retrospective Analysis for Research and Applications". *J Climate*. 24, pp. 3624–3648

[6] Richard Müller. et al. "Surface Solar Radiation Data Set - Heliosat (SARAH) - Edition 1". *Satellite Application Facility on Climate Monitoring*

[7] UK Meteorological Office. *MIDAS Land Surface Stations data* (1853-current) [Online]. British Atmospheric Data Centre, 2006 [2017], Available: http://badc.nerc.ac.uk/data/ukmo-midas

[8] Foyo-Moreno et al., 2014

A MULTIFUNCTIONAL LOW-COST SCALABLE FIELD MONITORING SYSTEM

C. Montes[1,2], O. González[1], G. Moncho[1], M. Padrón[1], J. Fernández[1], B. González-Díaz[2], C. Hernández-Rodríguez[3], J. Sanchiz[4], R. Guerrero-Lemus[3], J. Rodríguez[1], M. Friend[1] & M. Cendagorta[1].

[1]Instituto Tecnológico y de Energías Renovables, S. A. (ITER)
Pol. Industrial de Granadilla, s/n
38600 Granadilla de Abona, Spain.
[2]Departamento de Ingeniería Industrial.
[3]Departamento de Física and Instituto de Ciencia de Materiales.
[4]Departamento de Química and Instituto de Ciencia de Materiales.
Universidad de La Laguna.
Avenida Astrofísico Francisco Sánchez, s/n
38206 S/C de Tenerife. Spain

ABSTRACT: In this work, a new outdoor characterization system for photovoltaic solar modules is presented. The new low cost system has been designed to acquire the electrical parameters inside of each module through Modbus. Each module is connected to a supervisory system that collects and store the data. The supervisory system allows real-time control of the whole system. This characterization system has been fully designed, built and installed by the Instituto Tecnológico y de Energías Renovables S.A., in collaboration with the University of La Laguna.

Keywords: Characterisation, Energy Performance, Evaluation, Monitoring, Performance, Qualification and Testing

1 INTRODUCTION

The performance analysis and reliability of solar photovoltaic devices in outdoor conditions is an issue of vital importance in order to characterize the solar module parameters and to detect fabrication problems.

The analysis of the solar module electrical behavior is mainly carried out in solar simulators. The degradation analysis is usually evaluated in ageing chambers, where several cycles of UV light and moisture are applied.

According to the international standard IEC 60891, the electrical performance of the modules is translated to to the Standard Test Conditions (STC)

However, these techniques, even when procedures applied according to the normative, allow the characterization of the electrical performance, simulating outdoor conditions, are the best way to analyze the effect of several other possible exposure agents, such as wind impact or dust accumulation. The Instituto Tecnológico y the Energías Renovables, S.A. (ITER), together with the University of La Laguna (ULL), have developed a low cost new outdoor evaluation setup. The final system has been installed in ITER facilities, close to the Sahara Desert in South Tenerife (28º04'16.48"N 16º30'48.15"W), which provide environmental conditions that are unique in terms of very high sun irradiation, salinity, extreme thermal cycles, constant strong winds and atmospheric Saharan dust accumulations [1].

2 SENSOR PREPARATION

First, a preliminary batch of eight one-cell mini-modules was laminated at ITER's module fabrication facilities. Standard 156 mm x 156 mm multicrystalline silicon solar cells have been laminated in a conventional glass/EVA/TPT sandwich structure. Photovoltaic standard anodized aluminum profiles were used to frame each module. In order to preserve the integrity of the cell's circuit terminals and to provide accommodation for housing the electronic measurement components, large four-terminal class junction boxes were attached at their backside.

These mini-modules operate as the main sensors for the outdoor evaluation setup. As it can be observed in Fig. 1, the system is extremely versatile and can be modified on demand by simply laminating solar cells with different configuration, treatments or technology. The system has been designed to further extend its capabilities, in order to include studies of photon conversion or light concentration nature [2,3], adding films containing these materials on the glass of the mini-modules and/or on the encapsulants.

Figure 1: View of the one-cell modules after lamination (up-left), trimming operation (with exposed circuit terminals, up-right), frame (down-left) and junction box attachment (down-right).

3 SUPPORT STRUCTURE

To mount the mini-modules, a new support structure was developed, consisting of a 0.5 m stainless steel profile welded to a base plate, which can be fixed to the ground by using screws. Through inside of that pillar, a secondary pillar of 0.8m is introduced. This second pillar is made of anodized aluminum, which holds a tilting head with 10° steps, on which a flexible number of mini-modules can be installed.

Aluminum clamps fixed to aluminum beams of the required length allow different configuration in the solar modules mounting area (see Figure 2).

Finally, the height of the mounting structure can be changed from 0.8 up to 1.8 m by simply replacing the secondary pillar.

Figure 2: View of the support structure, showing the tilting head (up-left), exchangeable pillar (up-right) and mini-modules attached with clamps (down).

4 MEASURING DEVICES

A new kind of measuring device, consisting of an acquisition card, which can be assembled inside of the mini-module's junction box, in order to obtain better precision and to isolate the electronic device to other sources of external interferences, has been developed.

The sensors included in the electronic device acquire the electrical values. Also, every minute, an IV curve is acquired. The electronic device is implemented with a voltage regulator, controlled by a microcontroller, which also monitors the temperature radiated by the cell with infrared (IR) sensors. The data is acquired through 16-bit analog / digital converters and stored internally. Since the stored data is accessible via an RS485 interface by using Modbus RTU protocol, multiple cards of this kind can be connected in a linear, multidrop bus configuration (see Figure 3). Therefore, although the original configuration consisted of 8 mini-modules, the system is completely flexible in terms of the number of sensing units that can be connected.

Figure 3: View of a mini-module with the junction box opened showing the acquisition card assembled

5 ACQUISITION, STORAGE AND PRESENTATION OF THE DATA

A set of interface software which acquire, store and show on demand the data generated by the system has been developed. The setup for doing so can be summarized in the flowchart diagram shown at Figure 4.

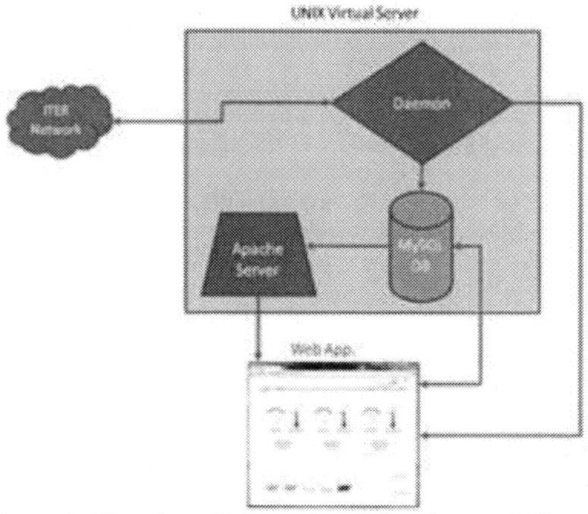

Figure 4: Flowchart diagram describing the acquisition, storage and presentation of the data.

As it can be observed, under a UNIX (Ubuntu) environment of a virtual dedicated server, an application developed in Java as a daemon operates by periodically interrogating the measurement system in real time via the MODBUS protocol. It also performs the necessary calculations and transfers them for their storage in a designed MySQL Data Base, as well as providing them to a web application (written in PHP and jquery) that is capable of displaying such data in different ways via a web server (Apache).

Although the system has been conceived for operating within ITER's supervisory control and data acquisition (SCADA) mainframe, it has been optimized for operating autonomously in lightweight computing hardware, such as single-board computers.

6 FIRST RESULTS AND DISCUSSION

After passing a test period, the system has been in operation since September 2016 at the ITER premises, to evaluate different cell technologies, as well as the effect on the power production of these cells resulting from trying different efficiency improvement strategies, such as applying down-converting films, as it can be observed in Figure 5.

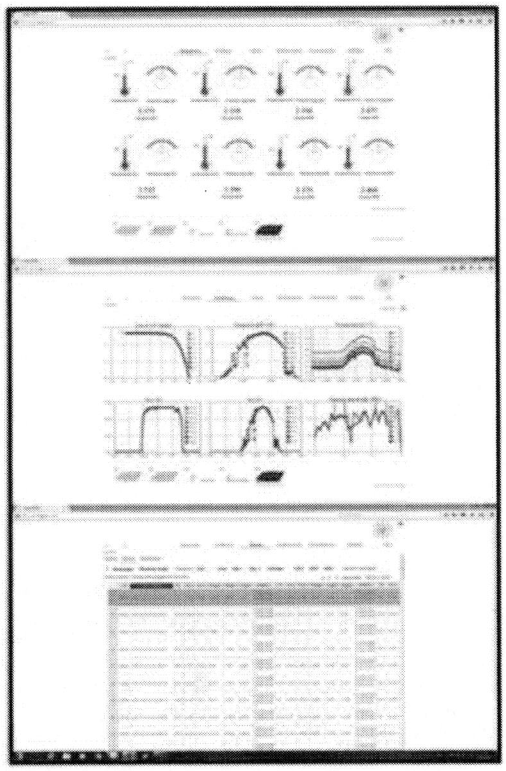

Figure 5: Screenshot captures of the web app. showing real time data acquired (up), processed graphically (middle) and raw data retrieval utilities (down) during the ongoing use of the system.

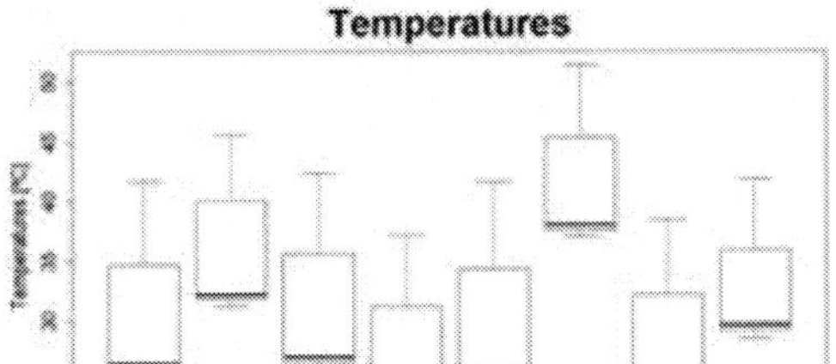

Figure 6: Evaluation of the Nominal Output Temperature (NOCT) from the eight panels under study.

The research results of the first test of minimodules with Down-Shifter layers have been recenty reported [4].

7 ACKNOWLEDGEMENTS

This work has been supported by the Ministerio de Ciencia e Innovación, Spain (Project ENE2013-41925R), co-supported by the European Social Fund and by the Fundación Cajacanarias (Project ENER10).

8 REFERENCES

[1] C. Montes, B. González-Díaz, A. Linares, E. Llarena, O. González, D. Molina, A. Pío, M. Friend, M. Cendagorta, J.P. Díaz and F.J. Expósito, "Effects of the Saharan dust in the performance of multi-MW PV grid connected facilities in the Canary Islands (Spain), 25th European Photovoltaic Solar Energy Conference and Exhibition, pp. 5046-5049, 6-10 September 2010, Valencia, Spain.

[2] T. Monzón-Hierro, J. Sanchiz, S. González-Pérez, B. González-Díaz, S. Holinski, D. Borchert, Cecilio Hernández-Rodríguez and Ricardo Guerrero Lemus, "A new cost-effective polymeric film containing an Eu(III) complex acting as UV protector and down-converter for Si-based solar cells and modules," Solar Energy Materials & Solar Cells vol. 136, pp. 187 - 192, 2015.

[3] S. González-Pérez, J. Sanchiz, B. González-Díaz, S. Holinski, D. Borchert and R. Guerrero-Lemus, "Luminescent polymeric film containing Eu(III) complex acting as UV protector and down-converter for Si-based solar cells and modules," Surface & Coatings Technology vol. 271, pp. 106-111, 2015.

[4] Benjamin Gonzalez-Diaz, Carlos Montes, Joaquin Sanchiz, Luis Ocana, Carlos Quinto, Oscar Gonzalez, Guillermo M. Bertolomeu, Miguel A. Padron-Gonzalez, Cecilio Hernandez-Rodriguez, Mari P. Friend, Manuel Cendagorta-Galarza, David Canadillas, Ricardo Guerrero-Lemus, "Evaluation Setup and First Power Production Results for Si Mini-Modules Covered by Eu3+-Based Down Converters in Extreme Outdoor Conditions", 2017 IEEE PVSC-44 | Washington, DC.

EVALUATION AND COMPARISONS OF THE MODELS TO CALCULATE SOLAR IRRADIATION ON INCLINED SOLAR PANELS FOR ANKARA

Talat Ozden[ad] , Abdullah Bugrahan Karaveli[bd], and Bulent G. Akinoglu[bcd]
[a]Electrical and Electronics Engineering, Gümüshane University, Gümüshane, 29100, Turkey
[b]Earth System Science Program, Graduate School of Natural and Applied Sciences, Middle East Technical University, 06800, Ankara, Turkey
[c]Department of Physics, Faculty of Arts and Sciences, Middle East Technical University, 06800, Ankara, Turkey
[d]Solar Energy Research Center (GUNAM), Middle East Technical University, 06800, Ankara, Turkey.

ABSTRACT: In the optimization and feasibility analysis of renewable energy systems, the accuracy of the simulation methods is extremely important. Thus, in the case of solar energy systems, estimating the monthly average daily solar irradiation incident on inclined solar panels is the main step of long-term feasibility analysis toward reaching the viable investments. In fact, most solar energy investor companies use silicon solar irradiation measuring instruments in Turkey although only a few of them also measure the solar irradiation using accurate instrumentation. Nevertheless, the need to estimate is still very essential and will continue to be so, mainly because of the fact that such on-land measurements are localized at single points on the Earth. In fact, the input, must be estimated or measured accurately to reach the efficiencies (output/input) of the systems. Consequently, this study analysis and compares the models to predict the solar irradiation on inclined surface in the most accurate way for Ankara. To this aim, the measurements have been taken since April 2016. Hourly isotropic model, two anisotropic models HDKR and Perez are used to estimate hourly solar irradiation on inclined surface and converted into monthly mean daily values. Meanwhile, isotropic sky model based on monthly average daily values is also used in the calculations, by utilizing a recent methodology developed by the authors to predict the diffuse component of solar irradiation falling on a horizontal surface. The analysis is carried out using the measurements of two Kipp and Zonen (CMP11) thermopile pyranometers and a new generation silicon pyranometer. The results show that HDKR is the best method. Another important result shows that the measurements of new generation silicon pyranometer has still considerable errors.
Keywords: Solar radiation, Pyranometer, Prediction of solar irradiation

1 INTRODUCTION

Accurate estimation of the input solar irradiation incident on solar energy systems must be used to reach their short and long term performances. Solar irradiation on tilted panels can be calculated using measured/estimated solar irradiation data on horizontal surface [1]. The calculation procedure mainly based on dividing the global solar irradiation values into its two components, namely beam and diffuse. There are various methodologies for this calculation and three of them are outlined in Duffie and Beckman [1].

This study aims to find out the best methodology to calculate solar irradiation falling on a tilted surface of solar panels for Ankara. In addition, we demonstrate the error that may be introduced in predicting the monthly mean **daily** values on tilted solar panels $\bar{H}_T$ using Liu and Jordan's [2] isotropic monthly average daily solar irradiation model directly rather than carrying out the calculations by well-known isotropic and two anisotropic sky **hourly** models of the literature. In the long term feasibility calculations of solar systems (for the last decade PV dominates all the other), monthly mean daily solar irradiation values are quite frequently utilized data [3] and thus the accuracy of these models are checked using measured and predicted monthly mean daily values $\bar{H}_T$ on tilted solar panels.

In the calculations of hourly diffuse solar irradiation on horizontal surface prior to reach the solar irradiation on tilted surface we used the well-known expression of Erbs et al. [1,4]. Then the monthly average solar irradiation on tilted surface $\bar{H}_T$ is calculated just by summing and averaging the hourly results.

There are mainly two types of hourly models to estimate the solar irradiation on inclined surfaces: isotropic and anisotropic sky models. A recent work [5] concluded that one of the anisotropic models – the Hay, Davies, Klucher, Reindl model (HDKR) – is better than another and also better than isotropic model [2] as given in Duffie and Beckman. In the present study, we used this isotropic model and two anisotropic models, namely HDKR and Perez models [1].

Lastly, we calculated the monthly average daily solar irradiation on tilted surface directly by the estimated/measured monthly mean daily diffuse component of solar irradiation ($\bar{H}_D$) on horizontal surface. In these calculations, to reach the diffuse component, we used a methodology based on a physical modeling approach [6] which has the best performance for Turkey as determined in our recent studies [7,8]. This methodology is monthly based and the coefficients used in the predictions are calculated for each month. The diffuse data taken by State Meteorological Service (which is 20 km away from the outdoor PV testing facility of GUNAM) is used to determine the monthly coefficients to be utilized in the prediction of $\bar{H}_D$.

Section 2 gives the materials and methods that we used. In Section 3 we present and discuss the results. Finally, last section is a summary and conclusions of our findings.

2 MATERIALS AND METHODS

2.1 Measured Data

The data that we measured is as the average of too many readings recorded at 10 minutes' intervals. Using this data, we calculated the hourly values and simply by summing we reached the daily values. Monthly averages of daily values are then calculated and utilized. Hourly data of total horizontal solar irradiation on horizontal surface (I) and on tilted surface (I_t) have been taken by using Kipp and Zonen pyranometer (CMP11) and Davis silicon pyranometer since April of 2016.

Figure 1: Tilted and Horizontal Pyranometers

In the present study we used one-year data between April, 2016 to March 2017. One of the Kipp and Zonen pyranometers was placed horizontally while the other was placed tilted at an angle of 32° (Fig. 1). The measuring systems are installed to the roof of Department of Physics within METU campus area. The site is a typical campus having moderate level buildings, the lands have parking lots together with 20-40 years old pine trees. The altitude is about 900 m. The climate is semi-arid continental.

2.2 Methods

As mentioned above, short and long term performances and feasibility studies of solar energy systems (essentially PV systems) are extremely important concepts toward the integration of the renewables. Accurate determination of the input to the systems, mainly the solar irradiation on tilted arrays is a must to reach truthful and feasible investments. Present study is to reach accurate determination of this input, which will be used in a software SOlAR TURnKEY: a software that is at the developing stage by the authors for techno-economic feasibilities and dynamic performances of the solar electricity investments to be installed in Turkey. Therefore, the accuracy that we can achieve in determining the input to the PV systems will essentially determine the performance (and the success) of the produced software.

The model of Erbs et al [4] that we used for the calculations of the **hourly diffuse** component is as follows:

$$\frac{I_d}{I} = \begin{cases} 1.0 - 0.09k_t & for\ k_T \leq 0.22 \\ 0.9511 - 0.1604k_t + 4.388k_t^2 - 16.638k_t^3 + 12.336k_t^4 & for\ 0.22 < k_T \leq 0.80 \\ 0.165 & for\ k_T > 0.80 \end{cases} \quad (1)$$

The first model that we used for the hourly solar irradiation on tilted surfaces is the isotropic sky model of Liu and Jordan [9]:

$$I_t = I_b R_b + I_d\left(\frac{1+\cos\beta}{2}\right) + I\rho_g\left(\frac{1+\cos\beta}{2}\right) \quad (2)$$

where I_b, I_d and I $(I = I_b + I_d)$ are hourly beam, diffuse and global solar irradiation on horizontal surface, respectively. R_b is the ratio of beam radiation on the tilted surface to that on the horizontal surface. β and ρ_g are tilt angle and ground albedo, respectively. We used a value of 0.20 for ρ_g, which is a typical average value [10].

The first **hourly** anisotropic sky model that we used was proposed by Hay and Davies [11]. It is later modified by Klucher and Reindl et al. [12], we refer this model as HDKR as named in Duffie and Beckman [1]:

$$I_t = (I_b + I_d A_i)R_b + I_d(1 - A_i)\left(\frac{1+\cos\beta}{2}\right)\left[1 + f\sin^3\frac{\beta}{2}\right] + I\rho_g\left(\frac{1-\cos\beta}{2}\right). \quad (3)$$

In this equation, A_i is an anisotropy index and f is a modulation factor.

Another anisotropic sky mode was proposed by Perez et al. [13]. They developed an expression which allows the calculation of four diffuse components: isotropic diffuse, the circumsolar diffuse, the horizon brightening and the ground reflected. The model was given as:

$$I_t = I_b R_b + I_d(1 - F_1)\left(\frac{1+\cos\beta}{2}\right) + I_d F_1\frac{a}{b} + I_d F_2 \sin\beta + I\rho_g\left(\frac{1-\cos\beta}{2}\right). \quad (4)$$

In this equation F_1 and F_2 are circumsolar and horizon brightening coefficients. a and b are terms to account for the cone of circumsolar radiation.

The last equation that we used is the **monthly** average daily model for tilted surfaces developed by Lui and Jordon [14] and later extended by Klein [15] as:

$$\bar{H}_t = \bar{H}_b\bar{R}_b + \bar{H}_d\left(\frac{1+\cos\beta}{2}\right) + \bar{H}\rho_g\left(\frac{1-\cos\beta}{2}\right) \quad (5)$$

The details on how to use these models to calculate the solar irradiation on the tilted surfaces are outlined in Duffie and Beckman [1]. In the estimation of monthly average **daily diffuse** component on horizontal surface is carried out using the aforementioned procedure that we developed based on a physical approach [6] and that we concluded to be the best [7].

3 RESULTS AND DISCUSSION

We carry out the same analysis both for the measured data with accurate thermopile pyranometers and the data taken by second generation silicon pyranometer. In the calculation of the levelized cost of energy (LCOE) and in reaching the life cycle assessment of PV systems, even small inaccuracies may lead to misleading conclusions [3]. Thus, it is extremely important to increase the accuracy of both measurements and the estimation procedures.

We carry out the same analysis both for the measured data with accurate thermopile pyranometers and the data taken by silicon pyranometer to determine the measurement accuracy of silicon pyronometer. The measured values for the 12 months starting from April 2016 till March 2017 are given in Table I as monthly mean daily values $\bar{H}$ and $\bar{H}_t$. Predicted values using the aforementioned methods are also given both for measurements of Kipp and Zonen and silicon pyranometer. The letters stand to indicate the following methodologies: classical **hourly** isotropic model is A; HDKR is B; Perez Model is C. The last column D is the isotropic **monthly average daily** model [1].

The measured values for the 12 months starting from April 2016 till March 2017 are given in Table I as monthly mean daily values $\bar{H}$ and $\bar{H}_t$. For the sake of getting a complete one year we extended our analysis to include 3 months from 2017. In Table I, predicted values using the aforementioned methods are also given both for measurements of Kipp and Zonen and silicon pyranometer. The letters stand to indicate the following methodologies: classical hourly isotropic model is A; HDKR is B; Perez Model is C. The last column D is the isotropic monthly average daily model [1].

Table I: Measured and calculated solar irradiation values

	Pyranometer (MJ m^{-2} d^{-1})						Silicon (MJ m^{-2} d^{-1})				
	Measured		Calculated $\bar{H}_t$				Measured	Calculated			
Months	$\bar{H}$	$\bar{H}_t$	A	B	C	D	$\bar{H}$	A	B	C	D
4,2016	21,67	23,33	22,29	23,22	23,04	21,75	20,81	21,33	22,27	22,09	20,95
5	21,21	20,71	20,04	20,79	20,46	19,49	20,45	19,23	19,98	19,63	18,78
6	26,23	26,07	23,89	24,45	24,28	22,53	25,63	23,04	23,63	23,44	21,97
7	27,39	28,84	25,63	26,26	26,22	24,18	27,40	25,23	25,80	25,82	24,19
8	23,91	24,92	23,69	24,50	24,39	23,55	23,25	22,99	23,80	23,68	22,93
9	14,38	16,55	15,83	16,68	16,56	17,43	18,70	20,98	22,17	22,10	21,49
10	12,20	16,07	15,55	16,69	16,50	16,06	12,79	15,98	17,18	17,03	16,62
11	8,84	13,58	12,32	13,39	13,25	12,34	8,18	11,30	12,38	12,30	11,72
12	6,00	9,43	8,03	8,72	8,60	8,04	5,48	7,65	8,48	8,39	7,55
1,2017	5,97	7,97	7,73	8,41	8,18	7,81	5,87	7,51	8,25	8,12	7,72
2	11,07	15,14	15,07	16,36	15,95	13,68	10,47	13,75	14,93	14,79	13,11
3	14,28	16,94	16,36	17,39	17,14	15,92	13,82	16,00	17,02	16,82	15,48

Table II and III give the results of the comparisons by two statistical errors: mean bias error (MBE) and root mean square error (RMSE), for the measurements of Kipp and Zonen and silicon pyranometer, respectively.

Table II: MBE and RMSE comparison of the models with respect to measurements of Kipp and Zonen

Methods	A		B		C		D	
Error	(MBE)	(RMSE)	(MBE)	(RMSE)	(MBE)	(RMSE)	(MBE)	(RMSE)
Error values	-1.09	1.38	-0.22	1.01	-0.42	1.01	-1.40	1.99

As seen from Table II, applying hourly calculations schemes for anisotropic sky model HDKR performs better (column B) when compared to the other hourly models and compared to the monthly mean daily model (D). Information about the difference in the accuracy of using hourly model HDKR and the isotropic daily can be deduced from Table II.

Table III: MBE and RMSE comparison of the models with respect to measurements of silicon pyranometer

Methods	A		B		C		D	
Error	(MBE)	(RMSE)	(MBE)	(RMSE)	(MBE)	(RMSE)	(MBE)	(RMSE)
Error values	-1,21	2,30	-0,31	2,11	-0,45	2.14	-1,42	2,74

Table III clearly reveals that with new generation silicon pyranometer the accuracy is considerably reduced. Further comparison to reveal the range of errors introduced in using the silicon pyranometers should be conducted. Continuing these measurements/analyses for other locations in Turkey and reaching the best estimation schemes for solar irradiation on inclined solar panels for different climates are our future prospects.

4 CONCLUSION

Present study is the evaluation of different inclined solar irradiation estimation procedures to conclude on their performances. The comparisons are carried out using one-year data recorded in Ankara, GUNAM's outdoor testing facility. Two Kipp and Zonen thermopile pyranometer and a silicon pyranometer are used in the evaluations. One of the thermopile pyranometer were horizontal and the other is tilted by an angle of 32° due South. The method developed by Hay and Davis model [11] which was later modified by Kulcher [16] and Reindl [12] (HDKR) were the best in 4 different methodologies.

The error introduced in using the measured horizontal solar irradiation values with the measurements of silicon pyranometer to estimate the monthly averages of daily solar irradiation on inclined surface was significantly large. It can reach up to even above 30% (Table I, 9th month) which is not tolerable to use in the feasibility studies of solar energy applications. Hence, except well-constructed and calibrated silicon pyranometers satisfying some standards, use of second generation silicon pyranometers should not be recommended in scientific researches. Nevertheless, it is within the knowledge of the authors that the second generation silicon photodetectors are still at their developing stage.

REFERENCES

[1] Duffie JA, Beckman WA., Solar engineering of thermal processes, 4th ed. New York: John Wiley and Sons; (2013).

[2] Liu BYH, Jordan RC., The long term average performance of flat plate solar energy collectors, Sol Energy 7 (1963) 53.

[3] Karaveli AB, Soytas U, Akinoglu BG., Comparison

of large scale solar PV (photovoltaic) and nuclear power plant investments in an emerging market, Energy 84 (2015) 656.
[4] Erbs DG, Klein SA, Duffie JA., Estimation of the diffuse radiation fraction for hourly, daily and monthly-average global radiation, Solar Energy 28 (1982) 293.
[5] Ozden T, Akinoglu BG, Turan R., Long term outdoor performances of three different on-grid PV arrays in central Anatolia – An extended analysis, Renew Energy 101 (2017) 182.
[6] Akınoğlu BG., Recent Advances in the Relations between Bright Sunshine Hours and Solar Irradiation, In: Badescu V, editor. Model. Sol. Radiat. Earth's Surf., Springer Berlin Heidelberg; (2008) 115.
[7] Karaveli AB, Akinoglu BG., Comparison of quadratic and linear models on the estimation of diffuse solar radiation for loction in Turkey, 9th Int. Conf. Sustain. Energy Environ. Prot. - SEEP2016, Kayseri (2016) 103.
[8] Karaveli AB, Akinoglu BG., Estimation of diffuse component for two locations in Turkey, Int. Conferance Sol. Energy Build. Ind. - Eurosun2016 (2016) 1467.
[9] Liu BYH, Jordan RC., The interrelationship and characteristic distribution of direct, diffuse and total solar radiation, Solar Energy 4 (1960) 1.
[10] Ineichen P, Guisan O, Perez R., Ground-reflected radiation and albedo. Solar Energy 44 (1990) 207.
[11] Davies JA, Hay JE., Calculation of the solar radiation incident on an inclined surface, Proc. First Can. Sol. Radiat. Data Work. (1980) 59.
[12] Reindl DT, Beckman WA, Duffie JA., Evaluation of hourly tilted surface radiation models, Solar Energy 45 (1990) 9.
[13] Perez R, Ineichen P, Seals R, Michalsky J, Stewart R., Modeling daylight availability and irradiance components from direct and global irradiance, Solar Energy 44 (1990) 271.
[14] Liu B, Jordan R., Daily insolation on surfaces tilted towards equator, ASHRAE J;(United States) (1961) 10.
[15] Klein SA., Calculation of monthly average insolation on tilted surfaces. Solar Energy 19 (1977) 325.
[16] Klucher TM., Evaluation of models to predict insolation on tilted surfaces, Solar Energy 23 (1979) 111.

A REVIEW OF DAILY GLOBAL SOLAR RADIATION MODELLING USING DIFFERENT STATISTICAL METHODS BASED ON SUNSHINE DURATION IN GRAN CANARIA ISLAND

F. Díaz, L. Mazorra Aguiar, F. Déniz*
University Institute for Intelligent Systems and Numerical Applications in Engineering.
University of Las Palmas de Gran Canaria, Edificio Central del Parque Tecnológico
* Electrical Department, University of Las Palmas de Gran Canaria
Campus Univ. Tafira +34928451989 - e-mail: felipe.diaz@ulpgc.es

ABSTRACT: Recently, solar photovoltaic and thermo electric systems have become one of the most important renewable energies. Global solar irradiation (GSR) is a very important variable for developing solar energy production in a region. GSR description and modeling requires solar radiation measurements in a wide range of location around this region. The high costs of solar radiation measurement stations avoid its installation in big areas around the world. Since years, many different GSR estimation models have been developed using several meteorological variables. Many papers can be found using sunshine duration, temperature, relative humidity and air pressure, which in fact are variables easier measured. In this paper, different Ångström models, based on sunshine duration, have been tested using measurement stations around Gran Canaria Island following previous works. The Island presents very different climatological behavior depending on the location because of its complicated orography and the influence of Trade Winds. In order to compare classical Angström models results, Artificial Neural Networks are proposed for modeling daily GSR using the day of the year and the sunshine duration as inputs. It is observed that for all measurement stations, neural networks outperform classical models using different network architectures.
Keywords: Solar Radiation, Sunshine duration, Neural networks, Ångström models

1 INTRODUCTION

Global Solar Radiation (GSR) is one the most important variable for estimating solar energy production in a region from photovoltaic and thermo electric systems. The percentage of solar energy generate in a region depends on the description of the solar radiation resource. GSR description and modelling requires solar radiation measurements in a wide range of location around this region. The cost of solar radiation measurements equipment may lead to a lack of stations in several areas around the world. However, different meteorological variables are easier to be measured and have been used for GSR modelling. Many papers can be found using sunshine duration, temperature, relative humidity and air pressure. In the present paper, the available data come from five measurement stations throughout the island. We establish a comparison between different classical Ångström models, based on sunshine duration following previous works, and an Artificial Neural Networks (ANN) proposed for modelling daily GSR using the day of the year and the sunshine duration as inputs.

This paper studies the capability of Artificial Neural Networks (ANN) for modelling daily GSR in regions with a complicated orography. During this work, several tests were estimated in order to find the optimal architecture for modelling GSR. Database was split into training and testing sets from measurement data obtained in different stations around the Island. In this case, we have focused on a common ANN architecture called Multilayer Perceptron (MLP) with only one hidden layer. MLP establish a relation between both input and output pairs of datasets. All neurons are connected by a few adjustable parameters, called weights, and affected by a hyperbolic tangent function. During the training process, these weights are optimized by minimizing the error between ANN output and measure data using an iterative scale conjugate gradient method.

Data used during this study were obtained in five measurement stations around Gran Canaria island kindly provide by Instituto Tecnológico de Canarias, I.T.C.: C0-Pozo Izquierdo (La: 27.8175N; Lo:15.4244W; h: 47m), C1-Las Palmas (La: 28.1108N; Lo: 15.4169W; h: 17m), C2-La Aldea (La: 27.9901N; Lo: 15.7907W; h: 197m), C5-Sta Brígida (La: 28.0337N; Lo: 15.4991W; h: 525m) and C6-Mogán (La: 27.8839N; Lo: 15.7216W; h: 300m).

2 SOLAR RADIATION MODEL DESCRIPTION

2.1 Ångström regression models

Ångström [1] first propose a relation between solar radiation and sunshine duration by simple linear model. The variables used in this model are the ratio between average daily global radiation and the corresponding value on a completely clear day and the ratio between average daily sunshine duration and the maximum sunshine duration. Later, Prescott proposed a model based on the top the atmosphere radiation instead of clear sky global irradiance [2] (1).

$$\frac{H}{H_0} = a + b * \frac{n}{N} \quad (1)$$

Where, H is daily GSR, H_0 is solar radiation on the top of the atmosphere at same location, n is measured sunshine duration and N is the maximum sunshine duration. While (a,b) are Ångström parameters, which are obtained through minimum square regression done with the available measurement stations data. Although linear model obtained quite good results taken into account the results error, it led us to negative daily GSR results with no physical meaning [3] [4] in Gran Canaria Island [5]. In the bibliography can be found different Ångström model variations [6] in order to outperform linear model physical meaning. In this paper, we have estimate daily GSR from sunshine duration using a linear model (1), Quadratic model (2), Third degree model (3), Logarithmic model (4) and an Exponential model (5).

$$\frac{H}{H_0} = a + b \cdot \frac{n}{N} + c \cdot \left(\frac{n}{N}\right)^2 \quad [7]\ (2)$$

$$\frac{H}{H_0} = a + b \cdot \frac{n}{N} + c \cdot \left(\frac{n}{N}\right)^2 + d \cdot \left(\frac{n}{N}\right)^3 \quad [8]\ (3)$$

$$\frac{H}{H_0} = a + b \cdot \ln\left(\frac{n}{N}\right) \quad [9]\ (4)$$

$$\frac{H}{H_0} = a \cdot e^{b \cdot \frac{n}{N}} \quad [5]\ (5)$$

In addition, in a previous work [5], it was proposed a compound model based on both exponential and linear ones. In this work, we compared these different models for modelling GSR using only sunshine duration for the same day. Linear model obtained good results for daily GSR data but this model leads to negative values for low sunshine duration data. Exponential model gave very good results and allows us to reproduce positive GSR results for the whole range of sunshine duration values.

2.2 Artificial Neural Networks

We used Artificial Neural Networks (ANNs) in this paper to improve solar radiation modelling using sunshine duration as input. ANNs are made of individual units called neurons and these neurons are connected by weights. One of the most common ANN architecture is the Multilayer Perceptron. In this paper we used one input layer, one hidden layer with no linear neurons and an output layer with no feedback connection between each layer. The general equation for the ANNs is as in (6). Firstly, weights are initialized randomly and optimized with a cost function during the training process. The average of square difference between estimated output and the measured data is the most common function (7).

$$y = y(x;w) = \sum_{j=0}^{h}\left[w_j f\left(\sum_{i=0}^{d} w_{ji} \cdot x_i\right)\right] \quad (6)$$

$$E(w) = \frac{1}{2}\sum_{i=1}^{N}(y_i - t_i)^2 \quad (7)$$

ANN architecture is one is one the most important issues to improved classical model results with ANNs. In this case we decided the number of hidden neurons and the optimal number of inputs. Networks with a high number of hidden units or inputs could obtain very good results with the training datasets but poor results when we used new testing dataset (overfitting problem). In the bibliography there are different regularization methods used to avoid overfitting and estimated the optimal ANN [10, 11]. During this work we used Bayesian regularization framework to control the architecture of the model [11]. To control the complexity Bayesian framework introduces different parameters, called hyperparameters, to the cost function (8). In Fig. 1 it is easily observed that Bayesian Framework control overfitting because training and testing datasets show similar model results.

$$S(w) = \frac{\beta}{2}\sum_{i=1}^{N}(y_i - t_i)^2 + \frac{\alpha}{2}\sum_{j=1}^{m} w_j^2 \quad (8)$$

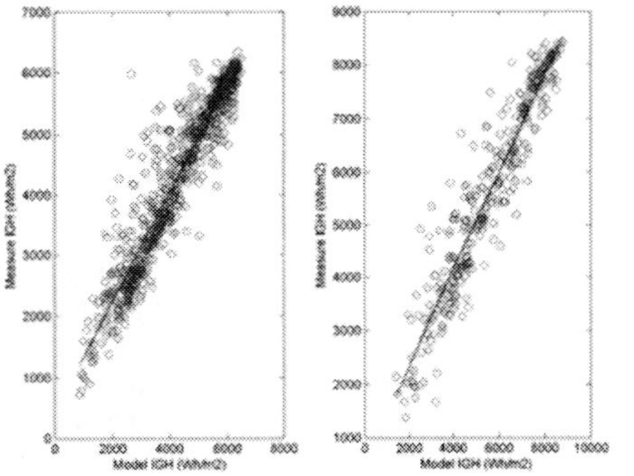

Figure 1.- ANN modelling with sunshine duration for C0 station data. Left: training set. Right: test set

The Bayesian framework gives us information for deciding the optimal model architecture. Automatic Relevance Determination (ARD) technique offers the possibility of determining the most important inputs during the training process assigning different regularization coefficient to each input. Moreover Bayesian Framework permits us to choose the number of hidden units by computing the probability of each model (evidence of the model). During the training process we calculated the model using different number of hidden units and the evidence of the model gives the optimal model architecture.

3 MODEL IMPLEMENTATION

To estimate de statistical models propose in this paper we should split first database in two different set, training and testing data. For all methods we establish the parameters and the architecture of the model using a learning process with the training dataset. The estimated model obtain during this process is then analyzed using new data for the model, testing dataset.

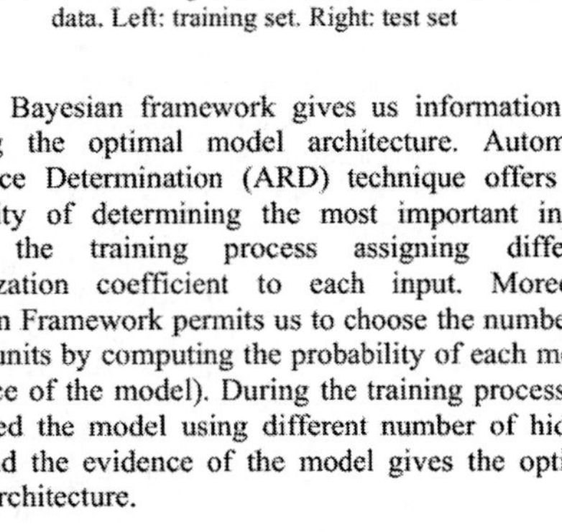

Figure 2.- Classical models daily GSR estimation.

Classical Ångström models use sunshine duration as input to obtain daily global solar radiation in each location. The model parameters are calculated using training dataset through minimum square regression. The composed exponential-linear model estimated model parameters using the same technique. In this case, for days that present lower sunshine duration data the GSR is estimated with exponential model. While for higher data we used linear model. During the training process it was established the optimal range of sunshine duration data for both methods. As seen in Fig. 2, all methods obtain good results. However, linear an logarithmic models would obtain negative GSR for lower sunshine data. In general, exponential-linear model outperforms classical models with an insignificant error difference.

The first advantage of using ANN is that it is possible to use different inputs at the same time for the training process while with the classical models we only used sunshine duration data for the same day. In this paper, we proposed only working with the number of the day and the sunshine duration for the same day and several days before as inputs for the ANN. During the training process, using the evidence of the model calculated with Bayesian Framework, we estimated the number of previous sunshine duration data it is possible to use to improve the model. So, input layer is composed by sunshine duration data from same and previous days and the output layer contains only the daily GSR for this day. During this work we compute different ANN architecture using different numbers of inputs in order to establish the optimal one. Specifically, we compute model results using the number of the day (d) and sunshine duration at same day (ANN 1Kn), sunshine duration from same day and one day before (ANN 2Kn) and sunshine duration from two days before (ANN 3Kn)..

During the training Bayesian framework also permits us to estimate the optimal number of hidden units for each case of different number of inputs. Finally, the best model was obtained using the number of the day, sunshine duration of same day and sunshine duration of one previous day. Using more number of previous sunshine duration data makes the calculation more complicated and increases the computational cost while not improving the results obtained previously. Fig. 3 shows the comparison between measured data and neural network optimal approach with two sunshine duration data (ANN 2Kn).

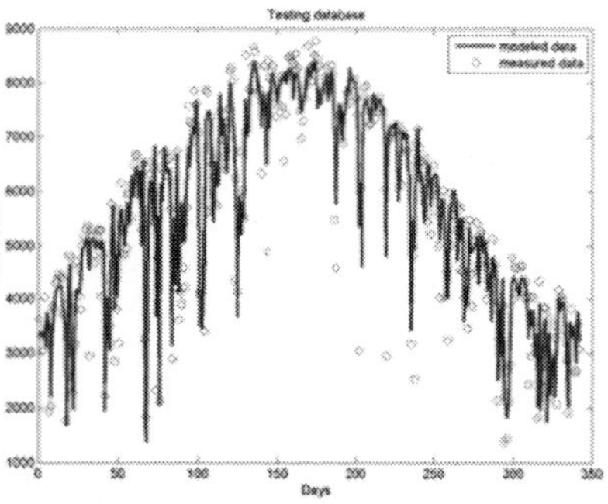

Figure 3.- Measured and modelled data using ANN for C0 station data.

4 SUMMARY AND CONCLUSIONS

As it is shown in Fig. 2 and Fig. 3 for Pozo Izquierdo's measurement station, classical Ångström models and ANNs fit measured data using only sunshine duration. MLP outperforms the rest of classical Ångström methods using sunshine duration data from the same and previous days, see Table I. The relative root mean square errors (%rRMSE) between daily GSR data obtained with ANN once it was trained and measured data are around 10% for southern stations and 15-20% for northern region. MLP outperforms classical Ångström around 1% in terms of %rRMSE for all the stations, except for C1-Las Palmas (in the North) where we can see more than 3%. The north of the island is affected by the Trade winds, which generate high cloudiness when they encounter the mountains.

Table I: Errors for three regression models using relative root mean square error (%RMSE)

% rRMSE	Linear	Exp.	ANN 2Kn
C0 Pozo Izquierdo	11.72	11.18	10.72
C1 Las Palmas	23.97	21.90	17.17
C2 La Aldea	12.57	12.64	11.88
C5 Sta. Brígida	17.62	17.48	16.93
C6 Mogán	14.08	14.18	12.92

5 ACKNOWLEDGMENTS

This work has been supported by the "*Convocatoria 2014 – Proyectos I+D+I – Programa Estatal de Investigación, Desarrollo e Innovación orientada a los retos de la sociedad*" of the Spanish Government, grant contracts CTM2014-55014-C3-1-R, as part of the project *"Integración de nuevas metodologías en simulación de campos de vientos, radiación solar y calidad del aire"*..

6 REFERENCES

[1] A. Ångström, «Solar and terrestrial radiation. Report to the international commission for solar research on actinometric investigations of solar and atmospheric radiation,» *Quarterly Journal of the Royal Meteorological Society,* vol. 50, nº 210, pp. 121-126, 1924.

[2] J. Prescott, "Evaporation from a water surface in relation to solar radiation," *Trans R Soc Sci Austr,* vol. 64, pp. 114-25, 1940.

[3] K. Hinrichsen, «The Angström formula with coefficients having a physical meaning,» *Solar Energy,* vol. 52, nº 6, pp. 491-495, 1994.

[4] A. Sahin and Z. Sen, "Statistical analysis of the Angström formula coefficients and application for Turkey," *Solar Energy,* vol. 62, no. 1, pp. 29-38, 1998.

[5] L. Mazorra Aguiar and F. Díaz, «Daily global solar radiation modeling for Gran Canaria Island,» *ENERGY SOURCES, PART A: RECOVERY, UTILIZATION, AND ENVIRONMENTAL EFFECTS,* vol. 38, nº 24, p. 3557–3564, 2016.

[6] J. Almorox and C. Hontoria, "Global solar radiation estimation using sunshine duration in Spain.," *Energy Conversion and Managment,* vol. 49, no. 9, pp. 1529-1535, 2004.

[7] B. Akinoglu and A. Ecevit, "Construction of a quadratic model using modified Angstrom coefficients to estimate global solar radiation," *Solar Energy,* vol. 45, pp. 85-92, 1990.

[8] C. Erketin and O. Yaldiz, "Comparison of some

existing models for estimating global solar radiation for Antalya (Turkey)," *Energy Conversion and Managment,* vol. 41, no. 4, pp. 311-330, 2000.

[9] D. Ampratwum and A. Dorvlo, "Estimation of solar radiation from the number of sunshine hours," *Applied Energy,* vol. 63, no. 3, pp. 161-167, 1999.

[10] C. Bishop, Neural Networks for Pattern Recognition, Oxford: Oxford University Press, 1995.

[11] P. Lauret, E. Fock, R. Randrianarivony and J. Manicom-Ramsamy, "Bayesian neural network approach to short time load forecasting," *Energy Conversion & Managment,* vol. 49, no. 5, pp. 1156-1166, 2008.

SOLAR ENERGY RESOURCE IN NEW ZEALAND

J. Ben Liley
National Institute of Water & Atmospheric Research (NIWA), Lauder
P B 50061 Omakau, 9352 New Zealand

ABSTRACT: To assist designers of solar energy systems and energy-efficient homes, researchers, and homeowners, NIWA has developed datasets, algorithms and web tools to provide information on available solar energy at any site in New Zealand. The NIWA Climate Database contains measurements, from over 100 sites, of global irradiance – the energy flux onto a horizontal surface. The data, at hourly or, for some stations, ten-minute intervals, are provided free alongside other climate data. We have used them to map solar energy flux nationwide, using satellite image analyses to allow for the effects of complex topography. To estimate solar energy flux onto or through surfaces of any orientation, we infer the diffuse and direct components of global irradiance. A web tool to deliver this information accounts for horizon shading and allows correction for local obstructions. Further refinement of information delivery is in development with a smartphone/tablet app, while use of satellite data with high-resolution reanalysis should give better initial datasets.
Keywords: Solar Radiation, System Performance, Resource Assessment

1 GLOBAL IRRADIANCE

1.1 NZ climate data

The New Zealand National Climate Database holds hourly data on global irradiance, ***G***, for over 200 climate stations around New Zealand. Of those stations, 181 are open (Fig. 1), and 100 have at least 10 years of data. Most sites have Licor LI200 silicon sensors, but 16 of them have Eppley PSP instruments. The former integrate only the 400-1100 nm range, rather than the full 300-3000 nm shortwave region, so the results are more variable under changing spectral composition. Appropriately scaled, they are nevertheless adequate for use in solar resource assessment.

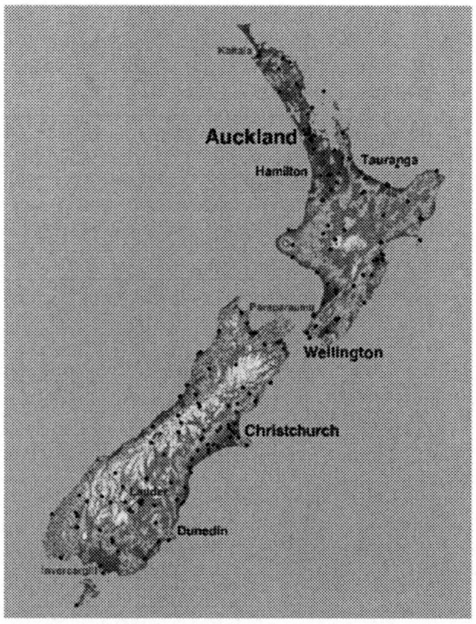

Figure 1: New Zealand sites measuring global irradiance (black dots, open; grey, closed) and global-diffuse-direct (red circles and labels), with cities of over 100,000 people labelled in black. Lauder (blue) is the best-instrumented radiation site in the southern hemisphere.

1.2 Mapping irradiance

New Zealand's mountainous topography straddling southern mid-latitudes creates complex meteorology and climatology. For example, the Central Otago region around Lauder is semi-arid, but 100 km west in Fiordland annual rainfall is around 6 metres. Spatial interpolation of just the measured irradiance would be misleading.

Analysis of five years' Advanced Very High Resolution Radiometer (AVHRR) radiance and texture data with a cloud classifying algorithm [1] has helped to elucidate the effects. Values of ***G*** from pyranometers are expressed as a fraction of clear-sky value, G_c, to quantify the radiative effect of cloud. They are supplemented by estimates from sunshine recorders via an Ångström relationship, and the result interpolated by thin-plate splines using average clear-sky frequency from the AVHRR analysis to estimate ***G*** at interstitial locations.

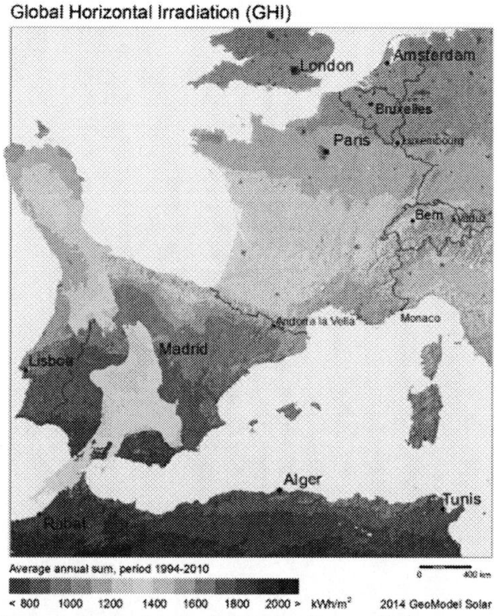

Figure 2: Interpolated global irradiance NZ compared with values for Europe from http://solargis.info.

1.3 Comparison with Europe

In Fig. 2, European data from SolarGIS are overlaid with New Zealand global irradiance, in the same units, projected through the earth. Annual average irradiance, especially in the North Island, is noticeably lower than at its antipodes in Spain.

The native Māori called the country *Aotearoa*, "Land of the Long White Cloud", which description accounts for lower average irradiance than at its antipodes in Spain and Portugal. The effect arises mostly from scattered or broken cloud, with high peak values due to low aerosol optical depth [2] and the orbital perihelion occurring in mid-summer. The same factors, plus lower ozone column amounts, make peak UV radiation in NZ up to 40% higher than at equivalent latitudes in both Europe and North America [3, 4].

2 DIRECT AND DIFFUSE

2.1 NZ data

As indicated in Fig. 1, four sites in NZ measure the separate direct normal incident flux, $\boldsymbol{R}$, and the diffuse irradiance, $\boldsymbol{F}$, as well as $\boldsymbol{G}$. Kaitaia, Paraparaumu, and Invercargill, labelled in red, are part of the NZ Climate Network. Lauder, at 45° S, is an atmospheric research station and one of 68 sites in the international Baseline Surface Radiation Network (BSRN).

The BSRN sites measure $\boldsymbol{G}$, $\boldsymbol{F}$, and $\boldsymbol{R}$ at high (1-3 minute) time resolution to the best available precision and consistency, in order to accurately quantify and monitor Earth's radiative balance. They provide data for satellite ground truth and validation of climate models. For greatest accuracy, the primary measure of $\boldsymbol{G}$ is $\boldsymbol{G_0} = \boldsymbol{F} + \boldsymbol{R}\cdot\cos\boldsymbol{\theta}$, where $\boldsymbol{\theta}$ is the solar zenith angle.

2.2 Inferring direct and diffuse

Analysis of hourly or 10-minute network and Lauder BSRN data, using the $\boldsymbol{G/G_c}$ fraction used for spatial interpolation, shows (Fig. 3) that in clean (low aerosol) conditions, $\boldsymbol{R}$ and $\boldsymbol{F}$ can be inferred from $\boldsymbol{G}$ with good precision. The correlation coefficients between inferred and measured hourly $\boldsymbol{R}$ values is 0.93, and it is 0.65 for $\boldsymbol{F}$.

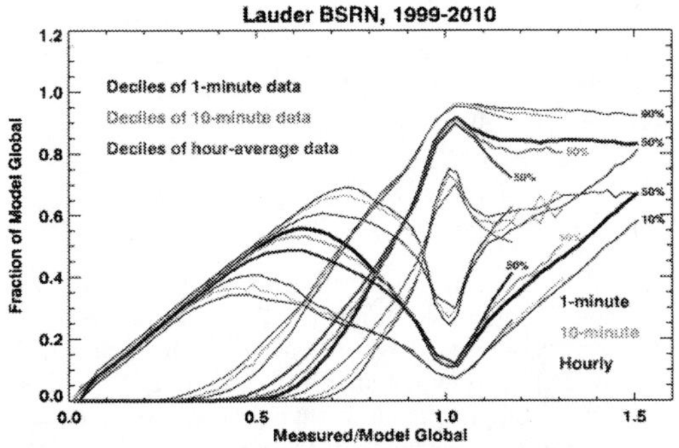

Figure 3: Deciles of $\boldsymbol{R}\cos\boldsymbol{\theta}/\boldsymbol{G_c}$ (red) and $\boldsymbol{F}/\boldsymbol{G_c}$ (blue) for different time averages show the same form, except that cloud enhancement ($\boldsymbol{G} > \boldsymbol{G_c}$) diminishes with longer averaging periods.

2.3 Time resolution

The relationships are notably stable against time resolution of the data, with almost the same median and decile range as 1-minute data are aggregated into 10-minute or hourly values. The exception is that the higher time resolutions show significant cloud enhancement when the sun is unobscured amid broken or scattered cloud that is brighter than blue sky. On those days, such values are countered by periods of sun obscuration in such a way that the mean prediction of $\boldsymbol{R}$ and $\boldsymbol{F}$ from $\boldsymbol{G}$ is very similar.

2.4 Tilted Surfaces

With a low and unobscured horizon, and sufficiently isotropic $\boldsymbol{F}$, solar panels, windows, or walls receive a tilted irradiance equal to

$$\max(\boldsymbol{R}\cdot\cos\boldsymbol{\varphi},0) + \boldsymbol{F}\cdot(1+\tau)/2 + \boldsymbol{G}\cdot\boldsymbol{\alpha}\cdot(1-\tau)/2 \qquad (1)$$

where $\boldsymbol{\varphi}$ is the solar angle to panel normal, $\boldsymbol{\tau}$ the cosine of tilt angle, and $\boldsymbol{\alpha}$ the ground albedo.

3 TOOLS FOR USERS

3.1 Solarview

We use the datasets and algorithms as above to provide an on-line tool for assessing available solar energy in New Zealand (http://solarview.niwa.co.nz). It uses a 100-m digital elevation model to calculate the local horizon, shows the solar path for summer and winter solstices, autumn equinox, and two shoulder months in spring, and gives cumulative solar energy on a panel of specified tilt and bearing. Data for all months, and hourly values for a Typical Meteorological Year [5], are tabulated for download from the web site.

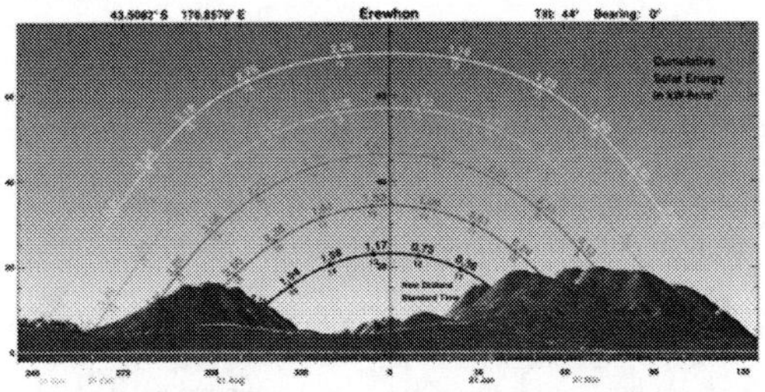

Figure 4: Estimates of average solar energy for a location in New Zealand, on a surface of arbitrary tilt and bearing (here tilted north to latitude angle). The plot shows solar results for 30 days centred on the listed dates.

3.2 Future developments

The design of Solarview was intended to support its use as a nomogram; users could sketch in other obstructions like buildings and trees, relative to the physical landscape, and read from the figures (at effective sunset minus sunrise) what fraction of the daily total was available.

We are developing a smartphone or tablet app to photograph the panorama, and register the image with orography from the digital model. This development will also explore whether the reflectance of surfaces affecting energy flux can be usefully inferred from the image, and how shading, reflection, and scatter from nearby objects should modify expression (1).

4 REFERENCES

[1] M.J. Uddstrom, *et al.*, Journal of Applied Meteorology, 40, (2001) 16.

[2] J.B. Liley, B.W. Forgan, Geophysical Research Letters, 36, (2009) L07811.

[3] G. Seckmeyer, *et al.*, Geophysical Research Letters, 22, (1995) 1889.

[4] R.L. McKenzie, *et al.*, Photochemical & Photobiological Sciences, 5, (2006) 343.

[5] J.B. Liley, *et al.*, NIWA, Lauder, Client Report LAU2007-02-JBL, (2007).

BUILDING ROOFTOP PHOTOVOLTAIC POTENTIAL IN MOUNTAINOUS REGIONS: A CASE STUDY FROM THE PYRENEES

Oriol Travesset-Baro, Guillem Francisco, Marc Vilella, Marc Pons

Observatori de la Sostenibilitat d'Andorra (OBSA), Sant Julià de Lòria, Andorra
{otravesset, gfrancisco, mvilella, mpons}@obsa.ad
Sustainability Measurement and Modeling Lab (SUMMLab), Universitat Politècnia de Catalunya, Terrassa, Spain
oriol.travesset@upc.edu

ABSTRACT: This paper describes a case study for the assessment of the energy generation potential of PV systems installed on building rooftops in two mountainous areas (Escaldes-Engordany in the Principality of Andorra and Tavascan in Catalonia). Their location, in the middle of the Pyrenees, means that the surrounding mountains generate shadows that can reduce significantly the PV potential. In this paper, a GIS-based 3D analysis is conducted to accurately represent urban areas and the surrounding topography. The results obtained show that the surrounding topography can reduce the PV potential significantly. Specifically, in our case study, this value ranges from 3.4% in Escaldes-Engordany to 12.8% in Tavascan. This study represents the first rooftop solar PV potential analysis conducted in Andorra and in the Pyrenees using high-resolution 3D models and establishes a methodology to conduct similar works in this and other mountainous regions.
Keywords: PV potential, mountainous areas, LiDAR, Andorra

1 INTRODUCTION

Nowadays, buildings account for 40% of the total final energy consumption and 36% of CO_2 emissions in the European Union (EU). The EU has set itself a long-term goal of reducing greenhouse gas emissions by 80–95% compared to 1990 levels by 2050 [1].

This context points to the need for countries to dedicate greater efforts in implementing effective policies that focus on minimising the energy-related CO_2 emissions of buildings. Solar energy technologies are expected to contribute to addressing this challenge in a near-future [2]. Solar PV rooftop systems have shown the capacity to convert residential and commercial neighbourhoods into distributed power generation areas [3], [4]. Urban areas offer a large amount of unused rooftops with an important energy generation potential. Major cities such as Barcelona, Boston or New York have characterised their building rooftops to estimate the PV energy generation potential and have published the results in a public web app to encourage citizens to install generation systems [5].

Many methods for estimating rooftop PV potential have been developed, ranging from simple multipliers of total building space to methods that employ complex geographic information systems (GIS) or three-dimensional (3D) models. Unlike simple methods, GIS-based 3D models are more detail-specific and replicable across multiple regions. However, these methods are time- and computer resource-intensive [6].

This paper describes a case study for the assessment of the energy generation potential of PV systems installed on rooftops in two mountainous areas (Escaldes-Engordany in the Principality of Andorra and Tavascan in Catalonia). The villages of Tavascan and Escaldes have around 100 and 15,000 inhabitants, respectively, and can be considered to be representative of small and medium-sized urban areas in the Pyrenees. Their location, in the middle of the Pyrenees, means that the surrounding mountains generate shadows that can reduce the PV potential significantly. In this paper, a GIS-based 3D analysis is conducted to accurately represent urban areas and the surrounding topography.

An estimate of rooftop suitability for PV systems is conducted using light detection and ranging data (LiDAR) and Solar Analyst, the ArcGIS extension for modelling solar radiation [7]. Although there is several studies analysing the rooftop PV potential using LiDAR data [8]–[10], most are developed in flat areas where topography does not influence the results. As far as we are currently aware of, only one study [11] has been conducted in a mountainous region. The main innovation of this study is the numerical quantification of the impact of the topography. The results obtained are specific for two zones in the Pyrenees, but can be used as a reference for other mountainous regions without the resources to conduct a GIS-based 3D analysis.

The paper is organised as follows. The methodology is presented in Section 2, Section 3 presents the results and the discussion. Finally, the conclusions and some prospects for future work are presented in Section 4.

2 METHODOLOGY

The assessment of the energy generation potential of PV systems installed on rooftops in Tavascan and Escaldes-Engordany consisted of five steps, which are detailed below and represented in the flowchart in Fig. 1. The impact of the surrounding topography on the received radiation was estimated by repeating part of the process.

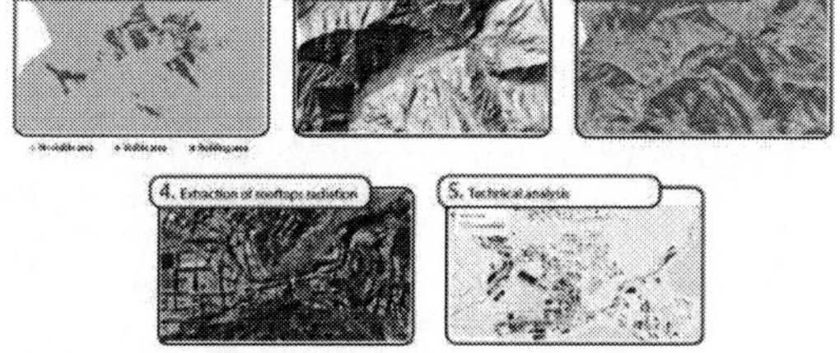

Figure 1: Flowchart of the methodology to estimate rooftop PV potential

2.1 Identification of the geographic area

The first step in the analysis was to identify the area of interest. It was represented using a 5x5 m digital elevation model (DEM) and the *Viewshed* [12] tool included in ArcGIS 3D Analyst package. This tool determines the raster surface locations visible to a set of observer features. In our case, visible locations determine the surrounding topography that may affect rooftop solar radiation. The DEM area had to be delimited to avoid excessive calculation time, however it needed to include all areas that could shadow the roofs. Although it is not of application in this study, some authors propose eliminating portions of the DEM that are not required or processing the data in batches to minimise computation data [9].

2.2 High-resolution DEM construction

The next step was to create a high-resolution 3D model to represent the buildings accurately. A new high-resolution DEM was built and merged with the previous DEM that represented the geographic area. Although a 5x5 m DEM is enough to represent the topography, urban areas require better resolution. In our model, we used LiDAR data with 1 m point spacing, the minimum to reflect accurately the characteristics of the rooftop, such as roof slope, roof orientation, and adjacent structures and vegetation [9]. Fig. 2 shows the high-resolution 3D model used to represent Escaldes-Engordany and its surrounding topography.

Figure 2: High-resolution 3D model of Escaldes-Engordany and surrounding topography

2.3 Solar radiation simulation

The global solar radiation was calculated using the Area solar radiation [13], included in ArcGIS Spatial Analyst toolbox. This tool derives incoming solar radiation from a raster surface. Table I shows the parameters considered in the analysis. It must be emphasized that the *Calculation directions* value was set to 32. Typically, a value of 8 or 16 is adequate for areas with gentle topography, whereas a value of 32 is adequate for complex topography. Increasing the number of directions increases the accuracy, but also increases the calculation time.

Table I: Parameters used in the solar radiation calculation

Parameter	Value
Day interval	14
Hour interval	1
Calculation directions	32
Zenith divisions	8
Azimuth divisions	8
Diffuse model type	UNIFORM_SKY
Diffuse proportion	0.3
Transmitivity	0.5

2.4 Extraction of rooftop radiation

The next step was to extract solar radiation values for each roof. The simplest method involved using the building's footprint. In both cases, Tavascan [14] and Escaldes-Engordany [15], we used topographical maps included in public databases, applying some manual corrections to represent the rooftop outline accurately.

2.5 Technical analysis

The last step was to determine the PV potential of each building rooftop by converting insolation falling in a given area into electricity. In this calculation, we considered 250 Wp; 1.7 m^2 PV modules, an overall system efficiency of 80%, and a useful area of 75% to exclude architectural and technical elements present on the roof that make it impossible to install PV systems.

2.6 Impact of the surrounding topography

In order to quantify the impact of the topography on the received radiation, we built a DEM of the urban area (similar to that in Section 2.2), but in this case, without including the surrounding topography. Then we repeated the steps explained in Sections 2.3, 2.4 and 2.5. Finally, we compared the results with those obtained with the topography.

3 RESULTS AND DISCUSSION

The areas under study represent small (Tavascan) and medium-sized (Escaldes) urban zones in the Pyrenees. In the sample, 112 buildings were identified in Tavascan, with a total footprint area of 11046 m^2. As a result, an average building footprint area of 98.6 m^2 was obtained. As shown in Table II, 1094 buildings were identified in Escaldes, with an average footprint area (263.1 m^2). The total solar energy received on the roofs was 8.9 GWh and 233.6 GWh in Tavascan and Escaldes, respectively, which means an average received radiation per building of 79.9 MWh and 224.2 MWh.

Table II: Main characteristics of the analysed areas

	Tavascan	Escaldes
Location (lat. / long.)	42.645 / 1.257	42.509 / 1.537
Number of buildings	112	1042
Total footprint area (m^2)	11046	274168
Avg. footprint area (m^2)	98.6	263.1
Received radiation (GWh)	8.9	233.6
Avg. received radiation (MWh)	79.9	224.2

Fig. 3 shows the histograms of building footprint areas in both zones. As it can be appreciated, rooftops in Escaldes-

Engordany are bigger than in Tavascan, including some buildings with footprint areas greater than 1000 m^2. On the other hand, Tavascan is mainly composed of building rooftops with a footprint area around 50 m^2, the largest footprint being under 500 m^2.

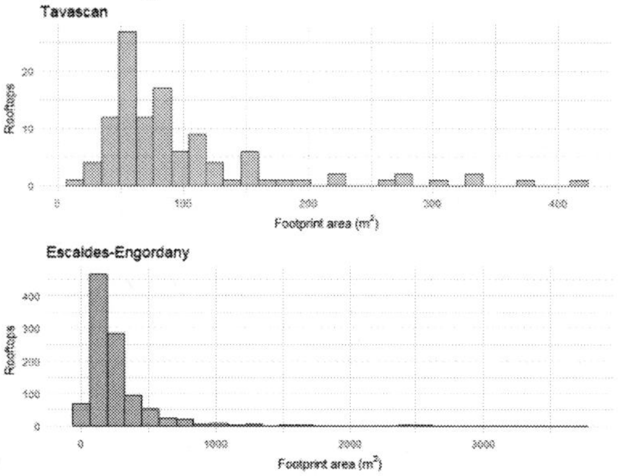

Figure 3: Histograms of building footprint areas in Tavascan and Escaldes-Engordany

The rooftop orientations in the areas analysed in this case study are represented in Fig. 4. It shows that in Tavascan, there is a higher prevalence of east, southeast and east orientations, whereas in Escaldes-Engordany, north and south are the most common orientations.

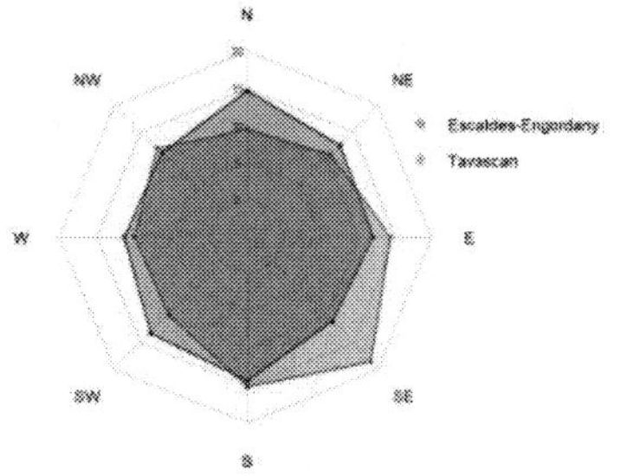

Figure 4: Rooftop orientations in Tavascan and Escaldes-Engordany (% of area)

3.1 Solar PV potential and reduction in emissions

The results obtained in the analysed areas show an overall PV roof potential of 21 GWh/year for an installed capacity of 29.2 MW in Escaldes-Engordany, and 780 MWh/year for an installed capacity of 1.2 MW in Tavascan. Considering the number of inhabitants and the average electricity consumption in Andorra and Catalonia, we can conclude that installing PV technology on rooftops in these areas would satisfy 21% and 127% of the demand in Escaldes and Tavascan, respectively.

The potential reduction in CO_2 emissions in the analysed areas was calculated using the current emission factor for electricity supply in both countries. In the case of Escaldes, this is calculated by taking into account imported electricity ratios from Spain and France, as more than 80% of Andorran electricity is imported. The Spanish emission factor is considered to be 302 gCO_2/kWh [16], whereas the French emission factor is 44 gCO_2/kWh [17]. Considering these factors and imports from each country, the emission factor for Andorran electricity is 181 gCO_2/kWh.

The overall reduction in emissions by installing solar PV panels in a 20-year period is 4710 tCO_2 in Tavascan and 74307 tCO_2 in Escaldes-Engordany.

3.2 Reduction of solar PV potential due to topography

As expected, the surrounding topography reduced the PV potential. The impact of this factor ranged from 3.4% in Escaldes-Engordany and 12.8% in Tavascan. The location of the latter in a closed valley could be the main reason for the higher reduction. Fig. 5 shows the cumulative histogram of the roofs as a function of the received radiation, comparing the simulations with and without topography. The higher difference between both simulations can be clearly seen in Tavascan, compared to Escaldes, due to the shadows caused by the surrounding topography.

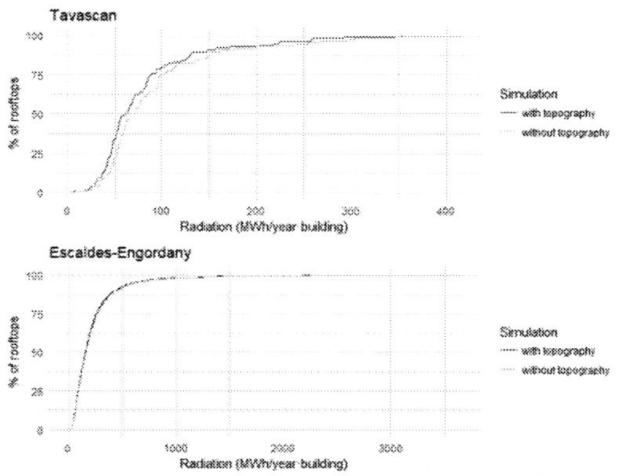

Figure 5. Cumulative histogram of the roofs as a function of received radiation (with and without topography)

4 CONCLUSIONS

The methodology used in this work is similar to that used in other papers [9] but we introduced the use of the *Viewshed* method to determine the geographic area of interest, which is fundamental in mountainous areas to include influent topography in the model. This study represents the first rooftop solar PV potential analysis conducted in Andorra and in the Pyrenees using high-resolution 3D models. Data availability and the quality of existing data are critical factors in developing this type of analysis. This paper established a methodology that was adapted to existing data in the Pyrenees and is applicable in other mountainous regions with similar data. In this way, further work could focus on expanding the study to other mountainous areas to extend the sample of results.

The next steps in this work will involve a more detailed GIS-based analysis of the roofs' architectural characteristics and the physical characteristics of the surrounding topography to determine their influence on received solar radiation. In addition, next efforts will concentrate on validating the results through ground-truth data from pyranometers or electricity-generation data from installed solar PV systems.

This type of study could be useful to encourage citizens to install PV solar generation systems. The results obtained through the simulations can be easily adapted and converted to valuable information to be provided to citizens through a platform where people can determine their rooftop solar potential, the economic profitability of a PV system or their environmental benefits. For this purpose, we have developed a web application that offers useful information adapted to the citizens of the areas analysed [18].

ACKNOWLEDGEMENTS

This research has been financially supported by the Government of Andorra as part of the framework of research grants for the Working Community of the Pyrenees (ACTP023-AND/2014). The first author also acknowledges a pre-doctoral grant from the Government of Andorra (ATCR012-AND 2014/2015).

REFERENCES

[1] European Commission, *Energy*. 2016. [Online]. Available at: https://ec.europa.eu/energy/en/topics/energy-efficiency/buildings. [Accessed: 06-Feb-2017].

[2] IRENA, *Roadmap For A Renewable Energy Future*. Abu Dhabi: 2016.

[3] James T, Goodrich A, Woodhouse M, Margolis R, and Ong S. *Building-Integrated Photovoltaics (BIPV) in the Residential Sector: An Analysis of Installed Rooftop System Prices*. National Renewable Energy Laboratory: Golden, Co, 2011.

[4] Barrier M and Irvine L. *Solar PV for Multifamily Affordable Housing in Seattle*. Northwest Sustainable Energy for Economic Development: Seattle, 2015.

[5] Mapdwell, *Solar system*. 2017. [Online]. Available at: https://www.mapdwell.com/en/solar/newyork. [Accessed: 27-Jul-2017].

[6] Melius J, Margolis R, and Ong S. *Estimating Rooftop Suitability for PV: A Review of Methods, Patents, and Validation Techniques*. National Renewable Energy Laboratory: Golden. December, 2013; 35.

[7] Fu P and Rich PM. "Design and Implementation of the Solar Analyst: An ArcView Extension for Modelling Solar Radiation at Landscape Scales", *Proceedings from the 19th Annual ESRI User Conference*, San Diego, USA. 1999; 1–24.

[8] Brito M, Gomes N, Santos T, and Tenedório J. "Photovoltaic potential in a Lisbon suburb using LiDAR data", *Solar Energy*, 2012; 86: 283–288.

[9] Kodysh JB, Omitaomu OA, Bhaduri BL, and Neish BS. "Methodology for estimating solar potential on multiple building rooftops for photovoltaic systems", *Sustainable Cities and Society*, 2013; 8: 31–41.

[10] Redweik P, Catita C, and Brito M. "Solar energy potential on roofs and facades in an urban landscape", *Solar Energy*, 2013; 97: 332–341.

[11] Mavromatidis G, Orehounig K, and Carmeliet J. "Evaluation of photovoltaic integration potential in a village", *Solar Energy*, 2015; 121: 152–168.

[12] Esri. *ArcGIS Desktop. Viewshed tool*. 2017. [Online]. Available at: http://pro.arcgis.com/en/pro-app/tool-reference/3d-analyst/viewshed.htm. [Accessed: 10-Jan-2017].

[13] Esri, *ArcMap. Area Solar Radiation 2017*. [Online]. Available at: http://desktop.arcgis.com/en/arcmap/10.3/tools/spatial-analyst-toolbox/area-solar-radiation.htm. [Accessed: 10-Jan-2017].

[14] Institut Cartogràfic i Geològic de Catalunya. *Topographic cartography*. 2016. [Online]. Available at: http://www.icgc.cat/en/Public-Administration-and-Enterprises/Downloads/Topographic-cartography. [Accessed: 18-Oct-2016].

[15] Andorra Government. *Andorra Spatial Data Infrastructure / Downloads*. 2016. [Online]. Available at: http://www.ideandorra.ad/geodades/. [Accessed: 18-Oct-2016].

[16] OCCC. *Nota informativa sobre la metodología del mix eléctrico por parte de la Oficina Catalana del Cambio Climático*. 2016.

[17] RTE France (French Transmission System Operator). *Key Figures 2016*. [Online]. Available at: http://www.rte-france.com/en/eco2mix/chiffres-cles-en. [Accessed: 07-Jun-2016].

[18] OBSA, *Potencial d'aprofitament solar en cobertes d'edificis del Pirineu*. 2017. [Online]. Available at: http://obsa.ad/solar/. [Accessed: 01-Aug-2017].

PERFORMANCE ANALYSIS OF THE DOMESTIC HOT WATER PRODUCTION WITH PV PANELS AND A HEAT PUMP

F.J. Aguilar, P.V. Quiles
[a]*Universidad Miguel Hernández, Departamento de Ingeniería Mecánica y Energía,*
Avenida de la Universidad s/n, Edificio Innova, 03202 Elche, Spain
faguilar@umh.es

ABSTRACT:

The current work describes an experimental and analytical work, carried out during an entire year, where a domestic hot water heat pump (DHW-HP) that is powered from photovoltaic panels (PV) and from the grid, has been analyzed.
The main aim of the project is focused on showing the real possibilities of using PV panels combined with a heat pump to supply the domestic hot water demand in dwellings. This system has been developed to have high energy efficiency and performance reliability. The experimental results have been used to validate the analytical model and to simulate the system behaviour under other working conditions such as: thermal demand, weather conditions, etc.

Keywords: Photovoltaic, Heat Pump, Renewable Energy, Domestic Hot Water

Nomenclature:

COP	Coefficient of Performance (-)
E	Electrical energy (kWh)
IRR	Solar radiation (W/m2)
P	Electrical power (W)
Q	Thermal energy (kWh)
SC	Solar Contribution (%)
SPF	Seasonal Performance Factor (-)
T	Temperature (°C)
η	Efficiency (-)
NOCT	Nominal Operation Cell Temperature (°C)

Subscripts

1	Compressor inlet (ref. thermodynamics cycle)
2	Compressor outlet (ref. thermodynamics cycle)
3	Condenser outlet (ref. thermodynamics cycle)
4	Evaporator inlet (ref. thermodynamics cycle)
C	Condenser
DHW	Domestic Hot Water
E	Evaporator
GRID	Energy consumptions from the grid
HP	Heat Pump
HT	Electrical heater
LOSS	Energy losses produced in the tank
O	Outdoor
PV	Photovoltaic
TOT	Total
SYST	System
U	Useful
W	Water

1 INTRODUCTION

Conventionally sourced energy is often obtained at great socio-political cost: it is unsustainable, fails to minimise environmental deterioration and fundamentally threatens human wellbeing. Thus, renewable and/or efficient energy systems play an increasingly important role in today's society. In addition to their popularly celebrated eco-friendly and economic benefits, their presence within the community is essential to sustainable growth, quality of life and even long term public health.

The Energy Performance of Buildings directive [1] was adopted in 2010 and focused on fomenting energy efficiency in buildings and streamlining existing previsions. The directive, which focuses in particular on air conditioning, insulation and renewable energy use, presents minimum performance requirements for buildings, energy performance calculation methods and energy certification criteria. The following notable objectives are set by the directive:

- From 2018 and on, all new governmental buildings must be Nearly Zero-Energy Buildings
- From 2020 and on, all new buildings must be Nearly Zero-Energy Buildings
- Renovations on existing buildings must improve the overall energy efficiency.
- Buildings and systems installed within them must be subject to regular inspection, especially heating and air conditioning systems.

Nearly Zero-Energy Buildings consume a very little amount of energy compared to current households. Said buildings have such high-energy performance and efficiency that they can be and are powered almost entirely by renewable sources.

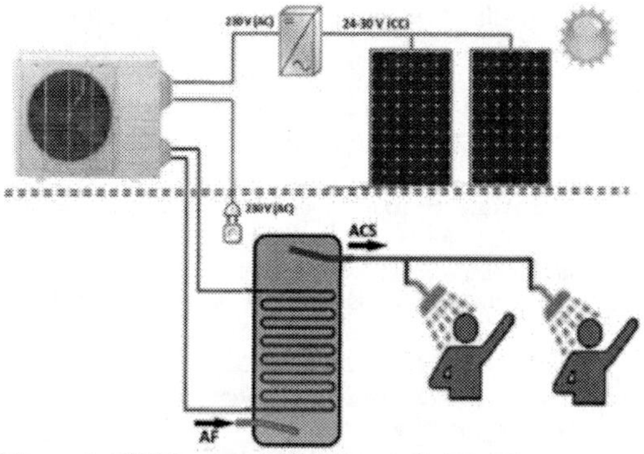

Figure 1: PV Heat Pump for Domestic Hot Water

Domestic hot water is one of the most important consumptions in dwellings. According to [2] in Spain it can be up to 22% of the total residential energy consumption and most of this consumption is still supported by conventional energy sources such as:

electricity or fossil fuels. This ratio will be higher when the standard nZEB of the Directive EPBD 2010 are applied (up to 50%), because the heating and cooling demand will decrease, but not the DHW demand.

Currently, the main alternative to conventional technologies that make use of renewable energy are the thermal solar collectors. However, the experiences in solar thermal systems in dwellings have not been very successful, because their reliability and energy savings are much lower than expected. Furthermore, stagnation and overheating are common problems in this kind of installation.

The current work describes an experimental and analytical work, carried out during an entire year, focused on a domestic hot water heat pump (DHW-HP) that is powered from photovoltaic panels (PV) and from the grid, simultaneously.

2 OBJECTIVES

The main aim of the project is focused on showing the real possibilities of using PV panels combined with a heat pump to supply the domestic hot water demand in dwellings. This system has been developed to have high energy efficiency and performance reliability. The experimental results have been used to validate the analytical model and to simulate the system behaviour under other working conditions such as: thermal demand, weather conditions, etc.

Several tasks have been carried oud to reach the aim previously described.

- To design and build the experimental facility.
- To measure and analyse the behavior of the system (DHW + PV) for a year.
- To analyse the experimental results obtained from the system.
- To develop an analytical model of the designed system capable of simulating its performance under different scenarios (climate conditions, thermal demand, etc.).

3. EXPERIMENTAL METHOD

The experimental part of the project is described in detail in [3]. Nevertheless, a brief description of this project has been included in this article.

3.1 Experimental facility

The heat pump analyzed is an ON/OFF equipment with a nominal heating capacity of 1.5 kW and a nominal electrical consumption of 470 W (nominal COP=3.19). Two photovoltaic panels with a total peak power of 470 Wp were connected to a micro-converter which was connected to the equipment at 230 Vac. The experimental set-up was installed on the roof of the University's research laboratory located in Elche (Spain). Figure 2 shows a photograph of the facility.

Table I shows the technical characteristics of the heat pump for DHW, while table II shows the technical characteristics of the photovoltaic panels.

Figure 2 : Photography of the Experimental Facility

Table I: Technical characteristics of the heat pump

Kaysun Compak KHP 15 190	Simb.	Unit	Eco
Thermal Capacity	Q_{HP}	W	1500
Compressor Power	$P_{E\text{-}COMP}$	W	470
COP	COP	%	3.19
Electrical Heater Power	Q_{ELEC}	A	2000
Refrigerant	---	---	R134a
Evaporator Fan Power	$P_{E\text{-}FAN}$	W	30
Tank Volume	V	L	190

Table II: Technical characteristics of the PV panels

Eurener 235	Simb.	Unit	Nom.
Nominal power	$P_{N,PV}$	W	235
Panel surface area	A_{PV}	m^2	1.67
Efficiency	Eff_{PV}	%	13.74
Short circuit current	I_{SC}	A	8.25
Open circuit voltage	V_{OC}	V	37.08
Nominal current	$I_{N,PV}$	A	7.66
Nominal voltage	$I_{N,PV}$	V	30.01
NOTC		°C	47
γ_{Pmax}		%/°C	-0.43

Tests were carried out under real climatic conditions, while a recirculation circuit was used to simulate water consumptions. The facility simulates 6 domestic-hot-water consumptions of 20-23 liters each one. The heat pump equipment heats the water to 55°C and it keeps the water temperature higher than 50°C. The water inlet temperature is between 12 and 15°C.

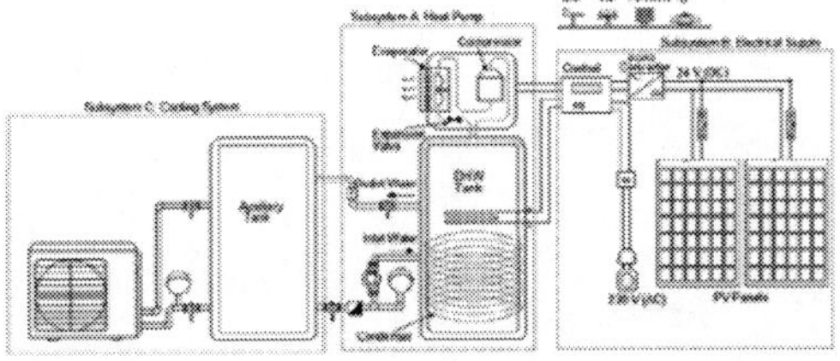

Figure 3 : Experimental setup

The total useful thermal energy produced ($Q_{U,TOT}$) is determined in [3] by two different methods simultaneously, in order to have more reliability in the results.

The first method is a "direct method", where the total heat produced by the system is calculated from measurements in the DHW tank. The total thermal energy can be separated into two components: useful energy for DHW (Q_{DHW}) production and tank losses (Q_{LOSS}).

The second method applied is an "indirect method", where the total useful thermal energy produced by the unit ($Q_{U,TOT}$) is calculated from the sum of two following values. The first one is thermal energy produced by the heat pump ($Q_{U,HP}$) that is calculated from measurements in the refrigerant thermodynamic cycle. The second one is the thermal energy produced by the electrical heater ($Q_{U,HT}$). This second value is determined using electrical measurements and applying a conversions factor of 1 between thermal and electrical energy.

3.2 Experimental results

During the experimental work more than 240 days were analyzed in detail (at least 20 days per month) in order to stablish seasonal results. Figures 4, 5 and 6 show an example day of the study. In figure 4the electrical power produced by the photovoltaic panels ($E_{PV,TOT}$), the electrical power consumed by the heat pump ($E_{TOT,HP}$) and the electrical energy consumed by the heat pump from the grid ($E_{GRID,HP}$) have been included.

It can be seen that the heat pump was working from 12 pm to 15 pm, when the maximum temperature of the water was reached.

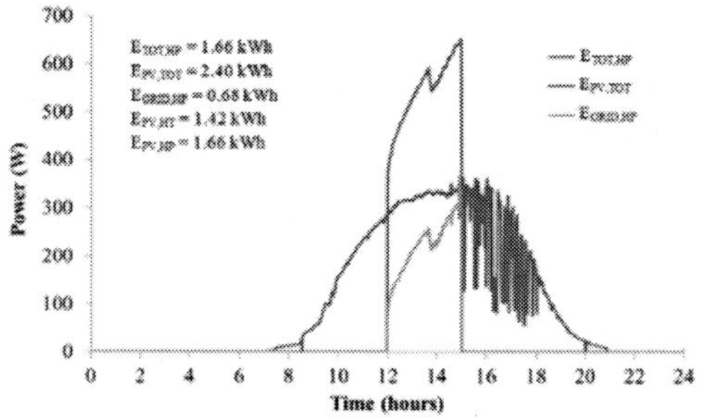

Figure 4 : Power consumed by the system (July 14th)

According to the control system of the installation when the PV energy is higher than the energy consumed by the heat pump this energy goes to an electrical heater installed in the water tank. However, if the energy produced by the PV panels is not enough to support the electrical energy demanded by the heat pump, it consumes energy from the grid.

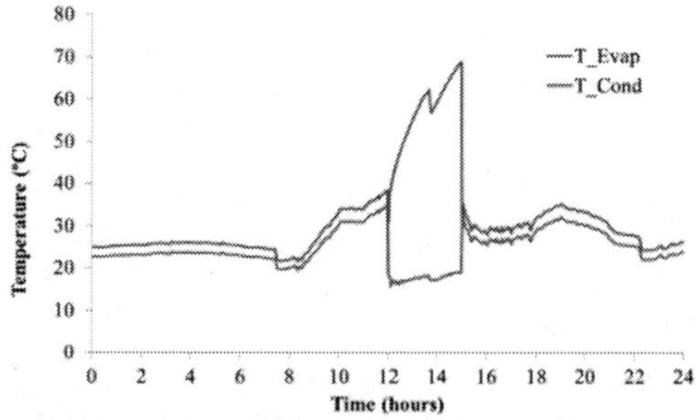

Figure 5 : Evaporator and Condenser temp. (July 14th)

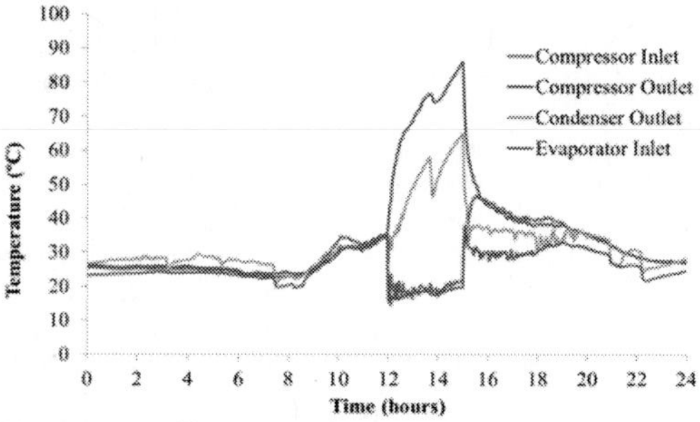

Figure 6 : Refrigerant cycle temperatures (July 14th)

Table III shows the experimental results from the example day, where the equations from [3] have been applied. Some of them are been summarized in the following equation.

$$COP_{HP} = \frac{E_{USEFUL,HP}}{E_{TOT,HP}} \quad (1)$$

$$COP_{HP,HT} = \frac{E_{USEFUL,HP}+E_{USEFUL,HT}}{E_{TOT,HP}+E_{PV,HT}} \quad (2)$$

$$COP_{SYST} = \frac{E_{USEFUL,HP}+E_{USEFUL,HT}}{E_{GRID}} \quad (3)$$

$$SC_{HP} = \frac{E_{PV,HP}}{E_{TOT,HP}} \quad (4)$$

$$SC_{SYST} = \frac{E_{USEFUL,HT}+E_{USEFUL,HP}\cdot SC_{HP}}{E_{USEFUL,HT}+E_{USEFUL,HP}} \quad (5)$$

Table III: Experimental results (July 14th)

	Units	Value
$E_{USEFUL,HP}$	kWh	6,40
$E_{USEFUL,HT}$	kWh	1,41
$E_{USEFUL,TOT}$	kWh	7,82
$E_{TOT,HP}$	kWh	1,66
$E_{PV,HT}$	kWh	1,41
$E_{PV,HP}$	kWh	0,98
$E_{PV,TOT}$	kWh	2,40
E_{GRID}	kWh	0,68
COP_{HP}	---	3,87
$COP_{HP,HT}$	---	2,55
COP_{SYST}	---	11,57
SC_{HP}	%	59,19%
SC_{SYST}	%	66,56%

The results from the experimental project demonstrated that the annual average efficiency of the heat pump (SPF_{HP}) could be close to 3.5 and the annual average efficiency of the whole system would be near 9, while the solar contribution would be higher than 60%.

4. ANALITICAL MODEL

4.1 Analytical model development

In this part of the project, the experimental results have been used to develop an analytical model capable of simulating the designed system under different working conditions. This study can help to select the optimum design parameters of the system in function of the DWH demand and climatic conditions in order to optimize the photovoltaic solar contribution and the energy efficiency.

The results when the performance of the system were as much stable as possible (steady work) were joined and collected in order to set an analytical model that illustrated the heat pump performance correctly. Thus, the aim was to remove non coherent results by-product of the transitory operation (mainly start-ups, breakdowns and moments of both hot water output and cold water input) when the heat pump performance changes sharply. And so, more than 300 valid points were collected.

The dependencies between the different variables were analyzed by using the Excel statistical analysis function.

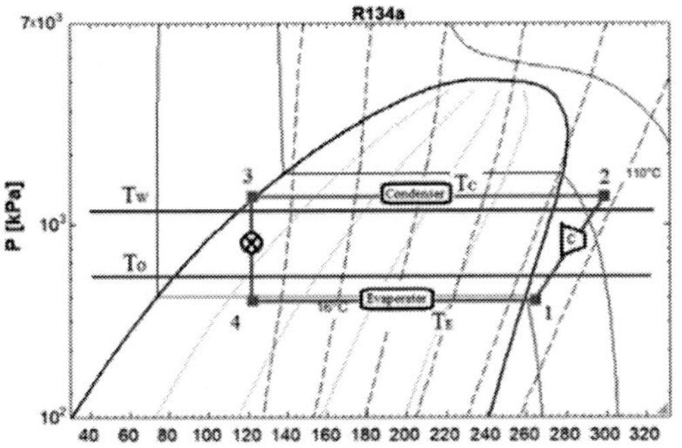

Figure 7 : P-h diagram (R134a)

It was stated that the useful thermal energy supplied by the heat pump to the water and the consumed electrical power depend on the evaporator temperature and the condenser temperature mainly (Figure 7). Also, the evaporation temperature will depend on the ambient temperature and the water temperature, while the condensation temperature is only function of the water temperature. The stratification effect of the tank will make the water temperature may considerably change between the bottom and the top of the tank. The water temperature needed to calculate the condensation temperature must be measured at the point where the coil is located (bottom part of the tank). It is described in [4] the numerical model uses to analyze the tank stratification. For further works the aim is to integrate this numerical model with the analytical model described in this article.

The following equations were defined by means an iterative process:

$$T_C[^{\circ}C] = 1.557 \cdot T_W - 20.09 \qquad (6)$$

$$T_E[^{\circ}C] = -14.46 + 0.645 \cdot T_O + 0.230 \cdot T_W \qquad (7)$$

$$Q_{U,HP}[W] = 3434.7 + 106.5 \cdot T_E - 66.4 \cdot T_C + 0.766 \cdot T_E^2 + 0.410 \cdot T_C^2 - 1.026 \cdot T_E \cdot T_C \qquad (8)$$

$$P_{E,HP}[W] = 310.7 - 0.097 \cdot T_E^2 + 0.510 \cdot T_C^2 + 0.097 \cdot T_E \cdot T_C \qquad (9)$$

The photovoltaic production was calculated by means the application of the equations given by IDAE in [5]. Some factors as cell temperature, temperature loss, inverter performance, etc. are considered in these equations.

$$T_{CELL}[^{\circ}C] = \frac{T_O + (NOTC - 20) \cdot IRR}{800} \qquad (10)$$

$$P_{DC,PV}[W] = \frac{P_{PV\,Panels} \cdot R \cdot [1 - g \cdot (T_{CELL})] \cdot IRR}{100} \qquad (11)$$

$$P_{AC,PV}[W] = P_{DC,PV} \cdot \eta_{converter} \qquad (12)$$

4.2 Validation of Analytical Model

In order to validate the defined model the described equations of several days of experimental work were applied, taking as input data the water temperatures at each moment [T_W], the ambient temperature [T_O] and the solar radiation IRR [W/m^2].

In the following figures the obtained results by means the experimental work and the ones obtained by means the model are compared. The high accuracy reached with the power consumed has to be highlighted, where the curves are almost equal.

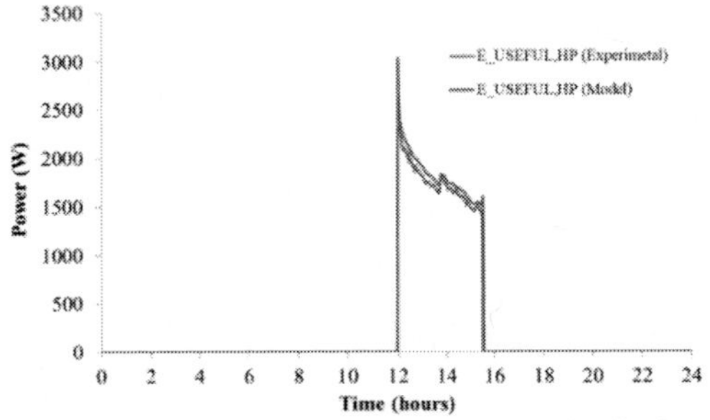

Figure 8 : Useful thermal energy produced by the heat pump. Experimental and analytical result (July 14th)

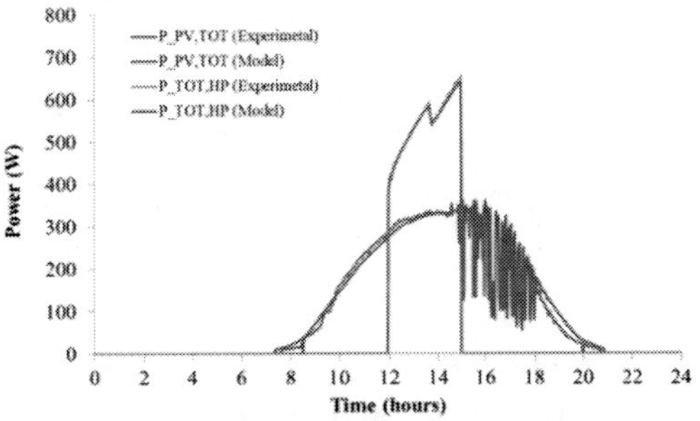

Figure 9 : Energy consumed by the heat pump and energy produced by the PV panels. Experimental and analytical results (July 14th)

Both graphic and table XX show that the differences are minimal. Errors below 6%.

Table IV: Experimental vs. analytical results (July 14th)

	Units	Experim.	Model	Error
$E_{USEFUL,HP}$	kWh	6.40	6.03	5.90%
$E_{USEFUL,HT}$	kWh	1.41	1.50	5.93%
$E_{USEFUL,TOT}$	kWh	7.82	7.52	3.77%
$E_{TOT,HP}$	kWh	1.66	1.65	0.36%
$E_{PV,HT}$	kWh	1.41	1.50	5.93%
$E_{PV,HP}$	kWh	0.98	0.96	1.88%
$E_{PV,TOT}$	kWh	2.39	2.46	2.73%
E_{GRID}	kWh	0.68	0.69	1.86%
COP_{HP}	---	3.87	3.65	5.57%
$COP_{HP,HT}$	---	2.55	2.39	6.15%
COP_{SYST}	---	11.57	10.93	5.52%
SC_{HP}	%	59.19%	58.28%	1.53%
SC_{SYST}	%	66.56%	66.58%	0.02%

Table V shows the comparative results for another studied day (June 2nd).

Table V: Experimental vs. analytical results (June 2nd)

	Units	Experim.	Model	Error
$E_{USEFUL,HP}$	kWh	6.19	6.50	5.02%
$E_{USEFUL,HT}$	kWh	1.26	1.33	5.47%
$E_{USEFUL,TOT}$	kWh	7.44	7.82	5.09%
$E_{TOT,HP}$	kWh	1.84	1.88	2.00%
$E_{PV,HT}$	kWh	1.26	1.33	5.47%
$E_{PV,HP}$	kWh	1.07	1.07	0.23%
$E_{PV,TOT}$	kWh	2.33	2.40	2.85%
E_{GRID}	kWh	0.77	0.81	4.31%
COP_{HP}	---	3.36	3.46	2.95%
$COP_{HP,HT}$	---	2.40	2.44	1.63%
COP_{SYST}	---	9.62	9.69	0.75%
SC_{HP}	%	58.29%	57.01%	2.18%
SC_{SYST}	%	65.34%	64.31%	1.58%

The same procedure has been repeated in more days with the result of similar differences to the ones shown at Tables IV and V. Thus, it can be assumed that the analytical defined model has correctly been validated.

4 CONCLUSIONS

- An experimental work where a heat pump for domestic hot water production (DHW-HP) that is powered from photovoltaic panels (PV) and from the grid has been analyzed.
- The results from the experimental work have been used to elaborate an analytical model of the designed system capable of simulating its performance under different scenarios (climate conditions, thermal demand, etc.).
- The analytical model has been validated by using the results from several days, where the errors have always been lower than 6%.

5 REFERENCES

[1] Directive 2010/31/EU of the European Parliament and of the Council of 19 May 2010 on the energy performance of buildings (recast); June 18, 2010.

[2] Energy Efficiency Indicators, International Energy Agency, 2016

[3] Aguilar, et al. Experimental study of the solar photovoltaic contribution for the domestic hot water production with heat pumps in dwellings, Volume 101 (2016), Pages 379-389

[4] S.A. Borman, Stratification within a Solar Assisted Thermal Storage Tank: Experimental Analysis & Numerical Model Development, 2017.

[5] Pliego de condiciones técnicas de instalaciones conectadas a red, IDAE, 2011.

INTEGRATION OF RENEWABLE ENERGY TECHNOLOGIES IN THE COMMUNITY OF THE AGRICULTURAL UNIVERSITY OF ATHENS

Christos Spyridon Karavas[a] and George Papadakis[a]*

[a]Agricultural University of Athens, Department of Natural Resources and Agricultural Engineering,
75 Iera Odos Street, Athens 11855, Greece,
*email: gpap@aua.gr

ABSTRACT: European Energy Union pursues for increased secure, sustainable, competitive, affordable and decarbonized energy. Increasing the share of renewable energy (RE) is an effective way to make energy supply more environmental friendly and energy sources more diverse. In this context, Iceland, Liechtenstein and Norway support energy efficiency programs in Greece. The programs aim to increase the share of RE and to meet EU targets on climate and energy. One of the ten projects which has been selected for funding is the project of the Agricultural University of Athens (AUA). The objective of this project is to demonstrate the application of RE at emblematic locations inside the AUA Campus. More specifically, four activities have been implemented; (a) An autonomous PV microgrid has been installed for the external lighting and also supply various electrical loads of AUA library, (b) Installation of efficient lighting systems in AUA's sports facilities powered by an autonomous PV microgrid, (c) A biomass boiler have be installed in AUA library in order to support the existing heating system and (d) an autonomous solar-driven charging station have be installed for supporting one electric mini-bus and ten electric bicycles. The project is expected to increase of RE consumption and to strengthen awareness and education in RE based solutions and to decrease the carbon footprint of AUA.
Keywords: Stand-alone PV Systems, Hybrid, Autonomous Microgrid

1 INTRODUCTION

The program of European Economic Area Financial Mechanism 2009-2014, (EEA grants of Iceland, Liechtenstein and Norway), provided funding to 16 EU countries in Central and Southern Europe. The Greek "GR03 – Renewable Energy" programme had a budget of €9,510,000. This program is co-funded by the Public Investment Program of Greece. The objective of this program is to increase the share of RE in energy production and to reduce the CO_2 emissions. The program also aims at mitigating social and economic disparities and strengthen bilateral cooperation between the donor countries and Greece. The ultimate target is the promotion and the effective implementation of a development framework which will create opportunities and prospects to the local economy and will familiarize the residents with sustainable energy technologies and improve the quality of their life. In particular, ten projects were selected for funding by this program in Greece. One of them, is the AUA's project.

The Agricultural University of Athens (AUA) has a long experience in developing and implementing Renewable Energy Systems (RES), as it has been active in research in the field of renewable energy and energy efficiency for the last 30 years [1-6]. Generally, Universities are recognized as ideal cases for demonstration and implementation of innovative interventions for RES internationally [7-12]. Therefore, AUA was selected as an appropriate site for RES installations due to its campus characteristics (25 ha of open space and 28 buildings at the centre of Athens), high visibility (over 5000 students and 500 staff members) and high research and educational impact.

The implementation of RES contributes to the University's carbon footprint mitigation in combination with lower energy consumption. On the other hand, the existence of such facilities will sensitize young researchers to be engaged with RE, academically and professionally, and increase research output of the university.

This project consists of four subprojects. Each subproject includes activities that have been implemented on the library building and the sport facilities of AUA. The library has been selected because it is the most visited building of AUA. Students from AUA and other universities of Athens visit its facilities for every day homework and studying. In addition, scientific events, conferences, seminars etc. are conducted very often. The sport facilities constitute a meeting point for students, employees, visitors etc. The four subprojects include the following installations:

(a) An autonomous PV microgrid for the external lighting and various electrical loads of AUA library.
(b) Lighting systems in AUA's sports facilities powered by an autonomous PV microgrid.
(c) A biomass boiler in AUA library.
(d) An autonomous solar-driven electric vehicle charging station and the purchase of one electric mini-bus and ten electric bicycles.

2 AUTONOMOUS MICROGRID

The library of AUA is a functional integral building while constitutes the heart of the University (Fig. 1).

Figure 1: AUA Library building

The objective of this subproject was to cover the lighting and various electrical loads of AUA library. An autonomous PV microgrid has been installed in order to secure the power supply. The microgrid includes energy resources and storage units. A PV array, rated at 100 kWp, has been installed on the library's roof (see Fig. 1) and a typical deep discharge solar OPzV battery bank (600 kWh at least) will secure the autonomy of the system for three days as a minimum. The autonomous mode of the microgrid requires the regulation of the voltage and frequency and the maintenance of a power balance in order to secure a stable operation. The microgrid is able to operate in connection with the main grid and use it as backup. The inverters and the batteries of the system are shown in Figure 3.

Figure 2: The PV installation in the library's roof

Figure 3: The inverters and the batteries of the autonomous microgrid of the AUA library

3 BIOMASS BOILER

This subproject aimed to assist the coverage of the library's thermal needs and demonstrate a sustainable technology of thermal power production. A pellets' boiler of nominal power of 30 kW has been installed and connected with the existing heating system of AUA library. The boiler features an automatic ignition and pellet delivery, with a robust burner. It works at high efficiency, up to 90%, and meets the European emissions standards. The biomass boiler is shown in Fig .4.

Figure 4: The biomass boiler

4 INTERCONNECTED AUTONOMOUS MICROGRID

The AUA provides sports facilities where both students and AUA's staff can exercise and participate in various sports activities. Among others, AUA offers four open field sport courts; (a) one basketball court, (b) one volleyball court and (c) two tennis courts.

The objective of this subproject was to cover the lighting needs of the above mentioned four sport courts with the use of sustainable energy technologies (Fig. 5). The lighting has been based on LED technology and the required energy is producing by photovoltaics. As it is presented in Figure 6, twelve LED floodlight towers have be installed in the courts. The power supply is handled by an autonomous AC microgrid. The microgrid consists of (a) a photovoltaic (PV) array of typical polycrystalline silicon panels, rated at 5 kWp, (b) a Sunny Island inverter for grid forming, (c) a Sunny inverter for the PV array, and (d) a typical deep discharge solar OPzV battery bank. The battery bank consists of 48 batteries. Each battery has a nominal voltage of 2 V and the DC bus has been set at 48 V. For this reason, 2 battery arrays were connected in parallel where each array includes 24 batteries in series. The PV array has been installed on a specially designed shelter (Fig. 7) that acts as well as a rest area for the athletes. Furthermore, this microgrid have be interconnected with the library's microgrid, in order to minimize the number of load shedding in the microgrid, providing an extra generation capacity (Fig. 8).

Figure 5: Lighting of the AUA's sport courts

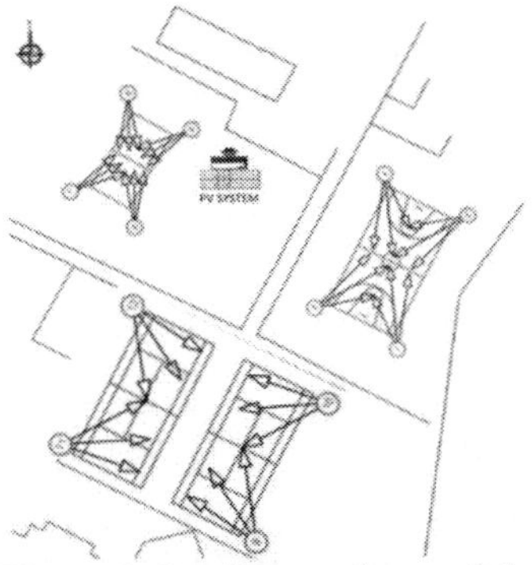

Figure 6: Installation scheme of the twelve floodlight towers

Figure 7: The PV shelter

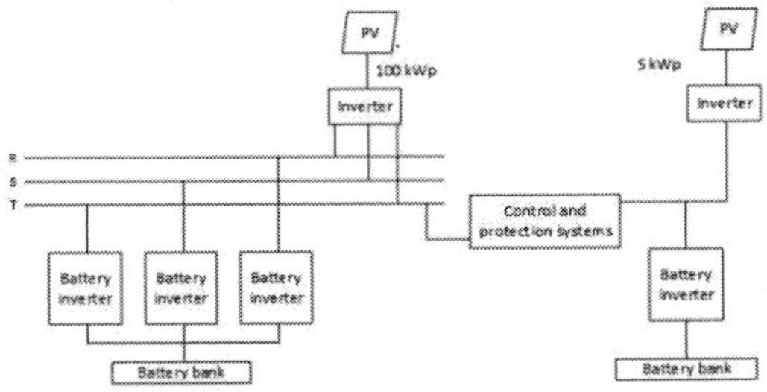

Figure 8: The interconnection of the two PV microgrids of 100 kWp and 5 kW.

5 ELECTRICAL VEHICLE CHARGING STATION

While the electric vehicles such as trains, trolleys and trams, are a safe and traditional mode of public transport in Greece, the penetration of the electric vehicles in the Greek market is very slow. AUA supports green projects on motion and the development of infrastructures in order to help students and staff to become familiar with electric vehicles.

This subproject refers to the study, the installation and the operation of an Autonomous Electric Vehicle Charging Station (EVCS). The EVCS aimed to cover the electrical needs of 10 electric bicycles and one electric vehicle (bus type) (Fig 9). Table I and Table II present the technical characteristics of the electric bicycles and electric vehicle, respectively.

Figure 9: The electrical bicycles and the electric vehicle

Table I: Technical characteristics of the electric bicycles

Type	
Battery Chemistry	Lithium-ion
Charge Time	6 hours (max)
Estimated Min Range	50 km
Top Speed	25 km/h

An autonomous microgrid power system based on SMA Sunny Island topology, was chosen for the power supply of the EVCS. The microgrid consists of: (a) a PV array of typical polycrystalline silicon panels, rated at 5 kWp, (b) three Sunny Island inverters for grid forming, (c) a Sunny inverter for the PV array, and (d) a typical deep discharge solar OPzV battery bank. The battery bank consists of 24 batteries, 2V each will be connected in series and the DC bus was set at 48 V. Furthermore, a photovoltaic parking shelter have been constructed (Fig. 10), for the parking of the electric vehicles.

Table II: Technical characteristics of the electric vehicle

Type	
Number of seats	16 plus driver
Motor Power	7.5 kW / AC
Motor Voltage	72 V / AC
Battery Chemistry	AGM
Number of Batteries	18 pieces at 8 V
Battery Capacity	160 Ah
Maximum Range	up to 120 km
Top Speed	45 km/h

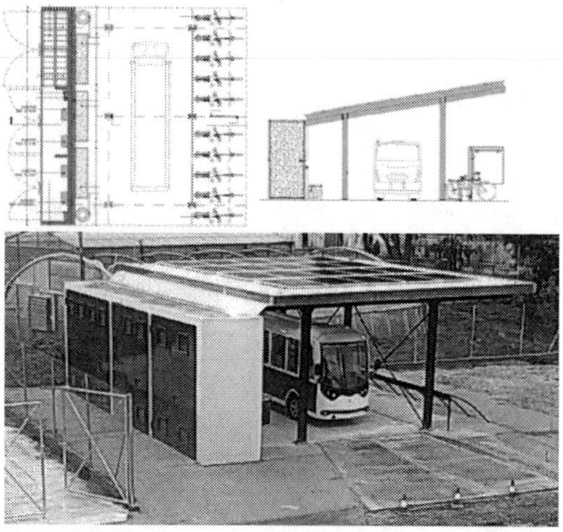

Figure 10: The photovoltaic parking shelter

6 DATA LOGGING SYSTEMS

Data recording systems have been installed, through which the production electricity data and any errors of the systems are recorded. Except of the safe operation of the systems, these data will be used for the implementation of future theses and systems' performance evaluation. Furthermore, three displays have been installed at each RES location in order to show the electric power, the produced energy and the avoided quantity of CO_2 emissions, on a daily basis.

7 CONCLUSIONS

The green character of the proposed activities in the community of AUA and the significant anticipated promotion of this project can act as an ideal developed environment in green utility. The transformation of the AUA community to "Green and Sustainable" will improve the quality of life at the campus. More in particular, the living conditions will be improved as the interventions in places where people work, will enhance the thermal comfort and upgrade the environmental sustainability performance of the university. The installation of PV systems will reduce the operation costs since 176 MWh/year will be produced and the biomass boiler will offer 14 MWh/year thermal produced energy. It is estimated that the exploitation of RES will reduce the AUA's carbon footprint at 1758 tnCO_2 per year. The electric vehicles will facilitate the access of students, staff and quests from the nearest metro station to AUA, with zero CO_2 emissions as the electric vehicles will be charged by solar energy. Finally, through the project activities, links between AUA and EU partners will be strengthened.

8 ACNOWLEDGMENT

The authors would like to acknowledge the funding authorities of the project, namely EEA Grants (85%), the Ministry of Economy and the Ministry of Environment and Energy of Greece (15%), as well as the Administration office and the Directorate of Technical Services of Agricultural University of Athens for their support on the project.

9 REFERENCES

[1] Kosmadakis, D. Manolakos, G. Papadakis. An investigation of design concepts and control strategies of a double-stage expansion solar organic Rankine cycle. International Journal of Sustainable Energy 2015, 34 (7), pp. 446-467.

[2] Karavas, C. S., Kyriakarakos, G., Arvanitis, K.G., Papadakis, G. A multi-agent decentralized energy management system based on distributed intelligence for the design and control of autonomous polygeneration microgrids. Energy Conversion and Management, 2015, 103, pp. 166-179.

[3] G. Kyriakarakos, D. D. Piromalis, K. G. Arvanitis, A. I. Dounis, G. Papadakis. On Battery-Less Autonomous Polygeneration Microgrids: Investigation of the Combined Hybrid Capacitors / Hydrogen Alternative. Energy Conversion and Management, 2015, 91, pp. 405-415

[4] Evangelos Dimitriou, Essam Sh. Mohamed, Christos Karavas and George Papadakis. Experimental comparison of the performance of two reverse osmosis desalination units equipped with energy recovery devices. Desalination and Water Treatment, 2015, 115, pp. 145-165.

[5] A.T. Balafoutis, E. Papageorgiou, Z. Dikopoulou, S. Fountas, G. Papadakis. Sunflower oil fuel for diesel engines: Experimental investigation and optimum engine setting evaluation using Multi-Criteria Decision Making approach. International Journal of Green Energy 2014, 11 (6), pp. 642-673

[6] D. Manolakos, G. Kosmadakis, S. Kyritsis, G. Papadakis. Identification of Behaviour and Evaluation of Performance of Small Scale, Low Temperature Organic Rankine Cycle Process coupled with RO Desalination Unit. Energy, Volume 34, Issue 6, June 2009, Pages 767-774.

[7] Cornell University: http://www.sustainablecampus.cornell.edu/: M.

[8] Harvard University: http://www.green.harvard.edu/

[9] Princeton University: http://www.princeton.edu/sustainability/

[10] University of California, Berkeley: http://sustainability.berkeley.edu/

[11] University of California, Davis: http://sustainability.ucdavis.edu/index.html

[12] University of Florida: http://www.sustainable.ufl.edu/

RELIABILITY AND DURABILITY OF COMPLETE POLYMER MATERIALS FOR BIPV APPLICATION

S. Boddaert[1], L. Bailly[2], C. Baguenard[2], S. Bourrigaud[3], M. Chaillou[4]

1 - CSTB, 290 route des lucioles, BP209, 06904 Sophia Antipolis CEDEX - France
2 - CANOE, 16 avenue Pey Berland, 33600 PESSAC – France
3 - ARKEMA – Lacq – France
4 - INNOVEOX, 14 Avenue Neil Armstrong, 33700 MERIGNAC - France

simon.boddaert@cstb.fr

ABSTRACT: Standards modules development focused on mass market and standards frame size and components, allowing to reach affordable prizes and large automation and optimized process. Unfortunately this standardization, limits the specific building applications and really adapted components to ensure building and envelope functions. To replace Glass cover is a fundamental step towards the definition of new components and new functionality. While the use of glass products do not just impact production, it also brings many strengths such as mechanical and fire resistance, durability and reliability. But it is weighty and reduces drastically integration possibilities. Moreover glass fabrication is time and energy consuming. Aim of this article is to promote/show research conducted during ISOCEL project, on the use and optimization of polymeric materials to produce photovoltaic modules specifically dedicated to building applications.
Keywords : Polymer, BIPV, Building code requirements, Energy.

1 INTRODUCTION

The use of polymeric materials is common but remains applicable to niche markets due to excessively high prices and manual manufacturing processes. Our approach consists of offsetting a higher price of the raw material by considerably reducing the manufacturing time of the modules and retaining the construction requirements of the standards modules. The ISOCEL project aims at defining a perfectly integration component (dedicated to building functions), but also on the ease of recycling these components. The advantages of recycling by the use of acrylics will not be presented in this article.

The dissemination of the standards 50583-1 [1] and 50583-2 [2] defining the requirements of the BIPV components and systems make it possible to better identify the implementation solutions in the building. Thus, the use of polymers allows modifications and adaptations of shape, thermal, optical and mechanical properties which the traditional components (glass) do not allow. Indeed, the use of polymer components permits an optimal architectural, technical and aesthetic adaptation without to deem to a secondary role the performance of power generation by a BIPV component.

1.1 Building requirement

What slows down the use of BIPV components is mainly due to geometric constraints or unacceptable cost levels for construction or renovation projects.

The work provided by NREL [3] and consolidated by the Architectural Solar Association, ASA [4], highlight the extra costs associated with the use of BIPV components is at least 27%. This cost can increase drastically if BIPV component have to ensure building material or structure component.

1.2 Reduction cost objectives

The ambition of ISOCEL project is not only to be able to offer suitable components but also has a more acceptable price/function ratio. The use of acrylic polymers makes it possible to use radically different crosslinking solutions based on UV and reducing curing time. The properties of the acrylic compounds are thus used, transparent to UV radiation.

The first tests carried out CANOE using photocrosslinkable components being positive, the optimization actions are continued with the aim of having adhesive properties identical to components such as EVA and a reticulation time divided by 10 compared with to a normal cure solution.

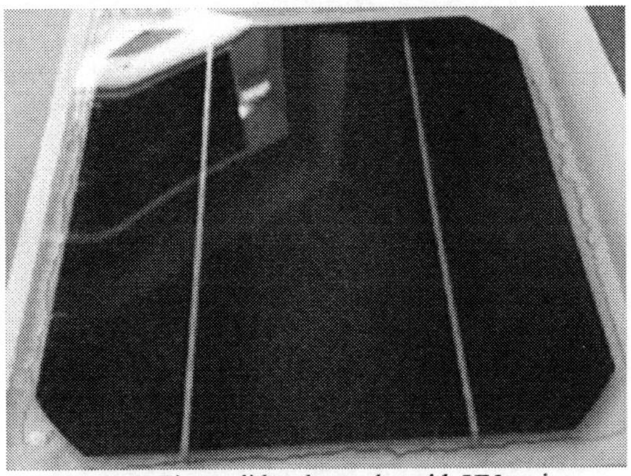

Figure 1: First validated samples with UV curing

1.3 Energy production

The technological choice also favors the stability or even the increase of electrical production. The acrylic components have an optical transmission of at least 0.92 for conventional components, which can be easily improved with specific coatings reducing reflection or haze effects. To promote the production of energy reduce the LCOE, critical indicator of the current BIPV components.

1.4 Weakness of polymers

Although chemical engineering now allows many combinations of component grades, it is still difficult to be able to surpass in all fields glass products (traditional module components), particularly in the fields of reliability and durability, making it possible to limit the LCOE.

2 FIELDS OF INVESTIGATION

The principle of photocrosslinking is based on statistical activation of all the crosslinking promoters during the manufacturing phase, which is a delicate point that has had to be treated. The current modules having 25 to 30 years of guarantee, it is essential to offer guarantees covering the same period. It is therefore essential to ensure an identical level of adhesion and to avoid all the delamination constraints.

Figure 2: Main problem encountered during the qualification phase – delamination.

The first part of this work consists of optimizing the formulation of the front and back barrier layers, notably on thermal aspects, scratch resistance, optical transmission. Acrylic formulation allows modifying and to select the compounds according to the criteria identified. Second part deals with UV resistance and fire resistance which are most challenging criteria to develop base acrylic BIPV components. We focus our investigations on these two elements, to aim for properties as close as possible to Glass/PVF PV modules.

2.1 Combined UV test and Fire test

The fire behavior is the greatest constraint, which today is still being optimized because whatever the flame retardant component it is particularly complex to achieve the properties of the glass. ARKEMA provides us with acrylic compounds using two different bases of fireproofing. Acrylics without charge are inert to UV radiation, but encapsulants components and fireproofing additives can greatly change this behavior

Fire and UV test are conducted to assess before and after UV accelerated aging. The aim is to determine the time resistance of the flame retardant components and the maintenance of the thermal and optical properties after aging

2.2 Fire test

Fire tests are carried out according to standard NF P 92-505 [5] and standard NF P 92-501 [6]. The aim being to target a product whose properties are more closely related to glass products. ARKEMA, supplied to us as a reference product of the Altuglas® V920T, which meets the optical requirements of the project. This is a Polymethyl Methacrylate Acrylic product (PMMA) that is widely available, endowed with interesting manufacturing properties and competitive prices. These components are used as the basic element; flame retardant grades are then produced and tested. The first flame-retardant tests are carried out with a fluoride base (usual solution as PVF or PVDF solution) and then based on different grades of phosphorus, specifically developed bay ARKEMA for this project.

2.3 UV test

These tests are carried out to measure the mechanical and optical characteristics, in the initial state and after simulated aging, using a WOM. Combined UV Cycle and damp heat test cycle are carried out on PMMA samples, but only on specimens with flame-retardant compound.

UV exposure is achieved with a WOM (ATLAS CI5000), equipped with a xenon arc lamp simulating the overall solar spectrum received at ground level and two borosilicate glass filters. The simulated aging tests are carried out according to the NF EN ISO 4892-2 method A cycle no. 1. Exposure time is 2000 hours to within 3%. Then UV and damp heat cycles are carried out with climatic chamber (CLIMATS EXCAL, 2211-HE) with an exposure bench according to Annex A of the standard NF EN ISO 12543-4: 2011. Mechanical and optical tests are carried out again after the 4000 hours total aging accelerated to determine the combined impact of UV and damp heat.

3 TEST and RESULTS

In view of the fact that these components must be used as BIPV components, it is therefore essential that their performance be as close as possible to that of traditional components (PV module), but they must also meet safety and durability criteria of the building.

3.1 Initial test

Altuglas® V920T is a Polymethyl Methacrylate Acrylic product, its optical performance are often take as a reference in polymers compounds and seems to be a perfect challenger to replace glass layer. All initial tests are carried out with this component as a reference element. The optimizations are carried out according to the results obtained and collaborative work with all project partners.

$$q = \frac{100\sum h}{t_i\sqrt{\Delta t}}$$

Equation 1: Fire index calculation, q .

The fire classification index, q, is the criterion for validating the choice of materials. The economic analysis is then carried out to validate the competitiveness of the solution.

Table I: Data for index calculation, conventional PMMA

Sample reference	ti	Δt	Σh	hmax
Reference CN	52	987	1737	100
Reference S.Up	59	1024	1568	98
Reference S.Up 2	62	982	1865	93
Reference CN3	55	925	1723	101

All these data make it possible to determine the classification index but above all they make it possible to determine the fire classification index M (described according to the standard NF P 92 501 -Radiation test), which is a necessary reference for the uses of components in the building. Traditional photovoltaic modules are generally classified M0, i.e. non-combustible. Thus, with the PMMA-based components on the front and rear of the photovoltaic modules, the classification obtained is shown in the table hereafter.

Table II: Fire index results

Sample reference	q	Mclass	$M_{class,mean}$
Reference CN	94,08	M4	M4
Reference S.Up	92,54	M4	M4
Reference S.Up 2	91,23	M4	M4
Reference CN3	96,2	M4	M4

The results obtained are globally bad for built-in materials (M4, that is qmean ≥ 50) because all grades of PMMA (standard, shielUp, optical coating) do not allow a fire classification allowing a safe use in the building (minimum M3 see below> M0). ARKEMA will provide us with many flame retardant product grades whose properties will also retain a durable and competitive product in terms of price.

3.2 Optimization test on fireproofing

Two solutions are particularly studied, the first one based on fluorine which is a component commonly used as a flame retardant, the second based on phosphor, cheaper but more delicate to use in solutions with copolymers because they are much more delicate to stabilize.

Table III: Fire index results with fire-retardant upgrade

Sample reference	q	Mclass	$M_{class,mean}$
Reference CN	94,08	M4	M4
Reference CN F_i	85,19	M4	M4
Reference CN F_f	62,20	M4	M4
Reference CN PO_i	40,18	M3	M3
Reference CN PO_f	28,20	M3	M3

Table III shows the results obtained with the new fireproofing materials and allows passing to class M3 (q_{mean} <50), thanks to the use of phosphorus-based product. These results are encouraging because the rate on the NC PO_i sample is only 9.5%. (Estimated at 14.5% on the NC PO_f sample).

A rate and an optimized formulation is currently being validated in order to reach a q_{mean} <15 index, which would allow an M2 ranking to be achieved. If we succeed in reaching this ranking, the use of PMMA could find a great deal in the building components, starting with the BIPV components.

3.3 Accelerating aging

The second part deals with the optimization of UV crosslinking and the management of the coefficients of adhesion and the proportions of elastomers. UV curing is particularly complex. It is necessary to determine the power, the exposure time the wavelengths and especially the exposure method for correct process. Under UV radiation, radicals are released and initiate the polymerization. Drying can be controlled and performed in several steps.

The UV tests carried out demonstrate that the crosslinking is not complete during the manufacturing phase. Indeed many tests with WOM but also in natural sunlight demonstrate that the reaction of crosslinking in the acrylic encapsulants is continued and includes a strong optical degradation.

Figure 3: UV aging effect – optical degradation

This phase is particularly difficult because it assumes that all UV promoters have reacted in addition to optical degradation, the adhesion of the components is greatly decreased, resulting in delamination or mechanical stress effects. The acrylic encapsulants having been modified in order to achieve a reaction of the 100% crosslinking promoters (optimization photocurable acrylic resin, FIG. 4), making it possible to obtain perfect samples.

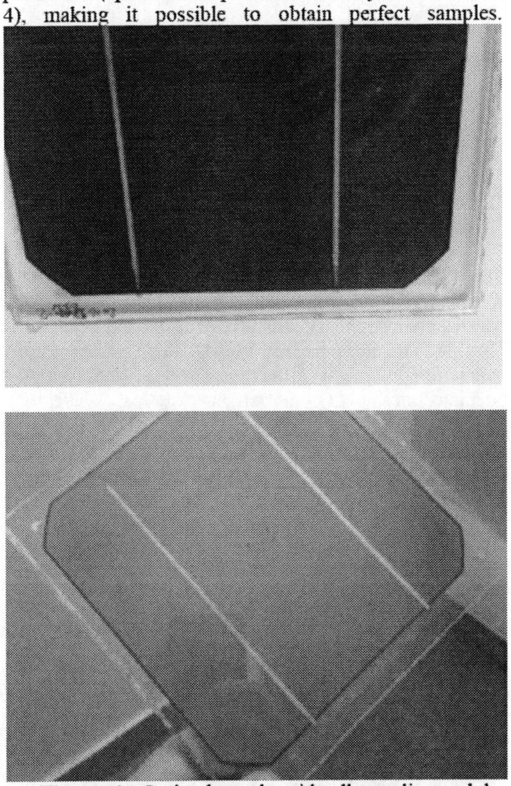

Figure 4: Optimal result with all acrylic module, (front layer, photocurable layer, back layer) – Front and back view.

The results obtained during the fire tests promote the use of flame-retardants to meet regulatory requirements. However, the chemical modification of the acrylics can lead to considerable modification during the manufacturing steps.

The UV tests and the damp heat tests are thus

preliminarily carried out on samples of flameproofed products of the CN PO_f type, being the best performers. The optical and mechanical characterizations are thus carried out before and after the WOM exposures in order to assess durability after accelerating aging.

Figure 5: Sample with flam-retardant in initial state.

In the initial state the component is as transparent as glass for a thickness of 3 mm. The latter is doped with the most efficient flame retardant solution based on phosphorus.

Table IV: Initial transmission and reflection factor (%)

Sample reference	τ_v^{nh}	τ_e^{nh}	ρ_v^{nh}	ρ_e^{nh}
Reference CN	92	84	7	7
Reference CN F_f	91	83	7	7
Reference CN PO_i	92	83	7	7
Reference CN PO_f	92	84	7	7

The NC PO_f component is the most effective element in achieving the ISOCEL project objectives. The thermal cycles are carried out in order to study the behavior of PMMA with fire-retardant. PMMA alone exhibits no UV alteration, which is different with this sample, which demonstrates an important optical alteration.

Figure 6: Sample after 126 damp heat cycle.

The results obtained on damp heat tests are particularly disappointing, the stability of the flame retardant compound is not good and a high migration on the faces of the component is identified. The surface is also altered and is no longer smooth.

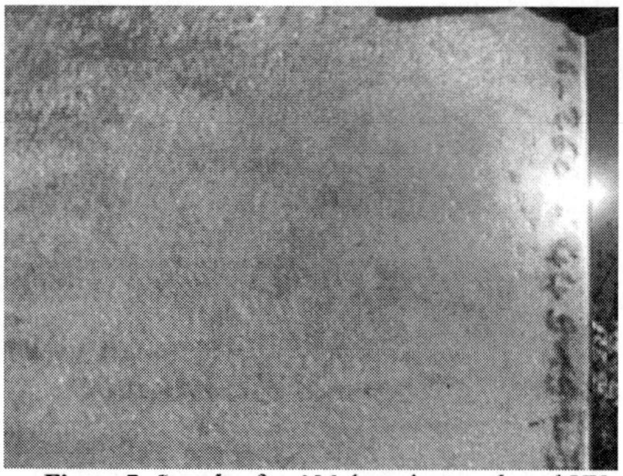

Figure 7: Sample after 126 damp heat cycle and UV aging.

The UV tests are also carried out in order to carry out the procedure in full. The optical degradations have increased considerably but a strong deformation of the component is also noted during the tests. These samples are then taken to the fire laboratory to assess whether the flame retardant component is still effective after the accelerated aging tests.

Table V: Final transmission and reflection factor (%)

Sample reference	τ_v^{nh}	τ_e^{nh}	ρ_v^{nh}	ρ_e^{nh}
Reference CN	92	84	7	7
Reference CN F_f	78	69	18	19
Reference CN PO_i	72	65	26	26
Reference CN PO_f	72	65	23	24

The optical results show a sharp decrease in optical performance with a sharp drop in optical transmission and a reflection that can't be used for solutions aimed at increasing power production.

3.4 Final fire test.

The aged components are then tested according to the same requirements as the non-aged components. The objective is to validate that despite the optical alterations, the behavior to the fire remains good.

Table VI: Fire index results with fire-retardant upgrade

Sample reference	q	Mclass	$M_{class,mean}$
Reference CN $PO_{f\text{-}init}$	28,20	M3	M3
Reference CN $PO_{f,aged}$	32,68	M3	M3

The fire results remain very good despite the optical alterations, which confirm that the flame retardant is always in a high proportion to protect the compounds.

It is also necessary to carry out mechanical tests to ensure that only optical degradations are observed.

4 MECHANICAL TESTS

As final validation, we perform mechanical behavior tests. The most important element for components of construction is to be able to resist despite the effects of aging.

Thus, some of the aged materials are mechanically characterized in order to ensure that the optical modifications do not lead to mechanical changes. It is acceptable to decrease the electricity production, but it is unfortunate to promote components so the mechanical proprieties can be dangerous for building and users. Mechanical characteristics in tension are determined in terms of the force and elongation of a test piece stretched along its main axis at a constant speed. The data are measured using an electromechanical machine (ZWICK-1494) equipped with a 2.5 kN force cell. The tests are carried out in accordance with NF EN ISO 527-1 [7].

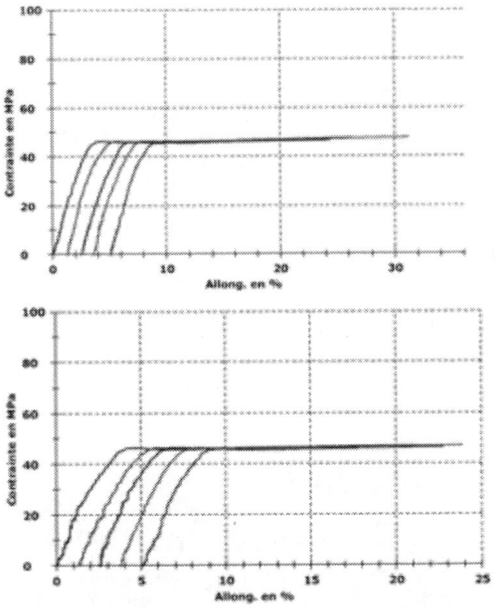

Figure 8: Elongation at tensile strength before and after aging tests.

Table VII: Traction characteristics.

Sample reference	E_t	σ_B	E_B
Mean	1577	47,0	23
Deviation	116	0,5	5
Dispersion	7	1	24
Sample aging	E_t	σ_B	E_B
Mean	1561	46,8	19
Deviation	122	0,3	1
Dispersion	8	1	5

With Et, elasticity module (MPa), σB, stress at break (MPa) and Eb deformation at break in %. We observe that the mechanical strength is almost constant despite the optical changes observed. It thus appears that the constitutions of the component are only slightly modified by aging.

5 CONCLUSION AND PERSPECTIVE

All these tests are carried out with the aim of defining the compatibility of the acrylic composites in the building. The BIPV components being always integrated with or near the construction members, it is imperative to validate the fire resistance, the mechanical and optical behavior of the BIPV components, which are likely to experience an important expansion in the years to come.

This work presents the testing process, which allowed validating the choices of the components used in order to propose a credible alternative to the traditional photovoltaic components.

There remains an important issue to be addressed, the optical reliability of flame retardant components. Indeed if in terms of safety the objectives seem to have been reached, the optical part leads to a strong alteration of the electrical production and therefore a strong increase in the LCOE specific to these components, which is not acceptable for these components to be massively used.

On the other hand the ease and the speed of realization of composites fully in acrylic presents an advantage that must be maintained because it is an important way of progress with a reduction of the manufacturing time which can reach 3 to 5 times the usual manufacturing time of the conventional modules.

Figure 9: 4-Cells acrylic module with JB and wire.

If all the work carried out today makes it possible to validate the entire process on small modules, important work still needs to be done to be able to produce modules of standard size. The current work deals with the validation of 24-cell modules, with the ultimate objective of achieving standard modules within 12 months. LCA analyzes will also be carried out to demonstrate the advantages of acrylic components in the recycling process.

6 REFERENCES

[1] EN 50583-1: 2016 Photovoltaics in buildings. BIPV modules
[2] EN 50583-2: 2016 Photovoltaics in buildings. BIPV systems
[3] NREL, U.S. Solar Photovoltaic System Cost Benchmark: Q1 2016
[4] Global business opportunities for architectural solar – Solar Plaza webinar / 22 February 2017
[5] NF – P 92-505 Test for fuse materials
[6] NF – P 92-501 Radiation test
[7] ISO 527-1, Plastics - Determination of tensile properties - Part 1 : general principles

ACKNOWLEDGEMENT

This research program was supported by French National Energy agency, ADEME, in Future Investment Program. All the partners of the project, industrialists and laboratories have been involved in this work to define a BIPV component combining aesthetics and performance.

GRAFFITI ON SOLAR NOISE BARRIERS
A CASE STUDY

Chris Tzikas[1], Minne de Jong[1], Lenneke Slooff[2], Michael Debije[3], Stijn Verkuilen[4], Wiep Folkerts[1]

[1]SEAC (Solar Energy Application Centre), Eindhoven, The Netherlands
[2]ECN (Energy research Centre of the Netherlands), Petten, The Netherlands
[3]TU/e Eindhoven University of Technology, Eindhoven, The Netherlands
[4]Heijmans, Rosmalen, The Netherlands

ABSTRACT: Infrastructure Integrated PhotoVoltaics (I2PV) is the next big step for large scale energy production. PV cells can be integrated in a wide range of infrastructure applications without requiring any additional space. In this way the infrastructure can have the additional functionality of producing electricity from sunlight. One of the promising I2PV applications are the PhotoVoltaic Noise Barriers (PVNBs). In this context the Solar Noise Barriers (SONOB) project was developed with the goal of studying the performance of PVNBs. In this case study, we study the effects of a common form of street pollution -graffiti- on PVNBs.
Keywords: Bifacial, System Performance, Graffiti, Shading, Infrastructure

1 INTRODUCTION

The aim of the SONOB project was the development of a breakthrough modular solar noise barrier concept. For this purpose, two PVNBs were constructed in the Living Lab, located at the Randweg in the South-West of 's Hertogenbosch, the Netherlands. The two barriers were placed in an orthogonal orientation to each other. One barrier faced South and North, the other faced East and West. The barriers inclined 15° away from the road: to the North and to the East, respectively. The Living Lab was officially opened on June 18th 2015, attracting national and international attention.

The design of the barrier was based on the Modulair GeluidsScherm (MGS) design directives of Rijkswaterstaat. It consisted of 4 cassettes of 5 x 1 m^2 equipped with, from bottom to top, monofacial polycrystalline c-Si, bifacial monocrystalline c-Si, an orange luminescent solar concentrator (LSC) and a Red LSC plate (see Figure 1) [1][2][3].

Figure 1: Depiction of the Living Lab prototype PVNB.

In this case study, we aim to investigate the effect that a common kind of street pollution -graffiti- has on the performance of a PVNB [4]. In this paper, the results comparing the effects of surface artwork on the monofacial and bifacial c-Si panels of the East/West oriented PVNB will be presented. Each panel was divided into two separate sub-panels consisting of 48 c-Si cells each (see Figure 2).

Prior to integration of the panels into the noise barrier prototype, all panels were flashed, showing an average rated power of 178 Wp for the monofacial panels, 207 Wp for the front side of the bifacial panels and 195 Wp for the rear side (bifaciality factor of ~94%). For the manufacture of the PVNB, both the monofacial and the bifacial c-Si cells were laminated between thick glass plates.

Figure 2: Closer depiction of the monofacial (bottom) and bifacial (top) c-Si cells of the PVNB.

To evaluate and analyse the electrical output of the PVNB, the Living Lab was equipped with measurement equipment for monitoring PV output (using IV tracers), solar irradiation, cell temperature and general weather data. The IV-tracer was controlled by a laptop computer and recorded measurements every 2 minutes for each of the panels, providing characteristic current *vs* voltage curves in order to study the effect of shading imposed by the graffiti. Every 24 hours, a summary of the measured data, consisting of irradiance data, V_{oc}, I_{sc}, V_{mpp}, I_{mpp}, cell temperatures and weather, data was automatically compiled and sent to the central SEAC server. In this way we were able to analyze the data remotely.

2 APPLICATION OF THE GRAFFITI

Graffiti is a potential risk for solar noise barriers, as it imposes a permanent, direct form of shading, blocking the incident light from reaching the PV cells and affecting the output of the PVNB. In the Living Lab we studied this effect in detail. Furthermore, we have treated a part of the barrier with anti-graffiti coating to study its effectiveness.

Figure 3: Artistic graffiti by Gart Smits on the East/West PVNB applied for research purposes.

Prior to the application of the street art, roughly half of the surface of the barrier was coated with a commercial anti-graffiti coating. The anti-graffiti coating was applied on the right half of the East/West PVNB while the left side remained uncoated for comparison reasons. The graffiti artwork represents a portrait of Nikola Tesla, a pioneer in electricity and the designer of our modern day electricity grid (see Figure 3). The portrait was produced by graffiti artist Gart Smits over a period of eight hours using commercial spray paints.

This experiment took place on the East/West PVNB while the South/North facing one remained undisturbed and acted as a reference for comparison purposes. The image effectively blocked incident sunlight from the West in the covered areas.

The purpose of the anti-graffiti coating application was to investigate the ease with which graffiti can be removed from a PVNB and the affect on panel performance after the removal of the artwork. A secondary analysis was performed to investigate the effect that this coating may have on the reflectance properties of the panel surface and thus to the energy production of the noise barrier.

3 PERFORMANCE RESULTS

The anti-graffiti coating application took place a few weeks before the application of the graffiti in order for its effect to be studied. In Figure 4 a single, clear day is chosen as representative for comparison between the coated and uncoated parts of the same panel.

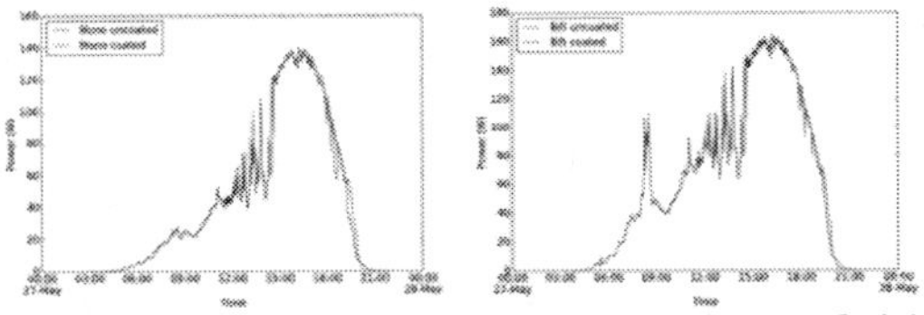

Figure 4: Power of the coated and uncoated mono-facial (left) and bi-facial (right) parts of the same panel of the East/West PVNB.

As shown in Figure 4, there is almost no negative effect on the power production of the anti-graffiti coated side of the panel which produces exactly the same power as the uncoated side. This result was consistent for the other days measured.

The graffiti remained on the PVNB surface for an entire summer month (June 2016) and its effect was studied using the DC Performance Ratio (PR) quality factor. The reason for selecting this factor was the fact that it is largely independent of the orientation, location, incident solar irradiation and conversion efficiency of a PV system [2]. It is a factor that describes the relationship between the actual and theoretical energy outputs of a PV system and gives an indication of how efficient a PV system is.

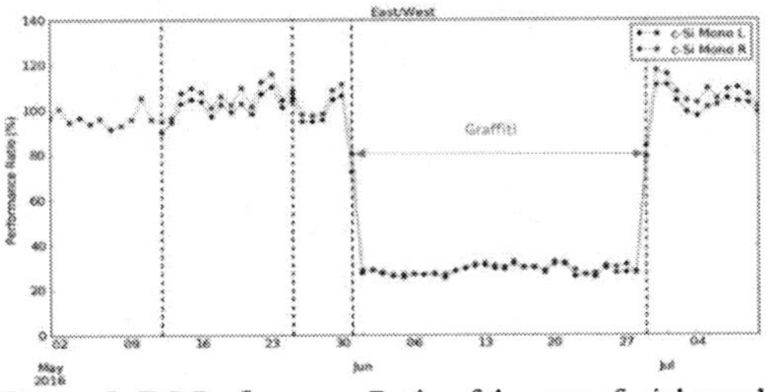

Figure 5: DC Performance Ratio of the monofacial panel as a function of time.

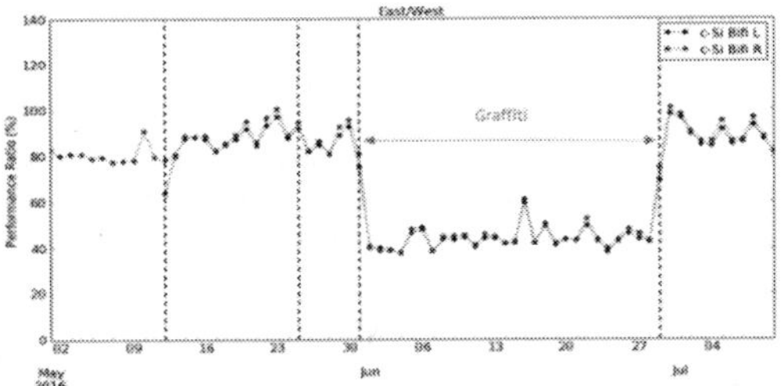

Figure 6: DC Performance Ratio of the bifacial panel as a function of time.

After the application of the graffiti the performance ratio of both the monofacial and bifacial modules dropped dramatically, as expected [5][6]. As can be seen in Figure 5 and Figure 6, the performance ratio of the monofacial panel dropped to below 30% and remained almost stable during all the days of the experiment, as both the direct and the diffuse light were blocked.

For the bifacial panel, the performance ratio was slightly higher as it dropped to just above 40%. This is primarily due to the fact that the rear side of the panel remained unaffected by the graffiti receiving light and slightly boosting the overall performance, although the coverage on this panel by the artwork was also somewhgat more restricted, which would also influence the PR.

4 REMOVAL OF THE GRAFFITI

One of the primary goals of the anti-graffiti coating application was to investigate the ease with which graffiti can be removed from a glass surface and whether the PV performance is recovered. For the removal of the graffiti, high-pressure warm water was used for the anti-graffiti coated part of the noise barrier, while application of a solvent was also needed for the uncoated part. The graffiti from the orange LSC plate was intentionally not removed.

After the removal of the graffiti the performance ratios of both panels were fully restored and appeared to be as high as before its application (see Figure 5 and Figure 6).

Following the completion a full year of field testing and the graffiti case study for the PVNB, the test field was dismantled according to the project plan in July 2016 and the different plates were transported for further studies.

Figure 7: Depiction of the PVNB after the removal of the graffiti.

5 CONCLUSIONS

After the application of the graffiti the DC performance ratio (PR) of both the monofacial and bifacial panels dropped dramatically, as expected. In the case of the bifacial panel, the PR was measured to be about 40% higher than the monofacial panel. This is partially due to the fact that the rear side of the panel remained unaffected by the graffiti, receiving light and slightly boosting the overall performance. After the removal of the graffiti the performance ratios of both panels were fully restored and measured to be as high as before its application.

An anti-graffiti coating was applied in the right half of the noise barrier. The performance comparison showed no negative effect in the coated area of the barrier in comparison with the uncoated one. For the removal of the graffiti, high-pressure warm water was used for the anti-graffiti coated part while also the application of a solvent was needed for the uncoated part. This was a very useful outcome for developers of commercial PVNBs as it constitutes an inexpensive solution that has the potential to significantly reduce the Operation & Maintenance costs of such an application.

6 REFERENCES

[1] M. Debije, P. Verbunt, *Adv. Energy Mater.* **2012**, *2*, 12–35.
[2] M. Kanellis, M. M. de Jong, L. Slooff, M. G. Debije, *Renew. Energy* **2016**, *103*, 647–652.
[3] M. G. Debije, C. Tzikas, V. A. Rajkumar, M. M. de Jong, *Renew. Energ.* **2017**, *113*, 1288–1292.
[4] C. McAuliffe, *J. Urban Aff.* **2012**, *34*, 189–206.
[5] A. Dolara, G. C. Lazaroiu, S. Leva, G. Manzolini, *Energy* **2013**, *55*, 466–475.
[6] J. Guerrero, Y. Muñoz, F. Ibáñez, A. Ospino, *IOP Conf. Ser. Mater. Sci. Eng.* **2014**, *59*, 12007.

ASSESSING FAÇADE-INTEGRATED PHOTOVOLTAICS: A METHODOLOGY FOR THEIR PRELIMINARY ASSESSMENT

Yanica Zammit and Simon Paul Borg
Department of Environmental Design, Faculty for the Built Environment, University of Malta, Msida, Malta
Yanica Zammit (yanicazammit@gmail.com); Simon P. Borg (simon.p.borg@um.edu.mt)

ABSTRACT: One strategy typically utilized to reduce the energy demand of buildings is the use of onsite renewables, typically photovoltaic systems. In this context, whereas the roof space allocated for photovoltaics is quickly reaching saturation, façade-integrated photovoltaics could be a good alternative. Unfortunately, however, compared to roof-installed photovoltaics, such systems require a higher design commitment, frequently accompanied with cost uncertainties, which are often cumbersome to assess in traditional building design. To address this the research being presented in this paper, describes a methodology which would enable building designers to preliminary assess their proposed design technical and economic potential in a quick and easy manner. The developed methodology was tested on one particular building, showing how the technical and economic performance varies depending of the inputs set by the building designer. Additionally, a section was also included to show the effect of façade orientation on the total monthly and annual energy yield, showing how façade-integrated photovoltaics which are not predominantly South facing should not be discarded *a priori* for not having an optimum orientation.
Keywords: *Façade-Integrated Photovoltaics; Energy Performance; Feasibility Analysis*

1 INTRODUCTION

In addressing the challenges posed in reducing the energy demand of buildings, two main strategies are typically utilised. A first strategy relies on reducing the energy demand of a building by improving its energy-efficiency [1-3]. A second strategy relies on utilising onsite renewable energy sources for energy generation [1, 4]. In this second strategy, although both wind and solar [4] have been considered as a potential renewable energy resource for use in the urban environment, solar driven renewable energy has been the technology which has been most successful in terms of market penetration [5, 6].

In most buildings, such a resource has so far most easily been exploited through roof-installed photovoltaics [7], as an add-on, away from the public eye. Space considerations, competing amenities and roof ownership issues can however limit the potential use of such a resource [8], especially in densely populated areas, where land uptake is at a premium. For this reason, building designers are now looking at building-integrated photovoltaics, more specifically, façade-integrated photovoltaics as an alternative for traditional façades [9, 10]. Such a design possibility is dictated not only from the fact that the façade is the centrepiece of a building, therefore an aspect which must be aesthetically prominent in nature, but also due to the fact that it is estimated that in urban areas, façades comprise 60-80% of a building surface [11].

Photovoltaic panels have, over time evolved greatly, in terms of their aesthetics, efficiency and customisation and a number of companies now provide innovative and customised products offering endless possibilities for building designers, to integrate such technologies in their building designs.

1.1 Façade-Integrated Photovoltaics

Façade-integrated photovoltaics have shown a high energy generation potential [12], however, compared to roof-installed photovoltaics, such systems require a higher design commitment, frequently accompanied with cost uncertainties, which are often cumbersome to assess in traditional building design.

Most of the research which has been carried out in this area currently presents only complex methods of analysing the problem, often leaving aside the cost issue [13], or if both aspects are considered, the results are then often very site specific [14]. This is of course a limitation for most professionals who often find the process of going into such detail a lengthy process.

In order to encourage the architectural integration of photovoltaics on façades, and ensure that such a way of designing buildings is seriously considered, there needs to be a simple, readily available methodology in place, to make sure that quick and easy decisions can be made by building professionals.

2 STUDY CARRIED OUT

2.1 Aim of Research

Based on the premise presented in the introduction, the primary aim of this study was to create a methodology which would enable building designers to preliminary assess their proposed design in a quick and easy manner, concurrently, investigating the technical and economic potential of façade-integrated photovoltaics. In the process, the impacts of various technical aspects, such as orientation and shading, were studied to determine optimal solutions. Cost uncertainties and economic aspects of façade-integrated photovoltaics were addressed through a purposely designed tool, assessing the system feasibility. With this tool the sensitivity of a particular project to varying fiscal incentives such as grants on capital investment and, or Feed-in-Tariffs (FIT) can be analysed.

The research which is hereby being presented utilises the particular case of façade-integrated photovoltaics on Maltese buildings as an example, however as discussed, the methodology is flexible and can be used anywhere, with the aim of encouraging designers and developers, to incorporate façade-integrated photovoltaics in their design.

2.2 Methodology

The methodology being presented here was split up into two main parts; a technical and an economic part, with the former feeding the latter.

For the purpose of testing out the methodology being proposed, and to show how the methodology can be applied in a design process, the parameters of a real life building in Malta (a free-standing building with a combination of glazed and cladded façades, facing various directions), were utilised.

The first part of the methodology aims at presenting the building designer with a series of step-by-step guidelines which can be used to technically assess the proposed project. Specifically, the guidelines are aimed at helping the designer analysing the energy generation potential of the project, based on site and project specific considerations such as latitude, orientation, shading from opposing buildings, usable frontage area (defined as the area which could potentially be covered by photovoltaics), tilt angle, rated maximum power and panel area of the proposed photovoltaic product.

The purpose of this part of the methodology is therefore to help the designer analyse the sensitivity of selected technical aspects on the energy performance of the proposed project, *e.g.* the sensitivity of energy generation potential to orientation for varying azimuth angles, as well as the effect of shading from opposing buildings. The latter was specifically modelled on the basis of an analysis carried out on different *façade height-to-(road) width ratios* of a typical urban/city scape and the resulting shading patterns, as shown in Figure 1. Table I shows the calculated *Shading Factors* (SF).

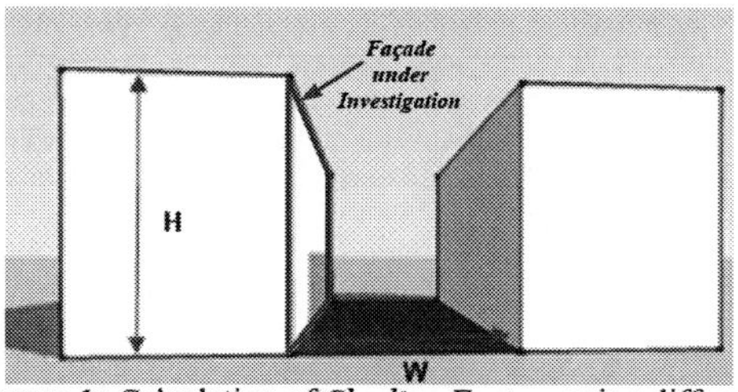

Figure 1: Calculation of *Shading Factors* using different façade height-to-(road) width ratios

Table I: *Shading Factors* for different façade height-to-(road) width ratios

Shading Factor (SF)	*Façade height-to-(road) width ratio*
1	1 >1:1 or for a property with an unobstructed frontage
0.8	>1:0.9 and <1:1
0.7	>1:0.7 and <1:0.9
0.55	>1:0.5 and <1:0.7
0.3	>1:0.3 and <1:0.5

An online tool, such as PVGIS [15], can then be used for simulation purposes to determine the potential annual energy generated.

The second part of the methodology makes use of a purposely designed tool, currently in the form of an excel sheet (but with the potential of becoming an interactive app), aimed at assessing the sensitivity of a proposed façade-integrated photovoltaic project to various fiscal parameters. The tool, feeds off the results produced in the technical part of the methodology to provide feasibility results in terms of the payback period of the project, the present worth (PW), specifically the Net Present Value (NPV) and the internal rate of return (IRR).

Project-dependent and independent variables can be introduced as input data for the model. These are financial elements which may or may not vary from one building to another. Project-independent variables include the inflation rate, the depreciated rate in energy generation, and the discount rate for future cash flow. Project-dependent variables include the usable frontage area and the grid electricity tariff.

The tool can consider both full-cost models, therefore excluding possible capital grants on the investment costs or FIT, and fiscally aided models, thus assessing the impact of fiscal incentives on the project's feasibility.

3 RESULTS

3.1 Initial Results: Creation of the Methodology

An initial result obtained through this research is as discussed the creation of a methodology which can be utilised for the preliminary technical analysis of façade-integrated photovoltaics. A high-level flowchart of the methodology being presented is shown in Figure 2.

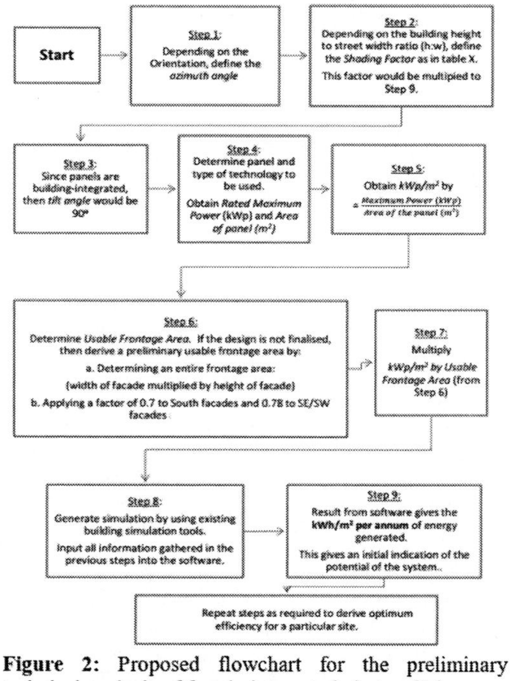

Figure 2: Proposed flowchart for the preliminary technical analysis of façade-integrated photovoltaics

The methodology provides a step-by-step approach on how to assess the preliminary technical parameters of façade-integrated photovoltaic technologies for any given project. This would allow the user to determine the optimal integration of façade-integrated photovoltaic for any building, with the aim of generating the highest possible energy generation levels. Also this way, designers can adopt a more design oriented approach, analysing what works best for their particular building, rather than applying general rules which do not necessarily apply for all case studies.

3.2 Specific Results from the Research - Energy

Figure 3 shows the building used to test the methodology - the newly built ICT Faculty at the University of Malta campus in Msida. It is a free-standing building with a combination of glazed and cladded façades, facing various directions

Figure 3: Educational/Office Building used to test the methodology (*Source*: Author)

The ICT Faculty was chosen as a case study in order to assess the potential of different types of photovoltaic systems, mainly crystalline photovoltaic panels and transparent photovoltaic glass. Different types of PV technologies were modelled and simulated in order to assess the feasibility of each different type of technology, with the usable frontage area necessarily varied to accommodate the different proposed designs. Each façade was modelled using its own azimuth angle, assuming a *Shading Factor* of 1. The existing façades are fully glazed, although with glazing units having different transparency levels. The lower and upper parts of the glazing for each floor are of a translucent nature, while the middle part is transparent to permit natural light ingress and outside views.

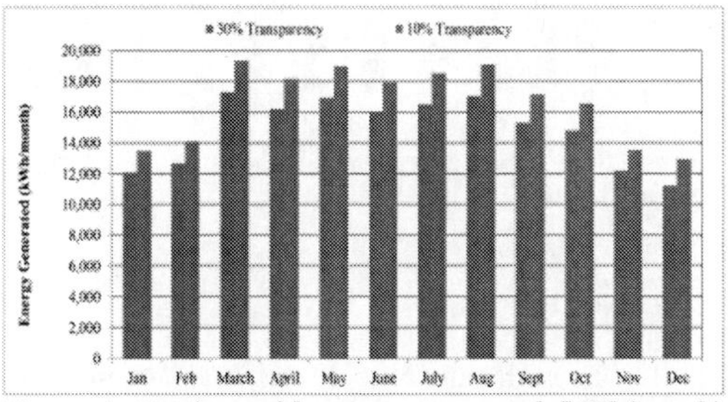

Figure 4: Total monthly energy generated (kWh/month) (*System 1:* Opaque crystalline and 30% transparent photovoltaic glass *vs. System 2:* Opaque crystalline and 10% transparent photovoltaic glass)

In the research being presented, the methodology was used to compare the energy performance of two types of façade-integrated photovoltaic projects. A first system, System 1 utilising 30% transparent photovoltaic glass, and a second system, System 2 utilising 10% transparent photovoltaic glass as a façade retrofit for the middle transparent part. The lower and upper glazed areas, requiring no transparency given their position were designed to be covered in both cases with opaque crystalline photovoltaic panels. The usable frontage area considered in both cases assumed an area of $1,170m^2$ covered by opaque crystalline photovoltaic panels and an area of $642m^2$ covered by transparent photovoltaic glass. Figure 4 shows the total monthly energy generated by the building for the two systems proposed.

3.2 Specific Results from the Research - Financial

Once the energy generation estimates are obtained for a proposed project, the tool designed to assess the financial performance of the proposed project, can then use this data to obtain a preliminary cost benefit analysis of the project.

As discussed earlier the methodology can cater for any type of financial scenario, including full cost models where the revenue stream is derived from the saved electricity costs (1), or fiscally aided models where FIT are utilised as financing mechanisms for grid fed electricity (2). Likewise, any capital grant (G) on the investment costs can be included or excluded.

$$NPV = -(1-G)I - \sum_{n=1}^{N} \frac{C\&M}{(1+r)^n} + \sum_{n=1}^{N} \frac{E \times c}{(1+r)^n} \quad (1)$$

$$NPV = -(1-G)I - \sum_{n=1}^{N} \frac{C\&M}{(1+r)^n} + \sum_{n=1}^{N} \frac{E \times FIT}{(1+r)^n} \quad (2)$$

Table II: Financial Parameters

Parameter	*Units*	*Rationale*
I	€	Installation costs *incl.* the cost of PV panels, mounting systems and installation
G	%	Percentage of capital investment financed as grant
C&M	€	Annual cost for cleaning & maintenance
c	€/kWh	Electricity cost
r	%	Discounted rate
FIT	€/kWh	Feed-in-Tariff rate
E	kWh	Energy generated
N	-	Number of years

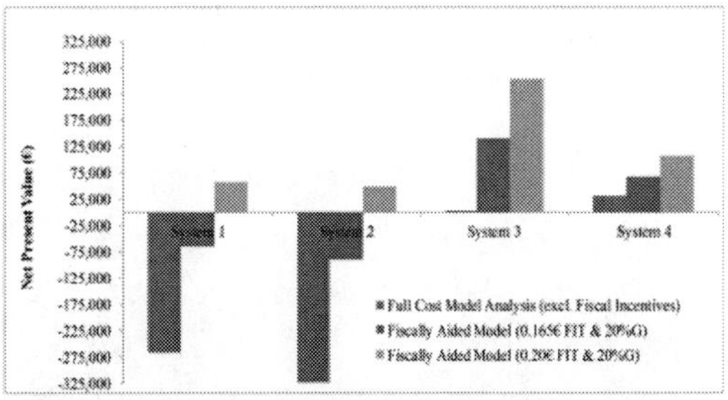

Figure 5: Net Present Value (€) for the different systems modelled for the Educational/Office building tested

Figure 5 shows the results for the financial return (in this case the Net Present Value) obtained for the two technical systems analysed for the educational/office building used to test the methodology. Three different fiscal scenarios were taken into consideration, namely:

- *Full Cost Model (excl. Fiscal Incentives)*;
- *Fiscally Aided Model (0.165€ FIT & 20%G)*; and
- *Fiscally Aided Model (0.20€ FIT & 20%G)*.

In all cases the discount rate for future cash in today's value flows was assumed as 2.2%, whilst a 30-year period was considered as the expected lifetime of the project. System 1 and System 2 describe, as per previous technical description, two different systems incorporating opaque crystalline photovoltaic panels and different types of transparent photovoltaic glass. System 3 and System 4 are two additional modelled systems. The former considers only the installation of opaque crystalline photovoltaic panels with a usable frontage area of 1,170m², without any transparent photovoltaic glass. The latter similarly considers only the installation of opaque crystalline photovoltaic panels without any transparent photovoltaic glass, but assumes a smaller usable frontage area of 341m².

3.3 Additional Results - Orienting Façades

In a study conducted by Pantic *et al.* [16], field and theoretical measurements were carried out using PVGIS for a vertically mounted PV panel and for three façade orientations: South, East and West, that is, 0°, -90° and 90° respectively. Results from this study showed that out of the three possibilities the South façade was the best in terms of energy generation. Results showed also that the energy generated by a façade facing either East or West was approximately half of that generated by the panels installed on a South facing façade. Pantic *et al.* did not however consider the intermediate angles between South and West and South and East. For buildings constrained by orientation and siting limitations, such azimuth angles are an important consideration for the integration of photovoltaics, as the overall applicable azimuths for which photovoltaic façades may be considered, increases from the dogmatic true South orientation.

For this reason, aside from the main research, a study was also carried to determine the effect of orientation on the performance of photovoltaic façades. Simulations on intermediate angles other than 0°, -90° and 90° were carried out using a photovoltaic clad façade sited on a four storey office block having a frontage of 6.4m (Simulated location Malta: Latitude ~ 35°). A *Shading Factor* of 1 and a usable frontage of 0.7 were assumed. Crystalline photovoltaic cells having a rated maximum power of 0.165kWp/m² were assumed for the simulations. The heat map shown in Figure 6 shows the total monthly energy generated for different orientations. Values are normalised on a per m² basis.

	90	60	45	30	15	0	-15	-30	-45	-60	-90
Jan	6.5	11.1	13.4	15.5	17.2	17.9	17.2	15.6	13.5	11.2	6.6
Feb	7.7	11.4	13.0	14.5	15.9	16.4	15.9	14.5	13.1	11.4	7.7
Mar	12.1	15.2	16.3	17.0	17.7	18.0	17.7	17.1	16.3	15.3	12.2
Apr	12.8	14.2	14.2	13.7	13.0	12.6	13.0	13.8	14.3	14.3	12.9
May	14.8	14.6	13.5	11.8	9.9	9.0	9.9	11.8	13.5	14.7	14.9
Jun	14.8	13.7	12.0	9.7	7.4	6.4	7.4	9.7	12.1	13.7	14.9
Jul	14.9	14.3	12.8	10.6	8.4	7.3	8.4	10.7	12.9	14.4	15.1
Aug	14.2	15.0	14.5	13.2	11.7	10.9	11.7	13.3	14.6	15.1	14.4
Sep	11.1	13.6	14.3	14.6	14.8	14.9	14.8	14.6	14.3	13.6	11.1
Oct	9.5	13.2	14.8	16.1	17.3	17.9	17.3	16.1	14.8	13.2	9.5
Nov	6.8	11.1	13.2	15.2	16.7	17.3	16.7	15.2	13.2	11.1	6.8
Dec	5.9	10.4	12.7	14.8	16.4	16.9	16.4	14.8	12.7	10.4	5.9

Figure 6: Heat map showing energy generation per month (kWh/m²) for varying azimuth angles (°)

As can be seen from the heat map, monthly energy generation is very susceptible to the azimuth angle. Counter intuitively to what one would expect, whereas for roof mounted systems the South orientation is practically always favoured, in the case of façades, results show that the monthly energy generated varies significantly. The reason behind these results can be attributed to the losses due to angular reflectance, where some of the energy falling on the photovoltaic façade is reflected (rather than absorbed) by the photovoltaic surface, proportionally to the angle of incidence. For predominately South facing façades (considered in this instance as being all angles between ±30°), during the summer months, the naturally occurring high angle of incidence between the solar radiation and the façade leads to high angular reflectance losses and hence in a decrease in the monthly energy generated. On the contrary during the autumn and winter months, the typical low angle of incidence leads to low angular reflectance losses and hence to an increase in the monthly energy generated, notwithstanding the lower magnitude of the solar radiation. For East and West facing façades the inverse is true. During summer the low angle of incidence typical of the daily early or the late afternoon hours, results in a significant energy yield.

On an annual basis (per m²) as can be seen in Figure 7 the best results are obtained where a compromise is reached between the reduction in angular radiation losses and the actual irradiation falling on the façade, resulting in the fact that façades which are not predominantly South facing (considered in this instance as being all angles between ±30° and ±45°) should not be discarded *a priori* for not having an optimum orientation.

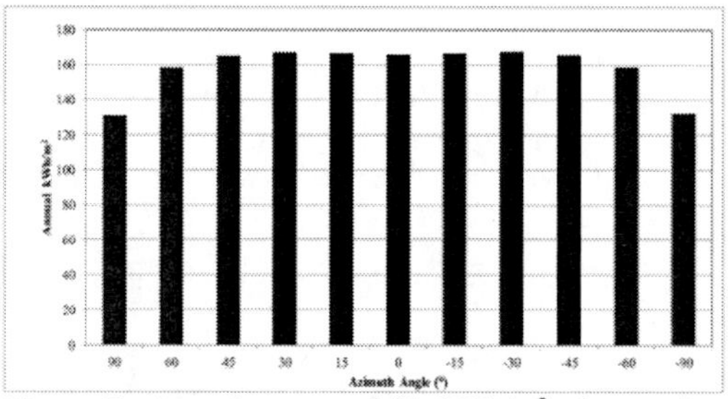

Figure 7: Annual energy yield (kWh/m²) for varying azimuth angles (°)

4 CONCLUSIONS

The research presented in this paper, presents the work done in the creation of a specific methodology for the preliminary assessment of façade-integrated photovoltaics. The methodology looks at both the technical and financial aspect of a proposed project and enables a building designer to quickly assess the feasibility of a proposed photovoltaic project based on a number of project related variables such as latitude, orientation, shading from opposing buildings, usable frontage area, rated maximum power for the technical part, and typical financial parameters for the cost benefit analysis part. The resulting methodology was then applied to a specific project relating to the possible retrofitting of a glazed building, with different photovoltaic systems, including a combination of opaque photovoltaic panels and transparent photovoltaic glass.

Additionally, a section was also included to show the

effect of façade orientation on the total monthly and annual energy yield. Results showed that contrary to roof installed photovoltaics, façade-integrated photovoltaics may be favourably affected from not having a perfect south facing orientation, and that façades which are not predominantly South facing (considered in this instance as being all angles between ±30° and ±45°) should not be discarded *a priori* for not having an optimum orientation.

5 REFERENCES

[1] Ferrante, A., Cascella, M.T., Zero energy balance and zero on-site CO_2 emission housing development in the Mediterranean climate, *Energy and Buildings* 2011; **43**: 2002–2010

[2] Stegou-Sagia, A., Antonopoulos, K., Angelopoulou, C., Kotsiovelos, G., The impact of glazing on energy consumption and comfort, *Energy Conversion and Management* 2007; **48**: 2844–2852

[3] Borg, S.P., Kelly, N.J., High resolution performance analysis of micro-trigeneration in an energy-efficient residential building, *Energy and Buildings* 2013; **67**: 153–165

[4] Marszal, A.J., Heiselberg, P., Jensen, R., Nørgaard, J., On-site or off-site renewable energy supply options? Life cycle cost analysis of a Net Zero Energy Building in Denmark, *Renewable Energy* 2012; **44**: 154–165

[5] Marszal A.J., Heiselberg P., Bourrelle J.S., Musall E., Voss K., Sartori I., Zero energy building - a review of definitions and calculation methodologies, *Energy and Buildings* 2011; **43**: 971–979

[6] Voss, K., Musall, E., *Net Zero Energy Buildings*, Munich: Detail Green Books; 2011

[7] Berger, W., Catalysts for the diffusion of photovoltaics - A review of selected programmes, *Progress in Photovoltaics: Research and Applications* 2001; **9**: 145–160

[8] Tymkow P., Tassou S., Kolokotroni M., Jouhara H., *Building Services Design for Energy Efficient Buildings*, New York: Routledge; 2013

[9] Scognamiglio, A., Røstvik, H.N., Photovoltaics and zero energy buildings: a new opportunity and challenge for design, *Progress in Photovoltaics: Research and Applications* 2013; **21**: 1319–1336

[10] Talal, S., Elias, K., Analysis of Building-Integrated Photovoltaic Systems: A case study of commercial buildings under Mediterranean Climate, *Procedia Engineering* 2015; **118**: 538–545.

[11] Esclapes, J., Ferreiro, I., Piera, J., Teller, J., A method to evaluate the adaptability of photovoltaic energy on urban façades, *Solar Energy* 2014; **105**: 414–427

[12] Bigaila, E., Rounis, E., Luk, P., Athienitis, A., A study of BIPV/T collector prototype for building façade application, *Energy Procedia* 2015; **78**: 1931–1936

[13] Argul, F.J., Castro, M., Delgado, A., Carpio, J., Peire, J., Software modelling of grid-connected photovoltaic buildings, *Progress in Photovoltaics: Research and Applications* 2003; **11**: 255–273

[14] Perez, M, Fthenakis, V., Kim, H., Pereira, A., Façade–integrated photovoltaics: A life cycle and performance assessment case study, *Progress in Photovoltaics: Research and Applications* 2012; **20**: 975–990

[15] Photovoltaic Geographical Information System (PVGIS), Joint Research Centre, Institute for Energy and Transport (IET), 2016

[16] Pantic, L., Pavlovic, T., Milosavljevic, D., Mijanic, D., Radonjic, I. & Radovic, M., Electrical Energy generation with differently oriented PV modules as façade elements, *Thermal Science* 2016; **20**: 1376–1386

PERFORMANCE OF A BUILDING INTEGRATED SEMITRANSPARENT PHOTOVOLTAIC FAÇADE ON A RESIDENTIAL HOUSE IN NORTHERN EUROPE

Andri Jagomägi (andri.jagomagi@gmail.com), Department of Materials and Environmental Technology, Tallinn University of Technology, Estonia
Andreas Wimmer, School of Engineering and Environmental Sciences, University of Applied Sciences Upper Austria, Wels 4600, Austria
Martin Thalfeldt, Department of Civil Engineering and Architecture, Tallinn University of Technology, Estonia

ABSTRACT: Semitransparent photovoltaic (PV) façade is a good alternative to the popular rooftop PV systems. In addition to the electricity generation the semitransparent PV facade regulates the indoor temperature and lighting conditions, alters the aesthetics, and acts as a noise barrier. In this paper we investigate the influence of semitransparent PV façade on a single family house in the Estonian climate. We compare the energy performance and the indoor environment of three modifications of the test house with IDA ICE simulation software: base version without window shadings, version with horizontal shades, and version with semitransparent PV façade. The results of the simulations show that the overheating of the base version can be equally well eliminated by the horizontal shades or the semitransparent PV façade. In contrast to the horizontal shades, the PV façade blocks the useful solar gain in wintertime. This results in higher heat demand by 60% and higher artificial lighting demand by 11%. Nevertheless, the extra energy demand is fully compensated by the PV electricity production, which makes the house a plus energy building. In addition to the energy consumption, the PV façade decreases the outdoor to indoor noise propagation by 32 dB.

Keywords: photovoltaic façade, semitransparent photovoltaics, building integrated photovoltaics, plus energy house, IDA ICE, noise reduction

1 INTRODUCTION

On the 4th of January 2003 the EU directive on the energy performance of buildings (EPBD) came into force and had to be implemented by the EU Member States at the latest on the 4th of January 2006 [1]. This directive requires member states to set certain limits to the energy demand of buildings and determines the energy performance certificates. It sets requirements for the energy consumption of buildings with the goal to reduce the $CO2$ emission in the building sector. In order to meet the requirements, not only the energy conservation has to be implemented, but also the passive and the active use of solar energy should be considered. However, without the onsite energy production the net zero energy house standard cannot be met in Northern Europe. Photovoltaics is one of the most promising small scale energy production technology in the residential areas. It is environmental friendly, long-lasting, and when properly installed, it does not disturb the local residential environment. Moreover, in the dense urban environments, where the land area is scarce, the installation of PV on the buildings is often the only option.

In addition to the electricity production, the building integrated photovoltaics (BIPV) can optimize the cost of construction by replacing the conventional building elements like façade skin or roof. The BIPV products can be used to control the inner climate of the building. The mounting method of semitransparent PV façades can include forced or natural cooling. Waste heat can be used for the space heating of the building when necessary. Also, the BIPV modify the aesthetic look of the building.

However, the BIPV is rather underused in urban environments. It is estimated that in 2013 the BIPV installations comprised only 3% of the total PV installations in Europe [2]. However, the potential share of PV on the buildings is 66% [3]. Personal communication indicates that one of the reasons for the low installation of BIPV products is the lack of knowledge of the technology among the architects who play the key role in the implementation of BIPV products on new buildings. Thus, in order to make the technology more transparent and reliable in the eyes of the architects, more case studies in terms of simulations, and demonstration projects are needed. It should be proven that the BIPV systems not only act as electricity producers, but their value is in-creased and payback period reduced as they provide some extra functions to the buildings.

Theoretical framework for the heat transfer analysis and thermal models of the flat PV façades have been published in several papers [3-14]. K.E. Park et al. designed the semitransparent PV glazing [3] and experimentally compared the temperature and the electrical performance of the samples with varying glass properties. The theoretical models of the heat transfer properties of naturally ventilated double skin PV façades have been analyzed by J.Cipriano et al. [4], M. Sandberg et al. [5], A. Guardo et al. [6] and G.E.Lau et al. [7]. The results of experimental studies of similar systems are reported by J.Peng et al. [8, 9, 10] and L.Gaillard [11]. These works help to quantify the cooling effect of the natural convec-tion on the PV facades. Comprehensive theoretical and experimental study on the thermal properties of double PV façade has been carried out by L.Mei et al. in the [12], where the space heating on the expense of waste heat of the PV modules is discussed. Also, the authors estimate the influence of such a system on the cooling and heating loads of the test house in three different climatic conditions. It was shown that compared to the brick wall, the semitransparent PV façade with space heating option slightly increase the cooling load and decrease the heating load. The same authors published simplified version of their thermal model in [13] and [14].

The influence of the semitransparent PV façade on the buildings' inner environment has been discussed in [15-30]. J.Han et al. [15] compare the effects of the semi-transparent BIPV façade with plain glass façade on the indoor temperature of a test room. The results of their paper show that the solar gain limiting effect of semi-transparent PV facade is very effective in reducing the cooling load of a building. The space heating application with air cooled opaque and semitransparent PV façades have been analyzed in [16, 17, 18] by K.Vats et al. and Kamthania et al. They provide detailed analytical model for the calculations of the performance of this type of systems in any climate. The same authors compare the electricity production and the space heating capabilities of semitransparent PV systems with different absorber materials [19, 20]. The theoretical framework of the PV façade with natural convection is presented by W. Sun et al. [21]. P.W.Wong et al. study the impact of semitransparent PV skin on the energy consumption of a residential building in Japan [22]. Their results show that with the optimized PV glazing in the cold climate conditions of Sapporo, the annual heating and cooling demand can be reduced by 5.3%. Another case study based on an office building was made by D.H.W.Li et al. [23]. The authors analyzed the energy performance and the financial im-pact of semitransparent PV skin to the office building in Hong Kong. It was found that the PV skin reduces well the cooling load of the building, also, it was estimated that the payback time of such a system is around 15 years.

G.Y.Yun et al. investigate the optimal ratio of the area of a transparent window to the area of a ventilated opaque PV facade on the buildings' envelope [24]. They study the effect of the ratio on the lighting, heating, and cooling loads of a test building in South-, Central-, and North-Europe. Semitransparent PV windows have been studied by P.K.Ng et al. [25]. This paper deals with the optimization of window to wall ratio and other window parameters in order to minimize the energy consumption of buildings in Singapore. Another window to wall ratio and the transparency optimization of semitransparent PV glazing on an office building in Tokyo is done by T. Miyazaki et al. [26]. L.Olivieri et al. report the measurement results of the electricity production of semitransparent PV glazing and its influence to the indoor environment in Madrid [27]. Another experimental paper was published by S.H.Yoo [28]. The author reports the field measurements and discusses the optimization of the electricity production and the shading properties of semi-transparent PV sunshades on a building in Korea. Recent review papers on the semitransparent photovoltaic fa-cades has been published by Lai et al. [29] and Quesada et al. [30].

The literature overview shows that not many studies have been conducted on the influence of semitransparent PV façade on the indoor environment and the energy balance of a private house in the Nordic climate. The main differences between the office and the private building space are in the usage hours, the availability of internal heat gain, and the illuminance requirements.

The goal of the present paper is to conduct the detailed simulation of the impact of semitransparent photovoltaic façade on the single family house in Estonia. In order to maximize the highest lighting levels and best visual field with the minimum heat loss, the typical architectural solutions in Northern Europe plan the largest windows to the south side of the house. Without the proper shading the large solar gains may create the overheating of the inner environment during the sunny season from the May to the August. Usually, the high solar gains are limited by the horizontal or tilted shades on top of the glass areas. In our study we investigate if the same effect can be achieved by photovoltaic façade. In our model the photovoltaic façade balances the energy consumption of the building by producing the electricity, it regulates the inner climate by reducing the solar gain, it provides extra space in terms of a winter garden, and it reduces the street noise inside the building. We conduct the detailed simulations of the effect of the façade on the yearly energy balance and on the inner climate of the building. The results are compared with the base model with plain glass façade, and the model with horizontal shades.

2 EXPERIMENTAL

In order to model the electrical performance of the façade laminated PV cells, we conducted the outdoor measurement campaign of a PV module that is made of similar PV cells that are used in our simulations. During the measurements we monitored the performance of the PV module during a half year from winter solstice to the summer solstice.

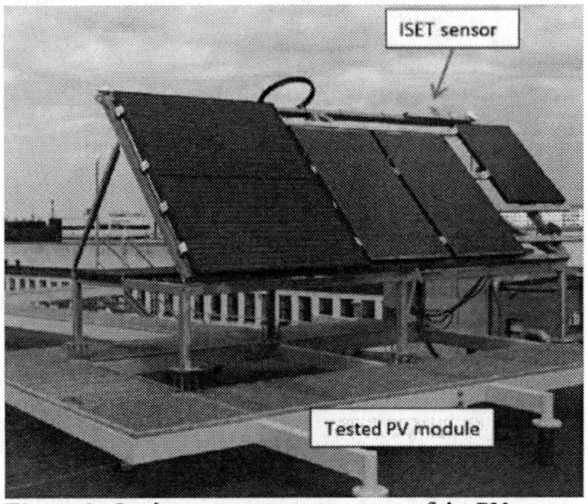

Figure 1: Outdoor measurement system of the PV modules in the Tallinn University of Technology.

The Figure 1 shows the photo of the experimental setup. The PV module was mounted on the frame that is facing south with 48 degrees of inclination angle from the horizon. The electrical characterization of the module is done via full fourth quadrant IV-curve tracing which is performed with the frequency of 1/min. The backside of the PV module is equipped with the RTD temperature sensor. The irradiance reference device is calibrated c-Si ISET sensor. The details of the measurement and data handling procedures are discussed in [31]. In our simulations we assumed that the PV façade comprises of the same monocrystalline silicon cells as the module we tested outdoors, thus, the outdoor measurement results were used to model the electrical performance of the PV façade.

3 DESCTIPTION OF THE MODEL

Our analysis is based on a two story single family house with a living space of 147 m2. In order to bring out the effect of the PV façade we defined three different versions of the model of the test house in the simulation software IDA Indoor Climate and Energy (IDA ICE). With this software, it is possible to calculate the hourly values of different indicators for every zone of the house, whereas every room of the house is defined as a single zone.

3.1 Version 1

The first version of the building shows the basic structure of the house. In this version, there is no shading and no photovoltaic system added. This version works as a reference and it helps to make Versions 2 and 3 comparable.

Figure 2: 3-D view of Version 1, which serves as a base for the simulations.

The structure of the house was chosen to reach low *U*-values and therefore to make the house energy efficient. Moreover, the approximately cubic form of the building leads to the good surface-to-volume-ratio. The 3D model of the Version 1 is depicted in the Figure 2. The detailed list of the materials of the building envelope are presented in the Table 1.

All the elements of the building envelope satisfy the legal requirements to the thermal transmittance (the *U*-value) stated in Ref. [33]. The thermal bridges of the building envelope are rated as 'good' and the infiltration rate 0.16 l/s per m^2 of the external surface are incorporated in the calculation model.

Table 1 Structure and the thermal transmittance of the building envelope

Thickness (cm)	Element
Roof (U = 0.106 W/m^{2}K)	
9.00	Substrate
1.00	Root protection layer
-	Separation layer (mat of fibers)
34.00	Heat insulation (extruded polystyrene)
0.50	Moisture proofing
-	Vapor pressure equalization
5.00 – 10.00	Incline concrete
25.00	Reinforced concrete ceiling
0.50	Ceiling plaster
Foundation (U = 0.135 W/m^{2}K)	
1.00	Surfacing
5.00	Screed
3.00	Footfall sound insulation
5.00	Expanded polystyrene
1.00	Moisture proofing
25.00	Reinforced concrete floor
-	Construction paper
16.00	Heat insulation (Extruded polystyrene)
10.00	Tamped concrete
-	Construction paper
30.00	Capillary breaching layer
-	Mat of fibers
Exterior walls (U = 0.141 W/m^{2}K)	
0.50	External plaster
50.00	Aerated concrete
0.50	Internal plaster

Table 2 Structure of the windows

Layer	Element
Pane	Low energy glazing
Gap	18mm Air (10%) / Argon (90%)
Pane	Clear glazing
Gap	18mm Air (10%) / Argon (90%)
Pane	Low energy glazing

Table 3 Orientation and size of the windows

Orientation	Width (cm)	Height (cm)
North	80	150
North	80	150
North	120	150
North	120	150
North	120	150
East	120	150
East	120	150
East	120	150
East	120	150
South	1100	260
South	341	260
South	328	260
South	402	260
West	120	150
West	80	150
West	120	150

In addition to the structure, the windows play important role in the building energy performance. In our simulations we used low *U*-value windows. The structure of the windows is listed in the Table 2. The *U*-value of the glazing and the frame are 0.55 W/m^{2}K and 0.60 W/m^{2}K respectively. The g-value of these windows is 0.453. The orientation and the sizes of the windows are summarized in the Table 3.

For the cooling purposes, the operation of the windows was realized with a macro that opens 20% of the glazing area when the indoor temperature reaches above 25 °C.

3.2 Version 2

The differences of the Version 2 compared to the Version 1 lie in the additional shading of the big glass area in the south façade and the shading of the windows on the east and west side (Figure 3). The shading of the south windows is designed as two awnings, one for each floor. The windows on the east and west side are shaded by vertical shading elements.

Figure 3: IDA ICE model of Version 2

The south side awnings are placed directly above the window frames and their width is equal to the height of the windows. The vertical shading elements of the east and the west windows have the same height and width as the respective windows. These vertical shading elements are mounted directly next to the window frame on the south side of the windows.

3.3 Version 3

Compared to the Version 1, the Version 3 has a semitransparent PV façade in the South side of the building (Figure 4). As in the Version 2, the shading of the west and east windows is provided by vertical shading elements, which are mounted directly next to the frame of the window. The extra space that the PV façade creates has the dimensions of 11 m, 2.5 m, and 5.85 m for the length, width, and the height respectively. We propose that this extra space can be used as a winter garden. The temperature of the winter garden is regulated by four 2.5mx1m windows: two in the west and two in the east. The Figures 5 and 6 show the renderings of the outside and the interior views of the house. The structure of the PV modules consists of two 4 mm glass layers and the monocrystalline silicon PV cells that are laminated in between the glass. This structure of the PV modules results in the U-value of 5.81 W/(m^{2}K). The modules are supported by the frame that takes 5% of the façade area and has the U-value of 2.0 W/(m^{2}K).
The nominal electrical power of the PV façade elements is calculated by considering the following parameters. The packing density of the cells in the PV glass is 0.8, which means that 80% of the glass is covered by the PV

Figure 4: IDA ICE model of Version 3

Figure 5. Rendering Facade Outside

cells and 20% is transparent. Also, we assumed that the conversion efficiency of the cells is 17.2%. Taking into account the frame area, the resulting efficiency of the façade area is 13.1%. The dimensions of the different parts of the PV façade and their nominal powers under standard test conditions (STC) are listed in the Table 4.

Figure 6. Rendering Facade Inside

Table 4 Areas of the PV façade and the nominal power under the Standard Test Conditions.

Orientation	Width (m)	Height (m)	Type	Nominal power (W)
South	11	5.85	PV	8412
West	2.5	3.6	PV	1176
West	2.5	1	Window	-
West	2.5	1	Window	-
East	2.5	3.6	PV	1176
East	2.5	1	Window	-
East	2.5	1	Window	-
Horizontal	2.5	11	PV	3595

In order to simulate the thermal and the optical properties of the PV façade in the IDA ICE software, the transmission and reflection coefficients of the structure must be defined. We assumed that the plain glazing part of the PV façade has the solar transmission coefficient of 0.9 and the solar reflection coefficient of 0.04. In addition, we assumed that 90% of the incident solar irradiance is absorbed by the PV cells. Taking into account the packing density of the cells we defined the PV façade in the IDA ICE software as one monolithic semitransparent layer with the transmission coefficient of 0.18 and the reflection coefficient of 0.088.

4 THEORETICAL FRAMEWORK

As a member of the European Union, Estonia had to implement the directive on the energy performance of buildings into their national regulations. Most relevant for this paper is the standard methodology for calculating the energy performance of buildings [33] and the minimum requirements for the energy performance [32].

4.1 Methodology for calculating the energy performance

The methodology includes six energy consumers of the building, which are the heating system, the cooling system, the ventilation system, the domestic hot water, the lighting, and the appliances. This energy is supplied by on site renewable energy sources or is delivered into the system from outside boundaries. In addition, the methodology takes into account solar gains, solar loads, heat transmission, internal heat gains, and internal heat loads.

4.2 Internal loads

The standard use of small residential buildings and the corresponding heat gain from appliances, lighting and occupants are provided in the Table 5. The table shows the building's number of occupied hours per day, the number of days of use per week and the heat gain from appliances as well as the heat gain from occupants during the building's occupied hours.

Table 5 Standard use of small residential buildings and corresponding values of heat gain [33].

Occupied time period	00:00-00:00
h/24h	24
d/7d	7
Usage rate	0.6
Appliances (W/m^2)	2.4
Occupants (W/m^2)	2
Occupants (m^2/person)	42.5

4.3 Minimum requirements for the energy performance

The regulation for the minimum requirements for the energy performance of buildings sets certain limits for the energy performance of buildings [32]. Moreover, it sets specific levels for the energy consumption and therefore separates the buildings into standard houses, low energy buildings, nearly zero-energy buildings, and net zero-energy buildings. This regulation is valid for the buildings to be constructed or to the buildings, which are undergoing major renovation and gives the maximum value of the building's total energy consumption. It also takes into account the type of the building and its technical parameters.
The key indicator is the energy performance indicator, which is the total consumed energy during the standard use of the building less the produced energy fed into the energy networks. This indicator is weighted, specific and it takes into account all the elements of the energy consumption of the building including the indoor climate control, the heating of hot water, and the operation of household appliances and other electrical equipment. In order to make the indicators of different buildings comparable it is given per square meter of the building's heated area. The Table 6 shows the maximum values of the energy performance indicator of residential buildings to be constructed in Estonia.

Table 6 Limits of the energy performance indicator of small residential buildings that will be newly constructed [32]

Standard house	160 kWh/m^2y
Low energy building	120 kWh/m^2y
Nearly zero-energy building	50 kWh/m^2y
Net zero-energy building	0 kWh/m^2y

4.4 Requirements for ventilation and indoor temperature

The Estonian legislation also determines the ventilation outdoor airflow rate and the heating and the cooling set points [32]. Both, the outdoor airflow rate and the set points are listed in the Table 7.

Table 7 Requirements for ventilation and indoor temperature for small residential buildings [32]

Outdoor air flow rate	0.42 l/(s m^2)
Heating set point	21 °C
Cooling set point	27 °C

The outdoor airflow rate is specific to one square meter of the building's heated area. Our energy calculations use the heating and cooling settings listed in the Table 7. In the case of partially heated rooms, we used the heating set point listed in the Table 7 as the indoor temperature.
Regarding the indoor comfort of residential buildings during the summer season, the cooling set point is not allowed to be exceeded by more than 150 degree hours (°Ch) annually. Passive cooling should be preferred to prevent overheating. It means that the cooling control should be executed as architectural and structural solutions, especially in the case of residential buildings. Measures such as the sun control, appropriately dimensioned and positioned glass surfaces, the orientation of the building with respect to the cardinal directions and other objects, and the use of massive structural elements should be preferred in any case [32]. In the case of residential buildings, the space cooling driven by the opening of the windows can be used.
In addition, we took into account that the house is equipped with the ventilation heat recovery unit with the efficiency of 80%.

4.5 Requirements for the building envelope

In order to provide proper energy performance and comfortable indoor temperature, to avoid the mold and the condensation on thermal bridges, inner surfaces or structural elements, the building envelope must provide good insulation and it must be permanently airtight [32]. The insulation characteristics of building envelope must meet the certain targets that are listed in the Table 8.

Table 8 Requirements for the thermal transmittance of building envelope [32].

Exterior walls	0.12 – 0.22 $W/(m^2K)$
Roofs and floors	0.1 – 0.2 $W/(m^2K)$
Windows and doors	0.6 – 1.1 $W/(m^2K)$

In total, the thermal transmittance of a building's envelope should not be higher than 0.5 $W/(m^2K)$.

4.6 Heating system

The energy consumption of the heating system in general depends on the outdoor temperatures. In our simulations we equipped the test house with the geothermal heat pump with the average coefficient of performance (COP) of 3.00. The electrical energy consumed by the heat pump is calculated by dividing the thermal energy produced with the COP.

4.7 Domestic hot water

For the calculations of the energy demand of the domestic hot water production the standard value of 25 $kWh/(m^2y)$ is considered in our calculations [33]. This figure assumes that in small residential building the specific use of hot water is 430 $l/(m^2y)$ and the temperature difference between the cold and the hot water is 50°C.

4.8 Lighting energy consumption

For the comparison of indoor natural light conditions between different house versions we counted the hours between 07:00 and 23:00 when the natural lighting level is lower than the recommended level in one year. The recommended levels were taken 500 lux for work area, 200 lux for bathrooms and living room, 150 lux for bedrooms, and 100 lux for hallways and service rooms [34]. The missing natural light in our simulations is compensated by the artificial lighting of modern LED lighting with the luminous efficacy of 100 lm/W.

4.9 Photovoltaic performance

In order to model the electrical performance of the PV façade we assumed that the PV modules consist of similar monocrystalline PV cells as the module that we used for the outdoor data collection.

The most important factor of the power output of photovoltaic modules is the solar irradiance. The second most important factor is the temperature of the PV cells, which in turn depends on the air temperature, wind speed, and the solar irradiance. The secondary effects that influence the performance of PV modules are the angle of incidence (AOI), and the spectrum of the irradiance. At the system level, additional losses are introduced by the module mismatch, cables, soiling, and the inverter.

There are many different models that are used to calculate the DC performance of PV modules. The model used in this work is a variant of King's model [35, 36] and is applied in the following form:

$$P_{DC} = P_{STC} G'(1 + C_0 \ln(G') + C_1 \ln(G')^2 + \gamma T') \quad (1)$$

where $G' = G/G_{STC}$ and $T' = T_{mod} - T_{STC}$.

The temperature constant $\gamma = -0.0043$ 1/K was taken from the datasheet of the PV module under the test. The nominal powers P_{STC} of different PV façade areas were taken from the Table 4. The empirical constants $C_0 = -0.0123$ and $C_1 = -0.0168$ quantify the nonlinear dependence of the photovoltaic system output power on the irradiance level and were fitted from the measurement data that was obtained from the experiment. The IDA ICA simulation software allows to model the temperatures of different components of the building. Thus, the hourly temperatures of the façade PV modules were taken from the results of the numerical simulations.

The Equation 1 assumes that the in-plane irradiance level is measured with the device that has a similar angular response and spectral sensitivity as the PV modules in the façade. In case the irradiance levels are measured with a device with the different properties adequate correction procedures must be applied. In our PV performance calculations we used the irradiance data measured with a broadband pyranometers [37], thus, we applied the corrections for the AOI effects. However, we did not take into account the temporal changes in solar spectrum since this input data is difficult to obtain. Accordingly, the irradiance levels at different façade surfaces were derived by translating the hourly horizontal diffuse and horizontal global irradiance data by using the Perez method [38] and then applying the AOI correction that is defined in Ref. [40].

DC side system losses that we took into account were the module mismatch loss MMF = 2%, cabling losses based on 4 mm^2 copper cables, and soling losses. In Ref. [40] it is described that in Germany, which has a rather similar climate to Estonia, the typical soiling loss of a PV system in a year is 2-6%. In our calculations we used 2% soiling loss for the vertical surfaces and 6% for the horizontal surfaces.

The DC/AC conversion of our PV façade is organized so that each surface is connected to the separate string inverter with the inverter sizing ratio of 1.3. The inverter efficiency [41] is calculated as

$$\eta_{inv} = \frac{p_0}{p_0 + (k_0 + k_1 p_0 + k_2 p_0^2)} \quad (2)$$

where p_0 is the ratio between AC output and the rated inverter power, and the constants k_i that represent different conversion loss mechanisms are calculated as

$$k_0 = \frac{5}{36\eta_{10\%}} - \frac{1}{4\eta_{50\%}} + \frac{1}{9\eta_{100\%}} \quad (3)$$

$$k_1 = \frac{5}{12\eta_{10\%}} + \frac{33}{12\eta_{50\%}} - \frac{4}{2\eta_{100\%}} - 1 \quad (4)$$

$$k_2 = \frac{5}{18\eta_{10\%}} - \frac{5}{2\eta_{50\%}} + \frac{20}{9\eta_{100\%}} \quad (5)$$

The experimental values $\eta_{10\%}$, $\eta_{50\%}$, and $\eta_{100\%}$ can be taken from the efficiency curve of the chosen inverter. In our calculations we assumed the following coefficients: $\eta_{10\%} = 94\%$, $\eta_{50\%} = 97\%$, and $\eta_{100\%} = 96\%$.

Based on the Equations 1-5 and also taking into account the additional DC side system losses, we calculated the AC performance of the different PV façade surfaces on hourly bases. The yearly electrical energy production is the sum of the hours.

5 RESULTS AND DISCUSSION

From the results of the IDA ICE simulations we were mainly interested in the indoor temperatures, lighting levels, energy balance, and the usage time of the winter garden.

5.1 Return air temperature

The overheating of a building can be quantified by degree hours, during which the indoor temperature was higher than the norm 27°C. The simulation results of the house Version 1 showed that with 581°Ch of the return temperature of the air handling unit (AHU), this version of the house is far beyond the norm of 150°Ch. This can be related to the large solar gains from unshaded south windows. The yearly temperature profile of the return air is plotted in the Figure 7. The temperature peaks appear from the June to the August, which corresponds to the hours from 3624 to 5832. Based on the return air over temperatures is it is clear that some sort of shading will be required. The two types of shadings are compared in Versions 2 and 3.

Due to the window shades of the Version 2, the over temperature degree hours are reduced significantly to 34. The three major peaks can be seen around 3000 hours (5th of May), the biggest one around 5100 hours (31st of July) and one around 6200 (15th of September) hours.

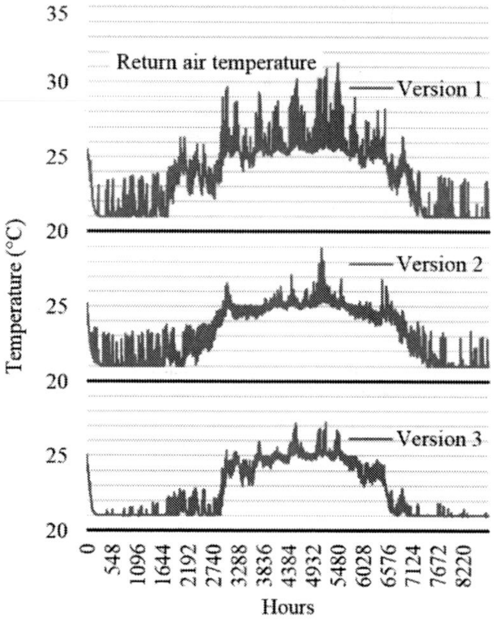

Figure 7. Return air temperatures of different versions.

5.2 Living room temperature

Because being on the south side of the house and directly bordering the winter garden in Version 3, the indoor temperature of the living room of different versions was chosen for the comparative analysis. The Figure 8 shows that the temperature of the living room is never below the heating set point 21°C, also, when necessary the cooling starts from the 25 °C by the window opening function. However, in the case of Version 1 the temperature still reaches up to 35 °C.

Compared to the Version 1, the living room of Version 2 never exceeds the temperatures above 30°C. The comfort temperature level 21°C during the wintertime is provided by the heating system and the summertime temperature is limited by opening of the windows.

The photovoltaic façade of Version 3 reduces the overheating of the living room even better than the shading elements in Version 2. The window cooling at 25°C is again visible as well as the action of the heating system. The peaks during the summertime are lower than the peaks in the Versions 1 and 2 and have a maximum level at 27.5°C.

5.3 Indoor lighting conditions

Due to the very big unshaded window area in the case of the Version 1, the artificial lighting demand is low. The results of the simulations show that the living room needs the artificial lighting 1707 hours per year. The total energy consumption of the lighting is 606 kWh/y. The simulations show that the required lighting level 200 lux in the living room area is satisfied already from the outdoor global horizontal irradiance level of 15 W/m^2.

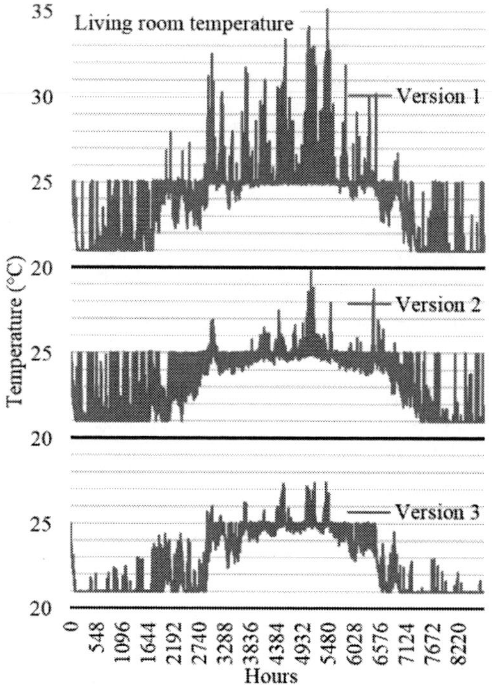

Figure 8. Living room temperatures of different versions.

In the case of Version 2 the direct sunlight during the summertime is greatly reduced by the horizontal shades above the south windows. However, due to the rather big window area the collection of diffuse light and the direct light during the low sun is still very good. The simulations show that the artificial lighting demand of the living room in the case of Version 2 is 1845 hours and the required lighting level is provided already by the outdoor horizontal irradiance of 20 W//m^2. The estimated annual electrical energy consumption of the artificial lighting of the Version 2 is 652 kWh.

The PV façade of the Version 3 reduces the natural lighting even more than the shades of Version 2. Nevertheless, the transparent glass area in the south direction is still 11.6 m^2. As a result the living room lighting demand is satisfied from the outdoor horizontal irradiance level of 50 W//m^2. The artificial lighting in the living room is required annually for 2392 hours and the total energy consumption by the artificial lighting of the Version 3 is 724 kWh/y.

5.4 Heating

The monthly heat energy that is needed to cover the heat losses by conduction, infiltration, and ventilation are seen in the Table 9.

The annual heating demand of the Versions 1 and 2 are similar, however, there are some monthly deviations that originate from the differences in the solar gain and infrared losses due to the shading elements of Version 2. Due to the reduced solar gain the Version 3 has the highest heating demand that is increased by 1700 kWh per year.

Table 9 Monthly heating demand of different versions.

Month	Version 1 (kWh)	Version2 (kWh)	Version 3 (kWh)
1	633	497	818
2	554	646	977
3	157	226	465
4	4	14	150
5	0	0	0
6	0	0	0
7	0	0	0
8	0	0	0
9	0	0	0
10	16	7	190
11	462	406	766
12	914	934	1 065
Total (kWh)	2 739	2 729	4 431
Total (kWh/m²/y)	18.6	18.6	28.9

5.5 PV façade performance

The AC electricity production of the photovoltaic façade for every orientation can be seen in the Table 10. The total annual electricity production is 10 691 kWh from which the majority of 6452 kWh is produced by the biggest South area of the PV façade.

Table 10 Electricity production of the photovoltaic facade

Orientation	Specific yield (kWh/kW)	Annual electricity (kWh)
South	767	6 452
East	605	711
West	582	684
Roof	791	2 844
Total	744	10 691

The specific yield of optimally oriented PV modules in Estonia is about 1000 kWh/kW. In the case of our PV façade the specific yield of South and Roof parts are 77-79% of the optimal surface. The specific yield of the East and the West areas are much lower 58-61% of the optimum. Thus, in order to optimize the cost efficiency of the electricity production the priority of the investment should be given to the Roof and South area of the façade. In the case of Roof and South PV installation only, the system would lose only 13% of the annual production compared to the full PV coverage.
As it is common to any solar energy system in the Northern Europe, the electricity production of the PV façade is very unevenly distributed over the year (Figure 9). The majority of 84% of the annual production is provided by the 7 months from March to the September. In contrast, the three darkest months from November to January produce only 6.6% of the annual total.

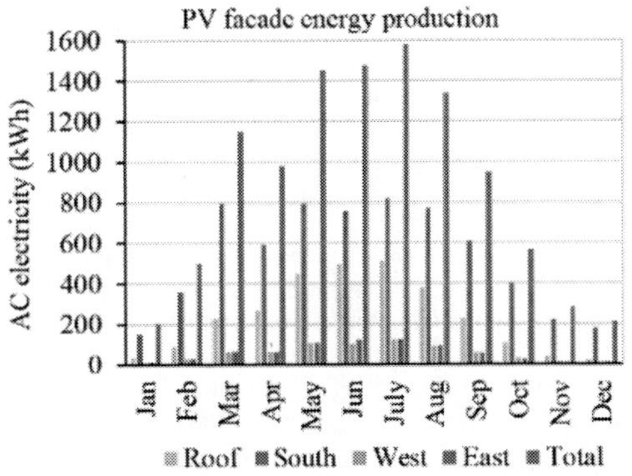

Figure 9. PV Energy Production

5.6 Total energy balance

The total electrical energy consumption of the building over one year is given in the Table 11. HVAC aux is the electrical energy consumed by the ventilation fans. The energy carrier conversion factor, which is the ratio of the primary energy use to the delivered energy, for electricity is 2 [32]. Accordingly, the primary energy demand of the Version 1 is 10 950 kWh per year, and the energy performance indicator is 74.5 kWh/m²/y. Not considering cooling, which is actually needed for the Version 1, the house would be a low energy building.
The total energy demand of the Version 2 is only slightly bigger than the demand of the Version 1. The only difference comes from the higher energy consumption of the artificial lighting. In sum, the energy performance indicator of the Version 2 is 75.5 kWh/m²/y, which makes it a low energy building. However, compared to the Version 1, the Version 2 would not require any cooling to satisfy the norms of the inner climate.
Due to the increased artificial lighting and heating demand the Version 3 consumes annually about 12% more electricity than the Versions 1 and 2. However, the whole consumption is covered by the on-site PV production. The comparison of the Tables 10 and 11 shows that the annual electricity consumption of Version 3 would be totally covered by the production of the South part of the façade only. In the case of all active façade parts, the surplus energy would be 4533 kWh/y, which makes the Version 3 a plus energy house.

Table 11 Annual energy consumption of different versions.

	Ver 1 (kWh)	Ver 2 (kWh)	Ver 3 (kWh)
Lighting, Facility	606	652	724
HVAC aux	888	888	888
Heating	913	910	1 477
Domestic Hot Water	1 225	1 225	1 225
Equipment, Tenant	1 843	1 843	1 843
Consumption total (kWh/y)	5 475	5 518	6 157
PV	0	0	-10 690
Balance (kWh/y)	5 475	5 518	-4 533
Primary energy (kWh/y)	10 950	11 035	-9 066
Energy performance indicator (kWh/m²/y)	74	75	-62

5.7 Soundproofing

The noise reduction properties of building elements can be quantified by the outside inside transmission class (OITC). The OITC characterizes the sound transmission loss of the exterior noise in the frequency range of 80-4000 Hz. The OITC rating of laminated glass depends on the thickness of the glass panes and the interlayer. The laminated structure glass/PVB/glass with corresponding thicknesses 3mm/.76mm/3mm has the OITC rating of 31 dB [42]. Similar structure with thicknesses 5mm/0.76mm/5mm has the OITC rating of 33 dB. Accordingly, we assume that our glass façade structure with 4 mm glass has the OITC rating of 32 dB. The triple glass structure that separates the winter garden from the living room has the estimated OITC rating of 31 [42]. In case the house is situated close to the busy road with the street noise of 75 dB, the noise level in the living room of the Version 1 and 2 would be 44 dB, which corresponds to the level of normal living room [43]. However, this is not the recommended level of the second room bedrooms that are under the same acoustic conditions. In the case of Version 3, the indoor noise is reduced to 13 dB, which is well below the typical bedroom level of 25 dB.

5.8 The usage time of the winter garden

The winter garden enveloped by the photovoltaic façade, provides additional living space or can be used for plants or other applications. This is only true when the temperatures are within a certain limit. For a temperature range between 18 °C and 30°C the winter garden is usable for 2041 hours per year, which is about 23 % of the year. The high fluctuations of the temperature are caused by several reasons. First, the winter garden has a low insulation value and therefore the indoor temperatures are highly dependent on the outdoor temperatures. Second, because of the transparent glass elements, the air inside the winter garden is heated up rapidly by the solar irradiation. Third, the winter garden has a very small heat storage capacity. The significant stops of the temperature rise at 25°C exist because of the window cooling.

On the other hand, the winter garden daily maximum temperatures reach above 18°C even at the coldest months. Due to the solar heating the temperatures reach up to 30°C above the outside ambient temperature. This implies that the winter garden can be used for recreational purposes almost all year around. There are 161 days when there is at least one hour during the day when the winter garden temperature is between 18°C and 30°C.

6 CONCLUSIONS

The observed photovoltaic façade has several advantages, but shows also some aspects which could be improved or analyzed more thoroughly.

For the reduction of the excess solar gain during the summer the semitransparent PV façade is a good alternative for the horizontal shades. The simulated façade with 80% of PV cell packing density reduces the summertime indoor temperature as well as the conventional shades. At the same time the transmitted daylight is still high enough to provide adequate indoor lighting conditions from the outdoor irradiance level of 50W/m^2. Nevertheless, compared to the simulations of the house with horizontal shades the living room annual artificial lighting demand is bigger by 547 hours, which results in 11% higher electricity consumption by the lighting system of the house.

Compared to the horizontal shading the semitransparent PV façade does not make difference between low and high sun when blocking the direct sunlight. Accordingly, the solar gain in the case of the PV façade is reduced equally in the summertime and winter. This results in 60% higher heating demand in the case of the PV façade. However, in the absolute units the increase in the primary energy consumption is not significant.

Despite of the higher primary energy demand by the heating and artificial lighting, the electricity production of the PV cells totally compensates the extra demand and makes the low energy building to a plus energy building.

In addition to the effects on the energy balance the semitransparent PV façade has aesthetic and architectural benefits. The façade provides additional space in the form of a winter garden, which has 23% of the year comfortable temperature and can be used during the daytime peak temperatures even at winter. Due to the visual isolation effect and the estimate noise reduction of 32 dB, the semitransparent PV façade may be an attractive solution for the residential buildings that are situated close to the roads with busy traffic.

Our simulations showed that semitransparent PV façades may be good alternatives for the popular rooftop PV systems. The following research could be made on the similar simulations with the variations in the PV cell packing density. Also, a detailed economic analysis of this kind of PV façades would bring out the real cost effectiveness of such solutions during the exploitation period of the buildings.

7 ACKNOWLEDGEMENTS

This work was supported by institutional research funding IUT (IUT19-28) of the Estonian Ministry of Education and Research, and by the European Union through the European Regional Development Fund project TK141.

8 REFERENCES

[1] European Parliament and the Council (2002) Directive 2002/91/EC on the energy performance of buildings.
[2] Y.B.Assoa, W.Sprenger, M.Misara, F.Roca, L.Mongibello, B.Kubicek, M.Wagner, S.Zamini, J.Merten, A.J.Carr, M.Machado. Thermal and Electri-cal Analysis of BIPV Systems in the Sophia Project. 29th EUPVSEC, 2014..
[3] K.E. Park, G.H. Kang, H.I. Kim, G.J. Yu, J.T. Kim, Analy-sis of thermal and electrical performance of semi-transparent photovoltaic (PV) module, Energy 35 (2010) 2681–2687.
[4] Jordi Cipriano, Guillaume Houzeaux, Daniel Chemisan, Chiara Lodi, Jaime Martí-Herrero, Numerical analysis of the most appropriate heat transfer correlations for free ventilated double skin photovoltaic façades, Applied Thermal Engineering 57 (2013) 57-68..
[5] M. Sandberg, B. Moshfegh, Buoyancy-induced air flow in photovoltaic facades Effect of geometry of the air gap and location of solar cell modules, Building and Environment 37 (2002) 211–218.
[6] A. Guardo, M. Coussirat, E. Egusquiza, P. Alavedra, R. Castilla, A CFD approach to evaluate the influence of construction and operation parameters on the perfor-

mance of Active Transparent Fac,ades in Mediterranean climates, Energy and Buildings 41 (2009).
[7] G.E.Lau, G.H. Yeoh, V.Timchenko, R.K.K.Yuen, Natural Convection in a PV-Integrated Double-Skin Façade using Large-Eddy Simulation, Procedia Engineering 14 (2011) 3277–3284.
[8] Jinqing Peng, Lin Lu, Hongxing Yang, Tao Ma, Compara-tive study of the thermal and power performances of a semi-transparent photovoltaic façade under different ventilation modes, Applied Energy 138 (2015) 572–583.
[9] Jinqing Peng, Lin Lu, Hongxing Yang, Jun Han, Investiga-tion on the annual thermal performance of a photovolta-ic wall mounted on a multi-layer façade, Applied Energy 112 (2013) 646–656.
[10] Jinqing Peng, Lin Lu, , Hongxing Yang, An experimental study of the thermal performance of a novel photovolta-ic double-skin facade in Hong Kong, Solar Energy 97 (2013) 293–304.
[11] Leon Gaillard, Christophe Ménézo, Stéphanie Giroux , Hervé Pabiou , Rémi Le-Berre, Experimental study of thermal response of PV modules integrated into naturally-ventilated double skin facades, Energy Procedia 48 (2014) 1254 – 1261.
[12] Li Mei, David Infield, Ursula Eicker, Volker Fux, Thermal modelling of a building with an integrated ventilated PV façade, Energy and Buildings 35 (2003) 605–617.
[13] David Infield, Ursula Eicker, Volker Fux, Li Mei, Jürgen Schumacher, A simplified approach to thermal perfor-mance calculation for building integrated mechanically ventilated PV facades, Building and Environment 41 (2006) 893–901.
[14] David Infield, Li Mei, Ursula Eicker, Thermal perfor-mance estimation for ventilated PV facades, Solar Ener-gy 76 (2004) 93–98.
[15] Jun Han, Lin Lu, Jinqing Peng, Hongxing Yang, Perfor-mance of ventilated double-sided PV façade compared with conventional clear glass façade, Energy and Build-ings 56 (2013) 204–209.
[16] K. Vats, G.N. Tiwari, Performance evaluation of a build-ing integrated semitransparent photovoltaic thermal sys-tem for roof and façade, Energy and Buildings 45 (2012) 211–218.
[17] Deepali Kamthania, G.N. Tiwari, Energy metrics analysis of semi-transparent hybrid PVT double pass facade con-sidering various silicon and non-silicon based PV module Hyphen is accepted, Solar Energy 100 (2014) 124–140.
[18] Deepali Kamthania, Sujata Nayak, G.N. Tiwari, Perfor-mance evaluation of a hybrid photovoltaic thermal dou-ble pass facade for space heating, Energy and Buildings 43 (2011) 2274–2281.
[19] Kanchan Vats, G.N. Tiwari, Energy and exergy analysis of a building integrated semitransparent photovoltaic thermal (BISPVT) system, Applied Energy 96 (2012) 409–416.
[20] Kanchan Vats, Vivek Tomar, G.N. Tiwari, Effect of pack-ing factor on the performance of a building integrated semitransparent photovoltaic thermal (BISPVT) system with air duct, Energy and Buildings 53 (2012) 159–165.
[21] Wei Sun, Jie Ji, Chenglong Luo, Wei He, Performance of PV-Trombe wall in winter correlated with south façade design, Applied Energy 88 (2011) 224–231.
[22] P.W. Wong, Y. Shimoda, M. Nonaka, M. Inoue, M. Mizuno, Semi-transparent PV: Thermal performance, power generation, daylight modelling and energy saving potential in a residential application, Renewable Energy 33 (2008) 1024–1036.
[23] Danny H.W. Li, Tony N.T. Lam, Wilco W.H. Chan, Ada H.L. Mak, Energy and cost analysis of semi-transparent photovoltaic in office buildings, Applied Energy 86 (2009) 722–729.
[24] Geun Young Yun, Mike McEvoy, Koen Steemers, Design and overall energy performance of a ventilated photovol-taic façade, Volume 81, Issue 3, March 2007, Pages 383–394.
[25] Poh Khai Ng, Nalanie Mithraratne, Harn Wei Kua, Energy analysis of semi-transparent BIPV in Singapore buildings, Energy and Buildings 66 (2013) 274–281.
[26] T. Miyazaki , A. Akisawa, T. Kashiwagi, Energy savings of office buildings by the use of semi-transparent solar cells for windows, Renewable Energy 30 (2005) 281–304.
[27] L. Olivieri, E. Caamaño-Martin, F .Olivieri b, J. Neila, Integral energy performance characterization of semi-transparent photovoltaic elements for building integration under real operation conditions, Energy and Buildings 68 (2014) 280–291.
[28] Seung-Ho Yoo, Simulation for an optimal application of BIPV through parameter variation, Solar Energy 85 (2011) 1291–1301.
[29] C.-M.Lai, S.Hokoi, Solar façades: A review, Building and Environment 91 (2015) 152-165.
[30] G.Quasada, D.Rousse, Y.Dutil, M.Badache, S. Hallé, A comprehensive review of solar facades. Transparent and translucent solar facades, Renewable and Sustainable Energy Reviews 16 (2012) 2643–2651.
[31] A. Jagomägi et al., "European Network of PV Outdoor Testing - Steps towards Harmonized Procedures", 24th EU-PVSEC, Hamburg, 2009, pp. 3432-3438..
[32] Regulation of the Estonian Government (27.01.2014), Minimum requirements for energy performance.
[33] Regulation of the Minister of Economic Affairs and Communications Estonia (09.01.2013), Methodology for calculating the energy performance of buildings.
[34] Marcus Steffen, 2014, Residential Lighting Design, Crowood Press, Wiltshire.
[35] D.L. King, J.A. Kratochvil, W.E. Boyson, W.I. Bower, Field experience with a new performance characterization procedure for photovoltaic arrays, in: Proceedings of the second World Conference and Exhibition on Photovoltaic Solar Energy Conversion, Vienna,.
[36] D.L. King, W.E. Boyson, J.A. Kratochvil, Photovoltaic Array Performance Model, SAND2004-3535, Sandia National Laboratories, 2004..
[37] T. Kalamees, J. Kurnitski, Estonian test reference year for energy calculations, in: Proceedings of the Estonian Academy of Sciences Engineering, Vol. 12, 2006, pp. 40–58..
[38] Perez R., Ineichen P., Seals R., Michalsky J., Stewart R., Modeling daylight availability and irradiance components from direct and global irradiance. Solar Energy, 44: 271-289, 1990..
[39] N. Martin, J.M. Ruiz (July, 2000) Solar Energy Materials & Solar Cells 25 – 38, Calculation of the PV modules angular losses under field conditions by means of an analytical model.

[40] IEA PVPS Task 7. Reliability Study of Grid Connected PV Systems - Field Experience and Recommended Design Practice. Report IEA-PVPS T7-08: 2002..
[41] M. Jantsch, H. Schmidt and J. Schmid. "Results of the concerted action on power conditioning and control". 11th E.C. Photovoltaic Solar Energy Conference. Montreux (1992)..
[42] Product datasheet. http://www.viracon.com/pdf/ViraconAcousticPerfDataTables.pdf.
[43] R.Probst, G.Grevers, H.Iro, 2006, Basic Otorhinolar-yngology: A Step-by-Step Learning Guide, Thieme, Stuttgart .
[44] Oskomera Solar Power Solutions B.V. (February, 2015) online, http://www.oskomera.com/prd47-semi-transparant-zonnepaneel.html.

AN IN-DEPTH COMPARISON OF PV MODULES IN A BIPV FACADE TEST SETUP

Jonathan Lehmann[1, 3], Wout Parys[1, 3], Juliana Gonçalves[1, 3], Jonathan Govaerts[4], Hans Goverde[4], Kris Baert[2, 3], Dirk Saelens[1, 3]
[1]Burgerlijke Bouwkunde, KU Leuven, Kasteelpark Arenberg 10, 3001 Heverlee-Leuven, Belgium
[2]ESAT-ELECTA, KU Leuven, Kasteelpark Arenberg 10, 3001 Heverlee-Leuven, Belgium
[3]EnergyVille, Thor Park 8310, 3600 Genk, Belgium
[4]imec, Kapeldreef 75, 3001, Heverlee-Leuven, Belgium

ABSTRACT: The purpose of this paper is to quantitatively compare different façade-integrated BIPV configurations comprising ventilated, unventilated façades and black or white rear encapsulant color. This paper introduces an experimental set-up and analyses the experimental results that were obtained on three representative days: two sunny days with different wind conditions and one cloudy day. Both temperatures and IV characteristics were continuously logged every 15 seconds. The electrical measurements have evidenced a strong impact of rear encapsulant color, with white modules performing up to 3% better. The thermal study has shown a strong cooling effect caused by cavity ventilation, with ventilated modules performing up to 6% better, as well as an important influence of meteorological conditions.
Keywords: Building Integrated PV (BIPV), c-Si, Façade, Module Integration, Performance, Thermal Performance

1 INTRODUCTION

An integral aspect of the policy of the European Union concerning the energy performance of buildings is the promotion of on-site energy production from renewable energy sources [1]. Integrating photovoltaic cells and modules in the building envelope allows to employ hitherto unused surfaces for this purpose, and this in an increasingly cost-efficient way [2]. So-called Building Integrated PV (BIPV) systems consist of solar photovoltaic cells and modules that are integrated in the building envelope as part of the building structure, thus replacing conventional building materials [3].

A BIPV building component needs to satisfy all typical building envelope component performance requirements, such as air and rain tightness, thermal insulation, aesthetics, fire resistance, durability and others [4], which can affect the PV panel performance.

The design parameters impacting the performance assessment of a BIPV building component can be roughly categorized into 3 levels: PV cell level, PV module level and building component integration level. Although PV parameters have been widely studied on a cell level, the authors have identified a need for a quantitative assessment of those parameters on a PV module and building component integration level in outdoor conditions, thus continuing earlier research [5].

In this work, we present the results of the experimental study of a module level parameter, namely rear encapsulant color, and an integration level parameter, namely back cavity ventilation.

The study is based upon an experimental measurement set-up of a façade BIPV mock-up, deployed in Leuven, Belgium.

This paper is structured as follows. The first section describes the experimental set-up, including the different BIPV component configurations and the measurement equipment, followed by the second and third section displaying and interpreting the measurement results.

2 DESCRIPTION OF THE EXPERIMENTAL SET-UP

The set-up consist of six glass-glass modules, each one composed of nine 6" full Al-BSF c-Si solar cell encapsulated in a front transparent 0.5mm EVA and rear black or white 0.5mm EVA and covered on both sides by a 2 mm glass (Figure 1a.). All modules are mounted in a BIPV structure with ventilated or not ventilated cavity (Figure 1b.).

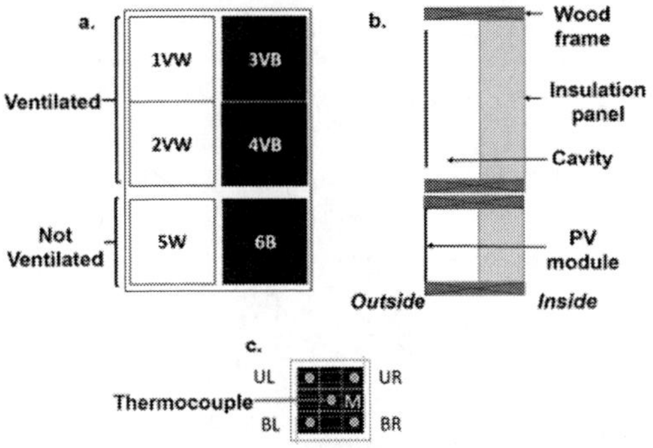

Figure 1: Schematic (a) front view, (b) section view of the setup and (c) thermocouples distribution at the back of each module

On the back of each module; we have installed five thermocouples (Figure 1c.) in order to measure the temperature. IV characteristics and temperatures of the PV modules are acquired every 15 seconds. Between each sweep, the PV modules are set at MPP value.

Acquiring IV curves instead of solely output power measurements allows to investigate the temperature-related open-circuit voltage [4] and irradiance-related short-circuit current [5] more in detail.

This, coupled with meteorological data measured separately will allow us to perform precise thermal analysis.

Finally, the modules are mounted in the vertical plane, on an S-W oriented façade of a KU Leuven test building with minimum shadowing. Next to the modules, a pyranometer has been installed to measure the illumination.

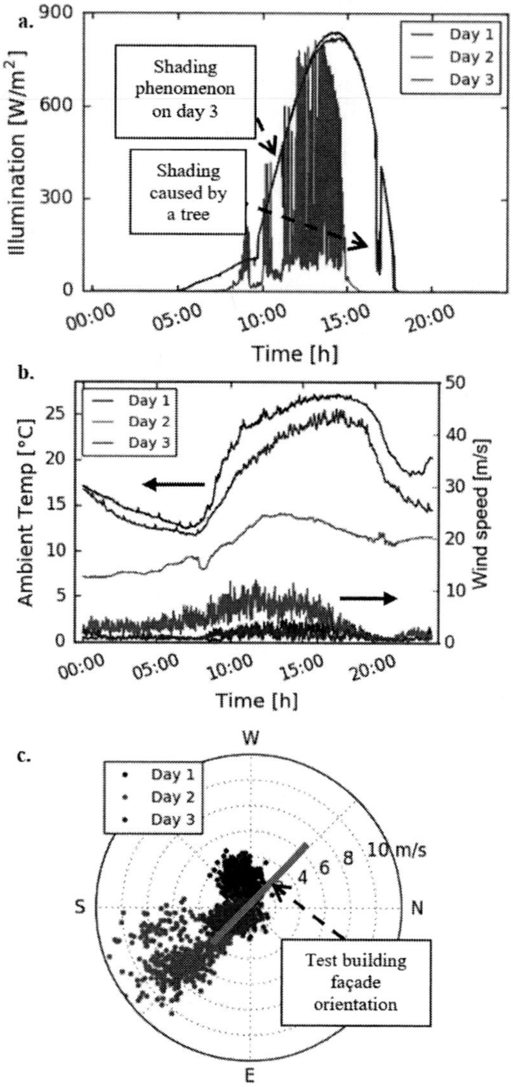

Figure 2: Day 1, 2, 3: (a) Illumination as a function of time (b) Ambient temperature and wind speed as a function of time (c) polar representation of wind direction between 9h00 and 20h00

After comparing the performances of black and white modules at standard test conditions, we will compare three different days. Day 1 and 3 are sunny days of September while day 2 is a cloudy winter day.

We observe almost identical illumination for day 1 and 3, with however a 2-3 degrees lower ambient temperature on day 3 (Figure 2a. & b.). Day 1 and 3 are subject to a moderate wind with different directions whereas day 2 is subject to a strong wind (Figure 2b. & c.).

3 ENERGY YIELD COMPARISON

3.1 Standard test condition performances

First, we performed a set of indoor electrical measurements under standard test conditions (Table 1). These results highlight a power difference of 5% at MPP between black and white modules. We observe a significant increase of Isc for white modules that explains the improved performances.

Table I: Average electrical performances of black and white module under standard test conditions

	Black module	White module
Short Circuit Current (A)	8.35	8.69
Open Circuit Voltage (V)	5.64	5.64
Power MPP (W)	35.81	37.19

3.2 Day 1: Sunny summer day

On day 1 (Figure 3a.), we observe higher energy yield for ventilated modules (except module 3VB) and higher energy yield for white modules compared to black modules.

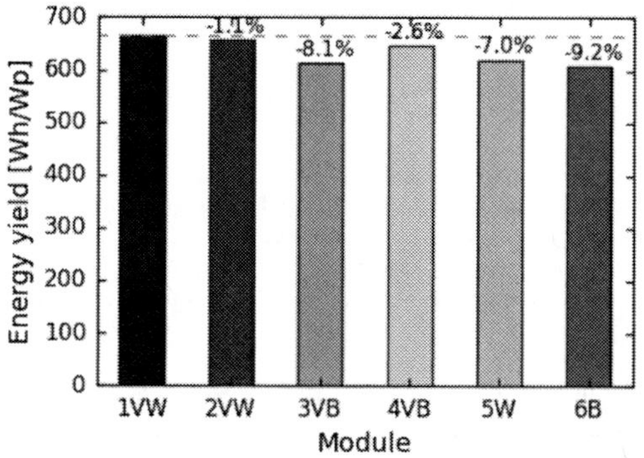

Figure 3: Energy yield of day 1 as a function of module number. On each bar, the energy yield difference in % compared to module 1 is written.

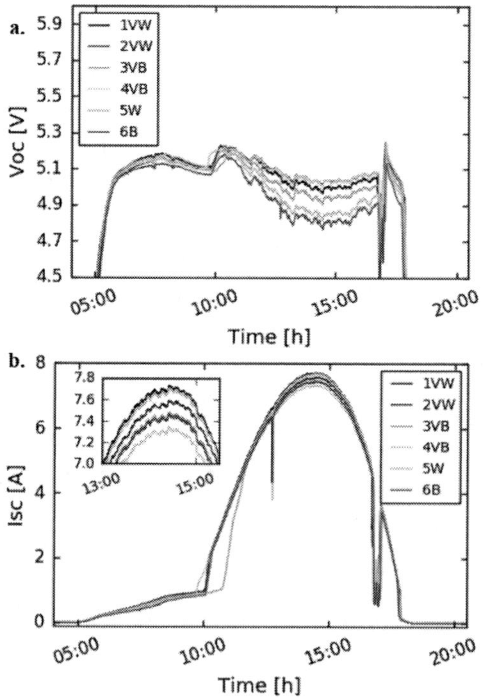

Figure 4: For day 1, (a) open circuit voltage (Voc) and (b) short circuit current (Isc) as a function of time

Throughout day 1, an important dispersion of Voc is observed (Figure 4a.). For module 3VB between 10h and 12h, a partial shading effect caused by a rain pipe is clearly noticeable (Figure 4b). This explains the lower energy yield observed on Figure 3. At peak illumination, we notice different behaviors, with ventilated modules exhibiting a higher Voc than unventilated modules (Figure 4a.) and white modules exhibiting a higher Isc than black modules (Figure 4b.).

3.3 Day 2: cloudy winter day

On day 2 (Figure 5), we obtain significant lower energy yield compared to day 1, around 200 Wh/Wp against 700 Wh/Wp. We can still observe higher energy yield for ventilated or white modules compared to not ventilated or black modules.

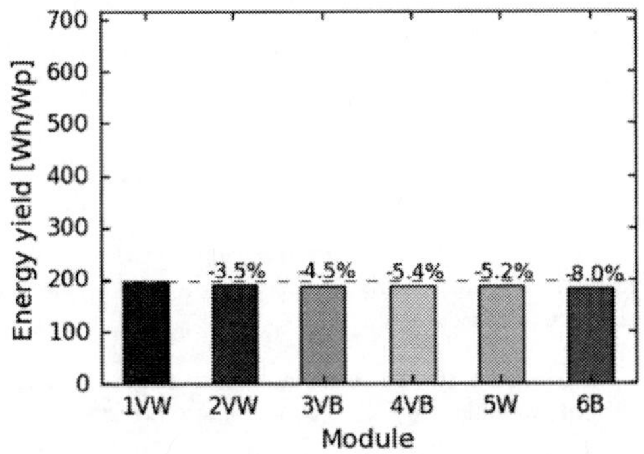

Figure 5: Energy yield of day 2 as a function of module number. On each bar, the energy yield difference in % compared to module 1 is written.

Because of temporal varying illumination, an important dispersion of Voc and Isc is observed (Figure 6). We measure comparable Voc for each module, showing that there is little temperature difference between them (Figure 6a.). The difference in temperature is however increasing after 12h00, when the illumination increases (Figure 2a.). We notice local dispersion of Isc on short periods (Figure 6b.), e.g. 9h00 and 15h00, evidencing a partial shading phenomenon caused by the cloudy weather.

a.

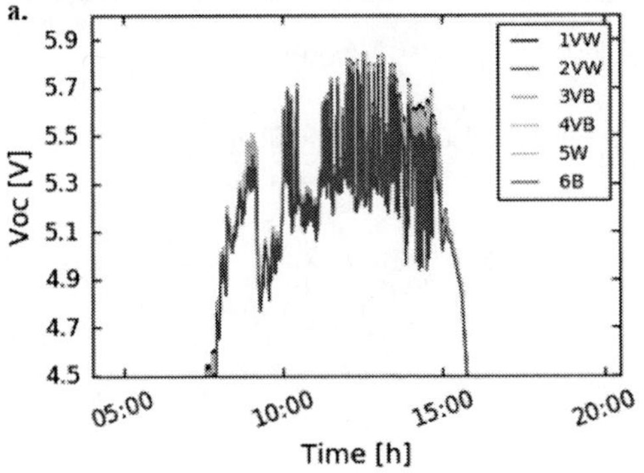

b.

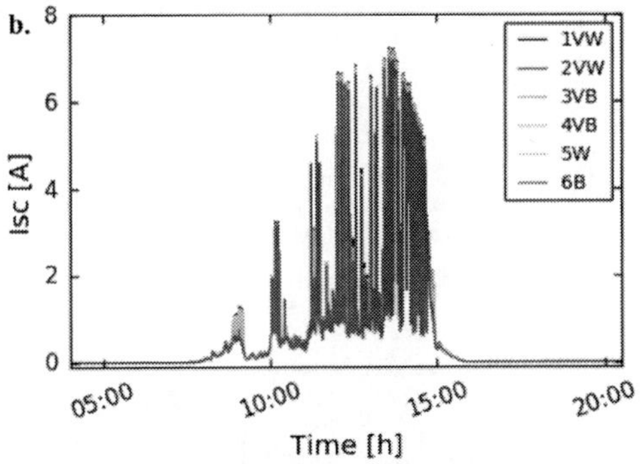

Figure 6: For day 2, (a) open circuit voltage (Voc) and (b) short circuit current (Isc) as a function of time

The energy yield comparison shows that the improved energy yield for white modules can be related to higher Isc whereas the improved energy yield for ventilated modules is caused by higher Voc. For day 3, similar results are obtained as for day 1.

4 MODULE TEMPERATURES

For each module, we measured the temperature distribution by a set of thermocouples mounted behind the glazing containing the solar cells. High temperatures are undesirable as they reduce the solar cell efficiency, will have an impact on the durability of the building components and may cause a hot thermal flux to the inside of the building. The legend entries 'U', 'B', 'R', 'M', and 'L' refer to the upper, bottom, right, middle and left temperature sensors respectively.

4.1 Effect of ventilation: module 1 and 5 on day 1

The temperature distribution for day 1 and module 1 (ventilated) and (b) module 5 (not ventilated) is depicted in Figure 7.

a.

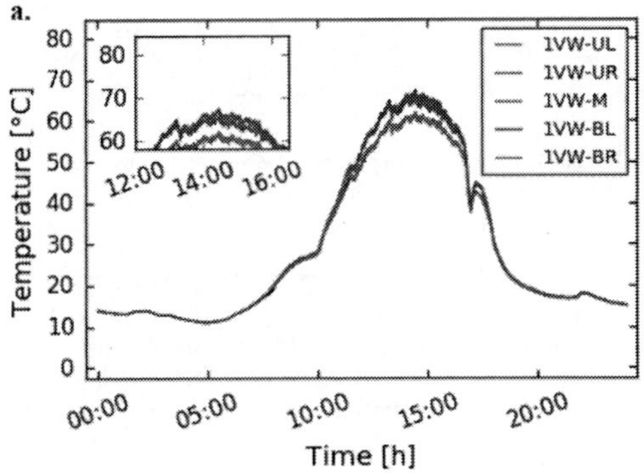

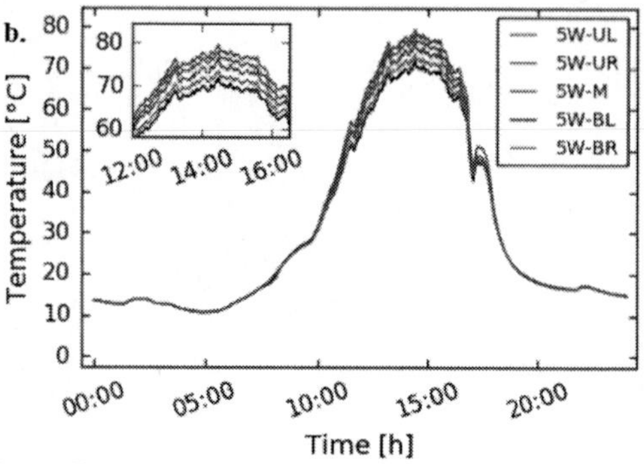

Figure 7: For day 1, thermocouple temperature of (a) module 1 (ventilated) and (b) module 5 (not ventilated) as a function of time

Compared to module 1, module 5 exhibits more dispersed temperatures that are 5-10°C higher (Figure 7). Because of cavity ventilation, on module 1, we observe a vertical temperature distribution with lower temperatures at the bottom of the cavity (Figure 7a.). Whereas on module 5, we observe an important edge effect resulting in a more elevated temperature in the middle of the module. If the module is mounted between façades with highly different temperatures, this edge effect can induce a difference between left and right temperatures of the module, as it is the case here (Figure 7b.).

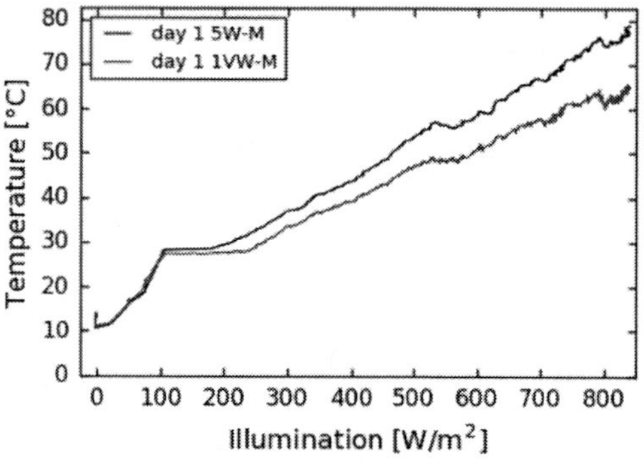

Figure 8: Middle of the module thermocouple temperature as a function of illumination between sunrise and maximum illumination for module 1, 5 on day 1

Figure 8 depicts the temperature of the thermocouple in the middle of the module (Figure 1c.) as a function of illumination between sunrise and maximum illumination. We observe significantly different slopes of heating for ventilated and un-ventilated modules. Due to cavity ventilation, module 1 shows a better cooling compared to module 5.

4.2 Effect of rear encapsulant color: module 5 and 6 on day 1

We observe similar temperatures for black and white modules at the bottom of the cavity (Figure 9), however for the white module the dispersion of temperatures is more important (Figure 9a.), leading to higher temperatures at the top of the cavity.

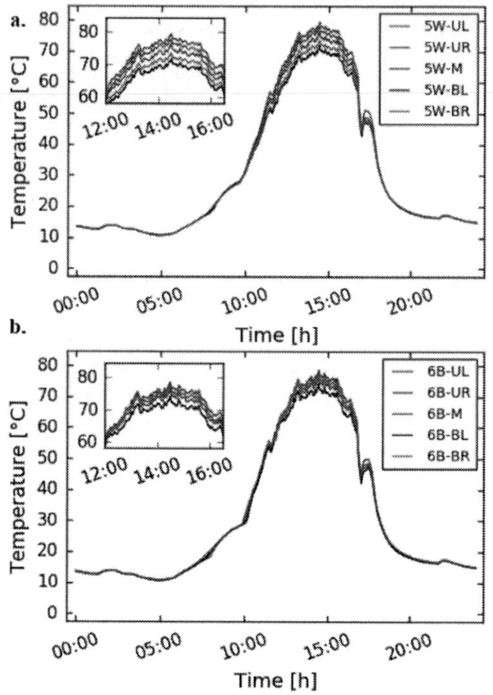

Figure 9: For day 1, thermocouple temperature of (a) module 5 and (b) 6 as a function of time

On Figure 10, we observe similar slopes of heating for both modules, therefore confirming this result.

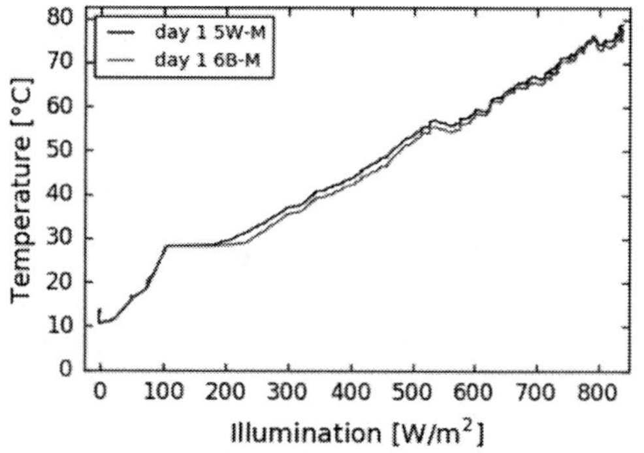

Figure 10: Middle of the module thermocouple temperature as a function of illumination between sunrise and maximum illumination for module 5, 6 on day 1

4.3 Effect of weather conditions: module 5 on days 1 and 3

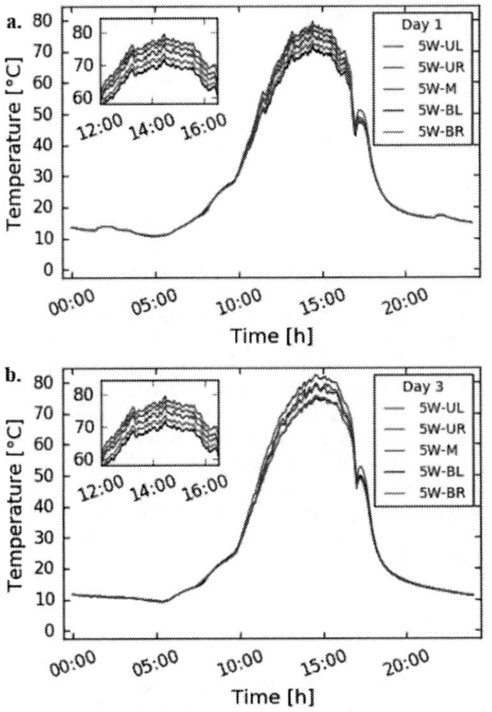

Figure 11: Thermocouple temperature of module 5 as a function of time on (a) day 1 and (b) day 3

On days 1 and 3 we have measured similar illumination (Figure 2a.). Therefore for un-ventilated module 5, we notice similar temperature dynamics for day 1 and 3 (Figure 11). On Figure 12 we observe that for day 1 before 500 W/m2 or 11h00, when the wind is low (Figure 2b.), temperatures are more elevated whereas at peak illumination, around 800 W/m2 or 15h00 when the wind is high, temperatures are more elevated on day 3.

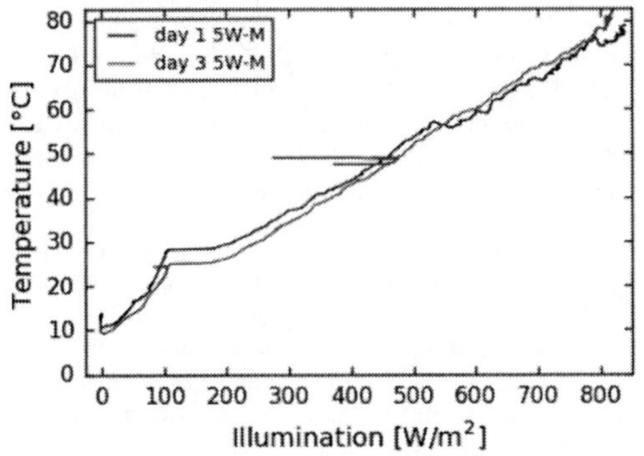

Figure 12: Middle of the module thermocouple temperature as a function of illumination between sunrise and maximum illumination for module 5 on day 1, 3

From these results, we can make the hypothesis that for low illumination, the PV module temperature difference between day 1 and 3 is mainly driven by the fact that ambient temperature is lower on day 3. However, after 11h00, when the wind is perpendicular to the façade and supporting module cooling on day 1, this cooling effect yields lower PV module temperatures on day 1.

In agreement with the electrical analysis, we have seen an important temperature difference between ventilated and not ventilated modules, caused by the cooling effect of ventilation. This is coupled to an important influence of meteorological conditions, mainly ambient temperature and wind. Finally, we have shown that rear encapsulant color has only a limited effect on the temperatures of the modules.

6 CONCLUSIONS

Through electrical and thermal measures, we have studied the impact of PV module structure and building component integration on the performances of a BIPV setup. The electrical measurements have evidenced a strong impact of rear encapsulant color, with white modules performing up to 3% better due to an increase of Isc. The thermal study has shown a strong cooling effect caused by cavity ventilation, with ventilated modules performing up to 6% better, as well as an important influence of meteorological conditions. As a conclusion, we have demonstrated that white rear encapsulant ventilated modules allow the highest energy yield due to highest currents and lowest temperatures.

7 ACKNOWLEDGEMENT

This work has been conducted within the EFRO SALK project, which receives the support of the European Union, the European Regional Development Fund ERDF, Flanders Innovation & Entrepreneurship and the Province of Limburg.

8 REFERENCES

[1] Directive 2010/31/EU of the European parliament and of the council of 19 May 2010 on the energy performance of buildings, *Official Journal of the European Union*, pp 153/13-35 2010

[2] G. Verberne, P. Bonomo, and F. Frontini, "BIPV Products for Facades and Roofs: A Market Analysis," in *European Photovoltaic Solar Energy Conference and Exhibition*, 2014.

[3] K. Sinapis and M. van den Donker, "BIPV REPORT 2013 - State of the art in Building Integrated Photovoltaics," 2013.

[4] P. Warren, "IEA ECBCS Annex 32 - Technical Synthesis Report: Integral Building Envelope Performance Assessment", 2003.

[5] G. Van den Broeck, W. Parys, H. Goverde, S. Van der Putten, J. Poortmans, J. Driesen, K. Baert, D. Saelens, "Experimental analysis of the performance of façade-integrated BIPV in different configurations" in *EU PVSEC*, 2016

[6] P. Löper, D. Pysch, A. Richter, M. Hermle, S. Janz, M. Zacharias, and S. Glunz. Analysis of the Temperature Dependence of the Open-Circuit Voltage. Energy Procedia, pages 135–142, 2012. doi:10.1016/j.egypro.2012.07.041.

[7] C. W. Hansen. Estimation of Parameters for Single Diode Models Using Measured IV Curves. IEEE 39th Photovoltaic Specialists Conference (PVSC), pages 223–228, 2013. doi:10.1109/PVSC.2013.6744135

EXPERIMENTAL INVESTIGATION AND CHARACTERIZATION OF BUILDING INTEGRATED PHOTOVOLTAIC/THERMAL ENVELOPE SYSTEM WITH THERMAL ENHANCEMENTS FOR ROOF AND CURTAIN WALL APPLICATIONS

E. D. Rounis, O. Kruglov, Z. Ioannidis, A. Athienitis, T. Stathopoulos
Dept. of Building, Civil and Environmental Engineering, Concordia University
1455 De Maisonneuve Blvd. W., Montreal, Quebec H3G 1M8, Canada

ABSTRACT: Integrating photovoltaic panels (PV) with the building envelope increases the cost-effectiveness of the installation by replacing common building materials, and the potential for generation of useful heat along with the electricity production. Large PV systems intended for building roof and/or facade applications require active cooling of the PV to avoid overheating and resulting reduction of the electrical efficiency, and decelerate the system's degradation. The heat collected by the cooling medium may be utilized in various ways such as to preheat fresh air, as a source for an air-source heat pump. Air-based systems are generally better suited for large installations, they are however, less efficient in extracting heat from the PV due to the thermo-physical properties of air and therefore need enhancements for their thermal performance. This study investigates ways of enhancing air-based Building Integrated Photovoltaic/Thermal (BIPV/T) systems, focusing on the use of multiple-inlets and presents the development and initial testing of a BIPV/T curtain wall prototype. The testing of the prototype showed superior performance for the double-inlet versus the single-inlet configuration, thermally (up to 18% higher thermal output), lower peak PV temperatures (reduced by up to 3.6°C) and a marginal increase in electrical output due to enhanced cooling.
Keywords: Building Integrated PV, Experimental Methods, Thermal Performance, Photovoltaic/Thermal

1 INTRODUCTION

Buildings are responsible for more than one third of the global energy consumption. In particular, the European building sector accounts for approximately 40% of the global consumption, and in North America for approximately 32% [1]. The need for net-zero energy buildings and the reduction of the energy footprint of existing ones is more prominent than ever and this is reflected in the new building codes and guidelines. The incorporation of solar technologies for electricity and useful heat generation provides an efficient means to that end.

A hybrid system that combines the properties of a photovoltaic system and a thermal collector is known as a photovoltaic thermal (PV/T) system. The photovoltaic layer of such a system has the role of the absorber and a circulating fluid removes part of the excess heat from it. The collected heat may then be utilized by various means, depending on the fluid's outlet temperature and delivery rate. PV/T systems have been found more efficient than PV and thermal collectors independently covering the same surface area [2].

Building-Integrated Photovoltaic/Thermal systems come from the integration of PV/T with the building envelope. This integration may have several advantages over using PV/T systems as stand-alone [3]. Apart from the architectural integration and the superior aesthetic result, the PV panels replace the outer shell materials (roof shingles, siding, concrete or masonry panels for façade etc.), increasing the cost effectiveness of the installation and the whole system functions as a building envelope element. This means that a BIPV/T system must also fulfill the requirements of the building envelope in terms of heat, air, water and moisture control [4]. Integration with the façade or the roof may offer a larger installation area, which would otherwise not be available for use, although BIPV/T systems cannot be optimally oriented and tilted as opposed to stand-alone PV/T systems.

The importance of the ventilation and cooling of the PV for building integrated systems has been stressed by several researchers [5, 6]. If the PV panels are directly set on the building surface, they will overheat since the wind-driven cooling is reduced, as opposed to racked PV panels. An air channel formed between the PV layer and the rest of the wall section for the ventilation of the system may serve two purposes: the cooling of the PV to prevent a decrease in electrical efficiency, and the collection of the preheated air for various purposes. Fan-assisted convection has been found to be more efficient in cooling the PV and extracting heat, especially in larger systems [7], in comparison to natural convection.

Depending on the coolant, BIPV/T systems are divided into water and air-based systems. Water based systems have greater heat exchange effectiveness. They require, however, special casing and piping, and may pose freezing issues in the winter, or steam formation in the summer. Air-based systems on the other hand are easier to install and maintain, they do not pose leakage or freezing issues and can be adapted to different kinds of facades and roofs. However, due to the thermo-physical properties of air they are less efficient in the extraction of excess heat from the PV. There have been several studies on ways to enhance the thermal performance of PV/T and BIPV/T systems, which are discussed in section 2.

A review on the existing literature shows that BIPV/T systems have not yet achieved widespread application, and full integration of the system is rarely achieved. Usually, a harmonic architectural integration with an existing façade or roof is enough to characterise the system as integrated. However, there are few examples of systems that replace the typical wall or roof structure of a building [8, 9]. This is mainly due to the fact that there is no current BIPV/T standard related to its integrated performance and function as a building envelope element.

In this study, thermal enhancement techniques proposed in the literature are reviewed and the lack of a BIPV/T performance design standard is addressed. Furthermore, the development and initial testing of a BIPV/T prototype, which was designed based on existing curtain wall practices and intended for full building integration is presented. The initial experimental test set-up focuses on the introduction of an additional air intake, however, several other thermal enhancement techniques will also be incorporated and studied in the future. The main purpose of this study is to provide insight on the design of fully integrated BIPV/T systems as well as the expected performance of various system configurations, which could lead to a BIPV/T performance standardization.

2 BIPV/T THERMAL ENHANCEMENTS

2.1 Air-based systems thermal enhancements

As mentioned earlier, air-based systems are generally more suitable for envelope integration and large-scale applications, but less efficient in heat extraction than water-based systems. A lot of research has been carried out on ways to enhance the thermal performance of air-based systems. The majority of the proposed enhancements were applied to a reference PV/T configuration consisting of the PV layer, the air channel and the insulated back surface (Fig. 1), while other refer to specific prototypes. Enhancement may only involve geometric and/or flow optimization, or the addition of material elements such as double glazing, fins or other porous media.

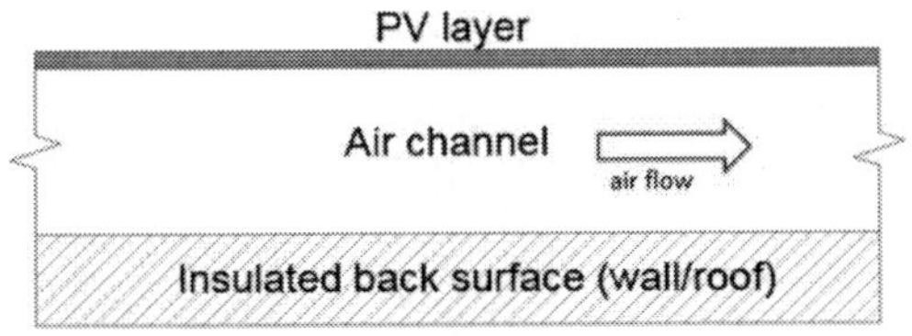

Figure 1: Typical PV/T configuration.

The criteria for enhancement vary among studies. Naturally ventilated systems with no heat collection are optimized for lower PV temperatures to increase the electrical performance and the systems durability. The optimization refers to the collector's aspect ratio (length over depth) for which the velocity of the buoyancy driven flow is maximized, but it may also involve the optimum tilt angle [10-14]. In general, studies on naturally ventilated systems agree on a 10-15 cm optimum channel gap and a 45°-55° tilt angle.

The enhancement of mechanically ventilated systems varies considerably, depending on whether the priority is given to either the electrical output, or the thermal output, or a balance between the two and whether the consumption of the cooling system is also considered. Several researchers have investigated the optimal aspect ratio of the collector length to the collector depth and the optimal flow for air-based PV/T systems [7, 15, 16]. In addition, some studies include the effect of pressure drop within the collector [15] and within the ducting that connects the PV/T to the HVAC system [17]. The suggested optimal flows vary between 0.026-0.05 kg/m2s, and the optimal channel depth between 0.025-0.09 m.

The practice of adding one or more glazing layers over the PV layer has been studied as a means to increase the thermal output of the system by reducing the wind induced heat losses [18-21]. The boosted thermal performance comes at the cost of the electrical, due to the higher PV temperature and the reduced solar radiation transmitted through the extra glazing (occasionally due to the formation of condensation on the outer glazing). The PV temperatures can be decreased with a double-pass configuration, by placing the PV layer between two air channels, independent or connected. However, unless the system is optimized for thermal output, glazed PV/T systems are not recommended, due to PV overheating and poor electrical performance [21].

Heat extraction from the PV layer can be further enhanced with the use of material elements inside the air channel. These elements could either be fins in contact with the absorber that act as heat sinks [22], or fins set on the back surface which induce turbulence and boost the convective heat transfer [23, 24]. Other elements can be porous media such as a metallic mesh [25, 26], honeycomb heat exchanger [27], and suspended thin metal sheet [24] which further increase the radiative and convective area. These techniques have been proven very effective in removing heat from the PV, as opposed to the reference design (Fig. 1). However, a major limitation is that they cannot be used for systems with long channels due to the added flow resistance in the air channel and consequent increase of fan consumption.

Most of the aforementioned studies has been based on small scale experimental models (collector length generally 1-1.5 m), the behaviour of which could be different in larger scale applications. This is primarily due to the formation of a thermal boundary layer and a consequent temperature rise of the PV layer along the collector, leading to non-uniform PV temperature distribution and lower performance of the collector.

One major issue that is raised is the lack of common criteria for the enhancement and optimization of the systems. Usually the enhancement refers to the PV/T system independently. This is due to the lack of a BIPV/T performance standard which would include the integration of the BIPV/T with the mechanical system of the building (heat pump, direct heat supply, heat exchanger, desiccant cooling etc.) and the corresponding requirements in terms of flow and fluid temperature, along with limitations for allowed PV temperatures. Furthermore, it is difficult to have a clear comparison between the various proposed systems due to the difference in dimensions and testing conditions, while the proposed models are usually case specific and the convective correlations are not generally applicable.

2.2 Multiple inlet BIPV/T

A more recent method of thermal enhancement is the introduction of more than one air intakes along the collector's channel. This technique is based upon the breaking of the thermal boundary layer and taking advantage of the introduced entrance effects. The convective heat exchange from the PV layer to the air stream is enhanced and more uniform PV temperatures can be achieved by regulation of the inlet flow distribution. The concept of a multiple-inlet BIPV/T evolved from the hybrid unglazed transpired collector (UTC) - PV/T system introduced by [8], which is installed at the mechanical penthouse of the John Molson School of Business (JMSB) building of Concordia University in Montreal. This system consists of custom PV modules attached onto a layer of UTC.

A double-inlet system was numerically investigated by Yang & Athienitis (2014) [26]. Convective heat transfer correlations were developed based on the testing of an experimental single-inlet prototype in an indoors solar simulator facility, which were used for the simulations of the performance of a two-inlet system. The addition of a second inlet increased the thermal efficiency of the system by 5%, reduced the maximum PV temperatures and marginally increased the electrical efficiency. Rounis et al (2016) [28] proposed a method of modelling multiple-inlet BIPV/T systems and simulated the performance of a single and a multiple-inlet BIPV/T system for a large-scale office building application. Results showed that the multiple-inlet system had the lowest and most uniform PV temperatures, as well as higher electrical and thermal efficiency. Mirzaei et al. (2014) [29] investigated experimentally the role of cavity flow on the performance of naturally ventilated BIPVs placed on inclined roofs. The experimental configurations included a flat and a stepped arrangement, which was essentially a multiple-inlet formation. It was found that the PV temperatures were significantly lower for the stepped configuration (multiple-inlet) cases, as opposed to the flat roof.

3 BIPV/T PROTOTYPE DEVELOPMENT

A prototype was designed and developed based on requirements regarding both the building envelope role of BIPV/T, as well as the electrical and thermal performance of the system. The purpose was to address issues of architectural and structural integration, but also to investigate and compare experimentally different configurations and performance enhancement methods using the same prototype, under the same experimental conditions. The objective of this effort would be to provide insight on the development of a BIPV/T design and performance standard.

3.1 Building envelope function

A BIPV/T system that is intended for full integration must satisfy the requirements of a building envelope in terms of heat, air, moisture and water control. From an architectural point of view, it must be structurally sound and should be seamlessly integrated with the building for a superior aesthetic result. Furthermore, a BIPV/T adapting to common building practices would be more cost effective and attractive to the market. These design considerations influenced the development process of the prototype and the curtain wall technique was chosen as the basis of the prototype's design (Fig. 2)

Figure 2: BIPV/T curtain wall concept rendering.

A typical metal curtain wall assembly usually consists of a glass pane for the visible section and a spandrel for the opaque, both fixed with pressure plates on vertical and horizontal metal mullions, which form the supporting frame. This technique relies on either a perfect seal approach for the control of air water and moisture, or pressure equalization for decreased rain penetration and the use of seals and a back-pan attached to the insulation as an air and moisture barrier. The heat transfer is regulated by insulation placed in the spandrel section and supported by a thin aluminum back-pan and double or triple-glazing at the visible section.

A BIPV/T curtain wall is intended for the opaque section of a building, although part of it could be modified to transmit solar radiation to the interior, as a double-skin façade [30, 31], or as a skylight section for roof applications. Figure 3 demonstrates the main aspects of the BIPV/T curtain wall design.

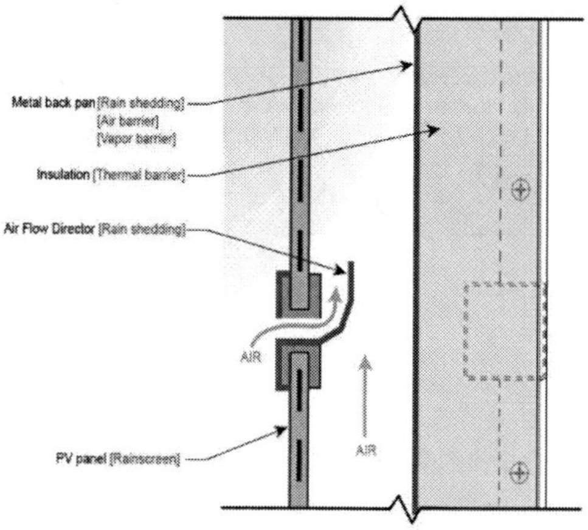

Figure 3: BIPV/T curtain wall cross section.

The main modifications from the original façade system include the following:

- The spandrel section extends for the full length of the system. The insulation is supported by an aluminum back-pan and an additional could be placed on top to form an extra absorber layer for the case of semitransparent PV.
- The horizontal structural elements (mullions) are shorter than the vertical, in order to facilitate a continuous air channel.
- The PV form the outermost layer of the section and are supported by either pressure plates or a combination of pressure plates and point supports, effectively replacing glazing.

Figure 4 demonstrates a plan view of the experimental BIPV/T curtain wall prototype. It consists of a metal frame, two frameless PV modules and a 2-inch rigid insulation sandwiched between aluminum sheets. The metal frame was fabricated by Unicel Architectural using commercial mullion extrusions. The frameless, semi-transparent PV modules were provided by Canadian Solar. Two sets of semi-transparent PV modules were used, one with 66 cells and one with 72 cells, in order to investigate the effect of different transparencies on the prototype's performance

A second inlet was introduced between the two PV modules, while the PV panels were placed horizontally to take advantage of the entrance effects of the second inlet flow. The prototype was designed with a reduced height to fit the test platform, but is considered as part of a story-high, or taller, system.

The panels were fixed with pressure plates at the side

and top mullions and a shorter middle mullion was used to reduce counter deflection and provide extra support for the dead load of the panels. A modification of the panel support was used for the roof BIPV/T presented in section 3.4.

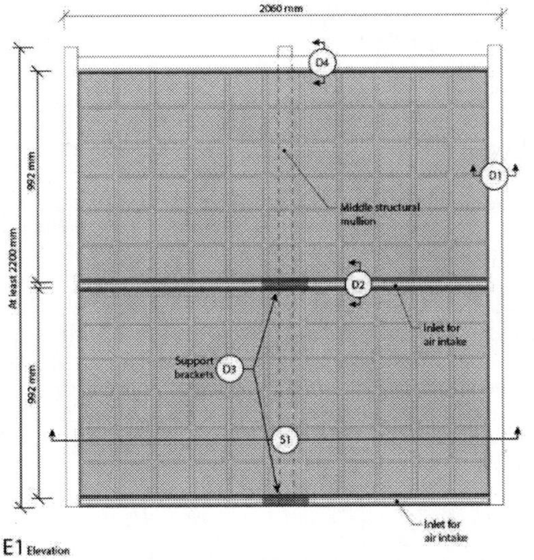

Figure 4: BIPV/T curtain wall prototype plan view.

3.3 Electrical and thermal performance

The prototype was also built to serve as a means of testing different configurations and thermal enhancement methods. For the first round of experiments discussed in this paper, the focus was on the comparison between the use of semitransparent PV modules with different transparencies, as well as the introduction of a second inlet between the modules. Further modifications will include the installation of fins and varying channel gap size.

3.4 Flow considerations

Regarding the flow of the air channel, several aspects were considered. The primary factor that dictated the overall flow design was the introduction of a second inlet between the PV panels. The opening of the inlet was roughly estimated by previous analysis [28] to approximately provide one third of the total flow when fully open. Furthermore, air flow directors were used (Fig. 3) to direct the air stream of the second inlet closer to the surface of the top PV module. This also acted as a measure against rain penetration. Finally, a custom manifold was built and connected to holes opened on the top mullion, at the top side of the collector. This consisted of four take-offs from the collector connected to the main duct of the air collection unit. The number of take-offs was a compromise between flow uniformity and reduction of the structural integrity of the top mullion.

3.4 Roof application

The roof of the row house designed and built by team Montreal for the Solar Decathlon China 2018 competition is a modification of the BIPV/T curtainwall prototype (Fig. 5). The BIPV/T is located at the front roof (Southern roof) of the house and measures 11.2m long and 5.4m wide.

Because of the row-house typology of the building it was imperative to provide the building with daylight, for which a skylight was implemented. The BIPV/T roof has a strip of glazing, which acts as a small solar collector and allows daylight to reach the skylight below.

Six longitudinal structural members separate the BIPV/T roof into 5 individual channels of approximately one meter (0.95m) width and a depth of 8 cm. This gap is between the PV panels and a layer of insulation enclosed with an aluminum back pan, which is discontinuoued at the glazing section above the skylight, replaced instead with a second layer of glazing to maintain the air channel continuity. At the end of each channel, a duct connects to the manifold where the heated air is collected before arriving at an air-to-water heat exchanger.

Figure 5: Rendering of the BIPV/T roof of Team Montreal's Solar Decathlon row-house with an included skylight (left), part of the actual fabricated system (right)

4 EXPERIMENTAL PROCESS AND PRELIMINARY RESULTS

The experimental prototype was tested in the Concordia University indoor Solar Simulator (Fig. 6). The Solar Simulator consists of a lamp field which emulates the sun light and a test bench upon which the test specimens are placed. The 8 metal halide lamps that comprise the lamp field can be moved individually in the horizontal and vertical axis and can produce irradiance intensities from 500 to 1200 W/m2 with a uniformity of up to 97% [32]. The spectral quality of the lamps fulfills the specifications of ISO 9806. Both the lamp field and the test platform can rotate between a horizontal and vertical orientation to accommodate testing cases varying between flat roofs, tilted roofs and vertical facades. The experiments carried out for the experimental process presented here was in the vertical position, corresponding to a façade integrated BIPV/T.

This paper includes the results from testing conditions similar to the Nominal Operating Cell Temperature (NOCT) conditions, namely:

- Average solar radiation, G=842 W/m2
- External wind velocity, Vwind=1 m/s
- Ambient temperature, Tamb=20°C

Two sets of PV modules were used (66 and 72 cell) and for each set, three configurations were tested: single inlet (with the middle inlet covered), double-inlet without the flow director and double inlet with the flow director installed. For each configuration, three total mass flow rates were investigated, namely, 400 kg/hr, 500 kg/hr and 600 kg/hr, with corresponding average channel velocities of 0.5 m/s, 0.62 m/s and 0.74 m/s. The most important results from this first set of experiments are presented in the following sections.

Figure 6: The BIPV/T prototype during testing in a vertical position at the Concordia University Solar Simulator.

4.1 Effect of transparency

The higher transparency of the 66-cell set of PV modules resulted in 10-11% lower electrical output versus the 72-cell set, which was consistent in all tested configurations and flow rates, reflecting the corresponding difference of PV cell coverage. The higher amount of radiation transmitted to the back absorber resulted in a 5-10% higher thermal output and slightly higher outlet temperature (up to 0.7°C). The peak PV temperatures for the 72-cell set were higher by 0.6°C - 3.6°C, due to the higher radiation absorbed by a larger PV cell area. These results agree with previous investigations [33,34]. The choice of packing factor in a real application would depend on the desired electrical and thermal output. Indicatively, Figure 7 demonstrates the electrical and thermal efficiencies of the two sets for all configurations at a collection rate of 600 kg/hr.

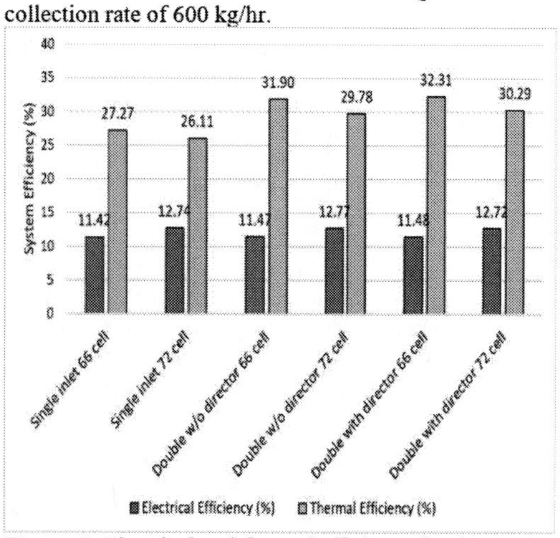

Figure 7: Electrical and thermal efficiency for the single-inlet, double-inlet and double-inlet with installed flow director, for the 66-cell and 72-cell PV module sets.

4.2 Effect of second inlet and flow director

The introduction of the second inlet may increase the thermal output by up to 18% and the outlet temperature by 1°C (for a collector of 2 m height). The electrical efficiency is marginally increased and the peak PV temperatures are lower by up to 1.6°C, depending on the flow rate. These results are in accordance with the experimental work by [26].

The use of an air flow director resulted in the highest differences in performance between the single and the double inlet configuration. The air flow entering from the second inlet was directed closer to the surface of the top PV module, extracting more heat and lowering further the temperature of the module. When the flow director was removed, the double-inlet configuration still performed better than the single-inlet in terms of electrical and thermal efficiency, as well as peak PV temperatures, however the difference in performance was less pronounced as for the case with the flow director installed.

Figure 8 presents the electrical and thermal efficiencies and the peak PV temperatures for the three configurations at the 600 kg/hr flow air collection rate. The results were similar for the other flow rates tested, as well as for the 66-cell set.

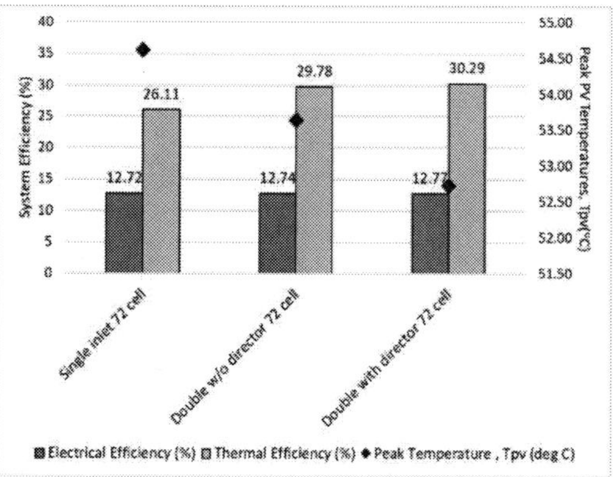

Figure 8: Electrical efficiency, thermal efficiency and peak PV temperature for the three configurations of the 72-cell set, at 600 kg/hr total mass flow rate (0.75 m/s average channel velocity).

5 CONCLUSION

This study presented an overview of the thermal enhancements that have so far been implemented in air-based hybrid PV/T collectors and the design and initial testing of a BIPV/T curtain wall prototype. The prototype has been developed to both provide design guidelines for fully integrated façade and roof BIPV/T systems and investigate the performance and applicability of various thermal enhancement techniques.

Regarding the literature, it has shown that there are numerous ways to enhance the thermal performance of air-based PV/T and BIPV/T systems. Regardless, the criteria of enhancement vary considerably among studies, while the majority of these studies refers to the optimization of small stand-alone units. There is a lack of a design and long-term performance BIPV/T standard, especially one that includes the integration with the envelope and with the HVAC system.

The initial testing of the prototype has shown that the

introduction of a second inlet may increase the thermal output by up to 18%, decrease the peak PV temperatures by up to 3°C and marginally increase the electrical output, as opposed to the case of a single-inlet system. The double-inlet system's performance was further boosted by an air flow director element, which directed the flow closer to the surface of the second PV module. The use of higher transparency PV modules increased the thermal output, by 5-10% and the outlet air temperature but the electrical output was decreased by 10-11%. The differences in performance for the single and double-inlet systems are expected to be more prominent in larger scale applications.

6 FUTURE WORK

Future work includes the investigation of more thermal enhancement techniques applied to the prototype, such as the use of fins and varying channel depth. A collective comparison of the tested methods will be carried at the end of the experimental procedure. Furthermore, models will be developed and validated according to the experimental results for the simulation of full scale systems and the suitability of already developed convective coefficient relations will be evaluated.

7 ACKNOWLEDGEMENTS

The authors would like to acknowledge the support of Unicel Architectural, Canadian Solar and the Hydro Quebec Industrial Chair.

8 REFERENCES

[1] Agathokleous, R. A., & Kalogirou, S. A. (2016). Double skin facades (DSF) and building integrated photovoltaics (BIPV): A review of configurations and heat transfer characteristics. *Renewable Energy*, *89*, 743–756.

[2] Kumar, R., & Rosen, M. A. (2011a). A critical review of photovoltaic-thermal solar collectors for air heating. *Applied Energy*, *88*(11), 3603–3614.

[3] Yang, T., & Athienitis, A. K. (2016). A review of research and developments of building-integrated photovoltaic/thermal (BIPV/T) systems. *Renewable and Sustainable Energy Reviews*, *66*, 886–912.

[4] Jelle, B. P., & Breivik, C. (2012). State-of-the-art building integrated photovoltaics. *Energy Procedia*, *20*(1876), 68–77.

[5] Bloem, J. J. (2008). Evaluation of a PV-integrated building application in a well-controlled outdoor test environment. *Building and Environment*, *43*(2), 205–216.

[6] Fossa, M., Menezo, C., & Leonardi, E. (2008). Experimental natural convection on vertical surfaces for building integrated photovoltaic (BIPV) applications. *Experimental Thermal and Fluid Science*, *32*(4), 980–990.

[7] Kaiser, A. S., Zamora, B., Mazon, R., Garcia, J. R., & Vera, F. (2014). Experimental study of cooling BIPV modules by forced convection in the air channel. *Applied Energy*, *135*, 88–97.

[8] Athienitis, A. K., Bambara, J., Neill, B., & Faille, J. (2011). Design and performance of a photovoltaic/thermal system integrated with transpired collector. *ASHRAE Transactions*, *117*(PART 2), 403–410.

[9] Chen, Y., Athienitis, A. K., & Galal, K. (2010). Modeling, design and thermal performance of a BIPV/T system thermally coupled with a ventilated concrete slab in a low energy solar house: Part 1, BIPV/T system and house energy concept. *Solar Energy*, *84*(11), 1892–1907.

[10] Brinkworth, B. (2000). Estimation of flow and heat transfer for the design of PV cooling ducts. *Solar Energy*, *69*(5), 413–420.

[11] Brinkworth, B. J., Marshall, R. H., & Ibarahim, Z. (2000). A validated model of naturally ventilated PV cladding. *Solar Energy*, *69*(1), 67–81.

[12] Brinkworth, B. J., & Sandberg, M. (2006). Design procedure for cooling ducts to minimize efficiency loss due to temperature rise in PV arrays. *Solar Energy*, *80*(1), 89–103.

[13] Brinkworth, B. J. (2006). Optimum depth for PV cooling ducts. *Solar Energy*, *80*(9), 1131–1134.

[14] Gan, G. (2009). Effect of air gap on the performance of building-integrated photovoltaics. *Energy*, *34*(7), 913–921.

[15] Bambrook, S. M., & Sproul, A. B. (2012). Maximizing the energy output of a PVT air system. *Solar Energy*, *86*(6), 1857–1871.

[16] Zogou, O., & Stapountzis, H. (2011). Experimental validation of an improved concept of building integrated photovoltaic panels. *Renewable Energy*, *36*(12), 3488–3498.

[17] Farshchimonfared, M., Bilbao, J. I., & Sproul, A. B. (2015). Channel depth, air mass flow rate and air distribution duct diameter optimization of photovoltaic thermal (PV/T) air collectors linked to residential buildings. *Renewable Energy*, *76*, 27–35.

[18] Hegazy, A. A. (2000). Comparative study of the performances of four photovoltaic / thermal solar air collectors. *Energy Conversion and Management*, *41*.

[19] Tiwari, A., Sodha, M. S., Chandra, A., & Joshi, J. C. (2006). Performance evaluation of photovoltaic thermal solar air collector for composite climate of India. *Solar Energy Materials and Solar Cells*, *90*(2), 175–189.

[20] Sopian, K., Liu, H. T., Kakac, S., & Veziroglu, T. N. (2000). Performance of a double pass photovoltaic thermal solar collector suitable for solar

drying systems. *Energy Conversion and Management*, *41*(4), 353–365.

[21] Tripanagnostopoulos, Y. (2007). Aspects and improvements of hybrid photovoltaic/thermal solar energy systems. *Solar Energy*, *81*(9), 1117–1131.

[22] Othman, M. Y., Yatim, B., Sopian, K., & Abu Bakar, M. N. (2007). Performance studies on a finned double-pass photovoltaic-thermal (PV/T) solar collector. *Desalination*, *209*(1-3 SPEC. ISS.), 43–49.

[23] Kumar, R., & Rosen, M. A. (2011b). Performance evaluation of a double pass PV/T solar air heater with and without fins. *Applied Thermal Engineering*, *31*(8-9), 1402–1410.

[24] Tonui, J. K., & Tripanagnostopoulos, Y. (2006). Improved PV/T solar collectors with heat extraction by forced or natural air circulation. *International Journal of Hydrogen Energy*, *31*(15), 2137–2146.

[25] Pantic, S., Candanedo, L., & Athienitis, A. K. (2010). Modeling of energy performance of a house with three configurations of building-integrated photovoltaic/thermal systems. *Energy and Buildings*, *42*(10), 1779–1789.

[26] Yang, T., & Athienitis, A. K. (2014). A study of design options for a building integrated photovoltaic/thermal (BIPV/T) system with glazed air collector and multiple inlets. *Solar Energy*, *104*, 82–92.

[27] Hussain, F., Othman, M. Y. H., Yatim, B., Ruslan, H., Sopian, K., Anuar, Z., & Khairuddin, S. (2015). An improved design of photovoltaic/thermal solar collector. *Solar Energy*, *122*, 885–891.

[28] Rounis, E. D., Athienitis, A. K., & Stathopoulos, T. (2016). Multiple-inlet Building Integrated Photovoltaic/Thermal system modelling under varying wind and temperature conditions. *Solar Energy*, *139*, 157–170.

[29] Mirzaei, P. A., Paterna, E., & Carmeliet, J. (2014). Investigation of the role of cavity airflow on the performance of building-integrated photovoltaic panels. *Solar Energy*, *107*, 510–522.

[30] Charron, R., & Athienitis, A. K. (2006). Optimization of the performance of double-façades with integrated photovoltaic panels and motorized blinds. *Solar Energy*, *80*(5), 482–491.

[31] Ioannidis, Z., Buonomano, A., Athienitis, A. K., & Stathopoulos, T. (2017). Double Skin Façades Integrating Photovoltaic Panels: A Comparative Analysis of the Thermal and Electrical Performance. *Energy & Buildings*. http://doi.org/10.1016/j.enbuild.2017.08.046

[32] Kapsis, K., Athienitis, A. K., & Harrison, S. J. (2016). Determination of solar heat gain coefficient for semi-transparent photovoltaic windows: an experimental study. *ASHRAE Transactions*, *125*.

[33] Bloem, J.J. (2008). Evaluation of a PV-integrated building application in a well-controlled outdoor test environment. *Building and Environment 43*, 205–216.

[34] Vats, K., Tomar, V., & Tiwari, G. N. (2012). Effect of packing factor on the performance of a building integrated semitransparent photovoltaic thermal (BISPVT) system with air duct. *Energy and Buildings*, *53*, 159–165.

ENERGY PERFORMANCE OF A BUILDING WITH SPLIT TANDEM PHOTOVOLTAIC WINDOWS

Marc JOBIN, Basile GRANDJEAN
hepia, University of Applied Sciences of Western Switzerland (HES-SO)
4 rue de la Prairie, 1202 Genève, SWITZERLAND
marc.jobin@hesge.ch, basile.grandjean@hesge.ch

ABSTRACT: We have made a 20cm x 20cm tandem photovoltaic (PV) window and developed a numerical model to investigate the energy performance of a building that would be equipped with such PV windows. The inside of the double-glazed window incorporates a tilted dichroic filter which is transparent for visible solar light and reflecting for near infra-red light. As a result, the transmitted visible photons are partially absorbed by a transparent organic solar cell while the reflected NIR photons are sent to silicon solar cells located in the frame of the window. To investigate the potential benefit of the technology, the net energy balance (NEB) simulation includes the choice of the location, the use of the building (residential or office), the percent coverage of the façade by the PV windows, etc. Several case studies are presented to emphasize the key points which influence the net global benefit of the window.
Keywords: Building integrated PV (BIPV), Tandem, Organic Solar Cell, Window

1 MOTIVATION

As it is getting always more challenging to improve the efficiency of solar cells, a general trend in photovoltaic industry is to use tandem technologies, that is, to combine multiple band gaps solar cells [1]. Another trend, in BIPV (building integrated photovoltaics), is to consider windows [2] and façades [3], as satisfactory and aesthetic solutions now exist for tiles [4].

The simplest solution when partially opaque windows are acceptable consists in covering 20-50% of the windows with opaque solar cells [5]. Nevertheless, for windows with 50%+ transparencies requirements and/or with constraints on colors, the situation is more challenging. New PV technologies such as OPV or DSSC can found a niche for these applications.

A second motivation of this study, apart from the prototype development, is to investigate the real economic potential of various PV windows technology, including the tandem we are proposing. We therefore have developed a simulation software which computes the net energy balance under realistic conditions for a building with PV windows.

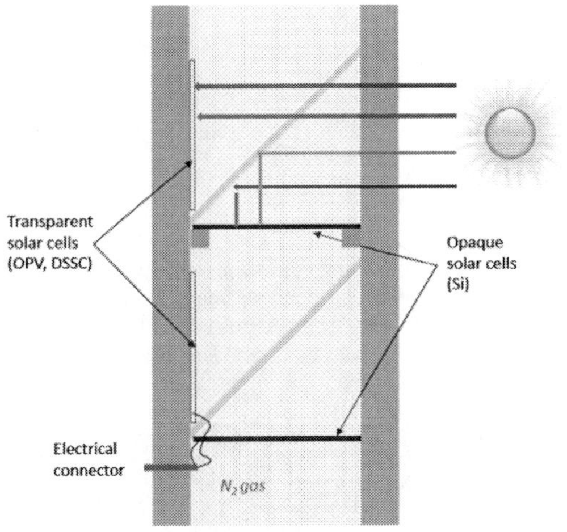

Figure 1 : Principle of the tandem PV window

2 CONCEPTION AND FABRICATION

2.1 Conception

The principle of the tandem window is depicted in Figure 1 . The incoming sun radiation is split by a tilted dichroic filter located between the two glasses of the doubled glazed window. The high-energy photons (VIS) are transmitted through the filter and arrive in the back window, on which a semi-transparent organic solar cell (OPV) is glued. The low energy photons (NIR) are reflected by the dichroic filter and sent to c-Si solar cells glued in the inside planes of the frame (see Fig. 3). We then take advantage of the transparency of OPV and the maturity and the performance of c-Si solar cells technology.

2.2 Dichroic filter

A key point of the tandem photovoltaic window is then the dichroic filter.

As we target for a low-cost photovoltaic window, we wanted the filter to be compatible with sol-gel technology, that is why we have investigated $[Al2O3/TiO2]^n$ stacks, as both can readily be prepared sol-gel. Note however that in the actual prototype, the layers have been deposited by atomic layer deposition. We have used the TFCalc software to optimize the layer thicknesses with the following target: minimum reflectivity for $\lambda(nm) \in$ [300;500] and maximum reflectivity for $\lambda(nm) \in$ [700;1200]. Restricting the number of stack to 3 (n=3) to keep low manufacturing costs, we found the optimized filter as described in Figure 2. The inset shows the thickness of each layer and the plot show the reflectance for 45° incidence angle.

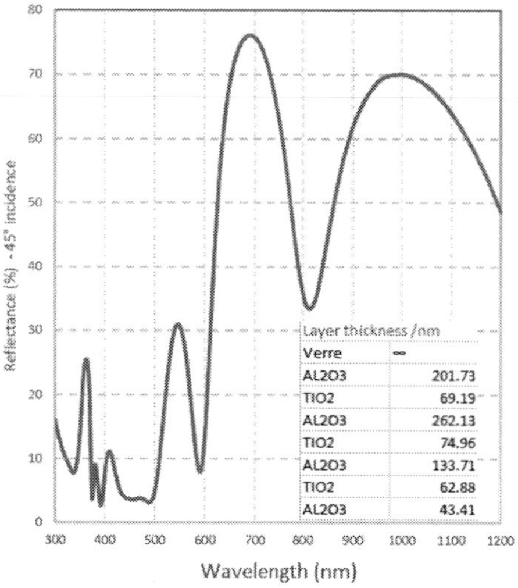

Figure 2 Computed optical reflectance of the optimised dichoiic filter at 45° incidence. Top : layer thickness

2.3 Fabrication

We have used commercial solar cells and integrated them directly on the window frame.

The thin (100 um) c-Si solar cells have been glued on three planes (out of the four available) of the frame inside the windows as shown in Figure 3 .

Figure 3 : Picture of the tandem PV window (showing the tilted dichroici filter, the c-Si inside the frame and the purple OPV solar cell on the back glass).

They are connected in parallel inside the windows to have only two electrical connectors. On one version of the window a gas connector was added to performed degradation studies under various atmospheres.

The organic solar cells have been purchased from OPVInfinity an placed on the inside plane of the back window. The external quantum efficiencies of both cells haves been measured with a our home-made system described in [6], but for the simulation we have used IQE's of standard c-Si cells and ITO/PEDOT:PSS/P3HT-PCBM/Al cells as found in the literature.

The dichroic filter is simply laid on a 3mm x 3mm rod of PMMA glued on the inner glass of the back window and it leans on the front window.

The window was assembled in a nitrogen glove box, ensuring a dry atmosphere in between the two glasses (and therefore a better degradation protection for the OPV cell)

3 ENERGY PERFORMANCE SIMULATION

3.1 Model's description

It has been established that the global energy impact of a semi-transparent building integrated photovoltaic (BIPV) modules not only depends on its production of electricity, but also on the impacts on the building's energy consumption for artificial lightning, heating, or cooling. Poh Khai Hg *et al.* [7] summarize the effects of a semi-transparent BIPV trough optical, thermal, and electrical impacts, and introduce the concept of net electricity benefit (NEB), as the sum of the PV production and potential the savings of electricity, minus the increase in electricity consumption due to the BIPV module. For example, in the case of a standard windows being replaced by a semi-transparent BIPV module, a part of the incoming flux of solar radiation is absorbed by the windows inducing a potentially lower indoor illuminance, hence an additional lightning cost. Furthermore, this absorption of radiance influences the cooling or heating loads because a part of the solar heat gains is partially obstructed by the BIPV.

We want here to establish the general conditions for which a PV windows makes really sense from a global energy (i.e economic) point of view, and we have developed a simulation tool to compare a reference non-photovoltaic window with the tandem PV window described above, and estimate the potential gains or losses. However, this comparison is not absolute but depends on the conditions of the simulation : sunshine illumination - therefore on the location and orientation of the building -, the use of the building (whether it is a residential or a commercial building, for which the operating hours are not the same), the windows to wall ratio (WWR), in others words the ratio of semi-transparent BIPV on the façade, and others parameters relatives to the geometry of the building, the electrical power of lightning, the average distance between the window and a workplace, or even the comfort threshold - corresponding to the illumination in lux below which we switch on the light in the building.

We had to choose simplifications in our model to bring out general principles for an optimal NEB, and avoid to focus on very specifics building. First, we only compute the indoor illuminance in terms of quantity of visible light transmitted, i.e. we do not take glaring problems into account. Also, we neglect the false color perception that could be induced by the absorption spectrum of the window. We also assume a comparison without blinds, because the results are strongly affected by the management of the blinds that not optimized for PV windows – this question of an optimal management of the blinds in the case of a PV windows deserve to be the topic of future work.

For the cooling and heating, we assume a factor of 70% including the regulation and the behavior of the users: for example, a unit of energy transmitted by the sun to the building only allows a reduction of 0.7 unit of the heating demand, the rest being lost for example in overheating effect due to the reaction of the regulation system, or even dissipated outside if a user feels to hot and opens the window.

Nevertheless, the direct comparison of the electricity generation and the thermal impact on heating in terms of energy units is inappropriate in most of the cases, as the exergetic value of thermal energy is much lower than for electrical energy. In addition, the thermal effects on the building do not only depend on the concept of PV windows by itself: for example it is possible to recover a part of this thermal energy and reinject it inside, for example in the case of a water-flow windows as studied in Ref. [8] Moreover, the energy for cooling or heating may be provided by renewable energy source, which make a general comparison even more complexe. In this cas, we simply choosed to calculate the results in CHF (swiss francs) in order to simplify this issue.

Figure 4a) is a snapshot of the simulation software, with the user entered parameters on the left part, the results on the center part and the data (sunshine illumination for the chosen location, quantum efficiencies of the solar cells,..) on the light part.

The simulation is run for each of eight façade orientations, and we compute :

(1) the energy (gain) produced by the PV window
(2) the energy (cost) for additional lightning
(3) the energy (cost) for additional heating demand
(4) the energy (gain) for savings in cooling demand
(5) the net energy balance (gain or cost).
(numbers refer to those of Figure 4b)).

The model also shows energy impact on lightning, in terms of extra duration in hours per day the artificial lightning has to be alight, and its average distribution for each hour of the day. Finally, we can choose to display the results for the tandem window describe above or the same window without the OPV cell.

3.2 Simulation's results

We choose the location to be Geneva (Switzerland) and the sunshine data (direct and indirect power for each of the 8760 hours of the year) are extract from the Swiss Meteo database. We choose a comfort threshold of 300 lux, and the operating hours of the building, here for an administrative building, from 8am to 6pm. We then play with the WWR, the orientation, and the period of time considered.

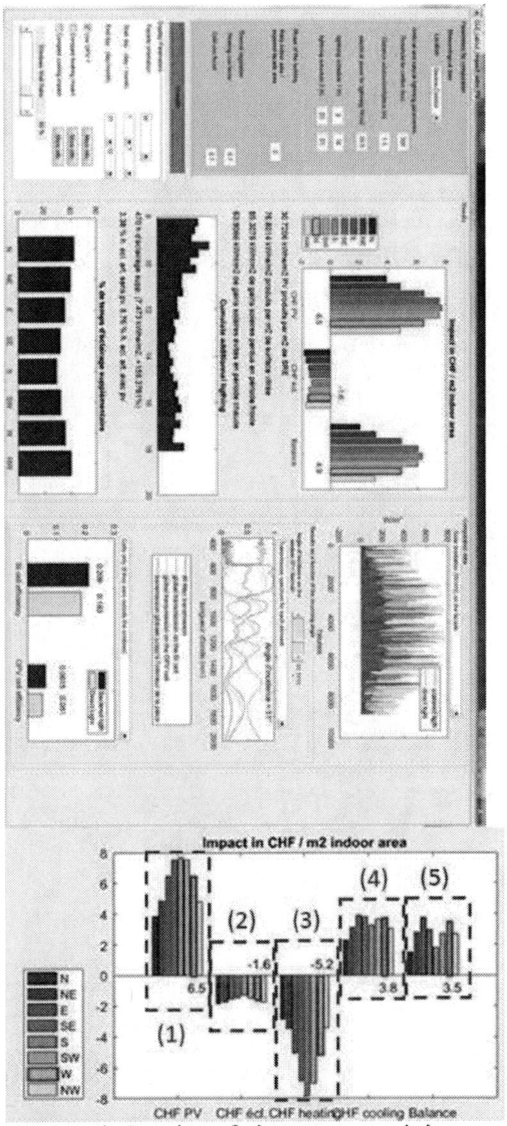

Figure 4 : a) Snapshot of the net energy balance calculation software b) NEB for each orientation

For example, in Geneva, with WWR = 80%, the amount of extra electricy for ligthning represent 17 % of the PV produced energy (or 14 % without the OPV). In termes of NEB, the tandem window present a gain of 14.3 % compare to the the same windows without the opv cell.

The tandem windows shows better NEB for high WWR, but decrease faster for small WWR than for a window without OPV cell. In Oslo, with more sun incident on the façade, the most of the extra needs for lightning are in winter. In summer the impacts on lightning is very low because of the direct radiation of the sun with a low angle all over the day into the building. For a location in Barcelona, the impacts on electricity consumption for lighting are very low for high WWR (order of 5% of the PV produced energy). Though, the highest PV production is not on the south orientation, but on the east and east sides.

The first conclusion we can draw if we play we this simulation tool is that the south orientation appears genrally to be the best compromize beteween the positives and negatis impacts on the net energy gain, tough it is not always the cases – east or west beeing more profitable in low latitude location. On the other hand, north exposed windows always seem to be the less effective orientation in terms of NEB.

As shown in Figure 5, we have seen that the NEB always increases with the WWR. For small PV windows, the extra energy required for lithnging is never compensated by the production of PV electricity, making then the NEB negative. In opposite, with a high WWR we have NEB>0 in most of the cases (whatever the other parameters). For a given set of parameters, we call the reference point for wich NEB=0 "NEB0", and compare it for different locations or condition (see Table 1).

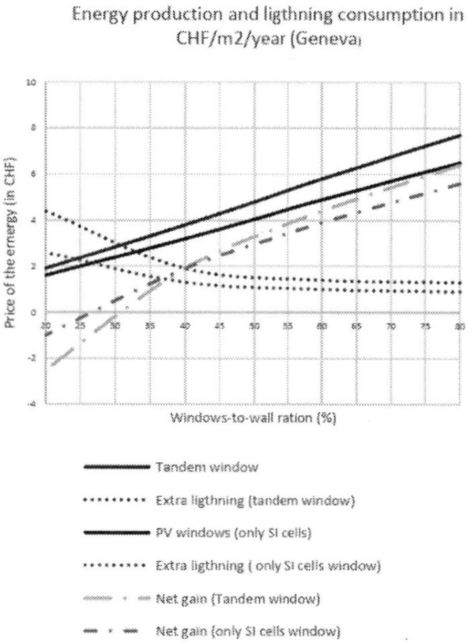

Figure 5 : : NEB in CHF as a function of coverage percentage for two kinds of PV windows. The tandem PV window shows slightly higher NEB for high WWR ratio. In the case of buildings with about 50% ot less, the glolal performance of a tandem windows is not better than a PV windows with infrared diffusion.

If we now estimate the potential increase of heating demand that correspond to the radiance obstructed by the PV windows and cannot be used as passive heating, NEB (equivalent in CHF) becomes close to 0. The impact of this loss of free heating being greater on the south, this leads to a slightly better global performance on the west or east orientation. Furthermore, the saving of energy for cooling turns out to be more important on east or west orientation, because of the position of the sun in summer. The better NEB is hence achieved for east or west orientation if we include our estimation of the thermal effects.

	Tandem window		
	WWR for NEB = 0	NEB WWR= 80% CHF/m2/year	Best orientation
Genève	31%	6.4	S
Oslo	33%	5.1	S
Barcelone	31%	7.7	E

	Only Silicium		
	WWR for NEB = 0	NEB WWR= 80% CHF/m2/year	Best orientation
Genève	26%	5.6	S
Oslo	29%	4.6	S
Barcelone	25%	6.7	E

Table 1: The point where NEB=0 depends on the location, the orientation, and the kinds of window. The table shows the results if we always choose the best orientation: the WWR for NEB=0 is higher for tandem windows in all the three locations.

Acknowledgment

This research is partially financed by "Energy District-2050" program from HES-SO, Switzerland.

References

[1] A. Polman and H. Atwater, Nature Materials 11(2012) 174.
For a recent development in perovskite-silicon tandem cell with potential industrial applications, see :
S. Albrecht *at al.*, Energy & Environ. Sci. 9 (2016), 81.
[2] For a review in BIPV including windows, see:
Akash Kumar Shukla, K. Sudhakar, Prashant Baredar, Energy and Buildings 140 (2017) 188.
[3] http://www.solaxess.ch/
[4] Among many other producers, see for example http://www.solarmassenergy.com/
[5] http://www.solarwall.ch/verre-photovoltaique/
[6] C. Pellodi, Master Thesis, HES-SO, 2013.
[7] Poh Khai Ng, Nalanie Mithraratne, Harn Wei Kua, Energy and Buildings 66 (2013) 274
[8] Pablo Sierra, Juan A. Hernandez, Energy and Buildings 15 (2017) 133

SMART WINDOWS BASED ON NANOPARTICLES SOLAR CONCENTRATORS

Paula Sánchez-Friera[1], Blas Puerto[1], Alberto Zapico[1], Helena Aguilar[2], João Gomes[3], Miguel Gonçalves[3], Mariana Ornelas[3], Catarina Costa[3], Ana Barros[3], Jimena Alarcón[4], Rodrigo García[4], Carlito García[5]

[1] Fundación PRODINTEC, Parque Científico-Tecnológico de Gijón, Av Jardín Botánico 1345, 33203 Gijón, Asturias, Spain.
[2] Nanolayer Coating Technologies, Lda, Rua Fernando Mesquita, 2785, 4760-034 Vila Nova de Famalicão, Portugal.
[3] CeNTI – Centre for Nanotechnology and Smart Materials, Rua Fernando Mesquita, 2785, 4760-034 Vila Nova de Famalicão, Portugal.
[4] Universidad del Bío-Bío, Avda. Collao 1202, 4051381, Concepción, Chile
[5] Universidad Nacional Evangélica, C Paseo de los Periodistas #54, Ensanche Miraflores, Santo Domingo, República Dominicana

ABSTRACT: This paper describes the main findings of the WINNER project, focused on the development of an innovative smart window based in the concept of luminescent solar concentrators (LSCs). Different luminophores have been used to realize semitransparent and colorless photovoltaic (PV) windows to convert the energy-passive façades of urban buildings into effective energy and, at the same time, promoting building's visual and thermal comfort. Here we used inorganic dyes with rare-earth elements such as Eu^{3+} and Dy^{3+}, and Nd^{3+}/Yb^{3+} lanthanide couple to down-shift the solar spectrum towards the NIR wavelength range where the solar cells have an optimal response. Experimental results demonstrate an optical power efficiency of 2.2% (inorganic dyes) and 1.3% (lanthanides) under AM1.5G conditions for 3.5x3.5 cm^2 tinted and fully transparent (optical transmittance ≈ 90%) LSCs. The design of the smart window was completed following a study of user and architectural requirements as well as market trends. The LSC system integrates not only the PV cells on the window frame but also all the basic electronic components for an autonomous mode of operation, including batteries and miniaturized balance-of-system elements. These results open the way for the realization of visible transparent solar energy harvesting clear glass windows.
Keywords: Building Integrated PV (BIPV), Smart windows, Luminophores, Performance, Balance-of-systems

1 INTRODUCTION

Nearly 40% of final energy consumption and 36% of greenhouse gas emissions takes place in buildings, according to the International Energy Agency. This consumption is growing due to the increasing number of electrical appliances in homes, growing population and growing urbanization. There is an urgent need for sustainable and efficient usage of energy in buildings to reduce CO_2 emissions, and this has led to governments launching initiatives and approving legislation towards this goal. In Europe, the building sector is central to the energy efficiency policy and it is considered one of the key sectors to achieve the 20/20/20 targets and beyond that, to meet the longer-term objectives laid down in the low carbon economy roadmap. According to the 2030 Communication published by the European Commission in July 2014, the building sector has the largest energy saving potential. In this context, the 2010 Energy Performance of Buildings Directive was launched to ensure that all new buildings are nearly zero energy buildings (NZEB) by 31 December 2020 and public buildings by 31 December 2018. In Latin American and Caribbean (LAC) countries, governments are also enforcing policies to implement incentives for renewable energies, as well as measures to promote energy efficiency and rational consumption.

The WINNER ("Smart windows for zero carbon energy buildings") project aims to contribute to these challenges by developing construction elements specifically designed for distributed energy generation and increase of energy efficiency in buildings (Fig. 1). The project received funds from the ERA-NET LAC program and it is implemented by an international consortium with partners both in Europe and in LAC regions.

The project's goal is the development of an innovative smart window system capable of generating photovoltaic energy for self-consumption at the same time as filtering out part of the IR radiation incident on the building, therefore reducing the energy demands for air conditioning, especially in hot climate areas. The project has also considered life cycle and economic analysis, paying attention to specific requirements set by different climates, geography and socioeconomic conditions in the EU and LAC regions.

2 CONCEPT

The integration of photovoltaic systems in buildings offers a solution to meet the NZEB set targets. The global solar photovoltaic energy technology has made enormous progress in the last decades reducing drastically the module costs and therefore bringing down electricity costs to values very close to those of conventional energy sources.

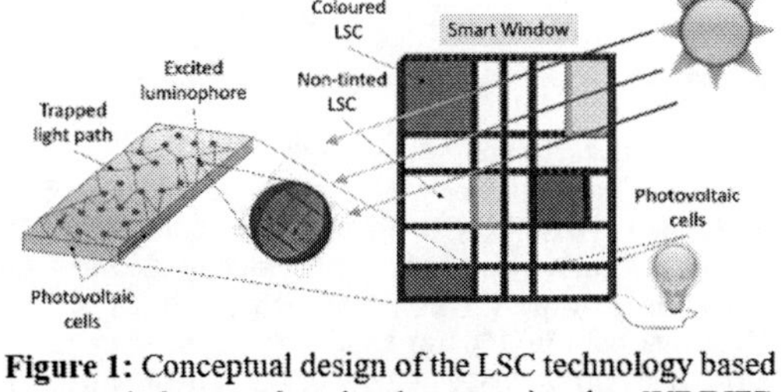

Figure 1: Conceptual design of the LSC technology based smart window under development in the WINNER project. The inset scheme highlights the wide possibility of coated tinted and colorless glass that can be used as part of the energy harvesting system.

In order to increase the efficiency of PV systems and reduce the use of materials, solar concentration has been used for decades. Concentration systems can be coupled to conversion elements to adapt the incident solar spectrum to the solar cell response. Luminescent solar concentrators combine concentration features with spectrum shifting. They consist normally on a highly transparent host material, in which the luminescent species are dispersed. These luminescent species absorb incident light and re-emit it at a shifted wavelength. By internal reflection, part of the light emitted by the luminescent materials travels to the concentrator perimeter, where solar cells can be placed to collect it and convert it into electricity.

In LSCs, both direct and diffuse light is concentrated. This allows the cost reduction of a PV system by substituting expensive semiconductor material in the solar cells by cheap concentrators. In addition, the system can provide additional advantages such as a smart climatization of the building, derived from the absorption of part of the incident radiation on the window panel. Previous developments have shown conversion efficiencies of LSC with different configurations in the range between 2.7% and 7.1% [1-3]. These systems use in most cases back reflectors that increase significantly the conversion efficiency of the LSCs, but low visible transmittance of the coatings still limits its application in smart windows because sunlight is unable to cross the window glass, which changes the visual perception of colors inside the building. The sizes of the reported high performing LSCs are in the order of 5x5 cm^2.

The concept developed in the WINNER project is based on planar LSCs consisting of an optically clear polymer matrix doped with inorganic dyes and lanthanide ions with emission spectra that matches the spectral peak of the external quantum efficiency (EQE, Fig. 2) of the crystalline solar cells installed in the LSC perimeter. The selected luminophores play only a partial role of the energy conversion functionality since embedded nanoparticles are responsible for multiple light scattering and dispersion effects in the polymeric waveguide.

In the WINNER approach, the color and degree of transparency of the LSC, which are determined by the type and density of the luminophores was also defined according to aesthetic and functional requirements of the smart window. This approach simultaneously enables electric energy generation and ensures a maximized visible range transparency. In addition, the window contains a double-glazing system, and infrared filters have been developed and coated on the inner glass for further optimization of the climatization conditions in the building.

3 METHODOLOGY

3.1 Solar cell selection

The solar cells selected for the 15x15 cm2 LSC prototypes were PERC p-type Cz-Si, with PECVD passivated emitter and rear, Ag front contacts, local Al-BSF and rear pads. They were manufactured at ISC Konstanz on 156x156 cm wafers and then laser cut to strips of 6 mm width. Table I shows the current-voltage (*I-V*) parameters of the selected Si solar cells before and after being cut into strips with appropriate dimensions to integrate the LSC-PV prototypes. V_{OC}, I_{SC}, FF, E_{ff} represent the open-circuit voltage (in V), the short-circuit current (in mA), the fill-factor and the efficiency (%) of the solar cell, respectively. The current and voltage at the maximum power point (P_{MPPT}) is given by I_{MPPT} and V_{MPPT}. It is clearly shown a significant decrease in the cells performance, mainly attributed to a sharp reduction in the generated photocurrent, which will ultimately have a strong impact in the global energy conversion efficiency of the LSC-PV system. This decrease is due partly to the much higher shading ratio of the metallization grid in the strips compared to the full cells.

To assemble 3.5x3.5 cm^2 LSC prototypes, IXOLAR™ SolarBITs monocrystalline solar cells with dimensions 22x7x1.8 mm were also used. Table I summarizes the *I-V* characteristics of the small size monocrystalline silicon solar cells measured under 1000 W/m^2 irradiance, AM1.5 G.

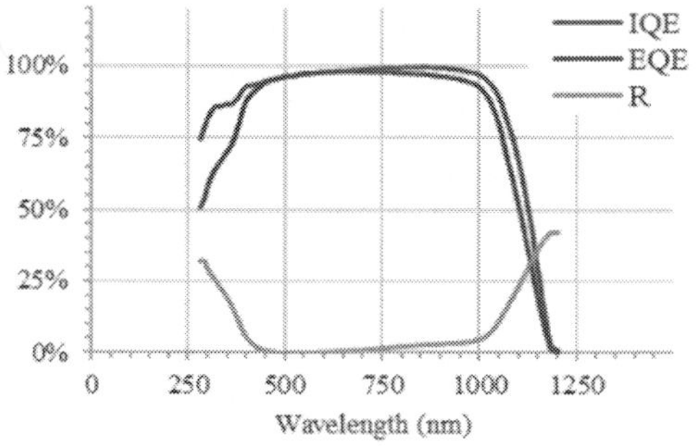

Figure 2: Internal (IQE), external (EQE) quantum efficiency and reflectance (R) of selected solar cells.

Table I: *I-V* parameters of PERC solar cells before cutting (full cells) and after cutting (strips), and of commercial SolarBITs cells.

PERC	V_{oc} (V)	FF (%)	E_{ff} (%)	Isc (A)	P_{MPPT} (W)	I_{MPPT} (A)	V_{MPPT} (V)	Area (mm^2)
Full cells	0.645	75.1	18.7	9.237	4.47	8.511	0.525	23 895
Strips	0.590	75.5	9.2	0.193	0.09	0.178	0.482	934
Solar BITs	V_{oc} (V)	FF (%)	E_{ff} (%)	Isc (mA)	P_{MPPT} (mW)	I_{MPPT} (mA)	V_{MPPT} (V)	Area (mm^2)
Full cells	0.644	0.670	21.3	76.08	25.55	51.10	0.500	120

3.2 LSC material selection

An extensive state-of-the-art review of materials to produce a LSC was performed. An initial selection of materials was done considering the requirements of: (i) matching the emission wavelength to the spectral peak of the external quantum efficiency (EQE) of the side-mounted PV cells; (ii) provide adequate color and degree of transparency, which are determined by the type and density of luminophores and the properties of the supporting polymeric medium; (iii) have high efficiency and stability; (iv) enable adequate processability with up-scalable techniques. Despite having the potential to revolutionize the future building architecture, the wide use of LSCs has so far been hindered by a lack of suitable emitters. Conjugated organic and organometallic fluorophores provide limited coverage of the solar spectrum, and suffer from significant optical losses associated with reabsorption phenomena. These deficiencies reduce the light-harvesting efficiency of LSCs and also lead to strong emission in the visible range of the spectrum (highly tinted glass), which imposes certain constraints on their use in architecture. On the other hand, colloidal or nanocrystal quantum dots (QDs) can help overcome these limitations and are typically more thermal and photostable. However, despite their great potential, LCSs fabricated using QDs suffer from incomplete coverage of the solar spectrum (large energy band gap of the material) which may also lead to strong coloring of devices. Additionally, the most interesting QDs, such as those with a core/shell structure typically (e.g. PbSe/CdSe quantum dots, which feature a lowered absorption onset and near-infrared luminescence in combination with a large effective Stokes shift) contain hazardous heavy metal ions that require very strict disposal/recycling protocols.

The WINNER project followed two different approaches:

1. Use of high-performance commercial luminophore substances embedded in a transparent colorless polymeric base (Polyvinyl alcohol - PVA). Some of the tested materials are inorganic dyes with **rare-earth elements such as Eu^{3+} and Dy^{3+}, and other fluorescent and phosphorescent materials** that present higher thermal and photostability than their organic counterparts. However, one main disadvantage of these dyes is the low quantum efficiency, some overlap of the absorption and emission spectra, and the fact they produce tinted window glasses (emission in the visible range).
2. Use of **lanthanide ions** (e.g. **Nd^{3+}/Yb^{3+} couple**; upon excitation of an Nd^{3+} ion with a UV-blue photon, Yb^{3+} ions emit two near-IR photons through an efficient two-step energy transfer from Nd^{3+} to Yb^{3+}; the wavelength of the emitted photons is close to the absorption peak of silicon based solar cells, so one can further increase the power output of the LSC devices). The conjugation of some lanthanide ions (Ln^{3+}) presents UV to NIR *downconversion*, rendering colorless LSC, while others are characterized by quantum-cutting luminescent transitions to the visible range of the solar spectrum resulting again in colored glass windows. In this approach, the selected materials are again embedded in an optically transparent polymeric base (PVA).

3.3 Material preparation

Two different dyes were incorporated in the PVA films to enhance the efficiency of the solar cells providing highly transparent colored LCs: Orange-Red and Yellow-Green.

To prepare the PVA films with dyes, luminescent particles at a mass loading of 10 % to the amount of PVA were used. Briefly, the luminescent particles were added to the PVA solution and vigorously stirred until a homogeneous dispersion was obtained before a silica sol based on tetraethoxysilane was added. After stirring the PVA/silica mixture at 60 °C for 5 hours, it was found that Yellow-Green degrades during the silica synthesis yielding a white and opaque film; therefore, only the polymeric film without silica was produced for this dye. Regarding the use of Orange-Red, both films (with and without silica) were deposited in glass slabs.

In parallel, lanthanide ions were used at a mass loading of 100 % to the amount of PVA in HCl 0.02 M and at a molar ratio of 1:10 (Nd^{3+}: Yb^{3+}). Films with and without silica were deposited in glass slabs by doctor-blade. The same procedure to incorporate the lanthanide ions was employed for both films: the lanthanide pair was added to the PVA aqueous solution ($pH \approx 2$) and vigorously stirred until a homogeneous dispersion was obtained. To prepare the PVA/silica film, the silica sol was added to the polymeric solution and the mixture was maintained under stirring for 5 hours at 60 °C.

3.4 Coating process development

Regarding the materials deposition technologies, a straightforward analysis was performed to select those able to provide the best performance of the LSC system without compromising scalability, sustainability and cost- and energy-effectiveness of the inherent production process with the future perspective of implementing this technology at industrial scale. Considering the properties of the polymeric waveguides and the chemical nature and morphology of the selected luminophores, the doctor blade technique was chosen as having several advantages over other coating technologies. This technique is highly versatile, does not require any special requirements and can be applied to standard window glass or any other flat surface made of an arbitrary material such as acrylic or PET. Very recently, the use of this inexpensive and highly scalable technique has been demonstrated for large area LSCs [5] and may in fact represent a practical route to real-life applications of LSCs in both semitransparent solar windows and high-concentration collectors of sunlight supplementing existing PV cells.

In the WINNER project, the glass size has been set to 3.5x3.5 cm^2 (lab-scale) and 15x15 cm^2 (prototype scale), which is larger than most of the LSCs previously reported in the literature. Considering the prototype dimensions, the coating process was also adapted for a roll-to-roll production platform looking towards future upscaling of the manufacturing process.

Figure 3: Doctor blade coating system integrated in roll-to-roll production platform at PRODINTEC.

3.5 Module assembly

For the characterization of the coating layers, the PV module with LSC was assembled following the configuration shown in Fig. 1 above. Several optical grade resins were used to attach the solar cells to the glass perimeter, including a UV curable resin, a 2-component silicone, a 2-component epoxy and PVA. All materials were proven apt for the application, although the 2-component epoxy was preferred in terms of processability due to fastest curing at room temperature. In the lab-scale system, one solar cell was fixed to the LSC edge on one side, and the remaining perimeter was covered with reflective aluminum tape to maximize light reflection and avoid deflected photons escaping through the LSC edges. In larger size prototypes, the solar cells were interconnected in strings and attached to the glasses in such a way that the active part of the cell was in contact with the active glass edge, whereas the part of the cell containing the busbar was outside the irradiated area, as shown in Fig. 4.

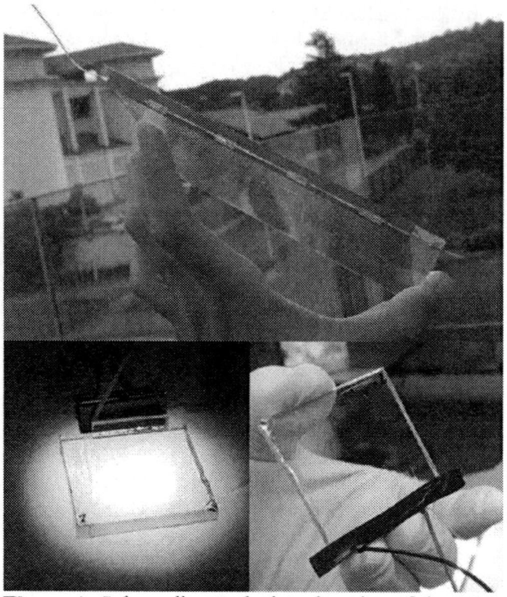

Figure 4: Solar cells attached to the edge of the LSC-coated glasses (above: 15x15 cm^2 prototypes; below: lab-scale 3.5x3.5 cm^2 LSC-PV samples).

4 RESULTS

4.1 Optical characterization

To perform an optical characterization of the films, several samples were prepared as described in section 3.2, using organic dyes and lanthanide ions.

The Orange-Red and Yellow-Green dyes used are insoluble in water and therefore produced a transparent and colored film but with solid particles dispersed (Fig. 5). However, in the PVA/silica hybrid film prepared with Orange-Red an obvious increase in the film transmittance was observed (Fig. 6).

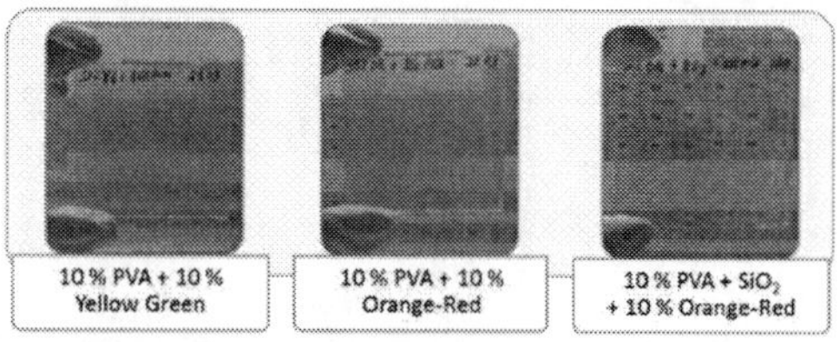

Figure 5: PVA films prepared with luminescent dyes (OR – orange red; YG – yellow green) and SiO_2.

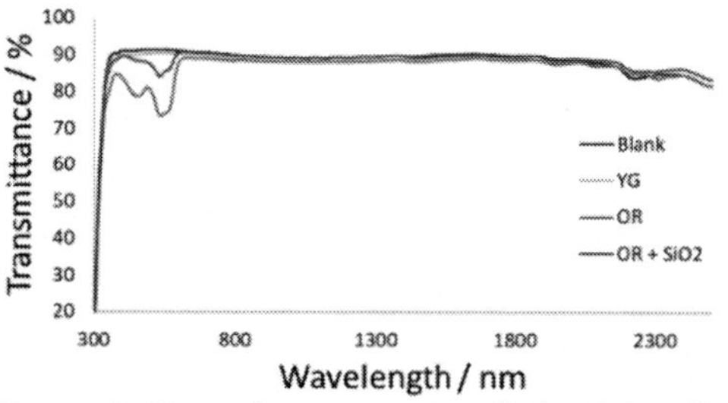

Figure 6: Transmittance spectra of the PVA films containing Yellow-Green, Orange-Red and Orange-Red and silica (the blank sample refers to an uncoated glass).

The films obtained with lanthanide ions were both transparent and colorless (Fig. 7). The transmittance spectra (Fig. 8) also confirms that the transmittance of the blank glass is not affected by the presence of the lanthanides nor the silica particles.

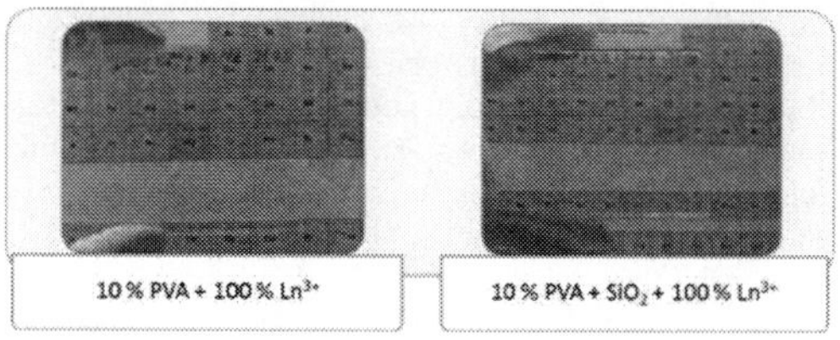

Figure 7: PVA films prepared with lanthanide ions, with and without SiO_2.

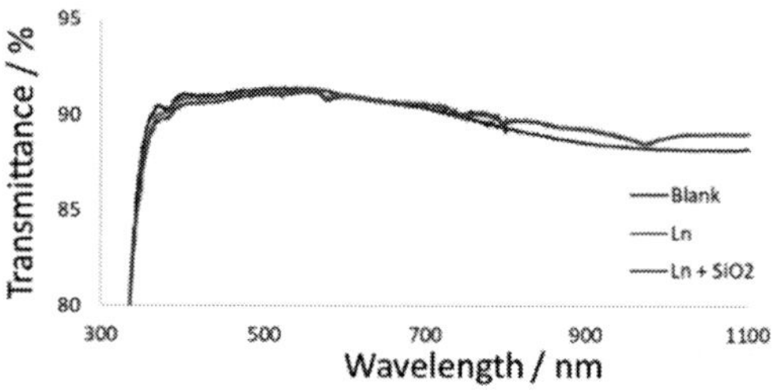

Figure 8: Transmittance spectra (UV-Vis NIR region) for the films prepared with the lanthanide pair Nd^{3+}/Yb^{3+}, with and without SiO_2.

4.2 Performance measurements

The LSC conversion efficiency is defined as the ratio between the power generated by the window (P_{out}) and the incident irradiance (P_{in}) over the full glass area, in standard conditions (1000 W/m^2 irradiance, 25 °C ambient temperature and AM1.5 spectrum). Indoor tests were performed using a solar simulator (Newport) to determine the global energy conversion efficiency of the lab-scale (3.5x3.5 cm^2) LSC-PV samples. Main results are shown in Table II. For large-scale prototypes, estimated efficiencies have been calculated by measuring outdoors the I_{SC} of the test samples and a reference module built with the same solar cells but with an uncoated glass. In this case, efficiencies in Table II are given as the ratio of the short-circuit photocurrent.

Table II: Estimated LSC efficiencies obtained from outdoor measurements at 770 W/m^2 for 15x15 cm^2 prototypes and results for indoor measurements of small scale samples under AM1.5G conditions.

	LSC η [%]	
Layer composition	15x15 cm^2 $\eta_1 = I_{SC} / I_{SCO}$	3.5x3.5 cm^2 $\eta_2 = P_{out} / P_{in}$
Blank glass	1.8	0.58
Ln^{3+} + SiO_2	1.4	1.06
Ln^{3+}	**5.2**	**1.32**
YG	2.9	0.75
OR	**6.2**	**2.18**
OR+SiO_2	3.7	1.72

Although enhancing light trapping performance in planar configuration LSCs, multiple scattering effects imparted by SiO_2 nanoparticles seem to contribute to optical losses within the concentrator, reducing the overall efficiency of the LSC-PV system.

Additionally, as the module has a double functionality as window and PV generator, the conversion efficiency is compromised as most of the radiation is still required to enter the building. The efficiency depends on the percentage of the light that is allowed to be captured, and also on the efficiency of the spectrum conversion process and the cell efficiency.

5 SYSTEM DEVELOPMENT

5.1 Design considerations

In parallel with the development of the LSC, a novel design of the smart window was completed following a study of user and architectural requirements as well as market trends. A conceptual design was developed considering thermal, acoustic, ergonomic and luminosity aspects, as well as an assessment of its use for zero-energy buildings. The smart window was decided to integrate not only the PV cells on the window frame, but also all the basic electronic components for an autonomous mode of operation, including batteries and miniaturized balance-of-system elements. Further details on the design and its implications for the goal to reach NZEB buildings are described in [4].

5.2 Balance of system

The electronic components of the system are integrated in the window frame, so that the window can operate as an autonomous element, thus simplifying installation costs. They have been chosen to enable the implementation of an independent low-voltage DC microgrid, which offers multiple advantages such as high safety and optimal adequacy for integration of LED lighting. High-density lithium-ion batteries are also integrated in each window to allow for an autonomous use of the smart windows.

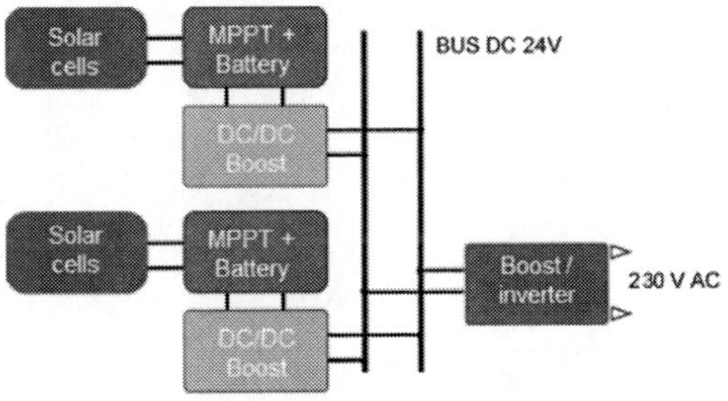

Figure 9: Electric scheme for the microgrid.

The system includes several components which are schematically shown in Figure 4. To maximize the energy generation, the system includes a maximum power point tracker (MPPT) which adjusts current demands to the optimal point. It also adapts power to charge the selected battery at the maximum allowed rate depending on the irradiance conditions. The battery is then connected to a DC/DC boost system where the voltage is elevated from typically 3-4.2 V to 24 V. All windows are then connected in parallel to a 24 V DC bus, which can already be used for lighting or small DC appliances. The grid can also include as an additional component a boost / inverter which transforms the 24 V available at the DC bus to a 230 V AC for standard home appliances.

The circuit described was designed and manufactured by the project partners, and all elements selected and integrated in the PCB as shown in Fig. 10.

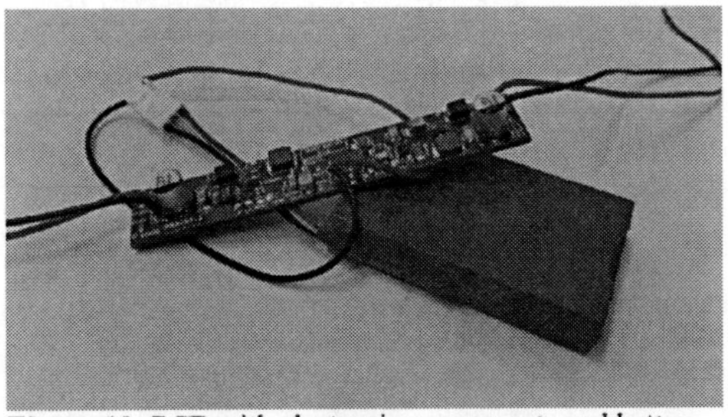

Figure 10: PCB with electronic components and battery.

5.3 Window full scale prototyping

Full-scale windows were manufactured using a production process similar to the one described in 3.5, but for the full solution with double glazing. In this case the first step was to assemble the glasses in pairs with the LSC in front and the glass with IR filter at the back. The assembly was done using a conventional process used in the double glazing industry, using a rigid separator and a two-system adhesive based on butyl and polysulphide, and including a dehydrating material to avoid condensation in the air chamber between the two glasses. The solar cells were interconnected by soldering a 1.1 mm wide ribbon, and they were attached to the glass perimeter using an epoxy as shown in Fig. 11. A protective thick silicone layer was deposited on the back of the cells.

Figure 11: Double-glass assemble showing light redirected on the LSC.

The final real scale smart window prototype is based on 12 units of 15x15 cm^2 and was manufactured using aluminum profiles painted in white. The active glasses were then introduced in the frame, interconnected and sealed. The junction box was built in polyamide using a 3D printing process and the electronics was allocated inside. It was then connected to the terminals and attached to the frame. The final prototype can be seen in Figure 12.

6 CONCLUSIONS

The WINNER project has addressed the development of a new energy generating system for buildings bringing the following main innovations: development of high transmittance nanomaterials for spectrum conversion and PV generation; development of innovative coating system for glass and integration of PV cells in window frames; design of sustainable smart window contributing to the promotion of NZEB buildings and development of smart embedded electronics.

We have proposed the use of inorganic spectrally selective scattering luminophores in conjunction with spectrally selective polymer coatings and Si solar cells to realize highly visible-transmissive (T ≈ 90% at 400-700 nm and NIR) energy harvesting solar concentrators with power conversion efficiencies up to 6.2% in outdoor conditions. Several prototypes and demonstrators were built and validated, demonstrating the potential of the smart window to generate electricity in buildings in a sustainable way.

Figure 12: Full window prototype with LSC and IR filters.

REFERENCES

[1] Currie M.J. et al (2008), High-efficiency Organic Solar Concentrators for Photovoltaics, Science 321, 226.
[2] Sloof L.H. (2008), A luminescent solar concentrator with 7.1% power conversion efficiency, Phys. Stat. sol. (RRL) 2, No. 6, 257-259.
[3] Goldschmidt J.C. et al (2009), Increasing the efficiency of fluorescent concentrator systems, Solar Energy Materials & Solar Cells 93, 176–182.
[4] Alarcón J et al (2017), Assessment of smart PV-windows for NZEB in Santiago of Chile, Proceedings of the 33rd European Photovoltaic Solar Energy Conference and Exhibition.
[5] Li H, et al. (2016). Doctor-blade deposition of quantum dots onto standard window glass for low-loss large-area luminescent solar concentrators. Nature Energy, 1(12), [16157]. DOI: 10.1038/nenergy.2016.157

ACKNOWLEDGEMENTS

This work is part of the WINNER project, funded under the programme ERA-NET LAC (Latin America, Caribbean and European Union), with project reference number ELAC2014/ESE-0146, and nationally in Spain by Ministerio de Economía y Competitividad (Acciones de Programación Conjunta Internacional, PCIN-2015-152), in Portugal by Ministry of Science, Technology and Higher Education via FCT- The Portuguese National Foundation for Science and Technology (Ref. ERANETLAC/0007/2014), in Chile by Comisión Nacional de Investigación Científica y Tecnológica (CONICYT) and in Dominican Republic by Ministry of Higher Education, Science and Technology (MESCYT).

The authors would also like to thank A. Halm, S. Schmitt and L. Popescu from ISC Konstanz for their support with the selection and characterization of the solar cells used in the prototypes. Professor Luiz Pereira from University of Aveiro is also recognized for his support with the *I-V* characterization test under AM1.5G.

ASSESSMENT OF SMART PV-WINDOWS FOR NZEB IN SANTIAGO OF CHILE

Jimena Alarcón-Castro[1], Rodrigo García-Alvarado[1], Alberto Zapico[2], Paula Sánchez-Friera[2], Fernando Berenguer[2], Helena Aguilar[3], João Gomes[4], Mariana Ornelas[4], Carlito García[5]
[1] Universidad del Bío-Bío, Avda. Collao 1202, 4051381, Concepción, Chile.
[2] Fundación PRODINTEC, Parque Científico-Tecnológico de Gijón, Av Jardín Botánico 1345, 33203 Gijón, Asturias, Spain.
[3] Nanolayer Coating Technologies, Lda, Rua Fernando Mesquita, 2785, 4760-034 Vila Nova de Famalicão, Portugal.
[4] CeNTI – Centre for Nanotechnology and Smart Materials, Rua Fernando Mesquita, 2785, 4760-034 Vila Nova de Famalicão, Portugal.
[5] Universidad Nacional Evangélica, C Paseo de los Periodistas #54, Santo Domingo, República Dominicana

ABSTRACT: This work aims to determine the contribution of smart photovoltaic (PV) windows with integrated luminescent solar concentrators (LSCs) to the energy performance of near-zero energy buildings (NZEB). The work focused a case study in Santiago of Chile, taking as a starting point local parameters such as the overall energy requirements, geometrical features and consumption patterns of office buildings in this location. Dynamic simulations were done to assess buildings features for energy efficiency, considering different estimated performances of the smart windows system. For rectangular-shaped buildings with windows mainly facing the sun, and considering daily occupation and the use of energy-efficient appliances together with high performance insulated walls, simulation results highlight a contribution of the smart window up to nearly one-half of the buildings energy demand. In usual office buildings this contribution is somewhat less expressive, but proper construction features can be regulated to meet the energy-efficiency goals stated by many countries, in line with the Paris Agreement for climate action.

Keywords: Building Integrated PV [BIPV], Photoluminescence, Façade, Evaluation, Modelling

1 ZERO-ENERGY BUILDINGS

The concept of zero-energy buildings has emerged in recent years, in the trend of sustainable development, as a goal of reducing resource consumption in operation of buildings, based on to combine actions of energy efficiency and integration of renewable generation systems. Recently to the United States Department of Energy proposed a definition of zero-energy building based on energy sources, considering that the annual balance between the regular demand is less than or equal to the renewable energy exported from the site [1].

The European Commission, in turn, called for the definition of Near Zero Energy Building (NZEB), through Directive 2010/31, by 2020, by committing it to all new and existing government property [2]. Defining that they must be very high-performance buildings with low energy requirements, largely assumed from renewable sources, which is calculated according to primary energy (in kilowatt hours per m2 of occupied area per year), regular services [3]. The European Federation of Professional Air Conditioning Organizations (REHVA) has attempted to standardize this definition, expressing the energy flows involved in the building [4], and to analyze the energy balance, through a graph of demand values versus generation in kWh/m2-y [5]. Although there are controversies in the establishment of metrics and scope of the concept [6, 7], as well as if elements external to the building or groups of buildings of the same property, management or location [8] are approached over the period of analysis, whether daily, monthly, yearly or full life cycle [7], and the relationship with building design characteristics.

In the United Kingdom, the concept of Zero-Carbon Buildings (ZCB: Zero-Carbon Building) has been disseminated, which includes measuring energy demand (kWh/m2-y), CO2 consumption (kg/m2-y) and execution. Another common meaning is the designation of Net-ZEB Buildings, promoted by North American institutions [9], which considers a reduction of energy needs compensated by the feeding of renewable technologies (own or external), linked to urban networks.

In Chile, since the 1980s, studies and normative suggestions on energy requirements in buildings have been carried out, and in 2000, for the first time in Latin America, a constructive legislation was introduced to improve the thermal performance of homes, and promote energy efficiency. In South America, Chile has been a pioneer on energy efficiency in buildings motivated by the lack of fuel sources of its own, a strong development of the building and construction sector, but also due to its more temperate and variable climates. In recent years, housing and sustainable building certifications have also been implemented, but on a voluntary basis and without performance targets. In 2012, a manual on energy management in public buildings was proposed [10], followed by a national strategy for sustainable construction [11], which proposes a 12% reduction in consumption by 2020 and a 10% increase in renewables. This is reinforced by Atmospheric Decontamination Plans, which provide funds for thermal reconditioning of housing by cities, and state development programs [12]. In addition, the national energy policy [13] proposes to implement, in all buildings, efficient building settings, improved comfort and intelligent management of energy, as well as a target of 70% of the overall national energy generation from renewable sources. These approaches, with smaller goals than European or North American ones and more prescriptive or administrative orientations, come from a more limited level of development, in which the application of restrictions usually produces rebound effects (increase energy consumption by lack of thermal comfort), although they are also convergent in energy efficiency and integration of general and on-site renewables. Therefore, a zero-energy building approach is pertinent to analyze in this context, to contribute to sustainable building policies in the continent.

The WINNER project of the ERANET-LAC 2014 European-LatinAmerican Call, is being carried out by the Fundacion Prodintec of Spain, the Center for Nanotechnology and Smart Materials (CeNTI) in Portugal, the National Evangelical University (UNEV) in the Dominican Republic and the Universidad del Bío-Bío (UBB) of Chile, with the aim to develop a system of photovoltaic windows with luminescent solar concentrators to provide zero-energy buildings (Figure 1). This work makes an evaluation of its application in an office building case study in Santiago of Chile, to determine the most appropriate characteristics and guide the design of the smart window. Beginning with a review of energy plans for buildings in Europe and Chile, and the definition of a study model in Santiago of Chile, the following parameters are evaluated by dynamic simulations: efficiency and energy generation with different construction settings, layout and efficiency of the windows.

For this work we consider that a zero-energy building should have a lower energy demand than traditional buildings with similar daily energy requirements and same location, as a result of better constructive schemes and more energy efficient household equipment, and that its annual balance includes the provision of energy from renewable sources, either generated *in-situ* or through grid-connected renewable energy power systems located nearby. Moreover, greenhouse gas (GHG) emissions, interaction with urban networks and energy costs depend on more circumstantial legislation or capacities, therefore for the present analysis their possible effect is omitted, as well as that of internal or collective energy management system. The simulation analysis concentrates on the demanded or required energy of the building itself. Likewise, it is considered an hourly equivalence for the energy supply / demand of the building, excluding costs of storage or charging for networks, and imbalances between urban sources.

Figure 1. WINNER project; development of LSC glasses (up), PV-electronic board (medium) and window design (below).

The application of photovoltaic windows with luminescent solar concentrators must fulfill functions of natural lighting, views and/or ventilation, adding solar energy generation features [14], unlike PV panels that typically overlap building façades and roofs, or are deployed in open areas exclusively dedicated to harvest solar energy. Photovoltaic generators integrated in roofs, walls, balconies or glazed areas (with embedded PV cells) have been installed in several buildings around the world [15], but the use of such technologies is not yet generalized [16]. A review of some thirty buildings that have implemented photovoltaic elements, some overlapping (e.g. in façades, roofs) and others integrated, have shown that a relevant number (about one-half) is able to cover the total energy demand of the building [17].

Some studies on the use of photovoltaics to achieve buildings near zero-energy (NZEB), highlight the relevance of the building typology and its location to achieve this goal [18]. That is, its proper implementation is dependent on the configuration, occupation and location of the building. Other work suggests the installation in offices and commercial centers located at medium latitudes with good solar exposure [19], but performance assessment was not presented. The SWING project of the Italian company ENEL and the Polytechnic of Milan has tested windows with integrated luminescent solar concentrators, but as autonomous units, without considering their participation in the global performance of the building [20]. The project assessed the effect of daylight, correlated color temperature, illuminance level and color rendering index and noted some limitations by the color range used.

2 DEVELOPMENT

2.1. Office Buildings in Santiago of Chile

The city of Santiago of Chile is one of the most powerful capitals in Latin America, with about 6,000,000 inhabitants and approx. 400,000,000 m2 of built surface, of which one third corresponds to buildings for services, industry, commerce and finance. These generate about a third of total energy consumption, around 20,000 GWh per year, estimating about 150 kWh /m2 with 3% of annual growth.

The office buildings in Santiago of Chile are characterized by having reinforced concrete structures, with heights from one to a hundred floors, and mixed façades with opaque enclosures and glazed cloths. The newer, taller and larger buildings, tend to have completely glazed façades. With an average height of 11 floors, total area of 10,000 m2 (with 6,000 m2 for offices), 4,000 m2 of façade and 2,000 m2 of window in the most frequent typologies [21]. There is evidence of higher energy consumption in buildings with a higher percentage of spans, due to overheating in summer and cooling requirements.

On the other hand, the city of Santiago of Chile has a warm temperate climate with winter rains and a long dry season. Its urban area is located at latitude 33°26'16 "S and longitude 70°39'01" W, with an average altitude of 570 m. and average annual radiation of 1,422 Kwh/m2.

Several studies have been carried out to characterize the energy performance of office buildings in Santiago of Chile. An engineering thesis [23], reviewing 150 representative buildings, determined an average annual electricity consumption of 95.4 kWh/m2, natural gas 34.6 kWh / m2 and drinking water of 1.43 m3/m2. With CO2 emissions of more than 50 kg/m2 per year, and annual costs in energy and potable water of US$ 14.5/m2, corresponding to 9% of the lease value (equivalent to US$ 16.7/m2 and for sale an average of 1,855 US$/m2). Recognizing a wide dispersion of values, the study then discriminates according to building heights. In low-rise buildings, it was reported an average annual electricity consumption of 86.5 kWh/m2 anually and 7.6 kWh/m2 per month, covering lighting, power, computing and air conditioning (mainly cooling).

In another study, a set of 51 office buildings, recently built in Santiago of Chile [23], were split into two groups, the first with partially glazed façades and the second one with totally glazed façades. The second group showed a heating demand of 3.9 kWh/m2 and a cooling rate of 62.5 kWh/m2, while the former needed 19.2 kWh/m2 for heating and 26.8 kWh/m2 for cooling, evidencing the seasonal thermal sensitivity and its relation with the vertical envelope.

Subsequently, another paper reviewed 10 cases [24], divided by shape and envelope characteristics, and correlated to their electrical consumption in January (summer) and July (winter). Considering the group of buildings with 55-64% of open façade spans, they reported an energy demand of 14.2 kWh/m2 in January and 12.8 kWh/m2 in July, which is close to the general average and recorded annual demand [23]. This is also in line with another study [23], carried out for office buildings in this area, reporting an average annual consumption per occupied area of 86.62 kWh/m2-y, based on energy consumption records that recognize a predominance of electrical requirements.

2.2. Case Study Morphology

To determine a building volume suitable for the energy contribution of photovoltaic windows, it is first considered that this typology usually presents a prismatic conformation (rectangular parallelepiped), with large variations of dimension, but with a similar horizontal partition (floors), in which daily energy-demanding activities are performed. The occupation must maintain a proximity to the vertical faces for natural illumination and ventilation, limiting the horizontal magnitude of the volumes (what are known as widths of cradles). Therefore, even if the magnitudes change, the relationship between the horizontal surfaces (which produces the consumption) and the façades (in which windows can be installed) is limited. Office buildings can range from a floor to tens, and a few meters in front or height, to more than a hundred meters, but its interior is usually divided into horizontal plants with 2.5 to 4 meters of height, and up to ten meters wide from an exterior façade. That is, each longitudinal meter of plant usually has 2.5 to 4 ml of façade, for 3 to 10 m2 of depth. Therefore from 0.3 to 1 m2 façade for each square meter of floor. The smaller proportion occurs in the larger and wider buildings, and the smaller, in the smaller and narrower buildings. As the floor area corresponds to the area of occupation that generates consumption, and the façade allows the integration of collector windows, for the study model it was considered a building of reduced size and extended form, which would allow a greater façade surface with respect to the plants.

On the other hand, in the façades are presented from 20% to 100% of windows (ratio between glazed and opaque surface). The lowest proportions correspond to buildings with small windows and some opaque sides, and the highest proportions, to buildings with completely glazed façades (called curtain walls). According to records of office buildings in Santiago [26], those with lower energy consumption have ratios of 24% to 44%, because higher glazed magnitudes have a lot of thermal loss at night and excessive daytime or summer solar uptake, requiring more heating and cooling. So, an efficient proportion is around 30%, or 0.3 m2 of window per square meter of façade. The windows are usually arranged in all façades, but in mid-latitudes (from 40 to 30°) the sunny side of a building receives about half of the solar radiation than the flat roofs, while the sides a quarter, and the posterior part of the building only one eighth. Therefore, in the study model, it was decided to concentrate the glazed ratio on the sunny side to get a building morphology that allows a solar energy uptake on windows surface that is about four times higher than what is obtained in an average building (equivalent to a completely glazed building), but with more efficiency (half of energy consumption).

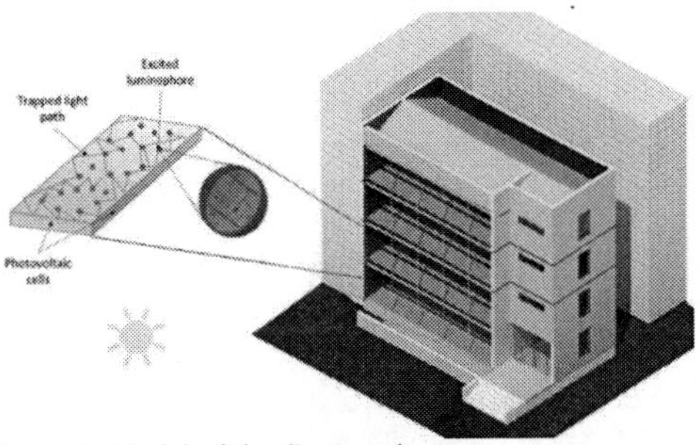

Figure 2. Model of the Case-study

The dimensions of the case study were then considered within the regular range of buildings, but of reduced magnitudes to limit their elaboration, and favorable proportions for a lower energy consumption and greater solar collection, with a usual configuration in this typology. The model contemplates a horizontal rectangular form of 8 x 18 m of extension (east-west), with four plants of 3 m of height each, on a base of one meter. In each plant, a smaller service area, staircase and elevators at one end, and most open work at desks (free floor), totaling 385 m2 of occupied area. The simulation considered small windows at the ends and on the sunniest façade (north) totaling 120 m2 of continuous windows distributed by four floors. Finally, the model assumed one contiguous building at one end and a nearby building at the rear side of the model building.

2.3. Simulations of Performance and Photovoltaic Generation

The software for energy performance simulations are a proven and standardized tool for the study of building behavior [25]. Considering the formation of a three-dimensional model, and a description of material characteristics, equipment and daily occupation, the model generates temperature profiles per zone, thermal flows and energy requirements.

The methods and tools for calculation of active solar generation are numerous [26, 27]. Procedures differ in solar radiation patterns, performance of PV panels and energy losses on facilities or surroundings, etc. In addition, different analysis strategies were proposed based on the energy demand, energy generation capacity of the PV system and facility configuration. Also the simulation procedures and PV systems usually have been verified for autonomous systems, but scarcely for those integrated in the building. Despite these limitations, simulation results are approximately consistent with experimental measurements.

For the energy performance simulations of the study case, the Design Builder software was used, together with the Energy Plus calculation tool, making a general model of the building and running a sequence of simulations set to different model parameters. The model considers a four floors building, each one divided into two zones (offices and surrounding walking areas). The first simulation run considered a regular configuration of the building and related equipment, and the occupation hours per day was set to reach an energy demand equivalent to the average consumption of this building typology for that same location; this procedure established a baseline situation. Subsequently different constructive improvements and more energy efficient equipment were tested, in line with recommendations for low energy consumption buildings implemented on the selected geographic location [21]. Models were then simulated for different locations.

The climate database for simulations includes: global and diffuse solar irradiation, percentage of cloud cover, dry bulb temperature, relative humidity, wind speed and direction. The data must be time-dependent, since the simulations are dynamic. The climate database used is epw (Energy Plus weather files).

The PVsyst software was used for the energy generation calculations, considering the available surface area in the sunniest façade of the study case as well as the climatic conditions of the selected location (Santiago of Chile). The model considers a smart window composed of a continuous glazing layer with overlapping glasses containing a layer of luminescent particles embedded in waveguide matrix capable of absorbing, converting and transmitting part of the incident radiation towards the perimeter of the window with photovoltaic cells that finally convert solar energy into electricity. The conversion efficiency is defined as the ratio between generated power and the irradiance striking the total glass collecting surface area, measured under standard conditions (1000 W/m2 irradiance, 25° C ambient temperature and AM1.5 spectrum). As the window has dual functionality, i.e. acts as building element and photovoltaic system, the conversion efficiency is compromised because most of the incoming solar radiation must be transmitted through the glass to reach the interior of the building.

Several research groups have published luminescent concentrator conversion efficiencies ranging from 2.7% to 7.1% [28, 29, 30]. Typically, these systems integrate a non-transmissive back reflector that manage to considerably increase the efficiency of the collector but prevents its use as a window glass because it blocks all incident light. Reported energy efficiencies are normally obtained for very small area collectors, in the order of 5x5 cm. In the WINNER project, developed by the authors, glass samples of 15x15 cm are considered. The estimated efficiencies range from 1% to 4%, depending on the percentage of light trapped inside the coated glass, the efficiency of the concentrator and the photovoltaic system optically coupled to it.

Based on this analysis, several simulations were performed by considering two new types of crystalline silicon photovoltaic modules with 1% and 4% efficiencies. The system was defined as having an available area of 120 m2 on the north façade of the building, and assuming different solar incidence angles. In each case, the most suitable inverter for the installation was chosen, and the possible shadows of contiguous buildings were neglected. Additional input parameters to the PVSyst software comprised energy losses due to different temperature conditions, irradiance, DC-AC energy conversion efficiency, and energy losses in the wiring (series resistance). An important feature of LSC-based solar windows is that, unlike other types of concentrators, they do not require the implementation of tracking systems to follow the sun throughout the day because they are capable of concentrating both direct and diffuse radiation.

3 RESULTS

3.1. Energy Demands of Case-study

Results from model simulations showed that office buildings set with standard features present a conventional energy demand profile. A relevant fraction of the energy requirements of office buildings originate from equipment, lighting, and plug loads determined by the building's intended use, its occupancy, and its scheduling, an energy demand that is quite stable throughout the year. An equivalent part is caused by air conditioning, both heating and cooling, in an inverted seasonal manner. The heating increases in winter periods and cooling in summer. To estimate the energy demand per unit area, the occupancy schedule was set from 10 AM to 4 PM, Monday to Friday, which means 30 hours per week (with a density of 0.1 p/m2). An annual energy demand of 86.5 kWh/m2-y was obtained (Figure 3), consistent with figures reported in the literature for the same building typology and geographic location [27, 28, 29]. The above working period is usual, considering that labor contracts rely on 44 effective working hours per week including outside activities, and office working hours are usually concentrated in central periods of the day.

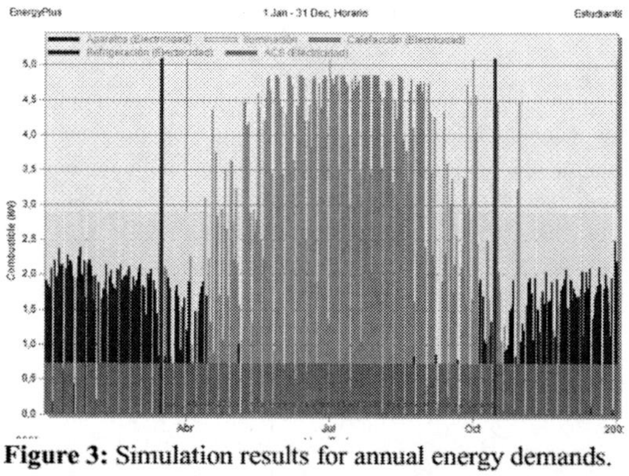

Figure 3: Simulation results for annual energy demands.

The simulations of the model accounting for different solar orientations, showed dissimilar energy demand values, due to different needs in air conditioning, since the demands of work equipment and lighting remained almost constant. The model considering buildings with most windows facing South, which is typically less sunny, presented the lowest total energy demand (67.42 kWh/m2-y), due to a lower cooling requirement in warm days. While the model with windows facing East delivered an intermediate energy demand (77.7 kWh/m2-y) and those facing West showed the highest value (87.42 kWh/m2-y), in both cases with slightly lower cooling variations in heating, reflecting the incidence of summer overheating, and in particular of dusk, by the concentration of glazed surface. As also the changing behavior shows the dynamic nature of the constructive features.

The models designed to include better construction schemes and energy efficient equipment presented significant variations, except for the change of traditional walls by CLT (counter-laminated tables) wood slabs, which showed a demand slightly higher (88.32 kWh/ m2-y) than the base model (86.5 kWh/ m2-y). Although this constructive alternative increases the thermal resistance and thickness of the vertical envelope, this result again reflects the variable nature of the thermal performance of the buildings. The energy demand for cooling is reduced due to lower losses of the envelope, but increase for heating, probably due to its lower thermal mass. However, in the model with panels in the windows there is a significant reduction of the global energy demand (61.86 kWh/m2-y), with a significant decrease of the cooling needs, apparently coming from partial shading effects of the plates that reduce the overheating; contrarily, heating needs increase due to lower thermal gains. The combined model of panels in windows with walls of CLT, delivers a slightly higher performance (64.63 kWh /m2-y), in which heating is reduced, but increases cooling. The simulation of the model with CLT plates and efficient equipment yields similar results (59.22 kWh/m2-y), although with an increased refrigeration demand. Not surprisingly, the model with CLT, window panels and efficient equipment achieves a relevant reduction in building's energy demand (42.33 kWh/m2-y), evidencing that the replacement of traditional equipment and lighting by better performance products has a significant contribution in the minimization of energy consumption.

3.2. Energy Supply of PV-Windows

The energy production of smart windows set in the model for the specific location in Santiago of Chile, calculated using PVSyst software, and considering PV panels with 1% and 4% conversion efficiency and several solar incidence angles are show in Figure 4.

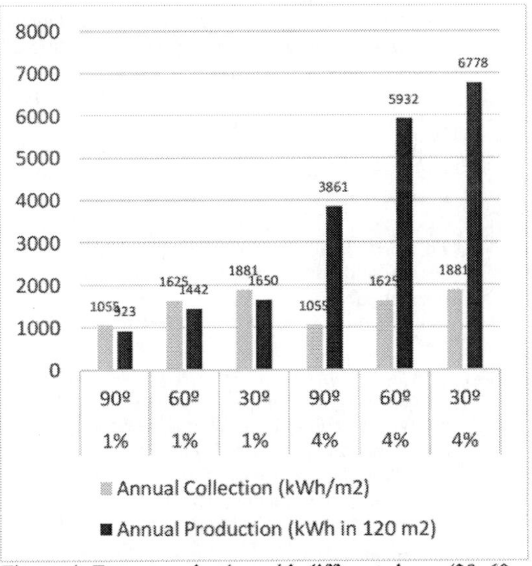

Figure 4: Energy production with different slopes (30, 60 and 90°) and PV efficiency rates (1 and 4%).

This production corresponds to 5.6% to 41.6% of the total energy demand (16,258.55 kWh), or also from 12.7% to 93.7% of the lighting requirements (7,230.3 kWh). That is, a provision relevant to the regular occupation of the building.

Regarding the energy conversion yields according to the tilt angle of the glass collector (with respect to the horizontal), there is no significant effect of the work equipment efficiency. At 30°, the greatest amount of energy is produced, decreasing substantially towards higher tilt angles. This angle is similar to the local latitude value (33°), which is consistent with recommendations from PV manufacturers.

According simulations the contribution of smart PV-windows is sensitive to solar orientation and amount in relation to building shape, but permanent to geographic location. The energy share is sensitive to building technology and occupation, and also is increased by the function of shading of smart PV-windows such reduce cooling requirements in amount close to the solar generation.

3.3. Energy balance.

According to the simulations, in the zero-energy balance graph (Figure 5), it can be observed that the building set to regular conditions, i.e. taking into account any performance add-ons, would have a demand of 86.52 kWh/m2-y. With improved construction characteristics, it would reach 59.22 kWh m2y, and with efficient equipment it would go down to 42.33 kWh/m2-y. The total contribution of the windows with LSC could represent a demand reaching values as low as 17.61 kWh/m2-y, i.e. managing to supply up to 41.6% of the annual energy demand of the improved building. Whereas, for a building with only improved construction schemes, this contribution represents less than a third (29.7%) and for a conventional building, about a fifth (20.35%). Changing the orientation of the façade, would decrease energy generation to something between less than a half and an eighth of the energy demand. It can be concluded that, under specific conditions of design, construction and equipment, PV-windows integrated at large-scale would be able to provide up to about one-half the energy required for a building, which is a relevant step for zero-energy balance.

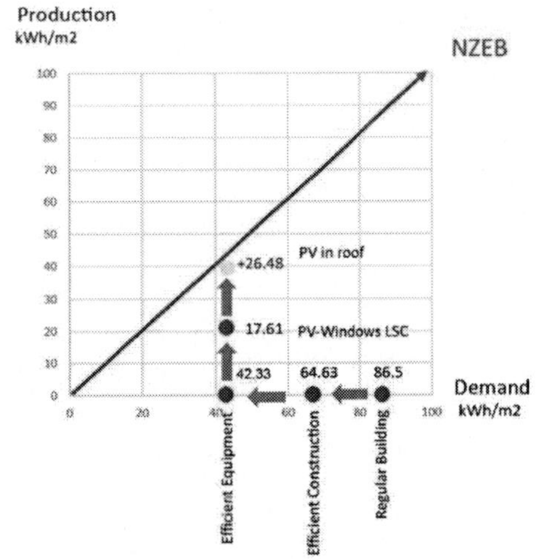

Figure 5: Energy balance.

An additional possibility is the installation of rooftop

photovoltaic panels, which according to the roof surface available in this building (100 m2) and its location (Lat 33°), with an optimum tilt angle and orientation (33 °-0°) and regular efficiency (11.98%), could reach a total annual energy production of 10,197 kWh-y (according to static calculations). This would represent 26.48 kWh-y/m2 that is more than half of the energy demand observed for conditioned buildings (according to previous simulations). Complementing this PV system set-up in rooftops with the collection of solar energy in smart windows will allow to provide the total energy required, reaching a zero balance. However, it must be considered that these contributions are usually synchronous, so the losses due to batteries or contributions to the network must be analyzed together.

4. CONCLUSION

This assessment is an innovative review of building features for PV-windows, to reduce energy demand and increase share of integrated solar generation, providing a new methodology to guide the development of BIPV products. The case-study represents an increasing building sector in emergent countries, and can be applied in other building types and locations. This work identifies the following novel conditions of PV-windows for NZEB:

- Geometrical relations of building shape and PV-windows to increase energy contribution.
- Parameters affecting the pathway to NZEB-goal in office buildings.
- Design conditions for smart PV-windows to be evaluated with prototypes and users.

In conclusion, the evaluation carried out in this work suggests that photovoltaic windows with LSC can make a relevant contribution to the objective of energy demand close to zero in certain buildings and geographic locations, as well as orient the design of windows and construction of buildings to increase contribution of BIPV to a more sustainable future

REFERENCES

[1] National Institute of Buildings Sciences (2015) A Common Definition for Zero Energy Buildings. Washington:US Department of Energy.

[2] EP-CEU (2010); Directive 2010/31 on the energy performance of buildings. The European Parliament and the Council of the European Union.

[3] Hermelink A.; Schimschar S.; Boermans T.; Pagliano L; Zangheri P.; Armani R.; Voss, K.; Musall E.; (2012) Towards nearly zero energy buildings Definition of common principles under the EPB, Ecofys, Koln.

[4] Kurnitski J.; Allard F.; Braham D.; Goeders G.; Heiselberg P.; Jagemar L.; Kosonen R.; Lebrun J.; Mazzarella L.; Railio J.; Seppänen O.; Schmidt M.; Virta M.; (2011) How to define nearly net zero energy buildings nZEB– REHVA proposal for uniformed national implementation of EPBD recast, REHVA-Journal, 6-12.

[5] Sartori I; Napolitano A. Marszal, Pless S, Torcellini P. and Voss; Criteria for Definition of Net Zero Energy Buildings, EuroSun Conference Graz 2010.

[6] Ayman M.; Ala H.; Kai S.; (2014) Fulfilment of net-zero energy building [NZEB] with four metrics in a single family house with different heating alternatives. Applied Energy, 114, 385.

[7] Deng S., Wang R.Z., Daa Y.J. (2014), How to evaluate performance of net zero energy building, a literature research, Energy 71 (2014) 1-16

[8] Scognamiglioa A., Adinolfia G., Graditia G., Saretta E. (2014); Photovoltaics in Net Zero Energy Buildings and Clusters: enabling the smart city operation, Energy Procedia, 61, 1171 – 1174

[9] Ares E. (2016), Zero Carbon Homes, Londres:House of Commons Library.

[10] Instituto de la Construcción (2012) Manual de Gestión de Energía en Edificios Públicos, Santiago: Innova-Corfo.

[11] MINVU (2013). Estrategia Nacional de Construcción Sustentable 2013-2020, Ministerio de Vivienda y Urbanismo [MINVU], Chile.

[12] MMA (2014). Planes de Descontaminación Atmosférica Estrategia 2014 – 2018, Ministerio del Medio Ambiente [MMA], Chile.

[13] MINER (2015). Política Energética de Chile 2050, Ministerio de Energía, Chile.

[14] Tsoi, S (2012) Structured luminescent solar energy concentrators: a new route towards inexpensive photovoltaic energy, Doctoral Thesis, Eindhoven: Tecnische Universitet Eindhoven.

[15] Hardy, D.A., Blekastad, S., Kerrouche, A., Roaf, S.C. and Richards, B.S., (2012) Creative use of BIPV materials: barriers and solutions. 27th European Photovoltaic Solar Energy Conference and Exhibition, Messe Frankfurt, Frankfurt, Germany.

[16] Strong S., (2011). Building Integrated Photovoltaics [BIPV]. [en línea]. Disponible en: https://www.wbdg.org/resources/bipv.php.

[17] Voss, K. y Musall, E. (2012). Net Zero Energy Buildings: International projects of carbon neutrality in buildings. Karlsruhe:ENob.

[18] Kim J-H; Kim H-R y Kim J-T (2015); Analysis of Photovoltaic Applications in Zero Energy Building Cases of IEA SHC/EBC Task 40/Annex 52, Sustainability, 7, 8782-8800; doi:10.3390/su7078782

[19] Gallo A., Téllez Molina B., Prodanovic M., González Aguilar J., Romero M. (2014); Analysis of net Zero-Energy Building in Spain. Integration of PV, solar domestic hot water and air-conditioning systems, Energy Procedia 48, 828 – 836

[20] Aste N., Tagliabue, L.C., Palladino P., Testa D. (2015), Integration of a luminescent solar concentrator: Effects on daylight, correlated color temperature, illuminance level and color rendering index, Solar Energy, 114, 174–182.

[21] Vasquez, C.; Encinas, F. y D'Alencon, R. (2015). Edificios de oficinas en Santiago: ¿Qué estamos haciendo desde el punto de vista del consumo energético? ARQ, 89, 50-61, http://dx.doi.org/10.4067/S0717699620150001000 08

[22] Pareja, P.A.,(2007). Desarrollo de índices de desempeño del consumo energético y de agua sanitaria en edificios de oficinas de Santiago. Tesis, Santiago: Universidad de Chile.

[23] Price Waterhouse Coopers (2007) Desarrollo de una línea base y metodología de medición del consumo energético en edificios, Santiago:Corfo.

[24] Bustamante W., De Herde, A., y Encinas (2011), Análisis de comportamiento térmico de edificios de oficinas en comunas de la Región Metropolitana, Chile. Revista de la Construcción. 10:1, 64-77

[25] Garcia-Alvarado, R., Gonzalez, A., Bustamante, W., Bobadilla, A. y Muñoz, C.,(2014). Características relevantes de la simulación energética de viviendas unifamiliares. Informes de la Construcción, 66- 533, DOI 10.3989/ic.12.108.

[26] Frontini F., Marzoli M., Doust N. (2013), Investigation of different simulation tools for solar photovoltaic modules, IBSPA-2013, 89-94

[27] Axaopoulos, P.J., Fylladitakis, E.D. & Gkarakis, K. (2014) Accuracy analysis of software for the estimation and planning of photovoltaic installations, Int J Energy Environ Eng 5:71

[28] Currie M.J. et al (2008), High-efficiency Organic Solar Concentrators for Photovoltaics, Science 321, 226.

[29] Sloof L.H. (2008), A luminescent solar concentrator with 7.1% power conversion efficiency, Phys. Stat. sol. [RRL] 2, No. 6, 257-259.

[30] Goldschmidt J.C. et al (2009), Increasing the efficiency of fluorescent concentrator systems, Solar Energy Materials & Solar Cells 93, 176–182.

ACKNOWLEDGEMENTS

This work is part of the WINNER project, funded under the programme ERA-NET LAC (Latin America, Caribbean and European Union), with project reference number ELAC2014/ESE-0146, and nationally in Spain by Ministerio de Economía y Competitividad (Acciones de Programación Conjunta Internacional, PCIN-2015-152), in Portugal by Ministry of Science, Technology and Higher Education via FCT- The Portuguese National Foundation for Science and Technology (Ref. ERANETLAC/0007/2014), in Chile by Comisión Nacional de Investigación Científica y Tecnológica (CONICYT) and in Dominican Republic by Ministry of Higher Education, Science and Technology (MESCYT).

PHOTOVOLTAIC ELECTROCHROMIC MODULE WITH UNIFORM COLOR CHANGE

Lee-May Huang, Cheng-Yu Peng, Chih-Hung Chen, Han-Chang Liu, Chorng-Jye Huang
Green Energy & Environment Research Laboratories, Industrial Technology Research Institute
195, Sec.4, Chung Hsing Rd., Chutung, Hsinchu, Taiwan 31057, R.O.C.
Corresponding author: Lee-May Huang, Phone: +886 (3)5912252, E-mail: lmhuang@itri.org.tw

ABSTRACT: A novel photovoltaic electrochromic (PV-EC) module for building integrated PV (BIPV) applications has been developed. The PV-EC module includes planarly distributed silicon thin film solar cells (Si-TFSCs) with successively deposited electrochromic thin films and stripe-shaped electrolyte layers. In addition to providing self-powered color changing, the series-integration of each Si-TFSC adds-up the photopotential of the Si-TFSCs and can supply electrical power as a general Si-TFSC module does. This dual functions PV-EC module provides an alternative solution to BIPV module, which demands not only power generation but also requires the balance of heat and light coming through the BIPV module.
Keywords: Photovoltaic, Electrochromic, BIPV, planar structured PV-EC

1 INTRODUCTION

Building integrated PV (BIPV) systems can provide crucial supplementary source of electrical energy from any surface available within the building's envelope. Glazing integrated photovoltaics are among the most promising solutions due to heating and cooling savings in addition to electricity production [1]. Electrochromic (EC) device has been developed for retrofitting existing windows, to dynamically control transmitted light, glare, and solar heat, while allowing light in and preserving view out [2]. The integration of PV module and EC device provides better efficiency in term of energy saving, for the PV-EC device can achieve color change in the EC layers from the sunlight converted electrical power (as described in figure 1). This self-powered smart glass can adjust the color of the EC device according to the intensity of sunlight irradiation to reduce indoor heat and to provide comfort.

Figure 1: The integration of see-through type BIPV module and EC smart glass.

Until recently, there are two types of solar powered EC devices available, the first one is tandem structure silicon (Si) based PV-EC device [3,4] and the second one is dye-sensitized solar cell (DSSC) based photoelectrochromic (PEC) device [5], both of which belong to National Renewable Energy Laboratory (NREL) technology. The silicon based PV-EC technology constitutes stacked layers of the photoabsorbers and EC active materials in one single device. However, this tandem-structure PV-EC device encounters problems such as low colored/bleached optical contrast, short circuit between the deposited layers and difficulty in producing large area device [3].

A photoelectrochromic technology separates the photoabsorbing layers of DSSC [6] and the EC layer to the anode and the cathode respectively for constituting a device. This PEC device can be described as having EC material inserted into DSSC, and has become the mostly studied photovoltaic integrated electrochromic technology. However, to apply such a structure to practical applications, many problems need to be overcome, such as long-term stability of the photo-absorbing layers and the possibility of developing devices having larger sizes.

Figure 2: Photograph showing the color changing of planar structure PV-EC device under exposure to sunlight, the active area of the PV-EC device is 5.5cm x 5cm, with each Si-TFSC stripe area=0.5cm x 5cm [7].

An innovative planar structure PV-EC device in which electrochromic materials are disposed between a semi-transparent silicon thin film solar cells (Si-TFSCs) and a transparent non-conductive substrate, was recently developed by our group [7-9]. The structure of the PV-EC device consists of planarly distributed semi-transparent Si-TFSCs with exposed transparent anode layers in-between the cathodes layers, configured in a superstrate structure, as shown in figure 2. The cathode of the Si-TFSCs are arranged in stripes, and the area outside the cathode blocks are anode which composed only of transparent conductive oxide (TCO).

In solution type planar structure PV-EC device, organic electrochromic solution is disposed between a planarly distributed Si-TFSC and a transparent non-conductive substrate [7]. Figure 3 shows the working principle of the solution type PV-EC device. When sunlight enters the PV-EC device, the Si-TFSCs generate electron-hole pairs. The photopotential difference at the anode and cathode of the Si-TFSC induces electrochemical redox reaction of the electrochromic solution. Simultaneously, cathodic and anodic electrochromic color change occurs at the cathode and anode of the Si-TFSC. The electronic current produced by the Si-TFSC is converted into an ionic current in the electrochromic solution. Thus, even though the anode and the cathode both contact the electrochromic solution when conducted, the problem of short circuit would not occur. Since the Si-TFSC cells are designed to be evenly distributed on a transparent substrate, the electric fields in the peripheral area and the center area are uniform, with a result that the color density in the peripheral and the center areas of the PV-EC device are the same, regardless of the size of the device.

When the intensity of the sunlight declines, the electrons and holes generated by the Si-TFSC are reduced and the photopotential and photocurrent of the solution type PV-EC device decreases gradually. Due to the self-erasable property of the solution type electrochromic material, the color of the electrochromic materials fade out gradually and resume to the original transparent state.

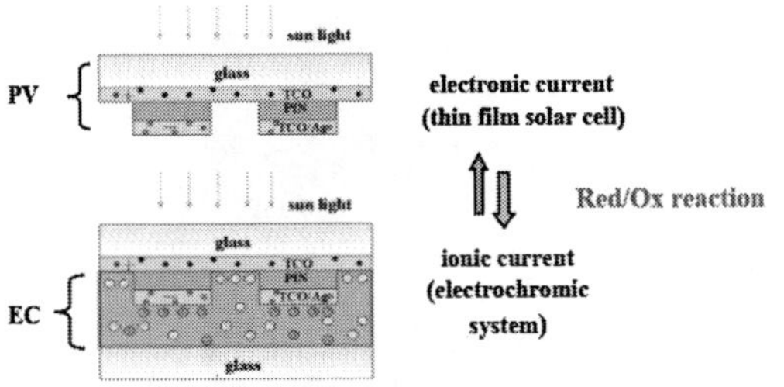

Figure 3: Working principle of solution type planar structure PV-EC device [7].

The overall transparency and the color contrast of this solution type PV-EC device are enhanced, as compared to tandem Si based PV-EC device [3,4]. However, this solution type PV-EC may cause erosion in Si-TFSC, and unless redox couples are added to the EC solution, the self-bleaching duration is too long for practical applications.

A solid-type planar structure PV-EC device which can solve the problem encountered by solution type PV-EC device is introduced [8]. Here, the electrochromic thin films (ECTFs) are successively deposited on the anodes of the Si-TFSCs. An electrolyte layer is then formed on the surfaces of the Si-TFSC cells to contact the ECTFs (EC(+) in figure 4) simultaneously. The electronic current generated by the Si-TFSCs is transformed to ionic current in the electrolyte layer that migrates transversely along the electrodes of the Si-TFSCs in order to maintain charge neutrality in the EC system. The electrolyte must be a good ionic current conductor, but it must be an isolator for electronic current generated by the Si-TFSCs.

In concordance with the structure of this solid type PV-EC device, the series-integration of each Si-TFSC adds-up the photopotential of the Si-TFSC, making large area PV-EC module fabrication feasible.

In this paper, the effects of two different electrolyte configurations: continuous electrolyte layer and stripe-shaped electrolyte layers, on the color change performance of sunlight irradiated PV-EC module are investigated.

2 EXPERIMENTAL

2.1 Semi-transparent Si-TFSC and module fabrication

A Si-TFSC (cell) was fabricated by sputter and plasma enhanced chemical vapor deposition processes. The structure of the Si-TFSC included a glass substrate; a ZnO:Al transparent conducting oxide layer which served as the anode; a P-type layer; a double junction a-Si:H/μc-Si:H thin films which served as the photoelectric conversion layers; a N-type layer; and a ZnO:Al layer and a back reflector metal layer together to constitute the cathode. A pulse laser of 532 nm was then used to remove a portion of the silicon thin film to form the Si-TFSC in stripes, the ratio of anode and cathode blocks was 1:1, respectively.

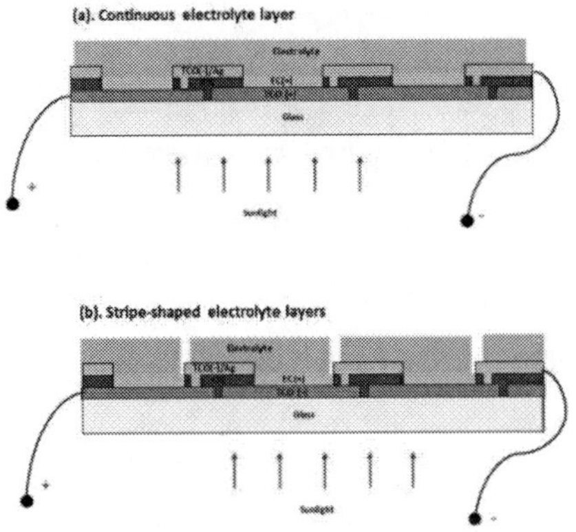

Figure 4: Planar structure PV-EC module with two types of electrolyte configurations: (a). a continuous electrolyte layer covering the whole PV-EC module, and (b) as striped electrolyte layers covering each pair of Si-TFSC forming an individual EC system.

A process flow of fabricating a PV-EC module was similar to that of conventional TFSC module]. A continuous ZnO:Al layer was preliminary formed on a glass substrate. Then, a first laser scribing step called P1 was conducted to remove the ZnO:Al layer within the areas P1 so as to form a plurality of anode layers, followed by a plasma-enhanced chemical vapor deposition (PECVD) of silicon layer. A scribing step, called P2 was then carried out to completely cuts through the silicon layer, forming a plurality of semiconductor thin film. Another ZnO:Al layer and (optionally a back reflector metal layer) were deposited successively, again followed by conducting a third laser scribing step P3 to remove the ZnO:Al layer and the metal layer, so as to form the cathode layers. When the processes were completed, adjacent Si-TFSCs were interconnected with each other in series (as shown in Figure 4). The transmittance of the Si-TFSC module was adjustable by controlling a P3 interval between the adjacent Si-TFSCs.

2.2 Preparation of the Prussian Blue electroplating solution

The preparation of Prussian Blue electroplating solution was as follows, 10mM of $K_3Fe(CN)_6$ was added into 50ml of DI-water, and 10mM of $FeCl_3$ and 10mM of KCl were added into 50ml of DI-water, so as to obtain two solutions. The two solutions were then mixed in a volume ratio of 1:1.

2.3 Electroplating of PBTF on the Si-TFSC

A Prussian Blue thin film (PBTF) was electrodeposited by traditional three electrode electrochemical method [8]. The Si-TFSC was put into a standard electroplating process set-up. During the electrodeposition process, the PBTF (EC(+) in figure 4) was deposited on a TCO layer, which had been isolated from direct contact with the anode of the Si-TFSC. The cathode area of Si-TFSC was masked to prevent contact with electroplating solution. An Autolab PGSTAT30 electrochemical analyzer with the TCO as working, platinum foil as counter and an Ag/AgCl as reference electrodes were utilized to carry out PBTF electrodeposition at a constant potential of 0.6V. The electrodeposition process was performed in dark room, in order to avoid additional photo-induced potential different of the Si-TFSC, which might disturb the uniformity of PBTF forming.

During the electroplating process, the color of the TCO layer gradually changed from transparent to light blue, indicating that a PBTF had been plated on the TCO surface. The thickness of the PBTF was approximately 200 nm. A reverse potential of -0.2V was then applied to the TCO layer to turn the PBTF into transparent colorless state.

2.4 Spreading electrolyte on Si-TFSC

0.1M of $LiClO_4$/PMMA/PC electrolyte was spread as: (a). a continuous layer covering the whole PV-EC module, and (b) as striped layers covering each pair of Si-TFSC, touching both the anode PBTF and the cathode of the Si-TFSC, forming an individual EC system, as shown in figure 4.

2.5 Measurements

The Spectrophotometry data of the anodic coloring PV-EC device was obtained *in-situ*, using a spectrophotometer (model UV-1601PC, Shimadzu, Japan) and a solar simulator. The IV measurement in simulated sunlight followed IEC60904-9 standard.

3 RESULTS AND DISCUSSION

When sunlight irradiates the PV-EC module, oxidation reaction occurs on the anode side of the EC system, and PBTFs start to change its tint from transparent colorless to light blue. However, figure 5 reveals that for the case of continuous electrolyte covering the Si-TFSCs (as shown in Figure 4(a)), as the duration of time exposure elapsed, the degree of color change of the four sets of the PBTFs are different. The non-uniform color change is expected to be induced by the potential difference of each set of the Si-TFSC connected in series, causing unequal redox reaction throughout the whole EC system. As a result, over-oxidation is likely to occur near the anode edge of the PV-EC module, and over-reduction appears near the cathode edge of the PV-EC module.

However, for the case of stripe-shaped electrolyte layer, the PV-EC module changes its tint uniformly. Here, the cations and anions within each pair of the cathode/anode of the Si-TFSCs can be confined inside a single EC system, as shown in figure 4(b) and figure 5. Since the stripe-shaped electrolyte is mutually independent, the electrolyte of one EC system is not in contact with the adjacent electrolyte. Charge imbalance caused by covering a continuous electrolyte layer on the serially connected Si-TFSC module can be avoided. Under sunlight irradiation, the PBTFs change color evenly, no over-oxidation or over-reduction reactions is observed after 20 minutes of continuous sunlight irradiation, as indicated in figure 6.

Although the electrical power generated by the PV-EC module takes into account the whole power output of the Si-TFSC module, the color/bleach of an ECTF is achieved by individually considering each pair of the anode and cathode of the Si-TFSC as an independent EC system [8]. This dual functions PV-EC module provides an alternative solution to building integrated photovoltaic (BIPV) application, which demands not only power generation but also requires the balance of heat and light coming through the BIPV module.

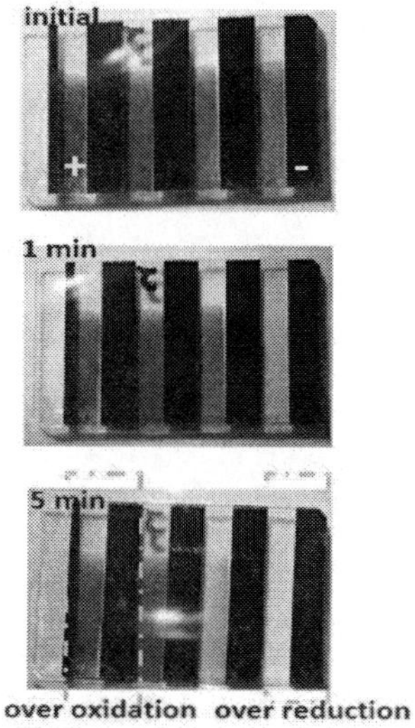

Figure 5: Non-uniform color change of sunlight irradiated PV-EC module with continuous electrolyte configuration.

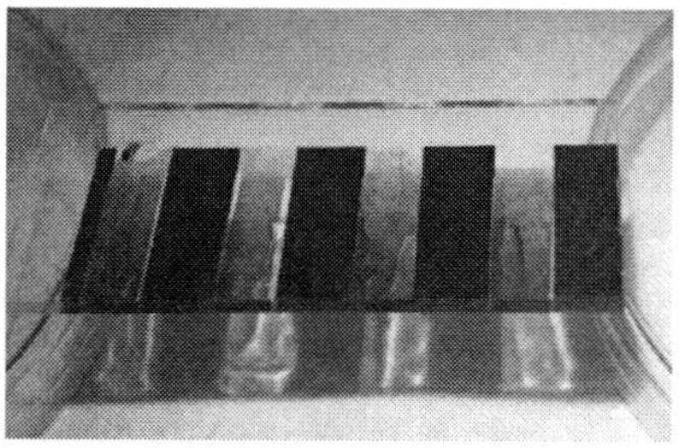

Figure 6 Uniform color change of sunlight irradiated PV-EC module with stripe-shaped electrolyte configuration

The photoelectric conversion and the electrochromic properties of a PV-EC device are presented in the following discussion. The IV characteristic of Si-TFSC and module shows insignificant changes in photoelectric properties after been subjected to Prussian Blue electrodeposition process as shown in table 1 [8]. For a-Si:H/μc-Si:H TFSC (cell), the measured Voc=1.32 V, Isc=27.65 mA, FF=61.94%, Pmax=22.65 mW and efficiency=9.44%. As for the a-Si:H/μc-Si:H TFSC module, the measured Voc=4.00 V, Isc=27.57 mA, FF=64.94%, Pmax=69.09mW and efficiency =5.23%.

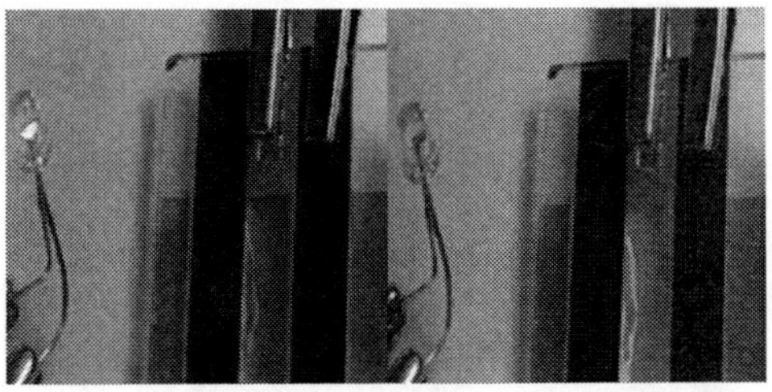

Figure 7: Photograph of a dual function PV-EC module, the colored and bleached state of a single PBTF in a PV-EC device can be switched independently even under light exposure, while the PV-EC module generates power to light LED bulb [8].

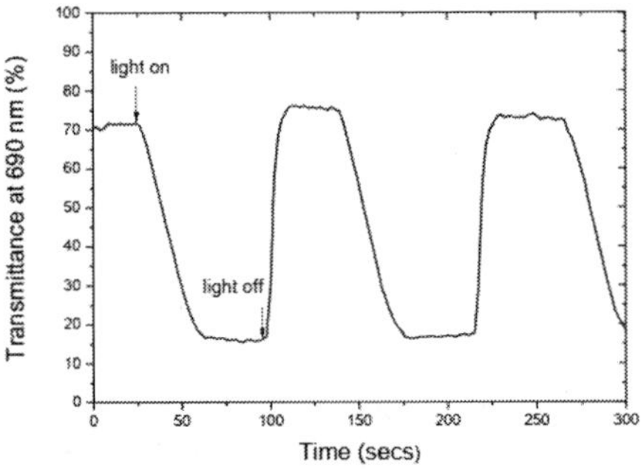

Figure 8: The spectral response of a PBTF in a solid type PV-EC device in the bleached and colored state at 690 nm wavelength [8].

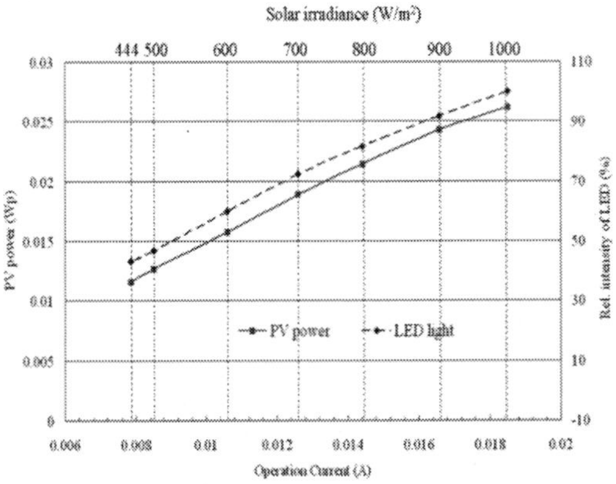

Figure 9: The dependence of a PV-EC module power generation and LED bulb illumination intensity to solar irradiance. The PV power and LED light vary proportionally with respect to sunlight irradiation [8].

Based on the circuit design described in previous study, the PBTF can be tinted or bleached under sunlight irradiation [8]. The optical contrast and spectral response of a PV-EC device are depicted in figure 8 and figure 9, respectively. The PBTF experiences anodic coloration from transparent to blue, the film darkens to its lowest transmission state in 30 seconds. The PBTF restores its transparency in 15 seconds. This result proves the feasibility of the concept of a tunable solid-state PV-EC device with stripe-shaped electrolyte structure.

Table 1: The characteristics of photoelectric conversion of a-Si:H/μc-Si:H-TFSC and module [8]

Type of solar cell	Voc (V)	Isc (mA)	FF (%)	Pmax (mW)	η (%)
a-Si:H/μc-Si:H TFSC	1.32	27.65	61.94	22.65	9.44
a-Si:H/μc-Si:H TFSC module	4.00	26.57	64.94	69.09	5.23*

* the efficiency of theSi-TFSC module is calculated by considering the actual occupied area of the TFSC cells not including the area of the laser scribed anode

The power generation of the PV-EC module is verified by connection to a LED bulb, as shown in Figure 9. The dependences of LED luminance intensity to various solar irradiances vary proportionally with the output power of the PV-EC module

4 CONCLUSIONS

The present study shows that planar structure PV-EC module having a plurality of stripe-shaped electrolyte with each single stripe individually covering each pair of Si-TFSC can effectively mitigate the phenomenon of charge imbalance, so as to have a better uniformity in color change during light irradiation.

An anodic coloring PV-EC module is built based on monolithic series-connected Si-TFSC, which fabrication method is compatible with the traditional Si-TFSC manufacturing process. The ECTFs are then simultaneously deposited on the anode of the Si-TFSC. Each pair of the Si-TFSC are covered with stripe-shaped electrolyte. The electrochromic behavior of each ECTF layer is achieved by individually considering each set of the anode and cathode of Si-TFSC (cell) as a single EC system. This PV-EC module proves for the first time that apart from generating power, Si-TFSC module is also capable to change its tint in response to the exposure to sunlight.

5 ACKNOWLEDGEMENT CONCLUSIONS

The financial support provided by Bureau of Energy (Grant No. 106-D0104) is gratefully acknowledged.

6 REFERENCES

[1] G. Gorgolis, D. Karamanis, Sol. Energy Mater. Sol. Cells 144 (2016) 559.

[2] P.M.S. Monk, R.J. Mortimer, D.R. Rosseinsky, Electrochromism and Electrochromic Devices, Cambridge University Press, Cambridge, 2007.

[3] K. Benson, H.M. Branz, Design goals and challenges for a photovoltaic powered electrochromic window covering, Sol. Energy Mater. Sol. Cells 39 (1995) 203-211.

[4] W. Gao, P. Liu, R.S. Crandall, S.H. Lee, D. K. Benson, Approaches for Large Area a-SiC:H Photovoltaic-Powered Electrochromic Window Coatings, J. Non-Cryst. Solids 266-269 (2000) 1140-1144.

[5] C. Bechinger, S. Ferrere, A. Zaban, J. Sprague, B.A. Gregg, Photoelectrochromic windows and displays, Nature 383 (1996) 608-610.
[6] B.O' Regan, M. Gratzel, A low-cost, high-efficiency solar-cell based on dye sensitized colloidal TiO2 films, Nature 353 (1991) 727-740.
[7] L.M. Huang, C.W. Hu, H.C. Liu, C.Y. Hsu, C.H. Chen, K.C. Ho, Sol. Energy Mater. Sol. Cells, 99 (2012) 154.
[8] L.M. Huang, C.K. Kung, C.W. Hu, C.Y. Peng, H.C. Liu, Sol. Energy Mater. Sol. Cells 107 (2012) 390.
[9] L.M. Huang, C.W. Hu, C.Y. Peng, C.H. Su, K.C Ho, Sol. Energy Mater. Sol. Cells, 145 (2016) 69.

DESIGN, FABRICATION AND TECHNO-ECONOMIC ANALYSIS OF SOLAR ENERGY CONVERSION SYSTEM BASED ON FLEXIBLE SOLAR PANELS

Mostafa Esmaeili Shayan[1], Gholamhassan Najafi[1*], Ahmmad Banakar[1]
1- Tarbiat Modares University, Tehran, Iran
g.najafi@modares.ac.ir, Tel.: +98(912)3664393, Fax: +98(21)48292200

ABSTRACT: In this study, photo-voltaic system based on flexible solar panels was designed, manufactured and evaluated on a scale of $1m^2$ and 56W. Conventional photo-voltaic systems cannot be used on curved surfaces. According to this, a flexible system was analyzed in real terms on a flat, cylindrical and hemispherical surface. Recognizing the importance of environmental variables, temperature sensing system to receive data and communicate online with LabVIEW software was designed and built, as well as economic analysis of the systems was carried out using the COMFAR software. Meteorological station data was obtained at the same time. Additionally, the Taguchi experimental design and quality specifications as "Bigger is better" was used to select the best system. Taguchi model showed: if the system of selection based on the optimal manufacturing process and variables is applied, then the system deployed on flat surface and standard condition takes the fill factor of 0.73 and maximum power production capacity equal to 46.71W with 7.09% efficiency. As well as, the system deployed on cylindrical and hemispherical surfaces have fill factor of 0.88 and 0.84, respectively. Their maximum power production capacities are 55.18W and 57.50W respectively with 7.45% and 7.45% efficiencies. Economic analysis indicated: the values of Internal Return Rate (IRR) with deployment on flat, cylindrical and hemispherical surfaces were 23.82%, 26.70% and 27.46%, respectively. And, their Net Present Value (NPV) were 6.747, 9.952 and 10.830 Million Rials. Economic justification for attracting investment projects were ranked respectively based on the hemispherical, cylindrical and flat surfaces.

Keywords: Photo-Voltaic, Flexible Solar Panels, LabVIEW, Taguchi.

1 INTRODUCTION

Solar energy is proposed as one of the most important renewable energy sources as both thermal and electrical power. European countries plan to provide close to 50% of their low-pressure heat demand by solar thermal sources by 2030 [1], but high-potential photo-voltaic technology can contribute up to 5% by 2030 and up to 11% by 2050 of electricity demand in Europe [2]. With the advancement of technology, nowadays, solar irradiance energy is converted directly into electricity, thereby this is a major step in gaining the power from the sun. Notwithstanding, only 1% of the total electrical power required by humans, equivalent to 18.400 TWh, is supplied still through solar cells [3, 4]. The importance of energy and its various sources of supply are now regarded as the main approaches of world's governments. Fossil fuels have accounted the bulk of energy requirement supply for themselves [5]. Energy sector poses the largest source of greenhouse gas emissions, and energy supply from fossil fuels has confronted the environmental with crises, emitting millions of greenhouse gases into the atmosphere [6]. A majority of the world's industries have designed their energy supply infrastructure with fossil fuel resources (coal, oil and natural gas). However, fossil fuels are exhaustible so that human beings must seek alternative resources to supply their required energy [5], while the energy intensity index[1] in Iran has risen from 100 to 300 from 1353 to 1373. The average energy intensity in the world is 0.4 and this figure is more than 0.6 in Iran [7]. Rising energy consumption, disadvantages and limitations of fossil fuels have increased the tendency of using renewable energy, especially solar energy [8], so that per capita energy production increased over 2000-2007 [9]. The increase is projected to rise more than 60% of consumption amount in 2000 by 2030. In providing renewable energy and given the potential of Iran, solar energy is environmentally friendly owing to not generating co_2; it also leads to maintain natural resources, to rehabilitate lands, to reduce power transmission lines, to increase regional autonomy in energy supply, to increase the speed of electricity transmission to rural areas, and to be cost-effective [10]. Research studies reveal that annually the number of sunny days and hours in Tehran is almost 313 days and 1742 hours, respectively. Its changing pattern in Iran is declining from the south-east to the north-west and is rising typically from the west to the east [11]. Using irradiance atlas and the application of mathematical irradiance models exhibit that the irradiance intensity in the clear air is potentially high in parts of Iran such as: Yazd, Kerman, Tabas, Birjand, Iranshahr, Chabahar, Shiraz, Bam, Bushehr and desert areas. But designing buildings usually does not allow the use of conventional solar systems with inflexible panels [11]. Flexible solar panels are capable of producing voltages over 50V. These circuits can supply the voltage of equipment requiring higher setup power in normal [12]. Mostly, the design of residential buildings, workshops, granary silos, biogas reservoirs and greenhouses in their optimal design own curves in order to provide energy and comfort conditions. Therefore, the usability of conventional crystal panels is not available. Conventional solar panels require metal structures to be installed, and usually are not compatible well with cylindrical and geometric shapes. To cope with these problems, one of the solutions is the design and fabrication of solar systems based on flexible panels, which, in addition to allowing the geometry of the desired structures and maintaining the previous capabilities, are economical and provide such structures with the energy requirements of electrical equipment.

2 LITERATURE REVIEW

Since 1973 oil shock, government of the developed countries has been spending much more costs on research and development of solar energy technologies. The

[1] . Energy intensity: Energy consumption is defined as ton equivalent of crude oil per 1,000 dollars of gross domestic production.

contribution of solar energy investment has been oriented relatively towards the advanced [13]. Research in the field of renewable energies was almost halted in the 1980s and 90s [14]. With the sharp fluctuations in oil and gas prices, as well as the severe risk of global warming due to the emission of greenhouse gasses induced by fossil fuels, renewable energy technologies such as solar energy were again become spotlighted. At present, more than 45% of the world's solar energy is generated in Japan, Europe is ranked second with 25% and the United States accounted for third with 19% [4]. More than 88% of the market demand in solar cells is dedicated to crystalline silicon [3]. Considering the world's energy basket, despite over 30% share of solar energy in the developed countries, its share in Iran is insignificant [15]

By the end of 2014, photo-voltaic capacity in the world reached 40,000 megawatts, countries such as America, Italy, Japan, Spain and the United States are the leading countries in the world's installed solar energy capacity. Rural electrification in Iran was started by using photovoltaics in 2006 first from Qazvin province and then in Gilan, Zanjan, Bushehr, Yazd and Kurdistan provinces. In the same vein, the photovoltaic electrification project for 634 rural households was planned in 2008, which the capacity of these power plants exceeds 386 kW [11].

Renewable power plants contribute almost insignificantly in the country's electricity supply. In the meantime, solar energy experiences the worst conditions. Among the three installed grid-connected renewable power plants (wind, biogas, and solar), solar power plants have the least capacity [2]. Nearly, all renewable electricity capacity exploited by the current 1- and 3-year period has been wind-type one [11].

In Hadjaissa et al. research conducted in the area of photovoltaic solar system in greenhouses, control and solar system monitoring algorithm based on fuzzy logic could pump daily $16m^3$ to a height of 8m with a power consumption of 55 to 250 watts [16]. Zamani Aghaei et al. used the Taguchi test analysis method in a solar thermal heater system with L_{16} (4^4) orthogonal array [11]. Murat worked with aerodynamic design on hybrid energy systems using flexible panels [14], also Gregg worked on the performance of amorphous silicon-based systems on a large scale and their cost-effectiveness. In a research on the economic justification of using photovoltaic systems relative to household fossil fuels, COMFAR software was utilized [17]. In a research, various analog converters and linking with different applications in MATLAB and LabVIEW were used to simulate and design a solar system [18]. Standard test analysis report for outdoor solar systems introduced the main performance-evaluating variables of the skin-driven photovoltaic system in electrical and thermal modes [19].

Considering the importance of deploying photovoltaic systems as the body of objects and figures, however, no studies have been conducted on photovoltaic systems deployment on cylindrical and spherical surfaces. In this research, we have tried to calculate and rate the actual cost of utilizing flexible solar systems on flat, hemispherical and cylindrical surfaces using techno-economic assessment methods and engineering economics, as well as the cost-effectiveness of using flexible solar systems is evaluated. To this end, NPV, IRR and PP criteria have been used.

3 MATERIALS AND METHODS

3.1 Technical analysis

In this research, a-si photovoltaic solar cells were used as a solar irradiance for electricity converter; flexible cells with proper thickness and strength could cover any structure and generate more power with a series connection. Imposing the test over flat, cylindrical and spherical surfaces, the acquisition of voltage and power data was carried out. For this purpose, testing surface was prepared by $1m^2$ and the design and fabrication of the cylindrical and spherical structures were carried out. Solar energy conversion system based on flexible panels according to Fig. 1 includes: geometric structure in cylindrical, hemispherical and flat surfaces, flexible solar module, battery bank, irradiance intensity sensor, temperature sensor, humidity sensor, anemometer, voltage meter, power amplifier, preparation circuits, battery fuses, analog-to-digital converter and computer.

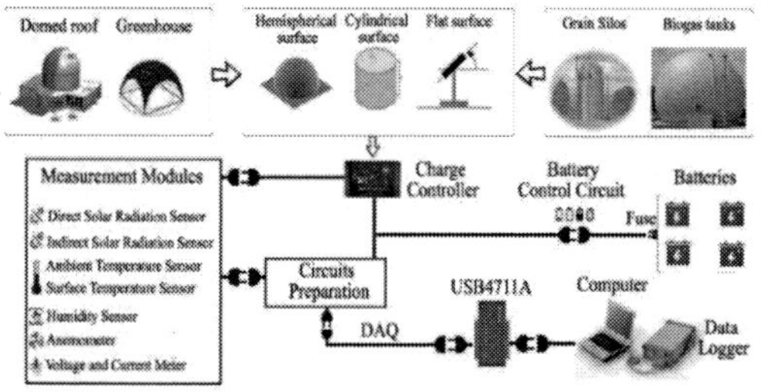

Figure 1: Flexible photovoltaic system and control equipment schematic

In current research, JNPsolar3W-12v flexible photovoltaic module was used with the specifications given in Table (1). JNsolar3W-12v film-thin solar silicon amorphous photovoltaic element with three connecting sites of deposition in a stainless polymer bed and a lock diode were employed to prevent the battery current discharge towards the solar panel in the module.

Table 1: JNsolar3W-12v flexible photovoltaic element Characteristics in standard conditions [14].

Characteristic	Unit	Value
Dimensions	mm	(290×210×5)
Power at maximum point	W	3.5
Open circuit voltage	V	14
Short circuit current	A	0.25
Voltage at maximum point	V	12
Performance	%	4-8

Relying on the dimensions of the rectangular-shaped panel, testing surface was chosen to be $1m^2$ [19]. The best fitting surface set by the selection of 1m solar panels will be $9744cm^2$, as well as the number of panels used will be 16. If the panel's marginal surface is deducted from the total surface, then the effective photovoltaic surface in the electrical design relations becomes $6996cm^2$ and the effective surface in geometric design relations of the structures will be $9744cm^2$. To connect the temperature module, RS232 serial port and LabVIEW software were used. This software benefits from a graphical programming language that contributes in working simplicity. Temperature was accurately regarded as the most important factor influencing the power generation of flexible photovoltaic systems. Microcontroller sends every 50 milliseconds the information of 16 temperature sensors via serial connection, and once received by LabVIEW, displays it and stores every 30 minutes in an Excel file in real-time. The maximum solar panel power is obtained from the product of voltage value (in open circuit mode) times current value (in short circuit mode) and fill factor coefficient. Fill factor is mathematically equivalent to the current-voltage coverage arc, and the more its value approaches one, the higher the quality of the solar system. Fill factor (FF) was calculated according to relation (1) on the surfaces [19].

$$FF = \frac{V_{MPP} \times I_{MPP}}{V_{oc} \times I_{sc}} \quad (1)$$

The efficiency of the photovoltaic system is one of the most important outdoor evaluation variables. Performance (or efficiency) or general performance (η) was calculated by the relationship (2) [19].

$$\eta = \frac{P/S_t}{I} \quad (2)$$

Where in the above relation, I is equal to the irradiance set on the array surface with unit W/m^2 and S_t is equal to the photovoltaic array surface with unit m^2 and P is the system power at the desired point with unit W. Using this indexation relation, calculating power dissipation in shadowing mode, temperature variations, dusts drop, light reflection and irradiance on the solar panel was undertaken. To draw the graphs related to the performance curve and voltage-current diagrams as well as the effect of variables on power, voltage and current, Excel 2013 version and Minitab 17.3 version were used; accordingly the variations of controllable and uncontrollable variables for each contact surface were assessed, separately. Ultimately, the optimal system was evaluated by Taguchi method. Taguchi orthogonal array design associated with the four-variable flexible solar system was eventually tested with 18 experiments. The orthogonal arrays of Taguchi experiments for the design $\{L_{18}\ (6^1 \times 3^3)\}$ evaluated the system performance over different surfaces.

The testing step is divided into three parts of flat, cylindrical and spherical test surfaces. Flexible panels are connected in the system in series and the system is separately designed from the local power network. The assessment step was carried out on the roof of a 7-story residential building in Tehran with 35.70 latitudes and 51.25 longitudes in three consecutive days for each surface. Local meteorological data near the system was measured and recorded by various devices. Irradiance data for each panel perpendicular to the surface was measured by the irradiance recorder. Meteorological data for the 9-day test was also obtained from Meherabad Meteorological Station (less than 5 km).

3.2 Economic Analysis

Evaluating economic performance of flexible solar systems was accomplished with an emphasis on engineering economics. Engineering economics, through project's economic analysis, addresses the study of different investment projects and choosing the most economical ones or making a decision to reject or accept the implementation of specific projects using mathematical techniques and quantitative evaluation criteria. In this research, we have used COMFAR software version 3. The engineering economics criteria used in the economic performance test are as follows.

3.2.1 Net present value (NPV)

This criterion, taking into account the monetarily time adjustment, creates an equilibrium between investment payments and incomes obtained from investment implementation. Assessing this equilibrium in relation to interest rate, it is a standard that design management has predetermined for investment and utilization of funds. This interest is referred to as "minimum attractive interest" or "capital cost ". The present value as a set of future cash fund flows is calculated by the relationship (3):

$$NPV = NCF_0 + \frac{NCF_1}{(1+i)} + \frac{NCF_2}{(1+i)^2} + \frac{NCF_t}{(1+i)^t} \quad (3)$$

In which, NPV is the net present value, NCF is equal to net cash funds, i is discount rate and t is equal to the financial period. NPV may be a negative number (not selecting any project) or a positive number (project selection). If NPV is equal to zero, then the designer will be indifferent to choosing whether to select a project, or not.

If the calculations show that by applying a specific discount rate as the investor's minimum interest rate, NPV criterion for a project is a positive value through inflation exertion, it indicates the project's justifiability; additionally the great gap between the project's output and input cash flows, as well as the magnitude of this indicator increase project's investment attractiveness.

3.2.2 Benefit-cost ratio

Benefit-cost ratio, actually, represents the ratio of the total present value in the revenues obtained from the project implementation to the total present value of its costs, which is another expression of the net present value of the projects.

3.2.3 Internal Rate of Return (IRR)

IRR is a well-known criterion in terms of projects' economic evaluation. This criterion specifies that greater IRR than cost of capital is the project's acceptance term. IRR stands for the discount rate based on which the NPV of the project is zero. If a project's NPV becomes positive, its IRR is greater than the return rate having been applied on investment. In NPV calculation, it is assumed that the discount rate is clear, while the NPV of the project ought to be determined. Besides in IRR calculation, project's NPV is set to zero and the discount rate, which is the same IRR of the project, is determined.

3.2.4 Payback Period (PP)

Using the criterion of payback period, the period in which the sum of annual revenues is broken even with the investment cost is specified. It is an approximate and simple way to deal with risks and the projects that have more revenue in the early years can be greatly benefited. In other words, the smaller the PP criterion, the more rapid the cash outflows are compensated by cash inflows; therefore, the project has more attractiveness for the investment. The payback period involves normal and dynamic payback periods. The concept of the normal payback period follows: the net cumulative cash flows of the project during the operation; coupled with the purpose of the dynamic payback period is that monetary time value is taken in payback period calculation into account and the calculations are performed based on discounted data. In general, the product produced by this project is electricity, which flexible solar systems generate electricity at nominal production capacity amount. As well as, project fabrication duration including the purchase and installation of equipment has been intended for one year. The operating or useful lifetime of the systems is 20 years.

The intended inflation rate in this section based on the average inflation rate in the last five years between 2011 and 2016 associating with goods category of water, electricity and fuel, on average, was 10% (Central Bank, 2017). The deposit interest rate (sight deposit) in Iran has recently decreased to 15%, but in the last 10 years it has averaged 16.7%, which is considered as the benchmark discount rate. According to the Iranian parliament's legislation, each kilowatt of renewable electricity is purchased from the consumer at 8,000 Rials, which this price is supposed to be 20 years of solar system's useful lifetime [7]. Therefore, performance-based flexible systems at their maximum power production conditions can have different revenues in deploying on different surfaces according to Table (2).

Table 2: List of solar power production costs in systems at different surfaces

Power Sales Revenue	Production	Total Revenue Per Year
Flat	81.369 kWh× Rial 8000	650952
cylinder	96.124 kWh× Rial 8000	768992
hemispherical	100.165kWh× Rial 8000	801320

3.2.5 Identifying costs and benefits of flexible solar systems

The first step in launching a solar energy conversion system based on flexible solar panels is to buy components of the system, which is actually a fixed investment cost. The cost (fixed price) of small-scale flexible photovoltaic panels is $ 0.8 per watt. Therefore, 56 watts is equal to 1568,000 Rials in 2016. The price growth of photovoltaic thin-film products rapidly declines. But in addition to the panel, other equipment are also used in the flexible photovoltaic systems. The total cost of a flexible photovoltaic system including purchase of panels, installation costs, etc. are estimated 5453350 Rials according to the standard in current photovoltaic projects.

Unforeseen costs included glue costs for reposing panels to surfaces, maintenance costs and soil cleaning. The total maintenance cost during a 20-year solar system useful lifetime was considered 10 million Rials.

4. RESULTS AND DISCUSSION

4.1 Technical analysis results

Flexible photovoltaic array consists of 16 flexible solar panels of JNPsolar3W-12V model serially connected on different surfaces. Fig. (2) Illustrates the analysis of flexible array power when deployed on different surfaces.

By understanding the amount of variance and its generating source, the best performance conditions in systems were simultaneously predicted, which this is one of the advantages of using Taguchi test method for the system. Table 3 represents the testing factors and surfaces. In this experiment, 4 variables (factor) were selected, the irradiance power variable in 6 levels, and the variables of temperature, wind speed, and system's average arc in three levels. The arc at level (1) is equivalent to the system test in flat mode, and level (2) relates to the system test on the hemisphere surface and level (3) associates with the deployment on the cylindrical surface.

The irradiance power variable are classified from sunrise to sunset from 0 to 1200 watts per m^2, which has been tested and analyzed in 6 levels. The temperature variable is involved in three levels of ambient temperature of 20 ° C to 50 ° C. The wind speed variable contributed in 3 levels from 0 to 3 m/s during the test.

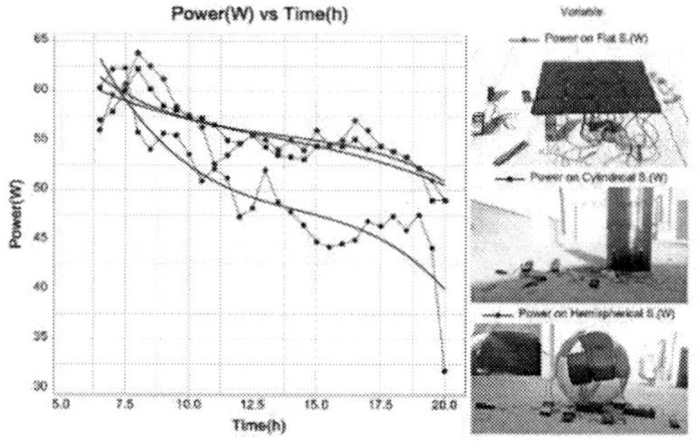

Fig. 2. Analysis of flexible array power when deploying on flat, cylindrical and spherical surfaces.

Table 3. Variables and test levels in Taguchi tests

Var. / Level	Irr. (W/m^2)	Temp. (°C)	w.s (m/s)	Flex. (°)
1	0-200	20-30	0-1	0-5
2	200-400	30-40	1-2	5-45
3	400-600	40-50	2-3	45-90
4	600-800	-	-	-
5	800-1000	-	-	-
6	1000-1200	-	-	-

If Taguchi test design analyzed based on the maximum power generation capability in Minitab, the output of the test analysis would be in accordance with Fig. 3.

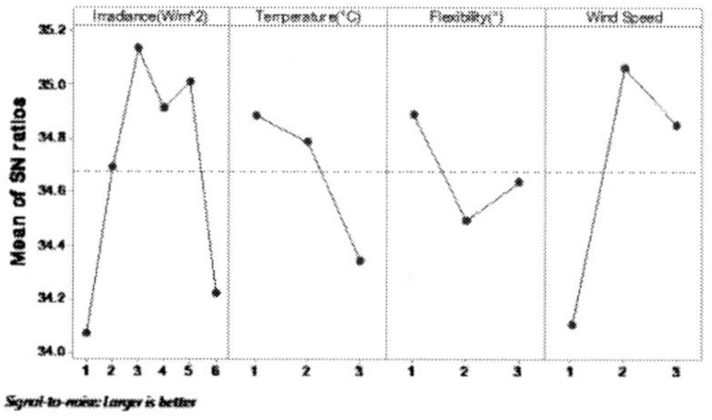

Fig. 3. Results of the average analysis of the Taguchi test impacts (signal-to-noise ratio) in the systems

In Fig. 3, all variables of irradiance, temperature, flexibility (angle) and wind speed are known. In the axis of the lengths, the graphs of the levels of variables are specified. The vertical axis represents the mean SNr. In this test, power has been evaluated based on "more, better". The vertical axis, in general, shows the larger the SNr, the closer the target, and no difference between choosing "more, better" and "or less, better" will exist. If the irradiance situates at level (1), more power is actually available. Irradiance will have a common property on systems up to the level (3), i.e., irradiance ranges 400-600

W/m^2; then, the effectiveness is a little diminished, but it is still optimal for the system and satisfies the objective. But in the irradiance power above the level (5), i.e., level (6) and above, the system is aimed in the opposite direction. With an overall glance to the irradiance power and SNr group, there can be seen that irradiance power could greatly influence the purpose of the potential of electric power generation by distancing away from the midline and high failure. The ambient temperature situated in the second group and temperature level (3) had the greatest impact on the system's purpose. Changes in wind speed from level (1) to level (2) were very effective at the system's aim and could alter the conditions in a way to generate power at high ambient temperatures. The flexibility or angle of the photovoltaic system showed that changing in this factor, namely, the use of the system in flat mode (level 1), or when deployed on a spherical surface (level 2) or on a cylindrical surface (level 3) can have an impact on generating maximum power objective. System's angle group and SNr revealed that: level (1) defined as using the system at the flat surface can have the greatest impact on the system. The application of the system on the level (2), meaning the application of the system on the spherical surface, despite its situation at the vicinity of the intermediate line of power production target, but it is more optimal than levels (1) and (3). Level (3) meant the application of the system over a cylindrical surface is more optimal than the flat surface, but the power production is maximum at level (2). Therefore, the use of the system in flat mode on the roof like the bitumen is not prioritized. As well as, if in a farm at the same time we have both grain storage silos and biogas reservoir, application of the system in these conditions will be preferable over a hemisphere surface, that's mean biogas reservoir.

Considering all the groups, the greatest influence on the system's purpose, that is, the maximum power of the solar system, is respectively related to the levels of irradiance, wind speed, temperature and application on different surfaces such as flat, cylindrical and spherical surfaces. Table 4 shows Taguchi signal to noise response in order to evaluate test variables and prioritize each one. Forecasting Taguchi's analysis in the present experiment showed that if the system were tested at selected levels based on the optimum production process and variables, it would be possible to expect a combined power equivalent to 49.9W with SNR = 33.929. If the test were calculated for the use of the system individually and on a completely flat surface, then the power of the system would be equivalent to 53.3W with SNr =34.4918. If the system were used to power prediction on the cylindrical surface, then the power of the system would be equal to 54.05W with SNR = 34.6368; for the hemispherical surface, the equivalent power of 55.7W with SNr= 34.8880 was computed.

Table 4: Mean signal-to-noise ratio in Taguchi experiments

Level	Irradiance	Temp.	Flexibility	Velocity
1	34.07	34.89	34.89	34.10
2	34.69	34.79	34.49	35.06
3	35.13	34.34	34.64	34.85
4	34.91	-	-	-
5	35.01	-	-	-
6	34.22	-	-	-
Division	1.07	0.54	0.40	0.96
Rating	1	3	4	2

4.2 Economic analysis results

The results of the economic analysis using COMFAR software imply that the net present value of the flexible system when deployed on the flat surface amounts to 6.747 million rials and its internal rate of return is 23.82% as well as the return period of capital is 9.24 years. Added to that, the net present values of the flexible system when deployed on the cylindrical and hemispherical surfaces are respectively equal to 9.952 rials and 10.830 rials, and their internal rates of return are 26.70% and 27.46%, as well as their return periods of capital are 8/24 and 8.02 years, respectively.

4.2.1 Net present value (NPV) analysis

The present value variable is one of the most important methods in engineering economics. Estimating the present value is a financial process, converting future value of all payments and receipts to the present value at the current time or origin of the project. If the net present value per the minimum attractive rate for a project is smaller than zero, then that project will be non-economic, indicating that the present value of the costs is lower than the present value of the revenues. And, if the net present value becomes positive, the present value of the costs is lower than the present value of the revenues and the project is economic. If the net present value is equal to zero, then the project is an economic one, since the minimum attractive rate has been provided to the investor. Fig. 4 shows the NPV of the designs for different discount rates at three fitting levels of flexible solar systems on flat, cylindrical, and hemispherical surfaces. Important highlights in the charts are:

1. With increasing discount rate, NPV for all three systems is reduced.
2. In the system deployed on a flat surface at the project's 23% discount rate, the present value of design's costs and revenues has been the same.
3. In the system deployed on a cylindrical surface at the project's discount rate of 26%, the current value of design's costs and revenues has been the same.
4. In the system deployed on a hemispherical surface at the project's discount rate of 27%, the current value of design's costs and revenues is equal.

By increasing the discount rate from 23% upwards at the flat surface, 26% upwards at the cylindrical surface and 27% upwards at the hemispherical surface, the present value of the costs gets higher than the one of revenues, in other words, the project becomes non-economic.

4.2.2 Internal Rate of Return (IRR) Analysis

When using this method, criterion for acceptance or rejection of a project is supposed a hypothetical one that balances revenues (annually) and costs (primary capital and annual costs). The rate at which the project's revenues and costs are balanced is called the "internal rate of return." The project's discounting indexes in the original state with a minimum attractive rate of 16.7% are described in Table (5).

Table 5: Discounting Indices in systems (Million Rials)

Power Plan	IRR	NPV
Flat Surface	23.82	6.747
Cylindrical Surface	26.70	9.952

Hemisphere Surface	27.46	10.830

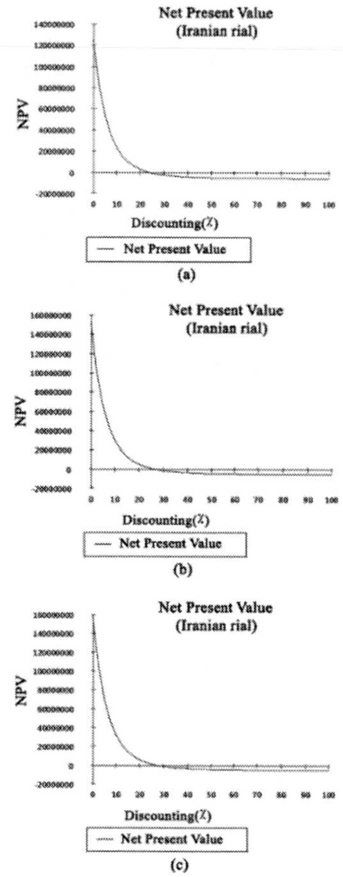

Figure 4: The results of net present value of the total capital in three systems a): deployed on a flat surface b): on a cylindrical surface c): on a hemispherical surface

4.2.3 Analysis of normal payback period graphs

Fig. (5) Displays cumulative net cash flows- normal payback period for the systems in three deployment modes. In fabrication duration and implementation of flexible solar systems, the net flows were negative and in the system when deployed on the flat, cylindrical and hemispherical surfaces, respectively from 9.24, 8.24 and 8.02 to the end of the operation period, cumulative net flows of designs were positive and steadily risen. Moreover, the normal PP of the capital in the system when deployed on the flat surface was 9 years, as well as it was 8 years in the case of deployment on the cylindrical and hemispherical surfaces (normal PP of the system on the hemispherical surface is equivalent to 8.07, so its year can be considered 7). That's being said, the paid initial capital will be re-acquired for the flat system in 1405, and for the cylindrical and hemispherical systems in 1404 (in solar calender), respectively.

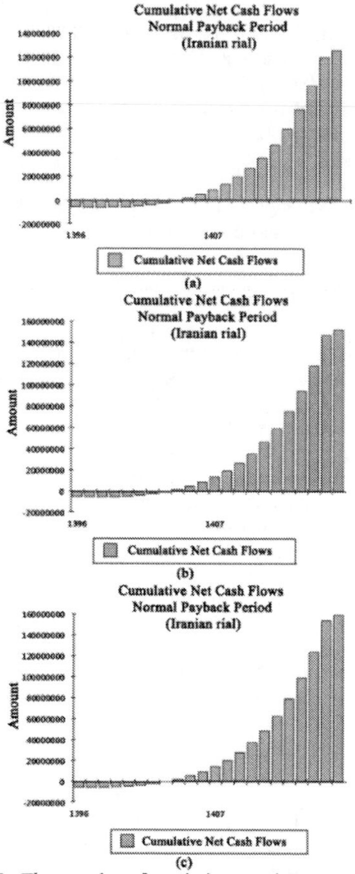

Figure 5: The results of capital normal PP- cumulative cash flow calculation in three systems a) deployment on the flat surface, b): deployment on the cylindrical surface c): deployment on the hemispherical surface

5 CONCLUSION

Given that the justification of designs is usually carried out in uncertainty, sensitivity analysis is an important criterion in the investment risk assessment. Indeed, in sensitivity analysis, with the repetition of financial calculations by changing the variables affecting the assessment results, the outcomes are compared with the initial results. If the changes made in the variables do not exclude the project from its justifiability, the investment will be done much confidently. In order to take into account factors not considered or the risk of agents endangering the return of the project, the sensitivity level of the project's indicators to these factors was estimated roughly. To this end, the sensitivity of project's indicators to the changes in revenues and costs is presented.

As such, in the flat-surface deployment system a 20% increase in project costs increased the internal rate by 21.30%; as well as a 20% drop in revenues will lead to a reduction in the project's internal rate of return by 22.99%. In addition, in a flexible solar system when deployed on cylindrical and hemispherical surfaces with an increase of 20% in project costs, the internal rate of return becomes 25.59% and 24.48%, respectively; coupled with, with a 20% drop in revenues, project's internal rate of return will reduce 25.79% and 26.53%, respectively. Therefore, the order of being economic in the systems will also be related to a flexible solar system when deployed on the hemispherical, cylindrical and flat surfaces. Summarizes the results of the techno-economic analysis test of the systems in table 6.

Table 6: Summary of systems' behavioral assessment for "more is better" power characteristic

	FF (%)	MP (W)	S.P (%)	OCV (V)	SCC (A)	IRR	NPV
F	73	46.7	7.1	185	0.34	23.8	6.7
C	88	55.2	7.4	179	0.35	26.7	9.9
H	84	57.5	7.4	178	0.36	27.4	10.8

The minimum fill factor (FF) relates to the flat surface (F) and the systems, when established on the cylindrical (C) and hemispherical (H) surfaces, have a fill factor of 0.88 and 0.84, respectively. The greater fill factor for systems allows the power controller to take more distribution of voltage and power. The maximum power (MP) is related to the system deployment on the hemispherical surface and is equivalent to 57.50W, while the minimum power of 46.71W is related to the system when deployed on the flat surface. The systems performance (S.P) under standard conditions (25°C and irradiance power of 1000 watts per m^2) for deploying on the cylinder and hemispherical surfaces has been computed 7.45%. The minimum performance was calculated 7.09% and related to the flat surface. The open-circuit voltage (OCV) and short circuit current (SCC) at the maximum state associated with the installation of a flexible system on the flat surface and were measured equal to 185.696 V and 0.3465amps, respectively. In addition, the maximum production current was achieved 0.36 amps at system deployment on the hemispherical surface. Therefore, the system in the case of deployment on the hemispherical surface by receiving diffusion irradiances in addition to vertical irradiances has optimal electrical and quality characteristics compared to other deployment surfaces. The second and third priorities are related to the cylindrical and flat surfaces, respectively. The economic analysis results of the systems exhibit that the internal rate of return in a flexible solar system when deployed on the flat, cylindrical and hemispherical surfaces are 23.82, 26.70 and 27.46%, respectively; subsequently the their net present value are respectively equal to 6.747 and 9.952 and 10.830 million Rials. Given that the net present value and appropriate rate of economic return are positive, the designs while having the economic justification for attracting the capital, are classified according to the application priority of flexible systems on the hemispherical, cylindrical and flat surfaces, respectively.

ACKNOWLEDGMENT

Thereby, the authors of the article heartily appreciate and acknowledge the professors of the Department of Architecture, Economics and Electrical Engineering of the Tarbiat Modares University who contributed to the research objectives advancement, as well as Iran Power Distribution Company and National Cartographic Center for providing the measurement tools.

REFERENCES

[1] ESTTP, Solar Heating and Cooling for a Sustainable Energy Future in Europe (Revised), European Solar Thermal Technology Platform (ESTTP), Belgium, (2017).

[2] IEA, Technology Roadmap (Solar Photovoltaic Energy), International Energy Agency, Paris, (2016).

[3] EPIA, Global Market Outlook for Solar Power 2017-2019, 9 June 2015. [Online]. Available: http://www.solarpowereurope.org.

[4] IEA, Snapshot of Global PV 1992-2014, 30 March 2017. [Online]. Available: http://www.iea-pvps.org/index.php?id=32.

[5] H. Kohjiro and A. Hironori, Handbook of Photovoltaic Science and Engineering, John Wiley & Sons, (2005).

[6] A. Ekrami, M. Sadeghi and B. Mahdavi, Economic evaluation of the development of solar power plants due to environmental considerations, Journal of Environmental Science and Technology, vol. No. 2, pp. 44-51, (2008).

[7] Saba, "Energy Consumption in Iran Iran Energy Efficiency Organization" 2017. [Online]. Available: www.saba.org.ir/saba_content/media/image/2009/03/133_orig.pdf

[8] Mahmoudi. M, Sadeghi. M, Farhadpour. F, Identification of two-phase current regimes and calculation of pressure drop in wells under gas jet, Oil Research Journal, No. 68. 70-57, (1390).

[9] M. Zohoori, Exploiting Renewable Energy Sources in Iran, Interdisciplinary Journal Of Contemporary Research In Business, vol. 4, pp. 849-862, (2012).

[10] Solangi, K. Islam, M. Saidur, R. Rahim, N. Fayaz, H. A Review on Global Solar Energy Policy, Journal of Renewable and Sustainable Energy Reviews, Vol. 15. PP. 2149-2163, (2011).

[11] Raufi Rad. M, Designing Solar Building Systems in Iran, First Edition, National Iranian Oil Company Publications, (2006).

[12] Miles, B. Jill, R. Lunt, J. Xu, A. Wang, C. Boyce, S. Vladimir, B. Karen, G Direct Monolithic Integration of Organic Photovoltaic Circuits on Unmodified Paper, Journal of Advanced Materials,vol.10 pp. 3500–3505, (2011).

[13] Harendra., S, Prashant. K, Tayal. S, Verma. S, the Third Generation Dye-Sensitized Solar Cell. International Journal of Advanced Technology in Engineering and Science (Ijates), Pp. 415-421, (2013).

[14] GMO, Global Market Outlook 2014, may 1 2016. [Online]. Available: www.pv-magazine.com/fileadmin/uploads/PDFs/Global_Market_Outlook_2016.pdf.

[15] Rabiee. M, Supplying 3% of the country's total energy with renewable resources, International Journal of Journalism Green Report Announcement and Announcement, No. 106, (1391).

[16] Hadjaissa. A, Ait. C, Ameur. K, Essounbouli. N, A Ga-Based Optimization of a Fuzzy-Based Mppt Controller For A Photovoltaic Pumping System (Case Study For Laghouat,Algeria), International Federation of Automatic Control (Ifac), Elsevier Ltd, Vol. 1, Pp. 692-697, (2016).

[17] Mahdi Adeli. M, Salimi Far. M, Qizilbash. A, Economic evaluation of solar energy (photovoltaic) and fossil fuels in domestic use (A Case Study of Three Units Complex in Mashhad city, Scientific Journal of Economic Policy, Year 6, No. 11-147-123, (1393).

[18] Koukouvaos. C, Kandris. D, Samarakou. M, Computer-Aided Modelling and Analysis of PV Systems: A Comparative Study, the Scientific World Journal,

Vol1.PP 17-29, (2014).
[19] Bloem. J, Lodi. C, Cipriano. J, Chemisa. D, an Outdoor Test Reference Environment for Double Skin Applications of Building Integrated Systems, Journal of Energy and Buildings, Elsevier, Vol. 50, Pp 63-73, (2012).

PERFORMANCE ASSESSMENT OF A NEW AIR-BASED BUILDING-INTEGRATED PHOTOVOLTAIC THERMAL SOLAR COLLECTOR

Véronique Delisle[1,*], Jun-Tae Kim[2], Jinhee Kim[2], Alexandre Gagné[1], Josef Ayoub[1]
[1] Natural Resources Canada, CanmetENERGY, Varennes, Québec, J3X 1S6, Canada
[2] Kongju National University, 1223-24 Cheonandae-Ro, Cheonan, Chungnam 31080, South Korea
*Corresponding author: tel: +1-450-652-7948, fax: +1-450-652-5177, veronique.delisle@canada.ca

ABSTRACT: Building-integrated photovoltaic (BIPV) technology is an active building envelope component that converts solar energy into electricity. When thermal energy is harvested, BIPV produces both thermal and electrical energy using the same building surface area. This article presents the performance of a new design of a BIPV module with thermal energy recovery that uses air as the heat recovery fluid (BIPVT/a). Two units of the newly developed prototype were tested indoors in a solar simulator under various operating conditions to obtain the steady-state thermal performance. At an irradiance level of 1005 W/m² and still air, a temperature rise of 54°C was measured at a flowrate of 55 kg/h. The BIPVT/a prototype was also mounted outdoors for a period of eight months to complete the electrical characterization and evaluate the thermal and electrical performance under transient conditions. With the measurements obtained, a model to predict the operating cells temperature from the ambient conditions and the fluid outlet temperature was developed. This model captures the interaction between the module thermal and electrical yield.
Keywords: Building Integrated PV (BIPV), Heat Recovery, Electrical Performance

1 INTRODUCTION

Building-integrated photovoltaic thermal (BIPVT) solar collectors produce both thermal and electrical energy simultaneously using the same surface area while acting as part of the building envelope. Compared to standard photovoltaic thermal (PVT) collectors, BIPVT technology has the added benefits of providing a more aesthetically pleasing architectural design and displacing the costs associated with traditional building envelope components. Thus, it has great potential to be integrated in buildings that have limited façade and roof areas with good solar exposure.

This article presents the performance of a new design of a BIPVT product that uses air as the heat recovery fluid (BIPVT/a). It also addresses one of the main challenges of BIPVT/a performance characterization by proposing a relationship between the collector thermal and electrical yield. The main objectives are:

- to characterize the thermal and electrical performance of a new collector design of a BIPVT/a product under both steady-state and transient conditions, and;
- to develop and validate a model to predict the photovoltaic cells temperature and capture the interaction between the collector thermal and electrical yield.

To fulfill these objectives, two units of the newly developed BIPVT/a prototype were tested in both indoor and outdoor conditions. The indoor testing was conducted in a large scale continuous solar simulator to characterize the collector thermal performance under steady-state conditions. The outdoor testing assessed the module daily thermal and electrical yield as well as completing its electrical characterization. A description of both experimental setups is provided in Section 2. The results on the collector thermal and electrical performance characterization are given in Sections 3 and 4, respectively. Section 5 provides a description of an empirical model to predict the cells operating temperature from the ambient and operating conditions and Section 6 presents the collector transient performance.

2 EXPERIMENTAL SETUP

2.1 BIPVT/a prototype

A schematic of the BIPVT/a prototype developed by Kongju National University is shown in Figure 1. It has an absorber consisting of both photovoltaic (PV) cells and thermal absorbers in alternate rows. The PV part of the absorber of each module consists of 30 monocrystalline silicon solar cells connected in series sandwiched between two layers of glass. The thermal absorber is made out of a thin metal sheet with highly selective blue coating. There are four thermal absorbers equally spaced in each module. They are located in the air cavity behind the glass-PV-glass assembly such that air can circulate above and below the thermal absorbers.

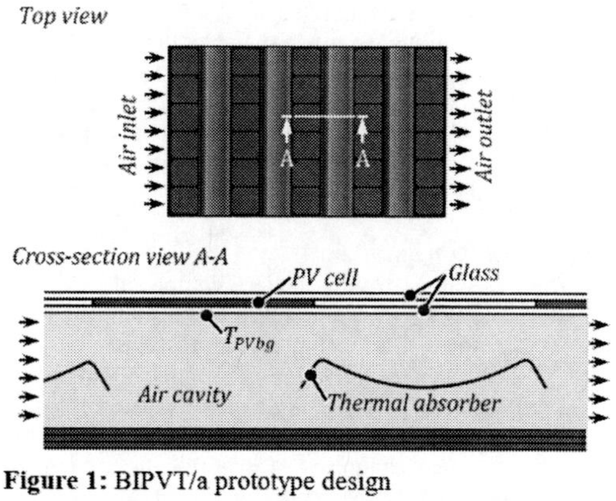

Figure 1: BIPVT/a prototype design

In both the indoor and outdoor experiments, two units of the prototype covering a total gross area of 3.27 m² were tested in series and in vertical position representing a building façade configuration. The back surface temperature of the glass-PV-glass assembly was measured with 10 surface thermocouples located in the middle of each PV row along the collector length. In addition, a distribution box was added at the second module outlet to promote flow uniformity across the cavity.

2.2 Indoor Experiment

The indoor testing was completed at the Concordia University Paul Fazio Solar Simulator and Environmental Chamber located in Montreal, Canada. The Solar Simulator consists of a set of 8 lamps with adjustable location and intensity that emulates a solar spectrum corresponding to an air mass of 1.5. The testing rig on which the units are mounted has a wind generator as well as a pyranometer and a hot-wire anemometer. These two instruments, mounted on a mobile scanning unit, measure the distribution and uniformity of irradiance level and wind speed across the specimen tested. As shown in Figure 2, the presence of an artificial sky located between the lamps and the testing rig minimizes the impact of long-wave irradiance.

Figure 2: Indoor experimental setup

The modules were connected to an air testing loop comprised of a variable-speed blower, a flowrate measuring station as well as pressure and temperature sensors. In this experiment, the air testing loop was located downstream of the collector with the blower operating in vacuum mode to circulate ambient air in the collector. The units were tested at an ambient temperature near 21°C and at three values of wind speeds (still air, 2.5 and 5 m/s), irradiance levels (818, 1005 and 1148 W/m²) and flowrates (55, 200 and 350 kg/h). The irradiance spatial non-uniformity was found to be at most 9.6%. This is less than the 15% non-uniformity required by the solar thermal collector testing standards ISO 9806 [1]. During testing, both modules were electrically connected in series to an adjustable load to maintain their operation at maximum power point to ensure that the thermal characterization is obtained under realistic electrical operating conditions.

2.3 Outdoor Experiment

The outdoor testing was completed at the CanmetENERGY PVT Testing Facility located in Varennes, Canada. This testing facility features:

- A thermal characterization system that circulates air in PVT or BIPVT/a modules at a temperature equal to or greater than outdoors to provide all the measurements required to quantify the amount of thermal energy recovered;
- An electrical characterization system that can maintain the PV module operation at its maximum power point while tracing current-voltage curves at regular time intervals;
- A testing rig with double-axis sun-tracking capabilities; and,
- A weather station including the following measurements: local wind speed, ambient temperature and relative humidity, ambient pressure, in-plane short-wave irradiance, in-plane long-wave irradiance and direct normal irradiance.

The two units were mounted on the testing rig shown in Figure 3 from January to September 2017. During this period, the modules were operated under both stagnation and heat recovery conditions at a slope of 90° either facing due south or in azimuth tracking mode. Under heat recovery conditions, the flowrate was fixed at 55 kg/h with air entering the collector at ambient conditions and the blower operating in vacuum mode. At all times, the system was maintained at maximum power point with current-voltage curves traced at every minute.

Figure 3: Outdoor experimental setup

3 THERMAL CHARACTERIZATION

3.1 Performance indicators

The thermal performance of BIPVT collectors can be characterized by the air temperature rise across the units, T_{rise} given by:

$$T_{rise} = T_o - T_i \qquad (1)$$

In Eq. (1), T_i and T_o are the air inlet and outlet temperatures, respectively. In both the indoor and outdoor experiments, the units were operated in open-loop such that T_i corresponds to the ambient air temperature, T_a. Also, the air flowrate was only measured at the outlet. Thus, it is assumed that the amount of air infiltrating the collector is negligible. As a result, the thermal energy recovered, Q_{th}, and the thermal efficiency, η_{th}, are calculated as follows:

$$Q_{th} = \int_{t_0}^{t_1} \dot{m}(h_o - h_i)\, dt \qquad (2)$$

$$\eta_{th} = \frac{Q_{th}}{A_g \int_{t_0}^{t_1} G\, dt} \qquad (3)$$

In Eq. (2) and (3), $\dot{m}$ is the air flowrate measured at the collector outlet, t is the time, G is the in-plane solar irradiance, A_g corresponds to the gross surface area of both modules (i.e., 3.27 m²) and h_i and h_o are the air specific enthalpy at the inlet and outlet, respectively.

3.2 Indoor steady-state thermal performance

The steady-state thermal performance characterization of the modules was completed with the indoor experiment. According to the standard ISO 9806 [1] on solar thermal collector testing methods, the determination of steady-state measurement periods should be done by comparing the deviation of 30-second averages of specific variables to their average value during that measurement period. In this study, measurement periods of 5 minutes were used. The maximum deviation allowed was set to that specified by ISO 9806 or stricter. In addition, each one of the glass-PV-glass assembly back surface temperature measurements, T_{PVbg}, were considered in the steady-state period identification. The temporal stability of irradiance and wind speed could not be evaluated since these measurements were only recorded at the beginning of the measurement period. Table I provides a summary of the variables considered and the maximum deviation used to identify steady-state measurement periods.

Table I: Criteria for steady-state measurement periods identification

Variable	Maximum deviation during measurement period	
	ISO 9806	Experiment
$\dot{m}$	2%	2%
T_{bg}	-	0.3°C
T_o	1.5°C	0.1°C
T_a	1.5°C	0.7°C
G	50 W/m²	Not applicable
Wind speed	0.5 m/s	Not applicable

The thermal efficiency as a function of air flowrate is shown in Figure 4 for an irradiance level of 1005 W/m². As it can be observed, the thermal efficiency increases with flowrate following a quadratic curve, but decreases with wind speed. This thermal performance dependency to wind speed indicates that the prototype behaves like an unglazed collector. This is further demonstrated in Figure 5 showing that at a flowrate of 55 kg/h, the temperature rise decreases from 54°C to 38°C when the wind speed increases from 0 m/s to 5 m/s.

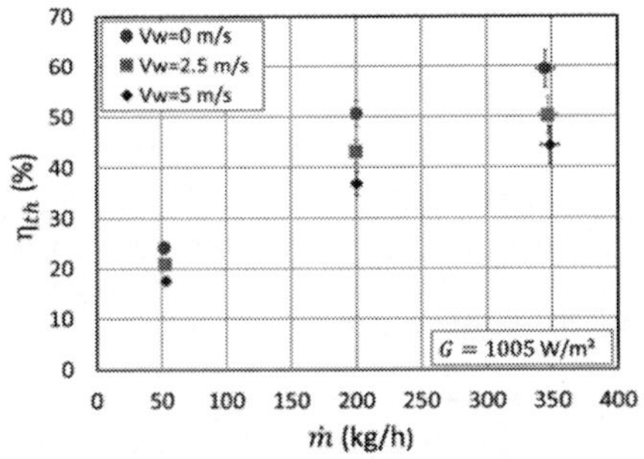

Figure 4: Thermal efficiency dependency to flowrate

Figure 6 presents the air temperature rise across the two units as a function of flowrate at a wind speed of 2.5 m/s. This graph shows that the air temperature rise increases with irradiance especially at the lowest flowrate of 55 kg/h where a difference of 12°C can be observed between the lowest and highest irradiance levels. The air temperature rise, however, decreases significantly with flowrate. At the medium irradiance level of 1005 W/m², the temperature rise is reduced by nearly 30°C when the flowrate changes from 55 kg/h to 350 kg/h.

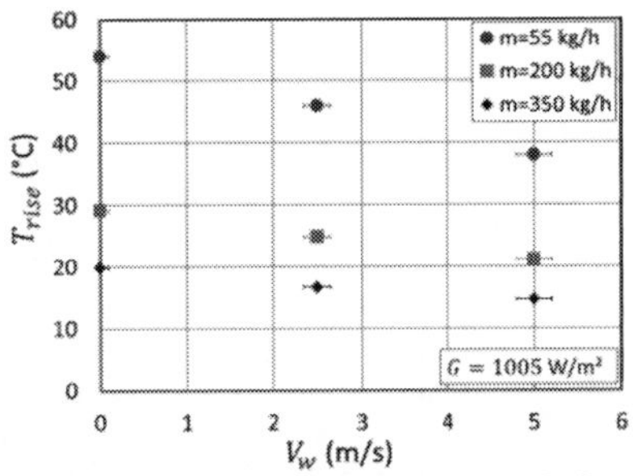

Figure 5: Temperature rise dependency to wind speed

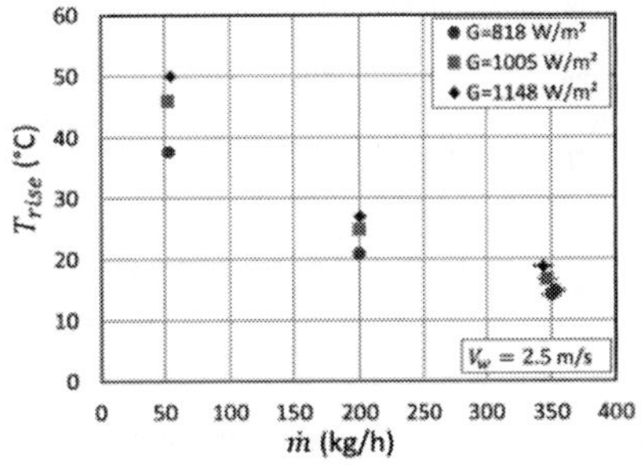

Figure 6: Temperature rise dependency to flowrate

3.3 Comparison between outdoor and indoor performance

Figure 7 presents a comparison between the thermal efficiency obtained indoors and outdoors as a function of wind speed. The outdoor data points correspond to measurement periods of 5 hours or more where the average irradiance level was between 850 and 985 W/m². The indoor data points shown on the graph are those obtained at irradiance levels of 818 W/m² and 1005 W/m². As it can be observed, both experiments provide comparable results in addition to showing a similar relation between thermal efficiency and wind speed.

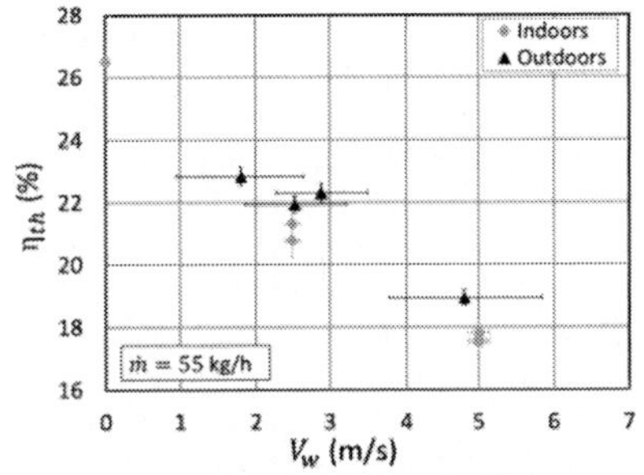

Figure 7: Indoor and outdoor thermal efficiency

4 ELECTRICAL CHARACTERIZATION

In PV and BIPVT modules, the electrical performance depends mainly on incident solar radiation, cell temperature and to a secondary order, on air mass and incidence angle. In a BIPVT module, however, the cells experience a much larger temperature gradient than in standard rack-mounted PV modules because of the presence of an air cavity. For example, at an irradiance level of 950 W/m² and ambient temperature of -10°C, maximum differences at the back surface of the glass-PV-glass assembly of 23°C and 14°C were measured along the collector length under heat recovery and stagnation conditions, respectively.

In order to get a better estimate of the module operating cell temperature, the equivalent cell temperature (ECT) is calculated. The ECT corresponds to the average temperature of the electronic junctions of the module and the temperature at which the module would operate if it were at a uniform temperature. This section presents the key parameters required to predict the ECT and the modules power production. The electrical analysis was done using outdoor data because the irradiance non-uniformity in the indoor test did not fulfill the 5% requirements of the IEC 61215 [2] standard on c-Si PV module design and qualification.

4.1 Equivalent cell temperature

In the standard IEC 60904-5 [3], the following relation is provided to calculate the ECT:

$$\mathrm{ECT} = T_{PV,ref} + \frac{1}{M_s\beta_{Voc}}\left[V_{oc} - V_{oc,ref} + DM_sN_s\ln\left(\frac{G_{ref}}{G}\right)\right] \quad (4)$$

In Eq. (4), V_{oc} is the open-circuit voltage, M_s is the number of modules in series, N_s is the number of cells in series per module, β_{Voc} is the open-circuit voltage temperature coefficient, D is the thermal diode voltage and T_{PV} is the cell temperature. The use of Eq. (4) involves finding the parameters β_{Voc} and D. It also requires that the open-circuit voltage ($V_{oc,ref}$) is known at a reference operating cell temperature ($T_{PV,ref}$) and irradiance level (G_{ref}).

4.2 Open-circuit voltage temperature coefficient

The open-circuit voltage temperature coefficient is obtained with the slope of the open-circuit voltage curve as a function of operating cell temperature. Since β_{Voc} is required to calculate the ECT, at this stage of the analysis T_{PV} has to be substituted by $T_{PVbg,avg}$ corresponding to the average of the ten T_{PVbg} measurements. To ensure that this temperature is close to T_{PV}, only data with all ten T_{PVbg} measurements within 7°C were kept for the analysis. This criteria was the most strict that would still provide a sufficient number of data for the analysis.

The open-circuit voltage as a function of $T_{PVbg,avg} - T_{PV,ref}$ is plotted in Figure 8 for fixed irradiance levels (± 20 W/m²) with $T_{PV,ref}$ taken as 25°C. The slope of these curves is on average -0.125 V/°C. Since this value is for two modules in series, β_{Voc} for a single module corresponds to -0.0625 V/°C.

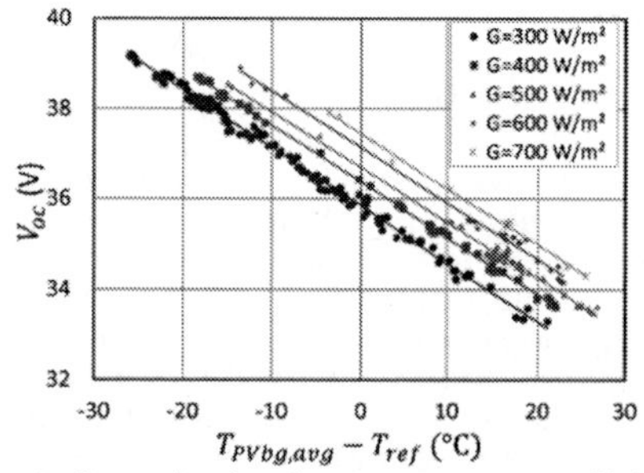

Figure 8: Open-circuit voltage temperature coefficient determination

4.3 Thermal diode voltage factor and open-circuit voltage at reference conditions

By substituting ECT with $T_{PVbg,avg}$ in Eq. 4, the following relation is obtained:

$$V_{oc} - \left(T_{PVbg,avg} - T_{PV,ref}\right)M_s\beta_{Voc} = V_{oc,ref} - DM_sN_s\ln\left(\frac{G_{ref}}{G}\right) \quad (5)$$

Figure 9 presents the left side of Eq. (5) as a function of $DM_sN_s\ln(G_{ref}/G)$ with G_{ref}=1000 W/m². According to this figure, D and $V_{oc,ref}$ correspond to 0.0286 V and 38.0 V, respectively.

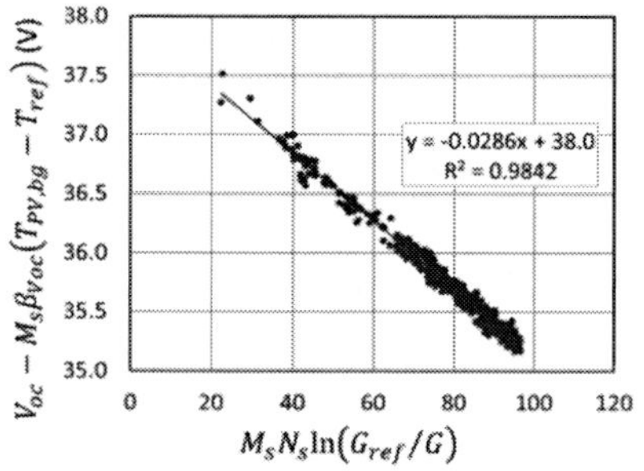

Figure 9: Thermal diode voltage factor and reference open-circuit voltage determination

4.4 Maximum power point temperature coefficient

The power temperature coefficient model is a simplified model able to estimate the production of PV modules at maximum power point, P_{mp}. It considers that P_{mp} is proportional to the level of irradiance and that the effect of cell temperature can be taken into account by applying a power temperature coefficient, $\gamma_{Pmp,rel}$:

$$P_{mp} = \frac{G}{G_{ref}}P_{mp,ref}\left[1 + \gamma_{Pmp,rel}\left(T_{PV} - T_{ref}\right)\right] \quad (6)$$

The maximum power dependency to cell temperature is shown in Figure 10 using ECT as T_{PV}. From this graph, it can be concluded that $P_{mp,ref}$ for both modules corresponds to 238.5 W and that $\gamma_{Pmp,rel}$=-0.0046 °C⁻¹.

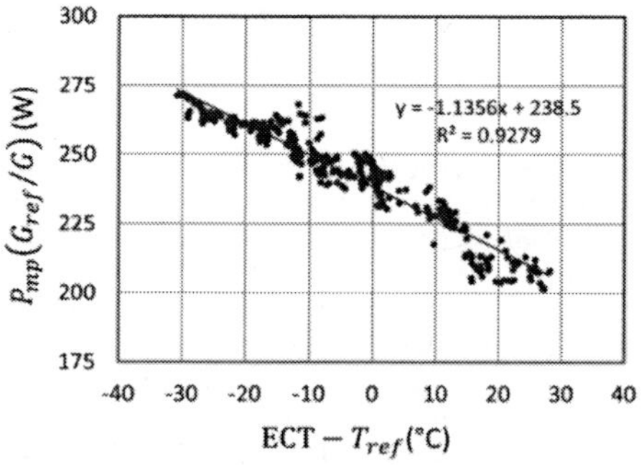

Figure 10: Reference maximum power point and power temperature coefficient

5 CELL OPERATING TEMPERATURE MODEL

5.1 Models

A common relation to estimate the cell operating temperature in a PV module is given by King et al. [4] as:

$$T_{PV} = T_a + G(e^{a+bV_w}) \quad (7)$$

In Eq. (7), V_w is the free stream wind speed, T_a is the ambient temperature and a and b are empirically-determined coefficients. In a BIPVT module under heat recovery conditions, this model cannot be applied because the flowrate, the inlet air temperature and ultimately, the outlet air temperature also affect the cell temperature. Since the fluid outlet temperature can be obtained from thermal performance curves, the following model is proposed to predict T_{PV} under open-loop heat recovery conditions:

$$T_{PV} = T_o + w(T_o - T_a) + kG(e^{a+bV_w}) \quad (8)$$

In Eq. (8), a, b and w are all empirically-determined coefficients. k is a constant corresponding to 1 (°C·m²)/W added to ensure the equation is dimensionally correct. Under stagnation conditions, the fluid outlet temperature is unknown. As a result, Eq. (8) cannot be used to predict T_{PV} and the model in Eq. (7) is used.

5.2 Parameters determination

The parameters in Eq. (7) and (8) were determined using 5 minute averages of the data collected during the outdoor experiment and using ECT as T_{PV}. Valid measurement periods were selected by comparing the deviation of specific variables to their average value during that measurement period. These variables and maximum allowable deviations were as follows:

- Flowrate within 2 kg/h
- Ambient temperature within 1°C
- Fluid outlet temperature within 1°C
- Irradiance stable within 25 W/m²
- Local wind speed within 1.5 m/s

The criteria for maximum allowable deviations were selected to be as strict as possible while still leaving a sufficient number of data for analysis.

The parameters and model root mean square error (RMSE), mean bias error (MBE) and coefficient of determination (R^2) are summarized in Table II.

Table II: Parameters and model performance at predicting ECT

	Heat Recovery	Stagnation
a	-4.34	-2.80
b (s/m)	-0.0473	-0.0322
w (°C⁻¹)	-0.321	-
RMSE	0.9°C	1.9°C
MBE	-0.1°C	-0.6°C
R^2	0.997	0.936

5.3 Link between thermal and electrical yield

The heat recovery temperature prediction model obtained provides a link between the module thermal and electrical yield. For example, at G=1005 W/m², T_a=21°C, $\dot{m}$ =55 kg/h and V_w= 0 m/s, Figure 5 indicates that a temperature rise of 54°C can be obtained corresponding to an outlet air temperature of 75°C. Using Eq. (8), the cell operating temperature is estimated at 70.8°C. At this temperature, Eq. (6) estimates an electrical power output of 189 W. Under the same operating conditions, but without any active heat recovery (i.e., under stagnation conditions), Eq. (7) predicts a cell operating temperature of 82.1°C corresponding to a power output of 177 W.

6 TRANSIENT PERFORMANCE

The module electricity production, Q_{el}, and electrical efficiency, η_{el}, over a period of time between t_0 and t_1 are calculated using the following relations:

$$Q_{el} = \int_{t_0}^{t_1} P_{mp}\, dt \quad (9)$$

$$\eta_{el} = \frac{P_{mp}}{A_g \int_{t_0}^{t_1} G dt} \quad (10)$$

Figure 11 and Figure 12 present the instantaneous thermal and electrical power produced by the two units under sunny and mildly cloudy conditions, respectively. For both measurement periods, the modules were under azimuth tracking mode and the flowrate was maintained at 55 kg/h ± 5%. The detailed conditions and thermal and energy produced during these periods are presented in Table III. As it can be observed, more thermal and electrical energy is produced during the sunny period. The thermal efficiency is lower, however, due to a combination of lower ambient temperature and higher wind speed.

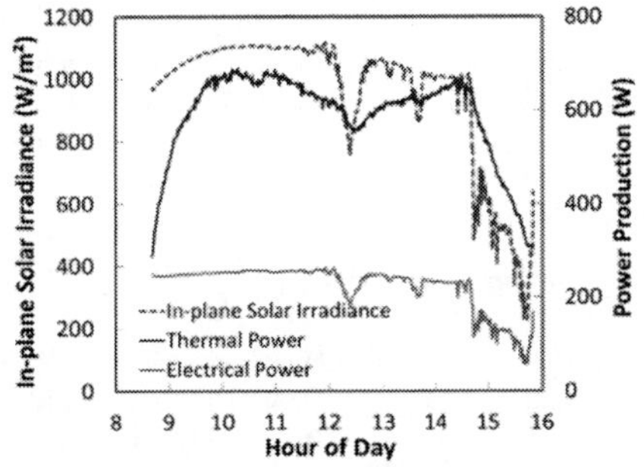

Figure 11: Thermal and electrical power production during a sunny day

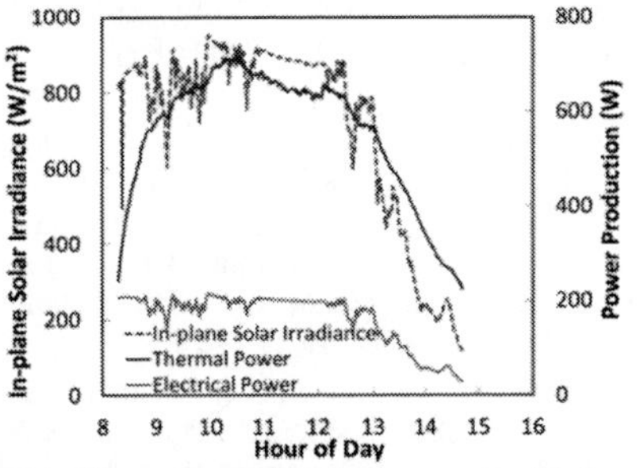

Figure 12: Thermal and electrical power production during a mildly cloudy day

Table III: Performance during measurement period

Day type	Sunny	Mildly cloudy
Duration (h)	7.3	6.4
Ambient temperature (°C)	-4.0	3.2
Total in-plane irradiance (MJ/m²)	24.8	16.4
Wind speed (m/s)	4.8	2.1
Q_{th} (MJ)	15.4	12.8
Q_{el} (MJ)	5.8	3.8
η_{th} (%)	18.9	23.9
η_{el} (%)	7.2	7.0

7 CONCLUSION

This article presented the steady-state and transient thermal and electrical performance of a new prototype of a BIPVT/a collector. The design of this product is unique because its absorber consists of a combination of both photovoltaic cells and thermal absorbers in alternate rows. Experimental measurements were taken in both outdoor and indoor conditions for two units in series operated in open-loop. Steady-state measurements performed indoors under a solar simulator and controlled environment showed that at a flowrate of 55 kg/h and an irradiance level of 1005 W/m², a temperature rise of 38°C and 54°C can be obtained at a 5 m/s and at sill air, respectively.

With the solar cells experiencing a large temperature gradient in the BIPVT module, the equivalent cell temperature calculated from the open-circuit voltage measurement was used to get a better estimate of the cell temperature. A model using data collected during the outdoor experiment was developed to estimate the equivalent cell temperature using the wind speed, ambient temperature, irradiance level and fluid outlet temperature as inputs. This model creates a link between the module thermal and electrical yield and can be used to estimate the module electrical production from its thermal performance.

8 ACKNOWLEDGEMENT

Financial support for this project was provided by an International Cooperation Grant (No. 20148520011270) from Korea Institute of Energy Technology Evaluation and Planning (KETEP) through Kongju National University for the international joint research project between Korea and Canada entitled "Development of air-type BIPVT system with modular methods", under Natural Resources Canada (NRCan) Contract Agreement #2015-046-CM-RND-411-PVTZEH, April 25th, 2015.

9 REFERENCES

[1] International Organization for Standardization, "ISO 9806 - Solar Energy - Solar thermal collectors - Test methods," International Organization for Standardization, Geneva, 2013.

[2] International Electrotechnical Commission, "IEC 61215 - Crystalline silicium terrestrial photovoltaic (PV) Modules - Design qualification and type approval," International Electrotechnical Commission, Geneva, 2005.

[3] International Electrotechnical Commission, "IEC 60904-5 Part 5, Determination of the equivalent cell temperature (ECT) for photovoltaic (PV) devices by the open-circuit voltage method," International Electrotechnical Commission, 2011.

[4] D. King, W. Boyson et J. Kratochvill, «Photovoltaic Array Performance Model,» Sandia National Laboratories, Albuquerque, 2004.

THERMAL PROPERTIES OF PHOTOVOLTAIC MODULES: THE DOUBLE FUNCTION OF BIPV SYSTEMS

Carlos Toledo [a], Rodolfo López-Vicente [a], Jose Abad [b] and Antonio Urbina [a]
[a] Department of Electronics, Technical University of Cartagena, Spain
[b] Department of Applied Physics, Technical University of Cartagena, Spain
Cartagena, 30202 Murcia, Spain

ABSTRACT: Building Integrated Photovoltaics (BIPV) play a crucial role to achieve the target of Net Zero Energy Building (NZEB). However, when a solar panel is fully integrated into the building envelope, its electrical and thermal behavior change compared to open-rack systems. BIPV projects tested and reproducing under real working conditions are essential to validate models and increase the penetration of BIPV systems. This article present the results of four simple squared structures with the same volume and with five faces (base excluded) made by PV modules from four different PV technologies: c-Si, two thin film (a-Si and CdTe) and an emergent organic photovoltaic (OPV). These structures, named PV-Cubes, are monitored with automated data acquisition of several thermal parameters. The results indicate that the highest measured temperature corresponds to the a-Si cube, however under diffuse conditions the a-Si panel has better performance from the thermal point of view. Furthermore, it is highlighted that the temperature differences of inner and outer faces of each panel (and for all face orientations) show a hysteresis when represented against incident irradiation, demonstrating a thermal inertia along a day, which has to be considered both for electrical power delivery and for thermal isolation in the building. The results of this research provide high quality experimental data and contributes to better understanding of BIPV systems from a dual-function perspective.
Keywords: Module integration; Monitoring; Thermal performance.

1 INTRODUCTION

Building Integrated Photovoltaics (BIPV) has evolved from a simple concept in which electricity generation was the only function of the photovoltaic system towards a more complex approach in which the photovoltaic module is part of the building structure, it is integrated into the building's design from an architectural point of view and therefore other functionalities have to be considered [1]. In particular, the thermal behavior has to be considered on top of the usual electrical power delivery by the photovoltaic modules. At least this double function has to be simultaneously considered when the photovoltaic modules are part of the building structure, usually acting as a skin of the building. Additionally, the thermal behavior also influences the photovoltaic function (via the thermal parameters of voltage and current photogeneration). This new vision of photovoltaic systems integrated in buildings is in line with European Union Directives [2], Horizon 2020 Program [3] and International Energy Agency programmes [4], all of which consider the use of PV in buildings essential to achieve the goal of Net Zero Energy Buildings (NZEB) contributing in this way to climate change mitigation. Recent research in BIPV systems highlights its environmental [5], [6] and aesthetics [7], [8] benefits. However, the relationship between PV and architecture continues to present a challenge.

BIPV systems, in order to achieve a multifunctional role, have to be considered not only focusing on energy generation, but also considering the module properties as a building structure; in particular, considering also the thermal properties of the photovoltaic module. When a solar panel is fully integrated into the building envelope, its electrical and thermal behavior change. There are currently only a few studies that simultaneously considers energy performance and thermal behavior. C. Young et al. developed a semi-transparent PV panel (Heat Insulation Solar Glass – HIGS) and investigated the potential thermal insulation based on the use of that panel [9]. Chatzipanagi et al. studied the impact of the operating cell temperature on standard modules overlapped on a ventilated façade and customized BIPV modules (double glass layer filled with Argon gas) into an low energy building prototype [10]. There are other studies deal the influence of the operating temperatures on the performance of PV modules but it was oriented towards the effect of airflow around BIPV systems [11], [12].

Therefore, there is a lack of specific studies about the performance of PV modules fully integrated in buildings under real test conditions considering simultaneously the study of thermal and electrical parameters with different orientations.

In this article, an experimental system has been designed and built in order to study the thermal properties of building integrated photovoltaic modules of four different PV technologies with different orientations under real test conditions as a first step to create models that improves the estimation of generated electricity in integrated systems and at the same time providing information about thermal isolation potential in each case.

2 EXPERIMENTAL SETUP

2.1 System description

Four structures, called "PV-Cubes" have been designed and built. They are the simplest model of a building in which all façades and roof are made of photovoltaic modules. The structure is simple: an aluminium frame and the photovoltaic modules; all the PV-Cubes have the same size (Fig. 1), in order to adapt the slight mismatch between top and lateral modules a small stripe of polycarbonate has been added to the cubes. Since the walls of the cubes are the modules themselves, they are considered as the "skin" of the simplified building represented by the cube. In this way, any model in which the thermal properties of the materials are considered will be mainly describing the thermal properties of the modules, with a small influence of the additional materials (which are exactly the same for all the modules: aluminium and polycarbonate).

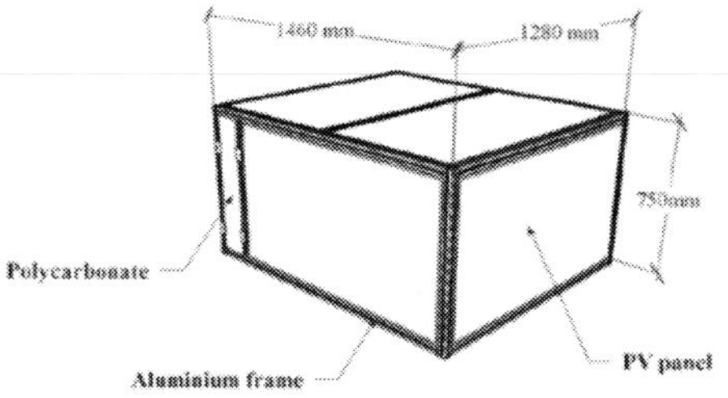

Figure 1: Structure design measurements

The cubes have been installed on the roof of one of the buildings at Murcia University (Spain). The latitude and longitude of the system location are 38° 1' 25'' N and 1° 10'33'' W respectively (Fig. 2).

Figure 2: Structures installed on the roof of a Murcia University building: four PV-Cubes of same size and different photovoltaic technology.

2.2 PV modules

Four technologies are considered: crystalline silicon (c-Si), cadmium telluride (CdTe) and amorphous silicon (a-Si) (from commercial manufacturers) and an organic photovoltaic module (OPV) (with P3HT:PCBM active layer) from Infinity PV, a start-up company from Technical University of Denmark. The modules installed and their sizes and Power Conversion Efficiency (PCE) are listed in table I.

Table I: PV modules used in the PV-Cubes (units in millimeters and kilograms).

PV tech.	c-Si	a-Si	CdTe	OPV
Model	SHS100	EPS80S24	FS397PLUS	Infinity
PCE	13%	11%	14%	3%
Length	1200	1205	1200	1200
Width	670	605	600	600
Thickness	35	35	6.8	12
Weight	8.2	12	12	1.2

2.3 Monitoring system

The temperature of each face is measured with a DS18B20 sensor attached to the backside and frontside of each PV module, the faces are oriented South, East, West, North and Horizontal (topside), additionally in the centre of each PV-Cube a temperature DS18B20 and humidity DHT11 sensors are installed (Fig. 3). At the same time, a weather station (Froggit, WH3080) and pyranometer BF5 Sunchine were installed to monitor *in situ* global and diffuse irradiance and other meteorological parameter such as ambient temperature and wind speed with a sampling data rate which measures all parameters every 5 minutes. Data logging is carried out by an Arduino card and sent by radio frequency communication with a XBee transmitter both installed within each cube (Fig.4) and, together with the meteorological station are collected by a receiver installed in a PC inside the building which stores all data.

Figure 3: Inside of OPV PV-Cube, showing the position of the electronic box and the temperature sensors at the back side of all photovoltaic modules.

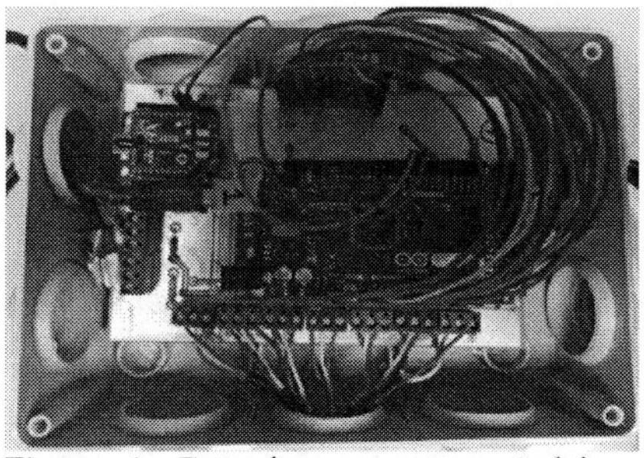

Figure 4: Data logger system. Arduino card and communication with XBee transmitters. All the electronic system is intalled in a sealed box inside the PV-Cubes.

3 RESULTS

The thermal evolution of the PV-Cubes has been continuously monitored, during day and night, for several months. Ambient parameters (outside), temperature within the cubes and temperature of the modules (inside and outside faces) are collected.

The highest temperature values, both ambient and panel temperatures, were recorded for a sunny day with wind speed less 5 m/s on August 3, 2017. Comparison of inside air temperature of each PV-Cube and outdoor ambient temperature for both sunny and cloudy day (August 29, 2017) during hot summer show overheating indoor environment of the structure (without airflow) which in the case of sunny day are up 20 °C (Figs. 5 and 6). A strong thermal effect on the air volume inside the PV-Cube demonstrates the importance of the thermal function of the PV panel when considered as a structural part of the building (the "cube" in this case). On the other hand, the heat input into the volume depends strongly on the orientation of each face (and the day time, i.e. solar orientation), creating an imbalance between the temperature of the air inside the PV-Cube and the temperature of the different faces. Furthermore, the temperature of the modules are different when measured on the outer or the inner faces, thus creating a gradient of temperature throughout the module. Globally, the thermal behavior of the system is complex and will have to be modelled considering several thermal coefficients (air, PV modules, auxiliary materials such as aluminum and polycarbonate). The thermal gradient throughout the polycarbonate and the aluminum are not considered, but they will be only small perturbations for the thermal behavior of the whole PV-Cube which is strongly dominated by the photovoltaic modules. During day time the temperature inside the PV-Cubes are higher than ambient temperature (up to more than 20 °C difference in an extreme sunny day, even for Murcia Region, with ambient temperature above 42 °C), but surprisingly, there is a thermal inversion since during night time, the air inside PV-Cubes is colder than ambient temperature.

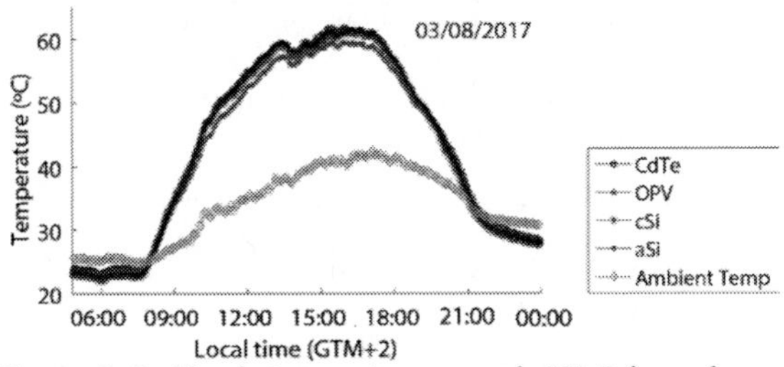

Figure 5: Inside air temperature on each PV-Cube and ambient temperature for a sunny day.

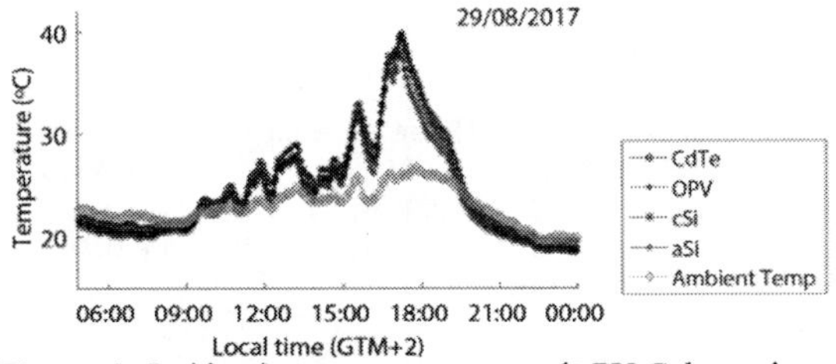

Figure 6: Inside air temperature on each PV-Cube and ambient temperature for a cloudy day.

The different orientations of the modules provide different thermal behavior for each PV technology considered along the day. At a given time, the irradiation on each face is very different (both global amount and direct/diffuse ratio), and is constantly changing as the day progresses. The higher recorded value is 77.12 °C for the horizontal inner surface in the a-Si structure. The results indicate that the horizontal surface has high impact on the total incident radiation in this time of the year since July and August are in general the months with the highest solar irradiance in Spain. Also, the high temperature reached by the PV module will have a strong negative influence on the electricity delivery by the module, since all faces will have a different thermal loss (and of course different incident irradiation), the power generated by each module should be managed at module level. In this case, each face has a single module, if considering larger structures, this power management should be made for the set of modules at the same orientation.

A detailed analysis of the thermal behavior of each face can be accomplished. As an example, the thermal difference between the rear and front horizontal face for all technologies is shown in Figures 7 and 8. The gradient temperature and therefore the thermal stress created within each module is large, up to 14ºC for the a-Si cube, and smaller (up to a maximum around 7ºC) for the other technologies. As expected, the thermal difference correlates well with the incident radiation.

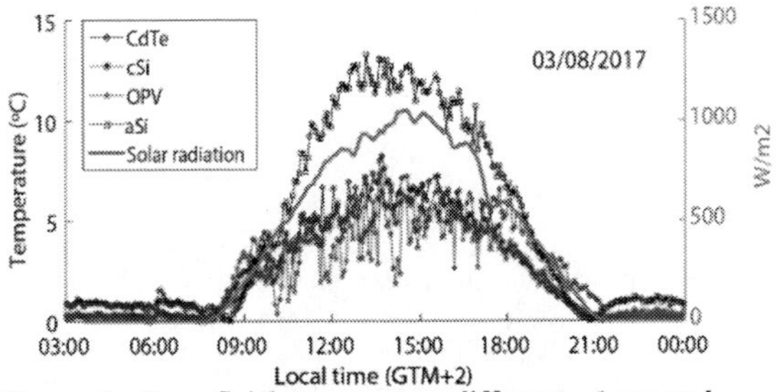

Figure 7: Superficial temperature difference (rear and front sensor) for a sunny day on the horizontal face.

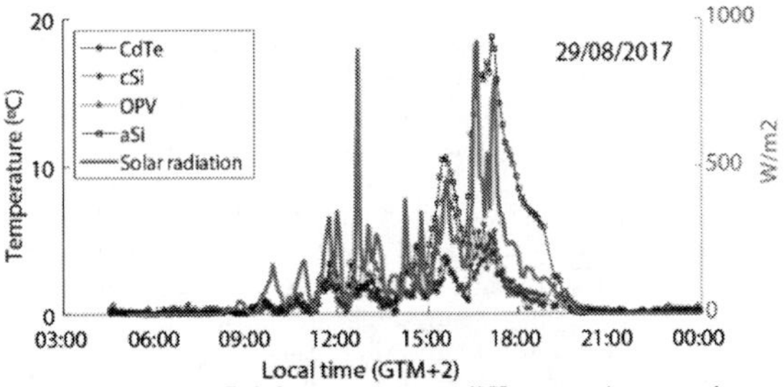

Figure 8: Superficial temperature difference (rear and front sensor) for a cloudy day on the horizontal face.

Finally, Figures 9 and 10 show the thermal evolution of the outer side of the horizontal face for all PV-Cubes versus the incident irradiance for diffuse conditions in the a-Si technology. For each irradiance value the plot provides two points, the "morning" point is when the Sun is raising and the "afternoon" point is when the Sun is setting. Each set of points creates a branch, therefore delivering a morning and an afternoon branch which differs by a significant temperature amount for all technologies. The results show hysteresis in all the cases (all technologies and all face orientations), the effect also depends on the ratio of global and diffuse irradiation, it is less prominent for the latter case (Figure 10).

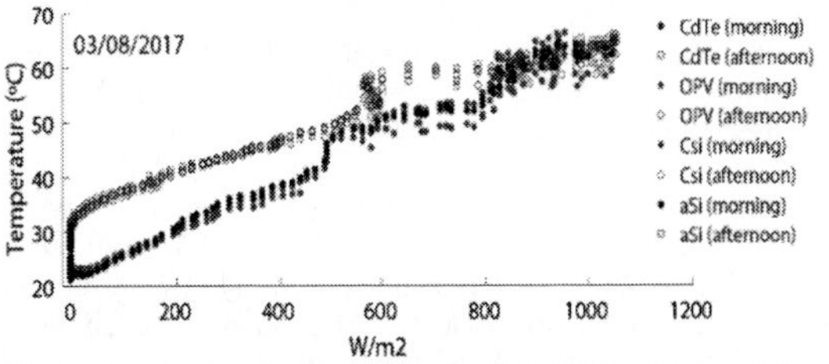

Figure 9: Temperature on the outer face of the horizontal PV panel for a-Si vs. horizontal global radiation in a sunny day.

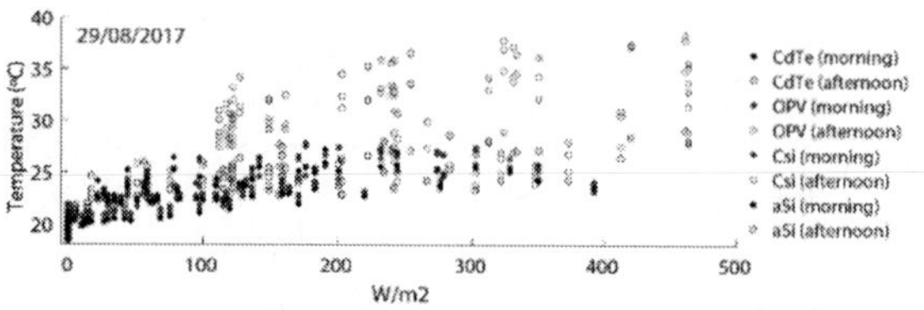

Figure 10: Temperature on the outer face of the horizontal PV panel for a-Si vs. horizontal global radiation in a cloudy day.

4 CONCLUSIONS

An experimental setup has been designed and built in order to provide information about the thermal behavior of photovoltaic modules under real working conditions for a BIPV prototype with four different PV technologies. The setup provides the measurement of meteorological parameters and 12 temperature sensors for each PV technology with a data sampling rate of 5 minutes. The system has been working continuously for several months.

An initial analysis of the experimental data, taking as an example some days that can be considered typical representatives of a "sunny" and a "cloudy" day, the following conclusions are obtained:

- The cube with a-Si photovoltaic modules reaches the highest temperatures (air temperature inside de PV-Cube and module temperatures, both inner and outer faces of the module) when compared with the other technologies (c-Si, CdTe, OPV). This high temperature will deteriorate the photovoltaic performance of the modules. However, under diffuse conditions the a-Si PV panel shows better performance from the thermal point of view.

- The air temperature differences between the interior and exterior of the structures are up to 20 °C, being slightly higher for a-Si PV technology.

- The temperature differences of inner and outer faces of each panel (and for all face orientations) may be as high as 14ºC, creating a thermal stress along the thickness of the PV modules. Additionally, the temperature of the modules (inner and outer faces) shows a hysteresis when represented against incident irradiation, demonstrating a thermal inertia of the PV-Cubes for all technologies.

The results of this research provide high quality experimental data which describe the thermal behavior of four photovoltaic technologies. As future work, detailed thermal models will be developed to simulate the observed thermal behavior and obtain empirical parameters which will characterize the thermal behavior of PV modules for integrated building applications. Additionally, continuous measurements of electrical parameters will be included in the experimental set up. They will be used to model the behavior of more complex buildings covered with photovoltaic modules that operate with a double function: electricity production and thermal insulation.

5 REFERENCES

[1] B. Petter Jelle, C. Breivik, H. Drolsum Røkenes, Sol. Energy Mater. Sol. Cells, Vol. 100 (2012), pp. 69–96.
[2] European Parliament and Council, Off. J. Eur. Union, Vol. L153 (2010), pp. 13–35.
[3] Commission European, (Online). Available: https://ec.europa.eu/easme/en/energy.
[4] International Energy Agency, (Online). Available: http://iea-pvps.org/index.php?id=task15.
[5] L. Serrano-Luján, R. García-Valverde, N. Espinosa, M. S. García-Cascales, J. M. Sánchez-Lozano, A. Urbina, Prog. Photovoltaics Res. Appl., Vol. 23 (2015), pp. 253–264.
[6] M. J. R. Perez, V. Fthenakis, H.-C. Kim, A. O. Pereira, Prog. Photovoltaics Res. Appl., Vol. 20 (2012), pp. 975–990.
[7] A. Scognamiglio, H. N. Røstvik, Prog. Photovoltaics Res. Appl., Vol. 21 (2013), pp. 1319–1336.
[8] C. Zomer, A. Nobre, P. Cassatella, T. Reindl, R. Rüther, Prog. Photovoltaics Res. Appl., Vol. 22 (2014), pp. 744–756.
[9] C.-H. Young, Y.-L. Chen, P.-C. Chen, Energy Build., Vol. 78 (2014), pp. 66–78.
[10] A. Chatzipanagi, F. Frontini, A. Virtuani, Appl. Energy, Vol. 173 (2016), pp. 1–12.
[11] P. A. Mirzaei, J. Carmeliet, Prog. Photovoltaics Res. Appl., Vol. 23 (2015), pp. 19–29.
[12] J. Peng, D. C. Curcija, L. Lu, S. E. Selkowitz, H. Yang, W. Zhang, Appl. Energy, Vol. 165 (2016), pp. 345–356.

Acknowledgements

The authors wish to acknowledge financial support from F. Séneca (Murcia, 19882/GERM/15) and MINECO (Spain, ENE2016-79282-C5-5-R).
C.T. is grateful to F. Séneca - CARM for grant (Exp. 19768/FPI/15).

A BUILDING-INTEGRATED SEMI-TRANSPARENT PV-GENERATOR ENDOWED WITH A MONO-AXIAL SOLAR TRACKER

Rosario Carbone
"D.I.I.E.S." Department - University "Mediterranea" of Reggio Calabria – Italy
rosario.carbone@unirc.it

ABSTRACT: In this paper we would like to describe constitutive characteristics of a photovoltaic generator conceived for being fully integrated in a building (BIPV-generator) also showing how it can be endowed by an unconventional and innovative "under-glass" mono-axial solar tracker. The semi-transparency special feature of our proposed BIPV-generator will be also underlined and valorized from the energy and architectural point of views. Some deepening about very low installation and maintenance costs of our BIPV-generator will be also given and the high reliability of its unconventional "under-glass" and mono-axial solar tracker will be underlined. Finally, we would like also to present and discuss the results of some first experimental measurements operated on a small-power prototype also introducing future tests to be performed on an additional prototype we are building "ad hoc" for performing "long-time" comparative tests among: (i) the on-field performances of the proposed BIPV-generator, (ii) the on-field performances of a conventional PV-generator built on the basis of the same quantity of PV-cells, with no solar-tracker and (iii) the on-field performances of a conventional PV-generator built on the basis of the maximum quantity of PV-cells which fully cover the available surface exposed to the sunlight (with no solar-tracker).
Keywords: Photovoltaic, Building Integration, Solar Tracker, Semi-transparent Photovoltaics.

1 INTRODUCTION

Recently, the technology of semi-transparent photovoltaics has received a great attention and different solutions have been introduced, tested and discussed by a lot of studies. Semi-transparent photovoltaic modules (PV-modules) have an obvious relevance in the field of greenhouses [1-6] while semi-transparent glass integrated PV-modules have been utilized on buildings as sunshade claddings or skylights or glass facade and so on, so significantly contributing to the development of building integrated photovoltaics (BIPV), where photovoltaic elements profitably replace other traditional building elements [7-9]. By this way, semi-transparent BIPVs clearly combine electrical energy production with other functional features and they can reduce the overall costs of investments on photovoltaics, by guaranteeing both a high energy efficiency and an aesthetically pleasing outlook of buildings. Furthermore, semi-trasparent BIPVs are able to practically avoid any environmental negative impact also overcoming the well known problem of the lack of installation space of photovoltaics.

One problem of semi-transparent PV-modules or, more in general, of BIPV-generators is that of their reduced generation of electricity caused by the reduced (or "non maximum") electrical PV-power installed per unit of surface area usefully exposed to the sunlight.

Undoubtedly, one way for catching more solar (and electrical) energy for a fixed quantity of PV-cells is that of introducing a solar tracking system; in fact, it is well known that solar trackers can improve the generated electricity up to 40% [10]. Nevertheless, solar trackers are affected by different issues; first of all, major initial costs and, then, also very low reliability and availability of their complex additional apparatus. In particular, availability of PV-generators endowed with a solar tracker can be particularly worsen because of continuous exposition of their additional servomechanisms to "aggressive" atmospheric agents as wind, dust, rain and also important variations of the temperature.

Remaining within the field of semi-transparent BIPVs, in the proposed paper we would like to show how, in our opinion, it is possible to design and realize an innovative BIPV-generator characterized by very low-costs and also by a high reliability of its included mono-axial solar tracker.

In a nutshell, a mono-axial solar tracker is suggested to be installed directly "under glass", on-board to a BIPV-generator specifically designed as described in the following, with some detail.

The basic idea has been already described and discussed by means and within an Italian Pending Patent Application [11]; anyway, in this paper, the designing procedure of our innovative semi-transparent BIPV-generator will be described in detail and its on-field performances will be tested and briefly discussed by referring to some first experimental results obtained by means of the installation and monitoring of a custom-designed low-power prototype, built on the basis of poly-crystalline (6x6) inch commercial PV-cells.

2 THE FIRST STEP: INTRODUCING AN INNOVATIVE MINI-PV-MODULE

To obtain a mono-axial solar tracker characterized, at the same time, by low costs, high efficiency and high reliability, first of all, the PV-generator has to be realized on the basis of innovative and unconventional mini-PV-modules (Fig. 1). Being these mini-PV-modules realized in the form of a single row of series-connected (conventional/commercial) PV-cells, they can be easily installed under a custom-designed transparent coverage (Fig.2), to be fully integrated in a building.

As it can be easily deduced from Figgs. 1-3, the (single-row) mini-PV-modules are immediately able to rotate around their central longitudinal axis, so simply making possible the realization of a mono-axial solar tracker. Furthermore, being the mini-PV-modules directly installed under the aforementioned building-integrated transparent coverage (practically "under glass"), all components of the PV-generator are fully protected from all potentially aggressive atmospheric agents, so ensuring a very high reliability and availability of all servomechanisms of the solar tracker system.

From Figgs. 2 and 3, it is also well evident that,

practically, only mini-PV-modules are expected to be rotated by the solar tracker servomechanisms while other heavy components of the semi-transparent BIPV-generator (e.g., the frames and the glass of the transparent coverage) remain fixed. As a consequence, solar tracker servomechanisms can have: i) a very low rated power, ii) negligible costs and iii) negligible electrical power consumption.

Figure 1 shows that each mini-PV-module is built on the basis of photovoltaic cells (preferably, of mono or poly crystalline silicon of the commercial type) which are sticked, in the form of a single row, on the flat surface exposed to the sun of a special and custom-designed elongated support, whose most important characteristics are graphically evidenced in Fig. 3.

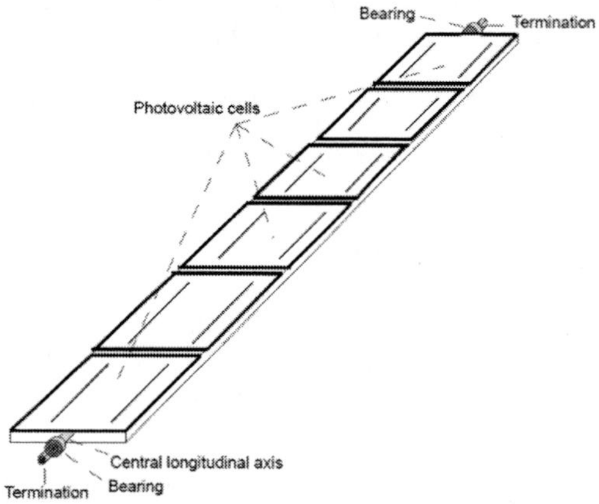

Figure 1: Graphical representation of one of our proposed single-row mini-PV-module.

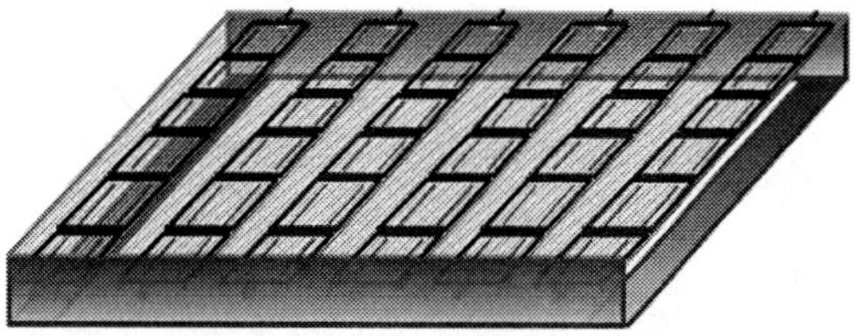

Figure 2: An example of a transparent coverage to be integrated in a building, under which some rows of our mini-PV-modules can be installed side by side.

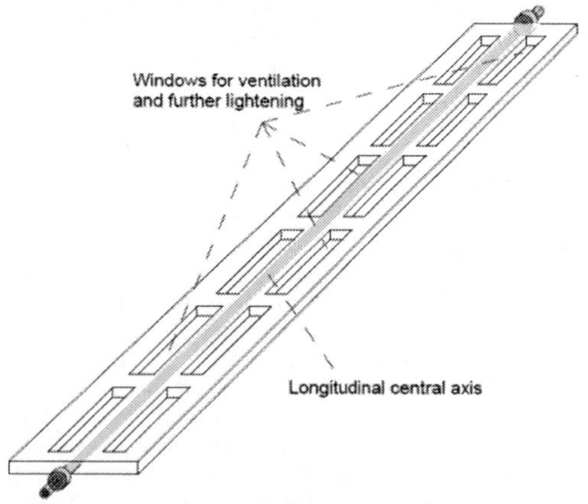

Figure 3: Graphical representation of our elongated (polystyrene) support, for conventional PV-cells.

First of all, please note the special elongated support is designed so that conventional PV-cells entirely cover its surface exposed to the sunlight.
Also, the aforesaid support is simply obtainable in the form of an elongated "strip", simply obtainable from a commercial sheet of a material which should be, at the same time, light, rigid, heat-resistant and electrically non-conductive, as it is, for example, the low-density expanded polystyrene.
Furthermore, the special support has a small thickness (between 1 and 3 cm at maximum, depending on its length and on the robustness of the material utilized) and has the form of a simple elongated "strip" provided with terminations to simply allow its rotation along its longitudinal central axis.
As clearly shown in Fig.3, the special support is also provided with some "windows", formed on both sides of its longitudinal axis, for ventilation and further lightening, so overall ensuring the construction of a mini-PV-module characterized by a total weight which is negligible if compared with the total weight of the prior art PV-modules including the same number and type of PV-cells. As an example, by using the low-density expanded polystyrene, it is possible to realize 10 mini-PV-modules each based on 6 conventional (6x6)" polycrystalline PV-cells (for a total of 60 PV-cells) whose total weight will result of about 1 kg, while, in comparison, a commercial PV-module based on the same 60 PV-cells (and ready for being installed on top of a conventional solar tracker) has a total weight of more than 20 kg.

Not of less importance, our proposed mini-PV-module is very cheap and, as a consequence, it is able to guarantee the proper trade-off between initial costs of the BIPV-generator to be realized and its relative payback-time.

3 THE BUILDING INTEGRATED SEMI-TRANSPARENT PV-GENERATOR WITH A MONO-AXIAL SOLAR TRAKER

As already illustrated in Fig.2, in order to avoid the mutual shading of the mini-PV-modules (that are placed side by side) during their complete daily rotation, they have to be necessarily inter-spaced.
As a consequence, on one hand, the installed electrical PV-power of our BIPV-generator can't be maximum with respect to the whole available surface profitably exposed to the sunlight.
However, on the other hand, once installed, the proposed BIPV-generator will profitably result semi-transparent.
The semi-transparency feature of our BIPV-generator can profitably guarantee also the partial lighting with natural sunlight of the interiors of the building in which the transparent coverage is integrated and, as a consequence, our BIPV-generator can be profitably utilized for obtaining buildings characterized by both aesthetic "delicious" architectural effects and high-values of their energy performance parameters [1-9].

Furthermore, it is within the objectives of the proposed paper to demonstrate that, by taking advantage of our unconventional, very cheap and high reliable mono-axial solar-tracker, the electrical energy generated by our proposed BIPV-generator can be maximized so much that it will result anyway equal or greater than the electrical energy which can be generated by a conventional PV-generator in which the whole available surface exposed to the sunlight is fully covered with a greater quantity of PV-cells (for obtaining the installation of the maximum electrical PV-power, obviously, with more spending and without obtaining the semi-transparency feature).

4 SOME FISRST EXPERIMENTAL TESTS

Once the main characteristics and properties of our proposed BIPV-generator have been underlined, in the following, an already tested small-power prototype (Fig. 4) will be briefly introduced for demonstrating, on the basis of some already done on-field tests, that our "under-glass" mono-axial solar-tracker is able to guarantee the same improvement (of about +30%) of the electrical energy generation typically guaranteed by a conventional (and well working) mono-axial solar-tracker, with respect of a fixed PV-generator based on the same quantity of PV-cells.

Experimental tests on the aforesaid PV-generator prototype have been performed recently, during an almost cloudless summer day, from 8.00 a.m. to 18:00 p.m.. The results are reported in Figgs 5-7, respectively, in terms of short circuit PV-generator current, $\mathbf{I_{cc}}$, open circuit PV-generator voltage, $\mathbf{V_o}$, and maximum generated power, $\mathbf{P_{max}}$, for both rotating and fixed constitutive mini-PV-modules.

From Fig.7, it is well evident a significant improvement of the daily mean power generated by the PV-generator when the solar tracker is activated; in fact, if rotating, the PV-generator delivers a daily mean power of about 86 W, against the about 64 W delivered by the fixed PV-generator.

This result confirms that, also with our proposed mono-axial "under-glass" solar tracker, an improvement of the generated power of about 34% it is possible, as it is typical for conventional mono-axial solar trackers.

In addition to aforesaid experiments, at moment, we are constructing a new and more complete small-power PV-generator prototype, specifically conceived for demonstrating and evidencing, in a more complete way, all the advantages of our BIPV-generator with respect not only to conventional PV-generator based on the same number of PV-cells exposed to the sunlight but also with respect to a conventional PV-generator whose electrical energy generation is maximized by utilizing a more quantity of PV-cells by entirely covering the whole available surface exposed to the sunlight.

Some more details about our project and intentions can be deduced also from the schematic representation of the new prototype (at the moment under construction) reported in Fig.8.

With some other words, we would like to demonstrate that, by properly designing the geometry and structure of the transparent coverage under which our rotating mini-PV-modules have to be installed, it is possible to realize a mono-axial solar tracker able to improve so much the generation of electricity of our (low-number) PV-cells that these last are able to generate a quantity of electricity which is equal (or greater) than that obtainable by using a conventional PV-generator based on a greater quantity of PV-cells fully covering the whole available surface exposed to the sunlight, which obviously suffers of higher costs and of the loss of the semi-transparency feature.

At this stage, please simply accept that, in principle, this could be right because the improvement of the electricity generated by our proposed mono-axial solar tracker depends from the maximum rotation angle guaranteed to the mini-PV-modules. In turn, this depends from the height of the "shoulders" which are on the basis of the construction of the transparent coverage and which are also necessary for installing the terminations of the mini-PVmodules (Figgs. 1, 2 and 3). Furthermore, the inter-distance among the different rows of mini-PV-modules (necessary to avoid mutual shading effects during their rotation) also depends from the rotation angle, together with the whole available surface exposed to the sunlight.

Figure 4: A picture of a first low-power rotating PV-generator prototype, already realized and tested.

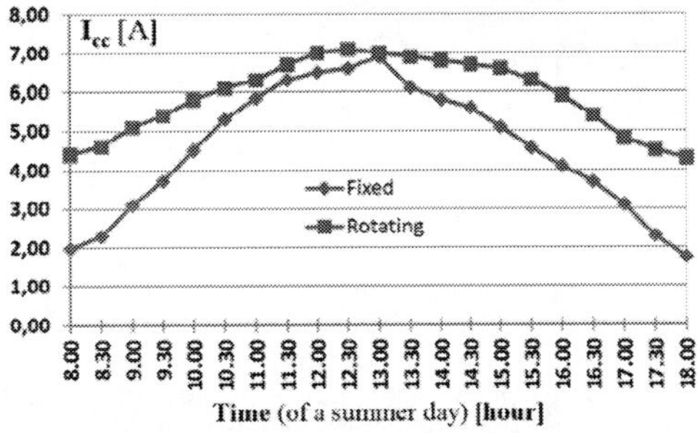

Figure 5: Short circuit current, $\mathbf{I_{cc}}$, of the PV-generator prototype, during an almost cloudless summer day.

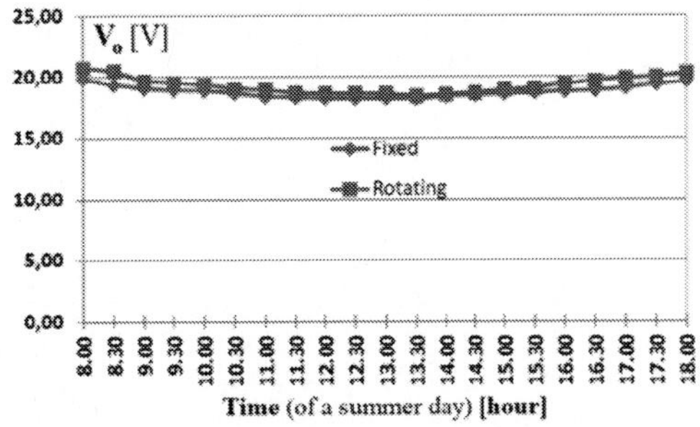

Figure 6: Open circuit voltage, $\mathbf{V_o}$, of the PV-generator prototype, during an almost cloudless summer day.

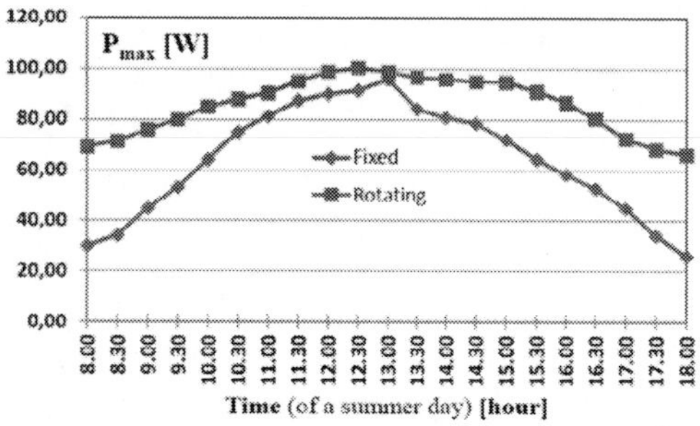

Figure 7: Maximum generated power, $\mathbf{P_{max}}$, of the PV-generator prototype, during an almost cloudless summer day.

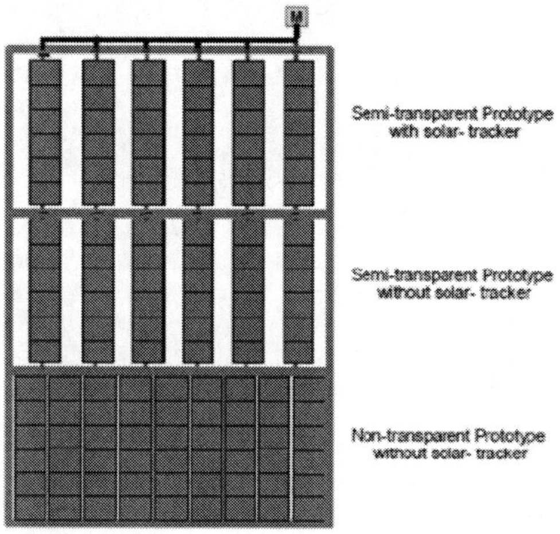

Figure 8: Schematic representation of the low-power prototype of the PV-generator under construction for performing more complete on-field measurements and comparative tests on the "long time".

7 CONLUSIONS

In this paper we have introduced and discussed an innovative "building integrated" and "semi-transparent" photovoltaic generator, also endowed by an innovative "under-glass" solar tracker, for maximization of the electrical power generation.

Some experiments, performed by using a first-step prototype seem to confirm the effectiveness and promised advantages of our PV-generator.

As future work, at the moment we are constructing a more complete prototype, for better testing on field performances of our proposed rotating and semi-transparent PV-generator with respect to performances of conventional fixed and non-transparent PV-generators, also on the "long-time".

8 REFERENCES

[1] R. Carbone, C. De Capua, R. Morello: "*PV-plants for Powering Greenhouses*". Proceedings of the IEEE International Conference "ICCEP 2011", 2011, Ischia, Italy. ISBN: 978 1 4244 8927 5X.

[2] M. Carlini, T. Honorati, S. Castellucci: "*Photovoltaic Greenhouses: Comparison of Optical and Thermal Behaviour for Energy Savings*". Mathematical Problems in Engineering, 2012. DOI: 10.1155/2012/743764.

[3] A. Marucci, D. Monarca, M. Cecchini, A. Colantoni, A. Manzo, A. Cappuccini: "*The Semitransparent Photovoltaic Films for Mediterranean Greenhouse: A New Sustainable Technology*". Mathematical Problems in Engineering, 14(2), 2012. DOI: 10.1155/2012/451934.

[4] Akira Yano, Mahiro Onoeb, Josuke Nakata: "*Prototype semi-transparent photovoltaic modules for greenhouse roof applications*". Elsevier International Journal on "Biosystems Engineering", Volume 122, June 2014, Pages 62–73.

[5] John Vourdoubas: "*Possibilities of Using Semi-Transparent Photovoltaic Modules on Rooftops of Greenhouses for Covering Their Energy Needs*". Journal of Agricultural Studies 4(1):90. December 2015. DOI: 10.5296/jas.v4i1.8694

[6] Marco Cossu, Akira Yano, Zhi Li, Mahiro Onoe, Hidetoshi Nakamura, Toshinori Matsumoto, Josuke Nakata: "*Advances on the semi-transparent modules based on micro solar cells: First integration in a greenhouse system*". In Applied Energy 162: 1042-1051. January 2016. DOI: 10.1016/j.apenergy.2015.11.002.

[7] Chehab O. "*The intelligent fac,ade photovoltaic and architecture*". *Renewable Energy 1994; 5(1):188–204.*

[8] Lam JC, Li DHW. "*An analysis of daylighting and solar heat for cooling-dominated office buildings*". Elsevier Solar Energy 1999; 65(4):251–62.

[9] Bodart M, De Herde A. "*Global energy savings in offices buildings by the use of daylighting*". Energy and Buildings 2002; 34(5): 421–9.

[10] Hassan Fathabadi: "*Comparative study between two novel sensorless and sensor based dual-axis solar trackers*". In Elsevier International Journal on Solar Energy, 138. November 2016. DOI: 10.1016/j.solener.2016.09.009.

[11] R. Carbone (designed inventor): "*Inseguitore solare fotovoltaico*". Italian Pending Patent, N. IT-RM-2015A000190, 30.04.2014. Applicant University "Mediterranea" of Reggio Calabria - Italy.

HAIL RESISTANCE OF COMPOSITE-BASED GLASS-FREE LIGHTWEIGHT MODULES FOR BUILDING INTEGRATED PHOTOVOLTAICS APPLICATIONS

Ana C. Martins[1*], Valentin Chapuis[1], Alessandro Virtuani[1], Laure-Emmanuelle Perret-Aebi[2], Christophe Ballif[1,2]

[1] École Polytechnique Fédérale de Lausanne (EPFL), Institute of Microengineering (IMT),
Photovoltaics and Thin Film Electronics Laboratory (PV-Lab), Rue de la Maladière 71b, 2000 Neuchâtel, Switzerland
[2] CSEM, PV-center, Jaquet-Droz 1, 2000 Neuchâtel, Switzerland
*contact person:ana.martins@epfl.ch

ABSTRACT: In this work, we manufacture lightweight (~5 kg/m^2) photovoltaic (PV) mini-modules for building integration replacing conventional glass sheets by a composite sandwich backsheet and a polymeric frontsheet. Our test devices are subjected to hail tests according to IEC 61215 and characterized electrically (to extract IV parameters) and by means of electroluminescence (EL) to visualize the hail-induced damages. The influence of both backsheet stiffness and frontsheet design on the hail test results are presented and discussed. Based on the results, two module design contributions to hail resistance are highlighted: i) a global energy dissipation, mostly related to the backsheet properties, is promoted by the deformation of the entire PV module structure and ii) a local energy dissipation occurs in the frontsheet under the hail impacts. These results clearly show that a balance between frontsheet design and backsheet stiffness has to be found in order to maximize hail resistance while minimizing the module's weight. By combining the ideal backsheet structure with the proper frontsheet layer, we show that a rigid glass-free medium-area module can easily pass relevant hail test with less than 5% power losses.

Keywords: BIPV, lightweight, composite sandwich structures, hail tests, qualification testing

1 INTRODUCTION

Photovoltaic (PV) technology has seen an increase of 50% over 2015, bringing the cumulative installed capacity worldwide to 303 GW in 2016 [1]. Of the global world's electricity demand, ~40% of the consumption is linked to buildings. Further, in several densely-populated countries, only little free-land is available for field PV installation [2]. Development of building integrated PV (BIPV) solutions is thus crucial to the growth of the total installed PV capacity as it enables electricity production with a minimal impact on free land. At European level, the main driver to massively promote the integration of PV into buildings is the Energy Performance in Buildings Directive (EPBD), which foresees that all new buildings in Europe need to comply at least with Nearly Zero-Energy-consumption Building (NZEB) standards by December 31st, 2020 [3]. The EPBD will favor the direct integration of decentralized energy sources in buildings, amongst which PV appears to be the most promising one, unlocking potentially a considerable market in Europe. To make an example, in countries like Switzerland already > 98% of solar installations take place in the built environment, and studies indicate that the use of solar energy in facades could provide up to 15 – 20% of the electricity production needed in urban areas [4]. However, current BIPV solutions suffer some limitations amongst which their relative high weight (>20 kg/m^2) due to the presence of glass as protective front and back layers [5, 6]. In many cases, especially old buildings, the weight loading capacity is limited. For some, rooftops in industrial parks, the limit is less than 10 kg/m^2, thus much lower than a typical PV module weight [7]. Moreover, most industrial buildings built before the 1997 Uniform Building Code (UBC) change are not suitable for conventional rooftop solar installation without costly structural improvements [6]. If lightweight PV modules were available, not only would this market segment be covered, but also other potentially interesting markets, such as that of temporary buildings and/or buildings in developing countries, where the load-bearing capacity of the building is limited, could benefit from this innovation.

Amongst the various types of PV modules, crystalline-silicon (c-Si) PV modules held close to 90% of the PV module market in 2016 [8]. Standard PV modules of c-Si solar cells are typically glass/backsheet modules having a weight of 12-16 kg/m^2, or glass/glass modules weighting 14-17 kg/m^2. For BIPV applications, glass/glass modules are generally preferred for the higher structural stability and for safety reasons. Lightweight PV modules based on glass-glass technology exist. For example, few companies propose a reduction of glass thickness from the standard 3.2 mm up to 0.8mm, achieving weight of 6.5kg/m^2. However, the cost of ultra-thin glass contributes to a considerable increase on PV module cost.

The commercially available lightweight PV modules on the market are predominantly flexible or semi-rigid solutions. The standard glass sheets are substituted by polymeric material as frontsheet and a glass fiber reinforced polymer as backsheet [9, 10] . These lightweight solutions have weights of 2.7-7 kg/m^2. From our laboratory experience, the drawback of these lightweight solutions is related to the long-term performance (reliability), even when the manufacturer claims compliance to IEC/UL certifications.

The aim of this work is to propose a BIPV module design that is contemporarily lightweight, rigid and resistant to hail, solving the problems of installation, supporting structure and reliability at the same time.

2 STRUCTURE OF THE LIGHWEIGHT SOLAR MODULES

The lightweight solar module is achieved by replacing the standard glass frontsheet by a thin transparent polymeric layer and by engineering the backsheet to replace the standard thin polymeric co-laminated PV backsheet by a composite sandwich structure. The presented lightweight module has a final weight of

5 kg/m². Figure 1 shows our lightweight design developed for BIPV applications. As one of the critical tests for glass-free solutions is the hail test, we show how the polymeric frontsheet design and composite backsheet stiffness influence the capability of the module to resist hail impacts. The composite sandwich element (based on glass fibers reinforced polymer and a honeycomb core) is intended to provide to the module its entire mechanical stability and is placed at the back of the module stack. Additionally, this lightweight solution is manufactured in a single lamination process, which we use to manufacture the composite backsheet and the full module in a single run. The main concern linked to this new module architecture, where the layer providing the mechanical stability is placed behind the cells and not in front, as in the case of glass/backsheet modules, is linked to its capability to still provide a sufficient protection against environmental degradations and especially against hail impacts. We show here how the composite backsheet and the polymeric frontsheet of our product must be engineered in terms of stiffness and thickness to provide optimal hail resistance.

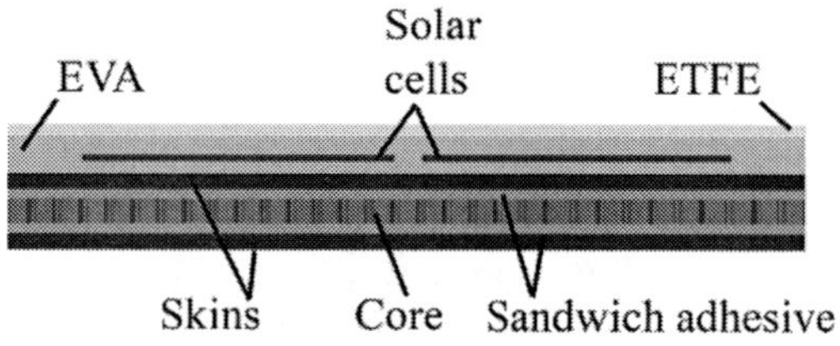

Figure 1: Sketch of lightweight PV module design for BIPV applications.

3 EXPERIMENTAL WORK

3.1 Mini-modules production

Two-cells (156 x 156 mm² Al-BSF c-Si solar cells with 3 busbars) mini-modules are produced according to Figure 1.

3.1.1 Influence of composite sandwich structure

To study the influence of the backsheet on the resistance to hail impact we prepare mini-modules using a constant frontsheet made of 100µm fluorine-based polymer (ETFE) and 900µm ethylene-vinyl acetate (EVA) and changing the composite backsheet materials or processing to tune its stiffness. The composite sandwich backsheet is made of glass fiber reinforced polymer (GFRP) skins (~0.7 mm thick) glued to a light honeycomb core (6 mm thick). The materials and process used to glue the skins to the core are varied to tune the final sandwich bending stiffness. The mini-modules composed of BS_1, BS_2 and BS_3 are processed in a vacuum-bag laminator in a single run by using conventional PV encapsulants. The mini-module composed of BS_4 is produced in a two-steps process. In a first step the solar cells are laminated to a single skin (see Figure 1) by using a simple lamination process. This proto-module composed of ETFE / EVA / solar cells / EVA / skin is then, in a second step, introduced in a vacuum bag to bond to the core and the other skin by using a liquid thermoset material (epoxy). This vacuum bag process takes 24h for a full adhesive crosslinking. Table I shows the bending stiffness of the composite sandwich structures selected for this study and measured by four-points bending test [11]. For each backsheet configuration, three mini-modules are prepared and subjected to a hail test.

Table I: Composite sandwich bending stiffness measured in four-points bending tests.

	Bending Stiffness, D [N.m²]
BS_1	4.8
BS_2	6.6
BS_3	8.4
BS_4	11.6

3.2.2 Influence of frontsheet thickness

In order to study the effect of the frontsheet on the hail resistance we evaluate four different frontsheet designs based on ETFE / EVA combinations. For one set of samples (FS_3), an additional glass scrim (GS) layer (17 g/m²) is introduced. Table II summarizes all test conditions. For this study, the backsheet is kept constant (composite backsheet with bending stiffness of 6.6 N.m²). For each condition, three identical samples are produced and tested using the same protocol as for the former study.

Table II: Conditions tested to evaluate the impact of the frontsheet design on hail resistance.

	Frontsheet layup			
	Layer 1 polymer	Layer 2 adhesive	Layer 3 polymer	Layer 4 adhesive
FS_1	100um	900um	-	-
FS_2	50um	450um	50um	450um
FS_3	100um	450um	GS-17/m²	450um
FS_4	200um	900um	-	-

3.2 Hail test setup

Hail test is performed according to IEC 61215-2:2016 [12]. Figure 2 represents the fixation system used and the impact spots selected. Ice balls of 25 mm diameter are shot at a velocity of 23.4 ± 1.5 m/s.

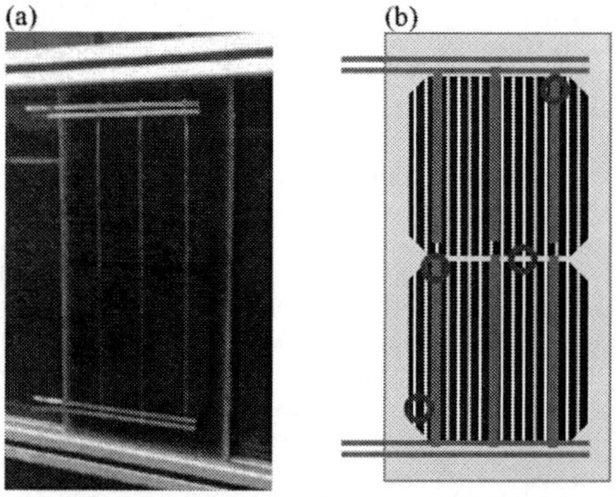

Figure 2: Hail test setup (a) example of fixation system used on two-cell mini-module during the hail test and (b) impact spots for the hail test.

The lightweight mini-modules are hit in four critical locations: (1) at the border of the solar cell, (2) at the electrical connection, (3) in between solar cells and (4) in

any point vulnerable to hail impacts.
The electrical characterization of the mini-modules was performed at standard test conditions (STC: AM1.5G, 25°C, 1000W/m^2) by means of current–voltage (IV) characteristics using a LED-halogen based sun simulator.

3.3 Drop-weight test

In a drop-weight test, a weight is dropped on the specimen under test from a given height. Knowing the mass and the height at the start position, we can easily control the impact energy at contact between the two objects. During the test, a weight of 5.5 kg is dropped on a specimen placed at the bottom of the system and the velocity before and after is measured by two light gates placed just above the impact spot. From the measured velocities, the absorbed energy can be obtained as follows:

$$E_{absorbed} = E_{impact} - E_{rebound} \quad (1)$$

The four types of frontsheet presented in Table II are manufactured on the top of a rigid aluminum plate. A replica of three samples per type of frontsheet is tested.

4 RESULTS

4.1 Influence of backsheet stiffness on hail resistance

Figure 3a shows the power and fill factor losses after hail test for each backsheet configuration and Figure 3b shows the corresponding EL image.

(a)

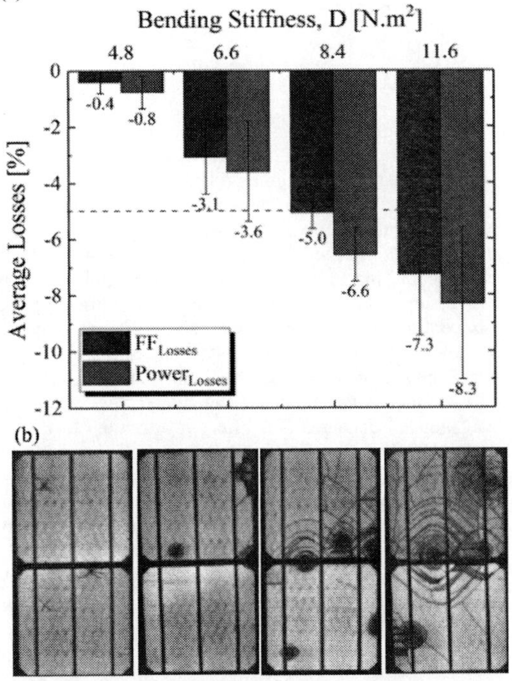

Figure 3: (a) Power and fill factor losses for our mini-modules after the hail test and (b) corresponding EL images. The test devices were manufactured using backsheets with different bending stiffness's.

None of the modules showed any visual damages after hail test (no crack visible to the naked eye). The results show that the power loss increases with the backsheet stiffness, with the maximum acceptable power loss of 5% being reached for a backsheet stiffness between 6.6 and 8.4 N.m^2. The EL images clearly show the change in damage patterns between the 6.6 N.m^2 backsheet for which the negative influence of the hail impact stays localized below the impact point and the 8.4 N.m^2 backsheet for which the damaged zone extends far away from the impact points.

Reducing the backsheet bending stiffness will enhance the resistance to hail of our lightweight PV module. However, it may not always be the ideal solution as it may lead to the need for a stronger support structure depending on the module size and depending on the loading (snow/wind load) it will experience during service lifetime. Therefore, it is equally important to evaluate the potential for hail damage reduction through energy dissipation in the frontsheet.

4.2 Influence of frontsheet thickness on hail resistance

The results summarized in Figure 4 show the fill factor and power loss after the hail test along with the corresponding EL images. As expected, reinforcing the frontsheet by increasing its volume (i.e. the thickness) leads to a decrease of the power loss after the hail test, as can be seen by comparing the results for FS_1 and FS_4. Using a drop-weight test we are able to quantify the energy absorbed by the different frontsheet.

(a)

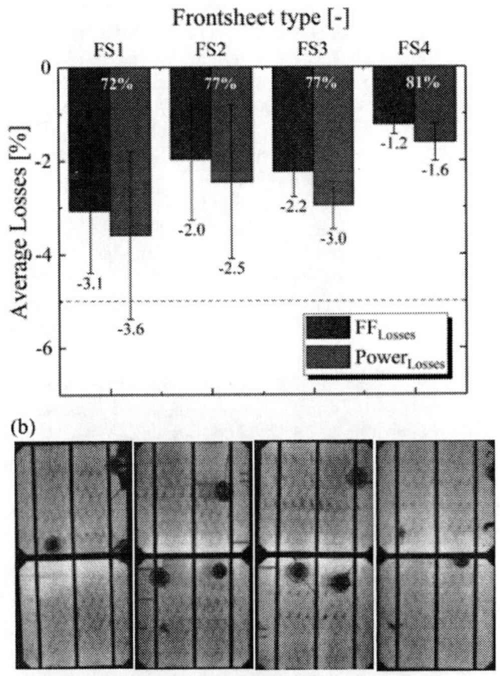

Figure 4: (a) Power and fill factor losses after the hail test for our 2-cells mini-modules with different frontsheet, and (b) the corresponding EL images after the ice-balls impact.

We observe that the frontsheet exhibiting the highest power loss (FS_1) is the one absorbing the less energy

(72%). Interestingly, keeping constant the volume of absorbing material but changing the materials layup (FS_1 vs FS_2) leads to a measurable improvement of the power loss by 35%. Again, we measured an increase of absorbed energy of 5% when the frontsheet is reinforced with two interlayers instead of a single thin layer. These results suggest that the presence of double interlayers contributes to the absorption of higher amount of energy during the impact. Moreover, we also see (FS_2 vs FS_3) that the glass scrim layer of 17 g/m^2 can be used to substitute one of the 50-μm fluorine-based polymer layer resulting in a thinner frontsheet showing equivalent hail resistance but improved thermal behavior thanks to the presence of the glass fibers. The FS_4 is the frontsheet that absorbed higher energy (81%) and consequently, is the one that show lower power loss.

4 UPSACALING: FROM MINI- TO MEDIUM-AREA MODULES

Targeting the upscaling of our mini-modules, we demonstrate that we are able to manufacture medium-area modules composed of 16 cell and we subject them to the IEC hail test as well. A medium-area module composed of a sandwich bending stiffness of 10 N.m^2 and a FS_1 is produced. Based on the IEC 61215-2 the hail test is performed after the damp-heat (DH) test (*Sequence E*). For this reason, before testing the resistance to hail of the medium-area module, we decided to aged it to meet the requirements presented in [12]. After lamination, the modules are characterized electrically and solar cells cracks identified by help of electroluminescence test. As a check, the leakage current test is performed as well before and after each test as required in IEC 61215 [12]. Following the initial characterization, the modules are introduced in DH for 1000h and subsequently to the hail test. The following impact spots are selected:

1. A corner of the module
2. Edge of the module
3. Edge of circuit (individual cell)
4. Close to the mounting structure
5. Any point which may prove especially vulnerable to hail
6. A point farthest from the above-selected points.

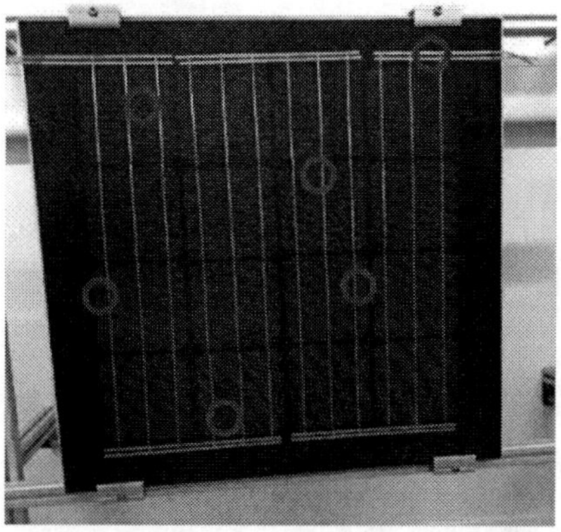

Figure 5: Medium-area module with indication of the impact spots used for the hail test.

Figure 5 shows the impacts locations considered and the fixation system. The test is performed considering the same test conditions as the ones used with the mini-modules (ice balls shot at 23.4 m/s with one-meter distance from canon). A moderate degradation is obtained after full *Sequence E*, being lower than the 5% pass/fail criterion set in the IEC standard. Moreover, the leakage current measure after both degradation tests is higher than the 40 MOhm·m^2 threshold to pass the test. Beside the good electrical performance, the EL images show small localized cracks due to the impact of the ice balls during the test (Figure 6). The presence of these small localized deformations can still be improved by using a frontsheet able to absorbed higher energy and thus reduce the amount of energy transfer to the solar cells. Moreover, the frontsheet used in the realization of the medium-area modules is the one exhibiting the weakest resistance to hail (FS_1), as can be seen in Figure 4.

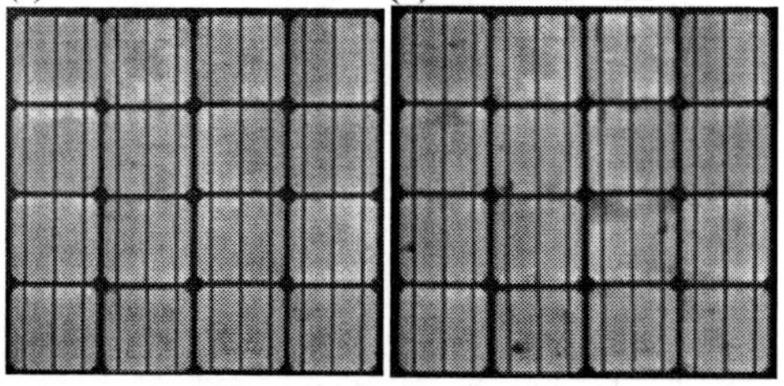

Figure 6: EL images of medium are modules composed of a stiff composite backsheet and a thin polymer frontsheet.

5 CONCLUSIONS

In this study, we show that we are able to manufacture a lightweight glass-free module with a rigid composite sandwich backsheet weighing 5 kg/m^2.

We show that the stiffness of the composite backsheet has a strong impact on the hail resistance, with a reduction of the hail induced damages for the low stiffness backsheet. This effect is attributed to a global energy dissipation of the hail impacts through the entire structure. Concerning the frontsheet, we show that not only the volume (thickness) of the polymeric frontsheet has a strong impact on resistance to hail, but that the distribution of materials in the thickness can further help mitigating hail induced damage. This effect is attributed to a local energy dissipation of the hail impact below the impact point.

Finally, we show that by using the correct trade-off between reinforcing the frontsheet and backsheet stiffness, it is possible to produce a lightweight glass-free module, with a thin polymer frontsheet and a rigid composite backsheet that passes *Sequence E* (DH+HT) of IEC 61215-2:2016.

ACKNOWLEDGMENTS

The authors gratefully acknowledge the financial support of the Swiss National Science Foundation (SNSF) through the National Research Programme (NRP) "Energy Turnaround" (NRP 70). Further information on NRP can be found at www.nrp70.ch. We further acknowledge the support of all staff at PV-Lab and CSEM, in particular Xavier Niquille.

REFERENCES

1. International Energy Agency Photovoltaic Power System Programme. Snapshots of global photovoltaic markets – 2016. Report IEA PVPS T1-31; 2017.

2. International Energy Agency. Energy Technology Perspectives 2017: Catalysing Energy Technology Transformations (Executive Summary). doi: http://dx.doi.org/10.1787/energy_tech-2017-en

3. Directive 2010/31/EU of the European Parliament and of the Council of 19 May 2010. The energy performance of building directive (2010). Official Journal of the European Union

4. Nowak S, Gutschner M, Ruoss D, et al. Potential for building integrated photovoltaics. Report IEA-PVPS T7-4 (2002)

5. Kajisa T, Miyauchi H, Mizuhara K, et al. Novel lighter weight crystalline silicon photovoltaic module using acrylic-film as a cover sheet. Japanese Journal of Applied Physics 2014; 53:092302. doi: http://dx.doi.org/ 10.7567/JJAP.53.092302

6. Nussbaumer H, Klenk M, Keller N. Small unit compond module: A new approach for light weight PV modules. In Proceedings of 32nd European Photovoltaic Solar Energy Conference (EU PVSEC); 2016, pp 56–60

7. Zhang F, Deng H, Margolis R, Su J. Analysis of distributed-generation photovoltaic deployment, installation time and cost, market barriers, and policies in China. Energy Policy 2015 81:43–55. doi: 10.1016/j.enpol.2015.02.010

8. Fraunhofer Institute for Solar Energy Systems, ISE (2017) Photovoltaics Report. Online at: https://www.ise.fraunhofer.de/content/dam/ise/de/documents/publications/studies/Photovoltaics-Report.pdf. Accessed 20 Sep 2017

9. DAS Energy GmbH Data sheet: Module DAS-Serie 240 – 250P.

10. SBM Solar Datasheet: Module SBM-165W Module.

11. Martins AC, Chapuis V, Virtuani A, Ballif C. Ultra-Lightweight PV module design for Building Integrated Photovoltaics. In Proceedings of 44th IEEE Photovoltaic Specialists Conference(IEEE); 2017

12. IEC 61215-2 Terrestrial photovoltaic (PV) modules - Design qualification and type approval - Part 2: Test procedures; 2016.

POWER LOSS THROUGH DECORATIVE ELEMENTS IN THE FRONT GLAZING OF BIPV MODULES

Max Mittag, Christoph Kutter, Matthieu Ebert, Helen Rose Wilson, Ulrich Eitner
Fraunhofer Institute for Solar Energy Systems ISE
Heidenhofstr. 2, 79110 Freiburg, Germany
max.mittag@ise.fraunhofer.de

ABSTRACT: Printed module front covers can be designed to mask the geometry of solar cells for Building-Integrated Photovoltaics (BiPV). Such prints reduce the transmittance of the module front layer, which decreases module power. We performed transmittance measurements with large-area and small-area measurement equipment (220 mm / 620 mm integrating sphere) on different samples and find measurements applying small-area illumination to result in lower transmittance ($\Delta = 3.1_{abs}\%$ at 40% coverage) due to lateral losses in the samples. By measuring the power of modules with different design prints we find that the results do not correspond to transmittance measurements of the glass covers alone, even if a large-area transmittance measurement setup is used ($\Delta = 16\%_{abs}$ at 40% coverage). We attribute the differences between optical and electrical measurements to module internal reflection, optical coupling of the solar cell, partial transmittance by coatings and diffuse scattering of decorative prints. Differences increase with the share of coated glass area. We perform electrical and LBIC measurements on modules with printed and unprinted reference glass covers and calculate the effective transmittance. Short circuit currents calculated by spatial integration of LBIC results are in good agreement with results from I_{SC} measurements. We predict the effective transmittance for arbitrary prints based on selected I_{SC} measurements and find them to be in good agreement to measurements. In conclusion, we find transmittance measurements on printed glass alone to be insufficient to predict the optical power losses as they overestimate the optical loss.

Keywords: Building Integrated Photovoltaic Module, BiPV, CTM, Cell-to-Module, Power Prediction

1 INTRODUCTION

The European directive 2010/31/EU (nearly zero energy buildings) encourages the large-scale deployment of Building-Integrated Photovoltaics (BiPV) [1]. Yet today, BiPV market penetration is still low. One reason is that the aesthetic appearance of the photovoltaic components does not satisfy the expectations of architectural design. To overcome this, many efforts are being undertaken to conceal the photovoltaic components such as the individual PV cells or cell interconnectors and to give the PV modules a homogeneous appearance [2]-[5]. A very promising concept that meets the aesthetic demands of building integrated photovoltaics (BiPV) is to completely coat the front cover material with optically active layers (Figure 1). Another option is the printing of patterns that only cover parts of the front interface (i.e. geometric patterns of dots etc. (Figure 2).

Figure 1: Full-area coating of module front glass cover [11]

The drawback of such printed decorative patterns is the reduction of transmittance, thus decreasing the nominal module power and finally increasing the financial barrier and return on investment.

Gains and losses in common photovoltaic modules are well understood and models exist to predict module power and cell-to-module losses under Standard Testing Conditions (STC) [6]-[9]. The effects of additional decorative elements have been investigated by Frontini [10], quantifying the difference between camouflage coverage and module power loss that result from wave-optical effects.

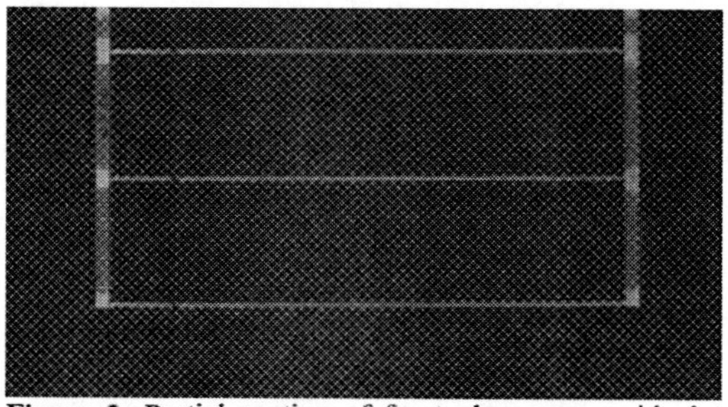

Figure 2: Partial coating of front glass cover with dot pattern

2 METHOD

Decorative elements applied to the module front cover (usually a glass pane) affect the module power by increasing reflection (1) and additional absorption (2), which both lead to shading losses (Figure 3).

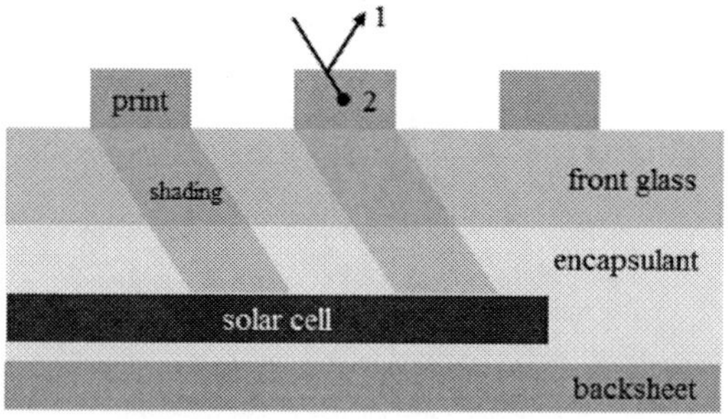

Figure 3: Schematic drawing of changes in module optics by application of decorative prints onto the module front cover

We analyze the effects of printed ceramic elements on the front glazing on module power. Transmittance measurements using samples with different printed patterns and coverage are performed. We compare the results of small-area and large-area transmittance measurement configurations. Furthermore, we characterize the ceramic print patterns using surface profilometry, microscope imaging and light-beam induced current (LBIC) measurements.

We compare the projected power of a module with coated glass based on geometrical coverage of the printed coating, based on transmittance measurement results as well as based on measurements using laser-beam induced current (LBIC) on manufactured module samples. In addition, we perform power measurements on the module samples. Results of the electrical and optical measurements are compared and discussed.

a. Geometric Coverage Ratio

Assuming complete opacity of the printed element, i.e. the print has zero transmittance; the ratio of the footprint of the design element to the total module is defined as the geometrical coverage ratio:

$$CR_{geometrical} = \frac{A_{decorative element}}{A_{module}} \quad (I)$$

While total opacity is a first-order assumption which is seldom valid in practice, the coverage ratio is nonetheless an important parameter as it represents a target value for production and dimensioning.

b. Optical Measurements of Glass

We perform optical measurements on front glasses and modules with different geometrical coverage ratios. White dots are printed onto a set of glass samples using a digital ink jet printer and ceramic based ink. The targeted geometrical coverage ratio is 10%, 20% and 40%. An unprinted reference glass pane is available (Figure 4).

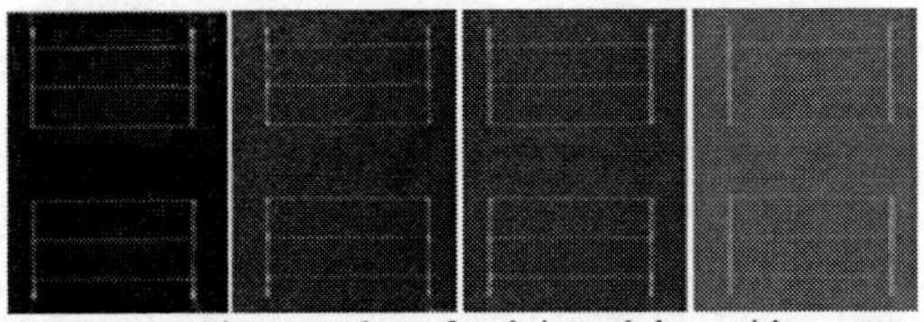

Figure 4: Photographs of mini-modules with covers made of digitally printed glass samples with different coverage ratios and reference glass pane. From left to right: Reference glass, 10% coverage ratio, 20% coverage ratio, 40% coverage ratio; small dots are printed

The glass samples are optically characterized using two different measurement setups. Method A uses a small-area light source and a small integrating sphere (diameter 220 mm). Method B uses a large integrating sphere (diameter 620 mm) with large-area illumination of the sample. It is widely accepted that a large aperture area and a large-area collimated radiation source are necessary to measure the transmittance of light-scattering, translucent materials accurately [12]-[14].

In Figure 5, the transmittance spectra of the different glass samples are shown. The effect of the two different characterization methods is reflected in the measured transmittance differences.

In Table 1 the measured transmittance values are shown, which are obtained by weighting the measured spectra by the AM1.5g spectrum and the spectral response of the solar cells that are used for prototype production.

The unprinted glass samples ($CR_{geometrical}$ = 0%) also features reflection and absorption. We calculate a relative transmittance ($T_{opt,\ rel}$) for both measurement setups which is relative to the transmittance of the unprinted glass sample.

The small-area illumination method (A) clearly underestimates the transmittance for higher coverage ratios, as expected. The differences between both measurement configurations increases with the coverage ratio (3.1$\%_{abs}$ at $CR_{geometrical}$ = 40%). For further consideration, the results from measurement B (large area) are therefore used.

The decrease of transmittance is not proportional to the increase of coverage. For example at 40% coverage, transmittance is still larger than 60%, showing that the geometrical coverage ratio alone is insufficient for predicting the power of a printed BiPV module.

Table 1: Average transmittance for different samples and characterization techniques, weighted with AM1.5g spectrum and spectral response of the solar cell used

$CR_{geometrical}$	0%	10%	20%	40%
$T_{opt,A}$	90.81%	82.32%	72.87%	58.32%
$T_{opt,A,rel}$	100%	90.7%	80.2%	64.2%
$T_{opt,B}$	90.56%	82.19%	74.35%	60.95%
$T_{opt,B,rel}$	100%	90.8%	82.1%	67.3%

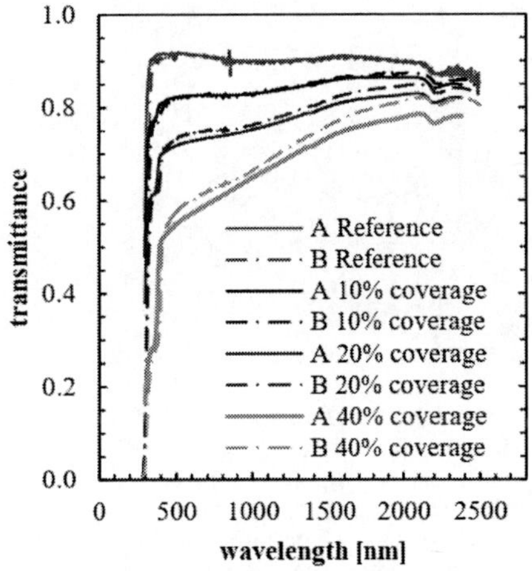

Figure 5: Transmittance spectra of glass samples covered with decorative elements (white dots), small (A) and large-area (B) illumination measurement setup, no additional weighting

Differences between geometrical coverage and transmittance measurement arise from a partial transparency of the decorative prints (3) as well as from scattering effects (4) as displayed in Figure 6. Shading of the solar cell (Figure 3) is replaced by reduced transmittance (Figure 6, 3 and 5). Scattering causes increased path lengths within the front glass that leads to additional attenuation of light. Within both optical measurement configurations, the transmittance of the glass samples is measured against air. Therefore light that is redirected by more than approx. 42° by scattering in the ceramic print undergoes internal (total) reflection at the glass/air boundary (5). As a result, increased attenuation as well as light redirection to the sides of the glass samples decreases the measured transmittance at the sensor. In the large-area illumination configuration, however, lateral losses are largely compensated by lateral gains.

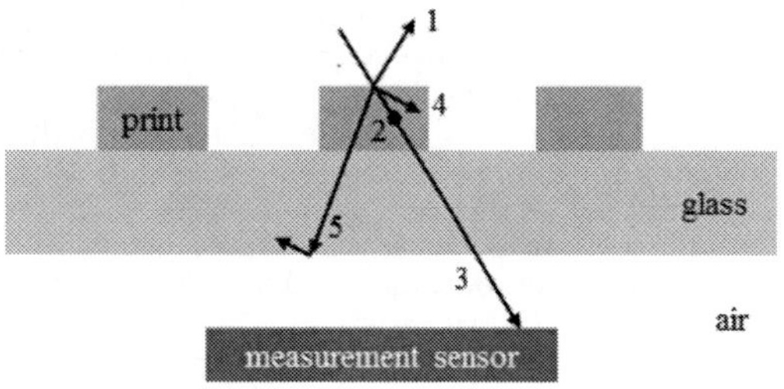

Figure 6: Schematic drawing of causes for differences between geometrical coverage and transmittance

c. Characterization of Printed Elements

We record microscope images of different coating elements and measured the size of the prints (example shown in Figure 7). The images show that the desired geometrical coverage does not equal the actual geometric coverage due to jagged edges and over-/underprinting.

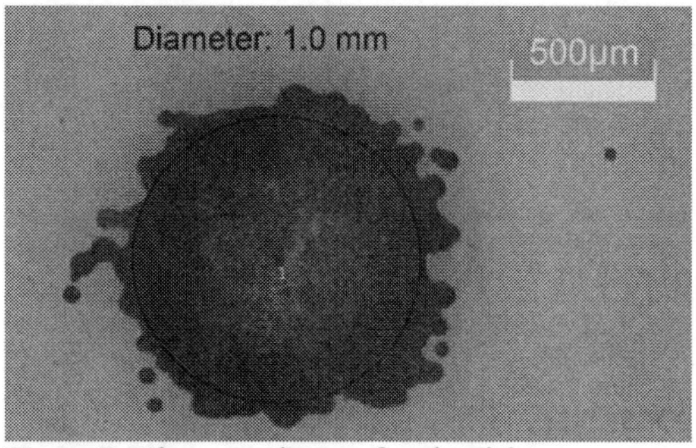

Figure 7: Microscope image of a printed dot

We perform a surface scan and find the thickness of printed elements to be inhomogeneous (Figure 8). Since the printed ink is partially transparent, the variation in thickness leads to spatially non-uniform transmittance of the printed element. We perform LBIC measurements on modules with printed glass and obtain spatially resolved information on module short circuit current. LBIC confirms the non-uniform transmittance of the printed elements (Figure 10).

Also the printed area features a rough surface, which leads to light scattering in transmission and reflection.

Figure 8: Surface profile of printed dots

We conclude that the geometrical coverage ratio is not a sufficient description of a print and that the coverage ratio may not be used as the sole basis for estimating the effective transmittance.

d. Power Measurements on Modules with Partially Printed Glass Covers

After measuring the transmittance of glass samples with different coverage ratios, we prepare mini-module samples (Figure 4) and measure electrical module parameters. For each sample, an IV measurement with a flash simulator is performed at the Fraunhofer ISE CalLab PV Modules under standard testing conditions (STC).

We find the short circuit current (I_{SC}) to deviate from the values which were estimated based on geometrical coverage. Current measurements also deviate from predictions based on transmittance measurements (Table 2). Increased coverage of the front glass cover does not translate into equivalently decreased current.

Table 2: Measured short circuit currents

$CR_{geometrical}$	0%	10%	20%	40%
$T_{opt,B,rel}$	100%	90.76%	82.10%	67.31%
$I_{sc,rel}$	100%	95.20%	89.20%	82.10%
$I_{sc,rel} - T_{opt,B,rel}$	0%	4.4%	7.1%	14.8%

Performing measurements of the module power, similar results are obtained (Table 3). The difference between the geometrical coverage ratio and expected power is even higher than for the short circuit current. The increased module power (P_{MPP}) in comparison to the short circuit current is caused by a better fill factor, due to a lower current. We find the difference between power and transmittance measurements to be significant. At 40% coverage a difference of $16\%_{abs}$ between transmittance and normalized power is found.

Table 3: Measured module power

$CR_{geometrical}$	0%	10%	20%	40%
$T_{opt,B,rel}$	100%	90.76%	82.10%	67.31%
$P_{MPP,rel}$	100%	95.90%	90.20%	83.60%
$P_{MPP,rel} - T_{opt,B,rel}$	0%	5.1%	8.1%	16.3%

In addition to our previous findings, we conclude that transmittance measurements on the glass samples alone are insufficient for predicting BiPV module power for increased coverage ratios. Differences between the setups for transmittance measurement (Figure 6) and module power measurement (Figure 9) arise from coupling gains [6] to the solar cell within the module (Figure 9, 7). Figure 9 also shows increased incidence of light on the cell surface through additional gains by internal reflection (6). Furthermore, the losses from reflection on position 2 of the front glass (inner interface) are reduced within the module (5) compared to the transmittance measurement, since the front glass and conventional encapsulation materials feature very similar refractive indices.

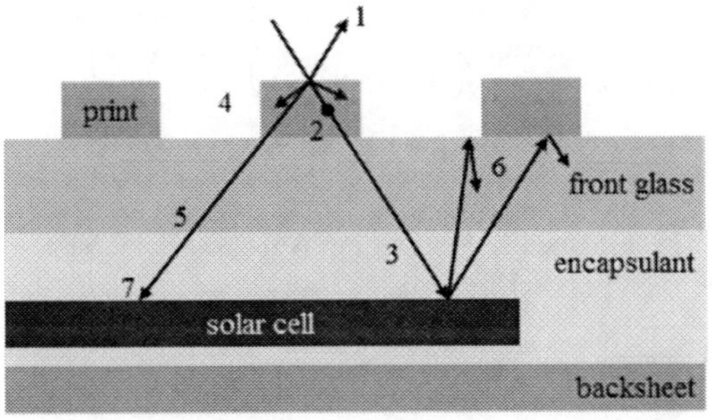

Figure 9: Schematic drawing of causes for differences between transmittance measurements and power measurements

In order to locally investigate the optical losses of the pattern with respect to cell current, we perform measurements on partially printed modules using laser-beam induced current (LBIC) and measure the external quantum efficiency (EQE) at five different wavelengths. On an area of 5x5 mm², 200x200 single spot measurements are performed. Samples with coverage ratios of 30% and 60% are measured. Using the EQE data from the LBIC measurements we compute the spatially resolved short circuit current (Figure 10).

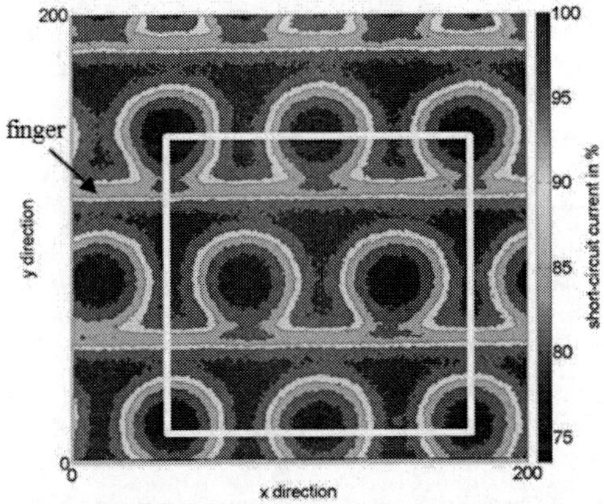

Figure 10: LBIC measurement results, spatially resolved relative short circuit current of a 1 mm white dot pattern. The area for spatial integration is outlined in white.

Horizontal lines in Figure 10 are due to the cell metallization (fingers).

We use a repeating area element (Figure 10, white square). By integrating over that area, we gain an effective value for short circuit current that includes the effects of scattering, and multiple reflections. It automatically weights the values for the printed and unprinted areas, analogous to the approach taken in [15, Annex C].

The integration over a defined area of the LBIC I_{SC} matrix is performed for the printed glass as well as for an unprinted reference. Both integrations are necessary to calculate the effective transmittance of the printed glass without additional effects of the cell metallization.

$$t_{eff,print} = \frac{\int_{x_{min,n}}^{x_{max,n}} \int_{y_{min,n}}^{y_{max,n}} I_{sc,norm} dxdy}{\int_{x_{min}}^{x_{max}} \int_{y_{min}}^{y_{max}} I_{sc,norm,Reference} dxdy} \quad \text{(II)}$$

We calculate the effective transmittance of printed front glass covers with different coverage ratios based on LBIC measurements performed on two different samples with coverage ratio of 30% and 60% (Figure 11, green square and red triangle, respectively). We find the calculated transmittance based on LBIC to be in good agreement with the transmittances based on the results of I_{SC} measurements ($\Delta < -0.37\%$abs.).

A prediction of the effective transmittance and thus of the module power for arbitrary coverage ratios is therefore possible based on two LBIC measurements.

Although a prediction of the effective transmittance based on LBIC measurements is possible, a prediction based on I_{SC} measurements is a simpler alternative, since both measurements require module prototypes and IV measurement equipment is widely available.

We calculate the effective transmittance of arbitrary coverage ratios (Figure 11, blue line) based on two I_{SC} measurements at 30 and 60% coverage (blue squares) and find them to be in good agreement with measured values (black squares; MSE = $0.59\%_{abs}$). While this procedure can be used to estimate the effective transmittance for different coverage ratios of the same print element, it requires module samples with two different coverage ratios of the same print element.

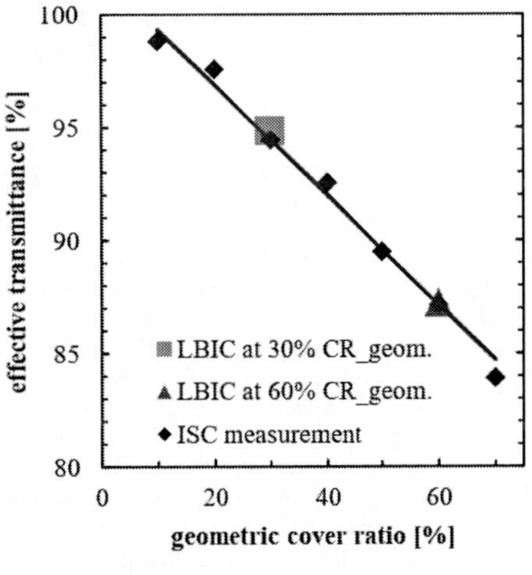

Figure 11: Calculation of the effective transmittance of printed front glass based on I_{SC} measurements on module samples with 30 and 60% coverage ratio versus measurements at other coverage ratios; effective transmittance calculated from LBIC measurements performed on samples with 30% (large green square) and 60% coverage ratio (large red triangle)

3 SUMMARY

Losses in common photovoltaic modules are well understood and models exist to predict module power. Effects of additional decorative elements have not yet been fully investigated. We analyze the effects in modules with printed front glass covers.

We perform electrical measurements on modules with different coverage ratios of printed elements. We find that the decrease in electrical power is not proportional to the increase in coverage ratio. The short circuit current is higher than expected, based on the degree of geometrical coverage alone.

We characterize the design prints by performing surface profilometry, microscope images and LBIC measurements. We find them to be non-homogeneous in thickness and also find their actual area to be different from the nominal size. LBIC images demonstrate that their transmittance is non-uniform.

We therefore conclude that the geometrical coverage ratio alone is not suitable for predicting module power or the effective transmittance of the printed glass.

We made transmittance measurements on glass that features design prints with different coverage ratios. Measurements were performed using illumination of small as well as large areas and with small (220 mm) as well as with large (620 mm) integrating spheres. We find the difference between the two measurement configurations to increase with the coverage ratio of the decorative print and the large-area measurements to result in higher transmittance values (Δ = 3.1$_{abs}$% at 40% coverage). While transmittance measurements of the glass panes alone include effects of imperfect prints and partial transparency, they obviously cannot take effects of optical coupling to solar cells in modules into account. Correspondingly, results of transmittance measurement and power measurements do not agree (Δ = 16%$_{abs}$ at 40% coverage). We conclude that the use of transmittance value of the glass alone is also insufficient to predict the module power.

We calculate the effective transmittance of glass panes with arbitrary coverage ratios based on selected LBIC measurements as well as on I_{SC} measurements and find them to be in good agreement with comparative IV measurements. A prediction of module power is therefore possible based on selected measurements on single module prototypes.

We made LBIC measurements on module samples with selected coverage ratios and calculated the effective transmittance of the printed glass in the module. We used this information to predict the transmittance for different coverage ratios. Transmittance calculated using spatially integrated LBIC is close to results from electrical measurements of the complete module and allows spatially resolved analysis of optical effects resulting from prints.

4 ACKNOWLEDGEMENTS

We would like to thank the German Ministry of Economic Affairs and Energy ("BiPVFab" FKZ 0324108A) for their funding.

5 REFERENCES

[1] European Parliament, "*Directive 2010/31/EU of the European Parliament and of the Council of 19 May 2010 on the energy performance of buildings*" http://data.europa.eu/eli/dir/2010/31/oj

[2] Verberne, G., et al, "*BIPV Products For Façades And Roofs: A Market Analysis*", 29th European PV Solar Energy Conference and Exhibition, 2014

[3] Scognamiglio, A., et al, "*Chapter 6 - Building-Integrated Photovoltaics (BIPV) for Cost-Effective Energy-Efficient Retrofitting*" Cost-Effective Energy Efficient Building Retrofitting: Woodhead Publishing, 2017, pp. 169–197

[4] Ferrara, C., et al, "*8 - Building-integrated photovoltaics (BIPV) A2 - Pearsall, Nicola,*" The Performance of Photovoltaic (PV) Systems, T1 - 8 - Building-integrated photovoltaics (BIPV) A2 - Pearsall, Nicola, Ed.: Woodhead Publishing, 2017, pp. 235–250

[5] Yang, T., et al, "*A review of research and developments of building-integrated photovoltaic/thermal (BIPV/T) systems*" Renewable and Sustainable Energy Reviews, vol. 66, pp. 886–912, 2016

[6] Haedrich, I., et al, "*Unified methodology for determining CTM ratios: Systematic prediction of module power*", Solar energy materials and solar cells 131 (2014), pp.14-23, ISSN: 0927-0248, SiliconPV, 2014

[7] Mittag, M., et al, "*Systematic PV module optimization with the cell-to-module (CTM) analysis software*", Photovoltaics International, vol. 36, 2017

[8] Mittag, M., et al, "*Electrical and Thermal Modeling of Junction Boxes*", 33rd European PV Solar Energy Conference and Exhibition, 2017

[9] Mittag, M., et al, "*Cell-to-Module (CTM) Analysis for Photovoltaic Modules with Shingled Solar Cells*", 44th IEEE Photovoltaic Specialists Conference, 2017

[10] Frontini, F., et al, "*Indoor And Outdoor Characterization Of Innovative Colored BIPV Modules For Façade Application*", 32nd European PV Solar Energy Conference and Exhibition, 2016

[11] Bläsi, B., et al, "*Morpho Butterfly Inspired Coloured BIPV Modules*", 33rd European PV Solar Energy Conference and Exhibition, 2017

[12] Chevalier, B., et al, "*Solar energy transmittance of translucent samples: A comparison between large and small integrating sphere measurements*" Solar Energy Materials and Solar Cells, vol. 54, no. 1–4, pp. 197–202, 1998

[13] Platzer, W. J., "*Directional-hemispherical solar transmittance data for plastic honeycomb-type structures*" Transparent insulation, vol. 49, no. 5, pp. 359–369, 1992

[14] Milburn, D., et al, "*An analysis of thick-sample effects in the measurement of directional-hemispherical transmittance*" Optics Communications, vol. 118, no. 1, pp. 1–8, 1995

[15] DIN EN 410:2011

PERFORMACE EVALUATION OF DIFFERENT ARCHITECTURAL FORMS AND ELECTRICAL TOPOLOGIES FOR BIPV PARKING LOTS

Cesar Biasi de Moura[1#], Sergio Shimura[2], Roberto Silva Simplicio[1], Rafael Herrero Alonso[1], Marcelo Knörich Zuffo[1]
[1]University of Sao Paulo, Polytechnic School, Department of Electronic Systems.
Av. Prof. Luciano Gualberto Trav.3 n. 158. CEP: 05508-900. Sao Paulo - SP – Brazil
[2]Federal Institute of Education, Science and Technology of Sao Paulo, Department of Electro-Electronic.
Rua Maria Cinto de Biaggi, n. 130. CEP: 18095-410. Sorocaba - SP – Brazil
[#]Phone: +55-11-30914248 – E-mail: cbmoura@lsi.usp.br

ABSTRACT: The objective of this study is to assess the impact on energy performance of different architectural forms of BIPV systems integrated into parking lots and the influence of the electrical connection configurations. We made the analysis of a 450 kWp photovoltaic installation installed in the parking lot at Candido Portinari Park, in Sao Paulo/SP. We analyzed four different architectural canopy shapes: inclined flat; half-circle arc; quarter-circle arc and a "V" shaped format. Additionally we considered the tilt angles of 6°, 10° and 20° for inclined flat and "V" shape formats. PVsyst V6.38 simulation software provided the results that were analyzed based on three metrics: the total energy produced in kWh, specific production and performance ratio (PR). The results indicate that the most appropriate shape is the inclined flat canopy with 6° tilt angle. However, the quarter-circle arc and "V" format with tilt angle between 6° to 10° and inclined flat with tilt angle of 10° also showed total energy produced close to the inclined flat canopy with 6° tilt angle with a difference of -3.3%, -1.5%, -2.3% e -0.2% respectively, demonstrating that these are feasible BIPV architectural forms for this parking lot.
Keywords: Photovoltaic power generation, Energy performance, BIPV

1 INTRODUCTION

Today the installed capacity of photovoltaic generation systems around the world already exceeded more than 130 GWp, reaching over 138.9 GWp in 2013. Projections indicate that in 2018 the installed capacity exceeds 400 GWp, be expected in American continent between 2.9 to 5.3 GWp of distributed generation installed [1].

BIPV systems integrated at parking lots offer an alternative to the installation of medium-sized photovoltaic systems in urban centers. They have the advantage of not occupying ground area as a conventional photovoltaic plant and be less affected by shading caused by typical obstacles of the urban environment as antennas, lightning rods, water towers, chimneys, etc., that may limit the useful area installation and reduce energy system performance [2].

In addition, other features like profile and values of solar insolation, operation temperature of the modules and the architectural form previously selected by the designer could indicate that the choice of a particular technology of photovoltaic modules, such as amorphous silicon modules, is more is more appropriate when compared to traditional silicon technologies (e.g. polycrystalline and monocrystalline). The designer can conclude that this choice is the most appropriate when considered, the generation costs for specific production (kWh / kWp), the installation location and the characteristics listed above [3].

Considering what it was previously described and presented studies on BIPV, this study shows relevant by approaching some of the choices that designers will face in the coming years when projects with BIPV systems integrated at parking lots becomes more and more present.

2 CHARACTERISTICS THAT AFFECT THE PERFORMANCE OF PHOTOVOLTAIC SYSTEMS

The performance of a photovoltaic system can be affected by different factors that affect the operation of the technology and/or limitations imposed by the installation site of project.

The performance of a photovoltaic system is inherent to technology when exposed to weather conditions of the site, essentially solar insolation, temperature and limits of the space for installation of the structure/support. Influencing the azimuth and tilt angle of the modules as well as the levels of shading in the presence of obstacles that cannot minimized. The inverter efficiency and distance between the PV array, cable connections, protection devices, soiling and degradation of modules also influence the performance of the system [4], [5] and [6].

Another important factor related to technology that affecting the yield of a photovoltaic system is the mismatch between the strings in same array. Several studies have presented their implications [5], [7], [8], [9] and [10].

The total energy generated is the amount of electrical energy injected into the utility power grid in kWh.

The specific production relates the total energy generated by the installed nominal peak power of the PV system, is an important way to measure performance between the different architectural form. The characteristics of the site where the project will installed affected this ratio, like as irradiation, temperature, wind speed, etc., is it useful for comparisons for different projects in the same location or for comparisons between different locations, but for the same project in kWh/kWp [4].

Already the performance ratio (PR) is a dimensionless number expressed as a percentage that relates the energy injected into the grid over time with the energy that would be injected into the grid to a system operating without losses, that is, constantly in conditions STC (Standard Test Conditions) [4] and [11].

For a photovoltaics system, studies indicate that the PR may vary from 60 to 90%. The literature considers a low PRs values below 70% and high PRs values above 80% [4], [12], [13], [14], [15] and [16]. However, there are already studies that indicate the possibility of photovoltaic systems with PR values above 90% in the near future [16].

3 METHODOLOGY

To evaluate the energy performance for the studied architectural forms, we realized a visit to the site of installation of the photovoltaic parking lot. During this visit, there was a distance measurement, position and approximate dimensions of obstacles obtained by CAD drawings, for field measurements with measuring tape and information obtained on the internet on the site of the present real estate projects surrounding the site, as well as photographic records with a reference object to validate the information obtained.

Based on this information, we built a 3D model of buildings and determined the shading potential in the installation site using the PVsyst V6.38 software. Due to absence of measured solarimetric data of the installation site at the time of this study was adopted Meteonorm 7.1 database.

The positioning of the modules on site are made to the photovoltaic parking lot accompany the existing layout of parking places to achieve the minimum power of 450 kWp.

For the architectural forms studied, "V" format and inclined flat canopy were studied the 6°, 10° and 20° angles. The angle of 6° was choice in the basic project of the constructor of photovoltaic parking lot. Already the angle of 10° was chosen because it is considered the lowest angle for self-cleaning of the modules [17], [18] and [19]. Finally, the angle of 20° for presenting the higher pure transposition factor for the location and azimuth if installation according the PVsyst V6.38.

For assessed architectural forms were simulated three different electrical topologies for inverters, central inverter with one and three MPPT(s), four inverters with one MPPT and string-inverter with one and two MPPT(s), in order to evaluate the energy performance for each of the 28 scenarios prepared.

In this study are considered the thermal losses by the way of installation of the modules, losses by Joule effect in AC/DC wiring, the losses from power transformer, losses by quality of modules, LID, electric mismatch, soiling, IAM and unavailability of the system.

3.1 Project presentation

The modules will installed in two sections of parking lot, with coordinates 23.5°S; 46.6°W and the modules used in the project is the SuryaVolt Series D SV-245D20 of DYA Solar with 245 Wp. This is a polycrystalline silicon module with 60 cells, 15.2% of efficiency and less of 4.5% of losses of efficiency at low irradiance. The Fig. 1 illustrates the modeled elements that may cause shading in the modules.

For the scenarios evaluated, there was the possibility of adopting the three different electrical topologies for inverters, considering the technical feasibility of the architectural format studied. For the simulation of the scenarios studied were chosen inverters Ingecon Sun 10 TL M (String-Inverter topology with 1 or 2 MPPTs), Ingecon Sun 110 TL B220 (Four inverters topology with 1 MPPT) and Ingecon Sun 410 TL M220 DC Indoor (Central topology with 1 or 3 MPPTs) manufactured by Ingeteam.

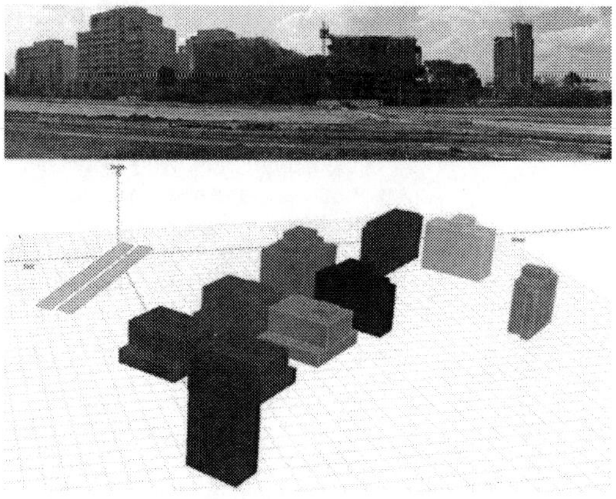

Figure 1: 3D model created from CAD drawings, field measurements, photographic records, etc.

To perform an evaluation of the influence of the number of MPPTs by inverter in the energy performance of the studied architectural forms, in the PVsyst V6.38 was enabling or not the "multi-MPPT" in order to make an exempt comparison between the results. It was not adopted in the study other inverter models with the same power and with only 1 MPPT for the simulations, as this could influence the results because of the difference in conversion efficiency values of the inverters, making it difficult to compare the simulations for the others scenarios.

3.2 Losses considered

As described in the methodology section, this study considers the various losses present in a photovoltaic system to evaluate this way the energy performance to the scenarios studied for the first year of operation.

For thermal and Joule losses of the system, was adopted the option “Free” mounted modules with air circulation in PVsyst V6.38 for systems in all scenarios for thermal losses. The losses by Joule effect on wiring between the inverter and transformer was used the value of 1.0% in STC, and these values chosen by recommendation on basic project of executor, such that the total losses in AC/DC wiring were less than 2.0%. We also considered losses in the transformer for this study. However, as this equipment is not specified in the basic project of executor, were used the values of 0.1% for the iron losses and 1.0% for resistive and inductive losses in the transformer, default values of PVsyst V6.38.

The losses due module quality and LID, were also used the default values stipulated by the PVsyst V6.38. For losses by quality of modules was adopted 2.5% because the tolerance on electrical parameters of module informed by the manufacturer is ±5%. Already for LID losses was used 2.0%, the default value on PVsyst V6.38 for silicon modules.

The mismatch losses for the different arrays were calculated using the PVsyst V6.38 and to perform the estimation of these losses has been adopted Gaussian distribution for an irradiance of 1000 W/m2 and 45° C of cell temperature, and was performed 500 random interactions. For the 28 scenarios studied, the estimation of mismatch losses with the PVsyst V6.38 resulting in a variation of 2.1% to 2.4%.

Table I: Accumulated monthly soiling losses considered

Month	Monthly soiling loss value
January	0.4%
February	0.9%
March	1.3%
April	1.7%
May	2.2%
June	2.6%
July	3.0%
August	3.5%
September	3.9%
October	4.3%
November	4.8%
December	5.2%

Already for soiling losses, it was adopted the value of 5.2% with approximately linear growth over the year, for simulations of the first year of operation (see Table I) [4]. Based on the identified values for annual soiling losses in Los Angeles, CA, city with features, size and problems with pollution of air similar to Sao Paulo, Brazil; we chose to adopt this value. The bibliographic survey also identified similar values in a photovoltaic plant in Cartagena, Spain [20] and [21].

For IAM losses was adopted ASHRAE model with "b_o" parameter equal to 0.05 in PVsyst V6.38.

Finally, for losses caused by unavailability was adopted an unavailability time fraction of 1.47% for central inverter topology, 1.42% for 4 inverters topology and 0.99% for string-inverter topology [22].

However, the number of inputs of unavailability on PVsyst V6.38 does not allow more than five periods per year, not allowing use the average number of failures per year and the average of downtime studied by [22] in simulations.

To enable in a more exempt way to compare the results of simulations without influence of other non-

Table II: Possible arrays layouts for each architectural form evaluated

Architectural Form	Electrical Topology	n. of Array	n. of Inverters	Size of Array	Tilt	Azimuth	Mismatch
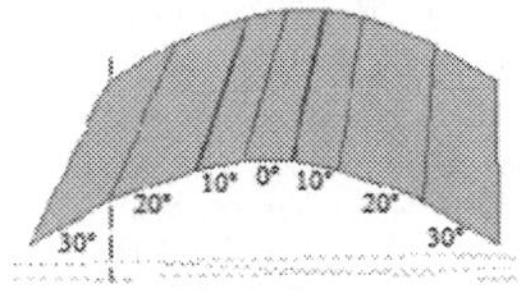 Half-Circle Arc	String-Inverter (1 or 2 MPPTs)	1	8	21x16	30.0°	-123.7°	2.4%
		2	8	21x16	20.0°	-123.7°	2.4%
		3	4	21x8	10.0°	-123.7°	2.4%
		4	4	21x8	0.0°	56.3°	2.4%
		5	4	21x8	10.0°	56.3°	2.4%
		6	8	21x16	20.0°	56.3°	2.4%
		7	4	21x16	30.0°	56.3°	2.4%
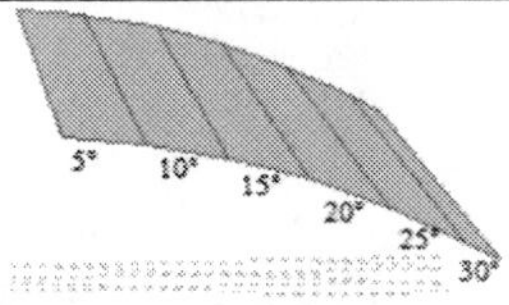 Quarter-Circle Arc	String-Inverter (1 or 2 MPPTs)	1	8	21x16	5.0°	56.3°	2.4%
		2	8	21x16	10.0°	56.3°	2.4%
		3	4	21x16	15.0°	56.3°	2.4%
		4	4	21x16	20.0°	56.3°	2.4%
		5	4	21x16	25.0°	56.3°	2.4%
		6	8	21x8	30.0°	56.3°	2.4%
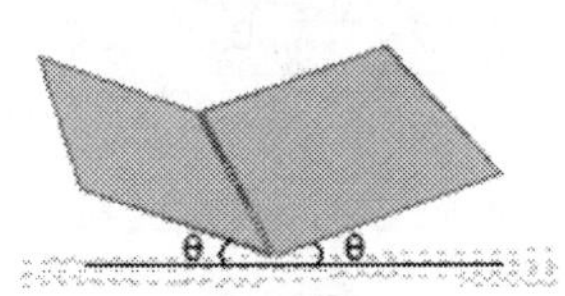 "V" Shaped Format	Four Inverters (1 MPPTs)	1	1	21x24	6°/10°/20°	56.3°	2.4%
		2	1	21x20	6°/10°/20°	-123.7°	2.5%
		3	1	21x20	6°/10°/20°	56.3°	2.5%
		4	1	21x24	6°/10°/20°	-123.7°	2.4%
	String-Inverter (1 or 2 MPPTs)	1	22	21x44	6°/10°/20°	56.3°	2.4%
		2	22	21x44	6°/10°/20°	-123.7°	2.4%
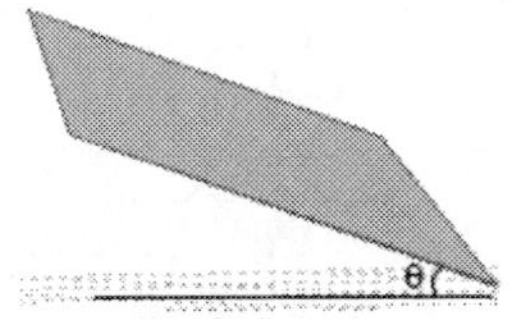 Inclined Flat Canopy	Central (1 MPPTs)	1	1	21x88	6°/10°/20°	56.3°	2.4%
	Central (3 MPPTs)	1	1	21x30	6°/10°/20°	56.3°	2.5%
		2		21x29	6°/10°/20°	56.3°	2.5%
		3		21x29	6°/10°/20°	56.3°	2.5%
	Four Inverters (1 MPPTs)	1	1	21x22	6°/10°/20°	56.3°	2.4%
		2	1	21x22	6°/10°/20°	56.3°	2.4%
		3	1	21x22	6°/10°/20°	56.3°	2.4%
		4	1	21x22	6°/10°/20°	56.3°	2.4%
	String-Inverter (1 MPPTs)	1	44	21x88	6°/10°/20°	56.3°	2.2%
	String-Inverter (2 MPPTs)	1	44	21x88	6°/10°/20°	56.3°	2.1%

controllable variables, we performed all simulations so that the five maximum downtime always occurs on the same days and schedules.
This way, we adopted in the simulations the following days and times: 03/09 at 08:00 am; 05/06 at 05:00 am, 06/02 at 04:00am, 09/30 at 08:00 am and 11/02 to 04:00 am.

3.3 Electrical topologies

How were evaluated four different architectural forms, for each of them was analyzed the possible electrical strings that could be adopted in the in a project of parking lot with photovoltaic generation, and all strings consist of 21 modules connected in series and positioned along the length of sections of parking lot. The modules are installed horizontally also along the length of the sections for the bypass diodes can operate properly, thereby reducing losses by partial shadings on modules caused by buildings in the surrounding of site and mutual shading caused by the sections themselves [3], [4] and [11].

The Table II shows the number of inverters, size of strings, and azimuth and tilt angle of modules for each string as well as their respective mismatch losses considered.

4 RESULTS AND DISCUSSION

The Table III shows the results of annual total energy produced, specific production and PR for the 28 simulated scenarios. Note that scenario n. 1 (half-circle arc with string-inverter 1 MPPTs topology) presented the worst energy performance when compared to the scenario n. 18 (inclined flat canopy with 6° tilt angle and string-inverter 2 MPPTs topology) the best performance. Although the PR for both scenarios have presented a variation of only 2.80%, scenario n. 1 showed a value of annual total energy produced -8.53% lower compared to the scenario n. 18.

The relative variation between the values of PR and annual total energy produced found between scenarios n. 1 and n. 18 is because the PR quantifies the percentage of utilization of the PV system in relation to the available solar resource, including all system losses. This manner we can conclude that scenario n. 1 had a utilization of almost as good as the available solar resource scenario n. 18, taking into account the chosen module.

See the Table III, we can see that the scenarios n. 3 and n. 4 (Quarter-Circle Arc) have a 5.6% higher performance approximately to annual total energy produced in relation to scenarios n. 1 and n. 2 (Half-Circle Arc). Despite these four scenarios presented near values of PR, with less than 0.5% difference between the scenarios n. 3 and n. 4 and scenarios n. 1 and n. 2, indicating that both formats can be used satisfactorily available solar energy.

This is due the fact that more than 45% of the modules in the Half-Circle Arc form are installed, seeking to follow the existing layout of parking places. This causes these modules are pointing to the south, as

Table III: Annual total energy, specific production and PR for the 28 scenarios evaluated

Scenario	Architectural Form	Tilt θ	Electrical Topology	Number of MPPTs by Inverter	Annual Total Energy Produced (kWh)	Specific Production (kWh/kWp)	PR
1	Half-Circle Arc	-	String-Inverter	1	458,823	1,013.39	73.20%
2				2	458,969	1,013.71	73.30%
3	Quarter-Circle Arc	-	String-Inverter	1	484,779	1,070.72	73.70%
4				2	484,926	1,071.04	73.70%
5	"V" Shaped Format	6°	Four Inverters	1	491,213	1,084.93	75.40%
6			String-Inverter	1	494,106	1,091.32	75.80%
7				2	494,124	1,091.36	75.80%
8		10°	Four Inverters	1	487,199	1,076.06	75.20%
9			String-Inverter	1	490,133	1,082.54	75.70%
10				2	490,156	1,082.60	75.70%
11		20°	Four Inverters	1	465,340	1,027.79	73.80%
12			String-Inverter	1	468,450	1,034.65	74.30%
13				2	468,593	1,034.97	74.30%
14	Inclined Flat Canopy	6°	Central	1	493,250	1,089.43	74.30%
15				3	492,799	1,088.43	74.60%
16			Four Inverters	1	495,911	1,095.31	75.10%
17			String-Inverter	1	501,044	1,106.64	75.90%
18				2	501,595	1,107.86	76.00%
19		10°	Central	1	492,148	1,087.00	74.30%
20				3	491,637	1,085.87	74.20%
21			Four Inverters	1	494,772	1,092.79	74.70%
22			String-Inverter	1	499,907	1,104.13	75.50%
23				2	500,545	1,105.54	75.60%
24		20°	Central	1	478,700	1,057.29	72.60%
25				3	477,996	1,055.74	72.60%
26			Four Inverters	1	480,855	1,062.05	73.00%
27			String-Inverter	1	486,357	1,074.20	73.80%
28				2	486,810	1,075.21	73.90%

the azimuth listed in Table II instead of the north, which is the recommended from the viewpoint of photovoltaic projects for the local latitude reducing considerably the global irradiance in the collector plane of modules for these arrays [3], [4] and [11].

As for the Inclined Flat Canopy form is possible to verify that the 6° tilt angle presented the best results for annual total energy produced and PR (scenarios n. 14 to n. 18), which contradicts the initial forecasts that pointed the 20° tilt angle (scenarios n. 24 to n. 28). Initial forecasts pointed the 20° angle as that present has a better performance due to a higher transposition factor, calculated using the PVsyst V6.38. It was verified through simulations that this resulted due to the increase in losses from near shading caused by buildings located in the surroundings, and the losses were approximately 40% higher for the 10° tilt angle (scenarios n. 19 to n. 23) and 200% for 20° tilt angle (scenarios n. 24 to n. 28).

Already the choice of electrical topology, it was verify that contrary to initial expectations, the central topology with 3 MPPTs had a lower annual total energy produced for the scenarios n. 15, n. 20 and n. 25. This was because the central topology with 3 MPPTs presents a loss by mismatch on arrays for each MPPT input bigger then the central topology with 1 MPPT (see Table II), also being observed through simulations that there was no reduction of electrical losses caused by shading even using a inverter with 3 MPPTs inputs.

Also with respect to the Inclined Flat Canopy form, each section of parking lot connects to a different input of the three MPPTs inputs available on inverter. The remaining strings (located at the ends both section of parking lot to the north), were connect to the third MPPT input of inverter.

In this case, the increase in MPPTs inputs to the central inverter showed no improvement to the value of annual total energy produced and PR therefore, the shading caused by buildings in the surroundings of site has a shading pattern with sizes smaller than the area occupied for each MPPT input of the inverter.

As the granularity of the shading pattern caused by buildings and verified through PVsyst V6.38 within the Construction/Perspective tool, are smaller than the area occupied by each of the three MPPTs inputs of the inverter, this results in the fact that central topology with 3 MPPTs cannot optimize the generation more than central topology with 1 MPPTs.

This fact was also confirmed by simulations with the four inverters topology because, the gain between 0.45% and 0.55% for annual total energy produced and from 0.4% to 0.8% approximately for PR of four inverters topology against central topology with 1 MPPTs.
This has as origin the reduction of losses by efficiency and availability on inverter that this topology presents, not by electrical losses by shading.

In this way, the string-inverter topology was the one presenting best results, in addition to having lower electrical losses caused by shading. It also has lower efficiency losses in the inverter and lower losses due to unavailability, what causes the string-inverter topology has a better performance approximately 1.7% for annual total energy produced and from 1.3% to 1.7% for PR against the central topology with 1 MPPT for the simulated angles.

Lastly, the "V" shaped format form presented results that are similar both to the half-circle arc form as to the inclined flat canopy form, depending on the analyzed angle. Just as the inclined flat canopy form, it we see that the 6° tilt angle presented the best results for annual total energy produced and PR. The results for 6° tilt angle presented closely values when compared to 10° tilt angle, which contradicts the same initial forecasts that pointed the 20° tilt angle as the angle with better performance due a higher transposition factor when calculated using the PVsyst V6.38.

As the inclined flat canopy form, in the "V" shaped format form the worst performance of the 20° tilt angle against the 6° tilt angle is also due to increase in losses with shading of the buildings located in the surroundings. These losses were approximately 24% higher for 10° tilt angle and 200% higher for the 20° tilt angle, however this is not only cause for "V" shaped format form with 20° tilt angle presents a difference of -5.17% in annual total energy produced and -1.5% in PR compared to the 6° tilt angle.

Additionally, to the problem of shading losses, the "V" shaped format form presented the same problem that half-circle arc format, with half of the modules also directed to southeast with -123.7° azimuth angle, to follow the existing layout of parking places. This has resulted in approximately 3.2% more losses by global irradiance in the collector plane of the modules when compared to the 6° tilt angle.

The fact that 50% of the modules was install to southeast also causes effects in 6° and 10° tilt angles to "V" shaped format form. However, these effects are attenuated due to the tilt angles are near to 0° (horizontal plane). This showed that in this case the attempt to increase the tilt angle of the modules with the intention of obtaining a better transposition factor causes a loss of yield for the system, when they are not considered if the architectural form adopted makes a considerable portion of the modules is pointed to south direction.

Already the adoption of the string-inverter topology against four inverters topology for the "V" shaped format form, presented the expected results, being the use of string-inverter topology with 2 MPPTs increase the annual total energy produced between 0.6% to 0.7% and 0.4% to 0.5% for the PR.

5 CONCLUSION

The impact study on the energy performance of BIPV systems is increasingly significant due to increase of generation energy costs, reduction of costs for the photovoltaic generation observed in recent years and the current context of Brazilian energy matrix. The results of this study show that, in the case of Villa Lobos Solar project, the architectural form with higher performance is the inclined flat canopy with 6° tilt angle and string-inverter topology with 2 MPPTs inputs. The difference in the results in each assessed scenario indicates that the choice of architectural form and electrical topology of the installation should be carefully analyzed, not meaning in any way the others architectural forms and electrical topologies are inadequate to this project and much less for projects located in other places, with near obstacles and positioning on terrain different from used in this study.

Despite the scenario with inclined flat canopy present better results, during the implementation of the project the architectural visual have a significant impact on the choice of the architectural form of parking lot.

In this way, we can use an architectural visual like the quarter-circle arc form or "V" shaped format form. This is shown by the results of scenarios n. 4, n. 7 and n. 10, who presented an approximate difference of -3.3%, -1.5% and -2.3% in annual total energy produced and PR values greater than 73%, 75% and 75% respectively when compared to the inclined flat canopy with 6° tilt angle (scenario n. 18), being that systems also have satisfactory energy performance.

Taking in consideration the results observed, the main reasons that made the half-circle arc form (scenarios n. 1 and n. 2) presents an approximate difference of -8.5% in annual total energy produced against the inclined flat canopy with 6° tilt angle (scenarios n. 14 to n. 18) is not the architectural form itself, but its positioning on site follow the existing layout of parking places as a project requirement.

This fact was the main cause of losses in this system because it led to losses in global irradiance on the collector plane of modules. Is possible well as verify by simulations, that the shading losses and electrical losses by shading were 1.6% higher in absolute value, thus explaining why such performance difference in the value of annual total energy produced.

It should be emphasized that the results observed for energy performance for the analyzes involving the 6°, 10° and 20° tilt angles for architectural forms "V" shaped format and inclined flat canopy are reflections of the conditions adopted in the simulations. This does not mean that the values of monthly losses for soiling in Table I and used for the three analyzed angles, accurately reflect the values that would be observed at the installation site.

Considering that the tilt angle has an influence on the values of soiling losses, this study has adopted a simplified premise and depending on the values that would be observed in the field, the final result of the simulations could eventually point the 10° tilt angle as the angle with larger annual total energy produced and PR values, if the soiling losses for 6° tilt angle show significantly higher for 10° tilt angle in the site, once the difference in annual total energy produced between the format inclined flat canopy with 6° and 10° tilt angles for string-inverter topology with 2 MPPTs inputs presented a difference of only 0.21%.

With respect the choice of electrical topology to be adopted in the system, the use of string-inverter topology with 2 MPPTs inputs may not necessarily be the most appropriate when taking into consideration other issues that were not the subject of this study, as the financial issue of price per kW for every power range of inverters and installation difficulty allied with system reliability.

Although it was not done an economic analysis for this study, it is important to note that topologies with central inverters tend to present a cost of USD/kWh lower than string-inverter topology, despite having 0.6% to 1.8% less of annual total energy produced. Thus, the choice of the correct topology will depend not only on energy performance, but additional factors such as the importance of the architectural visual of installation, system cost, availability, among other factors.

In the future, with the completion of Solar Villa Lobos project it will be possible compare the measured data to parking lot with photovoltaic generation with the simulated values and calculated in this study and, if necessary, reassess the methodology of this study, thus proposing its progressive and constant improvement.

6 ACKNOWLEDGE

The authors want to acknowledge the financial support by the Companhia Energetica de Sao Paulo (CESP) and others four utility companies, through the R&D public proposal from Brazilian government energy agency/ANEEL; the sponsors of Laboratory of Interdisciplinary Integrated Technologies (LSI-TEC); the project "Desenvolvimento e instalacao piloto de geracao fotovoltaica para modelo estrategico de referencia tecnologica, regulatoria, economica e comercial, inserindo esta energia na matriz energetica nacional" (ANEEL code: PE-0061-0034/2011).

7 REFERENCES

[1] EPIA 2014, "Global market outlook for Photovoltaics 2014-2018." European Photovoltaic Industry Association (2014) 60.

[2] R. Herrero, et al., Proceedings 29th European Photovoltaic Solar Energy Conference, Vol. I (2014) 2612.

[3] R. Rüther, Edifícios solares fotovoltaicos: o potencial da geração solar fotovoltaica integrada a edificações urbanas e interligada à rede elétrica pública no Brasil. 1st Ed, Florianópolis, UFSC/LABSOLAR (2004).

[4] H. Häberlin, Photovoltaics System Design and Practice, John Wiley & Sons (2012).

[5] B. Burger, R. Rüther, Inverter sizing of grid-connected photovoltaic systems in the light of local solar resource distribution characteristics and temperature. Solar Energy (2005).

[6] Zomer, et al., The balance between aesthetics and performance in building-integrated photovoltaics in the tropics. Progress in Photovoltaics: Research and Applications (2014).

[7] P. Bakas, A. Marinopoulos, B. Stridh, Impact of PV Module Mismatch on the PV Array Energy Yield and Comparison of Module, String and Central MPPT. 38th IEEE Photovoltaic Specialists Conference - PVSC (2012).

[8] H. Patel, V. Agarwal, MATLAB-Based Modeling to Study the Effects of Partial Shading on PV Array Characteristics. IEEE Transactions on Energy Conversion (2008).

[9] J. Webber, E. Riley, Mismatch Loss Reduction in Photovoltaic Arrays as a Result of Sorting Photovoltaic Modules by Max-Power Parameters. ISRN Renewable Energy (2013).

[10] R. Zilles, E. Lorenzo, An Analytical Model for Mismatch Losses in PV Arrays. International Journal of Solar Energy (1992).

[11] G. M. Masters, Renewable and Efficient Electric Power Systems, John Wiley & Sons (2014).

[12] B. Marion, et al., Performance Parameters for Grid-Connected PV Systems. 31st IEEE Photovoltaic Specialists Conference - PVSC (2005).

[13] C. D. Zomer, Método de Estimativa da Influência do Sombreamento Parcial na Geração Energética de Sistemas Solares Fotovoltaicos Integrados em Edificações. Universidade Federal de Florianópolis. Tese de Doutorado UFSC (2014).

[14] S. H. Ferreira de Oliveira, R. Zilles, Grid-Connected Photovoltaic Systems: The Brazilian Experience and The Performance of an Installation. Progress in Photovoltaics: Research and Applications (2001).

[15]R. Rüther, M. M. Dacoregio, Performance Assessment of a 2 kWp Grid-Connected, Building-Integrated, Amorphous Silicon Photovoltaic Installation in Brazil. Progress in Photovoltaics: Research and Applications (2000).
[16]N. H. Reich, B., et al., Performance Ratio Revisited: Is PR> 90% realistic?. Progress in Photovoltaics: Research and Applications (2012).
[17]A. Nobre, et al. 27th European Photovoltaic Solar Energy Conference, Vol. I (2012) 3763.
[18]A. Nobre, et al. 28th European Photovoltaic Solar Energy Conference, Vol. I (2013) 3673.
[19]J. T. Pinho, M. A. Galdino, Manual de Engenharia para Sistemas Fotovoltaicos. Rio de Janeiro, CEPEL – CRESESB (2014).
[20]J. T. Miller, J. Jackson, Impact of Soiling and Pollution on PV Generation Performance. CH2MHILL and Sandia National Laboratories Technical Memorandum (2009).
[21]E. Lorenzo, R. Moretón, I. Luque, Dust effects on PV array performance: in-field observations with non-uniform patterns. Progress in Photovoltaics: Research and Applications, v. 22, n. 6 (2014) 666.
[22]S. Shimura, et al. Production costs estimation in photovoltaic power plants using reliability. Solar Energy, Vol. 133 294-304 (2016).

BIPV AFFORDABILITY

Laura Maturi*[1], Jennifer Adami[1], Marco Lovati[1], Francesca Tilli [2], David Moser[1]
[1]Institute for Renewable Energy, EURAC Research, Viale Druso 1, 39100 Bolzano (BZ), Italy
[2]Gestore Servizi Energetici GSE Spa, Viale Maresciallo Pilsudski 92, 00197 Roma, Italy
*Phone: +39 0471 055633; Fax: +39 0471 055699; E-mail: laura.maturi@eurac.edu

ABSTRACT:
Several recent international surveys [1], [2] carried out among BIPV stakeholders reveals that one of the main barriers for the wide-spread of BIPV (building integrated photovoltaic) systems is the high cost.
The economic issue is still perceived as a barrier by architects and contractors, who are the main BIPV stakeholders.
On the other hand, the drastic cost breakdown of PV in recent years has enormously decreased BIPV prices leading to cost competitiveness with standard building materials.
It is thus essential to increase trust of architects, investors and financers stakeholders, showing business cases and real stories. Architects perception is highly influenced by "tangible examples" and real experiences more than by theoretical calculations.
This paper shows 16 realized BIPV projects as business case studies, providing information on their final user costs.
The case studies were selected among more than 40 examples collected in a local "call for case study", in order to represent "ordinary BIPV high quality" examples, including several kind of integration typologies representing both private and public sector. Our investigation field is not on "extraordinary, archistar-designed" BIPV projects, but on "ordinary BIPV high quality", meaning BIPV cases with high quality which have high replication potential overall Europe.
The economic matter is tackled from two different perspectives: the "PV" perspective –normalizing the cost to kWp, and the "building" perspective –normalizing the cost to m^2.
Results show that the cost of the analyzed BIPV systems, which construction year lays between 2004 and 2015, ranges (i) from 2.500 €/kWp to 8.300 €/kWp, with an average of around 5.500€/kWp and (ii) from 300 €/m^2 to 1.300 €/m^2, with an average of around 600 €/m^2. A clear decreasing trend is shown for the last decade (from 2004 to 2015) with values of ~8.000 €/kWp and ~950 €/m^2 in 2004 and of ~3.300 €/kWp and ~400 €/m^2 in 2015.
Looking in particular at the "building" perspective (which might be more interesting for architects), it is shown that the BIPV system capital cost lays in an acceptable range and it is even cheaper than some standard passive building materials [3] (e.g. glazed curtain walls, stone and others). This, without even considering the pay-back time period, which ranges from 4 to 11 years for the presented case studies and which is "infinite" for standard passive solutions (without taking into consideration energy savings).
These tangible examples demonstrate that, despite the economic issue is still perceived as a barrier for the wide spread of BIPV systems [1], [2], the use of PV in architecture is instead absolutely viable for many cases.

1 INTRODUCTION

Several definitions of BIPV have been adopted in literature and there is no univocal consensus within the PV community and the building sector about BIPV categorization and criteria (examples of criteria and definitions can be found in [4]–[6]). At international level, the recent standard EN 50583-1:2016 "Photovoltaics in buildings" [7], states that photovoltaic modules are considered to be building-integrated, if the PV modules form a construction product providing a function as defined in the European Construction Product Regulation CPR 305/2011 [8].
The meaning of the acronym BIPV in this publication, is intended in a broader meaning compared to the EN standard definition and in particular refers to a triple concept of integration: technology, aesthetic and energy integration. Technology integration is meant as the capability of the PV system to be "multifunctional" (as intended in the EN Standard) and aesthetic refers to the architectural appealing. The energy integration refers to the capability of a PV system to interact with the building and district energy system in order to maximize the local use of the produced electricity.
This is in our view a quite important aspect which is often not considered in the traditional BIPV definitions.
In fact, despite the existing "BIPV" definitions, we believe that in order to succeed in the BIPV system design, all three aspects must be considered. The selected BIPV cases studies are considered respondent to this triple concept.

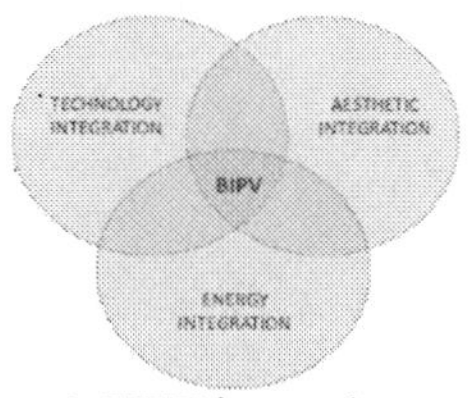

Figure 1: BIPV integration concept

2 CASE STUDIES SELECTION METHODOLOGY

In order to collect the most representative case studies for our investigation (i.e. ordinary BIPV high quality examples), a local "call for case studies" was launched by contacting most of the engineers, architects and professionals of Trentino Alto Adige region through

several channels (projects contact list, Engineers and Architect associations). This region has been very active in recent years in the BIPV field by boosting PV use and building energy efficiency through several measures: incentive schemes, dedicated policies, awareness raising, guidelines development and public engagement in the use of PV in public buildings.
Out of more than 40 collected cases, the best ones were selected through an internal workshop involving 19 Eurac collaborators, forming a "selection group". The selection was based on two main criteria, i.e. "overall impression/global approach" and "lesson learnt". The collaborators were asked to evaluate each case study by considering the following aspects:

- Overall impression / global approach, considering both technology and aesthetic integration, in particular:
  - the way PV integrated in the whole concept of the building and contributes to/preserve architectural quality;
  - the whole building concept;
  - surface texture, composition of visible materials, correlations of colours, details (especially successful new designs)/joints/fixing.
- Lesson learnt, also considering the energy integration, according to the following questions:
  - Is it an interesting experiment of PV integration?
  - Is there something to learn about it?
  - Are there innovative architectural concepts, original solutions, original applications?

Figure 2: Pictures of the workshop for the selection process

The cases getting a score higher than 20 marks were selected and included in this paper, meaning that the great majority of the "selection group" found the project interesting and inspirational for architects, as "ordinary BIPV high quality" examples (according to the BIPV definition given in the introduction).
In this way, 19 best case studies were selected.
For each of the selected case studies, a questionnaire was performed to collect information from at least one of the involved stakeholders (e.g. architect, owner, PV system electrical designer, installer, general contractor). Out of these 19, we could collect information and authorizations to publish on "cost issues" just from 16 of them.

3 CASE STUDIES OVERVIEW

The case studies were categorized according to the EN standard 50583:2016 "Photovoltaics in buildings" [9], as shown in Figure 3.

Figure 3: "Photovoltaics in buildings" categories according to [9]

The categories schematized in Figure 3 can be described as follows (from left to right):

- Category A: Sloped, roof-integrated, not accessible from within the building
- Category B: Sloped, roof-integrated, accessible from within the building
- Category C: Non-sloped (vertically) mounted not accessible from within the building
- Category D: Non-sloped (vertically) mounted accessible from within the building
- Category E: Externally integrated, accessible or not accessible from within the building

BIPV Case study	Architectural system	Standard category	Ownership
1	Semi-transparent roof	B	Private
2	Opaque tilted roof	A	Public
3	Semi-transparent roof	B	Private
4	External semi-transparent device	E	Private
5	Semi-transparent roof	B	Private
6	Opaque tilted roof	A	Private
7	External opaque device	E	Private
8	Opaque tilted roof	A	Private
9	Opaque cold façade	C	Private
10	External semi-transparent device	E	Public
11	Opaque tilted roof	A	Private
12	Opaque cold façade, semi-transparent façade	C, D	Public
13	Opaque cold façade	C	Public
14	Semi-transparent roof-façade	B, D	Private
15	Opaque cold façade	C	Public
16	Opaque tilted roof	A	Public

Table I: Case studies characterization according to their BIPV architectural system

Among the selected case studies, examples of different type of building typologies are presented, including office, residential, agricultural, industrial, community, religious, commercial and transportation buildings.
Several architectural integration are shown, including opaque and semi-transparent roof, warm, cold and double skin facades as well as external devices such as parapets and solar shading elements. The most predominant are façade and roof systems.
Different kind of ownerships (private and public) are also represented, giving an overview on different approaches to the BIPV matter, especially regarding the decision making related to economic issues.

Figure 4: Pictures of analyzed case studies

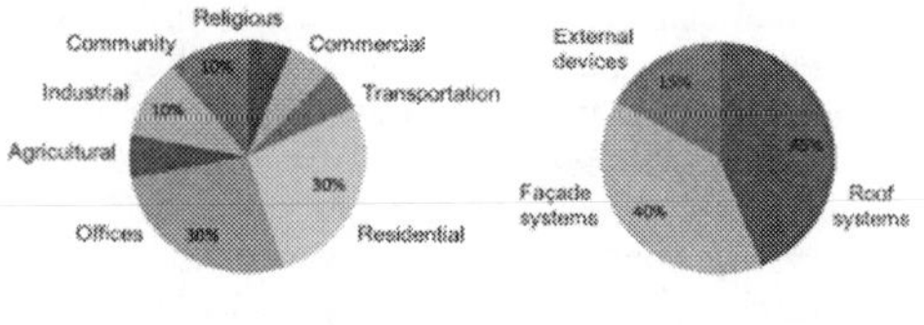

Figure 5: Building typologies and architectural integration types presented in the case studies

Regarding the PV modules, the crystalline technology is the mostly used, being applied in around 80% of the analysed case studies. Most of the systems are made using standard modules (only 30% are made with costume-made modules), showing that in many cases appealing BIPV systems can be realized without the need of customization.
A more detailed case studies description is insert in [12].

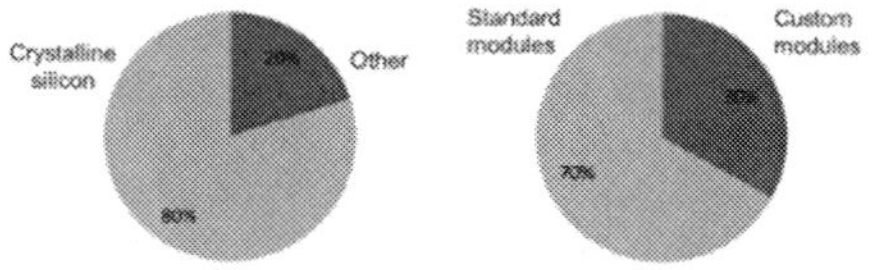

Figure 6: PV technologies and module types presented in the case studies

4 RESULTS

Figure 7 and Figure 8 show a summary of the cost of each BIPV system, looking at the same issue from two different perspectives (i.e. the "PV" perspective and the "BUILDING" perspective).
Figure 7 normalizes the BIPV systems cost to the nominal power (€/kWp), an indicator which is always used in the "PV sector". The cost of the analyzed BIPV systems ranges from 2.500 €/kWp to 8.300 €/kWp, with an average of around 5.500€/kWp, considering that the analyzed systems were built in the decade between 2004 and 2015.
This variation can be ascribed to several factors, such as the type of technological integration, type of components and, mostly important, the construction year, since the PV cost has seen an impressive decrease in the last few years.
In particular, looking at the technological integration types, the following average values are found:

- Opaque cold façade: ~7.900 €/kWp
- Semi-transparent roof-façade: ~5.100 €/kWp
- External device: ~4.900 €/kWp
- Opaque tilted roof: ~4.400 €/kWp

In order to look at the economic matter from another perspective, the cost has been normalized to the envelope covered surface (€/m^2), thus using an indicator which is normally used in the "BUILDING sector" (see Figure 8). The cost of the analyzed BIPV systems ranges from 300 €/m^2 to 1.300 €/m^2, with an average of around 600 €/m^2.
In particular, looking at the technological integration types, the following average values are found:

- Opaque cold façade: ~ 850 €/m^2
- Opaque tilted roof: ~ 600 €/m^2
- Semi-transparent roof-façade: ~ 500 €/m^2

- External device: ~ 500 €/m²

As the €/kWp index, the cost variation can be ascribed to several factors. In particular, this time a crucial role is played by the PV module efficiency. For this reason, looking at this indicator might be misleading but it is very useful to compare the BIPV system cost with standard building materials. It demonstrates that in fact, the BIPV system capital cost lays in an acceptable range and it is even cheaper than some standard passive building materials [3] (e.g. glazed curtain walls, stone and others). This, without even considering the pay-back time period, which ranges from 4 to 11 years for the presented case studies (this information was not available for all cases) and which is "infinite" for standard passive solutions (without taking into consideration energy savings).

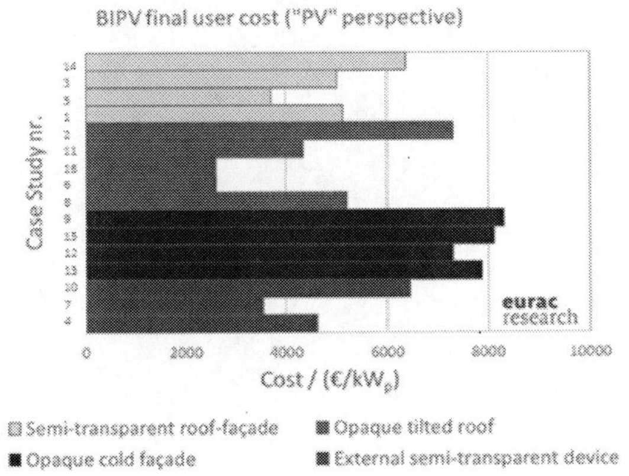

Figure 7: Final user costs of the analysed BIPV systems, normalized to the system nominal power

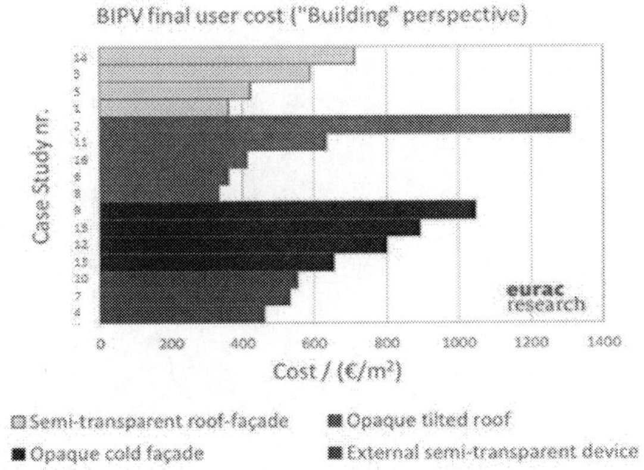

Figure 8: Final user costs of the analysed BIPV systems, normalized to the envelope covered surface

As mentioned above, the cost variation is very much influenced by the construction year, since the PV cost has seen an impressive decrease in the last few years. Figure 9 and Figure 10 show the trend over the years of the final user BIPV systems cost, considering the "PV" and "Building" perspective.

A clear decreasing trend is shown for the last decade (from 2004 to 2015) with values of ~8.000 €/kWp and ~950 €/m² in 2004 and of ~3.300 €/kWp and ~400 €/m² in 2015.

By comparing this data with standard not integrated PV systems, we might conclude that the "BIPV" trend cost in 2015, corresponding to 3.300€, is not too far from a ground-PV solution (considering a base line cost of around 2.500 €/kWp, typical of small plants in the last recent years).

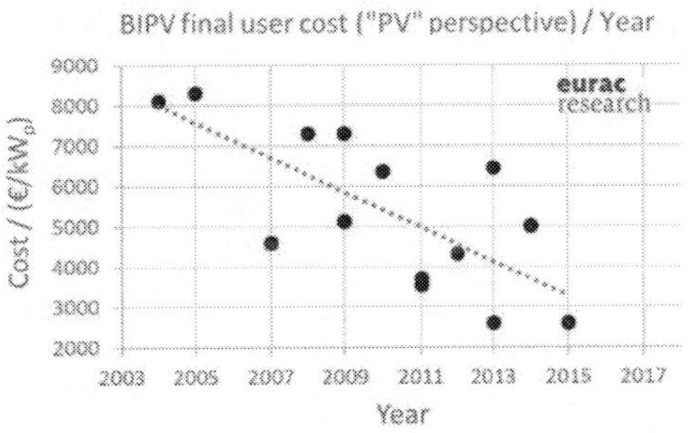

Figure 9: Final user costs of the analysed BIPV systems per construction year, normalized to the system nominal power

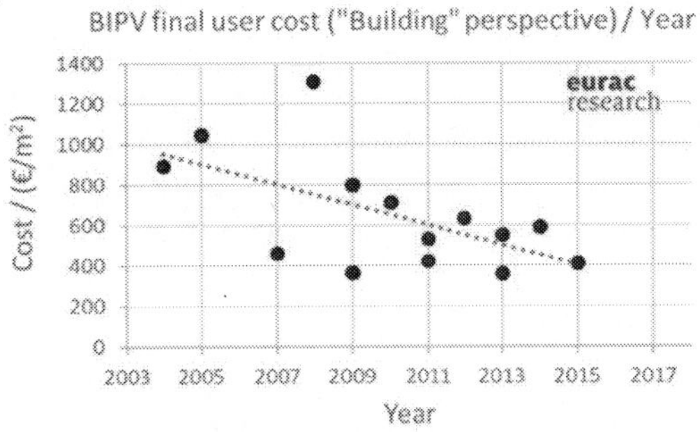

Figure 10: Final user costs of the analysed BIPV systems per construction year, normalized to the envelope covered surface

5 CONCLUSIONS

These tangible examples demonstrate that, despite the economic issue is still perceived as a barrier for the wide spread of BIPV systems [1], [2], the use of PV in architecture is instead absolutely viable from this point of view. A clear decreasing trend over the years during the last decade is shown in Figure 9 and Figure 10.

Even if some incentive schemes are over in Europe (e.g. the "Conto Energia" for Italy, which lasted from 2005 until 2013), they have paved the way to an irreversible process that cannot be stopped anymore. At Italian level, the current support schemes, relies mainly on two measures: tax credit, which allows to rescue 50% of the capital cost in 10 years, and the "net billing scheme" managed by GSE, which valorizes from an economic point of view delivered to the grid.

The economic viability of BIPV systems is thus preserved, even if we can somehow read a conceptual shift in the way to reward it: with the "Conto Energia" the formal and technological integration was mainly boosted (through a higher contribution foreseen for "innovative BIPV"), while, with the current schemes,

the energy integration is mostly pursued, in order to maximize the energy match between the produced and consumed energy.
This energy integration will become more and more important to cope with new ways to conceive buildings and their energy provision. In fact, also thanks to the EU policy oriented to promote the NZEB (nearly zero energy buildings) concept and RES (renewable energy sources) exploitation [10], [11], buildings are becoming more than just stand-alone units using energy from the grid. They are becoming micro energy hubs consuming, producing, storing and supplying energy, thus transforming the EU energy market, shifting from centralised, fossil-fuel based, national systems towards a decentralised, renewable, interconnected and variable system.
In this context, PV integration is irrevocably destined to play an essential role in the years to come and, learning from the experience gathered in realized projects, BIPV systems will have the opportunity to improve in all three aspects of its soul (technological, formal and energy).

6 ACKNOWLEDGMENTS

This work was produced in the context of collaboration between experts of the International Energy Agency (IEA) in the framework of the programme PVPS (IEA PVPS Task 15) under the title "Enabling Framework for BIPV acceleration", funded for Eurac by Stiftung Südtiroler Sparkasse.
The authors would like to thank all participants of the Task and, in particular, the ones of the Subtask A and the Italian participating team (GSE, ENEA, PoliMI).

7 REFERENCES

[1] PVSITES H2020 project, "BIPV market and stakeholder survey: Summary of results."
[2] IEA Task 41, "Building Integration of Solar Thermal and Photovoltaics – Barriers, Needs and Strategies," May-2012. [Online]. Available: http://task41.iea-shc.org/data/sites/1/publications/T41A1_Survey_FinalReport_May2012.pdf.
[3] G. Verberne *et al.*, "BIPV Products for Façades and Roofs: a Market Analysis." 2014.
[4] K. Farkas *et al.*, "Solar Energy Systems in Architecture - Integration Criteria and Guidelines," Internationa Energy Agency, IEA Task 41 T.41.A.2.
[5] T. Shoen, D. Prasad, and D. Ruoss, "Task 7 of the IEA PV Power Systems Program Achievements and outlook," presented at the 17TH EUROPEAN PHOTOVOLTAIC SOLAR CONFERENCE, 2001.
[6] B. Petter Jelle, C. Breivik, and H. Drolsum Røkenes, "Building integrated photovoltaic products: A state-of-the-art review and future research opportunities," *Sol. Energy Mater. Sol. Cells*, vol. 100, pp. 69–96, May 2012.
[7] "BS EN 50583-1:2016 - Photovoltaics in buildings. BIPV modules," 2016. [Online]. Available: http://shop.bsigroup.com/ProductDetail/?pid=000000000030266507. [Accessed: 15-Aug-2016].
[8] EU, "REGULATION (EU) No 305/2011 OF THE EUROPEAN PARLIAMENT AND OF THE COUNCIL of 9 March 2011 laying down harmonised conditions for the marketing of construction products and repealing Council Directive 89/106/EEC." Official Journal of the European Union, 2011.
[9] bsi, "Photovoltaic in buildings EN50583." .
[10] "Directive 2010/31/EU of the European Parliament and of the Council of 19 May 2010 on the Energy Performance of Buildings (EPBD)." 19-May-2010.
[11] European Parliament and European Council, "Directive 2009/28/CE on renewable energy sources (RES)." 23-Apr-2009.
[12] L. Maturi, J. Adami, Building Integrated Photovoltaic in Trentino Alto Adige, in press, 2018

THE CONTRIBUTION OF FAÇADES TO THE PV POTENTIAL FOR SITES WITH HIGH DIFFUSE FRACTION

S. Freitas[1], M.C. Brito[1]
[1]Instituto Dom Luiz – Faculdade de Ciências, Universidade de Lisboa
Faculdade de Ciências da Universidade de Lisboa, Ed. C8 (8.3.33), Campo Grande, 1749-016 Lisboa, Portugal.
e-mails: srefreitas@fc.ul.pt, mcbrito@fc.ul.pt

ABSTRACT: The recent but consistent reduction in the costs of PV systems and its technological improvement has prompted the installation of PV systems in non-optimal locations, such as building façades. This kind of application, although featuring higher levelized costs of electricity, has two main advantages: the large available area and the added value of widening the period of peak production throughout the day, thus allowing for a better match between demand and supply from PV. The solar electricity yield of vertical PV systems will depend on not only the absolute level of solar irradiation but on local climate conditions. The purpose of this work was to evaluate the effect of the fraction of diffuse irradiation on the contribution of façades to the solar PV potential of buildings in different locations in the world. It is shown that, for a given latitude, PV systems combining four façade orientations may generate up to 10% more electricity than a rooftop system in locations with diffuse fractions above 0.40.

Keywords: diffuse fraction, PVGIS, building façades, BIPV

1. INTRODUCTION

Traditionally, solar PV modules are installed in a way that the yearly electricity yield can be maximized. In the urban environment, building rooftops are usually the main candidates to host PV systems, due to tilts close to the optimum in gable rooftops or easy access and free space on horizontal ones.

However, as urban agglomerates tend to develop more vertically than horizontally, free areas in building rooftops become scarce: centralized air conditioning units, recreational structures, gardens/urban farms and even car parking are among the possible uses for a rooftop that might compete with the installation of PV systems. In this sense, the large available area in vertical façades should be considered, especially because the variety of orientations allows for a wider period of peak production and, therefore, a better match between demand and supply [1].

A vertical solar PV system will intersect solar irradiation differently from a horizontal or an optimally tilted system.

Lower solar elevations will lead to higher **direct radiation** reaching vertically tilted PV modules, whilst higher solar elevations favour more horizontal tilts. Hence, high latitudes are expected to be more favourable for vertical PV walls.

Diffuse radiation will often favour horizontal systems, due to higher sky view factors. On the other hand, **reflected radiation** from the surroundings might benefit vertical systems if the environment is highly reflective and/or there is a predominance of overcast days at the site. These impacts are related to the urban arrangement, surface materials and some weather impacts, such as snow cover and cloud cover variability.

The diffuse fraction [2] can be used to measure the variability in diffuse radiation associated to cloud cover at a certain location, and it is defined as:

$$k_D = \frac{DHI}{GHI} \quad (1)$$

where DHI is the diffuse horizontal irradiation [W/m^2] and GHI the global horizontal irradiation [W/m^2].

The goal of this work was to evaluate how the solar potential of façades depends on the amount of the diffuse solar radiation.

2. METODOLOGY

A routine to extract the fraction of diffuse over global radiation, i.e. diffuse fraction (k_D), from the PVGIS database [3] was written, following the indications in [4]. The map that covers Europe, Africa and most of Asia was queried with a latitude step of 2 degrees and a longitude step of 3 degrees (Fig.1).

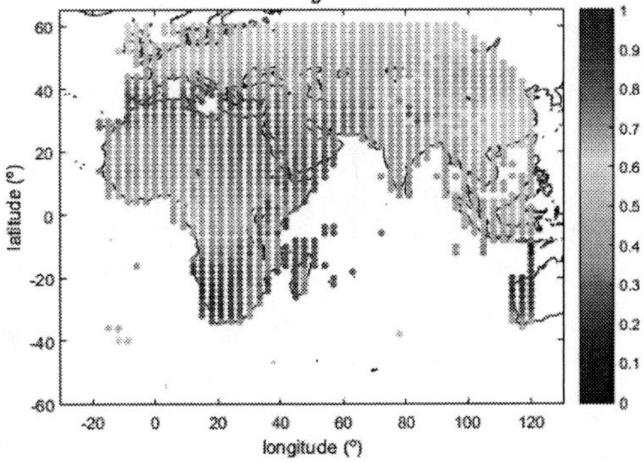

Fig.1 - Diffuse faction (k_D) values extracted from PVGIS database.

Annual electricity generation from PV systems was also obtained using this method. The query consisted on defining a conventional c-Si module 1kWp system with building integration and extracting the results for: horizontal rooftop (PV_{horiz}); optimally tilted and oriented rooftop (PV_{optim}); east- (PV_{east}); south- (PV_{south}); west- (PV_{west}) and north-facing façade (PV_{north}).
The system sizes were adjusted to the dimensions of a typical 3 storeys building: height=3m/storey, length=10m and width=10m (Fig.2). The assumed modules' efficiency leads to 150Wp/m^2.

PV_{horiz} PV_{optim} PV_{fac}

Fig.2 – Building archetype with 3 storeys used in the calculations.

The contribution from façades and rooftops can be compared using the ratios:

$$\alpha_{\text{horiz}} = \frac{PV_{fac}}{PV_{horiz}} \quad (2)$$

$$\alpha_{\text{optim}} = \frac{PV_{fac}}{PV_{optim}} \quad (3)$$

where PV_{fac}=$PV_{east} + PV_{south} + PV_{west} + PV_{north}$.

3. RESULTS

The results for α_{horiz} and α_{optim} are represented in Figure 3. One may note that both are always higher than unity, indicating that, for this very simplified building archetype, the solar potential of the facades exceeds that of the rooftop everywhere. It is important to recall that the analysis assumes that all four façades are exposed to solar radiation (hence no adjoining buildings are considered, which would, on average, halve the façade solar potential) and mutual shading between buildings is not considered. On the other hand, if taller buildings were considered, the contribution of façades could be much higher.
As expected lower latitudes translate into smaller α_{horiz} and α_{optim}, where, due to higher solar elevations, the optimal inclination for PV modules approaches zero and, therefore, vertical inclinations are strongly unfavoured.
For higher latitudes, the contribution of façades increases significantly. Furthermore, α_{horiz} becomes higher than α_{optim}. This is because steeper tilt angles are prefered where solar paths are mostly closer to the horizon, since sun rays will reach the module closer to the normal and winter snow will reflect more radiation.
In order to analyse the impact of diffuse fraction on the contribution of facades, this latitude bias has to be removed. For that effect, α_{horiz} and α_{optim} were normalized by fitting a cubic polynomial curve (Fig.4). Using the normalized values, denoted by g_{horiz} and g_{optim}, the positive impact of façades can be empirically quantified if written as a function of both latitude and diffuse fraction:

$$\alpha = f(lat) \times g(k_D) \quad (4)$$

where α stands for α_{horiz} or α_{optim}, the function $f(lat)$ is represented in Figure 4 and $g(k_D)$ can be retrieved from Figure 5 according to the diffuse fraction band.
It should be noted that, upon normalization, these coefficients no longer uphold their numerical significance; hence unity does not mean that the solar contribution of the façades equals that of the roof top. Instead, unity refers to the average α for that latitude band.

α_{horiz} α_{optim}

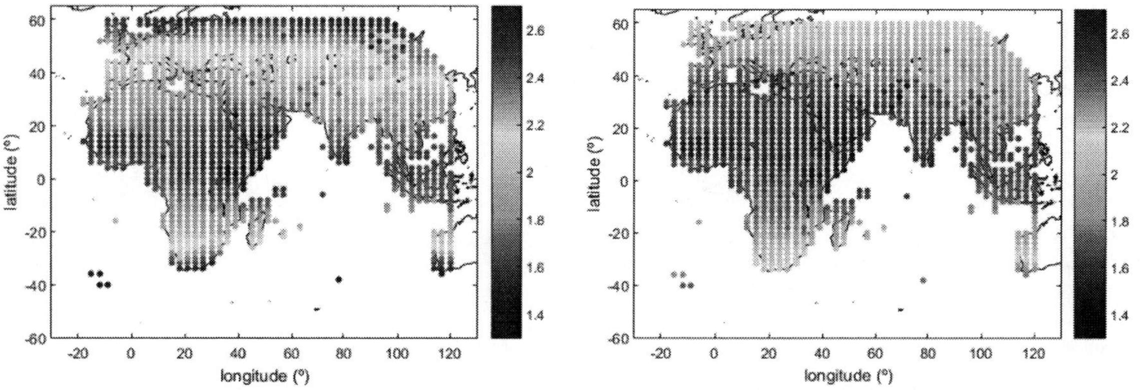

Fig.3 – Ratio between total façade PV production and: horizontal rooftop PV production (left) and optimally tilted and oriented rooftop PV production (right).

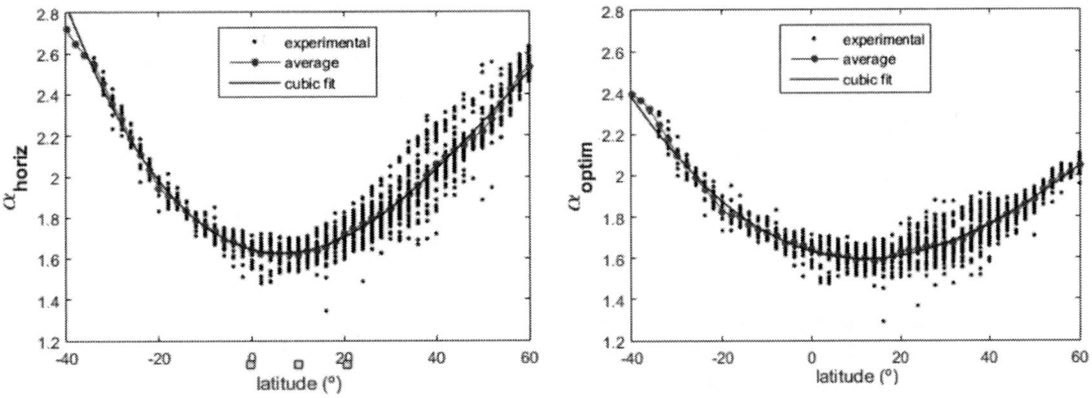

Fig.4 – Cubic fit to α_{horiz} (left) and α_{optim} (right) as a function of latitude:
$f(lat) = -2.6132 \times 10^{-7} lat^3 + 4.9471 \times 10^{-4} lat^2 - 5.8443 \times 10^{-3} lat + 1{,}6443$ (left) and
$f(lat) = -9.6113 \times 10^{-6} lat^3 + 2.7556 \times 10^{-4}. lat^2 - 6.1907 \times 10^{-3} lat + 1{,}632$ (right).

A general positive trend is visible in Figure 5, with diffuse fractions above 0.40 featuring g_{horiz} and g_{optim} higher values, although the variability is relatively high. These results show that, for a given latitude, the higher the fraction of diffuse radiation the more relevant the potential contribution of PV systems on façades. However, for g_{horiz} the improvement is not so clear. Whereas a horizontal panel gathers more direct radiation in the summer and a vertical panel in the winter, the former also features a higher sky view factor that might lead to higher irradiation for isotropic sky conditions. The vertical panel, on the other hand, benefits from the radiation that is reflected from its surroundings, although this quantity depends on the physical properties of the materials.

Given that the contribution from façades represents in average more 5% than rooftops' when the diffuse fraction is above 0.55, going up to 10%, it becomes clear that solar façades will generate more electricity at higher latitude sites with high fraction of diffuse radiation. Thus, one concludes that, as solar concentration works strictly with the direct component, technologies such as façade integrated CPV has very limited potential, although sun-tracking might improve this performance.

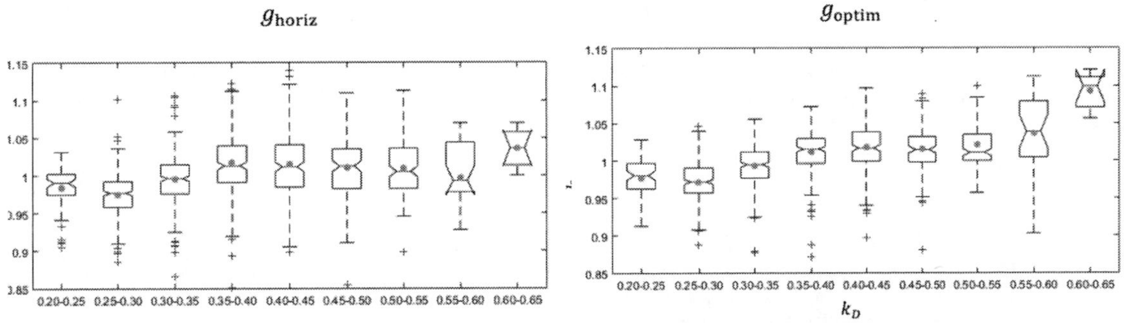

Fig.5 – Boxplot of g_{horiz} (top) and g_{optim} (bottom) for 0.05 by 0.05 bins of diffuse fraction (k_D). (The tops and bottoms of each "box" are the 25th and 75th percentiles, the red line in the middle of each box is the median, the green dots are the means, the black lines extending above and below each box are the extreme values and the red crosses are the outliers.)

In this work, a straightforward approach was taken, nonetheless, more locations in the world and more accurate results could have been obtained through more computationally demanding methodologies, such as the modelling of the urban context characteristic of each single location (through LiDAR data or other modelling techniques) and calculating solar irradiation in Radiance-based [5] software.

4. CONCLUSION

Diffuse fraction values and PV production estimations were retrieved from PVGIS database to evaluate the relative contribution of PV systems installed in façades. A simplified building archetype highlights that the contribution of facades may more than treble the solar potential of a building.

The relevance of PV façades to the overall solar potential of a building increases strongly with latitude, due to lower elevation angles for higher latitudes.

If this effect is removed, it is possible to test the impact of the diffuse fraction on the solar potential of PV façades.

Results show, for a given latitude, a general increase of facades' relative solar potential with diffuse fraction. Although facades feature lower sky view factors than horizontal roof tops, its higher areas lead to a relevant increase in irradiation for overcast conditions. This effect is more relevant when comparing irradiation on vertical façades with optimum tilted surfaces since the latter have lower sky view factors and higher footprints than horizontal modules on rooftops.

5. ACKNOWLEDGMENTS

Part of this work was supported by MIT Portugal Program for Sustainable Energy Systems and FCT grant SFRH/BD/52363/2013, Fundação Calouste Gulbenkian 2014 Award and by Projects MITP-TB/CS/0026/2013 (SUSCITY), PTDC/EMS-ENE/4525/2014 (PVCITY), UID/GEO/50019/2013 - Instituto Dom Luiz.

6. REFERENCES

[1] M. C. Brito, S. Freitas, S. Guimarães, C. Catita, and P. Redweik, "The importance of facades for the solar PV potential of a Mediterranean city using LiDAR data," *Renew. Energy*, vol. 111, pp. 85–94, Oct. 2017

[2] C. P. Jacovides, F. S. Tymvios, V. D. Assimakopoulos, and N. A. Kaltsounides, "Comparative study of various correlations in estimating hourly diffuse fraction of global solar radiation," *Renew. Energy*, vol. 31, no. 15, pp. 2492–2504, Dec. 2006

[3] JRC European Comission, "Photovoltaic Geographical Information System (PVGIS)," *Jt. Res. Cent. - Inst. Energy Transp.*, p. 10, 2014 [Online]. Available: http://re.jrc.ec.europa.eu/

[4] T. Huld and E. D. Dunlop, "PVGIS non-interactive interface, JRC's Institute for Energy and Transport - PVGIS - European Commission." [Online]. Available: http://re.jrc.ec.europa.eu/pvgis/automatic_interface.htm

[5] G. Ward, "The RADIANCE Lighting Simulation and Rendering System," in *4 SIGGRAPH conference*, 1994, pp. 459–72

MORPHO BUTTERFLY INSPIRED COLOURED BIPV MODULES

Benedikt Bläsi, Thomas Kroyer, Oliver Höhn, Martin Wiese, Claudio Ferrara, Ulrich Eitner, Tilmann E. Kuhn
Fraunhofer Institute for Solar Energy Systems ISE
Heidenhofstraße 2, 79110 Freiburg, Germany

ABSTRACT: In order to support acceptance and attractivity of building integrated PV (BIPV), there is a rapidly increasing demand for coloured PV modules. Especially, architects and building planners wish an individual colour choice, saturated colors, a homogeneous appearance for all angles of incidence and at the same time a high module efficiency. Furthermore, it is important that glare effects can be suppressed. State of the art concepts offer a restricted choice of colours, show a strong angular dependence, cause a high efficiency loss or do not allow for an efficient industrial production. We overcome these restrictions by using the Morpho butterfly effect, a bionic concept based on 3D photonic structures. Our novel technological realization of these structures is well suited for large scale production.
Keywords: Building Integrated PV (BIPV), Optical Properties, Module Manufacturing

1 INTRODUCTION

A detailed analysis of pathways for transforming the German Energy System by 2050 into a renewable energy system has shown that there a several alternatives of cost-effective energy systems and that these cost-effective energy systems include an installation of PV powerplants with 120 – 290 GW_p. This means that pv-modules with an area of 1200 – 2900 $km^2 \approx 15 - 36$ m^2/capita would have to be installed. The study considers all sectors and energy carriers. The model-based study investigates scenarios of system development and related costs to transform Germany's energy system in line with climate protection targets. [1].

Another recent study has shown that the technically and economically feasible area on roofs and facades of buildings is much larger than the required PV area for a renewable energy system [2].

We assume that the findings for the German Energy system are applicable also to many other countries and that there is therefore a huge economic potential for BIPV-systems. This potential can only be harvested with technically and architecturally well designed BIPV building components. This does not only include an architecturally pleasing appearance but also high electricity yield in order to achieve the targeted CO_2-reduction in the building stock.

Principally there are two approaches to achieve a pleasant appearance:

(1) "visible PV": the arrangement of the PV-cells and the interconnections can be used a design pattern.

(2) "invisible PV": the modules can be completely black, white or coloured with invisible PV-cells and interconnectors.

In this paper, we focus on concepts with "invisible PV". More precisely, coloured modules with high saturation, angle independent colour appearance and a minimized colour induced efficiency loss are targeted.

Different technologies can be used in order to hide the cell array with coloured layers at the front cover of the module. The coloured layer can be either inhomogeneous or homogeneous. Frit printed dots are a typical example of an inhomogeneous layer. The colour impression is stronger when the density of the dots is increased, but the module efficiency decreases at the same time since the frit printed area is almost opaque. Homogeneous coloured layers should have a high transmittance in the wavelength range with non-negligible spectral responsivity of the pv-cells. Spectral selectivity of the colour layer generally leads to higher transmittance and higher colour saturation but it is difficult to achieve neutral colours like grey or white and the colour often depends on the viewing angle. State of the art concepts offer a restricted choice of colours or a low colour saturation, show a strong angular dependence, cause a high efficiency loss or do not allow for straightforward industrial production [3–9].

This paper presents a novel technological approach for coloured modules with high efficiency, strongly improved visual appearance and the possibility of industrialized mass production.

2 THE MORPHO BUTTERFLY EFFECT

2.1 Basic understanding of the Morpho effect

Microscopic analyses of scales on the wings of Morpho butterflies revealed a 3D structure consisting of vertical ridges with horizontal lamellae attached to them. The horizontal lamellae are stacked almost equidistantly, while the arrangement of the ridges shows no strict regularity, but a defined disorder.

The basic principle of the Morpho effect is that thin film interference effects and structure effects interact in one three-dimensional photonic structure [10]. Due to the complex interactions in this functional layer, the angle dependence that is typical for planar multilayer stacks can be compensated.

In Fig. 1, the main constituents of the Morpho effect as described in [11,12] are sketched: The spectral reflectance peak is caused by a Bragg stack interference effect. The finite width of the ridges (on the real butterfly appr. 500 nm) leads to diffraction and, hence, to angular spread. Finally, the disorder caused by the height variation reduces or even destroys the spatial correlation between the ridges. This leads to suppressed interference effects between the ridges.

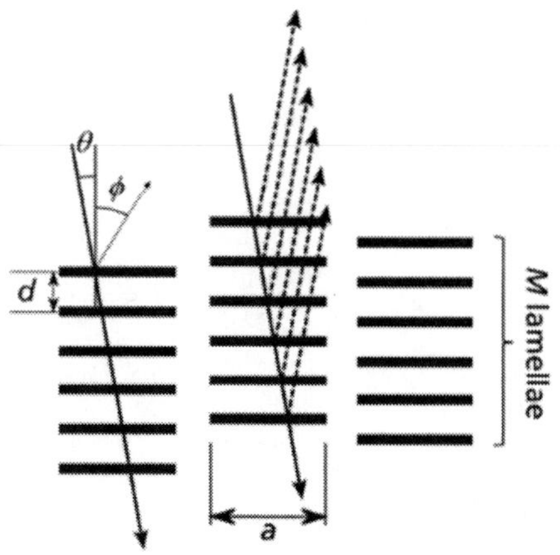

Figure 1: Main constituents of the Morpho effect: Interference from a finite Bragg stack (distance between lamellae d), diffraction from ridges of a given width a, and disorder caused by height variation. The angle of incidence is θ and the outgoing angle ϕ. (Sketches adapted from [11].)

2.2 From the 3D-Morpho to a corrugated thin film system

Structures as they can be found on the butterfly´s wing are extremely challenging to manufacture, especially if large area applications are aimed for. However, Chung et al [13] demonstrated that the Morpho effect can also be reproduced by a multilayer Bragg stack with alternating high and low refractive index (n_{high} and n_{low}) layers deposited on a substrate of defined roughness.

3 EXPERIMENTAL REALIZATION

The basic idea for the experimental realization has already been described in [14] and filed as a patent. However, in this earlier publication, only first qualitative results on glass samples could be shown, while in this paper PV module results are presented.

3.1 Module concept

For the integration in a PV module, the functional layer is applied to the rear surface of the module glass by a combination of structuring and sputter coating. In a standard laminated module it therefore is situated between glass and laminate (e.g. EVA). This means, that only the front glass has to be exchanged compared to standard module production. On the front side of the module glass, a texture may be applied in order to further enhance the colour tolerance and suppress glare effects (Fig. 2).

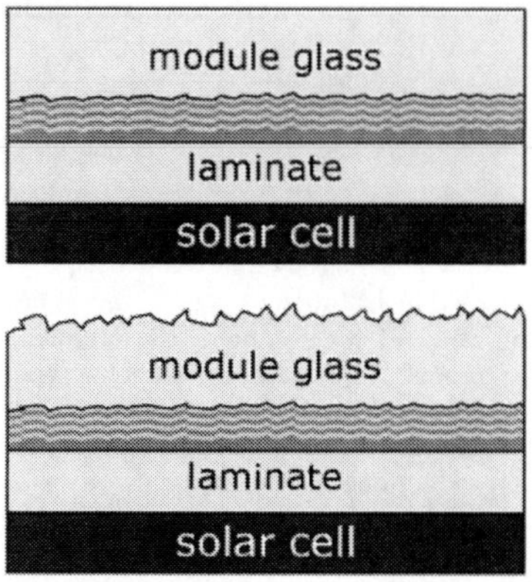

Figure 2: Sketch of the coloured module concept: the functional layer (green) is at the rear of the cover glass and in contact with the laminate. The front surface of the glass may be planar (top) or textured (bottom).

3.2 Morpho structures on Glass samples

To demonstrate the Morpho effect, glass samples were fabricated on small scale for the three colours red, green and blue. These glass samples were put on top of mini PV modules. The reflection at the intermediate interfaces (between glass sample and mini module) in the case of an air gap was eliminated by using a water film for approximate index matching. The photograph (Fig. 3) shows nicely the saturated colour. In the left part of the photograph, the module is not covered by the colour filter. So the dark grey solar cell is visible. In the central and right part, it can be seen that the bright colours only show up in the places where the cell is behind the filter. In the other areas with white background, high transmission is demonstrated. In a BIPV module, a homogeneous appearance with saturated colour will be achieved by using a black back sheet, black interconnectors and back junction cells (see below).

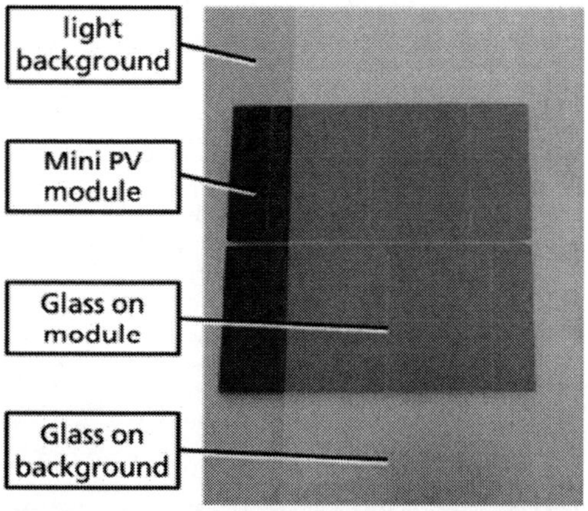

Figure 3: Photograph of a glass pane with colour layer applied to a mini module using a water film. In the left part of the photograph, the module is not covered by the colour filter. So the dark grey solar cell is visible. In the central and right part, it can be seen that the bright colours only show up in the places where the cell is behind the filter. In the other areas with light background, high transmission is demonstrated.

At these glass samples, angle dependent spectrometric reflectance measurements were performed. For the green coloured glasses, the spectra are shown in Fig. 4. One can clearly see the distinct and narrow reflectance peak at approximately 550 nm. Furthermore, the high angular stability of the peak position is evident.

It has to be noted that the spectra show additional reflection effects from rear interface (Morpho layer/air), which lead to a higher overall reflectance and a reduced saturation compared to the encapsulated module configuration.

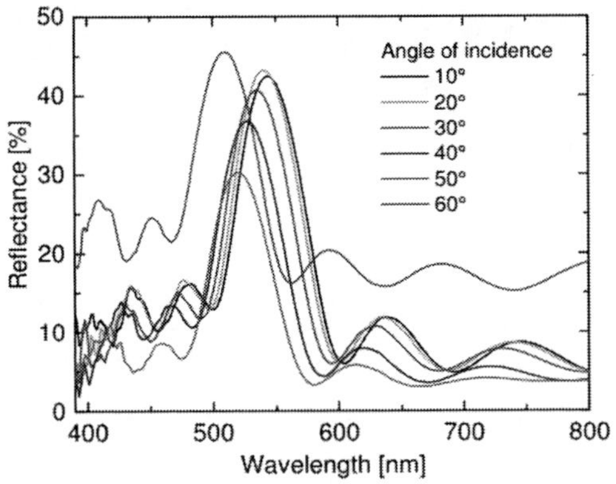

Figure 4: Angle dependent reflectance spectra, measured at the glass pane with green colour layer.

The measured reflectance spectra were used to deduce the colour appearance – via the representation of the spectra in RGB space (Fig. 5). Even though the RGB representation strongly depends on display and print properties, the high angular stability of the colours can be seen clearly.

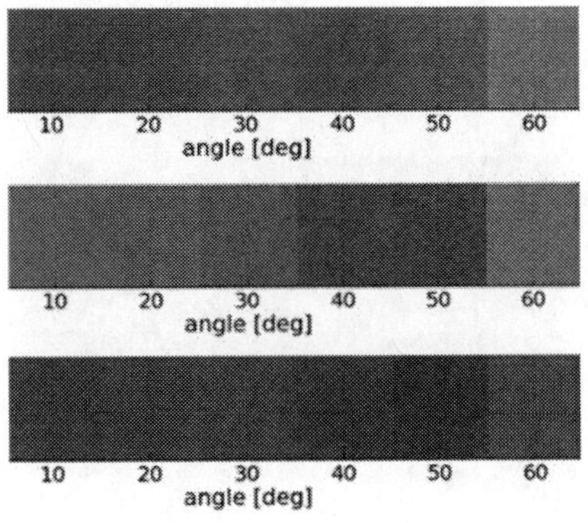

Figure 5: Representation of the measured reflectance spectra in RGB space.

In order to show the low angular dependence of the colours quantitatively, also the representation in the CIE chromacity diagram [15] is given (Fig. 6). Here x and y represent the normalized response of the blue and green receptors of the human eye. (The response of the red receptors is $z = 1 - x - y$.) The relatively large difference between the 50° and 60° values is caused by an enhanced overall reflectance at 60°, mainly caused by the rear interface. This effect can also be seen at the 60° spectrum in Fig. 4. In the laminated module, this effect will be suppressed.

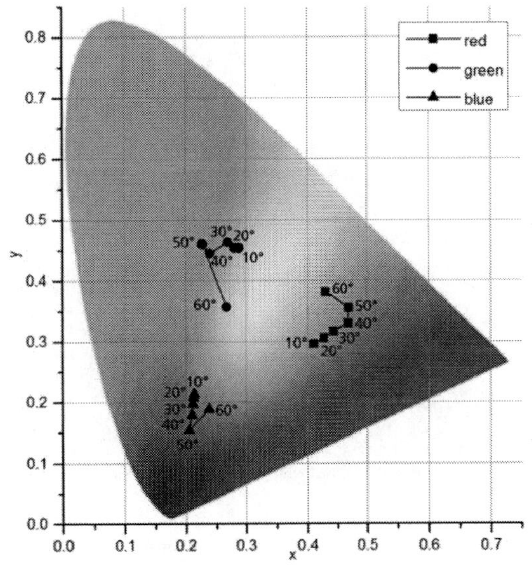

Figure 6: Representation of the measured reflectance spectra in the CIE chromacity diagram. The white point is at $x = y = 0.33$.

If one compares the reflectance spectrum with the external quantum efficiency (EQE) of a c-Si solar cell, it is evident that the reflectance is very low over the major part of the relevant spectral region. Hence, the module performance will not be strongly affected by the additional reflection losses. Since the Morpho layer/air interface leads to an enhanced reflection, the real module losses can be expected to be even lower. This is in line with the results shown in section 3.3.

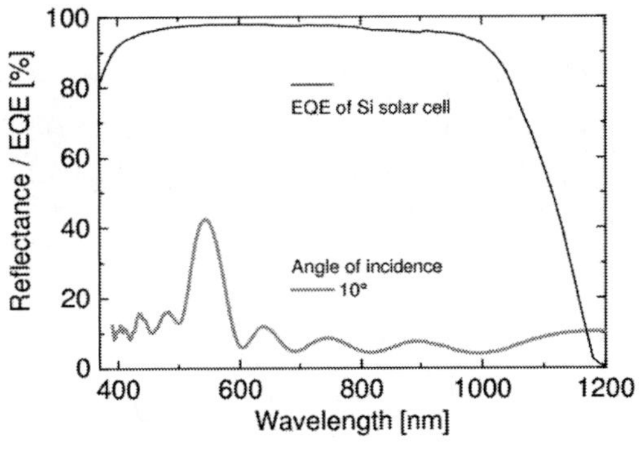

Figure 7: Comparison of measured reflectance spectrum (glass ample) and EQE of a c-Si solar cell.

3.3 Coloured modules

Demonstrator modules of the format 56 x 57 cm^2 (4 x 4 cells) were fabricated in the Module-TEC at Fraunhofer ISE. To show the effect of the functional layer, 8 modules with 3 different colours and 2 with uncoated glass were manufactured (Fig. 8). To reduce any disturbing internal reflection within the modules and maximize the effect of the filters the material selection was optimized. The used Sunpower Corp. back contact solar cells suit perfectly due to the missing metallization on the front side and thus appear homogeneously dark.

Figure 8: Photograph of the 56 x 57 cm^2 demonstrator modules with 4 different coating setups showing the colours red, green and blue; bottom right: uncoated glass.

To reduce the visibility of the electrical interconnection between the solar cells, we blackened the used connectors. For the rear side of the modules, a low reflective and deep dark backsheet foil was applied. (Fig. 9)

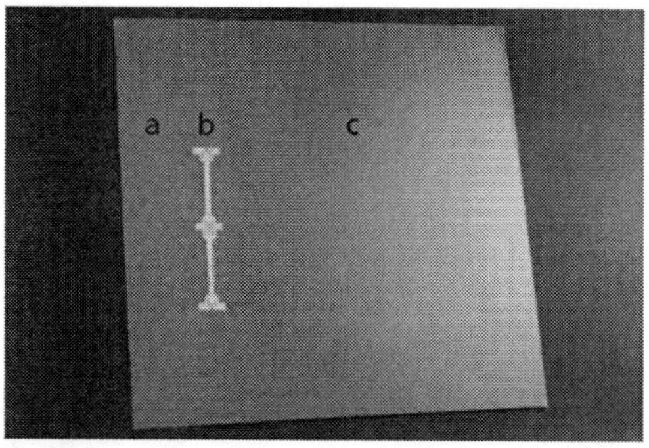

Figure 9: Photograph of a laminated test sample (20 x 20 cm^2) with green filter. Between cover glass and black backsheet, two interconnectors and one solar cell were laminated as demonstration.
Below a: blackened interconnector (invisible),
below b: standard interconnector,
below c: solar cell (invisible).
In a real module, only the narrow part of the interconnectors would be visible.

From a distance of only 1 m the modules appear highly homogenous and cells as well as connectors are invisible. Taking a closer look only the contours of the solar cell is visible due to the small mismatch in terms of colour and scattering between backsheet and solar cell. (Fig. 10)

Figure 10: Close up photograph of blue demonstrator. Electrical cell interconnection is matching to backsheet in terms of colour. Only contours of solar cells are slightly visible from a few distinct viewing angles.

The production of the modules themselves is equal to the manufacturing process of standard PV modules and thus should be transferable to any module production line as long as it is capable of interconnection of back contact solar cells. The module technology is also feasible with standard solar cells but with a slightly decreased overall homogeneity due to the high difference in terms of reflection between metallization and solar cell surface as well as the electric interconnectors.

For large area demonstration, the functional layer was applied to glass panes of the format 1.09 x 1.12 m^2 (7 x 7 cells). With these glass panes, PV module lamination was done at Ertex Solartechnik GmbH. As above, back contact solar cells from Sunpower Corp. and a black backsheet were used in order to achieve a homogeneous colour appearance with very high saturation (Fig. 11).

Power rating measurements at ISE CalLab PV Modules showed an extremely low colour induced efficiency loss. Compared to an uncoated cover glass, the relative loss of generated solar power for the coloured modules is appr. 7%. The power loss is in the same range for all three colours.

Figure 11: Photograph of the large area demonstration modules with the colours blue, green and red.

4 CONCLUSIONS

Using the Morpho effect, brightly coloured PV modules with high angular tolerance can be fabricated. We demonstrated the exemplary colours red, green and blue on large format BIPV modules.

With this technology, a large variety of colours can be achieved. A given colour appearance can be obtained by adapting the film and structure design of the photonic layer. Furthermore, mixed colours out of the RGB colour space can be composed out of functional layers for red, green and blue, respectively. With this approach a broad range of designs can be made accessible for architects. Since only the module glass is modified, standard solar cells can be used independent of the module design, making this approach also economically very promising. Similarly, the approach is applicable to solar thermal collectors.

5 ACKNOWLEDGEMENTS

Part of this work was funded by the German Federal Ministry of Economic Affairs and Energy under Contract Number 03ET1035A.

6 REFERENCES

[1] H.-M. Henning and A. Palzer, *What will the energy transformation cost? Pathways for transforming the German energy system by 2050*, 2015, www.ise.fraunhofer.de/content/dam/ise/en/documents/publications/studies/What-will-the-energy-transformation-cost.pdf.

[2] K. Fath, *Potenzial für Photovoltaik an Gebäuden in Deutschland. Entwicklung einer Methodik für großflächige solare Einstrahlungssimulationen und Hochrechnung mittels Data Mining*, Dissertation, KIT, 2017.

[3] A. Schüler, J. Boudaden, P. Oelhafen, E. de Chambrier, C. ROECKER, and J.-L. Scartezzini, "Thin film multilayer design types for colored glazed thermal solar collectors," Solar Energy Materials & Solar Cells **89**, 219–231 (2005).

[4] J. H. Selj, T. T. Mongstad, R. Søndenå, and E. S. Marstein, "Reduction of optical losses in colored solar cells with multilayer antireflection coatings," Solar Energy Materials & Solar Cells **95**, 2576–2582 (2011).

[5] F. Frontini, A. Scognamiglio, G. Graditi, C. Polo López, and M. Pellegrino, "From BIPV to Building Component," in *Proceedings of the 28th European Photovoltaic Solar Energy Conference and Exhibition (EUPVSEC)* (WIP, 2013).

[6] E. Klampaftis, D. Ross, and B. S. Richards, "Color, graphic design and high efficiency for photovoltaic modules," in *Proceedings of the 40th IEEE Photovoltaic Specialists Conference (PVSC)* (IEEE, 2014), pp. 25–29.

[7] V. Caer and A. Schuler, Laminated glazing with coloured reflection and high solar transmittance suitable for solar energy systems, US20150249424 A1 .

[8] C. Ballif, *Novel designs and materials for durable PV modules: applications on the ground, in cities and in the air* (2017).

[9] V. Neder, S. L. Luxembourg, and A. Polman, "Efficient colored silicon solar modules using integrated resonant dielectric nanoscatterers," Appl. Phys. Lett. **111**, 73902 (2017).

[10] S. Kinoshita, S. Yoshioka, and K. Kawagoe, "Mechanisms of structural colour in the Morpho butterfly: cooperation of regularity and irregularity in an iridescent scale," Proceedings of the Royal Society B: Biological Sciences **269**, 1417–1421 (2002).

[11] S. Kinoshita, Yoshioka S., Fujii Y., and Okamoto N., "Photophysics of Structural Color in the Morpho Butterflies," Forma **17**, 103–121 (2002).

[12] S. Kinoshita, S. Yoshioka, and J. Miyazaki, "Physics of structural colors," Rep. Prog. Phys. **71**, 76401 (2008).

[13] K. Chung, S. Yu, C.-J. Heo, J. W. Shim, S.-M. Yang, M. G. Han, H.-S. Lee, Y. Jin, S. Y. Lee, N. Park, and J. H. Shin, "Flexible, Angle-Independent, Structural Color Reflectors Inspired by Morpho Butterfly Wings," Adv. Mater. **24**, 2375–2379 (2012).

[14] B. Bläsi, T. Kroyer, O. Höhn, C. Ferrara, and T. E. Kuhn, "Coloured Module Glass for BIPV inspired by Morpho Butterfly," in *2016 OSA Light, Energy and the Environment Congress* (OSA, 2016), PW5A.2.

[15] E. C. Carter, Y. Ohno, M. R. Pointer, A. R. Robertson, R. Seve, J. D. Schanda, and K. Witt, *Colorimetry*, 3. ed. **15** (CIE, Central Bureau, 2004).

USE OF THE SLOPES OF THE CIRSURES' SANITARY LANDFILL FOR INSTALLATION OF PHOTOVOLTAIC PANELS: A PRELIMINARY ANALYSIS OF INITIAL PARAMETERS TO BE EVALUATED

Vitor De Brida, Fernando Soares dos Reis, Aline Cristiane Pan*
Solar Energy Technology Nucleus (NT-Solar), Pontifical Catholic University of Rio Grande do Sul, Av. Ipiranga, 6681, Porto Alegre, Zip code: 90619-900, RS, Brazil
Corresponding author*: aline.pan@pucrs.br

ABSTRACT: The present study investigates the initial parameters which should be assessed for the installation of photovoltaic panels in the area of the landfill of Cirsures, located in the municipality of Urussanga, Santa Catarina. The study conception was directly linked to analysis of the slopes of the landfill, its solid and electrical stability, sun incidence and shading. The proposed methodology for the study was obtained through photographic images taken on the spot and calculations based on characteristic data on the area. Thus, it was concluded that the slopes already consolidated sanitary landfill Cirsures, located in the municipality of Urussanga, Santa Catarina, meet the parameters analyzed for the future deployment of photovoltaic panels on them. The slopes are devoid of objects that can influence the assembly and efficiency of photovoltaic panels, in addition to presenting stability above the minimum required for your safety factor. The area has a great demand of sun incidence because of the staggered structure of the landfill, since the emission of solar radiation shows high rates between the months of November to February, leading to maximum efficiency of power generation.

Keywords: PV array, developing countries, grid-connect, sustainable.

1 INTRODUCTION

The development of strategies for energy production through renewable sources depends on a future vision on climate conditions that will be affected [1]. Among these strategies aimed at mitigating climate change, we relate to changes in sources of fossil fuels, incentive to use renewable energy sources and the use of energy-efficient appliances that reduce energy consumption [2].

Currently, the use of renewable energy sources has shown strong growth in the global electric chain, driven mainly by technological improvements that provided the reduction of production costs of photovoltaic panels, along with encouraging government policies to support renewable energy sources [3].

Among the alternative energy sources, solar energy is the most abundant renewable source of Earth, which leads the photovoltaic industry to show strong growth in terms of sustainability in the energy issue. Reduction result of the photovoltaic cell costs and the ability to link to the network, with an increase in power generation [4] [5] [6]. The increasing development of the photovoltaic market began in the 1980s with the application of multi-megawatt photovoltaic systems for energy production. The current photovoltaic market has a high growth rate from 30 to 40%. The use of photovoltaic solar energy is already considered popular in countries like the United States, Germany and Japan. Its applications in the world market are directly linked to remote industrial systems, developing countries and the systems connected to the network. Rural applications are the most present growth in the photovoltaic market, already the second largest is the sector of industries [7] [8].

Taking into account the population and new today's patterns of consumption growth, the final consequences related to this issue, is due to the large amounts of solid waste generated, which are usually disposed of in landfills because they are considered a simple process and low cost [9]. After the closure of landfills, we need a preventive monitoring for some years. This monitoring is necessary in order to avoid the formation and development of degradation processes in landfills, because after the discontinuance of landfills, they continue to have horizontal and vertical settlements, in addition to the continuous generation of leachate and gas, due to the chemical reactions of the organic materials constituting them. Already on the slopes of landfills, monitoring aims to monitor the instability of these caused by the liquidation of solid waste during the first five years after the consolidation of slopes [10].

With the development and viability of photovoltaic solar energy in Brazil, established a challenge to find suitable locations for capturing this energy. Landfills with their completed activities are considered local with viable potential for the installation of solar panels due to its geographical location [11] and its structure in staggered manner. The amount of solar radiation received by the landfill surface will depend on the slope surface orientation, as well as its geographical location [11].

This study aims to carry out an analysis of the initial parameters to be assessed for the future viability of the use of embankments consolidated a landfill located in the municipality of Urussanga, State of Santa Catarina, for the installation of photovoltaic panels. In this analysis the following parameters will be considered: geographic location, sunlight, shading, and solid and electrical stability of slopes, aiming to provide a more efficient future use of the landfill after its closure.

2 LANDFILL'S SLOPE

2.1 Geographical Location, Solar Radiation Incidence and Shading on Slopes

According to NBR 13896/1997, landfills, after its closure require a preventive monitoring for a few years, in order to minimize or prevent the release of liquids and / or gases to the water or the atmosphere. For embankments, monitoring is necessary in view of the possible instability of the same due to the liquidation of solid waste, incurred during the first five years after the consolidation of embankments [10].

It is essential to perform an initial analysis of the area that will be used to promote a site preparation where the solar panels will be installed. In addition to monitoring, the holding of local cleaning is required, or remove objects that may cause interference in the installation and operation of the panels [11].

Other parameters to be analyzed and considered are the sunlight at the location where the panels will be installed and shading along the sun's path, whereas the stepped structure of the landfill tends to form shadows of the slopes during the day. Through the realization of photographic images of the landfill Cirsures, municipality of Urussanga, Santa Catarina, we found the current situation found in the study area, with detail to the embankments already consolidated the landfill that are positioned to the north, as shown in Fig. 1 e Fig. 2.

Figure 1: Area that has the highest incidence of solar time at the landfill Cirsures by positioning account of embankments to the north, where you will be able to install photovoltaic solar panels.

Figure 2: Side view of the study area, with the main focus of the slopes of the landfill of Cirsures.

Using as a basis the photographic image of Fig. 1 was performed using software AutoCAD a future projection of the landfill after its closure. This projection was performed by grounding method for the formation of embankments, especially considering the area where there will be more sunlight in the landfill, due to the positioning of the slope to the north, as shown in Fig. 3.

2.2 Solid Stability of Slopes

An important factor that must be addressed in designing a project to install solar panels on consolidated slope landfills is the study of the stability thereof, to evaluate the safety, strength and viability. These studies aim to verify that the geometry of the slopes of the landfill meets the safety factor (SF), specified by the NBR 11862/2009 concerning the stability of slopes [12].

The safety factor to be adopted by the project should take into account the degree of certainty as to materials, environmental and human lives losses. Requiring a degree of security FS $\geq$ 1,5 [13]. This security level is associated with the carrying capacity of the slopes in order not to cause sinking or cracking weight account of solar panels as exemplified in Fig. 4. The result could be to not capture efficiency of sunlight, due to the change of its angle.

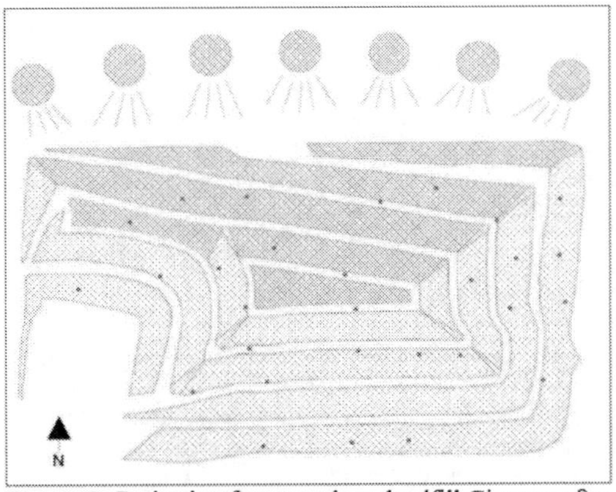

Figure 3: Projecting future sanitary landfill Cirsures after its closure, considering the area with higher solar incidence (northbound), using the *software AutoCAD*.

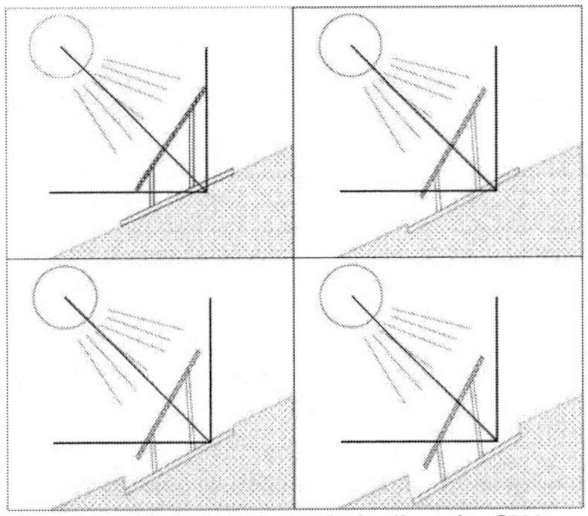

Figure 4: Outlining the change in the tilt angle of PV panels due to instability of the slope.

According to the Corona Engineering Ltd., company that conducted the study of stability in the landfill of Cirsures, the method used in the study of stability, for determining the minimum safety factor (SFmin) was by Bishop equation, based on the limit equilibrium theory. The results of this study indicate that the slopes of the landfill have a minimum safety factor of between 2,643 and 2,641, then, the slopes are safe as the breaks, thus ensuring the viability of the implementation of photovoltaic solar panels, without running the risk possible sags or cracking of the slopes. The determination of this safety factor performed by the company, based on geometry of the landfill, geotechnical parameters of the materials constituting the embankment and the foundation soil and groundwater in the normal position.

2.3 Electrical Stability of Slopes

An important point to be taken into account in Fig. 4 presented is the possibility of leakage of methane gas due to the sinking of photovoltaic solar panels, bringing the structure into contact with the layer of solid waste disposed there. What may cause consequently possible outbreaks of fires because of the emitted methane gas flammability, if any sparks in the solar photovoltaic system. The collection of biogas is considered mandatory because of the energy that biogas has and its danger of explosion due to methane gas potential [14].

For this, you must have a view to the completion of the internal monitoring of slopes already established, to verify whether it presents internal emissions of methane gas, because as mentioned above, the slopes continue to deliver liquids and gases in the first five years after consolidation of the same.

3 PHOTOVOLTAIC ENERGY

The solar radiation that reaches the Earth's surface is about 1 kilowatt per square meter (kW / m²) in clear conditions, and when the sun is near the zenith. This radiation is provided with two components: direct radiation coming directly from the Sun, and diffuse radiation, which is dispersed in the atmosphere indirectly. Solar radiation is an important parameter for life on the planet, in addition to being a renewable natural resource in abundance [6], [15]. For the calculation of the radiation intensity inclined surfaces varied chain arithmetic and models of spatial and temporal distribution of solar radiation is required.

The Solar Energy Laboratory of the Federal University of Rio Grande do Sul - UFRGS developed a software (Radiasol 2) able to perform these calculations quickly and directly. The Radiasol 2 uses internally mathematical models available in the literature, developed by other authors and / or members of the Solar Energy Laboratory. In software, calculations are made through routines that seek to determine the effect of the inclination of the receiving surface and the anisotropy of solar radiation in their direct and diffuse components. The radiation distribution models will be obtained immediately through charts or tables.

The input data used in the software to perform the calculations are based on actual data climatology of the city of interest to the project. The main data entry software for the calculations is the global solar radiation. As the municipality of Urussanga does not provide data for the global solar radiation it was necessary to conduct the radiation balance calculation through other meteorological elements (see Table I).

The estimated solar radiation can be taken from daily insolation held by Eq. (1) of Angstrom [17].

$$Q_g = Q_O\left(a + b\frac{n}{N}\right) \quad (1)$$

Where Qg: is the global solar radiation (kW / m².dia); Qo: solar radiation at the top of the atmosphere (kW/m².dia); a: constant for Urussanga - SC = 0.25; b: constant for Urussanga - SC = 0.50; n: average insolation (h/day); N: possible daily hours of sunshine. It is noteworthy that the constants a and b were obtained by contacting professionals specialized in climatology area of Agricultural Research and Rural Extension Company of Santa Catarina (EPAGRI) [18].

Applying the results in Eq. (1) in software Radiasol 2, it found the monthly values for global horizontal radiation, tilted, direct and diffuse (see Table II).

Table I: Monthly data used in Eq. (1) for the number of daily sunshine and radiation at the top of the atmosphere to the city of Urussanga, Santa Catarina [16].

	Number of Daily Sunshine (N)	Radiation at the Top of the Atmosphere (Q_o) (kW/m².dia)	Sunstroke Average (h/day)
Jan.	13,7	12,1	5,6
Feb.	13	11,3	5,9
Mar.	12,3	19,8	5,6
Apr.	11,4	7,9	4,8
May	10,7	6,4	4,8
Jun.	10,4	5,6	4,5
Jul.	10,6	5,9	4,8
Ago.	11,1	7,2	4,4
Sep.	11,9	8,9	4,7
Oct.	12,8	10,6	5,7
Nov.	13,5	11,8	5,7
Dec.	13,9	12,3	5,4

Table II: Results found in software Radiasol 2 for the global horizontal radiation, direct, diffuse and inclined in the area of the landfill of Cirsures [19].

	Global (kW/m². day)	Direct (kW/m². day)	Diffuse (kW/m². day)	Inclined (kW/m .day)
Jan.	5,51	2,23	3,14	5,4
Feb.	5,37	2,71	2,72	5,46
Mar.	4,64	2,48	2,57	5,08
Apr.	3,65	2,18	2,1	4,30
May	3,01	2,2	1,63	3,85
Jun.	2,62	1,81	1,65	3,48
Jul.	2,81	2,15	1,53	3,69
Ago.	3,18	2,06	1,92	3,86
Sep.	3,96	2,33	2,14	4,49
Oct.	5,00	2,42	2,81	5,25
Nov.	5,41	2,42	2,92	5,37
Dec.	5,47	2,23	3,01	5,27

4 ANALYSIS OF RESULTS

The analysis of the results achieved in the landfill of Cirsures showed that it meets the parameters evaluated, being necessary to carry out the monitoring of the generation of methane gas in the embankments already consolidated the landfill. According to Fig. 1 and Fig. 3 it can be seen that the landfill has a considerable area of solar radiation throughout the day, which promotes efficiency in the production of electricity. Already Fig. 2 shows that the slope of the landfill are clean, that is, not have objects that may interfere with the installation of photovoltaic panels and the production of electricity.

The data regarding the stability of the slopes show safety for the installation of photovoltaic panels, without there may be cracks and sinking which would cause a change in the inclination angle of the panels and thus the reduction in the solar radiation collection efficiency, as shown in Fig. 4. As for the electrical stability of the consolidated embankments, it is necessary to carry out continuous monitoring to verify whether they continue to emit methane gas. Thus avoiding possible fires due to sparks that may arise in the solar photovoltaic system.

The solar radiation in the area of the landfill, seen through the Table II, it was found that the high solar radiation rates are issued between the months of November to February, characteristic situation in the region where the landfill. Which leads to the production of electricity through photovoltaic panels operating at maximum efficiency. Among the months of June to August, the global solar radiation levels are low, leading photovoltaic panels operating less efficiently. However, it is known that the higher energy demand for this region, and in general for the country, occurs in the months where we have the highest efficiency of these panels (summer). Thus, the inclusion of these photovoltaic panels will serve as a solution to meet this demand and help maintain the stability of the distribution grid.

At the global level, some landfills have adopted or are considering such a measure of installing photovoltaic solar panels on their slopes already consolidated. In the United States for operating a photovoltaic plant of 1.00 MW in Atlanta, in Georgia. In New York, is being carried out the recovery of landfill Freshkills Park, which will be installed a photovoltaic solar power plant of 10 MW. Other American states, such as Massachusetts, New Jersey, Hawaii and Colorado are considering adopting these measures for future use of its landfills In Germany, the landfill Bremen has a photovoltaic plant of 0.84 MW and Japan are already underway encouraging studies for photovoltaics in landfills.

Comparing the locations where they are already operating solar plants in landfills with the South of Santa Catarina, as the city of Atlanta, the United States has the same levels of solar radiation, with an annual average of 4.5 kWh / m².dia. And solar power plant has been in operation since 2011 without reducing efficiency or interfere with the stability of the landfill. Already the town of Bremen, Germany, has an average solar radiation of 3.0 kWh / day, lower rate compared to the South of Santa Catarina.

For the design of a project of solar photovoltaic power generation in the landfill Cirsures, it is suggested the installation of photovoltaic panels multi crystalline on account of its current dominance in the world market. As the landfill is located in the south of Santa Catarina, the slope adopted for the panels would 24° North, for a higher uptake of solar energy in the months of November to February. As for the transmission of power generated by solar panels to the distribution networks, there would be difficulties, since the landfill is located close to power lines. From this it is seen that the landfill of Cirsures not find difficulties in the implementation of these panels.

5 CONCLUSIONS

Through this work, it was concluded that the embankments already established sanitary landfill Cirsures, located in the municipality of Urussanga, Santa Catarina, meet the parameters analyzed for the future deployment of photovoltaic panels on them. The slopes where there have been the disposal of waste and are consolidated, are devoid of objects that may cause interference in the installation and efficiency of photovoltaic panels. Therefore, the implementation of photovoltaic panels in landfills can be considered as an excellent alternative future for the development of photovoltaic solar energy. Since landfills, in most cases, they have areas that are no longer in operation and who have already undergone all appropriate measures for its closure safely, besides having areas with high solar radiation because of their staggered structure.

REFERENCES

[1] Wang, B. *et. al.* Renewable and Sustainable Energy Reviews, Vol. 40 (2014) 185-195.

[2] Abdullah, M. A., Agalgaonkar, A. P., Muttaqi, K. M., 2014. Renewable Energy, Vol. 66 (2014) 305-313.

[3] Cucchiella, F., D'Adamo, I. Rosa, P.. Renewable and Sustainable Energy Reviews, Vol. 47 (2015) 552-561.

[4] Cyrs, W. D. *et. al.* Energy Policy, Vol 68 (2014) 524-533.

[5] Herrera, C. M., Sánchez, F. H., Bañuelos, N. F. Energy Procedia, Vol. 57 (2014) 245-254.

[6] International Energy Agency. Technology Roadmap: Bioenergy for Heat and Power, OECD/IEA (2012)

[7] Kova, T. M. **et al.**, 2011. Solar Energy, Vol. 85, (2011). 1580–1608

[8] Makridesa, G. *et al.* Renewable and Sustainable Energy Reviews, Vol. 14 (2010) 754–762.

[9] Fernandes, A. *et. al.* Applied Catalysis B: Environmental, Vol. 176-177 (2015) 183-200.

[10] Lange, L. C. *et. al.* Project, operation and monitoring of landfills: guide of the professional in training, ReCESA (2008).

[11] Tansel, B., Kumar Varala, P., Londono, V., 2013. Sustainable Cities and Society, Vol. 8 (2013) 42-47.

[12] Corona Engineering Ltd. Enlargement of the Landfill: Analysis of Landfill Stability, Corona Engineering Ltd (2014).

[13] Brazilian Association of Technical Standards. NBR 11682: Stability of Landfills, ABNT(2009)

[14] Penteado, R. *et. Al.* Energy Policy, Vol. 42 (2012) 551-556.

[15] Park, J. K., Das, A, Park, J. H. Energy Conversion and Management, Vol. 101 (2015) 30-39.

[16] Biscaro, G. A. Meteorologia Agrícola Básica. UNIFRAF (2007).

[17] Vianello, R. L. Alves, A. R. Basic Meteorology and Applications. University Press (1991).

[18] Information Center of Environmental Resources and Hydrometeorology of Santa Catarina. Climate Standards of Urussanga, CIRAM (2015).

[19] Federal University of Rio Grande do Sul - UFRGS. Laboratory of Solar Energy: Radiasol 2, UFRGS. Available in <ftp://ftp.solar.ufrgs.br/progs/Radiasol2.zip>, (2016).

FROM PV SYSTEMS TO ENERGY SOLUTIONS PART II [1]
FROM THE CONCEPT TO REALITY

Thomas Nordmann • Ralph Lingel • Stephanie Fehling • Thomas Vontobel
TNC Consulting AG • General Wille-Str. 59 • CH-8706 Feldmeilen • Switzerland
P+41 44 991 5577• nordmann@tnc.ch • www.tnc.ch

ABSTRACT: Residential buildings are turning from energy consumers to prosumers using PV. The building as a system combines two or more of the following branches: energy efficiency, clever PV production, thermal energy management and electric mobility. Using a smart energy- and load management as a conductor or mastermind within the building allows controlling and optimizing power consumption according to PV production. Managing existing loads can help reduce the need for additional storage in form of batteries while optimizing self-consumption and grid-services.

2016 an existing building with PV and electric mobility was equipped with new heating technology (heat-pump), additional PV and the conductor/mastermind system and is since in operation. Production and consumption is monitored and evaluated. As of September 2017 a full heating-season as well as the intermediate seasons are measured, analysed and evaluated. First findings include: increased PV self-consumption, higher economic efficiency of the PV plant for post-FiT, individual mobility with PV (solar gas station), reduction of energy consumption for heat and general electricity with higher efficiency, lower energy costs, impact on CO_2 balance of buildings (decarbonisation) as well as monitoring of power production and energy consumption for inhabitants.

Keywords: Photovoltaic deployment, load management, self-consumption, storage, Grid-Services

AIM

In Switzerland about 1/3 of the overall final energy of 232.9 TWh (2015) is consumed in the transport sector, the largest part for individual mobility based on of fossil fuels. Another third of the Swiss energy is consumed by residential homes, whereof only a small part is direct electricity and the largest part is used for heating, again based on fossil fuels. The last third is used in industry and the service-industry.

For a decarbonised home, fossil fuels in individual mobility and heating have to be substituted. Heat-pumps and electric vehicles (EV) offer a chance to substitute these fossil fuels. At the same time this increases the overall demand for electricity. PV production in residential buildings with heat-pump and electric mobility therefor not only targets general electricity consumption but also primary energy demand for heating and mobility, multiplying possible fields for PV deployment and potential target market by factors. Todays energy-efficient residential buildings with PV can cover the energy demand, at least in the annual balance. Residential homes are becoming "prosumers", producers and consumers.

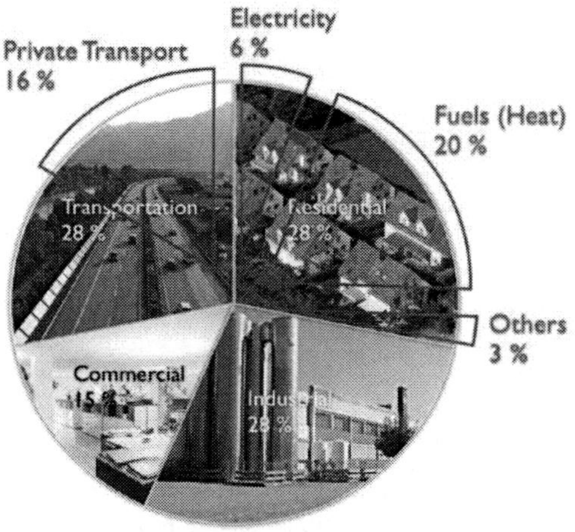

Fig. 1: Primary energy consumption for the different appliances. Red marked are shares for residential homes.

Mostly production and demand do not occur at the same time and are not coordinated. Further deployment of electric cars, heat-pumps and PV systems can lead to disturbing production and demand peaks and additional strains in power grids. In addition to these power grid aspects, it is also advantageous to coordinate production and

[1] Published at the 28th European Photovoltaic Solar Energy Conference and Exhibition (EU PVSEC 2013) 3. October 2013 Paris/France • PLENARY Paper: „From pv systems to energy solutions“ Thomas Nordmann • TNC Consulting AG • General Wille-Strasse 59 • CH-8706 Feldmeilen • Switzerland

consumption from an economic point of view by increasing self-consumption.

To coordinate the different components in a residential building, an electronic building automation “conductor” was developed and implemented, acting as the mastermind. It dynamically and continuously adapts the power demanding loads such as charging stations for EV’s and heat pumps to the available PV production while ensuring user needs. 2016 two pilot projects were realised to verify the capabilities of a conductor/ mastermind controller for Smart-Energy-Home.

The first pilot project is a low energy “Minergie” single-family house near Zurich, Switzerland. The house, built in 1999, was renovated with newest building technology including a 6 kWp PV system, a modulating brine-to-water heat pump (2.5 kWmax el.) and a controllable charging station for EV (11 kWmax) (Fig. 2). The charging station is primarily used for charging a car with 85 kWh battery capacity (Tesla S). Individual room temperature control and thermal management are also part of the system. “TNCALL Conductor” as mastermind for coordinating all these components is implemented in this project with a focus on optimizing PV self-consumption over all available loads. The project is equipped with a large number of sensors and a monitoring system, which continuously records the operating data.

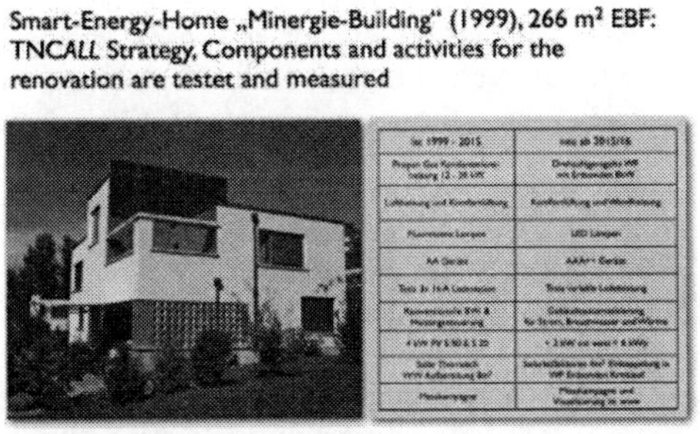

Fig. 2: Pilot project single-family house before and after renovation

The second pilot project is an office building equipped with a 12 kWp PV system and a controllable charging station for EV (11 kWmax). In this project, the focus is on charging the company's own electric car with 22 kWh battery capacity (Renault ZOE) with PV.

APPROACH

Targeting not only the general electrical consumption of buildings, but also the thermal energy (heat-pumps) and individual mobility (EV’s) multiplies the potential target markets for PV, while contributing to the transition to a renewable energy system and reducing CO_2 emissions on the road to decarbonisation. At the same time, this approach requires to consider buildings as energy systems with related and influencing components. With the use of an additional conductor or master-controller for coordinating and optimizing operation of existing controllers and optimizers for specific applications within buildings such as heat-pumps, charging stations for EVs, storage devices, etc. it is possible to operate a building with PV as an energy system and not only as the sum of energy components. This also allows for energy efficiency optimizations.

Possible operation modes can vary from optimization for PV self- consumption, self-sufficiency, economical optimization to power grid services and can be adapted based on software code without necessity of new hardware installations. This provides higher flexibility to adapt to changing requirements, political or economical frameworks.

The functionalities of the conductor are: the management of the different loads within a building such as charging electric cars, battery storage management, efficient heat pump operation for heating and hot water, operation of comfort ventilation systems, operation of domestic appliances. The direct PV yield is optimized in operation incl. the weather forecasts. The system provides monitoring of PV plant production and consumption of the different loads integrated for the inhabitants. Comfort requirements of the inhabitants have to be fulfilled at every point of time.

With the implementation of this system in pilot projects, first experiences in implementing and operating such a system are gained. This includes for example the integration of components with different interfaces for communication (bus-systems), shown in Fig. 3.

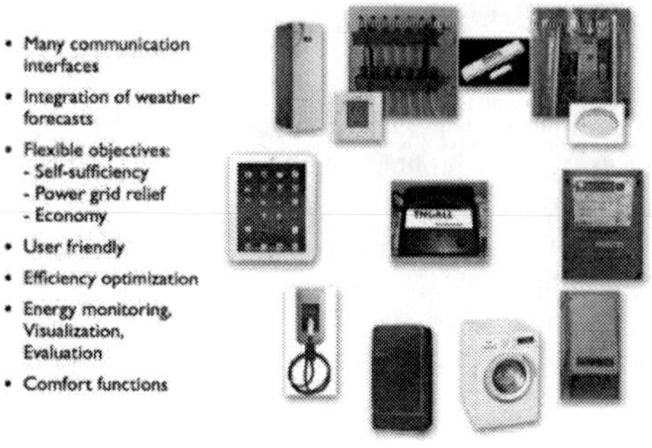

Fig. 3: TNCALL "Conductor" and the interfaces-hell

Monitoring of the components included in the system provides much needed and essential operational data. This data can not only be used for verifying operation as expected, but also to verify simulations or assumptions from existing models and models under evaluation for buildings.

Evaluation of operational data shows the potential and impact of using demand-side management and existing loads to reduce the need of additional storage for different purposes such as maximizing self-consumption or providing grid-services.

Measurements in the pilot projects before implementing the conductor show that the PV plant produces power at noon without enough loads to use the produced solar energy. At the same time, the electric vehicle is charging during the night, leading to high load peaks. (Fig. 4). Higher penetration rates of EV and PV in an uncoordinated system will lead to strains on the grid and high peak loads. Impact of the conductor on energy from grid and peak loads is evaluated.

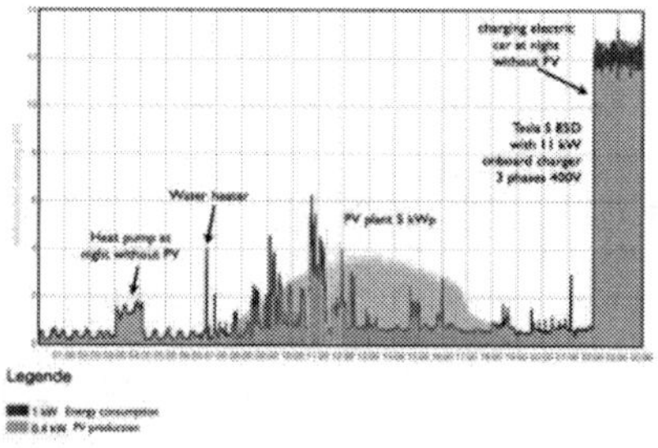

Fig. 4: Measured data of PV production and consumption in residential home without TNCALL

Fig. 5 shows that on the economical side the excess energy from the PV plant is redeemed at a low price by the grid operator. The electric vehicle is charged during the night and electricity must be bought at the higher end customer tariff. The obvious solution is to charge the car with the energy of the PV plant when available. The same can be done with other loads.

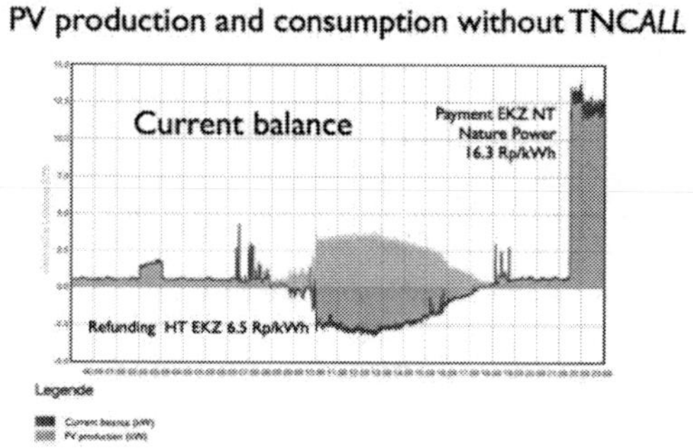

Fig 5: Current balance at grid connection point without TNCALL for residential home

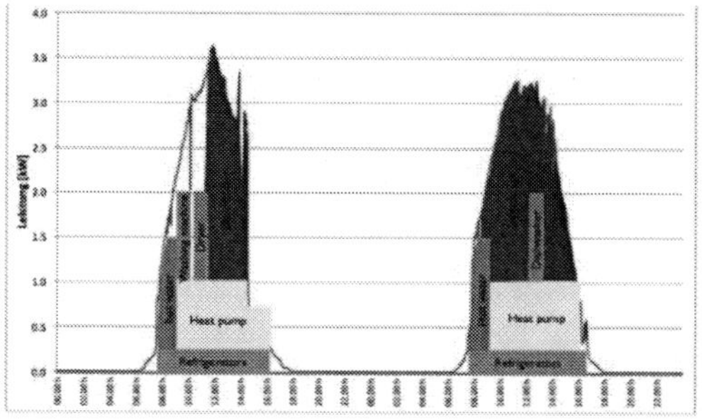

Fig. 6: Applied energy balance for a weekend with TNCALL

The aim is shown in Fig. 6. All loads should be moved to production times of the PV plant as far as possible. When the possibilities of the existing loads are exploited, a well-dimensioned stationary battery storage system can provide further optimization.

Results from operation will also allow to optimally size and layout future PV plants on buildings with the aim of providing a well-balanced energy system. Possible influencing factors besides nominal capacity of the PV plants can be the orientation and inclination of the PV modules, for example the forced implementation of PV on facades. At the same time the overall demand for electricity with the mandatory decarbonisation, especially for mobility and heating, will increase and lead to a multiplication of potential PV markets.

RESULTS

For the combination from PV with electric mobility first results from summer 2016 show, that with the implemented solution using the conductor, almost no energy from the PV plant is exchanged with the electricity grid.

In the residential house the car was mainly loaded on weekends with PV excess energy. This was sufficient, due to the large storage capacity of the vehicle and the rather low mileage driven during typical workdays. The system still provides the possibility to charge from the power grid, for example for longer road trips. When plugged in, almost all PV energy could be stored in the battery of the vehicle due to the large battery capacity and the rather small PV plant. (Fig. 7)

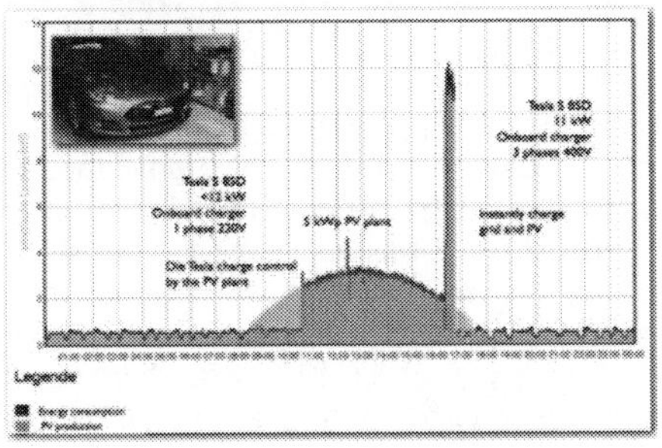

Fig. 7: Charging EV's with PV energy and optimally using the capacity of the EV's battery

In the office building the electric vehicle was almost exclusively charged with PV excess energy by the beginning of November 2016. Over a long period of time, a significantly greater mileage with PV energy for the vehicle would have been possible than the mileage of everyday use in real-life operating conditions demanded. This leads to the conclusion that additional EV's can be operated at this location or other consumers can be provided with PV energy in addition to the EV.

The heat pump of the residential home was operated with standard settings in the period January - June 2016. Starting from July 2016 the water heating system, also provided by the heat-pump, was adapted to the PV production. With the start of the heating season at the beginning of October 2016 and a PV-optimized control of the heat pump, it is shown that a self-consumption rate of nearly 100% can be achieved during the heating season (Fig. 8).

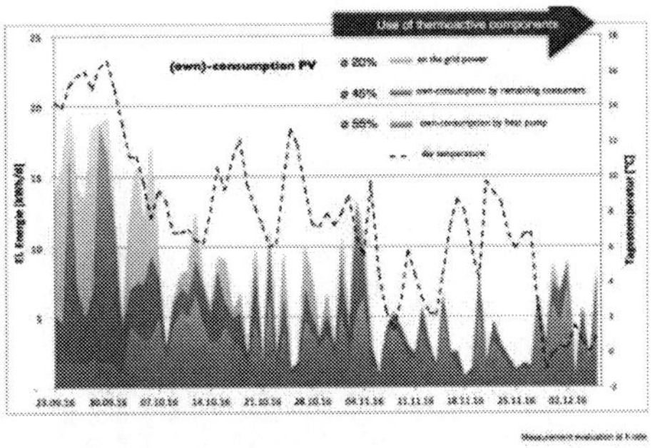

Fig. 8: Increasing self-consumption of 5.2 kWp PV plant and heat pump $P_{max\ el}$ 2.5 kW with TNCALL

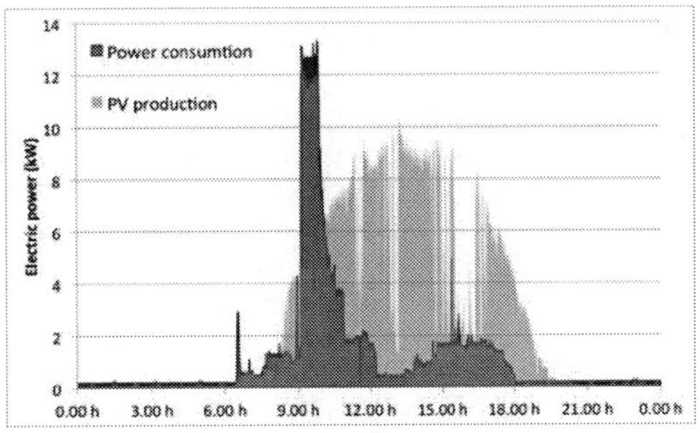

Fig. 8: Charging EV's and PV production on July 26th 2016 for pilot project 2 without optimized charging of EV

In pilot project 2 the focus lies on electro mobility and PV. Charging of the EV starts upon arrival of the EV at the work place with maximal charging power as shown in Fig. 8. Since the battery of the EV is not completely discharged with typical mileage, charging is done only during a short period of time.

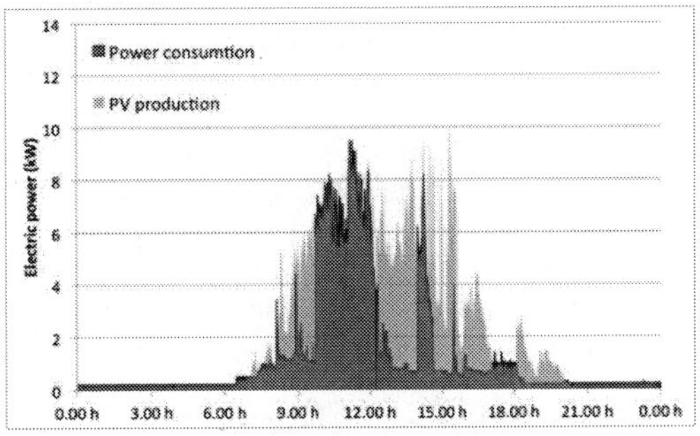

Fig. 9: Charging EV's and PV production on May 12th 2017 for pilot project 2 with optimized charging of EV

After installing the conductor, the EV is charged according to the available PV production. Charging takes a little longer but at lower power rates. Fig 9 shows that almost no energy has to be drawn from the grid with this optimized operation of the charging station.

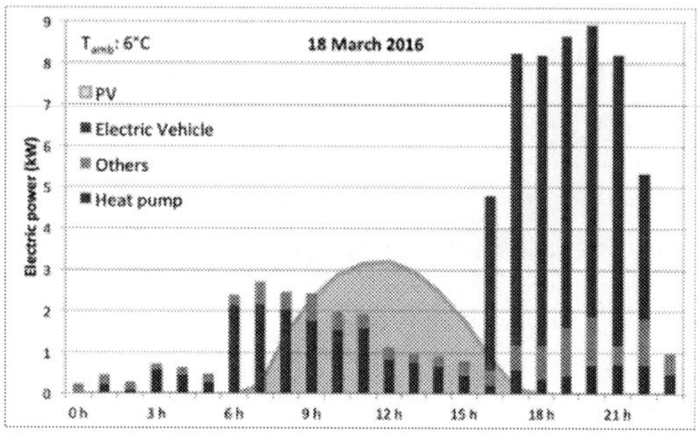

Fig. 10: Power consumption and PV production on March 18th 2016 for pilot project 1 without conductor to optimize loads according to production

More detailed analysis of power consumption on a day during heating period 2016 (ambient temperature is 6°C) for pilot project 1 shows heat pump operating for hot water during morning hours and EV being charged during evening hours with high charging power. Although PV yield was close to expected potential yield for a sunny day in spring, both the heat pump and the EV charging have drawn power from the grid.

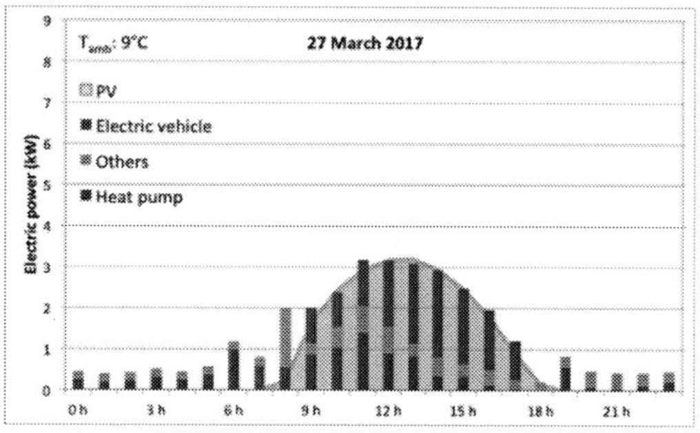

Fig. 11: Power consumption optimized to PV power production for pilot project 1 on March 27th 2017

March 27th 2017 was also a spring day during heating season (ambient temperature at 9°C) with nearly optimal PV yield and is therefor suited to compare with results shown in Fig. 10. Heat pump as well as EV charging were operated according to the available PV power production as shown in Fig 11. The conductor prioritised heat pump operation for hot water before charging the EV with increasing power. Power drawn from the grid is minimized and PV production is used in self-consumption with almost no excess PV energy fed into the grid at lower tariffs.

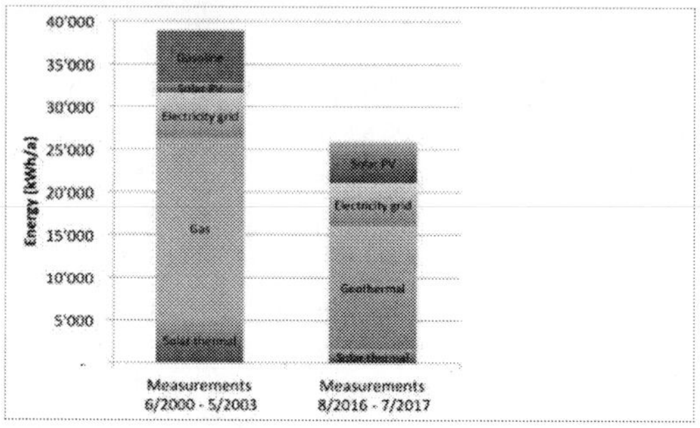

Fig. 12: Input energy for pilot project 1 with substitution of fossil fuels for mobility and heating/cooking in average over one year of operation.

Fig 12 shows the completed decarbonisation for a single family household. On the left side the input energy before transition is shown with fossil fuels for mobility, heating and cooking. Proof of concept for decarbonisation of a single family household is achieved, without reducing comfort for the inhabitants. The energy system of this house is now fully electrified. With this, it is basically possible to provide the energy system of this building including heating and individual mobility with renewable energy, such as PV. Solar thermal was reduced due to limited space available on the roof area, while PV nominal power was increased.

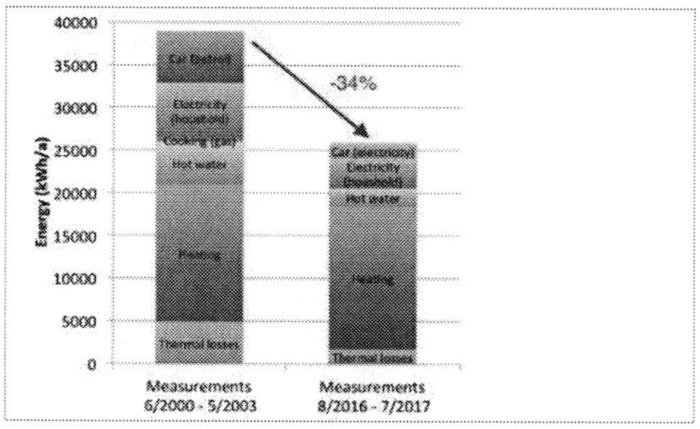

Fig. 13: Increase in energy efficiency combined with electrification and decarbonisation for a single family household over one year of operation.

Besides the switch to a fully electrified energy system and with this the decarbonisation for pilot project 1 an increase in efficiency of 34% was achieved. Main reasons for this higher efficiency are the higher efficiency of EVs for similar mileage and a reduction in thermal losses. Thermal insulation of the building was already good before the switch of energy production, so no additional insulation was applied. For average existing buildings this is not true, efficiency gains can be far higher with good thermal insulation. This should always be the first step before switching energy production to heat pump. Electrical appliances have been changed to state of the art efficient appliances. This includes highly efficient refrigerator and freezer, as well as washing machine and LED light. Standby losses of electrical appliances must be addressed in already efficient building.

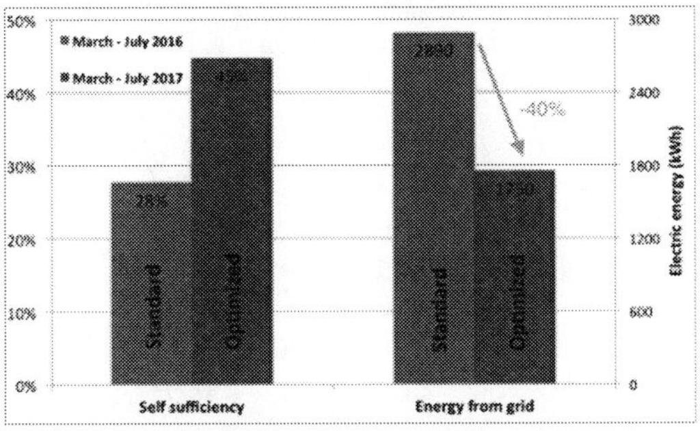

Fig. 14: Increase in self-sufficiency and decrease in energy from grid of 40% for pilot project 1.

Optimized EV charging and heat pump operation by the conductor/mastermind according to PV production increased overall self-sufficiency from 28% to 45% for pilot project 1 for the period of time evaluated. Energy consumption from the grid was reduced by 40%.

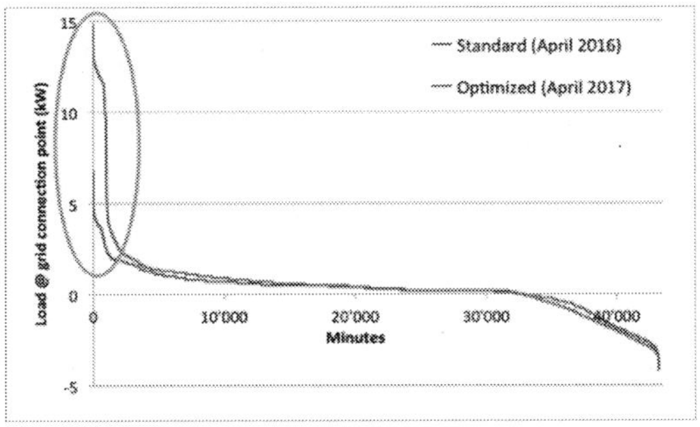

Fig. 15: Active load management reduces demand peaks drastically for period of time analysed.

Active load management reduces demand peaks drastically. Especially the high loads from EV charging are almost completely eliminated.

The combination of higher self-sufficiency, less energy necessary from the grid and reduced peak loads reduces strains on the power grid. These positive effects on the power grid and the additionally targeted energy markets from mobility and heating allow for far higher PV penetration rates and shares in overall energy production.

CONCLUSIONS

- We have to realize PV plants according to demand and not just from maximized production perspective.
- PV in buildings is part of the energy system of a building, together with the different loads.
- PV is not only used to cover general electricity consumption, but also heat pumps and electric mobility. This multiplies the potential target market of PV applications, contributes to a renewable energy system and reduces CO_2 emissions.
- The showcased pilot projects proof decarbonisation of residential buildings is possible today.
- Today's energy-related components in the building are not coordinated. The Smart-Energy-Home needs an electronic "conductor", providing load management and optimization on an energy system level. This concept is in operation.
- Evaluation of the pilot projects shows that energy storage is not only possible with stationary electrical storage (batteries), but also using the thermoactive building mass, the target application "Power to Heat", the time-shift of electrical consumers and the use of mobile storage capacities from electric mobility.
- The coordination and optimization of all loads and PV production in combination with efficiency measures reduces strains on the power grid (energy and peak load). Reduction of energy from grid of 40% has been shown, while efficiency was increased by 34% and peak loads have been reduced around 50%.
- Using an additional conductor/mastermind controller allows to adapt to different requirements and frameworks (economical and political) faster and easier by adapting the software (self-consumption, self-sufficiency, grid services, economical optimization, et.al.).
- The real-life solutions – verified by a monitoring system, measured data and analysis – are more challenging than the fast-printed high-gloss brochures.
- The conductor used by TNC provides possibilities to implement the Smart-Energy Home solution and evolve from PV system to energy solutions.

OUTLOOK

- Further operational data is monitored and evaluated
- Weather forecast is taken into account for optimization of self-consumption and demand-side management, especially thermal loads
- Further energy components are fully integrated into dispatch of PV energy available, especially stationary batteries
- Control and optimization algorithms are continuously improved
- Further products (heat pumps and EVs) are fully integrated to allow an even more flexible application of the conductor
- Further projects are developed, providing even more solidly based data for different operating conditions. A focus lies on apartment buildings with larger heat pumps and a multitude of EV charging stations. Impact of a conductor is expected to be higher for these cases.

IMPROVED DISTRIBUTION GRID HOSTING CAPACITY WITH OPTIMISED PV DEPLOYMENT

Magdalena Bledzinska[1], Grazia Barchi[2*], David Moser[2]
[1]Warsaw University of Technology- Faculty of Electrical Engineering, Pl. Politechniki 1,Warsaw, Poland
[2]EURAC Research - Institute for Renewable Energy Viale Druso 1, 39100 Bolzano (BZ), Italy
*Phone: +39 0471 055628; Fax: +39 0471 055699; E-mail: grazia.barchi@eurac.edu

ABSTRACT: Nowadays in Europe, the production from variable renewable energy sources (vRES) has seen a significant growth and the photovoltaic (PV) generation plays a central role. The unplanned proliferation of PV generators in different points of the distribution grid has caused different kind of disturbances that might compromise the correct operation and reliability of the grid. In particular, possible overvoltages caused by PV generation limit the amount of PV that can be installed in a certain node of a distribution system. In this respect, the aim of this work is twofold. First, the hosting capacity of a specific distribution network is analysed based on voltage buses, current branch and power at transformer level. Then, a suboptimal PV placement algorithm is proposed in order to reduce the overvoltage effect, thus mitigating the issues related to limited hosting capacity.

Keywords: *Phovoltaics, PV hosting capacity, power flow analysis, distribution grid*

1 INTRODUCTION

Installed global photovoltaic (PV) power capacity reached about 235 GW in 2015 and about 310 GW in 2016 [1]. This potential is able to cover 1.5% of the world electricity demand. Although the growth of renewable energy sources contributes to the decarbonisation of the energy production, high penetration of PV in the distribution grid may cause deviations from nominal operating conditions and lead to power quality issues. Thus, many studies are carried out to determine, to predict and to solve potential problems and challenges arising from such dynamic conditions. The impact of PV on the grid will depend on location, feeder parameters and nominal power installed. Different performance indicators concerning voltage, loading, protection and power quality are analysed in the literature in order to evaluate the maximum PV penetration [2]-[6]. Overvoltage at grid bus has a negative impact on customers' service as well as a possible cause of regulators failure. Moreover, high values of voltage and current have an impact on thermal limits that are important for safety reasons and insulation resistance, thus potentially disrupting the correct operation of protection devices. The aforementioned problems can be mitigated or solved by applying some solutions such as energy storage or reactive power control [4]. Reactive power control can be performed with smart inverters, on the basis of the active power related to power factor or the voltage magnitude in the point of common coupling [7]. Active transformers with tap changing are also considered as a positive asset to enable further PV growth [8]. As distinct from active methods mentioned above, hosting capacity can also be increased with some passive solutions. The most obvious and robust one is grid reinforcement. This however requires large investments and does not eliminate grid protection issues [8]. Although grid reinforcement mainly focuses on planning procedures of modernization and development, active control methods or optimal renewable energy source (RES) location scenarios may be considered as well, as a tool for minimizing the generation impact for distribution system operators.

This paper proposes an alternative method to mitigate the limit of hosting capacity based on PV sub-optimal placement. Indeed, in the literature, it is shown how optimal placement of PV arrays can minimize the power loss [9] or improve distribution grid performances through a sensitivity analysis [10]. In [11], both the voltage stability and losses indices are sensitive factors taken into account for optimal distributed generation (DG) placement. In [12], a genetic algorithm is used to find the best DG location under variable loads. However, no specific reference was found on the use of greedy algorithms to increase PV hosting capacity. The performed analysis is based on simulations using the open-source distribution grid software OpenDSS [13], which is used to compute the three-phase power-flow with high detail. It is coupled with the GridPV toolbox using a virtual COM port [14]. The proposed method is tested on the IEEE 13-bus Node Test Feeders [15] and on a larger European network.

2 METHODOLOGY

In this paragraph, first a definition of hosting capacity is given and the computational approach is presented. Then, a sub-optimal algorithm to mitigate the voltage deviations based on PV placement is described.

2.1 Hosting capacity

The hosting capacity term has become very popular and common among grid operators and researchers during the last ten years. The hosting capacity itself was already used before in completely different contexts such as networking, watermarking images or settlement of refugees [17]. In the context of distribution grids, the maximum amount of DG supported by an electric grid is defined as hosting capacity. The maximum quantity of DG (here we will refer only to PV generators) connected to the grid should guarantee reliability and voltage stability for customers and operators. The evaluation of different indicators may restrict the amount of PV that can be connected to the grid. For a given performance index, the hosting capacity is defined as the quantity of PV power that can be injected into a single grid point for which the chosen performance index does not exceed a specified limit, as shown in Fig.1.

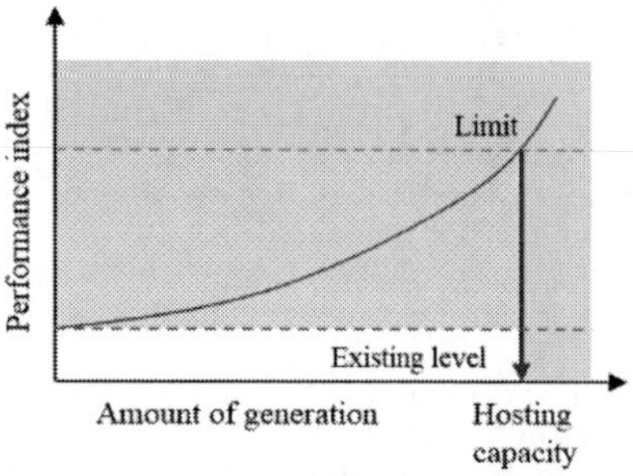

Fig. 1- Hosting capacity example with performance, which deteriorates with small PV generation [17].

There are several performance indices that can be taken into account to evaluate the quantity of power injected into a grid from a specific DG. The main ones are:
- Bus overvoltage and undervoltage levels;
- Line overcurrent level;
- Cable thermal limits (depending on the amount of current through the line);
- Maximum complex power to be transformed;
- Power quality indices (mainly due to harmonic distortion).

In this work, we evaluate the hosting capacity considering three of the abovementioned indices i.e.: the maximum or minimum bus voltage, ($V_{max-min}$), the maximum line current (I_{max}) and the maximum transformer power (S_{max}). Among them, PV generators are particularly sensitive to overvoltage (as confirmed by the simulation results reported in section III as well).
Assuming the two-node system model in Fig.2 , where the nominal phasor voltage at generator is $V_n e^{j\theta_n}$ with θ_n equal to 0° and V_n 1 p.u. (per-unit), Z is the line impedance composed by the R resistance and the reactance X and $V_g e^{j\theta_g}$ is the voltage phasor at the second node, the voltage deviation ΔV is given by:

$$\Delta V = V_g\text{-}V_n \qquad (1)$$

In the European Standard EN50160 (*"Voltage characteristics of electricity supplied by public distribution systems"* [18]), the magnitude voltage limits are equal to 5% of the nominal value at the slack bus, i.e. V_{max}=1.05 p.u. and V_{min}= 0.95 p.u. These are the values that will be considered in the following as overvoltage and undervoltage thresholds.
In order to evaluate how the chosen indices change as a function of the increasing amount of DG (e.g. Fig. 1), the power flows analysis is performed.
The power flow solutions give the steady-state operating conditions of an electrical grid, namely the voltage magnitude and phase at every bus as well as the real and reactive power flows through every line and transformer of the grid considered. If grid topology, generators, line parameters and loads are known, the power flow problem can be expressed as a system of non-linear algebraic equations that derive from the Kirchhoff's laws under

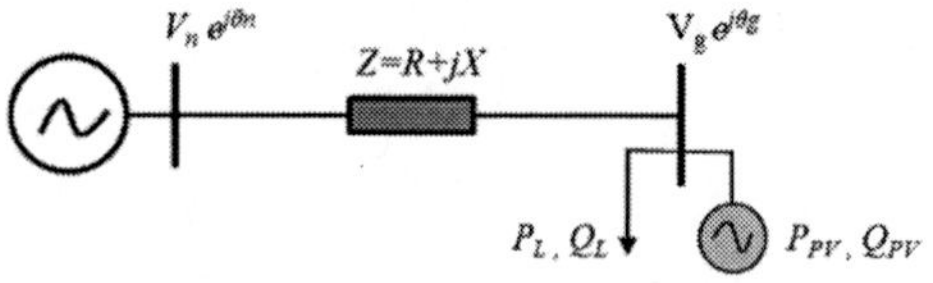

Fig. 2- Two-node system with PV generator.

proper constraints [19]. The most popular algorithms to solve the power flow problem are: Newton-Rapson and Gauss-Siedel [19]. However, the simulations performed in this paper are based on OpenDSS, which implements an algorithm based on harmonic analysis to compute the power flows [13]. This method has the advantage to be suitable not only for radial but also for mesh multi-phase distribution networks, which will be the most common topology in the next years [13].
If we define the PV penetration (in percentage) as:

$$PV_{\%} = \frac{\sum_{n=1}^{N_{PV}} P_{PV}(n)}{\sum_{l=1}^{N_{load}} \sqrt{P^2(l) + Q^2(l)}} \times 100 \qquad (2)$$

where n is the number of PV generators, l is the number of buses with active (P) and reactive (Q) power. The methods applied to evaluate the hosting capacity is shown in Fig.3. The first step is to model the three-phase (or single-phase) distribution network and to perform the power flow analysis without PV generators in order to have the baseline solution. After that, the PV generators are placed in some or all nodes and their power is increased by 1% steps according to (2). For each iteration, a new power flow solution is found and the results are compared with the limits for voltage, current and complex power. If at least one of these limits exceeds the threshold, the iteration stops and the maximum hosting capacity is achieved. For sure, the results strongly depend on the grid under test, e.g. on its topology, loads and admittance matrix. However, the method is quite general and can be applied to many different type of grids as shown in the literature [20].

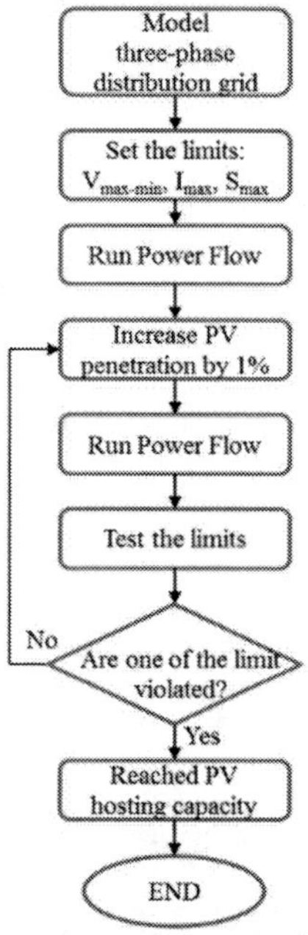

Fig. 3- Flow chart of the algorithm for hosting capacity computation.

2.3 PV placement through greedy algorithm
In the introduction, we have mentioned different kinds of methods to increase the PV hosting capacity, such as using reactive power control through smart inverters or grid line reinforcement. As mentioned in the Introduction, in

literature there are also some works which maximize the hosting capacity through optimal PV placement. This approach is also used in this paper. In particular, we try to mitigate the effect of PV generators, by properly placing them using greedy algorithms.
A greedy algorithm always makes the choice that looks best at the moment. It finds a sequence of locally optimal solutions, which usually leads to a globally optimal solution. However, this method not always converges to the global optimum. Nevertheless, this kind of algorithm is computationally efficient and fast enough to be suitable for effective PV placement even though the obtained solution is sometimes sub-optimal. The idea developed in this work is to install one PV generator at a time, finding the best bus location at every iteration. In particular, a PV generator is actually placed at the bus where the chosen cost function is minimum. Since, as previously mentioned, the voltage deviations are one of the main effect caused be high PV penetration, the cost function to minimize in the case at hand will be the root average mean square (RAMS), which is defined as:

$$RAMS_V = \sqrt{\frac{1}{N_B}\sum_{n=1}^{N_B}\left|V_n - \overline{V}_n\right|^2} \qquad (3)$$

In (3) N_B indicates the number of buses, while V_n and $\overline{V}_n$ are the nominal and actual value of the interested quantity at bus n respectively. In simulations, we will assume that the RAMS metrics will be applied to both bus voltage magnitude-only and phasors magnitude, which means to consider not only the amplitude but also the phase angle fluctuations.

3 SIMULATION RESULTS

This section presents some simulation results to evaluate the effectiveness of the algorithm shown in Fig. 3 and the validity of the greedy placement.

3.1 Evaluation of hosting capacity

In the first test, we use the IEEE 13-bus, three-phase distribution grid shown in Fig.4 to evaluate the PV hosting capacity using the method described in section 2.2.
Initially, we consider the operational case with PV generator installed at every three-phase bus, to evaluate to what extent PV penetration affects hosting capacity. Every PV generator has the same nominal power with a total penetration level given in (2). It is assumed to operate in steady-state conditions, i.e. active and reactive power at loads are approximately constant.

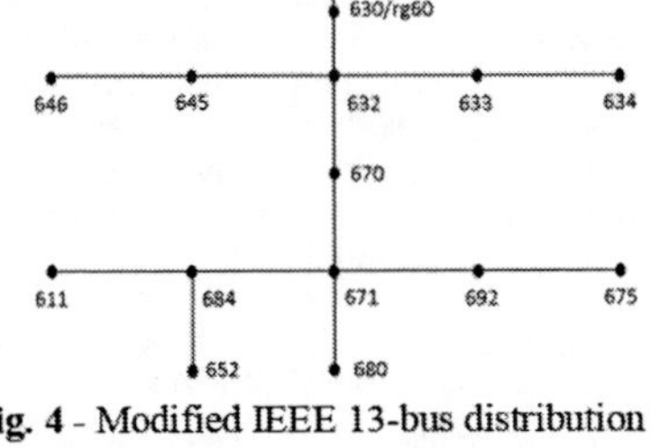

Fig. 4 - Modified IEEE 13-bus distribution feeder

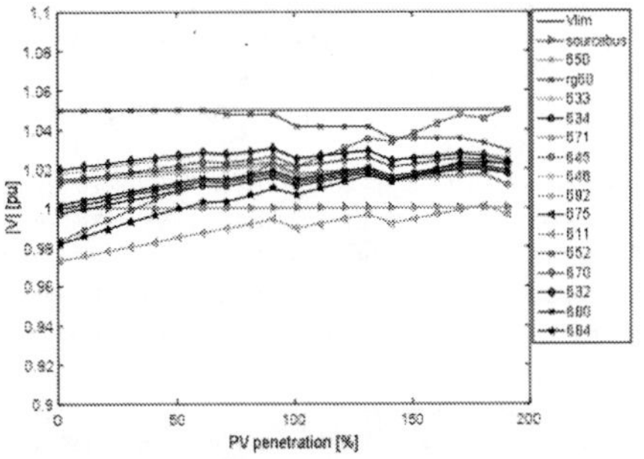

Fig. 5– Voltage magnitude for every bus by increasing PV penetration level

The penetration has been varied in the range [0%-200%] as defined in equation (2).
The limits for voltage deviations are ±5% of the nominal value as indicated in [18], while I_{max} equal to maximal line current and S_{max} equal to the maximum power at transformer. In the algorithm shown in Fig.3, after the baseline power flow computation, the PV penetration is increased iteratively by 1% steps. The limits are violated when PV penetration reaches 196%, because an overvoltage occurs at bus 652, as shown in Fig.5, which displays the average voltage magnitude of the three-phases as a function of the PV penetration at every bus. The red solid line indicates the upper -limit of the voltage (1.05 p.u. in this case).
Similar simulations (not reported for the sake of brevity) have been also performed using a smaller number of PV generators placed randomly. The results obtained in that case show that the IEEE 13-bus grid is robust to the current and power fluctuations, while the voltage limit is violated.

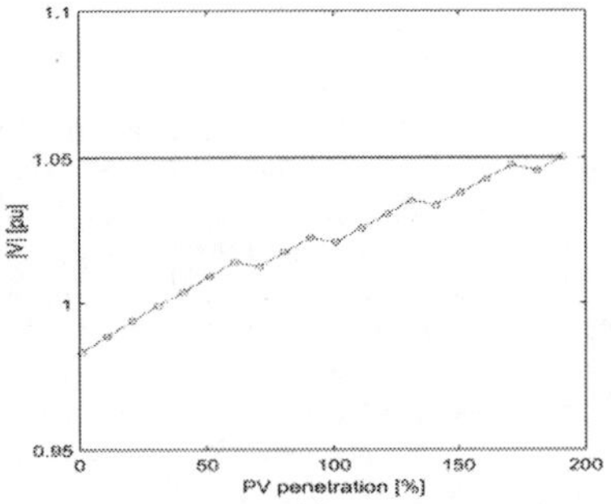

(a)

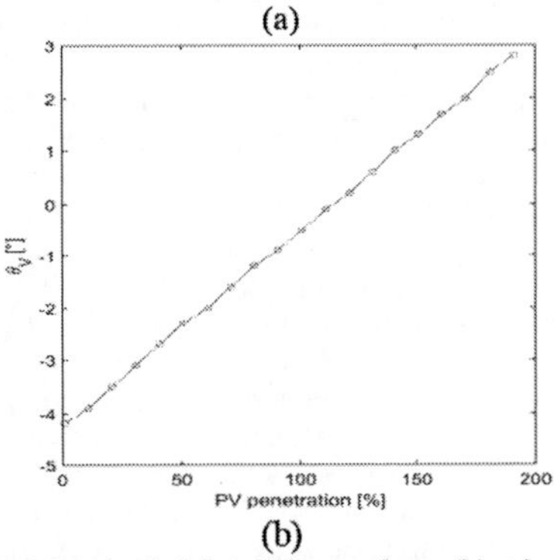

(b)

Fig. 6– Magnitude (a) and phase voltage (b) at bus 652, which exceed the hosting capacity limits

For this reason, the rest of the paper is focused on how to minimize voltage deviations only. It is worth noticing that as the amount of injected PV power increases, both voltage magnitude and angle at bus 652 grow linearly, as shown in Fig. 6(a)-6(b).
Therefore, even though currently just voltage magnitude deviations are taken into account in normal grid operation, in future active distribution grids, the increase of vRES, combined with time-variant loads (e.g. e-cars or storage) will require accurate bus voltage angle monitoring [16].

3.2 PV optimal placement
Previous simulations show that, in the case at hand, overvoltage is of primary concern for hosting capacity. Moreover, the linear growth of both voltage magnitude and angle shown in Fig. 6 is useful to choose the greedy algorithm optimization parameters. In particular, assuming to place one PV generator at a time in such a way that the RAMS metric at every iteration is minimum, we decide to compare the results assuming to consider either bus voltage phasors magnitude or bus voltage magnitude-only.
The corresponding percentage RAMS values are shown in Fig. 7(a)-(c), for three different PV penetration levels, i.e. 50% Fig. 7(a), 75% Fig. 7 (b) and 100% Fig. 7 (c).
Since in this case we assume to insert one PV generator at a time, the penetration percentage is equally distributed. This means that:

$$P_{PV} = \frac{PV_{\%}}{N_{PV}} \tag{4}$$

The red lines indicate the RAMS of voltage magnitude only obtained using sequential (solid line) or greedy (dashed line) placement, respectively. Here and in the following, for sequential placement we mean that PV generators are installed simply in the order as bus numbering. Dually, the blue lines are the RAMS values of voltage phasors magnitude for sequential (solid line) and greedy (dashed line) placement.
The obtained results are interesting for the following reasons.
First of all, it is evident that by increasing the amount of injected PV power the dispersion of bus voltage magnitude-only and phasors magnitude grow almost linearly with the number of generators installed (assuming that their nominal power is the same). The linear trend is particularly clear, when the greedy placement is used.
Moreover, for every considered penetration level, the red lines (voltage magnitude) are always below the blue lines (voltage phasor). This suggests that the phase fluctuations at different buses are significant in distribution grids and should be probably taken into account by Distribution System Operators (DSOs).
The greedy algorithm placement, as shown in Fig.7(a)-(c), generally returns better results (i.e. smaller fluctuations) than sequential placement. Further Monte Carlo simulations (not reported here for the sake of brevity) confirm that the greedy algorithm is highly likely to perform better than random placement.
An additional remark based on the results shown in Fig. 7 is that a small number of PV generators with high nominal power seem to have a greater impact on voltage phasor RAMS than many generators with low rated power. Indeed, for example considering Fig. 7(c) we have that even one PV with 100% of penetration caused a bus voltage deviations (phasor magnitude) higher than 13 PV with 50% of total penetration power.

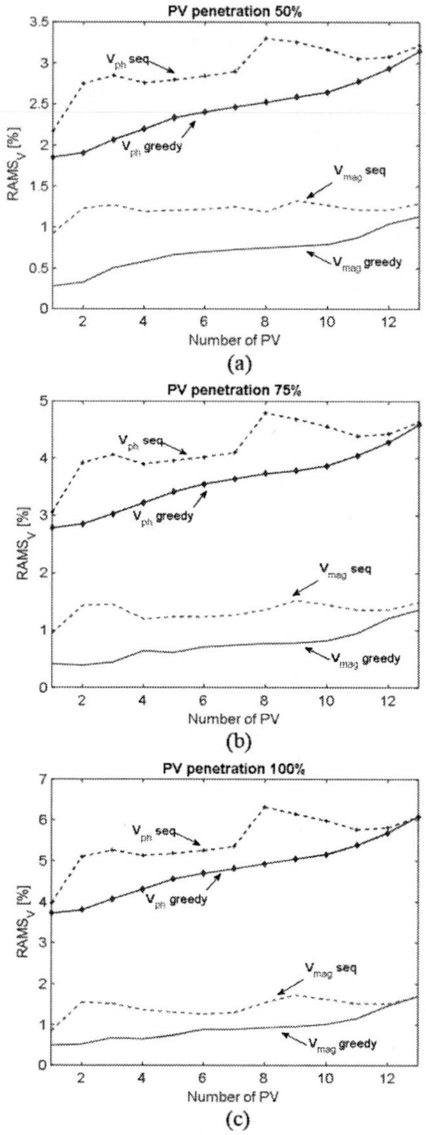

Fig. 7- Root average mean square values of bus voltage magnitude-only (red lines) and phasors magnitude (blue lines) for three different levels of PV penetration using sequential (solid line) and greedy (dashed line) placement.

In order to analyse the behaviour of the proposed method in a different context, similar simulations have been repeated using a modified version of the 130-bus LV European network available in [21]. In this case, only 30 PV generators are actually placed, varying their nominal power to provide 10%, 20% or 30% PV penetration. The RAMS values of bus voltage magnitude-only and phasors magnitude shown in Fig. 8(a)-(b) for sequential and greedy placement, respectively, essentially confirm all the previous considerations based on the results obtained using the IEEE test distribution grid.

4 CONCLUSIONS

In this paper, the impact of PV generator in terms of

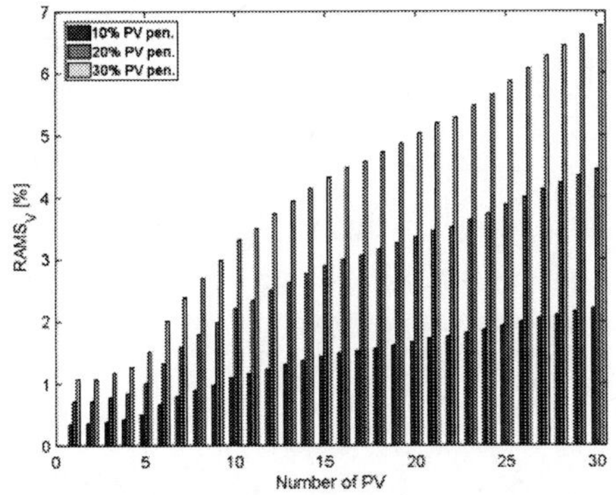

(a)

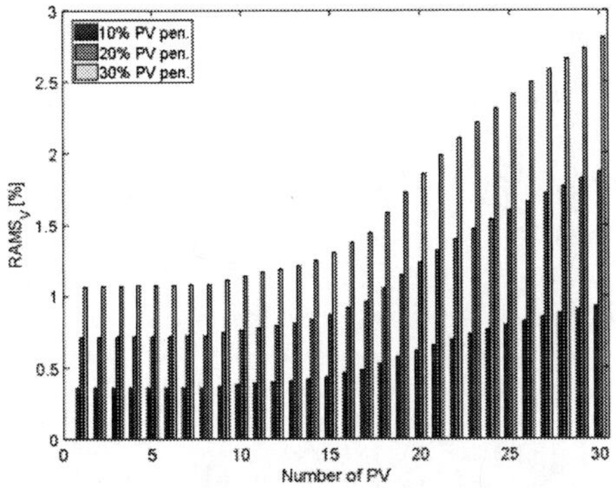

(b)

Fig. 8 – Root average mean square phasor voltage for different PV penetration using sequential (a) and greedy (b) algorithm applied to the European LV grid available in[21].

distribution grid hosting capacity has been analyzed and a method to mitigate the side effects of large PV penetration (particularly large voltage fluctuations) has been proposed. In particular, a PV generator placement algorithm based on a greedy approach has been used to minimize the root average mean squared (RAMS) values of either voltage magnitude only or voltage phasor magnitude from the respective nominal values at each bus. Multiple simulations show that the greedy algorithm performs better that both random and sequential placement. Moreover, it is demonstrated that PV penetration may have a significant impact not only on voltage magnitude, but also on voltage phase fluctuations. Indeed, the phasor magnitude RAMS values are always greater than those obtained in the magnitude-only case. For the considered distribution grids, it seems that a lower number of PV generators injecting a larger amount of power causes larger voltage fluctuations than many low-power PV generators. The results of the proposed PV placement applied to an European LV grid confirm those obtained using an IEEE test distribution grid.

In conclusion, we can said that the optimal placement based on greedy algorithm is a promising technique to mitigate the effect of high PV penetration in the distribution grid.

In future, we plan to analyze more complex distribution grids and to apply the optimal placement algorithm in the presence of distributed storage systems as well, in order to enhance grid hosting capacity.

5 ACKNOWLEDGEMENT

The research leading to these results has received funding from the European Union's Seventh Programme for research, technological development and demonstration under grant agreement No. 609019, Sinfonia project and from the from the European Funds for Regional Development (EFRE 2014-2020) South-Tyrol, INTEGRIDS project.

6 REFERENCES

[1] Arnulf Jäger –Waldau, PV Status Report 2016, JRC Science for Policy Report, October 2016

[2] CEER status review of regulatory approaches to smart electricity grids, 6.07.2011

[3] "Stochastic Analysis to Determine Feeder Hosting Capacity for Distributed Solar PV", EPRI, Palo Alto, CA: 2012.

[4] Xingtian Feng, Tongzhen Wei, "Study on Voltage Quality of Distribution Network with High Penetration of DG", International Conference on Power System Technology, 2010.

[5] National Renewable Energy Laboratory, J. Bank, B. Mather, J. Keller, M. Coddington, "High Penetration Photovoltaic Case Study Report".

[6] Ch. Bucher, G. Andersson, Lukas Küng, "Increasing the pv hosting capacity of distribution power grids – a comparison of seven methods", 28th European Photovoltaic Solar Energy Conference and Exhibition, 2013.

[7] O. Camacho Rascon; B. Schachler, J. Bühler; Matthias Resch; A. Sumper, Increasing the hosting capacity of distribution grids by implementing residential PV storage systems and reactive power control, 2016 13th International Conference on the European Energy Market (EEM)

[8] S. Hashemi and J. Østergaard, "Methods and strategies for overvoltage prevention in low voltage distribution systems with PV," in *IET Renewable Power Generation*, vol. 11, no. 2, pp. 205-214, 2 8 2017.

[9] M. Dixit, H. R. Jariwala, "Optimal placement of PV array in distribution system for power loss minimization considering feeder reconfiguration", IEEE 16th International Conference on Environment and Electrical Engineering (EEEIC), 2016

[10] S. Essallah, A. Bouallegue, A. Khedher, "Optimal Placement of PV-Distributed Generation units in radial distribution system based on sensitivity approaches", 16th International Conference on Sciences and Techniques of Automatic Control and Computer Engineering (STA), 2015.

[11] A. Parizad, A. Khazali and M. Kalantar, "Optimal placement of distributed generation with sensitivity factors considering voltage stability and losses indices," 2010 18th Iranian Conference on Electrical Engineering, Isfahan, Iran, 2010

[12] R. Shivarudraswamy; D. N. Gaonkar; Jayalakshmi N. S., GA based optimal location and size of the distributed generators in distribution system for different load conditions, IEEE 1st International

Conference on Power Electronics, Intelligent Control and Energy Systems, 2016

[13] http://smartgrid.epri.com/SimulationTool.aspx

[14] M. J. Reno and K. Coogan, "Grid Integrated Distributed PV (GridPV) Version 2," Sandia National Laboratories SAND2013-20141, 2014.

[15] http://ewh.ieee.org/soc/pes/dsacom/testfeeders/

[16] D. Macii, G. Barchi and D. Moser, "Impact of PMUs on state estimation accuracy in active distribution grids with large PV penetration," 2015 IEEE Workshop on Environmental, Energy, and Structural Monitoring Systems (EESMS) Proceedings, *Trento, 2015, pp. 72-77.*

[17] M. Bollen, S. K. Rönnberg, "Hosting Capacity of the Power Grid for Renewable Electricity Production and New Large Consumption Equipment ", Energies 2017, 10, pp. 1325.

[18] EN50160 – European Standard *"Voltage characteristics of electricity supplied by public distribution systems,* 2005.

[19] A. C. Santos, M. Nanni, M. R. Mansour, A. C. B. Delbem, J. B. A. London and N. G. Bretas, "A power flow method computationally efficient for large-scale distribution systems," *2008 IEEE/PES Transmission and Distribution Conference and Exposition: Latin America*, Bogota, 2008

[20] AlAlamat, Fadi. "Increasing the hosting capacity of radial distribution grids in Jordan." (2015).

[21] http://www.enwl.co.uk/sitemap/about-us/the-future/nia-lcnf-tier-1/lcnf-tier-1-projects/low-voltage-network-solutions

ESTIMATING THE TECHNICAL POTENTIAL OF GRID-CONNECTED PV SYSTEMS IN INDONESIA: A COMPARISON OF A METHOD BASED ON OPEN ACCESS DATA WITH A METHOD BASED ON GIS

Kun Kunaifi[1,2] and A.H.M.E. Reinders[1]

[1] University of Twente, Faculty of Engineering Technology, Department of Design, Production and Management, P.O. Box 217, 7500 AE Enschede, The Netherlands; k.kunaifi@utwente.nl; a.h.m.e.reinders@utwente.nl.
[2] UIN Suska Riau University, Faculty of Science and Technology, Department of Electrical Engineering, Jl. H.R. Soebrantas No. 115, Pekanbaru 28292, Indonesia: kunaifi@uin-suska.ac.id.

ABSTRACT: In this paper, we compare two methods for estimating the technical potential of grid-connected PV systems in Indonesia. One was a method developed by Veldhuis and Renders [1] and the other is a new method using Geographic Information System (GIS) and multi-criteria decision making (MCDM). The first method uses open access data from 2015 on a provincial level which can be applied to a complete country. Results from this method show that the total resource potential of grid-connected PV systems for all provinces in Indonesia in 2015 is 7,799 TWh which can be produced by 5,374 GWp total capacity of PV systems requiring 1.2% of the total land area of Indonesia. Because a detailed and accurate study on the technical potential of grid-connected PV systems in Indonesia is lacking, we propose another method which combines the capability of GIS and MCDM to develop a model which maps the potential of grid-connected PV systems in Indonesia. This model uses a digital elevation model (DEM) and thematic maps as inputs to produce a vector layer of calculable suitable areas for grid-connected PV systems. To validate this model, data has been collected during a field research which took place in Indonesia in Spring 2017. Also, a preliminary and more complex model for decision-making of grid-connected PV systems in Indonesia is introduced.

Keywords: potential, grid-connected PV systems, Indonesia, mathematical model, GIS, MCDM.

1 INTRODUCTION

In this paper, we compare two methods for estimating the technical potential of grid-connected PV systems in Indonesia. One method was developed by Veldhuis and Reinders [1] and has been implemented for estimating the technical potential of grid-connected PV systems in Indonesia. The other one is a new method which uses geographical information system (GIS) and-and multi-criteria decision making (MCDM). This method has not been implemented yet because it is now in the phase of concept development and it would become the first part of a full concept of developing a decision-making model for distributed solar photovoltaic (PV) systems which aims at increasing the electricity system's resilience in Indonesia. This work is part of the Ph.D. research currently undertaken at the Department of Design, Production and Management, University of Twente by the first author of this paper. Therefore, in this paper, the full preliminary and interdisciplinary concept mentioned above is also briefly introduced.

It is necessary to develop methods to estimate the technical potential of PV systems in Indonesia because - despite having a high irradiation and a favorable policy towards PV systems, the actual number of installations of PV systems in Indonesia is lagging behind the technical resource potential [4] [5]. By 2015, only 14 MWp of PV systems (both grid-connected and stand-alone systems) have been installed in Indonesia [4] which is small compared to a technical potential of 27 GWp estimated by Veldhuis and Reinders [1] using data from 2010.

Indonesia is an emerging economy with a large and is increasing energy demand. Its total area of 1.9 km^2 is divided into 34 provinces. Indonesia comprises of more than 13 thousand islands [6] in which about 4 thousand islands were inhabited in 2015 [7]. It is a tropical country which is located in Southeast Asia and Oceania (Figure 1). The population of Indonesia in 2015 was around 252

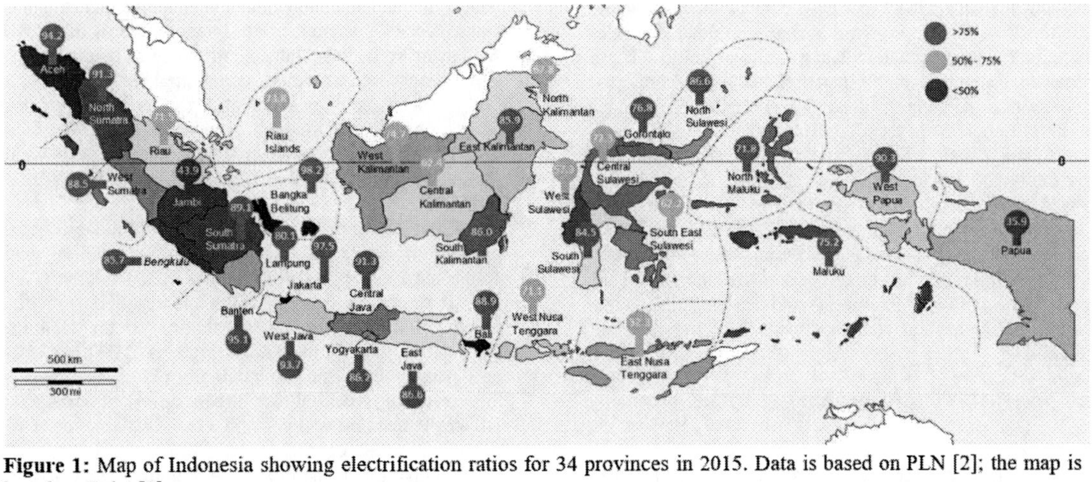

Figure 1: Map of Indonesia showing electrification ratios for 34 provinces in 2015. Data is based on PLN [2]; the map is based on Dalet [3].

million people which rank Indonesia as the fourth most populated country in the world [8].

However, Indonesia is still facing challenges in providing a sufficient, appropriately distributed, affordable, reliable, and cleaner electrical power supply to the whole population. In 2015, 11.7% of Indonesian people remained without electricity services [9] and the differences in electrification ratios (ER) inside the country are enormous (Figure 1). Also, Indonesia emits a significant and is increasing amount of CO_2 from the energy sector because the final energy consumption in Indonesia is dominated by fossil fuels [10]. Finally, the reliability of power grid in Indonesia is relatively low which is characterized by frequent blackouts and brownouts [9, 11-15].

Being located on the equator, a stable and high solar irradiation the whole year through is guaranteed. The annual average solar irradiation in Indonesia ranges from 4.4 kWh/m^2/day to 6.2 kWh/m^2/day [16] resulting in a total annual global irradiation in between 1,600 and 2,260 kWh/m^2/year depending on the location. Therefore, distributed solar PV systems are promising solutions to Indonesia's problems with the lack of electricity supply, low reliability of grid, and high CO_2 emissions. This is supported by favorable policy from the Indonesian Government which sets renewable energy (RE) target of 23% by 2020 and 31% by 2030 [17].

To stimulate the solar sector in Indonesia, it is, therefore, necessary to have a tool which could be used to carefully planning the grid-connected PV systems in Indonesia context. Such a tool should have a comprehensive approach. It should be capable to spatially locate the suitable points for grid-connected PV installations, estimate their technical potential and calculate the resulted improvement in grid reliability by considering technical and economic constraints. It should also take the social and regulatory aspects into consideration and estimate the environmental benefit from PV integration.

However, we are sure that there is no tool which has the afore-mentioned features available for Indonesia context. Most of the current studies in solar energy in Indonesia focus on off-grid PV applications. A few literatures about grid-connected PV systems in Indonesia are available. Tarigan et al. (2013) [18] evaluate the technical, economic and environmental aspects of grid-connected PV system in a typical residential in Surabaya, Indonesia using PVsyt and RETScreen software. Outhred and Retnanestri (2015) [19] evaluate the Australia's experience in developing household rooftop PV systems to provide insights to other countries including Indonesia contemplating similar scenarios. Those studies are appreciated, but issues related to grid-connected PV system potential studies and decision-making for Indonesia context are yet addressed.

In this paper, we will first introduce the method by Veldhuis and Reinders (Section 2.1) and its application based on data from 2015 (Section 2.2). In Section 3, the general concept of our new GIS-MCDM model for estimating grid-connected PV potential will be shown and a full preliminary concept of decision-making model for grid-connected PV in Indonesia will be briefly introduced. Finally, a discussion and conclusions are presented in Section 4.

2 METHOD VELDHUIS AND REINDERS

2.1 The Method

The method developed by Veldhuis and Reinders focused on estimating the resource and technical potential of grid-connected PV systems at provinces in Indonesia. This study filled the lack of literature in solar PV potential because previous studies did not distinguish between the potential of grid-connected and off-grid PV systems. Frost and Sullivan [20] reported that a total capacity of PV system for the whole Indonesia is more than 1,000 GWp. Further, a study from the Ministry of Energy and Mineral Resources of Indonesia [21] observed the PV system potential at province level with an average resolution of 300 km x 300 km.

In Veldhuis and Reinders [1] study, the mathematical model applied uses input data on a provincial level such as the population density, urbanization ratio, irradiation, electrification ratio and electricity demand. Both the resource potential and technical potential of grid-connected PV systems are determined for urban cores, suburbs and villages with population densities of respectively 5,000 persons/km^2, 1,000 persons/km^2 and 500 persons/km^2 (Figure 2). The full description of the mathematical operations in the methods of Veldhuis and Reinders can be found in their paper [1].

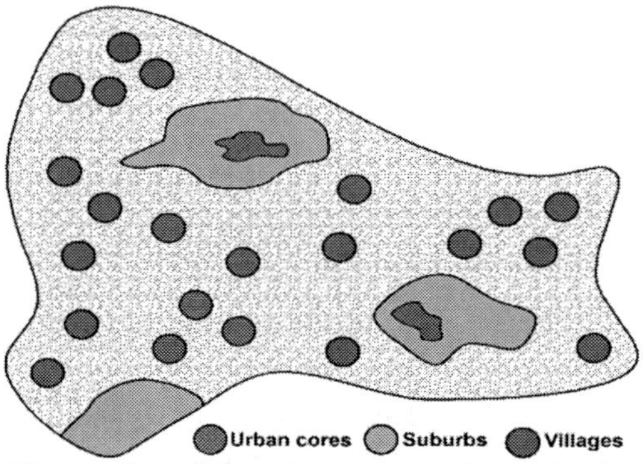

Figure 2: Schematic representation of the classification of the four areas inside a province: (1) urban core (red), (2) suburbs (orange), (3) villages (yellow), and (4) rural areas (green). Based on Veldhuis and Reinders [1].

2.2 The Results

Table I presents the results from the application of Veldhuis and Reinders's method using data from 2015. It shows the resource and technical potentials of grid-connected PV systems for all 34 provinces of Indonesia. As can be seen, the total potential of grid-connected PV systems in based on resource potential in 2015 is 5,374 GWp. The share of this potential for respectively urban cores, suburbs and village areas is 13%, 42%, and 45%. This amount of PV capacity would require 23,587 km^2 area which corresponds to 1.2% of the total land area of Indonesia.

If the above potential is limited to the electricity demand in 2015, the total technical potential of grid-connected PV systems in Indonesia in 2015, is 34 GWp. Fig. 1 shows an example of maps resulted from this method that presents the distribution of the technical potential of grid-connected PV systems (MWp) over 34 provinces in Indonesia using data from 2015. The share of this resource potential for urban cores, suburbs, and villages is resp. 30%, 39%, and 31%. This amount of PV system installed capacity would require 201 km^2 area which corresponds to 0.01% of the total land area of Indonesia.

Table I: Resource and technical potentials capacity of grid-connected PV systems for each province in Indonesia based on calculation method from Veldhuis and Reinders [1].

PROVINCE	PV SYSTEM CAPACITY (MWp) BASED ON RESOURCE POTENTIAL				PV SYSTEM CAPACITY (MWp) BASED ON TECHNICAL POTENTIAL			
	Urban core	Suburb	Village	Total	Urban core	Suburb	Village	Total
Aceh	0	54,817	129,691	184,508	0	37	87	124
Bali	0	66,257	26,769	93,026	0	369	149	519
Bangka Belitung	0	21,645	21,358	43,003	0	33	33	66
Banten	102,353	46,702	65,512	214,567	1,287	587	824	2,697
Bengkulu	0	20,586	39,707	60,293	0	14	27	41
Gorontalo	0	13,607	14,963	28,570	0	14	15	29
W. Papua	0	9,849	20,050	29,898	0	14	28	41
Jakarta	4,515	0	0	4,515	4,515	0	0	4,515
Jambi	0	26,760	11,259	38,020	0	74	31	105
W. Java	194,848	497,527	220,501	912,876	2,235	5,706	2,529	10,469
C. Java	38,804	232,018	269,446	540,268	231	1,380	1,602	3,213
E. Java	175,060	341,898	398,917	915,875	1,300	2,540	2,963	6,803
W. Kalimantan	18,280	30,575	68,072	116,927	24	40	90	154
S. Kalimantan	18,778	31,689	50,784	101,252	31	53	85	169
C. Kalimantan	0	26,105	20,888	46,993	0	42	34	76
E. Kalimantan	0	53,099	18,118	71,217	0	138	47	184
N. Kalimantan	0	7,525	238	7,763	0	113	4	116
Riau Islands	0	24,757	0	24,757	0	439	0	439
Lampung	0	80,643	167,224	247,867	0	132	274	406
N. Maluku	0	11,037	19,793	30,830	0	7	12	19
Maluku	0	19,776	21,965	41,741	0	18	20	38
W. N. Tenggara	11,281	45,178	36,632	93,090	8	32	26	67
E N. Tenggara	0	36,986	59,635	96,620	0	21	34	56
Papua	0	20,415	6,078	26,493	0	60	18	78
Riau	25,312	42,579	60,603	128,494	56	95	135	286
W. Sulawesi	0	9,929	16,913	26,841	0	5	9	14
S. Sulawesi	42,212	59,081	120,976	222,268	105	148	302	555
C. Sulawesi	0	26,967	50,665	77,632	0	14	27	42
S.E. Sulawesi	0	24,138	27,173	51,311	0	22	25	46
N. Sulawesi	0	34,713	29,072	63,785	0	56	47	103
W. Sumatra	0	71,225	80,850	152,075	0	218	248	466
S. Sumatra	0	98,558	161,061	259,620	0	204	334	538
N. Sumatra	62,641	132,609	159,569	354,819	213	450	542	1,205
Yogyakarta	8,650	44,056	13,141	65,847	35	181	54	270
INDONESIA	**702,733**	**2,263,306**	**2,407,621**	**5,373,660**	**10,041**	**13,255**	**10,653**	**33,948**

3 PRELIMINARY CONCEPT OF THE NEW MODEL

Using the method from Veldhuis and Reinders, the potential capacity of grid-connected PV systems in Indonesia can be easily estimated using open access data for the provincial or district level with the average resolution of around 100-300 km x 100-300 km. Further, it would be useful to estimate the accuracy of these studies and to increase their resolution by zooming into the level of cities with an average resolution of 15 km x 15 km or even better to the level of neighborhoods.

Because data for this purpose may not always be (made) available by local governments, it is difficult to use the method from Veldhuis and Reinders for a smaller area. Therefore, a more independent approach may be necessary in this case. Geographic Information Systems (GIS) and multi-criteria decision making (MCDM) could be a possible solution to tackle these issues and is part of our new approached that will be presented in Section 3. The GIS approach will be relevant because of ongoing efforts by government and private institutions in producing geospatial information such as digital elevation model (DEM), satelite imageries, maps, among others, for conductiong studies for the whole country.

3.1 GIS and MCDM approach

The GIS comprise sets of tools that can capture, store, retrieve, analyze and present various data which are spatially attributed to locations for a special set of purposes [22]. In short, GIS is embedded in a computer system that uses geographical data to create useful information. Maps have an essential role in GIS given its capability to store, retrieve, analyze, and present spatial data.

Based on several studies such as Khan (2014) [23], Lozano (2013) [24], Hafeznia (2017) [25], and Carrion (2008) [26], we summary the following main five steps to develop a GIS-MCDM model:

- identifying the criteria and input dataset,
- identifying appropriate software/tool that is capable of handling the previously-defined criteria,
- building the model in the GIS environment,
- analyzing the results.

1. Identifying the criteria and input dataset.

Due to multiple objectives that must be examined at the same time, making the most favorable decision could involve a complex process [24]. Some criteria are needed to ensure the judgment is measurable and quantified [26]. Like the GIS, MCDM has also been implemented in many fields [24] and therefore, they could be a good combination in this study. The MCDM is used due to its popularity in energy decision making [27] and also because it suits the complexity of factors to be considered in identifying suitable areas for grid-connected PV plants.

"MCDM refers to making decisions in the presence of multiple, usually conflicting, criteria" [28]. It is a decision support model which consists of a set of solutions, criteria and values [29]. The alternatives of the solution are ranked or sorted by the decision maker. The qualitative and quantitative of criteria are based on some predetermined indicators. The values of each alternative of corresponding criterion could be based on information from literature or an expert and stakeholder consultation.

MCDM applications in energy planning decision involve various methods. The methods are built based on priority setting, fuzzy principles, weighted averages, outranking, and their combinations [27]. In this study, we apply the weighted averages method using a set of criteria as shown in Figure 3.

2. Identifying appropriate software/tool that is capable of handling the defined criteria

Although some GIS software packages offer tremendous help in working with GIS and related tasks with their unified approaches, sometimes more than one tools are needed to tackle a problem. The selection of software depends on the objective of tasks to achieve. The general rule of thumb is that the software must be capable of creating, managing, analyzing, and visualizing geographic data. Therefore, the selected GIS software can be used as an environment to building the GIS model.

3. Analysis

The criteria previously described are used to conduct an analysis. An example from Carrion (2008 [26]) suggested that the criteria could be grouped into two classes: (i) supporting criteria, (ii) hindering criteria or restrictions. Supporting criteria are those that increase the suitability of an area for grid-connected PV plants while hindering criteria represents the negative factors that restrict an area for grid-connected PV installation.

In the next step, a final layer could be created showing areas with higher values only (have more supporting criteria), while areas with lower values (have more hindering criteria) will be removed or indicated low potential.

Figure 3 shows an example of a GIS-MCDM model which could be used for estimating the technical potential of grid-connected PV system for a city or neighborhood level in Indonesia context.

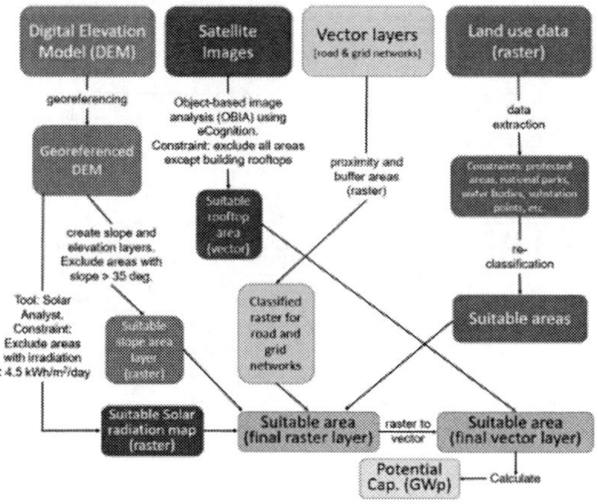

Figure 3: An example of GIS-MCDM model for estimating the potential capacity of grid-connected PV for a city or neighborhood level in Indonesia.

Initially, users define the inputs indicators for implementing the GIS and MCDM methods. Some examples of indicators are proximity to roads, unsuitable lands, harsh topography, and solar irradiation level. The model could be developed within various GIS packages such as ArcGIS from ESRI, eCognition from Trimble, QGIS from Quantum, and others. A digital surface model (DSM), vector and raster maps are examples of useful inputs

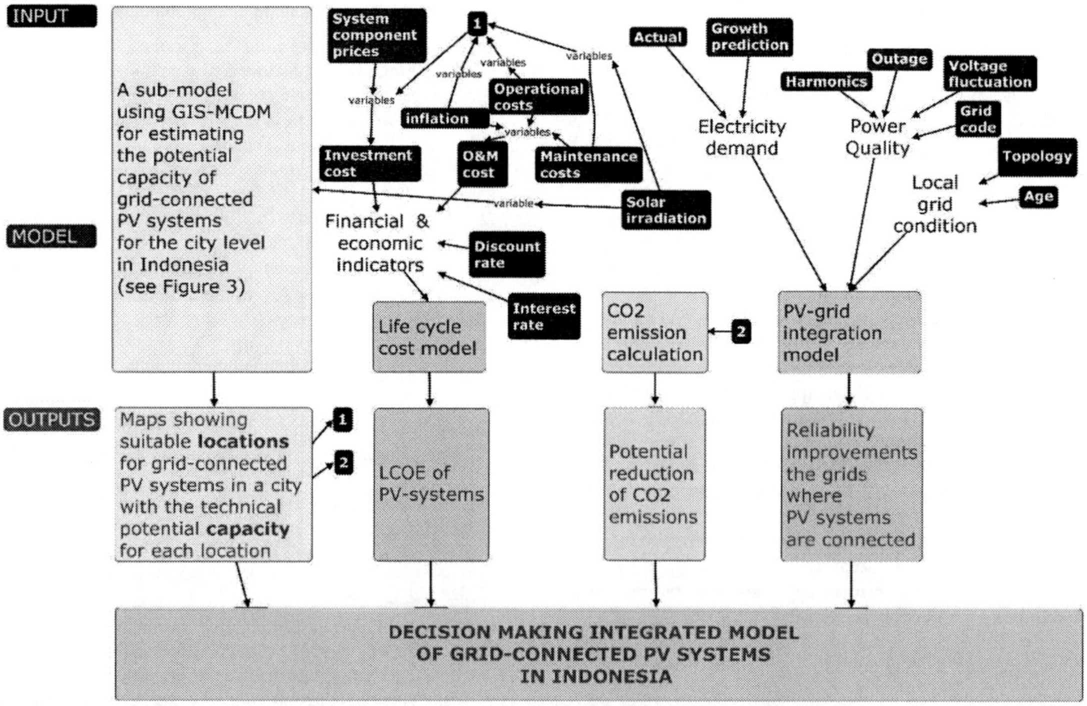

Figure 4: The full model proposed for decision-making of grid-connected PV systems in Indonesia.

for those tools that represent the area of a city which covers earth surface, buildings, roads, water bodies, open spaces, among others. Using the pre-defined indicators, a thematic layer of unsuitable areas could be created. In such a layer, only suitable areas of PV plants remain. Based on the knowledge about the suitable area and their distribution, the technical potential of grid-connected PV system could be calculated and mapped.

3.2 The Full Model

Besides estimating the potential capacity of grid-connected PV systems in Indonesia, as mentioned in Section 1, the definite function of this work is to develop a comprehensive decision-making model. The tasks of the model are to find the feasible locations for grid-connected PV plants and estimate their technical capacity, evaluate the reliability improvements on the grid to which the PV systems would be connected, and estimate the potential environmental benefit of distributed PV systems.

Figure 4 shows a general and preliminary concept of the proposed model. As can be seen, the first step of modeling is to create maps of suitable locations for grid-connected PV systems with their corresponding PV system capacity. This step could be conducted using the GIS-MCDM sub-model as shown in Figure 3. Once the capacities of PV systems are known, their values could be used to estimated the avoided CO_2 emissions using emission factors of different types of fossil fuels. Then, the levelized costs of energy (LCOE) could be calculated using life-cycle cost analysis (LCCA) based on various inputs. Finally, the improvements in the reliability of local grids by integrating PV systems could be estimated using a separate model.

4 DISCUSSION AND CONCLUSION

4.1. Grid-Connected PV potential estimation

Using data from 2015 as input for Veldhuis and Reinders method [1], it has been found that the total resource potential of grid-connected PV systems in Indonesia is about 5,374 GWp and the technical potential is about 34 GWp. The area required are 31,557 km^2 for the resource potential and 201 km^2 for the technical potential. A comparison of the results from the previous study based on data from 2010 [1] and the result from this study is described as follows: The resource potential changed by a factor of 5 and the economic potential increased by 26% following the growth of energy demand in the daytime from 2010 to 2015. The resource potential of electricity generation by PV systems in 2015 is 38 times the total electricity demand in 2015 based on data from the national utility PLN [2]. Logically the PV system capacity for resource potential increased nearly 5-fold from 1,100 GWp in 2010 to 5,374 GWp in 2015. This is due to growing population (at 1.5% per year) [30], increasing electricity demand (at 8.6% per year) [17] and increasing the PV module efficiency (from 15% in 2010 to 17% in 2015).

As shown in Figure 5, the relation of resource and technical capacities are not always linear like in Jakarta. For example, due to a large area, the resource potential in Lampung is 600 times its technical potential because of the small demand in the village and rural areas.

Table II shows a brief comparison of advantages and disadvantages of the Veldhuis and Reinders method and the GIS-MCDM model. By using the GIS-MCDM model proposed in this study, more accurate results could be expected because it can work with higher resolution input data. However, the GIS-MCDM model relies on the availability of DEM and other spatial information which is promising for the future considering the fast production of such data to date.

4.2. The New Decision-Making Model

Despite that this decision-making model offers a more comprehensive approach which considers many relevant variables, this is a preliminary concept that needs further studies. In particular, other influencing factors such should be included such as types of mounting methods, the balance of system (BOS), energy payback time, and existing grid capacity, among others. Later, after a complete development of the model, it needs to be tested and validated using more information which could take another one year to collect.

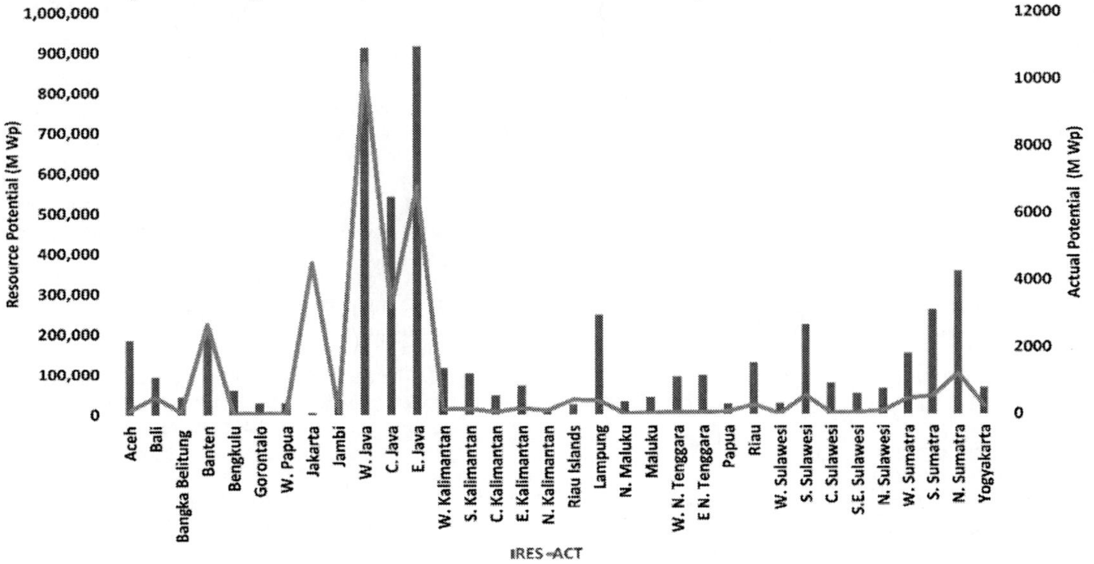

Figure 5: The technical potential of grid-connected PV systems in provinces in Indonesia based on the method from Veldhuis and Reinders.

Table II: Advantages and disadvantages of the Veldhuis and Reinders method in estimating the technical potentials of grid-connected PV systems for each province in Indonesia and its comparison to the GIS-MCDM method.

	Veldhuis and Reinders Model	GIS-MCDM model
Advantages	▪ Open access data for provincial and district levels can be used, ▪ The model is relatively simple, ▪ The model can be run on Ms. Excel, ▪ This is perhaps the first peer-reviewed work about PV potential estimation on provincial level in Indonesia.	▪ Literatures with examples from other counties are available, ▪ Higher level of accuracy and detail. It can show smaller area of cities or even neigbourhood, ▪ It could show spatial distribution of potential locations of grid-connected PV systems, ▪ It is dynamic, namely it could be used with different resolution as long as data is available
Disadvantages	▪ It cannot be used for smaller area than district level (resolution below 100 km x 100 km) because the data is not available, ▪ It does not include detail/local information, ▪ The spatial distribution of PV systems cannot be shown (lower aggregation level)	▪ Special data (not open access) are needed such as DEM, aerial imageries, maps of land use, etc. Data has not available for all locations in Indonesia yet, ▪ The model is quite complicated, requires special softwares (not for free)

ACKNOWLEDGEMENT

Funding for this study was provided by LPDP, which is the Indonesia Endowment Fund for Education. Expertise support is provided by the Advanced Research on Innovations in Sustainability and Energy (ARISE) of the University of Twente – The Netherlands.

REFERENCES

[1] A. J. Veldhuis and A. H. M. E. Reinders, "Reviewing the potential and cost-effectiveness of grid-connected solar PV in Indonesia on a provincial level," *Renewable and Sustainable Energy Reviews*, vol. 27, pp. 315-324, 2013.

[2] PLN, *Statistik PLN 2015*. Jakarta: PLN, 2016.

[3] D. Dalet, "Indonesia," *D-Maps*, Available: http://www.d-maps.com/

[4] BPPT, "Outlook Energy Indonesia 2016," BPPT, Jakarta2016.

[5] M. H. Hasan, T. M. I. Mahlia, and H. Nur, "A review on energy scenario and sustainable energy in Indonesia," *Renewable and Sustainable Energy Reviews*, vol. 16, pp. 2316–2328, 2012.

[6] Ministry of Marine Affairs and Fisheries, "Rekapitulasi Data Pulau di Indonesia," Available: http://www.ppk-kp3k.kkp.go.id/info-ppk/public_html/assets/uploads/files/11%20Selisih%20Jumlah%20Pulau%20Sebelum%20dan%20Sesudah%20Verifikasi.pdf

[7] "70% Pulau di Indonesia Tak Berpenghuni," ed: Liputan 6, 2014.

[8] "2016 Handbook of Energy & Economic Statistics of Indonesia," ESDM, Jakarta2016.

[9] *Statistik PLN 2015*. Jakarta: PLN, 2016.

[10] "Outlook Energy Indonesia 2016," BPPT, Jakarta2016.

[11] *Statistik PLN 2009*. Jakarta: PLN, 2010.

[12] *Statistik PLN 2011*. Jakarta: PLN, 2012.

[13] "Statistik PLN 2012," PLN, Jakarta2013.

[14] "Statistik PLN 2013," PLN, Jakarta2014.

[15] *Statistik PLN 2014*. Jakarta: PLN, 2015.

[16] NASA, "Surface meteorology and Solar Energy," NASA, Ed., 6.0 ed. Langley 2016.

[17] Government of Indonesia, "President Decree Regarding National Energy General Plan, 2016-2050," Government of Indonesia, Jakarta2016.

[18] E. Tarigan, Djuwaria, and F. D. Kartikasari, "Techno-Economic Simulation of a Grid-Connected PV System Design as Specifically Applied to Residential in Surabaya, Indonesia," *Energy Procedia*, vol. 65, no. 90 – 99, 2015.

[19] H. Outhred and M. Retnanestri, "Insights from the Experience with Solar Photovoltaic Systems in Australia and Indonesia," *Energy Procedia*, vol. 65, pp. 121 – 130 2015.

[20] Frost & Sullivan Asia Pacific, "2012 Indonesia Power Sector Outlook," Frost & Sullivan Asia Pacific, Jakarta, Presentation 2012.

[21] Government of Indonesia, "President Decree Regarding National Energy General Plan, 2017-2026," Government of Indonesia, Jakarta2017.

[22] I. Heywood, S. Cornelius, and S. Carver, *An Introduction to Geographical Information Systems*. New Jersey: Prentice Hall, 2011.

[23] G. Khan and S. Rathi, "Optimal Site Selection for Solar PV Power Plant in an Indian State Using Geographical Information System (GIS) " *International Journal of Emerging Engineering Research and Technology*, vol. 2, no. 7, pp. 260-266, 2014.

[24] J. M. S. Lozano, J. T. Solano, P. L. S. Elvira, and M. S. G. Cascales, "Geographical Information Systems (GIS) and Multi-Criteria Decision Making (MCDM) methods for the evaluation of solar farms locations: Case study in south-eastern Spain," *Renewable and Sustainable Energy Reviews*, vol. 24, pp. 544-556, 2013.

[25] H. Hafeznia, H. Yousefi, and F. R. Astaraei, "A novel framework for the potential assessment of utility-scale photovoltaic solar energy, application to eastern Iran," *Energy Conversion and Management*, vol. 151, pp. 240-258, 2017.

[26] J. A. Carrion, A. E. Estrellaa, F. A. Dols, M. Z. Toro, M. Rodrıguez, and A. R. Ridao, "Environmental decision-support systems for evaluating the carrying capacity of land areas: Optimal site selection for grid-connected photovoltaic power plants,"

Renewable and Sustainable Energy Reviews, vol. 12, pp. 2358–2380, 2008.

[27] S. D. Pohekar and M. Ramachandran, "Application of multi-criteria decision making to sustainable energy planning—A review," *Renewable and Sustainable Energy Reviews,* vol. 8, no. 4, pp. 365-381, 2004.

[28] L. Xu and J.-B. Yang, "Introduction to Multi-Criteria Decision Making and the Evidential Reasoning Approach " Manchester School of Management, University of Manchester Institute of Science and Technology, Manchester2001, vol. Working Paper No. 0106

[29] L. Krpan, V. Šteko, and Z. Koren, "Model for selecting locations for construction of solar power plants," *Građevinar,* vol. 64, no. 9, pp. 741-748, 2012.

[30] BPS-Statistics Indonesia, "Statistical Yearbook of Indonesia 2016," ed. Jakarta: BPS-Statistics Indonesia, 2016, p. 720.

On the development of long-term PV generation time series using PVGIS model for European power system analysis

Moustafelou I.*, Gonzalez-Aparicio I.*, Alves Dias P., Huld T. and Zucker A.

European Commission, DG-Joint Research Centre; Energy, Transport and Climate Directorate, Petten, Netherlands

* *Corresponding authors: ismini.moustafelou@gmail.com, iratxe.gonzalez-aparicio@ec.europa.eu*

Address: European Commission, PO Box 2 NL-1755 ZG Petten, The Netherlands

Abstract

The growing penetration of wind and solar renewable energy sources in the energy mix constantly increases the stochastic nature of the power system requiring a precise interpretation of the resource variability. This implies a better understanding on the accuracy of the approach to convert the weather data into a renewable energy sources dataset. As regards the solar generation, few studies focus on the sensitivity of the multi-criteria methods, the selection of the location to install solar panels and the assessment of implications on the total PV generation. This study quantifies the impact of applying different criteria in the selection of the land uses and the difference in the total annual generation at different aggregation levels (country, bidding zone and regional). For this purpose, PV hourly time series are calculated based on different assumptions using the PVGIS model. The results show the sensitivity of the impact on the total generation on different market zones under different assumptions and the relevance of defining an adequate rule based strategy for the development of future scenarios for power systems with high solar penetration.

1. Introduction

The European Commission's 2030 renewable energy targets suggest the transition towards a more decarbonised energy mix [1]. However, the growing penetration of intermittent renewable energy sources, (wind and solar) in the energy mix constantly increases the stochastic nature of the power system, potentially requiring more flexibility measures on both the supply and the demand side. To alleviate the drawbacks of increased shares of intermittent sources, modelling tools are deployed; but, to develop accurate models delivering results with lower uncertainties, adequate representation of the Renewable Energy Sources (RES) data is required and it is essential to estimate accurately the availability of RES sources at the hour level. The approach to convert weather data into a RES dataset is crucial for the accuracy of a dataset which should meet temporal and spatial requirements: data should be given at least in hourly time intervals and at different aggregation levels (country, bidding zone and regional) for modelling the European Power system with its different market areas. A consistent dataset should also keep a compromise between the geographical coverage accounting for European climate zones and a period long enough to capture the climate variability (a 30 year interval). Attention should also be given to the technical data of the wind and PV fleet such the orientation, spatial distribution, suitability of locations for installations and regulation of protected areas.

For the PV generation, existing studies are mainly focused on evaluating the theoretical, available and economical potential of solar energy in different locations of particular regions, based on multi criteria evaluation methods, commonly using Geographic Information Systems (GIS). For example, Sánchez-Lozano et. al assessed the suitable locations for solar farms in south-Eastern Spain, combining Multi Criteria Decision Making methods with GIS [2]. They evaluated the suitability based on restrictions/constraints and weighting factors or criteria. The optimum locations for the instalment of hybrid wind-solar power systems in Middle-East have been proposed by applying a Boolean model to GIS [3]. The annual technical, geographical and economical potential of solar energy has been explored at the regional level for Fujian Province in China implementing a GIS –based model [4]. Uyan ranked the suitability of Karapinar region for solar farms by establishing a multi criteria evaluation-GIS model [5]. In other studies, hourly data are applied to energy supply optimization model at regional scale. In particular, Killinger et al performed a technical optimization for PV systems, using as inputs the hourly generation of wind parks and several PV orientations in different regions of Germany [6].

Mainzer et al., explored the technical potential of solar PV in residential rooftops in two aggregation levels (federal state and municipality) in Germany [7]. The highest potentials were found in south and west regions; at municipality level the highest measured technical potential concerns Dortmund and Hannover. At state level North Rhine-Westphalia and Bavaria suggest the maximum solar power potential. A similar study for Switzerland was performed by Assouline et. al.; they estimated the solar energy technical potential at commune, canton and country basis, applying a GIS-Support Vector Machines (SVMs) method [8].

However, few are the studies analysing the relevance of the multi-criteria methods and the sensitivity in the total PV generation. To fill this gap, this analysis is devoted to quantify the PV potential under different land-use scenarios and the variation in the generation at different aggregation levels (country, bidding zone and regional) for modelling the European Power system with its different market areas.

2. Data and methodology

Two datasets of the hourly PV potential for 2015 (given by the maximum solar energy output (watt hour) for each kilowatt of installed capacity averaged over a region) are calculated based on different land-use assumptions using the PVGIS model [9]:

Assumption 1): The PVGIS model output is used for areas including all land-use types as available for the installation of solar panels.

Assumption 2): Some geographical areas are excluded for the installation of solar panels such as mine, dump and construction sites, artificial, non-agricultural vegetated areas, forests, glaciers and perpetual snow, wetlands and water bodies and any Natura 2000 protected areas. All the remaining suitable areas are merged creating the exclusion mask using GIS tools and then, it is implemented in the PVGIS model. To build the exclusion mask, different land-use databases and territorial administrative units are used: Nomenclature of Territorial Units for Statistics NUTS 2013 version, the 2006 CORINE Land Cover (CLC) and the 2015 version of Natura 2000 geographic layers sourced by, Eurostat [10], Copernicus [11], (and European Environmental Agency respectively [12].

Note that the selection of the excludable areas is rather conservative approach since Natura 2000 areas or artificial, non-agricultural vegetated areas could be permitted for solar PV installation under several circumstances, evaluating the land-use competition with other use [13] However, the scope of this study is to investigate the maximum bias of the generation between two extreme scenarios, in order to capture the sensitivity of the exclusion zones in the solar power generation only accounting for the availability of the land-use.

The PVGIS calculations were performed using hourly solar radiation data from the CM SAF SARAH solar radiation product [14] with a spatial resolution of 3 arc-minutes (~5km). Hourly PV power was calculated for each pixel taking into account the effects of shallow-angle reflectivity, temperature and low irradiance as well as cooling of the modules by wind [9]. The calculation was performed for PV arrays mounted on an open-rack mounting at 30°inclination south-facing. Finally, for each hour, the PV energy production was averaged over each region.

3. Results

3.1 Annual generation at different aggregation levels

Here it is shown the results of the PV potential difference between the two types of simulations at different aggregation levels. That is, the difference between the performances of the PVGIS model accounting for all the land-use types and with the exclusion criteria averaged over the NUTS2 regions. The difference is expressed either as absolute or as relative terms. The total generation in the non-exclusion scenario is estimated at 113.8 TWh. The results suggest that at the European level the difference in the total annual generation is not significant: The absolute difference is estimated at –59.15 GWh, while the relative difference at -0.052%. The maximum difference (-0.12%) occurs at 16:00h. At the regional scale, higher differences apply in different hours, ranging from -6.02 to 2.54%. This result shows that the land use availability, at continental level is not a constraint to reach the 27% goals of the EC.

At country level, the relative difference in the annual generation is negligible for most of the countries studied. The most notable differences concern Norway (-2.04%) and Switzerland (-1.41%), (Table 1). The maximum difference on a yearly basis (-48.2 GWh) is accounted for UK.

Table I: Solar Power Generation for each scenario and Country, and respective absolute and relative difference

Country	Generation Non-Exclusion (GWh)	Generation Exclusion (GWh)	Absolute Difference (GWh)	Relative Difference (%)
AT	839.89	842.03	-2.13	-0.25
BE	3151.31	3148.97	2.34	0.07
BG	1407.05	1406.40	0.66	0.05
CH	922.00	934.97	-12.98	-1.41
CZ	2268.53	2267.31	1.22	0.05
DE	42792.59	42782.26	10.34	0.02
DK	772.32	772.78	-0.46	-0.06
EE	5.56	5.56	0.00	-0.01
EL	3787.54	3784.43	3.11	0.08
ES	7710.52	7711.86	-1.34	-0.02
FI	9.03	9.03	0.00	-0.02
FR	8166.37	8172.71	-6.34	-0.08
HR	58.90	58.64	0.25	0.43
HU	36.40	36.40	0.00	-0.01
IE	0.94	0.94	0.00	-0.11
IT	28071.02	28076.75	-5.74	-0.02
LT	69.00	68.99	0.01	0.01
LU	124.15	124.07	0.08	0.07
LV	1.93	1.93	0.00	0.01
NL	1512.99	1513.23	-0.24	-0.02
NO	11.64	11.88	-0.24	-2.04
PL	95.62	95.60	0.02	0.02
PT	702.91	702.73	0.18	0.03
RO	1587.54	1587.58	-0.05	0.00
SE	93.36	93.42	-0.06	-0.07
SI	332.85	332.68	0.17	0.05
SK	624.05	623.77	0.28	0.04
UK	8616.67	8664.89	-48.22	-0.56

At NUTS2 level, the results indicate that many of the maximum differences between the two simulations occur at 12:00h, ranging between -6.02 % and 1.46%. However, the maximum difference of the PV potential at the European aggregation level (-0.12%) is noticed at 16:00. This is due to the larger number of the highly negative differences per NUTS2 region and therefore, smaller deviation between the NUTS2 regions (standard deviation of the difference at 16:00 and 12:00 equal to 0.47, and 0.64 respectively, see Table 2). The results indicate that the maximum standard deviation at the NUTS2 level during the peak hours (12-16) occurs at 12:00h (see Table 2). In addition this hour is considered to have the maximum difference in the simulation directly related to the power peaks and flexibility needs; hereinafter the hourly analysis is based at 12:00. The differences between the two simulations during this hour range from -6.02% to 1.46 %. Positive values indicate that the excluded areas in the second assumption are considered not very suitable for PV installations and then the averaged PV potential over the NUTS2 increases. On the contrary, negative values indicate that the excluded areas are suitable locations and then the averaged PV potential over the NUTS2 decreases (Figure 2). The dominant land type influencing the difference in the PV potential is the forests. Those differences become more significant when converting the PV potential into the total generation. By NUTS2, the maximum difference in total generation (absolute difference) is +3.03 (1.44% relative difference) and -12.5 GWh (-2.46% relative difference) and refers to Italian (ITC2) and UK (UKM2) territory, respectively. That means that for the case of Italy, the exclusion criterion followed reduced the PV potential and for UK increased the PV potential. Maximum differences are visible around 12:00 CET when the solar generation reaches the daily peak.

Table II: Maximum positive and negative relative differences and standard deviation at NUTS2 level during the solar power generation hours 4-21.

hour	NUTS2 max	Max (+) Difference (%)	NUTS2 min	Max (-) Difference (%)	Standard Deviation
4	CZ08	2.25	EL42	-2.29	0.36
5	CH05	2.45	UKM6	-3.96	0.63
6	CH01	2.54	NO04	-5.72	0.65
7	CH05	2.34	NO03	-4.45	0.47
8	CH05	1.01	NO04	-5.95	0.55
9	ITC2	0.83	NO04	-5.65	0.56
10	ITC2	1.01	NO03	-3.95	0.49
11	ITC2	1.20	NO03	-3.62	0.51
12	ITC2	1.46	NO04	-6.02	0.64
13	ITC2	2.07	NO04	-5.87	0.63
14	ITC2	1.89	NO04	-5.45	0.59
15	ITC2	2.30	NO04	-4.93	0.57
16	ITC2	1.90	NO04	-3.67	0.47
17	ITC2	1.29	NO04	-2.96	0.43
18	NO02	1.83	CH01	-2.92	0.41
19	HU32	2.03	CH01	-2.60	0.43
20	NO03	2.65	ES52	-2.40	0.46
21	UKE2	3.07	SE31	-2.09	0.35

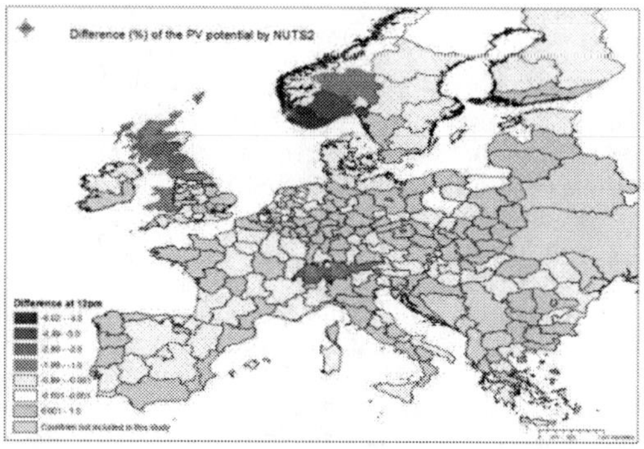

Figure 1: Difference (%) of the PV potential between the two types of simulations (all types of land-uses available – excluding areas) at NUTS 2 level

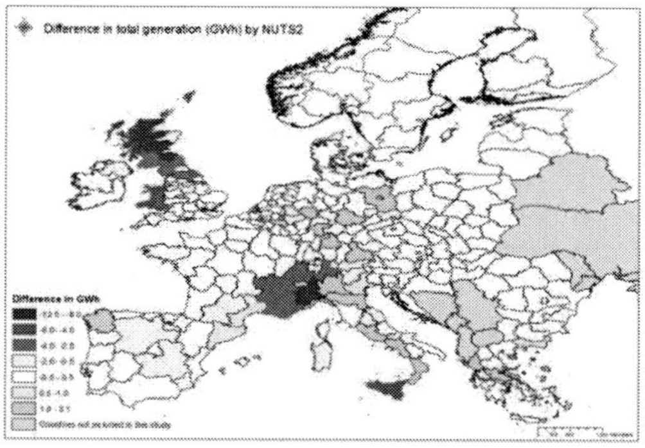

Figure 2: Difference (Total generation) of the PV potential between the two types of simulations (all types of land-uses available – excluding areas) at NUTS 2 level

In order to investigate the impact of the exclusion zones in the PV potential, the study is focused on the countries with the highest differences either at NUTS2 level or country level. For instance, in Switzerland (Figure 3), in some regions a maximum negative difference is noticed from 12pm to 15pm, while in other regions this difference is negligible. Note that the difference can reach up to 45% at some hours in the year (Figure 4). On the other hand, in Italy (Figure 5, top), both negative and positive peaks are observed during the same interval, depending on the region. The mean difference in solar power potential is negative during the hours 8pm to 7am in Swiss territory (shown by the red star in Figure 3), while for Italy (Figure 5, right) the difference is slightly positive from 7am to 7pm. In Norway the most negative differences are shown during the morning hours (6-12pm), while there is a decreasing trend of the relative difference (in absolute value) for five out of the seven NUTS 2 regions. The mean relative difference becomes the minimum (0.11 %) at 18:00 hours. In the UK, a negative trend is noticed between 6am and 8pm. The relative difference between the two assumptions shows the maximum value (-0.45%) at 12pm.

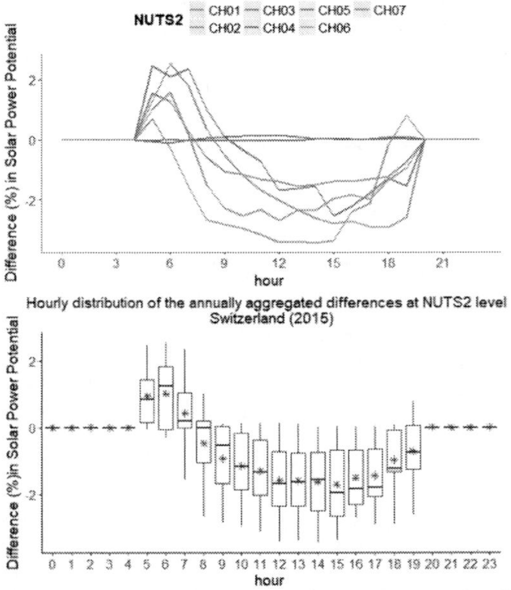

Figure 3: Difference (%) in the annually aggregated hourly Solar Potential (top) at NUTS2 level and its distribution (bottom) in Switzerland (2015). Median and mean values are illustrated by the black line and the labelled red star, respectively.

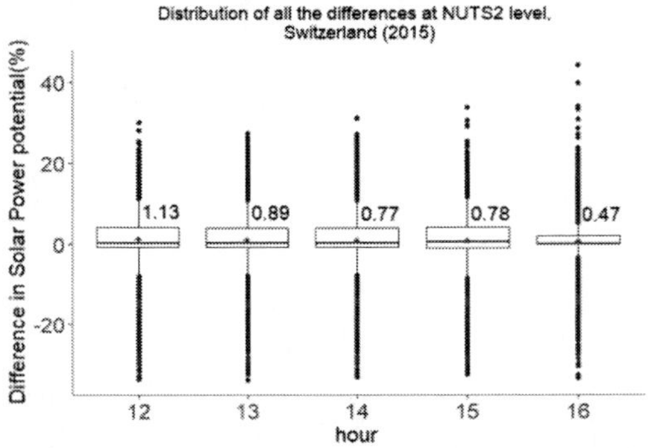

Figure 4: Distribution of all the differences in hourly solar potential in Switzerland in 2015 at NUTS2 level during peak hours (12-16h). Median and mean values are illustrated by the black line and labelled red dot, respectively.

To investigate the land type impact, as suggested by the abovementioned indications, on the solar power potential the four countries are further studied. The main class influencing the potential is the forests. This class is classified further to three subclasses; broad leaved, coniferous and mixed forest. The dominant classes in each part of Switzerland, Norway, Italy and United Kingdom are shown in Table 3.

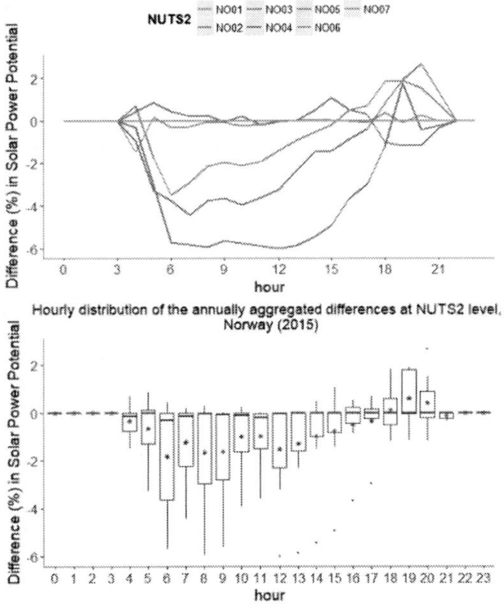

Figure 5: Difference (%) in the annually aggregated hourly Solar Potential (top) at NUTS2 level and its distribution (bottom) in Norway (2015). Median and mean values are illustrated by the black line and the red star, respectively.

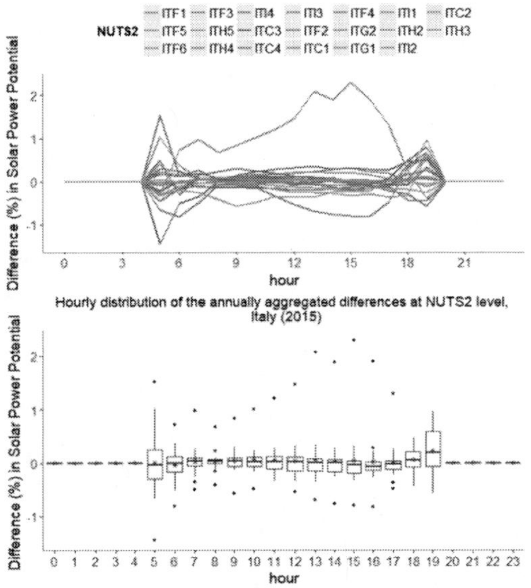

Figure 6: Difference (%) in the annually aggregated hourly Solar Potential (top) at NUTS2 level and its distribution (bottom) in Norway (2015). Median and mean values are illustrated by the black line and the red star, respectively.

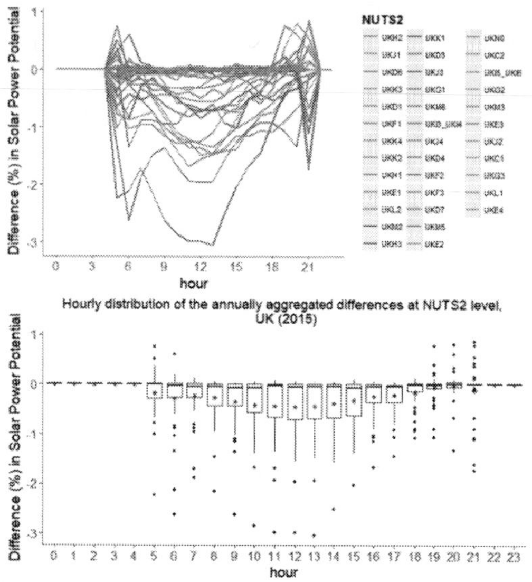

Figure 7: Difference (%) in the annually aggregated hourly Solar Potential (top) at NUTS2 level and its distribution (bottom) in UK (2015). Median and mean values are illustrated by the black line and the red star, respectively.

Table III: Dominant class in regions of Switzerland, Norway, Italy and United Kingdom

	CH	NO	IT	UK
Broad leaved forest		North, west	Central, south	South
Coniferous forest		South, Centre	North	North (Scotland)
Mixed forest	North			

3.2 Total annual available potential at European and country level

In order to assess the full annual potential that can be harvested if all available areas are covered by solar panels, under the two assumptions the following calculations were performed. PVGIS model data provide the potential as Wh per KW of installed capacity, capacity factors namely. To calculate the potential defined as the total power that can be generated in given available areas, it is assumed a power density of 25 MW/km^2 for solar farms [8]. Accounting with the same available area per NUTS2 level [15], the variable is the capacity factor of each scenario (see Table 4). Croatia (HR) is not included in the Eurostat source and therefore has been excluded from this analysis.

Since the results of the two simulations are very similar, the annual available solar potential by country under the assumption is selected to illustrate in Figure 8. Spain and France present the highest solar potential of 20338 and 17629 TWh per year, respectively. This is due both to their high capacity factor and large available areas. Other southern countries, such as Greece and Portugal present a moderate potential due to smaller territories compared to France and Spain. Portugal's total available potential is equal to 18% and 21% of that of Spain and France while Greece's potential stands for the 25% and 29%, respectively. Finland has a nearly 40% higher potential than Greece and Portugal due to its large available area (2.5 times the available area of Greece).

Table IV: Total Solar PV potential per Country, 2015

Country	Toal Potential Non-Exclusion Scenario (TWh)	Total Potential Exclusion Scenario (TWh)	Absolute Difference (TWh)
AT	2443.58	2449.81	-6.23
BE	827.35	826.70	0.64
BG	3727.03	3725.31	1.71
CH	1253.35	1270.67	-17.32
CZ	2163.25	2162.10	1.15
DE	9339.58	9337.29	2.28
DK	866.36	866.88	-0.52
EE	1047.43	1047.53	-0.10
EL	5092.30	5087.73	4.56
ES	20335.34	20337.99	-2.65
FI	6923.59	6924.83	-1.24
FR	17617.41	17628.66	-11.24
HU	416.75	416.78	-0.03
IE	1641.87	1643.83	-1.96
IT	10778.55	10780.37	-1.82
LT	9.24	9.23	0.00
LU	246.83	246.67	0.16
LV	224.63	224.61	0.02
NL	1099.08	1099.26	-0.18
NO	3463.28	3527.88	-64.60
PL	8571.93	8570.50	1.43
PT	3637.33	3636.30	1.03
RO	7554.46	7554.79	-0.33
SE	6080.88	6085.88	-5.00
SI	91.63	91.59	0.05
SK	1436.37	1435.70	0.67
UK	5812.48	5851.96	-39.49

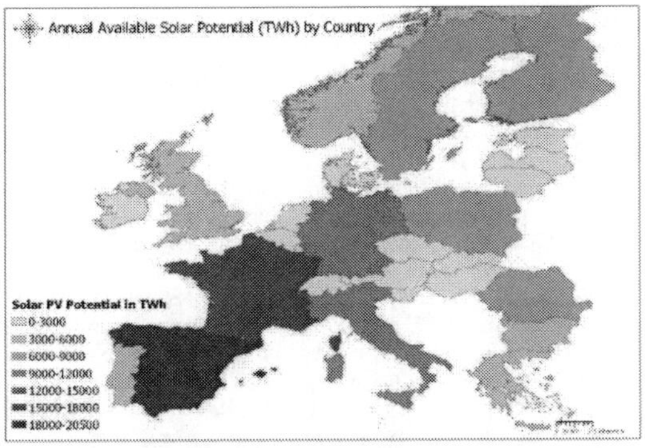

Figure 8: Annual Available solar potential (TWh) by country, using PVGIS time series for 2015

3.3 Complementarity between the capacity and the potential at continental scale

High technical potential in a region does not imply higher installed capacity per country. It also depends on other factors such as geomorphology, distance from transmission lines, from distribution stations or accessibility of the area and also subsidies, among others [16]. Here, the correlation between the two variables is calculated to observe the complementarity at EU level. At intracontinental scale, per NUTS2 region per country, the correlation of most of the countries is higher than 0.9 except for the Netherlands (0.831) and Germany (0.887). However, at European level the correlation factor is equal to 0.40, suggesting that there are countries of low potential with more installed capacity than others with higher potential.

National RES targets and policies affect the rate of investments. Hence the generous feed-in- tariff (FIT) support awarded from 2000 until 2017 along with the high RES-e target in electricity [17] (38.6% in 2020) in Germany provided incentives for deployment of RES-e. As of 2015 Germany had installed 39.2GW while France higher solar potential had installed only 6.2GW despite having similar power needs with Germany. The lower rate of deployment in France is attributed to the lower RES target for electricity (27% in 2020) [18]. United Kingdom has deployed 8.6GW of solar PV technology while Spain, the country with the highest solar potential has only installed 4.7GW of solar panels despite the high FIT rates during the past years. United Kingdom adopted a high RES-e target of 30% [19] in total power consumption and given the low solar potential had to install more solar capacity to meet the target. This is additional reason why the overall potential is less correlated with the technical one.

3.4 Case study on the Spanish power market

The last section is focused on a case study based on the Spanish power market, selected due to its high RES portfolio and resource availability (accounting for 15% of the EU-28 total capacity). Studying the relation between the actual RES-e generation and demand may provide useful insights for the flexibility needs of a power system with high RES-e capacity. The impact of wind (Figure and solar generation (Figure 10, top) on total load and day-ahead prices is correlated (Figure 10, bottom) using TSO data for 2015 [20]. Day-ahead prices are selected over intraday prices since most of the total load (87%) is traded in day-ahead market [21].

On an annual basis, although the aggregated wind and solar generation is weakly correlated with the load (R= 0.21); the correlation with the day ahead prices is higher with R=0.50 (Figure 10). This value gives an indication about the importance of the intermittency of the RES-e in the merit order curve. When there are available high amount of RES for the power system, they have priority to enter in the merit order at very low/zero/negative electricity prices decreasing the average price of electricity. However, the intermittency of the sources of the RES-e is not correlated with the load at daily cycle (Figure 10) so therefore, the need of strong flexibility measures [22].

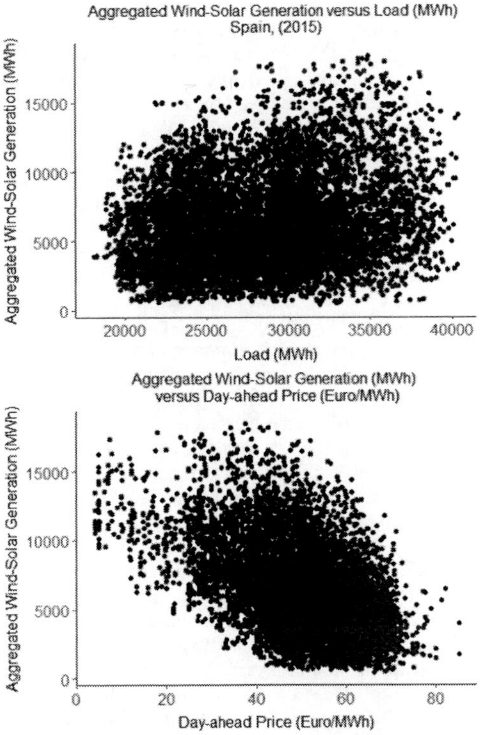

Figure 9: Aggregated wind and solar generation versus load (top) and day- ahead price (bottom), Spain (2015).

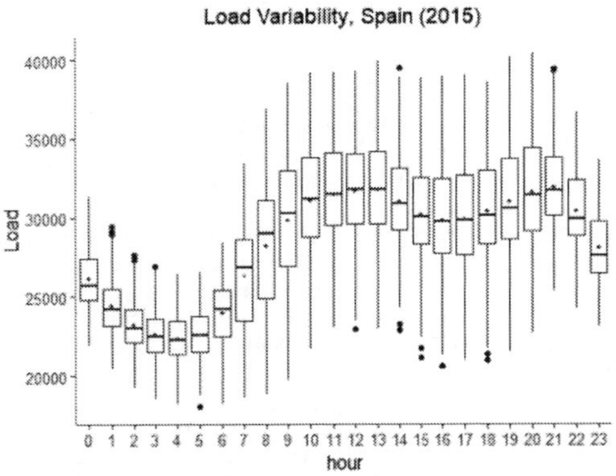

Figure 10: Hourly variability of load in Spain (2015). Median and mean values are illustrated by the black line and the red dot, respectively.

4. Conclusions

In the present study, the impact of the land use in the solar PV generation was evaluated. Two assumptions are formulated one in which all land-use types are available for solar PV installation and a second one in which some areas are excluded. At the annual scale, the land use has no significant effect in the potential of any of the three aggregation levels. At the country level, the maximum annually aggregated difference is noted in UK (-48.22GWh, Switzerland (-12.98GWh) and Germany (10.34GWh). The higher annually aggregated relative differences at the NUTS 2 regional level were observed in Italy (+1.44%) and UK (-2.46%). The main class influencing the difference in solar potential is the forests. On hourly scale, it is noticed that the difference between the two assumptions can vary significantly (-35% to 45%). The land use impact observed in these hours should be further studied since it might have an impact when studying the flexibility needs of a regional power system.

The correlation factor of capacity installed in 2015 with total available potential is calculated for each country. The distribution of solar capacity on the NUTS2 regions is highly optimized with the solar PV potential on the country level. However, countries with lower potential such as Germany and UK have deployed more solar capacity because of strong policy incentives and high RES-e targets. Therefore, on the European level the solar capacity is not distributed based on the resource availability contributing other factors to the variability.

Finally, the Spanish power market is studied to understand the flexibility needs of a power system with high RES-e integration characterised with high wind and solar capacities. Although the aggregated wind and solar power is weakly correlated with the load the prices are highly correlated indicating a strong dependence with the intermittency of the RES-e. Hence, flexibility needs for Spain are high and strategies should be adopted to decrease dependency from the wind and solar resources: deployment of storage systems by contributing in the flattening of the generation curves. For instance, compressed air energy storage or pump-hydro may store large output from RES-e while batteries may provide flexibility to end consumers [23]. Lastly with demand response programs consumers are incentivized to reduce or increase their consumption based on the power availability [24].

5. References

[1] European Commission 2014, Communication from the Commission to the European Parliament, the Council, the European Economic and Social Committee and the Committee of the Regions A policy framework for climate and energy in the period from 2020 to 2030

[2] Sánchez-Lozano, J. M., Teruel-Solano, J., Soto-Elvira, P. L., & García-Cascales, M. S. ,2013. Geographical Information Systems (GIS) and Multi-Criteria Decision Making (MCDM) methods for the evaluation of solar farms locations: Case study in south-eastern Spain. Renewable and Sustainable Energy Reviews, 24, 544-556

[3] Jahangiri, M., Ghaderi, R., Haghani, A., & Nematollahi, O., 2016. Finding the best locations for establishment of solar-wind power stations in Middle-East using GIS: A review. Renewable and Sustainable Energy Reviews, 66, 38-52.

[4] Sun, Y. W., Hof, A., Wang, R., Liu, J., Lin, Y. J., & Yang, D. W., 2013. GIS-based approach for potential analysis of solar PV generation at the regional scale: A case study of Fujian Province. Energy Policy, 58, 248-259.

[5] Uyan, M. 2013. GIS-based solar farms site selection using analytic hierarchy process (AHP) in Karapinar region, Konya/Turkey. Renewable and Sustainable Energy Reviews, 28, 11-17.

[6] Killinger, S., Mainzer, K., McKenna, R., Kreifels, N., & Fichtner, W. 2015. A regional optimisation of renewable energy supply from wind and photovoltaics with respect to three key energy-political objectives. Energy, 84, 563-574

[7] Mainzer, K., Fath, K., McKenna, R., Stengel, J., Fichtner, W., & Schultmann, F., 2014. A high-resolution determination of the technical potential for residential-roof-mounted photovoltaic systems in Germany. Solar Energy, 105, 715-731.

[8] Assouline, D., Mohajeri, N., & Scartezzini, J. L., 2017. Quantifying rooftop photovoltaic solar energy potential: A machine learning approach. Solar Energy, 141, 278-296

[9] Huld T. and Gracia Amillo A., 2015 Estimating PV Module Performance Over Large Geographical Regions: The Role of Irradiance, Air Temperature, Wind Speed and Solar Spectrum. Energies, 8, 5159-5181

[10] Eurostat. 2015, NUTS 2013, version 12/2015 http://ec.europa.eu/eurostat/web/nuts/local-administrative-units

[11] Copernicus-European Environment Agency. Corine Land Cover (CLC2006)- version 9/2016, http://land.copernicus.eu/pan-european/corine-land-cover/clc-2006/view

[12] European Environmental Agency. Natura 2000, version End 2015, http://www.eea.europa.eu/data-and-maps/data/natura-7/#tab-european-data

[13] Peschel, T., 2010 Solar parks–Opportunities for Biodiversity. A report on biodiversity in and around ground-mounted photovoltaic plants. Renews Special, 45, 6-14.

[14] Müller R, Behrendt T., Hammer A. And Kemper A., 2012 A New Algorithm for the Satellite-Based Retrieval of Solar Surface Irradiance in Spectral Bands. Remote Sensing, 4, 622-647

[15] Eurostat, 2017, Total and land area by NUTS 2 region http://ec.europa.eu/eurostat/estat-navtree-portlet-prod/BulkDownloadListing?file=data/tgs00002.tsv.gz

[16] Voivontas, D., Assimacopoulos, D., Mourelatos, A., & Corominas, J. (1998). Evaluation of renewable energy potential using a GIS decision support system. Renewable energy, 13(3), 333-344.

[17] Federal Republic of Germany, National Renewable Energy Action Plan in accordance with Directive 2009/28/EC on the promotion of the use of energy from renewable sources, http://ec.europa.eu/energy/sites/ener/files/documents/dir_2009_0028_action_plan_germany.zip

[18] Republique Francaise National action plan for the promotion of renewable energies 2009-2020 In accordance with Article 4 of European Union Directive 2009/28/EC http://ec.europa.eu/energy/sites/ener/files/documents/dir_2009_0028_action_plan_france.zip

[19] National Renewable Energy Action Plan for the United Kingdom Article 4 of the Renewable Energy Directive 2009/28/EC http://ec.europa.eu/energy/sites/ener/files/documents/dir_2009_0028_action_plan_united_kingdom.zip

[20] Red Electrica de Espana, 2017 https://www.esios.ree.es/en

[21] Ommie, 2016, Main results of the Electricity Market http://www.omie.es/files/mercado_electrico_ing.diptico_web_pdf.pdf

[22] Gonzalez-Aparicio I and Zucker A. (2015) Impact of wind power uncertainty forecasting on the market integration of wind energy in Spain. Applied Energy 159, 334-349.

[23] Palizban, O., & Kauhaniemi, K. (2016). Energy storage systems in modern grids—Matrix of technologies and applications. Journal of Energy Storage, 6, 248-259.

[24] Khajavi, P., Abniki, H., & Arani, A. B. (2011, May). The role of incentive based demand response programs in smart grid. In Environment and Electrical Engineering (EEEIC), 2011 10th International Conference on (pp. 1-4). IEEE.

APPLICATION OF BATTERY ENERGY STORAGE SYSTEM TO FACILITATE AND IMPROVE THE LV DISTRIBUTION NETWORK IN A COMMUNITY WITH PHOTOVOLTAIC SYSTEMS FOR A FUTURE LOAD SCENARIO

Ahmed Hosne Zenan, Dr. Edward Christopher, Professor Mark Sumner
University of Nottingham
University Park, Nottingham

ABSTRACT: This research aims to create a scenario where every household in the future will be susceptible to the consumption of larger loads due to electrification of heat and transport. Based on the notion of 'community energy', which advocates collective generation, management and sharing of energy from both grid and photovoltaic systems, the idea of this research is to be able to support the community function beyond its current capacity with little or no change to the core distribution network like underground cable layouts. This is to be implemented by integrating battery storage systems in the community based on optimum locations for maximum overall impact by maximizing energy utilisation 9within the community and minimizing energy loss under an acceptable rate of voltage fluctuation.

Keywords: Photovoltaic, Battery Storage and Control

1 RESEARCH METHODOLOGY

The primary platform for the modelling of the community and for the analysis of the eclectic components in the community is MATLAB. The electrical arrangements and components are executed mathematically using MATLAB programming to replicate a community having PV systems, regular domestic loads and provisions for higher electric loads, electric vehicle charging and battery storage. The programming uses Iterative ladder technique[1], a mathematical algorithm, to solve for currents and voltages at different nodes (houses), power and energy and losses in the low voltage distribution network(LVDN). The one-minute time step loading profile for all the houses was obtained from CREST model[2], a fully functional model developed by Loughborough University. The model uses real life data collected through surveys over a period of time and based on that a generic load profile is calculated for a household with at most 5 people for every day of the year. The CREST model also generates PV profile on a daily basis. For the electric vehicle (EV) modelling, a national transport survey data [3] was used to get an idea of the travel miles and duration of cars, for commute in and out of a community, to evaluate an approximate fuel consumption rate, which acted as the base for calculating electric car usage for every household in the modelling. In addition, every household has been limited to one EV for demonstration a future load scenario. The MATLAB programming also gives the user the flexibility to alter the number of houses, the penetration of PV systems, EV and Battery Storage Systems (BSS) in the community[4]. For now, the simulation can only be run for a day over 1440 mins and therefore all the data obtained from this modelling are per min based as well. The entire analysis is approached mathematically through MATLAB programming which gives the research base more flexibility

2 SCIENTIFIC INNOVATION

Ever since the breakthrough of the concept of alternative energy and its associated climate change policies have been established, there has been an increasingly popular quest for renewable sources like wind, hydro, solar, bio etc. Since then, not only big industries and farms, but also individuals have been taking the initiative to install renewable energy home systems to help generate electricity on their own as a secondary support besides the primary source, the grid. However, with alternative energy becoming increasingly popular and effective in many nations, there is a growing pressure on governments to encourage more penetration of green energy impacting communities as a whole. In UK, the government has been already encouraging individuals to work as a group to establish the concept of shared energy generation and distribution and therefore within the last five years, more than 5000 community led projects have been sprouting, exhibiting community energy schemes [5].Community led schemes can often overcome challenges more efficiently and more effectively than government alone thereby developing solutions to meet local needs that involve local people[6]. This will also give the customers more control over how they choose to manage energy in their house instead of simply switching energy suppliers and tariff rates. However, with an increase electrification of heat and transport and integration of embedded local generations, the future loading scenario in a regular community is expected to be higher. Thus, maximising the use of the available energy into the system and operating the network at its most efficient point at regulated voltages under a high load demand pressure on the entire LVDN and a community energy trading scheme, will become a major issue in the future. This research therefore attempts to comprehend the complexities that can arise due to the aforementioned scenario and aims to address the challenges while searching for an optimum, yet practical, solution. However, all the literature review done till date shows isolated attempts of mitigating problems arising from integrating distributed generators in the community, with very few literature taking into account any future loading scenarios as part of a community energy scheme. [7]. Therefore, this PhD research makes an attempt to merge all the possible future components and scenarios with a provision for flexibility to control network energy losses, currents and voltages by optimizing location of the storage systems in a scenario where the distribution network can operate to its maximum potential/capacity.

3 MODEL DETAILS AND ASSUMPTIONS

This section displays the basis of the research work i.e. the future scenario, on which, the solution of intelligently locating decentralised storage systems will be implemented. In essence, it shows the effects on the LVDN due to higher loads, distributed local generators and BSS.

3.1 Cable Length, Voltage Regulation and Thermal Rating

A voltage regulation criterion of -6%/+10% [8] for UK defines one of our network efficienct metric SSE(Sum of sqaured error). A statistical evaluation that indicates how deviated a value is from a stangard value. The closer the evaluated metric is to 0, the less fluctuated a parameter is, in this, the voltage at every node.The thermal limit has been determined by the current carrying capacity of the cables used in every sections of the distribution network. Pertinent to the modelling, the two main connection cables are mains cables (300 mm^2) and service cables (35 mm^2) [9] with a continuous per phase rating of 470A and 140A [10] respectively when laid directly [11]. For the calculation of cable length and thereby spacing between two houses, LV distribution network data for Meadows, Nottinham provided by Western Power has been used along with some meausrements from google map to have an approximate distance between each houses. Thisfacilitated the calculation of cable impedences between houses. An assumption for this modelling is that, all the houses are equally spaced, which might not be the case in practice.

3.2 ADMD and Transformer Rating

For an initial approach, the LVDN for a chosen community 'X' has been assigned to 18 households consisting of a maximum of 5 people per phase with around 33 kinds of appliances. Following the after diversity maximum demand (ADMD) power allocation for different combination of houses with different tariffs and loads, a transformer rating of 315kva was chosen.

3.3 Loading Profile

The loading profile is a one-minute time step power consumption for every appliance and for every household. The output data gives an aggregate load demand profile unique to each household from 0 minute to 1440 minute (1 day). To create a future load scenario, electric water storage heating, electric space heating, electric shower and electric vehicles were added on to the existing load profile.

3.4 PV and EV Penetration

The PV penetration in the modelling of the LVDN can vary from 0% to 100%, meaning either none of the houses will have a PV system installed or all 18 of the houses will have it installed respectively. But it is to be noted, that for the same PV penetration, the house the PV system is assigned. However, the EV penetration has been fixed to 100%, all 18 houses have one EV each. The EV is also allowed to charge in a different location apart from the household but that is beyond the boundary of the community.

3.5 Battery Storage System and Control

The battery storage system is kept relatively simple due to the existing complexity of the LV network with local embedded generation and reverse flower flows. The size of the battery is not a very significant factor yet, however, a generous 3kwh per battery, per house storage capacity was allowed. For a centralized battery storage system, it will be the sum of all the houses owning a PV system to allow for capturing maximum amount of excess energy in the community. A SUNNY ISLAND 3.0M / 4.4M battery inverter with a maximum rating of 3kw was chosen for modelling the battery control algorithm. However, this can be changed based on requirements. The battery is expected to be already charged by 50% at the beginning of the simulation i.e. 0th minute. Three times of control algorithm are being implemented to see how the community is affected collectively in terms of network energy efficiency and losses, producing an optimum energy management and trading solution among individual houses. The first control algorithm is very simple and divides all the power in excess coming from the community, or all the aggregate power demand from the whole community equally among batteries if there are multiple batteries. For a single decentralized battery or decentralized multiple batteries, the location of these BSS is also varied to see how position of the BSS affects the community's operations under shared energy schemes. The other algorithms will be based around the first simpler version but will however, will take more parameters into account while making decisions intelligently based on either load profiling or load and PV forecasting for example.

4 Results and Discussion

The research monitors the outcome pf the modelling at different phases and with the addition of each new element in the system while building up on the complexity gradually. The first phase focuses on forecasting a load profile from the future due to electrification of heat and transport and a more independent end user oriented energy network. The second phase introduces PV systems into the LVDN and monitors its impact on the energy network as a community in terms of clean energy generation, management, and voltage smoothing. The third phase primarily revolves around integrating centralized and decentralized BSS(S) in the LVDN and together with different basic and intelligent algorithm of battery control mechanism, assess the impact of such local embedded generations and storage in the community. Please note, all the simulations are carried out for a summer weekday. However, the modelling gives the flexibility to carry out the same tests for different seasons, days and cities. Moreover, the data representation on the graphs do not always start from 0 on the vertical axis for better understanding of the changes in values, however, in actual calculations all of the data are relative from 0 unit.

4.1 Effect of a higher load in the future

Traditional loads are a mixture of gas and electricity, with a fair percentage of gas used for heating. Electric vehicles are not an integral part of every household and as such results in a load profile with much lower average load and peak load. However, in the future, with an increase in demand for more energy, use of electricity for heating (space, water storage, electric shower etc.), cooling, and electrification of vehicles (Battery electric vehicle-BEV or EV), there will be a major upward shift in the existing average load and peak load paradigm. Figure 1 below, depicts a future load scenario for 18 houses per phase for a one-minute time step daily, aggregate load profile. The average load for all the houses of which stands at around 4kw and the peak load at around 12kw. It is also to be noted, EV being a constant power nodes, consumes a fixed amount of power for a long period (6-8 hour charging at 3kw from evening)

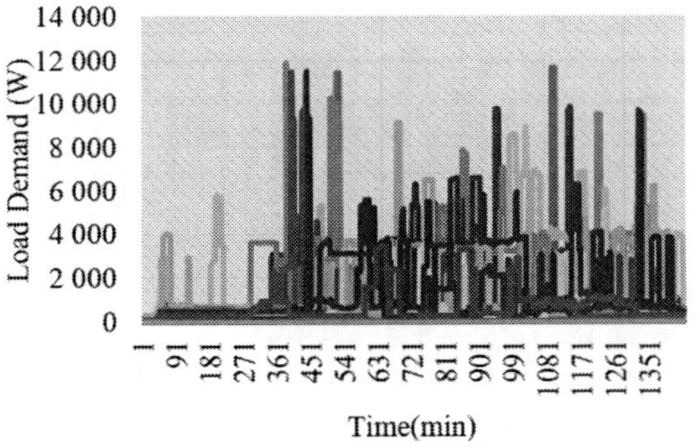

Figure 1: Load profile for 18 houses for 1 day(Each color represents a house)

4.2 Effect of PV Penetration

The model uses a PV system of 3.8 kw-peak for every house and assumed the same generation from all the PV systems in the community. Figure 2 shows the effect of increasing PV penetration from 0% to 100%. The location of the PV can also vary (could be random or fixed depending on the research specification) for the same penetration. Therefore, if the PV penetration is 20%, it means only 4 houses in the community will have a PV system installed and will facilitate collective self-consumption by dispatching energy to meet other households' demand, apart from its own. However, this raises the question, which four houses would have the PV installed. One approach could be simulating the model for different house combinations and taking the average in account or manually increasing the PV systems one house at a time and thereby calculating the PV penetration % to decrease the scope of variability in the model. The graph below in fig. 2 adopts the former approach. It can be seen, that with increase in penetration, the percentage loss (which is although not very significant now) decreases till it reaches 60% and starts picking up the curve upwards from there. The SSE (a measure of voltage variation) also shows the same results. Therefore, choosing a PV penetration of 60% for this particular load profile can be a wise decision in narrowing down the variables of the LVDN. For a different type of load profile, the model may yield a different result and as such, using that penetration will be a simpler decision. At the chosen PV penetration, the Eloss and SSE per phase (of a balanced 3 phase system) stand at approximately 0.16 and just below 0.002 percentage respectively. Figure 3, however, captures other network performance metrics such as how much the community is relying on the grid (grid utilization) or how much of the energy is locally generated or consumed (community utilization). The lower the grid utilization and the higher the community utilization, the better the LVDN performs in terms of facilitating a harmonized community energy system. The efficiency metric, which is essentially the same as the community utilization is calculated as the ratio of energy output to energy input and expressed as a percentage.

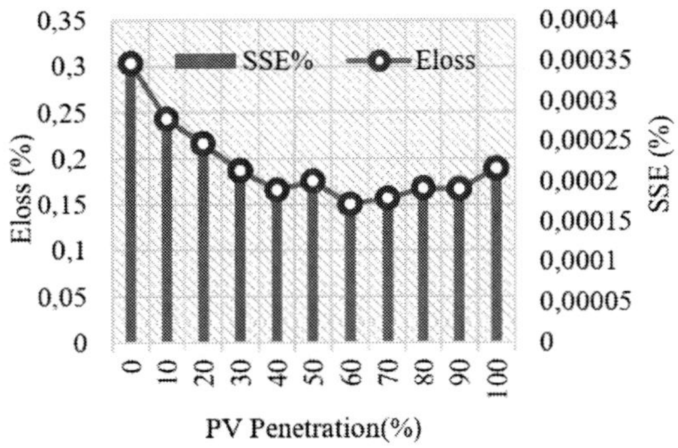

Figure 2: Effect of PV penetration on LVDN losses and SSE

From figure 3, it can be seen, that at 60% PV, the community utilization is the highest, with 50% much lower reliance on grid, therefore helping the community to work locally and independently.

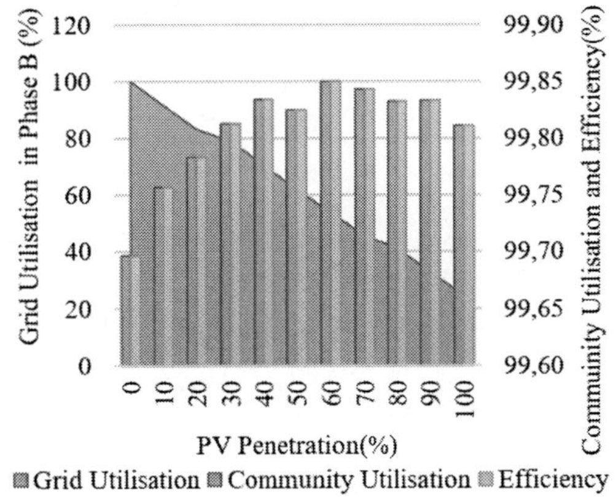

Figure 3: Effect of PV penetration on Grid utilisation and LVDN efficiency

4.3 Effect of a Single Centralized BSS

On reaching phase three of the research with a narrowed down option for variables to impact the LVDN, the community energy model starts integrating BSS. To start off, a centralized BSS is considered for the entire network with a 33kwh (approximate of a battery size of 3kwh per house multiplied by the number of houses having a PV system) system size and a max inverter power rating and of 3000W. Nevertheless, all these parameters are subject to change to test different kind of scenarios. The battery is initially assumed to be a charge of 50% at the beginning of the simulation and do not charge below 12% or above 90%. The charging or discharging of the BSS is prioritized over buying or selling energy from and to the grid. Figure 4, shows a BSS charging profile over one day in summer (21st June), discharging overnight, for EV charging mainly and space heating/cooling and may be water storage heating. During day time, when the PV is at its maximum generation, the excess is stored in the battery. Figure 5 and 6 show the impact a singular centralised battery system has on all the metrics as aforementioned. As it can be seen, the % range of the Eloss dropped to 0.15-0.14 from 0.16%, however, the SSE has increased very slightly to 0.0002.

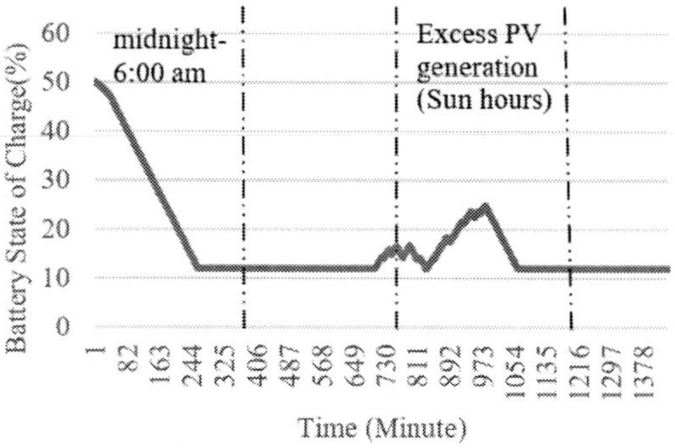

Figure 4:Battery Charging and Discharging for Centralised Approach at 60% PV penetration per phase

The simulation is run by position the battery at different nodes/houses starting from house 1 to house 18(In practice, all the houses will have their own BSS which will be switched off until asked by the intelligent control operation). According to figure 5, although the Eloss fluctiates in between 0.149 and 0.139% (owing to the MATLAB simulation error or 0.007 unit), it can be safe to assume, that the location of the battery does not matter at large for equally spaced houses. The SSE ,on the other hand stays constant at 0.0002% , which is negligible. Therefore, it will be more sensible to place the BSS where it will be most economical.

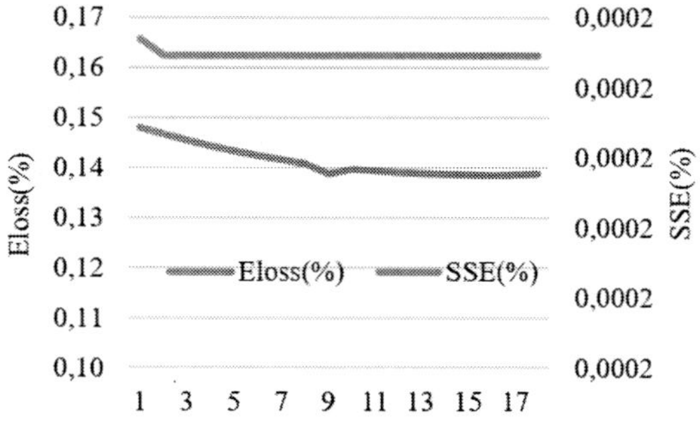

Figure 5:Effect of a single centralised BSS with variation in installed location of the system

Other metrics suggest a similar effects on the LVDN as depicted in figure 6, where the community's reliance on the grid has increased to appx. 51 %, ever after inlcuding one centralized storage and the community utilisation has decreased to 98.9% from a 99.85%. This is due to under-utilisation of a 33kwh BSS, with a 3000W inverter maximum power rating. A simple solution could be using smaller decntralised batteries for storage as discussed in section 4.2.

4.4 Effect of Multiple Batteries (Equal Share Algorithm)

As discussed before, different types of algorithms are to be deployed to test the community from different angles for the optimum result. To start off with, a very simple control mechanism has been implemented, coined as 'Equal Share Algorithm'. It is as the name suggests, equal participation of all the BSS(s) in the community for either storing cumulative excess PV generation in the community or for meeting the cumulative load demand-supply gap.

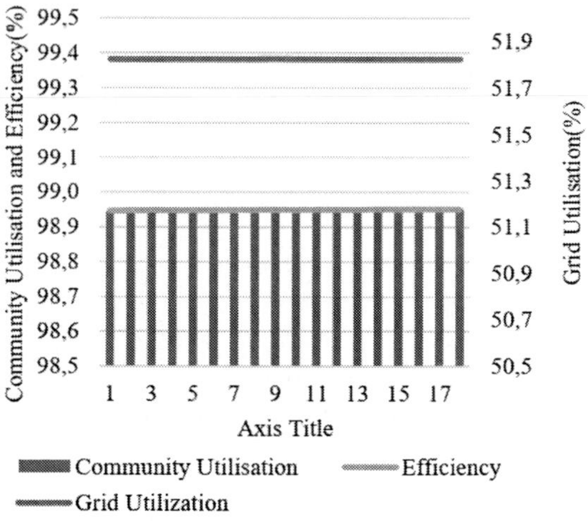

Figure 6: Effect of a single centralised BSS on positioning it from house 1 to house 18 in terms of grid utilisation, community utilisation and efficiency.

For example, if there is a total excess PV generation of X unit in the community, each BSS will simply store X/2 unit, regardless of other considerations.

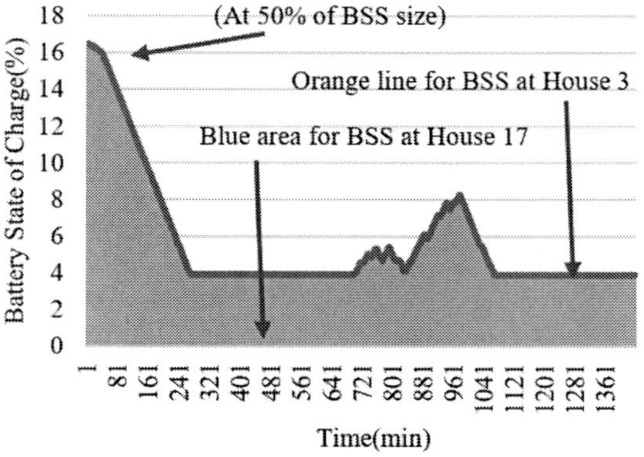

Figure 7: A snapshot of decentralised batteries, one at house 3 and the other at house 17 (for a 10% BSS penetration) at 60% PV penetration per phase

To implement this algorithm, a 10% BSS penetration i.e. 2 batteries for storage , was selected randomly. Figure 7 , therefore shows how the charging and discharging profile of both the batteries are exactly the same owing to the equal share control mechanism. The Eloss % is much smoother now and is around 0.14% for 22 different house combinations. For a community of 18 houses, a combination of 2 houses results in 151 possible solutions. However, figure 8 and figure 9, are therefore not true representatives of the sample and more simulations need to be run to reach a concrete conclusion. However, this gives a snapshot of what can be expected. As per expectations, for equally spaced housings, the locations of the batteries should not matter and this is, if not from a wide samples, is validated by the graphs below.

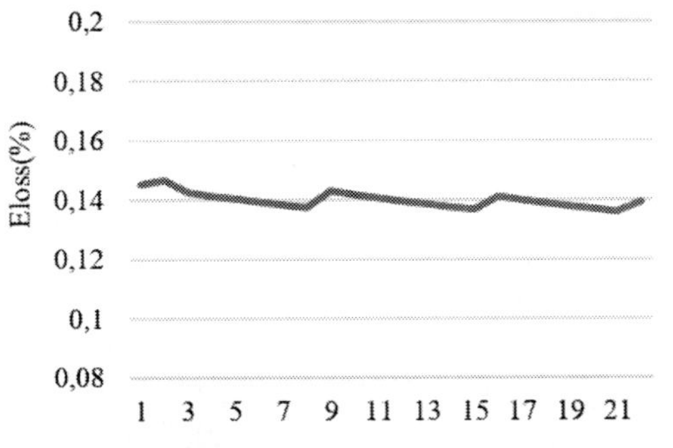

Figure 8: LVDN energy loss % for 22 different location combinations of two batteries of the same model

However, a decentralised BSS did contribute towards a lower reliance on grid(48%) and a higher community utilisation of around 99.99% (compared to than PV alone and PV and a centralised BSS).

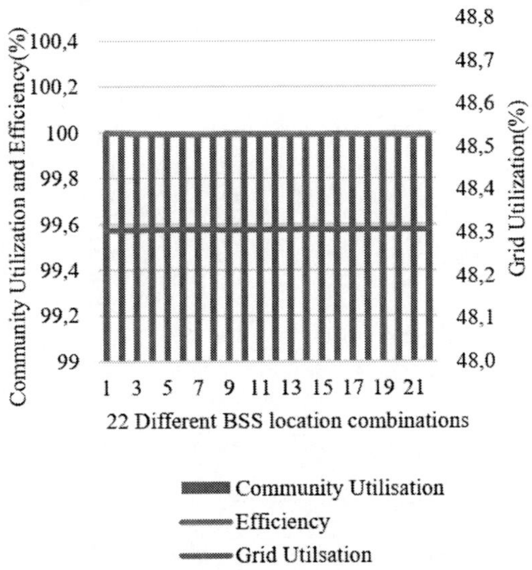

Figure 9: LVDN Performance evaluation metrics such as grid and community utilisation and efficiency for a 10% BSS penetration at 60% PV penetration , measured at 22 different possible locations

5 LIMITATIONS

The following are the limitations of the research for the current results but most of them can be overcome to improve accuracy:

- Lack of real-time data via smart metering
- Averaging of PV locations for the same penetration are subjected to only few samples
- EV profile is modelled using information gathered from petrol/diesel cars for office commutes from a national survey with the assumption, that in future, EVs will be used just like any other cars we use today.
- A constant MATLAB simulation error of 0.007% leading to misrepresentation of some data.
- Some power flow losses to and from the battery could not be calculated due to complexity of the iterative nature of the algorithm.
- The equal share algorithm uses a small sample from the population and may be not be a true representation of the population, however, it does how the expected outcome.

6 CONCLUSIONS

The contribution, of this research, will be towards developing an all-encompassing tool, for being able to put together a community with PV systems and BSS, catered for an inevitable market for electrificatiion of heat and transport.

6.1 Key Findings

The Key outcomes of the research work so far is summarized below:

- Future load Ranges from 3000W to 7000W on an average, with peaks reaching 12000W.
- Overall desired network performance increases with increase in PV penetration till 60%, and declines thereafter, with 60% penetration giving the best overall result.
- Single battery: Location does not matter for equally spaced houses
- Multiple batteries algorithm to be developed and analysed
- Robustness to be checked with multiple cities and seasons

6.2 General Contribution

This research will contribute significantly in the following areas as discussed as follows:

The contribution, of this research, will be towards

a) A more flexible and easily expandable network in terms of line capacity for future scenarios
b) A balanced network with eclectic combinations of PV and Storage Systems and all kinds of load demands (Regular Domestic, Electric Heating and Vehicles).
c) An optimized scenario with minimum over all distribution network energy loss and well-regulated voltages by controlling the battery storage system locations in the community
d) A holistic single tool, allowing users/consumers to plan the network ahead and analyse the network easily as per community demand

Besides the core benefits that this research contribution will bring about, it will not only create unity among the community members but will also eliminate any issues in the community pertinent to structural restrictions, ownership restrictions and government regulations that a single household can face as an individual, therefore resulting in a self sufficient independent community.

7 REFERENCES

[1] W. H. Kersting, Distribution System Modeling and

Analysis. Boca Raton, Florida, U.S.: CRC Press, 2001

[2] Richardson I. et al.., 2010. Domestic electricity use: A high-resolution energy demand model. Energy and Buildings, Vol. 42, pp 1878-1887

[3] UK Data Archive SN 5340 - National Travel Survey, 2002-2012

[4] MCKENNA, E., ... et al., 2013. Economic and environmental impact of lead-acid batteries in grid-connected domestic PV systems. Applied Energy, 104, pp. 239 – 249.

[5] Department of Energy and Climate Change, "Community Energy Strategy: People Empowering Change," Department of Energy and Climate Change, London, 2014.

[6] Department for Business, Energy & Industrial Strategy: Low carbon energy Published:25 March 2013Last updated: 26 January 2015,

[7] S. S,N. R. R.B. Hirematha, "Decentralized energy planning; modeling and application—a review," Renewable and Sustainable Energy Reviews, vol. 11, no. 5, pp. 729-752, 2007.

[8] T.Haggis, " Network Design Manual," E.0n Central Networks, 2006.

[9] S.Tucker, "Engineering Design Standard :11/6.6KV Secondary Distribution Network Design," UK Power Networks, 2016

[10] T. O. Directorate, "Engineering Recommendation G81 Part 3," Energy Networks Association, London, 2016.

[11] S.Cuddihey, "Engineering Design Standard: LV Network Design," UK Power Netwokrs, 2015.

STUDY AND ESTIMATION OF THE PHOTOVOLTAICS OPTIMUM SHARE IN MICROGRID BASED ON RENEWABLE ENERGY SOURCES FOR SMALL RURAL SETTLEMENTS IN CENTRAL EUROPEAN PART OF RUSSIA

Pavel P. Bezrukikh[1], Sergey M. Karabanov[2], Dmitry V. Suvorov[2], Pavel P. Bezrukikh (Jr.)[3], Andrey S. Karabanov[4]
1- G.M. Krzhizhanovsky Power Engineering Institute JSC, 19, Leninsky Prospekt, Moscow 119071, Russia
2- Ryazan State Radio Engineering University, 59/1 Gagarina St., Ryazan 390005, Russia
3- LUCOIL Ltd., 11, Sretensky Blvd, Moscow 101000, Russia
4- Helios Resource Ltd., 126 Proletarskaya St., Saransk 430001, Mordodia

ABSTRACT: The paper presents the research results of microgrid based on renewable energy sources for energy supply of small rural settlements in Central European part of Russia; the efficiency of renewable energy is estimated. The procedure of determining the optimal share of photovoltaics taking into account current and future prices of the equipment is given. The approaches to creation of microgrid providing continuous energy supply are considered.
Keywords: PV system, modelling, villages

1 INTRODUCTION

Russia has a high energy potential and a developed energy supply system. In 2016, the main energy indicators were as follows [1]:

- electricity generation: 1071.8 bln. kWh;
- electrical energy consumption: 1054.5 bln. kWh;
- installed capacity: 244.1 GW.

In general, the power system of the Russian Federation has high centralization. However, there are a large number of consumers (about 5 million people) who are not connected to the Unified Energy System.

The choice of new approaches to energy supply is caused by a number of factors:

1. Growth of tariffs for all types of energy;
2. 2. Price increase for connection to power grid.

In this regard, the use of renewable energy and microgrid becomes extremely challenging.

Russia has all kinds of renewable energy: solar energy, wind energy, hydropower, biomass, geothermal energy, etc. Table 1 shows the data on renewable energy potential in Russia [1].

Table I: The estimation of renewable energy sources potential in Russia

Resource	Gross potential, 10^6 t (c.e.)/ year	Technical potential, 10^6 t (c.e.)/ year	Economical potential, 10^6 t (c.e.)/ year
Wind power	44326	2216	11
Small hydropower	402	126	70
Solar energy	2205400	9695	3
Biomass energy	467	129	69
Geothermal energy (hydrothermal resources)	29200	11869	114
Low potential heat	563	194	53
TOTAL	**2251158**	**24229**	**320**

The data on the potential of renewable energy sources are the basis for making decision on their use in a particular region.

An important direction of energy development in Russia is the creation of distributed power generation systems, including microgrids. The Energynet program which defines the development of microgrid power industry has been established [2].

The paper presents the research results of stand-alone microgrid based on renewable energy sources for energy supply of small rural settlements in Central European part of Russia. The paper considers the issues of renewable energy efficiency. The analysis of electricity consumption from microgrid is carried out; mathematical model for calculating the optimal share of a photovoltaic power system in the microgrid structure based on a wind-solar power station, which provides full requirement in electricity for a settlement is presented; the evaluation of economic efficiency of microgrid use is carried out.

2 RENEWABLE ENERGY EFFICIENCY

The renewable energy efficiency is studied in paper [3]. The present paper considers efficiency as the ground for the creation of microgrid based on renewable energy sources, rather than on traditional energy sources. As is known, efficiency has a complex character and represents a set of a few efficiencies: energy efficiency, ecological efficiency, economic and social efficiency [3].

2.1 Energy efficiency

Energy efficiency of renewable energy has tremendous advantage over thermal power.

The energy efficiency is estimated according to the formula:

$$T_{epb} = E_{con} / E_y ,$$

where T_{epb} is the energy payback time, E_{con} is energy consumed for the production of a station, materials, equipment, operation, servicing; E_y – the electric energy annual production of a station. For a wind power station the energy payback time is less than one year (6 - 10 months), PV station c-Si T_{epb} is 2.2 years, PV station CdTe –9 months.

The picture is quite different for electric and thermal energy production at thermal and nuclear power plants. In terms of energy they are never repaid as during service life they constantly consume energy in the form of fuel and transform it to electric and/or thermal energy with the coefficient much less than unity. This is the difference of our point of view from many others according to which the fuel consumption during operation is not taken into account.

Therefore, using renewable energy sources the mankind can switch to effective electricity generation at wind power, PV power stations.

2.2 Ecological efficiency

The studies show that noxious emission from PV stations is less than from coal-fired power plants: for CO_2 – 35-55 times, NO_x – 44-100 times [3].

According to the research of World Steel Association in 2008: for 160 MW offshore wind power plant (eighty 2MW wind turbines) in Denmark in North Sea CO_2 emission was 7.6 g/kWh, according to [4] – 6.6 g/kWh, according to [5]– 8 g/kWh. So, CO_2 emission at wind power plants construction is two orders less than that of thermal stations.

2.3 Water consumption

Water requirement belongs to operational efficiency of wind and PV stations. Specific loss of water at a wind power station is 475 times less than that of nuclear power plant; 400 times less than at coal power stations and 275 times less than at gas-fired power plants. Specific loss of water at PV stations is 20 times less than at a nuclear power plant, 17 times less than at coal power stations and 14 times less than at gas-fired power plants (Table 2) [6, 7].

And it is apart from water requirement for a steam-water cycle of thermal and nuclear power plants and cooling water of turbine condensers, whereas at wind and PV stations the permanent water consumption is intended only for cleaning wind turbine blades and PV modules.

Table II: non-recoverable losses of water at the USA power plants

Power plant type	**NPP**	**Coal power station**	**Oil power plant**	**Gas power plant**	**Wind power plant**	**PV station**
Water specific loss, l/kWh	2.3	1.9	1.6	0.95	0.004	0.11

2.4 Economic efficiency

The potential of renewable energy technologies is not realized completely because of various barriers of the market, such as state support of traditional fuels [8]. According to the World Bank the annual state financing of gas industry in Russia amounts to 25 bln, power generating industry – USD 15 bln.

Obviously, by 2020 at PV stations (USD 1253/MWh) taking into account rapid unit cost reduction and efficiency increase of solar cells, levelized cost of electricity close to the cost of advanced coal and gas-turbine power plants will be achieved. In other words, the last barrier to broad-scale development of photovoltaics will be removed.

Table 3 shows the summarized results for 17 types of power plants. Levelized cost of electricity (LCOE) was used for economic efficiency evaluation.

Table III: Estimation of levelized cost of electricity of power plants in 2010 and power plants put into operation in 2018

Station type	**Levelized cost of electricity (LCOE), USD/MWh**					
	Minimum		**Average**		**Maximum**	
	2010	***2018***	***2010***	***2018***	***2010***	***2018***
Normal coal	90.1	89.5	***99.6***	**100.1**	116.3	118.3
Advanced coal	103.9	11.6	112.2	123.0	126.1	137.9
Advanced coal with CCS technology	129.6	123.9	140.7	135.5	162.4	152.7
Natural gas (standard)						
Normal with combined cycle	61.8	62.5	68.6	67.1	88.1	78.2
Advanced with combined cycle	58.9	60.0	65.5	65.6	76.1	76.1
Advanced with combined cycle with CCS technology	82.8	87.4	92.8	93.4	107.5	107.5
Normal gas-turbine	94.6	104.0	132.0	130.3	149.8	149.8
Advanced gas-turbine	80.4	90.3	105.3	104.6	119.0	119.0
Advanced nuclear	108.4	104.4	112.7	108.4	120.1	115.3
Geothermal	85.0	81.4	99.6	89.6	113.9	100.3
Biomass	101.5	98.0	120.2	111.0	142.8	130.8
Wind	78.2	73.5	**96.8**	**86.6**	114.1	99.8
Wind, offshore	307.3	183.0	330.6	221.5	350.4	294.7
Photovoltaic	122.2	112.5	156.9	144.3	245.6	224.4
PV thermal	182.7	190.2	251.0	261.5	400.7	417.6
Hydro	58.6	54.4	89.9	90.3	149.7	149.2

2.5 Social efficiency

Besides decrease of noxious emissions influencing negatively the population health, the significant social effect of renewable energy development is job creation for equipment production and operation of renewable energy facilities. Herewith workplaces directly relating to renewable energy (direct) and the ones in linked industries (indirect) are taken into account.

According to estimation of REN-21, 2011 the total number of workplaces in renewable energy sector in the world in 2010 was 3.5 million. The greatest contribution is made by biofuel production processes – 1.5 million workplaces [9].

However for Russia social efficiency of renewable energy is not settled by the quantity of workplaces.

The use of renewable energy sources for power supply allows to curtail drastically the electricity cost and to improve the population living conditions and development conditions of local production.

3 ANALYSIS OF MICROGRID ENERGY CONSUMERS

According to the Russian Statistical Yearbook (2015) [10], the rural population in Russia is 37 543 K people. In communities with the population of 51 – 100 people, 1006 K people live; in communities with the population of 101 – 200 people, 2133 K people live. These are the so-called small villages, which are the most affected by unstable electricity supply because of the extreme deterioration of the electrical network or even absence of power connection. So, the development of stand-alone microgrid is the most important for them. Out of 38 million of population in the Central, Northwest and Volga Federal Districts (the investigated regions) the rural population is 18 million people.

The present paper considers the issues of microgrid development for settlements with 100-200 people. Such population amounts to 1.5 million people

4 APPROACHES TO MICRORID DEVELOPMENT

The development of a microgrid is based on the following principles:

- providing comfortable living standard for the population served by a microgrid;
- maximum use of renewable energy sources for a given area;
- minimization of energy consumption for housing accommodation due to modern construction materials and technologies;
- control automation of electrogeneration, energy storage and power consumption;
- cost minimization for the development and operation of a microgrid.

A microgrid is to provide all basic vital requirements of people:

- lighting of a house and a plot of land, power supply of domestic electric appliances;
- cold and hot water supply;
- heating;
- outdoor lighting of a settlement.

The operation of the house power supply, water supply, outdoor lighting are to be completely automatic.

For each household the maximum power consumption based on output capacity of the feeding transformer and possible demand factor, minimum firm capacitance which is to be provided by a microgrid in all controlling cases are set.

Electricity metering of each household and public place, outdoor lighting should be done automatically on the basis of existing accounting systems.

A control system has to provide the accumulation data on electricity produced by all generating sources and on change of storage battery capacity.

5 DETERMINATION OF MICROGRID PARAMETERS

The determination of energy consumption and power is made on the basis of monitoring the energy consumption of a real house in the selected area during one year. The monthly and hourly electricity consumption was taken into account.

On the basis of the performed monitoring it was established:

- maximum daily power consumption is 3.75 kW (in September);
- the ratio of the maximum power consumption to the average is 3. This means that the estimated power consumption of one house should be 3 times higher than the average daily maximum power due to daily load curve irregularity, i.e. 3.75x3=11.25 kW (at further calculations, the power consumption was 10 kW).
- taking into account the fact that all houses can simultaneously consume maximum power, the demand factor is 0.5.
- based on the experimental data and accepted coefficients, the estimated value of the settlement power consumption is 150 kW, and the annual power consumption is 152 400 kWh.

The data were used for calculation of the generating capacity parameters.

6 MATHEMATICAL MODEL FOR CALCULATION OF THE PHOTOVOLTAICS OPTIMAL SHARE IN MICROGRID

Calculation of the monthly generation of electrical energy E_{pv} for a PV generator is determined by the method developed in [11].

Monthly electrical energy generation E_{pv} for a PV generator is determined by the equation:

$$E_{pv} = G_m(\beta) \cdot S \cdot \eta_{pv} \cdot (1 - \eta_{cable}) \cdot \eta_{mp} \quad (1),$$

where S is a PV generator area, η_{pv} – the conversion average efficiency of a PV module, η_{cable} – the cable loss (value 0.03), η_{mp} – the loss identified by absence of mpp tracking (value 0.8).

The coefficient η_{mp} is in the formula for calculation of the loss appearing when mpp tracking system is not available. The introduction of mpp tracking device in the system results in the necessity to calculate hourly solar radiation and also to have the data on average monthly temperatures.

For a battery the stored energy was calculate by:

$$E_{bat} = E_{pv} \cdot \eta_{bat} \quad (2),$$

where ηbat is the conversion average efficiency of a battery.

The battery capacitance was calculated by:

$$C_{bat} = \frac{E_{bat}}{DOD} \cdot (N+1) \qquad (3),$$

where N is the days without sun, DOD – the depth of the discharge.

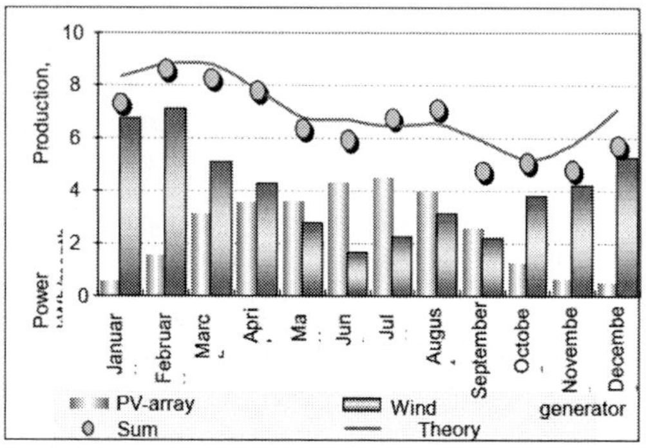

Figure 1: Monitoring data

To determine the optimum power ratio between wind and PV generators, a pilot plant was used [12]. It was used for monitoring solar-wind generation during a year. The research resulted in a good balance of calculations and monitoring data. The deviation of data was not more than 20%; the maximum deviation was observed in winter (Fig. 1).

The condition of the minimum power produced by a hybrid power plant is written as:

$$P_{nomPV\_n} + P_{nomwind\_n} = P_n = \min \quad (4),$$

where $P_{nomPV\ n}$ is the mean value of power generated by PV modules during n-month, $P_{nomwind\ n}$ - the mean value of power generated by a wind generator during n-month, P_n - the mean value of power generated by a microgrid during n-month (n-month is the number of a month in a year (n=1 ... 12)).

Below the most optimal proportions of nominal capacity of the wind and PV generators ($P_{nomwind}/P_{nomPV}$) are analyzed. The optimization was conducted by minimum irregularity of average monthly power production according to the formula:

$$\frac{\sum_{n} \left| (P_{average} - P_n) \right|}{P_{average}} = \min \qquad (5),$$

where $P_{average}$ - simple average of power production for the studied period (a year or a season), P_n- monthly power production (n=1 ... 12).

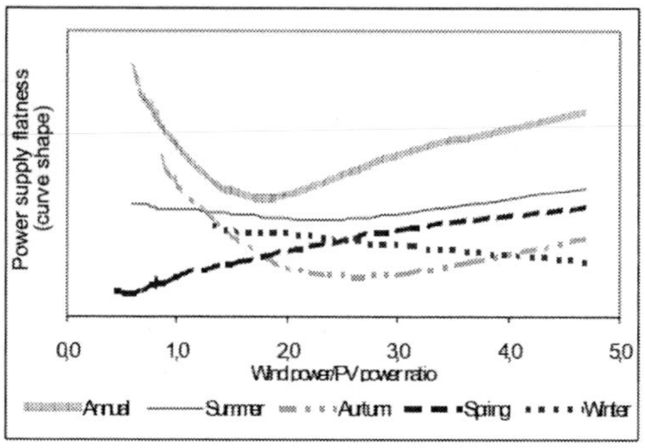

Figure 2: The curve shapes for different proportions of nominal capacities of wind and PV generators.

Fig. 2 shows the obtained curve shapes for different proportions of nominal capacities of wind and PV generators.

A flattened curve in winter and summer indicates noneffect of PV generator power increase on power supply character in winter and increase of the wind generator power in summer.

In spring, and especially in autumn, the variation of the power ratio is extreme. The optimal ratio for spring is 0.6, for autumn is 2.5–3. The optimal proportion for the whole year is 1.9.

The choice of proportions depends on the character of load consumption, its criticality to power supply irregularity in different seasons or during a year, and also on price relations per watt of wind and PV energy. In the experimental station the proportion was 2.35.

To provide all-year-round stand-alone power supply of customers in Central European part of Russia it is efficient to use hybrid power systems with the power ratio of 1 PV generator and 1.9 wind generator. At seasonal power supply this proportion is to be changed according to a season.

These results are valid only for the analyzed region and can be generalized with caution.

7 MICROGRID STRUCTURE

A microgrid consists of a wind power station, a PV station, a diesel generator, a control system, inverters, storage batteries, a microgrid automated control system (Fig. 3). To maintain the generation frequency, it is suggested to use a tank – a battery with hot water in each row. The settlement energy consumption is equal to 152 400 kWh per year, and the maximum power consumption at the demand factor of 0.4 is 120 kW.

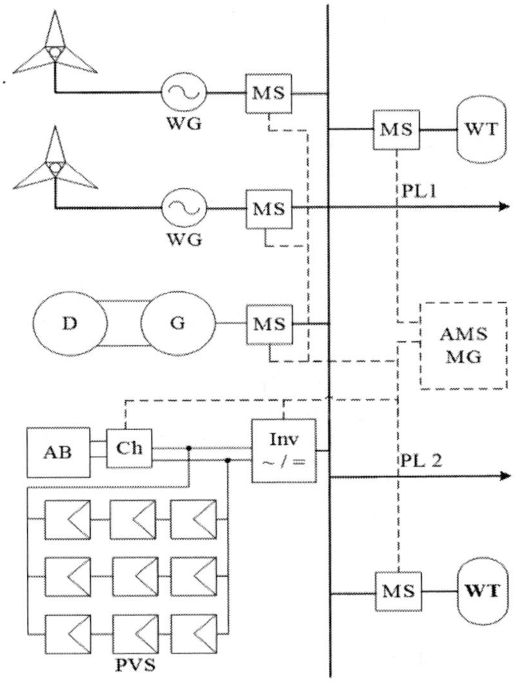

Figure 3: Microgrid structure: WG – Wind Generator, D – Diesel, G- Generator, AB – Accumulator Battery, Ch – Charger, PVS – Photovoltaic System, Inv – Invertor, WT – Water Tank, PL – Power Line, MS – Management System, AMS MG – Microgrid Automated Management System.

Based on the calculations, the main parameters of the generating devices are as follows:

- PV station peak power – 50 kW;
- installed capacity of a wind power station – 60 kW;
- diesel-generator power – 10 kW;
- storage battery capacity – 562 Ah.

The energy produced per year: a PV power station – 46 000 kWh; a wind power station – 92 000 kWh; a diesel-electric generator – 15 200 kWh. Microgrid is connected with houses by 0.4 kV power line.

Choice of types of wind generators, PV modules, inverters, etc. is carried out on the basis of economic calculations to minimize the costs.

8 ECONOMIC ANALYSIS OF USING MICROGRID

The economic analysis is based on calculation of design costs, the cost of microgrid equipment, the cost of facilities construction.

The costs of microgrid development amount to USD 370 000.

The costs of connecting to public electricity networks are highly dependent on the power line length. The cost of laying 1 km power lines is USD 17 000. With the power line length of more than 20 km, the cost of microgrid is less than that of the normal power line.

9 CONCLUSIONS

The studies show that reliable and efficient energy supply of small rural settlements may be carried out on the basis of high-tech equipment as a part of stand-alone microgrid made on the principles of combination of individual and collective systems. The optimal share of photovoltaics in microgrid is to be determined on the basis of the proposed procedure and a mathematical model.

10 REFERENCES

[1] S.M. Karabanov, M.V. Kuropov, and N.V. Kazakova, The prospects for Photovoltaics Development in Russia, Proceedings, 26th European Photovoltaic Solar Energy Conference, ISBN 3-936338-27-2.

[2] D. Kholkin, Pilot microgrid projects, Proceedings, 1st Interindustry International conference EnergyNet, The future of smart energy.

[3] S. Karabanov, P. Bezrukikh, P. Bezrukikh (Jr.), Renewable Energy Efficiency, Proceeding, 16th International Conference on Environment and Electrical Engineering (EEEIC), pp. 944-948.

[4] US Energy Information Administration, Annual Energy Outlook 2015, April 2015, DOE/EIA-0383.

[5] V. Fthenakis, H.C. Kim, R. Frichknecht, M. Rongei, P. Sinha and M. Stucki,. Photovoltaic power systems programme, Life cycle inventories and life cycle assessment of photovoltaic systems, IEA, October, 2011.

[6] P.P. Bezrukikh, P.P. Bezrukikh (Jr.), On indicators of energy status and renewable power efficiency under the conditions of economic crisis, Economic Questions, 2014, no. 8, pp. 92-105.

[7] P.P. Bezrukikh, S.M. Karabanov and P.P. Bezrukikh (Jr.), The efficiency of the world renewable energy, Energy, no. 12, 2015.

[8] A. Cabraal, S. Aqarwal, M. Takahashi, Rising to the challenge – the whys and whens of renewable energy, Renewable Energy World, 2007, vol. 10, no. 4, pp. 117–129.

[9] REN 21 Renewables. Global Status Report, 2012, 2013, 2014, 2015

[10] Russian statistical yearbook, 2015, Rosstat, Moscow, ISBN 978-5-89476-412-2.

[11] S.M. Karabanov, Y.V. Kukhmistrov, Calculation of ground photovoltaic systems by meteorological data, Technical Digest, 3rd World Conference on Photovoltaic Solar Energy Conversion, pp. 521-522.

[12] S.M. Karabanov, Y.V. Kukhmistrov, B.N. Sazhin, T.A. Shushkanova, Environmental tests of a wind-solar power station, Full Proceedings of World Renewable Energy Congress-VII, 29 June – 5 July, 2002, Cologne, Germany.

DEMAND SIDE MANAGEMENT USING PV, HEAT PUMPS AND BATTERIES – EFFECTS ON COMMUNITY AND BUILDING LEVEL

Luthander, Rasmus[1], Psimopoulos, Emmanouil [1,2], Widén, Joakim[1]
[1] Department of Engineering Sciences, Uppsala University, 751 21 Uppsala, Sweden
[2] Solar Energy Research Center (SERC), Dalarna University, 781 70 Borlänge, Sweden
Email: rasmus.luthander@angstrom.uu.se
Telephone: +46 18 471 62 15

ABSTRACT: This study examines how the energy management optimization on household level affects the maximum power flow in a community of houses and the contribution to load smoothening in the community. A detailed model of a single-family house with exhaust air heat pump and photovoltaic system is used in combination with high-resolution weather, electricity use and hot water use data. All five houses in the community are identical but the occupancy of the residents and their use of electric appliances and hot water differ. Results show no reduction of the maximum power delivered to the grid if the houses are operated to optimize the individual self-consumption and self-sufficiency. The highest aggregated power from the grid for the whole community occurred when the heat pumps were controlled by the PV electricity production but without any battery storage. This case also resulted in least smoothing of the aggregated household loads in the community. The conclusion of the study is that energy optimization for individual households in a community do not have to result in a reduction of the aggregated load and power production.
Keywords: Storage, Small grid-connected PV systems, Demand-Side, Self-consumption

1 INTRODUCTION

The energy use in buildings accounts for a large share of the total energy use in several developed countries. In Sweden, the residential and service sector accounts almost 40% of the total energy use [1]. Domestic hot water (DHW) and space heating (SH) accounted for more than half of the sector's total energy use in 2015, although the trend is reducing due to replacement of oil with electricity or district heating [1]. In Swedish single-family houses, electric heating is the most common option with a share of 45% of the total heat demand in the sector in 2014 [2]. For houses with electric heating, a heat pump combined with a hot water storage tank is an efficient way to lower the electricity demand for heating.

1.1 Demand side management in buildings

There are two ways to reduce the need of buying electricity in a household: either using solutions for energy savings or producing your own electricity. This paper focuses on the latter method to lower the need of buying electricity from a retailer by using photovoltaics (PV) for in-house electricity production.

With an electric heat pump more of the PV electricity can be used in the house in comparison to non-electric heated houses. This can be an advantage both from an economical point of view and from a power system's perspective. When buying prices of electricity exceed the selling prices, it is more profitable to replace bought electricity by own PV electricity than selling the excess production to the power grid. A higher use of the PV electricity also lowers the excess electricity from the PV system, which might reduce the stress on the power grid in areas with a high penetration of PV.

Electricity storage in batteries or load shifting of electric appliances can be used to increase the use of PV electricity in the house. The advantage of batteries is the possibility to store electricity for later use, while heat pumps can convert electricity into heat but not back again. Heat pumps do however have a coefficient of performance exceeding one, meaning that more heat power is delivered than electricity used.

1.2 Aim of study

The aim of this study is to investigate how the optimization of self-consumption and self-sufficiency for single households with rooftop mounted PV systems affects the aggregated electricity use and production on a community level. The energy management optimization includes batteries, heat pumps and thermal energy storage in domestic hot water (DHW) storage tanks. The detached houses are equipped with control algorithms to optimize the use of in-house PV electricity generation and lower the need of buying electricity. However, since the diversity of the load might reduce when the control algorithms are used, individual optimization of the energy use might not lead to an optimization for the whole community [3].

Studies of the aggregated effects by new energy control strategies of heat pumps and batteries can be helpful to assess how the power grid will be influenced. This is especially valid in a future of high penetration of PV-battery systems and demand side management.

2 METHODOLOGY AND DATA

This section describes the data and methodology used for the study.

2.1 Measures for evaluation

Several measures will be used to evaluate the demand side management in the community. On household scale, self-consumption and self-sufficiency, also called solar fraction, are commonly used to evaluate residential PV system [4]. Self-consumption is in this study defined as the share of PV generation that is consumed directly in the building. The share of the household load that can be supplied by the PV system is referred to as self-sufficiency.

Self-consumption and self-sufficiency can be defined mathematically by the instantaneous electric load for appliances and the heating system $L(t)$, the instantaneous power from the PV system $P(t)$. The in-house consumed

PV power $M(t)$ can be defined as

$$M(t) = \min\{L(t), P(t)\} \quad (1)$$

if no electric storage is used. When adding battery storage to the system, this can be extended to

$$M(t) = \min\{L(t), P(t) + S(t)\} \quad (2)$$

where $S(t) < 0$ denotes power to storage (charging) and $S(t) > 0$ denotes power from storage (discharging). With discrete time series, self-consumption SC and self-sufficiency SS can then be defined as:

$$SC = \frac{\sum_{t_1}^{t_2} M(t)\,dt}{\sum_{t_1}^{t_2} P(t)\,dt} \quad (3)$$

$$SS = \frac{\sum_{t_1}^{t_2} M(t)\,dt}{\sum_{t_1}^{t_2} L(t)\,dt} \quad (4)$$

The simultaneity of peak powers in houses will also be estimated with different control algorithms by using the load coincidence factor f_{load} [3]. A peak load in each house during the same period of time will lead to a high coincidence factor in the community. The coincidence factor gives a value of the smoothing effect of the maximum load for several buildings. The load coincidence factor is defined as

$$f_{load} = \frac{\max_t\left\{\sum_{n=1}^{i} L_{net,n}(t)\right\}}{\sum_{n=1}^{i} \max_t\{L_{net,n}(t)\}} \quad (5)$$

where i is the number of houses and $L_{net,n}$ is the net load in house n, i.e. power consumption minus PV power production. Similarly, the correlation of the PV power production can be calculated using the generation coincidence factor ($f_{generation}$), defined as

$$f_{generation} = \frac{\max_t\left\{\sum_{n=1}^{i} P_{net,n}(t)\right\}}{\sum_{n=1}^{i} \max_t\{P_{net,n}(t)\}} \quad (6)$$

where $P_{net,n}$ is the net power production in house n.

Net load and net power production in the equations above are defined as $L_{net}(t) = L(t) - M(t)$ and $P_{net}(t) = P(t) - M(t)$, respectively.

2.2 House model and cases

A detailed model of a single-family house with exhaust air heat pump, PV system and lithium-ion battery energy storage is used for the simulations. The heat pump, with help of an auxiliary heater during times with very high heat load, covers both domestic hot water and space heating. The house model is implemented in the simulation software TRNSYS 17 and described in detail in [6,7]. A control algorithm for the charging of the battery storage and operation of the heat pump is used to maximize the self-consumption from the PV system and minimize the need of buying electricity. Four cases presented in Tab. I will be examined, each of them with five households.

Table I: Control algorithms

	Battery storage	PV-controlled heat pump
Case 1	No	No
Case 2	Yes	No
Case 3	No	Yes
Case 4	Yes	Yes

In Case 1 and 2, the heat pumps are controlled solely by the temperature in the domestic hot water (DHW) storage tank and the need of space heating in the house. In Case 3 and 4, the heat pump can be switched on by excess PV power production to heat the DHW tank or for space heating. The control strategies were presented in [7]. The aim was to minimize the bought electricity (final energy) and maximize the self-consumption and self-sufficiency on a yearly basis for a single house. Much research on the optimization of self-consumption, self-sufficiency and final energy for individual households has been carried out. Studies of communities where all houses have individual optimization of the energy management is however more rare. This is important for system studies at power grid level since optimal control strategies vary depending on the boundary conditions for the optimization.

All simulations use same the house model but the electricity and hot water consumption series differ among the five households. The heat pumps and batteries are controlled individually in each house.

The results are valid for a community of five houses with parameters as in Tab. II. All houses in each case have the same sizes of batteries, PV systems and DHW tank. All houses are equipped with an exhaust-air heat pump of 2.05 kW electric power and an auxiliary heater for the DHW tank of 6.5 kW. The DHW tank temperatures are distributed equally between the houses in case 3 and 4. The set temperature in the rooms is not varied between the cases. See [7] for further details about the house model and boundary conditions.

Table II: Parameters for the study

	PVsize [kW_p]	Battery [kWh]	DHW tank [liters]	Set temp. in DHW tank [°C]
Case 1	6.2	0	180	48
Case 2	6.2	7.2	180	48
Case 3	6.2	0	180	44 - 52
Case 4	6.2	7.2	180	44 - 52

2.3 Data

High-resolution (one minute) meteorological data for Norrköping measured by the Swedish Meteorological and Hydrological Institute (SMHI) was used in the simulations. Synthetic activity sequences of individual household members were modeled using a Markov-chain model for activities, including occupancy, described in [8], with addition of two new states for DHW use (bathing and showering) identified from the same time-use data (TUD) as in the original model. The generated sequences are stochastic and thus represent well the random coincidence effects in real user behavior. Based on the activity and occupancy sequences, the internal heat energy gain from the residents, the electricity consumption and the domestic hot water use in each house can be calculated. Each of the five households has a different profile for domestic electricity and DHW load, generated by different runs of the load generator. This means that the total annuals loads are the same, but that the exact timings differ from house to house.

Details about how the electricity use is modeled from the activity sequences can be found in [8]. A new DHW model was added, in which tappings from the added bathing and showering activities are modeled along with a simple model for additional tappings, assumed to occur randomly with a fixed probability when residents are at home and awake. For showering a flow rate of 10 l/min during 4 minutes was assumed, and for bathing a flow rate of 16 l/min during 6 minutes. Additional tappings were assumed to occur at a given minute with a probability of 1%, with a flow rate of 4 l/min during 2 minutes.

The electricity use model has been validated previously against measurements [8]. The DHW model has not been validated directly against data, but flow rates and durations were chosen to comply with the DHW profiles developed within the IEA SHC Task 44, and the model performance was confirmed to be similar [9].

For the simulations in this paper, two adults and two children were assumed to live in each house, with the children simulated as one person.

3 RESULTS

The results for the four cases are presented in Figure 1-3 and Tab. III-V. Fig. 1 shows the distribution of the aggregated PV power production in the community, which will not differ from the distribution for a single house. This is due to the same azimuth and tilt angles of all PV systems. PV power in the night, i.e., 0 W, is omitted in the figure, since the distribution will be less distinguishable otherwise.

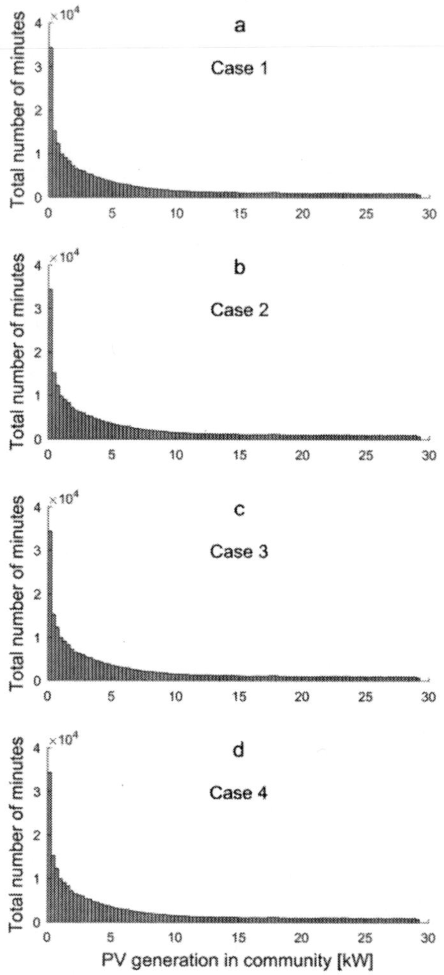

Figure 1: Distribution of aggregated PV generation in the community over one year with parameters as in Tab. II. PV generation of 0 Watt is omitted.

Fig. 2 shows the distribution of the power consumption in the community. The battery storage does not influence the power consumption distribution when the heat pump is operated independently of the PV power, i.e. Case 2 versus Case 1. The control algorithms including the heat pump (Case 3 and Case 4) result in more time with very low electricity use. A small peak is also seen for a power consumption of around 7 kW in Case 3 and Case 4.

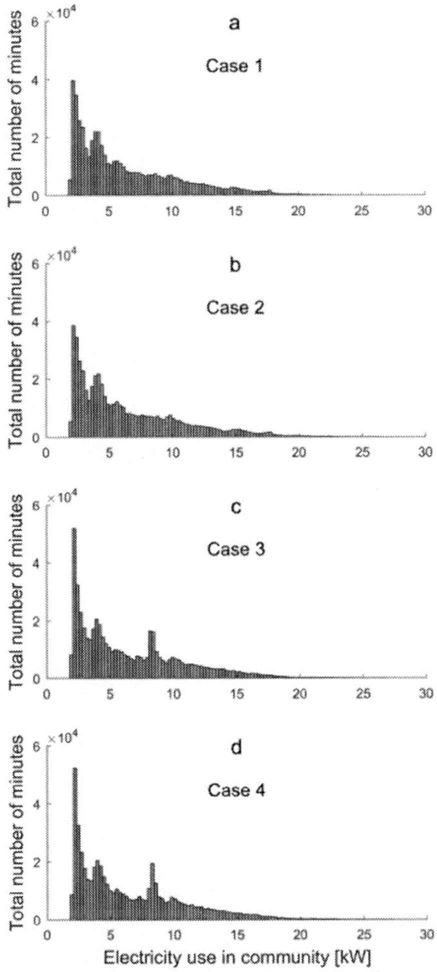

Figure 2: Distribution of the aggregated electricity consumption in the community over one year with parameters as in Tab. II.

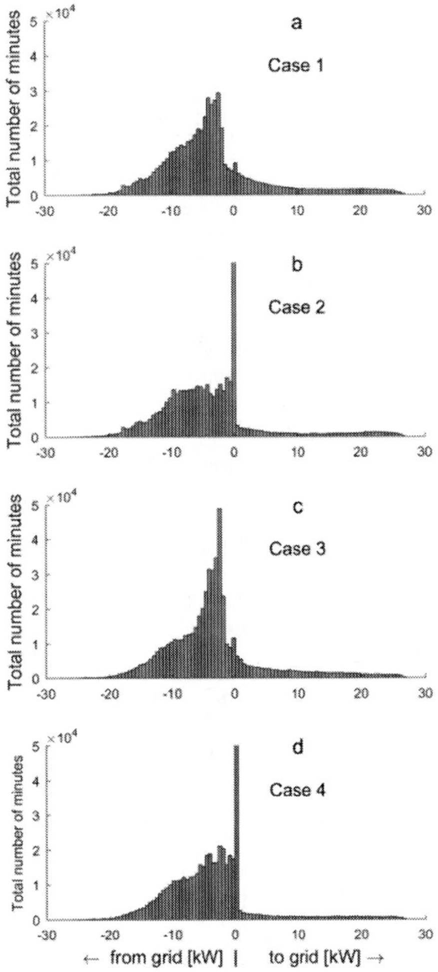

Figure 3: Distribution of the grid interaction in the community over one year with parameters as in Tab. II.

Fig. 3 shows the distribution of the net power consumption in the whole community. The figure shows a reduction in the share of the year with power to the grid (positive values in the figure) when using batteries (Case 2 and 4) compared to no batteries (Case 1 and 3). The time with zero gird interaction is significantly increased by the use of batteries. The control algorithm including the heat pump (Case 3 and 4) has a lower effect on the power flows than the control algorithm including batteries.

Tab. III shows the maximum power to and from the grid for the community. The number of minutes with very high power to or from the grid is not significantly affected neither by battery use of controlling of the heat pump. The maximum power to the grid is slightly affected by change of algorithm, where the highest power occurs for Case 3 with heat pump control but no battery.

Table III: Max power to and from grid for the whole community.

	Max power to grid [kW]	Max power from grid [kW]
Case 1	27.0	26.9
Case 2	27.0	26.4
Case 3	27.0	27.7
Case 4	27.0	25.8

Tab. IV shows the load and generation coincidence factors, self-consumption and self-sufficiency. The self-consumption and self-sufficiency are largely affected by the control algorithms in the different cases, whereas the generation coincidence factor is barely changed. The load coincidence factor is lowest for Case 2, i.e. with houses with batteries but without control algorithm regulating the operation of the heat pump. The highest load coincidence factor occurs in Case 3, where no batteries but a control of the heat pump operation are used. This shows that the control algorithms switch on the heat pumps at the same time when the total load is as highest.

The most of the increment in self-consumption and self-sufficiency occurs during times with moderate difference between PV generation and electricity use. During times with very high load or PV generation, the battery storages and DHW tanks are either at the highest or lowest charge level.

Table IV: Load coincidence factor (A), generation coincidence factor (B), self-consumption (C) and self-sufficiency (D) for the community. The latter two are mean values for the five houses in every case.

	A [%]	B [%]	C [%]	D[%]
Case 1	67.8	98.7	37.7	18.1
Case 2	67.0	98.7	63.0	29.1
Case 3	71.0	98.9	49.9	23.5
Case 4	67.6	98.9	72.2	33.1

Tab. V shows the occurrence of high power to or from the grid for the whole community. It is defined as power exceeding 95% of the maximum power in each case, see Tab. III. The total number of minutes during one year is 525,600. The number of minutes with high power from the grid, i.e. high aggregated overconsumption in the community, is much lower than the number of minutes with very high power to the grid.

Table V: Number of minutes with aggregated power to or from the grid exceeding 95% of maximum power.

	To grid [minutes]	From grid [minutes]
Case 1	1490	21
Case 2	1252	29
Case 3	1303	8
Case 4	1019	49

4 DISCUSSION

The results show that maximum power from grid (Tab. III) and load coincidence factor (Tab. IV) are highest for Case 3 with a control algorithm for the heat pump but without battery storage. Thus, energy use optimization in single household leads in this case to worse energy performance when it is used all houses in the community. However, the time of high power from the grid was short – only 8 minutes of the 525,600 minutes of the year with a power to the grid exceeding 95% of maximum. Even though the number of minutes with high power from the grid is low, the local grid needs to be dimensioned to handle it to avoid power failures.

There was a large difference in mean self-consumption and self-sufficiency for the cases without battery storage (Case 1 and Case 3) and with battery storage (Case 2 and Case 4). The improvement in self-consumption and self-sufficiency is larger when introducing batteries than heat pumps (Tab. IV). This is mainly due to the larger energy storage capacity in the batteries than in the hot water tanks, but also an effect of the lower heating demand in the summer when the PV generation is high then in the winter.

An important note is that the control algorithms were developed for energy optimization and not power optimization. This study however focuses on the power rather than the energy, which means that the results might change if the individual households are optimized for peak power reduction.

The main outcome of this study is that it is important to widen the boundaries and examine the system effects when a new energy management solution for households is proposed.

5. CONCLUSIONS

This study of a community of PV equipped households with individual energy use optimization shows that (*i*) the highest maximum power consumption in the community and higher load coincidence factor occurs when the heat pumps are controlled with the PV power production and no battery storage is used, and (*ii*) the lowest total power consumption occurs when battery storage is used in combination with heat pumps controlled by the PV power production.

REFERENCES

[1] Swedish Energy Agency. Energy in Sweden 2015. 2015.

[2] Swedish Energy Agency. Energistatistik för småhus 2014 [Energy statistics for detached houses 2014]. 2015.

[3] Strbac G. Demand side management: Benefits and challenges. Energy Policy 2008;36:4419–26. doi:doi.org/10.1016/j.enpol.2008.09.030.

[4] Luthander R, Widén J, Nilsson D, Palm J. Photovoltaic self-consumption in buildings: A review. Appl Energy 2015;142. doi:10.1016/j.apenergy.2014.12.028.

[5] Luthander R, Widén J, Nilsson D, Palm J. Photovoltaic self-consumption in buildings: A review. Appl Energy 2015;142:80–94. doi:10.1016/j.apenergy.2014.12.028.

[6] Persson T, Heier J. Småhusens framtida utformning : Hur påverkar Boverkets nya byggregler? [How do the new Swedish building codes affect detached houses of the future?]. Gävle, Sweden: Region Gävleborg,; 2010.

[7] Psimopoulos E, Leppin L, Luthander R, Bales C. Control algorithms for PV and heat pump system using thermal and electrical storage. Proc. EuroSun 2016 Conf., Palma de Mallorca, Spain: International Solar Energy Society; 2016. doi:10.18086/eurosun.2016.08.13.

[8] Widén J, Wäckelgård E. A high-resolution stochastic model of domestic activity patterns and electricity demand. Appl Energy 2010;87:1880–92.

[9] Haller MY, Dott R, Ruschenburg J, Ochs F, Bony J. The Reference Framework for System Simulations of the IEA SHC Task 44 / HPP Annex 38 Part A: General Simulation Boundary Conditions. 2013.

Modelling of PV Prosumers using a stationary battery, heat pump, thermal energy storage and electric vehicle for optimizing self-consumption ratio and total cost of energy

Dominik Keiner[1] and Christian Breyer[2]

[1] Ostbayerische Technische Hochschule Regensburg, Prüfeninger Str. 58, 93049 Regensburg, Germany
[2] Lappeenranta University of Technology, Skinnarilankatu 34, 53850 Lappeenranta, Finland, christian.breyer@lut.fi

ABSTRACT:
PV prosumers may be one of the most important market segments in the energy transition. By using one's own energy on the spot of production, PV prosumers have a significant possibility of supplying their own demands with cheap and pollution-free energy. This becomes most profitable by maximising self-consumption while avoiding large amounts of excess electricity being fed into the grid. Storage technologies, heat pumps and battery electric vehicles help to contribute to the highest possible self-consumption ratio for residential PV prosumer systems, which can reach grid-parity within the next decade in most regions of the world. The study provides a detailed overview of 8 regions on all continents and traces the benefits of PV prosumer systems for the world structured in 145 regions. A most noticeable threshold for the economical maximum battery capacity per installed PV capacity has been found, even independently of the inclusion of battery electric vehicles. Very high self-consumption ratios, demand cover ratios and heat cover ratios can be observed.

Keywords: PV; Prosumers; Households; Battery; Electric Vehicle; Energy cost; Heat Pump; Thermal Energy Storage; Vehicle-to-Home

1 INTRODUCTION

There is lot of interest in solar photovoltaic (PV) Producer and Consumers (Prosumers) all over the world, as the cost for PV systems are steadily decreasing and by the end of 2020 most residential electricity market segments will have reached grid-parity for solar PV [1] [2]. By optimizing the self-consumption ratio (SCR), this advantage can be exploited more. Utilising energy storage technologies, there is an opportunity for using PV electricity even at times when no energy is produced.

Some markets have already reached grid-parity for PV systems with battery storage [3]. Storing electricity in stationary batteries has different advantages. Stationary batteries are easy to use and relatively small, which enables them to be used in residential households. Furthermore, as the largest share of residential load appears in the evening and night hours, stationary batteries provide a way of using PV electricity produced during daytime for later use.

The characteristics of Lithium-Ion batteries enable stationary use for covering electricity demands of the household, in particular the electricity demand for heat by using an electric heat pump (HP). By completing the system with thermal energy storage (TES), this PV prosumer system is the basis of the simulated model.

Studies have shown a rapidly increasing stock of battery electric vehicles (BEVs) since 2010. The evolution of the global electric car stock reached nearly 2 million vehicles within six years from 2010 to 2016 [4]. The idea of SCR optimising can also be reached using BEVs and charging them with one's own produced PV electricity. It seems obvious that electric mobility, with an efficiency of 70 – 80% compared to a conventional combustion engine efficiency of 20 – 30% [5], will be one of the main drivers for the energy transition in the transport sector, above all in passenger transportation. But the high efficiency of BEVs can only be reached by charging them with electricity from renewable energy (RE) sources, such as wind, hydro or PV as the low efficiency of conventional energy generation would counteract the efficiency enhancement of BEVs [5]. The advantage of including BEVs in a residential energy system is the availability of cars during the long times when they are idle. Cars that are only used for short journeys and shopping trips can offer cheap electricity storage when vehicle-to-grid (V2G) service is made available. In this specific case, the vehicle may not always need to be grid connected to be of use to PV prosumers, instead offering a vehicle-to-home (V2H) service. This function is at its optimum if such BEVs are available at home during daytime when PV electricity is generated. However, not only 'usually at home' cars are of relevance for a residential energy investigation. Cars which are mostly used for work trips during the day have the constraint that charging stations are not available everywhere. From a residential point of view it would be very useful to charge this type of car at home and get it integrated into the residential energy system whenever possible, as explained later in more detail. Therefore, the 'fuel cost' for this car can also be reduced.

Space heating and cooling combined with hot water supply correspond to about a half of the energy consumption in buildings [6]. The European Union has a target share for RE of 20% of final energy consumption by 2020. This is estimated to include 4.9% of final energy from heat pumps and 2.9% from PV energy [6]. Several countries accept HPs as an RE application sourcing surface geothermal energy and support HP systems to enable the development of a higher share in buildings to replace conventional heating systems. The cost reduction of HPs is estimated to be about 30 – 40% for heat services by 2050 [6]. Furthermore, the technology of HPs is well known and well developed, even though there is this further cost reduction potential.

The benefits of PV and HP technologies can be best possibly utilised by combining them in residential energy systems. Past surveys show that PV + HP systems combined with TES can economically challenge conventional solar collector systems [7]. By combining PV, HP, TES and BEVs in a residential energy system, another basis for an optimised self-consuming home is built. However, system components are not sufficient to optimise the energy system. The aim of this work is to create a PV prosumer model for optimising PV energy use and total cost of energy for average residential homes for different regions worldwide. Further, the aim is to perform a more comprehensive investigation of the impact of the different system components and the development of the total cost of energy, self-consumption ratio (SCR), demand cover ratio (DCR) and heat cover ratio (HCR). This will include finding the least cost PV capacity and respectively optimised stationary battery sizes.

The results are presented for four different scenarios. Table 1 shows the usage of one 'usually away' car, from this point on called 'Car 1', and one 'usually at home' car, from now onwards called 'Car 2' for the four scenarios.

Table 1: *Usage of the BEV types in the modelled scenarios*

Scenarios	Car 1	Car 2 (V2H)
'Two Cars'	✓	✓
'Only Car 1'	✓	
'Only Car 2'		✓
'No Cars'		

2 METHODS AND DATA

The simulated model is based on the LUT Energy System model with an hourly resolution [8] [9] [10]. Therefore, 8760 hours are analysed for an entire year. To find the financially optimised (least cost) PV and stationary battery capacities, the simulation was performed by scanning PV capacities of 1 – 30 kW_p and stationary battery capacities of 1 – 50 kWh_{cap} in steps of 1 kW_p and 1 kWh_{cap}, respectively. Furthermore the simulation was carried out on a global scale, represented by 145 regions [9] [10], in 5-year steps from 2015 until 2050. A list of all regions with their abbreviations and including countries or areas can be found in the appendix (Table A1). The simulation programme MATLAB was used. In Figure 1, the entire setup with all system components is visualized. The black arrows represent electricity flows, and the orange arrows represent the flow of thermal energy in form of hot water.

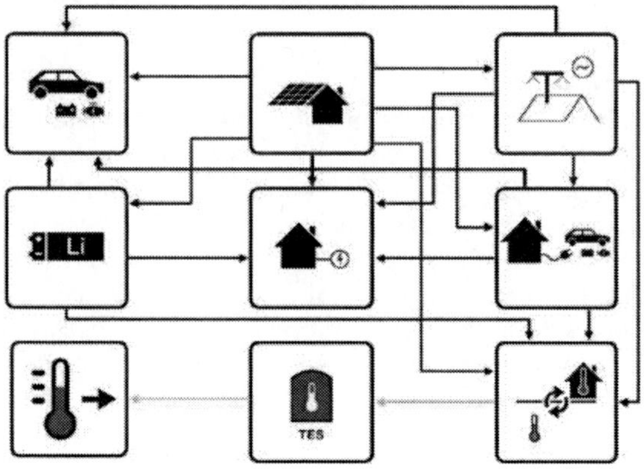

Figure 1: *Structure of the PV Prosumer model with all system components as it complies with the full 'Two Cars Scenario'.*

The prosumers are considered to be an average household for every region.

2.1 GDP per Capita and People per Household

Most of the data concerning were available on a national scale. With numbers for citizens per region, it is possible to calculate values per person. For 107 countries or areas, numbers for households are available from the United Nations database [11]. Using the Gross Domestic Product (GDP) per capita for those countries, a relation between GDP per capita and people per household (pph) can be used. The GDP projection is taken from [12]. To get the pph numbers for countries with no available household data, an interpolation is required.

For the development of pph values until 2050, a linear increase of GDP per capita was assumed. A further supposition was made for the impact of GDP per capita development on pph numbers. It was assumed that the effect on pph is a quarter of the GDP per capita development, which means a 100% increase of GDP per capita causes 25% decreasing of pph, as the pph numbers are highly dependent on GDP per capita values. In cases for regions with more than one country, the population weighted average of the included countries was taken. For countries which are split into more than one region, the same pph are assumed for all sub-regions. The development of pph numbers for all regions can be found in the appendix (Figure A1 and Table A2).

2.2 Solar Data

The input data for PV electricity generation were available for every region in full hourly resolution in the unit of kWh (kWp·h) according to [8] [9]. By multiplying those numbers with the installed PV capacity, the PV generation profile for a given capacity can be obtained.

2.3 Load Profiles

Residential electricity load profiles are available for Brazil, Italy, Japan, Senegal, Thailand, Germany, New Zealand and the United States [3]. The challenge is to create load profiles for all regions which represent the local behaviours of residential households. For this, the regions were divided in ten 'profile regions' – NA-US-SWMA (US South West and Middle America), NA-CA-US-NE (US North East and Canada), SA (South America), Europe-Nordic-RU (North Europe and Russia), Europe-CE-East (Central and East Europe), Europe-South-MENA (South Europe and MENA), SSA (Sub-Saharan Africa), Pacific (New Zealand and Australia), SEA (South East Asia) and NEA-Eurasia-CA (Central and East Asia). By combining all these profiles, an 'average human electricity consumption' load profile was built. Adapted with the abovementioned available profiles for each of the profile regions and allocating the regions to the profile types, it was possible to create load profiles containing area specific courses as well as average human behaviours (morning and evening peaks). The load profiles for the ten profile regions are shown in Figure 2. It can be seen that the profiles differ quite a lot during daytime, which has a great impact on Direct Self-Consumption (DSC) and therefore influences the results for energy use.

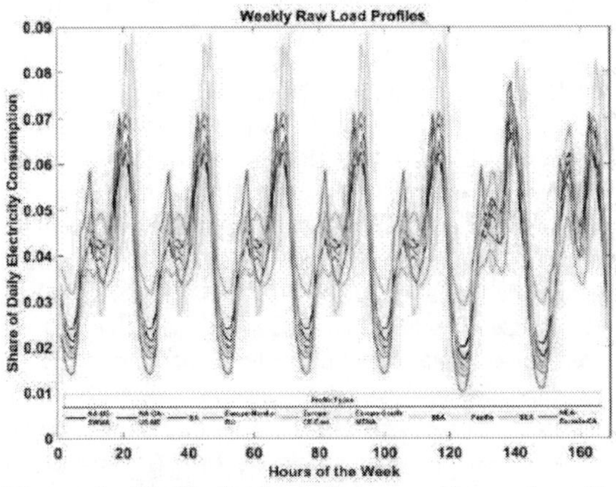

Figure 2: *Weekly load profiles as a share of weekly energy consumption for ten profile regions.*

Furthermore, the profiles are built as yearly factor profiles. As energy consumption reacts to seasonal changes, the linear course of the yearly profile had to be adapted, too. As national load profiles were available [12], factors compared to the yearly mean load were established. By combining weekly load profiles with the specific weekly factor using Equation 1 for every week of the year, the course for residential load profiles could be derived.

$$load_{real} = load_{raw} \cdot f_{week}$$

Equation 1: *Adapting of yearly load profiles to national load course. Abbreviations: $load_{real}$ – real load profile; $load_{raw}$ –raw load profile; f_{week} – factor of the investigated week compared to the annual national mean load.*

The real weekly load profiles were then set together for the yearly profile by repeating the introduced routine for all weeks of a year. Figure 3 shows the factorized yearly profile as an example for Germany. The load profiles are normalised to a yearly electricity consumption of 1 MWh. To get the real loads, the profiles have to be multiplied by the annual electricity consumption per household.

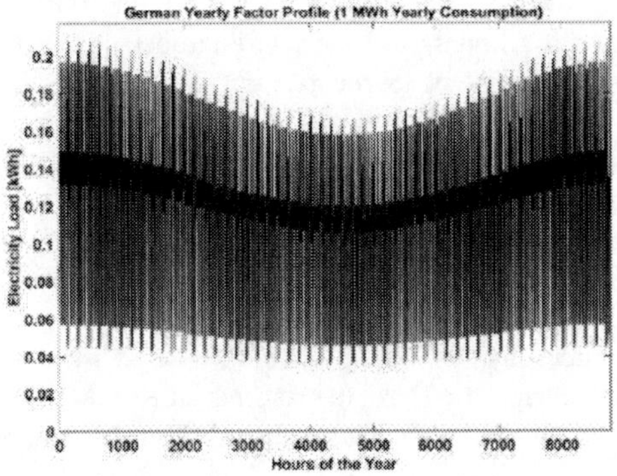

Figure 3: *German yearly factor profile, normalised on an annual electricity consumption of 1 MWh.*

With data for pph, total electricity consumption per regions [10] and share of residential electricity consumption for all regions, respectively, the annual residential electricity consumption per household can be calculated for each region. It was assumed that the share of residential electricity consumption will be the same for all years, which means that the increase of residential energy consumption is linked to the national increase of the total electricity consumption for each region, respectively.

Load profiles for Space Heating (SH) and Domestic Hot Water (DHW) demand were also available on a region-wide scale for all regions until 2050 [19]. It was possible to create SH and DHW load profiles with pph and population per region. Electricity consumption, SH and DHW for all regions and modelled years can be found in the appendix (Table A3-5).

2.4 Battery Electric Vehicles

The capacity of the BEV batteries was set to a specific value. Car 1 was assumed to have an 80 kWh_{cap} and Car 2 with a 60 kWh_{cap} battery. This can be explained with the availability of the cars for PV charging. As Car 1 is usually not at home during daytime and weekdays, it should be possible to charge as much energy as possible during weekends and evening hours. If needed, Car 1 has to receive its driving demand from the charged energy of the weekend for the whole following week. This could be compared to a filled tank of a conventional internal combustion engine (ICE) vehicle. As Car 2 is usually available for PV charging, but is used as a vehicle-to-home (V2H) device as well, it should be possible to charge as much PV electricity as possible for supplying household demands (house electricity and HP). Furthermore car 2 is also providing electricity for Car 1. The Depth of Discharge (DOD) for BEVs does not factor in, as a safety buffer of 25% of the storage capacity was included for emergency cases.

As already mentioned, Car 1 is used as a 'usually away' vehicle, which means it is away from home starting in the morning hours until the late afternoon to represent times for workplace journeys. Car 1 is also used for free time journeys on weekends. The decision whether Car 1 starts one of those free time journeys was reached by a randomised function within the simulation. This enables dynamic weekend journeys, independent of solar conditions on weekends. For Car 2 as a 'usually at home' vehicle, the length of the away time was limited to three hours per day, as Car 2 should be used mostly for shopping trips etc. Here, the starting time of the journey was chosen randomly. This programme enabled starting times of the journeys for Car 2 independent of solar conditions and load demands. The screening whether Car 2 just has one journey per day leads to leaving times mostly on late mornings. Data for leaving time of work journeys and shopping trips were taken from [13]. In cases of poor solar conditions, the BEVs have to be charged via the grid. The cars are charged via the grid in the early morning hours, when the house electricity demand is usually low (cf. Figure 2). To minimise grid charging it is checked if the State of Charge (SOC) of the vehicles is below the safety buffer plus the demand for one daily trip. The daily trip demand can be calculated by Equation 2 as following:

$$E_{Car,trip} = \frac{Yearly\ driving\ distance}{\sum journeys} \cdot E_{cons.}$$

Equation 2: *Calculation of electricity demand per trip to be applied for each car. Abbreviations:* $E_{car,trip}$ *– electricity demand per car and trip;* $E_{cons.}$ *– specific electricity consumption of BEV.*

The specific electricity consumption of the BEVs was set to 20 kWh/ 100 km [13], including energy losses and discharging efficiency. The discharging efficiency of Car 2 to supply household demands and Car 1 charging was set to 96.8%, which represents the charging and discharging efficiency of Li-Ion batteries of the LUT Energy System model [10]. Driving distances per year were set to 14,000 km/a for Car 1 and 10,000 km/a for Car 2.

2.5 Stationary Battery

As shown in Figure 1, the stationary battery is used to supply the electricity demand of the house and the HP as well as charging Car 1. In the model, charging the stationary battery has first priority after DSC. It is primarily envisaged for covering loads in the evening and night hours, but also for covering peaks during daytime when DSC is not sufficient, provided that the stationary battery is charged. In the morning, before grid charging for Car 1 is requested, it is checked whether electricity is left in the stationary battery. If yes, the electricity is transferred to Car 1. In this case it does not play a role whether load demands cannot be covered by the stationary battery, as charging Car 1 via the grid would cause the same cost. The advantage of this procedure is that the stationary battery is not liable to be fully charged when PV electricity is available again. This provides the possibility of using more PV electricity within the system. The type of the stationary batteries was chosen to be a Li-Ion battery, as they are increasingly common in residential households. The DOD was set to 95% and the charging and discharging efficiency to 96.8%.

If the stationary battery is discharged and Car 2 is available, Car 2 will step in and take over the tasks of the stationary battery. In the morning hours, when the SOC of Car 2 is higher than the safety buffer plus one daily demand, Car 2 also transfers as much electricity as possible to Car 1.

2.6 Heating System (Heat Pump, Heating Cartridge)

The HP type was chosen to be a brine/water (ground source) heat pump (GSHP), as it is known that the efficiency compared to air source HPs is much better, the cost is lower and availability compared to water source HPs is higher. Another advantage of GSHP's is the quite stable Coefficient of Performance (COP) over the whole year caused by the stored solar energy in the ground, even during wintertime with cold temperatures. The HP was modelled as a commercially available HP with the possibility of heating water up to 90°C. The rated power is 7 kW_{el}. For an easier simulation, the COP was set to a permanent value of 3.8 for operation at nominal value. For PV additional filling of the TES the COP is 3. The reason for two different operation modes is explained in the following section.

For regions with an SH demand less than 2 MWh_{th} per year and capita, an application of a HP with a TES of 800 litre capacity is not appropriate, as the investment cost for the system components are not reasonable for the heat demands. For that reason the modelled systems of the regions with a SH demand under the above-mentioned limit has been equipped with a heating cartridge instead of the HP with 7 kW_{th} output power. Figure 4 shows the regions indicating the respectively used heating systems. Only two regions change the system type during the energy transition period: Ecuador switches from a cartridge system to an HP system in 2045, and South Africa & Lesotho changes from an HP system to a cartridge system in 2050.

The European standards DIN EN 806-2 [14] and DIN EN 1717 [15] recommend a hot water temperature for potable water of minimum 60°C for the avoidance of legionella development. To be on the safe side, a hot water temperature for operation at a nominal value of 65°C was chosen. This consideration combines a water temperature which is as high as necessary for potable water quality and as low as possible for a most efficient HP operation.

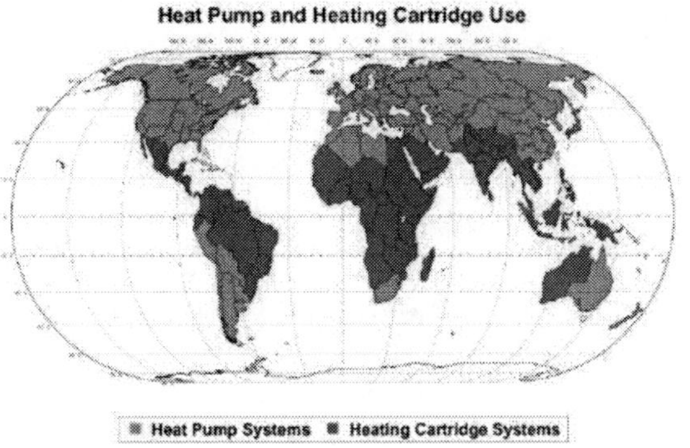

Figure 4: *All regions indicating the chosen heating system. Ecuador and South Africa & Lesotho are marked with their prevailing system.*

2.7 Thermal Energy Storage

The TES is assumed to be a tank-in-tank storage with a water capacity of 800 litres for HP systems and 200 litres for heating cartridge systems. This includes storage capacities for SH and DHW in one device, as they are usually used for solar thermal applications [5]. A standby share of 25% was set for the TES, which guarantees coverage of SH and DHW demands while refilling the storage. Therefore, occupants are able to take a bath or shower even when TES SOC is quite low and no refilling request was set off. This refilling request is triggered when the SOC of the TES falls under the 25% standby share. If this request comes during daytime when PV electricity is available, the demand is covered by DSC. If the opposite is the case, the demand gets covered by the stationary battery. If the stationary battery is discharged as well, in the 'Two Cars Scenario' and 'Only Car 2 Scenario', the demand gets covered by Car 2. If none of the energy storage options are able to cover the demand, the refilling demand is covered by the grid. In this case the TES gets filled by the HP with a COP of 3.8, as in section 2.6 already mentioned as operation at nominal value. The maximum TES capacity is therefore 800 litres at 65°C. Equation 3 describes the calculation of storable thermal energy of the TES.

$$E_{th,cap} = c_{p,water} \cdot V_{TES} \cdot dT$$

Equation 3: *Calculation of thermal energy storage capacity of TES. Abbreviations: E_{th} – thermal energy; $c_{p,water}$ – specific heat capacity of water (0.0016 $kWh_{th}/(kg{\cdot}K)$, derived from 4.19 $kJ/(kg{\cdot}K)$); V_{TES} – storage volume of TES; dT – temperature difference (45 K for nominal operation, 70 K for PV additional filling).*

The TES as well as the HP are designed for 90°C. Facing the situation that the stationary battery is fully charged, both cars are fully charged or not at home and the TES is charged with 800 litres at 65°C, the PV electricity surplus is normally fed into the grid, which would make no sense for a SCR optimisation, since there is still capacity in the TES available. Therefore the system has the possibility to fill the TES up to 89.6 kWh_{th}, which means full 800 litres at 90°C. Although the COP of 3 for PV additional filling is lower than for nominal operation, the possibility of charging more low cost energy balances the lower efficiency. With this consideration it is possible to have an added storage capacity of 32 $kWh_{th,cap}$, which can be used. Figure 5 visualises the TES filling conditions. The idea is to increase SCR at times with very good solar conditions. The PV additional filling feature is also available for heating cartridge systems with a 200 litre TES. For this system the same efficiency for the heating cartridge is assumed The nominal capacity for a 200 litre TES is 14.4 kW_{th} and the full capacity is 22.4 kWh_{th}.

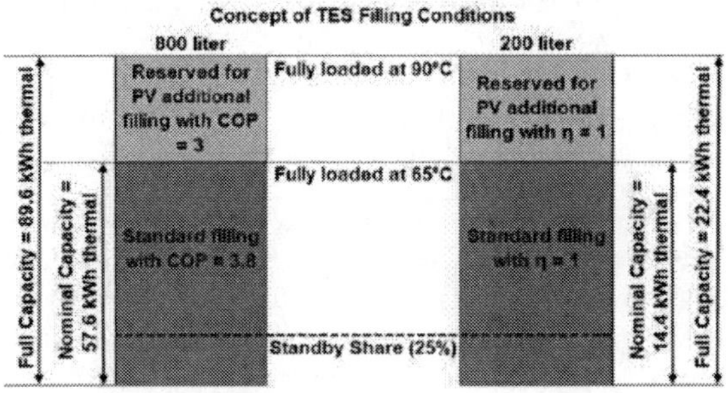

Figure 5: *Concept of TES filling conditions for the two operation modes of the HP.*

The losses of such thermal storage systems are given between 1.6 kWh_{th} and 2.5 kWh_{th} per 24h (cf. [16]). Applied to the 800 litre TES of the model, this results in thermal energy losses of about 0.15% per hour, considering a fully charged TES. For an easier simulation, this value was applied independently of the SOC and TES size. The efficiency assumptions for all relevant system components are listed in Table A8.

2.8 Financial Target Function

The yearly total cost of energy is calculated by applying Equation 4, which has to be minimised for the year.

$$Annual\ Energy\ Cost = \sum_i^{technology}(Capex_i \cdot crf_i + opex_{fix,i} + opex_{var,i} \cdot E_{throughput}) + cost_{grid} - income_{feed-in}$$

Equation 4: *Definition for annualised energy cost. Abbreviations: Capex – investment cost for technology, crf – annuity factor,* $opex_{fix}$ *– fixed operational expenditures,* $opex_{var}$ *– variable operational expenditures,* $E_{throughput}$ *– energy handling of component (e.g. discharged energy of the battery),* $cost_{grid}$ *– cost of remaining electricity supplied by the grid,* $income_{feedin}$ *– income for PV electricity fed-in to the grid.*

For a comparison of annual energy cost, Equation 5 was applied for a 100% grid supply of the energy system without a PV system and stationary battery. This has to be minimised for an entire year. For 100% grid supply, Car 2 is assumed to be a normal BEV, as a V2H application for grid supply may be not beneficial. The HP operates at nominal value for grid supply, to keep the yearly cost as low as possible.

$$Annual\ Grid\ Cost = \sum_i^{HP,TES}(Capex_i \cdot crf_i + opex_{fix,i} + opex_{var,i} \cdot E_{throughput}) + \left(\frac{E_{th,DHW} + E_{th,SH}}{COP_{nom}} + E_{el,house} + E_{el,car1} + E_{el,car2}\right) \cdot price_{grid}$$

Equation 5: *Definition for annualised cost for the energy system without PV and stationary battery and 100% grid supply. Abbreviations:* $E_{th,DHW}$ *– annual thermal energy for hot water demand;* $E_{th,SH}$ *– annual thermal energy for space heating;* COP_{nom} *– Coefficient of Performance of HP for operation at nominal value;* $E_{el,house}$ *– annual electricity consumption of household;* $E_{el,car}$ *– annual electricity demand for driving.*

Numbers for financial values of system components and grid prices are based on the LUT Energy System model and available (cf. [1], [2] and [10], Table A7). For the BEVs no storage costs are considered as it is assumed that the batteries are paid with the car anyway. For all regions a feed-in reimbursement of 0.02 €/kWh was used. Due to assumed grid limitations, the feed-in reimbursement was assumed to be paid for feed-in electricity up to 50% of generated electricity.

3 RESULTS

The following results are, if not mentioned specifically, selected for eight representative regions: FI (Finland), DE (Germany), JP-E (Japan East), AU-W (Australia West), ID-KL-SW (Indonesia East), KENUG (Kenya and Uganda), BR-SP (Brazil – Sao Paulo) and US-MW (United States – Midwest).

3.1 Total Cost of Energy and Grid-Parity

The results for annual total cost of energy for all scenarios are shown in Figure 6. With the modelled energy system and scenarios it is possible to reduce the annual costs from about 900 – 6000 €/a to approximately 550 – 5500 €/a for the shown regions. Notably, the achievable annual total cost of energy is independent of the simulated scenario. This can be explained by the fact that the electricity demand per year with BEVs is higher, but it is also possible to increase the SCR with BEV charging and therefore using cheaper PV electricity which compensates the expected additional costs. The additional costs can be clearly seen in Figure 6 for the 100% grid supply costs. The development of total energy cost for the PV systems decreases more or less linearly and follow the decreasing investment costs for the system components. Almost stagnating or slightly decreasing energy cost for 100% grid supply of some regions results from the decline of HP and TES costs, which are compensating increasing grid prices. Figure 7 shows the annual cost saving potential for all regions and scenarios in 2030 using a PV system compared to 100% grid supply. The results for 2050 are shown in Figure A2. The usage of SCR optimised residential energy systems is already profitable for most of the regions in 2015. Particularly regions with quite low grid electricity prices struggle with the high investment cost for stationary batteries. In Figure 8 the annual energy cost ratios, or rather respective system profitability for all regions, are shown for 2030. The stage of development, expressed by GDP per capita, has just a small impact on the level of profitability. The reason is a higher energy consumption of households in well developed countries, which makes an SCR optimisation more useful. Compared to full grid supply, SCR optimised energy systems may be able to reduce the annual total cost of energy up to 80% within the next decade. Until 2050 it is possible to save up to 90% of energy expenditures throughout all scenarios. The use of BEVs has a slight impact on the time of full regional grid-parity. For the 'Two Cars Scenario', the last region (Tajikistan & Kyrgizstan) will reach grid-parity between 2025 and 2030. For the 'Only Car 1 Scenario' and 'No Cars Scenario', the last region (Tajikistan & Kyrgizstan) will be able to get beyond grid-parity between 2035 and 2040. For the 'Only Car 2 Scenario', two regions (Tajikistan & Kyrgizstan and Kazakhstan) are able to reach grid-parity by the same time period in 2035. However, the heating system (HP and TES) is not optimised for every region. As the break-even point is very close to the grid-parity limit of a PV system cost / 100% grid supply cost ratio of 1, a slightly optimised TES size would be sufficient to shift the year of the last grid-parity event for all scenarios by five years earlier at minimum. Nevertheless, some regions are able to reduce their annual energy cost up to 90% compared to a 100% grid supply.

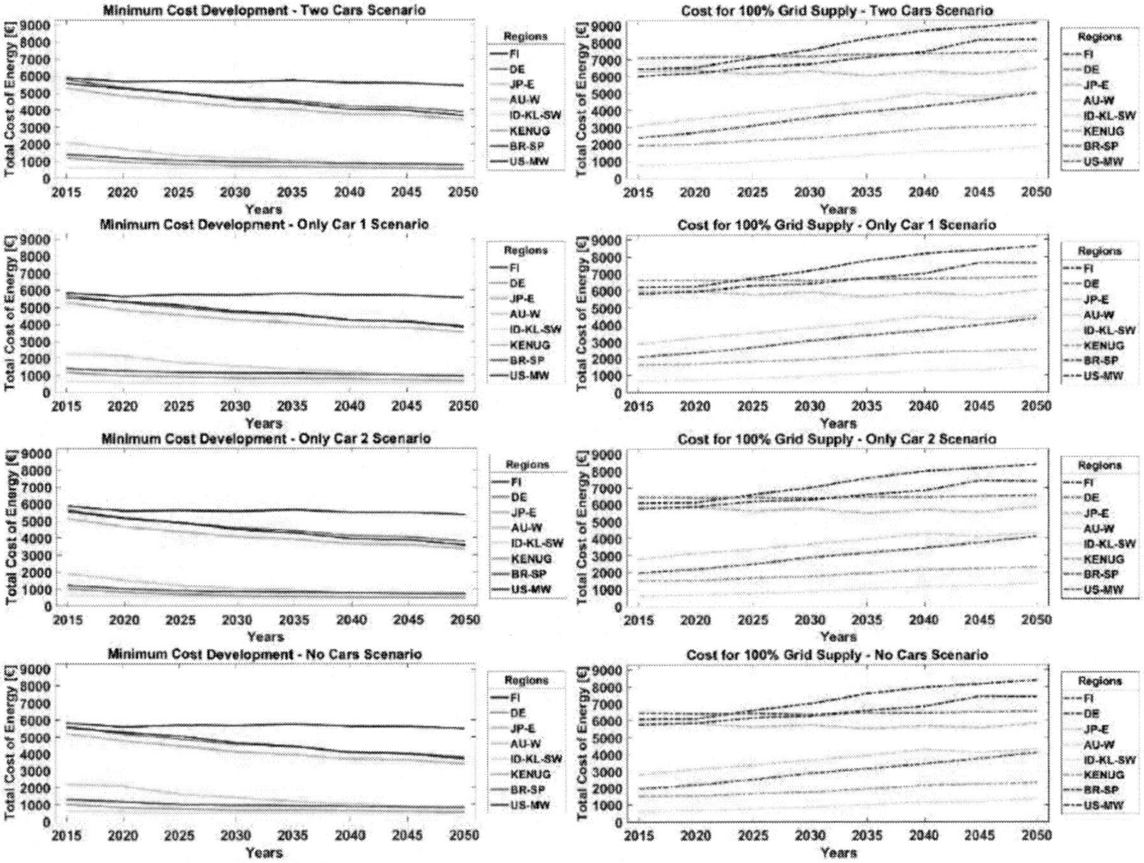

Figure 6: *Annual total cost of energy for least cost system (left) and full grid supply (right) for the six example regions for 'Two Cars Scenario' (top), 'Only Car 1 Scenario' (center top), 'Only Car 2 Scenario' (center bottom) and 'No Cars Scenario' (bottom).*

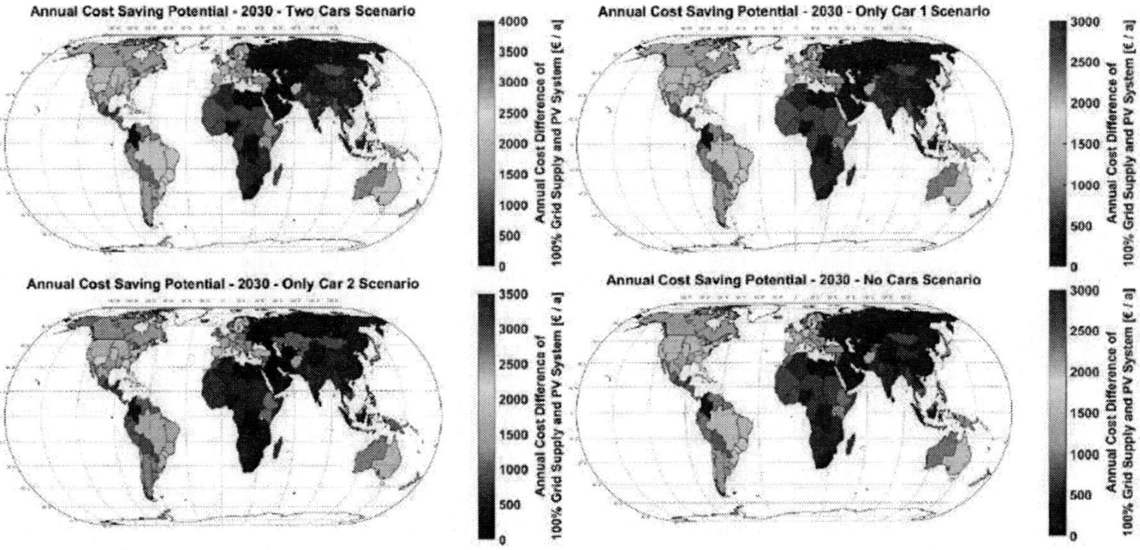

Figure 7 *Annual cost saving potential for all regions in 2030. Displayed are the potentials for 'Two Cars Scenario' (top left), 'Only Car 1 Scenario' (top right), 'Only Car 2 Scenario' (bottom left) and 'No Cars Scenario' (bottom right). Regions with no cost saving potential are marked grey.*

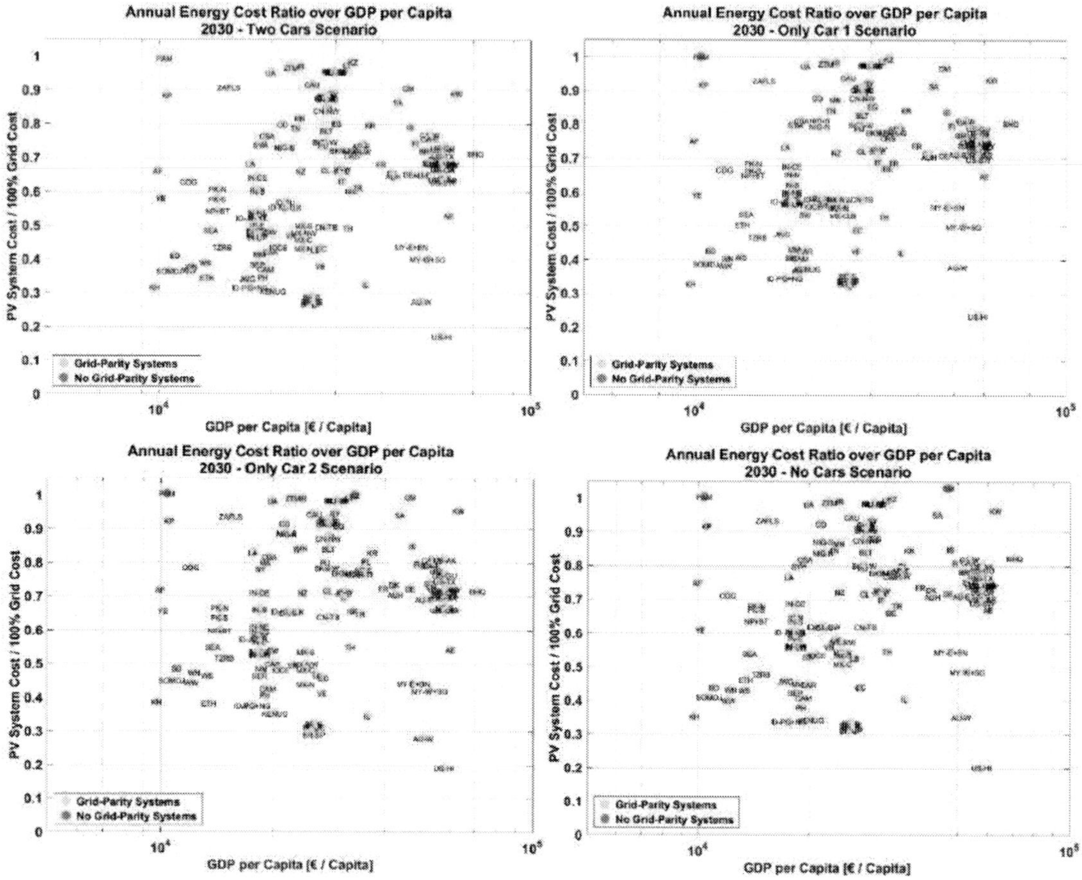

Figure 8: *System profitability of the last year with one or more regions which do not yet reach grid-parity over GDP per capita. Displayed are the years 2025 for the 'Two Cars Scenario' (top left), 2035 for the 'Only Car 1 Scenario' (top right), 2035 for the 'Only Car 2 Scenario' (bottom left) and 2050 for the 'No Cars Scenario' (bottom right)*

3.2 Self-Consumption Ratio

A real optimisation of SCR can be achieved partially. The substantial decline of PV system cost leads to the side effect that it makes large feed-in quantities economically practical. The SCRs will level off between about 30% - 60% until 2050 for all scenarios, as is shown in Figure 10. The difference between the scenarios is not very significant but noticeable. Comparing the 'Two Cars Scenario' and 'No Cars scenario', the difference is predictable. By using both cars the SCR can be improved by about 10 percentage points for some regions. The charging of Car 1 via the stationary battery causes a slightly higher possible SCR than using only one V2H BEV. The impact of the scenarios on the least cost system design is evaluated in section 2.5. There are several reasons for the mentioned difference between the 'Only Car 1 Scenario' and the 'Only Car 2 Scenario': it is useful for a good SCR that the stationary battery transfers as much energy as possible to Car 1 in the morning. Therefore, the SOC of the stationary battery is quite low when PV electricity is available again. This brings the opportunity to charge the stationary battery at a high level almost every day. This effect is lacking in the 'Only Car 2 Scenario'. The stationary battery gets beaten by Car 2 as a no cost storage. The availability of Car 2 during daytime is keeping the effort for recharging low. This leads to an always quite high SOC of Car 2. Furthermore, the lack of additional driving demand of Car 1, and therefore the transfer of electricity from Car 2 to Car 1, supports the development of SCR. The optimisation of storage capacity of the stationary battery has different effects on the difference between regions with very good and moderate solar conditions. For the 'Two Cars Scenario' the difference for Finland and Australia is merely insignificant, differing at about $15\%_{abs}$ at most. For the 'Only Car 1 Scenario' a spread of about $20\%_{abs}$ is expected between 2025 and 2045. In 2050 the SCRs will nearly match up. For the 'Only Car 2 Scenario' a difference of up to $15\%_{abs}$ until 2025 will be in the offing with similar SCRs in 2050. The impact on the SCR of using both cars, or either of them, compared to the 'No Cars Scenario' can be seen in Figure 9.

Comparing Figure 11 and 12 it can be seen that the SCR behaves in contrast to the annual energy costs. Furthermore, the simulation showed that the SCR is mostly dependent on the installed PV capacity in the 'Two Cars Scenario' and 'Only Car 2 Scenario'. For the 'Only Car 1 Scenario' and 'No Cars Scenario', the SCR increases for higher battery capacities for small to middle sized batteries. In a few of the reachable SCRs, a further improvement would make sense. For the example of Germany, a halving of the installed PV capacity, causing additional energy costs of under 200 €/a, can result in a SCR increase of approximately $25\%_{abs}$ (for the example of Figure 11, bottom left and Figure 12).

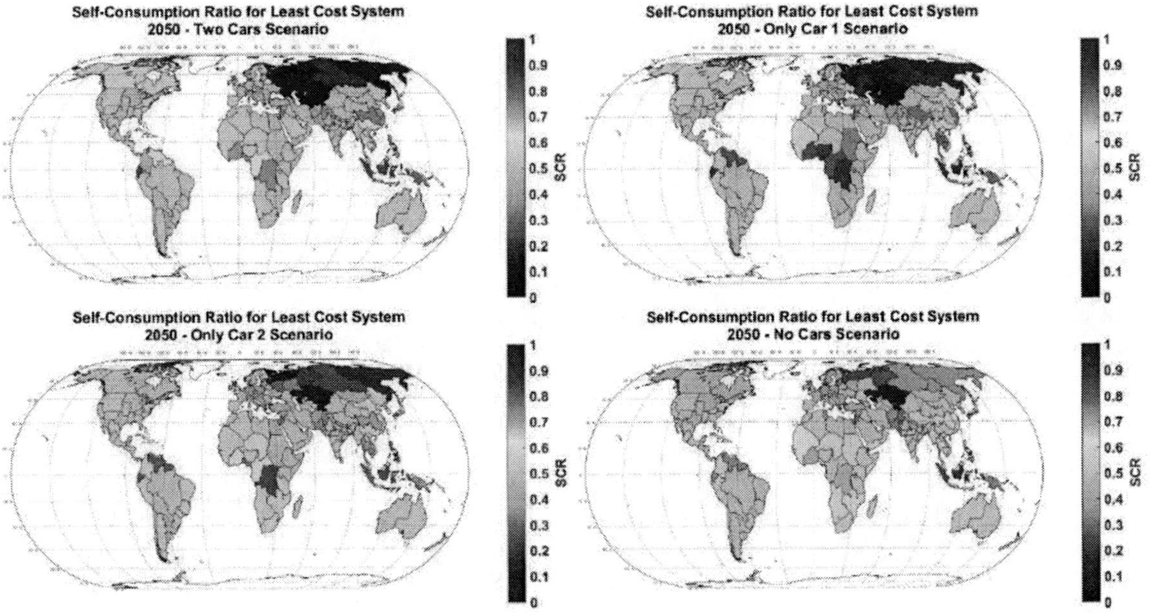

Figure 9: *Comparison of self-consumption ratios for all regions in 2050. Displayed are 'Two Cars Scenario' (left) and 'No Cars Scenario' (right).*

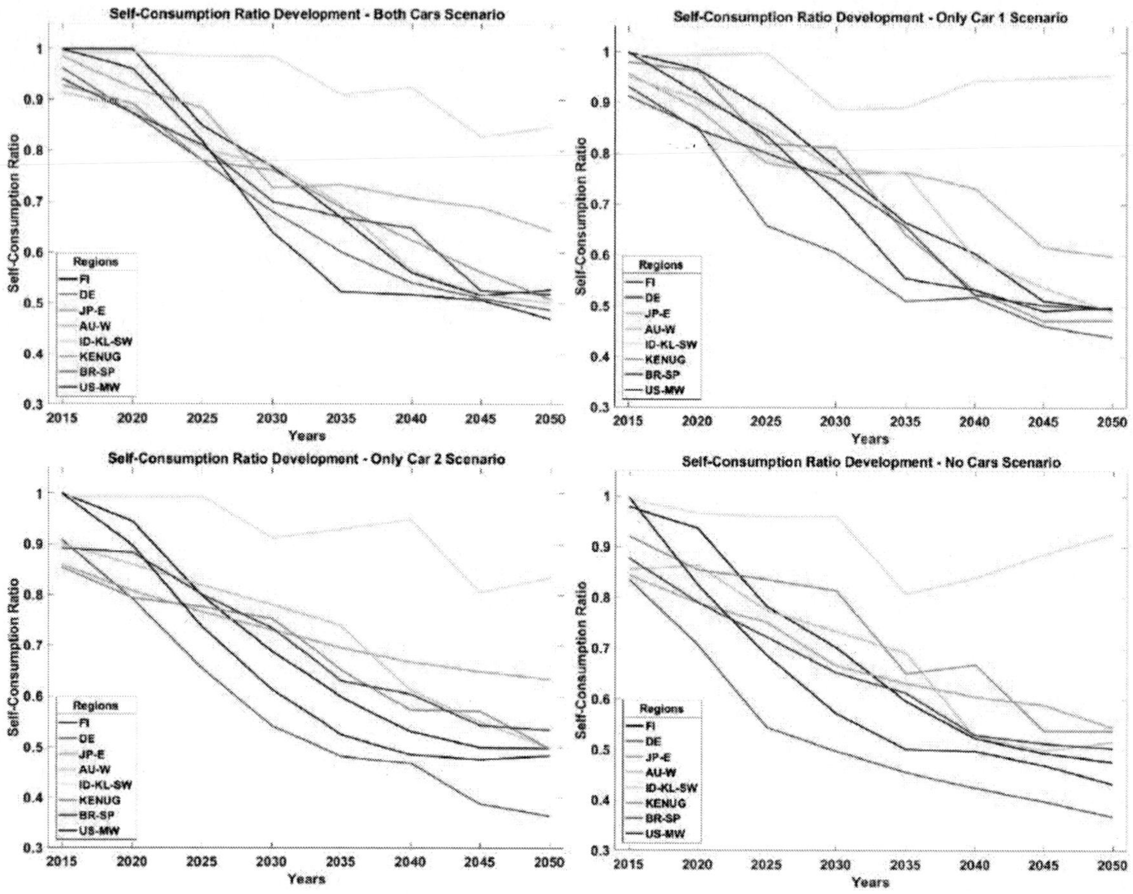

Figure 10: *Development of the Self-Consumption Ratio for the example regions for the 'Two Cars Scenario' (top left), the 'Only Car 1 Scenario' (top right), the 'Only Car 2 Scenario' (bottom left) and the 'No Cars Scenario' (bottom right).*

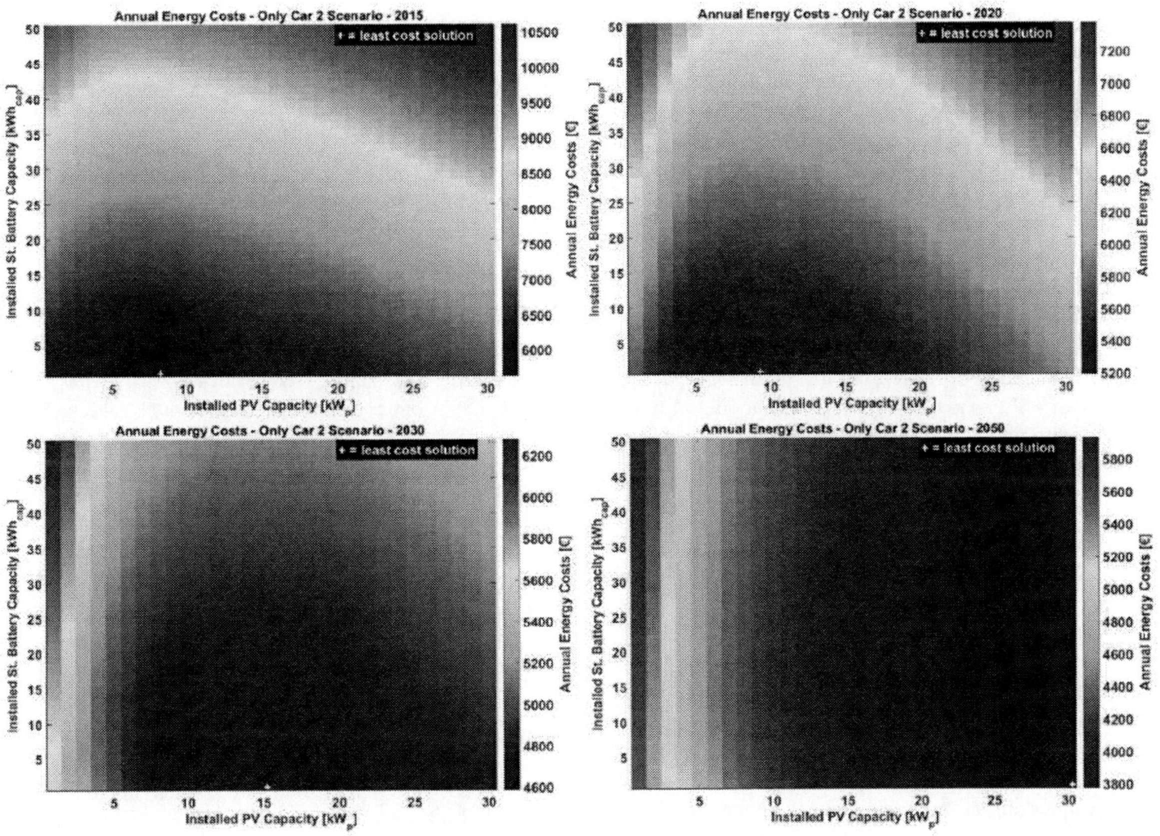

Figure 11: *Annual energy costs for Germany of the 'Only Car 2 Scenario' for 2015 (top left), 2020 (top right), 2030 (bottom left) and 2050 (bottom right).*

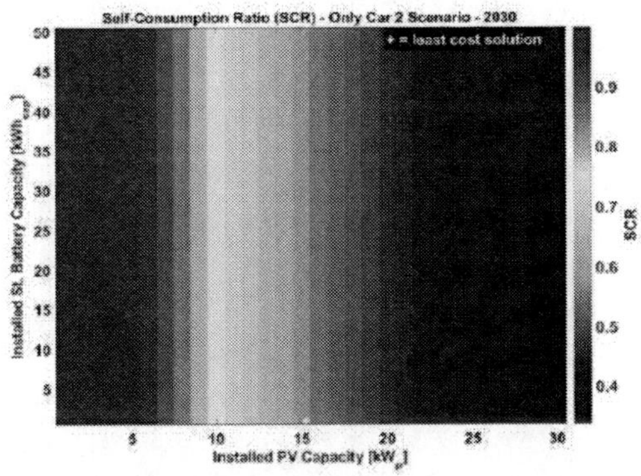

Figure 12: *Self-consumption ratio for Germany of the 'Only Car 2 Scenario' for 2030.*

3.3 Demand and Heat Cover Ratio

The differences for DCR and HCR are quite high among the scenarios. Regions with very good solar conditions will be able to cover theoretically up to 100% of their demands for electricity and heat in 2050 with self-produced PV electricity. Regions with balanced solar conditions (good solar conditions during summertime but moderate conditions in winter), such as the United States or Central Europe, are able to cover 90% - 95% of their electricity and 85% - 95% of their heat demand until 2050 on their own, among all scenarios. Regions with relatively low solar conditions and very high heat demands, such as Canada or Northern Europe, can cover 60% - 75% of their electricity demand and 60% - 70% of their heat demand. The best results are achieved for the 'Two Cars Scenario' and the worst for the 'No Cars Scenario'. Therefore, it can be said that widespread and affordable electric energy storage technologies are indispensable to effectively supply power and heat demands. Hereby, the course of the development through the years can be compared to a saturation with large changes in the early years and nearly stagnating developments from 2030 onwards for most regions. This can be explained by the fact that the added values for DCR and HCR of cost optimised systems are not very significant.

A real benefit for DCR can be noticed for scenarios without Car 2 at a range for small stationary batteries. For the majority of the regions, PV and stationary battery capacity increase have to coincide during the early years, resulting in significantly higher annual energy costs, whereas for later years when PV systems and batteries become more affordable and large PV capacities are part of the least cost system anyway, larger battery capacities can make the difference. For the example of Australia and the 'No Cars Scenario', the DCR for PV capacities from 7 kW_p – 30 kW_p are the same, but stationary batteries from 1 kWh_{cap} – 13 kWh_{cap} can increase the DCR up to $40\%_{abs}$ in 2050. For the case of Indonesia, this effect is even more significant. An increase from 1 kWh_{cap} to 4 kWh_{cap} causes a DCR improvement of $30\%_{abs}$. The course of HCR is quite similar to SCR but reversed in values. This means that a low SCR results in a high HCR. The explanation is the PV additional filling of the TES. As already mentioned, the SCR is mostly dependent on installed PV capacity. Very large PV capacities lead to a high availability of energy, with which the TES can be filled. The amount of energy which is necessary to fill the TES is comparably small to the amount of produced energy.

The differences of DCR and HCR over the regions and the impact of the BEVs are shown in Figure 13. It can be clearly seen that by using two cars, the DCR is slightly higher for most of the regions, compared to the 'No Cars Scenario'. The difference between the results for HCR of the 'Two Cars Scenario' and 'No Cars Scenario' is very clear for African and Asian regions. The lack of chargeable BEVs opens a great possibility for PV additional filling, which causes a large difference due to quite stable solar conditions throughout the whole year.

The range of least cost system designs differs a lot between regions in simulated scenarios. Figure 14 shows the least cost systems for the last years with one or more region or regions which do not reach grid-parity. The impact of Car 2 on the capacity of the stationary battery can be seen here as well. For the 'Two Cars Scenario', some regions which already have large PV capacities also have battery capacities up to 15 kWh_{cap} for their least cost system, except one region (Bahrain & Qatar). The majority of the regions have a resulting battery capacity of only 1 kWh_{cap} due to the availability of Car 2. The development for the 'Only Car 2 Scenario' is nearly the same, but with fewer regions with noticeable battery capacities (2050). The relevance of stationary batteries is much higher for the 'Only Car 1 Scenario' and 'No Cars Scenario'. Until 2050 the majority of the regions will have capacities greater than 1 kWh_{cap} for a least cost system design. For the two scenarios ('Only Car 1 Scenario' and 'No Cars Scenario') a dependence of battery capacities on PV capacities has to be expected. For high PV capacities, relatively high battery capacities will be necessary. Furthermore, the use of Car 1 with the possibility of energy supply via the stationary battery will have an impact on the least cost system design by increasing appropriate capacities. The least cost system designs for 2015, 2020, 2030 and 2050 can be found in the appendix (Figure A2-5).

3.4 Least Cost System Constructions

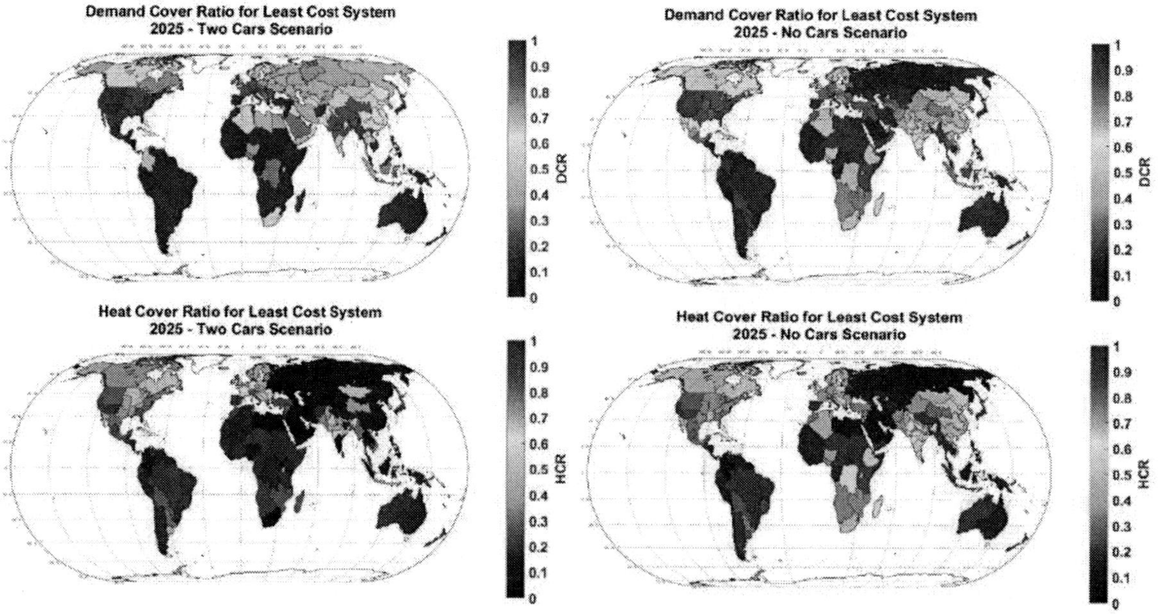

Figure 13: *Demand cover ratio for all regions in 2025 of the 'Two Cars Scenario' (top left) and 'No Cars Scenario' (top right) and heat cover ratio for all regions in 2025 of the 'Two Cars Scenario' (bottom left) and 'No Cars Scenario' (bottom right).*

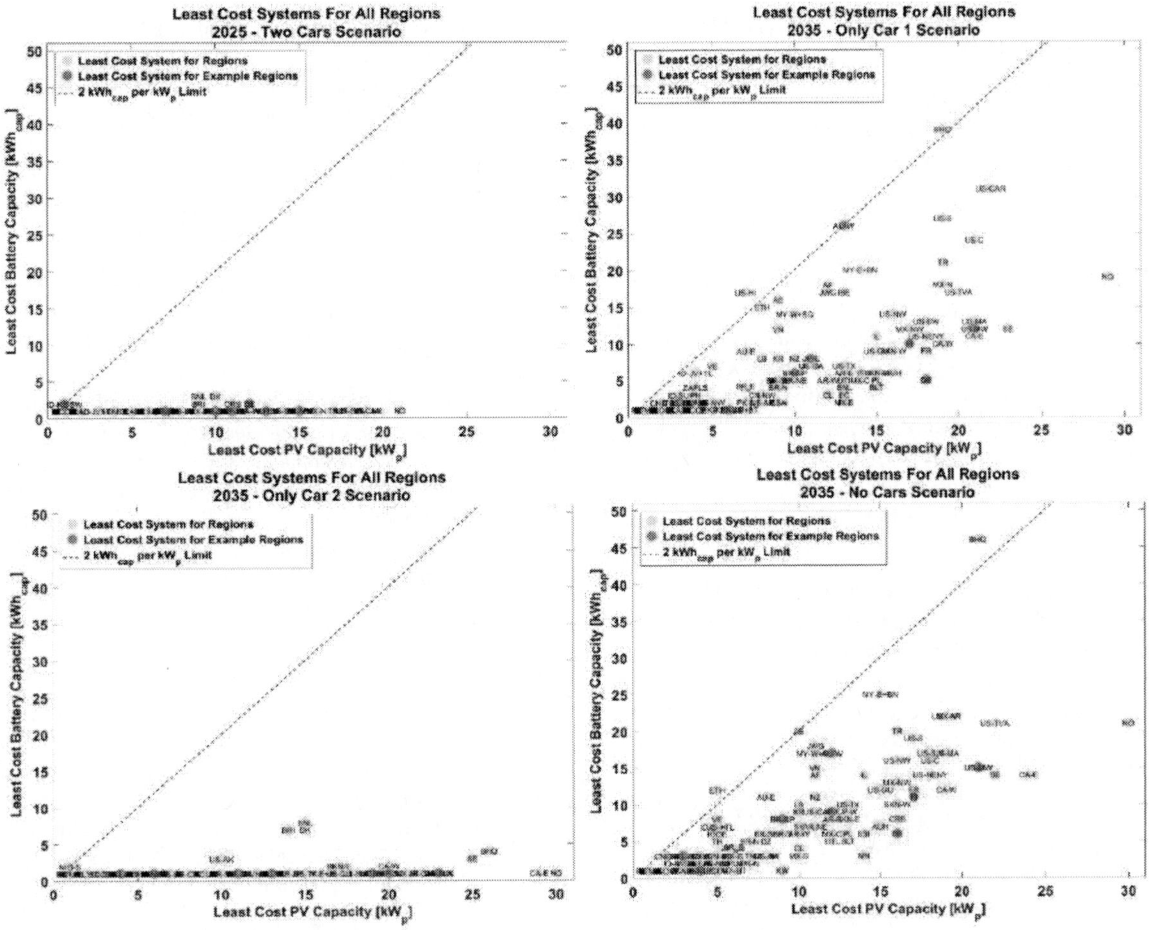

Figure 14: *Least cost system designs for the 'Two Cars Scenario' (2025 - top left), 'Only Car 1 Scenario' (2035 – top right), 'Only Car 2 Scenario' (2035 – bottom left) and 'No Cars Scenario' (2050 – bottom right). The example regions are marked green. Additionally, the 2 kWh_{cap}/kW_p line is marked as a suspected cost efficiency limit.*

For all scenarios the development of least cost systems seems to have a cost efficiency limit of a battery / PV capacity ratio of 2 kWh_{cap}/kW_p (Figure 14). With a few exceptions, most of the systems are below this limit. Through all scenarios it can be stated that very high battery capacities may not occur in residential energy systems for most regions. When not using a V2H vehicle, small and middle sized capacities will be most efficient. With such kinds of BEVs, only small battery capacities will be necessary or even superfluous. The PV capacities will vary through all sizes, depending on solar conditions and energy demand for electricity and heat.

Therefore, the optimization of energy system components is very beneficial for all regions, but can be broadened further, e.g. by including seasonal storage technologies in an intelligent residential energy system. The idea of this solution is to convert surplus electricity during summer times into hydrogen and store it. In winter times the hydrogen can be re-converted via fuel cells back into electricity and heat. Therefore, it would be able to increase the SCR and HCR at the same time. Residential seasonal storage solutions are already available [17].

4 CONCLUSION

The combination of PV systems with several energy storage technologies for electricity and heat will be highly beneficial for PV prosumers, particularly for declining annual energy cost. An absolute optimisation of annual energy cost and SCR at the same time is nearly impossible, so it is necessary to find a compromise. Although the sensitivity of energy costs is not high for most of the cases, a compromise is easy to find, at least for later years where the PV capacity does not play a big role in the whole system costs. A lot of energy fed into the grid could cause problems as already mentioned in [18]. Although, this may cause a very substantial potential for vehicle-to-home (V2H) services, which could improve the SCR.

Substantial PV prosumer activities may result in challenges for distribution grid management, but it can also be a source for least cost electricity supply for society. Overall, it can be said that there is not one least cost system design for all regions, and an optimisation of the system is very beneficial for all evaluated values. For some regions it may be practical to find an appropriate size for the TES and HP as well. For regions with moderate and low solar conditions, a TES of 800 litres and a HP of 7 kW_{el} seem to be good choices. Regions with good solar conditions and warm temperatures may not need such big system components to cover heat energy, as they mostly have to cover hot water demands and less space heating.

Including BEVs in a residential energy system is also very practical. Most of the energy for the driving demand can be covered by rooftop PV soon. Having an increasing cost for fuel in view, from a financial point of view the supply of driving demands via PV makes the most sense. Even an application of V2H will support a SCR and total energy cost optimisation, although the usage of a stationary battery and a V2H vehicle at the same time has to be rethought.

Compared to a 100% grid supply, an optimised residential energy system will be able to reduce the annual cost for energy up to $50\%_{abs}$, which can make a difference of several thousand euros per year and household, even when it could be more optimized by considering seasonal storage. Summarising the results, residential rooftop PV systems will be a very cost efficient possibility of delivering energy to households and for self-sufficiency of energy.

5 REFERENCES

[1] Gerlach A., Werner Ch., Breyer Ch., "Impact of financing costs on global grid-parity dynamics till 2030", 29th European Photovoltaic Solar Energy Conference, 2014.

[2] Breyer Ch., Gerlach A., "Global Overview on Grid-Parity", Progress in Photovoltaics: Research and Applications, 21, pp. 121-136, 2013.

[3] Werner Ch., Breyer Ch., Gerlach A., Beckel O., "Photovoltaic with energy storage: An overview on economics, system design and politics", 27th European Photovoltaic Solar Energy Conference, Franfurt 2012.

[4] International Energy Agency (IEA), "Global EV Outlook 2017", June 2017.

[5] Stadler I., Sterner M., "Energiespeicher – Bedarf, Technologien, Integration",Springer-Verlag GmbH Deutschland, ISBN 978-3-662-48892-8, 2nd edition, 2017

[6] International Renewable Energy Agency (IRENA), "Heat Pumps – Technology Brief", Energy Technology Systems Analysis Programme (ETSAP), January 2013

[7] Tjaden T., Krien U., Breyer Ch., "Simulation und techno-ökonomischer Vergleich von solarthermischen Heizungskonzepten und Photovoltaik-Wärmepumpen-Kombinationen im Wohnungssekor", 23. Symposium Thermische Solarenergie, Bad Staffelstein, April 2013.

[8] Bodganov D., Breyer Ch., "North-East Asian Super Grid for 100% renewable energy supply: Optimal mix of energy technologies for electricity, gas and heat supply options", Energy Conversion an Management 112 (2016) 176-190, January 2016.

[9] Breyer Ch., Bogdanov D., Gulagi A., Aghahosseini A., Barbosa L.S.N.S., Koskinen O., Barasa M., Caldera U., Afanasyeva S., Child M., Farfan J., Vainikka P., "On the role of solar photovoltaics in global energy transition scenarios", Progress in Photovoltaics: Research and Application 2017; 25:727–745, July 2017.

[10] Breyer Ch., Bogdanov D., Aghahosseini A., Gulagi A., Child M., Oyewo A.S., Farfan J., Sadovskaia K., Vainikka P., 2017. Solar Photovoltaics Demand for the Global Energy Transition in the Power Sector, Progress in Photovoltaics: Research and Applications, forthcoming

[11] United Nations Statistics Division (UNSD), "Occupied housing units by type of housing unit and urban/rural location", New York, 2017 www.data.un.org/Data.aspx?q=housing+units&d=POP&f=tableCode%3a278

[12] Toktarova A., Gruber L., Hlusiak M., Breyer Ch., 2017. Long-term load forecasting in high resolution for all countries globally, submitted

[13] Marwitz S., "Modellierung des Fahr- und Ladeverhaltens von Elektrofahrzeugen", Master Thesis, Karlsruhe Institute of Technology (KIT), September 2012.

[14] DIN Deutsches Institut für Normung e.V., "Specification for installations inside buildings conveying water for human consumption – Part 2: Design", German version EN 806-2:2005, Berlin, March 2005.

[15] DIN Deutsches Institut für Normung e.V., "Protection against pollution of potable water installations and general requirements of devices to prevent pollution by backflow", German version EN 1717:2000, Berlin, November 2000.

[16] Viessmann Werke GmbH & Co. KG, "Vitocell 100-B, Typ CVBB (750/950 l)", Data Sheet, Allendorf, Germany, 2017,

www.viessmann.de/de/wohngebaeude/warmwasserbereiter/warmwasserspeicher-bivalent/vitocell-100-b.html#

[17] HPS Home Power Solutions GmbH, Picea System, Berlin, Germany, 2017, www.homepowersolutions.de/en

[18] Hollinger R., Wille-Haussmann B., Erge T., Sönnichsen J., Stillahn T., Kreifels N., "Speicherstudie 2013", Fraunhofer ISE, Freiburg, 2013.

[19] Barbosa L.S.N.S and Breyer Ch., 2017. Heat demand in national-global resolution for the energy transition till 2050, internal working paper, Lappeenranta University of Technology

APPENDIX

Table A1: *Explanation of the modelled regions with abbreviations and including countries or splitted sub-regions.*

Region Number	Abbreviation	Including countries or splitted sub-regions
1	NO	Norway
2	DK	Denmark
3	SE	Sweden
4	FI	Finland
5	BLT	Baltic: Estonia, Latvia, Lithuania
6	PL	Poland
7	IBE	Iberia: Portugal, Spain, Gibraltar
8	FR	France, Monaco, Andorra
9	BNL	Belgium, Netherlands, Luxembourg
10	BRI	British Isles: Ireland, United Kingdom, Isle of Man, Guernsey, Jersey
11	DE	Germany
12	CRS	Czech Republic, Slovakia
13	AUH	Austria, Hungary
14	BKN-W	Balkan-West: Slovenia, Croatia, Bosnia & Hertzegovina, Serbia, Kosova, Montenegro, Macedonia, Albania
15	BKN-E	Balkan-East: Romania, Bulgaria, Greece
16	IT	Italy, San Marino, Vatican
17	CH	Switzerland, Liechtenstein
18	TR	Turkey, Cyprus
19	UA	Ukraine, Moldova
20	IS	Iceland
21	DZ	Algeria
22	BHQ	Bahrain and Qatar
23	EG	Egypt
24	IR	Iran
25	IQ	Iraq
26	IL	Israel
27	JWG	Jordan (incl. West Bank & Gaza Strip = State of Palestine)
28	KW	Kuwait
29	LB	Lebanon
30	LY	Libya
31	MA	Morocco
32	OM	Oman
33	SA	Saudi Arabia
34	TN	Tunisia
35	AE	United Arab Emirates
36	YE	Yemen
37	SY	Syria
38	RU-NW	Russia (Northwest)
39	RU-C	Russia (Center)
40	RU-S	Russia (South)
41	RU-V	Russia (Volga region)
42	RU-U	Russia (Urals)
43	RU-SI	Russia (Siberia)
44	RU-FE	Russia (Far East)
45	BY	Belarus
46	CAU	Armenia, Azerbaijan, Georgia
47	KZ	Kazakhstan
48	PAM	Tadjicistan, Kyrgizstan
49	ZU	Uzbekistan
50	TM	Turkmenistan
51	JP-E	Japan East
52	JP-W	Japan West
53	KR	South Korea
54	KP	North Korea
55	CN-NE	China (Northaast)
56	CN-N	China (North)
57	CN-E	China (East)
58	CN-C	China (Central)
59	CN-S	China (South)
60	CN-TB	China (Tibet)
61	CN-NW	China (Northwest)
62	CN-XU	China (Uigur region)
63	MN	Mongolia
64	NZ	New Zealand
65	AU-E	Australia East
66	AU-W	Australia West
67	ID-PG+NG	Indonesia-Papua + Papua New Guinea
68	ID-SU	Indonesia Sumatra
69	ID-JV+TL	Indonesia Java + Timor Leste
70	ID-KL-SW	Indonesia East
71	MY-W+SG	Malaysia West + Singapore
72	MY-E+BN	Malaysia East + Brunei
73	PH	Philippines
74	MM	Myanmar
75	TH	Thailand
76	LA	Laos
77	VN	Vietnam
78	KH	Cambodia
79	IN-E	India-East
80	IN-CE	India-Central East
81	IN-W	India-West
82	IN-CW	India-Central West
83	IN-N	India-North
84	IN-NW	India-North West
85	IN-UP	India-Uttar Pradesh
86	IN-S	India-South
87	IN-CS	India-South Central

88	IN-NE	India-Northeast
89	BD	Bangladesh
90	NP+BT	Nepal+Bhutan
91	PK-S	Pakistan-South
92	PK-N	Pakistan-North
93	AF	Afghanistan
94	LK	Srilanka
95	WW	Senegal,Gambia, Cape Verde Islands, Guinea Bissau, Guinea, Sierra Leone, Liberia, Mali, Mauritania, Western Sahara
96	WS	Ghana, Cote D'Ivoire, Benin, Burkina Faso (Upper Volta), Togo
97	WN	Niger, Chad
98	NIG-S	South Nigeria
99	NIG-N	North Nigeria
100	SER	Sudan, Eritrea
101	ETH	Ethiopia
102	SOMDJ	Djibouti, Somalia
103	KENUG	Kenya, Uganda
104	TZRB	Rwanda, Burundi, Tanzania
105	CAR	Central African Republic, Cameroon, Equatorial Guinea, Sao Tome and Principe, Congo, Republic of (Brazzaville), Gabon
106	COG	Congo, Democratic Republic (Kinshasha)
107	SW	Angola, Namibia, Botswana
108	ZAFLS	Republic of South Africa, Lesotho
109	SEA	Malawi, Mozambique, Zambia, Zimbabwe, Swaziland
110	IOCE	Comoros Islands, Mauritius, Mayotte, Madagascar, Seychelles
111	CAM	Panama, Costa Rica, Nicaragua, Honduras, El Salvador, Guatemala and Belize
112	CO	Colombia
113	VE	Venezuela, Guyana, French Guiana, Suriname
114	EC	Ecuador
115	PU	Peru
116	CSA	Bolivia and Paraguay
117	BR-S	Brazil South
118	BR-SP	Brazil São Paulo
119	BR-SE	Brazil Southeast
120	BR-N	Brazil North
121	BR_NE	Brazil Northeast
122	AR-NE	Argentina Northeast (includes Uruguay)
123	AR-E	Argentina East
124	AR-W	Argentina West
125	CL	Chile
126	CA-W	Canada-West
127	CA-E	Canada-East
128	US-NENY	US-New England & New York
129	US-MA	US-Mid-Atlantics
130	US-CAR	US-Carolinas
131	US-S	US-Southern
132	US-TVA	US-TVA
133	US-MW	US-Midwest
134	US-C	US-Central
135	US-TX	US-Texas
136	US-SW	US-Southwest
137	US-NW	US-Northwest
138	US-CA	US-California
139	US-AK	US-Alaska
140	US-HI	US-Hawaii
141	US-GU	US-Gulf
142	MX-NW	Mexico-Noroeste (Northwest)
143	MX-N	Mexico-Noreste (North)
144	MX-C	Mexico-Centro-Occidente (Center)
145	MX-S	Mexico-Centro & Sur-Sureste (South)

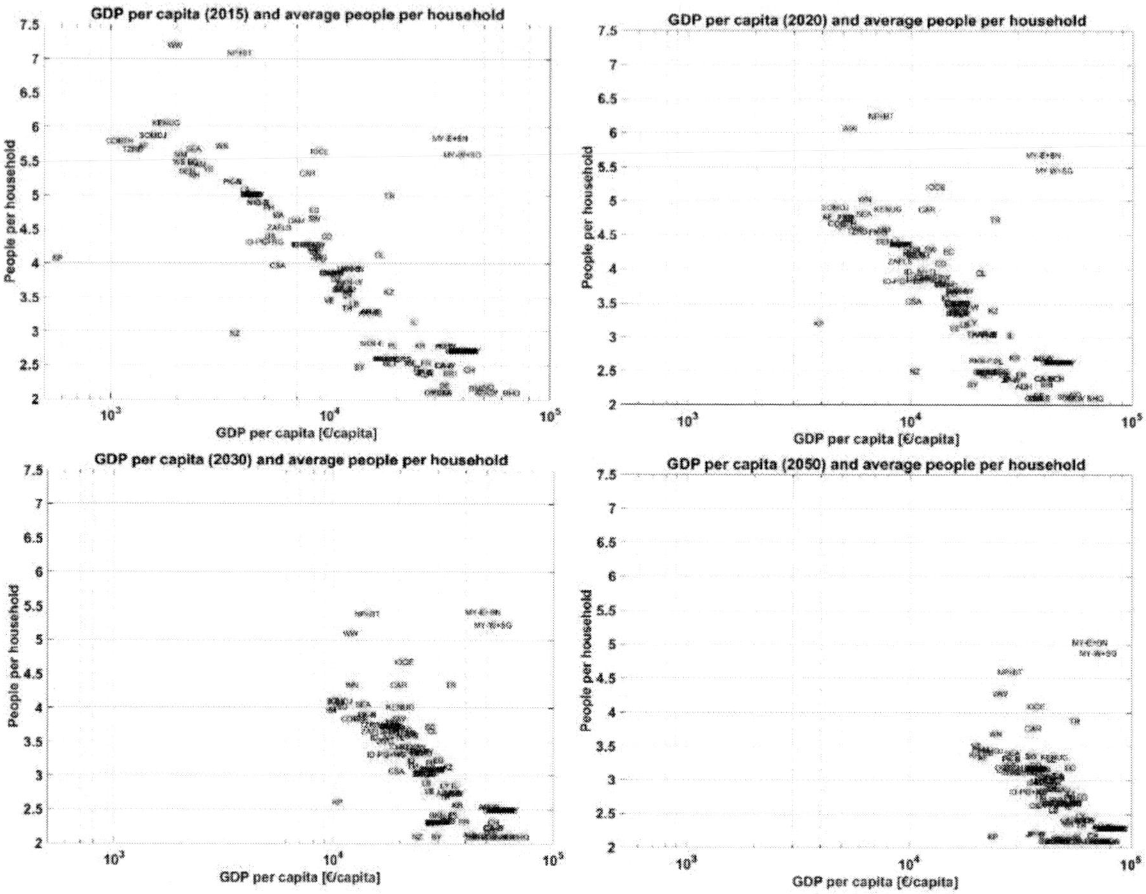

Figure A1: *Development for assumed GDP per capita and people per household for 2015 (top left), 2020 (top right), 2030 (bottom left) and 2050 (bottom right).*

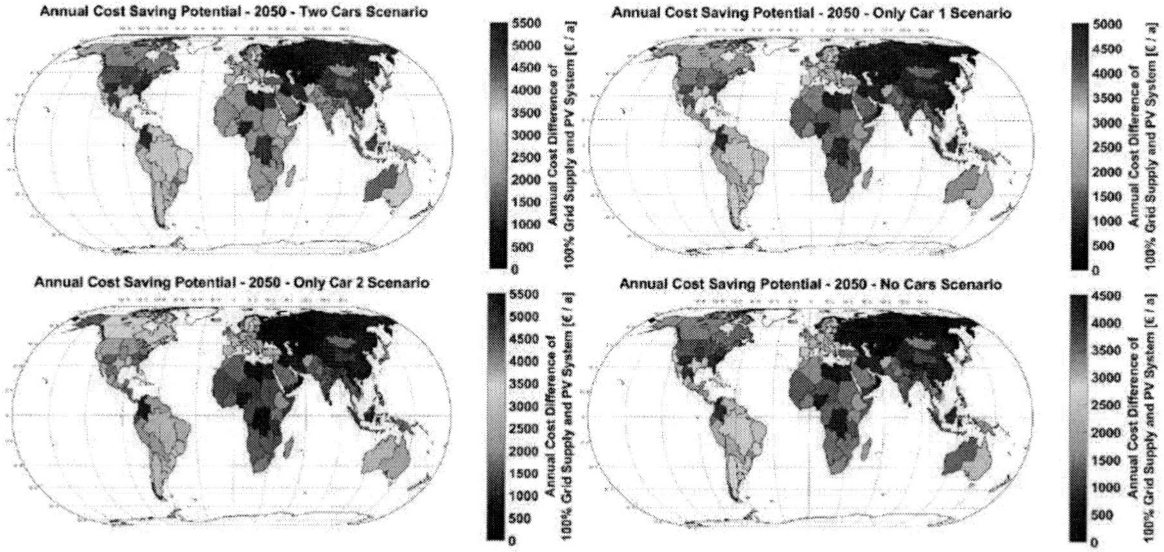

Figure A2: *Annual cost saving potential for all regions in 2050. Displayed are the potentials for 'Two Cars Scenario' (top left), 'Only Car 1 Scenario' (top right), 'Only Car 2 Scenario' (bottom left) and 'No Cars Scenario' (bottom right). Regions with no cost saving potential are marked grey.*

Table A2: *People per household for all regions and modelled years, structured to the 9 major regions Europe, Eurasia, MENA, Sub-Saharan Africa, India/ SAARC, Northeast Asia, Southeast Asia, North America and South America.*

Region	People per Household							
	2015	2020	2025	2030	2035	2040	2045	2050
				Europe				
NO	2.2	2.2	2.1	2.1	2.1	2.1	2.1	2.1
DK	2.1	2.1	2.1	2.1	2.1	2.1	2.1	2.1
SE	2.1	2.1	2.1	2.1	2.1	2.1	2.1	2.1
FI	2.1	2.1	2.1	2.1	2.1	2.1	2.1	2.1
BLT	2.5	2.4	2.4	2.3	2.2	2.2	2.1	2.1
PL	2.8	2.6	2.5	2.4	2.3	2.3	2.2	2.2
IBE	2.5	2.5	2.4	2.3	2.3	2.2	2.2	2.1
FR	2.5	2.5	2.4	2.3	2.3	2.2	2.2	2.1
BNL	2.2	2.1	2.1	2.1	2.1	2.1	2.1	2.1
BRI	2.4	2.3	2.2	2.2	2.1	2.1	2.1	2.1
DE	2.2	2.1	2.1	2.1	2.1	2.1	2.1	2.1
CRS	2.6	2.5	2.4	2.3	2.3	2.2	2.2	2.1
AUH	2.4	2.3	2.2	2.1	2.1	2.1	2.1	2.1
BKN-W	3.7	3.5	3.3	3.1	3.0	2.9	2.8	2.7
BKN-E	2.8	2.7	2.5	2.4	2.3	2.3	2.2	2.1
IT	2.4	2.4	2.4	2.3	2.3	2.2	2.2	2.2
CH	2.4	2.4	2.4	2.3	2.3	2.2	2.2	2.2
TR	5.0	4.7	4.5	4.3	4.2	4.1	4.0	3.9
UA	4.4	3.9	3.6	3.3	3.2	3.0	2.9	2.9
IS	2.1	2.1	2.1	2.1	2.1	2.1	2.1	2.1
				Eurasia				
RU-NW	2.6	2.5	2.4	2.3	2.2	2.2	2.1	2.1
RU-C	2.6	2.5	2.4	2.3	2.2	2.2	2.1	2.1
RU-S	2.6	2.5	2.4	2.3	2.2	2.2	2.1	2.1
RU-V	2.6	2.5	2.4	2.3	2.2	2.2	2.1	2.1
RU-U	2.6	2.5	2.4	2.3	2.2	2.2	2.1	2.1
RU-SI	2.6	2.5	2.4	2.3	2.2	2.2	2.1	2.1
RU-FE	2.6	2.5	2.4	2.3	2.2	2.2	2.1	2.1
BY	2.5	2.3	2.2	2.1	2.1	2.1	2.1	2.1
CAU	4.2	3.8	3.5	3.3	3.2	3.1	3.0	2.9
KZ	3.6	3.4	3.3	3.1	3.0	2.9	2.9	2.8
PAM	5.5	4.8	4.4	4.1	3.9	3.7	3.6	3.5
ZU	5.0	4.3	3.9	3.6	3.4	3.3	3.2	3.1
TM	3.7	3.5	3.3	3.1	3.0	2.9	2.8	2.8
				MENA				
DZ	3.8	3.4	3.2	3.0	2.9	2.8	2.7	2.6
BHQ	2.1	2.1	2.1	2.1	2.1	2.1	2.1	2.1
EG	4.3	3.8	3.5	3.2	3.1	3.0	2.8	2.8
IR	3.6	3.3	3.2	3.0	2.9	2.8	2.8	2.7
IQ	3.9	3.6	3.4	3.2	3.1	3.0	2.9	2.8
IL	3.1	3.0	2.9	2.9	2.8	2.7	2.7	2.6
JWG	4.1	3.9	3.7	3.5	3.4	3.3	3.2	3.1
KW	2.1	2.1	2.1	2.1	2.1	2.1	2.1	2.1
LB	3.4	3.2	3.0	2.9	2.8	2.7	2.6	2.6
LY	3.5	3.2	3.0	2.9	2.7	2.6	2.5	2.5
MA	4.7	4.2	3.9	3.7	3.5	3.3	3.2	3.1
OM	2.1	2.1	2.1	2.1	2.1	2.1	2.1	2.1
SA	2.1	2.1	2.1	2.1	2.1	2.1	2.1	2.1
TN	4.2	3.8	3.5	3.3	3.2	3.1	3.0	2.9
AE	2.1	2.1	2.1	2.1	2.1	2.1	2.1	2.1
YE	5.4	4.8	4.4	4.1	3.9	3.8	3.6	3.5
SY	5.0	4.4	4.0	3.7	3.5	3.4	3.3	3.2

				Sub-Saharan Africa				
WW	7.2	6.1	5.5	5.1	4.8	4.6	4.4	4.3
WS	5.5	4.6	4.1	3.9	3.6	3.5	3.3	3.2
WN	5.7	5.0	4.6	4.3	4.1	4.0	3.8	3.7
NIG-S	4.9	4.2	3.9	3.6	3.4	3.3	3.1	3.0
NIG-N	4.9	4.2	3.9	3.6	3.4	3.3	3.1	3.0
SER	5.4	4.4	4.0	3.7	3.5	3.3	3.2	3.1
ETH	5.8	4.7	4.2	3.8	3.6	3.4	3.3	3.2
SOMDJ	5.9	4.9	4.4	4.1	3.9	3.7	3.6	3.4
KENUG	6.1	4.9	4.4	4.0	3.8	3.6	3.5	3.3
TZRB	5.7	4.6	4.1	3.8	3.5	3.4	3.2	3.1
CAR	5.3	4.9	4.6	4.3	4.2	4.0	3.9	3.8
COG	5.8	4.7	4.2	3.8	3.6	3.4	3.3	3.2
SW	4.7	4.3	4.0	3.8	3.7	3.6	3.5	3.4
ZAFLS	4.5	4.1	3.9	3.7	3.5	3.4	3.3	3.2
SEA	5.7	4.8	4.4	4.1	3.8	3.7	3.5	3.4
IOCE	5.7	5.2	4.9	4.7	4.5	4.3	4.2	4.1
				India/ SAARC				
IN-E	5.0	4.4	4.0	3.7	3.6	3.4	3.3	3.2
IN-CE	5.0	4.4	4.0	3.7	3.6	3.4	3.3	3.2
IN-W	5.0	4.4	4.0	3.7	3.6	3.4	3.3	3.2
IN-CW	5.0	4.4	4.0	3.7	3.6	3.4	3.3	3.2
IN-N	5.0	4.4	4.0	3.7	3.6	3.4	3.3	3.2
IN-NW	5.0	4.4	4.0	3.7	3.6	3.4	3.3	3.2
IN-UP	5.0	4.4	4.0	3.7	3.6	3.4	3.3	3.2
IN-S	5.0	4.4	4.0	3.7	3.6	3.4	3.3	3.2
IN-CS	5.0	4.4	4.0	3.7	3.6	3.4	3.3	3.2
IN-NE	5.0	4.4	4.0	3.7	3.6	3.4	3.3	3.2
BD	5.5	4.7	4.3	4.0	3.8	3.7	3.5	3.4
NP+BT	7.1	6.2	5.7	5.4	5.1	4.9	4.7	4.6
PK-S	5.2	4.5	4.2	3.9	3.7	3.6	3.4	3.3
PK-N	5.2	4.5	4.2	3.9	3.7	3.6	3.4	3.3
AF	5.7	4.8	4.3	4.0	3.8	3.6	3.5	3.3
LK	4.2	3.8	3.5	3.3	3.2	3.1	3.0	2.9
				Northeast Asia				
JP-E	2.4	2.4	2.3	2.3	2.3	2.3	2.2	2.2
JP-W	2.4	2.4	2.3	2.3	2.3	2.3	2.2	2.2
KR	2.8	2.7	2.6	2.6	2.5	2.4	2.4	2.4
KP	4.1	3.2	2.8	2.6	2.5	2.3	2.2	2.2
CN-NE	3.9	3.5	3.3	3.1	3.0	2.8	2.8	2.7
CN-N	3.9	3.5	3.3	3.1	3.0	2.8	2.8	2.7
CN-E	3.9	3.5	3.3	3.1	3.0	2.8	2.8	2.7
CN-C	3.9	3.5	3.3	3.1	3.0	2.8	2.8	2.7
CN-S	3.9	3.5	3.3	3.1	3.0	2.8	2.8	2.7
CN-TB	3.9	3.5	3.3	3.1	3.0	2.8	2.8	2.7
CN-NW	3.9	3.5	3.3	3.1	3.0	2.8	2.8	2.7
CN-XU	3.9	3.5	3.3	3.1	3.0	2.8	2.8	2.7
MN	4.2	3.8	3.5	3.3	3.2	3.1	3.0	2.9
				Southeast Asia				
NZ	3.0	2.5	2.3	2.1	2.1	2.1	2.1	2.1
AU-E	2.8	2.7	2.6	2.5	2.5	2.4	2.4	2.3
AU-W	2.8	2.7	2.6	2.5	2.5	2.4	2.4	2.3
ID-PG+NG	4.3	3.8	3.5	3.3	3.2	3.0	2.9	2.8
ID-SU	4.3	3.9	3.6	3.4	3.3	3.2	3.1	3.0
ID-JV+TL	4.3	4.0	3.7	3.6	3.4	3.3	3.2	3.1
ID-KL-SW	4.3	3.9	3.6	3.4	3.3	3.2	3.1	3.0
MY-W+SG	5.6	5.5	5.3	5.2	5.1	5.0	5.0	4.9
MY-E+BN	5.8	5.7	5.5	5.4	5.3	5.2	5.1	5.0

PH	4.8	4.3	3.9	3.7	3.5	3.4	3.2	3.1
MM	5.6	4.6	4.1	3.8	3.6	3.4	3.3	3.2
TH	3.3	3.0	2.8	2.7	2.6	2.5	2.4	2.3
LA	5.1	4.4	4.0	3.8	3.6	3.4	3.3	3.2
VN	5.0	4.3	3.9	3.6	3.4	3.3	3.2	3.1
KH	5.3	4.6	4.2	4.0	3.8	3.6	3.5	3.4
				North America				
CA-W	2.5	2.4	2.3	2.2	2.2	2.1	2.1	2.1
CA-E	2.5	2.4	2.3	2.2	2.2	2.1	2.1	2.1
US-NENY	2.7	2.6	2.6	2.5	2.4	2.4	2.3	2.3
US-MA	2.7	2.6	2.6	2.5	2.4	2.4	2.3	2.3
US-CAR	2.7	2.6	2.6	2.5	2.4	2.4	2.3	2.3
US-S	2.7	2.6	2.6	2.5	2.4	2.4	2.3	2.3
US-TVA	2.7	2.6	2.6	2.5	2.4	2.4	2.3	2.3
US-MW	2.7	2.6	2.6	2.5	2.4	2.4	2.3	2.3
US-C	2.7	2.6	2.6	2.5	2.4	2.4	2.3	2.3
US-TX	2.7	2.6	2.6	2.5	2.4	2.4	2.3	2.3
US-SW	2.7	2.6	2.6	2.5	2.4	2.4	2.3	2.3
US-NW	2.7	2.6	2.6	2.5	2.4	2.4	2.3	2.3
US-CA	2.7	2.6	2.6	2.5	2.4	2.4	2.3	2.3
US-AK	2.7	2.6	2.6	2.5	2.4	2.4	2.3	2.3
US-HI	2.7	2.6	2.6	2.5	2.4	2.4	2.3	2.3
US-GU	2.7	2.6	2.6	2.5	2.4	2.4	2.3	2.3
MX-NW	3.9	3.7	3.5	3.4	3.3	3.2	3.1	3.0
MX-N	3.9	3.7	3.5	3.4	3.3	3.2	3.1	3.0
MX-C	3.9	3.7	3.5	3.4	3.3	3.2	3.1	3.0
MX-S	3.9	3.7	3.5	3.4	3.3	3.2	3.1	3.0
				South America				
CAM	4.6	4.2	3.9	3.7	3.5	3.4	3.3	3.2
CO	4.4	4.1	3.9	3.7	3.5	3.4	3.3	3.2
VE	3.5	3.1	2.9	2.8	2.7	2.6	2.5	2.4
EC	4.8	4.3	3.9	3.7	3.5	3.4	3.3	3.2
PU	4.1	3.7	3.4	3.2	3.0	2.9	2.8	2.7
CSA	4.0	3.5	3.3	3.1	2.9	2.8	2.7	2.6
BR-S	3.6	3.4	3.2	3.0	2.9	2.8	2.7	2.7
BR-SP	3.6	3.4	3.2	3.0	2.9	2.8	2.7	2.7
BR-SE	3.6	3.4	3.2	3.0	2.9	2.8	2.7	2.7
BR-N	3.6	3.4	3.2	3.0	2.9	2.8	2.7	2.7
BR_NE	3.6	3.4	3.2	3.0	2.9	2.8	2.7	2.7
AR-NE	3.3	3.0	2.9	2.7	2.6	2.6	2.5	2.4
AR-E	3.3	3.1	2.9	2.7	2.6	2.5	2.5	2.4
AR-W	3.3	3.1	2.9	2.7	2.6	2.5	2.5	2.4
CL	4.1	3.9	3.8	3.7	3.6	3.5	3.4	3.3

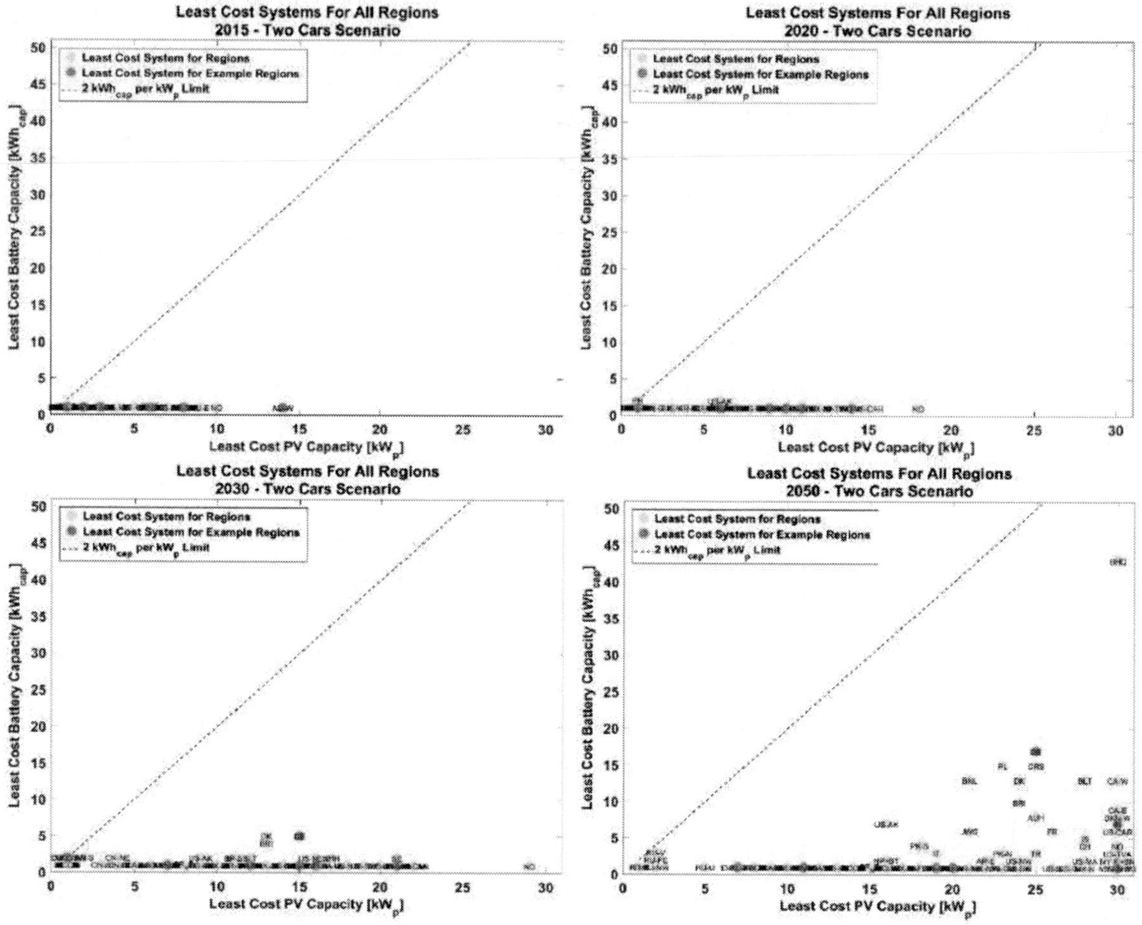

Figure A2: *Least cost system design for the 'Two Cars Scenario' in 2015 (top left), 2020 (top right), 2030 (bottom left) and 2050 (bottom right).*

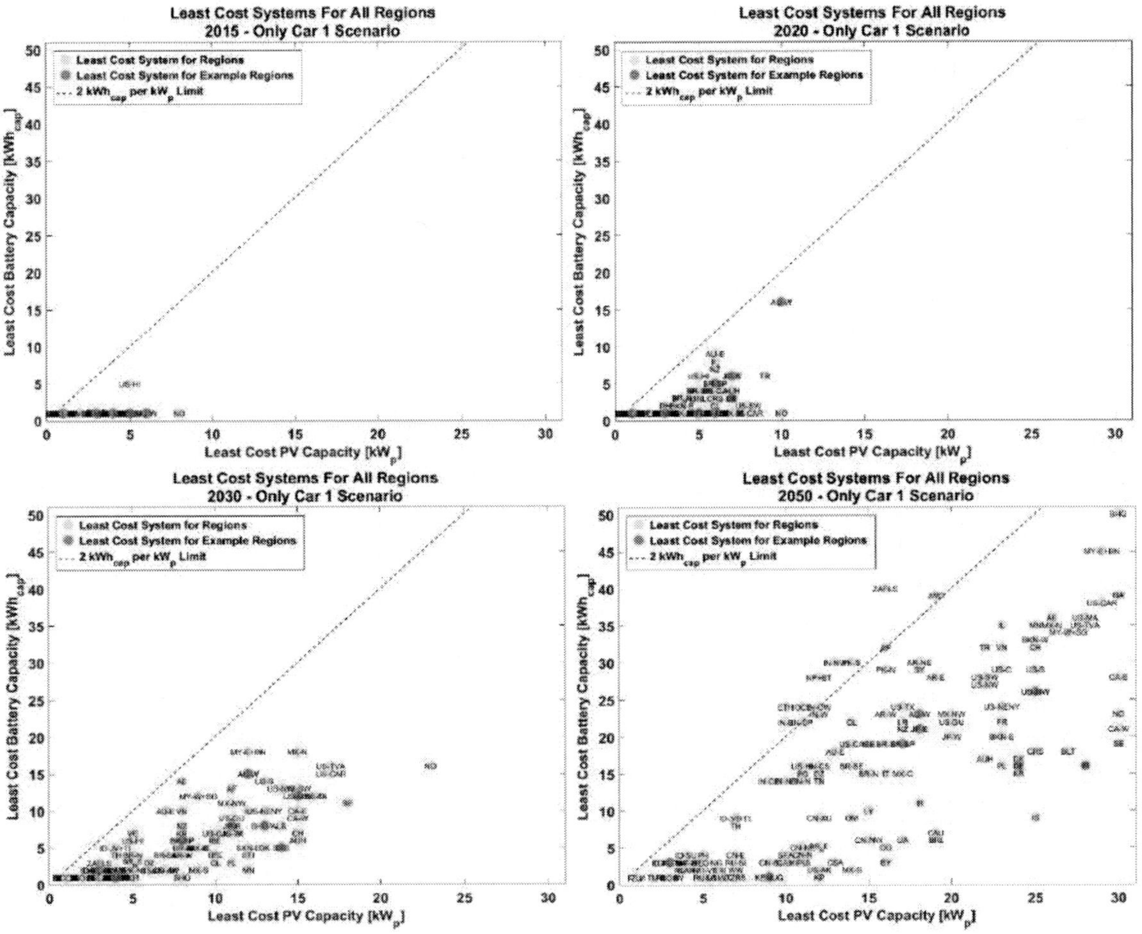

Figure A3: *Least cost system design for the 'Only Car 1 Scenario' in 2015 (top left), 2020 (top right), 2030 (bottom left) and 2050 (bottom right).*

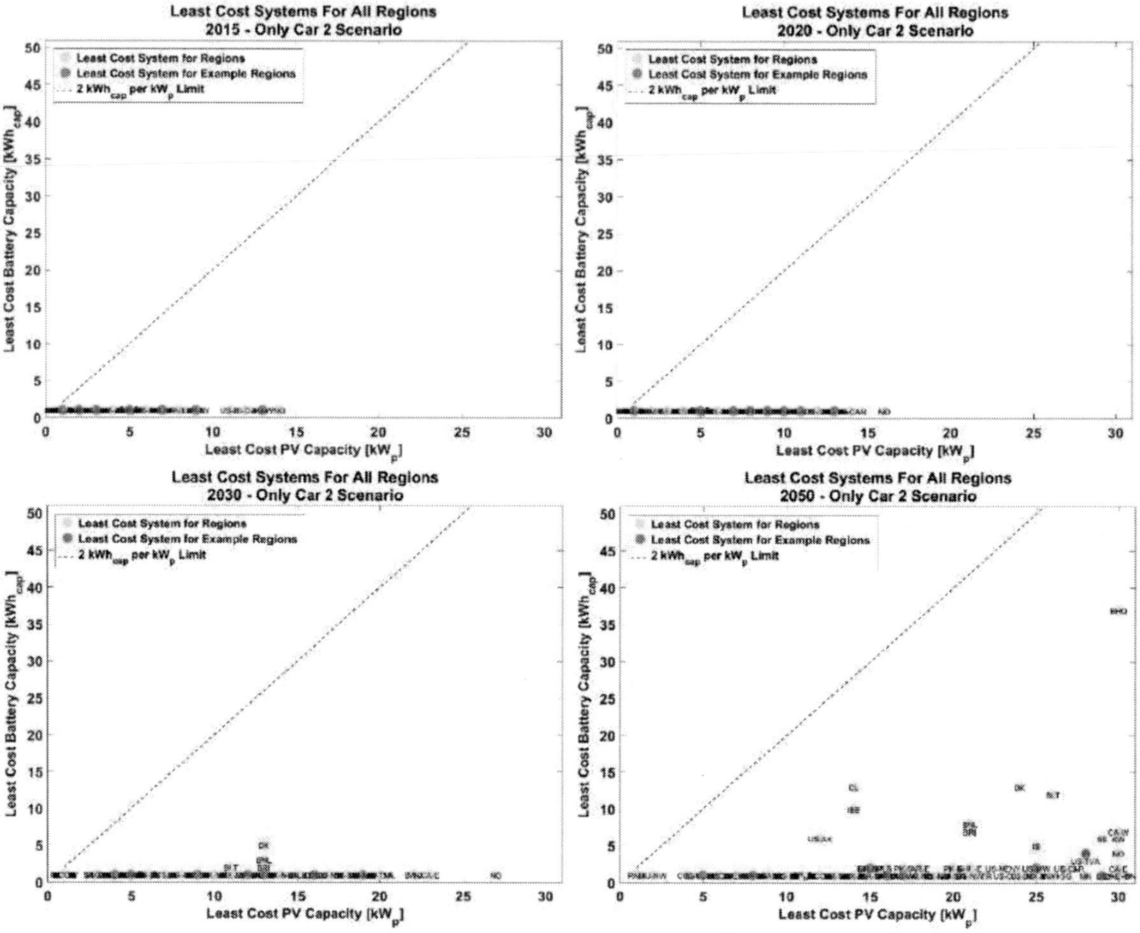

Figure A4: *Least cost system design for the 'Only Car 2 Scenario' in 2015 (top left), 2020 (top right), 2030 (bottom left) and 2050 (bottom right).*

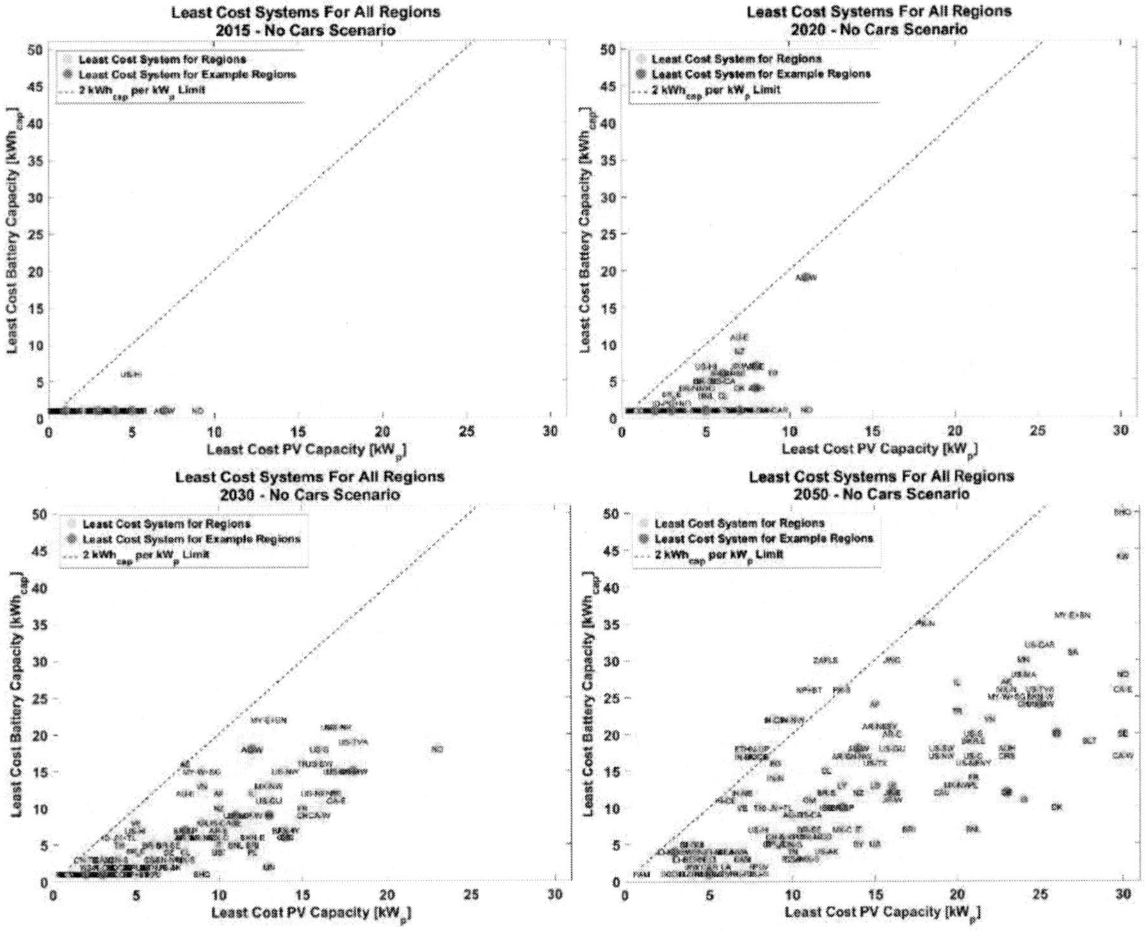

Figure A5: *Least cost system design for the 'No Cars Scenario' in 2015 (top left), 2020 (top right), 2030 (bottom left) and 2050 (bottom right).*

Table A3: *Annual electricity consumption per capita for all regions and modelled year, structured to the 9 major regions Europe, Eurasia, MENA, Sub-Saharan Africa, India/ SAARC, Northeast Asia, Southeast Asia, North America and South America.*

Region	Annual Electricity Consumption per Capita [MWh$_{el}$/a]							
	2015	2020	2025	2030	2035	2040	2045	2050
				Europe				
NO	8.24	8.27	8.39	8.53	8.73	8.97	9.24	9.53
DK	2.13	1.96	1.91	1.92	2.01	2.11	2.28	2.41
SE	5.02	5.13	5.08	5.01	5.19	5.20	5.36	5.46
FI	3.95	3.90	4.38	4.76	4.93	5.15	5.48	5.74
BLT	1.37	1.39	1.52	1.79	1.85	2.02	2.11	2.25
PL	1.00	1.13	1.19	1.24	1.31	1.41	1.55	1.64
IBE	2.07	2.13	2.26	2.37	2.47	2.55	2.67	2.76
FR	3.36	3.26	3.29	3.35	3.37	3.42	3.58	3.61
BNL	1.89	1.83	1.71	1.70	1.73	1.83	1.89	1.94
BRI	2.04	1.87	1.86	1.88	1.92	2.03	2.10	2.10
DE	1.86	1.83	1.82	1.87	1.90	1.97	2.02	2.12
CRS	1.63	1.69	1.76	1.87	2.07	2.20	2.33	2.44
AUH	1.71	1.71	1.82	1.92	1.97	2.08	2.23	2.35

BKN-W	1.82	1.90	1.97	2.07	2.18	2.37	2.52	2.72
BKN-E	1.53	1.65	1.72	1.86	2.01	2.19	2.39	2.58
IT	1.29	1.29	1.34	1.40	1.48	1.60	1.76	1.82
CH	2.62	2.67	2.73	2.79	2.88	2.98	3.10	3.23
TR	0.98	1.28	1.35	1.41	1.49	1.58	1.65	1.73
UA	0.80	0.87	0.94	1.01	1.10	1.20	1.32	1.45
IS	2.98	3.04	3.11	3.21	3.33	3.46	3.63	3.81
				Eurasia				
RU-NW	1.05	1.12	1.20	1.29	1.40	1.51	1.63	1.77
RU-C	0.71	0.75	0.81	0.87	0.94	1.02	1.10	1.19
RU-S	0.48	0.51	0.55	0.59	0.64	0.69	0.74	0.80
RU-V	0.87	0.92	0.99	1.07	1.15	1.25	1.35	1.45
RU-U	2.02	2.15	2.30	2.48	2.69	2.90	3.13	3.39
RU-SI	1.56	1.66	1.78	1.92	2.08	2.24	2.42	2.61
RU-FE	0.95	1.01	1.08	1.17	1.27	1.37	1.48	1.59
BY	0.75	0.80	0.86	0.94	1.02	1.11	1.22	1.34
CAU	1.59	1.61	1.65	1.73	1.82	1.93	2.07	2.23
KZ	0.58	0.58	0.59	0.61	0.64	0.66	0.69	0.72
PAM	0.54	0.51	0.50	0.48	0.47	0.47	0.47	0.48
ZU	0.31	0.31	0.31	0.31	0.32	0.33	0.35	0.37
TM	0.45	0.45	0.45	0.46	0.48	0.50	0.53	0.56
				MENA				
DZ	0.93	0.97	1.02	1.08	1.18	1.29	1.43	1.60
BHQ	11.82	12.39	13.24	14.38	15.70	17.24	19.09	21.30
EG	0.65	0.68	0.71	0.75	0.80	0.87	0.95	1.04
IR	1.50	1.60	1.74	1.91	2.11	2.33	2.60	2.94
IQ	1.19	1.16	1.13	1.12	1.11	1.11	1.12	1.15
IL	2.38	2.51	2.66	2.84	3.05	3.27	3.53	3.82
JWG	0.94	0.96	1.00	1.03	1.07	1.12	1.18	1.26
KW	8.98	9.23	9.72	10.38	11.21	12.11	13.20	14.52
LB	1.30	1.46	1.79	2.06	2.28	2.55	2.88	3.27
LY	2.02	2.11	2.25	2.44	2.68	2.97	3.31	3.75
MA	0.40	0.43	0.46	0.51	0.56	0.62	0.70	0.79
OM	2.13	2.25	2.42	2.66	2.95	3.27	3.61	3.94
SA	6.10	6.38	6.77	7.27	7.85	8.49	9.29	10.28
TN	0.41	0.45	0.49	0.54	0.61	0.68	0.78	0.89
AE	4.93	5.25	5.64	6.09	6.60	7.22	7.96	8.85
YE	0.21	0.21	0.21	0.21	0.21	0.22	0.23	0.25
SY	1.17	1.17	1.08	1.07	1.13	1.20	1.30	1.42
				Sub-Saharan Africa				
WW	0.06	0.07	0.09	0.10	0.13	0.16	0.21	0.26
WS	0.18	0.21	0.25	0.30	0.36	0.43	0.54	0.69
WN	0.03	0.03	0.03	0.04	0.04	0.05	0.06	0.07
NIG-S	0.14	0.19	0.23	0.30	0.37	0.46	0.61	0.80
NIG-N	0.14	0.18	0.23	0.30	0.37	0.46	0.61	0.81
SER	0.11	0.14	0.16	0.20	0.25	0.31	0.40	0.51
ETH	0.04	0.04	0.05	0.07	0.08	0.11	0.14	0.19
SOMDJ	0.03	0.04	0.04	0.05	0.06	0.07	0.09	0.11
KENUG	0.04	0.05	0.06	0.08	0.09	0.11	0.14	0.18
TZRB	0.04	0.04	0.05	0.06	0.07	0.08	0.10	0.13
CAR	0.11	0.13	0.15	0.17	0.20	0.24	0.29	0.34
COG	0.05	0.07	0.08	0.10	0.12	0.15	0.18	0.23
SW	0.29	0.32	0.36	0.41	0.47	0.54	0.63	0.74
ZAFLS	0.87	0.81	0.85	0.89	0.95	1.00	1.09	1.18
SEA	0.17	0.20	0.25	0.30	0.37	0.45	0.56	0.71
IOCE	0.13	0.15	0.18	0.20	0.25	0.31	0.38	0.47
				India/ SAARC				
IN-E	0.12	0.14	0.17	0.19	0.23	0.28	0.34	0.42

IN-CE	0.06	0.07	0.08	0.09	0.11	0.13	0.16	0.20
IN-W	0.24	0.28	0.32	0.38	0.45	0.54	0.66	0.81
IN-CW	0.23	0.27	0.31	0.36	0.43	0.52	0.64	0.79
IN-N	0.24	0.28	0.32	0.37	0.44	0.53	0.65	0.81
IN-NW	0.26	0.30	0.35	0.41	0.48	0.58	0.71	0.88
IN-UP	0.10	0.11	0.13	0.15	0.18	0.22	0.26	0.33
IN-S	0.21	0.25	0.29	0.33	0.40	0.48	0.58	0.72
IN-CS	0.21	0.24	0.28	0.33	0.39	0.47	0.57	0.71
IN-NE	0.06	0.07	0.08	0.10	0.11	0.14	0.17	0.21
BD	0.15	0.16	0.18	0.19	0.21	0.24	0.27	0.31
NP+BT	0.07	0.08	0.08	0.09	0.10	0.11	0.12	0.13
PK-S	0.37	0.40	0.42	0.44	0.48	0.52	0.56	0.62
PK-N	0.37	0.40	0.42	0.44	0.48	0.51	0.56	0.62
AF	0.09	0.08	0.08	0.08	0.09	0.09	0.10	0.11
LK	0.10	0.11	0.13	0.14	0.17	0.19	0.23	0.27
				Northeast Asia				
JP-E	2.29	2.36	2.44	2.53	2.65	2.77	2.90	3.04
JP-W	2.29	2.36	2.44	2.53	2.64	2.77	2.90	3.05
KR	1.50	1.72	1.99	2.31	2.70	3.17	3.79	4.57
KP	0.13	0.15	0.17	0.20	0.22	0.26	0.29	0.34
CN-NE	0.54	0.60	0.67	0.76	0.87	1.01	1.17	1.37
CN-N	0.54	0.60	0.67	0.76	0.87	1.01	1.17	1.38
CN-E	0.54	0.60	0.67	0.76	0.87	1.01	1.17	1.37
CN-C	0.54	0.60	0.67	0.76	0.87	1.01	1.17	1.38
CN-S	0.54	0.60	0.67	0.76	0.87	1.01	1.17	1.38
CN-TB	0.54	0.60	0.67	0.76	0.87	1.01	1.17	1.38
CN-NW	0.54	0.60	0.67	0.76	0.87	1.01	1.18	1.38
CN-XU	0.54	0.60	0.67	0.76	0.87	1.00	1.18	1.37
MN	0.55	0.60	0.65	0.71	0.79	0.89	1.01	1.15
				Southeast Asia				
NZ	3.08	3.04	3.00	2.97	2.97	3.00	3.03	3.08
AU-E	3.74	3.54	3.36	3.19	3.03	2.88	2.73	2.61
AU-W	5.34	5.06	4.80	4.55	4.32	4.12	3.92	3.73
ID-PG+NG	0.29	0.31	0.34	0.38	0.42	0.48	0.55	0.63
ID-SU	0.38	0.43	0.48	0.55	0.63	0.73	0.85	1.00
ID-JV+TL	0.67	0.77	0.88	1.02	1.19	1.38	1.61	1.87
ID-KL-SW	0.32	0.35	0.40	0.45	0.51	0.60	0.69	0.81
MY-W+SG	1.16	1.33	1.53	1.74	2.00	2.31	2.67	3.09
MY-E+BN	1.61	1.84	2.10	2.39	2.72	3.12	3.59	4.12
PH	0.32	0.36	0.40	0.44	0.49	0.56	0.65	0.75
MM	0.16	0.18	0.20	0.22	0.25	0.29	0.34	0.41
TH	0.59	0.71	0.85	1.03	1.25	1.54	1.92	2.40
LA	0.21	0.22	0.24	0.26	0.29	0.32	0.36	0.42
VN	0.72	1.12	1.56	2.17	2.52	2.98	3.55	4.24
KH	0.16	0.17	0.19	0.21	0.23	0.26	0.29	0.33
				North America				
CA-W	4.86	4.99	4.86	5.05	5.10	5.23	5.31	5.44
CA-E	4.47	4.47	4.43	4.66	4.78	4.92	5.00	5.01
US-NENY	3.53	3.56	3.59	3.61	3.64	3.66	4.01	4.00
US-MA	4.35	4.40	4.43	4.46	4.49	4.53	4.80	4.75
US-CAR	5.40	5.45	5.49	5.53	5.57	5.62	5.86	5.84
US-S	4.97	5.01	5.05	5.08	5.12	5.16	5.37	5.29
US-TVA	5.40	5.45	5.49	5.53	5.57	5.61	5.88	5.83
US-MW	4.12	4.16	4.19	4.21	4.25	4.28	4.53	4.49
US-C	4.27	4.31	4.34	4.37	4.40	4.44	4.62	4.55
US-TX	2.57	2.59	2.61	2.63	2.65	2.67	2.74	2.71
US-SW	4.28	4.32	4.35	4.37	4.41	4.45	4.56	4.49
US-NW	4.14	4.18	4.21	4.23	4.27	4.30	4.47	4.45

US-CA	2.20	2.22	2.24	2.25	2.27	2.29	2.45	2.42
US-AK	0.99	1.00	1.01	1.02	1.02	1.03	1.03	1.03
US-HI	1.84	1.86	1.87	1.88	1.90	1.92	1.94	1.90
US-GU	3.11	3.14	3.17	3.19	3.21	3.24	3.33	3.31
MX-NW	2.86	2.87	2.86	2.84	2.83	2.84	2.85	2.86
MX-N	4.67	4.68	4.65	4.62	4.61	4.62	4.64	4.66
MX-C	1.44	1.44	1.44	1.43	1.42	1.43	1.43	1.44
MX-S	0.65	0.65	0.65	0.65	0.65	0.65	0.65	0.65
				South America				
CAM	0.44	0.45	0.47	0.49	0.52	0.56	0.60	0.66
CO	0.65	0.69	0.75	0.81	0.89	0.97	1.08	1.21
VE	1.68	1.74	1.79	1.87	1.98	2.12	2.30	2.52
EC	0.44	0.46	0.48	0.49	0.52	0.55	0.60	0.66
PU	0.34	0.36	0.37	0.39	0.42	0.46	0.50	0.55
CSA	0.50	0.52	0.55	0.57	0.60	0.64	0.70	0.76
BR-S	1.03	1.10	1.18	1.27	1.39	1.53	1.70	1.90
BR-SP	1.24	1.32	1.41	1.52	1.66	1.83	2.04	2.28
BR-SE	0.90	0.97	1.03	1.11	1.22	1.34	1.49	1.67
BR-N	0.81	0.87	0.93	1.00	1.09	1.20	1.34	1.50
BR_NE	0.65	0.70	0.75	0.81	0.88	0.97	1.08	1.21
AR-NE	1.33	1.40	1.49	1.59	1.72	1.88	2.06	2.27
AR-E	1.52	1.59	1.68	1.77	1.90	2.06	2.25	2.47
AR-W	1.27	1.33	1.40	1.48	1.59	1.73	1.89	2.07
CL	0.68	0.67	0.66	0.66	0.67	0.68	0.70	0.72

Table A4: *Annual space heating demand per capita for every region and modelled year, structured to the 9 major regions Europe, Eurasia, MENA, Sub-Saharan Africa, India/ SAARC, Northeast Asia, Southeast Asia, North America and South America.*

Region	Annual Space Heating Demand per Capita [MWh_{th}/a]							
	2015	**2020**	**2025**	**2030**	**2035**	**2040**	**2045**	**2050**
				Europe				
NO	8.70	8.77	8.55	8.66	8.62	8.45	8.53	8.63
DK	5.96	6.01	5.87	5.96	5.95	5.86	5.94	6.03
SE	8.08	8.21	8.08	8.26	8.30	8.18	8.30	8.42
FI	9.32	9.47	9.33	9.54	9.61	9.49	9.64	9.80
BLT	8.30	8.26	7.94	8.84	9.27	9.68	10.56	11.32
PL	7.12	7.28	7.20	7.39	7.45	7.37	7.49	7.63
IBE	3.69	3.73	3.66	3.73	3.73	3.68	3.74	3.81
FR	5.31	5.38	5.27	5.37	5.39	5.32	5.40	5.49
BNL	5.33	5.36	5.24	5.31	5.31	5.23	5.30	5.38
BRI	5.56	5.66	5.59	5.74	5.78	5.70	5.79	5.87
DE	6.59	6.67	6.53	6.65	6.66	6.57	6.67	6.77
CRS	7.32	7.42	7.28	7.43	7.47	7.38	7.51	7.65
AUH	7.19	7.33	7.24	7.43	7.48	7.40	7.54	7.68
BKN-W	5.72	5.72	5.53	6.36	6.80	7.26	8.08	8.78
BKN-E	5.58	5.61	5.45	6.15	6.50	6.82	7.44	7.95
IT	4.97	4.99	4.86	4.92	4.91	4.83	4.91	5.00
CH	9.02	9.09	8.86	8.99	8.97	8.81	8.91	9.03
TR	4.42	4.51	4.45	4.57	4.62	4.59	4.69	4.79
UA	6.35	6.42	6.27	7.29	7.88	8.48	9.48	10.35
IS	9.01	9.13	8.97	9.15	9.18	9.05	9.19	9.34
				Eurasia				
RU-NW	9.03	8.96	8.60	9.81	10.43	11.07	12.28	13.31
RU-C	9.91	9.83	9.42	10.76	11.44	12.14	13.46	14.60
RU-S	9.13	9.06	8.68	9.91	10.54	11.19	12.41	13.46
RU-V	10.34	10.26	9.84	11.23	11.94	12.68	14.06	15.25

RU-U	9.88	9.80	9.40	10.73	11.41	12.11	13.43	14.57
RU-SI	10.38	10.30	9.87	11.27	11.98	12.72	14.11	15.30
RU-FE	8.90	8.83	8.47	9.67	10.28	10.91	12.10	13.12
BY	8.12	8.16	7.90	9.09	9.71	10.33	11.46	12.41
CAU	7.14	7.14	6.91	7.94	8.46	8.97	9.91	10.69
KZ	7.79	7.79	7.52	8.61	9.17	9.72	10.75	11.62
PAM	5.23	5.30	5.22	6.12	6.68	7.23	8.12	8.88
ZU	3.60	3.68	3.64	4.27	4.65	5.02	5.63	6.15
TM	3.52	3.53	3.41	3.91	4.18	4.46	4.95	5.37
				MENA				
DZ	4.34	4.49	4.56	4.02	4.22	4.22	3.90	3.57
BHQ	0.10	0.10	0.11	0.12	0.14	0.15	0.15	0.15
EG	1.80	1.88	1.94	1.73	1.83	1.83	1.69	1.54
IR	5.46	5.33	5.75	6.81	7.64	8.41	8.84	9.05
IQ	2.06	2.02	2.19	2.60	2.92	3.21	3.38	3.46
IL	0.75	0.73	0.79	0.93	1.03	1.13	1.18	1.20
JWG	1.12	1.07	1.12	1.29	1.43	1.55	1.61	1.63
KW	1.32	1.28	1.37	1.60	1.77	1.92	1.99	2.01
LB	1.53	1.50	1.63	1.93	2.17	2.40	2.52	2.58
LY	2.77	2.92	3.00	2.64	2.75	2.73	2.50	2.28
MA	2.79	2.90	2.96	2.61	2.75	2.76	2.56	2.36
OM	0.29	0.28	0.30	0.36	0.40	0.44	0.46	0.47
SA	0.64	0.62	0.67	0.78	0.87	0.95	0.99	1.00
TN	3.99	4.15	4.22	3.71	3.89	3.89	3.59	3.30
AE	0.11	0.11	0.12	0.14	0.15	0.17	0.17	0.18
YE	0.18	0.18	0.19	0.23	0.26	0.29	0.31	0.33
SY	3.11	3.09	3.36	4.03	4.57	5.08	5.38	5.54
				Sub-Saharan Africa				
WW	0.01	0.01	0.02	0.01	0.01	0.01	0.01	0.01
WS	0.01	0.01	0.01	0.01	0.01	0.01	0.01	0.01
WN	0.07	0.07	0.08	0.07	0.07	0.07	0.07	0.07
NIG-S	0.03	0.03	0.03	0.03	0.03	0.03	0.03	0.03
NIG-N	0.04	0.04	0.04	0.04	0.04	0.04	0.04	0.03
SER	0.11	0.11	0.12	0.10	0.11	0.11	0.10	0.10
ETH	1.21	1.29	1.35	1.22	1.31	1.34	1.26	1.18
SOMDJ	0.12	0.13	0.14	0.13	0.14	0.14	0.13	0.13
KENUG	0.56	0.58	0.59	0.52	0.56	0.57	0.53	0.49
TZRB	0.32	0.33	0.34	0.30	0.32	0.33	0.32	0.30
CAR	0.03	0.03	0.03	0.03	0.03	0.03	0.03	0.02
COG	0.05	0.05	0.06	0.05	0.05	0.06	0.05	0.05
SW	0.53	0.54	0.55	0.48	0.49	0.49	0.45	0.41
ZAFLS	2.76	2.47	2.48	2.16	2.26	2.25	2.08	1.91
SEA	0.47	0.49	0.50	0.44	0.47	0.48	0.45	0.42
IOCE	1.00	1.04	1.05	0.93	0.99	1.01	0.97	0.92
				India/ SAARC				
IN-E	0.79	0.75	0.75	0.78	0.76	0.73	0.80	0.87
IN-CE	0.83	0.78	0.78	0.82	0.79	0.76	0.83	0.91
IN-W	0.82	0.77	0.77	0.81	0.78	0.75	0.82	0.90
IN-CW	0.79	0.75	0.75	0.78	0.76	0.73	0.80	0.87
IN-N	0.85	0.81	0.81	0.84	0.82	0.79	0.86	0.94
IN-NW	0.82	0.77	0.77	0.81	0.78	0.76	0.83	0.90
IN-UP	0.83	0.78	0.78	0.82	0.79	0.76	0.84	0.91
IN-S	0.77	0.73	0.73	0.76	0.74	0.71	0.78	0.85
IN-CS	0.82	0.77	0.77	0.81	0.78	0.75	0.83	0.90
IN-NE	0.84	0.80	0.80	0.83	0.81	0.78	0.85	0.93
BD	0.12	0.13	0.12	0.12	0.12	0.13	0.13	0.13
NP+BT	1.03	1.17	1.05	1.06	1.10	1.17	1.21	1.21
PK-S	1.55	1.76	1.56	1.56	1.61	1.71	1.76	1.76

PK-N	1.67	1.90	1.68	1.68	1.74	1.85	1.90	1.90
AF	6.48	7.35	6.52	6.56	6.86	7.39	7.73	7.84
LK	0.00	0.00	0.00	0.00	0.00	0.00	0.00	0.00
				Northeast Asia				
JP-E	4.17	4.16	4.32	4.55	4.56	4.71	5.05	5.54
JP-W	3.87	3.85	4.01	4.21	4.23	4.37	4.68	5.14
KR	4.30	4.33	4.55	4.82	4.86	5.04	5.41	5.94
KP	7.09	8.22	7.47	7.68	8.18	8.92	9.36	9.48
CN-NE	2.61	2.63	2.77	2.85	3.06	3.19	3.34	3.61
CN-N	2.21	2.22	2.34	2.41	2.58	2.70	2.82	3.05
CN-E	2.06	2.07	2.19	2.25	2.42	2.52	2.64	2.85
CN-C	2.35	2.37	2.50	2.57	2.76	2.88	3.01	3.26
CN-S	2.21	2.22	2.35	2.42	2.59	2.71	2.83	3.06
CN-TB	0.10	0.10	0.10	0.11	0.11	0.12	0.12	0.14
CN-NW	5.75	5.78	6.11	6.28	6.74	7.04	7.34	7.95
CN-XU	2.31	2.32	2.45	2.53	2.71	2.83	2.95	3.20
MN	16.37	18.78	16.69	16.66	17.09	17.92	18.14	17.84
				Southeast Asia				
NZ	2.53	2.54	2.66	2.82	2.84	2.94	3.16	3.47
AU-E	2.11	2.09	2.14	2.19	2.13	2.12	2.18	2.28
AU-W	1.32	1.31	1.34	1.37	1.33	1.33	1.36	1.42
ID-PG+NG	0.21	0.24	0.22	0.22	0.23	0.25	0.26	0.26
ID-SU	0.00	0.00	0.00	0.00	0.00	0.00	0.00	0.00
ID-JV+TL	0.00	0.00	0.00	0.00	0.00	0.00	0.00	0.00
ID-KL-SW	0.00	0.00	0.00	0.00	0.00	0.00	0.00	0.00
MY-W+SG	0.00	0.00	0.00	0.00	0.00	0.00	0.00	0.00
MY-E+BN	0.00	0.00	0.00	0.00	0.00	0.00	0.00	0.00
PH	0.00	0.00	0.00	0.00	0.00	0.00	0.00	0.00
MM	0.17	0.20	0.18	0.18	0.19	0.21	0.21	0.21
TH	0.08	0.09	0.08	0.08	0.08	0.09	0.09	0.08
LA	0.32	0.37	0.33	0.33	0.34	0.36	0.37	0.37
VN	0.43	0.50	0.45	0.46	0.48	0.50	0.51	0.50
KH	0.01	0.01	0.01	0.01	0.01	0.01	0.01	0.01
				North America				
CA-W	10.73	10.93	10.80	10.68	10.82	10.71	10.67	10.61
CA-E	11.74	11.65	11.52	11.40	11.54	11.42	11.38	11.32
US-NENY	4.81	4.71	4.63	4.54	4.58	4.52	4.50	4.48
US-MA	6.99	6.84	6.71	6.60	6.65	6.57	6.53	6.50
US-CAR	6.42	6.28	6.17	6.06	6.10	6.03	6.00	5.97
US-S	5.79	5.67	5.56	5.46	5.51	5.44	5.41	5.38
US-TVA	6.18	6.05	5.94	5.83	5.88	5.81	5.78	5.75
US-MW	6.57	6.43	6.32	6.20	6.25	6.18	6.14	6.11
US-C	6.98	6.83	6.70	6.59	6.64	6.56	6.52	6.49
US-TX	6.40	6.26	6.15	6.04	6.09	6.01	5.98	5.95
US-SW	7.73	7.57	7.43	7.30	7.36	7.27	7.23	7.19
US-NW	6.40	6.26	6.15	6.04	6.09	6.01	5.98	5.95
US-CA	5.22	5.11	5.02	4.93	4.97	4.91	4.88	4.86
US-AK	5.29	5.18	5.09	5.00	5.04	4.97	4.95	4.92
US-HI	1.05	1.03	1.01	0.99	1.00	0.99	0.98	0.98
US-GU	6.69	6.55	6.43	6.32	6.37	6.29	6.25	6.22
MX-NW	1.60	1.61	1.60	1.60	1.64	1.65	1.66	1.67
MX-N	1.73	1.73	1.73	1.72	1.76	1.77	1.78	1.79
MX-C	1.86	1.86	1.86	1.85	1.90	1.90	1.92	1.93
MX-S	1.83	1.83	1.83	1.83	1.87	1.88	1.89	1.90
				South America				
CAM	0.11	0.13	0.15	0.17	0.21	0.24	0.27	0.31
CO	0.51	0.55	0.66	0.72	0.85	0.96	1.04	1.13
VE	0.05	0.06	0.07	0.07	0.08	0.10	0.11	0.12

EC	0.95	1.02	1.24	1.37	1.65	1.88	2.04	2.22
PU	2.34	2.54	3.09	3.42	4.08	4.63	5.01	5.43
CSA	3.18	3.43	4.16	4.59	5.50	6.29	6.90	7.59
BR-S	0.38	0.40	0.48	0.53	0.63	0.71	0.77	0.84
BR-SP	0.39	0.41	0.50	0.54	0.64	0.73	0.79	0.86
BR-SE	0.37	0.39	0.47	0.51	0.61	0.69	0.75	0.81
BR-N	0.36	0.39	0.47	0.51	0.60	0.68	0.74	0.81
BR_NE	0.34	0.36	0.44	0.48	0.57	0.64	0.69	0.75
AR-NE	2.79	3.01	3.65	4.02	4.78	5.39	5.82	6.28
AR-E	3.00	3.24	3.93	4.32	5.12	5.77	6.21	6.69
AR-W	3.13	3.38	4.10	4.50	5.34	6.02	6.48	6.98
CL	3.43	3.67	4.39	4.77	5.61	6.30	6.78	7.32

Table A5 *Annual hot water demand per capita for every region and modelled year, structured to the 9 major regions Europe, Eurasia, MENA, Sub-Saharan Africa, India/ SAARC, Northeast Asia, Southeast Asia, North America and South America.*

Region	Annual Hot Water Demand per Capita [MWh_{th}/a]							
	2015	**2020**	**2025**	**2030**	**2035**	**2040**	**2045**	**2050**
				Europe				
NO	0.68	0.68	0.69	0.69	0.70	0.70	0.71	0.71
DK	0.65	0.66	0.66	0.67	0.68	0.69	0.70	0.71
SE	0.67	0.69	0.71	0.73	0.74	0.75	0.76	0.77
FI	0.66	0.68	0.70	0.71	0.73	0.75	0.76	0.77
BLT	0.61	0.62	0.63	0.65	0.66	0.67	0.69	0.70
PL	0.62	0.65	0.67	0.69	0.71	0.73	0.74	0.75
IBE	0.64	0.66	0.67	0.68	0.69	0.70	0.71	0.72
FR	0.66	0.68	0.69	0.70	0.72	0.73	0.74	0.75
BNL	0.67	0.68	0.68	0.69	0.70	0.71	0.72	0.73
BRI	0.67	0.69	0.71	0.73	0.75	0.76	0.77	0.78
DE	0.68	0.69	0.70	0.72	0.73	0.74	0.75	0.76
CRS	0.63	0.64	0.66	0.67	0.69	0.70	0.72	0.73
AUH	0.64	0.66	0.68	0.70	0.72	0.73	0.74	0.75
BKN-W	0.55	0.57	0.59	0.62	0.64	0.67	0.69	0.71
BKN-E	0.59	0.61	0.63	0.66	0.68	0.70	0.71	0.73
IT	0.64	0.65	0.65	0.66	0.67	0.68	0.69	0.70
CH	0.72	0.72	0.73	0.74	0.75	0.75	0.76	0.77
TR	0.58	0.60	0.62	0.64	0.67	0.68	0.70	0.72
UA	0.49	0.52	0.56	0.60	0.63	0.67	0.70	0.72
IS	0.63	0.64	0.65	0.67	0.68	0.69	0.70	0.71
				Eurasia				
RU-NW	0.56	0.57	0.59	0.60	0.62	0.63	0.65	0.66
RU-C	0.62	0.63	0.64	0.66	0.68	0.69	0.71	0.73
RU-S	0.57	0.58	0.59	0.61	0.62	0.64	0.66	0.67
RU-V	0.64	0.66	0.67	0.69	0.70	0.72	0.74	0.76
RU-U	0.62	0.63	0.64	0.66	0.67	0.69	0.71	0.73
RU-SI	0.65	0.66	0.67	0.69	0.71	0.73	0.74	0.76
RU-FE	0.56	0.57	0.58	0.60	0.61	0.63	0.65	0.66
BY	0.57	0.60	0.63	0.65	0.67	0.69	0.71	0.72
CAU	0.70	0.74	0.78	0.81	0.84	0.87	0.89	0.91
KZ	0.62	0.65	0.67	0.69	0.71	0.72	0.74	0.75
PAM	0.38	0.43	0.48	0.54	0.59	0.64	0.68	0.71
ZU	0.43	0.48	0.52	0.57	0.61	0.65	0.68	0.70
TM	0.55	0.58	0.60	0.62	0.64	0.66	0.68	0.70
				MENA				
DZ	0.54	0.56	0.59	0.62	0.65	0.68	0.70	0.72
BHQ	0.46	0.47	0.47	0.47	0.47	0.47	0.47	0.47

EG	0.52	0.55	0.60	0.64	0.68	0.71	0.74	0.75
IR	0.57	0.58	0.60	0.62	0.64	0.66	0.68	0.70
IQ	0.55	0.57	0.59	0.62	0.64	0.66	0.68	0.70
IL	0.57	0.59	0.60	0.61	0.62	0.63	0.64	0.65
JWG	0.33	0.33	0.33	0.34	0.35	0.35	0.36	0.37
KW	0.64	0.64	0.65	0.65	0.66	0.66	0.67	0.67
LB	0.55	0.56	0.59	0.61	0.63	0.65	0.67	0.69
LY	0.56	0.61	0.64	0.68	0.70	0.72	0.73	0.75
MA	0.47	0.50	0.54	0.57	0.60	0.63	0.66	0.69
OM	0.67	0.68	0.70	0.72	0.74	0.75	0.77	0.77
SA	0.67	0.68	0.68	0.69	0.70	0.71	0.72	0.73
TN	0.51	0.53	0.56	0.59	0.62	0.65	0.67	0.69
AE	0.70	0.71	0.72	0.73	0.73	0.74	0.75	0.75
YE	0.39	0.41	0.43	0.45	0.49	0.53	0.58	0.62
SY	0.48	0.52	0.55	0.59	0.63	0.66	0.69	0.72
				Sub-Saharan Africa				
WW	0.29	0.32	0.35	0.40	0.45	0.50	0.55	0.60
WS	0.36	0.39	0.42	0.46	0.51	0.55	0.60	0.64
WN	0.31	0.35	0.40	0.45	0.51	0.56	0.61	0.65
NIG-S	0.42	0.45	0.49	0.53	0.57	0.60	0.63	0.65
NIG-N	0.48	0.52	0.56	0.61	0.66	0.70	0.74	0.77
SER	0.39	0.43	0.47	0.51	0.55	0.59	0.63	0.66
ETH	0.29	0.35	0.40	0.47	0.53	0.58	0.63	0.67
SOMDJ	0.23	0.30	0.37	0.43	0.50	0.55	0.60	0.63
KENUG	0.34	0.38	0.43	0.50	0.57	0.63	0.68	0.72
TZRB	0.32	0.35	0.39	0.44	0.50	0.56	0.61	0.66
CAR	0.36	0.38	0.41	0.44	0.49	0.54	0.59	0.63
COG	0.28	0.33	0.39	0.45	0.51	0.57	0.62	0.66
SW	0.48	0.51	0.54	0.57	0.60	0.63	0.65	0.68
ZAFLS	0.52	0.47	0.48	0.50	0.52	0.54	0.57	0.58
SEA	0.29	0.33	0.37	0.43	0.49	0.55	0.60	0.64
IOCE	0.25	0.26	0.28	0.30	0.34	0.39	0.45	0.51
				India/ SAARC				
IN-E	0.43	0.47	0.51	0.55	0.58	0.62	0.65	0.67
IN-CE	0.45	0.49	0.53	0.57	0.61	0.64	0.68	0.70
IN-W	0.44	0.48	0.53	0.57	0.60	0.64	0.67	0.69
IN-CW	0.43	0.47	0.51	0.55	0.58	0.62	0.65	0.67
IN-N	0.46	0.50	0.54	0.58	0.62	0.66	0.69	0.71
IN-NW	0.44	0.48	0.52	0.56	0.60	0.64	0.67	0.69
IN-UP	0.45	0.49	0.53	0.57	0.61	0.65	0.68	0.70
IN-S	0.42	0.46	0.50	0.53	0.57	0.60	0.63	0.65
IN-CS	0.44	0.48	0.53	0.57	0.60	0.64	0.67	0.69
IN-NE	0.46	0.50	0.55	0.59	0.63	0.66	0.70	0.72
BD	0.35	0.39	0.42	0.46	0.50	0.54	0.57	0.61
NP+BT	0.36	0.39	0.43	0.47	0.52	0.57	0.62	0.66
PK-S	0.40	0.42	0.45	0.48	0.52	0.56	0.59	0.63
PK-N	0.43	0.45	0.48	0.52	0.56	0.60	0.64	0.68
AF	0.31	0.33	0.37	0.41	0.46	0.52	0.58	0.63
LK	0.51	0.54	0.58	0.61	0.64	0.66	0.68	0.70
				Northeast Asia				
JP-E	0.66	0.66	0.66	0.66	0.67	0.68	0.68	0.69
JP-W	0.61	0.61	0.61	0.62	0.62	0.63	0.63	0.64
KR	0.64	0.66	0.67	0.68	0.69	0.70	0.71	0.72
KP	0.21	0.25	0.31	0.38	0.45	0.52	0.58	0.63
CN-NE	0.61	0.66	0.70	0.74	0.77	0.79	0.81	0.83
CN-N	0.52	0.56	0.60	0.63	0.65	0.67	0.69	0.70
CN-E	0.49	0.53	0.56	0.59	0.61	0.63	0.65	0.66
CN-C	0.56	0.60	0.64	0.67	0.70	0.72	0.74	0.75

CN-S	0.52	0.56	0.60	0.63	0.66	0.68	0.69	0.71
CN-TB	0.02	0.03	0.03	0.03	0.03	0.03	0.03	0.03
CN-NW	1.36	1.47	1.56	1.64	1.71	1.76	1.80	1.84
CN-XU	0.55	0.59	0.63	0.66	0.68	0.71	0.72	0.74
MN	0.52	0.56	0.59	0.62	0.65	0.68	0.70	0.72
				Southeast Asia				
NZ	0.62	0.63	0.64	0.65	0.67	0.68	0.68	0.69
AU-E	0.69	0.69	0.69	0.68	0.67	0.65	0.63	0.61
AU-W	0.43	0.43	0.43	0.43	0.42	0.41	0.39	0.38
ID-PG+NG	0.38	0.40	0.42	0.45	0.49	0.52	0.56	0.58
ID-SU	0.48	0.50	0.52	0.54	0.56	0.58	0.58	0.59
ID-JV+TL	0.45	0.48	0.51	0.54	0.57	0.58	0.59	0.59
ID-KL-SW	0.45	0.47	0.48	0.50	0.52	0.53	0.54	0.54
MY-W+SG	0.46	0.48	0.50	0.52	0.53	0.53	0.53	0.53
MY-E+BN	0.59	0.62	0.64	0.66	0.67	0.67	0.66	0.65
PH	0.45	0.48	0.51	0.55	0.59	0.62	0.65	0.68
MM	0.34	0.40	0.46	0.51	0.56	0.61	0.64	0.67
TH	0.56	0.59	0.62	0.66	0.69	0.71	0.73	0.74
LA	0.45	0.48	0.52	0.56	0.60	0.63	0.66	0.69
VN	0.44	0.49	0.54	0.59	0.63	0.66	0.69	0.71
KH	0.37	0.40	0.43	0.46	0.50	0.54	0.58	0.62
				North America				
CA-W	0.62	0.66	0.68	0.70	0.71	0.71	0.71	0.71
CA-E	0.68	0.71	0.73	0.75	0.76	0.76	0.76	0.75
US-NENY	0.54	0.56	0.57	0.58	0.58	0.58	0.58	0.57
US-MA	0.79	0.81	0.83	0.84	0.85	0.85	0.84	0.83
US-CAR	0.73	0.74	0.76	0.77	0.78	0.78	0.77	0.77
US-S	0.65	0.67	0.69	0.70	0.70	0.70	0.70	0.69
US-TVA	0.70	0.72	0.73	0.74	0.75	0.75	0.75	0.74
US-MW	0.74	0.76	0.78	0.79	0.80	0.80	0.79	0.79
US-C	0.79	0.81	0.83	0.84	0.85	0.85	0.84	0.83
US-TX	0.73	0.74	0.76	0.77	0.78	0.78	0.78	0.77
US-SW	0.87	0.90	0.92	0.93	0.94	0.94	0.93	0.92
US-NW	0.72	0.74	0.76	0.77	0.78	0.78	0.77	0.76
US-CA	0.59	0.61	0.62	0.63	0.63	0.63	0.63	0.63
US-AK	0.60	0.61	0.63	0.64	0.64	0.64	0.64	0.63
US-HI	0.12	0.12	0.12	0.13	0.13	0.13	0.13	0.13
US-GU	0.76	0.77	0.79	0.80	0.81	0.81	0.81	0.80
MX-NW	0.53	0.56	0.59	0.61	0.63	0.65	0.66	0.67
MX-N	0.56	0.59	0.62	0.64	0.67	0.68	0.70	0.70
MX-C	0.62	0.65	0.68	0.71	0.73	0.75	0.77	0.78
MX-S	0.61	0.64	0.67	0.70	0.72	0.74	0.76	0.77
				South America				
CAM	0.48	0.50	0.52	0.55	0.58	0.61	0.64	0.67
CO	0.54	0.56	0.58	0.60	0.63	0.65	0.68	0.70
VE	0.55	0.57	0.58	0.61	0.64	0.67	0.69	0.71
EC	0.52	0.54	0.57	0.61	0.65	0.68	0.71	0.73
PU	0.53	0.56	0.60	0.64	0.67	0.70	0.72	0.74
CSA	0.49	0.51	0.54	0.57	0.61	0.64	0.67	0.70
BR-S	0.55	0.57	0.59	0.62	0.64	0.67	0.69	0.71
BR-SP	0.58	0.60	0.63	0.65	0.68	0.70	0.73	0.75
BR-SE	0.55	0.57	0.59	0.62	0.64	0.67	0.69	0.71
BR-N	0.54	0.56	0.58	0.61	0.63	0.65	0.68	0.70
BR_NE	0.51	0.53	0.55	0.57	0.60	0.62	0.64	0.66
AR-NE	0.59	0.62	0.65	0.68	0.70	0.72	0.74	0.75
AR-E	0.57	0.60	0.63	0.66	0.68	0.70	0.72	0.73
AR-W	0.60	0.63	0.66	0.69	0.71	0.73	0.75	0.76
CL	0.59	0.61	0.62	0.64	0.65	0.67	0.68	0.70

Table A6: *Electricity prices for grid supply for all regions and modelled year, structured to the 9 major regions Europe, Eurasia, MENA, Sub-Saharan Africa, India/ SAARC, Northeast Asia, Southeast Asia, North America and South America.*

Region	Electricity Prices for Grid Covering [€/kWh$_{el}$]							
	2015	2020	2025	2030	2035	2040	2045	2050
				Europe				
NO	0.1590	0.1843	0.2137	0.2477	0.2655	0.2791	0.2791	0.2933
DK	0.2640	0.2775	0.2916	0.3065	0.3221	0.3386	0.3386	0.3558
SE	0.1670	0.1936	0.2244	0.2602	0.2735	0.2874	0.2874	0.3021
FI	0.1294	0.1501	0.1740	0.2017	0.2338	0.2606	0.2606	0.2739
BLT	0.1083	0.1338	0.1589	0.1874	0.2182	0.2511	0.2513	0.2699
PL	0.1420	0.1646	0.1908	0.2212	0.2565	0.2696	0.2696	0.2833
IBE	0.1655	0.1918	0.2224	0.2578	0.2716	0.2855	0.2855	0.3001
FR	0.1290	0.1495	0.1734	0.2010	0.2330	0.2649	0.2649	0.2784
BNL	0.1891	0.2192	0.2539	0.2703	0.2841	0.2986	0.2986	0.3138
BRI	0.1511	0.1754	0.2036	0.2341	0.2643	0.2777	0.2778	0.2920
DE	0.2330	0.2649	0.2784	0.2926	0.3075	0.3232	0.3232	0.3397
CRS	0.1460	0.1696	0.1967	0.2281	0.2608	0.2740	0.2740	0.2879
AUH	0.1901	0.2203	0.2551	0.2716	0.2855	0.3001	0.3000	0.3153
BKN-W	0.0915	0.1107	0.1338	0.1598	0.1859	0.2103	0.2088	0.2318
BKN-E	0.1146	0.1387	0.1626	0.1881	0.2182	0.2490	0.2491	0.2702
IT	0.2150	0.2492	0.2671	0.2808	0.2951	0.3101	0.3101	0.3260
CH	0.1320	0.1530	0.1774	0.2057	0.2384	0.2657	0.2657	0.2793
TR	0.1443	0.1673	0.1939	0.2248	0.2602	0.2735	0.2735	0.2874
UA	0.0215	0.0275	0.0351	0.0448	0.0571	0.0729	0.0729	0.0918
IS	0.1070	0.1340	0.1553	0.1800	0.2087	0.2419	0.2419	0.2645
				Eurasia				
RU-NW	0.0310	0.0396	0.0342	0.0436	0.0377	0.0481	0.0415	0.0530
RU-C	0.0310	0.0396	0.0342	0.0436	0.0377	0.0481	0.0415	0.0530
RU-S	0.0310	0.0396	0.0342	0.0436	0.0377	0.0481	0.0415	0.0530
RU-V	0.0310	0.0396	0.0342	0.0436	0.0377	0.0481	0.0415	0.0530
RU-U	0.0310	0.0396	0.0342	0.0436	0.0377	0.0481	0.0415	0.0530
RU-SI	0.0310	0.0396	0.0342	0.0436	0.0377	0.0481	0.0415	0.0530
RU-FE	0.0310	0.0396	0.0342	0.0436	0.0377	0.0481	0.0415	0.0530
BY	0.0730	0.0932	0.0805	0.1027	0.0887	0.1132	0.0978	0.1249
CAU	0.0489	0.0624	0.0539	0.0688	0.0594	0.0743	0.0655	0.0803
KZ	0.0150	0.0191	0.0165	0.0211	0.0182	0.0233	0.0201	0.0257
PAM	0.0109	0.0139	0.0120	0.0153	0.0132	0.0169	0.0146	0.0186
ZU	0.0240	0.0306	0.0265	0.0338	0.0292	0.0372	0.0322	0.0410
TM	0.0240	0.0306	0.0265	0.0338	0.0292	0.0372	0.0322	0.0410
				MENA				
DZ	0.0490	0.0625	0.0798	0.1019	0.1300	0.1507	0.1747	0.2026
BHQ	0.0258	0.0330	0.0421	0.0537	0.0685	0.0875	0.1095	0.1332
EG	0.0200	0.0255	0.0326	0.0416	0.0531	0.0677	0.0864	0.1103
IR	0.0133	0.0170	0.0217	0.0277	0.0353	0.0451	0.0576	0.0735
IQ	0.0130	0.0166	0.0212	0.0270	0.0345	0.0440	0.0562	0.0717
IL	0.1140	0.1400	0.1623	0.1882	0.2181	0.2529	0.2710	0.2849
JWG	0.1180	0.1422	0.1648	0.1910	0.2215	0.2568	0.2699	0.2836
KW	0.0160	0.0204	0.0261	0.0333	0.0425	0.0542	0.0692	0.0883
LB	0.0910	0.1161	0.1399	0.1622	0.1880	0.2180	0.2527	0.2709
LY	0.0120	0.0153	0.0195	0.0249	0.0318	0.0406	0.0519	0.0662
MA	0.1090	0.1365	0.1582	0.1834	0.2126	0.2465	0.2694	0.2831
OM	0.0090	0.0115	0.0147	0.0187	0.0239	0.0305	0.0389	0.0496
SA	0.0150	0.0191	0.0244	0.0312	0.0398	0.0508	0.0648	0.0827
TN	0.0810	0.1034	0.1319	0.1530	0.1773	0.2056	0.2383	0.2656

AE	0.0380	0.0485	0.0619	0.0790	0.1008	0.1287	0.1492	0.1729
YE	0.0409	0.0521	0.0665	0.0849	0.1084	0.1357	0.1573	0.1824
SY	0.0880	0.1123	0.1379	0.1599	0.1854	0.2149	0.2491	0.2670
Sub-Saharan Africa								
WW	0.1215	0.136	0.1561	0.1801	0.2028	0.2208	0.2354	0.2494
WS	0.102	0.1186	0.1428	0.1695	0.1938	0.2174	0.2409	0.2675
WN	0.1072	0.1274	0.154	0.1758	0.1989	0.2255	0.256	0.2776
NIG-S	0.023	0.028	0.0357	0.0455	0.0581	0.0742	0.0947	0.1208
NIG-N	0.023	0.028	0.0357	0.0455	0.0581	0.0742	0.0947	0.1208
SER	0.0987	0.1199	0.1442	0.1672	0.1938	0.2247	0.2597	0.273
ETH	0.1125	0.1341	0.1555	0.1803	0.209	0.2423	0.2648	0.2783
SOMDJ	0.1179	0.1379	0.1598	0.1853	0.2148	0.249	0.2696	0.2833
KENUG	0.1478	0.1683	0.1951	0.2179	0.2438	0.2698	0.2836	0.298
TZRB	0.0828	0.1007	0.1274	0.148	0.1719	0.1998	0.2315	0.2572
CAR	0.0906	0.1098	0.1349	0.1574	0.1834	0.2124	0.2456	0.2646
COG	0.037	0.045	0.0574	0.0733	0.0935	0.1193	0.1438	0.1667
SW	0.0615	0.0747	0.0928	0.1134	0.1366	0.1629	0.1912	0.2132
ZAFLS	0.0359	0.0435	0.0555	0.0707	0.09	0.1145	0.1374	0.1585
SEA	0.0652	0.0784	0.0988	0.1201	0.1441	0.1676	0.1933	0.221
IOCE	0.0709	0.0861	0.1099	0.1352	0.1525	0.1768	0.205	0.2356
India/ SAARC								
IN-E	0.069	0.0881	0.1124	0.138	0.16	0.1855	0.1855	0.2151
IN-CE	0.044	0.0562	0.0717	0.0915	0.1167	0.1406	0.1406	0.163
IN-W	0.056	0.0715	0.0912	0.1164	0.1403	0.1626	0.1626	0.1885
IN-CW	0.063	0.0804	0.1026	0.131	0.1518	0.176	0.176	0.2041
IN-N	0.039	0.0498	0.0635	0.0811	0.1035	0.1321	0.1321	0.1531
IN-NW	0.066	0.0842	0.1075	0.1346	0.156	0.1809	0.1809	0.2097
IN-UP	0.072	0.0919	0.1173	0.1413	0.1638	0.1899	0.1899	0.2201
IN-S	0.047	0.06	0.0766	0.0977	0.1247	0.1474	0.1474	0.1708
IN-CS	0.059	0.0753	0.0961	0.1227	0.145	0.168	0.168	0.1948
IN-NE	0.06	0.0766	0.0977	0.1247	0.1474	0.1709	0.1709	0.1981
BD	0.093	0.1187	0.143	0.1658	0.1922	0.2228	0.2228	0.2583
NP+BT	0.0636	0.0788	0.0919	0.1092	0.1324	0.1592	0.1596	0.1838
PK-S	0.051	0.0651	0.0831	0.106	0.1327	0.1539	0.1539	0.1784
PK-N	0.051	0.0651	0.0831	0.106	0.1327	0.1539	0.1539	0.1784
AF	0.142	0.1646	0.1908	0.2212	0.2565	0.2696	0.2696	0.2833
LK	0.051	0.0651	0.0831	0.106	0.1327	0.1539	0.1539	0.1784
Northeast Asia								
JP-E	0.1700	0.1971	0.1804	0.2091	0.1913	0.2218	0.2030	0.2353
JP-W	0.1700	0.1971	0.1804	0.2091	0.1913	0.2218	0.2030	0.2353
KR	0.0850	0.1085	0.0937	0.1196	0.1033	0.1319	0.1139	0.1399
KP	0.0690	0.0881	0.0761	0.0971	0.0839	0.1070	0.0925	0.1180
CN-NE	0.0690	0.0881	0.0761	0.0971	0.0839	0.1070	0.0925	0.1180
CN-N	0.0690	0.0881	0.0761	0.0971	0.0839	0.1070	0.0925	0.1180
CN-E	0.0690	0.0881	0.0761	0.0971	0.0839	0.1070	0.0925	0.1180
CN-C	0.0690	0.0881	0.0761	0.0971	0.0839	0.1070	0.0925	0.1180
CN-S	0.0690	0.0881	0.0761	0.0971	0.0839	0.1070	0.0925	0.1180
CN-TB	0.0690	0.0881	0.0761	0.0971	0.0839	0.1070	0.0925	0.1180
CN-NW	0.0690	0.0881	0.0761	0.0971	0.0839	0.1070	0.0925	0.1180
CN-XU	0.0690	0.0881	0.0761	0.0971	0.0839	0.1070	0.0925	0.1180
MN	0.0730	0.0932	0.0805	0.1027	0.0887	0.1132	0.0978	0.1249
Southeast Asia								
NZ	0.139	0.1611	0.1868	0.2166	0.251	0.2691	0.2828	0.2972
AU-E	0.111	0.1363	0.158	0.1832	0.2124	0.2462	0.2462	0.2691
AU-W	0.111	0.1363	0.158	0.1832	0.2124	0.2462	0.2462	0.2691
ID-PG+NG	0.1756	0.1906	0.208	0.2285	0.2527	0.2748	0.2748	0.298
ID-SU	0.047	0.06	0.0766	0.0977	0.1247	0.1474	0.1474	0.1708
ID-JV+TL	0.047	0.06	0.0766	0.0977	0.1247	0.1474	0.1474	0.1708

ID-KL-SW	0.047	0.06	0.0766	0.0977	0.1247	0.1474	0.1474	0.1708
MY-W+SG	0.0931	0.1059	0.1226	0.1392	0.1584	0.1756	0.1729	0.1946
MY-E+BN	0.0456	0.075	0.0945	0.1182	0.1436	0.1665	0.0197	0.193
PH	0.128	0.1484	0.172	0.1994	0.2312	0.2628	0.2628	0.2762
MM	0.096	0.1225	0.1448	0.1679	0.1946	0.2256	0.2256	0.2615
TH	0.059	0.0753	0.0961	0.1227	0.145	0.168	0.168	0.1948
LA	0.032	0.0408	0.0521	0.0665	0.0849	0.1084	0.1084	0.1357
VN	0.059	0.0753	0.0961	0.1227	0.145	0.168	0.168	0.1948
KH	0.151	0.1751	0.2029	0.2353	0.2622	0.2756	0.2756	0.2897
				North America				
CA-W	0.0785	0.1000	0.1254	0.1485	0.1724	0.1997	0.2293	0.2578
CA-E	0.0756	0.0953	0.1153	0.1391	0.1667	0.1932	0.2196	0.2420
US-NENY	0.1130	0.1391	0.1606	0.1854	0.2140	0.2451	0.2696	0.2833
US-MA	0.0972	0.1229	0.1464	0.1697	0.1967	0.2280	0.2593	0.2738
US-CAR	0.0975	0.1244	0.1470	0.1704	0.1976	0.2290	0.2603	0.2736
US-S	0.0931	0.1188	0.1430	0.1658	0.1922	0.2228	0.2558	0.2709
US-TVA	0.0807	0.1030	0.1315	0.1524	0.1767	0.2048	0.2375	0.2624
US-MW	0.0926	0.1178	0.1423	0.1651	0.1914	0.2219	0.2539	0.2709
US-C	0.0880	0.1123	0.1385	0.1606	0.1862	0.2159	0.2487	0.2696
US-TX	0.0868	0.1108	0.1361	0.1578	0.1829	0.2120	0.2458	0.2687
US-SW	0.0920	0.1167	0.1412	0.1637	0.1898	0.2200	0.2513	0.2721
US-NW	0.0762	0.0972	0.1214	0.1465	0.1698	0.1969	0.2274	0.2551
US-CA	0.1164	0.1402	0.1626	0.1885	0.2185	0.2533	0.2715	0.2853
US-AK	0.1964	0.2277	0.2588	0.2720	0.2859	0.3005	0.3158	0.3319
US-HI	0.3050	0.3206	0.3369	0.3541	0.3722	0.3912	0.4111	0.4321
US-GU	0.0754	0.0963	0.1222	0.1461	0.1694	0.1964	0.2276	0.2583
MX-NW	0.0472	0.0602	0.0768	0.0980	0.1251	0.1479	0.1714	0.1987
MX-N	0.0477	0.0609	0.0777	0.0991	0.1265	0.1495	0.1733	0.2009
MX-C	0.0482	0.0615	0.0786	0.1003	0.1280	0.1483	0.1720	0.1993
MX-S	0.0465	0.0594	0.0758	0.0967	0.1235	0.1459	0.1691	0.1961
				South America				
CAM	0.0972	0.1181	0.1400	0.1664	0.1953	0.2207	0.2435	0.2637
CO	0.0187	0.0237	0.0300	0.0377	0.0473	0.0595	0.0750	0.0946
VE	0.0800	0.1021	0.1303	0.1511	0.1751	0.2030	0.2354	0.2624
EC	0.0612	0.0781	0.0997	0.1272	0.1504	0.1743	0.2021	0.2343
PU	0.1080	0.1352	0.1567	0.1817	0.2107	0.2442	0.2669	0.2805
CSA	0.0586	0.1060	0.1273	0.1533	0.1795	0.2081	0.2331	0.2558
BR-S	0.1590	0.1843	0.2137	0.2477	0.2655	0.2791	0.2933	0.3082
BR-SP	0.1590	0.1843	0.2137	0.2477	0.2655	0.2791	0.2933	0.3082
BR-SE	0.1590	0.1843	0.2137	0.2477	0.2655	0.2791	0.2933	0.3082
BR-N	0.1590	0.1843	0.2137	0.2477	0.2655	0.2791	0.2933	0.3082
BR_NE	0.1590	0.1843	0.2137	0.2477	0.2655	0.2791	0.2933	0.3082
AR-NE	0.1070	0.1317	0.1564	0.1814	0.2103	0.2372	0.2662	0.2798
AR-E	0.0930	0.1187	0.1430	0.1658	0.1922	0.2228	0.2583	0.2714
AR-W	0.0930	0.1187	0.1430	0.1658	0.1922	0.2228	0.2583	0.2714
CL	0.1280	0.1484	0.1720	0.1994	0.2312	0.2628	0.2762	0.2903

Table A7: *Financial assumptions for the system technologies, applied to every region and all modelled years.*

	Unit	2015	2020	2025	2030	2035	2040	2045	2050
Financial Assumptions									
PV Rooftop									
Capex	[€/kW$_p$]	1360	1090	890	760	680	610	550	500
Opex fix	[€/(kW$_p$·a)]	20	16	13	11	10	9	8	8
Lifetime	[a]	30	30	35	35	35	40	40	40
Stationary Battery									
Capex	[€/kWh$_{cap}$]	600	300	200	150	120	100	85	75
Opex fix	[€/(kWh$_{cap}$·a)]	24	9	5	3.75	3	2.5	2.125	1.875
Opex var	[€/(kWh$_{through}$·a)]	0.0002	0.0002	0.0002	0.0002	0.0002	0.0002	0.0002	0.0002
Lifetime	[a]	15	20	20	20	20	20	20	20
Heat Pump									
Capex	[€/kW$_{el}$]	1600	1500	1500	1400	1400	1300	1300	1200
Opex fix	[€/(kW$_{el}$·a)]	20.6	19.5	19.5	17.9	17.9	17.4	17.4	16.9
Lifetime	[a]	20	20	20	20	20	20	20	20
Heating Cartridge									
Capex	[€/kW$_{el}$]	570	570	570	570	570	570	570	570
Opex fix	[€/(kW$_{el}$·a)]	10	10	10	10	10	10	10	10
Lifetime	[a]	30	30	30	30	30	30	30	30
Thermal Energy Storage									
Capex	[€/kWh$_{cap}$]	50	40	30	30	20	20	20	20
Opex fix	[€/(kWh$_{cap}$·a)]	0.75	0.6	0.45	0.45	0.3	0.3	0.3	0.3
Lifetime	[a]	25	25	25	25	30	30	30	30

Table A8: *Efficiency assumptions for the storage and energy conversion technologies, applied to every region and all modelled years.*

	Unit	Stationary Battery	TES	Car 1	Car 2	HP	Heating Cartridge
Efficiency Assumptions							
Charging	[%]	96.8	100	96.8	96.8	-	-
Discharging	[%]	96.8	100	-	96.8	-	-
Standby losses	[%/h]	0	0.15	0	0	-	-
Electricity-to-Heat conversion nominal COP	-	-	-	-	-	3.8	1
Electricity-to-Heat conversion PV additional filling COP	-	-	-	-	-	3.0	1

INTEGRATION OF SELF-SUPPLY ROOFTOP SOLAR SYSTEMS (PV & HOT WATER) WITH BATTERY STORAGE TO REDUCE GRID-BUY ELECTRICITY BY >96% AND ELIMINATE EVENING & MORNING ENERGY PEAKS: A CASE STUDY FOR RESIDENTIAL HAWAII

John Borland[1], Jay Moore[2], Corpuz Poncho[2], Takahiro Tanaka[3] and Harumi McClure[3]
[1]J.O.B. Technologies, 98-1204 Kuawa St, Aiea, Hawaii, 96701
[2]Poncho's Solar, 2669 Kilihau St, Honolulu, HI 96819
[3]Tabuchi Electric, 5225 Hellyer Ave, suite 150, San Jose, CA 95138
e-mail: JohnOBorland@aol.com

ABSTRACT: We report field data for the past 15 months starting with the June 1, 2016 installation of the rooftop solar system + battery storage and the steps we took to achieve >96% reduction in the daily grid-buy electricity from Hawaiian Electric going from 47kWh/day for April 2016 to 1.8kWh/day for July 2017.This required using multiple renewable energy storage techniques including electrical battery storage, hot thermal storage and cold thermal storage along with the optimized integration of TOU (time of use) for the key top energy consuming household appliances. We can now routinely achieve zero grid-buy satisfying 100% residential energy needs with renewable energy 6 days a week eliminating the Duck Curve problem by having the home energy management system in the standalone mode for off-grid operation.

Keywords: solar-PV, hot thermal storage, cold thermal storage, electrical battery storage, 100% renewable energy, off-grid

1 INTRODUCTION

The residential rooftop solar-PV market in Hawaii and California has reached the point that excess PV generation export/sold back to the grid with NEM (net energy metering) program has created what the industry calls the Duck Curve as shown in Fig.1, reported by Hawaiian Electric (HECO) [1]. The summer time back-feed of solar PV during peak generation time between 10AM to 2PM actually resulted in zero to negative load on the transformer and then a steep ramp from 3PM to evening peak at 8PM. For NEM programs, the economics and solar-PV electricity utility bill cost savings is quite different than for post-NEM. ROI (return on investment) is determined by generating the maximum amount of excess solar PV energy possible and selling/exporting it back to the grid so this can be used to off-set the costs of grid-buy energy used at night to achieve zero net metering. Since evening household appliance energy usage is usually much higher than daytime usage, over sizing rooftop solar-PV systems is required to compensate night time energy usage for zero net metering therefore resulting in the severe Duck Curve problem causing the utility (HECO) to end their NEM program so today in Hawaii with post-NEM no export/selling of excess rooftop solar PV generation back to the grid. In California to reduce the benefits of NEM, the peak time-of-use (TOU) rates are being shifted from daytime when solar-PV generation is abundant to evening when solar-PV generation drops off and residential energy demand is highest so with post-NEM the economics for electricity cost savings is very different. The key to fastest ROI in the post-NEM world is through reducing daily utility grid-buy energy to zero with properly sized rooftop solar-PV system with multiple energy storage (electrical battery and thermal) to reduce total system costs. Therefore we report steps taken to flatten residential energy usage curve eliminating the Duck Curve problem and achieve zero grid-buy for post-NEM residential Hawaii with an ROI of 2.7 years [2, 3, 4].

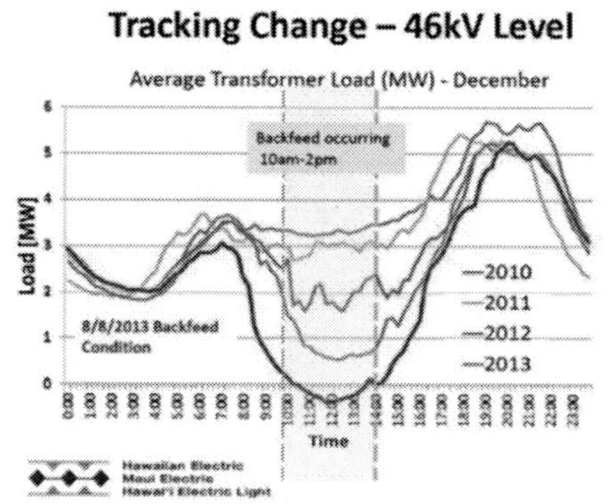

Fig.1: Hawaiian Electric utility load and severe Duck Curve formed by rooftop solar-PV back-feed.

2 SOLAR+BATTERY SYSTEM: 1st Month Results

On June 1, 2016 Poncho's Solar installed a 27-panel 7kWh rooftop solar-PV + 10kWh Panasonic Li-ion battery for storage connected to the Grid by the Tabuchi Electric 5.5kWh inverter control system in a post-NEM mode (no export back to the grid) and a single panel solar thermal hot water system with 80 gallons of storage as shown in Fig.2. After the 1st month the monthly utility bill from HECO was only reduced by ~50% or $180 still leaving a $180/month HECO bill which was 10x higher than NEM achieving zero net metering with a monthly minimum HECO bill of $18/month. This results in an ROI of 13-16 years for the $29.5K system due to battery replacement in year 11. The ROI is well beyond the 10 years battery warranty so a poor economic investment. Because of post-NEM operation with no export or selling of excess PV generation back to the grid PVWATTS.NREL.GOV calculation results showed about 35% PV energy generation loss or curtailment [2].

Fig.2: Poncho's Solar installed rooftop solar-PV and solar thermal hot water systems.

3 SOLAR+BATTERY SYSTEM: 2nd to 14th Month Results

Starting in July 2016 we spent the next 10 months and successfully figured out how best to operate (integrate and optimize) the solar system with multiple storage to achieve a reduction of 95% in grid-buy electricity going from 47kWh/day for April 2016 to 2.6kWh/day for April 2017 as shown in Fig.3 including 12 days with zero grid-buy and therefore 100% renewable energy as shown in Fig.4. Achieving zero grid-buy reduces ROI from 16 years to 7 years as shown in Fig. 5 or to 2.7 years if you include the state of Hawaii and federal government tax credits which reduces the net system costs to $13.5K.

As shown in Fig.3 grid-buy was first reduced by 50% to 22kWh/day after 1 month by installing rooftop solar PV with battery storage. Optimizing energy usage further reduced grid-buy by 70% to <15kWh/day from month 4 to 8 in Fig.3 which includes adjusting TOU of key household appliances to coincide with peak solar PV generation and super charging solar hot thermal storage peak temperature to >160°F. Additional hardware and software improvements to the solar and storage system between January to March 2017 resulted in the achievement of zero grid-buy and reduction of grid-buy by 95% to 2.6kWh/day for April 2017.

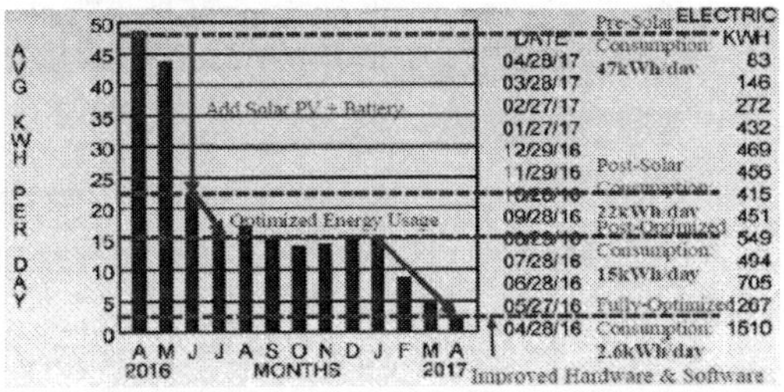

Fig.3: HECO home utility grid-buy monthly bill showing 53% reduction after 1 month.

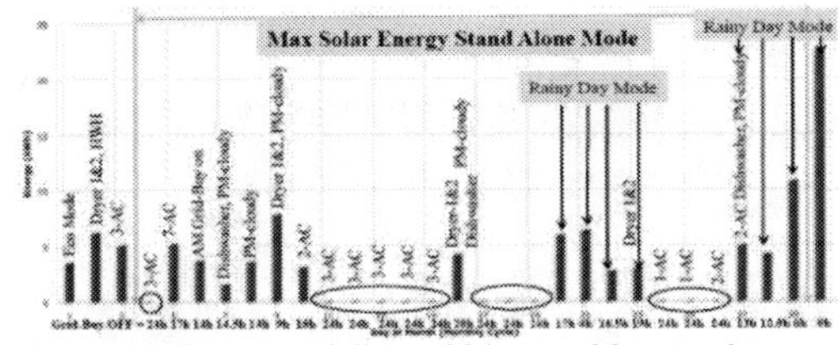

Fig.4: April 2017 daily grid-buy with 12 days at 0kWh/day.

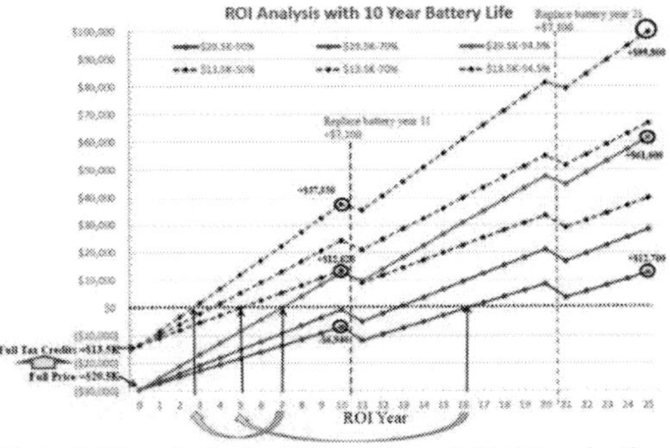

Fig.5: ROI analysis for 50%, 70% and 94.5% reduction in grid-buy.

To achieve 100% renewables we needed better monitoring of daily energy usage to look for areas to further reduce grid-buy by optimizing time of use (TOU) of key household appliances and identifying the root cause of each energy spike so we could find a renewable energy alternative. We used three different energy monitoring systems with 1 hour down to 3 seconds data collection resolution, Tabuchi Electric remote wall real time monitor with 3 second resolution, Laplace Solarlink web-site/App with 1 hour and 1 minute off-line monitoring and Hawaii's Blue Planet Foundation Bidgely web-site/App with 5 minute real time monitor. Fig.6 shows summer time energy usage for a hot sunny day (6/12/16) with total energy consumption/usage of 83.5kWh/day, solar PV generation of 39.5kWh/day and grid-buy energy of 42.3kWh/day. The high energy usage was caused by running 7-A/C (air conditioneer) units for 9 hours (~48kWh/day), hot water heating 6 times during the day (16kWh/day) and clothes dryer for 2.5 hours (12kWh/day) which adds up to 78kWh/day leaving 5kWh/day for other daily household appliance energy usage.

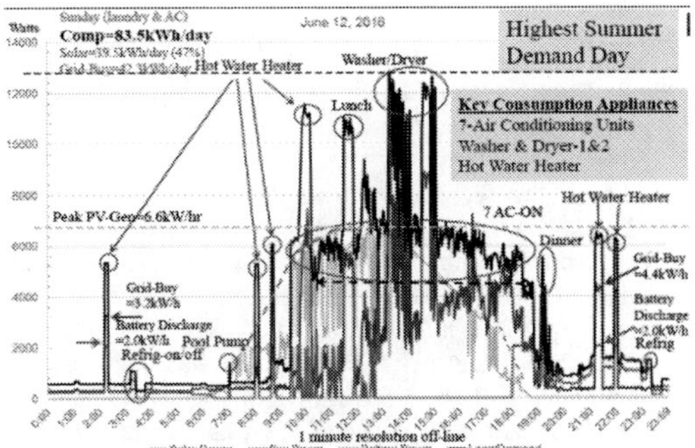

Fig.6: Summer time sunny day home energy usage.

In contrast, Fig.7 shows winter time energy usage for a rainy day (11/7/16) with total energy consumption of 34.9kWh/day, solar PV generation of 12.7kWh/day and grid-buy energy of 21.0kWh/day. Due to the rainy day the solar thermal hot water system peak water temperature only reached 105°F by 2PM requiring power to the electrical heater element to heat the water to 135°F 5 times during the day requiring multiple battery discharge and grid-buy energy shown in green and red totaling ~10kWh. The other grid-buy energy was used for cooking, refrigerator/freezer and pool pump. Fig.8 shows the flat grid-buy for a sunny winter day (1/19/17) and the elimination of Duck Curve morning and evening grid-buy energy peaks/spikes. The total energy consumption was 18.7kWh/day, solar PV generation of 10.4kWh/day and grid-buy of 8.4kWh. The grid-buy for hot water heating was 0.0kWh for this day, the hot thermal storage achieved super charging level by reaching 163°F for the solar thermal hot water system lasting 4 full baths of 40 gallons each or ~16kWh equivalent of energy storage.

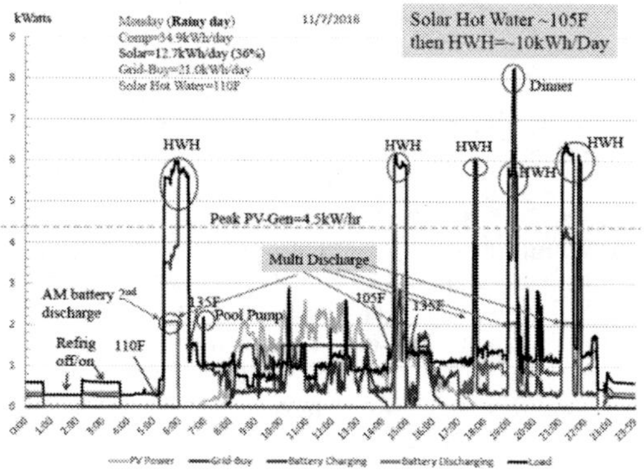

Fig.7: Winter time rainy day home energy usage.

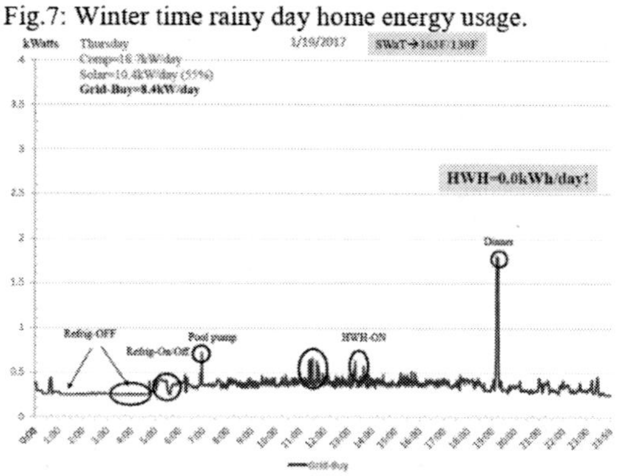

Fig.8: Elimination of Duck Curve morning and evening grid-buy energy peaks/spikes.

From the Pareto analysis the top energy usage appliance was A/C at 18-48kWh/day (2-5.5kWh peak energy) during summer months, #2 was the hot water heating at ~16kWh/day (4.5kWh peak energy), #3 was dryer at ~12.0kWh/day (5.5kWh peak energy), #4 was refrigerator/frezzer at 7.2kWh/day, #5 was pool pump at 6.5kWh/day and #6 was night lights at 2kWh/day. A break down showing household appliance energy usage differences between winter time, low summer time and high summer time are shown in Fig.9. Only A/C energy usage changes going from 60% during summer time high to 0% in winter.

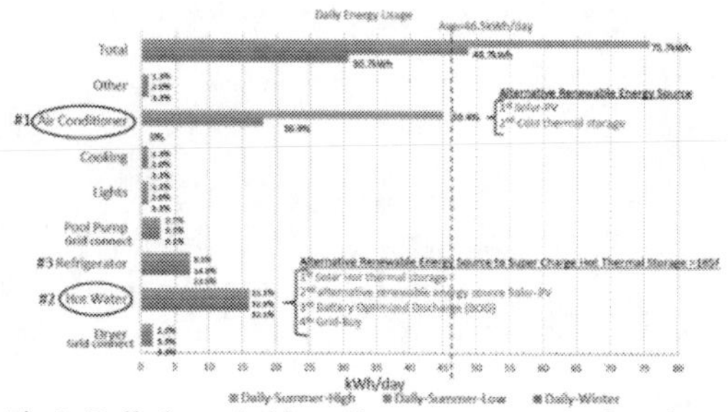

Fig.9: Daily household appliance energy usage for winter time versus summer time.

To achieve super charged >160°F hot thermal storage without grid-buy we made some hardware modifications for battery optimized discharge (BOD) at the end of Jan 2017 to give ~16kWh/day equivalent stored solar hot thermal energy. The Tabuchi Electric control system software was upgraded in mid-Feb 2017 to further reduce grid-buy by another 5kWh/day and the benefits for these new learning resulted in the continued drop in grid-buy from Feb 2017 to April 2017 achieving 2.6kWh/day shown in Fig. 10. For July 2017 grid-buy reached an all time low of 1.9kWh/day. A bar chart showing the renewable energy source usage is shown in Fig. 11 for winter time and summer time. Battery discharge stays at 6.6kWh/day year round but a 2nd battery was added on August 14, 2017 to increase total electrical battery storage capacity to 19.8kWh for >12kWh/day discharge. Solar PV generation goes from a low of 13kWh/day in winter to a high of 39kWh/day in summer, hot thermal storage goes from a low of 10kWh/day in winter to 16kWh/day in summer and cold thermal storage goes from zero in winter to a high of 10kWh/day in summer.

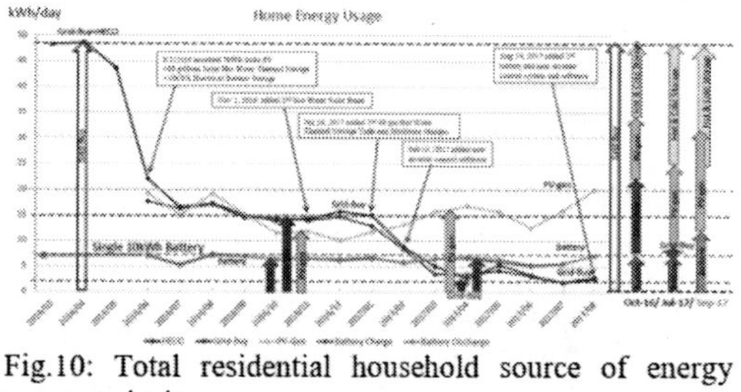

Fig.10: Total residential household source of energy usage analysis.

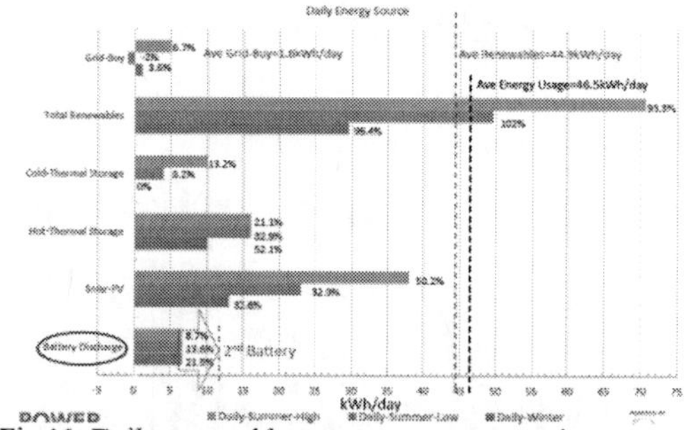

Fig.11: Daily renewable energy source comparison.

HEMS (home energy management system) communication between solar PV, battery charging/discharging and grid-buy with the integration of solar energy with multiple storage and optimized controllable TOU of key appliances are critical to realize these economic beneficial values with shortest ROI. Fig. 12 shows the household appliances that can be operated in the standalone (off-grid) mode using 100% renewable energy. The only appliances requiring grid-tie are dryer, stove/oven and pool pump using the upgraded Tabuchi Electric 16GU-2 inverter with dual battery for up to 3.3kWh energy demand in the standalone mode installed on August 14, 2017.

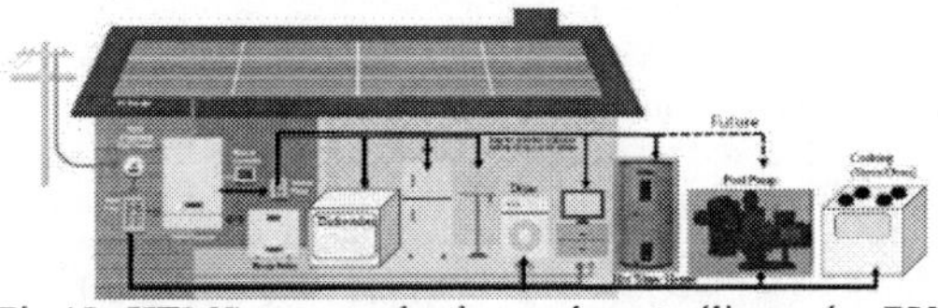

Fig.12: HEMS communicating and controlling solar PV generation, battery charging/discharging and grid-buy with up to 3.3kWh in standalone mode.

The benefits of the dual battery storage is shown for the results of September 5, 2017 a zero grid-buy standalone mode off-grid operation with multiple dual battery discharge of 13.7kWh/day in Fig.13. Note the triple AM morning BOD for hot water heating at 4AM, 6AM and 9AM due to the partly cloudy day before when solar hot thermal storage only achieved a super charged peak hot water temperature of 143°F resulting in the 4AM hot water temperature of only 115°F. Dual battery peak discharge of 4.1kWh and total discharge of 15.1kWh/day is shown in Fig.14 for Sunday September 3, 2017 to meet the total peak energy demand of 8-9kWh from the washer & dryer on this cloudy day. The HEMS was in the grid-tie mode for 3 hours from 10:30AM to 2:30PM to meet the high energy demand from the stove/oven and dryer resulting in a total grid-buy of only 2.8kWh/day.

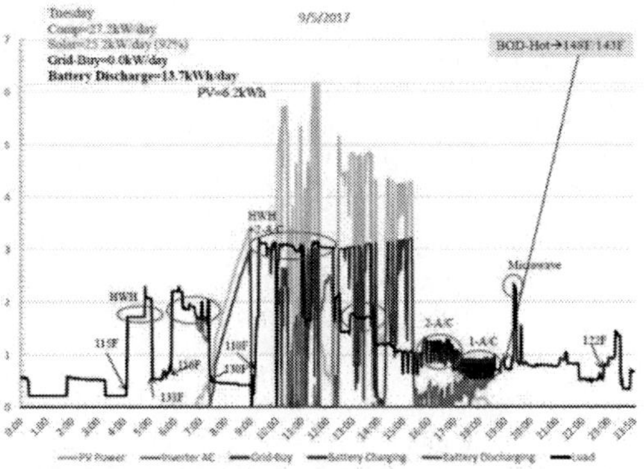

Fig.13: Dual battery with triple AM discharge in standalone mode with zero grid-buy.

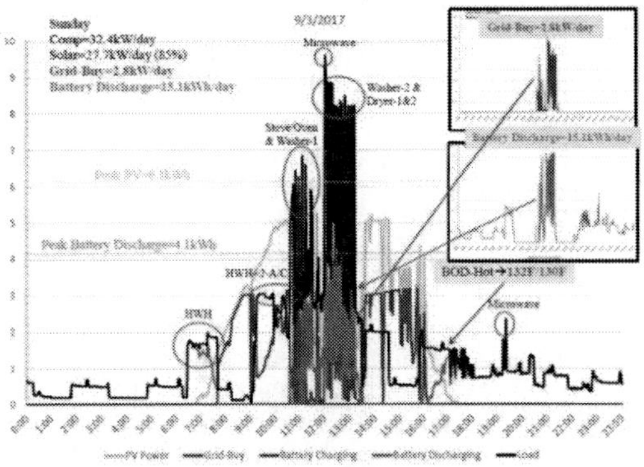

Fig.14: Dual battery with 4.1kWh peak batery discharge.

5 SUMMARY

For residential Hawaii customer self-supply in a post-NEM world, solar-PV + solar-thermal systems integrated with multiple storage techniques (electrical battery storage, hot thermal storage and cold thermal storage) and optimized energy usage of key household appliances with HEMS we reduced the utility grid-buy energy average daily usage from 47kWh/day for April 2016 to 1.9kWh/day for July 2017 with 12 days a month at zero grid-buy in the standalone mode for off-grid operation achieving 100% energy needs from renewables. This also eliminated the AM and PM grid-buy peaks causing the Duck Curve.

6 REFERENCES

[1] www.hawaiianelectric.com website for HECO Duck Curve.

[2] J. Borland et al., ">94%.5% Reduction in Grid-Buy Electricity and Elimination of AM & PM Energy Peaks/Spikes by Optimizing Energy Usage and Integration of Customer Self-Supply Rooftop Solar PV with Electrical & Thermal (Hot & Cold) Storage Batteries: A Case Study for Residential Hawaii", 44th IEEE-PVSC, paper #832, June 26-30, 2017.

[3] J. Borland et al., "Achieving 100 Percent Residential Renewables in Hawaii with Solar+Multi-Storage", Renewable Energy World, August 2, 2017.

[4] J. Borland et al., "Solar Energy (PV and Thermal) + Multi-Storage: Case Study For 100% Residential Hawaii Energy Needs", Solar Power International 2017, Technical Symposium paper 1.2.1, presentation material, Sep 11, 2017.

COMPARISON OF STRATEGIES FOR NET DEMAND FORECASTING IN CASE OF PHOTOVOLTAIC POWER PRODUCTION AND ELECTRICITY CONSUMPTION

Dennis van der Meer Joakim Widén Joakim Munkhammar

Built Environment Energy Systems Group (BEESG), Solid State Physics, Department of Engineering Sciences, The Ångström Laboratory, Uppsala University, Uppsala, SE-751 21, Sweden

ABSTRACT: This paper aims to investigate the relative difference in accuracy between forecasting net demand, i.e., electricity consumption less the photovoltaic (PV) power production, directly and indirectly, where the latter implies forecasting consumption and production separately before subtraction. Depending on the variability and penetration of PV power production, variability of the net demand time series is likely to increase as well, which may influence accuracy of the forecast. Therefore, we gradually increase the PV power production of a commercial building and quantify the accuracy. The well-known AutoRegressive Integrated Moving Average (ARIMA) model and AR neural network (ARNN) are employed to forecast the univariate time series. We show that the direct strategy leads to a forecast with higher accuracy in terms of root mean square error (RMSE), and is preferable for additional practical reasons as well, e.g., less computational burden.

Keywords: Forecasting, PV, Electricity consumption, Net demand

1 INTRODUCTION

The increasing penetration of photovoltaic (PV) systems and smart meters in the power system provides utilities and researchers with both challenges and opportunities. A striking example of a challenge is the increased strain on the electricity grid due to an increasingly variable net demand from buildings that produce and consume their own electricity through PV panels [1]. In this regard, net demand is considered to be electricity consumption less the PV power production and consequently, with increasing PV penetration, net demand will likely become more variable due to the variable nature of cloud movement over PV systems [1] [2].

An interesting opportunity is the increase of smart meter installations. Nowadays the time series of both PV power production and electricity consumption are on the same spatial and temporal resolution, which paves way to forecasting net demand directly, i.e., a forecast of consumption less production, as opposed to forecast net demand indirectly, i.e., forecast consumption and production separately and subsequently subtract the latter from the former [3].

Generally speaking, the typical day-ahead electricity consumption forecasting error on an aggregated level amounts to 3% [3], whereas the corresponding figure is 23% - 41% for irradiance forecasting with a one hour lead time, see e.g., [4]. Thus, net demand forecasting offers the potential advantage of a residual time series that may be less variable and can therefore be forecast with greater accuracy than the time series of PV power alone, assuming that consumption is higher than production. To the best of our knowledge there is one other study that investigated net demand forecasting, although they named it "net load forecasting" [5]. In their study, the authors compared two methods for net demand forecasting, i.e., an additive and integrated approach, of a microgrid with 33% annual solar energy penetration. The authors found that the additive model was outperformed by 10.69% in terms of root mean square error (RMSE), which is interesting since the integrated model utilized the PV power forecast as input to the load forecast. Unfortunately, the authors did not elaborate on this. Furthermore, the authors did not specify the forecast horizon, which we therefore assume to be 1 hour. Our study differs from [5] in three notable ways. Firstly, we compare forecasting electricity consumption and PV power production separately and subtract them afterwards, i.e., the indirect strategy, with subtracting the two time series directly and forecasting the residuals, i.e., the direct strategy. Secondly, we examine the stability of these strategies in case PV power penetration increases, and finally, we forecast 1 and 2 hours ahead to assess whether that affects accuracy.

A wide variety of data-driven models have been proposed for solar power and electricity consumption forecasting, of which a thorough review can be found in [6]. In case of solar power forecasting, these models include AutoRegressive Integrated Moving Average (ARIMA) models [4], [7] - [10], Artificial Neural Networks (ANNs) [11] - [16] and other regression models such as quantile regression (QR) [17] - [20]. Regarding electricity consumption forecasting, ARIMA and other regression models, such as multiple linear regression (MLR), are more common, see e.g., [21] - [27], although ANNs have been employed as well [28] - [30]. Since we aim to forecast both PV power production and electricity consumption by means of a direct and indirect strategy, we utilize both ARIMA and AutoRegressive Neural Network (ARNN) and implement these using the "forecast" package in R [31], due to its straightforward implementation of these models. Moreover, it is important to stress that the aim of this paper is not to create the most accurate model, but rather to study the effect of direct and indirect forecasting on accuracy.

The main contribution of this paper is therefore to establish what methodology will lead to the most accurate forecast in case net demand is considered: direct or indirect forecasting.

2 METHODOLOGY

2.1 Data

Production and consumption data of a commercial building in Uppsala, Sweden, are used to train two models, i.e., the ARIMA and ARNN models, and forecast with 1 and 2 hour horizons at a temporal resolution of 1 hour. It is perhaps interesting to note that the multiple step ahead forecast is performed in a recursive fashion. Furthermore, we utilize 70% of 2016 as training data, and 30% as testing data. Figure 1a presents the time series for a typical week in July 2016, from which at least three points of interest can be noted. Firstly, we observe that PV power production shows substantially more variability than electricity consumption, as anticipated. In fact, after normalizing the three respective time series on a scale of 0 to 1, we find that variances are 0.023, 0.046 and 0.022 for electricity consumption, PV power production and net demand, respectively. We can therefore conclude that variability does in fact decrease in case of net demand. Secondly, it can be seen that PV power production is relatively low compared to electricity consumption and finally that, because of the aforementioned, the net demand time series is similar to the original one of electricity consumption.

2.2 Approach

As stated above, investigating the accuracy of direct and indirect forecasting methodologies is the key element of this paper, inspired by [32], and is briefly explained here. The idea behind the direct forecasting strategy is that one subtracts the variable time series, i.e., PV power production, of the less variable time series, i.e., electricity consumption, in order to attain a relatively stable time series, i.e., net demand, which can be forecast more accurately. However, with increasing share of PV power into the net demand time series, variability of the latter will increase and it therefore seems plausible that there exists a limit after which the direct strategy might become counter effective. Examples of this phenomenon are presented in Figures 1a and b, where PV capacity is scaled with a factor of 4.5 and 9, respectively. In contrast to Figure 1a, the PV power production has a more substantial impact with increased capacity, which is especially visible for the case of net demand. There we can observe a few noteworthy points, e.g., that on sunny weekend days drops below zero and overall time series appears more variable. Numerically however, the variance of electricity consumption, PV power generation and net demand are found to be 0.023, 0.046 and 0.015, respectively, in case of Figure 1c. Therefore, we can conclude that in our case the increase of installed capacity has reduced the variance of net demand with 33%.

Consequently, we investigate what the effect on accuracy is when the share of PV power and the forecast horizon increase in terms of deterministic forecast characteristics. The following subsections introduce the forecast models and performance metrics utilized in this study, respectively.

2.3 Forecast models

2.3.1 Autoregressive integrated moving average

The ARIMA model is an established method to separate signal from noise in univariate time series modeling [33]. We take no exogenous variables into account in order to allow for fair comparison and reduce computational burden. The ARIMA model can be expressed in polynomial form as:

$$\phi(B)(1-B)^{\mathrm{d}}X_t = \theta(B)\varepsilon_t,$$

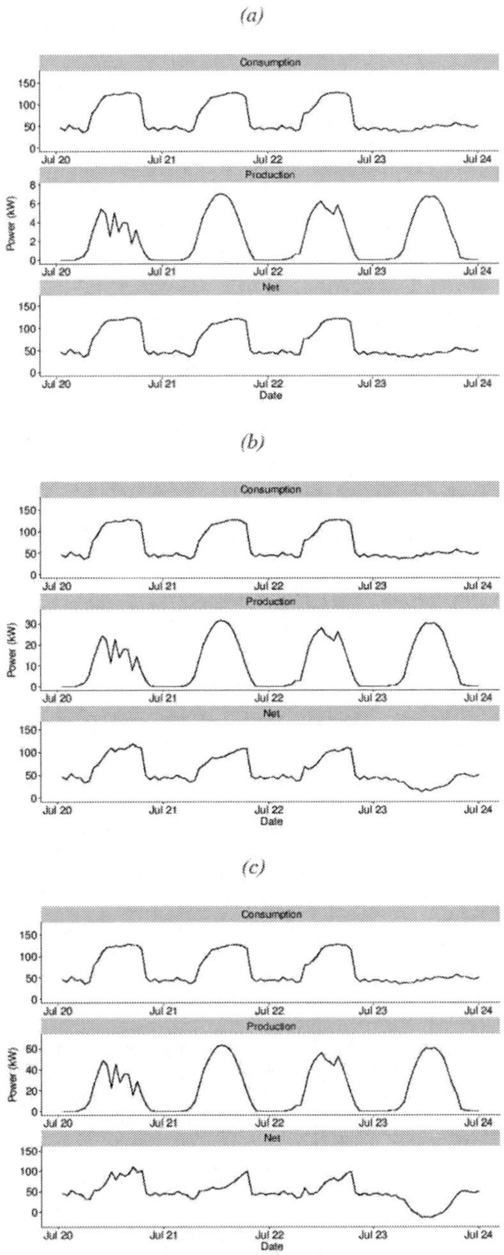

Figure 1: Time series of power production, consumption and net demand for six days in July, with increasing PV system capacity. More specifically, (a) presents the system in its original state, i.e., a 10 kW PV system. For figure (b), the PV system capacity is multiplied with 4.5 and for (c) it is multiplied with 9.

where $\phi(B) = 1-\phi_1 B-\phi_2 B^2-\cdots-\phi_{\mathrm{p}}B^{\mathrm{p}}$ describes the AR(p) process with $\phi_1,\ldots,\phi_{\mathrm{p}}$ as parameters, and $\theta(B) = 1+\theta_1 B+\theta_2 B^2+\cdots+\theta_{\mathrm{q}}B^{\mathrm{q}}$ the MA(q) process with $\theta_1,\ldots,\theta_{\mathrm{q}}$ as parameters. In addition, ε_t is a white noise term and B is the backward shift operator, defined as $BX_t = X_{t-1}$.

In this study, we use the "Arima" function of the "forecast" package [31] in R for the ARIMA predictions.

2.3.2 Autoregressive neural network

The ARNN utilized in this study is implemented with the "nnetar" function of the "forecast" package in R [31]. This particular neural network is a feed-forward neural network with a single hidden layer, that has lagged measurements as inputs. An exemplary ARNN is presented in Figure 2.

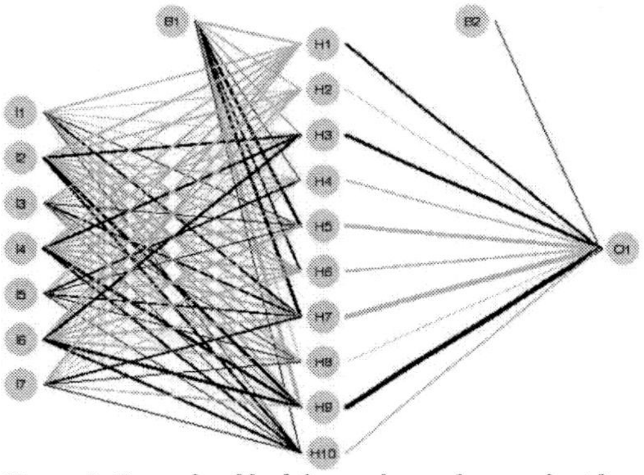

Figure 2: Example of feed-forward neural network with an input layer of 7 neurons and a hidden layer of 10 neurons. The thickness of the lines represents the weighting of input- to hidden layer.

Generally, it is understood that neural networks are not straightforward to apply, since many decisions regarding its parameterization ought to be made, e.g., architecture, learning rate, activation function, etc. [34]. Therefore, we view the neural network as a black-box for this study, by utilizing the "nnetar" function, and refer the interested reader to an elaborate description of NNs by Inman et al. [35].

2.4 Performance metrics

The deterministic error measures used in this study are the mean absolute error (MAE), mean absolute percentage error (MAPE), RMSE and normalized root mean square error (nRMSE), which are defined as follows:

$$\mathrm{MAE} = \frac{1}{N}\sum_{i=1}^{N}|\hat{y}_i - y_i|, \quad (1)$$

$$\mathrm{MAPE} = \frac{1}{N}\sum_{i=1}^{N}\frac{|\hat{y}_i - y_i|}{y_i}, \quad (2)$$

$$\mathrm{RMSE} = \sqrt{\frac{1}{N}\sum_{i=1}^{N}(\hat{y}_i - y_i)^2}, \quad (3)$$

$$\mathrm{nRMSE} = \sqrt{\frac{1}{N}\sum_{i=1}^{N}\left(\frac{\hat{y}_i - y_i}{y_i}\right)^2}, \quad (4)$$

where N is the length of the time series, $\hat{y}_i$ is the forecast value and y_i is the measured value.

3 RESULTS

In this section we present the results, which are achieved by gradually increasing the PV power capacity on the selected building and subsequently computing the performance metrics defined in eqs. (1) – (4). With regards to Figures 3a - 3d, we divide the result of the direct strategy by that of the indirect strategy, which implies that a result below unity favors the direct strategy, and vice versa.

3.1 ARIMA

Figure 3 presents the results of the forecasts using the AR and ARNN models with a forecast horizon of 1 and 2 hours. In particular, Figures 3a and b present the relative MAE and MAPE between the direct and indirect strategy, respectively. From these figures, we can observe that the accuracy of the ARIMA varies substantially among the two strategies, which is an observation that applies to the other metrics as well, albeit to a lesser extent. As a result, it is therefore difficult to point into the direction of a preferred strategy solely based on MAE or MAPE. However, Figures 3c and d do offer some additional pointers. The variation still exists but the overall consensus points towards the direct strategy. Since RMSE penalizes outliers more harshly and taking into account the results in Figure 3a and b, it can be deduced that the indirect strategy produces more outliers. At the highest penetration levels, there are substantial fluctuations between the two strategies and we come back to this in the next section. Overall however, we observe that increasing the installed capacity implies a downward trend in the relative performance metric at hand, and that the performance metric mainly determines the superior strategy.

In addition, Figures 3a - d present the comparison between two forecast horizons, i.e., 1 and 2 hours. In this comparison, there is general consensus that the relative measures of forecasts with 2 hour horizon lie below that of forecasts with 1 hour horizon. Especially in case one considers RMSE or nRMSE, the direct strategy is more accurate with increasing forecast horizon. In case of MAE or MAPE, the results remain spread out with varying installed capacity, and therefore no concrete statement can be made based solely on MAE or MAPE.

Finally, Figures 3e and f present the absolute RMSE and MAE, respectively, whilst nRMSE and MAPE are omitted for brevity. These figures confirm our earlier observations, i.e., that the direct strategy performs better with increasing penetration level, until it becomes too high and the indirect strategy is preferable. These figures therefore serve more as a comparison between the performance of the ARIMA and ARNN models, from which we can observe that the ARIMA outperforms the ARNN. Moreover, the degradation of the performance with increasing forecast horizon is substantially less in case of both RMSE and MAE.

3.3 ARNN

Similarly, Figure 3 presents the results of the ARNN model as well, for horizons of 1 and 2 hours. The results here contrast substantially with that of the ARIMA model in the sense that performance remains more constant when PV system capacity increases. In case of relative MAE, i.e., Figure 3a, we observe that both strategies perform quite similar, and that the forecast horizon determines the superior strategy, although the difference seems trivial. This similarity reduces slightly in case we look at relative MAPE, i.e., Figure 3b, since the performance difference between the strategies diverges with increasing PV system capacity. Figure 3c presents relative RMSE, which remains nearly constant with increasing installed capacity. Moreover, no noticeable difference between the different forecast horizons exists in this case.

A similar observation can be made in case of nRMSE, presented in Figure 3d, although at 90 kW installed capacity, the indirect strategy significantly outperforms the direct strategy. This behavior was observed in case of

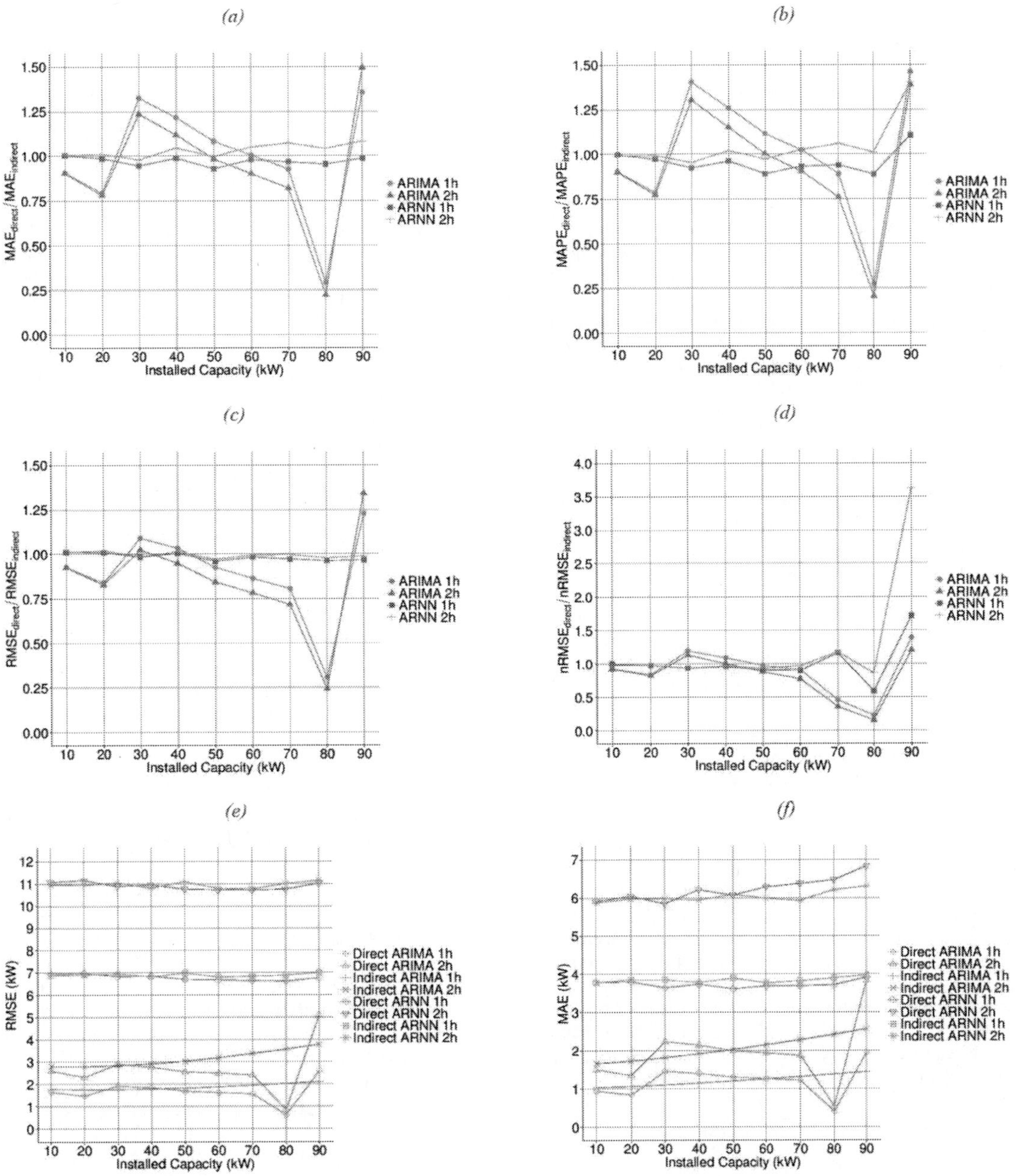

Figure 3: Results of the direct and indirect forecasting strategies for increasing PV system capacity and forecast horizon, in terms of MAE (a), MAPE (b), RMSE (c) and nRMSE (d). Note that the y-axes represent the ratio of the performance achieved by the direct and indirect strategies, respectively. Furthermore, figures (e) and (f) represent the absolute RMSE and MAE, respectively, whilst nRMSE and MAPE are omitted for brevity.

the ARIMA model forecasts as well, and is likely due to the occasional occurrence of negative values during the weekend, whereas negative values seldom occur when $P_{PV} \leq 80$ kW. With such level of PV power penetration, net demand can become negative during the weekend because the commercial building is not used. Due to the limited number of occurrences in the training data, it is particularly challenging to forecast.

Similarly, as mentioned in the previous section, we plot the absolute RMSE and MAE in Figures 3e and f, respectively, as well. The accuracy of the two strategies is highly similar, and also remains quite constant with increasing penetration level, which means that the nonlinearities introduced can be better explained by the ARNN. However, it still performs worse in absolute terms than the ARIMA, and there is a significant difference between performance at different forecast horizons.

4 CONCLUSION

In this study, we have compared two possible strategies one might utilize in case of forecasting net demand, i.e., electricity consumption minus photovoltaic (PV) power production, in combination with two forecast horizons, i.e., 1 and 2 hours. The direct strategy was defined in this paper as subtracting the aforementioned time series and forecasting the residuals. In contrast, the indirect strategy was defined as forecasting the two time series separately and subtracting the resulting forecasts afterwards. In order to achieve this, we utilized the AutoRegressive Integrated Moving Average (ARIMA) and AR neural network (ARNN) models.

The results revealed that overall, the strategies performed quite similarly and we can therefore conclude that although the variance reduces in case of net demand, this does not have substantial impact on forecast accuracy. However, this does not apply for low levels of PV power penetration, where the direct strategy clearly improves accuracy. Furthermore, we showed that, particularly in case of the ARIMA model but also to a lesser extent in case of the ARNN model, the direct strategy is preferable in case root mean square error (RMSE) is of importance. However, in case the PV system capacity is such that net demand sometimes, e.g., during weekends, drops below zero, it is more beneficial apply the indirect strategy.

In addition, our initial hypothesis that there exists some limit after which the direct strategy might become counter effective, seems to occur at 30 kW of installed PV capacity in our study. and only in case of the ARIMA model. However, further increasing of PV system capacity reduces the accuracy ratio again in favor of the direct strategy. Moreover, the results were inconclusive regarding the relationship between forecast horizon and strategy and therefore we will dedicate future work into establishing a solid conclusion on this aspect.

As a conclusion, since neither of the two strategies shows clear advantage over the other, it therefore seems preferable to select the direct strategy for at least three reasons. Firstly, the direct strategy has the advantage that one needs to train one model instead of two, which has obvious computational benefits. Secondly, the indirect strategy could complicate probabilistic forecasting in the sense that the predictive densities cannot straightforwardly be subtracted. And finally, one requires two data streams for the indirect strategy, one for electricity consumption and one for PV power production. Such data might not always be available.

ACKNOWLEDGEMENTS

This work was financially supported by SamspEL 2016 – 2020 in the project "Development and evaluation of forecasting models for solar power and electricity use over space and time", financed primarily by the Swedish Energy Agency.

Bibliography

[1] M. H. J. Bollen, Integration of distributed generation in the power system, Wiley, 2011.

[2] R. Luthander, J. Widén, D. Nilsson, and J. Palm, "Photovoltaic self- consumption in buildings: A review," *Applied Energy*, pp. 80-94, 2015.

[3] T. Hong and S. Fan, "Probabilistic Electric Load Forecasting: A Tutorial Review," *International Journal of Forecasting*, pp. 914-938, 2016.

[4] G. Reikard, "Predicting solar radiation at high resolutions: A comparison of time series forecasts," *Solar Energy*, pp. 342-349, 2009.

[5] A. Kaur, L. Nonnenmacher, and C. F. Coimbra, "Net load forecasting for high renewable energy penetration grids," *Energy*, pp. 1073-1084, 2016.

[6] D. van der Meer, J. Widén, J. Munkhammar, "Review on probabilistic forecasting of photovoltaic power production and electricity consumption," *Renewable Sustainable Energy Reviews*, pp. 1-29, 2016.

[7] P. Bacher, H. Madsen, and H. A. Nielsen, "Online short-term solar power forecasting," *Solar Energy*, pp. 1772-1783, 2009.

[8] A. Bracale, P. Caramia, G. Carpinelli, A. R. Di Fazio, and P. Varilone, "A bayesian-based approach for a short-term steady-state forecast of a smart grid," *IEEE Transactions on Smart Grids*, pp. 1760-1771, 2013.

[9] R. Bessa, A. Trindade, C. S. Silva, and V. Miranda, "Probabilistic solar power forecasting in smart grids using distributed information," *International Journal Electric Power Energy Systems*, pp. 16-23, 2015.

[10] M. David, F. Ramahatana, P. J. Trombe, and P. Lauret, "Probabilistic forecasting of the solar irradiance with recursive ARMA and GARCH models," *Solar Energy*, pp. 55-72, 2016.

[11] Y. Chu, M. Li, H. T. C. Pedro, and C. F. M. Coimbra, "Real-time prediction intervals for intra-hour DNI forecasts," *Renewable Energy*, pp. 234-244, 2015.

[12] H. Le Cadre, I. Aravena, and A. Papavasiliou, "Solar PV Power Forecasting Using Extreme Learning Machine and Information Fusion," *HAL*, pp. 22-24, 2015.

[13] S. Wang and C. Jia, "Prediction intervals for short-term photovoltaic generation forecasts," in *Proc. - 5th Int. Conf. Instrum. Meas. Comput. Commun. Control.* , 2016.

[14] J. Huang and M. Perry, "A semi-empirical approach using gradient boosting and k-nearest neighbors regression for GEFCom2014 proba- bilistic solar power forecasting," *International Journal of Forecasting*, pp. 1081-1086, 2016.

[15] F. Golestaneh, P. Pinson, and H. B. Gooi, "Very Short-Term Nonparametric Probabilistic Forecasting of Renewable Energy Generation; With Application to Solar Energy," *IEEE Transactions on Power Systems*, pp. 1-16, 2016.

[16] S. Chai, Z. Xu, and W. K. Wong, "Optimal Granule-Based PIs Construction for Solar Irradiance Forecast," *IEEE Transactions on Power Systems*, pp. 3332-3333, 2016.

[17] M. Zamo, O. Mestre, P. Arbogast, and O. Pannekoucke, "A benchmark of statistical regression methods for short-term forecasting of photo- voltaic electricity production. Part II: Probabilistic forecast of daily production," *Solar Energy*, pp. 804-816, 2014.

[18] M. P. Almeida, O. Perpiñán, and L. Narvarte, "PV power forecast using a nonparametric PV model," *Solar Energy*, pp. 354-368, 2015.

[19] G. I. Nagy, G. Barta, S. Kazi, G. Borbély, and G. Simon, "GEF- Com2014: Probabilistic solar and wind power forecasting using a generalized additive tree ensemble approach," *International Journal of Forecasting*, pp. 1087-1093, 2016.

[20] R. Juban, H. Ohlsson, M. Maasoumy, L. Poirier, and J. Z. Kolter, "A multiple quantile regression approach to the wind, solar, and price tracks of GEFCom2014," *International Journal of Forecasting*, pp. 1094-1102, 2016.

[21] B. Liu, J. Nowotarski, T. Hong, and R. Weron, "Probabilistic Load Forecasting via Quantile Regression Averaging on Sister Forecasts," *IEEE Transactions on Smart Grids*, 2015.

[22] J. Xie, T. Hong, T. Laing, and C. Kang, "On Normality Assumption in Residual Simulation for Probabilistic Load Forecasting," *IEEE Transactions on Smart Grids*, pp. 1-8, 2015.

[23] S. B. Taieb, R. Huser, R. J. Hyndman, and M. G. Genton, "Forecasting Uncertainty in Electricity Smart Meter Data by Boosting Additive Quantile Regression," *IEEE Transactions on Smart Grids*, pp. 2448-2455, 2015.

[24] A. Bracale, G. Carpinelli, P. D. Falco, R. Rizzo, A. Russo, A. Bracale, G. Carpinelli, P. D. Falco, and R. Rizzo, "New advanced method and cost-based indices applied to probabilistic forecasting of photovoltaic generation," *Journal of Renewable and Sustainable Energy*, 2016.

[25] J. Xie aand T. Hong, "GEFCom2014 probabilistic electric load forecasting: An integrated solution with forecast combination and residual simulation," *International Journal of Forecasting*, pp. 1012-1016, 2016.

[26] F. Ziel and B. Liu, "Lasso estimation for GEFCom2014 probabilistic electric load forecasting," *International Journal of Forecasting*, pp. 1029-1037, 2016.

[27] Y. He, Q. Xu, J. Wan, and S. Yang, "Short-term power load probability density forecasting based on quantile regression neural network and triangle kernel function," *Energy*, pp. 498-512, 2014.

[28] H. Quan, D. Srinivasan, and A. Khosravi, "Short-term load and wind power forecasting using neural network-based prediction intervals," *IEEE Transactions on Neural Networks Learning Systems*, pp. 303-315, 2014.

[29] H. Quan, D. Srinivasan, and A. Khosravi, "Uncertainty handling using neural network-based prediction intervals for electrical load forecasting," *Energy*, pp. 916-925, 2014.

[30] V. Almeida and J. Gama, "Prediction Intervals for Electric Load Forecast: Evaluation for Different Profiles," in *Intelligent Systems Applied to Power Systems*, 2015.

[31] R. J. Hyndman and Y. Khandakar, "Automatic Time Series Forecasting: The forecast Package for R (Version 8.1)," 2008.

[32] H. Shaker, H. Chitsaz, H. Zareipour, and D. Wood, "On comparison of two strategies in net demand forecasting using Wavelet Neural Network," in *North American Power Symposium*, 2014.

[33] G. E. P. Box, G. M. Jenkins, and G. C. Reinsel, Time series analysis: forecasting and control, Eagle Wood Cliffs, 1994.

[34] C. E. Rasmussen and C. K. I. Williams, Gaussian Processes for Machine Learning, MIT Press, 2006.

[35] R. H. Inman, H. T. Pedro, and C. F. Coimbra, "Solar forecasting methods for renewable energy integration," *Progress in Energy Combustion Science*, pp. 535-576, 2013.

ONLINE AND OFFLINE PV POWER FORECASTS FOR OPTIMAL CONTROL OF STORAGE SYSTEMS

James Barry and Jorge Thomas
Karlsruhe Institute of Technology
Hermann-von-Helmholtz Platz 1, 76344 Eggenstein-Leopoldshafen, Germany

ABSTRACT: Increasing penetration of solar photovoltaic installations coupled with batteries in household applications requires complex energy management software that includes reliable day-ahead forecasts of load and generation. The overall profitability of such systems depends to a large extent on the battery's lifetime, so that a charging regime that extends battery life increases the value of the investment. In this contribution two simple forecast models are incorporated into the control software of an AC-coupled storage system with a 32 kWh Li-ion battery and a 17 kWp solar array. The power forecasts are translated into SOC forecasts, allowing a calculation of the optimal charge power that ensures that the battery is only full at the end of the day. Examples of the application of both forecasts in the control algorithm are shown.
Keywords: Battery Storage and Control, Solar Home System, Solar Resource Forecasting

1 INTRODUCTION

The global shift away from baseload-dominated, dispatchable power generation towards non-dispatchable, renewable sources such as solar and wind has led to an increase in the variability of power supply, so that cost-effective and efficient energy storage technologies are indispensable. Li-ion batteries are the current market leader [1] for short-term (minutes to hours) storage applications, and are typically installed in combination with photovoltaic (PV) arrays in household or industry applications. In the household context, the economics of storage systems depend to a large extent on system lifetime and performance. The chosen charge/discharge regime is an important factor that influences battery life and ultimately the overall profitability of the system.

Apart from cyclic ageing due to charge and discharge cycles, it is well known that Li-ion batteries undergo so-called calendar ageing and that this is accelerated if they are stored at high temperatures and state-of-charge (SOC) levels [2]. Battery lifetime can thus be prolonged by controlling the system in an intelligent way, which necessitates the use of load and generation forecasts. A reliable day-ahead prediction of load requirements and PV power availability allows one to control charging in such a way that the battery is only full at the end of the day, thus minimising calendar ageing. In addition one should be able to reduce the level of grid feed-in of PV around the midday peak and at the same time have enough capacity to supply the user with power at times of high demand.

There are several ways to predict the solar irradiation and resulting PV power (see for example the reviews in Refs. [3], [4]); the different methods can be divided into those that use exogenous inputs ("online" forecasts [5]) and those that use only data collected on site ("offline" forecasts [6]. The former typically use numerical weather prediction (NWP) models, whereas the latter use a combination of different time series analyses methods. In addition one can combine several techniques to create a hybrid model and/or use an ensemble of forecasts to obtain an aggregated prediction [7], which has the added advantage of quantifying the uncertainty involved.

In the present work two forecast schemes are compared: an online prediction based on the global forecast system (GFS) numerical weather prediction model as well as a simple offline one based on data collected on site. These forecasts are incorporated into an intelligent charging algorithm that controls a 32 kWh AC-coupled Li-ion battery energy storage system, using standard household load curves as well as one of the arrays of the 1 MW solar-storage facility [8], [9] at Karlsruhe Institute of Technology (KIT) as PV generation input.

The test results allow an evaluation of forecasting methods in terms of the value they add to the storage system as a whole. Concurrent studies of commercially available storage systems (see [10] in these proceedings) have shown that intelligent battery management is only present in roughly 25% of storage systems on the market. In this respect it is useful to find simple and implementable software solutions to bring forecasting tools to the market. The unique combination of solar PV and storage facilities at KIT allows for in-depth studies of system control, both on the planning (forecast models and predictive analytics) and operational (control algorithms) level.

The next section describes the forecast models used, and provides some example data as well as error evaluation, the results of a test with the storage system are presented in section 3, and conclusions are given in section 4.

2 METHODOLOGY

2.1 Offline persistence forecast

The offline forecast is based on the commonly used assumption that today's weather will be roughly the same as yesterday's. It will be shown that in the storage system context this simple assumption suffices in most cases provided the forecast is updated often enough during the day.

A prediction is made at midnight for the day ahead in 5 minute intervals ($t \in [00{:}00, 24{:}00]$), using the predicted clear sky curve $\hat{P}_{\mathrm{PV}}^{\mathrm{CSC}}(t)$, i.e.,

$$\hat{P}_{\mathrm{PV}}^{\mathrm{offline}}(t) = \frac{\int_{\mathrm{yesterday}} P_{\mathrm{PV}}(t)\,dt}{\int_{\mathrm{today}} \hat{P}_{\mathrm{PV}}^{\mathrm{CSC}}(t)\,dt}\,\hat{P}_{\mathrm{PV}}^{\mathrm{CSC}}(t) \qquad (2.1)$$

where the correction factor scales the clear sky curve using the energy generated yesterday.

Figure 1 shows an example of the offline forecast for array 37 of the KIT 1 MW field, on 31. August 2017. The initial prediction at midnight is continually updated every hour during daylight hours, using Equation (2.1) with the numerator replaced by the integral over the past six hours.

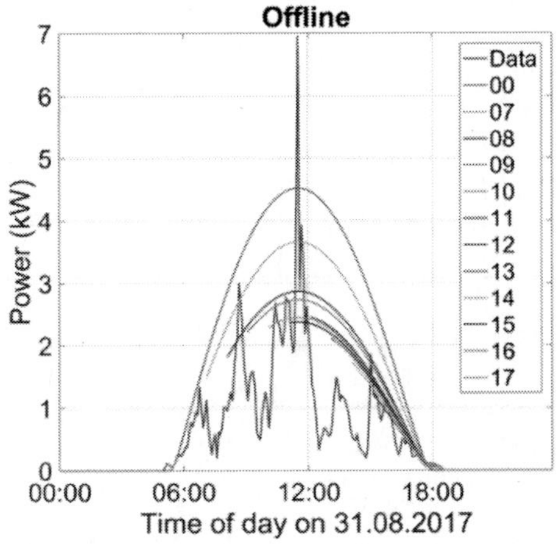

Figure 1: Example of offline persistence forecast with data from array 37 of the 1 MW PV field.

2.2 Online numerical weather prediction forecast

The online forecast uses radiation and temperature forecast data from the GFS model (downloaded from [11]) in order to predict the PV power expected from array 37. The GFS dataset provides a 72 hour forecast with 0.25° latitude resolution and a time resolution of one hour. The forecast is updated every six hours, and an example is shown in Fig. Figure **2**. Comparison of the forecasts shows that neither of them is able to predict the stochastic and highly fluctuating nature of solar radiation, as expected, but the online forecast is by nature better at predicting changes that happen between two days or between morning and afternoon.

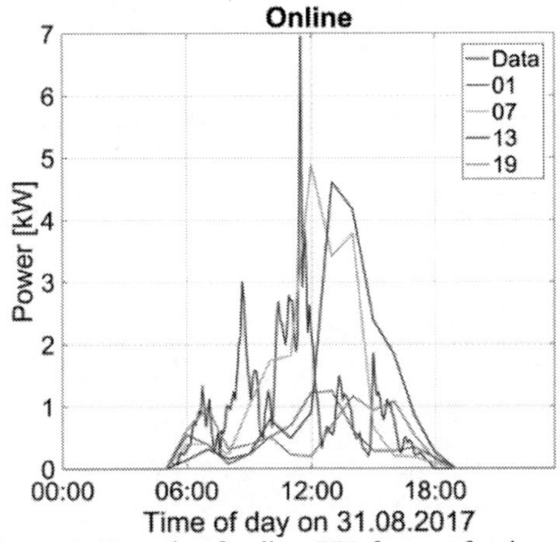

Figure 2: Example of online GFS forecast for the same day as in Fig. 1.

In order to compare the accuracy of the forecasts (before recalibration), data for one month (12.08.2017 to 11.09.2017) were compared with both offline and online forecast models. The error values shown in Table I show that there is overall little difference between the forecasts as far as energy goes. Note that in general the offline forecast tends to underestimate the yield, whereas the online one overestimates it.

Table I: Forecast errors calculated for one month of data, from 12.08.2017 to 11.09.2017

Forecast / Errors	$\hat{P}_{PV}^{offline}$ (00:00)	$\hat{P}_{PV}^{online}$ 01:00	07:00	13:00	19:00
RMSE (kW)	1.182	1.186	1.193	1.197	1.171
MAE (kW)	0.636	0.640	0.650	0.655	0.634

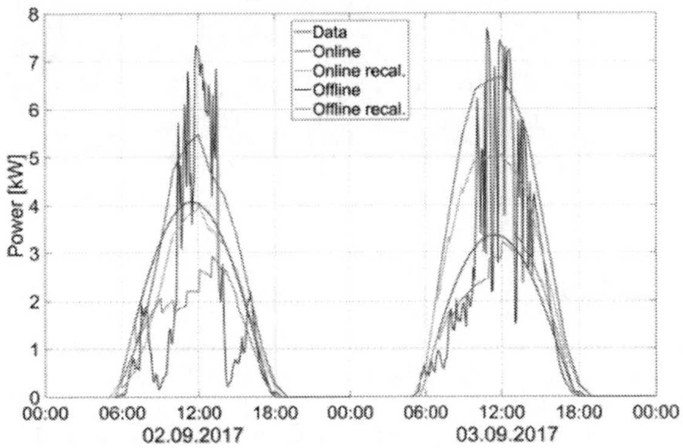

Figure 3: Recalibrated hourly forecasts (both offline and online), for two days in September 2017

The recalibration procedure described above for the offline forecast can also be used for the online one, so that data from the past six hours are used to recalculate the forecast for the rest of the day. Recalibrated hourly forecasts for both variants are shown in Fig. Figure **3** for two days with relatively pronounced PV fluctuations. These examples show the typical variability that can occur in autumn and spring, and it is evident that the two models show different results. The recalibrated version of the forecast will be employed in the storage system control algorithm.

2.3 AC-coupled storage system

The forecast models described above were integrated into the control algorithm of an AC-coupled storage system on KIT's north campus (see the diagram in Fig. Figure **4** and the photographs in Fig. Figure **5**).

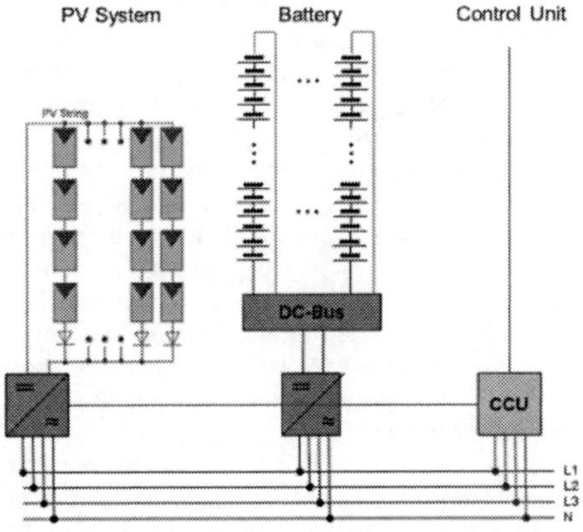

Figure 4: Schematic diagram of AC-coupled storage system.

Figure 5: Photos showing an aerial view of the PV system at KIT as well as the Li-ion battery system.

The predicted AC power from the PV modules (the power from array 37 was multiplied by a factor of 2) is translated to the effective DC power for charging the battery, using the inverter efficiency curve shown in Fig. Figure **6**. Note that the storage system shows a round-trip efficiency of around 80% above 15 kW charge/discharge power, which includes losses from the battery as well as the inverter.

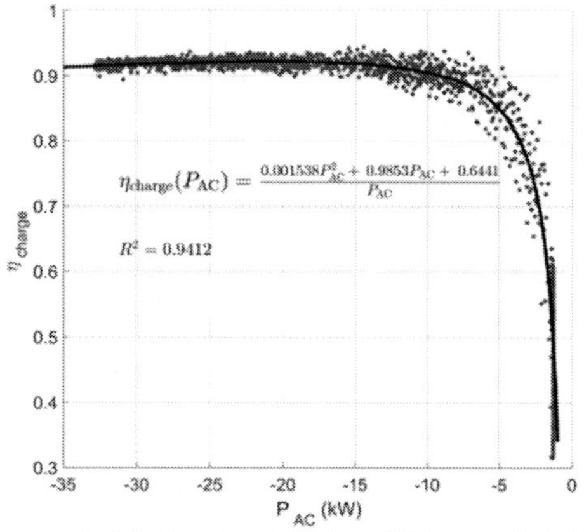

Figure 6: Fit function for the efficiency of AC-DC conversion for battery charging.

2.3 Predictive control algorithm

The forecasts were incorporated into a custom-designed control algorithm running at a frequency of 200 ms (operational control). Figure Figure **8** provides an overview of the implementation of the predictive module, where the SOC prediction is recalculated every hour, $\Delta t = 5$ minutes and $E_{\text{tot}} = 26$ kWh is the (net) battery capacity. In this case η is a predefined function for the conversion efficiency (cf. Fig. Figure **6**), and the algorithm is designed to iteratively reduce battery charging if the predicted SOC reaches 95% before the load-PV crossover point, i.e. the point just before sunset where load becomes larger than generation.

In order to test the algorithm the system was run with the inputs shown in Table II, where the load data was taken from the summer week of households 2, 8 and 17 from the ADRES data set [12]. In order to forecast the battery power one has to combine the PV forecasts described above with a forecast of the load. In this case a simple 5 minute average of the load was used as a forecast (see Fig.Figure **7**).

Table II: Source data for storage system test

Input	Source	Res.	Prediction
Load	ADRES HH 2+8+17 (summer)	1 s	Average
PV	$2 \times P_{Array\ 37}$ (KIT 1 MW field)	200 ms	Offline/online

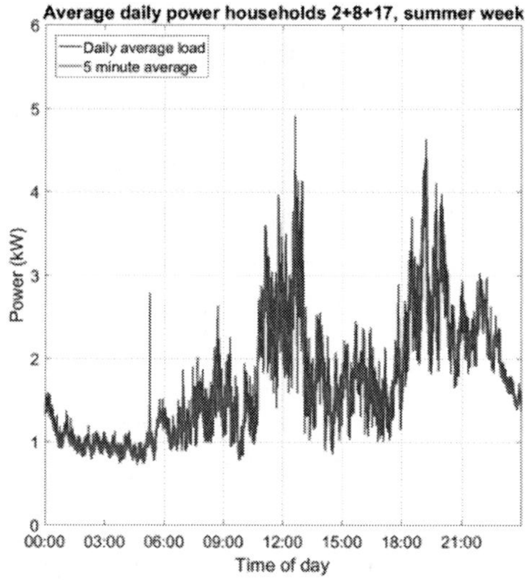

Figure 7: Average daily load data for the summer week of the sum of ADRES household 2, 8 and 17.

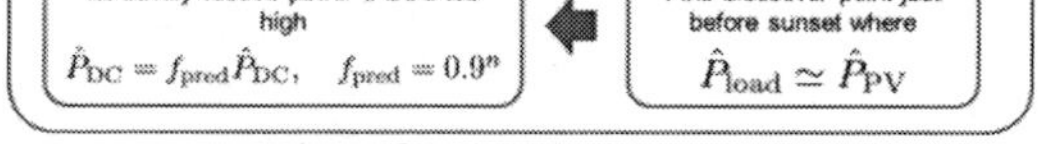

Figure 8: Overview of the predictive control algorithm

3 RESULTS AND DISCUSSION

The preliminary results show that the algorithm successfully controls battery charging (note that discharging is not allowed during the day). Two examples are shown in Figs. Figure **9** and Figure **10**, for the offline and online forecast, respectively. Due to the iterative nature of the process it is often better to underestimate production so that the battery is charged quicker at first, since one can always slow down the charging process later in the day. This can be seen in Figure 9 , where the offline forecast grossly underestimates production in the morning. However, by the end of the day the recalibration

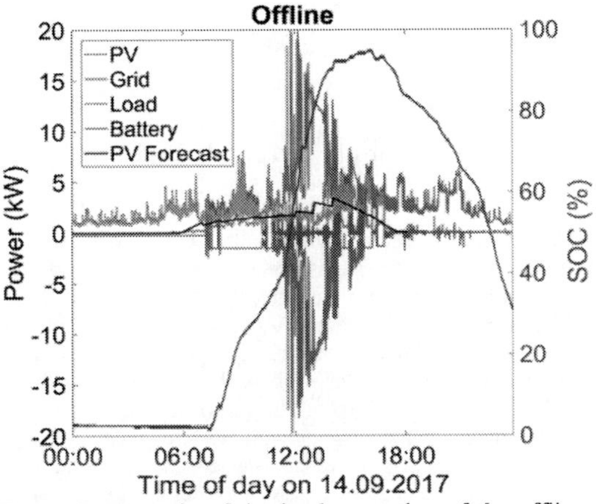

Figure 9: Example of the implementation of the offline forecast in charging the battery, where the SOC is shown in magenta (right axis).

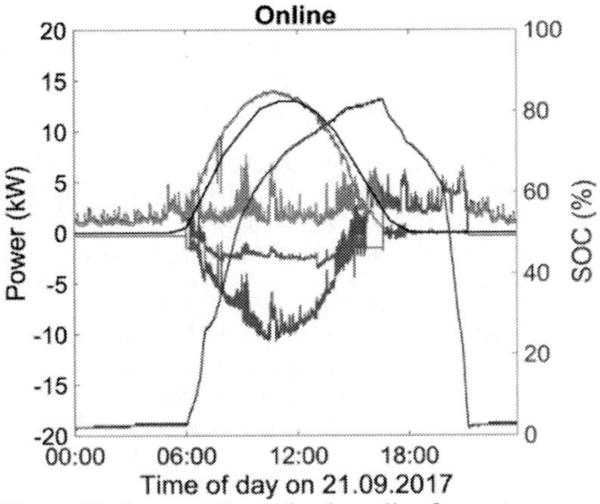

Figure 10: Same as Fig. 9, for the online forecast.

compensates for this, so that in this context the simple persistence forecast is sufficient, provided it does not grossly overestimate the yield (this could happen on foggy days, for instance). If one overestimates production one may throttle the charging too much, which in the end leads to a reduction in self-sufficiency – the operator is forced to buy electricity at night to compensate for the battery not being fully charged. Another aspect that has not been taken into account is the overall conversion efficiency, which suffers when charging the battery at partial load, see also [10]. By reducing the charge power one runs into the danger of operating at low powers, which leads one to consider other methods of slowing down charging such as simply delaying it until later in the day. A complete analysis of the effect on self-sufficiency is still to be carried out.

4 SUMMARY AND CONCLUSIONS

Two different forecast models have been incorporated into the control algorithm of an AC-coupled Li-ion battery storage system, in order to control the daily charge cycle of the battery. It has been shown that both models perform similarly when it comes to a day-ahead forecast of PV yield, and that the ability of the software to perform hourly calibration means that initial errors can be compensated for. Further analysis on the effect of the control scheme on self-sufficiency and overall system economics are still to be carried out.

5 ACKNOWLEDGEMENTS

The authors thank Nina Munzke and Bernhard Schwarz for helpful discussion.

6 REFERENCES

[1] D. U. Sauer, "Lithium-ion batteries become the benchmark for stationary applications – Markets, players, prices, technology," in 10th International Renewable Energy Storage Conference, 2016.

[2] P. Keil, S. F. Schuster, J. Wilhelm, J. Travi, A. Hauser, R. C. Karl, and A. Jossen, "Calendar Aging of Lithium-Ion Batteries: I. Impact of the Graphite Anode on Capacity Fade," J. Electrochem. Soc., vol. 163, no. 9, pp. A1872–A1880, Jan. 2016.

[3] J. Antonanzas, N. Osorio, R. Escobar, R. Urraca, F. J. Martinez-de-pison, and F. Antonanzas-torres, "Review of photovoltaic power forecasting," Sol. Energy, vol. 136, pp. 78–111, 2016.

[4] R. H. Inman, H. T. C. Pedro, and C. F. M. Coimbra, "Solar forecasting methods for renewable energy integration," Prog. Energy Combust. Sci., vol. 39, no. 6, pp. 535–576, 2013.

[5] E. Lorenz, J. Kühnert, D. Heinemann, K. P. Nielsen, J. Remund, and S. C. Müller, "Comparison of global horizontal irradiance forecasts based on numerical weather prediction models with different spatio-temporal resolutions," Prog. Photovoltaics Res. Appl., vol. 24, pp. 1626–1640, 2016.

[6] H. T. C. Pedro and C. F. M. Coimbra, "Assessment of forecasting techniques for solar power production with no exogenous inputs," Sol. Energy, vol. 86, no. 7, pp. 2017–2028, 2012.

[7] Y. Ren, P. N. Suganthan, and N. Srikanth, "Ensemble methods for wind and solar power forecasting -- A state-of-the-art review," Renew. Sustain. Energy Rev., vol. 50, pp. 82–91, 2015.

[8] J. Barry, N. Munzke, and J. Thomas, "Short-term power fluctuations in densely clustered PV-battery systems," 32nd Eur. Photovolt. Sol. Energy Conf. Exhib., 2016.

[9] J. Thomas and N. Munzke, "A Static Solar-Tracking Configuration for PV Power Plant to Redistribute the Daily Power Supply," in 31st European Photovoltaic Solar Energy Conference and Exhibition, 2015.

[10] N. Munzke, B. Schwarz, and J. Barry, "Evaluation of the Performance of Household Li-Ion Battery Storage Systems and Their Impact on Profitability," in 33rd European Photovoltaic Solar Energy Conference and Exhibition, 2017.

[11] National Centers for Environmental Prediction, "Global Forecast System FTP Download." [Online]. Available: ftp://ftp.ncep.noaa.gov/pub/data/nccf/com/gfs/prod/.

[12] A. Schuster, D. Wertz, A. Frohner, and C. Karner, "Energie der zukunft: ADRES-Concept," Vienna, 2012.

COMPUTATIONALLY INEXPENSIVE PV SYSTEM MODEL AS A SIMULATION AGENT FOR LARGE SCALE INTEGRATION ANALYSIS

Colin Levis, Martin Hill
Cork Institute of Technology, Dept. of Electrical Engineering, Bishopstown, Cork, Ireland.
E-mail: colin.levis@mycit.ie, martin.hill@cit.ie

ABSTRACT: In recent years the proliferation of renewable energy systems, especially photovoltaic (PV), generating into the low voltage (LV) distribution network has been increasing and these systems are poised to contribute a significant portion of the energy supply. This poses a challenge for maintaining grid stability, which is traditionally reliant on large centralised synchronous machines. For large-scale integration of distributed PV generation, PV systems must be capable of providing grid support services (GSS) emulating the traditional system. This paper presents an accurate, reduced order, computationally inexpensive agent based model (ABM) of a three-phase PV inverter that will provide a platform for analysing advanced control strategies in a multi-agent LV distribution network. The ABM is designed to simulate the actions and interactions of autonomous/decentralised PV systems with GSS for voltage and frequency stability using droop controllers. The developed ABM is validated by comparing results from three independent step changes [Irradiation, frequency & voltage] with a full switching model in MATLAB/ Simscape Power Systems™. These resuts show similar system dynamics and a computational burden reduction of 135:1, hence, providing a simulation agent for large scale integration analysis.

Keywords: State space, Linearized, Computationally inexpensive, Agent based model, Photovoltaic

1 INTRODUCTION

PV systems are now playing a significant role in achieving a renewable power system, and with the reduction in PV module prices, large-scale grid integration is happening around the world. In 2016, a total of 76.6 GW of PV was connected to the grid resulting in a present global accumulative capacity of 306.5GW [1]. There are, however, power system stability concerns with the proliferation of PV within the grid especially countries like Germany, Greece and Italy with penetration levels over 7% [2]. As penetration levels increase conventional synchronous generators are gradually being dispalced reducing the inherent system synchronous inertia created by traditional rotating machines resulting in frequency stability implications throughout the power system in the event of a sudden generation and load imbalance [3][4]. Another area of concern is voltage stability, in contrast to conventional synchronous generators connected in a centralised manner at transmission level, PV generators tend to be small and geographically located connected at distribution level, particularly rooftop PV as seen in Australia that account for 85% of installations [5]. If the distributed PV generation now exceeds the local demand, the reverse power flow can cause voltage variations exceeding regulated limits [6] [7].

To allow for large scale penetration of distributed PV and other variable renewable enegy sources, these nonsynchronous systems must provide grid support services in the form of frequency response and voltage support. Recent results from the EU REserviceS project indicate that allowing GSS from variable renewable resources such as wind and PV can be economical and stable with penetration levels of up to 50%, with the greatest benefits from fast frequency response (FFR), especially downward frequency containment reserve (FCR), with delivery times of less than five minutes [8]. Similiarly, fast voltage support can be provided at the point of connection by regualting reactive power, therefore correcting voltage variations due to the variable nature of the active power delivered to the grid [9].

Current inverter technology can support this development by utisling droop controllers emulating tradional power system responses for both active and reactive power [10][11]. However, there is still a major challenge in the coordination between centralised power plants and distributed PV. To meet these demands new techniques are required for collaborative interactions with PV generators [12][13]. Evaluation of these new techniques through simulation of high penetration levels on low voltage networks will require hundreds of interacting distributed PV systems as shown in Fig. 1. For such simulations using detailed switching models would not be feasible due to the computational resources required. Alternatively, to analyse GSS and introduce techniques such as game theory in the LV distribution network, an accurate, reduced order, computationally inexpensive agent based model is required as shown in Fig.1.

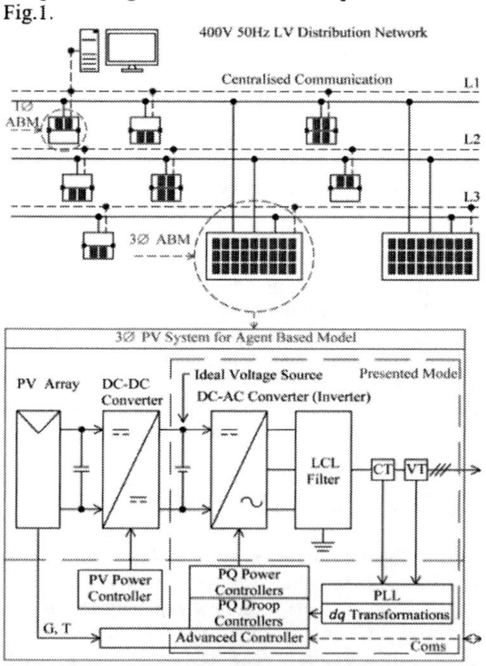

Figure 1: Multi-Agent based LV Distribution Network

Since a PV system is nonlinear and time varying, the approach taken in the development of the ABM is to create a time invariant, linearised small signal system model operating about a typical large scale operating point. Current literature uses similiar techniques [14], however, the addition of this paper is to include a slow centralised communication structure as shown in Fig. 1. It is intended that all agents are autonomous, capable of functioning without the need for centralised communication with the added benefit of advanced control strategies. The communication ensures that each agent provides data on its previous performance to the central server and can receive updated droop control characteristics from the central server for future events.

The aim of this paper is to develop a complete small signal model for three-phase PV inverter that will provide a platform for analysing advanced control strategies with increased penetration levels of PV within the LV distribution network. The developed ABM is validated by comparing results from three independent step changes [Irradiation, frequency & voltage] with a full switching model in MATLAB/ Simscape Power Systems™.

2 SYSTEM OVERVIEW

The three phase PV inverter and control structure is presented in Fig. 2. This is a voltage source current controlled grid connected inverter (VSI CC). The inverter converts the input DC power into an AC output, an LCL filter is used to attenuate the switching frequency harmonics to an acceptable level created from the inverter. By using the Park transformation, the three-phase filtered voltage and currents are transformed into a time invariant system creating a direct and quadrature (dq) axis. Classical cascaded proportional integral (PI) controllers independently control both active and reactive power. The active power reference depends on irradiation, temperature and the desired reserve capacity for frequency response which in turn is dependent on the frequency-watt droop controller. The reactive power reference is dependent on the volt-var droop controller. For this model the nominal reactive power is zero. To synchronise the output current from the inverter to the grid a phase locked loop (PLL) is used. The PLL also provides an input to the frequency-watt droop controller to implement GSS. The PLL and filter dynamics are critical to system performance in the ability of providing grid support services.

3 SYSTEM MODEL

In this section, the three phase PV inverter as shown in Fig. 2 is used to derive the mathematical ABM. Each section can be represented as a set of linear/nonlinear differential equations that describe the systems dynamics. Linearizing these nonlinear system differential equations around an optimum operating point provides a small signal model suitable for large-scale integration analysis.

3.1 Phase-Locked Loop

Phase-locked loops (PLLs) provide power inverters the ability to synchronize and monitor the instantaneous state of the grid at the point of common coupling (PCC) as shown in Fig. 2. This is an adaptive process, allowing both the grid and power inverter to work together. In three phase systems the synchronous-reference frame (SRF) PLL is a basic solution to maintain synchronism[15]. A SRF PLL transforms the PCC grid voltages using Parks transformation to create quadrature time invariant voltages (v_{od}, v_{oq}) as in Fig. 3. To achieve grid voltage synchronism, this PLL uses a PI controller with gains $(k_p^{PLL} \& k_i^{PLL})$ acting on the alignment error angular frequency $(\Delta\omega)$ (1) until $v_{oqF} = 0$.

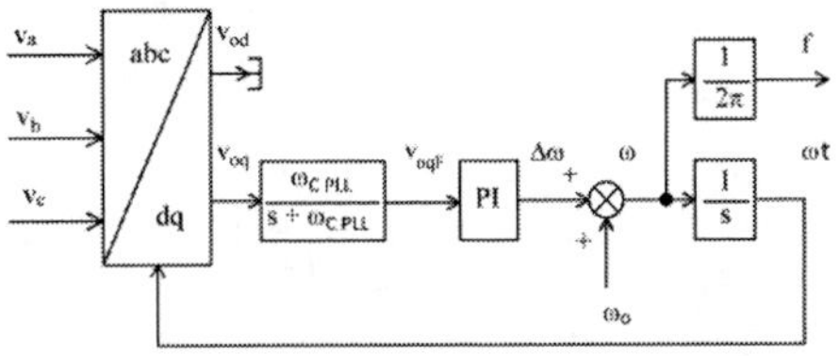

Figure 3: Synchronous Reference Frame PLL

The addition of the alignment error and reference angular frequency (ω_0) gives the estimated system angular frequency (ω) (2). Integrating the estimated system angular frequency then gives the transformation angle (ωt) necessary to achieve synchronism. The PLL provides the PCC instantaneous grid transformation angle ωt and the system frequency required for the regulation of active

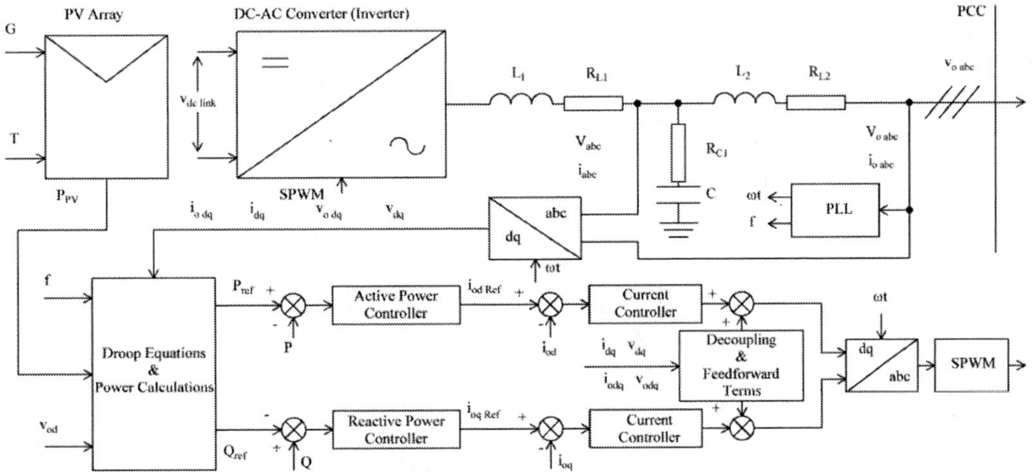

Figure 2: PV Inverter System Model

and reactive power from the power inverter to the grid. The state variables for the PLL are given in equations (3)(4).

$$\Delta\omega = k_p^{PLL} v_{oqF} + k_i^{PLL} \int v_{oqF}\, dt \quad (1)$$

$$\omega = \omega_0 + \Delta\omega \quad (2)$$

$$\frac{dv_{oqF}}{dt} = v_{oq}\omega_{cPLL} - v_{oqF}\omega_{cPLL} \quad (3)$$

$$\frac{d\mu}{dt} = v_{oqF} \quad (4)$$

3.2 Power Calculation

The instantaneous active (P) and reactive (Q) power from the inverter at the PCC can be calculated using the *dq* transformation currents and voltages (5)(6). Since the PLL dynamically sets v_{oq} to zero, v_{od} is then equal to the peak amplitude phase voltage, therefore active power is directly proportional to i_{od} and reactive power is proportional to i_{oq}.

$$P = \frac{3}{2}\left(v_{od} i_{od} + v_{oq} i_{oq}\right) \quad (5)$$

$$Q = \frac{3}{2}\left(v_{oq} i_{od} - v_{od} i_{oq}\right) \quad (6)$$

The calculated instantaneous powers P & Q are then passed through low pass filters with a cutoff frequency of ω_{cp} & ω_{cq} before entering the controller (7)(8).

$$P_F = P \cdot \frac{\omega_{cp}}{s+\omega_{cp}} \Rightarrow \frac{dP_F}{dt} = P\omega_{cp} - P_F\omega_{cp} \quad (7)$$

$$Q_F = Q \cdot \frac{\omega_{cq}}{s+\omega_{cq}} \Rightarrow \frac{dQ_F}{dt} = Q\omega_{cq} - Q_F\omega_{cq} \quad (8)$$

3.3 Active Power Reference

As the PV array and DC/DC converter dynamics are not prsented in this paper the available power (P_{av}) from a PV array at a given irradiance and temperature is given by [16](9).

$$P_{av} = \gamma \cdot P_{rated}\left(\frac{G}{Gn}\right)[1 + \alpha(T - T_n)] \quad (9)$$

Where γ is the de-rating factor in percentage (%), P_{rated} is the PV array rated power (*W*), *G* solar irradiation (*W/m²*) , *Gn* solar irradiation for standard test conditions (STC) (1000*W/m²*), α *temperature* coefficient of power (*%/°C*), *T* cell temperature (*°C*) and T_n cell temperature for STC (25*°C*). A 12kWp PV array based on CIT's RES2020 micro-grid is used for the power source in this paper. The array consists of 16 series Hyundai IMG 250Wp Multi-crystalline panels with 3 parallel strings where α = -0.0043 *%/°C*.

3.4 Inverter & LCL Filter

Averaging the switching states over a single switching period provides an average inverter model based on the principle that the fundamental frequency (*f*) is significantly smaller than the inverter switching frequency (f_{sw}). Fig. 4, shows the equivalent *dq* time invariant circuit of the LCL filter. The non-linear time invariant model differential equations are given in equations (10-15).

$$L_1 \frac{d}{dt} i_d = d_d v_{dc} 0.5 - i_d R_{L1} - v_d + \omega L_1 i_q \quad (10)$$

$$L_1 \frac{d}{dt} i_q = d_q v_{dc} 0.5 - i_q R_{L1} - v_q - \omega L_1 i_d \quad (11)$$

$$L_2 \frac{d}{dt} i_{od} = v_d - i_{od} R_{L2} - v_{od} + \omega L_2 i_{oq} \quad (12)$$

$$L_2 \frac{d}{dt} i_{oq} = v_q - i_{oq} R_{L2} - v_{oq} - \omega L_2 i_{od} \quad (13)$$

$$C \frac{d}{dt} v_{cd} = i_d - i_{od} + \omega C v_{cq} \quad (14)$$

$$C \frac{d}{dt} v_{cq} = i_q - i_{oq} - \omega C v_{cd} \quad (15)$$

where

$$v_d = \left(i_d - i_{od} + \omega C v_{cq}\right) R_c + v_{cd}$$

$$v_q = \left(i_q - i_{oq} - \omega C v_{cd}\right) R_c + v_{cq}$$

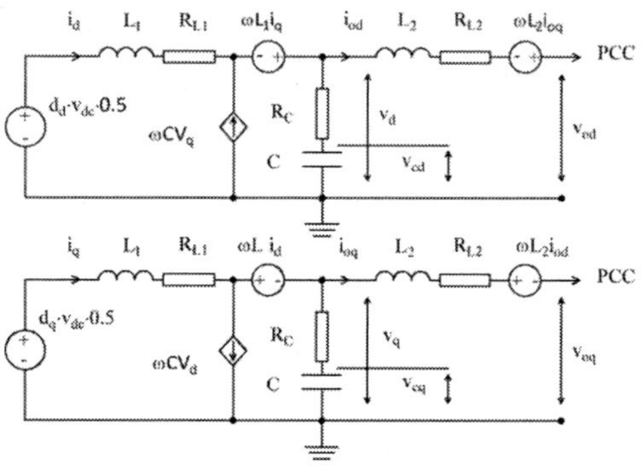

Figure 4: Average Inverter and LCL Filter *dq* Circuit

3.5 Power Controllers

To achieve the reference powers set from the droop characteristics a classical cascaded PI controller ensures zero steady state error. Implementation of the PI controllers are in the *dq* rotating reference frame with an angular frequency (*ω*). To eliminate the cross coupling created from the *dq* transformation as shown in Fig. 4, decoupling terms are included in both controllers as shown in Fig. 5. The addition of both cascade PI controllers creates four extra state variables ($\emptyset_P, \emptyset_Q, \gamma_P \& \gamma_Q$)(16-19).

$$\frac{d\emptyset_P}{dt} = P_{ref} - P_F \quad (16)$$

$$\frac{d\emptyset_Q}{dt} = -Q_{ref} + Q_F \quad (17)$$

$$\frac{d\gamma_P}{dt} = k_p^P\left(P_{ref} - P_F\right) + k_i^P \emptyset_P - i_{od} \quad (18)$$

$$\frac{d\gamma_Q}{dt} = k_p^Q\left(-Q_{ref} + Q_F\right) + k_i^Q \emptyset_Q - i_{oq} \quad (19)$$

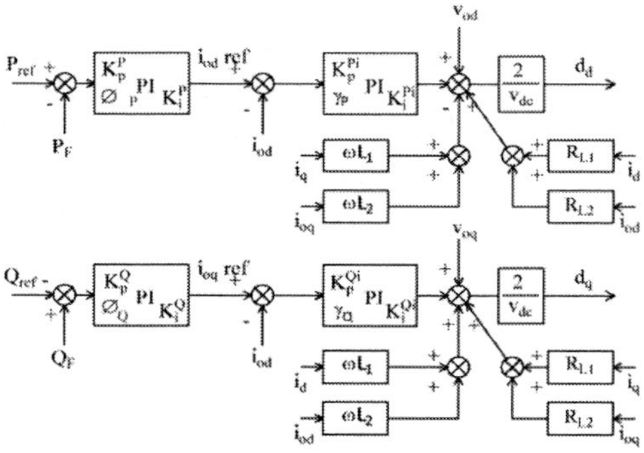

Figure 5: Cascaded PQ Power Controllers

3.6 Linearization

Linearization of the nonlinear system differential equation uses the Taylor's series expansion about a point x_0 expressing the function as an infinite sum of the function derivatives (20).

$$f(x) = f(x_0) + \left.\frac{df}{dx}\right|_{x=x_0} \frac{(x-x_0)}{1!} + \left.\frac{d^2f}{dx^2}\right|_{x=x_0} \frac{(x-x_0)^2}{2!} + \dots + \left.\frac{d^nf}{dx^n}\right|_{x=x_0} \frac{(x-x_0)^n}{n!} \quad (20)$$

For small perturbations around the steady state

operating point x_0, the first order terms yield a linear relationship between $f(x)$ and the excursion from x_0 resulting in a small signal model (21).

$$f(x) - f(x_0) \approx \left.\frac{df}{dx}\right|_{x=x_0} \tag{21}$$

The initial steady state operating points chosen for the 12kW system are 6kW and 0kVar (P_e, Q_e) which are mid-range operating powers. Alternatively, depending on location (solar irradiation & temperature) and system capacity a statistically optimal annual/monthly operating point can be chosen. From these initial operating points, the entire system steady state operating points can be found by solving the differential equations ($\dot{x} = 0$). The system equilibrium points as presented in Table I and are quantified from simulation using MATLAB/Simscape Power Systems™ as shown in Fig. 6 with the component values from Table II.

Table I: System Equilibrium Points (6kW, 0kVar)

Parameter	Value	Parameter	Value
P_e	6kW	Q_e	0kVar
V_{de}	331.6V	V_{qe}	3.883V
V_{ode}	325.5V	V_{oqe}	0V
I_{ode}	12.29A	I_{oqe}	0A
I_{de}	12.29A	I_{qe}	2.295A
D_{de}	0.6989	D_{qe}	0.07563
V_{dce}	950V		

3.7 Complete Simulation Model

A general state-space representation of the linearized first order differential equations describing the systems dynamics is in the form (22).

$$\begin{aligned} \dot{\tilde{x}} &= A\tilde{x} + B\tilde{u} \\ \tilde{y} &= C\tilde{x} + D\tilde{u} \end{aligned} \tag{22}$$

In total, the system contains 14 state variables and 6 state inputs with active and reactive power as the desired outputs (23).

$$\begin{aligned} \tilde{x} &= [\widetilde{i_d}; \widetilde{i_{od}}; \widetilde{i_q}; \widetilde{i_{oq}}; \widetilde{v_{cd}}; \widetilde{v_{cq}}; \widetilde{v_{oqF}}; \widetilde{P_F}; \widetilde{Q_F}; \widetilde{\emptyset_P}; \widetilde{\emptyset_Q}; \widetilde{\gamma_P}; \widetilde{\gamma_Q}; \tilde{\mu}]^T \\ \tilde{u} &= [\widetilde{P_{ref}}; \widetilde{Q_{ref}}; \widetilde{v_{od}}; \widetilde{v_{oq}}; \widetilde{v_{dc}}; \tilde{\omega}]^T \\ \tilde{y} &= [\tilde{P}; \tilde{Q}]^T \end{aligned} \tag{23}$$

Table II: System Component Values

Parameter	Value	Parameter	Value
V_{dc}	950V	L_1	8mH
L_2	1mH	L_3	50µH
R_{L1}	0.5Ω	R_{L2}	0.5Ω
R_{L3}	0.1Ω	R_C	0.1Ω
C	22µF	Z	1Ω
f_{cPLL}	100Hz	f_{cP}	50Hz
f_{cQ}	50Hz	$V_{g\,L\text{-}L}$	437V
k_p^P	1.30E-05	k_i^P	0.25
k_p^{Pi}	3.8	k_i^{Pi}	2.30E-04
k_p^Q	2.00E-05	k_i^Q	0.5
k_p^{Qi}	5	k_i^{Qi}	2.30E-04
k_p^{PLL}	1	k_i^{PLL}	2

4 DROOP CHARACTERISTIC EQUATIONS

In a power system the control of active power is related to frequency and the control of reactive power is related to voltage, hence they can be controlled separately. For stable load power sharing between generators connected in parallel a droop power characteristic must be provided, which allows proportional load sharing according to their power capacity [17].

4.1 Active Power

In order for the PV system to provide GSS in the form of frequency response a frequency-watt droop characteristic is implemented as shown in Fig. 7 [10]. To participate in under frequency events the PV system must have a frequency containment reserve (P_{res}) from either an energy storage system or de-loading the system by the desired reserve power. The frequency-watt droop characteristic curve is represented by the piecewise function (24).

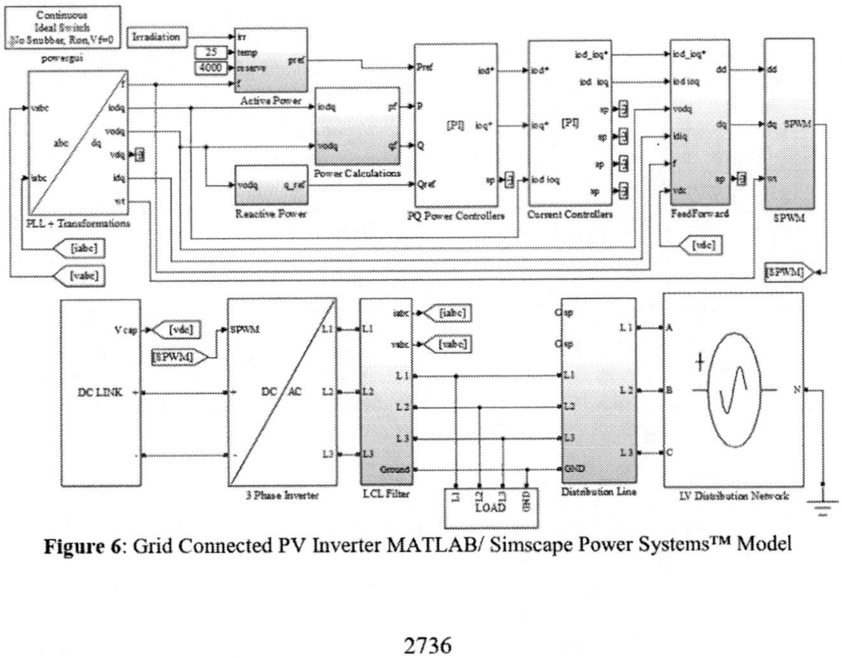

Figure 6: Grid Connected PV Inverter MATLAB/ Simscape Power Systems™ Model

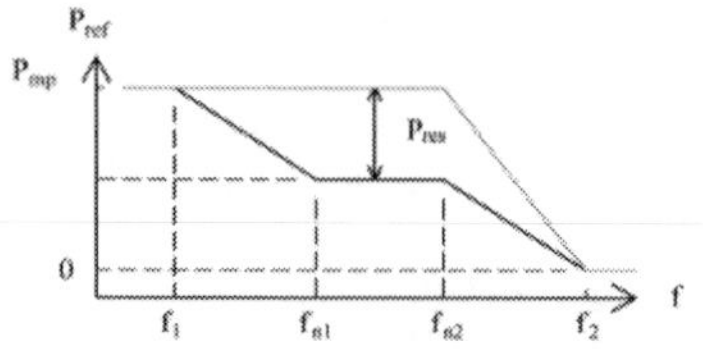

Figure 7: Frequency-Watt Droop Characteristic

$$P_{ref} = \begin{cases} P_o & f_{n1} \geq f \leq f_{n2} \\ P_o + \frac{P_{res}}{f_{n1}-f_1}(f_{n1} - f), & f_{n1} < f > f_1 \\ P_o + \frac{P_o}{f_2 - f_{n2}}(f_{n2} - f), & f_{n2} > f < f_2 \end{cases}$$

$$where\ P_o = P_{av} - P_{res} \quad (24)$$

4.2 Reactive Power

Similarly, for voltage support a volt-var droop characteristic is implemented as shown in Fig. 8. The amount of reactive power that can be delivered or consumed which is dependent on the maximum power capacity of the inverter and can be adjusted depending on the amount of active power being supplied [7]. The volt-var droop characteristic curve is represented by the piecewise function (25).

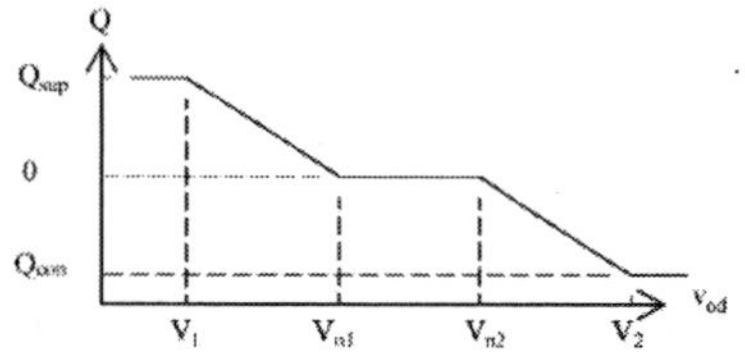

Figure 8: Volt-Var Droop Characteristic

$$Q_{ref} = \begin{cases} 0 & v_{n1} \geq v_{od} \leq v_{n2} \\ \dfrac{Q_{sup}}{V_{n1} - V_1}(V_{n1} - v_{od}), & v_{n1} < v_{od} > v_1 \\ \dfrac{Q_{con}}{V_{n2} - V_2}(V_{n2} - v_{od}), & v_{n2} > v_{od} < v_2 \end{cases} \quad (25)$$

5 VALIDATION

The linearized mathematical ABM of the PV inverter presented in this paper is validated by comparing results with simulations in MATLAB/ Simscape Power Systems™. The reference model is a 3ϕ/400V/50Hz/ 12kW full switching inverter model as shown in Fig. 6. The inverter is connected to a low voltage distribution network including distribution line resistance & reactance with a local load connected at the PCC of the PV system. An ideal voltage source represents the dc link as a constant voltage. Fig. 9, represents the ABM in state-space using Simulink for comparison simulations to include three separate events:

1) Step change in irradiation
2) Step change in system frequency
3) Step change in system voltage

The step changes used to compare the switching model and state space ABM are unlikely to occur during normal operation but instead serve as a severe test condition scenario that can anticipated in the LV grid.

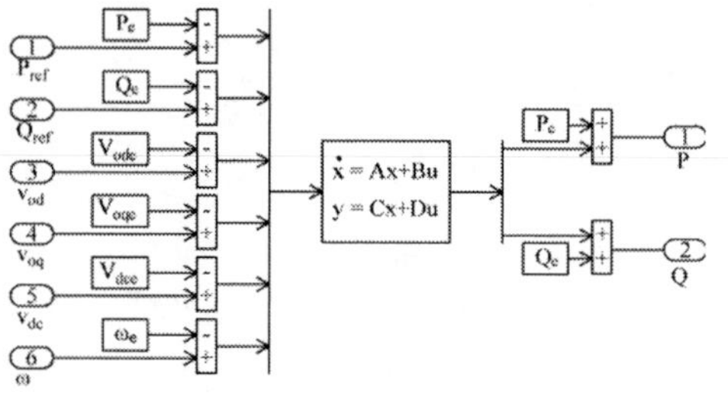

Figure 9: Agent Based Model in Simulink

6 RESULTS

The results presented in this section have the PV inverter operating with a constant reserve capacity of 4kW for under frequency support.

6.1 Irradiation Step

System reference tracking performance is evaluated using step changes in irradiation [400W/m^2 from 0-100ms, 800W/m^2 from 100-150ms, 1000W/m^2 from 150-200ms and 400W/m^2 from 200-250ms]. Since the PV system is operating with a constant reserve power of 4kW, the system must operate at 4kW below the maximum power available for any given irradiance above 333W/m^2, otherwise the output power would be zero. The target powers for each irradiation step [0.8kW (4.8kW MPP) from 0-100ms, 5.6kW (9.6kW MPP) from 100-150ms, 8kW (12kW MPP) from 150-200ms and 0.8kW (4.8kW MPP) from 200-250ms]. The results shown in Fig. 10 show fast reference tracking characteristics delivering the target set points in 35ms with similar system dynamics produced from both the switching model and ABM.

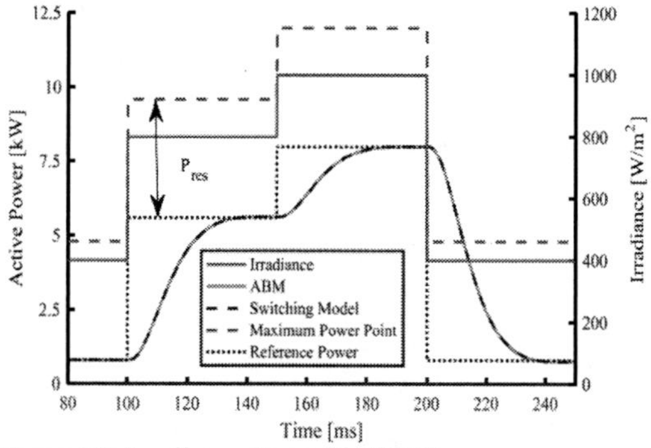

Figure 10: Irradiance Step with 4kW Reserve

6.2 Frequency Step

A step change in frequency is used for system frequency response evaluation [-1.2Hz at 100ms and the return to nominal conditions 50Hz at 200ms] as shown in Fig. 11. The results show fast frequency response characteristics delivering 4kW in 40ms with near equivalent dynamics produced from both the switching model and ABM. The dynamics of the PLL contribute significantly to the step change in frequency taking 8ms to reach the new steady state power reference. This highlights the importance of the PLL for system dynamics.

6.3 Voltage Step

A step change in distribution network voltage is used for voltage support system evaluation [-21.85V at 100ms and the return to nominal conditions at 200ms] as shown in

Fig. 12. The results show fast voltage support response characteristics delivering 2.15kVar in 35ms.

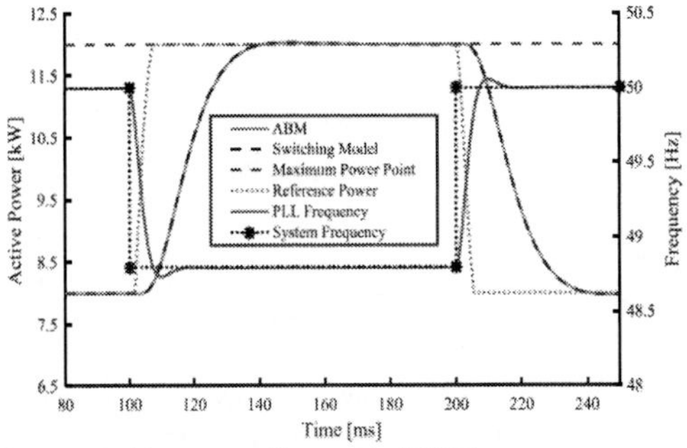

Figure 11: Frequency Step with 4kW Reserve

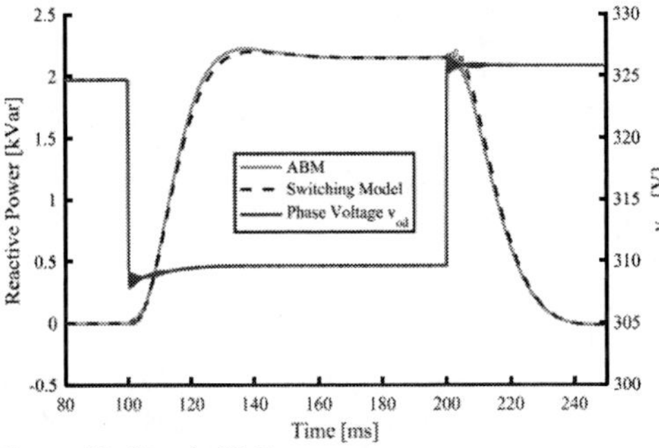

Figure 12: Step in Voltage

7 DISCUSSION

The results show similar system dynamics between the full switching PV inverter model and the ABM with a computational burden reduction of 135:1. These results demonstrate accuracy and computational burden reduction of the ABM suitable for LV distribution network analysis. However, the ABM presented in this paper only shows the dynamics of the PV inverter only. To develop a full PV system model the PV array and DC/DC converter must be included as shown in Fig. 1. Also the frequency and voltage validation tests do not include any interaction with the distribution network. In future work a full PV system ABM will be developed and analysed in an interactive distribution network. This will allow for PV multi-agent analysis and the evaluation of GSS's with advanced control strategies such as game theory within a reasonable simulation time.

8 CONCLUSION

This paper demonstrates an accurate, reduced order, computationally inexpensive ABM of a three-phase PV inverter that will provide a platform for analysing advanced control strategies in a multi-agent LV distribution network. The developed ABM is validated by comparing results from three independent step changes [Irradiation, frequency & voltage] with a full switching model in MATLAB/ Simscape Power Systems™. The results show that the PV system has an inherently fast response where the PLL contributes significantly to the dynamic response and a computational burden reduction of 135:1 is achieved from the derived ABM.

9 REFERENCES

[1] Schmela Michael (SolarPower Europe), "Global Market Outlook For Solar Power/2017-2021," 2017.

[2] IEA, "Snapshot of Global Photovoltaic Markets," 2015.

[3] P. Tielens and D. Van Hertem, "The relevance of inertia in power systems," *Renew. Sustain. Energy Rev.*, vol. 55, pp. 999–1009, 2016.

[4] D. S. G. ENTSOE, "Frequency Stability Evaluation Criteria for the Synchronous Zone of Continental Europe," 2016.

[5] (APV) Austrailian PV Institute, "Solar PV Status," *funded by the Australian Renewable Energy Agency.* [Online]. Available: http://pv-map.apvi.org.au. [Accessed: 01-Aug-2017].

[6] L. Collins and J. K. Ward, "Real and reactive power control of distributed PV inverters for overvoltage prevention and increased renewable generation hosting capacity," *Renew. Energy*, vol. 81, pp. 464–471, Sep. 2015.

[7] P. Jahangiri and D. C. Aliprantis, "Distributed Volt/VAr control by PV inverters," *IEEE Trans. Power Syst.*, vol. 28, no. 3, pp. 3429–3439, 2013.

[8] F. Van Hulle, P. Wilczek, and I. Pineda, "Economic grid support services by wind and solar PV," *REserviceS*, 2014.

[9] C. Schauder, "Advanced Inverter Technology for High Penetration Levels of PV Generation in Distribution Systems Advanced Inverter Technology for High Penetration Levels of PV Generation in Distribution Systems," *National Renewable Energy Laboratory (NREL)*, no. March, p. 35, 2014.

[10] J. Neely, J. Johnson, J. Delhotal, S. Gonzalez, and M. Lave, "Evaluation of PV frequency-watt function for fast frequency reserves," in *Applied Power Electronics Conference and Exposition - APEC*, 2016, vol. 2016–May, pp. 1926–1933.

[11] M. Fazeli, J. B. Ekanayake, S. Member, and P. M. Holland, "Exploiting PV Inverters to Support Local Voltage — A Small-Signal Model," *IEEE Trans. Energy Convers.*, vol. 29, no. 2, pp. 453–462, 2014.

[12] H. A. Shayanfar and S. Malek, "Photovoltaic Microgrids Control by the Cooperative Control of Multi-Agent Systems," in *30th Power System Conference (PSC2015)*, 2015, no. November, pp. 287–293.

[13] N. Quijano *et al.*, "The Role of Population Games and Evolutionary Dynamics in Distributed Control Systems," *IEEE Control Systems Magazine*, vol. 37, no. 1, pp. 70–97, 2017.

[14] B. Wen, "Stability Analysis of Three-phase AC Power Systems Based on Measured D-Q Frame Impedances," PhD, Virginia Polytechnic Institute and State University, 2014.

[15] R. Teodorescu, M. Liserre, and P. Rodríguez, *Grid Converters for Photovoltaic and Photovoltaic and.* A John Wiley & Sons, Ltd, 2011.

[16] F. Brihmat, "PV Cell Temperature / PV Power Output Relationships Homer Methodology Calculation," *IJSTR Int. J. Sci. Technol. Res.*, vol. 2, no. 1, 2014.

[17] Kundur.P, *Power System Stability and Control.* Mc Graw Hill. Inc, 1994.

USE OF LOAD PROFILES TO OPTIMIZE MICROGRIDS OF NON-RESIDENTIAL ENVIRONMENTS

Jesus da Costa Fernandes, Michael Schmidt, Thomas Feldmann, Elmar Bollin
Hochschule Offenburg – University of Applied Sciences / INES – Institute for Energy System Technology
24, Badstraße, 77652, Offenburg, Germany
Phone: +49 781 205 354, Fax: +49 781 205 45354
E-mail: costa@hs-offenburg.de
Internet: fgnet.hs-offenburg.de, ines.hs-offenburg.de, www.hs-offenburg.de

Three real-lab trigeneration microgrids are investigated in non-residential environments (educational, office/administrational, companies/production) with a special focus on domain-specific load characteristics. For accurate load forecasting on such a local level, à priori information on scheduled events have been combined with statistical insight from historical load data (capturing information on not explicitly-known consumer behavior). The load forecasts are then used as data input for (predictive) energy management systems that are implemented in the trigeneration microgrids. In real-world applications, these energy management systems must especially be able to carry out a number of safety and maintenance operations on components such as the battery (e.g. gassing) or CHP unit (e.g. regular test runs). Therefore, energy management systems should combine heuristics with advanced predictive optimization methods. Reducing the effort in IT infrastructure the main and safety relevant management process steps are done on site using a Smart & Local Energy Controller (SLEC) assisted by locally measured signals or operator given information as default and external inputs for any advanced optimization. Heuristic aspects for local fine adjustment of energy flows are presented.
Keywords: Smart Grid Operation, microgrids, trigeneration microgrids

1 INTRODUCTION

Microgrids play an increasingly important role in future power systems. They allow integrating higher shares of distributed renewable sources like photovoltaics (PV) and wind power by balancing the local intermittent generation with local demand by means of load management and storage. Reduced transportation and distribution losses lead to a high level of efficiency as well as– in case of tri-generation microgrids (power, heating & cooling) – the use of heat of combined heat and power units (CHP) near to the source. If microgrids are able to perform grid-stabilizing functions or can even switch from grid-connected mode to islanded mode they can effectively help to increase overall system stability and reliability.

Intensive research was devoted to the reliable and efficient operation of microgrids over the last decades [3, 8, 14, 15, 16, 17, 22]. Many authors propose the use of mathematical optimization and control methods like linear optimization or model predictive control [7, 10, 14, 21, 22, 24, 26]. Typically, such optimization methods need forecasts of supplied loads and of the renewable power generation in order to generate operation schedules for switchable and controllable units like storages, conventional generators or devices and appliances.

The accurate forecast of loads respectively the generation of representative load profiles for different consumer types and consumer situations is therefore crucial for the success of optimization efforts of energy systems. This holds for small spatial scales like microgrids as well as for larger scales like distribution grids and beyond. Intensive research was therefore also devoted to load forecasting [1, 2, 3, 4-10, 26, 27].

Operating tri-generation microgrids in the real world, the proposed theoretical methods often can only be applied with major modifications. For instance, mathematical optimization schemes and forecasting technologies typically ignore component requirements like the regular gassing of a battery. Therefore, this paper takes a strongly application oriented point of view. It presents operating experience with focus on load profiles and automated operation of three microgrids operated and monitored as real-labs in non-residential environments:

1. Educational environment - A cogeneration micro grid (with 1.2 kW PV peak power) on the main campus of the University of Applied Sciences in Offenburg primarily supplies a students' learning area with power (cf. Fig 1). The heat of the natural gas driven 15 kW CHP-unit is distributed to other lab areas and a third of the power is used to charge a lead acid battery when renewable power sources are not sufficient.

Figure 1: Real-lab cogeneration microgrid with rooftop installed renewable energy sources (left) and students' earning corner (right) in university building B in Offenburg/Germany [12].

2. Office and administrational buildings - A tri-generation microgrid with 6.5 kW PV peak power at the Institute for Energy System Technology (INES) on the northern campus of the Offenburg university supplies an office area, an electro-vehicle and some lab facilities with power (cf. Fig. 2). An extension to supply heat and cooling is ongoing integrating a CCHP unit to the microgrid.

Figure 2: Front view of the INES-Campus North with office building and outdoor installations in Offenburg.

3. Companies and production facilities - A tri-generation grid (with 500 kW heat and up to 160 kW power) supplies a poultry farm with eggs and noodle production with power and different levels of heat and cold (cf. Fig. 3).

Figure 3: Overview of the company real-lab poultry farm in Schönberg near Gengenbach/Germany [13, 15].

The remainder of this paper is organized as follows. In Section 2 we will give a short overview of theoretical work on microgrid optimization and load and generation forecasting with a special focus on contributions of the authors. In Section 3 the focus is set on the use of heuristics in microgrid operation in order to account for important events changing optimization conditions and additional requirements like battery gassing, state-of-charge-calibration etc. In Section 4 the real-world microgrids mentioned above will be presented in more detail giving a practical insight on the generation and interpretation of load profiles for these cases. Finally, we present our conclusions and outlook in Section 5.

2 OVERVIEW OF LITERATURE AND PRELIMINARY WORK

The theoretical foundation and practical realization of (trigeneration) microgrid operation is an extensive field of research. In this section, we focus on the state of the art and preliminary work related to predictive optimization and management, load forecasting and the role of different storage options.

2.1 Predictive optimization and management

Forecasting of solar power generation is an important ingredient for predictive optimization at the management of microgrids. In many cases, the solar power forecast is based on irradiation forecast. Irradiation forecast stem in many cases from numerical weather predictions, often in combination with ground measurements and satellite data. Therefore the spatial resolution of the forecast is typically limited to several kilometers of regional dimensions and can therefore not take into account specific conditions of the located site of the microgrid.

Down to 3 % accuracy can often be achieved for larger areas mainly in periods with weak precipitation. Local weather conditions lead to strong variations in accuracy of over 20 % and strongly ask for additional tools to improve forecasting quality. Schmelas, Feldmann, da Costa Fernandes et al. of the research group sustainable energy technology (FGNET) at INES presented in [11] that this range in forecast accuracy is realistic also for microgrids. An improvement of the power forecast can be achieved by taking into account shading conditions and the knowledge of local weather conditions like precipitation forecast, snow coverage in winter or cloud movement. This can be done by merging non-local weather forecast information with local information e.g. with Artificial Neural Networks (ANN) in order to obtain an improved local forecasting.

In a second step the load profile forecast should achieve at least the same accuracy compared with the forecasting of the energy supply, which is typically a mix of dynamic and unsteady sources as well as of highly stable energy sources. Energy storage plays a crucial role to balance supply and demand in the presence of forecast errors.

In the preliminary work the research group FGNET at INES has developed energy management and optimization solutions for many different applications over the last years. Already at the REHVA Clima Conference 2005 in Lausanne, first developments of data communication infrastructures for energetic optimization have been presented with the intelligent dynamic building (idB) operation[23]. First results to validate an optimization process applying the idB functions were presented with the simulation supervised operation of a grid connected PV system in Offenburg by Eicker et al. [24]. In the same context a building operation was supported by online simulation with the optimized night ventilation of the Solar Info Center Building SIC in Freiburg by Bollin, da Costa F. and Feldmann [25]. An algorithm using a Multi-Linear Regression (MLR) was able to adjust the intensity and duration of night ventilation in a very efficient way using the forecasted average ambient temperature for the day ahead. The forecast was included to the local automation system using a weather forecast webservice developed by the FGNET in cooperation with a weather forecaster.

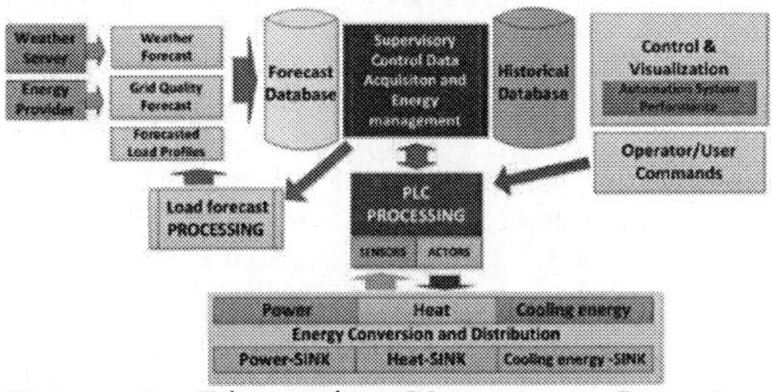

Figure 4: Trigeneration Management Concept – Overview of the functional components for the smart microgrid [13]

Schmelas et al. [26] continued the development work and extended the MLR-Concept to an Adaptive MLR algorithm for thermo-activated building components. Hong [7] applied MLR to load forecasting for demand side management. Actual works on optimization continue with MPC (Model Predictive Control) approaches by several colleagues at INES mostly in cooperation with the University of Freiburg. The application potential can be stated as wide open with development focuses on building integration of renewable energy sources, energy efficiency and optimized operation of smart microgrids like the INES Smart Grid, cf. Fig. 2.

Currently, the concepts of predictive optimization which have been applied to the operation of single components are transferred to trigeneration microgrids delivering the necessary hardware and communication infrastructures, cf. Fig. 4.

2.2 Load forecasting

Accurate load forecasting is an important prerequisite for an efficient grid operation for many years, and intensive research was dedicated over the last decades to improve forecasting methods. The management of load structures on the middle and low voltage level made it compulsory to continuously analyze and predict the state and performance of power grids in order to run them economically efficient. Recent developments show a strong activity tending to smart grids and smart homes. A significant role is given since 2012 to the Global Energy Forecasting Competition (GEFCom2012) [4] stimulating a huge number of publications in energy optimization directions and methods. The most important approaches to mention in the case of short term forecasting are based on the use of similar-day, time series, regression, fuzzy logic, expert systems and neural networks. Various hybrid concepts and algorithms that add external information or statistical analysis have been published and validated. A good survey on the various concepts is given by Hernandez et al [10].

While grid operators typically need a forecast of the aggregated load of grid sections, microgrid operators have to accurately predict few local loads. They may have the advantage of additional knowledge on scheduled events, but cannot rely on averaging effects for non-planned loads. Therefore, it is especially important to capture domain-specific knowledge on the load characteristics and to merge statistical methods for load-profiles with à priori knowledge on scheduled events.

One attempt to capture the different characteristics of local load profiles has been presented by Chen, Canizares and Singh [5]. They compose the load profile of four functions representing a normal load (L_n), weather sensitive load (L_w), special event load (L_s) and random load (L_r),

$$L = L_n + L_w + L_s + L_r \qquad (2.1)$$

Typically, the variation of the total load in time during a day is hence not completely predictable, e.g. since the consumers only partly follow a given time schedule or specific use pattern. This finding was already stated by Jardini et al. [6] and can therefore be sufficiently described with a daily course of the mean and standard deviation curves for forecasting purposes.

As in the case of generation forecast errors, energy storage is a crucial tool to deal also with demand forecast errors.

2.3 Role of storages

As seen in the previous subsections, energy storage play a crucial role in balancing energy supply and demand in general, but especially also to compensate for forecasting errors.

In trigeneration systems, electrical storages as well as storages for heat and cold can help to balance demand and supply of the different media. Here we give a short overview of the main concepts.

Thermal energy storages (TES)

As the most simple thermal energy storage concept was used the thermosiphon system to store solar heat in a water tank. Sophisticated developments led to heat pumps, geothermal heat exchangers, seasonal storages or ice storages mostly combined with water or frost resistant media. A wide range of technical applications are available on the market and therefore considered as already part of engineering. Charging and discharging at the same time or today's concepts of thermal layer storages require relatively complex system models and also more sophisticated controls.

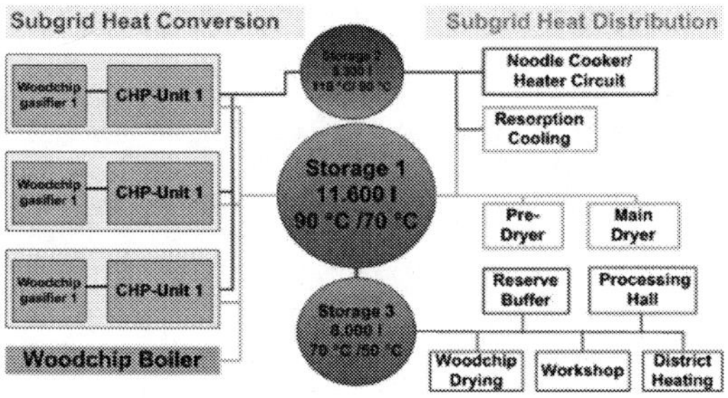

Figure 5: Heat Conversion and Storage concept of the real-lab noodle-production. The resorption chiller connected to Storage 1 and 2 additionally supplies the ice storage and the chilling circuits.

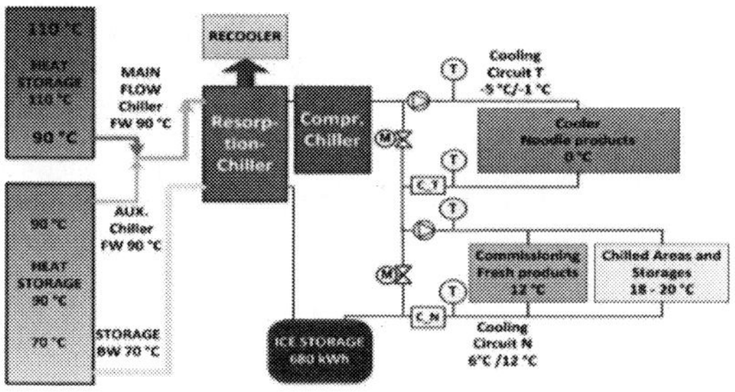

Figure 6: Cold storage and chilling circuits of the real-lab noodle production.[15]

Electrical energy storages

In the research history back to the 90´s electrical energy storages were mainly dominated by lead acid batteries as used in decentral power supplies and backup systems. Scientific modelling was done for example with the focus on PV based remote power systems. A large number of research and demonstration projects were supported by the European Energy Research Funding Initiatives like JOULE, THERMIE or ALTENER. However, in the following years, most installed PV systems have been connected to the grid without need for electrical storage. Only in the last years, with PV grid share constantly increasing, the combination of PV and battery storage in order to increase self-consumption and lower the effect on the distribution grids, the combination of PV and local battery storage has gained new dynamics. This tendency is even more increased due to newest smart grid concepts that would allow for a grid-friendly use of local storage units.

Viewed of a modeling and control point batteries still present some major research and development challenges, like an accurate estimation of the state of charge (SOC) or state of health (SOH). In real world operation, calibration events are necessary in order to obtain accurate SOC and SOH estimation on the basis of respective estimation algorithms. Such calibration events, as well as health-improving operations like gassing for lead-acid batteries, have to be triggered by energy management systems and should, ideally, be taken into account by optimization algorithms.

Discussing electrical storage devices, electric vehicles play a special role. If a bidirectional use of the Li-Ion-accumulators is allowed by the manufacturers and the charging infrastructure is capable of it, the typically large-sized batteries of electric vehicles could be used for the energy balancing within microgrids or within the utility grid. Possible conflicts of interests between the user of the car and the microgrid or distribution grid operator are here a special research and development challenge. The effect of this connection is still a strong focus of research activity in projects like E-mobility [27] analyzing load profiles of e-vehicles in the company environment.

Power-to-Gas is another interesting electrical storage option, possibly serving as long-time and seasonal storage, but also offering high energy densities like conventional fossil fuels. Especially regenerative surplus of power that cannot be consumed during the moment of generation is converted to gases like hydrogen or methane and fed to the gas distribution or filled directly or liquidized to tanks. Important works were done in the context of technical potentials of chemical storages and how to include new power-to-gas solutions to economic reality [20].

3 METHODS AND APPROACH

In this section, we present some aspects of load forecasting and heuristics-based control that appear very powerful in practical applications but which are not so widely covered in the literature often focusing on theoretical methods.

3.1 Practical approaches to improved load forecasting

3.1.1 Use of historical data to identify day-specific patterns

Based on historical data of the system monitoring database a calendar evaluation can identify system specific patterns that are day-specific and allows days-ahead forecasting. Sometimes, standard load profiles as published by the BDEW (German Association of Energy Providers) can be matched with or adapted to measured load profiles. This works best for pooled consumers (like a city quarter) with low shares of renewables and local energy management, but is typically not directly applicable to single self-managed entities.

3.1.2 Use of additional planning information (e.g. occupancy or production schedules)

Another very valuable source of information is the use of operation plans for machines, occupancy schedules for rooms, production plans for whole production lines etc. Especially up-to-date information on deviations of regular schedules is crucial for an optimized operation of the energy system. If this planning data is merged with the statistical analysis of historical data, significant improvements of the load forecast accuracy can be achieved.

3.2 Improving controls including heuristic features

The operation of components like batteries or CHP units requires certain maintenance actions that are typically not taken into account by classical energy management algorithms that are based on the mathematical optimization of cost functions, i.e. such maintenance events are typically out-of-scope of the mathematical optimization. Nevertheless are they crucial for a smooth and safe system operation and therefore triggered by the energy management system. Another category of out-of-scope events is related to the manual interaction of energy system users. We present now some examples for such out-of-scope events that are typically not taken into account by the optimization algorithms respectively by the load forecasting.

3.3.1 Battery calibration events

Most algorithms that continuously calculate the state of charge (SOC) of a battery are drifting because of the difficulty in modelling the chemical complexity of batteries. Temperature, acid density, current and voltage dependence are four key characteristics that have to be accurately captured by such models.

In conjunction with renewable energy sources like wind and solar, batteries may have to deal with rapidly changing charging and discharging currents as well as with fast changes in the temperature conditions, which makes it even more difficult to estimate the correct SOC. As a consequence there is significant influence on the state of health (SOH), which is also difficult to estimate. Therefore SOC and SOH algorithms typically need to recalibrate in regular time intervals. A simple method for SOC calibration is to force the battery to the end-of-charge (EOC) point. In addition, an overlapping with the gassing sequence of the battery charger can be implemented. Such a strategy was first implemented with a lead acid battery within the real-lab Energieinsel in 2014.

3.3.2 Manual Interaction by User

Any user/occupancy dependent load profile is strongly influenced by the acceptance of the technical systems the user should rely on and how the human behavior could be included [18]. The role of comfort, occupant satisfaction and acceptance in office buildings have been evaluated by Gossauer [19] within the German federal EnOB initiative. A significant role on load profiles was revealed related to humans using a technical infrastructure. Manual overrides of automation and technical devices should be included in short term load forecasting as well as manually activated faults by operation commands. Here a crucial role is given to the performance of the local automation system and the person(s) in charge of the implementation. A special importance must be given to this aspect if only a reduced automation is performed due to economic reasons.

We propose to realize necessary maintenance actions on the basis of heuristics that overrun the operation schedules generated by an optimization algorithm using load and generation forecasts with cost functions. Similarly users will regularly cause deviations of these operation schedules by not following load forecasts, e.g. by overrunning automated use profiles for heaters, air conditioning, ventilation appliances etc.

Ideally this change by respectively overrun of the classical schedules should be fed back as hard constraints to the classical optimization routine in order to allow an update of optimization that now considers the necessary maintenance events or user interactions.

A solution discussed among experts is the theory of agent based control systems which can ease the fine adjustment of specific grid components respecting the overall management on a higher communication layer [1, 2, 3, 14, 15,21, 23, 24, 26]. A management of events and calibration routines to readjust the optimization algorithm can be integrated using flags or state signals written to the database as loop-back information of the system. These vectors can trigger the update rate of forecasts and the choice of subroutines according to their need. The optimization priority can be controlled if an event occurs causing a breakdown in communication or even forces a fallback to a local save mode.

4 NON-RESIDENTIAL REAL LABS

In this section we describe three non-residential trigeneration microgrids in more detail, with a special focus on load profiles, forecast-based energy management and the parallel use of heuristics in order to ensure necessary maintenance events. All three projects are equipped with an advanced energy monitoring and energy management system, and the development of new energy management features and algorithms is ongoing.

4.1 Educational environment: PV-hybrid system with cogeneration

The "Energieinsel" (Energy Island) is a PV-hybrid system with cogeneration located in and on a building at the main campus of the Offenburg University of Applied Sciences and represents an educational environment. It is PLC-controlled and has a Panel-PC as HMI. Its main loads are the data acquisition and control units as base or normal load and a student' learning corner. The loads of the learning corner are lighting and small appliances like phone and notebook chargers. The lighting of workplaces can be adjusted according to personal requirements of users via Panel-PC. If a motion sensor (passive infrared, PIR) detects that there are no users the lighting intensity is adapted to daylight intensity or switched off.

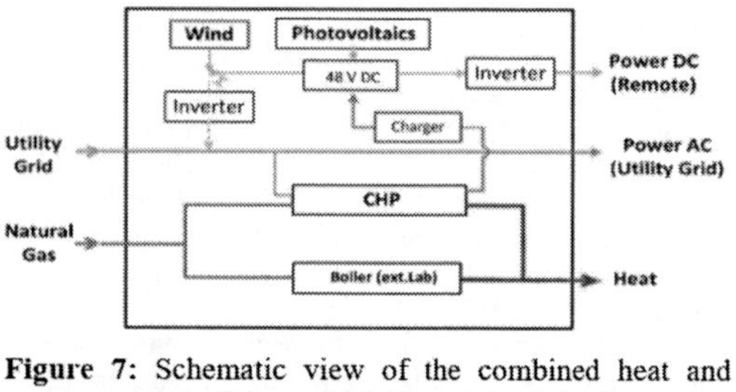

Figure 7: Schematic view of the combined heat and power concept of the real-lab Energieinsel (main campus of the Offenburg University of Applied Sciences).

The power is supplied by different generators on the roof of the building. The photovoltaic generator and the small wind turbine of Figure 1 are basic components of the hybrid system running the learning corner. The energy supply concept of Figure 7 shows a scenario which is representative for many municipal building pools or an educational campus. Within this concept and proved by measurement campaigns on the main campus and on the north campus wind generators play a minor role in Offenburg and many other cities across Germany. Wind turbines can only be stated as moderate sources in many urban environments.

With respect to characteristic load profiles, the analysis offers good empirical findings to combine approved forecasting methods with site-specific knowledge gained for local correction features.

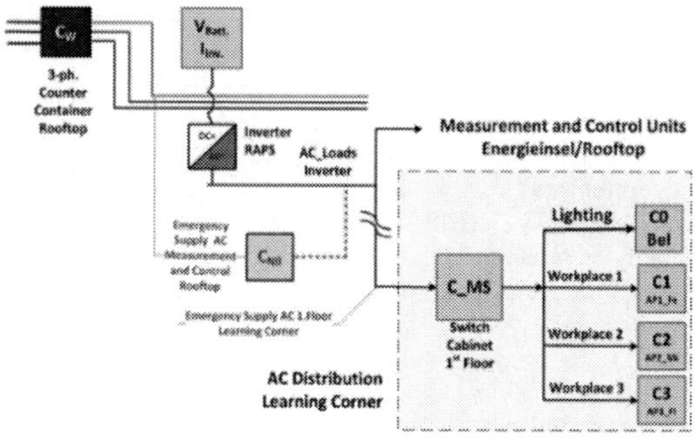

Figure 8: Load distribution of the real-lab Energieinsel and the learning corner.

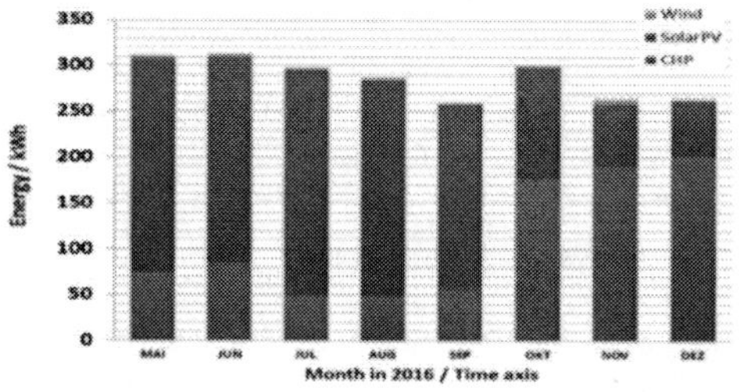

Figure 9: Energy shares of sources CHP, solar PV and wind to supply the power demand of the learning corner.

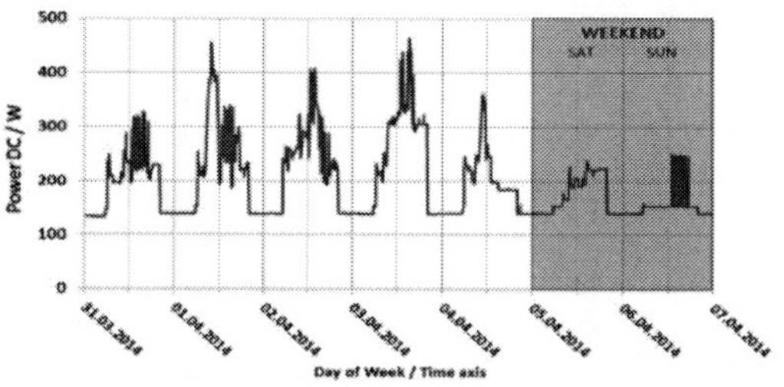

Figure 10: Load profile of the learning corner during a peak period of the semester. In peak periods a Saturday is often a workday for students and the normal load is given with roundabout 130 W.

Calendar analysis for improved load forecasting

We now present how a calendar analysis with day types and seasonal characteristics can help to identify repetitive patterns. The example of the learning corner adds to the daytime/night change three semester seasonal factors:

- Period of no occupancy (outside lecture period, low intensity)
- Regular occupancy period (during lecture period, typical use)

- Occupancy before and during critical activity phases (before and during exams, high intensity)

The power load profiles of the learning corner and the identified patterns are representative for educational and seminar buildings, as can be seen of a comparison with a measured load profile of a public school in Offenburg (see Figure 16). The performance check reveals typical patterns which can be used for forecasting a day applying an adequate weight for the expected demand. The identified patterns are also very helpful for a suitable dimensioning of storages when optimizing the system and its operation.

There are three load intensities to be identified classifying the lecture period, weekends and holiday as further periods. If these periods cannot be identified by a given time schedule the calendar evaluation can offer highly probable preset values for the demand forecast. Mainly office and administrational buildings tend to similar patterns of occupancy. Here the example of the office area at the northern Campus of the university in Offenburg represents the typical load profile of small groups in companies or municipal units working in defined spaces. Scaling in this case is easily possible as well as the energy share for air-conditioning and space heating.

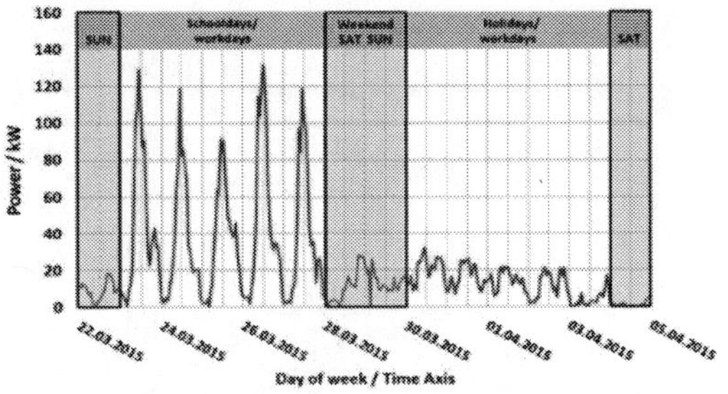

Figure 16: Measured load profile of a secondary school in Offenburg/ Germany [15] with significant patterns.

Heuristics for energy maintenance

We now present a practical example for improving the battery management on the basis of heuristics, realized and validated in the real-lab Energieinsel.

The CHP operation to supply AC Loads and to recharge the lead-acid-accumulator up to a float level can be started by the solar charger or by the PLC when the following conditions are satisfied:

Due to the course of solar and wind power as main energy sources together with the consumption of loads a steady decrease of the battery level can be stated in autumn, winter and spring. Even assuming a typical energy demand the remaining power is very often not sufficient to realize a significant raise of the battery level which is preliminary for a gassing to improve the acid density distribution of the electrolyte. To improve the SOC of the battery a complementary use of the energy sources should help to lift up the cycling level. The following conditions were defined to initiate the raising procedure in a weekly interval on every Sunday morning:

If the battery voltage level never reached the EOC during the last six days the control sequence will start the CHP on Monday night at latest 2 o'clock until the CHP-stop level were passed. With the following solar power in the morning renewable energy should be sufficient to bring the battery to a significantly higher SOC-level and closer to gassing conditions.

For the longer term a higher cycling of the accumulator allows a more efficient gassing and reduces operation of the CHP with the hysteresis of the backup-function given for a deeper cycling.. The winter period is affected by three ambient conditions: the higher need of artificial light and heating demand together with missing sunlight as energy source. In winter a higher share of CHP energy is quite inevitable but the use of the heat is guaranteed in the early morning time and overlaps with the battery support. Solar power is mostly too weak to bring the battery to gassing. Measured data of the real-lab Energieinsel confirmed the effect of the forced CHP operation in summer but confirmed also that the actual consumption level is still too high for the given renewable energy sources.

4.2 Office and administrational buildings: A multimode microgrid

At campus north of the Offenburg University, a three-phase microgrid with diverse generation, storage and load options is operated in conjunction with a trigeneration system. The design of this smart microgrid allows an autonomous, grid supported and grid interactive mode. For several projects the lab infrastructure is best suited to develop and validate optimization algorithms on the low voltage level of utility grids [9]. The significant components of the microgrid can be listed with a 6.5 kW peak photovoltaic generator, a 4.5 kW wind turbine, an e-vehicle with 26 kWh Li-Ion-battery, a KOH-electrolyzer of up to 6 kW to produce hydrogen and hydrogen PEM fuel cell of 1.7 kW to feedback hydrogen power. A meteorological station was recently installed to measure ambient conditions allowing local fine adjustments of predictive management and control tools.

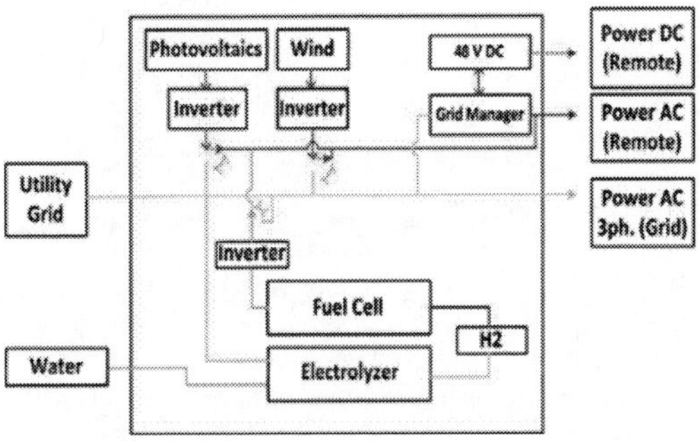

Figure 10: Simplified energy supply concept of the INES Smart Grid ISG on Campus North in Offenburg.

The office area contains 4 offices with two workplaces each. The working times are usually scheduled between 8 a.m. and 5:30 p.m. Due to varying occupancy (vacation, training, internships, changing work places) the daily energy demand is fluctuating for all appliances related to the workplaces like IT devices, lighting, water use or heating demand in the winter period. Thermal aspects are mainly defined by water heating realized by small electric boilers in the washing rooms. Additionally, lighting and kitchen appliances like a refrigerator, a water boiler and a coffee machine cause additional consumption. The scenario can be considered as representative for many office areas in administrational or non-residential buildings.

First load profiles were taken with a short term measurement campaign. Already the first data sets revealed the difficulties in getting accurate load forecasting if schedules about occupancy were not kept up-to-date. Lighting in common areas is mainly controlled by presence detection.

Figure 10: INES Smart Grid - R&D test environment on Campus North of the Offenburg University of Applied Sciences in Offenburg/Germany.

Figure 11 reveals that occupancy is not strongly following a schedule and causes random effects. Additionally to a base load and an activity determined offset there is at least a third share being difficult to forecast in a satisfying way. A pattern definition is possible with days of activity and days of no-activity, nighttime and daytime offset. Peaks of the specific weekly profile are mainly caused by electric boiler units while the base load is kept very low.

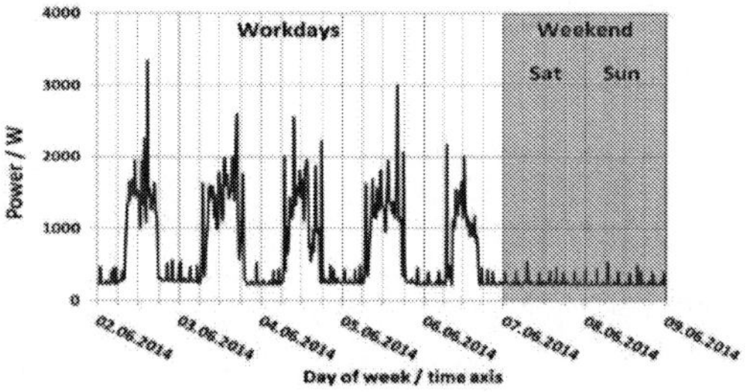

Figure 11: Measured week profile of the power demand in the office area at Campus North.

4.3 Companies and production facilities: A trigeneration microgrid

A tri-generation supplies a poultry farm with egg and noodle production with power and different levels of heat and cold (cf. Fig. 3).

The engineering concept of the smart trigeneration grid is schematically described in Figure 7 and works with three lignite gas fueled CHP units with a total electric power of 135 kW and 360 kW of process heat , an 18 kW peak PV generator, a 135 kW boiler and two chiller units with up to 120 kW of cooling power.

Thermal storages for three heat and two cooling power levels complete the smart energy configuration. The energy consumption of the grid automation is kept as low as possible. An internet based management concept works on the experience of previous projects and should be extended to the predictive energy management developed for the Energie-Plus-Haus in Berlin [14].

The load profile of Figure 15 was measured after the substitution of a major part of thermal consumption units which are now covered by the thermal heat distribution.

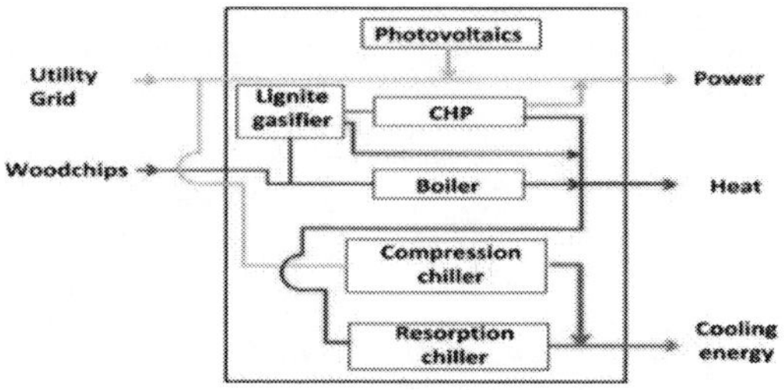

Figure 13: Simplified Energy supply concept of a poultry farm as company type.

The remaining power demand achieves peak values up to 140 kW. While the resorption chiller is a thermal consumer the backup compression chiller is power driven and had to be installed to secure the reliability of the cold storages where fresh eggs and noodle products are stored before expedition.

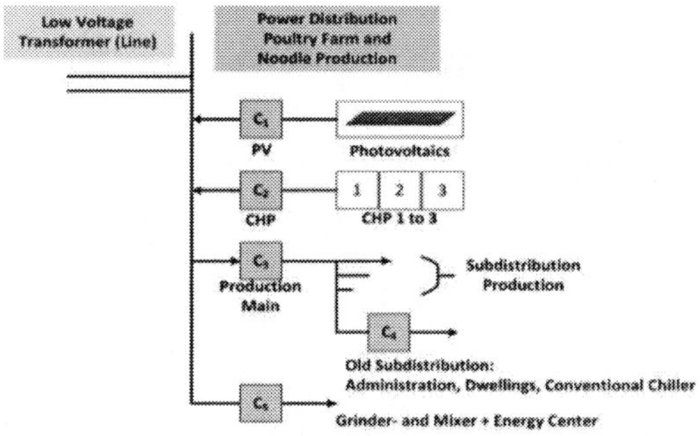

Figure 14: Power distribution of the trigeneration grid with functional areas noodle production, poultry farm with administration and dwellings.

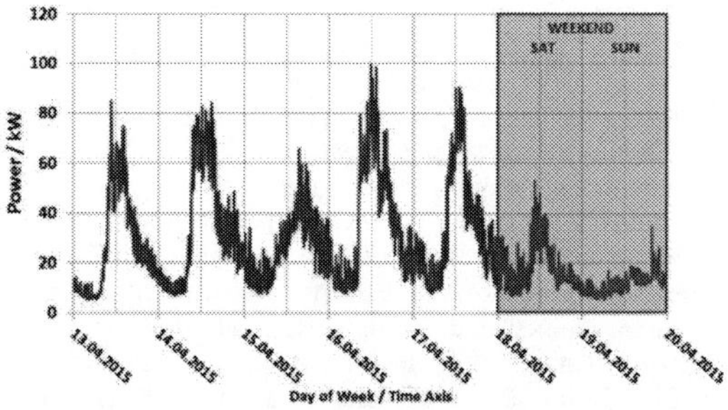

Figure 15: Measured week profile of the power demand of the noodle production at the poultry farm.

The monitoring and setup of fundamental energy management functions of all real-labs were finished in 2016 and are described in detail in a final report of the project "Energy grid management" supported by the innovation fund of a local energy provider under the ID 2012-09 [15].

5 CONCLUSIONS AND OUTLOOK

Load forecasting and generation forecasting play a crucial role in modern energy management systems that apply mathematical and predictive optimization in order to achieve optimal system operation with respect to a user-defined cost function. Many sophisticated load forecasting methods are available and practical experience shows that especially the evaluation of historical monitoring data with respect to day-specific patterns and the combination with extra information like occupancy or production plans is very efficient.

The experience and analysis of monitoring data of three real-world microgrids in non-residential environments show that such load forecasting is challenging for different reasons: Averaging effects that apply to larger pools of aggregated consumers do not fit here. A part of the loads is strongly related to foreseen use profiles (occupancy or production schedules) which are in principle easy to forecast. Unfortunately deviations of these schedules are often not communicated to the energy management system. Users can overrun scheduled usage of components and hence cause deviations of the load forecast. Operation and maintenance events must be realized but the information about is often not included in the load and generation forecasts serving as inputs of higher-level energy management algorithms.

Triggering of such operation and maintenance events can be reliably implemented using heuristics that overrun the higher-level energy management. Ideally the information about the triggering is fed back to the higher level optimization layer initiating a new optimization run taking into account the new constraints.

In the future, the development of hybrid control strategies is planned. These strategies shall merge the use of mathematical optimization tools with heuristics for operation and maintenance, and shall also focus on the known sources of load forecast deviations described above.

REFERENCES

[1] Winkler G., Meisenbach C., Hable M., Meier P.: Intelligent energy management of electrical power systems with distributed feeding on the basis of forecasts of demand and generation, CIRED 2001, Amsterdam/NL, 18. - 21. June 2001.

[2] Krost G., Beckhaus P.: Intelligent forecasting energy management for hybrid photovoltaic systems; Proc. of IEEE ISAP2001 Conference, Budapest/Hungary, June 18-21, 2001.

[3] Beckhaus P.: Simulation und Anlagenmanagement für dezentrale Energieversorgungssysteme; Dissertation, Universität Duisburg, 2002.

[4] Lloyd J.R.: Load forecasting at GEFCom2012 and overviews on, 2014, website: www.gefcom.org

[5] Chen H., Canizares C.A., Singh A. et al.: ANN-Based Short-Term Load Forecasting in Electricity Markets. Proc.of the IEEE Power Eng. Society Transmission and Distribution Conference, 2:411–415, 2001.

[6] Jardini, Tahan, Gouvea et al.: Transactions on Power Delivery, Vol. 15, No.1/ January 2000.

[7] Hong T., Gui M., Baran E. M., Willis, H.L.: Modeling and Forecasting Hourly Electric Load by Multiple Linear Regression with Interactions, 978-1-4244-6551-4/1, IEEE 2010.

[8] Harb H. et al.: Optimal design of energy conversion units for residential houses, Conference Paper, RWTH Aachen, Report Nr. RWTH-2016-04264.

[9] Gobmeier Th. et al.: Simulationsgestützte Prognose des elektrischen Lastverhaltens, Endbericht der Forschungsstelle für Energiewirtschaft (FfE), June 2012.

[10] Hernandez L., Baladrón C., Aguiar M. J. et al.: A Survey on Electric Power Demand Forecasting: Future Trends in Smart Grids, Microgrids and Smart Buildings, IEEE Communications Surveys & Tutorials, VOL. 16, No. 3, 3rd Quarter 2014

[11] Schmelas M., Feldmann T., da Costa Fernandes J., Bollin E.: Photovoltaics Energy Prediction Under Complex Conditions for a Predictive Energy Management System, Journal of Solar Energy Engineering, June 2015, Vol. 137 / 031015-3.

[12] Da Costa Fernandes J.: Forecast based energy optimization for solar photovoltaic and hybrid energy systems, Session 6AV.4.6 at 29th European Photovoltaic Solar Energy Conference and Exhibition, ISBN: 3-936338-34-5, p.3648-3653, Amsterdam/NL, 2014.

[13] da Costa Fernandes J., Eid M.: Forecast based energy management for trigeneration subnets in smart grids", 7th Int. Conference on PV Hybrids and Mini Grids, Bad Hersfeld, April 10/11, 2014.

[14] Feldmann T., Schmelas M., da Costa Fernandes J., Bollin E.: Prädiktives Energiemanagement für ein Energie-Plus-Haus mit Versorgung von Elektrofahrzeugen, Forschungsinitiative „Zukunft Bau", Aktenzeichen: II3-F20-11-1-, Berlin 2013.

[15] da Costa Fernandes J., Zapf G., Schneble H.-J., Bollin E.: Final Report of Project, "Energienetzmanagement dezentraler, wärmegeführter Blockheizkraftwerke im speichergestützten KWKK-Betrieb diverser Verbraucher-strukturen", Hochschule Offenburg, Innovationsfonds badenova, Project No. 2012-09, 06/2017.

[16] Farhangi H.: The path of the smart grid, IEEE Power & Energy Magazine, Jan. / Feb. 2010.

[17] Savola T.: Modelling biomass-fueled small-scale CHP plants for process synthesis optimization, Dissertation, Helsinki Univ. of Technology, 2007.

[18] Gigerenzer G., Todd P.M. et al.: Simple heuristics that make us smart, Behavioral and Brain Sciences (2000) 23, 727–780, Oxford University Press, New York, 1999.

[19] Gossauer E.: Nutzerzufriedenheit in Bürogebäuden, Dissertation, Universität Karlsruhe, 2008.

[20] Lietz, F.: Rechtlicher Rahmen für die Power-to-Gas-Stromspeicherung, Dissertation Universität Göttingen, ISBN 9783848742677, NOMOS, 2017.

[21] Hirsch, C.: Fahrplanbasiertes Energiemanagement in Smart Grids, Dissertation Karlsruhe Institut für Technologie (KIT) Karlsruhe, 2015.

[22] Krost G., Matics J.: Mikro-Blockheizkraftwerke zur Gebäudeversorgung Flexibles autoadaptives Betriebs-management, eBWK Nr. 60, 2008 Nr.3.

[23] da Costa Fernandes J., Feldmann T., Bollin E.; idB-Simulation and Automation assisted Information and Control Tool for Sustainable Building Operation, Technical Sessions at Clima 2005, 8th REHVA World Congress in Lausanne, 2005.

[24] Eicker U., Pietruschka D., Schumacher J., da Costa Fernandes J., Feldmann T., Bollin E.; Improving the Energy Yield of PV Power Plants Through Internet Based Simulation, Monitoring and Visualization, Proceedings of the 20th European Photovoltaic Solar Energy Conference and Exhibition in Barcelona, Barcelona, 2005.

[25] Bollin E., da Costa Fernandes J., Feldmann T.: Energie optimiertes Bauen, Teilkonzept 3: Langzeitmonitoring des Neubauvorhabens Solar Info Center Freiburg, Phase II Monitoring/ Betrieb, Abschlußbericht FKZ: BMWi 0335007U, Hochschule Offenburg, 2008.

[26] Schmelas M., Feldmann T., Wellnitz P., Bollin E.: Adapt ive predictive control of thermo-active building systems (TABS) based on a multiple regression algorithm: First practical test. *Energy Build.* 2016, *129*, 367–377.

[27] Meßmer J., Bollin E., Hochberg U.: Final Report "E-Mobility meets urban Services", Hochschule Offenburg, Innovationsfonds der EW Mittelbaden, 2015.

BIDIRECTIONAL ELECTRIC VEHICLES STORES PV ENERGY AND CREATES NEW BUSINESS CASES FOR PV - CAN THIS REPLACE THE STATIONARY BATTERIES?

Urs Muntwyler, Benjamin Ulrich, Eva Schüpbach
Bern University of Applied Sciences (BFH), Jlcoweg 1, CH-3400 Burgdorf
phone: +41 (0)34 426 68 37, fax: +41 (0)34 426 68 63/ urs.muntwyler@bfh.ch

ABSTRACT: While many research projects mention or simulate using electric cars as moveable batteries, practical tests and measurements are rare. To better understand the current state of art and to measure the round-trip efficiency, the authors built a test system and performed various measurements. This paper describes the challenges, the performed measurements, the results and the potential for the technology.
Keywords: Energy Performance, Grid Integration, Interfaces, System Performance

1 INTRODUCTION

The marketing of PV electricity increasingly becomes a difficult challenge as the feed-in tariff schemes are blocked by overbooking or political actions against a higher contribution of PV electricity. In Switzerland, there is no chance to get the feed-in-tariff (Swiss-FIT called "kostendeckende Einspeisevergütung KEV"). In the end of 2016, 12'684 PV-Installations with 592 MWp were accepted in the Swiss-FIT. But 35'028 installations with 2'105MWp were on the waiting list. This is quite a lot in a country with about 8 million inhabitants. Hence, the self-consumption of PV electricity is crucial. Replacing grid electricity by PV electricity gives us a value of 0.25€/ kWh as an example. This works well with all electric consumers. But most consumers in a house are quite small compared to the production of the PV installation [1].

Figure 1: PV installation for the Tour de Sol 1987 – enough energy for 6 light electric vehicles from Horlacher AG (the installation is still in use).

Figure 2: Light electric vehicles from the Tour de Sol 1987.

Mobility as a new consumer of PV electricity is therefore an interesting market for PV. This idea is not new and was introduced first in the "solar car races" of the Tour de Sol, the first solar car race in the world. In the 3rd race, the Tour de Sol 1987, a new category for "grid connected EVs" was created. This allowed the race participants to share a 3 kWp PV plant on the roof of the company "Horlacher AG", a producer of light-weight electric vehicles in Switzerland, see Figure 1 and Figure 2. It was the starting point for private decentralized PV installations. Now the dreams of the Tour de Sol-organizers are coming true. For the PV market, EVs powered with PV is a "low hanging fruit" and can bring the potential of using the energy from nearly 2kWp per car. For Switzerland with its nearly 5 million cars with internal combustion engines (ICE) this is an additional 8-10 GWp market.

2 EVS ARE MORE EFFICIENT AND WILL BE CHEAPER THAN CARS WITH INTERNAL COMBUSTION ENGINES

Using PV electricity to eliminate the use of fossil fuel in inefficient processes gives a leverage effect. We can avoid the huge losses of these processes by replacing them with a highly efficient electric driven process. PV electricity has then an even higher value. This is especially the case for consumers of large amounts of electricity such as the heating and hot water systems with heat pumps. But the heating system is mainly used in winter times. In this season, the PV systems have less overproduction and thus the best new consumer with the leverage effect is the electric car. An electric car with a consumption of 15 kWh/100 km (for example the Opel "Ampera") needs a total of 2'250 kWh to run 15'000 km a year. With an electricity price of 0,25 €/ kWh this costs € 562.50 in Switzerland. A traditional gasoline car needs about 900 litres of gasoline (6 litre/ 100 km) for the same distance. This costs in Switzerland (price summer 2017: 1,5€/ litre) and so the user pays € 1'350. Here we have the first price advantage of the EV-user! Further cost advantages are lower service costs, eventually less insurance costs and in some countries subsidies (not in Switzerland). In the future, EVs will also be cheaper to buy. This is a function of the mass market and the "economy of scale". It can be expected that EVs will be cheaper around 2020. The functionality of modern EVs is similar or better compared to most conventional cars. The "fun to drive" is already better right now. Therefore, EVs will replace cars with combustion engines in most of the applications in the next 20 years.

3 ENERGY FOR THE EV WILL MAINLY COME FROM PHOTOVOLTAICS

EVs are a new consumer of electricity. There is often a confusion about the source of electricity. This is used by the lobby against EVs to calculate "ridiculous" CO_2 emission to worsen the environmental balance of EVs. In practice, most of the additional electricity will be produced by wind or photovoltaics. In our case, we are looking mainly on PV. This allows us to exploit additional benefits of these two technologies. The capacity of PV (and wind) installations is not a problem. Both technologies are growing much faster than the EV market. The lifetime of a PV installation is 25-30 years. The lifetime of a car is about 8-10 years. Therefore, a kWp PV sees three cars in its lifetime. This means for Switzerland that - with only 300 MWp/year of new PV capacity - all ICE can be replaced by EVs after 25-30 years. This is about the annual market in Switzerland. Quicker expansion of the PV capacity is no problem, as the world market was 75GWp in 2016.

4 SOLAR POWERED EVS HAVE ADDITIONAL POTENTIALS

With EV's powered by PV, we do not reduce emissions, we eliminate them. The elimination on local level is total. The motor emits no emissions. Only particles from the tyres are produced. This will allow local authorities to ban ICE's from inner city traffic. It solves also the CO_2 problem and allows national government to ban diesel cars, as EVs are much more efficient and emit no CO_2. This gives the EV with PV (and wind) by far the best rating of all motor fuels [2]. For the owner of a PV installation there are more benefits, such as, e.g., an increase of the "own consumption" of the PV plant. In this case, the energy costs the owner only the amount that would be obtained for feeding the electricity into the grid. The Swiss utility company Bernische Kraftwerke AG (BKW AG) pays 4 cent€/ kWh. In that case, the 2'250 kWh would cost only € 90 per year. In practice, not the entire 2'250 kWh will come from "overproduction". But from spring to fall, the EV will mainly enhance the "own consumption". This situation can be improved with special chargers, which switch on the EV in situation of overproductions. Such an installation was developed by two students at Bern University of Applied Sciences BFH in Burgdorf in 2016 [3]. A similar installation will be installed on the solar carport at BFH in Burgdorf.

5 PV INSTALLATIONS WITH RESIDENTIAL BATTERIES

Due to restrictions from utilities and governments, stationary PV batteries are more and more used to store a part of the overproduction of PV electricity. In practice, these installations are not very economical. A fair price for the PV overproduction would be a better solution. But this is not very realistic. The installed batteries are usually in the range of 1 kWh/kWp. This means that usually the batteries are below 10 kWh for single homes. Such batteries are still quite expensive to buy and to install. And while prices for the battery elements will drop in a similar way how PV module prices dropped, the cost of installation will continue to be high. These systems need expensive cabinets to host the battery elements and the power electronics. This and the installation costs will stay even if the battery elements are below 100€/kWh (which is to be expected in the next five years from now). In addition to these costs, space is needed to place the battery system.

6 EV BATTERY IN BIDIRECTIONAL USE – A PRACTICAL EXAMPLE WITH A NISSAN LEAF

6.1 The potential of EV battery to replace the stationary battery

The electric car offers even an additional benefit. The battery can replace the stationary batteries of PV installation. This offers low costs, as the EV battery is huge compared to stationary batteries in PV installations. In the past, most EVs had only relative small battery sizes of 10-20 kWh (with the exception of the Tesla 85-100 kWh). Such batteries offer not a big capacity for additional use. With the new 40-60 kWh batteries of the new Nissan Leaf/ Zoe and Ampera-e, more EVs have big capacities compared to the stationary PV batteries. In the present quality, these batteries will not see their "end of life" in the car. An Opel Ampera-e with 400-500 km and 1'000 cycles would reach 400-500'000 km, a distance that hardly an Opel will see in its lifetime. This gives room for further use of the battery. In a PV installation, only one daily charge is to be expected. In wintertime, there will be no charge at all due to the low production of the PV installation. Therefore, more than 200 full cycles in a year are not realistic. But even this would mean that the full charge could be used daily in the house. So, if we assume that only 50% of a 40 kWh battery are used, then even the 20 kWh would be too much as this energy must be used in the night, when no charge occurs from the PV installation. In a single house with 4 persons, the daily energy consumption for electricity is about 4'000 kWh [4]. The daily consumption per night is less than 5 kWh or 1'000 kWh per year [5]. Therefore, the lifetime of the EV-battery will not be affected by this additional service. Only if we can serve additional users we can use the full 20kWh. This could be expected in a parking situation with many other cars, where the EV supports other EVs or the user of the parking get access to the EVs to offer grid services.

6.2 Bidirectional EV (V2G - vehicle to grid)

To use this benefit, a bidirectional EV is needed. As these EVs are not on the market yet, the simplest way is an EV with a CHAdeMO DC-charging facility, a Japanese charging standard that charges the EV battery on the DC-side. The Photovoltaic Laboratory (PV LAB) at BFH Burgdorf hence bought a Nissan Leaf (shown in Figure 3) offering this possibility. Using this car, the PV LAB carried out a study on "business cases" of bidirectional use. The car has an AC-charger too. The schema for this is shown in Figure 4. The BFH PV LAB then developed an electronic device for the Nissan Leaf to inject the energy of the battery into the grid. The PV LAB also tried to cooperate with the car import company or the garage, but our requirements were outside of their expertise. Below, details on the technology and the potential of bidirectional EVs to replace stationary PV-batteries are given.

6.3 Bidirectional charging unit for CHAdeMO-EVs

The Nissan Leaf has a CHAdeMO DC-charging possibility, see Figure 5. This DC-access can be used to give the power of the battery to the outer world. We developed an access unit to switch the battery to the CHAdeMO outlet. This uses a CAN-bus and a respective protocol. To ensure safety, we developed a software that includes all the needed safety features. We are now able to install a small standard inverter used for wind applications, to feed the DC energy of the battery into the grid. The power was not very high as the study aim was to mainly support household applications.

Figure 3: Nissan Leaf of the PV LAB at Bern University of Applied Sciences BFH in the solar carport.

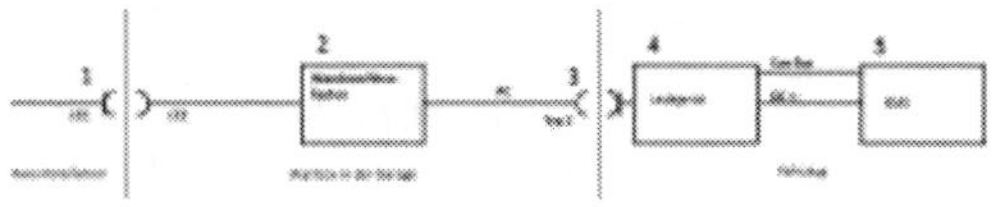

Figure 4: The charging scheme of the Nissan LEAF: AC charging of the Nissan Leaf.

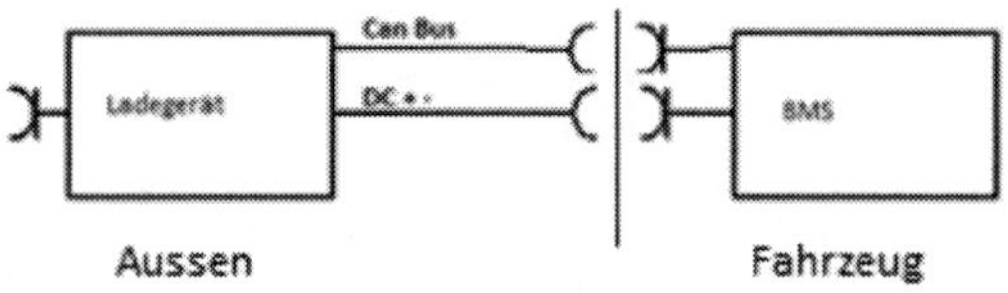

Figure 5: CHAdeMo DC-fast charging of the Nissan Leaf.

The CHAdeMo-plug (Figure 6) is massive, as the charging power is 50 kW. This results in currents up to 150 ADC. The CHAdeMo DC-plug gives the access to the battery of the Nissan Leaf (Figure 8). The communication is done with a CAN-Bus. This is visualised in the flowchart in Figure 9.

Figure 6: CHAdeMo DC-plug.

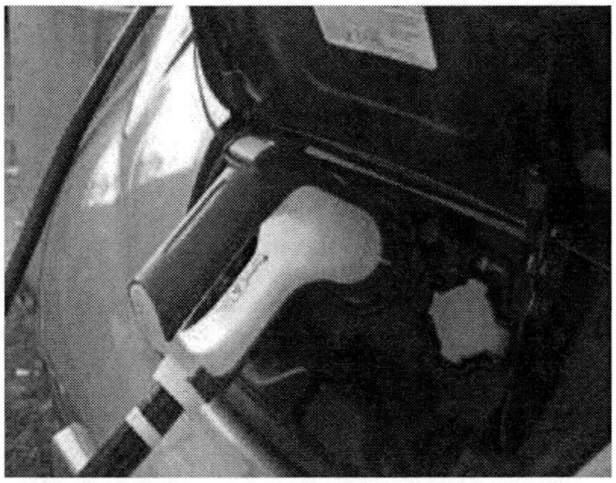

Figure 7: CHAdeMO-plug in the Nissan Leaf.

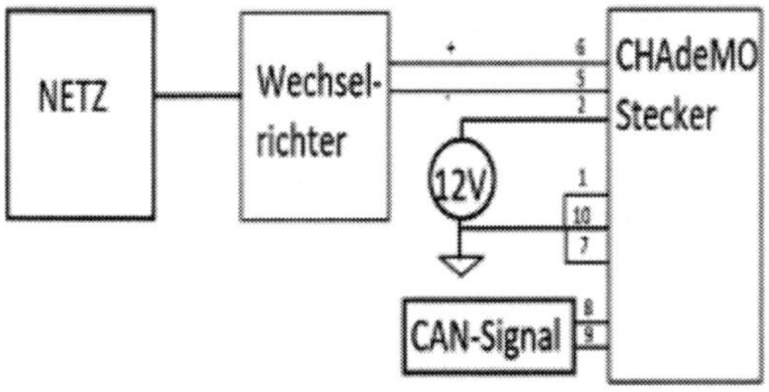

Figure 8: Technical schematics for the discharge of the car battery via CHAdeMO access.

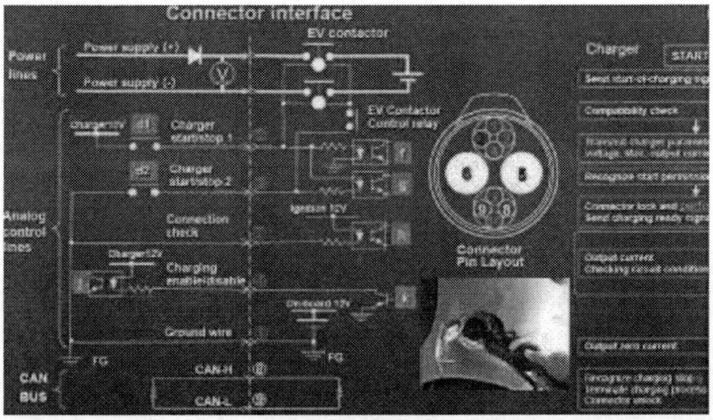

Figure 9: Flowchart diagramm of the CHAdeMO access of the Nissan Leaf.

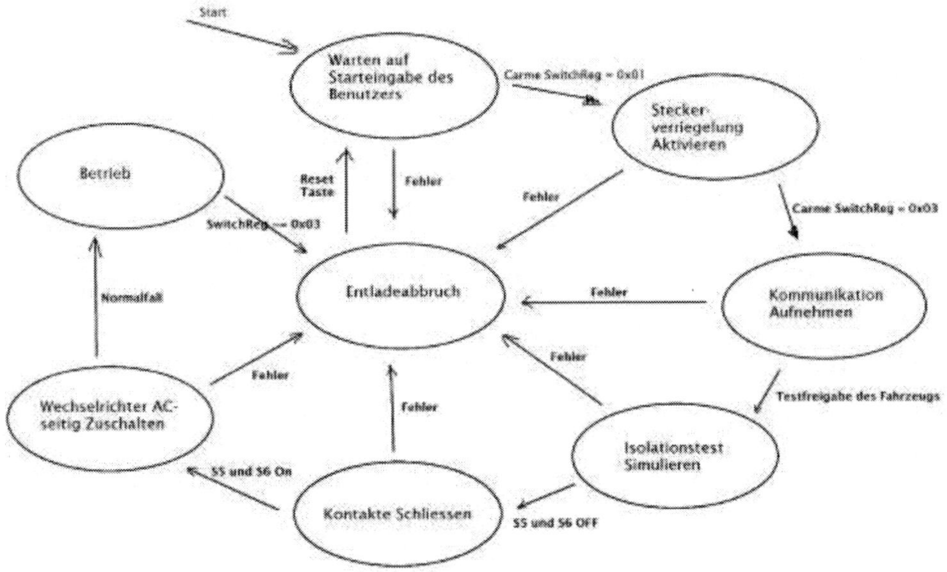

Figure 10: Flow chart of the control.

To understand the CAN-protocol, we visited a commercial charging facility and copied the CAN-protocol, see Figure 11. This allowed us to build up the charging regulation with the inverter, as shown in Figure 12. The software for the switch-on of the inverter was programmed following the flow chart in Figure 10.

Figure 11: Copy of the CAN-protocol on a commercial charging station

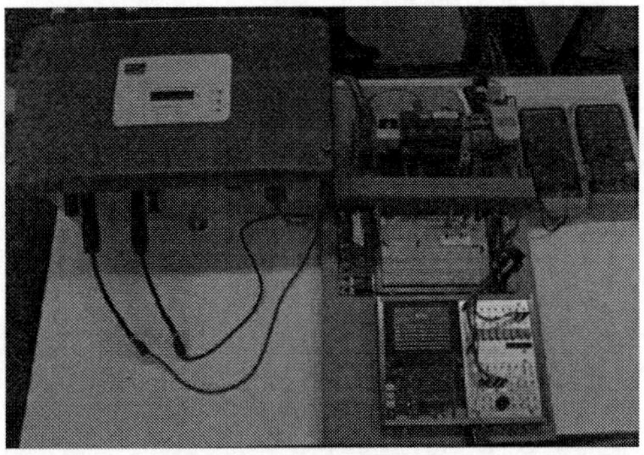

Figure 12: Discharge configuration with standard-inverter (for wind generator) and the measurement-, signal-conversion-, communication- and safety devices.

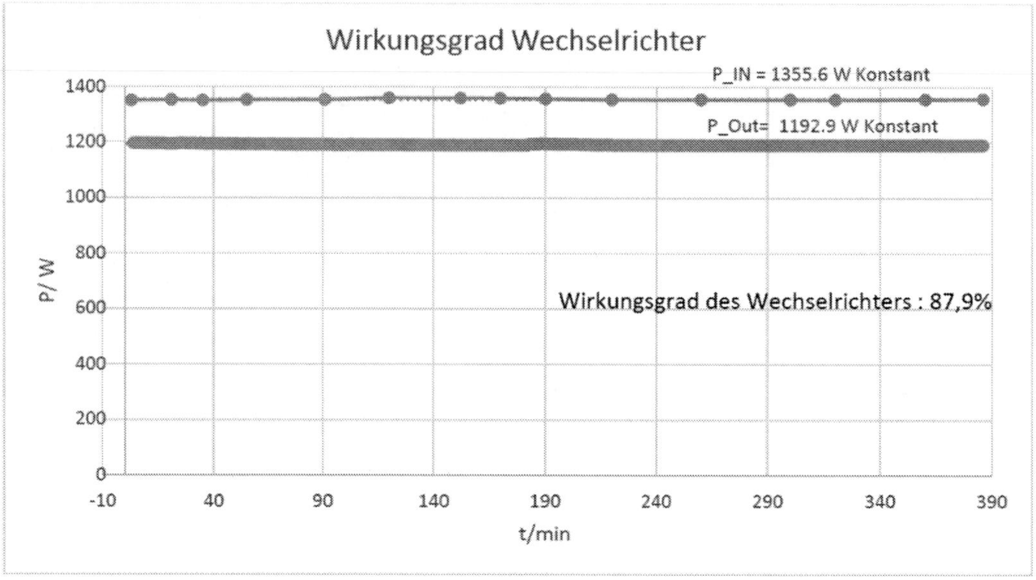

Figure 13: Efficiency of the inverter.

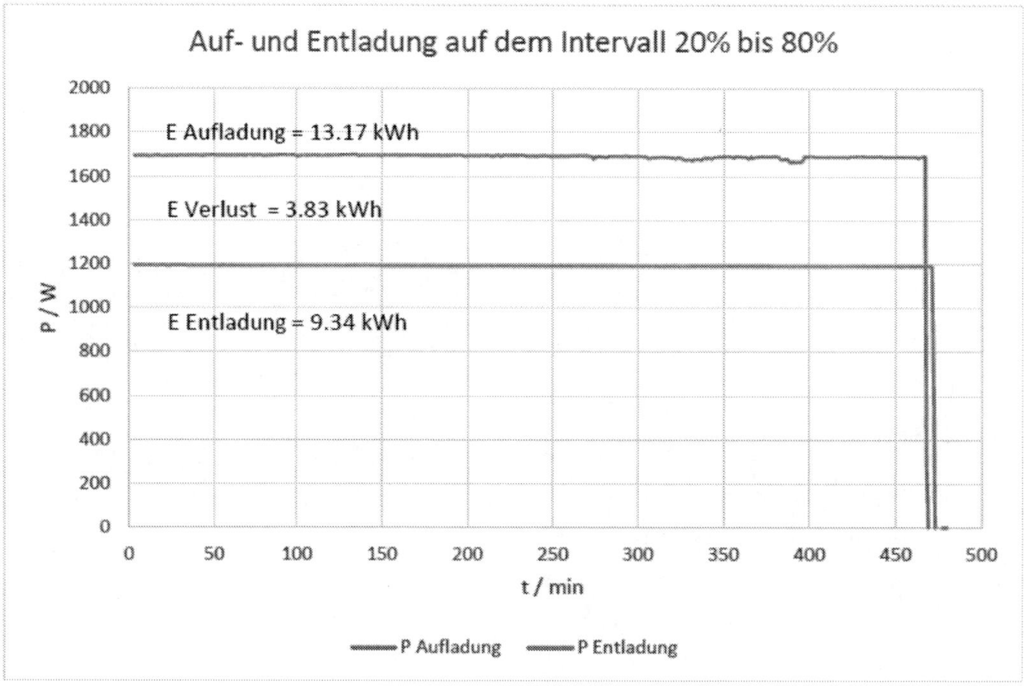

Figure 14: Energy use during charging and energy yield during discharge of the battery.

With that procedure, we got access to the battery and first measurements could be done. We could measure the voltage and the state of charge SOC of battery as shown in Figure 15.

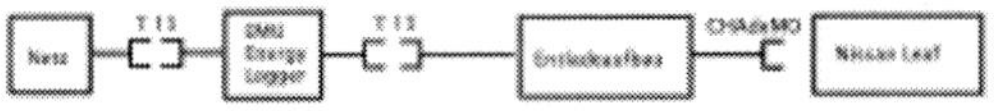

Figure 15: Measurement configuration of the discharge of a Nissan Leaf.

The efficiency of the inverter (galvanic separated) could then be measured. The efficiency of 87,9% is not very high. With optimized devices, similarly to PV inverters, efficiencies of about 98% will be possible. We could measure the efficiency of the charge- and discharge between 20% to 80% of the state of charge SOC of the battery. The total round trip efficiency was 70.9%.

Like the discharge, the total efficiency can be improved and 90-95% total efficiency will be possible.

6.4 Future application

In the future, all these devices can be implemented in a car. The detour over the CHAdeMO can be avoided with an integrated charge-/ discharge unit. This allows the EV user to implement further business cases as "load management"/ "grid services" etc. We identified several business cases which vary strongly with the local conditions.

6.5 Minor use of the EV battery

In an Opel "Ampera-e" with 400 km range, 500 full cycles represent 200'000 km. For a typical user with 12-15'000 km a year, this represents a total life of the car of about 15 years. The use of 200 cycles at 15% discharge a year represents therefore about additional 30 full cycles per year. As these cycles will occur by very low currents and the systems will be not stressed very much.

We can, therefore, expect very low prices per stored kWh and is anticipated below the 100-150 US$/ kWh for the full battery of the EV. But we need to pay the additional injection unit for the EV. This could be in the range of under 1'000US$ based on current figures for a 3 kWp PV inverters. This would add costs of about 4 cent/ kWh for the stored energy, if we assume 200 cycles at 15% of a 60 kWh EV battery in 15 years. This is far below all PV battery systems.

7 COLLABORATION WORK ON THE "PV AND EV" - TOPIC – PLANNED TASK OF THE TECHNICAL COLLABORATION PROGRAM "PHOTOVOLTAICS" AND "HYBRID- AND ELECTRIC VEHICLES"

This new application makes it very attractive to speed up the development with international collaboration. In the Technical Collaborations Programs TCP of the Internationale Energy Agency IEA, currently two efforts are planned in this regard:

7.1 Task 28 – V2X TCP HEV

The vehicle to grid V2G procedure is the topic of the Task 28 "V2X" of the Transport Collaboration Group „Hybrid- and ElectricVvehicles" (HEV) of the International Energy Agency IEA that works on all aspects of V2G (vehicle to grid) or V2X (vehicle to something). The main target is the international collaboration. Research institutes and companies from Europe, USA, Canada and South Korea are involved [6]. The PV LAB at BFH Burgdorf in Switzerland is a Member of his Task 28. It is during one of the Task meetings that the PV LAB at BFH generated the idea to conduct this study on the bidirectional Nissan Leaf.

7.2 Planned task "PV and EV" for 2018

The New Energy Development Organization NEDO currently prepares a new Task in the Technical Collaboration Program "Photovoltaics". Here, the focus is mainly on the "EV and PV"-topic. The task is expected to start in 2018. First workshops were held in Washington and Vienna in 2017.

8 RESULTS AND CONCLUSION

The Nissan Leaf can be used as bidirectional charging/ discharging unit. The needed devices in hard- and software can be improved in the future for a more practical use. It shows that the bidirectional use of an EV battery offers new advantages and cost benefits using an EV together with PV. An EV in use with PV electricity is much cheaper than an ICE. A 2.5 kWp PV installation produces the "gasoline" for 30 years of driving. This is equivalent to 27'000 litres of gasoline. The current higher price for the EV will go down in the next years. Finally, EVs will be cheaper than ICE's. With the new 40-60 kWh batteries of the new models, the initial conditions for the bidirectional use are given. This allows the PV and EV-user to replace the stationary battery with his bidirectional EV. For the car industry, this bidirectional mode is yet another big challenge. Eventually, the dream of the "solar pioneers" of the 1980ies have come true, and PV powered EVs will reach the mass market in the next years!

Acknowledgements
This study was carried out in the frame of the Swiss Centre for Competence in Energy Research on the Future Swiss Electrical Infrastructure (SCCER-FURIES) and with Swiss industry partners. We gratefully acknowledge financial support from the Swiss Commission for Technology and Innovation (CTI - SCCER program) and from Bern University of Applied Sciences BFH.

References

1. Urs Muntwyler (2017) Labor für Photovoltaiksysteme. https://www.pv-test.ch
2. Urs Muntwyler (2016) Electric vehicles powered with PV electrcity as a new driver for photovoltaics EUPVSEC 2016.
3. Michael Briggeler (2016) Analyse und PV-optimierte Ladestrategie für Solarcarport der BFH. BSc-Diplomarbeit, Berner Fachhochschule.
4. Urs Muntwyler (2016, 2017), Plushaus-workshop energy-cluster.ch
5. Benjamin Ulrich (2017) Bidirektionales Laden von Elektrofahrzeugen. BSc-Diplomarbeit, Berner Fachhochschule.
6. IEA TCP HEV Annual report 2016, Task 28 – Home grids and V2X Technologies: page 115-123.

POTENTIAL APPLICATIONS OF A LOAD-MANAGING PHOTOVOLTAIC SYSTEM

Joseph A. Azzolini and Meng Tao
School of Electrical, Computer and Energy Engineering, Arizona State University
551 E Tyler Mall, Tempe, AZ 85281, USA

ABSTRACT: Conventional photovoltaic (PV) systems are designed to compensate for the intermittent nature of solar energy, and therefore require multiple power managing components like batteries and power converters. Each of these components adds an additional cost and has an associated power loss. In this paper, we present a new topology for photovoltaic systems. Instead of managing the power from the PV array, our system manages the loads and matches them to the instantaneous maximum power available from the PV array. This results in a far simpler system than conventional systems with a much higher efficiency and much lower cost. Several potential applications of this system are discussed including electric vehicle charging and industrial electrolytic processes.
Keywords: Photovoltaic System, Load Management, Electric Vehicle, Electrolysis, Hydrogen

1 INTRODUCTION

Conventional photovoltaic (PV) systems are designed to compensate for the intermittent nature of solar energy, and therefore require multiple power-managing components like batteries and power converters. Each time the PV power is conditioned or managed, there is an associated cost and power loss incurred. For commercial-scale PV systems, these power management devices account for 10–35% of the total system cost depending on the complexity of the system [1], and lead to a typical system efficiency of 77% [2].

In this paper we report a new topology for PV systems and its potential applications. Instead of managing the power from the PV array, our system manages the loads and matches them to the instantaneous maximum power available from the PV array. This results in a far simpler system with an efficiency approaching 100%. The simultaneous lower cost and higher efficiency promises a 15–40% lower levelized cost of electricity (LCOE) than conventional PV systems. The load-managing system can be configured to handle a large number of loads, without a significant cost increase. The system efficiency also increases with the number of loads being managed. The load-managing system can be operated as a stand-alone system or be backed by a secondary power source, such as the electric grid. Although a load-managing PV system is not a general purpose concept, it enables the use of solar power in several impactful applications such as electric vehicle (EV) charging, electrolytic processes, and remote power applications.

2 SYSTEM OVERVIEW

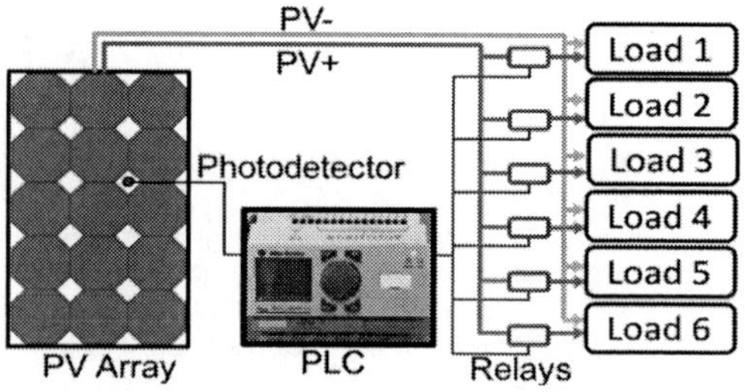

Figure 1: Stand-alone load-managing PV system architecture with DC loads

The core of the load-managing system is shown in Fig. 1. The components of the system include a PV array, a photodetector, a programmable logic controller (PLC), a set of relays, and a number of electrical loads to manage. During a typical day, the solar irradiance will increase throughout the morning. As the available power increases with the solar irradiance, the PLC determines the number of loads it can power based on the information provided by the photodetector. When the output power from the PV array reaches its maximum around midday, the PLC will connect the maximum number of loads to the PV array. In the afternoon when the solar irradiance decreases, the PLC begins to disconnect loads from the PV array until the sun sets. The system will also respond to any weather conditions throughout the day and adjust the number of loads connected accordingly. Overall, this system utilizes the maximum available power from the PV array while mitigating the amount of power losses without a conventional maximum power point tracker. This load-managing system can also be backed by a secondary power source, such as the electric grid, for increased reliability. The grid-backed load-managing system still utilizes the maximum available power from the PV array, but takes supplemental power from the grid if necessary. It will be discussed later in this paper.

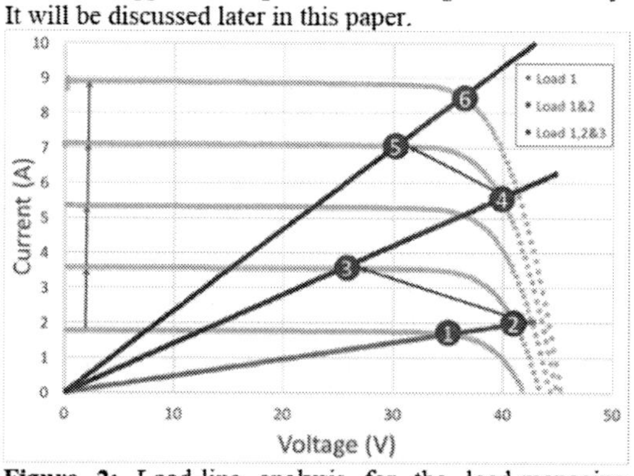

Figure 2: Load-line analysis for the load-managing system with three equal loads

The load-line analysis in Fig. 2 provides more detail about how the system operates for optimal efficiency. Several current-voltage (I-V) curves of a commercial 72-cell silicon panel at different levels of irradiance are plotted. As the irradiance increases throughout the morning, the I-V curve gradually shifts up. To simplify the illustration, we assume that there are only three loads with a fixed, equal resistance. The fixed-resistance loads appear as straight lines in Fig. 2. The slope of each line is equal to the reciprocal of the combined resistance of the loads, and the system can only operate along these lines. At point 1, the first load is connected to the PV array. As the irradiance increases, the operating point will move along the load 1 line, until it reaches point 2. The system will switch from point 2 to point 3 by connecting load 2 in parallel with load 1. This critical point is referred to as a switch point (SP). The optimal switch point occurs when the power at point 3 is equal to the power at point 2. The system will then operate along the load 1&2 line until it reaches the next switch point, and so on. In the afternoon, this process will happen in reverse as the irradiance decreases.

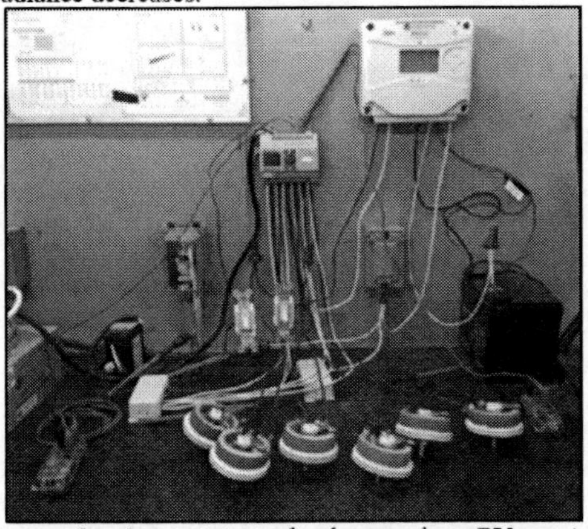

Figure 3: A prototype load-managing PV system constructed at ASU

A prototype stand-alone load-managing PV system has been implemented to explore the practice challenges of this system topology, pictured in Fig. 3. This system includes a PV panel, photodetector, PLC with six built-in relays, and six variable resistors to represent the loads. Data-logging digital multimeters take measurements of the total power delivered to the loads throughout the day. The solar charge controller and battery in the photo were only used to take initial measurements of the solar panel, and thus are not part of the load-managing system. The photodetector circuit (not pictured) outputs a 0-10V signal to the PLC based on the measured short-circuit current of a small silicon solar cell. Analysis was performed that proved a linear relationship between short-circuit current and maximum power of a solar cell under ideal conditions. Therefore, the voltage signal can estimate the maximum available power from the PV array. Overall, this system allows for manipulation of number of loads, resistance per load, and the switch point of each load.

4 RESULTS AND DISCUSSION

Fig. 4 shows the experimental performance of the system in terms of power delivered to the loads as a function of time. There are four loads in this experiment, each set to 12 Ω, with switch points of 1.25V, 4.00V, 6.75V, and 8.20V. The switch points (SP1 through SP4) are visualized four times in the morning. As the solar irradiance increases throughout the morning, the PLC subsequently connects loads to the PV panel based on the voltage signal received from the photodetector. A dip in power is observed near SP4, which is attributed to a sub-optimal switch point setting. Optimal switch point settings result in a continuous power curve, like at SP2 and SP3. By noon, all four loads are powered by the PV panel. As the solar irradiance decreases throughout the afternoon, the loads are disconnected from the PV panel accordingly. The load-managing system also responds to changes in weather conditions in real time. At about 3:30 p.m., the system briefly experienced cloud coverage. The photodetector responded to the drop in irradiance, and the rest of the system reacted accordingly.

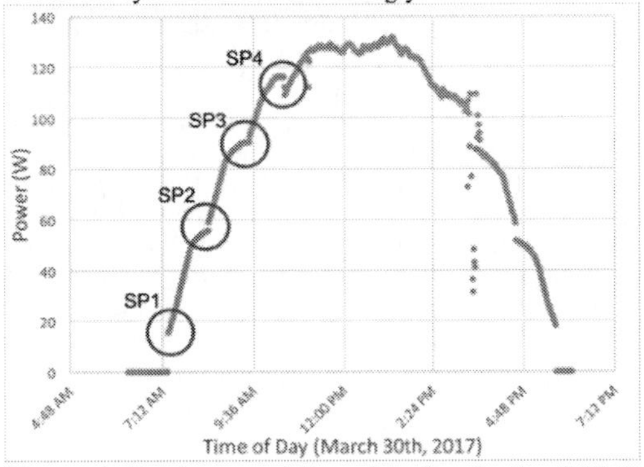

Figure 4: Total power delivered to four loads of 12 Ω each

An estimate of the system efficiency can be obtained by first encompassing the data in Fig. 4 with a sine function. The sine curve represents the maximum available power from the solar panel during the day. The daily system efficiency is then the ratio of the area under the data to the area under the sine curve:

$$\eta = \frac{\textit{Power Delivered to Loads}}{\textit{Total Power Available}} \times 100 \qquad (1)$$

The performance of the load-managing system was modeled in MATLAB. The MATLAB program uses an ideal solar panel (infinite shunt resistance and zero series resistance) to calculate the efficiency of the system when managing any number of loads with equal resistance. The results below in Fig. 5 show the theoretical efficiency as a function of number of loads being managed. The system efficiency reaches above 99% when managing just eight loads. The program can also be used to calculate the optimal switch points for each of the loads as well.

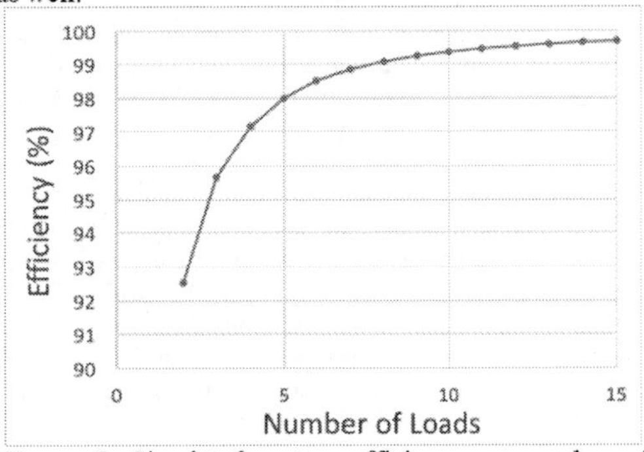

Figure 5: Simulated system efficiency vs. number of loads

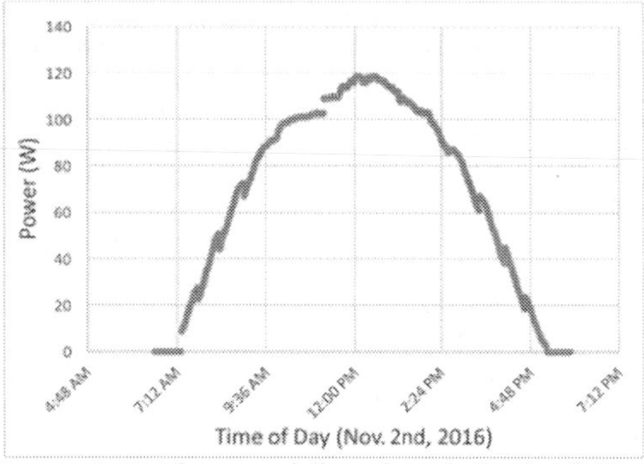

Figure 6: Total power delivered to six loads of 21 Ω each

More measurements have been performed. In Fig. 6, the number of loads being managed by the system is increased to six, with each load set to 21 Ω. This change causes the data to fit more closely under a sine function, showing that the efficiency increases with an increase in number of loads being managed. In Fig. 7, the four loads being managed are set to different resistances to represent uneven loads, with first, second and fourth loads at 17.75 Ω and third load at 8.875 Ω. Again a close to sine curve is observed for the total power delivered.

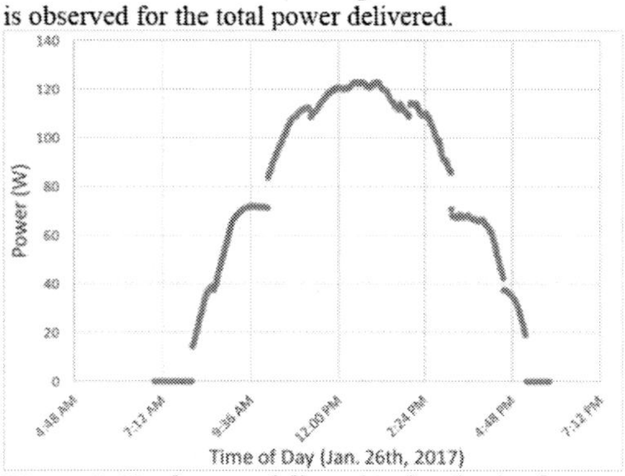

Figure 7: Total power delivered to four loads of unequal resistance values

5 APPLICATIONS

A load-managing PV system is not a general purpose system, and therefore the ideal application would have to meet several requirements. First, the total electrical load should be able to be divided into multiple sub-loads. The system would manage these sub-loads and connect them to the PV array in such a way as to extract the maximum available power throughout the day. These sub-loads should be able to tolerate intermittent power and the possibility of turning on and off frequently.

It is important to note that these requirements can be relaxed by adding modifications to the core of the system, without affecting the system's inherent benefits. For instance, AC loads can be managed by adding a DC-AC inverter for each load, DC loads with strict voltage requirements can be managed by adding a DC-DC converter for each load, and most importantly, the load-managing system can be backed up by the electric grid, eliminating its intermittency. If power converters are utilized for any of the loads, they would only have to be rated for the power requirements of the load and not the entire system, reducing their cost and improving system scalability.

The complexity of the system's control can be modified as well. For example, multiple photodetectors can be used for a large PV array. The outputs of these photodetectors can be used to make a more accurate estimation of available power and can help to compensate for partial-shading conditions. Variable-resistance loads such as batteries can also be managed by adding additional feedback signals. Lastly, a microprocessor or computer can replace the PLC to handle more complex control algorithms.

There are several key applications that meet the requirements discussed above including electric vehicle charging, industrial electrolytic processes, and remote power applications.

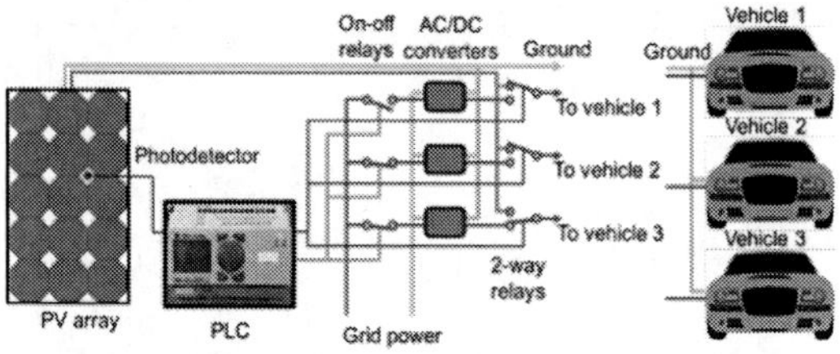

Figure 8: Load-managing PV system for charging electric vehicles

5.1 Electric Vehicle Charging

Cars and trucks contribute ~22% to total carbon emissions in the U.S. and ~17% globally [3]. The electric vehicle (EV) is one potential solution to this problem. However, the environmental benefit of EVs is arguable when charged by the grid, as ~67% of grid electricity is derived from fossil fuels in the U.S. [4]. Significant carbon reduction can be achieved if electric vehicles are charged with solar electricity.

For this application, a grid-backed load-managing system as shown in Fig. 8 would be implemented. By adding a secondary power source, a full charge can be guaranteed to each vehicle by the end of the day. To illustrate the benefits of this system, let us examine the example of EVs charging at the workplace. Most vehicles will stay until about 5:00 p.m. and therefore, not every vehicle needs to start charging as soon as they arrive in the morning. The load-managing system will connect the maximum amount of EVs to the PV array throughout the day. In the afternoon, when solar irradiance begins to decrease, the load-managing system can disconnect vehicles from the PV array and connect them to the grid instead. Therefore, any vehicles not being charged by the PV array are still being charged. In Fig. 8, the PLC has determined there is enough solar power to charge vehicles 1 and 2, while vehicle 3 is charged by the grid. If a driver needs to leave at 10 a.m., the on-off relay can be manually switched on to bypass the load-managing system and connect that vehicle to the grid in early morning. Overall, this system utilizes the maximum available power from the PV array and takes supplemental power from the grid if necessary, ensuring all vehicles are fully charged by the end of the day.

5.2 Industrial Electrolytic Processes

Electrolysis is widely used in the chemical and metallurgical industries for the production and purification of many raw materials that enable our modern society. Examples of electrowon or electrorefined materials include chlorine (Cl_2), hydrogen (H_2), and caustic soda (NaOH) by the chloralkali process from an aqueous sodium chloride (NaCl) solution [5], which produces about 66,000,000 tonnes/yr of NaOH [5]. The electricity required to produce 1 kg of NaOH is about 2.5 kWh/kg [5], translating into a total electricity consumption of about 1.65×10^{11} kWh/yr for the electrolysis. In addition, aluminum (Al) by the Hall–Heroult process from aluminum oxide (Al_2O_3) dissolved in molten salt [5], which is produced at about 50,000,000 tonnes/yr [6]. The electricity required to produce 1 kg of Al is about 14 kWh/kg, resulting in a total electricity consumption of about 6.12×10^{11} kWh/yr for the electrolysis [6].

Hydrogen fuel cell vehicles have also been proposed as an alternative to gasoline/diesel powered vehicles. These vehicles use an electrochemical process to convert compressed hydrogen gas into electricity and water. Hydrogen gas can be produced from water through electrolysis. This process can be powered by the load-managing PV system. A stack of hydrogen cells serves as a load and there are multiple stacks of hydrogen cells in the system to be managed.

Given the simultaneous benefit of improved efficiency and reduced cost, the stand-alone and grid-backed load-managing PV systems described in this paper have the potential to increase the consumption of solar electricity in these industrial processes. The system could use grid power for backup or just use solar power if so desired.

5.3 Remote Power Applications

Stand-alone PV systems are used for a variety of applications in remote areas where access to electricity is not available. Direct-coupled PV systems are the cheapest and simplest to implement in these situations as they require no storage or power conversion. However, these systems typically perform well below the rated efficiency as the maximum power is not being tracked. Direct-coupled PV systems can be combined with DC pumps for water circulation and irrigation purposes as in [7]. PV systems can also be directly-coupled to PV powered refrigerators which take advantage of DC permanent magnet motors, as seen in [8,9]. A load-managing PV system can be modified to accommodate and improve upon any of the applications above by significantly increasing the efficiency without much of a cost increase, when multiple pumps or multiple refrigerators are being managed.

6 CONCLUSIONS

The load-managing PV system presented here has several key advantages over conventional PV systems. By managing the loads instead of the power, the system cost is significantly reduced while the system efficiency improves. The simultaneous benefit of reduced cost and improved efficiency leads to a 15-40% lower levelized cost of electricity as compared to conventional PV systems. The load-managing PV system also has an excellent scalability as efficiency increases with the number of loads being managed. The core of the system contains a PV array, a photodetector, a programmable logic controller, and a set of relays. The system performs maximum power point tracking by varying the number of loads in parallel to match the combined resistance of the loads with the characteristic resistance of the PV array. With a system efficiency approaching 100%, and the option of grid-backing, this system offers a cost effective way to incorporate solar electricity into several impactful applications.

7 ACKNOWLEDGMENT

This work was partially supported by National Science Foundation under grant number CBET-1336297.

8 REFERENCES

[1] D. Chung, C. Davidson, R. Fu, K. Ardani, and R. Margolis, U.S. photovoltaic prices and cost breakdowns: Q1 2015 benchmarks for residential, commercial, and utility-scale systems, NREL, 2015, http://www.nrel.gov/docs/fy15osti/64746.pdf.

[2] K. Ardani and R. Margolis, 2010 Solar technologies market report, NREL, 2011.

[3] E. Roston, Electric cars can help clean up the grid - Electricity needs to come from renewable sources in order to arrest climate change, http://www.bloomberg.com/news/articles/2016-04-29/electric-cars-can-help-clean-up-the-grid.

[4] US Energy Information Administration, What is U.S. electricity generation by energy source? https://www.eia.gov/tools/faqs/faq.cfm?id=427&t=3.

[5] G. G. Botte, Electrochemical manufacturing in the chemical industry, The Electrochemical Society Interface, 23(3), pp. 49-55 (2014).

[6] International Aluminum Institute, Current IAI Statistics, 2014. http://www.world-aluminium.org/statistics/.

[7] Q. Kou, S. A. Klein, and W. A. Beckman, A method for estimating the long-term performance of direct-coupled PV pumping systems, Solar Energy, 64(1–3), pp. 33–40 (1998).

[8] M. de Blas, J. Appelbaum, J. L. Torres, A. García, E. Prieto, and R. Illanes, A refrigeration facility for milk cooling powered by photovoltaic solar energy, Progress in Photovoltaics: Research and Applications, 11(7), pp. 467–479 (2003).

[9] R. Foster, B. Jensen, B. Dugdill, B. Knight, A. Faraj, J. K. Mwove, and W. Hadley, Solar milk cooling: Smallholder dairy farmer experience in Kenya, International Solar Energy Society, 2015, https://doi.org/10.18086/swc.2015.03.05.

COMPENSATION OF FORECAST ERROR IN LARGE PV PLANTS WITH BATTERY STORAGE: ASSOCIATED STRATEGIES

J. Marcos *, I. de la Parra, M. Muñoz, M. García, L. Marroyo

Department of Electrical and Electronic Engineering – Institute of Smart Cities
Public University of Navarre (UPNa)
Campus Arrosadía s/n, Edificio de los Pinos, Pamplona 31006, Spain.
E mail: javier.marcos@unavarra.es Tel.: +34-948-169-277.

*Author for correspondence.

ABSTRACT: High penetration rates of utility-scale photovoltaics (PV) plants can force the system operator (TSO) to allocate significant levels of energy reservoirs, and as a consequence, an increase of the operation costs can occur. For this reason, TSO traditionally applies sanctions to every generator which fails to deliver the scheduled power. This paper presents two novel strategies which allow a large PV plant, in conjunction with an energy storage system (EES), to completely compensate the forecast errors and fulfil the programme referred. The general validity of these algorithms was confirmed through simulations carried out with both proposed strategies and real operational PV power output data taken every 5s in the course of two year 38.5 MW PV power plant of Amareleja (Portugal).

Keywords: Large Grid-connected PV systems, storage, forecast.

1 INTRODUCTION

As renewables energies increase their presence in the power generation mix, the TSO must allocate energy reserves in order to balancing the variable power generated by intermittent sources like wind and PV generators. These energy reservoirs can lead to an increase in system inefficiency and thus, impacting in the electricity price [1, 2]. Hence, transmission system operators force wind-power energy producers to present a production schedule with over twenty-four hours in advance to participate in the daily wholesale electricity market. Corresponding forecast errors can mean economic penalties to the wind farm owner [3]. Nowadays, this is different for large PV plants owners in the great majority of electric markets. However, everything would seem to indicate that large PV plants will participate in the same way that wind farms and then, to receive some kind of economic penalty depending of the forecast error. Hence, there are two ways of improvement in order to minimize these errors: using a battery energy storage system (BESS) to balance out the prediction errors [4, 5] or increasing the accuracy of forecast methods [6].

In this paper, two different BESS control strategies are presented in order to minimize storage requirements to totally compensate the forecast error. Taking advance of the several 1 hours intra-diary market sessions (i.e. Spanish future wholesale electricity market), the PV plant owner can correct the prediction for the next hour and hence, open the way to a state of charge control to avoid large energy requirements. The first strategy injects into the grid the mean power of the previous 1 hours, while the second strategy modifies the forecast PV power for each next 1 hour period according to the previous BESS state of charge (SOC). In consequence, the first strategy can achieve less energy requirements, at the cost of increasing the BESS cycle degradation.

These strategies have been successfully simulated and validated with one real 5-s data recorded at all the 70 inverters of 550 kW located at the Amareleja PV plant (Portugal), owned by Acciona Energia S.A. during two years (Oct'12-Sept'14). The one day ahead PV production prediction was obtained by a parametric method [6] which transforms 1-hour irradiance and ambient temperature forecast data provided Meteologica S.A. into PV power data.

A first analysis quantifies the energy requirements without any kind of SOC control, i.e. we analyze the error bias between forecast data and production data (charging or discharging tendency). This can be seen as the theoretical maximum possible energy requirements. At the other end, and subsequently, we estimate the theoretical minimal energy requirements Emin that any strategy can achieve, that is, the maximum 1 hour energy error. Finally both strategies are developed ensuring that Emin is the ultimate objective, as well as a low cycling degradation.

2 DATABASE

2.1 Large PV plant power data

The experimental data for this work is taken from the Amareleja (South Portugal) PV plant. This plant occupies an area of 250 Ha and includes 2520 solar trackers with a rated output of 17.7 to 18.8 kWp, up to a total peak power of 45.6 MWp. The corresponding inverter power, P^*, is 38.5 MW. The trackers are one-vertical axis models, with the receiving surface tilted 45° from the horizontal. The plant is divided into 70 units, each comprising 36 tracking systems connected to a 550 kW DC/AC inverter. Thanks to extensive monitoring, 5 sec synchronized records of the output power of all the inverters are available from during two years (Oct'12-Sept'14).

Figure 1. Amareleja PV plant.

2.2 Forecast PV power data

Likewise, the one day ahead PV production prediction was obtained by a parametric method [6]. This model provides the AC power at the output of the PV plan using as input variables predicted ambient temperature and predicted global horizontal irradiance, both obtained from open sources. It has been evaluated and validated using historical production data of a fixed generator (2,24kWp) and a large PV power plant (Amareleja PV plant also) and historical forecast data from the open source Meteogalicia. In addition, a toolbox to implement this parametric methodology has been built upon the Matlab environment. The toolbox has been developed to provide the PV plant production one day in advance.

The parametric model methodology has two parts:

1. Transform global horizontal irradiance into irradiance in the plane of the PV array (G_I) and both ambient temperature and irradiance in the plane of the PV module into cell temperature (T_C).
2. Simulate the losses in each element of the PV installation.

Figure 2 displays a diagram for the general configuration of a simulated grid-connected PV system.

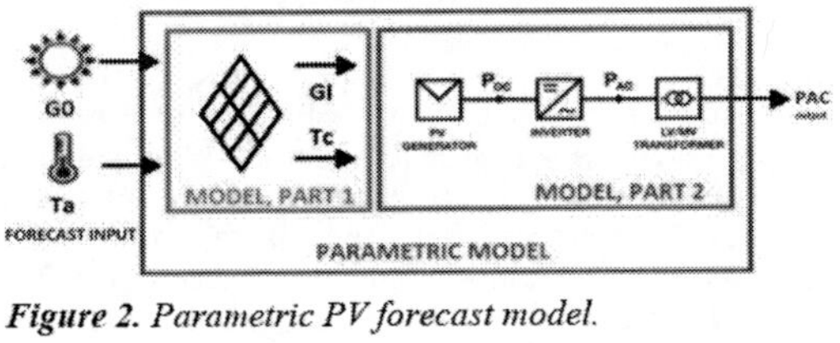

Figure 2. Parametric PV forecast model.

Considering this model, the greater deviation is due to the prediction of G_0 and T_a from official the meteorological services (close to 18%). If the model is fed with accurate meteorological variables, the error of the model is less than ± 4%. In addition, the first part of the model, which transforms variables into the plane of the array introduces more deviation than the PV plants components model (around ±2%).

3 CONTROL STRATEGIES

3.1 Energy requirements without any kind of SOC control

First, we analyse the error bias between forecast data and production data (charging or discharging tendency). This can be seen as the theoretical maximum possible energy requirements. Obviously:

$$P_{BAT}(t) = P_{FOR}(t) - P_{PV}(t) \qquad (1)$$

Then, the corresponding power excess or shortage is either taken from ($P_{BAT} > 0$) or stored into ($P_{BAT} < 0$) the battery. The energy stored at the battery, E_{BAT}, is given by the integral of P_{BAT} over time. In this way, the behavior of the whole system can be easily simulated for any time series of $P_{PV}(t)$ and $P_{FOR}(t)$. For the sake of simplicity, any potential battery and associated electronic converter losses are disregarded here. The evolution of the energy in the battery during the whole period (Oct'12-Sept'14) has been calculated. Figure 3 shows this energy transformed to storage time, calculated as E_{BAT}/P^*.

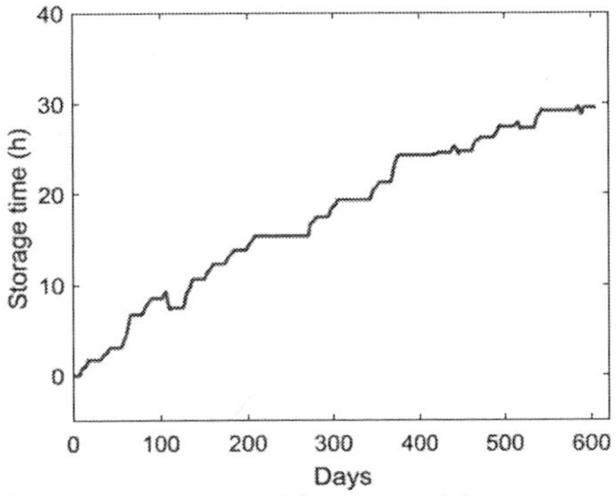

Figure 3. Parametric PV forecast model.

In addition, we have analysed the worst hour storage time required (h), that in our case is 0.7h. This can be seen as the minimum theorical storage needed in order to compensate the forecast error. Any strategy developed would not be able to reduce this storage time.

As the example shown in Figure 4, the tendency of the battery to charge continuously affects the entire two years period. This means that, typically, $P_{PV}(t) > P_{FOR}(t)$ and then, the battery charging tendency leads to excessive battery capacity requirements, in the order of some hours. Hence, a SOC control strategy can be implemented in order to reduce the storage requirements.

3.2 Strategy 1: Hourly Moving Average

This first strategy does not use forecast data in order to complete fulfil the scheduled power. Every hour, the power injected into the grid and scheduled for the next hour $P_G(H+1)$ is the mean-average of the previous hour $P_{PV}(H)$, that is:

$$P_G(H+1) = mean(P_{PV}(H)) \qquad (2)$$

Due to mean average properties, there is no need for any type of SOC control to prevent the battery charge /discharge. The flowchart of the strategy is, Figure 4:

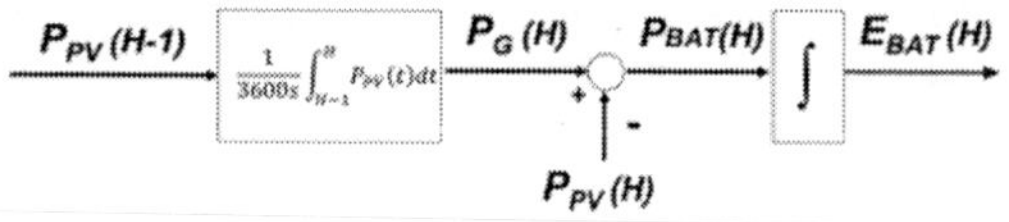

Figure 4. *Strategy 1: flowchart.*

However, the strategy does not take advantage of forecast data and introduces a complete cycle in the battery per day. Figure 5 shows an example of simulation this strategy during 2nd October 2012.

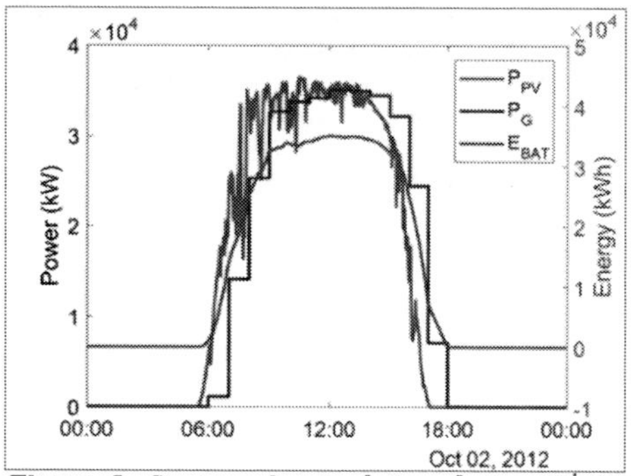

Figure 5. *Strategy 1: simulation during 2nd October 2012.*

3.3 Strategy 2: Forecast and SOC control

The second strategy modifies the forecast PV power for each next 1 hour period according to the previous SOC value. In particular, the power injected into the grid and scheduled for the next hour $P_G(H+1)$ hour is calculated by the sum of:

- The PV power forecast for the next hour $P_{PV,FOR}(H+1)$.
- The mean error between the PV power forecast $P_{PV,FOR}(H)$ and PV power during the last hour $P_{PV}(H)$, in order to correct the bias of the prediction.
- A SOC control set point, in order to maintain the BESS SOC near the reference and then, reduce the storage requirements.

In other words, Eq. (3):

$$P_G(H) = P_{PV,FOR}(H) + ME_{FOR}(H-1) + e_{BAT}(H) \qquad (3)$$

The flowchart of the strategy is, Figure 6:

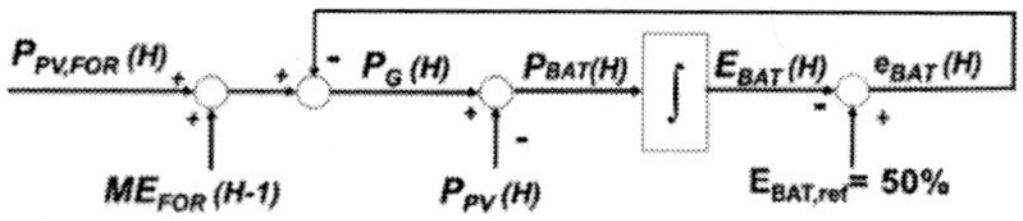

Figure 6. *Strategy 2: flowchart.*

Figure 7 shows an example of simulation this strategy during the same previous day 2nd October 2012. As it can be checked, the SOC control acts properly with the purpose of return the SOC to its initial state. At the same time, the forecast bias error is also correct.

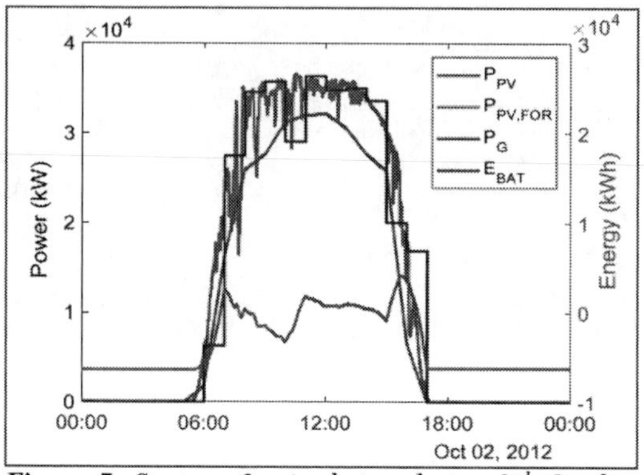

Figure 7. *Strategy 2: simulation during 2nd October 2012.*

4 RESULTS

Both strategies have been successfully simulated during a two years period. It is important to remark that during this time, SOC remains controlled and the storage requirements are considerable reduced. The strategies have been analysed against Storage Time and Cycling Degradation merit indexes (Table 1). This last parameter has been calculated using Rainflow - Counting Algorithm [7].

SUMMARY OF THE VALIDATION RESULTS	STRATEGY 1	STRATEGY 2
Worst hour storage time required (h)	0.7	
Annual storage time required (h)	0.98	1.31
Cycling degradation (%)	13.46	3.32

Table 1. *Two years simulation results.*

Due the annual storage time required is so close to the worst day requirements, we can affirm that both strategies achieve a good SOC control. Likewise, despite Strategy 1 provide slightly better storage requirements, Strategy 2 shows that the use of forecast extends considerably the BESS lifetime, over 4 times more. The reason lies in the fact that this strategy must act every day, regardless of whether there are any forecast error or not, whilst the first strategy only acts in the event of an error between the predicted PV power and the real one-

5 CONCLUSIONS

These strategies have demonstrated that with roughly 1 hour of storage it is possible to fully compensate the forecast errors in a large PV plant, according to future market regulations. Likewise, the use of open source meteorological forecast can extend the battery life in more than 4 times compared with not having it. Although present BESS prices can be high enough to balance economic penalties, there are no technical barriers in order to make a large PV plant completely full predictable and hence, reduce the cost of TSO energy reserves.

6 AKNOWLEDGMENTS

The authors would like to thank ACCIONA for authorizing measurements at its PV plants and for their staff helpful collaboration. This work has been supported by the Spanish State Research Agency (AEI) and FEDER-UE under grants DPI2016-80641-R and DPI2016-80642-R.

8 REFERENCES

[1] Kreikebaum, F., Moghe, R., Prasai, A., & Divan, D. (2011, September). Evaluating the application of energy storage and day-ahead solar forecasting to firm the output of a photovoltaic plant. In Energy Conversion Congress and Exposition (ECCE), 2011 IEEE (pp. 3556-3561). IEEE.

[2] Wan, C., Zhao, J., Song, Y., Xu, Z., Lin, J., & Hu, Z. (2015). Photovoltaic and solar power forecasting for smart grid energy management. CSEE Journal of Power and Energy Systems, 1(4), 38-46.

[3] Cai, Z., Bussar, C., Stöcker, P., Moraes, L., Magnor, D., Leuthold, M., & Sauer, D. U. (2015). Application of Battery Storage for Compensation of Forecast Errors of Wind Power Generation in 2050. Energy Procedia, 73, 208-217.

[4] J. Miettinen, et al., "Minimizing wind power producer's balancing costs using electrochemical energy storage," the Nordic Conference on Electricity Distribution Management and Development Stockholm, Sweden September 8–9, 2014.

[5] F. Luo, et al., "Coordinated operational planning for wind farm with battery energy storage system," Sustainable Energy, IEEE Transactions on vol.6 , pp. 253 - 262, 2014.

[6] M. P. Almeida, M. Muñoz, I. de la Parra, & O. Perpiñán (2017). Comparative study of PV power forecast using parametric and nonparametric PV models. Solar Energy, 155, 854-866.

[7] Dufo-López, R.; Lujano-Rojas, J.M.; Bernal-Agustín, J.L. Comparison of different lead-acid battery lifetime prediction models for use in simulation of stand-alone photovoltaic systems. Appl. Energy 2014, 115, 242–253.

SIZING OF URBAN DISTRIBUTION TRANSFORMERS IN A NEIGHBOURHOOD WITH PV GENERATION AND ENERGY STORAGE

S. Freitas[1], T. Santos[2], M.C. Brito[1]

[1]Instituto Dom Luiz – Faculdade de Ciências, Universidade de Lisboa, Campo Grande, 1749-016 Lisboa, Portugal.

[2]Interdisciplinary Centre of Social Sciences - Faculty of Social Sciences and Humanities, Universidade Nova de Lisboa, 1069-061, Lisboa, Portugal

e-mails: srefreitas@fc.ul.pt, teresasantos@fcsh.unl.pt, mcbrito@fc.ul.pt

ABSTRACT:
The goal of this work is to understand the power balance on urban distribution transformers, and the changes in power capacities that may be needed in a large-scale PV deployment scenario. Hourly load demand by power transformer was estimated for a suburb of Lisbon, Portugal, from reference profiles published by the distribution grid operator, and hourly PV production was calculated using irradiance simulations from SolarAnalyst extension for ArcGIS, for building rooftops, and the LadyBug plug-in for Grasshopper/Rhinoceros, for building façades. The PV potential greatly surpasses the transformers' capacity, suggesting that network upgrades will be unavoidable in the future. Additionally, a storage strategy for maximization of self-consumption was investigated to smooth power ramps in the grid. It is shown that without proper management strategy, small storage capacity might lead to higher ramps, whilst larger capacity might smooth ramps when the PV production peaks.

Keywords: building façades, urban BIPV, distribution transformers, storage

1. INTRODUCTION

The increase in popularity of building integrated photovoltaics (BIPV) raises concerns on its impacts on the electricity supply grid at different times of the day. Thus, it becomes relevant to assess the possible outcomes of high PV deployment scenarios. One of the challenges is the urban distribution transformers' power capacity that must be large enough to accommodate all the surplus electricity from PV systems that is being fed into them. This might not be the case for all the transformers, especially if the PV potential from façades is added to the rooftops'- which is expected to occur in the medium term, when PV costs reach low enough levels to enable PV deployment in less favourable conditions.
A power distribution transformer (PT) is expected to have an operational life of around 30 years [1][2], thus the standard sizing practice contemplates the installation of oversized equipment to ensure a lower load level on the transformer until additional capacity is required to allow for future demand growth. In this sense, the potential of distributed generation with BIPV might represent a means of supplementing network capacity, postponing substantial investment on expansions or upgrades of the distribution systems, such as acquiring new power transformers [3].
Given that the installation of PV systems in buildings in urban environments is growing relatively fast, excessive electricity feed-in might represent a challenge to the stability of the grid due to power ramps, and thus a challenge for the network operator. Therefore, the local distribution grid ought to consider PV generated electricity input when sizing new urban distribution transformers, or in the upgrade of current assets.
The main goal of this work is to understand the power balance on urban distribution transformers and the changes in power capacities that may be needed if the total PV potential of the buildings in a suburb is considered. Additionally, the impacts of aggregated storage in power ramps is analyzed.

2. METODOLOGY

This study considers the area of Alvalade in the city of Lisbon (Portugal, 38.71°N 9.14°W). The data available includes the power capacity of all transformers as well as georeferenced polygons representing the building footprints (Fig. 1A) including information on the heights and number of floors of the respective buildings. In the absence of information of the grid network data describing the interconnections between buildings and transformers, it was assumed that each transformer has its own influence zone. The zones of influence (Fig. 1B) were computed as the area delimited by Thiessen polygons using the points that correspond to the location of the transformers (red dots in Fig.1 B). Then, the building polygons whose centroids fall inside an influence zone are assigned to the respective transformer. Using the centroids of the building polygons assures that each one has only one connection and, therefore, belong to a unique influence zone.

Fig. 1 – Delimitation of the study area (blue area) and location of the transformers (red dots) (A), Thiessen polygons depicting the influence zones of all transformers (B)

To accurately investigate the transformer spare capacity for accommodating all potential PV electricity generation in the case study, one needs to know when the latter is expected to peak. As the peak depends on azimuth and slope of the building surfaces, different transformers will feature peak electricity injection at different times of the day. The existence of large façade areas ought to further shift peak production. The estimation of PV potential consisted on an hourly time step solar irradiance simulation, allowing for the detection of the hour of highest injection of PV generated electricity into the grid.
The SolarAnalyst [4] extension for ArcGIS was employed for the hourly solar irradiance simulations on the rooftops. Using a 1m² resolution LiDAR derived digital surface model (DSM) of the area and monthly averages of diffuse radiation fraction and atmospheric transmissivity, the total amount of incoming solar radiation was calculated for each location of the DSM, following the approach reported in [5]. Since the SolarAnalyst extension is not capable of handling full-3D solar irradiance simulations, another tool had to be used to estimate the PV potential on façades. The building footprints were imported into the CAD software Rhinoceros3D [6] and extruded to their respective heights (Fig. 2) using the integrated graphical algorithm editor Grasshopper. Then, employing the plug-in LadyBug [7], the hourly solar irradiance in the vertical surfaces was computed. This step had to be performed inside a loop that cycled over a total of 550 buildings and a nested loop for the hourly irradiance calculation, otherwise the software would crash due to the huge amount of building geometries.
For simplicity, four reference days were considered in the hourly simulations: the 21st of March, June, September and December.
The irradiance results were input into Eq. 1 to estimate the PV potential, assuming standard 1m² panels with 15% efficiency.

$$PV_z(t) = I_h(t) \times \eta \times PR\,, \tag{1}$$

where PV means the PV generation potential [kW] for a transformer influence zone z, at hour t, I_h is the total hourly solar irradiance received by considered building surfaces in the influence area [kW], η is the efficiency of the panels (0.15) and PR is a performance ratio (0.85).
In the absence of real metered data, an estimate for the power demand by transformer influence zone at a given hour was derived using the reference consumption profiles for low voltage clients in class C (contracted power below 13.8kVA) for 2017 [8] and knowing that average yearly electricity consumption in the city of

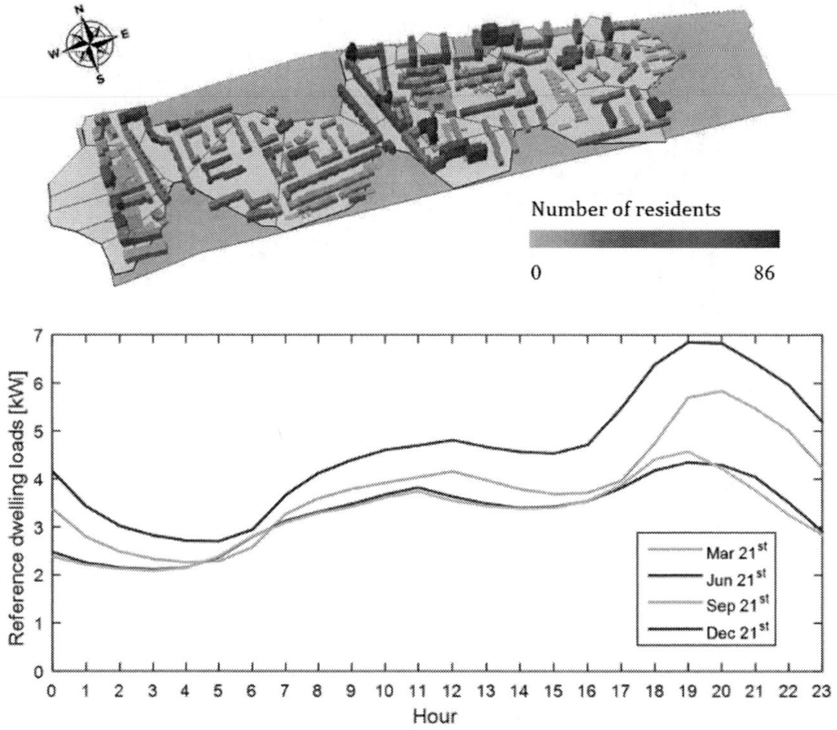

Fig. 2 – 3D model of the buildings inside the transformer influence zones (after the extrusion of building footprints) and the number of residents per building (top) and the seasonal reference electricity load for single dwellings (bottom). (For interpretation of the references to colour in this figure, the reader is referred to the web version of this paper.)

Lisbon is around 1300 kWh/year/person [9]. The profiles are estimated and published by the national transportation grid operator every year for billing purposes, and are applied to customers who do not own a net metering device.

$$P_{load,b}(t) = N_{p,b} \times P_{load,ref}(t)\,, \qquad (2)$$

where $P_{load,b}(t)$ is the electricity consumption [kW] in building b at hour t, $N_{p,b}$ is the number of residents in building b (Fig. 2, top), $P_{load,ref}(t)$ is the value of the reference electricity demand at hour t [kW] (Fig. 2, bottom).

Li-ion batteries with capacities equal to 0, 0.25, 0.5, 1, 2 and 4 kWh/kW$_p$ were considered and a self-consumption maximization strategy was employed for the four days:

1) if the PV production exceeds the load demand, the surplus is used to charge up storage and the remaining production, if any, is sold to the grid;
2) the same as 1, but all the surplus is sold if the storage is already at maximum charge;
3) if the PV production is lower than the electricity demand, the deficit is drawn from the storage bank;
4) same as 3, but if the storage hits minimum level, the remaining demand is purchased from the grid.

Lastly, the hourly power balance at each PT ($P_{balance}$) was obtained through:

$$P_{balance}(t) = P_{PT} + P_{load}(t) + P_{storage} - P_{PV}(t) \qquad (3)$$

where P_{PT} stands for the transformer power capacity, P_{load} is the local power demand, $P_{storage}$ is the total storage capacity and P_{PV} generated PV electricity.

3. RESULTS

The transformers will suffer the highest PV electricity injection at 10am of September 21st (Fig. 3). The addition of façades widens the period of electricity production and shifts the peak towards the morning period, except in December when this peak is shifted towards the afternoon. The PV potential greatly surpasses the electricity consumption from 8am to 3pm, which would be more difficult to achieve if PV was installed solely on the rooftops.

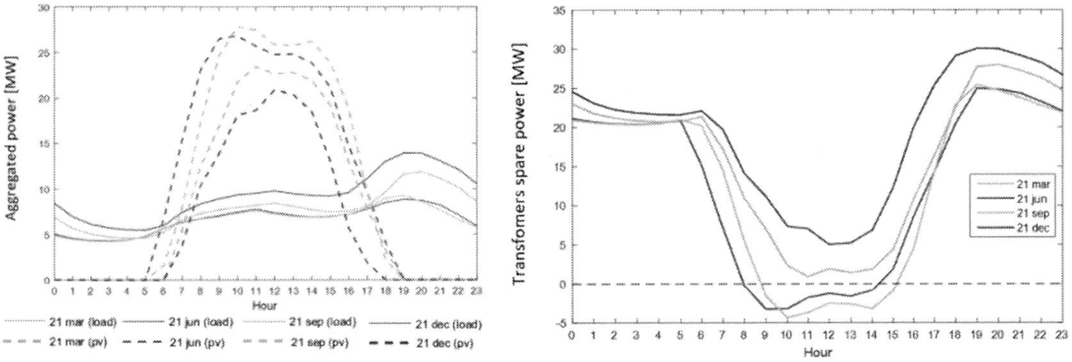

Fig. 3 – Aggregated hourly electricity load and PV production (left) and transformers spare power (right) for the 21st of March, June, September and December.

The maximum hourly PV potential of all rooftops reaches about 10MW and with façades becomes 28MW, twice the whole transformer power capacity, and greatly surpassing the electricity consumption. There are however some sub-areas that can still accommodate the PV production (Fig. 4), where there are relatively fewer buildings with low energy consumption and/or smaller rooftops. The overall transformer grid has no spare capacity to accommodate such high electricity production, thus the upgrade of almost all the transformers would be essential, regardless of few changes in neighbour transformer interconnections.

Given that high power ramps threaten grid stability, the potential of aggregated electricity storage was investigated to increase the transformers' availability to supply/receive what prosumers generate/demand. In Figure 5, boxplots for no storage (0 kWh/kW$_p$) show that a power capacity upgrade of 4 times would be required. These impacts are further enhanced, for most sub-areas, when the storage capacity exists but is relatively small (between 0.25 and 1kWh/kW$_p$). This occurs either because there is a moment when the batteries reach their maximum capacity and a high amount of electricity is suddenly injected into the grid, or the reverse, i.e. when batteries discharge completely and the demand must be assured by the grid. For storage capacities above 2 kWh/kW$_p$ seem to produce a relief impact on the grid, in particular in the summer.

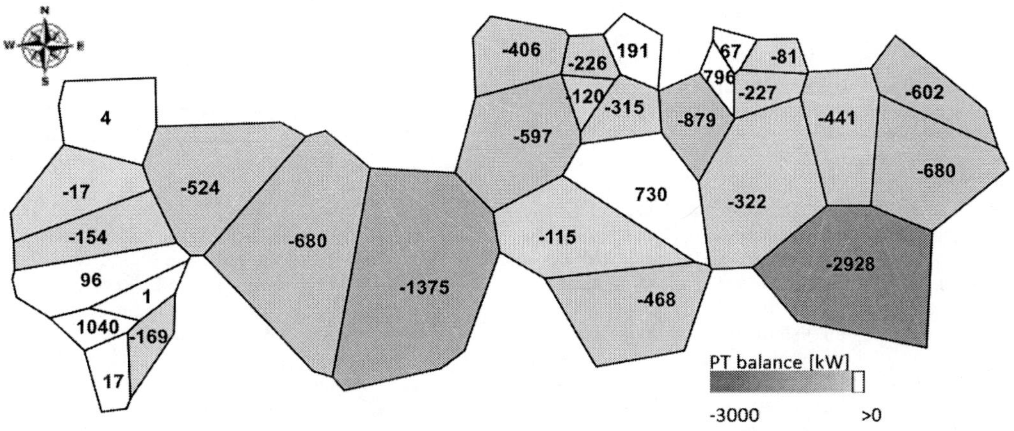

Fig. 4 - P_{GAP} at each transformer influence zone considering rooftop and façade PV generation. Negative values indicate failure at the respective transformer.

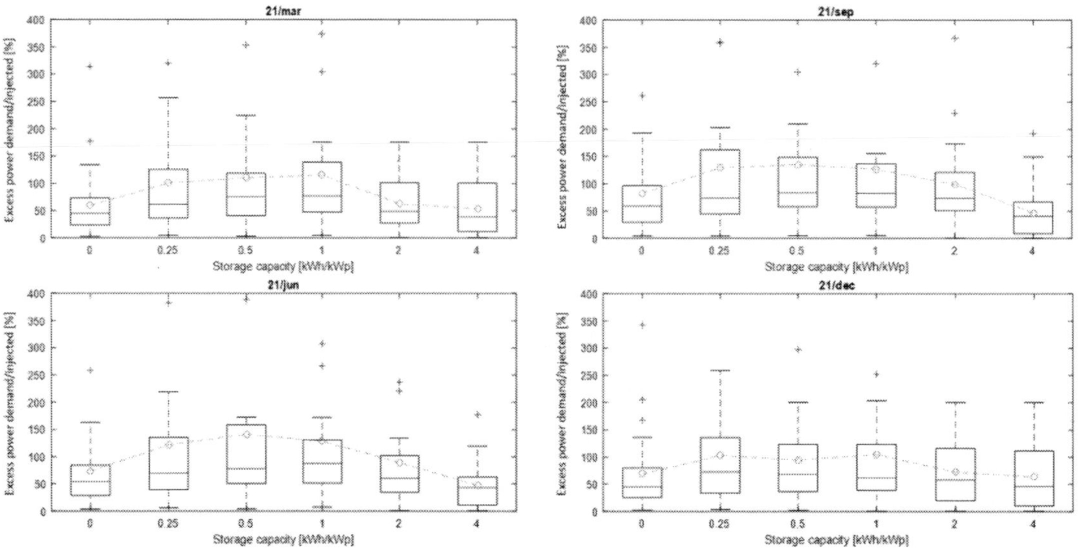

Fig. 5 – Boxplots representing the distribution of the maximum ramp value, as a percentage of the transformers' capacity, for different storage capacities in all hours of the typical days analysed. (The tops and bottoms of each "box" are the 25th and 75th percentiles, the red line in the middle of each box is the median, the green dots are the average, the black lines extending above and below each box are the extreme values and the red crosses are the outliers.)

4. CONCLUSION

The solar PV potential of rooftops and building façades in a small area was computed to study the power balance of the distribution grid in a future scenario with very high PV penetration. It is shown that the grid becomes locally less resilient with massive PV deployment and therefore the sizing of new urban power transformers ought to anticipate future installation of solar photovoltaics in buildings. As for current infrastructure, since upgrading transformers is a costly solution, one might contemplate rearrangement of interconnections among neighbouring transformers, so that all/some of the surplus PV electricity from one transformer could be fed into another with spare power capacity.

Perhaps unexpectedly, the introduction of local storage further enhances the need for higher grid capacity, in particular for smaller battery systems for PV self-consumption. Well managed large storage capacities on the other hand may be very useful for low grid impact of large PV production.

Aside from the sizing of transformers, the management of storage must be thought carefully along with proper deployment of PV systems, which should consider the demand profiles and the variety of surface tilts and aspects for achieving a better demand-supply match.

5. ACKNOWLEDGMENTS

The authors would like to thank EDP Distribuição for the distribution transformers data set.

This publication was partially supported by CICS.NOVA, UID/SOC/04647/2013, with the financial support of FCT/MCTES through National funds, MIT Portugal Program for Sustainable Energy Systems, FCT grant SFRH/BD/52363/2013, Projects MITP-TB/CS/0026/2013 (SUSCITY), PTDC/EMS-ENE/4525/2014 (PVCITY), UID/GEO/50019/2013 - Instituto Dom Luiz and Calouste Gulbenkian Foundation award 2014.

6. REFERENCES

[1] K. D. Harden, "Optimizing energy efficiency standards for low voltage distribution transformers," Fort Wayne, Indiana: Purdue University, 2011.

[2] A. R. A. Manito, A. Pinto, and R. Zilles, "Evaluation of utility transformers' lifespan with different levels of grid-connected photovoltaic systems penetration," *Renewable Energy*, vol. 96, pp. 700–714, Oct. 2016.

[3] A. Piccolo and P. Siano, "Evaluating the Impact of Network Investment Deferral on Distributed Generation Expansion," *IEEE Transactions on Power Systems*, vol. 24, no. 3, pp. 1559–1567, Aug. 2009.

[4] P. Fu and P. Rich, "Design and implementation of the solar analyst: an arcview extension for modeling solar radiation at landscape scales," in *19th annual ESRI User Conference*, 1999.

[5] N. M. P. Gomes, "Integração de dados LIDAR com imagens de muito alta resolução espacial para determinação de áreas urbanas com potencial solar," Faculdade de Ciências Sociais e Humanas, Universidade Nova de Lisboa, 2011.

[6] McNeel, "Rhinoceros 5.0," 2016. [Online]. Available: https://www.rhino3d.com/

[7] M. S. Roudsari and M. Pak, "Ladybug: a parametric environmental plugin for grasshopper to help designers create an environmentally-conscious design," in *13th IBPSA*, 2013.

[8] ERSE, "Perfis de perdas, perfis de consumo e de autoconsumo, e perfis de produção," 2016: http://www.erse.pt/pt/electricidade/regulamentos/acessoasredesaasinterligacoes/Paginas/PerfishorariosdeperdasedeconsumoemBTEBTNeIP.aspx?master=ErsePrint.master

[9] M. C. Brito, N. Gomes, T. Santos, and J. A. Tenedório, "Photovoltaic potential in a Lisbon suburb using LiDAR data," Solar Energy, vol. 86, no. 1, pp. 283–288, Jan. 2012.

PV AS MAIN ENERGY SOURCE FOR THE ENERGY SUPPLY OF URBAN RESIDENTIAL DISTRICTS IN CENTRAL EUROPE

J.-S. Telle[1], R. Völker[1], T. Kilper[1], I. Hauer[2], K. v.Maydell[1]
[1]DLR Institute of Networked Energy Systems
Carl-von-Ossietzky-Straße 15, 26129 Oldenburg, Germany
[2] Department of Electric Energy Systems, Otto-von-Guericke University of Magdeburg
Universitätsplatz 2, 39106 Magdeburg, Germany

ABSTRACT: This case study investigates the technical and economic potential of photovoltaic (PV) as main energy source of urban residential districts, considering sector coupling and a centralized district storage system (DSS). Therefore, load and PV generation profiles are modeled for a sector-coupled energy supply. Sector coupling includes the integration of heat pumps for the implementation of a power-to-heat concept and battery electric vehicles (BEV) for power-to-vehicle applications. Simulation results show the benefit of sector coupling and the need of DSS to optimize the self-consumption. The DSS supports the decentralization and self-sufficiency of the districts energy supply and provides flexibility by different operating strategies. The results of economic analysis point out that, under the investigated circumstance, LCOE (Levelized Costs of Electricity) of PV are in a competitive position with the current of price of electricity. Compared to the current EEG remuneration, the business model of "quarter electricity" (germ. Mieterstrom) provides the opportunity to increase the margin for providers and prosumers of PV. Additionally customers of urban electricity could benefit from lower prices of electricity and increase the acceptance of PV within urban residential districts. Moreover, the technical and economic potential for providers, consumers and prosumer of PV is shown.
Keywords: Photovoltaic, Sector Coupling, District Storage System, Economic Analysis, Quarter Electricity

1 INTRODUCTION

Global growth of PV systems leads to the fact that PV is no longer only an ecologically favorable technology. It is also an economically attractive power generation solution. LCOE of PV in Germany has been decreased since 2008 [1], however the expansion of PV has been stagnated and the German government targets has not been achieved. In order to increase the share of renewable energies, especially of the PV, sector coupling could play a key role. The share of renewable energies in the sector of electricity is about 30 %; however, in the whole sector of end-use energy it sums up to only 12.6 % [2]. The electrification of heating, cooling and mobility sector introduces opportunities for PV and other renewable energy technologies. Based on a case study of a planned urban residential district, the potential of the PV at household level is analyzed. Besides the PV, a DSS, heat pumps and battery electric vehicle (BEV) were investigated. The district storages were applied to disable the delay of load and energy generation and to provide flexibility options. Aiming to address the economic advantage of the PV and the energy supply through sector coupling, investment calculation, the LCOE and the business model of quarter energy were analyzed.

The first part of this case study consists of modelling and simulation of a sector-coupled energy supply of a new projected settlement. The load and generation profiles were formed for one year in a 1-minute resolution. Simulation with different parameters were performed and technically and economically evaluated.

2 MODEL DESCRIPTION

2.1 Topology

The base of this investigation is an urban district with a high penetration of PV. The district consists of 4 one-family houses (OFH) and 8 multi-family houses (MFH), overall of 108 residential units. The houses are built on a high energy standard according to the KfW 40 Standard from the EnEv 2009 [3]. The energy supply will be implemented according to an energy concept through sector coupling (see figure 1). The demand of space heating and domestic hot water (DHW) will be covered by a power-to-heat concept. For every OFH and MFH, air-water heat pumps will be used in combination with buffer storages. For a power-to-vehicle concept inside the district, it is assumed that charging stations are available for the inhabitants.

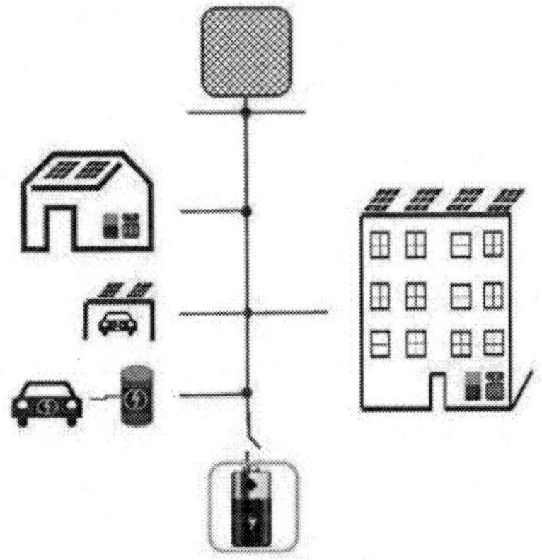

Figure 1: Conception of energy supply of the case study

2.2 Load and PV-generation profiles in a high resolution

Firstly, the load and PV-generation profiles are created using different sources for a simulation period of one year in a 1-minute resolution.

The load profiles for space heating and DHW are developed considering the norm VDI4655 [4], the EnEV 2009 and indications about consumptions from [5]. The number of people living in every household is specified with data from [6] (see [5] for more details about electricity load profiles).

Initially, the annual energy demand of space heating is determined at 20 kWh/m²·a and 1,000 kWh/WE (MFH) or rather 500 kWh/Pers (OFH) for DHW.

In order to know the demand of the heat sector, "Typical-days" (see VDI 4655) are created. At the beginning, 10 different Typical-days are created and seasons are distinguished as follows: summer (S), winter (W), transition period (Ü); working day (W) and Sunday and holiday (S); cloudiness degree: cloudy (B) and bright (H). Ten Typical-days are created using the combination of these factors (in summer there are only two days). With the help of the meteorological data, each day of the year is assigned to one Typical-day. Using the Typical-day factors from VDI 4655 and the annual demand, the number of each Typical-day and the daily demand of space heating and DHW are determined. To provide thermal energy, water-air heat pumps in combination with buffer storages are utilized. Assuming that all households have underfloor heating systems and a similar consumer behavior in relation to the DHW consumption, profiles in 1-minute resolution are determined for the whole year.

Load profiles of electricity consumption are modelled by using the 74 load profiles (1-minute resolution) from [7]. To define the annual demand of electricity for each household, a standard distribution of the German average consumption from 2014 [8] and the distribution of district inhabitants are calculated. In order to create 108 profiles, 58 out of the 74 profiles are used twice, but were shifted plus/minus 15 minutes. Subsequently, the annual demand is assigned through a standard derivation (see [5] for more details about electricity load profiles).

Taking the mobility into account for a sector coupling concept, a simple model of power-to-vehicle is implemented. It is assumed that a part of the district mobility is BEV, being charged on an existing charging infrastructure within the district. To generate charging profiles in a 1-minute resolution, the following assumptions are made: 25 BEV, 39,1 km of daily mileage, 3.6 kW of charging capacity and two charging times per day: at 6 p.m. and 10 p.m. [5], [9].

The PV is installed as rooftop system with a south alignment and an inclination angle of 40° for the OFH and an east-west alignment with a 13° inclination angle for the MFH and some carports. Also, a centralized district storage system (DSS) will be integrated in the district energy supply. The 1-minute resolution PV generation profiles are modeled as follows: The first step choosing a type of module and inverter. For this case study, a monocrystalline module type (η_{Module}=18.7 %) and a power inverter with an efficiency of η_{Inv}=97 % are chosen. The efficiency of the system is determined as $\eta_{Sys} = 18\,\%$. in addition, the alignment, the inclination angle, the meteorological data (irradiance and temperature) and the module area must be implemented as input parameters for the simulation [10], [11]. The output parameter is the power output of the PV in a 1-minute resolution. To have a look on the potential of the PV as main energy source, the number of installed modules or rather the module area was varied (see [5] for more details to the modelling of the PV-system).

2.3 District storage system and operating strategies

Based on [12], the centralized alternative DSS is chosen in opposite to single household storage systems. It is assumed a chemical and technical design of commercial lithium-ion cells. In the model, ageing processes and efficiencies are not implemented. A C-Rate of one is defined.

To optimize the self-consumption, the use of a DSS with two different operating strategies is examined.

In the first strategy *Fast charge*, the storage responds to the energy balance from the district's PV-generation and the energy demand. A positive balance means that there is a higher generation than demand and the storage can store the surplus of PV production. The other way around implies that the storage has to be discharged and provide energy.

The second strategy *scheduled* deals with the assumption of a perfect forecast. Besides the target of optimizing the self-consumption, grid relief is another goal. Thereby, the energy balance of the next day is forecasted and an operating schedule with the objective of capping the peaks of energy surplus is determined (see figure 2 and [5] for more details to the modelling of the DSS operating strategies).

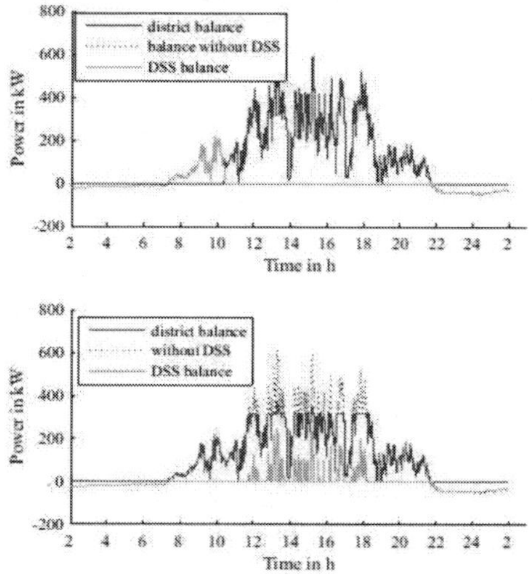

Figure 2: Compare of the DSS operating strategies, exemplary summer day with unsettled weather

2.4 Evaluation parameters and simulation scenarios

The most important parameters to evaluate the technical potential of the PV and the benefits of a DSS are the district energy balance, the self-consumption rate (SCR) and the self-sufficiency rate (SSR).

The energy balance of the district is defined as the difference between annual PV-generation and annual sector-coupled household load. A positive balance means an annual surplus of PV-generation.

The SCR indicates the share of the annual self-consumption in relation to the annual PV generated energy.

The degree of district self-sufficiency is indicated by the SSR (self- sufficiency rate). It is calculated from the relation of annual self-consumption to annual district load. A rate of 100 % means a complete self-sufficiency of the district.

The main simulation scenarios are defined as follows: without DSS (0), Fast charge (1) and scheduled (2). sub-scenarios are determined by the variation of the installed PV-Power (1-3) or the used roof area. In addition, the following subcases are considered: sector coupling (a) and without sector coupling (b) (Table I).

Table I: Simulation scenarios

Installed PV In kW_{peak}	Without DSS	DSS Fast charge	DSS scheduled
673	S.0.1a/b	S.1.1a	S2.1a/b
369	S.0.2a/b	S.1.2a	S2.2a/b
268	S.0.3a/b	S.1.2a	S2.3a/b

2.5 LCOE and business model

The first step to point out the economic potential of the PV in urban residential districts, considering a sector coupling, is an investment calculation. Therefore, the technologies of the PV and DSS are assessed with the annuity method according to the [5]. The investment amount is split in annually equal parts during an assessment period, dividing the net present value (NPV) by the annuity factor. Thereof, the equal parts resulting in the equivalent annual cost (EAC) or annuity.

The EAC considers the annuity of capital-related costs, like the investment amount or procured replacement, and the operation-related costs, for instance costs for maintenance, servicing or inspection.

Over an assessment period of 15 years, with an interest factor of 1.035 [5] and without price changes, an annuity factor of a = 0.0868 is obtained. The assessment of the capital- and operation-related costs from the PV is based on [1] and [13] assumes an operating lifetime of 20 years. According to [10], the capital-related costs are calculated with1, 150 €/kW_{Peak} and the operating-related costs with 1 % of the investment amount.

The capital-related costs of the DSS are valued at the capacity specific price of 600 €/kWh, the power-electronic specific price of 150 €/kW and in addition to 100 €/kW for other costs (e.g. cooling system and transformers) [14]. The operating-related costs are valued at 2 % of the investment amount. The annuity can hence be calculated. To rate the investment afterwards, the annuity of the revenues has to be considered. The total annuity of the whole annual cash flows is calculated by the sum of the annuity of revenues and costs. The LCOE and LCOS (levelized cost of storage) indicate the costs of one kilowatt-hour generated from the district PV or stored with the DSS. The LCOE of the PV is calculated as the annuity of the costs (capital and operating costs) divided by the annual energy output, whereas the LCOS is calculated as the quotient of the annuity of the costs by the annual discharged energy in addition to the PV LCOE.

To figure out the economic potential of the PV, the business model of quarter electricity is investigated. Quarter electricity is not a new concept, however the circumstances have changed. The EEG-remuneration decreases, the LCOE of PV decreases and the model of self-consumption become more attractive. To be a self-consumer in the legal sense and to get every advantage of ceasing grid fees (only 40 % of the EEG-apportionment has to be paid), the owner of the PV-system has to be the consumer as well as the operator in the same natural or legal person. For the model of quarter electricity, it is possible to several persons. The advantages remain with the exception of the EEG-apportionment which has to be paid to 100 %. Therefore, communities of tenants, lessors, prosumers, contractors or utilities can offer quarter electricity. The quarter electricity respect to the new EEG 2017 considers a possibly reduced EEG-remuneration. The price for electricity within the district is developed with the actual legal framework for quarter electricity and compared through the average price of energy at household level (29 ct/kWh [15]). The condition for this model is the spatial link between PV and point of consumption and no transit through the public grid.

3 RESULTS AND DISCUSSION

3.1 District energy balance and dimensioning of the DSS

The simulation scenarios without DSS result in the following simulation outputs (see Table II).

Table II: Simulation results for scenarios without DSS

Scenario	0.1a	0.1b	0.2a	0.2b	0.3a	0.3b
PV-gen. in MWh/a	627		348		255	
Load in MWh/a	376	270	376	270	376	270
SCR in %	21,7	17,9	34,8	29,6	43,6	37,8
SSR in %	36,4	41,5	32,4	38,2	29,8	35,8

It shows that both evaluation parameters (SCR and SSR) had different behaviors relating to the annual PV generation and load. Considering the high PV penetration and the low level of household load, it leads to a surplus of energy or rather a low SCR and high SSR. The scenarios considering sector coupling (a) have a quite higher SCR. The proportion of energy surplus and the amount of withdrawal from the grid can be extracted from Figure 2. The scenarios 0.1 a and b show that the PV system is over-dimensioned related to the load with a high feed-in into the public grid. Whether it makes sense to use the whole area for PV installation depends on the operator(s) and owner(s) of the PV-system. Overall, it becomes clear that the PV has the potential to increase the share of renewable energies within the final energy mix.

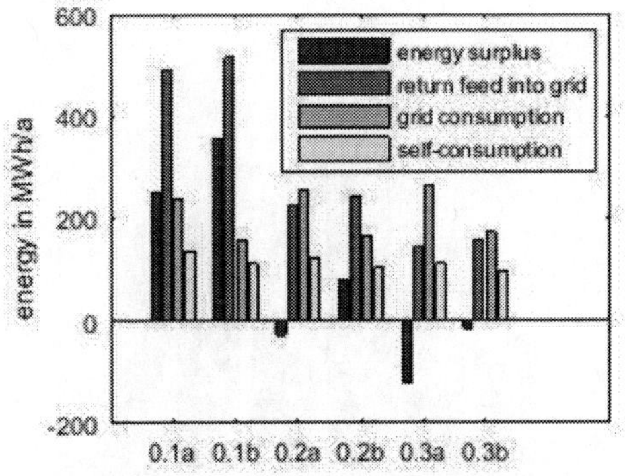

Figure 3: District energy balances without DSS

Figure 3 shows the potential for a DSS: store the surplus (dark blue) to increase the self-consumption (yellow) and decreases the power withdrawal (green).

To find the optimum size of the DSS, the SCR and SSR can be indicators for the technical dimensioning. Besides the technical parameters, the economic parameters also play a key role for the dimensioning. This case study focuses on the technical efficiency. Figure 3 shows the SCR with increasing capacity of the DSS (left) and the rate of increasing SCR over the capacity (right). It shows that there is a point between

200 and 250 kWh where the rate of increase approaches zero.

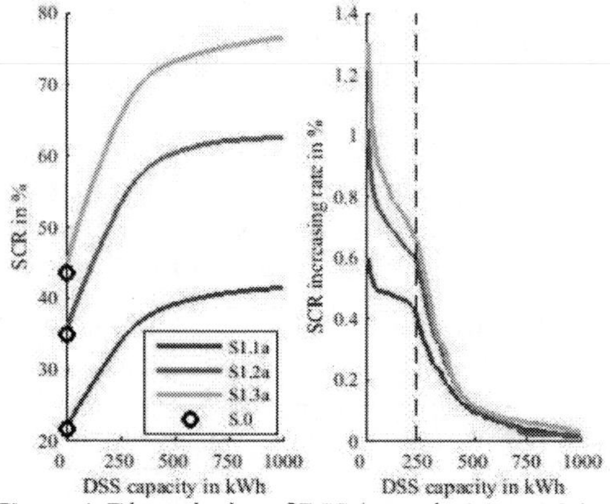

Figure 4: Dimensioning of DSS (scenario 1.1a – 1.3a)

At this point, the capacity and power for the DSS are chosen. To simplify further calculations, the same capacity of 240 kWh is chosen for all scenarios with sector coupling (see Figure 4) or rather 190 kWh for all scenarios without sector coupling.

3.2 Optimized self-consumption and flexibility options

Figure 5 shows how the SCR and SSR increase with the chosen DSS.

The larger the energy surplus of the district, the smaller the increase of the SCR is because of the DSS. From a technical point of view, a smaller PV-system can be more efficient and a DSS brings higher SCR benefit.
Besides the optimization of self-consumption with the DSS, the possibility as flexibility option is also investigated. The DSS operating strategy scheduled offers the possibility to relieve the public grid providing ancillary service. Another possibility is to integrate the DSS into a pool which provides control reserve.

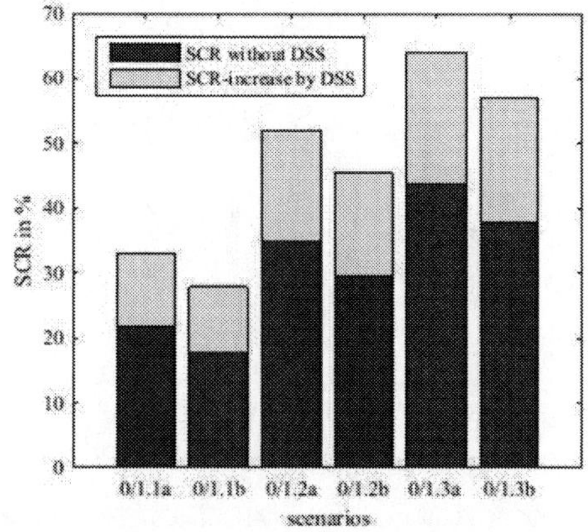

Figure 5: Share of SCR increase by the DSS

3.3 LCOE and district energy price

From the annuity method results and simulation results, the LCOE of PV is calculated as 10.5 ct/kWh up to 10.7 ct/kWh for an assessment period of 20 years. For the two different sizes of DSS and depending on the scenario, the LCOS values range between 40.0 ct/kWh and 52.7 ct/kWh. Due to very low differences between the simulation results of both DSS-scenarios, there is no differentiation for the LCOS.

Table III: Levelized Cost of Electricity and Storage

	1a	1b	2a	2b	3a	3b
LCOE ct/kWh	10.7	10.7	10.6	10.6	10.5	10.5
LCOS Ct/kWh	41.5	40.1	47.5	43.6	52.7	45.5

Particularly results high prices for stored energy compared to the current mean price of electricity in Germany [15]. It shows that the smallest PV system has the smallest LCOE. But in consideration to the amount of self-consumed energy and the residual need from the grid there is no disadvantage for the over-dimensioned PV system.

It results in high prices for the stored energy, compared to the current mean price of electricity in Germany [15]. However, considering to the amount of self-consumed energy and the residual need from the grid, there is no disadvantage for the over-dimensioned PV system.

The development of the price of electricity for the district is based on the described business model of quarter electricity. Assuming a EEG-remuneration from 11.01 ct/kWh [16], EEG apportionment from 6.88 ct/kWh, price for billing 3.2 ct/kWh and a VAT rate of 19 %, it results in the margin of 4.06 ct/kWh, compared to the 29.16 ct/kWh [13] ,[15]. To get a EEG-remuneration for quarter electricity but also to advise an attractive price to the costumers, the price has to be reduced in comparison to the base price of the local utility. Considering a price discount of 2 ct/kWh and the LCOE PV for scenario 1.1a, a margin of 2 ct/kWh is derived (Figure 6).

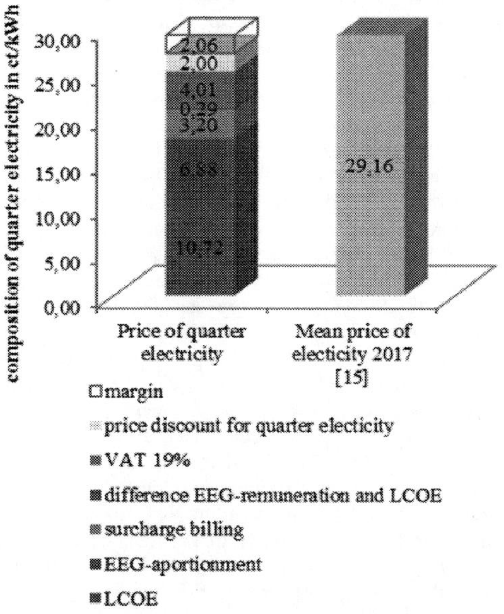

Figure 6: Price composition of quarter electricity

The conventional marketing with 11.01 ct/kWh EEG-remuneration results in a margin of 0.29 ct/kWh. Comparing the margin from the conventional case with the margin from the quarter electricity, the benefit of the quarter electricity is shown. It should be kept in mind that every household in the district could freely choose its energy utility. However, considering that the SSR was below 100 %, there remains a rest requirement of electricity which the utility can provide from its portfolio and which can therefore generate another margin. Depending on the provider of quarter electricity, delivery of the rest need of electricity is an important part.

4 CONCLUSION

Prior case study underlines the technical and economically potential of PV in urban residential districts. It was found that there is a high technical potential of PV considering a sector coupling and integrated DSS. Integration of DSS demonstrated the possibility to increase the SCR and SSR to become more independent from global energy prices. It provides also the opportunity of flexibility options for the relief of the grid.

Including sector coupling to the energy supply of a urban residential district shows approaches to implement power to heat or power to vehicle concepts and increases consumption of renewable energy solutions.

For an optimized energy supply of an urban residential district, it could be beneficial to over-dimension the PV. Therefore, the self-consumption and autarky could be maximized.

Quarter electricity provides the opportunity to develop a beneficial business model for PV as main energy source, optimizes the local self-consumption and includes all participants (owner, operator and consumer) in lower energy costs.

The determined PV-LCOE and the price of quarter electricity indicate the economic potential regardless of EEG-remuneration and could be an incentive for the stagnating expansion of PV in Germany. The economic analysis also shows that the LCOS has to be decreased to be competitive with the current prices of electricity.

5 REFERENCES

[1] D. H. Wirth, "Recent Facts about Photovoltaics in Germany," Frauenhofer ISE, Germany, 2017.

[2] Umweltbundesamt, "Primärenergieverbrauch," 3 2017. [Online]. Available: https://www.umweltbundesamt.de/daten/energie-als-ressource/primaerenergieverbrauch#textpart-1. [Accessed 9 2017].

[3] Bundesmindesterium für Justiz und Verbraucherschutz, "Verordnung über energiesparenden Wärmeschutz und energiesparende Anlagentechnik bei Gebäuden," 10 2015. [Online]. Available: https://www.gesetze-im-internet.de/enev_2007/.

[4] Verein Deutscher Ingenieure e. V., *VDI 4655 2008-05 - Reference load profiles of single,* Germany, 2008.

[5] T. J.-S., *Master thesis : "Potentialanalyse verschiedener Betriebsstrategien und Geschäftsmodelle eines Quartierspeichers innerhalb einer PV-Klimaschutzsiedlung unter Berücksichtigung einer Sektorkopplung",* Germany: University of Magdeburg, 2017.

[6] German Federal Statistical Office, ""Wirtschaftsrechnungen"," 2013. [Online]. Available: https://www.destatis.de/DE/Publikationen/Thematisch/EinkommenKonsumLebensbedingungen/Konsumausgaben/EVS_AufwendungprivaterHaushalte2152605139004.pdf?__blob=publicationFile. [Accessed 9 2017].

[7] Hochschule für Technik und Wirtschaft HTW Berlin, *„Repräsentative elektrische Lastprofile für Einfamilienhäuser in Deutschland auf 1-sekündiger Datenbasis", Datensatz,* Berlin, Germany, 2015.

[8] Federal Ministry for the Environment, Nature Conversation,, "Stromspiegel für Deutschland 2014," Germany, 2017.

[9] Kraftsfahrtsbundesamt, "Kurzbericht Gesamtkilometer," 2016. [Online]. Available: https://www.kba.de/DE/Statistik/Kraftverkehr/VerkehrKilometer/2016_vk_kurzbericht_pdf.pdf;jsessionid=BCB582E7338D7F1DD8659022C9287A7C.live21304?__blob=publicationFile&v=11. [Accessed 9 2017].

[10] Meteotest, *Meteonorm 7, Software,* Bern, Switzerland.

[11] National Technology and Engineering Solutions of Sandia, LLC., *PV_LIB Toolbox (Matlab), Software.*

[12] R. Völker et al., "Analysis of stationary electrical storage solutions for residential districts with high photovoltaic penetration," 32nd European Photovoltaic Solar Energy Conference and Exhibition, Munich, Germany, 2016.

[13] Dr. H. Will, Urban Energie GmbH, München & Fabian Zuber, 1°energy | local markets, Berlin, ""Geschäftsmodelle mit PV-Mieterstrom" - PV Financing Project," Germany, 2016.

[14] J. Fleer et al., "Model-based economic assessment of stationary battery systems providing primary control reserve," 10th International Renewable Energy Storage Conference, IRES 2016, Düsseldorf, Germany, 2016.

[15] BDEW Bundesverbrand der Energie und Wasserwirtschaft e.V., "BDEW-Strompreisanalyse Februar 2017," Berlin, Germany, 2017.

[16] Bundesnetzagentur, "Photovoltaic installations data submissions and EEG-supported feed-in tariffs," 2017. [Online]. Available: https://www.bundesnetzagentur.de/DE/Sachgebiete/ElektrizitaetundGas/Unternehmen_Institutionen/ErneuerbareEnergien/Photovoltaik/DatenMeldgn_EEG-VergSaetze/DatenMeldgn_EEG-VergSaetze_node.html. [Accessed 9 2017].

SUN HUB – ENERGY HUB FOR OUTDOOR TABLES

Peter B. Poulsen[1], Gisele A. dos Reis Benatto[1], Nicholas Riedel[1], Sune Thorsteinsson[1], Adrian Alejo Santamaria Lancia[1], Joanna Karolina Symonowicz[1], Kris Retoft[2] and Ib Mogensen[2]

[1]Department of Photonics Engineering, Technical University of Denmark, Frederiksborgvej 399, 4000, Roskilde, Denmark, ppou@fotonik.dtu.dk
[2]Outsider, Kigkurren 8A, Islands Brygge, Copenhagen, Denmark

ABSTRACT: Solar cells integrated into products are attracting more and more attention especially due to the dramatically declining cost of solar cells. Furthermore, we are getting more dependent on portable units like mobile phones, tablets and PCs which has to be charged to be of any use. Especially on festivals where people camps for several days it can be hard to have your portable units charged. In this this work we report a solar powered hub, as an add-on to a table in the urban environment for charging mobile phones and tablets and other handheld devices through USBs, charging laptops through AC connections, providing opportunity to stream music via Bluetooth and play it from a handheld device to the table and lastly to provide LED lighting on the table during the dark hours. 3 prototypes of the system was built and tested at the Roskilde Festival 2017. Electrical logger units were built into the 3 Sun Hubs to monitor the overall energy system and the consumption of each functionality in the table.
Keywords: Stand-alone PV Systems

1 INTRODUCTION

Solar cells integrated into products are attracting more and more attention especially due to the dramatically declining cost of solar cells. Furthermore, the cost of batteries is being driven down by the increasing market for electrical cars being a major driver with e.g. Tesla investing heavily in both. Lithium based secondary batteries are now approaching prices for Lead Acid, which gives some major advantages in solar cell product design, since lighter and more compact form factors can be used without introducing a major extra costs. Moreover, Lithium battery chemistry is getting to a level where it can survive so many cycles that it in many cases don't need to be changed in in the lifetime of the product which earlier on has been one of the major disadvantages of solar products. Especially when addressing the western countries markets, where labor cost is high and battery change is often a showstopper. The recent year's development in the solar and battery market mentioned above has made major difference for the feasibility of solar powered product that will just become more and more feasible in the future.

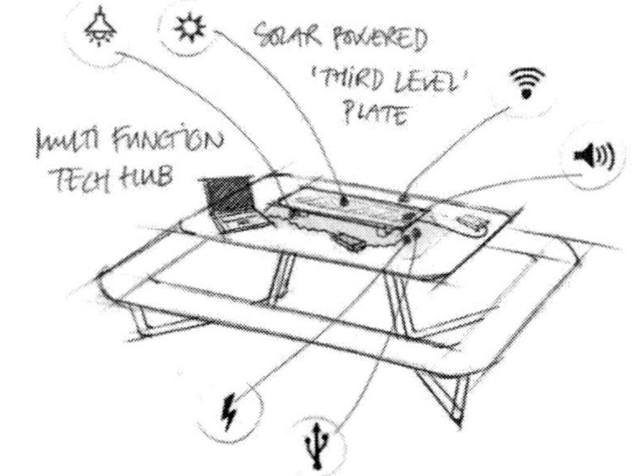

Figure 1: Sketch of the Sun Hub concept on top of the Plateau table produced and sold by the company outsider in Denmark.

Solar products has earlier on been more primitive, basically powering products from the sun to save on cables which in e.g. a lighting installation in Denmark represent around the same cost as the cost of the lamp being installed. In this study we report a solar powered hub, as an add-on to a table in the urban environment for charging mobile phones and tablets and other handheld devices through USBs, charging laptops through AC connections, providing opportunity to stream music via Bluetooth and play it from a handheld device to the table and lastly to provide LED lighting on the table during the dark hours.

Figure 2: The Plateau table without Sun Hub

2 METHODOLOGY

A sketch of the Sun Hub can be seen on Figure 1. For dimensioning, some assumptions were made to both fit the dimensions of the Plateau table where the hub can be mounted as an add-on feature, and to satisfy the need for the different functions the Sun Hub makes available to the user. A mock-up was made to find the best size of the hub leaving enough space around it for working with also a relatively large laptop. 3D software models were made to extend this study virtually and an image from this process can be seen in Figure 3.

Figure 3: 3D drawing of the Plateau table with the Sun Hub and some different sizes of laptops.

When the best compromise between size of solar panel for energy production and space for the user to work around the hub was reached, the active components for the hub could be chosen. The major components are listed below.

- 2x 30 Wp Monocrystalline silicon panels for shade tolerance (Voc: 25.65V, Isc: 1.48A for each panel)
- 2x Panasonic LC-CA1212P 12V Lead Acid Battery 12Ah.
- 18 W LED flex band (3014 Sidelight SMD LED - IP 67) to be dimmed down.
- Charge regulator: GenaSun GV-5 (Lead Acid)
- 12V Inverter Hama Safety 100W
- LeaningTech 50 W 4.0 Wireless Digital Bluetooth with Audio Receiver Amplifier Board
- 4x 2.5" full range speakers
- 2x PIR motion sensors looking 180 degrees out of each of the ends.

A rendering of the Sun Hub with the above components embedded is shown below on figure 4.

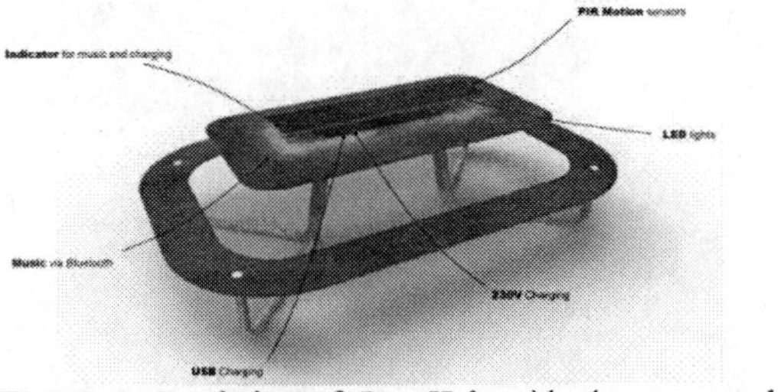

Figure 4: Rendering of Sun Hub with the suggested components implemented in the commercial Plateau table.

In figure 5 a diagram of the electrical components in the systems is shown schematically. Due to the limited resources in the project it was chosen to find commercial components on the market and connect them by a custom electronical driver system controlling them. More cost effective solutions would be to build the electronics into one printed circuit board which will be a task for the future.

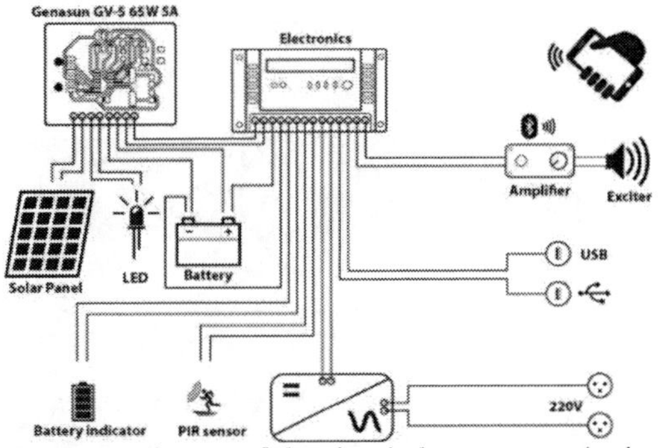

Figure 5: Diagram of the electrical components in the Sun Hub.

An electronic mock-up was made to test the component and especially the losses in the system. Since, as stated above, the project did not have financing for developing dedicated and optimized electronics, e.g. a low power commercial inverter needed to be used being very inefficient and having a 6W standby consumption. The charge efficiency was found to around 80% when charging close to 100W and much lower when charging at lower wattages. Also, a commercial board with Wireless Digital Bluetooth with Audio Receiver Amplifier was used having about 0.6 W standby consumption. The energy consumption while playing music was for the selected board around 3W at the maximum volume. Since the Bluetooth will be searching for a phone to pair with it was decided to turn this function on only, when PIR was activated opening a window of 10 minutes (being renewed every time PIR is activated again). So when no PIR was activated the power to the whole audio board was turned off.

The hub has 2 sockets for USB charging and 2 sockets for AC charging. The latter is sharing the 100 W inverter, so only 100W can be drawn from the 2 sockets in total. This means that very powerful laptops cannot charge at maximum rate, but will have a longer charging time which is a compromise that has to be made. A rather simple microcontroller based controller PCB was made to control the individual components and with the measured losses in the system, a mathematical model could be made. Assuming that the Sun Hub are placed in a not shadowed environment in Denmark in a typical meteorological year (TMY), an energy balance could be made (see figure 6) based on the measurements of the charge efficiencies and standby consumptions of the components.

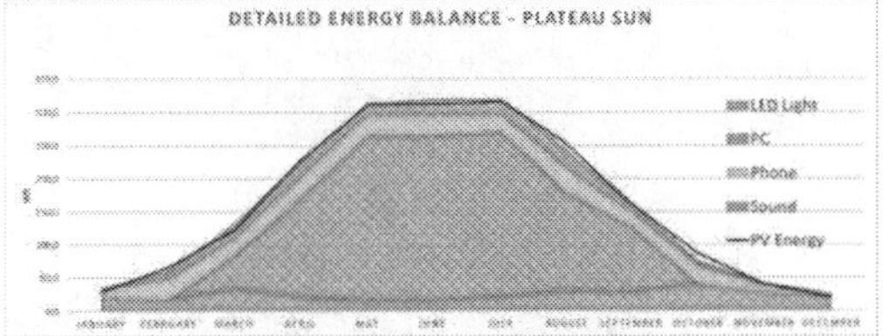

Figure 6: Energy balance of the functions offered by the Sun Hub.

The power consumption is very different for the different functions, and the availability is controlled by the microprocessor prioritizing:

1. USB charging
2. Streaming music
3. AC charging of laptops

Figure 7: Energy prioritizing for the individual functions

The microcontroller turns off the individual functionality at different programmed thresholds changing the indicator from green to red. The table always leaves energy for lighting to the night – even when all indicators are red. 3 prototypes of the Sun Hub were built and can be seen on Figure 8.

Figure 8: 3 prototypes of the Sun Hub.

On figure 9 it can be seen how the electronics is mounted and fitted into the hub.

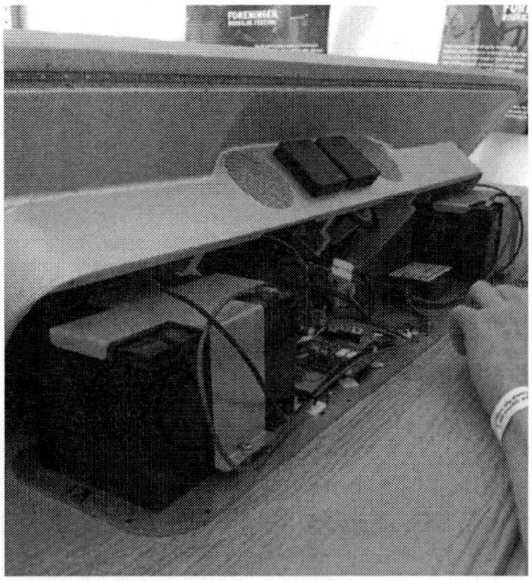

Figure 9: Inside look of the Sun Hub.

The 3 prototypes was field tested at the Roskilde Festival 2017. A photo can be seen on Figure 10.

Figure 10: Photo of field test of a Sun Hub integrated into a Plateau table at the Roskilde Festival 2017.

3 RESULTS AND DISCUSSION

The 3 prototype hubs were observed by a group of 6 students from the Technical University of Denmark. The aim of this was to identify if the user interfaces has been designed in and intuitive way. Furthermore an anthropology study was made to identify how the user felt about the table, its functions, its user interface etc.

Figure 11: A Sun Hub with LEDs turned on and all the other functions off during a rainy day at the Roskilde Festival 2017.

Some interesting observations were that many users used the 230V AC socket for charging their mobile devices. That is seen from the table a very inefficient way of using the energy since it is converted from 12V DC to 230 V

AC and back to DC by the portable unit's charger. A lot of loss will come in each conversion step, and the DC/AC inverter is very inefficient at low power consumption. A technical challenge was the high stand-by consumption of the DC/AC inverter. A magnet relay was built into the lid of the 230VAC socket to only turn on the inverter when the lid was opened (expectedly to insert a plug for charging). This magnet relay was not really working well leaving the inverter on unintentionally and draining the battery even when no charging was going on. In the software of the Sun Hubs controller electronics a threshold was made turning off the 230VAC charger and preventing this feature for being used by when the battery was under a chosen state of charge as shown on Figure 7. This saved some energy for the other functionality of the Sun Hubs. In the user survey the users was asked how they felt about not having all functionalities available at all times e.g. due to bad weather or heavy use of the table earlier in the day. The users really didn't complain as they saw it as the nature of a solar product that its services vary with the weather. This is a really important attitude of the users for securing satisfied customers to the product. The user interfaces seem to have been made successfully intuitive. The indicators of service (as shown on Figure 7) though gave rise to some confusion and some users tried to press them as a button to try to make the service available that was automatically had been disabled. The energy system was dimensioned well providing the services very similar to the calculated in the energy balance (Figure 6). The problem with the magnet switch was creating the biggest deviation from the calculations and that problem has to be solved before making it into a product.

4 CONCLUSIONS

A solar powered hub for charging laptops, phones, streaming/playing music and providing light for an outdoor table was researched, developed, prototyped and tested. 3 prototypes was constructed and tested at the Roskilde Festival 2017. The energy system was dimensioned well and providing the calculated services to the user when using solar irradiation data for a TMY in Denmark end of June where the hubs was tested at the Roskilde Festival 2017. The user survey showed that people accepted that the hubs services were not always available due to bad weather (the Roskilde Festival 2017 had very few sun hours this year). The 12V DC to 230V AC inverter had very high stand-by consumption and a rather low efficiency and a good solution for only turning this on when a connection is made to the plug is still to be found.

5 ACKNOWLEDGEMENTS

The work was supported by Innovation Fund Denmark under the Innobooster programme. The project was granted support for developing the Sun Hub solution in one grant and a special grant was giving to support testing it at the Roskilde Festival 2017. The Sun Hub is designed by Holscher Design, who was not a partner in the project but a very active subcontractor. The project partners will also like to thank the 6 students from the Technical University of Denmark who made the observations during the festival.

AFFORDABLE AND CLEAN ENERGY: ADDRESSING PROJECT DEVELOPMENT CHALLENGES OF UTILITY-SCALE SOLAR PV PLANTS

Simon Benmarraze[1], Carlos Ruiz[1], Roland Roesch[1]
[1]International Renewable Energy Agency (IRENA)
IRENA Innovation and Technology Center, Robert Schuman Platz 3, D-53175 Bonn, Germany
Corresponding author contact: sbenmarraze@irena.org

ABSTRACT: Despite a rise in installed Solar PV generation capacity worldwide, the deployment of utility-scale solar PV power plants can be challenging because some projects may not meet certain required standards to obtain financial support. Furthermore, there is a general perception of high technology risk, cumbersome administrative procedures, insufficient transparency, as well as limited access to financing instruments. In some regions, smaller project developers are especially exposed since the recent experience of developing utility-scale solar projects still largely was with large companies and utilities. To mitigate project failure, project development guidelines must become more easily available. This paper emphasizes the importance of the early stages of a Solar PV project development and how they are connected to reducing the risk of project failure. The paper addresses the key issues encountered during a typical development of a utility-scale solar PV project and provide key recommendations to develop a bankable solar PV project proposal.

Keywords: Renewable Energy, Solar PV, Project Development, Implementation, Stand-alone PV Systems, Plant, Performance, Economic Analysis, Marketing, PV Market, Project Financing, International Renewable Energy Agency, IRENA, Project Navigator

1 INTRODUCTION

The Paris Agreement adopted at the United Nations Climate Change Conference in December 2015 puts renewables at the forefront of the required energy transformation towards a low-carbon future. Among all renewable energy technologies, solar photovoltaic (solar PV) is a commercially available and reliable technology with a total installed capacity of 290 GW worldwide in 2016 and more than 70 GW added in 2016 alone [1].

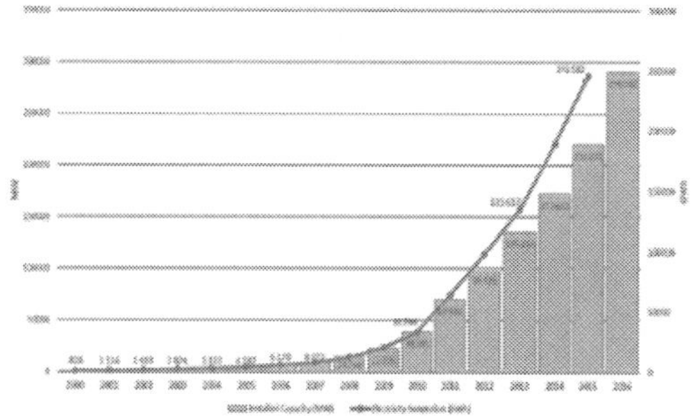

Figure 1: Global solar PV deployment chart

Even though solar PV is a mainstream renewable energy technology, many projects face obstacles during their development due to the unavailability of transparent guidelines, complicating the path of many project developers towards financial close and commercial operations. In its mission to promote the widespread adoption and sustainable use of solar PV energy technology, the International Renewable Energy Agency (IRENA) has developed the Project Navigator, an online platform providing comprehensive, easily accessible, and practical information, tools and guidance to assist in the development of bankable renewable energy projects.

The platform supports developers throughout the project development process by providing knowledge, tools and best practices to develop a successful utility-scale solar PV project, which generally can be identified by its size and equipment – typically starting at 5 megawatts (MW) with large inverters and a connection to a high- or medium-voltage grid. The project development process presented in this paper is based on the IRENA Project Navigator Technical Concept Guidelines for Utility-Scale Solar PV [2] composed of nine phases which range from the identification of a business opportunity to the execution of a given solar PV project:

- **Identification**: Analysis of local constraints by evaluating potential issues, opportunities as a basis for project development kick-off.
- **Screening**: Systematic protocol used to identify characteristics of various alternatives within a defined project scope.
- **Assessment**: Determination of the value of a given project and analysis of activities required to implement it.
- **Selection**: Decision-support process that benchmarks project ideas and selects the project configuration with the highest potential or priority.
- **Pre-development**: Establishment of project requirements and concept documentation, providing the basis for design and evaluation of the final project configuration.
- **Development**: Critical phase where requirements and activities are finalized in order to approach financial close with investors and lenders.
- **Construction**: On-site equipment and systems design, procurement and installation with qualified contractors. This phase demands high quality standards and extensive expertise in order to meet project performance targets.

- **Operation and maintenance**: After commissioning, the plant is operated and managed to maintain the highest level of availability. Preventive and correction maintenance protocols are applied in a quality environment to prevent any outages or failures.
- **Decommissioning**: At the plant's end-of-life, equipment is dismantled and eventually prepared for removal and recycling.

2 UTILITY SCALE SOLAR PV PROJECT DEVELOPMENT

The IRENA Project Navigator introduces a project lifecycle process structured in nine distinct phases designed to support the progressive development of renewable energy projects. These phases become increasingly more involved and detailed, commensurate with the requirements at that phase. The main aspects of each phase are described in detail, along with the minimum requirements for the development of a bankable project. In addition, practical frameworks, checklists, forms, models and various templates are provided. Each section provides insights on the key conditions and requirements to be fulfilled, mainly in terms of the technical suitability and bankability of the project to achieve financial close.

Phase 1. Identification Phase

The first step in the development of a solar PV power plant project is the identification phase. In this phase, potential locations are identified for the installation of the plant in a certain region or country.

Project developers must be aware of and have a clear overview of the different key stakeholders involved in a typical solar PV project. The key stakeholders that take part in a large-scale solar PV project under a power purchase agreement (PPA) model are represented in figure 1:

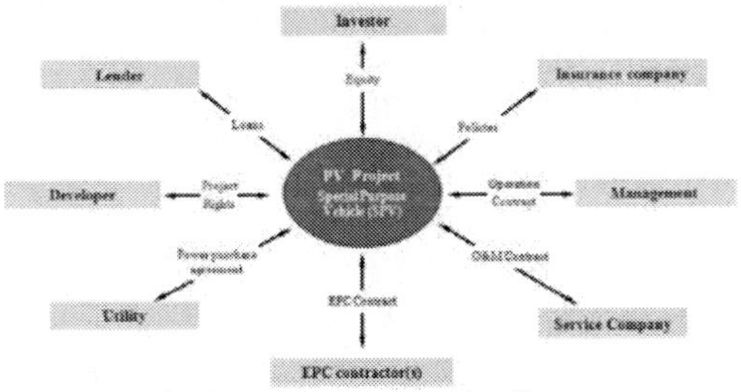

Figure 1: Solar PV project stakeholder diagram

Project developers will need to identify different sites and check their availability and suitability against a set of constraints, from the local up to the national level, including sensitive installations (e.g., environmental constraints, archaeological restrictions, military installations, etc.), which includes identifying the best site from both regulatory and socio-economic standpoints.
The project developer should be aware of the potential restrictions for the pre-selected sites (governmental, environmental and regulatory), investigate whether the sites generally look suitable from a technical point of view (e.g., size, topography, slope, ground conditions), clarify who are the land owners and whether they are willing to lease/sell their land, and identify all relevant stakeholders for each project opportunity. At the end of this stage, the project developer should have completed a list of potential project sites including location/co-ordinates, size, contact information of land owner. A template is available in the guidelines.

The methodology for identifying suitable sites proceeds under general criteria, such as:

- Governmental guidelines of the specific region or country must be considered.
- General identification and selection of sites for the implementation of large-scale solar PV power plants should be closely coordinated with the regulatory authorities and grid operator.
- Agricultural and former agricultural areas typically are suitable for solar PV power plants as these require substantially less preparation works. Nevertheless, other categories are possible (former industrial land, depolluted land, etc.) if no permit restrictions are in place. Unprotected forests also can be considered as potential locations for PV plants, notwithstanding the need of additional work for site preparation.
- Horizontal and level terrain is most suitable for solar PV installations, as uneven terrain is more prone to shading effects. Furthermore, uneven terrain requires more extensive consolidation and structure installation works. The slope profile of the terrain must not exceed 10%, and often a maximum of 6% is specified as an identifier for suitable areas.

Phase 2. Screening

In the second phase, all potential sites selected during the identification phase will be broadly examined with the objective of discarding options which are unlikely to be successful due to practical limitations. For this phase, the project developer can use a site screening checklist template available in the platform.

The key action of this stage is the screening of the identified sites in more detail. The project developer should identify potential "showstoppers" and how they would be resolved. At the end of this stage, the project developer should have completed site screening checklist completed for each site.

Project developers should be aware of certain situations, known as "showstoppers", that can lead to a stop in the development of a project once they are identified. These can happen throughout the entire development process, including the early stages. Potential showstoppers include the following:

- The land owner changes his or her mind and is no longer willing to sign an agreement for the land that was pre-selected for the solar PV project.
- The land which was initially identified turns out to be unsuitable for the selected technology or

- planned capacity (e.g., due to site complexity, slope, soil conditions, access to the site, etc.).
- The land initially identified has new large "exclusion areas" that the developer was not aware of at the beginning of the project. This can include archaeological discoveries, protected species on the site, protected areas in the vicinity, or other developments such as residential houses, streets or airports, from which a minimum distance is prescribed.
- The local authority does not approve the site for "industrial use" when it was classified previously as agriculture land.
- The geotechnical desk study shows that the site is located within a flood-prone area with periodical flooding.
- The cost of grid connection is prohibitive due to lack of planning (e.g., the need to cross streets, rivers or other obstacles, and/or long lead times for relevant permits), or lack of available capacity at the nearest substation (the cost of a new build and/or extending an existing substation could be high).
- The government changes the regulatory framework (e.g., lowering the feed-in tariff, new application requirements such as for expensive bank guarantees, tax, self-consumption).
- Unforeseen political instability in the region/country.

Phase 3. Assessment

In the third step, the assessment phase, the various areas or sites are compared and evaluated under technical and socio-environmental aspects. The project developer should assess whether the proposed site match technical viability criteria, and assess whether the proposed site match environmental criteria. At the end of this stage, the project developer should have completed a matrix of technical and socio-environmental site assessment for each site. For example, the resource potential of any given site should be assessed early on in the development of a solar PV project. The long-term average annual global horizontal irradiation (GHI) is of most interest to solar PV project developers as they indicate the expected future solar irradiation at a site.

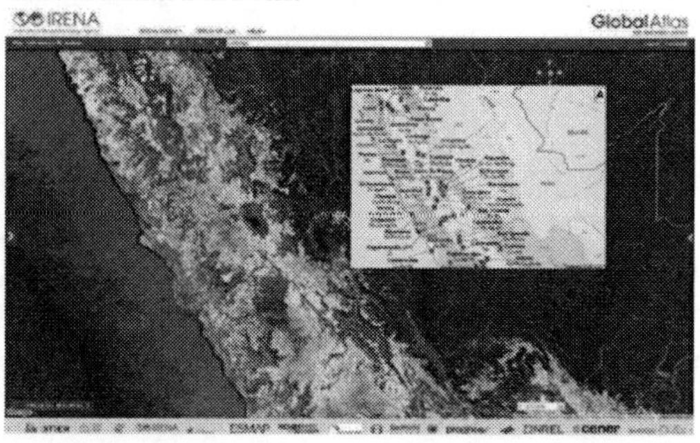

Figure 2: Example of map-based solar PV regional analysis using IRENA Global Atlas

With regard to the solar resource assessment, several tools and data available on the Internet can be applied in a first step. The following sources are free of cost and give an impression of the solar irradiation at site level:

- IRENA Global Atlas for Renewable Energy (http://globalatlas.irena.org/)
- US National Aeronautics and Space Administration (NASA) solar database (https://eosweb.larc.nasa.gov/sse/)
- Solar PVGIS: Photovoltaic Geographical Information System for Africa, Asia and Europe (http://re.jrc.ec.europa.eu/pvgis/)
- Solar Atlas published by several countries.

In addition to the technical suitability of the different sites, environmental and social issues should be considered. Therefore, some aspects should be considered to detect "red flag" issues that could put project implementation at risk or require stringent mitigation measures.

Phase 4. Selection

Based on the results from the assessment phase, the pros and cons of each area should be examined and ultimately the area or site with the lowest social environmental impact and the best technical viability should be preferred.

At this stage, the usual objective of site selection is oriented towards maximum output of the power plant at minimal cost and minimum socio-environmental impact. It also is possible that the objective is not the highest output but the highest installed capacity. This depends on the business model and is project-specific. It can be, for example, that an investor is paying the developer for a certain installed capacity and not for a maximized plant performance. With more installed capacity at a given site, the overall MWh produced per year will be higher, but the specific yield (kWh/kWp) will be lower. The project developer should make sure that the site selection is objective and independent, define final selection criteria list for site selection (e.g., maximum yield vs. maximum capacity), investigate any showstoppers, and evaluate whether the project development team has the required resources for further development. At the end of this stage, the project developer should have completed a table with sites for final project development.

Phase 5. Pre-Development

During this phase, major engineering studies are performed, including the basic design of the solar PV power plant. These design activities form the basis to prepare applications for permits, licenses and authorizations. There are many trade-offs that need to be made to achieve the optimum balance between performance and cost. Solar PV module technologies, technical design and factors, as well as degradation and warranty-related aspects.

Figure 3 provides an example of technical and financial activities to be performed during pre-development phase:

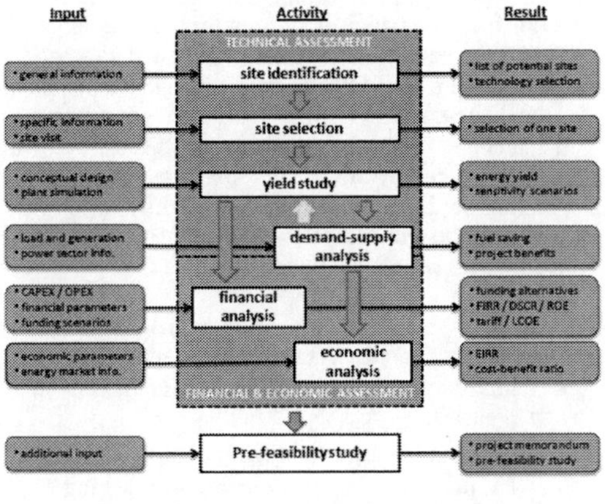

Figure 3: Project development activity flow diagram

The key activities of this stage are the evaluation of relevant power technologies (i.e. PV technology costs per module type and per supplier, suitability of primary components with regard to site-specific climatic and environmental conditions), the development of a sound plant technical design (i.e. preparation of grid connection assessment study, preparation of solar resource and yield study, execution of geotechnical and topographic surveys), the evaluation of the plant performance (i.e. detailed yield study highlighting assumptions and uncertainties based on verified location-specific solar resource files with calibration of satellite-derived meteorological data interpolated with ground-based measurements of at least one year) and the control of all permitting processes (i.e. local consultations with community stakeholders, follow up, tendering and execution of environmental studies).

For the technology evaluation, the project developer should obtain datasheets for PV module, inverter and other balance of system (BOS) equipment, confirm whether plant power ratings, efficiencies and losses in line with industry standards, obtain favorable commercial terms for major equipment in terms of guarantees, warrantees, verify major solar PV module characteristics (performance, voltage control, tolerance, temperature coefficients, degradation, low-light profiles), confirm characteristics with key suppliers (MPPT, voltage efficiencies), and produce technical design study confirming that all selected equipment are suitable for the project site's climatic conditions.

Poly-Si solar cells generally attain lower efficiencies in comparison to mono-Si cells. Mono-Si cells reach up to 25.6% conversion efficiencies, whereas poly-Si cells achieve 21.3% efficiencies on a laboratory scale [3]. Typically, poly-Si and mono-Si modules are fabricated with a rated power of 210-345 W, and their dimensions vary between 0.7 m^2 and 1.6 m^2. Solar PV modules suffer power losses because of different ageing processes occurring over time, such as delamination of layers, failures in cell interconnections and water intrusion. A degradation effect which is particular to crystalline modules occurs after several hours of sun exposure. This effect is called light-induced degradation (LID). For poly-Si, the power losses related to this effect are usually around 1% of the nominal power, while typical LID losses for mono-Si are around 2%.

Recently, a new kind of degradation has been seen, referred to as "potential induced degradation" (PID). PID is caused by leakage current in the solar PV module, also known as high-voltage stress (HVS). This is a loss of up to 80% of the nominal module power caused by leakage current at high voltages. In case the modules are not PID-free, it is highly recommended to adopt the use of inverters that allow (solid) grounding of the negative pole of the DC system to avoid or minimize the effect of PID.
Estimates for degradation figures found in the market for installation at moderate climatic conditions for c-Si modules are 0.25% to 0.8% per year after LID. On average, for current industry practice, annual degradation of the nominal power of crystalline solar PV modules is accepted as 0.5%, or 12.5% over the 25 years of the expected lifetime of solar PV systems. In a financial model, an average yearly power degradation of 0.5% is recommended for the lifetime of the solar PV power plant.

When selecting solar PV modules, the following key aspects should be considered to meet bankability requirements defined by investors and lenders:

- Appropriate performance objective of solar PV modules in terms of efficiency and supply cost with regard to land and associated BOS equipment requirements.
- Module power tolerance should be quoted "positive only", for example 0/+5 W and no longer the former market standard of +/-3%.
- The power warranty quoted by manufacturers should be linear over 25 years and not a stepwise guarantee.
- Product warranties of at least 5, 10 or even 12 years must be given by the manufacturer to cover product failures.
- Modules should be PID-free.
- If installed near the sea, the modules should have a salt mist corrosion test certificate (IEC 61701).
- Evaluation of the actual performance of a sample solar PV module should be performed by a third-party accredited laboratory.

The inverter may be transformer-less or include a transformer to step up the voltage. Inverters with transformers are galvanically isolated and can be used for grounding of certain solar PV modules that require such grounding, for example thin-film modules. Where transformer-less string inverters are installed, the efficiency is higher and the size and weight are reduced as well as cost, even if this is partially offset by the need for additional protective equipment. When operating in hot climates, inverters suffer de-rating at temperatures above 45-55 °C. As the temperature rises further, power output drops, for instance at 2% per °C. This means that the inverter temperature must be monitored and cooling might be required.

For the technical design, the project developer should produce technical design in line with best practices for tilt angle, orientation, inter-row distance, shading, string optimization, and civil works and security, discuss plant design with an independent engineer, public authorities and potential insurance providers, confirm that all requirements from the grid code for the inverter selection and sizing are met, confirm that plant design takes into account horizon and near-shading obstacles for the plant layout, identify potential restrictions on land availability and use together with local authorities, confirm availability of water supply and the associated regulatory requirements and costs, and double check that geotechnical studies cover all relevant aspects related to foundations and soil compaction.

The design typically should consider the following aspects:

- row distance to minimize internal shading and shading losses
- optimized low- and medium-voltage cable runs to minimize electrical losses
- internal access roads within the plant to assure efficient movement of vehicles for maintenance
- module inclination to optimize energy generation for the specific site by minimizing shading and optimizing area usage (in some projects the aim also can be to maximize the installed capacity, accepting somewhat more shading in this specific case)
- module orientation to maximize energy generation. Usually, this is to the south (azimuth = 0) in the northern hemisphere and to the north in the southern hemisphere.
- inverter voltage limits, MPPT range as well as safety and grid requirements.

For example, the spacing between module rows is a compromise to reduce inter-row shading while keeping the required area of the solar PV power plant within reasonable limits. Inter-row shading can practically never be reduced to zero: in the morning and in the afternoon, the shadows are very long. Figure 4 illustrates the different angles that must be considered in the design process.

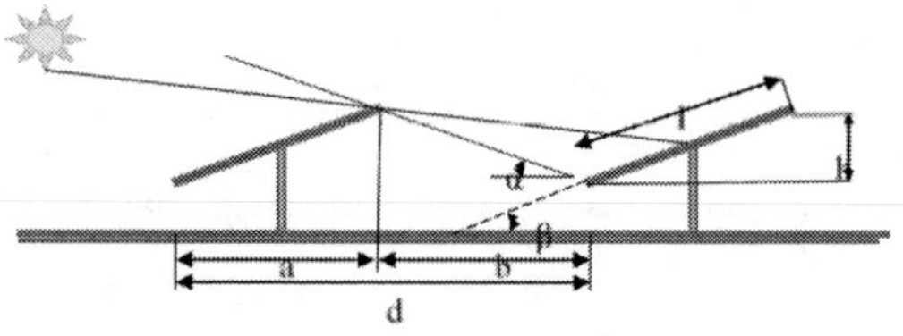

Figure 4: Shading angle diagram

The shading limit angle α is the solar elevation angle beyond which there is no inter-row shading on the modules. If the elevation of the sun is lower than α, then a part of the module will be shaded with an associated yield loss.

For the performance evaluation, ground-based solar radiation measurement is important for the development of solar PV projects because solar radiation data generated only by satellite sources introduce a bias in the estimation of the site-specific solar resource in terms of instantaneous, average, frequency distribution and statistical representability of solar radiation values. Depending on the approach, the uncertainty of long-term averages can be reduced below 5% for GHI using ground-measured data coupled with satellite data.

To increase the accuracy of the calculation, other information and parameters are needed, such as:

- long-term weather data with GHI and ambient temperature: TMY
- technical datasheets of main components: solar PV module, inverter, transformer, tracking system (if applicable)
- system layout and design
- loss factors: e.g., degradation, shading losses, cable losses, transformer losses, soiling losses, own/auxiliary consumption.

The key parameter for evaluating the performance and quality of a solar PV power plant is the performance ratio (PR). The PR is the ratio of total energy yield to the theoretically available energy, and it gives an indication of the quality of the installation and the weather conditions.

$$PR = \frac{AC\ Yield\ [kWh]}{Installed\ Capacity[kWp]x\dfrac{Irradiance\ on\ module\ plane\ \left[\frac{kWh}{m^2}\right]}{STC\ Irradiance\ \left[\frac{kW}{m^2}\right]}}$$

The PR is determined by measuring the solar irradiation on the module plane and the AC energy yield. Usually the ambient and module temperature also are logged to allow temperature adjustment of the PR calculation. During project development the PR is predicted using different modelling software. Usual PR values for new solar PV systems vary from less than 75% in summer to more than 82% in winter. This parameter identifies the overall impact of losses on the rated output. As it is normalised with respect to irradiation, this allows comparison of solar PV power plants in different locations. The higher the PR, the more efficient the solar PV power plant is, which means basically that more energy is obtained from solar irradiation.

The project developer should validate quality and accuracy of the meteorological data and make sure it includes long-term weather patterns at the site, use industry-standard algorithms to convert GHI values to tilted values at the surface of the solar PV modules, confirm that all components and the system layout are captured in the plant performance model, verify that all losses and uncertainties have been taken into account in the performance model including cable, soiling, auxiliary and grid availability losses, certify that prices in the cost estimation are realistic and complete, document all relevant investment and operations costs associated with the agreed performance level, confirm that the key parameters in the financial model reflect local market price and best practices, and design a financial model that covers the full life cycle of the project.

A thorough environmental study is required to obtain almost all authorizations for the development of a solar PV power plant. A distinction should be made between the need for an Environmental Impact Assessment (EIA) or an Environmental and Social Impact Assessment (ESIA), as per applicable national environmental legislation of the project country and the requirements of the international financing institutions, which may differ from the legislative requirements.

Therefore, the project developer should be aware of all requirements of an environmental impact assessment study, discuss with local authority to confirm the exact site land use (i.e., agricultural/industrial/landfill), obtain verifiable documentation on land ownership, and be prepared to provide any relevant supporting files for permit applications (including environmental studies, layout and preliminary drawings).

At the end of this stage, the project developer should have completed a list of selected equipment including datasheets, certifications and commercial terms, a set of project studies including topological, geotechnical, civil engineering fields with relevant plant and grid connection layouts, a validated solar resource assessment study and associated files, a detailed yield study with ratios and performance metrics, a draft financial model, all environmental studies including addendum requests by regulatory authorities to process permit applications.

The project developer also should discuss with stakeholders about project contingencies that are unknown during pre-development due to a lack of complete project definition and engineering. Estimating these contingencies relies on developers' and contractors' experience because they cannot be determined at the time the cost estimates are prepared. They typically include:

- Change in scope (i.e., adding more modules or change in foundation type)
- Labor productivity (i.e., accounting for seasonal effects or contracting approach)
- Procurement delays due to design changes or shipping issues
- Unforeseen cost escalation.

Phase 6. Development

During the final project development phase, all activities previously performed under multiple tasks and in different steps will be finalized with the aim of reaching financial close. The development phase, as analyzed in detail in the previous sections and provided as a guide, is the crucial part of the whole process during which final decisions are taken affecting the solar PV power plant performance.

The key activities of this stage are the finalization of all bankability requirements (i.e. completion of bankability checklist, coordination of technical and legal due diligence with independent advisors representing the financing parties), project contracting approach finalization (i.e. analysis of contracting terms with regard to project structure such as performance guarantee requirements imposed by lenders), contractual performance (i.e. negotiation of contractual agreements), financial modeling (i.e. validation of detailed financial model with financing parties) and risk management (i.e. finalization of insurance policies for construction and operation).

For the bankability concerns, the project developer should investigate whether there are any new requirements (permits, environmental, grid) to be considered, identify potential financial or economic incentives with relevant community stakeholders, agree with the lender's advisors on the final yield report and address any potential issues, and discuss with investors and lenders on the project bankability levels (feedback from technical and legal due diligence).

For the contracting final approach, the project developer should identify all risks and rewards of each contracting alternative, be comfortable with the complexity of the chosen contracting approach, and confirm that the contracting approach in line with your partners' expectations. At the end of this stage, the project developer should have completed Report on contract approach recommending the best approach to work with contractors

To manage contractual performance, the project developer should issue all tendering documentation to potential contractors with clear technical definitions, clarify contracts' scope of work been clarified with your contractors, agree with contractors and project lenders and investors on the level of performance bonds or guarantees, confirm project milestones (commissioning, Provisional Acceptance Certificate (PAC), Final Acceptance Certificate (FAC), etc.) well-defined and understood by the contractors.

For example, the definition of the project milestones and the completion deadline is a crucial part of the EPC contract. Project milestones usually are linked with interim payments that the plant owner has to pay to the contractor during the construction phase [4]. Upon milestone payment, the property of the corresponding part of the works is transferred to the owner, while the risk usually remains with the contractor until provisional acceptance of the plant. Figure 5 indicates the typical duration, milestones and guarantees of an EPC contract.

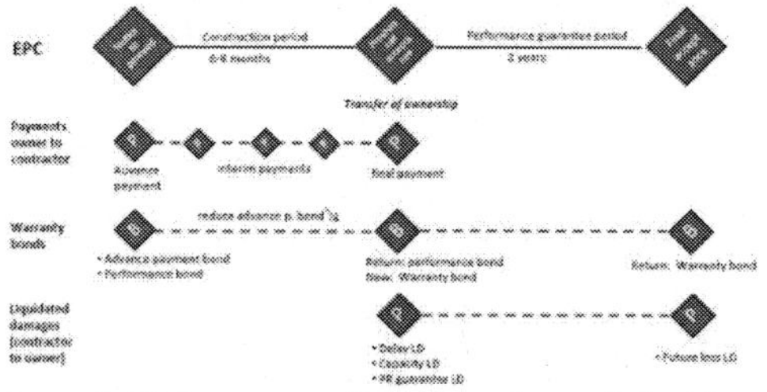

Figure 5: EPC contract milestones and completion dates

In addition, there is a need to verify that test protocols are well-defined and reflect industry best practices including performance ratio (PR) test protocols at different project milestones, confirm that contractor's liquidated damages are line with financial partners' expectations in case of non-performance, agree on a defects warranty period covering at least the period between provisional and final acceptance, consider potential consequences of delays in the project and are you prepared to cope with/manage them if necessary, confirm that the PPA payment terms are in line with the level of debt service required by lenders, review all provisions included in the PPA regarding minimum annual electricity generation, payment schedule and penalties in case of unavailability of the grid and subsequent curtailment [5], and align contractual terms across all agreements (loan, equity, debt, offtake).

To finalize the project financial model, the project developer should finalize the financing structure of the project, escalate discussions with investors and lenders regarding the final financial model version, confirm that all financial models converging to the same values, and provide to investors and lenders a clear sensitivity analysis for key parameters to be able to perform stress tests with P90/P99 profiles.

These profiles are produced based on simulation results, assumptions for module degradation, and uncertainty analyses. The results of the statistical analysis are the so-called P-cases (probability cases). The P50 value is the yield without considering and deducting any uncertainty. In this case the likelihood is 50% that the expected value will be exceeded, as shown in figure 6 (1 500 kWh/kWp). In the P75 case it is 75% probable that the value will be reached (and only 25% that it will not be reached). Consequently, the P75 value is lower than the P50 value.

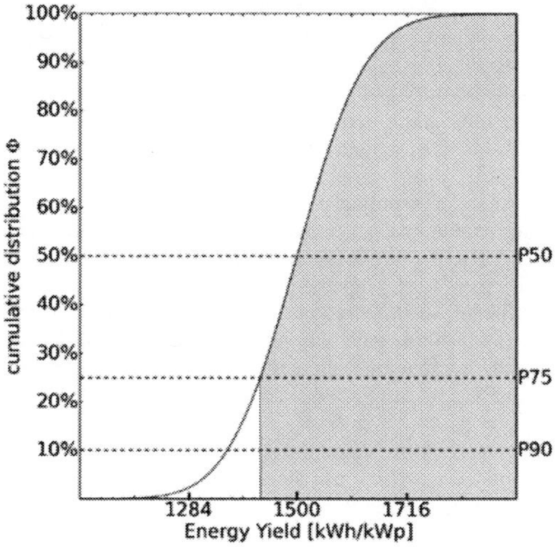

Figure 6: Probability cases cumulative distribution diagram

There are no general requirements for a minimum energy yield from lenders, but their minimum expectations for a return must be at least met or exceeded. The usual P-cases considered by lenders are P50, P75, P90 and P99. In addition to these P-cases, investors and lenders also apply so-called downside scenarios to consider potential overestimation of solar irradiation and electricity generation.

At the end of this stage, the project developer should have completed the final version of all engineering studies, permitting, authorization and licensing processes, a strong bankability report, a virtual data room with all required information from investors and lenders, project contracts and/or term sheet signed and all terms and conditions pre-agreed and a detailed financial model.

Phase 7. Construction

The construction phase demands high quality standards and extensive expertise to meet the main performance targets. The key milestones of this stage are the finalization of detailed technical documentation completion of master project planning production of project quality plan production of project health, safety and environment (HSE) plan setting up of project management structure application for any temporary permitting preparation of all procurement and shipping documentation clearance of all access and internal roads execution of project implementation plan and commissioning of the PV plant.

The project developer should obtain all required technical documentation for the installation of the solar PV plant project, supervise the construction schedule with contractors, and plan well in advance the final activities of the construction phase, i.e., tests and commissioning.
During the construction stage, the project team should have produced a project master plan, a project implementation report, monthly project progress reports, a provisional and final commissioning report, and prepared the receipt of provisional and final acceptance certificates.

Phase 8. Operations

It is important to define the conditions and parameters for O&M of a solar PV power plant during its service life. Proper maintenance of the solar PV power plant is essential for safe, efficient and reliable long-term operation. Consequently, a yearly O&M fee is to be considered in the financial model. There are two kinds of maintenance: preventive (or scheduled) maintenance and corrective (or unscheduled) maintenance. Preventive maintenance usually adheres to the requirements and recommendations of the component suppliers, whereas corrective maintenance is done in case of an unexpected failure in the solar PV power plant:

- Preventive maintenance – this is the result of activities scheduled before starting operation with the purpose of preventing faults or underperformance.
- Corrective maintenance – this is the result of activities performed during operation of the solar PV power plant due to unexpected failures.

Under the O&M contract, an "availability warranty" should be given by the O&M contractor to assure the performance of the solar PV power plant above an agreed value. The project developer should agree on sufficient liquidated damages to overcome any significant outage period, agree with the O&M contractor on an optimized approach to spare parts stock and replacement, agree on clear maintenance procedures following industry standard practices (ISO, VDI etc.), and review all provisions related with preventive maintenance activities to optimize the availability period of the plant. During this period, the project team should have produced and updated operation & maintenance procedures and manuals.

Phase 9. Decommissioning

End-of-life disposal of solar PV modules is a serious environmental issue. Because of the very long life span of solar PV modules (25-30 years), most modules installed today have not yet reached the disposal stage. Therefore, little experience with knowledge of solar PV waste management and recycling is available. In general, decommissioning involves three activities. The first is planning of the general dismantling and removal process, which covers the two other activities of waste management and recycling.

A conservative approach to consider the cost of decommissioning would be to set aside a part of the revenue of the last months of plant operation before decommissioning starts. Some developers have a riskier approach and do not explicitly require this in the financial model. They rely on the expected high value of materials because they can be recycled and sold.

The project developer should allocate an accrued reserve account for decommissioning, investigate all regulations in place regarding the dismantlement and recycling of each equipment of the plant, inform all relevant authorities of the works planning, establish transparent procedures for the dismantlement and removal of waste, and agree on who will earn the proceeds of any sale of materials or equipment.

3 REFERENCES

[1] IRENA (2017), Renewable capacity statistics 2017, International Renewable Energy Agency (IRENA), Abu Dhabi

[2] IRENA (2016), IRENA Project Navigator – Technical Concept Guidelines for Utility-Scale Solar PV, International Renewable Energy Agency, Abu Dhabi.

[3] Green, M. et al. (2016). Solar Cell Efficiency Tables (Version 47), Progress in PV: Research and Applications.

[4] Kerf, M. et al. (1998). Concessions for Infrastructure – A Guide to Their Design. World Bank Technical Paper No. 399. World Bank with Inter-American Development Bank. Washington, DC, pp. 160, 161, 163.

[5] World Bank (2008). Power Purchase Agreement. Public-Private Partnership in Infrastructure Resource Center for Contracts, Laws and Regulations (PPPIRC). World Bank. Washington, DC.

4 ACKNOWLEDGEMENTS

The authors would like to thank Fichtner GmbH & Co KG for their contribution to the preparation of the IRENA Project Navigator Technical Concept Guidelines for Utility-Scale Solar PV.

IMROVING THE COMPETITIVENESS OF SOLAR PV WITH ELECTRICITY STORAGE

Eero Vartiainen
Fortum Growth Oy
POB 1, FI-00048 Fortum, Finland, eero.vartiainen@fortum.com

Gaëtan Masson
Becquerel Institute
Rue d'Arlon 69-71, Brussels 1040, Belgium, g.masson@becquerelinstitute.org

Christian Breyer
Lappeenranta University of Technology
Skinnarilankatu 34, FI-53850 Lappeenranta, Finland, christian.breyer@lut.fi

David Moser
Eurac Research
Via G.Di Vittorio 16, I-39100 Bolzano, Italy, david.moser@eurac.edu

ABSTRACT: Solar photovoltaics (PV) is becoming the cheapest form of electricity generation in many sunny countries around the world. In some European countries with moderate annual solar irradiation, true competitiveness of solar PV has not yet been reached in all market segments. However, this competitiveness might be enhanced by using electricity storage. The challenge for PV generation is that it does not always match the electricity consumption. Especially in residential PV installations, a large share of the self-produced solar electricity cannot be immediately consumed locally, and is fed to the grid with very low remuneration. With storage, the share of PV self-consumption can be increased which could lead to a better business case for the prosumer. Currently, the average price of battery storage is still too high in most cases to make a PV system with storage competitive against PV without storage. However, in countries with high retail electricity price, like Italy and Germany, PV with storage is already becoming competitive. The continuing cost decrease of battery storage will soon make it a feasible option in more countries and will allow a higher share of PV generation in the energy system.

Keywords: LCOE, IRR, electricity storage, grid parity, prosumer, PV self-consumption

1 INTRODUCTION

Solar photovoltaics (PV) is already becoming the cheapest form of electricity generation in the sunny parts of the world, manifested in the record low PV power purchase agreements (PPAs) in several countries around the world [1]. In many European countries, PV electricity is already cheaper than electricity bought from the grid in all consumer segments [2]. However, in some European countries with moderate annual solar irradiation, true competitiveness of solar PV has not yet been reached. The main problem with PV generation is that it does not always match the electricity consumption. Especially in residential PV installations, a large share of the self-produced solar electricity cannot be immediately consumed locally, and needs to be fed in to the grid. Often the remuneration for the surplus feed-in electricity is very low compared with the full value of retail electricity including energy cost, distribution fees and taxes. This leads to a non-optimal business case for the prosumer. But with electricity storage, the self-consumption ratio of PV could be increased. Whether this increase will enhance the prosumer business case is the topic of this paper.

To determine the true competitiveness of PV without storage is relatively straight-forward. It can be done by comparing the levelised cost of electricity (LCOE) of PV with the average PV electricity value [3]. PV electricity value is the weighted average of the variable retail price of electricity which directly replaces consumption and the value of the surplus PV electricity fed into the grid. Only the variable part of the retail electricity price is considered, since for example the fixed annual grid fee cannot be avoided by self-consumption, as long as the prosumer is grid-connected.

The situation becomes more complex when the PV system is connected to an electricity storage. The main driver for the storage is the price difference between the retail electricity and feed-in electricity. This price difference has to cover the extra cost of the storage system in order to the storage investment be profitable. In essence, it becomes a sizing issue where the PV and the storage system must be optimised in relation to the electricity consumption.

For example, an unprofitable PV system with a low self-consumption ratio could become profitable with a storage system if the self-consumption increases a lot. On the other hand, a profitable PV system with relatively high self-consumption could become unprofitable if the increase of self-consumption with storage is too little. The form of the consumption profile also has an effect: if most of the electricity is consumed around midday, the storage has less impact compared with a situation where morning or evening consumption dominates.

To determine which is the most suitable system depends on which economic metric is used. One can, for example, maximise autarchy, minimise the combined LCOE of PV + storage, maximise the revenue relative to one's electricity bill, or maximise net present value (NPV) or internal rate of return (IRR). The IRR is a discount rate which makes the NPV zero. IRR is very commonly used and is especially useful in sensitivity analysis. There is also a difference between making a new PV + storage system investment or just adding a storage to an already existing PV system.

In the previous works by the authors [2-3], the year of true grid parity, i.e. when PV LCOE equals the average value of PV electricity was determined for various countries and market segments. This was done with the help of historical learning curves and future projections for the PV system component prices. The same approach could be used here for the PV + storage system. However, the battery storage learning curve is still at a relatively early stage and there are big uncertainties regarding the exact rate of the price reduction.

In this paper, IRR is used as the main metric for a PV system investment with and without storage. If the IRR, in real terms, is positive, the investment is considered competitive compared with a situation where it is not made and all electricity consumed is bought from the grid. Additionally, the IRRs for the PV and PV + storage system investments are compared to determine which is more competitive. As explained earlier, there can be many rationales for making such investments. The main focus of this paper is to analyse how much various parameters affect the profitability of each investment.

2 INPUT DATA

2.1 Electricity consumption and price

The main consumer segment studied here is the residential. A standard consumption profile for a German household (Standardlastprofil H0 - Haushalt) [4] by German Association of Electricity Industry is used. This is a daily profile with 15 minute time steps with slightly different values for winter, summer and spring/autumn days. Because the simulation model used here for the analysis has hourly time steps, the 15 minute values were averaged hourly. The standard household profile gives a 1 MWh annual electricity consumption which in turn is scaled up according to the value used in the simulation. Most hourly simulation cases were made for Berlin in Germany, but the same profile was used also for sensitivity analysis in Rome in Italy, Brussels in Belgium and Helsinki in Finland.

For the commercial or industrial case, a real measured hourly time series from a shopping mall in Sicily, Italy, with about 2500 MWh annual consumption was used. In addition, a constant load profile representing a continuous manufacturing industry was used for the sensitivity analysis.

To assess the value of the PV and stored electricity, average retail electricity prices gathered by Eurostat [5] were used. Retail electricity includes both energy, grid, and taxes and fees components. However, the fixed annual and monthly fees were deducted from the total retail price because they cannot usually be saved by own consumption [2]. Peak load shaving and subsequent savings in the fixed power fees were not considered here.

For the feed-in surplus electricity, 2016 average hourly spot market electricity prices were used with a typical 10% deduction for the administrative fee which is usually charged for the feed-in. No special feed-in tariffs or net-metering schemes were considered here, apart from hourly net-metering. The electricity prices used in this study can be seen in Table I.

Table I: Electricity prices used in the study, derived from Eurostat 2015 data

€/kWh	Residential	Industrial	Feed-in
Germany	0.273	0.105	0.026
Italy	0.266	0.149	0.039
Belgium	0.178	0.090	0.033
Finland	0.119	0.058	0.029

The reasoning behind country selection was that for moderate annual irradiation countries, Germany had high, Belgium average and Finland low electricity retail prices, whereas Italy represents a country with high irradiation and electricity price.

2.2 PV system

All input data related to the PV system is according to the recent ETIP PV report [2]. For example, the module prices have been projected according to the historical learning curve [6,7] and base case volume growth [2]. Table II summarises the CAPEX and OPEX prices used for all countries. It is true that these prices may differ from country to country according to the market situation, but in the long term they should converge in Europe.

Table II: CAPEX and OPEX prices for residential and industrial PV systems

	CAPEX (€/Wp)		OPEX(€/kWp/a)
	Residential	Industrial	
2017	1.35	0.81	19
2020	1.17	0.68	18
2025	0.97	0.55	16
2030	0.83	0.46	14

PV system lifetime is 30 years and power degradation 0.5% per year. For PV generation profiles, European Commission Joint Research Centre's EMHIRES dataset [8] was used. This data includes an hourly time series of capacity factors for years 1986-2015 in country, bidding zone, NUTS-1 and NUTS-2 level for most European countries. The country level dataset gave extremely and unrealistically low values for the Northern European countries like Finland. For this reason, the NUTS-2 level data was used here. For each city, the closest NUTS-2 area was looked up and then a year with an annual yield close to the average of years 1986-2015 was selected for each area. Table III shows the annual yields and comparison with Solargis [9] irradiation data for a 30 degrees tilted and south-oriented PV system for each city. It can be seen that the Performance Ratio (PR) is between 79% and 84% for each city which seems quite realistic [10].

Table III: Annual Solargis irradiation, EMHIRES NUTS-2 level yield and Performance Ratio for a south-oriented 30 degrees tilted PV system

	Irradiation (kWh/m^2)	Yield (kWh/kWp)	PR
Berlin	1240	1037	83.6%
Rome	1860	1496	80.4%
Brussels	1240	1022	82.4%
Helsinki	1160	924	79.7%

2.3 Storage system

The biggest uncertainty in this analysis concerns the

future development of the storage system prices. The market is still immature and currently there is not one price for a certain storage system. For example, the price range for battery systems in Germany in early 2016 was from 750 €/kWh to 2500 €/kWh [11]. In early 2017, Bloomberg reported that the global average for small-scale battery systems was $1175/kWh [12] which is close to 1000 €/kWh with current exchange rate. This is used as the starting point of the price development 2017 here.

The price learning curve for battery systems, especially Li-ion batteries, has been a topic of hot debate in the recent years. One of the first estimates made by Hoffmann [13] was a learning rate (LR) of 15-20% for Li-ion batteries. Recently, it has been established [14] that LR for residential Li-ion battery systems has been 12% ± 4%. On the other hand, LR for consumer electronics batteries has been as high as 30% ± 3% according to the same study. A learning rate of 15% is used here for the base case.

For the future price development curve, a market volume growth projection is also needed since the price will decrease 15% every time the global cumulative volume doubles. There is no consensus opinion on the volume growth or even the current market size, but one can always make an educated guess. A good example would be the growth in solar PV installations for the past decade. According to the data collected by SolarPower Europe [15], the global annual PV market had an average compound annual growth rate (CAGR) of 48% from 2006 to 2016. For the cumulative PV installations, the average CAGR for the same period was 47%. Therefore, it is not inconceivable if the average CAGR for cumulative stationary Li-ion battery systems were 50% at least until 2025. From 2025 to 2030, it is assumed that the average CAGR would be 30%, from 2030 to 2035 20%, from 2035 to 2040 10%, and after 2040 5%.

Assuming that the global cumulative stationary Li-ion battery capacity was around 5 GWh at the end of 2016, with the above described growth rates the cumulative shipments would grow to about 25 GWh in 2020, 200 GWh in 2025, 700 GWh in 2030 and 2900 GWh in 2040. It is important to distinguish between shipments and installed capacity because lifetime of the battery storage is relatively short and by 2040 a significant share of the market would be battery replacement investments.

The projected cumulative battery shipments would be about 0.5 kWh storage capacity per kWp of installed PV capacity by 2040 projected by the ETIP PV report [2] and the ratio would stay at about the same level after that. This sounds feasible when considering that an economic optimum for this ratio seems to be considerably higher [16].

Fig. 1 shows the price development for a residential battery storage system according to the above described volume growth and learning rate of 10%, 15% and 20%. As said before, 15% LR is the base case here.

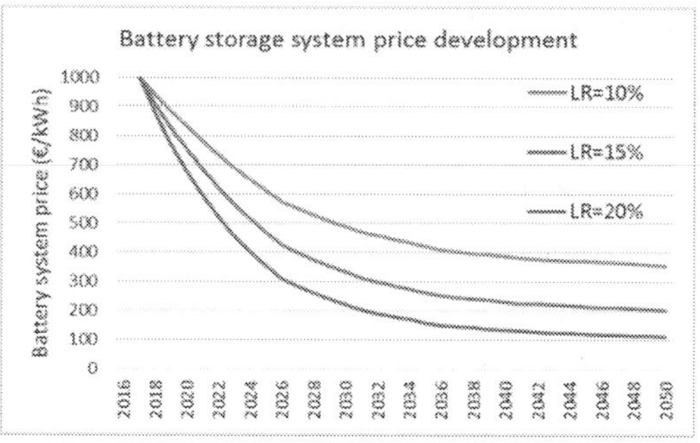

Figure 1: Residential Li-ion battery system price development projection for 2017-2050 with three different learning rates, prices in 2017 real money

As can be seen, the price would be about 50% higher with a 10% LR than with a 15% LR by 2030, whereas 20% LR would give almost 40% lower price. Of course, there is an added uncertainty from the volume growth projections. A 50% lower or 100% higher cumulative volume at a given time would obviously mean a difference of one volume doubling and therefore, about ±15% difference in price with a 15% learning rate.

In addition to price, the battery system lifetime is a very important parameter influencing the profitability of the storage investment. It is assumed here that the battery calendar life is 15 years. This means that the battery needs to be renewed once during the lifetime of the PV system. In all simulated cases, the storage was used for less than 4000 full cycles or 5000 80% deep cycles, since the average maximum usable capacity during the lifetime of the battery was set to 80% of the nominal capacity. This means that the end of calendar life is more probable than the end of cycle life for the battery systems simulated here. In addition, the round-trip efficiency of the storage was set to 80%, leading to a less than 5% loss of the PV generation in all simulated cases. And finally, the OPEX price for the battery system was set at 5 €/kWh per nominal storage capacity in 2017, and decreasing with the same rate as the PV system OPEX over time.

3 RESULTS

The base case for the analysis is a residential 5 kWp PV system in Berlin, Germany with a 5 kWh battery storage and 5 MWh annual electricity consumption. This base case is compared with a similar system with different PV system sizes ranging from 3 kWp to 10 kWp. Fig. 2 shows the comparison with current PV and storage CAPEX prices in 2017. The blue column gives the IRR for the PV system investment without the storage, red column for a PV + storage investment and gray column for a storage only investment to an existing PV system.

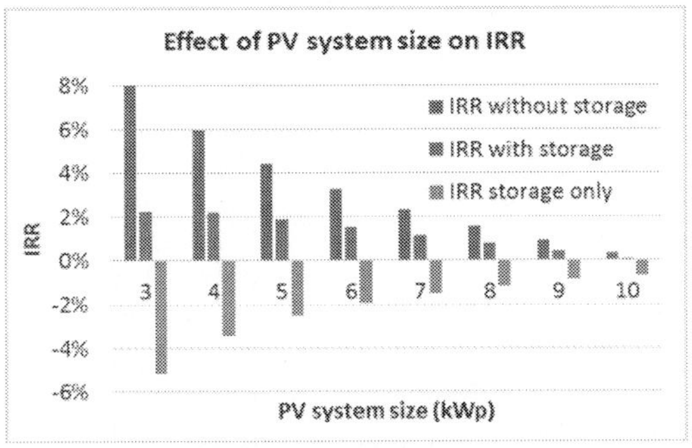

Figure 2: Effect of PV system size on the IRR for a residential PV system investment in Germany 2017 with and without a 5 kWh storage, electricity consumption 5 MWh/a, storage CAPEX 1000 €/kWh + VAT

It can be seen that the investment is profitable, i.e. the IRR is positive, for all PV system sizes from 3 to 10 kWp with or without storage. However, the IRR without storage is higher than with storage in all cases. Therefore, if one wishes to maximise the profit, a PV system without a storage is the better investment. If there already is an existing PV system, adding a storage gives a negative IRR which means that the investment is not profitable with current CAPEX and OPEX prices.

There is a big uncertainty concerning the current price and it would be interesting to know at which price the storage investment becomes more competitive than the PV investment. Fig. 3 shows the same comparison for a system with a storage CAPEX of 460 €/kWh + VAT, i.e. at about half the price of the base case. Now the 5 kWp PV system with and without storage gives the same IRR and the IRR for storage only investment is also the same. For larger than 5 kWp systems, the investment with storage is more profitable than without it. Moreover, all investments in storage only would now be profitable.

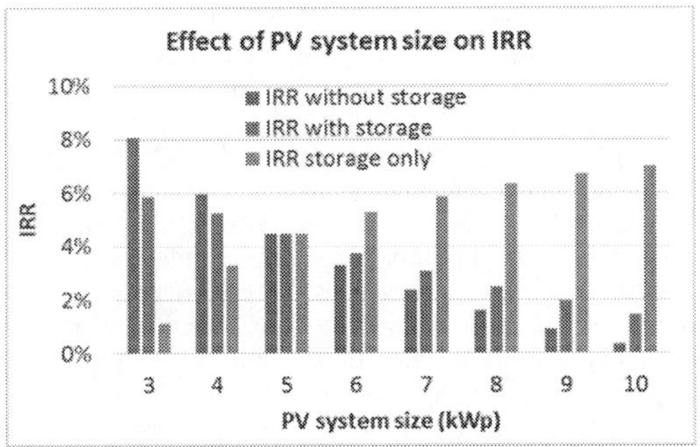

Figure 3: Effect of PV system size on the IRR for a residential PV system investment in Germany 2017 with and without a 5 kWh storage, electricity consumption 5 MWh/a, if storage CAPEX were 460 €/kWh + VAT

Unfortunately, a small-scale battery system turnkey CAPEX price below 500 €/kWh is not yet quite realistic. With 1000 €/kWh current price and 15% learning rate, 500 €/kWh price level would be reached around year 2025. However, by this time the price for a PV system will also have decreased, which changes the comparison a little bit. Fig. 4 shows the comparison with projected 2025 CAPEX and OPEX prices.

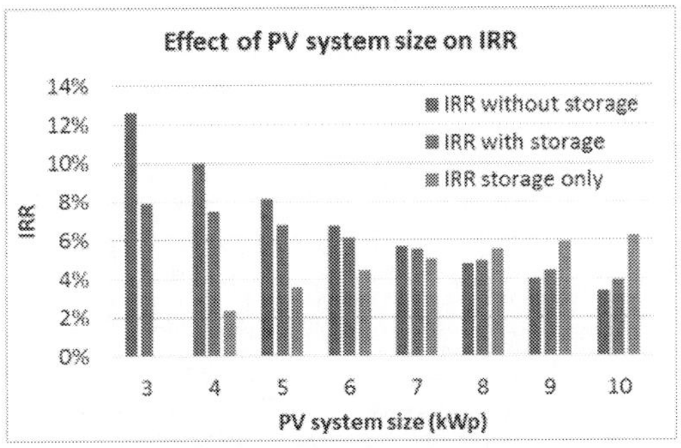

Figure 4: Effect of PV system size on the IRR for a residential PV system investment in Germany 2025 with and without a 5 kWh storage, electricity consumption 5 MWh/a, storage CAPEX 460 €/kWh + VAT

Now the investment with storage becomes more profitable than without it for larger than 7 kWp PV system sizes. When the PV generation increases, more surplus is created and the storage can be used more frequently. The same applies for the storage only investment. There is a slight decrease of storage only IRR compared with Fig. 3 because the storage replacement investment cost in Fig. 3 was halved from Fig. 2, whereas in Fig. 4 the replacement investment decreases according to the 15% LR and is thus a little higher than in Fig. 3.

Fig. 5 shows the comparison for a similar system in Rome, Italy. The variable retail electricity price is quite similar to Germany, but the PV yield in Rome is almost 50% higher than in Berlin. Therefore, more surplus PV generation is created with the same system size and the storage only investment starts to be profitable for larger than 5 kWp systems. However, PV without storage is still more profitable than with storage for all PV system sizes.

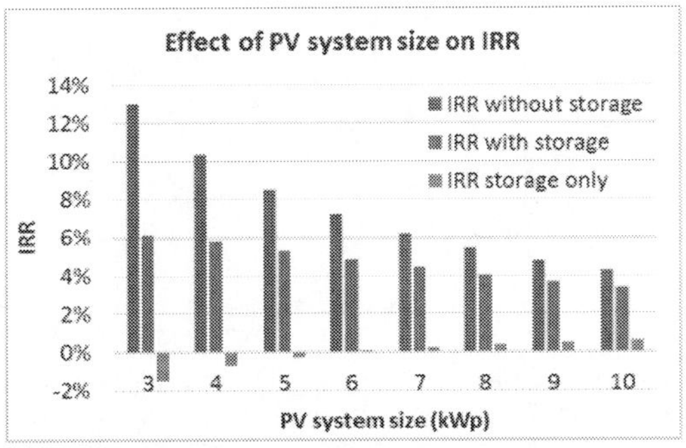

Figure 5: Effect of PV system size on the IRR for a residential PV system investment in Italy 2017 with and without a 5 kWh storage, electricity consumption 5 MWh/a, storage CAPEX 1000 €/kWh + VAT

Fig. 6 shows the comparison for Brussels, Belgium. PV investment without storage is already profitable with system sizes smaller than 7 kWp. However, PV+storage or storage only investment is not yet profitable. The variable residential retail electricity price in Italy and Germany is about 50% higher than in Belgium, which means that the storage system should be much cheaper in Belgium in order to compensate for the reduced revenue for the prosumer.

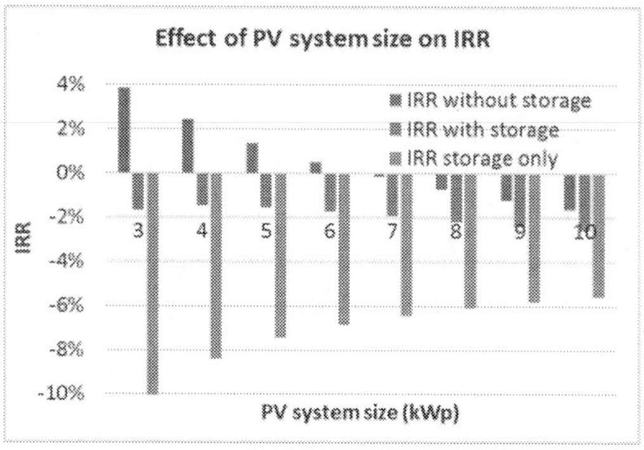

Figure 6: Effect of PV system size on the IRR for a residential PV system investment in Belgium 2017 with and without a 5 kWh storage, electricity consumption 5 MWh/a, storage CAPEX 1000 €/kWh + VAT

Fig. 7 shows the comparison for Helsinki, Finland. The storage only IRRs have been left out because they are more than 10% negative for all PV system sizes. Only PV systems smaller than 2.5 kWp and without storage are currently profitable in Finland with an annual 5MWh electricity consumption. Even that is arguable, since the PV system price per unit increases for the smaller systems and the base case price here is for 5 kWp. The variable residential retail electricity price in Finland is less than half compared with Germany or Italy. Therefore, it would take a very long time or much faster learning rate for the storage investments to become profitable in Finland.

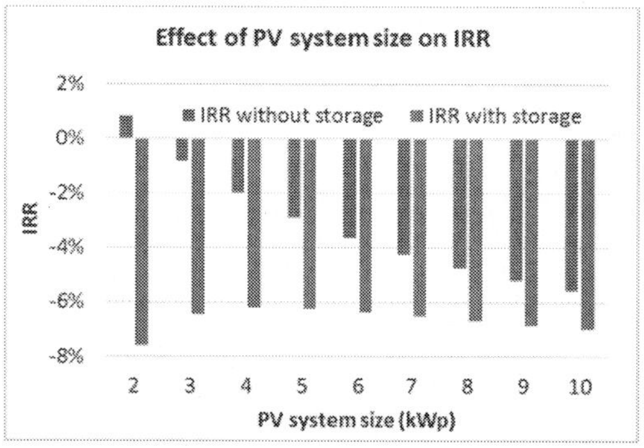

Figure 7: Effect of PV system size on the IRR for a residential PV system investment in Finland 2017 with and without a 5 kWh storage, electricity consumption 5 MWh/a, storage CAPEX 1000 €/kWh + VAT

4 SENSITIVITY ANALYSIS

Instead of varying the PV system size in relation to the storage size and electricity consumption, it is also possible to vary the other two main parameters. Fig. 8 shows the base case 2017 for Germany with fixed 5 kWp PV system size and 5 MWh annual electricity consumption with different storage system sizes. It can be seen that the PV+storage system is profitable for all battery sizes below 10 kWh. However, storage investment alone is not yet profitable with current prices.

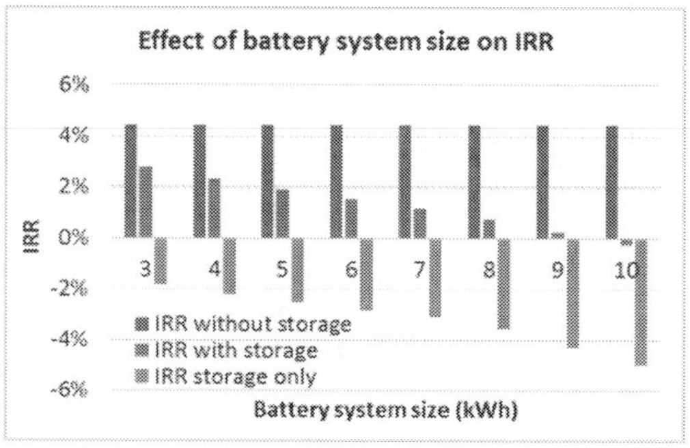

Figure 8: Effect of storage system size on the IRR for a residential 5 kWp PV system investment in Germany 2017 with or without the storage, electricity consumption 5 MWh/a, storage CAPEX 1000 €/kWh + VAT

Fig. 9 shows the comparison for the 2017 base case in Germany with fixed 5 kWp PV system and 5 kWh storage system size with different annual electricity consumption levels. PV+storage system is a profitable investment for higher than 3 MWh annual consumption, but storage investment alone is not profitable.

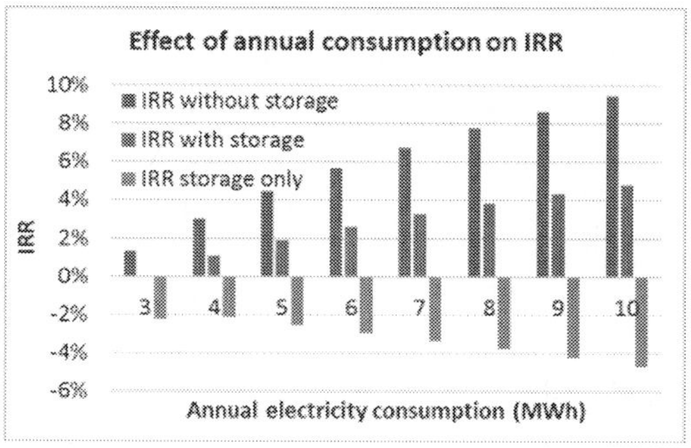

Figure 9: Effect of annual electricity consumption on the IRR for a residential 5 kWp PV and 5 kWh storage system investment in Germany 2017 with or without the storage, storage CAPEX 1000 €/kWh + VAT

Apart from the annual electricity consumption, the consumer segment and consumption profile also has some effect. Fig. 10 shows the comparison between different PV system sizes for an hourly consumption profile of a shopping mall in Italy with a 1 MWh storage capacity and 1 GWh annual electricity consumption. Because the storage is a commercial installation of industrial size, it is estimated at half the unit price of the residential system and without VAT. On the other hand, the variable commercial or industrial electricity price is also lower than the residential price.

It can be seen that the IRR with storage is quite close to the IRR without it for larger than 2 MWp systems, and storage investment alone starts also to be competitive with larger PV system sizes. However, when comparing a residential 5 kWp/5 kWh storage per 5 MWh consumption with the commercial 1 MWp/1 MWh storage per 1 GWh consumption, the storage alone investment is slightly more profitable for the residential case. A commercial system like a shopping mall has relatively more electricity consumption during daytime and thus benefits less from the storage.

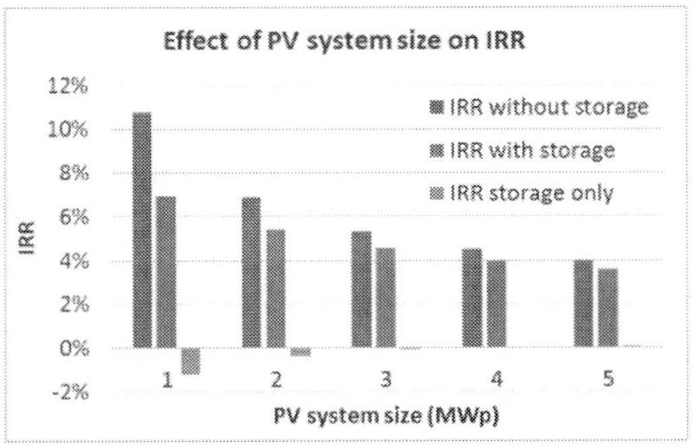

Figure 10: Effect of PV system size on the IRR for a large commercial PV system investment in Italy 2017 with and without a 1 MWh storage, electricity consumption 1 GWh/a, storage CAPEX 500 €/kWh

Fig. 11 shows the same case as Fig. 10 with a constant load profile which could represent for example a continuous factory manufacturing. As can be seen, there is not a big difference to Fig. 10 which means that the impact of the consumption profile could be smaller than one might think.

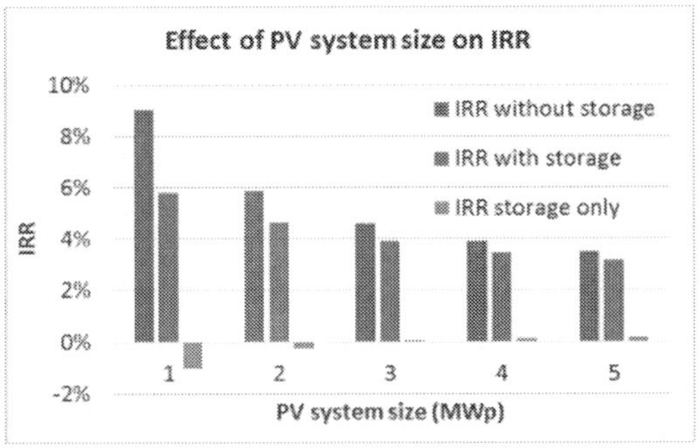

Figure 11: Effect of PV system size on the IRR for an industrial PV system investment in Italy 2017 with and without a 1 MWh storage, electricity consumption 1 GWh/a with constant load, storage CAPEX 500 €/kWh

Other parameters which have an effect on the IRR include the electricity retail and feed-in prices. In order to make a sensitivity analysis on the effect of various parameters on the IRR, the 5 kWp/5 kWh storage per 5 MWh consumption of Fig. 3 is used here. This base case for the sensitivity analysis, shown in Fig. 12, with 460 €/kWh + VAT storage CAPEX gives the same IRR for both PV, PV+storage and storage alone investments.

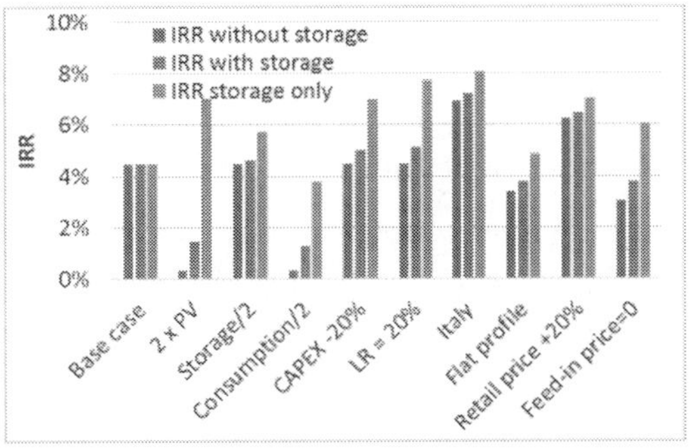

Figure 12: Effect of various parameters on the IRR for a residential PV system investment in Germany with and without a storage, PV system size in the base case (left) 5 kWp, storage capacity 5 kWh, electricity consumption 5 MWh/a, storage CAPEX 460 €/kWh + VAT

Doubling the PV system size obviously increases the surplus generation and thus makes the PV investment less profitable, but the storage investment becomes relatively more profitable. Halving the annual consumption has similar effect, but the absolute IRR for the storage alone investment decreases slightly because of less stored electricity. Halving the storage capacity makes more effective use of it and therefore increases the IRR.

Storage CAPEX has a big impact as does the learning rate. Using a 20% LR instead of 15% would reduce the CAPEX by about 25% by 2025. Increasing the yield increases the surplus generation, which means that the storage investment would be more profitable in Italy than in Germany if the same CAPEX and electricity prices and hourly load profile are used. Using a flat electricity profile instead of the standard German hourly household profile will shift the load more to night-time and therefore, improves the storage investment. On the other hand, the IRR for PV investment without the storage decreases because the standard profile has a better match with the PV generation profile.

Increasing the retail electricity price clearly improves the investment with or without the storage. If the value for the surplus feed-in electricity were zero, the IRR for storage alone investment increases because the difference between retail and feed-in prices increases. However, the PV investment in this case becomes less profitable because surplus generation has no value.

5 CONCLUSIONS

A residential PV system with battery storage is already a profitable investment in countries with high electricity price like Germany and Italy if the PV and battery capacity is sized sensibly in relation to the electricity consumption. In most cases, the PV system investment without a storage still has a higher IRR than a system with storage. Adding a storage to an existing PV system would already be profitable in Italy with relatively large PV systems generating a lot of surplus electricity.

Li-ion battery storage prices are still relatively high and must clearly come down in order to become competitive in more countries. With a moderate battery volume growth and 15% learning rate the battery CAPEX price should be at least halved by 2025. However, there is big uncertainty about the battery price and volume development because the market is still immature. Even current prices vary a lot depending on the project. It is feasible that the volume growth will be much faster than projected here, especially if one considers the Li-ion batteries needed for the growing electric vehicle market. Moreover, Li-ion batteries for consumer electronics have shown a much higher learning rate, up to 30% [14]. It is also possible that some new technology, e.g. flow batteries, would eventually become a more competitive storage option.

Internal rate of return (IRR) was selected here as the metric for analysing the profitability of the investment. One could also use other metrics, like net present value (NPV) or levelised cost of electricity (LCOE). Other than economic metrics such as autarchy or self-consumption ratio could also be used by the investor. The main benefit of electricity storage is probably not to improve the IRR of a PV investment but to allow larger PV systems and thus a larger share of renewable electricity in our energy system.

6 ACKNOWLEDGEMENTS

The work for this paper has been carried out under the framework of the EU PV Technology and Innovation Platform (ETIP PV). We also gratefully acknowledge Solargis for the irradiation data, European Commission Knowledge Management Unit, Directorate C Energy, Transport and Climate, Joint Research Centre for the dissemination of EMHIRES hourly capacity factors, Chris Werner for providing the German consumption profiles, and Benjamin Wilkin from APERE for providing Belgian consumption data.

7 REFERENCES

[1] IEA PVPS, Snapshot of Global Photovoltaic Markets 2016, Report IEA PVPS T1-31:2017.
[2] E. Vartiainen, G. Masson, Ch. Breyer, The True Competitiveness of Solar PV – a European Case Study, European PV Technology and Innovation Platform (2017).
[3] E. Vartiainen, G. Masson, Ch. Breyer, The True Competitiveness of Solar PV – a European Case Study, 32nd European PV Solar Energy Conference and Exhibition (2016).
[4] Verband der Elektrizitätswirtschaft e.V., Standardlastprofil H0 – Haushalt
[5] Eurostat database, ec.europe.eu/Eurostat/data (2016).
[6] ITRPV, International Technology Roadmap for Photovoltaic, Eighth Edition, 2016 Results, March 2017.
[7] I. Theologitis, G. Masson, Potential for Cost Reduction of PV Technology – Impact of CHEETAH Research Innovations, 31st European PV Solar Energy Conference and Exhibition (2015).
[8] I. Gonzalez-Aparicio, T. Huld, F. Careri, F. Monforti, A. Zucker, EMHIRES dataset - Part II: Solar power generation. European Meteorological derived HIgh resolution RES generation time series for present and future scenarios, Part II: PV generation using the PVGIS model; EUR 28629 EN; doi: 10.2760/044693, JRC Science for Policy report (2017).
[9] Solargis database (2017)
[10] N.H. Reich, B. Müller, A. Armbruster, K.Kiefer, W.G.J.H.M. van Sark & Ch. Reise, Performance ratio revisited: Are PR>90% realistic?, Proceedings of the 26th European Photovoltaic Solar Energy Conference and Exhibition (2011).
[11] C. Werner, Private communication on market overview of battery prices in Germanybased on real offers, Chris Werner Energy Consulting, June 2, 2016.
[12] PV magazine 3/2017.
[13] W. Hoffmann, Importance and evidence for cost effective electricity storage, Proceedings of the 29th European Photovoltaic Solar Energy Conference and Exhibition (2014).
[14] O. Schmidt, A. Hawkes, A. Gambhir, I. Staffell, The future cost of electrical energy storage based on experience rates, Nature Energy, Vol. 2, Article Nr 17110 (7/2017).
[15] SolarPower Europe, Global Market Outlook For Solar Power 2017-2021 (2017)
[16] J. Weniger, T. Tjaden, V. Quaschning, Sizing of residential PV battery systems, Energy Procedia, 46, 78-87 (2014).

CROWDFUNDRES: A NEW OPPORTUNITY FOR FINANCING RENEWABLE ENERGY PROJECTS

TOPIC 7: PV ECONOMICS, MARKETS AND POLICIES
SUB-TOPIC 7.1: PV ECONOMICS AND MARKETS

Author for correspondence
Silvia Caneva, WIP – Renewable Energies, Sylvensteinstr. 2, 81369 Munich, Germany
Phone: +49 89 720 12 733, E-mail: silvia.caneva@wip-munich.de

Co-Authors
I. Weiss, M. Papapetrou & P. Alonso[1)], O. Gajda & K. Kohl[2)], A. Bergmann & B. Burton[3)], T. Aschenbeck-Florange, A. Dlouhy & T. Drefke[4)], A. de Ferrari & M. Martinoli[5)], J. Wahlmüller & S. Egger [6)], T. Harwood, R. van Maaren & K. Harder [7)], S. Müller-Windisch & V. Daoud Henderson[8)], A. Raguet & M.-V. Gauduchon [9)], L. Pulles, M. de Jong, E. Hünewaldt & S. van Beurden [10)] C. Arnaud, M. Papoutsi & A. Roesch [11)], S. Wannop, A. Gregory &D. Crockford 12), D. Maguire, L. Clifford & R. Kelly [13)] and C. Rumolino & F. Petit [14)]

[1)] WIP – Renewable Energies, Germany, [2)] European Crowdfunding Network, [3)] University of Dundee, [4)] Osborne Clarke, [5)] youris.com, [6)] GLOBAL 2000, [7)] Abundance, [8)] Green Crowding, [9)] Lumo, [10)] Oneplanetcrowd, [11)] Solar Power Europe, [12)] Renewable Energy Generation, [13)] BNRG Renewables and [14)] VALOREM

ABSTRACT

We are currently seeing a deceleration of renewable energy growth in Europe. This is partly attributed to the challenges for financing renewable energy projects. Reduced access to conventional financing options over the past few years has triggered innovative financing schemes to emerge, with crowdfunding attracting a lot of attention. The project CrowdFundRES recognises the vast potential of crowdfunding for financing renewable energy projects and brings together the three following target groups: renewable energy project developers, citizen investing their savings in renewable energy projects and crowdfunding platforms acting as intermediaries between the public and the project developers.

The overall objective of the CrowdFundRES project is to contribute to the acceleration of the renewable energy growth in Europe by unleashing the potential of crowdfunding for financing renewable energy projects. In order to achieve this, the work has been structured for achieving the following objectives:

- Gain a deep understanding of the public's perception of crowdfunding
- Analyse the challenges faced by the application of crowdfunding for renewable energy projects in Europe
- Develop guidelines that support accepted practices in crowdfunding renewable energy projects
- Apply the guidelines and review them based on practical experience
- Develop policy recommendations to improve the market and regulatory framework
- Promote the crowdfunding concept and its advantages among those who could contribute or raise funds

The CrowdFundRES project is a European project, co-financed by the European Commission under H2020 Programme. The project started on 1st February 2015 and will end on 31th January 2018. It involves 14 partners across 6 different European Countries representing the relevant stakeholders in the field of crowdfunding for renewable energy projects. The European Crowdfunding Network and four of the most active crowdfunding platforms in RES project in Europe are involved: Abundance, Green Crowding, Lumo and OnePlanetCrowd. 3 project developers REG, BNRG Renewables and Valorem, are also involved in the project. The consortium is completed by Solar Power Europe, the University of Dundee, Osborne Clarke (multinational law firm expert in Crowdfunding), GLOBAL 2000 (association which represents developers of projects with high untapped crowdfunding potential and a network of citizens sensitive to environmental issues) and youris.com, leader in the dissemination and communication of the leading-edge European innovation via TV media and the web. The project is coordinated by WIP – Renewable Energies.

The EU PVSEC will represent a unique opportunity to get informed about the results achieved during the CrowdFundRES project in terms of innovative business models for projects financing and, at the same time, providing instruments to increase the awareness and public acceptance of renewable energy projects.

Detailed information on the CrowdFundRES project is available at the project website: http://www.crowdfundres.eu

Keywords: public acceptance, awareness campaigns, financing, innovative business models, crowdfunding, policies

1. PURPOSE OF THE WORK AND APPROACH

The overall aim of the CrowdFundRES project is to contribute to the acceleration of the renewable energy growth in Europe by unleashing the potential of crowdfunding for financing renewable energy projects.

> *"Crowdfunding can be defined as a collective effort of many individuals who network and pool their resources to support efforts initiated by other people or organizations. This is usually done via or with the help of the Internet. Individual projects and businesses are financed with small contributions from a large number of individuals."*
>
> Source: K. De Buysere, O. Gajda, R. Kleverlaan, D. Marom, "A Framework for European Crowdfunding"

In order to achieve the overall objective, the work has been structured for achieving the following objectives:

1. Gain a deep understanding of the public's perception of crowdfunding
2. Analyse the challenges faced by the application of crowdfunding for renewable energy projects in Europe
3. Develop guidelines that support accepted practices in crowdfunding renewable energy projects
4. Apply the guidelines and review them based on practical experience
5. Develop policy recommendations to improve the market and regulatory framework
6. Promote the crowdfunding concept and its advantages among those who could contribute or raise funds

This paper provides a detailed description of the results achieved until now in the project and on the results to be achieved in the upcoming months and to be presented at the EU PVSEC in Amsterdam, The Netherlands, in September 2017.

2. SCIENTIFIC INNOVATION AND RELEVANCE

The growing difficulties of renewable energy project developers to finance their project using the conventional avenues have led to the deceleration of renewable energy growth. This is putting significant pressure on independent RES developers, from which only the well-capitalised ones are able to maintain control of their projects at the pre-construction stage, most often by increasing their own equity participation in their projects[1]. There is a need for making use of innovative financing options, such as crowdfunding, that will cover some part of the increased funding requirements. This alternative financing method shows a vast potential for financing renewable energy projects and can help bridging the existing funding gap.

[1]Taylor Wessing, "Bridging the Funding Gap: The financing challenge for European cleantech and renewable energy", 2010

According to the 2nd European Alternative Finance Industry Report, the European alternative finance market as a whole, covering a range from equity-based crowdfunding to peer-to-peer business lending and from reward based crowdfunding to debt-based securities, grew from €2,833m in 2014 to €5,431m in 2015[2].

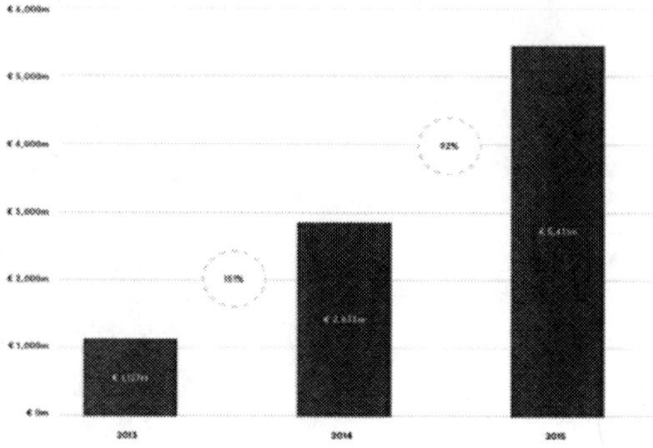

European Online Alternative Finance Market Volumes

The fast development of crowdfunding in Europe has attracted the interest of the European Commission and the European Parliament in this relatively new form of financing. The Parliament provided an overview of the current status of crowdfunding in Europe in the briefing 'Crowdfunding in Europe: Introduction and state of play' published in January 2017. Moreover, the Commission published a working document 'Crowdfunding in the EU Capital Markets Union', which discloses the potential of crowdfunding as a key source of financing for SMEs over the long term. However, the report states that crowdfunding remains relevant in a national domain with limited cross-border activity despite the fast growth of the crowdfunding sector at EU level. Even so, given the upward trend, the European Commission recognises the need of monitoring the market and regulatory developments in the crowdfunding sector. For that purpose, the Commission will maintain regular dialogue with European Supervisory Authorities, Member States, and the crowdfunding sector to promote convergence, sharing of best practice and keep developments under review.

The revision of the energy EU regulatory framework is being currently discussed after the publication of the Clean Energy for all Europeans package. The revision of the renewable energy directive, and the electricity regulation and directive will notably impact the further development of renewables and the way citizens and communities will engage in the energy transition. Such revision will have an impact on the business cases, regulatory stability, capital costs, and administrative burdens for renewables projects in the next decade.

Crowdfunding can play an important role in supporting the effective renewable energy growth in Europe.

[2]B. Zhang, R. Wardrop, T. Ziegler, A. Lui, J. Burton, A. James, K. Garvey, "Sustaining Momentum: The 2nd European Alternative Finance Industry Report", September 2016

Key to further support this trend is harnessing the fast-growing crowdfunding market where projects and businesses are financed through contributions from a large number of sources rather than through just a few or one.

The idea of CrowdFundRES is to exploit the potential of crowdfunding for filling part of the funding gap that RES project developers have been experiencing by bringing them closer to citizens who are willing to participate as contributors through the use of online crowdfunding platforms. While the idea of reaching out to the crowd for solutions or for collecting funds isn't new, the phenomenon of using the internet to directly connect with funders has emerged only recently. However, this approach is becoming more relevant as information technologies have been developing and market structures have been evolving.

CrowdFundRES aims to exploit the potential of crowdfunding for financing renewable energy projects by bringing together crowdfunding platforms, renewable energy project developers and the general public, identifying and addressing limiting factors and promoting more effective practices among them. This would support further market integration of renewable energy by securing, at acceptable terms, the financing for renewable energy projects that might have difficulties in using conventional financing options; for example when the profile, expected returns or scale of the project do not match the portfolio requirements for bank financing, or when the applied technology is relatively new/complicated and the perceived risk is too high.

Moreover, when members of the public become investors in renewable energy projects the sense of ownership grows and it contributes to changing mentalities, reducing public opposition that is sometimes associated with renewable energy developments.

For that reason, despite project developers often having access to conventional financing at good terms, they still choose to cover part of the investment through crowdfunding, usually targeted at the local population.

There have been several other projects with a focus ranging from community-owned energy projects with the involvement of cooperatives and communities to the creation of an online aggregating platform for renewable energy projects. CrowdFundRES brings together for the first time all kinds of project developers (incl. traditional RES developers, cooperatives, ESCOs, venture capitalists, etc.) with all citizens in a broader sense (e.g. without excluding those from urban areas who have the largest potential as contributors or even citizens from other countries). Also, the project works rather as an enabler for the creation of new crowdfunding platforms and the use of existing ones for RES project development than just an aggregator of projects hosted in existing platforms.

3. RESULTS AND CONCLUSIONS

Six main objectives have been identified in order to achieve overall aim of the CrowdFundRES project in contributing to the acceleration of the renewable energy growth in Europe. For each of these objectives an overview on the results achieved until now or on the results to be achieved in the upcoming months is provided.

Gain a deep understanding of the public's perception of crowdfunding

CrowdFundRES has performed a qualitative and quantitative analysis through surveys offering a unique insight on the elements of crowdfunding that are positively or negatively received by the public. It has also highlighted how the results of this analysis change depending on geographical or other socioeconomic classifications of the public. More specifically, the surveys analyzed perceptions about crowdfunding for renewable energy with the first one focusing on citizens' perceptions as potential contributors, the second one examining the views of crowdfunding platforms and the third one those of RES project developers. The findings provided a detailed picture of the current state of opinion across the EU regarding the present state and future potential of the sector. It was done by conducting a more detailed examination of contributors' preferences on types of investments. Geographical or socioeconomic aspects in the surveys were further analyzed where the sample size allowed.

The surveys have fed into the formulation of guideline insights for platforms and developers as well as have contributed to the policy recommendations report, published in July, 2017. They will be open until the end of the project when the additional collected input will be studied to see if there are any useful conclusions that can be drawn.

Analyse the challenges faced by the application of crowdfunding for renewable energy projects in Europe

All practical experience of crowdfunding for RES projects has been evaluated in close cooperation with the involved parties in order to identify challenges and limiting factors. Also within the project, crowdfunding of new RES developments has be undertaken and close monitoring provided useful input to the analysis on the challenges and their classification according to technologies, size of projects and differences in the regulatory and market frameworks. The work done until now has helped to analyze the challenges faced by the application of crowdfunding for renewable energy projects in Europe from a regulatory, market, socioeconomic and technical perspective

Develop guidelines that support accepted practices in crowdfunding renewable energy projects

The improved understating of the public's perspectives and the identification of challenges for RES developers and platforms have be used to develop three sets of guidelines that will support the wider and more effective development of crowdfunding.

The set of guidelines includes: guidelines for RES project developers interested in crowdfunding, guidelines the crowdfunding platforms interested in hosting RES projects and guidelines for investors in Clean Energy Projects via Crowdfunding.

Apply the guidelines and review them based on practical experience

The guidelines are being used within CrowdFundRES for crowdfunding new RES projects and the practical experiences of applying the guidelines will provide feedback for improving them. Additional experience will be gained by using the guidelines in supporting the creation of new platforms.

Develop policy recommendations to improve the market and regulatory framework

The framework conditions in all EU-28 Member States has been reviewed and the project has developed policy recommendations that will help shape the regulatory and market framework to foster crowdfunding for renewable energies. The improvement expected by supporting these measures will include: common power market design to overcome cross-border barriers, legal definition for crowdfunding models and market actors, promotion of increased integration of renewable electricity generation into markets, advance energy efficiency, cleanliness and performance to achieve EU climate goals and Implementation of Energy Union governance rules.

In this context, aimed at overcoming cross-border barriers for investments, Lumo and Oneplanetcrowd have successfully developed a solution that gives French and Dutch investors alike the opportunity to co-invest in a solar park in Torreiles (France). The success of this unprecedent collaboration shows that EU citizens are enthusiastic about the new investment opportunities offered by cross-border crowdfunding.

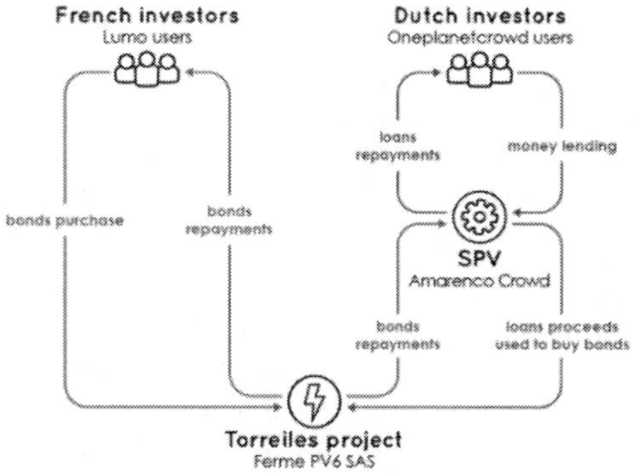

Torreiles project crowdfunding scheme

Regulatory authorities and policy makers around Europe can therefore play a decisive role in establishing a harmonized legal framework amongst EU that enables cross-border investments. Supporting cross-border crowdfunding would help boost renewable energy projects and bring us a step closer to a more united Europe.

Promote the crowdfunding concept and its advantages among those who could contribute or raise funds

More than 42.000 stakeholders including crowdfunding experts, different types of renewable energy developers and professionals, NGOs, environmental organisations, policy makers and the wider public have been reached through the projects dissemination activities in the first half of the project. More specifically this was achieved by press releases, newsletters, leaflets, visits to the website, stakeholders reached on LinkedIn through the "Crowdfunding for Renewable Energy" and "Energy Crowdfunding" groups, users viewed the project videos on YouTube, events where the project was represented, stakeholders and citizens reached the surveys. If the 136.879 accounts reached through the #CrowdFundRES twitter discussion group are also taken into account, more than 170.000 individuals have been reached through the promotion campaign and the number of our reach is expected to increase in the second half of the project.

ACKNOWLEDGEMENT

The CrowdFundRES project is a European project, co-financed by the European Commission under the H2020 Programme. The project started on 1st February 2015 and will end on 28th February 2018 .The logo of each partner is shown below.

DISCLAIMER

The sole responsibility for the content of this publication lies with the authors. It does not necessarily reflect the opinion of the European Commission. The European Commission is not responsible for any use that may be made of the information contained therein.

GROWTH REGIONS IN PHOTOVOLTAICS 2016
UPDATE ON LATEST GLOBAL SOLAR MARKET DEVELOPMENT

Ch. Werner[1], A. Gerlach[2], Ch. Breyer[3], G. Masson[4]

[1] Chris Werner Energy Consulting, 06846 Dessau, Germany
Phone: +49 (0) 176 64093993, e-mail chris.werner@cw-energyconsulting.de
[2] Alexander Gerlach New Energy Consulting, 99755 Ellrich, Germany,
[3] Lappeenranta University of Technology, Skinnarilankatu 34, 53850 Lappeenranta, Finland
[4] Becquerel Institute, 69-71 Rue D'Arlon, 1040 Brussels, Belgium

ABSTRACT
The installed capacity of photovoltaics (PV) shows a constant growth. The major part of PV is installed in highly developed countries. Further, there is an increasing number of emerging markets that show high growth rates of cumulative and annual PV installations. However, reliable installation rates for PV are available only for a small number of countries. Not only in emerging markets, even for some developed countries exact numbers of cumulative PV installation are rarely available. For the end of 2016 IRENA reports 2900.98 GWp of globally installed PV capacity, giving clear national specific data for 147 countries. IEA PVPS provides a number of 303.00 GWp installed by providing detailed data for 30 countries.
This paper gives an overview of installed PV capacities for all countries in the world based on the examination of publically accessible data. Furthermore, an analysis of the development of cumulative PV installations in recent years is given. Resulting from this evaluation, PV installations are localized in 196 countries, representing 305.78 GWp.

Keywords
PV Markets, Installed Capacity, PV Installation, Market Growth

1 MOTIVATION

The global photovoltaic (PV) installations constantly show high growth rates [1]. A variety of countries has reached large cumulative PV installations contributing relevant shares to their annual power generation. Further, numerous markets that today are not among the leading countries per cumulative or annual PV installations, show a high potential to become the new growth markets of photovoltaics. Since reliable and officially traced data on PV installations are only available for a small number of countries, public reports do not usually show any data for installed capacities in those new and emerging PV markets. These markets are usually named as Rest of World (RoW). However, they often show a constant growth of installed PV capacity. With rising photovoltaic capacities and growing economies in these markets, they are about to become important photovoltaic markets in the years to come. Thus, RoW markets have to be better understood.

As a result, an analysis of cumulative installed PV capacity for all countries in the world is strongly required. This paper provides an analysis of photovoltaic capacities and growth rates ranging from 2009 to today for every market globally. Further data starting in 2001 are available, but exceed the scope of this paper. On this basis, an estimation of future growth rates and PV capacities expected to be installed in all markets can be calculated.

2 EXAMINATION METHOD

For numerous PV markets no reliable estimation of installed PV is available due to a lack in public authority monitoring. This is especially valid for emerging markets, but a number of developed countries as well. To enhance a reliable valuation of the photovoltaic markets of all countries, the examination of several sources is necessary to ensure a credible estimation.

A basic source is the international customs database, monitored by the 'Market Analysis and Research' section of the International Trade Centre, which is an agency of UN's World Trade Organization. This database contains the customs data of all countries worldwide since the year 2000, receiving the data from the particular national customs authorities. The database provides the opportunity to evaluate the total value of imported and exported products for each country per year. Furthermore, a possibility of evaluating bilateral trades for a specific product is given. Products are classified in several specific product numbers, so-called HS Codes, obligatory for more than 200 countries, customs and economic unions, representing more than 98% of world's trade [2]. HS Code group 854140 represents 'photosensitive semiconductors devices, photovoltaic cells and light emitting diodes'. More specific descriptions with detailed determination of codes between light emitting diodes and PV are available for approximately 95% of the data. Otherwise, an experienced ratio of 80% PV and 20% light emitting diodes is used. Surprisingly, a large number of developing countries' customs agencies have reported detailed data. It is valid that: *'imports A from B'* = *'exports B to A'*. Data have been assigned for 190 countries in a period of 2000 until 2016. The basis of the calculation of PV capacities per country is the following assumption: *'import A'* – *'exports A'* + *'production A'* = *'market A'*. The production volumes refer to the publications of the magazine 'Photon' on the topic of production output [3]. From this, attention is paid to the fact that approximately 25% of annual production will be installed in the following year as significant quantities of

PV products are being either in transit to their destination or being warehoused before their installation.

Since the data include the monetary value of the products, a conversion to PV capacities is necessary. For this purpose a reliable estimation of PV prices per Wp for every country is needed. These price data refer to a worldwide annual average PV module price per Wp [4-6]. Since prices depend on countries' locations and market sizes, there is a significant ratio of divergence. Therefore, all countries are classified in five main groups with one of them having two subgroups, each representing a group of countries that are nearly comparable in market size and economic conditions. The world's average price is multiplied by a factor quoting the ratio of countries' prices for each of these groups. The ratio is based on experience and statements of several countries' experts that have been contacted in the past [7-9]. This has been validated in previous publications [10-15]. Since the world's given average annual PV prices are selling prices of modules' producers, a factor of 10% is added considering distributor margins and shipping, etc. Therewith, prices for the year 2016 vary between 0.55 USD/Wp for advanced PV market countries and 1.36 USD/Wp for small PV applications in African and Asian low-income countries as can be seen in Table 1.

To lower the probability of errors, further sources with installation rates estimated by IRENA, International Energy Agency Photovoltaic Power Systems Programme (IEA PVPS), journal du photovoltaique, Solar Power Europe and public resources are used [1, 16-45]. Some countries report PV installation data by giving DC capacities while others refer to their installed AC capacities. Thus a small probability of mismatch between report data and those number calculated by using the DC capacity according to the customs database is possible. However numbers referring to DC capacities are favored.

Table 1: PV module price for conversion of customs values into PV [4-6]. Abbreviations: North America (NA) and Latin America (LA).

[USD/Wp]	**2000**	**2001**	**2002**	**2003**	**2004**	**2005**	**2006**	**2007**	**2008**	**2009**	**2010**	**2011**	**2012**	**2013**	**2014**	**2015**	**2016**
average world price:	5.17	4.96	4.36	4.13	4.24	4.55	5.07	4.57	4.38	2.53	2.05	1.65	1.21	0.89	0.85	0.74	0.62
countries >100 MW/year	4.66	4.47	3.92	3.72	3.82	4.09	4.56	4.11	3.94	2.28	1.85	1.49	1.09	0.80	0.76	0.66	0.55
countries >10 MW/year	5.69	5.46	4.79	4.55	4.67	5.00	5.57	5.02	4.82	2.78	2.26	1.82	1.33	0.98	0.93	0.81	0.68
countries >5 MW/year	7.76	7.44	6.54	6.20	6.36	6.82	7.60	6.85	6.57	3.79	3.08	2.48	1.82	1.34	1.27	1.11	0.92
countries <5 MW/year	9.31	8.93	7.84	7.44	7.63	8.19	9.12	8.22	7.89	4.55	3.70	2.97	2.18	1.60	1.52	1.33	1.11
countries <1 MW/year	10.35	9.93	8.71	8.27	8.48	9.10	10.13	9.14	8.76	5.06	4.11	3.30	2.42	1.78	1.69	1.47	1.23
Africa & Asia	11.38	10.92	9.59	9.10	9.33	10.01	11.14	10.05	9.64	5.57	4.52	3.63	2.66	1.96	1.86	1.62	1.36
Europe & NA & LA	9.83	9.43	8.28	7.86	8.06	8.64	9.62	8.68	8.33	4.81	3.90	3.14	2.30	1.69	1.61	1.40	1.17

Furthermore, all sources are weighted in a certain order: For countries that have their own cell or module production, the arithmetic mean of all sources except customs data is taken for estimation of installation rates, due to production margins and local PV value chain structures influencing the reliability of the customs database. In case of countries having no production, the average of all sources is used.

Due to statistical impacts of additional databases, it is possible in individual cases that estimated installed PV capacities at the end of 2016 can be less than those estimated at the end of previous years.

3 OVERVIEW ON PV INSTALLATIONS 2016

Within the use of this examination, the world's total installed PV capacity is estimated to be up to 305.78 GWp at the end of 2016. It is expected that there is a further 487 MWp that had been installed before the year 2000. PV installations are found in almost all countries. Appendix Table 1 includes the results for PV installation rates for all countries in the world at end of the year 2016.

As can be seen in Figure 1, China, with a share of 25% of total global installed PV, is again the country with leading PV installations even having increased that share from 19% by end of 2015. Further five markets show more than 10 GWp of PV capacity installed. Another 20 markets claim more than one GWp of PV capacity by the end of 2016. Thus, for the first time more than 25 countries are among those with more than one GWp deployed. These countries represented by China (77.69 GWp), Japan (42.55 GWp), Germany (41.27 GWp), the United States (41.14 GWp), Italy (19.14 GWp), the United Kingdom (11.43 GWp), India (9.31 GWp), France (7.12 GWp), Australia (5.90 GWp), Spain (5.21 GWp), the Republic of Korea (4.27 GWp), Belgium (3.37 GWp), Canada (2.93 GWp), Greece (2.61 GWp), Thailand (2.25 GWp), Czech Republic (2.18 GWp), the Netherlands (2.00 GWp), South Africa (1.73 GWp), Switzerland (1.63 GWp), Chile (1.62 GWp), Romania (1.40 GWp), Taiwan (1.24 GWp), Austria (1.08 GWp), Israel (1.05 GWp), Bulgaria (1.04 GWp) and Pakistan (1.02 GWp) in total amount to 95% of global PV installations. Chile, Taiwan, Austria, Israel and Pakistan are new representatives of this group of markets. Furthermore, there are 29 additional countries claiming more than 100 MWp per country (Ukraine, the Philippines, Denmark, Turkey, Slovakia, Portugal, Honduras, Mexico, Russia, Algeria, Malaysia, Brazil, Slovenia, the United Arab Emirates, Jordan, Morocco, Bangladesh, Hungary, Luxembourg, the French overseas department of Reunion, Peru, Indonesia, Saudi Arabia, Poland, Puerto Rico, Sweden, Singapore, Senegal, the Dominican Republic). Remarkable is the highly increasing share of emerging markets among those countries from Latin America and Southeast Asia. Further other emerging and developing countries show a significant installed PV base, in particular Iran (95

MWp), Nigeria (94 MWp), or Panama (91 MWp). Even countries that belong to the group of least income countries show respective numbers of PV like Ghana (89 MWp), Yemen (64 MWp), Kenya (61 MWp) or Sri Lanka (51 MWp). Especially in least developed low electrified countries in Sub-Saharan Africa and South Asia like Bangladesh, Tanzania or Myanmar high ratios of installed PV capacity are realised as off-grid systems, i.e. 5 MWp solar home systems with an average size of 50 Wp represent a solar power solution for 100,000 families [46]. On a regional perspective Asia-Pacific and Europe, by far, show highest cumulative PV installations, with 146.43 GWp and 105.45 GWp, respectively. Thus, for the first time regions surpassed the threshold of more than 100 GWp deployed.

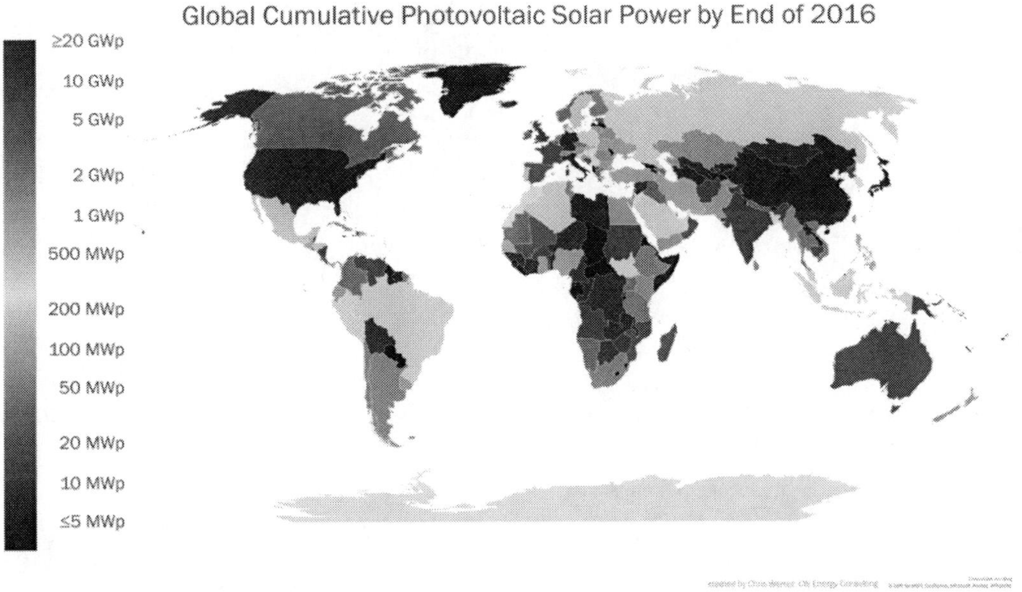

Figure 1: Total installed PV capacity per country by end of 2016.

Approximately 61% of global installed PV capacity (187.71 GWp) is installed in countries that have a gross domestic product (GDP) per capita higher than 30,000 USD per year. Low income countries (GDP/capita <10,000 USD) claim 101.09 GWp, with a high share represented by China. Countries with a GDP per capita between 10,000 USD and 30,000 USD represent only 15.06 GWp. Especially the two groups low and medium income countries with a GDP/capita below 10,000 USD and between 10,000 and 20,000 USD show high growth rates with having increased their cumulative installed capacity by 68% each [47, 48].

For three years in a row, the majority of 47% of total PV at end of 2016 is located in countries with medium irradiation (1,500-2,000 kWh/(m²·a)) [49]. Countries with less than 1,500 kWh/(m²·a), which are mainly highly developed countries represent 24% of total installed PV at low growth rates. Those countries predestined to use PV due to their very high insolation (>2,000 kWh/(m²·a)) claim 5.9% of global PV being installed, but show highest growth rates of country groups evaluated.

Figure 2 shows the PV installations per capita per country. Germany (505 Wp/cap.), Luxembourg (439 Wp/cap.), Japan (332 Wp/cap.), Italy (321 Wp/cap.), Belgium (298 Wp/cap.), the French overseas department of Reunion (283 Wp/cap.), Australia (247 Wp/cap.), Greece (232 Wp/cap.), Malta (215 Wp/cap.), the Czech Republic (205 Wp/cap.), Switzerland (195 Wp/cap.), the United Kingdom (174 Wp/cap.), the French overseas department of Martinique (168 Wp/cap.), Guam (158 Wp/cap.), Denmark (157 Wp/cap.) and the French overseas department of French Guiana (156 Wp/cap.) are the ones with more than 150 Wp/capita. Another ten countries show more than 100 Wp/capita (the French overseas department of Guadeloupe, Bulgaria, Slovenia, Israel, the United States, Austria, the Netherlands, Spain, France, Slovakia). Especially, highly developed countries and island states show significantly high PV installations per capita. In total, 15 of the top 50 countries ranking in PV per capita are small islands or states with large parts of their territory located on small islands. This might be due to PV being a clear cost competitive energy source that enables those remote island countries to be decoupled from high prices for conventional electricity generation by diesel power plants [50, 51]. As a world average, every human claims 41.5 Wp of PV.

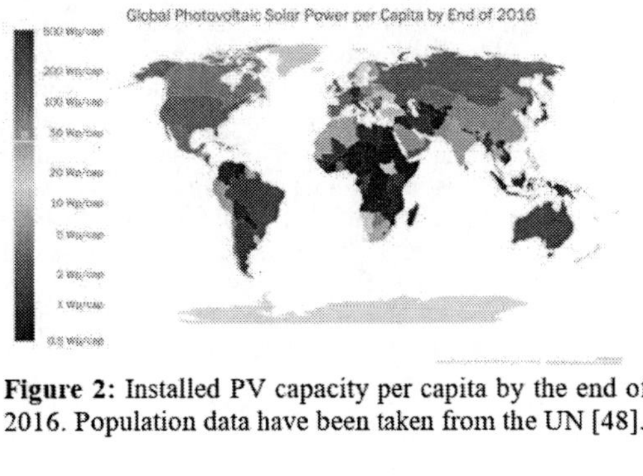

Figure 2: Installed PV capacity per capita by the end of 2016. Population data have been taken from the UN [48].

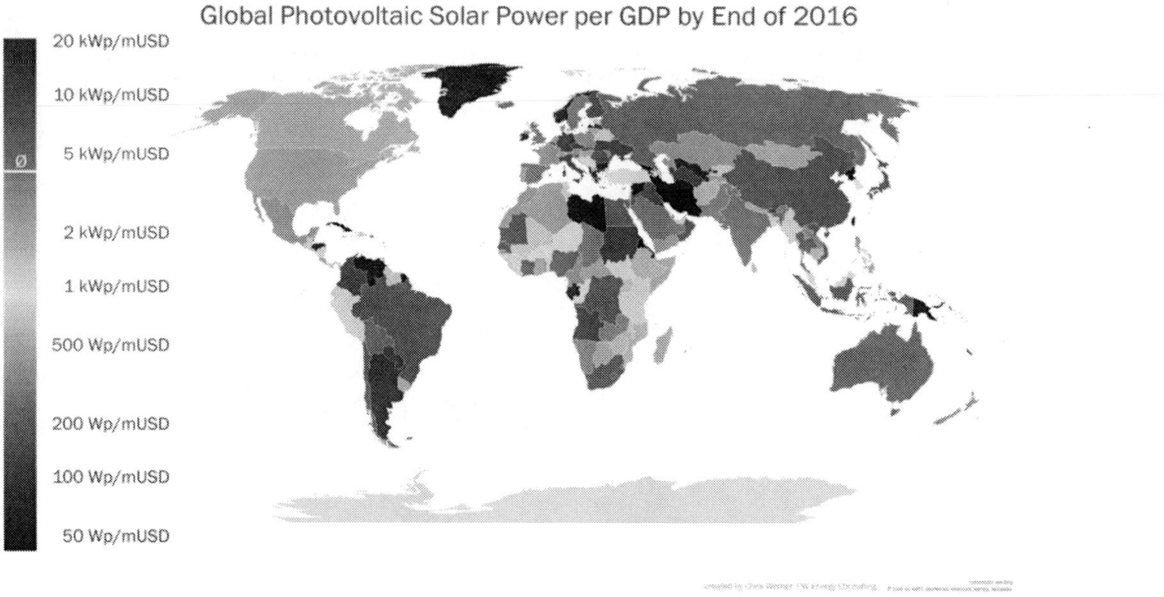

Figure 3: Global overview on cumulative PV installations per million USD of GDP by the end of 2016. Data for GDP have been taken from the World Bank [49].

An indicator of how much of a country's financial capabilities are invested relatively in PV is recognizable in installed PV capacities per countries' GDP (Figure 3). With the use of this comparative parameter, markets are able to be described as overcoming different economic possibilities. Some countries have reached significant ratios of PV installations per one million USD of GDP. Honduras (20.0 kWp/mUSD), Bulgaria (19.8 kWp/mUSD), Kiribati (13.8 kWp/mUSD), Greece (13.4 kWp/mUSD), Germany (11.9 kWp/mUSD), the Czech Republic (11.3 kWp/mUSD), Italy (10.3 kWp/mUSD) and Ukraine (10.2 kWp/mUSD) are those countries with more than 10 kWp installed per million USD of GDP. A large number small countries that often are rarely recognized due to having installed only a few MWp can be distinguished. Thus, countries like Malta (8.4 kWp/mUSD), Mauritania (7.2 kWp/mUSD), Senegal (7.2 kWp/mUSD), Micronesia (5.9 kWp/mUSD), Barbados (4.0 kWp/mUSD), Gambia (3.7 kWp/mUSD), Togo (3.4 kWp/mUSD), the Republic of Palau (3.2 kWp/mUSD) or Namibia (3.2 kWp/mUSD) invest more of their often limited economic power in PV than the bulk of highly developed countries. In comparison to this, highly developed countries such as Norway (0.069 kWp/mUSD), Ireland (0.089 kWp/mUSD), Latvia (0.112 kWp/mUSD) still have a vast potential of investing more capital in PV although some of them may not have the need for large PV investments due to already claiming high shares of renewable electricity. Especially, emerging markets, like the African ones or island states, have identified PV as being able to improve significantly people's living conditions and save enormous financial resources in the mid- to long-term. Due to a fast amortization of PV systems in rural areas of developing countries [52-53], these markets have no need for subsidies and make off-grid PV countries foremost in proportion of PV expenditures to GDP.

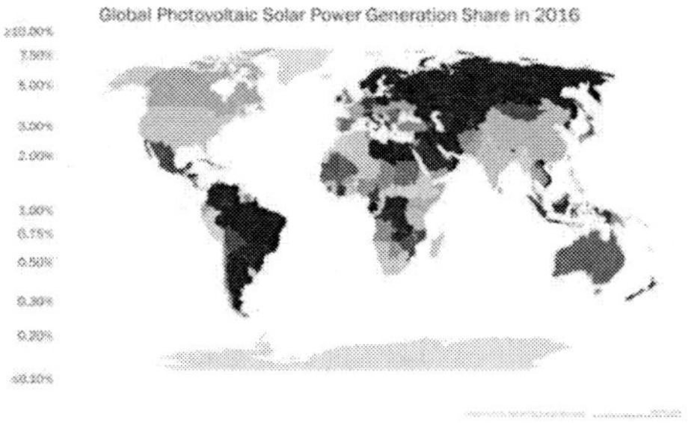

Figure 4: Global overview of PV share in the respective countries total power generation [55].

As a consequence of high ratios of PV to GDP, numerous countries have reached a relevant share of PV in their total power generation (Figure 4). In 2016 14.4% of Kiribati's power generation capacity is complemented by PV. Honduras, with 12.8%, as well claims more than 10% of its power generated by solar PV. Further Italy (8.6%), the island of St. Helena (8.1%), Cape Verde (7.6%), Germany (7.5%), Comoros (7.3%), Guinea-Bissau (6.8%), Greece (6.8%), Malta (6.3%), the Solomon Islands (6.2%), Equatorial Guinea (5.9%), Sierra Leone (5.5%) and Japan (5.2%) saw more than 5% of their power generation provided by PV. In total 83 countries show more than 1% of PV share in their power production with a global average of 1.72% of all global power generated by solar PV. Especially, European, Western African, Asian countries and island states show high ratios of PV in their power generation. While regarding the African and insular countries that represent high ratios of PV in their power generation, it is obvious that they are the ones using a bulk of diesel power plants for generation of electricity. Countries leading in usage of diesel power, due to their remote location usually lead in shares of PV in their total power generation. Thus, in addition to small off-grid appliances, PV is also used as a sustainable alternative for larger solutions as applied in hospitals, schools, supply of commericial appliances and in more and more cases as an upgrade for existing diesel/ oil fired power plants. This is especially prevailing in the environment of high diesel prices and remote areas to be supplied with affordable electricity [51-54].

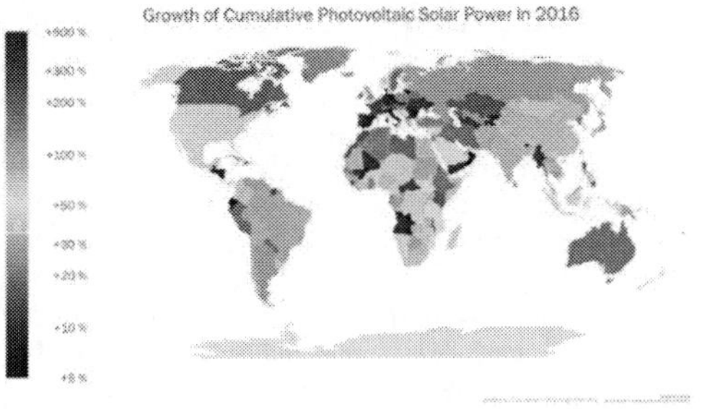

Figure 5: Ratio of added PV capacity in 2016 to cumulative PV installations in 2015, representing growth in PV markets' cumulative capacity.

Figure 5 gives an overview of PV market growth in 2016. Regularly, some countries show huge growth rates in a short period of time. In Yemen, 2016 affected by civil war, the cumulative installed PV capacity has grown by more than 775% from 7.3 MWp to 63.9 MWp, enhanced by low availability of local power grids triggering investments in small scale off-grid PV. Further, other emerging countries have shown high growth rates in 2016. In El Salvador, PV capacity by the end of 2016 was more than 580% higher than the end of 2015, reaching 44.9 MWp installed. The Philippines added some 683 MWp in 2016 reaching 898 MWp of cumulative capacity, equivalent to a growth rate of 317%. Comparable growth rates can be identified for Malawi (+316%), Morocco (+294%), Azerbaijan (+200%), Lebanon (+194%) or Senegal (+176%). This indicates largely growing PV markets especially in emerging countries all over the world. In Europe, Hungary tops PV growth rates with adding 189% last year reaching 257 MWp. In addition Belarus (+173%), Ireland (+171%), Turkey (+141%), Poland (+135%) and Finland (+105%) have also more than doubled their installed PV. With Italy (+1.2%), Slovakia (+1.0%), Bulgaria (+1.0%), Spain (+0.6%) and the Czech Republic (+0.4%) some of the countries with least growing cumulative PV installations are located in Europe. In contrast, further African markets like Ethiopia, Kenya or Nigeria showed high growth of their installed capacity per country. In 2016, South Africa added another 400 MWp of PV helping to overcome power shortages. Chinese PV installations remain at huge growth by adding 33.3 GWp in just one year, equivalent to a growth of 74% and thus has been the first time ever market to surpassed the 30-GWp-threshold of annual new installations. Apart from Eastern Asia, some South and Southeast Asian markets show significant annual installations. The MENA region, led by Morocco, Jordan, Israel and Algeria, after years of low growth, finally seems to start increasing in their PV capacity at measurable scale with countries like Yemen, Iraq and Egypt starting to expand PV investments. In the Americas, Latin American countries like El Salvador, the Bahamas, Jamaica, Cuba, Barbados and Chile are the ones with high relative growth rates. Chile even ranks second in 2016 annual installations with 773 MWp deployed, ahead of Canada (400 MWp) and behind the dominant US market that added some 15,150 MWp in 2016. Even the great potentials in Brazil and Mexico start to get used by each adding more than 100 MWp in 2016, after years of low installation rates.

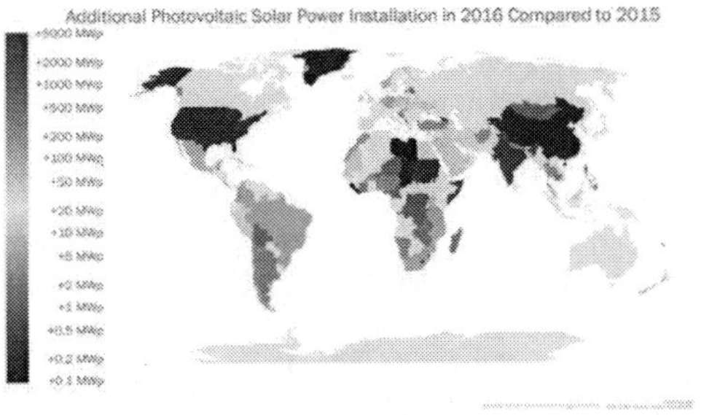

Figure 6: Additional new annual installations in 2016 compared to the annual installations of 2015. Countries marked grey show a decline of annual installs from 2015 to 2016.

Figure 6 shows results of a new research on the development of annual installations on a year-over-year comparison, trying to give an answer on the question what the significantly larger 75.9 GWp 2016 global annual installations compared to the 52.3 GWp during 2015 is based on. With its 33.3 GWp installed in 2016, China installed 17.3 GWp more than the 15.9 GWp installed during 2015. Comparable trends can be identified for 106 other countries that increased their annual installations from 2015 to 2016. The US increased the annual installs by 7.5 GWp on top of the 7.6 GWp installed during 2015 and thus reached 15.1 GWp of newly installed PV during 2016. Germany that installed 1,450 MWp during 2015 slightly increased its installations by 130 MWp to 1,580 MWp in 2016. It is remarkable that especially emerging markets in Latin American, African and MENA region markets have grown their annual PV installations in 2016 compared to 2015. With these countries adding 30.3 GWp (13.0 GWp without China) more during 2016 that they added in 2015, they easily compensated the loss of 6.8 GWp of annual installations in 85 other countries (marked grey in Figure 6).

The development of PV in emerging markets could be more different than in the markets where PV has developed until now. PV starts to progress in several emerging countries based on calls for tenders for large-scale PV systems, rather than rooftop PV. This development will be even stronger in places where neither the population, nor companies will have the ability to finance PV systems for self-consumption. This trend is already visible since 2014 with the share of large-scale PV increasing in all regions except in Europe. In parallel, small off-grid applications are starting to develop for the electrification of households with very low electricity consumption, as for instance Bangladesh. This development implies the possible cohabitation in emerging countries between Pico PV and Solar Home Systems and utility-scale plants, leaving the rooftop market aside.

Figure 7 shows a segmentation of PV markets. Further, an attractiveness of PV markets is represented. For this, a portfolio analysis chart is applied to PV markets with the use of a logarithmic order. 'Cash cows' are mature markets that claim significant installed PV capacities (more than 0.1% of global installed PV at the end of 2016) with relatively low growth rates, but still remaining the most important driver of the global PV industry. Germany, Italy, Belgium, France, Portugal, Denmark, Slovakia, Japan, Canada or Australia are typical representatives of this PV market segment. The United Kingdom, the Republic of Korea, Taiwan, Pakistan, Honduras, Russia and Algeria are new members of that group, having been categorized as 'rising stars' in 2015. This segment claims 36% of total installed PV capacity. Even with very low growing markets, such as Germany (4% in 2016) or Italy (1%), standing for annual installation of a few 100 MWp due to having an installed PV capacity base of a few GWp, this is the most important PV market segment. Furthermore, 'cash cows' mostly are countries with high incomes. In total 30 countries belong to this group, whereas 2016 it have been 28.

'Rising stars' are probably the most interesting market segment, as respective installed PV capacities (>0.1% of global PV in 2016) and growth rates are above the world's long term average (>46%) and led to significant installations in the relevant year. Brazil, China, Chile, India, Mexico, the Philippines, Turkey and the United States are those that represent this market segment in 2016. Therefore, this segment stands for 60% of total global PV installations. An interesting question is whether these markets will stay in this segment. In the short- to mid-term, 'rising stars' will become 'cash cows', as such high growth rates will not be able to keep up with increasing cumulative PV capacities in the years to come. It is remarkable that the United States and China managed to increase their annual PV installations in 2016 at a such high scale, which resulted in them moving from 'cash cows' to 'rising stars' again.

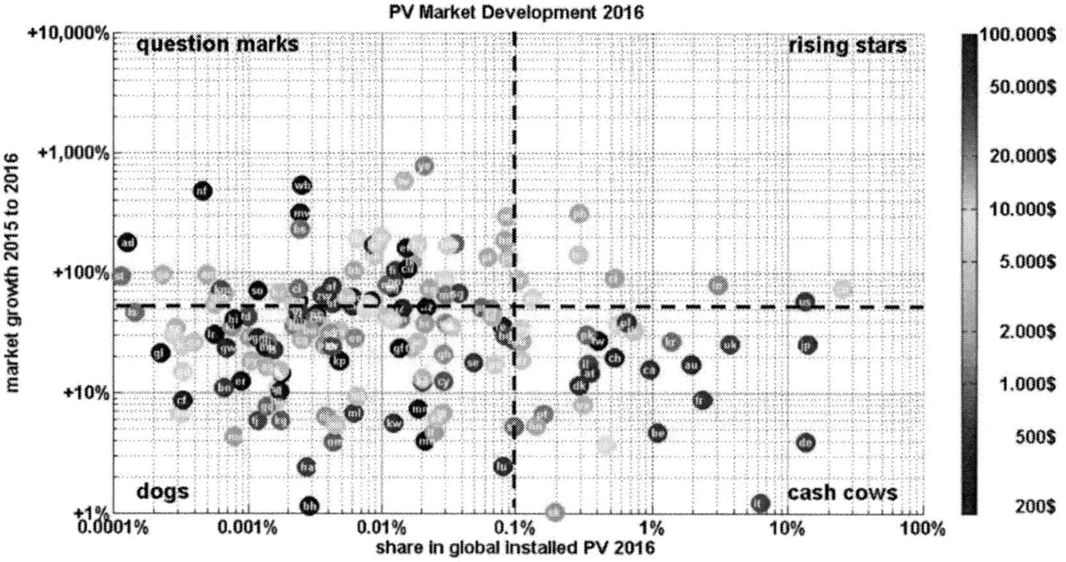

Figure 7: Cumulative installed PV by end of 2016 (x-axis) and PV capacities growth rate in 2016 (y-axis). Every bubble represents one country. The color code of the bubbles refers to the GDP/capita (left axis). The code inside of the bubble represents the internet-domain of the relevant country.

Markets that show high growth rates, but relatively low installed PV can be named as 'question marks'. Among 46 others, this segment includes markets such as Jordan, Morocco, Hungary, Saudi Arabia, Poland, Puerto Rico, Singapore, Senegal, Iran, Malta, Panama, Eqypt or New Zealand that show relatively high growth rates, but because of their low installed cumulative PV capacity, they are so far marginal in the global PV perspective. If these markets keep their high growth rates for several years, becoming a 'rising star' can be a valid opportunity. Most interesting in this segment are markets close to the threshold of becoming a 'rising star'. Jordan, Hungary, Poland and Morocco are among those that are expected to become new 'rising stars' in the years to come.

At first view 'dogs' seem to be markets of less interest since only 1.1% of total PV is installed in those countries. But especially these are non-feed-in tariff driven markets which often claim a high share of off-grid PV applications, such as Tanzania, Kenya or Zambia. In 2016 for the first time markets that before could be categorized as 'cash cow' moved back to being a 'question mark' resulting from growth rates significantly below average.

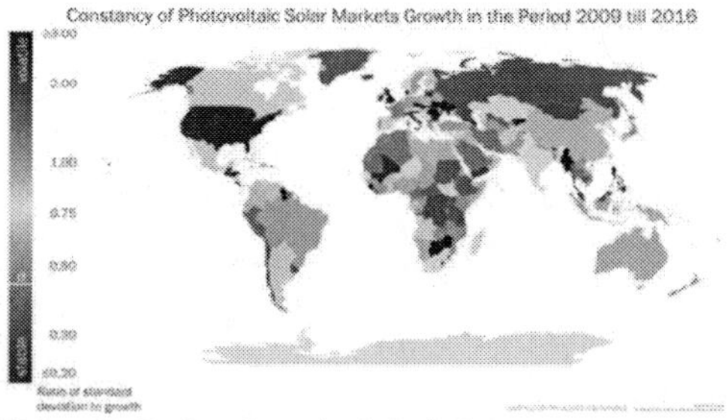

Figure 8: Ratio of standard deviation to average growth of PV capacity in the period from 2009 to 2016. The blue color refers to very unstable PV markets, while red refers to relatively constant growing photovoltaic capacities.

High growth rates of 'rising star' or 'question mark'

countries are not always an indicator for a sustainable and constant growth. Deviations in PV markets' annual growth rates are significant, due to markets often being driven by politically motivated feed-in tariffs. Thus, average annual growth rates have to be regarded in the context of annual growth rate standard deviation. Thus, sustainable growth can be defined as significant but constant growth. Very constant growing PV markets can be identified by the ratio of standard deviation to average growth.

As can be seen in Figure 8, such significant but constant growth in recent years can be identified especially in the United States, Ireland, Tanzania, Saudi Arabia, Japan, New Zealand, Singapore or Portugal and a variety of further emerging markets. All of them show a ratio of standard deviation to average growth of less than 0.50 (refer to Figure 8). Portugal represents an average annual growth of its cumulative PV of 26% in the period from 2009 to 2016. In 2010 Tanzania's PV capacity grew by 81%, in 2011 by 86%, in 2012 by 78%, in 2013 by 50%, in 2014 by 33%, in 2015 by 51% and in 2016 again by 51% to 43.4 MWp at the end of the year. This is an example of a very constant growth in PV and can definitely be regarded as sustainable. If this constant and high growth continues, Tanzania could reach 65 MWp by the end of 2017. Another example is the United States, the PV capacity grew by 53% in 2010, 65% in 2011, 78% in 2012, 65% in 2013, 51% in 2014, 41% in 2015 and 58% in 2016, which leads to an average annual growth rate of 58% with a relatively low standard deviation of only 11%. If this constant growth goes on steadily, 60 GWp might be cumulatively installed by the end of 2017 in the US. Markets such as Denmark, Ukraine, Slovakia or even the United Kingdom show high annual growth rates, but high standard deviations, due to variable growth with boom years being followed by years of very little installations [12-15]. Thus, they show ratios of standard deviation to average growth of more than two. By following this approach, the most volatile markets are among others Cape Verde (ratio of 18.4), Kyrgyzstan (ratio of 14.0), Belize, Botswana, Slovakia, Denmark and Zimbabwe showing ratios of more than five.

With the use of this evaluation of average annual growth, the constancy of growth rates, the help of publications and the knowledge of six years of analysis of cumulative PV capacities [11-15], a rough forecast of cumulative PV capacities for the years 2017 and 2018 is possible. In 2017, the global cumulative installed PV capacity is likely to increase by 30% to 397.50 GWp by the end of 2017 (new installations of about 91.70 GWp in 2017) and by the end of 2018, 496.80 GWp (new installations of about 99.30 GWp in 2018) of cumulative installed PV capacity can be expected.

4 CONCLUSIONS

In total, 305.78 GWp can be allocated to 196 countries by end of 2016. Out of those 75.93 GWp have been newly installed. The global trade volume of PV modules reached 31.9 bnUSD in 2016, after 33.5 bnUSD in 2015.

The most significant amounts of PV are still installed in mature markets in Europe, East Asia and North America. However, a relevant paradigm shift to developing and emerging markets as the new growth regions can be seen. Some low and middle income countries already show remarkable numbers of newly installed PV.

In the years to come, new growth markets in Central and South America (Brazil, Mexico, Argentina, Caribbean, Panama, Cuba, Peru), Asia-Pacific (Indonesia, Singapore, New Zealand, Vietnam), MENA (Iran, Jordan, Iraq, Morocco, Egypt) and Sub-Saharan Africa (Nigeria, Senegal, Ghana) will increasingly play a major role within the next years of global PV markets development. In Europe Poland, Hungary, Turkey and Russia will expand their growth perspectives on the short to mid-term, while the Balkan countries and the Caucasus states show interesting developments on the long term.

One is able to assess that every world region might represent a great PV market potential. Enormous growth rates in specific regions of the world show that PV is on the way to becoming an important part of global electricity supply. Some developing countries are on the road to solar electrification, others use PV as an opportunity to decrease suffering from lacking access to electricity in remote locations. Lower energy costs enabled by reduced dependence on energy imports will contribute to leading these countries to sustainable development and stable growth.

REFERENCES

[1] [IRENA] – International Renewable Energy Agency, 2017. Renewable Capacity Statistics 2017, Dubai

[2] International Trade Centre Market Analysis Tool, 2017. Product Map, Market Analysis and Research Section, International Trade Centre, Geneva, www.trademap.org

[3] Photon International, 2012. Zellproduktion 2011, Photon Magazin, 2012(4), [plus respective data for the years 2000-2011 either of Photon Magazin or Photon International]

[4] Breyer Ch., Birkner C., Meiss J., Goldschmidt J.C., Riede M., 2013. A top-down analysis: determining photovoltaics R&D investments from patent analysis and R&D headcount, Energy Policy, 62, 1570-1580

[5] Photon, 2016. Factory Gate c-Si Module Prices, Photon Consulting, Aachen

[6] Hanwha Q CELLS Market Intelligence, 2011-2016. Intern Price Snapshot, Hanwha Q CELLS, Bitterfeld-Wolfen

[7] Werner Ch., Gerlach A., Adelmann P., Breyer Ch., 2011. Global Overview on Cumulative Installed Photovoltaic Power, Proceedings 26th European Photovoltaic Solar Energy Conference, Hamburg

[8] Szewzcyk W., 2017. private communication, PV insights for Poland, May 31

[9] Horn S., 2017. private communication, PV insights for Malaysia and South East Asia, March 29

[10] Werner Ch., Gerlach A., Adelmann P., Breyer Ch., 2011. Global Overview on Cumulative Installed Photovoltaic Power, 2nd Symposium Small PV-Applications, Ulm

[11] Werner Ch., Gerlach A., Adelmann P., Breyer Ch., 2011. Global Overview on Cumulative Installed Photovoltaic Power, Proceedings 26th European Photovoltaic Solar Energy Conference, Hamburg
[12] Werner Ch., Gerlach A., Adelmann P., Breyer Ch., 2013. Analysis and Forecast of Global Installed Photovoltaic Capacity and Identification of Hidden Growth Markets, Proceedings 28th European Photovoltaic Solar Energy Conference, Paris
[13] Werner Ch., Gerlach A., Breyer Ch., Masson G., 2014. Global Installed Photovoltaic Capacity and Identification of Hidden Growth Markets, Proceedings 29th European Photovoltaic Solar Energy Conference, Amsterdam
[14] Werner Ch., Gerlach A., Breyer Ch., Orlandi S., Masson G., 2015. Latest Developments in Global Installed Photovoltaic Capacity and Identification of Hidden Growth Markets, Proceedings 31st European Photovoltaic Solar Energy Conference, Hamburg
[15] Werner Ch., Gerlach A., Breyer Ch., Orlandi S., Masson G., 2016. Global Photovoltaics in 2016 – Analysis of Current and Future Solar Energy Markets, Proceedings 32nd European Photovoltaic Solar Energy Conference, Munich
[16] [PVMA] – PV Market Alliance, 2016. The PVMA Global PV Market Report, Brussels
[17] [IEA PVPS] - International Energy Agency Photovoltaic Power Systems Programme, 2017. Snapshot of Global PV Markets 2016, Paris
[18] Journal des Énergies Renouvelables, 2017, Photovoltaic Barometer, EurObserv'er, 2017 (April), Paris
[19] [REN21], 2017. Renewables 2016 – Global Status Report, REN21, Paris
[20] Solar Power Europe, 2017. Global Market Outlook – For Solar Power 2017-2021, Brussels
[21] Bellini E., 2017. French Development Agency finances solar project in Bolivia with €11.5 million, PV Magazine, Berlin, https://www.pv-magazine.com/2017/06/15/french-development-agency-finances-solar-project-in-bolivia-with-e11-5-million/
[22] Morias L., 2017. Brazil's solar installed capacity reaches 99 MW in April, Renewables Now, Sofia, https://renewablesnow.com/news/brazils-solar-installed-capacity-reaches-99-mw-in-april-571599/
[23] Bellini E., 2017. Work begins on 4.4 MW PV plant in Cuba, PV Magazine, Berlin, https://www.pv-magazine.com/2017/06/07/work-begins-on-4-4-mw-pv-plant-in-cuba/
[24] Bellini E., 2017. Cyprus sees share of large-scale solar rise, PV Magazine, Berlin, https://www.pv-magazine.com/2017/05/12/cyprus-sees-share-of-large-scale-solar-rise/
[25] Bellini E., 2017. France adds 78 MW of PV capacity in Q1 2017, PV Magazine, Berlin, https://www.pv-magazine.com/2017/06/06/france-adds-78-mw-of-pv-capacity-in-q1-2017/
[26] Zaripova A., 2017. US company to build 50 MW PV plant in the republic of Georgia, PV Magazine, Berlin, https://www.pv-magazine.com/2016/04/25 /us-company-to-build-50-mw-pv-plant-in-the-republic-of-georgia_100024274/
[27] Roselund Ch., Lopez B.D., 2017. Breaking new ground, PV Magazine, Berlin, https://pv-magazine-usa.com/2017/04/23/breaking-new-ground/
[28] Bellini E., 2017. Hungarian PV market sees robust growth, PV Magazine, Berlin, https://www.pv-magazine.com/ 2017/04/03/hungarian-pv-market-sees-robust-growth/
[29] Bellini E., 2017. Iran's PV market expected to see strong growth over next 12 months, PV Magazine, Berlin, https://www.pv-magazine.com/2017/03/24/irans-pv-market-expected-to-see-strong-growth-over-next-12-months/
[30] Bellini E., 2017. Lebanon issues RfP for 180 MW of solar, PV Magazine, Berlin, https://www.pv-magazine.com/2017/05/31/lebanon-issues-rfp-for-180-mw-of-solar/
[31] Kenning T., 2017. Malaysia installed 32MW solar under FiTs in 2016, PV-Tech, London, https://www.pv-tech.org/news/malaysia-installed-32mw-solar-under-fits-in-2016
[32] Parkinson G., 2017. Norfolk Island has "too much" solar, now it wants storage, REneweconomy, Mullumbimby, http://reneweconomy.com.au/nor folk-island-has-too-much-solar-now-it-wants-storage-58159/
[33] Bellini E., 2017. Panama's Social Security Fund seeks to invest in solar, PV Magazine, Berlin, https://www.pv-magazine.com/2017/06/26/pana mas-social-security-fund-seeks-to-invest-in-solar/
[34] Crisostoma L., 2017. Philippines installs over 900 MW of PV solar power capacity in 2016, Solenergy Energy Philippines, Makati City, http://solenergy.com.ph/philippines-installs-over-900-mw-of-pv-solar-power-capacity-in-2016/
[35] Ahola J., 2017. National Survey Report of PV Power Applications in Finland 2015, IEA PVPS Task 1, Paris
[36] Bellini E., 2017. Portugal: renewable energy sector wants first auction for large-scale solar and wind, PV Magazine, Berlin, https://www.pv-magazine.com/2017/05/26/portugal -renewable-energy-sector-wants-first-auction-for-large-scale-solar-and-wind/
[37] Bellini E., 2017. Romania's PV capacity grew by 70 MW in 2016, PV Magazine, Berlin, https://www.pv-magazine.com/2017/03/08/roman ias-pv-capacity-grew-by-70-mw-in-2016/
[38] EMA, 2017. Singapore PV Statistics , Energy Market Authority Singapore, Singapore, https://www.ema.gov.sg/cmsmedia/Publications_an d_Statistics/Statistics/31RSU.pdf
[39] Bellini E., 2017. PV in Slovakia is kept alive by residential segment, PV Magazine, Berlin, https://www.pv-magazine.com/2017/03/21/pv-in-slovakia-is-kept-alive-by-residential-segment
[40] Bellini E., 2017. Slovenia's PV capacity tops 259 MW, net-metering expected to support more growth, PV Magazine, Berlin, https://www.pv-magazine.com/2017/03/10/slovenias-pv-capacity-tops-259-mw-net-metering-expected-to-support-more-growth/
[41] Bellini E., 2017. South Africa's PV capacity reaches 1.47 GW, new PV installations for 2016 total 509 MW, PV Magazine, Berlin, https://www.pv-magazine.com/2017/03/17/south-

africas-pv-capacity-reaches-1-47-gw-new-pv-installations-for-2016-total-509-mw/
[42] Bellini E., 2017. Sweden tops 140 MW of installed PV capacity, PV Magazine, Berlin, https://www.pv-magazine.com/2017/04/04/sweden-tops-140-mw-of-installed-pv-capacity/
[43] Bellini E., 2017. Switzerland installed 250 MW of new PV systems in 2016, PV Magazine, Berlin, https://www.pv-magazine.com/2017/03/15/switzerland-installed-250-mw-of-new-pv-systems-in-2016/
[44] Bellini E., 2017. Tunisia launches tender for 70 MW of solar, PV Magazine, Berlin, https://www.pv-magazine.com/2017/05/19/tunisia-launches-tender-for-70-mw-of-solar/
[45] Bellini E., 2017. Uruguay's installed PV capacity tops 88.4 MW, PV Magazine, Berlin, https://www.pv-magazine.com/2017/06/23/uruguays-installed-pv-capacity-tops-88-4-mw/
[46] Breyer Ch., Werner Ch., Rolland S., Adelmann P., 2011. Off-Grid Photovoltaic Applications in Regions of Low Electrification: High Demand, Fast Financial Amortization and Large Market Potential, Proceedings 26th European Photovoltaic Solar Energy Conference, Hamburg
[47] World Bank, 2017. World Development Indicators, Gross Domestic Product 1960-2016, World Bank, New York
[48] [UN] – United Nations, 2017. World Population Prospects: The 2017 Revision, Department of Economic and Social Affairs, United Nations, New York
[49] Breyer Ch. and Schmid J., 2010. Population Density and Area weighted Solar Irradiation: Global Overview on Solar Resource Conditions for Fixed Tilted, 1-Axis and 2-Axis PV Systems, 25th EU PVSEC, Valencia, DOI: 10.4229/25thEUPVSEC2010-4BV.1.91
[50] Breyer Ch., Gerlach A., Schäfer D., Schmid J., Fuel-Parity: New Very Large and Sustainable Market Segments for PV Systems, IEEE EnergyCon, Manama, 2010, December 18–22
[51] Blechinger P., Cader C., Bertheau P., Huyskens H., Seguin R., Breyer Ch., 2016. Global Analysis of the Techno-Economic Potential of Renewable Energy Hybrid Systems on Small Islands, Energy Policy, 98, 674-687
[52] Cader C., Bertheau P., Blechinger P., Huyskens H., Breyer Ch., 2015. Global cost advantages of autonomous solar-battery-diesel systems compared to diesel-only systems, Energy for Sustainable Development, 31, 14-23
[53] Werner Ch. and Breyer Ch., 2012. Analysis of Mini-grid Installations: An Overview on System Configurations, Proceedings 27th European Photovoltaic Solar Energy Conference, Frankfurt, DOI: 10.4229/27thEUPVSEC2012-5DO.9.6
[54] Bertheau P., Cader C., Huyskens H., Blechinger P., 2015. The Influence of Diesel Fuel Subsidies and Taxes on the Potential for Solar-Powered Hybrid Systems in Africa, Resources, 4, 673-691
[55] [CIA], 2016. The World Factbook. Central Intelligence Agency, Washington

Appendix Table 1: Alphabetical list of all countries in the world and their cumulative installed PV capacity at the end of 2016. Data are compiled as documented in the paper and based on IRENA, IEA PVPS and further publications plus ITC data [11-45].

	PV installation by end of 2016 [MWp]	PV installation by end of 2015 [15] [MWp]	PV installation by end of 2014 [14] [MWp]	PV installation by end of 2013 [13] [MWp]	PV installation by end of 2012 [12] [MWp]	PV installation by end of 2011 [11] [MWp]	PV installations per capita by end of 2016 [Wp/cap]	PV installations per GDP by end of 2016 [Wp/mUSD]	PV share in power generation 2016 [%]
Afghanistan	12.90	7.30	5.30	4.60	3.10	1.60	0.38	662.6	0.45%
Albania	1.70	1.10	0.99	0.30	0.14	0.10	0.58	142.5	0.03%
Algeria	339.00	285.00	78.00	17.60	17.60	15.20	8.50	2,172.0	1.16%
Andorra	0.39	0.14	0.12	0.07	0.05	0.05	5.00		0.00%
Angola	13.70	13.00	9.80	9.50	6.70	3.60	0.49	152.8	0.46%
Antigua and Barbuda	4.20	3.60	0.51	0.29	0.17	0.12	42.03	2,898.2	0.00%
Argentina	57.00	45.00	35.20	32.70	29.90	14.90	1.31	104.4	0.07%
Armenia	1.50	0.76	0.37	0.26	0.20	0.14	0.51	142.2	0.03%
Australia	5,900.00	5,020.00	4,080.00	3,200.00	2,380.00	1,380.00	247.90	4,897.8	3.75%
Austria	1,080.00	940.00	794.00	639.00	420.00	184.00	124.44	2,794.8	1.63%
Azerbaijan	30.40	10.10	1.10	0.52	0.24	0.19	3.16	803.2	0.15%
Bahamas	7.50	2.26	1.59	1.33	1.16	0.09	19.39	829.0	0.50%
Bahrain	8.80	8.70	6.80	5.70	4.90	0.39	6.41	276.2	0.00%
Bangladesh	258.00	198.00	172.00	140.00	83.60	58.50	1.60	1,165.2	0.84%
Barbados	18.80	9.20	6.00	3.90	2.20	0.38	66.15	4,098.0	0.00%
Belarus	56.90	20.80	13.10	7.60	6.60	2.50	6.00	1,199.6	0.10%
Belgium	3,370.00	3,220.00	3,120.00	2,990.00	2,620.00	2,030.00	298.55	7,226.1	3.87%
Belize	5.20	3.10	1.30	1.10	0.97	0.03	14.47	2,945.7	0.99%
Benin	7.30	4.90	4.40	1.60	1.10	0.76	0.69	850.5	1.02%
Bermuda	4.10	3.30	2.40	1.40	0.97	0.33	66.13		0.00%
Bhutan	0.25	0.24	0.22	0.20	0.19	0.17	0.32	111.8	0.02%
Bolivia	8.80	6.60	5.60	4.50	4.10	3.20	0.82	260.3	0.19%
Bosnia-Herzegovina	16.40	10.10	5.70	4.00	1.90	0.54	4.64	990.4	0.13%
Botswana	9.10	9.10	5.70	4.90	4.40	3.30	4.12	595.8	0.52%
Bouvet Island	0.00	0.00	0.00	0.00	0.00	0.00			
Brazil	326.00	174.00	154.00	109.00	78.90	39.90	1.58	181.5	0.08%
Brunei	2.00	1.80	1.30	0.95	0.14	0.01	4.79	175.4	0.08%
Bulgaria	1,040.00	1,030.00	1,020.00	1,000.00	920.00	149.00	144.90	19,849.2	4.66%
Burkina Faso	13.10	8.40	6.90	5.80	4.90	3.50	0.72	1,081.3	1.89%
Burma	57.90	53.90	4.00	2.00	1.30	0.74	1.10	858.7	1.12%
Burundi	2.40	1.70	1.40	1.20	0.97	0.33	0.24	798.1	1.05%
Cambodia	12.80	10.30	9.30	7.70	5.60	3.80	0.82	639.5	0.50%
Cameroon	9.80	6.90	5.40	3.30	2.50	1.40	0.43	404.9	0.23%
Canada	2,930.00	2,530.00	1,910.00	1,320.00	877.00	628.00	81.50	1,915.3	0.65%
Cape Verde	13.10	10.50	10.20	9.90	9.20	8.50	24.58	8,099.1	7.55%
Central African Republic	1.00	0.92	0.34	0.28	0.25	0.19	0.22	569.4	0.93%
Chad	3.00	2.10	1.40	1.00	1.00	0.68	0.21	312.5	2.38%
Chile	1,620.00	847.00	404.00	134.00	13.00	4.60	91.20	6,558.0	3.31%
China	77,690.00	44,420.00	28,470.00	18,310.00	6,890.00	3,250.00	55.61	6,937.1	1.44%

Colombia	36.70	26.20	20.00	17.60	12.10	7.20	0.76	129.9	0.09%
Comoros	1.90	1.10	0.50	0.25	0.19	0.17	2.44	3,081.1	6.86%
Congo, Dem. Rep.	7.40	5.10	3.80	3.10	1.90	1.40	0.10	211.4	0.13%
Congo, Republic	7.80	6.10	4.10	3.60	2.30	2.00	1.56	995.7	1.23%
Costa Rica	24.20	15.60	11.50	8.50	6.10	1.20	5.03	421.3	0.31%
Croatia	64.80	47.00	31.80	23.60	12.30	11.90	15.30	1,285.1	0.42%
Cuba	47.60	22.80	9.40	8.00	3.20	2.80	4.15		0.35%
Cyprus	89.00	79.10	69.40	34.00	23.70	11.30	76.66	4,494.6	3.51%
Czech Republic	2,180.00	2,170.00	2,170.00	2,140.00	2,060.00	1,960.00	205.59	11,299.8	3.60%
Denmark	894.00	801.00	654.00	573.00	376.00	17.80	157.15	2,920.2	2.73%
Djibouti	4.30	3.40	1.80	1.80	1.80	1.60	4.64		2.29%
Dominica	1.80	1.10	0.81	0.13	0.08	0.07	24.60	3,425.8	3.08%
Dominican Republic	102.00	75.30	31.00	21.50	11.50	4.50	9.69	1,424.9	1.19%
Ecuador	79.20	74.60	34.00	14.50	10.20	4.50	4.91	809.8	0.54%
Egypt	69.60	40.00	26.60	23.20	11.00	6.00	0.74	207.0	0.07%
El Salvador	44.90	6.60	4.40	3.10	1.10	0.41	7.11	1,675.5	0.80%
Equatorial Guinea	4.20	3.90	2.70	2.40	2.10	0.68	3.57	412.6	5.69%
Eritrea	2.70	2.40	1.60	1.40	0.63	0.46	0.56		1.59%
Estonia	19.20	14.90	12.80	8.50	4.80	3.40	14.60	829.8	0.22%
Ethiopia	46.60	17.70	15.00	13.40	12.30	9.00	0.47	643.9	1.08%
Fed. St. of Micronesia	1.90	1.90	1.40	0.51	0.15	0.15	18.19	5,900.6	1.72%
Fiji	5.30	4.60	3.20	2.20	1.30	0.65	5.94	1,144.3	1.01%
Finland	39.00	19.00	17.00	12.60	9.34	6.92	7.11	164.7	0.03%
France	7,120.00	6,540.00	5,620.00	4,690.00	3,990.00	2,850.00	110.46	2,887.9	1.75%
French Guiana	42.00	34.00	34.00	34.00	0.00	0.00	156.31		
Gabon	1.20	0.95	0.67	0.47	0.40	0.35	0.62	84.4	0.09%
Gambia	3.60	2.80	2.00	1.40	0.94	0.75	1.82	3,732.1	2.49%
Georgia	0.71	0.36	0.36	0.36	0.36	0.36	0.18	49.5	0.01%
Germany	41,270.00	39,700.00	38,250.00	35,680.00	32,360.00	24,760.00	505.09	11,904.5	7.33%
Ghana	88.80	73.20	23.80	15.70	5.00	1.80	3.22	2,080.1	1.13%
Greece	2,610.00	2,610.00	2,590.00	2,570.00	1,530.00	624.00	232.67	13,415.0	6.34%
Greenland	0.68	0.56	0.48	0.21	0.06	0.06	12.06		0.00%
Grenada	1.00	0.87	0.35	0.24	0.10	0.10	9.36	984.1	0.97%
Guadeloupe	67.10	67.10	0.00	0.00	0.00	0.00	148.97		
Guam	25.60	25.60	0.13	0.13	0.13	0.13	158.22		2.89%
Guatemala	88.30	82.70	31.40	3.60	2.00	1.80	5.43	1,284.1	1.67%
Guinea	7.70	7.70	5.10	3.20	2.20	1.80	0.64	1,222.4	1.40%
Guinea - Bissau	2.10	1.70	1.20	0.99	0.82	0.62	1.19	1,864.8	6.83%
Guyana	4.50	4.50	4.30	4.00	2.90	2.70	5.86	1,305.8	1.15%
Haiti	8.50	8.30	5.80	4.10	2.50	1.40	0.79	1,059.5	3.19%
Honduras	431.00	409.00	73.60	5.00	2.80	1.70	48.10	20,030.8	12.89%
Hong Kong	2.50	2.50	2.50	2.50	0.90	0.33	0.35	7.8	0.01%
Hungary	257.00	88.90	20.40	14.40	3.90	3.40	26.27	2,066.9	0.54%
Iceland	4.80	3.90	2.50	1.50	0.90	0.53	14.53	239.4	0.00%
India	9,310.00	5,180.00	3,270.00	2,610.00	1,460.00	470.00	7.11	4,113.1	1.36%
Indonesia	207.00	145.00	83.00	46.70	34.70	13.30	0.80	222.0	0.15%

Iran	95.00	34.96	9.14	1.43	1.43	1.43	1.20		0.05%
Iraq	28.30	10.40	7.20	4.50	4.00	3.10	0.78	165.0	0.06%
Ireland	26.30	9.70	4.00	2.00	1.00	0.50	5.60	89.4	0.06%
Israel	1,050.00	895.00	690.00	455.00	243.00	158.00	130.20	3,294.2	2.92%
Italy	19,140.00	18,910.00	18,320.00	17,780.00	16,300.00	12,810.00	321.66	10,346.1	8.63%
Ivory Coast	7.30	4.20	2.80	2.30	1.60	1.20	0.32	201.9	0.18%
Jamaica	26.50	11.00	6.60	2.70	0.43	0.43	9.23	1,889.2	1.06%
Japan	42,550.00	33,870.00	23,420.00	13,500.00	7,230.00	4,910.00	332.49	8,614.4	5.24%
Jordan	274.00	117.00	31.70	4.80	1.10	0.79	29.91	7,088.4	2.26%
Kazakhstan	62.10	54.80	25.20	12.80	4.70	1.30	3.50	464.6	0.10%
Kenya	61.70	54.90	31.60	25.40	17.80	12.30	1.31	874.8	1.49%
Kiribati	2.30	1.70	0.94	0.51	0.34	0.20	20.46	13,875.1	14.41%
Korea, DPR	14.70	12.40	7.70	5.30	3.30	2.40	0.58		0.13%
Korea, Rep.	4,270.00	3,360.00	1,900.00	1,350.00	974.00	786.00	84.40	3,025.7	1.12%
Kuwait	37.50	35.50	15.70	9.30	6.00	3.50	9.53		0.12%
Kyrgyzstan	5.40	5.10	5.00	4.10	1.20	0.02	0.92	824.3	0.08%
Laos	5.60	5.60	3.00	2.70	1.80	0.68	0.84	352.1	0.29%
Latvia	3.10	2.40	2.30	1.50	1.20	0.68	1.56	112.0	0.04%
Lebanon	20.00	6.80	4.10	2.10	0.74	0.74	3.42	420.7	0.18%
Lesotho	0.44	0.30	0.29	0.26	0.18	0.12	0.20	200.0	0.00%
Liberia	1.70	1.30	0.92	0.87	0.67	0.49	0.38	809.1	0.77%
Libya	5.50	4.80	4.20	3.90	2.80	2.50	0.88		0.03%
Lithuania	75.50	72.10	60.70	37.00	8.80	0.74	25.75	1,766.5	0.78%
Luxembourg	249.00	243.00	233.00	104.00	50.90	33.30	439.35	4,153.6	0.00%
Macedonia	19.80	18.10	15.70	11.20	6.50	2.90	9.52	1,816.6	0.37%
Madagascar	18.20	11.20	5.90	5.70	3.50	2.30	0.75	1,821.7	1.31%
Malawi	7.50	1.80	1.40	1.30	1.00	0.55	0.43	1,378.2	0.38%
Malaysia	336.00	250.00	229.00	73.70	61.90	13.50	10.94	1,133.8	0.35%
Maldives	6.80	4.40	4.20	2.30	1.80	1.20	16.25	1,893.6	0.00%
Mali	18.80	17.60	16.20	15.50	13.80	10.60	1.08	1,338.5	3.60%
Malta	92.00	55.60	32.80	21.60	14.60	9.60	215.15	8,402.5	6.27%
Martinique	65.00	62.50	0.00	0.00	0.00	0.00	168.46		
Mauritania	33.70	18.70	15.60	12.60	7.80	1.70	8.06	7,271.4	4.80%
Mauritius	23.20	15.50	12.10	11.00	4.10	1.60	18.42	1,907.2	1.31%
Mexico	410.00	254.00	187.00	125.00	51.70	35.30	3.26	392.0	0.24%
Moldova	2.40	2.30	1.20	0.91	0.63	0.20	0.59	355.6	0.06%
Mongolia	7.00	5.00	4.50	3.50	3.50	3.30	2.35	627.2	0.22%
Morocco	265.00	67.10	57.40	36.90	23.70	18.00	7.61	2,612.3	1.07%
Mozambique	18.20	11.90	11.80	9.60	3.50	2.50	0.65	1,652.3	0.22%
Namibia	33.40	23.20	15.90	8.50	3.20	2.80	13.77	3,253.1	1.26%
Nepal	37.70	20.40	6.70	6.20	3.80	3.70	1.32	1,783.0	1.56%
Netherlands	2,000.00	1,440.00	1,050.00	695.00	262.00	139.00	118.07	2,594.6	1.46%
New Caledonia	9.90	6.80	4.30	3.80	3.40	3.20	36.79		0.52%
New Zealand	66.60	43.70	30.10	23.50	14.50	7.50	14.43	360.0	0.18%
Nicaragua	11.60	10.90	7.50	4.70	3.60	2.30	1.91	876.7	0.51%
Niger	7.30	4.60	3.30	2.00	1.80	1.80	0.37	972.2	1.22%

Nigeria	93.60	68.10	45.60	37.80	28.70	23.30	0.52	231.1	0.52%
Norfolk Island	1.40	0.24	0.24	0.24	0.24	0.17			
Norway	25.60	16.10	13.00	11.10	8.40	7.80	4.92	69.1	0.02%
Oman	13.30	12.80	5.90	4.70	2.20	1.60	3.17	200.6	0.11%
Pakistan	1,020.00	780.00	121.00	45.50	18.20	10.80	5.39	3,595.9	1.91%
Panama	90.60	48.30	19.90	8.90	6.00	1.50	22.83	1,641.7	1.34%
Papua New Guinea	11.40	9.10	6.40	4.20	3.00	1.20	1.44		0.48%
Paraguay	3.20	2.70	1.30	0.74	0.37	0.18	0.48	116.6	0.06%
Peru	215.00	183.00	163.00	100.00	49.20	14.80	6.85	1,119.2	0.89%
Philippines	898.00	215.00	109.00	21.00	16.20	14.30	8.83	2,945.2	1.34%
Poland	192.00	81.60	27.60	7.90	5.80	2.90	5.02	408.9	0.10%
Portugal	492.00	461.00	405.00	284.00	231.00	166.00	47.22	2,405.1	1.56%
Puerto Rico	171.00	113.00	96.00	96.00	1.90	1.90	46.55		1.31%
Qatar	41.00	41.00	30.80	13.10	5.00	2.40	16.52	268.9	0.22%
Republic of Palau	0.96	0.90	0.32	0.19	0.14	0.14	45.10	3,276.5	
Reunion	245.00	180.00	0.00	0.00	0.00	0.00	283.77		
Romania	1,400.00	1,350.00	1,230.00	1,050.00	69.60	7.70	70.43	7,499.0	3.32%
Russia	340.00	268.00	83.00	13.00	10.00	7.00	2.36	265.0	0.03%
Rwanda	11.80	11.80	11.20	2.60	2.10	1.20	1.01	1,408.8	4.73%
Sao Tome and Principe	0.35	0.18	0.14	0.08	0.02	0.01	1.79	997.0	0.00%
Saudi Arabia	205.00	137.00	81.60	55.20	33.40	21.30	6.50	317.1	0.14%
Senegal	107.00	38.70	27.60	19.10	15.60	10.10	7.14	7,246.6	4.79%
Serbia	14.60	11.00	10.40	5.70	3.50	1.30	1.65	386.8	0.06%
Seychelles	2.00	1.20	0.85	0.53	0.16	0.10	21.34	1,401.2	0.94%
Sierra Leone	5.30	4.80	4.40	4.00	2.80	0.40	0.73	1,444.6	5.57%
Singapore	115.00	68.40	37.00	14.30	6.40	4.70	20.78	387.3	0.00%
Slovakia	595.00	589.00	576.00	559.00	531.00	470.00	109.39	6,644.2	2.12%
Slovenia	300.00	285.00	283.00	270.00	208.00	87.20	144.59	6,819.6	2.67%
Solomon Island	3.00	3.00	2.40	2.20	1.30	0.55	5.11	2,495.6	5.93%
Somalia	3.60	2.10	1.10	0.32	0.16	0.11	0.26	579.1	1.63%
South Africa	1,730.00	1,260.00	915.00	252.00	118.00	68.00	31.29	5,867.6	1.22%
Spain	5,210.00	5,180.00	5,180.00	5,020.00	4,790.00	4,420.00	112.29	4,228.6	3.22%
Sri Lanka	51.40	23.00	18.00	15.30	10.10	8.30	2.48	632.1	0.53%
St. Helena	0.35	0.35	0.02	0.01	0.01	0.01	86.76		8.09%
St. Lucia	1.00	0.80	0.33	0.07	0.07	0.07	5.64	725.4	0.00%
St. Vincent & Grenadines	0.85	0.65	0.13	0.06	0.06	0.06	7.77	1,102.8	1.07%
Sudan	12.50	9.50	6.80	5.80	4.20	2.70	0.32	130.8	0.35%
Suriname	5.10	4.60	3.30	1.10	0.79	0.56	9.22	1,408.6	0.46%
Swaziland	0.89	0.66	0.60	0.54	0.50	0.46	0.67	238.8	0.09%
Sweden	150.00	127.00	71.50	40.40	21.90	14.60	15.36	293.5	0.10%
Switzerland	1,630.00	1,360.00	1,050.00	748.00	414.00	214.00	195.92	2,470.3	3.02%
Syria	7.00	4.60	2.00	1.70	1.50	1.20	0.37		0.04%
Taiwan	1,240.00	975.00	590.00	342.00	207.00	66.00	52.80		0.64%
Tajikistan	3.60	3.40	2.80	2.50	1.70	0.70	0.42	517.9	0.04%
Tanzania	43.40	28.80	19.10	14.40	9.60	5.40	0.81	915.0	1.30%
Thailand	2,250.00	1,690.00	1,400.00	926.00	439.00	213.00	32.77	5,530.4	1.92%

Togo	15.20	15.20	7.50	3.30	2.00	1.40	2.05	3,454.5	2.41%
Trinidad and Tobago	4.50	4.50	3.90	3.30	1.80	0.38	3.31	214.4	0.09%
Tunisia	51.10	41.30	28.30	25.00	15.90	9.80	4.53	1,214.9	0.53%
Turkey	880.00	364.00	70.70	49.40	33.40	13.90	11.24	1,025.9	0.46%
Turkmenistan	5.00	4.60	2.60	1.70	0.85	0.50	0.90	138.2	0.06%
Uganda	36.90	20.90	13.20	9.40	6.90	5.00	0.92	1,445.5	1.62%
Ukraine	957.00	886.00	855.00	692.00	416.00	193.00	21.43	10,260.5	0.65%
United Arab Emirates	298.00	233.00	182.00	118.00	118.00	44.00	32.55	854.5	0.51%
United Kingdom	11,430.00	9,080.00	5,260.00	3,370.00	1,810.00	955.00	174.78	4,364.5	2.90%
United States	41,140.00	25,990.00	18,370.00	12,200.00	7,400.00	4,160.00	128.59	2,215.5	1.26%
Uruguay	91.80	66.40	22.40	6.00	4.50	2.40	26.75	1,751.2	1.23%
Uzbekistan	5.60	5.60	3.50	2.70	1.40	0.33	0.18	83.3	0.02%
Venezuela	14.40	14.40	12.90	10.10	6.60	4.40	0.46		0.02%
Vietnam	42.50	30.00	15.00	11.80	11.80	9.70	0.45	209.8	0.04%
Virgin Islands	2.40	2.40	0.00	0.00	0.00	0.00			0.00%
West Bank	7.70	1.20	0.00	0.00	0.00	0.00	1.65	574.8	0.14%
Western Sahara	0.25	0.25	0.25	0.25	0.25	0.25	0.48		0.53%
Yemen	63.90	7.30	6.60	1.30	0.73	0.50	2.37	2,339.2	1.70%
Zambia	6.80	5.00	4.80	4.00	2.70	1.80	0.42	347.8	0.12%
Zimbabwe	11.00	6.70	4.30	3.90	3.30	3.00	0.70	675.3	0.23%

ARENA's Large Scale Solar funding impact on utility-scale solar in Australia

Bert Herteleer[1], Olivia Boyd[2], Anthony Dobb[2], Steven Rodgers[2], Lyndon Frearson[1]

[1]Ekistica, Desert Knowledge Precinct, Alice Springs, 0870, Northern Territory, Australia,
bert.herteleer@ekistica.com.au
[2]Australian Renewable Energy Agency (ARENA), 2 Phillip Law St NewActon, Canberra City, 2601, Australian Capital Territory, Australia

ABSTRACT: Over 2015-16, the Australian Renewable Energy Agency ran a competitive reverse auction process to provide up to AU$100M grant funding for utility-scale solar photovoltaic projects in Australia, aimed at stimulating the utility-scale solar market in the country, building local capacity, and reducing long-term costs. Over an 18-month period, ARENA evaluated more than 75 project proposals with a large spread in $/W costs, and used multiple selection criteria and phases to arrive at 12 projects that were provided funding for a total of 492 MW_{ac}. The use of multiple selection criteria (rather than a pure focus on cost) ensured that the long-term financial viability of the projects were considered alongside cost-competitiveness. As a result, all funded projects reached financial close by May 2017, less than 18 months after the competitive round was launched, and started construction shortly after. Drivers for success for similar reverse auctions are discussed, highlighting the benefits of an established network of due diligence providers, previous internal experience with similar projects and a market-driven approach. The funding round design and previous ARENA experience reduced the need for ARENA grant support from an expected AU$0.5/W (and AU$1.6/W in 2014) to AU$0.18/W for successful applicants under the round. The utility-scale solar market in Australia has since demonstrated the ability to secure financing and reduce total system costs without any further need for grant funding support.

Keywords: reverse auction, utility-scale solar, Knowledge Sharing

1 INTRODUCTION

Historically, the uptake of utility-scale solar power has lagged the residential sector in Australia by a significant margin. By the end of 2015, only 4 solar photovoltaic (PV) plants with over 5 MW_{ac} capacity were active on Australia's grids, whereas a total of 4683 MW_p of non-utility scale solar had been installed, with most of this capacity in the residential sector [1].[1] Since its inception, the Australian Renewable Energy Agency (ARENA) and its predecessor organisations and initiatives have funded a few utility-scale solar installations, ranging from 1 MW (Uterne, 2011) to 102 MW (Nyngan, 2014) in Australia.

While the successive efforts by ARENA and its predecessor agencies had provided funding for utility-scale solar in Australia, the rate and trend of uptake was slow, compared to the domestic residential market and worldwide utility-scale developments. Australian utility-scale PV system costs (in US$/W and AU$/W) and the Levelised Cost of Electricity (LCOE) historically exceeded similar projects worldwide, due to a variety of reasons, many of which were a direct result of the immaturity of the Australian utility scale solar market, ranging from supply chain issues, labour, and shipping costs, to difficulties of obtaining local financing. Early utility scale solar projects funded by ARENA had a levelised cost ranging between $AU152-182/MWh (US$137-164/MWh based on 2014 exchange rates), and had needed up to AU$1.6/W of ARENA grant support.

Throughout 2014-2015, ARENA observed an uptick of individual applications for funding for utility-scale solar projects. With ARENA's mandate and focus on stimulating innovation and lowering costs of renewable power in Australia, a process of consultations with stakeholders (developers, Engineering, Procurement and Construction (EPC) firms and financiers) was undertaken to develop a funding approach that would best serve the large-scale solar market, and deliver improved value for the ARENA-supplied grant funding.

From these consultations, and the experience ARENA had gained on its previous funding for renewable projects (solar and non-solar) – e.g. AGL's Nyngan and Broken Hill solar farms and the FRV's Moree Solar Farm through the Solar Flagships program, and a range of off-grid and fringe-of-grid solar farms through the Regional Australia's Renewables Program [2] – ARENA announced the Large-Scale Solar PV (LSS) competitive funding round on 14 July 2015 [3], and opened the funding round to Expressions of Interest (EOIs) from industry on 9 September 2015 [4]. ARENA's announcement of up to AU$100M of grant funding available coincided with the Clean Energy Finance Corporation (CEFC) announcing that it would provide approximately AU$250M of debt financing for interested and approved projects. ARENA's expectation was that this funding would result in approximately 200 MW_{ac} of added capacity, at an average grant cost to ARENA at AU$0.5/W. The LSS competitive funding round was run as a reverse auction, which provided ARENA (and subsequently, the market) with a price discovery benefit.

To qualify for funding, projects had to meet the following criteria:

- All electricity must be generated from solar PV;
- PV system size greater than 5 MW_{ac};
- A validated LCOE below AU$135/MWh (≈US$95/MWh, ≈€90/MWh)[2] using ARENA's LCOE calculator;

[1] The Large Scale Solar PV Competitive Round defines 'large scale' solar as > 5 MW_{ac}.

[2] Average exchange rate at time of announcement.

- A maximum grant size of AU\$30M for a single project;
- The project must be connected to the National Electricity Market (NEM) or the South-West Interconnected System (SWIS), the two regulated electricity markets in Australia which are deemed to be on-grid according to Australian regulations;
- The projects must not involve behind the meter elements.

The LSS funding round shares common ground with the Regional Australian Renewables (RAR) programme (more details in [2]), which was launched by ARENA in 2013. To achieve the goals of ARENA's Knowledge Sharing mandate, applicants who chose to participate in the LSS competitive PV funding round must provide ARENA and the market[3] with:

- Standard project information (system size, orientation, fixed or 1-axis tracking, …);
- Financial models used, highlighting the impact of ARENA's grant on the project standard metrics such as the Internal Rate of Return (IRR);
- A capital expenditure (CAPEX) breakdown of the project cost completion;
- Operational data post completion for a period of 5 years, to demonstrate that the Key Knowledge Sharing components of the project are met;
- Lessons learnt (Knowledge Sharing) reports for external distribution to the market.

By giving the market access to a level of granularity of data not otherwise accessible, ARENA intended to establish benchmarks that could drive further competition in utility scale solar, help to increase market understanding of solar yield and performance, and enabling a more objective comparison of Australian utility scale solar performance to international standards. Using this data, ARENA has provided the market with the (typically) aggregated information in a variety of formats, ranging from workshops, blog posts to lessons learnt reports on ARENA's website. Once the projects are operational, the performance data will also allow future similar projects to benchmark their expected yields, which is important given the wide range of climatic conditions present in Australia.

2 LSS FUNDING ROUND PHASES AND EVALUATIONS

2.1 EOI phase

ARENA received 77 eligible EOIs for funding between 9 September and 2 November 2015, for a total capacity exceeding 2.8 GW. The competitiveness of the respective bids was then assessed on the basis of five merit criteria [5]:

1. Contribution to the funding program objectives: cost reduction and competitiveness (measured as required ARENA grant funding per MWh of expected generation, transparency and price discovery, demonstrated ability to deliver large-scale solar PV at a cost below AU\$135/MWh;
2. Applicant capability and capacity to deliver the project, including demonstrated level of experience and expertise of the applicant/consortium. Appropriateness of the proposed allocation of roles and responsibilities within the project team/consortium, the quality and completeness of project documentation, and project team/consortium's ability to reach strict program deadlines;
3. Activity design, methodology, risk and compliance: The focus was on the quality of the project plan and other supporting information, the resources in place to deliver the project as per the application specifications, the quality and completeness of the risk management plan, and the overall risk profile of the project;
4. Financial viability and co-funding commitment: The focus was on whether a project had firm financing commitments in place on which due diligence had already been performed, a demonstrated sound approach to managing risks that might impact the financial viability of the project and project revenues, and a demonstrated understanding of how the financing process may affect other areas of the project (e.g. contracting);
5. Knowledge sharing: The focus was on the degree to which the applicant agreed to comply with the terms of the model LSS Knowledge Sharing Plan, which sets out all data to be shared, reports to be produced and other standardised knowledge sharing requirements.

The assessment process involved parallel assessments by ARENA staff and external due diligence (including legal, technical and financial service providers), whose reports were combined and analysed. Further assessments were obtained from ARENA's Advisory Panel (AAP), which consists of recognised experts on renewable energy who provide advice to ARENA on an ongoing basis. The various project assessments were subsequently ranked and moderated, with the top-rated projects (according to the merit assessment criteria) invited to proceed to the full application.

In March 2016, ARENA published aggregated and anonymized information on the EOI applications [6], covering factors such as typical Operations & Maintenance Expenses (OPEX) cost per MWh, or the Capital Expenditures (CAPEX) costs for systems in \$/W_{ac} and \$/W_{dc}. Media coverage and interviews undertaken by ARENA stuff in April 2017, which included financiers in some of Australia's 'big four' banks and due diligence providers, demonstrated that this initial data release provided crucial detailed benchmark information on utility-scale PV, in an immature market that had previously had a vacuum of such benchmark data both in the public domain and in private firms' databases. Anecdotal evidence suggests that this early data release may have played a role in driving further competition and accelerating the de-risking of utility scale solar PV in Australia.

[3] The information provided to the market is typically in aggregated/anonymised form, with some exceptions on standard descriptions, such as system size, total cost and ARENA funding amount.

2.2 Full application phase

A shortlist of 22 projects was publicly announced on 14 January 2016 [7], who were invited to submit a full application, with the deadline on 15 June 2016.

The round included a provision that 'material changes' in the project from the project plan and design submitted in a shortlisted application would result in exclusion from funding. 'Material changes' were defined as a change in project location, a deviation project AC power rating by over 10%, or an upward revision of the project LCOE in excess of 5%. This was an important feature of the round design to ensure continuity in the project proposals: projects were required to remain generally similar to the original proposal.

As in the EOI phase, projects were subjected to detailed due diligence by ARENA staff and external parties, which also included AAP assessments. The assessments from the different sources were combined, and recommendations were issued on the project proposals. The project proposals recommended for ARENA grant funding were then delivered to the ARENA board, which, upon approving the proposals, resulted in the projects being issued an Offer of Funding / Offer to Negotiate.

ARENA facilitated transfer of knowledge from the first wave of utility scale solar PV developers in Australia to the shortlisted applicants, in an effort to improve the quality of all final bids. In an ARENA-hosted closed-door workshop with the 22 short-listed applicants that took place in April 2016, project managers from the ARENA-funded Moree, Broken Hill and Nyngan plants shared their key project development lessons learnt with shortlisted applicants, the CEFC provided lessons learnt on securing financing, and ARENA worked with the short-listed applicants to identify future data and knowledge sharing priorities [8]. This was followed by another workshop where other ARENA-funded utility scale solar plants shared lessons learnt on challenges associated with reaching financial close and permitting, planning and construction with the final successful applicants [9].

On 8 September 2016 ARENA announced that 12 out of the 22 projects had been approved for ARENA grant funding. This marked a significant acceleration of the deployment of utility scale solar PV in Australia, exceeding even ARENA's original expectations for the round. In total 482 MW of utility-scale solar projects were selected to proceed towards financial close (compared to an expected 200 MW at the outset of the round), with an average ARENA grant at AU$0.19/W, for a total support from ARENA at AU$91.7M [10].

3.3 Financial close phase

Previous experience of ARENA funding for large scale renewable energy projects for emerging technologies had demonstrated the risk that long lag times between signing a funding agreement for successful projects and satisfying the conditions necessary for financial close could endanger the project's cost competitiveness and viability, particularly when technology costs were falling rapidly.

Bid bonds were a crucial design feature of the Large Scale Solar PV Competitive Round, and were built into the round in order incentivise swift progress by a project towards meeting the conditions necessary for ARENA to commit to project financial close. Projects that accepted an Offer of Funding from ARENA were then required to lodge a bid bond calculated at AU$20k/$MW_{ac}$, with a minimum value of AU$250k and a maximum of AU$500k. The projects which subsequently reached financial close on time would have their bid bonds returned, thus further incentivizing project proponents towards project delivery.

Throughout the financial close negotiations, ARENA reserved the right to review the grant funding to be provided to projects if the project economics changed materially to the benefit of proponents. In the context of a rapidly evolving market (and rapidly evolving input costs), this ensured that ARENA, as a taxpayer-funded entity, could continue to support renewable project developments without excessive profits for the awardee or causing significant market distortion.

By May 2017, all 12 projects approved for ARENA funding had reached financial close, with further changes to the total project costs, system capacities (the total changed from 482 MW to 492.5 MW) and required amount of ARENA funding [11]. Figure 1 shows the locations and power ratings of the LSS funded projects.

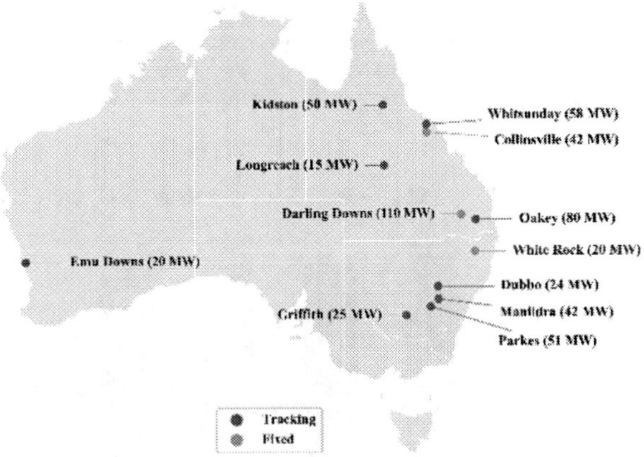

Figure 1: Locations and AC power ratings of the LSS-funded utility-scale solar systems in Australia. Tracking refers to single-axis tracking systems.

3 RESULTS AND DISCUSSION

The price discovery aspects of the LSS funding round are evident in Figure 2, where the impact of the design choices and the merit criteria used by ARENA are also evident: while low $/W values were of importance (and that highest $/W CAPEX values allowed easier project exclusion from further consideration), other aspects such as the proponents' capacity to successfully deliver the project, willingness to participate in knowledge sharing, or ability to replicate the results and economics of a winning project across the industry in the future were given equal consideration.

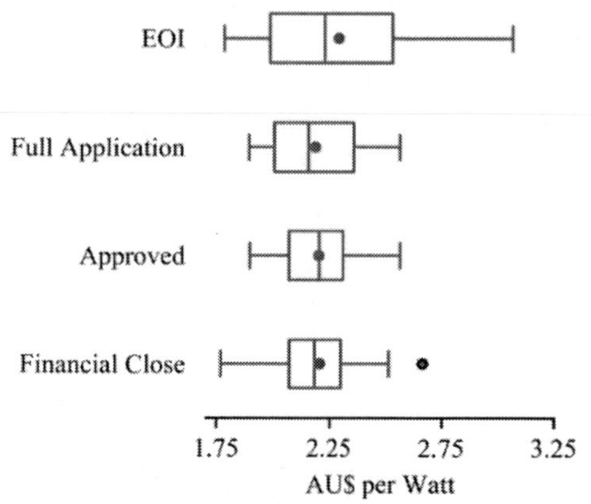

Figure 2: Boxplots for the four main phases in the LSS competitive funding round, illustrating the spread in CAPEX costs for the projects. The dot inside the box indicates the mean, and the line the median value. Values calculated and shown on a per-project basis.

This resulted in a marked reduction in the min-max spread of project costs, nearly halving from AU$1.28/W at the EOI stage to AU$0.67/W in subsequent LSS phases.

Figure 3 shows how the structure of the funding agreements and the due diligence performed over the different phases allowed ARENA to reduce the average grant funding to projects (as calculated from the total funding amounts), even as some projects' total cost increased from approval to financial close. In general, proponents achieved $/W improvements over the financial close phase.

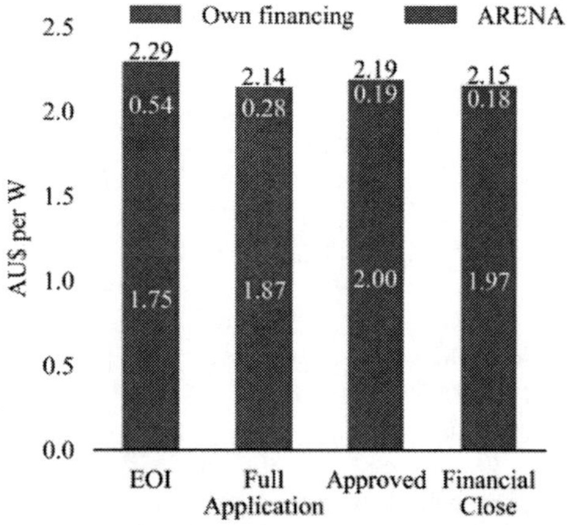

Figure 3: Capacity-weighted average system CAPEX costs over the LSS phases, showing the reduction in the need for ARENA funding over time (EOI – Financial Close: approximately 18 months).

In general terms from ARENA's perspective, the results of the LSS competitive funding round have been a resounding success on multiple criteria.

- Market interest and capacity: the EOI phase saw a combined total of 2.8 GW from 77 eligible projects.
- Price, and price discovery: the majority of projects achieved an LCOE below AU$135/MWh, and many went below this by a significant margin. Of the approved projects, using ARENA's standard LCOE calculator [12] and the Full Application's June 2016 average exchange rate [13], the highest LCOE was AU$132.5/MWh (€90/MWh, US$101.5/MWh), and the lowest LCOE was AU$97.8/MWh (€66.4/MWh, US$74.9/MWh).
- Value for (taxpayer) money and market stimulus: Comparing the historical amount of ARENA support needed to non-LSS utility-scale solar projects, the average support was AU$1.25/$W_{ac}$ for 246 MW_{ac}. The selection criteria and iterative process of the LSS round resulted in projects being able to reduce their need for ARENA funding from the EOI phase to financial close, from an average of approximately AU$0.54/$W_{ac}$ to less than AU$0.18/$W_{ac}$ for 492 MW_{ac}.
- Self-sufficiency: the Australian utility-scale market has taken the momentum and interest from the LSS round further, with new projects continuing to being announced with further cost declines. The Australian Energy Market Operator (AEMO) reporting on announced solar projects totalled more than 1700 MW in 2016, while (solar-only) connection requests have increased to 11,043 MW by 1 July 2017 [14].
- Project momentum: All projects approved for LSS grant funding were able to secure financing and reach financial close, and were in various stages of construction in September 2017. Contrasting this with reverse auctions held in other countries (e.g. in Brazil [15]), timeliness and completion rates for reverse auctions are not always 100%. Auctions that focus on cost only as a selection and decision criterion have had projects delay start of construction [16], or failing to reach financial close, as the resultant bidding was too aggressive to deliver projects on the ground in a limited timeframe. Very low winning bids in international reverse auctions focused only on price have also raised concerns that winning projects may encounter performance issues over the standard 25-year lifespan for a solar PV plant.

4 DRIVERS FOR SUCCESS OR FAILURE

While the LSS competitive funding round has been remarkably successful in providing the much-needed push to the utility-scale solar industry in Australia, not all of this success can be ascribed to ARENA, and entities wishing to replicate this model need to be cognisant of the drivers for successful, and unsuccessful, reverse auctions. In particular, many drivers or levers that influence the success of reverse auctions cannot be directly controlled by agencies such as ARENA.

Among the aspects that ARENA was able to control or influence were:

The definition of success: ARENA's primary goals were related to stimulating competition and strengthening the necessary supply chains (both technological and know-how) to enable the utility-scale solar industry to be self-sufficient. ARENA's focus on continuous innovation has resulted in no more funding to be needed for LSS-like solar projects in Australia, and if such projects apply for funding in the future, they would need to demonstrate a strong need for such support, e.g. by combining utility-scale solar power with other technologies that improve integration of renewable energy into a secure and reliable grid, such as storage or co-location with wind farms.

The LSS funding round was *not* focused on funding the lowest cost projects. Rather it was focused on increasing the competitiveness of the large scale solar industry in Australia overall, including building market capacity, de-risking utility scale solar to ensure uptake of lower-cost private debt financing, and, ultimately, the growth of the utility scale solar PV sector without further need for government subsidies. This unwillingness to simply accept the lowest-cost bid was more than compensated by subsequent solar market development. All LSS projects under the round achieved financial close less than a year after they submitted shortlisted applications, and new utility scale solar farms without grant-funding from ARENA continue to be announced as the sector scales up to achieve the Renewable Energy Large-Scale Target of 33,000GWh of power generation from renewable energy by 2020. The funding boundaries imposed under the round (e.g. maximum AU$30M for a single project) also resulted in more projects being proposed. Fewer, larger projects with a higher magnitude of grant funding may not have benefited the market to the same extent.

The combination of grant funding from ARENA and the potential for CEFC debt funding for utility-scale solar provided greater certainty for solar developers, EPCs and financiers that participation in the LSS funding round would have a high enough probability of success, and that the financial incentive of participation was large enough. Similarly, the shortlisted project proponents who failed to obtain ARENA funding had the ability to claim up to 50% of their verifiable third-party costs up to a cost of AU$2M, if they could prove that the project would otherwise have continued, would have reached financial close within 6 months of approval, continued to meet the LCOE benchmark and changes not be materially significant [5].

The LSS funding round design with good due diligence processes (and established networks of trusted companies to provide these services to ARENA) struck a balance between providing funding to projects, while ensuring that ARENA could objectively determine whether benefits were correct. Previous ARENA experience with providing grant funding to large, utility-scale, high risk projects for emerging and relatively un-tested renewable energy technologies (at least in the Australian context) also provided critical internal lessons learnt that influenced the design of the LSS round.

ARENA's LSS LCOE calculator enabled a more objective comparison of the different projects (especially at the EOI phase), and projects were required to link their financial models to the standardised LCOE calculator. This provided an extra objective metric for inclusion or exclusion of projects, and made analyses across the portfolio of submissions more manageable. However, the (significantly) lower PPAs struck between retailers and solar farm developers also point to the limitations of the LCOE as a PPA-forecasting metric, particularly given the many project-specific factors that may enable more favourable long-run marginal costs.

The consultations with the market allowed ARENA to develop a reverse auction that was fit for purpose, and answered and unlocked latent market demand. Once these consultations indicated the need and benefit of a utility-scale solar reverse auction, ARENA could move quickly and announce the LSS competitive funding round. However, the ability to move quickly was there because ARENA had laid the ground work in the previous years by establishing the necessary company networks for due diligence work and had the internal capacity to design and evaluate the funding round.

Other aspects that contributed to the success of the LSS funding round, yet are rarely guaranteed to be in place are the following:

The existence of latent demand and capacity to deliver in Australia: large engineering firms in Australia looked to diversify in the years after the Australian mining boom had ended, and invested in developing the capacity to deliver EPC services in the renewable energy sector. Moreover, a limited number of solar farms had been developed in parallel to ARENA-funded projects with Australian State Government support (such as the Australian Capital Territory's reverse renewables auction), which increased the capacity for the market to deliver. Together with worldwide CAPEX improvements in PV system hardware, many of the conditions for rapid uptake were present. The timely boost of funding from ARENA and the CEFC delivered through a focused, competitive process was able to harness and accelerate the impact of these trends.

The utility scale solar PV sector built upon previous experience that financing institutions in Australia had built up in the wind energy sector (which had historically developed at a better pace than the solar sector). This contributed to a marked drop in financing costs for solar projects in Australia over the course of the LSS funding round, with observed (base case) Debt Service Coverage Ratios (DSCRs) for Power Purchase Agreements (PPAs) as low as 1.25 [17].

Policy uncertainty surrounding carbon pricing and the Renewable Energy Target (RET) in Australia slowed investment and built up latent demand, with wind and solar developers delaying planned investments. The impact of these delays on the Federally legislated Renewable Energy Target (RET) and its target of 33,000 GWh from renewable energy sources by 2020 have pushed up prices of Large Generation Certificates for systems that are active before 2020. Given the short window of opportunity to 2020 to build sufficient capacity to meet the RET, the shorter project development timeframes for utility scale solar versus wind farms has further stimulated interest in the solar industry in Australia. ARENA's LSS funding round can therefore not claim all of the impact on the utility-scale solar market in Australia, however, the LSS funding round has had a marked role in pushing forward investments and cost declines in utility-scale solar by up to 3 years, compared to the previously observed trends [17].

5 CONCLUSIONS

ARENA's competitive funding round for large-scale solar projects has provided a critical stimulus to the utility-scale solar industry in Australia, providing funding for 12 projects whose combined capacity of 492 MW_{ac} is almost double the previously installed utility-scale solar capacity. All 12 LSS-funded projects have successfully achieved financial close and are expected to be operational from 2018 onwards.

Multiple selection criteria that integrated cost competitiveness with project viability, and design features to incentivise the delivery of low-risk projects on time and on budget has provided ARENA and the Australian solar market with price discovery benefits. This has reduced the need for ARENA grant funding from average values of more than AU$0.5/W to an average LSS value of AU$0.18/W (-64%) for the successful projects under the round. The LSS funding round and associated interest has stimulated the utility-scale solar market in Australia to such an extent that it no longer needs ARENA funding.

The design of the LSS funding round has balanced economic returns for the participating projects while ensuring that ARENA's support was limited to the required amount. ARENA was able to negotiate with the project proponents on the strength of its internal and external due diligence network, and this allowed ARENA to select and support projects which could best contribute to the longer-term health of the utility-scale solar industry in Australia. The deliberate choice by ARENA by selecting project proposals for quality over cost only in the LSS reverse auction has been validated by the subsequent utility-scale solar PV market development.

The Knowledge Sharing requirements for project proponents has allowed ARENA to provide early information to project proponents and the market on price discovery, and the performance data that will be gathered from the supported projects will allow future projects that incorporate utility-scale solar to be benchmarked in Australia, and further inform decision-makers on investments to be made.

ACKNOWLEDGEMENTS
The authors wish to thank Lachlan McLeod and George Dickeson for the graphic inputs, and ARENA staff in the development and delivery of the round (including Matt Walden, Chris Twomey and Dan Sturrock, who reviewed an earlier draft of this paper), as well as the LSS projects for their participation in the LSS funding round. Ekistica is ARENA's Knowledge Sharing partner and data handler under Work Order CEAP-WO-14/15-003.

REFERENCES

[1] W. Johnston and R. Egan, "PV in Australia in 2015," 18 August 2016. [Online]. Available: http://www.iea-pvps.org/index.php?id=93&eID=dam_frontend_push&docID=3109. [Accessed 15 June 2017].

[2] B. Herteleer, O. Boyd, A. Dobb, S. Rodgers and L. Frearson, "Identifying risks, costs and lessons from ARENA-funded off-grid renewable energy projects in regional Australia," in *33rd EU PVSEC*, Amsterdam, 2017.

[3] ARENA, "ARENA announces new priorities," 14 July 2015. [Online]. Available: https://arena.gov.au/news/arena-announces-new-priorities/. [Accessed 15 September 2017].

[4] ARENA, "$350m to kick start new large-scale solar projects," 9 September 2015. [Online]. Available: https://arena.gov.au/news/350m-to-kick-start-new-large-scale-solar-projects/. [Accessed 15 September 2017].

[5] ARENA, "Updated Funding Announcement," 02 2017. [Online]. Available: https://arena.gov.au/assets/2017/02/LSSPV-Funding-Announcement-updated-1.pdf. [Accessed 18 September 2017].

[6] ARENA, "Large-scale solar PV competitive round – EOI application data," March 2016. [Online]. Available: https://arena.gov.au/funding/programs/advancing-renewables-program/large-scale-solar-photovoltaics-competitive-round/eoi-application-data/. [Accessed 18 September 2017].

[7] ARENA, "ARENA selects 22 large-scale solar projects to take next step," 14 January 2016. [Online]. Available: https://arena.gov.au/news/arena-selects-22-large-scale-solar-projects-to-take-next-step/. [Accessed 15 September 2017].

[8] J. Martin II, "Large-scale solar developers get a leg up through ARENA knowledge sharing," 4 May 2016. [Online]. Available: https://www.solarchoice.net.au/blog/news/large-scale-solar-developers-leg-up-through-arena-knowledge-sharing-030516. [Accessed 21 September 2017].

[9] ARENA, "Why sheep could be good for your solar farm," 17 November 2016. [Online]. Available: https://arena.gov.au/blog/why-sheep-could-be/. [Accessed 21 September 2017].

[10] ARENA, "Historic day for Australian solar as 12 new plants get support," 8 September 2016. [Online]. Available: https://arena.gov.au/news/historic-day-for-australian-solar-as-12-new-plants-get-support/. [Accessed 19 September 2017].

[11] ARENA, "Large Scale Solar PV Projects," June 2017. [Online]. Available: https://arena.gov.au/assets/2017/06/AU21509_ARENA_Large_Scale_Solar_Brochure_v6-1.pdf. [Accessed 18 September 2017].

[12] ARENA, "ARENA LCOE Calculator," February 2017. [Online]. Available: https://arena.gov.au/assets/2017/02/LCOE-calculator-1.xlsx. [Accessed 18 September 2017].

[13] Australian Taxation Office, "Monthly exchange rates for 1 July 2015 to 30 June 2016," 20 December 2016. [Online]. Available: https://www.ato.gov.au/Tax-professionals/TP/1-July-2015-to-30-June-2016/. [Accessed 19 September 2017].

[1 AEMO, "Electricity Statement of Opportunities," September 2017. [Online]. Available:

[14] https://www.aemo.com.au/-/media/Files/Electricity/NEM/Planning_and_Forecasting/NEM_ESOO/2017/2017-Electricity-Statement-of-Opportunities.pdf. [Accessed 20 September 2017].

[15] A. Viane, "Reverse Auctions to Scale Renewable Energy: Brazilian Approach," 6 June 2017. [Online]. Available: https://d2oc0ihd6a5bt.cloudfront.net/wp-content/uploads/sites/837/2017/06/4_Auctions_Renewables_Brazilian_Approach.pdf. [Accessed 18 September 2017].

[16] IRENA and CEM, "Renewable Energy Auctions – A Guide to Design," 24 June 2015. [Online]. Available: http://www.irena.org/DocumentDownloads/Publications/Renewable_Energy_Auctions_A_Guide_to_Design.pdf. [Accessed 14 September 2017].

[17] M. Walden, "LSS Round, challenges experienced and lessons learned from the process and trends we have observed," Clean Energy Council, 11 May 2017. [Online]. Available: https://www.cleanenergycouncil.org.au/dam/cec/events/SIF-17/matt-walden-presentation/Matt%20Walden.pdf. [Accessed 18 September 2017].

TECHNICAL AND ECONOMIC POTENTIAL OF PV IN LEBANON AND JORDAN AIMING FOR REGIONAL READINESS LEVEL DEVELOPMENT

M. Haidar[1,2] P. Baliozian[1], A. Mustafa[2], S. Mourad[2], and R. Preu[1]
[1] Fraunhofer Institute for Solar Energy Systems (ISE), Heidenhofstraße 2, 79110 Freiburg, Germany
[2] Albert-Ludwigs-Universität Freiburg, Zentrum für Erneuerbare Energien (ZEE), Tennenbacher Straße 4, 79106 Freiburg, Germany
Contact email: mohamad.haidar@ise.fraunhofer.de

ABSTRACT:
This paper provides an approach to calculate the technical and economic potential for photovoltaic (PV) power generation in Lebanon and Jordan as an initial attempt to regionally quantify energy. The technical potential was estimated using analytical and *GIS* (*Geographic Information System*) based methods having sets of constraints and assumptions. Moreover, the economic potential is also estimated using *GIS*. The results of this study indicate a calculated technical potential with a total of 21330 TWh/a in both Lebanon and Jordan compared to the analytically calculated technical potential of 20941 TWh/a. As for the economically feasible energy production, the expected total electricity production is 596 TWh/a. The *GIS* method provides a regional visualization of the potentials on the map showing regions with high and low potentials. Having projects in the PV sector can clearly cover a significant share of the demand, knowing that the economic potential in this approach shows an energy production with an order of magnitude higher than the total electricity demand of 26.7 TWh/a. The concepts of technology readiness level (TRL) together with the results obtained from the study motivate the introduction of a new concept in standardizing PV installed capacity maturities called Regional Readiness Levels (*RRL*). The *RRL* notion ranks regions according to parameters based on market potential of PV, aiming to overcome communication barriers within the PV community.
Keywords: Photovoltaic, technical potential, economic potential, GIS, Regional Readiness Level

1 INTRODUCTION

With today's environmental challenges and increasing energy demand, renewable energies have become inevitable to secure a sustainable future. Photovoltaic (PV) based solar energy conversion systems play a significant role in the clean energy transition. Today, solar PV installations worldwide exceed 300GW[1]. According to the International Technology Roadmap for Photovoltaics, the installed PV capacity is expected to increase in a logistic growth curve, reaching a yearly annual growth of 70 GW_p [2]. The aggregate penetration of photovoltaic electricity in the energy market requires technological advancements achieving higher efficiencies coupled with decreasing costs, along with progress in careful resource planning.

The main driver of PV systems' power generation is the solar irradiation that varies by a factor of four from a location to another on earth's surface [3]. Meaning that regional studies are necessary to estimate resources and study the technical and economic feasibility as an effort towards creating market targeting incentives.
Lebanon and Jordan are the countries that witnessed a high flux of refugees due to the Syrian war that started in 2011, reaching around 1.7 million refugees [4]. It is overt that a sudden increase of population requires a positive response in increasing electricity production to cover the increase of demand, especially if the population of the host country is of the same order of magnitude. Additionally, Lebanon's electricity infrastructure was severely damaged in the civil war (1975-1990) and was never properly reconstructed for many reasons. The electricity demand exceeds the supply, and Lebanese residents are subjected to daily blackouts reaching 13 hr/day [5]. The availability of solar energy, growth of the PV sector and the urgent need of electricity generation directs the focus towards this particular region.

The overall aim of this research paper is to assess the PV situation in Lebanon and Jordan, and calculate the technical and economic potentials of PV power generation within the vicinity of the two countries as a region of interest. In addition to that, a PV Regional Readiness Level (RRL) is introduced, with an approach of placing regions in a scale of PV market maturity. This study not only calculates an approximate PV electricity potential of Lebanon and Jordan; it also approaches the topic of standardizing regional PV installation maturities in RRLs as a possible method of establishing a common language within the PV community. In section 2, a brief description of the electricity production in the two countries Lebanon and Jordan is portrayed. Section 3 defines the PV potentials, and narrows down to the method of calculation used in obtaining the technical and economic potential of the region. In section 4, the results of the calculations are shown in addition to the introduction of the Regional Readiness Level table. The methodology used is illustrated in figure 1 below.

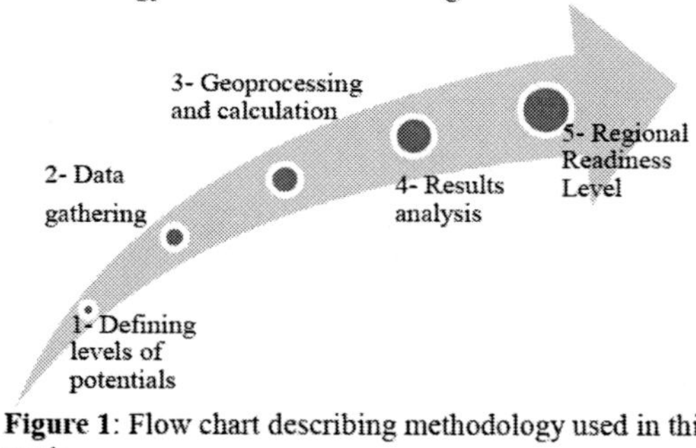

Figure 1: Flow chart describing methodology used in this work.

2 CURRENT ELECTRICITY PRODUCTION OVERVIEW

2.1 Lebanese electricity production

The main supplier of electricity in Lebanon is Electricity of Lebanon (*EDL*), which is a public institution linked to the Ministry of Energy and Water. *EDL* covers around 67 % of the electricity demand, leaving the rest for the households to manage. Two electricity bills are then paid, one for *EDL* and another expensive one for the private diesel generator owners that provide the back-up self-generated electricity during *EDL*'s blackouts.

In 2015, around 94% of *EDL*'s produced electricity was fossil fuel based, leaving around less than 6% to hydropower and imports. Solar PV electricity generation provides around 0.11% of *EDL*'s electricity generation, around 0.014 TWh out of the total 12.41 TWh of electricity generated [6]. Lebanon's official renewable energy target is attaining 12% renewable electricity and heat generation by 2030 [7]. The electricity generation and distribution sector in Lebanon needs to improve both in covering the demand as well as increasing the share of PV to meet the target. The Lebanese government recently approved plans to implement 180 MW of solar PV by 2020[8].

2.2 Jordanian electricity production

The Jordanian power sector is now regulated by the Electricity Regulatory Committee (*ERC*). The generation companies include the privately-owned and state-owned Electric Power Generating Company. The electricity generated from generators is sold to the state-owned National Electric Power Company (*NEPCO*) which operates the transmission systems. The electricity is then distributed to consumers through three privately-owned distribution companies [9]. 99.9% of the population has access to landline electricity. The total nominal capacity of the generation system is about 3,300 MW.

Jordan is heavily dependent on the import of foreign energy resources with 96% of its energy resources have to be imported. In 2012, around 14.3 TWh of electricity was consumed in Jordan, with a steady increase of around 5% per year over recent years [10]. Jordan has an official target of covering 10% of its primary energy demand using renewables by 2020 [7]. To meet its targets, the Jordanian energy sector has already implemented 77 MW of solar PV with another 540 MW currently under execution and 200 MW to be awarded in 2017. [11]

3 THE PV POTENTIALS, CALCULTION METHODS AND ASSUMPTIONS

3.1 Defining the potentials of PV

Different sources of renewable energy present different opportunities for utilization. It is then necessary to evaluate and assess the opportunities or potential presented by each one of them. Combining different sets of information, assumptions and constraints will yield different levels of potential. Lopez *et al.* define four main levels of potential. These levels are built upon each other, providing a clearer and more comprehensive understanding of the potential presented by a given source of renewable energy [12]. A quick overview of the different levels of potential is shown below in Figure 1.

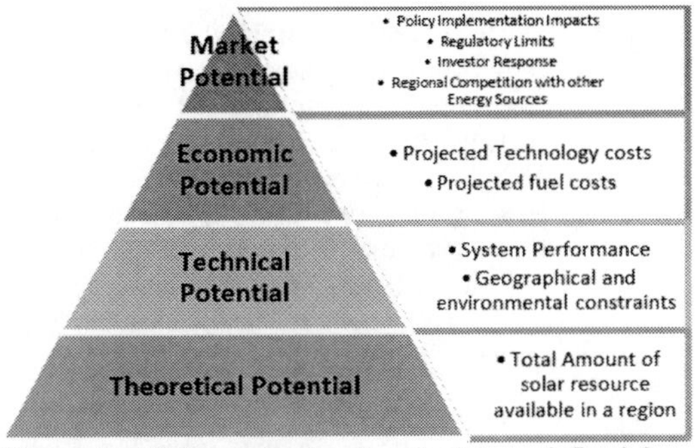

Figure 2: Different levels of potential as defined by Lopez *et al.*, adapted from [12].

3.2 Assumptions and calculation method of the PV potentials in Lebanon and Jordan

After defining the different potentials of PV system installations, Lebanon and Jordan are chosen as a region of interest for the potential study. The *Yingli* YL 260 W multi-crystalline solar panel is a common module used in Jordan. [13] It was selected for the study as standard module with an efficiency of 16%.

Analytical calculations using equations 1 and 2 and a set of assumption and givens from references:

$$RP = A_{total} * G_{avg} \quad (1)$$

RP: resource potential in GWh/a

A_{total}: total area of country or region of interest in km^2

G_{avg}: Yearly average Global Horizontal Irradiation (GHI) for the country/region of interest in $\frac{kWh}{m^2}/a$

$$TP = A * G * \eta * PR \quad (2)$$

TP: Technical potential in GWh/a

A: Available land area is determined by land use constraints, km^2

- Exclude all forests and protected areas
- Exclude all agricultural areas (arable land and permanent crops)
- Exclude all water areas

G: Yearly average GHI for the constricted region of interest in $\frac{kWh}{m^2}/a$

η: Standard module efficiency is the module efficiency under standard testing conditions (Module Temperature T= 25°C, Irradiance G= 1000W/m^2)

PR: Performance Ratio as defined in [14].

Assumptions:

- All PV panels are installed horizontally
- Performance Ratio of PV system is 0.8 for both Lebanon and Jordan since in Jordan PRs were found to be around 0.82 for Amman and around 0.77 for Aqaba [10].

- Module efficiency of selected module is 0.16
- Only 10% of urban areas can be used for PV: 20% of roof space is considered to be available and only 50% of the rooftops suitable for PV installation. This is a conservative assumption; however it is very common in Lebanon and in Jordan to have water storage tanks on the roof, Heating Ventilation and Air conditioning (HVAC) equipment and multiple satellite dishes.

As for the economic potential:

The levelized cost of PV electricity over 20 years varies depending on site, system price and operation and maintenance (O&M) costs. The table 1 below provides a comparison of PV and fossil based generation. The PV prices vary depending on system type, location, and O&M costs while fossil-based depends heavily on fossil fuel prices. Grid connected utility scale systems are cheaper and off-grid systems with battery storage are the most expensive [15]. Many PV systems have achieved grid parity and some systems are cheaper than fossil based electricity. Systems including storage remain more expensive. Therefore, the economic potential of PV is estimated to be at a penetration rate of 20-40% of total electricity demand. Above this penetration rate, PV will lose its economic value and become more expensive due to the requirement of storage caused by the intermittent nature of solar energy [16]. The fact that annual demand and annual solar irradiation curves follow the same pattern helps to allow a higher penetration rate due to direct consumption.

Table 1: LCoE of PV compared to fossil fuels

Country	Jordan	Lebanon
PV LCoE (EUR/kWh)	>0.057 [11]	0.046-0.485 [15]
Average cost of production and distribution from fossil fuels (EUR/kWh)	0.146 [10]	0.14 [5]

Geographic Information System (*GIS*) based calculation:

The calculation of the potential can also be computed and visualized on *GIS*. Visualizations might help policy makers pass more effective and region specific legislation. The original equation is modified in the following manner:

$$\boldsymbol{TP = A_{cell} * N_{cells} * G_{cell,Avg.} * \eta * PR} \quad (3)$$

TP: technical potential in kWh/a
A_{cell}: Raster Cell Area in m^2
N_{cells}: Number of cells
$G_{cell,avg.}$: Average GHI/cell, calculated using Zonal Statistics in *GIS* once all exclusions and assumptions taken into account in $\frac{kWh}{m^2}/a$.

Excluded zones consist of:

- Zones with slopes higher than 5%
- Land used for Agriculture (Permanent Crop areas + Arable Land)
- Protected and Forest Areas
- Lakes and rivers
- Only 10% of Urban areas can be used for PV, urban areas are included regardless of slope. Other methods to identify available urban space as explained by Gagnon *et al.* in [17] require either unavailable light detection or ranging (LIDAR) sensor data or work intensive selection rendering them inapplicable for large areas.

The same formula can be used for the economic potential. Estimations can be made by adding the following assumptions and constraints to the study:

- Neglect PV module degradation rate and penetration rate constraint.
- Exclude all regions with below average radiation levels
- Exclude all regions not located within 15 km buffer radius of electrical substations
- Exclude all regions above 1500 meters in Altitude
- Include urban areas with a 1/3 ratio of urban technical potential: In other words, utility scale PV is more economic than residential/urban PV installations. [13][15].

The necessary data for the study was collected from different sources. Table 2 below is listing all data and sources:

Table 2: *GIS* data sources

Data Content	Data Type	Data Source
Urban Areas	Vector	Natural Earth
Water	Vector	Open Street Maps by GeoFabrik
Land use (Agriculture)	Vector	Open Street Maps by GeoFabrik
Protected Areas and Forests	Vector	WPDA consortium
Digital Elevation Model	Raster	ASTER Global DEM v2 from EarthData Nasa
Global Horizontal Irradiation	Raster	Solar-Med-Atlas
Direct Normal Irradiation	Raster	Solar-Med-Atlas

4 RESULTS AND DISCUSSION

The results are shown in table 3. The *GIS* processing produced the required results, visible in figures 3 to 5.

With the current state of technology and applying many geographical and environmental constraints, the technical potential obtained is huge for both countries. It is three orders of magnitude larger than demand for Lebanon and 4 for Jordan.

Table 3: Analytical results

Country	Unit	Lebanon	Jordan	Total	Ref.
Total Area	km^2	10452	89342	99794	[20] [21]
G_{avg}	$kWh/m^2/a$	1900	2100	-	[18] [19]
RP	TWh/a	19858	187618	207476	-
Land area	km^2	10230	88802	-	[20] [21]
Forest area	%area	13	1.1	-	[20] [21]
Agricultural land	%area	25	11.4	-	[20] [21]
Built-up area	%area	23.3	4.5	-	[20] [21]
$TP_{analytic}$	TWh/a	1021	19919	20941	-
TP_{GIS}	TWh/a	730	20600	21330	GIS
EP_{GIS}	TWh/a	518	78	596	GIS

The *GIS* technical potential of both Lebanon and Jordan returned similar results to the analytical estimation. As for Lebanon, the technical potential is lower than the analytically derived one. This is mainly because many highly sloped areas were excluded during *GIS* processing. Lebanon has large mountainous regions that can explain this exclusion. The variations can also be explained by the fact that, each time an area is excluded a new average for GHI has to be calculated which is done in the GIS but not in the analytic calculations.

The estimated economic potential is very high, considering the assumptions taken. In fact, it is 36 times demand for Jordan and 6 times for Lebanon. However the real economic potential would be at a penetration rate of around 20% to 40% as discussed above. Therefore, the current methodology does not attempt to project the level of PV generation that might be deployed in the future, but rather as a tool to identify the regions where PV projects are economically feasible and interesting. More advanced economic studies can be done using LCoE, which would require unavailable detailed country wide project costs data. Such studies exist on a regional European level as shown in Ossenbrink *et al.* [22]. The current analysis can be further refined to select the most economic locations enough to meet a certain PV penetration rate. A first approximation or judgement is that the central part of Jordan is best suited for PV as it is not far away from the demand and has very high economic potential. For Lebanon, the eastern region known as the Beqaa valley is suited for economic PV generation

A political and social analysis will further decrease the potential. However, it is very hard to define constraints to analyze this potential analytically or using *GIS*. That is where the concept of RRL becomes helpful.

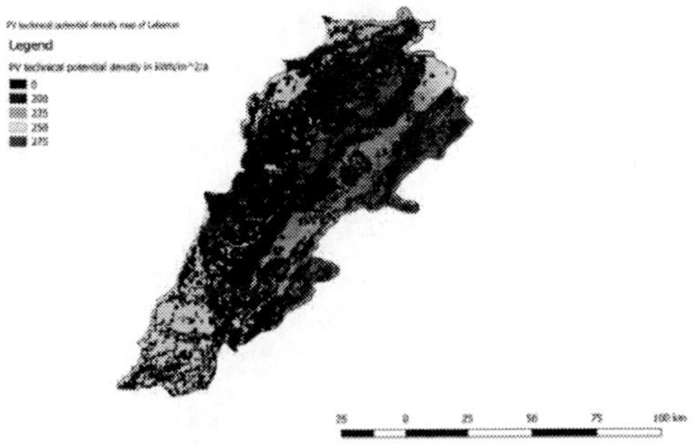

Figure 3: PV technical potential of Lebanon as estimated using the implemented methodology.

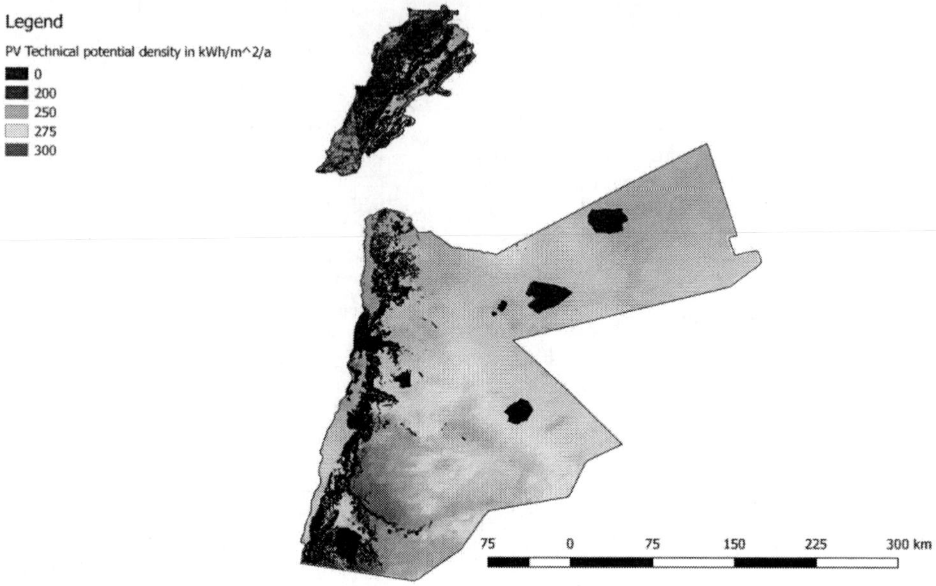

Figure 4: PV technical potential of Lebanon and Jordan as estimated using the implemented methodology.

5 REGIONAL READINESS LEVELS

A comparison of this assessment to other similar studies is difficult since different methodologies, definitions and assumptions are used. The idea of an RRL or country readiness level comes from the Technology Readiness Level *NASA Goddard Space Flight center* first proposed in the 1970s as a method to scale technology maturity from the concept observation stage to space flight readiness maturity [23]. A scale of 9 levels was created to rate the different maturities on a checklist format. The *European Commission* within the *Horizon 2020* program adopted such an approach to create 9-level TRLs as well as 10-level Manufacturing Readiness Levels (*MRLs*) [24]. Moreover, *Fraunhofer ISE* created a 10-level TRL stages table for PV cell production maturity [25].

The ambitious goal of *RRLs* differs from the previous concepts by assessing the readiness of implementing PV in a certain country instead of assessing a specific technology readiness/development. The RRL will characterize the maturity of a specific region by considering technical, economic and socio-political aspects, thus aiming for a standardized simplified language of discussion among renewable energy stakeholders. The RRL in table 4 below begins from conceptual aspects of PV planning to project execution to mass deployment, including the different sides in a single simple scale. The RRL is to be updated yearly.

Having a Regional Readiness Level scale provides a standard way of placing regions in levels of PV market maturity. In other words, it provides a qualitative and quantitative overview of the actual development stage in the PV market depending on parameters affecting such a development. The following table is a suggestion of such a scale. A more detailed RRL table can be created by including additional relevant factors affecting the advancement of PV in a specific region. After having a holistic scale agreed upon in the PV community, regions such as EU, Lebanon and Jordan could be marked

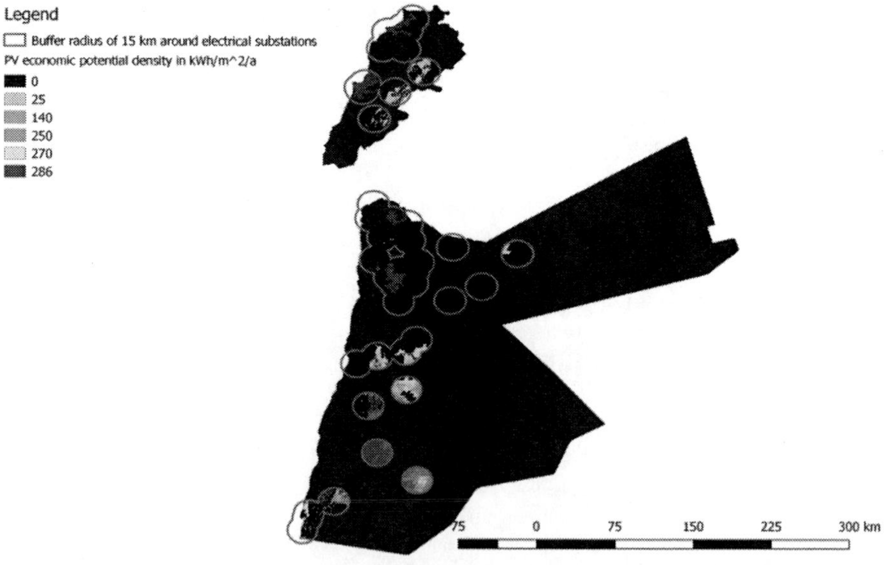

Economic Potential of PV for Lebanon and Jordan as estimated using current methodology

Figure 5:

according to their respective RRLs. If the notion of RRLs develops to a future PV project standard, it would be a keyword used by scientists, engineers, investors and policy makers, thus closing the communication gap between these stakeholders.

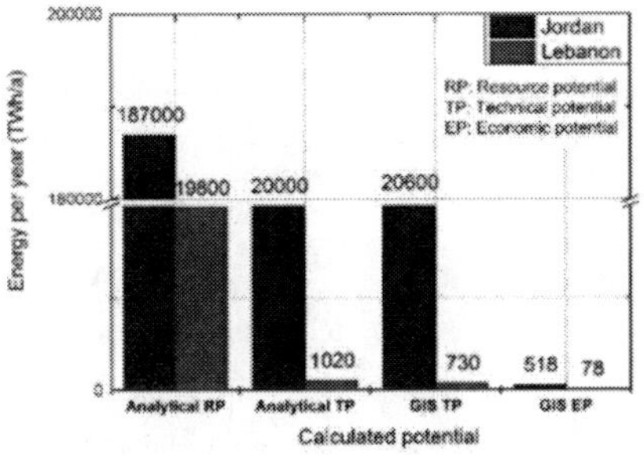

Figure 6: Results for the different calculated PV potentials in Lebanon and Jordan

Table 4: Regional Readiness Levels for PV

RRL0	• Basic Resource Potentials Observed • Official PV target exists
RRL1	• PV energy action plan • Market entry mechanism defined (1st come 1st served, tenders....)
RRL2	• Legal Framework for Renewables exists • Financial and regulatory incentives (Feed-in tariffs, tax reductions)
RRL3	• Institution responsible for tracking PV progress exists • Forecast for PV installation exists • Project subject to milestones or development timelines
RRL4	• Project mobilization and tracking through above mentioned milestones • Infrastructure proved to withstand PV installation
RRL5	• Local manufacturing of PV or part of the value chain • Local installation of PV
RRL6	• Peak shaving of electricity demand by PV in some areas at certain times
RRL7	• Peak shaving of electricity demand by PV in several areas for most of the time
RRL8	• Significant amount of electrical energy of the regional demand is supplied by PV
RRL9	• PV is a dominating source to supply the regions electrical demand

Lebanon can be placed currently at the RRL level of 2 based on the above overview of its electricity sector, since new plans have been announced. Policy makers in Lebanon would have to offer investors more incentives and plan ahead as well. Jordan is more advanced, with an RRL level of 6, since local manufacturing exists and the amount of PV currently operational allows peak shaving in some areas at certain times [26]. Germany is placed at an RRL level of 8, since PV currently represents a significant portion of the energy supply. [27]
The RRL is then a good tool to guide policy towards changes that will be beneficial for proper implementation of PV.

6 CONCLUSION AND OUTLOOK

In conclusion, Lebanon and Jordan have shown a high technical and economic potential compared to demand. Jordan has the larger potential between the two countries. The assumption based *GIS* study enables us to visualize the different potentials on a regional level. This can be beneficial for different stakeholders.

Developing a regional readiness scale simplifies the analysis even further for these stakeholders and presents the results in a unified standard format, hence closing the gap between the stakeholders during discussions. The concept of RRL is therefore introduced in this paper. Such a scale helps to compare the PV capabilities of different countries on a regional level and is an important tool for policy makers to assess the current situation.

The accuracy of this study is limited to the accuracy of the *GIS* data. For more reliable studies, assumptions should be further reduced by different models, while constraints and parameters increased. The need for more reliable and sensor measured data is urgent as well. The RRL could be developed to incorporate more detailed factors affecting the PV installed capacity on a regional level. Regions like EU can be included and compared to Lebanon and Jordan.

7 ACKNOWLEDGEMENTS
The authors would like to thank Heinz Ossenbrink for valuable input during the 32nd EUPVSEC session he chaired, which shaped the idea behind this study. The authors would also like to thank Julian Fleischmann from the University of Freiburg, Rami Fakhoury, Reem Irany and Patil Mesrobian from the Lebanese Center for Energy Conservation (LCEC), for their valuable information and references.

8 REFERENCES

[1] J. Dobson, C. Schlufter, T. Knoblich, A. Stiehler, Longer Term Investments, Renewables, UBS, 2017

[2] VDMA (2017). International Technology Roadmap for Photovoltaics - www.*ITRPV*.net.

[3] Hoogwijk, Monique Maria (2004): On the global and regional potential of renewable energy sources.

[4] United Nations High Commissioner for Refugees (UNHCR). "UNHCR Syria Regional Refugee Response." UNHCR Syria Regional Refugee Response. N.p., n.d. Web. 18 Jan. 2017.

[5] Fardoun *et al.* (2012) Electricity of Lebanon: Problems and Recommendations, Energy Procedia 19 (310-320).

[6] Solar PV Status Report for Lebanon (2016), UNDP,

Ministry of Energy and Water.

[7] Pan-Arab Renewable Energy Strategy 2030 – Roadmap of Actions for Implementation. International Renewable Energy Agency (2014).

[8] Bellini, Emiliano. "Lebanon issues RfP for 180 MW of solar." Pv magazine International. N.p., 31 May 2017. Web. 07 June 2017.

[9] Åberg, Emma, *et al.* "Up-scaling Solar PV for Self-Consumption in the Jordanian Market." TO THE FUTURE (2014): 72.

[10] Enabling PV in the MENA Region The Emerging PV Market in Jordan, GIZ, 2014

[11] Middle East Solar Outlook 2017. Rep. Middle East Solar Industry Association, 2016. Web. 9 June 2017.

[12] Lopez *et a.l,* U.S. Renewable Energy Technical Potentials: A GIS Based Analysis, NREL, July 2012

[13] Jordan Solar Power. "P.V Solar Panel." P.V Solar Panel. N.p., n.d. Web. 23 Jan. 2017. <http://jordansolarpower.net/main/index.php/products/p-v-solar-panel>.

[14] Smets, Arno H. M., Klaus Jäger, Olindo Isabella, René ACMM Van Swaaij, and Miro Zeman. Solar energy: the physics and engineering of photovoltaic conversion, technologies and systems. Cambridge, England: UIT Cambridge Ltd., 2016. Web.

[15] El Khoury, Pierre, Memorandum Ref: M13-6-V 5, Lebanese Center for Energy Conservation (LCEC), 2016

[16] Brown *et al.* "Estimating Renewable Energy Economic Potential in the ...", NREL, n.d. Web. 23 Jan. 2017.

[17] Gagnon et al. "Rooftop Solar Photovoltaic Technical Potential in the ...", Technical Report NREL/TP-6A20-65298 NREL. N.p., Jan. 2016. Web. 18 Jan. 2017.

[18] GroundWork for the development of Renewable Energy Action Plan. Rep. Association Libanaise pour la maîtrise de l'énergie et pour l'environnement, Dec. 2015. Web. 9 June 2017.

[19] Etier, Issa, Anas Tarabsheh, and Mohamad Ababneh. "Analysis of Solar Radiation in Jordan." Jordan Journal of Mechanical and Industrial Engineering 4.6 (2010): 733-38. Web. 9 June 2017.

[20]"The World Factbook: LEBANON." Central Intelligence Agency. Central Intelligence Agency, 12 Jan. 2017. Web. 18 Jan. 2017.

[21]"The World Factbook: JORDAN." Central Intelligence Agency. Central Intelligence Agency, 12 Jan. 2017. Web. 18 Jan. 2017.

[22] Ossenbrink *et al.* (2015), Towards a European Union of Photovoltaic Solar Electricity. 31st European Photovolatic Solar Energy Conference and Exhibition (EUPVSEC)

[23] Mankins, John C. "Technology readiness assessments: A retrospective." Acta Astronautica 65.9 (2009): 1216-1223.

[24] European Commission, Appendix G of Horizon 2020 – Work Program 2014-2015.

[25] Baliozian, P *et al.* (2016), Photovoltaic Development Standardizing Based on Roadmaps and Technology Readiness Levels (TRLs). 32nd European Photovoltaic Solar Energy Conference and Exhibition (EUPVSEC).

[26] "Why Philadelphia Solar?" Why Philadelphia Solar - Philadelphia Solar, philadelphia-solar.com/why-philadelphia-solar/. Accessed 3 Sept. 2017.

[27] Wirth, Harry, "Recent facts about photovoltaics in Germany", Fraunhofer ISE, n.d. Web 20 Sep 2017.

EVALUATING THE EFFECTIVENESS OF PAST AND FUTURE FEED-IN TARIFF POLICY IN GREAT BRITAIN USING AN AGENT-BASED MODEL

Phoebe Pearce[1], Raphael Slade[2]
[1] Blackett Laboratory, Department of Physics, Imperial College London, SW7 2AZ, U.K.
[2] Centre for Environmental Policy, Imperial College London, SW7 2AZ, U.K.

ABSTRACT: A new agent-based model was used to investigate the effect of feed-in tariffs on the adoption of small-scale solar photovoltaic arrays by households in Great Britain. Both historical (2010-2016) and future (2016-2021) scenarios are modelled, allowing comparison with actual installation data and near-future projections. Agent-based models provide a natural modelling paradigm to include population heterogeneity, non-optimization decision-making and the effect of a social network. Observed installation patterns in the period 2010-2016 can be modelled accurately by considering decision-making based on four factors: household income, the payback period of the investment in the array, the number of adopters in a household's social network, and the capital cost of the installation. The level of feed-in tariffs significantly affects the adoption rate, so a realistic historical subsidy scenario was compared with several alternative scenarios. This indicates that an improved strategy for reducing feed-in tariffs over time could have given higher deployment at lower cost. The model is also used to project future deployment considering the effect of different feed-in tariff levels, electricity prices, and the cost of photovoltaics.
Keywords: Funding and Incentives, Strategy, Modelling

1 INTRODUCTION

Since 2010, feed-in tariffs (FiTs) for small-scale renewable energy installations, including solar photovoltaics (PV), have been available to individuals, communities, and industrial and commercial organisations in Great Britain, as part of the UK's climate change mitigation strategy. The majority of FiT installations (over 99%) are solar PV [1].

The total peak capacity of small-scale (defined here to mean up to 10 kW_p) PV systems installed with support from the FiT scheme now exceeds 2 GW [1], [2]. Supporting this generation comes at significant cost to utilities and thus their customers, and costs continue to rise as the scheme remains open to registrations while payments to existing installations are guaranteed for 20 to 25 years. Given the scheme's significant cost as well as the importance of increasing the share of renewable energy in electricity generation, a review of the implementation of the FiT scheme, both over the period 2010-2016 and planned policy for the next few years, is relevant. Specifically, investigating if FiT policy encouraged installation in an effective manner in the period 2010-2016, and predicting the outcome of future policy in the short term (up to 2022) can provide relevant insights for policy-makers. To assess FiT policy, this work uses a new agent-based model (ABM) constructed specifically to investigate the adoption of small-scale PV by households in Great Britain (feed-in tariffs are not available in Northern Ireland and thus do not cover all of the UK). While this model focuses on the effect of FiTs, it also includes other economic factors and the effect of a social network on adoption decisions.

2 POLICY BACKGROUND

Great Britain's FiT scheme was set out in the 2008 Energy Act and took effect from April 2010, supporting electricity generation from renewable technologies including PV [3]. The FiT scheme is intended for installations under 5 MW.

The scheme operates by requiring electric utility companies to pay owners of registered renewable energy installations for the electricity produced by a PV system. The feed-in tariff is actually composed of two tariffs: a

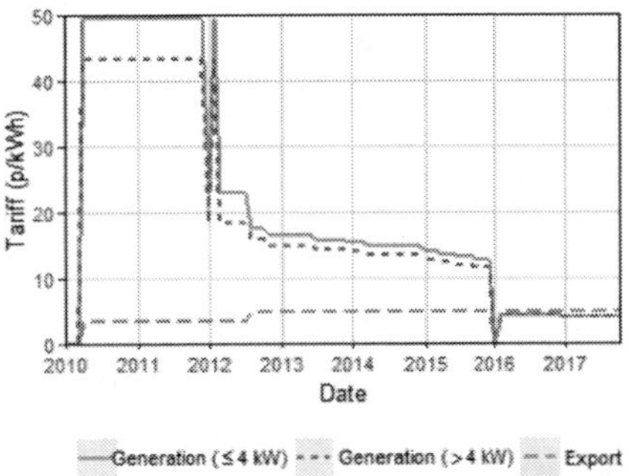

Figure 1: Generation and export tariff levels (on a monthly basis) for small scale PV installations (≤ 10 kW) in pence (£0.01) per kWh. These are the 2016 levels of the FiT [4], which is adjusted annually according to the Retail Price Index.

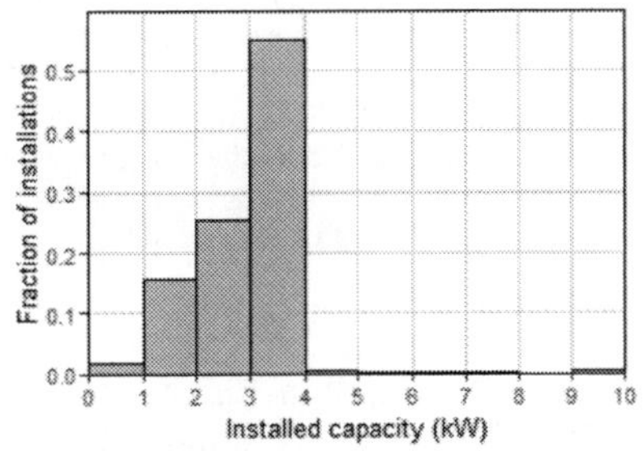

Figure 2: Capacity distribution of small-scale, domestic PV systems registered under the FiT scheme. Data from [1].

generation tariff, which is paid for all the electricity generated by a system, and the export tariff, which is paid for the energy exported (or deemed to have been exported) to the grid. Feed-in tariffs (specifically the generation tariff) have varied significantly since the introduction of the scheme in 2010, as shown in Fig. 1, with various policy consultations, legal challenges, and changes of government leading to sudden decreases in the FiT. For most of the FiT scheme's existence (2010-2015), the generation tariff paid was dependent on the peak capacity installed, with systems ≤ 4 kW receiving higher payment per unit of electricity. This appears to have led to a clear incentive to install less than 4 kW, as shown in Fig. 2. Most generators under 30 kW capacity do not have an export meter installed (since this is not required for small systems), and their electricity exports are deemed to be half of the total production [5], [6]. The model assumes this "deemed export" for all installations, since it focuses exclusively on ≤ 10 kW systems. When considering the installation of small-scale systems, it is valid to focus on adoption by households exclusively; Fig. 3 shows domestic and non-domestic FiT-registered, small-scale PV capacity in GB. Clearly, domestic installations make up the majority of all small-scale installations.

Following a consultation on the FiT scheme in 2015 [8], the scheme was paused from 15 January to 8 February 2016, after which FiTs were reduced. In addition, they are no longer capacity-dependent for installations ≤ 10 kW. The scheme now includes deployment caps which limit the amount installed under each tariff level and the total capacity which can receive FiT incentives [9], [10]. The aim of the deployment caps is to keep the additional cost of supporting new systems installed between 2016 and 2019 below £100 million/year [11]. For every quarter, a feed-in tariff and deployment cap are set [9], [10]. If the deployment cap is reached before the end of the quarter, this triggers degression, which is a reduction in the FiT to the next lowest tariff. Excess capacity is carried forward. Even if the cap is not reached, the FiT will reduce at the start of the next quarter. Deployment caps and FiTs are currently set until 31 March 2019, although the total available capacity (700 MW) available under the deployment caps may run out before this date.

3 MODEL SPECIFICATION

ABMs provide an intuitive framework to take into account explicit characteristics of both technology and human behaviour. The basic modelling elements are agents (which may represent e.g. individual people, households, or a government agency), and the collective actions of these agents leads to emergent behaviour [12, Ch. 2]. ABMs also address the issue of control; large-scale optimization models (e.g. [13], [14]) implicitly assume there is some centralised control over the energy system, which is often not the case, especially in the case of small-scale, privately-owned technologies such as solar PV.

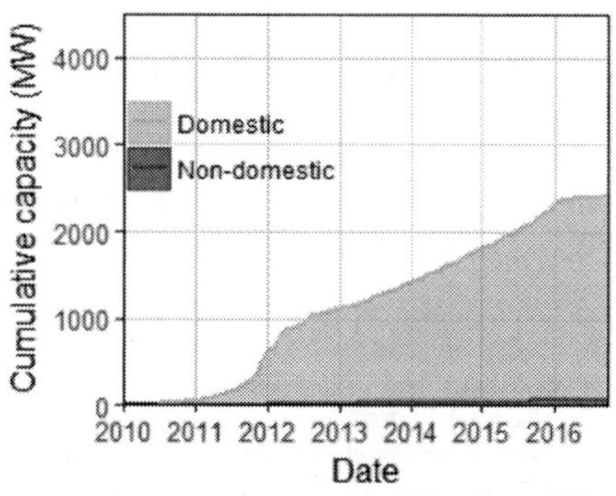

Figure 3: Small-scale PV systems which are registered under the FiT scheme, divided into capacity registered as domestic or otherwise (non-domestic). From [7].

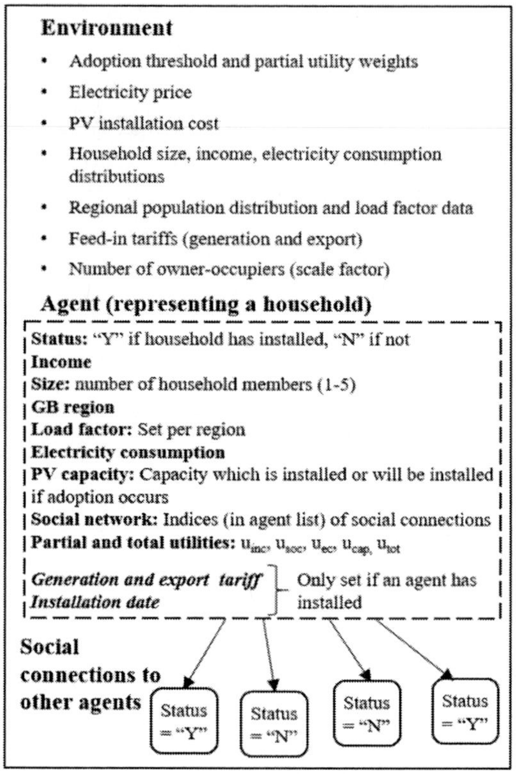

Figure 4: Schematic representation of the ABM's constituent parts and their interactions. The agent exists in the environment; an agent's social connections can be considered a part of its environment.

This model aims to predict adoption decisions (if, when, and how much PV is installed) and translate the information on individual adoptions into cumulative installed capacity and production data, while tracking associated subsidy and capital costs. The model assigns realistic agent and environment characteristics relevant to decision-making, implements a decision-making strategy, and scales the results to the GB population. In order to provide useful projections and alternative historical outcomes, the model was calibrated to observed PV deployment in the period 2010-2016. The model was implemented in the R language. The basic modelling elements are the agents, which in this model represent individual households. The model generates a population of agents of the desired size, assigning each agent relevant and realistic characteristics as summarised in Fig. 4. The characteristics are assigned randomly, based on probability distributions describing the various agent characteristics; this means each agent population is unique. The model population size is scaled to the number of households occupying properties they own ("owner-occupiers") in Great Britain (around 17 million).

3.1 Agent decision-making

The decision strategy used by agents is a "utility function" made up of four components which influence the adoption decision, similar to previous work in the field [15]–[17]. The four partial utilities, the factors which determine them, and their functional forms are:

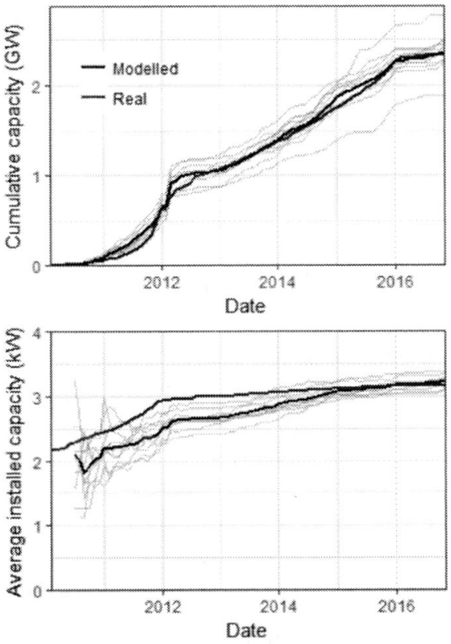

Figure 5: Real PV capacity data [1] and the output of the calibrated model running under a realistic scenario with FiTs and other parameters set equal to their actual 2010-2016 values. The thin grey lines show ten individual model runs; the overall result (black line) is their mean. Top: cumulative installed capacity. Bottom: average installation capacity (cumulative capacity divided by number of adopters). From [7].

1. u_{inc}: An income factor, based on household income I_k of agent k and mean national household income $\bar{I}$.

$$u_{inc,k} = \frac{1}{1+\exp\left(\frac{\bar{I}-I_k}{5000}\right)}$$

2. u_{ec}: Economic attractiveness of the investment, based on the simple payback period pp_k.

$$u_{ec,k}(t) = \frac{20-pp_k(t)}{19}$$

3. u_{soc}: Social environment, based on how many adopters A_k an agent is linked compared to its total number of links L_k.

$$u_{soc,k}(t) = \frac{1}{1+\exp\left(1.2\times\left[\frac{L_k}{4}-A_k(t)\right]\right)}$$

4. u_{cap}: Capital cost barrier, based on household income I_k and the capital cost C_k of installing a PV system.

$$u_{cap,k}(t) = \frac{1}{1+\exp\left(-0.0007\times\left[\frac{I_k}{5}-C_k(t)\right]\right)}$$

The linear and logistic equations which describe partial utilities 1-4 relate the characteristics considered to a number in the range [0, 1]. The weighted sum of these partial utilities is compared to some threshold t; an agent (household) k will adopt if

$$u_{tot,k} = \sum_i w_i u_{i,k} = w_{inc}u_{inc,k} + w_{ec}u_{ec,k} + w_{soc}u_{soc,k} + w_{cap}u_{cap,k} \geq t$$

Where the partial utility weights w sum to 1, and all partial utilities lie in the range [0, 1], so that the total utility $u_{tot,k}$ is always ≤ 1. This means the threshold t must also lie in the range [0, 1].

3.2 Model calibration

The partial utility weights and threshold are set globally and were calibrated so the model output matched observed installation when FiTs, electricity prices and PV costs, etc. in the model were set to realistic values over the period 2010-2016. This calibration was done through an exhaustive search covering the w and t parameter space. The results of the calibration are shown in Fig. 5. The cumulative capacity as predicted by the model matches observation very well, which is unsurprising as this is the objective of the calibration. The trends in average installed capacity predicted by the model match those observed in reality, though this was not required by the calibration.

3.3 Environment variables

The market for rooftop solar PV in the UK is currently dominated by mature technologies, predominantly crystalline silicon [18]. The model does not differentiate between different PV technologies, assuming one average installation cost per kW of peak power installed, and the same load factor for all installations in the same region (GB is divided into 12 regions for which average load factor data is reported [19]). Like the available FiTs, the cost of PV and electricity price change over time (e.g. the total system cost of a small-scale PV system has fallen from around £5000/kW to £1600/kW since 2010). For historical simulations, electricity and PV prices are based on real data, but for projections, predictions of these parameters have to be made.

3.4 Model operation

The model steps through the time frame under consideration in monthly intervals. For each time step, the model updates the environmental parameters and thus the corresponding agent parameters and utility functions. It then checks if any new agents will adopt under these conditions. The population generated for each model run is different, meaning they will produce slightly different results (see Fig. 5). For each policy scenario, the time

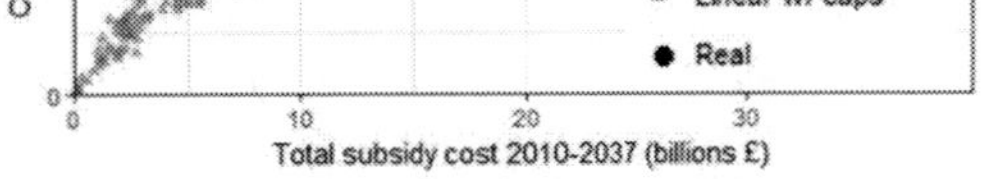

Figure 6: Cumulative installed capacity by 1 October 2016 with total expected subsidy cost, for installations installed during the period April 2010-September 2016. These are the results of around 850 independent simulations using different FiT degression strategies. The shaded area corresponds to scenarios which lead to equal or higher cumulative capacity for lower or equivalent total expected subsidy costs as compared to the simulated realistic scenario (•). Two alternative policy scenarios A and B are labelled.

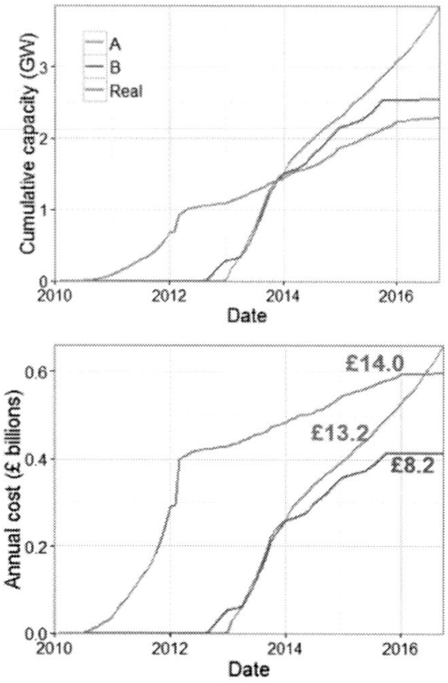

Figure 7: Top: cumulative capacity with time in the realistic scenario and alternatives A and B. Bottom: annual subsidy cost commitment with time for the five scenarios. The labels are the expected total subsidy cost to the end of the guarantee period for systems installed during the simulation period.

evolution is repeated 10 times, each time using a newly generated population of 5000 agents. Taking the mean of the outcomes (e.g. deployment, electricity production, subsidy costs) of all model runs at each time step gives an averaged result.

There were two distinct modes of model operation; one for historical simulations over the period January 2010 - September 2016, and one for projections (October 2016 - December 2021). While historical simulations start with a new agent population with no adopters, projections require pre-generated populations reflecting the fact that around 4% of the agent population has adopted by October 2016.

4 RESULTS

4.1 Historical simulations (2010-2016)

To compare real FiT policy in the period 2010-2016 to alternative scenarios utilising various degression strategies, the effects of a large number of alternative policy scenarios were modelled. The modelled cumulative capacity by October 2016 varying with total expected subsidy cost based on expected electricity production for these alternative scenarios is shown in Fig. 6. This shows that there are a number of scenarios using any of the alternative degression strategies which are predicted to have a better outcome than the realistic (calibration) scenario. The main issue with FiT policy was that initially, there was no mechanism in place for reducing tariffs, which led to rapid cost escalation in late 2011 and early 2012. Alternative scenarios with lower initial feed-in tariffs and/or pre- planned tariff reduction strategies avoid

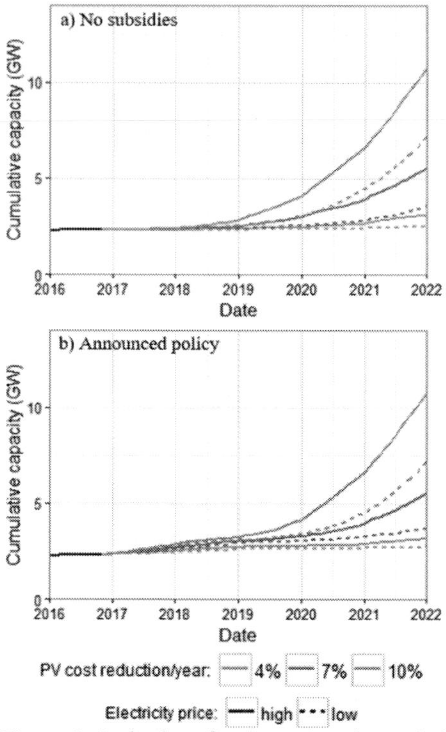

Figure 8: Projections for the projected cumulative PV capacity up to the end of 2021, depending on electricity and PV cost trends, for the two different future scenarios: a) no subsidies and b) policy as announced.

this cost escalation. Two alternative policy scenarios are marked in Fig. 6, and further details of the evolution of cumulative capacity over the simulation period and concurrent expected cost are shown in Fig. 7.

4.2 Projections (2016-2022)

The results for projections under two different viable policy scenarios are shown in Fig. 8. These scenarios are *a)* no subsidy from October 2016 onwards, and *b)* deployment cap policy as currently in place and planned up to March 2019. Fig. 8 compares total installed capacity and total expected subsidy cost under these policy scenarios as well their dependence on various trends for the price of electricity and PV installation. Clearly, the rate at which total PV system cost reduces is of critical importance to predicting the installation rate. Fig. 8 also shows that the continued existence of the FiT scheme in its currently planned form (scenario *b*) has little effect on installed capacity by January 2022 when compared with scenario *a*. Scenario *b* leads to slightly earlier adoption during the period in which subsidies are available (up to 2019), but installed capacity by January 2022 is very similar to scenario *a*, in which no further subsidies are offered.

The stated aim of implementing deployment caps in February 2016 was limiting the additional cost of supporting post-February 2016 installations to £100 million/year. The maximum additional cost incurred in scenario *b*, with high electricity and low PV cost, is £42 million/year. Small-scale PV installations account for 58% of the total solar capacity registered under the FiT scheme, and solar technologies collectively make up the bulk of installations; this indicates that costs will remain below £100 million/year.

5 CONCLUSIONS

It is possible to construct an ABM which, using demographic data to generate agent populations and simulating agent decision-making based on four factors, can be calibrated to replicate the observed installation patterns of small PV systems well. Using this model to investigate both historical and future policy shows that using a reasonable degression strategy, it would have been possible to achieve a higher installed capacity by 2016 at a lower expected subsidy cost. Using the model to project future installed capacity shows the current FiTs are so low that they have little effect in encouraging adoption, with PV cost reduction being a far more important driver. However, although the FiT does not have much effect on PV installation by 2022, it is also meeting the policy goal of keeping the cost of subsidising new FiT installations below £100 million/year.

The full model code is available at github.com/phoebe-p/FiTABM.

REFERENCES

[1] Ofgem, 2016. [Online]. Available: https://www.ofgem.gov.uk/publications-and-updates/feed-tariff-installation-report-30-september-2016. [Accessed: 01-Nov-2016].

[2] Department for Business Energy & Industrial Strategy, *Feed-in Tariff statistics*, 2016. [Online]. Available: https://www.gov.uk/government/statistics/solar-photovoltaics-deployment. [Accessed: 04-Jan-2016].

[3] Parliament of the United Kingdom, 2008.

[4] Ofgem, 2016. [Online]. Available: https://www.ofgem.gov.uk/system/files/docs/2016/04/version_2-_feed-in_tariff_scheme_tariff_table_1_april_2016_-_31_march_2017_pv_only.pdf. [Accessed: 08-Jul-2016].

[5] Energy Saving Trust, [Online]. Available: http://www.energysavingtrust.org.uk/sites/default/files/reports/Measuring the electricity you export.pdf. [Accessed: 19-Jul-2016].

[6] DECC, 2015. [Online]. Available: https://www.gov.uk/government/statistics/national-energy-efficiency-data-framework-need-report-summary-of-analysis-2015.

[7] P. Pearce and R. Slade, *under peer review in Energy Policy*, 2017.

[8] DECC, 2015. [Online]. Available: https://www.gov.uk/government/uploads/system/uploads/attachment_data/file/467074/Consultation_on_a_review_of_the_feed-in_tariff_scheme.pdf.

[9] Ofgem, 2016. [Online]. Available: https://www.ofgem.gov.uk/system/files/docs/2016/07/010416-300616_quarterly_deployment_caps_report.pdf.

[10] Ofgem, 2016. [Online]. Available: https://www.ofgem.gov.uk/system/files/docs/2016/03/caps_monthly_report_february_2016.pdf.

[11] DECC, 2015. [Online]. Available: https://www.gov.uk/government/uploads/system/uploads/attachment_data/file/487300/FITs_Review_Govt__response_Final.pdf.

[12] K. H. van Dam, I. Nikolic, and Z. Lukszo, Dordrecht: Springer, 2013.

[13] S. Messner and L. Schrattenholzer, *Energy*, vol. 25, no. 3, pp. 267–282, 2000.

[14] A. S. Manne and C. O. Wene, Upton, NY, Feb. 1992.

[15] V. Iachini, A. Borghesi, and M. Milano, 2015, pp. 136–148.

[16] J. Palmer, G. Sorda, and R. Madlener, *Technol. Forecast. Soc. Change*, vol. 99, pp. 106–131, 2015.

[17] J. Zhao, E. Mazhari, N. Celik, and Y.-J. Son, *Simul. Model. Pract. Theory*, vol. 19, no. 10, pp. 2189–2205, 2011.

[18] DECC, 2013. [Online]. Available: https://www.gov.uk/government/uploads/system/uploads/attachment_data/file/249277/UK_Solar_PV_Strategy_Part_1_Roadmap_to_a_Brighter_Future_08.10.pdf.

[19] DECC, 2015. [Online]. Available: https://www.gov.uk/government/statistics/quarterly-and-annual-load-factors.

This is a pre-peer review version of the paper submitted to the journal Progress in Photovoltaics.
It has been selected by the Executive Committee of the EU PVSEC 2017 for submission to Progress in Photovoltaics.

THE SOCIAL PROFITABILITY OF PHOTOVOLTAICS IN GERMANY

Javier López Prol[a,b,*], Karl W. Steininger[a,b]
[a] Wegener Center for Climate and Global Change, University of Graz, Brandhofgasse 5, A-8010 Graz, Austria.
[b] Department of Economics, University of Graz, Universitaetsstr. 15, A-8010 Graz, Austria.
* Corresponding author: javier.lopez-prol@uni-graz.at

ABSTRACT: While Germany has led the market in photovoltaic (PV) implementation throughout the last decade, there has been increasing criticism of PV support policies due to their high cost. Although declining, the levelized cost of electricity (LCOE) from PV is still above the German wholesale electricity price. However, using LCOE as an evaluation yardstick falls short in at least two respects: it neither takes into account integration costs rising with PV penetration (i.e. undervaluing its actual cost), nor avoided externalities of replacing conventional for renewable generation (social cost overvaluation). We thus calculate the social profitability of PV in Germany by including not only private costs and benefits, but also integration costs and avoided externalities, using the internal rate of return and the profitability index as indicators. Our results show that when these factors are considered, PV investments - up to a penetration level of at least 15% - have a social profitability above 10% at the lower bound of the social cost of carbon (150 €/tCO_2), and one that is at least positive up to that penetration level if the social cost of carbon were only 50 €/tCO_2. Results also show the level of private profitability if all externalities were internalized, and assert that subsidies are justified in order to align private and social profitability. The proposed method could be used as a complementary indicator to private profitability by public institutions, development banks and companies with social responsibility values.

Keywords: economic analysis, photovoltaic, sustainable

1 INTRODUCTION

Germany has led the photovoltaics (PV) implementation market throughout the last decade, with the highest installed capacity per capita and the second highest in absolute value after China [1]. However, although PV costs have been declining during the last three decades at a learning rate (i.e. cost decline per each doubling in installed capacity) between 20 and 23% [2–4], PV support policies have been under increasing criticism due to their high costs.

While PV is quickly achieving grid parity in many parts of the world [5], its levelized cost (LCOE) is still above wholesale electricity prices in Germany [4]. Additionally, increasing PV penetration causes rising integration costs to the electricity system – due to its variability, uncertainty and location specificity [6,7]. Finally, all generation technologies also cause external costs not included in their market prices [8], causing social damages which are usually ignored in traditional cost-benefit analyses. Therefore, any comprehensive cost-benefit analysis of generation technologies should take into account not only private costs and benefits, but also the integration costs caused to the electricity system and the avoided or additional externalities of switching between technologies.

We develop a method to calculate the "social profitability" of photovoltaics by including integration costs and avoided externalities alongside the assessment of private costs and benefits, applying our approach to Germany. By including social costs and benefits we perform a more comprehensive analysis to inform policy and investment decisions. First, our method demonstrates the competitiveness of the technology when social costs and benefits are taken into account. Second, it can justify taxes/subsidies when social profitability is lower/higher than private profitability in order to arrive at welfare-optimal investment decisions. Third, it indicates private profitability levels if all externalities were to be internalized. Finally, it is a useful indicator to complement private profitability for investment decisions of public institutions, development banks, and companies with social responsibility values.

We calculate "social profitability" through two widely used indicators, the internal rate of return and the profitability index, and according to the two approaches available in the literature: considering integration costs either as an additional cost or as a lower value of PV electricity [7]. Aside from this calculation, we present a wide range of results regarding the main parameters to cope with the uncertainty concerning actual values of these parameters and their future evolution.

The remaining of the paper is structured as follows: section 2 frames the discussion regarding the profitability and competitiveness of variable renewables. Section 3 presents the method for the calculation of the social profitability of PV and the input data to the different approaches. Section 4 presents the main results, first as a function of the social cost of carbon, and then as a function of PV installation cost and electricity yield. Section 5 summarizes insights derived from the previous sections' analysis and its relevance.

2 PROFITABILITY AND COMPETITIVENESS OF VARIABLE RENEWABLES

Competitiveness of different electricity-generating technologies is usually assessed and compared through the levelized cost of electricity (LCOE), which measures the lifecycle costs of a technology per kWh of electricity generated during the lifetime of the system, discounted at a determined discount rate [9–11]. However, this indicator has increasingly been criticized for its inability to capture a range of features divergent across technologies, and other indicators have been proposed, such as the Levelized Avoided Cost of Electricity (LACE), which accounts for how much it would cost the grid to generate the electricity otherwise displaced by the new generation project [12]; and the System LCOE, which includes the cost of integrating the new generation into the existing electricity system, moving beyond the traditional LCOE [6].

This is a pre-peer review version of the paper submitted to the journal Progress in Photovoltaics.
It has been selected by the Executive Committee of the EU PVSEC 2017 for submission to Progress in Photovoltaics.

From a levelized costs perspective, competitiveness or "grid parity" is achieved when a generation technology reaches the costs of conventional technologies [5,13]. Equivalently, competitiveness can be assessed from the point of view of profitability. In this sense, competitiveness would be achieved when a technology is profitable in the market without subsidies [see 15,16 for PV]. However, the existence of market failures may entail the departure of private profitability from social profitability, which would not be the case in the presence of perfect and complete markets. Only in the latter case would private and social profitability be equal and the amount of investment be welfare optimal. In the presence of incomplete and/or imperfect markets, however, private investment decisions may yield socially sub-optimal outcomes. In this case, the government can subsidize/tax activities with positive/negative externalities such that private and social profitability align again [16].

PV profitability has been widely studied in regards to evaluating the impact and evolution of Feed-in Tariffs and other incentives [17–22]. However, on the one hand, private profitability does not capture the integration costs caused to the electricity system in situations with higher PV penetration and on the other, the avoided externalities when PV displaces conventional generation. On this basis, we calculate the social profitability of PV by taking into account not only private costs and benefits, but also integration costs imposed on the electricity system due to higher PV penetration as well as net avoided external costs. Since we want to evaluate the social profitability of the technology to displace conventional generation we do not include corrective incentives such as Feed-in Tariffs or investment subsidies in our calculations, since they are interventions designed to correct for externalities themselves and would therefore result in double counting. Positive social profitability would entail that the technology is competitive when all factors are accounted for. Likewise, social profitability above/below private profitability would justify subsidies/taxes to that technology in order to adjust for social benefits/costs. Finally, social profitability shows how much private profitability would be if all externalities were internalized.

3 METHOD AND DATA

3.1 Social profitability index and social rate of return

We calculate social profitability by computing the internal rate of return and its equivalent profitability index, including not only private costs and benefits, but also integration costs and avoided externalities. We assume that new PV generation displaces non-renewable generation, and we do not consider other spill-over effects or broader macro policy objectives such as energy security, job creation, etc.

The internal rate of return is the discount rate (d[%]) at which the Net Present Value (NPV[€]) equals zero. The NPV is the sum of the discounted cash flows (costs and benefits) triggered by the investment during its lifetime (n). The first addend of equation 1 represents the benefits derived from the electricity generated (annual electricity generated per kWp EPV [$\text{kWhy}^{-1}/\text{kWp}$] multiplied by the wholesale electricity price P [€/kWh] and the value factor of PV electricity at each penetration level (VF_ρ[proportion]). Since – as penetration (ρ[%]) increases – the variability, uncertainty and location-specificity of PV generation are captured as either a lower value of the electricity generated (value approach; VF_ρ) or equivalently as a higher integration cost (cost approach; IC_ρ[€/kWh]) at the aggregated level, we do not need high temporal resolution of PV generation and wholesale electricity prices. This simple model allows us to compute social profitability through one of the two approaches by either setting integration costs to zero and adjusting the value factor (in the value approach) or choosing a constant value factor of 1 and acknowledging integration cost (cost approach).

$PVin$ [€/kWp] and $PVom$ [€/kWp] are the private costs of the system: installation and operation and maintenance, respectively. Due to the quantitative importance and uncertainty related to climate change, we differentiate between climate change externalities (avoided carbon emissions multiplied by the respective social cost of carbon (ACE [$\text{tCO}_2\text{e/kWh}$] $*$ SCC [€/tCO_2e]), and all other avoided externalities (EXT[€/kWh]). Finally, IC_ρ[€/kWh] represents the integration costs for each penetration level. The parameter ∂_x (eq. 2) represents the discount factor for each of the addends (a through e, see eq. 3 to 7), capturing the annual escalation rates of each of the variables we are concerned with (ε_y[%]), the annual rate at which the PV system loses efficiency (dg[%]) and the discount rate (d).

$$NPV(d) = EPV * P * VF_\rho * \partial_a - PVin - PVom * \partial_b + EPV * ACE * SCC * \partial_c + EPV * EXT * \partial_d - EPV * IC_\rho * \partial_e = 0 \quad (1)$$

$$\partial_x = \frac{K_x * (1 - K_x{}^n)}{1 - K_x} \quad (2)$$

$$K_a = \frac{(1 + \varepsilon_P) * (1 + \varepsilon_{VF}) * (1 - dg)}{1 + d} \quad (3)$$

$$K_b = \frac{(1 + \varepsilon_{PVom})}{1 + d} \quad (4)$$

$$K_c = \frac{(1 + \varepsilon_{ACE}) * (1 + \varepsilon_{SCC}) * (1 - dg)}{1 + d} \quad (5)$$

$$K_d = \frac{(1 + \varepsilon_{EXT}) * (1 - dg)}{1 + d} \quad (6)$$

$$K_e = \frac{(1 + \varepsilon_{IC}) * (1 - dg)}{1 + d} \quad (7)$$

We calculate the Internal Rate of Return (IRR) as the discount rate at which the Net Present Value (NPV) equals zero. Likewise, we calculate the Profitability Index ($PI(d)$) as the ratio between the Net Present Value at a determined discount rate ($NPV(d)$) and the investment cost ($PVin$; eq. 8) [23,24]. The social internal rate of return therefore represents the annual social yield of a PV investment at a certain penetration level, and the social profitability index represents the present value (at a predetermined discount rate) of the investment as a proportion of the investment cost. Both indicators are equivalent such that a breakeven 0% internal rate of return corresponds to a profitability index of zero at zero discount rate. Likewise, 2 $PI(0)$ is equivalent to 10% IRR (see figure A1 in the appendix for a detailed equivalence between both indicators and across different discount rates).

$$PI(d) = \frac{NPV(d)}{PVin} \quad (8)$$

Finally, we can calculate the breakeven social cost of carbon (SCC), installation cost ($PVin$) and annual electricity yield (or equivalent solar irradiation) (EPV) at a certain discount rate by simply clearing the interest parameter in eq. 1 as shown, respectively, in eq. 9-11:

$$SCC(d) = \frac{PVin+PVom*\partial_b+EPV*IC_\rho*\partial_e-EPV*P*VF_\rho*\partial_a-EPV*EXT*\partial_d}{EPV*ACE*\partial_c} \quad (9)$$

$$PVin(d) = EPV*P*VF_\rho*\partial_a - PVom*\partial_b + EPV*ACE*SCC*\partial_c + EPV*EXT*\partial_d - EPV*IC_\rho*\partial_e \quad (10)$$

$$EPV(d) = \frac{PVin+PVom*\partial_b}{P*VF_\rho*\partial_a+ACE*SCC*\partial_c+EXT*\partial_d-IC_\rho*\partial_e*} \quad (11)$$

3.2 Cost vs. Value approach to integration costs

Increasing the penetration of variable renewable energy (VRE) technologies causes integration costs to the electricity system [25–27]. Integration costs arise due to the uncertainty, variability and location-specificity of this type of technology, which are usually higher than those of conventional dispatchable generation alternatives. Integration costs can be measured and conceptualized in a "cost approach" as the marginal cost of increasing VRE penetration, or in a "value approach" as the declining marginal value of VRE electricity at higher VRE penetration [7]. We use both approaches to calculate the social profitability of PV in Germany.

Integration costs can be decomposed into three sub-categories according to the respective feature of the VRE causing them [6,7]. The first, balancing costs, arise from the uncertainty regarding the amount of potential electricity generation at any given time, with the latter depending on climatic conditions, known ahead of time with only some degree of certainty. Grid costs derive likewise from the locational specificity of VRE, which must be deployed where the resource is available and electricity then transmitted and distributed to consumption points, which entails costs related to grid use and expansion. Finally, profile costs are caused by the profile of generation (including daily and seasonal cycles) not perfectly coinciding with load profiles, and can be likewise divided into three sub-categories: (i) backup costs derived from the low capacity credit of VRE; i.e. the need to have idle capacity to cover the intermittency and seasonality of VRE; (ii) full load hour reduction referring to the decrease of operation hours of thermal generators, which increases their cost per kWh; and (iii) overproduction costs arising when generation at a given time is higher than demand. This is particularly relevant for PV as its generation is rather concentrated in a few hours of the day and months of the year.

While there is a rich literature regarding the quantification of integration costs for wind (see [6,7] for an overview), the literature on PV integration costs estimates is more scarce. The input parameters of our model – and therefore our results– are a rough estimation of shape and order of magnitude at different penetration rates, rather than a final and exact calculation. Input data are extracted from the literature and refers specifically to PV in European (profile and grid costs) and American (balancing costs) thermal systems. We assume balancing costs to be a linear function of penetration starting from 2 €/MWh at 0% [28] up to 6 €/MWh at 30% penetration [29]. According to [7], grid costs are within the single digit range in terms of €/MWh. Therefore, we consider the most pessimistic shape of grid costs suggested by [30] (an inverted U curve peaking at 20% penetration) at an assumed 10 €/MWh to be in the high end of the range. Profile costs, derived from the variability of the PV generation, entail the most important share of integration costs (Figure 1). Within profile costs, backup costs are most prominent at low penetration levels, reduction of load hours for conventional generators prevails at penetrations of 5% to 20%, and overproduction costs rocket to dominance beyond that point [6]. Figure 1 illustrates the input data for the cost model, which is derived from the aforementioned literature (also see sources given with Figure 1) and constitutes a quantification of the shape and order of magnitude rather than an exact calculation [6].

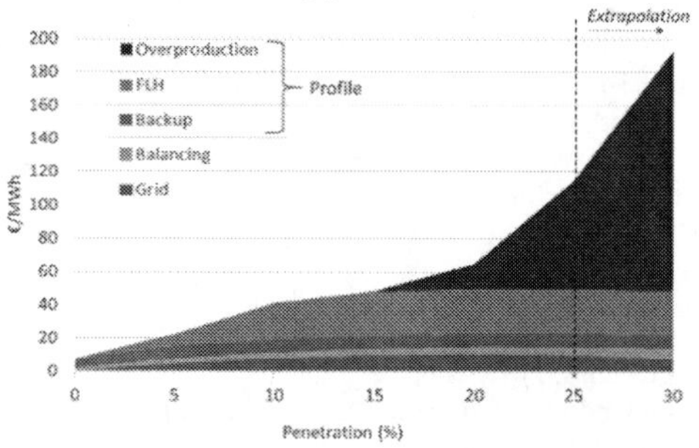

Figure 1: PV integration costs across market penetration level. Sources: [6,7,28–30]. Note: profile costs are extrapolated beyond 25% penetration.

Likewise, increasing penetration of almost-zero marginal cost electricity technologies cause the decline of wholesale electricity prices by shifting the supply curve to the right in the so-called "merit-order effect" [31–34]. This effect is particularly relevant for PV since the price drop is stronger at times of high PV generation. Therefore, the market value of PV electricity declines as its penetration in the market increases [35–38]. In the presence of perfect and complete markets, the decline in the value of PV electricity would be equal to the decrease of its value factor in the wholesale electricity market, where the value factor represents the ratio of the unit revenue of PV (i.e. the generation-weighted average price) and the time-weighted average wholesale electricity price [39,40]. Therefore, integration costs can also be understood as the decline in the market value of PV electricity, this approach being equivalent to the cost perspective [6,7].

The market value of PV electricity can be estimated empirically *ex-post* from market data, or theoretically *ex-ante* through dispatch or investment and dispatch models. [39,40] provide a comprehensive overview of the concept and estimations of the market value of PV electricity. First, they review estimates found in the literature for different countries and penetration levels. Second, they estimate *ex-post* value factors from the German wholesale electricity market. Finally, they run the EMMA dispatch and investment model to estimate the *ex-ante* long term value factors for penetration levels up to 15%. Their average results are illustrated in Figure 2 (labelled as "review", "empirical" and "model" respectively), with respective

This is a pre-peer review version of the paper submitted to the journal Progress in Photovoltaics.
It has been selected by the Executive Committee of the EU PVSEC 2017 for submission to Progress in Photovoltaics.

extrapolations (linear for the "review" and "empirical" and power for the "model" estimates) up to 30% penetration.

We use the model benchmark estimates at 0 €/tCO_2 social cost of carbon (SCC) (since we account for the SCC separately) and its respective power extrapolation (see appendix for more details on extrapolations) as input data. The estimates at 0 €/tCO_2 show a higher solar value at low penetration but a stronger decline as penetration increases. These estimates are derived from the partial equilibrium EMMA model, which is a dispatch and investment model representing the Northwestern European power system (Germany, Belgium, The Netherlands, France and Poland) and does not account for internal grid constraints, so it only partially captures grid integration costs, the total of which are relatively small according to [6] (see Fig. 1).

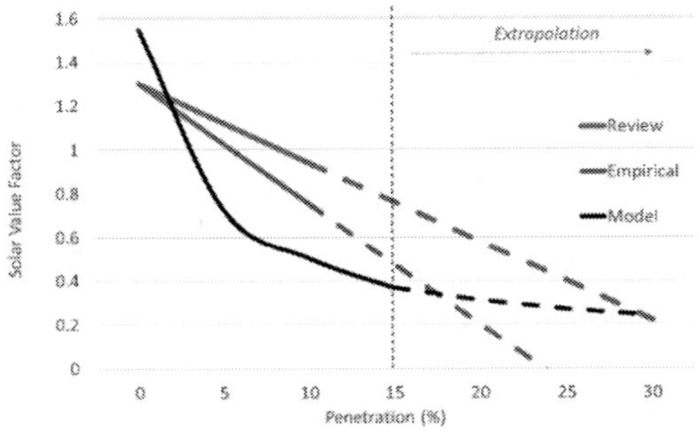

Figure 2: Solar value factor trends according to literature estimates (Review), *ex-post* estimation (Empirical) and *ex-ante* estimation (Model). Note: see sources [6,7] for a more detailed illustration.

In conclusion, both cost and value approaches are suitable and theoretically equivalent to account for the integration costs of PV or any other VRE. However, the input data for both approaches are estimated differently, which explains the differences in our results for each of the approaches. The cost approach input data is less consistent, since the different components come from different studies and are only rough estimations of shape and order of magnitude, with balancing and grid cost interpolation and profile cost extrapolation between 25-30% penetration. The value approach input data is more recent and consistent, however, it only partially captures grid costs, and as extrapolation starts at 15% penetration, so it is unlikely to capture the overproduction costs arising from this penetration level. Although both approaches include endogenous adaptation (i.e. optimal response of the electricity mix to increasing VRE penetration), the cost approach shows the "worst case scenario" without any additional adaptation, while the value approach applied beyond 15% shows the "best case" if all overproduction costs could be avoided with additional adaptation measures such as storage, demand side adjustment or intercontinental interconnections.[1]

Our model allows for the computation of both cost and value approaches. Table 1 summarizes the input values for each of the approaches, taking into account that integration costs would be 0 in the value approach, and likewise the value factor would be 1 in the cost approach. We calculate our model in real terms, and assume the real integration costs and value factor are unchanging due to reasons other than the market penetration level.[2]

Penetration	Integration cost	Value Factor
[%]	IC [€/MWh]	VF
0	7	1.546
5	22.3	0.722
10	41.0	0.502
15	48.2	0.366
20	64.7	0.258*
25	114.6	0.184*
30	192.7*	0.131*
Source	[6,7,28–30]	[6,7]
ε_y[%]	0	0

Table 1: Cost vs. value approach (see figures 1 and 2 respectivelly). Note: *extrapolation.

3.3 Climate change and other externalities

Externalities can be defined as the costs or benefits caused by an economic activity to a third party not involved in the transaction. Since external costs or benefits are not included in market prices, they lead to a welfare-suboptimal allocation of resources [16]. In the case of energy, virtually all generation technologies impose external costs not included in their price, such as human toxicity, depletion of resources or climate change, among others [8]. Reaching a welfare-optimal allocation of resources requires the integration of externalities into market processes by, for instance, taxing/subsidizing activities with negative/positive externalities.

Due to the complexity and uncertainty regarding the climate change (CC) externality caused by the release of CO_2 and other greenhouse gas (GHG) emissions, externalities are divided in our model into CC externalities, valued at the social cost of carbon (SCC), and other externalities, for which we use the estimations provided by Ecofys [8]. We include the net positive externalities of PV electricity by subtracting the negative externalities of PV itself from the per kWh external cost of non-renewable German electricity generation that PV substitutes for, as shown in Figure 3.

[1] Indeed, the value approach results beyond 15% penetration show a similar shape as produced by the cost approach when overproduction costs are excluded (see Figure A2 in the appendix).

[2] More specifically this results in a profitability index or IRR for a specific penetration level. If penetration increases further in the future, this may induce changes to the integration costs or value factor, also for all earlier investments. As we neglect this potential impact, our results are valid for the respective point in time and a specific penetration level, but profitability and SRR could decline for such investment, when penetration is further enhanced later – in the same way as the new profitability index at the higher penetration level will decline.

This is a pre-peer review version of the paper submitted to the journal Progress in Photovoltaics.
It has been selected by the Executive Committee of the EU PVSEC 2017 for submission to Progress in Photovoltaics.

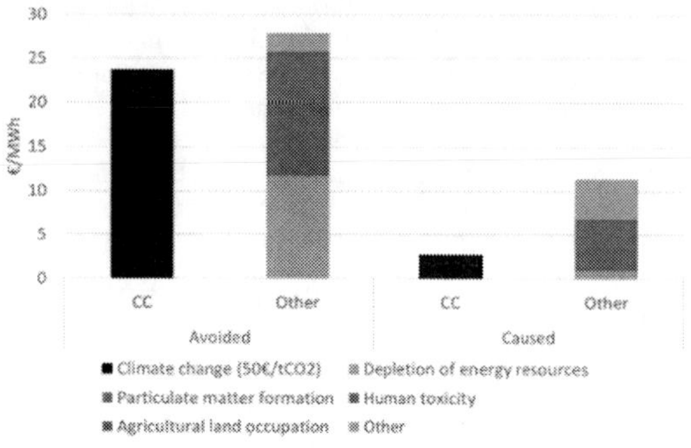

Figure 3: Figure 3. Average external cost per MWh of the non-renewable German electricity generation (avoided) and of PV (caused). Source: [8] and Eurostat (non-renewable generation).

Since we assume that the new PV generation displaces conventional generation, the avoided carbon-equivalent emissions are given by the CO_2-equivalent emissions intensity of non-renewable generation in Germany, provided by Eurostat and valued at the SCC. The SCC is defined as the marginal cost of carbon dioxide-equivalent emissions caused by climate change damages with respect to a business-as-usual scenario. The SCC, estimated through Integrated Assessment Models (IAM), is uncertain. The most comprehensive review of the SCC estimates has been done by the US Interagency Working Group, which analyses the three main models (PAGE, FUND and DICE) and the five most relevant scenarios to reach a central estimate of 36 $\$_{2007}/tCO_2$ in 2015 [41]. Another review of more than 200 estimates determines an average value[3] of 41 $\$_{1995}/tCO_2$ [42], and the most recent update of the DICE model suggests a value of 31 $\$_{2010}/tCO_2$.

These average estimates, however, conceal large uncertainties. For instance, [43] show the large range of estimates provided by the DICE model by simply exploring the plausible values of four types of uncertainties: (i) climate sensitivity, climate damages at (ii) low and (iii) high temperatures and (iv) the social discount rate, obtaining values between 28 and 892 $\$/tCO_2$ in 2010. Likewise, Pindyck criticizes the pretension of knowledge provided by IAMs and argues that climate policies should not be based on average outcomes, but on the possibility of extreme events, and urges the implementation of carbon prices to internalize the climate change externality [44–47].

In this sense, [48] argue that the average estimates from IAMs are gross underestimates: first, because they include both low and high social discount rates, which undermine the present value of future damages, and second, because they ignore factors such as (i) uncertainty about greenhouse gas (GHG) concentrations, (ii) costs of a large rise in temperature, (iii) overall higher climate damages, (iv) low-probability/high-impact climate change risks, and (v) risk aversion. According to this meta-analysis, the lower bound of the SCC should be at least 125 $\$_{1995}/tCO_2$ (i.e. about 150 $€_{2015}/tCO_2$)[4]. The fact that IAMs do not generally take into account the existence of tipping points and positive climate change feedback loops, which is also likely to be a source of undervaluation, in addition to the latest developments in earth and climate sciences, which point at higher climate sensitivity than expected [49], permafrost tipping point risks [50], higher than expected sea level rise [51,52], and even climate-driven polar motion [53], lead us to expect that the SCC might be even higher than the lower bound of 150 $€_{2015}/tCO_2$ suggested by [48]. We still use the value of 150 $€/tCO_2$ as a benchmark lower bound of the SCC but report a wide range between 0 and 300 $€/tCO_2$ to account for the uncertainty surrounding its actual value and the (more likely than not) upwards evolution of future estimations acknowledging risk aversion.

3.4 Other data and assumptions

Since we calculate our model in real terms and assume that nominal operation and maintenance costs, electricity prices and integration costs (the latter over time, but at identical penetration levels) rise at the same rate as inflation, their respective real escalation rates are equal to zero. The SCC increases over time because greenhouse gas concentrations increase over time, with future emissions thus producing larger incremental damages [41]. Although its real escalation rate is also uncertain (between 1.2% and 4.4% depending on the assumptions and the timeframe [41]), we assume an intermediate value of 2.5%. We assume an installation cost of 1,000 €/kWp, the wholesale electricity price at its average value between years 2005-2010, and a system lifetime of 25 years (see Table 2 for a summary of the input data).

---

[3] Originally reported in \$ per ton of carbon [42]: 151 \$ per ton of carbon is equivalent to 41.2 \$ per ton of CO_2.

[4] [46] derives the lower bound of the SCC of 125 $\$_{1995}/tCO_2$ by applying a 206% surcharge on the average value indicated by [40] (personal communication to authors). We apply the US GDP deflator (Federal Reserve data base) for 1995-2015 (1.46) and the 2015 average US\$-€ exchange rate (1.11) as well as perform a sensitivity analysis with the long-term (1990-2015) average exchange rates (1.11 to 1.23) to account for the real depreciation of the US\$ in the last years, together resulting in a range between 148 and 164 $€_{2015}/tCO_2$, out of which we consider a conservative value of 150 $€_{2015}/tCO_2$ as a benchmark for the lower bound of the SCC.

This is a pre-peer review version of the paper submitted to the journal Progress in Photovoltaics.
It has been selected by the Executive Committee of the EU PVSEC 2017 for submission to Progress in Photovoltaics.

Parameter	Installation cost	Operation and maintenance cost	Electricity price (2005-2010 av.)	Electricity yield	Net avoided carbon emissions	Net avoided external cost
Notation [unit]	PV_{IN} [€/kWp]	$PVom$	P [€/MWh]	EPV [kWhy^{-1} /kWp]	ACE $[\frac{tCO_2e}{MWh}]$	EXT [€/GWh]
Value	1,000	10	48.375	938	0.701	16.54
Source	[11]	[54]	Epex Spot	[55]	Eurostat	[8]
ε_y[%]	-	0	0	-0.08*	0	0

Table 2: Data and assumptions. Note: *Degression rate at which the PV system losses efficiency

4 RESULTS AND DISCUSSION

We now present our model results, comparing the cost and the value approaches. While their interpretation up to 15% penetration is the same, it differs beyond that point. Both approaches include endogenous adaptation of the electricity system to higher VRE generation. However, while the cost approach includes overproduction costs arising beyond 15% penetration, these are not covered in the value approach, since input data are extrapolated beyond the point at which overproduction start to arise. Therefore, the cost and value illustrations below can be considered as worst and best cases respectively, considering either no additional adaptation measures beyond endogenous electricity system capacity optimization (cost approach), or perfect adaptation to remove all overproduction costs beyond 15% penetration (value approach).

The cost of capital is not included in the model so it is under the discretion of the reader to judge the required profitability level to cover the opportunity cost of capital. However, if we assume that PV deployment is financed by the government to correct electricity market externalities, it is reasonable to consider the 25-year German bond yield as a plausible cost of capital for PV subsidies. The interest rates on 25-year German bonds have ranged between 1% (in 2017) and 5% in the last ten years, so the white area between the solid and the dashed lines in Figures 4-6 is a plausible range for covering the "social cost of capital".

4.1 Social profitability as a function of the social cost of carbon

Figure 4 shows the results for the two approaches as a function of PV penetration and the social cost of carbon (SCC). The color gradient represents the social profitability index (SPI) at a 0% discount rate and the solid line shows the social profitability breakeven (i.e. social rate of return equal to 0%). The dashed and dotted lines show social rates of return (SRR) of 5% and 10% respectively (see Fig. A.1. for a detailed relation between the social profitability index and the social rate of return and between the social profitability index at different discount rates). Both cost and value approaches agree that PV profitability at current penetration (7%) is above 10% at the lower bound of the SCC (150 €/tCO_2), and around 5% at a SCC as low as 50 €/tCO_2. Indeed, PV would still be socially profitable at that level of the SCC and double the current penetration (50 €/tCO_2 and 14% penetration).

While the primary axes provide the basic information to interpret Figure 4, we provide illustrative secondary axes to help the reader evaluate their own dynamic scenarios. Thus, the secondary horizontal axis shows the years in which different penetration levels were achieved in Germany (e.g. 7% in 2016), as well as different projections regarding the future evolution of PV penetration for Germany (IRENA Reference and ReMap scenarios for 2030 [56] and UBA projection for the 100% renewables scenario in 2050 [57]); and the EU-27 (Energy [R] evolution scenario [58]). Since the SCC increases over time (because future emissions occur at a higher GHG concentration level and thus produce larger incremental damages), the secondary vertical axis illustrates the years in which the corresponding values of the SCC are reached according to the "high impact" (or risk acknowledging) scenario of the Interagency Working Group [41]. This scenario considers a SCC of 107 €$_{2015}$/tCO_2 in 2015 rising up to 215 €$_{2015}$/tCO_2 by 2050[5].

The wide range of parameters provided in Figure 4 allows the reader to evaluate their own dynamic scenarios according to the assumed value and evolution of the social cost of carbon, the expected evolution of PV penetration and the required social profitability. For instance, if we assume a desired social profitability at 5% and the evolution of the SCC given by the "high impact" scenario of the Interagency Working Group, the dashed line in figure 4 panel A would represent a diffusion pathway such that the maximum socially desired PV penetration in 2015 would be 20%, 25% in 2030 and almost 30% in 2050 at current installation costs (see section 4.2 for different installation costs).

[5] Using the US GDP deflator between 2007-2015 (1.128) and the US$/€ average exchange rate in 2015 (1.11), provided by the Federal Reserve.

This is a pre-peer review version of the paper submitted to the journal Progress in Photovoltaics.
It has been selected by the Executive Committee of the EU PVSEC 2017 for submission to Progress in Photovoltaics.

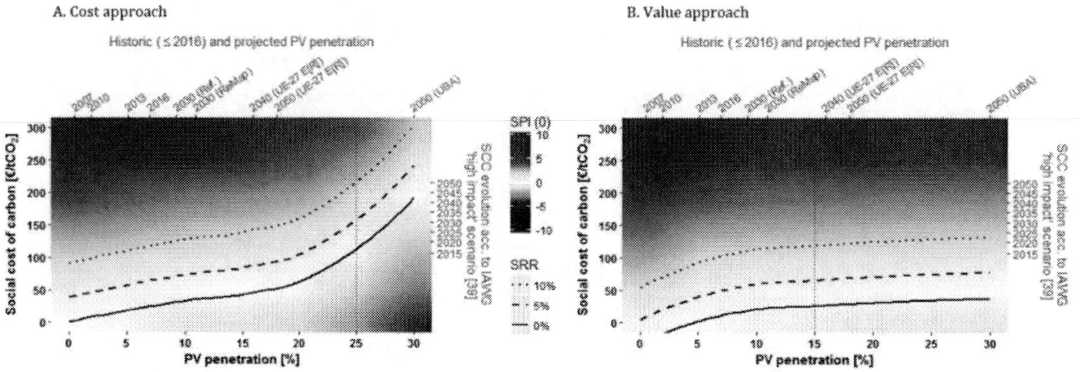

Figure 1: Social profitability index at 0% discount rate ($SPI(0)$) and social rate of return (SRR) as a function of the PV penetration [%] and the Social Cost of Carbon (SCC [€/tCO_2]). Note: the secondary vertical axis illustrates the year in which the corresponding SCC is reached according to the "high impact" scenario of the Interagency Working Group [41]. The secondary horizontal axis depicts the evolution of PV penetration in Germany up to 2016, and different projections for Germany (IRENA Reference and ReMap scenarios for 2030 [56] and UBA projection for the 100% renewables scenario in 2050 [57]), and the EU-27 (Energy [R] evolution scenario [58]). Input data to the right of the vertical dotted lines are extrapolations.

Both cost and value approaches show similar results up to 15% PV penetration, which suggests the accuracy of the input data. Beyond that point, the cost approach represents the worst case scenario given current conditions, while the value approach would represent the potential for PV social profitability if overproduction costs could be totally removed with additional adaptation measures, such as demand side adjustment, storage or intercontinental interconnections [57].

The positive social profitability of PV entails that the technology is competitive when factoring in both integration costs and the external cost of conventional generation. This can, on one hand, justify subsidies for this technology such that private profitability aligns with social profitability. On the other hand, this method could be used to compare different investment alternatives in terms of social profitability rather than only private profitability, for which more transparency and data on externalities valuation is needed to make this kind of analysis possible and accurate. Finally, it shows how profitable PV would be if all externalities were internalized.

4.2 Social profitability as a function of the installation cost

We now extend our analysis to cover a wide range of possibilities regarding PV installation costs and PV yield per kWp, which deepen the dynamic interpretation of the results and allow the reader to evaluate different scenarios according to assumptions on the evolution of different parameters. Since at the lower bound of the SCC (150 €/tCO_2) PV is socially profitable for the entire assessed penetration range, we present the results for half that value (75 €/tCO_2) to better depict the relationship between these variables and PV penetration (see Fig. A3 and A4 in the appendix for results with SCC of 0, 50 and 150 €/tCO_2).

Figure 5 shows the social profitability of PV at increasing penetration depending on the installation cost, with the SCC set at 75 €/tCO_2. Both approaches agree that (i) the lower the installation cost, the higher the spread of the social profitability across penetration levels; and (ii) at the current installation costs and SCC at half its lower bound, PV penetration would be socially profitable up to 20%. Beyond that penetration level, the cost approach shows a strong and rapid decline in social profitability and a virtual penetration barrier beyond 20-25%. This suggests that PV will only be profitable at high penetration levels if climate change damages (and therefore the SCC) are high. Exogenous adaptation, however, could substantially improve the social profitability of PV as suggested by the value approach. Figure 5 also shows the forecasted evolution of installation costs in Germany by the Fraunhofer institute [11] between 2020 and 2050. According to their average estimations, and in the worst case (represented by the cost approach with SCC at only half its lower bound value), PV would achieve 10% social profitability at 10% penetration in 2030 (see appendix for results at different SCC values).

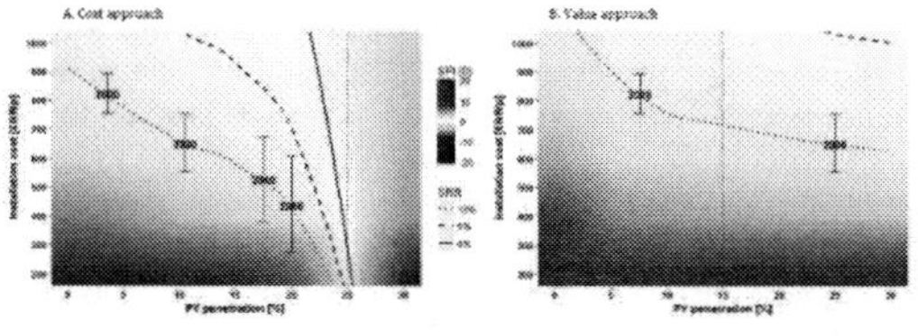

Figure 5. Social profitability index at 0% discount rate ($SPI(0)$) and social rate of return (SRR) at 75 €/tCO_2 social cost of carbon as a function of PV penetration [%] and installation cost ($PVin$ [€/kWp]). Note: error bars represent the forecasted evolution of PV installation costs by the Fraunhofer institute [11]. Input data to the right of the vertical dotted line are extrapolations.

4.3. Social profitability as a function of the PV yield per Wp

Finally, we present the social profitability of PV as a function of its penetration and its potential yield per kWp, again with a SCC at only half its lower bound (75 €/tCO_2). The potential PV yield per kWp is taken from the PV GIS project [55], which assumes a 75% performance rate. Figure 6 illustrates the position of several European countries as a function of their respective PV penetration and average PV potential. Due to the lack of country-specific integration cost estimations, these results can provide a useful illustration about their approximate position. However, since avoided carbon emissions is a key variable, countries with ± 10% of the German level are depicted in black, countries with levels between 10-50% lower emissions intensity of their non-renewable generation are depicted in blue (meaning that their social PV profitability is actually lower than their position implies due to the lower abatement benefits), and countries with emissions intensity at levels between 10-50% higher are depicted in red. Countries beyond ± 50% German levels are not depicted at all.

All selected countries show positive social profitability and a potential to increase penetration up to at least 20% at half the lower bound of the SCC. Greece and Poland, although differing on their solar irradiation and penetration level, both show PV potential due to their high emissions intensity. Portugal and Italy, although having lower potential abatement benefits due to their lower emissions intensity, have high social PV profitability thanks to their high insolation. Social profitability of PV would be around 10% in Romania and Bulgaria, according to the conservative scenario depicted in Fig. 6A.

Figure 6. Social profitability index at 0% discount rate

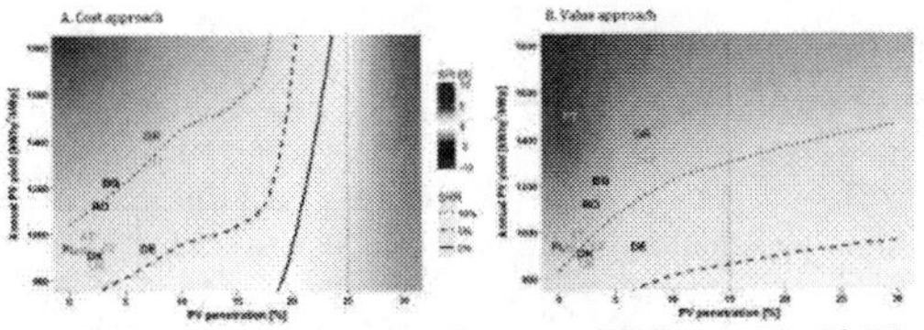

($SPI(0)$) and social rate of return (SRR) at 75 €/tCO_2 social cost of carbon as a function of PV penetration [%] and annual PV yield (EPV [$\text{kWhy}^{-1}/\text{kWp}$]). Note: country codes written in black have ± 10% of the German emissions intensity of the non-renewable generation, countries in blue between 10-50% lower, and countries in red between 10-50% higher emissions intensity of non-renewable generation. Input data to the right of the vertical dotted line are extrapolations.

5 CONCLUSIONS

We have calculated the social profitability of photovoltaics (PV) in Germany by including not only private costs and benefits, but also integration costs of higher PV penetration in the electricity system and avoided external costs of displacing non-renewable generation. We have computed social profitability from a cost and value perspective. Both approaches agree that at the lower bound of the social cost of carbon (SCC 150 €/tCO_2), PV social profitability is above 10% at current German penetration rate (7%), and it is still positive up to at least double the current penetration level even at only 50 €/tCO_2.

This entails that PV is competitive when both integration costs and externalities are included in the analysis, and shows the level of private profitability if all the external costs and benefits were internalized. Therefore, subsidies to PV are economically justified in these ranges in order to align private and social profitability and reach a welfare-optimal allocation of resources.

Although uncertainties are large for penetration levels beyond 15%, the comparison between cost and value approaches suggests the high potential of of exogenous adaptation (such as storage, demand side adjustment and intercontinental interconnections) to boost PV social profitability, also in situations beyond 20% penetration. In the worst case, if no additional adaptation measures are implemented, PV will only be competitive beyond 20% penetration in order to abate high climate change damages at exponentially increasing costs due to overproduction. On the contrary, if all overproduction costs could be totally removed at zero cost, PV would be socially profitable at half the lower bound of the SCC, independently of its penetration.

The method presented in this article could be used as a complementary indicator to evaluate investment alternatives by public institutions, investment banks and companies with social responsibility values. Additionally, it would be interesting to apply this method to other technologies and types of investments to be able to compare them with PV and have a more comprehensive understanding of the role of different abatement technologies. For instance, it could well be that other technologies or types of investments (e.g. wind installations or efficiency measures) have higher social profitability than PV in certain circumstances. It is also important to further explore (and improve) the data related to external costs as well as their estimation, so that this type of analysis can be generalized and extended to other technologies and sectors.

ACKNOWLEDGEMENTS

This research has been possible thanks to the funding of the Austrian Science Fund under the research grant W 1256-G15: Doctoral Program Climate Change – Uncertainties, Thresholds and Coping Strategies, and thanks to the support of the Wegener Center for Climate and Global Change and the University of Graz.

REFERENCES

[1] IRENA - International Renewable Energy Agency 2016 *Renewable Energy Statistics 2016*

[2] IEA I E A 2014 *Technology Roadmap. Solar Photovoltaic Energy.pdf*

[3] Feldman D 2014 Photovoltaic (PV) pricing trends: historical, recent, and near-term projections

[4] Fraunhofer ISE 2017 *Recent Facts about Photovoltaics in Germany*

[5] Breyer C and Gerlach A 2013 Global overview on grid-parity: Global overview on grid-parity *Progress in Photovoltaics: Research and Applications* **21** 121–36

[6] Ueckerdt F, Hirth L, Luderer G and Edenhofer O 2013 System LCOE: What are the costs of variable renewables? *Energy* **63** 61–75

[7] Hirth L, Ueckerdt F and Edenhofer O 2015 Integration costs revisited – An economic framework for wind and solar variability *Renewable Energy* **74** 925–39

[8] Ecofys 2014 *Subsidies and costs of EU energy. Final report*

[9] Branker K, Pathak M J M and Pearce J M 2011 A review of solar photovoltaic levelized cost of electricity *Renewable and Sustainable Energy Reviews* **15** 4470–82

This is a pre-peer review version of the paper submitted to the journal Progress in Photovoltaics.
It has been selected by the Executive Committee of the EU PVSEC 2017 for submission to Progress in Photovoltaics.

[10] Fraunhofer ISE 2013 *Levelized Cost of Electricity Renewable Energy Technologies. Study November 2013*

[11] Fraunhofer ISE 2015 *Current and future cost of photovoltaics. Long-term scenarios for market development, system prices and LCOE of utility-scale PV systems*

[12] EIA - Energy Information Administration 2017 *Levelized Cost and Levelized Avoided Cost of New Generation Resources in the Annual Energy Outlook 2017*

[13] EPIA E P I A 2011 *Solar photovoltaics competing in the energy sector. On the road to competitiveness*

[14] Cucchiella F, D'Adamo I and Gastaldi M 2016 A profitability assessment of small-scale photovoltaic systems in an electricity market without subsidies *Energy Conversion and Management* **129** 62–74

[15] Lang T, Ammann D and Girod B 2016 Profitability in absence of subsidies: A techno-economic analysis of rooftop photovoltaic self-consumption in residential and commercial buildings *Renewable Energy* **87** 77–87

[16] Pigou A C 2013 *The Economics of Welfare* (Palgrave Macmillan)

[17] Talavera D L, Nofuentes G and Aguilera J 2010 The internal rate of return of photovoltaic grid-connected systems: A comprehensive sensitivity analysis *Renewable Energy* **35** 101–11

[18] Talavera D L, Muñoz-Cerón E, Ferrer-Rodríguez J P and Nofuentes G 2016 Evolution of the cost and economic profitability of grid-connected PV investments in Spain: Long-term review according to the different regulatory frameworks approved *Renewable and Sustainable Energy Reviews* **66** 233–47

[19] Chiaroni D, Chiesa V, Colasanti L, Cucchiella F, D'Adamo I and Frattini F 2014 Evaluating solar energy profitability: A focus on the role of self-consumption *Energy Conversion and Management* **88** 317–31

[20] Spertino F, Di Leo P and Cocina V 2013 Economic analysis of investment in the rooftop photovoltaic systems: A long-term research in the two main markets *Renewable and Sustainable Energy Reviews* **28** 531–40

[21] Danchev S, Maniatis G and Tsakanikas A 2010 Returns on investment in electricity producing photovoltaic systems under de-escalating feed-in tariffs: The case of Greece *Renewable and Sustainable Energy Reviews* **14** 500–5

[22] De Boeck L, Van Asch S, De Bruecker P and Audenaert A 2016 Comparison of support policies for residential photovoltaic systems in the major EU markets through investment profitability *Renewable Energy* **87** 42–53

[23] Chabot B 1998 From costs to prices: economic analysis of photovoltaic energy and services *Progress in Photovoltaics Research and Applications* **6** 55–68

[24] Nofuentes G, Aguilera J and Muñoz F J 2002 Tools for the Profitability Analyses of Grid-Connected Photovoltaics.pdf *Prog. Photovolt: Res. Appl.* 555–70

[25] IPCC - Intergovernmental Panel on Climate Change 2011 Integration of Renewable Energy into Present and Future Energy Systems *IPCC Special Report on Renewable ENergy Sources and Climate Change Mitigation* (Cambridge University Press)

[26] Energy G E 2010 *Western wind and solar integration study* (Citeseer)

[27] Milligan M, Ela E, Hodge B-M, Kirby B, Lew D, Clark C, DeCesaro J and Lynn K 2011 Integration of Variable Generation, Cost-Causation, and Integration Costs *The Electricity Journal* **24** 51–63

[28] Luoma J, Mathiesen P and Kleissl J 2014 Forecast value considering energy pricing in California *Applied Energy* **125** 230–7

[29] Gowrisankaran G, Reynolds S S and Samano M 2016 Intermittency and the value of renewable energy *Journal of Political Economy* **124** 1187–1234

[30] MIT M I of T 2015 *The Future of Solar Energy*

[31] Clò S, Cataldi A and Zoppoli P 2015 The merit-order effect in the Italian power market: The impact of solar and wind generation on national wholesale electricity prices *Energy Policy* **77** 79–88

[32] Cludius J, Hermann H, Matthes F C and Graichen V 2014 The merit order effect of wind and photovoltaic electricity generation in Germany 2008–2016: Estimation and distributional implications *Energy Economics* **44** 302–13

[33] Sensfuss F, Ragwitz M and Genoese M 2008 The merit-order effect: A detailed analysis of the price effect of renewable electricity generation on spot market prices in Germany *Energy policy* **36** 3086–3094

[34] Woo C K, Moore J, Schneiderman B, Ho T, Olson A, Alagappan L, Chawla K, Toyama N and Zarnikau J 2016 Merit-order effects of renewable energy and price divergence in California's day-ahead and real-time electricity markets *Energy Policy* **92** 299–312

[35] Borenstein S 2008 The market value and cost of solar photovoltaic electricity production *Center for the Study of Energy Markets*

[36] Borenstein S 2012 The Private and Public Economics of Renewable Electricity Generation *Journal of Economic Perspectives* **26** 67–92

[37] Baker E, Fowlie M, Lemoine D and Reynolds S S 2013 The economics of solar electricity *Annual Review of Resource Economics* **5** 387–426

[38] Clò S and D'Adamo G 2015 The dark side of the sun: How solar power production affects the market value of solar and gas sources *Energy Economics* **49** 523–30

[39] Hirth L 2013 The market value of variable renewables *Energy Economics* **38** 218–36

[40] Hirth L 2015 Market value of solar power: Is photovoltaics cost-competitive? *IET Renewable Power Generation* **9** 37–45

[41] Interagency Working Group on Social Cost of Greenhouse Gases and , United States Government 2016 *Technical Support Document: Technical Update of the Social Cost of Carbon for Regulatory Impact Analysis Under Executive Order 12866*

[42] Tol R S 2009 The economic effects of climate change *The Journal of Economic Perspectives* **23** 29–51

[43] Ackerman F and Stanton E A 2012 Climate Risks and Carbon Prices: Revising the Social Cost of Carbon *Economics: The Open-Access, Open-Assessment E-Journal* **6** 1

[44] Pindyck R S 2013 Climate Change Policy: What Do the Models Tell Us? *Journal of Economic Literature* **51** 860–72

[45] Pindyck R S 2013 The Climate Policy Dilemma *Review of Environmental Economics and Policy* **7** 219–37

[46] Pindyck R S 2015 *The use and misuse of models for climate policy* (National Bureau of Economic Research)

This is a pre-peer review version of the paper submitted to the journal Progress in Photovoltaics.
It has been selected by the Executive Committee of the EU PVSEC 2017 for submission to Progress in Photovoltaics.

[47] Pindyck R S 2013 Pricing carbon when we don't know the right price *Regulation* **36** 43

[48] van den Bergh J C J M and Botzen W J W 2014 A lower bound to the social cost of CO2 emissions *Nature Climate Change* **4** 253–8

[49] Tan I, Storelvmo T and Zelinka M D 2016 Observational constraints on mixed-phase clouds imply higher climate sensitivity *Science* **352** 224–7

[50] Chadburn S E, Burke E J, Cox P M, Friedlingstein P, Hugelius G and Westermann S 2017 An observation-based constraint on permafrost loss as a function of global warming *Nature Climate Change* **7** 340–4

[51] DeConto R M and Pollard D 2016 Contribution of Antarctica to past and future sea-level rise *Nature* **531** 591–7

[52] Hansen J, Sato M, Hearty P, Ruedy R, Kelley M, Masson-Delmotte V, Russell G, Tselioudis G, Cao J, Rignot E, Velicogna I, Tormey B, Donovan B, Kandiano E, von Schuckmann K, Kharecha P, Legrande A N, Bauer M and Lo K-W 2016 Ice melt, sea level rise and superstorms: evidence from paleoclimate data, climate modeling, and modern observations that 2 °C global warming could be dangerous *Atmospheric Chemistry and Physics* **16** 3761–812

[53] Adhikari S and Ivins E R 2016 Climate-driven polar motion: 2003-2015 *Science Advances* **2** e1501693–e1501693

[54] Talavera D L, Nofuentes G and Aguilera J 2010 The internal rate of return of photovoltaic grid-connected systems: A comprehensive sensitivity analysis *Renewable Energy* **35** 101–11

[55] Šúri M, Huld T A, Dunlop E D and Ossenbrink H A 2007 Potential of solar electricity generation in the European Union member states and candidate countries *Solar Energy* **81** 1295–305

[56] IRENA - International Renewable Energy Agency 2015 *REmap 2030. A Renewable Energy Roadmap. Renewable Energy Prospects: Germany*

[57] Klaus T, Vollmer C, Werner K, Lehmann H and Müschen K 2010 Energieziel 2050: 100% Strom aus erneuerbaren Quellen *Dessau: Umweltbundesamt*

[58] Teske S, Rahlwes R, Wijhoven J, Pregger T, Simon S, Naegler T, O'Sullivan M, Schmid S, Graus W H J, Zittel W and others 2013 Energy [R] evolution-a sustainable Netherlands energy outlook. Report 2013 Netherlands energy scenario

LITHUANIAN SMART SPECIALIZATION AND CLUSTERING ACTIVITIES IN PHOTOVOLTAIC SECTOR

Juras Ulbikas
Applied Research Institute for Prospective Technologies
Vismaliuku str. 34, LT-10243, Vilnius, Lithuania, juras.ulbikas@protechnology.lt

ABSTRACT: Paper is presenting developments of Lithuanian PV sector in the last two programming periods demonstrating efficient collaboration between national RTD system players with competencies in material research and businesses willing to develop innovative PV business. Paper presents bottom up approach of PV cluster development integrating RTD and industrial partners with further introduction of the open access dedicated to PV RTD infrastructure for testing and industrial scale research. Paper is discussing impact of governmental support through country smart specialization priorities for PV sector competitiveness and ability to survive market pressure and is presenting examples of successful implementation cluster based activities leading to novel competitive products introduction.
Keywords: see enclosed list of keywords

1 AIM AND APPROACH USED

PV manufacturing industry in Lithuania emerged as a new high tech sector promoted by Lithuanian government in previous programming period taking into account R&D competencies available in the country. Support for PV industry was strengthened in recent EU Financial Framework (2014-2020) by putting PV high on priority list of Lithuanian Smart specialization areas. Photovoltaic technologies are integrated under categories[1] 1 and 4 identified for the Lithuania's Operational Programmes of the EU Structural Funds for the period 2014-2020:

- *Solar energy equipment and technologies for its use for the production of electricity, heat and cooling* under priority area 'Energy and sustainable environment': Strive for energy independence in Lithuania, which highly depends on imports from Russia, resulted in strong promotion of renewable energy generation as well as production of such generators. PV manufacturing industry emerged as part of RES promotion policies understanding that R&D competencies available in the country will ensure stability and growth of this industry.
- *Photonic technologies and functional materials and coatings* under priority area "New product processes, materials and technologies, transport". High level of material research has long lasting tradition in Lithuania when country was known as Silicon Valley for Soviet Union. Knowledge available in research environment is essential to ensure permanent PV technology performance growth and technology upgrade. On the other hand, photonic technologies and lasers in particular are one of the leading R&D areas in Lithuania with successful fast-growing and export-driven industry providing high level competencies also for PV manufacturing industry, where laser based processes are increasingly replacing traditional solutions.

2 SCIENTIFIC INNOVATION AND RELEVANCE

From the very beginning of emerging PV manufacturing industry in Lithuania it was clear that worldwide PV market challenges and pressure from third countries producers can be met only by competitiveness of local PV products which can be ensured by fluent flow of innovations into manufacturing business. In PV manufacturing business to stay competitive means permanent upgrade of technology keeping in line with fast technological progress. It calls for systemic industrial scale research where close collaboration of businesses with research community is a necessity. Industrial research is specific research in high 7-9 levels of technology readiness and typically is internal company's activity, but needs also deep knowledge and analytic equipment available in research centres.

Lithuanian PV industry introduced sustainable mechanism for manufacturing industry competitiveness support by means of research intensive cluster integrating most of PV businesses and research entities active in PV sector. Applied Research Institute for Prospective Technologies (PROTECH) was appointed as cluster administrator and dedicated cluster research infrastructure with focus of its activities on industrial research in photovoltaic technologies. Lithuanian PV Technology Cluster, founded in 2008, is non-profit organization operating under Joint Venture Agreement and has 35 members, i.e. 26 SME members and 8 research institutions. Manufacturing facilities were introduced in the segments of value chain where competencies are available locally first of all cSi based Solar cells and cSi based PV modules. Initiatives of local investors to start PV business triggered interest in other sectors of relevance to PV therefore today Lithuania PV sector is also moving into new market segments with higher profit margins such as BIPV and power supply for devices requiring nonstandard solutions.

[1] The priorities fall under 6 categories, namely 1) Energy and sustainable environment, 2) Health technologies and biotechnologies, 3) Agro-innovation and food technologies, 4) New product processes, materials and technologies, transport, 5) Transport, logistics and information and communication technologies, 6) Inclusive and creative society. The priorities were approved by the Government of the Republic of Lithuania on 14 October 2013. The documents available for consultation at http://www.mosta.lt

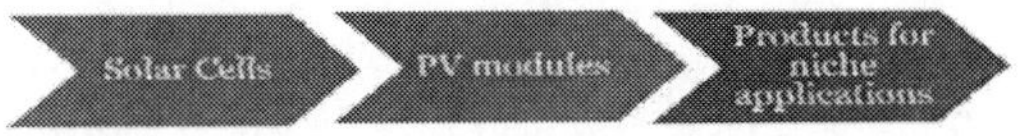

Figure 1: Availability of manufacturing capacities in Lithuania along value chain.

Supported by ERDF funding PV Cluster invested into dedicated PV industrial research infrastructure consisting today of 5 laboratories:

- Solar Cell Materials Testing Laboratory;
- Solar Cell Manufacturing Processes Research Laboratory;
- Solar Modules Testing Laboratory;
- Photovoltaic Applied Research Laboratory;
- Photovoltaic Demonstration Laboratory.

The objective of the Cluster is to consolidate Lithuanian companies and research institutions operating in PV sector in order to increase sustainability and international competitiveness of Lithuanian PV producers. Cluster provides these support services:

- Access to public support (regional/national programmes, innovation vouchers, etc.);
- Access to technology services;
- Facilitation of collaboration between members.

PV Cluster actively participates in national and EU Support Programmes: FP7, H2020, INTERREG, EIT KICs, European Union Structural Funds. It has already successfully implemented cluster focused projects PV4B, PVPLUS, PVPLUS+, SOLARROK, SMART-FLEX.

3 PROTECH APPROACH IMPLEMENTATION EXAMPLES

1. PERC technology development and production line upgrade

In 2011 discussions about next generation PV cell technology which can be suitable for PV manufacturers in Lithuania were started and decision to adopt PERC cell concept was made. On the basis of Joint Venture Agreement two projects were launched:

Infrastructure development: Implementation of PERC cell cluster industrial research infrastructure in SOLITEK (project PVPLIUS+) manufacturing environment covering purchase and integration of industrial scale research equipment for main PERC solar cells technological steps: thin films deposition (SINGULAR), chemical etching and polishing (LINEA) and laser ablation (INOLAS).

Technology development: Research project with participation of SOLITEK, PROTECH, SINGULUS, ALTECHNA and PREMET SC aiming at implementation of PERC cell on industrial research infrastructure. For that semi-finished solar cells from SOLITEK manufacturing line were shifted to PERC technological equipment and returned back into the line to get final product. Project resulted in achieving significantly higher solar cell efficiency (by ~20%) and providing recommendations to equipment manufacturer for improving their equipment.

Figure 2: PERC cell industrial research laboratory in SOLITEK premises.

2. Cost competitive BIPV solutions

In 2012 discussions with PV Cluster partner UAB VIASOLIS hosting glass-glass PV module manufacturing line was launched about possible novel products suitable for their technology. As solution for growing pressure from China manufacturers for mainstream products, focus on niche BIPV market (building integrated PV) was proposed. Agreement resulted in launching demonstration project SMART-FLeX (*http://www.smartflex-solarfacades.eu*) under which novel BIPV product for glass facades was proposed.

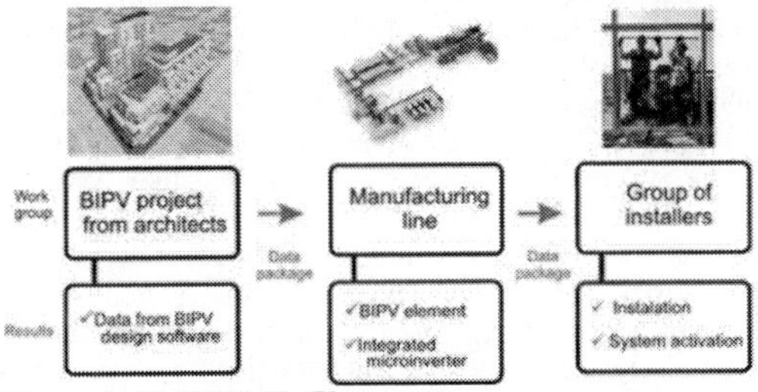

Figure 3: SMART-FLeX concept.

Project is covering modification of production line to increase flexibility of the PV product and demonstration of the proposed solution in real environment. Project was closed on May 2017 demonstrating flexibility and cost reduction on all levels of PV value chain starting from architectural design through manufacturing line and simplified integration. It is expected that project will ensure strong competitive position for module manufacturer UAB VIASOLIS in growing in EU market for BIPV (building integrated PV) products.

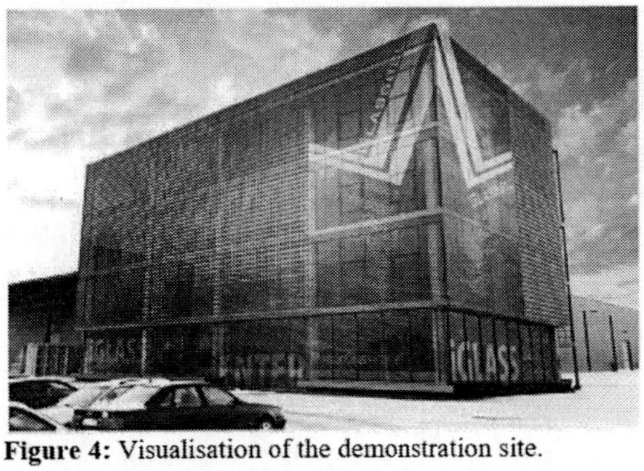

Figure 4: Visualisation of the demonstration site.

Figure 5: Realization of demonstration site.

PROTECH currently has 38 employees, including 5 PhDs, 1 postdoc and 2 doctoral researchers.

4 ADDITIONAL COMPONENTS

Today Lithuanian PV cluster is among strongest manufacturing clusters in Lithuania while PROTECH is recognized as leading national PV competence centre able to provide relevant support for manufacturing industries. Methodology of PROTECH approach toward industry relevant research implementation can be summarised as follow:

- PROTECH is providing PV cluster businesses with studies and insights about technological trends in PV technologies and in joint discussions with cluster companies promising and attractive for development solutions are elaborated.
- Decisions for developing particular technology or product are formalised by Joint Venture agreements indicating the need for research activities and relevant research infrastructure.
- Agreed activities are implemented as separate projects resulting in new cluster research infrastructure and appropriate research performed. Depending on scope, projects are searching for appropriate funding in FP/H2020 or other support schemes (ERDF, BSR etc.) or can be implemented as internal company projects.

Successful implementation of research projects with strong industry involvement triggered interest of private investors to work with PROTECH developing businesses on the basis of technologies developed under PROTECH. As a result of such collaboration spin-off companies up-taking PROTECH technologies were established: UAB MODERN E-TECHNOLOGIES adopting sol-gel technology for solar cells doping, UAB PRECIZIKA-MET SC for sol-gel based solar cell technologies developed under FP7 SELFLEX project, UAB "PRELAS" implementing novel laser ablation equipment.

Lithuanian PV sector managed to survive through the years of downturn of EU PV manufacturing businesses due to its focus onto innovative solutions, superior quality and competitive pricing. Finding niche products and novel applications which are providing customers with added value comparing to mainstream solutions are the main advantages resulting from research intensive PV cluster activities targeting at long lasting competitiveness of PV sector in the country.

5 ACKNOWLEDGMENT

The presentation of this paper was supported by the European Regional Development Fund project "Photovoltaic Technology Cluster Innovation Support System Development Enforcing Competitiveness of PV Sector (InoPV)", contract No 01.2.1-LVPA-K-833-01-0004.

SOLAR-ERA.NET
ERA-NET ON SOLAR ELECTRICITY FOR THE IMPLEMENTATION OF THE SOLAR EUROPE INDUSTRY INITIATIVE

Stefan Nowak, NET Nowak Energy & Technology, Waldweg 8, 1717 St. Ursen, Switzerland,
T: +41 26 494 00 30, F: +41 26 494 00 34, stefan.nowak@netenergy.ch
Marcel Gutschner, NET Nowak Energy & Technology, marcel.gutschner@netenergy.ch
Thomas Biel, NET Nowak Energy & Technology, thomas.biel@netenergy.ch
Stefan Oberholzer, Swiss Federal Department for Environment, Transports, Energy and Communication,
stefan.oberholzer@bfe.admin.ch
Christoph Hünnekes, Forschungszentrum Jülich – PtJ, ch.huennekes@fz-juelich.de
Herman Bastek, Forschungszentrum Jülich – PtJ, h.bastek@fz-juelich.de
Dorothea Brockmann, Forschungszentrum Jülich – PtJ, d.brockmann@fz-juelich.de
Melanie Schulte, Forschungszentrum Jülich – ETN, me.schulte@fz-juelich.de
Stefan Rabe, CEF-NRW, rabe@cef.nrw.de
Joachim Kutscher, Forschungszentrum Jülich – ETN, jo.kutscher@fz-juelich.de
Karin Wikman, TEKES Teknologian Ja Innovaatioiden Kehittaemiskeskus, Karin.Wikman@tekes.fi
Manon Gerbaud, Agence de l'Environnement et de la Maîtrise de l'Energie (ADEME), manon.gerbaud@ademe.fr
Tristan Carrere, Agence de l'Environnement et de la Maîtrise de l'Energie (ADEME), tristan.carrere@ademe.fr
José Herrero, Centro de Investigaciones Energéticas, Medioambientales y Tecnológicas (CIEMAT),
jose.herrero@ciemat.es
Severino Falcón Morales, Ministerio de Economía y Competitividad (MINECO), severino.falcon@mineco.es
Laurence Polain, Service Public de Wallonie (SPW), laurence.polain@spw.wallonie.be
Nicolas Delsaux, Service Public de Wallonie (SPW), nicolas.delsaux@spw.wallonie.be
Elsie De Clercq, Agentschap Innoveren Ondernemen (VLAIO), elsie.declercq@vlaio.be
Jan Osiński, Narodowe Centrum Badan i Rozwoju (NCBR), jan.osinski@ncbr.gov.pl
Kaan Karaösz, Turkiye Bilimsel ve Teknolojik Arastirma Kurumu (TUBITAK), kaan.karaoz@tubitak.gov.tr
Riza Seymen, Turkiye Bilimsel ve Teknolojik Arastirma Kurumu (TUBITAK), riza.seymen@tubitak.gov.tr
Otto Bernsen, Ministerie van Economische Zaken, Landbouw en Innovatie / RVO, otto.bernsen@rvo.nl
Stathis Tselepis, Kentro Ananeosimon Pigon ke Exikonomisis Energeias (CRES), stselep@cres.gr
Christian Inglis, InnovateUK, Christian.Inglis@innovateuk.gov.uk
Amy Oswin, InnovateUK, Amy.Oswin@innovateuk.gov.uk
Leonidas Antoniou, Research Promotion Foundation (RPF), lanto@research.org.cy
Ioanna Loizou, Research Promotion Foundation (RPF), iloizou@research.org.cy
Adriana Agrimi, Regione Puglia, a.agrimi@regione.puglia.it
Carlo Gadaleta Caldarola, ARTI - Agenzia Regionale per la Tecnologia e l'Innovazione,
c.gadaletacaldarola@arti.puglia.it
Dario Tornabene, Regione Sicilia, dario.tornabene@regione.sicilia.it
Theodor Zillner, Bundesministerium für Verkehr, Innovation und Technologie (BMVIT), theodor.zillner@bmvit.gv.at
Elvira Lutter, Klima- und Energiefonds (KLIEN), elvira.lutter@klimafonds.gv.at
Gernot Wörther, Klima- und Energiefonds (KLIEN), gernot.woerther@klimafonds.gv.at
Anita Hipfinger, Österreichische Forschungsförderungsgesellschaft (FFG), Anita.Hipfinger@ffg.at
Pierre-Jean Rigole, Swedish Energy Agency (SWEA), pierre-jean.rigole@energimyndigheten.se
Tobias Walla, Swedish Energy Agency (SWEA), tobias.walla@energimyndigheten.se

ABSTRACT: SOLAR-ERA.NET is a network that brings together more than 20 RTD and innovation programmes in the field of solar electricity technologies in the European Research Area. The network of national and regional funding organisations has been established in order to increase transnational cooperation between RTD and innovation programmes and to contribute to achieving the objectives of the Solar Europe Industry Initiative (SEII) through dedicated transnational activities (especially transnational calls). SOLAR-ERA.NET was an EU funded FP7 project running from 2012 to 2016. Through the support of the funding organisations, some 60 MEUR will have been mobilised for transnational RTD and innovation projects through four joint calls carried out. Results are presented for the transnational SOLAR-ERA.NET calls in this paper. Within the H2020 Cofund ERA-NET scheme, the network continues to work as SOLAR-ERA.NET Cofund.

Keywords: R&D and Demonstration Programmes, Photovoltaic, Funding and Incentives, Strategy

1 INTRODUCTION AND CONTEXT

SOLAR-ERA.NET is a European network of national and regional funding organisations and RTD and innovation programmes in the field of solar electricity generation, i.e. photovoltaics (PV) and solar thermal electricity (STE) and concentrating solar power (CSP). SOLAR-ERA.NET shall contribute to reaching the objectives of the former Solar Europe Industry Initiative (SEII), replaced by the European Technology & Innovation Platforms (ETIPs) in 2015, by carrying out the coordination and support actions between national and regional R&I programmes.

The SEII was a joint initiative of the industry sector, EC and member states and was embedded in the European Strategic Energy Technology Plan (SET-Plan) which aims to increase, coordinate and focus EU support on key low-carbon energy technologies in order to achieve Europe's 2020 and 2030 energy and climate objectives in the future.

2 NETWORK AND MISSION

2.1 Network

As the largest ERA-NET (network in the European Research Area) ever in the solar power field, SOLAR-ERA.NET involves more than 20 national and regional RTD and innovation programmes dealing with PV and CSP. This high level of involvement of most relevant stakeholders provides excellent outreach and allows for a solid coordination needed for an efficient and coherent approach in the highly diverse and versatile RTD landscape.

The network is composed of 19 organisations being programme owners and / or programme managers (see list of authors). Countries and regions participating in the SOLAR-ERA.NET consortium and / or in the trans-national calls are identified in Figure 1. SOLAR-ERA.NET is in principle open to other countries and programmes wishing to join in transnational calls.

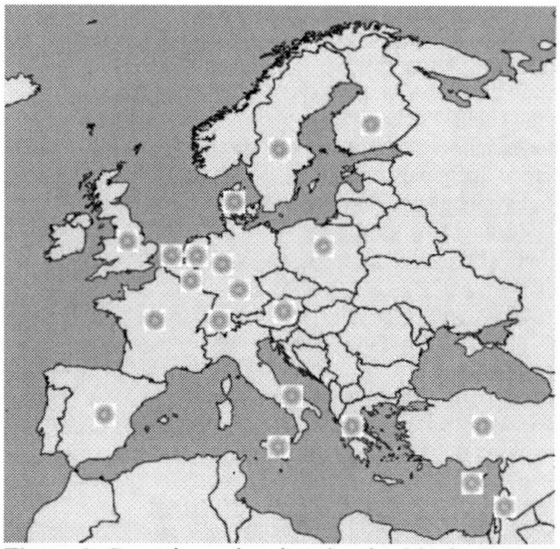

Figure 1: Countries and regions involved in the SOLAR-ERA.NET consortium and / or in the transnational calls

2.2 Missions and goals

SOLAR-ERA.NET has two fundamental missions. As a network supported by the EC within the ERA-NET scheme, the mission is to improve the coordination and cooperation between national and regional RTD programmes. In the context of the Strategic Energy Technology (SET-) Plan, the mission of the network is to implement central parts of the SET plan on a transnational level and thus contribute to achieving the goals defined in the SET plan.

The more specific and essential goals and activities of SOLAR-ERA.NET are i) to launch joint calls for RTD proposals by national and regional RTD and innovation programmes and ii) to define and support the best joint activities, strategic information exchange and use of implementation tools.

By identifying and choosing priority topics based on the Implementation Plans for PV respectively CSP for transnational calls, the SOLAR-ERA.NET network selects and funds industrially relevant transnational RTD and innovation projects in the field of solar electricity technologies.

In quantitative, financial terms, SOLAR-ERA.NET shall result in a total funding volume by the participating national and regional programmes of up to 50 MEUR for innovative projects.

3 ACTIVITIES, FACTS AND FIGURES

3.1 Set up of transnational calls

The network started in November 2012. Four sets of transnational joint calls have been launched so far in March 2013, January 2014, December 2014 and December 2015 with the participation of up to 18 countries covering up to 8 PV and 4 CSP topics per call.

Call topics in photovoltaics (PV) and concentrating solar power (CSP) are commonly defined by the SOLAR-ERA.NET consortium and adopted by the participating programmes and countries according to their national / regional priorities. The topics can be found in Table 1.

The transnational call is based on a 2-step/stage-procedure with i) a preproposal and ii) a full proposal stage. Preproposals are checked according to the national / regional regulations and then discussed in the SOLAR-ERA.NET consortium setting up lists of preproposals recommended respectively not recommended for stepping on to the full proposal phase.

Once the full proposals are evaluated by independent international experts as well as by national experts, a list of projects suggested for funding is established. The ultimate funding decisions are however taken by the national agencies / ministries.

The transnational SOLAR-ERA.NET joint calls are each allocated with a total public funding budget of approximately 12 million euros provided by the participating national and regional programmes and agencies. The joint call consortium is mainly composed of SOLAR-ERA.NET partners but can be extended by associate partners (like Denmark and Israel). Applicants may also come from other countries (like Ireland and Norway) but they do have to provide their own funding for their participation in the projects.

3.2 Participation in transnational calls

The four transnational SOLAR-ERA.NET calls found good interest in the solar power industry sector and research community. Some 180 preproposals were submitted with a total budget of 230 MEUR (of which 144 MEUR funding requested) involving 774 partners respectively contributions from 18 countries.

A slight majority (54%) of the organisations involved in the preproposals submitted belong to the categories of small and medium sized enterprises and large enterprises; research organizations have a share of 43%. Figure 2 shows the share of different types of organizations among the partners being involved in the preproposals submitted.

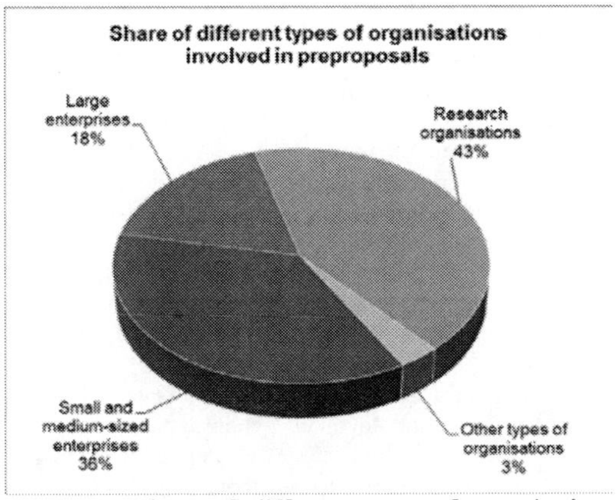

Figure 2: Share of different types of organisations involved in preproposals

83 full proposals were submitted in the four joint calls. Figure 3 shows the number of coordinators and partners in the full proposals submitted by call as well as by country / region.

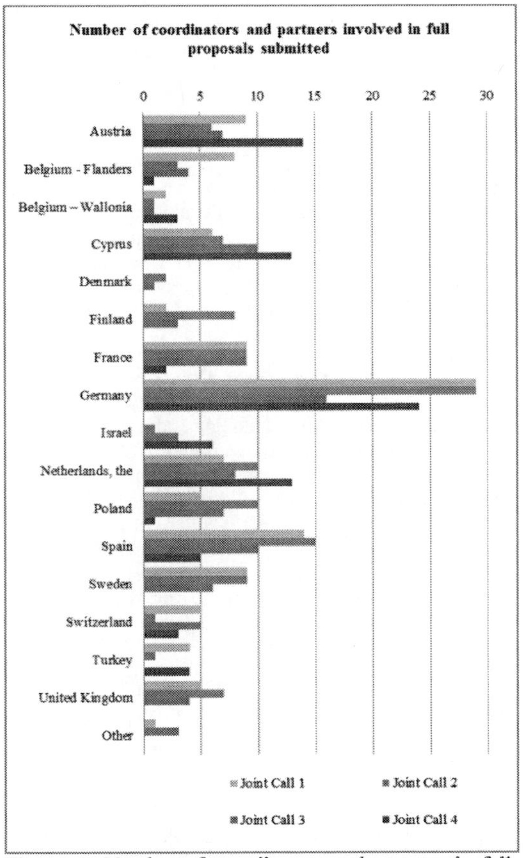

Figure 3: Number of coordinators and partners in full proposals submitted according to the countries / regions participating in the four transnational SOLAR-ERA.NET calls

4 RESULTS AND OUTLOOK

4.1 Selected results

The transnational SOLAR-ERA.NET calls found good interest and participation of the sector:

- 180 preproposals submitted with total project costs of 230 MEUR and funding requested of 160 MEUR
- 774 partners / contributions involved in preproposals
- 18 countries and regions participating in the joint call consortium
- 83 full proposals submitted
- 38 projects funded (8 projects still in negotiation phase)
- Funding committed so far (38 projects): approx. 32 MEUR public funding for a total project volume of around 49 MEUR

Table 1 shows the number of full proposals suggested for funding per call topic. Some topics have been added in the second and third call.

Table 1: Four transnational SOLAR-ERA.NET calls - topics and numbers of full proposals suggested for funding (number of full proposals per call and topic as well as total sum)

Transnational SOLAR-ERA.NET topics in calls I, II, III, IV and all four together	I	II	III	IV	Σ
PV1 Innovative processes for inorganic thin-film cells & modules	2	2	1	2	7
PV2 Dedicated modules for BIPV design and manufacturing	3	2	1	0	6
PV3 Grid integration and large-scale deployment of PV	2	1	1	3	7
PV4 High-efficiency PV modules based on next generation c-Si solar cells	1	5	3	2	11
PV5 Solar glass and encapsulation materials	1	2	0	1	4
PV6 Concentrator PV technology		0	2	0	2
PV7 Si feedstock, crystallization and wafering		1	0	0	1
PV8 Organic solar cells and other emerging concepts			2	0	2
CSP1-4 All four CSP topics together	1	2	1	2	6
Total	*10*	*15*	*11*	*10*	*46*

The 38 projects (33 PV projects, 5 CSP projects) funded are:

- SLAGSTOCK: Low-cost Sustainable Thermal Energy Storage Systems Made of Recycled Steel Industry Waste

- LIMES: Light Innovative Materials for Enhanced Solar Efficiency
- INTESEM: Intelligent Solar Energy Management Pipeline from Cell to Grid
- NOVACOST: Non vacuum based strategies for cost efficient low weight chalcogenide photovoltaics
- NovaZOlar: All-non-vacuum processed ZnO-based buffer and window layers for CIGS solar cell technology
- InnoModu: Leadfree modules with low silver content and innovative busless cell grid
- AER II: Industrialization and System Integration of the Aesthetic Energy Roof Concept
- PV4FACADES: Photovoltaics for high-performance building-integrated electricity production using high-efficiency back-contact silicon modules
- BLACK: Black silicon and defect engineering for highly efficient solar cells and modules
- PV2GRID: A next generation grid side converter with advanced control and power quality capabilities
- THESEUS: Tandem High Efficiency Solar Cells Utilizing III-V Semiconductors on Silicon
- APPI: Atmospheric Pressure Processing for Industrial Solar Cells
- SNOOPI: Smart Network Control with Coordinated PV Infeed
- SPRINTCELL: Sulfide-based Ink for Printable Earth Abundant Solar Cell
- IPERMON: Innovative Performance Monitoring System for Improved Reliability and Optimized Levelized Cost of Electricity
- ACCESSCGIS: Atmospheric Cost Competitive Elemental Sulfo Selenization for CIGS
- PV me: organic PhotoVoltaic systems integrated in manufactured building elements
- U-light: Ultra lightweight PV modules and their applications in innovative PV systems achieving lowest levelized cost of electricity (LCOE)
- inGRID: High efficiency PV modules with cells interconnected by a novel conducting encapsulation layer
- MONOSCRIBE: Roll-to-Roll Monolithic Interconnection of Customizable Thin-film Solar Modules
- HighCast: High Performance Silicon Casting and Wafering
- EDITOR: Evaluation of the Dispatchability of a Parabolic Trough Collector System with Concrete Storage
- FunGlass: Multi-Functional Glass for PV Application
- SITEF: Silicone Fluid Test Facility
- PROOF: Competitive Industrialized Photovoltaic Roofing
- SolFieOpt: Optimal Heliostat Fields for Solar Tower Power Plants
- SPRINTCELL: Sulfide-based Ink for Printable Earth-Abundant Solar Cell
- DINAMIC: Dilute Nitride Based Concentrator Multijunction Solar Cells, With Efficiencies Over 47%
- IPERMON: Innovative Performance Monitoring System for Improved Reliability and Optimized Levelized Cost of Electricity
- HVolt-PV: High voltage IBC photovoltaic i-Cells and modules
- CNT-PV: Carbon nanotube hole-transporting and collecting layers for semi-transparent, flexible and low-cost solid-state photovoltaic cells
- HIPPO: High-efficiency poly-Si passivated contact solar cells and modules
- Liquid Si 2.0: Polysilane materials for liquid phase deposition of functional silicon layers for cost-effective high efficiency solar cells
- FrontCIGS: Re-designing front window in flexible CIGS modules for cost-effective moisture protection
- SIMON: Silicone Fluid Maintenance and Operation
- DURACIS: Advanced global encapsulation solutions for long term stability in industrial flexible Cu(In,Ga)Se2 photovoltaic technology
- Enhance: Enhanced rooftop PV integration through kinetic storage and wide area monitoring
- PEARL TF-PV: Performance and Electroluminescence Analysis on Reliability and Lifetime of Thin-Film Photovoltaics

4.2 Outlook

SOLAR-ERA.NET will likely result in some 46 projects with a total project volume of around 60 MEUR of which around 40 MEUR comes from public funding. This means that the EC funding for SOLAR-ERA.NET (1,2 MEUR) will be multiplied by a factor of around 50 resp. 30.

The successful projects demonstrate i) potential commercial impact / relevance to industrial and market needs / contribution to the Solar Europe Industry Initiative and added transnational value, ii) scientific and technological excellence and iii) high quality and efficiency of the implementation and the management. Further information on running projects from the transnational joint calls can be found on the solar-era.net website.

Within the Horizon 2020 Cofund ERA-NET, the network continues its activities and launched another joint call - allocated with a funding budget of 20 MEUR - in December 2016. The SOLAR-ERA.NET Cofund call addresses 4 topics, i.e. innovative and low-cost PV manufacturing, advanced PV products and applications, PV system integration and CSP cost reduction and system integration, and support successful projects aiming at Technology Readiness Levels of 5 to 7. By the time of writing this text, full proposals are being evaluated and ranked by independent international experts.

News on the future SOLAR-ERA.NET Cofund, will be channelled through the solar-era.net and funding agencies involved.

REFERENCES

Further information is available on the project website www.solar-era.net.

ACKNOWLEDGEMENTS

The support of the SOLAR-ERA.NET joint activities by the EC under grant agreement N° 321571 within the 7th Framework Programme and SOLAR-ERA.NET Cofund under grant agreement N° 691664 within Horizon 2020 is gratefully acknowledged.

HIGHLIGHTS FROM THE FP7 PROJECT ON PHOTOVOLTAICS CHEETAH: MORE POWER WITH LESS MATERIAL

Jan M. Kroon[1], Franco Roca[2], Iver Lauermann[3], Karsten Bittkau[4], Maté Heisz[5], Kris Van Nieuwenhuysen[6], Adrien Danel[7], Paul Sommeling[1], Martina Schmidt[3,8], Suren Gevorgyan[9] and on behalf of the rest of the CHEETAH consortium
[1]ECN Solar Energy, [2]ENEA (Italy), [3]HZB (Berlin), [4]FZ Jülich (Germany), [5]Solar Power Europe (Belgium), [6]IMEC (Belgium), [7]CEA/INES, [8]University of Duisburg (Germany),[9]DTU (Denmark)
P.O. Box 1, 1755 ZG Petten, The Netherlands
E-mail: j.kroon@ecn.nl, tel. +31885154734

ABSTRACT. CHEETAH is an FP7 integrated research project on photovoltaics funded by the European Commission and was initiated by EERA PV. This project is structured into two type of activities: Coordinative and Support actions (CSA) and Joint Research activities (JRA). The CSA focuses on the creation of a long term collaborative platform by developing tools for knowledge sharing, e-learning platforms for training and education, mobility between researchers, efficient use of infrastructures and promoting best practices and standards. The JRA focuses on developing new concepts and technologies for wafer-based crystalline silicon PV (modules with thin cells < 100 micron), thin-film PV (advanced light management) and organic PV (very low-cost barriers), resulting in reduced cost of environmentally benign/abundant/non-toxic materials and increased module performance. This paper gives a summary of the main achievements of this project for the two categories of activities.

Keywords: Cost reduction, PV materials, Education and training

1 INTRODUCTION

CHEETAH (Acronym for *Cost-reduction through material optimisation and Higher EnErgy outpuT of solAr pHotovoltaic modules - joining Europe's Research and Development efforts in support of its PV industry*) [1] is a 4 year combined collaborative project and coordination and support action funded under the European Commission's 7th Framework programme and runs since January 2014. The project consortium consists of 34 partners from research institutes and universities.

CHEETAH's aim is to solve specific R&D issues in coherence with the EERA-PV Joint Program and SET plan, to overcome fragmentation of European PV R&D in Europe and intensify the collaboration between R&D providers and industry to accelerate the industrialization of innovations.

The main objectives in this project are threefold:

- **More power with less materials**: developing new concepts and technologies for wafer-based crystalline silicon PV (modules with thin cells < 100 micron), thin-film PV (advanced light management) and organic PV (very low-cost barriers), resulting in (strongly) reduced cost of environmentally benign/abundant/non-toxic materials and increased module performance.
- **Fostering long-term European cooperation in the PV R&D sector**, by sharing knowledge, organizing workshops, exchange and training researchers inside and outside Europe, efficient use of infrastructures, promoting best practices and standards
- **Accelerating the implementation of innovative technologies in the PV industry** by the involvement of Solar Power Europe and EIT InnoEnergy in this program.

This paper will describe the consortium (in section 2) and the main achievements of the project which are divided into two main categories: coordinative and supportive actions, and Joint Research activities (in section 3). Some concluding remarks and perspectives of the results of this integrated research project are given in Section 4.

2 DESCRIPTION OF THE CONSORTIUM

The CHEETAH project brings together a consortium of 34 partners spread around Europa, consisting of 31 research centers (universities and research institutes) of excellence in PV R&D, complemented by Solar Power Europe (formely known as EPIA) and Innoenergy (formerly known as KIC Innoenergy) as the associations with the aim to shape the regulatory environment and enhance business opportunities for solar power in Europe and consultancy company Ayming for project management support (see Figure 1 and appendix).

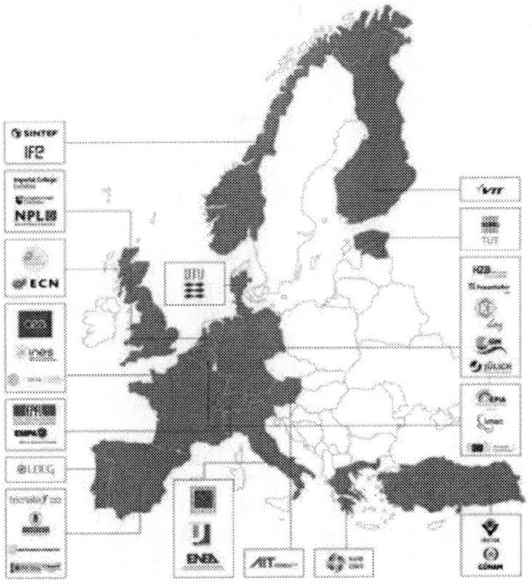

Figure 1. Geographical spreading of the 34 CHEETAH consortium members

3 DESCRIPTION OF THE PROJECT

The CHEETAH project is structured around two blocks of activities:

- **Coordination activities (CSA)**: developing the basis for long-term research and giving access to the knowledge to researchers and industry (WP2-5)
- **Joint research activities (JRA)**: improving the services provided and fill in the gaps: more power with less materials (WP6-10)

A Pert diagram of the project structure of CHEETAH is shown in Figure 2.

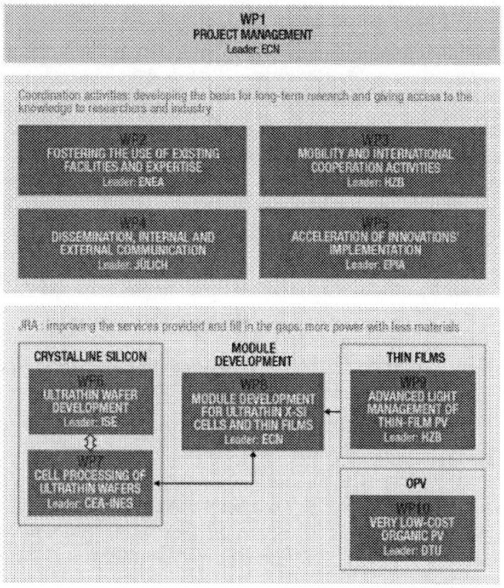

Figure 2. Pert diagram of the CHEETAH workpackage structure

In the following subsections 3.1 and 3.2 the main highlights of the achievement within the CSA and JRA activities will be presented.

3.1 Coordination (CSA) activities

The main aim of these cluster of coordination activities was to create a framework for cooperation by creating tools for knowledge exchange like a web-based knowledge exchange area portal, e-learning platform available for training courses, workshop lectures, as well to organize mobility actions by organizing research exchanges within the consortium for training and educational purposes of student and young researchers.

Some examples of important achievements within this coordinative cluster are listed in subsections 3.1.1 and 3.1.2.

3.1.1 Cheetah knowledge exchange web area portal

The CHEETAH Knowledge Exchange Web Area Portal (KEAP) is an online information source to promote knowledge exchange among experts and trainees on photovoltaic (PV) solar energy research in Europe. It is developed in the context of CHEETAH and open to non-CHEETAH partners and the wider photovoltaic community and is accessible via www.cheetahexchange.eu. KEAP supports cooperation among the CHEETAH partners and within the broader PV research community in Europe by facilitating knowledge and information sharing, identifying technical and scientific needs and making available a wide range of information on expertise, infrastructures, equipment and PV technologies.

This portal improves the state-of-the art of knowledge exchange thanks to a dynamic database, a powerful build-in search engine and a dedicated e-learning platform for on-line meetings with internal and external stakeholders, webinars, on-line tests and experiments. A lot has been done to improve the accessibility and impact of KEAP by advertising via all public channels (youTube. Linkedin, ResearchGate, Twitter, etc). KEAP counts 33 CHEETAH and 13 non CHEETAH corporate descriptions, 205 expert and 45 infrastructure profiles of both CHEETAH and non-CHEETAH organizations. 46 webinars have been organized since January 2014 by the platform with more than 1010 registered users, 930 "live" participants and hundreds accessing via streaming.

3.1.2 Joint support actions

As a continuation of the Round Robin trials, organized within the FP7 Sophia project, a number of Round Robin and inter comparison campaigns have been initiated in CHEETAH. They focused on characterization of thin film Silicon based tandem cells and perovskite cells with the aim to test and validate measurement protocols and research infrastructure. The outcomes of these campaigns are best practice documents that will be used as input for new pre-normative test protocols and standards. These documents will be shared and disseminated *via* peer reviewed papers and/or the KEAP.

3.2 Joint research (JRA) activities

In the current PV landscape, the PV market is dominated (>90 %) by crystalline Silicon (x-Si) PV and very strong cost driven inhibiting adoption of novel innovations. Thin film PV like CIGS has a much lower market share (< 10 %) but possesses a very low cost potential and able to compete with x-Si if module efficiencies are comparable. Emerging technologies like Perovskite/Organic PV could widen applicability of PV by introducing new assets like flexibility, free-form and colour.

For the three mentioned PV technologies, costs are highly dominated by the materials, i.e. for x-Si the wafer, for thin films the semiconductor like including indium, which is present in CIGS and some TCO's (ITO) and for emerging technologies the encapsulation. These were selected as the drivers for the JRA carried out within CHEETAH. A few illustrations of the progress that has been made in CHEETAH are shown in the following subsections.

3.2.1 Cost reduction strategy in Wafer-based crystalline Si

The JRA activities were organized as follows:

1. Fabricate (ultra)thin wafers by
   - Epitaxial growth `kerf free` (EPI) foils ranging from 40-80 micron
   - Wire sawn wafers (high TRL) chemically thinned down to 80 micron
2. Development of a process for the fabrication of advanced solar cells and modules

Develop new technology bricks for ultra-thin wafers (40 µm)

- **front side textures** → optical light trapping, Jsc improvement
- **High surface passivation level** → Voc improvement
- **Advanced metallization** → low breakage rate, Jsc and FF improvement
- **Develop and/or redesign industrially compatible cell and module fabrication processes** for thin cells
  - Maximize yield, maintain performance

Fabrication of kerf free EPI foils from 40-80 micron

Processes were developed to prepare release layers, epitaxially thicken the Si foil and release layer resulting in thin epitaxially (EPI) grown Si wafers varying in thickness from 40 up to 80 µm featuring electronic and mechanical properties to allow for >20% solar cells. For the epitaxial thickening and layer release the standard foil size increased from 5 x 5 cm^2 to 12.5 x 12.5cm^2 and several batches of 40 µm foils were made for cell processing with good detachment yield and lifetimes up to 100 µs on average.

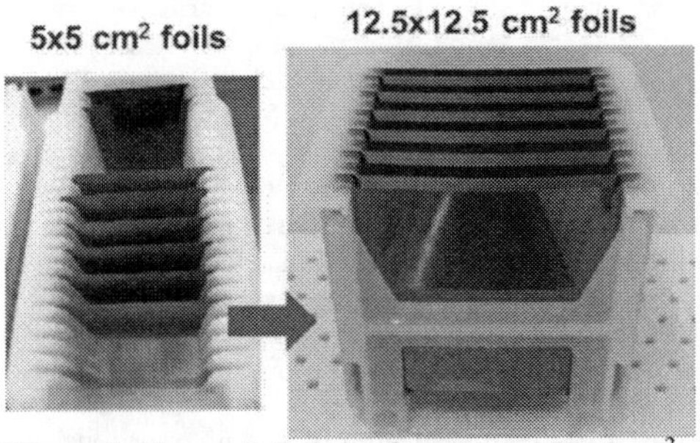

Figure 3. Change in foil size from 25 to 156.25 cm^2

Cell processing of ultra-thin EPI-wafers

Table I shows the I-V characteristics of a successful heterojunction (SHJ) cell process integration run performed on 50µm 125x125 mm^2 Epifoils (4x4 cm^2 cells) compared to Cz based 50 and 180µm SHJ cell processed in the same run. Optimized efficiencies up to 17 % were achieved, with further room for improvement for all the I-V parameters.

Table I. J-V characteristics for SHJ cells on thin, thick Cz wafers and thin EPI foils (*data courtesy imec*)

	count	Jsc	Voc	FF	eta
Thick Cz	12	35.8	722	70.6	18.3
180 µm best		36.2	730	77.2	20.3
Thin Cz	20	33.9	723	71.6	17.6
50 µm best		34.7	744	74.4	18.8
Epifoil	17	32.9	656	72.1	15.6
50 µm best		34.1	693	75.5	17.0

Cell processing of thin wire sawn wafers

Heterojunction (HJ) cells down to 80 µm were manufactured using the semi-industrial pilot line in automatic mode with minor process adjustments and low breakage rates compared to standard thickness. A maximum efficiency of **22 %** has been achieved to date for a ~90 µm four busbar bifacial SHJ cell. Wafers were all chemically thinned from 160-180µm commercially available sawn wafers. Complementary tests were also conducted on 120µm sawn wafers provided by CHEETAH partner SINTEF demonstrating further the compatibility of the industrialization line with such thin incoming wafers.

In addition, Interdigitated Back Contact (IBC) cells down to 80 µm were processed using a pre-pilot process flow starting from the same batch of 120 µm sawn wafers. Both cell types (SHJ and IBC) were used as input for module integration for which different interconnection technologies are used as is illustrated in the next paragraphs.

Cell integration into modules

Functional mini (4 cells) IBC and full sized (60 cells) HJ modules based on cells between 80 and 100 µm were successfully fabricated.

Figure 4 shows an IBC mini module (left) and an EL image (middle) revealing no cell breakage/cracks and FF > 74 %. The modules are manufactured in a mini-pilot line using foil based interconnection technology where cells are contacted to a conductive back sheet via electrically conductive adhesives [5]. The right picture demonstrates feasibility of cell handling down to 80 µm via pick an place using industrial module equipment.

Figure 4. Left: an IBC mini module manufactured at ECN; centre: an EL image and right: industrial equipment showing feasibility of handling thin wafers

In addition, three 60 cells SHJ modules were fabricated, using standard glass, conductive glue, encapsulant and lamination process. A set of 93µm record SHJ cells were integrated in a 60-cell glass/backsheet and glass/glass module configuration, leading to very promising final module output power of 313Wp (cell to module power ratio – CTM - of 99% and 0.98Wp per g Si) and 297Wp (CTM of 96%) respectively, without specific breakage observed after PL inspection. This demonstrates the full compatibility of thin wafers with current module configuration.

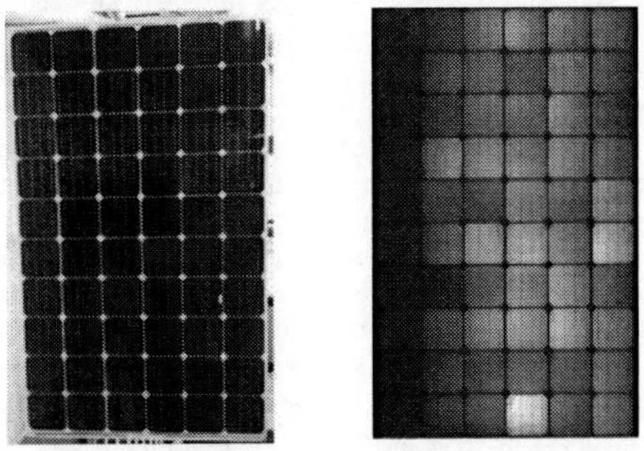

Figure 5. A first 60 cell HJ module + EL image, with HJ 4 bus bar cells (HJ) having an average thickness of 93µm (from 86 to 99µm), manufactured at CEA-INES.

3.2.2 Cost reduction strategy in thin film PV

The main driver for the JRA activities on thin film PV was to achieve higher efficiencies of thin-film solar cells (thin film Si, $Cu(In,Ga)(S,Se)_2$ and Kesterites) with less solar cell base materials by enhancing todays maximum efficiency using light management strategies and an envisaged cost reduction of 20 %. Moreover, the supply risks of In and Ga were simultaneously addressed for $Cu(In,Ga)(S,Se)_2$ cells. The consortium worked on two innovative approaches:

Light management strategies for thin film Silicon produced by liquid phase crystallization

Light management strategies like nano-patterning/-structuring of thin film silicon cells that are prepared via liquid phase crystallization or by introducing grating structures are applied to improve the optical absorption of the solar cell absorber without any electrical losses of the device. Work is in progress to reach efficiencies of ≥ 15% in the project with the ultimate goal to reach the potential of competitive efficiencies > 20%.

Micro concentrator approach

The approach is based on restriction of absorber material to a few ten to hundred microns diameter and use concentrated sunlight by a factor of 50 - 500 onto absorbers to compensate for the loss in light absorption.
In a top-down proof of concept study, close to 30% relative efficiency increase was achieved by light concentration onto planar CIGS solar cells of sub-mm^2 size patterned by lithography [3] and inkjet-printed CIGSSe.

Working CISe micro cells were made using a novel bottom-up approach for local chalcopyrite growth from indium islands. Corresponding millimeter-sized lenses are fabricated from PMMA by a casting process. Finally, an angular splitting concentrator exploiting both direct and diffuse light components by combining chalcopyrite and kesterite absorber material was proposed and are in the early stages of research. A more detailed description of the innovative approaches followed can be found in reference [4].

3.2.3 Cost reduction strategy in materials research for Organic PV

The main aim of the OPV JRA activities is to identify materials and develop device structures which are intrinsically stable towards water and oxygen so that barrier layer requirements can be significantly reduced. This will substantially reduce the costs of flexible OPV since the barrier is considered as one of the most important cost drivers for flexible OPV.

The consortium worked on the approach to establish a link between intrinsic and extrinsic stability of devices and effect of different barrier materials on device lifetime. Large intercomparison lifetime testing campaigns have been executed so far using various device designs and organic polymers. A wide range of active materials has been investigated in the cells with the purpose of improving the intrinsic lifetime of devices. The studies revealed that while the best device architectures could only improve the lifetime of an unprotected device by a factor of 2 to 3, application of a simple PET foil improved the device lifetime by two orders of magnitude. Although the use of ultrabarriers with UV filter could improve the lifetime by another order of magnitude compared to PET, the price for such barriers is too high. Thus, the conclusion of the study was that the best is to focus on developing cheap methods for improving barrier properties of cheap foil materials like PET and adhesives. Currently, deposition of barriers directly onto devices or PET via sputtering methods are investigated.

New organic materials (amongst others a non-fullerene acceptor) were additionally identified that both increased performance and thermal stability.

4 CONCLUDING REMARKS AND OUTLOOK

The technology developments within CHEETAH should lead to the realization of innovative and competitive PV concepts with a significant reduction in cost of materials and increase of the overall performance. These innovative developments should finally lead to a contribution to an accelerated implementation in the European PV industry, so that Europe can regain and build up own manufacturing capacity in all parts of the value chain in due time. In addition, the establishment of an effective collaborative platform for the PV R&D sector established via the CSA activities within CHEETAH will help Europe in realizing these goals.

The short term impact of the various technical achievements in terms of exploitability is not easy to judge in general. An important aim of many JRA in CHEETAH was to prove technical feasibility and bring it to TRL levels up to 7 so that will generate more interest from the industry. The production of thin cells down to 80 micrometer in an industrial environment as well as the technical capabilities to handle thin cells in a module line are examples of CHEETAH results that have the potential to be exploited by industry on a shorter timeframe as soon as silicon becomes more expensive. These are important achievements and demonstrate that we are far ahead of roadmap predictions by for instance ITRPV [6]. Silicon wafer production with Kerf free solutions as investigated in CHEETAH also attracts the attention of start-ups since this can ultimately result in significant cost savings for Silicon. For thin film PV and emerging technologies like Organic and Perovskite PV, the followed approaches within CHEETAH are still in its early infancy and more time is needed to bring these concepts up to higher TRLs(6-7) after which they can be scaled up and truly evaluated for their manufacturability and cost reduction potential.

Despite the fact that the PV industry is again in a consolidation phase and has other priorities at this moment to survive, there is certainly consensus that longer-term innovations like those investigated in CHEETAH are required to further reduce the costs to levels that will make PV a very low cost and competitive electricity source compared to others.

5 REFERENCES

[1] CHEETAH website: www.cheetah-project.eu
[2] S. Harrison et al., EUPVSEC Proceedings 2016, 358-362
[3] E. Lotter et al., EUPVSEC Proceedings 2016, 1156-1160
[4] B. van Aken. Photon International, Vol. 33, October 2016, 97-105
[5] M. Schmidt et al., EUPVSEC Proceedings 2017
[6] ITRPV Eighth Edition 2017, http://itrpv.net/Reports/

6 ACKNOWLEDGEMENT

The European Union is acknowledged for funding the CHEETAH project through the Seventh Framework Programme (FP7/2007-2013) under grant agreement n° 609788 (CHEETAH)

7 APPENDIX

List of beneficiaries

Nr.	Short name	Country
1	ECN	Netherlands
2	CEA/INES	France
3	Fraunhofer ISE	Germany
4	DTU	Denmark
5	HZB	Germany
6	FZ Julich	Germany
7	AIT	Austria
8	ENEA	Italy
9	EPFL	Switzerland
10	IFE	Norway
11	IKZ	Germany
12	IMEC	Belgium
13	NPL	United Kingdom
14	SINTEF	Norway
15	Tallinn University	Estonia
16	ZWS	Germany
17	LNEG	Portugal
18	Tor Vergata	Italy
19	METU	Turkey
20	TECNALIA	Spain
21	UPM	Spain
22	CIEMAT	Spain
23	KAPE-CRES	Greece
24	Loughbourough Univ	United Kingdom
25	EMPA	Switzerland
26	Imperial	United Kingdom
27	JRC-EC	Belgium
28	TUBITAK	Turkey
29	VTT	Finland
30	UPVLC	Spain
31	UNIMIB	Italy
32	Solar Power Europe	Belgium
33	Innoenergy	Netherlands
34	Ayming CG	France

CAFÉ AU LIGHT: HOW TO IMPROVE GUINEAN PEOPLE'S LIVES BY COMBINING COFFEE AND PHOTOVOLTAICS

J. Cárabe[1], N.N. Malo[2], A. Bautista[3], L. Barrios[3] & M. Loua[4]
[1] CIEMAT. Division of Renewable Energies. Avda. Complutense 40. E-28040 Madrid. Spain. julio.carabe@ciemat.es.
[2]UDECOM. BP 523, Boma, Samoé, N'Zérékoré. Guinée
[3]CLEANERGETIC SEERS SOLUTIONS SL. c/ Agosto 6, loft 16. E-28022 Madrid. Spain
[4]Ambassade de la République de Guinée en Espagne. Madrid. Spain.

ABSTRACT: Photovoltaics developed for a long time as a field clearly focused on stand-alone applications, in which the lack of any grid in certain places justified the installation of PV systems in spite of the high costs they used to have decades ago. Fortunately the successful development of new technologies, combined with a spectacular market deployment (based on grid-connected systems), has reduced the cost of PV electricity below grid-parity on a large scale. The need for electricity in many countries is still, however, a main concern and PV is now in an optimum position to offer technically and economically efficient solutions thereto. *Café au Light* is a project aimed at improving life conditions and stimulating economic development in a number of villages in the N'Zérékoré prefecture of Guinea by means of the installation of PV microgrids that will provide electricity both for public services and private homes. Funding is not obtained through grants but through a significant increase in the productivity of the main economic resource of the people living in these towns: coffee. With this, the undertaking is twofold, having to deploy efforts on two areas: coffee and photovoltaics. The initiative is not conceived as a development-aid project but as an endeavour in which business is equally profitable for all parts on the basis of two essential goals: net benefit for the people and sustainability.
Keywords: Developing Countries, Education and Training, Economic Analysis, Financing, Rural Electrification, Stand-alone PV Systems, Villages.

1 INTRODUCTION

Photovoltaic solar energy is a very competitive and clean energy source which is already contributing in a significant way to an impressive change in the energy model of industrialised countries, providing effective solutions to the three main energy & environment challenges of the European Union: sustainability, security of supply and lowest environmental impact. Yet, other, stand-alone, more traditional applications of photovoltaics should be considered equally important, because they can play a critical role in a process of huge improvement of the conditions of life of millions of people in countries having a high economic potential but relatively little technological development.

The Republic of Guinea, in the heart of western Africa is an 11-million people country with large natural resources. Its economy is largely dependent on agriculture and mineral production. It is the world's second largest producer of bauxite and has rich deposits of diamond and gold. Large areas of land are dedicated to cultivating coffee, cocoa and palm. In spite of such an appreciable potential, only about 26% of the population has access to electricity (53% of the people living in cities and 11% of those living in rural areas). This restriction does not only affect directly their private life, but also limits their economic and technological development. Photovoltaic solar energy can be used as a powerful tool for the economic development of Guinean people, who may have this way a unique opportunity to improve very noticeably their living standards, with particular emphasis on access to water, health services and basic welfare.

This initiative, called *Café au Light*, was launched by UDECOM, the University for Community Development, in N'Zérékoré, a prefecture in the inner region of Guinea. UDECOM is conceived as a tool aimed at providing not just academic knowledge, but the know-how which can be useful for the people who live there, taking into account their real needs and environment. UDECOM decided to promote the installation of photovoltaic systems in a number of towns of the N'Zérékoré prefecture by establishing an agreement with a reference research and technology centre having a large expertise in the field (CIEMAT - Spain) and by providing the necessary infrastructure for the education and training of local technicians and engineers who would be able to guarantee a proper maintenance of the PV systems installed. Soon, CIEMAT established collaboration for this project with a company focused on the installation of photovoltaic systems (Cleanergetic Seers Solutions - Spain). These contacts originated the present *Café-au-Light* consortium.

The project has a technical photovoltaic dimension which runs in parallel with the commercial component. In principle, the idea was just to design and install a number of PV systems, to request funding and to establish a mechanism for the reimbursement of the loans through the payments by the end users. However, it soon became evident that the economic capacity of these end users was well below both the potential offered by their resources and the solvency required for funding the PV installations. In other words, it seemed unreasonable to improve the capacity of consumption of the people without trying to enhance at the same time their capacity of production of goods (which is clearly susceptible of huge improvements without great efforts).

2 BACKGROUND

Thirteen years ago, in 2004, a Guinean PhD now living in Canada, Dr. Nianga Malo, decided to found in N'Zérékoré, the second largest city in the country, located in the innermost region thereof, a University in which Guinean young people could acquire knowledge and learn techniques being useful for their society. The main goal was to generate technicians having skills fitting the social reality of the country. This way

UDECOM (*Université pour le Développement Communautaire*, i.e. University for Communal Development) was created.

More recently, with the aim of generating economic activity and addressing the serious problem of the lack of electric energy affecting not only the inhabitants of the N'Zérékoré prefecture but in fact most Guineans, Dr. Malo conceived the possibility to install electric-energy generation systems based on photovoltaic modules in some villages, starting with three of them (Samoé, Guéla and Kouénala), which add up a population of about 7000 inhabitants. In Dr. Malo's proposal a number of considerations were clearly taken into account, such as the facts that photovoltaic solar energy is one of the cleanest and cheapest sources of electricity, requires simple infrastructures and moderate investments, is fully renewable and is probably the best option not only for producing clean and cheap electric energy in grid-connected facilities, but also for electrifying areas being far from any electrical grid.

In 2015 UDECOM, by way of Dr. Malo and with the support of the Embassy of the Republic of Guinea in Madrid, on the basis of their confidence on CIEMAT as a centre of reference on energy research and development, requested the support of their Photovoltaic Unit for the initiative. CIEMAT accepted the proposal with high interest and invited CLEANERGETIC SEERS SOLUTIONS, a consultancy, engineering and PV-system integration company with extensive previous collaboration, to join in order to provide the installation capacity needed for the project.

3 TECHNICAL DEFINITION OF THE PV SYSTEM

Designing, installing and starting a PV facility to provide electricity to a number of isolated towns does not involve any major technical difficulty for a centre specialised in energy R&D and a company with a broad experience on photovoltaics.

Three mini-grids, one per village, have been planned. Individual small PV systems could be exceptionally envisaged in cases of buildings far from the population nuclei. The components of each of the micro-grids are:

- Electric generator: photovoltaic modules which, in case of need, might be supported by alternatives such as wind turbines, a mini-hydraulic system or even a diesel generator, for instance.
- Energy management system, which transforms the energy provided by the generator into electricity in a form (voltage, frequency) directly usable by the customer. It also provides battery charging and consumption.
- Batteries, which store the spare energy produced in order to make it available whenever it is needed.
- Distribution box, providing protection elements and a basic framework to the system.
- One individual electricity meter per end user (home or public building or service) providing automatic limitations of power and energy and even a warning subsystem which can indicate that either of these limits has been surpassed or is being reached and even switch off electricity. Control and readings may be remote.
- Customer management system, which rationalises the energy use and the billing.

The following electricity requirements have been identified:

- Lighting.
- Cooling.
- Water pumping.
- Computers, printers, mobile-phone chargers and electronic equipment in general.
- Domestic appliances.

The photovoltaic power potential in the N'Zérékoré area can be obtained from the Guinea Solar-Resource Map published by the World-Bank Group [1]. As shown on Fig. 1, a reasonable value is 1500 kWh/(kW_p·year).

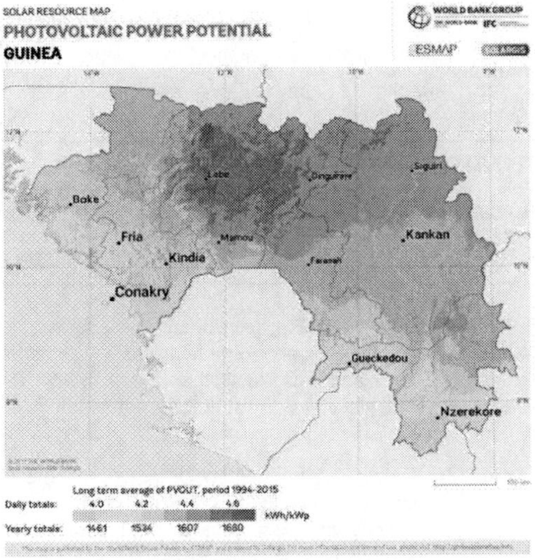

Fig. 1: Photovoltaic power potential of the different regions of Guinea (source World Bank Group [1].

The population and the number of dwellings to be endowed with PV electricity are shown on Table I.

Table I: Basic demographic parameters of the towns.

Town	Inhabitants	Dwellings
Samoé	4 531	1 133
Guéla	1 068	267
Kouénala	1 416	354
TOTAL	**7 015**	**1 754**

Each dwelling will be provided with 3 12-W, 1300-lm LED lamps, which will be active 5 h/day, plus a plug for small appliances (mobile phones, radio, etc.). In addition, each dwelling will have an extra LED lamp which will contribute to public lighting. With these parameters, the domestic energy consumption per dwelling will be 0.37 kWh/day or 135 kWh/year. The corresponding overall figures for the whole PV system, broken down per town and including not only private-home electricity but also public services, are expressed on Table II.

Table II: Calculated PV-electricity consumption per town in kWh/day (total also in kWh/year), covering both home- and public services.

	Samoé	Guéla	Kouénala	TOTAL
Home use	413.55	97.46	129.21	**640.21**
Health centre	42.43	10.00	13.26	**65.69**
Food refrigerators	8.49	2.00	2.65	**13.14**
Community building	42.43	10.00	13.26	**65.69**
School	42.43	10.00	13.26	**65.69**
TOTAL:				
kWh/day	549	129	172	**850**
kWh/year	200 498	47 251	62 647	**310 396**

The sizing of the photovoltaic systems of the respective microgrids are obtained from the above calculated consumptions and the above-cited 1500 kWh/(kW$_p$·year) (see please Table III).

Table III: Calculated nominal peak power of the PV systems needed and corresponding capital expenditure assuming a total system cost (including BOS) of 2.5 €/W$_p$.

	Nominal Peak Power (kW$_p$)	Capital expenditure (€)
Samoé	134	334 164
Guéla	32	78 752
Kouénala	42	104 412
TOTAL	**207**	**517 328**

This first system design contemplates photovoltaic modules for basic services, including essential electricity for domestic use, night lighting of main streets and public buildings with emphasis on health- and education centres, having in mind that the ebola virus affected this area only three years ago. The estimated rate to be paid per dwelling is 3 to 4 €/month. Higher nominal power and energy consumption figures are foreseen in further phases, once the initial investments have been partly amortised.

Inverters will have, in total, approximately the same nominal power as the modules, i.e. around 200 kW overall. Output will be 230 V, 50 Hz. Batteries will be sized to be able to provide electricity for up to two days of autonomy. Their total capacity (considering the sum for the three microgrids) will thus be about 1700 kWh or about 35000 Ah, assuming 48-Volt batteries.

4 FINANCIAL ASPECTS AND COFFEE

Although the initial approach covered only the design and installation of the PV-based mini-grids, it soon became evident that the project would only be feasible if financial issues were suitably addressed within an action plan contemplating both technical and financial considerations. A strategy aimed at getting funding was immediately deployed. Grant schemes are scarce and often focused on small self-limited projects, usually based on charity schemes making them non scalable. In most cases they are implemented through utilities, which hoard up most of the benefits of the project. The typical result is one or several photovoltaic systems financed by a public entity, installed by a non-local company and forgotten once the project has finished. In such conditions, the impact on local people, economic development, employment and other essential factors is extremely low. These considerations, combined with the impressions received during the first visits to N'Zérékoré, mainly concerning the need for much broader-scoped initiatives aimed at addressing not just energy needs but economic expansion and the subsequent struggle to attain social welfare, configured a totally new financing scheme. It soon became clear that the only realistic option for a long-term, scalable, sustainable project was to establish a scheme according to which the electricity end users would themselves provide the funding while improving their economic capacity by introducing a number of improvements in their productive activity.

In February 2016 a CLEANERGETIC engineer, Luis Barrios, visited Guinea accompanied by Dr. Malo. During that visit, both the Guinean authorities and the people from the villages in question were contacted for the first time. The support by the former was as clear as the enthusiasm by the latter. A first report was elaborated covering the starting point, the water and energy needs and the economic potential. On the basis of these data the PV-system design data shown in the previous section were generated.

As reflected on this report, most of the inhabitants of the target villages are peasants who have coffee production as their main source of revenues. An analysis of their activity clearly showed that the economic efficiency of their activity was very low. They basically did not take care of their plantations. Once a year they just harvested their coffee in one single run, without selecting mature cherries, and sold their crop without any further treatment to a local trader who paid for it amounts well below the standard prices of average-quality coffee in the international markets. With this working scheme, coffee quality, quantity and economic performance were well below standards and, therefore, left a lot of room for improvement. Such a diagnostic indicated that the project had to be twofold and cover both photovoltaics and coffee production. For this reason, it was named *Café au Light*.

Along the year 2016 the activity of the working team, made of CIEMAT, CLEANERGETIC and UDECOM, was deployed towards looking for technical support in the cultivation and treatment of coffee and establishing contacts in order to find import- and commercialisation ways in Europe. This way, a strategy was developed for the transformation of the cultivation, storage, treatment and transport of coffee in Guinea, while at the same time contact was established with several coffee companies who showed their interest in establishing trade contracts and even providing technical support. This way, a basis was established in order to go from a situation in which the Guinean farmers sold at low rate a non-selected coffee to local traders, to a new scenario in which they will obtain a much better product and, through a society incorporated by CLEANERGETIC and UDECOM, sell it in the European market at standard prices. The additional income will be used to amortise the new coffee equipment, increase the farmers' revenues and finance the PV systems.

In March 2017 CLEANERGETIC CEO Dr. Bautista, and UDECOM founder Dr. Malo organised the second visit to Guinea. Official links to the Guinean authorities were established, the project was officially launched and more field work was done in order to collect more data. An inventory list of the coffee plantations was elaborated. A more detailed diagnostic of the whole growth, harvesting, transport and commercialisation of the coffee in the area was achieved. During this second visit, 40 towns submitted their application to participate in a second phase of the project.

A Guinean-based company, *Café au Light Group Sarl*, was incorporated in last July, having UDECOM and CLEANERGETIC as its partners. A piece of land was bought for the installation of a coffee-processing plant, drying patios and a warehouse. The machinery, with a process capacity of 1 Ton per hour, has already been purchased, is being manufactured, and is expected to be operative on site before the end of 2017. An agreement with Aguipex (*Agence Guinéene de Promotion des Exportations* – Guinean Agency for the promotion of exports) has been drafted and will soon be signed in order to facilitate access to public and corporate financing. A vehicle has been acquired for the transport to the plant of the fresh coffee cherries provided by the 713 farmers who have signed an agreement with the company and who are presently (September 2017) collecting their coffee crops. The total amount of coffee from the three towns in the present year is 240 Tons, to be shipped in 12 20-ton containers.

In October, the Spanish sister company, *Café au Light Group SL*, will be incorporated in Madrid. The possible undertaking of a third phase in which towns from virtually the whole country would participate is clearly supported by the Guinean Government and seriously envisaged by the project consortium.

5 TRAINING AND EDUCATION

Both the installation and maintenance of the photovoltaic systems and the adoption and maintenance of totally new procedures in the coffee activity require the active participation of the people from the towns affected. Without their implication and without a plan for their sustained training and education, the project would have a significant probability of failure. This aspect is considered of the highest importance, not only because it guarantees the medium- and long-term sustainability of the endeavour, but because it also provides an effective mechanism for its integration in the social substrate.

A training-and-education programme, based on the UDECOM capacities and infrastructure and on the technical support provided by CIEMAT and CLEANERGETIC in the field of photovoltaics, is being elaborated and will be launched soon. The coffee expertise available from the collaborations established, including a consultancy service, will be the basis for UDECOM to implement the training programmes in the coffee area.

The photovoltaic training programme will cover the knowledge needed to participate in the installation of the minigrids and to take care of the running systems in order to be able to provide a basic on-site technical support and maintenance service. CIEMAT researchers and CLEANERGETIC engineers will be teachers in these courses. The possibility to facilitate the creation of a number of technical jobs in the area is considered a very important aspect of the project.

6 SUMMARY

In summary, a project initially conceived as purely technical, focused on the design of photovoltaic systems, has turned out to cover a significantly broader scope and require the effective addressing of three aspects: (a) the design and installation of photovoltaic systems according to the requirements of the towns affected, (b) the implementation of completely new processes for the growth, harvesting, storage, treatment and commercialisation of coffee, and (c) the elaboration of training programmes on both coffee production and photovoltaics.

Café au Light was officially presented in an event held at CIEMAT, in Madrid, on June 12, 2017. The initiative is far different from conventional development-aid projects. Its focus is not just installing a photovoltaic facility, but generating economic development, welfare, training, clean energy, employment and health by means of an approach which is basically well-designed all-parties-winning business in the framework of a really paradigmatic and visionary project. The whole work will be performed together with the Guinean people, looking for a built-in integration of the project in the local society, and having in mind long-term technical, economic, environmental and social sustainability as absolute priorities.

The endeavour is based on an agreement signed by CIEMAT, UDECOM and CLEANERGETIC. In this arrangement, UDECOM provide the knowledge of the local reality, the infrastructure needed to guarantee an adequate training programme, and full support for all local operations. CLEANERGETIC offer a valuable know-how on the design and installation of photovoltaic systems, their training capacity and an entrepreneurial structure allowing to undertake investment actions required for the implementation of the project and hire technical-support services in the coffee area. CIEMAT give full technical and institutional support to the project, collaborate in maximising its social impact and provide their full training competence.

The authors of the present paper live this experience with enthusiasm, because the project combines development, solidarity, environment protection, and other aspects being of primary importance for the lives of many people. If the project is successful –as we are fully convinced it will be- we will have done much more than installing a few photovoltaic minigrids. We will have furthermore contributed to trigger an action plan which results in benefits for all parties and which can likely be replicated in other places.

7 REFERENCES

[1] Photovoltaic power potential of Guinea. World Bank Group. Esmap. Solargis. http://globalsolaratlas.info.

VALUE ADDED BY PV INSTALLATIONS IN THE NETHERLANDS (2015): A MODELING APPROACH

K. Heinbach[2], S.Salecki[2], C. Olson*[1], F. Lenzmann[1], L. Beurskens[1], M. Sonne[1], , B. Hirschl[2],

[1] Energy research Centre of the Netherlands, P.O.Box 1, 1755 ZG Petten, The Netherlands
[2] Institut für ökologische Wirtschaftsforschung (IÖW), Potsdamer Str. 105, D-10785 Berlin. Germany

ABSTRACT: Renewable energy installations stimulate the economy through the monetary flows of salaries, profits and taxes, as well as job creation. The WeBEE model created by the Institute for Ecological Economy Research (IÖW) to quantify value added and employment effects generated by RE technologies in Germany was adapted to reflect the Dutch economy. It was then used to calculate the value-added to the Dutch economy due to the photovoltaic (PV) in 2015, the most recent year for which complete economic data was available. To understand how the value added is distributed, it is broken down into components on two levels: the adminstrative level, i.e., the municipal, provincial and national level; as well as on the level of value chain elements, such as manufacturing, planning or installation.

Keywords: Economic Analysis, Sociological, PV System, Modelling

1 INTRODUCTION

In 2010, the Institute for Ecological Economy Research (IÖW) developed a model to quantify value added and employment effects generated by renewable energy (RE) technologies in Germany (WeBEE model) [1]. Since its creation, this model has been revised, extended and applied in a variety of research projects in Germany. The model delivers employment (in FTE) as well as the economic value added (in €) in the following three components:

- After tax company profits
- Net employee income
- Tax revenues

This conference contribution reports on a recent collaboration between ECN and IÖW to 1) adapt the WeBEE model to reflect the Dutch economy, and 2) present initial results of the value added by residential and commercial rooftop PV systems in 2015, the most recent year for which economic data across all the relevant sectors of the Dutch economy was available.

The adaption of the model from the German to the Dutch economy entails revising the model's economic parameters to reflect the profitability and productivity figures for companies, wage levels and the tax system specific to the Netherlands.

Within the scope of this conference paper, the methodological approach of the WeBEE model for the Netherlands is adapted for two selected RE value chains, representing the added PV capacity in 2015 typical residential roof-top systems (2,8 kWp) and commercial - agricultural or cooperative - roof-top systems (170 kWp).

Initial results for specific value added and employment effects are presented and discussed.

2 METHODOLOGY

2.1 Basic description of the model

The primary basis for assessing value added within the model is an analysis of the specific turnovers generated along the value chain. To this end the value chain is broken down into "value added steps". Examples of value added steps are poly-Si production, wafer production, solar cell production etc. or installation work, service/repair work. Every step is then labeled with a classification code that characterizes the type of its economic activity. We used the NACE (Nomenclature statistique des Activités économiques dans la Communauté Européenne = NACE) codes from the statistical classification of economic activities in Europe. [2] For example, the value added step "module production" is labeled by NACE code 26 (Manufacture of computer, electronic and optical products).

The breakdown into value added steps by economic activity (NACE code) allows for the appropriate parameters for the turnover analysis to be identified, i.e.:

- Operating profit margin,
- Labor productivity
- Gross average employee income
- Applicable taxes (which depend on the legal status of value step actors)

The first three of these parameters can be obtained from economic statistics tables that are available as a function of NACE code [3]. The tax parameters need to be extracted from the tax system [4].

On the basis of this breakdown and with the appropriate parametrization, the three value added components (after tax company profits, net employee income, and taxes) as well as employment can now be calculated for every value added step as exemplified in Figure 1.

So, this method allows to generate a value added and employment profile for every value added step. The results are often aggregated into larger clusters, for example "value chain phases", e.g., PV system manufacture, planning & installation, operations & maintenance and PV system owner. Another interesting way to view the distribution of the results is broken down into administrative levels, i.e. municipal, provincial, and national level.

The analysis only refers to turnovers which are directly relevant for RE technologies. Thus, this model depicts only direct economic effects related to the level of the RE operator and direct up-stream suppliers. Further up-stream suppliers are defined as indirect stages of the overall RE value chain and are not part of the WeBEE model. However, the inclusion of these indirect effects in the economic evaluation can be achieved through an input-output approach which can be interlinked with the

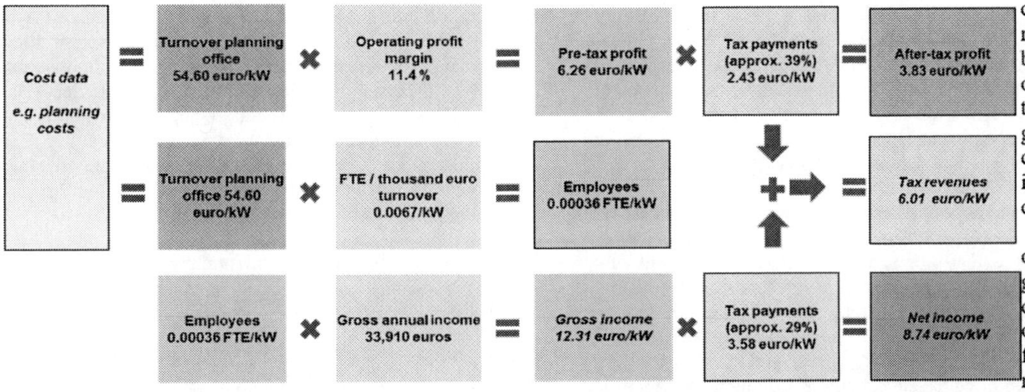

Figure 1. Example of calculation scheme for value added components and employment effects. Source: IÖW WeBEE model. [1]

2.2 Cost structures for RE technologies

Because about 70% of the ~500 MWp of photovoltaic systems installed in 2015 were residential systems of on average 2.8 kWp capacity (11 panels of 260 Wp), this size system was chosen as a representative residential roof-top system. [5] The majority of the remaining 30% of PV systems installed in 2015 ranged between 50 to 500 kWp. Only 4% of the systems had a capacity of >500 kWp. There were only 74 ground mounted systems in the NL, making up a cumulative capacity of 52 MW (out of a cumulative capacity at the end of 2015 of 1.5 GW). On the other hand, systems between 100 and 250 kWp, a typical size for agricultural and cooperative systems, made up a full 50% of systems greater than 50 kWp. Therefore, the second system representative of the 2015 Dutch market was taken to be a 170 kWp rooftop system (654 panels at 260 Wp). [5]

2.2.1 Residential rooftop PV investment costs

The total investment for a 2.8 kWp roof-top photovoltaic system in 2015 was 5,208 euros, corresponding to a specific investment cost of 1,860 €/kW. [6] Since the installers of small household systems mostly obtain the modules and inverters from a wholesaler for photovoltaic systems, we assumed that the investment costs for these key components include a trade margin of 22 %. Since small household systems are usually partly financed by borrowed capital, we assumed a typical consumer ratio of borrowed capital of 80 percent, a credit period of 10 years and an average interest on borrowed capital of 6 percent.

The pre-tax profits of the plant operator of the small household system were calculated by using an average return on equity. For this, we assumed a value of 5.5 % for the average return on equity and an equity ratio of 20%. [6] The typical legal form of plant operators of small roof-top photovoltaic systems is that of a 'natural person', equivalent to a non-incorporated company.

2.2.2 Commercial rooftop PV investment costs

The total investment for the reference system (with an installed capacity of 170 kWp) amounts to around 183,850 euros, corresponding to specific investment costs of 1,081 €/kW. [6] The grid connection costs include service personnel costs as well as material costs. It was assumed that the share of these costs are similar to those in the German market, in which material expenses were considered to be 35% of the total grid connection costs. The operating costs consist of costs for repair and maintenance, rooftop lease, commercial and technical management, insurance, network fees and financing over a system lifetime of 25 years, and amount to about 4% of the investment costs. For the financing based on borrowed capital, we assumed a ratio of borrowed capital of 90 percent, a credit period of 15 years and an average interest on borrowed capital of 3.75 % [8]. Regarding the lease payments for the roof surface, it was assumed that 6 percent of the lease payments go to public real estate owners (municipalities) and 94 percent go to private real estate owners, which was based on the percentage of public real estate in the Netherlands.[9]

The pre-tax annual profits of the system operator of the large photovoltaic system were calculated by using an average return on equity of 10% and an equity ratio of 10%.[8] The typical legal form of plant operators of commercial roof-top PV systems is an incorporated company.

2.3 Calculating value added and employment effects

2.3.1 Company profits and salaries

Pre-tax profits are calculated using the branch-specific operating profit margin before taxes on the turnover. For the case of the financing of power plants we allocated the branch of "Financial service activities" (NACE-Code 64) and used the credit volume as a reference for the operating margin instead of the turnover.[10]

2.3.2 After-tax profits, incomes and taxes

Since the taxation of profits is dependent on the corporate structure of companies, the calculation of the tax load and the after-tax profits have to be done separately for incorporated and non-incorporated companies. Information on the share of incorporated or non-incorporated companies in each step of the value chain was found in the Dutch Central Bureau of Statistics. [3]

The tax rate on incorporated companies is 20% up to a taxable profit of 200,000 euros, and 25% for profits above this limit. [4a] Because there was insufficient information available on the average absolute taxable profit, we used a mean value of 22.5%. The after-tax profits were calculated by subtracting the tax payments from the pre-tax profits.

With insufficient information on the average absolute pre-tax profit in the different branches, we could not include possible tax deductions such as the investment deduction. Furthermore, there was not enough information available for the calculation of the property tax. And, as there was insufficient information available on the composition of different kinds of shareholders for incorporated companies, we did not calculate income taxes on shareholder level and assumed no pay out of company profits.

Non-incorporated companies typically pay a tax-rate of 42%. [4b] This tax applies for the average employment income in the Netherlands of 38,000 euros (CBS 2017) which was used to estimate the average capital income of shareholders of non-incorporated companies. The same average income was used to calculate the general tax rebate and the work rebate, which both reduce the calculated income tax payment for the shareholders. The pre-tax income was lowered by the reduced income tax and the long-term care tax (5%), resulting in the after-tax income of the shareholders of the non-incorporated companies.

The undistributed after-tax profits on company level and the after-tax incomes of the shareholders both for the incorporated and non-incorporated companies sum up to the after-tax profits of the participating enterprises as one of the value added components.

A different method was applied to assess the profits of the PV system owner. For the assessment of the profits in this stage, pre-tax earnings were calculated with the help of average return on equity . [7][8]

The turnover which was calculated for each of the value chain steps may also be translated into employment effects. Using the economic statistics of the Dutch economy, the employment intensity, (FTE/million euros turnover), was calculated for the relevant value chain branches, and used to estimate the number of jobs (FTEs) created per kW of installed capacity.[3] The average gross annual income of employees (per FTE) for the relevant branches, obtained from the same source, were multiplied with the employment intensity to calculate the gross income per euro turnover.

Tax rates on employee income were found to be between 42 - 52%. Part of the income tax are the social security contributions, which account for 28 percent of the employee income at a maximum of 9,455 euros. [4c] We interpreted the income tax rate minus the social security contributions as a residual income tax rate since social security contributions and income taxes are distributed differently on administrative levels. The resulting residual income tax is reduced by the general tax rebate and the work tax rebate, both depending on yearly income. Taking into account the social security contributions, the residual income tax and the income tax rebates we calculated the after-tax income for employees.

Employers have to pay social security contributions at a rate of 18.4 percent of the employee income. [4c] It does not affect the after-tax income of the employees but accounts as part of the fiscal revenue.

2.3.3 Tax allocation

The tax payments on profits and employee's income are calculated for each type of tax. As most of the relevant tax payments are revenues for the national level and redistributed to the provinces and municipalities we used official distribution key figures to calculate the tax revenues of different administrative levels. Table I presents the shares of total taxes on the administrative levels including the taxes collected by provinces and municipalities. As the social security contributions are not redistributed and the provinces and municipalities collect some taxes on their own, we adjusted these figures to calculate the shares of redistributed taxes for the administrative levels (last column).

Table I: Tax allocation: total breakdown (including social security contributions), and breakdown of redistributed taxes (separate from social security). [10]

Administrative level	Share of total taxes (%)	Share of redistributed taxes (%)
Supra-National level	1.00	1.80
Central Government	46.60	83.75
Provinces	1.52	0.93
Municipalities	10.32	13.52
Social Security	39.60	

The tax revenues on the different administrative levels are added up to calculate the value added on each level.

3 RESULTS & DISCUSSION

Figure 2 provides visual information about the value added on the municipal, provincial and national levels respectively. The largest part of the value added, roughly two thirds of it, is generated on the municipal level resulting mainly from employee income and company profits, along with a small amount of municipal tax revenues. On the provincial level, an insignificant amount is added resulting from provincial tax revenues. The national level benefits by further, more significant tax revenues, which represent another third of the total value added. These national tax revenues primarily come from taxes on salaries and profits.

Looking at the relative contributions to the value added by the two considered market segments (residential, commercial), it is striking that the value added is generated predominantly by the residential systems. An important reason for that (but not the only one) is simply that the added capacity in 2015 for the residential systems is larger by a factor of almost 3. Furthermore, the relatively larger demand on installation services and larger trade margin for system components (due to smaller system scale) and the larger profitability for system owners due to net metering (which is not applicable to commercial systems) play a role. So, installation, trade and system ownership are significant additional factor for this difference.

Another perspective on the results is given in figure 3, which shows a breakdown of the total value added (163 mln. €) into different value chain phases. In the order of decreasing significance we find, installation (47%), trade (31%), system owner (13%), operations & maintenance (9%).

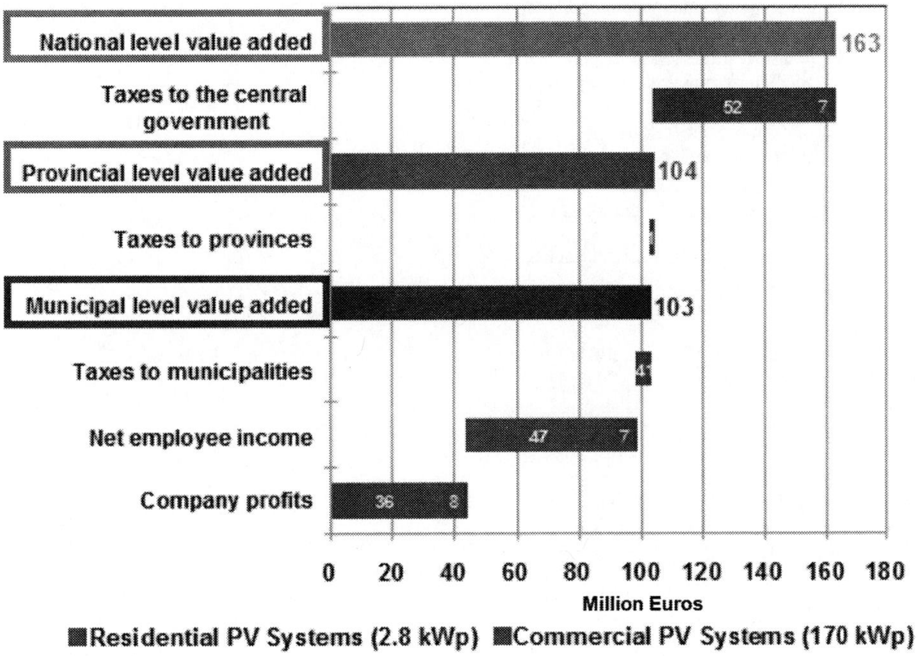

Figure 2. Value added due to PV installations in 2015 in the Netherlands, by administrative level.

According to our model, a total amount of ~2000 jobs was present in the PV sector of the Netherlands in 2015, predominantly related to installation work (> 70%) and to a smaller extent to trade.

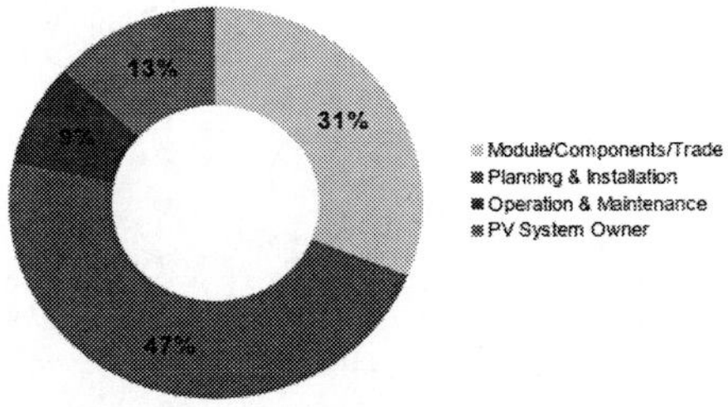

Figure 3. Breakdown of value added due to PV installations in 2015, by value chain element

CONCLUSIONS

We applied a modeling approach (WeBEE model), which allows to determine value added from economic and other input data such as turnover, productivity and tax rates to analyze the distribution of the value added and employment effects by PV installations in the Netherlands.

REFERENCES

[1] Hirschl, Bernd, Katharina Heinbach, Andreas Prahl, Steven Salecki, André Schröder, Astrid Aretz und Julika Weiß (2015): Wertschöpfung durch Erneuerbare Energien. Ermittlung der Effekte auf Länder- und Bundesebene. Schriftenreihe des IÖW 210 (Dezember). https://www.ioew.de/fileadmin/user_upload/BILDER_und_Downloaddateien/Publikationen/Schriftenreihen/IOEW_SR_210_Wertsch%C3%B6pfung_durch_erneuerbare_Energien_auf_Landes-_und_Bundesebene.pdf.

[2] Eurostat (2008): NACE Rev. 2 - Statistical classification of economic activities in the European Community. http://ec.europa.eu/eurostat/documents/3859598/5902521/KS-RA-07-015-EN.PDF/dd5443f5-b886-40e4-920d-9df03590ff91?version=1.0.

[3] Centraal Bureau voor de Statistiek (CBS) (2017): Statline. Statline. http://statline.cbs.nl/Statweb/?LA=nl (Zugriff: 21. August 2017).

[4] Belastingdienst (a): Tarieven voor de vennootschapsbelasting in 2015. https://www.belastingdienst.nl/wps/wcm/connect/bldcontentnl/belastingdienst/zakelijk/winst/vennootschapsbelasting/veranderingen_vennootschapsbelasting_vorige_jaren/veranderingen_vennootschapsbelasting_2015_sitegebied/tarieven_vennootschapsbelasting_2015 (Zugriff: 21. August 2017).
Belastingdienst (b): Tarieven voor personen die de AOW-leeftijd nog niet hebben bereikt. https://www.belastingdienst.nl/wps/wcm/connect/bldcontentnl/belastingdienst/prive/werk_en_inkomen/bijzondere_situaties/middeling_sterk_wisselende_inkomens/belastingteruggaaf_berekenen/u_bent_het_hele_middelingstijdvak_jonger_dan_de_aow_leeftijd/tarieven_voor_personen_die_de_aow_leeftijd_nog_niet_hebben_bereikt (Zugriff: 21. August 2017).
Belastingdienst (c): Sociale verzekering https://www.belastingdienst.nl/wps/wcm/connect/bldcontentnl/belastingdienst/prive/werk_en_inkomen/sociale_verzekeringen/

[5] Polder PV, Peter Segaar, http://www.polderpv.nl/nieuws_PV137.htm#24

dec2016_CBS_maatwerktabel_PV_NL_tm_2015;
http://www.polderpv.nl/nieuws_PV137.htm#24dec2016_CBS_maatwerktabel_PV_NL_tm_2015;
http://www.polderpv.nl/PV_projecten_overzicht_22dec2016.htm#PV_projecten_segmentatie_NL_provincies ;

[6] van Sark, Wilfried und Tony Schoen (2016): Inventarisatie PV markt Nederland. Status februari 2016. http://www.zonnestroomnl.nl/wp-content/uploads/2016/07/SMZ_201602_inventarisatiemarkt.pdf.

[7] Milieu Centraal, Kosten en opbrengst zonnepanelen, https://www.milieucentraal.nl/energie-besparen/zonnepanelen/zonnepanelen-kopen/kosten-en-opbrengst-zonnepanelen/

[8]Zuijlen, Christine van und Sander Lensink (2015): Eindadvies basisbedragen SDE+ 2016. ECN and DNV GL en TNO. https://www.ecn.nl/publications/PdfFetch.aspx?nr=ECN-E--15-052.

[9] Kadaster, https://www.kadaster.nl/

[10] Ministerie van Financiën (2017): Open Data. Open Data. Website: http://opendata.rijksbegroting.nl/ (Zugriff: 21. August 2017).

POTENTIAL IMPACTS OF SOLAR ENERGY INTEGRATION ON FUEL-MIX STRATEGIES IN QATAR

Moiz Bohra, Nasreddine El Dehaibi, Antonio Sanfilippo, Marwan Khraisheh
{mbohra, neldehaibi}@qf.org.qa, {asanfilippo, mkhraisheh}@hbku.edu
Qatar Environment and Energy Research Institute, Hamad Bin Khalifa University, Qatar Foundation
P.O. Box 5825, Doha, Qatar

ABSTRACT: This study evaluates the cost utility of various fuel-mix strategies in Qatar to maximize savings while meeting the country's solar energy adoption targets. We use nonlinear programming (NLP) to optimize the relative shares of oil and gas to be displaced by solar energy with reference to savings. In view of plummeting costs for solar photovoltaic (PV) energy, we find that displacing domestic use of oil and natural gas with solar energy in Qatar is cost-effective in most hydrocarbon price scenarios. Moreover, we find that it is more economical to displace more oil when oil prices are higher (above $88/bbl) and to displace more gas at lower oil prices. Based on our findings, we recommend that Qatar curtail electricity subsidies by aligning investments with return opportunities based on oil prices. Moreover, we recommend avoiding further expansions in gas-fired power plants and instead investing in electric vehicle infrastructure to continue promoting the growth of solar energy.
Keywords: fuel-mix, photovoltaic, modeling, economic analysis.

1 PURPOSE OF WORK

Hydrocarbons are the leading source of Qatar's export revenue while acting as virtually the entire energy source for the country [1]. It is therefore advantageous for the country to displace its consumption of hydrocarbons with renewable energy, both from an economic and environmental perspective – i.e. increase natural gas export and reduce CO_2 emissions.

As part of the Qatar National Vision 2030, Qatar has mandated energy goals where 2% of electricity will be generated from solar by 2020, and 20% by 2030 [2]. Using electricity generation forecasting methodology [3], the energy targets for Qatar can therefore be estimated as shown in Table I.

Table I: Qatar electricity gen. forecasts for 2020 and 2030

	2020	2030
Total electricity generated (TWh forecasted)	57.5	61.9
Qatar's national target for solar energy adoption [2]	2%	20%
Electricity generated from solar (TWh)	1.15	12.38

Introducing solar energy into the fuel mix opens interesting opportunities for Qatar. Firstly, solar energy can be used to displace natural gas by generating electricity for the residential, commercial, or industrial sectors. Secondly, solar energy can be used to displace oil in the transportation sector through the deployment of hybrid or electric vehicles.

This study aims to evaluate the cost utility of various fuel-mix strategies in Qatar to maximize savings while meeting the country's solar energy adoption targets. The following research questions are addressed:

1. How much electricity production can be satisfied through solar energy in 2020 and 2030?
2. What is the optimal distribution of solar PV energy in replacing oil and gas in terms of money saved per megawatt hour (MWh)?
3. What are the relevant policy insights that emerge from the cost-optimal use of solar PV energy to replace oil and gas, and the policy undertakings that such insights support?

2 SCIENTIFIC INNOVATION AND RELEVANCE

Several studies have developed mathematical models to analyze fuel-mix strategies.

Torvanger [4] uses a Divisia index approach [5] to evaluate the reduction of CO_2 emissions in nine countries belonging to the Organisation for Economic Co-operation and Development (OECD).

Jafar et al [6] use a Leontief's input-output framework to estimate pollutant emission scenarios in Malaysia for 2020, concluding that CO_2, sulfur dioxide (SO_2), and nitrogen oxide (NO_x) emissions would increase significantly based on the Fuel Diversification Strategy incorporated in the Malaysian Energy Policy in 1999.

Roques et. al [7] use a Mean-Variance Portfolio theory [8] to identify optimal generation portfolios as characterized by fuel type, CO_2 emissions, and electricity price risk in liberalized electricity markets. While the study focuses on fossil and nuclear fuels, the approach could in principle be extended to include renewables.

Marrero et al [9] also use Mean-Variance Portfolio theory to assess the natural gas and renewables strategy proposed for the electricity industry in the 2006 energy plan of the Canaries, considering the average cost and the risk associated with the alternatives for electricity generation. The use of this approach for the problem addressed in this chapter would have to take into consideration that Qatar as a major gas exporter may have an advantage in using natural Marrero et al gas for electricity generation.

Atwa et al [10] uses mixed-integer nonlinear programming (NLP) to optimize the allocation of different types of distributed renewable energy resources in the distribution system to minimize energy losses. The objective function aims to minimize energy losses in the distribution system subject to network system constraints.

Our approach differs from past efforts, in that it aims to establish the relative percentages of natural gas and oil that should be replaced by solar energy to maximize savings. In terms of mathematical modeling methods, we follow Atwa et al [10] in using NLP as an optimization technique.

3 APPROACH

We use NLP as an optimization technique to develop

a function that measures the economic utility of diverse fuel-mix scenarios emerging from the replacement of oil and natural gas with solar energy. The NLP objective function optimizes the percentages of oil and gas to be replaced by solar energy in diverse scenarios determined by variables that are either under or outside Qatar's control. These variables include oil and gas price, oil and gas subsidies, gas levelized cost of operation and maintenance (LCO&M), and a potential carbon tax. NLP is used in this study since most of the constraints, especially gas LCO&M and the carbon tax, exhibit non-linear behavior.

The objective function takes into account three potential sources savings: subsidy returns, carbon tax savings, and electricity generation savings, that are maximized in terms of the relative share of oil and gas that are displaced by solar energy.

The savings from subsidy returns are derived from increased exports of oil and gas and is expressed in equation 1.

$$\textit{subsidy returns} = \frac{\text{oil replaced} * (\text{oil price} - \text{subsidized oil price}) + \text{gas replaced} * (\text{gas price} - \text{subsidized gas price})}{\text{solar pv adopted}} \quad (1)$$

Carbon tax savings are derived from a hypothetical carbon tax that is used to curtail CO_2 emissions as shown in equation 2.

$$\text{carbon tax savings} = \frac{\begin{bmatrix}(\text{oil replaced} * \text{CO2 emissions from oil}) + \\ (\text{gas replaced} * \text{CO2 emissions from gas})\end{bmatrix} * \text{carbon tax}}{\text{solar PV target}} \quad (2)$$

Electricity generation savings are derived from costs saved due to reduced natural gas production. Fuel costs are considered in subsidy returns. These savings are expressed in equation (3).

$$\text{electricity generation savings} = \frac{\text{replaced gas} - \text{powered electricity} * \text{gas LCO\&M}}{\text{solar pv target}} \quad (3)$$

The total savings is therefore expressed in (4) as the summation of (1), (2), and (3).

$$\text{total savings} = \text{subsidy savings} + \text{carbon tax savings} + \text{electricity generation savings} \quad (4)$$

The savings are determined for a diverse set of plausible scenarios that encompass a wide range of variable constraints. Ranges for oil and gas subsides were determined using the International Energy Agency (IEA) Price-Gap approach [11] along with historical data on energy subsidies in Qatar [12]. Considering that the introduction of a carbon tax in Qatar is currently unlikely, modest ranges were chosen using the European Emissions Trading System as a reference [13]. Oil price ranges were derived from historical data of Brent Crude prices [14] and a partial correlation was applied to gas prices using Henry Hub spot prices [15] and Pearson's correlation [16]. Finally, EIA projections for power plants expected to enter service in 2022 were used to determine ranges for LC+O&M of natural gas [17].

Scenarios are generated as the transitive closure of all variable combination within their available value ranges, which are shown in Table II. The relative share of oil and gas to be displaced is then optimized for each scenario to maximize returns.

Table II: Description of variables and variable constraints

Variables	Value range constraints	Value interval constraints	Units
Within Qatar's control			
Oil/gas split	$0 < x_1 < 1$	-	real
Oil subsidy	$0 < x_4 < 56$	5	percentage
Gas subsidy	$0 < x_5 < 41$	5	percentage
Carbon tax	$0 < x_6 < 10$	5	$/tCO$_2$
Outside Qatar's control			
Gas LC+O&M Cost	$1.42 < x_7 < 4.36$	2	c/kWh
Oil price	$10 < x_2 < 140$	14	$/bbl
Gas price	$1.78 < x_3 < 14.76$	13	$/MMBtu

4 RESULTS

To quantify the relative impact of the various parameters in Table II on cost savings, we performed 'one-factor-at-a-time' sensitivity analysis. We first set a base scenario with fixed parameter values, as shown in Table III, and then test the scenario with each parameter set at its lowest and the highest values (Table II), while keeping the other parameters fixed, to verify the ensuing changes on savings/returns.

Table III: Base scenario with parameter values for sensitivity analysis

Parameter	Value	Unit
Oil price	50	$/bbl
Gas price	6	$/MMBtu
Oil subsidy	30	%
Gas subsidy	30	%
Carbon tax	5	$/tCO2
Gas-powered electricity capital cost	0.03	$/kWh
Base returns (output)	47.6 (replacing gas with solar energy) 31.3 (replacing oil with solar energy)	$/MWh

Sensitivity analysis shows that the model is highly sensitive to changes in oil price. We found that replacing oil with solar energy yields higher savings when oil prices are $88/bbl or greater, but replacing gas with solar energy is better at lower oil prices. Oil subsidy shows a smaller effect, with savings ranging from $48/MWh of solar energy replacing gas at 0% oil subsidy, and $55/MWh of solar energy replacing oil at 56% oil subsidy. At a low gas price of $2/MMBtu, gas is replaced by solar energy to return $37/MWh. At a high gas price of $14/MMBtu, gas is replaced by solar energy to return $68/MWh. These are significant savings in all scenarios. Gas subsidies varying from 0% to 41% of market price return \$32/MWh to \$53/MWh of solar energy replacing gas. The trade-off curve that indicates the optimal fuel-replacement strategy in terms of oil and gas prices is illustrated in Figure 1.

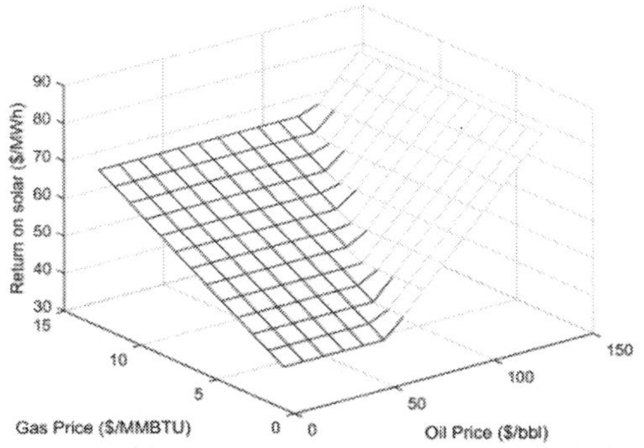

Figure I: Trade-off Curve for Optimal Fuel Displacement Strategy. Gas and oil prices vary, while other variables are fixed. The blue region indicates higher savings when solar energy displaces gas, while the yellow region indicates higher savings when solar energy displaces oil.

We found that savings from carbon tax were insignificant unless carbon taxes were set very high (above \$50/tCO2e). Finally, the cost of gas-fired electricity generation (capital, operating and maintenance costs) can also effect savings. At \$0.0142/kWh, the bottom range for gas-fired electricity cost (see LC+O&M in Table II), the optimal base return is \$32/MWh of solar energy replacing gas, while at the top range of \$0.0436/kWh, the optimal base return is \$61/MWh.

5 FINAL REMARKS

The abundance of hydrocarbon resources and adequate gas-fired electricity capacity has slowed the adoption of renewable energy in Qatar. However, the adoption of solar energy in Qatar, as well as other countries with rich fossil fuel resources, can help meet growing domestic demand and open up additional hydrocarbon resources for export, with ensuing economic benefits. For example, our analysis shows that due to the continuously falling costs of solar energy and the subsidy structures for hydrocarbons, investing in solar power makes economic sense for a range of hydrocarbon prices.

Our analysis suggests that a gradual reduction in subsidies for electricity generated from natural gas in order to level the field for solar energy may provide the opportunity for cost savings through the replacement of oil and gas with solar energy. Cost savings from solar energy may be reinvested to incentivize the adoption of renewable and efficient energy technologies, creating additional opportunities to replace oil and gas to achieve further savings. Finally, by diverting funding from the creation of new natural gas power plants to the development of solar energy capabilities, Qatar can avoid locking in expensive infrastructure for decades to meet its growing electricity demand in an environmental friendly manner.

6 REFERENCES

[1] Center for International Development at Harvard University (no date) The Atlas of Economic Complexity. http://www.atlas.cid.harvard.edu. Accessed 30 Aug 2016

[2] REN21 (Renewable Energy Policy Network for the 21st Century) (2016) Renewables 2016 Global Status Report. REN21 Secretariat, Paris. http://www.ren21.net/status-of-renewables/global-status-report/

[3] Sanfilippo A, Pederson L (2016) Impacts of PV Adoption in Qatar on Natural Gas Exports to the Far East. In: Lester L, Efird B (eds) Energy Relations and Policy Making in Asia: The benefits of mutual interdependence. Palgrave Macmillan.

[4] Torvanger A (1991) Manufacturing sector carbon dioxide emissions in nine OECD countries, 1973–87: A Divisia index decomposition to changes in fuel mix, emission coefficients, industry structure, energy intensities and international structure. Energy Economics 13(3):168-186

[5] Divisia F (1926) L'indice monétaire et la théorie de la monnaie. Revue d'Économie Politique LX (1)49-81

[6] Jafar AH, Al-Amin AQ, Siwar C (2008) Environmental impact of alternative fuel mix in electricity generation in Malaysia. Renewable Energy 33(10):2229-2235

[7] Roques FA, Newbery, DM, Nuttall, WJ (2008) Fuel mix diversification incentives in liberalized electricity markets: A Mean–Variance Portfolio theory approach. Energy Economics 30(4):1831-1849

[8] Markowitz HM (1952) Portfolio Selection. The Journal of Finance 7(1):77–91

[9] Marrero GA, Ramos-Real FJ (2010) Electricity generation cost in isolate system: the complementarities of natural gas and renewables in the Canary Islands. Renewable and Sustainable Energy Reviews 14(9):2808-2818

[10] Atwa YM, El-Saadany EF, Salama MMA et al (2010) Optimal renewable resources mix for distribution system energy loss minimization. IEEE Transactions on Power Systems 25(1):360-370

[11] IEA (International Energy Agency) (no date) Fossil-fuel subsidies – methodology and assumptions: The Price Gap Approach. International Energy Agency and World Energy Outlook. http://www.worldenergyoutlook.org/resources/energysubsidies/methodology/. Accessed 31 Aug 2016

[12] IEA (International Energy Agency) Data Services (no date). http://wds.iea.org. Accessed 30 Aug 2016

[13] European Energy Exchange (EEX) Emission Allowances (no date). https://www.eex.com/en/market-data/emission-allowances/spot-market/european-emission-allowances#!/. Accessed 31 Aug 2016

[14] Indexmundi Brent Crude Oil Prices. http://www.indexmundi.com/commodities/?commodity=crude-oil-brent. Accessed 31 Aug 2016

[15] Indexmundi Henry Hub Natural Gas Spot Prices. http://www.indexmundi.com/commodities/?commodity=natural-gas. Accessed 31 Aug 2016

[16] Alterman S (2012) Natural Gas Price Volatility in the UK and North America. The Oxford Institute for Energy Studies. https://www.oxfordenergy.org/wpcms/wp-content/uploads/2012/02/NG_60.pdf

[17] EIA (2017) Levelized Cost and Levelized Avoided Cost of New Generation Resources in the Annual Energy Outlook 2017. https://www.eia.gov/forecasts/aeo/pdf/electricity_generation.pdf. Accessed 10 Mar 2017

A METHODOLOGY TO EVALUATE THE POTENTIAL OF USING PVRO DESALINATED BRACKISH WATER IN IRRIGATION ON LARGE FARMING SCALE: APPLICATION TO SALINE AREA IN RHAMNA PROVINCE (MOROCCO)

Youness Ettayeb[1], Ndiaga Mbodji[2], Safaa Hajji[3], Toky A. A. Arisily[4], Ali Hajji[5]
[3] Al Akhawayn University, Ifrane, Morocco
[1,2,4,5] Process Engineering and Environment Research Unit,
Institut Agronomique et Vétérinaire Hassan II, Rabat, Morocco
ahajji61@hotmail.com

ABSTRACT: Photovoltaic-powered reverse osmosis (PVRO) desalination can potentially be the solution to two major global issues: water and energy. This technique not only increases water supply but also decreases the dependency on fossil fuels. Many countries like Morocco today face the problem of water stress, which has become a crucial issue for the government. The brackish underground water reserves are estimated at about 500 million m^3 with a salinity of 0.5 to 16 g L^{-1}.
This paper presents a computerized approach based on operations research methods for evaluating the potential of using PVRO desalinated brackish water in agriculture, as part of investigating solutions to the problem of abandoned or poorly exploited arable land due to freshwater shortages. It is based on a 4-step process: (1) Selection and collection of information on the study area, (2) Determination of possible crops, (3) Selection of energy source and calculation of water production specific cost, and (4) Determination of the optimal operating plan and evaluation of the gain potential.
The methodology was applied to Jaidate County (Rhamna Province, Morocco), located in a saline geological zone, where 206 hectares of land is available for farming. The results showed that the PVRO desalinated brackish water in agriculture (1,528 m^3 per day) requires an investment of 17,500,000 MAD (about 1,564,000 €) with a water specific cost of 5.1 MAD m^{-3} (about 0.46 € m^{-3}). The payback period is between 5.8 to 8.3 years. The profitability is 96 to 141% with solar energy according to the financing scenario, 56 to 96% with utility grid and 42 to 65% with electricity produced by a diesel generator. The PVRO unit will generate a gain potential of 2,500,000 MAD (about 223,425 €) or 1,700,000 MAD (about 151,930 €) with corn and oat as crops, according to the financing scenario.
Keywords: Desalination water, Reverse osmosis, Brackish, Agriculture, Methodology, Photovoltaic, Morocco

1 INTRODUCTION

Morocco is one of the countries most threatened by water stress, placed as "extremely risky" [1]. Scientists have already been alarming for several years regarding this water scarcity. Faced with such a problem, two solutions are recommended, finding new sources, and reducing our consumption.

Firstly, desalination of seawater and brackish water offers the possibility of using other sources of near-inexhaustible water, knowing that Morocco has an enormous brackish groundwater potential estimated at 500 million m^3 with a salinity of 0.5 to 16 g L^{-1} [2]. Secondly, 80% of water use concerns agriculture [3]. It is therefore in this field that rationalization should be the most important. The farms must change their irrigation mode, converting the current practice called gravity mode, with which most water is lost, by the micro-irrigation method.

The availability of poor quality water resources, such as brackish groundwater, is now an important factor, particularly for irrigated agriculture in arid and semi-arid zones in Morocco. Land is being abandoned due to the unavailability of fresh water sources. Thus, the use of saline water for irrigation is one of the possible answers to adapt to the growing shortage of water resources [4].
Large-scale desalination plants can be built to provide adequate water supply. The energy required to supply these plants is often a limiting factor and presents a challenge because desalination technologies are very energy-intensive, particularly when water is transported to land, but water desalination technologies are emerging in the world thanks to the progress made in reducing the energy consumption of equipment and the emergence of new sources of renewable energies, which helps to reduce costs. Reverse Osmosis (RO) is currently the leading desalination technology, holding 60% of the total installed capacity in the world [5]. According to a report published by Hemping [6], on the economic analysis of RO desalination, the energy cost of RO desalination of both brackish and seawater varies between 37-43% [6]. However, if the desalination techniques are well controlled, their implementation will be a plus for Morocco. Therefore, it is necessary to have new tools for the design, sizing and financial analysis of desalination units.

The main objective of this work is to develop a methodology for the evaluation of the potential use of desalinated brackish water in agriculture, using Photovoltaic-powered Reverse Osmosis (PVRO) and operational research for profit maximization.
Section 2 below presents the methodology to evaluate the potential of using desalinated brackish water in agriculture, which was applied to Jaidate County in Morocco, whereas Section 3 gives the main results using the net present value (NPV), the internal rate of return (IRR), the payback period, the profitability as well as the optimal operating plan and the gain potential.

2 STEPS OF THE METHODOLOGY

The methodology for the evaluation the potential of using desalinated brackish water for agricultural purpose consists of 4 steps, and was applied to Jaidate County.

Step 1: Selection and collection of information on the study area

The most important part of the project feasibility study is the choice of a study area that produces only brackish water unsuitable for human consumption and agriculture, with an availability of land. In our case, the study will be carried out in Jaidate County, located on the edge of a saline geological zone (Figure 1), in central region of Morocco. It is located in Rhamna Province, with 12,000 inhabitants and is 40 km from Marrakech. The land extends over 206 ha and belongs to an association of farmers [7].

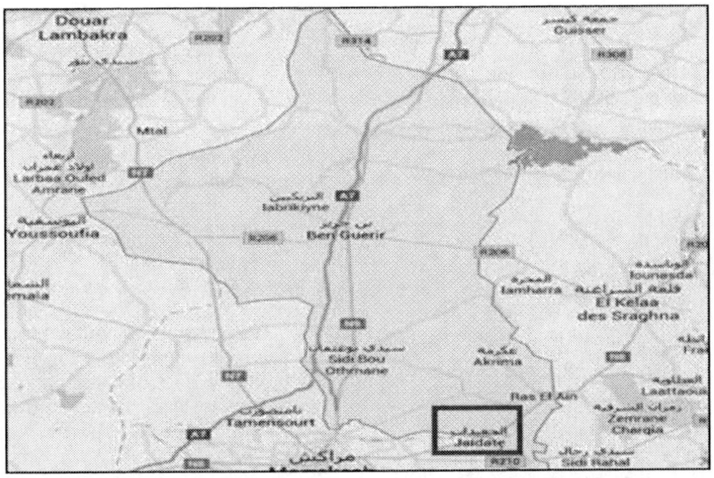

Figure 1 : Map showing the saline geological area of Jaidate County

After selecting the study area, the next step is the grouping of information on the study area. Table I gives the main information collected for Jaidate County which is characterized by a superficial groundwater, a semi-arid climate, a brown isohumic soil and a silty texture. Existing crops are cereals, citrus, cactus, apricot and olive.

Table I: Information on the study area

Characteristic	Value	Unit
Potential	75	Mm^3
Salinity	2 à 6	$g\ L^{-1}$
Average daily radiation	5.8	$kWh\ m^{-2}$
Annual rainfall	200-300	mm
Assimilable phosphorus	56.5	$mg\ kg^{-1}$
Potassium	585.8	$mg\ kg^{-1}$
Organic material	2.15	%

Step 2: Determination of possible crops

The water requirements of different crops are evaluated using dynamic soil water balance methods, including meteorology, evolution of the phenological stage of the crop and direct measurements of soil moisture status [8]. In other words, knowledge of plant parameters, climatic and soil data is required.

Crops selection criteria vary according to market or customer demand, but the most important are soil type, water availability and climatic conditions.

The average annual consumption of cereals in Morocco is almost 200 kg per capita, of which soft wheat represents 70% in urban areas and 66% in rural areas [9]. In the world, the average annual consumption is around 152 kg per capita, where the importance of cereals [9]. According to forecasts by the Ministry of Agriculture, local demand for cereals could reach 137.5 million quintals by 2020 [9].

The main possible crops are the following cereals:

- Soft wheat;
- Durum wheat;
- Barley;
- Corn;
- Oat.

Step 3: Selection of the energy source and calculation of water production specific cost

The choice of the energy source remains crucial in order to greatly reduce the RO desalinated water specific cost. Reverse osmosis is a process of separation in liquid phase by permeation through semi-selective membranes under the effect of a pressure gradient and consists mainly of a water intake with a feed pump, pretreatment, pressurization, separation membranes, post-treatment, energy recovery system and brine disposal [5]. Possible options are solar energy, utility grid, diesel generator, or a combination of these various options. Details on the PVRO process and on the specific cost calculation are discussed in [5].

In our case, the annual water requirement is approximately 371,200 m^3 (1,528 m^3 per day). In the first analysis, solar energy is chosen as source. The implementation of the project of desalination of brackish water requires an investment of about 17,500,000 Moroccan Dirhams (MAD). Operating costs are estimated at about 2,000,000 MAD, without energy costs. The specific cost of desalinated water is therefore 5.1 MAD m^{-3}.

Step 4: Determination of the optimal operating plan and evaluation of the gain potential

This step consists of determining the optimal operating plan and calculating the maximization of the gain potential using the simplex method (operational research), with a determination of the decision variables, the objective function (OF) and the constraints.

$$OF = \sum_{i=1}^{n} [P_i^S R_i S_i - (P_i^G G_i S_i + P_i^W W_i S_i + P_i^F F_i S_i + P_i^P P_i S_i + P_i^O N J1 S_i + P_i^L J2 S_i + P_i^M S_i + P_i^C C_i S_i + P_i^T G_i S_i)] \quad (1)$$

- **Decision variables**

The use of the surface areas as decision variables makes it possible to know the surface areas available for each crop and thus determine the operating cost per hectare.

The variables that will be used in our case are: (S1, S2, S3, ... Sn) positive:
[S1≥0, S2≥0, S3≥0 ... Sn≥0.].
With:

- S1: Soft wheat surface area;
- S2: Durum wheat surface area;
- S3: Barley surface area;
- S4: Corn surface area;
- S5: Oat surface area.

- **Objective function**

Once the variables have been chosen, we can formulate the problem with the construction of a mathematical function composed of the decision variables which represents the modeled physical model, with the aim of maximizing OF under constraints.

i =1	Culture 1
i =2	Culture 2
..	..
i = n	Culture n

The maximization of the objective function OF makes it possible to assign a value to each different configuration (see equation (1)).

With:

$$R_i = \frac{\text{Harvested quantities (t)}}{\text{Surface areas (ha)}} \quad (2)$$

$$G_i = \frac{\text{Grain quantities (t)}}{\text{Surface areas (ha)}} \quad (3)$$

$$W_i = \frac{\text{Water quantities (t)}}{\text{Surface area (ha)}} \quad (4)$$

$$F_i = \frac{\text{Fertilizer quantities (t)}}{\text{Surface area (hectare)}} \quad (5)$$

$$P_i = \frac{\text{Pesticide quantities (t)}}{\text{Surface areas (ha)}} \quad (6)$$

$$C_i = \frac{\text{Combustible (kWh)}}{\text{Surface areas (ha)}} \quad (7)$$

With:

P_i^S	Harvest selling price	MAD t^{-1}
P_i^G	Grain purchase price	MAD t^{-1}
P_i^W	Desalinated water price	MAD t^{-1}
P_i^F	Fertilizer price	MAD t^{-1}
P_i^P	Pesticide price	MAD t^{-1}
P_i^O	Load per worker per day	MAD wo^{-1} day^{-1}
P_i^L	Land rental price	MAD ha^{-1} day^{-1}
P_i^M	Machine rental price	MAD ha^{-1}
P_i^C	Fuel price	MAD kWh^{-1}
P_i^T	Transport price	MAD t^{-1}
N	Number of workers per hectare	Wo ha^{-1}
J1	Number of working days for i	Day
J2	Number of culture days for i	Day

The objective function coefficients are calculated from equation (1):

$$OF = 3{,}654\, S_1 + 5{,}088\, S_2 + 2{,}494\, S_3 + 8{,}601\, S_4 + 4{,}935\, S_5 \quad (8)$$

- **Constraints**

The optimization problem consists in maximizing (or minimizing) an objective function of n decision variables subjected to a set of constraints expressed in the form of equations which represent the limitations on the variables.

$$\sum_i^n S_i \leq S_{Total} \leq 206 \text{ ha} \quad (9)$$

$$\sum_i^n W_i S_i \leq W_{Total} \leq 75\ 10^6 \text{ t} \quad (10)$$

$$Gi\, S_i \leq CF_i \quad (11)$$

$$R_i\, S_i \leq DM_i \quad (12)$$

The quantities of seed must be given by the suppliers (CF_i) and the quantities of crops must be absorbed by the market (DM_i).
Table II gives the constraint coefficients used in our case.

Table II: Constraint coefficients

Crops	W_i	Gi	CF_i	R_i	DM_i
Soft wheat	980	0.16	40	7.3	11,943
Durum wheat	880	0.16	40	7.2	4,593
Barley	660	0.16	40	6.5	2,297
Corn	5,100	0.16	40	10.0	919
Oat	780	0.16	30	3.5	439

- **Results**

The different equations have been solved using the Excel Solver depending on the objective function and the decision variables, while satisfying the constraints listed. For a farm of 206 ha with crops (soft wheat, durum wheat, barley, corn, oats), determining the distribution of the optimal operating plan gives corn and oat as crops yielding more gain with the following proportions:

- Corn: 92 ha;
- Oat: 114 ha.

The value of the objective function or the maximum gain (difference between the selling price and the operating costs) is estimated at 2,500,000 MAD.

$$OF = \text{Selling price} - \text{Operating costs} \quad (13)$$

When the gain potential is judged to be too low, returning to Step 2 makes it possible to change the energy source and to recalculate the water specific cost and thus calculate a new gain potential.

3. FINANCIAL STUDY AND PROJECT SENSITIVITY ANALYSIS

There are two financing scenarios for the installation of the reverse osmosis unit:

- Financing from the various farmers belonging to the association (case 1) according to their water requirements;
- Private financing (case 2), in which case farmers will buy desalinated brackish water via a private investor who will add a profit of 1.7 MAD to the water specific cost.

Different sources of energy are also studied for each case:

- Option 1: The reverse osmosis plant is powered by PV solar energy without a battery bank;
- Option 2: The locality is considered to be connected to the utility grid, in which case the connection is direct and the average cost of electricity is set at 1.0 MAD kWh^{-1} using the prices of the National Office of Electricity and Water (ONEE, French acronym);
- Option 3: It is considered that the locality is not connected to the utility grid. In this case, it is necessary to add to the electricity price the connection costs which are estimated at 5,000,000 MAD. The costs of the lines and the transformer station are 180,000 MAD km^{-1} and 1,500 MAD kW^{-1}, respectively. Table III gives the details for 20 km and 500 kW.

Table III: Connection costs

Designation	Value (MAD)
Lines	3,600,000
Transformer station	750,000
ONEE penalties and care	600,000
Total	4,950,000

- Option 4: A diesel generator is used to operate the reverse osmosis system. The price of the electricity produced by the generator is 2.7 DH kWh^{-1}. The investment cost of the AC diesel generator is 1,200,000 MAD using 2,400 MAD kW^{-1}.

In order to analyze whether the investment will prove to be profitable or not, the net present value, the internal rate of return, the payback period of a reverse osmosis unit, the profitability which is defined as the ratio of net profit to the sum of production and investment costs, are calculated as well as the optimal operating plan and the gain potential (see Table IV). The operating costs also include the energy costs.

Table IV: Summary of different cases

Financing	Case 1	Case 2	Case 1	Case 2	Case 1	Case 2	Case 1	Case 2
Energy source	PVRO without battery bank		Utility Grid-powered Reverse Osmosis (UGRO)		UGRO + Connection costs		Diesel Generator-powered Reverse Osmosis (DGRO)	
Investment (10^6 MAD)	17.5	17.5	8.6	8.6	13.6	13.6	9.8	9.8
Operating costs per year (10^6 MAD)	2.0	2.0	2.4	2.4	2.7	2.7	4.1	4.1
Energy costs per year (10^6 MAD)	-	-	1.0	1.0	1.0	1.0	2.7	2.7
Specific cost of water (MAD m^{-3})	5.1	6.8	5.0	6.7	5.5	7.2	8.3	10.0
Cash-flow (10^6 MAD)	30.0	39.9	19.8	29.7	23.0	33.0	24.8	34.8
NPV (10^6 MAD)	12.5	22.4	11.2	21.1	9.4	19.3	15.0	25.0
IRR (%)	18.0	24.0	24.0	36.0	18.0	26.0	27.0	37.0
Payback period (year)	8.3	5.8	5.8	3.7	8.7	5.6	5.3	3.6
Profitability (%)	96.0	141.0	56.0	96.0	52.0	86.0	42.0	65.0
Crop types (surface area (ha))	S4(92) S5(114)	S4(92) S5(114)	S4(92) S5(114)	S4(92) S5(114)	S4(92) S5(114)	S2(114) S4(92)	S2(206) -	S2(206) -
Gain potential (10^6 MAD)	2.5	1.7	2.3	1.5	2.0	1.3	0.6	0.6

The results showed that the specific costs of desalinated brackish water using PVRO without battery bank (5.1 or 6.8 MAD m^{-3}) and UGRO brackish water (5.0 or 6.7 MAD m^{-3}) are comparable regardless of the funding scenario. The first option has the advantage of having a lower operating cost (2,000,000 MAD) and a higher gain potential (2,500,000 MAD) while the second option has a lower investment cost (8,600,000 MAD). The payback period of desalinated brackish water by PVRO is between 5.8 to 8.3 years. When connection costs are integrated in UGRO or when a diesel generator is used, the gain potential is less interesting than the first two. Profitability with the first option is estimated at 96% on the farmers' side and 141% on the private investor's side. These values are estimated, respectively, at 56% and 96% when electricity comes from the utility grid, at 42% and 65% when electricity is produced from a diesel generator.

4. CONCLUSIONS

The computerized approach to evaluating the potential of using PVRO desalinated brackish water in agriculture was developed and used for Jaidate County (Rhamna Province) to desalinate a daily capacity of 1,528 m^3. Different cases of financing and energy supply options were studied to determine the lowest specific cost and the highest gain potential. Desalination of brackish water by PVRO requires a total investment of 17,500,000 MAD (about 1,564,000 €) with a specific water cost of 5.1 or 6.8 MAD m^{-3}. The payback period is between 5.8 to 8.3 years. The corresponding profitabilities are respectively 96% and 141% on the farmers' side and on the private investor's side. The results of the study showed that for a farm of 206 ha, the farmer should plant corn (92 ha) and oat (114 ha) in order to have a maximum profit estimated at 2,500,000 (about 223,425 €) or 1,700,000 MAD (about 151,930 €) according to the financing scenario. Land could now be used for agricultural production and water would no longer be a constraint that would restrict crops.

The methodology developed is interesting, general and can be applied in areas with agricultural land abandoned due to poor water quality without any source of supply. It is quantitative (determination of the gain potential) because it allows having a very clear idea of the project on farmer's side and on private investor's side.

The use of desalinated brackish water allows the increase of agricultural land, consequently a generation of economic activities and the stabilization of the water stress by the mobilization of brackish groundwater.

References

[1] T. Luo, R. Young, and P. Reig, Aqueduct Projected Water Stress Country Rankings, Technical Note World Resources Institute, (2015).

[2] Ministère de l'Energie, des Mines, de l'Eau et de l'Environnement, Eaux Souterraines Saumâtres au Maroc, Potentialités en tant que Ressources Alternatives, (2008).

[3] M. P. Chambost, Un Nouveau Rapport Alarme sur le Stress Hydrique au Maroc, Telquel, (2014).

[4] V. Renaudin, Le Dessalement de l'Eau de Mer et des Eaux Saumâtres, Culture Sciences Chimie, (2003).

[5] S. Hajji, A. Bentamy, N. Mbodji, and A. Hajji, Design and Cost Optimization of Small-Scale PV-Powered Reverse Osmosis Desalination (Case Study), 32 European PV Solar Energy Conference (EU PVSEC), June 20–24, Munich-Germany, pp. 2801-2807, (2016).

[6] A. Hemping, Economic Analysis of Reverse Osmosis Desalination of Water for Agricultural Irrigation, California Polytechnic State University, (2011).

[7] P. Lherbier, Accès à l'Eau Potable dans la Commune de Jaidate, (2010).

[8] K. Kadi, M. Chergui, N. Malkia, S. Hamli, S. Boukeria, and A. Yahia, Estimation des Besoins en Eau de la Culture de l'Ail par Différentes Formules, European Scientific Journal, Vol. 10 N°2, pp. 378-396, (2014).

[9] A. Tahhar, Organisation de la Chaine Logistique dans la Filière Céréales au Maroc : Etat des Lieux et Perspectives, Thèse Présentée en Vue de l'Obtention du Diplôme de Hautes Etudes du CIHEAM - Master of Science, (2015).

PHOTOVOLTAIC SYSTEM AND COMPONENTS PRICE DEVELOPMENT IN THE NETHERLANDS

Wilfried G.J.H.M. van Sark[1,2,*], Tony Schoen[3]
[1]Stichting Monitoring Zonnestroom (SMZ), Utrecht, The Netherlands
[2]Copernicus Institute for Sustainable Development, Utrecht University, Heidelberglaan 2, 3584CS Utrecht, the Netherlands, [*]T: +31 30 253 7611, E: w.g.j.h.m.vansark@uu.nl
[3]New-Energy-Works (NEW), Utrecht, the Netherlands

ABSTRACT: Market inventories have been made since fall 2011 for modules, inverters, and systems that are available for purchase in the Netherlands to provide the fast growing Dutch market with proper market information. The most recent inventory was made in December 2016. After the drop in module price in 2012 the average selling price of modules has been found to stabilize, while inverter prices continue to decrease. PV system prices have decreased considerably by ~20% since 2011, with some fluctuations. December 2016 average selling prices are 1.109 €/Wp and 0.29 €/Wp for modules and inverters, respectively, with 10-15% lower price for modules from China. System prices have seen a temporary increase mid 2013 and 2016. Average installation costs amount to 0.34 €/Wp. Grid parity is obvious: for a residential 2.5 kWp system the levelized cost of electricity (LCOE) is calculated to be 0.133 €/kWh for a system price of 1.77 €/Wp (including installation), based on an energy yield of 900 kWh/kWp, 25 years system lifetime, 3% discount rate, and 1% operation and maintenance (O&M) cost.
Keywords: PV modules, inverters, system price, kWh price, grid parity, LCOE, market analysis

1 INTRODUCTION

The Dutch photovoltaic (PV) market has been growing very fast in the past years, with 2 GW installed capacity at the end of 2016, see Fig. 1. Market research to assess the development of the cost of modules, inverters, and complete systems has been performed since October 2011 [1-3]. The availability of price data is of importance to consumers as potential investors in PV systems. We have found that the average selling price of modules, inverters, and systems decreased considerably, predominantly in the year 2012. In this paper an update (status end 2016) of the market prices for modules, inverters and complete systems is presented.

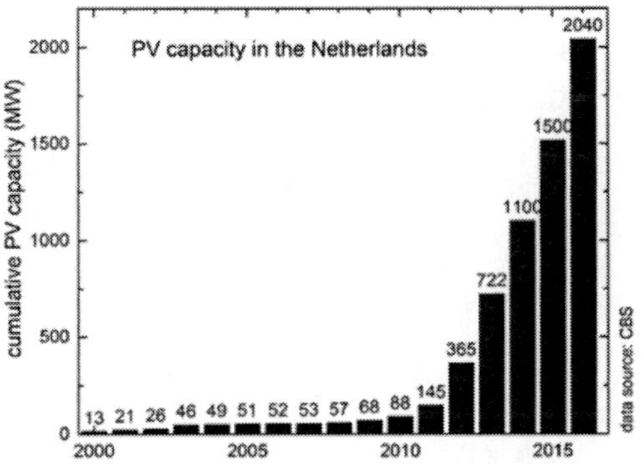

Figure 1. Development of cumulative installed PV capacity in the Netherlands (data source: CBS [4]).

We continued our effort in making further inventories of the PV market in the Netherlands. As before, we collected price data on PV modules, inverters, other system components including installation and consultancy, and complete systems. We also determined the market size and prices of modules of non-European origin in the Netherlands. Finally, for a number of different typically sized PV systems, performance and cost calculations were performed. This allowed for the determination of the levelized cost of electricity of PV systems in order to assess consumer grid parity.

2 METHODOLOGY

The collection of market data was performed using extensive Internet searches that identified the relevant retailers of PV modules, inverters, complete systems, and other Balance of System (BOS) components. We have identified the country of origin for most of the modules. Note that all prices in this paper are quoted *including* 21% VAT, as this is relevant for consumers.

The Levelized Cost of Electricity (LCOE) was calculated using

$$LCOE = \frac{\alpha I + OM}{E} , \qquad (1)$$

where α is the capital recovery factor, I the initial investment, OM the operation and maintenance cost, and E the annual electricity production. The capital recovery factor is defined as

$$\alpha = \frac{r}{1-\left(1+r\right)^{-L}} \qquad (2)$$

with r the discount rate, and L the lifetime of the system.

In the Netherlands a large number of 4-6 module systems are installed, and still are offered, mostly for historical reasons: system size then can be as small as about 600 Wp, and installation is straightforward do-it-yourself, complying to electricity regulations. A typical household roof system that contributes substantially to a typical annual electricity demand of 3500 kWh is sized at 2.5 - 3 kWp. Present legislation in force until 2023 allows for unlimited net metering. We therefore also include a 5 kWp system in our calculations. Larger systems for commercial buildings and offices range from 10-50 kWp. Larger systems as solar parks are excluded here, as information on prices are not available. We have chosen four typical sizes for the LCOE calculation: 0.6, 2.5, 5, and 50 kWp. We used interest rates of 3%, 6% and 8% Further, energy ratings of systems in the Netherlands are between 800 and 1000 kWh/kWp, depending on correct and optimal installation. System lifetime is 25 years, and O&M cost is taken as 1%.

3 RESULTS

3.1 PV modules

In the first inventory of PV modules at the end of October 2011, 165 different types were collected. Only mono and multicrystalline silicon modules were found. The average selling price was 2.28 €/Wp; the average capacity was 144 Wp/m^2. In the inventory taken at the end of December 2016 much more modules were collected (1348 different types, also some thin film silicon and CdTe modules, ~5% of total). In December 2016, the average price per rated power was 1.109 €/Wp, which is 48% lower than at the end of October 2011, as illustrated in Fig. 2. It is clear that the largest price reduction occurred in 2012. In fact, a slight increase is seen for 2016, while the large error bars must be noted.

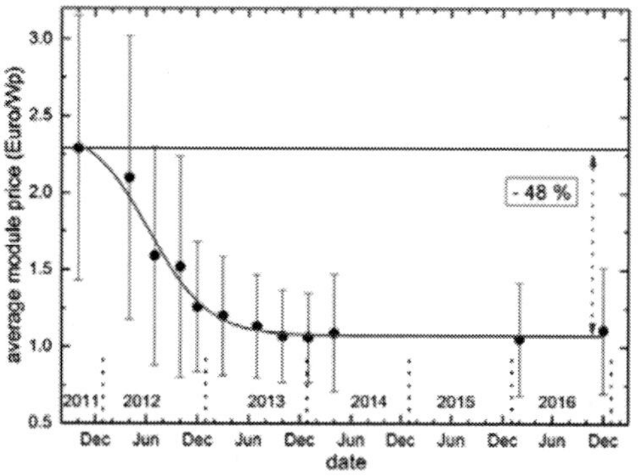

Figure 2. Development of PV module price (including tax) between October 2011 and December 2016.

Our analysis of the country of origin of the modules revealed that 40.2% of the modules originate from China, as shown in Fig. 3. Figure 4 shows that modules from China have a 10-15% lower average price compared to the average price of all modules.

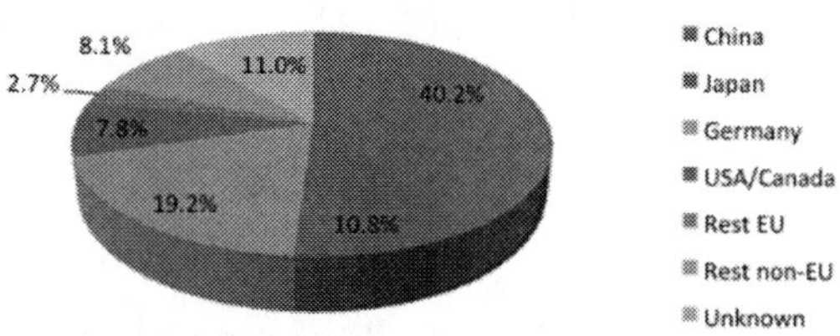

Figure 3. Country of origin of the modules (status December 2016).

3.2 Inverters

At the end of October 2011, the inverter inventory contained 98 unique ones, with average selling price of 0.45 €/Wp. The December 2016 update showed a price decrease to 0.287 €/Wp presently (40.2% decrease) as shown in Fig. 5, with the December 2016 update containing 777 inverters.

3.3 Complete systems

Since the first time that the inventory for systems was made in April 2012, the number of systems for tilted (flat) roofs in the inventory rose dramatically from 1557 (1477) in December 2012 to over 24000 in December 2016. The average selling price for tilted systems

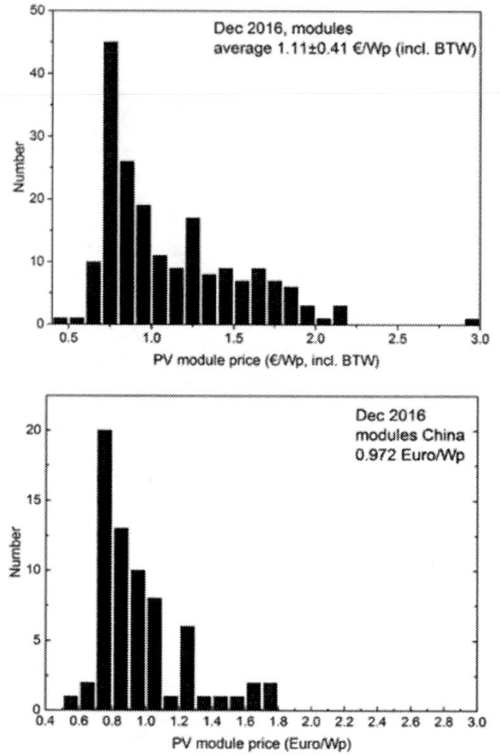

Figure 4. Distribution of PV module price for (a) all modules (top), and modules from China (bottom).

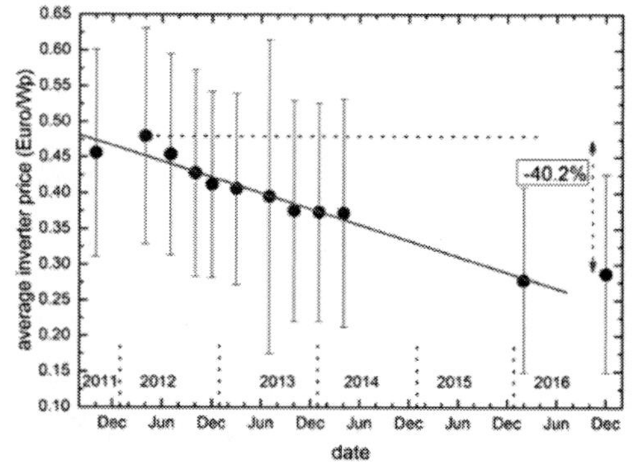

Figure 5. Development of inverter price (including tax) between October 2011 and December 2016.

(excluding installation) decreased by 20% between April 2012 and December 2016, from 1.63 to 1.32 €/Wp for tilted systems, and from 1.67 to 1.39 €/Wp for flat systems. Early 2013 and in 2016 the system price experienced an increase. The 2013 price increase was due to anticipated EU import duties for Chinese made panels, while the 2016 price increase possibly is due to market conditions.

Average installation cost is 0.34 €/kWp, and varies with installation size, as can be seen in Figure 7.

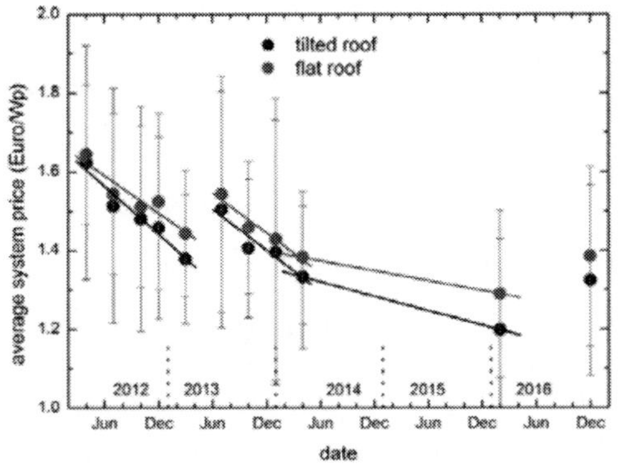

Figure 6. Development of system price (including tax, excluding installation) between April 2012 and December 2016.

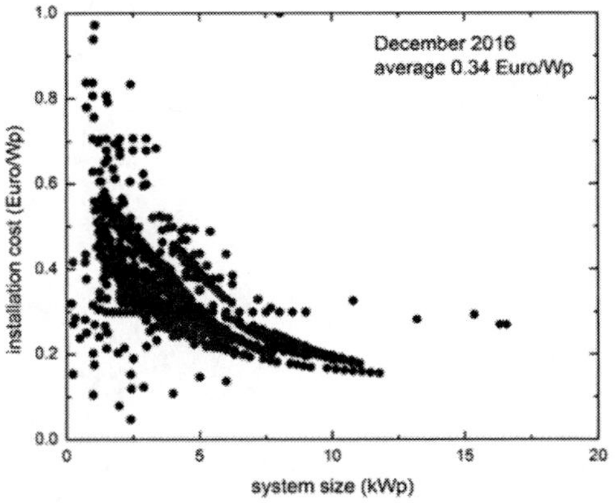

Figure 7. System installation cost as a function of system size, December 2016.

The average prices for three system size ranges have been determined from the inventory data, i.e., for 0-1 kWp, 1-5 kWp, and 5-25 kWp. This is also performed for installation cost, in order to calculate levelized cost of electricity. Table I shows the results, while Fig. 8 shows the development of system price including installation.

Table I: Prices of four typically sized PV systems

system size kWp	price €/Wp	installation €/Wp	total €/Wp
0.6	1.50	0.60	2.10
2.5	1.37	0.40	1.77
5	1.26	0.30	1.56
50	1.14	0.20	1.34

3.5 Levelized cost of electricity

For the four system sizes the LCOE is calculated for three different values of the interest rate (3, 6 and 8%), including installation cost. Results are shown in Table II; only for the smallest system and highest interest rate values are higher than the current price of electricity (0.23 €/kWh) that utilities charge to customers. Clearly, grid parity continues for soft and mortgage type loans, i.e., the cost of electricity is lower than that charged by utilities. This clearly holds for systems of 2.5-5 kWp, and depends on energy yield as well. With energy rating values of 1000 or 800, LCOE values will be 11% lower or 11% higher, respectively.

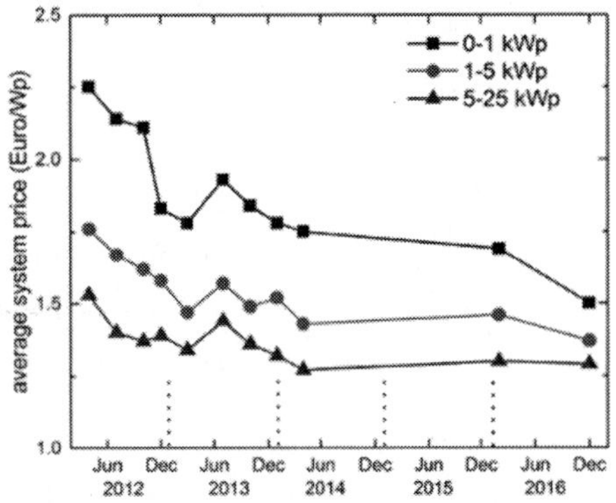

Figure 8. Development of average system price, including installation.

Table II: Levelized cost of electricity for the four typically sized PV systems. Data for two yield values are shown, 850 and 900 kWh/kWp. Data illustrate that grid parity has been reached.

850 kWh/kWp

system size	price	interest rate		
kWp	€/Wp	3%	6%	8%
0.6	2.10	0.167	0.218	0.256
2.5	1.77	0.140	0.184	0.216
5	1.56	0.124	0.162	0.190
50	1.34	0.106	0.139	0.163

900 kWh/kWp

system size	price	interest rate		
kWp	€/Wp	3%	6%	8%
0.6	2.10	0.157	0.206	0.242
2.5	1.77	0.133	0.174	0.204
5	1.56	0.117	0.153	0.180
50	1.34	0.100	0.131	0.154

4 CONCLUSION

A continued market analysis is performed which shows the development of prices on the market for PV modules, inverters, and complete systems. With the data available, we find that PV module price is stabilizing, while inverter price is steadily decreasing. For PV systems the prices have decreased, although with some fluctuations up and down. Modules from companies based in China are cheaper than others.

Our method of acquiring data is based on the availability of price data on websites that sell PV components and systems. We increasingly encounter inactive websites, or websites at which price information cannot be downloaded directly, as a quote for a system is required. This has lead to a lower amount of data for our analysis, which may have affected our results.

5 ACKNOWLEDGEMENTS

We would like to thank TKI Urban Energy for financial support. Parts of these results have been reported in the Dutch National Solar Trend Report 2017 [5].

6 REFERENCES

[1] W.G.J.H.M. van Sark, P. Muizebelt, J. Cace, A. de Vries, P. de Rijk, *Grid parity reached in the Netherlands for consumers*, Proceedings of the 27th European Photovoltaic Solar Energy Conference, 2012, pp. 4010-4014.

[2] W.G.J.H.M. van Sark, P. Muizebelt, J. Cace, A. de Vries, P. de Rijk, *Photovoltaic market development in the Netherlands*, Proceedings of the 28th European Photovoltaic Solar Energy Conference, 2013, pp. 4723-4726.

[3] W.G.J.H.M. van Sark, P. Muizebelt, J. Cace, G. Rutten, A. de Vries, P. de Rijk, *Photovoltaic market development in the Netherlands – 2013, the year of price stabilization,* Proceedings of the 29th European Photovoltaic Solar Energy Conference, 2014, pp. 4059-4061.

[4] CBS Statline, http://statline.cbs.nl/statweb/, last access date 20/09/2017.

[5] Solar Solutions, *National Solar Trend Report 2017*, http://www.solarsolutions.nl/download-solar-trendrapport-2017/, last access date 20/09/2017.

EVALUATING THE FACTORS AFFECTING THE BREAK-EVEN COST OF ON-SITE PV GENERATION AT INDUSTRIAL UNITS

Michael Papapetrou [1,2], George Kosmadakis [1,4], Pablo Frías [3], Tomas Gómez [3], Lorenzo Simons [3],
1) Wirtschaft und Infrastruktur GmbH & Co Planungs-KG (WIP Renewable Energies), Phone: +49 89 720 12 712, E-mail: pmp@wip-munich.de
2) Dipartimento di Ingegneria dell'Innovazione Industriale e Digitale, Università degli Studi di Palermo.
3) Institute for Research in Technology (IIT), of Universidad Pontificia Comillas
4) Department of Natural Resources and Agricultural Engineering, Agricultural University of Athens, Iera Odos 75, 11855, Athens, Greece

ABSTRACT: This paper studies the way in which the regulatory framework and market rules affect the feasibility of on-site PV generation for large industrial units. In most European markets net metering and feed-in tariffs for self-consumed electricity are not possible or are being phased out, providing an incentive to the industry for becoming more flexible in the way electricity is consumed in order to maximise the percentage of the variable electricity generated on-site that is self-consumed. The electricity cost for the industry is the benchmark for PV or any other on-site generation technology and in general on-site solar energy is competitive with that. However, as the regulation develops further, the exemptions of paying the regulated charges for the electricity that is self-consumed are phased out. Also the cost of flexibility required to self-consume all variable on-site generation has to be added to the LCOE of solar electricity, moving it further away from the competitiveness benchmark. Still, as the LCOE of solar electricity reduces continuously mostly due to the reduction of PV system costs, it becomes competitive for more and more users in more and more target markets.

Keywords: Economic Analysis, Funding and Incentives, Grid Integration, On-site generation, Self-generation, Flexibility, Regulatory framework

1 INTRODUCTION

Currently energy policy in Europe is in a transition and faces fundamental choices. Among others, the following two topics have been at the top of the agenda: (a) The cost-effective integration of variable renewable electricity into the power systems of Europe; (b) The rising cost of electricity and its effects on the competitiveness of the European Industry.

On-site generation through PV when combined with flexible self-consumption minimises excess electricity exported to the grid. This facilitates further growth and integration of variable renewable energy, while it can reduce the industrial electricity costs. Thus, on-site generation through PV combined with flexible self-consumption can help addressing at the same time both issues at the top of the European energy policy agenda.

The European rules suggested in 2014 regarding renewable energy support schemes, foresee increased market exposure and balancing responsibilities for renewable energy [1]: *"The following conditions apply from 1 January 2016 to all new aid schemes and measures: (a) Aid is granted as a premium in addition to the market price (premium) whereby the generators sell its electricity directly in the market; (b) Beneficiaries are subject to standard balancing responsibilities, unless no liquid intra-day markets exist; (c) Measures are put in place to ensure that generators have no incentive to generate electricity under negative prices."*

This trend, leads also the renewable energy generator to look for applications like on-site generation with flexible consumption, where there is no dependence from subsidies.

This study focuses specifically on the issue of on-site generation for the industrial sector. It is relevant to examine separately this sector as its access to resources, demand flexibility and large roof spaces make it ideal candidate for self-generation; however the low energy prices that the industries have access to, make it less attractive to adopt this new trend and the market design issues become more critical.

2 METHODOLOGY

That paper highlights the regulatory and market elements affecting the feasibility of on-site renewable energy generation for industrial users, aiming to enable wider adoption of on-site generation through regulatory adjustments and the use of business models that maximise returns under certain conditions.

As part of the IndustRE project [2], we performed a systematic analysis of the regulatory framework and the market rules in main European markets (Belgium, France, Germany, Italy, Spain and the UK), assessing the impact they have on the feasibility of on-site generation and consumption [3]. This analysis is presented in sections 3.1 and 3.2. Then, we reviewed the cost of electricity for small, medium and large industrial users, in our effort to define the benchmark that on-site generation levelised cost of electricity (LCOE) has to meet for becoming competitive. This work was performed as part of the RED Heat-to-Power project [4] and is presented in section 3.3. Finally, the demand flexibility emerges as a key factor affecting the feasibility of on-site generation and therefore section 3.4 was added addressing this issue.

The work focused on the markets in Belgium, France, Germany, Italy, Spain and the UK. These countries have important industrial production and together they represent more than 65% of the EU population and almost 80% of all the installed wind and PV capacity. These figures allow us to balance between a manageable effort of working with six target countries and still having an important impact on a European level.

3 RESULTS AND DISCUSSION

3.1 Electricity procurement and self-generation options

Traditional business models are defined as industrial electricity users signing a contract with a supplier, who in turn charges them for the consumed energy. Usually the supplier also charges them the network charges, other regulated charges and taxes according to options provided by the existing regulation and the agreed conditions.

However, large industrial consumers that are connected to the medium or high voltage grid can choose to follow an alternative process and purchase energy directly from the wholesale electricity market or through bilateral contracts with generators. In that case they pay separately the use of system charges, or network charges, other regulated charges, and taxes as defined by the regulation according to their level of voltage and power/energy consumed.

Own generation behind the consumer's meter, is allowed in all target countries under certain circumstances. The way that the self-consumed and the excess energy fed into the grid are treated, follow one (or a combination) of the three approaches listed below:

a) A feed-in-tariff/premium approach on all instantly self-consumed and/or excess energy injected into the grid.

b) Net metering approach, which is a billing system by which the excess electricity injected into the grid can be used to offset net consumption, not instantly but cumulatively throughout a whole billing period, e.g. a month or a year.

c) Market value approach, by which the electricity that is not self-consumed but injected (instantly) into the grid would be rewarded at the market price of the time of injection.

Consumers under any of these schemes may be exempted from, or asked to contribute to the network and other regulated charges on self-generated energy/power.

In section 3.2 more details are provided regarding the way that self-generation is regulated in the six target countries.

3.2 Target country review

The analysis of the situation in the target countries [3] did take into account the good overview included in the relevant Commission Staff Working Document [5], which accompanied the Communication of the European Commission on 2015 about the new energy market design [6]. This analysis is presented next for the target countries/markets.

Belgium: Consumers with renewable energy installations under 10 kW can benefit from a net metering system. This system entails that the amount of renewable electricity produced annually is deducted directly from the general electricity bill of the consumer, while there is no other remuneration for the part of the electricity injected in the grid. For renewable energy installations over 10 kW, typically two meters are installed: (i) A meter which measures the renewable generation, which is used to calculate the subsidy. (ii) A bi-directional grid connection meter: typically there is still a single grid connection but with separate meter readings for injection and consumption. Consumption and injection are subject to different grid costs and additional charges.

This regulation creates an incentive for medium sized companies to become more flexible in the way the use electricity, following the generation patterns and minimizing the exchange of electricity with the grid.

It has to be noted though, that from July 2015, in the Flemish region, a specific tariff for prosumers was launched to make prosumers contribute to the distribution network costs. This tariff ranges from 67 to 106 €/kW installed.

France: Self-consumption is allowed in France [7]. This foresees the option to consume on-site all self-generated electricity and to be exempted from paying a series of taxes, such as VAT, the Renewable and Social surcharge (CSPE) and the Municipal tax (TICFE). This regulation encourages self-consumption, but it is still possible to feed any excess generation into the grid and be remunerated under the feed-in tariff scheme if the user complies with the terms established in the feed-in tariff policy.

Germany: The review of the German Renewable Energy Act (EEG) in 2012 has introduced a limiting factor for grid injection which is favouring direct consumption. For the excess energy injected into the grid, the consumer is paid the market price. Regarding grid tariffs and other charges, since 01/08/2014, self-consumed energy is totally exempted if < 10 kWp and < 10 MWh/year. Above these levels, a reduced renewable energy surcharge is applied on self-consumption. This surcharge increased gradually, representing always just a fraction of the total surcharge: 30% by 2015, 35% by 2016 and 40% by 2017.

Italy: For generation units < 500 kWp owned by the consumer, a net-billing system in used as an alternative to net metering. According to this system, the value of the excess energy injected into the network is calculated at the wholesale price. This can be used as credit for subsequent periods or is paid to the customer. For larger renewable energy systems, up to 20 MW, there is a scheme in place called SEU20 ("Sistemi Efficienti di Utenza"), by which large consumers can self-consume or sell the excess energy at a market price. Since 2014, under the SEU scheme, industries with on-site generation are exempted from paying 95% of the volumetric part of the general system charges on the self-consumed electricity, encouraging self-consumption. SEU requires a private connection between a VRE plant and a single industrial consumer.

Spain: Self-consumption is allowed in Spain and is regulated by the Real Decreto 900/2015 of 10th October 2015. According to this regulation, there are two potential self-generation types of consumers: i) consumers with on-site renewable energy generation capacity under 100 kWp for self-consumption, where any excess energy fed into the grid is not remunerated; ii) consumers with officially registered generation facilities for self-consumption, who could optionally receive economic remuneration for the energy injections to the grid, according to the specific regulation in place for the generation technology in question. In any case, installed generation capacity must be below the consumption contracted capacity. Under both modalities, the consumer with self-consumption must pay for a series of regulated charges, such as: (i) Network access tariffs, including a volumetric charge and a capacity charge, applied respectively on the hourly net demand supplied from the grid and on the contracted capacity. (ii) Charges associated to the electricity system regulated costs, applied on hourly consumption, i.e. the electricity withdrawn from the grid plus the self-consumed energy. (iii) Charges for other services of the system, or "support charge", which is a specific charge determined by the Ministry of Industry and energy aimed at recovering the costs incurred by the system to support the connection of self-consumption, which would not be necessary if the prosumer were not connected to the grid. This charge would be applied on hourly self-consumed energy. There is a transitional period until the charges associated to the electricity system regulated costs are defined, after which the full charges mentioned above will be introduced.

United Kingdom: A feed-in-tariff approach is implemented in the UK, by which small-scale PV and wind systems (<50 kWp) eligible to receive a feed-in-tariff are given not only a generation tariff for the energy production (that is self-consumed), but also a bonus for up to 50% of the excess electricity fed into the grid. For an installed capacity > 50 kWp and < 5 MWp, there is only a feed-in-tariff paid to the energy generation. Regarding grid and system costs, prosumers are exempted from paying any of them for the part of the electricity that is self-consumed, incentivizing self-consumption among industrial consumers. The tariff structure further incentivizes industries to self-consume, so they will try to forecast peak demand periods and manage their injection/withdrawals during those hours (either by using on site generation or by reducing their consumption).

We can see that only Belgium among the target countries actually presents a net metering policy, but it is limited to very small units (below 10 kW), so large industrial consumers cannot benefit from this policy.

It can also be observed that, in most cases self-consumption is incentivized to some extent for large industrial consumers. This is done by exempting them from paying certain regulated charges on the self-consumed energy, such as a series of taxes (e.g. in France), part of the network charges (e.g., the exemption to pay for 95% of the volumetric part of the system charges under the SEU scheme in Italy, and in Germany the requirement to pay only for a RES-surcharge on self-consumed energy that corresponds to a reduction of the normal grid tariff and other regulated rates) or any grid and system charges at all, as in the UK and Belgium. This creates an important incentive for self-consumption from on-site generation. This is also an incentive for flexibility in order to maximise the percentage of the on-site generated electricity that is self-consumed.

Overall, as the net metering and the feed-in tariffs for self-generated electricity are phased out, the price that the prosumer pays for electricity supply from the grid becomes the benchmark for the cost of the renewable electricity generated on-site. Therefore, in section 3.3 we review the industrial electricity prices.

On the contrary, grid tariff exemptions on self-consumed electricity are gradually being cut down or eliminated in other places, e.g. in Spain prosumers are required to pay the network access tariffs on the net consumption plus an additional charge for grid and system costs on the self-consumed electricity and in the Flemish region of Belgium a specific distribution network tariff for prosumers was launched in July 2015. These developments make self-consumption a less attractive option, as the benchmark LCOE for on-site generation moves lower.

3.3 Industrial electricity prices

In some countries favourable prices are available, especially for large industrial sites with annual electricity consumption of over 70 GWh. The electricity prices (including all taxies and levies) for small/medium and for large industrial sites for the period 2012-2016 are given in Fig. 1 for some representative EU Member States, according to 2016 data from Eurostat [8].

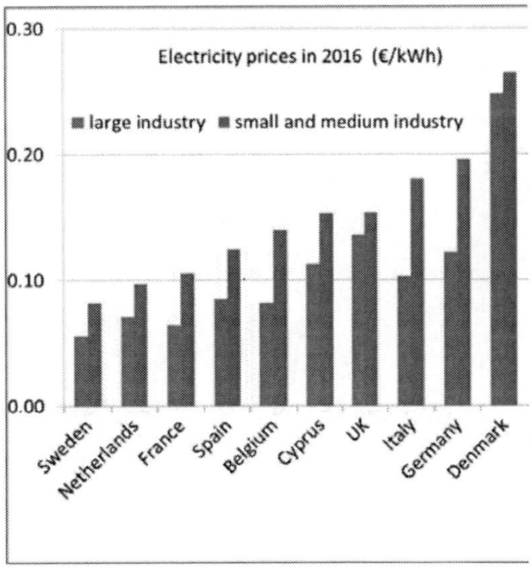

Figure 1: Electricity prices for small/medium and for large industrial sites in EU countries for the year 2016 (Eurostat data)

Average electricity prices for small/medium industrial sites are about 0.15 €/kWh, reducing to 0.10 €/kWh for large ones. Most countries are not too far from the average European prices, while Denmark is an exception as it has much higher prices and Sweden, Netherlands and France have lower prices.

In case large industries procure electricity from the wholesale market, the prices vary in time and are generally lower. An example is provided in Fig. 2 from the German/Austrian energy market for the period June 2016-June 2017 [9].

There are moments with negative prices (in total 3 days during the last year), due to overproduction of electricity. However, there are even more moments with electricity price over 0.05 €/kWh, which for the last year were about 40 days (10 days during December 2016, and from 10th January up to 10th February 2017). On the other hand, similar market prices for France according to EEX data [9] are shown in Fig. 3 for the same duration.

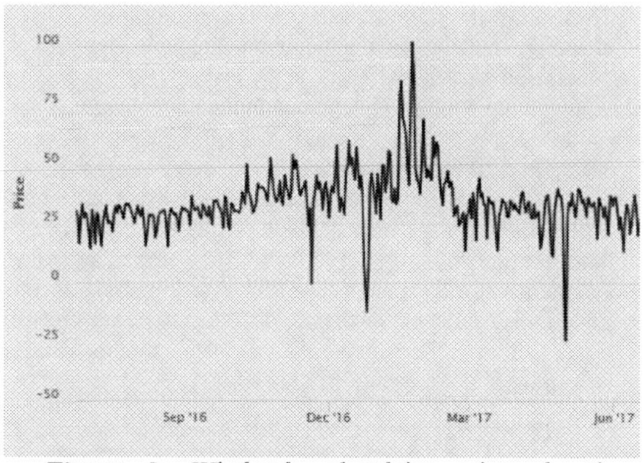

Figure 2: Wholesale electricity prices in the German/Austrian market for the period June 2016-June 2017, price in €/MWh (EEX data)

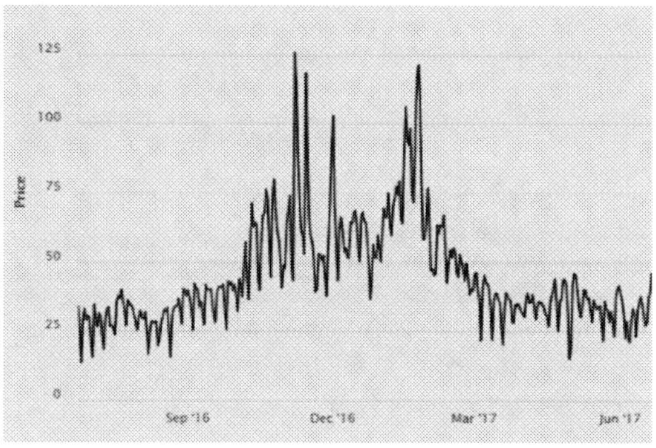

Figure 3: Wholesale electricity prices in the French market for the period June 2016-June 2017, price in €/MWh (EEX data)

In France the prices of the wholesale market are more variable and we observe a longer period with prices over 0.05 €/MWh (about 4 months, October 2016-January 2017).

Overall we have seen that the prices that small and medium industry pays for electricity in most European countries are in the region of 0.10 to 0.20 €/kWh. On the other hand the levelised cost of electricity of on-site solar energy ranges between 0.05 to 0.14 €/kWh according to the latest data [10]. The very wide range of LCOEs in that study stems from the fact that the systems considered range from very small (5 kWp) to large rooftop systems (1 MWp), while different climates are also considered. Still, in all cases, the overall conclusion is that on-site PV generation can be competitive.

The benchmark for large industry is lower, especially if they have access to wholesale electricity prices. However, the solar energy costs can still be competitive and as the solar energy costs decrease further, they will become even more competitive; offering also cost stability, as opposed to the price fluctuations of the wholesale energy market.

Finally, according to the available data, electricity prices for industrial users have experienced a small increase about 2-3% in average in EU during the last decade [8], revealing the potential of introducing promising energy measures in industries.

3.4 Industrial demand flexibility

The price that the industry pays for electricity is a benchmark only if the industry can consume all electricity generated on-site. This is the case if the capacity of the on-site renewable energy system is small compared to the industrial demand. If we want to accommodate larger systems, the flexibility of the on-site generation and/or consumption becomes of interest. In the case of solar energy, the generation becomes more flexible by adding a battery, which increases also the electricity costs. The flexibility of the industrial demand on the other hand can be an interesting case, which we discuss in section 3.4.

Having flexibility to shift electricity demand over time can be beneficial as this can be used to increase the percentage of the self-generated electricity that is self-consumed. The amount of flexibility that is available to be activated by industrial users varies depending on the kind of industrial activity, the technical installations, the production processes and the objective perceptions, preferences and requirements of the industrial plant owner. In order to quantify the available flexibility, the research group of VITO has developed a process called Demand Response Audit (DRA) as part of the IndustRE project [11]. This process was further developed as part of the IndustRE project and it was applied to several case studies. The main conclusion was that there is often more flexibility available than the plant operators think. However, the cost of activating this flexibility is case specific and can be determined only by the industrial plant owners/operators.

4 CONCLUSIONS

We have seen that in most of the European markets studied (Belgium, Germany, Italy, Spain, France and the UK) net metering and feed-in tariffs for self-consumed electricity are not possible or are being phased out. This provides an incentive to the industry for becoming more flexible in the way electricity is consumed in order to maximise the percentage of the variable electricity generated on-site that is self-consumed.

The electricity cost for the industry is the benchmark for PV or any other on-site generation technology. This consists of the cost of energy (market-based price), the network charges and any other taxes, levies and surcharges that may apply (regulated costs). In the vast majority of cases (considering size of systems, solar irradiation and tariffs available to the end-user) the LCOE of solar energy is already competitive. However, as the regulation develops further, the exemptions of paying the regulated charges for the electricity that is self-consumed are phased out. This lowers the benchmark, taking some of the large energy users that have access to low energy tariffs out of the zone of on-site solar energy competitiveness, especially in countries where solar irradiation is not so high. For, the cases where the industry changes its consumption patterns in order to self-consume all electricity generated on-site, the cost for activating this flexibility has to be added to the LCOE of solar energy, moving it further away from the competitiveness benchmark. Still, as the LCOE of solar energy reduces continuously, it becomes competitive for more and more users in more and more target markets.

5 ACKNOWLEDGEMENTS

The work presented in this paper has been carried out within the projects IndustRE and RED Heat-to-Power. Both are co-financed by the European Commission under the H2020 Programme (Grant Agreement number IndustRE 646191; RED Heat-to-Power 640667). Detailed information on the IndustRE and the RED Heat-to-Power projects is available at the project websites: http://www.industre.eu/, http://www.red-heat-to-power.eu/

6 DISCLAIMER

The sole responsibility for the content of this publication lies with the authors. It does not necessarily reflect the opinion of the European Commission. The European Commission is not responsible for any use that may be made of the information contained therein.

7 REFERENCES

[1] EC Guidelines on State aid for environmental protection and energy 2014-2020

[2] IndustRE project,: http://www.industre.eu/, co-financed by the European Commission under the H2020 Programme (Grant Agreement number 646191)

[3] Mercedes Vallés, Tomás Gómez and Pablo Frías, Business models and market barriers, IndustRE D2.4, March 2016, http://www.industre.eu/downloads/download/business-models-and-market-barriers

[4] RED Heat-to-Power project, http://www.red-heat-to-power.eu/ , co-financed by the European Commission under the H2020 Programme (Grant Agreement number 640667

[5] European Commission SWD(2015) 141 final "Best practices on Renewable Energy Self-consumption", SWD(2012) 141

[6] European Commission Delivering a New Deal for Energy Consumers {COM(2015) 339 final}

[7] Decree law n° 2008-386 - 23rd April 2008

[8] Eurostat 2016, Electricity prices for household consumers - bi-annual data (from 2007 onwards), http://ec.europa.eu/eurostat/web/energy/data/database

[9] European Power Exchange, Market Data, https://www.eex.com/en/market-data/power/spot-market/auction#!/2017/06/24

[10] European PV Technology and Innovation Technology Platform, The True Competitiveness of Solar PV, A European Case Study, March 2017, http://www.etip-pv.eu/fileadmin/Documents/ETIP_PV_Publications_2017-2018/LCOE_Report_March_2017.pdf

[11] Annelies Delnooz, Davy Geysen, Daan Six, Jef Verbeeck and Ana Virag, Adapted methodology for optimal valorization of Flexible Industrial Electricity Demand, IndustRE D3.2, June 2016, http://www.industre.eu/downloads/download/adapted-methodology-for-optimal-valorization-of-fl

AN OVERVIEW OF PATENT APPLICATION DATA IN EUROPE IN THE FIELD OF PHOTOVOLTAICS

Alberto Visentin[1], B. Erol Sağol[1], Mauro Boero[2], and Christian Königstein[2]
[1,2]European Patent Office, Directorate General 1
[1] Gitschinerstrasse 103, D-10969 Berlin
[2]Postbus 5818, NL-2280 HV Rijswijk, The Netherlands
e-mail: avisentin@epo.org, bsagol@epo.org, mboero@epo.org, ckoenigstein@epo.org

ABSTRACT: The present study is aimed at giving an update on patent activities in the photovoltaics (PV) field covering the past 45 years. In this study, the number of patent applications published during this time period, per year and for the whole period, in several of the most relevant solar cell technologies is analysed. Patent filing figures and the origin of patent applications are shown for the different types of solar cells including polymer and organic solar cells, and dye sensitized solar cells.

Keywords: PV patent activity

1 INTRODUCTION

The European Patent Office (EPO) is the executive body of the European Patent Organisation, an intergovernmental organisation that was created with the entry into force of the European Patent Convention (EPC) on 7 October 1977. In 2013 the EPO celebrated the 40th anniversary of the signature of the EPC, which forms its legal basis. The Organisation currently comprises 38 member states [1], two extension states – Bosnia and Herzegovina and Montenegro – and two validation states [6] – Morocco and Moldavia - which recognise the validity of European patent applications and patents upon request by the patentee.

Figure 1: Member states of the EPO (shown in red)

Patent applications at the EPO (which numbered over 290.000 in 2016) are assessed by the EPO's patent examiners *inter alia* on novelty and inventiveness. The examiner's search for prior art – to see if the invention is really novel or already known – requires a huge documentation in a large number of technical fields. The EPO's documentation comprises an internal database [5], containing 78.7 million published patents (corresponding to about 20 million patent family members) and 24.5 million technical literature papers. In 2016 alone, 3.73 million patent documents and about 728 000 literature documents were added. The documents are ordered using a new classification scheme called the Cooperative Patent Classification (CPC), which has been developed jointly by the EPO and the United States Patent and Trademark Office [2]. The classification scheme is built up in a hierarchical manner and based on the International Patent Classification (IPC) [3], allowing one to narrow down searches to a certain set of specifications and technical features describing the invention. The CPC classification scheme contains about 260.000 entries, allowing the precise classification of the documents – compared to about 73.400 entries in IPC. Every year a few 1000 new entries are added to the CPC scheme. The latest scheme update entered into force on August 1, 2017 [7]. In the field of photovoltaics (PV), our database contains at present (May 2016) 125.000 entries, including about 7.250 non-patent literature documents. The 117.000 patent documents correspond to about 37.000 patent families.

This database also allows for statistical evaluations of the information supplied for each application, such as publication date and country of origin of the application. Due to the recently increasing contribution of organic solar cells to the field of PV and the growing number of companies producing such cells, an overview of the patent application history in the field is interesting.

2 METHODOLOGY

The presented statistics are extracted from our database using the CPC scheme to specify the fields of interest and cover the years 1970 - 2016. The numbers are based on a single count for every patent family, wherein the data are based on the priority date of the priority application, i.e. the date of first filing. Since patent applications are published 18 months after their first filing date, applications having a priority date after 1.1.2016 are not yet available and, therefore, cannot be considered. For example, an application having a priority date of 31/12/2015 will be published on 1/7/2017. Therefore, filing numbers are presented up until priority year 2015.

The geographical analyses are based on the priority patent, i.e. the country of first filing. The EPO member states [1] are taken into account as well as the US, Japan, Australia, Republic of Korea, Canada, Province of China Taiwan, China, Israel and New Zealand [4]. In addition, direct European ("EP") and international Patent Cooperation Treaty (PCT) first filings ("WO") are identified separately. When referring to patent applications in Europe, all patent applications filed with the EPO or with one of its member states are taken into account. However, when referring to EP Applications or EP Patents, only those applications are considered that have been filed with the EPO (as first or second filing). The data for PCT applications for the last two years have to be seen as preliminary, because of the 30 month delay for entering into the European phase.

3 RESULTS

The development of the total field of PV - including inorganic, organic and dye sensitized solar cells and modules - is shown in figure 2.

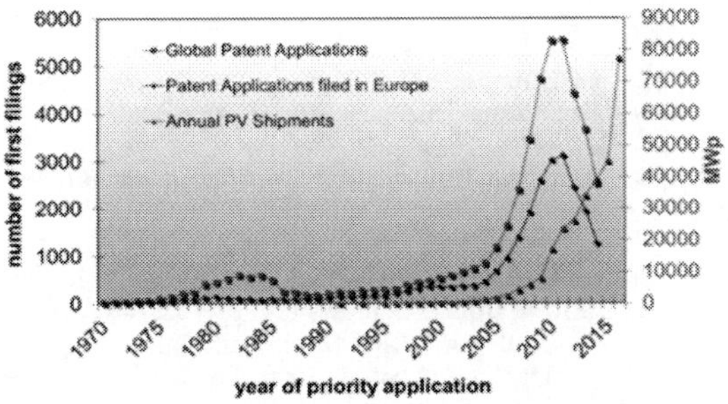

Figure 2: Annual Patent Applications (first filings) in PV

The graph shows a more or less general increase in the number of filings in the European states up to 2012. For the global number of filings - including patent applications that had been filed only outside Europe - a big temporary increase can be seen for the early 1980s, during the period of the second oil crisis. This increase is mainly attributed to US and Japanese patent applications that were not filed in Europe. Over the last four years a decrease in the number of filings has been observed – both globally and in Europe. The decrease for the last two years is partly due to a bias linked to a lack of data available, in particular for international applications where applicants have up to 30 months before entering a regional or national phase, plus the time it takes for us to receive the data and update the collection. Nonetheless, a decrease in the patent activity after 2012 has to be acknowledged.

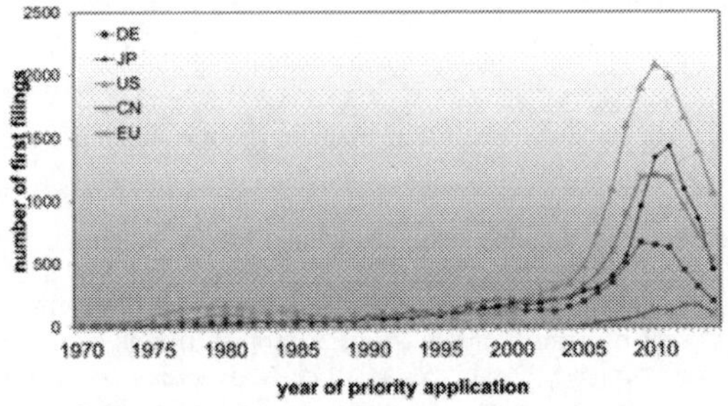

Figure 3: Annual Patent Applications (first filings) in PV for US, JP, EU (=Europe), DE, and CN first filings

While the maximum of patent applications was reached around 2010 in Europe and especially in Germany, which amounts to half of the European applications, the activity in Asia – in Japan and especially in China – continued to be strong for some more years (see Figure 3).

When comparing patent activity in the field of PV with activity in wind energy and nuclear energy (see Figure 4), a clear increase can be shown for the field of renewable energies (PV and wind) up to 2012, whereas nuclear energy (nuclear fission and nuclear fusion) did not show any substantial increase in activity.

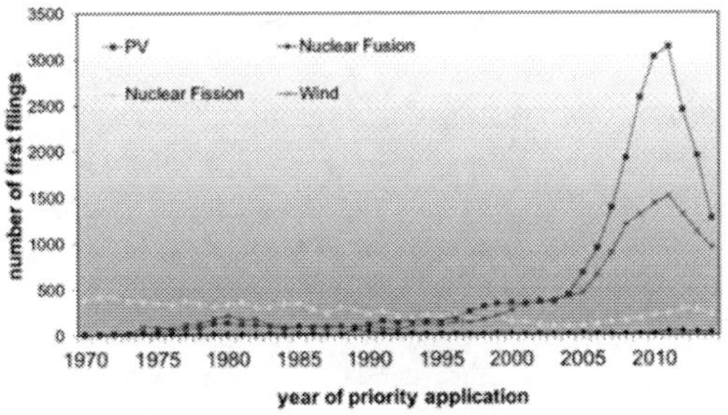

Figure 4: Patent Applications filed in Europe for PV, wind energy, nuclear fission and nuclear fusion

From 2012 onwards, we see again a downward trend in the number of filings for PV, wind energy and nuclear fission.

Looking at patent activity in the field of PV (including all types of solar cells and PV modules), 60% of the patent applications filed in Europe, i.e. applications filed with the EPO or with one of its member states [1], originate from Germany, Japan and the US (Figure 5), with Germany still having the largest share of 25 %.

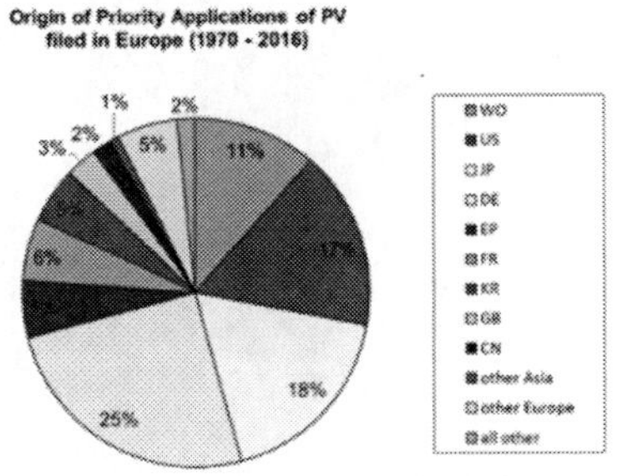

Figure 5: Origin of Patent Applications Filed in Europe - All Types of Solar Cells

A slightly different picture emerges when only those patent applications are taken into account, which are filed at the EPO (Figure 6).

In that case, Germany has a much smaller share (8% of EP applications compared with 26% of all European applications, including national applications), while the shares of Japan (14% vs. 18%) and the US (17% vs. 17%) remain almost unchanged. This can be explained by the fact that most of the German patent applications are only filed nationally (i.e. in Germany), but not with the EPO. More than one third of the applications (38%) received by the EPO arrive via the international (PCT) route (WO). For those applications the country of origin is not accessible. Direct EP first filings amount to 9%, again the origin of the applicant is not directly accessible.

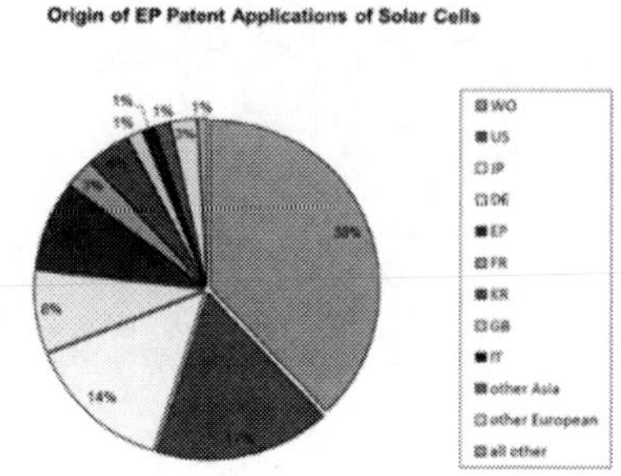

Figure 6: Origin of Patent Applications Filed at the EPO - All Types of Solar Cells

When looking at granted EPO patents (Figure 7), it can be seen that Germany has a larger share than in the patent applications (12% of the grants vs 8% of the applications). The US (17% of applications vs 13% of grants) have a larger share of applications and Japan has a larger share of grants than applications (16% of grants, 14% of applications).

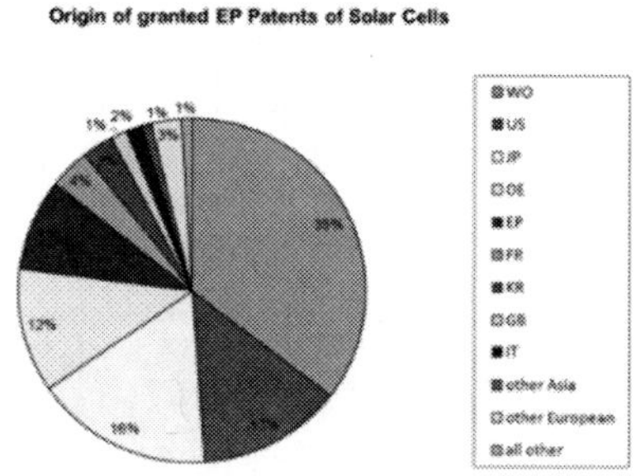

Figure 7: Origin of Granted EPO Patents - All Types of Solar Cells

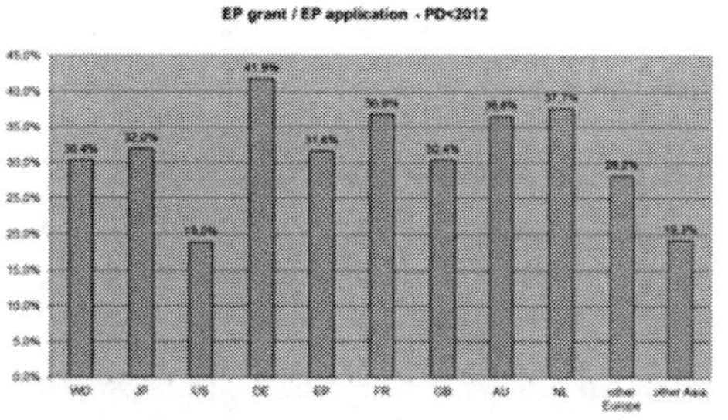

Figure 8: Granted EP Patents in % of EP Applications - All Types of Solar Cells

Patent activity as presented in this study is usually based on patent applications. However, not every application becomes a valid (granted) patent in the end. At the EPO, on average, 50% of EP patent applications in the field of PV are granted, i.e. become a valid European patent. EP applications originating from Japan, Germany, France, Australia and the Netherlands are amongst those that have a higher success rate than the average (Figure 8).

Up to now, filing figures have been considered for the whole period of 1970 – 2015. When analysing the filing figures for different decades (Figure 9), the evolution of the PV industry in different regions of the world can be seen. While in Germany and Japan a continuous increase in patent activity can be seen, France, the UK ("GB"), and the US showed a dip in the period 1984 – 1993.

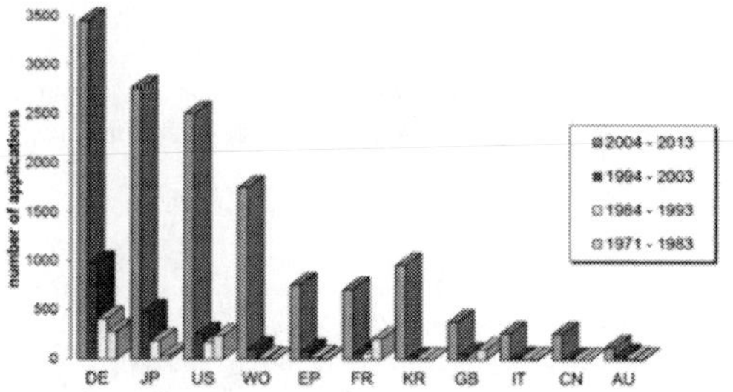

Figure 9: Patent Applications Filed in Europe for different decades

Asian countries other than Japan (i.e. China & Korea), showed significant filing figures only for the last decade (2004 – 2013). Both China and Korea had almost no patent activity in the field of PV before 2000. For the last decade, Korea has surpassed France, the UK and Italy, which had much higher filing figures than Korea in the decades before.

Apart from inorganic solar cells, organic solar cells and dye sensitized solar cells – also called second generation solar cells – are becoming more and more important. When looking at the development of filings in organic PV, it is interesting to compare the number of filings in dye sensitized solar cells with that of other organic solar cells, including layered and bulk hetero junction cells (Figure 10).

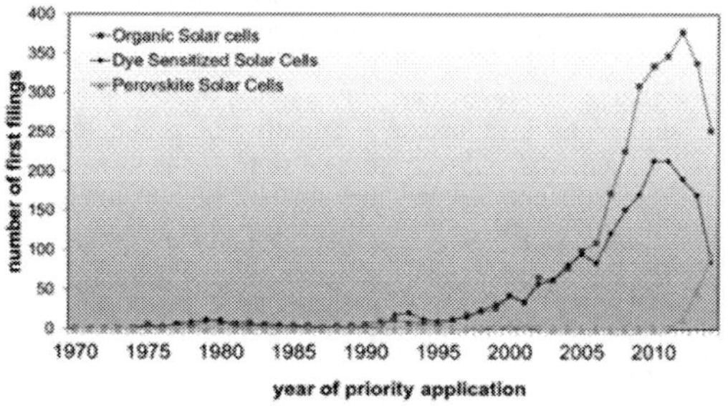

Figure 10: Patent Applications Filed in Europe - Organic and Dye Sensitized Solar Cells

Both fields develop similarly from 1990 on, although the number of filings in dye sensitized cells was outgrown by filings in other organic cells for the first time in 2004. Since 2012 (year of priority application), a drop in filings could be observed for dye sensitized solar cells, whereas the filing figures of organic solar cells increased further, however, at a lower rate. For the last two years, also a decline of applications in organic solar cells can be observed. This decrease, however, needs to be confirmed in the future.

Looking at the filing figures in the field of dye sensitized solar cells, as presented in more detail in Figure 10, the first significant growth can be noted in the early 1990s. The increase in patent activity seems to be related to the work of Michael Grätzel et al. on TiO_2 cells using transition metal complexes as sensitizing dye (EP 0 333 641 A1).

In the field of organic solar cells, growing patent activity can be observed starting in the late 1990s, after the publication of the patent application by Sariciftci and Heeger (WO 94/05045) describing a bulk hetero junction solar cell based on a mixture of a polymer and a fullerene.

Recently a new type of solar cells emerged, which uses a metal organic perovskite (typically $CHNH_3PbI_{3-x}Br_x$) as active material. For this type of devices, steep increase of patent activity has been observed.

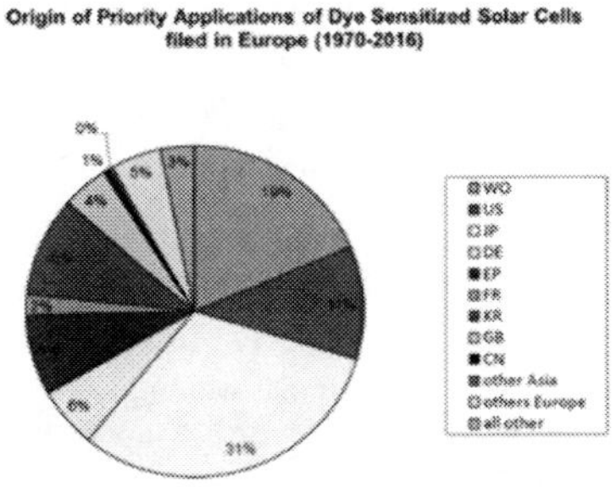

Figure 11: Origin of Patent Applications in Europe – Dye Sensitized Solar Cells

Whereas patent activity in the field of dye sensitized solar cells is dominated by Japan (Figure 11), patents in the field of organic solar cells originate from the US and Japan at an almost equal share (Figure 12).

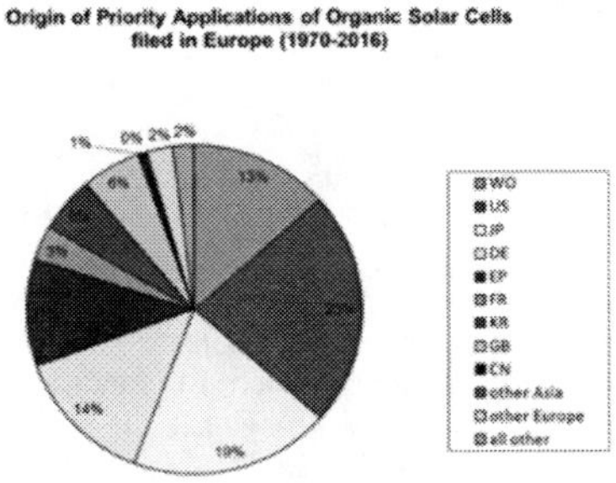

Figure 12: Origin of Patent Applications in Europe - Organic Solar Cells

As shown in Figure 12, by far the greatest part of filings in organic solar cells comes from the US (23%), followed by Japan (19%) and Germany (14%). Other major players are the UK (6%) and Korea (6%). Direct PCT filings ("WO") (13%) and direct EP filings (11%) cannot be attributed to a specific country. Both Japan and Korea are much stronger in the area of dye sensitized solar cells with 31% and 10%, respectively. In the area of dye sensitized solar cells, other big players are the US (11%) and Germany (6%) – again with major contributions from direct EP first filings (8%) and PCT filings (19%).

A different picture is obtained for inorganic solar cells (see Figure 13), where Germany has the biggest share (28%) followed by the US (17%) and JP (16%).

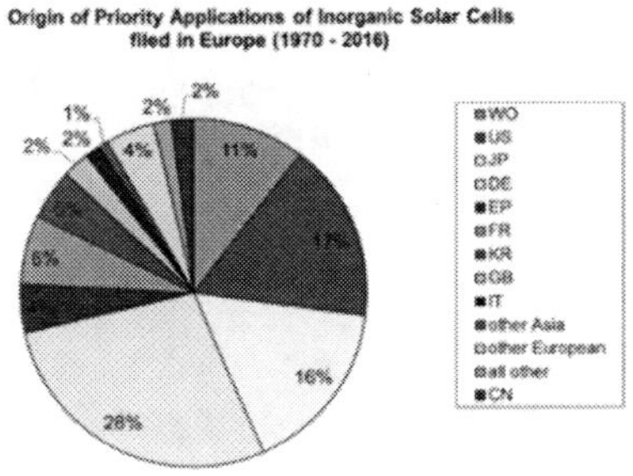

Figure 13: Origin of Patent Applications in Europe - Inorganic Solar Cells

6 SUMMARY

The tremendous success of PV is reflected in the increasing number of patent applications in Europe, which peaked in 2012. Since then, a decrease of applications can be observed. The origin of patent applications filed in Europe is nearly equally distributed between Europe, Japan, the US and other countries. Within Europe, most of the patent applicants come from Germany.

The number of applications relating to organic (polymer) solar cells and dye sensitized solar cells increased considerably over the past 15 years. Perovskite solar cells are expected to become the future boom field in PV patent applications with a doubling in application figures over the past two years.

References

[1] EPO member states: AL = Albania, AT = Austria, BE = Belgium, BG = Bulgaria, CH = Switzerland, CY = Cyprus, CZ = Czech republic, DE = Germany, DK = Denmark, EE = Estonia, ES = Spain, FI = Finland, FR = France, GB = United Kingdom, GR = Greece, HR = Croatia, HU = Hungary, IE = Ireland, IS = Iceland, IT = Italy, LI = Liechtenstein, LT = Lithuania, LU = Luxembourg, LV = Latvia, MC = Monaco, MK = Former Yugoslav Republic of Macedonia, MT = Malta, NL = Netherlands, NO = Norway, PL = Poland, PT = Portugal, RO = Romania, SE = Sweden, SI = Slovenia, SK = Slovakia, SM = San Marino, RS = Serbia, TR = Turkey; Extension States: BA = Bosnia and Herzegovina, ME = Montenegro

[2] Information on the Cooperative Patent Classification scheme is available at www.cpcinfo.org; A search using the classification scheme can be performed using Espacenet: http://worldwide.espacenet.com/

[3] Information on the International Patent Classification is available at http://web2.wipo.int/ipcpub/

[4] US = USA, JP = Japan, CN = China, AU = Australia, KR = Korea, TW = Taiwan, NZ = New Zealand, CA = Canada, IL = Israel, WO = PCT Patent Application.

[5] https://www.epo.org/searching-for-patents.html

[6] http://www.epo.org/about-us/organisation/validation-states.html

[7] Information on the Cooperative Patent Classification: http://www.cooperativepatentclassification.org/

CURRENT AND FUTURE ESTIMATES OF THE LCOE FOR A 10-MW GROUND-MOUNT SOLAR PLANT DEPENDING ON LOCATION AND COMPARING WITH LOCAL PPAs

Julie Cren[1], Robert Thomas[1]
[1]Univ. Grenoble Alpes, F-38000 Grenoble, France
CEA, F-38054 Grenoble Cedex 09, France

ABSTRACT: 2016 was another record year for the solar industry, with yearly installations reaching 75 GWp, a global capacity installed exceeding 300 GWp, and power-purchase-agreement (PPA) prices beneath the symbolic value of $30/MWh. The levelized cost of energy (LCOE) generation is becoming a key metric to understand the economic and technological evolutions of solar energy. It has been the subject of numerous studies and is used by developers and investors to assess project worth. This paper presents our LCOE model for a 10-MWp ground-mount solar plant, using the results of interviews conducted with the stakeholders of the field during 2016-2017. Representative values for the capital investment, operating costs and means of financing were obtained, and the study focused on showing the impact of local factors, not only climatic, but also local economic and regulatory constraints. The results obtained for the LCOE varied between 30 and 70 €/MWh according to location showing good agreement with prices announced as part of 2016-2017 PPAs. In 2030, we estimate that this range will continue to decrease to reach values between 20 and 40 €/MWh with improvements due to increases in module efficiency, the increasing use of trackers and innovations such as bifacial modules.

Keywords: Photovoltaic, Levelized cost, LCOE, Power purchase agreement, PPA, Grid parity

1 AIM AND APPROACH USED

The LCOE approach to calculate the cost of energy from different sources, including renewables, is a classical one. What has become more interesting in the last few years is our ability to compare LCOE calculations in different parts of the world with the actual price data coming from the calls for tender (power purchase agreements) that are now almost universally prevalent for ground-mount solar plants. For example, how do the figures from PPAs compare with model estimates considering the differences in the level of solar irradiance by location? Which local factors other than irradiation can explain why the latest average PPA price in France is 16% higher than in Denmark, when Denmark receives 30% less sunlight? What evolution in the cost of solar energy should we expect by 2030 given future reductions in capital expenditure and further increases in module efficiency? Based on current capital expenditure levels, what effect does module efficiency have on the value of the LCOE and can high efficiency compensate for higher module cost?

The approach has been to detail each element of the LCOE, such as each component that forms part of the capital expenditure 'capex' or operational expenditure 'opex'. In order to ensure up-to-date parameters, interviews were conducted with key industrial actors covering the full value chain of a ground-mount PV plant, including equipment suppliers, developers, investors and operators. This approach allowed us to build a model for 2016 that has fine granularity and can be used to estimate the potential to further reduce prices, to measure the impact of expected incremental and disruptive innovations on the LCOE or else to assess the incidence of local factors including not only climate but policy and regulation.

2 SCIENTIFIC INNOVATION AND RELEVANCE

This study builds upon a body of work that aims at estimating the cost of energy production of solar PV technologies. However, as solar technology matures and observed differences in the cost of both module and balance of system (BOS) hardware reduce, a finer level of analysis is possible with respect to the benefit of certain emerging technologies (such as the use of trackers) and specific conditions by region or country. Examples include the impact of land-use policies (discrimination in favour or against the installation of solar on agricultural land), policies regarding interconnection and the effect of grid reinforcement charges, and the importance that supportive national policy frameworks can have on prevailing risk assessments and the cost of capital. In addition, a sensitivity analysis allowed us to look at the impact that module-efficiency improvements can have on plant BOS costs.

3 RESULTS

The enormous impact that we have begun to see from solar technology has its origin in the phenomenal and persistent learning rate of solar technology: as we produce more units, module prices decline as industry's combined experience increases. For silicon, this learning rate has meant a reduction in module prices by a factor of 40 in 30 years.

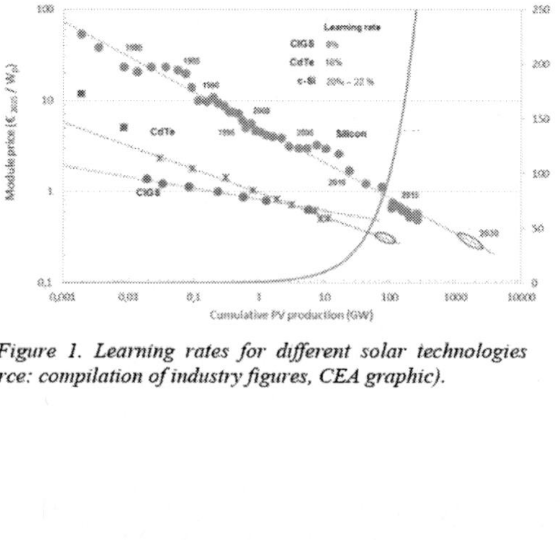

Figure 1. Learning rates for different solar technologies (Source: compilation of industry figures, CEA graphic).

This decline in module prices is not simply the function of lower-cost production, but of improvements in module performance. Thin-film technologies CIGS and CdTe have made enormous strides in the last 5 years, with efficiency improvements close to +1% per year, closing the gap on mainstream p-type Al:BSF silicon. But this longstanding silicon architecture is beginning to give way to advanced designs based on the PERC family of technologies that have shown efficiency advantages in the 1%-2% range. We find that efficiency improvement rates of commercial modules (0.3-0.4% for silicon) are in line with ITRPV [1] projections.

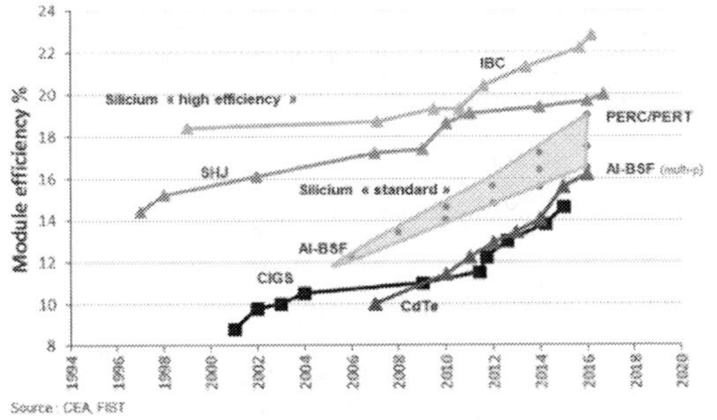

Figure 2. Evolution of commercial module efficiencies for different cell technologies (Source: compilation of industry figures, CEA graphic).

While module prices are largely standardised in Europe and the United States (our reference cost is 400 €/kW), BOS capex levels show some variations, especially for soft costs such as the cost of interconnection/grid reinforcement and labour. The dominant BOS hardware categories remain the inverter/transformer skid or box, the electrical components (BOS elec) and the support and mounting structure and industry feedback suggests these costs are quite generalised in Europe and the United States. Across all BOS categories, we find that France shows somewhat higher costs than Germany, Italy or the United States. Considerable variations are still possible in BOS costs depending on land prices, structural components to cope with particular local climatic or soil conditions and grid interconnection distances. Overall, we estimate total capital investment levels (including the module) between 700 €/kW and 900 €/kW, with an estimated reference value of 800 €/kW.

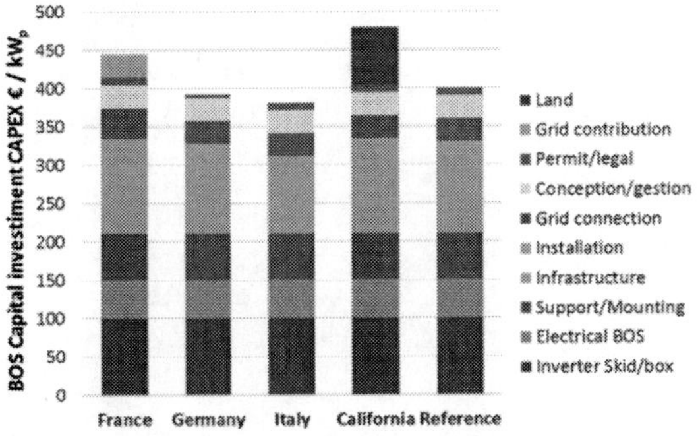

Figure 3. Variation of BOS costs by country (Source: CEA).

Though most plant developers rent land for solar plant development, the continuing federal tax credits in the United States favour land acquisition rather than rental, which would increase plant capex and correspondingly reduce operating expenditure.

Opex costs are determined by a range of factors, not just the well-known maintenance charges, but also insurance, land rental, management charges (including billing) and in some cases taxes on energy providers.

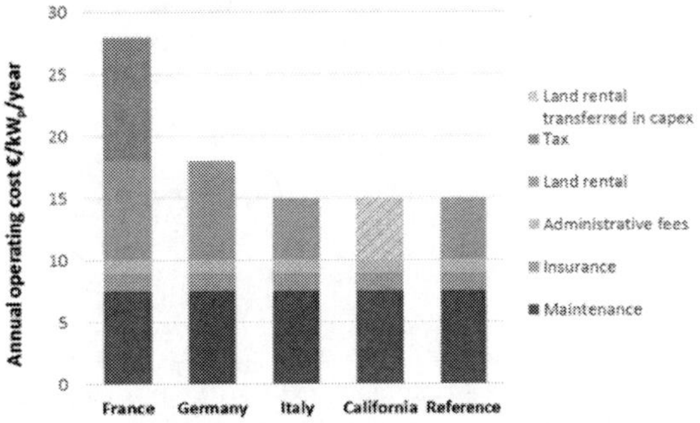

Figure 4. Operating costs for a ground-mount solar plant across four locations (Source: CEA).

The cumulative effects of these capex and opex figures, together with plant energy output and the weighted average cost of capital are used to calculate the LCOE by location (figure 5). Whereas the values for Dresden and San Diego are close to the reference values, Toulouse, France and Naples, Italy are penalised by different additional country-specific cost factors. In the case of France, it is the cost of grid interconnection/reinforcement, land costs and taxes and for Italy it is the higher cost of capital (8% instead of the reference level of 4%), due to the absence of supportive policy frameworks. In addition, we find good agreement between our estimates for 2016 and PPA contract prices announced in the different regions in 2016.

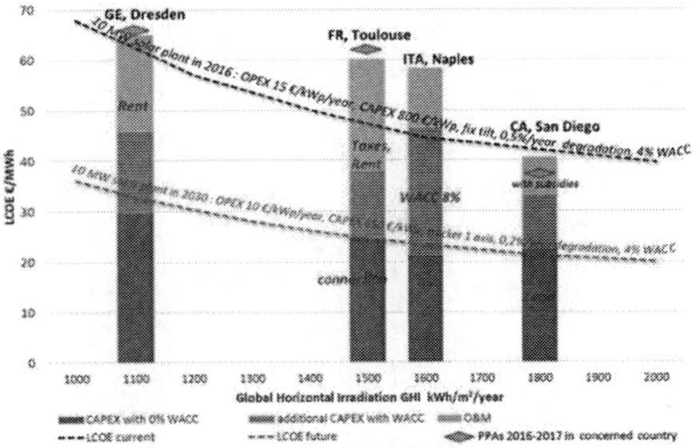

Figure 5. Current and projected (2030) curves of LCOE according to GHI. LCOE breakdown of four locations (Source: CEA).

A sensitivity analysis allows us to understand the critical elements affecting the value of the LCOE. Figure 6 shows the effects on the value of the LCOE in Toulouse (reference value 60 €/MWh) due to changes in the cost of capital, the level of insolation and so on. Unsurprisingly, the level of irradiance is an important factor and moving from Toulouse to Lille or Dresden increases the LCOE by 28%. Equally future capex reductions feed quite directly through to the LCOE, with our estimates for 2030 of 600 €/kW reducing the LCOE by 20%. Perhaps less evident is the importance of the WACC: an increase in the WACC from 4% to 8% (the value estimated for

Italy) is like moving a solar plant from southern France to northern France or northern Germany. Inverter replacement frequency and module degradation have less impact than might be supposed.

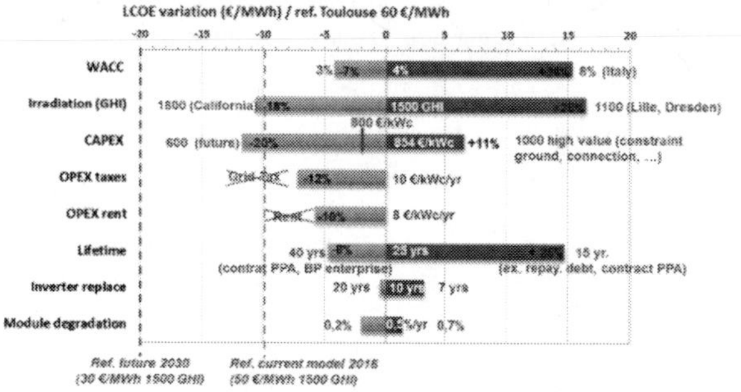

Figure 6. Tornado graph of sensitivity study on LCOE parameters with Toulouse LCOE (60 €/MWh) as a reference (Source: CEA).

The LCOE analysis can be generalised globally and compared with the values obtained from recent PPAs (see figure 7). In general, we find good agreement between theory and actual contract prices, including the impact of using one-axis trackers that are now becoming much more prevalent. One particular comparison that we find stands is between the recent PPAs announced in Denmark and Abu Dhabi, two locations with vastly different levels of irradiance. Both plants are unique in their own ways, show industry-leading values for the contract price and probably favourable local conditions regarding land availability, grid interconnection and taxes. Despite large differences in plant size (5x10 MW in Denmark) and 350 MW in Abu Dhabi, 90% of the difference in the contract price (a factor of 2.4) is due to the difference in irradiance (a factor of 2.2).

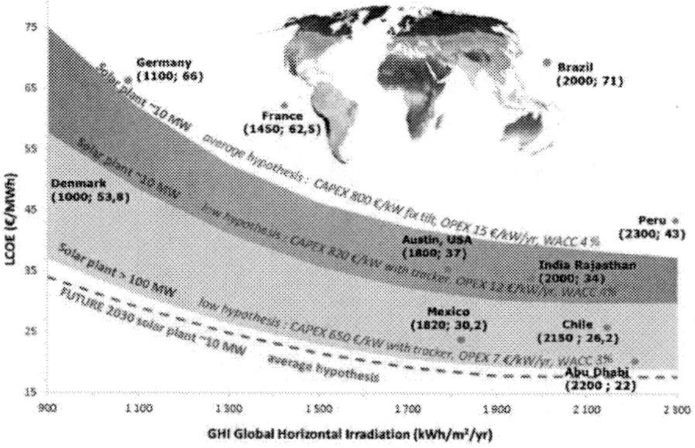

Figure 7. Reference curves of LCOE and PPAs in 2016 according to GHI (Source: CEA).

4 CONCLUSIONS

The LCOE model developed for 2016 suggests solar energy prices between 30 and 70 €/MWh (GHI between 2200 and 900 kWh/m^2/an). This range should continue to reduce. Combining our estimates for reductions in capex with increases in module efficiency suggests a value of the LCOE of between 20 and 40 €/MWh in 2030, and in favourable regions in Europe (GHI > 1300 kWh/m^2/an) the figure should tend toward 30 €/MWh.

At such price levels, solar power will cement its position as the lowest energy technology globally, a position that it now already enjoys in many specific locations in the world. To echo the views the REN21 network [2], solar will become viewed as an "ordinary" energy resource alongside current conventional sources. Of course, challenges remain in terms of grid integration and intermittence, but in just a few years solar energy has demonstrated its competitiveness as an energy investment without subsidy.

5 ACKNOWLEDGMENTS

The authors would like to acknowledge support from the French National Alliance for the Coordination of Energy Research (CVT ANCRE) that commissioned this study.

6 REFERENCES

[1] International Technology Roadmap for Semiconductors (ITRPV), Seventh edition, March 2016.

[2] REN21 Renewable Energy Policy Network for the 21st Century, Renewables Global Status Report, 'http://www.ren21.net/status-of-renewables/global-status-report/

SUPPORTING MARKET UPTAKE OF BUILDING-INTEGRATED PHOTOVOLTAIC TECHNOLOGIES WITH THE PVSITES PROJECT

M. Machado(1), S. Challet(2), I. Weiss(2), E. Román(1), J.M. Espeche(3), F. Noris(3), T. Reijenga(4), E. Rico(5), I. Huerta(5), Y.B. Assoa(6), F. Burgun(6), J.C Esteban(7), J. Escribano(7), P. Alamy(8), V.K. Nguyen(8)

(1) TECNALIA RESEARCH & INNOVATION
Donostia-San Sebastián, Spain
maider.machado@tecnalia.com
eduardo.roman@tecnalia.com

(2) WIP RENEWABLE ENERGIES
München, Germany
ingrid.weiss@wip-munich.de
simon.challet@wip-munich.de

(3) R2M SOLUTION
Pavia, Italy
juan.espeche@r2msolution.com
federico.noris@r2msolution.com

(4) BEAR-iD
Rotterdam, The Netherlands
tjerk@bear.nl

(5) ONYX SOLAR ENERGY
Avila, Spain
erico@onyxsolar.com
ihuerta@onyxsolar.com

(6) CEA INES
Le Bourget-du-Lac, France
ya-brigitte.assoa@cea.fr
françoise.burgun@cea.fr

(7) ACCIONA CONSTRUCCION
Madrid, Spain
josecarlos.esteban.matias@acciona.com
jorge.escribano.troncoso@acciona.com

(8) CADCAMATION
Onex, Switzerland
palamy@cadcamation.ch
vknguyen@cadcamation.ch

ABSTRACT: In the framework of increasingly demanding legislation related to energy performance in buildings, the integration of photovoltaic systems in the building skin offers many possibilities. However, a continuous development of the building-integrated photovoltaics (BIPV) market in the next years necessarily demands the compliance with strict requirements in terms of design flexibility, aesthetics, durability, cost reduction, grid integration, testing standards and operation & maintenance. Within this context, a European consortium of 15 partners is developing the H2020 PVSITES project, "BIPV technologies and systems for large-scale market deployment". The project will demonstrate an ambitious set of BIPV solutions, specially tailored to provide a comprehensive and robust answer to market demands. This paper focuses on some of the major results achieved at the time of writing. These include a detailed BIPV market and regulatory framework analysis and recommendations for the standardization of BIPV products. Novel developments of glazed crystalline silicon modules for BIPV integration are presented, with both opaque and see-thru variations. Also included are the selection of a battery technology best suited for inclusion in a custom-made building energy management system (BEMS), as well as a BIM-compatible software tool to assist the design of BIPV projects.
Keywords: Building Integrated PV (BIPV), Cost reduction, R&D and Demonstration Programmes

1 PURPOSE OF THE WORK

1.1 Context for European BIPV

BIPV is currently a steadily growing market. Market analysts estimate a compound annual growth rate (CAGR) of 18.7% and a total of 5.4 GW installed worldwide between 2013 and 2019 [1]. Currently the BIPV market holds a market share of around 2% of the overall PV market. However, the "BIPV Technologies and Markets: 2015-2022" report forecasts that BIPV penetration will rise to a 13% penetration by 2022 [2].

The EU currently leads the worldwide BIPV market, with 41% of total installations, followed by the USA with 27%. One of the main drivers for BIPV market growth in the EU is the increasingly demanding legislation related to energy performance in buildings.

Despite this favorable framework, it is a fact that estimations of BIPV market growth have been subsequently overestimated in the past few years. A series of demands from the stakeholders that have not been properly addressed by the BIPV value chain are probably the cause for this deviation. These key requirements are mainly related to:

- Flexibility in design
- Aesthetical considerations
- Lack of digital tools integrating PV and building performance at the earliest stage of a project to provide best bioclimatic design and financial options
- Lack of demonstration of long-term reliability of the technology
- Compliance with legal regulations
- Smart interaction with the grid
- Cost effectiveness

1.2 Project objectives

Within this context, the objective of the ongoing EU funded PVSITES project is to drive BIPV technology to a large market deployment by a combined virtual and real demonstration of an ambitious portfolio of building-integrated solar technologies and systems, giving a forceful, reliable answer to the market requirements identified by the industrial members of the consortium in their day-to-day activity. As a consequence, high impact demonstration and dissemination actions will be accomplished in terms of cost-effective renewable energy generation, aesthetics in solar architecture, reduction of energy demands and smart energy management.

The ambition of this project is, therefore, to pave the way towards a BIPV global market uptake led by the European industry and the emergence of a robust supply chain generating economic growth and job opportunities. An industrial consortium including partners with international reputation in the photovoltaics, building-integrated photovoltaics, construction and energy service sectors, supported by several RTOs and BIM experts for Energy Modeling, has joined efforts to reach this challenging objective. A wide range of BIPV solutions are being virtualized, developed and demonstrated in terms of design, building integration, performance, energy management, operation and maintenance, deconstruction, recycling and cost effectiveness.

This demonstration activity is carried out in 6 representative buildings across Europe plus experimental buildings and outdoor test benches, covering different climates, building uses, architectural implementations and energy management models.

1.3 Scientific innovation and relevance

PVSITES has a large innovation potential based on the generation and demonstration of flexible, robust, and cost-effective BIPV module and energy management technologies, together with associated software tools to support the design process.

- The BIPV solutions based on high efficiency crystalline silicon (c-Si) BIPV modules will substantially reinforce the aesthetics, flexibility of design and high performance considerations. Different strategies based on occultation of metallic contacts, high efficiency back-contact solar cells, glass treatments and coatings for improved passive properties and framing systems compatible with XL format glazing units. In addition to this, a low-concentration solar control BIPV system will be demonstrated, providing improved performance in terms of electricity generation and reduction of cooling demands, giving place to an unprecedented case in BIPV.
- The BIPV solutions based on high efficiency flexible lightweight CIGS thin film solar modules are opening a new paradigm in cost effective BIPV applications. Monolithically interconnected flexible lightweight CIGS solar modules produced with a roll-to-roll manufacturing process can provide low production costs of solar modules (< € 0.50/W for a 100 MW plant) and open novel opportunities for applications on roofs and facades in energy efficient buildings with lower carbon footprint.
- The combination of innovative power conversion and energy management technologies will increase BIPV electricity utilization (self-consumption). Novel BIPV generation and local electrical demand forecasting tools will reduce storage system costs by means of a practical selection and sizing of electrical storage system for each use case. The required storage capacity will also be reduced by means of active load management tools. Overall the achieved effect will be a cost reduction of the self-consumed electricity, coupled to an increase in the efficiency, flexibility and robustness of power conversion systems.
- The software tool for the coupled simulation of BIPV elements and building energy performance is, as of today, one of the only tools in the market offering this integrated approach. BIM and Energy Modeling will support and guarantee architectural and engineering workflows, as data will be added and exchanged through virtual workspaces, and even directly connected to the manufacturers' databases.

2 FIRST RESULTS

The PVSITES project started in January 2016 and will reach its conclusion in June 2019. This section summarizes a selection of results from completed or still ongoing tasks related to the project, as of September 2017. More detailed information can be consulted in a series of public reports made available on the project website [3].

2.1 BIPV market and regulatory framework analysis

Every analysis on the current BIPV market situation shows that there are a series of demands from the stakeholders that have not been properly addressed by the industry. The PVSITES project was conceived from its early stages to address these critical aspects with a joint action bringing together the different actors of the BIPV value chain.

A detailed market assessment has been one of the first accomplished targets of the project. The current BIPV markets, trends and regulatory framework have been analysed by project partners and mainly R2M Solutions and Acciona, together with the stakeholders' needs, from the global figures to a more detailed analysis of BIPV in Europe and a final focus on the 4 countries in which PVSITES demonstration activities are taking place (France, Switzerland, Belgium and Spain). The markets and stakeholders involved in BIPV-related industries have also been part of the study. To support information gathering a stakeholder survey was developed within the consortium and distributed to key market actors through different channels. It was for instance sent online to members of photovoltaic associations and of the IEA PVPS Task 15 on BIPV, and circulated at *ad hoc* events such as the EUPVSEC 2016.

The market analysis showed that the penetration of BIPV has been overestimated by most analysts in the past years, mainly due to an unforeseen price drop of regular PV which made it a more profitable business than BIPV. It is nevertheless expected that the global BIPV market will grow in the coming years. This growth will be more conservative in the period 2016-2018 and will likely increase faster in the following years. It must be noted that within the PV market, the BIPV technologies will probably remain a niche market with no more than 8% of the total PV shipments by 2020, reaching 5 GW by 2020 [4].

Europe and the USA are foreseen as the BIPV market leaders, with approximately 3.5B€ invested in each market every year. Regulations toward near zero energy buildings, better incentives schemes in some countries and citizen consciousness are the main drivers for the optimistic forecast for the BIPV economy in Europe [5].

Among the targeted countries, France and Switzerland are in a better position for a faster and higher penetration of BIPV. Specific feed-in-tariffs and tenders for BIPV along with a higher social awareness regarding environmental issues increase the opportunities for investments in BIPV, although a recent decrease in French feed-in tariffs has reduced financial profitability for future installations. On the other hand, Spain has faced a deep crisis in the solar renewable energy field which led to almost zero PV installations in the past years. A recovery is now to be expected, with the recent tendering of large PV installations to be completed in 2019-2020 and the adoption of new regulation favoring self-consumption of renewable energy in buildings, although this will depend on political and economic stability in the longer term. Belgium has known a similar but less drastic situation than Spain with a significant decrease in annual installed capacity for three consecutive years in 2012-2014. The yearly installation volume has slowly recovered since then despite a lack of specific incentives for BIPV, and future trends remain uncertain.

At the European level a planned update of the Energy Performance of Buildings Directive (EPBD), expected be voted in October 2017, will set more stringent requirements for member states to accelerate the transition towards nearly-zero-energy buildings (NZEBs), with a specific focus on renovating the existing building stock. Once implemented this new legislative framework is expected to be favorable to BIPV installations.

The worldwide BIPV market can be divided into three sub-markets: glass, non-glass roofing and non-glass walling. The non-glass roofing market is currently the biggest one and it will continue to lead the market in the future, reaching 10B€ per year in 2021. Therefore, this market is and should remain the most attractive, almost doubling the investments in the other two by 2020.

The stakeholder analysis performed within the PVSITES project further shows the need for better integration between BIPV actors to reduce complexity and costs in the value chain. Contractors, architects and BIPV manufacturers are the main stakeholders and a close collaboration among them will foster their synergies, bringing benefits to the rest of the actors involved in the BIPV market. The main obstacles to the uptake of BIPV in Europe identified by market actors are related to the high costs of BIPV solutions, combined with a perceived uncertainty on the energy performance and lifetime of the products, as shown in Figure 1. Design flexibility, product aesthetics and compliance with local regulations are also major requirements.

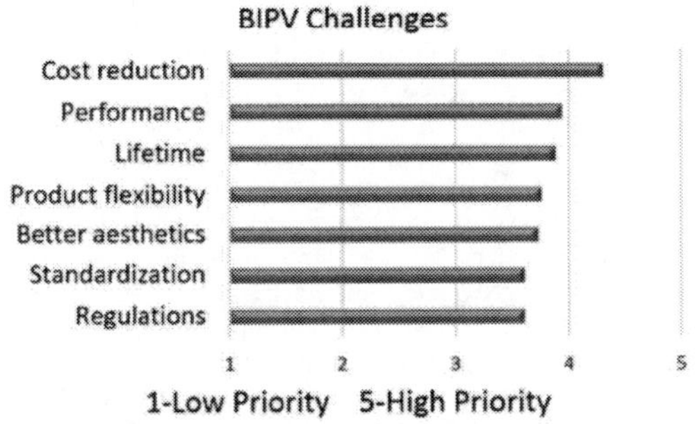

Figure 1: Ranking of most important challenges to the development of BIPV in Europe as perceived by the 35 respondents to a PVSITES questionnaire [4].

2.2 Standardization of BIPV elements

In order to be suitable for building integration, BIPV products have to demonstrate compliance with the regulations from the PV and construction sectors. BIPV modules must therefore comply with the electro-technical requirements as set out in the Low Voltage Directive 2006/95/EC [6] and, as construction products, the requirements from the Construction Products Regulation CPR 305/2011 [7] have to be fulfilled. The still ongoing development of a specific and complete standardization context for BIPV elements, however, is one of the barriers affecting the market uptake of the technology.

The recently developed standard EN 50583 addresses the standardization needs for BIPV modules and systems used as construction products. It establishes three hierarchic levels of requirements for BIPV modules. The top level includes the relevant requirements for all kinds of PV modules, regardless of their technology and composition and the requirements for all technologies but depending on whether the modules do contain glass or not. The second level provides the requirements for glass and non-glass based modules, regardless of their location in the building but as a function of their backsheet material for the non-glazed ones. The third level applies only to glazed modules, and establishes differentiated requirements as a function of their location within the building.

BIPV products must also comply with the specific standards and regulations in the country of use. Depending on the considered country, insurance applications, town planning rules (zoning regulation regarding colors, building shape...) and national building codes (fire, acoustic, thermal...) must also be met.

Taking all these aspects into account, the PVSITES consortium conducted an in-depth analysis of the standardization needs applicable to the project developments [8]. The results have shown that the EN 50583 standard provides a good reference though not all relevant requirements are covered and, in some cases, adaptations of tests or proposals of complementary standard tests are necessary. As an example, for low concentration BIPV products and curved glass/glass products in the PVSITES project, specific recommendations were proposed in order to adapt standard tests conditions to their configuration and properties. To ensure performance and long term reliability, standard test sequences were defined for each BIPV product and inverter developed within the project.

2.3 Crystalline silicon BIPV products

A large portfolio of products based on crystalline silicon (c-Si) and flexible thin film CIGS technologies is undergoing development within the PVSITES project. This paper will focus on the presentation of c-Si products as the CIGS products are expected to be fully developed at later stages.

One of the main constrains for BIPV market acceptability constrains is that most c-Si photovoltaic solutions show visible cell busbars and string L-interconnections. These metallic strips are necessary to harvest the electrical production occurring in the individual solar cells (busbars) and to connect the cells to each other within the modules (L-interconnections), but alter the otherwise homogenous aspect of the products. PVSITES partner Onyx Solar has therefore followed different strategies for the development of two generations of opaque c-Si glass-glass modules with hidden bus bars and L-interconnections [9]. Two product generations have been developed, as shown in Figure 2:

- 1st generation: Use of black plastic sheets compatible with glass lamination to hide the L- interconnections and use of black frit patterned rear glazing.
- 2nd generation: Use of black plastic sheets to hide the L-interconnections, implementation of black conductive ribbons and use of black frit patterned rear glazing to obtain a fully opaque solution.

Figure 2: Glazed c-Si modules with hidden bus bars and L interconnections (1st and 2nd generations shown left and right, respectively) [9].

For specific building solutions such as skylights and curtain walls, the decision maker is looking for certain light transmission values. For these cases, a compromise between cell density providing the see-thru degree, efficiency and aesthetical added value should be found. Attending to these needs, Onyx Solar has implemented back-contact solar cells within the photovoltaic glazing design [10].

Onyx Solar has also analysed potential glazing treatments compatible with BIPV technology, such as self-cleaning, mass coloured glass, frit patterned glass, reflective coatings and low-emissivity (low-E) coatings, in order to understand their effects on the BIPV glazing and the improvement generated on the passive properties of c-Si BIPV units [11]. A set of prototypes was manufactured based on this work, as shown in Figure 3.

The cost competitiveness of these newly-developed glazed c-Si modules has been analysed by the manufacturer [9,10]. Onyx Solar estimates end-user final prices in the range of 275-315 €/m² once commercial production starts. These price levels enable payback times of complete BIPV systems on the order of 5-12 years, depending on local contexts and project specifics.

Figure 3: Samples of c-Si based products with colour reflective (top), mass coloured glass and low-E treatments (bottom) for improved passive properties [11].

Framing systems suitable for the integration of extra-large and large thickness BIPV units in curtain walls, vertical facades and skylights have furthermore been defined and prototyped [12]. All above mentioned-products will be subjected to indoor validation testing to ensure their safety and reliability. A few of them will then be installed at the project's demonstration sites (see Section 2.5).

2.4 New inverter and BEMS technology

At system level, a combination of flexible and high efficiency grid interface and new building energy management strategies is under preparation. This work includes the demonstration of:

- A new planner tool to optimize the use of selected batteries for each use case, reducing the cost per unit of energy served (€/kWh that passes through the batteries).
- New power conversion designs (low-cost and high efficiency PV inverter with DC coupled storage system and a low-cost and robust PV inverter based on Silicon Carbide - SiC - technology), to enhance the conversion efficiency of the whole system.
- A novel low cost and reliable 24-hours-ahead forecasting tool, to develop innovative energy management strategies increasing overall economic performance.
- The integration of active load management tools, to reduce required storage capacity.

Work on the building energy management systems (BEMS) started with an analysis of commercially available electricity storage technologies and the recommendation of the most suitable one for the project's objectives. This documented comparison involved a series of tests by project partner CEA-INES [13]. The tested batteries were based on the aqueous sodium-ion (SI), lithium-iron-phosphate / graphite (LFP/G) and lithium-iron-phosphate / lithium-titanate (LFP/LTO) technologies. The investigated battery parameters included energy density, charge and discharge rates at different temperatures, cycle life testing under different charge/discharge conditions and thermal stability (for lithium-ion batteries). Lithium-iron-phosphate batteries were identified as the most suitable for electricity storage in BIPV applications due to superior discharge rates. Among LFP batteries, the testing campaign recommended the choice of LFP/G over LFP/LTO due to lower costs (370 €/kWh for LFP/G compared to 2055 €/kWh for LFP/LTO).

2.5 Demonstration activities

The developed BIPV products, inverters and BEMS will be implemented and tested over six demonstration sites and two test sites, representing a variety of building types and located in different European locations and climates. Six different building types are covered with the objective of testing the PVSITES products under as much different architectural integrations as possible and with different electrical requirements and load balance situations. This will considerably increase the use cases and strategies in which the products will be validated. Monitoring systems have been installed at each demonstration site to precisely track performance before and after the implementation of the PVSITES products. This includes both the energy production or storage and the improvements brought as a constructive element. The International Performance Measurement and Verification Protocol (IPVMP) methodology is being used to ensure the quality of the measurement and validation of the results. In the next page Table I summarizes preliminary information about the expected demo sites, as the exact implementation plans are still under definition. Implementation at the demonstration sites will be paired with training courses for installers, while guided visits will be offered after the commissioning phase.

Table I: Preliminary overview of the PVSITES demonstration sites and respective BIPV installations.

	Single House	Educational building	Carport

(1)	FormatD2	Flisom	Flisom
(2)	Stambruges Belgium	Geneva Switzerland	Zürich Switzerland
(3)	Roofing shingles (CIGS on steel) 	Large tiles on façade (CIGS on metal substrate)	Roof tiles (CIGS on metal sheets)
(4)	Flisom	Flisom	Flisom
(5)	SSW	E + W	Horizontal
(6)	75 m²	136 m²	180 m²
(7)	8 kWp	4 + 8 kWp	16 kWp
	Industrial building	**Apartment building**	**Office building**

(1)	Cricursa	Vilogia	Tecnalia
(2)	Barcelona Spain	Wattignies France	San Sebastián Spain
(3)	Large roofing shingles (CIGS on metal substrate) 	Ventilated façade (c-Si modules with hidden bus bars)	Ventilated façade (Glass-glass back contact c-Si cells)
(4)	Flisom	Onyx Solar	Onyx Solar
(5)	Horizontal	SSE	SSE
(6)	225 m²	130 m²	170 m²
(7)	19 kWp	20 kWp	10 + 10 kWp

(1) PVSITES partner responsible for the demonstration site
(2) Location
(3) BIPV product demonstrated
(4) Manufacturer
(5) Orientation of BIPV modules
(6) Surface area of BIPV modules
(7) Installed PV power

2.6 Integrated simulation software tools

Finally, aimed at the earliest design phases of BIPV projects, a BIM-compatible and bioclimatic simulation software for the joint prediction of photovoltaic and building energy performance is being finalized by project member CADCAMation, as illustrated in Figure 4. BIM technology makes possible the early virtualization of all BIPV products created during the project into the architectural design of buildings. 3D geometry properties coupled with optical and thermal modeling will feed-in the engineering workflow with contextual data, optimizing the technical integration into the building skin and the coupling with thermal comfort simulations.

The work-in-progress software will answer a series of well-defined customer needs and strengthens decision processes, by providing the following features:

- A specific design tool for PVSITES products integration into the building skin, as a specialized plugin into BIM ready BIPV software.
- Libraries of standardized BIPV products and related PVSITES solutions, turned into BIM components.
- Compliance of BIPV products with legal requirements demonstrated from the earliest feasibility study through "BIM constraints check" services.
- Bridging the gap between architects, engineers and PV manufacturers as BIM is first and foremost a collaborative methodology generating collaborative workspaces.

A pre-commercial version of the user-friendly software was released in September 2017 and can be accessed on the project website [3].

Figure 4: 3D visualisation of yearly solar irradiance on a building façade with the PVSITES software tool [3].

3 CONCLUSIONS

Between January 2016 and September 2017, the activities of the PVSITES consortium have contributed towards a greater market uptake of BIPV technologies by fulfilling some of the most relevant demands from stakeholders. The first project results presented in this paper will be refined in the months ahead and together with the rest of the work in progress, will culminate in the demonstration of an ambitious portfolio of BIPV systems in real buildings by the summer 2019. The expected impacts are classified in several categories:

- Market impacts, through a relevant increase in total BIPV installed capacity in the period 2020-2024 and with a payback time of 5-7 years at maximum.
- Increased competitiveness of EU industry with a very relevant associated job creation.
- Environmental impacts from additional installed power due to PVSITES, with a reduction of overall CO2 emissions 12% higher than in the business-as-usual case.
- Early integration of BIPV elements in building projects through BIM-based software modeling.
- Smart management of the building energy demand and of the electricity injection into the grid, thus reducing potential grid stress and the need to reinforce the grid.
- Increased stakeholder awareness and acceptance of BIPV as a consequence of high impact demonstration and targeted dissemination actions.

Regular project news can be consulted on the project website [3].

4 REFERENCES

[1] Transparency Market Research, "Building Integrated Photovoltaics (BIPV) Market: Global Industry Analysis, Size, Share, Growth, Trends and Forecast, 2013 – 2019", 2014.

[2] n-tech Research, "BIPV technologies and markets: 2015-2022," 2015.

[3] PVSITES project website, www.pvsites.eu
[4] PVSITES Report, "BIPV market and stakeholder analysis and needs", 2016.
[5] PVSITES Report, "European regulatory framework for BIPV", 2016.
[6] Official Journal of the European Union, "Directive 2014/35/EU of the European Parliament and of the Council", 2014.
[7] Official Journal of the European Union, "Regulation (EU) No 305/2011 of the European Parliament and of the Council", 2011.
[8] PVSITES Report, "Standardization needs", 2016.
[9] PVSITES Report, "Opaque photovoltaic glazing solutions based on crystalline silicon with hidden busbars and L interconnections", 2017.
[10] PVSITES Report, "See-through photovoltaic glazing solutions based on back contact solar cells", 2017.
[11] PVSITES Report, "Samples of crystalline silicon based photovoltaic glazing products for indoor validation tests", 2017.
[12] PVSITES Report, "Prototypes of framing systems compatible with XL format and large thickness photovoltaic glazing units", 2017.
[13] PVSITES Report, "Recommendations on storage systems for BIPV systems", 2017.

6 ACKNOWLEDGMENTS

This project has received funding from the European Union's Horizon 2020 research and innovation programme under grant agreement No 691768. Neither the European Commission nor any person acting on behalf of the Commission is responsible for the use which might be made of the following information. The views expressed in this publication are the sole responsibility of the authors and do not necessarily reflect the views of the European Commission.

The PVSITES project is a collaboration between the following 15 European organisations:

- Tecnalia, Spain – Project coordinator
- Acciona Construcción, Spain
- BEAR-iD, The Netherlands
- CADCAMation, Switzerland
- CEA INES, France
- Cricursa, Spain
- CTCV, Portugal
- Film Optics, United Kingdom
- Flisom, Switzerland
- Format D2, Belgium
- Nobatek, France
- Onyx Solar, Spain
- R2M Solution, Italy
- Vilogia, France
- WIP Renewable Energies, Germany

FP7-CHEETAH KNOWLEDGE EXCHANGE PLATFORM: RESULTS AND THEIR EXPLOITATION

Francesco Roca[1]*, David Casaburi[1], Francesco Beone[1], Claudia Diletto[1], Ilaria Falcone[1], Anna De Girolamo[1], Riccardo Miscioscia[1] Karsten Bittkau[8], Iver Lauerman[9], Máté Heisz[27], Suren A. Gevorgyan[5], Ivan Gordon[6], Alexandre Roesch[27], Carlos del Cañizo[22], Martina Schmid[9], Adrien Danel[4], Paul Sommeling[2], Kris Van Nieuwenhuysen[6], Jan Kroon[2]

Simona Binetti[24], Torsten Boeck[15], Francesca Brunetti[12], Jake Bowers[17], Stephan Buecheler[13], Julio Cárabe[10], Aldo Di Carlo[11], Maarja Grossberg[21], Georgios Halambalakis[12], Jukka Hast[25], Antonio Joyce[16], Rannveig Kvande[19], Erwin Lotter[26], Franziska Ringleb[15], Eduardo Román[20], Rasit Turan[18], Juan F. Trigo[10], Guillermo Sánchez-Plaza[23], Nicolas Wyrsch[14], Sjoerd Veenstra[2], Shokufeh Zamini[3]

[1]ENEA- Agenzia Nazionale per le Nuove Tecnologie; [2]ECN-Stichting Energieonderzoek Centrum Netherland; [3]AIT Austrian Institute of Technology; [4]CEA-INES Commissariat à l'Energie Atomique Aux Energies Alternatives France; [5]DTU-Danmarks Tekniske Universitet Danmark; [6]IMEC-Interuniversity Microelectronics Centre Belgium ; [7]FH-ISE Fraunhofer-Instituts für Solare Energiesysteme Germany; [8]IEK5–Forschungszentrum Jülich Germany; [9]HZB-Helmholtz-Zentrum Berlin für Materialien und Energie GmbH Germany; [10]CIEMAT-Centro de Investigaciones Energeticas Medioambientales y Tecnologicas; [11]CHOSE - University of Rome "Tor Vergata. Italy; [12]CRES-Center for Renewable Energy Resources and Saving ; [13]EMPA Swiss Federal Laboratories for Materials Science and Technology [14]IMT-Institute of Microengineering- Ecole Polytechnique de Lausanne Switzerland; [15]IKZ-Leibniz-Institut für Kristallzüchtung Germany; [16]LNEG-Laboratório Nacional de Energia e Geologia Lisboa Portugal ; [17]Loughborough University- School of Electronic Electrical and Systems Engineering Electrical and Systems Engineering, United Kingdom; [18]METU-Middle East Technical University Turkey; [19]SINTEF Materials and Chemistry, Norway; [20]Tecnalia, Spain; [21]TUT-Tallinn University of Technology, Estonia; [22]UPM-IES Instituto de Energía Solar Universidad Politécnica de Madrid, Spain; [23]UPVLC-Universitat Politècnica de València, Spain; [24]UMIB-University of Milano- Bicocca, italy; [25]VTT Technical Research Centre of Finland Ltd; [26]ZSW-Zentrum für Sonnenenergie- und WasserstoffForschung Baden-Württemberg Germany ; [27]SolarPower Europe Belgium;

* corresponding author franco.roca@enea.it;+38-081-7723270

ABSTRACT: FP7-CHEETAH is a combination of a collaborative project (CP) and a coordination and support action (CSA), receiving funding from the European Union under grant agreement No 609788. The project aims at solving specific R&D issues to overcome fragmentation of European PV R&D and to accelerate the industrialization of innovations by intensifying the collaboration between R&D providers and industry. This proceeding reports on the strategy and key tools brought by the CHEETAH project to improve the state of the art in knowledge exchange on PV RTD. Its CHEETAH Knowledge Exchange Platform (KEP) features a dynamic database, a powerful build-in search engine, the dedicated e-learning platform for on-line meetings with internal and external stakeholders, webinars, on-line tests and experiments. The portal profits from the best practice in more efficient social and professional networks web portal by representing a significant step forward in knowledge exchange for the European photovoltaics community to support training, share knowledge and research infrastructures and foster collaboration opportunities at EU scale.

Keywords: e-Learning, Education and Training, Dissemination, Qualification and Testing, Stability, PV Technologies, Social and professional Networks

1 INTRODUCTION

FP7-CHEETAH "*Cost-reduction through material optimization and higher Energy output of Solar Photovoltaic Modules - joining Europe's research and development efforts in support of its PV Industry*" coordinated by ECN, NL [1,2,3] is a combined collaborative project (CP) and coordination and support action (CSA) funded under the European Commission's 7th Framework program with the aims to solve specific R&D issues and to overcome fragmentation of European PV R&D by intensifying the collaboration between R&D providers and industry to accelerate the industrialization of innovations. The project description and the Pert diagram are shown in Fig. 1. This project is also tightly linked to the EERA-PV Joint Program and for this reasons its outputs will also be continued under the EERA-PV Joint Program at the end of project in December 2017 [7].

The CHEETAH Knowledge Exchange Area Portal (KEP), in parallel to the project web site and other dissemination activities (newsletter, communication, etc.), constitutes the pillar of the project to bring information from different sources in a uniform and simple way to any interested CHEETAH partner and external organization on availability of infrastructures, equipment, expertise, technical documents [4,5,6] (fig. 2).

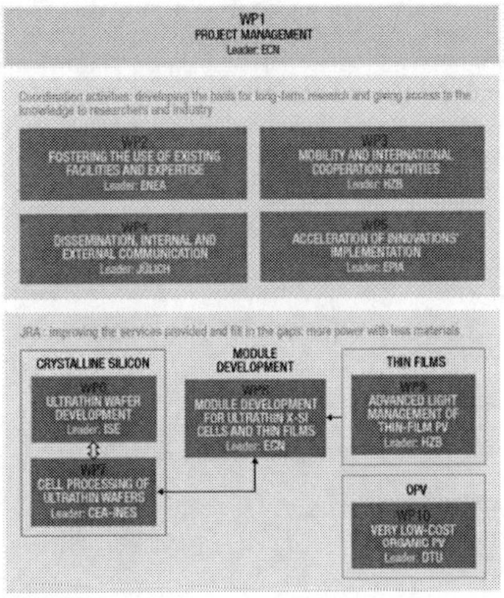

Figure 1: Description of CHEETAH Project and Pert diagram of the workpackage structure

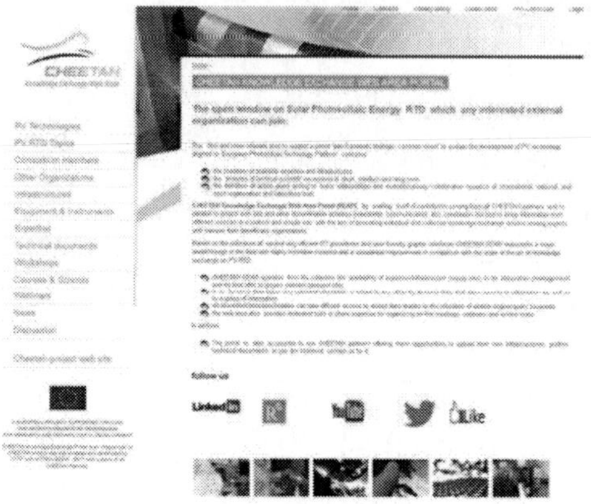

Figure 2: CHEETAH Knowledge Exchange Platform (KEP) web site https://www.cheetah-exchange.eu

It operates from collection in its database of expertise/infrastructure (supply site), to its elaboration (management) and its final offer to project partners (demand site) by utilizing user-friendly dedicated tools having a very effective potential to share expertise (see fig. 3)

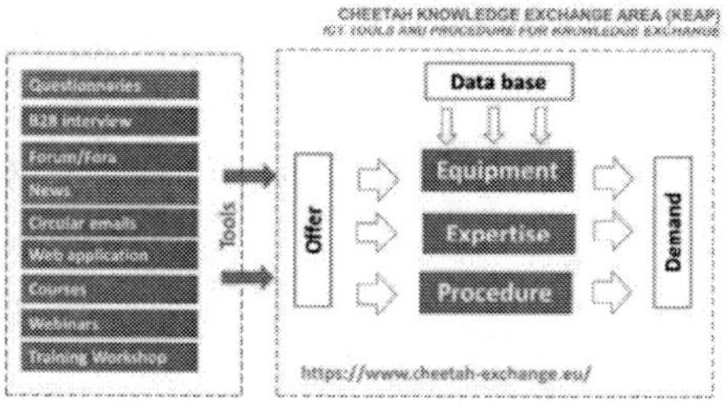

Figure 3: CHEETAH Knowledge Exchange Platform (KEP) ICT Tools and its rationale.

The web portal operates by an Operational database management systems (also referred to as OLTP On Line Transaction Processing databases). Any data is updated in real-time and KEP allows us to do more than simply view archived data: any uploaded information is dynamically linked to all others by logic connections that allow access to any individual information as well as to information already uploaded by improving the informative content (available equipment? expertise? where are they located? who could I contact? Etc.).

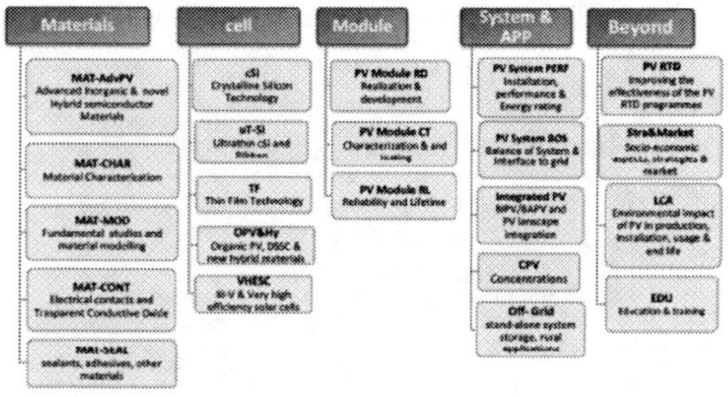

Figure 4: CHEETAH Knowledge Exchange Platform (KEP) cataloguing criteria

It is also based on the utilization of structured cataloguing criteria applied to PV technologies/PV RTD topics and organizations involved in CHEETAH (see fig. 4).

From this point of view, the procedure to collect and offer information to CHEETAH partners and other project consortia represents a major breakthrough in the field with highly innovative content and a substantial improvement in comparison with the state of the art of knowledge exchange on PV RTD and it makes CHEETAH KEP unique in its approach.

CHEETAH KEP offers the following opportunities:

- All interested browsers/readers can have efficient access to stored data thanks to the utilization of search engine/query keywords used in connection with the more diffuse and efficient ICT procedures and user friendly graphic interfaces.
- Offer/demand and needs of infrastructures, equipment, expertise, technical documents, test procedures
- Open access on-line lectures, courses , workshop by widely using webinars based on the implementation of the fruitful results and experiences of SOPHi@Webinar, the internal e-learning platform of FP7-SOPHIA project
- Public access in reserved web-area (documents, data, reports, etc.)
- Restricted area for specific technical documents (for user-partners only)
- On-line forum/fora for internal, external and technical/scientific discussions on specific themes
- On-line questionnaires & ICT tools to optimize submissions and collection of specific information
- For the first time the platform is also moving to propose live experiments followed remotely by other scientists.

CHEETAH KEP is highly secure and user-friendly. It is the portal for the European photovoltaics community to support training, share knowledge and research infrastructures and foster collaboration opportunities at EU scale.

2 CHEETAH KEP Rationale and organization

The strategic goal of CHEETAH is on the one hand, to identify currently running technical-scientific needs of the European PV RTD sector and, on the other hand, to establish the best way of taking profit from the strength owned by each CHEETAH partner in terms of facilities and expertise, in order to improve collaboration inside the European PV Research and Industrial community. This goal is also a practical translation of an objective of EERA.

CHEETAH KEP defines the facilities and expertise needed to achieve those goals based on a multistep process

2.1 Collection of information

Information on partners knowledge needs is realized by circulating questionnaires, business-to-business interviews, discussion among partners by fora/circular emails, and news. It is further elaborated in connection with other parallel initiatives launched by the European and international photovoltaic community such as PV road maps [3,4,5,6]

2.2 Knowledge necessities: demand side
The CHEETAH knowledge exchange necessities inventory is based on assuming that for each of the indicated topics (PV Technologies, PV RTD topics, Equipment) the interest shown by each respondent is strongly influenced by his/her role, typology and level of expertise, and clearly involvement in the project. This imposes that the category to which the respondent corresponds needs to be identified at least for statistic motivations to know better and more efficiently about the knowledge needs.

The following categories of respondent are defined

i. *Organization Contact points*: they are the respondents for their own organization and they assign the relevance showed by each organization in the specific topic.
ii. *WP/task leaders*: they offer evaluation specific for their own field of expertise within the CHEETAH project.
iii. *Group leader*: they offer an evaluation of relevance of their own laboratory/group of activities, and the elaboration will be based on the coherence of the group they belong to. Just as example it could be the relevance assigned to a specific equipment by a group of laboratories having the same objective.
iv. *Student/younger researcher:* they express the interest of researchers, on specific topic manly at an initial stage of their career.

The elaboration of received questionnaires offered different points of view on the relevance assigned to each topic depending on the role of respondents and it will be also a very useful tool to be shared towards a global European strategic vision of different categories.

Furthermore, the respondents express actually the intensity of his/her personal interest or on behalf of his/her organization on the specific PV RTD technology/topic/ and/or on the equipment utilization by indicating a score raging from no-interest (=0) , moderately involved (=1), normally involved (=2), widely involved (=3) ; to strongly involved (=4).

The collected information takes also in consideration the evolution in time of the specific interest of each researcher/scientist. In fact, the respondents assign also a score expressing interest on today, but also in perspective for the short, medium and long term. This offers opportunities to also draw some road map on the evolution of the interest in each specific PV RTD technology/topic.

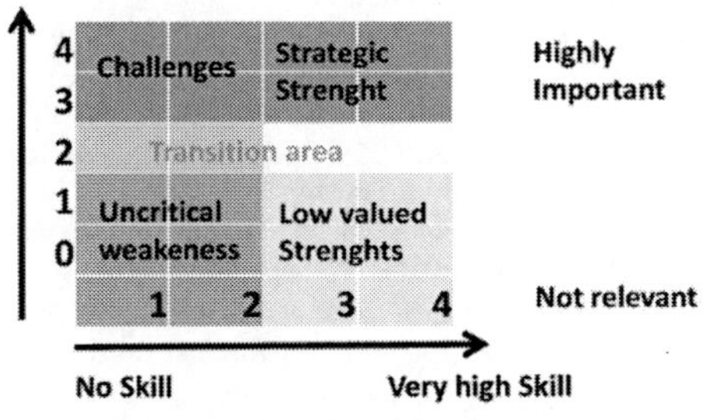

Figure 5 Skill assigned by each interviewee to any PV RTD technology/PV RTD Topic/PV RTD Equipment Level of importance assigned by each Interviewee to any PV RTD technology/PV RTD Topic/PV RTD Equipment

Four areas of different needs are determined in this way;

Uncritical weakness:

an area where for each specific PV technology / PV RTD topic and equipment a both not prepared / not-relevant condition is verified for the specific organization;

Low-valued strength:

an area where for each specific PV technology / PV RTD topic and equipment a well prepared / not-relevant condition is verified for the specific organization;

Challenge area:

where needs should be satisfied with targeted actions (both not-prepared / highly important verified)

Strategic Strength:

an area where for each specific PV technology / PV RTD topic and equipment both well prepared / and relevant conditions are verified for the specific respondent organization

The opportunity to extend the questionnaires outside of the CHEETAH consortium, mainly to improve the population for statistic evaluation and to serve the PV RTD community with useful tools for evaluation of knowledge exchange needs has also been evaluated.

2.2 Knowledge necessities: offer side
Based on the FP7-SOPHIA experience, this step focuses on the available inventory of infrastructures and expertise and technical/scientific documentation (papers, reports, strategic position documents, etc.). It also concerns the development of tools and procedures to optimize the exchange process in order to adequately sustain demand/supply of expertise and infrastructures.

The large number of outstanding research organizations working together via the CHEETAH project can avoid the useless replication of a large number of small efforts by sharing information on the availability of knowledge. The CHEETAH Knowledge Exchange Portal (KEP) aims at widely distributing information among partners on available expertise, infrastructures, equipment and technical documents, to provide the scientific community with a common data base/catalogue. A list of useful information is associated to each CHEETAH organization (expertise, infrastructures, technical documents, courses, webinars, discussion, news, etc.) and currently uploaded on the CHEETAH KEP web site.

2.3 Data elaboration
The detailed and statistic elaboration of information on the main interest in PV technologies, PV RTD topics and equipment of CHEETAH partners is fundamental for several reasons:

I. In this way the expertise and the experimental facility needs can be evaluated to be shared with the full services available
II. It can be offered at the European level and the indication of different facilities/equipment could be used for different technological topics
III. Experts, students stages and courses/webinar can focus on the field of main interest to optimize depending on constraints on financial budgets and time availability of the experts and participants

2.2 Data Management

The core of data management is represented by the CHEETAK KEP web site. It uses the same graphic frame as the main project web site, but it is based on a different ICT organization. In fact the content is periodically revised and implemented on a daily/weekly/monthly basis to adapt the web portal to new necessities of information on PV RTD fields/topic cataloguing/search criteria.

The updating process isn't just an updating of the information content but it deeply involves the development of new procedures and new and advanced ICT tools. For that reason CHEETAH KEP cannot be managed by utilizing the main project web site, but it needs to exploit a dynamic ICT environment that is totally different from any conventional htlm web site approach for the following motivations:

- All key participants are directly involved in the process and in real time, as it usually happens in exchange web sites like Linked-in, Research-gate, Google-Scholar, etc.
- It offers a more user-friendly shorthand approach . The technical/scientific content among WP/task leaders, webmaster and end-users and the scientists/ researchers/experts/students interested in accessing the information are defined by focusing mainly on technical aspects and their needs, more than formal procedures to approve amendments in web site content, as it is typically requested by any "official" project Web Site,

The utilization of a different area also reduces any technical ICT difficulty which could be caused to the main project web site by the very frequent revision/maintenance/implementation requested by very frequent periodical updating.

3 CHEETAH KEP WEB SITE DESCRIPTION

In the following paragraph we summarize information on the main facilities offered by the CHEETAH Exchange Area web portal by leaving to the browser/reader the opportunities to directly access them by browsing on the web site

3.1 CHEETAH KEP PV Technologies

Photovoltaic (PV) technology steadily progresses thanks to both research and development at the laboratory level and the technological deployment and experience from the market place. Research and Development – "R&D" – is crucial to sustain in mid and long-terms the further development of PV technology. Performing joint research addressing well-chosen issues plays an important role in achieving the critical mass and effectiveness required to meet the sector's ambitions for definitive technology implementation and industry competitiveness,

Following indications of the "Strategic Research Agenda for Photovoltaic Solar Energy Technology" EU PV Technology Platform and indications of CHEETAH experts, a range of technologies and topics has been determined the development of which is expected to significantly contribute to reach the overall targets and can be the objective of knowledge exchange:

- cSi Wafer Based Technology
- Si Ultrathin Wafer development & ribbon
- TFSi-Thin Film Silicon
- CIGSS- Copper indium gallium diselenide
- CdTe-Cadmium telluride
- Emerging/Novel Inorganic, Hybrid PV Materials & Nanotechologies (Perovskite, Kesterite, quantum dot/quantum wire SC, Intermediate band SC, etc)
- Organic Photovoltaics
- DSSC- Dye Sensitized Solar Cells
- Material and Device Characterisation
- Materials, devices, system modelling
- PV Module realization & Development
- PV Module qualification & testing
- PV Components, Systems & Interface to grid
- CPV Concentration Photovoltaics
- BIPV Building Integration Photovoltatics
- Education & training
- Socio-economic aspects & market
- Environmental impact, waste reduction and recycling
- PV RTD Networks, coordination of research efforts, strategy and PV RTD projects management

The CHEETAH Knowledge Exchange portal uses them as the base to catalogue the offer/demand among consortium partners of requested expertise, infrastructures, equipment, courses, technical documents, by promoting further development of knowledge in the same field of interest also outside of the consortium.

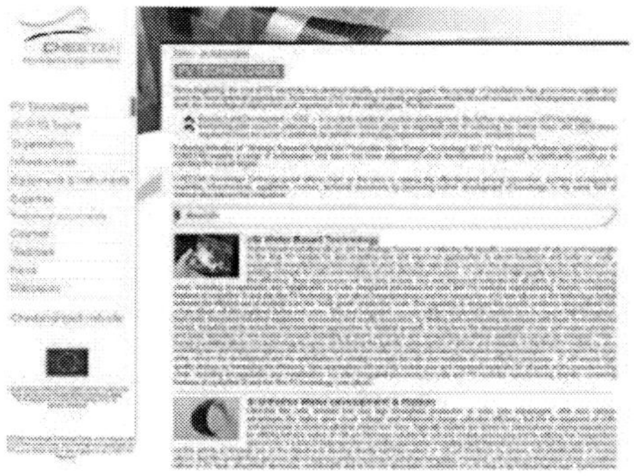

Figure 6 - PV RTD TECHNOLOGIES web page http://www.cheetah-exchange.eu/pv_technologies.asp

3.2 PV RTD TOPICS

Based on the CHEETAH catalogue of main PV RTD Technologies, a list of relevant R&D topics is proposed for the main existing technologies. The main R&D topics per technology area are summarized in the CHEETAH web site and consist of around 120 topics ranging from basic science to applications.

It is a tentative list that is dynamically updated following the indication of CHEETAH experts, and that we would deeply optimize in the final release of CHEETAH KEP. The PV RTD Topic can be sorted by utilizing different criteria. To each specific PV RTD topic is associated, when applicable, the list of available CHEETAH experts, infrastructures, equipment, courses, technical documents with the aim of intensifying the collaboration and the knowledge sharing among CHEETAH R&D providers.

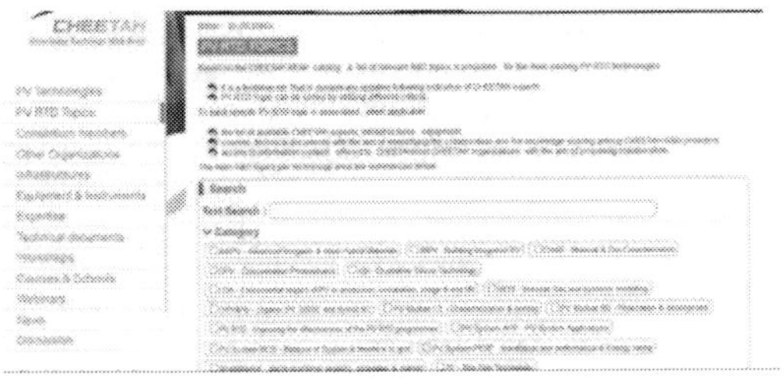

Figure 7: PV RTD TOPICS web page http://www.cheetah-exchange.eu/tags.asp

3.3 CHEETAH KEP: Expertise and infrastructures

The main goal of dissemination of the information on existing infrastructures of the CHEETAH partners is to make available some of the existing top-class PV Research Infrastructures and scientists for the benefit of the whole European photovoltaic community.

This is proposed on the basis of the establishment of a reciprocal collaboration framework within funded European Community collaborative projects or in perspective of some projects based on agreements between parties by business-to-business approach or bilateral/multilateral public/private funded project.

The extended database with the description of CHEETAH experts is available to comment/interact on a wide range of PV RTD topics.

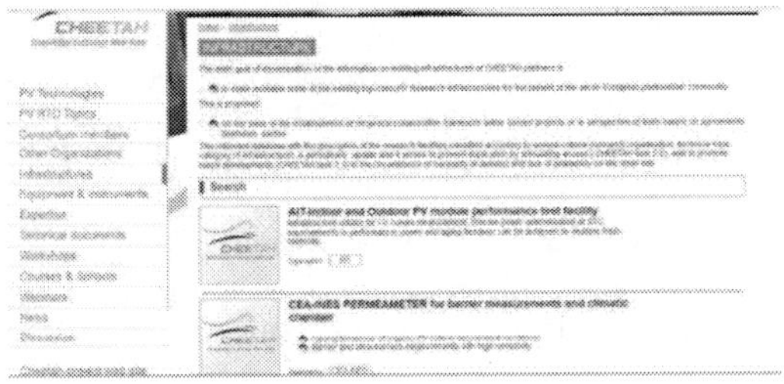

Figure 8 - CHEETAH Infrastructures web pages http://www.cheetah-exchange.eu/infrastructure.asp

A list is offered in alphabetical order and searching criteria can be utilized by typing names or keywords, such as PV Technology/ PV RTD/Topic/ Involved Organization/,

The research facilities are classified according to several criteria (research organisation, technical topic, category of infrastructure) and they are periodically updated. It serves to prevent duplication by stimulating access and to promote future developments in the circumstance of necessity of demand and lack of availability on the other side.

215 experts and 45 infrastructure profiles were uploaded as Sept. 2017 and they are growing day-by-day

4 CHEETAH WEBINARS

Based on the success and the fruitful results and experiences of the SOPHi@Webinar, the internal e-learning platform of FP7-SOPHIA, CHEETAH KEP is also proposing a set of interactive training courses and webinars planned in collaboration with all Joint Research Activity leaders, mobility and international cooperation activities and dissemination, internal-external communication responsibles.

Electronic conferencing is very powerful for knowledge exchange. It requires no more than a set of interconnected computers with suitable software and hardware. Participants can connect with conferences and workshops at their convenience, regardless of their time zone and geographical location [4,5]-

Web conferencing is time saving, it reduces travel costs, video conferences enhance the possibilities for all to interactively review a subject, allowing them to share efficiently ideas, documents, conclusions and concerns.

There are no limitations on event location and it can be realized by laptops, tablets and mobile phones and can be followed from everywhere.

The involvement of participants can focus on the topic issues of direct interest, saving time from topics in which they are not involved or interested. Storage and streaming of produced output is very useful also to review the informative content. In fact the acquired material (i.e. slides and video presentations) can be easily distributed by webcasting them even after the event

The targeted audience ranges from 5-10 to a maximum of 20 participants, for short on-line courses, in order to enable a large number of them to effectively interact with experts, up 100-200 participants for lectures.

Very soon the platform will also offer on-line live experiments: scientists/researchers operating in a test site or by remote equipment can interact live with other researchers interested in the characterization/ measurements.

CHEETAH is proposing different categories of webinars:

- Lectures and short on-line courses focused on best practice in the utilization of infrastructures, equipment, realization of experiments and test procedures (WP2);
- Lectures on specific technical-scientific aspects and dissemination of CHEETAH results/output (WP3 and WP4);
- Webinars concerning strategy in PV RTD, market and interaction with industries (WP4 and WP5)

The full list of material is available @ https://www.cheetah-exchange.eu/webinars.asp

- 46 webinars have been organized since January 2014 by the platform
- With more than, 930 "different" users
- They registered to participate 1450 times, and several hundred are accessing via streaming.

A modular series of web-courses targeted to researchers and PhD students will be very soon proposed.

This new MOOC-Massive Open Online Courses "learning PV at CHEETAH speed" lead by UPM will be organized as courses. Each one will ask for a student dedication of 5-20 hours and will provide tools for the students to address specific PV-related topics that may help in their work.

Trainers will try to go one step beyond "Fundamentals of PV" and tackle specific PV-related topics and tools that may be of help in their research work: for example,:

- introduction to solar cell simulation,
- extracting information from I-V and QE measurements,
- calculating limiting efficiencies of solar cells.

UPM will be responsible to launch the first series to serve as an example, by leading its organization and by also involving other CHEETAH partners.

Furthermore CHEETAH partners are invited to organize focused on line/onsite training courses on the utilization of infrastructures by also including practice sessions. The webinar platform will be provided with tools assuring opportunity to follow live experiments/tests

5 WEB AREA ACCESS TO NON-CHEETAH EXTERNAL PARTNERS

The CHEETAH Knowledge Exchange web area Portal is an open source portal open on request to any interested non-CHEETAH organization. Any public, private research centres and universities actively involved in PV energy research and located in the European Union or associated countries can propose its access to CHEETAH KEP but the elective members are EERA-PV and EU-PV platform partners. Non-CHEETAH partners could be also interested to participate in project activities .

A set of useful information can be uploaded on the web site which is also associated to each external organization (expertise, infrastructures, technical documents, courses, webinars, discussion, news).

Access of the portal to external non-CHEETAH partners has imposed that some special areas of CHEETAH KEP can be reserved to project partners only. For instance, information concerning maintenance, or some relevant results or information related to the project management.

Access to CHEETAH KEP is offered by a registration procedure by user name (email) /password. Three different access right levels are arranged depending on the project necessities to access to the information content. The registration procedure offers also the opportunity to collect statistical data (role, gender, nationality, etc.) on registered users and the procedure includes also the opportunity to recover forgotten passwords.

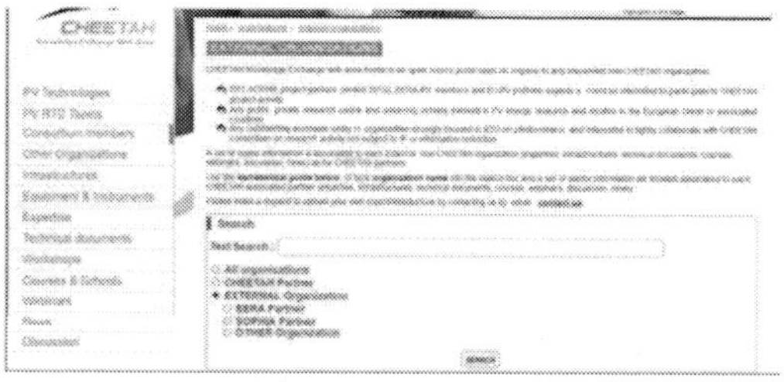

Figure 9. CHEETAH External partners web Log-in area

6 CHEETAH KEP: ACHIEVED RESULTS

We warmly invite interested readers to directly browse within the platform web pages to learn more about the potential of the platform.,

The main achieved results (as Sept 2017) are the following:

- a collaborative spirit has been created within the project partners and external organization as demonstrated by the numerous actions in Knowledge Exchange.
- A more efficient "ranking list" of the most demanded technologies / Topics / Equipment extended also to EERA-PV Members has been produced and it will be objective of another upcoming report.
- More than 60.000 accesses by its dynamic links moved interested users to 33 CHEETAH and 13 non CHEETAH corporate descriptions
- 215 experts and 45 infrastructure profiles uploaded are growing day-by-day
- 46 webinars with 930 "different users" registered to participate 1450 times, and many hundreds accessing via streaming.

7 CHEETAH RESULTS in PERSONNEL AND TEST PROCEDURE EXCHANGE

During the first 3 years of the project, there were 49 visits of researchers to project partner laboratories ranging from 1 day to months. A total of 13 summer universities and workshops was associated with the project. An international coordination board was formed with high-ranking experts from 5 countries.

As an additional activity within WP3 supported by WP2 and joint Research Work packages the following round robin test procedure were launched:

- RR Thin Si Tandem Solar Cells
  Organizer JRC-ESTI & Jülich
  Participants: DTU; CREST; TÜBİTAK; AIT; UNIMIB; IFE; ECN; CEA-INES; ENEA; ISE; HZB
- RR & common test procedures on determining the accurate composition in CIGS absorber layers
  Participants: EMPA, ZSW, HZB
- RR on OPV Solar Cells (ISOS-L-3)
  Participants: DTU, FhG-ISE, ECN, ENEA,
- RR Perovskites Interlaboratory Studies (ISOS-D3 life testing) in collaboration with EERA

CONCLUSION

The CHEETAH Knowledge Exchange Portal constitutes a very useful tool to bring information from different sources in a uniform and simple way to all CHEETAH and non-CHEETAH partners, as permanent channel fostering the use of existing facilities and Expertise.

The CHEETAH Knowledge Exchange Platform (KEP) represents a significant step forward in the PV RTD knowledge exchange taking profit from the best practice in more diffused social, scientific and professional networks.

It is very useful to identify the current European PV RTD Community's technical-scientific needs and efficiently establish and promote channels and procedures to transfer information among each partner in order to enforce potential and effectiveness of RTD activities held by each organization.

The portal is also accessible to non CHEETAH partners offering them information on the availability of infrastructures, equipment, expertise, technical documents to widely foster the interaction and collaboration among organizations involved.

In addition, the web area provides dedicated tools to share expertise by organizing also on-line meetings, webinars and on-line tests. Such tools have been realized with the aim of promoting individual and collective knowledge exchange actions among experts and trainees from beneficiary organizations. CHEETAH KEP represents a major breakthrough in the field with highly innovative contents and a substantial improvement in comparison with the state of the art of knowledge exchange on PV RTD.

Acknowledgement

We would like to thank all CHEETAH members for their continuous support and their contribution in helping to enrich the web site of information and technical – scientific content to make the CHEETAH KEP initiative possible and successful.

We also thank as of September 2017 all CHEETAH KEP browsers who for more than 60.000 times have accessed to our web pages and several of them have reacted with enthusiasm and interest by supporting us with comments and amendments, they are essential for the development of the portal.

We acknowledge the technical staff of ENEA, which has consistently supported both the development and the use of applications and David Casaburi, ENEA and Francesco Beone for Web site development and the Adobe Connect Corporate e-learning platform, respectively and their technical help.

Finally, but not least, we would thank the European Commission that for the financial support of 7FP7-CHEETAH (grant agreement n° 609788), which is also supporting the CHEETAH KEP initiative.

REFERENCES

[1] CHEETAH project web site http://www.cheetah-project.eu

[2] Highlights from the FP7 Project on Photovoltaics CHEETAH: More Power with Less Materials J.M. Kroon ECN, Petten, Netherlands 7EO.3.5 This Conference

[3] CHEETAH Knowledge Exchange Platform web site : https://www.cheetah-exchange.eu

[4] CHEETAH Webinars web pages: https://www.cheetah-exchange.eu/webinars.asp

[5] Innovation brought by FP7-CHEETAH Project in Management of Knowledge Exchange for PV RTD F. Roca, et al.proc. 32nd European Photovoltaic Solar Energy Conference and Exhibition EUPVSEC 2016 20-24 June 2016, Munich, Germany

[6] FP7-CHEETAH project knowledge exchange portal: an advanced tool to efficiently bring information to the European photovoltaic RTD community F. Roca, et al. proc. 31st EU PVSEC 14-18 September 2015 CCH - Congress Center Hamburg, Hamburg, Germany

[7] EERA success stories fiches: Knowledge for a common vision: the CHEETAH KEAP e-learning platform presented during EERA Annual assembly 2016

[8] CHEETAH first public event Symposium. 16th September 2015 (EUPVSEC1015 parallel event) EU PV platform. A Strategic Research

[9] SOPHi@webinar: The E-Learning Platform of the FP7-Sophia Project: Obtained Results and Perspective for its future Exploitation F. Roca, et al. et al. proc. 31st EU PVSEC 14-18 September 2015 CCH - Congress Center, Hamburg, Germany

[10] SOPHi@Webinar the advanced e-learning platfom of the Photovoltaic European Research Infrastructure FP7-SOPHIA Project, F.Roca et Al. 29th European Photovoltaic Solar Energy Conference and Exhibition, 7DV.4.46; EU PVSEC 2014 Amsterdam, NL Sept 22-26, 2014

[11] SOPHIA Symposium on European PV Research Infrastructures .- Chambery, France, 22 January 2015.

[12] Strategic Vision on PV Research Infrastructure in Europe. J Kroon et Al. 29th European Photovoltaic Solar Energy Conference and Exhibition, EU PVSEC 2014 Amsterdam, NL Sept 22-26, 2014

[13] Photovoltaic Solar Energy Technology Edition II (2011)

DEVELOPMENT OF BIPV COURSEWARE FOR STUDENTS AND PROFESSIONALS ON THE NEED FOR EDUCATION IN THE BIPV SECTOR: THE DEM4BIPV PROJECT

Wilfried van Sark[1], Sofía Arancón[2], Ingrid Weiss[2], Momir Tabakovic[3], Hubert Fechner[3], Atse Louwen[1], George Georghiou[4], George Makrides[4], Maria Hadjipanayi[4], Eliza Loucaidou[5], Monica Ioannidou[5]
[1]Utrecht University, Copernicus Institute, Heidelberglaan 2, 3584 CS, Utrecht, The Netherlands
[2]WIP-Renewable Energies, Sylvensteinstrasse 2, D-81369 Munich, Germany
[3]FH Technikum Wien, ENERGYbase, Giefinggasse 6, 1210 Wien, Austria
[4]University of Cyprus, Panepistimiou 1 Avenue, P.O. Box 20537, 1678 Nicosia, Cyprus
[5]Deloitte Ltd, 213 Arch. Makariou III Avenue, 3030 Limassol, Cyprus
Email: w.g.j.h.m.vansark@uu.nl ; Phone: +31 (0) 30 253 7611/7600

ABSTRACT: As lack of education is seen as one of the barriers for BIPV deployment, the Dem4BIPV project (www.dem4bipv.eu/) aims to develop innovative training material for higher education in building integrated photovoltaics (BIPV). This paper presents the results of a survey that was conducted among BIPV stakeholders with the aim to identify major knowledge gaps in education and target audiences as well as teaching goals. From that, potential courses dedicated to the needs of each target audience related to BIPV are developed. It aims to provide modular content and a schedule for a 10 week, halftime course (7.5 ECTS).

Keywords: BIPV, building-integrated, photovoltaics, training, education, Erasmus+

1 PURPOSE OF THE WORK

Building-integrated photovoltaics (BIPV) are photovoltaic elements that are used to replace conventional building materials in parts of the building envelope (roofs, or facades). They are increasingly being incorporated into the construction of new buildings as a principal or ancillary source of electrical power, rather than added on later, although existing buildings may be retrofitted with similar technology. The advantage of integrated PV systems over more common non-integrated systems is that the initial cost can be offset by reducing the amount spent on building materials and labor that would normally be used to construct the part of the building that the BIPV modules replace. These advantages make BIPV the optimal method of installing renewable energy systems in urban, built-up areas where undeveloped land is both scarce and expensive. In fact a wide variety of BIPV systems are available in today's markets. In both new projects and renovations, BIPV is proving to be an effective building energy technology in residential, commercial, industrial, and institutional buildings and structures. It is generally expected that in the next century, photovoltaics will be able to contribute substantially to the mainstream power production, and that through their widespread commercialization, BIPV systems will become the backbone of the nearly zero energy building (NZEB) European target for 2020 [1].

Despite technical promise, social barriers to widespread use have been identified, i.e. conservative culture of the building industry and integration with high-density urban design. In fact there is a notable disparity between the progress made in terms of the technology and the knowledge and skills of the professionals (architects, engineers, designers, planners) who are ultimately responsible for the integration of BIPV systems. PV can be included in building projects only if the professionals involved in the development have sufficient knowledge about PV technologies and appropriate design tools to assist them. High level of knowledge and skills is required for successful BIPV systems planning since the projects realized in the past show that a successful BIPV system design is based heavily on technical knowledge and competences. Poorly designed systems usually have to be redesigned or repaired, consequently swelling maintenance costs and lowering system efficiency rate.

In an effort to respond to the growing industry demand for a specialised BIPV workforce, a consortium of leading European experts and Higher Education Institutions (HEIs) launched the Dem4BIPV project in 2015. The general objective of HEIs is to provide high quality education utilizing modern learning methods. In particular, for engineering departments, learning has to be in line with the needs of the industry and the new trends (such as renewable sources of energy, sustainability and PV for example). Meeting the needs of the EU industry in general and the local industry in particular and forging sustainable relationships with the industry are important issues. For the industry the objective is to be competitive and help enhance their presence on the market in the field of BIPV. An important aspect is skilled staff with up to date knowledge, something which this project is aiming at. For other partners and stakeholders their objective is to help enhance knowledge in the field and keep up to date with the technology and adoption of it.

2 PROJECT OBJECTIVES

The objective of the Dem4BIPV project is to develop an innovative and multidisciplinary training scheme that will enable graduated architects and engineers to acquire the technical knowledge and competencies required for the successful implementation of BIPV projects. It will be implemented at the postgraduate level and will be part of Master programmes in the field of Sustainable Energy in a number of leading universities in Europe, i.e. the Utrecht University in The Netherlands, the University of Cyprus and the FH Technikum Wien in Austria.

Dem4BIPV is based on the principle of European cooperation through which innovative educational material utilizing ICT will emerge on the topic of BIPV, which is of crucial importance for the future development and penetration of the PV market in Europe with a potential significant contribution in meeting Europe's energy challenges. Education and training are crucial for both economic and social progress, and aligning skills with labour market needs plays a key role in this. This project has been designed and structured so as to meet the real needs of the PV market and contributes positively to EU benchmarks for 2020 in relation to education. It also indirectly tackles fast-rising youth unemployment, as it places emphasis on delivering the right skills for employment in the BIPV industry and increasing the efficiency of higher education in the field of sustainable energy and on working collaboratively with all relevant stakeholders.

The project will contribute to:

- Raising awareness of the job prospects offered within the renewable energy sector in the field of BIPV
- Increasing awareness, sensitivity, and promotion of BIPV technologies among key target groups within the local economy.

3 PROJECT ACTIVITIES

The Dem4BIPV project has five main activities:

- Analysis of the existing and future market needs in terms of BIPV system integration and education needs in this field, as well as an identification of the Best Practices in Europe mainly but not exclusively, which will result in the definition of the framework and the actual requirements of the course component.
- Development of high-quality didactic content on BIPV for higher education
- Development of a virtual learning environment to support the practical aspects of the course (i.e. laboratory work of experimental nature).
- Deployment of remote laboratories
- Pilot testing of the course and refinement

The project's tangible and intangible outcomes include the below:

I. Development of appropriate and up to date educational material for BIPV.
II. The creation of a Virtual Learning Platform and Remote Labs for BIPV education and training.
III. Accreditation (in parallel but beyond the scope of the project).

4 PROJECT RESULTS

The Dem4BIPV project started in September 2015 and will reach its conclusion in August 2018. This section summarizes a selection of results from achieved or still ongoing tasks related to the project, as of September 2017. More detailed information can be consulted on the project website at www.dem4bipv.eu [2].

4.1 Stakeholder survey in the BIPV sector

Before educational material was developed a thorough analysis of educational needs was performed using a survey among identified stakeholders, i.e. professionals from the building sector (architects, building contractors, building planners, manufacturers of building envelope products, electrical engineers), BIPV manufacturers (BIPV producers, mounting systems producers, façade manufacturers), installers (PV installers, electricians, façade installers), investors, research organisations. In total 100 completed questionnaires from 15 different countries have been analysed.

Groups and topics covered in the questionnaire are shown in Table I. The educational needs for many of the topics were found to be strong or very strong (see Figure 1). In fact, for 30% of the topics the sum of 'strongly needed' and 'very strongly needed' was larger than 75%. The topic 'aesthetics' scored highest with 92% combining 'strongly needed' and 'very strongly needed', followed by BIPV software /tools, market maturity, and building envelope material properties.

Table I. Groups and topics used in the questionnaire.

Group	Topic
Constructional integration	Combination with conventional building materials
	Building envelope material properties
	Moisture protection
	Orientation/inclination
Energetic integration	Electrical interconnection
	Electrical protection
	Effects of integration type on energy performance
Design integration	Glass (optical, colours,..)
	Sun protection
	Different BIPV /products / materials / technologies
	Aesthetics (variety of colours, sizes and types)
	BIPV software /tools for early integration in the design process
Regulatory barriers	Building law (EPBD - Energy)
	Heritage law
	Noise regulation
	Heat regulations
	Renewable Energy Directives
Economical barriers	Price of the BIPV systems
	Payback time
	Maintenance of the modules
	Feed-in tariffs
Standards	EC (low voltage directive)
	CENELEC
	prEN 50583
Maintenance and recycling	Recycling
	Exchange of the Modules
Manufacturing barriers	Customization vs. Standardization
Market driven barriers	Market maturity
Cultural barriers	Cultural barriers
Lack of information	Best practice examples/demonstration examples

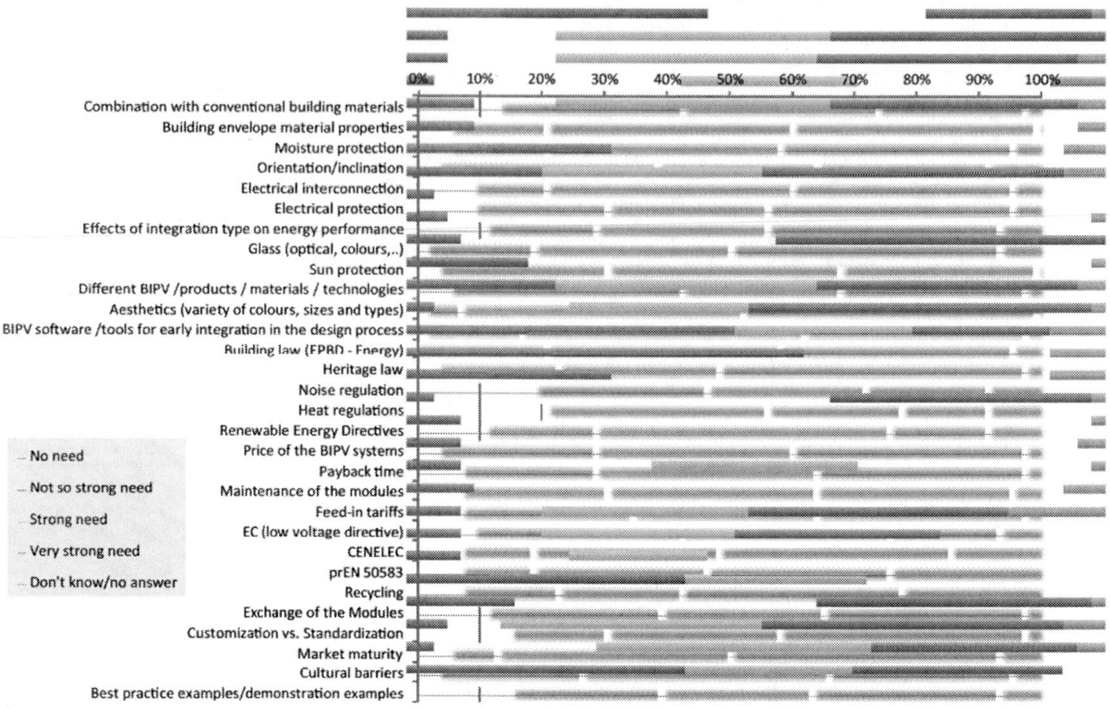

Figure 1: Analysis of the questionnaires

BIPV brings different sectors of professionals together, which did not work together in the past. The results of the analysis show that the new players in the BIPV arena have to enhance their knowledge base and skills in relation to BIPV systems. This primarily pertains to architects, civil and electrical engineers, followed by mechanical and environmental engineers according to the survey result. The main group for educational need is identified as the professionals (i.e. Architects, Engineers, Planners etc.), followed by building contractors and post-graduate students (MSc) in relevant fields (i.e. architectural, engineering, etc.).

The results of this analysis are collected in the report 'Framework and Requirements' Analysis' [3] which is available for download from the Dem4BIPV website [2]. In addition, the report reviews the present status and outlook of the BIPV market on a global and European scale. In particular, it provides a comprehensive review of the market situation and the future trends for Austria, Cyprus, France, Germany, Italy and the Netherlands until the year 2020.

4.2 Didactic content for students

The results of the stakeholder survey allowed identifying major knowledge gaps in education and teaching goals. From that, potential courses dedicated to the needs of each target audience related to BIPV are developed.

The course outline is developed by Utrecht University (UU) with the UU students from the Energy Science and Sustainable Development master programmes in mind. The course will be implemented in a modular manner with four building blocks:

I) Introduction to BIPV
II) BIPV design and engineering
III) BIPV performance
IV) BIPV and its environmental footprint
V) BIPV market, barriers, incentive schemes

Modules can be combined and customized to design a BIPV course for different target groups. Hands-on work will be included, by using the virtual labs.

I) Introduction to BIPV

The learning goals of this module are to familiarize the students with BIPV technology, e.g. to show what the definition of the technology is, what its pros and cons are, how the BIPV market has developed and what kind of aspects are related to the application of BIPV in the urban environment. Optional in this course module is a general introduction into photovoltaics, which could precede the BIPV introduction, depending on the pre-existing knowledge of the student (see Fig. 2).

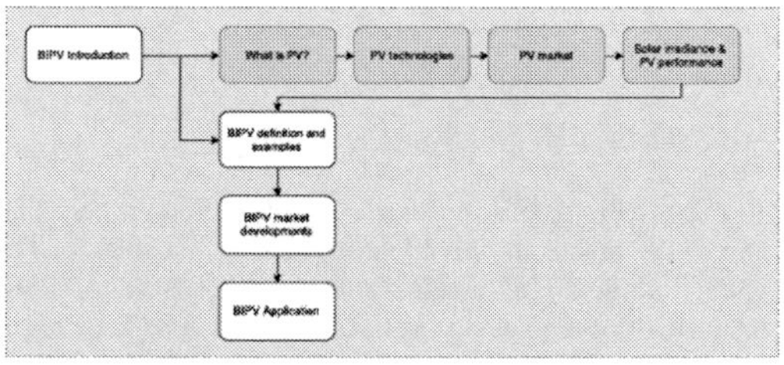

Figure 2: Example: BIPV intro module set-up

II) BIPV design and engineering

In this module aspects of design and engineering are treated to students. This includes physical aspects such as thermal isolation, construction of components, layout and design examples of buildings and components, testing procedures. Design software such as SketchUp [4] can be used, as well als Building Information Modeling (BIM) [5]

III) BIPV performance

This module will inform the students about the performance of BIPV systems, and will thus address general PV performance from basic physics to system operation, but will focus on BIPV specific aspects (see

Fig. 3). The learning goals are to gain knowledge on which BIPV specific factors affect performance, and what this means in terms of BIPV system yields compared to conventional PV systems. Also, a goal is to familiarise students with some tools that allow for irradiance and (BI)PV performance modelling, like the new BIPV software tool developed in the European project PVSITES [6].

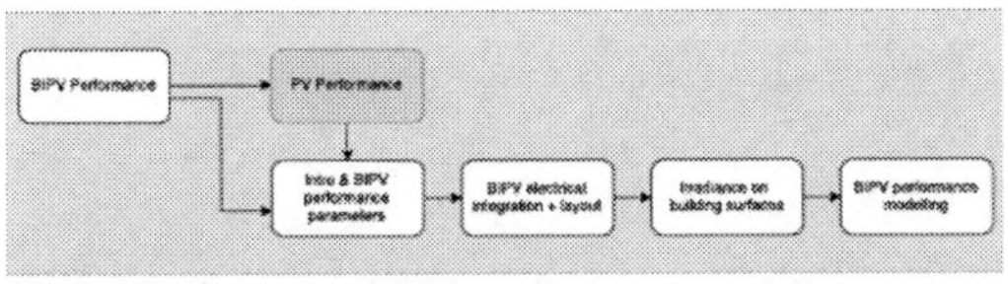

Figure 3: Example: BIPV performance module set-up

IV) BIPV and its environmental footprint: LCA and recycling

This course module will address the environmental impact of BIPV, and will thus address general topics such as how environmental impact is established using life cycle analysis (LCA), what the environmental impact of BIPV is (results from literature) and how this compares to conventional PV and other electricity sources. The module will also focus on showing which BIPV specific factors affect the environmental performance, and if BIPV specific characteristics are suitably accounted for in LCA standards, as allocation of the impact to the multiple functions of BIPV can have a very significant effect on e.g. the impact of the building vs. the impact of the generated electricity from the BIPV system. Furthermore, the module will address PV and BIPV recycling or end-of-life treatment, showing the contribution of recycling to the environmental impact, and analysing whether BIPV specific characteristics affect the ability to perform recycling on end-of-life systems.

V) BIPV market: barriers, incentive schemes

In this course module, we focus on the BIPV market, and assess the barriers in the market that possibly are preventing large scale deployment of BIPV. The module addresses the development of BIPV costs over time, makes the comparison with conventional PV and other renewable electricity sources, and analyses the current BIPV market to assess the dynamics and maturity of this market. Furthermore, different renewable energy incentive schemes will be shown and compared, and we address whether these schemes are applicable and functional for BIPV systems. The course module will also aim to assess business cases for different BIPV options.

4.3 Virtual learning environment (VLE) and remote laboratories

The project also foresees the development of a VLE i.e. a web-based platform for the support and delivery of developed course modules particularly focusing on the practical aspects of the course. The VLE platform will be accessible via the Dem4BIPV website and will link remote users to the didactic content and the remote labs in each participating university.

Furthermore, UCY, UU and FHTW have designed and are currently developing a remote lab each to be hosted in their respective laboratories. The remote labs are real physical experimental setups that will be accessed remotely by students for the performance of different experimental exercises. Each remote lab will be fully integrated within the VLE through web-services and will also be integrated into the BIPV course developed. The remote labs currently being developed are:

I) Remote laboratory for outdoor testing of BIPV modules (UCY): This comprises of a small-scale experimental setup for testing BIPV modules in real conditions. The lab will serve the purpose of measuring in real time among others BIPV electrical parameters, meteorological parameters, I-V and thermal behavior measurements.

II) Multidisciplinary functional integration of PV power systems into buildings and grids (FHTW): The lab will allow for comparative research of PV systems performance in terms of different geographic irradiation conditions, module orientation/azimuth/elevation, and different design techniques. Moreover, the lab will enable the study of the effect of aggregation of volatile energy resources (i.e. virtual power plant) and relevant methods of information modelling and communication.

III) Virtual laboratory using real data for simulations (UU): The lab will enable students to combine measured data from UU's outdoor PV test facility with different advanced modelling approaches for predicting PV performance of different PV module technologies on different building surfaces.

5 CONCLUSIONS

This paper has presented the progress achieved in the Dem4BIPV project, starting with a brief overview of the results of the "Dem4BIPV Framework and Requirements' Analysis". The aim of the analysis was to identify the key players in the BIPV sector and to provide a clear picture of the BIPV education needs in Europe.

The survey shows that architects, building contractors, building planners, façade and mounting systems producers, as well as investors play an important role in the future deployment of BIPV. According to the survey, roofs and façade systems will be the main application of BIPV. The main requirements that must be addressed by a BIPV system in the building are its cost, its durability and maintenance, as well as the aesthetical component. Also design, constructional integration and regulatory issues have to be addressed with high priority.

Based on the outcomes of this analysis, Utrecht University with the other partners is developing a curriculum for potential courses dedicated to the needs of each target audience related to BIPV. It aims to provide modular content and a schedule for a 10 week, halftime course (7.5 ECTS). This information will be used for the development of course material including didactic content for students, the development of a virtual learning environment and the deployment of remote laboratories.

The first pilot Dem4BIPV courses will be implemented by the Utrecht University (The Netherlands), the University of Cyprus (Cyprus) and the FH Technikum Wien (Austria) in 2018.

6 REFERENCES

[1] European Commission. Directive 2010/31/EU of the European Parliament and of the Council of 19 May 2010 on the energy performance of buildings, Official Journal of the European Union L 153, 18.6.2010, p. 13–35.

[2] Dem4BIPV project website, www.dem4bipv.eu/

[3] Dem4BIPV report, 'Framework and Requirements' Analysis' 2016.

[4] SketchUp, https://www.sketchup.com

[5] Autodesk, https://www.autodesk.com/solutions/bim/architecture

[6] PVSITES project, www.pvsites.eu

7 ACKNOWLEDGMENTS

This project is funded by the KA2 Strategic Partnerships for higher education programme of Erasmus+ under contract 2015-1-NL01-KA203-008882

Co-funded by the Erasmus+ Programme of the European Union

THE PILOT TENDER FOR PV IN GREECE WITHIN 2016. RESULTS AND CONCLUSIONS.

Dr. Dionysios Papachristou[1], Panagiota Kapetana[1], Panagiotis Daliouris[1], Thanassis Petmezas[2]
[1]Regulatory Authority for Energy, 132, Pireos Str., GR-11854, Athens, Tel.: +30 210 3727409
Fax: +30 210 3464407, E-Mail: papachristou@rae.gr , kapetana@rae.gr , daliouris@rae.gr [2]cosmoONE Hellas
Marketsite S.A., 26 Olympias Str, GR-141 22 Attika, Tel.: 210 2723810, Fax: 210 2723820, E-Mail:
thanassis.petmezas@cosmo-one.gr

ABSTRACT: In line with the "Guidelines on State aid for environmental protection and energy (2014-2020)" (2014/C 200/01), Greece run a pilot tender in 2016 for a 5% of the total Renewable Energy Sources' (RES) capacity which was to be installed within the period 2015-2016. The purpose of this paper is to present the procedure and the results of the pilot tender for photovoltaic installations which took place in Greece in 2016. A specific innovative electronic platform was developed to simplify the procedure, secure its accuracy and transparency and increase the competitiveness with the use of an electronic auction. The main goal is to adopt the procedure as a best practice in the future tenders for RES installations.
Keywords: Photovoltaic, PV Market, RES Tenders

1 NEW SUPPORT SCHEME FOR RES

In accordance with the Greek legal framework and specifically Law 4414/2016 "*New support scheme for renewable energy sources power plants and high efficiency combined heat and power plants. - Provisions on the legal and functional separation of the supply and distribution branches in the natural gas market and other provisions*" ("Law"), Greece implemented a new RES support mechanism, in compliance with the Guidelines on State aid for environmental protection and energy (2014-2020) (Communication from the Commission, EC C200/28.6.2014).

One of the main purposes of the Law was the development of an adequate support scheme for power generation from RES in the context of a common European Member States' policy on fighting climate change, reducing greenhouse gas emissions and achieving the target of RES participation in gross final energy consumption. In this legislation framework, specific provisions took place for the planning, organization and execution of a pilot tender. The legislation stated that "*RAE shall be responsible for carrying out competitive bidding processes and certifying their results*".

2 CONDUCT OF THE PILOT COMPETITIVE BIDDING PROCESS

2.1 Procedure

RAE was obliged, according to the Law, to conduct a pilot tender on photovoltaic power plants within the year 2016, which was also used as an opportunity for collecting information that would facilitate the introduction of a generalized tendering scheme for all RES technologies that will take place in the following years.

According to the Law, Article 7, par. 9, the pilot tender on photovoltaic plants should have the following main characteristics:

- the tendered capacity should be at least 40 MW,
- the PV-Plants should be divided into two Categories, according to the installed capacity of each installation (Category I for PV-Plants ≤ 1MW and Category II for PV-Plants > 1MW), whereas the tendered capacity of Category I could not exceed 20% of the overall auctioned power.

The maximum allowed bidding price in accordance to the legislation was set at 104 €/MWh for Category I and 94 €/MWh for Category II. In compliance to the main legislative guidelines, two competitive bidding processes took place online on the 12th of December 2016, in cooperation with a Greek company specialized in B2B online auctions. Instead of the traditional competition procedure with the submission of heavy hard copy application files and an unnecessary load of paperwork, the online application procedure allowed the participants to upload all the necessary technical and legal documents that had to be sent for the participation in the tender, minimizing workload and simplifying the procedures in hole, with both, participants and RAE benefiting from the simplicity.

A **40% rule of a minimum level of competition** was decided by the Greek regulator, ensuring the competition between the participants in each category. This rule determined that each category's capacity sum must be equal or exceed 140% of the auctioned capacity.

A rule, for the letter of guarantee at the level of 1% of the total investment cost per project was enforced to all the participants in order to participate in the auction procedure. In addition, after the success in each category, each participant had the obligation to cover **a rule of "4% guarantee of proper performance".** This means that each participant must submit to the Greek Regulator an additional letter of guarantee, only if the sum of all letters of guarantee which were submitted to all the operators (Grid or System operator) does not equal or exceed the amount defined by the **4% rule** of the total investment cost for each project.

2.2 Electronic Platform

A specific electronic platform was developed in order to serve all the procedures of the pilot tender. The algorithm was also custom developed for the electronic auctions which took place. Important steps were added in the platform's algorithm for improvement and innovation of the whole process, as well as for the proper implementation of the procedural part of the competition, i.e. the posting, processing, submission, evaluation, and local storage of supporting documents via an online platform.

In this way, the applications of the interested parties, as well as the supporting legal and technical documents, were submitted online. Each company had to create a unique profile through which all necessary files could be uploaded. Then, RAE evaluated all the applications and the projects which passed successfully through the phase A (administrative check) had the opportunity to participate in the competitive tendering procedure (phase B) using this specific electronic platform. The pattern used for the execution of the process is shown in the following Figure.

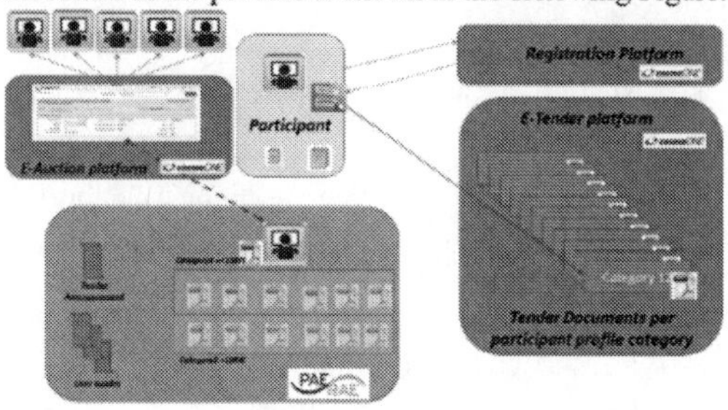

Figure 1: The procedure of the participants' registration and application onto the online platform

User guides including presenting all the important steps via visual examples for registering in the platform, applying and taking part in the tenders were available online. In addition, all the participants were called upon training through dry runs in order to make them acquainted with the platform hosting the tenders. In this way, an innovative auction, which was never used before in the field of RES plants, became accessible and understandable to all the participants, independent of the profile of the company - varying from full S.A. to individuals.

2.3 Online Auctions

The final tenders took place online on the same electronic platform with an absolute success for the participants, for RAE and for the end-consumers.

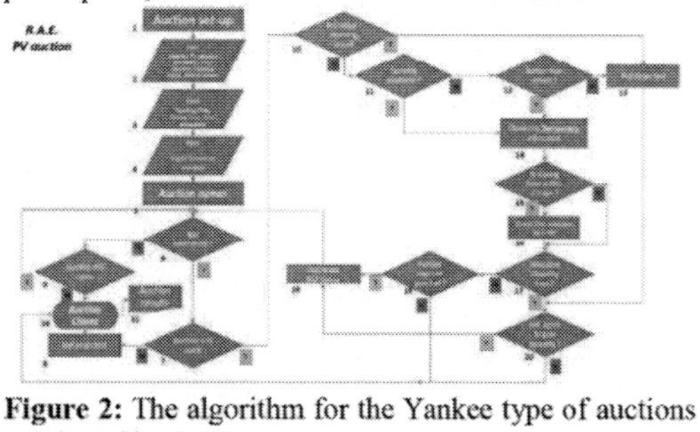

Figure 2: The algorithm for the Yankee type of auctions conducted by RAE

The auctions used were a variation of the Yankee auction type. The basic "RULE" of this auction was the following: all accepted bids are registered in ascending order with criterion the submitted price. In case of equal prices, the bid which relates to the smaller quantity of PV prevails. In case of bids with equal price and quantity, the bid that was submitted earlier prevails. In that way, it is never possible to have in the system two totally equal bids. The allocation of the financial aid starts from top to bottom, till it exceeds the available PV quantity. The algorithm followed during these procedures is shown in Figure 2. It consists of the following steps:

1. Auction set up

This step includes all necessary actions that need to be done, in order to set up the YANKEE type of auction, according to the specific requirements of the RAE pilot tenders.

2. Set up of bidders' info

The auction Service provider (cosmoONE) as administrator of the platform, "invites" in the system the eligible bidders by constructing a list of companies that can have access to the specific auction. Companies that are not invited have neither access to the auction nor awareness that the specific auction exists and takes place.

3. Set up of PV capacities and related info

The auction Service provider inputs the initial values of the installed PV capacity per participating company/user in the auction, the total available capacity, the opening value per Unit of measurement and the bid increment.

4. Set up of Time and date related info

The auction Service provider inputs the date and time of the auction opening, the duration of the auction, the provision of time extensions, the duration of the extensions, the number of extensions, the rule of extensions etc. There were no extensions in the case of the pilot tender, as decided by RAE.

5. Auction Opens

The auction opens and the time clock starts counting down the time. Any bidder can submit his bid as long as the auction is still open.

6. Based on the market situation, bids can be submitted or not.

7. If no bid is submitted till the time of auction closing, the auction closes with no bids (path 5-6-9-6-9-….-30).

8. If a bid is submitted, there is the first control point and the system checks if the auction is still open. In case that the duration (either the initial planned or with time extensions) of the auction has passed, the bid is rejected, the auction changes status to 'close' and no more bids are accepted (path 5-6-7-8-30).

9. The auction results are automatically generated at closing time and the system produces an analytical list of all bids from all bidders with their outcome (accepted, quantity, value, rejected, reason for rejection etc.) and the winners' list that indicated the final allocated quantities and the respected values on the measure of unit (path 30-31).

10. In case that a bid is submitted and the auction is still open, there is the second control point and the system checks the value (per unit of measure) against the opening price (path 6-7-10).

11. In case that the submitted value is greater than the opening price, the bid is rejected and the system allows the bidder to improve his bid (path 10-13-20).

12. In case that the submitted value is equal or less than the opening price, the system performs the third check that concerns the quantity that is tight with the specific bidder, against the available (not already temporary allocated) quantity.

If the available quantity is equal or greater than the requested, the system grands the bid and temporary allocates this quantity to the bidder (path 10-11-14).

13. If the available quantity is less than the requested, the system makes a comparison of the bid against the existing bids that have already allocated quantity.

In case that the bid is "better" that the existing accepted bids according to the "RULE", the system grabs the quantity from one or more bidders with the lowest current ranking and temporary allocates the quantity to the specific bidder (path 10-11-12-14). The bidders who lose their quantity are immediately notified by the system.

In case that the bid is "worse" that the existing accepted bids according to the "RULE", the bid is not granting quantity to the bidder, it is registered in the system, and the system allows the bidder to improve his bid (path 10-11-12-13).

14. Check for time extensions.

Whenever a bid is accepted, the system checks if the bid has been submitted within the time window, where the time extensions can be activated. In case that the submission time is within this period, the system grands a time extension (path 14-15-16).

15. (Temporarily) allocated quantity is lost.

During the auction, every bidder can lose his quantity due to the submission of a better bid by another bidder. This is indicated in his screen and the bidder has the right to bid again to regain his quantity.

16. Quantity is lost and the bidder does not want to improve his bid and bid again.

The bidder does not submit any further bids and waits till the auction closes, as he can still gain his quantity under certain conditions that derive from the "RULE" (path 17-20-30-31).

17. Quantity is lost and the bidder wants to improve his bid and bids again.

The bidder improves his bid by decreasing his bid value and submits the bid again (path 17-20-19-6).

18. Quantity is not lost, but the bidder wants to improve his bid.

If the bidder wants to strengthen his position in the auction and be in a better position against the other bidders according to the "RULE", he is eligible to submit a new bid. To do so, he must improve his bid value and submit the bid (path 17-18-19-6).

The thirty (30) minute processes were observed in real time by the committee of RAE specially called upon the evaluation of the applications and the completion of the tenders. The bids followed a different pattern in each auction, as shown in the following Figures.

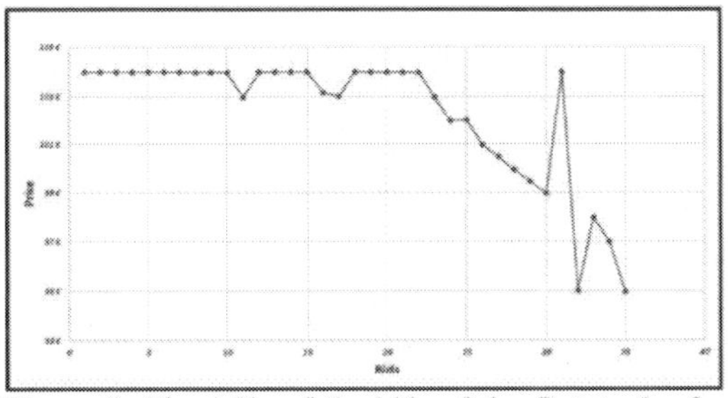

Figure 3: Timetable of the bids of the first tender for Category I (installed capacity ≤ 1MW)

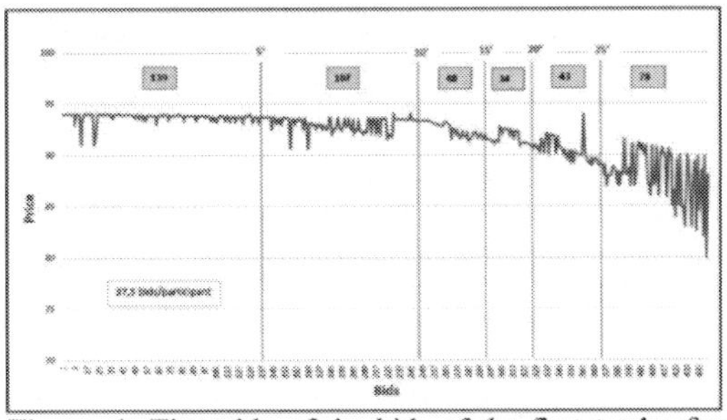

Figure 4: Timetable of the bids of the first tender for Category II (installed capacity > 1MW)

In Category I, the minimum price was 94,97 €/MWh and the maximum price was 104 €/MWh, as indicated by the Law 4414/2016 (**Figure *3*:** Timetable of the bids of the first tender for Category I (installed capacity ≤ 1MW). In Category II, the minimum price was 79,97 €/MWh and the maximum was 88 €/MWh (**Figure *4***). The participants were informed in every step through the way whether their bid was valid or whether they had to make a new price offer.

After the procedure was determined, the participants had the possibility of raising objections to the results. Since the procedure was objective, described in great detail in the invitation to the tender, also conducted by RAE, and clarified through the dry runs and the users' guides available to all the participants, there were no objections which were not clarified by the publication of RAE's decision on the final results, which was accepted by RAE's Board.

3 RESULTS

The results and the conclusions of the pilot tender were analyzed and presented in detail in this paper. The participants in the **first Category I (≤ 1 MW)** were fourteen (14) and the photovoltaic power plants that successfully achieved a price in the tender were nine (9). The **weighted average price of the successful bids was 98,78 €/MWh**. As for the second **Category II (> 1 MW)**, fourteen (14) participants applied and seven (7) acquired a price; the **weighted average price of the successful bids was 83,3 €/MWh.** In conclusion, there was a decrease of the starting price of about 15%, an important outcome for the end-consumers whose benefit was one of RAE's prime goals.

The main outcomes of the procedure were the following:

- Zero reported problems, in both understanding and use of the three relevant applications for registration, document submission and bidding.
- There was 100% success rate in the participants' registration process, application and submission of documents to the system.
- Among the 488 accepted bids, just one was rejected by the system, as not complying with the bidding rule.
- The allocation of the available quantities reached 100%.

4 CONCLUSIONS

The approach of using all tools provided by our digital

era was Innovative for this market in Greece and especially for the energy sector, setting a benchmark for all future auctions in the domestic and European market alike.

The objectives of the RES committee for transparency, neutrality, accuracy and speed of execution were fully met through the adjustment of the platform's algorithm.

Public Sector and Regulatory Authorities can lead innovation to their markets with Public-Private Cooperation schemes, when they jointly carefully plan and flawless execute new e-tendering procedures.

5 REFERENCES

[1] RAE (Regulatory Authority for Energy of Greece), www.rae.gr

[2] IPTO (Independent Power Transmission Operator), www.admie.gr

[3] LAGIE SA (Operator of Electricity Market), http://www.lagie.gr/nc/en/home/

[4] HEDNO SA (Hellenic Electricity Distribution Network Operator S.A, http://www.deddie.gr

[5] Bundesnetzagentur, BNetzA https://www.bundesnetzagentur.de/EN/Home/home_node.html;jsessionid=D327D59E9B4ABECFCB2ED90D437DE8A9

[6] Guidelines on State aid for environmental protection and energy 2014-2020 - (2014/C 200/01).

[7] EU Decision C (2016) 7272 final, Brussels, 16.11.2016: State Aid Support Mechanism, SA.44666 – Greece. New operating aid scheme for the production of electricity from RES and HECHP. http://ec.europa.eu/competition/elojade/isef/index.cfm.

[8] Law 4414/2016 "*New support scheme for renewable energy sources power plants and high efficiency combined heat and power plants. - Provisions on the legal and functional separation of the supply and distribution branches in the natural gas market and other provisions*" (Series No 149/09.08.2016 Government Gazette of the Hellenic Republic).

[9] IPTO (Independent Power Transmission Operator): Electricity Generation Adequacy Report 2017-2027, Athens, May 2017. www.admie.gr

[10] IPTO (Independent Power Transmission Operator): Ten Years Network Development Plan, 2017-2026, 4 Oct. 2016. www.admie.gr

[11] Dr. Dionysios Papachristou, Workshop on the Pilot Tender on Photovoltaic Plants, National Technical University of Athens, 21.10.2016.

[12] Dr. Dionysios Papachristou, Workshop on the Competitive auction procedure for RES; National Technical University of Athens, 10.02.2017.

[13] Dr. Dionysios Papachristou, Workshop for Competitive auction procedures in Wind Sector and latest RES reforms, ELETAEN, 03.03.2017, www.eletaen.gr

PHOTOVOLTAIC POWER PRODUCTION IN GREECE: HISTORY, CURRENT STATUS AND NEW POLICIES FOR FUTURE DEPLOYMENT

John S. Anagnostopoulos
Greek Regulatory Authority for Energy
Pireos 132, Athens 11854; E-mail: anagno@rae.gr

ABSTRACT: This article starts with the presentation of the first development stages of PV production in Greece and the corresponding evolution and changes in the regulatory framework in the last years, along with the effects on the electricity systems and the consumers. The new policy and regulatory framework for RES deployment that has been recently developed in the country is then analyzed, within the framework of the energy market transformation and the radical modifications towards the EU target model. The organization, implementation and results of the first pilot auction for PV plants are then briefly discussed, followed by the regulatory framework for the regular competitive auctions for new projects that is being prepared, in order to be able to start this procedure early the next year 2018, based on a sliding feed-in-premium concept. The paper closes with the prospects of the PV production and corresponding market in Greece the next decades.

Keywords: Photovoltaic; PV Market; National Program; Pilot Auction; Competitive Tenders;

1 HISTORY OF PV TECHNOLOGY IN GREECE

The installation of the first PV power plant in Greece is dated back in 1983, when a 100 kWp pilot plant is constructed in the island of Kythnos, which was one of the first such units in Europe. This unit, combined with a 100 kW wind farm installed in the same island one year before, and with a Diesel generator, in a hybrid power production system, was successfully operated for some years, and the system was covering up to 75% of the electricity needs of the island. However, this technology was not extended to other locations in the country, mainly due to the high capital cost, and the PV installations remained negligible for almost 25 years (740 kWp until 2007), in spite of the great solar potential available throughout the country (Fig. 1).

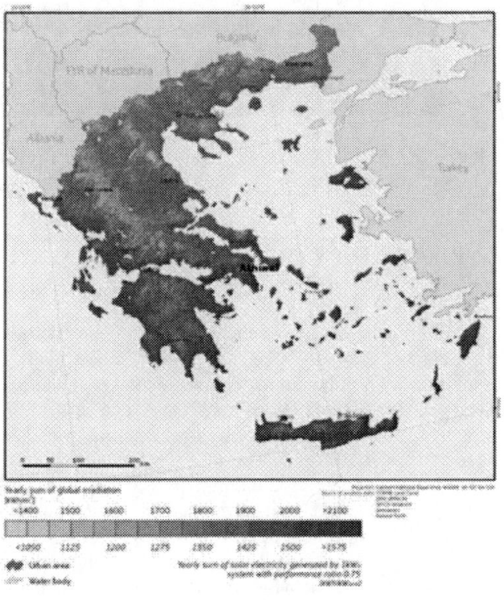

Figure 1: Solar radiation and photovoltaic electricity potential in Greece for optimally-inclided PV modulus (Source: PVGIS © European Union [1-2])

Then, the government by the Law 3468/2006 set strong incentive measures by extremely high FiTs especially for PVs, ranged from 400 €/MWh for interconnected units above 100 kW, to 500 €/MWh for non-interconnected units below 100 kW. Also, a simplified licensing procedure was adopted for PV plants below 150 kW.

A few years later, in 2009, a special program for small rooftop PV installations was launched for units up to 10 kW, with an even higher FiT, at 550 €/MWh, and very small tax rate. The licensing procedure for RES projects was further simplified by the Law 3851/2010, while the FiTs were not changed despite the rapid decrease of PVs cost, thus resulting in greatly increased applications in that year 2010 to the HEDNO: more than 25000, with a cumulative power of almost 5 GW [3].

In the same year (Oct. 2010), the national action plan for RES development (MD 19598/11.10.2010) envisaged a significant increase of RES share in the gross energy consumption of the country in order to reach 40% renewable electricity by the year 2020. Deployment of 7.5 GW Wind and 2.5 GW PV units was stipulated in that plan. This policy entailed a boom in PV sector development, and an expansion of installed power to 2.6 GW up to 2013 (Fig. 2), exceeding even the target for year 2020.

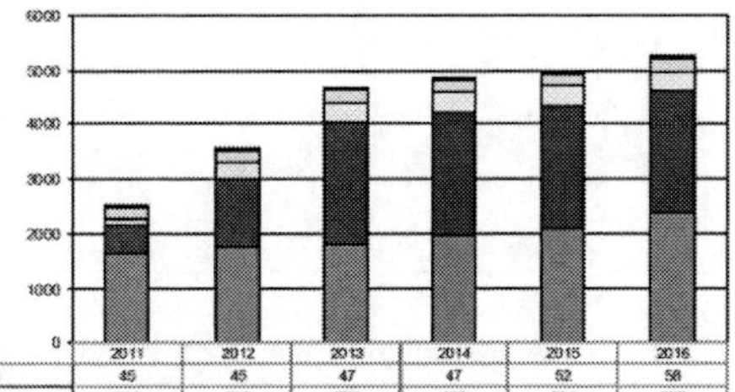

	2011	2012	2013	2014	2015	2016
Biomass	45	45	47	47	52	58
SHP	206	213	220	220	224	223
PV rooftop	95	298	373	375	376	375
PV	522	1338	2210	2221	2229	2230
WG	1640	1753	1810	1978	2089	2370

Figure 2: Installed power of RES in Greece by technology, for the period 2011 to 2016 [3-5]

At the same time (January 2012) the government and the RAE initiated plans to rationalize the tariffs for new

PV plants, starting by a 15% reduction for new units (Ministerial Decision (MD) 2262/2012), followed by a further 28% reduction later in the year (MD 2301/16933/2012). A solidarity charge of 25-30% was then foreseen by the Law 4093/2012 for PV plants, while the guaranteed FiT was as the valid one at the time the PV plant was set in operation. Later, in May 2013, the FiT for PVs was further decreased by 25% for future plants (MD 1288/9011/2013), while the Law 4152/2013 increases the solidarity surcharge up to 42%.

However, the effect of the above regulation was minor, since it was not applied to the existing PV plants, while the Special Account for Renewables, funded largely by final consumers, was accumulating a significant deficit. For this reason, eventually the government was forced to impose a retroactive haircut of the tariffs of PVs in operation, the so-called "new-deal", based on the activation date and the installed power of each particular unit (Law 4254/2014), repealing the special solidarity surcharge for all RES producers. The smallest PV tariffs of this classification were still quite high, in the range of 150 to 270 €/MWh. Also, the Sale Agreements for new plants were extended by a period of 7 years during which the tariff is defined equal to 90 €/MWh or as determined by the Market Operator.

The injection of large quantities of very expensive PV energy in the electricity system, in conjunction with the reduction of the demand due to the economic recession, created soon an uncontrollable deficit in the special fee for RES, which since then remains problematic, in spite of the significant surcharge of the final consumers' bills and the considerable haircut of the PVs feed-in-tariffs in 2014.

After that, as also due to the financial turmoil in the years 2014 and 2015, the investment initiatives were postponed and several large PV projects were stalled (Fig. 3), in contrast to the wind projects that showed a remarkable recovery in the last three years. The realization of new PV projects in the country was completely terminated in 2015, when a new legislation dropped the tariffs for such units to the opposite edge, by correlating them with the quite low system marginal price, and hence leaving numerous production licenses inactive.

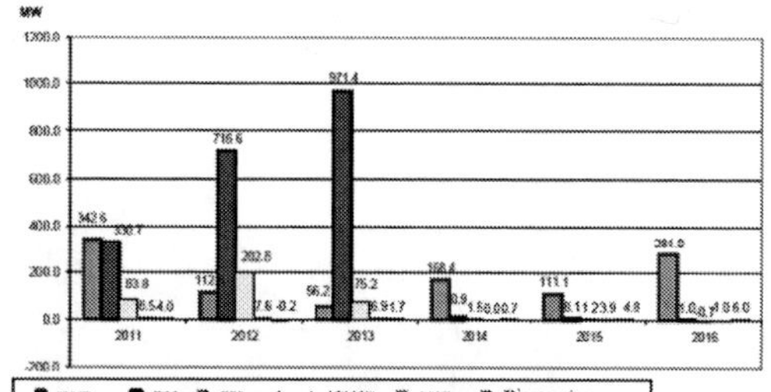

Figure 3: Incorporation of new RES units in the period 2011-2016 [3-5]

The aggregated power of RES projects with production license approaches today 30 GW, exceeding many times the average electricity power needs of the country (about 5 GW today). The electricity production mix in the interconnected system of Greece in 2016 is shown in Fig. 4. The share of PV farms and rooftop installations exceeds 7%, whereas the total contribution of RES approaches 26%. The latter increases to 28% at national level, when the autonomous island systems are also included.

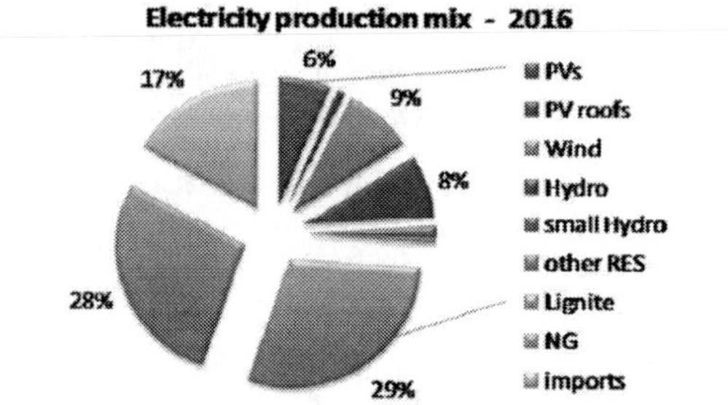

Figure 4: Percentage electric energy production in the interconnected system of Greece

The distribution of all Photovoltaic units in the country, including licensed, operating, or upon consideration for approval, can be find in the RES geo-information created by RAE (Fig. 5), which is available at www.rae.gr/geo/.

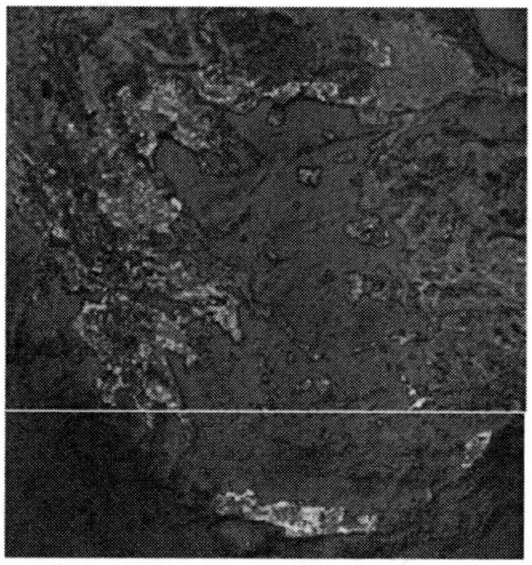

Figure 5: Mapping of PV projects in Greece (Source: RAE, www.rae.gr/geo/)

2 REGULATORY FRAMEWORK AND RESENT LEGISLATION

Important changes of the regulatory framework for RES took place in 2016, in harmonization with the "Guidelines on State aid for environmental protection and energy 2014-2020 (2014/C/200/01). A new institutional support mechanism for RES was introduced by the Law 4414/2016, "State Aid SA.44666 - Greece New operating aid scheme for the production of electricity from RES and HECHP", which then approved by the Competition Commission authorities in November 2016 [6-7].

The new law, which was approved by the vast majority of Greek parliament members, abandons the FiT policy and introduces the feed-in-premiums scheme, along with competitive tenders and virtual net metering, aiming at the gradual integration and participation of high RES and CHP production in the electricity market, in an optimal and cost-benefit way for the society and the final consumers.

The new framework supports RES based on differential incremental status and a Sliding Feed-in-Premium scheme, in which the new RES plants will participating in the wholesale energy market and will be given a variable premium, on top of the standard market price for the generated green power. The total compensation cannot exceed an upper threshold, the Reference Price (RP). Hence, if the market price is higher than this RP threshold, the producers will have to return their excess income to the system (Fig. 6). The FiP will be valid for 20 years, however solar thermal plants will receive it for 25 years. RAE will recommend every year to the Ministry for possible readjustment of the RP for units licensed after 1/1/2016. Only small PV plants up to 500 kW will continue to receive operating aid on a fixed price basis. Exceptions were provided only for non-interconnected islands, until the full opening of the energy market in them [8-9].

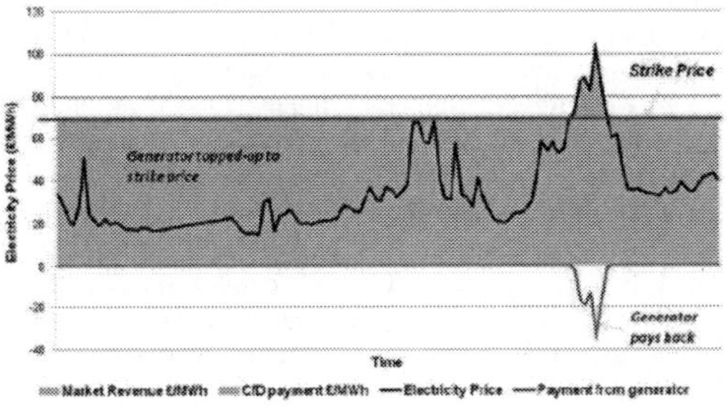

Figure 6: Schematic of the new slide Feed-in-Premium mechanism for RES

Also, the operating projects and those that have signed power purchase agreement before 2016 do not have to participate in the energy market and can continue to apply under the old fixed feed-in tariff system. However, the precondition is for wind projects to be connected to the grid by the end of June, 2018, and for all other projects by the end of year 2017 [9].

From 1/1/2017 the new scheme to approve new renewable energy capacity is based on competitive tenders. Following recommendations of the energy Regulator, the Minister will be able to call on a tender for specific capacities and technologies, and possibly for particular regions. Technology based tenders are at first considered for Greece, because PV projects are much more cost-competitive than wind power plants. However, this policy is subjected to the permission of the EU (DGComp), the decision of which is expected soon. The RP will be determined through this competitive procedure, in order to reduce the cost of electricity from RES to the consumers. This operational support will be given only to successful candidates.

Finally, the new legislation expands the net-metering schemes to ground-mounted PV systems, and allows for the first time the realization of virtual net metering projects by governmental entities (municipalities and municipal enterprises, education institutes, hospitals, etc.) or private not-for-profit or public benefit organizations, in order to facilitate these organizations to become autonomous and cut down significant expenses for electricity [9].

Virtual net metering (VNM) is also allowed now for farmers and farming associations, which moreover can combine in this account more consumption points being in different locations. The above projects refer to any type of RES technologies, including biomass/biogas. The regulatory framework for VNM has recently completed by the Greek Regulator.

3 PILOT COMPETITIVE BIDDING PROCESS

The first movement towards a next deployment period was made in October 2016, when the Regulatory Authority for Energy organized and carried out a pilot auction only for PV projects, in accordance with the legal framework of Law 4414/2016, with the following characteristics [9]:

- Tendered capacity 40 MW.
- Type: "Pay as Bid". In case of a tie, priority is given to the smallest quantity.
- PV-Plants divided in two Categories:

  Category I of PV plants with installed capacity up to 1 MW, which are exempted from the obligation to issue a generation license (ref), and Category II of PV plants with installed power > 1 MW, which are subject to the issuance of a generation license.
- Maximum allowed bidding price 104 €/MWh for Category I and 94 €/MWh for Category II.
- Category I could not exceed 20% of the auctioned power.
- Maximum allowable power per bid in Category II: 10 MW.
- All selected candidates of both Categories have to comply with the rule of 4% Performance Guarantees
- connection activation of the PV plants with an awarded price from the pilot tender (in commercial or trial operation) should take place within:
  - 18 months for PV installations in Category I
  - 24 months for PV installations in Category II

In order to secure a minimum competition, the Regulator set as precondition the realization of each tender the cumulative capacity of the applications in each category should be at least 40% higher than the corresponding auctioned amount. This term introduced a risk in the pilot process, but it was successfully fulfilled.

Special care was devoted by the RAE to secure the simplicity and transparency of this innovative procedure, as also to provision of plenty of information and assistance to all potential participants. More specifically, the following measures were used:

- Development of a specific electronic platform, by which almost all the necessary documents of the applicants were submitted via the internet. Significant load of paperwork is avoided by this procedure.
- Cooperation with a Greek company specialized in conducting online tenders and creation of a specific electronic platform.
- Organization of a workshop to discuss all components of the procedure with potential participants, and essential adjustments according to the bidders' input.
- Continuous support of participants before and during the preparation of the tender procedure by an established group of engineers, IT experts and lawyers.
- Virtual training auctions for the participants to be familiarized with the environment of the electronic auctions platform.

The following Fig. 7 presents the final auctions bids for Category II, showing a remarkable competition. The successful participants who managed to get a price from the two tender processes were sixteen (16).

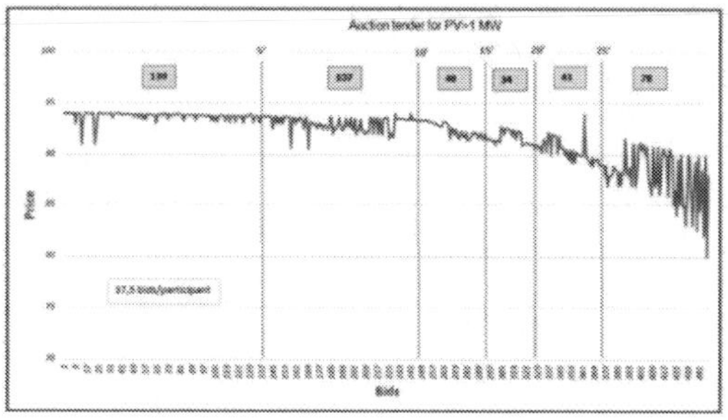

Figure 7: Overview of the auction tender for Category II

The procedure was successfully completed, and a number of sixteen (16) participants managed to get a price from the two tender processes, while the feedback from the applicants and the market was very positive. More specifically:

- o There were no problems of understanding and using the applications during the users training neither any mistakes in the registration process, procedure of application, and participation.
- o Approximately 100% of the auctioned Capacity was covered.
- o The application of the "40% Rule" was beneficial for the Consumer. Remarkable competition achieved in both Categories and resulted in remarkable price reduction:
  - Category I: weighted average price: 98,78 € / MWh (minimum: 94,97 € / MWh)
  - Category II: weighted average price: 83,3 € / MWh (minimum: 79,97 € / MWh)
- o The requirements for transparency, neutrality, simplicity, accuracy, reliability, and speed of execution are fully met.

4 FUTURE PROSPECT FOR PV DEPLOYMENT

The regular competitive bidding process is currently under development by RAE and the Ministry of Energy, based on a sliding feed-in-premium concept. Several critical questions are being considered to this purpose, like the advantages or disadvantages of technology-specific or regional auctions against neutral auctions, and the setting of several other design elements and rules (frequency, auction type, evaluation criteria etc.).

A first conclusion is that the organization of separate auctions for different RES technologies is preferable, for the following reasons:

- Different Weighted Production Cost of Electricity, thus competition between technologies not being possible, e.g. a solar thermal plant cannot compete with a PV or a wind station.
- In most regions of the mainland country PVs are more competitive than wind farms.
- Different way of operating stations (stations of controlled and non-controlled production).
- Achievement of other benefits (domestic value added, jobs, environmental benefits).
- Achievement of a better functioning of the system with the inclusion of all RES technologies (gradual independence on imported fossil fuels).
- Improve the electricity grid stability.

This proposal has been notified to the Commission and it is expected that will be approved, following the approval of a similar request of Germany, on December 2016 (http://europa.eu/rapid/press-release_IP-16-4471_en.htm).

The regular tenders for new RES plants are planned to be started early next year, while the auctioned power will be determined separately for each tender, based on the directions of the ongoing national energy planning, the development and stability of the mainland transmission grid, the difference in RES potential of various regions of the country, and the progress of the interconnection projects of the Aegean Islands.

In the mean time, RAE is considering the organization of a neutral technology base pilot tender for a specific region of Greece of high wind potential, in which wind and solar technologies can be competitive.

A significant number of licenses for PV plants are expected to be activated in the next years, since the existing grid infrastructure can provide the required flexibility and transmissibility for at least 3-4 GW of new RES plants. According to the RES deployment scenario used by the Greek IPTO to conduct the Electricity Generation Adequacy Report earlier this year 2017 [10], as also the forecasts of other experts, at least half of this new RES power will be given by new PV units (Fig. 8). This penetration will become even more enhanced after the interconnections of Crete and other islands of the Aegean Sea, which are planned by the IPTO to be completed in the next years [11-12].

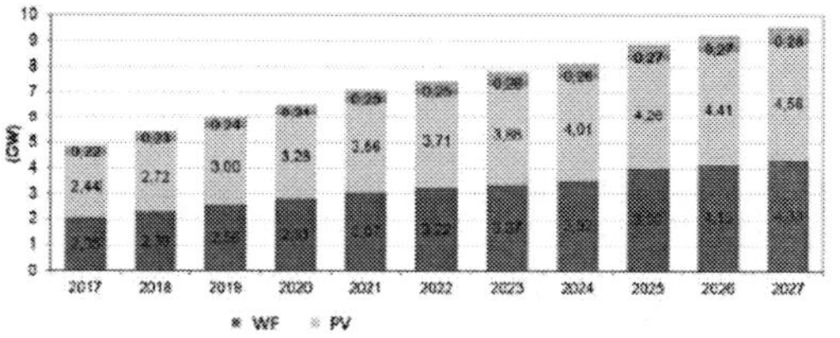

Figure 8: RES deployment scenario (Source: IPTO, 2017 [10])

The local investors are waiting for the final regulatory framework and the permanent tenders conditions for new PV plants. It must be noted however, that the current tariff levels are still quite high compared to other European countries. For example, the resulting price in the recent pilot tender for large PVs (above 10 MW) was 85 €/MWh on the average, whereas in Germany the corresponding prices are approaching 70 €/MWh [13]. Moreover, one must take into account that the capacity factor for PVs in Germany varies between 10 and 13%, whereas in Greece between 19.4 and 24.2%. The above numbers indicate that the amortization period for such investments in Greece is less than half than in Germany. Also, the per capita installed power for both PV plants in Greece is less than half than in Germany. Therefore, there are much greater prospects and returns for such investments today in Greece than in the northern Europe.

The development of the National Energy Roadmap to 2030 has announced on September 2017 by the Minister of Energy, and will be carried out by the Ministry and the RAE, with the aid of an energy planning committee composed by representatives from the energy system operators, the energy producers, the involved public authorities, and specialized scientists and institutions, as well as based on the results of public consultations and of specialized energy studies. The target is to conclude, with the maximum possible consensus, to certain quantitative targets for the main energy quantities (e.g. energy production mixture, energy savings, emissions reduction, etc.) in the next 3-4 months, and then to design a realistic and effective action plan for their achievement.

Finally, in the framework of the Management Code of Non-Interconnected Islands (www.rae.gr, 2014), and aiming at the maximization of RES share in the energy production mix, RAE has recently carried out a public consultation for the possibility of using storage systems in PV and wind units and the results are being considered for the preparation of a regulatory framework supporting the exploitation of this technology that will be announced in the next months. A key point under consideration is the possibility of setting increased tariffs for the controllable production coming from the storage units, which may depend on the specific technology (batteries or pumped storage), on the grid size, or on the current production cost in an island, or on a combination of the above. The possibility of future interconnection of an island should be also taken into account in such storage contracts.

5 CONCLUSIONS

The exploitation of RES in Greece enters a second development phase, in which both the state and the investors have gained adequate experience and maturity to avoid the mistakes and distortions of the past, and to establish a health competition climate to the benefits of all parts, and mainly the Consumer.

The experience and the gained knowledge of the first pilot tender for PV plants showed that Public Sector and Regulatory Authorities are able to plan and flawless execute new e-tendering procedures. Key factor for this success is the wide and transparent information of all stakeholders, which moreover, can contribute to the creation of a trustfulness and confidence climate in the energy market.

Today, the Greek electricity market is being reformed fast towards the European Target model, and to the complete deliberation of the electricity market. In this new environment, which is planned to be activated by the end of 2018, the electricity market rules will become clearer and more fair, securing sales transparency and boosting the participation of more players.

Photovoltaic production, in conjunction with wind energy will be the major RES technologies that will be significantly deployed in the next decades in Greece. Competition is expected to be strong, with solar potential being significant across the entire country, while the wind potential predominates in the Aegean islands.

REFERENCES

[1] M. Šúri, T.A. Huld, E.D. Dunlop, H.A. Ossenbrink, Potential of solar electricity generation in the European Union member states and candidate countries. Solar Energy, 81 (2007) 1295–1305. http://re.jrc.ec.europa.eu/pvgis/.
[2] T. Huld, R. Müller, A. Gambardella, A new solar radiation database for estimating PV performance in Europe and Africa. Solar Energy, 86 (2012) 1803-1815.
[3] HTSO (Hellenic Transmission System Operator). http://www.desmie.gr
[4] LAGIE (Operator of Electricity Market), http://www.lagie.gr/nc/en/home/
[5] HEDNO (Hellenic Electricity Distribution Network Operator S.A, http://www.deddie.gr
[6] EU Decision C (2016) 7272 final, Brussels, 16.11.2016: State Aid Support Mechanism, SA.44666 – Greece. New operating aid scheme for the production of electricity from RES and HECHP. http://ec.europa.eu/competition/elojade/isef/index.cfm.
[7] Law 4414/2016. Establishing a new support scheme for RES and High Efficiency Cogeneration of Heat and Power (HECHP) power plants, Ministry of Environment and Energy. www.ypeka.gr
[8] RAE (Regulatory Authority for Energy), Report of Activities in the year 2016. www.rae.gr
[9] D. Papachristou, The Regular competitive tenders for RES in Greece, Workshop of the Regulatory Authority for Energy, NTUA, Athens, February 10, 2017.
[10] IPTO (Independent Power Transmission Operator): Electricity Generation Adequacy Report 2017-2027, Athens, May 2017. www.admie.gr
[11] IPTO (Independent Power Transmission Operator): Ten Years Network Development Plan, 2017-2026, 4 Oct. 2016. www.admie.gr
[12] RAE (Regulatory Authority for Energy), Update of the strategic study for the interconnection of autonomous island networks to the mainland power system. www.rae.gr
[13] A. Muller, Germany – Experiences with implementing RES E auctions, Federal Ministry for Economic Affairs in Energy, 2016. www.bmwi.de

PROMOTING A SUSTAINABLE DIFFUSION OF SOLAR PV ELECTRICITY IN AFRICA: RESULTS OF THE CODEV PROJECT

Eleonora Annigoni[1*], Alessandro Virtuani[1], Ababacar Ndiaye[2], Nicolas Wyrsch[1], Mamadou Lamine Ndiaye[2], Christophe Ballif[1], Cheikh Mouhamed Fadel Kebe[2]
[1] École Polytechnique Fédérale de Lausanne (EPFL), Institute of Microengineering (IMT), Photovoltaics and Thin Film Electronics Laboratory (PV-Lab), Rue de la Maladière 71b, 2000 Neuchâtel, Switzerland
[2]Ecole Supérieure Polytechnique de Dakar (ESP-Sénégal), Centre International de Formation et de Recherche en Energie Solaire (CIFRES)
BP:5085 Dakar (Fann), Senegal

*Tel: +41 21 69 54274, eleonora.annigoni@epfl.ch

ABSTRACT: In this work we present the result of a collaboration between the Polytechnic Schools of Dakar (ESP) and Lausanne (EPFL) on the testing and monitoring of solar photovoltaic modules. The collaboration has involved the exchange of knowledge, methodologies and data, and, in particular, the analysis of the aging of PV modules exposed to the hot semi-arid climate of Dakar for eight years.
With the aim of promoting a focus on quality and reliability, the long-term goal of the collaboration would be to set-up a testing laboratory for PV modules and systems in Dakar and a training center. The testing laboratory will be working in close collaboration with the University and should potentially have a "lean" and easy-replicable structure.
The implementation of a third-party institution able to assess independently the quality of components and support system developers and installers in the design, commissioning and maintenance of PV projects is crucial to promote a "sustainable" diffusion of solar electricity in Africa, particularly when considering the residential and commercial/industrial rooftop PV market segment. By minimizing risk, focus on quality should promote a virtuous cycle leading to: (1) mitigation of financing costs of solar projects, therefore, considerably reducing the overall costs of this technology, (2) increase positive perception and awareness about PV.
Keywords: reliability, testing, international cooperation, emerging markets.

1 INTRODUCTION

Affordable and Clean Energy (i.e. ensuring access to affordable, reliable, sustainable and modern energy for all) is one of the 17 Sustainable Development Goals (SDGs), officially known as *Transforming our world: the 2030 Agenda for Sustainable Development*, set by the United Nations in 2015 [1]. At today's prices, and with a steadily declining price trend, solar photovoltaic (PV) electricity is increasingly perceived as one of the key-enabling technologies potentially leading to the achievement of this single target in those parts of the word where the access to electricity is still extremely limited. As for example in sub-Saharan Africa where 70% of the population (i.e. 600 million person) has no access to this form of energy [2].

In recent years, however, with a total (all Africa) cumulative capacity reaching over 2 GW in 2015, the market for new solar PV installations in Africa is expanding very rapidly [3]. Most of the new additions in capacity are covered by the utility-scale segments (e.g. South-Africa and Algeria in 2015), but rooftop (i.e. residential, industrial and commercial) as well as micro/pico-solar PV installations are experiencing similar growth rates, though official statistics covering these market segments are by far less reliable.

Nevertheless, the diffusion of rooftop PV installations in developing countries is often hindered by: (1) a poor perception about the reliability of this technology, and (2) by the relatively high up-front capital and financing costs, combined with a difficult access to credit [4, 5]. A promotion and sustainable diffusion of solar electricity in developing countries cannot take place if adequate control mechanism and quality checks (at component and system level) are not set in place.

The paper presents the results of a collaboration between ESP and EPFL that has involved the exchange of knowledge, methodologies, and data, and, in particular, the analysis of the aging of PV modules exposed to the hot semi-arid climate of Dakar for eight years.

As a follow-up of this initial collaboration, the paper addresses the issue (and relevance) of establishing a testing laboratory for PV in Senegal with the aim of promoting a focus on quality and promote a smooth diffusion of solar electricity in the region, particularly when considering the solar rooftop market segment.

2 EXPERIMENTAL WORK

A set of modules from different manufacturers has been exposed for eight years on the premises of ESP in Dakar (see Fig. 1), see [3] and [4] for a description of the testing facility and measurements methods at ESP. Dakar has a *hot semi-arid* climate (Köppen climate classification: *BSh*), with a short rainy season and a lengthy dry season. Other climates, which may be encountered in Senegal are mostly *hot-desert* (BWh) in the North and *tropical savannah* (Aw) in the South.

Figure 2 shows a map of Global Tilted Irradiation, (GTI) and solar electricity potential of Senegal for optimally-inclined modules [7]. With a yearly average of cumulative irradiation higher than 2000 kWh/m^2/y for all the country, the map clearly indicates the extremely good availability of solar resources in Senegal, and the large potential for solar electricity.

A characterization of a sub-set of modules has been performed at ESP after four years of operation and at EPFL in late 2016 (after 8 years). This has included visual inspection, Infra-Red (IR) and Electro-Luminescence (EL) imaging, and power (IV) measurements at STC of the modules exposed in Dakar for 8 years.

Fig.1: Researchers of ESP and EPFL on the outdoor test facility at ESP in Dakar.

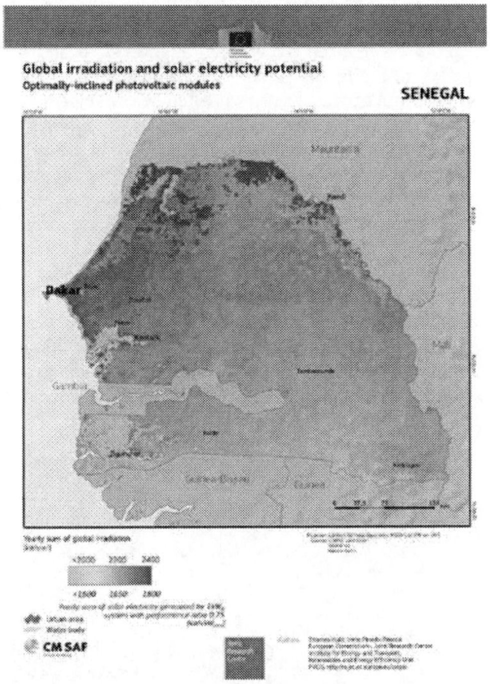

Fig.2: Map of Global Irradiation and solar electricity potential of Senegal for modules installed at optimal tilt [8].

At ESP, IV curve measurement were performed under natural sunlight with a portable IV tracer (IV HT400, see Ref. [6]), and at EPFL using a LED-Halogen based pulsed solar simulator that replicates the AM1.5 spectrum. EL and IR imaging were performed by applying a current corresponding to the short-circuit current (I_{sc}) to the modules by means of a power supplier.

3 RESULTS

For one specific module, representative of the whole set, the results of IV curve measurements are reported in Table I, showing the measured PV parameters before the modules installation and after 4 years of exposure, performed at ESP. The results of the measurements performed at EPFL after 8 years are shown as well.

In Table I, one can observe that the degradation trend is not consistent if comparing EPFL and ESP measurements, which proves the additional difficulty of assessing degradation rates when the measurement procedures (and possibly the calibration of instruments and sensors) are not constant during the years. It also highlights the necessity to have a good alignment and traceability to SI metrology standards of both measurement set-ups.

Table I: Degradation of PV parameters after exposure to the climate of Dakar for four and eight years, respectively.

	P_{mpp} [W]	V_{oc} [V]	I_{sc} [A]	FF [%]
Initial measurements at ESP	113.15	21.76	7.41	70.18
Measurements at ESP after 4 years	98.56	21.73	7.19	63.06
Measurements at EPFL after 8 years	110.69	21.51	7.49	68.7
Initial vs 4 yrs (%)	-12.89	-0.18	-2.90	-10.13
Initial vs 8 yrs (%)	-2.18%	-1.17	1.14%	-2.09%

For the same module, Fig. 3 shows results of visual inspection, EL and IR imaging performed on the modules after 8 years of exposure. The EL image highlights the presence of cracked cells, and the IR one the presence of hot-spot correlated with the presence of localized shunting paths.

Fig.4 shows the picture of a cell from the same module, exhibiting a pattern common to several of the modules installed on the roof (including modules from different manufacturers): i.e. the presence of decorated micro-cracks with burnmarks and discoloration of finger grids, in combination with a possible delamination of the encapsulant.

Finally Fig. 5 and 6 show how soiling can be a major issue in Dakar leading to current (and power) losses as high as -60%, particularly in the dry season. In particular, Fig. 6 shows an open junction box (J-box), belonging to one of the exposed modules, claiming IP65 protection against dust and water ingress, where the first and second digit refer to dust (6 = dust tight) and water protection (5 = protected against water jets).

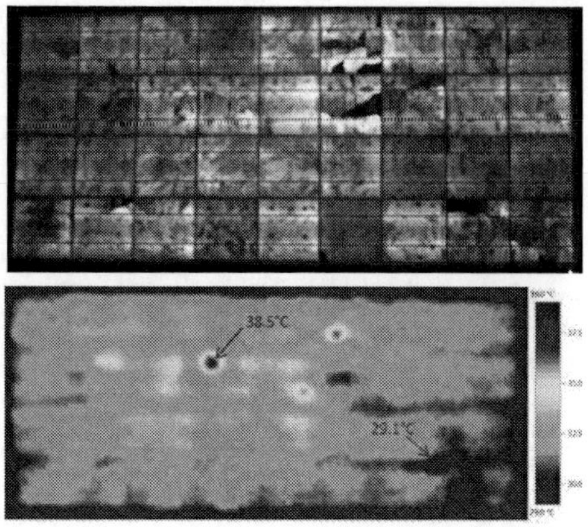

Fig. 3: From top to bottom: visual inspection, EL and IR imaging of a PV module exposed in Dakar for 8 years. Measurements were performed at EPFL in 2016.

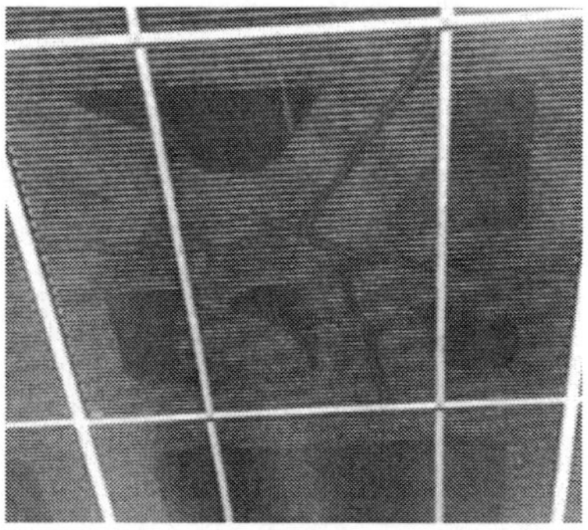

Fig. 4: Particular of one of the cells of the module shown in Fig. 3, highliting the presence of decorated micro-cracks with burnmarks and discoloration of finger grids, in combination with a possible delamination of the encapsulant.

Fig. 5: Soiling can account up to 60% current and power losses.

Fig. 6: Open junction box, belonging to one of the exposed modules, claiming IP65 protection against dust and water ingress.

4 THE QUEST FOR QUALITY

As can be inferred by viewing Figures 3, 4 and 6 the quality of the modules (and other components) available in several African countries is often below standard, and, often, no compliancy to the meaningful international qualification standards (e.g. IEC, UL) can be proven. Most of the PV modules available for individual users or small installers are distributed through retailers with often a very limited knowledge about PV, who cannot offer any traceability about the origin of the products. Fig. 7 shows the picture of PV module available for sale on the street of a local market (Marché Sandaga) in Dakar. This combined with the sometimes extremely harsh climatic conditions that can be experience in several parts of the continent (from desert to hot and humid climates) can pose a serious threat to the long-term performance of the PV plants realized using these modules.

Fig. 7: PV modules available for sale at a local market (Marché Sandaga) in Dakar.

Another major barrier for a "sustainable" diffusion of solar PV electricity in Africa, is the level of training of system designers and installers, which should be supported by adequate training and qualification schemes. Professionals should therefore learn how to select the right components, how to design a system based on the project-specific requirements and constraints (e.g.

presence of shading, siting, budget, etc.). Similarly, adequately skilled professionals should be in charge of the installation, commissioning and operation and maintenance (O&M) phases.

By focusing on some of these aspects (quality of components and training), the long-term aim of the collaboration between ESP and EPFL is to set up a PV component test laboratory in Dakar which should be able to qualify modules available on the Senegalese market and certify minimal quality standards. This center should also serve as a hub for teaching and technical support between the University users, installers and project leaders as well as public authorities. Within the collaboration between ESP and EPFL, module test sites and dedicated monitoring equipment are being deployed in several University sites in Senegal. We thereby aim at further studing degradation and lifetime of modules in various types of sub-Saharan climates. In this context the presence of a test laboratory should offer additional access to data and experiences.

We believe that, if successful and adequately replicable in other contexts, the implementation of such a testing and training laboratory could really contribute in setting up a virtuous cycle, bringing to a "sustainable" diffusion of solar electricity in the whole continent.

Adequately trained professionals, combined with the possibility of verifying the quality of the components directly in the country, will prospectively have a strong impact on the bankability of small/medium projects, and therefore contribute in reducing the cost of financing of the projects. Cost of financing that, due to the large up-front costs of solar PV projects, constitute per-se a significant barrier to the diffusion of this technology. Another, positive side effect of would be that contributing in increasing the awareness and the positive perception of solar PV among the population.

5 CONCLUSIONS

Harsh climatic conditions such as the ones prevailing in Senegal seriously threat long-term module efficiency. This problem is even more acute given the often poor quality of the products available on the market. A collaboration between the Polytechnic schools of Dakar in Senegal and Lausanne in Neuchâtel to investigate PV module degradation and failure modes related to such climatic environments was started with the help of the Cooperation and Development Center of EPFL (CODEV).

As a follow-up of this initial collaboration, a testing laboratory for PV in Senegal is being set in Dakar with the aim of promoting a focus on quality and a smooth diffusion of solar electricity in the region, particularly when considering the solar rooftop market segment. This test laboratory should become an important actor of the PV value chain in Senegal and should offer, besides testing services, also education and technical advices. The center is also expected to become a valuable partner for future collaboration projects between EPFL and ESP to contribute to the understanding of module degradation in African countries and to establish technical guidelines for improved lifetime and quality.

6 ACKNOWLEDGEMENTS

This work was funded by the Cooperation and Development Center of EPFL (CODEV). The creation of a quality test center in Dakar is supported by the REPIC (www.repic.ch) platform. We gratefully acknowledge the contribution of all PV-team members at EPFL and ESP.

REFERENCES

[1] Sustainable Development Goals: United Nations Resolution A/RES/70/1 of 25 September 2015, par. 54;

[2] WEO-2014, Africa Energy Outlook, International Energy Agency (IEA) report, 2014;

[3] SOLAR PV IN AFRICA: COSTS AND MARKETS, International Renewable Energy Agency (IRENA) report 2016;

[4] Ivan Nygaard, Ulrich Elmer Hansen, Gordon Mackenzie & Mathilde Brix Pederse, *Measures For Diffusion Of Solar PV In Selected African Countries*, Intern. Jour. Sust. En. 36, 707-721, 2015;

[5] Joseph Amankwah-Amoah, Solar Energy in Sub-Saharan Africa: The Challenges and Opportunities of Technological Leapfrogging, Thunderbird International Business Review, 2014;

[6] Ababacar Ndiaye, Cheikh M.F. Kébé, Abdérafi Charki, Papa A. Ndiaye, Vincent Sambou, Abdessamad Kobi, *Degradation evaluation of crystalline-silicon photovoltaic modules after a few operation years in a tropical environment,* Solar Energy 103, 70–77, 2014;

[7] JRC – PV-Gis maps: http://re.jrc.ec.europa.eu/pvgis/cmaps/afr_new.htm

DISTRIBUTED PV GENERATION: CHALLENGES AND SOLUTIONS FOR ITS EXPANSION AND INTEGRATION IN THE BRAZILIAN GRID WITH CASE STUDY OF THE IMPACT OF THE WHITE TARIFF

SILVA, Vinícius Oliveira da[1*]; BERNHARD, Diego Biasseto [2]; RELVA, Stefania Gomes[1]; GIMENES, André Luiz Veiga[1]; UDAETA, Miguel Edgar Morales[1]; SALLES, Maurício Barbosa de Camargo.
[1]Energy Group of the Electric Energy and Automation Engineering Department of the Polytechnic Scholl of the University of São Paulo.
[2]Flue8 Industries.
[3]Electric Energy and Automation Engineering Department of the Polytechnic Scholl of the University of São Paulo.
Avenida Luciano Gualberto, travessa 3, n° 158, Sala AN-16 do prédio Anexo, Zip code: 05508-010.
vinicius.oliveira.silva@usp.br

ABSTRACT: The objective of this work is to study and evaluate the impacts and challenges of the insertion of distributed generation (DG) linked to photovoltaic (PV) technology in Brazil and to verify its impact in a residence energy bill. The Brazilian panorama of the generation PV, the obstacles for its implantation, as well as its technical and regulatory solutions are analyzed. Since 2012 there was an increase in the installation of DG due to regulatory changes. The exemption from taxes such as PIS/COFINS and ICMS caused a growth of 233% in PV installations between 2015-2016. In relation to tariff changes, the appearance of the white tariff in 2018 may negatively influence the growth of PV photovoltaic, because, for those who adopt the tariff, it would reduce the off-peak energy offset credit, being in this period of the day the biggest PV generation. Thus, it will reduce the savings of its monthly bill in relation to the conventional tariff. The white tariff comes as a financial stimulus to the consumers change their load curve, shifting their maximum demand out of the peak hours, however it reduces the economy generated by the DG PV, because the reduction of the residential energy bill with the installation of DG PV is 90% in relation to the conventional tariff, however, the reduction of the energy bill with the change of habits added to DG FV and with the white tariff is 78%. Thus, the implementation of DG PV is directly linked to its cost and the consumer who choose to install DG PV should remain with the conventional tariff.
Keywords: Photovoltaic system, Distributed Generation (DG), Electricity tariff, White Tariff, Conventional tariff.

1 INTRODUCTION

Distributed generation (DG) can be understood as a power system allocated close to the consumption center without the need to deploy extensive transmission networks [1]. Currently, it refers to a small generation system, installed in a dispersed way and connected to the local utility grid [2]. The module 1 of PRODIST also defines distributed generation as: Power generating plants of any power, with installations connected directly to the electrical distribution system or through consumer installations, being able to operate in parallel or in isolation and dispatched - or not - by ONS. In mature markets GD is considered as an alternative to the expansion of large and centralized parks [2], being used in energy planning as a rule and not as an exception [1].

The main factors that contribute to the implementation of DG are: (i) alternative to expansion or use of transmission lines, (ii) increasing demand for reliable and quality electricity, (iii) stabilization or flexibilization of electricity prices, (iv) support for ancillary services, (v) concerns about climate change, (vi) liberalization of the energy market, and (vii) development of small scale electricity generation technologies [1] such as gas turbines, microturbines, positive displacement, photovoltaic modules, wind turbines.

In Brazil, there is an increase in the number of DGs (Fig. 1), mainly PV systems (Fig. 2) [3], which have recently undergone profound changes related to the reduction of technology costs, improved regulation, taxation and sources of financing [4], making the environment more competitive and, for in turn, expanding its implantation, mainly in the residential sector [5].

In this sense, the objective of this work is to study the insertion of the DG linked to photovoltaic technology in Brazil, seeking to understand the obstacles inherent to its implementation and expansion process, by means of the survey of the norms, laws and directives in force in Brazil. primary focus on the change in tariffs linked to the residential consumption class.

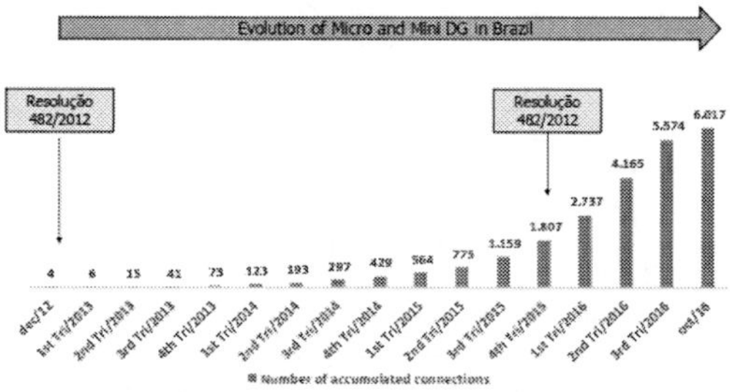

Figure 1. Evolution of micro and mini DG in Brazil [5].

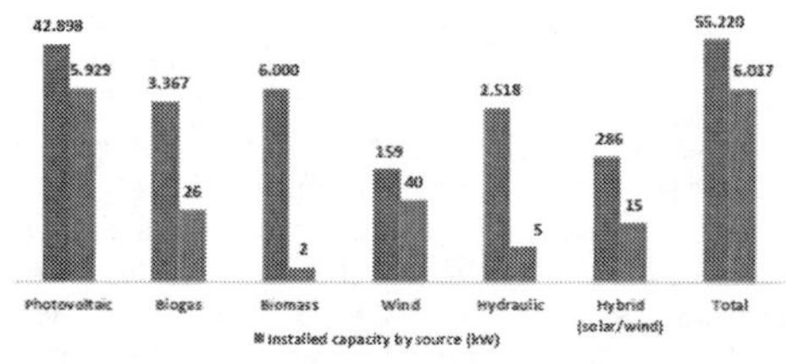

Figure 2. Installed capacity of DG in Brazil by source [5].

2 OVERVIEW OF DG IN BRAZIL

Since the publication of the Resolution No 482 in 2012 up to October 2016, 6,017 mini and micro power stations of GD have been installed, totaling 55.2 MW, of which 5,929 (99%) are from solar photovoltaic (PV), 40 wind power, 15 hybrid (solar/wind), 26 of biogas, 2 to biomass and 5 to hydraulics [3], which represents less than 0.04% of the Brazilian controlled power [6]. In 2024, ANEEL predicts that there will be about 1.2 million consumer units producing its own energy, totaling 4.5 GW [7], reaching 2.2% of the installed power [7].

2.1. Characterization of micro and mini DG and energy compensation

Normative Resolution No 687/2015 [8]:

- Micro and mini DG are small generating plants that use renewable sources of electricity or qualified cogeneration connected to the distribution network, through installations of consumer units.
- Micro DG with installed power P <= 75 kW and Mini DG 75 kW <P <= 5 MW (3 MW for the water source.
- Credits expire in 60 months and can reduce the consumption of the same owner units in different locations within the same distributor area.
- Allows the installation of DG in condominiums with sharing among condominium owners in defined%.
- It allows a consortium or cooperative to install a micro or mini DG and use the energy to cut down on the consortium or cooperative invoices.

In relation to the Brazilian taxation for DG [8]:

- Law No. 13,169 / 2015 removes the incidence of PIS and COFINS from the positive difference between the energy consumed and the energy injected by the consumer unit with micro or mini DG. Serves all states.
- CONFAZ published ICMS Agreement 16/2015, repealing ICMS Agreement 6/2013, which authorizes the States to grant exemption in internal operations related to the circulation of electric energy.

Until November 2016, 21 States joined the ICMS 16/2015 [9] (Fig. 3).

Figure 3. DG connection by State of Brazil.

About the Brazilian tariff modality and the rules of the white tariff see Table 1, Table 2, Table 3 and Fig. 4.

Table 1. Tariff modality by group of consumption of Brazil [10].

Tariff Modality	Group	Tariff Rule
Conventional monomial	B	Constant tariffs for electric energy consumption, regardless of hours of use throughout the day.
White hour	B[a]	Differential rates of electricity consumption, according to hours of use throughout the day

a. except for the subgroup B4 and for the sub-classes Low Income of subgroup B1

Table 2. Tariff hour type in group B [10].

Tariff hour type	Tariff Rule
On-peak	Three consecutive daily hours defined according to the daily load curve of the distributor area, except for Saturdays, Sundays and national holidays.
Intermediary	Conjugated to the peak tariff position, being one hour immediately before and another immediately after, applied to Group B.
Off-peak	Set of consecutive and complementary daily hours, those defined as being on-peak an intermediate for Group B.

Table 3. Entry into force of the white tariff [10].

Implementation	Average Consumption Class
2018	kWh/month ≥500
2019	500>kWh/month≥ 250
2020	class of group B[a]

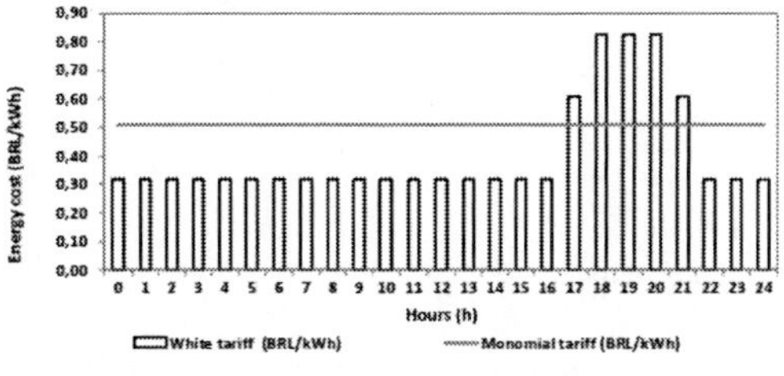

Figure 4. Electricity price by consumption hour [11].

3 CASE STUDY

The case study considers a residence with (Table 4):

i. A PV system of 5.0 kWp installed;
ii. At a site with a capacity factor (CF) of 16% [12];
iii. Average monthly consumption at the on-peak of 42% and off-peak of 58% [13];
iv. Tariff according to Figure 4.

Considering these premises, the monthly saving or loss for six scenarios consists of the comparison of:

i. Conventional monomial (CM) with PV system (DG);
ii. CM with White Tariff (WT);
iii. CM with WT and changes in consumption habits (WT+CH), in this case part of the consumption and moved to hours outside the on-peak;
iv. WT with WT+CH;
v. WT with WT+DG+CH; and
vi. WT with WT+CH.

4 RESULTS

The white tariff comes as a financial stimulus, in the case the economy generated, for consumers to change their load curve, shifting their peak demand out of peak hours. From the economic point of view, the micro and mini DG, combined with the white tariff and the tariff flags, can affect the billing of the distribution companies. In addition, the tariff policy directly affects the payback of the installation of photovoltaic systems.

On the other hand, consumers who adopt DG PV would be discouraged from adhering to the white tariff because they would reduce their credit gains in the generation because it would be almost totally out of point, where it would have less value.

The results show that the best scenario is with CM and DG charging, when there is 90% savings over conventional payment. On the other hand, the worst-case scenario is that of the white tariff without a change of habit in which there is a 5% increase over the conventional payment (Table 5).

5 CONCLUSIONS

1. From the economic point of view, the micro and mini DG, combined with the white tariff and the tariff banners, can affect the sales of distribution companies.
2. The white tariff comes as a financial stimulus, in the case the economy generated, so that consumers change their load curve, shifting their maximum demand out of peak hours, however it reduces the savings generated by DG FV.
3. The load curve of the consumer unit is changed with the distributed generation.
4. Load curve changes can be controlled with active demand management using energy
5. storage.
6. Changes in power factor can be regulated by inverter control.
7. The change in load curve reduces the load on distribution transformers and
8. dramatically increases their life.
9. Photovoltaic generation can raise grid voltage levels.
10. The grid voltage can be moderated by inverter injection control.

Thus, the scale model architecture of a cPVPP is effective and possible to be reproduce in scale. If there is expansion of this type of system in conjunction with HPP it is possible to increase the bulk energy in a commercial HPP and store water in its dam.

Table 4. Premises of Generation and Consumption of electricity.

Installed Capacity (kWp)	CF	Monthly Electricity Generation (kWh)	Monthly Consumption (kWh)		
			Total	On-peak	Off-peak
5.0	16%	576	640	269	371
			100%	42%	58%

Table 5. Consolidated results of the case study.

Base	CM	CM	CM	WT	WT	WT
Cases	DG	WT	WT+CM	WT+DG	WT+CH	WT+DG+CH
Saving	293.76	-15.80	52.90	184.56	68.70	253.02
Saving/base	90%	-5%	16%	57%	21%	78%

RERERENCES

[1] G. Pepermans, J. Driesen, D. Haeseldonckx, R. Belmans, W. D'haeseleer. "Distributed generation: definition, benefits and issues." Energy Policy, vol. 33, pp. 787-798, Abr. 2005

[2] R. R. Sgyaçabia, "Avaliação do Impacto da Geração Distribuída sobre o Sistema de Proteção de Sobrecorrente de uma Rede de Distribuição Operando em Ilhamento Intencional" Dissertação (Mestrado), Departamento de Engenharia Elétrica da Escola de Engenharia de São Carlos, Universidade de São Paulo, Brasil, 2015.

[3] EPE (2014), Nota Técnica DEA19/14: Inserção da Geração Fotovoltaica Distribuída no Brasil – Condicionantes e Impactos, Brasil, p. 60.

[4] Aneel (2016). Acesso: http://www2.aneel.gov.br/scg/gd/gd.asp

[5] Aneel (2015). Resolução Normativa nº 687, de 24 de novembro de 2015. Acesso: http://www2.aneel.gov.br/cedoc/ren2015687.pdf

[6] Aneel (2016). Acesso: http://www2.aneel.gov.br/aplicacoes/capacidadebrasil/capacidadebrasil.cfm

[7] EPE (2015). "Plano Decenal de Expansão de Energia 2024". Ministério de Minas e Energia. Empresa de Pesquisa Energética. Brasília: MME/EPE, p. 468

[8] Aneel (2016). Micro e minigeração distribuída: sistema de compensação de energia elétrica. 2. ed, Brasília, p. 31.

[9] CONFAZ (2016). Conselho Nacional de Política Fazendária. Acesso: https://www.confaz.fazenda.gov.br/legislacao/convenios/2015/cv016_15.

[10] Aneel (2016). Acesso: http://www2.aneel.gov.br/area.cfm?idArea=43&idPerfil=2&idiomaAtual=0

[11] Aneel (2016). Micro e minigeração distribuída: sistema de compensação de energia elétrica. 2. ed, Brasília, p. 31

[12] da Silva, V. , Udaeta, M. , Gimenes, A. and Linhares, A. (2017) Improving the Performance of Photovoltaic Power Plants with Determinative Module for the Cooling System. Energy and Power Engineering, 9, 309-323. doi: 10.4236/epe.2017.95021.

[13] A. A. Francisquini. "Estimação de curvas de carga em pontos de consumo e em transformadores de distribuição". Dissertação (Mestrado), Universidade Estadual Paulista, Ilha Solteira Brasil, 2015.

TRAINING THE NEXT GENERATION OF PV RELIABILITY EXPERTS – NEW MARIE-SKLODOWSKA CURIE (MSCA) PROJECT SOLAR-TRAIN

K.-A. Weiss[1], S. Saile[1], A. Keiner[1], L. Pitta Bauermann[1], G. Oreski[2], R. Gottschalg[3], D. Moser[4], M. Topic[5], A. R. Lagunas[6], P. Chiantore[7], M. van Iseghem[8]

[1] Fraunhofer Institute for Solar Energy Systems ISE, Freiburg, Germany
[2] Polymer Competence Center Leoben, Leoben, Austria
[3] Loughborough University, Loughborough, United Kingdom
[4] EURAC Research, Bolzano, Italy
[5] University of Ljubljana, Ljubljana, Slovenia
[6] Centro Nacional de Energías Renovables, Sarriguren, Spain
[7] BayWa r.e. Operation Services, Rome, Italy
[8] Electricité de France, Moret-sur-Loing, Franc

Karl-Anders Weiss, Fraunhofer Institute for Solar Energy Systems (ISE), Heidenhofstr. 2, 79110 Freiburg, Germany
karl-anders.weiss@ise.fraunhofer.de, Phone +49-761-4578-5474

The service life estimation in the photovoltaic (PV) industry is yet in its infancy, requiring both underpinning science and trained personnel to reduce costs of energy. An unmet industrial requirement is an accelerated and operating environment specific, service life time and energy yield assessment. SOLAR-TRAIN qualifies early stage researchers (ESRs) in the field of PV durability as part of a highly innovative, multi-disciplinary project meeting industry requirements. The research objective is to develop novel and validated models for the service life time and energy prediction of PV modules and systems. The elements to this puzzle are researched in the frame of 14 PhD projects with individual areas of focus, ranging from investigations on climatic degradation factors, through to research on material level, simulations and modelling of energy yield and life time. The research projects are enhanced by an innovative training scheme in close cooperation with industry. This inter-sectoral approach provides excellent theoretical and technical background as well as immersion in different business sectors and career mentoring.

Keywords: PV, reliability, durability, life time estimation, modelling, yield assessment, encapsulation, polymers

1 MOTIVATION

Service life estimations are standard in the majority of industries but crucially not in photovoltaics. It is typically measured in hours of life-time. Not only has this to be realized for PV to ensure commercial viability but also the performance over time needs to be predicted as the value of functional materials depends also on the level of performance throughout the life-time.

SOLAR-TRAIN delivers an essential contribution to this problem and significantly advances state of the art of service life prediction for PV. This will decide the viability of companies, as devices and systems may be over-engineered or not meet the predicted 20-25 year lifetime. Low failure and ageing rates ensure financial viability of investments. This has been recognised widely.

In the framework of the Energy Roadmap 2050, PV is seen as a key technology for decarbonizing energy supply. The focus of the PV industry has been largely to produce cheaper devices (modules) and increase efficiencies, to reduce initial system costs. It has been shown that the biggest uncertainty in cost of energy generated from PV is system degradation.

Supposedly identical modules show very different degradation rates in the field. The variance appears to be larger for newer systems which has an effect on the financial return of a PV system. This is the biggest uncertainty in energy generation cost and is largely independent of solar cell technology. This is mainly driven by the weathering behavior of the encapsulation, which is a multi-layer polymer system.

It is not possible to predict either aging rate or lifetime today. All estimations are done based on standardized laboratory tests, taken from other industries and, as shown e.g. by Wohlgemuth et al [1], often accelerating inappropriate failure mechanisms. Yet, there is still a large gap in understanding this multi-disciplinary problem.

2 CROSS-SECTORAL, MULTI-DISCIPLINARY APPROACH

2.1 14 Individual Research Projects

To counter these issues SOLAR-TRAIN investigates module performance along the entire value chain. The work packages are achieved by the collaborative effort of 14 ESRs with individual research projects. The different topics for ESR theses are innovative elements within the research framework. ESRs will carry out detailed root-cause analysis, site meteorology, device meteorology, data-mining and elemental modelling of material, device and system behavior in relation to module degradation.

Starting in summer 2017, commercial and test samples are produced and tested in distributed measurement campaigns and deployed in different operating environments. They are exposed to state-of-the-art and to-be-developed stress cycles to allow a validated link of degradation to stresses, production methods, materials and methods of deployment. Accelerated and lower cost test cycles for the assessment of innovative materials and module developments will be delivered.

SOLAR-TRAIN's research questions are investigated by a multi-disciplinary team of chemists, physicists, mathematicians and electronic engineers. They will develop new insights in all subtopics of PV reliability and durability such as (a) climatic degradation factors, (b) system analytics, (c) material (polymer) parameters, (d) service life & energy models, (e) linking production to performance and (f) performance enhancement by improved O&M.

Table I: ESR research projects

ESR	Topic (host institution)
ESR 1	Investigation of the correlation of different climatic degradation factors as a basis for typical load definition (Fraunhofer ISE, Germany)
ESR 2	Analysis of material and module parameters and correlation to degradation modes (Fraunhofer ISE, Germany)
ESR 3	Development of service life model for PV modules (Fraunhofer ISE, Germany)
ESR 4	Simulating realistic ageing conditions (Loughborough University, UK)
ESR 5	Influence of manufacturing processes on durability (Loughborough University, UK)
ESR 6	Statistical analysis of reliability and durability of fielded systems (Loughborough University, UK)
ESR 7	Potential, temperature and humidity induced degradation of PV module energy yield. (University of Ljubljana, Slovenia)
ESR 8	Outdoor performance monitoring, technology related energy yield / uncertainty analysis and statistical techniques for the evaluation degradation (University of Ljubljana, Slovenia)
ESR 9	Investigation of PV module degradation mode indicators (CENER, Spain)
ESR 10	Influence of encapsulant - backsheet combinations on PV module degradation modes (PCCL, Austria)
ESR 11	Characterization of degradation behaviour and material interactions in PV modules (PCCL, Austria)
ESR 12	Uncertainty framework for the PV energy yield to link failures, outdoor effects and degradation with financial indicators (EURAC; Italy)
ESR 13	Service life modelling of a PV plant using a physical approach based on climatic data and failure mode analysis (EDF; France)
ESR 14	Causes of degradation and performance improvement in a complete PV system (BayWa r.e., Italy)

The individual research projects will be merged with the aim to improve the state-of-the-art in service life prediction of PV modules and systems to enhance the industry's economic security and predictability.

2.2 Knowledge beyond mere academia

At the core of SOLAR-TRAIN is the intersection between research and industry. Throughout the entire project duration, ESRs are obliged to move across sectors and borders for so-called Secondments, i.e. scientific stays at the projects beneficiaries or one of its 10 partner organizations. Each will support the career development of ESRs by training-on-the-job and courses on economic issues, such as impact assessment and entrepreneurship.

For a most effective cross-sectoral training, beneficiaries and partners present the entire value chain, from material developers and manufacturers, through to operators and insurance companies, allowing ESRs deep insights into every aspect of PV manufacturing and deployment and finally also to build up a sustainable professional network across Europe.

3 INNOVATIVE TRAINING SCHEME

The academic part of the project is supplemented by an innovative training scheme composed of eight different elements including summer schools and workshops fostering interdisciplinary understanding, interaction and cooperation by individually selected soft skill modules. Action centered learning and mentoring rounds off the education part of the project with the aim to obtain a PhD.

3.1 Basic Training

Each beneficiary and host institution has different regulations. The ESRs will have to follow the prescribed training approach of their nominated host partner to ensure compliance with local regulations. All institutions have a long track record and have training courses tailored to their individual framework. These are good practice and sufficient training to pass a PhD. The basic training consists of the following elements.

- Laboratory and health safety
- Academic research and writing: Literature survey, scientific writing, documentation, giving presentations, metrology and data evaluation
- PhD as project: time management and planning
- Work flow and organization of each beneficiary
- Managing your supervisor

In addition, there is the project work plan and related questions such as how to ensure that the various beneficiaries will be reached and how to deliver the distributed work program. All these issues are subject to the Basic Training.

3.2 Beginners' Week

The Beginners' Week is designed to equip ESR's of different disciplines with the equal foundations to be able to work effectively in multi-disciplinary environment. Parts of the curriculum are fundamentals in photovoltaics, module design and reliability assessment. The teaching modules are designed and taught by renowned PV and reliability experts.

Beyond developing an understanding of the technological framework, this basics seminar is a platform for a first face-to-face interaction of SOLAR-TRAIN's ESRs. These personal contacts are essential for the future successful cooperation. To ensure a successful social integration, the schedule includes time slots for social events and activities.

3.3 Summer Schools

Consecutive to the basics seminar, a series of three Summer Schools is envisaged. Summer Schools are obligatory for all ESR's who have to present their work and progress. Each presentation will be followed by a peer discussion of the presented research and joint planning of next steps.

The second half of the summer school will provide advanced courses on SOLAR-TRAINs fields of research as well as coaching on exploitation and entrepreneurship. The planned topics are listed in Table II.

Table II: SOLAR-TRAIN Summer Schools

#	Topic
1	**State-of-the-art in PV Module Technology** (advanced course) at the University of Loughborough, UK
2	**PV Module Lifetime & Reliability** (advanced course) at the University of Ljubljana, Slovenia
3	**Industrial Soft Skills & Entrepreneurship** (advanced course) at PCCL, Austria

The Summer School series deepens the understanding of cell technologies, PV module composition, components and materials, standardization and standardization and proceeds to the specifics of life time and reliability testing, performance monitoring and life time modelling. The final Summer School places these insights into the economic context, focusing on aspects of quality management, Intellectual Property Rights (IPR) management, ramp up processes, product development and project management in industry.

3.4 Online Seminars

To ensure a continuous exchange between ESRs in the periods between the Summer Schools there are monthly online seminars. Each month half the ESRs will report on their progress. For each meeting a chair of the ESRs is defined who will have to organize the meeting, take minutes of discussions, collect presentations and archive them in the project software.

Typically, presentations are followed by questions/discussions and then a round of formative feedback on how to improve the presentation. The seminars will be essential to track progress and also for the cohort feeling as the ESRs 'meet' regularly. The rotating chair and the minute taking teaches essential skills for future scientific careers.

3.5 Individual Training Modules

For the transfer of the acquired knowledge into academia or industry, soft skills and methodologies for an effective work and time management are paramount. In SOLAR-TRAIN, extracurricular training activities that prepare ESRs for the work are given particular consideration as they are embedded as individual training modules (ITM) throughout the complete ESR period of three years.

Offered are two modules per year, i.e. a total of six of courses providing oral presentation skills, grant proposal writing, project management for scientists, leadership skills for scientists, acquisition and negotiation skills and intercultural communication. The requirement is to attend a minimum of three during the course of the ESR period.

Individual training modules are held by selected trainers and coaching experts from different companies. This ensures a change of perspective and input for every seminar as well as learning methodologies suitable for the individual preferences for each ESR.

3.6 Action Centered Learning

Action Centered Learning evolves in annual training events, so-called 'Challenge Calls' or competitive group projects. The supervisors of the hosting institutions specify small research tasks, contributing towards the deliverables of the WPs which require interactions of ESRs. The ESRs are split into three teams. The team selects a team-leader as well as other team roles as required. The team has to write a project application as it would do for an external client, i.e. including business plan and deliverables.

The teams are required to complete the tasks within their budgets, both in terms of time as well as financial and deliver the promised deliverables. The projects are then submitted and a 'winner' is chosen based on scientific quality, cost effectiveness and timely delivery.

The aim of these actions is to train the ESRs in skills such as team work, project planning, team leadership and time management.

3.7 Mentoring

Each ESR is mentored by a senior staff from industry on the basis of research focus and interest. An individual training plan is created and specific training needs will be discussed throughout the project duration and possibly also beyond.

Mentors act as career coaches or additional sources of feedback on the wider issues of the project. It is also anticipated that students are co-supervised by industry as far as possible.

3.8 Intersectoral Secondments

Each ESR undertakes at least one Secondment within a different business sector, i.e. academic-based ESRs will stay at a company and vice versa. This strategy ensures a multi-sectoral approach and will enhance career prospects of ESRs.

4 CONCLUSION

SOLAR-TRAIN is a Marie Sklodowska-Curie (MSCA) Innovative Training Network (ITN) funded by the European Union (Figure I). It started in September 2016 with a total duration of four years. The international early stage researchers active in the project were selected in a global recruiting process in January 2017 and took up their work in spring and summer 2017. They will stay with their host institutions for three years.

Their aim is to develop novel and validated models for service life time and energy prediction of PV modules and systems to ultimately enable a scientific assessment of the triangle quality, durability and costs. The project progress and publications are recorded on www.solar-train.eu.

Figure I: Official SOLAR-TRAIN logo representing the cooperation of the eight hosting institutions.

5 ACKNOWLEDGMENT

SOLAR-TRAIN has received funding in the framework of the European Union's H2020 MSCA program under Grant Agreement No. 721452.

6 REFERENCES

[1] J.H. Wohlgemuth, M.D. Kempe, "Equating damp heat testing with field failures of PV modules," in Proc. 39th IEEE PVSC, 2013, pp. 126-131.
[2] https://solar-train.eu/partners

SCHOOL SUSTAINABLE: ECOLOGICAL FARMING, SOLAR ENERGY AND RAINWATER CAPTURE AS ELEMENT EDUCATOR IN COUNTRY GRAVATAÍ / RS - BRAZIL

Aline Cristiane Pan[1,*], Andreia Golembieski Machado[2], Letícia Paranhos Menna de Oliveira[2], Lilian Alves Schmitt[2], Anderson Antunes De Paulo[3], Luis Fabiano Ribeiro Gomes[4], Rosane Souza da Silva[2]
[1]Solar Energy Technology Nucleus (NT-Solar), [2]Institute of Environment and [3]Engineering Faculty of the Pontifical Catholic University of Rio Grande do Sul
Av. Ipiranga, 6681, Porto Alegre, 90619-900, RS, Brazil, e-mail*: aline.pan@pucrs.br
[4]Padre Nunes State High School
Rua Belcindo, 25, Planaltina, Gravataí , RS, Brazil

ABSTRACT: The project Sustainable School proposed by evaluating the environmental performance of the State High School Padre Nunes, in the City of Gravataí, State of Rio Grande do Sul, the preparation of a report containing suggestions for action in the management, teaching and school extension to provide support taking decisions by the school administration for the purpose of reducing the environmental impact of their actions and the creation of a map bringing together all the proposed changes in the school environment, including sustainable structures as educators elements. Building a tank with a capacity of 16 thousand liters of rainwater was carried out, construction of a solar thermal panel as water heaters with a capacity of 146 liters of clean water for use in school kitchen, installation of a photovoltaic module supply the pressurize for non-potable water distribution network supplied by the rainwater tank, several interventions to revitalize the school yard as educational mosaics and paintings, conducting agro-ecological planting structures as vertical gardens, herb spiral, compost, among others. In addition, training workshops were carried out with the scholarship students and the courses of different levels.
Keywords: education and training, sustainable, photovoltaic, thermal performance.

1 INTRODUCTION

The compulsory nature of environmental education in schools is an advance [1], but it brings new challenges for teachers, who find it difficult to insert it into school planning. The lack of practical elements often means that the education carried out is "on the environment" by addressing only conceptual aspects and making the subject only treated by science teachers, opposing the character of the environmental education that must be [2] [3]. Social technologies applied at school make the environmental education plan into action, in a practical way and involving students in projects that respect the environment where it is possible to think, plan and execute alternatives that integrate the man and his way of life with the other species [4] [5] [6].

The Sustainable School project was launched in 2011 with the objective of promoting environmental management activities in Porto Alegre schools through the improvement of technical-administrative, pedagogical and community extension procedures. The team is multidisciplinary and composed of undergraduate students and professors from the Pontifical Catholic University of Rio Grande do Sul (PUCRS). The main lines of action of the project are water, energy, waste, suppliers, raw materials, school and education and community extension. In the Water axis, we seek the valuation of the water resource, optimization of the resource through improvements in the infrastructure. In the Energy axis, we seek the discussion about the various forms of energy, optimizing the use of energy through adaptations in the infrastructure. As for the raw material, we seek to identify the main materials acquired in the school routine, search for sustainable alternatives and optimize the use of materials through changes in the administrative routine. In the Suppliers axis we emphasize the importance of choosing more sustainable suppliers (local companies, certified and linked to the solidarity economy). In the axis that we call Waste we help in the identification of the main waste produced, and in the definition of segregation and correct disposal (through the realization and delivery of the "Disposal Map" where the main collection points near the school are), besides contributing in campaigns of awareness. In the axis that surrounds the School Area, we evaluate the possibilities of insertion of sustainable niches in the school environment (identification of trees, trails, minhocary, vertical / horticultural gardens / tires, cisterns, herbal spirals, lakes, animal husbandry, solar energy thermal and photovoltaic). Finally, in the Teaching and Extension area, we assist in the identification of pedagogical and extension activities related to environmental sustainability (lectures, workshops, gymnastics, films). In addition to the teachers and content taught, the physical space also acts in a direct and symbolic way on the students, which can facilitate or inhibit behaviors. The tendency of schools, mainly in urban centers, is to pave floors and have few green areas [7], but a courtyard that does not have elements such as water, land, plants and animals may make it harder to implement a curriculum the workshops carried out by the project aim to promote the revitalization of the school space with sustainable forms of exploitation of natural resources, providing teachers with the continuous development of teaching and extension activities with the students and making it possible to continue the project. Therefore, the main objective of this study was to assist the Padre Nunes State School of Education (EEEM Padre Nunes) and the local community of the Gravataí-Rio Grande do Sul (RS), metropolitan region of Porto Alegre capital, in the quest for sustainability environmental, social and economic improvement through the improvement of technical-administrative, pedagogical and community extension procedures, with emphasis on ecological agriculture in the school yard, use of solar energy and rainwater harvesting.

2 METHODOLOGY EMPLOYED

2.1 Object

The city of Gravataí has more than 270 thousand

inhabitants and is the 73rd city in the ranking of the human development index of Rio Grande do Sul, Brazil. E. E. E. M Padre Nunes is located in Bairro Planaltina and serves 1100 students of basic education. The school before the intervention of the Sustainable School Project was totally paved and almost monochromatic, not offering an environment that aroused the pleasure of being in the school environment and not stimulating the taste for environmental preservation, as well as the critical analysis of the use of natural resources by current generation.

2.2 Work Team

The work team of the Sustainable School project, made up of professors and students from PUCRS, worked together with a professor and six students from the EEEM Padre Nunes, who were scholarship holders of the Project approved by the Public Announcement Initiation in Sciences, Mathematics, Engineering, Creative Technologies and Literature (PICMEL), Call Notice n° 015/2013 and Project n° 0328-2551 / 14-4, of the Foundation for Research Support of the State of Rio Grande do Sul (FAPERGS).

2.3 Survey of the School's Environmental Performance

The survey was carried out in loco through a specific form filled through interviews with the administration, consultation of the school records and observation of the places of interest of the physical structure of the School. The selected themes for the evaluation are: (1) Environmental Programs; (2) Buildings; (3) Hazardous Wastes; (4) Common Waste; (5) Water; (6) Energy; (7) Pedagogical Activities; (8) Administrative Activities; (9) Community Activities; (10) Suppliers; (11) Green Area and (12) Power. From the evaluation of the school environmental performance, suggestions were selected from a stock bank that is continuously updated. The suggestions were presented to the school managers through a detailed report and a presentation to the managers and other employees. Among the various suggestions are the special workshops.

2.4 Selecting the Location of Sustainable Structures

The choice of location for the implementation of sustainable structures occurred together with the teachers and students involved in this work through walking in the school and socialization of knowledge about the local reality.

2.5 Team building

The promotion of the deepening of ecological agriculture in the schoolyard, solar thermal system for water heating, photovoltaic system and ferrocement cistern with the project grant holders through workshops with the Institute of Environment and Natural Resources of the Nucleus of Technology in Solar Energy and through guided trails in PUCRS (School of Education) and in the Piratini Palace.

2.6 Workshops

The activities with teachers and students were carried out through analysis of the selected site, projection of the activity through drawings made by the students and execution of the practical activity of implementation of sustainable structures related to the use of solar energy, rainwater through ferrocement tank and the incorporation of elements of agroecology.

Cistern → Fertilizer cistern for collecting rainwater. The ferrocement tank is considered a social technology because it is a simple, replicable process that meets a community need. The ferrocement cistern can be constructed collectively, being possible the participation of children in some stages. The use of cisterns in schools brings to light the theme of rainwater as a natural resource of still underestimated value. Training activities and workshops with students included discussions about the water problem, presentation of a model of the tank, as well as, in the execution of the construction of the cistern, the students' guided participation in some stages. The painting of the cistern with motifs related to the preservation of water resources. The catchment area of the cistern is 100 m², a rainfall index of 1,440 mm / year has been estimated (data for the metropolitan area of Porto Alegre), which means that it has a catchment capacity of 9,600 L / month (100 m² x 120 mm / month x 0.8 - loss coefficient). The cistern was built with a radius of 1.30 m and 2 m in height, totaling a volume of 16 thousand liters of total capacity. The water is collected in the gutters installed on the roof of the school cafeteria, then taken to the separator of the first waters and then directed to the cistern. The water outlet is provided by a 50 mm pipe installed in the center of the cistern directed to a register.

Agroecology → The elements of agroecology introduce the school community in the questions of rational use of land, approach to nature, through the reduction of the area and use of native plants [8]. Workshops were held on the concepts of agroecology and permaculture, as well as practical activities for the preparation of vertical gardens on pallets and the spirals of herbs, compost, ferns and fruit plants

Thermal Solar Energy → The initiative was to undertake concrete elements of appreciation of new ways of building within the educational space without this presenting costs or impacts, replacing existing technology based on the use of solar energy for heating water inside the school kitchen. Using solar panels to heat water even on cloudy days significantly reduces energy consumption for heating and still be a cost-effective option [9] [10].

The context of the social reality of the metropolitan region of Porto Alegre shows us a harsh reality, that there are not enough resources to warm rustic dwellings, or even hot water for personal hygiene, given the economic condition presented by the inhabitants of the towns or favelas of so many communities. The intention is to consider the needs of these poor communities and democratize the knowledge developed by providing a flow of socialization practices environmental education and sustainability to meet some of their real needs.

Crossing the academic experiences and popular knowledge in the community was a challenge put in evidence and prejudices. The paradigm of the use of solar energy during the winter, in the southernmost Brazilian state, and of lower solar incidence should be broken convincing of the energy efficiency existing in such experiment was fundamental. Putting this novelty with a small experiment in the classroom, fact outdated the moment was the definitive construction of the equipment.

The idea of reuse as an educator element of the team was always transverse, which was promptly attended to in the collection of materials that could be readily delivered and operated. The PET bottles, along with the Tetra Pack packs were very useful in the activity. With the material collection gymkhana, it was possible to accumulate a significant amount of these materials and thus to review the viable and complete sustainability intentionality. In reviewing sources of consultation, it was observed the possibility of assembling a solar thermal collector with low cost already presenting significant results, consisting of a PVC plastic pipe coil, found in trade of construction material for hydraulic installations and of easy manipulation and welding.

Following this, an estimate was made of the daily use of hot water by the school kitchen, proposed as a replacement challenge in the energy source, replacing the electric faucet connected to the general grid. Due to its flow rate and the time used, the determination of 130 liters was achieved. Based on the diameter of the black painted PVC pipes involved in the construction of the solar thermal collector, 25 mm, we included a heating coil, with 31 bars of 2m length each, and a larger reservoir, using a 50 mm pipe with 6m in length, which resulted in a set of sufficient 147 liters of water heated by solar energy (Fig. 1).

Figure 1: Details of the construction of the solar thermal collector for water heating.

The temperature gain provided by the solar thermal system was measured at 6.3 ° C in the Winter Season (measurement taken at Solstice - June 22) and at the beginning of the school year, at the end of February we reached 11.5 ° C thermal measurements in common tap connected to the distribution network in Gravataí. In order to protect from the typical meteorological weather of Rio Grande do Sul, we again resorted to materials based on the reuse of PET bottles, making it interesting to protect, making it more efficient at heating from 2 ° C to 4 ° C more, not only as a "greenhouse" barrier to the rays already trapped around the PVC pipe. At the end of the construction, the Tetra Pack was used as a form of thermal insulation for the winter since the cold roof decreased the amount of heat that was transferred to the asbestos of the tile by up to 3.5 ° C, and also becoming an important reflector for the interior rear area of the serpentine giving a gain of plus 2.1 ° C.

Photovoltaic Solar Energy → The installation of a 140 W photovoltaic module of the company *Yingli* was carried out to feed the non-potable water supply pressurizer supplied by the cistern based on [11]. The cistern that had a capacity of 16,000 liters of rainwater needed to be pumped into the water box and then into the taps of the school yard. This water needed to have minimum conditions of pressure, because it is used for the external and internal cleaning of the school. A Jabsco 2.9 GPM 12 V pressurizing pump of DC was chosen so that it was not necessary to use an inverter. A stationary, Freedom brand battery of 185 Ah / 170 Ah was also installed to store the produced and unused electrical energy. Also, a load controller, brand Epsolar Landstar, was introduced to the system. The module was installed in the orientation of 30 ° North under a support developed for the same in the Laboratory of the Faculty of Engineering of PUCRS. The whole system was made apparent, because the objective is that the student and the whole school community can understand the complete operation of the water cycle and the photovoltaic module. This insertion is very valuable, since it can cover the needs of this school inexhaustible and can serve as a basis for sustainable development.

3 RESULTS AND DISCUSSION

During the development of this work, scholars, managers, technicians and teachers were trained in educational elements through the use of social technologies focused on agroecology, solar energy and rainwater. Within the agroecology axis, the activities of vertical and horizontal vegetable gardens, herb spiral and compost were developed. Within the solar energy axis was developed a solar-thermal water heater and installed a photovoltaic module to supply the pressurizer for non-potable water network supplied by the tank. Within the rainwater axis was built a ferrocement tank. The main steps of the activities carried out in E. E. M. Padre Nunes and their methodologies are described in Tab. 1.

Through Tab. 1 we observed that the objectives of this work were fully achieved and it was possible to develop the proposed activities with students, teachers and technical-administrative personnel. The fellows of the team participated actively in the training, which reflected in the quality of the workshops. Nevertheless, it is necessary to improve the utilization of the educational structures implemented in the form of inclusion of curricular activities, which are being studied.

During the period of execution of this work (1 year), 03 volunteers and 03 fellows of Scientific Initiation, graduates and graduates of PUCRS, respectively, and 06 fellows of Junior Scientific Initiation of E.E.E.M. Father Nunes participated in the training and execution activities. The consolidation of the executed project model will serve for the implementation in other schools and will consist of publication for the dissemination of this approach in a school environment. E.E.E.M. Father Nunes, we believe that this will serve as a model to be followed by other schools and will have the opportunity to develop innovative pedagogy by inserting elements of environmental education transversally into the school curriculum using the elements implemented in the schoolyard.

More pragmatically, E.E.E.M. Padre Nunes will be able to overcome the challenges of lack of water during the hot months, since the cistern of 16 thousand liters can supply water for class cleaning and irrigation of the gardens, saving drinking water.

The various interventions with elements of agroecology bring to school the approximation of the school community to nature, providing a link between them and aligning a parsimonious and fruitful coexistence.

The idea of revising the energy matrix as an element of environmental education is of primordial action in our school practices. Installing equipment such as photovoltaic modules with immediate energy application (without external assistance), changing paradigms in heating drinking water with serpentine structures coated with PET bottles and Tetra Pack packaging is an innovation still needed and never seen by the professionals involved or even school community present in the environment, envisioning an innovation and reinforcing the intention of the educational element and the use of alternatives, rather than present.

Table I: Stages of the activities carried out in E. E. M. Padre Nunes and their methodologies used.

Proposed / Performed Steps		Methodology
1 - Survey of the environmental performance of the school		Predefined questionnaire; Technical visits.
2 - Actions and Suggestions		Report containing suggestions for the reduction of the environmental impact in each axis.
3 - Choice of location for workshops		Participatory mapping with local students, teachers and scholars to implement interventions in the school yard; Photographic records made by the fellows about possible places to be revitalized in the schoolyard; Student activity on photo printing in order to give space to students' imaginations and desires on the school yard.
4 - Group discussions on works		Training of scholars on the theme of rational use of water, energy and the benefits of agroecology; Visit of the scholarship holders in the case of the Piratini Palace (cistern, gardens and compost); Visit of fellows to the trail of sustainable elements of the central campus and Nucleus of Technology in Solar Energy of PUCRS.
5 - Workshops	Cistern	Construction of cistern of ferrocement with execution by specialized outsourced service and interventions of the scholarship and staff in the orientation of the students and teachers for the learning of the technique, and the use of this as an educating element in the water theme.
	Garden on pallet	Creation of 4 vertical gardens reusing pallets and carpets. The location of the gardens was defined according to the microclimate being used species of succulents and native flowers.
	Palette Bench	Two benches were made by reusing pallets. The seats were upholstered and arranged in the schoolyard.
	Mosaic on table and reels	A school table was revitalized through the mosaic technique with the reuse of old tiles. This technique was also used to compose benches reusing wooden reels. These banks were intended for the second-year primary school students.
	Herb Spiral	A spiral of herbs was placed in the central courtyard using stones for the structure of the spiral, broken tiles and soil for substrate and seedlings of various spices.
	Composter	A composter was made for the access of the school kitchen staff. The compost was made with pallets and reused tiles and received soil and straw / newspaper substrate.
	Thematic garden	In an environment with low solar illumination the planting of pteridófitas was realized, creating a thematic place for the study of these plants.
	Revitalization of the bike rack	The bike rack was painted and decorated using the stencil technique which reuses X-ray sheets.
	Agroforestry	Areas of the courtyard were revitalized with the planting of native trees (fruit trees) and flower beds
	Solar Energy	The non-potable hydraulic network was designed to distribute the water pressurized by photovoltaic solar energy. Where a photovoltaic module connected to a battery was installed to feed the pressurizer to the non-potable tank network (Fig. 2). As a result, three non-potable water hoses were installed to be used to clean the school yard, irrigate the gardens, among others. It was established the most adequate measure for the construction of a solar thermal collector for heating potable water with a view to attending the school kitchen, sending its confection with material collected from the school community consisting of PET bottles and tetrapack containers (Fig. 3) Two commercial lighting posts were installed in LED lamps, powered by photovoltaic solar cells

Figure 2: Installed photovoltaic module and pressurizer that is powered by the same.

Figure 3: Solar thermal collector for heating installed on the roof of the School.

The application of solar energy to the workshop is carried out in a workshop. It is clear that we must disseminate and socialize our society in its simplest applications. When teachers and students, even in needy regions, fail to realize the immediate application of really direct and even simple techniques of appropriation of this energy source. We must make constant our efforts to present for teaching purposes all possible ways of revising our way of seeking and using sources of energy, from conventional sources.

Table 2 presents a comparison of the measures taken of the assembled equipment and its improvements developed in the school, using a TDS3 digital thermometer, based on a heating time of 12 h of the collector.

Table II: Stages Comparison of the water temperatures obtained by the solar thermal panel

Date	Water Temperature (°C)				
	Red	Tube without paint	Tube paint	Tube PET Coated	Tube PET and Tetra Pack
June	16.1	22.4	22.4	24.4	26.5
Febru ary	21.0	33.5	33.5	37.5	41.0

4 CONCLUSIONS

The benefits achieved by the project extend much more than the implanted structures to review our energy matrix and reuse of solar heat. Much more to the students who participated in this process as building agents of a school, and consequent education, quality and application of their knowledge. They become authors of their process of social intervention, multipliers of healthy experiences and placing them as promoters of science. Their self-esteem, their innovative character, the capacity for initiative and understanding of change in the environment makes them references and leaderships for the school community, stimulating and discussing the role of the school and the intervention to the environment occupied by them.

It is well known the difficulty presented by many communities in the sense of having equipment that brings comfort and convenience to their financial needs. If initiatives to socialize certain technologies for their immediate implantation and seen the reutilization of simple materials increasing the efficiency, we see that we have reached the proposed objectives and that each time solar energy should be popularized in its use. We need means of dissemination that present the population with these means that reduce costs and bring enormous benefits to our environment, a clear proposal of true sustainability.

Acknowledgements

This work was performed as part of the Universal Project 2014, n° 458145/2014-9, of the National Council for Scientific and Technological Development (CNPq). In same way, by Research Program for Students of PUCRS (BPA/2017). It is of great importance the confidence shown by the professors, employees and especially by Prof. Silvia Bittello of E. E. E. Padre Nunes.

REFERENCES

[1] Ministry of Education, Resolution n° 2, (2012).
[2] Capra, F. A teia da vida: uma nova compreensão científica dos sistemas vivos, São Paulo, Cultrix (1996)
[3] World Commission on Environment and Development. Our common future, 2. ed., Rio de Janeiro: Fundação Getúlio Vargas (1992)
[4] Cavalcanti, C. (Org.). Environment, Sustainable Development and Public Policies, São Paulo: Cortez (1999).
[5] Dias, R. Gestão ambiental: responsabilidade social e sustentabilidade, São Paulo, Atlas (2010).
[6] Trajber, R., Sato, M. Revista Eletrônica do Mestrado em Educação Ambiental, Rio Grande – RS, (2010) 1517-1256.
[7] Legan, L. Criando habitats na Escola Sustentável, Imprensa Social (2010).
[8] Vieira, E. F. Geografia Física e Vegetação - Rio Grande do Sul, Editora Sagra (1984).
[9] Chigueru, T. *et al.* Atlas Solarimétrico do Brasil, Recife, Editora Universitária da UFPE (2014).
[10] Bezerra, A. M. Aplicações térmicas da energia solar, João Pessoa, UFPB, 4ª ed. (2011).
[11]Rüther, R. Edifícios solares fotovoltaicos: o potencial da geração solar fotovoltaica integrada a edificações urbanas e interligada à rede elétrica pública no Brasil, Florianópolis, LABSOLAR (2004).

AUTHOR INDEX

AUTHOR INDEX

AUTHOR INDEX

AUTHOR INDEX

AUTHOR INDEX

AUTHOR INDEX

AUTHOR INDEX

AUTHOR INDEX

AUTHOR INDEX

AUTHOR INDEX

AUTHOR INDEX

AUTHOR INDEX

AUTHOR INDEX

AUTHOR INDEX

AUTHOR INDEX

AUTHOR INDEX

AUTHOR INDEX

AUTHOR INDEX

AUTHOR INDEX

AUTHOR INDEX

AUTHOR INDEX

AUTHOR INDEX

AUTHOR INDEX

AUTHOR INDEX